South America
Central America
and the
Caribbean
2006

# South America Central America and the Caribbean 2006

14th Edition

Routledge
Taylor & Francis Group
LONDON AND NEW YORK

**First published 1985**
**Fourteenth Edition 2006**

**© Routledge 2005**
Haines House, 21 John Street, London WC1N 2BP, United Kingdom
*(A member of the Taylor & Francis Group)*

All rights reserved. No part of this
publication may be photocopied, recorded,
or otherwise reproduced, stored in a retrieval
system or transmitted in any form or by any
electronic or mechanical means without the
prior permission of the copyright owner.

ISBN 1-85743-312-2
ISSN 0258-0661

Editor: Jacqueline West
Regional Organizations Editors: Catriona Appeatu Holman, Helen Canton
Statistics Editor: Philip McIntyre
Technology Editor: Ian Preston
Assistant Editors: Eleanor Baynes, Camilla Chew, Nicola Gollan, Michael Grayer, Kirstie Macdonald, James Middleton, Patrick Raleigh, Daniel Ward
Production Co-ordinator: Andreas Gosling
Contributing Editor (Commodities): Simon Chapman
Editorial Clerical Assistant: Charley McCartney
Series Editor: Joanne Maher

Typeset in New Century Schoolbook

Typeset by Unwin Brothers Limited, The Gresham Press, Old Woking, Surrey
Printed and bound by Polestar Wheatons, Exeter

# FOREWORD

The 14th edition of SOUTH AMERICA, CENTRAL AMERICA AND THE CARIBBEAN provides a survey of the political and economic life both of the region and of the 48 countries and territories within it. The region, as suggested by the title, contains three distinct geopolitical areas. This volume also contains three distinct, though complementary, areas. Part One consists of eight introductory articles, covering a variety of subjects of regional significance, be it the political shift to the left in Latin America; crime and reform in the Caribbean; trade agreements in Latin America; US anti-drugs policy in the Andean region post-Plan Colombia; or trade preferences in the Caribbean. In Part Two all of the chapters in the region comprise a geography section, historical and economic articles written by experts, a statistical survey, directory and, where applicable, a bibliography. Part Three contains information on the region's major primary commodities, as well as sections on international organizations, research institutes, books and periodicals relevant to the region.

The year since the publication of the last edition of SOUTH AMERICA, CENTRAL AMERICA AND THE CARIBBEAN has seen much change. In early 2005 there were fears that Bolivia was moving towards civil war, as left-wing groups demanded nationalization of the hydrocarbons industry while pro-business forces insisted on greater autonomy for the country's oil-producing regions. The conflict resulted in the resignation of President Carlos Mesa Gisbert in June. In Ecuador, popular protest against apparent government interference in the Supreme Court forced President Lucio Gutiérrez Borbua out of office in April. In October 2004 the leftward shift in Latin America continued with the election of Tabaré Ramón Vázquez Rosas to the presidency in Uruguay. Earlier, in August, Venezuela's controversial head of state, Hugo Chávez Frías, had survived an opposition-sponsored referendum to force his recall from office. The consolidation of his position, as well as high international oil prices, allowed Chávez in 2005 to accelerate his so-called 'Bolivarian revolution', and to seek alliances with other left-of-centre governments in the region. Meanwhile, in Brazil, President Luiz Inácio ('Lula') da Silva's Partido dos Trabalhadores was embroiled in a corruption scandal that threatened to bring down his Government in mid-2005. In Mexico, where a presidential election was scheduled to be held in 2006, hopeful candidates from President Vicente Fox's party were trailing in popularity behind the potential left-wing nominee, Andrés Manuel López Obrador. Argentina continued to recover from the financial crises of earlier in the decade, although negotiations with the IMF remained tense. Meanwhile, the continuing political breakdown and humanitarian crisis in Haiti showed little prospect for improvement, even in the event that elections were able to proceed as scheduled in November 2005.

The Editors are grateful to all the contributors for their articles and advice, and to the numerous governments and organizations that provided statistical and other information.

August 2005

# HEALTH AND WELFARE STATISTICS: SOURCES AND DEFINITIONS

**Total fertility rate** Source: WHO, *The World Health Report* (2005). The number of children that would be born per woman, assuming no female mortality at child-bearing ages and the age-specific fertility rates of a specified country and reference period.

**Under-5 mortality rate** Source: UNICEF, *The State of The World's Children* (2005). The ratio of registered deaths of children under 5 years to the total number of registered live births over the same period.

**HIV/AIDS** Source: UNAIDS. Estimated percentage of adults aged 15 to 49 years living with HIV/AIDS. < indicates 'fewer than'.

**Health expenditure** Source: WHO, *The World Health Report* (2005).
*US $ per head (PPP)*
International dollar estimates, derived by dividing local currency units by an estimate of their purchasing-power parity (PPP) compared with the US dollar. PPPs are the rates of currency conversion that equalize the purchasing power of different currencies by eliminating the differences in price levels between countries.
*% of GDP*
GDP levels for OECD countries follow the most recent UN System of National Accounts. For non-OECD countries a value was estimated by utilizing existing UN, IMF and World Bank data.
*Public expenditure*
Government health-related outlays plus expenditure by social schemes compulsorily affiliated with a sizeable share of the population, and extrabudgetary funds allocated to health services. Figures include grants or loans provided by international agencies, other national authorities, and sometimes commercial banks.

**Access to water and sanitation** Source: WHO/UNICEF Joint Monitoring Programme on Water Supply and Sanitation (JMP) (Mid-Term Assessment, 2004). Defined in terms of the percentage of the population using improved facilities in terms of the type of technology and levels of service afforded. For water, this includes house connections, public standpipes, boreholes with handpumps, protected dug wells, protected spring and rainwater collection; allowance is also made for other locally defined technologies. Sanitation is defined to include connection to a sewer or septic tank system, pour-flush latrine, simple pit or ventilated improved pit latrine, again with allowance for acceptable local technologies. Access to water and sanitation does not imply that the level of service or quality of water is 'adequate' or 'safe'.

**Human Development Index** (HDI) Source: UNDP, *Human Development Report* (2004). A summary of human development measured by three basic dimensions: prospects for a long and healthy life, measured by life expectancy at birth; knowledge, measured by adult literacy rate (two-thirds' weight) and the combined gross enrolment ratio in primary, secondary and tertiary education (one-third weight); and standard of living, measured by GDP per head (PPP US $). The index value obtained lies between zero and one. A value above 0.8 indicates high human development, between 0.5 and 0.8 medium human development, and below 0.5 low human development. Countries with insufficient data were excluded from the HDI. In total, 177 countries were ranked for 2002.

# ACKNOWLEDGEMENTS

The editors gratefully acknowledge the co-operation, interest and enthusiasm of all the authors who have contributed to the volume. We are also indebted to the many organizations connected with the region, particularly the national statistical offices. We owe special thanks to a number of embassies and ministries. We are also grateful to the University of Southampton Cartographic Unit for supplying the maps included in the Geography sections.

We are most grateful for permission to make extensive use of material from the following sources: the UN's statistical databases and *Demographic Yearbook, Statistical Yearbook, Industrial Commodity Statistics Yearbook, International Trade Statistics Yearbook* and *Monthly Bulletin of Statistics*; UNESCO's *Statistical Yearbook*; the UN Food and Agriculture Organization's statistical database; the UN Economic Commission for Latin America and the Caribbean's *Statistical Yearbook*; the ILO's statistical database and *Yearbook of Labour Statistics*; the World Bank's statistical database and *World Bank Atlas, Global Development Finance, World Development Report* and *World Development Indicators*; the World Tourism Organization's *Yearbook of Tourism Statistics*; and the IMF's statistical database, *Government Finance Statistics Yearbook* and *International Financial Statistics*. We are also grateful to the International Institute for Strategic Studies, Arundel House, 13–15 Arundel Street, London WC2R 3DX, for the use of defence statistics from *The Military Balance 2004–2005*.

# EXPLANATORY NOTE ON THE DIRECTORY SECTION

The Directory section of each chapter covering a major country is arranged under the following, or similar, headings, where they apply:

**THE CONSTITUTION**

**THE GOVERNMENT**
- HEAD OF STATE
- CABINET/COUNCIL OF MINISTERS
- MINISTRIES

**LEGISLATURE**

**POLITICAL ORGANIZATIONS**

**DIPLOMATIC REPRESENTATION**

**JUDICIAL SYSTEM**

**RELIGION**

**THE PRESS**

**PUBLISHERS**

**BROADCASTING AND COMMUNICATIONS**
- TELECOMMUNICATIONS
- RADIO
- TELEVISION

**FINANCE**
- CENTRAL BANK
- STATE BANKS
- DEVELOPMENT BANKS
- COMMERCIAL BANKS
- FOREIGN BANKS
- BANKING ASSOCIATIONS
- STOCK EXCHANGES
- INSURANCE

**TRADE AND INDUSTRY**
- GOVERNMENT AGENCIES
- DEVELOPMENT ORGANIZATIONS
- CHAMBERS OF COMMERCE
- INDUSTRIAL AND TRADE ASSOCIATIONS
- EMPLOYERS' ASSOCIATIONS
- MAJOR COMPANIES
- UTILITIES
- TRADE UNIONS

**TRANSPORT**
- RAILWAYS
- ROADS
- SHIPPING
- CIVIL AVIATION

**TOURISM**

**DEFENCE**

**EDUCATION**

# THE CONTRIBUTORS

**Luciano Baracco.** Researcher on Central America at the University of Bradford.

**Andrew Bounds.** Former Central America correspondent of the *Financial Times*.

**Dr Ed Brown.** Lecturer in Human Geography at the University of Loughborough.

**Dr Julia Buxton.** Senior Research Fellow at the Centre for Co-operation and Security, Department of Peace Studies at the University of Bradford.

**Prof. Peter A. R. Calvert.** Professor of Comparative and International Politics at the University of Southampton.

**Prof. Paul Cammack.** Professor in the Department of Government at the University of Manchester.

**Greg Chamberlain.** Journalist specializing in Caribbean affairs.

**Luis Eduardo Fajardo.** Researcher in the Department of Economics at the University of Rosario.

**Dr James Ferguson.** Writer specializing in Caribbean affairs.

**Prof. David Fleischer.** Professor of Political Science at the University of Brasília.

**Lila Haines.** Economic historian and business journalist specializing in the Cuban economy.

**Canute James.** Senior Lecturer at the Caribbean Institute of Media and Communication at the University of the West Indies.

**Melanie Jones.** Writer and researcher specializing in Cuban affairs.

**Prof. Adrian McDonald.** Professor of Environmental Management at the University of Leeds.

**Duncan McDonald.** Researcher at the Centre for Research into Environmental Health at the University of Aberystwyth.

**James McDonough.** Editor and Publisher at EPIN Publishing, including *The Puerto Rico Report* newsletter.

**Sandy Markwick.** Writer and researcher specializing in Latin American affairs.

**Canon Prof. Kenneth N. Medhurst.** Professor Associate, University of Sheffield, Senior Research Associate, the Von Hugel Institute, St Edmund's College, University of Cambridge, and Canon Theologian, Anglican Diocese of Bradford.

**Sir Keith Morris.** Former British Ambassador to Colombia.

**Katharine Murison.** Writer specializing in Latin American and African affairs.

**Philip J. O'Brien.** Senior Lecturer in the Department of Sociology at the University of Glasgow.

**Dr Gabriel Palma.** Lecturer in Economics at the University of Cambridge.

**Dr Francisco Panizza.** Senior Lecturer in Latin American Politics at the London School of Economics and Political Science.

**Prof. Jenny Pearce.** Professor of Latin American Politics at the University of Bradford.

**Nicola Phillips.** Hallsworth Research Fellow at the University of Manchester, Editor of *New Political Economy* journal and Co-Editor of the *International Political Economy Yearbook*.

**Rod Prince.** Journalist specializing in Caribbean affairs.

**Diego Sánchez-Ancochea.** Lecturer in Economics at the Institute for the Study of the Americas at the University of London.

**Helen Schooley.** Writer specializing in Latin American affairs.

**Dr Ken Shadlen.** Lecturer at the Development Studies Institute at the London School of Economics and Political Science.

**Dr Rachel Sieder.** Senior Lecturer in Politics at the Institute for the Study of the Americas at the University of London.

**Colin Smith.** Editor of the *Catholic Standard* and British Broadcasting Corporation correspondent in Guyana.

**Prof. A. E. Thorndike.** Former Senior Lecturer at the Graduate School of International Studies at the University of Birmingham.

**René van Dongen.** Geographer and Vice-Principal of the School of the Nations, Georgetown, Guyana.

**Phillip Wearne.** Writer and researcher specializing in Latin American affairs.

**Mark Wilson.** Writer and researcher specializing in Caribbean affairs.

# CONTENTS

| | |
|---|---|
| The Contributors | page viii |
| Abbreviations | xiii |
| International Telephone Codes | xvi |
| Maps | xvii |

## PART ONE
# General Survey

| | |
|---|---|
| Latin America's Turn to the Left  Dr FRANCISCO PANIZZA | 3 |
| Debt, Finance and the IMF: Three Decades of Debt Crises in Latin America  Dr KEN SHADLEN | 8 |
| Crime in the Caribbean: Reform and Progress  CANUTE JAMES | 13 |
| An Environmental Portrait of Latin America and the Caribbean  Prof. ADRIAN McDONALD and DUNCAN McDONALD | 17 |
| Trade Agreements in the Americas  Dr NICOLA PHILLIPS | 23 |
| 'Plan Colombia' and Beyond: US Anti-Drugs Policy in the Andean Region  LUIS EDUARDO FAJARDO | 28 |
| Trade Preferences in the Caribbean  MARK WILSON | 32 |
| Religion in 21st Century Latin America  Canon Prof. KENNETH N. MEDHURST | 37 |

## PART TWO
# Country Surveys

See page vii for explanatory note on the Directory section of each country.

**ANGUILLA** 45
| | |
|---|---|
| Geography and Map | 45 |
| History  MARK WILSON | 45 |
| Economy  MARK WILSON | 46 |
| Statistical Survey | 47 |
| Directory | 48 |

**ANTIGUA AND BARBUDA** 52
| | |
|---|---|
| Geography and Map | 52 |
| History  MARK WILSON | 53 |
| Economy  MARK WILSON | 54 |
| Statistical Survey | 55 |
| Directory | 56 |
| Bibliography | 61 |

**ARGENTINA** 62
| | |
|---|---|
| Geography and Map | 62 |
| History  Prof. PETER CALVERT | 63 |
| Economy  Prof. PETER CALVERT | 69 |
| Statistical Survey | 74 |
| Directory | 80 |
| Bibliography | 98 |

**ARUBA** 99
| | |
|---|---|
| Geography and Map | 99 |
| History | 99 |
| Economy | 101 |
| Statistical Survey | 102 |
| Directory | 103 |
| Bibliography | 107 |

**THE BAHAMAS** 108
| | |
|---|---|
| Geography and Map | 108 |
| History  MARK WILSON | 109 |
| Economy  MARK WILSON | 111 |
| Statistical Survey | 115 |
| Directory | 116 |
| Bibliography | 123 |

**BARBADOS** 124
| | |
|---|---|
| Geography and Map | 124 |
| History  MARK WILSON (Based on an earlier article by Prof. A. E. THORNDIKE) | 124 |
| Economy  MARK WILSON | 126 |
| Statistical Survey | 129 |
| Directory | 130 |
| Bibliography | 136 |

**BELIZE** 137
| | |
|---|---|
| Geography and Map | 137 |
| History  HELEN SCHOOLEY (Revised by the editorial staff) | 138 |
| Economy  HELEN SCHOOLEY (Revised by the editorial staff) | 141 |
| Statistical Survey | 143 |
| Directory | 145 |
| Bibliography | 150 |

**BERMUDA** 151
| | |
|---|---|
| Geography and Map | 151 |
| History | 151 |
| Economy | 152 |
| Statistical Survey | 153 |
| Directory | 154 |

**BOLIVIA** 159
| | |
|---|---|
| Geography and Map | 159 |
| History  Prof. PETER CALVERT | 160 |
| Economy  Prof. PETER CALVERT | 164 |
| Statistical Survey | 168 |
| Directory | 173 |
| Bibliography | 182 |

**BRAZIL** 183
| | |
|---|---|
| Geography and Map | 183 |
| History  Prof. DAVID FLEISCHER (Based on an earlier text by Dr PAUL CAMMACK) | 185 |
| Economy  Dr GABRIEL PALMA | 191 |
| Statistical Survey | 198 |
| Directory | 205 |
| Bibliography | 228 |

**THE BRITISH VIRGIN ISLANDS** 230
| | |
|---|---|
| Geography and Map | 230 |
| History  MARK WILSON | 230 |
| Economy  MARK WILSON | 231 |

# CONTENTS

| | |
|---|---|
| Statistical Survey | 232 |
| Directory | 233 |
| **THE CAYMAN ISLANDS** | **237** |
| Geography and Map | 237 |
| History | 237 |
| Economy | 238 |
| Statistical Survey | 239 |
| Directory | 240 |
| **CHILE** | **245** |
| Geography and Map | 245 |
| History   PHILIP J. O'BRIEN | 246 |
| Economy   PHILIP J. O'BRIEN | 250 |
| Statistical Survey | 255 |
| Directory | 260 |
| Bibliography | 276 |
| **COLOMBIA** | **278** |
| Geography and Map | 278 |
| History   Sir KEITH MORRIS | 279 |
| Economy   Sir KEITH MORRIS | 284 |
| Statistical Survey | 289 |
| Directory | 294 |
| Bibliography | 307 |
| **COSTA RICA** | **308** |
| Geography and Map | 308 |
| History   DIEGO SÁNCHEZ-ANCOCHEA (Based on an earlier article by Prof. JENNY PEARCE) | 309 |
| Economy   DIEGO SÁNCHEZ-ANCOCHEA (Based on an earlier article by ANDREW BOUNDS) | 312 |
| Statistical Survey | 317 |
| Directory | 321 |
| Bibliography | 330 |
| **CUBA** | **331** |
| Geography and Map | 331 |
| History   MELANIE JONES | 332 |
| Economy   LILA HAINES | 338 |
| Statistical Survey | 344 |
| Directory | 348 |
| Bibliography | 361 |
| **DOMINICA** | **362** |
| Geography and Map | 362 |
| History   MARK WILSON | 362 |
| Economy   MARK WILSON | 364 |
| Statistical Survey | 365 |
| Directory | 366 |
| Bibliography | 370 |
| **THE DOMINICAN REPUBLIC** | **371** |
| Geography and Map | 371 |
| History   Dr JAMES FERGUSON | 372 |
| Economy   Dr JAMES FERGUSON | 375 |
| Statistical Survey | 379 |
| Directory | 383 |
| Bibliography | 393 |
| **ECUADOR** | **394** |
| Geography and Map | 394 |
| History   SANDY MARKWICK | 395 |
| Economy   SANDY MARKWICK | 400 |
| Statistical Survey | 405 |
| Directory | 411 |
| Bibliography | 421 |
| **EL SALVADOR** | **422** |
| Geography and Map | 422 |
| History   ANDREW BOUNDS (Revised by DIEGO SÁNCHEZ-ANCOCHEA) | 423 |
| Economy   ANDREW BOUNDS (Revised by DIEGO SÁNCHEZ-ANCOCHEA) | 427 |
| Statistical Survey | 431 |
| Directory | 435 |
| Bibliography | 443 |
| **THE FALKLAND ISLANDS** | **444** |
| Geography and Map | 444 |
| History | 444 |
| Economy | 445 |
| Statistical Survey | 446 |
| Directory | 446 |
| **FRENCH GUIANA** | **449** |
| Geography and Map | 449 |
| History   PHILLIP WEARNE (Revised by Dr JAMES FERGUSON) | 449 |
| Economy   PHILLIP WEARNE (Revised by Dr JAMES FERGUSON) | 451 |
| Statistical Survey | 452 |
| Directory | 454 |
| Bibliography | 457 |
| **GRENADA** | **458** |
| Geography and Map | 458 |
| History   MARK WILSON | 458 |
| Economy   MARK WILSON | 459 |
| Statistical Survey | 460 |
| Directory | 462 |
| Bibliography | 466 |
| **GUADELOUPE** | **467** |
| Geography and Map | 467 |
| History   PHILLIP WEARNE (Revised by Dr JAMES FERGUSON) | 468 |
| Economy   PHILLIP WEARNE (Revised by Dr JAMES FERGUSON) | 470 |
| Statistical Survey | 471 |
| Directory | 473 |
| Bibliography | 476 |
| **GUATEMALA** | **477** |
| Geography and Map | 477 |
| History   Dr RACHEL SIEDER | 478 |
| Economy   Prof. JENNY PEARCE (Revised by the editorial staff) | 481 |
| Statistical Survey | 486 |
| Directory | 490 |
| Bibliography | 499 |

# CONTENTS

| | |
|---|---|
| **GUYANA** | 501 |
| Geography and Map | 501 |
| History COLIN SMITH | |
| (Based on an earlier article by JAMES McDONOUGH) | 502 |
| Economy COLIN SMITH | |
| (Based on an earlier article by JAMES McDONOUGH) | 507 |
| Statistical Survey | 513 |
| Directory | 514 |
| Bibliography | 521 |
| **HAITI** | 522 |
| Geography and Map | 522 |
| History GREG CHAMBERLAIN | 523 |
| Economy GREG CHAMBERLAIN | 526 |
| Statistical Survey | 529 |
| Directory | 532 |
| Bibliography | 539 |
| **HONDURAS** | 540 |
| Geography and Map | 540 |
| History HELEN SCHOOLEY | |
| (Revised by SANDY MARKWICK) | 541 |
| Economy PHILLIP WEARNE | |
| (Revised by SANDY MARKWICK) | 544 |
| Statistical Survey | 547 |
| Directory | 551 |
| Bibliography | 558 |
| **JAMAICA** | 559 |
| Geography and Map | 559 |
| History Dr JAMES FERGUSON | 559 |
| Economy Dr JAMES FERGUSON | 562 |
| Statistical Survey | 566 |
| Directory | 570 |
| Bibliography | 578 |
| **MARTINIQUE** | 580 |
| Geography and Map | 580 |
| History PHILLIP WEARNE | |
| (Revised by Dr JAMES FERGUSON) | 581 |
| Economy PHILLIP WEARNE | |
| (Revised by Dr JAMES FERGUSON) | 582 |
| Statistical Survey | 583 |
| Directory | 585 |
| Bibliography | 588 |
| **MEXICO** | 589 |
| Geography and Map | 589 |
| History SANDY MARKWICK | 590 |
| Economy SANDY MARKWICK | 597 |
| Statistical Survey | 603 |
| Directory | 609 |
| Bibliography | 629 |
| **MONTSERRAT** | 630 |
| Geography and Map | 630 |
| History MARK WILSON | 631 |
| Economy MARK WILSON | 632 |
| Statistical Survey | 633 |
| Directory | 634 |
| **THE NETHERLANDS ANTILLES** | 637 |
| Geography and Map | 637 |
| History RENÉ VAN DONGEN | |
| (Revised by the editorial staff) | 638 |
| Economy RENÉ VAN DONGEN | |
| (Revised by the editorial staff) | 641 |
| Statistical Survey | 644 |
| Directory | 645 |
| Bibliography | 652 |
| **NICARAGUA** | 653 |
| Geography and Map | 653 |
| History ED BROWN | |
| (Based on an earlier article by LUCIANO BARACCO | |
| and revised by the editorial staff) | 654 |
| Economy PHILLIP WEARNE | |
| (Revised by the editorial staff) | 660 |
| Statistical Survey | 665 |
| Directory | 669 |
| Bibliography | 677 |
| **PANAMA** | 678 |
| Geography and Map | 678 |
| History HELEN SCHOOLEY | |
| (Revised by KATHARINE MURISON) | 679 |
| Economy PHILLIP WEARNE | |
| (Revised by the editorial staff) | 683 |
| Statistical Survey | 687 |
| Directory | 692 |
| Bibliography | 700 |
| **PARAGUAY** | 701 |
| Geography and Map | 701 |
| History Prof. PETER CALVERT | 702 |
| Economy Prof. PETER CALVERT | 706 |
| Statistical Survey | 709 |
| Directory | 713 |
| Bibliography | 720 |
| **PERU** | 721 |
| Geography and Map | 721 |
| History SANDY MARKWICK | 722 |
| Economy SANDY MARKWICK | 728 |
| Statistical Survey | 734 |
| Directory | 739 |
| Bibliography | 751 |
| **PUERTO RICO** | 752 |
| Geography and Map | 752 |
| History Prof. PETER CALVERT | |
| (Based on an earlier article by JAMES McDONOUGH) | 753 |
| Economy Prof. PETER CALVERT | |
| (Based on an earlier article by JAMES McDONOUGH) | 756 |
| Statistical Survey | 759 |
| Directory | 762 |
| Bibliography | 769 |
| **SAINT CHRISTOPHER AND NEVIS** | 770 |
| Geography and Map | 770 |
| History MARK WILSON | 770 |
| Economy MARK WILSON | 772 |
| Statistical Survey | 772 |
| Directory | 774 |

# CONTENTS

| | |
|---|---|
| **SAINT LUCIA** | 779 |
| Geography and Map | 779 |
| History  MARK WILSON | 779 |
| Economy  MARK WILSON | 780 |
| Statistical Survey | 781 |
| Directory | 782 |
| **SAINT VINCENT AND THE GRENADINES** | 788 |
| Geography and Map | 788 |
| History  MARK WILSON | 789 |
| Economy  MARK WILSON | 790 |
| Statistical Survey | 791 |
| Directory | 792 |
| **SOUTH GEORGIA AND THE SOUTH SANDWICH ISLANDS** | 797 |
| **SURINAME** | 798 |
| Geography and Map | 798 |
| History  JAMES McDONOUGH (Revised by the editorial staff) | 799 |
| Economy  JAMES McDONOUGH (Revised by the editorial staff) | 803 |
| Statistical Survey | 807 |
| Directory | 808 |
| Bibliography | 814 |
| **TRINIDAD AND TOBAGO** | 815 |
| Geography and Map | 815 |
| History  MARK WILSON (Based on an earlier article by ROD PRINCE) | 816 |
| Economy  MARK WILSON | 820 |
| Statistical Survey | 824 |
| Directory | 829 |
| Bibliography | 837 |
| **THE TURKS AND CAICOS ISLANDS** | 838 |
| Geography and Map | 838 |
| History | 839 |
| Economy | 839 |
| Statistical Survey | 840 |
| Directory | 841 |
| **THE UNITED STATES VIRGIN ISLANDS** | 845 |
| Geography and Map | 845 |
| History | 846 |
| Economy | 846 |
| Statistical Survey | 847 |
| Directory | 848 |
| **URUGUAY** | 851 |
| Geography and Map | 851 |
| History  HELEN SCHOOLEY (Revised by KATHARINE MURISON) | 852 |
| Economy  HELEN SCHOOLEY (Revised by the editorial staff) | 856 |
| Statistical Survey | 860 |
| Directory | 864 |
| Bibliography | 873 |
| **VENEZUELA** | 874 |
| Geography and Map | 874 |
| History  Dr JULIA BUXTON | 875 |
| Economy  Dr JULIA BUXTON | 881 |
| Statistical Survey | 889 |
| Directory | 895 |
| Bibliography | 909 |

## PART THREE
## Regional Information

| | |
|---|---|
| **REGIONAL ORGANIZATIONS** | 913 |
| The United Nations in South America, Central America and the Caribbean | 913 |
| Permanent Missions to the UN | 913 |
| United Nations Information Centres/Services | 914 |
| Economic Commission for Latin America and the Caribbean—ECLAC | 915 |
| United Nations Development Programme—UNDP | 916 |
| United Nations Environment Programme—UNEP | 921 |
| United Nations High Commissioner for Refugees—UNHCR | 925 |
| United Nations Peace-keeping | 927 |
| World Food Programme—WFP | 928 |
| Food and Agriculture Organization of the United Nations—FAO | 929 |
| International Bank for Reconstruction and Development—IBRD (World Bank) | 933 |
| International Development Association—IDA | 936 |
| International Finance Corporation—IFC | 937 |
| Multilateral Investment Guarantee Agency—MIGA | 939 |
| International Fund for Agricultural Development—IFAD | 940 |
| International Monetary Fund—IMF | 941 |
| United Nations Educational, Scientific and Cultural Organization—UNESCO | 945 |
| World Health Organization—WHO | 948 |
| Other UN Organizations active in the Region | 954 |
| Andean Community of Nations | 956 |
| Caribbean Community and Common Market—CARICOM | 959 |
| Central American Integration System | 964 |
| The Commonwealth | 967 |
| European Union—Regional Relations | 974 |
| Inter-American Development Bank—IDB | 977 |
| Latin American Integration Association—LAIA | 980 |
| North American Free Trade Agreement—NAFTA | 981 |
| Organization of American States—OAS | 982 |
| Southern Common Market—MERCOSUR/MERCOSUL | 987 |
| Other Regional Organizations | 990 |
| **MAJOR COMMODITIES OF LATIN AMERICA** | 998 |
| **RESEARCH INSTITUTES** | 1038 |
| **SELECT BIBLIOGRAPHY (BOOKS)** | 1048 |
| **SELECT BIBLIOGRAPHY (PERIODICALS)** | 1054 |
| **INDEX OF REGIONAL ORGANIZATIONS** | 1061 |

# ABBREVIATIONS

| | | | |
|---|---|---|---|
| Abog. | Abogado | Commdt | Commandant |
| AC | Acre | Commr | Commissioner |
| Acad. | Academician; Academy | Confed. | Confederation |
| Adm. | Admiral | Cont. | Contador |
| admin. | administration | Corpn | Corporation |
| AG | Aktiengesellschaft (Joint Stock Company) | CP | Case Postale; Caixa Postal (Post Box) |
| Ags | Aguascalientes | Cres. | Crescent |
| a.i. | ad interim | CSTAL | Confederación Sindical de los Trabajadores de América Latina |
| AID | (US) Agency for International Development | | |
| AIDS | acquired immunodeficiency syndrome | CT | Connecticut |
| AK | Alaska | CTCA | Confederación de Trabajadores Centro-americanos |
| AL | Alabama, Alagoas | | |
| ALADI | Asociación Latino-Americana de Integración | Cttee | Committee |
| Alt. | Alternate | cu | cubic |
| AM | Amazonas; Amplitude Modulation | cwt | hundredweight |
| amalg. | amalgamated | | |
| AP | Amapá | DC | District of Columbia, Distrito Central |
| Apdo | Apartado (Post Box) | DE | Delaware, Departamento Estatal |
| approx. | approximately | Dec. | December |
| Apt | Apartment | Del. | Delegación |
| Apto | Apartamento | Dem. | Democratic; Democrat |
| AR | Arkansas | Dep. | Deputy |
| asscn | association | dep. | deposits |
| assoc. | associate | Dept | Department |
| asst | assistant | devt | development |
| Aug. | August | DF | Distrito Federal |
| auth. | authorized | Dgo | Durango |
| Ave | Avenue | Diag. | Diagonal |
| Av., Avda | Avenida (Avenue) | Dir | Director |
| AZ | Arizona | Div. | Division |
| | | DN | Distrito Nacional |
| BA | Bahia | Dr(a) | Doctor(a) |
| BCN | Baja California Norte | Dr. | Drive |
| BCS | Baja California Sur | dpto | departamento |
| Bd | Board | dwt | dead weight tons |
| Blvd, Blvr | Boulevard | | |
| b/d | barrels per day | E | East, Eastern |
| Bldg | Building | EC | Eastern Caribbean; European Community |
| BP | Boîte Postale (Post Box) | ECCB | Eastern Caribbean Central Bank |
| br.(s) | branch(es) | ECLAC | (United Nations) Economic Commission for Latin America and the Caribbean |
| Brig. | Brigadier | | |
| BSE | bovine spongiform encephalopathy | Econ. | Economist |
| BTN | Brussels Tariff Nomenclature | ECOSOC | (United Nations) Economic and Social Council |
| | | ECU | European Currency Unit |
| C | Centigrade | Ed.(s) | Editor(s) |
| c. | *circa*; cuadra(s) (block(s) ) | Edif. | Edificio (building) |
| CA | California; Compañía Anónima | edn | edition |
| CACM | Central American Common Market | EEC | European Economic Community |
| CAFTA | Central American Free Trade Agreement | EFTA | European Free Trade Association |
| Camp. | Campeche | e.g. | exempli gratia (for example) |
| cap. | capital | eKv | electron kilovolt |
| Capt. | Captain | eMv | electron megavolt |
| CARICOM | Caribbean Community and Common Market | Eng. | Engineer; Engineering |
| CCL | Caribbean Congress of Labour | Ens. | Ensanche (suburb) |
| Cdre | Commodore | ES | Espírito Santo |
| CE | Ceará | Esc. | Escuela; Escudos; Escritorio |
| Cen. | Central | esq. | esquina (corner) |
| CEO | Chief Executive Officer | est. | established; estimate, estimated |
| cf. | confer (compare) | etc. | et cetera |
| Chair. | Chairman | eV | eingetragener Verein |
| Chih. | Chihuahua | EU | European Union |
| Chis | Chiapas | excl. | excluding |
| Cia, Cía | Companhia, Compañía | exec. | executive |
| Cie | Compagnie | Ext. | Extension |
| c.i.f. | cost, insurance and freight | | |
| C-in-C | Commander-in-Chief | F | Fahrenheit |
| circ. | circulation | f. | founded |
| CIS | Commonwealth of Independent States | FAO | Food and Agriculture Organization |
| cm | centimetre(s) | Feb. | February |
| CMEA | Council for Mutual Economic Assistance | Fed. | Federation; Federal |
| Cnr | Corner | FL | Florida |
| CO | Colorado | FM | frequency modulation |
| Co | Company | Fri. | Friday |
| Coah. | Coahuila | fmrly | formerly |
| Col | Colonel | f.o.b. | free on board |
| Col. | Colima, Colonia | Fr | Father |
| Comm. | Commission | Fr. | Franc |
| Commdr | Commander | ft | foot (feet) |

xiii

# ABBREVIATIONS

| | | | |
|---|---|---|---|
| g | gram(s) | Maj. | Major |
| GA | Georgia | Man. | Manager; managing |
| GATT | General Agreement on Tariffs and Trade | MD | Maryland |
| GDP | gross domestic product | MDG | Millennium Development Goal |
| Gen. | General | ME | Maine |
| GM | genetically modified | mem. | member |
| GmbH | Gesellschaft mit beschränkter Haftung (Limited Liability Company) | MEV | mega electron volt |
| | | Méx. | México |
| GMT | Greenwich Mean Time | mfrs | manufacturers |
| GNP | gross national product | MG | Minas Gerais |
| GO | Goiás | Mgr | Monseigneur; Monsignor |
| Gov. | Governor | MHz | megahertz |
| Govt | Government | MI | Michigan |
| Gro | Guerrero | Mich. | Michoacán |
| grt | gross registered tons | Mlle | Mademoiselle |
| GSP | Global Social Product | mm | millimetre(s) |
| Gto | Guanajuato | Mme | Madame |
| gWh | gigawatt hours | MN | Minnesota |
| | | MO | Missouri |
| ha | hectares | Mon. | Monday |
| HE | His (or Her) Eminence; His (or Her) Excellency | Mor. | Morelos |
| | | MP | Member of Parliament |
| hg | hectogram(s) | MS | Mato Grosso do Sul; Mississippi |
| Hgo | Hidalgo | MSS | Manuscripts |
| HGV | Heavy goods vehicle | MT | Montana |
| HI | Hawaii | MW | megawatt(s); medium wave |
| HIPC | heavily indebted poor country | MWh | megawatt hour(s) |
| HIV | human immunodeficiency virus | | |
| hl | hectolitre(s) | N | North, Northern |
| HM | His (or Her) Majesty | n.a. | not available |
| Hon. | Honorary (or Honourable) | NAFTA | North American Free Trade Agreement |
| HQ | Headquarters | Nat. | National |
| HRH | His (or Her) Royal Highness | NATO | North Atlantic Treaty Organisation |
| | | Nay. | Nayarit |
| IA | Iowa | NC | North Carolina |
| ibid. | ibidem (in the same place) | NCO | Non-Commissioned Officer |
| IBRD | International Bank for Reconstruction and Development (World Bank) | ND | North Dakota |
| | | NE | Nebraska |
| ICC | International Chamber of Commerce | NGO | Non-governmental organization |
| ICFTU | International Confederation of Free Trade Unions | NH | New Hampshire |
| | | NJ | New Jersey |
| ID | Idaho | NL | Nuevo León |
| IDA | International Development Association | NM | New Mexico |
| IDB | Inter-American Development Bank | NMP | net material product |
| i.e. | id est (that is to say) | No(.) | number, número |
| IGAD | Intergovernmental Authority on Development | Nov. | November |
| IL | Illinois | nr | near |
| ILO | International Labour Organisation | nrt | net registered tons |
| IMF | International Monetary Fund | NV | Naamloze Vennootschap (Limited Company); Nevada |
| IN | Indiana | | |
| in (ins) | inch (inches) | NY | New York |
| Inc | Incorporated | | |
| incl. | including | OAS | Organization of American States |
| Ind. | Independent | Oax. | Oaxaca |
| Ing. | Engineer | Oct. | October |
| Insp. | Inspector | OECD | Organisation for Economic Co-operation and Development |
| Inst. | Institute, Instituto | | |
| Int. | International | OECS | Organisation of Eastern Caribbean States |
| IRF | International Road Federation | Of. | Oficina |
| irreg. | irregular | OH | Ohio |
| Is | Islands | OK | Oklahoma |
| ISIC | International Standard Industrial Classification | OPEC | Organization of the Petroleum Exporting Countries |
| Jal. | Jalisco | op. cit. | opere citato (in the work quoted) |
| Jan. | January | opp. | opposite |
| Jr | Junior | OR | Oregon |
| Jt | Joint | Org. | Organization |
| | | ORIT | Organización Regional Interamericana de Trabajadores |
| kg | kilogram(s) | | |
| kHz | Kilohertz | oz | troy ounces |
| km | kilometre(s) | | |
| KS | Kansas | p. | page |
| kW | kilowatt(s) | PA | Pará, Pennsylvania |
| kWh | kilowatt hours | p.a. | per annum |
| KY | Kentucky | Parl. | Parliament(ary) |
| | | PB | Paraíbo |
| LA | Louisiana | PC | Privy Counsellor |
| lb | pound(s) | PE | Pernambuco |
| LIBOR | London Inter-Bank Offered Rate | Perm. Rep. | Permanent Representative |
| Lic. | Licenciado | PI | Pianí |
| Licda | Licenciada | pl. | place |
| LNG | liquefied natural gas | PLC | Public Limited Company |
| LPG | liquefied petroleum gas | PMB | Private Mail Bag |
| Lt, Lieut | Lieutenant | POB | Post Office Box |
| Ltd, Ltda | Limited, Limitada | pp. | pages |
| m | metre(s) | PR | Paraná |
| m. | million | PREF | Poverty Reduction and Growth Facility |
| MA | Maranhão; Massachusetts | | |

# ABBREVIATIONS

| | |
|---|---|
| Pres. | President |
| Prin. | Principal |
| Prof. | Professor |
| Propr | Proprietor |
| Prov. | Province; Provincial |
| Pte | Private |
| Pty | Proprietary |
| p.u. | paid up |
| publ. | publication; published |
| Publr(s) | Publisher(s) |
| Pue. | Puebla |
| Pvt. | Private |
| | |
| Q. Roo | Quintaya Roo |
| QC | Queen's Counsel, Québec |
| q.v. | quod vide (to which refer) |
| Qro | Querétaro |
| | |
| Rd | Road |
| reg., regd | register; registered |
| reorg. | reorganized |
| Rep. | Republic; Republican; Representative |
| Repub. | Republic |
| res | reserve(s) |
| retd | retired |
| Rev. | Reverend |
| RI | Rhode Island |
| RJ | Rio de Janeiro |
| Rm | Room |
| RN | Rio Grande do Norte |
| RO | Rondônia |
| RR | Roraima |
| RS | Rio Grande do Sul |
| Rt | Right |
| | |
| S | South; Southern; San |
| SA | Société Anonyme, Sociedad Anónima (limited company) |
| SARL | Sociedade Anônima de Responsabilidade Limitada (Joint Stock Company of Limited Liability) |
| Sat. | Saturday |
| SC | Santa Catarina, South Carolina |
| SD | South Dakota |
| SDR(s) | Special Drawing Right(s) |
| Sec. | Secretary |
| Sen. | Senior; Senator |
| Sept. | September |
| Sgt | Sergeant |
| Sin. | Sinaloa |
| SITC | Standard International Trade Classification |
| SJ | Society of Jesus |
| SLP | San Luis Potosí |
| s/n | sin número (no number) |
| Soc. | Society |
| Son. | Sonora |
| SP | São Paulo |
| Sq. | Square |
| sq | square (in measurements) |
| Sr | Senior; Señor |
| Sra | Señora |
| St(s) | Saint(s); Street(s) |
| Sta | Santa |
| Ste | Sainte |
| subs. | subscriptions; subscribed |
| | |
| Suc. | Sucursal |
| Sun. | Sunday |
| Supt | Superintendent |
| | |
| Tab. | Tabasco |
| Tamps | Tamaulipas |
| Tce | Terrace |
| tech., techn. | technical |
| tel. | telephone |
| Thurs. | Thursday |
| TJ | tetrajoule |
| Tlax. | Tlaxcala |
| TN | Tennessee |
| TO | Tocatins |
| Treas. | Treasurer |
| Tue. | Tuesday |
| TV | television |
| TX | Texas |
| | |
| u/a | unit of account |
| UEE | Unidade Ecónomica Estatal |
| UK | United Kingdom |
| UN | United Nations |
| UNCED | United Nations Conference on Environment and Development |
| UNCTAD | United Nations Conference on Trade and Development |
| UNDP | United Nations Development Programme |
| UNESCO | United Nations Educational, Scientific and Cultural Organization |
| UNHCHR | United Nations High Commissioner for Human Rights |
| UNHCR | United Nations High Commissioner for Refugees |
| Univ. | University |
| USA (US) | United States of America (United States) |
| USAID | United States Agency for International Development |
| USSR | Union of Soviet Socialist Republics |
| Urb. | Urbanización (urban district) |
| UT | Utah |
| | |
| VA | Virginia |
| VAT | Value-Added Tax |
| v-CJD | new variant Creutzfeldt-Jakob disease |
| Ver. | Veracruz |
| VHF | Very High Frequency |
| VI | (US) Virgin Islands |
| viz. | videlicet (namely) |
| vol.(s) | volume(s) |
| | |
| W | West; Western |
| WA | Washington |
| WCL | World Confederation of Labour |
| Wed. | Wednesday |
| WFTU | World Federation of Trade Unions |
| WHO | World Health Organization |
| WI | Wisconsin |
| WTO | World Trade Organization |
| WV | West Virginia |
| WY | Wyoming |
| | |
| yr | year |
| Yuc. | Yucatán |

# INTERNATIONAL TELEPHONE CODES

To make international calls to telephone and fax numbers listed in *South America, Central America and the Caribbean*, dial the international code of the country from which you are calling, followed by the appropriate country code for the organization you wish to call (listed below), followed by the area code (if applicable) and telephone or fax number listed in the entry.

| | Country code | + GMT* |
|---|---|---|
| Anguilla | 1264 | –4 |
| Antigua and Barbuda | 1268 | –4 |
| Argentina | 54 | –3 |
| Aruba | 297 | –4 |
| Bahamas | 1242 | –5 |
| Barbados | 1246 | –4 |
| Belize | 501 | –6 |
| Bermuda | 1441 | –4 |
| Bolivia | 591 | –4 |
| Brazil | 55 | –3 to –4 |
| British Virgin Islands | 1284 | –4 |
| Cayman Islands | 1345 | –4 |
| Chile | 56 | –4 |
| Colombia | 57 | –4 |
| Costa Rica | 506 | –6 |
| Cuba | 53 | –4 |
| Dominica | 1767 | –4 |
| Dominican Republic | 1809 | –4 |
| Ecuador | 593 | –4 |
| El Salvador | 503 | –6 |
| Falkland Islands | 500 | –4 |
| French Guiana | 594 | –3 |
| Grenada | 1473 | –4 |
| Guadeloupe | 590 | –4 |
| Guatemala | 502 | –6 |
| Guyana | 592 | –4 |
| Haiti | 509 | –4 |
| Honduras | 504 | –6 |
| Jamaica | 1876 | –4 |
| Martinique | 596 | –4 |
| Mexico | 52 | –6 to –7 |
| Montserrat | 1664 | –4 |
| Netherlands Antilles | 599 | –4 |
| Nicaragua | 505 | –6 |
| Panama | 507 | –4 |
| Paraguay | 595 | –4 |
| Peru | 51 | –4 |
| Puerto Rico | 1787 | –4 |
| Saint Christopher and Nevis | 1869 | –4 |
| Saint Lucia | 1758 | –4 |
| Saint Vincent and the Grenadines | 1784 | –4 |
| Suriname | 597 | –3 |
| Trinidad and Tobago | 1868 | –4 |
| Turks and Caicos Islands | 1649 | –4 |
| US Virgin Islands | 1340 | –4 |
| Uruguay | 598 | –3 |
| Venezuela | 58 | –4 |

*Time difference in hours – Greenwhich Mean Time (GMT). The times listed compare the standard (winter) times. Some countries adopt Summer (Daylight Saving) Times—i.e. + 1 hour for part of the year.

South America

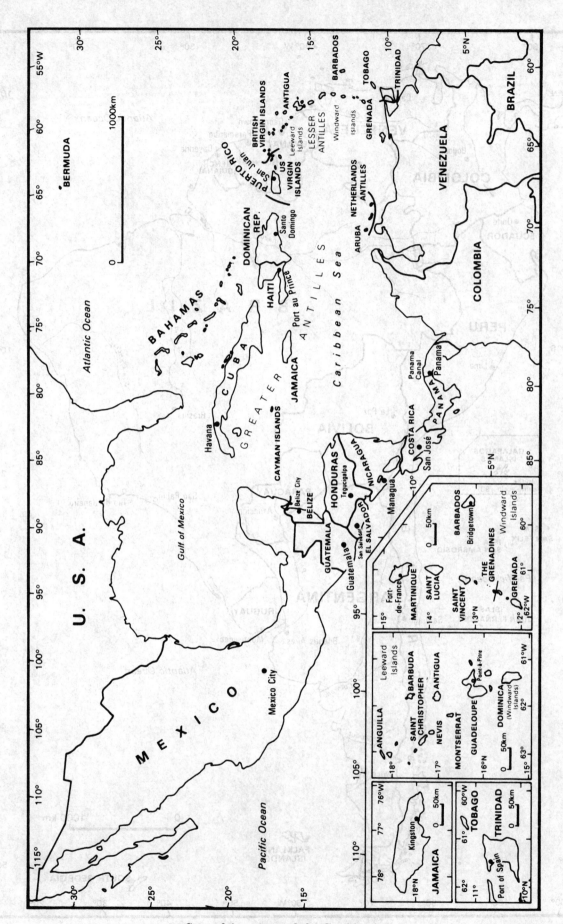

Central America and the Caribbean

# PART ONE
# General Survey

# LATIN AMERICA'S TURN TO THE LEFT

## Dr FRANCISCO PANIZZA

**INTRODUCTION**

After being ruled by centre-right governments in the 1990s, Latin America appears to be turning to the left in the 21st century. Left-wing and left-of-centre governments are in office or have been in office in Argentina, Bolivia, Brazil, Chile, Ecuador, Uruguay and Venezuela, and in Mexico the leftist former mayor of Mexico City, Andrés Manuel López Obrador, is the leading potential contender in the 2006 presidential election. What accounts for this political shift? In part, the political advances of the left are a reaction to the failures of the twin processes of democratization and free-market reforms that dominated the region in the last two decades of the 20th century. When they took off in the 1980s, democratization and economic reform brought with them high hopes for the people of Latin America. In the words of President Raúl Alfonsín of Argentina (1983–89) during the heyday of transition to democracy, 'with democracy the people lead better lives'.

By the early 1990s Latin America appeared to be on the verge of leaving behind its history of political and economic turbulence. A mix of democracy, economic liberalization and regional integration gained widespread consensus among the region's political élites. Moreover, the formula had considerable popular support, as shown by the electoral victories of economic reformers such as Carlos Menem in Argentina in 1989 and 1995, Ernesto Zedillo Ponce de León in Mexico in 1994, Fernando Henrique Cardoso in Brazil in 1994 and 1998 and Alberto Fujimori in Peru in 1995. The mid-1990s was also a period in which relations between the USA and Latin American countries were at one of their warmest levels in the region's troubled history. US President Bill Clinton (1993–2001) offered the prospect of a US leader in tune with the political and economic changes in the region, including a commitment to democracy and human rights unimpeded by cold war politics, and market-led economic growth no longer limited by outdated economic protectionism. A manifestation of the rapprochement between Latin American nations and their powerful northern neighbour was the inclusion in the Declaration of the First Summit of the Americas in Miami, Florida, in 1994 of a commitment to conclude negotiations 'no later than 2005' on a Free Trade Area of the Americas (FTAA), in which barriers to trade and investment were to be eliminated progressively.

Just over a decade later, optimism about the benefits of democracy and free-market reforms has given way to disenchantment and disappointment. Democracy has survived in the region and arguably it has been furthered by the election of President Vicente Fox Quesada in Mexico in 2000. However, political instability has overshadowed the survival of democracy. Between 1995 and 2005 seven South American presidents have been forced to resign or been impeached, while President Lt-Gen. (retd) Hugo Chávez Frías of Venezuela (1999–) was deposed briefly in a military coup in April 2002 before being reinstated to office. Significantly, in Argentina in 2001, in Ecuador in 2000 and 2005 and in Bolivia in 2003 and 2005, democratically elected presidents were forced to resign by popular protests. Meanwhile, market reforms have failed to deliver the sustained growth and job creation that are necessary though not sufficient conditions for the alleviation of poverty. According to figures from the Economic Commission for Latin America and the Caribbean (ECLAC), gross domestic product per head in the region expanded by an annual average of just 0.4% between 1995 and 2003. Urban unemployment rose from 7.3% in 1990 to an average of 10.5% in 2002–04. In 2002 there were 221m. poor people in the region (44.0% of the total population), of whom 97m. (19.4%) were living in extreme poverty or indigence.

Surveys indicate that the majority of Latin Americans are disappointed with the way that democracy and markets work in the region. In 2004 a region-wide survey by the polling organization Latinobarómetro found that 65% of interviewees were dissatisfied with the working of democracy in their country. Parliaments and political parties are among the less-trusted institutions region-wide. Of particular significance is the perception that democratic regimes are unwilling or unable to uphold the rule of law. Political factors, particularly corruption, have contributed to popular dissatisfaction with the working of democracy. However, the inability of the region's democratic regimes to live up to President Alfonsín's promise of a better life lies behind popular disenchantment with democracy and market reforms: according to the survey, satisfaction with the working of the market economy declined from 24% in 2002 to 19% in 2004.

Disenchantment with democracy and free-market economics may have created the conditions for the rise of the new wave of left-of-centre governments, but it is also pertinent to analyse how the advances of democracy in the region, as much as its flaws, have contributed to the successes of the left. Despite its many weaknesses, most Latin American countries are experiencing the longest uninterrupted period of democratically elected governments in their history. Dissatisfaction with the work of democracy does not equal the rejection of democracy as a political regime: according to Latinobarómetro, in 2004 71% of Latin American citizens believed that while democracy may have its problems, it was none the less the best system of government. Elections are now the path to office for the left, as revolution has lost its appeal. With the exception of the Fuerzas Armadas Revolucionarias de Colombia—Ejército del Pueblo (FARC—EP) and the Ejército de Liberación Nacional (ELN) in Colombia, the Zapatistas in Mexico and a rump Sendero Luminoso guerrilla movement in Peru, the left has abandoned its Cuban-inspired armed insurgency strategy to become engaged fully in democratic politics.

Democratization has allowed left-wing parties to participate in elections: hitherto many of these parties had been proscribed. As part of the democratic process, left-wing parties won elections at municipal and provincial level, granting them their first taste of holding public office. A number of the region's main cities, such as Buenos Aires, São Paulo, Mexico City, Bogotá and Montevideo, are controlled or have been controlled by parties of the left. President Tabaré Vázquez of Uruguay (2005–) was mayor of Montevideo before winning the presidency, and Andrés Manuel López Obrador, a potential future President of Mexico, was, until mid-2005, mayor of Mexico City. Governing at provincial or municipal level has allowed parties of the left to gain policy-making and administrative experience, test policy initiatives and showcase their competence in office.

Democracy has also helped strengthen civil society, which has benefited the left. Some of the more successful left-wing parties, such as the Partido dos Trabalhadores (PT) in Brazil, the Encuentro Progresista—Frente Amplio in Uruguay and the Movimiento al Socialismo (MAS) in Bolivia, have strong roots in civil society. Social movements were principal actors in the transition to democracy, as exemplified by their role in the 'Diretas já' campaign for direct elections in Brazil in 1984 and in the plebiscites in Uruguay in 1980 and Chile in 1988. The return to electoral politics has resulted in social movements losing their political centrality, while neoliberal economic policies have weakened the trade unions. But overall civil society is today stronger than ever before throughout the region. Of particular importance has been the resurgence of indigenous peoples' social and political movements in countries such as Ecuador, Bolivia and Peru, as well as new social organizations associated with the contemporary anti-globalization lobby.

If the flaws of democracy in the region should not detract from the implications of its resilience for the advances of the left, claims about the failure of neoliberalism should be qualified at least by the very different attitudes adopted by left-of-centre governments towards the market economy. Opposition to neoliberalism has been a unifying factor for left and left-of-centre groups of otherwise different political leanings. However, as will be shown in more detail below, for all their anti-neoliberal rhetoric, most left-of-centre governments have been surprisingly cautious in departing from the policy prescriptions of the so-called 'Washington Consensus'. And while the 2004 Latinobarómetro survey may have found that more than 71% of Latin Americans were unhappy with the working of the market economy in their own countries, 72% believed that a market economy was the only economic system for their countries to become developed.

Unstable yet resilient electoral democracies that fail to uphold the rule of law, and economic reforms that have not delivered sustained economic growth do not necessarily produce left-wing governments, although they may precipitate political change. The resurgence of the left has as much to do with the national political and economic contexts in which left-wing parties operate as with general socio-economic trends in the region. Without entering into a theological debate about the meaning of the left, it is possible to combine politico-ideological and functional criteria to identify three main actors that represent the main faces of the left in contemporary Latin America: radical-populist, grassroots and reformist-left movements.

## THE RADICAL-POPULIST LEFT

Latin America has a long tradition of populism. A contentious concept for which there is no agreed definition, populism has been associated with a number of traits, such as a personalist leadership and an anti-establishment appeal that bypasses political institutions. The anti-status quo element of populism makes it a typical occurrence in politically fragmented societies when the political system fails to address its citizens' grievances. In these circumstances the populist leader appears as a unifying force, able to reconstitute symbolically the unity of the nation by rallying the people against their political and economic enemies. And yet populist politics are not necessarily left-wing politics. In Latin America populism has been associated with both left- and right-wing leaders. The latest wave is linked to popular backlashes against the inequalities and economic insecurities of neoliberalism in a context of weakly institutionalized and exclusionary democracies.

The rise to power of Lt-Col (retd) Chávez in Venezuela is a textbook example of the politics of contemporary left-wing populism. As the imprisoned leader of a junior officers' rebellion against the unpopular administration of President Carlos Andrés Pérez (1988–93), Chávez gained nation-wide popularity by rising against a corrupt political establishment that had colonized the state through clientelism and patronage. In the 1988 election he made full use of his image as a political outsider by campaigning against the 'partidocracia'—the patronage-based party system—that had dominated Venezuelan politics for the previous four decades. Once in power, Chávez followed his anti-establishment rhetoric with the promotion of radical political reforms that eroded the support-base of Venezuela's traditional parties, while creating parallel political institutions to consolidate his position. The most important reform was that of the Constitution, approved by national referendum in December 1999. The new Constitution reduces the role of political parties and provides mechanisms for direct political participation, such as recall elections and referendums.

President Chávez created his own political party, the Movimiento V República, and developed links with civil society organizations. He has sought to mobilize the lower sectors of society against opponents of his 'Bolivarian Revolution', which include, among others, business, most of the middle classes, the financial sector, the media, the hierarchy of the Catholic Church and the US Government. Political turmoil in a context of social and political polarization has resulted in highly charged street demonstrations by pro- and anti-Chávez followers. An unsuccessful coup attempt in April 2002 and a general strike that lasted 10 weeks in 2002–03 evidenced the breakdown of traditional forms of political mediation. However, as even the US Government has been forced to admit, for all his political radicalism, Chávez has remained within the boundaries of democratic politics. Street mobilizations and the deinstitutionalization of Venezuela's traditional party system have been paralleled by the recourse to elections as the ultimate source of political legitimacy: between 1997 and 2004 eight nation-wide electoral contests were held, including a recall referendum, won by Chávez in August 2004.

Economic change has only partially matched political change. President Chávez has been a strong critic of neoliberalism and even of capitalism. He has used his Government's petroleum wealth to finance a number of social programmes. More recently he has stated that his ultimate goal is the construction of a socialist society of a yet undefined nature. The Venezuelan state has become more interventionist and protectionist. In late 2001 the Government enacted a package of 49 economic reform laws. Legislation included agrarian reform, the state control of oil ventures, state support for workers' co-operatives and the re-nationalization of the social security system. Other heterodox measures have since followed, including raising levies on petroleum companies and imposing capital and exchange controls. In February 2005 the Government announced that it would review all contracts signed with foreign companies in the country's non-oil sector to ensure that firms 'act in line with the national interest'. Even so, on the whole economic reforms to date have been moderate, more akin to traditional developmentalism than to state socialism. The new Constitution guarantees the right to private property and assigns the state the role of promoting private initiative. Venezuela has honoured its external debt and the authorities have stated that current agreements with foreign companies would be honoured, the revision of contracts notwithstanding. Even the agrarian reform allows for land expropriation only under highly restricted conditions. Whether economic policy will take a more radical turn, as some recent rhetoric from Chávez appears to suggest, remains to be seen, as 'Chavismo' is divided increasingly between moderates and radicals, with the President playing a mediating role.

While President Chávez has been the outrider of radical-left populism in Latin America, the Government of President Néstor Carlos Kirchner (2003–) in Argentina can also be included within the populist tradition. Similarly to Chávez, Kirchner came to office in a country experiencing political and economic turmoil. Following the collapse of the Argentine economy in late 2001 that forced the resignation of President Fernando de la Rúa, the political élite became utterly discredited, as politicians were regarded as responsible for the hardship that affected not just the urban poor but also significant sectors of the middle classes. Social mobilization by groups such as pensioners and the piqueteros (a disparate group of demonstrators) created an environment of political unrest that was compounded by high levels of common crime. 'Que se vayan todos' ('All politicians out') was the slogan that expressed the mood of the Argentine people at the time. And yet, unlike in the past, democratic institutions held in Argentina. The April 2003 election became a contest between two strands of Justicialismo, the movement of 1950s populist leader Juan Domingo Perón: the free-market, pro-US policies represented by former President Carlos Menem, and the more nationalist one embodied by Kirchner.

Elected with just 22% of the vote and without the full control of the Peronist (Justicialista) party, once in office Kirchner sought to reassert his authority over both his party and the country. To this end, he has retrieved elements of Peronism's national-popular tradition and sought to broaden his political support-base by forging alliances with left-of-centre forces and social movements sympathetic to his policies, such as the human rights movement and sectors of the piqueteros. He has sought to present himself as a defender of the Argentine people against powerful domestic and foreign interests. He has clashed with the IMF, with the owners of the privatized public utilities, with the country's external debt holders and even with the Roman Catholic Church. While he has been an uncompromising critic of the policies of his Peronist predecessor, Menem, Kirchner has not been averse to concentrating power in the Executive. He has maintained the anti-institutional bias that characterized the Menem administration by showing little regard for checks and balances. In the first year of his mandate he issued 67 Decrees of Necessity and Urgency—laws issued by the Executive without parliamentary approval—compared with 64 in the same period of the first Menem administration. He has purged the Supreme Court of Menem's cronies but nominated politically friendly substitutes.

Economically, the administration of President Kirchner has been characterized by nationalist rhetoric and an emphasis on social justice. He has appealed to the IMF to acknowledge its responsibility for the misguided economic policies that led to 2001–02 crisis. When early in 2005 the petroleum company Shell raised petrol prices he called for a boycott of the company, accusing it of acting against the interests of the Argentine people. He adopted an uncompromising position in the negotiation of the country's external debt, resulting in a reduction in capital payments of around 70%. He has resisted demands from the privatized utilities for tariffs adjustments to compensate for the devaluation of the peso in 2002. And he has shown little concern for microeconomic issues and stable rules of the game necessary for attracting foreign investment.

However, there are clear limits to President Kirchner's populism. Politically, Kirchner has much less control over the country's political institutions than President Chávez in Venezuela. He has been unable to either bypass or reshape the country's political institutions like Chávez's 'Bolivarian Revolution'. Argentina remains today a more institutionalized, more pluralist and less politically polarized society than Venezuela. Kirchner's nationalist, anti-neoliberal economic rhetoric has been accompanied by a highly orthodox fiscal policy that has delivered the largest primary budget surpluses of the past two decades. He has paralleled his anti-IMF tirades with repaying the country's obligations to the Fund, and has stated his disposition to enter negotiations with the Fund with a view to signing a new agreement later in 2005 or 2006. Although the privatized utilities have suffered significant losses because of the Government's refusal to adjust their tariffs, there are no plans for their renationalization. Wages have been kept low and the US dollar high, allowing business to reconstitute profit margins and exports to benefit from high commodity prices (although profits on petroleum and agricultural exports have been offset somewhat by taxation).

## THE GRASSROOTS LEFT

Political developments in Bolivia and Ecuador share some of the features associated with the rise of radical populism elsewhere in Latin America, but there are significant differences. Among the common elements are the loss of legitimacy of political systems and popular discontent with neoliberal reforms, leading to backlashes against political institutions. Both countries have high levels of socio-economic inequality compounded by the social, political and economic exclusion of their indigenous populations, while nationalism has played a significant role in framing popular protests against the economic order. They also have a history of populist governments, both from the left and the right. The rise to office of Col Lucio Gutiérrez Borbua (2003–05) in Ecuador has a strong resemblance to that of President Chávez in Venezuela. Both are former military officers jailed for short periods of time for rising against democratically elected governments, and subsequently they won elections with the support of radical political parties.

However, there are significant differences in the political situations of Bolivia and Ecuador compared to those of Venezuela and Argentina. In Bolivia and Ecuador the process of deinstitutionalization of democracy has gone much further. While in Venezuela and Argentina Presidents Chávez and Kirchner have emerged as strong political leaders able to command considerable popular support and at least partially bridge the gap between the state and the citizens, no such figures exist in the two Andean republics. Unlike Chávez, once in office Gutiérrez reneged on his left-wing populist electoral platform and fostered strong ties with the IMF, thereby alienating him from his left-wing supporters without ingratiating himself with the country's powerful conservative forces; this double failure ultimately cost him the presidency.

In Bolivia and Ecuador the political scenario is today one of weak states and political fragmentation in which the political game is increasingly being played outside formal institutions. Popular grassroots organizations played a crucial role in the street protests that led to the fall of democratically elected presidents in both countries and continue to be of central importance in the highly unstable scenarios that followed. Indigenous peoples' organizations have been at the forefront of these movements, although protests have also incorporated significant sectors of the urban population.

Bolivia has a history of political instability and a long tradition of social protest, frequently operating outside the political system. This pattern began to change in the mid-1980s and 1990s, a period in which Bolivia enjoyed the longest period of political stability in the country's history. A political system based on inter-party pacts and coalitions ensured political stability, while a programme of economic reforms that ended chronic hyperinflation and led to the closure of loss-making state-owned mines brought economic stability. A by-product of the closure of inefficient mines was the weakening of trades unions in the industrial sector, particularly the powerful miners' federation, the Federación Sindical de Trabajadores Mineros de Bolivia. Nevertheless, civil society organizations remained strong in the countryside, particularly among the sindicatos (peasant organizations) and the ayllu (rural community organizations). There was also a considerable rise in ethnic awareness and organization among indigenous groups that constitute the majority of the country's population. Of particular significance was resistance to the drive to eradicate coca plantations in the Chapare (Tropic of Cochabamba) region, intensified under 'Plan Dignidad' during the Governments of Hugo Bánzer Suárez and Jorge Quiroga Ramírez (1997–2002). Resistance to the military-led attempt to eradicate the coca crop resulted in violent clashes with peasants and accusations of human rights violations perpetrated by the military. The leader of the coca growers, Evo Morales, became the presidential candidate of a new party, the MAS, established by the sindicatos, which became the second largest party in the 2002 general election.

During 2000–03 a wave of popular protests spread from the traditional peasant heartlands of the Tropic of Cochabamba to the sprawling urban dwellings of El Alto, near La Paz, including ever-growing numbers of popular organizations. Protests were triggered by different grievances, such as the rise in the price of water by a foreign-owned utility in the so-called 'water war' in Cochabamba, land conflicts in Santa Cruz, the campaign to establish a minimum living pension and opposition to an increase in income tax. Although they were mostly peaceful, the protests caused widespread disruption, in particular the tactic of blocking highways and marching on the capital. Social protests culminated in the so-called 'gas war' in October 2003. In its origins the protest was against the export of liquefied natural gas to the USA and Mexico through a Chilean port, but it soon turned into a generalized complaint against the Government of President Gonzalo Sánchez de Lozada. After a bloody attempt at repressing the protesters at the cost of some 80 dead, the President was forced to resign on 17 October and flee the country. His Vice-President, Carlos Mesa Gisbert, assumed the presidency, but he in turn was forced to resign in June 2005 after continued protests and demonstrations. Eduardo Rodríguez Veltzé, hitherto President of the Supreme Court, succeeded him in an acting capacity.

A crisis of governability also affected Ecuador in 2005, following the resignation of President Gutiérrez in April. Gutiérrez was forced to resign after a series of popular protests, triggered by his replacement of Supreme Court judges and the new Court's subsequent decision to allow former President Abdalá Bucaram Ortiz (1996–97) to return to the country. As in Bolivia, the protests that led to Gutiérrez's resignation were part of a longer wave of popular mobilizations in which resurgent indigenous organizations and the urban poor have played, and continued to play, a leading role. The new President, former Vice-President Alfredo Palacio González, presides over a weak Government trying to perform an impossible balancing act between the demands of a highly mobilized popular sector, local élites and foreign interests.

## THE REFORMIST LEFT

Unlike radical populism, in which power is personalized and concentrated at the top, and grassroots-left movements, in which power is fragmented and deinstitutionalized, reformist-left governments operate in institutionalized political systems in which power is exercised through the checks and balances characteristic of liberal democracies. Three left-of-centre governments fall within this category: the Concertación por la

Democracia coalition in Chile, headed by Socialist President Ricardo Lagos (2000–); the EP–FA Government of President Tabaré Ramón Vázquez Rosas (2005–) in Uruguay; and the PT-led administration of President Luiz Inácio 'Lula' da Silva in Brazil (2003–). Chile, Uruguay and Brazil are among the most consolidated democracies in South America. They also have comparatively strong states, with a tradition of rule of law (particularly Chile and Uruguay) and powerful legislatures. Party systems are well-established in Chile and Uruguay, and while this is less so the case in Brazil, the party system has become consolidated progressively in the second half of the 1990s and the early 2000s, with the PT and the centrist Partido da Social Democracia Brasileira emerging as the country's main political forces.

The three countries' main left-wing parties, the Partido Socialista (PS) in Chile, the EP–FA in Uruguay and the PT in Brazil, commenced on the radical left and moved progressively to the centre left at different times in the course of their history. The shift was the result of political learning, institutional constraints and electoral strategies. Crucial in the case of Chile were the lessons drawn by the PS leadership from the failure of the Government of President Salvador Allende (1970–73), in which political polarization and economic mismanagement were perceived as having contributed to the breakdown of the country's democratic order. As a result, in the 1980s the PS abandoned its historical links with the Communists and forged alliances with the Christian Democrats and other centrist parties. The alliance originally came together in the plebiscite that defeated the Government of General Augusto Pinochet in 1988 and it has since dominated Chilean politics, held together by the ideological convergence of the allied parties, political success and an electoral system devised by the outgoing dictatorship that favoured coalition politics.

In Brazil three electoral defeats convinced the PT leader Lula da Silva of the need to broaden the party's appeal by moderating its electoral platform and presenting himself as a responsible candidate. The PT's shift to the centre became evident in the 'letter to the Brazilian people', published in the early stages of the 2002 electoral campaign, in which da Silva stated his commitment to fiscal responsibility and honouring the country's external obligations. A similar process occurred in Uruguay's 2004 electoral campaign, in which the EP–FA candidate, Tabaré Vázquez, won the presidency at his third attempt. At the beginning of the campaign Vázquez made a well-publicized trip to Washington, DC, where he met representatives of the IMF and the World Bank to reassure international investors; he also announced the name of a moderate centre-left economist, Danilo Astori, as his future Minister of the Economy. Moderation was required not only to appeal to moderate voters but also to build electoral alliances with centre parties, which became governing coalitions after the elections. In Brazil the PT won less than 23% of the seats in Congress. To secure a congressional majority the Government formed a heterogeneous coalition of left-wing and centre parties, precariously held together more by traditional patronage than by policy agreements. In Uruguay the EP–FA was itself a coalition of no fewer than 21 groups ranging from the radical left to the centre, of which eight were represented in the legislature.

If the political moderation of reformist-left governments can be accounted for largely by the politico-institutional contexts in which they operate, the conditions of the three countries' economies largely explain these governments' tendency towards economic moderation. In contrast to the economic dislocations that characterized the countries in which radical-populist and grass-roots forces are in the political ascendancy, Chile, Brazil and Uruguay have relatively stable economies. In Chile the Concertación inherited the most successful Latin American economy of the 1980s. In its 15 years in office the Concertación has continued to run the country on strong free-market economic principles with considerable economic success. In Brazil the Real Plan, enacted in 1994, ended years of hyperinflation. The Government of President da Silva made economic stability a key goal and, after a difficult first year in 2003, the economy grew at a healthy pace in 2004 and 2005. Uruguay was badly affected by the 2001–02 Argentine crisis, forcing a massive devaluation of the currency. However, the economy made a strong recovery in 2004 and the new administration of President Vázquez has sought to maintain the economic momentum by adopting the same macroeconomic policies of the preceding centre-right Government.

While distancing themselves rhetorically from neoliberalism, these centre-left governments have embraced the market economy with varying degrees of ideological conviction and pragmatic calculation. They have encouraged foreign investment and maintained friendly relations with international financial agencies. However, they have attempted to make market economies more sensitive to the needs of the poor by enacting a number of targeted social programmes. In Brazil and Uruguay the state has been more interventionist and protectionist than in Chile, but unlike Argentina and Venezuela their governments have sought to work with the business sector rather than against it. Consensus on the benefits of the market economy is much stronger in Chile than in Brazil or Uruguay, where some left-wing critics have voiced disquiet about the continuation of the preceding Government's economic policy.

## CONCLUSIONS

An overview of the rise of the left in Latin America shows that while there may be some common causes that explain the left's ascendancy, left-wing governments appear to have more differences than commonalities. The advances and shortcomings of democracy in the region are the political ground in which parties and movements of the left have flourished. The degree of consolidation of each country's political and economic institutions appears to be the main variable in the left's attitudes towards liberal democracy and free-market economies. Left-wing forces have benefited from real or perceived political and economic failures of neoliberal economic reforms in the region. Resurgent nationalism has led to the questioning of the benefits of foreign investment in the Andean countries, while demands for a more socially aware and interventionist state have gained ground throughout the region. Nevertheless, there are considerable differences in the extent to which left-wing governments have departed from the principles of free-market economics. Chile has the most free-market economy in the region, while the oil-funded 'Bolivarian Revolution' in Venezuela has not matched its anti-neoliberal, anti-capitalist rhetoric with equally radical policy changes.

Perhaps the rationale for the cautious approach to economic change taken by left-wing governments is that for all their criticisms they have yet to find an alternative model to the 'Washington Consensus' that dominated Latin American economies in the 1990s. In October 2003 Presidents Kirchner and da Silva signed the so-called 'Buenos Aires Consensus'. It advocated fighting hunger and poverty. It proposed strengthening the role of the state and the implementation of development policies tailored to the region's reality. It urged the strengthening of Mercosur (Mercado Común del Sur/Mercado Comum do Sul, the Southern Common Market) and supported the efforts of developing nations to advance their interests in the 'Doha round' of negotiations in the World Trade Organization. But, as shown by the different economic policies of the Argentine and Brazilian Governments, the 'Buenos Aires Consensus' is little more than a collection of loose principles that is far from constituting an alternative set of policies to the limited but specific propositions of the 'Washington Consensus'.

Finally, what are the implications of the rise of the left in Latin America for relations with the USA? After the 11 September 2001 terrorist attacks on the USA, Latin America's relations with the USA have been distinctively chillier than during the Clinton years. The Administration of George W. Bush (2001–) has largely neglected relations with Latin America, which ill fits its anti-terrorist global agenda. The US-led invasion of Iraq was hugely unpopular in the region, as it reminded the Latin American people of a long history of US military interventions in the name of democracy, and of human rights abuses perpetrated by the US-trained military in the name of fighting domestic subversion. Negotiations on the FTAA have stalled. The US Government has been concerned particularly by the increasingly belligerent anti-US rhetoric of President Chávez of Venezuela and the danger of political disintegration in Bolivia and Ecuador. However, paradoxically, the reformist-left governments may offer the USA the best way for securing its paramount interest in the region, namely political stability. The Government of Brazil has thus become of primary importance to the USA. For all the Brazilian Government's south-south political and economic activism, the administration of President da Silva is perceived as a stabilizing force in Latin America. And given Brazil's significance in the region, the future of South America may well depend on the domestic and international success of its left-of-centre Government.

## BIBLIOGRAPHY

Alvaredo, F., and Benedetti, P. 'Left of Centre Governments in Argentina: Deeds vs Facts'. Paper presented to the Workshop Left of Centre Governments in Latin America: Current Practices and Future Prospects. London, Institute for the Study of the Americas, 18 March 2005.

Comisión Económica para América Latina y el Caribe. *Anuario Estadístico de América Latina y el Caribe 2004*. Washington, DC, 2005.

Crabtree, J. *Patterns of Protest. Politics and Social Movements in Bolivia*. London, Latin America Bureau, 2005.

Ellner, S. 'Revolutionary and Non-Revolutionary Paths of Radical Populism: Directions of the Chavista Movement in Venezuela', in *Science & Society*. New York, NY, April 2005.

Panizza, F. E. 'Unarmed Utopia Revisited. The Resurgence of Left of Centre Politics in Latin America', in *Political Studies*. London, forthcoming December 2005.

# DEBT, FINANCE AND THE IMF: THREE DECADES OF DEBT CRISES IN LATIN AMERICA

Dr KEN SHADLEN

## INTRODUCTION*

The financial crisis that engulfed Argentina from late 2001 was the latest in a series of debt crises that has affected Latin America since the early 1980s. In fact, both the 1980s, recalled as the 'lost decade', and the 1990s were marked by recurrent debt and financial crises throughout the region. By the start of the 21st century, Latin American external debt amounted to some US $782,900m., the largest share of any region in the developing world. In addition to its quantity, a remarkable feature of this debt was that more than three-quarters of it was owed to private, commercial creditors. While indebtedness was not limited to Latin American countries, in other regions of the developing world (e.g. Africa) the debt was owed principally to foreign governments or international institutions, such as the IMF or the World Bank. Thus, Latin American debt stands out on account of its longevity (the fact that many countries were still dealing with the effects of debt accumulated decades ago), its size, and the fact that the lenders were for-profit institutions whose own health depended on being repaid.

Understanding Latin American debt is critical for understanding Latin American economic history. The debt crises that are the subject of this chapter began in August 1982, when Mexico announced its inability to make the payments that were coming due. Soon after, finance officials in Latin America's other largest—and most indebted—economies found themselves in similar predicaments. The Latin American debt crisis had begun. The debt crisis would prove to be a watershed event, provoking fundamental and lasting changes in Latin American countries' relationship with international financial markets, and the global economy more generally.

This chapter provides an overview of the key issues and events, and in doing so provides an important background for the country analyses that follow in Part Two. We begin with an examination of how such large debts were accumulated in the years prior to Mexico's near default, placing the borrowing and lending processes in the context of broader changes experienced in the world of international finance. Upon considering the factors that converted the massive borrowing and lending extravaganza of the 1970s into the crisis of the 1980s, we then examine the changing strategies for management of international debt in the 1980s, a period in which capital only trickled to the region as most countries struggled to escape from overwhelming debt-servicing burdens. Indeed, this was a period that defied the logic of international finance, as net flows of capital went *from* developing *to* developed countries. By the early 1990s, however, international finance returned to Latin America. After quickly contrasting the dominant forms of lending in the 1970s and 1990s, we examine the vulnerability generated by the new international financial context of development, vulnerability which led to yet more debt crises in the 1990s and early years of the 21st century. In the conclusion, we consider some of the broader implications for economic policy and development of Latin America's changing relationship with the international financial system.

---

*Some of the data used in this article come from the following sources: Cohen, B. J. *In Whose Interest? International Banking and American Foreign Policy*. Yale, CT, Yale University Press, 1986; Devlin, R. *Debt and Crisis in Latin America: The Supply Side of the Story*. Princeton, NJ, Princeton University Press, 1989; Stallings, B. *Banker to the Third World: U.S. Portfolio Investment in Latin America, 1900–1986*. Berkeley, CA, University of California Press, 1987.

## THE BORROWING AND LENDING EXTRAVAGANZA OF THE 1970s

What factors brought about the accumulation of such enormous debt in the years prior to the crisis? The standard explanation for the borrowing and lending in this period, rooted in the recycling of so-called 'petro-dollars,' goes as such: the 'oil shock' of 1973 engendered a massive transfer of money from oil consumers throughout the world to a small number of petroleum exporters; the exporting countries, now with massive trade surpluses, deposited their revenues in Western banks; financial intermediaries, which by definition take deposits from savers and lend to borrowers, proceeded to lend this money, principally to developing countries that had unprecedented import bills on account of the increase in oil prices. In a sense, then, the oil crisis produced a marriage of convenience between borrowers and creditors, creating both the demand for capital inflows, in the form of developing countries that sought to take on debt to continue to afford imported oil, and the supply of capital, in the form of deposits that banks would seek to recycle.

The petro-dollar narrative certainly contributes to understanding the accumulation of debt in the 1970s, but it cannot alone explain the events of this period. This is because the standard narrative understates the dynamics that contributed to both increased demand for loans on the part of borrowers and the supply of loans on the part of creditors. It also overlooks a set of important institutional changes that facilitated such a massive expansion of commercial debt. Indeed, a careful examination of lending from international banks to developing countries reveals that the trend began prior to 1973. The oil crisis accelerated this, but the new process of international financial mediation had already begun.

### Borrowers

With regard to demand, the 1970s featured heavy borrowing by both oil importers and also oil exporters. To be sure, net petroleum importers, such as Argentina and Brazil, accumulated large debts in this period, but so too did net oil exporters, such as Mexico and Venezuela. This suggests that the demand for foreign loans was driven by more than the need to pay for oil imports. In some countries borrowing fuelled investment drives, while in other cases investment tended to support consumption. Investing or consuming, the most salient differences among developing countries in the 1970s were not their profiles as oil importers or exporters, but simply the size and structure of the manufacturing sector: middle-income industrializing countries had an insatiable demand for inflows of foreign capital, and borrowing from banks appeared to be an easy and attractive way to access abundant quantities with few conditions attached.

Indeed, to the extent that countries could attract capital inflows via international bank loans, they could circumvent many of the restrictions on both the amounts and use of loans that were imposed by multilateral lenders such as the IMF and the World Bank. Private bank loans also allowed countries to avoid the tensions that had accompanied the expansion of multinational enterprises after 1945, such as tensions over investment strategies and the repatriation of profits. As an indicator of how much Latin American countries came to demand this form of capital inflow, consider the expanding contribution of commercial bank lending to net financial inflows to Latin America, from a mere 1.6% in 1961–65, to 8.1% in 1966–70, 42.4% in 1971–75, and amounting to 58.3% in 1976–80.

### Creditors

A second limitation of the petro-dollar narrative regards the supply of capital. Net international bank lending in this period far exceeded the amount of money deposited in banks by the OPEC countries. Clearly, there was another source of capital, beyond petro-dollars. Key to the growing supply of funds was the expanding pool of dollars—largely a function of persistent US trade deficits in the post-Second World War era—that were deposited in international banks since the early 1960s. The net size of the 'eurocurrency market', which includes US dollars deposited in banks outside the USA, expanded almost 15-fold in 1965–75, and then more then doubled again between 1975 and 1980. It was this generalized increase in global liquidity, reflected by the expansion of eurocurrency markets, in conjunction with the petro-dollars that produced such abundant supply of money to be lent in the 1970s.

Importantly, 'offshore' banks operating in eurocurrency markets were largely unregulated. Unlike their national counterparts, for example, offshore banks were not obliged to hold specified amounts of cash relative to their outstanding loans. And because they were unburdened by requirements to diversify their portfolios, they could continue lending to the same set of countries. Indeed, by the early 1980s some of the larger US banks had lent as much as 200% of their capital to a handful of Latin America's largest debtors.

**Mobilizing Resources and Minimizing Risk**

Even given developing countries' increased demand for loans and the banks' increased supply of capital, we still need to understand why international banks were both able and willing to engage in such risky behaviour. After all, they could have found more conservative destinations for their money than Third World debt. Yet international banks devised a set of mechanisms that allowed them to mobilize massive sums of capital for international lending and to minimize (though not eliminate) the risk of doing so.

One factor that facilitated lending was a significant growth of overseas bank branches within developing countries during the 1960s and 1970s, which increased contacts and assisted in establishing customer bases among local authorities and companies. In 1960 only eight US banks had overseas branches; by 1980 this figure had increased to 139 banks (although not all of these were in the developing world).

Banks also minimized their risk by lending on adjustable interest rates, with borrowers' payment obligations linked to the London Inter-Bank Offer Rate (LIBOR). Adjustable (or floating) rates placed the burden of risk largely on the borrower, especially during a period, such as the mid-1970s, when real interest rates were low (and, at times, even negative). Were real interest rates to increase, as indeed they did, so too would countries' debt-servicing obligations. (As a note, the fact that Latin American countries accepted the banks' insistence on floating rates illustrates both the asymmetrical bargaining strength between lenders and borrowers and the extent of these countries' demand for commercial bank loans.)

A crucial mechanism on the part of the large 'money centre' banks for minimizing risk was through the creation of loan 'syndicates'. These were large lending networks, including as many 50 banks participating in the same loan. The pyramidal shape of the 'tombstones' announcing the loans in the financial press, for example, illustrated this nicely: the main organizing bank at the top, followed by other well-known banks at the second level, then lesser-known banks, and, finally, a plethora of smaller banks. Syndicated loans increased the amount of banks involved, as they provided a means for smaller banks that lacked their own international lending departments to participate in the lending boom, and thus increased the amount of money available to be lent. However, perhaps more importantly, syndicates also constituted a mechanism for larger banks to spread the risk of international banking.

With these mechanisms in place, the 1970s witnessed a revolution in international financial intermediation. International banking was not new in this period, not by any stretch of the imagination, but these changes allowed it to assume a more important role in the operations of more banks. For the 10 largest US banks, for example, earnings from international operations (not just in developing countries) as a percentage of total earnings more than tripled, from 17.5% in 1970 to 54.7% in 1982.

To summarize, the 1970s did indeed witness a marriage of convenience between borrowers and lenders, but the events of the decade were much more complex than suggested by the standard narrative of recycling petro-dollars. Increased demand for loans, increased supply of capital available to be lent, and a set of mechanisms that facilitated the expansion of new forms of international financial intermediation produced a borrowing and lending extravaganza.

## DEBT CRISES IN THE 1980s

A confluence of events turned the borrowing and lending extravaganza of the 1970s into the debt crisis of the 1980s. First, real interest rates rose dramatically in the early 1980s, as central banks in developed countries attempted to combat inflation. The LIBOR, the relevant rate for international loans, which had been negative, in real terms, in 1974–77, turned positive in 1978 and reached 6% by 1981. Since Latin American countries had borrowed on floating rates, the increase in the LIBOR produced serious financial shocks, as highly indebted countries faced increasingly large debt-servicing obligations. At the same time as the increase in interest rates, a decline in commodity prices reduced developing countries' export revenues. Subsequently, total debt service as a share of exports increased throughout Latin America and Caribbean, from 26.8% in 1977 to 46.9% in 1982. In some countries the increase was even more spectacular (e.g. Venezuela's debt-servicing to export ratio almost quadrupled).

Many Latin American governments were also unable to repay their foreign debts on account of increased capital flight. Significant amounts of money were reinvested by nationals in real estate and financial instruments outside Latin America. Although this problem was particularly acute in countries such as Argentina, Mexico, and Venezuela, where comparatively open capital accounts made the private export of capital relatively simple, the phenomenon was widespread throughout the region. Among the many effects of capital flight, the most relevant fact to the present discussion is that the constant conversion of local currency into dollars to move abroad placed a drain on a country's foreign reserves. Finally, a new sense of fear and skittishness on the part of creditors reduced the ability of developing countries to borrow fresh money to make payments on loans coming due. As debt-servicing obligations increased, Latin American countries began to appear less creditworthy, and as a consequence bankers became reluctant to extend additional loans.

The confluence of these four factors left many developing countries in crisis. While a spike in real interest rates increased developing countries' debt-servicing obligations, declining commodity prices, capital flight, and bankers' gradual turn away from the region reduced debtor countries' ability to meet the increased obligations through either export revenues, foreign reserves, or new financing.

**Mexico as Trigger and Template**

In August 1982 the Mexican finance minister, Jesús Silva Herzog, travelled to the US capital, Washington, DC, to explain his country's inability to make its upcoming payment. Silva Herzog met with officials from the US Federal Reserve, the US Treasury and the IMF. The Mexican crisis triggered a panic. It was obvious that Mexico was not unique, either in terms of level of indebtedness to commercial banks or in its (in)capacity to pay. It was merely the first country to acknowledge the gravity of the situation. Moreover, many came to fear a chain reaction, with default in Mexico followed by default in the other large Latin American debtors, particularly Argentina, Brazil, and Venezuela (these four countries accounted for nearly 80% of Latin America's total external debt). Furthermore, fear of additional default exacerbated the third and fourth conditions noted above, accelerating capital flight and further reducing banks' eagerness to lend fresh money.

From the perspective of the creditors and creditor governments, the situation was alarming for the simple reason that the largest banks' high levels of exposure left them exceptionally vulnerable to default. Were the USA's largest banks to go bankrupt, which many feared was possible given the amount they had lent through their unregulated 'offshore' branches, the US economy, if not the entire international financial system, was at risk. In response, the US Treasury and Federal Reserve, working with central banks from Organisation for Economic Co-operation and Development member countries, quickly put together an emergency rescue package of nearly US $5,000m. so that Mexico was able to make the payments that were coming due and thus avert default. The rescue package was obviously not a solution. The USA and others were simply lending money to Mexico that was to be used to repay the banks. It was designed as a bridging loan to avert default and calm the markets until a strategy for dealing with international debt could be developed.

### Return of the IMF

After August 1982 the IMF would become a major player, arranging emergency financing packages with both Mexico and Mexico's creditors. To Mexico, the Fund guaranteed fresh inflows of capital in exchange for wrenching domestic adjustment—sharp reductions in public spending and reductions in imports. However, the IMF would not release the money until Mexico's private creditors also agreed to loan fresh funds, a strategy that came to be labelled 'involuntary lending'. Hence, a new template was established: debtors would enter into agreement with the IMF to implement domestic austerity measures and the IMF would provide emergency financing conditional on the involuntary lending of private creditors.

It is important to note how the 1982 crisis ushered in a new era for the IMF. The Fund had played only a marginal role in the process of international financial intermediation for most of the period leading up to the Latin American debt crisis, when commercial bank loans became the dominant source of foreign capital. Those countries that could borrow from international banks, such as middle-income countries in Latin America, could bypass standard conditionality imposed by the IMF. After 1982, with commercial banks now resolutely uninterested in extending additional money, the Fund came to play a more pivotal and central role. Countries would no longer be able to escape IMF conditionality, not just to gain access to IMF funding, but also to secure private bank loans. The IMF's new position in the centre of the emerging debt-management arrangements would give it unprecedented leverage over developing countries' ability to tap into international credit markets.

The Mexico crisis was also a turning point in terms of the IMF's relationship with creditors, for this was the first time that, in negotiating an agreement with a debtor country, the IMF had stipulated the size of the new contribution from private lenders. Certainly, the banks were reluctant to make new loans, as many regarded this as 'throwing good money after bad'. Moreover, each bank, wishing to contribute as little new money as possible but receive interest payments on outstanding loans, had an incentive to 'free ride' on the actions of the others. The creditors' collective action problem was compounded by the sheer number of banks involved and their very different levels of exposure. For example, many of the smaller banks that were brought into international lending through syndicates typically had lent less money relative to their whole lending portfolio and, unlike the heavily exposed 'money centre' banks, could feasibly write off these loans and simply cut their losses. The IMF helped the creditors achieve the necessary co-ordination. Because the Fund quickly established close relations with the debtors, it gained information that it could share with the banks. Furthermore, under the IMF's tutelage, the major lenders formed steering committees to negotiate collectively with debtor countries and prevent any of the smaller creditors from declaring default.

## CHANGING STRATEGIES FOR MANAGING LATIN AMERICAN DEBT

There can be no denying that the emergency financing strategy of the early 1980s was designed with the highest priority given to averting default. The strategy, quite simply, was to reschedule loans that were immediately due, mobilize the bare minimum of new resources to enable orderly debt servicing, and reduce 'absorption' in the borrowing countries via fiscal and monetary retrenchment. The underlying expectation was that reduced economic activity would dampen demands for imports and subsequently free resources to service foreign debt, and that once the threat of default passed, capital flows would return to the region.

To the extent that the debt crisis was a crisis facing overexposed banks, the response in the early 1980s worked. The crisis was solved: countries did not default and the banks continued to receive interest payments. However, within two years it was clear that the debt crisis was far from over. Most debtor countries plunged into recession without recovery. Once the initial fears of international financial collapse were gone, it quickly became evident that most countries were still in no position to meet their debt-servicing obligations. In fact, after oil prices collapsed in the mid-1980s many countries, such as Mexico and Venezuela, appeared to be in even worse conditions.

### The Baker Plan

In 1985 the US Secretary of the Treasury, James Baker, announced a new approach to the debt crisis in Latin America. The Baker Plan proposed US $20,000m. in new lending by the commercial banks, another $9,000m. in additional lending by both the IMF and the World Bank, and the introduction of structural-adjustment policies on the part of participating countries. This third aspect was of critical importance, for in contrast to freeing resources by reducing absorption, the Baker Plan aimed to generate new resources, and thus repayment capacity, through so-called 'growth-oriented adjustment'. Thus, the sorts of policies prescribed in the mid-1980s—a package of 'neoliberal' measures such as trade liberalization, deregulation of foreign investment, and privatization—were different from, and more extensive than, the straightforward austerity measures advocated in the early years of the debt crisis.

The principle limitation of the Baker Plan was that the private banks still had little interest in lending new money. The banks were receiving interest, and after the scare of 1982–83 most preferred to reduce their exposure to the region. In fact, virtually all of the money lent by private banks in this period was done so grudgingly. This involuntary lending was part of the IMF-supervised rescheduling operations. However, if the banks would not lend more money, and if the USA and IMF would only mobilize small amounts of occasional involuntary lending to refinance upcoming interest payments, the result was that countries were typically paying out more in the form of interest payments than they were receiving in the form of loans and investment. From 1985–87, for example, accumulated net resource transfers to international financial markets from Latin America's four largest borrowers (Argentina, Brazil, Mexico, and Venezuela) amounted to nearly US $40,000m., notwithstanding the calls for new money contemplated by the Baker Plan. Although the hope of the Baker Plan was that structural reforms would lead to growth and thus make debtor countries more attractive to the banks, there was reluctance to lend to highly indebted countries that appeared to have little prospects of growth in the immediate future. Latin American countries' large accumulated debts—their 'debt overhang'—led to their experiencing net negative resource transfers (NNRT), which impeded investment and growth.

One lesson of the Baker Plan, thus, was that an effective strategy for debt management would have to deal with the problems of debt overhang and NNRT. Growth-oriented adjustment, whatever the merits of the policies being prescribed, could not offer a solution to the debt crisis, for the simple reason that crushing debt burdens and the net outflow of capital inhibited investment. The critical challenge was in working out how to reduce the problem of debt overhang, although neither the USA nor the IMF was prepared to compel private banks to cancel or reduce debts. At the banks' insistence—and the banks did insist—creditor participation in any programmes of debt reduction would have to be voluntary.

### The Brady Plan

A new strategy for managing international debt emerged in 1989, once again proposed and named after the incumbent Secretary of the Treasury, Nicholas Brady. In some regards, the Brady Plan was similar to the Baker Plan. In both schemes the structural-adjustment conditionality attached to participation amounted to adoption of neoliberal economic policies, and both relied on the voluntary participation of creditors. However, the Brady Plan (actually the brainchild of the Japanese finance minister) marked a fundamental change in debt management in that it included a mechanism to reduce debt overhang. The Plan called for creditors either to lend new money, reduce the amount of outstanding debt, or to reduce the interest on outstanding debt. Importantly, the second and third options would be executed by converting loans into bonds. To entice otherwise reluctant creditors to participate voluntarily, money from the IMF, the World Bank, and the Japanese Government were used to purchase US Treasury bonds that served as collateral for the converted debt. In the years after 1990, virtually all Latin American countries reached new agreements with their creditors under the framework of the Brady Plan. Throughout the region, bank loans from the 1970s were converted into bonds, so-called Brady Bonds.

The Brady Plan would have critical effects on Latin American economies. In the first regard, by offering a potential solution to the problem of NNRT, the Brady Plan provided strong incentives for developing countries to undertake the deep structural reforms that the IMF, the World Bank, and the US Treasury had been advocating since the mid-1980s. This is a key point worth underscoring: countries could only qualify for the benefits promised by the Brady Plan if they agreed to undertake significant neoliberal economic-policy changes. During the early 1990s most Latin American countries underwent fundamental changes in their strategies of economic development and, while the conditions attached to the Brady Plan were not wholly responsible for these changes, it was clear that such changes were the price to be paid for potentially escaping from the suffocating burden of debt overhang.

The Brady Plan also contributed to the reintroduction of voluntary capital inflows to Latin America. One reason for this was that the policy changes implemented by participating countries created new opportunities for foreign investment. The privatization of state enterprises, removal of restrictions on foreign investment and liberalization of domestic financial markets all made Latin American countries very attractive sites for foreign investors. A second way that the Brady Plan contributed to the new inflow of capital was that bonds, unlike bank loans, have deep secondary markets. Latin American debt, thus, became attractive and entered into the portfolios of entirely new sets of investors who buy and sell international bonds. And as the debtors regained creditworthiness, they could issue new bonds as well. Both of these new investment opportunities dovetailed with a larger change that was occurring at the same time—the growth of new institutional investors in the developed world (such as managers of mutual funds and pensions) who eagerly sought higher-risk investments to increase their returns and diversify their portfolios. Thus, in the early 1990s capital flows resumed to Latin America. In fact, by 1995 the level of capital flows exceeded that of 1980. Finance, which had flooded into the region in the 1970s and left, en masse, in the 1980s, had returned. Although the Brady Plan 'resolved' the debt crisis of the 1980s, the renewed eagerness of creditors to lend to Latin American countries meant that indebtedness would remain an important and explosive issue.

## NEW FINANCIAL FLOWS, NEW VULNERABILITY

Importantly, the 1990s marked not just a return of capital flows, but the emergence of a new form of capital flows. If the 1970s was the era of the bank loan, the 1990s was the era of the portfolio investment, such as purchase of government bonds or equity. From 1977–82, roughly three-quarters of the net capital flows to Latin America and the Caribbean were in the form of commercial bank loans. From 1990–96, however, portfolio investments amounted to approximately two-thirds of net capital flows to the region.

The return of finance and, in particular, prominence of portfolio investments, created a new international financial context for development. One of the key features was the prevalence of short-term and highly mobile capital flows. Bonds tend to be issued with relatively short maturities, significantly shorter then bank loans. The effect of this is that a change in investor sentiment—if either the outlook in the borrowing country or region looks worse or the outlook in an alternative destination looks better—can lead to rapid withdrawal of the stock of outstanding bonds. Likewise, equity investments can leave overnight. Indeed, for many institutional investors the ability to make rapid withdrawals and rebalance portfolios is essential, for it allows them to compete for high returns while always being able to meet customers' demands to adjust or liquidate their accounts. It is this spectacular mobility and volatility that earned the new form of capital flows the label 'hot money'.

Latin America's dependence on volatile capital flows made the region's economies more vulnerable in the 1990s. To be sure, if one looks throughout the developing world, then by comparative standards, Latin American countries have historically been highly integrated into international financial markets. This condition was amplified in the 1990s, owing both to the economic policy changes that included augmented financial opening and to the significant increases in capital mobility and financial integration on a global scale. Although increased integration facilitated the expansion of new capital flows, it also increased countries' vulnerability to external shocks. Massive inflows tend to cause currencies to appreciate, which leads to trade deficits, which increase dependence on continued inflows in order to finance the trade deficits. However, money that enters the country can also leave it, generally just as fast; and when the direction of flows reverse, a simple phenomenon given the high degree of mobility of 'hot money', crisis ensues.

In a stylized fashion, this scenario of 'hot money' entering and then leaving, for whatever reason, aptly summarizes the financial crises experienced by Mexico in 1994–95, Brazil in 1998–99, and Argentina in 2001–02 (as well as the financial crises of the late 1990s in East Asia, Russia, and Turkey). The country chapters deal with the Latin American cases in detail. For the present purposes it is worth underscoring that the massive inflows of capital that these countries received in the aftermath of the arrangements made under the framework of the Brady Plan, while important for fuelling economic recovery, clearly did not reduce these countries' vulnerability. They each suffered major crises, with wrenching adjustment and social dislocation, in the 'post-debt crisis' financial environment.

To a certain extent, the international responses to the more recent round of debt crises (now called 'financial crises') were similar to what transpired in the 1980s. The international financial community, led by the USA and IMF, mobilized resources to avert massive default (significantly more energetically and enthusiastically in the cases of Mexico and Brazil than Argentina), and in exchange for this international support the debtors underwent periods of extensive economic adjustment. In both periods, external lending increased vulnerability to crisis, external lending was followed by crisis, and debtors bore the burden of adjustment.

Ultimately, responding to financial crises in the current environment presents new challenges. In the first regard, the sheer size and expense of contemporary financial crises makes the recurrent mobilization of public resources increasingly difficult. The rescue plans organized by the USA and IMF in the 1980s, although at the time unprecedented in terms of the amounts of money available, were insignificant compared with what is necessary in the face of contemporary financial crises. Moreover, the sheer number and diversity of the actors involved generates new complexities. The creditors are no longer simply banks, large and small. Rather, a key attribute of the new international financial environment is that developing country debt is held by so many different types of institutional (and individual) investors, all of which have dramatically different relationships with borrowers and depositors. While reaching agreements on rescheduling syndicated loans in the first round of debt crises was no easy feat, the difficulties in achieving agreement with, and co-ordination among, the more recent and more heterogeneous group of creditors in the new environment are much greater. It was in response to these challenges and difficulties that the IMF's First Deputy Managing Director, Anne Krueger, proposed the creation of Sovereign Debt Restructuring Mechanism (SDRM) in 2001. The core of the proposal was to establish a centralized and simplified process for countries which lack ability to service their debts to negotiate with creditors, under the supervision of the Fund, and, crucially, for the agreement reached to become binding on all creditors. However, creditors and creditor governments (particularly, but not exclusively, the USA) reacted coolly to Krueger's proposal, concerned that any process that simplified debt restructuring would in effect sanction and encourage default. Creditors were also wary of sacrificing their legal rights to pursue litigation against debtors, and warned that an SDRM would greatly reduce investors' interest in developing country debt. This latter concern was then echoed by many representatives of developing countries. Finally, the SDRM was also assailed by both creditors and debt-relief campaigners for strengthening the Fund's role in the process of debt restructuring: private creditors raised the possibility that the Fund could impose greater levels of private debt reduction in those countries where the Fund itself had lent significant amounts of money and was heavily exposed; debt-relief campaigners feared that the Fund's role in the SDRM would instill a bias against minimal debt relief. On account of the widespread opposition from various quarters, the project for the SDRM did not come to fruition, though the underlying motivations and concerns have not gone away. Indeed, regardless of what one thinks of the particularities of Krueger's proposals, the on-going saga of Argentina and the countries' creditors engaging in three years of protracted negotiations serves as a useful reminder of the inadequacies of the current arrangements for managing sovereign debt. Indeed, even as negotiations between Argentina and creditors appeared to have successfully led to a restructuring agreement at the beginning of 2005, dissenting investors were able to delay the entire process and prevent conclusion of the deal—precisely something that would be impossible under a binding process such as that contemplated with the SDRM.

## CONCLUSION: DEBT, FINANCE AND DEVELOPMENT IN THE GLOBAL ECONOMY

The 1970s witnessed a massive credit boom, as commercial banks in the developed countries lent significant amounts of money to developing countries, particularly in Latin America. As already discussed, changed international conditions in the early 1980s brought the lending and borrowing extravaganza to an abrupt halt. The ensuing debt crisis brought about a new relationship between Latin America, the international financial system and the IMF and, ultimately, a new relationship was established between Latin America and the international economy. In many ways, the shift to the neoliberal economic model in the 1980s and 1990s has its roots in the debt crisis and the subsequent strategies for managing international debt. In the 1990s, capital returned to the region, in abundant fashion, but the new forms of capital flows are marked by high levels of volatility. In important ways, the first and last chapters of the story of Latin American debt since the 1970s are similar, in that euphoric lenders provided money to enthusiastic borrowers, with neither side appropriately considering the risks involved, only for changed international conditions to turn inflows into outflows and generate debt and financial crises. Latin America regained access to voluntary capital markets in the 1990s, but vulnerability and susceptibility to crisis remained intensely problematic features of this relationship. Furthermore, as long as countries need to borrow and so long as investors are anxious to lend, not just debt but debt crises will remain critical aspects of the global political economy. Devising a mechanism to address debt crises in an orderly and equitable manner is a major challenge for the 21st century.

At a more general level, the new form of capital inflows has serious and potentially perverse effects on policy-making in Latin America. To attract the interest of institutional investors and retain mobile capital, countries face pressures to raise interest rates, reduce spending, and increase taxes (particularly consumption taxes). Thus, while the purported benefit of capital inflows is to increase investment and growth, countries often need to implement austere monetary and fiscal policies that dampen investment and growth so as to attract the capital. This dilemma is heightened during periods of capital outflow, as the volatility of capital flows tends to encourage pro-cyclical policies: policy-makers fear stimulating the economy, either through monetary or fiscal policy, out of concern that such measures will cause further capital outflows. Moreover, the fact that money can enter and leave so rapidly can provoke significant movement in the exchange rate, movements that might dampen trade and investment. Thus, countries typically gather large amounts of foreign reserves as a mechanism to protect against and minimize exchange-rate volatility. However, the opportunity costs of hoarding foreign reserves must be considered: money set aside to ward off exchange-rate volatility is money is not being used for infrastructure, education, health care and poverty reduction.

Put together, these effects generate serious concerns as to whether the new international financial context creates an environment conducive for growth and development. Latin American countries were among the highest-growing countries in the developing world in the 1960s and 1970s, but fell behind thereafter. Average growth rates in both the 'lost decade' of the 1980s and the post-Brady Plan 'emerging market' era of the 1990s were significantly lower then elsewhere in the developing world—notwithstanding the massive inflows of capital in the latter period. However, extended periods of low growth is not a luxury the region can afford, given the high levels of poverty and indigence. Furthermore, the region's massive inequalities with regard to the distribution of income and wealth are only exacerbated during periods of crisis. The fear, then, is that, in addition to compounding vulnerability to debt crises, the new form of integration into international financial markets creates a low-level equilibrium of low growth, high unemployment and persistent poverty and inequality.

# CRIME IN THE CARIBBEAN: REFORM AND PROGRESS
## CANUTE JAMES

### INTRODUCTION

After years of uncertainty and indecision, in early 2005 several Caribbean countries made a tentative move towards legal and judicial reform with the establishment of a regional court, the Caribbean Court of Justice (CCJ). Based in Port of Spain, Trinidad and Tobago, this new body was to arbitrate disputes that were expected from the establishment of the Caribbean Single Market and Economy (CSME), scheduled to be inaugurated by the end of 2005. The nature of the Court was not exactly what the countries of the Caribbean Community and Common Market (CARICOM) had originally planned. In removing the Privy Council (based in London, United Kingdom) as the final appellate court for most of the former British colonies in the region, and replacing it with the CCJ, representatives of these countries had wanted to make a fundamental change to the justice system in the region, and, in a wider sense, to the perception of the Caribbean internationally.

The creation of the CCJ is just one aspect of the reform that the islands of the Caribbean Basin have begun to implement in the 21st century. The region was faced with increasingly complex problems of public security, reflected in rising levels of crime, and persistent concerns, particularly from the USA, that the financial services sector of many jurisdictions were open to abuse by international criminals, such as money-launderers and terrorist-financiers. These issues were partly owing to the islands' position as an entrepôt in the global narcotics trade: drugs were smuggled from South America mainly to Haiti, the Dominican Republic, Puerto Rico and Jamaica, from where they were distributed to North America and Europe.

### BACKGROUND

In 2001 10 members of CARICOM—Antigua and Barbuda, the Bahamas, Barbados, Belize, Grenada, Guyana, Jamaica, Saint Christopher and Nevis, Saint Lucia and Saint Vincent and the Grenadines—agreed to establish a regional court as the final court of appeal for those countries. This new court would replace the Privy Council, which had been retained as the ultimate appellate court since independence by all of the 10 except Guyana.

This new regional court was intended to deal with trade disputes expected from the implementation of the common market, as well as with criminal matters hitherto heard by the Privy Council in London. The CCJ was mired in debate across the region from its inception. Its detractors argued that it was not necessary, and that the Privy Council should be retained. Some claimed that the cost of establishing the regional court was exorbitant, while others questioned the quality of jurisprudence in the region. Questions were also raised about the ability of the planned court to insulate itself from influence by politicians and other interests. There were also claims that the regional court was an attempt by governments to bypass the frequent Privy Council rulings halting the execution of convicted murderers. Meanwhile, supporters of the CCJ countered that the Caribbean countries were among the few that continued to have links with the Privy Council, and that such an arrangement was out of step with the political status of most of the countries in the region. Furthermore, the Court was central to the proper functioning of the proposed CSME.

Moves towards the establishment of the CCJ were delayed in early 2005. In late July 2004 the Jamaican Parliament had approved legislation seeking to replace the Privy Council with the CCJ as the country's final appeal court. Several other countries in the region had planned to follow Jamaica's move. However, in February 2005 the Privy Council itself upheld an appeal by the main opposition Jamaican Labour Party and human rights organizations that the legislation was unconstitutional. The Privy Council ruled that the proposed CCJ 'lacks protection', since its status could be enacted or abolished by a simple majority in the Jamaican Parliament. It further ruled that to establish the CCJ as Jamaica's final appellate court without a constitutional amendment to that effect would undermine the rights of the Jamaican people as entrenched in the Constitution. This ruling forced hasty reconsideration in the region. It caused a delay in plans by the Governments of Barbados, Jamaica and Trinidad and Tobago to sign an agreement in mid-February creating an economic union. The other CARICOM members were to sign up to the Court later, after making legislative and administrative changes. In the event, the CCJ was inaugurated on 16 April, when the Governments of Barbados, Jamaica and Trinidad and Tobago signed an agreement making it the arbiter solely on trade issues in the region's economic union. Although this move severely limited the CCJ's role, it also removed much of the criticism of the institution.

In July 2003, when planning the regional court, CARICOM leaders agreed to create a Regional Judicial and Legal Services Commission to determine the composition of the bench, which was to comprise nine justices. The Commission's members were to be appointed by regional bar associations, civil society organizations and legal educators. This Commission was inaugurated in August. The Caribbean Development Bank (CDB) raised US $100m. to finance the Court's operations: some $12m. of this was allocated to the costs of setting up the CCJ, while the rest was placed in a trust fund, to be managed by the CDB. Operational costs of the Court were to be met from the $5m. that the fund was anticipated to yield annually. Several heads of government and the CCJ President, Michael de la Bastide, concurred that the method of appointing judges and the financing of the Court would insulate it from the influence that some critics claimed was a threat to its independence. Speaking at the inauguration of the CCJ in April 2005, de la Bastide, a retired Chief Justice of Trinidad and Tobago, said:

'When one is dealing with a regional court it is not so easy for an individual government to put pressure on it. It is less likely that a court of this sort is going to be subject to local influences. For the persons who can amend the agreement to agree to do so they must have some unity of purpose for this and it is difficult to see the circumstances which will make this possible.'

De la Bastide went on to state that the way the Court was funded provided 'a very effective shield against influence' from interested parties.

Political leaders also claimed that the Court was necessary to the region's economic development, and they rejected the arguments of the critics. Kenny Anthony, Prime Minister of Saint Lucia and holder of the CARICOM portfolio for justice and governance, believed that the creation of the regional court indicated 'confidence in our ability to continue the journey which the first step implies'. He rebutted suggestions that the Court would bring about a reduced quality of justice, claiming that the legal profession in the Caribbean was among the strongest in the Commonwealth.

'The Caribbean is not a fledgling state approaching tentatively the threshold of the rule of law...We laid the foundations of a strong legal profession and legal and judicial institutions many decades ago, both at home and abroad. The Caribbean Court of Justice is not a leap into the dark. It is a leap of enlightenment. This region has the most sturdy credentials for creating a regional court of appeal that can respond, and respond with finality, to the most rigorous standards of the rule of law.'

Patrick Manning, Prime Minister of Trinidad and Tobago, affirmed that although not all regional governments had, by 2005, arranged to have the CCJ as the final court of appeal, 'there could be no doubt that it is merely a matter of time before we all complete the journey, which will confirm our status as truly independent people of the Caribbean.' By mid-2005, however, only Barbados and Guyana had adopted the CCJ as their final appellate court. Edwin Carrington, Secretary-General of CARICOM, claimed that the CCJ was 'the court of the Caribbean people, by the Caribbean people, for the Caribbean people.' Carrington believed that the new court was 'a critical pillar' of the planned CSME, which was 'the most effective means by which the economies of the region can be successfully integrated into the proposed new hemispheric and global economy.'

## VIOLENT CRIME IN THE REGION

Advocates of the CCJ have insisted that there was no merit to charges that the institution was intended to be a 'hanging court' that would expedite the execution of convicted criminals. Executions, usually by hanging, have been rare in the region in recent years. Appeals to the Privy Council in London against planned executions of murderers generally have been upheld. Claims by opponents to the CCJ that the Court would be used to speed execution have been fuelled by indications of annoyance from several Governments in the region at the rulings of the Privy Council on the treatment of convicts awaiting the death penalty. Officials spoke frequently of the Privy Council's rulings 'tying the hands' of judicial systems in dealing with the increasing crime and violence in the region.

This increasing level of criminal violence in the region was reflected in the rise in the number of murders and kidnappings over the past five years, most noticably in the Dominican Republic, Haiti, Jamaica, Guyana and Trinidad and Tobago. Police sources attributed the upward trend to the increase in narcotics-trafficking in those countries. Drugs cartels fighting to gain control of the trade in the region, often had deadly results. Security ministers and police chiefs across the region contended that this problem was being compounded by the thousands of criminal deportees that were sent back from the USA each year. Most of these were returned to the Dominican Republic, Haiti and Jamaica. The Jamaican police asserted that deportees were involved in most of the murders, armed robberies and shootings, while the Guyanese authorities claimed that the violent crimes that were common in the country by the early 21st century— 'drive-by' shootings, kidnappings and carjackings—were rare before the influx of deportees began in 1997. Government sources there claimed that, in most instances, these deportees returned to a region with little or no family connections, but with a sophisticated knowledge of the methods of modern crime. The USA, the United Kingdom and Canada frequently issued travel advisories, warning their nationals against travelling to several countries in the region whenever there was an increase in the violent crime rate. Such warnings obviously had adverse effects on those economies dependent on tourism. The US Department of State's Overseas Security Advisory Council in 2005 designated the Dominican Republic a 'high crime post', adding that it saw 'no reason to believe that the crime rate will decline in the near future.' It reported that in 2004 there was an increase in all crimes, from murder, assaults, armed robberies and home invasions, to petty thefts and pickpocketing.

In May 2005, during a two-day meeting of ministers responsible for national security in Port of Spain, Patrick Manning, Trinidad and Tobago's Prime Minister, who also held the CARICOM crime and security portfolio, maintained that the profile of crime and security in the Caribbean had changed significantly over the last two decades:

'Today the trafficking and proliferation of illegal firearms, the emergence of new forms of crime, cyber attacks on financial institutions, and the threat of terrorism dictate that we address crime as a major concern for the region.'

He added that governments in the region were concerned that over the next three years the potential for further criminal and security threats could emerge, as the Caribbean moved towards establishing its economic union, and as the planned Free Trade Area of the Americas took shape. However, Manning believed that the problems created by criminal deportees affected all countries in the region and that the issue could only be addressed by engaging the Caribbean's 'collective strength', and by re-examining the issue according to an agreed new strategy. At the same meeting, CARICOM Secretary-General Carrington stated that all Caribbean states were concerned over security as the increase in criminal violence was eroding public safety and security, and would deter investors.

'There isn't a day we do not have reports of ghastly crimes and shocking violence in our region...Gone are the days when citizens of our region would hear or read news of violent crime elsewhere in the world with some measure of reassurance that 'that can't happen here.'

Some police chiefs, such as Ausbert Regis, the Police Commissioner of Saint Lucia, believed that drastic action was needed to deal with violent crime, raising the question of extending capital punishment. In March 2005 he stated: 'If this means that we revisit the execution of persons who commit a particular class of crime, then so be it. But we need to ensure that persons will not commit crime with no fear of consequences.'

## DRUGS-TRAFFICKING AND MONEY-LAUNDERING

The region's key position in the global narcotics trade, and the efforts by some countries to broaden the bases of their economies, brought another problem to the Caribbean in recent years. The rise in drugs-trafficking was of increasing international concern, but particularly a US concern. Following the terrorist attacks on New York and Washington, DC, of 11 September 2001, the USA moved to limit the ability of terrorists and terrorist organizations to finance their activities. The financial services sector of the Caribbean was considered by the US Administration to be a likely source of terrorist-financing. The US Department of the Treasury adopted the stance that increasing the transparency and security of the financial system in each country, and in all its vulnerable sectors, in addition to targeting the sources of terrorist support, was an essential element of the fight against terrorism and financial crimes. (For further details, see Testimony of Juan Carlos Zarate, Assistant Secretary, Terrorist Financing and Financial Crimes, US Department of the Treasury Before the House Financial Services Subcommittees on Domestic and International Monetary Policy, Trade and Technology And Oversight and Investigations, 30 September 2004.)

As a result, the Caribbean's fledgling 'offshore' financial services sectors, already under pressure from the industrialized nations, became the focus of concerns about the financing of terrorist organizations. The efforts to deal with terrorist-financing further intensified in early 2004 after US officials and agencies suggested that the methods being used for tracking funds were not proving very effective. According to the USA, small financial services jurisdictions, such as those in the Caribbean, were vulnerable to abuse by terrorists and their financers because regulation of the sector was concentrated on large sums. Smaller sums could be transacted in the smaller jurisdictions, and consequently evade scrutiny.

Long hesitant about accepting direction from industrialized countries about how to manage their 'offshore' sectors, and with the threat of punishment if they did not meet standards on information exchange and transparency intended to fight tax criminals, money-launderers and, now significantly, terrorist-financiers, Caribbean governments were forced into implementing changes that many had strongly resisted. These jurisdictions had been facing increasing pressure from two separate, but similar, fronts. The Organisation for Economic Co-operation and Development (OECD), based in Paris, France, and consisting of the world's richest nations, had, in June 2000, issued a list of 35 tax-favourable jurisdictions, or so-called tax 'havens', claiming that these countries and territories were providing 'harmful' tax competition. The jurisdictions, which offered a no tax, or low tax, environment to attract a range of international financial business, included 15 in the Caribbean region: Anguilla, Antigua and Barbuda, Aruba, the Bahamas, Barbados, the British Virgin Islands, Dominica, Grenada, Montserrat, the Netherlands Antilles, Saint Christopher and Nevis, Saint Lucia, Saint Vincent and the Grenadines, the Turks and Caicos Islands and the US Virgin Islands. OECD demanded greater levels of co-operation from these countries and territories, through exchange of information on their 'offshore' financial sectors. It demanded also that levels of transparency in the administration of these sectors be increased. If the jurisdictions did not agree to the reforms, various sanctions would be imposed against them, the nature of which was not revealed. OECD's intention was to ascertain the levels of tax evasion occurring, and the ability of the tax-favourable jurisdictions to attract corporate and individual financial business through a preferential environment.

The other source of pressure for the 'offshore' jurisdictions came from the Financial Action Task Force on Money Laundering (FATF), which, like OECD, was based in Paris. In 1990 the FATF had issued a report containing 40 'recommendations', which it encouraged all countries to adopt. These recommendations provided a comprehensive blueprint of the measures needed to address the problem of money-laundering. They included the criminalization of the laundering of the proceeds of serious crimes, and the enactment of laws to seize and confiscate the proceeds of crime. Financial institutions in the jurisdictions were obliged to identify all clients, including any beneficial owners of property, and to keep appropriate records. They were also required to report suspicious transactions to the competent national authorities and to implement a comprehensive range of internal control measures, through systems for the monitoring and supervision of financial institutions. The FATF also declared that there was a need to enter into international treaties or agreements and to pass national legislation to allow countries to provide prompt and effective international co-operation at all levels. Failure to implement the FATF's recommendations would lead to the 'naming and shaming' of the delinquent jurisdictions, with attendant damage to that country's reputation and a reduction in its ability to attract legal international business.

The positions of OECD and the FATF alarmed those jurisdictions with 'offshore' financial sectors. While some of the larger and more established sectors, such as those in the Cayman Islands and the Bahamas, had by then developed into significant pillars of their economies, others, such as those in the eastern Caribbean, were newcomers to international financial services sector. The Bahamas' financial services sector, for example, was created to meet the needs of the country's earliest tourists, some of whom spent as much as six months at a time in the islands.

The development of financial services in some jurisdictions was encouraged by their location between North America and Europe, which meant they could take advantage of the time differences. Furthermore, the changing demands in Europe and North America for a range of financial services and the expansion of the Eurodollar market gave a fillip to the business done by the jurisdictions. New demands in global insurance stimulated further development, as some jurisdictions promoted a range of special-purpose vehicles to attract increasing volumes of business. Gradually, 'offshore' centres in the Caribbean region became an integrated part of global finance. They were accepted in the industrialized countries as a way to accelerate capital formation and to create a market for capital. Demand grew for their services, and the jurisdictions expanded. Long-standing 'offshore' financial centres such as Bermuda, the Bahamas and the Cayman Islands, were joined by newer rivals such as Antigua and Barbuda, Aruba and the British Virgin Islands. The sector increased in importance in many of the region's economies: for example in 2004 financial services contributed some 25% to the gross domestic product (GDP) of the Cayman Islands and an estimated 20% to the GDP of the Bahamas.

Over the past five years all the Caribbean jurisdictions originally included on the FATF's 'blacklist' of 'non-co-operative' jurisdictions were removed, with Saint Vincent and the Grenadines being the last to be taken off in June 2003, after pledging to undertake reform of the financial sector. However, despite increasing attempts to reform the administration of its financial services sectors, and the strengthening of regulatory oversight to meet the demands of OECD and the FATF, the region remained a source of concern to the US Administration. In its report on money-laundering and terrorist-financing in 2004, the US Department of State placed several Caribbean countries and territories on its list of 'jurisdictions of primary concern'. According to that Department, the financial institutions in these countries engaged in 'currency transactions involving significant amounts of proceeds from international narcotics-trafficking'. The report indicated particular US concern about the United Kingdom's Overseas Territories, warning that Anguilla, Bermuda, the British Virgin Islands, the Cayman Islands, Montserrat and the Turks and Caicos Islands should continue to strengthen their regulations to prevent and inhibit money-laundering and terrorist-financing. According to the report, although Anguilla's financial sector was small in comparison with other jurisdictions in the Caribbean, the ability to register companies there online and the use of bearer shares made the island vulnerable to money-laundering. Bermuda had improved the integrity of its financial regulatory system, was co-operating with the USA and the international community to counter money-laundering and terrorist-financing and continued to update its legislation. However, the report suggested that measures to detect and monitor cross-border transportation of cash and monetary instruments should be enacted. The British Virgin Islands remained vulnerable to money-laundering, despite improved legislation, and needed to strengthen its regulatory oversight by implementing legislation to counter terrorist-financing. The Department of State concluded that the threat in the Cayman Islands was mainly owing to the significant size of that territory's financial services sector. Although the administration there had improved protection against money-laundering, it should move towards criminalizing terrorist-financing. Montserrat, a smaller jurisdiction, was said to be attractive to money-launderers because of the weak regulation in place there. The report pointed specifically to the absence of exchange controls on transactions of less than US $100,000, and suggested that the Government criminalize laundering related to domestic drugs-trafficking and improve regulatory oversight to deter criminal and terrorist organizations. The Turks and Caicos Islands was alleged to be a transhipment point for narcotics-traffickers, and was vulnerable to money-laundering because of its large financial services sector. This was compounded by bank and corporate secrecy laws and internet gaming activities. The Department of State recommended increased co-operation efforts with foreign law enforcement agencies and administrative authorities, as well as further supervisory oversight provision, to ensure that criminal or terrorist organizations did not abuse the financial sector there.

The US Department of State's 2004 report also listed Antigua and Barbuda, the Bahamas and the Dominican Republic as jurisdictions of 'primary concern' in the Caribbean. These countries' financial institutions engaged in 'transactions involving significant amounts of proceeds from serious crime'. It also listed Barbados, Grenada, Jamaica, Saint Christopher and Nevis, Saint Lucia and Saint Vincent and the Grenadines as countries 'of concern'. Despite having 'comprehensive legislation' for its financial sector, Antigua and Barbuda was susceptible to money-laundering, particularly through its 'offshore' sector. The purchase of real estate, large vehicles and jewellery, and the processing of money through a complex national or international network of legitimate businesses and 'shell' companies (companies registered in the jurisdiction but with no staff or offices) were the main money-laundering methods used in the Bahamas, according to the Department of State. It noted that large quantities of cash were being kept in properties that were considered 'safe' houses, following legislation that reduced launderers' ability to deposits large sums of cash. The report said that the clandestine movement of large amounts of cash by couriers and wire transfer remittances were the primary methods of transferring illicit funds from the USA into the Dominican Republic, making that country a major transit post for drugs, mostly cocaine and heroin, en route to Europe and the USA. It claimed that the Government of the Dominican Republic had the legislative framework to combat money-laundering and terrorist-financing, but that insufficient implementation left the country vulnerable to criminal financial activity and abuse. The Bahamas was among the jurisdictions that moved early to reduce the possibilities of terrorist-financing. The Government enacted anti-terrorism legislation in December 2004. At that time, Alfred M. Sears, the Attorney-General, announced that the country had adopted the FATF's recommendations relating to the financing of terrorism. The new laws would work in conjunction with existing legislation, including the Proceeds of Crime Act that gave the police, customs officials and the courts powers in relation to money-laundering, and the search, seizure and confiscation of the proceeds of crime. The Bahamas Financial Services Board claimed that the new legislation enabled the Government to authorize financial institutions to 'freeze' the accounts of persons identified as terrorists by the UN, under UN resolutions relating to terrorist-financing.

Despite the concerns of the industrialized nations about the level of regulatory oversight of Caribbean financial services sectors, and the US Department of State's listing of potential abuse by those financing terrorism, administrators of Caribbean centres claimed that significant improvements in protecting the financial sector had been implemented. The British Overseas Territories in the region were further subject to increased disclosure of its operations through a European Union directive on the management of savings held in institutions in the jurisdictions. Although they supported moves to reduce the possibilities of terrorist-financing, Caribbean finance ministers were, however, worried that efforts by OECD and the FATF, and the concerns expressed in the US Department of State's 2004 report could result in the end of a source of financing that was helping to keep afloat many of the region's pressured economies. The concern was fuelled by the Department of State's reference to the Dominican Republic as a jurisdiction of 'primary concern' for money-laundering, and the identification of wire transfer remittances as one of the primary methods for moving illicit funds from the USA to the Dominican Republic. According to figures from the Inter-American Development Bank, nationals from the Dominican Republic, Jamaica, Haiti, Guyana, Suriname and Trinidad and Tobago working abroad sent remittances worth over US $5,000m. in 2004. Most of this was sent from the USA. The funds—$2,400m. to the Dominican Republic, $1,500m. to Jamaica and $1,000m. to Haiti—were an important source of income for these struggling economies. Finance authorities in the sending countries were being forced increasingly to tighten controls to prevent remittances being exploited by those financing terrorism, a move about which Caribbean governments were, naturally, wary.

## CONCLUSION

The pace and nature of the institutional reform taking place in the Caribbean in the mid-2000s would determine significantly how the region confronted and overcame the problems it faced as a consequence of its geography and lingering social problems.

Situated on the main route of the global trade in narcotics, the Caribbean inevitably would continue to suffer from organized and violent crime perpetrated by the proxies of the drugs cartels. These problems were compounded by the two other elements. The first was the impact of criminal deportees from North America and Europe, who, according to regional police authorities, were ready recruits to organized crime. The second element was one that the region would find difficult to resolve for many years. Caribbean countries, faced with the need to protect their small economies from the impact of globalization, feared that current high levels of poverty would not be reduced rapidly enough to curb levels of violent and organized crime.

Financially pressured governments were being forced to reform their law enforcement capacities. Police forces across the region were implementing co-operation schemes and many were seeking foreign assistance in order to address the rising rates of murder, kidnapping, arms- and people-smuggling, money-laundering and terrorist-financing. The challenges were not helped by the widespread corruption among law enforcement and other public officials. In the Dominican Republic, human rights organizations were critical of the police force there whose officers shot and killed 360 people in 2004, 160 more than in the previous year. There was similar criticism of the actions of the police in Jamaica.

Other and more difficult security considerations were also confronting the region. In Trinidad and Tobago, the Government was working with the USA to protect the country's gas installations from possible terrorist attack. Trinidad and Tobago was the source of 70% of US imports of liquefied natural gas, and the installations were considered a likely terrorist target. The region's ability to deal effectively with these problems would continue to be determined by the perceived national interests of the USA. The Bush Administration stated in early 2005 that it would expand its 'third border' (Caribbean) initiative and improve relations with the region, strengthening the security and legal aspects of the relationship. The region's police authorities believed that the increased co-operation with US law enforcement agencies had resulted in further drugs interdictions and arrests, which made the region, and the USA, safer. Nevertheless, the more fervent advocates of the inevitability and the value of the CCJ, and supporters of the reforms undertaken in the areas of law enforcement and regulatory oversight of financial services, were unlikely to see these reforms as sufficient to bring about early and satisfactory solutions to the rising levels of crime in the Caribbean.

# AN ENVIRONMENTAL PORTRAIT OF LATIN AMERICA AND THE CARIBBEAN

## Prof. ADRIAN MCDONALD and DUNCAN MCDONALD

### INTRODUCTION

Reviewing the environments and environmental issues to be found in Latin America and the Caribbean, a very diverse and significant part of the global land surface, is a challenging task and more so if the material is to be expressed in a few pages. In previous editions of *South America, Central America and the Caribbean* we have considered the key issues and the role of El Niño, or, to give it its proper meteorological term, El Niño Southern Oscillation (ENSO), in creating dramatic variations in climate patterns. We have also examined the resulting economic and personal burden on people. El Niño and its counterpoint La Niña, remain a key physical influence on the environmental variability in much of Latin America. In this article, however, we shall present three elements: first, a review of the main environmental sectors raising concern; second, an assessment of what appear to be the key current issues attracting attention from domestic and international pressure groups; and third, a statistical portrait of Latin America and the Caribbean. All three approaches cover material that will be subject to change, particularly the current issues. In contrast, some elements of such a statistical portrait will remain constant in practical time horizons; elements such as the land take of individual countries. Other elements, however, such as natural cover, land use, the fisheries and the political and legal responses and agreements are more responsive to global policy and global markets. Change, in particular the rate and direction of change, and commitments made to the international community, are the most important and interesting elements of this environmental portrait and it is on these aspects that this article concentrates. The UN Environment Programme's (UNEP) publication *One Planet, Many Peoples: Atlas of Our Changing Environment* shows the environmental changes that have been captured in the era of satellite imagery. The publication brings into sharp focus the importance of the changes in Latin America and the Caribbean. One of the more alarming images shows the loss of mangrove and coastal wetland in Honduras as the land-water margin is converted to shrimp farming developments, making these areas significantly more vulnerable to sea surges.

### ENVIRONMENTAL SECTORS OF CONCERN

It is not possible to address all the environmental concerns of a continent, thus, below some of the key sectors that raise concern are considered separately. Although addressed as separate entities, these various environmental issues are, however, highly inter-related. For example, the mining can affect air quality, water quality and land, in addition to the direct resource depletion. The Millennium Development Goals (MDG) set by the UN Development Programme will clearly have a strong influence on the rate and direction of environmental initiatives over the next decade. The challenging water supply and sanitation objectives (globally, for a further 700m. people to have ready access to water, where ready access is defined as a community tap per 250 people) have a real prospect of delivery in Latin America, where substantial progress has been made over the last 30 years. However, although promised, a regional, integrated vision of the attainment of the MDGs is still unrealized. The MDGs are likely to influence all of the sectors discussed below.

#### Mining

Latin America contains a wealth of mineral resources. The US Geological Survey (USGS) states that in 2000 the region was responsible for 43% of world copper production, 41% of world tin production, 39% of world silver output, 26% of world bauxite output and 24% of iron ore production. The USGS estimated that total tonnages of these minerals would continue to rise until 2007. Foreign mining companies, predominantly from Canada, the USA and Europe, have been moving into Latin America in strength since the 1990s, although indigenous, much older, mining activity continued. Increased opposition to the presence of these foreign companies and an internationalization of the protest were anticipated.

The appetite for foreign investment in the majority of Latin American governments' policies have led to favourable legal conditions on opening and operating mines in the region, certainly in comparison with mines in the USA or Canada, where environmental, health and safety and planning regulations are stricter. Indeed, a problem for implementing new environmental regulations in certain countries, including Mexico and Guatemala, is that trade agreements allow investors to sue for damages if their operations are affected by changes in environmental regulations (see section on waste management). It is this remarkable legal approach that creates conflict between resource development and environmental management.

The steady rise in many mineral prices, as well as the introduction of modern, large-scale, industrial methods of strip mining and cyanide-based refining, have made more mine sites financially viable. However, the reliance on the global market price for metals also means that the long-term viability (and thus employment) of such sites is insecure.

Modern open-cast mining practices carry serious environmental risks and issues: the large scale of the operations removes entire hillsides with the associated risks to flora and fauna, while the pulverized rock exposes dangerous elements that can contaminate groundwater; the use of chemicals such as cyanide or mercury to extract precious metals can lead to releases of these dangerous substances into the environment; and the mining operation itself consumes substantial quantities of energy and water. The lack of environmental regulation and enforcement has led to several high-profile environmental incidents, for example, the release, in 1996, of 400,000 metric tons of toxic waste into the Pilcomayo River in Paraguay, which devastated downstream fisheries, wildlife, as well as affecting indigenous communities who were not informed of the incident.

In recent years opposition to poorly controlled or illegal mining has been increasing both from Latin American governments and from international and local pressure groups. Several protests to mining proposals in Mexico, Guatemala and Peru have become violent (e.g., at the Marlin gold mine in western Guatemala in early 2005 and at the Peruvian La Zanja mine and another site near Cajamarca in November 2004). Honduran and Costa Rican courts both cancelled mining concessions in 2004 and Costa Rica imposed a moratorium on granting new mining rights from 2002.

#### Waste and Industrial Effluents

Deposition of untreated sewage and industrial wastes into rivers, streams and lakes continued to be a major problem across much of the region. In the late 1990s, according to the Pan American Health Organization, only about 13% of Latin America's collected sewage received any treatment and large numbers of people were unconnected to any sewage collection system.

Agriculture accounts for a large proportion of economic activity across the region, with ranching dominating the sector. Such agricultural activity is associated with the use of fertilizers, herbicides and pesticides. Fertilizer consumption rose by 42% between 1990 and 1998 (according to the UN Economic Commission for Latin America and the Caribbean—ECLAC, and UNEP). It is anticipated that eutrophication (excessive aquatic nutrient-enrichment) problems will increase in Latin America, as will the release of diverse pesticides to the food chain.

Mining is also a major source of water pollution: the large-scale operations described above historically have had a poor environmental record, with several accidental releases of toxic material into the environment. Mining-related contamination is derived from both small, indigenous and large-scale, international activity. Small-scale, independent gold-mining operations in Brazil (by *garimpeiros*) has led to fish in many streams exceeding the World Health Organization regulations for mercury. Heavy metal waste such as lead, cadmium and mercury will be retained in the sediments of the river systems and will become a pollution legacy for the future.

### Air Quality

Air quality is a major concern to many Latin American governments, particularly in the large and rapidly growing city environments where air pollution poses a public health risk. However, it is also an issue in rural and remote areas where mining and logging operations impact on air quality. Between 1990 and 1999 regional emissions of 'greenhouse' gases rose by 37%, largely owing to the increased use of cars and electricity. Other pollutants also reported to be growing in the same period were sulphur dioxide (by 22%), nitrogen oxides (41%), hydrocarbons (45%) and carbon monoxide (28%). Specific targeted initatives have improved air quality in certain locations in recent years, but it is estimated that air pollution permanently affects the health of more than 80m. people in the region (according to ECLAC and UNEP). By 2010, again according to UNEP, some 85% of the population was likely to be living in urban areas and so addressing urban air pollution was a priority.

### Biodiversity

The tropical, sub-tropical and temperate habitats of Latin America and the Caribbean are exceptionally rich in biodiversity, containing up to 40% of the plant and animal species on the planet (see UNEP's *Global Environmental Outlook 2000*). Major risks to biodiversity are: population growth and unsustainable economic development; introduced invasive species; mining and oil exploration; colonization politics; logging; subsistence farming; and the wildlife trade and habitat fragmentation. The Amazon region alone holds an estimated 10%–17% of the world's species (see Gibson, A. C. (Ed.). *Neotropical Biodiversity and Conservation*. Mildred E. Mathias Botanical Garden, University of California, Los Angeles, 1996).

Fragmentation of habitats through largescale burning of forests in both Central and South America create islands of forest surrounded by agricultural or burned land. Indigenous species in these areas encounter several problems: animal species requiring a large area to roam are disadvantaged, and encroachment of non-indigenous species from the surrounding areas alter the balance of competition. As is common around the globe, human encroachment on these areas in the form of road and transport links, create more agricultural and industrial pressures on the landscape.

## LAND AND LAND USE

### Deforestation

Deforestation is probably the single most important environmental degradation taking place in the early 21st century. There are two reasons for this. The first relates to the biodiversity of the tropical forests and the second lies in the carbon stored in the tropical rainforests. We will consider these two factors in reverse order. The planet would not be capable of supporting life (at least as we know it) without an atmosphere that controls the global heat balance. Shortwave radiation from the sun passes through the atmosphere and warms the earth's surface. The warm surface radiates heat back towards space, but this is longwave radiation (the wavelength depends on the temperature of the emitting body: the hotter the body, the shorter the wavelength) and this energy is trapped, in part, by carbon dioxide in the atmosphere. Too much carbon dioxide causes too much radiation to be captured and so global warming occurs. The tropical forests of Latin America are a major store of carbon. While these forests exist that carbon is not released into the atmosphere. Deforestation causes a release of carbon into the atmosphere, causing global warming and thus, potentially, catastrophic global change.

Of course, this analysis could be wrong but, should the world wait until it is proven correct, it will be too late to remedy. For this reason, all governments in countries with substantial stocks of forest are under major pressure to control deforestation. Whether such political pressure is being translated into ecologically responsible policies is considered below. However, first, the significance of biodiversity needs to be captured. The environment has always fluctuated. There are eras of higher and lower temperatures, rainfall, ice advances, etc., as well as catastrophic events such as meteorite impacts; however, the natural environment has been able to adapt to such changes. The key to such adaptability is the diversity of the life on the planet: there are always species available that can cope with the new conditions and expand to fill the place of species that could not survive. The tropical rainforests of Latin America and the Caribbean are the major repository of this vital biodiversity. Put more starkly, the resilience of the planet is housed, in large part, in Latin America. So does the rate of deforestation show any sign of stopping, or at least slowing?

Forested areas in Latin America and Caribbean are shown in Table 1. This table is derived from the World Resources Institute's (2004) tabulations, which are drawn primarily from UN reports. The dominant contributor to the forest cover is Brazil with more than one-half of the total forest cover for South America, Central America and the Caribbean. Peru, Mexico, Bolivia, Colombia and Venezuela, in descending order, are the next most significant forest-cover countries, each contributing between 49m. ha and 64m. ha. Globally, forest cover declined by an average of 0.2% per year in 1990–2000. However, in South America the annual average percentage change over this period was double this figure and in Central America and the Caribbean in the same 10 years the decline in forest cover averaged 1.1% per year. The highest annual declines in forest cover were achieved in Haiti, where an annual average rate of 5.7% was recorded, in El Salvador, with 4.6%, and in Nicaragua, where forest cover decreased by 3% per year. Only in Uruguay and Cuba was an increase in forest cover per year over the last decade recorded.

It is the absolute significance of forest cover and the steady decline in this cover that has created international pressure on these countries to reduce the forest harvest. (Perhaps cull would be a more appropriate term, as harvest implies a degree of sustainability with a suggestion that there will be further annual harvests, whereas the exploitation of the tropical rainforest is often associated with a change in use to urban, grassland or cropland.) This pressure exerted by the international communities and international environmental groups is not entirely fair, since most of Europe, for example, has in the past, exploited its forest cover far more severely than Latin America. Nevertheless, in much of the developed world, timber must now be certified and the most widely recognized certification scheme is that produced by the Forest Stewardship Council, an independent, non-governmental organization (NGO) based in Bonn, Germany. Just over 5m. ha of forest land have been certified in Latin America from a total forest cover of just short of 1,000m. ha. While this is a meagre land area relative to the total forest area, it starts to be a significant contribution to controlling the annual exploitation of timber, particularly in the light of the very rapid increase in certified timber holdings achieved in the 1998–2003 period, which amounted to an annual increase in certified forest area of approximately 30%. Superficially, this suggests in very broad terms, that in 20–30 years the certified plantation forest area will have grown to sufficient size to arrest the decline in natural forest cover. This, however, makes the huge assumption that the certified forest areas can directly substitute for the timber exploitation of the old growth, largely tropical rainforest.

To explore this assumption it is necessary to consider the harvest in a little more detail. Harvest statistics are given in Table 2. South America, with 337.6m. cu m of roundwood removals in 2003, according to the UN's Food and Agriculture Organization, is the most significant harvest area. In contrast, Central America and the Caribbean produced a timber harvest of 94.0m. cu m in the same period. In terms of individual countries, Brazil, with the largest forest cover, is the most significant harvest country, producing 238.5m. cu m of roundwood in 2003. The key, however, to understanding the environmental policy significance of these statistics is the balance

**Table 1: Forest Area in Latin America and the Caribbean**

|  | Total forest | | Natural forest | | Plantations | |
| --- | --- | --- | --- | --- | --- | --- |
|  | Area ('000 ha) 2000 | Annual % change 1990–2000 | Area ('000 ha) 2000 | Annual % change 1990–2000 | Area ('000 ha) 2000 | Annual % change 1990–2000 |
| **Central America and the Caribbean** | 78,737 | −1.1 | 76,556 | −1.2 | 1,295 | −0.5 |
| Belize | 1,348 | −2.3 | 1,345 | −2.4 | 3 | 3.6 |
| Costa Rica | 1,968 | −0.8 | 1,790 | −1.4 | 178 | 9.6 |
| Cuba | 2,348 | 1.3 | 1,867 | 0.1 | 482 | 7.6 |
| Dominican Republic | 1,376 | — | 1,346 | −0.3 | 30 | — |
| El Salvador | 121 | −4.6 | 107 | −6.1 | 14 | — |
| Guatemala | 2,850 | −1.7 | 2,717 | −2.2 | 133 | — |
| Haiti | 88 | −5.7 | 68 | −7.6 | 20 | 5.1 |
| Honduras | 5,383 | −1.0 | 5,335 | −1.1 | 48 | — |
| Jamaica | 325 | −1.5 | 317 | — | 9 | — |
| Mexico | 55,205 | −1.1 | 54,938 | −1.1 | 267 | — |
| Nicaragua | 3,278 | −3.0 | 3,232 | −3.2 | 46 | 14.3 |
| Panama | 2,876 | 1.6 | 2,836 | −1.8 | 40 | 17.3 |
| Trinidad and Tobago | 259 | −0.8 | 244 | — | 15 | — |
| **South America** | 885,618 | −0.4 | 875,163 | −0.5 | 10,455 | 6.7 |
| Argentina | 34,648 | −0.8 | 33,722 | −1.1 | 926 | — |
| Bolivia | 53,068 | −0.3 | 53,022 | −0.3 | 46 | 3.7 |
| Brazil | 543,905 | −0.4 | 538,924 | −0.4 | 4,982 | 3.2 |
| Chile | 15,536 | −0.1 | 13,519 | −0.8 | 2,017 | 5.5 |
| Colombia | 49,601 | −0.4 | 49,460 | −0.4 | 141 | 6.2 |
| Ecuador | 10,557 | −1.2 | 10,390 | −1.3 | 167 | 2.4 |
| Guyana | 16,879 | −0.3 | 16,867 | — | 12 | — |
| Paraguay | 23,372 | −0.5 | 23,345 | −0.5 | 27 | 11.3 |
| Peru | 66,215 | −0.4 | 64,575 | −0.5 | 640 | 15.2 |
| Suriname | 14,113 | — | 14,100 | 0.0 | 13 | 0.8 |
| Uruguay | 1,292 | 5.0 | 670 | 0.0 | 622 | 16.3 |
| Venezuela | 49,506 | −0.4 | 48,643 | −0.5 | 863 | 8.7 |

Source: mainly World Resources Institute.

between industrial roundwood (tree trunks) and fuel wood. In both South America and in Central America and the Caribbean the wood fuel harvests *exceed* that harvested for industrial use. Indeed in only two countries, Argentina and Chile, is the majority of the forest harvest as industrial roundwood. What does this mean for the effectiveness of environmental policy? Firstly, forest destruction is strongly influenced by domestic indigenous exploitation as well as by multinational timber companies. Secondly, as a consequence of this, control by international treaty and by certification of Western timber sales is unlikely to be successful as the means of controlling tropical deforestation. It is necessary to influence the harvest at community and family levels and to provide an economically and socially acceptable fuel substitute.

According to figures based on satellite imagery from the Brazilian Ministry of the Environment, between August 2003 and August 2004 an area of 26,130 sq miles (67,677 sq km) of forest was cleared, the second highest figure recorded since annual surveys began in 1989. Average annual loss of forest in 1980–90 was 7.4m. ha, and, in 1990–95, annual deforestation averaged 5.8m. ha. Almost all (95%) of this loss was in tropical regions (according to ECLAC and the UN's Food and Agriculture Organization). Nevertheless, the region still has the highest percentage of the world's closed woodland at 32%. There is, however, some room for optimism in these declining figures: UNEP's *One Planet, Many Peoples* publication clearly indicates that, while the loss of forest in Brazil and Paraguay in recent years has been enormous, the losses in Argentina, where environmental regulation is much more effective, was much smaller. Thus, it can be concluded that much of the failure to control deforestation is owing to lack of political will.

The direct causes of deforestation are expansion of agricultural activity and forest fires. Forest fires can form part of the natural eco-cycle, but in the Latin American region and, particularly, in Central America and tropical South America, forest fires are started to clear land for agriculture and to promote grass growth for grazing livestock. These practices lead to a permanent change in the predominant ecosystem. Largescale burning and deforestation threatens air quality over large regions.

**Table 2: Total Roundwood Removals, 2003**
('000 cu m)

|  | Industrial roundwood | Wood fuel |
| --- | --- | --- |
| **Central America and the Caribbean** | 12,826.6 | 81,178.4 |
| Bahamas | 17.0 | — |
| Barbados | 5.0* | — |
| Belize* | 61.6 | 126.0 |
| Costa Rica | 1,687.0 | 3,453.8* |
| Cuba* | 808.0 | 1,828.0 |
| Dominican Republic* | 6.3 | 556.0 |
| El Salvador | 682.0 | 4,147.1 |
| Guadeloupe* | 0.3 | 15.0 |
| Guatemala | 509.0 | 15,551.9* |
| Haiti* | 239.0 | 1,985.2 |
| Honduras | 801.0 | 8,703.4* |
| Jamaica* | 282.4 | 577.1 |
| Martinique* | 2.0 | 10.0 |
| Mexico | 7,420.0 | 38,090.4* |
| Nicaragua | 93.0 | 5,865.4* |
| Panama | 153.0 | 1,233.5* |
| Trinidad and Tobago | 60.0 | 35.3* |
| **South America** | 148,615.0 | 188,942.8 |
| Argentina | 5,335.0 | 3,972.0* |
| Bolivia | 650.0 | 2,205.6* |
| Brazil | 102,994.0 | 135,542.5* |
| Chile | 27,491.0 | 12,712.4* |
| Colombia | 2,068.0 | 7,890.5 |
| Ecuador | 913.0 | 5,349.5* |
| French Guiana* | 60.0 | 89.2 |
| Guyana | 292.0 | 869.6* |
| Paraguay* | 4,044.0 | 5,842.5 |
| Peru | 1,192.0 | 9,073.5 |
| Suriname | 155.0 | 44.0* |
| Uruguay | 2,132.0 | 1,607.0* |
| Venezuela | 1,289.0 | 3,744.5* |

*FAO estimate(s).

Source: FAO.

## Desertification

The general perception of Latin America is of a land of rainforest. However, between one-quarter and one-third of the land area is desert or dryland. Although this is a very minor component, for some countries the percentage of dryland is significant: Mexico, Argentina, Ecuador, Paraguay, Peru, Venezuela and Jamaica all have more than 30% of their land area as drylands and are all therefore sensitive to the global threat of desertification. In absolute dryland area, Mexico is dominant, but Brazil, by virtue of its vast land area, has almost the same dryland area designations. The UN Convention to Combat Desertification characterizes the dry regions of Latin America thus:

'The deserts of Latin America's Pacific coast stretch from southern Ecuador along the entire Peruvian shoreline and well into northern Chile. Further inland, lying at 3,000–4,500 metres in altitude, the high, dry plains (Altiplano) of the Andean mountains cover large areas of Peru, Bolivia, Chile, and Argentina. To the east of the Andes, an extensive arid region extends from the Chaco's northern reaches in Paraguay to Patagonia in southern Argentina. Northeast Brazil contains semi-arid zones dominated by tropical savannah. Most of Mexico is arid and semi-arid, especially in the north. The Caribbean states of the Dominican Republic, Cuba, Haiti and Jamaica, amongst others, also contain arid zones; erosion and water shortages are noticeably intensifying in many East Caribbean islands.' (www.unccd.int)

The vulnerable drylands support about 25% of the population and the issues (crop vulnerability, water limitations, and soil loss) are accentuated in island communities. Desertification (anthropogenic processes that produce desert-like conditions in drylands) is strongly associated with socio-economic characteristics of the population. Typically, the people live in extreme poverty and, because the land is relatively unproductive and so cannot support a large population, there is out-migration, often by males to the cities to seek work. This leaves a female-dominated population in the drylands. The Inter-American Development Bank is supporting initiatives to bring integrated management to the drylands in Argentina, the Dominican Republic, El Salvador, Guatemala, Honduras and Nicaragua.

Perhaps more encouraging, given the international scale of the problem, are the international initiatives implemented in the early 21st century. A programme to address the common problems encountered in the Gran Chaco Americano region, the largest dryland ecosystem in Latin America, is under way in Argentina, Bolivia and Paraguay. The forests of the Gran Chaco, although structurally very different from those of the Amazon, are being lost at a rate greater than that experienced in the rainforest regions. Initiatives have been launched to improve regulations, education and health, as have joint-investment programmes, supported by a working groups representing farmers, the community, local governments, local NGOs and the private sector. A broadly similar effort, known as the Puna Programme, in the Andes region of Argentina, Bolivia, Chile, Ecuador and Peru, focuses on the promotion of sustainable development.

## LOCAL REACTION TO RESOURCE DEVELOPMENT

When existing resource use is successful and substantial, new, typically larger scale activities that threaten the viability of the existing enterprises are frequently opposed at the local level. For example, in Peru the area of Tambogrande successfully sustains lemon, papaya and mango farming, partially owing to internationally funded irrigation and land-distribution projects, and provides sustainable livelihoods for some 20,000 farmers. In 2002, in a community referendum, over 90% of local people voted against a proposal by the Manhattan Mineral Corporation to develop a gold mine, fearing that contamination of water supplies by mining activities would threaten their community.

The simple scale of new developments and, often, the environmental impact of associated developments (secondary processing and transport, for example) raise significant concerns. These concerns can be based on the expected direct impact or on the possibility that the regulatory framework will be insufficient to cope with the large, often international, scale and diversity of the development. The Alumysa project raises such concerns. The project, a venture by Canadian mining company Noranda to create an aluminium plant, three hydroelectric dams and a new port in the Aisén region of Chile, was suspended in August 2003 following local and international protest at the environmental impacts of the proposed development and associated facilities. However, neither the Chilean government nor Noranda have cancelled the project. According to the environmental organization Friends of the Earth Chile the operation involves a US $2,750m. aluminium production plant, which will necessitate construction of a new port for the import of raw alumina and other materials. Furthermore, 'the three planned hydroelectric dams that will supply the requisite 758 MW of energy for the plant will flood an area of 9,600 ha. The project also includes 95 km of new roads, 79 km of power lines, a wharf and the expansion of two nearby towns.'

Environmental concerns about the plant centred not only around the impact on the biodiversity of the area caused by the flooding of a large area of land, but also around the increase in shipping required to transport the raw materials and products and the large volumes of waste produced by plants of this scale, estimated at over 1m. metric tons annually. It was feared that this waste would pollute local rivers, streams and lakes and impact on the flora, fauna and the smallscale fishing enterprises that generate employment for local people. Friends of the Earth Chile also claimed that the project offered limited employment opportunities for the region's inhabitants, with only 10% of the 1,100 jobs at the plant likely to go to local people.

Honduran environmental groups highlight the impact that timber-export, banana, mining and tobacco companies have on the local environment. Deforestation owing to timber-export and large mono-crop activities has led to a reduction in biodiversity. Banana plantations cause water pollution through their use of non-biodegradable plastic bags that dress the banana clusters, which allegedly contain pesticides and dioxins.

Many vital resources such as water supplies are inadequate both in quality and quantity, and in sites of rapid urbanization the resulting problems can be acute. However, interventions designed to remedy such inadequacies may simply provide a product that is financially unattainable by a significant proportion of the local population, and will result in the loss of the inadequate but attainable previous resource provision system. One example of this was in the Bolivian city of Cochabamba in late 1999 and early 2000, when more than 100,000 people protested against a sharp increase in water charges by the newly privatized water authority, owned by Aguas del Tunari, a joint venture between the US-based Bechtel and the Italian company Edison. Before its privatization, the local water company was debt-ridden and had a reputation for mismanagement. Furthermore, only 60% of Cochabamba residents had access to water. Aguas del Tunari improved the hours of supply and the water pressure, but the concomitant increase in water charges, by 200% according to some local sources, was unpopular. The Bolivian Government responded to the public protests by suspending the 40-year lease agreement with Aguas del Tunari, prompting the water company to threaten to sue the Government.

Environmental groups, both local and international, continue to raise the issue of the deforestation of the Amazon basin. One method is through the creation of systems of certification such as the Forest Stewardship Council accreditation, which guarantees that the timber is sourced from renewable and well-managed forests. However, the market value of the product generates pressure to circumvent the regulations. Over 26,000 sq miles of forest were lost in the Amazon basin in 2004, the second largest annual loss ever. According to figures released by Brazil's National Space Sciences Institute (Instituto Naciona de Pesquisas Espaciais), deforestation in the Amazon in the 12 months from August 2003 increased significantly compared with the previous 12 months. It was estimated that around 17% of the original natural vegetation in the Brazilian Amazon has been destroyed. Global conservation organizations, such as the World Wide Fund for Nature, have criticized the Brazilian Government for encouraging real-estate speculation within forest areas in order to expand cattle-ranching and industrial-scale farming. In June 2005, nevertheless, following an extensive operation, police in the Mato Grosso region arrested over 80 people in connection with organized illegal logging activities.

Environmental concerns also frequently relate to the incidental adverse effects of a development. In southern Peru, near the city of Arica, for example, proposals for a copper mine in one of the driest regions in the world led to environmental concerns over the availability of water. Friends of the Earth report that the mining operation will use 700 litres of water per second. Locals are concerned about the use of groundwater. Chilota, one of the sources from which the mining company, Minera Quellaveco, plans to extract water, is a biologically diverse wetland that also provides grazing for alpacas and lamas, allowing local communities a subsistence income. The use of Chilota groundwater may well lower the area's water table and change the microclimate. Farmers in nearby Tala claim they will be impacted by the dust from the mining operation, and that the planned diversion of the nearby Asana river into a small riverbed running through the area will flood their productive lands. In this case then, are concerns over loss of water resources, reduced biodiversity, loss of grazing, change in microclimate, atmospheric pollution and flooding.

## ENVIRONMENTAL EXPLOITATION AND PROTECTION

According to the Convention on International Trade in Endangered Species of Wild Fauna and Flora (CITES) 2004, in 2000 South America was a net exporter of live animals, mainly lizards, of which there were some 200,575 net exports in that year. South America is a net exporting region for all elements of species trade with the exception of snake skins. Central America and the Caribbean is a net exporter of live animals, but a strong net importer of skins, 710,492 in 2000. Skin imports to the region were totally dominated by Mexico, which, in 2000, registered net imports of 932,624 skins (net crocodile skin imports numbered 275,647, lizard skins 238,011 and snake skins 180,969). As such specialist leathers are all directly used in manufacture (cases, shoes, clothing, etc.), this suggests that Mexico, which has also stood apart from many other countries in other aspects of this analysis, is industrializing and, despite a Latin language basis, is physically and economically turning towards the USA and Canada rather its southern Latin American community.

The picture thus far then is of a group of countries rapidly exploiting their important forest resources and exporting both live and dead natural faunal products. To combat this, many Latin American countries have started to take seriously conservation and protection of species and habitats. In recognition of the numbers of species under threat of extinction, most countries have set aside land and marine reserves under the criteria set out in the IUCN—The World Conservation Union's Management Categories I–VI. (These categories are: strict nature reserve or wilderness area; national park; natural monument; habitat/species management area; protected landscape/seascape; managed resource protected area.) Overall, in Central America and the Caribbean, a total of 15.1% of land is protected, while in South America this figure is lower, at 10.6%. These figures would bear comparison with, for example, the United Kingdom or Canada. This figure ranges for individual countries from 0.3% in Guyana and Uruguay to 84.6% in Jamaica, but, of course, these percentages have to be viewed in the context of the total land mass of these countries. While Brazil has one of the smaller holdings of land under IUCN protection (6.7%), it has by far the largest area (125.0m. ha in 2002) of biosphere reserves (a less specific categorization protecting entire habitats).

Perhaps the international manifestation of commitment to environmental protection is in the agreements to which a country has put its name. In part, membership is driven by geographical location and so only those countries that have part of their land mass 'close' to a direct oceanic connection to the Antarctic have opted to sign for these more regionally focused Antarctic initiatives. Several environmental treaties— Biodiversity, Climate Change Convention, Desertification, Endangered Species and Ozone Layer Protection—have had an almost universal take up rate among the Latin American countries. The popularity of such agreements can be driven by the political perceptions of the time or by self-interest, either because the subject matter is particularly significant for that country or because the perception is that the agreement can be delivered without cost to the government of the country. In broad terms, however, as a topic becomes more specific with identified deliverables or limitations the rate of ratification declines.

## CURRENT ISSUES

Drawing a line between pure environmental issues and the social and urban environment is problematic. Certainly, the management of waste, the issues of urbanization and the consequences of rural to urban migration and the consequential failure of water and sewerage infrastructures, if they were ever sufficient, are current issues that will not be resolved in this or the next generation. (In the 20 years since the international water supply and sanitation decade around 1980, little progress has been made and in 2004 there remained 1,000m. people without ready access to clean water and effective sanitation.) In this section we will deal with the rural natural environment, rather than the urban and semi-urban environment.

**Table 3: Adapted Land Use in Latin America and the Caribbean**
(sq km)

|  | Grasslands | Croplands | Urbanized and built-up areas |
|---|---:|---:|---:|
| **Central America and the Caribbean** | 333,104 | 403,755 | 3,385 |
| Belize | 1,011 | 6,263 | 1 |
| Costa Rica | 52 | 8,406 | 25 |
| Cuba | 7,507 | 48,201 | 406 |
| Dominican Republic | 5,750 | 9,743 | 127 |
| El Salvador | 102 | 6,862 | 180 |
| Guatemala | 5,665 | 27,948 | 213 |
| Haiti | 5,173 | 11,565 | 15 |
| Honduras | 2,272 | 32,262 | 68 |
| Jamaica | 1,255 | 1,243 | 42 |
| Mexico | 301,448 | 193,863 | 1,884 |
| Nicaragua | 336 | 29,603 | 42 |
| Panama | 603 | 18,030 | 78 |
| Trinidad and Tobago | 25 | 11 | 80 |
| **South America** | 1,101,213 | 1,218,853 | 16,625 |
| Argentina | 540,646 | 368,279 | 3,292 |
| Bolivia | 65,840 | 16,880 | 116 |
| Brazil | 116,212 | 584,276 | 9,832 |
| Chile | 86,521 | 14,185 | 628 |
| Colombia | 44,792 | 49,648 | 528 |
| Ecuador | 16,987 | 23,702 | 259 |
| Guyana | 1,514 | 809 | 9 |
| Paraguay | 10,912 | 6,684 | 293 |
| Peru | 133,590 | 15,970 | 184 |
| Suriname | 192 | 105 | 12 |
| Uruguay | 66,098 | 37,973 | 268 |
| Venezuela | 17,794 | 10,316 | 1,202 |

Current environmental issues in Latin America depend to a great extent on the nature of the country concerned. For example, the majority of the Caribbean countries, being islands with relatively short rivers and therefore relatively rapid runoff, accentuated by seasonal rainfall and a seasonal tourist population that is not coincident with the higher rainfall periods, are concerned with water shortages and water resources. Authorities in Anguilla, Antigua and Barbuda, the British Virgin Islands, the Cayman Islands, the Dominican Republic, the Turks and Caicos Islands and the US Virgin Islands have all reported concerns over freshwater resources, either because of the limited nature of the fundamental resource or because the availability of the resource was becoming compromised by other environmental degradations, such as deforestation. Several countries (Barbados, Belize, Costa Rica, El Salvador, Guatemala, Honduras, Mexico, Trinidad and Tobago) have reported water pollution problems ascribed to careless agriculture and mining activities. These water pollution circumstances compromise the availability of the limited freshwater resources. A smaller number of countries (Costa Rica, the Dominican Republic, Jamaica, Nicaragua, Saint Vincent and the Grenadines) have reported that the pollution is concentrated in the coastal environment and thus threatens the loss of the coral environment and tourism. The Global Water Partnership, an NGO based in Stockholm, Sweden, that supports countries in the sustainable management of their water resources, offers a water-related vision for water in the 21st century.

In contrast, the 'continental' Latin American countries are less concerned with water and marine issues. Here, the current environmental issues are reported as waste disposal, deforestation, desertification, overgrazing and soil erosion, as well as the consequences of urbanization. These environmental problems do not arise in isolation. Deforestation is likely to lead to soil erosion. This occurs because in tropical regions the bulk of the nutrients are locked in the trees and not in the soil. When the trees are removed, the nutrients are also removed and the successor vegetation is nutrient-poor and thus vulnerable. This vegetation also lacks the foliage cover to protect against intense rainfall. Land degradation owing to deforestation, desertification or overgrazing leads to a reduced capacity to support people who then are forced to migrate to cities, thus generating the problems associated with rapid urbanization. Urbanized areas are provided in Table 3.

## CONCLUSIONS

Latin America and the Caribbean is the guardian of many of the world's most important habitats. It holds the key to the preservation or otherwise of huge tracts of tropical rainforest and associated fauna and flora likely to be the world's greatest terrestrial 'gene bank', the vast majority of which is unknown, or at least unassessed. Failure to preserve the rainforest is likely to have a significant effect on global climate. Furthermore, and perhaps perversely given the continent's tropical habitats, Latin America and the Caribbean is closest to many Antarctic sites and holds many of the small island nations that will be most at risk in the event of sea-level changes and altered storm tracks.

The vast majority of Latin American and Caribbean governments are engaging with international environmental treaties at least to the same degree as some economically wealthy countries. Not all jurisdictions are at the same stage in the signing and ratification process, however, and it remains to be seen how effective the implementation of such treaties will be promoted in each country.

## BIBLIOGRAPHY

Central Intelligence Agency (CIA). *CIA National Statistics Handbook*. Washington, DC. www.cia.gov/cia/products/hies.

Convention on International Trade in Endangered Species of Wild Fauna and Flora (CITES), 2004. www.cites.org.

UN Environment Programme. *Global Environment Outlook 2000*. Nairobi, 2005.

*One Planet, Many Peoples: Atlas of Our Changing Environment*. Nairobi, 2005.

World Resources Institute. Earthtrend tabulations, 2004. www.earthtrends.wri.org.

# TRADE AGREEMENTS IN THE AMERICAS

## Dr NICOLA PHILLIPS

### INTRODUCTION

Following the *de facto* collapse of the negotiations towards a Free Trade Area of the Americas (FTAA) in late 2003, the politics and substance of trade agreements in the region have undergone a striking shift. Until that time, the emphasis since the early 1990s had been firmly on hemispheric integration and the negotiation of a free trade area stretching 'from Alaska to Tierra del Fuego', alongside the establishment or overhaul of the subregional blocs associated with the North American Free Trade Agreement (NAFTA), the Southern Common Market (Mercado Común del Sur/Mercado Comum do Sul—Mercosur/Mercosul), the Caribbean Community and Common Market (CARICOM), the Central American Common Market (CACM) and Andean Community of Nations (Comunidad Andina de Naciones—CAN). In 2004–05, however, the USA's focus, and indeed that of other key countries of the region, has been the negotiation of new, bilateral trade agreements within the region and outside it. To a large extent, this shift has been a direct response to the stalemate and paralysis of the FTAA negotiations and the similar deadlock that has for some time prevailed in the multilateral arena. In other words, bilateralism appears to have been firmly established as the format for trade negotiations in the Americas at a time when the FTAA process has lost all impetus and the prospects for successful conclusion of the World Trade Organization's (WTO) multilateral 'Doha round' of negotiations seemed, at best, uncertain.

The primary aim of this article is to examine these recent developments in the patterns and politics of trade in the Americas, and to evaluate the implications of both the shift to bilateralism and the specific agreements that are in train. It will first survey the nature of the FTAA project and sketch out the contours of the diverse and divergent interests that have shaped the politics of the negotiations and, ultimately, precipitated the stalemate that led to their collapse in late 2003. It will then turn, in this light, to the reasons for the reformulation of US trade strategies to emphasize bilateral negotiations with a range of partners in the region, paying particular attention to the Central American Free Trade Agreement (CAFTA) that, in late July 2005, received congressional approval in the USA. It will then outline briefly the ways in which this new bilateralism is also favoured by many other countries in the region and, in these cases, is not particularly 'new': countries such as Canada, Chile and Mexico have long prioritized bilateral trade agreements alongside their subregional and hemispheric commitments. The final section will offer concluding thoughts on the political, economic and developmental implications of this new template for trade negotiations in the region.

### THE FTAA PROJECT

The FTAA project grew out of the initial impetus given to a notion of hemispheric integration by the Enterprise for the Americas Initiative (EAI), announced by US President George Bush, Sr (1989–93) in 1990. This set in train a broader hemispheric process known as the 'Summitry of the Americas', the first summit meeting of which was convened in Miami, Florida, USA, in 1994. It was at this meeting that the cornerstone of this new 'multilateralism' in the Americas—the project of a hemispheric free trade area—was first given formal expression, with the goal of achieving an agreement to this end between all of the countries of the Americas (except Cuba) by the beginning of 2005. What was especially interesting about the FTAA process, in this sense, was that it signalled decisively the abandonment by the US state of its traditional reticence in matters of regional integration, preferring instead to adhere to a longstanding ideological preference for multilateralism. The greater openness to regionalism in the late 1980s and early 1990s was largely a defensive reaction against regionalist developments elsewhere, particularly the deepening of integration in the European Union (EU), the proliferation of regionalist initiatives elsewhere, and the sorts of economic competition with Asia (especially, at the time, with Japan) that had been both characteristic of the 1980s and intimately connected with perceptions in the USA of its declining global hegemony. The result of this combination of trends, as well as the significant movements towards neo-liberalism and democracy across Latin America and the Caribbean, was the signing of the NAFTA by Canada, Mexico and the USA in 1993 and the process of wider hemispheric integration encapsulated in the EAI and the FTAA.

Yet, just as the NAFTA process was driven primarily by insistent pressure from Mexico, the FTAA process was propelled forward largely by the interest of Latin American and Caribbean governments and a number of other (especially business) actors. Following the announcement of the EAI, there was something of a hiatus until the Administration of Bill Clinton (1993–2001) instigated negotiations some years later. Even then, it is striking that trade and economic matters were largely second-order issues on the agenda for the First Summit of the Americas (Weintraub, 2000, see Bibliographical References), achieving later salience not as a result of US interest but rather as a result of sustained advocacy from Latin American governments, as well as a result of vocal support from regional business interests. Indeed, one of the defining characteristics of the FTAA process throughout the 1990s was the absence of serious political leadership from the USA—or indeed any other country—and the concentration of activity and discussion in the technical working groups rather than in political arenas. What was also notable for much of the 1990s was that the FTAA hardly made an appearance in domestic political debates across the region—particularly not in the USA—and that the negotiating process was shrouded in secrecy, in as much as the negotiations were treated as confidential and lacked input or participation from civil society. It was not until the late 1990s that information about the FTAA was made more freely available to the public, largely as a result of sustained pressure from civil society organizations, and that serious political debate thus began to emerge within, and indeed between, the countries of the region.

Once the FTAA process was underway, the immediately contentious issues centred around the form that the negotiations should take, the format in which they should be conducted, and their desirable scope. Many of the debates and tensions around these issues were only slowly resolved. It was largely agreed from the beginning of the process that the FTAA was to be conceived as a 'single undertaking'; that is, that there would be a single agreement encompassing all of the various negotiating areas, which would apply equally to all of the contracting countries, notwithstanding some support from US negotiators for allowing a so-called 'early harvest' should some areas of negotiation advance more quickly than others. Beyond this, preferences regarding the shape of the negotiations were distinctly different. The initial vision of the FTAA that prevailed in the USA was of a 'hub-and-spoke' regionalist arrangement, in which the countries of the region would be brought into the North American orbit through essentially bilateral negotiations with the USA. Furthermore, the project was conceived in the USA as a process of enlarging the NAFTA, in effect a 'NAFTA on steroids', as it was dubbed, in which the existing North American agreement would constitute the template for the wider hemispheric agreement. Both these notions were vetoed early in the negotiations by Latin America and Caribbean participants. What emerged instead, and was enshrined at the WTO's Third Trade Ministerial meeting in Belo Horizonte, Brazil (following a first meeting in Denver, Colorado, USA, in June 1995 and a second in Cartagena, Colombia, in March 1996), was a format in which the FTAA would be the result of negotiation between subregional blocs. With the exception of the NAFTA bloc, the member countries of each of the existing subregional units in the Americas thus undertook a strategy of collective negotiation or 'bloc bargaining', designed explicitly to enhance their own leverage in the negotiations and, by extension, to dilute the unilateral dominance of the agenda by the USA.

The Fourth Trade Ministerial in San José, Costa Rica, in March 1998 was the point at which the structure of the negotiating process was agreed. It had been established at the outset that the negotiations would proceed largely within a range of technical working groups, each dedicated to a specific issue area. Initial proposals from US negotiators concerning the structure of these groups, however, garnered some controversy as they excluded the two areas of key importance to the rest of the participants—agriculture and trade remedies (the range of remedial mechanisms capable of deployment in instances of 'unfair' commercial practices by trade partners, such as anti-dumping measures, countervailing duties, safeguards and quotas). This omission reflected the categorical US insistence that neither area was to be opened for negotiation in a regional arena; rather, these areas could only be negotiated in the multilateral forum of the WTO. The eventual inclusion of agriculture and trade remedies in the working group structure was eventually conceded at the equally trenchant insistence of Latin American and Caribbean participants (Feinberg, 1997). The structure agreed in San José thus comprised a total of nine technical working groups: market access; agriculture; services; investment; government procurement; intellectual property rights; subsidies/anti-dumping/countervailing duties; competition policy; and dispute settlement. The areas of labour and environmental issues, on the one hand, and special and differential treatment for smaller and poorer economies, on the other, did not have separate forums within this structure, but rather remained to be indirectly filtered into the activities of the nine working groups by interested negotiators.

Finally, the desirable scope of the negotiations was linked not only to the remit established by the identification of the nine negotiating areas, but also to the level and depth of agreement to which negotiators should aspire. In this regard, preferences again diverged. The US and Canadian positions, along with those of some other countries such as Chile, were consistently that the FTAA should aspire to a coverage and depth that exceeded those of existing multilateral provisions. That is, the FTAA should be 'WTO-plus'. The majority of Latin American and Caribbean positions stipulated conversely that the FTAA should aim, at least in the first instance, to be merely 'WTO-compatible'. Such preferences arose from widespread perceptions that the provisions agreed in the so-called 'Uruguay round' of multilateral trade negotiations had yet fully to be implemented, particularly in the key area of market access. The principle of 'WTO-plus', in addition, was held to be a selective one for the USA: its ongoing refusal to open the key areas of agricultural subsidies and trade remedies for negotiation, as well as reticence in granting significant concessions on market access in strategic sectors such as steel and textiles, were deemed by other participants to compromise the legitimacy of the US government's insistence on 'WTO-plus' as the template for a 'comprehensive' agreement. The principle of 'WTO-plus' was, nevertheless, formally accepted as the template for the negotiations, but not until November 2002, at the Seventh FTAA Ministerial in Quito, Ecuador (a fifth meeting was held in Toronto, Canada in November 1999 and a sixth was convened in the Argentine capital, Buenos Aires, in April 2001). Even so, Latin American and Caribbean negotiators' central concerns to ensure fuller implementation of Uruguay round market access agreements continued to form the foundation of their negotiating strategies. In other words, the broad stance remained one based on 'WTO-compatibility', seen as yet to be achieved, emphasizing at the same time that a 'WTO-plus' format must be genuinely that and reach across the full range of negotiating areas, including those unilaterally excluded from the agenda by the USA.

## DIVERGENCES AND INEQUALITIES IN THE FTAA PROCESS

These controversies surrounding what the FTAA process should look like arose in a fundamental sense from the wide divergences in the interests that motivated participation in an FTAA project and informed the priorities and preferences that were brought to the negotiating table. While Latin American and Caribbean interests cannot be considered in any way homogenous, and some of the individual countries' positions will be discussed below, nevertheless there are a range of ways in which the faultlines are mapped out largely between Latin America and the Caribbean, on the one hand, and the USA (and, in the main, Canada), on the other. Latin American and Caribbean interest in the process was, and remains, dictated primarily by market access to the US economy. For much of the region, this is, in part, as a result of the pronounced levels of economic dependence on the USA: around 80% of Canadian and Mexican trade, some 50% of Central American trade and about 40% of Caribbean and Andean countries' trade is with the USA.

In other part, Latin American and Caribbean interest stems from a generalized concern over the high tariff and non-tariff barriers in the USA to the entry of products of key importance to many Latin American and Caribbean economies. While average tariff levels in the USA and Canada remain considerably lower than in many parts of the region (around 4.5%, compared with 14% in Brazil and 16% in Mexico, for example), sectoral peaks remain highest in the USA (reaching 350%) and, moreover, the US economy remains protected by a formidable array of non-tariff barriers. These include frequent deployment of the armoury of trade remedies available to US policy-makers, extensive resort to *ad valorem* duties and other contingency measures, and the maintenance of significant government subsidies and other support mechanisms in a range of sectors. Agriculture (in which the USA accounts for around one-fifth of total global agricultural subsidies), textiles, steel and softwood lumber are examples of sectors in which US protectionist measures have been highly contentious. Indeed, under the presidency of George W. Bush (2001–) they have become more so, following measures implemented by his Administration to enhance this protection, notably the introduction of the Farm Security and Rural Investment Act of 2002 and the imposition of extensive steel quotas, the latter subsequently deemed illegal by the WTO. The structure of trade in the region is thus one in which barriers to the US market are greater and considerably more diverse than those encountered by US exports to the rest of the region.

The interest of Latin American and Caribbean governments, exporters and companies in a regional free trade initiative is thus largely self-explanatory. What are less obvious are the motivations of the USA. Outside the NAFTA area, the Americas region is only modestly significant in the overall profile of US trade, and indeed this modest significance extends only to the parts of the region clustered geographically around the USA itself. US interests in the region decline steadily as one moves south, becoming, on the whole, minimal by the time one reaches the Southern Cone economies. In contrast with the figures noted earlier for dependence on the US market in the rest of the region, the US accounts for only about 20% of the Mercosur countries' (Argentina, Brazil, Paraguay and Uruguay) aggregate trade. Moreover, given the widespread processes of unilateral trade liberalization that occurred across Latin America and the Caribbean during the 1980s and 1990s, market access-related benefits from a free trade agreement would accrue predominantly in the Latin American and Caribbean region, as it is in the US economy, not in Latin American and Caribbean economies, that barriers to trade and investment remain most concentrated.

There are certainly a range of market access benefits to be derived by the USA from hemispheric free trade, particularly in the area of the relatively high tariffs on US exports of manufactured products that continue to apply in many countries of the region. This, indeed, was one of the primary reasons for US interest in CAFTA. The expansion of trade in services was also a key motivating factor in the FTAA negotiations, as well as in subsequent bilateral agreements. However, the set of interests shaping US trade strategies have been, and remain, defined by a range of other economic disciplines associated with the 'new trade agenda'. Most notable among these are investment, intellectual property, competition policy and government procurement, as well as, to an extent, environmental and labour standards. Trade, in summary, is not only about trade. Rather, US interests are dictated primarily by the prospects of achieving binding regional agreements across a range of other economic policy areas, and trade is the mechanism by which the USA has chosen to pursue these priorities (Phillips, 2006). This is clear in the ways in which the FTAA was articulated, from its inception as a 'single undertaking': market access and trade benefits come only with agreements on investment, competition policy, intellectual property, etc. The structure of the technical working groups also reflects these priorities and, consequently, significant US leverage over the terms on which the negotiations have been conducted. The FTAA project was thus of particular relevance to US trade strategies in as much as it offered significant opportunities for implanting in the region the range of economic disciplines central to contemporary US economic policy, thereby circumventing the ponderous and gridlocked multilateral process.

In some respects, then, the divergences of interests in the FTAA negotiations followed a fairly straightforward faultline between the USA (and often Canada) and Latin America and the Caribbean. Yet the positions of Latin American and Caribbean participants in the FTAA process also featured important differences. The interest of the Mercosur countries in an FTAA was always rather less pronounced than that of many others given that, at an aggregate level, the subregional trade structure and export profile was considerably diversified. The marked Brazilian reticence in the hemispheric process stemmed in important part from the consequently much closer identification of its interests with the multilateral trading system and, indeed, although for different reasons, the subregional Mercosur project. Moreover, it came from a steadfast opposition to the US vision of the FTAA project and to a regional economic order thus dominated by peculiarly US interests. The negotiations were, in this sense, dominated by a stand-off between the USA and Brazil, and the respective 'coalitions' that each came to lead in the final stages of the process.

Mexican interest in a putative hemispheric agreement was also lukewarm, largely in view of the potential consequences of an FTAA for its existing preferential relationship with the USA. Conversely, successive Canadian Governments, while enjoying the same preferential relationship with the USA, have consistently advocated as comprehensive as possible an FTAA. To an extent, this can be attributed to an historically entrenched predisposition towards multilateralism in Canada, and the fact that Canada has an extensive investment relationship with many economies in Latin America and the Caribbean. Yet the primary reason for Canadian commitment to a genuinely hemispheric undertaking related to the implications for this preferential North American relationship of expanded access to the US market under the terms of an FTAA, and, thus, a need to retain influence over the shape that this access should take. Interestingly, this same concern came to prevail in Mexico as progress in the FTAA negotiations began visibly to falter over 2002 and 2003 and the USA became much more proactive at a bilateral level with various countries in the region.

The positions in the negotiations of the smaller countries of the Caribbean and Central America, along with some of the Andean countries, were dominated by concern with the profound adjustments required for effective participation in an FTAA. Across the board, this led to a focus in the negotiations on the single issue of special and differential treatment. Interestingly, however, the positions of Caribbean countries were considerably more 'sceptical' than those of either the Central American or the Andean countries, and the members of CARICOM aligned themselves much more closely with the Brazilian position in the final stages of the negotiations.

## THE COLLAPSE OF NEGOTIATIONS AND THE 'FTAA-LITE'

The result of these divergences in the negotiations was a steady loss of momentum in the FTAA process, displayed, in the first instance, as a questioning of the 2005 deadline by Brazil and some Caribbean countries. Over 2002 and 2003 it took the much more fundamental form of a fracturing of commitment to the single undertaking. A coalition emerged in favour of what came to be called an 'FTAA-lite', composed of the Mercosur countries, CARICOM countries and Venezuela. The aims of this grouping were set out in the Brazilian Government's announcements, some months before the pivotal Eighth FTAA Ministerial in Miami in November 2003, of its intentions to restructure Brazilian negotiating strategies with regard to the FTAA. It proposed to negotiate the key issues of market access, services and investment in a bilateral Mercosur-US format, leaving only 'basic elements' such as dispute settlement, trade facilitation and special and differential treatment to be negotiated multilaterally. These proposals gained some (very guarded) support from some Mercosur partners and CARICOM, but were rejected emphatically by the USA.

Yet the outcome of this fragmentation was an unlikely alliance between Brazil and the USA. It was formed on the basis of their agreement, brokered behind closed doors before the Miami meetings, to abandon the single undertaking in favour of a 'buffet-style' arrangement, in which countries could select the undertakings they wished to make in an FTAA. The 'Miami Declaration', after considerable rancour, set out the new format of an FTAA that operated effectively at two levels, or speeds: the first comprising a 'common and balanced set of rights and obligations applicable to all countries', the second 'additional liberalization and disciplines' that countries might 'choose to develop'. Significantly also, the Declaration reflected the abandonment of the 'WTO-plus' format, referring merely to a commitment to make FTAA provisions 'consistent' with multilateral provisions. This new 'FTAA-lite' format was opposed trenchantly by a grouping of 13 countries, led by Canada, Chile and Mexico, each of which, for different reasons, saw its interests as being served only by a comprehensive and reciprocal agreement among all of the countries of the region. Canadian and Chilean officials sought together to respond by assembling a draft text of an alternative plan. The principle of tailoring the pace of liberalization in accordance with levels of development remained intact in this draft, but the text introduced the idea that benefits from an FTAA should be made conditional upon the extent of obligations assumed under the agreement. While this was accepted in principle by then US Trade Representative Robert Zoellick, he also indicated a likely 'pragmatic' approach to dealing with the problem in future negotiations rather than a strict sliding scale of access to US markets dependent on how many of the FTAA obligations countries chose to adopt.

Governments opposed to such a reduction in the scope of the FTAA subsequently agreed to acquiesce as it became clear that it was not, according to the US Trade Representative and Brazilian trade negotiators, open for discussion. As a result, the Miami meetings ended a day early, as negotiators declared themselves to have exhausted topics for discussion. At the Special Summit of the Americas, held in Monterrey, Mexico, in January 2004, however, recriminations persisted and disputes over the foundations of an FTAA were still clearly in evidence. The Mercosur members continued to press for a 'gradual' FTAA, and raised trenchant objections to US aspirations to greater hemispheric-level liberalization in the areas of services and investment. Those countries also rejected the creation of extranational tribunal systems, such as those established under the investor-state provisions of NAFTA, and demanded a system of compensations for the adverse competitive impact of government supports to US farmers. No agreement was reached on the agriculture issue, and the talks were deemed largely to have failed, stymied by trenchant differences between the Mercosur countries and the group of countries led by the USA. Since that time, there has been virtually no activity nor substantive progress, and there has emerged a general perception across the region, including in the USA, that the FTAA project has slipped into dormancy, at least in the short term.

## THE NEW BILATERALIST IMPETUS

The counterpart of the move to an 'FTAA-lite' was the shift to a more sustained focus in US trade policy on bilateral trade negotiations. In the context of the Americas, bilateralism represents, in essence, a political response by the US Administration to the political difficulties encountered in realizing its particular vision of the FTAA and the regional economic governance agenda. Bilateralism provided a way of obtaining access to services markets in the region and of shaping economic disciplines such as investment rules and intellectual property in exchange for concessions on market access for a range of goods, while, at the same time, witholding significant concessions on agricultural liberalization or modification of domestic legislation on trade remedies. Crucially also, bilateralism offered a way of reviving the principle of 'WTO-plus' in new trade agreements following the collapse of this aspiration as the foundation for an FTAA. Without exception, the bilateral agreements that US trade officials refer to as 'state of the art' conform to a 'WTO-plus' template. Notably, however, the terms of 'WTO-plus' are the same as those that prevailed in the FTAA negotiations; that is, 'WTO-plus' is not universal across the various areas of negotiation and, as in the hemispheric negotiations, it does not encompass agreements on trade remedies, agricultural subsidies or various strategic and politically sensitive sectors.

However, the pursuit of bilateral agreements has also been deemed useful as a mechanism for increasing the incentives of other partners (notably Brazil) to engage in similar negotiations, or else for increasing their interests in the success of the FTAA negotiations and thus encouraging a softening of negotiating positions. The 'incentive' has been invoked consistently by US trade officials in the FTAA process. Similar pressures in the multilateral arena were brought to bear following the collapse of the Fifth WTO Ministerial meeting in Cancún, Mexico, in September 2003, when US Trade Representative Zoellick declared his determination not to wait for the 'won't do' countries in the multilateral system and to undermine the emerging Brazil-led coalition of developing countries. The early defection from this grouping of such countries as Colombia, El Salvador, Costa Rica and Peru was directly a consequence of US trade officials' rebukes and warnings that trade agreements with the USA could be threatened by participation in this alternative grouping. The US overtures to Argentina in mid-2005, muted but, nevertheless, apparently serious, played directly, in similar fashion, on the rather tense condition of the Argentine-Brazilian relationship and augured an emphatic undermining of the Brazilian Government in its own subregion.

In summary, it is through the progressive prioritization of bilateral negotiations that US influence over the architecture of the region has been most easily asserted. Indeed, the bilateralist emphasis facilitates the construction of precisely the hub-and-spoke regional arrangements and the extension of the NAFTA that the USA initially envisaged in the FTAA. This is manifest in the substance of the two major bilateral agreements into which the USA has entered—with Chile in 2002 and with the Central American countries in 2004—as well as in the ongoing negotiations with the group of Andean countries. The template, as with the 'WTO-plus' format of the FTAA, has been to use each successive agreement as the baseline for subsequent negotiations, with the expectation that the provisions of each new agreement will exceed those of the preceding one. Yet, while bilateralism has emerged as the preferred strategy for the pursuit of US trade and economic interests in the region, it is notable that it has encountered serious trouble in the domestic political arena in the USA. This is particularly clear, most recently, in the case of CAFTA: the politics of congressional ratification indicate that, in many ways, the US Trade Representative's strategies have been somewhat out of kilter with large sections of prevailing political and public opinion in the USA, indicating a more troubled panorama for US bilateral strategies than is often recognized.

## THE POLITICS OF CAFTA

The putative CAFTA is, in many ways, the archetypal example of an agreement that bears all of the hallmarks of US preferences and US dominance in the negotiating process. It provides for the elimination of US tariffs on most of its agricultural products within 15 years (slightly longer than in the Chilean agreement), but, again like the Chilean agreement, excludes any concessions on subsidies and other non-tariff barriers. It eliminates tariffs and quotas on textiles products that comply with rules of origin provisions, that is, that they are made using cloth that has been dyed and finished in the USA, and in this sense CAFTA augurs an expansion of trade in textiles. It also amends the rule of origin to include some fabrics from Canada and Mexico, intended to foster the integration of North American textiles and apparel industries. Yet these provisions intersect with existing multilateral arrangements for the global liberalization of textiles and apparel quotas at the start of 2005. Taken together, this multilateral elimination of quotas and the terms of the CAFTA agreement have been calculated to signify a 50% reduction in the expansion of Central American textiles and clothing exports to the USA (Hilaire and Yang, 2003). The CAFTA agreement resembles the Chilean agreement in its explicit incorporation of a range of provisions that go beyond trade and impinge directly on domestic legal structures. In the agreement, the USA achieves its stated goals of compelling reforms of the domestic legal and business environment, which include greater transparency, of strengthening the rule of law and of enacting much more extensive protection and enforcement of intellectual property rights. CAFTA also establishes stringent additional provisions in matters of intellectual property, notably in the area of pharmaceuticals, in the form of a five–10-year waiting period on the production of generic drugs competing with brand names unless the former have undergone the same sort of extensive clinical trials as the latter. This provision obliged the Guatemalan Government to revoke legislation approved in 2004 allowing the marketing of generic products.

In matters of labour and environmental standards, CAFTA goes beyond the equivalent provisions in the Chilean agreement. The environmental chapter provides for the benchmarking of environmental co-operation activities and input from international organizations (the agreement with Chile also carries such provisions, but these were negotiated only after the original agreement). The provisions on labour standards require the full implementation of legislation in each country, but this requirement is supplemented by a range of co-operative schemes to improve these laws and their enforcement, as well as financial and technical assistance measures as part of a broader commitment to provide what has been called 'trade capacity-building assistance' to developing countries. In the context of CAFTA, a US $6.75m., four-year grant was extended by the USA in 2003 to support 'good labor conditions' in its Central American trade partners, funded by the US Department of Labor. It is oriented to the implantation of inspection systems, the education of employers and workers on matters of labour laws and the construction of industrial dispute settlement systems.

The politics of CAFTA in the USA, however, have been fraught and have generated significant opposition from a range of quarters. The majority of Democrats and labour unions have taken issue with the provisions on labour standards, arguing that, in establishing only a requirement to implement domestic legislation, they stop short of fostering genuine improvement in labour rights in Central American countries and, moreover, dilute the leverage that the USA can exercise in this respect. Opposition from sugar producers to the increased quota for exports to the USA from Central America has also been trenchant, and indeed has induced some Republicans to declare intentions to vote against ratification. The textiles industry has been the other protagonist in the constellation of forces opposing the Agreement, and similarly has wielded enormous weight within Congress. At the same time, CAFTA is supported by business organizations, although not by textiles associations. More broadly, public attitudes to CAFTA have been consistent with the generalized decline in support for trade since the beginning of the decade, and at the start of the second Bush Administration in 2005 trade was undoubtedly more controversial in US politics than it had been for some time. Much of this stemmed from the dominant concern about the impact of trade on the US labour market, particularly in view of the perceived threat from the People's Republic of China, as well as from countries such as those of Central America.

Nevertheless, the early stages of the US congressional passage of CAFTA indicated that it would eventually obtain congressional approval, although, as it indeed turned out, by the narrowest of margins. On 14 June 2005 the Senate's Finance Committee agreed (by 11 votes to nine) to send the proposed treaty to Congress for approval, with only minimal changes, and this was subsequently approved, by 54 votes to 45, in the full Senate vote on 30 June. On the same day the Agreement found majority support (24 votes to 11) in the House Ways and Means Committee. However, it was generally accepted that the greatest difficulty would be encountered when the legislation reached the House floor in the congressional session beginning on 11 July; nevertheless, on 28 July the House of Representatives ratified CAFTA, by 217 votes to 215. The Agreement was signed into law by President Bush on 2 August. In light of the political manoeuvring necessary to secure its legislative approval, what the politics of the CAFTA process has shown clearly is that trade in the Americas is increasingly dominated by a US political landscape that is more hostile to new trade agreements than it has been for some time, particularly to those which are perceived to carry implications for the US labour market. The strategies of the US Trade Representative to put in place a regional trading order that is fundamentally oriented to the particular preferences and interests of the USA are, by extension, hostage not only to intra-regional politics, but also to increasingly hostile political opinion within the USA itself.

## BILATERALISM IN REGIONAL TRADE STRATEGIES

Although bilateralism has become most visibly the foundation for US trade strategies in the Americas, it constituted a long-standing dimension of the trade strategies in a range of countries in the region. Canada, Mexico and Chile are salient examples. In this respect, the USA has consistently lagged behind the trend, a fact frequently invoked by US trade officials as justification for both regional and bilateral agreements. Bilateralism has also come to be more widely favoured as the best means of pursuing strategic priorities in trade negotiations given the difficulties encountered in both the hemispheric and the multilateral arenas. The important point is that it has found most robust expression in the particular area of market access; that is, in the defining pillar of Latin American and Caribbean interests in hemispheric trade negotiations. In this sense, US attempts to prioritize bilateral negotiations have been facilitated by the ways in which bilateralism connects with the key negotiating priorities of Latin American and Caribbean governments, even while the more recent bilateral agreements with the USA have gone far beyond the more traditional deals that centred on market access arrangements. Despite residual preferences for a more encompassing agreement in many parts of the region, in this sense, the USA has, therefore, encountered receptive responses to its overtures to 'bilateral' agreements.

At the same time, reactions to the USA's forceful bilateral agenda has also spurred other governments, either to seek to pursue the same sorts of negotiations with trading partners, or else to extend their existing bilateral strategies in the region and outside it. Particularly notable have been the ways in which countries such as Brazil, Mexico and Chile have prioritized consolidation of their positions in Asia and have responded to the emergence of Chinese economic power by negotiating bilateral arrangements with a range of countries in that region, including Japan and the People's Republic of China itself.

## CONCLUSIONS

These shifts in the patterns and politics of trade agreements in the Americas carried a range of consequences that augured an entrenchment of many of the structures of inequality—of both power and development—that continued to characterize the region. What was undoubtedly clear was that the NAFTA template for trade agreements and a distinctly hub-and-spoke form of regionalism, both of which were favoured by the USA at the start of the negotiations, were indeed those that came to prevail as the FTAA process unfolded. Furthermore, the collapse of those negotiations facilitated a shift to bilateralism that presaged even greater likelihood of an entrenchment of hub-and-spoke arrangements in the region. The initiation and completion of a range of bilateral agreements with countries in the region facilitated US dominance of both the negotiating process and the substance of economic and trading relationships, in a way that had not been possible in recent hemispheric and multilateral processes. Indeed, the gross disparities of political, economic and negotiating power between the USA and its partners meant that these agreements were concluded relatively quickly, and with few additional concessions on the part of the USA, although the passage through the US Congress of CAFTA was significantly more tortuous than that of the US agreement with Chile, as a result of the perceived implications for the US labour market.

The shape of the new 'state of the art' bilateral agreements, in addition, reflects the ways in which the regional trade agenda has become one of the primary vehicles through which a neo-liberal economic regime, widely deemed to have been detrimental to the developmental performance of Latin American and Caribbean countries over the 1990s, has been advanced in the region. At a time when many governments in Latin America and the Caribbean have moved away from orthodox neoliberal strategies and rejected the precepts of the so-called 'Washington Consensus', the leverage afforded by bilateral trade negotiations has been deployed to foster the adoption of a range of economic disciplines, particularly relating to intellectual property and investment rules, that are an inherent part of a neoliberal economic order in the region. Together with the ways in which these trade agreements make limited or no concessions in the areas of greatest developmental importance to the majority of Latin American and Caribbean economies (agricultural subsidies, trade remedies and key sectors such as textiles, steel, footwear and sugar), the efficacy of bilateral trade relations with the USA as a development strategy were distinctly dubious. Given also the adverse reaction against trade in the USA, which permeated congressional politics throughout 2005, the medium-term prospects for the construction of a regional economic and trading order that would address many of the key inequalities and developmental obstacles in the region seemed decisively slim.

## BIBLIOGRAPHICAL REFERENCES

Feinberg, R. E. *Summitry in the Americas: A Progress Report*. Washington, DC, Institute of International Economics, 1997.

Hilaire, A., and Yang Yongzheng. 'The United States and the New Regionalism/Bilateralism'. IMF Working Paper, 03/206. Washington DC, International Monetary Fund, October 2003.

Phillips, N. 'US Power and the Politics of Economic Governance in the Americas', in *Latin American Politics and Society*, p. 48. Boulder, CO, Lynne Rienner Publrs, 2006.

Weintraub, S. *Development and Democracy in the Southern Cone: Imperatives for US Policy in South America*. Washington, DC, CSIS Press, 2000.

Zoellick, R. 'Trading in Freedom: The New Endeavour of the Americas', in *Economic Perspectives: An Electronic Journal of the US Department of State*, pp. 6–12. Washington, DC, 2002.

# 'PLAN COLOMBIA' AND BEYOND: US ANTI-DRUGS POLICY IN THE ANDEAN REGION

## LUIS EDUARDO FAJARDO

### INTRODUCTION

US policy towards the Andean countries during the Administration of President George W. Bush (2001–) has been dominated by the 'Plan Colombia', and its regional extension, the Andean Regional Initiative. Plan Colombia was due to formally end in 2005, but US involvement would continue for the foreseeable future, in spite of controversy around its results. Plan Colombia began towards the end of the Administration of Bill Clinton (1993–2001), when the US Government pledged to commence a large anti-drugs aid initiative in Colombia, beginning in 2000. Soon after taking office in 2001, the new Bush Government announced the expansion of the programme, to provide military and economic aid to six other Latin American countries apart from Colombia. Besides confronting drugs-trafficking in this part of Latin America, the Andean Regional Initiative was also intended to counteract the political instability and violence generated by the illegal drugs trade in Colombia and in Bolivia, Brazil, Ecuador, Panama, Peru and Venezuela. In subsequent years the Andean Regional Initiative directed US resources into the region with mixed and controversial results.

The latest US Department of State report on anti-drugs activities in 2004 followed the same broad lines as previous reports: Colombia continued to show a substantial increase in eradication of coca fields, while results for Peru and particularly Bolivia were less promising. None the less, drug exports to the USA continued unabated.

Drugs-fuelled political instability and violence continued to be a problem in the Andean region in the early 21st century. Again, Colombia showed the greatest improvements in restoring order through a strong internal effort backed by US aid. However, in spite of a growing sense of confidence in the Colombian Government's capacity to restore order in many parts of the country previously affected by guerrilla and paramilitary activity, there remained serious concern about the long-running civil war, in which rebel armies such as the Marxist Fuerzas Armadas Revolucionarias de Colombia—Ejército del Pueblo (FARC—EP) and the Ejército de Liberación Nacional (ELN), as well as the right-wing Autodefensas Unidas de Colombia (AUC) paramilitary organization, continue to use drugs-originated profits to fund their wars against the Government. Official US sources have estimated that the FARC—EP and the AUC obtain as much as 70% of their income from the drugs trade.

Moreover, drugs-fuelled violence has continued to target US interests in the Andean region. The US Department of State's annual report on the 'Global Patterns of Terrorism', pointed out that, in 2000, nearly 85% of terrorist attacks on US interests abroad occurred in Latin America, most of these in Colombia. In the last few years these incidents have included the kidnapping and murder of US citizens (both civilians and government agents involved in anti-narcotics activities), sabotage against US economic interests (particularly oil-related), and a terrorist attack against the US embassy in Lima, Peru.

### BACKGROUND: FROM PLAN COLOMBIA TO THE ANDEAN REGIONAL INITIATIVE

Even though the Andean Regional Initiative was conceived as a regional programme, Colombia is clearly at the centre of US concerns in the region. In the last three decades, US-Colombian relations have been largely shaped by problems relating to drugs-trafficking. In spite of its traditional stance as a staunchly pro-US democracy, Colombian relations with the USA deteriorated sharply after 1994 when US authorities claimed that Colombian President Ernesto Samper Pizano's (1994–98) electoral campaign had benefited from financial support from the Cali drug cartel. During Samper's term in office, Colombia's international isolation coincided with a slowdown in economic growth. These problems were compounded by a corresponding increase in guerrilla activity and by the widespread belief that the FARC—EP and the ELN were now obtaining large amounts of funds from their involvement in drugs-trafficking. This led to greater anxieties over the Colombian Government's capacity to counter the flourishing narcotics trade, and prompted worries of increased political instability in Colombia among US policymakers. The election to the presidency of the strongly pro-US candidate Andrés Pastrana Arango in 1998 was followed shortly after by the launching of a major US aid initiative, which came to be known as Plan Colombia.

During President Pastrana's first state visit to the USA in October 1998, President Clinton announced an increase in military aid to Colombia and pledged to mobilize US and international support for the peace process. By August 1999 the USA offered a substantial increase in US assistance if Colombia devised a comprehensive anti-drugs strategy. The Colombian Government responded with Plan Colombia, unveiled formally in September 1999, which proposed expenditures of US $7,500m. in the fight against the cultivation of illegal narcotics. Colombia would contribute $4,000m. of this total, and would hope to raise in the next months, with international solidarity, the other $3,500m.

In July 2000 Congress approved a $1,300m. package which focused on helping Colombian security forces undertake a 'push into southern Colombia', where drugs cultivation and guerrilla activity were most intense, as well as providing support for other interdiction efforts by the Colombian Government and police force. It was also intended to fund institutional reform programmes in the Colombian judiciary, and to provide support for alternative development schemes subsidizing legal crops in order to make coca and poppy cultivation less attractive. However, the vast majority of the US resources were approved for military aspects of the drugs-eradication campaign. US law initially forbade this assistance from directly supporting counter-insurgency efforts. It was to focus instead on anti-drugs operations.

The election of George W. Bush to the US presidency in late 2000 led to expectations of increased and deepened US involvement in the efforts to eradicate drugs cultivation in Latin America. This took the form of the Andean Regional Initiative, a US $880m. programme, launched at the Third Summit of the Americas in Québec, Canada in April 2001, that was intended to expand US assistance to Colombia's neighbours.

The election in 2002 of the hardline Alvaro Uribe Vélez to the presidency of Colombia led to even stronger support from the USA. While many other countries in the region had begun voting into office left-leaning and occasionally anti-US leaders, President Uribe quickly established himself as the staunchest US ally in the region. He shared with the Bush Administration a commitment to take a strong stance against rebels, framing Colombia's counter-insurgency campaign as part of a global 'war on terror' that commenced following the terrorist attacks on the USA of 11 September 2001.

### US INVOLVEMENT IN COLOMBIAN COUNTER-INSURGENCY

One of the most visible signs of increased support for Colombia in the post-11 September climate was the decision, taken by the US Congress in July 2002, to remove legal limitations prohibiting US military aid being employed directly in counter-insurgency activities (as opposed to strictly drugs-control operations). In the first phases of Plan Colombia, the US Army shared intelligence with the Colombian armed forces, trained Colombia's counter-narcotics brigade, aviation personnel and a marine brigade, and provided support for counter-narcotics infrastructure, including radar and airstrips. Following the lifting of legal barriers, in late 2002 US Army Special Forces arrived in Colombia to train a batallion deployed in protecting the strategic Cano Limón-Covenas oil pipeline from sabotage by FARC—EP and ELN guerrillas. In October 2004 the US Congress approved an additional $109.3m. in funding for continued pipeline protection and the training of special Colombian army units who would be sent to apprehend prominent rebel leaders in the southern jungles.

In 2004–05 the US-backed military in Colombia seemed better equipped to fight the insurgent threat than it had been in the late 1990s, when FARC—EP seemed to have strategic advantage. During 2004 the Colombian armed forces established an institutional presence in 158 municipalities that previously had been controlled by guerrillas. In that year the national homicide rate decreased by 15% and that of kidnapping by 35%, while terrorist attacks were reported to have fallen by 42%. Furthermore, in May 2004 US aid allowed the military to announce the implementation of 'Plan Patriota', the largest ever military offensive against the FARC—EP rebels. The initiative involved deploying several thousand soldiers in the southern Amazon jungle, the rebels' main stronghold, in pursuit of the movement's leaders. Co-operation with the USA also improved in other fronts in 2004. In a break with precedent, in March and December of that year, respectively, the Colombian Government extradited to the USA two major guerrilla leaders, Omaira Rojas Cabrera, known as 'Sonia', and Ricardo Palmera-Pineda, known as 'Simon Trinidad'. Rojas Cabrera faced drugs-trafficking charges while Palmera-Pineda was arraigned on drugs-trafficking and kidnapping charges.

A large element of the controversy in the USA surrounding Plan Colombia and the Andean Regional Initiative focused on fears that the USA would become increasingly involved in the Colombian civil war, as happened in Viet Nam in the 1960s. An amendment to the Foreign Operations Appropriations Act of 2002, known as the Byrd Amendment, limited the official US presence in Colombia to 400 military and 400 civilian personnel in the field at any given time. However, in October 2004 the US Congress voted, as part of the United States Defense Department Authorization Act of 2005, to increase to 800 the limit on US military presence in Colombia and to increase to 600 the number of civilian personnel. The increases were in support of the Plan Patriota and the then head of the US Southern Command, Gen. James T. Hill, insisted that US troops would provide only logistical support to Colombian soldiers fighting the guerrillas. Nevertheless, on several occasions US personnel have found themselves in harm's way, most notoriously in March 2003 when a US government contractor on an anti-drugs mission was murdered by FARC—EP guerrillas, and a further three contractors kidnapped, after the aeroplane in which they were travelling went down in rebel-held territory. In early 2005 US servicemen in Colombia again attracted controversy when several US soldiers were arrested in a series of incidents involving alleged drugs- and weapons-trafficking. Instead of facing trial in Colombia, they were repatriated to the USA under the terms of a treaty granting US personnel judicial immunity from Colombian authorities. Such incidents added to the controversy over the risk of increased US involvement in the Colombian armed conflict.

As the USA increased its military and economic aid to Colombia, it also became involved in domestic political controversies relating to security issues. While the Bush Administration enthusiastically supported President Uribe's strong military challenge to the left-wing guerrillas, US government officials, however have expressed concerns about the possible effects of some of his policies on the right-wing AUC militias. The Colombian Government continued military operations to defeat the AUC, but also sought dialogue with its leaders, hoping for an agreement that would lead to the group's eventual demobilization. Some AUC leaders have commented that if these dialogues were to succeed, they should lead to the reduction or abandonment of criminal charges against them in exchange for a ceasefire. The US ambassador to Colombia, William Wood, has been critical of the fact that negotiations were being conducted with AUC leaders allegedly still involved in the international drugs trade, some of whom US authorities wanted to extradite.

## DRUGS ERADICATION IN COLOMBIA

By 2005 there was a broad consensus that Plan Colombia was successful in destroying record amounts of coca and poppy fields in Colombia. However, it was still not clear that these successes had diminished the Colombian drugs trade in a definitive way. In 2004, according to official sources, Colombian government forces captured 178 metric tons of cocaine, eradicated through aerial spraying a record 136,555 hectares (ha) of coca fields and 3,060 ha of poppy fields, as well as a further 10,991 ha of coca and 1,497 ha of poppy plants through manual eradication. The US Department of State estimated that in 2004 Colombia's coca production potential fell by 21% compared with the previous year. Such results were consistent with the earlier years of Plan Colombia. While in 2000 the potential harvest of coca was estimated at 136,200 ha, in 2003 it had fallen to 113,850 ha, the lowest figure since 1998. The total acreage of eradicated fields was four times larger in 2004 than in 2000. By 2005 the USA had spent more than US $3,000m. on anti-drugs aid to Colombia since 2000, and in April 2005, the Bush Administration requested congressional approval of a further $600m. for the 2006 fiscal year. Colombia was considered to be running the most successful anti-drugs programme in the Andean region. Nevertheless, the country continued to be the world's largest producer of cocaine base, leading some US law-makers to question the wisdom of continuing large-scale aid to Colombia when coca-growers seemed to be able to replace coca fields almost as fast as they were destroyed.

US military aid to Colombia is contingent on the US Department of State annually certifying that there has been improvement on the often-criticized human rights record of the Colombian security forces. The US Government continued to certify Colombian government co-operation in curbing human rights abuses in the armed forces. Nevertheless, several human rights organizations, including Amnesty International and the Office of the UN High Commissioner for Human Rights, continued to denounce alleged abuses by Colombian government agents. Starting in 2003, efforts by the Colombian Government to conduct a peace process with right-wing paramilitary groups included discussions on offering leniency to many of their members accused of participating in human rights violations, in exchange for their demobilization. Many human rights groups were vigorously opposed to this prospect.

## DRUGS CULTIVATION IN BOLIVIA

If US anti-drugs policy was controversial in Colombia, it led to even greater doubts in other Andean countries. Bolivia, which previously had been used as an example of the viability of supply-side anti-drugs policies, owing to an earlier successful eradication campaign in the Chapare province, experienced a resurgence in coca cultivation from 2003. This was, in part, owing to the increasing political influence of organizations representing small-scale coca farmers, which, predictably, opposed the new eradication campaigns. According to the US Department of State, in 2004 Bolivia eradicated 8,000 ha of coca fields in the Chapare region, but with cultivation increasing in other regions such as Yungas, overall coca cultivation in Bolivia increased by 6% in 2004. The net area under coca cultivation, which reached 14,600 ha in 2000, was estimated at 24,600 ha in 2004. Moreover, coca growers' organizations, who were strongly opposed to the forced eradication programmes in the Chapare region, continued to enjoy high levels of popular support throughout the country, and were instrumental in the resignation of President Carlos Mesa in June 2005. In that year, following the disappointing results of the anti-narcotics war in the country, the USA announced it was considering reducing drugs-eradication funding to Bolivia for the 2006 fiscal year.

## IMPROVEMENTS IN PERU

By 2003 Peru was the second largest cocaine producer in the world but had made some progress in its US-supported eradication efforts. During 2004 Peruvian authorities eradicated around 10,500 ha of coca, of which nearly 20% was achieved through voluntary eradication programmes. According to US Department of State figures, net coca cultivation area stood at 31,150 ha in 2003, a fall from 51,000 ha in 1998 and much lower than the peak figure of 115,300 ha in 1995.

US Department of State figures also indicated that about two-thirds of the Peruvian crop was grown in the Upper Huallaga and Apurimac valley regions, where anti-drugs efforts were traditionally concentrated. However, in a worrying development for the Peruvian Government, new cultivation areas recently appeared in the the regions of Loreto and Puno. Another worrying trend was the appearance of stronger links between drugs-trafficking organizations and the Maoist insurgency grouping Sendero Luminoso. The group, which was thought eliminated by the mid-1990s, reappeared with military actions in coca-growing regions such as Huánuco province. It was obtaining new funding through drugs-activities, prompting fears of a repetition of the Colombian so-called 'narco-guerrilla' phenomenom.

## UNCERTAINTY IN ECUADOR

Ecuador has been adversely affected by the internal conflict in neighbouring Colombia. The ongoing civil war there resulted in an influx of refugees into Ecuador's northern provinces, while guerrillas and paramilitaries made frequent border incursions. At the same time, transnational drugs-trafficking continued at high levels. From 2003 to his overthrow in April 2005, President Lucio Gutiérrez Borbua, a nationalist former army colonel, maintained a co-operative discourse with its neighbours and with the USA.

Such co-operation efforts were highlighted in early January 2004 when Ecuadorean police, apparently acting on information received from Colombian and US authorities, captured Ricardo Palmera-Pineda, a leading figure in the Colombian FARC—EP rebel group (see above), and within days deported him to Colombia. Since 1999 Ecuador has allowed US military personnel to operate an air base at Manta, for anti-naroctics operations. US funds helped to sustain a US $465m. economic and social development programme, launched in 2000, in the country's northern border area, aimed at preventing the expansion of illicit crop cultivation in the area.

President Gutiérrez's successor, Alfredo Palacio Gonzáles, had yet to signal a major change of direction in Ecuador's collaborative efforts with overall US anti-drugs policy, although initial declarations by the new Government's spokesmen seemed to indicate less willingness to involve Ecuador in Colombia's war against narcotics-dealers and guerrillas, who continued to use Ecuador as a strategic rear guard.

## THE WAR ON DRUGS IN OTHER COUNTRIES

The President of Brazil, Luiz Inácio (Lula) da Silva, who took office in January 2003, pledged to continue support of regional anti-drugs operations, in spite of his nationalist stance. In 2005 the Brazilian Government continued an increased security presence along its northern borders. 'Operation Cobra', which had begun in 2000 as an effort to control illegal activities along the Colombian–Brazilian border, was expanded in 2003 to include surveillance of borders with Peru, Venezuela and Bolivia.

Panama also received a small but steady number of Colombian war refugees and suffered incursions from Colombian guerrillas and paramilitaries into its territory. In 2005 the USA continued to offer technical and financial support to Panamanian law enforcement agencies. President Martín Torrijos Espino, the eldest son of former nationalist dictator Omar Torrijos, who took office in September 2004, maintained a co-operative stance with the USA, announcing, among other initiatives, his intention of seeking a free trade agreement with that country.

In the early 2000s Venezuela was the country most critical of US policies in the Andean region and maintained an increasingly confrontational stance with neighbouring Colombia. Venezuela's drugs cultivation remained very small, but cocaine seizures in the first half of 2004 equalled the record figures for all of 2003, according to the US Department of State, largely reflecting that country's status as a transhipment point for drugs originating in neighbouring Colombia. President Hugo Chávez Frías survived severe political difficulties in 2004, including a recall referendum. Nevertheless, his Government began 2005 in a much stronger position, as record oil prices bolstered state revenues and allowed him to engage in social spending. Among the many criticisms directed against Chávez, domestically and internationally, was the accusation that he undermined regional efforts to contain drugs-financed subversive groups, through his alleged affinity towards the Colombian FARC—EP rebels, who claimed ideological sympathies with Chávez's left-wing 'Bolivarian Revolution'.

Regional security was adversely affected in 2004–05 by an increasing hostility between the Governments of Colombia and Venezuela. This reached a nadir in January 2005 following the arrest, in the previous month, of Rodrigo Granda Escobar, the alleged international spokesperson of the FARC—EP. The Venezuelan Government alleged that that Colombian police had been involved in an illegal operation to seize Granda Escobar from Venezuela, where he was living, and covertly return him to Colombia to be arrested. The accusations were denied by the Colombian Government. President Chávez retaliated by recalling the Venezuelan ambassador to Colombia and suspending economic co-operation projects. Although a diplomatic solution was eventually reached in February, relations with Colombia remained tense. In the meantime, Chávez repeatedly accused the USA of plotting to destabilize his Government, and in May he threatened to suspend diplomatic ties with the Bush Administration completely. In 2005 Venezuela was the most serious opponent of US policy in the Andean region, and seemed likely to remain as such for the next few years. The Chávez regime was counting on its status as a main provider of petroleum to the USA in a time of fuel scarcity to avoid any major retaliations by the USA to its contentious actions. However, the US Government seemed determined to prevent President Chávez from establishing stronger links with other countries in the region, and was likely to continue diplomatic efforts across the continent to present Venezuela as a threat to regional stability.

## THE ANDEAN TRADE PREFERENCE ACT

Although most US non-military aid to the region under the Andean Regional Initiative was geared towards helping specific groups of impoverished farmers in drugs-producing areas, a related programme, the expansion and renewal of the Andean Trade Preference Act (APTA), through the approval of the Andean Trade Promotion and Drug Eradication Act (ATPDEA), was expected to bring more widespread benefits to the economies of Colombia, Bolivia, Ecuador and Peru. By removing tariffs for Andean exports to the USA, it hoped to create further opportunities for industries other than coca production. It would also create benefits for many previously established legal industries in the four targeted countries.

The ATPDEA received presidential approval on 31 October 2002. This legislation renovated and widened the previous ATPA, which had been originally granted to the Andean countries in 1991, and which had expired in December 2001. The new APTA was to remain in operation until 31 December 2006.

The APTA provides free access to the US market of any product not specifically excluded. Colombian exports to the US market recorded a real increase in value and as a percentage of total Colombian exports every year from 1993. According to Colombian government estimates, between 1992 and 1999 the ATPA programme generated more than 140,000 new jobs. In contrast, trade between the USA and ATPA countries grew by less than 1% in 2002, compared with the previous year, following the expiry of the programme.

In 2005 the trend towards expanded commercial interaction in the Andean region continued as Colombia, Peru and Ecuador conducted negotiations with the USA towards the eventual signing of a free trade agreement. Such an agreement would indefinitely extend many of the trade benefits conceded to the Andean countries in the last few years as part of the USA's anti-drugs policy in the region. However, the viability of trade pacts with Ecuador might be threatened if the continuing political uncertainty there in 2005 led to a more nationalist stance being adopted by the new Government.

## CONCLUSIONS AND PROSPECTS: BEYOND PLAN COLOMBIA

The Andean Regional Initiative and, in general, US policy towards the Andean region in the 2000s cannot be understood exclusively as an anti-coca-cultivation programme. Although that was the rationale used to justify it in US government and legislative circles, it is a regional security initiative as much as it is a programme to curb drugs-trafficking from Latin America to the USA. The Initiative was a response to increasing US concerns over the combined destabilizing effects of drugs profits and guerrilla groups on Colombia, and over the possibility of civil strife and large-scale violence spreading across the region into other Andean democracies.

The Andean Regional Initiative and its predecessor, Plan Colombia, did appear to achieve strong coca crop-eradication success in Colombia and, to a lesser extent, in Peru over the last five years, although neither succeeded effectively to disrupt the overall supply of drugs to the USA. In the early 2000s Colombia was considered by the US Administration as having improved its anti-narcotics efforts. This was in striking contrast with the mid-1990s, when US Drug Enforcement Agency officials openly described Colombia as a 'narcodemocracy'.

The hardening of Colombian attitudes towards guerrillas and an increased government drive for military results against drugs-trafficking was not just the result of US influence, however; it also reflected domestic political developments. In the midst of generalized public disillusionment over previous negotiation attempts with the FARC—EP, Colombian President Alvaro Uribe Vélez maintained very high popularity levels as he reached the third year of his term in office, owing, to a large extent, to his hardline stance against the guerrillas; his attempts to seek re-election in 2006 were based on the promise of these policies continuing until 2010. A cornerstone of this strategy was the denial of one of the guerrillas' largest funding mechanisms through the eradication of coca and poppy fields in areas of rebel influence.

Anti-drugs efforts in other Andean countries produced more mixed results, however. In Peru there were worrying signs of an emerging alliance between drugs-traffickers and the Sendero Luminoso, indicating that narcotics-fuelled political instability remained a risk in that country. As part of its global 'war on terror' policy, the USA might well use this development to justify further anti-drugs aid to Peru as a means of preventing Sendero Luminoso regaining the strength it enjoyed in the 1980s. In Bolivia, the focus of coca cultivation merely shifted, from Chapare to Yungas. Bolivia, and to a lesser extent Peru, also experienced a strengthening in the political force of small-scale coca growers, who became increasingly organized and vociferous. Ecuador and Panama, while not major drugs-producing countries, faced new challenges in the early 2000s, owing to the effect of events in Colombia. The abrupt change of government in Ecuador in April 2005 brought renewed doubts over the continuation of the high level of co-operation with the USA and Colombia in drugs-interdiction efforts.

In Bolivia, the coca-growers' movements electoral triumphs and the bucking of the trend in coca-acreage reduction prompted fresh doubts as to the success of supply-side-based anti-drugs policies there. The success of Bolivia in the late 1990s in substantially reducing its drugs production was often used as an argument to prove that, in spite of constant, or growing, demand for cocaine from Western markets, drugs-eradication programmes could come close to terminating production from Andean countries. While coca production levels in the last five years in Bolivia were nowhere near as high as those experienced in the 1990s, the increase in cultivation observed there highlighted the difficulties involved in attempting to find a permanent solution to drugs-trafficking in the face of steady demand in the USA and Europe. It also generated further scepticism as to the long-term feasibility of voluntary eradication programmes, which were a dominant feature of earlier anti-drugs strategies, particularly in Bolivia, and which were followed by many coca farmers demanding the right to return to coca cultivation. The enormous political uncertainty in Bolivia in 2005, with nationalist forces gaining strength and coca growers' leader Evo Morales emerging again as a strong contender to the presidency, might lead, in the medium term, to a radical departure from collaboration with the USA in its regional security and anti-drugs policies.

This 'radical populism' was also evident in Venezuela, which emerged in the early 2000s as the most serious challenger to US Andean policy. With oil-related windfalls increasing the political confidence of the Chávez regime, in 2005 the Government increased its militant anti-US rhetoric and cemented an alliance with Cuba. While Venezuela was not a very significant drugs-production centre, there were continued fears that the country could provide sanctuary for the FARC—EP's military operations against the Colombian state, or even provide the guerrillas with weapons.

US policy in the Andean region would continue to be influenced by the global political climate created by the international 'war on terror'. Following the terrorist attacks on the USA of 11 September 2001, a much more assertive USA removed previous limitations on its financial and military involvement in the Colombian domestic conflict. The flow of financial aid explicitly aimed at counter-insurgency activities, and the direct involvement of US government officials and contractors in operations in the field in Colombia, pointed to greater US intervention in the region. In spite of growing military commitments in the Middle East and elsewhere, the Bush Administration continued to expand its military presence in Colombia and seemed set to maintain a presence there, particularly if President Uribe, one of the USA's strongest allies in a continent increasingly shifting towards more nationalist and left-leaning governments, secures a second term in office in 2006.

As bilateral free trade negotiations with Andean countries advance, US policy towards such pacts might eventually become less intensely focused on drugs and security issues. However, in the foreseeable future, these would remain the central points of interest in US relations with the region. Colombia would most likely remain committed to a US-backed military offensive against FARC—EP for several years to come. The USA's 'war on terror' did not determine this offensive, but it certainly made it easier for the USA to justify large-scale military aid to this country. Across the rest of the region, US efforts to ensure friendly governments, free trade and anti-drugs efforts were likely to face increasing difficulties, owing to local political conditions. An increasingly hostile and self-confident Venezuelan regime, as well as growing political instability and anti-US sentiments in Ecuador and Bolivia were just three of the many potential sources of trouble in the near future for the US strategy of containing political violence and curtailing drugs-trafficking in the Andean nations.

## BIBLIOGRAPHY

Executive Office of the President, Office of National Drug Control Policy. Press Statement regarding the Latest Estimate for Poppy Cultivation in Colombia. Washington, DC, 9 May 2003. (www.state.gov/g/inl/rls/prsrl/ps/20774.htm).

Tickner, A. 'La Guerra contra las drogas: las relaciones Colombia-EU durante la administración Pastrana', in *Plan Colombia: Ensayos críticos*. Bogotá, DC, Universidad Nacional de Colombia, 2001.

US Department of State, Bureau for International Narcotics and Law Enforcement Affairs. *International Narcotics Control Strategy Report*. Washington, DC, 2005.

White House, Office of the Press Secretary. *Fact Sheet, Andean Trade Preferences Act*. Washington, DC, March 2002. (www.whitehouse.gov/news/releases/2002/03/20020323-10.html).

# TRADE PREFERENCES IN THE CARIBBEAN

## MARK WILSON

On 16 April 2004 the first round of negotiations began in Jamaica towards an Economic Partnership Agreement (EPA) between the European Union (EU) and the Caribbean Forum (CARIFORUM) grouping of Caribbean states.[1] By September 2005 the aim is to reach agreement on priorities for support of Caribbean integration and negotiating targets, with a draft partnership agreement to follow in 2006, and the main negotiations reaching a conclusion by the beginning of 2008. The talks followed an extended period during which the special arrangements made for former Caribbean colonies as exporters of traditional agricultural products—principally bananas, sugar, rice and rum—have been eroded. The remaining market privileges are due, in most cases, to disappear altogether within a few years. At the same time, the small states of the Caribbean have been concerned about more general trade liberalization initiatives—the World Trade Organization (WTO) agreements, and the proposed Free Trade Area of the Americas.

The new EPA will be negotiated under the 20-year Cotonou Agreement, signed in that city in June 2000 by heads of state and government from African, Caribbean and Pacific (ACP) countries and EU members. The Cotonou Agreement, in turn, followed a succession of five-year Lomé Conventions, the first of which came into force on 1 April 1976. These linked 77 of Europe's former ACP colonies with the European Community and its successor, the EU. Each convention combined an aid package, export earnings stabilization schemes, special market access arrangements for most traditional exports, and one-way duty-free status for most ACP manufactured exports to Europe. In contrast to earlier agreements, however, the new EPAs will not cover the full ACP grouping, but will be separately negotiated for regions such as the Caribbean or West Africa. They will also be reciprocal, with two-way market access covering EU exports to the ACP members, as well as their exports to Europe. They will also cover services, as well as trade in goods. To give an opportunity for gradual adjustment they will come into force in their final form only after a transition period that may exceed 10 years.

In spite of the extended adjustment period, the proposals for new trading arrangements are not viewed with any enthusiasm by the Caribbean governments or private sector. In a wide range of trade negotiations, Caribbean countries have emphasized the need for special and differential treatment for small states, with continuing one-way trade privileges and protection for domestic industries over an extended period. The security afforded by tariffs and barriers to entry for small and medium-scale manufacturers and service providers of varying efficiency is emphasized to the exclusion of the benefits to consumers and businesses of access to more competitive markets with a wider choice of price, product type, and quality. Some Caribbean countries, particularly those such as the Bahamas where there is no income tax, have attributed to trade liberalization the need to reform their fiscal systems, replacing import duties, for example, with a value-added tax (VAT).

Reforming the trade regime is not simply a matter of choice. The former Lomé arrangements are incompatible with WTO rules, and an existing waiver lasts only until 2007. Europe has also made commitments to other partners, which conflict with the interests of the Caribbean jurisdictions. The EU's 'Everything but Arms' initiative, announced in February 2001, gives duty-free and quota-free access to 48 of the world's poorest countries. This dilutes the privileges of the Caribbean, and opens the European market to much larger and lower-cost producers such as Bangladesh and Viet Nam. Again, a transition period is intended to mitigate the effects on the Caribbean. There will be phased reductions in existing tariffs, extending to January 2006 for bananas, September 2009 for rice, and July 2009 for sugar.

In some respects, the former trading arrangements were a success. Traditional agricultural producers lived in a fairly secure world. Small farmers could grow bananas, rice and, in some cases, sugar cane, sure of a steady export market, and usually a predictable price. The same was true for private and state-owned companies producing sugar and rum. However, the cost of production remained high, while there was much less success in stimulating non-traditional exports. Indeed, the proportion of EU imports originating in the wider ACP group crept steadily down, from 6.7% in 1976 to only 3.0% in 2000. Furthermore, of that modest flow, 10 products made up 60% of the total.

Producers of traditional commodities in the Caribbean are a diverse group. Most attention has been given internationally to small farmers, a vulnerable group whose problems have direct implications not only for the economy but for social and political stability. These farmers were the traditional mainstay of the Windward Islands' banana industry, and they also produced a significant proportion of rice in Guyana and Suriname, and of sugar cane grown in certain countries, for example, Trinidad and Tobago.[2] A second broad group are the larger private-sector players; these assumed a major role, for example, in Jamaica's sugar and banana industries, as well as in rice growing and processing. These are rural employers of great local importance. Rum distillers such as Demerara Distillers in Guyana or Wray and Nephew in Jamaica are more closely integrated in their countries' corporate sector, where in some cases they play a leading role. A third group of commodity producers are the large state-owned companies, which in some cases originated with the nationalization of foreign-owned sugar estates following independence. These include the Guyana Sugar Corporation (Guysuco), Trinidad and Tobago's former agricultural conglomerate, Caroni Ltd, and, with rather different origins, Surinamese companies such as the former banana producer Surland, now restructured and marked for privatization as the Stichting Behoud Bananensector Suriname (SBBS), and the rice grower Stichting voor Machinale Landbouw (Mechanised Agriculture Foundation), also recently restructured.

Although Caribbean governments are rightly keen to defend the interests of their traditional agricultural producers, the role of this sector in most regional economies declined steadily in importance during the second half of the 20th century. By the early 21st century the main source of foreign exchange on most Caribbean islands was tourism. There are exceptions: in Trinidad and Tobago the energy sector is the driving force of the economy; in Jamaica and Suriname, and to a lesser extent in Guyana, bauxite plays a major role. However, traditional agriculture remains of great importance as a source of employment, and has until now underpinned the social and economic stability of large parts of the rural Caribbean.

## BANANAS

Of the traditional agricultural sectors, banana cultivation has, until recently, played the most important social and economic role, and its problems have therefore attracted the most international attention. In the Windward Islands[3] the colonial authorities fostered banana growing in the 1950s as a response to the demise of the sugar industry on some islands and owing to the difficulties faced by some other commercial crops. Small farmers, many with holdings of less than half a hectare (ha), sold fruit to island-wide growers' associations, which in turn supplied Geest Bananas Ltd, a British company that shipped the produce to the United Kingdom. In contrast to most tropical crops, bananas provide a year-round cash income, as harvesting is continuous. In contrast to tree crops, the time-lag from planting to harvest is measured in months, not years. Additionally, production is possible in a wide variety of conditions, allowing even farmers with marginal land to participate in the industry, albeit at low yields. This pattern of very small farms was not universal, however. Most banana farms in Belize measure 40 ha–200 ha, while in Jamaica the role of small farmers was supplemented by a few large commercial estates. In Suriname a state-owned company, the former Surland, has been the only banana exporter.

Dependence on the banana industry was greatest in the small economies of the Windward Islands, and in particular in Dominica, an island with only 70,200 people in 2003 and a per-head income among the lowest in the English-speaking Caribbean. In 1992 bananas comprised some 55% of merchandise exports and, in 1990 30% of all foreign exchange earnings, still exceeding tourism. More important, most rural households were in some way directly involved with the banana industry. Together, the 10,000 small producers in the Windward Islands in 1992 supplied 45% of the British banana market. In Jamaica, with a similar total production, the economic importance of the crop was much smaller, at 3.8% of merchandise exports in 1992 and 1.7% of foreign exchange earnings in the same year.

Until 1993 the British market was preserved for the Caribbean banana exporters[4], with a small quantity of so-called 'dollar bananas'[5] from Latin America permitted only when supplies fell short. With the single European market in operation from 1992, this system could no longer operate. To protect the interests of the traditional producers, a complex system of licences, tariffs and quotas was developed. The USA, as well as several Latin American banana producers, used the machinery of the WTO to challenge this new EU banana import regime; in 1997 they obtained a ruling broadly in their favour, which held that there should be changes in the quota and licence system. The EU moved to a tariff for non-ACP bananas, with a duty-free quota for ACP producers; this in turn was challenged and was, in April 1999, found to violate WTO rules. After a brief 'trade war', in which the USA imposed punitive tariffs on selected European imports, a new licensing system was agreed. This would run from 2001, with a transition to a tariff-only system from 2006, which until 2008 would incorporate residual privileges for ACP producers. This new regime necessitated a WTO waiver, which was forthcoming in November 2001.

In a free-market situation, the British landed price of South and Central American bananas is around 30% below that of Windward Islands fruit. This is in part because of greater scale economies on the very large estates that operate there, and in part because agricultural wages are much lower; in Ecuador, salaries are around one-quarter of the levels prevailing in the Eastern Caribbean. There is also a quality issue, although it should be made clear that importers and supermarkets use this word in a rather specialized sense, to denote a bright colour, uniform size and unblemished skin, rather than taste.

The various EU import regimes in force since 1992 were designed to protect the interests of Caribbean and other traditional producers, but succeeded only to a limited extent. At times, for example in 1998–2000, illegally imported fruit, in excess of quotas, has flooded the European market (so-called 'banana-laundering'). Partly because of the limited success of these import regimes, and partly because of underlying imbalance between demand and supply, prices have moved unevenly downwards, creating uncertainty for the growers and leading many to abandon the industry. In 1990–2000 the wholesale price of bananas in the United Kingdom fell by 14%. There was a further 30% fall in prices by October 2003, as intensified competition between British supermarkets helped drive down the price of the fruit; changes in exchange rates and final market prices left the price paid to Caribbean producers broadly unaltered thereafter. Lower export volumes have also led to higher overheads and, at times, to crippling 'dead freight' payments. Agronomic problems have also damaged the industry. Black Sigatoka disease, hitherto absent from the Caribbean, affected Jamaica in the 1990s, necessitating expensive spraying programmes. There was a severe outbreak of Leaf Spot disease in Saint Lucia in 2003. There have also been problems with drought, and with storm damage, while Belize has suffered directly from several hurricanes, most recently 'Hurricane Iris', which devastated the banana-growing district in 2001. Exports have declined across the region. In Dominica, for example, the value of banana exports in 2003 was US $4.7m., compared with $30.1m. in 1992; a recovery to $6.0m in 2004 did not signal a reversal of the underlying trend. In Jamaica, banana exports fell from $39.5m. in 1992 to just $12.8m in 2003.

The direct economic impact of the decline in banana exports can be seen most clearly in Dominica, which has only a small tourism industry and few other exports. GDP grew slowly in the 1990s, then contracted by 4.9% in 2001 and 2.7% in 2002, largely because of a further sharp fall in banana exports. Unemployment in 2003 was put at an estimated 25.0% by the World Bank. The Government's fiscal problems were intensified by heavy debt, some of it contracted at commercial rates for ill-conceived development projects. While the banana industry has not been neglected in official policy-making, recent attempts to restructure the economy have tended to focus on other sectors, mainly tourism.

There has been substantial EU assistance to Caribbean banana producers. Part has been from payments under the former Stabilisation of Export Earnings (Stabex) scheme, triggered by decreasing banana earnings and totalling €200.9m for the Windward Islands alone for the crop years 1995–99, with an additional €189.0m from a Special Programme of Assistance for the banana industry.

Funds have been directed in part to assistance for the banana industry, including, but not limited to, agronomic improvements such as 'trickle and drip' irrigation, or tissue culture to provide high quality planting material. As a result, yields for some farmers have substantially increased, while those on land that is too dry or too steeply sloping for successful banana cultivation have, in most cases, moved out of the industry. With innovations such as field packing, improved transport infrastructures and inland reception centres, the percentage of fruit meeting export quality standards increased sharply in the late 1990s. In the Windwards, banana growers' associations have been privatized; in Saint Lucia, growers can now choose from three competing buyers. These developments have clearly not been enough to stem the decline in banana production; however, the industry's proponents argue that they will enable it to survive in a slimmed-down form.

Outside the banana industry, EU assistance has also been channelled to economic diversification, including, but not limited to, new agricultural products; and to provision of a social safety net, including retraining and employment opportunities for the rural population. However, cumbersome project development and approval procedures have made the disbursement of funds in some cases painfully slow; for example, disbursement of Stabex funds compensating for 1995 crop shortfalls began only in 1998.

Banana growers in the Windward Islands have attempted to increase their share of the final supermarket price by moving into international transport, ripening and distribution. The West Indies Banana Development Company (Wibdeco)—owned jointly by four island governments and banana growers' associations—in 1996 entered into a joint venture with an Irish company, Fyffes Group, to buy the shipper and distributor Geest Bananas Ltd; the purchase was financed with loans totalling £20m from the Allied Irish Bank (AIB). Direct sales contracts were negotiated with British supermarkets. From January 2001 Wibdeco subsequently took on sole responsibility for marketing and distribution of Windwards bananas, with Geest retaining responsibility for freight; at this point Wibdeco took on a £4m working capital loan from Fyffes, to be repaid over two years. This new commitment was met, but with export volumes below expectation, Wibdeco had difficulty in meeting earlier commitments to the AIB. Although a final payment was due in July 2003, a balance of £3.2m. remained due in the first half of 2004. The Governments of Saint Lucia, Saint Vincent and the Grenadines and Grenada therefore in 2004 guaranteed borrowing of US $10m. from Citibank's Trinidad and Tobago affiliate.

In late 2004 and in 2005 interest focused on the tariff structure to be adopted when quotas are finally phased out at the beginning of 2006. The EU proposed a tariff of €230 per metric ton for Latin American and other non-traditional producers. CARICOM continued to lobby for a tariff of €275 per ton, but with little chance of success. Latin American producers argued for a much lower tariff of €70 per ton, claiming that a higher rate would cost them some of their existing market share, which was 63% in the former 15-member EU and close to 100% in the 10 new member states. On 30 March Honduras, Colombia, Costa Rica, Panama and Guatemala referred the issue to the WTO for arbitration, a relatively quick process, with a ruling expected by August 2005.

There have been efforts to increase earnings by exploiting niche markets. There is a premium price for smaller fruit, which is now packaged and sold separately. Most Windwards fruit meets employment and environmental 'Fair Trade' standards, and is sold with a 'social premium', which is paid to community groups for infrastructural or social development projects. There is also a market for certified organic fruit, which sells at a premium of 25%–40%; however, the islands which have been most heavily involved in traditional banana production will have greatest difficulty in identifying land that has been chemical-free for the required period.

More emphasis has been placed also on the substantial regional market, where CARICOM has a punitive tariff on Latin American bananas. Windwards bananas are exported to Trinidad and Tobago and Barbados. Jamaicans eat their way through an estimated 120,000 metric tons per year. Exports decreased from 205,000 tons in 1966 to 43,000 tons in 2003. Falling prices and the cost of spraying against Black Sigatoka disease, which arrived in the island only in the 1990s, have driven smaller domestic producers out of production, allowing the commercial producers to devote two-thirds of their production to the home market.

Restructuring has been most far-reaching in the case of the state-owned Surinamese producer Surland, which closed in April 2002 after sustaining heavy losses and accumulating Sf45,000m. in debt, laying off 2,000 staff. With an €18m. EU assistance programme and US $7.3m. in working capital from the Inter-American Development Bank, the company was then restructured as the SBBS, and a full replanting exercise using tissue culture, new irrigation systems, and cableways for transport of fruit was undertaken. Exports were resumed in March 2004, and privatization of the company is now planned.

An exercise of this sort is not possible with the large number of relatively small, private-sector growers who remain active in the Windward Islands, or even with the somewhat larger farmers active in Belize and Jamaica. In mid-2004 the survival in some form of a commercial banana industry in these economies appears probable; however, a rural way of life centred on small scale cultivation of fruit for export by independent farmers is already receding into the past.

## SUGAR

Sugar was the economic mainstay of most of the Caribbean from the 18th century until the 1950s. In contrast to bananas, the industry was dominated by large estates, with heavy capital investment in factories producing raw sugar from cane. After independence, large estates and sugar factories were in most cases nationalized, as in Guyana, Trinidad and Saint Kitts, so that the fortunes of the industry became an important component of the public-sector financial balance. Loss of trade preferences does not therefore carry quite the same emotional charge as for bananas, but does, nevertheless, have major social and economic implications.

Physical conditions in most of the Caribbean are not ideal for cane production: problems include low rainfall and uneven terrain in Barbados, field layouts unsuited to mechanization in Guyana, and the small scale to which the industry is constrained in Saint Kitts. Productivity is generally low, while wage levels are much higher than in producers such as Brazil, although of course lower than in others such as Australia. Factory equipment is, in many cases, antiquated, while Barbados in particular was plagued by a many-layered and costly management structure for what, by international standards, have been small production units. For these and other reasons, costs are extremely high by international standards.

The Caribbean sugar industry, which dominated the world market until the early 19th century, is no longer of great international significance. The CARICOM producers account for close to 0.5% of world sugar exports. Within most Caribbean economies, sugar is now overshadowed. In Barbados, for example, sugar made up 5% of GDP in 1980, but only 1% in 2001, in which year it accounted for only 2% of foreign exchange earnings and 2% of employment. In Guyana, however, sugar in 2003 still employed 7% of the labour force, produced 26% of total exports, and comprised 15% of GDP. Even in the economies where sugar has least overall importance, it remains significant as an employer of unskilled rural labour, which would be difficult for other industries to absorb. In Barbados, a tourism-based economy where sugar still dominates the rural landscape, closure of the industry would have far-reaching environmental and visual implications.

The sugar protocol of the Lomé Conventions allocates Caribbean producers export quotas to the EU, at a guaranteed price of €523.70 per metric ton. In recent years this has been approximately three and a half times higher than the fluctuating free market price for raw sugar. This lucrative market has absorbed almost all of CARICOM's sugar exports, and indeed some producers, such as Barbados, have imported lower-cost sugar so as to use their own crop to supply the EU quota. However, even at this price, many Caribbean sugar industries have been heavily in loss, notably those of Trinidad, Barbados and Saint Kitts. In the case of Saint Kitts, the industry's losses in recent years have been equivalent to 4.0% of GDP, a heavy burden on the public purse. In Guyana, however, the state-run Guysuco has, in most years, made a substantial contribution to central government finances, through dividends, taxes and other payments, with costs at around US $400 per ton.

The EU was in the process of reassessing its internal sugar regime. Three alternatives were put forward to the Commission in September 2003: an extension to the existing regime beyond 2006 with reduced quotas and prices; a gradual elimination of production quotas, with the EU internal price adjusting itself to the price of non-preferential imports; and a complete liberalization of the current sugar regime with integration of sugar producers into the Single Farm Payment system. In June 2004 the EU Farm Commissioner, Franz Fischler, proposed a 20% reduction in sugar prices in 2005, followed by a further cut in 2007, bringing the cumulative reduction to 33%. Although compensatory aid would be granted to those affected by the changes, the announcement, nevertheless, was opposed by Caribbean producers. Since then the proposed 2005 price reduction has been postponed, with implementation now scheduled for July 2006, a date by which most Caribbean producers will have harvested and sold that year's crop, with the second cut to follow in July 2007. Externally, meanwhile, Brazil, Thailand and Australia in 2002 challenged the EU sugar regime through the mechanisms of the WTO; in August 2004 a WTO panel ruled, on a preliminary basis, that the EU subsidies were unlawful.

Although the Caribbean is likely to continue to enjoy some form of support in the European market, this is unlikely to be as beneficial as the present system, and those countries that already subsidize their sugar industries will face increasing losses. In 2003 Trinidad and Tobago closed its state-owned sugar company Caroni Ltd. For the 9,200 staff, this closure was cushioned by redundancy payments totalling TT $724m., with an additional TT $300m. pension fund enhancement. A successor, the Sugar Manufacturing Company, will continue to buy cane from independent farmers, but its future too appears uncertain, with output well below target in both 2004 and 2005. Saint Christopher and Nevis announced in December 2004 (two months after a general election) that the sugar industry would be closed on completion of next annual crop, in mid-2005. Sugar workers were to be retrained, and much of the land used for tourism developments. The Governments of Barbados and Jamaica have difficult decisions to make. In the two mainland producers, Guyana and Belize, there is a realistic prospect of continuing profitability. Guyana in particular proposes to invest US $135m. in the industry, in order to grow an additional 66,000 metric tons of sugar on 88 sq km of land to be converted for cane growing. A new sugar factory will also be built by October 2007, as will a co-generation plant to produce electricity for the public power supply and, probably, a refinery for further processing of raw sugar. Finance will come from the People's Republic of China, the World Bank and the Caribbean Development Bank, as well as from retained earnings. This initiative is expected to reduce costs to around US $265 per ton. Proposals were also under consideration in early 2005 for a US $35.2m. loan from India to fund a modernization programme at three other existing sugar factories in Guyana. Guyana and Belize will enjoy a continuing advantage in the Caribbean market, which charges a Common External Tariff of 40% on sugar and some other agricultural products, and where there is a significant saving in freight costs over extra-regional producers. Local earnings from sugar exports can also be increased through sales of packaged and branded sugar; in 2003 Guysuco launched its Demerara Gold product in selected markets. There has also been a pilot project to produce organic sugar.

## RICE

Suriname and Guyana are the only ACP countries that export rice to Europe. They are small producers by international standards, with exports in 2001 totalling 260,000 metric tons, of which 80% was from Guyana. However, rice has been the backbone of both countries' agricultural sectors, and has provided a livelihood for several thousand small farmers, as well as for a smaller number of large agriculturalists, millers and processors.

Following a deep decline in the national economy in the 1980s, the Guyanese rice industry in particular prospered in the 1990s, with improved plant varieties and technology, better irrigation, duty-free concessions for inputs, a more favourable pricing system, and privatization of state-owned rice mills. Production peaked in 1999; since then, however, a changing economic environment has brought problems.

Until 1997 Caribbean and other ACP rice could be exported duty- and quota-free to Europe if it was processed in one of the EU's overseas countries and territories. In the Caribbean there were rice mills in Curaçao (part of the Netherlands Antilles), Saint Vincent and Montserrat, none of which grow rice. From 1997 this route into the EU market was limited to 35,000 metric tons. An additional quota of 145,000 tons for direct exports to the EU is subject to a tariff, and to regulations on timing and security deposits, which make it difficult for local exports to exploit the quota's full potential benefit. To compound these problems, local weather patterns were disrupted by a drought brought about by El Niño and then by flooding caused by La Niña. As a result, there was a decline in world prices. The export price of Surinamese rice fell by 40% from 1997 to 2001. According to local industry sources, prices were by the end of this period one-third below production costs.

During the rice boom Guyanese millers and farmers had borrowed extensively from local banks, while Surinamese farmers were forced to borrow owing to their country's severe macroeconomic difficulties. Lower prices after the late 1990s left many producers in both Suriname and Guyana unable to service their debts. This was a problem not only for the rice industry, but also for those countries' financial sectors. In Guyana total rice industry debt had, by 2001, reached G $12,000m. in principal with G $4,600m. in accumulated interest, owed by 1,380 farmers, and necessitating G $6,000m. in loan loss provision by commercial banks.

By 2005 world rice prices had recovered, rising by 40% in 2001–04. In February 2002 the Guyanese Government brokered a debt forgiveness scheme, writing off accumulated interest and 25% of principal owed for debts of less than G $10m., with repayment of the remaining amount rescheduled over 10 years. There has also been some assistance from the EU, in the shape of a €24m. aid package for Caribbean rice, agreed in 2003. Of this aid, €11.7m. was allocated to Guyana and €9.3m. to Suriname for appropriate credit facilities, market and product research, institution and capacity building, research and development, as well as drainage and irrigation support for the low-lying coastal farmlands of both countries. These funds, while useful, fell far short of meeting the full restructuring needs of the rice industry. In Guyana a debt relief programme for small farmers was agreed in principle in February 2002, although implementation has been unsteady. However, a long term strategy was needed to resolve the long term problems of the industry.

An alternative to the EU market is that of CARICOM, with imports of over 170,000 metric tons per year. Guyana supplies almost one-half of this market, and also exports rice to Suriname. However, the tariff on extra-regional imports, at 25%, is lower than for sugar, while islands such as Jamaica import substantial quantities of subsidized US rice. Regional trading arrangements remain problematic, although the Caribbean Court of Justice, inaugurated in April 2005, will provide a legally binding mechanism for the enforcement of CARICOM rules, to the advantage of Guyana and Suriname. While the industry is likely to survive in some form, stagnation appears more likely than a renewed period of growth.

## RUM

Rum is of course not directly an agricultural product, but is intimately linked in its history and development to the sugar industry. Most Caribbean countries make rum, including several which no longer have a sugar industry, and use imported molasses. Caribbean brands have for many years had strong local visibility, but until recently the mainstay of the Caribbean rum industry was production of bulk rum for blending overseas. Trinidad and Tobago's rum company, Angostura Holdings Ltd, for example, produced spirit for Bacardi, whose Bahamian blending and bottling plant exported to Europe under the Lomé Conventions. A quota system for Caribbean rum was in force until 2000, giving producers little incentive to invest in larger-scale production facilities, but assuring duty-free access for a steady quantity of product. Rum producers are not in the same social need category as small banana growers, but do impact significantly on Caribbean economies.

Under a 'zero-for-zero' agreement on white spirits, the EU and the USA phased out most tariffs on imported rum and certain other spirits from most sources from July 1997 to January 2003. While Caribbean rums retained a very slight tariff advantage, this agreement opened the EU market to lower-cost third-country producers with much greater economies of scale. The best-adapted regional rum producers have moved from bulk exports to developing their own brands in export markets. This requires considerable capital investment; it can take US $30m. over five to 10 years to gain recognition in a major market. Some companies, such as Demerara Distillers of Guyana, have successfully promoted brands in selected international niche markets, gaining a small share of a business carrying much higher margins than bulk rum. Others are affiliated with international majors. Mount Gay in Barbados is majority owned by Remy-Cointreau, which in turn is part of the Maxxium alliance. As with traditional agricultural products, the EU has agreed a €70m. assistance programme over four years for institutional capacity-building, plant modernization, pollution control and, crucially, overseas distribution and marketing. In some countries, rum exports have increased as sugar has declined. In Barbados, for example, rum comprised 10.2% of domestic exports in 2003, up from 3.6% in 1991, growing from Bds $8.8m. to Bds $33.6m. in absolute terms.

## PROSPECTS

Each of the Caribbean's traditional export commodities seems likely to survive in some form. However, survival is unlikely for some groups of producers, such as the smaller banana growers, and some national industries, such as sugar production on those islands where the industry is most heavily in loss. Where traditional industries do survive, they will employ less labour, and smaller independent businesses will play a sharply reduced role. This implies job loss, mainly among unskilled and semi-skilled staff in rural areas, who are not well placed for retraining or redeployment. Financing of social 'safety nets' is important, but will prove only a temporary palliative. More important is the growth of alternative export industries and securely based domestic production to absorb labour released by the traditional sectors, and replace them as a source of foreign exchange. Some new industries appear promising. Shrimp and fisheries exports have increased in countries such as Belize, Guyana and Suriname, and the lucrative EU market appears to provide additional opportunities. Some telemarketing initiatives also appear potentially lucrative. Nevertheless, the most promising exchange earner on a region-wide basis remains tourism.

## FOOTNOTES

1. The Caribbean Forum (CARIFORUM) includes the 13 independent English-speaking Caribbean countries: Antigua and Barbuda, the Bahamas, Barbados, Belize, Dominica, Grenada, Guyana, Haiti, Jamaica, Saint Lucia, Saint Christopher and Nevis, Saint Vincent and the Grenadines

and Trinidad and Tobago. It also includes Haiti and Suriname, which, along with their English-speaking regional neighbours, are members of the Caribbean Community and Common Market (CARICOM); and the Dominican Republic, which is not a CARICOM member. The scope of this article extends to the English-speaking CARICOM members and Suriname, but not to Haiti nor to the Dominican Republic, where issues relating to the export of agricultural commodities are different in nature.

2. Trinidad and Tobago and Saint Christopher and Nevis are used when referring to the Governments of that country. Trinidad and Saint Kitts are used when referring to the individual islands on which sugar is grown. No sugar is grown on the islands of Tobago or Nevis.

3. The Windward Islands are composed of Dominica, Grenada, Saint Lucia and Saint Vincent and the Grenadines.

4. At the same time, the French market was preserved for its former colonies and Overseas Departments, with similar historic arrangements in place for some other European importers, and a low-cost free-market system in operation in Germany.

5. This term originated in the 1950s, when the United Kingdom gave preference to trade with the 'sterling area', comprising mainly Commonwealth countries keeping a proportion of their currency reserves in British pounds, as against the 'dollar area', which included the USA and Latin America.

# RELIGION IN 21ST CENTURY LATIN AMERICA

## Canon Prof. KENNETH N. MEDHURST

### THE LATIN AMERICAN CHURCH'S GLOBAL SIGNIFICANCE

At the beginning of the third millennium Christianity's global future depends significantly on developments in Latin America. That region contains the world's largest single concentration of Roman Catholics. Approximately 45% of all Roman Catholics are Latin American. It is also the scene of unprecedented Protestant growth that was presenting the Catholic Church with novel forms of competition. That competition's outcome could have major implications for the future shape of world-wide Christianity. In recent times, new radical expressions of Latin American Catholicism have served as something of a model for co-religionists in such diverse countries as the Philippines and South Africa. In that context, the outcome of the Roman Catholic Church's own efforts to restrain such radicalism, and its responses to fresh Protestant challenges, has a global importance.

### THE HISTORICAL CONTEXT[1]

The contemporary Latin American Church's position has to be viewed in the context of its history. For centuries the Roman Catholic Church enjoyed an ecclesiastical monopoly that had been introduced into the region in the 16th century under the auspices of the Spanish and (in Brazil's case) Portuguese colonizing power. In practice, the papacy conceded control over the Church to the Spanish and Portuguese crowns. Bishops were royal appointees and, despite some questioning of imperial authority by such early missionaries as the Dominican Bartolomé de las Casas, ecclesiastical leaders generally acted as state servants. Equally, the Church became a privileged beneficiary of a hierarchically structured socio-economic order constructed by colonists. It was, for example, a major landowner. For its part, the state underwrote the Church's ecclesiastical monopoly. In principle, Church and society were coterminous. In practice, mass, and sometimes enforced, 'conversions' meant that official understandings of Catholicism frequently failed to impinge significantly upon the local culture.

Political independence initially did little to change this situation. Post-colonial political élites sought to take over control of the Church from imperial predecessors. They also continued to preside over largely unreconstructed societies. However, these élites contained liberal, secularizing elements, whose reform programmes contained demands for the diminution of church influence. The outcome was a series of intra-élite conflicts partly concerned with Catholicism's future status. Such conflicts cemented alliances between the Church and society's more conservative and, especially, landowning interests.

The outcome of such conflicts varied according to local circumstances. In Colombia a Concordat agreed by the state with the Vatican assured the Church a privileged position. At the opposite end of the spectrum, Mexico's Church, during early 20th century's revolutionary upheavals, was, for a while, persecuted. For a long time after it remained politically marginalized. Most countries reached accommodations at a variety of intermediate points between these two polarized positions. In general, the Church was left with few incentives actively to re-consider inherited positions. Only in the atypical Uruguayan case did it confront a patently secularized society that obviously limited its influence.

### 20th CENTURY SOCIO-ECONOMIC CHANGES

Socio-economic changes of the 20th century disturbed such complacency. Urbanization and a measure of industrialization profoundly changed the mainly rural social order within which the Church had become rooted. The institution lacked the flexibility needed quickly to adjust to the new situation. Many of its lower-strata constituents, when freed from the constraints of rural society, were seen to owe it little significant allegiance. It frequently became evident that religion was of the syncretic variety, owing much to pre-Columbian or, in the case of those descended from slaves, to African influences. In many other instances commitment was to a pervasive 'cultural Catholicism' that owed relatively little to official church teaching. The institution was exposed as a primarily upper or middle class affair.[2]

Most notably in Chile, this changing environment provided the Church with new forms of competition that offered previously absent incentives to re-evaluate traditional strategies.[3] The competition, in Chile, came from early Pentecostal experiments and, perhaps more seriously, from a Marxist-inspired political movement. In response, some European-trained clergy sought to organize alternative middle-sector, working-class and even peasant groups designed to preserve traditional clerical control. They were essentially defensive in concept and not intended to encourage serious structural change.

In the 1950s the obvious limitations of this approach led some Chilean middle class professionals to establish a Christian Democratic party.[4] Under the influence of such progressive European Catholic thinkers as Jacques Maritain they offered a 'middle way' between traditional Conservatism and the secular left. It was not tied to the church hierarchy and was even open to non-Catholics. It aimed, however, to deploy Catholic credentials in the building of a broad electoral coalition. The intention was to reach out beyond established middle-class interests and to previously unmobilized elements amongst shanty-town dwellers and peasants. These groups were offered a programme of significant structural reform.

Initially, the party enjoyed spectacular success. In 1964, within six years of its foundation, it came to power under the leadership of President Eduardo Frei (1964–70). For a while it seemed to offer a successful model for the whole region. In practice, however, Frei's administration presided over a process of political polarization that led to the election of Salvador Allende Gossen's Marxist-inspired 'Government of Popular Unity' and then, in 1973, to the coup which established Gen. Augusto Pinochet Ugarte's military-based dictatorship.

The process was partly the result of Christian Democrats raising unfulfillable expectations. It was also a reflection of the fragmentation of their original coalition. Many supported them as the best available way of blocking the left and subsequently returned to the traditional right. Frustrated radicals turned to Allende.

### THE INTERNATIONAL CONTEXT

By the time of Allende's election the Latin American Church had moved on. This was partly owing to developments within the region, but also to a changed international context. In the process, Chile's vanguard role was taken over by progressive groups within the Brazilian Church.[5]

In the 1950s and early 1960s Brazil was caught up in a major social and political crisis. Catholic radicals, along with others, became involved in an unprecedented process of mass political mobilization entailing competition for the allegiance of hitherto mainly unorganized lower-strata interests. The process was largely caused by the dynamics of Brazil's political system, but also owed something to the Cuban revolution's influence. The Catholic student and workers' groups embroiled in the competition owed much to a small but well organized section of the Brazilian episcopate, led by Archbishop Dom Helda Camara of Recife. Such support partly sprang from the realization that amid a period of upheaval the Church's institutional interests might be best served by sinking deeper roots among society's less advantaged sectors. It also sprang from a major theological reassessment that pointed toward strengthened commitments to the cause of social justice.

That re-assessment was initially inspired by unfolding Brazilian conflicts. It was also substantially encouraged by changes in the international Church associated with the Second Vatican Council (Vatican II—1962–65). That gathering sanctioned an 'updating' of the Church's internal life and provided for openness to democratic, secular and even left-wing political forces. The result was to legitimate and to strengthen forces for change already working in the Latin American and, especially, the Brazilian Church.

Brazilian mass political mobilization, in 1964, precipitated a military coup. It was the forerunner of several similar military interventions that, in the next two decades, were to frustrate the Latin American left's hopes and were to provide the political framework within which the Church was constrained to function. The resulting dictatorship initially placed Brazil's Catholic radicals on the defensive but they were subsequently to exercise a major influence within the wider Latin American Church.

In particular, they were prominent among the expert advisers who influenced the historic conference of Latin American bishops held in Medellín (1968).[6] The Medellín Conference was intended to apply Vatican II's conclusions to Latin America's particular circumstances. In practice, it went further than many of the participating bishops may have initially intended. Certainly, it represented a watershed in the regional Catholic Church's history. It embraced an analysis of Latin American realities that attributed the area's socio-economic and political circumstances to unjust international and national structures. Challenging these was seen as a major feature of Christian commitment on the grounds that they perpetuated unacceptably dehumanizing inequalities. Within this analysis traditional Catholic understandings of sin were revised. The idea of structural sin was incorporated. This theological shift pointed toward a potentially massive and historically unprecedented move on the Church's part towards leftward-inclined political allegiances.

## 'LIBERATION THEOLOGY'[7]

One especially important outcome of resulting debates was the articulation of 'liberation theology'. Its first internationally known exponent was the Peruvian priest Gustavo Gutiérrez, who, significantly, was in attendance at Medellín. He and his successors frequently differed over their precise prognoses and prescriptions. Nevertheless, they shared certain preoccupations and general approaches. Thus, it is intimated that the Judeo-Christian tradition can be seen to place a premium upon humanity's 'liberation' from all that impedes the realization of its God-given potential. The delivery of the ancient Jewish people from slavery in Egypt is viewed as one particularly important paradigm. Emphasis is placed upon political expressions of faith entailing the challenging of dehumanizing structures. There is talk of 'praxis', whereby faith is seen as a matter of creative collective engagement with mundane political realities, rather than adherence to abstract doctrinal propositions.

This general approach has been influential within the worldwide Church as well as within Latin America itself. It has also been controversial. Many priests, nuns and lay people have been inspired to commit themselves to self-sacrificial struggles on behalf of the poor. Such commitment has been made evident in consciousness-raising work among marginalized urban and rural dwellers. It was also evident in, for example, the work of the Chilean 'Christians for Socialism', who first allied themselves with Allende and then joined forces with other sectors of the Church in opposition to Pinochet's regime. Radicalized Catholics similarly inspired opposition to Brazil's military dictatorship.

One particular significant phenomenon, associated with such developments, was the emergence of Comunidades Eclesiales de Base (CEBs, or base communities).[8] These were conceived as sub-parochial entities able to facilitate lay participation in church life and to equip members for participation in the wider society's affairs. They were, in part, a response to a clergy shortage which had left Latin America's Church substantially dependent on foreign priests. They were also seen as harbingers of a more popular, declericalized and decentralized Church—a Church consistent with Vatican II's notion of 'a Pilgrim People'.

During the years of military rule Brazil and other countries in the region had thousands of such entities. For many they offered the possibility of a renewed faith, a sense of community amid otherwise atomized societies, and for a chance to acquire skills deployable in the Church's service and in political struggles. Sometimes there was an 'overspill' effect, whereby community members were enabled to share in broad-based campaigns ranging from efforts to improve municipal services to the challenging of dictators. It is significant, for example, that Brazil's post-military opposition party, the Partido dos Trabalhadores (Workers' Party), emerged from such a background.

For many, CEBs offered a creative new way of 'being the Church'. It was also one of several things that made liberation theology highly controversial. Thus, from an episcopal perspective, CEBs could be seen as potentially divisive centrifugal forces liable to escape from hierarchical controls. Similarly, there was an anxiety that liberation theology, more generally, signified an undue politicization of faith. A particular concern was the extent to which Marxist categories were sometimes imported into theological discourse. Not least, there was resistance to any possible espousal of class warfare at odds with the idea of a Church open to all. Equally, there was a suspicion that the Christian idea of 'the Kingdom of God' could, in practice, become conflated with Marx's concept of a classless society. Above all, there were problems concerning the issue of violence.

Even before the Medellín Conference a Colombian priest, Camilo Torres, controversially raised the issue by joining a Marxist rural guerrilla organization. His subsequent death made him, for some, an emblematic martyr figure. The majority of even radical Catholics, however, opposed such an option. Pope Paul VI (1963–78) condemned violence and Archbishop Helda Camara echoed his views. Nevertheless, the issue remained alive. It was especially significant in Central America.[9] In El Salvador, Catholic base communities were involved in opposition activities that spiralled out of church control and helped to trigger a protracted civil war. This entailed massive human rights infractions and claimed many lives, including those of radical Catholics. In particular, there was the case of Archbishop Oscar Romero y Galdames, who was assassinated in 1980 for his advocacy of justice in the face of military opposition. In Nicaragua, radicalized Catholics, frequently linked to base communities, were involved in armed Sandinista-led opposition to Somoza's dictatorship. The resulting Government similarly received some active Catholic support. Priests even held ministerial office despite papal and official Nicaraguan church disapproval.

Such histories tended to give credence to those who, for ethical reasons, consistently opposed violence. In El Salvador and elsewhere, militancy in the medium term seemed more likely to provoke violent reaction than to promote structural change. Nicaragua could be seen as a partial exception, but the advent of the Sandinistas did not bring an end to civil war or assure their long term hold on power.

Divisions in the Nicaraguan Church pointed, in particularly acute form, to the tensions more widely generated within the Latin American Church in the post-Medellín period.[10] In, for example, Brazil and Chile, portions of the hierarchy initially welcomed military intervention as relief from political turbulence. They were influenced by the expectation that military rule would be short-lived. There were other church leaders, however, who, from the outset, went into opposition. Over time, the circumstances of protracted dictatorship induced a closing of ranks, albeit one that obscured the persistence of underlying divisions. Faced with human rights abuses and economic policies that heaped the costs of change on the poor, conservative leaders moved in the direction of more radical colleagues. The resulting alliances entailed significant episcopal support for working class, peasant, human rights and other movements pledged to ending dictatorship. This was both a matter of seizing opportunities to broaden the Church's social base and of taking principled stands. Sometimes church leaders, as in Paraguay, used their influence to promote dialogue between government and opposition, but, when faced by intractable rulers, the overall movement was towards opposition. The general effect was to undermine any claims to legitimacy on the part of dictators. Equally, there was a tendency for the Church to emerge as the chief focus for broadly based opposition movements. The institution offered an umbrella under cover of which others could assemble. In Chile, for example, the Catholic Church, in association with other religious bodies, offered protection to victims of human rights abuses irrespective of the loyalties of those concerned. In doing so, much goodwill was gathered from once alienated groups. Chilean ecclesiastical agencies similarly helped to orchestrate the anti-Pinochet vote which, in a 1988 plebiscite, precipitated the dictator's resignation and democracy's restoration.

Only the Argentine hierarchy remained wedded to a reliance upon private dealings with incumbent élites. They, atypically, stayed generally loyal to military rulers.[11] This has to be understood against the background of Argentina's specific historical circumstances. Thus, in the 1920s the Church and the military forged an alliance as a consequence of shared opposition to secularizing democratic groups.

That alliance was reinforced in the early 1950s when the incumbent populist President, Gen. Juan Domingo Perón, turned on the Church as part of a strategy for shoring up his anti-clerical working-class base. The alliance only came to be seriously questioned during the 1990s and in the wake of Argentina's last disastrous experience of military government.

## THE RETURN TO DEMOCRACY[12]

The restoration of democratic regimes that characterized the 1980s and 1990s, drew attention once more to those internal church tensions that had been partially submerged during the years of dictatorship. Situations varied according to local circumstances and the balance of forces within particular local hierarchies.

The general trend, however, was for official church leaders to place some distance between themselves and the political arena. They simultaneously sought to reassert traditional hierarchical control. With the resumption of competitive party politics the perceived need to be routinely engaged in politics seemed to recede. Many church leaders welcomed the opportunity to give priority to traditionally understood pastoral and evangelistic tasks. In some cases there may have been a belated reaction against the heated ideological climate that had first precipitated military intervention. To the same general end steps were taken to discipline or restrain radical church activists perceived as threatening traditional understandings of ecclesiastical authority.

On the one hand, there was a general presumption in favour of support for newly restored democratic governments. (In Chile the advent of a centrist Christian Democratic-led administration did much to build the necessary trust.) On the other hand, ecclesiastical patronage tended to be withdrawn from radical politically activist elements. In particular, there was pressure for base communities to focus on traditional liturgical activities if they were not completely to disappear. Sometimes specific local circumstances inhibited the process of political disengagement. In Brazil, most notably, post-military Governments pursued economic policies so reminiscent of the preceding dictatorship and remained so tainted by corruption that the episcopate felt constrained to sustain a critical running commentary and to offer continuing support to agencies concerned with such matters as land reform, environmental protection, human rights and support for indigenous peoples.

## THE SIGNIFICANCE OF CONSERVATIVE REACTION

The Brazilian situation makes it plain that, despite everything, impulses emanating from the Medellín Conference remain strong. Nevertheless, the last 20 years or so have seen a regrouping and strengthening of those ecclesiastical forces concerned to impede, if not reverse, some of the more radical innovations associated with that change of course. One important milestone was reached when opponents of liberation theology, led by Colombian bishops, took control of the bureaucracy of the Consejo Episcopal Latinoamericano (CELAM—Latin American Episcopal Conference).[13] They replaced those who had set the Medellín agenda. Consequently, at the next CELAM gathering, held in Puebla, Mexico, in 1979, a renewed emphasis was placed upon evangelism rather than on political activism.[14]

On the other hand, the Puebla conference's findings retained a strong commitment to social justice. The same was true of the subsequent assembly held in 1992 in the Dominican Republic. This testifies to some continuing capacity on the part of radical political activists to influence debate. It also suggests that it may sometimes be misleading to define churchmen in purely political terms and without reference to theologically driven categories. Ecclesiastical conservatives need not be socially conservative.

Within most Latin American hierarchies there remain probably minority groups who are essentially pre-Vatican II in outlook and, consequently, are both ecclesiastically and politically conservative. Some of them may even have a preference for authoritarian rule. They are likely to be associated with the very conservative Opus Dei organization, which exercises significant influence in several Latin American countries. In Peru for example, seven of the country's 52 bishops belong to that body. Most official church leaders, however, are now, to varying degrees, committed to the cause of social justice and change. The real controversy concerns the priority to be given to political engagement and the extent to which engagement implies a more decentralized or declericalized Church.

In that controversy, John Paul II's papacy (1978–2005) was clearly associated with traditional hierarchical understandings of ecclesiastical authority. In Latin America, as elsewhere, an appointments policy has been pursued whereby leaders in the mould of Archbishop Helda Camara have been steadily replaced by known supporters of the Church's 're-Romanization'. The Brazilian archdiocese of São Paulo, widely perceived to be the progressive Church's most important centre was, in 1989, divided into five parts. The predictable result was an undermining of the influence of Cardinal Archbishop Evaristo Arns, a once potent patron of radical causes. Similarly, such prominent liberation theologians as the Brazilian Leonardo Boff were subjected to Vatican inquiries or bans. The Vatican, moreover, issued its own condemnatory critique of liberation theology. The process has continued in more recent times. Thus, 2005 saw the retirement of Dom Pedro Casaldáliz who, as Bishop of São Félix de Araguaiu and President of the Church's Pastoral Land Commission, gave a high priority to socioeconomic issues. The nature of his possible successor was indicated by the fact that out of four new Brazilian episcopal appointments, two were members of Opus Dei and one of the ideologically similar Communione e Liberazione organization.

Despite all this, the late Pope John Paul II regularly used official pronouncements and visits to offer unusually powerful critiques of capitalism and of neoliberal economic assumptions. Human rights abuses have been similarly condemned. Thus, even if liberation theology's exponents are beleaguered they can still claim a certain legitimacy. Equally, the Church's 'mainstream', though now clearly conservative in its ecclesiology, still generally gives a priority to the cause of the poor, and to social justice, which runs strongly counter to once prevalent pre-Vatican II and pre-Medellín traditions. Helda Camara's death in 1995 and Evaristo Arns' retirement in 1999 apparently symbolized the end of an era and a high water mark for the forces of ecclesiastical conservatism. Nevertheless, the Latin American Church remains very different from the one that preceded the developments of the 1960s.

## THE ROMAN CATHOLIC CHURCH'S STANDING: AN EVALUATION[15]

Continuing debates concerning the Catholic Church's future raise questions about its general standing in the early 21st century. Amid conflicting ecclesiological understandings the underlying issue concerns the institution's long term strength, the extent of its ongoing influence and the depth of its roots in Latin American society. Experience of military rule created an expectation that the Church might have significantly widened its social base. Certainly, political engagement during that era earned the institution goodwill among many hitherto on its margins. The tendency, even in the Colombian and Argentine cases, to renounce reliance on formalized church–state partnerships was similarly seen as conducive to the fostering of popular trust (as well as perhaps signifying a greater institutional self-confidence). Equally, credence was given to the assumption, found among exponents of liberation theology, that it was possible to bridge the divide separating the official Church from those mainly lower-strata groups which share in a popular, if not syncretic, religion. There is evidence from several countries suggesting that such notions are by no means entirely ill-founded. In a number of societies there are signs that the Church remains a significant and relatively prestigious institution capable of commanding substantial trust across social divides. During the crisis surrounding President Alberto Fujimori's contested re-election in 2000, Peruvian Church leaders were commonly regarded as credible arbitrators. There were doubts concerning the allegiances of the Archbishop of Lima, Cardinal Juan Luis Cipriani Thorne, a member of Opus Dei and perceived to have been unduly close to Fujimori; however, in 2003, significantly enough, episcopal colleagues replaced Cipriani as President of their national Bishops' Conference and he remains a divisive figure. Similarly, the Bolivian Church's leadership continues to demonstrate moral authority amid destabilizing conflicts over ownership of the country's huge natural gas reserves and the political role of the country's hitherto largely disempowered indigenous majority. Most parties to such disputes continue to look to the Church as a mediator and the only national institution commanding substantial respect. A perverse measure of the Colombian Church's standing is that paramilitary forces and drugs cartels have assassinated ecclesiastical leaders who have denounced corruption, opposed violence or urged national reconciliation. The killing of Cali's Archbishop, Isías Duarte Cancino, in 2002, was but the most clearly dramatic case. At the same time, both government and guerrilla groups, albeit reluctantly, were constrained to use the Church's good offices in their attempts to sustain a measure of dialogue. Perhaps still more strikingly, ecclesiastical intervention in the 2004 presidential election in the Dominican Republic apparently pre-empted the then incumbent regime's attempt to falsify the result.

One apparently minor change of fortune concerns the Argentine Church. The stand taken by some of its leaders on behalf of the victims of their country's recent socio-economic and political upheavals has contributed to a significant increase in public confidence. Work done at local level for the relief of suffering and with a view to nurturing popular grass-roots participation has worked to the same end. Opinion polls carried out in 2003 and 2004 suggest that amid generally discredited national institutions the Church stood out as the one retaining most public trust.

The contrasting cases of Venezuela and Cuba offer a less clear cut, but still positive, picture. Thus, the Venezuelan Church, though divided by long-running conflicts within the wider society between supporters and opponents of President Hugo Chávez Frías (1999–), and despite much state-inspired anti-Church rhetoric, retains higher levels of public confidence than the Government. In Cuba an internationally beleaguered regime has seen some need to look for Catholic support. (In April 2005 Cuban leader Fidel Castro Ruz attended a mass celebrated in honour of the late Pope John Paul II.) For its part, the local Church has gained considerable goodwill by combining opposition to external intervention in Cuban affairs with unofficial but significant support for defenders of human rights.

There is evidence, however, pointing in different directions. Thus when, following democracy's restoration, local hierarchies diminished their support for base communities, these were sometimes shown to lack strong durable support. Some survived as politically relevant entities, but others continued minus their original political vocation or even disintegrated. In the 1980s and 1990s it became evident that many such bodies may have previously been unduly dependent on purely clerical leadership, episcopal protection and external resourcing. Sometimes the result, at the grass-roots level, has been something of a vacuum.

This latter situation may partly have been an expression of macro-level socio-economic and political realities over which the Church can have no immediate control. Thus, in the context of an apparently triumphant international capitalism and neo-liberal market economics, the Latin American left, with whom significant sections of the Church have been allied, became much enfeebled. Equally, the end of the cold war meant that revolutionary impulses once associated with Cuba and Nicaragua became much less relevant. In Nicaragua the Catholic left found itself very much on the defensive in face of a Catholic hierarchy sometimes perceived as unduly close to post-Sandinista Governments (and even implicated in allegations of political corruption). Similarly, in the Central American republics of El Salvador and Guatemala, church representatives have sometimes found it difficult to obtain justice for past human rights abuses.

Within this much changed environment church leaders have sometimes so reordered their priorities that they may now tend to find allies among their most traditional supporters. Thus, in Argentina, Brazil and Chile, to cite only three cases, an emphasis on strict Roman Catholic orthodoxy in such matters as divorce, birth control and abortion brought the Church into collision with public authorities and pushed it into alliance with those minority and generally privileged sectors of lay Catholic opinion disposed strongly to campaign on behalf of traditional positions. Secularizing tendencies have left many at variance with such teaching and supportive of legal liberalization. (Not least, the Church finds itself at odds with a mainly middle class, but still growing, feminist movement.) For example, in 2004, most Chileans supported government-sponsored legislation that, for the first time, introduced divorce. That legislation was approved in spite of official Church opposition. Its chief supporters came from diminished groups of those upper class conservatives that society's least advantaged sectors have good reason to distrust. This could bode ill for the Church's long-term standing among the disadvantaged majority.

In the tussle between the Church's more conservative and radical elements much could depend upon developments in Brazil. Because of its sheer size that country and its Church seem bound to have a major impact on relationships between 'the religious' and 'the political' throughout Latin America. Within the Brazilian Church, and despite conservative 're-Romanizing' tendencies, radical forces have been significantly, and perhaps irreversibly, entrenched. They have remained associated with community development, educational and other grass-roots projects conducted on behalf of indigenous peoples, peasant groups and other marginalized sectors. To those affected they have represented powerful signs of hope and signified, from the grass roots upwards, growth of a more vibrant civil society. Not least, they have sometimes been able to derive encouragement from episcopal critiques of prevailing socio-economic orthodoxies.

The election, in 2002, of Luíz Inácio ('Lula') da Silva to the Brazilian presidency, seems to offer further, quite substantial, hope to radicals, both Catholic and secular. President da Silva's Partido dos Trabalhadores (PT) has always owed much to the participation of Catholic activists and to unofficial, but significant, encouragement from the Church. In its lengthy quest for electoral success, the PT had moderated its initial programme in order to reassure both middle-sector Brazilians and international financial institutions. Nevertheless, its continuing commitment to greater equality and structural change signifies a substantial departure from the general thrust of post-military Brazilian politics. Moreover, the party's programme substantially mirrors official church objectives. A Catholic priest has even been put in charge of the new Government's poverty-eradication programme. Further evidence of sensitivity to Catholic opinion was supplied in February 2005 when, following the murder of a North American nun who had opposed land-owning and business interests responsible for the destruction of large swathes of Amazonian jungle territory, Brazil's federal Government was constrained to place new legal limits on the activities of such interests. On such a basis, significant expectations have been aroused throughout Latin America concerning the possibility of a serious challenge to existing socio-economic and political models. The outcome of this fresh experiment is therefore likely to have far-reaching consequences both for the region's politics and for the terms upon which the region's Catholic Church will, in the long run, engage with the political process.

## PROTESTANTISM[16]

It is partly against such a background that Protestantism's dramatic growth has to be placed. Its expansion may, to some extent, represent a critical commentary upon the still majority Roman Catholic Church.

Latin American Protestantism first arrived in the wake of 19th and 20th century immigration from Europe and under the auspices of Lutheran or other 'historic' Churches. The early 20th century, most notably in Chile, saw the subsequent growth of sometimes US-inspired Pentecostal Churches. The last approximately three decades, however, have witnessed a further and unprecedented massive expansion.

Sometimes, particularly in Central America, North American influence has again been involved. Indeed, some commentators have sensed an officially inspired attempt to neutralize the influence of a changed Catholic Church. The new religious right's links to Republican Administrations and to co-religionists in, most notably, Guatemala, have given credence to such interpretations. The adhesion to a California-based Protestant sect of the former and obviously repressive Guatemalan military dictator, Gen. José Efraín Ríos Montt (1982–83), has worked to the same end. In reality, however, recent growth has been mainly of indigenous inspiration.

The growth, principally involving Pentecostal Churches, has been so spectacular that in Brazil, for example, there are now more Protestant pastors than Catholic priests. Even more strikingly, there are now more Protestant Church members than members of base communities. In other words, growth of Protestantism has occurred mainly among those very lower-strata groups to whom liberation theology's exponents were concerned to appeal. A measure of the situation is that according to official censuses the Catholic population of Brazil declined from 89% of the total in 1980 to 74% in 2000. Current projections raise the possibility that, by 2025 Latin America's entire Roman Catholic population might represent no more than 50% of the total.

This helps to explain why the Protestant challenge represents a significant unifying factor within the Roman Catholic Church. Catholic traditionalists are disturbed by an invasion of their own once guaranteed space by forces threatening to inherited cultural identities. Radicals react against what is viewed as an apolitical or socially conservative form of religion, generally supportive of the *status quo*. They respond particularly negatively to an ethos which owes a good deal to North American models and which frequently seems to stress the possibility of individual and even material improvement.

The short-term political impact of Protestant growth lends some weight to those latter views. Although for example in Chile, some Pentecostal Churches joined with Roman Catholics in opposition to military rule, the evidence often points in a contrary direction. Other Chilean Pentecostalists offered Gen. Pinochet support or eschewed political involvement. In such a context it seems especially noteworthy that during Chile's years of military rule approximately 14% of army officers embraced Pentecostalism.

In the post-military realm of electoral politics Protestant interventions have frequently seemed marked by opportunism rather than considered attempts to promote major change. Leading Brazilian pastors, for example, have operated traditional clientelistic forms of politics whereby they mobilize denominational members in order to secure their own election to public office. The small political parties they have formed frequently seem chiefly to serve as vehicles for personal ambition. They often lack any clearly defined programme beyond promoting denominational interests. In Peru, by partial contrast, an apparently well-organized Protestant constituency promised to have a major impact on national politics. Evangelical Protestants were a significant factor in Peruvian President Fujimori's initial election. Protestant leaders, however, were unable to translate this into lasting political influence. Fujimori was subsequently to see greater political profit in seeking out supporters amongst the Catholic hierarchy.

These manoeuvrings are confined to small Protestant élites. Their activities leave open deeper questions concerning the underlying causes of Protestant growth and its general significance.

The clue seems to lie in the immediacy of the responses that Pentecostal Churches seem to offer to the pressing personal needs of their disadvantaged constituents. At a structural level, they enjoy a flexibility that often enables them more readily to draw close to members than is the case with Roman Catholics.

Their long-established institutions can appear remote, forbidding and unduly clericalized. Even base communities frequently seem, in practice, to presuppose learned clerical leadership. At the spiritual and psychological levels, Pentecostals offer theologically direct and immediately appealing messages that contrast with the cerebral and long-term strategic approaches often associated with liberation theology. Such messages sometimes seem more readily to tap into the region's popular religious culture than is the case with even revised Catholic liturgies. Not least, Pentecostal congregations often seem quick to supply some sense of community to the socially isolated or marginalized. Such communities may even serve to introduce important elements of order into disorganized lives, an order that can positively impact on family life, to the particular benefit of women. The possibility of a modest material security may follow.

In the short run, all of this suggests a capacity to confer meaning, purpose and identity upon lives caught up in disorientating processes of mainly urban, socio-economic and political change. In the long run, some commentators more controversially contemplate the possibility of value shifts in Latin American culture ultimately supportive of economic development and democratic politics. The evidence concerning this is mixed and the final outcome necessarily uncertain.

## CONCLUSION

The great transformations of recent times clearly co-exist with some underlying continuities. The Catholic Church's structures and the surrounding society's hierarchies even now bear some of the hallmarks of their colonial origins. Against that background the great historic watershed for Latin American Catholicism was the Medellín Conference. That gathering produced a sea change in the institution's own life and in its attitudes toward society and politics. In the process, an institution more or less uniformly conservative in theology and politics became divided along a variety of axes. Resulting controversies even now, in 2005, remain unresolved. Nevertheless, and despite some recent shifts in the balance of forces, the institution's overall centre of gravity seems permanently to have moved. There is now a widespread, albeit less than complete, consensus concerning the general claims of social justice. The most fundamental divisions concern the way the Church is most appropriately to be conceived and how the demands of justice may best be met. The currently dominant model is of a rigidly centralized Church generally reluctant to become routinely enmeshed in political controversy, albeit particularly disposed to act as society's tutor in matters of sexual morality.

However, doubts remain concerning generally prevalent modes of ecclesiastical governance. Protestantism's success suggests the desirability of greater decentralization and participation. Equally, in the political realm, prevailing attitudes seem unduly to discount the need for democratic institutions to be underpinned by a vibrant civil society conducive to popular participation and able to constrain the politically or economically powerful—a society to whose building the Church could make a major contribution. Recent developments in Argentina offer further examples. Promising, government-backed innovations in Brazil offer still more signs of hope. The Chilean Church's role in building support for the idea of holding Gen. Pinochet to account in his own country is striking. There is more, however, that might be done to assist in the building of a deeply rooted, democratic, political culture.

More than ever the Church's role has to be seen in a global context.[17] This is true for Catholics and Protestants. In the case of Catholics, the 2005 papal election served as a powerful reminder of the extent to which they are caught up in a complex international ecclesiastical system. On that occasion at least three Latin American cardinals were cited as 'papables': Cláudio Hummes of São Paulo; Jorge Mario Bergoglio of Buenos Aires; and Oscar Andrés Rodríguez Maradiaga of Tegucigalpa. Although, in the event, a European, Benedict XVI, was chosen to be Pope, the very fact of, for the first time, seriously entertaining the idea of a Latin American pontiff testified to the place that the region now occupies on the international Catholic stage.

Similarly, Latin America's 'Protestant explosion', must be seen within a broader context. It is part of a wider and, more especially, developing, world movement. Moreover, it may be that, within Latin America, changes in both the Catholic and Protestant communities represent separate yet complementary processes of renewal: processes partially prompted by shared experiences of the cultural, socioeconomic and political upheavals attendant upon globalization. Equally, it may be that spiritual and cultural encounters arising from the co-existence of these two considerable forces may have major implications, not only for Latin America, but also for global Christianity's future. It could even have implications for Christianity's capacity fruitfully to influence the global political agenda.

## FOOTNOTES

1. For an important historical study of the Latin American Church, see Mecham, J. L. *Church and State in Latin America*. Revised Edn, Chapel Hill, NC, University of North Carolina Press, 1966.
2. On the general subject of 'popular religion', see Levine D H, (Ed.). *Religion and Political Conflict in Latin America*. Chapel Hill, NC, University of North Carolina Press, 1986. See also Bruneau, T. C., et al (Eds). *The Catholic Church and Religion in Latin America*. Centre for Developing Area Studies, Montréal, QC, 1985.
3. On the Chilean Church, See Smith, B. H. *The Church and Politics in Chile: Challenge to Modern Catholicism*. Princeton, NJ, Princeton University Press. 1982.
4. On the subject of Christian democracy, see Williams, E. J. *Latin American Christian Democratic Parties*. Knoxville, TN, University of Tennessee Press, 1967. Also, Fleet, M. *The Rise and Fall of Chilean Christian Democracy*. Princeton, NJ, Princeton University Press, 1985.
5. On the Brazilian Church, see Bruneau, T. C. *The Political Transformation of the Brazilian Catholic Church*. London, Cambridge University Press, 1984, and, by the same author, *The Church in Brazil. The Politics of Religion*. Austin, TX, University of Texas Press, 1982. See also Mainwaring, S. *The Catholic Church and Politics in Brazil*. London, Oxford University Press, 1986.
6. On the Medellín Conference, see Poblate, R. 'From Medellín to Puebla: Notes for Reflection', in Levine D. H. (Ed.), *Churches and Politics in Latin America*. London, Sage Publications, 1979.
7. On the subject of 'liberation theology' much has been written. For an introduction to the subject, see Kee, A. *Domination or Liberation*. London, SCM, 1986. Also the seminal work, Gutiérrez, G. *A Theology of Liberation, History, Politics and Salvation*. Mary Knoll, Orbis Books, 1973; Gibellini, G. *Frontiers of Theology in Latin America*. London, SCM, 1975; Smith, C., *The Emergence of Liberation Theology. Radical Religion and Social Movement Theory*. Chicago, IL, Chicago University Press, 1991; finally, Levine, D. H. *Popular Voices in Latin American Catholicism*. Princeton, NJ, Princeton University Press, 1992. For a more recent discussion, see Batsone, D., et al (Eds), *Liberation Theologies, Postmodernity and the Americas*. London, Routledge, 1997.
8. On base communities, see Levine, D. H. (Ed.). *Religion and Political Conflict in Latin America*, Chapel Hill, NC, University of North Carolina Press, 1986. For a more recent account, see Vázquez, M. A. *The Brazilian Popular Church and the Crisis of Modernity*. Cambridge, Cambridge University Press, 1998.
9. On Central America, see Berryman, P. *The Religious Roots of Rebellion*. London, SCM, 1984.
10. For a general country by country survey of the Church during this period, see the articles by Medhurst, K. N., in Mews, S. (Ed.). *Religion in Politics*. London, Longman, 1989.
11. On the general subject of the Argentinean Church, see Medhurst, K. N., and Gutiérrez, M. C. 'The Roman Catholic Church in Argentina'. A paper presented to the congress of the International Political Science Asscn, Berlin, 1994.
12. For a discussion of recent events, see Cleary, E. L., and Stewart-Gambino, H. *Conflict and Competition. The Latin American Church in a Challenging Environment*. London, Lynne Reinner Publrs, 1992. See also Gill, A. *Rendering Unto Caesar: the Catholic Church and the State in Latin America*. Chicago, IL, University of Chicago Press, 1998; Keogh, D. (Ed.), *Church and Politics in Latin America*. London, Macmillan, 1990; Lehmann, D. *Democracy and Development in Latin America. Economics, Politics and Religion in the Post-War Period*. Cambridge, Polity Press, 1990; Medhurst, K. N. *Christianity and Democracy in Argentina and Chile;* Dobson, A., and Stanyer, J. (Eds). *Contemporary Political Studies*, 1998, Vol. 2, pp. 1004–15. Nottingham, Political Studies Asscn, 1998; Smith, C., and Prokopy, J. (Eds). *Latin American Religion in Motion*. London, Routledge, 1999.
13. On the Colombian Church, see Levine, D. H. *Religion and Politics in Latin America; The Catholic Church in Venezuela and Colombia*. Princeton, NJ, Princeton University Press, 1981, and Medhurst, K. N. *The Church and Labour in Colombia*. Manchester, Manchester University Press, 1984.
14. On the Puebla Conference, see Berryman, P. 'What Happened at Puebla', in Levine, D. H. (Ed). *Churches and Politics in Latin America*. London, Sage Publications, 1979.
15. See note 12.
16. On Protestantism, see Martin, D. *Tongues of Fire. The Explosion of Protestantism in Latin America*. Oxford, Blackwell Publrs, 1993; Stoll, D. *Is Latin America Turning Protestant?* Berkeley, CA, University of California Press, 1992 . For a comparative study of Catholicism and Protestantism, see Lehmann, D. *Struggle for the Spirit. Religious Transformation and Popular Culture in Brazil and Latin America*. Cambridge, Polity Press, 1994. For a comparative study of Pentecostalism extending beyond Latin America, see Preston, P. *Evangelicals and Politics in Asia, Africa and Latin America*. Cambridge, Cambridge University Press, 2001.
17. For a discussion of Latin America in a global setting, see Davie, G. *Europe the Exceptional Case: Parameters of Faith in the Modern World*. London, Darton, Longman and Todd, 2002, especially pp. 54–83.

# PART TWO
# Country Surveys

# ANGUILLA

## Geography

### PHYSICAL FEATURES

The United Kingdom Overseas Territory of Anguilla is in the north-eastern Caribbean and is the most northerly part of the Leeward Islands in the Lesser Antilles. The territory includes the islet of Sombrero, the pivot of the Lesser Antilles, between the main arc of the archipelago running south-eastwards and the Virgin Islands running westwards. The British Virgin Islands lie some 40 km (25 miles) to the west of the territory, but the nearest neighbour is only 8 km to the south—the French (northern) part of the island of Saint Martin (Sint Maarten), which is under the jurisdiction of Guadeloupe (and, therefore, part of France). Anguilla itself was previously part of the federation of Saint Christopher (Kitts)-Nevis-Anguilla, but seceded and reverted to British colonial status (Saint Kitts is over 110 km to the south-east). Anguilla comprises over 96 sq km (37 sq miles) of territory, the main island itself consisting of 91 sq km. This makes the colony the smallest territory in the Caribbean.

The main island, aligned roughly south-west to north-east, is long and narrow, which is why the French named it for an eel (Anguilla), echoing the Carib name, which meant a sea serpent (Malliouhana). It is a low, coral-and-limestone island, about 26 km in length and never wider than 5 km. The highest point on Anguilla is at Crocus Hill (65 m or 213 ft). Most of the more rugged terrain is at the north-eastern end of the island, where it faces into the Atlantic weather, sheltering some denser vegetation than the usual scrub of the arid interior behind 30-m cliffs. There are some areas of wetland along the 60 km or so of coastline, favoured by the island's varied bird life, but it tends to be the clear seas favoured by coral and the many wide, sandy beaches that draw more lucrative visitors.

Apart from the detached islets of Scrub Island, at the north-eastern tip of the main island, and the even smaller Anguillita, at the south-western tip, the territory also comprises a number of other islands and cays. To the north-west of the mainland is Dog Island and, just to the east of that, the Prickley Pear Cays and Seal Island. Further to the north-west, almost 50 km from Anguilla itself, is the sea-washed rock of Sombrero, uninhabited since its lighthouse was automated. The light, 51 m (166 ft) above the sea, serves shipping using the Anegada Passage from the Atlantic Ocean into the Caribbean Sea. The island, just over 1.5 km long, not even 0.5 km wide and around 10 m above sea level, is particularly rich in bird life.

### CLIMATE

The climate is a subtropical one, tempered by the north-eastern trade winds off the Atlantic. The lack of altitude means Anguilla often misses the rains, but also that it is prone to flooding, particularly during the June–September hurricane season.

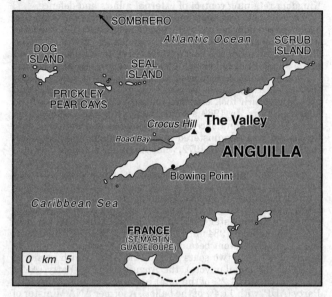

Annual rainfall is 36 ins (914 mm) per year and falls mainly between September and December. The mean temperature is 80°F (27°C), varying little over the course of the year.

### POPULATION

There are some white and some mulatto (mixed-race) people native to Anguilla, but most of the population is black. Everyone speaks English, the official language, and most are adherents of one or other of numerous Christian denominations. According to figures from the 1990s, 40% of the population was nominally Anglican and 33% Methodist, with the Seventh-day Adventists, Baptists and Roman Catholics the next largest groups.

Official estimates put the total population at 12,200 at the end of 2003, with about 1,400 living in and around the capital, The Valley, which is located in the centre of the island, near the northern coast. Across the island, to the south-west, is the ferry terminal of Blowing Point, although the main anchorage is at Road Bay, directly north of that port. The ferry gives easy access to the French and Dutch island of Saint Martin/Sint Maarten, encouraging tourism, but also helpful for emigration, which was a major demographic problem until the end of the 20th century. Greater prosperity means net emigration has ceased. Indeed, over 1,000 workers from abroad were employed in Anguilla at the beginning of the 2000s.

## History

### MARK WILSON

Anguilla is a United Kingdom Overseas Territory. A Governor, who is the representative of the British monarch, has important reserve powers, including responsibility for national security and defence, the civil service, the judiciary and certain financial matters. A Chief Minister is responsible to the Legislative Council, similar in function to a parliament, which contains a majority of elected members. The Governor presides over an Executive Council, similar in function to a cabinet, which includes the Chief Minister and the other ministers. A constitutional review was in progress in 2005.

A few traces remain of the original Amerindian inhabitants. The first known European visitors were French, and named the island for its eel-like shape; however, the first European settlers were British, arriving in 1650. With a dry climate and thin soil, the island was not a major centre for plantation agriculture, and from 1825 it was tied increasingly closely in its administration to that of Saint Christopher and Nevis (St Kitts). From 1871 to 1956 Anguilla formed part of the Leeward Islands Federation, as part of the presidency of Saint Christopher. With the Leeward Islands Federation disbanded in 1957, Saint Christopher-

Nevis-Anguilla joined the Federation of the West Indies in 1958 along with nine other British colonies. Following the departure of Jamaica and Trinidad and Tobago in 1962, the Federation collapsed, and an attempt to unite the remaining colonies as the 'little eight' was unsuccessful. Along with its neighbours, Saint Christopher-Nevis-Anguilla became a British Associated State in 1967, responsible for its own internal affairs, with the United Kingdom retaining control of external affairs and defence.

However, this arrangement was fiercely resisted by Anguilla, which feared domination by its larger neighbour, and in particular by Robert Llewellyn Bradshaw, leader of the St Kitts-Nevis-Anguilla Labour Party, which, in spite of its name, never had a substantial following outside the larger island.

Ronald Webster, leader of the People's Progressive Party (PPP), led a movement to break away from the three-island grouping, which took effective control of many aspects of the island's affairs. In 1969 a detachment from the United Kingdom's Metropolitan Police Force was landed, and from that date the island was administered under a resident Commissioner; the British police left in 1972, when Anguilla established its own force.

From 1980 Anguilla was formally established as a British Dependent Territory under a Governor, and from 1998, as a United Kingdom Overseas Territory. In local politics, office alternated between the PPP and the Anguilla National Alliance (ANA) of Emile (later Sir Emile) Gumbs, who held office in 1977–80 and from 1984 until his retirement in 1994. Thereafter, the two-party system became less clear. A general election in March 1994 gave two seats each to the ANA (which garnered 35.7% of the popular vote), the Anguilla United Party (AUP, which attracted 31.2% of the vote), and the Anguilla Democratic Party (ADP, with 11.4% of the ballot). A former ANA Minister of Finance, Osbourne Fleming, took the seventh legislative seat. Hubert Hughes, also a former ANA minister, but now leader of the AUP, formed a coalition with Victor Banks of the ADP, and became Chief Minister. In May controversy arose over the proposed appointment by the Governor of a nominated member to the Legislative Council, David Carty, who had failed to win a legislative seat in the March election. Hughes objected to Carty's nomination, and alleged that he had not been consulted over the appointment, a charge denied by the Governor. Despite a ruling in Carty's favour by the Constitutional Court, the Speaker refused to swear him in as a member of the House of Assembly in December 1995.

Anguilla was severely damaged by 'Hurricane Luis' in October 1995. Hughes was critical of what he saw as insufficient support from the United Kingdom. The announcement by the British Government in January 1997 that it was considering the extension of its powers in the Dependent Territories of the Caribbean attracted further criticism from Hughes, as did the proposed reintroduction of the Governor's reserve powers, whereby the Governor (with the consent of the British Government) can amend, veto or introduce legislation without the consent of the local legislature.

In a legislative election held on 4 March 1999, the AUP held its two seats, securing 1,579 votes, as did the ADP, with 704 votes. Hughes therefore continued in office as leader of a coalition Government. The ANA, with Fleming now a member, also retained its three seats and attracted 2,053 votes. However, in June the ADP leader and Minister of Finance, Banks, left the governing coalition following a dispute about the ADP's role in government. Hughes therefore no longer enjoyed the support of a majority of the elected members; however, with several appointed members also sitting in the Legislative Council, the constitutional position was unclear. To force an eventual resolution, Banks and the three ANA members withdrew from the House of Assembly, thus denying it a quorum. This left the Government unable, in December 1999, to introduce a budget for 2000 or implement any policy. Thus, fresh elections were called for 3 March 2000.

The election of March 2000 left the ANA still with three parliamentary seats, the AUP still with two seats, while Banks (representing the ADP) and Edison Baird (an Independent) each held one. Fleming was appointed Chief Minister (and Minister of Home Affairs, Tourism, Agriculture, Fisheries and Environment) with support from four of the seven elected members. Banks became Minister of Finance, Economic Development, Investment and Commerce.

In February 2000 the island's Governor, Robert Harris, who had not enjoyed good relations with the Hughes Government, and who had reportedly been extremely frustrated by the political crisis, departed Anguilla. He was succeeded by Peter Johnstone, who, in turn, was replaced by Alan Huckle in May 2004. In the same month Edison Baird replaced Hubert Hughes as Leader of the Opposition after Albert Hughes resigned from the AUP and transferred his support to Baird.

In June 2001 the Government officially approved the draft of its National Telecommunications Policy, which would liberalize the telecommunications sector. Cable & Wireless, the territory's sole telecommunications provider, in April 2003 signed an agreement with the Government to open the market for competition. In August the Government commenced the sale of 6m. shares at US $1 each in the Anguilla Electricity Company, a profit-making public utility, in order to raise funds for the EC $49.2m. expansion and reconstruction of Wallblake Airport. The proceeds of the sale were primarily to go towards lengthening the airport's runway in order to accommodate larger aircraft.

At the most recent general election, which was held on 21 February 2005, there were some variations in party names or affiliations, but the same members were returned to Government as in the previous ballot. The Anguilla United Front, an alliance comprising the ADP and the ANA, led by Osbourne Fleming, won four seats, with 38.9% of the votes cast, while the Anguilla National Strategic Alliance secured two seats and the Anguilla United Movement (as the AUP had been renamed) attained one. Some changes were to be expected, however, by the time of the next election, with some of the existing representatives likely to retire from active politics.

In May 2002 the British Overseas Territories Act, having received royal assent in the United Kingdom in February, came into force and granted British citizenship to the people of its Overseas Territories, including Anguilla. Under the new law Anguillans are able to hold British passports and work in the United Kingdom and anywhere else in the European Union.

# Economy

## MARK WILSON

Anguilla is a United Kingdom Overseas Territory in the eastern Caribbean, with an area of 96 sq km and a population, according to the Government, of some 12,200 at the end of 2003. Immigration resulting from economic prosperity resulted in an average annual population growth of 2.6% in 1990–2001. Anguilla is an associate member of the Caribbean Community and Common Market, or CARICOM, which in principle is attempting to develop a single market by 2006. It is also a member of the Organisation of Eastern Caribbean States, which links nine of the smaller Caribbean territories, while the Eastern Caribbean Central Bank, headquartered in Saint Christopher and Nevis, supervises its financial affairs.

The island has a prosperous middle-income economy, with a per-head gross domestic product (GDP) of US $9,662 in 2003, and overall GDP increased at an average annual rate of 3.8% in 1994–2004. Moreover, unemployment is low by regional standards—the rate was 7.8% in 2002, and has fallen since, with labour shortages a major concern by 2005. None the less, the economy is vulnerable to tropical storms and GDP contracted by 4.1% in 1995 after 'Hurricane Luis' devastated the island, and

# ANGUILLA

by 0.3% in 2000, when major hotels were unable to open as a result of the effects of the previous year's 'Hurricane Lenny'. A downturn in international demand for Caribbean tourism in 2001–02 resulted in an economic contraction and fiscal problems. However, the economy has since recovered dramatically, with GDP growth of 13% in 2004, led by tourism and hotel investments. Government revenue in that year was 25% above the budget projection, with a recurrent surplus equivalent to 6% of GDP.

Tourism is the mainstay of the economy; the main attractions are the island's tranquillity, the clear surrounding waters, low rainfall, and white sandy beaches. There were 53,987 stop-over visitors in 2004, a 15.1% increase over the previous year. Of these, 66.2% were from the USA; the ratio of tourists to local population is therefore higher than on such islands as Antigua and Barbuda, Barbados, or Jamaica. Most tourists stay in luxury accommodation; their high spending power per head is of further economic benefit. The island is not a port of call for cruise ships, which helps preserve its pleasant atmosphere. A EC $49.2m. improvement programme at Wallblake Airport was completed in December 2004, after several delays, to accommodate mid-range jets, operated by American Airlines, the main international carrier.

Residential, commercial, public-sector and tourism-related investment has resulted in a high level of construction activity, which contributed 14.7% of GDP in 2004. With tourism showing some dynamism, the construction sector will remain strong. Tourism-related investments proposed or in progress, and scheduled for completion by 2012, total US $1,050m., equivalent to more than nine times the annual level of GDP. There is no significant agricultural sector, but a few small-scale farmers keep livestock and grow food crops. There has also been some recent development of commercial fishing. There is no manufacturing industry.

There is a small 'offshore' financial sector, which has been reasonably well regulated. Following pressure from the Organisation for Economic Co-operation and Development (OECD), which, in June 2000, included Anguilla on a list of so-called tax 'havens', efforts were made to improve the transparency of the sector. Legislation to control money-laundering was strengthened, and the Government made a commitment to move towards the international exchange of information in criminal and civil tax investigations. As a result of these efforts, in 2002 OECD removed Anguilla from its blacklist. Online company registration through an approved agent is quick and cost-effective; a total of 3,041 International Business Companies had been registered by the end of 2003. The Government intends to develop e-commerce, while one potential sideline is the sale of internet addresses ending in the island's registered suffix, '.ai' (although this meant that sites ending in this suffix might contain unsuitable material, particularly as 'ai' means 'love' in Chinese). However, an earlier proposal from a US company based in Texas to develop a rocket-launching site on the uninhabited dependency of Sombrero did not come to fruition.

Anguilla earned an unexpected economic 'windfall' worth EC $15m. per year in 1998–99 from European Union (EU) customs duty paid to the island on aluminium ingots exported by Brazil to Europe. By paying duty while in Anguillan waters, where shipping costs produced a lower cost, insurance and freight value than at the final destination, the exporters benefited from a small saving. This activity was ended by an EU ruling at the end of 1999.

The island is in the heart of the hurricane belt, and has been damaged by several storms in recent years, including 'Luis' in October 1995, 'Georges' in September 1998, and 'Lenny' in 1999. There is no volcanic risk.

# Statistical Survey

Source: Government of Anguilla, The Secretariat, The Valley; tel. 497-2451; fax 497-3389; e-mail stats@gov.ai; internet www.gov.ai.

## AREA AND POPULATION

**Area** (sq km): 96 (Anguilla 91, Sombrero 5).
**Population:** 11,561 (males 5,705, females 5,856) at census of 9 May 2001. *2003* (official estimate): 12,200 at 31 December 2003.
**Density** (at 31 December 2003): 127.1 per sq km.
**Principal Towns** (population at 2001 census): South Hill 1,495; North Side 1,195; The Valley (capital) 1,169; Stoney Ground 1,133. *Mid-2003* (UN estimate, incl. suburbs): The Valley 1,380 (Source: UN, *World Urbanization Prospects: The 2003 Revision*).
**Births, Marriages and Deaths** (2003): Registered live births 139; Birth rate 11.4 per 1,000; Registered marriages 75; Marriage rate 6.1 per 1,000; Registered deaths 65; Death rate 5.3 per 1,000.
**Expectation of Life** (official estimates, years at birth): 78.88 (males 76.52; females 81.11) in 2003.
**Economically Active Population** (persons aged 15 years and over, census of 9 May 2001): Agriculture, fishing and mining 183; Manufacturing 135; Electricity, gas and water 81; Construction 830; Trade 556; Restaurants and hotels 1,587; Transport, storage and communications 379; Finance, insurance, real estate and business services 433; Public administration, social security 662; Education, health and social work 383; Other community, social and personal services 164; Private households with employed persons 164; Activities not stated 871; *Total employed* 5,644 (males 3,014, females 2,630); Unemployed 406 (males 208, females 198); *Total labour force* 6,050 (males 3,222, females 2,828). *July 2002:* Total employed 5,496 (males 3,009, females 2,487); Unemployed 465 (males 204, females 261); Total labour force 5,961 (Source: ILO).

## HEALTH AND WELFARE

**Under-5 Mortality Rate** (per 1,000 live births, 1997): 34.0.
**Physicians** (per 1,000 head, 2003): 1.2.
**Health Expenditure** (% of GDP, 1998): 4.92.
Sources: Caribbean Development Bank, *Social and Economic Indicators 2004* and Pan American Health Organization.
For definitions, see explanatory note on p. vi.

## AGRICULTURE, ETC.

**Fishing** (FAO estimates, metric tons, live weight, 2003): Marine fishes 180, Caribbean spiny lobster 60, Stromboid conchs 10; Total catch 250. Source: FAO.

## INDUSTRY

**Electric Energy** (million kWh): 53.60 in 2001; 55.33 in 2002; 58.39 in 2003.

## FINANCE

**Currency and Exchange Rates:** 100 cents = 1 Eastern Caribbean dollar (EC $). *Sterling, US Dollar and Euro Equivalents* (31 May 2005): £1 sterling = EC $4.909; US $1 = EC $2.700; €1 = EC $3.329; EC $100 = £20.37 = US $37.04 = €30.04. *Exchange Rate:* Fixed at US $1 = EC $2.70 since July 1976.

**Budget** (EC $ million, 2002, preliminary): *Revenue:* Tax revenue 63.8 (Taxes on domestic goods and services 28.9, Taxes on international trade and transactions 34.4); Non-tax revenue 20.1; Capital grants 2.0; Total 85.9. *Expenditure:* Current expenditure 81.0 (Personal emoluments 41.1, Other goods and services 33.8, Transfers and subsidies 2.8, Interest payments 3.4); Capital expenditure 3.3; Total 84.3. Source: Eastern Caribbean Central Bank, *Report and Statement of Accounts*.

**Cost of Living** (Consumer Price Index; base: 2000 = 100): 101.5 in 2002; 108.5 in 2003; 114.0 in 2004.

**Expenditure on the Gross Domestic Product** (EC $ million at current prices, 2003): Government final consumption expenditure 56.71; Private final consumption expenditure 262.55; Gross fixed capital formation 94.26; *Total domestic expenditure* 413.52; Export of goods and services 203.55; *Less* Import of goods and services 298.79; *GDP in purchasers' values* 318.28. Source: Eastern Caribbean Central Bank.

# ANGUILLA

**Gross Domestic Product by Economic Activity** (EC $ million at current prices, 2003): Agriculture, hunting, forestry and fishing 7.23; Mining and quarrying 2.88; Manufacturing 4.03; Electricity, gas and water 13.80; Construction 29.71; Wholesale and retail 19.02; Hotels and restaurants 70.95; Transport 16.72; Communications 21.24; Banks and insurance 36.37; Real estate and housing 8.46; Government services 45.59; Other services 5.95; *Sub-total* 281.95; *Less* Imputed bank service charge 29.01; *GDP at factor cost* 252.94. Source: Eastern Caribbean Central Bank.

**Balance of Payments** (EC $ million, 2003): Export of goods f.o.b. 11.49; Imports of goods f.o.b. –182.42; *Trade balance* –170.93; Exports of services 192.06; Imports of services –116.37; *Balance on goods and services* –95.24; Other income received 4.98; Other income paid –19.82; *Balance on goods, services and income* –110.08; Current transfers received 23.28; Current transfers paid –23.50; *Current balance* –110.30; Capital account (net) 16.64; Direct investment (net) 76.66; Portfolio investment (net) –0.97; Other investment (net) 23.36; Net errors and omissions 13.72; *Overall balance* 19.11.

## EXTERNAL TRADE

**Principal Commodities** (EC $ million, 2004): *Imports:* Food and live animals 41.1; Beverages and tobacco 17.2; Mineral fuels, lubricants, etc. 26.5; Chemicals and related products 19.9; Basic manufactures 50.6; Machinery and transport equipment 86.3; Miscellaneous manufactured articles 27.4; Total (incl. others) 276.6. *Exports* (incl. re-exports): Beverages and tobacco 1.0; Mineral fuels, lubricants, etc. 1.6; Basic manufactures 1.0; Machinery and transport equipment 6.8; Miscellaneous manufactured articles 4.8; Total (incl. others) 15.5.

**Principal Trading Partners** (EC $ million, 2004): *Imports:* Barbados 3.9; Canada 4.0; Guadeloupe 5.0; Netherlands Antilles 20.5; Puerto Rico 21.1; Trinidad and Tobago 19.2; United Kingdom 15.4; USA 140.7; US Virgin Islands 8.9; Total (incl. others) 276.6. *Exports* (incl. re-exports): British Virgin Islands 0.2; Guadeloupe 0.6; Guyana 0.6; Netherlands Antilles 7.4; Saint Christopher and Nevis 0.3; Saint Lucia 1.3; USA 2.4; US Virgin Islands 1.6; Total (incl. others) 15.5.

## TRANSPORT

**Road Traffic** (motor vehicles licensed, 2003): Private cars 3,198; Hired cars 830; Buses/trucks/jeeps/pickups 2,002; Motor cycles 53; Tractors 5; Heavy equipment 168; Other 7.

**Shipping:** *Merchant Fleet* (registered at 31 December 2004): 3; Total displacement 701 grt. Source: Lloyd's Register-Fairplay, *World Fleet Statistics*.

## TOURISM

**Visitor Arrivals:** 104,974 (Stop-overs 47,965, Excursionists 57,009) in 2001; 111,118 (Stop-overs 43,969, Excursionists 67,149) in 2002; 109,282 (Stop-overs 46,915, Excursionists 62,367) in 2003.

**Visitor Arrivals by Country of Residence** (2003): Canada 4,296; Caribbean 10,044; Germany 1,424; Italy 2,115; United Kingdom 5,081; USA 56,992; Total (incl. others) 109,282.

**Tourism Receipts** (estimates, US $ million): 61.0 in 2001; 55.3 in 2002; 61.7 in 2003.

## COMMUNICATIONS MEDIA

**Radio Receivers** (1997): 3,000 in use.

**Television Receivers** (1999): 1,000 in use.

**Telephones** (2002): 5,796 main lines in use.

**Facsimile Machines** (1993): 190 in use.

**Mobile Cellular Telephones** (2000): 15,000 subscribers.

**Internet Connections** (2002): 1,391.

Sources: partly UN, *Statistical Yearbook*; International Telecommunication Union.

## EDUCATION

**Pre-primary** (2003): 11 schools; 38 teachers; 499 pupils.

**Primary** (2003): 8 schools; 92 teachers; 1,438 pupils.

**General Secondary** (2002/03): 1 school; 91 teachers; 1,102 pupils.

**Adult Literacy Rate** (UNESCO estimates): 95.4% (males 95.1%; females 95.7%) in 1995. Source: UNESCO, *Statistical Yearbook*.

# Directory

## The Constitution

The Constitution, established in 1976, accorded Anguilla the status of a British Dependent Territory. It formally became a separate dependency on 19 December 1980, and is administered under the Anguilla Constitution Orders of 1982 and 1990. British Dependent Territories were referred to as United Kingdom Overseas Territories from February 1998 and draft legislation confirming this change and granting citizens rights to full British citizenship and residence in the United Kingdom was published in March 1999. The British Overseas Territories Act entered into effect in May 2002. The British Government proposals also included the requirement that the Constitutions of Overseas Territories should be revised in order to conform to British and international standards. The process of revision of the Anguillan Constitution began in September 1999.

The British monarch is represented locally by a Governor, who presides over the Executive Council and the House of Assembly. The Governor is responsible for defence, external affairs (including international financial affairs), internal security (including the police), the public service, the judiciary and the audit. The Governor appoints a Deputy Governor. On matters relating to internal security, the public service and the appointment of an acting governor or deputy governor, the Governor is required to consult the Chief Minister. The Executive Council consists of the Chief Minister and not more than three other ministers (appointed by the Governor from the elected members of the legislative House of Assembly) and two *ex-officio* members (the Deputy Governor and the Attorney-General). The House of Assembly is elected for a maximum term of five years by universal adult suffrage and consists of seven elected members, two *ex-officio* members (the Deputy Governor and the Attorney-General) and two nominated members who are appointed by the Governor, one upon the advice of the Chief Minister, and one after consultations with the Chief Minister and the Leader of the Opposition. The House elects a Speaker and a Deputy Speaker.

The Governor may order the dissolution of the House of Assembly if a resolution of 'no confidence' is passed in the Government, and elections must be held within two months of the dissolution.

The Constitution provides for an Anguilla Belonger Commission, which determines cases of whether a person can be 'regarded as belonging to Anguilla' (i.e. having 'belonger' status). A belonger is someone of Anguillan birth or parentage, someone who has married a belonger, or someone who is a citizen of the United Kingdom Overseas Territories from Anguilla (by birth, parentage, adoption or naturalization). The Commission may grant belonger status to those who have been domiciled and ordinarily resident in Anguilla for not less than 15 years.

## The Government

**Governor:** ALAN HUCKLE (sworn in 28 May 2004).

### EXECUTIVE COUNCIL
(July 2005)

**Chief Minister and Minister of Home Affairs, Gender Affairs, Immigration, Labour, Lands and Physical Planning and Environment:** OSBOURNE FLEMING (AUF).

**Minister of Finance, Economic Development, Investment, Tourism and Commerce:** VICTOR F. BANKS (AUF).

**Minister of Education, Health, Sports and Youth and Culture:** EVANS MCNEIL ROGERS (AUF).

**Minister of Infrastructure, Communications, Public Utilities, and Housing:** KENNETH HARRIGAN (AUF).

**Attorney-General:** RONALD SCIPIO.

**Deputy Governor:** HENRY MCCRORY.

**Parliamentary Secretary with responsibility for Water, Agriculture and Fisheries:** ALBERT HUGHES.

### MINISTRIES

**Office of the Governor:** Government House, POB 60, The Valley; tel. 497-2622; fax 497-3314; e-mail govthouse@anguillanet.com.

# ANGUILLA

**Office of the Chief Minister:** The Secretariat, The Valley; tel. 497-2518; fax 497-3389; e-mail chief-minister@gov.ai.

All ministries are based in The Valley, mostly at the Secretariat (tel. 497-2451; internet www.gov.ai).

## Legislature

### HOUSE OF ASSEMBLY

**Speaker:** DAVID CARLY.

**Clerk to House of Assembly:** Rev. JOHN A. GUMBS.

**Election, 21 February 2005**

| Party | % of votes | Seats |
|---|---|---|
| Anguilla United Front (AUF) | 38.9 | 4 |
| Anguilla United Movement (AUM) | 19.4 | 1 |
| Anguilla National Strategic Alternative (ANSA) | 19.2 | 2 |
| Anguilla Progressive Party (APP) | 9.5 | — |
| Total (incl. others) | 100.0 | 7 |

There are also two *ex-officio* members and two nominated members.

## Political Organizations

**Anguilla National Strategic Alliance (ANSA):** The Valley; Leader EDISON BAIRD.

**Anguilla Progressive Party (APP):** The Valley.

**Anguilla United Front (AUF):** The Valley; internet www.unitedfront.ai; f. 2000; Leader OSBOURNE FLEMING; alliance comprising:

**Anguilla Democratic Party (ADP):** The Valley; f. 1981 as Anguilla People's Party; name changed 1984; Leader VICTOR F. BANKS.

**Anguilla National Alliance:** The Valley; f. 1980 by reconstitution of People's Progressive Party; Leader OSBOURNE FLEMING.

**Anguilla United Movement (AUM):** The Valley; f. 1979; revived 1984; previously known as the Anguilla United Party—AUP; conservative; Leader HUBERT B. HUGHES.

## Judicial System

Justice is administered by the High Court, Court of Appeal and Magistrates' Courts. During the High Court sitting, the Eastern Caribbean Supreme Court provides Anguilla with a judge.

## Religion

### CHRISTIANITY

#### The Anglican Communion

Anglicans in Anguilla are adherents of the Church in the Province of the West Indies, comprising nine dioceses. Anguilla forms part of the diocese of the North Eastern Caribbean and Aruba.

**Bishop of the North Eastern Caribbean and Aruba:** Rt Rev. LEROY ERROL BROOKS, St Mary's Rectory, POB 180, The Valley; tel. 497-2235; fax 497-3012; e-mail brookx@anguilla.net.com.

#### The Roman Catholic Church

The diocese of St John's-Basseterre, suffragan to the archdiocese of Castries (Saint Lucia), includes Anguilla, Antigua and Barbuda, the British Virgin Islands, Montserrat and Saint Christopher and Nevis. The Bishop resides in St John's, Antigua.

**Roman Catholic Church:** St Gerard's, POB 47, The Valley; tel. 497-2405.

#### Protestant Churches

**Methodist Church:** South Hill; Minister Rev. LINDSAY RICHARDSON. The Seventh-day Adventist, Baptist, Church of God, Pentecostal, Apostolic Faith and Jehovah's Witnesses Churches and sects are also represented.

## The Press

**Anguilla Life Magazine:** Caribbean Commercial Centre, POB 109, The Valley; tel. 497-3080; fax 497-2501; 3 a year; circ. 10,000.

**The Light:** POB 1373, Herbert's Commercial Centre, The Valley; tel. 497-5058; fax 497-5795; e-mail thelight@anguillanet.com; f. 1993; weekly; newspaper; Editor GEORGE C. HODGE.

**Official Gazette:** The Valley; tel. 497-5081; monthly; government news-sheet.

**What We Do in Anguilla:** POB 1373, Herbert's Commercial Centre, The Valley; tel. 497-5641; fax 497-5795; e-mail thelight@anguillanet.com; f. 1987; monthly; tourism; Editor GEORGE C. HODGE; circ. 50,000.

## Broadcasting and Communications

### TELECOMMUNICATIONS

**Cable & Wireless Anguilla:** POB 77, The Valley; tel. 497-3100; fax 497-2501; internet www.cwwionline.com/buhome.asp?bu=Anguilla; Gen. Man. SUTCLIFFE HODGE.

**Cingular Wireless:** Hanna-Waver House, The Valley; tel. 497-8322; e-mail rhon.rogers@an.cingular.com; owned by Digicel Ltd (Bermuda); fmrly AT&T Wireless; to commence operations in 2005.

### BROADCASTING

#### Radio

**Caribbean Beacon Radio:** POB 690, Long Rd, The Valley; Head Office: POB 7008, Columbus, GA 31908, USA; tel. 497-4340; fax 497-4311; f. 1981; privately owned and operated; religious and commercial; broadcasts 24 hours daily; Pres. Dr GENE SCOTT; CEO B. MONSELL HAZELL.

**Heart Beat Radio:** The Valley; tel. 497-3354; fax 497-5995; e-mail info@hbr1075.com; internet hbr1075.com; f. 2001; commercial.

**Kool FM:** North Side, The Valley; tel. 497-0103; fax 497-0104; e-mail kool@koolfm103.com; internet www.koolfm103.com; commercial.

**Radio Anguilla:** Dept of Information and Broadcasting, Treasury Dept, The Valley; tel. 497-2218; fax 497-5432; e-mail radioaxa@anguillanet.com; internet www.radioaxa.com; f. 1969; owned and operated by the Govt of Anguilla since 1976; 250,000 listeners throughout the north-eastern Caribbean; broadcasts 17 hours daily; Dir of Information and Broadcasting KENNETH HODGE; News Editor WYCLIFFE RICHARDSON.

**ZJF FM:** The Valley; tel. 497-3157; f. 1989; commercial.

**Voice of Creation Station:** The Valley; internet www.voiceofcreation.com; f. 2000; religious station with devotional and gospel music.

**ZJF FM:** The Valley; tel. 497-3157; f. 1989; commercial.

#### Television

**Anguilla Television:** terrestrial channels 3 and 9; 24-hour local and international English language programming.

**Caribbean Cable Communications (Anguilla):** POB 336, George Hill, The Valley; tel. 497-3600; fax 497-3602; e-mail alisland@anguillanet.com; internet www.caribcable.com.

## Finance

(cap. = capital; res = reserves; dep. = deposits; m. = million; amounts in EC dollars)

### CENTRAL BANK

**Eastern Caribbean Central Bank:** Fairplay Commercial Complex, POB 1385, The Valley; tel. 497-5050; fax 497-5150; e-mail eccbaxa@anguillanet.com; internet www.eccb-centralbank.org; HQ in Basseterre, Saint Christopher and Nevis; bank of issue and central monetary authority for Anguilla, Antigua and Barbuda, Dominica, Grenada, Montserrat, Saint Christopher and Nevis, Saint Lucia and Saint Vincent and the Grenadines; Gov. Sir K. DWIGHT VENNER.

### COMMERCIAL BANKS

**Caribbean Commercial Bank (Anguilla) Ltd:** POB 23, The Valley; tel. 497-3917; fax 497-3570; e-mail service@ccb.ai; internet

# ANGUILLA

www.ccb.ai; f. 1976; Chair. OSBOURNE B. FLEMING; Man. Dir PRESTON B. BRYAN.

**FirstCaribbean International Bank Ltd:** The Valley; e-mail care@firstcaribbeanbank.com; internet www.firstcaribbeanbank.com; f. 2002 following merger of Caribbean operations of Barclays Bank PLC and CIBC; Exec. Chair. MICHAEL MANSOOR; CEO CHARLES PINK.

**National Bank of Anguilla Ltd:** POB 44, The Valley; tel. 497-2101; fax 497-3310; e-mail info@nba.ai; internet www.nba.ai; f. 1985; 5% owned by Govt of Anguilla; cap. 28.6m., res 10.1m., dep. 309.1m. (March 2002); Chair. JOSEPH N. PAYNE; CEO E. VALENTINE BANKS.

**Scotiabank Anguilla Ltd:** Fairplay Commercial Centre, POB 250, The Valley; tel. 497-3333; fax 497-3344; e-mail scotia@anguillanet.com; internet www.scotiabank.com; Man. WALTER MACCALMAN.

There are 'offshore', foreign banks based on the island, but most are not authorized to operate in Anguilla. There is a financial complex known as the Caribbean Commercial Centre in The Valley.

## TRUST COMPANIES

**Barwys Trust Anguilla Ltd:** Caribbean Suite, The Valley; e-mail jbw@barwys.com; Man. BART WISJMULLER.

**Codan Trust Co (Anguilla) Ltd:** c/o Intertrust Offices, POB 147, The Valley; tel. 498-4126; fax 498-8423; e-mail anguilla@cdp.bm; internet www.cdp.bm.

**Financial Services Co Ltd:** POB 58, The Valley; tel. 497-3777; fax 497-5377; e-mail firstanguilla@anguillalaw.com.

**First Anguilla Trust Co Ltd:** Mitchell House, POB 174, The Valley; tel. and fax 498-8800; e-mail information@firstanguilla.com; internet www.firstanguilla.com; Mans JOHN DYRUD, PALMAVON WEBSTER.

**Geneva Trust Corporation:** National Bank Corporate Bldg, Airport Rd, The Valley; e-mail dekokera@soa.wits.ac.za; Man. AWYN DE KOKER.

**Hansa Bank and Trust Co Ltd:** Hansa Bank Bldg, POB 213, The Valley; tel. 497-3802; fax 497-3801; e-mail hansa@attglobal.net; internet www.hansa.net.

**Intertrust (Anguilla) Ltd:** POB 1388, The Valley; tel. 497-2189; fax 497-5007; e-mail anguilla@intertrustgroup.com; internet www.intertrust-group.com.

**HWR Services (Anguilla) Ltd:** Harlaw Chambers, POB 1026, The Valley; tel. 498-5000; fax 498-5001; e-mail heather.wallace@harneys.com; internet www.harneys.com.

**Renaissance International Trust:** POB 687, The Valley; tel. 498-7878; fax 498-7872; e-mail renaissance@anguillanet.com.

**Sinel Trust (Anguilla) Ltd:** Sinel Chambers, POB 1269, The Valley; tel. 497-3311; fax 497-5659; e-mail arichardson@sineltrust.com; internet www.sineltrust-anguilla.com.

**Sterling Trust (Anguilla), Ltd:** National Bank of Anguilla Bldg, St Mary's Rd, The Valley; tel. 497-2189; fax 497-5007; e-mail hwoltz@sterlinggroup.bs; Man. HOWELL W. WOLTZ.

## REGULATORY AUTHORITY

**Anguilla Financial Services Commission:** The Secretariat, POB 60, The Valley; tel. 497-3881; fax 497-5872; e-mail lanston_c@anguillafsd.com; internet www.anguillafsc.com; f. 2004 to replace the Financial Services Department of the Ministry of Finance, Economic Development, Investment and Commerce; Chair. JOHN LAWRENCE.

## STOCK EXCHANGE

**Eastern Caribbean Securities Exchange:** based in Basseterre, Saint Christopher and Nevis; e-mail info@ecseonline.com; internet www.ecseonline.com; f. 2001; regional securities market designed to facilitate the buying and selling of financial products for the eight member territories—Anguilla, Antigua and Barbuda, Dominica, Grenada, Montserrat, Saint Christopher and Nevis, Saint Lucia and Saint Vincent and the Grenadines; Gen. Man. TREVOR BLAKE.

## INSURANCE

**A-Affordable Insurance Services Inc.:** Old Factory Plaza, POB 6, The Valley; tel. 497-5757; fax 497-2122.

**Barbados Mutual Life Assurance Society:** Herbert's Commercial Complex, The Valley, POB 492; tel. 497-3712; fax 497-3710.

**British American Insurance Co Ltd:** Herbert's Commercial Centre, POB 148, The Valley; tel. 497-2653; fax 497-5933; e-mail britam@anguillenet.com.

**Caribbean General Insurance Ltd:** POB 65, South Hill; tel. 497-6541.

**D-3 Enterprises Ltd:** Herbert's Commercial Centre, POB 1377, The Valley; tel. 497-3525; fax 497-3526; e-mail d-3ent@anguillanet.com; Man. CLEMENT RUAN.

**Gulf Insurance Ltd:** Blowing Point; tel. 497-6613; fax 497-6713.

**Malliouhana Insurance Co Ltd:** Herbert's Commercial Centre, POB 492, The Valley; tel. 497-3712; fax 497-3710; e-mail maico@anguillanet.com; Man. MONICA HODGE.

**Nagico Insurance:** POB 79, The Valley; tel. 497-2976; fax 497-3303; e-mail fairplay@anguillanet.com.

**National Caribbean Insurance Co Ltd:** Herbert's Commercial Centre, POB 323, The Valley; tel. 497-2865; fax 497-3783.

**Nem West Indies Insurance Ltd (Nemwil):** Old Factory Plaza, POB 6, The Valley; tel. 497-5757; fax 497-2122.

# Trade and Industry

## DEVELOPMENT ORGANIZATION

**Anguilla Development Board:** POB 285, Wallblake Rd, The Valley; tel. 497-3690; fax 497-2959.

## CHAMBER OF COMMERCE

**Anguilla Chamber of Commerce and Industry:** POB 321, The Valley; tel. 497-2701; fax 497-5858; e-mail acoci@anguillanet.com; internet www.anguillachamber.com; Pres. JOHN BENJAMIN; Exec. Dir CALVIN BARTLETT.

## INDUSTRIAL AND TRADE ASSOCIATION

**Anguilla Financial Services Association (AFSA):** POB 1071, The Valley; tel. 497-8367; fax 497-3096; e-mail pwebster@websterdyrud.com; internet online.offshore.com.ai/afsa; Pres. PAM WEBSTER; Exec. Dir ODELL MCCANTS, Jr.

## UTILITIES

### Electricity

**Anguilla Electricity Co Ltd:** POB 400, The Valley; tel. 497-5200; fax 497-5440; e-mail info@anglec.com; internet www.anglec.com; f. 1991; operates a power-station and generators; Chair. EVERET ROMNEY; Gen. Man. NEIL MCCONNIE.

# Transport

## ROADS

Anguilla has 140 km (87 miles) of roads, of which 100 km (62 miles) are tarred.

## SHIPPING

The principal port of entry is Sandy Ground on Road Bay. There is a daily ferry service between Blowing Point and Marigot (St Martin).

**Link Ferries:** Little Harbour; tel. 497-2231; fax 497-3290; e-mail fbconnor@anguillanet.com; internet www.link.ai; f. 1992; daily services to Julianna International Airport (St Martin) and charter services; Owner FRANKLYN CONNOR.

## CIVIL AVIATION

Wallblake Airport, 3.2 km (2 miles) from The Valley, has a bitumen-surfaced runway with a length of 1,100 m (3,600 ft). In 1996 Anguilla signed an agreement with the Government of Aruba providing for the construction of a new jet airport on the north coast of the island, with finance of some US $30m. from the Aruba Investment Bank. In May 2004 reconstruction and expansion of Wallblake Airport commenced. The project was to cost EC $49.2m., most of which was intended for the extension of the runway to a length of 1,470 m (4,900 ft).

**Air Anguilla:** POB 110, Wallblake; tel. 497-2643; fax 497-2982; scheduled services to St Thomas, St Martin/St Maarten, Saint Christopher and Beef Island (British Virgin Islands); Pres. RESTORMEL FRANKLIN.

**American Eagle:** POB 659, Wallblake Airport; tel. 497-3131; fax 497-3502; operates scheduled flights from Puerto Rico three times a day (December to April) and once daily (May to November).

ANGUILLA

**Caribbean Star Airlines:** Airport, POB 1628W; tel. 497-8690; fax 497-8689; e-mail ceo@flycaribbeanstar.com; internet www.flycaribbeanstar.com.

**Tyden Air:** POB 107, Wallblake Airport; tel. 497-3419; fax 497-3079; charter company servicing whole Caribbean.

## Tourism

Anguilla's sandy beaches and unspoilt natural beauty attract tourists and also day visitors from neighbouring St Martin/St Maarten. In November 1999 'Hurricane Lenny' badly affected Anguilla, forcing the temporary closure of the island's two largest hotels. Tourism receipts totalled some US $61.7m. in 2003 and there were 759 hotel rooms on the island; in the same year 52.5% of visitors were from the USA, 9.2% were from other Caribbean countries, while most of the remainder were from the United Kingdom, Canada, Italy and Germany. Visitor numbers totalled 109,282 in 2003.

**Anguilla Tourist Board:** POB 1388, The Valley; tel. 497-2759; fax 497-2710; e-mail atbtour@anguillanet.com; internet www.anguilla-vacation.com.

**Anguilla Hotel and Tourism Association:** POB 1020, Coronation Ave, The Valley; tel. 497-2944; fax 497-3091; e-mail ahta@anguillanet.com; internet www.ahta.ai; f. 1981; Exec. Dir MIMI GRATTON.

## Defence

The United Kingdom is responsible for the defence of Anguilla.

## Education

Education is free and compulsory between the ages of five and 16 years. Primary education begins at five years of age and lasts for six years. Secondary education, beginning at 11 years of age, lasts for a further six years. There are six government primary schools and one government secondary school. A 'comprehensive' secondary school education system was introduced in 1986. Post-secondary education is undertaken abroad. According to the 2003 budget address, government expenditure on education was to total EC $13.28m. in that year, equivalent to 14.2% of proposed recurrent expenditure.

# ANTIGUA AND BARBUDA

## Geography

### PHYSICAL FEATURES

Antigua and Barbuda is in the Leeward Islands, in the northeastern Caribbean, the Atlantic Ocean spreading to the east. The country's nearest neighbour is the British dependency of Montserrat, 43 km (27 miles) to the south-west of Antigua island, but only 24 km south-east of Redonda, the uninhabited western outpost of Antigua and Barbuda. Guadeloupe, a part of France, lies to the south and Saint Christopher (Kitts) and Nevis to the west. The islands that continue the chain of the Lesser Antilles to the north-west are variously parts of the Dutch and French Antilles. Antigua and Barbuda has a total surface area of 442 sq km (171 sq miles), Antigua (280 sq km) being larger than Barbuda (161 sq km). Redonda covers only 1.6 sq km.

Most of the country consists of a flat, coral or limestone terrain, but there are higher areas anciently formed by volcanic activity (southern Antigua and Redonda). This contributes to an irregular shoreline of many beaches and harbours on the main island (in total, the country has 153 km of coastline). Antigua was the largest of the British Leeward Islands and, historically, the site of an important port (English Harbour) for the Royal Navy in the West Indies. The arid island is about 23 km long (east–west) and 18 km wide, and its complex coast is girdled by reefs and islets. In the south-west the largely treeless highlands culminate in Boggy Peak (402 m—1,319 ft).

By contrast, Barbuda, 42 km north of Antigua, is an entirely low-lying island (the highest point above sea level, in the north-eastern, ambitiously named Highlands, reaches only 38 m), with smoother shores and only one harbour, the large Codrington Lagoon on the west. The Lagoon is formed by the south-westward jutting Palmetto Point on the central western coast, and a long, narrow spit of land heading northwards from there and culminating in Cedar Tree Point, which runs in a north-easterly direction to form the narrow sea entrance. The other side of this northern entrance to the lagoon is formed by Goat Island (which is actually connected to the mainland by a narrow isthmus). South of the Lagoon is the world's largest colony of frigate birds, and the island is home to many other birds and to turtles.

The rocky, scrubby cone of Redonda, 55 km west-south-west of Antigua, lies in the western or inner of the two chains of islands that the Lesser Antilles split into here, between Montserrat and Nevis. It is just over 1.5 km long (north–south) and barely 0.5 km wide. Redonda has achieved some notoriety as a putative kingdom, established by the Irish Shiell family from Montserrat seven years before the United Kingdom formally annexed the island in 1872 (placing it under the jurisdiction of the Antiguan authorities). Those authorities never bothered to dispute the royal title, which became particularly noted in British literary circles after the Second World War, when the poet John Gawsworth ('King Juan') promoted its court with the creation of an 'intellectual aristocracy'. The title is now disputed, but sovereignty of Redonda is firmly vested in the Crown of Antigua and Barbuda, while actual possession by goats and birds is seldom challenged.

### CLIMATE

Antigua and Barbuda is prone to hurricanes and droughts, the low altitudes drawing little of the moisture carried by the constant Atlantic trade winds. Antigua receives more rainfall

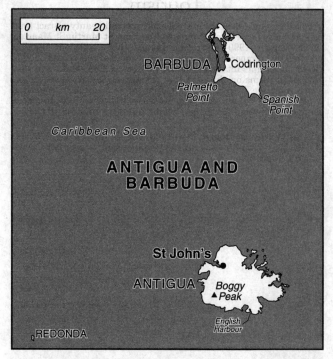

than Barbuda, at some 45 ins (1,143 mm) annually. Most rainfall is in September–November, at the end of the hurricane season. Average monthly temperatures range between 73°F and 85°F (23°–29°C).

### POPULATION

Most people are black, but there are some native whites (traditionally of British or Portuguese descent), as well as more recent communities from Syria and Lebanon. Moreover, as much as 10% of the population is reckoned to have emigrated from the Dominican Republic. Such groups often retain use of their own languages, but English is the official tongue. An English patois is also widely spoken. Most people are Christian, the main denomination being the Anglican Communion, as represented by the Church in the Province of the West Indies, although there are also Roman Catholic, Moravian and Protestant communities.

There were 77,426 people in the country at the time of the May 2001 census (over 98% of them on Antigua). There is a large immigrant population, notably from the Dominican Republic and, since 1995, from Montserrat—some 3,000 are reckoned to have fled the volcano to Antigua. Saint John's, in the north, is the national capital, the largest city and the main port. It had an estimated population of 27,516 people in mid-2003, while the chief town of Barbuda, Codrington, is home to most of the island's population of some 1,500. For administrative purposes, Barbuda is separate from Antigua (Redonda is also a separate unit), and the main island is divided into six parishes.

# History

## MARK WILSON

Antigua and Barbuda is a constitutional monarchy within the Commonwealth. Queen Elizabeth II is Head of State, and is represented in Antigua and Barbuda by a Governor-General. There is a bicameral legislature, Parliament, with an elected chamber.

Few traces remain of the islands' original Amerindian inhabitants. Christopher Columbus landed and named the main island Santa María de la Antigua in 1493, but the Spanish took little interest in the islands, and the first permanent European settlement was by English colonists in 1632. The inlet of English Harbour provided shelter from hurricanes, and was a major British naval base in the 18th and 19th centuries. Antigua was captured by France only once, and very briefly, in 1666. Sugar was grown from 1674, and cultivation of this crop and of cotton was the dominant economic activity for most of the colonial period, with plantations worked until 1834 by slaves of African origin, and then by their descendants as free, but badly paid, labourers. Following emancipation there was a high rate of emigration to other Caribbean countries, and later to Great Britain and North America. Antigua was a separate British colony until 1871, when it became the seat of government for the Leeward Islands Federation.

The Antigua Trades and Labour Union (ATLU) was founded in 1939, organizing the low-paid sugar workers and other manual employees. Vere C. Bird, Sr, became President of the ATLU in 1943 and, in 1946, founded the Antigua Labour Party (ALP), which won a decisive election victory in the first universal-suffrage elections in 1951. Bird, an impressive public speaker with a determined political style, was in power during a period in which Antigua and Barbuda, along with the other British possessions in the Caribbean, enjoyed steady improvements in living standards, education and social services. As a result, he won the fierce loyalty of most lower-income Antiguans.

Following the dissolution of the Leeward Islands Federation in 1957, Antigua and Barbuda joined the Federation of the West Indies in 1958, along with nine other British colonies. Following the departure of Jamaica and Trinidad and Tobago in 1962, the Federation collapsed and an attempt to unite the remaining colonies as the 'little eight' was unsuccessful. Along with its neighbours, Antigua and Barbuda became a British Associated State in 1967, responsible for its internal affairs, with the United Kingdom retaining control of external affairs and defence.

Vere Bird's ALP remained in government until February 1971, when George Walter's Progressive Labour Movement (PLM) began a troubled period in office. Regaining power in 1976, the ALP presided over the transition to independence on 1 November 1981.

After 1976 the ALP won six successive election victories. The Bird family retained tight control on the party machinery, and continued to enjoy loyal popular support. Almost 29% of the labour force is employed directly by the state, and many public employees felt, and indeed in some cases still feel, a close and personal bond with the ALP's leadership. Several small and large private-sector businesses had close links with ALP government ministers. Until November 2000 no radio or television station deemed to be critical of the Government was licensed to operate.

However, the Government's political credibility was damaged over an extended period by a series of scandals. Of these, the best documented came to light in 1990. A Colombian diplomatic note dated 3 April 1990 and delivered to the Antiguan permanent representative at the Organization of American States (OAS) stated that Israeli-manufactured weapons found on the property of a known Colombian narcotics trafficker had been traced to a delivery made to the Antigua Defence Force. The Prime Minister's eldest son, Vere Bird, Jr, was Minister of National Security at the time. The Governor-General appointed a British Queen's Counsel, Sir Louis Blom Cooper, to undertake a Commission of Inquiry. His conclusions included a recommendation that Vere Bird, Jr, 'should not hold any public office again'. Acting on the recommendations of the Inquiry, Bird, Jr, was dismissed from his post and banned for life from holding government office. The head of the Defence Force was also dismissed.

The Inquiry's conclusions helped to settle a succession struggle within the Bird family. Vere Bird, Sr, retired as leader of the ALP in September 1993, at the age of 84, and was succeeded by another son, Lester Bryant Bird, who led the party into an election on 8 March 1994. The campaign was hard-fought. Three small opposition parties had merged in 1992 to form the United Progressive Party (UPP), led by Baldwin Spencer, a senior official of the Antigua Workers' Union. At the election, the UPP increased its parliamentary representation from one seat to five. The ALP's share of the popular vote fell to 54.4%, from 63.8% in 1989, but the party retained 11 seats, and Lester Bird became Prime Minister. In May 1996 Vere Bird, Jr, was controversially appointed to the post of Special Adviser to the Prime Minister.

At the next election, held on 9 March 1999, the ALP's share of the popular vote was reduced slightly further, to 52.6%, but the party gained an additional seat from the UPP by a margin of eight votes. A Commonwealth observer team criticized the electoral process, while independent observers noted that large numbers of voters had been given generous gifts of food, or had been allowed to import vehicles duty free in what the Prime Minister referred to as an exercise in 'poverty alleviation'. Vere Bird, Jr, was appointed Minister of Agriculture, Lands and Fisheries, on the basis that he had been elected to Parliament by voters who were aware of the findings of the Blom Cooper Commission of Inquiry.

The attention of the Government was, from mid-2001, absorbed by the country's fiscal difficulties, which were heightened by fierce resistance within the business community to a 2% levy on gross turnover. A serious controversy also developed over the management of the Government's Medical Benefits Scheme. This prompted the dismissal, in May 2001, of the Attorney-General, Errol Cort, and the Leader of Government Business in the Senate, George 'Bacchanal' Walker. The Government reluctantly agreed to appoint a Commission of Inquiry into the management of the Scheme. In April 2002, following fraud allegations arising from the Inquiry, a minister resigned. The report of the Inquiry, which was published in July 2002, detailed serious and systematic mismanagement, and recommended that the Director of Public Prosecutions should consider prosecuting 14 public officials, including two former health ministers. Some of these have since been charged.

Also in April 2002 a further public dispute emerged, which prompted the Prime Minister to initiate a libel suit against leading members of the UPP, including Baldwin Spencer, as well as against an independent newspaper and radio station and a 15-year-old girl. In June the Prime Minister abandoned an initial attempt to obtain a court injunction against further repetition of the allegations, paying costs of EC $26,000 to his opponents. The libel case was scheduled to open in February 2003, but Bird abandoned the suit against the newspaper and radio journalists before proceedings began. The case against the 15-year-old girl continued, however, although she, in turn, initiated civil proceedings against the Prime Minister, before dropping the case in October. In May 2002 an ALP Member of Parliament, Sherfield Bowen, called for the Prime Minister to demit office along with his Chief of Staff, to allow an independent investigation of allegations made against both of them; in May 2003 he tabled a motion of 'no confidence' in the Prime Minister, which was expected to lead to a vigorous parliamentary debate. At the same time, the standing of the Government was weakened by its repeated difficulties in meeting financial commitments, including salary payments. In June an internal party dispute temporarily cost the ALP its parliamentary majority as four dissidents declared themselves independents. The return of Longford Jeremy to the party, after being rewarded with a ministerial portfolio, restored the Government's majority; however, party feuds continued and intense public criticism damaged its political standing.

In advance of the 2004 general election a completely new electoral register was, with the assistance of the Electoral Office of Jamaica, prepared for the first time since 1975. Elimination of the names of the deceased and non-residents reduced the list by more than one-fifth, while voters were issued with identity cards including a photograph and fingerprint, decreasing the risk that illegitimate votes would be cast. At the election, held on 23 March, the UPP increased its share of the vote to 55.2%, securing 12 of the 17 seats. The ALP took only 41.7% of the vote and four seats; both Lester Bird and Vere Bird, Jr, were defeated. Voter turn-out, at 91.2%, was extremely high, and the election was widely seen as a watershed in national politics. Baldwin Spencer became Prime Minister, and committed his Government to a pragmatic reformist programme and the control of corruption and mismanagement. None the less, the ALP retained a strong core of support, particularly in the civil service. Immediately before the election, a large number of files at the Prime Minister's office were destroyed, making it difficult for the incoming Government to establish the true state of public finances, debt or contracts with the private sector. In mid-2005 the Government was attempting to restabilize the economy, and had established friendly relations with major investors, including key ALP allies. In addition, there were attempts to investigate allegations against members of the former ALP administration, but no charges had been brought by early July.

The island of Barbuda has maintained an independent political tradition, in spite of its small population, and there have been several complex disputes between the islanders and the central Government over land development and other matters. An elected Council manages some aspects of the island's affairs. In the 2004 general election, the UPP-aligned Barbuda People's Movement (BPM) and the ALP-aligned Barbuda People's Movement for Change (BPMC) each won 400 votes. A re-run election on 20 April resulted in a narrow win for the BPM, with 408 votes to 394 for the BPMC. The new Member of Parliament, Trevor Walker, was appointed Minister of State with Responsibility for Barbuda Affairs, thus integrating the smaller island into the national decision-making process. In March 2005 the BPM also won the island council election, which ended a deadlock produced by the BPM and the BPMC each winning four seats at the 2003 election.

# Economy

## MARK WILSON

Antigua and Barbuda is the third smallest country in the Western hemisphere, in terms of population, with some 78,600 inhabitants living on its 442 sq km in 2003. However, the islands have developed a relatively prosperous middle-income economy, with a per-head gross domestic product (GDP) of US $9,685 in 2003. GDP grew at an annual average rate of 1.5% in 2001 and 2.1% in 2002, with the key tourism industry stagnating, before registering 5.5% growth in 2003. Unemployment is fairly low, estimated at 8.1% in 2001, and wages are high enough to attract immigrants from less prosperous Caribbean countries such as Dominica, the Dominican Republic and Guyana. There is concern about the extent of government debt; according to the 2004 budget speech, total foreign and domestic public debt was equivalent to EC $2,813m. (or 137% of GDP), much of it private borrowing at commercial rates, and as much as 29% of the labour force is employed directly by the Government, while wages and salaries account for 66% of recurrent revenue. The previous (Antigua Labour Party—ALP) regime found increasing and obvious difficulty in paying salaries on time or in meeting regular financial commitments. The 2005 budget, the first presented for a full year by the new United Progressive Party (UPP) Government, projected an overall deficit close to 11% of GDP, but aimed to narrow the gap with new taxes, including an income tax for the highest-paid 25% of the population, with a value-added tax to be introduced from 2006.

Antigua and Barbuda is a member of the Caribbean Community and Common Market (CARICOM), which is attempting to develop a single market, in principle by 2006. It is also a member of the Organisation of Eastern Caribbean States, which links nine of the smaller Caribbean territories (Antigua and Barbuda, Dominica, Grenada, Montserrat, Saint Christopher (St Kitts) and Nevis, Saint Lucia, Saint Vincent and the Grenadines, with Anguilla and the British Virgin Islands as associate members), while the Eastern Caribbean Central Bank, headquartered in neighbouring Saint Christopher and Nevis, supervises its financial affairs. Antigua is of much greater economic significance than Barbuda, which is two-thirds the area of the main island, but which has a population of fewer than 2,000. A third island, Redonda, is much smaller and is uninhabited.

The main source of foreign-exchange revenue is the service sector, principally tourism. Net earnings from services were US $241.3m. in 2003, which was not, however, enough to cover the $309.4m. deficit on merchandise trade. The islands' main attractions are their white sandy beaches, of which there are reputed to be 365. There are also fine historic naval sites at English Harbour, Nelson's Dockyard and Shirley Heights in the southeast of the island. There were 245,456 stop-over tourist arrivals in 2004, of whom 29% came from the USA, 46% from Europe (mostly the United Kingdom) and 4% from Canada. There were also 522,753 cruise-ship passengers, although, with much lower per-head spending, they made a smaller contribution to the economy. Yachting and pleasure-boating is also an important activity, centred on English Harbour; Antigua's annual sailing week in April attracts several hundred yachts from across the Caribbean and world-wide. Barbuda has a small tourism industry, with three luxury hotels. Visitor expenditure was EC $812.7m. in 2003.

The tourism industry has been somewhat stagnant in recent years. In 2003 hotels and restaurants accounted for 9.3% of GDP, down from a peak value of 15.6% in 1994. Stop-over visitors only in 2004 surpassed the 1997 peak of 232,141. High labour costs have been a particular problem for the sector. Relations of the former ALP Government with established investors were not always good. The industry suffered in 2001–03 from a downturn in the international economy, and from the fear in some markets of possible terrorist attacks on tourists, although a recovery was in progress in 2004. In that year too, a US entrepreneur of Texan origin who acquired Antiguan citizenship, Allen Stanford, announced plans for an EC $3,000m. resort development, to employ 1,000 on completion, on a site previously earmarked under the ALP Government in 1997 for a Malaysian initiative that failed to materialize.

Antigua serves as a hub for airline services to the smaller neighbouring islands, and is an important operational centre for two regional airlines. These are: LIAT (1974) Ltd, a well-established airline headquartered in Antigua which has experienced financial difficulties in recent years, but employs a staff of close to 1,000; and Caribbean Star Airlines, established in 2000 by Stanford, who in 2003 moved his airline's administrative headquarters to St Kitts. The wholly owned Stanford group played an important role in the local economy, with total employment of 800 and stated investments of EC $130m., including domestic and 'offshore' financial institutions, a daily newspaper and a property development company, which in 2003 had plans for an EC $256m. development on the airport perimeter. The group had close relations with the ALP administration, and was believed to have provided significant loan finance when required, including a US $40m. loan made in 2002. An agreement announced in November 2004 with the incoming UPP administration provided for further new funding, as well as a US $18.5m. debt cancellation.

There is an 'offshore' financial services sector, which has at times been a source of international concern. In 1999 the USA, the United Kingdom and other countries instructed their financial institutions to apply 'enhanced scrutiny' to transactions with Antigua and Barbuda; however, legislation to control money-laundering and financial fraud has since been more strictly enforced, and most restrictions were lifted by 2001. The USA signed a Tax Information Exchange Treaty with Antigua and Barbuda in December 2001. Although the country was never included on the Financial Action Task Force's list of 'non-co-operative' jurisdictions in its fight against money-laundering, in late 1999 a money-laundering operation was uncovered in Antigua and Barbuda, allegedly operated by Pavlo Lazarenko, a former Prime Minister of Ukraine. The country was placed on a list of so-called tax 'havens' by the Organisation for Economic Co-operation and Development (OECD, based in Paris, France) in June 2000, but was removed from this list in 2002 after the Government signed a commitment to move towards the exchange of information with civil and criminal tax investigators by 2005. Antigua and Barbuda withdrew from this agreement in 2004, citing the failure of some OECD members to make similar commitments. At the beginning of 2004 there were 15 licensed 'offshore' banks, of which 11 were 'shell' banks with no physical presence; more than 30 other banks had been closed for regulatory reasons since 1999. There were also approximately 13,500 International Business Corporations, as well as 26 internet gambling sites, supervised in principle by the Directorate of Overseas Gaming, to which the owners of these sites paid a substantial annual licence fee. Antigua in 2003 challenged US restrictions on internet gambling through the structures of the World Trade Organization, and in April 2005 received a ruling that was interpreted as partly in its favour.

The Government places emphasis on the importance of new technology, and telemarketing is seen as a promising area of activity. However, the islands have lagged behind their neighbours in liberalizing the telecommunications industry. Domestic landline services are provided by the state-owned Antigua Public Utilities Authority, and the local Cable & Wireless affiliate controls international telecommunications services; however, there is limited competition in cellular telephony.

Since the closure of the sugar industry, most of the land has been uncultivated. Agriculture and fishing accounted for only 3.5% of GDP in 2003. The state owns most rural land. This has given it close control over residential, commercial and tourism-orientated development projects. There is a very small manufacturing sector, which contributed 2.1% of GDP in 2003, compared with 3.4% in 1990. It includes a brewery and a paints company, which cater mainly for the domestic market.

The islands are at serious risk from hurricanes, and were severely damaged by 'Hurricane Luis' in 1995. Other recent hurricanes to affect the island were 'Marilyn', also in 1995, 'Georges' in 1998, and 'José' and 'Lenny' in 1999.

# Statistical Survey

Source (unless otherwise stated): Ministry of Finance, Economic Development and Planning, Govt Office Complex, Parliament Dr., St John's; tel. 462-5015; fax 462-4860; e-mail budget@candw.ag

## AREA AND POPULATION

**Area:** 441.6 sq km (170.5 sq miles).

**Population:** 62,922 at census of 28 May 1991; 77,426 (provisional result, males 37,002, females 40,424) at census of 28 May 2001. *Mid-2003* (estimate): 78,600 (Source: Caribbean Development Bank, *Social and Economic Indicators*).

**Density** (mid-2003): 178.0 per sq km.

**Principal Town:** St John's (capital), population 22,342 at 1991 census. *Mid-2003* (UN estimate, incl. suburbs): St John's 27,516 (Source: UN, *World Urbanization Prospects: The 2003 Revision*).

**Births, Marriages and Deaths** (registrations, 2001): Live births 1,366 (birth rate 18.04 per 1,000); Marriages 1,784 (marriage rate 23.55 per 1,000); Deaths 457 (death rate 6.03 per 1,000).

**Expectation of Life** (WHO estimates, years at birth): 72 (males 70; females 75) in 2003. Source: WHO, *World Health Report*.

**Employment** (persons aged 15 years and over, census of 28 May 1991): Agriculture, forestry and fishing 1,040; Mining and quarrying 64; Manufacturing 1,444; Electricity, gas and water 435; Construction 3,109; Trade, restaurants and hotels 8,524; Transport, storage and communications 2,395; Finance, insurance, real estate and business services 1,454; Community, social and personal services 6,406; Activities not adequately defined 1,882; Total employed 26,753 (males 14,564, females 12,189) (Source: ILO). *2001:* Total active labour force 46,440. *Mid-2003* (estimates in '000): Agriculture, etc. 8; Total labour force 34 (Source: FAO).

## HEALTH AND WELFARE
### Key Indicators

**Total Fertility Rate** (children per woman, 2003): 1.6.

**Under-5 Mortality Rate** (per 1,000 live births, 2003): 12.

**Physicians** (per 1,000 head, 1999): 0.78.

**Hospital Beds** (per 1,000 head, 1998): 3.88.

**Health Expenditure** (2002): US $ per head (PPP): 527.

**Health Expenditure** (2002): % of GDP: 4.8.

**Health Expenditure** (2002): public (% of total): 68.6.

**Access to Water** (% of persons, 2002): 91.

**Access to Sanitation** (% of persons, 2002): 95.

**Human Development Index** (2002): ranking: 55.

**Human Development Index** (2002): value: 0.800.

For sources and definitions, see explanatory note on p. vi.

## AGRICULTURE, ETC.

**Principal Crops** (FAO estimates, '000 metric tons, 2003): Vegetables 2.2; Melons 0.8; Mangoes 1.4; Other fruits 8.5. Source: FAO.

**Livestock** (FAO estimates, '000 head, year ending September 2003): Asses 1.6; Cattle 14.0; Pigs 5.5; Sheep 18.6; Goats 35.5; Poultry 100. Source: FAO.

**Livestock Products** (FAO estimates, '000 metric tons, 2003): Beef and veal 0.5; Cows' milk 5.4. Source: FAO.

**Fishing** (metric tons, live weight, 2003): Groupers and seabasses 352; Snappers and jobfishes 341; Grunts and sweetlips 212; Parrotfishes 254; Surgeonfishes 196; Triggerfishes and durgons 143; Carangids 85; Caribbean spiny lobster 243; Stromboid conches 469; Total catch (incl. others) 2,587. Source: FAO.

## INDUSTRY

**Production** (estimates, 1988): Rum 4,000 hectolitres; Wines and vodka 2,000 hectolitres; Electric energy (estimate, 2001) 99m. kWh. Source: UN, *Industrial Commodity Statistics Yearbook*.

## FINANCE

**Currency and Exchange Rates:** 100 cents = 1 Eastern Caribbean dollar (EC $). *Sterling, US Dollar and Euro Equivalents* (31 May 2005): £1 sterling = EC $4.909; US $1 = EC $2.700; €1 = EC $3.329; EC $100 = £20.37 = US $37.04 = €30.04. *Exchange rate:* Fixed at US $1 = EC $2.700 since July 1976.

**Budget** (EC $ million, preliminary, 2003): *Revenue:* Tax revenue 381.5 (Taxes on income and profits 62.6, Taxes on Property 11.1, Taxes on domestic goods and services 75.4, Taxes on international transactions 232.4); Other current revenue 44.4; Capital revenue 1.1; Total 427.0, excluding grants received (3.2). *Expenditure:* Current expenditure 514.8 (Personal emoluments 250.1, Employment contributions 32.5, Other goods and services 89.1, Interest payments 92.3, Transfers and subsidies 50.8); Capital expenditure and net lending 81.2; Total 596.0. Source: IMF, *Antigua and Barbuda: Statistical Appendix* (November 2004).

# ANTIGUA AND BARBUDA

**International Reserves** (US $ million at 31 December 2004): IMF special drawing rights 0.01; Foreign exchange 120.12; Total 120.13. Source: IMF, *International Financial Statistics*.

**Money Supply** (EC $ million at 31 December 2004): Currency outside banks 113.34; Demand deposits at deposit money banks 389.14; Total money 502.48. Source: IMF, *International Financial Statistics*.

**Cost of Living** (Consumer Price Index; base: 2000 = 100): 98.7 in 2001; 99.8 in 2002; 103.2 in 2003. Source: IMF, *Antigua and Barbuda: Statistical Appendix* (November 2004).

**Expenditure on the Gross Domestic Product** (preliminary, EC $ million at current prices, 2003): Government final consumption expenditure 525.74; Private final consumption expenditure 657.51; Gross capital formation 1,055.61; *Total domestic expenditure* 2,238.86; Exports of goods and services 1,281.77; *Less* Imports of goods and services 1,465.71; *GDP in purchasers' values* 2,054.92. Source: Eastern Caribbean Central Bank.

**Gross Domestic Product by Economic Activity** (preliminary, EC $ million at current prices, 2003): Agriculture, hunting, forestry and fishing 65.16; Mining and quarrying 29.81; Manufacturing 40.10; Electricity and water 50.59; Construction 243.85; Trade 175.50; Restaurants and hotels 173.43; Transport and communications 348.41; Finance, insurance, real estate and business services 291.86; Government services 322.03; Other community, social and personal services 126.69; *Sub-total* 1,867.43; *Less* Imputed bank service charges 120.25; *GDP at factor cost* 1,747.18. Source: Eastern Caribbean Central Bank.

**Balance of Payments** (preliminary, EC $ million, 2003): Exports of goods f.o.b. 129.38; Imports of goods f.o.b. –964.86; *Trade balance* –835.48; Exports of services 1,152.39; Imports of services –500.85; *Balance on goods and services* –183.94; Other income received 34.60; Other income paid –129.47; *Balance on goods, services and income* –278.81; Current transfers received 64.13; Current transfers paid –49.92; *Current balance* –264.60; Capital account (net) 18.51; Direct investment from abroad (net) 255.44; Portfolio investment assets –0.16; Other investment assets (net) 10.78; Net errors and omissions 50.66; *Overall balance* 70.63. Source: Eastern Caribbean Central Bank.

### EXTERNAL TRADE

**Total Trade** (EC $ million): *Imports f.o.b.:* 913.4 in 2000; 847.2 in 2001; 887.0 in 2002 (preliminary figure). *Exports f.o.b.:* 44.1 in 2000; 46.1 in 2001; 47.0 in 2002 (preliminary figure). Source: Eastern Caribbean Central Bank, *Report and Statement of Accounts*.

**Principal Commodities** (US $ million, 1999): *Imports:* Food and live animals 61.5 (Meat and meat preparations 12.3, Vegetables and fruit 14.1); Beverages and tobacco 14.6 (Beverages 13.5); Crude materials (inedible) except fuels 11.4; Mineral fuels, lubricants, etc. 37.4 (Refined petroleum products 34.7); Chemicals and related products 21.8; Basic manufactures 54.8; Machinery and transport equipment 114.7 (Telecommunications and sound equipment 16.6, Road vehicles 52.9); Miscellaneous manufactured articles 38.1. Total (incl. others) 356.0. *Exports:* Food and live animals 0.6; Crude materials (inedible) except fuels 0.5; Chemicals and related products 1.7 (Pigments, paints, varnishes, etc. 1.2); Basic manufactures 3.0 (Textiles, yarn, fabrics, made-up articles, etc. 0.7, Iron and steel 1.4); Machinery and transport equipment 5.9 (General industrial machinery and equipment 0.6, Telecommunications and sound equipment 1.8, Road vehicles 1.1); Miscellaneous manufactured articles 2.7 (Photographic equipment, optical goods and watches, etc. 0.8); Total (incl. others) 15.0. Source: UN, *International Trade Statistics Yearbook*.

**Principal Trading Partners** (US $ million, 1999): *Imports:* Barbados 6.7; Canada 9.4; Japan 36.2; Korea, Republic 5.1; Netherlands Antilles 19.6; Saint Vincent and the Grenadines 4.6; Trinidad and Tobago 21.4; United Kingdom 22.3; USA 176.1; Venezuela 14.7; Total (incl. others) 356.0. *Exports:* Anguilla 0.2; Barbados 0.5; British Virgin Islands 0.2; Canada 0.2; Dominica 1.0; France 1.1; Germany 0.2; Grenada 0.2; Jamaica 0.9; Montserrat 2.0; Netherlands Antilles 0.5; Saint Christopher and Nevis 1.7; Saint Lucia 0.8; Saint Vincent and the Grenadines 0.3; Trinidad and Tobago 0.8; United Kingdom 1.1; USA 3.2; Total (incl. others) 15.0. Source: UN, *International Trade Statistics Yearbook*.

### TRANSPORT

**Road Traffic** (registered vehicles, 1998): Passenger motor cars and commercial vehicles 24,000. Source: UN, *Statistical Yearbook*.

**Shipping** (international freight traffic, '000 metric tons, 1990): Goods loaded 28; Goods unloaded 113 (Source: UN, *Monthly Bulletin of Statistics*). *Merchant Fleet* (registered at 31 December): 1,028 vessels (total displacement 6,914,568 grt) in 2004 (Source: Lloyd's Register-Fairplay, *World Fleet Statistics*).

**Civil Aviation** (traffic on scheduled services, 1999): Kilometres flown (million) 11; Passengers carried ('000) 1,371; Passenger-km (million) 276; Total ton-km (million) 26. Source: UN, *Statistical Yearbook*.

### TOURISM

**Visitor Arrivals** ('000 persons): 510.3 (198.1 stop-over visitors, 312.2 cruise-ship passengers) in 2002; 607.8 (224.0 stop-over visitors, 383.8 cruise-ship passengers) in 2003; 768.3 (245.5 stop-over visitors, 522.8 cruise-ship passengers) in 2004. Source: Caribbean Development Bank, *Social and Economic Indicators*.

**Tourism Receipts** (US $ million, excl. passenger transport): 272 in 2001; 274 in 2002; 301 in 2003. Source: World Tourism Organization.

### COMMUNICATIONS MEDIA

**Radio Receivers** (1997): 36,000 in use*.

**Television Receivers** (1999): 33,000 in use*.

**Telephones** (2002): 38,000 main lines in use†.

**Facsimile Machines** (year ending 31 March 1997): 850 in use‡.

**Mobile Telephones** (2002): 28,200 subscribers†.

**Internet Users** (2002): 10,000†.

**Daily Newspapers** (2003): 2.

**Non-daily Newspapers** (1996): 4*.

\* Source: UNESCO, *Statistical Yearbook*.
† Source: International Telecommunication Union.
‡ Source: UN, *Statistical Yearbook*.

### EDUCATION

**Pre-primary** (1983): 21 schools; 23 teachers; 677 pupils.

**Primary** (2000/01): 55 schools; 525 teachers; 10,427 students.

**Secondary** (2000/01): 14 schools; 361 teachers; 5,794 students.

**Special** (2000/01): 2 schools; 15 teachers; 61 students.

**Tertiary** (1986): 2 colleges; 631 students.

**Adult Literacy Rate:** 86.6% in 2000 (Source: Organization of Eastern Caribbean States).

# Directory

## The Constitution

The Constitution, which came into force at the independence of Antigua and Barbuda on 1 November 1981, states that Antigua and Barbuda is a 'unitary sovereign democratic state'. The main provisions of the Constitution are summarized below

### FUNDAMENTAL RIGHTS AND FREEDOMS

Regardless of race, place of origin, political opinion, colour, creed or sex, but subject to respect for the rights and freedoms of others and for the public interest, every person in Antigua and Barbuda is entitled to the rights of life, liberty, security of the person, the enjoyment of property and the protection of the law. Freedom of movement, of conscience, of expression (including freedom of the press), of peaceful assembly and association is guaranteed and the inviolability of family life, personal privacy, home and other property is maintained. Protection is afforded from discrimination on the grounds of race, sex, etc., and from slavery, forced labour, torture and inhuman treatment.

### THE GOVERNOR-GENERAL

The British sovereign, as Monarch of Antigua and Barbuda, is the Head of State and is represented by a Governor-General of local citizenship.

### PARLIAMENT

Parliament consists of the Monarch, a 17-member Senate and the House of Representatives composed of 17 elected members. Senators

# ANTIGUA AND BARBUDA

are appointed by the Governor-General: 11 on the advice of the Prime Minister (one of whom must be an inhabitant of Barbuda), four on the advice of the Leader of the Opposition, one at his own discretion and one on the advice of the Barbuda Council. The Barbuda Council is the principal organ of local government in that island, whose membership and functions are determined by Parliament. The life of Parliament is five years.

Each constituency returns one Representative to the House who is directly elected in accordance with the Constitution.

The Attorney-General, if not otherwise a member of the House, is an *ex-officio* member but does not have the right to vote.

Every citizen over the age of 18 is eligible to vote.

Parliament may alter any of the provisions of the Constitution.

## THE EXECUTIVE

Executive authority is vested in the Monarch and exercisable by the Governor-General. The Governor-General appoints as Prime Minister that member of the House who, in the Governor-General's view, is best able to command the support of the majority of the members of the House, and other ministers on the advice of the Prime Minister. The Governor-General may remove the Prime Minister from office if a resolution of no confidence is passed by the House and the Prime Minister does not either resign or advise the Governor-General to dissolve Parliament within seven days.

The Cabinet consists of the Prime Minister and other ministers and the Attorney-General.

The Leader of the Opposition is appointed by the Governor-General as that member of the House who, in the Governor-General's view, is best able to command the support of a majority of members of the House who do not support the Government.

## CITIZENSHIP

All persons born in Antigua and Barbuda before independence who, immediately prior to independence, were citizens of the United Kingdom and Colonies automatically become citizens of Antigua and Barbuda. All persons born outside the country with a parent or grandparent possessing citizenship of Antigua and Barbuda automatically acquire citizenship as do those born in the country after independence. Provision is made for the acquisition of citizenship by those to whom it would not automatically be granted.

# The Government

**Monarch:** HM Queen ELIZABETH II (succeeded to the throne 6 February 1952).

**Governor-General:** Sir JAMES BEETHOVEN CARLISLE (took office 10 June 1993).

## CABINET
(July 2005)

**Prime Minister and Minister of Foreign Affairs, International Transportation, Trade, National Security, Telecommunications, Broadcasting, Information, Labour, Public Administration, Empowerment, Ecclesiastics and Barbuda Affairs:** BALDWIN SPENCER.

**Deputy Prime Minister and Minister of Public Works:** WILMOTH DANIEL.

**Minister of Finance, Economic Development and Planning:** Dr ERROL CORT.

**Attorney-General:** JUSTIN SIMON.

**Minister of Justice and Legal Affairs:** COLIN DERRICK.

**Minister of Health, Youth Affairs and Sports:** JOHN MAGINLEY.

**Minister of Tourism:** HAROLD LOVELL.

**Minister of Education, Human Development and Culture:** BERTRAND JOSEPH.

**Minister of Agriculture, Lands, Marine Resources and the Environment:** CHARLESWORTH SAMUEL.

**Minister of Housing, Culture and Social Transformation:** HILSON BAPTISTE.

**Minister of State in the Ministry of Agriculture, Lands, Marine Resources and the Environment:** Sen. JOANNE MASSIAH.

**Minister of State in the Ministry of Health, Youth Affairs and Sport, with Responsibility for Youth Affairs and Sport:** WINSTON WILLIAMS.

**Minister of State in the Ministry of Housing, Culture and Social Transformation, with Responsibility for Culture:** ELESTON ADAMS.

**Minister of State in the Office of the Prime Minister, Responsible for Public Administration:** JACQUI QUINN-LEANDRO.

**Minister of State in the Office of the Prime Minister, Responsible for Telecommunications, Broadcasting and Information:** Dr EDMOND MANSOOR.

**Minister of State with Responsibility for Barbuda Affairs:** TREVOR WALKER.

## MINISTRIES

**Office of the Prime Minister:** Queen Elizabeth Highway, St John's; tel. 462-4956; fax 462-3225; e-mail cora.richards@antigua.gov.ag; internet www.antigua.gov.ag.

**Ministry of Agriculture, Lands, Marine Resources and the Environment:** Queen Elizabeth Highway, St John's; tel. 462-1543; fax 462-6104.

**Ministry of Barbuda Affairs:** Govt Office Complex, Parliament Dr., St John's.

**Ministry of Ecclesiastics:** Govt Office Complex, Parliament Dr., St John's.

**Ministry of Education, Human Development and Culture:** Govt Office Complex, Queen Elizabeth Highway, St John's; tel. 462-4959; fax 462-4970; e-mail doristeen.etinoff@ab.gov.ag.

**Ministry of Finance, Economic Development and Planning:** Govt Office Complex, Parliament Dr., St John's; tel. 462-5015; fax 462-4260; e-mail budget@candw.ag.

**Ministry of Foreign Affairs:** Queen Elizabeth Highway, St John's; tel. 462-1052; fax 462-2482; e-mail minforeign@candw.ag; internet www.antiguabarbuda.net/external.

**Ministry of Health:** St John's St, St John's; tel. 462-1600; fax 462-5003.

**Ministry of Housing, Culture and Social Transformation:** St John's.

**Ministry of Information:** St John's.

**Ministry of International Transportation:** St John's St, St John's; tel. 462-0894; fax 462-1529.

**Ministry of Justice and Legal Affairs, and Office of the Attorney-General:** Hadeed Bldg, Redcliffe St, St John's; tel. 462-6037; fax 562-1879; e-mail legalaffairs@candw.ag; internet www.antiguabarbuda.net/ag.

**Ministry of Labour, Public Administration and Empowerment:** Cnr of Nevis St and Friendly Alley, St John's; tel. 462-3331; fax 462-1595; e-mail minlabour@antigua.gov.ag.

**Ministry of National Security:** Govt Office Complex, Parliament Dr., St John's.

**Ministry of Public Administration:** St John's.

**Ministry of Public Works:** St John's.

**Ministry of Tourism:** Queen Elizabeth Highway, St John's; tel. 462-4625; fax 462-2836; e-mail mintourenv@candw.ag.

**Ministry of Trade:** Redcliffe St, St John's; tel. 462-1543; fax 462-5003.

**Ministry of Youth Affairs and Sports:** St John's.

# Legislature

## PARLIAMENT

### Senate

**President:** Dr EDMOND MANSOOR.

There are 17 nominated members.

### House of Representatives

**Speaker:** D. GISELLE ISAAC.

**Ex-Officio Member:** The Attorney-General.

**Clerk:** SYLVIA WALKER.

ANTIGUA AND BARBUDA

**General Election, 23 March 2004\***

| Party | Votes cast | % | Seats |
|---|---|---|---|
| United Progressive Party | 21,892 | 55.2 | 12 |
| Antigua Labour Party | 16,544 | 41.7 | 4 |
| Barbuda People's Movement | 408 | 1.0 | 1 |
| Barbuda People's Movement for Change | 394 | 1.0 | — |
| Independents and rejected ballots | 391 | 0.9 | — |
| Total | 39,629 | 100.0 | 17 |

\* The table reflects the final election results after a re-run ballot on 20 April 2004 in the constituency of Barbuda following a tie between the Barbuda People's Movement and the Barbuda People's Movement for Change.

## Political Organizations

**Antigua Freedom Party (AFP):** St John's; f. 2003.

**Antigua Labour Party (ALP):** St Mary's St, POB 948, St John's; tel. 462-2235; f. 1968; Leader LESTER BRYANT BIRD; Chair. MOLWYN JOSEPH.

**Barbuda People's Movement (BPM):** Codrington; campaigns for separate status for Barbuda; Parliamentary Leader THOMAS HILBOURNE FRANK; Chair. FABIAN JONES.

**Barbuda People's Movement for Change (BPMC):** Codrington; f. 2004; effectively replaced Organisation for National Reconstruction, which was f. 1983 and re-f. 1988 as Barbuda Independence Movement; advocates self-government for Barbuda; Pres. ARTHUR SHABAZZ-NIBBS.

**Democratic People's Party (DPP):** St John's; f. 2003; Leader SHERFIELD BOWEN.

**First Christian Democratic Movement (FCDM):** St John's; f. 2003; Leader EGBERT JOSEPH.

**National Labour Party (NLP):** St John's; f. 2003.

**National Movement for Change (NMC):** St John's; f. 2002; Leader ALISTAIR THOMAS.

**New Barbuda Development Movement:** Codrington; linked with the Antigua Labour Party.

**Organisation for National Development:** Upper St Mary St, St John's; f. 2003 by breakaway faction of the United Progressive Party; Sec.-Gen. VALERIE SAMUEL.

**United Progressive Party (UPP):** Nevis St, St John's; tel. 462-1818; fax 462-5937; e-mail upp@candw.ag; f. 1992, by merger of the Antigua Caribbean Liberation Movement (f. 1979), the Progressive Labour Movement (f. 1970) and the United National Democratic Party (f. 1986); Leader BALDWIN SPENCER; Dep. Leader WILMOTH DANIEL; Chair. LEON SYMISTER.

## Diplomatic Representation

### EMBASSIES AND HIGH COMMISSION IN ANTIGUA AND BARBUDA

**China, People's Republic:** Cedar Valley, POB 1446, St John's; tel. 462-1125; fax 462-6425; e-mail chinaemb_ag@mfa.gov.cn; Ambassador REN XIAOPING.

**United Kingdom:** British High Commission, Price Waterhouse Coopers Centre, 11 Old Parham Rd, POB 483, St John's; tel. 462-0008; fax 562-2124; e-mail britishc@candw.ag; High Commissioner JOHN WHITE (resident in Barbados).

**Venezuela:** Cross and Redcliffe Sts, POB 1201, St John's; tel. 462-1574; fax 462-1570; e-mail embaveneantigua@yahoo.es; Ambassador JOSÉ LAURENCIO SILVA MÉNDEZ.

## Judicial System

Justice is administered by the Eastern Caribbean Supreme Court, based in Saint Lucia, which consists of a High Court of Justice and a Court of Appeal. One of the Court's Puisne Judges is resident in and responsible for Antigua and Barbuda, and presides over the Court of Summary Jurisdiction on the islands. There are also Magistrates' Courts for lesser cases.

**Chief Justice:** DENNIS BYRON.
**Solicitor-General:** LEBRECHT HESSE.
**Attorney-General:** JUSTIN SIMON.

## Religion

The majority of the inhabitants profess Christianity, and the largest denomination is the Church in the Province of the West Indies (Anglican Communion).

### CHRISTIANITY

**Antigua Christian Council:** POB 863, St John's; tel. 462-0261; f. 1964; five mem. churches; Pres. Rt Rev. DONALD J. REECE (Roman Catholic Bishop of St John's-Basseterre); Exec. Sec. EDRIS ROBERTS.

#### The Anglican Communion

Anglicans in Antigua and Barbuda are adherents of the Church in the Province of the West Indies. The diocese of the North Eastern Caribbean and Aruba comprises 12 islands: Antigua, Saint Christopher (St Kitts), Nevis, Anguilla, Barbuda, Montserrat, Dominica, Saba, St Martin/St Maarten, Aruba, St Bartholomew and St Eustatius; the total number of Anglicans is about 60,000. The See City is St John's, Antigua.

**Bishop of the North Eastern Caribbean and Aruba:** Rt Rev. LEROY ERROL BROOKS, Bishop's Lodge, POB 23, St John's; tel. 462-0151; fax 462-2090; e-mail dioceseneca@candw.ag.

#### The Roman Catholic Church

The diocese of St John's-Basseterre, suffragan to the archdiocese of Castries (Saint Lucia), includes Anguilla, Antigua and Barbuda, the British Virgin Islands, Montserrat and Saint Christopher and Nevis. At 31 December 2003 there were an estimated 15,322 adherents in the diocese. The Bishop participates in the Antilles Episcopal Conference (whose Secretariat is based in Port of Spain, Trinidad).

**Bishop of St John's-Basseterre:** Rt Rev. DONALD JAMES REECE, Chancery Offices, POB 836, St John's; tel. 461-1135; fax 462-2383; e-mail djr@candw.ag.

#### Other Christian Churches

**Antigua Baptist Association:** POB 277, St John's; tel. 462-1254; Pres. IVOR CHARLES.

**Methodist Church:** c/o POB 863, St John's; Superintendent Rev. ELOY CHRISTOPHER.

**St John's Evangelical Lutheran Church:** Woods Centre, POB W77, St John's; tel. 462-2896; e-mail lutheran@candw.ag; Pastor M. HENRICH, Pastor J. STERNHAGEN, Pastor T. SATORIUS.

There are also Pentecostal, Seventh-day Adventist, Moravian, Nazarene, Salvation Army and Wesleyan Holiness places of worship.

## The Press

**Antigua Sun:** 15 Pavilion Dr., Coolidge, POB W263, St John's; tel. 480-5960; fax 480-5968; e-mail editor@antiguasun.com; internet www.antiguasun.com; twice weekly; published by Sun Printing and Publishing Ltd; Editor and Business Man. PATRICK HENRY.

**Business Expressions:** POB 774, St John's; tel. 462-0743; fax 462-4575; e-mail chamcom@candw.ag; monthly; organ of the Antigua and Barbuda Chamber of Commerce and Industry Ltd.

**Daily Observer:** Fort Rd, POB 1318, St John's; fax 462-5561; internet www.antiguaobserver.com; independent; Publr SAMUEL DERRICK; Editor WINSTON A. DERRICK; circ. 4,000.

**The Nation's Voice:** Public Information Division, Church St and Independence Ave, POB 590, St John's; tel. 462-0090; weekly.

**National Informer:** St John's; weekly.

**The Outlet:** Marble Hill Rd, McKinnons, POB 493, St John's; tel. 462-4410; fax 462-0438; e-mail outletpublishers@yahoo.com; f. 1975; weekly; publ. by the Antigua Caribbean Liberation Movement (founder member of the United Progressive Party in 1992); Editor CONRAD LUKE; circ. 5,000.

**The Worker's Voice:** Emancipation Hall, 46 North St, POB 3, St John's; tel. 462-0090; f. 1943; twice weekly; official organ of the Antigua Labour Party and the Antigua Trades and Labour Union; Editor NOEL THOMAS; circ. 6,000.

ANTIGUA AND BARBUDA                                                                                                          *Directory*

## Publishers

**Antigua Printing and Publishing Ltd:** POB 670, St John's; tel. 462-1265; fax 462-6200.

**Wadadli Productions Ltd:** POB 571, St John's; tel. 462-4489.

## Broadcasting and Communications

### TELECOMMUNICATIONS

Most telephone services are provided by the Antigua Public Utilities Authority (see Trade and Industry).

**Antigua Public Utilities Authority Personal Communications Services (APUA PCS):** St John's; internet www.apuatel.com; f. 2000; digital mobile cellular telephone network; Man. JULIAN WILKINS.

**Cable & Wireless (Antigua and Barbuda) Ltd:** 42–44 St Mary's St, POB 65, St John's; tel. 480-4000; fax 480-4200; internet www.cwantigua.com; owned by Cable & Wireless PLC (United Kingdom).

**Cingular Wireless Antigua and Barbuda:** St John's; internet www.cingularwireless.com; joint venture between SBC Communications and BellSouth; scheduled to sell its operations and licences in the Caribbean to Digicel in 2005; Pres. and CEO STANLEY T. SIGMAN.

### RADIO

**ABS Radio:** POB 590, St John's; tel. 462-3602; internet www.cmattcomm.com/abs.htm; f. 1956; subsidiary of Antigua and Barbuda Broadcasting Service (see Television); Programme Man. D. L. PAYNE.

**Caribbean Radio Lighthouse:** POB 1057, St John's; tel. 462-1454; fax 462-7420; e-mail lighthouse@candw.ag; internet www.radiolighthouse.org; f. 1975; religious broadcasts; operated by Baptist Int. Mission Inc (USA); Dir CURTIS L. WAITE.

**Caribbean Relay Co Ltd:** POB 1203, St John's; tel. 462-0994; fax 462-0487; e-mail cm-crc@candw.ag; jtly operated by British Broadcasting Corpn and Deutsche Welle.

**Observer Radio:** f. 2001; independently owned station; Gen. Man. WINSTON A. DERRICK.

**Radio ZDK:** Grenville Radio Ltd, POB 1100, St John's; tel. 462-1100; f. 1970; commercial; Programme Dir IVOR BIRD; CEO E. PHILIP.

**Sun FM Radio:** St John's; commercial.

### TELEVISION

**Antigua and Barbuda Broadcasting Service (ABS):** Directorate of Broadcasting and Public Information, POB 590, St John's; tel. 462-0010; fax 462-4442; scheduled for privatization; Dir-Gen. HOLLIS HENRY; CEO DENIS LEANDRO.

**ABS Television:** POB 1280, St John's; tel. 462-0010; fax 462-1622; f. 1964; Programme Man. JAMES TANNY ROSE.

**CTV Entertainment Systems:** 25 Long St, St John's; tel. 462-0346; fax 462-4211; cable television co; transmits 33 channels of US television 24 hours per day to subscribers; Programme Dir K. BIRD.

## Finance

(cap. = capital; res = reserves; dep. = deposits; m. = millions; brs = branches)

### BANKING

The Eastern Caribbean Central Bank, based in Saint Christopher, is the central issuing and monetary authority for Antigua and Barbuda.

**ABI Bank Ltd:** High St and Corn Alley, POB 1679, St John's; tel. 480-2700; fax 480-2750; e-mail abib@abifinancial.com; internet www.abifinancial.com; f. 1990 as Antigua Barbuda Investment Bank Ltd; ABI Financial Group comprised of ABI Bank Ltd, AOB (Antigua Overseas Bank) Ltd, ABIT (ABI Trust Antigua) Ltd, ABID (ABI Development Co) Ltd and ABII (ABI Insurance Co) Ltd; cap. EC $11.7m., res EC $8.4m., dep. EC $447.2m. (Sept. 2003); Chair. EUSTACE FRANCIS; Man. Dir MCALLISTER ABBOTT.

**Antigua and Barbuda Development Bank:** 27 St Mary's St, POB 1279, St John's; tel. 462-0838; fax 462-0839; f. 1974; Man. S. ALEX OSBORNE.

**Antigua Commercial Bank:** St Mary's and Thames Sts, POB 95, St John's; tel. 462-1217; fax 462-1220; f. 1955; auth. cap. EC $5m.; Man. JOHN BENJAMIN; 2 brs.

**Bank of Antigua:** 10 Pavilion Dr., POB 315, Coolidge; tel. 462-4282; fax 462-4718; internet www.bankofantigua.com; f. 1981; total assets EC $340m. (Dec. 2003); Chair. ALLEN STANFORD; 2 brs.

**RBTT Bank Caribbean Ltd:** 45 High St, POB 1324, St John's; tel. 462-4217; fax 462-8575; e-mail hollings@candw.ag; Country Man. KATHRYNE ARMSTRONG-HOLLINGSWORTH.

### Foreign Banks

**Bank of Nova Scotia** (Canada): High St, POB 342, St John's; tel. 480-1500; fax 480-1554; Man. LEN WRIGHT.

**FirstCaribbean International Bank (Barbados) Ltd:** St John's; internet www.firstcaribbeanbank.com; adopted present name in 2002 following merger of Caribbean operations of CIBC and Barclays Bank PLC; Exec. Chair. MICHAEL MANSOOR; CEO CHARLES PINK.

**Royal Bank of Canada:** High and Market Sts, POB 252, St John's; tel. 480-1151; fax 480-1190; offers a trustee service.

In November 2004 there were 16 registered 'offshore' banks in Antigua and Barbuda.

### Regulatory Body

**Financial Services Regulatory Commission (FSRC):** Patrick Michael Bldg, Lower Nevis St, POB 2674, St John's; fmrly known as International Financial Sector Regulatory Authority, adopted current name in 2002; Chair. LEBRECHT HESSE; Admin. LEROY KING.

### STOCK EXCHANGE

**Eastern Caribbean Securities Exchange:** e-mail info@ecseonline.com; internet www.ecseonline.com; based in Basseterre, Saint Christopher and Nevis; f. 2001; regional securities market designed to facilitate the buying and selling of financial products for the eight member territories—Anguilla, Antigua and Barbuda, Dominica, Grenada, Montserrat, Saint Christopher and Nevis, Saint Lucia and Saint Vincent and the Grenadines; Gen. Man. TREVOR BLAKE.

### INSURANCE

Several foreign companies have offices in Antigua. Local insurance companies include the following:

**General Insurance Co Ltd:** Upper Redcliffe St, POB 340, St John's; tel. 462-2346; fax 462-4482.

**Sentinel Insurance Co Ltd:** Coolidge, POB 207, St John's; tel. 462-4603.

**State Insurance Corpn:** Redcliffe St, POB 290, St John's; tel. 462-0110; fax 462-2649; f. 1977; Chair. Dr VINCENT RICHARDS; Gen. Man. ROLSTON BARTHLEY.

## Trade and Industry

### DEVELOPMENT ORGANIZATIONS

**Barbuda Development Agency:** St John's; economic development projects for Barbuda.

**Development Control Authority:** St John's; internet www.antiguagov.com/dca.

**Industrial Development Board:** Newgate St, St John's; tel. 462-1038; fax 462-1033; f. 1984 to stimulate investment in local industries.

**St John's Development Corporation:** Heritage Quay, POB 1473, St John's; tel. 462-2776; fax 462-3931; e-mail stjohnsdevcorp@candw.ag; internet www.firstyellow.com/stjohnsdev; f. 1986; manages the Heritage Quay Duty Free Shopping Complex, Vendors' Mall, Public Market and Cultural and Exhibition Complex.

### CHAMBER OF COMMERCE

**Antigua and Barbuda Chamber of Commerce and Industry Ltd:** Cnr of North and Popeshead Sts, POB 774, St John's; tel. 462-0743; fax 462-4575; e-mail chamcom@candw.ag; f. 1944 as Antigua Chamber of Commerce Ltd; name changed as above in 1991, following the collapse of the Antigua and Barbuda Manufacturers' Asscn; Pres. CLARVIS JOSEPH; Exec. Dir HOLLY PETERS.

### INDUSTRIAL AND TRADE ASSOCIATIONS

**Antigua Cotton Growers' Association:** Dunbars, St John's; tel. 462-4962; Chair. FRANCIS HENRY; Sec. PETER BLANCHETTE.

**Antigua Fisheries Corpn:** St John's; e-mail fisheries@candw.ag; partly funded by the Antigua and Barbuda Development Bank; aims to help local fishermen.

**Antigua Sugar Industry Corpn:** Gunthorpes, POB 899, St George's; tel. 462-0653.

**Private Sector Organization of Antigua and Barbuda:** St John's.

### EMPLOYERS' ORGANIZATION

**Antigua Employers' Federation:** Upper High Street, POB 298, St John's; tel. and fax 462-0449; e-mail aempfed@candw.ag; f. 1950; 117 mems; Chair. PEDRO CORBIN; Exec. Sec. HENDERSON BASS.

### UTILITIES

**Antigua Public Utilities Authority (APUA):** St Mary's St, POB 416, St John's; tel. 480-7000; fax 462-2782; generation, transmission and distribution of electricity; internal telecommunications; collection, treatment, storage and distribution of water; Gen. Man. PETER BENJAMIN.

**Caribbean Power Ltd:** supplies Barbuda with water and electricity services.

### TRADE UNIONS

**Antigua and Barbuda Meteorological Officers Association:** c/o V. C. Bird International Airport, Gabatco, POB 1051, St John's; tel. and fax 462-4606; Pres. LEONARD JOSIAH.

**Antigua and Barbuda Public Service Association (ABPSA):** POB 1285, St John's; tel. 463-6427; fax 461-5821; e-mail abpsa@candw.ag; Pres. JAMES SPENCER; Gen. Sec. ELLOY DE FREITAS; 550 mems.

**Antigua and Barbuda Union of Teachers:** c/o Ministry of Education, Human Development and Culture, Church St, St John's; tel. 462-4959; Pres. COLIN GREENE; Sec. FOSTER ROBERTS.

**Antigua Trades and Labour Union (ATLU):** 46 North St, POB 3, St John's; tel. 462-0090; fax 462-4056; e-mail atandlu@candw.ag; f. 1939; affiliated to the Antigua Labour Party; Pres. WIGLEY GEORGE; Gen. Sec. NATALIE PAYNE; about 10,000 mems.

**Antigua Workers' Union (AWU):** Freedom Hall, Newgate St, POB 940, St John's; tel. 462-2005; fax 462-5220; e-mail awu@candw.ag; f. 1967 following split with ATLU; not affiliated to any party; Pres. MAURICE CHRISTIAN; Gen. Sec. KEITHLYN SMITH; 10,000 mems.

## Transport

### ROADS

There are 384 km (239 miles) of main roads and 781 km (485 miles) of secondary dry-weather roads.

### SHIPPING

The main harbour is the St John's Deep Water Harbour. It is used by cruise ships and a number of foreign shipping lines. There are regular cargo and passenger services internationally and regionally. At Falmouth, on the south side of Antigua, is a former Royal Navy dockyard in English Harbour. The harbour is now used by yachts and private pleasure craft.

**Antigua and Barbuda Port Authority:** Deep Water Harbour, POB 1052, St John's; tel. 462-4243; fax 462-2510; f. 1968; responsible to Ministry of Finance, Economic Development and Planning; Chair. LLEWELLYN SMITH; Port Man. LEROY ADAMS.

**Joseph, Vernon, Toy Contractors Ltd:** Nut Grove St, St John's.

**Parenzio Shipping Co Ltd:** Nevis St, St John's.

**Vernon Edwards Shipping Co:** Thames St, POB 82, St John's; tel. 462-2034; fax 462-2035; e-mail vedwards@candw.ag; cargo service to and from San Juan, Puerto Rico.

**The West Indies Oil Co Ltd:** Friars Hill Rd, POB 230, St John's; tel. 462-0140; fax 462-0543; e-mail wiocfs@candw.ag.

### CIVIL AVIATION

Antigua's V. C. Bird (formerly Coolidge) International Airport, 9 km (5.6 miles) north-east of St John's, is modern and accommodates jet-engined aircraft. There is a small airstrip at Codrington on Barbuda.

Antigua and Barbuda Airlines, a nominal company, controls international routes, but services to Europe and North America are operated by American Airlines (USA), Continental Airlines (USA), Lufthansa (Germany) and Air Canada. Antigua and Barbuda is a shareholder in, and the headquarters of, the regional airline, LIAT. Other regional services are operated by BWIA (Trinidad and Tobago) and Air BVI (British Virgin Islands).

**LIAT (1974) Ltd:** POB 819, V. C. Bird Int. Airport, St John's; tel. 480-5600; fax 480-5625; e-mail li.sales.mrkting@candw.ag; internet www.liatairline.com; f. 1956 as Leeward Islands Air Transport Services, jtly owned by 11 regional Govts; privatized in 1995; shares are held by the Govts of Antigua and Barbuda, Montserrat, Grenada, Barbados, Trinidad and Tobago, Jamaica, Guyana, Dominica, Saint Lucia, Saint Vincent and the Grenadines and Saint Christopher and Nevis (30.8%), BWIA (29.2%), LIAT employees (13.3%) and private investors (26.7%); scheduled passenger and cargo services to 19 destinations in the Caribbean; charter flights are also undertaken; Chair. WILBUR HARRIGAN; CEO GARRY CULLEN.

**Carib Aviation Ltd:** V. C. Bird Int. Airport; tel. 462-3147; fax 462-3125; e-mail caribav@candw.ag; charter co; operates regional services.

**Caribbean Star Airlines:** Coolidge Industrial Estate, Airport Rd, POB 1628W, St George; f. 2000; operates regional services; Pres. and CEO PAUL MOREIRA.

## Tourism

Tourism is the country's main industry. Antigua offers a reputed 365 beaches, an annual international sailing regatta and Carnival week, and the historic Nelson's Dockyard in English Harbour (a national park since 1985). Barbuda is less developed, but is noted for its beauty, wildlife and beaches of pink sand. In 1986 the Government established the St John's Development Corporation to oversee the redevelopment of the capital as a commercial duty-free centre, with extra cruise-ship facilities. In 2004 there were 245,456 stop-over visitors and 522,753 cruise-ship passengers. In the same year 29% of stop-over visitors came from the USA and in 2003 expenditure by all visitors totalled US $301m.

**Antigua and Barbuda Department of Tourism:** c/o Ministry of Tourism, Queen Elizabeth Highway, POB 363, St John's; tel. 462-0029; fax 462-2483; e-mail deptourism@antigua.gov.ag; internet www.antigua-barbuda.org; Dir-Gen. SHIRLENE NIBBS.

**Antigua Hotels and Tourist Association (AHTA):** Island House, Newgate St, POB 454, St John's; tel. 462-3703; fax 462-3702; e-mail ahta@candw.ag; internet www.antiguahotels.org; Chair. ANDREW V. MICHELIN; Exec. Dir CYNTHIA G. SIMON.

## Defence

There is a small defence force of 170 men (army 125, navy 45). The US Government leases two military bases on Antigua. Antigua and Barbuda participates in the US-sponsored Regional Security System. The defence budget in 2004 was estimated at EC $12.0m.

## Education

Education is compulsory for 11 years between five and 16 years of age. Primary education begins at the age of five and normally lasts for seven years. Secondary education, beginning at 12 years of age, lasts for five years, comprising a first cycle of three years and a second cycle of two years. In 2000/01 there were 55 primary and 14 secondary schools; the majority of schools are administered by the Government. In the same year some 10,427 primary school pupils and 5,794 secondary school pupils were enrolled. Teacher-training and technical training are available at the Antigua State College in St John's. An extra-mural department of the University of the West Indies offers several foundation courses leading to higher study at branches elsewhere. Current government expenditure on education in 1993/94 was projected at EC $37.1m., equivalent to 12.8% of total budgetary expenditure.

# Bibliography

For works on the Caribbean generally, see Select Bibliography (Books)

*Antigua and Barbuda Foreign Policy and Government Guide*, 2nd edn. USA International Business Publications, 2000.

Coram, R. *Caribbean Time Bomb: The United States' Complicity in the Corruption of Antigua.* New York, NY, William Morrow & Co, 1993.

Dyde, B. *The Unsuspected Isle: A History of Antigua.* Oxford, Macmillan Caribbean, 2003.

Lazarus-Black, M. *Legitimate Acts and Illegal Encounters: Law and Society in Antigua and Barbuda.* Washington, DC, Smithsonian Institution Press, 1994.

Lowes, S., et al. *Antigua and Barbuda (World Bibliographical Series).* Santa Barbara, CA, ABC-CLIO, 1995.

Nicholson, D. V. *The Story of the Arawaks in Antigua and Barbuda.* Antigua Archaeological Society, Fort Collins, CO, Linden Press.

# ARGENTINA

## Geography

### PHYSICAL FEATURES

The Argentine Republic is the second largest country in Latin America, after Brazil, and occupies the broad territories east of the Andes in the tapering southern half of South America. Along the Andes, from north to south, Argentina and Chile share the second longest land border in the Americas, at 5,150 km (3,198 miles) and it continues from north to south across the Isla Grande de Tierra del Fuego (the south-western city of Ushuaia is the most southerly city in the world). To the north is Bolivia (beyond an 832 km border) and to the north-east Paraguay (1,880 km). Also in the north-east, thrust up between south-eastern Paraguay and southern Brazil, Argentina extends an arm along the Mesopotamia between the Paraná and Uruguay rivers; Brazil lies to the north, east and south-east of this region (the border is 1,224 km in length). The border with Uruguay (579 km) continues along the River Uruguay to the sea—Uruguay lies east across the river, but also to the north-east where the Argentinean capital, Buenos Aires, faces it across the great Río de la Plata (River Plate) estuary. In all, Argentina covers 2,780,403 sq km (1,073,519 sq miles), although it also claims a further 28,202 sq km of territory in the South Atlantic (the two British dependencies of the Falkland Islands and of South Georgia and the South Sandwich Islands) and in Antarctica (where its claims partly overlap those of Chile and, again, of the United Kingdom).

Argentina lies between the converging lines drawn by the Continental Divide and the eastern coast of South America, but also includes the eastern part of the main island of Tierra del Fuego, the tip of the eastward-curling tail of the continent, which ends in a broken mass of islands (mainly held by Chile). Argentina includes the Islas de los Estados just to the east (further east, and a little north, are the Falklands, which Argentina claims as the Islas Malvinas). The country's coastline, bulging in the north-east, then bitten by broad gulfs or bays southwards, is 4,989 km in length. Northern Argentina does not have a coast, but thrusts into the centre of the continent, extending the country's maximum length to some 3,330 km (it is 1,384 km at its widest), making north and south very different prospects, depending on latitude. Moreover, the terrain not only varies between the tropical rainforest of the north through fertile plains to the bleak landscape of Patagonia, but in the essential contrast between the plains and the high Andes. Indeed, Argentina contains both the highest and the lowest places in South America: Cerro Aconcagua (6,960 m or 22,843 ft), the highest mountain in the world outside the great ranges in the middle of Asia, is west of Mendoza; and the Salinas Chicas (–40 m) is on the Valdés peninsula in northern Patagonia. The country can be divided into the Andean highlands, the northern lowlands, the central plains of the Pampas (*pampa*) and the windswept Patagonian steppe.

The lower Patagonian Andes in the south (seldom exceeding 3,600 m) mount northwards into the main Andean cordillera, where some peaks rise above 6,400 m. Parallel ranges and spurs extend the mountainous terrain, which is generally inhospitable, except in some of the broader valleys, deep into north-western Argentina. Just south of here, in central Argentina, is the only other highland of significance, the Sierra de Córdoba (less than 3,000 m). The plains beneath these heights consist of the southern, Argentine, part of the Gran Chaco in the north, the Pampas stretching south for about 1,600 km and Patagonia in the narrower south. From the far north to the Colorado many rivers disappear into sinks or empty into marshes, and the northern and central plains are dotted with lakes and swampy wetlands. In the north and north-east subtropical and tropical conditions vary the landscape with rainforest, notably the tannin-rich quebracho trees (which thrive on the peculiarly saline soil of the Chaco), but generally grasslands dominate the vast, sometimes gently undulating, but largely treeless, plains.

Forests cover only 19% of Argentina, while pasturelands cover 52%. Vegetation also includes the pine forests of the Andes and of Tierra del Fuego, the hardy shrubs and brambles of Patagonia and the cacti and thorny bushes of the arid, mountainous north-west—but grasslands dominate. The Pampas proper, the main region of fertile farmland and rich grazing for livestock, falls gently from about 600 m at the base of the Andean system in the west to sea level. Patagonia, semi-arid and with a more tortured terrain, has sharp contrasts between heights of over 1,500 m and depressions deeper than 30 m below sea level. Its desolate plains end in a lake district of waters fed by glaciers and icecaps (Argentina has 30,200 sq km of inland waters in total). The main rivers of the south, the Negro and the Colorado, are further north, while the most important rivers in the north are those that drain into the Plate—the Paraguay, the Paraná and the Uruguay. The fauna native to this varied land ranges from, for instance, the parrots, jaguars, tapirs and monkeys of the north, through the rheas (American ostriches), armadillos, martens and deer of the Pampas, to the Andean condors, llamas, alpacas and guanacos.

### CLIMATE

Apart from a small tropical area in the north-east and the subtropical Chaco (the Tropic of Capricorn passes through northernmost Argentina), most of the country has a temperate climate, although it is arid in the north-west, and the dry south gradually tends to the subarctic. Most of the country is shadowed by the Andes and most rainfall is in the east. Altitude also affects the temperature. The lower Patagonian Andes leave the south exposed to the prevailing westerlies, which descend less humid from the mountains. The semi-arid Pampas ranges from cool to humid subtropical. The Pampas and the north-east can be subjected to violent windstorms, known as pamperos, while flooding and earthquakes can add to the natural hazards of the country. Average annual rainfall is at its height in the north (over 1,500 mm or some 60 ins), becoming drier to the west and south—Buenos Aires, on the north-east coast, but at a central latitude, receives 950 mm (37 ins). The average minimum and maximum temperatures in Buenos Aires range from winter's

8°C and 15°C (46°–50°F) in July to mid-summer 20°–30°C (67°–86°F) in January. To the west, in the foothills of the Andes, the extremes can be more pronounced, while in the far north there have been summer temperatures of 45°C (113°F) recorded and the average winter temperature in western (inland) Patagonia is 0°C (32°F).

## POPULATION

Most Argentines are of European origin (85%), especially of Spanish and Italian extraction (but also British, French, German and Russian, for instance). Although most of the rest of the population are Mestizos, of mixed descent, these are relatively few compared to other Latin American countries and mainly originate from elsewhere in South America. Only about 1% of the population is Amerindian. Other ethnic groups are also well represented, such as the 1m. or so of Arab descent, and the Jews that form one of the largest Jewish communities in the world outside Israel. This rich racial diversity does not compromise the predominance of the Roman Catholic Church (nominally 92% of the population—but with less than 20% practising), but it means that other Christian denominations (2%) and faiths (Jews 2%, others 4%—including Muslims or native traditions) are represented. The official language is Spanish, although some minorities also retain use of their own tongues for domestic purposes, and English, Italian, German and French are not uncommon. The three remaining Amerindian languages are Tehuelche, Guaraní and Quechua, most speakers of native tongues living in the north and west of the country.

According to official mid-year estimates, in 2004 the total population of Argentina was 38.2m., which was relatively small compared with the vast landscape they inhabited. Moreover, almost two-fifths of the total lived in the greater metropolitan area of Buenos Aires, the capital. In all, about 86% of the population is classed as urban. After Buenos Aires (the city proper had a population of almost 2.8m. at the time of the 2001 census), the chief cities of Argentina are Córdoba (1.3m.—northwest of Buenos Aires, its old rival, and mid-way to the Chilean border) and Rosario (0.9m.—upriver from Buenos Aires, on the Paraná). Argentina is a federal republic constituted of 23 provinces and one autonomous city (the Federal Capital, Buenos Aires).

# History
## Prof. PETER CALVERT

In colonial times Argentina lay on the furthest edges of the Spanish Empire. The first Spanish settlers in what is now Argentina came from Peru. Buenos Aires, though founded in 1580 and a fine natural port, stagnated while all trade had to be channelled through Lima and, until 1776 (when the Viceroyalty of the River Plate—Río de la Plata was established), the region remained backward and neglected. Direct trade between Buenos Aires and Spain, in hides and in silver from what is now Bolivia, stimulated the growth of the town. After 1776 it began to grow rapidly. With Spain itself in the hands of the French Emperor Napoleon Bonaparte's forces, the city freed itself from Spanish rule in 1810 and its Cabildo (town council) governed on behalf of the Spanish king, Ferdinand VII, even though he was, at the time, a captive of Napoleon. It was only in 1816, when the Spanish moved to recapture Buenos Aires, that the independence of the United Provinces of South America was declared at Tucumán.

Much of Argentine history in the 19th century focused on the problem of constitutional organization, owing to the rivalry between Buenos Aires and the other Provinces. The former favoured a centralized structure, with Buenos Aires dominant, while the latter wanted a federal structure with provincial autonomy. Buenos Aires derived its revenues from the port, while the prices of provincial manufactures, such as textiles, were undercut by cheaper imports. Buenos Aires did not share its wealth with the other Provinces and, moreover, grew more European in its outlook and amenities, while the Provinces remained backward, dominated by autocratic, often savage, rulers (caudillos). The struggle between federalists and centralists was halted after 1835 by the dictatorship of Juan Manuel de Rosas, caudillo of Buenos Aires. He ignored the national problem, giving the provincial rulers complete freedom of action in return for their recognition of him as national leader. Paradoxically, this helped create the national unity he opposed. In 1852 he was deposed by a coalition of his political opponents.

In 1853 a federal Constitution was created for the new Argentine Republic. Buenos Aires seceded from the federation, but in 1859 was defeated in a military confrontation with the other Provinces. Then, in 1861, it joined the union in order to dominate it, and, with its economic strength, finally triumphed in 1880, when the city of Buenos Aires replaced Rosario as the national capital. A new capital for the old Province of Buenos Aires was built at La Plata. The provincial caudillos made the transition to being more conventional politicians, though it was the great landowners, the estancieros, who dominated national life.

The next four decades were years of economic transformation as the combined impact of British investment, European immigration, the expansion of the railways and exploitation of corn and grain of the Pampas made Argentina by far the most advanced of the Latin American states. These developments initially benefited the landowners (cattle barons and commercial agriculturists) who dominated politics, but new classes of professionals (bankers, brokers and lawyers) and an urban working class emerged with the rapid growth of cities and began to challenge the hold of the ruling classes.

The first challenge came in the 1890s with the foundation of the Radical and Socialist parties, but fraudulent elections kept the oligarchy of landowners, merchants and bankers in power. In 1912, however, President Roque Sáenz Peña insisted on the adoption of a law introducing secret ballots, to reduce electoral corruption. As a result, in 1916 Hipólito Yrigoyen, nephew of the founder of the Radical party, the Unión Cívica Radical (UCR—Radical Civic Union), became Argentina's first popularly elected President and began a six-year term of office. The economy was still growing strongly and manufacturing industry developed under Yrigoyen's Radical successor, Marcelo T. de Alvear. The Radicals now dominated government and in 1928 Yrigoyen was elected to a second term, causing a split in the party. Before the effects of the 1929 depression were felt, Yrigoyen's reclusiveness made it possible for a small band of armed cadets to seize power in 1930, led by a retired general, José E. Uriburu.

From 1932 to 1943, a period known as the 'Infamous Decade', the oligarchy resumed power in the form of a loose coalition (the Concordancia) of Conservatives and anti-personalist Radicals, supported by the Armed Forces. It was their friends and supporters who benefited from the economic recovery. The Second World War divided Argentine society further. Some leaders were strongly pro-Allied, but others, not least in the Armed Forces, pro-Axis.

## THE PERONATO

The turning point in modern Argentine history came in 1943, when it seemed that the civilian politicians would install a pro-Allied president, and the Army intervened again. Col Juan Domingo Perón, secretary of the army lodge that planned and executed the coup, became Minister of War and Secretary for Labour and Social Welfare in the military Government. He promoted labour reforms and encouraged unionization, becoming immensely popular with the masses, though not with the oligarchy. In 1946, in a free election, he won the presidency decisively. In 1949 he amended the Constitution to permit his immediate re-election and held power until 1955.

The regime had many of the marks of a dictatorship—control of the media, suppression of dissent, interference in universities. However, it rested on an alliance with trade unions and the popular support of the urban underprivileged, the *descamisados* (shirtless ones). Large welfare programmes brought real benefits to the poor and were publicized by Perón's charismatic wife, Eva Duarte de Perón ('Evita'), who came to be regarded virtually as a saint.

A staunch nationalist, Perón bought out the British-owned railways and other public utilities, greatly accelerated industrialization under strong government control, and increased the role of the state in the economy. In his foreign policy he sought a 'Third Position', later to be termed non-alignment, and the leadership of South America. However, he neglected the agricultural sector, formerly the basis of Argentina's export trade. Rural migration increased and serious economic imbalances developed. As inflation rose and agricultural output fell, the economy's growth slowed down. Eva Perón died in July 1952, depriving her husband of his strongest ally with the masses. However, though the military had been alienated by her prominence, and some officers were already tired of Perón himself, support for the regime remained strong.

During his second term, from 1951, after surviving an attempted military coup, Perón tempered his policies. He resisted workers' wage demands, supported the farmers and, from 1954, encouraged foreign capital in the petroleum industry. Such changes alienated many former supporters. Discontent grew with both the repressive nature of the regime and with the large and over-zealous bureaucracy. Attacks by his supporters on the Roman Catholic Church compounded Perón's problems. Finally, in September 1955, the Armed Forces intervened and Perón went into exile in Spain. However, his legacy and his political movement survived, to form the fundamental divide in Argentine politics for the next three decades. The critical factor was the antagonism between the Armed Forces and the Peronists, the former trying for 18 years to exclude both Perón and his supporters from national politics.

## THE MILITARY INTERVENE, 1955–83

Between 1955 and 1983 Argentina's political history was very turbulent. The leader of the coup of 1955, Gen. Lonardi, who was prepared to work with the Peronists, was soon deposed by the more uncompromising Gen. Pedro Aramburu. For three years (1955–58) he attempted to suppress all vestiges of Peronism. Elections from which the Peronists were barred returned a left-wing Radical, Arturo Frondizi (1958–62), under whom economic development accelerated and inflation became an endemic problem. When Frondizi proposed to allow the Peronists (though not Perón himself) to stand for election, he was deposed by the Army. New elections were held in 1963 and Dr Arturo Illia, another Radical, was elected. However, in 1966 the military ousted him also, claiming that he had been ineffective. The new military Government headed by Juan Carlos Onganía (1966–70) produced a viable, but austere, recovery plan.

Meanwhile, under the influence of the Cuban Revolution and its aftermath, younger Peronists had adopted revolutionary tactics. Two main urban guerrilla movements emerged, a pro-Cuban organization called the Ejército Revolucionario del Pueblo (ERP—Revolutionary Army of the People) and a Peronist group known as the Montoneros. The Montoneros kidnapped and murdered former President Aramburu, thereby undermining Onganía, whose colleagues deposed him in June 1970. An unknown general, Roberto Levingston, was appointed to head the Government, but was replaced in March 1971 by Gen. Alejandro Lanusse, the organizer of the 1970 coup, who took over the presidency himself.

### The Peróns, 1973–76

Lanusse inherited an impossible situation. With guerrilla violence growing and with the economy suffering from such frequent political changes that long-term, consistent policies could not be implemented, he took the ultimate gamble of holding fresh elections in March 1973 and allowing the Peronists to take part for the first time in 20 years. Perón's candidacy was disallowed (he had now been in exile for 18 years), but his proxy, Héctor Cámpora, was allowed to stand, and was duly elected. However, Cámpora's presidency was short; having offered freedom to captured guerrillas, he resigned in order to force Perón's return. The military, internally divided, yielded to the demand for new elections. Perón's return to the country in June 1973 was marred by a gun-battle at Ezeiza Airport in which many died. In September, nevertheless, he was returned to the presidency, with his third wife, María Estela ('Isabelita') Martínez de Perón, as Vice-President. Inevitably, the conflicting hopes of his supporters were disappointed. Perón, now a sick man of 78, was unable to meet the many conflicting demands made on him during his short third term as President, though much was said and planned along the social democratic lines which he had seen working well in Europe. He tried to distance himself from the leftist fringe that had infiltrated the movement during the last years of military proscription, only to find that political violence originating from both the left- and right-wing increased.

At his death on 1 July 1974, his widow Isabelita assumed the presidency in a situation of increasing chaos, becoming Latin America's (and, indeed, the world's) first woman executive president. The Peronist movement was now not only divided, its extreme wings were, in fact, at war, and rightist 'death squads' appeared, controlled by Isabelita Perón's chief confidant and Minister of Social Welfare, José López Rega. The country slid towards anarchy; inflation rose to 364% in 1976, violence spread and the Government did little to stop it. In March 1976 the Armed Forces again seized power.

### The Military Junta, 1976–83

The governing junta of service chiefs chose the commander of the Army, Gen. Jorge Videla, as the new President. Under their leadership there began what the Government termed euphemistically the 'Process of National Reorganization'. The period has since become better known abroad as the 'dirty war', *la guerra sucia*.

The Process was a concerted attempt to eradicate terrorism by the use of terror. Tens of thousands of 'suspects' were arrested and, in many cases, tortured and murdered. There were reports that people were arrested simply to fulfil the quotas imposed on government agencies. The most conservative estimates put the number of people killed, or who 'disappeared', at between 10,000 and 15,000. Such a wholesale purge inevitably included some genuine terrorists and by 1978 the capacity for disruption by the Montoneros and the ERP had been drastically reduced by the death, exile or imprisonment of their known leaders.

Meanwhile, massive borrowing and the overvaluation of the currency led to a rapid growth in consumer spending, giving the middle classes a false sense of prosperity. Videla, who had retired from the Army and junta in August 1978, was succeeded as President, in a quasi-constitutional fashion, by Gen. Roberto Viola in March 1981, just at the start of economic recession. Ill-health and the opposition of military conservatives to his relatively conciliatory policies forced Viola out of office in November. A right-wing nationalist, Gen. Leopoldo Fortunato Galtieri, took over in December as President and head of the junta.

At the beginning of 1982 there was a series of labour demonstrations and strikes, culminating, on 30 March, in a violent confrontation between demonstrators and government forces in Buenos Aires. From the moment of his seizure of power, however, Galtieri had prepared a plan which, he believed, would guarantee his position. On 2 April Argentine forces seized the British-ruled Falkland Islands (Islas Malvinas) in the South Atlantic, title to which had been disputed by Argentina since British occupation in 1833. The tiny British garrison was rapidly overwhelmed by 10,000 Argentine soldiers and repossession of sovereignty was declared. Initially, the seizure was a resounding success for the Government—even the Peronists supported it. However, the final defeat of the Argentine forces by British troops on 14 June was a catastrophe and a national humiliation. The military Government of Galtieri, inept politically and economically, had failed in its own professional field. Galtieri was abruptly replaced, and a retired general, Reynaldo Bignone, selected to head an interim Government under cover of which the military could retreat from power.

# THE RETURN TO DEMOCRACY

Bignone, having established a dialogue with the political parties, called elections for 30 October 1983, in an atmosphere of deep economic crisis and national confusion, fuelled by growing civilian demands for the investigation of human rights abuses committed by the services. The UCR candidate, Raúl Alfonsín, who had courageously opposed the war and had a record of defence of human rights, obtained a massive victory over the Peronist Italo Luder, who lacked charisma and whose party was deeply divided.

## The Presidency of Raúl Alfonsín, 1983–89

Inaugurated as President in December 1983, Alfonsín faced massive problems. Of a foreign debt of US $40,000m. left by the military government, one-quarter at least had been wasted and another quarter was never traced. Meanwhile, 'hyperinflation' and economic stagnation combined to disappoint the hopes that Alfonsín's victory had aroused. During the first year of Alfonsín's presidency, after protracted negotiations, he managed to renegotiate the foreign debt, maintain political stability and reach some accord with the Peronists. Equally significantly, he finalized with Chile a long-standing dispute over possession of three islets in the Beagle Channel, the settlement of which had almost brought the two countries to war in 1978.

Two issues dominated Alfonsín's Government: relations with the military and the resuscitation of the economy. In theory committed to eradicating the military from Argentine politics and to establishing a working democratic system, Alfonsín soon had to temper his policies with reality. The Falklands/Malvinas débâcle provided the opportunity to restructure the military high command; Alfonsín removed anti-democratic senior officers and replaced them with more co-operative ones, and the defence budget was drastically reduced, for economic as well as for political reasons.

## The 'Dirty War' and Relations with the Military

Complicating the issue was the popular demand for the trial of service personnel for gross abuses of human rights during the 'dirty war'. Ostensibly an asset to Alfonsín, public pressure on this issue was, in fact, a two-edged weapon. National revulsion at evidence of military atrocities came up against the claims of the military that the 'dirty war' was a 'just war', and that a general amnesty should be enacted. Alfonsín attempted to resolve this conflict in December 1986, when the Congreso (Congress) approved a law, known popularly as the 'Punto Final' (Full Stop) Law, which established a 60-day deadline for new indictments. The Government had expected that only some 70 such cases might be presented. However, by March 1987, owing to the zeal of the civil courts, over 250 indictments had been accepted, and, for the first time, serving officers, as well as retired ones, were accused.

In April 1987 military rebellions broke out in Córdoba and later at the Campo de Mayo itself. Although there were popular demonstrations in support of democracy, concessions were made to the military, despite the ensuing controversy. A new army Chief of Staff was appointed and Alfonsín submitted to the Congreso legislative proposals, which came to be known as the 'Obediencia Debida' (Due Obedience) Law, whereby most military officers accused of human rights violations were to be absolved of their crimes on the grounds that they had simply obeyed orders. Out of some 370 members of the Armed Forces due to be tried for human rights offences, only between 30 and 50 were now left to face charges. Even this number was too many for some sectors of the military and there were further insurrections in January and December 1988. These were outwardly firmly suppressed by the Government, although again there were accusations that Alfonsín subsequently made concessions to the military which resembled some of the rebels' demands.

## Economic Crisis

The second major imponderable for Alfonsín was the Argentine economy. After pursuing gradualist policies to reactivate the economy in 1984 and the first half of 1985, the Government turned instead to the fashionable doctrine of the 'heterodox shock' in June 1985. This involved the introduction of austerity measures, under the 'Austral Plan', named after the new currency, which was intended to end inflation, then running at an estimated annual rate of 1,129%. The Government's action led to labour unrest, with the Peronist-dominated trade unions holding a series of one-day strikes. Hence, despite its initial promise, the Plan was effectively abandoned while the mid-term elections took place.

With the elections out of the way and inflation again soaring, in February 1987 a new economic programme, the second Austral Plan (the 'Australito') was implemented. Again the Government's attempts at achieving economic stabilization were unsuccessful; the deficit on the public-sector account worsened and inflation continued to spiral upwards. In early 1989 the World Bank, which until that time had, like most other multilateral lending agencies, supported Alfonsín, suspended all its financing in Argentina. Negotiations with international creditors were deferred indefinitely. Despite the efforts of two new finance ministers in as many months, no credible economic strategy emerged and control over inflation was abandoned while the elections of May 1989 were held.

## President Menem's First Term, 1989–95

Victory went to the Peronist candidate, Carlos Saúl Menem, the flamboyant former Governor of the inland Province of La Rioja. In his campaign Menem captured the popular vote and a Peronist majority in the Senado (Senate) with the promise of a 'production revolution' based on wage increases and significant aid to industry. Yet long before he was scheduled to take office, the seriousness of the economic crisis led Alfonsín to try to reach an agreement with the Peronists on economic strategy. Menem wisely declined this, and all attempts at accommodation between the incoming and outgoing Governments resulted in failure. Instead, food riots, looting and bombings in several Argentine cities forced Alfonsín to impose a state of siege and to hand over presidential power to Menem five months early, on 8 July, in order to avoid a total breakdown of public authority.

Menem's succession was the first time since 1928 that an elected President had handed over to his successor without military pressure. Democracy, even in the midst of the greatest economic crisis of the century, was immensely popular. It was immediately clear, moreover, that Menem's economic strategy, to turn Argentina into what he called a 'popular market-capitalist country', was much closer to that of the Radicals than the campaign had suggested. The intention was not to revolutionize the system, but to bring together a corporate capitalist coalition, containing businessmen, foreign creditors, the unions, the Armed Forces and the Church. This policy marked an immediate departure from the interventionist approach associated with early Peronism, though it was closer to the many pragmatist elements in Perón's thought and policies. Meanwhile, a widespread lack of confidence was reflected in the collapse of the currency and a second hyperinflationary wave.

Menem's relaxed use of presidential authority, divisions in the Cabinet and reluctant support for policies so different to traditional Peronism increased the difficulty in reducing state expenditure and in selling state-owned enterprises. With regard to the military, President Menem, who had spent the entire period of the military regime under house arrest, from the outset advocated conciliation, despite strong resistance from among his own supporters. After weeks of rumour a series of pardons affecting 277 individuals, some of them guerrillas, and including members of the military junta responsible for the Falklands débâcle, was issued in October 1989. In other ways President Menem was firmer; the leaders of the rebellions against Alfonsín were dismissed and, following a further rebellion on 3 December, any question of negotiation was speedily rejected, and the leaders charged with insurrection. However, Menem issued the irreversible decision on 29 December to pardon Gen. Jorge Videla and others convicted for human rights crimes during the 'dirty war'. At the same time the relative strength of the Armed Forces continued to decline. In February 1990 the President resumed diplomatic relations with the United Kingdom, while postponing discussion of the sovereignty of the disputed islands.

# TRANSFORMING THE ECONOMY

Meanwhile, in January 1991 Domingo Cavallo was appointed economy minister and proceeded to implement a far-reaching programme of economic stabilization. This had three main aims. The first was the so-called 'dollarization' of the economy, whereby the Argentine economy was linked to the US economy by the establishment of a new currency, the peso, worth 10,000 australes, at parity with the US dollar. The second was an ambitious programme of privatization, reversing 40 years of Peronist policy. The third aim was to improve government finances by raising revenue and eliminating tax evasion, by means such as the 'fiscal pact' with an agricultural development association, the Sociedad Rural Argentina (SRA—Argentina Rural Society). By the time of President Menem's state visit to the USA in November, he was celebrated as Latin America's leading free-market reformer and US ally. As a result, in March 1992 the President secured a promise of debt reduction under the Brady Initiative (a plan for debt relief originally proposed by Nicholas Brady, the US Secretary of the Treasury, in 1989).

The Peronists consolidated their control of the Cámara de Diputados (Chamber of Deputies) in partial legislative elections of 3 October 1993. In the same month they also gained senate approval for the reform proposal and announced their intention to conduct a national referendum on constitutional reform, which would include an amendment to the Constitution to permit Menem, like Perón before him, to be re-elected for a consecutive term. In December President Menem declared his intention to seek re-election in 1995.

Elections to a 305-seat constituent assembly, which was to draft and approve the proposed constitution, took place on 10 April 1994. The Peronists failed to gain the majority that would guarantee the passing of the constitutional amendments. On 22 August the assembly approved the new Constitution, which allowed for the possibility of presidential re-election for a second term. It came into force on the following day. Other constitutional amendments included: the reduction of the presidential mandate from six years to four; delegation of some presidential powers to a Chief of Cabinet; a run-off election for presidential and vice-presidential candidates when neither obtained 45% of the votes cast, or 40% when the nearest candidate gained less than 30% of the ballot; the establishment of an autonomous government in the city of Buenos Aires with a directly elected mayor; the extension of congressional sessions to nine months; an increase in the number of senators from each province from two to three; and the creation of a Council of Magistrates and other judicial reforms.

Presidential elections were scheduled for 14 May 1995. The UCR presidential candidate was Horacio Massaccesi, Governor of Río Negro. Menem also faced a strong challenge from José Octavio Bordón, of the Frente por un País en Solidaridad (Frepaso—the name chosen in December 1994 for the Frente Grande), which gained fresh support from dissident Peronists and some Radicals. In the event, however, the election was overshadowed by the impact on the economy of the Mexican economic crisis (the 'tequila effect'), caused by the dramatic fall in the value of the Mexican peso in December 1994. In March 1995 the Government undertook an economic consolidation programme, with the aid of US $7,000m. in international and domestic credits. Public spending was drastically reduced and the unemployment rate increased from 12.5% to 18.6% of the economically active population in April–July. An employment initiative costing $1,500m. was announced by President Menem in August, but there was still widespread social unrest in major cities throughout the year.

In spite of the worsening economic situation and the Government's continuing austerity programme, in May 1995 President Menem became only the third Argentine president to win re-election, gaining 49.9% of the votes cast in the first round of voting and thereby avoiding a second ballot. At concurrent provincial and legislative elections the ruling Partido Justicialista (PJ) won nine of the 14 gubernatorial elections and increased their representation in the enlarged 257-seat Cámara de Diputados to an overall majority. Frepaso, however, won the largest share of the 130 legislative seats contested, largely at the expense of the UCR.

### President Menem's Second Term, 1995–99

The second term, almost inevitably, was an anticlimax. Following his victory, President Menem announced the composition of a largely unaltered Cabinet, led by Eduardo Bauzá, in the new post of Chief of Cabinet. However, tensions within the new Government soon became evident. President Menem showed increasing irritation at the power of Domingo Cavallo, who had been reappointed to the economy ministry, and who complained publicly of resistance to reform among entrenched interests in government. President Menem established an Economic and Social Council, consisting of trade-union and business leaders, which was to assist in determining economic policy. The move was opposed by Cavallo, who accused the President of attempting to employ interventionist policies. He was dismissed in July and was replaced as economy minister by Roque Fernández, hitherto President of the Central Bank. Fernández, an ally of Cavallo, continued his predecessor's policies.

Nevertheless, the effects on the political situation were marked. In March 1997 Cavallo formed his own party, Acción por la República (AR—Action for the Republic), and in July he and Gustavo Béliz, the former Minister of the Interior and leader of the political grouping Nueva Dirigencia (ND—New Leadership), announced their electoral alliance. They immediately launched a mordant attack on alleged governmental corruption. Meanwhile, throughout 1997 the Government's popularity was further undermined by widespread and sometimes violent social and industrial unrest, caused by discontent with the Government's economic austerity measures and reports of corruption, and in the mid-term elections of October 1997 the PJ lost its overall majority in the Cámara de Diputados. The popularity of the Peronists remained low in 1998 as unemployment persisted and President Menem and Governor Duhalde battled publicly for control of their party. The President launched an unsuccessful campaign to amend the Constitution again, in order that he might stand for a third term, but, as expected, Governor Duhalde won the Peronist nomination.

At the same time, hopes intensified of holding senior military officers to account for their role in the 'dirty war'. In 1997 Judge Alfredo Bagnasco began an investigation into the alleged systematic theft of as many as 300 infants from jailed political opponents of the junta. Most of these children had then been adopted by military couples and the details of their true parentage remained obscure. Though many 'dirty war' criminals had been convicted and jailed for other crimes in 1985, and subsequently pardoned by President Menem, Judge Bagnasco argued that the pardons did not cover crimes against children, as these crimes were ongoing in effect. Former junta Presidents, Gens Videla and Bignone, were arrested in June 1998 and January 1999, respectively, and others summoned to give evidence included Gen. Galtieri and the former head of the Navy, Adm. Emilio Massera. Vice-Adm. Rubén Oscar Franco and Gen. Guillermo Suárez Mason were arrested in 1999, after former naval captain Jorge Acosta had surrendered voluntarily. Although, in January 2000, a federal judge refused their extradition, the arrests of Franco and Suárez Mason, as well as the discovery in Lomas de Zamora of some bodies of the 'disappeared', proved that the issue was still salient. In July 2002 Gen. Galtieri was arrested for the alleged kidnap, torture and murder of left-wing militants during the 'dirty war'. A further 44 senior retired police and military officers, including Gen. Suárez Mason, were charged in connection with the same offence.

### The Presidency of Fernando de la Rúa, 1999–2001

Elections to the presidency were held on 24 October 1999. Fernando de la Rúa of the UCR, the conservative, 62-year-old Mayor of the Federal District of Buenos Aires, was the candidate of the unified opposition Alianza para el Trabajo, la Justicia y la Educación (ATJE). His two main opponents were Eduardo Duhalde, whose nomination by the PJ was endorsed by Menem, and Domingo Cavallo of the AR. De la Rúa won a decisive victory, gaining 49% of the votes cast, compared to Duhalde, who secured 38% of the ballot, and Cavallo, who came third with 10% of the votes. Carlos Ruckauf, the outgoing Vice-President, was elected to succeed Duhalde as Governor of the Province of Buenos Aires, the nation's second most powerful political position, in concurrently held gubernatorial elections. The ATJE was also successful in partial elections to the Cámara de Diputados, also held on the same day, winning 63 of the 130 seats to be renewed, and thereby increasing the party's total number to 127, just two short of an absolute majority. In contrast, the Peronists won 50 seats, reducing their legislative representation to 101. Cavallo's AR secured nine seats.

The new President was sworn in on 10 December 1999, and immediately faced a serious economic crisis. He appointed

Rodolfo Terragno as his head of Cabinet and José Luis Machinea as Minister of the Economy. Later that month the new Congreso approved an austerity budget which reduced public expenditure by US $1,400m., as well as a major tax-reform programme and a federal revenue-sharing scheme. In April 2000 the Senado approved a major revision of employment law. The legislation met with public criticism and led to mass demonstrations by public-sector workers and, subsequently, to two 24-hour national strikes organized by the Confederación General del Trabajo (CGT). In May the ruling ATJE confirmed its position when its candidate, Aníbal Ibarra, was elected Mayor of the Federal District of Buenos Aires. However, later that year the Government came under intense pressure after it was alleged that some senators had received bribes from government officials to approve the controversial employment law. In September the Congreso voted to end the immunity that protected law-makers, judges and government ministers from prosecution, and thus allow an investigation into the corruption allegations. However, in October the Government was further weakened after the resignation of the Vice-President and leader of Frepaso, Carlos 'Chacho' Alvarez, who was angered by the President's refusal to replace ministers implicated in corruption allegations.

In September 2000 the Government proposed to reduce the fiscal deficit in 2001 from US $5,300m. to $4,100m., mainly by reducing government expenditure by $700m. The Alliance in the Congreso then declared an economic emergency and passed an anti-evasion law and a law to reduce corporate taxes. On 14 November, in a televised address, the President announced that the country faced 'a veritable catastrophe'. The IMF did agree to emergency aid, amounting to more than $7,000m., and an agreement to this effect was signed in January 2001; however, there were severe conditions attached. These included abolition of the existing state pension scheme and the 'freezing' of federal transfers to the provinces. Feeling that he did not have the necessary political backing for his austerity programme, on 2 March the Minister of the Economy, José Machinea, resigned. President de la Rúa replaced him with an orthodox economist, Ricardo López Murphy, who promised to maintain both the dollar 'peg' and convertibility, but when he announced further spending cuts of $2,000m. in 2001 and $2,500m. in 2002, two UCR ministers and a Frepaso minister resigned. In the ensuing cabinet reshuffle, Domingo Cavallo, who since 1996 had been isolated as leader of the right-wing AR, was reappointed Minister of the Economy.

In early June 2001 former President Carlos Menem was arrested for illegal arms-trafficking to Croatia and Ecuador during the 1990s. Although he strongly denied the allegations, he was held under house arrest for four months, but no evidence was brought against him. Although constitutionally ineligible for immediate re-election, he nevertheless contested the presidency in 2003.

## FINANCIAL CRISIS
### Five Presidents in 14 Days, 2001–02
The appointment of the 54-year-old Domingo Cavallo, who as Minister of the Economy in 1991–96 had carried through the successful dollarization of the economy, briefly stabilized the economic situation. However, although the dollar 'peg' was credited with producing a decade of financial stability in Argentina, it was at the price of economic growth. From 1999 it had become increasingly clear that continuing deflation and the overvaluation of the currency was leading to escalating levels of Argentine debt. However, successive Governments proved either unwilling or unable to adhere to their spending plans. Cavallo chose tax increases rather than spending cuts as the method to attempt to reduce the 2001 budget deficit by US $3,000m. In late March the Senado (by 50 votes to four) passed a 'Competitiveness Law' that delegated special powers to the Executive, permitting it to alter tax rates and undertake administrative reforms. There were three significant exceptions to these powers, however: the Executive was not allowed to dismiss public-sector workers, cut their salaries or reduce state retirement benefits. Confidence that the situation would improve began to ebb when, on 17 April, Cavallo sent a proposal to the Congreso to replace the dollar 'peg' for one based on both the US dollar and the euro. In May the Government successfully swapped $29,500m. in bonds, due to mature by 2006, for long-term bonds with maturity dates between 2006 and 2031. This postponed payment on the maturing debt, but at 'ruinous' (according to Cavallo) interest rates of 15.3%.

Following an agreement with 14 powerful provincial Governors, the Senado finally passed a 'zero deficit' law in late July 2001, which prohibited budgetary overspending by provincial governments. The legislation persuaded the IMF to allow the early disbursement of US $1,200m. (due in September) in August and to lend the Government a further $8,000m. This was in spite of the strong reservations of the new US Administration of George W. Bush and clear signs that the economic situation was rapidly deteriorating.

At the mid-term elections on 14 October 2001 the electorate demonstrated their anger with the Government by giving the opposition Peronists a majority in both houses of the Congreso. The end came for the UCR Government when, in response to a massive run on the banks on 30 November, the Government imposed limits both on cash withdrawals and on capital movements (known as the *corralito*), effectively ending convertibility on 1 December. The IMF destabilized the situation further when it withheld a US $1,260m. disbursement, which was due to be paid in December. On 15–16 December public order began to break down as desperate citizens looted supermarkets and crowds mounted a mass demonstration to demand the Government's resignation, banging empty pots in frustration as they marched on the presidential palace, the Casa Rosada, in the Plaza de Mayo. Heavy-handed police intervention claimed 27 lives and injured hundreds more. Finally, on 20 December President de la Rúa asked the Peronists to join a government of national unity. They prudently refused. De la Rúa resigned the same day and escaped from the roof of the Casa Rosada by helicopter, while Cavallo, in fear for his life, fled to Patagonia. (In April 2002 Cavallo was charged with involvement in illegal arms shipments to Croatia and Ecuador during the 1990s, but he was released from gaol after a federal court ruled there was insufficient evidence to support the charges.)

Ramón Puerta, the Peronist President of the Senado, was immediately appointed interim President while the Congreso met to choose someone to hold office until fresh presidential elections could be held in March 2002. Puerta's first, and indeed only, act as President was to ratify the state of emergency imposed by his predecessor. After a 15-hour session, the Congreso elected Adolfo Rodríguez Sáa as President. Sáa was a charismatic Peronist who had served as Governor of San Luis since 1983. He reaffirmed that he would maintain both convertibility and the dollar 'peg'. However, the empty-pots protests (*cacerolazos*) resumed at the news that the affluent had spirited some US $20,000m. out of the country while the actual value of the peso was rapidly depreciating. The new Minister of the Economy, Jorge Capitanich, proposed the introduction of a third currency, the argentino, to run parallel with the US dollar and the peso. This was immediately interpreted as a form of disguised devaluation. As further rioting broke out, Rodríguez Sáa summoned a conference of provincial Governors to demonstrate support for the new Government. However, only five Governors attended, and on 30 December 2001, Rodríguez Sáa resigned the presidency. The following day, pleading ill health, Ramón Puerta yielded the presidency to Eduardo Camaño, the Peronist President of the Cámara de Diputados. Finally, on 1 January 2002, the Congreso appointed Eduardo Duhalde, the Peronist candidate in the 1999 presidential election, to serve out the remaining part of de la Rúa's term, the fifth man to wear the presidential sash in a fortnight.

### The Presidency of Eduardo Duhalde, 2002–03
In early 2002 the Minister of the Economy in the new Duhalde Government, Jorge Remes Lenicov, wrestled unsuccessfully with the task of reconciling internal political demands with the requirements of the IMF. On 3 January Argentina officially defaulted on its loan repayments, reportedly the largest ever sovereign debt default, and on 6 January the Senado authorized the Government to set the exchange rate, thus officially ending the 10-year-old parity between the US dollar and the peso. In early February the Supreme Court ruled (by a margin of six to three) that the restrictions imposed on bank withdrawals (the

*corralito*) were unconstitutional. Some accounts were freed from the restrictions but, in order to forestall the complete collapse of the financial system, the Government announced, by decree, a six-month ban on legal challenges to the remainder of the bank withdrawal regime. Numerous bank holidays were also decreed to prevent another run on the banks and a further devaluation of the currency. A constitutional crisis ensued as the Congreso initiated impeachment proceedings against the unpopular Supreme Court Justices. Later that month the Government signed a new tax-sharing pact with the provincial Governors, linking the monthly amount distributed to the provinces to tax collections, as recommended by the IMF. However, in late April Lenicov resigned following an uncompromising message from the IMF that no further help would be available unless drastic reforms were forthcoming and the Senado's refusal to support an emergency plan to exchange 'frozen' bank deposits for government bonds. He was replaced by Roberto Lavagna (the sixth Argentine economy minister in 12 months).

Nevertheless, public anger against the Government and at the state of the economy did not subside. In late June 2002 thousands of anti-Government protesters marched in front of the Congreso while teachers and public-sector workers went on strike. On 3 July President Duhalde announced that a presidential election would take place six months earlier than planned, on 30 March 2003, and that a new administration would take office on 25 May. In November, following negotiations between representatives of provincial and federal government, it was announced that a presidential election would be held earlier than planned on 27 April 2003, with a presidential run-off, if necessary, on 18 May. It was hoped that this move would reduce political pressures on the Government during its ongoing negotiations with the IMF.

Meanwhile, a series of court rulings challenging the compulsory conversion of US dollars into pesos, at various unequal rates, increased the pressure on the Government in mid-2000. The compensation ordered left the banking system in debt to some US $20,000m. and on the verge of collapse. On 26 July a federal appeals court overrode a decree banning legal challenges to the *corralito* and in mid-August another appeals court upheld a judgment that the *corralito* was unconstitutional. This decision was relaxed somewhat on 30 September, but, to the fury of depositors, withdrawals could still only be made in pesos. Meanwhile, on 22 August the Supreme Court ruled that the 13% pay cut imposed on public-sector workers by the UCR administration in July 2001 was also unconstitutional, a decision that would cost the Government an additional $1,400m. On 1 October an appeals court in Buenos Aires effectively suspended proposed price increases for public utilities. Efforts to persuade the IMF to allow it to defer all payments to the Fund until the end of 2003 were unsuccessful, and in November the Government, seeking to pressurize the Congreso to approve much-needed legislation, deliberately failed to meet its next payment.

Delinking the peso from the US dollar proved disastrous for the millions of Argentines who had lost money, and particularly for the professional middle class. Unexpectedly, though, inflation remained relatively subdued, and although it reached 40% by the end of 2002, hyperinflation did not ensue. Much of the credit for this was due to economy minister Lavagna, who successfully reined in government expenditure, generating a month-on-month fiscal surplus. None the less, violence throughout the country again increased and became endemic. In greater Buenos Aires there were on average seven murders per day in 2002, compared with four per day in 2001.

## THE PRESIDENCY OF NÉSTOR KIRCHNER

In January 2003, following an entire year of intermittent negotiations between the interim Duhalde Government and the IMF, a provisional agreement on debt-restructuring was finally reached. This would cover the period to the end of August, by which time the new Government would have taken office. In the first months of the year an economic recovery, which appeared to have begun in November 2002, continued, and by election day, 27 April 2003, the peso had risen from its low of 3.6 to the US dollar to 2.8. However, with more than 60% of the population unofficially below the poverty line, violence remained endemic amid a general sense of hopelessness throughout the country.

Meanwhile, IMF delegations held negotiations with the prospective presidential candidates, while belated informal talks began with the country's private creditors, over the more than US $55,000m. held by them, which there was no obvious way of repaying.

Infighting among the ruling Peronists was so intense that for the first time the party failed to unite behind a single candidate for the 2003 presidential election. No primaries were held, and in January 2003 the PJ congress agreed to allow all three Peronist aspirants to contest the ballot. Meanwhile, the UCR was so discredited that two of its members chose to run as independents. In the months preceding the presidential election 72-year-old former President Carlos Saúl Menem (representing the Frente por la Lealtad/Unión del Centro Democrático faction of the PJ) held a narrow lead in the opinion polls. Despite the widespread feeling that it was his Government that had brought about the economic catastrophe, Menem laid all the blame on his successors, and gained a good deal of support from those who remembered the prosperity they had enjoyed under his rule.

Menem attracted the largest share of the first-round presidential ballot, on 27 April 2003, with 24.34% of the votes cast, followed by Néstor Kirchner (representing the Frente para la Victoria faction of the PJ), with 21.99% of the ballot. Kirchner, whose 12 years as Governor of the sparsely populated Province of Santa Cruz had been marked by exceptional probity and sound financial management, was the favoured candidate of President Duhalde. Ricardo López Murphy of the electoral alliance Movimiento Federal para Recrear el Crecimiento, a former UCR Minister of Defence and (briefly) the economy minister under President de la Rúa, came third, with 16.35% of the ballot. The third Peronist candidate, former President Adolfo Rodríguez Sáa, of the Frente del Movimiento Popular/Partido Unión y Libertad, attracted 14.12% of the votes. He was narrowly beaten by Elisa Carrió of the Alternativa por una República de Iguales, who won 14.15% of the votes. As no candidate had won a clear majority of the votes, a run-off ballot between the two leading candidates was to be held on 18 May. However, when four days beforehand it became clear that Kirchner was set to win some 63% of the second-round votes, Menem withdrew his candidacy.

Although thus denied the legitimacy of a massive popular vote in his favour, Kirchner was sworn in as the new President on 25 May 2003. He retained Roberto Lavagna as Minister of the Economy and granted the Central Bank autonomy in fixing rates. By the end of 2003 the economy had started to recover from its poor performance between 1998 and 2002. Unrest continued, however, both in protest against the steep rise in the rate of unemployment and against the Government's autocratic style and apparent lack of interest in the social consequences of its policies. With more than one-half of the population below the poverty line, the crime level had risen significantly. However, in September 2003 the IMF finally approved a loan of US $12,500m. over three years, and talks began on the rescheduling of some $90,000m. of foreign debt. Meanwhile, in a popular move, the Senado approved legislation that would allow the annulment of the two amnesty laws passed in the 1980s, which had hitherto protected military officers accused of human rights offences. (In mid-June 2005 the Supreme Court ruled, by seven votes to one with one abstention, that the two laws were unconstitutional, thereby opening the way for an estimated 500–1,000 further prosecutions for human rights abuses committed under military rule.)

In legislative and gubernatorial elections, held between August and November 2003, the ruling Peronists performed strongly, winning 127 of the 257 seats in the Cámara de Diputados and 41 seats in the 72-seat Senado. Their nearest rivals, the UCR, won only 46 and 14 seats, respectively, and five of the 22 governorships contested, while the Frente Grande retained the governorship of the Federal Capital. In fact, the multiplicity of small parties and the large number of independent candidates made it very difficult to obtain a majority coalition in the Cámara. By February 2004, therefore, when Lavagna again formally met representatives of the IMF, there was already concern over whether the Government would be able to fulfil its promises of fiscal reform.

At the first anniversary of his presidency, in May 2004, President Kirchner himself remained very popular, but his

Government less so, and the press was uniformly gloomy about the country's future. However, Kirchner's own popularity continued to increase as his Government continued its policy of aggressively seeking more favourable terms to the repayment of the national debt. The acceptance of these terms in February 2005 (see Economy) enabled him further to strengthen his position by carefully considered popular gestures, such as calling for a boycott of the petroleum company Royal Dutch/Shell after a fuel price increase and dismissing publicly the further demands of the IMF. 'There is life after the IMF, and it's a very good life,' Kirchner said in Munich, Germany, in April, on the last day of his state visit to that country.

## FOREIGN POLICY

In the 1990s Argentina emerged as a close ally of the USA and an active participant in multinational peace-keeping operations, signalling the country's intention to adopt a more prominent role in international affairs, despite economic constraints that forced reductions in the military budget and the strength of the Armed Forces. During an official visit in 1997 US President Bill Clinton (1993–2001) commended Argentina's participation in more than 12 UN missions of the preceding decade, including those in Bosnia and Herzegovina, Cyprus and Haiti, and announced that he would seek non-NATO (North Atlantic Treaty Organization) ally status for the country. This would allow Argentina access to certain military funding and to a wider range of surplus US and NATO weaponry. The announcement drew protests from neighbours, particularly Brazil and Chile, who claimed that it could lead to a regional imbalance. By mid-2005 relations with the USA remained good, with Kirchner praising the US Administration's 'positive and prudent' attitude towards Argentina. The US Government, meanwhile, did not obstruct Argentina in its negotiations to restructure its external debt, and publicly expressed its appreciation of Argentina's moderating influence on the left-wing Government of President Lt-Col (retd) Hugo Chávez Frías in Venezuela.

Although Argentina restored full diplomatic relations with the United Kingdom in February 1990, and agreements were subsequently concluded on the protection of fish stocks and the reduction of military restrictions in the South Atlantic region, the question of the Falkland Islands' disputed sovereignty was not resolved. The new Constitution of August 1994 reiterated Argentina's claim to sovereignty over the Islands. Nevertheless, discussion of the issue was suspended while tense, long-running negotiations on lucrative fishing rights in the region continued. In September 1995 Argentina and the United Kingdom signed a comprehensive agreement on joint oil exploration in the region. In January 1997 President Menem reasserted both Argentina's claim to the Malvinas and his policy of seeking a peaceful resolution to the dispute. Meanwhile, relations between the two countries continued to improve and this *rapprochement* was demonstrated by an official visit by Menem to the United Kingdom in October 1998. Two months later the United Kingdom announced the relaxation of its arms embargo against Argentina. In July 1999 Argentina and the United Kingdom agreed to end the ban on Argentine citizens visiting the Falkland Islands and to re-establish direct flights there from October. In August 2001 Tony Blair became the first serving British Prime Minister to visit Argentina. The sovereignty issue, however, was not discussed and President Duhalde confirmed his intention to maintain existing policy towards the Islands, to seek their 'recovery', but solely by peaceful means, shortly after his appointment in January 2002.

President Néstor Kirchner followed suit at his inauguration in July 2003, and raised the Falklands question that same month during a meeting in London with Tony Blair. Thereafter Argentina resumed its traditional policy of exploiting international meetings and mobilizing support from third parties for bilateral talks on sovereignty with the United Kingdom. In May 2005 the inclusion of the Falkland Islands as an Overseas Territory of the United Kingdom in the draft European Union (EU) constitution provoked vociferous complaints from the Kirchner Government.

Following treaties, signed in 1985 and 1991, respectively, which resolved sovereignty disputes involving the Beagle Channel islands and territory in the Antarctic region, relations with Chile were further stabilized in 1996 when an Arbitration Court settled the last remaining mainland boundary question (the 'southern glaciers' issue). In August 1998 the two countries held joint naval exercises for the first time and in December the two countries' heads of state signed a further agreement on the Southern Glaciers question, which was ratified by the Argentine Senado in June 1999, thus ending a century of border disputes in the Southern Andes. Relations between the two countries were strained in mid-2004, however, following President Kirchner's decision to reduce exports of gas to Chile by some 25% in order to meet a critical shortfall in stocks for domestic consumption. The Chilean Government claimed that this violated an agreement signed in 1995. In September tensions between the two countries were exacerbated following the appointment of Ignacio Walker Prieto as the Chilean Minister of Foreign Affairs. In a newspaper editorial prior to his appointment Walker had criticized both President Kirchner for opportunism and Peronism in general for displaying 'authoritarian' and 'fascistic' traits.

Relations with Brazil, traditionally Argentina's largest trading partner, remained broadly positive in mid-2005, owing principally to ideological affinities between the countries' Governments. In May 2005 President Kirchner attended talks in Brasília with the Presidents of Brazil and Venezuela aimed at increasing economic, infrastructural, social and energy co-operation in the region. In the same month the Brazilian Government expressed its support for Argentina's protest against the inclusion of the Falkland Islands as a British Overseas Territory in the EU's draft constitution. However, co-operation between Argentina and Brazil has continually been hindered by Argentine concerns at the perceived desire of Brazil to assume a role of regional leadership. In October 2004 the Argentine Government reiterated its deep misgivings at Brazil's attempts to gain a permanent seat on the UN Security Council. Also, restrictions on Brazilian imports were imposed from mid-2004 by the Kirchner Government in response to what it considered to be disadvantageous imbalances in bilateral trade that stifled Argentina's industrial development. Brazil disputed the imposition of trade restrictions, claiming they obstructed the economic integration promoted by the regional trade bloc, the Mercado Común del Sur (Mercosur, Southern Common Market), of which Argentina is a member.

# Economy

## Prof. PETER CALVERT

In 2005 Argentina was on the road to recovery. A second year of strong economic growth in 2004 allowed for the successful renegotiation of much of the country's foreign debt in early 2005, although the purchasing power of wages and salaries was still some 15% below the level of 2001. The sudden collapse of the economy in that year and the consequent default in January 2002 had been the first major economic crisis of the 21st century, with important implications for Latin America generally and for the wider world economic system. Previously, Argentina had been seen as a model for its 'free-market' reforms, its social security and education and its relative affluence. Subsequently, things were much more uncertain. At the height of the crisis popular anger was diverted from the Government to the banks closed by government decree in order to avert a total collapse of the system. In 2004 the British bank Lloyds TSB, which had operated in Argentina for more than 140 years, was chief among those deciding to end operations in the country.

There were two main causes for the crisis. The 'Convertibility Plan', introduced in 1991, had 'pegged' the peso to the US dollar at par, thereby ending inflation virtually overnight. This move was so successful, however, that even after recession began in 1997, governments were reluctant to contemplate any adjustment. At the same time, falling revenues and the inability of provincial governments to control expenditure created a widening fiscal gap. This could only be filled by increasing borrowing. At the end of 2001 the Government was bankrupt, some 28.5% of the population were unemployed and an estimated 37% were living in poverty. This led to an acute social and political crisis, and to a new Government, which defaulted on its international debt obligations at the beginning of 2002. While there were those within the international multilateral lending agencies who held that no help should be given to Argentina, as an example to other countries, more cautious counsels prevailed, and early in 2003 a temporary IMF agreement was concluded (though agreement with the country's overseas creditors was not reached until February 2005—see below). By that time, the economy had already stabilized, although at a lower level than in the late 1990s. Growth resumed, while the rates of both inflation and unemployment continued to fall.

With a land area of 2,780,403 sq km (excluding areas of Antarctica and the Falkland Islands, claimed by Argentina), Argentina is the eighth largest country in the world by area. However, though in population it ranks fourth in Latin America, in the world it ranks only 29th, with an estimated 39.5m. inhabitants in early 2005. Population density was low, at 13.7 per sq km in mid-2004, yet over one-third of the population lived in Greater Buenos Aires (population some 13.8m. in 2001) and 86% of the country was regarded as urban. The country's average annual rate of population growth in 2005 was 0.98%. The labour force had increased by an average of 2.2% in 1997–2003 and 97.0% of the adult population was literate in 2002. In 2005 the crude birth rate was estimated at 16.9 per 1,000, the crude death rate at 7.6 per 1,000 and infant mortality at 16.2 per 1,000. Expectation of life at birth was 75.9 years (80.0 for females and 72.2 for males). According to World Bank figures, Argentina had a gross national income (GNI) of US $122,310m., measured at 2001–03 prices, in 2003. Gross domestic product (GDP) per head was $3,650 in that year (or $10,920 on an international purchasing-power parity basis), ranking Argentina as an upper middle-income country. On the 2003 Human Development Index of the UN Development Programme (UNDP), Argentina ranked 34th in the world.

Geographically, Argentina comprises four very diverse regions. The Andean area adjoins the entire western boundary with Chile, and with Bolivia in the north-west. In the north the forested flat lands of the Chaco and the flood plains of Mesopotamia (between the Paraná and Uruguay rivers) border Paraguay and Brazil. The Pampas (*pampa*), the vast plain which is the heart of the country, stretches from Buenos Aires on the Atlantic to the Andes and, north to south, from the Chaco to the Río (River) Colorado. This is the heartland of the country, the location of both its rich agricultural resources and its industry. To the south of the Río Colorado lies the fourth region, Patagonia, characterized by windswept plateaux pastured by sheep. The climate varies from subtropical in the north to subantarctic in the far south. Ushuaia, in Tierra del Fuego, is the most southerly city in the world. Argentina also claims as its national territory a substantial sector of Antarctica and several groups of islands in the South Atlantic (notably the Falkland Islands, or Islas Malvinas as they are known in Argentina, the subject of dispute with the United Kingdom—see below).

## HISTORY

The modern economic history of Argentina began with unification of most of the country under constitutional rule in 1853. In 1860 Argentina had a population of some 2m. and no more than 80 km of railway track. However, stable government, good communications and good economic opportunities attracted foreign migrants, capital and enterprise. Between 1857 and 1930 more than 6m. immigrants, mostly from Italy and other Southern European countries, entered Argentina. By 1914 the railway network, much of it built by British capital and engineering but much of it also by the state, had grown to 32,000 km, centred on Buenos Aires and Rosario. The key to this growth lay in the exploitation of the rich soil of the Pampas, as a small landowning élite and their employees, with improved breeds of cattle and sheep, turned Argentina into the world's greatest grazing ground and one of its principal granaries. The exploitation of the refrigerator ship from 1876 made possible an export trade in chilled and frozen meat. By 1914 Argentina was responsible for one-half of the world's beef exports, one-sixth of its mutton exports and one-10th of its wool exports, as well as providing vast exports of wheat, maize and linseed to growing consumer markets in Europe and North America.

The Federal Capital, Buenos Aires, mirrored that growth. Its population grew fivefold between 1880 and 1914, to 1.5m. As well as being the hub of the railway network, it was the major port for burgeoning exports and increasing imports of manufactured goods. Foreign capital, skills and personnel were crucial in transforming country and capital alike, with British enterprise predominant.

After the world depression that followed the First World War, the economy resumed its growth, with manufacturing industry developing from the 1920s. However, the financial crisis of 1929 severely damaged the economy of Argentina. Recovery in the 1930s was based on import substitution for manufactured goods, and by the 1940s the contribution of industry to GDP was higher than that of the agrarian sector, which had benefited from Argentina's neutrality in the Second World War. With the advent to power of Col Juan Domingo Perón in 1946, the trend towards industrialization was accentuated and the growth of a powerful trade union movement under his leadership allowed the masses a larger share of the national wealth. However, later Governments found this labour movement unmanageable, while the neglect of agriculture (the traditional basis of export earnings) diminished its competitiveness in world markets, though for a long time this fact was concealed by the enormous fertility of the Pampas.

The military Governments of 1976–83 adopted free-market policies. In theory they were seeking to introduce domestic efficiency by allowing liberal external competition, encouraging foreign investment and reducing the state's role in economic management. In practice, however, the overvaluation of the currency created an illusion of prosperity among their middle-class supporters (the *plata dulce*), encouraged imports and accelerated inflation. Furthermore, military authoritarianism actually extended the role of the state and, with it, the powers and numbers of the bureaucracy. Foreign borrowing covered the huge budget deficits, so that debt servicing grew enormously. With the return to civilian government in 1983, successive finance ministers struggled to overcome the legacy of their predecessors but without notable success until the early 1990s, when Argentina's GDP began to show respectable annual growth, after almost two decades of stagnation or decline. It was only towards the end of the century that the structural weaknesses that had plagued the economy for 50 years began to reassert themselves.

## AGRICULTURE, FORESTRY AND FISHING

It was the Pampas that created modern Argentina, and its products—agricultural and pastoral—remain vital to the economy. It comprises only one-fifth of the total national territory, but contains two-thirds of the population, including that of the capital, Buenos Aires. In 2002, of Argentina's 278m. ha, 33.7m. were arable, 1.3m. under permanent crops and 142m. permanent pasture. According to the 2001 census, agriculture (including forestry and fishing) employed only 910,996 people, some 8.3% of the active work-force, and accounted for just an estimated 4.5% of GDP in the same year, though by 2004 it was estimated to have risen again to 10.6%. The main cash crops are wheat, maize, sorghum and soya beans. The country is completely self-sufficient in basic foodstuffs, and the majority of most agricultural products are available for export. Food and live animals accounted for some 23.7% of exports by value in

2001. During 1990–2003 agricultural GDP increased at an average annual rate of 2.8%.

The 2004 cereal crop totalled only 32.5m. metric tons, a further slight fall from the previous year, and well down on 1996/97, when it had reached a record 52m. tons. According to the UN's Food and Agriculture Organization (FAO), in 2004 Argentina produced 14.5m. tons of wheat, 15.0m. tons of maize, 2.7m. tons of sorghum, 34.8m. tons of soybeans and 7.9m. tons of oilcrops—all (with the exception of sorghum) an improvement on the previous year. In 2002 exports of wheat and flour totalled 11.5m. tons, valued at US $1,132.1m. Exports of soybeans totalled 6.2m. tons in the same year, valued at $1,118.8m.

Other crops and fruits include sugar cane, rice, linseed, potatoes, tomatoes, cotton, tea and grapes. In 2003 an estimated 19.3m. metric tons of sugar cane was harvested, the same level as the previous year. Most of Argentina's internationally known wines are grown in the Province of Mendoza; some from the valley of Cafayate, south of Salta. Although only about 2% of the average production of over 22m. hectolitres (hl) per year was exported in the early 1990s, quality increased sharply thereafter and successful efforts were made to expand traditional export markets and to find new ones, with such success that in 2004 the value of wine exports totalled US $168m.

Although Argentina is no longer the world's principal exporter of beef, livestock production and meat sales abroad remained highly significant in both the internal and external economy. The cattle population was an unofficial 51.0m. in 2003. In 2000 meat and meat preparations exports were valued at US $790m. However, in 2001, owing to an outbreak of foot-and-mouth disease, this figure had fallen to an estimated $365m. Despite improved internal controls, the USA and Japan still imposed a ban on imports of Argentine fresh and frozen meat. However, in 2002 exports of meat and meat preparations rose again, to $576m., and reached $735m. in 2004. Domestic consumption accounted for as much as 90% of total beef production.

The sheep population, which fell sharply during the 20th century—from about 66m. head in 1900 to only 12.5m. in 2003—still accounted for a significant proportion of economic activity in Patagonia. Wool (greasy) output in 2003 was an estimated 72,000 metric tons. Pig and poultry products are not significant export items, although the home markets have in the past expanded with the rising cost of living, and goat meat is an essential element of the Argentine asado (barbecue). In 2003 FAO recorded an estimated 4.3m. pigs, 4.2m. goats and 110.7m. chickens. An estimated 3.7m. horses attested to both their value for transport in the more remote areas where paved roads are unknown and to Argentina's world-wide pre-eminence on the polo field.

Forestry was comparatively neglected. By the early 1990s woodland was being lost at an average rate of about 0.2% per year. Although forests covered 18.6% of the land area in 1995, according to FAO, later statistics are unreliable. Total roundwood production in 2003 amounted to 9.3m. cu m, most of which was consumed by the domestic market. Fishing, too, was underexploited, although offshore resources were large and the vessels of many other nations were active in the South Atlantic. However, by the beginning of the 21st century the value of fishing exports had risen considerably, from US $86.2m. in 1985, to an estimated $868m. by 2003, in which year the total catch was estimated at 916,200 metric tons.

## MINING AND POWER

The chief metallic ore used is iron, with 5.1m. metric tons of crude steel being produced in 2004. Aluminium production stood at 272,048 tons in 2004 and zinc at 37,807 tons. In 2003 mining output stood at 12,100 tons of lead, 199,000 tons of copper, 133,917 kg of silver and 29,744 kg of gold. Uranium reserves were estimated at some 25,000 tons in the early 1990s but the ore was of low quality and production ceased in 2000. In 1997 the Bajo de la Alumbrera open-pit gold and copper mine, the country's largest, became fully operational, increasing substantially total mineral output.

Argentina was almost fully self-sufficient in energy and likely to continue to be so. According to the World Bank, in 2002 it consumed 1,543.2 kg of oil equivalent per head. In that year natural gas accounted for 50.8% of total primary energy consumption. The major oilfield, Golfo San Jorge, discovered in 1907, is at Comodoro Rivadavia in Patagonia. It provides over 30% of national output. The other major onshore fields are Neuquén-Río/Negro-La Pampa, with 28% of total production, and the Cuyana field near Mendoza, in the foothills of the Andes west of Buenos Aires, providing some 25%. Argentina had proven reserves of petroleum amounting to 2,900m. barrels in 2004, the third largest in Latin America. Production averaged 755,000 b/d in 2004, and consumption only 486,000 b/d. The depletion of land-based reserves had encouraged interest in increased offshore exploration on the continental shelf. Coal production has revived in the face of high fuel costs, but was still modest. Some 210,000 metric tons were produced in 2004, although 640,000 tons were consumed.

The operating losses and huge overseas debt of the state petroleum company, Yacimientos Petrolíferos Fiscales (YPF), led to the Government allowing private-sector companies, including Esso Standard Oil Company and Royal Dutch/Shell Group, to operate in Argentine oilfields in the late 1980s, and in 1991 the industry was deregulated. In January 1999 the Government sold a 15% stake in YPF to Repsol of Spain. In June Repsol completed its purchase of the remaining 85% of the company (of which the Government itself still owned 5.3%, valued at over US $800m.), to form the world's 10th largest oil and gas corporation, controlling 4,100m. barrels of petroleum reserves and, in 2003, accounting for some 39% of Argentina's oil production, according to the US Energy Information Administration (EIA). The next two production companies by size after Repsol YPF are Pérez Compañc (owned by Petrobrás of Brazil) and Petrolera Argentina San Jorge (a member of the US Chevron Group).

Natural gas reserves were estimated at 768,000m. cu m in 2004, the third largest in Latin America, and production was an estimated 37,150m. cu m in the same year. It was hoped that the discovery, in 1999, of extensive gas and oil deposits off the coast of Tierra del Fuego would eventually eliminate the need to import gas from Bolivia. Other deposits were at Neuquén in the far south of Patagonia. Thousands of kilometres of pipelines linked these fields, those of Mendoza and the southern Bolivian gas field with Buenos Aires and other main urban centres. Production nearly doubled between 1993 and 1998, to 38,631m. cu m. Distribution was largely in the care of Transportadora de Gas del Sur, SA, the largest pipeline company in South America. At present rates of consumption Argentina's known gas reserves would last some 20 years; however, production was planned to expand at an annual rate of 3.6% in the decade to 2010. In mid-2004, however, a critical shortfall arose in gas supply to meet rapidly increasing domestic consumption, caused primarily by government-imposed restrictions on the price of gas (see History). To avoid future shortages the Government of President Néstor Kirchner committed itself to future liberalization of the gas market. In June 2005 the Presidents of the four Mercosur (Mercado Común del Sur—Southern Common Market) member countries (Argentina, Brazil, Paraguay and Uruguay) signed an agreement creating an 'energy ring' (anillo energético), which would facilitate the supply and storage of natural gas from the Camisea field in Peru.

Total installed electrical generating capacity in 2002 was some 27 GW, according to the EIA. Electricity production in 2004 was 89,014m. kWh. Of this, about 52% was provided by conventional thermal generation, 41% by hydroelectric power and 7% from nuclear power (Argentina was the leading nuclear-power producer in Latin America). The hydroelectric potential of Argentina was estimated at some 30,000 MW and has increasingly been exploited since the 1970s, when Argentina increased its hydroelectric output 10-fold. Concerns about the long-term effects of silting spurred work on the Yacyretá dam, a joint project with Paraguay on the River Paraná, with an installed generating capacity of 3,200 MW (scaled down from the planned 4,050 MW). Construction was delayed by financial shortages and the plant, which came on line in 1998, has been operating considerably below capacity. In August 2001 an agreement was concluded to raise the water level in the impoundment to bring output closer to design capacity, but it met stiff resistance from indigenous interests. The Kirchner Government sought funding from the World Bank; the report of the inspection team was discussed by the Bank in May 2004, but no decision was reached.

A hydroelectric complex with an expected capacity of 2,000 MW, which was under construction on the Limay river, was expected to provide a further 9% of the country's energy requirements. However, the joint Argentine-Paraguayan Corpus project on the River Paraná, with an anticipated generating capacity of 6,900 MW, was also delayed following public opposition. The interconnector with the Brazilian grid, between Paraná and São Paulo, Brazil, has been in commercial operation since June 1999 and the seond stage was close to completion in 2005. The 1,890-MW Salto Grande project was operated jointly by Argentina and Uruguay. In 2002 Argentina exported 2,818,000m. kWh of electric energy and imported 8,775,000m. kWh.

## MANUFACTURING

Argentine industrialization proceeded apace from the 1930s, stimulated by import-substitution policies. As early as the 1940s the value of industrial production exceeded that of the agricultural and pastoral sectors combined. However, the 'free market' policies of the military regimes of 1966–73 and 1976–83 allowed foreign competition to increase and reduced industrial output. Industry accounted for 34.7% of GDP in 2003, and manufacturing for 23.9%.

The key industrial and manufacturing sectors are the motor industry, consumer goods (especially 'white' goods such as refrigerators, washing machines and television receivers), pharmaceuticals, cosmetics, electronic equipment, fibres, cement, rubber and paper and other wood products. Argentina is self-sufficient in basic manufactures, producing 5.2m. metric tons of cement (an increase of 33.4% on 2002) and an estimated 1.4m. tons of paper in 2003. Argentina was bound to depend essentially on its primary exports in the foreseeable future, but world markets were weak in the early 2000s, and only a continued restructuring of the industrial sector to increase manufactured exports was likely to compensate for this. The country has comparative advantages in such fields as petrochemicals and agro-industries, but has long lacked a stable economic climate in which they could prosper. However, 546,869 tons of polyethylene and 170,770 tons of polyvinylchloride (PVC) were produced in 2003.

In 2003 Argentine factories produced 109,364 cars, 50,799 commercial motor vehicles and 9.8m. motor vehicle tyres, and 357,220 washing machines. In the same year 332,200 television sets and 138,110 domestic refrigerators were produced.

The Government of Carlos Menem liberalized the industrial sector in the early 1990s, selling off a wide-ranging series of state concerns. The Government claimed that liberalization resulted in the creation of many new jobs, but between 1980 and 1993 the proportion of those employed in manufacturing fell from 40% to 30%, while there, as elsewhere, part-time labour increased, especially for women. The unemployment rate peaked at 28.5% in November 2001, but fell back to 19.7% in 2002. In December 2004 unemployment was still high, at some 16%.

## TRANSPORT AND COMMUNICATIONS

Argentina long had the most developed communications system in Latin America, stemming from its earlier economic development and the establishment of a rail network from the mid-19th century. Privatization of the rail sector began in 1991 with the franchising of the Rosario–Bahía Blanca grain line to a Techint–Iowa State Railroad consortium. The franchising strategy was complicated by protests from the provincial governments over payments introduced to subsidize loss-making passenger services. Furthermore, in March 1993 12 provinces lost their passenger-train services, although Córdoba's service later won a reprieve. Following privatization, railway use increased significantly. In 2004 there were approximately 31,091 km of track. In 2003 Argentine railways moved 378m. passengers (up from 355m. in 2002) and 20.5m. metric tons of freight.

In 2004 there were 214,471 km of roads, of which 63,648 km (29.7%) were paved. National bus services are good, with interconnection to other Latin American states (the longest, from Buenos Aires to Caracas, Venezuela, carries 3m. passengers per year a distance of 9,660 km). More than 5m. passenger cars and almost 1.5m. commercial vehicles were in use in 1998. The leasing of highways proved to be the most controversial, problematic and politically unpopular part of the Menem Government's privatization programme. Nevertheless, in 1998 the network carried 87m. vehicles, which paid US $366.8m. in tolls.

In the Federal Capital, metropolitan bus services were complemented by an efficient and spacious underground network. The Subterráneo (Subte), the oldest system in South America, was privatized in 1993. Shipping is not well developed, considering Argentina's large export trade and access to 10,950 km of navigable waterways. There are 13 major ports, of which Buenos Aires is the largest and most significant. The major shipping firm is Empresas Líneas Marítimas Argentinas. In 2004 there were 507 vessels, amounting to 436,700 gross registered tons, under the national flag.

The major international airport at Buenos Aires is at Ezeiza. The smaller Aeroparque Jorge Newbery lies at the heart of the city and, though it operates essentially as an internal entrepôt, also caters for a limited number of flights to neighbouring Latin American countries. There are eight other international airports in the country and some 1,342 airports in total. Aircraft departures in 2004 exceeded 90,000.

Argentina's telephone service has been much improved in recent years. In 2002 there were 8m. fixed lines (219 per thousand population) and 6.5m. mobile cellular telephones in use. In the same year there were 4.1m. internet users, an average of 109 per thousand. By 2003 there were 14 internet service providers in the country, a steep fall from 33 in 2000.

Tourism is still comparatively underdeveloped, partly because of the distance from Europe and the USA and the considerable distances between the major centres of tourist interest, and partly because, until the devaluation of the Argentine peso in January 2002, the country was a prohibitively expensive location. Tourist arrivals in 2003 were 3.4m., an increase of 21.4% on the previous year, encouraged by the devaluation of the peso. Receipts from tourism amounted to some 8% of total exports in 2003.

## INVESTMENT AND FINANCE

The policies pursued by the successive military Governments from 1976 to 1983 emphasized international competitiveness and a reduction in the rate of inflation. Devaluation and obstruction of external trade, caused by the Falklands conflict, created a fall of 45% in the value of goods purchased abroad in 1982, compared with 1981.

The Government of Raúl Alfonsín attempted to stimulate economic growth while increasing taxation and reducing public expenditure, broadly in line with the IMF's austerity proposals. However, investment remained depressed and inflation reached 434% in the year to December 1983. The Austral Plan, introduced in June 1985, was the first example in Latin America of the so-called 'heterodox shock'. It sought to eliminate inflation at once by a complete wage and price 'freeze' for 90 days, while at the same time breaking the inflationary spiral by instituting a new currency, the austral, worth 1,000 pesos argentinos. However, inflation continued to rise and the wages and prices 'freeze', which was a necessary part of the Plan, was breached by the Government.

The economic team of President Menem introduced a new stabilization plan in July 1989. The rate of inflation quickly fell to under 10% per month, allowing Menem to agree on a US $1,400m. stand-by loan from the IMF in October. However, the recovery was short-lived. Effective stabilization of the currency only came as a result of the measures introduced in January 1991, by the new economy minister, Domingo Cavallo. The key to his Convertibility Plan was the so-called 'dollarization' of the economy, linking the Argentine currency to the US dollar, with a fixed exchange rate, supported by foreign exchange or gold reserves at the Central Bank. This had the immediate effect of restraining the rise in consumer prices, although inflation to the year end was, nevertheless, 84%. In January 1992 another currency, the nuevo peso argentino (equal to US $1) replaced the austral as a visible symbol of the change.

In July 1993 industrial output reached record levels, the peso was rated the third strongest currency in the world, after the Japanese yen and the Singapore dollar, and the economy, as measured by GDP at constant prices, grew by an annual rate of 6.0%. However, Argentina's recovery suffered a sharp reverse from the effects of the Mexican peso crisis of December 1994 (the 'tequila effect'). Following the Mexican devaluation, in early 1995 severe restrictions were placed on public spending, in an attempt to support the Argentine peso. Widespread tax evasion continued to limit the capacity of the Government to raise revenue. The economy contracted in 1995, but economic growth resumed thereafter, reaching 8.1% in 1997. At the same time, inflation remained under control. However, in 1998, owing to the economic crisis in Brazil, economic growth slowed, and in 1999 the economy contracted.

In December 1999 the incoming Government of President Fernando de la Rúa inherited an economy already in crisis. The new Government arranged a US $7,400m. stand-by facility with the IMF. Throughout 2000 it struggled with the Congreso to implement an austerity programme. The situation continued, however, to worsen and in December a further $14,000m. loan agreement was negotiated. In spite of the loan agreement, in March 2001 the first-quarter fiscal target was missed because of falling tax revenues. At this point Domingo Cavallo was again appointed Minister of the Economy and in May he successfully negotiated a supplementary agreement with the IMF that would give the country continued access to a $40,000m. package of financial aid, even though the fiscal deficit at the year end was expected to exceed the IMF-agreed target. (In the event, the 2001 fiscal deficit was $9,090m., 39.8% higher than the IMF target.) However, with confidence already eroded, there were continuing fears that he would not be able to avoid both devaluation and default. The economy contracted by 4.4% in that year. In June the Senado approved the introduction of a new exchange-rate regime, which would fix the peso at mid-point between the US dollar and the euro as soon as the euro reached parity with the dollar. However, the de la Rúa Government collapsed in December before it could be implemented. During the brief term of President Rodríguez Sáa a new currency, the argentino, was proposed, but the new administration of Eduardo Duhalde was immediately forced to abandon the fixed exchange-rate system and devalue the Argentine peso by about 30% in early January 2002. Once 'unpegged' from the US dollar, the peso continued to fall in value, leading to the intermittent closure of the banks in the first half of 2002, in a desperate attempt to stem the outflow of currency and its conversion into US dollars. In the latter part of 2002, however, the exchange rate stabilized, rising to approximately 2.84 pesos to the US dollar. Controls on bank deposits were gradually relaxed. Nevertheless, GDP contracted by 10.9% in 2002 as a whole, with disastrous effects on the country's living standards. As a result, and contrary to expectations, hyperinflation was averted. Although the rate of inflation in 2002 rose to 25.8%, it fell back to 13.5% in 2003 and to 8.8% over the 12 months to April 2005, despite some relaxation in wage policy. A positive economic growth rate of 8.8% in 2003 and an estimated 9.0% in 2004 was meanwhile attributed partly to the effects of the devaluation and partly to higher commodity prices.

## FOREIGN TRADE, BALANCE OF PAYMENTS AND DEBT

In the 1980s Argentina generally had a favourable balance of trade but an unfavourable balance of payments. However, in the 1990s the trade balance turned to deficit, owing to a continued rise in the value of imports. The trade deficit reached US $3,099m. in 1998. This deficit decreased to $795m. in the following year, and in 2000 a trade surplus of $2,454m. was registered, which increased to $7,385m. in 2001, mainly owing to a decrease in the cost of imports. In 2002 a steep fall in imports gave an even larger surplus of $17,178m., but this fell back to $5,473m. in 2004 as the economy improved.

There was steady progress in the 1990s in extending intra-regional trade. From July 1990 Argentina actively participated in the creation of Mercosur. On 1 January 1995 the common external tariff was established, as well as an open frontier between Argentina and its partners. A major trade dispute with Brazil over creating a balanced regime for trade in motor vehicles was ended by agreement in March 2000, at which a number of minor points of dispute were also shelved. However, restrictions on Brazilian imports were imposed from mid-2004 by the Kirchner Government in response to what it considered to be disadvantageous imbalances in bilateral trade that stifled Argentina's industrial development (see History).

Exports were valued at an estimated US $29,566m. in 2003, rising to an estimated $34,554m. in 2004. Edible oils, fuels, cereals, feed and motor vehicles constituted the main export items. By the 1990s the European Union (EU) was established as Argentina's largest trading partner, but Brazil was the largest single-country trading partner. Global figures disguised interesting variations. Figures for 2003 again confirmed Brazil to be Argentina's largest export market, the destination for 15.8% of exports by value in that year, followed by Chile (12.0%), the USA (10.6%), the People's Republic of China (8.7%) and Spain (4.7%). In the same year Brazil accounted for 34.0% of imports, the USA 16.4%, Germany 5.6%, the People's Republic of China 5.4% and Italy 3.2%. Total imports, however, which had fallen to only $8,990m. in 2002, rose again to $22,445m. in 2004. The main import items were machinery and equipment, motor vehicles, chemicals, metal manufactures and plastics.

Apart from the vertiginous rise and fall of inflation (Argentina is the only country in the world to have experienced continuous triple-digit inflation for more than 15 years), the most dramatic feature of the economy at the end of the last century was the massive increase in foreign debt. This grew from US $35,700m. in 1981 to $74,473m. in 1993. Rescheduling negotiations between Argentina, the IMF and creditor banks became a permanent feature of financial policy and, on occasions, were difficult. The first part of the privatization programme, begun in 1990, recovered more than $5,000m. in foreign debt, as payment was made largely by debt-for-equity 'swaps'. A further substantial reduction followed in 1991, and in 1992 agreement was reached with foreign private creditor banks to consolidate $30,000m. of debt and so, reduce the country's capital obligation to some $48,000m. Under the principles of the US Brady Initiative (an initiative on debt relief originally proposed by Nicholas Brady, the US Secretary of the Treasury, in 1989), the agreement included a substantial debt write-off.

Although Argentina was able to withstand the effects of the Mexican financial crisis in late 1994 and early 1995 and maintain convertibility, the Government, as a consequence, was forced to resume borrowing. In February 1998 it signed a new three-year accord with the IMF under the Extended Finance Facility, for US $2,800m. As the recession deepened, borrowing increased, and by the end of 1999 the World Bank classified Argentina as a severely indebted middle-income country. At the end of 2000, with the country sinking deeper into recession, Argentina's total external debt stood at $146,172m. This figure gave a debt-to-GDP ratio of 51.4%, and debt service as 71.3% of exports of goods and services. The total cost of debt servicing in 2000 was $25,500m. The administration of Fernando de la Rúa was faced with the unpalatable prospect of negotiating a voluntary restructuring of the national debt at a time when it was unlikely to have any access to international capital markets for months to come. Public-sector pension and salary cuts were insufficient to improve the rapidly deteriorating economic situation. Other austerity measures included the introduction of one-year bonds, known as 'patacones', as payment to 160,000 public-sector workers. The refusal of the IMF to disburse emergency funds to Argentina and the introduction of the corralito (government 'freeze' on bank deposits) in December 2001 led to the collapse of the de la Rúa Government and to the inevitable official default on the country's debt in January 2002.

Negotiations between the new Duhalde Government and the IMF continued throughout 2002, culminating in a temporary agreement early in January 2003 that would last until August. Meanwhile, however, on 14 November 2002, Argentina had become the first country ever to default on a loan from the World Bank. At the end of 2003 the total external debt stood at US $166,207m. compared with $149,989m. in 2002. Debt servicing in 2003 cost $14,009m., representing 37.9% of exports of goods and services, compared with 16.6% in 2002. The Government received a favourable report and some conditional support from the IMF in 2003. Meanwhile, the Argentine Government was counting on a new agreement with the Fund in order to extend debt maturing in 2004, and, with the support of those who wished to see Argentina taking a positive role in the extension of free trade in the region, it was therefore able to maintain its policy of aggressively seeking better terms for the restructuring of some $82,000m. of debt. Under the plan, Argentina offered its creditors new bonds, most of which were worth about 33 US cents for each dollar of the old bonds. Minister of the Economy Roberto Lavagna defended the new terms, saying that international financial institutions shared the blame for Argentina's economic problems by pressuring indebted countries to produce restructuring plans that satisfied creditors, rather than making sustainable economic growth the main priority, thereby keeping the countries on 'artificial respiration'.

In February 2005 76.6% of bondholders accepted the new terms. The IMF, however, then demanded that the Government show 'good faith' by allowing the remaining 24.4% (including the so-called 'vulture funds') who initially rejected the Government's offer and who held some US $20,000m. in defaulted debt, to be given the same terms. At the same time, the Fund called for further structural reform of the Argentine economy, including a higher primary budget surplus. It also recommended that the Government respect property rights of the foreign-owned utility and oil companies that entered the country during the privatization programme of the 1990s. In April 2005, however, President Kirchner was able publicly to dismiss these demands.

In 2005 central government expenditure was forecast at US $77,530.8m. Of this, $44,953.8m. (58.0%) was spent on social services, $10,052.3m. (13.0%) on debt servicing, $5,612.2m. (7.2%) on defence and security, and $5,040.0m. (6.5%) on general administration.

## CONCLUSION

Argentina's resources—material and human—are large, and the country has made good progress in surmounting the crisis of 2001. Sadly, however, it was not the first time that economic crisis has occurred in Argentina. The disastrous cycle of military intervention, economic crisis and ineffective civilian government that, in the period 1930–90, took Argentina from being the seventh richest country in the world to being the 77th, was temporarily broken only at the beginning of the 1990s. The 'dollarization' of the economy virtually eliminated inflation and brought financial stability and the repatriation of capital for productive investment. However, the problems of trade and fiscal deficits, high levels of unemployment and an overvalued currency recurred in the late 1990s. The Government of President de la Rúa failed to take effective action to rein in government expenditure. However, to its credit, the interim Government of President Duhalde did successfully restrain central government expenditure and persuade the provincial Governors that they had no alternative but to accept fiscal discipline.

Since taking office in 2003, President Kirchner has successfully built on returning confidence in order to reconcile the pressures of multilateral lending institutions, financial institutions and creditors with the demands of the severely impoverished and militant Argentine people. The first steps were to address the persistent shortages of gas and electricity, which were impeding a return to economic growth, by the creation in May 2004 of a new state energy company, Energía Argentina, SA, to direct energy policy, and to restore co-operation with Brazil within the framework of Mercosur. Kirchner's greatest success thus far has been to persuade bondholders to accept up to a 70% loss on their investment in an attempt to restructure the US $102,600m. in bad debt. Despite complaints, especially in Italy (where some 450,000 small investors lost their savings on Argentine bonds), about his Government's tough negotiating style, the majority of bondholders did accept the terms offered, which would restore Argentina's rating with the credit rating agency Standard & Poor's only to B– (equal to the worst in Latin America), but give the country access once more to the world's capital markets.

# Statistical Survey

Sources (unless otherwise stated): Instituto Nacional de Estadística y Censos, Avda Julio A. Roca 609, 1067 Buenos Aires; tel. (11) 4349-9200; fax (11) 4349-9601; e-mail ces@indec.mecon.gov.ar; internet www.indec.mecon.ar; and Banco Central de la República Argentina, Reconquista 266, 1003 Buenos Aires; tel. (11) 4348-3955; fax (11) 4334-5712; e-mail sistema@bcra.gov.ar; internet www.bcra.gov.ar.

## Area and Population

### AREA, POPULATION AND DENSITY

| | |
|---|---:|
| Area (sq km) | 2,780,403* |
| Population (census results)† | |
| 15 May 1991 | 32,615,528 |
| 17–18 November 2001 | |
| Males | 17,659,072 |
| Females | 18,601,058 |
| Total | 36,260,130 |
| Population (official estimates at mid-year) | |
| 2002 | 37,515,632 |
| 2003 | 37,869,730 |
| 2004 | 38,226,051 |
| Density (per sq km) at mid-2004 | 13.7 |

* 1,073,519 sq miles. The figure excludes the Falkland Islands (Islas Malvinas) and Antarctic territory claimed by Argentina.

† Figures exclude adjustment for underenumeration, estimated to have been 0.9% at the 1991 census.

### PROVINCES
(2001 census)

| | Area (sq km) | Population | Density (per sq km) | Capital |
|---|---:|---:|---:|---|
| Buenos Aires— | | | | |
| Federal District | 203 | 2,776,138 | 13,679.6 | |
| Buenos Aires— | | | | |
| Province | 307,571 | 13,827,203 | 45.0 | La Plata |
| Catamarca | 102,602 | 334,568 | 3.3 | San Fernando del Valle de Catamarca |
| Chaco | 99,633 | 984,446 | 9.9 | Resistencia |
| Chubut | 224,686 | 413,237 | 1.8 | Rawson |
| Córdoba | 165,321 | 3,066,801 | 18.6 | Córdoba |
| Corrientes | 88,199 | 930,991 | 10.6 | Corrientes |
| Entre Ríos | 78,781 | 1,158,147 | 14.7 | Paraná |
| Formosa | 72,066 | 486,559 | 6.8 | Formosa |
| Jujuy | 53,219 | 611,888 | 11.5 | San Salvador de Jujuy |
| La Pampa | 143,440 | 299,294 | 2.1 | Santa Rosa |

# ARGENTINA

*Statistical Survey*

| — continued | Area (sq km) | Population | Density (per sq km) | Capital |
|---|---|---|---|---|
| La Rioja | 89,680 | 289,983 | 3.2 | La Rioja |
| Mendoza | 148,827 | 1,579,651 | 10.6 | Mendoza |
| Misiones | 29,801 | 965,522 | 32.4 | Posadas |
| Neuquén | 94,078 | 474,155 | 5.0 | Neuquén |
| Río Negro | 203,013 | 552,822 | 2.7 | Viedma |
| Salta | 155,488 | 1,079,051 | 6.9 | Salta |
| San Juan | 89,651 | 620,023 | 6.9 | San Juan |
| San Luis | 76,748 | 367,933 | 4.8 | San Luis |
| Santa Cruz | 243,943 | 196,958 | 0.8 | Río Gallegos |
| Santa Fe | 133,007 | 3,000,701 | 22.6 | Santa Fe |
| Santiago del Estero | 136,351 | 804,457 | 5.9 | Santiago del Estero |
| Tucumán | 22,524 | 1,338,523 | 59.4 | San Miguel de Tucumán |
| *Territory* | | | | |
| Tierra del Fuego | 21,571 | 101,079 | 4.7 | Ushuaia |
| **Total** | **2,780,403** | **36,260,130** | **13.0** | **Buenos Aires** |

## PRINCIPAL TOWNS
(population at 2001 census)*

| | | | |
|---|---|---|---|
| Buenos Aires (capital) | 2,776,138 | Malvinas Argentinas† | 290,691 |
| Córdoba | 1,267,521 | Berazategui† | 287,913 |
| La Matanza† | 1,255,288 | Bahía Blanca† | 284,776 |
| Rosario | 908,163 | Resistencia | 274,490 |
| Lomas de Zamora† | 591,345 | Vicente López† | 274,082 |
| La Plata† | 574,369 | San Miguel† | 253,086 |
| General Pueyrredón† | 564,056 | Posadas | 252,981 |
| San Miguel de Tucumán | 527,150 | Esteban Echeverría† | 243,974 |
| Quilmes† | 518,788 | Paraná | 235,967 |
| Almirante Brown† | 515,556 | Pilar† | 232,463 |
| Merlo† | 469,985 | San Salvador de Jujuy | 231,229 |
| Salta | 462,051 | Santiago del Estero | 230,614 |
| Lanús† | 453,082 | José C. Paz† | 230,208 |
| General San Martín† | 403,107 | Guaymallén | 223,365 |
| Moreno† | 380,503 | Neuquén | 201,868 |
| Santa Fe | 368,668 | Formosa | 198,074 |
| Florencio Varela† | 348,970 | Godoy Cruz | 182,563 |
| Tres de Febrero† | 336,467 | Escobar† | 178,155 |
| Avellaneda† | 328,980 | Hurlingham† | 172,245 |
| Corrientes | 314,546 | Las Heras | 169,248 |
| Morón† | 309,380 | Ituzaingó | 158,121 |
| Tigre† | 301,223 | San Luis | 153,322 |
| San Isidro† | 291,505 | San Fernando† | 151,131 |

* In each case the figure refers to the city proper. At the 2001 census the population of the Buenos Aires agglomeration was 13,827,203.
† Settlement within the Province of Buenos Aires.

**Mid-2003** (UN estimate, incl. suburbs): Buenos Aires 13,047,115 (Source: UN, *World Urbanization Prospects: The 2003 Revision*).

### BIRTHS, MARRIAGES AND DEATHS

| | Registered live births | | Registered deaths | |
|---|---|---|---|---|
| | Number | Rate (per 1,000) | Number | Rate (per 1,000) |
| 1995 | 658,735 | 18.9 | 268,997 | 7.7 |
| 1996 | 675,437 | 19.2 | 268,715 | 7.6 |
| 1997 | 692,357 | 19.4 | 270,910 | 7.6 |
| 1998 | 683,301 | 18.9 | 280,180 | 7.8 |
| 1999 | 686,748 | 18.8 | 289,543 | 7.9 |
| 2000 | 701,878 | 19.0 | 277,148 | 7.5 |
| 2001 | 683,495 | 18.2 | 285,941 | 7.6 |
| 2002 | 694,684 | 18.3 | 291,190 | 7.7 |

**Marriages:** 130,553 (marriage rate 3.5 per 1,000) in 2001.
Source: mainly UN, *Demographic Yearbook* and *Population and Vital Statistics Report*.

**Expectation of life** (WHO estimates, years at birth): 74 (males 71; females 78) in 2003 (Source: WHO, *World Health Report*).

## ECONOMICALLY ACTIVE POPULATION
(persons aged 14 years and over, census of November 2001)

| | Males | Females | Total |
|---|---|---|---|
| Agriculture, hunting and forestry | 805,293 | 92,228 | 897,521 |
| Fishing | 11,843 | 1,632 | 13,475 |
| Mining and quarrying | 35,068 | 2,911 | 37,979 |
| Manufacturing | 966,056 | 279,488 | 1,245,544 |
| Electricity, gas and water | 76,428 | 13,737 | 90,165 |
| Construction | 621,598 | 16,968 | 638,566 |
| Wholesale and retail trade; repair of motor vehicles, motorcycles and personal and household goods | 1,287,017 | 624,364 | 1,911,381 |
| Hotels and restaurants | 168,929 | 132,755 | 301,684 |
| Transport, storage and communication | 629,188 | 88,385 | 717,573 |
| Financial intermediation | 111,717 | 74,796 | 186,513 |
| Real estate, renting and business activities | 456,005 | 255,746 | 711,751 |
| Public administration and defence; compulsory social security | 649,790 | 319,490 | 969,280 |
| Education | 200,538 | 730,387 | 930,925 |
| Health and social work | 197,282 | 394,824 | 592,106 |
| Other community, social and personal services | 266,782 | 186,202 | 452,984 |
| Private households with employed persons | 85,213 | 701,219 | 786,432 |
| Extra-territorial organizations and bodies | 1,084 | 917 | 2,001 |
| Activities not adequately described | 243,200 | 184,107 | 427,307 |
| **Total employed** | **6,813,031** | **4,100,156** | **10,913,187** |
| Unemployed | 2,212,776 | 2,138,820 | 4,351,596 |
| **Total labour force** | **9,025,807** | **6,238,976** | **15,264,783** |

**May 2003** (labour force survey of 31 urban agglomerations, '000 persons aged 10 years and over): Total employed 8,570.8 (males 4,887.7, females 3,683.1); Unemployed 1,583.6; Total labour force 10,154.4 (males 5,836.7, females 4,317.7).

# Health and Welfare

## KEY INDICATORS

| | |
|---|---|
| Total fertility rate (children per woman, 2003) | 2.4 |
| Under-5 mortality rate (per 1,000 live births, 2003) | 20 |
| HIV (% of persons aged 15–49, 2003) | 0.7 |
| Physicians (per 1,000 head, 1998) | 3.01 |
| Hospital beds (per 1,000 head, 1996) | 3.29 |
| Health expenditure (2002): US $ per head (PPP) | 956 |
| Health expenditure (2002): % of GDP | 9.5 |
| Health expenditure (2002): public (% of total) | 50.2 |
| Access to water (% of persons, 2000) | 79 |
| Access to sanitation (% of persons, 2000) | 85 |
| Human Development Index (2002): ranking | 34 |
| Human Development Index (2002): value | 0.853 |

For sources and definitions, see explanatory note on p. vi.

# Agriculture

**PRINCIPAL CROPS**
('000 metric tons)

|  | 2001 | 2002 | 2003 |
|---|---|---|---|
| Wheat | 15,428 | 12,300 | 14,530 |
| Rice (paddy) | 859 | 713 | 718 |
| Barley | 573 | 570* | 549 |
| Maize | 15,365 | 15,000 | 15,040 |
| Rye | 81 | 80 | 37 |
| Oats | 645 | 500 | 348 |
| Sorghum | 2,909 | 2,847 | 2,685 |
| Potatoes | 2,505 | 2,133 | 2,150* |
| Sweet potatoes | 309 | 310* | 315* |
| Cassava (Manioc)* | 165 | 165 | 170 |
| Sugar cane* | 19,050 | 19,250 | 19,250 |
| Dry beans | 263 | 278 | 216 |
| Soybeans (Soya beans) | 26,864 | 30,180 | 34,800 |
| Groundnuts (in shell) | 563 | 517 | 314 |
| Olives* | 90 | 93 | 95 |
| Cottonseed | 257† | 104 | 111† |
| Sunflower seed | 3,043 | 3,843 | 3,714 |
| Artichokes* | 85 | 86 | 88 |
| Tomatoes | 648 | 668 | 670* |
| Pumpkins, squash and gourds | 291 | 291* | 295* |
| Chillies and green peppers* | 122 | 123 | 125 |
| Dry onions | 576 | 642 | 645* |
| Garlic | 135 | 126 | 127* |
| Carrots* | 220 | 225 | 230 |
| Other vegetables* | 640 | 645 | 648 |
| Watermelons* | 125 | 126 | 126 |
| Cantaloupes and other melons | 78 | 79* | 79* |
| Bananas* | 175 | 180 | 180 |
| Oranges | 918 | 767 | 687 |
| Tangerines, mandarins, clementines and satsumas | 501 | 464 | 381 |
| Lemons and limes | 1,218 | 1,313 | 1,236 |
| Grapefruit and pomelos | 199 | 205 | 184 |
| Apples | 1,429 | 1,157 | 1,307 |
| Pears | 585 | 537 | 639 |
| Peaches and nectarines | 258 | 212 | 256 |
| Plums | 106 | 106 | 151 |
| Grapes | 2,244 | 2,360 | 2,370* |
| Tea (made) | 63 | 63 | 64 |
| Mate | 310 | 285 | 285 |
| Tobacco (leaves) | 98 | 125 | 118 |
| Cotton (lint) | 167† | 62 | 65† |

* FAO estimate(s).
† Unofficial estimate.
Source: FAO.

**LIVESTOCK**
('000 head, year ending September)

|  | 2001 | 2002 | 2003 |
|---|---|---|---|
| Horses* | 3,600 | 3,650 | 3,655 |
| Mules* | 180 | 180 | 185 |
| Asses* | 95 | 95 | 98 |
| Cattle | 48,851 | 48,100 | 50,869† |
| Pigs* | 4,000 | 3,500 | 3,100 |
| Sheep | 13,500* | 12,400 | 12,450* |
| Goats | 3,387 | 4,000 | 4,200 |
| Chickens* | 110,000 | 110,500 | 110,700 |
| Ducks* | 2,300 | 2,350 | 2,355 |
| Geese* | 130 | 135 | 140 |
| Turkeys* | 2,800 | 2,850 | 2,900 |

* FAO estimate(s).
† Unofficial figure.
Source: FAO.

**LIVESTOCK PRODUCTS**
('000 metric tons)

|  | 2001 | 2002 | 2003 |
|---|---|---|---|
| Beef and veal | 2,461 | 2,493 | 2,621 |
| Mutton and lamb* | 50 | 50 | 52 |
| Pig meat | 198 | 165 | 150 |
| Horse meat* | 55 | 56 | 56 |
| Poultry meat | 993 | 742 | 781 |
| Cows' milk | 9,769 | 8,793 | 8,197 |
| Butter† | 53 | 55 | 55 |
| Cheese† | 420 | 370 | 350 |
| Hen eggs | 302 | 255 | 263 |
| Honey | 80 | 83 | 75 |
| Wool: greasy | 56 | 65 | 72 |
| Cattle hides (fresh)* | 348 | 345 | 375 |
| Sheepskins (fresh)* | 24 | 23 | 23 |

* FAO estimates.
† Unofficial figures.
Source: FAO.

# Forestry

**ROUNDWOOD REMOVALS**
('000 cubic metres, excl. bark)

|  | 1999 | 2000 | 2001 |
|---|---|---|---|
| Sawlogs, veneer logs and logs for sleepers | 3,547 | 2,091 | 2,274 |
| Pulpwood | 2,944 | 3,794 | 2,962 |
| Other industrial wood | 161 | 120 | 99 |
| Fuel wood | 3,950 | 3,965 | 3,972 |
| **Total** | 10,602 | 9,970 | 9,307 |

**2002–2003:** Production as in 2001 (FAO estimates).
Source: FAO.

**SAWNWOOD PRODUCTION**
('000 cubic metres, incl. railway sleepers)

|  | 1999 | 2000 | 2001 |
|---|---|---|---|
| Coniferous (softwood) | 686 | 619 | 1,630 |
| Broadleaved (hardwood) | 722 | 202 | 500 |
| **Total** | 1,408 | 821 | 2,130 |

**2002–03:** Production as in 2001 (FAO estimates).
Source: FAO.

# Fishing

('000 metric tons, live weight)

|  | 2001 | 2002 | 2003 |
|---|---|---|---|
| Capture* | 944.7 | 958.5 | 914.6 |
| Southern blue whiting | 54.3 | 42.5 | 44.6 |
| Argentine hake | 249.4 | 358.8 | 334.1 |
| Patagonian grenadier | 111.8 | 98.7 | 97.8 |
| Argentine red shrimp | 78.8 | 51.4 | 52.9 |
| Patagonean scallop | 39.0 | 51.0 | 49.5 |
| Argentine shortfin squid | 230.3 | 177.3 | 140.9 |
| Aquaculture | 1.3 | 1.5 | 1.6 |
| **Total catch*** | 946.0 | 959.9 | 916.2 |

* FAO estimates.
Note: The data exclude aquatic plants (metric tons, capture only): 3 in 2000. Also excluded are aquatic mammals, recorded by number rather than by weight. The number of La Plata River dolphins caught was 160 in 2001; 215 in 2002; 893 in 2003.

Source: FAO.

# ARGENTINA

## Mining

('000 metric tons, unless otherwise indicated)

|  | 2001 | 2002 | 2003 |
|---|---|---|---|
| Crude petroleum ('000 42-gallon barrels)* | 285,381 | 275,894 | 270,349 |
| Natural gas (million cu metres, gross)* | 45,994 | 45,819 | 50,664 |
| Lead ore† | 12.3 | 12.0 | 12.1‡ |
| Zinc ore† | 39.7 | 37.3 | 29.8‡ |
| Silver ore (kilograms)† | 152,802 | 125,868 | 133,917‡ |
| Copper ore† | 191.7 | 204.0 | 199.0‡ |
| Gold ore (kilograms)† | 30,630 | 32,506 | 29,744‡ |

* Source: US Geological Survey.
† Figures refer to the metal content of ores and concentrates.
‡ Estimates.

## Industry

**SELECTED PRODUCTS**
('000 metric tons, unless otherwise indicated)

|  | 2001 | 2002 | 2003 |
|---|---|---|---|
| Wheat flour | 3,537 | 3,550 | 3,793 |
| Beer (sales, '000 hectolitres) | 12,390 | 11,990 | 12,600 |
| Wine (sales, '000 hectolitres) | 12,036 | 11,986 | 12,291 |
| Cigarettes (million units) | 1,739 | 1,812 | 1,990 |
| Paper | 1,229 | 1,208 | 1,394 |
| Aluminium | 248 | 269 | 272 |
| Rubber tyres for motor vehicles ('000) | 8,037 | 9,093 | 9,758 |
| Portland cement | 5,545 | 3,910 | 5,217 |
| Urea | 934 | 1,120 | 1,300 |
| Ammonia (estimates) | 732 | 771 | 884 |
| Motor spirit (petrol, '000 cu metres) | 30,241 | 29,040 | 30,405 |
| Passenger cars ('000 units) | 170 | 111 | 109 |
| Washing machines ('000 units, estimates) | 295 | 162 | 357 |
| Home refrigerators ('000 units, estimates) | 246 | 162 | 138 |

**Electric energy** (million kWh): 89,014 in 2000; 90,189 in 2001; 84,430 in 2002 (Source: UN Economic Commission for Latin America and the Caribbean, *Statistical Yearbook*).

## Finance

**CURRENCY AND EXCHANGE RATES**

**Monetary Units**
100 centavos = 1 nuevo peso argentino (new Argentine peso).

**Sterling, Dollar and Euro Equivalents** (31 May 2005)
£1 sterling = 5.2052 new pesos;
US $1 = 2.8630 new pesos;
€1 = 3.5304 new pesos;
100 new pesos = £19.21 = $34.93 = €28.33.

**Average Exchange Rate** (new pesos per US $)
2002   3.0633
2003   2.9006
2004   2.9233

Note: From April 1996 to December 2001 the official exchange rate was fixed at US $1 = 99.95 centavos. In January 2002 the Government abandoned this exchange rate and devalued the peso: initially there was a fixed official exchange rate of US $1 = 1.40 new pesos for trade and financial transactions, while a free market rate was applicable to other transactions. In February, however, a unified 'floating' exchange rate system, with the rate to be determined by market conditions, was introduced.

**BUDGET**
(million new pesos, forecasts)*

| Revenue | 2003 | 2004 | 2005 |
|---|---|---|---|
| Current revenue | 61,597.1 | 61,345.0 | 81,096.2 |
| Tax revenue | 46,665.8 | 46,588.2 | 63,548.8 |
| Direct | 7,960.3 | 11,946.1 | 16,475.3 |
| Indirect | 38,705.5 | 34,642.1 | 47,073.5 |
| Social security contributions | 11,719.9 | 10,586.9 | 13,284.1 |
| Sale of public goods and services | 281.8 | 382.1 | 464.1 |
| Property income | 1,107.5 | 1,087.0 | 1,174.5 |
| Other non-tax revenue | 1,522.3 | 1,643.5 | 1,863.7 |
| Current transfers | 299.9 | 1,057.4 | 761.0 |
| Capital revenue | 482.3 | 669.0 | 1,009.4 |
| **Total** | **62,079.4** | **62,014.0** | **82,105.7** |

| Expenditure | 2003 | 2004 | 2005 |
|---|---|---|---|
| Current expenditure | 58,725.3 | 55,205.4 | 68,344.5 |
| General administration | 3,630.8 | 4,042.5 | 5,040.0 |
| Legislative | 345.6 | 362.1 | 393.9 |
| Judicial | 849.9 | 903.2 | 1,067.4 |
| Foreign relations | 608.8 | 592.4 | 663.2 |
| Interior | 1,299.6 | 1,637.0 | 2,023.1 |
| Fiscal | 159.1 | 88.1 | 21.3 |
| Defence and security | 4,265.2 | 4,790.7 | 5,612.2 |
| Defence | 2,315.9 | 2,365.7 | 2,654.0 |
| Internal security | 1,533.6 | 1,818.7 | 2,210.1 |
| Penal system | 187.4 | 249.2 | 351.4 |
| Intelligence services | 228.3 | 357.0 | 396.8 |
| Social services | 35,666.2 | 38,078.8 | 44,953.8 |
| Health | 3,644.3 | 2,294.0 | 2,648.3 |
| Social welfare | 2,193.1 | 2,937.3 | 4,254.6 |
| Social security | 22,187.4 | 24,730.4 | 28,756.0 |
| Education and culture | 3,093.4 | 3,548.7 | 4,611.8 |
| Labour | 3,806.9 | 3,729.1 | 3,673.3 |
| Economy | 1,188.8 | 1,399.6 | 2,686.0 |
| Energy, fuel and mining | 302.4 | 292.4 | 1,116.2 |
| Communications | 103.0 | 169.4 | 156.2 |
| Transport | 297.4 | 400.6 | 483.8 |
| Agriculture | 230.8 | 262.4 | 334.1 |
| Industry | 100.8 | 108.9 | 385.6 |
| Debt-servicing | 13,974.3 | 6,893.9 | 10,052.3 |
| Capital expenditure | 3,033.2 | 4,506.6 | 9,186.4 |
| **Total** | **61,758.5** | **59,712.0** | **77,530.8** |

* Budget figures refer to the consolidated accounts of the central Government.
Source: Oficina Nacional de Presupuesto, Secretaría de Hacienda, Ministerio de Economía, Buenos Aires.

**INTERNATIONAL RESERVES**
(US $ million at 31 December)

|  | 2002 | 2003 | 2004 |
|---|---|---|---|
| Gold* | 3 | 4 | 769 |
| IMF special drawing rights | 94 | 1,008 | 877 |
| Foreign exchange | 10,395 | 13,145 | 18,007 |
| **Total** | **10,492** | **14,157** | **19,653** |

* National valuation.
Source: IMF, *International Financial Statistics*.

**MONEY SUPPLY**
(million new pesos at 31 December)

|  | 2002 | 2003 | 2004 |
|---|---|---|---|
| Currency outside banks | 16,430 | 26,649 | 33,872 |
| Demand deposits at commercial banks | 11,843 | 16,292 | 22,052 |
| **Total money** | **28,273** | **42,940** | **55,923** |

Source: IMF, *International Financial Statistics*.

# ARGENTINA

*Statistical Survey*

## COST OF LIVING
(Consumer Price Index for Buenos Aires metropolitan area; annual averages; base: 1999 = 100)

|  | 2001 | 2002 | 2003 |
|---|---|---|---|
| Food and beverages | 97.2 | 130.8 | 155.9 |
| Clothing | 92.6 | 125.7 | 152.9 |
| Housing | 99.2 | 105.8 | 114.3 |
| All items (incl. others) | 98.8 | 124.3 | 141.1 |

## NATIONAL ACCOUNTS
(million new pesos at current prices)

### National Income and Product

|  | 2001 | 2002 | 2003* |
|---|---|---|---|
| Gross domestic product (GDP) at market prices | 268,697 | 312,580 | 375,909 |
| Net primary income from abroad | -7,744 | -19,402 | -22,429 |
| Gross national income (GNI) | 260,953 | 293,178 | 353,480 |
| Net current transfers | 281 | 1,311 | 1,319 |
| Gross national disposable income | 261,234 | 294,490 | 354,799 |

* Preliminary figures.

### Expenditure on the Gross Domestic Product

|  | 2001 | 2002 | 2003 |
|---|---|---|---|
| Final consumption expenditure | 223,201 | 231,727 | 281,044 |
| Households* | 185,164 | 193,482 | 238,047 |
| General government | 38,037 | 38,245 | 42,997 |
| Gross capital formation | 41,953 | 34,311 | 54,690 |
| Gross fixed capital formation | 38,099 | 37,387 | 56,913 |
| Changes in inventories | 3,854 | -3,076 | -2,223 |
| Total domestic expenditure | 265,154 | 266,038 | 335,734 |
| Exports of goods and services | 30,977 | 86,552 | 93,878 |
| *Less* Imports of goods and services | 27,434 | 40,010 | 53,381 |
| Gross domestic product (GDP) in market prices | 268,697 | 312,580 | 376,232 |
| GDP at constant 1993 prices | 263,997 | 235,236 | 256,024 |

* Including non-profit institutions serving households (NPISHs).

Source: IMF, *International Financial Statistics*.

### Gross Domestic Product by Economic Activity

|  | 2001 | 2002 | 2003 |
|---|---|---|---|
| Agriculture, hunting and forestry | 11,565 | 30,482 | 37,267 |
| Fishing | 711 | 1,422 | 1,558 |
| Mining and quarrying | 6,657 | 18,674 | 20,528 |
| Manufacturing | 43,242 | 63,603 | 84,530 |
| Electricity, gas and water supply | 6,332 | 5,352 | 6,127 |
| Construction | 11,597 | 7,888 | 11,531 |
| Wholesale and retail trade | 32,831 | 33,691 | 41,183 |
| Hotels and restaurants | 7,309 | 6,681 | 8,232 |
| Transport, storage and communications | 22,873 | 23,116 | 29,975 |
| Financial intermediation | 13,762 | 16,457 | 13,683 |
| Real estate, renting and business activities | 42,697 | 39,885 | 42,195 |
| Public administration and defence* | 17,117 | 16,867 | 18,846 |
| Education, health and social work | 24,052 | 22,717 | 24,263 |
| Other community, social and personal service activities† | 14,232 | 11,807 | 13,455 |
| Sub-total | 254,976 | 298,641 | 353,375 |
| Value-added tax | 16,233 | 16,468 | 22,022 |
| Import duties | 1,575 | 1,308 | 2,289 |
| *Less* Imputed bank service charge | 4,087 | 3,837 | 1,774 |
| GDP in market prices | 268,697 | 312,580 | 375,909 |

* Including extra-territorial organizations and bodies.
† Including private households with employed persons.

## BALANCE OF PAYMENTS
(US $ million)

|  | 2002 | 2003 | 2004 |
|---|---|---|---|
| Exports of goods f.o.b. | 25,651 | 29,566 | 34,453 |
| Imports of goods f.o.b. | -8,473 | -13,118 | -21,185 |
| Trade balance | 17,178 | 16,448 | 13,267 |
| Exports of services | 3,178 | 4,048 | 4,937 |
| Imports of services | -4,739 | -5,496 | -6,670 |
| Balance on goods and services | 15,617 | 15,000 | 11,534 |
| Other income received | 3,020 | 3,083 | 3,533 |
| Other income paid | -10,507 | -11,291 | -12,742 |
| Balance on goods, services and income | 8,130 | 6,792 | 2,324 |
| Current transfers received | 772 | 897 | 1,072 |
| Current transfers paid | -219 | -319 | -368 |
| Current balance | 8,682 | 7,370 | 3,029 |
| Capital account (net) | 406 | 70 | 45 |
| Direct investment abroad | 627 | -774 | -319 |
| Direct investment from abroad | 2,149 | 1,887 | 4,254 |
| Portfolio investment assets | 477 | -95 | -77 |
| Portfolio investment liabilities | -5,117 | -7,663 | -9,129 |
| Other investment assets | -8,896 | -4,403 | -3,085 |
| Other investment liabilities | -9,925 | -4,581 | -1,970 |
| Net errors and omissions | -1,805 | -888 | 55 |
| Overall balance | -13,402 | -9,077 | -7,197 |

Source: IMF, *International Financial Statistics*.

# External Trade

## PRINCIPAL COMMODITIES
(US $ million)

| Imports c.i.f. | 2001 | 2002 | 2003* |
|---|---|---|---|
| Mineral fuels, mineral oils and products of their distillation | 798 | 426 | 478 |
| Organic chemicals | 1,322 | 987 | 1,291 |
| Pharmaceutical products | 639 | 393 | 475 |
| Plastics and articles thereof | 1,036 | 544 | 861 |
| Nuclear reactors, boilers, machinery and mechanical appliances | 3,521 | 1,281 | 2,273 |
| Electrical machinery and equipment and parts; sound and television apparatus, parts and accessories | 2,562 | 598 | 1,104 |
| Vehicles other than railway or tramway rolling-stock, and parts and accessories | 1,877 | 722 | 1,520 |
| Total (incl. others) | 20,320 | 8,990 | 13,833 |

| Exports f.o.b. | 2001 | 2002 | 2003* |
|---|---|---|---|
| Fish, crustaceans, molluscs, etc. | 939 | 711 | 868 |
| Cereals | 2,448 | 2,127 | 2,307 |
| Oil seeds and oleaginous fruits; miscellaneous grains, seeds and fruit; industrial or medicinal plants; straw and fodder | 1,401 | 1,288 | 1,993 |
| Animal or vegetable fats, oils and waxes | 1,637 | 2,087 | 2,832 |
| Residues and waste from food industries; prepared animal fodder | 2,628 | 2,790 | 3,500 |
| Mineral fuels, mineral oils and products of their distillation | 4,475 | 4,419 | 5,120 |
| Raw hides and skins (other than furskins) and leather | 803 | 685 | 712 |
| Plastics | 629 | 642 | 696 |
| Nuclear reactors, boilers, machinery and mechanical appliances | 797 | 684 | 641 |
| Vehicles other than railway or tramway rolling-stock, and parts and accessories | 1,978 | 1,603 | 1,428 |
| Total (incl. others) | 26,543 | 25,651 | 29,566 |

* Provisional figures.

# ARGENTINA

**PRINCIPAL TRADING PARTNERS**
(US $ million)*

| Imports c.i.f. | 2001 | 2002 | 2003 |
|---|---|---|---|
| Brazil† | 5,278 | 2,518 | 4,708 |
| Chile† | 506 | 177 | 290 |
| China, People's Republic‡ | 1,113 | 342 | 743 |
| France (incl. Monaco) | 736 | 262 | 320 |
| Germany | 1,052 | 554 | 769 |
| Italy (incl. San Marino) | 839 | 312 | 444 |
| Japan | 767 | 314 | 396 |
| Korea, Republic | 410 | n.a. | n.a. |
| Mexico | 437 | 158 | 238 |
| Paraguay | 303 | 255 | 295 |
| Spain | 712 | 311 | 392 |
| Taiwan | 230 | n.a. | n.a. |
| United Kingdom | 407 | 195 | 215 |
| USA | 3,781 | 1,804 | 2,264 |
| Uruguay† | 329 | 122 | 164 |
| **Total** (incl. others) | 20,320 | 8,990 | 13,833 |

| Exports f.o.b. | 2001 | 2002 | 2003 |
|---|---|---|---|
| Belgium | 290 | 250 | 232 |
| Bolivia | 269 | 299 | 242 |
| Brazil† | 6,188 | 4,848 | 4,663 |
| Canada | 225 | 186 | 216 |
| Chile† | 2,849 | 2,959 | 3,536 |
| China, People's Republic‡ | 1,224 | 1,177 | 2,576 |
| Egypt | 345 | n.a. | n.a. |
| France (incl. Monaco) | 257 | 313 | 330 |
| Germany | 455 | 605 | 721 |
| India | 440 | n.a. | n.a. |
| Iran | 417 | 339 | 47 |
| Italy (incl. San Marino) | 853 | 851 | 931 |
| Japan | 351 | 370 | 344 |
| Korea, Republic | 396 | n.a. | n.a. |
| Malaysia | 294 | n.a. | n.a. |
| Mexico | 485 | 670 | 796 |
| Netherlands | 802 | 1,038 | 1,094 |
| Paraguay | 500 | 345 | 445 |
| Peru | 394 | 445 | 414 |
| Southern African Customs Union§ | 312 | 292 | 334 |
| Spain | 1,093 | 1,135 | 1,388 |
| Thailand | 304.2 | n.a. | n.a. |
| United Kingdom | 291 | 380 | 384 |
| USA | 2,884 | 2,980 | 3,134 |
| Uruguay† | 746 | 546 | 543 |
| Venezuela | 235 | 148 | 140 |
| **Total** (incl. others) | 26,543 | 25,651 | 29,566 |

\* Imports by country of origin; exports by country of destination.
† Including free-trade zones.
‡ Including Hong Kong and Macao.
§ Comprising South Africa, Namibia, Botswana, Lesotho and Swaziland.

# Transport

**RAILWAYS**
(traffic)

| | 2000 | 2001 | 2002 |
|---|---|---|---|
| Passengers carried (million) | 478 | 432 | 355 |
| Freight carried ('000 tons) | 16,265 | 16,960 | 17,469 |
| Passenger-km (million) | 8,939 | 7,975 | 6,586 |
| Freight ton-km (million) | 8,696 | 8,989 | 9,444 |

**2003:** Passengers carried (million) 378; Freight carried ('000 tons) 20,544.

**ROAD TRAFFIC**
('000 motor vehicles in use)

| | 1998 | 1999 | 2000 |
|---|---|---|---|
| Passenger cars | 5,047.8 | 5,056.7 | 5,386.7 |
| Commercial vehicles | 1,094.0 | 1,029.0 | 1,004.0 |

Source: UN, *Statistical Yearbook*.

**SHIPPING**

**Merchant Fleet**
(registered at 31 December)

| | 2002 | 2003 | 2004 |
|---|---|---|---|
| Number of vessels | 481 | 497 | 507 |
| Total displacement ('000 grt) | 422.9 | 433.9 | 436.7 |

Source: Lloyd's Register-Fairplay, *World Fleet Statistics*.

**International Sea-borne Freight Traffic**
('000 metric tons)

| | 1996 | 1997 | 1998 |
|---|---|---|---|
| Goods loaded | 52,068 | 58,512 | 69,372 |
| Goods unloaded | 16,728 | 19,116 | 19,536 |

Source: UN, *Monthly Bulletin of Statistics*.

**Total maritime freight handled** ('000 metric tons): 118,965 in 2000; 125,685 in 2001; 112,944 in 2002 (Source: Dirección Nacional de Puertos y Vías Navegables).

**CIVIL AVIATION**
(traffic on scheduled services)

| | 1999 | 2000 | 2001 |
|---|---|---|---|
| Kilometres flown (million) | 170 | 160 | 107 |
| Passengers carried ('000) | 9,192 | 8,904 | 5,809 |
| Passenger-km (million) | 14,024 | 15,535 | 8,330 |
| Total ton-km (million) | 1,559 | 1,751 | 883 |

Source: UN, *Statistical Yearbook*.

# Tourism

**TOURIST ARRIVALS BY REGION**
('000)

| | 2001 | 2002 | 2003 |
|---|---|---|---|
| Europe | 370.9 | 323.7 | 456.0 |
| USA and Canada | 179.8 | 152.6 | 224.5 |
| South America | 1,828.2 | 2,094.5 | 1,970.6 |
| Bolivia | 113.2 | 119.1 | 59.7 |
| Brazil | 333.0 | 345.0 | 350.3 |
| Chile | 520.3 | 749.0 | 767.8 |
| Paraguay | 469.2 | 518.3 | 429.8 |
| Uruguay | 392.5 | 363.0 | 363.1 |
| **Total** (incl. others)* | 2,620.5 | 2,820.0 | 2,995.3 |

\* Excluding nationals residing abroad.

**Tourism receipts** (US $ million, incl. passenger transport): 2,756 in 2001; 1,716 in 2002; 2,397 in 2003.

Source: World Tourism Statistics.

ARGENTINA

## Communications Media

|  | 2000 | 2001 | 2002 |
|---|---|---|---|
| Television receivers ('000 in use) | 11,500 | 11,800 | n.a. |
| Telephones ('000 main lines in use) | 7,894.2 | 8,108.0 | 8,009.4 |
| Mobile cellular telephones ('000 subscribers) | 6,050.0 | 6,974.9 | 6,500.0 |
| Personal computers ('000 in use) | 2,560 | 2,000 | 3,000 |
| Internet users ('000) | 2,600 | 3,000 | 4,100 |

**Radio receivers** ('000 in use): 24,300 in 1997.

**Facsimile machines** (estimate, '000 in use): 87 in 1998.

**Book production:** 13,148 titles in 2001.

**Daily newspapers:** 34 in 1998.

Sources: mainly UNESCO, *Statistical Yearbook*, and International Telecommunication Union.

## Education
(2000)

|  | Institutions | Students | Teachers |
|---|---|---|---|
| Pre-primary | 16,000 | 1,246,597 | 78,386 |
| Primary | 22,283 | 4,668,006 | 306,210 |
| Secondary | 21,114 | 3,419,901 | 182,863 |
| Universities* | 36 | 1,196,581 | 117,596 |
| Colleges of higher education | 1,754 | 440,164 | 12,825 |

* 2001 figures for state universities only.

Source: former Ministerio de Cultura y Educación.

**Adult literacy rate** (UNESCO estimates): 97.0% (males 97.0%; females 97.0%) in 2002 (Source: UN Development Programme, *Human Development Report*).

# Directory

## The Constitution

The return to civilian rule in 1983 represented a return to the principles of the 1853 Constitution, with some changes in electoral details. In August 1994 a new Constitution was approved, which contained 19 new articles, 40 amendments to existing articles and the addition of a chapter on New Rights and Guarantees. The Constitution is summarized below:

### DECLARATIONS, RIGHTS AND GUARANTEES

Each province has the right to exercise its own administration of justice, municipal system and primary education. The Roman Catholic religion, being the faith of the majority of the nation, shall enjoy state protection; freedom of religious belief is guaranteed to all other denominations. The prior ethnical existence of indigenous peoples and their rights, as well as the common ownership of lands they traditionally occupy, are recognized. All inhabitants of the country have the right to work and exercise any legal trade; to petition the authorities; to leave or enter the Argentine territory; to use or dispose of their properties; to associate for a peaceable or useful purpose; to teach and acquire education, and to express freely their opinion in the press without censorship. The State does not admit any prerogative of blood, birth, privilege or titles of nobility. Equality is the basis of all duties and public offices. No citizens may be detained, except for reasons and in the manner prescribed by the law; or sentenced other than by virtue of a law existing prior to the offence and by decision of the competent tribunal after the hearing and defence of the person concerned. Private residence, property and correspondence are inviolable. No one may enter the home of a citizen or carry out any search in it without his consent, unless by a warrant from the competent authority; no one may suffer expropriation, except in case of public necessity and provided that the appropriate compensation has been paid in accordance with the provisions of the laws. In no case may the penalty of confiscation of property be imposed.

### LEGISLATIVE POWER

Legislative power is vested in the bicameral Congreso (Congress), comprising the Cámara de Diputados (Chamber of Deputies) and the Senado (Senate). The Chamber of Deputies has 257 directly elected members, chosen for four years and eligible for re-election; approximately one-half of the membership of the Chamber shall be renewed every two years. Until October 1995 the Senate had 48 members, chosen by provincial legislatures for a nine-year term, with one-third of the seats renewable every three years. Since October 1995 elections have provided for a third senator, elected by provincial legislatures. In 2001 the entire Senate was renewed; one-third of the Senate was to be renewed every two years from 2003.

The powers of Congress include regulating foreign trade; fixing import and export duties; levying taxes for a specified time whenever the defence, common safety or general welfare of the State so requires; contracting loans on the nation's credit; regulating the internal and external debt and the currency system of the country; fixing the budget and facilitating the prosperity and welfare of the nation. Congress must approve required and urgent decrees and delegated legislation. Congress also approves or rejects treaties, authorizes the Executive to declare war or make peace, and establishes the strength of the Armed Forces in peace and war.

### EXECUTIVE POWER

Executive power is vested in the President, who is the supreme head of the nation and controls the general administration of the country. The President issues the instructions and rulings necessary for the execution of the laws of the country, and himself takes part in drawing up and promulgating those laws. The President appoints, with the approval of the Senate, the judges of the Supreme Court and all other competent tribunals, ambassadors, civil servants, members of the judiciary and senior officers of the Armed Forces and bishops. The President may also appoint and remove, without reference to another body, his cabinet ministers. The President is Commander-in-Chief of all the Armed Forces. The President and Vice-President are elected directly for a four-year term, renewable only once.

### JUDICIAL POWER

Judicial power is exercised by the Supreme Court and all other competent tribunals. The Supreme Court is responsible for the internal administration of all tribunals. In April 1990 the number of Supreme Court judges was increased from five to nine.

### PROVINCIAL GOVERNMENT

The 22 provinces, the Federal District of Buenos Aires and the National Territory of Tierra del Fuego retain all the power not delegated to the Federal Government. They are governed by their own institutions and elect their own governors, legislators and officials.

## The Government

### HEAD OF STATE

**President of the Republic:** Néstor Carlos Kirchner (took office 25 May 2003).

### CABINET
(July 2005)

**Cabinet Chief:** Alberto Fernández.

**General Secretary to the Presidency:** Oscar Parrilli.

**Minister of the Interior:** Aníbal Domingo Fernández.

**Minister of Foreign Affairs, International Trade and Worship:** Rafael Bielsa.

**Minister of Education, Science and Technology:** Daniel Fernando Filmus.

**Minister of Defence:** José Juan Bautista Pampuro.

**Minister of the Economy:** Roberto Lavagna.

**Minister of Labour, Employment and Social Security:** Carlos Tomada.

**Minister of Planning, Public Investment and Services:** Julio Miguel de Vido.

**Minister of Health and the Environment:** Dr Ginés Mario González García.

# ARGENTINA

**Minister of Justice, Security and Human Rights:** ALBERTO IRIBARNE.

**Minister of Social Development:** Dr ALICIA MARGARITA KIRCHNER DE MERCADO.

## MINISTRIES

**General Secretariat to the Presidency:** Balcarce 50, 1064 Buenos Aires; tel. (11) 4344-3662; fax (11) 4344-3789; e-mail secgral@presidencia.net.ar.

**Ministry of Defence:** Azopardo 250, 1328 Buenos Aires; tel. (11) 4346-8800; e-mail mindef@mindef.gov.ar; internet www.mindef.gov.ar.

**Ministry of the Economy:** Hipólito Yrigoyen 250, 1310 Buenos Aires; tel. (11) 4349-5000; e-mail ministrosecpriv@mecon.gov.ar; internet www.mecon.gov.ar.

**Ministry of Education, Science and Technology:** Pizzurno 935, 1020 Buenos Aires; tel. (11) 4129-1000; e-mail info@me.gov.ar; internet www.me.gov.ar.

**Ministry of Foreign Affairs, International Trade and Worship:** Esmeralda 1212, 1007 Buenos Aires; tel. (11) 4819-7000; e-mail web@mrecic.gov.ar; internet www.mrecic.gov.ar.

**Ministry of Health and the Environment:** 9 de Julio 1925, 1332 Buenos Aires; tel. (11) 4381-8911; fax (11) 4381-2182; e-mail consultas@msal.gov.ar; internet www.msal.gov.ar.

**Ministry of the Interior:** Balcarce 50, 1064 Buenos Aires; tel. (11) 4339-0800; fax (11) 4331-6376; e-mail secretariaprivada@mininterior.gov.ar; internet www.mininterior.gov.ar.

**Ministry of Justice, Security and Human Rights:** Sarmiento 329, 1041 Buenos Aires; tel. (11) 4328-3015; internet www.jus.gov.ar.

**Ministry of Labour, Employment and Social Security:** Leandro N. Alem 650, 1001 Buenos Aires; tel. (11) 4311-2913; fax (11) 4312-7860; e-mail consultas@trabajo.gov.ar; internet www.trabajo.gov.ar.

**Ministry of Planning, Public Investment and Services:** Buenos Aires.

**Ministry of Social Development:** Avda 9 de Julio 1925, 14°, 1332 Buenos Aires; tel. (11) 4379-3648; e-mail desarrollosocial@desarrollosocial.gov.ar; internet www.desarrollosocial.gov.ar.

## President and Legislature

### PRESIDENT

Election, First Round, 27 April 2003*

| Candidates | Votes | % of votes cast |
|---|---|---|
| Carlos Saúl Menem (PJ) | 4,677,213 | 24.34 |
| Néstor Carlos Kirchner (PJ) | 4,227,141 | 21.99 |
| Ricardo López Murphy (MFRC†) | 3,142,848 | 16.35 |
| Elisa M. A. Carrió (ARI) | 2,720,143 | 14.15 |
| Adolfo Rodríguez Sáa (PJ) | 2,714,760 | 14.12 |
| Others | 1,737,982 | 9.04 |
| Total‡ | 19,220,087 | 100.00 |

*A second round of voting between the two leading candidates, Carlos Menem and Néstor Kirchner, was scheduled to be held on 18 May 2003; however, on 14 May Menem withdrew his candidacy from the ballot. The presidency was thus won by default by Kirchner.

† The Movimiento Federal para Recrear el Crecimiento, an electoral alliance comprising the Partido Recrear para el Crecimiento and several provincial parties.

‡ Excluding 535,082 invalid votes.

### CONGRESO

#### Cámara de Diputados
(Chamber of Deputies)

**President:** EDUARDO OSCAR CAMAÑO.

The Chamber has 257 members, who hold office for a four-year term, with approximately one-half of the seats renewable every two years.

Distribution of Seats by Party, December 2004

| | Seats |
|---|---|
| Partido Justicialista (PJ) | 129 |
| Unión Cívica Radical (UCR) | 45 |
| Alternativa por una República de Iguales (ARI) | 11 |
| Frente del Movimiento Popular | 7 |
| Partido Socialista | 6 |
| Frepaso (Frente País Solidario) | 5 |
| Autodeterminación y Libertad | 4 |
| Frente Popular Bonaerense (Frepobo) | 4 |
| Movimiento Popular Neuquino | 4 |
| Fuerza Republicana | 3 |
| Partido Unidad Federalista (PaUFe) | 3 |
| Compromiso para el Cambio | 2 |
| Frente Cívico y Social | 2 |
| Frente Nuevo | 2 |
| Fuerza Porteña | 2 |
| Nuevo Buenos Aires | 2 |
| Renovador de Salta | 2 |
| Other parties | 21 |
| Vacant | 3 |
| **Total** | **257** |

#### Senado
(Senate)

**President:** DANIEL OSVALDO SCIOLI.

The Senate has 72 directly elected members, three from each province. One-third of these seats are renewable every two years.

Distribution of Seats*, December 2004

| | Seats |
|---|---|
| Partido Justicialista (PJ) | 40 |
| Unión Cívica Radical (UCR) | 15 |
| Frente Cívico y Social de Catamarca | 2 |
| Fuerza Republicana | 2 |
| Movimiento Popular Neuquino | 2 |
| Cruzada Renovadora de San Juan | 1 |
| Frente Cívico Jujeño | 1 |
| Frente Grande | 1 |
| Frepaso (Frente País Solidario) | 1 |
| Justicialista 17 de Octubre | 1 |
| Partido Socialista | 1 |
| Radical Independiente | 1 |
| Radical Rionegrino | 1 |
| Renovador de Salta | 1 |
| Vecinalista-Partido Nuevo | 1 |
| Vacant | 1 |
| **Total** | **72** |

*The table shows the number of seats by legislative bloc.

### PROVINCIAL ADMINISTRATORS
(July 2005)

**Mayor of the Federal District of Buenos Aires:** ANÍBAL IBARRA.

**Governor of the Province of Buenos Aires:** Dr FELIPE SOLÁ.

**Governor of the Province of Catamarca:** Dr EDUARDO BRIZUELA DEL MORAL.

**Governor of the Province of Chaco:** Dr ROY NIKISCH.

**Governor of the Province of Chubut:** Dr MARIO DAS NEVES.

**Governor of the Province of Córdoba:** Dr JOSÉ MANUEL DE LA SOTA.

**Governor of the Province of Corrientes:** Dr HORACIO RICARDO COLOMBI.

**Governor of the Province of Entre Ríos:** Dr JORGE PEDRO BUSTI.

**Governor of the Province of Formosa:** Dr GILDO INSFRÁN.

**Governor of the Province of Jujuy:** Dr EDUARDO FELLNER.

**Governor of the Province of La Pampa:** Dr CARLOS VERNA.

**Governor of the Province of La Rioja:** Dr ANGEL EDUARDO MAZA.

**Governor of the Province of Mendoza:** JULIO CÉSAR CLETO COBOS.

**Governor of the Province of Misiones:** CARLOS EDUARDO ROVIRA.

**Governor of the Province of Neuquén:** JORGE OMAR SOBISCH.

**Governor of the Province of Río Negro:** Dr MIGUEL SAIZ.

**Governor of the Province of Salta:** Dr JUAN CARLOS ROMERO.

**Governor of the Province of San Juan:** Dr JOSÉ LUIS GIOJA.

ARGENTINA

**Governor of the Province of San Luis:** Dr ALBERTO RODRÍGUEZ SAÁ.

**Governor of the Province of Santa Cruz:** SERGIO ACEVEDO.

**Governor of the Province of Santa Fe:** Dr JORGE OBEID.

**Governor of the Province of Santiago del Estero:** GERARDO ZAMORA.

**Governor of the Province of Tucumán:** JOSÉ JORGE ALPEROVICH.

**Governor of the Territory of Tierra del Fuego:** Dr MARIO JORGE COLAZO.

## Political Organizations

**Acción por la República (AR):** Buenos Aires; e-mail accionrepublica@geocities.com; internet www.ar-partido.com.ar; f. 1997; right-wing; Pres. CÉSAR ALBRISI; Sec.-Gen. ALFREDO CASTAÑON.

**Alternativa por una República de Iguales (ARI):** Buenos Aires; e-mail ; internet www.ari.org.ar; f. 2001; progressive party; Leader ELISA CARRIÓ.

**Corriente Patria Libre:** Humberto I 542, Buenos Aires; tel. (11) 4307-3724; e-mail sitio@patrialibre.org.ar; internet www.patrialibre.org.ar/local.htm; f. 1987; nationalist and left-wing.

**Frente para el Cambio:** Buenos Aires; Leader ALICIA CASTRO.

**Frepaso** (Frente País Solidario): Buenos Aires; tel. (11) 4370-7100; e-mail frepaso@sion.com; internet www.frepaso.org.ar; f. 1994; centre-left coalition of socialist, communist and Christian Democrat groups; Leader CARLOS ALVAREZ.

**Movimiento por la Dignidad y la Independencia (Modin):** Buenos Aires; e-mail modin@causa.zzn.com; internet www.geocities.com/CapitolHill/Parliament/9268/modin.htm; f. 1991; right-wing; Leader Col ALDO RICO.

**Movimiento de Integración y Desarrollo (MID):** Buenos Aires; e-mail midinterior@mid.org.ar; internet www.mid.org.ar; f. 1963; Leader CARLOS ZAFFORE; 145,000 mems.

**Movimiento al Socialismo (MAS):** Chile 1362, 1098 Buenos Aires; tel. (11) 4381-2718; fax (11) 4381-2976; e-mail masarg@arnet.com.ar; internet www.mas.org.ar; Leaders RUBÉN VISCONTI, LUIS ZAMORA; 55,000 mems.

**Nueva Dirigencia (ND):** Buenos Aires; internet www.nuevadirigencia.org.ar; f. 1996; centre-right; Leader GUSTAVO BÉLIZ.

**Partido Comunista de Argentina:** Buenos Aires; e-mail central@pca.org.ar; internet www.pca.org.ar; f. 1918; Leader PATRICIO ECHEGARAY; Sec.-Gen. ATHOS FAVA; 76,000 mems.

**Partido Demócrata Cristiano (PDC):** Combate de los Pozos 1055, 1222 Buenos Aires; tel. (11) 4305-1229; fax (11) 4306-8242; e-mail democracia_cristiana@fibertel.com.ar; internet www.pdca.org.ar; f. 1954; Leader ROBERTO PABLO MEYER; 85,000 mems.

**Partido Demócrata Progresista (PDP):** Chile 1934, 1227 Buenos Aires; internet www.demoprogresista.org.ar; Gen. Sec. Dr ALBERTO NATALE; c. 100,000 mems.

**Partido Federal:** Buenos Aires; internet www.federal.org.ar; f. 1973; Pres. Dr MARTÍN BORELLI.

**Partido Intransigente:** Buenos Aires; e-mail nacional@pi.org.ar; internet www.pi.org.ar; f. 1957; left-wing; Leaders Dr OSCAR ALENDE, LISANDRO VIALE; Sec. MARIANO LORENCES; 90,000 mems.

**Partido Justicialista (PJ):** Buenos Aires; internet www.pj.org.ar; Peronist party; f. 1945; Pres. EDUARDO FELLNER; Leader ALBERTO FERNÁNDEZ; 3m. mems; three factions within party contested the 2003 presidential election.

   **Frente por la Libertad/Unión del Centro Democrático:** Presidential Candidate CARLOS SAÚL MENEM.

   **Frente del Movimiento Popular/Partido Unión y Libertad:** Presidential Candidate ADOLFO RODRÍGUEZ SÁA.

   **Frente para la Victoria:** Presidential Candidate NÉSTOR CARLOS KIRCHNER.

**Partido Memoria y Movilización Social:** Buenos Aires; f. 2003; left-wing; mem. of the Mesa Coordinadora para un Nuevo Proyecto Nacional; Pres. EDUARDO LUIS DUHALDE.

**Partido Nacional de Centro:** Buenos Aires; f. 1980; conservative; Leader RAÚL RIVANERA CARLES.

**Partido Nuevo Triunfo (PNT):** Buenos Aires; e-mail pnt@libreopinion.com; internet pnt.libreopinion.com; f. 1990 as Partido de los Trabajadores; illegal extremist right-wing group; Pres. ALEJANDRO CARLOS BIONDINI; Vice-Pres. JORGE CASÓLIBA.

**Partido Obrero:** Ayacucho 444, Buenos Aires; tel. (11) 4953-3824; fax (11) 4953-7164; internet www.po.org.ar; f. 1982; Trotskyist; Leaders JORGE ALTAMIRA, CHRISTIAN RATH; 61,000 mems.

**Partido Popular Cristiano (PPC):** Entre Ríos; internet www.partidopopularcris.com.ar; Leader JOSÉ ANTONIO ALLENDE.

**Partido Recrear para el Crecimiento (RECREAR):** Buenos Aires; internet www.recrearargentina.org; contested 2003 elections as Movimiento Federal para Recrear el Crecimiento, in alliance with several provincial parties; Pres. RICARDO LÓPEZ MURPHY; Vice-Pres. JOSÉ MARÍA LLADÓS; Sec. RICARDO URQUIZA.

**Partido de la Revolución Democrática (PRD):** Buenos Aires; internet www.bonassoprd.com.ar; f. 2001; left-wing; mem. of the Mesa Coordinadora para un Nuevo Proyecto Nacional; Leader MIGUEL BONASSO.

**Partido Socialista:** Buenos Aires; f. 2003, following merger of the Partido Socialista Democrático and the Partido Socialista Popular.

**Partido Socialista Auténtico:** Sarandí 56, Buenos Aires; tel. and fax (11) 4952-3103; e-mail psacomitenacional@espeedy.com.ar; internet webs.psa.org.ar; f. 1896; Sec.-Gen. MARIO MAZZITELLI.

**Partido Unidad Federalista (PaUFe):** Buenos Aires; e-mail mlisa@unidadfederalista.org.ar; internet www.paufe.org.ar; Leader LUIS ABELARDO PATTI.

**Polo Social:** Echeverría 441, 1878 Quilmes, Buenos Aires; tel. and fax (11) 4253-0971; e-mail polosocial@cscom.com.ar; internet polosocial.agruparnos.com.ar; f. 1999; mem. of the Mesa Coordinadora para un Nuevo Proyecto Nacional; Leader FRANCISCO ('BARBA') GUTIÉRREZ.

**Unión del Centro Democrático (UCeDé):** Buenos Aires; f. 1980 as coalition of eight minor political organizations; Leader CARLOS CASTELLANI.

**Unión Cívica Radical (UCR):** Alsina 1786, 1088 Buenos Aires; tel. and fax (11) 4375-2000; e-mail info@ucr.org.ar; internet www.ucr.org.ar; moderate; f. 1890; Pres. ANGEL ROZAS; Sec.-Gen. GERARDO MORALES; 2.9m. mems.

**Unión para la Nueva Mayoría:** Buenos Aires; f. 1986; centre-right; Leader JOSÉ ANTONIO ROMERO FERIS.

The following political parties and groupings also contested the 2003 legislative elections: Alianza Nuevo Espacio Entrerriano, Autodeterminación y Libertad, Compromiso para el Cambio, Cruzada Renovadora de San Juan, Demócrata Mendoza, Demócrata Progresista, FISCAL—Mendoza, Frente Alternativa Pampeana, Frente Cívico Jujeño, Frente Cívico y Social de Catamarca, Frente Grande, Frente Nuevo, Frente Popular Bonaerense (Frepobo), Frente Renovador, Frente de Unidad Provincial, Fuerza Portena, Fuerza Republicana, Izquierda Unida, Movimiento Multisectorial de Trabajo y la Autodeterminación, Movimiento Popular Neuquino, Nuevo Buenos Aires, Proyecto Corrientes, Radical Independiente, Radical Rionegrino, Renovador de Salta, Unión por la Argentina and Vecinalista-Partido Nuevo.

### OTHER ORGANIZATIONS

**Asociación Madres de Plaza de Mayo:** Hipólito Yrigoyen 1584, 1089 Buenos Aires; tel. (11) 4383-0377 ; fax (11) 4954-0381; e-mail madres@satlink.com; internet www.madres.org; f. 1979; formed by mothers of those who 'disappeared' during the years of military rule, it has since become a broad-based grouping with revolutionary socialist aims; Founder and Leader HEBE MARÍA PASTOR DE BONAFINI.

**Mesa Coordinadora para un Nuevo Proyecto Nacional:** Buenos Aires; f. Oct. 2004; umbrella org. comprising various left-wing parties and *piquetero* groupings, including:

   **Federación de Tierra y Vivienda (FTV):** e-mail ftv@cta.org.ar; moderate grouping; 150,000 mems; Leader LUIS D'ELIA.

   **Frente Transversal Nacional y Popular:** Leader EDGARDO DEPETRI.

   **Movimiento Barrios de Pie:** moderate grouping; Leader JORGE CEBALLOS.

   **Movimiento de Trabajadores Desocupados 'Evita' (MTD Evita):** Leader EMILIO PÉRSICO.

   **Partido Memoria y Movilización Social:** see Political Organizations.

   **Partido de la Revolución Democrática (PRD):** see Political Organizations.

   **Polo Social:** see Political Organizations.

Other *piquetero* groupings include:

**Bloque Piquetero Nacional.**

**Corriente Clasista y Combativa (CCC):** e-mail info@cccargentina.org.ar; internet www.cccargentina.org.ar; radical grouping; Leader CARLOS 'PERRO' SANTILLÁN.

# ARGENTINA

**Movimiento Independientes de Jubilados y Descocupados (MIJD):** radical grouping; Leader Raúl Castells.

**Movimiento Social de Trabajadores Teresa Rodríguez:** Leguizamón 2570, Mar de Plata; tel. (223) 487-9931; e-mail noticias@elteresa.org.ar; internet www.elteresa.org.ar; f. 1996, Movimiento de Trabajadores Desocupados; adopted present name in 1997; extreme left-wing.

**Movimiento de Trabajadores Desocupados 'Aníbal Verón' (MTD Aníbal Verón):** tel. (11) 5489-1374.

**Movimiento sin Trabajo—Teresa Vive (MST—TV):** radical grouping; Leader Gustavo Giménez.

**Polo Obrero:** Leader Néstor Pitrola.

## Diplomatic Representation

### EMBASSIES IN ARGENTINA

**Albania:** Avda del Libertador 946, 4°, 1001 Buenos Aires; tel. (11) 4812-8366; fax (11) 4815-2512; e-mail ambasada.bue@fibertel.com.ar; Ambassador Edmond Trako.

**Algeria:** Montevideo 1889, 1021 Buenos Aires; tel. (11) 4815-1271; fax (11) 4815-8837; Ambassador Nourredine Ayadi.

**Angola:** Buenos Aires; Ambassador Fernando Dito.

**Armenia:** Avda Roque S. Peña 570, 3°, 1035 Buenos Aires; tel. (11) 4816-8710; fax (11) 4812-2803; e-mail armenia@fibertel.com.ar; Ambassador Ara Aivazian.

**Australia:** Villanueva 1400, 1426 Buenos Aires; tel. (11) 4779-3500; fax (11) 4779-3581; e-mail info.bageneral@dfat.gov.au; internet www.argentina.embassy.gov.au; Ambassador Peter Hussin.

**Austria:** French 3671, 1425 Buenos Aires; tel. (11) 4802-1400; fax (11) 4805-4016; e-mail embajada@austria.org.ar; internet www.austria.org.ar; Ambassador Dr Gudrun Graf.

**Belarus:** Cazadores 2166, 1428 Buenos Aires; tel. (11) 4788-9394; fax (11) 4788-2322; e-mail argentina@belembassy.org; Ambassador Vadim Ivanovich Lazerko.

**Belgium:** Defensa 113, 8°, 1065 Buenos Aires; tel. (11) 4331-0066; fax (11) 4311-0814; e-mail buenosaires@diplobel.org; internet www.diplobel.org/argentina; Ambassador Koenraad Rouvroy.

**Bolivia:** Avda Corrientes 545, 2°, 1043 Buenos Aires; tel. (11) 4394-6042; fax (11) 5217-1070; Ambassador Arturo Liebers Baldivieso.

**Bosnia and Herzegovina:** Miñones 2445, Buenos Aires; tel. (11) 4896-0284; fax (11) 4896-0351; Ambassador Duško Ladan.

**Brazil:** Cerrito 1350, 1010 Buenos Aires; tel. (11) 4515-2400; fax (11) 4515-2401; e-mail embras@embrasil.org.ar; internet www.brasil.org.ar; Ambassador Mauro Vieira.

**Bulgaria:** Mariscal A. J. de Sucre 1568, 1428 Buenos Aires; tel. (11) 4781-8644; fax (11) 4786-6273; e-mail embular@sinectis.com.ar; internet www.sinectis.com.ar/u/embular; Ambassador Atranas I. Budev.

**Canada:** Tagle 2828, 1425 Buenos Aires; Casilla 1598 C1000WAP Correo Central, Buenos Aires; tel. (11) 4808-1000; fax (11) 4808-1111; internet www.dfait-maeci.gc.ca/bairs; e-mail bairs-webmail@dfait-maeci.gc.ca; Ambassador Yves Gagnon.

**Chile:** Tagle 2762, 1425 Buenos Aires; tel. (11) 4802-7020; fax (11) 4804-5927; e-mail data@embajadadechile.com.ar; internet www.embajadadechile.com.ar; Ambassador Luis Maira Aguirre.

**China, People's Republic:** Avda Crisólogo Larralde 5349, 1431 Buenos Aires; tel. (11) 4543-8862; fax (11) 4545-1141; e-mail embchinaargentina@hotmail.com; internet ar.chineseembassy.org/esp; Ambassador Ke Xiaogang.

**Colombia:** Carlos Pellegrini 1363, 3°, 1010 Buenos Aires; tel. (11) 4325-0258; fax (11) 4322-9370; e-mail ebaires@minrelext.gov.co; internet www.embajadacolombia.int.ar; Ambassador Rodrigo Ernesto Holguín Lourido.

**Congo, Democratic Republic:** Callao 322, 2°, Buenos Aires; tel. (11) 4373-7565; fax (11) 4374-9865; Chargé d'affaires a.i. Yemba Lohaka.

**Costa Rica:** Avda Santa Fe 1460, 3°I, 2023 Buenos Aires; tel. (11) 4815-0072; fax (11) 4814-1660; e-mail embarica@ifibertel.com.ar; Ambassador Eduardo Francisco Otoya Boulanger.

**Croatia:** Gorostiaga 2104, 1426 Buenos Aires; tel. (11) 4777-6409; fax (11) 4777-9159; e-mail embajadadecroacia@velocom.com.ar; Ambassador Rikard Rossetti.

**Cuba:** Virrey del Pino 1810, 1426 Buenos Aires; tel. (11) 4782-9049; fax (11) 4786-7713; e-mail argoficemb@ecuargentina.minrex.gov.cu; Ambassador Alejandro González Galiano.

**Czech Republic:** Junín 1461, 1113 Buenos Aires; tel. (11) 4807-3107; fax (11) 4807-3109; e-mail buenosaires@embassy.mzv.cz; internet www.mfa.cz/buenosaires; Ambassador František Padělek.

**Dominican Republic:** Avda Santa Fe 830, 7°, 1059 Buenos Aires; tel. (11) 4312-9378; fax (11) 4894-2078; Ambassador Rafael Calventi.

**Ecuador:** Presidente Quintana 585, 9° y 10°, 1129 Buenos Aires; tel. (11) 4804-0073; fax (11) 4804-0074; e-mail embecuador@embecuador.com.ar; Ambassador Dr José Rafael Serrano Herrera.

**Egypt:** Virrey del Pino 3140, 1426 Buenos Aires; tel. (11) 4553-3311; fax (11) 4553-0067; e-mail embegypt@fibertel.com.ar; Ambassador Hazem Mohamed Taher Elsaed.

**El Salvador:** Esmeralda 1066, 7°, 1059 Buenos Aires; tel. (11) 4311-1864; fax (11) 4314-7628; Ambassador Rafael Alfonso Quiñónez Meza.

**Finland:** Avda Santa Fe 846, 5°, 1059 Buenos Aires; tel. (11) 4312-0600; fax (11) 4312-0670; e-mail sanomat.bue@formin.fi; internet www.finembue.com.ar; Ambassador Risto Kaarlo Veltheim.

**France:** Cerrito 1399, 1010 Buenos Aires; tel. (11) 4515-2930; fax (11) 4515-0120; e-mail ambafr@abaconet.com.ar; internet www.embafrancia-argentina.org; Ambassador Francis Lott.

**Germany:** Villanueva 1055, 1426 Buenos Aires; tel. (11) 4778-2500; fax (11) 4778-2550; e-mail administracion@embajada-alemana.org.ar; internet www.embajada-alemana.org.ar; Ambassador Dr Rolf Schumacher.

**Greece:** Arenales 1658, 1061 Buenos Aires; tel. (11) 4811-4811; fax (11) 4816-2600; e-mail gremb.bay@mfa.gr; Ambassador Cogevinas Alexios.

**Guatemala:** Avda Santa Fe 830, 5°, 1059 Buenos Aires; tel. (11) 4313-9160; fax (11) 4313-9181; e-mail embargentina@minex.gob.gt; Ambassador (vacant).

**Haiti:** Avda Figueroa Alcorta 3297, 1425 Buenos Aires; tel. (11) 4802-0211; fax (11) 4802-3984; e-mail embahaiti@interar.com.ar; Ambassador Edris Saint Armand.

**Holy See:** Avda Alvear 1605, 1014 Buenos Aires; tel. (11) 4813-9697; fax (11) 4815-4097; Apostolic Nuncio Most Rev. Adriano Bernardini (Titular Archbishop of Faleri).

**Honduras:** Avda del Libertador 1146, 1112 Buenos Aires; tel. (11) 4804-6181; fax (11) 4804-3222; e-mail honduras@ciudad.com.ar; Ambassador Napoleón Alvarez Alvarado.

**India:** Córdoba 950, 4°, 1054 Buenos Aires; tel. (11) 4393-4001; fax (11) 4393-4063; e-mail indemb@indembarg.org.ar; internet www.indembarg.org.ar; Ambassador Rinzing Wangdi; Ambassador Pramathesh Rath (designate).

**Indonesia:** Mariscal Ramón Castilla 2901, 1425 Buenos Aires; tel. (11) 4807-2211; fax (11) 4802-4448; e-mail emindo@tournet.com.ar; internet www.indonesianembassy.org.ar; Ambassador Max Pangemanan.

**Iran:** Avda Figueroa Alcorta 3229, 1425 Buenos Aires; tel. (11) 4802-1470; fax (11) 4805-4409; Chargé d'affaires a.i. Muhammad Ali Tabatabaei Hasan.

**Ireland:** Avda del Libertador 1068, 6°, 1112 Buenos Aires; tel. (11) 5787-0801; fax (11) 5787-0802; e-mail info@irlanda.org.ar; internet www.irlanda.org.ar; Ambassador Kenneth Thompson.

**Israel:** Avda de Mayo 701, 10°, 1084 Buenos Aires; tel. (11) 4338-2500; fax (11) 4338-2624; e-mail info@buenosaires.mfa.gov.il; internet buenosaires.mfa.gov.il; Ambassador Rafael Eldad.

**Italy:** Billinghurst 2577, 1425 Buenos Aires; tel. (11) 4802-0071; fax (11) 4804-4914; e-mail ambitalia@ambitalia-bsas.org.ar; internet www.ambitalia-bsas.org.ar; Ambassador Roberto Nigido.

**Japan:** Bouchard 547, 17°, 1106 Buenos Aires; tel. (11) 4318-8200; fax (11) 4318-8210; e-mail taishikan@japan.org.ar; internet www.ar.emb-japan.go.jp; Ambassador Shinya Nagai.

**Korea, Republic:** Avda del Libertador 2395, 1425 Buenos Aires; tel. (11) 4802-8865; fax (11) 4803-6993; e-mail argentina@mofat.go.kr; internet www.embcorea.org.ar; Ambassador Yang-Boo Choe.

**Kuwait:** Uruguay 739, 1015 Buenos Aires; tel. (11) 4374-7202; fax (11) 4374-8718; e-mail info@kuwait.com.ar; internet www.kuwait.com.ar; Ambassador Ali Hussain As-Sammak.

**Lebanon:** Avda del Libertador 2354, 1425 Buenos Aires; tel. (11) 4802-4492; fax (11) 4802-2909; e-mail embajada@ellibano.com.ar; Ambassador Hicham Salim Hamdan.

**Libya:** 3 de Febrero 1358/62, 1426 Buenos Aires; tel. (11) 4788-3760; fax (11) 4784-9895; Chargé d'affaires a.i. Dr Ali Y. Guima ben Guima.

**Lithuania:** Gelly 3530, Buenos Aires; tel. (11) 4801-6007; fax (11) 4805-8238; e-mail amb.ar@urm.lt; also responsible for relations with Chile, Colombia, Ecuador, Uruguay and Venezuela; Chargé d'affaires Arvydas Naujokaitis.

# ARGENTINA

**Malaysia:** Villanueva 1040-1048, 1062 Buenos Aires; tel. (11) 4776-0504; e-mail mwbaires@ciudad.com.ar; Ambassador Dato MOHD ROZE'RADZI ABDUL RAHMAN.

**Mexico:** Arcos 1650, Belgrano, 1426 Buenos Aires; tel. (11) 4789-8800; fax (11) 4789-8836; internet www.embamex.int.ar; Ambassador MARÍA CRISTINA DE LA GARZA SANDOVAL.

**Netherlands:** Edif. Porteño II, Olga Cossenttini 831, 3°, 1107 Buenos Aires; tel. (11) 4338-0050 ; fax (11) 4338-0060; e-mail bue@minbuza.nl; internet www.embajadaholanda.int.ar; Ambassador ROBERT JAN VAN HOUTUM.

**New Zealand:** Carlos Pellegrine 1427, 5°, 1010 Buenos Aires; tel. (11) 4328-0747; fax (11) 4328-0757; e-mail kiwiargentina@datamarkets.com.ar; internet www.nzembassy.com/argentina; Ambassador CARL ROBINSON WORKER.

**Nicaragua:** Callao 1564, 1°, Of. A, 1024 Buenos Aires; tel. (11) 4807-0260; fax (11) 4806-1221; e-mail embanic@fibertel.com.ar; Ambassador SILVIO AVILEZ GALLO.

**Nigeria:** Rosales 2674, Casilla 2100, 1636 Olivos, 1000 Buenos Aires; tel. (11) 4790-7565; Ambassador Dr M. A. WALI.

**Norway:** Esmeralda 909, 3°B, 1007 Buenos Aires; tel. (11) 4312-2204; fax (11) 4315-2831; e-mail emb.buenosaires@mfa.no; internet www.noruega.org.ar; Ambassador GUNNAR H. LINDEMAN.

**Pakistan:** Gorostiaga 2176, 1426 Buenos Aires; tel. (11) 4775-1294; fax (11) 4776-1186; e-mail parepbaires@sinectis.com.ar; Ambassador MOHAMMAD NISAR.

**Panama:** Avda Santa Fe 1461, 1°, 1060 Buenos Aires; tel. (11) 4811-1254; fax (11) 4814-0450; e-mail epar@fibertel.com.ar; internet www.embajadadepanama.com.ar; Ambassador MANUEL ORESTES NIETO.

**Paraguay:** Avda Las Heras 2545, 1425 Buenos Aires; tel. (11) 4802-3826; fax (11) 4801-0657; Ambassador (vacant).

**Peru:** Avda del Libertador 1720, 1425 Buenos Aires; tel. (11) 4802-2000; fax (11) 4802-5887; e-mail embperu@arnet.com.ar; Ambassador Dr MARTÍN BELAÚNDE MOREYRA.

**Philippines:** Mariscal Ramón Castilla 3075/3085, 1425 Buenos Aires; tel. (11) 4807-3334 ; fax (11) 4804-1595; e-mail phba@peoples.com.ar; Ambassador GEORGE B. REYES.

**Poland:** Alejandro M. de Aguado 2870, 1425 Buenos Aires; tel. (11) 4802-9681; fax (11) 4802-9683; e-mail polemb@datamarkets.com.ar; Ambassador SLAWOMIR RATAJSKI.

**Portugal:** Maipú 942, 17°, 1340 Buenos Aires; tel. (11) 4312-0187; fax (11) 4311-2586; e-mail embpor@embajadaportugal.org.ar; Ambassador ANTÓNIO CARLOS CARVALHO DE ALMEIDA RIBEIRO.

**Romania:** Arroyo 962-970, 1007 Buenos Aires; tel. and fax (11) 4326-5888; fax 4322-2630; e-mail embarombue@rumania.org.ar; internet www.roemb.com.ar; Ambassador ALEXANDRU VICTOR MICULA.

**Russia:** Rodríguez Peña 1741, 1021 Buenos Aires; tel. (11) 4813-1552; fax (11) 4812-1794; Ambassador YEVGENII M. ASTAKHOV.

**Saudi Arabia:** Alejandro M. de Aguado 2881, 1425 Buenos Aires; tel. (11) 4802-3375; fax (11) 4806-1581; e-mail embasaudita@fibertel.com.ar; internet www.embajadasaudi.org; Ambassador ADNAN BIN ABDULLAH BAGHDADI.

**Serbia and Montenegro:** Marcelo T. de Alvear 1705, 1060 Buenos Aires; tel. (11) 4812-9133; fax (11) 4812-1070; e-mail yuembaires@ciudad.com.ar; Ambassador IVAN SAVELJIĆ.

**Slovakia:** Avda Figueroa Alcorta 3240, 1425 Buenos Aires; tel. (11) 4801-3917; fax (11) 4801-4654; e-mail embsl@fibertel.com.ar; Ambassador VLADIMIR GRACZ.

**Slovenia:** Santa Fe 846, 6°, 1059 Buenos Aires; tel. (11) 4894-0621; fax (11) 4312-8410; e-mail vba@mzz-dkp.gov.si; Ambassador BOJAN GROBOVŠEK.

**South Africa:** Marcelo T. de Alvear 590, 8°, 1058 Buenos Aires; tel. (11) 4317-2900; fax (11) 4317-2951; e-mail embasa@ciudad.com.ar; internet www.sudafrica.org.ar; Ambassador Prof. MLUNGIS WASHINGTON MAKALIMA.

**Spain:** Mariscal Ramón Castilla 2720, 1425 Buenos Aires; tel. (11) 4802-6031; fax (11) 4802-0719; Ambassador MANUEL ALABART FERNÁNDEZ-CAVADA.

**Sweden:** Casilla 3599, Correo Central 1000, Buenos Aires; tel. (11) 4329-0800; fax (11) 4342-1697; e-mail ambassaden.buenos-aires@foreign.ministery.ce; Ambassador MADELEINE STRÖJE-WILKENS.

**Switzerland:** Avda Santa Fe 846, 10°, 1059 Buenos Aires; tel. (11) 4311-6491; fax (11) 4313-2998; e-mail vertretung@bue.rep.admin.ch; internet www.eda.admin.ch/buenosaires_emb/s/home.html; Ambassador DANIEL VON MURALT.

**Syria:** Calloa 956, 1023 Buenos Aires; tel. (11) 4813-2113; fax (11) 4814-3211; Ambassador RIYAD SNEIH.

**Thailand:** Federico Lacroze 2158, 1426 Buenos Aires; tel. (11) 4774-4415; fax (11) 4773-2447; e-mail thaiembargen@fibertel.com.ar; Ambassador ASIPHOL CHABCHITRCHAIDOL.

**Tunisia:** Ciudad de la Paz 3086, 1429 Buenos Aires; tel. (11) 4544-2618; fax (11) 4545-6369; e-mail atbuenosaires@infovia.com.ar; Ambassador HAMIDA MRABET LABIDI.

**Turkey:** 11 de Setiembre 1382, 1426 Buenos Aires; tel. (11) 4788-3239; fax (11) 4784-9179; e-mail turquia@fibertel.com.ar; Ambassador SUKRU TUFAN.

**Ukraine:** Lafinur 3057, 1425 Buenos Aires; tel. (11) 4552-0657; fax (11) 4552-6771; e-mail embucra@embucra.com.ar; internet www.embucra.com.ar; Ambassador OLEKSANDR NYKONENKO.

**United Kingdom:** Dr Luis Agote 2412, 1425 Buenos Aires; tel. (11) 4808-2200; fax (11) 4808-2274; e-mail askinformation.baires@fco.gov.uk; internet www.britain.org.ar; Ambassador Dr JOHN HUGHES.

**USA:** Avda Colombia 4300, 1425 Buenos Aires; tel. (11) 5777-4533; fax (11) 5777-4240; internet buenosaires.usembassy.gov; Ambassador LINO GUTIÉRREZ.

**Uruguay:** Avda Las Heras 1907, 1128 Buenos Aires; tel. (11) 4803-6030; fax (11) 4807-3050; e-mail webmaster@embajadauruguay.com; internet www.embajadadeluruguay.com.ar; Ambassador FRANCISCO BUSTILLO BONASSO.

**Venezuela:** Virrey Loreto 2035, 1428 Buenos Aires; tel. (11) 4788-4944; fax (11) 4784-4311; e-mail embaven@fibertel.com.ar; internet www.la-embajada.com.ar; Ambassador FREDDY BALZÁN MORREL.

**Viet Nam:** 11 de Septiembre 1442, 1426 Buenos Aires; tel. (11) 4783-1802; fax (11) 4782-0078; e-mail sovnartn@teletel.com.ar; Ambassador DIEN NGUYEN NGOC.

# Judicial System

## SUPREME COURT

**Corte Suprema**

Talcahuano 550, 4°, 1013 Buenos Aires; tel. (11) 4370-4600; fax (11) 440-2270; e-mail albinogomez@cjsn.gov.ar; internet www.csjn.gov.ar.

The nine members of the Supreme Court are appointed by the President, with the agreement of at least two-thirds of the Senate. Members are dismissed by impeachment.

**President:** ENRIQUE SANTIAGO PETRACCHI.

**Vice-President:** AUGUSTO CÉSAR BELLUSCIO (until 1 Sept. 2005).

**Justices:** EUGENIO RAÚL ZAFFARONI, AUGUSTO CÉSAR BELLUSCIO (until 1 Sept. 2005), ENRIQUE SANTIAGO PETRACCHI, ANTONIO BOGGIANO (suspended), RICARDO LORENZETTI, JUAN CARLOS MAQUEDA, CARLOS S. FAYT, ELENA I. HIGHTON DE NOLASCO, CARMEN ARGIBAY.

## OTHER COURTS

Judges of the lower, national or further lower courts are appointed by the President, with the agreement of the Senate, and are dismissed by impeachment. From 1999, however, judges were to retire on reaching 75 years of age.

The Federal Court of Appeal in Buenos Aires has three courts: civil and commercial, criminal, and administrative. There are six other courts of appeal in Buenos Aires: civil, commercial, criminal, peace, labour, and penal-economic. There are also federal appeal courts in: La Plata, Bahía Blanca, Paraná, Rosario, Córdoba, Mendoza, Tucumán and Resistencia. In August 1994, following constitutional amendments, the Office of the Attorney-General was established as an independent entity and a Council of Magistrates was envisaged. In December 1997 the Senate adopted legislation to create the Council.

The provincial courts each have their own Supreme Court and a system of subsidiary courts. They deal with cases originating within and confined to the provinces.

**Attorney-General:** ESTEBÁN RIGHI.

# Religion

## CHRISTIANITY

At 31 December 2003 there were an estimated 34,400,446 Catholics in Argentina.

**Federación Argentina de Iglesias Evangélicas** (Argentine Federation of Evangelical Churches): José María Moreno 873, 1424 Buenos Aires; tel. and fax (11) 4922-5356; e-mail presidencia@faie.com.ar; internet www.faie.org.ar; f. 1938; 29 mem. churches; Pres.

# ARGENTINA

Rev. EMILIO MONTI (Methodist Evangelical Church); Exec. Sec. Rev. FLORENCIA HIMITIAN.

### The Roman Catholic Church

Argentina comprises 14 archdioceses, 50 dioceses (including one each for Uniate Catholics of the Ukrainian rite, of the Maronite rite and of the Armenian rite) and three territorial prelatures. The Archbishop of Buenos Aires is also the Ordinary for Catholics of Oriental rites, and the Bishop of San Gregorio de Narek en Buenos Aires is also the Apostolic Exarch of Latin America and Mexico for Catholics of the Armenian rite.

**Bishops' Conference:** Conferencia Episcopal Argentina, Suipacha 1034, 1008 Buenos Aires; tel. (11) 4328-0993; fax (11) 4328-9570; e-mail seccea@cea.org.ar; internet www.cea.org.ar; f. 1959; Pres. Mgr EDUARDO VICENTE MIRÁS (Archbishop of Rosario).

#### Armenian Rite

**Bishop of San Gregorio de Narek en Buenos Aires:** VARTAN WALDIR BOGHOSSIAN, Charcas 3529, 1425 Buenos Aires; tel. (11) 4824-1613; fax (11) 4827-1975; e-mail exarmal@pcn.net; internet fast.to/exarcado.

#### Latin Rite

**Archbishop of Bahía Blanca:** GUILLERMO JOSÉ GARLATTI, Avda Colón 164, 8000 Bahía Blanca; tel. (291) 455-0707; fax (291) 452-2070; e-mail arzobis@arzobispadobahia.org.ar.

**Archbishop of Buenos Aires:** Cardinal JORGE MARÍA BERGOGLIO, Rivadavia 415, 1002 Buenos Aires; tel. (11) 4343-0812; fax (11) 4334-8373; e-mail arzobispado@arzbaires.org.ar.

**Archbishop of Córdoba:** CARLOS JOSÉ NÁÑEZ, Hipólito Irigoyen 98, 5000 Córdoba; tel. and fax (351) 422-1015; e-mail info@arzobispado.org.ar.

**Archbishop of Corrientes:** DOMINGO SALVADOR CASTAGNA, 9 de Julio 1543, 3400 Corrientes; tel. and fax (3783) 422436; e-mail arzctes@arnet.com.ar.

**Archbishop of La Plata:** HÉCTOR AGUER, Calle 14, 1009, 1900 La Plata; tel. (221) 425-1656; e-mail cancillerialp@infovia.com.ar; internet www.arzolap.org.ar.

**Archbishop of Mendoza:** JOSÉ MARÍA ARANCIBIA, Catamarca 98, 5500 Mendoza; tel. (261) 423-3862; fax (261) 429-5415; e-mail arzobispadomza@supernet.com.ar.

**Archbishop of Mercedes-Luján:** RUBÉN HÉCTOR DI MONTE, Calle 22, No 745, 6600 Mercedes, Buenos Aires; tel. (2324) 432-412; fax (2324) 432-104; e-mail arzomerce@yahoo.com.

**Archbishop of Paraná:** MARIO LUIS BAUTISTA MAULIÓN, Monte Caseros 77, 3100 Paraná; tel. (343) 431-1440; fax (343) 423-0372; e-mail arzparan@arzparan.org.ar.

**Archbishop of Resistencia:** CARMELO JUAN GIAQUINTA, Bartolomé Mitre 363, Casilla 35, 3500 Resistencia; tel. and fax (3722) 441908; fax (3722) 434573; e-mail arzobrcia@arnet.com.ar.

**Archbishop of Rosario:** EDUARDO VICENTE MIRÁS, Córdoba 1677, 2000 Rosario; tel. (341) 425-1298; fax (341) 425-1207; e-mail arzobros@uolsinectis.com.ar; internet www.delrosario.org.ar.

**Archbishop of Salta:** MARIO ANTONIO CARGNELLO, España 596, 4400 Salta; tel. (387) 421-4306; fax (387) 421-3101; e-mail arzobisposalta@infovia.com.ar.

**Archbishop of San Juan de Cuyo:** ALFONSO ROGELIO DELGADO EVERS, Bartolomé Mitre 250 Oeste, 5400 San Juan de Cuyo; tel. (264) 422-2578; fax (264) 427-3530; e-mail arzobispadosanjuan@infovia.com.ar.

**Archbishop of Santa Fe de la Vera Cruz:** JOSÉ MARÍA ARANCEDO, Avda General López 2720, 3000 Santa Fe; tel. (342) 459-1780; fax (342) 459-4491; e-mail curia@arquisantafe.org.ar.

**Archbishop of Tucumán:** LUIS HÉCTOR VILLALBA, Avda Sarmiento 895, 4000 San Miguel de Tucumán; tel. (381) 422-6345; fax (381) 431-0617; e-mail arztuc@arnet.com.ar.

#### Maronite Rite

**Bishop of San Charbel en Buenos Aires:** CHARBEL MERHI, Eparquía Maronita, Paraguay 834, 1057 Buenos Aires; tel. (11) 4311-7299; fax (11) 4312-8348; e-mail mcharbel@hotmail.com.

#### Ukrainian Rite

**Bishop of Santa María del Patrocinio en Buenos Aires:** Rt Rev. MIGUEL MYKYCEJ, Ramón L. Falcón 3950, Casilla 28, 1407 Buenos Aires; tel. (11) 4671-4192; fax (11) 4671-7265; e-mail pokrov@ciudad.com.ar.

### The Anglican Communion

The Iglesia Anglicana del Cono Sur de América (Anglican Church of the Southern Cone of America) was formally inaugurated in Buenos Aires in April 1983. The Church comprises seven dioceses: Argentina, Northern Argentina, Chile, Paraguay, Peru, Bolivia and Uruguay. The Primate is the Bishop of Argentina.

**Bishop of Argentina:** Rt Rev. GREGORY VENABLES, 25 de Mayo 282, 1002 Buenos Aires; Casilla 4293, Correo Central 1000, Buenos Aires; tel. (11) 4342-4618; fax (11) 4331-0234; e-mail diocesisanglibue@arnet.com.ar.

**Bishop of Northern Argentina:** Rt Rev. MAURICE SINCLAIR, Casilla 187, 4400 Salta; tel. (387) 431-1718; fax (387) 431-2622; e-mail sinclair@salnet.com.ar; jurisdiction extends to Jujuy, Salta, Tucumán, Catamarca, Santiago del Estero, Formosa and Chaco.

### Protestant Churches

**Convención Evangélica Bautista Argentina** (Baptist Evangelical Convention): Virrey Liniers 42, 1174 Buenos Aires; tel. and fax (11) 4864-2711; e-mail ceba@sion.com; f. 1909; Pres. CARLOS A. CARAMUTTI.

**Iglesia Evangélica Congregacionalista** (Evangelical Congregational Church): Perón 525, 3100 Paraná; tel. (43) 21-6172; f. 1924; 100 congregations, 8,000 mems, 24,000 adherents; Supt Rev. REYNOLDO HORSTT.

**Iglesia Evangélica Luterana Argentina** (Evangelical Lutheran Church of Argentina): Ing. Silveyra 1639-41, 1607 Villa Adelina, Buenos Aires; tel. (11) 4766-7948; fax (11) 4766-7948; f. 1905; 30,000 mems; Pres. WALDOMIRO MAILI.

**Iglesia Evangélica del Río de la Plata** (Evangelical Church of the River Plate): Mariscal Sucre 2855, 1428 DVY Buenos Aires; tel. (11) 4787-0436; fax (11) 4787-0335; e-mail presidente@ierp.org.ar; f. 1899; 25,000 mems; Pres. FEDERICO HUGO SCHÄFER.

**Iglesia Evangélica Metodista Argentina** (Methodist Church of Argentina): Rivadavia 4044, 3°, 1205 Buenos Aires; tel. (11) 4982-3712; fax (11) 4981-0885; e-mail iema@iema.com.ar; internet www.iema.com.ar; f. 1836; 6,040 mems, 9,000 adherents, seven regional superintendents; Bishop ALDO M. ETCHEGOYEN; Exec. Sec.-Gen. DANIEL A. FAVARO.

## JUDAISM

**Delegación de Asociaciones Israelitas Argentinas (DAIA)** (Delegation of Argentine Jewish Associations): Pasteur 633, 7°, 1028 Buenos Aires; tel. and fax (11) 4378-3200; e-mail daia@daia.org.ar; internet news.daia.org.ar; f. 1935; there are about 250,000 Jews in Argentina, mostly in Buenos Aires; Pres. Dr GILBERT LEWI; Sec.-Gen. Dr JULIO TOKER.

# The Press

## PRINCIPAL DAILIES

### Buenos Aires

**Ambito Financiero:** Avda Paseo Colón 1196, 1063 Buenos Aires; tel. (11) 4349-1500; fax (11) 4349-1505; e-mail correo@ambito.com.ar; internet www.ambitoweb.com.ar; f. 1976; morning (Mon.–Fri.); business; Dir JULIO A. RAMOS; circ. 115,000.

**Buenos Aires Herald:** Azopardo 455, 1107 Buenos Aires; tel. (11) 4342-8476; fax (11) 4331-3370; e-mail info@buenosairesherald.com; internet www.buenosairesherald.com; f. 1876; English; morning; independent; Editor-in-Chief ANDREW GRAHAM-YOOLL; circ. 20,000.

**Boletín Oficial de la República Argentina:** Suipacha 767, 1008 Buenos Aires; tel. (11) 4322-3982; fax (11) 4322-3982; internet www.boletinoficial.gov.ar; f. 1893; morning (Mon.–Fri.); official records publication; Dir JORGE EDUARDO FEIJOO; circ. 15,000.

**Clarín:** Piedras 1743, 1140 Buenos Aires; tel. (11) 4309-7500; fax (11) 4309-7559; e-mail cartas@claringlobal.com.ar; internet www.clarin.com; f. 1945; morning; independent; Dir ERNESTINA L. HERRERA DE NOBLE; circ. 616,000 (daily), 1.0m. (Sunday).

**Crónica:** Avda Juan de Garay 40, 1063 Buenos Aires; tel. (11) 4361-1001; fax (11) 4361-4237; f. 1963; morning and evening; Dirs MARIO ALBERTO FERNÁNDEZ (morning), RICARDO GANGEME (evening); circ. 330,000 (morning), 190,000 (evening), 450,000 (Sunday).

**El Cronista:** Honduras 5663, 1414 Buenos Aires; tel. (11) 4778-6789; fax (11) 4778-6727; e-mail cronista@sadei.org.ar; f. 1908; morning; Dir NÉSTOR SCIBONA; circ. 65,000.

**Diario El Popular:** Vicente López 2626, Olivarría, 7400 Buenos Aires; tel. (22) 8442-0502; fax (22) 8442-0502; e-mail diario@elpopular.com.ar; internet www.diarioelpopular.com.ar; f. 1974; morning; Dir ALBERTO ALBERTENGO; circ. 145,000.

**La Gaceta:** Beguiristain 182, 1870 Avellaneda, Buenos Aires; Dir RICARDO WEST OCAMPO; circ. 35,000.

# ARGENTINA

**La Nación:** Bouchard 551, 1106 Buenos Aires; tel. (11) 4319-1600; fax (11) 4319-1969; e-mail cescribano@lanacion.com.ar; internet www.lanacion.com.ar; f. 1870; morning; independent; Pres. JULIO SAGUIER; circ. 184,000.

**Página 12:** Avda Belgrano 671, 1092 Buenos Aires; tel. (11) 6772-4400; fax (11) 4334-2335; e-mail lectores@pagina12.com.ar; internet www.pagina12.com.ar; f. 1987; morning; independent; Dir ERNESTO TIFFEMBERG; Editor FERNANDO SOKOLOWICZ; circ. 280,000.

**La Prensa:** Azopardo 715, 1107 Buenos Aires; tel. (11) 4349-1000; fax (11) 4349-1025; e-mail laprensa@interlink.com; internet www.interlink.com.ar/laprensa; f. 1869; morning; independent; Dir FLORENCIO ALDREY IGLESIAS; circ. 100,000.

**La Razón:** Río Cuarto 1242, 1168 Buenos Aires; tel. and fax (11) 4309-6000; e-mail larazon@arnet.com.ar; internet www.larazon.com.ar; f. 1992; evening; Dir OSCAR MAGDALENA; circ. 62,000.

**El Sol:** Hipólito Yrigoyen 122, Quilmes, 1878 Buenos Aires; tel. and fax (11) 4257-6325; e-mail autalan@elsolquilmes.com.ar; internet www.elsolquilmes.com.ar; f. 1927; Editor and Dir LUIS ANGEL AUTALÁN; Pres. RODRIGO GHISANI; circ. 25,000.

**Tiempo Argentino:** Buenos Aires; tel. (11) 428-1929; Editor Dr TOMÁS LEONA; circ. 75,000.

## PRINCIPAL PROVINCIAL DAILIES

### Catamarca

**El Ancasti:** Sarmiento 518, 4700 Catamarca; tel. and fax (3833) 431385; e-mail lector@elancasti.net.ar; internet www.elancasti.com.ar; f. 1988; morning; Dir ROQUE EDUARDO MOLAS; circ. 8,000.

### Chaco

**Norte:** Carlos Pellegrini 744, 3500 Resistencia; tel. (3722) 428204; fax (3722) 426047; f. 1968; Dir MIGUEL A. FERNÁNDEZ; circ. 14,000.

### Chubut

**Crónica:** Namuncurá 122, 9000 Comodoro Rivadavia; tel. (297) 447-1200; fax (297) 447-1780; e-mail diariocronica@arnet.com.ar; internet www.diariocronica.com.ar; f. 1962; morning; Dir HERMINIA M. PRESAS; Chief Editor HÉCTOR RODRÍGUEZ; circ. 15,000.

### Córdoba

**Comercio y Justicia:** Mariano Moreno 378, 5000 Córdoba; tel. and fax (351) 422-0040; e-mail redaccion@comercioyjusticia.info; f. 1939; morning; economic and legal news with periodic supplements on architecture and administration; Pres. EDUARDO POGROBINKI; Dir JAVIER ALBERTO DE PASCUALE; circ. 5,800.

**La Voz del Interior:** Monseñor P. Cabrera 6080, 5008 Córdoba; tel. (351) 475-7000; fax (351) 475-7247; e-mail lavoz@lavozdelinterior.com.ar; internet www.lavozdelinterior.com.ar; f. 1904; morning; independent; Dir Dr CARLOS HUGO JORNET; circ. 68,000.

### Corrientes

**El Litoral:** Hipólito Yrigoyen 990, 3400 Corrientes; tel. and fax (3783) 422227; internet www.corrientes.com.ar/el-litoral; f. 1960; morning; Dir CARLOS A. ROMERO FERIS; circ. 25,000.

**El Territorio:** Avda Quaranta 4307, 3300 Posadas; tel. and fax (3752) 452100; e-mail info@territoriodigital.com; internet www.territoriodigital.com.ar; f. 1925; Dir GONZALO PELTZER; circ. 22,000 (Mon.–Fri.), 28,000 (Sunday).

### Entre Ríos

**El Diario:** Buenos Aires y Urquiza, 3100 Paraná; tel. (343) 423-1000; fax (343) 431-9104; e-mail info@eldiarioentrerios.com.ar; internet www.eldiario.com.ar; f. 1914; morning; Dir Dr LUIS F. ETCHEVEHERE; circ. 25,000.

**El Heraldo:** Quintana 42, 3200 Concordia; tel. (345) 421-5304; fax (345) 421-1397; e-mail administracionelheraldo@infovia.com; internet www.elheraldo.com.ar; f. 1915; evening; Editor Dr CARLOS LIEBERMANN; circ. 10,000.

### Mendoza

**Los Andes:** San Martín 1049, 5500 Mendoza; tel. (261) 449-1280; fax (261) 449-1217; e-mail havila@losandes.com.ar; internet www.losandes.com.ar; f. 1982; morning; Dir ELVIRA CALLE DE ANTEQUEDA; Gen. Man. MIGUEL A. BAUZA; circ. 60,000.

### Provincia de Buenos Aires

**El Atlántico:** Bolívar 2975, 7600 Mar del Plata; tel. (223) 435462; f. 1938; morning; Dir OSCAR ALBERTO GASTIARENA; circ. 20,000.

**La Capital:** Avda Champagnat 2551, 7600 Mar del Plata; tel. (223) 478-8490; fax (223) 478-1038; e-mail diario@lacapitalnet.com.ar; internet www.lacapitalnet.com.ar; f. 1905; Dir FLORENCIO ALDREY IGLESIAS; circ. 32,000.

**El Día:** Avda A. Diagonal 80, 817-21, 1900 La Plata; tel. (221) 425-0101; fax (221) 423-2996; e-mail redaccion@eldia.com; f. 1884; morning; independent; Dir RAÚL E. KRAISELBURD; circ. 54,868.

**Ecos Diarios:** Calle 62, No 2486, 7630 Necochea; tel. (2262) 430754; fax (2262) 424114; e-mail ecosdiar@satlink.com; internet www.ecosdiarios.com.ar; f. 1921; morning; independent; Dir GUILLERMO IGNACIO; circ. 6,000.

**La Nueva Provincia:** Rodríguez 55, 8000 Bahía Blanca; tel. (291) 459-0000; fax (291) 459-0001; e-mail info@lanueva.com; internet www.lanueva.com.ar; f. 1898; morning; independent; Dir DIANA JULIO DE MASSOT; circ. 22,000 (Mon.–Fri.), 30,000 (Sunday).

**El Nuevo Cronista:** Mercedes 619, 5°, 6600 Buenos Aires; tel. (11) 2324-4001; e-mail info@nuevocronista.com.ar; internet www.nuevocronista.com.ar.

**La Voz del Pueblo:** Avda San Martín 991, 7500 Tres Arroyos; tel. (2983) 430680; fax (2938) 430682; e-mail redaccion@lavozdelpueblo.com.ar; f. 1902; morning; independent; Dir ALBERTO JORGE MACIEL; circ. 8,500.

### Río Negro

**Río Negro:** 9 de Julio 733, 8332, Gen. Roca, Río Negro; tel. (2941) 439300; fax (2941) 430517; e-mail comentario@rionegro.com.ar; internet www.rionegro.com.ar; f. 1912; morning; Editor NÉLIDA RAJNERI DE GAMBA.

### Salta

**El Tribuno:** Avda Ex Combatientes de Malvinas 3890, 4412 Salta; tel. (387) 424-6200; fax (387) 424-6240; e-mail redaccion@eltribuno.com.ar; internet www.eltribuno.com.ar; f. 1949; morning; Dir ROBERTO EDUARDO ROMERO; circ. 25,000.

### San Juan

**Diario de Cuyo:** Mendoza 380 Sur, 5400 San Juan; tel. (264) 429-0038; fax (264) 429-0063; e-mail direcciondc@diariodecuyo.com.ar; internet www.diariodecuyo.com.ar; f. 1947; morning; independent; Dir FRANCISCO B. MONTES; circ. 20,000.

### San Luis

**El Diario de La República:** Lafinur 924, 5700 San Luis; tel. and fax (2623) 422037; e-mail administracion@grupopayne.com.ar; internet www.eldiariodelarepublica.com; f. 1966; Dir-Gen. FELICIANA RODRÍGUEZ SAA; circ. 12,000.

### Santa Fe

**La Capital:** Sarmiento 763, 2000 Rosario; tel. (341) 420-1100; fax (341) 420-1114; e-mail elagos@lacapital.com.ar; f. 1867; morning; independent; Dir CARLOS MARÍA LAGOS; circ. 65,000.

**El Litoral:** Avda 25 de Mayo 3536, 3000 Santa Fe; tel. (342) 450-2500; fax (342) 450-2530; e-mail litoral@litoral.com.ar; internet www.litoral.com.ar; f. 1918; morning; independent; Dir GUSTAVO VITTORI; circ. 37,000.

### Santiago del Estero

**El Liberal:** Libertad 263, 4200 Santiago del Estero; tel. (385) 422-4400; fax (385) 422-4538; e-mail liberal@teletel.com.ar; internet www.sdnet.com.ar; f. 1898; morning; Exec. Dir JOSÉ LUIS CASTIGLIONE; Editorial Dir Dr JULIO CÉSAR CASTIGLIONE; circ. 20,000.

### Tucumán

**La Gaceta:** Mendoza 654, 4000 San Miguel de Tucumán; tel. (381) 431-1111; fax (381) 431-1597; e-mail redaccion@lagaceta.com.ar; internet www.lagaceta.com.ar; f. 1912; morning; independent; Dir ALBERTO GARCÍA HAMILTON; circ. 70,000.

## WEEKLY NEWSPAPER

**El Informador Público:** Uruguay 252, 3°F, 1015 Buenos Aires; tel. (11) 4476-3551; fax (11) 4342-2628; f. 1986.

## PERIODICALS

**Aeroespacio:** Avda Dorrego 4019, Buenos Aires; tel. and fax (11) 4514-1561; e-mail info@aeroespacio.com.ar; internet www.aeroespacio.com.ar; f. 1940; every 2 months; aeronautics; Dir CARLOS GUSTAVO RINALDI; circ. 24,000.

**Billiken:** Azopardo 579, 1307 Buenos Aires; tel. (11) 4346-0107; fax (11) 4343-7040; e-mail vzumbo@atlantida.com.ar; f. 1919; weekly; children's magazine; Dir JUAN CARLOS PORRAS; circ. 240,000.

# ARGENTINA

**Casas y Jardines:** Sarmiento 643, 1382 Buenos Aires; tel. (11) 445-1793; f. 1932; every 2 months; houses and gardens; publ. by Editorial Contémpora SRL; Dir Norberto M. Muzio.

**Chacra y Campo Moderno:** Editorial Atlántida, SA, Azopardo 579, 1307 Buenos Aires; tel. (11) 4331-4591; fax (11) 4331-3272; f. 1930; monthly; farm and country magazine; Dir Constancio C. Vigil; circ. 35,000.

**Claudia:** Avda Córdoba 1345, 12°, Buenos Aires; tel. (11) 442-3275; fax (11) 4814-3948; f. 1957; monthly; women's magazine; Dir Ana Torrejón; circ. 150,000.

**El Economista:** Avda Córdoba 632, 2°, 1054 Buenos Aires; tel. (11) 4322-7360; fax (11) 4322-8157; internet www.eleconomista.com.ar; f. 1951; weekly; financial; Dir Dr D. Radonjic; circ. 37,800.

**Fotografía Universal:** Buenos Aires; monthly; circ. 39,500.

**Gente:** Azopardo 579, 3°, 1307 Buenos Aires; tel. (11) 433-4591; f. 1965; weekly; general; Dir Jorge de Luján Gutiérrez; circ. 133,000.

**El Gráfico:** Paseo Colón 505, 2°, 1063 Buenos Aires; tel. (11) 5235-5100; fax (11) 5235-5137; e-mail elgrafico@elgrafico.com.ar; internet www.elgrafico.com.ar; f. 1919; weekly; sport; Editor Martin Mazur; circ. 127,000.

**Guía Latinoamericana de Transportes:** Florida 8287, esq. Portinari, 1669 Del Viso (Ptdo de Pilar), Provincia de Buenos Aires; tel. (11) 4320-7004; fax (11) 4307-1956; f. 1968; every 2 months; travel information and timetables; Editor Dr Armando Schlecker Hirsch; circ. 7,500.

**Humor:** Venezuela 842, 1095 Buenos Aires; tel. (11) 4334-5400; fax (11) 411-2700; f. 1978; every 2 weeks; satirical revue; Editor Andrés Cascioli; circ. 180,000.

**Legislación Argentina:** Talcahuano 650, 1013 Buenos Aires; tel. (11) 4371-0528; e-mail jurispru@lvd.com.ar; f. 1958; weekly; law; Dir Ricardo Estévez Boero; circ. 15,000.

**Mercado:** Rivadavia 877, 2°, 1002 Buenos Aires; tel. (11) 4346-9400; fax (11) 4343-7880; e-mail gacetillas@mercado.com.ar; internet www.mercado.com.ar; f. 1969; monthly; business; Dir Miguel Angel Diez; circ. 28,000.

**Mundo Israelita:** Pueyrredón 538, 1°B, 1032 Buenos Aires; tel. (11) 4961-7999; fax (11) 4961-0763; f. 1923; weekly; Editor Dr José Kestelman; circ. 15,000.

**Nuestra Arquitectura:** Sarmiento 643, 5°, 1382 Buenos Aires; tel. (11) 445-1793; f. 1929; every 2 months; architecture; publ. by Editorial Contémpora SRL; Dir Norberto M. Muzio.

**Para Ti:** Azopardo 579, 1307 Buenos Aires; tel. (11) 4331-4591; fax (11) 4331-3272; e-mail redaccionparati@atlantida.com.ar; internet www.parati.com.ar; f. 1922; weekly; women's interest; Dir Aníbal C. Vigil; circ. 104,000.

**Pensamiento Económico:** Avda Leandro N. Alem 36, 1003 Buenos Aires; tel. (11) 4331-8051; fax (11) 4331-8055; e-mail cac@cac.com.ar; internet www.cac.com.ar; f. 1925; quarterly; review of Cámara Argentina de Comercio; Dir Dr Carlos L. P. Antonucci.

**La Prensa Médica Argentina:** Junín 845, 1113 Buenos Aires; tel. (11) 4961-9793; fax (11) 4961-9494; e-mail presmedarg@hotmail.com; f. 1914; monthly; medical; Editor Dr P. A. López; circ. 8,000.

**Prensa Obrera:** Ayacucho 444, Buenos Aires; tel. (11) 4953-3824; fax (11) 4953-7164; f. 1982; weekly; publication of Partido Obrero; circ. 16,000.

**La Semana:** Sarmiento 1113, 1041 Buenos Aires; tel. (11) 435-2552; general; Editor Daniel Pliner.

**La Semana Médica:** Arenales 3574, 1425 Buenos Aires; tel. (11) 4824-5673; f. 1894; monthly; Dir Dr Eduardo F. Mele; circ. 7,000.

**Siete Días Ilustrados:** Avda Leandro N. Alem 896, 1001 Buenos Aires; tel. (11) 432-6010; f. 1967; weekly; general; Dir Ricardo Cámara; circ. 110,000.

**Técnica e Industria:** Buenos Aires; tel. (11) 446-3193; f. 1922; monthly; technology and industry; Dir E. R. Fedele; circ. 5,000.

**Visión:** French 2820, 2°A, 1425 Buenos Aires; tel. (11) 4825-1258; fax (11) 4827-1004; f. 1950; fortnightly; Latin American affairs, politics; Editor Luis Vidal Rucabado.

**Vosotras:** Avda Leandro N. Alem 896, 3°, 1001 Buenos Aires; tel. (11) 432-6010; f. 1935; women's weekly; Dir Abel Zanotto; circ. 33,000; monthly supplements:

   **Labores:** circ. 130,000.

   **Modas:** circ. 70,000.

## NEWS AGENCIES

**Agencia TELAM, SA:** Bolívar 531, 1066 Buenos Aires; tel. (11) 4339-0330; fax (11) 4339-0353; e-mail telam@telam.com.ar; internet www.telam.com.ar; Gen. Man. Haedo L. Lazzaro.

**Diarios y Noticias (DYN):** Avda Julio A. Roca 636, 8°, 1067 Buenos Aires; tel. (11) 4342-3040; fax (11) 4342-3043; e-mail info@dyn.com.ar; internet www.dyn.com.ar; f. 1982; Pres. Carlos Laria.

**Noticias Argentinas, SA (NA):** Suipacha 570, 3°B, 1008 Buenos Aires; tel. (11) 4394-7522; fax (11) 4394-7648; e-mail infogral@noticiasargentinas.com; internet www.noticiasargentinas.com; f. 1973; Dir Luis Fernando Torres.

### Foreign Bureaux

**Agence France-Presse (AFP):** Avda Corrientes 456, 4°, Of. 41–42, 1366 Buenos Aires; tel. (11) 4394-0872; fax (11) 4393-9912; e-mail afp-baires@tournet.com.ar; internet www.afp.com; Bureau Chief Frédéric Garlan.

**Agencia EFE** (Spain): Avda Moreau de Justo 1720, 1°, 1107 Buenos Aires; tel. (11) 4312-5721; fax (11) 4815-8691; e-mail efebas@ciudad.com; Bureau Chief Agustín de Gracia.

**Agenzia Nazionale Stampa Associata (ANSA)** (Italy): San Martín 320, 6°, 1004 Buenos Aires; tel. (11) 4394-7568; fax (11) 4394-5214; e-mail ansabaires@infovia.com.ar; Bureau Chief Antonio Cavallari.

**Associated Press (AP)** (USA): Avda Leandro N. Alem 712, 4°, 1001 Buenos Aires; tel. (11) 4311-0081; fax (11) 4311-0083; e-mail bcormier@ap.org; Bureau Chief William Cormier.

**Deutsche Presse-Agentur (dpa)** (Germany): Buenos Aires; tel. (11) 4311-5311; e-mail msvgroth@ba.net; Bureau Chief Dr Hendrik Groth.

**Informatsionnoye Telegrafnoye Agentstvo Rossii—Telegrafnoye Agentstvo Suverennykh Stran (ITAR—TASS)** (Russia): Avda Córdoba 652, 11°E, 1054 Buenos Aires; tel. (11) 4392-2044; Dir Isidoro Gilbert.

**Inter Press Service (IPS)** (Italy): Buenos Aires; tel. (11) 4394-0829; Bureau Chief Ramón M. Gorriarán; Correspondent Gustavo Capdevilla.

**Magyar Távirati Iroda (MTI)** (Hungary): Marcelo T. de Alvear 624, 3° 16, 1058 Buenos Aires; tel. (11) 4312-9596; Correspondent Endre Simó.

**Prensa Latina** (Cuba): Buenos Aires; tel. (11) 4394-0565; e-mail prela@teletel.com.ar; Correspondent Víctor Carriba.

**Reuters** (United Kingdom): Avda Eduardo Madero 940, 25°, 1106 Buenos Aires; tel. (11) 4318-0600; fax (11) 4318-0698; Dir Carlos Pía Mangione.

**Xinhua (New China) News Agency** (People's Republic of China): Tucumán 540, 14°, Apto D, 1049 Buenos Aires; tel. (11) 4313-9755; Bureau Chief Ju Qingdong.

The following are also represented: Central News Agency (Taiwan), Interpress (Poland), Jiji Press (Japan).

## PRESS ASSOCIATION

**Asociación de Entidades Periodísticas Argentinas (ADEPA):** Chacabuco 314, 3°, 1069 Buenos Aires; tel. and fax (11) 4331-1500; e-mail adepa@adepa.org.ar; internet www.adepa.org.ar; f. 1962; Pres. Lauro F. Laino.

# Publishers

**Aguilar, Altea, Taurus, Alfaguara, SA de Ediciones:** Avda Leandro N. Alem 720, 1001 Buenos Aires; tel. (11) 4119-5000; fax (11) 4119-5021; e-mail info@alfaguara.com.ar; internet www.alfaguara.com.ar; f. 1946; general, literature, children's books; Pres. Esteban Fernández Rosado; Gen. Man. David Delgado de Robles.

**Aique:** Valentín Gómez 3530, Buenos Aires; tel. and fax (11) 4865-5152; e-mail editorial@aique.com.ar; internet www.aique.com.ar; f. 1976; division of Aique Grupo Editor, SA.

**Amorrortu Editores, SA:** Paraguay 1225, 7°, 1057 Buenos Aires; tel. (11) 4816-5812; fax (11) 4816-3321; e-mail info@amorrortueditores.com; internet www.amorrortueditores.com; f. 1967; academic, social sciences and humanities; Man. Dir Horacio de Amorrortu.

**Az Editora, SA:** Paraguay 2351, 1121 Buenos Aires; tel. (11) 961-4036; fax (11) 961-0089; e-mail info@az-editora.com.ar; internet www.az-editora.com.ar; f. 1976; educational, social sciences and medicine; Pres. Dante Omar Villalba.

**Biblioteca Nacional de Maestros:** c/o Ministerio de Educación, Ciencia y Tecnología, Pizzurno 935, planta baja, 1020 Buenos Aires; tel. (11) 4129-1272; fax (11) 4129-1268; e-mail gperrone@me.gov.ar; internet www.bnm.me.gov.ar; f. 1884; Dir Graciela Perrone.

# ARGENTINA

**Carlos Lohlé, SA:** Tacuarí 1516, 1139 Buenos Aires; tel. (11) 4427-9969; f. 1953; philosophy, religion, belles-lettres; Dir FRANCISCO M. LOHLÉ.

**Club de Lectores:** Avda de Mayo 624, 1084 Buenos Aires; tel. (11) 4342-6251; f. 1938; non-fiction; Man. Dir MERCEDES FONTENLA.

**Cosmopolita, SRL:** Piedras 744, 1070 Buenos Aires; tel. (11) 4361-8925; fax (11) 4361-8049; f. 1940; science and technology; Man. Dir RUTH F. DE RAPP.

**Crecer Creando Editorial:** Avda Callao 1225, 14°B, 1023 Buenos Aires; tel. and fax (11) 4812-4586; e-mail info@crecercreando.com.ar; internet www.crecercreando.com.ar; educational.

**De Los Cuatro Vientos Editorial:** Avda Balcarce 1053, Local 2, CP 1064, Buenos Aires; tel. and fax (11) 4300-0924; e-mail info@deloscuatrovientos.com.ar; internet www.deloscuatrovientos.com.ar; Dir and Editor PABLO ALBORNOZ.

**Depalma, SA:** Talcahuano 494, 1013 Buenos Aires; tel. (11) 4371-7306; fax (11) 4371-6913; e-mail info@ed-depalma.com; internet www.ed-depalma.com; f. 1944; periodicals and books covering law, politics, sociology, philosophy, history and economics; Man. Dir ROBERTO SUARDIAZ.

**Edicial, SA:** Rivadavia 739, 1002 Buenos Aires; tel. (11) 4342-8481; fax (11) 4343-1151; e-mail edicial@ssdnet.com.ar; internet www.ssdnet.com.ar/edicial; f. 1931; education; Man. Dir J. A. MUSSET.

**Ediciones Atril:** Hortiguera 1411, Buenos Aires; tel. (11) 4924-3003; e-mail atril@interlink.com.ar.

**Ediciones La Aurora:** José María Moreno 873, 1°, 1424 Buenos Aires; tel. and fax (11) 4922-5356; f. 1925; general, religion, spirituality, theology, philosophy, psychology, history, semiology, linguistics; Dir Dr HUGO O. ORTEGA.

**Ediciones de la Flor SRL:** Avda Gorriti 3695, 1172 Buenos Aires; fax (11) 4963-5616; e-mail edic-flor@datamarkets.com.ar; internet www.edicionesdelaflor.com.ar; f. 1966; fiction, poetry, theatre, juvenile, humour and scholarly; Co-Dirs ANA MARÍA T. MILER, DANIEL DIVINSKY.

**Ediciones Librerías Fausto:** Avda Corrientes 1316, 1043 Buenos Aires; tel. (11) 4476-4919; fax (11) 4476-3914; f. 1943; fiction and non-fiction; Man. RAFAEL ZORRILLA.

**Ediciones Macchi, SA:** Alsina 1535/37, 1088 Buenos Aires; tel. (11) 446-2506; fax (11) 446-0594; e-mail info@macchi.com.ar; internet www.macchi.com; f. 1947; economic sciences; Pres. RAÚL LUIS MACCHI; Dir JULIO ALBERTO MENDONÇA.

**Ediciones Nueva Visión, SAIC:** Tucumán 3748, 1189 Buenos Aires; tel. (11) 4864-5050; fax (11) 4863-5980; e-mail ednuevavision@ciudad.com.ar; f. 1954; psychology, education, social sciences, linguistics; Man. Dir HAYDÉE P. DE GIACONE.

**Ediciones del Signo:** Julián Alvarez 2844, 1°A, 1425 Buenos Aires; tel. (11) 4861-3181; fax (11) 3462-5031; e-mail info@edicionesdelsigno.com.ar; internet www.edicionesdelsigno.com.ar; f. 1995; philosophy, psychoanalysis, politics and scholarly.

**Editorial Abril, SA:** Moreno 1617, 1093 Buenos Aires; tel. (11) 4331-0112; f. 1961; fiction, non-fiction, children's books, textbooks; Dir ROBERTO M. ARES.

**Editorial Acme, SA:** Santa Magdalena 635, 1277 Buenos Aires; tel. (11) 4328-1508; f. 1949; general fiction, children's books, agriculture, textbooks; Man. Dir EMILIO I. GONZÁLEZ.

**Editorial Albatros, SACI:** Torres Las Plazas, J. Salguero 2745, 5°, 1425 Buenos Aires; tel. (11) 4807-2030; fax (11) 4807-2010; e-mail info@albatros.com.ar; internet www.albatros.com.ar; f. 1945; technical, non-fiction, social sciences, sport, children's books, medicine and agriculture; Pres. ANDREA INÉS CANEVARO.

**Editorial Argenta Sarlep SA:** Avda Corrientes 1250, 3°, Of. F, Buenos Aires; tel. and fax (11) 4382-9085; e-mail argenta@millic.com.ar; internet www.editorialargenta.com; f. 1970; literature, poetry, theatre and reference.

**Editorial Bonum:** Avda Corrientes 6687, 1427 Buenos Aires; tel. and fax (11) 4554-1414; e-mail ventas@editorialbonum.com.ar; internet www.editorialbonum.com.ar; f. 1960; religious, educational and self-help.

**Editorial Don Bosco:** Avda Don Bosco 4069, 1206 Buenos Aires; tel. (11) 4883-0111; fax (11) 4883-0115; e-mail administracion@edb.com.ar; internet www.edb.com.ar; f. 1993; religious and educational; Gen. Man. GERARDO BENTANCOUR.

**Editorial Catálogos SRL:** Avda Independencia 1860, 1225 Buenos Aires; tel. and fax (11) 4381-5708; e-mail catalogos@ciudad.com.ar; internet www.catalogossrl.com.ar; religion, literature, academic, general interest and self-help.

**Editorial Claretiana:** Lima 1360, 1138 Buenos Aires; tel. (11) 427-9250; fax (11) 427-4015; e-mail editorial@editorialclaretiana.com.ar; internet www.editorialclaretiana.com.ar; f. 1956; Catholicism; Man. Dir DOMINGO ANGEL GRILLIA.

**Editorial Claridad, SA:** Viamonte 1730, 1°, 1055 Buenos Aires; tel. (11) 4371-5546; fax (11) 4375-1659; e-mail editorial@heliasta.com.ar; internet www.heliasta.com.ar; f. 1922; literature, biographies, social science, politics, reference, dictionaries; Pres. Dra ANA MARÍA CABANELLAS.

**Editorial Columba, SA:** Sarmiento 1889, 5°, 1044 Buenos Aires; tel. (11) 445-4297; f. 1953; classics in translation, 20th century; Man. Dir CLAUDIO A. COLUMBA.

**Editorial Contémpora, SRL:** Sarmiento 643, 5°, 1382 Buenos Aires; tel. (11) 445-1793; architecture, town-planning, interior decoration and gardening; Dir NORBERTO C. MUZIO.

**Editorial Difusión, SA:** Sarandí 1065–67, Buenos Aires; tel. (11) 4941-0088; f. 1937; literature, philosophy, religion, education, textbooks, children's books; Dir DOMINGO PALOMBELLA.

**Editorial Errepar:** Paraná 725, 1017 Buenos Aires; tel. (11) 4370-8025; fax (11) 4383-2202; e-mail clientes@errepar.com; internet www.errepar.com; legal texts.

**Editorial Glem, SACIF:** Avda Caseros 2056, 1264 Buenos Aires; tel. (11) 426-6641; f. 1933; psychology, technology; Pres. JOSÉ ALFREDO TUCCI.

**Editorial Grupo Cero:** Mansilla 2686, planta baja 2, 1425 Buenos Aires; tel. (11) 4328-0614; internet www.editorialgrupocero.com; fiction, poetry and psychoanalysis.

**Editorial Guadalupe:** Mansilla 3865, 1425 Buenos Aires; tel. (11) 4826-8587; fax (11) 4823-6672; e-mail ventas@editorialguadalupe.com.ar; internet www.editorialguadalupe.com.ar; f. 1895; social sciences, religion, anthropology, children's books, and pedagogy; Man. Dir LORENZO GOYENECHE.

**Editorial Heliasta, SRL:** Viamonte 1730, 1°, 1055 Buenos Aires; tel. (11) 4371-5546; fax (11) 4375-1659; e-mail editorial@heliasta.com.ar; internet www.heliasta.com.ar; f. 1944; literature, biography, dictionaries, legal; Pres. Dra ANA MARÍA CABANELLAS.

**Editorial Hispano-Americana, SA (HASA):** Alsina 731, 1087 Buenos Aires; tel. (11) 4331-5051; f. 1934; science and technology; Pres. Prof. HÉCTOR OSCAR ALGARRA.

**Editorial Inter-Médica, SAICI:** Junín 917, 1°, 1113 Buenos Aires; tel. (11) 4961-9234; fax (11) 4961-5572; e-mail info@inter-medica.com.ar; internet www.inter-medica.com.ar; f. 1959; medicine and veterinary; Pres. SONIA MODYEIEVSKY.

**Editorial Inter-Vet, SA:** Avda de los Constituyentes 3141, Buenos Aires; tel. (11) 451-2382; f. 1987; veterinary; Pres. JORGE MODYEIEVSKY.

**Editorial Juris:** Moreno 1580, 2000 Rosario; tel. (341) 426-7301; fax ; e-mail editorialjuris@arnet.com.ar; internet www.editorialjuris.com; f. 1952; legal texts.

**Editorial Kier, SACIFI:** Avda Santa Fe 1260, 1059 Buenos Aires; tel. (11) 4811-0507; fax (11) 4811-3395; e-mail info@kier.com.ar; internet www.kier.com.ar; f. 1907; Eastern doctrines and religions, astrology, parapsychology, tarot, I Ching, occultism, cabbala, freemasonry and natural medicine; Pres. HÉCTOR S. PIBERNUS; Mans SERGIO PIBERNUS, OSVALDO PIBERNUS.

**Editorial Losada, SA:** Moreno 3362/64, 1209 Buenos Aires; tel. (11) 4373-4006; fax (11) 4375-5001; e-mail administra@editoriallosada.com; f. 1938; general; Pres. JOSÉ JUAN FERNÁNDEZ REGUERA.

**Editorial Médica Panamericana, SA:** Marcelo T. de Alvear 2145, 1122 Buenos Aires; tel. (11) 4821-5520; fax (11) 4825-5006; e-mail info@medicapanamericana.com.ar; internet www.medicapanamericana.com.ar; f. 1962; medicine and health sciences; Pres. HUGO BRIK.

**Editorial Mercosur:** Dean Funes 923/25, 1231 Buenos Aires; tel. (11) 4822-4615; e-mail info@editorialmercosr.com; internet www.editorialmercosur.com; self-help and general interest.

**Editorial del Nuevo Extremo:** Juncal 4651, 1425 Buenos Aires; tel. (11) 4773-3228; fax (11) 473-8445; e-mail editorial@delnuevoextremo.com; internet www.delnuevoextremo.com; general interest.

**Editorial Plus Ultra, SA:** Avda Callao 575, 1022 Buenos Aires; tel. (11) 4374-2953; f. 1964; literature, history, textbooks, law, economics, politics, sociology, pedagogy, children's books; Man. Editors RAFAEL ROMÁN, LORENZO MARENGO.

**Editorial Santillana:** Buenos Aires; f. 1960; education; Pres. JESÚS DE POLANCO GUTIÉRREZ.

**Editorial Sigmar, SACI:** Belgrano 1580, 7°, 1093 Buenos Aires; tel. (11) 4383-3045; fax (11) 4383-5633; e-mail editorial@sigmar.com.ar; f. 1941; children's books; Man. Dir ROBERTO CHWAT.

# ARGENTINA

**Editorial Sopena Argentina, SACI e I:** Maza 2138, 1240 Buenos Aires; tel. and fax (11) 4912-2383; f. 1918; dictionaries, classics, chess, health, politics, history, children's books; Dir MARTA A. J. OLSEN.

**Editorial Stella:** Viamonte 1984, 1056 Buenos Aires; tel. (11) 4374-0346; fax (11) 4374-8719; e-mail admin@editorialstella.com.ar; internet www.editorialstella.com.ar; general non-fiction and textbooks; owned by Asociación Educacionista Argentina.

**Editorial Sudamericana, SA:** Humberto 545, 1°, 1103 Buenos Aires; tel. (11) 4300-5400; fax (11) 4362-7364; e-mail info@edsudamericana.com.ar; internet www.edsudamericana.com.ar; f. 1939; general fiction and non-fiction; Gen. Man. OLAF HANTEL.

**Editorial Troquel, SA:** Pichincha 967, 1219 Buenos Aires; tel. (11) 4941-7943; e-mail troquel@ba.net; f. 1954; general literature and textbooks; Pres. GUSTAVO A. RESSIA.

**Editorial Zeus SRL:** Rosario; tel. (341) 449-5585; fax (341) 425-4259; e-mail editorialzeus@citynet.net.ar; internet www.editorial-zeus.com.ar; legal texts; Editor and Dir GUSTAVO L. CAVIGLIA.

**Emecé Editores, SA:** Alsina 2062, 1090 Buenos Aires; tel. (11) 4954-0105; fax (11) 4953-4200; f. 1939; fiction, non-fiction, biographies, history, art, essays; Pres. ALFREDO DEL CARRIL.

**Espasa Calpe Argentina, SA:** Avda Independencia 1668, 1100 Buenos Aires; tel. (11) 4382-4043; fax (11) 4383-3793; f. 1937; literature, science, dictionaries; publ. *Colección Austral*; Dir GUILLERMO SCHAVELZON.

**EUDEBA** (Editorial Universitaria de Buenos Aires): Rivadavia 1573, 1033 Buenos Aires; tel. (11) 4383-8025; fax (11) 4383-2202; e-mail eudeba@eudeba.com.ar; internet www.eudeba.com.ar; f. 1958; university text books and general interest publications; Pres. Dr LUIS YANES.

**Fabril Editora, SA:** Buenos Aires; tel. (11) 4421-3601; f. 1958; non-fiction, science, arts, education and reference; Editorial Man. ANDRÉS ALFONSO BRAVO; Business Man. RÓMULO AYERZA.

**Galerna:** Lambaré 893, 1185 Buenos Aires; tel. (11) 4867-1661; fax (11) 4862-5031; e-mail contacto@galerna.net; internet www.galernalibros.com; fiction, theatre, poetry and scholarly.

**Gram Editora:** Cochabamba 1652, 1148 Buenos Aires; tel. (11) 4304-4833; fax (11) 4304-5692; e-mail grameditora@infovia.com.ar; internet www.grameditora.com.ar; catholic.

**Gránica Editorial:** Lavalle 1634, 3°G, 1048 Buenos Aires; tel. (11) 4374-1456; fax (11) 4373-0669; e-mail granica.ar@granicaeditor.com; internet www.granica.com; juvenile and fiction.

**Kapelusz Editora, SA:** San José 831, 1076 Buenos Aires; tel. (11) 4342-7400; fax (11) 4331-8020; e-mail empresa@kapelusz.com.ar; internet www.kapelusz.com.ar; f. 1905; textbooks, psychology, pedagogy, children's books; Vice-Pres. RAFAEL PASCUAL ROBLES.

**Plaza y Janés, SA:** Buenos Aires; tel. (11) 4486-6769; popular fiction and non-fiction; Man. Dir JORGE PÉREZ.

**Schapire Editor, SRL:** Uruguay 1249, 1016 Buenos Aires; tel. (11) 4812-0765; fax (11) 4815-0369; f. 1941; music, art, theatre, sociology, history, fiction; Dir MIGUEL SCHAPIRE DALMAT.

## PUBLISHERS' ASSOCIATION

**Cámara Argentina de Publicaciones:** Reconquista 1011, 6°, 1003 Buenos Aires; tel. (11) 4311-6855; f. 1970; Pres. AGUSTÍN DOS SANTOS; Man. LUIS FRANCISCO HOULIN.

# Broadcasting and Communications

**Secretaría de Comunicaciones:** Sarmiento 151, 4°, 1000 Buenos Aires; tel. (11) 4318-9410; fax (11) 4318-9432; internet www.secom.gov.ar; co-ordinates 30 stations and the international service; Sec. MARIO GUILLERMO MORENO.

**Subsecretaría de Planificación y Gestión Tecnológica:** Sarmiento 151, 4°, 1000 Buenos Aires; tel. (11) 4347-9970; Under-Sec. Ing. ALEJANDRA CABALLERO.

**Subsecretaría de Radiocomunicaciones:** Sarmiento 151, 4°, 1000 Buenos Aires; tel. (11) 4311-5909.

**Subsecretaría de Telecomunicaciones:** Sarmiento 151, 4°, 1000 Buenos Aires; tel. (11) 4311-5909.

**Comité Federal de Radiodifusión (COMFER):** Suipacha 765, 9°, 1008 Buenos Aires; tel. (11) 4320-4900; fax (11) 4394-6866; e-mail mlagier@comfer.gov.ar; internet www.comfer.gov.ar; f. 1972; controls various technical aspects of broadcasting and transmission of programmes; Insp. JULIO DONATO BÁRBARO.

## TELECOMMUNICATIONS

**Cámara de Informática y Comunicaciones de la República Argentina (CICOMRA):** Avda Córdoba 744, 2°, 1054 Buenos Aires; tel. (11) 4325-8839; fax (11) 4325-9604; e-mail cicomra@cicomra.org.ar; internet www.cicomra.org.ar; f. 1985; Pres. NORBERTO CAPELLÁN.

**Comisión Nacional de Comunicaciones (CNC):** Perú 103, 9°, 1067 Buenos Aires; tel. (11) 4347-9242; fax (11) 4347-9244; internet www.cnc.gov.ar; f. 1996; Pres. Dr ROBERTO CATALÁN; Insp. FULVIO MARIO MADARO.

**Cía Ericsson SACI:** Güemes 676, 1°, 1638 Vicente López PCIA, Buenos Aires; tel. (11) 4319-5500; fax (11) 4315-0629; e-mail infocom@cea.ericsson.se; internet www.ericsson.com.ar; Dir-Gen. Ing. ROLANDO ZUBIRÁN.

**Movicom BellSouth:** Ingeniero Enrique Butty 240, 1001 Buenos Aires; tel. (11) 5321-0000; fax (11) 4321-0334; e-mail rree@movi.com.ar; internet www.movi.com.ar; telecommunications services, including cellular telephones, trunking, paging and wireless access to the internet; acquired by Telefónica Móviles in Jan. 2005; Pres. Lic. MAURICIO E. WIOR.

**Telcosur, SA:** Don Bosco 3672, 5°, 1206 Buenos Aires; tel. (11) 4865-9060; e-mail telcosur@tgs.com.ar; internet www.telcosur.com.ar; f. 1998; 99% owned by Transportador de Gas del Sur (TGS); Operations Man. EDUARDO VIGILANTE; Commercial Man. EDUARDO MARTÍN.

**Telecom Argentina:** Alicia Moreau de Justo 50, 1107 Buenos Aires; tel. (11) 4968-4000; fax (11) 4968-1420; e-mail inversores@intersrv.telecom.com.ar; internet www.telecom.com.ar; provision of telecommunication services in the north of Argentina; Pres. AMADEO R. VÁZQUEZ.

**Telefónica de Argentina, SA (TASA):** Tucumán 1, 17°, 1049 Buenos Aires; tel. (11) 4345-5772; fax (11) 4345-5771; e-mail gabello@telefonica.com.ar; internet www.telefonica.com.ar; subsidiary of Telefónica (Spain); provision of telecommunication services in the south of Argentina; Pres. MARIO EDUARDO VÁSQUEZ.

**Unifón:** Corrientes 655, 1043 Buenos Aires; tel. (11) 4130-4000; internet www.unifon.com.ar; 98% owned by Telefónicas Móviles, SA (Spain); operates mobile telephone network.

## RADIO

There are three privately owned stations in Buenos Aires and 72 in the interior. There are also 37 state-controlled stations, four provincial, three municipal and three university stations. The principal ones are Radio Antártida, Radio Argentina, Radio Belgrano, Radio Ciudad de Buenos Aires, Radio Excelsior, Radio Mitre, Radio El Mundo, Radio Nacional, Radio del Plata, Radio Rivadavia and Radio Splendid, all in Buenos Aires.

**Radio Nacional Argentina (RNA):** Maipú 555, 1006 Buenos Aires; tel. (11) 4325-4590; fax (11) 4325-4313; e-mail direccion@radionacional.gov.ar; internet www.radionacional.gov.ar; five radio stations: Nacional; Nacional Folklórica; Nacional Clásica; Nacional Faro; and RAE (q.v.); Exec. Dir ADELINA OLGA MONCALVILLO.

**Radiodifusión Argentina al Exterior (RAE):** Maipú 555, 1006 Buenos Aires; tel. (11) 4325-6368; fax (11) 4325-9433; e-mail rae@radionacional.gov.ar; f. 1958; broadcasts in seven languages to all areas of the world; Dir MARCELA CAMPOS.

**Asociación de Radiodifusoras Privadas Argentinas (ARPA):** Juan D. Perón 1561, 8°, 1037 Buenos Aires; tel. (11) 4382-4412; f. 1958; an association of all but three of the privately owned commercial stations; Pres. DOMINGO F. L. ELÍAS.

## TELEVISION

There are 42 television channels, of which 29 are privately owned and 15 are owned by provincial and national authorities. The national television network is regulated by the Comité Federal de Radiodifusión (see above).

The following are some of the more important television stations in Argentina: Argentina Televisora Color LS82 Canal 7, LS83 (Canal 9 Libertad), LV80 Telenueva, LU81 Teledifusora Bahiense SA, LV81 Canal 12 Telecor SACI, Dicor Difusión Córdoba, LV80 TV Canal 10 Universidad Nacional Córdoba, and LU82 TV Mar del Plata SA.

The Argentine Government holds a 20% stake in the regional television channel Telesur (q.v.), which began operations in May 2005 and is based in Caracas, Venezuela.

**Asociación de Teleradiodifusoras Argentinas (ATA):** Avda Córdoba 323, 6°, 1054 Buenos Aires; tel. (11) 4312-4208; fax (11) 4315-4681; e-mail info@ata.org.ar; internet www.ata.org.ar; f. 1959; asscn of 21 private television channels; Pres. CARLOS FONTAN BALESTRA; Sec. JORGE RENDO.

**América:** Fitzroy 1650, Buenos Aires; internet www.america2.com.ar; Programme Man. LILIANA PARODI.

**Canal 13:** Lima 1261, 1147 Buenos Aires; tel. (11) 4305-0013; fax (11) 4307-0315; e-mail prensa@artear.com; internet www.canal13

.com.ar; f. 1989; leased to a private concession in 1992; Programme Man. ADRIÁN SUAR.

**LS82 TV Canal 7:** Avda Figueroa Alcorta 2977, 1425 Buenos Aires; tel. (11) 4802-6001; fax (11) 4802-9878; e-mail info@canal7argentina.com.ar; internet www.canal7argentina.com.ar; state-controlled channel; Controller RICARDO PALACIO; Dir of News ANA DE SKALON; Dir of Drama LEONARDO BECHINI.

**LS83 TV Canal 9:** Dorrego 1708, Buenos Aires; tel. (11) 777-2321; fax (11) 777-9620; e-mail noticias@azultv.com; internet www.infobae.com/canal9/interiores/home.php; f. 1960; private channel; Dir-Gen. DANIEL HADAD; Gen. Man. SANDRO SCARAMELLI.

**Telefé (Canal 11):** Pavón 2444, 1248 Buenos Aires; tel. (11) 4941-9549; fax (11) 4942-6773; e-mail prensa@telefe.com; internet www.telefe.com.ar; private channel; Programme Man. CLAUDIO VILLARRUEL.

# Finance

(cap. = capital; res = reserves; dep. = deposits; m. = million; amounts in nuevos pesos argentinos—$, unless otherwise stated)

## BANKING

In September 2001 there were two public banks, 11 municipal banks, 33 domestic private banks, 39 foreign private banks and two co-operative banks.

### Central Bank

**Banco Central de la República Argentina:** Reconquista 266, 1003 Buenos Aires; tel. (11) 4348-3500; fax (11) 4348-3955; e-mail sistema@bcra.gov.ar; internet www.bcra.gov.ar; f. 1935 as central reserve bank; it has the right of note issue; all capital is held by the state; cap. and res $3,837.5m., dep. $992.3m. (Dec. 1999); Pres. MARTÍN REDRADO.

### Government-owned Commercial Banks

**Banco de la Ciudad de Buenos Aires:** Florida 302, 1313 Buenos Aires; tel. (11) 4329-8600; fax (11) 4112-098; e-mail bcdad39@sminter.com.ar; internet www.bancociudad.com.ar; municipal bank; f. 1878; cap. $169.5m., res $654.3m., dep. $4,463.0m. (Dec. 2002); Chair. ROBERTO J. FELETTI; Gen. Man. HUGO HARLEY ARBARELLO; 42 brs.

**Banco de Inversión y Commercio Exterior, SA (BICE):** 25 de Mayo 526, 1002 Buenos Aires; tel. (11) 4317-6900; fax (11) 4311-5596; e-mail info@bice.com.ar; internet www.bice.com.ar; f. 1991; cap. $489.2m., res $17.8m., dep. $121.8m. (Dec. 2000); Pres. and Chair. Dr ARNALDO MÁXIMO BOCCO; Vice-Pres. JOSÉ ALBERTO SPOSATO.

**Banco de la Nación Argentina:** Bartolomé Mitre 326, 1036 Buenos Aires; tel. (11) 4347-6000; fax (11) 4347-6316; e-mail gerencia@bna.com.ar; internet www.bna.com.ar; f. 1891; national bank; dep. $18,766.0m. (Dec. 2002); Chair. HORACIO E. PERICOLI; Gen. Man. JUAN CARLOS FÁBREGA; 645 brs.

**Banco de la Pampa SEM:** Carlos Pellegrini 255, 6300 Santa Rosa; tel. (2954) 433008; fax (2954) 433196; e-mail cexterior@blp.com.ar; internet www.blp.com.ar; f. 1958; cap. $8.8m., res $153.2m., dep. $656.2m. (June 2002); Pres. LUIS E. ROLDÁN; Gen. Man. OSWALDO LUIS DADONE; 112 brs.

**Banco de la Provincia de Buenos Aires:** Avda San Martín 137, 1597 Buenos Aires; tel. (11) 4347-0000; fax (11) 4347-0229; e-mail gsaldana@bpba.com.ar; internet www.bpba.com; f. 1822; provincial govt-owned bank; cap. $1,250.0m., res $244,137m., dep. $9,918,116m. (Sept. 2003); Pres. JORGE EMILIO SARGHINI; Gen. Man. JAVIER ARTACHO; 370 brs.

**Banco de la Provincia de Córdoba:** San Jerónimo 166, 5000 Córdoba; tel. (351) 420-7200; fax (351) 422-9718; f. 1873; provincial bank; cap. $47.6m., res $142.8m., dep. $1,036.1m. (Dec. 1994); Pres. LUIS ENRIQUE GRUNHAUT; 150 brs.

**Banco de la Provincia del Neuquén:** Avda Argentina 41/45, 8300 Neuquén; tel. (299) 4496618; fax (299) 4496622; internet www.bpn.com.ar; f. 1960; dep. $232.8m., total assets $388.6m. (March 1995); Pres. FÉLIX RACCO; Vice-Pres. CARLOS ALBERTO SANDOVAL; 22 brs.

**Banco Social de Córdoba:** 27 de Abril 185, 1°, 5000 Córdoba; tel. (351) 422-3367; dep. $187.3m., total assets $677.3m. (June 1995); Pres. Dr JAIME POMPAS.

**Banco de Tierra del Fuego:** Maipú 897, 9410 Ushuaia; tel. (2901) 441600; fax (2901) 441601; e-mail entradas@bancotdf.com.ar; internet www.bancotdf.com.ar; national bank; cap. and res $21.8m., dep. $54.2m. (June 1992); Pres. GUSTAVO LOFIEGO; Gen. Man. MIGUEL LANDERRECHE; 4 brs.

**Nuevo Banco de la Rioja, SA:** Rivadavia 702, 5300 La Rioja; tel. (3822) 430575; fax (3822) 430618; f. 1994; provincial bank; Pres. ELIAS SAHAD; Gen. Man. CLAUDIA L. DE BRIGIDO; 2 brs.

**Nuevo Banco de Santa Fe, SA:** 25 de Mayo, 3000 Santa Fe, Rosario; tel. (342) 429-4200; fax (342) 455-4543; e-mail contactobc@bancobsf.com.ar; internet www.bancobsf.com.ar; f. 1847 as Banco Provincial de Santa Fe, adopted current name in 1998; provincial bank; cap. $60.0m., res $31.1m., dep. $643.5m. (Dec. 1998); Chair. JOSÉ ENRIQUE ROHM; Exec. Dir MARCELO MUIÑO; 103 brs.

### Private Commercial Banks

**Banco BI Creditanstalt, SA:** Bouchard 547, 24° y 25°, 1106 Buenos Aires; tel. (11) 4319-8277; fax (11) 4319-8296; e-mail bicreditanstalt.com.ar; internet www.bicreditanstalt.com.ar; f. 1971 as Banco Interfinanzas, name changed as above in 1997; cap. and res $126,689m., dep. $749.7m. (Dec. 2002); Pres. Dr MIGUEL ANGEL ANGELINO; Gen. Man. RICARDO RIVERO HAEDO.

**Banco CMF, SA:** Macacha Güemes 555, Puerto Madero, 1106 Buenos Aires; tel. (11) 4318-6800; fax (11) 4318-6812; f. 1978 as Corporación Metropolitana de Finanzas, SA; adopted current name in 1999; cap. $49.0m., res $111.2m., dep. $254.0m. (Dec. 2002); Pres. JOSÉ P. BENEGAS LYNCH.

**Banco COMAFI:** Avda Roque Sáenz Peña 660, 1035 Buenos Aires; tel. (11) 4347-0400; fax (11) 4347-0404; e-mail contactenos@comafi.com.ar; internet www.comafi.com.ar; f. 1984; assumed control of 65% of Scotiabank Quilmes in April 2002; total assets US $410.1m. (June 2001); Pres. GUILLERMO CERVIÑO; Vice-Pres. EDUARDO MASCHWITZ.

**Banco Comercial Israelita, SA:** Bartolomé Mitre 702, 2000 Rosario; tel. (341) 420-0557; fax (341) 420-0517; f. 1921; cap. $4.3m., res $20.5m., dep. $232.0m. (June 1998); Pres. Ing. DAVID ZCARNY; 4 brs.

**Banco de Corrientes:** 9 de Julio 1099, esq. San Juan, 3400 Corrientes; tel. (3783) 479300; fax (3783) 479283; internet www.elbancodecorrientes.com.ar; f. 1951 as Banco de la República de Corrientes; adopted current name in 1993, after transfer to private ownership; cap. and res $38.5m., dep. $126.7m. (June 1995); Pres. ALEJANDRO RETEGUI; 33 brs.

**Banco Finansur, SA:** Avda Corrientes 400, 1043 Buenos Aires; tel. (11) 4324-3400; fax (11) 4322-4687; e-mail bafin@bancofinansur.com.ar; f. 1973 as Finansur Compañía Financiera; adopted current name in 1993; Pres. JORGE SÁNCHEZ CÓRDOVA; 5 brs.

**Banco de Galicia y Buenos Aires SA:** Juan D. Perón 407, Casilla 86, 1038 Buenos Aires; tel. (11) 4394-7080; fax (11) 4393-1603; e-mail bancogalicia@bancogalicia.com.ar; internet www.bancogalicia.com.ar; f. 1905; cap. $468.7m., res $2,244.3m., dep. $19,004.7m. (Dec. 2002); Pres. JUAN MARTÍN ETCHEGOYHEN; CEO EDUARDO ARROBAS; 281 brs.

**Banco Macro Bansud, SA:** Sarmiento 447, 1041 Buenos Aires; tel. (11) 5222-6500; fax (11) 5222-6624; e-mail international@macrobansud.com.ar; internet www.macrobansud.com; f. 1995 as Banco Bansud after merger of Banesto Banco Shaw, SA, and Banco del Sud, SA; adopted current name in 2002 following merger with Banco Macro (f. 1988); cap. $64.4m., res $1,132.0m., dep. $1,056.3m. (Dec. 2002); Pres. JORGE HORACIO BRITO; 149 brs.

**Banco Mariva, SA:** Sarmiento 500, 1041 Buenos Aires; tel. (11) 4321-2200; fax (11) 4321-2222; e-mail info@mariva.com.ar; internet www.mariva.com.ar; f. 1980; cap. $30.0m., res $10.0m., dep. $206.3m. (Dec. 1994); Pres. RICARDO MAY.

**Banco Patagonia Sudameris Argentina, SA:** Juan D. Perón 500, 1038 Buenos Aires; tel. (11) 4132-6055; fax (11) 4132-6059; e-mail international@patagoniasudameris.com.ar; internet www.patagoniasudameris.com.ar; f. 1912 as Banco Sudameris; adopted current name in 2003 following merger with Banco Patagonia; cap. $89.0m., res $11.5m., dep. $977.2m. (Dec. 1998); Chair. JOSÉ MARÍA DAGNINO PASTORE.

**Banco Regional de Cuyo, SA:** Avda San Martín 841, 5500 Mendoza; tel. (261) 449-8800; fax (261) 449-8801; internet www.bancoregional.com.ar; f. 1961; Pres. and Gen. Man. JOSÉ FEDERICO LÓPEZ; 12 brs.

**Banco Río de la Plata, SA:** Bartolomé Mitre 480, 1036 Buenos Aires; tel. (11) 4341-1000; fax (11) 4342-8962; internet www.bancorio.com.ar; f. 1908; cap. $346.7m., res $1,988.9m., dep. $13,008.2m. (Dec. 2002); Pres. JOSÉ LUIS ENRIQUE CRISTOFANI; 266 brs.

**Banco de San Juan:** Rivadavia 86, 5400 San Juan; tel. (264) 429-1000; fax (264) 421-4126; internet www.bancosanjuan.com.ar; f. 1943; 25% owned by provincial govt of San Juan; 75% privately owned; Pres. ENRIQUE ESKENAZI; Gen. Man. RAÚL RIOBÓ; 8 brs.

**Banco Santiago del Estero:** Avda Belgrano 529 Sur, 4200 Santiago del Estero; tel. (385) 422-2300; e-mail info@bse.com.ar;

ARGENTINA *Directory*

internet www.bse.com.ar; dep. $69.9m., total assets $101.6m. (Jan. 1995); Pres. Américo Daher.

**Banco de Valores, SA:** Sarmiento 310, 1041 Buenos Aires; tel. (11) 4323-6900; fax (11) 4334-1731; e-mail info@banval.sba.com.ar; internet www.bancodevalores.com; f. 1978; cap. $10.0m., res $40.5m., dep. $142.2m. (Dec. 2002); Pres. Julio A. Macchi; 1 br.

**Banex:** San Martín 136, 1004 Buenos Aires; tel. (11) 4340-3000; fax (11) 4334-4402; e-mail infobcra@banex.com.ar; internet www.banex.com.ar; f. 1998 by merger of Exprinter Banco with Banco San Luis, SA; cap. $29.1m., res $18.8m., dep. $196.6m. (Dec. 1999); Pres. Julio Patricio Supervielle; Gen. Man. Gabriel Coqueugniot; 49 brs.

**HSBC Bank Argentina, SA:** Avda de Mayo 701, 27°, 1084 Buenos Aires; tel. (11) 4344-3333; fax (11) 4334-6679; internet www.hsbc.com.ar; f. 1978 as Banco Roberts, SA; name changed to HSBC Banco Roberts, SA, in 1998; adopted current name in 1999; cap. $512.4m., res $1,068.4m., dep. $5,694.3m. (June 2003); Chair. and CEO Michael Smith; 68 brs.

**Nuevo Banco de Entre Ríos SA:** Monte Caseros 128, 3100 Paraná; tel. (343) 423-1200; fax (343) 421-1221; e-mail exterior .cambios@nuevobersa.com.ar; internet www.nuevobersa.com.ar; f. 1935; provincial bank; transferred to private ownership in 1995; cap. $20.4m., dep. $624.5m. (Dec. 2002); Pres. Ricardo Matías Taddeo; 73 brs.

**Nuevo Banco Industrial de Azul, SA:** Avda Córdoba 675, 1054 Buenos Aires; tel. (11) 4311-4666; fax (11) 4315-8113; e-mail info@bancoazul.com; internet www.bancoazul.com; f. 1971; Pres. and Gen. Man. Alberto Meta; 2 brs.

**Nuevo Banco Suquía, SA:** 25 de Mayo 160, 5000 Córdoba; tel. (351) 420-0200; fax (351) 420-0443; internet www.bancosuquia.com.ar; f. 1961 as Banco del Suquía, SA; adopted current name in 1998; dep. $2,051.0m. (June 2000); Pres. José P. Porta; Gen. Man. Raúl Fernández; 100 brs.

### Co-operative Bank

**Banco Credicoop Cooperativo Ltdo:** Reconquista 484, 1003 Buenos Aires; tel. (11) 4320-5000; fax (11) 4324-5891; e-mail credicoop@bancocredicoop.coop; internet www.credicoop.com.ar; f. 1979; cap. $0.7m., res $546.5m., dep. $2,324.9m. (June 2003); Pres. Raúl Guelman; Gen. Man. Carlos Heller; 227 brs.

### Foreign Banks

**ABN Amro Bank NV** (Netherlands): Florida 361, Casilla 171, 1005 Buenos Aires; tel. (11) 4320-0600; fax (11) 4322-0839; f. 1914; cap. and res $37.7m., dep. $94.5m. (June 1992); Gen. Man. César A. Deymonnaz; 7 brs.

**Banca Nazionale del Lavoro, SA (BNL)** (Italy): Florida 40, 1005 Buenos Aires; tel. (11) 4323-4400; fax (11) 4323-4689; internet www.bnl.com.ar; cap. $272m., dep. $2,306m. (June 1999); took over Banco de Italia y Río de la Plata in 1987; Pres. Ademaro Lanzara; Gen. Man. Niccolo Pandolfiu; 136 brs.

**Banco do Brasil, SA** (Brazil): Sarmiento 487, Casilla 2684, 1041 Buenos Aires; tel. (11) 4394-0939; fax (11) 4394-9577; f. 1960; cap. and res $33.9m., dep. $2.0m. (June 1992); Gen. Man. Hélio Testoni.

**Banco do Estado de São Paulo** (Brazil): Tucumán 821, Casilla 2177, 1049 Buenos Aires; tel. (11) 4325-9533; fax (11) 4325-9527; cap. and res $11.7m., dep. $6.7m. (June 1992); Gen. Man. Carlos Alberto Bergamasco.

**Banco Europeo para América Latina (BEAL), SA:** Juan D. Perón 338, 1038 Buenos Aires; tel. (11) 4331-6544; fax (11) 4331-2010; e-mail bealbsa@interprov.com; f. 1914; cap. and res $60m., dep. $121m. (Nov. 1996); Gen. Mans Jean Pierre Smeets, Klaus Krüger.

**Banco Itaú, SA** (Brazil): 25 de Mayo 476, 2°, 1002 Buenos Aires; tel. (11) 4325-6698; fax (11) 4394-1057; internet www.itau.com.ar; fmrly Banco Itaú Argentina, SA, renamed as above following purchase of Banco del Buen Ayre, SA, in May 1998; cap. and res $20.2m., dep. $1.2m. (June 1992); Pres. Olavo Egydio Setubal; 78 brs.

**Banco Société Générale, SA** (France): Reconquista 330, 1003 Buenos Aires; tel. (11) 4329-8000; fax (11) 4329-8080; e-mail info@ar.socgen.com; internet www.ar.socgen.com.ar; f. 1887 as Banco Supervielle de Buenos Aires, SA, adopted current name in 2000; cap. $50.5m., res $56.8m., dep. $820.6m. (Dec. 2001); Chair. and Gen. Man. Marc-Emmanuel Vives; 64 brs.

**Bank of Tokyo-Mitsubishi, Ltd** (Japan): Avda Corrientes 420, 1043 Buenos Aires; tel. (11) 4348-2001; fax (11) 4322-6607; f. 1956; cap. and res $20m., dep. $81m. (Sept. 1994); Gen. Man. Kazuo Omi.

**BankBoston NA** (USA): Florida 99, 1005 Buenos Aires; tel. (11) 4346-2000; fax (11) 4346-3200; f. 1784; cap. $456.8m., dep. $4,012m., total assets $8,677m. (Sept. 1998); Pres. Ing. Manuel Sacerdote; 139 brs.

**Banque Nationale de Paris, SA** (France): 25 de Mayo 471, 1002 Buenos Aires; tel. (11) 4318-0318; fax (11) 4311-1368; f. 1981; cap. and res $29.4m., dep. $86.7m. (June 1992); Gen. Man. Chislain de Beaucé.

**BBVA Banco Francés, SA:** Reconquista 199, 1003 Buenos Aires; tel. (11) 4346-4000; fax (11) 4346-4320; internet www.bancofrances.com; f. 1886 as Banco Francés del Río de la Plata, SA; changed name to Banco Francés, SA, in 1998 following merger with Banco de Crédito Argentino; adopted current name in 2000; cap. $368.1m., res $2,543.3m., dep. US $10,654.1m. (Dec. 2002); Chair. Gervásio Collar Zavaleta; CEO and Gen. Man. Antonio Martínez-Jorquera; 308 brs.

**Chase Manhattan Bank** (USA): Arenales 707, 5°, 1061 Buenos Aires; tel. (11) 4319-2400; fax (11) 4319-2416; f. 1904; cap. and res $46.3m., dep. $12,387m. (Sept. 1992); Gen. Man. Marcelo Podestá.

**Citibank, NA** (USA): Colón 58, Bahía Blanca, 8000 Buenos Aires; tel. (11) 4331-8281; f. 1914; cap. and res $172.8m., dep. $660.8m. (June 1992); Pres. Ricardo Angles; Vice-Pres. Guillermo Stanley; 16 brs.

**Deutsche Bank, SA** (Germany): Tucumán 1, 14°, 1049 Buenos Aires; tel. (11) 4590-2800; fax (11) 4590-2882; f. 1998; cap. $125.5m., res $182.6m., dep. $197.1m. (Dec. 2002); Pres. Patricio Eduardo Kelly; 47 brs.

**Morgan Guaranty Trust Co of New York** (USA): Avda Corrientes 411, 1043 Buenos Aires; tel. and fax (11) 4325-8046; cap. and res $29.1m., dep. $11.6m. (June 1992); Gen. Man. José McLoughlin.

**Republic National Bank of New York** (USA): Bartolomé Mitre 343, 1036 Buenos Aires; tel. (11) 4343-0161; fax (11) 4331-6064; cap. and res $17.2m., dep. $13.6m. (March 1994); Gen. Man. Alberto Muchnick.

### Bankers' Associations

**Asociación de Bancos del Interior de la República Argentina (ABIRA):** Avda Corrientes 538, 4°, 1043 Buenos Aires; tel. (11) 4394-3439; fax (11) 4394-5682; f. 1956; Pres. Dr Jorge Federico Christensen; Dir Raúl Passano; 30 mems.

**Asociación de Bancos Públicos y Privados de la República Argentina (ABAPPRA):** Florida 470, 1°, 1005 Buenos Aires; tel. (11) 4322-6321; fax (11) 4322-6721; e-mail info@abappra.com.ar; internet www.abappra.com; f. 1959; Pres. Enrique Olivera; Man. Luis B. Bucafusco; 31 mems.

**Federación de Bancos Cooperativos de la República Argentina (FEBANCOOP):** Maipú 374, 9°/10°, 1006 Buenos Aires; tel. (11) 4394-9949; f. 1973; Pres. Omar C. Trillo; Exec. Sec. Juan Carlos Romano; 32 mems.

## STOCK EXCHANGES

**Mercado de Valores de Buenos Aires, SA:** 25 de Mayo 367, 8°–10°, 1002 Buenos Aires; tel. (11) 4313-6021; fax (11) 4313-4472; e-mail merval@merval.sba.com.ar; internet www.merval.sba.com.ar; f. 1929; Pres. Héctor J. Bacqué.

There are also stock exchanges at Córdoba, Rosario, Mendoza and La Plata.

### Supervisory Authority

**Comisión Nacional de Valores (CNV):** 25 de Mayo 175, 1002 Buenos Aires; tel. (11) 4342-4607; fax (11) 4331-0639; e-mail webadm@cnv.gov.ar; internet www.cnv.gob.ar; monitors capital markets; Pres. Guillermo Harteneck.

## INSURANCE

In March 2001 there were 222 insurance companies operating in Argentina, of which 121 were general insurance companies. The following is a list of those offering all classes or a specialized service.

### Supervisory Authority

**Superintendencia de Seguros de la Nación:** Avda Julio A. Roca 721, 5°, 1067 Buenos Aires; tel. (11) 4331-8733; fax (11) 4331-9821; f. 1938; Supt Dr Ignacio Warnes.

### Major Companies

**La Agrícola, SA:** Buenos Aires; tel. (11) 4394-5031; f. 1905; associated co La Regional; all classes; Pres. Luis R. Marco; First Vice-Pres. Justo J. de Corral.

**Aseguradora de Créditos y Garantías, SA:** Avda Corrientes 415, 4°, 1043 Buenos Aires; tel. (11) 4394-4037; fax (11) 4394-0320; e-mail acgtias@infovia.com.ar; internet www.acg.com.ar; f. 1965; Pres. William A. Franke; CEO Horacio G. Scapparone.

**Aseguradora de Río Negro y Neuquén:** Avda Alem 503, Cipolletti, Río Negro; tel. (299) 477-2725; fax (299) 477-0321; f. 1960; all classes; Gen. Man. ERNESTO LÓPEZ.

**Aseguradores de Cauciones, SA:** Paraguay 580, 1057 Buenos Aires; tel. (11) 4318-3700; fax (11) 4318-3799; e-mail directorio@caucion.com.ar; internet www.caucion.com.ar; f. 1968; all classes; Pres. JOSÉ DE VEDIA.

**Aseguradores Industriales, SA:** Juan D. Perón 650, 6°, 1038 Buenos Aires; tel. (11) 4326-8881; fax (11) 4326-3742; f. 1961; all classes; Exec. Pres. Dir LUIS ESTEBAN LOFORTE.

**La Austral:** Buenos Aires; tel. (11) 442-9881; fax (11) 4953-4459; f. 1942; all classes; Pres. RODOLFO H. TAYLOR.

**Colón, Cía de Seguros Generales, SA:** San Martín 548–550, 1004 Buenos Aires; tel. (11) 4320-3800; fax (11) 4320-3802; f. 1962; all classes; Gen. Man. L. D. STSCK.

**Columbia, SA de Seguros:** Juan D. Perón 690, 1038 Buenos Aires; tel. (11) 4325-0208; fax (11) 4326-1392; f. 1918; all classes; Pres. MARTA BLANCO; Gen. Man. HORACIO H. PETRILLO.

**El Comercio, Cía de Seguros a Prima Fija, SA:** Avda Corrientes 415, 3° y 5°, 1043 Buenos Aires; tel. (11) 4394-1300; fax (11) 4393-1311; internet www.bristolgroup.com.ar; f. 1889; all classes; Pres. HORACIO SCAPPARONE; Vice-Pres. EDUARDO MARTELLI.

**Cía Argentina de Seguros de Créditos a la Exportación, SA:** Corrientes 345, 7°, 1043 Buenos Aires; tel. (11) 4313-3048; fax (11) 4313-2919; f. 1967; covers credit and extraordinary and political risks for Argentine exports; Pres. LUIS ORCOYEN; Gen. Man. Dr MARIANO A. GARCÍA GALISTEO.

**Cía Aseguradora Argentina, SA:** Avda Roque S. Peña 555, 1035 Buenos Aires; tel. (11) 430-1571; fax (11) 430-5973; f. 1918; all classes; Man. GUIDO LUTTINI; Vice-Pres. ALBERTO FRAGUIO.

**La Continental, Cía de Seguros Generales SA:** Avda Corrientes 655, 1043 Buenos Aires; tel. (11) 4393-8051; fax (11) 4325-7101; f. 1912; all classes; Pres. RAÚL MASCARENHAS.

**La Franco-Argentina, SA:** Buenos Aires; tel. (11) 430-3091; f. 1896; all classes; Pres. Dr GUILLERMO MORENO HUEYO; Gen. Man. Dra HAYDÉE GUZIAN DE RAMÍREZ.

**Hermes, SA:** Edif. Hermes, Bartolomé Mitre 754/60, 1036 Buenos Aires; tel. (11) 4331-4506; fax (11) 4343-5552; e-mail hermes@mbox.servicenet.com.ar; f. 1926; all classes; Pres. DIONISIO KATOPODIS; Gen. Man. FRANCISCO MARTÍN ZABALO.

**India, Cía de Seguros Generales SA:** Avda Roque S. Peña 728/36, 1035 Buenos Aires; tel. (11) 4328-6001; fax (11) 4328-5602; f. 1950; all classes; Pres. ALFREDO JUAN PRIESSE; Vice-Pres. Dr RAÚL ALBERTO GUARDIA.

**Instituto Italo-Argentino de Seguros Generales, SA:** Avda Roque S. Peña 890, 1035 Buenos Aires; tel. (11) 4320-9200; fax (11) 4320-9229; f. 1920; all classes; Pres. ALEJANDRO A. SOLDATI.

**La Meridional, Cía Argentina de Seguros SA:** Juan D. Perón 646, 1038 Buenos Aires; tel. (11) 4909-7000; fax (11) 4909-7274; e-mail meridi@starnet.net.ar; f. 1949; life and general; Pres. GUILLERMO V. LASCANO QUINTANA; Gen. Man. PETER HAMMER.

**Plus Ultra, Cía Argentina de Seguros, SA:** San Martín 548–50, 1004 Buenos Aires; tel. (11) 4393-5069; f. 1956; all classes; Gen. Man. L. D. STSCK.

**La Primera, SA:** Blvd Villegas y Oro, Trenque Lauquén, Prov. Buenos Aires; tel. (11) 4393-8125; all classes; Pres. ENRIQUE RAÚL U. BOTTINI; Man. Dr RODOLFO RAÚL D'ONOFRIO.

**La Rectora, SA:** Avda Corrientes 848, 1043 Buenos Aires; tel. (11) 4394-6081; fax (11) 4394-3251; f. 1951; all classes; Pres. PEDRO PASCUAL MEGNA; Gen. Man. ANTONIO LÓPEZ BUENO.

**La República Cía Argentina de Seguros Generales, SA:** San Martín 627/29, 1374 Buenos Aires; tel. (11) 4314-1000; fax (11) 4318-8778; e-mail ccastell@republica.com.ar; f. 1928; group life and general; Pres. JOSÉ T. GUZMAN DUMAS; Gen. Man. EDUARDO ESCRIÑA.

**Sud América Terrestre y Marítima Cía de Seguros Generales, SA:** Florida 15, 2°, Galería Florida 1, 1005 Buenos Aires; tel. (11) 4340-5100; fax (11) 4340-5380; f. 1919; all classes; Pres. EMA SÁNCHEZ DE LARRAGOITI; Vice-Pres. ALAIN HOMBREUX.

**La Unión Gremial, SA:** Mitre 665/99, 2000 Rosario, Santa Fe; tel. (341) 426-2900; fax (341) 425-9802; f. 1908; general; Gen. Man. EDUARDO IGNACIO LLOBET.

**La Universal:** Buenos Aires; tel. (11) 442-9881; fax (11) 4953-4459; f. 1905; all classes; Pres. RODOLFO H. TAYLOR.

**Zurich-Iguazú Cía de Seguros, SA:** San Martín 442, 1004 Buenos Aires; tel. (11) 4329-0400; fax (11) 4322-4688; f. 1947; all classes; Pres. RAMÓN SANTAMARINA.

### Reinsurance

**Instituto Nacional de Reaseguros:** Avda Julio A. Roca 694, 1067 Buenos Aires; tel. (11) 4334-0084; fax (11) 4334-5588; f. 1947; reinsurance in all branches; Pres. and Man. REINALDO A. CASTRO.

### Insurance Associations

**Asociación Argentina de Cías de Seguros (AACS):** 25 de Mayo 565, 2°, 1002 Buenos Aires; tel. (11) 4312-7790; fax (11) 4312-6300; e-mail secret@aacsra.org.ar; internet www.aacsra.org.ar; f. 1894; 32 mems; Pres. ROBERTO F. E. SOLLITTO.

**Asociación de Entidades Aseguradoras Privadas de la República Argentina (EAPRA):** Esmeralda 684, 4°, 1007 Buenos Aires; tel. (11) 4393-2268; fax (11) 4393-2283; f. 1875; asscn of 12 foreign insurance cos operating in Argentina; Pres. Dr PIERO ZUPPELLI; Sec. BERNARDO VON DER GOLTZ.

## Trade and Industry

### GOVERNMENT AGENCIES

**Agencia de Desarrollo de Inversiones:** Avda Pte Julio A. Roca 651, 5°, Sector 20, 1067 Buenos Aires; tel. (11) 4349-3442; fax (11) 4349-3453; e-mail adi@mecon.gov.ar; internet www.inversiones.gov.ar; promotion of investment in Argentina; supervised by the Secretaría de Industria, Comercio y de la Pequeña y Mediana Empresa.

**Cámara de Exportadores de la República Argentina:** Avda Roque S. Peña 740, 1°, 1035 Buenos Aires; tel. (11) 4394-4351; fax (11) 4328-1003; e-mail contacto@cera.org.ar; internet www.cera.org.ar; f. 1943 to promote exports; 700 mems; Pres. Dr ENRIQUE S. MANTILLA.

**Consejo Federal de Inversiones:** San Martín 871, 1004 Buenos Aires; tel. (11) 4313-5557; fax (11) 4315-1639; e-mail cfi@cfired.org.ar; internet www.cfired.org.ar; federal board to co-ordinate domestic and foreign investment and provide technological aid for the provinces; f. 1959; Sec.-Gen. Ing. JUAN JOSÉ CIÁCERA.

**Dirección de Forestación (DF):** Avda Paseo Colón 982, anexo jardín, 1063 Buenos Aires; tel. (11) 4349-2124; fax (11) 4349-2102; e-mail bfores@mecon.gov.ar; assumed the responsibilities of the national forestry commission (Instituto Forestal Nacional—IFONA) in 1991, following its dissolution; supervised by the Secretaría de Agricultura, Ganadería, Pesca y Alimentación; maintains the Centro de Documentación e Información Forestal; Library Man. NILDA E. FERNÁNDEZ.

**Instituto de Desarrollo Económico y Social (IDES):** Araoz 2838, 1425 Buenos Aires; tel. (11) 4804-4949; fax (11) 4804-5856; e-mail ides@ides.org.ar; internet www.ides.org.ar; f. 1960; investigation into social sciences and promotion of social and economic devt; 700 mems; Pres. ADRIANA MARSHALL; Sec. LUIS BECCARIA.

**Oficina Nacional de Control Comercial Agropecuario (ONCCA):** Avda Paseo Colón 922, Buenos Aires; tel. (11) 4349-2034; fax (11) 4349-2005; e-mail infooncca@sagpya.minproduccion.gov.ar; oversees the agricultural sector; supervised by the Secretaría de Agricultura, Ganadería, Pesca y Alimentación.

**Organismo Nacional de Administración de Bienes (ONABE):** 1302 Avda J. M. Ramos Mejia, 3°, Of. 300, 1104 Buenos Aires; tel. (11) 4318-3458; responsible for overseeing privatization of state property.

**Secretaría de Agricultura, Ganadería, Pesca y Alimentación:** Avda Paseo Colón 922, 1°, Of. 146, 1063 Buenos Aires; tel. (11) 4349-2000; fax (11) 4349-2292; e-mail comunica@sagpya.minproduccion.gov.ar; internet www.sagpya.mecon.gov.ar; f. 1871; undertakes regulatory, promotional, advisory and administrative responsibilities on behalf of the meat, livestock, agriculture and fisheries industries; Sec. MIGUEL SANTIAGO CAMPOS.

### DEVELOPMENT ORGANIZATIONS

**Instituto Argentino del Petróleo y Gas:** Maipú 645, 3°, 1006 Buenos Aires; tel. (11) 4325-8008; fax (11) 4393-5494; e-mail informa@iapg.org.ar; internet www.iapg.org.ar; f. 1958; established to promote the devt of petroleum exploration and exploitation; Pres. Ing. JORGE OSCAR FERIOLI.

**Secretario de Programación Económica:** Hipólito Yrigoyen 250, 8°, Of. 819, Buenos Aires; tel. (11) 4349-5710; fax (11) 4349-5714; f. 1961 to formulate national long-term devt plans; Sec. Dr JUAN JOSÉ LACH.

**Sociedad Rural Argentina:** Florida 460, 1005 Buenos Aires; tel. (11) 4324-4700; fax (11) 4324-4774; e-mail prensa@ruralarg.com.ar; internet www.ruralarg.org.ar; f. 1866; private org. to promote the devt of agriculture; Pres. Dr LUCIANO MIGUENS; 9,400 mems.

# ARGENTINA — Directory

## CHAMBERS OF COMMERCE

**Cámara Argentina de Comercio:** Avda Leandro N. Alem 36, 1003 Buenos Aires; tel. (11) 5300-9000; fax (11) 5300-9058; e-mail gerencia@cac.com.ar; internet www.cac.com.ar; f. 1924; Pres. CARLOS RAÚL DE LA VEGA.

**Cámara de Comercio, Industria y Producción de la República Argentina:** Florida 1, 4°, 1005 Buenos Aires; tel. (11) 4331-0813; fax (11) 4331-9116; f. 1913; Pres. JOSÉ CHEDIEK; Vice-Pres Dr FAUSTINO S. DIÉGUEZ, Dr JORGE M. MAZALAN; 1,500 mems.

**Cámara de Comercio Exterior de Rosario:** Avda Córdoba 1868, 2000 Rosario, Santa Fe; tel. and fax (341) 425-7147; e-mail ccer@commerce.com.ar; internet www.commerce.com.ar; f. 1958; deals with imports and exports; Pres. JUAN CARLOS RETAMERO; Vice-Pres. EDUARDO C. SALVATIERRA; 150 mems.

**Cámara de Comercio de los Estados Unidos en la República Argentina (AMCAM):** Viamonte 1133, 8°, 1053 Buenos Aires; tel. (11) 4371-4500; fax (11) 4371-8100; e-mail amcham@amchamar.ar; internet www.amchamar.com.ar; f. 1918; US Chamber of Commerce; Mem. Man. GONZALO LERGA.

Similar chambers are located in most of the larger centres and there are many other foreign chambers of commerce.

## INDUSTRIAL AND TRADE ASSOCIATIONS

**Asociación de Importadores y Exportadores de la República Argentina:** Avda Belgrano 124, 1°, 1092 Buenos Aires; tel. (11) 4342-0010; fax (11) 4342-1312; e-mail aiera@aiera.org.ar; internet www.aiera.org.ar; f. 1966; Pres. HÉCTOR MARCELO VIDAL; Man. ADRIANO DE FINA.

**Asociación de Industriales Textiles Argentinos:** Buenos Aires; tel. (11) 4373-2256; fax (11) 4373-2351; f. 1945; textile industry; Pres. BERNARDO ABRAMOVICH; 250 mems.

**Asociación de Industrias Argentinas de Carnes:** Buenos Aires; tel. (11) 4322-5244; meat industry; refrigerated and canned beef and mutton; Pres. JORGE BORSELLA.

**Asociación Vitivinícola Argentina:** Güemes 4464, 1425 Buenos Aires; tel. (11) 4774-3370; f. 1904; wine industry; Pres. LUCIANO COTUMACCIO; Man. Lic. MARIO J. GIORDANO.

**Cámara de Sociedades Anónimas:** Libertad 1340, 1016 Buenos Aires; tel. and fax (11) 4000-7399; e-mail camsocanon@camsocanon.com; internet www.camsocanon.com; Pres. Dr ALFONSO DE LA FERRERE; Man. CARLOS ALBERTO PERRONE.

**Centro de Exportadores de Cereales:** Bouchard 454, 7°, 1106 Buenos Aires; tel. (11) 4311-1697; fax (11) 4312-6924; f. 1943; grain exporters; Pres. RAÚL S. LOEH.

**Confederaciones Rurales Argentinas:** México 628, 2°, 1097 Buenos Aires; tel. (11) 4261-1501; Pres. ARTURO J. NAVARRO.

**Coordinadora de Actividades Mercantiles Empresarias:** Buenos Aires; Pres. OSVALDO CORNIDE.

**Federación Agraria Argentina (FAA):** e-mail comunicacion@faa.com.ar; internet www.faa.com.ar; f. 1912; oversees the interests of small and medium-sized grain producers; Pres. EDUARDO BUZZI.

**Federación Lanera Argentina:** 25 de Mayo 516, 4°, 1002 Buenos Aires; tel. (11) 4878-8800; fax (11) 4878-8804; e-mail info@flasite.com; internet www.flasite.com; f. 1929; wool industry; Pres. RICHARD VON GERSTENBERG; Sec. GEORGE J. LEFEBVRE; 80 mems.

## EMPLOYERS' ORGANIZATION

**Unión Industrial Argentina (UIA):** Avda Leandro N. Alem 1067, 11°, 1001 Buenos Aires; tel. (11) 4313-4474; fax (11) 4313-2413; e-mail uia@uia.org.ar; internet www.uia.org.ar; f. 1887; re-established in 1974 with the fusion of the Confederación Industrial Argentina (CINA) and the Confederación General de la Industria; following the dissolution of the CINA in 1977, the UIA was formed in 1979; asscn of manufacturers, representing industrial corpns; Pres. ALBERTO ALVAREZ GAIANI; Sec. JUAN CARLOS SACCO.

## MAJOR COMPANIES

**Aceros Bragado, SACIF:** Bernardo de Yrigoyen 190, 1072 Buenos Aires; tel. (11) 4385-952; fax (11) 4112-068; f. 1969; foundry, mill rolls, bearing trucks, laminating; Pres. BERNARDO ABEL COLL; 1,670 employees.

**Acindar Industria Argentina de Aceros, SA:** Estanislao Zeballos 2739, Beccar, Buenos Aires; tel. (11) 4719-8300; fax (11) 4719-8501; e-mail sac@acindar.com.ar; internet www.acindar.com.ar; f. 1942; production of iron and steel; Commercial Man. ROQUE MONASTERIO; 3,922 employees.

**Alpargatas, SAIC:** Avda Regimiento de Los Patricios 1142, 1265 Buenos Aires; tel. (11) 4303-0041; fax (11) 4303-2401; f. 1885; textile and footwear manufacturers; Pres. GUILLERMO A. GOTELLI; 8,000 employees.

**ALUAR** (Aluminio Argentino, SAIC): Pasteur 4600, Victoria, 1644 Buenos Aires; tel. (11) 4725-8060; fax (11) 4725-8091; internet www.aluar.com.ar; f. 1970; aluminium production; Pres. DOLORES QUINTANILLA DE MADANES; 1,700 employees.

**Agrometal, SA:** Misiones Monte Maiz 1974, 2659 Córdoba; tel. (34) 6847-1311; fax (34) 6847-1804; internet www.agrometal.com.ar; manufacture of agricultural machinery; Chair. ROSANA MARÍA NEGRINI.

**Alto Palermo, SA:** Hipólito Yrigoyen 476, Buenos Aires; tel. (11) 4344-4600; e-mail finanzas@altopalermo.com.ar; internet www.altopalermo.com.ar; land and real estate, finance; Chair. EDUARDO SERGIO ELSZTAIN, Sr; CEO ALEJANDRO G. ELSZTAIN.

**Astra, Compañía Argentina de Petróleo, SA:** Tucumán 744, 19° y 20°, 1049 Buenos Aires; tel. (11) 4324-0096; fax (11) 4329-0019; e-mail sfidalgos@repsol-ypf.com; f. 1912; petroleum services and products; owned by Repsol-YPF; Pres. Dr J. M. RANERO DÍAZ; 725 employees.

**Atanor, SA:** Albarellos 4914, Munro, Buenos Aires; tel. (11) 4721-3408; fax (11) 4721-3400; e-mail dircomex@atanorsa.com.ar; internet www.atanorsa.com.ar; f. 1943; producers of chemicals and petrochemicals; Pres. MIGUEL ANGEL GONZÁLEZ; 768 employees.

**BAESA** (Buenos Aires Embotelladora, SA): Roque Saenz Peña 308, San Isidro Buenos Aires; tel. (11) 4747-8317; fax (11) 4747-3846; f. 1989; makers of canned and bottled soft drinks; owned by Quilmes, SA; Chair., Pres. and CEO OSVALDO H. BAÑOS; 5,500 employees.

**Bagley, SA:** Avda Montes de Oca 169, 1270 Buenos Aires; tel. (11) 4300-0202; fax (11) 4341-4013; f. 1887; manufacturers of crackers, snacks, biscuits, chocolate products and wines and spirits; Pres. Dr ERNESTO O'FARRELL; 5,011 employees.

**Bayer Argentina, SA:** Ricardo Gutiérrez 3652, Munro, 1605 Buenos Aires; tel. (11) 4762-7000; fax (11) 4762-7100; internet www.bayer.com.ar; f. 1911; production of chemicals, agrochemicals and pharmaceuticals; parent co Bayer AG, Germany; Pres. and Gen. Man. HELMUT FLETCHNER; 1,176 employees.

**BGH, SA:** Brasil 731, 1154 Buenos Aires; tel. (11) 4309-2001; fax (11) 4362-6690; e-mail info@bgh.com.ar; internet www.bgh.com.ar; electronic appliances manufacturers; f. 1913 as Boris Garfunkel e Hijos; Gen. Man. EDUARDO SCARPELLO; 530 employees.

**Boldt, SA:** Aristóbulo del Valle 1257, 1295 Buenos Aires; tel. (11) 4309-5400; fax (11) 4361-3435; e-mail contact@boldt.com; internet www.boldt.com.ar; information technology, telecommunications, land and leisure management; Chair. ANTONIO ANGEL TABANELLI; 932 employees.

**Borax Argentina, SA:** Tincalayu, Salta; e-mail boraxargentina@borax.com; internet www.borax.com; owned by Rio Tinto; mining of borates.

**CANALE, SA:** Avda Martín García 320, 1165 Buenos Aires; tel. (11) 4307-4000; fax (11) 4307-3003; f. 1975; manufacturers of biscuits; Gen. Man. CARLOS MOLLINERI; 1,688 employees.

**Celulosa Argentina, SA:** Dardo Rocha 3278, Martínez, 1640 Buenos Aires; tel. (11) 4717-6077; e-mail contacto.comercial@celulosaargentina.com.ar; internet www.celulosaargentina.com.ar; f. 1929; manufacturers of paper and paper products; Pres. RICARDO ZERBINO CAVAJANI; 1,000 employees.

**Cervecería Maltería Quilmes, SAICA:** 12 de Octubre y Gran Canaria s/n, 1878 Quilmes, Buenos Aires; tel. (11) 4394-1700; fax (11) 4326-0026; internet www.quilmes.com.ar; f. 1888; beer and malt producers; Pres. CARLOS MIGUENS; 1,601 employees.

**Colorín, SA** (Colorín Industria de Materiales Sintéticos, SA): Juramento 5853, Munro, 1605 Buenos Aires; tel. (11) 462-7700; fax (11) 4762-7796; e-mail catcli@colorin.com; internet www.colorin.com; paint producers; Gen. Man. MARTÍN RODRÍGUEZ; 223 employees.

**Compañía Azucarera Concepción, SA:** San Martín 662, 5°, 1004 Buenos Aires; tel. (11) 4311-3444; fax (11) 4312-0418; manufacturers of sugar cane and alcohol; Pres. HORACIO GARCÍA GONZÁLEZ; 3,289 employees.

**COVIASA, SA:** Avda Castañares 1581/95, 1406 Buenos Aires; tel. (11) 4921-5794; e-mail coviasa@com.ar; internet www.coviasaherrajes.com.ar; mining co.

**Disco, SA:** Larrea 847, 1°, 1117 Buenos Aires; tel. (11) 4964-8000; fax (11) 4964-8039; e-mail feedback@disco.com.ar; internet www.disco.com.ar; f. 1961; supermarket chain; Chair. LEITZIA VEJO DE PEIRANO; 5,100 employees.

**Dycasa, SA:** Avda Leandro N. Alem 986, 4°, 1001 Buenos Aires; tel. (11) 4318-0200; fax (11) 4318-0230; e-mail info@dycasa.com.ar; internet www.dycasa.com; construction; Pres. ENRIQUE TOMÁS HUERGO.

**Esso SA Petrolera Argentina:** Carlos María Della Paolera 297, 1001 Buenos Aires; tel. (11) 4319-1400; fax (11) 4319-1163; f. 1911; active in all spheres of the petroleum industry; subsidiary of Exxon Corpn, USA; Man. JENS DREYER; Public Relations Man. ALEJANDRO MUENTE; 2,000 employees.

**Ferrum, SA:** España 496, Avellaneda, 1870 Buenos Aires; tel. (11) 4222-1500; fax (11) 4222-3464; e-mail info@ferrum.com; internet www.ferrum.com; f. 1911; building material manufacturers; Pres. IMGENIERO GUILLERMO VIEGENER; Man. Dir DANIEL H. CALABRÓ; 1,200 employees.

**Fiplasto, SACI:** Alsina 756, 1087 Buenos Aires; tel. (11) 4331-2518; fax (11) 4331-2136; e-mail info@fiplasto.com.ar; internet www.fiplasto.com.ar; f. 1945; hardwood and veneer manufacturers; Pres. MAXIMO FEDERICO LELOIR; 300 employees.

**Ford Argentina, SA:** Henry Ford/Ruta Panamericana s/n, Ricardo Rojas, 1617 Buenos Aires; tel. (11) 4756-9000; fax (11) 4756-9001; internet www.ford.com; f. 1913; manufacture of motor vehicles; owned by Ford Motor Co, USA; Pres. THEODORE CANNIS; 5,200 employees.

**GATIC, SAICFIA:** Eva Perón 2535, San Martín, 1650 Buenos Aires; tel. (11) 4724-7200; fax (11) 4724-7676; f. 1953; rubber footwear makers; Pres. EDUARDO IEVART BAKCHELLIAN; 4,666 employees.

**General Electric Technical Services Co, Inc.:** Avda Leandro N. Alem 619, 9°, 1001 Buenos Aires; tel. (11) 4313-2880; fax (11) 4313-2880; f. 1920; sales of industrial equipment; engineering services; Pres. ALEJANDRO BOTTAN; subsidiary of International General Electric Co, USA; 900 employees.

**Grafex, SAGCIF:** Crisólogu Larralde 3414, 1872 Sarandí, Provincia de Buenos Aires; tel. (11) 4206-6478; fax (11) 4207-4720; e-mail ventas@grafex.com.ar; internet www.grafex.com.ar; f. 1894; manufacturers of inks, varnishes and lacquers for the printing industry, and distributors of printing supplies; Pres. ALDO BLARDONE; 95 employees.

**Grimoldi, SA:** Florida 251, Buenos Aires; tel. (11) 4489-6400; fax (11) 4627-7763; internet www.grimoldi.com.ar; footwear retailers; Pres. ALBERTO LUIS GRIMOLDI.

**IBM Argentina, SA:** Ing. Enrique Butty 275, 1300 Buenos Aires; tel. (11) 4898-4898; fax (11) 4313-2360; e-mail ibm_directo@ar.ibm.com; internet www.ibm.com.ar; f. 1923; computer hardware and software; owned by IBM Corpn, USA; Gen. Man. for Latin America JUAN FERNÁNDEZ OLIVA; Pres. MARCELO LEMA; 1,200 employees.

**ICI Argentina SAIC:** Paseo Colón 221, 5°, 1063 Buenos Aires; tel. (11) 4343-2011; fax (11) 4331-1185; e-mail industrial@ici.com.ar; internet www.ici.com.ar; f. 1928 as Duperial; adopted present name in 1995; manufacturers of wine chemicals, alcohol, grapeseed oil; 200 employees.

**Ingenio y Refinería San Martín del Tabacal, SA:** Leandro N. Alem 986, 1001 Buenos Aires; tel. (11) 4576-7710; fax (11) 4576-7720; e-mail ingenio@tabacal.com.ar; internet www.tabacal.com.ar; f. 1920; sugar and alcohol production and citrus fruit cultivation; owned by Seaboard Corpn of the USA; Man. Dir RANDOLPH I. FLEMING; 1,200 employees.

**Inversiones y Representaciones, SA** (IRSA): Moreno 877, 1066 Buenos Aires; tel. (11) 4344-4600; fax (11) 4344-4611; internet www.irsa.com.ar; f. 1943; land and property development; Chair. EDUARDO SERGIO ELSZTAIN.

**Kraft Foods, SA Argentina:** Avda Int Francisco Rabanal 3220, 1437 Buenos Aires; tel. (11) 4630-8000; fax (11) 4924-3003; internet www.kraft.bumeran.com.ar; f. 1933; chocolate, sweets and frozen confectionery; owned by Altria; Chair. and CEO JUAN PEDRO MUNRO; 600 employees.

**Ledesma, SAAIC:** Avda Corrientes 415, 13°, 1043 Buenos Aires; tel. (11) 4378-1555; fax (11) 4325-7666; e-mail adiciancio@ledesma.com.ar; internet www.ledesma.com.ar; f. 1908; sugar producers; Pres. Dr CARLOS PEDRO BLAQUIER; 3,970 employees.

**Loma Negra, SA:** Bouchard 680, Buenos Aires; tel. (11) 4319-3003; fax (11) 4319-3000; e-mail info@lomanegra.com.ar; internet www.lomanegra.com.ar; f. 1926; cement and building materials manufacturing; Pres. MARÍA AMALIA SARA LACROZE DE FORTABAT; Dir-Gen. VÍCTOR SAVANTI; 2,200 employees.

**Massalin Particulares, SA:** Avda Leandro N. Alem 466, 9°, 1003 Buenos Aires; tel. (11) 4319-4100; fax (11) 4319-4150; f. 1980; cigarette and tobacco producers; Pres. RAFAEL ARGÜELLES; 2,500 employees.

**Mercedes Benz Argentina, SACIFIM:** Avda del Libertador 2424, 1425 Buenos Aires; tel. (11) 4801-0061; fax (11) 4808-8701; internet www.mercedes-benz.com.ar; f. 1951; manufacturers of trucks, buses and engines; subsidiary of Daimler Benz AG, Germany; Pres. KARL HEINZ HARTMANN; 1,600 employees.

**Juan Minetti, SA:** Ituzaingó 87, 5000 Córdoba; tel. 0800 777-6463; fax (351) 426-7551; e-mail conexion@grupominetti.com.ar; internet www.juanminetti.com.ar; f. 1932; manufacturers of hydraulic cement; Pres. MANUEL AUGUSTO JOSÉ BALTAZAR FERRER; Vice-Pres. JUAN JAVIER NEGRI; 830 employees.

**Molinos Río de la Plata, SA:** Uruguay 4075, Victoria, 1644 Buenos Aires; tel. (11) 4340-1100; e-mail atconsum@molinos.com.ar; internet www.molinos.com.ar; f. 1931; manufacturers of flour and grain products; part of the Pérez Companc group; COO GUILLERMO GARCÍA; 3,000 employees.

**Morixe Hermanos, SA:** Federico García Lorca 210–250, 1405 Buenos Aires; tel. (11) 4431-1281; fax (11) 4431-4079; e-mail info@morixe.com.ar; internet www.morixe.com.ar; flour and grain processing; f. 1923; Pres. JORGE JERONIMO DE ACHÁVAL; 215 employees.

**Nestlé Argentina:** Avda del Libertador 1855, Vicente López, 1638 Buenos Aires; tel. (11) 4329-8100; fax (11) 4329-8200; internet www.nestle.com.ar; manufacturers of condensed milk, instant coffee, milk powder and confectionery; subsidiary of Nestlé, SA, Switzerland; f. 1930; Dir-Gen. CLAUDIO BARTOLINI; 3,400 employees.

**Nobleza-Piccardo, SAICF:** San Martín 645, San Martín, 1650 Buenos Aires; tel. (11) 4724-8444; fax (11) 4313-2499; internet www.noblezapiccardo.com; f. 1898; cigarette and tobacco manufacturers; owned by British American Tobacco; Pres. and Gen. Man. MARK M. COBBEN; 1,000 employees.

**PBBPolisur SA:** Eduardo Madero 900, 7°, 1106 Buenos Aires; tel. (11) 4319-0100; e-mail fbepoli@dow.com; internet www.dow.com/polisur; wholly owned by Dow Chemical since February 2005; petrochemicals.

**Peugeot Citroën Argentina, SA:** Juan Domingo Perón 1001, Villa Bosch, 1682 Provincia de Buenos Aires; tel. (11) 4734-3005; fax (11) 4734-3007; internet www.peugeot.com.ar; f. 1965 as Sevel Argentina, SA; automobile manufacturers; Chair. FRANCISCO NACRI; 7,500 employees.

**Philips Argentina, SA:** Vedía 3892, 1430 Buenos Aires; tel. (11) 4546-7777; fax (11) 4546-7600; internet www.philips.com.ar; f. 1935; manufacturers of electrical equipment; subsidiary of Philips Golampenfabrieken NV, Netherlands; CEO JUAN LARRAÑAGA; 1,780 employees.

**Pirelli Cables, SAIC:** Cervantes 1901, Merlo, 1722 Buenos Aires; tel. (11) 4489-6000; fax (11) 4489-6603; internet www.pirelli.com.ar; Pres. ELVIO BALDINELLI; f. 1948; 1,900 employees.

**Plavinil Argentina, SA:** Timoteo Gordillo 5490, 1439 Buenos Aires; tel. (11) 4605-4042; fax (11) 4605-8061; e-mail plavinil@infovia.com.ar; internet www.plavinil.com.ar; industrial plastic textiles; f. 1947; Pres. and Gen. Man. CARLOS MARCELO AGOTE; 150 employees.

**Pluspetrol Exploración y Producción, SA:** Edif. Pluspetrol Lima 339, 1073 Buenos Aires; tel. (11) 4340-2222; fax (11) 4340-2215; e-mail rrhh-cv@pluspetrol.com.ar; internet www.pluspetrol.net; f. 1977; oil and gas exploration and production; Chair. and Pres. LUIS ALBERTO REY; 497 employees.

**Quest International Argentina, SA:** Ruta 9, Panamericana Km 36, Colectora Oeste, 1619 Garín; tel. (3327) 456-800; fax (3327) 456-927; internet www.questintl.com; subsidiary of ICI International; produces fragrances, flavours, food ingredients.

**Química Estrella, SACII:** Avda de los Constituyentes 2995, 1427 Buenos Aires; tel. (11) 4254-8100; fax (11) 4522-3022; e-mail info@quimicaestrella.com.ar; f. 1906; pharmaceutical manufacturers; Pres. JOSÉ A. MARTÍNEZ DE HOZ; 600 employees.

**Renault Argentina, SA:** Fray Justo María de Oro 1744, 1414 Buenos Aires; tel. (11) 4778-2000; fax (11) 4778-2023; e-mail src-renault.argentina@renault.com; internet www.renault.com.ar; f. 1955 as Ciadea, SA; motor vehicle manufacturers; Pres. MANUEL FERNANDO ANTELO; 2,211 employees.

**Repsol YPF (Yacimientos Petrolíferos Fiscales), SA:** Avda Roque Saénz Peña 777, 1035 Buenos Aires; tel. (11) 4329-2000; fax (11) 4329-2001; e-mail lrejonp.ir@repsolypf.com; internet www.ypf.com.ar; f. 1922; petroleum and gas exploration and production; state-owned until 1992; Pres. ANTONI BRUFAU NIUBÓ; Dir-Gen. for Argentina, Bolivia and Brazil ENRIQUE LOCUTURA; 9,750 employees.

**Rigolleau, SA:** Lisandro de la Torre 1651, Berazategui, 1009 Buenos Aires; tel. (11) 4256-2010; fax (11) 4256-2544; e-mail info@rigolleau.com.ar; internet www.rigolleau.com.ar; f. 1882; makers of glass and glass products; Pres. ENRIQUE CATTORINI; 1,485 employees.

**Roggio e Hijos Benito, SA:** Blvd Las Heras 402, 5000 Córdoba; tel. (351) 420-2202; fax (351) 445-1717; e-mail webmaster@roggio.com.ar; internet www.roggio.com.ar; f. 1955; group of construction companies; Pres. VITO REMO ROGGIO; 2,156 employees.

**Shell Compañía Argentina de Petróleo SA:** Avda Roque Saénz Peña 788, 1035 Buenos Aires; tel. (11) 4130-2000; e-mail

# ARGENTINA

nuevoinforme@shell.com; internet www.shell.com.ar; f. 1922; active in all spheres of the petroleum industry; owned by Royal Dutch Shell; Pres. JORGE BREA.

**Techint:** Carlos María Della Paolera 299, 17°, 1001 Buenos Aires; tel. (11) 4018-4100; fax (11) 4018-1000; e-mail info@techint.com; internet www.techint.com; f. 1945; steel and petroleum extraction and refining; Dir SERGIO EINAUDI.

**Siderar, SA:** Avda Leandro N. Alem 1067, 23°, 1001 Buenos Aires; tel. (11) 4018-2100; fax (11) 4318-2460; e-mail aparej@siderar.com; internet www.siderar.com; f. 1962; manufacturers of steel; Pres. PAOLO ROCCA; Dir.-Gen. FREDY CAMEO; 5,695 employees.

**Texaco Argentina, SA:** Virgilio 4501, Lomas de Zamora, Provincia de Buenos Aires; tel. (11) 4285-1116; fax (11) 4285-4664; e-mail deloar@chevrontexaco.com.

**Xerox Argentina, ICSA:** Casilla de Correo 1664, Jaramillo 1595, 1429 Buenos Aires; tel. (11) 4703-7700; fax (11) 4703-7701; e-mail webmaster-xrx-ar@xerox.com; internet www.xerox.com.ar; f. 1967; document processing and office equipment hire; subsidiary of Xerox Corpn, USA; Pres. EDUARDO GABRIEL LIJTMAER; 500 employees.

## UTILITIES

### Regulatory Authorities

**Energía Argentina, SA (ENARSA):** Neuquén; f. 2004; state-owned; Dir EXEQUIEL ESPINOSA.

**Ente Nacional Regulador de la Electricidad (ENRE):** Avda Eduardo Madero 1020, 10°, 1106 Buenos Aires; tel. (11) 4314-5805; fax (11) 4314-5416; internet www.enre.gov.ar; Vice-Pres. Ing. RICARDO MARTÍNEZ LEONE.

**Ente Nacional Regulador del Gas (ENARGAS):** Suipacha 636, 10°, 1008 Buenos Aires; tel. (11) 4325-9292; fax (11) 4348-0550; internet www.enargas.gov.ar; Pres. Ing. HÉCTOR ENRIQUE FORMICA.

### Electricity

**CAPEX:** Melo 630, Vicente López, 1638 Buenos Aires; tel. (11) 4796-6000; e-mail info@capex.com.ar; internet www.capex.com.ar; f. 1988; electricity generation; Chair. ENRIQUE GÖTZ; Vice-Chair. ALEJANDRO GÖTZ.

**Central Costanera, SA (CECCO):** Avda España 3301, 1107 Buenos Aires; tel. (11) 4307-3040; fax (11) 4307-1706; generation, transmission, distribution and sale of thermal electric energy; Chair. JAIME BAUZÁ BAUZÁ.

**Central Puerto, SA (CEPU):** Avda Tomás Edison 2701, 1104 Buenos Aires; tel. (11) 4317-5000; fax (11) 4317-5099; electricity generating co; CEO ANTONIO BÜCHI BUĆ.

**Comisión Nacional de Energía Atómica (CNEA):** Avda del Libertador 8250, 1429 Buenos Aires; tel. (11) 4704-1384; fax (11) 4704-1176; e-mail freijo@cnea.edu.ar; internet www.cnea.gov.ar; f. 1950; scheduled for transfer to private ownership; nuclear energy science and technology; Pres. JOSÉ PABLO ABRIATA.

**Comisión Técnica Mixta de Salto Grande (CTMSG):** Avda Leandro N. Alem 449, 1003 Buenos Aires; operates Salto Grande hydroelectric station, which has an installed capacity of 650 MW; joint Argentine-Uruguayan project.

**Dirección de Energía de la Provincia de Buenos Aires:** Calle 55, 570, La Reja, 1900 Buenos Aires; tel. (11) 4415-000; fax (11) 4216-124; f. 1957; electricity co for province of Buenos Aires; Dir AGUSTÍN NÚÑEZ.

**Empresa Distribuidora y Comercializadora Norte, SA (EDENOR):** Azopardo 1025, 1107 Buenos Aires; tel. (11) 4348-2121; fax (11) 4334-0805; e-mail ofitel@edenor.com.ar; internet www.edenor.com.ar; distribution of electricity; Pres. FERNANDO PONASSO.

**Empresa Distribuidora Sur, SA (EDESUR):** San José 140, 1076 Buenos Aires; tel. (11) 4381-8981; fax (11) 4383-3699; internet www.edesur.com.ar; f. 1992; distribution of electricity; Gen. Man. JOSÉ MARÍA HIDALGO.

**Entidad Binacional Yacyretá:** Avda Eduardo Madero 942, 21°–22°, 1106 Buenos Aires; tel. (11) 4510-7500; e-mail rrpp@eby.org.ar; internet www.eby.org.ar; operates the hydroelectric dam at Yacyretá on the Paraná river; owned jointly by Argentina and Paraguay; completed in 1998, it is one of the world's largest hydroelectric complexes, consisting of 20 generators with a total generating capacity of 2,700 MW; Exec. Dir OSCAR ALFREDO THOMAS.

**Hidronor Ingeniería y Servicios, SA (HISSA):** Hipólito Yrigoyen 1530, 6°, 1089 Buenos Aires; tel. and fax (11) 4382-6316; e-mail info@hidronor.com; internet www.hidronor.com.ar; fmrly HIDRONOR, SA, the largest producer of electricity in Argentina; responsible for developing the hydroelectric potential of the Limay and neighbouring rivers; Pres. CARLOS ALBERTO ROCCA; transferred to private ownership in 1992 and divided into the following companies.

**Central Hidroeléctrica Alicurá, SA:** Avda Leandro N. Alem 712, 7°, 1001 Buenos Aires.

**Central Hidroeléctrica Cerros Colorados, SA:** Avda Leandro N. Alem 690, 12°, 1001 Buenos Aires.

**Central Hidroeléctrica El Chocón, SA:** Suipacha 268, 9°, Of. A, Buenos Aires.

**Hidroeléctrica Piedra del Aguila, SA:** Avda Tomás Edison 1251, 1104 Buenos Aires; tel. (11) 4315-2586; fax (11) 4317-5174; Pres. Dr URIEL FEDERICO O'FARRELL; Gen. Man. IGNACIO J. ROSNER.

**Transener, SA:** Avda Paseo Colón 728, 6°, 1063 Buenos Aires; tel. (11) 4342-6925; fax (11) 4342-7147; energy transmission co.

**Petrobrás Energía, SA:** Maipú 1, 22°, 1084 Buenos Aires; tel. (11) 4344-6000; fax (11) 4331-8369; internet www.petrobrasenergia.com; f. 1946 as Pérez Compancs, SA; petroleum interests acquired by Petrobrás of Brazil in 2003; operates the hydroelectric dam at Pichi Picún Leufu; Chair. JORGE GREGORIO PÉREZ COMPANC.

### Gas

**Asociación de Distribuidores de Gas (ADIGAS):** Diagonal Norte 740, 5°B, Buenos Aires; tel. (11) 4393-8294; e-mail consultas@adigas.com.ar; internet www.adigas.com.ar; f. 1993 to represent newly privatized gas companies; Gen. Man. CARLOS ALBERTO ALFARO.

**Distribuidora de Gas del Centro, SA:** Avda Hipólito Yrigoyen 475, 5000 Córdoba; tel. (351) 4688-100; fax (351) 4681-568; state-owned co; distributes natural gas.

**Gas Natural Ban, SA:** Isabel la Católica 939, 1269 Buenos Aires; tel. (11) 4303-1380; internet www.gasnaturalban.com.ar; f. 1992; distribution of natural gas; Gen. Man. ANTONI PERIS MINGOT.

**Metrogás, SA:** Gregorio Aráoz de Lamadrid 1360, Buenos Aires; tel. (11) 4309-1434; fax (11) 4309-1025; f. 1992; gas distribution; Chair. RICK LYNN WADDELL; Dir LUIS AGUSTO DOMENECH.

**Transportadora de Gas del Norte, SA:** Don Bosco 3672, 3°, 1206 Buenos Aires; tel. (11) 4959-2000; fax (11) 4959-2242; internet www.tgn.com.ar; f. 1992; distributes natural gas; Gen. Man. FREDDY CAMEO.

**Transportadora de Gas del Sur, SA (TGS):** Don Bosco 3672, 5°, 1206 Buenos Aires; tel. (11) 4865-9050; fax (11) 4865-9059; e-mail totgs@tgs.com.ar; internet www.tgs.com.ar; processing and transport of natural gas; f. 1992; Gen. Dir EDUARDO OJEA QUINTANA.

### Water

**Aguas Argentinas:** Buenos Aires; internet www.aguasargentinas.com.ar; distribution of water in Buenos Aires; privatized in 1993; Dir-Gen. JEAN-LOUIS CHAUSSADE.

## TRADE UNIONS

**Central de los Trabajadores Argentinos (CTA):** Avda Independencia 766, 1099 Buenos Aires; tel. (11) 4307-3829; e-mail internacional@cta.org.ar; internet www.cta.org.ar; dissident trade union confederation; Gen. Sec. VÍCTOR DE GENNARO; Int. Sec. PEDRO WASIEJKO.

**Confederación General del Trabajo (CGT)** (General Confederation of Labour): Azopardo 802, 1107 Buenos Aires; tel. (11) 4343-1883 ; e-mail secgral@cgtra.org.ar; internet www.cgtra.org.ar; f. 1984; Peronist; represents approx. 90% of Argentina's 1,100 trade unions; Sec.-Gen. HUGO MOYANO.

**Movimiento de Trabajadores Argentinos (MTA):** Buenos Aires; dissident trade union confederation.

# Transport

**Comisión Nacional de Regulación del Transporte (CNRT):** Maipú 88, 1084 Buenos Aires; tel. (11) 4819-3000; e-mail msenet@mecon.gov.ar; internet www.cnrt.gov.ar; regulates domestic and international transport services.

**Secretaría de Obras Públicas y Transporte:** Hipólito Yrigoyen 250, 12°, 1310 Buenos Aires; tel. (11) 4349-7254; fax (11) 4349-7201; Sec. Ing. ARMANDO GUIBERT.

**Secretaría de Transporte Metropolitano y de Larga Distancia:** Hipólito Yrigoyen 250, 12°, 1310 Buenos Aires; tel. (11) 4349-7162; fax (11) 4349-7146; Under-Sec. Dr ARMANDO CANOSA.

**Secretaría de Transporte Aero-Comercial:** Hipólito Yrigoyen 250, 12°, 1310 Buenos Aires; tel. (11) 4349-7203; fax (11) 4349-7206; Under-Sec. FERMÍN ALARCIA.

**Dirección de Estudios y Proyectos:** Hipólito Yrigoyen 250, 12°, 1310 Buenos Aires; tel. (11) 4349-7127; fax (11) 4349-7128; Dir Ing. JOSÉ LUIS JAGODNIK.

## RAILWAYS

Lines: General Belgrano (narrow-gauge), General Roca, General Bartolomé Mitre, General San Martín, Domingo Faustino Sarmiento (all wide-gauge), General Urquiza (medium-gauge) and Línea Metropolitana, which controls the railways of Buenos Aires and its suburbs. There are direct rail links with the Bolivian Railways network to Santa Cruz de la Sierra and La Paz; with Chile, through the Las Cuevas–Caracoles tunnel (across the Andes) and between Salta and Antofagasta; with Brazil, across the Paso de los Libres and Uruguayana bridge; with Paraguay (between Posadas and Encarnación by ferry-boat); and with Uruguay (between Concordia and Salto). In 2002 there were 34,463 km of tracks. In the Buenos Aires commuter area 270.4 km of wide-gauge track and 52 km of medium-gauge track are electrified.

Plans for the eventual total privatization of Ferrocarriles Argentinos (FA) were initiated in 1991, with the transfer to private ownership of the Rosario–Bahía Blanca grain line and with the reallocation of responsibility for services in Buenos Aires to the newly created Ferrocarriles Metropolitanos, prior to its privatization.

In 1993 central government funding for the FA was suspended and responsibility for existing intercity passenger routes was devolved to respective provincial governments. However, owing to lack of resources, few provinces have successfully assumed the operation of services, and many trains have been suspended. At the same time, long-distance freight services were sold as six separate 30-year concessions (including lines and rolling stock) to private operators. By late 1996 all freight services had been transferred to private management, with the exception of Ferrocarril Belgrano Cargas, SA, which was in the process of undergoing privatization. In the mid-1990s the FA was replaced by Ente Nacional Administrador de los Bienes Ferroviarios (Enabief), which assumed responsibility for railway infrastructure and the rolling stock not already sold off. The Buenos Aires commuter system was divided into eight concerns (one of which incorporates the underground railway system) and was offered for sale to private operators as 10- or 20-year (subsidized) concessions. The railway network is currently regulated by the Comisión Nacional de Regulación del Transporte (CNRT—see above).

**Ente Nacional de Administración de Bienes Ferroviarios (ENABIEF):** Avda Raqmos Mejía 1302, 6°, Buenos Aires; tel. (11) 4318-3594.

**Ferrocarriles Metropolitanos, SA (FEMESA):** Bartolomé Mitre 2815, Buenos Aires; tel. (11) 4865-4135; fax (11) 4861-8757; f. 1991 to assume responsibility for services in the capital; 820 km of track; Pres. MATÍAS ORDÓÑEZ; concessions to operate services have been awarded to the following companies.

**Ferrovías:** Avda Ramos Mejía 1430, 1104 Buenos Aires; tel. (11) 4314-1444; fax (11) 3311-1181; operates northern commuter line in Buenos Aires; Pres. B. G. ROMERO.

**Metropolitano:** Avda Santa Fe 4636, 1425 Buenos Aires; tel. (11) 4778-5800; fax (11) 4778-5878; e-mail eltren@metropolitano.co.ar; f. 1993; operates three commuter lines; Pres. J. C. LOUSTAU BIDAUT.

**Metrovías (MV):** Bartolomé Mitre 3342, 1201 Buenos Aires; tel. (11) 4959-6800; fax (11) 4866-3037; e-mail info@metrovias.com.ar; internet www.metrovias.com.ar; f. 1994; operates subway (Subterráneos de Buenos Aires, q.v.) and two commuter lines; Pres. A. VERRA.

**Trenes de Buenos Aires, SA (TBA):** Avda Ramos Mejía 1358, 1104 Buenos Aires; tel. (11) 4317-4400; fax (11) 4317-4409; e-mail prensa@tbanet.com.ar; took over operations of two commuter lines from state in 1995; Pres. S. C. CIRIGLIANO.

**Cámara de Industriales Ferroviarios:** Alsina 1609, 1°, Buenos Aires; tel. (11) 4371-5571; private org. to promote the devt of Argentine railway industries; Pres. Ing. ANA MARÍA GUIBAUDI.

The following consortia were awarded 30-year concessions to operate rail services, in the 1990s:

**Ferrobaires (Unidad Ejecutora del Programa Ferroviario Provincial) (UEPFP):** General Hornos 11, 4°, 1084 Buenos Aires; tel. (11) 4305-5174; fax (11) 4305-5933; f. 1993; operates long-distance passenger services; Pres. G. CRESPO.

**Ferrocarril Buenos Aires al Pacífico/San Martín (BAP):** Avda Santa Fe 4636, 3°, Buenos Aires; tel. (11) 4778-2486; fax (11) 4778-2493; operates services on much of the San Martín line, and on 706 km of the Sarmiento line; 6,106 km of track; bought by Brazil's América Latina Logistica, SA, in 1999; Pres. N. SILVA.

**Ferrocarril Belgrano Cargas, SA (FCGB):** Maipú 88, 1084 Buenos Aires; tel. (11) 4343-7220; fax (11) 4343-7229; f. 1993; scheduled for privatization; operates freight services; 7,300 km of track and 120 locomotives; Pres. Dr IGNACIO A. LUDVEÑA.

**Ferrocarril Mesopotámico General Urquiza (FMGU):** Avda Santa Fe 4636, 3°, 1425 Buenos Aires; tel. (11) 4778-2425; fax (11) 4778-2493; operates freight services on the Urquiza lines; 2,272 km of track; bought by Brazil's América Latina Logistica, SA, in 1999; Pres. N. SILVA.

**Ferroexpreso Pampeano (FEPSA):** Bouchard 680, 9°, 1106 Buenos Aires; tel. (11) 4318-4900; fax (11) 4510-4945; operates services on the Rosario–Bahía Blanca grain lines; 5,193 km of track; Pres. H. MASOERO.

**Ferrosur Roca (FR):** Bouchard 680, 8°, 1106 Buenos Aires; tel. (11) 4319-3900; fax (11) 4319-3901; e-mail ferrosur@impsat1.com.ar; operator of freight services on the Roca lines since 1993; 3,000 km of track; Gen. Man. SERGIO DO REGO.

**Nuevo Central Argentino (NCA):** Avda Alberdi 50, 2000 Rosario; tel. (341) 437-6561; fax (341) 439-2377; operates freight services on the Bartolomé Mitre lines since 1993; 5,011 km of track; Pres. M. ACEVEDO.

Buenos Aires also has an underground railway system:

**Subterráneos de Buenos Aires:** Bartolomé Mitre 3342, 1201 Buenos Aires; tel. (11) 4862-6844; fax (11) 4864-0633; f. 1913; became completely state-owned in 1951; fmrly controlled by the Municipalidad de la Ciudad de Buenos Aires; responsibility for operations was transferred, in 1993, to a private consortium (Metrovías) with a 20-year concession; five underground lines totalling 36.5 km, 63 stations, and a 7.4 km light rail line with 13 stations, which was inaugurated in 1987; Pres. A. VERRA.

## ROADS

In 2002 there were 215,434 km of roads, of which 29.5% were paved. Four branches of the Pan-American highway run from Buenos Aires to the borders of Chile, Bolivia, Paraguay and Brazil. In 1996 9,932 km of main roads were under private management. Construction work on a 41-km bridge across the River Plate (linking Punta Lara in Argentina with Colonia del Sacramento in Uruguay) was scheduled to begin in the late 1990s; however, by 2005 the agreement with Uruguay to build the bridge still had not been ratified by the Congreso.

**Dirección Nacional de Vialidad:** Avda Julio A. Roca 378, Buenos Aires; tel. (11) 4343-2838; fax (11) 4343-7292; controlled by the Secretaría de Transportes; Gen. Man. Ing. ELIO VERGARA.

**Asociación Argentina de Empresarios Transporte Automotor (AAETA):** Bernardo de Irigoyen 330, 6°, 1072 Buenos Aires; e-mail aaeta@sei.com.ar; internet www.aaeta.org.ar; f. 1941; Pres. JUAN CARLOS VÁZQUEZ.

**Federación Argentina de Entidades Empresarias de Autotransporte de Cargas (FADEEAC):** Avda 25 de Mayo 1370, 3°, 1372 Buenos Aires; tel. (11) 4383-3635; fax (11) 4383-7870; e-mail fadeeac@fadeeac.org.ar; Pres. LUIS A. MORALES.

There are several international passenger and freight services, including:

**Autobuses Sudamericanos, SA:** Buenos Aires; tel. (11) 4307-1956; fax (11) 4307-1956; f. 1928; international bus services; car and bus rentals; charter bus services; Pres. ARMANDO SCHLECKER HIRSCH; Gen. Man. MIGUEL ANGEL RUGGIERO.

## INLAND WATERWAYS

There is considerable traffic in coastal and river shipping, mainly carrying petroleum and its derivatives.

**Dirección Nacional de Construcciones Portuarias y Vías Navegables:** Avda España 221, 4°, Buenos Aires; tel. (11) 4361-5964; responsible for the maintenance and improvement of waterways and dredging operations; Dir Ing. ENRIQUE CASALS DE ALBA.

## SHIPPING

There are more than 100 ports, of which the most important are Buenos Aires, Quequén and Bahía Blanca. There are specialized terminals at Ensenada, Comodoro Rivadavia, San Lorenzo and Campana (petroleum); Bahía Blanca, Rosario, Santa Fe, Villa Concepción, Mar del Plata and Quequén (cereals); and San Nicolás and San Fernando (raw and construction materials). In 2003 Argentina's merchant fleet totalled 497 vessels, with a combined aggregate displacement of 433,909 grt.

**Administración General de Puertos:** Avda Ing. Huergo 431, 1°, Buenos Aires; tel. (11) 4343-2425; fax (11) 4331-0298; e-mail institucionales@puertobuenosaires.gov.ar; internet www.puertobuenosaires.gov.ar; f. 1956; state enterprise for direction, administration and exploitation of all national sea- and river-ports; scheduled for transfer to private ownership; Gen. Sec. Dr CARLOS FERRARI.

# ARGENTINA

**Capitanía General del Puerto:** Avda Julio A. Roca 734, 2°, 1067 Buenos Aires; tel. (11) 434-9784; f. 1967; co-ordination of port operations; Port Capt. Capt. PEDRO TARAMASCO.

**Administración General de Puertos (Santa Fe):** Duque 1 Cabacera, Santa Fe; tel. (42) 41732.

**Consorcio de Gestión del Puerto de Bahía Blanca:** Avda Dr Mario M. Guido s/n, 8103 Provincia de Buenos Aires; tel. (91) 57-3213; Pres. JOSÉ E. CONTE; Sec.-Gen. CLAUDIO MARCELO CONTE.

**Terminales Portuarias Argentinas:** Buenos Aires; operates one of five recently privatized cargo and container terminals in the port of Buenos Aires.

**Terminales Río de la Plata:** Buenos Aires; operates one of five recently privatized cargo and container terminals in the port of Buenos Aires.

**Empresa Líneas Marítimas Argentinas, SA (ELMA):** Avda Corrientes 389, 1327 Buenos Aires; tel. (11) 4312-9245; fax (11) 4311-7954; f. 1941 as state-owned org.; transferred to private ownership in 1994; operates vessels to northern Europe, the Mediterranean, west and east coasts of Canada and the USA, Gulf of Mexico, Caribbean ports, Brazil, Pacific ports of Central and South America, Far East, northern and southern Africa and the Near East; Pres. PABLO DOMINGO DE ZORZI.

Other private shipping companies operating on coastal and overseas routes include:

**Antártida Pesquera Industrial:** Moreno 1270, 5°, 1091 Buenos Aires; tel. (11) 4381-0167; fax (11) 4381-0519; Pres. J. M. S. MIRANDA; Man. Dir J. R. S. MIRANDA.

**Astramar Cía Argentina de Navegación, SAC:** Buenos Aires; tel. (11) 4311-3678; fax (11) 4311-7534; Pres. ENRIQUE W. REDDIG.

**Bottacchi SA de Navegación:** Buenos Aires; tel. (11) 4392-7411; fax (11) 411-1280; Pres. ANGEL L. M. BOTTACCHI.

**Maruba S. en C. por Argentina:** Maipú 535, 7°, 1006 Buenos Aires; tel. (11) 4322-7173; fax (11) 4322-3353; Chartering Man. R. J. DICKIN.

**Yacimientos Petrolíferos Fiscales (YPF):** Avda Roque S. Peña 777, 1364 Buenos Aires; tel. (11) 446-7271; privatization finalized in 1993; Pres. NELLS LEÓN.

## CIVIL AVIATION

Argentina has 10 international airports (Aeroparque Jorge Newbery, Córdoba, Corrientes, El Plumerillo, Ezeiza, Jujuy, Resistencia, Río Gallegos, Salta and San Carlos de Bariloche). Ezeiza, 35 km from Buenos Aires, is one of the most important air terminals in Latin America. More than 30 airports were scheduled for transfer to private ownership.

**Aerolíneas Argentinas:** Bouchard 547, 9°, 1106 Buenos Aires; tel. (11) 4317-3000; fax (11) 4320-2116; internet www.aerolineas.com.ar; f. 1950; transfer to private ownership initiated in 1990; 97.9% stake acquired by the Spanish company AirComet Marsans in 2001; services to North and Central America, Europe, the Far East, New Zealand, South Africa and destinations throughout South America; the internal network covers the whole country; passengers, mail and freight are carried; Pres. ANTONIO MATA.

**Austral Líneas Aéreas (ALA):** Corrientes 485, 9°, 1398, Buenos Aires; tel. (11) 4317-3600; fax (11) 4317-3777; internet www.austral.com.ar; f. 1971; domestic flights linking 27 cities in Argentina; Pres. MANUEL CASERO.

**Aerovip:** Ricardo Rojas 401, 5°, Buenos Aires; tel. (11) 4312-6954; fax (11) 4312-7080; f. 1999; domestic scheduled and charter flights.

**Air Plus Argentina:** Juncal 858, 3°c, Buenos Aires; tel. (11) 4393-9935; fax (11) 4328-3609; f. 1999; charter flights between Argentina and the USA; Pres. LUIS LÚPORI; Gen. Man. MIGUEL MAGGI.

**Líneas Aéreas del Estado (LADE):** Perú 710, 1068 Buenos Aires; tel. (11) 4361-7071; fax (11) 4362-4899; e-mail director@lade.com.ar; f. 1940; Dir GUILLERMO JOSÉ TESTONI.

**Líneas Aéreas Federales, SA (LAFSA):** Buenos Aires; f. 2003; founded following the suspension of operations of Líneas Aéreas Privadas Argentinas and Dinar Líneas Aéreas; co-operation accord signed with Southern Winds Líneas Aéreas in September 2003 (annulled by the Government in February 2005); signed a co-operation accord with Línea Aérea Nacional de Chile in March 2005, with a view to the creation of Línea Aérea Nacional de Argentina in mid-2005.

**Southern Winds Líneas Aéreas:** Quaglia 262, Local 13, Buenos Aires; internet www.fly-sw.com; f. 1996; co-operation accord signed with Líneas Aéreas Federales, SA in Sept. 2003 (annulled by the Govt in Feb. 2005); Pres. JUAN MAGGIO.

**Transporte Aéreo Costa Atlántica (TACA):** Bernardo de Yrigoyen 1370, 1°, Ofs 25–26, 1138 Buenos Aires; tel. (11) 4307-1956; fax (11) 4307-8899; f. 1956; domestic and international passenger and freight services between Argentina and Bolivia, Brazil and the USA; Pres. Dr ARMANDO SCHLECKER HIRSCH.

**Transportes Aéreos Neuquén:** Diagonal 25 de Mayo 180, 8300 Neuquén; tel. (299) 4423076; fax (299) 4488926; e-mail tancentr@satlink.com.ar; domestic routes; Pres. JOSÉ CHALÉN; Gen. Man. PATROCINIO VALVERDE MORAIS.

**Valls Líneas Aéreas:** Río Grande, Tierra del Fuego; f. 1995; operates three routes between destinations in southern Argentina, Chile and the South Atlantic islands.

## Tourism

Argentina's superb tourist attractions include the Andes mountains, the lake district centred on Bariloche (where there is a National Park), Patagonia, the Atlantic beaches and Mar del Plata, the Iguazú falls, the Pampas and Tierra del Fuego. Tourist arrivals in Argentina in 2004 totalled an estimated 2.4m. In that year tourism receipts were an estimated US $2,500m.

**Secretaría de Turismo de la Nación:** Suipacha 1111, 20°, 1368 Buenos Aires; tel. (11) 4312-5611; fax (11) 4313-6834; e-mail info@turismo.gov.ar; internet www.sectur.gov.ar; Sec. CARLOS ENRIQUE MEYER.

**Asociación Argentina de Agencias de Viajes y Turismo (AAAVYT):** Viamonte 640, 10°, 1053 Buenos Aires; tel. (11) 4325-4691; fax (11) 4322-9641; e-mail secretaria@aaavyt.org.ar; internet www.aaavyt.org.ar; f. 1951; Pres. TOMÁS PATRICIO RYAN; Gen. Man. GERARDO BELIO.

## Defence

In August 2004 Argentina's Armed Forces numbered an estimated 71,400: Army 41,400, Navy 17,500 (including Naval Air Force), Air Force 12,500. There were also paramilitary forces numbering 31,240. In April 1995 conscription was ended and a professional (voluntary) military service was created in its place.

**Defence Expenditure:** An estimated 4,800m. new pesos in 2004.

**Chair. of the Joint Chiefs of Staff:** Brig.-Gen. JORGE CHEVALIER.

**Chief of Staff (Army):** Brig.-Gen. ROBERTO FERNANDO BENDINI.

**Chief of Staff (Navy):** Adm. JORGE OMAR GODOY.

**Chief of Staff (Air Force):** (vacant).

## Education

Education from pre-school to university level is available free of charge. Education is officially compulsory for all children at primary level, between the ages of six and 14 years. Secondary education lasts for between five and seven years, depending on the type of course: the normal certificate of education (bachillerato) takes five years, whereas a course leading to a commercial bachillerato lasts five years, and one leading to a technical or agricultural bachillerato takes six years. Technical education is supervised by the Consejo Nacional de Educación Técnica. Non-university higher education, usually leading to a teaching qualification, is for three or four years, while university courses last for four years or more. There are 36 state universities and some 48 private universities. The total enrolment at primary and secondary schools in 1996 was estimated at 99.4% and 67.2% of the school-age population, respectively. Government expenditure on education, culture, science and technology in 2001 was 2,632.4m. new pesos (5.7% of total public expenditure).

# Bibliography

For works on South America generally, see Select Bibliography (Books)

Alonso, P. *Between Revolution and the Ballot Box: The Origins of the Argentine Radical Party*. Cambridge, Cambridge University Press, 2000.

Arceneaux, C. L. *Bounded Missions: Military Regimes and Democratization in the Southern Cone and Brazil*. University Park, PA, Penn State University Press, 2001.

Auyero, J. *Poor People's Politics: Peronist Survival Networks and the Legacy of Evita*. Durham, NC, Duke University Press, 2001.

Barton, R., Tedesco, L. *The State of Democracy in Latin America: Post-Transitional Conflicts in Argentina and Chile*. London, Routledge, 2004.

Blustein P. *And the Money Kept Rolling in (and Out): Wall Street, the IMF, and the Bankrupting of Argentina*. London, Public Affairs, 2005.

Brown, J. C. *A Brief History of Argentina*. Facts on File, Inc, www.factsonfile.com, 2002.

Calvert, S. and P. *Argentina: Political Culture and Instability*. London, Macmillan, 1989.

Corrales, J. *Presidents Without Parties: The Politics of Economic Reform in Argentina and Venezuela in the 1990s*. University Park, PA, Penn State University Press, 2002.

Dominguez, J. I., and Shifter, M. (Eds). *Constructing Democratic Governance in Latin America (An Inter-American Dialogue Book)*. Baltimore, MD, Johns Hopkins University Press, 2003.

Escude, C. *Foreign Policy Theory in Menem's Argentina*. Gainesville, FL, University Press of Florida, 1997.

Ferradas, C. A. *Power in the Southern Cone Borderlands: An Anthology of Development Practice*. Westport, CT, Bergin & Garvey, 1998.

Fuentes, C. *Contesting the Iron Fist: Advocacy Networks and Police Violence in Democratic Argentina and Chile*. London, Routledge, 2004.

Gibson, E. *Class and Conservative Parties: Argentina in Comparative Perspective*. Baltimore, MD, Johns Hopkins University Press, 1996.

Goñi, U. *The Real Odessa: How Perón Brought the Nazi War Criminals to Argentina*. London, Granta, 2000.

Heinz, W. S., and Frühling, H. *Determinants of Gross Human Rights Violations by State and State-sponsored Actors in Brazil, Uruguay, Chile and Argentina*. The Hague, Martinus Nijhoff Publrs, 1999.

Helmke, G., *Courts Under Constraints: Judges, Generals, and Presidents in Argentina (Cambridge Studies in Comparative Politics)*. Cambridge, Cambridge University Press, 2005.

Karush, M. B. *Workers or Citizens: Democracy and Identity in Rosario Argentina, 1912–1930*. Alberquerque, NM, University of New Mexico Press, 2002.

Keeling, D. J. *Contemporary Argentina*. Boulder, CO, Westview Press, 1998.

Levine, L. W. W., Levine, L. W. and Ortiz, F. *Inside Argentina from Peron to Menem: 1950–2000 from an American Point of View*. Ojai, CA, Edwin House Publishing, 2001.

Levitsky, S. *Transforming Labor-Based Parties in Latin America: Argentine Peronism in Comparative Perspective*. Cambridge, Cambridge University Press, 2003.

Lewis, D. K. *The History of Argentina*. Westport, CT, Greenwood Publishing Group, 2001.

Lewis, P. H. *The Crisis of Argentine Capitalism*. Chapel Hill, NC, University of North Carolina Press, 1990.

*Guerrillas and Generals: The 'Dirty War' in Argentina*. Westport, CT, Greenwood Publishing Group, 2001.

Llanos, M. *Privatization and Democracy in Argentina: An Analysis of President-Congress Relations*. Boston, MA, St Martin's Press, 2002.

Marchak, P. *God's Assassins: State Terrorism in Argentina in the 1970s*. Montréal, QC, McGill-Queens University Press, 2002.

Middlebrook, M. *The Fight for the Malvinas*, London, Pen & Sword Books, 2003.

Mussa, M. *Argentina and the IMF: From Triumph to Tragedy*. Washington, DC, Institute for International Economics, 2002.

Norden, D., and Russell, R. *The United States and Argentina: Changing Relations in a Changing World*. London, Routledge, 2002.

Osiel, M. J. *Mass Atrocity, Ordinary Evil and Hannah Arendt: Criminal Consciousness in Argentina's Dirty War*. New Haven, CT, Yale University Press, 2002.

Peralta Ramos, M., and Waisman, C. H. (Eds). *From Military Rule to Liberal Democracy in Argentina*. Boulder, CO, Westview Press, 1987.

Podalsky, L. *Specular City: The Transformation of Culture, Consumption and Space after Peron*. Philadelphia, PA, Temple University Press, 2002.

Powers, N. *Grassroots Expectations of Democracy and Economy: Argentina in Comparative Perspective*. Pittsburgh, PA, University of Pittsburgh Press, 2001.

Robben, A. C. G. M., *Political Violence and Trauma in Argentina*. Philadelphia, PA, University of Pennsylvania Press, 2005.

Rock, D. *Authoritarian Argentina: The Nationalist Movement, Its History and Its Impact*. Berkeley, CA, University of California Press, 1995.

*State Building and Political Movements in Argentina, 1860–1916*. Stanford, CA, Stanford University Press, 2002.

Romero, J. L. *Las Ideas Políticas en Argentina*. Buenos Aires, Fondo de Cultura Económica Argentina, 2002.

Romero, L. A. *A History of Argentina in the Twentieth Century*. University Park, PA, Penn State University Press, 2002.

Sabato, H. *The Many and the Few: Political Participation in Republican Buenos Aires*. Stanford, CA, Stanford University Press, 2001.

Sawers, L. *The Other Argentina: The Interior and National Development*. Boulder, CO, Westview Press, 1998.

Spektorowski, A. *The Origins of Argentina's Revolution of the Right*. Notre Dame, IN, University of Notre Dame Press, 2001.

Tedesco, L. *Democracy in Argentina*. Ilford, Essex, Frank Cass & Co Ltd, 1999.

Teichman, J. A. *The Politics of Freeing Markets in Latin America: Chile, Argentina, and Mexico*. Chapel Hill, NC, University of North Carolina Press, 2001.

Tulchin, J. S., and Garland, A. M. *Argentina: The Challenges of Modernization*. Wilmington, DC, Scholarly Resources Inc, 1998.

# ARUBA

## Geography

### PHYSICAL FEATURES

Aruba is a constituent of the tripartite Kingdom of the Netherlands, together with the metropolitan country in Europe and the other Dutch Caribbean islands, grouped in the Netherlands Antilles (including Aruba, until it gained *status aparte* in 1986). The island of Aruba is one of the Lesser Antilles, lying in the southern Caribbean, the most westerly of that part of the chain paralleling the South American coast. Indeed, the island lies only some 25 km (16 miles) north of mainland Venezuela (the Paraguná peninsula). It is 68 km west of Curaçao, the chief island of the Netherlands Antilles. With Curaçao and Bonaire, Aruba constitutes what the Dutch confusingly call the 'Leeward Islands' (*Benedenwindse Eilands*). They are more familiarly called the 'ABC islands'. Aruba covers an area of 193 sq km (74.5 sq miles).

Aruba is the smallest of the three Dutch islands in the southern Caribbean. It is about 32 km at its longest (running from the south-east to the north-west) and almost 10 km at its widest. The island tapers fairly evenly towards the south-east, but the northerly facing weather coast extends further than the other, gentler shore, as, to the north-west of the capital, Oranjestad, the coast turns abruptly towards the north-east, curving into a western coastline that arcs up to the pointing north-western tip of Aruba. Most of the main towns and tourist resorts are on the leeward, reef-fringed western and southern shores. There are over 68 km of seashore. The interior (*cunucu*) of the dry island is naturally covered by scrub, cacti and wind-bent divi divi (*watapana*) trees, and little land is farmed. The lack of trees is owing to human exploitation of the scarce wood resources, although the more endangered native species are now protected, and there are replanting programmes and initiatives designed to keep goats out of vulnerable areas. Bird life is rich, particularly during November–January, when migratory species swell the local avian population. The terrain is generally flat, although there are some hills, the highest being Jamanota (189 m or 620 ft). There are no rivers.

### CLIMATE

Aruba has an even, tropical marine climate, with minimal seasonal temperature variation—the average is fairly constant at 27°C (81°F), seldom going below 26°C or above 32°C. August, September and October are the hottest months, while it is slightly cooler than the rest of the year in December–February. The island is outside the Caribbean hurricane belt (although it is constantly cooled by the trade winds) and is very dry. There are only an average of 510 mm (20 ins) of rainfall per year, mostly falling in October–December.

### POPULATION

The main ethnic group (80%) is of mixed white and Amerindian (Arawak) race (there have been no full-blooded Amerindians

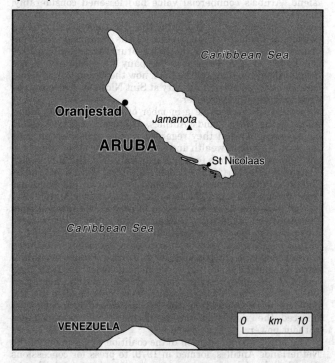

since the late 19th century), but, as a long-established and cosmopolitan trading centre, Aruba has attracted a rich diversity of communities and racial groups. As in the other southern Dutch Antilles, this varied background has given rise to a widely spoken Creole language, Papiamento, of mixed Portuguese, Spanish, Dutch and English descent. The official language, however, is Dutch, although English and Spanish are also widely spoken. The historic influence of Latin America on this part of the Dutch Caribbean is revealed by estimates of religious affiliation—the main faith on Aruba is the Roman Catholic denomination of Christianity, to which more than 80% of the population adhere. A further 8% are Protestant Christians, while other groups represented include Hindus, Muslims, Jews and Confucians.

The total population in mid-2004 was officially estimated at 97,518, although during the course of the typical year well over 600,000 more visit as stopping tourists and almost that number again as passing cruise-ship passengers. Oranjestad (with a population of some 28,817 in mid-2003) is at the more northerly end of the south-western coast, with the 'oil town' of Sint Nicolaas at the southern end. Inland, near the western shore is Noord and in the centre of the island is Santa Cruz.

## History

The six islands of the Netherlands Dependencies—comprising Aruba and the Netherlands Antilles (the 'Antilles of the Five', see separate chapter)—formed part of the once-powerful Dutch trading empire, and still fulfil that role. The two island groups were administered as Curaçao and Dependencies between 1845 and 1948. Having been promised independence by Queen Wilhelmina of the Netherlands during the Second World War, in 1954 Aruba and the Antilles of the Five were granted autonomous federation status, and, along with the metropolitan Netherlands, declared an integral part of the Kingdom of the Netherlands.

Although Aruba was claimed by the Spanish in 1499, and settled by them, the Spanish claim to the island was successfully challenged by the Dutch, who seized it in a series of raids between 1630 and 1640, mostly through the entrepreneurial activity of the Dutch West India Company. The Treaty of Munster eventually recognized Dutch claims in 1648. The trading importance of the island was emphasized by the fact that the

Dutch West India Company ruled it until 1828. Before being established as entrepôt trading centres, the Leeward Islands had been smuggling havens, serving Terra Firma (now Venezuela). In addition, salt pans were developed, which were of great commercial importance to the Dutch, owing to Spain's monopoly on supplies from its South American possessions. However, by the time the Dutch Crown assumed control of the island, Aruba's commercial value had lessened considerably. Slavery was abolished in 1863 and the island suffered an economic decline that lasted until the discovery of significant petroleum deposits in and around Lake Maracaibo, in Venezuela, at the beginning of the 20th century. As a result, in 1929 the Lago Oil and Transport Company (a subsidiary of the Standard Oil Company of the USA, now the Exxon Corporation) established a petroleum refinery at Sint Nicolaas. The economy expanded rapidly thereafter.

In 1954 Aruba became a member of the autonomous federation of the Netherlands Antilles. However, many islanders came to resent what they regarded as the excessive demands made upon Aruban wealth and resources by the other islands in the federation. Aruba's principal political party, the Movimentu Electoral di Pueblo (MEP—People's Electoral Movement), led by Gilberto (Betico) Croes, campaigned for Aruban independence from its foundation in 1971. In a referendum held in March 1977 82% of participants were in favour of Aruba's withdrawal from the Antillean federation. Although the other islands were opposed to proposed independence, the Netherlands itself had begun, from the early 1960s, to encourage secession. The Dutch Government was responsible for providing extensive administrative and financial aid to its dependencies. In 1969 high levels of unemployment provoked serious riots and looting in Willemstad, Sint Maarten, which were only quelled by the deployment of troops from the Netherlands. The incident reinforced the view that the Netherlands had responsibility without power.

The MEP used its position in the coalition Government of the Netherlands Antilles, formed in 1979, to press for concessions from the other islands towards early independence for Aruba. As a result, there were frequent clashes with other members of the Government and on two occasions the MEP withdrew from the coalition. A major issue was the distribution of offshore petroleum revenue. Whereas the Federal Government wanted it to be shared among the islands, the MEP insisted that it should accrue to the island concerned. Following negotiations with the Dutch, in 1983 it was agreed that Aruba would leave the federation on 1 January 1986 and assume *status aparte*. It would proceed to independence 10 years later. The only stipulation was that Aruba would form a co-operative union with the Antilles of the Five in economic and monetary affairs.

In November 1985 the MEP lost power in Aruba, to a four-party coalition led by the Arubaanse Volkspartij (AVP—Aruban People's Party). Aruba gained, as planned, its *status aparte*, and the AVP leader, Jan Hendrik Albert (Henny) Eman, became its first Prime Minister.

At first the Aruban Government and people were confident that the Dutch would yield on the question of independence, which few islanders now actually wanted, following a downturn in the economy precipitated by the closure in 1985 of the Exxon refinery. However, by 1989, following an improvement in the economy, the popular mood was less certain. Extensive tourist development and associated construction work was reinforced by the development of an 'offshore' finance facility, a resurgence in the petroleum storage and transhipment industry and plans to reopen parts of the Lago refinery in 1991. The pro-independence MEP, by this time led by Nelson Oduber, was returned to power in the January 1989 election. The AVP remained highly critical of the new Government's plans on independence and proposed a referendum on the issue. Eventually, the Netherlands Government proposed a new constitutional settlement between the three 'Kingdoms' and, in 1990, ceased to demand Aruban secession by 1996.

In March 1994, at a meeting in The Hague, the Governments of Aruba, the Netherlands and the Netherlands Antilles decided to abandon plans for Aruba's transition to full independence. The possibility of a passage to full independence at a later date was not excluded, but was not considered a priority, and would require the approval of the Aruban people, by referendum, as well as the support of a two-thirds' majority in the Staten. Instead, the aim came to be an increase in autonomy, or *separashon*, in the federation, which would bind Aruba more closely to the Antilles of the Five, with whom relations had improved after 1986. Moreover, in the general election in July the AVP won the greatest number of seats in the parliament and formed a coalition Government with the Organisacion Liberal Arubano (OLA—Liberal Aruban Organization). This coalition lasted until September 1997, when the OLA withdrew from government and the Staten was dissolved. A general election was held in December, in which all political groupings retained their number of parliamentary seats. Following protracted negotiations, the AVP and the OLA renewed their coalition in mid-1998, and a new Council of Ministers, headed by Eman, was appointed.

In June 2001 the governing coalition again collapsed, following the withdrawal of the OLA's support for the AVP's plan to privatize the Aruban Tourism Authority. As a result, the legislative elections that had been scheduled to be held in December were brought forward to 28 September. The MEP defeated the incumbent AVP in the elections, securing 52.5% of the votes cast and 12 seats in the Staten. The AVP's legislative representation was reduced to six seats. The three remaining seats were shared between the PPA (two) and the OLA (one). Oduber was once again appointed Prime Minister and a new Government took office in November.

In early 2004, after several months of discussion, the Aruban and Dutch Governments agreed to appoint Fredis Refunjol, hitherto Minister of Education, as Governor. The Dutch Government objected to the fact that there was only one candidate for the position and argued that the appointment was overtly political. Nevertheless, since Refunjol's candidature had strong cross-party support in the Staten he was duly sworn in on 7 May. Meanwhile, the erstwhile President of the Staten, Francisco Walfrido Croes, assumed the vacant education portfolio. Meanwhile, in January of that year the Staten voted against becoming an Ultra Periphery Area of the European Union (EU). The island thus remained an Overseas Territory of the EU, although the Aruba Trade and Industry Association argued that an issue of such importance should be decided by the result of a referendum.

In recent years Aruba's relations with the metropolitan Netherlands have been dominated by the latter's pressure for more control to be exercised over the large amount of aid that it gives to Aruba, and the issue of independence. Perhaps the key issue was the future arrangements for the island's security, since Aruba's geographical proximity to the South American mainland was considered to make it particularly susceptible to use as a drugs-trafficking base. In 1990 Aruba announced that it was to adopt the 1988 UN Convention on measures to combat trade in illegal drugs, and a joint Dutch and Aruban team was formed to conduct investigations. Nevertheless, in 1996 the USA included Aruba on its list of major drugs-producing or transit countries. As a result, new legislation was introduced in October 1997 to facilitate the extradition of suspected drugs-traffickers and money-launderers. In May 1999 the US navy and air force began patrols from a base in Aruba in an effort to counter the transport of illicit drugs. In November 2003 Aruba signed an agreement with the USA to exchange tax information in order to combat illegal financial activities, such as money-laundering, that are associated with international terrorism and drugs-trafficking. The issue of illegal immigration again arose in late 2004: in the first 10 months of the year Aruba deported 864 Colombians, most of whom arrived on the island from Venezuela. It was feared that instability in Venezuela could prompt a further increase in illegal arrivals.

# Economy

As part of the Kingdom of the Netherlands, Aruba is classed as an Overseas Territory in association with the European Union. It forms a co-operative union with the Antilles of the Five in monetary and economic affairs. Aruba also has observer status with the Caribbean Community and Common Market (CARICOM).

The secession of Aruba in 1986 from the Netherlands Antilles federation was marked by political rancour and economic disruption. Although the island's economy showed remarkable resilience in the late 1980s and 1990s, neither Aruba nor the 'Antilles of the Five' anticipated the extent of 'capital flight' (withdrawal of funds and investments) that occurred, mostly by local citizens. In spite of this, at the beginning of the 21st century Aruba still enjoyed a relatively prosperous economy, substantially based on tourism and commerce. Compared with other islands in the Dutch Caribbean, Aruba was able to capitalize on its remarkable beaches and offshore attractions to develop a mass tourism industry very rapidly (and maintain low levels of unemployment even once the period of rapid construction activity was over). Another alleviating factor was the generous levels of aid, both official and unofficial, through welfare payments and remittances, from the metropolitan Netherlands. The administration of Jan Hendrik Albert (Henny) Eman of the Arubaanse Volkspartij (Aruban People's Party), in office in 1997–2001, demanded the establishment and acceptance by the Antilles (and of Curaçao in particular) of the Aruban guilder, on a par with the Antillean guilder. There was also much argument over the division of the former federation's assets between the Antilles of the Five and Aruba, at the agreed ratio of 70:30. Gold and foreign reserves were allocated on the ratio of 63:27, with 10% to be held in a reserve fund.

Between 1987 and 1993 the Aruban economy grew at an average annual rate of 10%. However, since this growth was entirely attributed to unexpectedly high levels of investment in tourism, as hotel accommodation reached saturation point and construction eased in the mid-1990s, so too did the rate of economic expansion. Tourism receipts fell because of excessive discounting, causing the Government to ease monetary restrictions in March 1995. Money supply and consumer demand increased, prompting the IMF to warn of the risk of higher rates of inflation. In late 1995 strict credit restrictions were imposed and average inflation for that year was 3.4%, down from 6.3% the year before. Thereafter, the increase in consumer prices continued to fall, reaching 1.9% in 1998. However, by 2000 the annual inflation rate had increased to 3.5%, before decreasing to 2.9% in 2001. In 2002 the increase in consumer prices stood at 3.3%, falling to 2.2% in 2003, before increasing again to 2.8% in 2004. In 2000 gross domestic product (GDP) increased by 3.6% in real terms, but, following the terrorist attacks on the USA in September 2001, which badly affected the tourism and construction sectors, GDP decreased by 0.7% in 2001 and by 2.6% in 2002. Despite the US-led military campaign in Iraq in 2003 and its potentially negative effect on the Aruban economy, growth of 1.6% was achieved in that year after a modest revival in tourism and an expectedly sharp increase in public and private investment. Strong growth of 5.1% was achieved in 2004, buoyed by developments in the tourism sector and related growth in construction. In real terms, the economy increased by an annual average of 2.5% during 1996–2003. Some 8.0% of the work-force was unemployed at the end of 2003.

There is little agricultural activity in Aruba, owing to the poor quality of the soil, combined with water shortages (large desalination plants provide most of the water used, although the cost of this process is high). The major crop is aloes (used in the manufacture of cosmetics and pharmaceuticals). There is a small fishing industry and some livestock is raised. In 2000 agriculture engaged only 0.6% of the active labour force.

For most of the 20th century the refining and transhipment of imported petroleum and petroleum products formed the basis of Aruba's relative economic prosperity. The Lago Oil and Transport Company (a subsidiary of the Standard Oil Company of the USA, now the Exxon Corporation) established a petroleum refinery at Sint Nicolaas in 1929. Most of the crude petroleum for the refineries came from the nearby Venezuelan fields, and the refined products were mainly shipped to the USA. At its peak, in the mid-1950s, the Lago plant was processing up to 600,000 barrels per day (b/d). In the 1970s the industry diversified into petroleum storage. A huge terminal was built to handle supertankers carrying African and Middle Eastern petroleum and petroleum products. Since US ports on the eastern seaboard could not handle tankers over 60,000 dwt, the deep-water facilities of the island were ideal. However, in the early 1980s the sector suffered a sudden, dramatic decline following the US authorities' decision to allow construction of large offshore terminals on the US Gulf of Mexico and eastern coast. Furthermore, these new US terminals fed into US refineries.

As a consequence of this and of the reduction in supplies from Venezuela, owing to the imposition of production quotas by the Organization of the Petroleum Exporting Countries (OPEC), the Lago plant ceased operations in 1985. The Government lost nearly US $80m. per year in tax revenues, or about 60% of the island's income. Following riots in protest at the closure, the Dutch Government agreed to emergency aid, on the condition that emergency financial austerity measures were put into effect. Production reached an estimated 202,000 b/d in 1999 and, following a further $250m. renovation in 2000, increased to 280,000 b/d. In March 2004 the refinery was purchased by the US-based Valero Energy Corporation, which, in 2005, announced that it was considering a US $6,000m. expansion of the plant in order to increase production to as much as 800,000 b/d by about 2012. There is a large petrochemical transhipment terminal on Aruba and a small petrochemicals industry. In 1995 a $100m. advanced-technology coker plant opened, to supply liquefied petroleum gas, largely for export to the USA. There are believed to be exploitable reserves of hydrocarbons within the territory, and Aruba also has reserves of salt.

Service industries, particularly the financial and tourism sectors, became Aruba's principal economic activity from the mid-1980s. The financial sector was well established and attempts were made to encourage the growth of a data-processing industry, aimed at US companies in particular. Aruba also established an 'offshore' company facility, aimed particularly at the Latin American market. In June 2000, however, the Organisation for Economic Co-operation and Development (OECD) urged the Government to improve further the accountability and transparency of the financial services it provided, or face economic sanctions. In June 2001 the Aruban Government agreed to conform to OECD's guide-lines on the elimination of harmful tax practices and was subsequently removed from OECD's so-called 'blacklist'.

Aruba's main source of income was tourism. This sector was responsible for Aruba's dramatic economic recovery following the closure of the Lago refinery. Liberal, but closely monitored, casino laws were promoted, as well as the exceptionally good beaches, clear waters and reefs. The number of hotel rooms increased from 2,078 in 1986 to 6,912 in 2003. Aggressive marketing and new air services resulted in stop-over arrivals increasing to 728,157 in 2004, from 181,200 in 1986. Cruise-ship passengers totalled 576,320 in 2004 and receipts from tourism amounted to A. Fl. 1,872m. in the same year. The expansion of Aruba's airport, to increase its passenger-handling capacity to 2.6m. per year by 2010, was in progress in 2005. In May 2005, however, the IMF warned the Government that diversification of the economy away from tourism was necessary to reduce Aruba's vulnerability to external influences; this vulnerability was especially illustrated by Aruba's economic recession in 2001–02, largely caused by a decline in the tourism market following the terrorist attacks on the USA in September 2001. In mid-2005 the unpredictability of the industry again threatened to undermine the rest of the economy when the disappearance of a US teenager on the island, coupled with an apparently poorly managed police investigation, generated large amounts of negative publicity in North America.

# ARUBA

In 2004 the budget deficit was A Fl. 326.5m. In the same year, Aruba recorded a visible trade deficit of A Fl. 485.2m., but there was a small surplus on the current account of the balance of payments of A Fl. 17.9m. Despite an improvement in the trade balance (from A. Fl. 965.4m. in 2003), Aruba is obliged to import most of its requirements, particularly machinery and transport equipment, basic manufactures and foodstuffs. In 2004 the principal source of imports, excluding the petroleum sector and the 'free zone', was the USA (58.8% of the total); other major sources were the Netherlands, Venezuela and the Netherlands Antilles. The principal market for exports in 2004 was also the USA (accounting for 35.6% of the total), followed by the Netherlands Antilles, the Netherlands and Venezuela. At the end of 2004 total government debt was A Fl. 1,718.2m. (equivalent to 45.0% of GDP), of which 49.0% was owed to external creditors, primarily the Governments of the USA and the Netherlands. In spite of Aruba's economic prosperity, it seemed clear that dependence on the Dutch would continue into the 21st century.

# Statistical Survey

Sources (unless otherwise stated): Department of Economic Affairs, Commerce and Industry (Direktie Economische Zaken, Handel en Industrie), Sun Plaza Bldg, L. G. Smith Blvd 160, Oranjestad; tel. 5821181; fax 5834494; e-mail deaci@setarnet.aw; internet www.arubaeconomicaffairs.aw; Centrale Bank van Aruba, J. E. Irausquin Blvd 8, POB 18, Oranjestad; tel. 5822509; fax 5832251; e-mail cbaua@setarnet.aw; internet www.cbaruba.org.

## AREA AND POPULATION

**Area:** 193 sq km (74.5 sq miles).

**Population:** 66,687 (males 32,821, females 33,866) at census of 6 October 1991; 97,518 in mid-2004 (official estimate).

**Density** (mid-2004): 505.3 per sq km.

**Principal Towns** (population estimates, 2002): Oranjestad (capital) 20,700; Sint Nicolaas 17,400. Source: Stefan Helders, *World Gazetteer* (internet www.world-gazetteer.com). *Mid-2003* (UN estimate, incl. suburbs): Oranjestad 28,817 (Source: UN, *World Urbanization Prospects: The 2003 Revision*).

**Births and Deaths** (2001): Registered live births 1,266 (birth rate 13.8 per 1,000); Registered deaths 477 (death rate 5.2 per 1,000).

**Expectation of Life** (years at birth, 2000): Males 70.0; Females 76.0.

**Economically Active Population** (persons aged 15 years and over, 2000): Agriculture, hunting and forestry 251; Manufacturing electricity, gas and water 2,940; Construction 3,892; Wholesale and retail trade, repairs 7,112; Hotels and restaurants 7,651; Transport, storage and communications 2,905; Financial intermediation 1,485; Real estate, renting and business activities 3,722; Public administration, defence and social security 3,573; Education 1,431; Health and social work 1,986; Other community, social and personal services 2,776; Private households with employed persons 1,870; Other 324; Total employed 41,918; Unemployed 3,118; Total labour force 45,036.

## HEALTH AND WELFARE

**Under-5 Mortality Rate** (per 1,000 live births, 1996): 4.1.

**Physicians** (per 1,000 head, 1999): 1.28.

**Hospital Beds** (per 1,000 head, 1995): 37.0.

**Health Expenditure** (% of GDP, 1998): 2.5.

**Access to Water** (% of persons, 2002): 100.

**Access to Sanitation** (% of persons, 1995): 100.
Source: partly Pan American Health Organization.
For definitions, see explanatory note on p. vi.

## FISHING

**Total catch** (FAO estimates, metric tons, live weight, 2003): 150 (Groupers 15, Snappers and jobfishes 45, Wahoo 50, Other marine fishes 40). Source: FAO.

## INDUSTRY

**Electric Energy** (million kWh, 2002): 824.6.

## FINANCE

**Currency and Exchange Rates:** 100 cents = 1 Aruban gulden (guilder) or florin (A Fl.). *Sterling, Dollar and Euro Equivalents* (31 May 2005): £1 sterling = A Fl. 3.254; US $1 = A Fl. 1.790; €1 = A Fl. 2.207; A Fl. 100 = £30.73 = $55.87 = €45.31. Note: The Aruban florin was introduced in January 1986, replacing (at par) the Netherlands Antilles guilder or florin (NA Fl.). Since its introduction, the currency has had a fixed exchange rate of US $1 = A Fl. 1.79.

**Budget** (A Fl. million, 2004): *Revenue:* Tax revenue 707.1 (Taxes on income and profits 321.9, Taxes on commodities 255.4, Taxes on property 43.0, Taxes on services 60.7, Foreign exchange commission 26.2); Other current revenue 77.8; Total 784.9, excluding grants received (31.1). *Expenditure:* Wages 286.2; Employers' contributions 184.3; Wage subsidies 122.7; Goods and services 191.1; Interest payments 85.2; Development fund spending 32.9; Investments 34.2; Other expenditure 205.8; Total 1,142.5.

**International Reserves** (US $ million at 31 December 2004): Gold 48.66; Foreign exchange 295.42; Total 344.08. Source: IMF, *International Financial Statistics*.

**Money Supply** (A Fl. million at 31 December 2004): Currency outside banks 130.83; Demand deposits at commercial banks 826.71; Total money (incl. others) 960.99. Source: IMF, *International Financial Statistics*.

**Cost of Living** (Consumer Price Index at December; base: September 2000 = 100): 107.7 in 2002; 110.1 in 2003; 113.2 in 2004.

**Gross Domestic Product** (A. Fl. million at current prices): 3,421 in 2002; 3,599 in 2003; 3,819 in 2004 (preliminary).

**Expenditure on the Gross Domestic Product** (A. Fl. million at current prices, 2004, preliminary): Final consumption expenditure 3,024; Gross capital formation 1,129; *Total domestic expenditure* 4,153; Exports of goods and services 2,454; *Less* Imports of goods and services 2,788; *GDP in purchasers' values* 3,819.

**Gross Domestic Product by Economic Activity** (A. Fl. million at current prices, 2000): Agriculture, hunting, forestry and fishing, and mining and quarrying 14; Manufacturing 91; Electricity, gas and water (incl. petroleum refining) 212; Construction 202; Trade 440; Restaurants and hotels 355; Transport, storage and communications 287; Finance, insurance, real estate and business services 877; Government services 390; Other community, social and personal services 365; *Sub-total* 3,234; *Less* Imputed bank service charges 163; Indirect taxes, *less* subsidies 255; *GDP in purchasers' values* 3,326. Source: UN, *National Accounts Statistics*.

**Balance of Payments** (A Fl. million, 2004): Exports of goods f.o.b. 4,860.2; Imports of goods f.o.b. –5,345.4; *Trade balance* –485.2; Exports of services 2,218.9; Imports of services –1,425.4; *Balance on goods and services* 308.3; Other income received 65.1; Other income paid –168.5; *Balance on goods, services and income* 204.9; Current transfers received 71.8; Current transfers paid –258.8; *Current balance* 17.9; Capital account (net) 32.9; Direct investment abroad 1.4; Direct investment from abroad 233.8; Portfolio investment assets –32.3; Portfolio investment liabilities 114.0; Other investment (net) –359.1; Net errors and omissions 2.6; *Overall balance* 11.2.

## EXTERNAL TRADE

**Principal Commodities** (A Fl. million, 2004): *Imports c.i.f.:* Live animals and animal products 94.8; Food products 169.4; Chemical products 158.7; Base metals and articles thereof 104.6; Machinery and electrical equipment 234.3; Transport equipment 134.7; Total (incl. others) 1,478.9. *Exports f.o.b.:* Live animals and animal products 2.6; Machinery and electrical equipment 5.4; Transport equipment 3.5; Art objects and collectors' items 7.6; Total (incl. others) 40.5. Note: Figures exclude transactions of the petroleum sector and those of the Free Trade Zone of Aruba.

**Principal Trading Partners** (A Fl. million, 2004): *Imports c.i.f.*: Japan 36.9; Netherlands 205.1; Netherlands Antilles 44.5; USA 869.8; Venezuela 50.5; Total (incl. others) 1,478.9. *Exports f.o.b.*: Colombia 1.7; Netherlands 6.9; Netherlands Antilles 8.2; USA 14.4; Venezuela 3.7; Total (incl. others) 40.5. Note: Figures exclude transactions of the petroleum sector and those of the Free Trade Zone of Aruba.

### TRANSPORT

**Road Traffic** (motor vehicles in use, December 2002): Passenger cars 42,802; Lorries 804; Buses 391; Taxis 398; Rental cars 3,324; Other cars 549; Motorcycles 960; Total 49,228.

**Shipping** (2003): *Arrivals:* Oil tankers 464; Cruise ships 315.

**Civil Aviation:** *Aircraft Landings:* 17,283 in 2001; 16,874 in 2002; 15,642 in 2003. *Passenger Arrivals:* 821,454 in 2001; 759,085 in 2002; 761,085 in 2003.

### TOURISM

**Tourist Arrivals:** 1,224,822 (642,627 stop-over visitors, 582,195 cruise-ship passengers) in 2002; 1,184,233 (641,906 stop-over visitors, 542,327 cruise-ship passengers) in 2003; 1,304,477 (728,157 stop-over visitors, 576,320 cruise-ship passengers) in 2004.

**Tourism Receipts:** A. Fl. 1,872.0m. in 2004.

### COMMUNICATIONS MEDIA

**Radio Receivers** (1997): 50,000 in use.
**Television Receivers** (1997): 20,000 in use.
**Telephones** (2001): 37,132 main lines in use.
**Facsimile Machines** (1996): 3,600 in use.
**Mobile Cellular Telephones** (2001): 53,000 subscribers.
**Internet Users** (2001): 7,912.
**Daily Newspapers** (1996): 13 titles (estimated circulation 73,000 copies per issue).
Sources: mainly UNESCO, *Statistical Yearbook*; International Telecommunication Union; UN, *Statistical Yearbook*.

### EDUCATION

**Pre-primary** (2000/01): 23 schools; 2,737 pupils; 105 teachers.
**Primary** (2000/01): 33 schools; 8,849 pupils; 415 teachers.
**General Secondary** (2000/01): 10 schools; 4,251 pupils; 242 teachers.
**Technical-Vocational** (2000/01): 2 schools; 3,237 pupils; 263 teachers.
**Community College** (1999/2000): 1 school; 1,187 pupils; 106 teachers.
**University** (1999/2000): 1 university; 208 students; 24 tutors.
**Teacher Training** (2000/01): 1 institution; 180 students; 25 teachers.
**Special Education** (2000/01): 4 schools; 272 pupils; 56 teachers.
**Private, Non-aided** (1999/2000): 4 schools; 553 pupils; 58 teachers.
**International School** (2000/01): 154 pupils; 25 teachers.
**Adult Literacy Rate** (official estimates, 2000): Males 97.6%; Females 97.1%.

# Directory

## The Constitution

On 1 January 1986 Aruba acquired separate status (*status aparte*) within the Kingdom of the Netherlands. The form of government is similar to that for the Netherlands Antilles, which is embodied in the Charter of the Kingdom of the Netherlands (operational from 20 December 1954). The Netherlands, the Netherlands Antilles (Antilles of the Five) and Aruba each enjoy full autonomy in domestic and internal affairs, and are united on a basis of equality for the protection of their common interests and the granting of mutual assistance. In economic and monetary affairs there is a co-operative union between Aruba and the Antilles of the Five, known as the 'Union of the Netherlands Antilles and Aruba'.

The Governor, who is appointed by the Dutch Crown for a term of six years, represents the monarch of the Netherlands in Aruba. The Government of Aruba appoints a minister plenipotentiary to represent it in the Government of the Kingdom. Whenever the Netherlands Council of Ministers is dealing with matters coming under the heading of joint affairs of the realm (in practice mainly foreign affairs and defence), the Council assumes the status of Council of Ministers of the Kingdom. In that event, Aruba's Minister Plenipotentiary takes part, with full voting powers, in the deliberations.

A legislative proposal regarding affairs of the realm and applying to Aruba as well as to the metropolitan Netherlands is sent, simultaneously with its submission, to the Staten Generaal (the Netherlands parliament) and to the Staten (parliament) of Aruba. The latter body can report in writing to the Staten Generaal on the draft Kingdom Statute and designate one or more special delegates to attend the debates and furnish information in the meetings of the Chambers of the Staten Generaal. Before the final vote on a draft the Minister Plenipotentiary has the right to express an opinion on it. If he disapproves of the draft, and if in the Second Chamber a three-fifths' majority of the votes cast is not obtained, the discussions on the draft are suspended and further deliberations take place in the Council of Ministers of the Kingdom. When special delegates attend the meetings of the Chambers this right devolves upon the delegates of the parliamentary body designated for this purpose.

The Governor has executive power in external affairs, which he exercises in co-operation with the Council of Ministers. He is assisted by an advisory council which consists of at least five members appointed by him.

Executive power in internal affairs is vested in a nominated Council of Ministers, responsible to the Staten. The Aruban Staten consists of 21 members, who are elected by universal adult suffrage for four years (subject to dissolution), on the basis of proportional representation. Inhabitants have the right to vote if they have Dutch nationality and have reached 18 years of age. Voting is not compulsory.

## The Government

### HEAD OF STATE

**Queen of the Netherlands:** HM Queen BEATRIX.
**Governor:** FREDIS J. REFUNJOL (took office 7 May 2004).

### COUNCIL OF MINISTERS
(July 2005)

**Prime Minister and Minister of General Affairs and Utilities:** NELSON ORLANDO ODUBER.
**Deputy Prime Minister and Minister of Social Affairs and Infrastructure:** MARISOL J. TROMP.
**Minister of Education:** FRANCISCO WALFRIDO CROES.
**Minister of Finance and Economic Affairs:** NILO J. J. SWAEN.
**Minister of Justice:** HYACINTHO RUDY CROES.
**Minister of Public Health and the Environment:** CANDELARIO A. S. D. (BOOSHI) WEVER.
**Minister of Sports, Culture and Labour:** TAI FOO RAMON LEE.
**Minister of Tourism and Transportation:** EDISON BRIESEN.
**Minister Plenipotentiary and Member of the Council of Ministers of the Realm for Aruba in the Netherlands:** ALICIA A. TROMP-YARZAGARAY.
**Minister Plenipotentiary of the Realm for Aruba in Washington, DC (USA):** HENRY BAARH.

### MINISTRIES

**Office of the Governor:** Plaza Henny Eman 3, Oranjestad.
**Office of the Prime Minister:** Government Offices, L. G. Smith Blvd 76, Oranjestad; tel. 5880300; fax 5880024.
**Ministry of Education:** L. G. Smith Blvd 76, Oranjestad; tel. 5830937; fax 5828328.
**Ministry of Finance and Economic Affairs:** L. G. Smith Blvd 76, Oranjestad; tel. 5880269; fax 5880347; e-mail minfin.ecaffairs@setarnet.aw.

# ARUBA

**Ministry of General Affairs and Utilities:** L. G. Smith Blvd 76, Oranjestad; tel. 5839022; fax 5838958.

**Ministry of Justice:** L. G. Smith Blvd 76, Oranjestad; tel. 5839131; fax 5825388.

**Ministry of Public Health and the Environment:** L. G. Smith Blvd 76, Oranjestad; tel. 5834966; fax 5835082.

**Ministry of Social Affairs and Infrastructure:** L. G. Smith Blvd 76, Oranjestad; tel. 5880700; fax 5880032; e-mail minszi@setarnet.aw; internet www.minszi.aw.

**Ministry of Sports, Culture and Labour:** L. G. Smith Blvd 76, Oranjestad; tel. 5839695; fax 5835985.

**Ministry of Tourism and Transportation:** L. G. Smith Blvd 76, Oranjestad; tel. 5839035; fax 5835084.

**Office of the Minister Plenipotentiary for Aruba:** R. J. Schimmelpennincklaan 1, 2517 JN The Hague, Netherlands; tel. (70) 3566200; fax (70) 3451446; e-mail mail@arubahuis.nl; internet www.arubahuis.nl.

## Legislature

### STATEN

**President:** MARLON WERLEMAN, Staten, L. G. Smith Blvd 72, Oranjestad.

**General Election, 28 September 2001**

| Party | % of votes | Seats |
| --- | --- | --- |
| Movimentu Electoral di Pueblo | 52.5 | 12 |
| Arubaanse Volkspartij | 26.6 | 6 |
| Partido Patriótico Arubano | 9.6 | 2 |
| Organisacion Liberal Arubano | 5.7 | 1 |
| Aliansa Democratico Arubano | 3.5 | — |
| Conscientisacion y Liberacion Arubano | 1.1 | — |
| Acción Democratico Nacional | 1.1 | — |
| **Total** | **100.0** | **21** |

## Political Organizations

**Acción Democratico Nacional (ADN)** (National Democratic Action): Oranjestad; f. 1985; Leader PEDRO CHARRO KELLY.

**Aliansa Democratico Arubano** (Aruban Democratic Alliance): Oranjestad; Leader ROBERT FREDERICK WEVER.

**Arubaanse Volkspartij (AVP)** (Aruba People's Party): Oranjestad; tel. 5833500; fax 5837870; f. 1942; advocates Aruba's separate status; Leader MICHIEL GODFRIED EMAN.

**Conscientisacion y Liberacion Arubano (CLA)** (Concentration for the Liberation of Aruba): Oranjestad; Leader MARIANO DUVERT BLUME.

**Movimentu Electoral di Pueblo (MEP)** (People's Electoral Movement): Santa Cruz 74D, Oranjestad; tel. 5854495; fax 5850768; e-mail mep@setarnet.aw; internet www.mep.aw; f. 1971; socialist; 1,200 mems; Pres. and Leader NELSON ORLANDO ODUBER.

**Organisacion Liberal Arubano (OLA)** (Aruban Liberal Organization): Oranjestad; f. 1991; Leader GLENBERT FRANCOIS CROES.

**Partido Patriótico Arubano (PPA)** (Patriotic Party of Aruba): Oranjestad; f. 1949; social democratic; opposed to complete independence for Aruba; Leader BENEDICT (BENNY) JOCELYN MONTGOMERY NISBETT.

## Judicial System

Legal authority is exercised by the Court of First Instance. Appeals are heard by the Joint High Court of Justice of the Netherlands Antilles and Aruba.

**Attorney-General of Aruba:** RUUD ROSINGH.

**Courts of Justice:** J. G. Emanstraat 51, Oranjestad; tel. 5822294; fax 5821241; e-mail griffiekopie@setarnet.aw.

## Religion

Roman Catholics form the largest religious community, numbering more than 80% of the population. The Anglicans and the Methodist, Dutch Protestant and other Protestant churches have a total membership of about 6,500. There are approximately 130 Jews.

### CHRISTIANITY

#### The Roman Catholic Church

Aruba forms part of the diocese of Willemstad, comprising the Netherlands Antilles and Aruba. The Bishop resides in Willemstad (Curaçao, Netherlands Antilles).

**Roman Catholic Church:** J. Yrausquin Plein 3, POB 702, Oranjestad; tel. 5821434; fax 5821409.

#### The Anglican Communion

Within the Church in the Province of the West Indies, Aruba forms part of the diocese of the North Eastern Caribbean and Aruba. The Bishop is resident in The Valley, Anguilla.

**Anglican Church:** Holy Cross, Weg Seroe Pretoe 31, Sint Nicolaas; tel. 5845142; fax 5843394; e-mail holycross@setarnet.aw.

#### Protestant Churches

**Baptist Church:** Aruba Baptist Mission, SBC, Paradera 98-C; tel. 5883893.

**Church of Christ:** Pastoor Hendrikstraat 107, Sint Nicolaas; tel. 5848172.

**Dutch Protestant Church:** Wilhelminastraat 1, Oranjestad; tel. 5821435.

**Evangelical Church:** C. Huygenstraat 17, POB 272, Oranjestad; tel. 5822058.

**Faith Revival Center:** Rooi Afo 10, Paradera; tel. 5831010.

**Iglesia Evangelica Pentecostal:** Asamblea di Dios, Reamurstraat 2, Oranjestad; tel. 5831940.

**Jehovah's Witnesses:** Guyabastraat 3, Oranjestad; tel. 5828963.

**Methodist Church:** Longfellowstraat, Oranjestad; tel. 5845243.

**New Apostolic Church:** Goletstraat SA, Oranjestad; tel. 5833762.

**Pentacostal Apostolic Assembly:** Bernhardstraat 185; tel. 5848710.

**Seventh-day Adventist:** Weststraat, Oranjestad; tel. 5845896.

### JUDAISM

**Beth Israel Synagogue:** Adriaan Laclé Blvd, Oranjestad; tel. 5823272; fax 5823534.

### BAHÁ'Í FAITH

**Spiritual Assembly:** Bucutiweg 19, Oranjestad; tel. 5823104.

## The Press

### DAILIES

**Amigoe di Aruba:** Patriastraat 13, POB 323, Oranjestad; tel. 5824333; fax 5822368; e-mail amigoearuba@setarnet.aw; internet amigoe.com; f. 1884; Dutch; Man. RICKY BEAUJOHN; circ. 12,000 (in Aruba and Netherlands Antilles).

**Aruba Today:** Weststraat 22, Oranjestad; tel. 5827800; fax 5827093; e-mail news@arubatoday.com; internet www.arubatoday.com; Editor-in-Chief JULIA C. RENFRO.

**Bon Dia Aruba:** Weststraat 22, Oranjestad; tel. 5827800; fax 5827044; e-mail noticia@cspnv.com; internet www.bondia.com; Dir JOHN CHEMALY, Jr.

**Diario:** Engelandstraat 29, POB 577, Oranjestad; tel. 58826747; fax 58828551; e-mail diario@setanet.aw; internet www.diarioaruba.com; f. 1980; Papiamento; morning; Editor/Man. JOSSY M. MANSUR; circ. 15,000.

**Extra:** Dominicanessenstraat 17, Oranjestad; tel. 58834034; fax 5821639; Papiamento; Dir C. FRANKEN.

**The News:** Italiestraat 5, POB 300, Oranjestad; tel. 5824725; fax 5826125; e-mail thenewsaruba@setarnet.aw; f. 1951; English; Publr GERARDUS J. SCHOUTEN; Editor BEN BENNET; circ. 6,900.

**Nobo:** Dominicanessenstraat 17, Oranjestad; tel. 5834034; fax 5827272; Papiamento; Dir ADRIAAN ARENDS.

**La Prensa:** Bachstraat 6, POB 566 Oranjestad; tel. 5821199; fax 5828634; e-mail laprensa@laprensacur.com; internet www.laprensacur.com; f. 1929; Papiamento; Editor THOMAS C. PIETERSZ.

ARUBA

### NEWS AGENCIES

**Algemeen Nederlands Persbureau (ANP)** (The Netherlands): Caya G. F. (Betico) Croes 110, POB 323, Oranjestad; tel. 5824333; fax 5822368.

**Aruba News Agencies:** Bachstraat 6, Oranjestad; tel. 5821243.

## Publishers

**Aruba Experience Publications NV:** Verbindingsweg 2, POB 634, Oranjestad; tel. 5834467; fax 5384520; e-mail info@arubaexperience.com; internet www.arubaexperience.com; Gen. Man. MICHEL J. M. JANSSEN.

**Caribbean Publishing Co (CPC) Ltd:** L. G. Smith Blvd 116, Oranjestad; tel. 5820485; fax 5820484.

**De Wit Stores NV:** L. G. Smith Blvd 110, POB 386, Oranjestad; tel. 5823500; fax 5821575; e-mail dewitstores@setarnet.aw; f. 1948; Man. Dir R. DE ZWART.

**Gold Book Publishing:** L. G. Smith Blvd 116, Oranjestad; tel. 5820485; fax 5820484; internet www.caribbeanhotels.org; a division of the Caribbean Hotel Asscn, based in the Cayman Islands.

**Oranjestad Printing NV:** Italiestraat 5, POB 300, Oranjestad; Man. Dir GERARDUS J. SCHOUTEN.

**ProGraphics Inc:** Italiestraat 5, POB 201, Oranjestad; tel. 5824550; fax 5822526; e-mail vadprinting@setarnet.aw; f. 2001; fmrly VAD Printers Inc.

**Publicidad Aruba NV:** Emanstraat 110, POB 295, Oranjestad; tel. 5835139.

**Publicidad Exito Aruba SA:** Domenicanessenstraat 17, POB 142, Oranjestad; tel. 5822020; fax 5824242; f. 1958.

**Rozenstand Publishing Co:** Cuquisastraat 1, Oranjestad; tel. 5824482.

**Van Dorp Aruba NV:** Caya G. F. (Betico) Croes 77, POB 596, Oranjestad; tel. 5823076; fax 5823573.

## Broadcasting and Communications

### TELECOMMUNICATIONS

**Digicel:** POB 662, Oranjestad; e-mail customercare@digicelaruba.com; internet www.digicelaruba.com; f. 2003; owned by an Irish consortium; Chair. DENIS O'BRIEN.

**Servicio di Telecomunicacion di Aruba (SETAR):** Seroe Blanco z/n, POB 13, Oranjestad; tel. 5251576; fax 5836970; e-mail setar@setarnet.aw; internet www.setar.aw; f. 1986; Man. Dir PATRICIO NICOLAS.

### BROADCASTING

#### Radio

**Canal 90 FM Stereo:** Van Leeuwenhoekstraat 26, Oranjestad; tel. 5828952; fax 837340; e-mail info@canal90fm.aw; internet www.canal90fm.aw.

**Cristal Sound 101 7 FM:** J. G. Emanstraat 124A, Oranjestad; tel. 5827726; fax 5820144.

**Radio 1270:** Bernardstraat 138, POB 28, Sint Nicolaas; tel. 5845602; fax 5827753; commercial station; programmes in Dutch, English, Spanish and Papiamento; Dir F. A. LEAUER; Station Man. J. A. C. ALDERS.

**Radio Carina FM:** Datustraat 10A, Oranjestad; tel. 5821450; fax 5831955; commercial station; programmes in Dutch, English, Spanish and Papiamento; Dir-Gen. ALBERT R. DIEFFENTHALER.

**Radio Caruso Booy FM:** G. M. de Bruynewijk 49, Savaneta; tel. 5847752; fax 5843351; e-mail radiocarusobooy@hotmail.com; internet www.geocities.com/carusobooy; commercial station; broadcasts for 24 hrs a day; programmes in Dutch, English, Spanish and Papiamento; Pres. HUBERT ERQUILLES ANTONIO BOOY; Gen. Man. SIRA BOOY.

**Radio Galactica FM:** J. G. Emanstraat 120, Oranjestad; tel. 5830999; fax 5838999; e-mail radiogalactica@hotmail.com; internet www.galactica999fm.aw; f. 1990; Dir MODESTO J. ODUBER; Station Man. MAIKEL J. ODUBER.

**Radio Kelkboom:** Bloemond 14, POB 146, Oranjestad; tel. 5821899; fax 5834825; e-mail radiokelkboom@setarnet.aw; f. 1954; commercial radio station; programmes in Dutch, English, Spanish and Papiamento; Owners CARLOS A. KELKBOOM, E. A. M. KELKBOOM; Dir EMILE A. M. KELKBOOM.

*Directory*

**Radio Victoria:** Washington 23, POB 5291, Oranjestad; tel. 5873444; fax 5873444; e-mail radiovictoria@setarnet.aw; internet www.setarnet.aw/users/radiovictoria; f. 1958; religious and cultural FM radio station owned by the Radio Victoria Foundation; programmes in Dutch, English, Spanish and Papiamento; Pres. N. J. F. ARTS.

**Voz di Aruba** (Voice of Aruba): Van Leeuwenhoekstraat 26, POB 219, Oranjestad; tel. 5824134; commercial radio station; programmes in Dutch, English, Spanish and Papiamento; also operates Canal 90 on FM; Dir A. M. ARENDS, Jr.

#### Television

**ABC Aruba Broadcasting Co NV:** Royal Plaza Suite 223, POB 5040, Oranjestad; tel. 5838150; fax 5838110; e-mail 15atv@setarnet.aw.

**Tele-Aruba NV:** Pos Chiquito 1A, POB 392, Oranjestad; tel. 5857302; fax 5851683; e-mail telearuba@hotmail.com; internet www.telearuba.aw; f. 1963; fmrly operated by Netherlands Antilles Television Co; commercial; govt-owned; Gen. Man. M. MARCHENA.

## Finance

(cap. = capital; res = reserves; dep. = deposits; m. = million; brs = branches; amounts in Aruban florin, unless otherwise stated)

### BANKING

#### Central Bank

**Centrale Bank van Aruba:** J. E. Irausquin Blvd 8, POB 18, Oranjestad; tel. 5252100; fax 5252101; e-mail cbaua@setarnet.aw; internet www.cbaruba.org; f. 1986; cap. 10.0m., res 130.7m., dep. 329.8m. (Dec. 2003); Pres. A. R. CARAM; Exec. Dirs K. A. H. POLVLIET, J. R. FIGAROA-SEMELEER.

#### Commercial Banks

**Aruba Bank NV:** Caya G. F. (Betico) Croes 41, POB 192, Oranjestad; tel. 5821550; fax 5829152; e-mail customersupport@arubabank.com; internet www.arubabank.com; f. 1925; acquired Interbank Aruba NV in Dec. 2003; total assets US $260m. (Dec. 2004); Man. Dir and CEO ILDEFONS D'ANDELO SIMON; 5 brs.

**Banco di Caribe NV:** Vondellaan 31, POB 493, Oranjestad; tel. 5232000; fax 5832422; e-mail bdcaua@setarnet.aw; internet www.bancodicaribe.com; f. 1987; Gen. Man. EDUARDO A. DE KORT; 1 br.

**Caribbean Mercantile Bank NV:** Caya G. F. (Betico) Croes 53, POB 28, Oranjestad; tel. 5823118; fax 5824373; e-mail executive_office@cmbnv.com; internet www.cmbnv.com; f. 1963; cap. 4.0m., res 43.0m., dep. 835.3m. (Dec. 2003); Pres. L. CAPRILES II; Man. Dir W. G. CARSON; 6 brs.

**RBTT Bank Aruba NV:** Italiestraat 36, Sasakiweg, Oranjestad; tel. 5833221; fax 588211756; e-mail firstet@setarnet.aw; internet www.rbtt.co.tt; f. 2001; fmrly First National Bank of Aruba NV (f. 1985 and acquired by Royal Bank of Trinidad and Tobago Ltd in 1998); total assets US $125.3m. (Dec. 2000); Chair. PETER J. JULY; Pres. EDWIN L. TROMP; 6 brs.

#### Investment Bank

**AIB NV:** Wilhelminastraat 34–36, POB 1011, Oranjestad; tel. 5827327; fax 5827461; e-mail aib@setarnet.aw; f. 1987 as Aruban Investment Bank; name changed as above in April 2004; Pres. P. C. M. VAN DER VOORT VAN ZIJP.

#### Mortgage Banks

**Fundacion Cas pa Comunidad Arubano (FCCA):** Sabana Blanco 66, Oranjestad; tel. 5238800; fax 5836272; e-mail mail@fcaa.com.

**OHRA Hypotheek NV:** Caya G. F. (Betico) Croes 85, Oranjestad; tel. 5839666; fax 5839498; e-mail info@ohrabank-aua.com.

#### 'Offshore' Banks

**Inarco International Bank NV:** Punta Brabo z/n, Arulex Bldg, Oranjestad; tel. 5822138; fax 5832363; e-mail fred.aarons@citicorp.com.

**Citibank NA:** J. G. Emanstraat 61, Oranjestad; tel. 5822138; fax 5832363.

### INSURANCE

There were eight life insurance companies and 14 non-life insurance companies active in Aruba in December 2004.

#### Association

**Insurance Association of Aruba (IAA):** L. G. Smith Blvd 160, Oranjestad; tel. 5821111; fax 5826138.

## Trade and Industry

### DEVELOPMENT ORGANIZATION

**Department of Economic Affairs, Commerce and Industry** (Direktie Economische Zaken, Handel en Industrie): Sun Plaza Bldg, L. G. Smith Blvd 160, Oranjestad; tel. 5821181; fax 5834494; e-mail deaci@setarnet.aw; internet www.arubaeconomicaffairs.aw; Acting Dir MARIA DIJKHOFF-PITA.

### CHAMBER OF COMMERCE AND INDUSTRY

**Aruba Chamber of Commerce and Industry:** J. E. Irausquin Blvd 10, POB 140, Oranjestad; tel. 5821120; fax 5883200; e-mail secretariat@arubachamber.com; internet www.arubachamber.com; f. 1930; Pres. JOHN G. EMAN; Exec. Dir LORRAINE C. DE SOUZA.

### TRADE ASSOCIATION

**Aruba Trade and Industry Association:** ATIA Bldg, Pedro Gallegostraat 6, POB 562, Oranjestad; tel. 5827593; fax 5833068; e-mail atiaruba@setarnet.aw; internet www.atiaruba.org; f. 1945; Pres. SERGE MANSUR.

### MAJOR COMPANIES

**ACM & Industries NV:** Barcadera 5B, POB 2197; cultured marble/onyx products; Dir JAN SJAUW KOEN FA.

**AIISCO NV:** Barcadera 5A, POB 503, Oranjestad; tel. 5855912; fax 5855120; manufactures industrial detergents and disinfectants; Dir SAMUEL JONKHOUT.

**Antilles Industrial Gases NV:** Balashi, POB 387, Oranjestad; tel. 5822173; fax 5822823; f. 1963; industrial-gas producers; Man. Dir STANLEY DE MARCHENA; 15 employees.

**Antilliaanse Handel Maatschappij NV:** Fergusonstraat 7, Oranjestad; tel. 5824040; construction materials; Man. Dir HERMAN STEENHUIZEN; 81 employees.

**Aruba Aloe Balm NV:** Sabana Blanco 41, POB 360, Oranjestad; tel. 5883222; fax 5826081; e-mail aausa@arubaaloe.com; internet www.arubaaloe.com; aloe-based skin-care products; Dir JACOBUS VEEL.

**Arubaanse Verffabriek NV:** L. G. Smith Blvd 144, POB 297, Oranjestad; tel. 5822519; fax 5827225; manufactures paints and fillers; Dir HANS HENRÍQUEZ.

**Aruba Candle Co NV:** Franlinstraat z/n, POB 240, Oranjestad; tel. and fax 5821958; candles, disinfectants; Dir MICHAEL SALADIN.

**Aruba Handelmaatschappij NV:** Nassaustraat 14, Oranjestad; groceries; Man. Dir RAOUL C. HENRÍQUEZ; 120 employees.

**Aruba Mariott Resort and Stellaris Casino:** L. G. Smith Blvd 101, Oranjestad; tel. 5869000; fax 5860649; internet marriott.com/property/propertypage/AUAAR; Gen. Man. DAVID SHAHRIARI; 670 employees.

**Aruba Trading Company:** Weststraat 15–17, Oranjestad; tel. 5823950; fax 5832156; e-mail info@arubatrading.com; internet www.arubatrading.com; f. 1928; distribution of general goods, incl. food and drink, clothing and household items; Man. Dir ANDREW L. BARBER; 80 employees.

**Barcadera Cement Aruba:** c/o J. G. Emanstraat 118A, POB 614, Oranjestad; tel. 5837286; fax 5831545; cement producers; Dir CANDELARIO THIEL.

**Brouwerij Nacional Balashi NV:** POB 5317, Balashi; tel. 5854805; fax 5854785; f. 1997; brewery.

**Carex Paper Products NV:** Belgiëstraat 5, Oranjestad; tel. 5821404; fax 5831114; manufactures toilet paper, kitchen towels; Man. MARLON JACOBS.

**Caribbean Paint Factory NV:** Fergusonstraat 57D, POB 273, Oranjestad; tel. 5825339; fax 5837063; manufactures paint; Dir ANTON KAMERMANS.

**Hoori NV:** Balashi 70, POB 1176, Oranjestad; tel. 5856400; fax 5851466; manufactures disposable plastic materials; Dir MOON CHIU CHAN.

**Hyatt Regency Aruba Resort:** J. A. Irausquin Blvd 85, Palm Beach; tel. 5861234; fax 5865478; internet www.aruba.hyatt.com; f. 1987; Gen. Man. BARRY KAPLAN; 670 employees.

**R. J. van der Sar NV:** Barcadera 9, POB 299, Oranjestad; tel. 5850631; fax 5850645; internet www.vandersarnv.com; manufactures disinfectants, soaps, detergents; Dir ROBERT NIEUW.

**Valero Energy Corporation:** 5 Lago Weg, Sint Nicolaas; internet www.valero.com; 775 employees.

### UTILITIES

#### Electricity and Water

**Utilities Aruba NV:** govt-owned holding co.

**Electriciteit-Maatschappij Aruba (ELMAR) NV:** Wilhelminastraat 110, Oranjestad; tel. 5823700; fax 5828991; e-mail elmar.aruba@setarnet.aw; internet www.elmararuba.com; independently managed co, residing under Utilities Aruba NV; electricity distribution; Man. Dir ISMAEL W. F. WEVER.

**Water en Energiebedrijf Aruba (WEB) NV:** Balashi 76, POB 575, Oranjestad; tel. 5254600; fax 5857681; e-mail info@webaruba.com; internet www.webaruba.com; f. 1991; independently managed co, residing under Utilities Aruba NV; production and distribution of industrial and potable water, and electricity generation; Gen. Dir JOSÉ LACLÉ.

#### Gas

**Aruba Gas Supply Company Ltd (ARUGAS):** Barcadera z/n, Oranjestad; tel. 5851198; fax 5852187; e-mail arubagas@setarnet.aw.

**BOC Gases Aruba:** POB 387, Oranjestad; tel. 5852173; fax 5852823; e-mail bocaruba@mail.setarnet.aw; internet www.boc.com.

### TRADE UNIONS

**Federashon di Trahadornan di Aruba (FTA)** (Aruban Workers' Federation): Bernardstraat 23, Sint Nicolaas; tel. 5845448; fax 5845504; e-mail federacion@hotmail.com; f. 1964; independent; affiliated to World Confederation of Labour; Sec.-Gen. JOSÉ RUDOLF GEERMAN.

There are also several unions for government and semi-government workers and employees.

## Transport

There are no railways, but Aruba has a network of all-weather roads.

**Arubus NV:** Sabana Blanco 67, Oranjestad; tel. 5827089; fax 5828633; state-owned company providing public transport services.

### SHIPPING

The island's principal seaport is Oranjestad, whose harbour can accommodate ocean-going vessels. There are also ports at Barcadera and Sint Nicolaas, the latter administered by the Coastal Aruba Refining Company.

**Aruba Ports Authority NV:** L. G. Smith Blvd 23, Oranjestad; tel. 5826633; fax 5832896; e-mail aruports@setarnet.aw; f. 1981; responsible for the administration of the ports of Oranjestad and Barcadera; Man. Dir ERWIN O. CROES.

**Coastal Aruba Refining Co NV:** Seroe Colorado, POB 2150, Sint Nicolaas; tel. 5894904; fax 5849087; f. 1989; petroleum refinery, responsible for the administration of the port of Sint Nicolaas; Gen. Man. DAVID LAM.

#### Principal Shipping Companies

**Beng Lian Shipping S. de R. L. A. V. V.:** Dominicanessenstraat 22, Oranjestad.

**Magna Shipping Co:** Koningstraat 52, Oranjestad; tel. 5824349.

**Rodoca Shipping and Trading SA:** Parkietenbos 30, Barcadera Harbour; tel. 5850096; fax 5823371; fmrly Aruba Shipping and Chartering Co NV.

**Windward Island Agencies:** Heyligerweg, POB 66, Oranjestad.

### CIVIL AVIATION

The Queen Beatrix International Airport (Aeropuerto Internacional Reina Beatrix), about 2.5 km from Oranjestad, is served by numerous airlines, linking the island with destinations in the Caribbean, Europe, the USA and Central and South America. After renovation and expansion, the airport was expected to be able to handle 2.6m. passengers per year by 2010. In November 2000 the national carrier, Air Aruba, was declared bankrupt.

## Tourism

Aruba's white sandy beaches, particularly along the southern coast, are an attraction for foreign visitors, and tourism is a major industry. The number of hotel rooms increased from 2,078 in 1986 to 6,912 in 2003. In 2004 most stop-over visitors came from the USA (73.5%), Venezuela (8.1%) and the Netherlands (5.2%). In 2004 728,157 stop-over visitors and 576,320 cruise-ship passengers visited Aruba. Receipts from tourism totalled A. Fl. 1,872.0m. in 2004.

**Aruba Tourism Authority (ATA):** L. G. Smith Blvd 172, Oranjestad; tel. 5823777; fax 5834702; e-mail ata.aruba@aruba.com; internet www.aruba.com; f. 1953; Man. Dir MYRNA JANSEN-FELICIANO.

**Aruba Hotel and Tourism Association (AHATA):** L. G. Smith Blvd 174, POB 542, Oranjestad; tel. 5822607; fax 5824202; e-mail info@ahata.net; internet www.ahata.com; f. 1965; 85 mems; represents some 6,000 guest rooms; CEO JORGE PESQUERA.

**Cruise Tourism Authority—Aruba:** POB 5254, Suite 227, Royal Plaza Mall, L. G. Smith Blvd 94, Oranjestad; tel. 5833648; fax 5835088; e-mail int1721@setarnet.aw; internet www.arubabycruise.com; f. 1995; non-profit government organization; Exec. Dir KATHLEEN ROJER.

## Defence

The Netherlands is responsible for Aruba's defence, and military service is compulsory. The Dutch-appointed Governor is Commander-in-Chief of the armed forces on the island. A Dutch naval contingent is stationed in the Netherlands Antilles and Aruba. In May 1999 the USA began air force and navy patrols from a base on Aruba as part of efforts to prevent the transport of illegal drugs.

## Education

A Compulsory Education Act was introduced in 1999 for those aged between four and 16. Kindergarten begins at four years of age. Primary education begins at six years of age and lasts for six years. Secondary education, beginning at the age of 12, lasts for up to six years. The main language of instruction is Dutch, but Papiamento (using a different spelling system from that of the Netherlands Antilles) is used in kindergarten and primary education and in the lower levels of technical and vocational education. Papiamento is also being introduced onto the curriculum in all schools. Aruba has two institutes of higher education: the University of Aruba, comprising the School of Law and the School of Business Administration, which had 208 students in 1999/2000; and the Teachers' College, which had 180 students in 2000/01. There is also a community college. The majority of students, however, continue their studies abroad, generally in the Netherlands. Government expenditure on education in 1999 was planned at A Fl. 113.8m., 12.6% of total spending and equivalent to an estimated 3.4% of GDP.

# Bibliography

Croes, R. R. *Anatomy of Demand in International Tourism: The Case of Aruba*. Assen, Van Gorcum Ltd, 2000.

Haanappel, P., et al (Eds). *The Civil Code of the Netherlands Antilles and Aruba*. Kluwer Law International, 2002.

*The Kingdom of the Netherlands—Aruba: 2005 Article IV Consultation*. Washington, DC, IMF Staff Country Report, 2005.

Schoenhals, K. *Netherlands Antilles and Aruba*. Oxford, Clio Press, 1993.

# THE BAHAMAS

## Geography

### PHYSICAL FEATURES

The Commonwealth of the Bahamas is in the West Indies, but, as it is located to the north of the Greater Antilles and east of Florida (USA), the archipelago lies in the Atlantic Ocean, not the Caribbean Sea. The island chain begins some 80 km (50 miles) off the coast of south-eastern Florida, arcing south and east to tail off in the Turks and Caicos Islands, a British dependency which lies just over 60 km south-east of Mayaguana and to the west of Great Inagua, the southernmost of the Bahamas. The country ends here, just to the north of the Windward Passage between the Antillean islands of Cuba and Hispaniola, giving the Bahamas two more international neighbours less than 100 km from its shores—Haiti to the south-east and Cuba to the south. The Tropic of Cancer bisects the country, crossing Long Island. The myriad islands of the Bahamas cover 13,939 sq km (5,382 sq miles), of which 3,870 sq km are enclosed or inland waters.

There are almost 700 islands and some 2,000 cays (keys) and rocky islets in the Bahamas, giving the country coastlines that total 3,542 km in length. Most of the islands are coralline, usually flat, many of the larger ones being long and thin. Low hills relieve the landscape in places, the highest point in the country being Mt Alvernia (63 m or 207 ft) on Cat Island (once believed to be the San Salvador where the Italian navigator Christopher Columbus first set foot in the Americas). This terrain, as well as the dry climate, does not support much arable land. In the north of the archipelago is Grand Bahama, the main island nearest to Florida, while to its east, marking the other edge of the Little Bahama Bank, are Little Abaco and Great Abaco. To the south of Grand Bahama is the Great Bahama Bank, marked above sea level by the landmass of Andros, the largest island in the country, as well as Bimini and Berry and Williams Islands. East of Andros, beyond the continuation of the Northeast Providence Channel known as the Tongue of the Ocean (which plunges some 6,000 m beneath the surface), rises the relatively small, but densely populated, island of New Providence, the location of Nassau, the national capital. East of New Providence Eleuthera continues the chain of long, thin islands heading south from Great Abaco, with Cat Island and Long Island following. East of Cat Island and Long Island is, among others, the island now called San Salvador (formerly Watling Island) while to the west is Great Exuma, another long, thin stretch of territory. By now heading more east than south, the remaining large islands of the Bahamas include Crooked and Acklins Islands and, beyond the Mayaguana Passage, Mayaguana itself and, to its south, Little Inagua (where Bahamian territory comes closest to the Turks and Caicos Islands—West Caicos) and Great Inagua, the latter dominated by Lake Rosa at its heart. Particularly to the west of the main chain of islands and islets, numerous reefs and cays dot the ocean, extending the country's territorial waters to cover well over 0.25m. sq km. The Andros Barrier Reef is the third largest such reef in the world and the largest after that of Belize in the Americas.

### CLIMATE

The climate is tropical marine, moderated by the Gulf Stream and the Atlantic trade winds. The islands are prone to hurri-

canes, which can be particularly devastating given the low-lying terrain. Precipitation is not profuse, and the annual average is 1,360 mm (53 ins), falling mostly in June and then the four months thereafter. Seasonal variations in temperature are slight, with the weather always being warm—the month with the highest average daily maximum temperature is August (89.3°F or 31.8°C), while the lowest daily minimum is in January (62.1°F or 16.7°C).

### POPULATION

The population is predominantly black (about 85% of African descent, according to the 1990 census), but there is a substantial white population (12%), which, statistically, does not include those of Hispanic descent (3%). The people of the Bahamas are generally Christian, the largest denomination being the Baptists (31%), but there are also numerous other Protestant denominations, as well as sizeable Roman Catholic (16%) and Anglican (16%) communities. Apart from the Creole spoken by some Haitian immigrants, the main language in use, as well as the official one, is English.

According to official estimates at mid-2004, the total population of the Bahamas was 320,700, compared with the 1990 figure of 255,095. In 1990 just over two-thirds of the population lived on New Providence, which is dominated by the country's capital and largest city, Nassau. By mid-2003 the city and its suburbs housed 222,163. The other islands, of which about 29 are inhabited, are known as the Out Islands or Family Islands. The more populous ones are Grand Bahama, where the second-largest city, Freeport, is located, Eleuthera, the original site of settlement by the British, and Andros. The country is divided into 21 administrative districts.

# History

MARK WILSON

## INTRODUCTION

The nearly 700 islands and 2,000 uninhabited cays that make up the Bahamian archipelago stretch in a 1,220-km arc towards the northern edge of the Caribbean from a point some 80 km off the coast of Florida (USA). One of the islands, San Salvador or Watling Island, is widely believed to have been the navigator Christopher Columbus's first landfall in the New World, in October 1492. The Spanish are thought to have deported and enslaved the original Lucayan inhabitants, but otherwise took little interest in the dry and somewhat barren islands. The British also found little to attract them to the islands, although a royal charter permitting their exploitation was granted to Sir Robert Heath in 1629. The first British settlers were Puritans from Bermuda, who arrived on Eleuthera in 1647. Other migrants from Bermuda also came, to seek salt. New Providence became the site of the capital, Nassau, from 1666. The other islands became collectively known as the Out Islands and, more recently, as the Family Islands. The soil and climate were not suitable for commercial agriculture and piracy became the basis of the economy, until its eradication from 1719 by the British Governor, Capt. Woodes Rogers.

Population growth was slow. As recently as 1782 there were only 4,000 inhabitants, of whom some 43% were white. However, with the ending of the American War of Independence at this time, loyalist settlers who had been expelled from the former British colonies on the mainland arrived. New Englanders settled some of the smaller islands as fishermen, while southerners brought slaves and established cotton plantations on some of the larger islands. Within four years, the population had grown to 8,950, of whom 67% were black. However, the cotton plantations were soon abandoned, owing to insect pests and soil exhaustion. In total, fewer than 10,000 slaves were landed in the colony and many were freed from bondage long before emancipation in 1834, when the population was an estimated 21,000. Then, as now, whites formed a significant minority. A further minority group originated directly from Africa, without any experience of New World slavery: several Bahamian villages were constructed on land granted to Africans freed by the British navy from captured Spanish vessels, which were intercepted *en route* to Cuba, following the abolition of the slave trade. Poor soil and a dry climate continued to prevent the development of plantation agriculture. The economy was based successively on plundering wrecks, gun-running during the American Civil War (1861–65), the cultivation of citrus and pineapples, sponge fishing and the smuggling of rum and whisky during the US 'Prohibition' of alcoholic liquors in 1919–33.

Although there was some miscegenation, a rigid colour bar prevented black advancement. Accordingly, when a representative House of Assembly was established in 1729, it was an exclusive preserve of the white settlers and merchants. Despite being briefly suspended in 1776, when the colony was captured by the rebel American colonists, and again in 1782 when it surrendered to the Spanish, the House of Assembly remained a permanent feature of Bahamian politics.

Although free black property owners were able to vote from 1807, and the Assembly had four non-white members by as early as 1834, political and economic life was controlled, to all intents and purposes, by a white merchant élite, the so-called 'Bay Street Boys' (named after Nassau's main commercial street). This oligarchy practised blatant electoral bribery in small Out Island constituencies, where the secret ballot was not introduced until 1949. The élite's electoral power was challenged by just a small number of black and mixed-race members and by Sir Etienne Dupuch's 'Nassau Tribune'.

## POLITICAL AWAKENING

Black political awakening can be traced to the so-called Burma Road riots in 1942. By this time, after centuries of extreme variations in the island's fortunes, the economy was beginning to expand, with tourism growing steadily from the 1920s and the construction of US military bases required by the Second World War. The riots erupted over the issue of differential wages paid during the construction of a US Air Force base, to foreign and white workers on the one hand and to black Bahamians on the other. Two people were killed and 25 injured in the riots, which were followed by significant pay increases for black workers. The colony's first political party, the Progressive Liberal Party (PLP) was founded by a group of mixed-race professionals in 1953. In 1956 Sir Etienne Dupuch successfully advocated legislation to outlaw racial discrimination in public places. The 'Bay Street Boys' responded to the formation of the PLP by creating the United Bahamian Party (UBP). Lynden (later Sir Lynden) Pindling, a newly qualified black lawyer, later took over the leadership of the PLP, breaking with an older generation of mainly mixed-race PLP politicians. A tourism industry strike in 1958 was followed by universal male suffrage and the creation of four seats in New Providence, to reduce under-representation of the most densely populated island. The UBP benefited from the continuing over-representation of the less populous islands and won a majority of seats in the 1962 election, with only 36% of the popular vote, less than the 45% polled by the PLP. However, reports of the privately owned Grand Bahama Port Authority, which had recently been granted a casino licence, paying large consultancy fees to cabinet ministers severely damaged the reputation of the oligarchy.

## INDEPENDENCE

In the historic January 1967 general election vigorous PLP campaigning resulted in each party winning 18 seats in the 38-seat House of Assembly. The sole representative of the Labour Party held the balance of power, and pledged his support for the PLP, which was therefore able to form a Government; Pindling, hailed as the 'Black Moses', became premier. The racial strife and economic collapse confidently predicted by the UBP never materialized and a series of constitutional reforms was soon introduced. Because of the insistence by the PLP on black advancement, local control over immigration was regarded as especially important, and the import of mainly white, expatriate labour became progressively restricted.

The PLP won the 1968 election, securing 29 legislative seats; and, after further endorsement at a general election in 1972, the PLP led the country to independence in 1973. There was, however, growing middle-class resentment at Pindling's dictatorial style of leadership, while allegations of corruption became widespread. Dissident members resigned from the PLP and in 1972 merged with the remnants of the UBP to form a new opposition party, the Free National Movement (FNM).

By the mid-1970s drugs-trafficking and money-laundering had become significant activities in the Bahamas. In 1984 the total value of the illegal drugs trade passing through the islands was conservatively estimated at some US $800m. Worse still was a virtual epidemic in the local use of drugs. Violent crime increased dramatically as addiction began to affect members of all social classes across the Bahamas. Initially, Pindling resisted pressure from the US Administration, notably the Drug Enforcement Agency (DEA). However, in 1983 he was publicly accused by a US television network of personal involvement and was forced to establish a Royal Commission to investigate the issue.

The Commission reported in December 1984. Although no evidence was published implicating Pindling, the same could not be said of several other ministers. However, a minority report noted a prominent Bahamian businessman, Everette Bannister, had made substantial payments to Pindling, enabling the Prime Minister to spend more than eight times his official salary (in early 1989 Bannister was indicted in the USA, on charges of conspiring with the Colombian Medellín cartel to transport drugs through the Bahamas). However, few dismissals followed and Pindling was re-endorsed as party leader at the October 1985 PLP convention. Two cabinet ministers, Hubert Ingraham and Perry Christie, who also shared a legal practice, resigned in protest. These two later both served as Prime Minister, with Ingraham leading an FNM Government for 10 years from 1992, and Christie succeeding him in 2002, as leader of a PLP Government.

The damage caused to Bahamian–US relations was serious. The US Administration's concern went beyond trafficking: the strict banking secrecy laws, under which US and other 'offshore' banks in the islands operated, were also criticized because of their misuse by criminals. Moreover, US companies and individuals who used Bahamian banks deprived the US Internal Revenue Service (IRS) of tax income. Consequently, the Bahamas was excluded from the US Caribbean Basin Initiative and tax concessions were not granted to US insurance companies wishing to exploit Bahamian 'offshore' facilities, nor to US corporations holding conventions in the islands. The Bahamas Government denounced US demands that the DEA and IRS be allowed access to bank records and be permitted to search for drugs as an 'imperialist' infringement of sovereignty.

Aware of increasing public concern and international pressure, Pindling realized that co-operation with the USA was essential. A compromise was reached: drugs searches in the Bahamas could be made by the US Coastguard and the DEA, but only with the Bahamian police involved. The first such joint operation took place as early as April 1985, in Bimini, when the authorities seized 34,000 metric tons of marijuana, 2,500 kg of processed cocaine and US $1.4m.-worth of aircraft and ships. Expenditure on coastguard and defence forces was increased and several notable Colombian smugglers were arrested, including Carlos Lehder Rivas. Other measures followed: free passage and diplomatic immunity were granted to DEA agents; a joint drugs-interdiction force was formed; and in 1989 both countries signed a Mutual Legal Assistance Treaty (M-LAT), providing for collaboration in the investigation of criminal allegations within the 'offshore' industry. A similar agreement was also signed with Canada. However, drugs-trafficking continued to pose a problem, despite the limited success of anti-smuggling measures.

At the end of the 1980s Pindling's confidence was such that he was dismissive of accusations (made by Lehder at his trial in Miami, Florida, USA) that Lehder had given him US $400,000. In 1991 elected representatives were required to state their assets for the first time; 10 of the 13 cabinet ministers and 25 members of the Assembly declared more than $1m. Local opinion was, if anything, surprised at the modesty of the amounts reported. One PLP minister, Kendal Nottage, who was dismissed in 1985, but reappointed after the 1987 elections, reported assets of more than $11m. He was indicted in the USA in 1989 on charges of conspiring to conceal the illegal profits of a major Colombian drugs dealer.

### DOMESTIC POLITICS AND THE RISE OF THE FNM

The PLP remained the dominant political force in the 1980s, accusing the FNM of subservience to the USA and to local white Bahamian interests. However, the party lost some ground in the 1987 election, winning 31 seats to the FNM's 16 in an enlarged 49-seat House of Assembly. Christie and Ingraham were elected as independents. The FNM leader, Kendal Isaacs, subsequently resigned and was succeeded by his deputy, Cecil Wallace-Whitfield, who, however, died in 1990. Hubert Ingraham, who had joined the FNM one month earlier, became leader of the party. Perry Christie soon afterwards rejoined the PLP.

Prospects for the FNM improved as the economy worsened. Measures taken under US pressure against cocaine smuggling and money-laundering reduced the free flow of funds into the economy from the mid-1980s. A consequent fall in consumer demand induced a recession in the formal sector of the economy, followed by an 11% decline in tourist arrivals in 1989–92. As a result, Pindling was forced to introduce three successive austerity budgets and restraints on credit. In addition, the Government was damaged by new scandals, this time over widespread corruption in state-owned businesses. A vigorous campaign by the FNM resulted in the defeat of the PLP in the general election of 19 August 1992, with the new Government taking 32 of the 49 seats and 55.3% of the popular vote. Ingraham became Prime Minister and the extent of economic mismanagement and of corrupt practices in several public corporations under the PLP regime only then began to become apparent.

The FNM's confident campaign prior to the March 1997 general election culminated in a clear victory, with the party winning 34 of the 40 seats in a reduced House of Assembly and taking 58.0% of the popular vote. Sir Lynden Pindling, diminished by a series of scandals and suffering from the effects of treatment for cancer, resigned as PLP leader in April, and later announced his retirement from Parliament, after 41 years. Perry Christie succeeded him as party leader. In September the PLP's parliamentary strength was reduced to only five seats when the Government won a by-election in Pindling's old seat, South Andros. The PLP was further damaged in January 2000 by the resignation of its deputy leader, Bernard Nottage, who subsequently formed a rival party, the Coalition for Democratic Reform.

In spite of the FNM's 10-year record of economic success, and a dramatic improvement in the standard of government, the party was increasingly perceived as élitist, autocratic and subservient to the interests of foreign investors. From the late 1990s expansion of large hotels such as Sun International's Atlantis resort on Paradise Island gave rise to concerns over beach access. There was vigorous trade-union opposition in 2000 to the proposed privatization of the Bahamas Telecommunications Company (BaTelCo). From late 2000 sections of the financial community and the legal profession were disturbed by the Government's willingness to co-operate with international moves by the Financial Action Task Force on Money Laundering (FATF, based in Paris, France), and by the Organisation for Economic Co-operation and Development's (OECD, also based in Paris) initiative against tax evasion (see below).

These considerations laid the ground for a sharp turnaround in the fortunes of the two main parties, which gathered force from mid-2001. In line with Ingraham's long-standing commitment to step down as premier after two terms in office, on 16 August the FNM held a convention to elect a leader-designate. Following a hard-fought contest, the Minister of Tourism, Orville Alton Thompson ('Tommy') Turnquest, emerged as the victor, with 192 of the 381 votes cast. Turnquest was widely accused of having used improper methods to influence delegates, and was seen as a significantly less substantial figure than Ingraham. His father, Sir Orville Alton Thompson Turnquest, was Governor-General, but, in order to prevent a possible conflict of interest, he was replaced in this office by Dame Ivy Dumont, a respected former Minister of Education.

An FNM proposal to amend the Constitution, originally put forward before the 1997 election, suddenly gathered pace at the beginning of 2002. Five proposed amendments were put to a referendum on 27 February. These included: the creation of an independent Director of Public Prosecutions; the establishment of an independent body to oversee teachers' employment; the appointment of an independent Parliamentary Electoral Commissioner; and the creation of a Boundaries Commission. The PLP voted for the amendments in Parliament, but campaigned for a 'no' vote in the referendum, on the grounds that the reforms had been rushed and ill-prepared. To the consternation of the Government, the population rejected all five proposals by large margins.

### THE 2002 ELECTIONS

A general election was held on 2 May 2002, close to the constitutional deadline. The PLP increased its share of the popular vote to 52%, taking 29 of the 40 seats. The FNM's legislative representation was reduced to seven seats (they attracted 41% of the votes), with the four remaining seats and votes taken by independent candidates. Only one FNM cabinet minister was re-elected to Parliament; Turnquest lost his seat, but was nominated by his party to the Senate and remained party leader. However, the position of Leader of the Opposition, which, according to the Consitition, must be taken by a member of the the lower house, was given to Alvin Smith, a former junior education minister.

Perry Christie was appointed Prime Minister. Key cabinet appointments included Cynthia Pratt, a former youth worker, who was made Deputy Prime Minister and Minister of National Security, Frederick Mitchell, who assumed the foreign affairs portfolio, and James Smith, Ambassador for Trade under the previous Government, who was appointed Minister of State for Finance. Christie himself assumed the post of Minister of Finance.

The new Government pledged the establishment of a national health-insurance system, as well as constitutional reform; commissions were subsequently established to develop detailed proposals with regard to these commitments. The commission on health insurance reported in May 2004, but without costed proposals or a timetable for implementation, while no firm recommendations had been made by mid-2005 for constitutional reform, a topic that appeared to have slipped off the Government's agenda.

## REGIONAL ISSUES

In the late 1990s there was a resurgence in drugs-trafficking in the Bahamas, as the Mexican route for Colombian cocaine became more difficult and Haiti, just south of the Bahamas, increased its importance as a transhipment point, as did Jamaica. A report published in March 2004 by the US Department of State suggested that 12% of cocaine imports to the USA were transhipped through the Jamaica–Cuba–Bahamas corridor. Moreover, an earlier analysis of balance-of-payments statistics by the US embassy in Nassau suggested that cocaine and marijuana smuggling resulted in foreign-exchange inflows of US $200m.–$300m. annually, a sum which would be equivalent to 11%–17% of formal-sector merchandise imports, or 4%–7% of gross domestic product. On this basis, the report suggested drugs transhipment might rival, or even surpass, the Bahamian banking industry in economic impact. The USA included the Bahamas on a list of 23 major drugs source or transhipment countries, and, although the USA acknowledged progress in reducing narcotics activity, strong differences over counter-narcotics policy surfaced in December 2002 when Bahamian representatives walked out of a meeting of a joint task force in a protest against opening remarks by the erstwhile US Ambassador to the Bahamas, J. Richard Blankenship. Blankenship had called for a clear national drugs-control strategy on the islands; the appointment of a 'drugs tsar;' an enquiry into the 1992 'Lorequin' incident, in which cocaine from a US-controlled drugs shipment reportedly went astray; and the reform of the Defence Force, which was blamed for the Lorequin incident, and with which the USA refused to share sensitive law-enforcement information. The USA appeared embarrassed by the tenor of Blankenship's remarks, and he resigned as Ambassador in June 2003; however, most of the concerns noted in his speech were later addressed. An inquiry into the Lorequin incident was completed in September 2004, criticizing defence force officers but not recommending prosecutions, while a $3m. national counter-narcotics plan was made public in June, and a new maritime agreement with the USA was signed in the same month.

Two important drugs-trafficking organizations were defeated in 2001, with eight major participants arrested. Partly as a result of these successes, there was a dramatic decrease in the murder rate, which fell sharply to 3.2 per month in 2001, from 6.2 per month in the previous year; the rate increased to 4.3 per month in 2002. There were further breakthroughs, with two further drugs-trafficking organizations dismantled in December 2002 and June 2004; others, however, remained in operation. Extradition from the Bahamas to the USA was difficult, however. Aggregate cocaine seizures in 2003 totalled 4.0 metric tons.

There was increasing international concern from 1999 over the standard of regulation of international financial centres. In June 2000 the Bahamas was included by the FATF on a 'blacklist' of jurisdictions considered to be failing in attempts to combat money-laundering. As a result, in the second half of 2000 the Government devoted considerable political attention to reform of the 'offshore' financial sector. New legislation established a Financial Intelligence Unit in October, improved the mechanism for international co-operation in criminal proceedings and required international business companies to keep records showing their beneficial ownership. A further law was passed increasing the regulatory capacity of the Central Bank. As a result of the measures, in June 2001 the FATF removed the Bahamas from its list of non-co-operative countries. An agreement to exchange information on tax with the USA was signed in January 2002 and came into effect a year later. However, several aspects of the new regulatory regime were challenged in the courts on constitutional grounds, and the new PLP Government proposed a review of recent financial legislation and international agreements.

Another ongoing security concern was illegal migration from Haiti. The census of 2000 reported that mainly legal migrants from Haiti made up 7% of the resident population; however, in 1995–2003 more than 35,000 Haitian illegal migrants and refugees were deported from the Bahamas, with a peak figure of 7,589 in 2001. There was also a substantial volume of illegal migration from Jamaica and Cuba, while illegal migrants from the People's Republic of China also used the Bahamas as a transit point for entry to the USA.

# Economy

### MARK WILSON

The economy of the Bahamas is one of the most prosperous of the Caribbean and Latin American nations, with a per-head gross domestic product (GDP) of some US $17,883 in 2004, a higher figure than any independent country in Latin America and the Caribbean (the Bahamian dollar is at par with the US dollar). The Bahamas' A3 foreign-debt rating with the Moody's investor agency was also high in comparison with other regional economies, as was the Standard & Poor's A– long-term sovereign credit rating. The country was rated 51st of 177 countries in the UN Development Programme's Human Development Index for 2003 (seventh in the Latin America and Caribbean region after Barbados, which was placed 29th, Argentina (34th), Saint Christopher and Nevis (39th), Chile (43rd), Costa Rica (45th) and Uruguay (46th)).

Following a recession in 1988–93, from 1993 there was continuous economic growth, which in 1996–2000 averaged 4.3% per year. Foreign direct investment has been very high since the late 1990s and was equivalent to an estimated 6.3% of GDP in 2004. Recovery was prompted largely by new hotel investment, much of it stimulated by the sale to new investors of the dilapidated properties of the Government's Hotel Corporation, and by the increased hotel occupancy of stop-over visitors. However, in 2001 the economy grew by only 0.8%, as the downturn in the US economy and security concerns after the terrorist attacks on the USA in September 2001 slowed the vital tourism sector. This was followed by growth of only 1.4% in 2002 and 1.9% in 2003, with the rate of unemployment rising to 10.8% in the latter year (compared with a rate of around 8% in 2001). However, growth had recovered to an estimated 3.0% in 2004, with the unemployment rate improving slightly to 10.2%.

There is no personal or corporate income tax in the Bahamas. Duties on imports made up 54.4% of the Government's recurrent revenue in 2002/03 (or 60.8% of tax revenue). As a result, retail prices were relatively high, although the inflation rate declined from an annual average of 4.4% in 1985–95, to an average annual rate of 1.6% in 1995–2004. (Consumer prices rose by

0.8% in 2004.) Taxes directly related to tourism (hotel occupancy tax, departure tax and gaming tax) made up a further 11.6% of government revenue in 2003/04. However, the small tax base posed problems. In spite of successive tax increases, government revenue was equivalent to only 17.1% of GDP in 2003/04, much lower than the proportion prevailing in other Caribbean islands, such as Barbados, where government revenue was equivalent to 34.3% of GDP in the same year. The Bahamas was likely to join the World Trade Organization within the next few years, a move which would necessitate a reduction in import duties. However, membership of the Caribbean Community and Common Market (CARICOM) single market and economy was not likely, while there remained vocal opponents in the Bahamas to membership of these organizations or of the proposed Free Trade Area of the Americas. To protect the tax base while avoiding the need for an income tax, it was predicted that a value-added tax would be introduced within the next few years, although the May 2005 budget presentation ruled out any immediate moves in this direction.

The fiscal deficit remained low from the mid-1990s until 2001/02, with the exception of 1996/97, when the deficit increased sharply to reach 3.6% of GDP, partly as a result of spending commitments made before the 1997 general election. The fiscal deficit fell back to 1.6% of GDP in 1997/98 and was down to 0.3% in 1999/2000 and 2000/01. With the recurrent fiscal account in surplus, the Government aimed to eliminate its overall deficit in 2000/01, an objective which it narrowly failed to achieve. However, both recurrent and overall fiscal accounts moved sharply into deficit from the last quarter of 2001, with revenue adversely affected by the downturn in tourism and in domestic consumer demand, reaching 3.1% in 2001/02, 3.4% in 2002/03, and 2.9% for 2003/04, with a projected 2.4% for 2004/05. Total public and government-guaranteed debt increased from 38% of GDP in 2002 to 44% in 2004; however, the latter was considered to be a fairly manageable figure, and remained low by Caribbean standards. Interest payments on debt were US $123.3m. in 2004/05, equivalent to 12.0% of the estimated recurrent revenue of $1,030m. The foreign debt-service ratio, however, was an estimated 3.6% of the estimated value of exports of goods and services in 2004. Salaries and related expenses accounted for 51.4% of total recurrent expenditure in 1993/94, but only 44.4% in 2003/04, marking a gradual decrease in public-sector employment. Most government-owned hotels were sold in the mid-1990s. However, plans by the Free National Movement (FNM) Government to privatize the Bahamas Telecommunications Corporation (BaTelCo) failed to bear fruit, as the Progressive Liberal Party (PLP) Government, elected in May 2002, failed to reach agreement with any of three potential strategic partners shortlisted in 2003 to purchase a 49% stake in the restructured company, now known as the Bahamas Telecommunications Company, or BTC. Suggestions for the privatization of the Bahamas Electricity Corporation were at an early stage of conceptualization in mid-2005. However, the Government hoped during the third quarter to develop firm proposals for divestment of the national airline Bahamasair, which has a history of heavy financial losses.

In 1955 an area of 603 sq km (233 sq miles), approximately one-third of the island of Grand Bahama, was granted to the privately owned Grand Bahama Port Authority (GBPA), with important tax concessions under the Hawksbill Creek Agreement. Freeport, which owed its existence entirely to that Agreement, developed considerably. The GBPA aimed originally to construct a manufacturing centre. The focus then shifted to tourism and residential development. Within Freeport, the operation of the port and airport, land development and the management of the water and electricity supply were private-sector activities. Under successive PLP Governments Grand Bahama was regarded as an opposition stronghold, and was consequently neglected by the central administration. The petroleum refinery and cement plant were closed, and tourism stagnated. In 1993, with the FNM in office, the Hawksbill Creek Agreement was extended to 2054, with tax exemptions running to 2015. This contributed to renewed economic development, with a Hong Kong-owned company, Hutchison Whampoa, the principal investor.

The Out Islands, or Family Islands, varied enormously in the degree of economic development achieved. In the northern Bahamas, activities such as tourism and fishing brought prosperity to a number of islands, including Abaco, Spanish Wells, Eleuthera and Bimini. In contrast, some islands in the southern Bahamas, such as Mayaguana, Acklins and Crooked Island had small populations and a very limited range of economic activity.

## TOURISM

Tourism was the basis of prosperity in the Bahamas, and remained by far the most important sector of the economy. The islands are closer to the USA than are most other destinations in the Caribbean region. For this reason, commercial tourism developed comparatively early, with the construction of the Royal Victoria Hotel in Nassau commencing just before the American Civil War of the 1860s. Luxury winter tourism developed further between 1920 and 1940, and from the 1950s jet travel opened the islands as a mass-market destination. The number of stop-over tourists increased from 32,000 in 1949 to 1.5m. in 2004. Such a marked increase led to rapid growth in commercial activity and a significant increase in living standards.

Following a fall in tourist arrivals in the early 1990s, a vast increase in tourism-related investment prompted a recovery in the sector. This investment totalled US $1,500m.–$2,000m. over five years and involved hotel construction and refurbishment, an extension to Nassau International Airport and new air links. As a result, room occupancy in Nassau increased from 52.0% in 1992 to 76.0% in 1999, in spite of a greatly increased room stock, before decreasing to 69.0% in 2004. Meanwhile, tourism receipts rose to $1,855m. in 2004. The opening of 'private island' facilities by cruise lines and a refurbished cruise-port facility in Nassau contributed to an increase in cruise-passenger arrivals, which totalled 3.4m. in 2004. The Bahamas received more cruise-ship passengers than any other Caribbean destination, although with much lower per-head spending ($54 in 2002) than stop-over tourists ($1,061).

The increase in tourism investment in the late 1990s involved the reconstruction of most of the existing hotels in Nassau, followed, in 1999–2001, by the major properties on Grand Bahama. This activity involved the extensive development, at a cost of US $450m., of the Sun International (later Kerzner International) Atlantis resort on Paradise Island. After completion of its Phase II, the project included 2,395 rooms, up from 1,100 in 1993, as well as a marina, a casino and a conference facility, and had 5,600 staff, equivalent to about 4% of national employment. In contrast to some of its rivals, the resort maintained high occupancy and profits throughout the tourism downturn of 2001–03. Gross revenue was $520m. in 2003, with occupancy at over 80%. Following the Government's decision to invest substantially in airport, road, water and electricity improvements, and its commitment to a further 11 years of tax concessions and up to $4m. per year in marketing assistance over five years, in May 2003 Kerzner International announced a further $600m. Phase III expansion, increased in scope to $1,000m. per year later. This was to include 1,500 additional hotel rooms for completion by 2007, and was to increase staff numbers to 9,000. After completion of Phase III, Kerzner International is expected to account for almost one-half of stop-over tourism in the Bahamas, giving the company a dominant role in the national economy. In 2004 Kerzner purchased the neighbouring Club Méditerranée property on Paradise Island for $40m., and in 2005 the company bought the island's Hurricane Hole marina. Another very large tourism investment was proposed in 2005. The Baha Mar Development Company, owned by a locally resident investor, agreed in March of that year to buy the Wyndham Nassau Resort, Crystal Palace Casino and Nassau Beach Hotel, and on 4 May finalized the purchase of the Radisson Cable Beach Resort, the last remaining major property of the state-owned Hotel Corporation, along with a substantial area of government-owned land. For a renovation and construction investment estimated at $1,200m., the project aimed to develop two 1,000-room hotels, a 300-room and a 400-room hotel, a 0.7 ha casino, a golf course, some 1 ha of convention space, residential units and a marina, and to employ a staff of 5,000 on completion in 2009.

Most of Nassau's hotel properties were large, and many belonged to international chains, a great asset for US marketing, which relied heavily on branding. On Grand Bahama, much of the available accommodation was previously owned by the Hotel Corporation and had, by the 1990s, become rather dilapidated. In 1997 the largest properties were sold to Hutchison Whampoa, which undertook a US $400m. redevelopment and reopened as the Our Lucaya complex, with 1,350 rooms in late 2000. The island remained a relatively weak area for Bahamian tourism. The Princess Resort and Casino was sold in May 2000 and reopened after a $42m. renovation as the 965-room Resort at Bahamia; renamed the Royal Oasis, it was closed after 'Hurricane Frances' in September 2004, with substantial debts to the Government, staff and local businesses. In spite of their excellent beaches and wildlife resources, the Family Islands remained relatively undeveloped. There were few direct air links to the north-eastern USA, and travel via Nassau could be inconvenient. Airport improvements in Nassau and the Family Islands formed an important component of the Government's economic strategy, and large-scale developments were by 2005 built or planned on several islands, including Eleuthera, Abaco, San Salvador and Exuma, where a $140m. resort was completed in late 2003. In 2005 proposals were under development for large-scale resorts (with airports to serve them) on smaller and more remote islands, such as Rum Cay, which currently has only 70 inhabitants; the Government's aim was for each island to have at least one 'anchor' property to underpin its development.

Agriculture and manufacturing were poorly developed, and tourist operators purchased little locally. Food, furnishings, even souvenirs, were generally shipped in from Miami (Florida, USA); some construction projects imported pre-assembled and pre-fitted hotel rooms, thus minimizing the need for local labour. For this reason, retained earnings were poor. In addition, many major hotel projects were given important tax concessions, reportedly equivalent in value to 20% of the development cost for the Baha Mar and Atlantis Phase III projects.

Visitor arrivals by air in 2004 were 2.0% below their 2000 peak value, but were 11.1% above the 1998 figure. Cruise-passenger numbers, however, were at an all-time high, and had increased by 93% since 1998. In 2003 the hotels and restaurants sector directly contributed an estimated 9.7% of GDP, while in 2004 travel receipts comprised 72.4% of earnings from goods and services, and were 2.5% greater in value than the cost of all goods imported to the Bahamas. Owing to the proximity of the USA, the European market was less important to the Bahamian tourism industry than it was to that of the rest of the Caribbean. In 2003 some 86.4% of stop-over arrivals were estimated to have travelled from the USA. In the same year 6.2% of visitors came from Europe and 4.2% came from Canada.

## 'OFFSHORE' FINANCE AND REGISTRATION

Nassau is a major international financial centre, with banking and insurance accounting for 11.5% of GDP in 2003. In 2004 the Minister of Financial Services and Investment, Allyson Maynard-Gibson, stated that financial institutions in the Bahamas had combined balance sheet assets of US $300m., with $100,000m. under administration, and a combined annual payroll of $203m. Despite this, by the 1990s the Bahamas had lost its former dominant position as an 'offshore' financial centre. From being one of the world's best known 'offshore' centres, dating from the 1920s, the financial services industry in the Bahamas declined almost every year from the early 1980s until 1991. Whereas in 1976 some 49% of Organisation for Economic Co-operation and Development (OECD) 'offshore' assets in developing countries were administered from Nassau, by 1998 its market share was only 10%, with the Cayman Islands taking 31% and Hong Kong 25%. In the world-wide ranking of the financial centres of metropolitan and developing countries, Nassau declined from third place in 1976 to 15th by 1998. Nassau, however, lost market share partly because there was no longer a tax advantage in booking large international loans 'offshore', leading to loss of business, which had in any event accounted for relatively little value added. However, in the more economically significant business of private banking and trust management, the reputation of the Bahamas was severely damaged by extensive money-laundering activity. Suffering from the country's image of sleaze and corruption and with many financial institutions having to defend themselves against the persistent enquiries of overseas law-enforcement and tax authorities (mainly from the USA), a number of institutions reduced their operations or chose alternative centres, notably the Cayman Islands. Aggressive marketing of company registration and other services by newer financial centres, such as the British Virgin Islands, also allowed them to capture a share of the market which they did not thereafter relinquish.

Resolute action was taken to cleanse the industry of illegal funds and to remove the minority damaging the industry's reputation. The reforms began in 1989, with more rapid progress being made after the 1992 elections and the change of government. There was much on which to base a recovery: the accumulation of expertise was probably unrivalled in the Caribbean and, in spite of the Mutual Legal Assistance Treaty signed with the USA in 1989, under most conditions, legal protection for banking secrecy remained. By 1991 the industry was once more expanding. Legislation was updated on international business companies and mutual funds in 1995, on money-laundering in 1996 and on trusts in 1998.

From April 1998 the Financial Services Promotion Board handled overseas marketing of financial services. However, the captive insurance sector was less developed than in Bermuda, Barbados and other jurisdictions that, unlike the Bahamas, enjoyed tax concessions granted by the US or Canadian authorities.

In June 2000 the intergovernmental body charged with combating money-laundering, the Financial Action Task Force on Money Laundering (FATF, based in Paris, France) included the Bahamas in a list of 15 'non-co-operative' jurisdictions. The FATF criticized: the practice of issuing bearer shares, which made beneficial ownership of financial assets impossible to trace; long delays and restricted responses to requests for judicial assistance; and rules which allowed intermediaries to avoid revealing the names of their clients. Along with the other jurisdictions cited, in June the Bahamas was the subject of an advisory statement from the US Treasury, which asked US financial institutions to apply 'enhanced scrutiny' to transactions with the country. In response, the Bahamian authorities accelerated their efforts to bring the jurisdiction into line with international requirements, and 11 new financial-reform measures were passed by Parliament by the end of the year. As a result, the Bahamas was removed from the FATF 'blacklist' in June 2001. In January 2002 the Bahamas concluded a tax information-exchange agreement with the USA.

The Bahamas International Securities Exchange opened in May 2000 and in June 2005 was trading the shares of 19 local companies. However, with market turnover low, at US $24.3m. for 2004, commission income was not sufficient to cover running costs. (In 2002 the Exchange had been in severe financial difficulty and had required government financial support.)

Local expenditure by 'offshore' companies was US $134.0m. in 2004, an increase from $106.0m. in 2003, and equivalent to 10.0% of the visible trade deficit. Total employment in banking was approximately 4,300, close to 3% of the employed labour force. Although expatriates continued to play an important role, by the start of the 21st century Bahamian nationals held a high proportion of professional positions.

During the 1990s company registration, in particular, witnessed spectacular growth, following the International Business Company Act of 1990. Owing to simplified automated procedures and tax concessions, there was an estimated total of 102,000 registered International Business Companies (IBCs) at the beginning of 2001. Financial-sector professionals in the Bahamas have always insisted that most 'offshore' business is legitimate. As a result of new requirements that beneficial owners should be identifiable, the number of new IBCs registered in 2003 was less than one-fifth of the new registrations recorded in 2000, while many existing IBCs failed to renew their registration. The total number of IBCs accordingly fell to 39,984 in 2003. The licences of 183 'shell' banks (banks registered in the jurisdiction but with no staff or offices there) were revoked in 1999–2003 as the result of a new requirement that 'offshore' banks must maintain a physical presence in the Bahamas. With other licences lapsing later, 262 'offshore' banks remained in operation in March 2005. However, the requirement to maintain an office and employ staff meant that the remaining banks each had a more significant economic impact.

Legal challenges were made against various aspects of the new financial legislation, but by mid-2005 no major challenge had succeeded. The PLP Government had pledged to review the recent financial legislation, but had not in practice relaxed controls significantly. Levels of compliance with new requirements were uneven, and the deadline for bank customers to provide documentary proof of identity was extended from December 2001 to December 2003 and then to March 2004. A number of criminal investigations and troubling incidents in 2002–05 underlined the need for continuing regulatory vigilance, while the FATF in 2004 expressed concern that the Government was lax in responding to regulatory requests. On a more positive note, legislation approved in February 2003 provided a legal basis for e-commerce, while legislation approved in May 2004 extended the civil law concept of investment foundations to the Bahamas, established Purpose Trusts, and allowed IBCs to establish segregated accounts to protect assets against creditors.

The other area of 'offshore' activity that continued to perform well into the 2000s was that of ship registration. The Bahamas ship registry, relaunched after the adoption of the Bahamas Maritime Authority Act in 1995, was the world's third in terms of gross registered tonnage, behind Panama and Liberia. It was managed by the Bahamas Maritime Authority, which had offices in London, United Kingdom, Nassau and New York, USA. Registered displacement was some 35.4m. grt by December 2004. In 1997 Hutchison Whampoa opened a US $78m. international container-transhipment terminal at Freeport, Grand Bahama; after completion of a $71m. expansion in 1999 it was capable of handling 950,000 container units per year and was to act as an intercontinental hub port serving North America, the Caribbean and South America. A further $75m. expansion in 2003 increased the capacity to 1.5m. container units, and an additional expansion to 1.9m.-unit capacity was planned. A $30m. airport-improvement programme was completed in 2004, and a 320 ha industrial park was also planned.

## INDUSTRY

Industrial development was a relatively late arrival in the Bahamas, dating from 1960 and the formation of the Grand Bahama Development Company, based in the free-trade zone in Freeport, itself only established by the 1955 Hawksbill Creek Agreement. Freeport was at first dominated by ship bunkering and petroleum refining. However, proposals to develop Grand Bahama as a major industrial centre did not come to fruition; refining operations at the Bahamas Oil Refining Company ceased in 1985 and a cement plant built to serve the US market also closed. There remained some interest in petroleum and natural-gas exploration; drilling by a US company, on Grand Bahama Bank, continued from 1986, but with no positive results. A further exploration programme was agreed with US investors Kerr McGee in 2003, with exploratory drilling expected in 2005 following encouraging seismic-survey results in 2004.

In the 1990s, however, high labour and utilities costs, and the small size of the local market, combined to limit severely the development of industrial activity, although a number of manufacturers produced for the local market, among them a brewery, soft-drinks companies, paper converters and printers. Bacardi operated a rum refinery in Nassau, which made use of imported molasses to produce light rums for export to Europe under the Lomé Convention (which expired in February 2000 and was replaced by the Cotonou Agreement). Rum exports fluctuated: they fell from US $33.4m., or 27% of domestic export earnings, in 1991 to only $2.9m. (3.1% of domestic exports) in 1995, before recovering to $38.2m., or 10.1%, in 2001.

Of greater significance were a number of enclave industries located on Grand Bahama that benefited from the island's tax concessions and port facilities. A Freeport company had facilities for the repair of containers, fabrication of steel structures and instrumentation system maintenance. There is also a plant producing expandable polystyrene. The US $75m. Grand Bahama Shipyard cruise-ship repair facility was established in 2000. A 320 m dry dock, the largest in the Americas, was in operation from early 2002, by which time employment had increased to 423. Although engineering skills shortages held the number of local staff to 206 at this date, a training centre was opened by the end of the year. A separate operation included the world's largest covered yacht-repair facility. In 2002–03 the Venezuelan state oil company, Petróleos de Venezuela (PDVSA), expanded its petroleum-storage facility at the former Bahamas Oil Refining Company (BORCO) on Grand Bahama, from a capacity of 12m. barrels to 20m. barrels, at a cost of $38m. Three separate proposals were made in 2001 to establish a Liquefied Natural Gas (LNG) regasification terminal and handling facilities in either Grand Bahama or the small island of Ocean Cay. The proposed facilities would serve Florida via an underwater pipeline. With local environmental concerns strong, no final decision had been made on any of these proposals by mid-2005; however, there appeared to be a possibility that the Ocean Cay scheme would proceed. In 1998 a 13.5 MW electricity generating plant, constructed at a cost of $17.5m., opened in Freeport. On the remote island of Inagua in the southern Bahamas, the main employer was the US-owned Morton Salt Company, which produced salt through the evaporation of sea water. Exports of salt in 2001 were worth $13.5m.

Industry (including construction and utilities) employed 16.6% of the working population in 1999 (construction accounted for 11.4%). Manufacturing and mining contributed some 5.2% of GDP in 2003, with construction contributing a further 7.0% and electricity, gas and water 3.6%. However, high wages and costs continued to be serious disadvantages.

## AGRICULTURE AND FISHING

Agriculture never played a leading role in the Bahamian economy and in 2003 it was estimated to account for only 0.5% of GDP. More than 80% of food was imported. Agricultural resources are severely limited since rainfall is low, few areas have groundwater for irrigation and much of the land consists of bare limestone rock with only scattered patches of soil.

In 2002 agricultural production accounted for only 1% of total land area. There were small-scale farmers on many of the Family Islands, who produced livestock, fruit and vegetables in limited quantities for local markets. Larger commercial farms on New Providence, the most populous island, produced eggs and poultry, constituting one-half of agricultural production in 2001; however, the largest of these was forced to close in 2002 as a result of reduced protection from imports. Feed and other supplies were imported. Ornamental plants and flowers accounted for another 12.7% of agricultural production. Following government efforts in the early 1990s to make land available to local and foreign investors for fruit and vegetable farming, by 1995 over 8,000 ha were devoted to citrus-fruit cultivation. To protect local producers, imports of some crops, such as bananas, were restricted. A number of sizeable agro-businesses in the larger islands of the northern Bahamas operated as enclave industries, growing citrus and other fruit and vegetable crops for export. Some farms used crushed limestone rock as a growing medium and nutrients were sometimes added to groundwater used for irrigation. Total exports of fruit and vegetables amounted to US $7.6m. in 2001. However, commercial citrus production experienced a serious setback in 2004–05, when all trees on two major farms covering a total of 2,500 ha were uprooted after an outbreak of citrus canker disease.

Commercial fishing remained an important economic activity on some of the smaller islands, such as Spanish Wells, making up 2.2% of national GDP, and supplied 19.1% of domestic exports in 2001. Exports of crawfish (spiny lobster) amounted to US $67.7m. in 2001, with another $4.4m. from other fisheries exports. Such primary economic activities, however, remained peripheral in their contribution to the national wealth of the Bahamas.

THE BAHAMAS

# Statistical Survey

Source (unless otherwise stated): Central Bank of the Bahamas, Frederick St, POB N-4868, Nassau; tel. 322-2193; fax 322-4321; e-mail cbob@centralbankbahamas.com; internet www.centralbankbahamas.com.

## AREA AND POPULATION

**Area:** 13,939 sq km (5,382 sq miles).

**Population:** 255,095 at census of 2 May 1990; 303,611 (males 147,778, females 155,833) at census of 1 May 2000; 320,700 at mid-2004. *By Island* (1990): New Providence 172,196 (including the capital, Nassau); Grand Bahama 40,898; Andros 8,187; Eleuthera 10,586. Source: partly Caribbean Development Bank, *Social and Economic Indicators.*

**Density** (mid-2004): 23.0 per sq km.

**Principal Town** (population, UN estimate, incl. suburbs): Nassau (capital) 222,163 in mid-2003. Source: UN, *World Urbanization Prospects: The 2003 Revision.*

**Births, Marriages and Deaths** (1999, unless otherwise indicated): Registered live births 6,367 (birth rate 21.4 per 1,000); Registered marriages (1996) 2,628 (marriage rate 9.3 per 1,000); Registered deaths 1,567 (death rate 5.3 per 1,000). *2001:* Registered live births 5,353 (birth rate 17.3 per 1,000); Registered deaths 1,609 (death rate 5.6 per 1,000). Sources: UN, *Demographic Yearbook, Population and Vital Statistics Report,* and Caribbean Development Bank, *Social and Economic Indicators.*

**Expectation of Life** (WHO estimates, years at birth): 72 (males 69; females 75) in 2003. Source: WHO, *World Health Report.*

**Economically Active Population** (sample survey, persons aged 15 years and over, excl. armed forces, April 1999): Agriculture, hunting, forestry and fishing 5,835; Mining, quarrying, electricity, gas and water 1,745; Manufacturing 5,910; Construction 16,540; Wholesale and retail trade 19,955; Hotels and restaurants 23,300; Transport, storage and communications 10,305; Finance, insurance, real estate and business services 13,350; Community, social and personal services 47,780; Activities not adequately defined 630; *Total employed* 145,350 (males 77,245, females 68,105); Unemployed 12,290 (males 4,955, females 7,335); *Total labour force* 157,640 (males, 82,200, females 75,440) (Source: ILO Caribbean Office). *2002* (figures rounded to nearest 10 persons): Total employed 152,690 (males 78,410, females 74,280); Unemployed 15,290 (males 7,560, females 7,730); Total labour force 167,980 (males 85,970, females 82,010).

## HEALTH AND WELFARE

### Key Indicators

**Total Fertility Rate** (children per woman, 2003): 2.3.

**Under-5 Mortality Rate** (per 1,000 live births, 2003): 14.

**HIV/AIDS** (% of persons aged 15–49, 2003): 3.0.

**Physicians** (per 1,000 head, 1998): 1.06.

**Hospital Beds** (per 1,000 head, 1996): 3.94.

**Health Expenditure** (2002): US $ per head (PPP): 1,074.

**Health Expenditure** (2002): % of GDP: 6.9.

**Health Expenditure** (2002): public (% of total): 48.6.

**Access to Water** (% of persons, 2002): 97.

**Access to Sanitation** (% of persons, 2002): 100.

**Human Development Index** (2002): ranking: 51.

**Human Development Index** (2002): value: 0.815.
For sources and definitions, see explanatory note on p. vi.

## AGRICULTURE, ETC.

**Principal Crops** (FAO estimates, '000 metric tons, 2003): Roots and tubers 1.0; Sugar cane 55.5; Tomatoes 4.0; Other vegetables 19.7; Bananas 3.4; Lemons and limes 8.5; Grapefruit and pomelos 12.5; Other fruits 3.6. Source: FAO.

**Livestock** (FAO estimates, '000 head, year ending September 2003): Cattle 0.7; Pigs 5.0; Sheep 6.5; Goats 14.0; Poultry 2,975. Source: FAO.

**Livestock Products** (FAO estimates, '000 metric tons, 2003): Chicken meat 8.0; Cows' milk 0.7; Goats' milk 1.0; Hen eggs 0.9. Source: FAO.

**Forestry** (FAO estimates, '000 cubic metres, 2003): *Roundwood Removals (excl. bark):* Sawlogs and veneer logs 17 (output assumed to be unchanged since 1992); *Sawnwood Production (incl. railway sleepers):* Coniferous (softwood) 1 (output assumed to be unchanged since 1970). Source: FAO.

**Fishing** (metric tons, live weight, 2003, estimates): Capture 12,611 (Nassau grouper 422; Snappers 700; Caribbean spiny lobster 10,378; Stromboid conchs 620); Aquaculture 42; *Total catch* 12,653. Source: FAO.

## MINING

**Production** (estimates, '000 metric tons, 2003): Unrefined salt 900; Aragonite 1,200. Source: US Geological Survey.

## INDUSTRY

**Production** (estimate, million kWh, 2002): Electric energy 1,826.

## FINANCE

**Currency and Exchange Rates:** 100 cents = 1 Bahamian dollar (B $). *Sterling, US Dollar and Euro Equivalents* (31 May 2005): £1 sterling = B $1.8181; US $1 = B $1.0000; €1 = B $1.2331; B $100 = £55.00 = US $100.00 = €81.10. *Exchange Rate:* Since February 1970 the official exchange rate, applicable to most transactions, has been US $1 = B $1, i.e. the Bahamian dollar has been at par with the US dollar. There is also an investment currency rate, applicable to certain capital transactions between residents and non-residents and to direct investments outside the Bahamas. Since 1987 this exchange rate has been fixed at US $1 = B $1.225.

**General Budget** (preliminary, B $ million, 2002): *Revenue:* Taxation 796.0 (Taxes on international trade and transactions 506.6; Taxes on tourism 88.2; Taxes on property 76.6; Taxes on companies 61.2); Other current revenue 89.7; Total 885.6. *Expenditure:* Wages and salaries 475.5; Goods and services 227.4; Interest payments 100.8; Subsidies and transfers 91.4; Capital expenditure and net lending 131.1; Total 1,026.2 (Source: IMF, *Bahamas: Statistical Appendix* (July 2003)). *2003* (preliminary, B $ million): Total revenue and grants 901.8; Total expenditure (excl. net lending, 43.1) 1,046.5 (Current expenditure 962.7, Capital expenditure 83.8). *2004* (approved estimates, B $ million): Total revenue and grants 991.5; Total expenditure (excl. net lending, 25.9) 1,088.6 (Current expenditure 968.9, Capital expenditure 119.7).

**International Reserves** (US $ million at 31 December 2004): Reserve position in IMF 9.7; Foreign exchange 664.7; Total 674.4. Source: IMF, *International Financial Statistics.*

**Money Supply** (B $ million at 31 December 2004): Currency outside banks 177; Demand deposits at deposit money banks 860; Total money (incl. others) 1,037. Source: IMF, *International Financial Statistics.*

**Cost of Living** (consumer price index; base: 2000 = 100): 104.3 in 2002; 107.4 in 2003; 108.3 in 2004. Source: IMF, *International Financial Statistics.*

**Gross Domestic Product** (B $ million at current prices): 5,400.1 in 2002; 5,502.2 in 2003 (estimate); 5,734.6 in 2004 (preliminary). Source: Caribbean Development Bank, *Social and Economic Indicators.*

**Expenditure on the Gross Domestic Product** (B $ million at current prices, 2004, preliminary): Government final consumption expenditure 810.5; Private final consumption expenditure 3,991.2; Increase in stocks 166.5; Gross fixed capital formation 1,681.5; *Total domestic expenditure* 6,649.7; Exports of goods and services 2,471.4; *Less* Imports of goods and services 3,290.9; Statistical discrepancy –95.6; *GDP in purchasers' values* 5,734.6.

THE BAHAMAS

**Gross Domestic Product by Economic Activity** (B $ million at current prices, 2003, estimates): Agriculture, hunting, forestry and fishing 151.8; Mining and quarrying 58.6; Manufacturing 232.3; Electricity, gas and water 199.6; Construction 393.5; Wholesale and retail trade 730.7; Restaurants and hotels 541.0; Transport, storage and communications 504.2; Finance, insurance, real estate and business services 1,670.5; Government services 328.5; Other community, social and personal services 793.1; *Sub-total* 5,603.8; *Less* Financial intermediation services indirectly measured and other indirect taxes 803.8; *Gross value added in basic prices* 4,800.0; Taxes on products 634.6; *Less* Subsidies on products 52.4; Statistical discrepancy 120.0; *GDP in purchasers' values* 5,502.2.

**Balance of Payments** (US $ million, 2003): Exports of goods f.o.b. 424.8; Imports of goods f.o.b. –1,629.5; *Trade balance* –1,204.7; Exports of services 1,979.9; Imports of services –1,079.5; *Balance on goods and services* –304.3; Other income received 48.4; Other income paid –211.5; *Balance on goods, services and income* –467.4; Current transfers received 48.8; Current transfers paid –11.2; *Current balance* –429.8; Capital account (net) –37.4; Direct investment from abroad (net) 145.0; Other investment assets 46,576.8; Other investment liabilities –46,462.1; Net errors and omissions 317.5; *Overall balance* 110.0. Source: IMF, *International Financial Statistics*.

### EXTERNAL TRADE

**Principal Commodities** (US $ million, 2001): *Imports c.i.f.*: Food and live animals 298.1 (Meat and preparations 78.3); Mineral fuels, lubricants, etc. 301.2 (Refined petroleum products, etc. 278.3); Chemicals 146.8; Basic manufactures 551.4 (Non-metallic mineral manufactures 297.7); Machinery and transport equipment 536.1 (General industrial machinery and equipment 105.0; Road vehicles 151.8); Miscellaneous manufactured articles 385.9 (Articles of apparel and clothing accessories 143.2); Total (incl. others) 2,370.1. *Exports f.o.b.*: Food and live animals 80.8 (Fish, crustacean and molluscs, and preparations thereof 72.1); Beverages and tobacco 40.8 (Alcoholic beverages 39.6); Crude materials (inedible) except fuels 27.8; Refined petroleum products 68.8; Chemicals 97.8 (Organic chemicals 14.4; Artificial resins and plastic materials, and cellulose esters, etc. 71.7); Machinery and transport equipment 45.5 (General industrial machinery and equipment 14.2); Total (incl. others) 378.2.

**Principal Trading Partners** (US $ million, 2001): *Imports c.i.f.*: USA 2,032.5; Venezuela 107.9; Total (incl. others) 2,370.1. *Exports f.o.b.*: France 21.5; Germany 14.5; Spain 12.5; United Kingdom 12.0; USA 293.1; Total (incl. others) 378.2.
Source: UN, *International Trade Statistics Yearbook*.

### TRANSPORT

**Road Traffic** (vehicles in use, 1998): 67,400 passenger cars; 16,800 commercial vehicles. Source: Auto and Truck International (Illinois), *World Automotive Market Report*.

**Shipping:** *Merchant fleet* (displacement, '000 grt at 31 December): 35,798 in 2002; 34,752 in 2003; 35,388 in 2004 (Source: Lloyd's Register-Fairplay, *World Fleet Statistics*). *International sea-borne freight traffic* (estimates, '000 metric tons, 1990): Goods loaded 5,920; Goods unloaded 5,705 (Source: UN, *Monthly Bulletin of Statistics*).

**Civil Aviation** (2001): Kilometres flown (million) 7; Passengers carried ('000) 1,626; Passenger-km (million) 391; Total ton-km of freight (million) 48. Source: UN, *Statistical Yearbook*.

### TOURISM

**Visitor Arrivals** ('000): 3,979.9 (1,428.2 by air, 2,551.7 by sea) in 2001; 4,205.0 (1,402.9 by air, 2,802.1 by sea) in 2002; 4,398.8 (1,428.6 by air, 2,970.2 by sea) in 2003; 4,810.0 (1,450.0 by air, 3,360.0 by sea) in 2004. Source: Caribbean Development Bank, *Social and Economic Indicators*.

**Tourism Receipts** (B $ million): 1,763 in 2002; 1,782 in 2003; 1,855 in 2004.

### COMMUNICATIONS MEDIA

**Radio Receivers** (1997): 215,000 in use.
**Television Receivers** (1999): 73,000 in use.
**Telephones** (2003): 131,700 main lines in use.
**Facsimile Machines** (1996): 500 in use.
**Mobile Cellular Telephones** (2003): 116,300 subscribers.
**Internet Users** (2003): 84,000.
**Daily Newspapers** (1996): 3 titles (total circulation 28,000 copies). Sources: UN, *Statistical Yearbook*; UNESCO, *Statistical Yearbook*; International Telecommunication Union.

### EDUCATION

**Pre-primary** (1996/97): 20 schools; 76 teachers; 1,094 pupils.
**Primary** (1996/97): 113 schools; 1,792 teachers; 42,839 pupils (2001).
**Secondary:** 37 junior/senior high schools (1990); 2,097 teachers (1996/97); 23,401 students (2001).
**Tertiary** (1987): 249 teachers; 5,305 students. In 2002 there were 3,463 students registered at the College of the Bahamas.
Sources: UNESCO, *Statistical Yearbook*; UN, Economic Commission for Latin America and the Caribbean, *Statistical Yearbook*; Caribbean Development Bank, *Social and Economic Indicators 2001*.

**Adult Literacy Rate** (UNESCO estimates): 95.5% (males 94.6%; females 96.3%) in 2002. Source: UN Development Programme, *Human Development Report*.

# Directory

## The Constitution

A representative House of Assembly was first established in 1729, although universal adult suffrage was not introduced until 1962. A new Constitution for the Commonwealth of the Bahamas came into force at independence, on 10 July 1973. The main provisions of the Constitution are summarized below.

Parliament consists of a Governor-General (representing the British monarch, who is Head of State), a nominated Senate and an elected House of Assembly. The Governor-General appoints the Prime Minister and, on the latter's recommendation, the remainder of the Cabinet. Apart from the Prime Minister, the Cabinet has no fewer than eight other ministers, of whom one is the Attorney-General. The Governor-General also appoints a Leader of the Opposition.

The Senate (upper house) consists of 16 members, of whom nine are appointed by the Governor-General on the advice of the Prime Minister, four on the advice of the Leader of the Opposition and three on the Prime Minister's advice after consultation with the Leader of the Opposition. The House of Assembly (lower house) has 40 members. A Constituencies Commission reviews numbers and boundaries at intervals of not more than five years and can recommend alterations for approval of the House. The life of Parliament is limited to a maximum of five years.

The Constitution provides for a Supreme Court and a Court of Appeal.

## The Government

### HEAD OF STATE

**Monarch:** HM Queen Elizabeth II (succeeded to the throne 6 February 1952).
**Governor-General:** Dame Ivy Dumont (took office 1 January 2002).

### THE CABINET
(July 2005)

**Prime Minister and Minister of Finance:** Perry G. Christie.
**Deputy Prime Minister and Minister of National Security:** Cynthia A. Pratt.
**Attorney-General and Minister of Justice and of Education:** Alfred M. Sears.
**Minister of Agriculture, Fisheries and Local Government:** V. Alfred Gray.
**Minister of Financial Services and Investments:** Allyson Maynard-Gibson.
**Minister of Foreign Affairs and Public Service:** Frederick A. Mitchell.
**Minister of Health and Environmental Services:** Dr Marcus C. Bethel.

# THE BAHAMAS

**Minister of Housing and National Insurance:** D. SHANE GIBSON.
**Minister of Labour and Immigration:** VINCENT A. PEET.
**Minister of Public Works and Utilities:** BRADLEY B. ROBERTS.
**Minister of Social Services and Community Development:** MELANIE S. GRIFFIN.
**Minister of Tourism:** OBEDIAH 'OBIE' H. WILCHCOMBE.
**Minister of Trade and Industry:** LESLIE O. MILLER.
**Minister of Transport and Aviation:** GLENNYS M. E. HANNA-MARTIN.
**Minister of Youth, Sports and Culture:** NEVILLE W. WISDOM.
**Minister of State in the Ministry of Finance:** Sen. JAMES H. SMITH.

## MINISTRIES

**Attorney-General's Office and Ministry of Justice:** Post Office Bldg, East Hill St, POB N-3007, Nassau; tel. 322-1141; fax 356-4179.
**Office of the Prime Minister:** Sir Cecil V. Wallace-Whitfield Centre, West Bay St, POB CB-10980, Nassau; tel. 327-5826; fax 327-5806; e-mail info@opm.gov.bs; internet www.opm.gov.bs.
**Office of the Deputy Prime Minister:** Churchill Bldg, Bay St, POB N-3217, Nassau; tel. 356-6792; fax 356-6087.
**Ministry of Agriculture, Fisheries and Local Government:** Levy Bldg, East Bay St, POB N-3028, Nassau; tel. 325-7502; fax 322-1767.
**Ministry of Education:** Thompson Blvd, POB N-3913, Nassau; tel. 502-2704; fax 322-8491.
**Ministry of Finance:** Cecil V. Wallace-Whitfield Centre, West Bay St, POB N-3017, Nassau; tel. 327-1530; fax 327-1618; internet www.bahamas.gov.bs/finance.
**Ministry of Financial Services and Investments:** Sir Cecil V. Wallace-Whitfield Centre, West Bay St, POB CB-10980, Nassau; tel. 327-5826; fax 327-5806.
**Ministry of Foreign Affairs and Public Service:** East Hill St, POB N-3746, Nassau; tel. 322-7624; fax 328-8212; e-mail mfabahamas@batelnet.bs.
**Ministry of Health and Environmental Services:** Royal Victoria Gardens, Shirley St, POB N-3730, Nassau; tel. 322-7425; fax 322-7788.
**Ministry of Housing and National Insurance:** National Insurance Bldg, Baillou Rd, POB N-7508, Nassau; tel. 502-1500; fax 323-3048.
**Ministry of Labour and Immigration:** Post Office Bldg, East Hill St, POB N-3008, Nassau; tel. 323-7814; fax 325-1920.
**Ministry of National Security:** Churchill Bldg, Bay St, POB N-3217, Nassau; tel. 356-6792; fax 356-6087.
**Ministry of Social Services and Community Development:** Frederick House, Frederick St, POB N-3206, Nassau; tel. 356-0765; fax 323-3883.
**Ministry of Public Works and Utilities:** John F. Kennedy Dr., POB N-8156, Nassau; tel. 322-4830; fax 326-7344.
**Ministry of Tourism:** British Colonial Hilton Hotel, Bay St, POB N-3701, Nassau; tel. 322-7500; fax 328-0945; e-mail tourism@bahamas.com; internet www.bahamas.com.
**Ministry of Trade and Industry:** Manx Bldg, West Bay St, POB N-4849, Nassau; tel. 328-2700; fax 328-1324.
**Ministry of Transport and Aviation:** Pilot House Complex, POB N-10114, Nassau; tel. 394-0445; fax 394-5920.
**Ministry of Youth, Sports and Culture:** 7th Floor, Post Office Bldg, East Hill St, POB N-4891, Nassau; tel. 322-6250; fax 322-6546.

## Legislature

### PARLIAMENT

#### Senate

**President:** SHARON R. WILSON.
There are 16 nominated members.

#### House of Assembly

**Speaker:** OSWALD INGRAHAM.
The House has 40 members.

### General Election, 2 May 2002

| Party | Seats |
|---|---|
| Progressive Liberal Party (PLP) | 29 |
| Free National Movement (FNM) | 7 |
| Independents | 4 |
| **Total** | **40** |

## Political Organizations

**Bahamas Freedom Alliance (BFA):** Nassau; formed Coalition Plus Labour alliance with CDR and PLM to contest 2002 legislative elections; Leader D. HALSTON MOULTRIE.
**Coalition for Democratic Reform (CDR):** Nassau; f. 2000; centrist; formed Coalition Plus Labour alliance with BFA and PLM to contest 2002 legislative elections; Leader BERNARD NOTTAGE.
**Free National Movement (FNM):** Mackey St, POB N-10713, Nassau; tel. 393-7853; fax 393-7914; e-mail fnm@coralwave.com; internet www.freenationalmovement.org; f. 1972; Leader ORVILLE ALTON THOMPSON (TOMMY) TURNQUEST; Chair. CARL W. BETHEL.
**People's Labour Movement (PLM):** Nassau; formed Coalition Plus Labour alliance with BFA and CDR to contest 2002 legislative elections.
**Progressive Liberal Party (PLP):** Sir Lynden Pindling Centre, PLP House, Farrington House, POB N-547, Nassau; tel. 325-5492; fax 328-0808; e-mail eightmr@bahamiansfirst.com; internet www.progressiveliberalparty.com; f. 1953; centrist party; Leader PERRY G. CHRISTIE; Chair. RAYNARD S. RIGBY.

## Diplomatic Representation

### EMBASSIES IN THE BAHAMAS

**China, People's Republic:** 3 Orchard Terrace, Village Rd, POB SS-6389, Nassau; tel. 393-1415; fax 393-0733; e-mail chinaemb_bs@mfa.gov.cn; Ambassador LI YUANMING.
**Haiti:** Sears House, Shirley St, POB N-666, Nassau; tel. 326-0325; fax 322-7712; Chargé d'affaires JOSEPH J. ETIENNE.
**USA:** Mosmar Bldg, Queen St, POB N-8197, Nassau; tel. 322-1181; fax 328-7838; e-mail embnas@state.gov; internet nassau.usembassy.gov; Ambassador JOHN D. ROOD.

## Judicial System

The Judicial Committee of the Privy Council (based in the United Kingdom), the Bahamas Court of Appeal, the Supreme Court and the Magistrates' Courts are the main courts of the Bahamian judicial system.

All courts have both a criminal and civil jurisdiction. The Magistrates' Courts are presided over by professionally qualified Stipendiary and Circuit Magistrates in New Providence and Grand Bahama, and by Island Administrators sitting as Magistrates in the Family Islands.

Whereas all magistrates are empowered to try offences that may be tried summarily, a Stipendiary and Circuit Magistrate may, with the consent of the accused, also try certain less serious indictable offences. The jurisdiction of magistrates is, however, limited by law.

The Supreme Court consists of the Chief Justice, two Senior Justices and six Justices. The Supreme Court also sits in Freeport, with two Justices.

Appeals in almost all matters lie from the Supreme Court to the Court of Appeal, with further appeal in certain instances to the Judicial Committee of the Privy Council.

#### Supreme Court of the Bahamas

Parliament Sq., POB N-8167, Nassau; tel. 322-3315; fax 323-6895.
**Chief Justice** Sir BURTON HALL.

**Court of Appeal:** POB N-8167, Nassau; tel. 322-3315; fax 325-6895; Pres. Dame JOAN SAWYER.
**Magistrates' Courts:** POB N-421, Nassau; tel. 325-4573; fax 323-1446; 15 magistrates and a circuit magistrate.
**Registrar of the Supreme Court:** ESTELLE G. GRAY EVANS, POB N-167, Nassau; tel. 322-4348; fax 325-6895; e-mail estelleevans@bahamas.gov.bs.

**Attorney-General:** ALFRED M. SEARS.

**Office of the Attorney-General:** 3rd Floor, Post Office Bldg, East Hill St, POB N-3007, Nassau; tel. 322-1141; fax 356-4179; Dir of Legal Affairs RHONDA BAIN; Dir of Public Prosecutions BERNARD TURNER.

**Registrar-General:** ELIZABETH E. M. THOMPSON, 50 Shirley St, POB N-532, Nassau; tel. 322-3316; fax 322-5553; e-mail registra@batelnet.bs.

## Religion

Most of the population profess Christianity, but there are also small communities of Jews and Muslims.

### CHRISTIANITY

According to the census of 1990, there were 79,465 Baptists (31.2% of the population), 40,894 Roman Catholics (16.0%) and 40,881 Anglicans (16.0%). Other important denominations include the Pentecostal Church (5.5%), the Church of Christ (5.0%) and the Methodists (4.8%).

**Bahamas Christian Council:** POB N-3101, Nassau; tel. 326-7114; f. 1948; 27 mem. churches; Pres. Rev. WILLIAM THOMPSON.

#### The Roman Catholic Church

The Bahamas comprises the single archdiocese of Nassau. At 31 December 2003 there were an estimated 41,077 adherents in the Bahamas. The Archbishop participates in the Antilles Episcopal Conference (whose Secretariat is based in Port of Spain, Trinidad). The Turks and Caicos Islands are also under the jurisdiction of the Archbishop of Nassau.

**Archbishop of Nassau:** Most Rev. LAWRENCE ALOYSIUS BURKE, The Hermitage, West St, POB N-8187, Nassau; tel. 322-8919; fax 322-2599; e-mail rcchancery@batelnet.bs.

#### The Anglican Communion

Anglicans in the Bahamas are adherents of the Church in the Province of the West Indies. The diocese also includes the Turks and Caicos Islands.

**Archbishop of the West Indies, and Bishop of Nassau and the Bahamas:** Most Rev. DREXEL GOMEZ, Bishop's Lodge, POB N-7107, Nassau; tel. 322-3015; fax 322-7943.

#### Other Christian Churches

**Bahamas Conference of the Methodist Church:** POB SS-5103, Nassau; tel. 393-3726; fax 393-8135; e-mail bcmc@bahamas.net.bs; Pres. KENRIS L. CAREY.

**Bahamas Conference of Seventh-day Adventists:** Harrold Rd, POB N-356, Nassau; tel. 341-4021; fax 341-4088; internet www.bahamasconference.org.

**Bahamas Evangelical Church Association:** Carmichael Rd, POB N-1224, Nassau; tel. 362-1024.

**Greek Orthodox Church:** Church of the Annunciation, West St, POB N-823, Nassau; tel. 322-4382; f. 1928; part of the Archdiocese of North and South America, based in New York (USA); Priest Rev. THEOPHANIS KULYVAS.

**Methodist Church Conference in the Bahamas (MCCA):** POB N-3702, Nassau; tel. 373-1888; Conference Pres. Rev. LIVINGSTON MALCOLM.

Other denominations include African Methodist Episcopal, the Assemblies of Brethren, Christian Science, the Jehovah's Witnesses, the Salvation Army, Pentecostal, Presbyterian, Baptist, Lutheran and Assembly of God churches.

### OTHER RELIGIONS

#### Bahá'í Faith

**Bahá'í National Spiritual Assembly:** Arianna House, Dunmore Lane, Nassau; tel. 326-0607; e-mail nsabaha@mail.com.

#### Islam

There is a small community of Muslims in the Bahamas.

**Islaamic Centre:** Carmichael Rd, POB N-10711, Nassau; tel. 341-6612.

**Islaamic Centre Jamaat Ul-Islam:** 13 Davies St, Oakes Field, POB N-10711, Nassau; tel. 325-0413.

#### Judaism

Most of the Bahamian Jewish community are based on Grand Bahama.

**Bahamas Jewish Congregation Synagogue:** POB CB-11003, Cable Beach Shopping Centre, Nassau; tel. 327-2064.

## The Press

### NEWSPAPERS

**The Bahama Journal:** Media House, POB N-8610, Nassau; tel. 325-3082; fax 356-7256; internet www.jonesbahamas.com; daily; circ. 5,000.

**The Freeport News:** Cedar St, POB F-40007, Freeport; tel. 352-8321; fax 352-3449; internet freeport.nassauguardian.net; f. 1961; daily; Gen. Man. DORLAN COLLIE; Editor ROBYN ADDERLEY; circ. 6,000.

**The Nassau Guardian:** 4 Carter St, Oakes Field, POB N-3011, Nassau; tel. 302-2300; fax 328-8943; e-mail editor@nasguard.com; internet www.thenassauguardian.com; f. 1844; daily; CEO CHARLES CARTER; Man. Editor ANTHONY CAPRON; circ. 12,277.

**The Punch:** POB N-4081, Nassau; tel. 322-7112; fax 323-5268; twice weekly; Editor IVAN JOHNSON; circ. 25,000.

**The Tribune:** Shirley St, POB N-3207, Nassau; tel. 322-1986; fax 328-2398; e-mail tribune@100jamz.com; f. 1903; daily; Publr and Editor EILEEN DUPUCH CARRON; circ. 13,500.

### PERIODICALS

**The Bahamas Financial Digest:** Miramar House, Bay and Christie Sts, POB N-4271, Nassau; tel. 356-2981; fax 356-7118; e-mail michael.symonette@batelnet.bs; f. 1973; 4 a year; business and investment; Publr and Editor MICHAEL A. SYMONETTE; circ. 15,890.

**Bahamas Tourist News:** Baypearl Bldg, Parliament St, POB N-4855, Nassau; tel. 322-4528; fax 322-4527; e-mail starpub@batelnet.bs; f. 1962; monthly; Editor BOBBY BOWER; circ. 371,000 (annually).

**Nassau City Magazine:** Miramar House, Bay and Christie Sts, POB N-4824, Nassau; tel. 356-2981; fax 326-2849.

**Official Gazette:** c/o Cabinet Office, POB N-7147, Nassau; tel. 322-2805; weekly; publ. by the Cabinet Office.

**What's On Magazine:** Woodes Rogers Wharf, POB CB-11713, Nassau; tel. 323-2323; fax 322-3428; e-mail info@whatsonbahamas.com; internet www.whatsonbahamas.com; monthly; Publr NEIL ABERLE.

## Publishers

**Bahamas Free Press Ltd:** POB CB-13309, Nassau; tel. 323-8961.

**Etienne Dupuch Jr Publications Ltd:** 51 Hawthorne Rd, POB N-7513, Nassau; tel. 323-5665; fax 323-5728; e-mail dupuch@bahamasnet.com; internet www.dupuch.com; f. 1959; publishes *Bahamas Handbook*, *Trailblazer* maps, *What To Do* magazines, *Welcome Bahamas*, *Tadpole* (educational colouring book) series and *Dining and Entertainment Guide*; Dirs ETIENNE DUPUCH, Jr, S. P. DUPUCH.

**Media Enterprises Ltd:** 31 Shirley Park Ave, POB N-9240, Nassau; tel. 325-8210; fax 325-8065; e-mail info@bahamasmedia.com; internet www.bahamasmedia.com; Pres. and Gen. Man. LARRY SMITH; Publishing Dir NEIL E. SEALEY.

**Printing Tours and Publishing:** Miramar House, Bay and Christie Sts, POB N-4846, Nassau; tel. 356-2981; fax 356-7118.

**Sacha de Frisching Publishing:** POB N-7776, Nassau; tel. 362-6230; fax 362-6274; children's books.

**Star Publishers Ltd:** POB N-4855, Nassau; tel. 322-3724; fax 322-4537; e-mail starpub@bahamas.net.bs; internet www.supermaps.com.

## Broadcasting and Communications

### TELECOMMUNICATIONS

The telecommunications sector was in the process of being opened to private participation in 2005.

**Bahamas Telecommunications Co (BTC):** POB N-3048, John F. Kennedy Dr., Nassau; tel. 302-7000; fax 326-7474; e-mail info@batelnet.bs; internet www.batelnet.bs; f. 1966, fmrly known as BaTelCo; state-owned; intended sale of 49% stake to private

# THE BAHAMAS — Directory

investors failed in 2003; privatization remained pending in 2005; Chair. RENO BROWN; Gen. Man. MICHAEL SYMONETTE.

**Cable Bahamas Ltd:** POB CB-13050, Nassau; tel. 356-2200; fax 356-8997; e-mail info@cablebahamas.com; internet www.cablebahamas.com; f. 1995; provides cable television and internet services; Pres. RICHARD PARDY.

## BROADCASTING

### Radio

**Broadcasting Corporation of the Bahamas:** 3rd Terrace, POB N-1347, Centreville, New Providence; tel. 322-4623; fax 322-6598; f. 1936; govt-owned; operates the ZNS radio and television network; Chair. MICHAEL D. SMITH; Gen. Man. ANTHONY FOSTER.

**Radio Bahamas:** internet www.univox.com/radio/zns.html; f. 1936; broadcasts 24 hours per day on four stations: the main Radio Bahamas ZNS1, Radio New Providence ZNS2, which are both based in Nassau, Radio Power 104.5 FM, and the Northern Service (ZNS3—Freeport); f. 1973; Station Man. ANTHONY FORSTER; Programme Man. TANYA PINDER.

**Cool 96 FM:** POB F-40773, Freeport, Grand Bahama; tel. 353-7440; fax 352-8709.

**Love 97 FM:** Bahamas Media House, East St North, POB N-3909, Nassau; tel. and fax 356-2555; e-mail twilliams@jonescommunications.com; internet www.love97fm.com.

**More 95.9 FM:** POB CR-54245, Nassau; tel. 361-2447; fax 361-2448; e-mail morefm94.9@batelnet.bs.

**Tribune Radio Ltd** (One Hundred JAMZ): Shirley and Deveaux St, POB N-3207, Nassau; tel. 328-4771; fax 356-5343; e-mail michelle@100jamz.com; internet www.100jamz.com; Gen. Man. STEPHEN HAUGHEY; Programme Dir ERIC WARD.

### Television

**Broadcasting Corporation of the Bahamas:** see Radio.

**Bahamas Television:** f. 1977; broadcasts for Nassau, New Providence and the Central Bahamas; transmitting power of 50,000 watts; full colour; Programme Man. CARL BETHEL.

US television programmes and some satellite programmes can be received. Most islands have a cable-television service.

# Finance

The Bahamas developed into one of the world's foremost financial centres (there are no corporation, income, capital gains or withholding taxes or estate duty), and finance has become a significant feature of the economy. At 31 December 2003 there were 284 banks and trust companies operating in the Bahamas; 169 banks and trust companies had a physical presence in the islands.

## BANKING

(cap. = capital; res = reserves; dep. = deposits; m. = million; brs = branches)

### Central Bank

**The Central Bank of the Bahamas:** Frederick St, POB N-4868, Nassau; tel. 322-2193; fax 322-4321; e-mail cbob@centralbankbahamas.com; internet www.centralbankbahamas.com; f. 1973; bank of issue; cap. B $3.0m., res B $92.2m., dep. B $273.4m. (Dec. 2003); Gov. and Chair. JULIAN W. FRANCIS; Dep. Gov. WENDY M. CRAIGG.

### Development Bank

**The Bahamas Development Bank:** Cable Beach, West Bay St, POB N-3034, Nassau; tel. 327-5780; fax 327-5047; e-mail info@bahamasdevelopmentbank.com; internet www.bahamasdevelopmentbank.com; f. 1978 to fund approved projects and channel funds into appropriate investments; Chair. K. NEVILLE ADDERLEY; Man. Dir GEORGE E. RODGERS; 2 brs.

### Principal Bahamian-based Banks

**Bank of the Bahamas International:** Claughton House, Shirley St, POB N-7118, Nassau; tel. 326-2560; fax 325-2762; f. 1970 as Bank of Montreal (Bahamas and Caribbean); name changed as above in 2002; jtly owned by Govt and Euro Canadian Bank; 50% owned by Govt, 50% owned by c. 4,000 Bahamian shareholders; cap. B $12.0m., res B $9.0m., dep. B $294.8m. (June 2003); Chair. HUGH G. SANDS; Man. Dir PAUL McWEENEY; 10 brs.

**Commonwealth Bank Ltd:** 610 Bay St, POB SS-5541, Nassau; tel. 394-7373; fax 394-5807; e-mail cbinquiry@combankltd.com; internet www.combankltd.com; f. 1960; total assets B $17m. (Dec. 2002); Pres., CEO and Dir WILLIAM BATEMAN SANDS, Jr; 9 brs.

### Principal Foreign Banks

**Banque Privée Edmond de Rothschild Ltd** (Switzerland): 51 Frederick St, POB N-1136, Nassau; tel. 328-8121; fax 328-8115; internet www.lcf-rothschild.com; f. 1997; cap. 67.4m. Swiss francs, res 4.2m. Swiss francs, dep. 7.0m. Swiss francs (Dec. 2002); Chair. ROBERT MISRAHI; Dir CLAUDE MESSULAM.

**BNP Paribas (Bahamas) Ltd:** Beaumont House, 3rd Floor, Bay St, POB N-4883, Nassau; tel. 326-5935; fax 326-5871; internet www.bnpgroup.com; Man. Dir ANDRÉ LAMOTHE.

**BSI Overseas (Bahamas) Ltd** (Italy): Bayside Executive Park, West Bay St, POB N-7130, Nassau; tel. 702-1200; fax 702-1250; f. 1990; wholly owned subsidiary of BSI AG (Switzerland); cap. US $10.0m., res US $66.1m., dep. US $1,103.4m. (Dec. 2001); Chair. RETO KESSLER; Man. Dir MARTIN HUTTER.

**Canadian Imperial Bank of Commerce (CIBC)** (Canada): 4th Floor, 308 East Bay St, POB N-8329, Nassau; tel. 393-4710; fax 393-4280; internet www.cibc.com; Area Man. TERRY HILTS; 9 brs.

**Citibank NA** (USA): 4th Floor, Citibank Bldg, Thompson Blvd, Oakes Field, POB N-8158, Nassau; tel. 302-8859; fax 302-8625; internet www.citibank.com; Gen. Man. ALISON JOHNSTON; 2 brs.

**Crédit Suisse (Bahamas) Ltd** (Switzerland): 4th Floor, Bahamas Financial Centre, Shirley and Charlotte Sts, POB N-4928, Nassau; tel. 356-8100; fax 326-6589; f. 1968; subsidiary of Crédit Suisse Zurich; portfolio and asset management, 'offshore' company management, trustee services; cap. US $12.0m., res US $20.0m., dep. US $500.7m. (Dec. 1997); CEO and Man. Dir EUGEN SCHAEFLI; Chair. JOACHIM STRÄHLE.

**Eni International Bank Ltd:** 1st Floor, British American Insurance House, Marlborough St, POB SS-6377, Nassau; tel. 322-1928; fax 323-8600; e-mail eib-info@enibank.eni.it; cap. US $50.0m., dep. US $260.0m., total assets US $324.3m. (Dec. 2003); Pres., Chair. and Man. Dir PAOLO CARCMOSINO.

**FirstCaribbean International Bank Ltd:** Charlotte House, Shirley St, POB N-3221, Nassau; tel. 325-7384; fax 322-8267; internet www.firstcaribbeaninternational.com; f. 2002 following merger of Caribbean operations of Barclays Bank PLC and CIBC; Exec. Chair. MICHAEL MANSOOR; CEO CHARLES PINK.

**Guaranty Trust Bank Ltd:** Lyford Manor Ltd, Lyford Cay, POB N-4918, Nassau; tel. 362-7200; fax 362-7210; e-mail guaranty@gtblbah.com; f. 1962; cap. US $18.0m., res US $0.1m., dep. US $106.1m. (Jan. 2004); Chair. Sir WILLIAM C. ALLEN; Man. Dir JAMES P. COYLE.

**HSBC Private Banking (Bahamas) Ltd** (Switzerland): 3rd Floor, Maritime House, Frederick St, POB N-10441, Nassau; tel. 328-8644; fax 328-8600; e-mail hfccfint@bahamas.net.bs; f. 1971 as Handelsfina Int., name changed as above in 2002; cap. US $5.0m., res US $15.5m., dep. US $581.3m. (Dec. 1999); Chair. MARC DE GUILLEBON; CEO JEAN-JACQUES SALOMON.

**National Bank of Canada (International) Ltd:** Ground Floor, Goodman's Bay Corporate Centre, POB N-3015, Nassau; tel. 502-8100; fax 323-8086; e-mail info@nbclintl.com; f. 1978; 100% owned by Natcan Holdings International Ltd; cap. US $20.0m., dep. US $156.2m., total assets US $537.6m. (Oct. 2002); Pres. and Man. Dir JACQUES LATENDRESSE.

**Overseas Union Bank and Trust (Bahamas) Ltd** (Switzerland): 250 Bay St, POB N-8184, Nassau; tel. 322-2476; fax 323-8771; f. 1980; cap. US $5.0m., res US $6.2m., dep. US $97.9m. (Dec. 1997); Chair. Dr CARLO SGANZINI; Gen. Man. URS FREI.

**Pictet Bank and Trust Ltd** (Switzerland): Building No. 1, Bayside Executive Park, West Bay St and Blake Rd, POB N-4837, Nassau; tel. 302-2222; fax 327-6610; e-mail pbtbah@bahamas.net.bs; internet www.pictet.com; f. 1978; cap. US $1.0m., res US $10.0m., dep. US $126.2m. (Dec. 1995); Chair. CHRISTIAN MALLET; Pres. and Man. YVES LOURDIN.

**Private Investment Bank Ltd:** Devonshire House, Queen St, POB N-3918, Nassau; tel. 302-5950; fax 302-5970; f. 1984 as Bank Worms and Co International Ltd; renamed in 1990, 1996 and 1998; in 2000 merged with Geneva Private Bank and Trust (Bahamas) Ltd; cap. US $3.0m., res US $12.0m., dep. US $151.3m. (Dec. 2003); Chair. and Dir FRANÇOIS ROUGE.

**Royal Bank of Canada Ltd** (Canada): 4th Floor, Royal Bank House, East Hill St, POB N-7549, Nassau; tel. 356-8500; fax 328-7145; internet www.royalbank.com; f. 1869; Chair. ALLAN TAYLOR; Vice-Pres. MICHAEL F. PHELAN; 16 brs.

**Scotiabank (Bahamas) Ltd** (Canada): Scotiabank Bldg, Rawson Sq., POB N-7518, Nassau; tel. 356-1400; fax 322-7989; e-mail scotiabank@batelnet.bs; internet www.scotiabank.com; Man. Dir ANTHONY C. ALLEN; Pres. BRUCE BIRMINGHAM; 18 brs.

# THE BAHAMAS

**SG Hambros Bank and Trust (Bahamas) Ltd** (United Kingdom): SG Hambros Bldg, West Bay St, POB N-7788, Nassau; tel. 302-5000; fax 326-6709; e-mail dominique.lefevre@sghambros.com; internet www.sghambros.com; f. 1936, above name adopted in 1998; cap. B $2.0m., res B $10.3m., dep. B $461.3m. (Dec. 2003); Chair. WARWICK NEWBURY; Man. Dir DOMINIQUE LEFEVRE.

**UBS (Bahamas) Ltd** (Switzerland): UBS House, East Bay St, POB N-7757, Nassau; tel. 394-9300; fax 394-9333; internet www.ubs.com/bahamas; f. 1968 as Swiss Bank Corpn (Overseas) Ltd, name changed as above 1998; cap. US $4.0m., dep. US $420.2m. (Dec. 1997); Chair. MARTIN LIECHTI; CEO ERIC TSCHIRREN.

### Principal Bahamian Trust Companies

**Ansbacher (Bahamas) Ltd:** Ansbacher House, Bank Lane, POB N-7768, Nassau; tel. 322-1161; fax 326-5020; e-mail info@ansbacher.bs; internet www.ansbacher.com; incorporated 1957 as Bahamas International Trust Co Ltd, name changed 1994; cap. B $1.0m., res B $9.7m., dep. B $190.3m. (Sept. 1998); Man. Dir IAN D. TOWELL.

**Bank of Nova Scotia International Ltd:** Scotiabank Bldg, Rawson Sq., POB N-7545, Nassau; tel. 356-1517; fax 328-8473; e-mail bahamas@scotiatrust.com; internet www.scotiabank.com; wholly owned by the Bank of Nova Scotia; Vice-Pres. and Gen. Man. STEPHEN J. GRAINGER.

**Chase Manhattan Trust Corpn:** Shirley and Charlotte Sts, POB N-3708, Nassau; tel. 356-1305; fax 325-1706; Gen. Man. KEN BROWN; 4 brs.

**Leadenhall Bank and Trust Co Ltd:** IBM Bldg, East Bay and Church Sts, POB N-1965, Nassau; tel. 325-5508; fax 328-7030; e-mail drounce@leadentrust.com; f. 1976; Man. Dir DAVID J. ROUNCE.

**Oceanic Bank and Trust:** POB AP-59203, Nassau; tel. 502-8822; fax 502-8840; f. 1969; Man. Dir BRUCE BELL.

**Winterbotham Trust Co Ltd:** Bolam House, King and George Sts, POB N-3026, Nassau; tel. 356-5454; fax 356-9432; e-mail nassau@winterbotham.com; internet www.winterbotham.com; cap. US $3.8m.; CEO GEOFFREY HOOPER; 2 brs.

### Bankers' Organizations

**Association of International Banks and Trust Companies in the Bahamas:** POB N-7880, Nassau; tel. 356-3898; Chair. ANDREW LAW.

**Bahamas Financial Services Board (BFSB):** 4th Floor, Euro-Canadian Centre, POB N-1764, Nassau; e-mail info@bsfb-bahamas.com; internet www.bfsb-bahamas.com; f. 1998; jt govt/private initiative responsible for overseas marketing of financial services; CEO and Exec. Dir WENDY C. WARREN; Chair. MICHAEL PATON.

**Bahamas Institute of Bankers (BIB):** POB N-3203, Nassau; tel. 325-4921; fax 325-5674; e-mail bibroyalbah@netbahamas.com; Pres. KIM BODIE.

### STOCK EXCHANGE

**Bahamas International Securities Exchange (BISX):** British Colonial Centre of Commerce, 1 Bay St, POB EE-15672, Nassau; e-mail info@bisxbahamas.com; internet www.bisxbahamas.com; f. 2000; 19 local companies listed at January 2005; Chair. IAN FAIR; CEO KEITH DAVIES (acting).

### INSURANCE

The leading British and a number of US, Canadian and Caribbean companies have agents in Nassau and Freeport. Local insurance companies include the following:

**Allied Bahamas Insurance Co Ltd:** 93 Collins Ave, POB N-1216, Nassau; tel. 326-5439; fax 356-5472; general, aviation and marine.

**Bahamas First General Insurance Co Ltd:** 93 Collins Ave, POB N-1216, Nassau; tel. 326-5439; fax 326-5472; Man. Dir PATRICK G. W. WARD.

**Colina Insurance Co Ltd:** 12 Village Rd, POB N-4728, Nassau; tel. 393-2224; fax 393-1710; Colina Insurance Co merged with Global Life Assurance Bahamas in July 2002; operates under above name.

**Commonwealth General Insurance Co Ltd:** POB N-4200, Nassau; tel. 322-8210; fax 322-5277; Man. Dir ALBERT ARCHER.

**Family Guardian Insurance Co Ltd** (FamGuard): East Bay St, POB SS-6232, Nassau; tel. 393-1023; fax 394-1631; f. 1965.

**Royal and Sun Alliance Insurance (Bahamas) Ltd:** POB N-4391, Nassau; tel. 328-7888; fax 325-3151; Gen. Man. STEVEN WATSON.

**Security and General Insurance Co Ltd:** POB N-3540, Nassau; tel. 326-7100; fax 325-0948; Gen. Man. MARK SHIRRA.

**Summit Insurance Co Ltd:** POB SS-19028, Nassau; tel. 394-2351; fax 394-2353; Gen. Man. COLIN JONES.

### Association

**Bahamas General Insurance Association:** POB N-860, Nassau; tel. 323-2596; fax 328-4354; e-mail bgia@bahamas.net.bs; Co-ordinator ROBIN B. HARDY.

# Trade and Industry

## DEVELOPMENT ORGANIZATIONS

**Bahamas Agricultural and Industrial Corpn (BAIC):** Levy Bldg, East Bay St, POB N-4940, Nassau; tel. 322-3740; fax 322-2133; e-mail baicorp@batelnet.bs; internet www.baic.gov.bs; f. 1981 as an amalgamation of Bahamas Development Corpn and Bahamas Agricultural Corpn for the promotion of greater co-operation between tourism and other sectors of the economy through the development of small and medium-sized enterprises; Chair. MICHAEL HALITIS.

**Bahamas Financial Services Board (BFSB):** see Finance, Bankers' Organizations.

**Bahamas Investment Authority:** Cecil V. Wallace-Whitfield Centre, POB CB-10980, Nassau; tel. 327-5970; fax 327-5907; e-mail info@opm.gov.bs; internet www.opm.gov.bs/bia.php; Deputy Dir PHILIP MILLER.

**Bahamas Light Industries Development Council:** POB SS-5599, Nassau; tel. 394-1907; Pres. LESLIE MILLER.

**Nassau Paradise Island Promotion Board:** Dean's Lane, Fort Charlotte, POB N-7799, Nassau; tel. 322-8381; fax 325-8998; f. 1970; Chair. GEORGE R. MYERS; Sec. MICHAEL C. RECKLEY; 30 mems.

## CHAMBER OF COMMERCE

**Bahamas Chamber of Commerce:** Shirley St and Collins Ave, POB N-665, Nassau; tel. 322-2145; fax 322-4649; e-mail bahamaschamber@coralwave.com; internet www.bahamasb2b.com/bahamaschamber; f. 1935 to promote, foster and protect trade, industry and commerce; Exec. Dir PHILIP SIMON; 450 mems.

## EMPLOYERS' ASSOCIATIONS

**Bahamas Association of Land Surveyors:** POB N-10147, Nassau; tel. 322-4569; Pres. DONALD THOMPSON; Vice-Pres. GODFREY HUMES; 30 mems.

**Bahamas Boatmen's Association:** POB ES-5212, Nassau; f. 1974; Pres. and Sec. FREDERICK GOMEZ.

**Bahamas Contractors' Association:** POB N-8170, Nassau; Chair. GODFREY E. LIGHTBOURN; Sec. ROBERT E. MYERS.

**Bahamas Employers' Confederation:** POB N-166, Nassau; tel. 328-5719; fax 322-4649; e-mail becon@bahamasemployers.org; internet www.bahamasemployers.org; f. 1963; Pres. BRIAN NUTT.

**Bahamas Hotel Employers' Association:** SG Hambros Bldg, West Bay St, POB N-7799, Nassau; tel. 322-2262; fax 502-4221; e-mail bhea4mcr@hotmail.com; f. 1958; Pres. J. BARRIE FARRINGTON; Exec. Vice-Pres. MICHAEL C. RECKLEY; 16 mems.

**Bahamas Institute of Chartered Accountants:** Star Plaza, Mackey St, POB N-7037, Nassau; tel. 394-3439; fax 394-3629; f. 1971; Pres. L. EDGAR MOXEY.

**Bahamas Institute of Professional Engineers:** Nassau; tel. 322-3356; fax 323-8503; Pres. ANTHONY DEAN.

**Bahamas Motor Dealers' Association:** POB N-3919, Nassau; tel. 328-7671; fax 328-1922; Pres. HARRY ROBERTS.

**Bahamas Real Estate Association:** Bahamas Chamber Bldg, POB N-8860, Nassau; tel. 325-4942; fax 322-4649; Pres. PATRICK STRACHAN.

## MAJOR COMPANIES

### Construction

**Bahamas Marine Construction Co Ltd:** Bay and Victoria Sts, POB N-7512, Nassau; tel. 325-1654; fax 326-5127; e-mail bmcmail@mosko.com; internet www.mosko.com/bmc.htm; f. 1980; focuses on the construction of ports, docks and marinas; Chair. GEORGE MOSKO; Man. Dir JAMES GEORGE MOSKO; 111 employees.

**Freeport Concrete Co Ltd (FCC):** POB F-42647, Freeport; tel. 352-7511; fax 351-3669; internet www.fccbahamas.com; f. 1995; ready-mix concrete manufacturers; Pres., Dir and CEO RAY SIMPSON; Chair. HANNES BABAK.

**Mosko's United Construction Ltd:** Bay St and Victoria Ave, POB N-641, Nassau; tel. 322-2825; fax 325-2571; e-mail moskomail@mosko.com; internet www.mosko.com; f. 1958; Pres. GEORGE MOSKO; 567 employees.

# THE BAHAMAS

## Food and Beverages

**Aquapure Water Ltd:** POB SS-6244, Nassau; tel. 393-1904; fax 393-1936; f. 1975; produces purified and deionized bottled water; Dir JOHN MCSWEENEY.

**Bacardi and Co Ltd:** Millar Rd, POB N-838, Nassau; tel. 362-1412; fax 323-2371; f. 1960; distilling and rum manufacture; Pres. MANUELA J. CUTILLAS JUSTIZ; 120 employees.

**Bahama Palm Groves Ltd:** Don Mackely Blvd, POB 20096, Marsh Harbour; tel. 367-3086; fax 367-2223; f. 1986; fruit sellers; Man. Dir RANDY KEY; 287 employees.

**Bahamas Supermarkets Ltd:** East West Highway, POB N-3738, Nassau; tel. 393-2830; fax 393-1232; f. 1968; Pres. C. WINGE; Chair. JAMES KUFELDT; 700 employees.

**Butler and Sands Co Ltd:** John F. Kennedy Dr., POB N-51, Nassau; tel. 322-7586; fax 326-6655; f. 1949; distribution of alcoholic beverages; Pres. MARK A. G. FINLAYSON; CEO GARET O. FINLAYSON; 150 employees.

**Grand Bahamas Food Co:** Grand Bahama Highway, POB F-2540, Freeport; tel. 352-9801; fax 352-8870; f. 1968; food wholesalers and retailers; Pres. MINAS VARDAOULIS; 245 employees.

**Pepsi Cola (Bahamas) Bottling Co Ltd:** Soldier Rd, POB N-3004, Nassau; tel. 364-5640; fax 393-5744; f. 1991; soft-drinks manufacturer; Man. MARCIO RAMOS; 21 employees.

**Super Value Food Stores Ltd:** Golden Gates Shopping Centre, POB N-3039, Nassau; tel. 361-5220; fax 361-5583; f. 1965; Pres. RUPERT WINER ROBERTS, Jr; Vice-Pres M. A. ROBERTS, CANDY KELLY; 567 employees.

## Pharmaceuticals

**PFC Bahamas Ltd:** West Sunrise Highway, POB F-42430, Freeport; tel. 352-8171; fax 352-6950; f. 1967; manufacturer of pharmaceutical products; owned by Roche (Switzerland); Pres. JAMES WILSON; Chair. PAUL FREIMAN; 10,000 employees.

## Miscellaneous

**Baha Mar Development Co Ltd:** c/o Wyndham Nassau Resort and Crystal Palace Casino, West Bay Street, POB N-8306, Nassau; tel. 327-6200; purchased Wyndham Nassau Resort, Crystal Palace Casino, Nassau Beach Hotel and Radisson Cable Beach Resort in 2005; Pres. ROBERT HELLER.

**Bahamas Realty Ltd:** POB N-1132, Nassau; tel. 393-8168; fax 393-0326; e-mail brealty@bahamasrealty.bs; internet www.bahamasrealty.bs; f. 1978 as Caribbean Management and Sales Ltd; real estate; Pres. ROBIN B. BROWNING.

**BORCO Bahamas Oil Refining Co:** POB F-2435; West Sunrise Highway, Freeport; tel. 352-9811; fax 352-4029; e-mail jvargas@borcoltd.com; f. 1964; petroleum distribution; owned by Petróleos de Venezuela (PDVSA); Pres. JAIME VARGAS; 108 employees.

**Grand Bahama Development Co Ltd:** Pioneer Way, POB F-2666; Nassau; tel. 352-6711; fax 352-9864; f. 1961; holding co; Chair. JACK HAYWOOD; 300 employees.

**H. G. Christie Ltd:** Millar's Court, POB N-8164, Nassau; tel. 322-1041; fax 326-5642; e-mail sales@hgchristie.com; internet www.hgchristie.com; f. 1922; real estate service; Vice-Pres. JOHN CHRISTIE.

**Hutchison Port Holdings (HPH):** Headquarters Bldg, Container Port Rd, Queen's Highway, POB F-42465, Freeport; tel. 350-8000; fax 351-8044; internet www.hph.com.hk; port holding co; owned by Hutchison Whampoa of Hong Kong; subsidiaries include Grand Bahama Airport Co, Freeport Container Port and Freeport Harbour Co; Man. Dir JOHN E. MEREDITH; Exec. Dirs RICHARD PEARSON, JAMES S. TSIEN.

**J. S. R. Real Estate Ltd:** POB F-40093, Freeport; tel. 352-7201; fax 352-7203; e-mail jsrreal@batelnet.bs; f. 1957; real estate, developments, condominiums, residency and investing.

**Kerzner International Ltd:** c/o Holiday Inn, POB N-4777, Paradise Island; tel. 363-3000; South African co, owns the Atlantis Resort and the Holiday Inn on Paradise Island; Chair. SOL KERZNER; CEO BUTCH KERZNER.

**Morton Salt Co (Bahamas):** Gregory St, Matthew Town, Inagua; tel. 339-1300; internet www.mortonsalt.com; f. 1954 when West Chemical Co was amalgamated into Morton Salt Co.

**Polymers International Ltd:** POB F-42684, Freeport; tel. 352-3506; fax 352-2779; e-mail gebelhar@polymersintl.com; f. 1997; manufacturer of polystyrene products; Plant Man. JOSEPH GREGORY EBELHAR, III; 50 employees.

**Resort at Bahamia:** Mall St Sunrise, POB F-40207, Freeport; tel. 350-7000; fax 350-7003; f. 1973, renamed 2000; management of hotels; Man. TYRONE THURSTON; 500 employees.

**Resorts Int. (Bahamas) Ltd:** POB N-4777, Nassau; tel. and fax 363-3703; f. 1967; holding co involved in management and operation of hotels; Man. Dir GABRIEL SASTRE; 2,900 employees.

**Taylor Industries Ltd:** 111 Shirley St, POB N-4806, Nassau; tel. 322-8941; fax 328-0453; f. 1945; electrical appliances and supplies; Pres. and Gen. Man. DEREK TAYLOR; 87 employees.

**Templeton Global Advisers Ltd:** POB N-7759, Lyford Cay, Nassau; tel. 362-4600; fax 361-4308; f. 1986; investment consultants and security brokers; Chair. CHARLES JOHNSON; 400 employees.

## UTILITIES

### Electricity

**Bahamas Electricity Corpn (BEC):** Big Pond and Tucker Rds, POB N-7509, Nassau; tel. 325-4101; fax 323-6852; internet www.bahamaselectricity.com; f. 1956; state-owned; transfer to private ownership pending in 2005; provides 70% of the islands' power-generating capacity; Chair. AL JARRETT; Gen. Man. BRADLEY S. ROBERTS.

**Freeport Power Co Ltd:** Mercantile Bldg, Cedar St, POB F-888, Freeport; tel. 352-6611; f. 1962; privately owned; Pres. ALBERT J. MILLER.

### Gas

**Caribbean Gas Storage and Terminal Ltd:** POB N-9665, Nassau; tel. 327-5587; fax 362-5006; e-mail info@caribbeangas.com; internet www.caribbeangas.com; f. 1992; Man. Dir PETER T. NEWELL.

**Tropigas:** Nassau; tel. 322-2404.

### Water

**Bahamas Water and Sewerage Corpn (WSC):** 87 Thompson Blvd, POB N-3905, Nassau; tel. 302-5500; fax 328-3896; e-mail wscomplaints@wsc.com.bs; internet www.wsc.com.bs; f. 1960; Chair. ABRAHAM BUTLER.

## TRADE UNIONS

All Bahamian unions are members of one of the following:

**Commonwealth of the Bahamas Trade Union Congress:** Congress House, 3 Warwick St, POB CB-10992, Nassau; tel. 394-6301; fax 394-7401; e-mail tuc@bahamas.net.bs; Pres. OBIE FERGUSON Jr; 12,500 mems.

**National Congress of Trade Unions:** Horseshoe Dr., POB GT-2887, Nassau; tel. 356-7459; fax 356-7457; Pres. LEROY (DUKE) HANNA; Gen. Sec. KINGSLEY L. BLACK; 20,000 mems.

The main unions are as follows:

**Bahamas Airport, Airline and Allied Workers' Union:** Workers' House, Harold Rd, POB N-3364, Nassau; tel. 323-4491; fax 323-7086; f. 1958; Pres. FRANKLYN CARTER; Gen. Sec. PATRICIA TYNES; 550 mems.

**Bahamas Brewery, Distillers Union:** POB N-838, Nassau; tel. 362-1412; fax 362-1415; f. 1968; Pres. JOSEPH MOSS; Gen. Sec. RAFAEL HOLMES; 140 mems.

**Bahamas Communications and Public Officers' Union:** Farrington Rd, POB N-3190, Nassau; tel. 322-1537; fax 323-8719; e-mail prebcpou@batelnet.bs; f. 1973; Pres. D. SHANE GIBSON; Sec.-Gen. ROBERT A. FARQUHARSON; 2,100 mems.

**Bahamas Doctors' Union:** Nassau; Pres. Dr EUGENE NEWERY; Gen. Sec. GEORGE SHERMAN.

**Bahamas Electrical Workers' Union:** East West Highway, POB GT-2535, Nassau; tel. 393-1838; fax 356-7383; Pres. CHARLES ROLLE; Gen. Sec. PATRICIA JOHNSON.

**Bahamas Hotel Catering and Allied Workers' Union:** Harold Rd, POB GT-2514, Nassau; tel. 323-5933; fax 325-6546; f. 1958; Pres. PATRICK BAIN; Gen. Sec. LEO DOUGLAS; 6,500 mems.

**Bahamas Maritime Port and Allied Workers' Union:** Prince George Docks, POB FF-6501, Nassau; tel. 322-2049; fax 322-5445; Pres. FREDERICK N. RODGERS; Sec.-Gen. LEON WALLACE.

**Bahamas Musicians' and Entertainers' Union:** Horseshoe Dr., POB N-880, Nassau; tel. 322-3734; fax 323-3537; f. 1958; Pres. LEROY (DUKE) HANNA; Gen. Sec. PORTIA NOTTAGE; 410 mems.

**Bahamas Public Services Union:** Wulff Rd, POB N-4692, Nassau; tel. 325-0038; fax 323-5287; f. 1959; Pres. JOHN PINDER; Sec.-Gen. SYNIDA DORSETT; 4,247 mems.

**Bahamas Taxi-Cab Union:** Nassau St, POB N-1077, Nassau; tel. 323-5818; fax 326-2919; Pres. FELTON COX; Gen. Sec. ROSCOE WEECH.

**Bahamas Union of Teachers:** 104 Bethel Ave, Stapledon Gardens, POB N-3482, Nassau; tel. 323-4491; fax 323-7086; f. 1945; Pres. KINGSLEY L. BLACK; Sec.-Gen. HELLENA O. CARTWRIGHT; 2,600 mems.

**Bahamas Utility Services and Allied Workers' Union:** Nassau; Pres. HUEDLEY MOSS.

**Eastside Stevedores' Union:** POB N-2167, Nassau; tel. 322-4069; fax 364-7437; f. 1972; Pres. CURTIS TURNQUEST; Gen. Sec. TREVOR CAREY.

**Grand Bahama Construction, Refinery, Maintenance and Allied Workers' Union:** 33A Kipling Bldg, POB 42397, Freeport; tel. 352-2476; fax 351-7009; f. 1971; Pres. NEVILLE SIMMONS; Gen. Sec. MCKINLEY JONES.

**United Brotherhood of Longshoremen's Union:** Wulff Rd, POB N-7317, Nassau; f. 1959; Pres. JOSEPH MCKINNEY; Gen. Sec. DEGLANVILLE PANZA; 157 mems.

## Transport

### ROADS

There are about 966 km (600 miles) of roads in New Providence and 1,368 km (850 miles) in the Family Islands, mainly on Grand Bahama, Cat Island, Eleuthera, Exuma and Long Island. In 1999 57.4% of roads were paved.

### SHIPPING

The principal seaport is at Nassau (New Providence), which can accommodate the very largest cruise ships. Passenger arrivals exceed 2m. annually. The other main ports are at Freeport (Grand Bahama), where a container terminal opened in 1997, and Matthew Town (Inagua). There are also modern berthing facilities for cruise ships at Potters Cay (New Providence), Governor's Harbour (Eleuthera), Morgan's Bluff (North Andros) and George Town (Exuma).

The Bahamas converted to free-flag status in 1976. The fleet's aggregate displacement was 35,388,244 grt in December 2004 (the third largest national fleet in the world).

There is a weekly cargo and passenger service to all the Family Islands.

**Bahamas Maritime Authority:** POB N-4679, Nassau; tel. 323-3130; fax 323-2119; internet www.bahamasmaritime.com; e-mail nassau@bahamasmaritime.com; f. 1995; promotes ship registration and co-ordinates maritime administration.

**Freeport Harbour Co Ltd:** POB F-42465, Freeport; tel. 352-9651; fax 352-6888; e-mail fhcol@batelnet.bs; Gen. Man. MICHAEL J. POWER.

**Nassau Port Authority:** Prince George Wharf, POB N-8175, Nassau; tel. 326-7354; fax 322-5545; e-mail aja@batelnet.bs; regulates principal port of the Bahamas; Port Dir ANTHONY ALLENS.

#### Principal Shipping Companies

**Bahamas Fast Ferries:** Bahamas Ferry Services Ltd, POB N-3709, Nassau; tel. 323-2166; fax 322-8185; e-mail info@bahamasferries.com; internet www.bahamasferries.com; services Spanish Wells, Harbour Island, Current Island and Governors Harbour in Eleuthera, Morgan's Bluff and Fresh Creek in Andros and George Town in Exuma; Chair. CRAIG R. SYMONETTE.

**Cavalier Shipping:** Arawak Cay, POB N-8170, New Providence; tel. 328-3035.

**Dockendale Shipping Co Ltd:** Dockendale House, West Bay St, POB N-10455, Nassau; tel. 325-0448; fax 328-1542; e-mail dockship@dockendale.com; f. 1973; ship management; Man. Dir L. J. FERNANDES; Tech. Dir K. VALLURI.

**Eleuthera Express Shipping Co:** POB N-4201, Nassau.

**Grand Master Shipping Co:** POB N-4208, Nassau.

**Grenville Ventures Ltd:** 43 Elizabeth Ave, POB CB-13022, Nassau.

**HJH Trading Co Ltd:** POB N-4402, Nassau; tel. 392-3939; fax 392-1828.

**Gladstone Patton:** POB SS-5178, Nassau.

**Pioneer Shipping Ltd:** Union Wharf, POB N-3044, Nassau; tel. 325-7889; fax 325-2214.

**Teekay Shipping Corporation:** TK House, Bayside Executive Park, West Bay St & Blake Rd, POB AP-59212, Nassau; tel. 502-8820; fax 502-8840; internet www.teekay.com; petroleum transporting; Chair. C. SEAN DAY; Pres. and CEO BJORN MOLLER.

**Tropical Shipping Co Ltd:** POB N-8183, Nassau; tel. 322-1012; fax 323-7566.

**United Shipping Co Ltd:** POB F-42552, Freeport; tel. 352-9315; fax 352-4034; e-mail info@unitedship.com; internet www.unitedship.com.

### CIVIL AVIATION

Nassau International Airport (15 km—9 miles—outside the capital) and Freeport International Airport (5 km—3 miles—outside the city, on Grand Bahama) are the main terminals for international and internal services. There are also important airports at West End (Grand Bahama) and Rock Sound (Eleuthera) and some 50 smaller airports and landing strips throughout the islands. A B $30m. airport improvement programme was completed in 2004.

**Bahamasair Holdings Ltd:** Coral Harbour Rd, POB N-4881, Nassau; tel. 377-8451; fax 377-7409; e-mail astuart@bahamasair.com; internet www.bahamasair.com; f. 1973; state-owned, scheduled for privatization by Sept. 2005; scheduled services between Nassau, Freeport, destinations within the USA and 20 locations within the Family Islands; Chair. ANTHONY MILLER; Man. Dir PAT ROLLE (acting).

## Tourism

The mild climate and beautiful beaches attract many tourists. In 2004 tourist arrivals increased by an estimated 1.5%, compared with the previous year, to some 1,450,000 (excluding 3,360,000 visitors by sea). The majority of stop-over arrivals (86.4% in 2003) were from the USA. Receipts from the tourism industry stood at B $1,855m. in 2004. In September 1999 there were 223 hotels in the country, with a total of 14,080 rooms.

**Ministry of Tourism:** British Colonial Hilton Hotel, Bay St, POB N-3701, Nassau; tel. 322-7500; fax 328-0945; e-mail tourism@bahamas.com; internet www.bahamas.com; Dir-Gen. VINCENT VANDERPOOL-WALLACE.

**Bahamas Hotel Association:** Dean's Lane, Fort Charlotte, POB N-7799, Nassau; tel. 322-8381; fax 326-5346; e-mail bhainfo@batelnet.bs; Pres. JEREMY MACVEAN COX; Exec. Vice-Pres. BASIL H. SMITH.

**Bahamas Tourism and Development Authority:** POB N-4740, Nassau; tel. 326-0992; fax 323-0993; e-mail linkages@batelnet.bs; f. 1995; Exec. Dir FRANK J. COMITO.

**Hotel Corporation of the Bahamas:** West Bay St, POB N-9520, Nassau; tel. 327-8395; fax 327-6978; Chair. GEOFFREY JOHNSTONE; Chief Exec. WARREN ROLLE.

## Defence

The Royal Bahamian Defence Force, a paramilitary coastguard, is the only security force in the Bahamas, and numbered 860 (including 70 women) in August 2004. The defence budget includes expenditure on the 2,300-strong police force and is mostly used to finance the campaign against drugs-trafficking.

**Defence Budget:** B $30m. in 2004.

**Commodore:** DAVID ROLLE.

## Education

Education is compulsory between the ages of five and 16 years, and is provided free of charge in government schools. There are several private and denominational schools. Primary education begins at five years of age and lasts for six years. Secondary education, beginning at the age of 11, also lasts for six years and is divided into two equal cycles. In 2000 98% of children in the relevant age-group were enrolled at primary level. In the same year 89% of children in the relevant age-group were enrolled at secondary level. The University of the West Indies has an extra-mural department in Nassau, offering degree courses in hotel management and tourism. The Bahamas Hotel Training College was established in 1992. The Bahamas Law School, part of the University of the West Indies, opened in 1998. Technical, teacher-training and professional qualifications can be obtained at the two campuses of the College of the Bahamas.

Government expenditure on education in 2004/05 was budgeted at B $201.8m. (or 17.1% of total spending from the General Budget).

# Bibliography

For works on the Caribbean generally, see Select Bibliography (Books)

Block, A. A. *Masters of Paradise: Organized Crime and the Internal Revenue Service in the Bahamas*. Somerset, NJ, Transaction Publishers, 1997.

Craton, M., and Saunders, G. *Islanders in the Stream: A History of the Bahamian People: From the Ending of Slavery to the Twenty-First Century*, Vol. I. Athens, GA, University of Georgia Press, 1998.

*Islanders in the Stream: A History of the Bahamian People: From the Ending of Slavery to the Twenty-First Century*, Vol II. Athens, GA, University of Georgia Press, 2000.

Culmer Jenkins, O., and Saunders, G. *Bahamian Memories*. University Press of Florida, Gainesville, FL, 2000.

Eneas, G. *Agriculture in the Bahamas: Historical Development 1492–1992*. Nassau, Media Publishing Ltd, 1998.

Howard, R. *Black Seminoles in the Bahamas*. Gainesville, FL, University Press of Florida, 2002.

Johnson, H. *Bahamas—Slavery to Servitude*. Gainesville, FL, University Press of Florida, 1997.

Johnson, W. B. *Race Relations in the Bahamas 1784–1834: The Nonviolent Transformation from a Slave to a Free Society*. Fayetteville, AK, University of Arkansas Press, 1999.

Keegan, W. F. *The People Who Discovered Columbus*. Gainesville, FL, University Press of Florida, 1992.

Kelly, R. C., Ewing, D., Doyle, S., and Youngblood, D. *Country Review, Bahamas, 1998/1999*. Commercial Data International, 1998.

McCulla, P. E. *Bahamas*. Broomall, PA, Chelsea House Publrs, 1998.

# BARBADOS

## Geography

### PHYSICAL FEATURES

Barbados lies in the Lesser Antilles, between the Atlantic Ocean and the Caribbean Sea. It is the most easterly of the West Indian islands, and its nearest neighbour is Saint Vincent and the Grenadines, about 160 km (100 miles) to the west. The island nation has an area of 430 sq km (166 sq miles).

The wider southern part of the island of Barbados continues in the north-west, tapering northwards. The terrain is generally flat, limestone scored by deep, vegetation-filled gullies, gently rising into a central highland area, especially in the north-east, where Mt Hillaby reaches 336 m (1,103 ft). The 97 km (156 miles) of coast are more rugged in the east and north, protecting the island from the worst of the oceanic weather and sheltering the gentler waters and beaches of the west. Reefs surround much of Barbados. There are rivers in the north-eastern Scotland District, where the ancient limestone cap has been eroded to expose even older rocks. The well-watered island is fertile, and two types of flora warrant particular mention: the typical local fig trees, with their aerial roots, which are believed to be the origin of the island's name (Los Barbados, 'the bearded ones', of Pedro a Campos of Portugal in 1536); and the grapefruit, which originates on Barbados. Wildlife includes the Barbados green monkey (vervet monkeys originally transported from West Africa) and the red-footed tortoise.

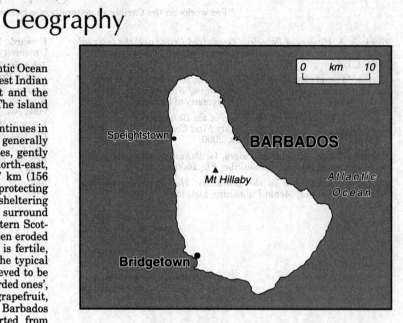

### CLIMATE

The climate is tropical, but exposure to the Atlantic alleviates the extremes. Indeed, when the wind is strong off the ocean, certainly in the highlands it can get chilly. Generally, the average temperature is 27°C (81°F). The rainy season is between June and October, when there are likely to be hurricanes, although Barbados is on the edge of the region prone to them.

### POPULATION

The people of the island, Barbadians (Bajans), are now predominantly black (89% at the 1990 census), but there is still a white community (3%), with the balance comprising mixed-race and Asian groups. Most of the people are Christians, the largest denomination being the Anglican Church, claiming the adherence of 40% of the population, followed by Pentecostalists with 8%, Methodists 7% and all the other Protestant groups 12%. Roman Catholics account for 4% of the islanders. More than 140 religions and sects are represented on the island, although an unusually high (for the West Indies) proportion of the population, 17%, professes no religion at all. There are small Muslim and Jewish communities—the synagogue is built on the site of one of the first two built in the Americas. English is the universal and official language.

The total population was estimated at 269,000 in mid-2004, making Barbados one of the most densely populated countries in the Caribbean. The capital and main port, Bridgetown, is on the south-western coast. On the coast in the north-west, Speightstown is the chief commercial centre for the north of the island. There are currently 11 parishes, and it is being debated whether to give the capital separate parish status.

## History

### MARK WILSON

Based on an earlier article by Prof. A. E. Thorndike

The Amerindians who settled Barbados from around AD 350 left the island during the 16th century, so that the first British settlers, who arrived in 1627, found no indigenous inhabitants. Barbados remained under British sovereignty until political independence in 1966, thereby earning itself the sobriquet 'Little England', and Barbadians played an important role in the settlement and administration of Britain's other Caribbean possessions. The first British settlers were smallholders growing tobacco and other crops, using the labour of indentured servants. However, fundamental change came with the introduction in the 1640s of sugar cane, by Dutch merchants, who brought plants from the Dutch settlements in Brazil. Sugar production required considerable labour and capital for the manufacturing process, a large work-force and extensive acreage, so large estate-owners supplanted the smallholders and, increasingly, slaves (of African origin) replaced the European servants. Although the first slaves arrived in 1627, they were few in number until the 1640s. By 1655 slaves formed 47% of a population of 43,000 and by 1712 they formed 77% of a total population of some 54,500. Many whites moved on to other British settlements in the Caribbean or on to the American mainland. Those who stayed, unless they were landowners, became craftsmen, overseers or merchants, or in some cases led a marginal and socially isolated existence as 'poor whites'. The black slave population was harshly treated and there were attempted slave revolts in 1675, 1692, 1702 and 1816. Slavery was eventually abolished in 1834, but its legacy was a highly stratified class-based society, still based, to some extent, on gradations of colour.

The British settlers established a House of Assembly in 1639 to represent their interests. Based on the 'representational system', the franchise was strictly limited by a property qualification. *De facto* power was exercised by the House of Assembly through its control of the public purse; hence it was able to hinder any attempt at reform by successive Governors. In 1876 the British Governor, John Pope Hennessey, proposed the establishment of a confederation to link Barbados and the Windward Islands. This suggestion was resisted by the Assembly, but was seen by many blacks as a partial solution to their difficulties. Eight people were killed in the ensuing Confederation Riots, and Pope Hennessey was subsequently transferred to Hong Kong. The first mixed-race member of the Assembly, Samuel Jackman Prescod, was elected in 1843, and the franchise was significantly widened in 1884. However, even the reformed property qualification continued to exclude the majority of blacks from the franchise. In 1856 the Assembly introduced district medical officers, and a Board of Education was formed in 1878, under the influence of the Anglican bishop, John Mitchinson. A non-white professional middle-class emerged during the 19th century; Sir Conrad Reeves, a mixed-race politician and lawyer, was Chief Justice from 1886 to 1902. However, further political and social advance was to wait until the rise of the labour movement in the 1930s.

Charles Duncan O'Neal founded the Democratic League, influenced by Fabian principles, in 1924. Its first member, Chrissie Brathwaite, was elected to the Assembly in the same year, and the Workingmen's Association was founded in 1926. The poor economic climate and the impoverished condition of most Barbadians (Bajans) led, as was also the case in most of the other British West Indian territories, to labour disturbances. In July 1937 14 people were killed and 47 injured in island-wide riots. A later commission of inquiry expressed no surprise at the disturbances, once the inequalities in Barbadian society at the time were revealed. The Barbados Progressive League was founded in 1938, its leaders including Grantley (later Sir Grantley) Adams. It gained five seats in the House of Assembly in 1940 and was strengthened considerably by an alliance with the Barbados Workers' Union (BWU), founded by Adams and Hugh Springer. In 1943 the League successfully campaigned for an extension of the franchise. In the 1944 general election the League won seven seats, with eight won by Wynter Crawford's more radical West Indian National Congress party and eight by the traditionalist Electors' Association, established by the landowning and merchant élite. Adams and other elected members subsequently joined the Executive Committee, the principal policy-making instrument. The League was renamed the Barbados Labour Party in 1946, and made gains at the expense of both other parties in the elections held in 1946 and 1948.

Universal adult suffrage was introduced in 1950 and in the general election held in 1951 the BLP won 16 of the 24 seats. Ministerial government was introduced in 1954 and Adams was appointed Premier. He subsequently became Prime Minister of the West Indies Federation, from January 1958 until its dissolution in 1962, and was succeeded as Premier of Barbados by Dr Hugh Cummins.

Following the 1951 election victory, those who favoured a more socialist approach, such as Errol Barrow, became disenchanted with those, like Adams, who favoured gradualist policies. In 1955 a small group, later named the Democratic Labour Party (DLP), led by Barrow, split from the BLP and joined forces with former members of the Congress Party. The BLP won 15 seats in the subsequent 1956 election, with the DLP and the Progressive Conservative Party each obtaining four. In October 1961 full internal self-government was granted and, in the ensuing general election, the DLP won 16 seats to the BLP's five, with one Independent seat and four for the traditionalist Barbados National Party. Britain tried to promote an association of Barbados and the neighbouring Leeward and Windward Islands, following the collapse of the West Indies Federation in 1962. However, this attempt was not successful, and Barrow led Barbados to separate independence in November 1966 and became its first Prime Minister. Thereafter, a two-party system, based on the two Labour parties, prevailed.

The DLP was ousted from power in 1976 and the BLP leader, J. M. G. M. 'Tom' Adams, son of Sir Grantley Adams, became Prime Minister. The BLP also won the 1981 election, securing 17 of the 27 seats in the newly enlarged House of Assembly. Adams played a leading role in support of the US military intervention in Grenada in 1983. He died suddenly in 1985 and was succeeded by Bernard St John, under whose leadership the BLP was heavily defeated by the DLP in the May 1986 election, when the BLP won only three seats. St John lost his seat and a former Minister of Foreign Affairs, Henry (later Sir Henry) Forde, assumed the BLP leadership. Errol Barrow once again became Prime Minister. The underlying reason for the Government's defeat lay in the country's past history. The issue of racism and 'white power' was never far from the surface; the BLP had become too closely identified, in many people's view, with the light-coloured business élite, from which it received substantial funds. Barrow, on the other hand, appealed to the black population and promised tax reforms to aid the black middle class. There was also dissatisfaction with the BLP's strong identification with US policy in the region, which offended nationalist sensibilities. Barrow's stand on this and other regional issues made him an imposing political force. His sudden death in June 1987 was, however, not entirely unexpected, given his refusal of medical advice pertaining to his strenuous work schedule. He was succeeded by his one-time deputy, Erskine Sandiford.

Sandiford's first political test was the resignation, in September 1987, of the Minister of Finance, Dr Richard (Richie) Haynes, who saw himself as a possible future Prime Minister. At the same time, the economic situation began to deteriorate, partly as a result of Haynes' tax measures and partly as a result of the closure in 1986 of the island's main manufacturing operation, a semi-conductor plant. In 1989 Haynes formed the National Democratic Party (NDP), with three other DLP parliamentarians, thus displacing Forde as the official Leader of the Opposition. In the general election of January 1991 the BLP won 10 seats to the DLP's 18. The fortunes of the BLP were restored with those of the two-party system; all four NDP members lost their seats. Support for the Sandiford regime declined rapidly thereafter, as austerity measures were introduced as a condition of IMF assistance to the economy, made necessary by a serious economic crisis. This situation was exploited to great effect by a rejuvenated BLP, led by a dynamic new leader, Owen Arthur. At the general election of September 1994 the BLP won 19 of the 28 seats, while the DLP retained only eight and the NDP regained one, although the DLP and the NDP together secured more votes than the BLP. Arthur became Prime Minister.

Arthur, a former professional economist, promoted economic recovery and international competitiveness in order to reduce the high levels of unemployment in the country, and economic growth resumed from 1993. He proved himself capable of populist gestures, passing a constitutional amendment forbidding future reductions in pay for public employees. He also significantly broadened the BLP's support base, retaining its links with the business community while at the same time attracting a considerable number of nationalist intellectuals and trade unionists. In 1997 a public holiday to mark Emancipation Day was declared, and a National Heroes' Day was introduced on the anniversary of the birth of Sir Grantley Adams. The anniversary of the 1937 Confederation Riots was declared a 'day of national significance', and a pilot project was commenced towards the teaching of black studies in primary and secondary schools.

In December 1998 a constitutional commission, chaired by former premier Forde, recommended the replacement of the British monarch as head of state with a ceremonial President. It also proposed changes in the composition of the Senate, provision for peoples' initiatives to introduce new legislation and stronger constitutional support for the Auditor-General and the Public Accounts Committee.

Support from across the political spectrum and a rapidly growing economy helped the BLP to gain an unprecedented victory in the general election of 20 January 1999, receiving 64.8% of the total votes cast and winning 26 of the 28 seats in the House of Assembly. The opposition DLP emerged from the election severely weakened, gaining 35.1% of the votes cast and only two seats; the party was further weakened from September 2001 by a dispute between its two members of parliament over the party leadership. An appointed senator, Clyde Mascoll, was chosen as party president, to keep the peace between the two elected members.

In August 2000 Arthur announced that a referendum would be held on the replacement of the monarch with a republic, a proposal which had the support of all political parties. However, a series of sharp political controversies in Trinidad and Tobago over the constitutional powers of the President in 2000–02 led to an enhanced appreciation of the need for careful consideration of the relationship between an elected Government and a ceremonial President. In early 2002 the Deputy Prime Minister and Minister of Foreign Affairs and Foreign Trade, Billie Miller, announced that new constitutional legislation would be drafted by the end of the year; however, the only change made was an amendment approved in September to override human rights judgments by the Privy Council, making it easier to make use of the death penalty.

A general election was held on 21 May 2003, one year ahead of the consitutional deadline. With a strong economic record, in spite of a recent downturn in tourism, Owen Arthur led the BLP to a third successive electoral victory. However, the DLP increased its share of the popular vote to 44% and its parliamentary strength to seven; moreover, the Government held a further four seats only by narrow margins. Some voters were concerned over allegations of serious mismanagement in tourism, waste disposal and other public-sector projects, and were at the same time concerned to temper the perceived arrogance of the BLP Government, while expecting it to remain in office. Mascoll was appointed opposition leader, ending the DLP's leadership dispute, and allowing it to present itself as a more credible alternative. In August Reginald Toppin, the Minister of Commerce, Consumer Affairs and Business Development, resigned. Senator Lynette Eastmond was appointed as successor to Toppin, who gave no reason for his abrupt departure from government.

Constitutional questions moved back onto the agenda in 2005. In February the Prime Minister again promised a referendum on the monarchy. Meanwhile, legislation enacted in May made Barbados one of only two countries to make the new Caribbean Court of Justice (inaugurated in the previous month in Trinidad and Tobago) its final court of appeal, replacing the Privy Council in London, United Kingdom.

Barbados played a leading role in the early 2000s in moves by 'offshore' financial centres to block initiatives by the Organisation for Economic Co-operation and Development (OECD, based in Paris, France) against tax 'havens', and by the intergovernmental Financial Action Task Force on Money Laundering (FATF, also based in Paris—see Economy). In January 2001 Barbados hosted a joint OECD-Commonwealth conference on eliminating harmful tax regimes. Barbados has an active anti-money-laundering policy. The country co-chaired a joint working group on money-laundering, established as a result of an initiative at the Commonwealth Finance Ministers' meeting in Malta in September 2000. This was a considerable diplomatic success for Barbados, both regionally and internationally.

A dispute with Trinidad and Tobago over the delimitation of the Exclusive Economic Zone and fishing rights for Barbadian vessels within Tobago waters escalated in early 2004, after almost 10 years of intermittent negotiations. Although fisheries issues have attracted most attention, considerable importance is also attached to geological structures close to the probable boundary line between the countries that may contain significant quantities of oil and natural gas. Barbados in February 2004 imposed a temporary licensing requirement on some Trinidad and Tobago exports, and referred the maritime dispute to the International Tribunal for the Law of the Sea in Hamburg, Germany. Although this forum was expected to take some years to resolve the dispute, bilateral relations with Trinidad and Tobago had been normalized in other respects by the second half of 2004.

# Economy

## MARK WILSON

Barbados is more economically and socially developed than most of the English-speaking Caribbean, with a per-head gross domestic product (GDP) of US $10,325 in 2004, among the highest in the region. The island functions as an air-transport hub for the eastern Caribbean and is the site of the headquarters of several regional organizations. With high standards of education, infrastructure and health care, in 2004 the island was ranked higher than any other economy in Latin America and the Caribbean by the Human Development Index, produced by the UN Development Programme. Barbados ranked 29th worldwide, behind the members of the Organisation for Economic Co-operation and Development (OECD), but ahead of any other country in the region. This was all the more remarkable because the country had one of the largest population densities in the world (some 625.6 per sq km in mid-2004). A US $213m. programme for curriculum reform and the introduction of technology into education, known as Edutech 2000, was scheduled for completion by 2010.

GDP grew by an average of 1.0% per year in 1990–2003; the economy contracted by 0.5% in 2002 as a result of a sharp decline in the tourism sector, but following a revival in tourism demand growth resumed in 2003 and 2004, when GDP increased by 2.0% and 3.7%, respectively. With the foreign-reserves position strong, the Government was able to continue with a counter-cyclical infrastructural investment programme to deal with the downturn, which included major investments in the port and airport. A sewage system to serve the south coast was also completed. In addition, the Government privatized a number of public-sector companies, including the Barbados National Bank and the Insurance Corporation of Barbados.

As a result of the economy's underlying strength, the recession of 2001–02 was barely noticed by most Barbadian consumers, and the recovery was rapid. In December 2003 the unemployment rate was 10.0%, significantly below the March 1993 figure of 26.2%, and only slightly above the 9.3% rate recorded at the beginning of 2001; by the end of 2004, the rate was down to 8.7%. Although some consumer prices are high by international standards, the underlying rate of inflation was modest from 1992. Prices rose sharply at the start of 1997, when the introduction of a value-added tax (VAT) increased the annual inflation rate to 7.7%. Thereafter, however, annual inflation averaged 1.3% per year until 2003; inflation in 2004 remained at the same level.

In the 1990s successive Governments took a reasonably cautious approach to budgeting and overall fiscal deficits remained under control, with the current account largely in surplus. A 1996 deficit equivalent to 3.0% of GDP was reduced in 1997 to 1.1% with the introduction of the VAT. The deficit remained low in 1998–2000, but increased sharply in 2001 and 2002, to 3.6% and 5.9%, respectively, decreasing (according to preliminary estimates) to 4.5% in 2003 and 2.4% in 2004; the Central Bank projected a 2.7% deficit for 2005. The IMF noted, however, in May 2004 that off-balance-sheet spending on improvements to the airport and on the new Hilton hotel was equivalent to 4.0% of GDP, thus substantially increasing the reported deficit. The ratio of public debt to GDP rose to 76% in January 2005 compared with a figure of 63% in 2001; the increase would have been greater but for the 2004 sale of the state's remaining majority holding in the Barbados National Bank to Republic Bank (of Trinidad and Tobago). This rise was owing to a slower than expected increase in tax revenue as a result of a downturn in the local economy. With the reserves position strong, the Government chose to continue with its capital programmes in 2001–03, in order to maintain employment and complete infrastructural improvements in time for the expected phase of renewed growth in the tourism sector. The planned sale of a 51% stake in the Insurance Corporation of Barbados to BF&M Ltd of Bermuda for Bds $66m. was expected to impact positively on the debt balance and the balance-of-payments capital account.

The balance-of-payments current account was in surplus from 1992 to 1996, with high and rising earnings from tourism and other services compensating for a permanent negative balance on merchandise trade. Continuing economic growth from 1997, however, brought about a rapid increase in imports of consumer and capital goods, resulting in a current-account deficit, which rose from US $49.3m. in 1997 to $147.6m. in 1999, before falling back slightly to $110.6m. in 2001 with reduced domestic demand, and then increasing to an estimated $277.8m. in 2004, a level equivalent to 9.9% of GDP and considered by the Government to be unsustainable. Investment in tourism-related construction projects, as well as official borrowing, brought about strong capital inflows, with a surplus on the capital account of the balance of payments averaging $222.4m. per year in 1999–2002, rising to $302.1m. in 2003 before decreasing to $104.7m. in 2004.

From the 1990s net international reserves increased strongly. From a low point of Bds $38.9m. in 1991, reserves reached a high of Bds $1,510.2m. in May 2002. Tourism receipts and precautionary borrowing on international financial markets of US $100m. in June 2000 and US $150m. in December 2001 helped the healthy growth. Reserves reached a new all-time high of Bds $1,539.3m. in March 2004, falling back to Bds $1,159.4m. at the end of the year as seasonal tourism inflows weakened and import levels remained high; this last reserves figure included Bds $183.4m. in 'second tier' reserves, held overseas by Barbados financial institutions on condition that they will be repatriated if necessary at the request of the Central Bank. In spite of its apparent strength, there are underlying concerns over the health of the economy. The Standard and Poor's agency in August 2004 reduced its rating for Barbados to BBB+ for long-term foreign-currency debt and A- for long-term domestic-currency debt. Concerns include rising debt ratios, the high level of off-balance-sheet spending and the weakness of the sugar and manufacturing sectors.

The period of positive economic growth that began in 1993 had been preceded by economic stagnation from the mid-1980s, which had forced the Government to seek the assistance of the IMF. This downturn had followed two decades of political stability and economic prosperity. In 1991, as a precondition of an IMF agreement, the Government was forced to introduce a severe austerity programme, including a stabilization tax and an 8% reduction in pay for public-service workers. This coincided with a period of severe difficulties for the sugar industry (until the 1970s the principal foreign-exchange earner), which recorded low production. A downturn in tourism (which from the 1970s had replaced the sugar industry as the principal economic sector) was also experienced, associated with a recession in North America. With stricter macroeconomic policies, there were sharp increases in interest rates and in unemployment. While the stabilization programme provided immediate relief to the economy, recovery began only in 1993, when economic conditions improved in Barbados' main tourist and export markets. In 1993 the Government, the private sector and the trade unions agreed a tripartite two-year protocol covering wages, prices and conditions of employment, which reflected a high degree of consensus regarding social and economic goals. New protocols agreed in 1995, 1998 and 2002 had considerable symbolic importance in cementing the national traditions of consensus and compromise; in 2003 the Government proposed to entrench this consultative process in the Constitution.

## TOURISM

Barbados has a well-established tourism industry, employing 10.6% of the working population in 2001. Following a decline in the sector in the early 1990s, brought about by poor market conditions, a fall in charter-plane services and an increasingly poor reputation for value for money, the industry entered a period of strong growth; stop-over tourist arrivals increased by 34.3% between 1992 and 2000, to reach 545,027 in the latter year, the highest number ever recorded. As a result of uncertain market conditions in the USA and a decrease in the local purchasing power of sterling for tourists from the United Kingdom, the number of stop-over arrivals decreased to 497,899 in 2002, but recovered to a new high of 551,502 in 2004. Tourism receipts accounted for 12.4% of GDP in 2004 and in 2003 contributed 54.4% of foreign-exchange earnings from goods and non-factor services. Barbados is increasingly important as a home port for cruises in the eastern Caribbean, and consequently cruise traffic brings more benefits to the local economy than to some other islands. The number of cruise-ship passengers rose from 99,168 in 1984 to 721,270 in 2004 and helped to propel tourism receipts to an estimated US $767m. in 2003.

Labour costs in Barbados were much higher than in some competing destinations and the landscape less immediately striking. However, the island enjoyed a high percentage of repeat visitors. All-inclusive holiday packages accounted for a relatively small market share, thus encouraging spending outside hotel premises. The closure of the Barbados Hilton for reconstruction in 1999 meant that no major international chains were represented on the island for some years, and there were no properties with more than 300 rooms; however, a new Bds $117m. 350-room Hilton was opened in May 2005. Although several other major projects were planned or under construction by mid-2005, in 2003 two mid-market hotels with a total of 350 rooms closed because of financial difficulties, temporarily reducing the island's available room stock. Barbados' tourism has been particularly successful in the United Kingdom, which in 2004 accounted for some 38.8% of tourist arrivals. The lack of large brand-name hotels was partly responsible for Barbados' poor performance in North America, which also impeded the development of conference and incentive tourism. However, there was some reversal in this trend in 2002–03, following special discounting promotions in the US market. As a result, US tourists comprised 23.5% of the total in 2004 (compared with 20.6% in 2000).

Tourist accommodation was, from the 1950s, concentrated on the sheltered south and west coasts, which were highly urbanized. By the 1990s further expansion depended largely on the development of inland or east-coast sites, with the use of sports facilities to compensate for the distance from the sea, or on the redevelopment of sites which were already used for tourism or other urban purposes, such as the reconstruction of the Barbados Hilton on the south coast, along with the planned redevelopment of the adjacent former Mobil petroleum refinery site for tourism. Attempts were made to attract a major international chain for this project. Several major proposals were under development in mid-2005, including a mixed tourism development in central Bridgetown, golf-based developments in eastern and northern Barbados, a retirement village (also in northern Barbados) and a hotel proposal for the former Cunard Paradise Beach property on the west coast, which had been empty since the late 1990s.

## FINANCIAL AND INFORMATION SERVICES

Barbados has a well-developed local banking and insurance industry. The largest regional commercial bank, FirstCaribbean, is headquartered in Barbados. With assets of US $8,650m. in 2003, it was formed in 2002 through a merger of the Caribbean interests of the Canadian Imperial Bank of Commerce and Barclays Bank. Sagicor, a major insurance company with interests throughout the region and assets in 2004 of US $1,570m., was also formed in 2002 through the merger of the two largest Barbados-based insurers, followed by a demutualization exercise. There was a small local securities exchange, on which 20 locally based companies had a market capitalization of Bds $10,782m. in June 2005, with one additional company listed on the so-called 'small cap' market. In addition, four Trinidadian and Jamaican companies had cross-listings on the Barbados exchange, bringing total capitalization to Bds $21,332m.

The 'offshore' financial sector, mainly specializing in insurance, was encouraged by the negotiation of double taxation agreements with other countries, although the sector did not operate on the same scale as that of the Bahamas, Bermuda or the Cayman Islands. However, it was a significant employer and foreign-exchange earner; under Barbados regulations a high proportion of 'offshore' entities maintained an office and employed local staff, rather than existing simply on paper. Barbados has an extensive network of double taxation and tax-information exchange treaties, and a fairly high standard of financial regulation. There is an active anti-money-laundering regime, which was further strengthened by new legislation in 1998 and 2001, while bank supervision was reinforced in 2002. Active regulation discouraged more dubious clients, but attracted businesses such as captive insurance companies. An International Business Companies (IBCs) Act, passed in 1991, resulted in the registration of 4,635 IBCs by November 2004;

additionally, there were 413 exempt insurance companies and 53 'offshore' banks with total assets of US $32,000m.

International developments at the turn of the 21st century placed all 'offshore' financial services sectors under increasing pressure. In July 1999 a World Trade Organization (WTO) dispute panel ruled in favour of a complaint by the European Union (EU) that tax exemptions for US foreign sales corporations (FSCs) amounted to an export subsidy worth US $2,000m. annually; legislation providing specifically for FSCs was repealed in 2000. The USA was instructed to reform its regulations by October 2000. This was of particular concern to Barbados, which is one of a very small number of 'offshore' centres approved by the USA for FSC registration; the future of these corporations was still unclear in mid-2005. The island's inclusion on the OECD list of 'non-co-operative tax havens' in June 2000 received a hostile reaction from the Barbadian Government, which played a leading role in international lobbying by 'offshore' centres against OECD's initiative and that of the Financial Action Task Force on Money Laundering (FATF, based in Paris, France). Barbados was removed from the OECD list in early 2002, after the Organisation modified its criteria for the definition of a tax 'haven'. Although, unlike many other jurisdictions included on OECD's blacklist, Barbados did not pledge to reform its financial sector, the Government did introduce legislative changes intended to improve financial transparency, as well as demonstrating a willingness to enter into tax information exchange agreements with other countries.

The informatics industry originated in the early 1980s with data-entry operations. Employment in the sector reached 2,972 in December 1997, but fell subsequently, to 1,128 by December 2001, recovering to 1,442 by the end of 2003. By the mid-1990s routine work was handled by scanning systems, and work that was still done manually could be sourced more cheaply from countries with lower labour and telecommunications costs. Informatics costs in Barbados were cheaper than in the USA and the island had a higher standard of education and a more adaptable English-speaking labour force than many of its competitors and was also in the same time zone as the eastern USA. Although these factors meant that Barbados retained a comparative advantage in some high-end work, by the early 2000s the island's comparative advantage appeared to be diminishing even at the top end of the market, although call-centre employment continued to increase. With new cellular-service providers in operation, to be followed by competitive fixed-wireless, international and finally landline services, liberalization of the telecommunications sector was well advanced in 2005.

## MANUFACTURING

The manufacturing sector was in a depressed state from the mid-1980s and the restraints on credit and consumer spending, imposed progressively from 1991, resulted in bankruptcies and factory closures. Economic recovery in the late 1990s failed to benefit most manufacturing industries, because of the gradual erosion of protective barriers against imports. The main manufacturing industries employed over 15,000 people in 1986, but only 9,700, representing 7.3% of total employment, at the end of 2002. However, Barbados had a surprisingly wide range of manufacturing industries, producing consumer products for the local and regional market, although labour costs were higher than in most Caribbean islands and power costs much higher than in Trinidad and Tobago. The closure of the small petroleum refinery in 1998 was a gain for the economy, as refinery products could be imported more cheaply from Trinidad and Tobago, and a valuable site was released for tourist development.

The removal of trade barriers caused difficulties for some manufacturers in the 1990s. The clothing and wooden furniture industries virtually disappeared. However, the more efficient producers in other branches of manufacturing managed to survive and in some cases to expand their export sales. In 2001 manufactured exports (excluding sugar) earned US $124.9m.

The remaining heavy industrial plant was the Arawak cement plant in the north of the island, owned by a Trinidadian company, which used local limestone and low-cost orimulsion fuel from Venezuela. After having been closed for some years, the kiln reopened in 1997 and, owing to the activity of the construction industry in Barbados and throughout the eastern Caribbean, worked close to full capacity.

Petroleum production was another industrial activity. Crude petroleum production began in 1973, and rose to 679,000 barrels in 1985 before declining to 328,000 barrels in 1997 as wells were depleted. However, an onshore drilling programme begun in that year increased production dramatically from 1998. Total extraction reached 708,500 barrels in 1999, equivalent to one-third of local consumption. With reduced drilling, output fell to 365,000 barrels in 2003; however, a new drilling programme in 2004–05 was expected to increase production. Natural gas production stood at 31.0m. cu m in 2003. There is a piped supply to most urban and suburban areas, while some electric generation capacity transferred to natural gas from 1999. With local gas resources limited in extent, proposals were sketched to import gas from Trinidad and Tobago, either by pipeline or by barge as compressed natural gas. A US-based company, Conoco Inc., operated an offshore petroleum and gas exploration programme in partnership with the French concern Elf Aquitaine (now part of Total); initial seismic surveys showed promising deep-water geological prospects. The first exploratory well, drilled in December 2001, produced disappointing results. Further drilling was expected to take place, but was likely to await delimitation of the Exclusive Economic Zone boundary with Trinidad and Tobago.

## SUGAR AND AGRICULTURE

Sugar was the mainstay of the economy until the rise of the tourism industry in the mid-20th century. In 1946 the sugar industry accounted for 37.8% of GDP and 55% of foreign-exchange earnings, employing some 25,100 people. By 2002, however, the industry accounted for only 2.0% of GDP, 1.4% of foreign-exchange earnings and an estimated 2.5% of total employment.

In spite of a guaranteed price for exports to Europe, under the Lomé Convention (which expired in February 2000 and was replaced by the Cotonou Agreement in June), the sugar industry suffered from severe economic problems from the 1980s. These related mainly to high production costs and inefficiencies associated with the traditional management system. Farmers left many sugar farms uncultivated, in order to capitalize upon the high prices that could sometimes be obtained for land set aside for urban development. By 1992 a high proportion of estates had borrowed heavily from the state-owned Barbados National Bank and had accumulated debts greater than the value of their assets. Barbados Sugar Industries Ltd (BSIL), which owned and operated the factories, was forced into receivership. The factories and the most heavily indebted estates were placed under the management of the Barbados Agricultural Management Company, which from 1994 was run on a management contract by a British company, Booker-Tate. Amid considerable political controversy, Booker-Tate insisted on a restructuring of the entire sugar industry, with one-third of the debt to be cancelled, one-third to be converted into equity in a new company and the remaining one-third to be serviced as an active loan. The Government also guaranteed a loan of US $11.5m. from the Caribbean Development Bank. In return, a plan to increase production to 75,000 metric tons of raw sugar within five years was put into operation. In early 1993 the IMF approved the plan. The initial contract expired in June 1998 and after several temporary extensions, Booker-Tate's direct involvement was phased out. In an attempt to reduce costs, one of the three sugar factories was closed in June 2002, with another to follow in 2005, leaving one refurbished factory in operation. Another proposal was the development of high-value niche markets for speciality branded sugars; 3,000 tons of high-value yellow-crystal sugar was exported in 2002.

As a result of falling production, Barbados in 2003 surrendered part of its valuable EU quota of 54,000 metric tons of sugar (worth some US $24m. in foreign exchange), which was not fulfilled from 1991 to 1995, nor after 2001; the new figure is 35,000 tons. No use is now made of the US quota, which was reduced from 14,239 tons to 8,000 tons in 1995. From 1990 the rum industry was forced to import most of its rising annual requirement of molasses. Production reached a low point of 38,500 tons in 1995, partly as a consequence of a severe drought

# BARBADOS

in the previous year. Thereafter output recovered to a high of 64,600 tons in 1997 but had decreased to a new low of 34,357 tons in 2004. Earnings meanwhile suffered from the declining value of the euro against the US dollar. With production costs estimated at Bds $1,493 per ton, sugar subsidies, which totalled Bds $100m. in the 1990s, threatened to spiral out of control. Although the recovery in the euro raised the export price to Bds $1,225 by mid-2003, the medium-term outlook for the industry remained poor, with the EU announcing a cumulative price cut of 33% in 2006–07, and the structure of the EU import regime under challenge within the WTO from Brazil and Australia.

The soil held nutrients well, but was thin. Rainfall was generally adequate, but there were occasional severe droughts. Groundwater supplies were not sufficient for large-scale irrigation. These problems impeded the development of other agricultural activities. There was some commercial production of vegetables and root crops. The island was virtually self-sufficient in poultry products and in fresh milk, activities which depended to a significant extent on imported inputs. There was some pig farming and lamb was produced from the local Black Belly short-haired sheep. However, despite some self-sufficiency, agricultural production (including forestry and fishing, but excluding sugar) contributed only an estimated 3.6% of GDP in 2004, and (including sugar) employed 4.5% of the working population in 2003.

# Statistical Survey

Sources (unless otherwise stated): Barbados Statistical Service, National Insurance Bldg, 3rd Floor, Fairchild St, Bridgetown; tel. 427-7841; fax 435-2198; e-mail barstats@caribsurf.com; internet www.bgis.gov.bb/stats; Central Bank of Barbados, Tom Adams Financial Centre, Spry St, POB 1016, Bridgetown; tel. 436-6870; fax 427-1431; e-mail cbb.libr@caribsurf.com; internet www.centralbank.org.bb.

## AREA AND POPULATION

**Area:** 430 sq km (166 sq miles).

**Population:** 252,029 (males 119,665, females 132,364) at census of 12 May 1980; 257,082 (provisional) at census of 2 May 1990; 269,000 at mid-2004 (Source: UN, *World Population Prospects: The 2004 Revision*).

**Density** (mid-2004): 625.6 per sq km.

**Ethnic Groups** (*de jure* population, excl. persons resident in institutions, 1990 census): Black 228,683; White 8,022; Mixed race 5,886; Total (incl. others) 247,288.

**Principal Town:** Bridgetown (capital), population 5,928 at 1990 census. *Mid-2003* (UN estimate, incl. suburbs): Bridgetown 139,854 (Source: UN, *World Urbanization Prospects: The 2003 Revision*).

**Births, Marriages and Deaths** (provisional, 2000): Live births 3,762 (birth rate 14.1 per 1,000); Marriages (1997) 3,377 (marriage rate 12.7 per 1,000); Deaths 2,367 (death rate 8.8 per 1,000). Source: partly UN, *Population and Vital Statistics Report*.

**Expectation of Life** (WHO estimates, years at birth): 75 (males 71; females 78) in 2003. Source: WHO, *World Health Report*.

**Economically Active Population** (provisional and rounded figures, labour force sample survey, '000 persons aged 15 years and over, excl. armed forces, 2001): Agriculture, forestry and fishing 5.4; Manufacturing 9.2; Electricity, gas and water 2.0; Construction and quarrying 14.2; Wholesale and retail trade 18.9; Tourism 13.7; Transport, storage and communications 5.9; Financing, insurance, real estate and business services 10.4; Community, social and personal services 48.8; Activities not adequately defined 0.2; *Total employed* 128.7 (males 67.6, females 58.5); Unemployed 14.1 (males 6.1, females 7.5); *Total labour force* 142.8 (males 73.7, females 66.0) (Source: ILO Caribbean Office). *2002:* Total employed 128.7; Unemployed 14.5; Total labour force 143.2 (Source: IMF, *Barbados: Statistical Appendix* (May 2004)).

## HEALTH AND WELFARE
### Key Indicators

**Total Fertility Rate** (children per woman, 2003): 1.5.

**Under-5 Mortality Rate** (per 1,000 live births, 2003): 13.

**HIV/AIDS** (% of persons aged 15–49, 2003): 1.5.

**Physicians** (per 1,000 head, 1999): 1.21.

**Hospital Beds** (per 1,000 head, 1996): 7.56.

**Health Expenditure** (2002): US $ per head (PPP): 1,018.

**Health Expenditure** (2002): % of GDP: 6.9.

**Health Expenditure** (2002): public (% of total): 68.4.

**Access to Water** (% of persons, 2002): 100.

**Access to Sanitation** (% of persons, 2002): 99.

**Human Development Index** (2002): ranking: 29.

**Human Development Index** (2002): value: 0.888.

For sources and definitions, see explanatory note on p. vi.

## AGRICULTURE, ETC.

**Principal Crops** ('000 metric tons, 2003): Sweet potatoes 2.6; Yams 1.2; Other roots and tubers 0.6*; Sugar cane 364.6; Pulses 1.0*; Coconuts 1.7*; Cabbages 0.6; Tomatoes 1.2; Cucumbers 2.0; Chillies and peppers 1.3; String beans 0.9*; Carrots 1.0; Okra 1.5; Green corn 0.7*; Other vegetables 4.6*; Bananas 0.6*; Other fruits 3.0*.
*FAO estimate.

**Livestock** (FAO estimates, '000 head, year ending September 2003): Horses 1.1; Mules 2.0; Asses 2.1; Cattle 13.5; Pigs 16.6; Sheep 27.0; Goats 5.0; Poultry 3,450.

**Livestock Products** ('000 metric tons, 2003): Beef and veal 0.4; Pig meat 2.2 (FAO estimate); Poultry meat 11.1; Cows' milk 7.0; Hen eggs 1.6.

**Forestry** (FAO estimates, '000 cubic metres): Roundwood removals 5 in 2001; 5 in 2002; 5 in 2003.

**Fishing** (FAO estimates, metric tons, live weight, 2003): Total catch 2,500 (Yellowfin tuna 116; Flyingfishes 1,590; Common dolphinfish 550).
Sources: FAO.

## MINING

**Production** (estimates, 2003): Natural gas 31m. cu m; Crude petroleum 365,000 barrels. Source: US Geological Survey.

## INDUSTRY

**Selected Products** (2002, unless otherwise indicated): Raw sugar 36,325 metric tons (2003); Rum 11,000,000 litres; Beer 6,800,000 litres; Cigarettes 65m. (1995); Batteries 17,165 (official estimate, 1998); Electric energy 827m. kWh. Sources: partly UN, *Industrial Commodity Statistics Yearbook*, and IMF, *Barbados: Statistical Appendix* (May 2004).

## FINANCE

**Currency and Exchange Rates:** 100 cents = 1 Barbados dollar (Bds $). *Sterling, US Dollar and Euro Equivalents* (31 May 2005): £1 sterling = Bds $3.6362; US $1 = Bds $2.0000; €1 = Bds $2.4662; Bds $100 = £27.50 = US $50.00 = €40.55. *Exchange Rate:* Fixed at US $1 = Bds $2.000 since 1986.

**Budget** (Bds $ million, year ending 31 March 2003): *Revenue:* Tax revenue 1,620.6 (Direct taxes 731.6, of which Personal income tax 358.3, Corporate taxes 220.3, Taxes on property 95.7; Indirect taxes 889.0, of which Value-added tax 494.5, Excises 114.2, Import duties 160.1, Other indirect taxes 106.9); Non-tax revenue 114.6; Total 1,735.2. *Expenditure:* Current 1,688.3 (Wages and salaries 683.6, Other goods and services 215.3, Interest payments 268.7, Transfers and subsidies 520.7); Capital (incl. net lending 2.8) 316.0; Total 2,004.3. Note: Budgetary data refer to current and capital budgets only and exclude operations of the National Insurance Fund and other central government units with their own budgets.

**International Reserves** (US $ million at 31 December 2004): IMF special drawing rights 0.06; Reserve position in IMF 8.00; Foreign exchange 571.80 Total 579.86. Source: IMF, *International Financial Statistics*.

**Money Supply** (Bds $ million at 31 December 2004): Currency outside banks 398.7; Demand deposits at commercial banks 1,247.7; Total money (incl. others) 1,650.3. Source: IMF, *International Financial Statistics*.

**Cost of Living** (Index of Retail Prices; base: 2000 = 100): 102.7 in 2002; 104.4 in 2003; 105.8 in 2004. Source: IMF, *International Financial Statistics*.

**Expenditure on the Gross Domestic Product** (Bds $ million at current prices, 2004): Government final consumption expenditure 1,176; Private final consumption expenditure 4,009; Increase in stocks 482; Gross fixed capital formation 598; *Total domestic expenditure* 6,265; Exports of goods and services 2,790; *Less* Imports of goods and services 3,430; *GDP in purchasers' values* 5,625. Source: Caribbean Development Bank, *Social and Economic Indicators*.

**Gross Domestic Product by Economic Activity** (estimates, Bds $ million at current prices, 2001): Agriculture, hunting, forestry and fishing 199; Mining and quarrying 28; Manufacturing 256; Electricity, gas and water 146; Construction 243; Wholesale and retail trade 645; Hotels and restaurants 471; Transport, storage and communications 467; Finance, insurance, real estate and business services 798; Government services 758; Other community, social and personal services 211; *GDP at factor cost* 4,223; Indirect taxes, *less* subsidies 875; *GDP in purchasers' values* 5,098.

**Balance of Payments** (US $ million, 2003): Exports of goods f.o.b. 264.2; Imports of goods f.o.b. –1,065.6; *Trade balance* –801.4; Exports of services 1,165.5; Imports of services –518.8; *Balance on goods and services* –154.7; Other income received 75.1; Other income paid –182.0; *Balance on goods, services and income* –261.6; Current transfers received 126.5; Current transfers paid –34.3; *Current balance* –169.4; Direct investment abroad –0.5; Direct investment from abroad 58.3; Portfolio investment assets –22.9; Portfolio investment liabilities 84.1; Other investment assets –83.1; Other investment liabilities 166.5; Net errors and omissions 34.4; *Overall balance* 67.4. Source: IMF, *International Financial Statistics*.

### EXTERNAL TRADE

**Principal Commodities** (preliminary, US $ million, 2003): *Imports c.i.f.*: Food and beverages 159; Motor cars 37; Fuels 114; Chemicals 52; Construction materials 74; Total (incl. others) 1,025. *Exports f.o.b.*: Sugar and molasses 21; Rum 16; Other food and beverages 21; Chemicals 19; Electrical components 12; Other manufactures 39; Total (incl. other exports and re-exports) 229. Source: IMF, *Barbados: Statistical Appendix* (May 2004).

**Principal Trading Partners** (US $ million, 2002): *Imports:* Brazil 16.7; Canada 37.1; China, People's Republic 26.6; Demark 10.0; France 19.1; Germany 17.8; Guyana 10.6; Italy 10.4; Japan 45.0; Mexico 10.4; Netherlands 18.6; New Zealand 11.5; Suriname 6.7; Switzerland 7.3; Trinidad and Tobago 116.2; United Kingdom 78.6; USA 439.8; Total (incl. others) 996.6. *Exports:* Antigua and Barbuda 7.2; Bahamas 2.1; Belize 2.5; Canada 4.5; Dominica 4.1; Grenada 6.7; Guyana 8.5; Jamaica 15.0; Saint Christopher and Nevis 5.5; Saint Lucia 11.4; Saint Vincent and the Grenadines 9.1; Spain 2.9; Suriname 5.3; Trinidad and Tobago 23.7; United Kingdom 25.7; USA 35.6; Total (incl. others) 259.3 (Source: UN, *International Trade Statistics Yearbook*). 2003 (preliminary, US $ million): Total imports 1,025; Total exports 229 (Source: IMF, *Barbados: Statistical Appendix* (May 2004)).

### TRANSPORT

**Road Traffic** (motor vehicles in use, 2002): Private cars 81,648; Buses and coaches 736; Lorries and vans 6,962; Motor cycles and mopeds 1,856; Road tractors 921. Source: International Road Federation, *World Road Statistics*.

**Shipping** (estimated freight traffic, '000 metric tons, 1990): Goods loaded 206; Goods unloaded 538 (Source: UN, *Monthly Bulletin of Statistics*). Total goods handled ('000 metric tons, 1997): 1,095 (Source: Barbados Port Authority). *Merchant Fleet* (vessels registered at 31 December 2004): Number of vessels 85; Total displacement 580,262 grt (Source: Lloyd's Register-Fairplay, *World Fleet Statistics*).

**Civil Aviation** (1994): Aircraft movements 36,100; Freight loaded 5,052.3 metric tons; Freight unloaded 8,548.3 metric tons.

### TOURISM

**Tourist Arrivals** ('000 persons): *Stop-overs:* 497.9 in 2002; 531.2 in 2003; 551.5 in 2004. *Cruise-ship Passengers:* 523.3 in 2002; 559.1 in 2003; 721.3 in 2004. Source: Caribbean Development Bank, *Social and Economic Indicators*.

**Tourist Arrivals by Country** ('000 persons, 2003): Antigua and Barbuda 6.5; Canada 49.6; Germany 7.6; Grenada 6.0; Guyana 15.1; Jamaica 7.0; Saint Lucia 13.7; Saint Vincent and the Grenadines 11.8; Trinidad and Tobago 27.5; United Kingdom 202.6; USA 129.3; Total (incl. others) 531.2.

**Tourism Receipts** (US $ million, incl. passenger transport): 706 in 2001; 666 in 2002; 767 in 2003.
Source: World Tourism Organization.

### COMMUNICATIONS MEDIA

**Radio Receivers** (1999): 175,000 in use.
**Television Receivers** (2000): 83,000 in use.
**Telephones** (2003): 134,000 main lines in use.
**Facsimile Machines** (year ending 31 March 1997): 1,800 in use.
**Mobile Cellular Telephones** (2003): 140,000 subscribers.
**Personal Computers** (2002): 28,000 in use.
**Internet Users** (2003): 100,000.
**Newspapers:** *Daily* (1996): 2 (circulation 53,000). *Non-daily* (1990): 4 (estimated circulation 95,000).
Sources: partly UNESCO, *Statistical Yearbook*, UN, *Statistical Yearbook*, and International Telecommunication Union.

### EDUCATION

**Pre-primary** (1995/96): 84 schools; 529 teachers; 4,689 pupils.
**Primary** (2002): 109 schools; 1,823 teachers; 29,502 pupils.
**Secondary** (2002): 32 schools; 1,389 teachers; 21,436 pupils.
**Tertiary** (2002): 4 schools; 339 teachers; 11,226 students.
Source: Ministry of Education, Youth Affairs and Sport.

**Adult Literacy Rate** (UNESCO estimates): 99.7% (males 99.7%; females 99.7%) in 2002. Source: UN Development Programme, *Human Development Report*.

# Directory

## The Constitution

The parliamentary system has been established since the 17th century, when the first Assembly sat, in 1639, and the Charter of Barbados was granted, in 1652. A new Constitution came into force on 30 November 1966, when Barbados became independent. Under its terms, protection is afforded to individuals from slavery and forced labour, from inhuman treatment, deprivation of property, arbitrary search and entry, and racial discrimination; freedom of conscience, of expression, assembly, and movement are guaranteed.

Executive power is nominally vested in the British monarch, as Head of State, represented in Barbados by a Governor-General, who appoints the Prime Minister and, on the advice of the Prime Minister, appoints other ministers and some senators.

The Cabinet consists of the Prime Minister, appointed by the Governor-General as being the person best able to command a majority in the House of Assembly, and not fewer than five other ministers. Provision is also made for a Privy Council, presided over by the Governor-General.

Parliament consists of the Governor-General and a bicameral legislature, comprising the Senate and the House of Assembly. The Senate has 21 members: 12 appointed by the Governor-General on the advice of the Prime Minister, two on the advice of the Leader of the Opposition and seven as representatives of such interests as the Governor-General considers appropriate. The House of Assembly has (since 2003) 30 members, elected by universal adult suffrage for a term of five years (subject to dissolution). The minimum voting age is 18 years.

The Constitution also provides for the establishment of Service Commissions for the Judicial and Legal Service, the Public Service,

# BARBADOS

the Police Service and the Statutory Boards Service. These Commissions are exempt from legal investigation; they have executive powers relating to appointments, dismissals and disciplinary control of the services for which they are responsible.

## The Government

### HEAD OF STATE

**Monarch:** HM Queen ELIZABETH II (succeeded to the throne 6 February 1952).

**Governor-General:** Sir CLIFFORD HUSBANDS (appointed 1 June 1996).

### THE CABINET
(July 2005)

**Prime Minister and Minister of Finance, Economic Affairs and the Civil Service:** OWEN S. ARTHUR.

**Deputy Prime Minister, Attorney-General, Leader of Government Business in the House of Assembly and Minister of Home Affairs:** MIA AMOR MOTTLEY.

**Minister of Agriculture and Rural Development:** Sen. ERSKINE GRIFFITH.

**Minister of Commerce, Consumer Affairs and Business Development:** Sen. LYNETTE EASTMOND.

**Minister of Education, Youth Affairs and Sport:** REGINALD FARLEY.

**Minister of Energy and Public Utilities:** ANTHONY P. WOOD.

**Minister of Foreign Affairs and Foreign Trade:** Sen. BILLIE A. MILLER.

**Minister of Health:** Dr JEROME WALCOTT.

**Minister of Housing, Lands and the Environment:** ELIZABETH THOMPSON.

**Minister of Industry and International Business:** DALE MARSHALL.

**Minister of Labour and Social Security:** RAWLE C. EASTMOND.

**Minister of Public Works:** GLINE A. CLARKE.

**Minister of Social Transformation:** HAMILTON F. LASHLEY.

**Minister of Tourism and International Transport:** NOEL A. LYNCH.

**Minister of State in the Office of the Prime Minister and the Ministry for the Civil Service:** Sen. JOHN WILLIAMS.

**Minister of State in the Ministry of Foreign Affairs and Foreign Trade:** KERRIE SYMONDS.

**Minister of State in the Ministry of Education, Youth Affairs and Sport:** CYNTHIA FORDE.

### MINISTRIES

**Office of the Prime Minister:** Government Headquarters, Bay St, St Michael; tel. 436-6435; fax 436-9280; e-mail info@primeminister.gov.bb; internet www.primeminister.gov.bb.

**Ministry of Agriculture and Rural Development:** Graeme Hall, POB 505, Christ Church; tel. 428-4150; fax 420-8444; internet www.barbados.gov.bb/minagri.

**Ministry for the Civil Service:** Roebuck Plaza, 20–23 Roebuck St, Bridgetown; tel. 426-2390; fax 228-0093.

**Ministry of Commerce, Consumer Affairs and Business Development:** Government Headquarters, Bay St, St Michael; tel. 427-5270.

**Ministry of Education, Youth Affairs and Sport:** Elsie Payne Complex, Constitution Rd, Bridgetown; tel. 430-2700; fax 436-2411; e-mail mined1@caribsurf.com; internet www.edutech2000.gov.bb.

**Ministry of Energy and Public Utilities:** Government Headquarters, Bay St, St Michael.

**Ministry of Finance and Economic Affairs:** Government Headquarters, Bay St, St Michael; tel. 436-6435; fax 429-4032.

**Ministry of Foreign Affairs and Foreign Trade:** 1 Culloden Rd, St Michael; tel. 436-2990; fax 429-6652; e-mail barbados@foreign.gov.bb; internet www.foreign.gov.bb.

**Ministry of Health:** Jemmott's Lane, St Michael; tel. 426-5080; fax 426-5570.

**Ministry of Home Affairs:** General Post Office Bldg, Level 5, Cheapside, Bridgetown; tel. 228-8961; fax 437-3794; e-mail mha@caribsurf.com.

**Ministry of Housing, Lands and the Environment:** National Housing Corpn Bldg, 'The Garden', Country Rd, St Michael; tel. 431-7800; fax 435-0174; e-mail nurses@gob.bb.

**Ministry of Industry and International Business:** The Business Centre, Upton, St Michael; tel. 430-2200; fax 429-6849; e-mail mtbbar@caribsurf.com.

**Ministry of Labour and Social Security:** Docklands Pl. and Cavans Lane, Bridgetown; tel. 427-7550; fax 426-8959; internet www.labour.gov.bb.

**Ministry of Public Works:** The Pine, St Michael; tel. 429-2191; fax 437-8133; internet www.publicworks.gov.bb.

**Ministry of Social Transformation:** Nicholas House, Broad St, Bridgetown; tel. 228-5878.

**Ministry of Tourism and International Transport:** Sherbourne Conference Centre, Two Mile Hill, St Michael; tel. 430-7500; fax 436-4828; e-mail barmot@sunbeach.net; internet www.barmot.gov.bb.

**Office of the Attorney-General:** Sir Frank Walcott Bldg, Culloden Rd, St Michael; tel. 431-7750; fax 228-5433; e-mail attygen@caribsurf.com.bb.

## Legislature

### PARLIAMENT

#### Senate

**President:** Sir FRED GOLLOP.
There are 21 members.

#### House of Assembly

**Speaker:** ISHMAEL ROETT.

**General Election, 21 May 2003**

| Party | % of votes | Seats |
|---|---|---|
| Barbados Labour Party (BLP) | 55.8 | 23 |
| Democratic Labour Party (DLP) | 44.1 | 7 |
| **Total** | 100.0 | 30 |

## Political Organizations

**Barbados Labour Party:** Grantley Adams House, 111 Roebuck St, Bridgetown; tel. 429-1990; e-mail hq@blp.org.bb; internet www.blp.org.bb; f. 1938; moderate social democrat; Leader and Chair. OWEN S. ARTHUR; Gen. Sec. WILLIAM DUGUID.

**Democratic Labour Party (DLP):** 'Kennington', George St, Belleville, St Michael; tel. 429-3104; fax 427-0548; e-mail dlp@sunbeach.net; internet www.dlpbarbados.org; f. 1955; Pres. Sen. FREUNDEL STUART; Leader CLYDE MASCOLL.

## Diplomatic Representation

### EMBASSIES AND HIGH COMMISSIONS IN BARBADOS

**Brazil:** Sunjet House, 3rd Floor, Fairchild St, Bridgetown; tel. 427-1735; fax 427-1744; e-mail brembarb@sunbeach.net; internet www.brazilbb.org; Ambassador ORLANDO GALVÊAS OLIVEIRA.

**Canada:** Bishops Court Hill, Pine Rd, POB 404, St Michael; tel. 429-3550; fax 429-3780; e-mail bdgtn@international.gc.ca; internet www.dfait-maeci.gc.ca/barbados; High Commissioner MICHAEL C. WELSH.

**China, People's Republic:** 17 Golf View Terrace, Rockley, Christ Church; tel. 435-6890; fax 435-8300; e-mail chineseembbds@caribsurf.com; Ambassador YANG ZHIKUAN.

**Costa Rica:** Omega Bldg, Suite 4, 1st Floor, Dayreels Rd, Christ Church; tel. 431-0250; fax 431-0261; e-mail embc@sunbeach.net; Ambassador EMILIA SALAZAR BORBÓN.

**Cuba:** Palm View, Erdiston Dr., St Michael; tel. 435-2769; fax 435-2734; e-mail embajadadecuba@sunbeach.net; Ambassador JOSÉ JOAQUÍN ALVAREZ PORTELA.

**United Kingdom:** Lower Collymore Rock, POB 676, St Michael; tel. 430-7800; fax 430-7813; e-mail britishhcb@sunbeach.net; internet www.britishhighcommission.gov.uk/barbados; High Commissioner JOHN WHITE.

# BARBADOS

**USA:** Canadian Imperial Bank of Commerce Bldg, Broad St, POB 302, Bridgetown; tel. 436-4950; fax 429-5246; internet bridgetown.usembassy.gov; Ambassador MARY ELIZABETH KRAMER.

**Venezuela:** Hastings, Main Rd, Christ Church; tel. 435-7903; fax 435-7830; e-mail jesusmachin@sunbeach.net; Ambassador MARÍA CORINA RUSSIAN FRONTADO.

## Judicial System

Justice is administered by the Supreme Court of Judicature, which consists of a High Court and a Court of Appeal. Final appeal lies with the Judicial Committee of the Privy Council, in the United Kingdom. There are Magistrates' Courts for lesser offences, with appeal to the Court of Appeal.

**Supreme Court:** Judiciary Office, Coleridge St, Bridgetown; tel. 426-3461; fax 246-2405.

**Chief Justice:** Sir DAVID SIMMONS.

**Justices of Appeal:** G. C. R. MOE; ERROL DA COSTA CHASE; COLIN A. WILLIAMS.

**Judges of the High Court:** FREDERICK A. WATERMAN; MARIE A. MACCORMACK; E. GARVEY HUSBANDS; CARLISLE PAYNE; SHERMAN MOORE; LIONEL DACOSTA GREENIDGE.

**Registrar of the Supreme Court:** SANDRA MASON.

**Office of the Attorney-General:** Sir Frank Walcott Bldg, Culloden Rd, St Michael; tel. 431-7750; fax 228-5433; e-mail attygen@caribsurf.com.bb; Dir of Public Prosecutions CHARLES LEACOCK; e-mail cbleacock@inaccs.com.bb.

## Religion

More than 100 religious denominations and sects are represented in Barbados, but the vast majority of the population profess Christianity. About 40% of the total population were Anglican, while the Pentecostal (8%) and Methodist (7%) churches were next in importance. There are also small groups of Hindus, Muslims and Jews.

### CHRISTIANITY

#### Barbados Christian Council

Caribbean Conference of Churches Bldg, George St and Collymore Rock, St Michael; tel. 426-6014.

#### The Anglican Communion

Anglicans in Barbados are adherents of the Church in the Province of the West Indies, comprising eight dioceses. The Archbishop of the Province is the Bishop of Nassau and the Bahamas, resident in Nassau, the Bahamas. In Barbados there is a Provincial Office (St George's Church, St George) and an Anglican Theological College (Codrington College, St John).

**Bishop of Barbados:** Rt Rev. RUFUS THEOPHILUS BROOME, Leland, Philip Dr., Pine Gardens, St Michael; fax 426-0871; e-mail mandeville@sunbeach.com.

#### The Roman Catholic Church

Barbados comprises a single diocese (formed in January 1990, when the diocese of Bridgetown-Kingstown was divided), which is suffragan to the archdiocese of Port of Spain (Trinidad and Tobago). At 31 December 2003 there were an estimated 10,750 adherents in the diocese. The Bishop participates in the Antilles Episcopal Conference (currently based in Port of Spain, Trinidad and Tobago).

**Bishop of Bridgetown:** Rt Rev. MALCOLM PATRICK GALT, St Patrick's Presbytery, Jemmott's Lane, POB 1223, Bridgetown; tel. 426-3510; fax 429-6198; e-mail rcbishopbgi@caribnet.com.

#### Other Churches

**Baptist Churches of Barbados:** National Baptist Convention, President Kennedy Dr., Bridgetown; tel. 429-2697.

**Church of God (Caribbean Atlantic Assembly):** St Michael's Plaza, St Michael's Row, POB 1, Bridgetown; tel. 427-5770; Pres. Rev. VICTOR BABB.

**Church of Jesus Christ of Latter-Day Saints (Mormons)—West Indies Mission:** Bridgetown; tel. 435-8595; fax 435-8278.

**Church of the Nazarene:** District Office, Eagle Hall, Bridgetown; tel. 425-1067.

**Methodist Church:** Bethel Church Office, Bay St, Bridgetown; tel. and fax 426-2223; e-mail methodist@caribsurf.com.

**Moravian Church:** Roebuck St, Bridgetown; tel. 426-2337; Superintendent Rev. ERROL CONNOR.

**Seventh-day Adventists (East Caribbean Conference):** Brydens Ave, Brittons Hill, POB 223, St Michael; tel. 429-7234; fax 429-8055; e-mail president@eastcarib.org; internet wwww.eastcarib.org; Pres. JAMES F. DANIEL.

**Wesleyan Holiness Church:** General Headquarters, Bank Hall; tel. 429-4864.

Other denominations include the Abundant Life Assembly, the African Orthodox Church, the Apostolic Church, the Assemblies of Brethren, the Berean Bible Brethren, the Bethel Evangelical Church, Christ is the Answer Family Church, the Church of God the Prophecy, the Ethiopian Orthodox Church, the Full Gospel Assembly, Love Gospel Assembly, the New Testament Church of God, the Pentecostal Assemblies of the West Indies, the People's Cathedral, the Salvation Army, Presbyterian congregations, the African Methodist Episcopal Church, the Mt Olive United Holy Church of America and Jehovah's Witnesses.

### ISLAM

In 1996 there were an estimated 2,000 Muslims in Barbados.

**Islamic Teaching Centre:** Harts Gap, Hastings; tel. 427-0120.

### JUDAISM

**Jewish Community:** Nidhe Israel and Shaara Tzedek Synagogue, Rockley New Rd, POB 651, Bridgetown; tel. 437-1290; fax 437-1303; Pres. RACHELLE ALTMAN; Sec. SHARON ORAN.

### HINDUISM

At the census of 1980 there were 411 Hindus on the island.

**Hindu Community:** Hindu Temple, Roberts Complex, Government Hill, St Michael; tel. 434-4638.

## The Press

**Barbados Advocate:** POB 230, St Michael; tel. 467-2000; fax 434-1000; e-mail news@barbadosadvocate.com; internet www.barbadosadvocate.com; f. 1895; daily; Editor (vacant); circ. 11,413.

**The Broad Street Journal:** Plantation Complex, St Lawrence Main Rd, Christ Church; tel. 420-6245; fax 420-5477; e-mail bsj@sunbeach.net; f. 1993; weekly; business; Editor RICHARD COSTAS.

**The Nation:** Nation House, Fontabelle, POB 1203, St Michael; tel. 430-5400; fax 427-6968; e-mail nationnews@sunbeach.net; internet www.nationnews.com; f. 1973; daily; also publishes the *Sun on Saturday*, the *Sunday Sun* (q.v.) and the *Visitor* (a free publ. for tourists); Pres. and Editor-in-Chief HAROLD HOYTE; circ. 23,144 (weekday), 33,084 (weekend).

**Official Gazette:** Government Printing Office, Bay St, St Michael; tel. 436-6776; Mon. and Thur.

**Sunday Advocate:** POB 230, St Michael; tel. 467-2000; fax 434-1000; e-mail news@sunbeach.net; internet www.barbadosadvocate.com; f. 1895; Editor REUDON EVERSLEY; circ. 17,490.

**The Sunday Sun:** Nation House, Fontabelle, St Michael; tel. 430-5400; fax 427-6968; e-mail sundaysun@sunbeach.net; internet www.nationnews.com; f. 1977; Dir HAROLD HOYTE; circ. 48,824.

**Weekend Investigator:** POB 230, St Michael; tel. 434-2000; circ. 14,305.

### NEWS AGENCIES

**Caribbean Media Corporation (CMC):** Unit 1B, Bldg 6A, Harbour Industrial Estate, St Michael; tel. 467-1000; fax 429-4355; e-mail admin@cananews.com; internet www.cananews.com; f. by merger of Caribbean News Agency (CANA) and Caribbean Broadcasting Union; COO GARY ALLEN.

### Foreign Bureaux

**Inter Press Service (IPS)** (Italy): POB 697, Bridgetown; tel. 426-4474; Correspondent MARVA COSSY.

**Xinhua (New China) News Agency** (People's Republic of China): Christ Church; Chief Correspondent DING BAOZHONG.

Agence France-Presse (AFP) and Agencia EFE (Spain) are also represented.

## Publishers

**The Advocate Publishing Co Ltd:** POB 230, St Michael; tel. 434-2000; fax 434-2020.

**Business Tutors:** POB 800, St Michael; tel. 428-5664; fax 429-4854; e-mail pchad@caribsurf.com; business, management, computers.

**Carib Research and Publications Inc:** POB 556, Bridgetown; tel. 438-0580; f. 1986; regional interests; CEO Dr Farley Braithwaite.

**Nation Publishing Co Ltd:** Nation House, POB 1203, Fontabelle, St Michael; tel. 430-5400; fax 427-6968.

# Broadcasting and Communications

### TELECOMMUNICATIONS

**Cable & Wireless (Barbados) Ltd:** POB 32, Wildey, St Michael; tel. 292-6000; fax 427-5808; e-mail bdsinfo@caribsurf.com; internet www.candwbet.com.bb; f. 1984; fmrly Barbados External Telecommunications Ltd; became Cable & Wireless BET Ltd, which merged with Barbados Telephone Co Ltd (BARTEL), Cable & Wireless Caribbean Cellular and Cable & Wireless Information Systems in 2002; provides international telecommunications services; owned by Cable & Wireless PLC (United Kingdom); Chair. Stephen Emtage; Pres. Donald Austin; Gen. Man. Vincent Yearwood.

**Cingular Wireless Barbados:** St Michael; internet www.cingularwireless.com; joint venture between SBC Communications and BellSouth; scheduled to sell its operations and licences in the Caribbean to Digicel in 2005; Vice-Pres. and Gen. Man. Jonathan Koshar.

**Digicel Barbados Ltd:** The Courtyard, Hastings, Christ Church; tel. 434-3444; fax 426-3444; e-mail customercare@digicelgroup.com; internet www.digicelbarbados.com; f. 2003; awarded licence to operate cellular telephone services in March 2003; owned by an Irish consortium; Dir Ralph 'Bizzy' Williams.

**Sunbeach Communications:** 'San Remo', Belmont, St Michael; tel. 430-1569; fax 228-6330; e-mail customerservice@sunbeach.net; internet www.sunbeach.net; f. 1995; awarded licence to operate cellular telephone services in March 2003; Man. Dir Michael Wakley.

### BROADCASTING

#### Radio

**Barbados Broadcasting Service Ltd:** Astoria St George, Bridgetown; tel. 437-9550; fax 437-9203; e-mail action@sunbeach.net; f. 1981; operates BBS FM and Faith FM (religious broadcasting).

**Barbados Rediffusion Service Ltd:** River Rd, Bridgetown; tel. 430-7300; fax 429-8093; f. 1935; public company; runs the commercial stations HOTT FM (f. 1998), Voice of Barbados (f. 1981) and YESS Ten-Four FM (f. 1988); Gen. Man. Vic Fernandes.

**Caribbean Broadcasting Corporation (CBC):** The Pine, POB 900, Wildey, St Michael; tel. 429-2041; fax 429-4795; e-mail cbc@caribsurf.com; internet www.cbcbarbados.bb; f. 1963; Chair. Folzo Brewster; operates three radio stations.

**CBC Radio 900 AM:** tel. 434-1900; f. 1963; spoken word and news.

**Liberty 98.1 FM:** tel. 429-5522; f. 1984; popular music.

**Quality 100.7 FM:** international and regional music, incl. folk, classical, etc.

**Voice of Barbados 92.9 FM:** Bridgetown; internet vob929.com; f. 1981; owned by the Starcom network; current affairs, sport, music.

#### Television

**CBC TV:** The Pine, POB 900, Wildey, St Michael; tel. 429-2041; fax 429-4795; e-mail cbc@caribsurf.com; internet www.cbcbarbados.bb; f. 1964; part of the Caribbean Broadcasting Corpn (q.v.); Channel Eight is the main national service, broadcasting 24 hours daily; a maximum of 115 digital subscription channels will be available through Multi-Choice Television; Gen. Man. Melba Smith; Programme Man. Hilda Cox.

# Finance

By November 2004 there were 4,635 international business companies and 53 'offshore' banks registered in Barbados. In addition, there were 2,981 foreign sales corporations in December 2001.

### BANKING

(cap. = capital; auth. = authorized; res = reserves; dep. = deposits; brs = branches; m. = million; amounts in Barbados dollars unless otherwise indicated)

#### Central Bank

**Central Bank of Barbados:** Tom Adams Financial Centre, Spry St, POB 1016, Bridgetown; tel. 436-6870; fax 427-9559; e-mail cbb.libr@caribsurf.com; internet www.centralbank.org.bb; f. 1972; bank of issue; cap. 2.0m., res 10.0m., dep. 690.2m. (July 2004); Gov. Marion V. Williams; Deputy Gov. Carlos A. Holder.

#### Commercial Banks

**Caribbean Commercial Bank Ltd (CCB):** Lower Broad St, POB 1007c, Bridgetown; tel. 431-2500; fax 431-2530; f. 1984; purchased by RBTT Financial Holdings, Trinidad and Tobago, in 2004; cap. 25.0m., res 6.0m., dep. 258.5m. (Dec. 2002); Chair. David C. Shorey; Pres. and CEO Mariano R. Browne; 4 brs.

**FirstCaribbean International Bank Ltd:** Warrens, POB 503, St Michael; tel. 367-2500; fax 424-8977; e-mail firstcaribbeanbank@firstcaribbeanbank.com; internet www.firstcaribbeanbank.com; f. 2002; previously known as CIBC West Indies Holdings, adopted present name following merger of CIBC West Indies and Caribbean operations of Barclays Bank PLC; cap. and res 1,898.1m., dep. 14,512.2m. (Oct. 2003); Chair. Michael K. Mansoor; CEO Charles Pink; 9 brs.

**Mutual Bank of the Caribbean Inc:** Trident House, Lower Broad St, Bridgetown; tel. 436-8335; fax 429-5734; 4 brs.

#### Regional Development Bank

**Caribbean Development Bank:** Wildey, POB 408, St Michael; tel. 431-1600; fax 426-7269; e-mail info@caribank.org; internet www.caribank.org; f. 1970; cap. US $155.7m., res US $14.1m., total assets US $756.2m. (Dec. 2003); Pres. Dr Compton Bourne.

#### National Bank

**Barbados National Bank Inc (BNB):** Independence Sq, POB 1002, Bridgetown; tel. 431-5700; fax 426-2606; internet www.bnbbarbados.com; f. 1978 by merger; cap. 48.0m., res 56.4m., dep. 1,161.7m. (Sept. 2003); partially privatized; Chair. Ronald F. D. Harford; Man. Dir and CEO Louis A. Greenidge; 6 brs.

#### Foreign Banks

**Bank of Nova Scotia (Canada):** Broad St, POB 202, Bridgetown; tel. 431-3000; fax 228-8574; e-mail peter.vanschie@scotiabank.com; f. 1956; Gen. Man. Peter F. van Schie; 6 brs.

**Royal Bank of Canada:** 2nd Floor, Bldg 2, Chelston Park, Collymore Rock, POB 986, St Michael; tel. 431-6680; fax 436-9675; e-mail roycorp@caribsurf.com; f. 1911; 1 br.

#### Trust Companies

**Bank of Nova Scotia Trust Co (Caribbean) Ltd:** Bank of Nova Scotia Bldg, Broad St, POB 1003b, Bridgetown; tel. 431-3120; fax 426-0969.

**Barbados International Bank and Trust Co:** Bissex House, Bissex, St Joseph; tel. 422-4629; fax 422-7994; e-mail bdosintlbank&trustco@caribsurf.com; f. 1981; Chair. Douglas Leese.

**Clico Mortgage & Finance Corporation:** C. L. Duprey Financial Centre, Walrond St, Bridgetown; tel. 431-4719; fax 426-6168; e-mail cmfc@sunbeach.net.

**Concorde Bank Ltd:** The Corporate Centre, Bush Hill, Bay St, POB 1161, St Michael; tel. 430-5320; fax 429-7996; e-mail concorde@sunbeach.net; f. 1987; cap. US $2.0m., res US $2.0m., dep. US $16.1m. (June 2003); Pres. Gerard Lussan; Man. Marina Corbin.

**Ernst & Young Trust Corporation:** Bay St, POB 261, St Michael; tel. 430-3900; fax 435-2079.

**FirstCaribbean International Trust and Merchant Bank (Barbados) Ltd:** Warren, St Michael; tel. 367-2324; fax 421-7178; internet www.firstcaribbeanbank.com; known as CIBC Trust and Merchant Bank until 2002.

**Royal Bank of Canada Financial Corporation:** 2nd Floor, Bldg 2, Chelston Park, Collymore Rock, POB 986, St Michael; tel. 431-6580; fax 429-3800; e-mail roycorp@caribsurf.com; Man. N. L. Smith.

**St Michael Trust Corpn:** c/o PriceWaterhouseCoopers, Collymore Rock, St Michael; tel. 436-7000; fax 429-3747; e-mail bb@fiscglobal.com; internet www.fiscglobal.com; f. 1987.

### STOCK EXCHANGE

**Barbados Stock Exchange (BSE):** Carlisle House, 1st Floor, Hincks St, Bridgetown; tel. 436-9871; fax 429-8942; internet www

.bse.com.bb; f. 1987 as the Securities Exchange of Barbados; in 1989 the Govts of Barbados, Trinidad and Tobago and Jamaica agreed to link exchanges; cross-trading began in April 1991; reincorporated in 2001; Gen. Man. MARLON YARDE.

### INSURANCE

The leading British and a number of US and Canadian companies have agents in Barbados. By November 2004 there were 413 exempt insurance and insurance management companies were registered in the country. Local insurance companies include the following:

**Barbados Fire & Commercial Insurance Co:** Beckwith Pl., Broad St, POB 150, Bridgetown; tel. 431-2800; fax 426-0752; e-mail info@bfc-ins.com; internet www.bfc-ins.com; f. 1996 following merger of Barbados Commercial Insurance Co Ltd and Barbados Fire and General Insurance Co; f. 1880; Man. Dir DAVID A. DEANE.

**Insurance Corporation of Barbados (ICB):** Roebuck St, Bridgetown; tel. 427-5590; fax 426-3393; e-mail icb@icbbarbados.com; internet www.icbbarbados.com; f. 1978; partially privatized in 2000; cap. Bds $3m.; Chair. Dr JOHN MAYERS; Man. Dir WISMAR GREAVES; Gen. Man. MONICA SKINNER.

**Sagicor:** Sagicor Financial Centre, Lower Collymore Rock, St Michael; tel. 467-7500; fax 436-8829; e-mail info@sagicor.com; internet www.sagicor.com; f. 1840 as Barbados Mutual Life Assurance Society (BMLAS), changed name as above in 2002 after acquiring majority ownership of Life of Barbados (LOB) Ltd; Chair. COLIN G. GODDARD; Pres. and CEO DODRIDGE D. MILLER.

**United Insurance Co Ltd:** United Insurance Centre, Lower Broad St, POB 1215, Bridgetown; tel. 430-1900; fax 436-7573; e-mail united@caribsurf.com; f. 1976; Man. Dir DAVE A. BLACKMAN.

#### Association

**Insurance Association of the Caribbean Inc:** IAC Bldg, Collymore Rock, St Michael; tel. 427-5608; fax 427-7277; e-mail info@iac-caribbean.com; internet www.iac-caribbean.com; regional asscn; Dir (Barbados) DAVIS BROWNE.

## Trade and Industry

### GOVERNMENT AGENCY

**Barbados Agricultural Management Co Ltd:** Warrens, POB 719c, St Michael; tel. 425-0010; fax 425-0007; e-mail bamc@cariaccess.com; Chair. LINDSAY HOLDER; Gen. Man. E. LEROY ROACH.

### DEVELOPMENT ORGANIZATIONS

**Barbados Agriculture Development and Marketing Corpn:** Fairy Valley, Christ Church; tel. 428-0250; fax 428-0152; f. 1993 by merger; programme of diversification and land reforms; Chair. TYRONE POWER; CEO E. LEROY ROACH.

**Barbados Investment and Development Corpn:** Pelican House, Princess Alice Highway, Bridgetown; tel. 427-5350; fax 426-7802; e-mail bidc@bidc.org; internet www.bidc.com; f. 1992 by merger; facilitates the devt of the industrial sector, especially in the areas of manufacturing, information technology and financial services; offers free consultancy to investors; provides factory space for lease or rent; administers the Fiscal Incentives Legislation; CEO ANTON NORRIS.

**Department for International Development in the Caribbean:** Chelsea House, Chelsea Rd, POB 167, St Michael; tel. 430-7900; fax 430-7959; division of the British Govt's Dept for Int. Devt.

### CHAMBER OF COMMERCE

**Barbados Chamber of Commerce and Industry:** Nemwil House, 1st Floor, Lower Collymore Rock, POB 189, St Michael; tel. 426-2056; fax 429-2907; e-mail bdscham@caribsurf.com; internet www.bdscham.com; f. 1825; 205 mem. firms; some 300 reps; Pres. MARK A. B. THOMPSON.

### INDUSTRIAL AND TRADE ASSOCIATIONS

**Barbados Agricultural Society:** The Grotto, Beckles Rd, St Michael; tel. 436-6683; fax 435-0651; e-mail heshimu@sunbeach.net; Pres. TYRONE POWER.

**Barbados Association of Medical Practitioners:** BAMP Complex, Spring Garden, St Michael; tel. 429-7569; fax 435-2328; e-mail bamp@sunbeach.net; Pres. Dr MARGARET O'SHEA.

**Barbados Association of Professional Engineers:** POB 666, Bridgetown; tel. 425-6105; fax 425-6673; f. 1964; Pres. GLYNE BARKER; Sec. PATRICK CLARKE.

**Barbados Hotel and Tourism Association:** Fourth Ave, Belleville, St Michael; tel. 426-5041; fax 429-2845; e-mail bhta@maccs.com.bb; internet www.funbarbados.com/bhta; Pres. ALAN BANFIELD; First Vice-Pres. SUSAN SPRINGER.

**Barbados Manufacturers' Association:** Bldg 1, Pelican Industrial Park, St Michael; tel. 426-4474; fax 436-5182; e-mail bmex-products@sunbeach.net; internet www.bma.org.bb; f. 1964; Pres. PETER MILLER; Exec. Dir WENDELL CALLENDER; 100 mem. firms.

### EMPLOYERS' ORGANIZATION

**Barbados Employers' Confederation:** Nemwil House, 1st Floor, Collymore Rock, St Michael; tel. 426-1574; fax 429-2907; e-mail becon@sunbeach.net; internet www.barbadosemployers.com; f. 1956; Pres. Dr HENSLEY SOBERS; Exec. Dir HARRY HUSBANDS; 235 mems (incl. associate mems).

### MAJOR COMPANIES

#### Construction and Cement

**Arawak Cement Co Ltd:** Checker Hall, St Lucy; tel. 439-9880; fax 439-7976; e-mail arawak@arawakcement.com.bb; internet www.arawakcement.com.bb; f. 1981; manufacture and marketing of cement and lime (quicklime and hydrated lime); 100% owned by Trinidad Cement Ltd; sales US $30m. (2003); Chair. WALTON JAMES; CEO Dr ROLLIN BERTRAND; Gen. Man. FRED BROOME-WEBSTER; 300 employees.

**Dacosta Mannings:** POB 103, Bridgetown; tel. 431-8700; fax 421-2014; e-mail info@dacostamannings.com; internet www.dacostamannings.com; f. 1969; trade in building materials, furniture and hardware products; Man. Dir WAYNE KIRTON; 276 employees.

**Edghill Associates Ltd:** Websters Industrial Park, Wildey, St Michael; tel. 427-2941; fax 426-5958; e-mail edghill@caribsurf.com; heavy construction; Dir RICHARD EDGHILL; 300 employees.

**C. O. Williams Construction Ltd:** Lears, St Michael; tel. 436-3910; fax 427-5336; e-mail info@cow.bb; f. 1960; Chair. Sir CHARLES OTHNEIL WILLIAMS.

#### Food and Beverages

**ADM Barbados Mills Ltd:** Spring Garden Highway, POB 260, St Michael; tel. 427-8880; fax 427-8886; internet www.admworld.com; f. 1977; manufacturer of flour and other grain-derived products; Gen. Man. CARLOS BELGRAVE; 93 employees.

**Banks Barbados Breweries Ltd:** Wildey, St Michael; tel. 429-2113; e-mail info@banksbeer.com; internet www.banksbeer.com; f. 1961; brewery; sales Bds $146m. (2002/03); 100 employees.

**Barbados Agricultural Management Co Ltd:** POB 719, Bridgetown, St Michael; tel. 425-2211; fax 425-3505; e-mail bamc@cariaccess.com; f. 1993; growing and processing of sugar cane; sales Bds $43.6m. (2000); Chair. R. CARL SYLVESTER; Exec. Chair. ATTLEE BRATHWAITE; Gen. Man. E. LEROY ROACH; 1,285 employees.

**Barbados Bottling Co Ltd:** POB 226, Bridgetown; tel. 420-8881; fax 428-4095; manufacturer of soft drinks; f. 1944; Chair. ALAN FIELDS; Man. Dir RICHARD COZIER; Gen. Man. DAN B. STOUTE; 113 employees.

**Barbados Dairy Industries Ltd:** Pine Hill Dairy, POB 56B, Bridgetown; tel. 430-4100; fax 429-3514; e-mail cgibson@banksholdings.com.bb; f. 1966; manufacturer of dairy and related products; sales US $29m. (2003); Chair. ALLAN FIELDS; Man. Dir CLYDE GIBSON; 200 employees.

**BICO Ltd:** Harbour Industrial Park, Bridgetown; tel. 430-2100; fax 426-2198; e-mail admin@bicoltd.com; f. 1901; manufacturer and distributor of ice cream; operators of public cold-storage facilities; sales Bds $13.1m. (2000); Chair. F. EDWIN THIRLWELL; 115 employees.

**A.S. Bryden and Sons (Barbados) Ltd:** Barbarees Hill, St Michael; tel. 431-2600; fax 426-0755; e-mail barbados@brydens.com; internet www.brydens.com; f. 1898; Manufacturers' representative and distributor for food and beverages, pharmaceuticals, photographic supplies, personal care and household cleaners; sales US $40m.; CEO ANDREW LEWIS; 123 employees.

**Hanschell Inniss:** Goddard's Complex, Kensington, Fontabelle, St Michael; tel. 426-0720; fax 427-6938; internet www.goddardenterprisesltd.com; manufacture and distribution of food and soft drinks; owned by Goddard Enterprises Ltd since 1973; Chair. PATRICK MAYERS; 117 employees.

**Mount Gay Distilleries Ltd:** POB 208, Brandon's Gap, Deaco's Rd, POB 208, Bridgetown; tel. 425-9897; fax 425-8770; e-mail info@mountgay.com; internet www.mountgay.com; f. 1955; subsidiary of McKesson Corpn, USA; rum distilling; Man. Dir PATRICK DUSSOSSOY MARSHALL; 55 employees.

**R. L. Seale and Co Ltd:** Clarence House, Tudor Bridge, POB 864, St Michael; tel. 426-0334; fax 436-6003; e-mail rseale@caribsurf

# BARBADOS

.com; f. 1926; manufacture and distribution of rum, sale of food; sales US $55m. (2003); Chair. and Man. Dir Sir DAVID SEALE; 275 employees.

### Miscellaneous

**Barbados Shipping and Trading Co Ltd:** see section on Shipping.

**BRC West Indies Ltd:** Cane Garden, St Thomas; tel. 425-0687; fax 425-2941; e-mail brc@caribsurf.com; internet www.brcwestindies.com; f. 1979; manufacturer of wire mesh and steel products; sales Bds $24m. (1998); Chair. R. S. WILLIAMS; Man. Dir JOHN FRANCIS; 42 employees.

**Collins Ltd:** Warrens Industrial Park, POB 203, Bridgetown, St Michael; tel. 425-4550; fax 424-9182; e-mail colcar@caribsurf.com; distributor of pharmaceuticals, hospital supplies, toiletries, confectionery, canned and snack foods throughout the Caribbean; Man. Dir PETER F. BOURNE; 185 employees.

**Goddard Enterprises Ltd:** Mutual Bldg, 2nd Floor, Lower Broad St, Bridgetown; tel. 430-5700; fax 436-8934; e-mail gelinfo@goddent.bb; internet www.goddardenterprisesltd.com; f. 1921; rum production, meat processing, bakery production, in-flight and airport-terminal catering, duty-free sales, lumber and building supplies, air conditioning and electrical contracting, insurance and financial services, shipping agent, automotive agency; sales Bds $354m. (1997/98); Chair. C. G. GODDARD; Man. Dir JOSEPH N. GODDARD; 2,593 employees.

**Mico Garment Factory Ltd:** Harbour Industrial Park, POB 621, Bridgetown; tel. 426-2941; fax 429-7267; e-mail info@misons-group.com; internet www.misons-group.com; f. 1964; manufacture of clothing; Chair. MOHAMMED IBRAHIM JUMAN; Man. Dir APHZAL JUMAN; 143 employees.

**Roberts Manufacturing Co Ltd:** POB 1275, Lower Estate, Bridgetown, St Michael; tel. 429-2131; fax 426-5604; e-mail roberts@rmco.com; internet www.rmco.com; f. 1944; subsidiary of Barbados Shipping and Trading Co Ltd; (see under Shipping); manufacturers of shortening, margarine, edible oils and animal feeds; sales Bds $58m. (1999); Chair. ALAN C. FIELDS; Gen. Man. M. A. CLARKE; 160 employees.

**St James Beach Hotels PLC:** Warrens, St Michael; tel. 438-4690; fax 438-4696; f. 1969; holding co in the hotel industry; Chair. RAY HORNEY; 720 employees.

## UTILITIES

### Electricity

**Barbados Light and Power Co (BL & P):** POB 142, Garrison Hill, St Michael; tel. 436-1800; fax 429-6000; f. 1911; electricity generator and distributor; operates three stations with a combined capacity of 209,500 kW; Chair. F. O. MCCONNEY; Man. Dir ANDREW GITTENS.

### Gas

**Barbados National Oil Co Ltd (BNOCL):** POB 175, Woodbourne, St Philip; tel. 423-0918; fax 423-0166; e-mail ronhewitt@bnocl.com; internet www.bnocl.com; f. 1979; extraction of petroleum and natural gas; state-owned, scheduled for privatization; Chair. HARCOURT LEWIS; Gen. Man. RONALD HEWITT; 166 employees.

**National Petroleum Corporation (NPC):** Wildey, St Michael; tel. 430-4020; fax 426-4326; gas production and distribution; Chair. Dr WILLIAM DUGUID; Gen. Man. KEN LINTON.

### Water

**Barbados Water Authority:** The Pine, St Michael; tel. 426-4134; fax 426-4507; e-mail bwa@caribsurf.com; f. 1980; Gen. Man. DENNIS YEARWOOD.

## TRADE UNIONS

Principal unions include:

**Barbados Secondary Teachers' Union:** Ryeburn, Eighth Ave, Belleville, St Michael; tel. and fax 429-7676; e-mail bstumail@caribsurf.com; f. 1949; Pres. PHIL PERRY; Gen. Sec. PATRICK FROST; 375 mems.

**Barbados Union of Teachers:** Merry Hill, Welches, POB 58, St Michael; tel. 427-8510; fax 426-9890; e-mail but4@hotmail.com; internet www.butbarbados.org; f. 1974; Pres. UNDENE WHITTAKER; Gen. Sec. HERBERT GITTENS; 1,800 mems.

**Barbados Workers' Union (BWU):** Solidarity House, Harmony Hall, POB 172, St Michael; tel. 436-3492; fax 436-6496; e-mail bwu@caribsurf.com; internet www.bwu-bb.org; f. 1941; operates a Labour College; Pres.-Gen. HUGH ARTHUR; Gen. Sec. Sir LEROY TROTMAN; 25,000 mems.

**National Union of Public Workers:** Dalkeith House, Dalkeith Rd, POB 174, St. Michael; tel. 426-1764; fax 436-1795; e-mail nupwbarbados@sunbeach.net; f. 1944; Pres. CECIL W. DRAKES; Gen. Sec. JOSEPH E. GODDARD; 6,000 mems.

# Transport

## ROADS

**Ministry of Public Works:** The Pine, St Michael; tel. 429-2191; fax 437-8133; internet www.publicworks.gov.bb; maintains a network of 1,600 km (994 miles) of roads, of which 1,578 km (981 miles) are paved; Chief Tech. Officer C. H. ARCHER.

## SHIPPING

Bridgetown harbour has berths for eight ships and simultaneous bunkering facilities for five. In October 2003 the Government announced a 10-year plan to expand the harbour, at an estimated cost of US $101m. The plan was to include the construction of a new sugar terminal and a new cruise-ship pier.

**Barbados Port Authority:** University Row, Bridgetown Harbour; tel. 430-4700; fax 429-5348; e-mail administrator@barbadosport.com; internet www.barbadosport.com; f. 1979; Chair. LARRY TATEM; Gen. Man. EVERTON WALTERS; Port Dir Capt. H. L. VAN SLUYTMAN.

**Shipping Association of Barbados:** 2nd Floor, Trident House, Broad St, Bridgetown; tel. 427-9860; fax 426-8392; e-mail shasba@caribsurf.com.

### Principal Shipping Companies

**Barbados Shipping and Trading Co Ltd (B. S. and T.):** 1st Floor, The Auto Dome, Warrens, St Michael; tel. 417-5110; fax 417-5116; e-mail info@bsandtco.com; internet www.bsandtco.com; f. 1920; Chair. ALLAN C. FIELDS; CEO G. ANTHONY KING.

**Bernuth Agencies:** Bridgetown; tel. 431-3343.

**Carlisle Shipping Ltd:** Carlisle House, Bridgetown; tel. 430-4803.

**DaCostas Ltd:** Carlisle House, Hincks St, POB 103, Bridgetown; tel. 426-3451; fax 429-5445; shipping company; Man. Dir JOHN WILKINSON.

**T. Geddes Grant Bros:** White Park Rd, Bridgetown; tel. 431-3300.

**Hassell, Eric and Son Ltd:** Carlisle House, Hincks St, Bridgetown; tel. 436-6102; fax 429-3416; e-mail info@erichassell.com; internet www.erichassell.com; shipping agent, stevedoring contractor and cargo fowarder.

**Maersk:** James Fort Bldg, Hincks St, Bridgetown; tel. 430-4816.

**Tropical Shipping Kensington:** Fontabelle Rd, St Michael; tel. 426-9990; fax 426-7750; internet www.tropical.com.

## CIVIL AVIATION

The principal airport is Grantley Adams International Airport, at Seawell, 18 km (11 miles) from Bridgetown. A US $100m. contract to build a new arrivals terminal was awarded in late 2001. The project was scheduled for completion by late 2005.

# Tourism

The natural attractions of the island consist chiefly of the warm climate and varied scenery. In addition, there are many facilities for outdoor sports of all kinds. Revenue from tourism increased from Bds $13m. in 1960 to some $1,534m. in 2003. The number of stop-over tourist arrivals was 551,502 in 2004, while the number of visiting cruise-ship passengers was 721,270. There were some 7,000 hotel rooms on the island in 2003.

**Barbados Hotel and Tourism Association (BHTA):** POB 711C, Bridgetown; tel. 426-5041; fax 429-2845; e-mail info@bhta.org; internet www.bhta.org; f. 1952 as the Barbados Hotel Asscn; adopted present name in 1994; non-profit trade asscn; Pres. JON MARTINEAU.

**Barbados Tourism Authority:** Harbour Rd, POB 242, Bridgetown; tel. 427-2623; fax 426-4080; e-mail btainfo@barbados.org; internet www.barbados.org; f. 1993 to replace Barbados Board of Tourism; offices in London, New York, Los Angeles, Miami, Munich, Paris and Caracas; Chair. HUDSON HUSBANDS; Pres. and CEO STUART LANE.

## Defence

The Barbados Defence Force was established in April 1978. It is divided into regular defence units and a coastguard service with armed patrol boats. The total strength of the armed forces at August 2004 was an estimated 610, comprising an army of 500 members and a navy of 110. There was also a reserve force of 430 members.

**Defence Budget (2004):** Bds $26.0m. (US $13m.).

**Chief of Staff:** Col ALVIN QUINTYNE.

## Education

Education is compulsory for 12 years, between five and 16 years of age. Primary education begins at the age of five and lasts for seven years. Secondary education, beginning at 12 years of age, lasts for six years. In 2002 enrolment of children in the primary age-group was 99.1% (males 99.8%, females 98.5%), and in the secondary-school age-group enrolment was 98% (males 97.4%, females 98.6%). Tuition at all government schools is free. In the same year enrolment at tertiary level was some 48% of the relevant age-group (males 38.5%; females 57.5%), with 11,226 students registered. Degree courses in arts, law, education, natural sciences and social sciences are offered at the Cave Hill campus of the University of the West Indies. A two-year clinical-training programme for medical students is conducted by the School of Clinic Medicine and Research of the University, while an in-service training programme for teachers is provided by the School of Education. The Government announced in September 2000 that it expected to realize its ambition of providing universal pre-school education by the academic year 2002. In 2002 80% of children aged three and four attended school. Current expenditure on education by the central Government in 2001/02 totalled some Bds $375m. (equivalent to about 19.5% of total expenditure). According to the 2002 budget address, government expenditure on education in 2002/03 would equal Bds $396m., representing 19.8% of total expenditure. A US $200m. programme for curriculum reform and the introduction of technology into education, known as Edutech 2000, was scheduled for completion by 2010.

## Bibliography

For works on the Caribbean generally, see Select Bibliography (Books)

Beckles, H. *A History of Barbados from Amerindian Settlement to Nation State*. Cambridge, Cambridge University Press, 1990.

Broberg, M. *Barbados*. New York, NY, Chelsea House Publications, 1998.

Carmichael, T. A. *Barbados: Thirty Years of Independence*. Kingston, Ian Randle Publrs, 1996.

Drummond, I., and Marsden, T. *The Condition of Sustainability*. London, Routledge, 1999.

Girvan, N. (Ed.). *Poverty, Empowerment and Social Development in the Caribbean*. Bridgetown, Canoe Press, 1997.

Ligon, R. *A True and Exact History of the Island of Barbadoes*. London, Frank Cass Publrs, 1998.

Schomburg, R. *History of Barbados*. London, Frank Cass Publrs, 1998.

# BELIZE

## Geography

### PHYSICAL FEATURES

Belize is on the north-eastern shores of Central America. Mexico lies to the north, beyond a 250-km (155-mile) border, and Guatemala to the west and south, beyond a 266-km border, with Honduras in the south-east, across the Gulf of Honduras. Belize is a Central American, Commonwealth country, which was known as British Honduras until 1973, when it was a dependent territory of the United Kingdom. It became independent only in 1981, the last country on the American mainland to do so, but its history and culture have made it more usually associated with the anglophone West Indian states than with its Spanish-speaking neighbours (it is the only Central American country not to have a Pacific coast). These neighbours also have territorial claims on Belize, with Guatemala going so far as only to recognize the country's independence in September 1991. The Organization of American States (OAS) is currently mediating the disputes over Guatemala's territorial claims and its rights of maritime access to the Caribbean. There are also problems associated with the 2000 agreement on managing disagreements within the 'Lines of Adjacency' (1 km either side of the Belize–Guatemala border), which attempt to limit illegal immigration (by 'squatters') coming into Belize. Honduras claims the Sapodilla Cays. The current total area of Belize covers 22,965 sq km (8,867 sq miles), making it about the size of Wales (United Kingdom) but with a population less than that of Cardiff, the Welsh capital.

The territory of Belize, which includes 160 sq km of inland waters and a maritime littoral of 386 km, is a flat, swampy coastal plain, with low mountains in the south. A resurvey of the Maya Mountains recently superseded Victoria Peak, in the south-east, with the nearby peak of Doyle's Delight (1,174 m or 3,853 ft) as the highest point in the country. The low-lying north of the country was once the bed of the sea, and supports scrubby vegetation or dense tropical hardwood forest. Particularly in this area, the landscape is typified by jungles laced with a seasonally navigable river network. Central Belize has sandy soil, supporting savannah grasslands, to the south the land rising into the lofty Mountain Pine Ridge area and, thence, the Maya Mountains, which continue west into Guatemala. Here rainfall fuels the many streams, such as the Macal (which, with the Mopan, becomes the River Belize). South of the watershed is a more precipitous landscape, with short, fast streams carrying fertile soils and detritus to the coast, which permits not only a flourishing agriculture but also the longer-established tropical rainforest. Belize's shores are guarded by a coral barrier-reef system that is second in size only to the Great Barrier Reef of Australia. This extends the territory of the country to include a number of islands and cays (mangrove cays and island cays) off shore, in the Caribbean Sea. At almost 300 km in length, it is certainly the longest reef in the Americas. Two-fifths of the country, including the marine environment, is protected by parks and reserves.

### CLIMATE

The climate is tropical and very hot and humid, despite the prevailing winds off the Caribbean. The country has an annual mean temperature of 79°F (26°C), with maximums seldom above 96°F (36°C) or below 60°F (16°C), even at night. The rainy season is in May–November, with hurricanes likely from June. The coast is prone to flooding, particularly in the south. The dry season is in February–May. Average annual rainfall ranges from 50 ins (1,270 mm) in the north to 170 ins (4,320 mm) in the south. Complications in this pattern have been observed in recent years (noticeably owing to the El Niño phenomenon and global warming).

### POPULATION

The original racial balance in Belize has changed since independence, and particularly during the 1980s, mainly owing to immigration. The growing mixed Maya-Spanish or Mestizo population accounts for 46% of the total, while the previously dominant, black Creole population accounts for 28%. The autochthonous Amerindians consist mainly of the Maya (10%), although there are also immigrant, mixed-race Garifuna ('Black Carib'—6%) peoples. Other groups include those of European descent (including German Mennonites), 'East' Indians and Chinese. This changing balance of population has not had a discernible effect on the pattern of religious adherence, with Roman Catholics in 2003 accounting for about three-quarters of the population. Other Christian denominations command the faith of the majority of the remainder of the population (mainly Pentecostalists, Anglicans, Seventh-day Adventists, Mennonites, Methodists and Jehovah's Witnesses). English is the official language, but immigration (often, originally, illegal) from the neighbouring Hispanic countries means that Spanish is now widely spoken. There is also a Creole dialect in use and native speakers of Amerindian tongues such as Maya, Garifuna and Ketchi.

The total population was 282,600, according to estimates for mid-2004, making Belize the smallest country of Central America in terms of population (El Salvador is slightly smaller in extent). The capital since 1972 has been Belmopan, in the centre of the country, although, according to official estimates, it still had a population of only 12,300 at mid-2004. The old capital on the coast, Belize City, remains the largest urban centre (59,400), and both Orange Walk in the north and San Ignacio near the central western border have larger populations than the capital. The chief town of the south is Punta Gorda. The country is the least urbanized in Central America, with only about one-10th of the population living in urban areas. Belize is divided into six districts.

# History

## HELEN SCHOOLEY

Revised for this edition by the editorial staff

### INTRODUCTION

Belize lies on the Caribbean coast of the Central American isthmus. It was the centre of the Mayan empire, which flourished between AD 300 and 600, but by the time the Spanish arrived in the 16th century there were very few Mayans left, and the Mayan buildings were already in ruins. As Spanish colonial rule here was only tenuous, the earliest European settlers were British buccaneers, who began arriving in the mid-17th century and made use of the many sheltered anchorage points along the coast for raids on Spanish shipping. Spain mounted repeated attacks against the British settlers, but in 1763 granted them rights to fell logwood, and in 1802 recognized British sovereignty by the Treaty of Amiens. Under the name British Honduras, the area was governed by British administrators from 1786. It became subject to British law in 1840 and was designated a crown colony in 1862. The significance of the colony to the United Kingdom (which had trading interests all along the Caribbean coast) was primarily economic; it had the best deepwater port facilities in the region and in the early 19th century up to 95% of Central American trade passed through the settlement.

The country's ethnic structure reflects successive waves of immigration, encouraged because of the country's underpopulation. There were relatively few original Amerindian (chiefly Mayan) inhabitants, and not many British settlers stayed permanently. From the early 17th century many Africans came as slaves (slavery in the British colonies was abolished progressively between 1833 and 1838), and many early immigrants came from the West Indies. According to the 1980 census, Creoles (of mainly African descent) accounted for 40% of the population, Mestizos (of mixed Mayan and European descent) for 33%, Garifuna or 'Black Caribs' (of mixed Amerindian and African blood, deported from Saint Vincent in 1797) for 8% and Amerindians, the ethnic descendants of the original inhabitants, for 7%. However, the arrival in the 1980s of an estimated 30,000 Central American refugees into a population of around 200,000, caused a major change in the country's ethnography, and meant that Spanish speakers soon outnumbered English speakers, causing some resentment. By 2001 Mestizos accounted for 46% of the population, Creoles for 28%, Amerindians for 10% and Garifuna for 6%, with a further 10% of the population being more recent immigrants, notably from Asia and the Middle East. In 2001 an estimated 70,000 Belizean citizens were resident in the USA.

After gaining independence from Spain in 1821, Guatemala began to lay claim to the area (as did Mexico for a limited period). In 1859 Guatemala recognized the colony's boundaries in exchange for a British undertaking to construct a road from Guatemala City to the Caribbean coast. However, the road was never built, as the commercial importance of British Honduras declined markedly with the completion of the Panama railway in 1855, moving the focus of Central American trade to the Pacific seaboard. Guatemala did not actively pursue the matter of the unbuilt road until it abrogated the treaty in 1945, claiming that the United Kingdom had failed to fulfil its obligations.

### THE RISE OF THE PUP

In 1949 the colony's first major internal political issue arose, when a scandal occurred over a substantial devaluation of the British Honduras dollar. The controversy led to the formation of the People's United Party (PUP) in 1950, led by George Price, which assumed an anti-British stance and demanded independence. A general election with universal suffrage was held in 1954, in accordance with a new Constitution. Of the nine seats in the new Legislative Assembly, the PUP won eight. The party won all nine seats at the 1957 election and all 18 seats in the enlarged, and renamed, House of Representatives in 1961, elected under a revised Constitution that established ministerial government with a bicameral legislature. The post of First Minister was assumed by Price.

The colony was granted self-government in 1964 and, after the 1965 elections, George Price became Premier. In January 1972 the city of Belmopan, built to replace Belize City, which had been devastated by a hurricane in 1961, was declared the new capital, and in June 1973 British Honduras was officially renamed Belize (the name is variously described as originating from the Mayan description of the 'muddy water', *belix*, of the Belize river, or from the Spanish pronunciation of Wallace, the name of the buccaneer who established the first British settlement, in 1638).

### INDEPENDENCE

The granting of self-government prompted Guatemala to renew its claim to the territory, and this claim has been the subject of a series of protracted negotiations ever since. Talks collapsed in 1972, when the United Kingdom decided to station a permanent military garrison in the colony, and again in February 1975, when Guatemala proposed the cession of about one-quarter of Belize's territory. In response to the impasse, in 1975 the PUP began an international campaign for independence, and gained the support of the Caribbean Community and Common Market (CARICOM), which Belize had joined in 1974, and the Non-aligned Movement. In 1980 the UN General Assembly also issued a resolution in favour of Belizean independence—139 countries (including, for the first time, the USA) in favour, seven abstaining and Guatemala refusing to vote. In 1976 Panama had become the first of the Central American states to oppose Guatemala's claim to the territory and by 1980 only Honduras and El Salvador still supported Guatemala.

Unlike the PUP, Belize's other main political organization, the United Democratic Party (UDP), advocated a gradual process of independence. Its members organized anti-Government demonstrations in March 1981 when a tripartite agreement was reached, incorporating a broad commitment to Belizean independence and respect for its territorial integrity, while providing Guatemala with access to the Caribbean Sea. The so-called Heads of Agreement proposed a formal treaty, to be concluded before a constitutional conference on independence held in April. When the constitutional conference was held first instead, Guatemala protested and suspended consular relations with the United Kingdom. The accord was also unpopular in the colony, but, with the British finally conceding security guarantees, it was concluded that conditions were sufficient for Belize to proceed to independence.

Belize attained full independence, within the Commonwealth, on 21 September 1981 and with George Price as Prime Minister. The UDP boycotted the celebrations and Guatemala refused to recognize the state's sovereignty. Meanwhile, the United Kingdom agreed to continue its defence commitment to Belize for 'an appropriate period'. During the 1980s negotiations foundered on Guatemala's refusal to recognize any delegate from Belize, and it was not until 1991 that the Guatemalan Government officially recognized 'the right of the people of Belize to self-determination', and decided to establish diplomatic relations with the country. In return for this concession, Belize agreed to a redefinition of its maritime boundaries, allowing Guatemala access to ports on its Caribbean coast. The Maritime Areas Bill was approved in 1992.

The 30-year domination by Price and the PUP of national politics had ended in December 1984, when the UDP gained an overwhelming victory in legislative elections, and its leader, Manuel Esquivel, became Prime Minister. The PUP returned to power in September 1989, still under Price's leadership, but lost in 1993. The announcement of the departure of the British garrison and criticism of the accord with Guatemala (see below), as well as economic issues—in particular, the fear of devaluation in order to meet election-campaign tax promises—resulted in the PUP's poor performance in the June general election. Although the PUP won more than 51% of the votes, the UDP gained 16 seats in the 29-seat House of Representatives, and Esquivel became Prime Minister. Meanwhile, George Price had left the premiership for the final time; his successor as leader of the PUP was Said Musa.

## THE INTERNATIONAL RAMIFICATIONS OF DOMESTIC POLITICS

Guatemala was not the only foreign country to have an intimate connection with Belizean affairs. Relations with Guatemala (see below) could dominate the political agenda, and the installation of a new UDP Government in Belize saw the suspension of the 1993 agreement's ratification process and the renewal of Guatemalan territorial claims against the Caribbean country. However, in the second half of the 1990s another issue attracted more international attention than the long-running dispute with Guatemala.

In January 1995 the UDP Government announced the latest in a series of controversial 'economic citizenship' programmes, under which foreign nationals could apply for a Belizean passport in return for an application fee of US $50,000 per family unit. The scheme was designed to attract wealthy Hong Kong citizens, and a similar programme had first been introduced by the PUP Government of 1979–84. Information offices were opened in many countries world-wide, although the scheme came under scrutiny from mid-1995, with reports that Belize was being used as a channel for illegal immigration, particularly from the People's Republic of China and from Taiwan (recognized by Belize as the Republic of China), to the USA. In domestic terms, however, the illegal immigration causing concern was that by Guatemalan peasants, which in mid-1994 Prime Minister Esquivel claimed was being encouraged by the Government of Guatemala. In an atmosphere of economic uncertainty, the increase in the Spanish-speaking population was causing some disquiet among the once dominant, English-speaking Creoles, so the passport programme and other efforts to encourage immigration from the Far East and the Middle East were seen as a balance. With the re-election of a PUP Government in 1998 though, the scheme was doomed. In March 2000 the Political Reform Commission, established in the previous year, recommended that the policy of economic citizenship be abolished. The Government undertook to end the practice from February 2002, but it had not yet lost the ability to attract controversy. In mid-2002 it was revealed that the sale of Belizean passports to foreign nationals had continued until July, despite the official termination of the policy, and Prime Minister Musa dismissed the Minister of Home Affairs, Maxwell Samuels, as a result. (Samuels subsequently returned to the Cabinet.)

Increasing dissatisfaction with the UDP Government had resulted in an overwhelming victory for the PUP in the general election of August 1998. The PUP won more than 59% of the votes cast, gaining 26 seats in the 29-seat House of Representatives, and its new leader, Said Musa, became Prime Minister. One of the contributing factors to the economic decline that had fuelled the unpopularity of the UDP was the withdrawal of the British garrison. The PUP, meanwhile, enjoyed the support of a less official British connection.

The PUP victory in 1998 immediately focused attention on the political influence of Michael Ashcroft (Lord Ashcroft from 2000), a businessman with dual Belizean and British citizenship, with extensive financial interests in Belize. He was widely reputed to have donated major support to the PUP election campaign. Ashcroft was chairman and chief executive officer (CEO) of the Belize Holdings (BHI) Corporation (known as Carlisle Holdings Ltd from 1999). In 1990 the previous PUP administration had granted BHI a 30-year tax exemption, which the Esquivel Government had attempted, unsuccessfully, to overturn.

An investigation by the British Government into the country's financial structure, carried out in 1999–2000, drew attention to the activities of Carlisle Holdings, and in early 2001 authorities in both the United Kingdom and the USA alleged that Ashcroft's financial and offshore enterprises might have been used by criminal interests for money-laundering. The British Government agreed to retract its allegations of criminal activity at Carlisle Holdings Ltd and to pay some £500,000 in damages, following a British High Court action by Ashcroft in June. Meanwhile, concern about the country being used by drugs traffickers and money launderers remained, although it was removed from the list of nations deemed by the USA to be not 'fully co-operating' with anti-drugs efforts. Also, a negative report in 2000 by an international task force against money laundering prompted some reform of the 'offshore' financial sector.

Although the international controversy over Ashcroft abated in the 2000s, he remained an influential figure in Belizean domestic politics. Nevertheless, Ashcroft's relations with the PUP Government were strained in late 2001, when the Government failed to renew the exclusive 15-year licence held by his company, Belize Telecommunications Ltd (BTL). The company sharply increased call charges in August 2002 and sought to have contracts that had been awarded to a competitor annulled in a judicial review. The Ministry of Communications and Public Utilities described BTL's actions as arrogant and, in November, the Supreme Court rejected BTL's case. Relations with Ashcroft deteriorated to such an extent that the Government instead sought a deal with a US businessman and entrepreneur, Jeffrey J. Prosser, who had significant telecommunications interests in the US Virgin Islands. He assumed control of BTL in April 2004. However, the deal unravelled one year later, when the Government accused Prosser of failing to pay as promised and resumed control of the company, incurring significant expenses, including a controversial contempt fine by a US court based in Florida, and the renewed interest of Lord Ashcroft, who again came to an arrangement with the Government.

In 2002 the country faced further criticism from abroad, when it was claimed that the system of allowing foreign ships to be registered with the Belizean authorities under a 'flag of convenience' was being abused by vessels fishing illegally. In September the Government considered measures to tighten the regulation of fishing boats, including the satellite tracking of vessels. In the following month the Attorney-General also had to defend before the Inter-American Commission on Human Rights a proposal to establish the Belize Court of Appeal as the highest appellate court in the judicial system, thereby preventing prisoners sentenced to death from appealing to the Privy Council (based in the United Kingdom). A group of prisoners claimed that the amendment would be incompatible with the Government's obligations under an Organization of American States (OAS) treaty on human rights.

Among the more serious influences from beyond the borders are the tropical storms and hurricanes from the Caribbean, which particularly affect the south. In October 2000 'Hurricane Keith' destroyed power and communications lines and caused damage worth about US $200m., and in October 2001 'Hurricane Iris' left 20 people dead and at least 13,000 homeless, prompting the Government to designate the south of the country a disaster zone. Such events have not helped the programme to improve living conditions in the south (introduced in January 1998). The south is home to many Honduran and Guatemalan refugees, and also to about one-half of the country's Mayan population, who oppose a number of regional development projects, notably challenging the Government's right to grant concessions to fell logwood on their ancestral land. The most significant development project to incur local and international opposition, however, was the Chalillo Dam, on the Macal river in central Belize (see Economy).

## THE 2003 GENERAL ELECTION

In a general election held on 5 March 2003 the PUP and Said Musa were returned to power, with the party securing 53.2% of votes cast; however, its parliamentary representation was reduced from 26 seats to 22. Nevertheless, the party had secured the first consecutive term in office of any Government since independence. It also still controlled five of the eight municipalities (down from seven in 2000). The PUP campaign focused on its record of increased spending on health, housing and education, although the opposition alleged that some of this expenditure had been biased in favour of the families of PUP party members. Nevertheless, despite Government problems with both international and domestic criticism of its financial arrangements and institutions, of shipping registration and of the sale of passports, the PUP was re-elected. The first years of office had been disturbed by changes in the Government, but a stable Cabinet since August 2002 had reassured the public (the Government was only reorganized in January 2004). However, many of the problems reappeared soon after the general election, not helped by fractures within the administration.

The most immediate domestic issue of Musa's second term was the country's rising levels of crime, notably of violent and organized crime. The trend, which was being experienced elsewhere in Central America, was fuelled by Belize's role as a transit point in drugs-trafficking. In addition to the crime wave, there were three other sources of pressure on the Government's continued popularity. Corruption remained a concern; during 2003 there were allegations that a number of government appointments had been on the basis of family connection. In December the highly controversial construction of the Chalillo Dam was referred to the Privy Council. The resulting judgment in January 2004 failed to halt the operation. Nevertheless, claims that the project would not fulfil its anticipated generating capacity and would pose an enormous and irreversible threat to the environment continued. The third issue was the state of the national economy, which had grown in the previous term of office, consequently raising expectations and increasing the fiscal deficit (see Economy).

On 16 August 2004 concerns about macroeconomic management within the Government were made dramatically public, when seven ministers, including the Deputy Prime Minister, John Briceño, resigned. They expressed particular concern about the growth in external debt and the manner in which government funds had been used in support of private ventures. Although an accommodation kept the group (sometimes referred to as the 'G7') in the Government and in a position to exercise some fiscal oversight, attention had been drawn to the parlous state of public finances and to some dubious practises in the administration. Some reforms were instituted, but by December, on the eve of budget preparations, Prime Minister Musa and his ally, Ralph Fonseca, the Minister of Home Affairs and Investment, were ready to reassert their control. Separate deals ensured that the solidarity of the 'G7' was split and, on 29 December, three of the ministers were dismissed—Mark Espat, Eamon Courtenay and Cordel Hyde (the former two being the principal representatives of the south)—and PUP support in the legislature consequently weakened.

Musa had regained control of the Government only to find popular protests a more significant challenge in 2005. The budget proposals in January, which both introduced tax increases and withheld a final instalment of an agreed public-sector pay rise, ignited public demonstrations that descended into violence in Belmopan. Widespread protests by political and popular opposition, uniting business and labour, prompted the Government to make some concessions in early February. The budget was revised and an inquiry into the collapse of the Development Finance Corporation was promised (it first sat in May), as was a reform of the procedures for awarding of state contracts. The Government agreed to limit public debt and increase oversight. Effectively independent budgets for certain public officials were proposed (notably for the judiciary and the Auditor-General), while certain appointments were to require approval by a two-thirds' majority in the Senate. A review of legislation that affected basic constitutional rights was ordered. However, the tax rises were not withdrawn and the Government refused to permit the pay increase. The tax rises took effect in March and in April the latest twist in the BTL fiasco sparked off further protests and rioting in Belize City, with communications for the whole country being severely disrupted. The strike by BTL workers was followed by industrial action by teachers, but a general strike did not materialize and the complete slowdown of January was not achieved. The Government accused the opposition of fomenting trouble, but the leak in May of an IMF document recommending further tax rises indicated that unrest was likely to continue.

## INTERNATIONAL RELATIONS

Although Belize traditionally identified itself as a Caribbean country, independence from the United Kingdom and accession to the OAS in 1991 (with recognition by Guatemala) heralded an increasing involvement in Latin American affairs. The influx during the 1980s of thousands of Central American refugees and settlers caused major changes in the population, with racial, political and social implications. By the 1991 census Spanish speakers outnumbered English speakers; during the 1990s the difference between the number of Mestizos and of Creoles widened, but official encouragement of immigration from Asia and the Middle East acted as a counterbalance to the Spanish influx.

On 13 May 1993 the United Kingdom announced the withdrawal of the British garrison in Belize, on the grounds that it would no longer be needed, given Belize's improved relations with Guatemala and the ending of civil wars in Nicaragua and El Salvador. Although the base continued to be used for regular military exercises, the gradual reduction of British military personnel, from 1,350 to 180, had pronounced effects on the national economy and in June 2000 led to an unsuccessful appeal by Belize for a return to the pre-1993 level of military aid. By the 2000s Belize was cultivating US military assistance and the first major deployment of US troops in the country occurred under the 'New Horizon' joint military exercise, held in February–May 2000. The exercise involved 3,500 US troops, working in relays of small teams, on social infrastructure projects, including the establishment of a programme of basic medical services. The USA is one of Belize's largest sources of aid, although an important donor is Taiwan, which Belize recognizes as the legitimate Government of China.

In February 1996 the Government ratified an agreement with Guatemala, El Salvador and Honduras to co-operate in the conservation and tourism promotion of Mayan archaeological sites. In March the Inter-American Investment Corporation, part of the Inter-American Development Bank (IDB), announced the allocation of funds for development projects in Belize. On 1 December 2000, following nine years of observer status, Belize was given full membership of the Sistema de la Integración Centroamericana (SICA—Central American Integration System), of which it occupied the presidency during the second half of 2003. In 2001 Belize formally applied for membership of the organization's Banco Centroamericano de Integración Económica (BCIE—Central American Bank for Economic Integration). The country was already a participant in the Caribbean Community and Common Market (CARICOM), which remained the main bloc for the purposes of economic policy.

Following Guatemalan recognition of Belize, in April 1993 Belize and Guatemala signed a non-aggression pact. However, relations between the two countries were jeopardized by the ousting of the Guatemalan President, Jorge Serrano Elías, on 1 June, and by the surprise defeat of the PUP in early legislative elections held later in the same month. The new UDP Government suspended ratification of the agreement with Guatemala, although the new Guatemalan President, Ramiro de León Carpio, announced that Guatemala would continue to respect Belize's independence. In March 1994 Guatemala formally reaffirmed its territorial claim against Belize. In mid-1994 Prime Minister Esquivel made his accusation about Guatemala employing destabilizing tactics by encouraging its peasants to occupy and settle in areas of Belizean forest.

Despite continued border tension, relations with Guatemala improved in the late 1990s, particularly with the return to power of the PUP. In 1998 there was an agreement to establish a joint commission to deal with immigration and cross-border traffic. However, bilateral negotiations scheduled for February 2000 were overshadowed by renewed tension. In January the Guatemalan Government announced that it was increasing its military presence near to the disputed territory, precipitating a series of border and diplomatic incidents. Bilateral talks resumed in May, when a panel of negotiators was installed at the headquarters of the OAS, in Washington, DC, USA. There were further border incidents in late 2000, despite an agreement in November designed to stabilize diplomatic and trade relations and to increase communications between the Belizean and Guatemalan armed forces. In February 2001 the Pan-American Institute of Geography and History, in Mexico City, Mexico, issued a report determining the location of the 'Lines of Adjacency' between Belize and Guatemala. The Institute's ruling was accepted by both countries, without prejudice on the sovereignty issue, as a result of which 45 Guatemalan families living in the disputed territory were relocated. In November three Guatemalan nationals were shot and killed during a confrontation with Belizean security forces in Toledo. Following a government investigation into the incident, the Prime Minister announced that the families of the victims were to be compensated.

In June 2002 the Minister of Foreign Affairs, Assad Shoman, and his Guatemalan counterpart, Gabriel Orellana Rojas, met for OAS-mediated discussions. These resulted in an agreement that Guatemala would recognize Belize's land boundary in exchange for extra maritime territory and proposals for the development of the border area. Plans to hold public referendums on the agreement in both countries by the end of the year were disrupted by a further border incident in October. The eventual signing, in February 2003, of a co-operation agreement, which provided for a 'transition process' and 'confidence-building measures', was hailed as a major landmark in the negotiation process. However, Guatemala continued to express reservations and in August rejected the OAS framework accord, preferring to take the case to the International Court of Justice. Belize emphasized the binding nature of the framework and of the subsequent transition agreement. The Guatemalan Government reiterated its rejection in January 2004, citing constitutional difficulties with a referendum. In February a new administration confirmed that decision but conceded that confidence-building measures remained desirable and that negotiations under the transition agreement should continue. Tensions over border incidents continued, notably in April and in early 2005, but co-operative initiatives also proceeded. The most important development was the move to liberalize trading links. In mid-2004 Guatemala offered discussion of a free trade agreement with Belize, which was not included in Central American developments in that area and, in November, the terms of reference of a 'partial scope' agreement were drawn up. Formal negotiations began in February 2005, proceeding with every likelihood of imminent resolution into June.

# Economy

## HELEN SCHOOLEY

Revised for this edition by the editorial staff

British settlers were originally attracted to Belize by its forests, for dyes and timber. Following the country's decline as a trading base in the mid-19th century, forestry became the chief economic activity until it was overtaken by agriculture in the 1960s. In the late 1950s some 40% of the land was held in large estates (one-third by the Belize Estates and Produce Co) and only 3% was under cultivation; by 2002 some 6.7% of the total land area was under production, although this represented less than one-fifth of the possible acreage. Sugar became the main crop, especially after the arrival of the British company Tate and Lyle in 1963. In order to address the shortage of manpower, successive governments encouraged immigration, and, increasingly, migrant workers from Central America settled in the country.

The economy expanded rapidly in the late 1980s, chiefly as a result of programmes of diversification and privatization. In the early 1990s the rate of gross domestic product (GDP) growth contracted sharply, but grew again in the second half of the decade. The development of agricultural processing compensated for problems in the timber industry and sluggish performance in construction and garment assembly. The economy revived in 2000, when GDP growth was put at 12.3%, but revealed its vulnerability to natural disaster in the huge amount of damage caused by 'Hurricane Keith' in 2000 and 'Hurricane Iris' in 2001. The GDP growth rate steadied at 4.9% and 4.3% in 2001 and 2002, respectively, before rising again in 2003 when increased activity in the fishing and tourism sectors contributed to GDP growth of 9.4%. The economy expanded by a more modest 4.2% in 2004 and the growth rate was expected to decline to 3% in 2005. Dependence on agriculture has been eased by development of the manufacturing and service sectors. The Government has actively sought foreign investment in food-processing and other agricultural processing ventures, and has promoted the country's financial services and tourism potential.

## AGRICULTURE

Agriculture, forestry and fishing contributed 15.9% of GDP in 2003 and employed some 22.6% of the working population. In 2001 the sector contributed some 89.0% of export earnings. Traditionally, the main crops were sugar cane, bananas and citrus fruits (notably oranges and grapefruit), and since 2000 citrus fruits and related products have rivalled sugar as the country's leading export commodity. In June 2001 the Government approved the Sugar Industry Bill, designed to reduce the state's role in the sector and encourage outside investment. Other significant crops were maize, red kidney beans, rice, tobacco and marine products. Since independence there have been initiatives to diversify the agricultural sector, including the cultivation of papayas, mangoes, peanuts, pineapples and soya beans, and also to revitalize timber production. Dependence on agriculture made the economy vulnerable to the fluctuations of international markets, disease and the country's location on a hurricane belt. (In 1931 a hurricane almost totally destroyed Belize City and, after 'Hurricane Hattie' struck in October 1961, it was decided to build a new capital city at Belmopan, at a cost of over £4m., most of which was provided by the United Kingdom.) As a member of the Commonwealth, Belize benefited from low tariffs on its exports to the European Union (EU) under the Cotonou Agreement (the successor to the Lomé Convention, which expired in 2000), although this was scheduled to end by 2008 (see below), and tariff-free access to the USA under the Caribbean Basin Initiative.

New areas planted with citrus trees resulted in increased yields from the early 1990s, and the amount of land under rice cultivation was also extended at this time. In the banana industry, which expanded markedly after its privatization in 1985, the adoption of a new variety was expected to raise the annual crop by between 20% and 50% in the early 2000s. Bananas contributed BZ $65.8m. in export earnings (excluding re-exports) in 2000, but the 2001 figure was cut by one-third as a result of hurricane damage. The subsequent recovery in banana exports was slow, and the sector contributed just BZ $53.0m. in export earnings in 2004. Following the resolution of an EU–USA dispute on banana imports, the EU preferential market for Caribbean bananas ended in April 2001, and a transitional system, issuing licences according to historical trade patterns, was adopted. The definitive tariff-only system was scheduled to be in place by 1 January 2006, and was expected to reduce Belize's banana revenue. A similar preferential trade agreement with the EU for Caribbean sugar was scheduled to be phased out by 2009. An initial reduction of 20% in the preferential rate paid by the EU was to be implemented in 2005, with the full reduction scheduled to take effect in 2007/08. On a much smaller scale, the cultivation of papaya expanded from the late 1990s, to yield export earnings of BZ $22.8m. in 2004, and the further development of dried papaya production was expected to be still more lucrative. Another new crop in the 1990s was hot peppers, grown for the US market. Coffee was first planted on a commercial scale in 1989, and by 2003 some 229 metric tons of coffee were produced. The coffee was promoted as an organic crop, some for local consumption, with exports going chiefly to the USA and Singapore.

Encouragement of cultivation of new land and development of the timber industry brought the agricultural sector into conflict with environmental concerns. A 1996 plan to redistribute land and increase the country's profitable citrus-fruit production resulted in the destruction of some of the country's rainforest. One of the most prized Belizean timber products was mahogany, which in 1995 was listed internationally as an endangered species. Between 1973 and 1997 the Government granted 17 logging concessions in the Toledo district, in the south of the country, to foreign companies (a number of which were from South-East Asia) requiring mahogany. In late 1996 controversy arose over the Government's proposals to upgrade the road through the mountains to Punta Gorda on the coast, linking it to the Central American Highway through Guatemala. A Mayan delegation successfully appealed to the Inter-American Development Bank (IDB) to base funding on greater consideration of the project's cultural and environmental impact. The finance was eventually approved in January 1998.

Belize has rich fishing grounds, yielding both white fish and shellfish, and in 2004 marine products (excluding aquarium products) were the country's principal export earner, contributing an estimated 26.0% (BZ $106.7m.) of export earnings (excluding re-exports). The fishing sector expanded markedly during the 1990s, with further growth of 39% over the period 1999–2001. Following a slight contraction in 2002, in 2003 the sector expanded by 111%, largely owing to an expansion in farmed shrimp production. This sector grew 10-fold in the 1990s and accounted for BZ $85.2m. in export revenue in 2004. Initially, the majority of the fish exported went to the USA, but further development of the shrimp industry, financed by US interests, was directed at the European market. The next most significant fishery products were lobster, with exports of BZ $15.1m. in 2004, and conch at $5.8m. Other potential growth areas were sharks, mangrove oysters, crabs and scallops and aquarium fish.

## MANUFACTURING

Two significant areas of manufacturing to emerge were food-processing and garment assembly. In 2003 the manufacturing sector employed 7.6% of the active work-force and contributed an estimated 9.2% of GDP, while garments accounted for around 9.0% of export revenue in 2004. Successive governments offered (often controversial) tax concessions for foreign agribusiness investors, including Tate and Lyle and the US chocolate concern Hershey.

The two prime processed-food exports in 2004 were sugar, which provided some 19.9% of export revenue, and orange and grapefruit concentrates, which provided an estimated 19.3%. Exporters of orange and grapefruit concentrates had duty-free access to the US and EU markets and to the Caribbean Community and Common Market (CARICOM). New ventures included the manufacture of hot pepper sauces, worth US $1.1m. in 2004, meat-processing and dried-fruit production, and there was small-scale production of items in wood and conch shell for the tourist market. Longer-established sectors were cement and garment manufacture. The latter faced competition from Mexican assembly plants after the implementation of the North American Free Trade Agreement (NAFTA) from 1994. Belizean investment promotion focused on the country's proximity to Mexico and the favourable trade agreements with Canada, the EU and the USA. The establishment of NAFTA prompted Belize to build a series of 'in-bond' manufacturing plants under joint ventures with Mexico, in free trade zones situated along the northern border with Mexico. These plants concentrated on light manufacturing and telecommunications equipment. It was too early to determine what effect the advent of the Central American Free Trade Agreement (CAFTA), signed in May 2004 between the USA and Costa Rica, El Salvador, Guatemala, Honduras and Nicaragua, would have on Belize's assembly-plant sector. In April 2005 a second round of negotiations were held for a limited bilateral trade agreement with Guatemala.

## MINING AND ENERGY

A number of mineral deposits were located in Belize, including bauxite and gold, but none proved to be commercially viable, with the exception of limestone quarrying. Mining employed 0.4% of the labour force and accounted for 0.6% of GDP in 2003. In 1988 a Mines and Minerals Law was passed, designed chiefly for petroleum production, awarding all mineral rights to the Government. Exploration had begun in 1938, concentrated on shore, in the north of the country, and off shore in the south, in the Bay of Campeche. During the 1980s Belize became particularly attractive to foreign petroleum companies for investment, in comparison with its more turbulent Central American neighbours. The first well was drilled in 1955, and the largest deposit discovered had an estimated yield of between 10 and 20 barrels per day, while national requirements ran to around 600,000 barrels per year. In 2004 imports of mineral fuels and lubricants accounted for an estimated 15.0% of Belize's total import costs.

The Government also encouraged the development of other forms of energy, and, from 1991, removed all duties on solar- and wind-energy devices. During the 1990s the Molléjon hydroelectric station was constructed on the Macal river, and in 2001 a total of US $15m. was allocated to current schemes to supply electricity to rural areas. Although many infrastructure projects were scheduled, including the development of facilities at Belmopan's Fort Point river port and the installation of the country's second underground fibre-optic telecommunications link, the Government was also faced with replacing services destroyed by natural disasters. There was controversy over the proposed construction of the Chalillo dam, also on the Macal river, by the Canadian company Fortis. In 2001 Fortis had acquired both the Molléjon dam and a monopoly on the generation and supply of power in Belize. Supporters of the project argued that it would reduce the country's imports of Mexican electricity, while opponents claimed the water flow would be insufficient to meet targets. The principal objection, however, was the environmental damage caused by flooding an area of natural rainforest containing several endangered plant and animal species, notably the tapir, Belize's national animal and a natural food source for jaguars, and the scarlet macaw. In March 2002 the non-governmental organization Belize Alliance of Conservation (BACONGO) began legal action to reverse government approval of the dam. Although construction work had already begun in 2003 (involving many drafted Asian workers) BACONGO took the case to the Privy Council in London, United Kingdom, in July; the ruling given in January 2004 found in favour of proceeding with the construction of the dam.

## TOURISM AND ENVIRONMENT

By the late 1990s the tourism sector had become the country's second largest foreign-exchange earner, and in 2002 it was worth over US $132m. The chief attractions were the country's Caribbean coast, its Mayan heritage and its natural environment, particularly its rainforests and the barrier reef, the second largest in the world. The majority of visitors came from the USA, and in 2003 tourist arrivals reached an estimated 686,591, according to the country's tourism board (compared with the 2001 figure of 244,072). In 1999 the Government passed the Gaming Control Act regulating new gambling ventures, designed chiefly for the tourist market, both in a casino and also in electronic form. The sector employed about 10.5% of the working population in 2003 and accounted for 22% of GDP in 2000. Expansion of the sector concentrated on diving, low-impact tourism, conservation of the country's varied flora and fauna, and sustainable-development projects, as well as cruise tourism. There were also moves to diversify into retirement centres and hospitality schools.

The development of tourism highlighted the importance of the country's environment. During the 1990s legislation was introduced to protect the nation's forestry reserves and the mangrove belts along the coast, which formed a natural filtration system for the coastal fisheries, especially the shellfish beds. As a result, nearly 40% of land in Belize has been granted some form of protected status, but reports suggested that much of the legislation was rarely enforced. Another major environmental concern was the effect of chemical products used in the cultivation of sugar cane, which ran off the fields into the water system, affecting the fisheries. In addition, the chemicals curbed the rat population, thereby reducing the number of snakes available for export, either live, or dead and dried for medicinal use.

In 2000 scientists at the Smithsonian Institution in Washington, DC, USA, expressed concern regarding the destruction of the central barrier reef off the coast of Belize. The destruction, by 'bleaching' of the coral, was a result of rising sea temperatures caused by global warming.

## BANKING AND INVESTMENT

From the early 1990s the Government promoted Belize as an 'offshore' financial centre, stressing its political stability. The first 'offshore' banks opened in 1996 and by 2003 there were over 25,000 'offshore' companies registered in Belize. The Financial Institutions Act of January 1996 tightened the regulations on the country's banking system and aligned them with those of other CARICOM member countries. In 1999 the Government established an International Financial Commission to regulate the sector, and in July 2000, in order to increase transparency, the Commission ended the practice of dealing in bearer shares. The banking and finance sector was dominated by Michael Ashcroft (see History), whose interests were regrouped into Carlisle Holdings Ltd in 1999. The Carlisle Holdings company also owned the Belize Bank and operated the sale of passports under the 1995 'economic citizenship' programme.

In mid-2004 it was discovered that the Social Security Board (SSB) had contracted a number of unsecured loans, principally to the housing and tourism sectors and to the Development Finance Corporation (DFC), incurring substantial financial losses. In December it was announced that the SSB was to be prohibited from agreeing any further loans pending the implementation of new investment regulations, and that the DFC was to be dissolved.

## FINANCES

After enjoying budget surpluses from the 1980s, in the mid-1990s, as a result of an economic downturn, Belize began to record budget deficits. The scaling down of the British garrison from 1994 resulted in a significant loss of income in Belize. The presence of the garrison was estimated to have contributed 15%–20% of the country's GDP. Austerity measures were subsequently implemented, and the renewal of the controversial 'economic citizenship' programme helped to reduce the budget deficit, but increased government spending by the People's United Party (PUP) Government took the deficit to US $38.8m. in 2001. In its second term, which began in 2003, the Government of Said Musa reduced expenditure on non-essential infrastructure and, from November 2004, implemented measures to improve the collection of taxes, particularly land tax, and customs duties. Further tax measures introduced in the 2004/05 budget targeted the business sector, notably banks, real estate agencies and casinos. The largest share of government revenue came from import duty (37% in 2005), with indirect taxes contributing 27% and income tax the relatively low figure of 20%.

The trade deficit widened steadily in the late 1990s owing to a sharp increase in imports, to a peak of US $213.7m. in 2001. The deficit, which stood at US $209.3m. in 2003, was expected to decline in 2004, largely owing to increased tourism receipts. However, the deficit on the current account of the balance of payments increased from US $77.5m. in 1999 to $181.0m. in 2003 (equivalent to 18.3% of GDP).

From 1976 the currency was pegged to the US dollar at the rate of US $1.00 equal to BZ $2.00. The average annual rate of inflation dropped from 6.4% in 1996 to less than 1.0% in 1997, and the country posted negative rates in 1998 and 1999. Growth resumed in 2000, rising slowly from 0.6% in 2000 to an estimated 2.5% in 2003 and to 3.1% in 2004; the rate was anticipated to rise further, to more than 4% in 2005. Total external debt stood at US $1,057.6m. at the end of 2003 (equivalent to 26.8% of GDP); the country's bilateral debt was owed mainly to the USA and the United Kingdom, and much of it was held on concessional terms. Increased government spending in the first Musa administration further inflated the debt (both in real terms and as a ratio of GDP), and in May 2003 the country's debt rating was lowered accordingly. In April 2004 the IMF warned that a reduction in government expenditure and measures to refinance the debt burden were imperative to avoid an unmanageable deficit on the balance of payments. The budgetary deficit was reduced from 8.1% of GDP in 2003/04 to an estimated 4.2% in 2004/05. The 2005/06 budget aimed to further reduce the fiscal deficit to 2.8% of GDP.

## CONCLUSION

In 2005 the country faced competition as a result of the gradual cessation of the EU preferential market for Caribbean bananas and sugar. Nevertheless, the trade deficit was beginning to decrease, and the rate of unemployment had declined from around 14% in the mid-1990s to 11.6% in 2004 and was expected to remain stable in 2005.

In 2001 the terrorist attacks on the USA caused a sudden fall in visitor numbers and in October 'Hurricane Iris' damaged about one-fifth of the country's hotels. Although the tourism sector proved its ability to recover relatively quickly between seasons, the damage highlighted the limited nature of the country's infrastructure, such as fewer than 500 km of paved roads. Furthermore, in the 2004/05 budget expenditure on infrastructure was restricted in an attempt to reduce the fiscal deficit.

The question of scale has proved crucial to the Belizean economy. A small domestic market, the exploitation of only a small proportion of its cultivable land, inadequate infrastructure, the closure of some roads during the rainy season, the burden of fuel and machinery imports, and the country's relatively high labour and energy costs have all tended to discourage major economic development. These limitations and the vulnerability of the economy to natural disasters and international fluctuations would restrict government attempts to reduce the high levels of foreign debt and budget deficit.

# Statistical Survey

Sources (unless otherwise stated): Central Statistical Office, Cnr Culvert Rd and Mountain View Blvd, Belmopan; tel. 822-2207; fax 822-3206; e-mail info@cso.gov.bz; internet www.cso.gov.bz; Central Bank of Belize, Gabourel Lane, POB 852, Belize City; tel. 223-6194; fax 223-6226; e-mail cenbank@btl.net; internet www.centralbank.org.bz.

## AREA AND POPULATION

**Area:** 22,965 sq km (8,867 sq miles).

**Population:** 189,774 at census of 12 May 1991; 240,204 (males 121,278, females 118,926) at census of 12 May 2000; 282,600 at mid-2004 (official estimate).

**Density** (mid-2004): 12.3 per sq km.

**Districts** (official estimates at mid-2004, rounded figures): Belize 84,200; Cayo 63,900; Orange Walk 43,800; Corozal 35,000; Stann Creek 28,900; Toledo 26,800.

**Principal Towns** (official estimates at mid-2004, rounded figures): Belize City (former capital) 59,400; San Ignacio/Santa Elena 16,100; Orange Walk 15,000; Belmopan (capital) 12,300; Dangriga (fmrly Stann Creek) 10,400; Corozal 8,600; San Pedro 7,600; Benque Viejo 6,700; Punta Gorda 4,900.

**Births, Marriages and Deaths** (provisional figures, 2003): Registered live births 7,440 (birth rate 27.3 per 1,000); Registered marriages 1,713 (marriage rate 6.3 per 1,000); Registered deaths 1,277 (death rate 4.7 per 1,000).

**Expectation of Life** (WHO estimates, years at birth): 68 (males 65; females 71) in 2003. Source: WHO, *World Health Report*.

# BELIZE

**Economically Active Population** (2003): Agriculture 17,164; Forestry 864; Fishing 2,120; Mining and quarrying 360; Manufacturing 6,739; Electricity, gas and water 778; Construction 7,475; Wholesale and retail trade and repairs 14,687, Tourism (incl. restaurants and hotels) 9,400; Transport and communications 3,297; Financial intermediation 1,518; Real estate and renting 1,770; General government services 10,309; Community, social and personal services 10,779; Work abroad and in commercial free zone 1,676; Other 286; *Total employed* 89,222; *Total unemployed* 13,219; *Total labour force* 102,441. *April 2004* (labour force survey): Total employed 95,911 (Primary sector 19,566, Secondary sector 17,456, Services 58,889); Total unemployed 12,580; Total labour force 108,491.

## HEALTH AND WELFARE
### Key Indicators

**Total Fertility Rate** (children per woman, 2003): 3.1.

**Under-5 Mortality Rate** (per 1,000 live births, 2003): 39.

**HIV/AIDS** (% of persons aged 15–49, 2003): 2.4.

**Physicians** (per 1,000 head, 2000): 1.05.

**Hospital Beds** (per 1,000 head, 1996): 2.13.

**Health Expenditure** (2002): US $ per head (PPP): 300.

**Health Expenditure** (2002): % of GDP: 5.2.

**Health Expenditure** (2002): public (% of total): 47.3.

**Access to Water** (% of persons, 2002): 91.

**Access to Sanitation** (% of persons, 2002): 47.

**Human Development Index** (2002): ranking: 99.

**Human Development Index** (2002): value: 0.737.

For sources and definitions, see explanatory note on p. vi.

## AGRICULTURE, ETC.

**Principal Crops** ('000 metric tons, 2003): Rice (paddy) 13.2; Maize 33.6; Sorghum 9.2; Sugar cane 1,073.2; Dry beans 5.0; Fresh vegetables 7.9 (FAO estimate); Bananas 73.5; Plantains 28.0 (FAO estimate); Oranges 165.5; Grapefruit and pomelos 39.2; Other fresh fruit 21.6 (FAO estimate).

**Livestock** ('000 head, year ending September 2003): Horses 5 (FAO estimate); Mules 5 (FAO estimate); Cattle 58; Pigs 21; Sheep 5; Chickens 1,450 (FAO estimate).

**Livestock Products** (FAO estimates, '000 metric tons, 2003): Beef and veal 2.2; Chicken meat 13.4; Pig meat 1.0; Cows' milk 3.8; Hen eggs 1.5.

**Forestry** (1995): *Roundwood Removals* ('000 cubic metres, excl. bark): Sawlogs, veneer logs and logs for sleepers 62, Fuel wood 126, Total 188. *Sawnwood* ('000 cubic metres, incl. railway sleepers): Coniferous (softwood) 5, Broadleaved (hardwood) 30, Total 35. *1996–2003*: Annual production as in 1995 (FAO estimates).

**Fishing** ('000 metric tons, live weight, 2002): Capture 24.8 (Sardinellas 5.0; European pilchard 2.0; European anchovy 5.0; Jack and horse mackerels 5.4; Chub mackerel 3.7; Caribbean spiny lobster 0.5; Stromboid conchs 1.4); Aquaculture 4.4 (Whiteleg shrimp 4.4); Total catch 29.2. *2003*: Capture 5.2; Aquaculture 10.2; Total catch 15.4.
Source: FAO.

## INDUSTRY

**Production** (2002): Raw sugar 107,209 long tons; Molasses 35,633 long tons; Cigarettes 84 million; Beer 2,388,000 gallons; Batteries 4,307; Flour 26,078,000 lb; Fertilizers 27,775 short tons; Garments 1,183,000 items; Citrus concentrates 4,351,000 gallons; Soft drinks 3,684,000 cases. Source: IMF, *Belize: Selected Issues and Statistical Appendix* (April 2004).

## FINANCE

**Currency and Exchange Rates**: 100 cents = 1 Belizean dollar (BZ $). *Sterling, US Dollar and Euro Equivalents* (31 May 2005): £1 sterling = BZ $3.6362; US $1 = BZ $2.0000; €1 = BZ $2.4662; BZ $100 = £27.50 = US $50.00 = €40.55. *Exchange rate*: Fixed at US $1 = BZ $2.000 since May 1976.

**Budget** (BZ $ million, year ending 31 March 2005, estimates): *Revenue*: Taxation 444.4 (Taxes on income, profits 102.0; Domestic taxes on goods and services 135.0; International trade 187.4); Other current revenue 47.0; Capital revenue 13.0; Total 504.3, excl. grants (16.0). *Expenditure*: Current expenditure 424.4 (Wages and salaries 214.1, Pensions 23.9, Goods and services 70.5, Interest payments 82.6, Subsidies and current transfers 33.3); Capital expenditure 129.7; Total 554.1.

**International Reserves** (US $ million at 31 December 2004): IMF special drawing rights 2.55; Reserve position in the IMF 6.58; Foreign exchange 39.12; Total 48.25. Source: IMF, *International Financial Statistics*.

**Money Supply** (BZ $ million at 31 December 2004): Currency outside banks 115.31; Demand deposits at commercial banks 321.98; Total money (incl. others) 439.93. Source: IMF, *International Financial Statistics*.

**Cost of Living** (Consumer Price Index; base: 1990 = 100): All items 122.9 in 2002; 126.0 in 2003; 129.9 in 2004.

**Expenditure on the Gross Domestic Product** (BZ $ million at current prices, 2003): Government final consumption expenditure 284.3; Private final consumption expenditure 1,527.1; Increase in stocks 10.8; Gross fixed capital formation 394.6; *Gross domestic expenditure* 2,216.9; Exports of goods and services 1,101.2; *Less* Imports of goods and services 1,316.9; Statistical discrepancy –24.3; *GDP at market prices* 1,976.8.

**Gross Domestic Product by Economic Activity** (BZ $ million at current prices, 2003): Agriculture, forestry and fishing 279.4; Mining and quarrying 9.8; Manufacturing 161.3; Electricity and water 60.7; Construction 77.7; Wholesale and retail trade, restaurants and hotels 396.0; Transport, storage and communications 183.1; Financial intermediation 128.6; Real estate, renting and business services 137.9; Community, social and personal services 120.0; General government services 202.8; *Sub-total* 1,757.5; Taxes, less subsidies, on products 272.6; *Less* Imputed bank service charges 53.3; *GDP at market prices* 1,976.8; *GDP at constant 2000 prices* 1,989.8.

**Balance of Payments** (BZ $ million, 2003): Exports of goods f.o.b. (incl. re-exports) 632.6; Imports of goods f.o.b. –1,051.2; *Trade balance* –418.6; Exports of services 425.9; Imports of services –278.6; *Balance on goods and services* –271.3; Other income received 8.4; Other income paid –188.0; *Balance on goods, services and income* –450.9; Current transfers received 94.0; Current transfers paid –4.9; *Current balance* –361.9; Capital account (net) –96.1; Direct investment abroad –0.7; Direct investment from abroad 57.6; Portfolio investment assets –0.3; Portfolio investment liabilities 151.1; Financial derivatives assets 1.1; Other investment assets –17.1; Other investment liabilities 192.4; Net errors and omissions 13.8; *Overall balance* –60.1.

## EXTERNAL TRADE

**Principal Commodities** (BZ $ million, 2004): *Imports c.i.f.*: Food and live animals 109.2; Mineral fuels and lubricants 154.7; Chemicals and related products 76.3; Manufactured goods 136.9; Miscellaneous manufactured articles 81.8; Machinery and transport equipment 175.9; Commercial free zone 156.6; Export processing zone 113.9; Total (incl. others) 1,028.2. *Exports f.o.b.*: Marine products (excl. aquarium fish) 106.7; Sugar 81.5; Orange concentrate 55.5; Grapefruit concentrate 23.8; Bananas 53.0; Garments 37.1; Papaya 22.8; Total (incl. others) 410.1.

**Principal Trading Partners** (BZ $ million, 2004): *Imports c.i.f.*: USA 398.1; Mexico 106.0; United Kingdom 23.9; Canada 12.5; Total (incl. others) 1,028.2. *Exports f.o.b.* (excl. re-exports): USA 226.2; United Kingdom 80.6; Mexico 5.5; Total (incl. others) 410.1.

## TRANSPORT

**Road Traffic** ('000 vehicles in use, 1998): Passenger cars 9,929; Buses and coaches 416; Lorries and vans 11,339; Motorcycles and mopeds 270 (Source: IRF, *World Road Statistics*). *2000* ('000 vehicles in use): Passenger cars 21.5; Commercial vehicles 3.9 (Source: UN, *Statistical Yearbook*).

**Shipping** (sea-borne freight traffic, '000 metric tons, 1996): Goods loaded 255.4; Goods unloaded 277.1. *Merchant Fleet* (vessels registered at 31 December 2004): Number of vessels 1,019; Total displacement 1,687,460 grt (Source: Lloyd's Register-Fairplay, *World Fleet Statistics*).

**Civil Aviation** (2001): Passenger arrivals 174,201. Source: IMF, *Belize: Statistical Appendix* (October 2002).

## TOURISM

**Tourist Arrivals**: 450,689 (cruise-ship passengers 271,737, stopover visitors 178,952) in 2002; 686,591 (cruise-ship passengers 488,917, stop-over visitors 197,674) in 2003; 744,677 (cruise-ship passengers 569,642, stop-over visitors 175,035) in 2004.

**Tourism Receipts** (US $ million): 120.5 in 2001; 132.8 in 2002; 156.2 in 2003.
Source: Belize Tourist Board.

## COMMUNICATIONS MEDIA

**Radio Receivers** (1997): 133,000 in use*.

**Television Receivers** (2000): 44,000 in use†.

**Telephones** (2003): 33,300 main lines in use†.
**Facsimile Machines** (1996): 500 in use‡.
**Mobile Cellular Telephones** (2003): 60,400 subscribers†.
**Personal Computers** (2002): 35,000 in use†.
**Internet Users** (2002): 30,000†.
**Book Production** (1996): 107 titles*.
**Non-daily Newspapers** (1996): 6 (circulation 80,000)*.
*Source: UNESCO, *Statistical Yearbook*.
†Source: International Telecommunication Union.
‡Source: UN, *Statistical Yearbook*.

## EDUCATION

**Pre-primary** (2002/03): 103 schools, 214 teachers, 3,773 students.
**Primary** (2003/04): 275 schools, 2,618 teachers, 62,074 students.
**Secondary** (2003/04): 43 schools, 1,074 teachers, 15,344 students.
**Higher** (1997/98): 12 institutions, 228 teachers, 2,853 students.
Source: Ministry of Education, Youth and Sports.

**Adult Literacy Rate** (UNESCO estimates): 76.9% (males 77.1%; females 76.7%) in 2002. Source: UN Development Programme, *Human Development Report*.

# Directory

## The Constitution

The Constitution came into effect at the independence of Belize on 21 September 1981. Its main provisions are summarized below:

### FUNDAMENTAL RIGHTS AND FREEDOMS

Regardless of race, place of origin, political opinions, colour, creed or sex, but subject to respect for the rights and freedoms of others and for the public interest, every person in Belize is entitled to the rights of life, liberty, security of the person, and the protection of the law. Freedom of movement, of conscience, of expression, of assembly and association and the right to work are guaranteed and the inviolability of family life, personal privacy, home and other property and of human dignity is upheld. Protection is afforded from discrimination on the grounds of race, sex, etc., and from slavery, forced labour and inhuman treatment.

### CITIZENSHIP

All persons born in Belize before independence who, immediately prior to independence, were citizens of the United Kingdom and Colonies automatically become citizens of Belize. All persons born outside the country having a husband, parent or grandparent in possession of Belizean citizenship automatically acquire citizenship, as do those born in the country after independence. Provision is made which permits persons who do not automatically become citizens of Belize to be registered as such. (Belizean citizenship was also offered, under the Belize Loans Act 1986, in exchange for interest-free loans of US $25,000 with a 10-year maturity. The scheme was officially ended in June 1994, following sustained criticism of alleged corruption on the part of officials. A revised economic citizenship programme, offering citizenship in return for a minimum investment of US $75,000, received government approval in early 1995, but was ended in 2002.)

### THE GOVERNOR-GENERAL

The British monarch, as Head of State, is represented in Belize by a Governor-General, a Belizean national.

### Belize Advisory Council

The Council consists of not less than six people 'of integrity and high national standing', appointed by the Governor-General for up to 10 years upon the advice of the Prime Minister. The Leader of the Opposition must concur with the appointment of two members and be consulted about the remainder. The Council exists to advise the Governor-General, particularly in the exercise of the prerogative of mercy, and to convene as a tribunal to consider the removal from office of certain senior public servants and judges.

### THE EXECUTIVE

Executive authority is vested in the British monarch and exercised by the Governor-General. The Governor-General appoints as Prime Minister that member of the House of Representatives who, in the Governor-General's view, is best able to command the support of the majority of the members of the House, and appoints a Deputy Prime Minister and other Ministers on the advice of the Prime Minister. The Governor-General may remove the Prime Minister from office if a resolution of 'no confidence' is passed by the House and the Prime Minister does not, within seven days, either resign or advise the Governor-General to dissolve the National Assembly. The Cabinet consists of the Prime Minister and other Ministers.

The Leader of the Opposition is appointed by the Governor-General as that member of the House who, in the Governor-General's view, is best able to command the support of a majority of the members of the House who do not support the Government.

### THE LEGISLATURE

The Legislature consists of a National Assembly comprising two chambers: the Senate, with eight nominated members; and the House of Representatives, with 29 elected members. The Assembly's normal term is five years. Senators are appointed by the Governor-General: five on the advice of the Prime Minister; two on the advice of the Leader of the Opposition or on the advice of persons selected by the Governor-General; and one after consultation with the Belize Advisory Council. If any person who is not a Senator is elected to be President of the Senate, he or she shall be an *ex-officio* Senator in addition to the eight nominees.

Each constituency returns one Representative to the House, who is directly elected in accordance with the Constitution.

If a person who is not a member of the House is elected to be Speaker of the House, he or she shall be an *ex-officio* member in addition to the 29 members directly elected. Every citizen older than 18 years is eligible to vote. The National Assembly may alter any of the provisions of the Constitution.

## The Government

### HEAD OF STATE

**Sovereign:** HM Queen ELIZABETH II (succeeded to the throne 6 February 1952).

**Governor-General:** Sir COLVILLE YOUNG (appointed 17 November 1993).

### THE CABINET
(July 2005)

**Prime Minister and Minister of Finance, National Development and the Public Service:** SAID MUSA.

**Deputy Prime Minister and Minister of Natural Resources, Local Government and the Environment and Minister of State in the Ministry of Finance:** JOHN BRICEÑO.

**Attorney-General and Minister of Education, Youth, Sports and Culture:** FRANCIS FONSECA.

**Minister of Home Affairs and Public Utilities:** RALPH FONSECA.

**Minister of Health, Labour and Defence:** VILDO MARIN.

**Minister of Foreign Affairs, Foreign Trade and Tourism, NEMO and Information:** GODFREY SMITH.

**Minister of the Works, Transport and Communications and Minister of State in the Ministry of Finance:** JOSÉ COYE.

**Minister of Human Development and Housing:** SYLVIA FLORES.

**Minister of Agriculture and Fisheries:** MIKE ESPAT.

**Minister without Portfolio in the Office of the Prime Minister:** MARCIAL MES.

**Minister without Portfolio in the Office of the Deputy Prime Minister:** SERVULO BAEZA.

### MINISTRIES

**Office of the Prime Minister:** New Administration Bldg, Belmopan; tel. 822-0399; fax 822-3323; e-mail pmbelize@btl.net.

**Ministry of Agriculture and Fisheries:** 2nd Floor, West Block Bldg, Belmopan; tel. 822-2241; fax 822-2409; e-mail mafpaeu@btl.net.

BELIZE — *Directory*

**Ministry of the Attorney-General and of Education, Youth, Sports and Culture:** West Block Bldg, Belmopan; tel. 822-2380; fax 822-3389; e-mail info@moes.gov.bz; internet www.moes.gov.bz.

**Ministry of Finance, National Development and the Public Service:** New Administration Bldg, Belmopan; tel. 822-2169; fax 822-3317; e-mail finsecmof@btl.net.

**Ministry of Foreign Affairs, Foreign Trade and Tourism, NEMO and Information:** NEMO Bldg, POB 174, Belmopan; tel. 822-2167; fax 822-2854; e-mail belizemfa@btl.net; internet www.mfa.gov.bz.

**Ministry of Health, Labour and Defence:** Water Reservoir Area, Belmopan; tel. 822-0589; fax 822-2942; e-mail healthbze@btl.net.

**Ministry of Home Affairs and Public Utilities:** New Administration Bldg, Belmopan; tel. 822-2218; fax 822-2195; e-mail investment@btl.net.

**Ministry of Human Development and Housing:** West Block, Independence Hill, Belmopan; tel. 822-2161; fax 822-3175; e-mail mhd@btl.net.

**Ministry of Natural Resources, Local Government and the Environment:** Market Sq., Belmopan; tel. 822-2226; fax 822-2333; e-mail info@mnrei.gov.bz; internet www.mnrei.gov.bz.

**Ministry of Works, Transport and Communications:** New 2 Power Lane, Belmopan; tel. 822-2136; fax 822-3282; e-mail works@btl.net.

## Legislature

### NATIONAL ASSEMBLY

#### The Senate

**President:** PHILIP ZUNIGA.
There are eight nominated members.

#### House of Representatives

**Speaker:** ELIZABETH ZABANEH.
**Clerk:** JESUS KEN.
**General Election, 5 March 2003**

|  | Votes cast | % of total | Seats |
| --- | --- | --- | --- |
| People's United Party (PUP) | 52,934 | 53.2 | 22 |
| United Democratic Party (UDP) | 45,415 | 45.6 | 7 |
| Others | 1,211 | 1.2 | — |
| Total | 99,560 | 100.0 | 29 |

## Political Organizations

**People's United Party (PUP):** 3 Queen St, Belize City; tel. 223-2428; fax 223-3476; f. 1950; based on organized labour; merged with Christian Democratic Party in 1988; Leader SAID MUSA; Exec. Chair. EIDEN SALAZAR, Jr; Deputy Leaders MAXWELL SAMUELS, JOHN BRICEÑO.

**United Democratic Party (UDP):** South End Bel-China Bridge, POB 1898, Belize City; tel. 227-2576; fax 227-6441; e-mail unitedd@btl.net; internet www.udp.org.bz; f. 1974 by merger of People's Development Movement, Liberal Party and National Independence Party; conservative; Leader DEAN BARROW; Chair. DOUGLAS SINGH.

## Diplomatic Representation

### EMBASSIES AND HIGH COMMISSION IN BELIZE

**China (Taiwan):** 20 North Park St, POB 1020, Belize City; tel. 227-8744; fax 223-3082; e-mail embroc@btl.net; Ambassador CHARLES ERH-HUANG TSAI.

**Costa Rica:** 10 Unity Blvd, POB 288, Belmopan; tel. 822-1582; fax 822-1583; e-mail fborbon@btl.net; Ambassador FERNANDO BORBÓN.

**Cuba:** 6087 Manatee Dr., Buttonwood Bay, POB 1775, Belize City; tel. 223-5345; fax 223-1105; e-mail embacuba@btl.net; Ambassador REGALA CARIDAD DÍAZ HERNÁNDEZ.

**El Salvador:** 49 Nanche St, POB 215, Belmopan; tel. 823-3404; fax 823-3569; e-mail embasalva@btl.net; Ambassador MANUEL ANTONIO VÁSQUEZ MEZA.

**Guatemala:** 8 'A' St, King's Park, POB 1771, Belize City; tel. 233-3150; fax 235-3140; e-mail embbelice1@minex.gob.gt; Ambassador TÉLLEZ MIRALDA.

**Honduras:** 114 Bella Vista, POB 285, Belize City; tel. 224-5889; fax 223-0562; e-mail embahonbe@sre.hn; Ambassador JOSÉ RIGOBERTO ARRIAGA CHINCHILLA.

**Mexico:** 18 North Park St, POB 754, Belize City; tel. 223-0193; fax 227-8742; e-mail embamexbze@btl.net; internet www.sre.gob.mx/belice; Ambassador JOSÉ ARTURO TREJO NAVA.

**Nicaragua:** 124 Barrack Rd, Belize City; tel. and fax 223-2666; e-mail embanicbelize@btl.net; Ambassador JAVIER WILLIAMS SLATE.

**United Kingdom:** Embassy Sq., POB 91, Belmopan; tel. 822-2146; fax 822-2761; e-mail brithicom@btl.net; internet www.britishhighbze.com; High Commissioner ALAN JONES.

**USA:** 29 Gabourel Lane, POB 286, Belize City; tel. 227-7161; fax 223-0802; e-mail embbelize@state.gov; internet www.usembassy.state.gov/belize; Ambassador ROBERT J. DIETER.

**Venezuela:** 19 Orchid Garden St, POB 49, Belmopan; tel. 822-2384; fax 822-2022; e-mail embaven@btl.net; Ambassador AMINTA GUACARAN TORREALBA.

## Judicial System

Summary Jurisdiction Courts (criminal jurisdiction) and District Courts (civil jurisdiction), presided over by magistrates, are established in each of the six judicial districts. Summary Jurisdiction Courts have a wide jurisdiction in summary offences and a limited jurisdiction in indictable matters. Appeals lie to the Supreme Court, which has jurisdiction corresponding to the English High Court of Justice and where a jury system is in operation. From the Supreme Court further appeals lie to a Court of Appeal, established in 1967, which holds an average of four sessions per year. Final appeals are made to the Judicial Committee of the Privy Council in the United Kingdom.

**Supreme Court:** Supreme Court Bldg, Belize City; tel. 227-7256; fax 227-0181; e-mail chiefjustice@btl.net; internet www.belizelaw.org/supreme_court/chief_justice.html; Registrar MICHELLE ARANA; Chief Justice Dr ABDULAI OSMAN CONTEH.

**Magistrate's Court:** Paslow Bldg, Belize City; tel. 227-7164; Chief Magistrate HERBERT LORD.

## Religion

### CHRISTIANITY

Most of the population are Christian, the largest denomination being the Roman Catholic Church (an estimated 76% of the population in 2003). The other main groups were the Pentecostal, Anglican, Seventh-day Adventist, Methodist and Mennonite churches.

**Belize Council of Churches:** 149 Allenby St, POB 508, Belize City; tel. 227-7077; f. 1957 as Church World Service Committee, present name adopted 1984; nine mem. Churches, four assoc. bodies; Pres. Rev. LEROY FLOWERS; Gen. Sec. SADIE VERNON.

#### The Roman Catholic Church

Belize comprises the single diocese of Belize City-Belmopan, suffragan to the archdiocese of Kingston in Jamaica. At 31 December 2003 it was estimated that there were 208,949 adherents in the diocese. The Bishop participates in the Antilles Episcopal Conference (whose secretariat is based in Port of Spain, Trinidad and Tobago).

**Bishop of Belize City-Belmopan:** OSMOND PETER MARTIN, Bishop's House, 144 North Front St, POB 616, Belize City; tel. 227-2122; fax 223-1922; e-mail episkopos@btl.net.

#### The Anglican Communion

Anglicans in Belize belong to the Church in the Province of the West Indies, comprising eight dioceses. The Archbishop of the Province is the Bishop of the North Eastern Caribbean and Aruba, resident in St John's, Antigua and Barbuda.

**Bishop of Belize:** Rt Rev. SYLVESTRE DONATO ROMERO-PALMA, Bishopthorpe, 25 Southern Foreshore, POB 535, Belize City; tel. 227-3029; fax 227-6898; e-mail bzediocese@btl.net; internet www.belize.anglican.org.

#### Protestant Churches

**Methodist Church in the Caribbean and the Americas (Belize/Honduras District) (MCCA):** 88 Regent St, POB 212,

Belize City; tel. 227-7173; fax 227-5870; f. 1824; c. 2,620 mems; District Pres. Rev. DAVID GOFF.

**Mennonite Congregations in Belize:** POB 427, Belize City; tel. 823-0137; fax 823-0101; f. 1958; an estimated 3,575 members live in eight Mennonite settlements, the largest of which is Altkolonier Mennotitengemeinde with 1,728 mems; Bishop AARON HARDER.

Other denominations active in the country include the Presbyterians, Baptists, Moravians, Jehovah's Witnesses, the Church of God, the Nazarene Church, the Assemblies of Brethren and the Salvation Army.

### OTHER RELIGIONS

There are also small communities of Hindus (106, according to the census of 1980), Muslims (110 in 1980), Jews (92 in 1980) and Bahá'ís.

## The Press

**Amandala:** Amandala Press, 3304 Partridge St, POB 15, Belize City; tel. 202-4476; fax 222-4702; e-mail editor@amandala.com.bz; internet www.amandala.com.bz; f. 1969; 2 a week; independent; Publr EVAN X. HYDE; Editor RUSSELL VELLOS; circ. 45,000.

**Ambergris Today:** San Pedro Town, Ambergris Caye; tel. 226-3462; e-mail ambertoday@btl.net; internet www.ambergristoday.com; weekly; independent; Editor DORIAN NUÑEZ.

**The Belize Times:** 3 Queen St, POB 506, Belize City; tel. 224-5757; fax 223-1940; e-mail belizetime@btl.net; internet www.belizetimes.bz; f. 1956; weekly; party political paper of PUP; Editor-in-Chief ANDREW STEINHAUER; circ. 6,000.

**Belize Today:** Belize Information Service, East Block, POB 60, Belmopan; tel. 822-2159; fax 822-3242; monthly; official; circ. 17,000.

**Government Gazette:** Print Belize, 1 Power Lane, Belmopan; tel. 822-0194; fax 822-3367; e-mail admin@printbze.bz; internet www.printbelize.bz; f. 1871; official; weekly; CEO LAWRENCE J. NICHOLAS.

**The Guardian:** Ebony St and Bel-China Bridge, POB 1898, Belize City; tel. 207-5346; fax 227-5343; e-mail info@guardian.bz; internet www.guardian.bz; weekly; party political paper of UDP; Editor JOHN AVERY; circ. 5,000.

**The Reporter:** 147 Allenby St, POB 707, Belize City; tel. 227-2503; fax 227-8278; e-mail editor@belizereporter.bz; internet www.reporter.bz; f. 1967; weekly; Editor ANN MARIE WILLIAMS; circ. 6,500.

**The San Pedro Sun:** POB 35, San Pedro Town, Ambergris Caye; fax 226-2905; e-mail sanpedrosun@btl.net; internet www.sanpedrosun.net; weekly; Editors RON SNIFFIN, TAMARA SNIFFIN.

### NEWS AGENCY

**Agencia EFE** (Spain): c/o POB 506, Belize City; tel. 224-5757.

## Publishers

**Angelus Press Ltd:** 10 Queen St, POB 1757, Belize; tel. 223-5777; fax 227-8825; e-mail angelgm@btl.net; Gen. Man. AMPARO M. NOBLE.

**Cubola Productions:** Montserrat Duran, 35 Elizabeth St, Benque, Viejo del Carmen; tel. 932-0853; fax 932-240; e-mail cubolabz@btl.net; internet www.belizebusiness.com/cubola.

**Print Belize Ltd:** 1 Power Lane, Belmopan; tel. 822-0194; fax 882-3367; e-mail admin@printbze.bz; internet www.printbelize.bz; f. 1871; responsible for printing, binding and engraving requirements of all govt depts and ministries; publications include annual govt estimates, govt magazines and the official Government Gazette.

## Broadcasting and Communications

### TELECOMMUNICATIONS

In September 2001 the Government announced that the telecommunications sector was to be liberalized.

**Public Utilities Commission (PUC):** see Utilities—Regulatory Body; Dir of Telecommunications Sector ROBERTO YOUNG.

**Belize Telecommunications Ltd (BTL):** Esquivel Telecom Centre, St Thomas St, POB 603, Belize City; tel. 223-2868; fax 223-1800; e-mail prdept@btl.net; internet www.btl.net; f. 1987; owned by Innovative Communication Corpn; taken over by the Govt in Feb. 2005; CEO GASPAR AGUILAR.

### RADIO

**Love FM:** 33 Freetown Rd, Belize City; tel. 223-2098; fax 223-0529; e-mail lovefm@btl.net; internet www.lovefm.com; f. 1992; purchased Friends FM in 1998; Man. Dir RENE VILLANUEVA.

**Radio Krem Ltd:** 3304 Partridge St, POB 15, Belize City; tel. 227-5929; fax 227-4079; e-mail kremwub@hotmail.com; internet www.krem.bz; commercial; purchased Radio Belize in 1998; Man. EVA S. HYDE.

Other private radio stations broadcasting in Belize include: Estereo Amor, More FM, My Refuge Christian Radio, Radio 2000 and Voice of America.

### TELEVISION

In 1986 the Belize Broadcasting Authority issued licences to eight television operators for 14 channels, which mainly retransmit US satellite programmes, thus placing television in Belize on a fully legal basis for the first time.

**Baymen Broadcasting Network (CTV-Channel 9):** 27 Baymen Ave, Belize City; tel. 224-4400; fax 223-1242; commercial; Man. MARIE HOARE.

**Channel 5 Belize:** POB 679, Belize City; tel. 227-3146; fax 227-4936; e-mail gbtv@btl.com; internet www.channel5belize.com; f. 1991.

**Tropical Vision (Channel 7):** 73 Albert St, Belize City; tel. 227-3988; fax 227-5602; e-mail tvseven@btl.net; internet 7newsbelize.com; commercial; Man. NESTOR VASQUEZ.

## Finance

(cap. = capital; res = reserves; dep. = deposits; brs = branches; amounts in BZ $, unless otherwise indicated)

### BANKING

#### Central Bank

**Central Bank of Belize:** Gabourel Lane, POB 852, Belize City; tel. 223-6194; fax 223-6226; e-mail cenbank@btl.net; internet www.centralbank.org.bz; f. 1982; cap. 10m., res 12.9m. (2004); Gov. SYDNEY CAMPBELL; Dir and Chair. JORGE MELITON AUIL.

#### Development Bank

**Development Finance Corporation:** Bliss Parade, Belmopan; tel. 822-2360; fax 822-3096; e-mail info@dfcbelize.org; internet www.dfcbelize.org; f. 1972; ceased to finance loans following a govt review in Dec. 2004; issued cap. 10m.; Chair. GLENN GODFREY; CEO TROY GABB; 5 brs.

#### Other Banks

**Alliance Bank of Belize Ltd:** 106 Princess Margaret Dr., POB 1988, Belize City; tel. 223-6783; fax 223-6785; e-mail alliance@btl.net; internet www.alliancebank.bz; f. 2001; dep. 6,876m., total assets 113.7m.; 3 brs.

**Atlantic International Bank Ltd:** Cnr Freetown Rd and Cleghorn St, POB 481, Belize City; tel. 223-3152; fax 223-3528; e-mail banking@atlabank.com; internet www.atlanticibl.com; f. 1971; offshore banking; dep. 161.0m., total assets 191.9m. (2001); Chair. Dr GUILLERMO BUESO; 8 brs.

**Bank of Nova Scotia (Scotiabank)** (Canada): Albert St, POB 708, Belize City; tel. 227-7027; fax 227-7416; e-mail bns.belize@scotiabank.com; Man. Dir PATRICK ANDREWS; 6 brs.

**Barclays Bank PLC** (United Kingdom): c/o FirstCaribbean Bank Ltd, 21 Albert St, POB 363, Belize City; tel. 227-7211; fax 227-8572; e-mail barclaysbz@btl.net; Man. TILVAN KING; 5 brs.

**Belize Bank Ltd:** 60 Market Sq., POB 364, Belize City; tel. 227-7132; fax 227-2712; e-mail bblbz@belizebank.com; internet www.belizebank.com; cap. 4.3m., res 5.9m., dep. 639.3m. (2003); Chair. Sir EDNEY CAIN; Senior Vice-Pres. LOUIS ANTHONY SWASEY; 12 brs.

**Provident Bank and Trust of Belize:** 35 Barrack Rd, POB 1867, Belize City; tel. 223-5698; fax 223-0368; e-mail services@providentbank.bz; internet www.providentbelize.com; f. 1998; cap. US $6.0m., res US $1.5m., dep. US $121.3m. (2003); Chair. JOY VERNON GODFREY; Pres. ALVARO ALAMINA.

There is also a government savings bank. In late 2001 the Government amended the exchange-control regulations to allow foreign-currency exchange bureaux.

## INSURANCE

The insurance sector is regulated by the Office of the Supervisor of Insurance, part of the Ministry of Finance.

**Atlantic Insurance Company Ltd:** Atlantic Bank Bldg, 3rd Floor, Cnr Cleghorn St and Freetown Rd, Belize City; tel. 223-2657; fax 223-2658; e-mail info@atlanticinsurancebz.com; internet www.atlanticinsurancebz.com; f. 1990; Chair. Dr GUILLERMO BUESO; Gen. Man. MARTHA GUERRA.

**Belize Insurance Centre:** 212 North Front St Belize City; tel. 227-7310; fax 227-4803; e-mail info@belizeinsurance.com; internet www.belizeinsurance.com; insurance broker; Gen. Man. CYNTHIA AWE.

**F & G Insurance:** 6 Fort St, POB 438, Belize City; tel. 227-7493; fax 227-8617; e-mail fandg@btl.net; internet www.fandginsurance.com.

**Regent Insurance Company Ltd:** 81 North Front St, POB 661, Belize City; tel. 227-3744; fax 227-2022; e-mail regent@btl.net; internet www.regentinsurance.com; f. 1982; CEO ANTHONY FLYNN.

# Trade and Industry

## STATUTORY BODIES

**Banana Control Board:** c/o Dept of Agriculture, West Block, Belmopan; management of banana industry; in 1989 it was decided to make it responsible to growers, not an independent executive.

**Belize Marketing Board:** 117 North Front St, POB 633, Belize City; tel. 227-7402; fax 227-7656; f. 1948 to encourage the growing of staple food crops; purchases crops at guaranteed prices, supervises processing, storing and marketing intelligence.

**Belize Sugar Cane Board:** 7 Second St South, Corozal Town; tel. 422-2005; fax 422-2672; f. 1960 to control the sugar industry and cane production; includes representatives of the Government, sugar manufacturers, cane farmers and the public sector.

**Citrus Control Board:** c/o Dept of Agriculture, West Block, Belmopan; tel. 822-2199; f. 1966; determines basic quota for each producer, fixes annual price of citrus.

## DEVELOPMENT ORGANIZATIONS

**Belize Reconstruction and Development Corporation:** 1 Bliss Promenade, POB 92, Belize City; tel. 227-7424; e-mail recondev@btl.net; Gen. Man. ALOYSIUS PALACIO.

**Belize Trade and Investment Development Service (BELTRAIDE):** 14 Orchid Garden St, Belmopan; tel. 822-3737; fax 822-0595; e-mail beltraide@belizeinvest.org.bz; internet www.belizeinvest.org.bz; f. 1986 as a joint government and private-sector institution to encourage export and investment; Exec. Dir LOURDES SMITH.

## CHAMBERS OF COMMERCE

**Belize British Chamber of Commerce (BBCC):** Embassy Sq., Ring Rd, Cayo District, POB 91, Belmopan; tel. 822-2146; fax 822-2761; internet www.belizebritishchamber.com; Chair. BILL HOWIE; Sec. KATHY ESQUIVEL.

**Belize Chamber of Commerce and Industry:** 63 Regent St, POB 291, Belize City; tel. 227-3148; fax 227-4984; e-mail bcci@btl.net; internet www.belize.org; f. 1920; Pres. ARTURO VASQUEZ; Sec. CYNTHIA AWE; 300 mems.

## EMPLOYERS' ASSOCIATIONS

**Banana Growers' Association:** Big Creek, Independence Village, Stann Creek District; tel. 622-001; fax 622-112; e-mail banana@btl.net.

**Belize Cane Farmers' Association:** 34 San Antonio Rd, Orange Walk; tel. 322-005; fax 323-171; f. 1959 to assist cane farmers and negotiate with the Sugar Board and manufacturers on their behalf; Chair. PABLO TUN; 16 district brs.

**Belize Livestock Producers' Association:** 47.5 miles Western Highway, POB 183, Belmopan; tel. 823-202; fax 823-886; e-mail blpa@btl.net; Chair. PETE LIZARRAGA.

**Citrus Growers Association:** Mile 9, Stann Creek Valley Rd, POB 7, Dangriga, Stann Creek District; tel. 522-3585; fax 522-2686; e-mail cga@belizecitrus.org; internet www.belizecitrus.org; f. 1966; CEO BRIDGET CULLERTON.

## MAJOR COMPANIES

**Belize Brewing Co Ltd:** 1 King St, POB 1068, Belize City; tel. 227-7031; fax 227-2399; e-mail belikin@bowenbz.com; f. 1968; subsidiary of Bowen & Bowen Ltd; producers of malt liquors; Pres. KEVIN BOWEN; 120 employees.

**Belize Estate Co Ltd:** Slaughterhouse Rd, POB 151, Belize City; tel. 223-0641; fax 223-1367; e-mail bec@bowen.com.bz; internet www.kiarental.com; f. 1875; subsidiary of Bowen & Bowen Ltd; importers and dealers of alcoholic drinks, shipping agents, main Ford dealers, operation of tourist enterprises; Pres. BARRY M. BOWEN; Man. Dir W. F. BOWMAN; 94 employees.

**Belize Sugar Industries Ltd:** Tower Hill, POB 29, Orange Walk Town; tel. 322-2150; fax 322-3247; e-mail bsires@btl.net; f. 1935; public co; raw sugar manufacturers; Man. Dir JOSÉ MONTALVO; Contact JOHN MACFARLANE; 650 employees.

**Belize Timber Ltd:** 2 Mapp St, POB97, Belize City; tel. 224-4193; fax 223-3552; e-mail bzetimber@btl.net; logging and forestry.

**Carlisle Holdings Ltd:** 60 Market Sq., POB 1764, Belize City; tel. 227-2660; fax 227-5854; e-mail info@carlisleholdings.com; internet www.carlisleholdings.com; holding co with banking and financial services operations in Belize; investments in infrastructure development and agro-processing and distribution in Central America and the Caribbean region; Chair. and CEO Lord ASHCROFT; Deputy Chair. DAVID B. HAMMOND; 437 employees.

**Deloitte and Touche Belize:** 40A Central American Blvd, POB 1235, Belize City; tel. 227-3020; fax 227-5792; e-mail deloitte@deloittebelize.com; internet www.deloittebelize.com; accountancy, management consultancy, offshore services; Man. Partner JULIÁN CASTILLO.

**Esso Standard Oil SA Ltd:** Caesar Ridge Rd, POB 328, Belize City; tel. 227-7323; fax 227-7726; e-mail ecgss@btl.net; petroleum exploration and distribution; Man. RUFINO LIN.

**Femagra Industries Ltd:** POB 65, Belize City; tel. 223-3430; fax 223-3097; e-mail vivian@femagra.com; internet www.femagra.com; f. 1984; chemical and food processing, water purifying.

**Fresh Catch Belize Ltd:** Democracia Village, Belize District; f. 2001; tilapia fish farm; Man. EMILE MENA.

**Hofius Ltd:** 19 Albert St, POB 226, Belize City; tel. 227-7231; fax 227-4751; e-mail hofiusace@btl.net; f. 1892; hardware and home products, real-estate sales, boats and marine fittings, food distribution; CEO JOHN CRUMP; 50 employees.

**KPMG Corporate Services (Belize) Ltd:** Jasmine Court, 35A Regent St, Suite 210, Belize City; tel. 271-755; fax 271-117; e-mail kpmg-csbl@btl.net; assurance, management consultancy and offshore company formation services; Senior Partners STANLEY ERMEAV, LISA ZAYDEN.

**Madisco:** C42 Leghorn St, POB 34, Belize City; tel. 224-4158; fax 224-1797; e-mail sales@madisco.bz; internet www.madisco.bz; marketing and distribution.

**Netkom Internet Solutions:** 18 Leslie St, Kings Park, POB 855, Belize City; tel. 223-3274; e-mail support@netkombelize.com; internet www.netkombelize.com; internet consultancy and provider of digital media production services.

**Northern Fishermen Cooperative Society:** 49 North Front St, POB 647, Belize City; tel. 224-4448; fax 223-0978; e-mail norficoop@btl.net; producers, processors and exporters of seafood products; Gen. Man. ROBERT USHER.

**Nova Companies Belize Ltd:** 12.5 Miles Northern Highway, POB 1360, Belize City; tel. 225-2301; fax 225-2309; e-mail novaco@btl.net; aquaculture; f. 1989; Gen. Man. CHARLES HYDE, Jr.

**Prosser Agrotec and Fertilizer Co:** Mile 8 Western Highway, POB 566, Belize City; tel. 223-5392; fax 222-5548; e-mail prosser@btl.net; manufacture of industrial chemicals and fertilizers; CEO SALVADOR ESPAT.

**Sol Belize Ltd:** 2.5 miles Northern Highway, Belize City; tel. 223-0406; fax 223-0704; e-mail joe.habet@solpetroleum.com; f. 1938; subsidiary of Interamericana Trading Ltd (Barbados); marketing and distribution of petroleum, petroleum products and chemical products; Gen. Man. JOE HABET; 14 employees.

**Texaco Belize Ltd:** 4.5 miles Western Highway, POB 627, Belize City; tel. 224-340; fax 224-355; subsidiary of Texaco Inc, USA; exploration and production of petroleum and gas, refining and distribution of petroleum and gas products; Man. HECTOR LÓPEZ.

## UTILITIES

### Regulatory Body

**Public Utilities Commission (PUC):** 63 Regent St, Belize City; tel. 227-1176; fax 227-1149; e-mail consumeraffairs@puc.bz; internet www.puc.bz; regulatory body, headed by seven commissioners, replaced the Offices of Electricity Supply and of Telecommunications; Chair. Dr GILBERT CANTON.

## Electricity

**Belize Electricity Co Ltd (BECOL):** 115 Barrack Rd, POB 327, Belize City; tel. 227-0954; fax 223-0891; e-mail bel@btl.net; owned by Fortis Inc (Canada); operates Mollejón 25 MW hydroelectric plant, which supplies electricity to Belize Electricity Ltd (BEL—see below); Chair. NESTOR VASQUEZ; Pres. and CEO LYNN YOUNG.

**Belize Electricity Ltd (BEL):** 2.5 miles Northern Highway, POB 327, Belize City; tel. 227-0954; fax 223-0891; e-mail pr@bel.com.bz; internet www.bel.com.bz; fmrly Belize Electricity Board, changed name upon privatization in 1992; Govt held 51% of shares until 1999; 68% owned by Fortis Inc (Canada); Pres. and CEO LYNN R. YOUNG; Vice-Pres. RENE BLANCO; Chair. ROBERT USHER; 242 employees.

### Water

**Belize Water Services Ltd:** POB 150, Central American Blvd, Belize City; tel. 222-4757; fax 222-4759; e-mail bws_ceosec@btl.net; f. 1971 as Water and Sewerage Authority (WASA); changed name upon privatization in March 2001; 82.7% owned by Cascal, BV (United Kingdom/Netherlands); CEO MARTIN R. GREENHALGH.

## TRADE UNIONS

**National Trade Union Congress of Belize (NTUCB):** POB 2359, Belize City; tel. 227-1596; fax 227-2864; e-mail ntucb@btl.net; Pres. HORRIS PATTEN; Gen. Sec. GEORGE FRAZER.

### Principal Unions

**Belize Communications Workers' Union (BCWU):** POB 1291, Belize City; tel. 223-4809; fax 224-4090; e-mail bcwu@btl.net; Pres. PAUL PERRIOTT; Gen. Sec. CHRISTINE PERRIOTT.

**Belize Energy Workers' Union:** c/o Belize Electricity Ltd, 2.5 miles Northern Highway, POB 327, Belize City; tel. 227-0954; Pres. MARK BUTLER; Gen. Sec. FLOYD HERRERA.

**Belize National Teachers' Union:** 10 Eve St, POB 382, Belize City; tel. 227-2857; fax 223-5233; e-mail bntu@btl.net; Pres. ANTHONY FUENTES; Gen. Sec. LOIS BARBER; 1,000 mems.

**Belize Workers' Union:** Tate St, Orange Walk Town; tel. 822-2327; Pres. HORRIS PATTEN.

**Christian Workers' Union:** 107B Cemetery Rd, POB 533, Belize City; tel. and fax 227-2150; e-mail cwu@btl.net; f. 1962; general; Pres. ANTONIO GONZÁLEZ; 1,000 mems.

**Public Service Union of Belize:** 2 Mayflower St, POB 458, Belmopan; tel. 822-0282; fax 822-0283; e-mail belizepsu@btl.net; f. 1922; public workers; Pres. DYLAN RENEAU; Sec.-Gen. FRANCISCO G. ZUNIGA; 1,600 mems.

**United Banners Banana Workers' Union:** Dangriga; f. 1995; Pres. MARCIANA FUNEZ.

# Transport

**Department of Transport:** Power Lane, Belmopan; tel. 822-2435; fax 822-3317; e-mail transcom@btl.net; Commissioner PHILLIP BRACKETT.

### RAILWAYS

There are no railways in Belize.

### ROADS

There are 2,710 km of roads, of which 2,210 km (1,600 km of gravel roads, 300 km of improved earth roads and 310 km of unimproved earth roads) are unpaved. In 2000 the World Bank approved a loan of US $13m. for road construction. In 2002 the European Union approved a grant of BZ $3.6m. towards the construction of a bridge over the New Sibun river. The Government provided BZ $1m. in funding and the project was completed in late 2004.

### SHIPPING

There is a deep-water port at Belize City and a second port at Commerce Bight, near Dangriga (formerly Stann Creek), to the south of Belize City. There is a port for the export of bananas at Big Creek and additional ports at Corozal and Punta Gorda. Nine major shipping lines operate vessels calling at Belize City, including the Carol Line (consisting of Harrison, Hapag-Lloyd, Nedlloyd and CGM. A proposal to develop a cruise-ship port at Port Loyola at a cost of BZ $963.5m. was submitted to parliament in October 2004.

**Belize Ports Authority:** Caesar Ridge Rd, POB 633, Belize City; tel. 227-2480; fax 227-2500; e-mail bzportauth@bpa.org.bz; internet bpa.org.bz; f. 1980; Chair. FRED HUNTER; Commr of Ports LLOYD JONES.

**Marine & Services:** 95 Albert St, POB 611, Belize City; tel. 227-2112; fax 227-5404; e-mail shipping@marineservices.bz; internet www.marineservices.bz; f. 1975; shipping and cargo services, cruise line agent; Gen. Man. JOSÉ GALLEGO.

**Port of Belize Ltd:** Caesar Ridge Rd, POB 2674, Belize City; tel. 227-2439; fax 227-3571; e-mail portbz@btl.net; operates the main port facility; CEO FRANCINE WAIGHT.

### CIVIL AVIATION

Phillip S. W. Goldson International Airport, 14 km (9 miles) from Belize City, can accommodate medium-sized jet-engined aircraft. The runway was extended in 2000. There are 37 airstrips for light aircraft on internal flights near the major towns and offshore islands.

**Belize Airports Authority (BAA):** POB 1564, Belize City; e-mail aviation@btl.net; tel. 225-2045; fax 225-2439; e-mail bzeaa@btl.net; CEO PABLO ESPAT.

**Maya Island Air:** Municipal Airport, POB 458, Belize City; tel. 223-1403; fax 223-0576; e-mail mayair@btl.net; internet www.mayaairways.com; f. 1997 as merger between Maya Airways Ltd and Island Air; operated by Belize Air Group; internal services, centred on Belize City, and charter flights to neighbouring countries; CEO EUGENE ZABANEH; Gen. Man. FERNANDO TREJOS.

**Tropic Air:** San Pedro, POB 20, Ambergris Caye; tel. 226-5671; fax 226-2338; e-mail reservations@tropicair.com; internet www.tropicair.com; f. 1979; operates internal services and services to Guatemala; Chair. CELI MCCORKLE; Man. Dir JOHN GREIF, III.

# Tourism

The main tourist attractions are the beaches and the barrier reef, diving, fishing and the Mayan archaeological sites. There are nine major wildlife reserves (including the world's only reserves for the jaguar and for the red-footed booby), and government policy is to develop 'eco-tourism', based on the attractions of an unspoilt environment and Belize's natural history. The country's wildlife also includes howler monkeys and 500 species of birds, and its barrier reef is the second largest in the world. However, in May 2000 scientists reported that the high sea temperatures recorded in 1998, caused by the El Niño phenomenon and global warming, had resulted in extensive damage to the coral population. There were some 437 hotels in Belize in 2002. In 2004 there were an estimated 744,677 tourist arrivals and in 2003 tourism receipts totalled an estimated US $156.2m. In February 1996 the Mundo Maya Agreement was ratified, according to which Belize, El Salvador, Guatemala, Honduras and Mexico would co-operate in the management of Mayan archaeological remains. In June 2000 the Inter-American Development Bank approved a loan of US $11m. towards the Government's tourism development plan.

**Belize Tourism Board:** Level 2, New Central Bank Bldg, POB 325, Gabourel Lane, Belize City; tel. 223-1913; fax 223-1943; e-mail info@travelbelize.org; internet www.travelbelize.org; f. 1964; fmrly Belize Tourist Bureau; eight mems; Chair. THERESE RATH; Dir TRACY TAEGAR-PANTON.

**Belize Tourism Industry Association (BTIA):** 10 North Park St, POB 62, Belize City; tel. 227-5717; fax 227-8710; e-mail info@btia.org; internet www.btia.org; promoyes sustainable tourism; Exec. Dir ANDREW GODOY.

# Defence

The Belize Defence Force was formed in 1978 and was based on a combination of the existing Police Special Force and the Belize Volunteer Guard. Military service is voluntary, but provision has been made for the establishment of National Service, if necessary, to supplement normal recruitment. In August 2004 the regular armed forces totalled approximately 1,050 and there were some 700 militia reserves. In 1994 all British forces were withdrawn from Belize, with the exception of about 30 troops who remained to organize jungle-warfare training.

**Defence Budget:** an estimated BZ $38m in 2004.

**Belize Defence Force Commandant:** Brig.-Gen. LLOYD GILLETT.

# Education

Education is compulsory for all children between the ages of five and 14 years. Primary education, beginning at five years of age and lasting for eight years, is provided free of charge, principally through subsidized denominational schools under government control. In

2003/04 there were 62,074 pupils enrolled at 275 primary schools. Secondary education, beginning at the age of 13, lasts for four years. There were 15,344 students enrolled in 43 general secondary schools in 2003/04. In 1999 the combined enrolment of primary, secondary and tertiary education was an estimated 73% (males 73%; females 72%).

In 1997/98 there were 2,853 students enrolled in 12 higher educational institutions, which included technical, vocational and teacher-training colleges. The University College of Belize was established in 1986 and there is also an extra-mural branch of the University of the West Indies in Belize. In early 2000 it was announced that a new University of Belize was to be formed through the amalgamation of five higher education institutions, including the University College of Belize and Belize Technical College. In the financial year 2004/05 budget expenditure on education was envisaged at BZ $152.6m., representing 26.7% of total spending by the central Government.

# Bibliography

For works on the Caribbean generally, see Select Bibliography (Books)

Barry, T., and Vernon, D. *Inside Belize*. London, Latin America Bureau, 1995.

Belize News Network. www.belizenews.net.

Cutlack, M. *Belize, Ecotourism in Action*. London, Macmillan, 1993.

Kroshus Medina, L. *Negotiating Economic Development: Identity Formation and Collective Action in Belize*. Tucson, AZ, University of Arizona Press, 2004.

Moberg, M. *Myths of Ethnicity and Nation: Immigration, Work and Identity in the Belize Banana Industry*. Knoxville, TN, University of Tennessee Press, 1997.

Norton, N. *Belize*. London, Cadogan Press, 1997.

Phillips, M. D. *Belize*. Lanham, MD, University Press of America, 1996.

Roessingh, C. *The Belizean Garifuna: Organization of Identity in an Ethnic Community in Central America*. Amsterdam, Rozenberg, 2002.

Simmons, D. C. *Confederate Settlements in British Honduras*. Jefferson, NC, McFarland & Co, 2001.

Sutherland, A. *The Making of Belize*. Westport, CT, Bergin & Garvey, 1998.

Thomson, P. *History of Belize*. Oxford, Macmillan Caribbean, 2005.

Turner, B. L., and Harrison, P. D. (Eds). *Pulltrouser Swamp: Ancient Maya Habitat, Agriculture and Settlement in Northern Belize*. Salt Lake City, UT, University of Utah Press, 2000.

# BERMUDA

## Geography

### PHYSICAL FEATURES

Bermuda lies in the North Atlantic, about 900 km (560 miles) east of Cape Hatteras in North Carolina (USA), and is an Overseas Territory of the United Kingdom (indeed, it is the oldest British colony). Although geographically part of North America, Bermuda shares many of the features of the West Indian islands, and is generally included in that region—it is north-west of the Bahamian archipelago and north of the Virgin Islands. Bermuda, located on what were called the Somers Islands, has an area of only 53.3 sq km, making it the smallest territory in the Western hemisphere.

Bermuda is built of coral perching on the southern rim of the summit of an underwater volcanic mountain. The 103 km (64 miles) of coastline define some 138 islands and islets, which are strung out across 35 km, running south-westwards from St George's Island to hook around the Great Sound in the south and taper off in a more northerly direction. There are many surrounding reefs, banks and islets. About 20 of the islands are inhabited and seven are linked by bridges and causeways. South of St George's is St David's Island, then Great Bermuda or Main Island (23 km in length). The landscape is richly vegetated, although there is now little land used for farming (over two-fifths of the territory is kept rural and undeveloped), with low hills separated by fertile depressions. The highest point is in the far south, at Gibbs Hill (78 m or 256 ft), itself surmounted by a 36-m, cast-iron lighthouse.

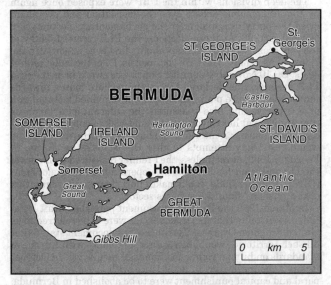

### CLIMATE

The subtropical climate is influenced by the Gulf Stream, which enables Bermuda to consist of the most northerly coral islands in the world. The weather is mild, but humid, with strong winds common in winter. Hurricanes can occur between June and November. The good, year-round rainfall (averaging some 1,500 mm—59 ins—annually) makes up for the scarcity of natural freshwater resources. The annual average temperature is 76°F (24°C), ranging between an average minimum of 59°F (15°C) in February and an August maximum of 85°F (29°C).

### POPULATION

Bermuda, named for the Spanish sailor who visited it in the early years of the 16th century, was only inhabited in the 17th century, by the British. The black population forms the majority (61% of the total in 2000), but whites still comprise 39%. Most people are Christian, with the main (of many) denominations being the Anglican Church (23%) and the African Methodist Episcopalian Church. English is the official language. There is a reasonably significant Portuguese-speaking community.

The total population at the time of the 2001 census was 65,545, giving a population density of some 1,237 per sq km—Bermuda is the most densely populated territory of the Americas. The capital is at Hamilton, near the centre of Main Island and on the north-eastern shores of the Great Sound. It is one of two municipalities in the territory, the other being that of St George's, the original settlement and old capital in the north. The latter municipality actually has the larger population (1,648 at the census of 1991). There are also three villages on the islands, while for administrative purposes there are nine parishes.

## History

Bermuda is a Crown Colony of the United Kingdom. The British monarch is represented in the islands by a Governor.

The 1968 Constitution gave Bermuda (the oldest British colony, established in 1684) internal self-government. The 1968 elections were won by the moderate, multiracial United Bermuda Party (UBP). However, Bermudian society was riven by racial tensions and in the early 1970s there were considerable levels of violence, including the assassination of the Governor of the Colony. At the 1976 election the UBP remained in power, although it lost seats to the mainly black, pro-independence, left-wing Progressive Labour Party (PLP). The PLP gained 46% of the vote, but only 14 of the 40 seats in the House of Assembly. In 1980 the UBP only narrowly retained power, but regained the seats it had lost in the following election, in 1983. Meanwhile, in 1982, John (later Sir John) W. Swan became party leader and Premier. The UBP won a decisive victory in the 1985 elections, but in 1989 it lost seats, although it maintained its majority. The UBP remained in government after the general election of October 1993 (the first to be held since the voting age was lowered from 21 to 18 years), having won 22 seats (compared with 18 seats secured by the PLP).

On 16 August 1995 a referendum was held on independence from the United Kingdom, despite the strenuous opposition of the PLP, which stated that the issue should be decided in a general election, and urged its supporters to boycott the poll. In the referendum, which had a relatively low turn-out for Bermuda of 58.8%, some 73.6% of participants registered their opposition to independence. The debate polarized opinion within the ruling party and Swan subsequently resigned as Premier and as leader of the UBP. The erstwhile finance minister, Dr David Saul, succeeded him in both posts. Saul remained neutral on the independence issue, and the divisions within the UBP deepened under his leadership. A motion of censure against him, prompted by Saul's decision to authorize the establishment of a foreign-owned franchise restaurant in Bermuda (in contravention of a ruling by the Bermuda Monetary Authority), was approved, with the support of five UBP members, in the House of Assembly in June 1996. Having failed to restore party unity,

Saul resigned as Premier and UBP leader in March 1997. Pamela Gordon replaced him.

In August 1996 the Leader of the Opposition, Frederick Wade, died. At a by-election in October Wade's former seat was retained by the PLP, and in the following month the party's Deputy Leader, Jennifer Smith, was elected to the leadership of the organization.

The deep divisions within the UBP were exposed once again by the acrimonious debate over the legislation and overshadowed the party's campaign for the parliamentary elections of 9 November 1998. At these elections, the PLP received 54.2% of the votes cast, winning 26 seats and its first ever majority in the House of Assembly. The UBP obtained 44.1% of the ballot and 14 seats. On 10 November Jennifer Smith was sworn in as the first PLP Premier. She declared that no immediate moves towards independence were planned, although it remained a stated aim of the party. She also reassured the international business community that her Government would seek to enhance Bermuda's attraction as an international business centre, and that she would resist any attempts to alter the island's tax status. In February 1999, nevertheless, the Premier reassured members of the Organisation for Economic Co-operation and Development (OECD) that Bermuda would make efforts to improve regulation of the 'offshore' financial services sector.

In March 1999 the British Government published draft legislation confirming that its Dependencies were to be referred to as United Kingdom Overseas Territories; the document stated that all such territories would be required to comply with European standards on human rights and financial regulation. In October it was announced that in accordance with these reforms, corporal and capital punishment were to be abolished in Bermuda.

Following OECD's investigation into tax 'havens', in June 2000 Bermuda pledged to conform to international standards on financial transparency before the end of 2005. This was likely to include the abolition of the 60/40 rule, by which all businesses aimed at the local market had to be 60%-owned by a Bermudian individual or entity. The Government had, however, pledged to maintain its existing tax system, which included no income tax. On 2 November 2000, in an internal PLP election, Smith was re-elected leader of the party, with 81 of the 116 votes cast.

Aside from the worrying decline in the tourism industry and the increase in the value of drugs seizures by Bermudian customs, the most important issue the Government faced during the first six months of 2001 was that of constitutional change. The Government and the opposition UBP fundamentally disagreed about proposed changes to the voting system and electoral boundaries. Racial issues were at the heart of the conflict, with the ruling PLP claiming the present system favoured the traditionally 'white' UBP. In December 2000 the House of Assembly approved a motion requesting the British Government to approve the establishment of a boundaries commission, which would recommend the size of the island's constituencies. In January 2001 the UBP submitted an 8,500-signature petition to the British Government, demanding a constitutional conference or referendum before any changes were made. Gordon also claimed that some Bermudians had refused to sign the petition fearing recriminations. The British Foreign and Commonwealth Office (FCO), however, stated that a constitutional conference was unnecessary, and in April it began consultations over the proposed changes. The commission recommended reducing the number of seats in the House of Assembly by four, to 36. In addition, whereas deputies had previously been elected from 20 two-member constituencies, under the proposed scheme each member of parliament would be elected by a separate constituency. The FCO approved the changes before the July 2003 general election.

In November 2001 Sir John Vereker succeeded Sir Thorold Masefield as Governor of Bermuda.

In May 2002 the British Overseas Territories Act, having received royal assent in the United Kingdom in February, came into force and granted British citizenship to the people of its Overseas Territories, including Bermuda. Under the new law Bermudians would be able to hold British passports and work in the United Kingdom and anywhere else in the European Union.

In mid-2002 the issue of financial transparency and corporate governance in Bermuda re-emerged following the disastrous collapse of the US energy concern Enron and several other multinational firms in late 2001 and early 2002. The indictment on tax-evasion charges of Dennis Kozlowski, the former CEO of Tyco International Ltd, a manufacturer of electronic security systems with a nominal headquarters on the island, fuelled a growing campaign in the US media and the US Congress against the perceived lack of financial regulation and scrutiny in Bermuda. Since 1997, it was reported, nine major US companies had reincorporated in Bermuda and the Cayman Islands.

Elections to the smaller 36-seat House of Assembly were held on 24 July 2003 and were dominated by resentment towards wealthy foreign workers and criticism of the personal style of Smith, regarded by many as uncommunicative and aloof. The PLP retained its majority in parliament, securing 22 seats, but its share of the popular vote was reduced to 51.6% (compared with 48.0% for the UBP). Smith, however, resigned as Premier after it emerged that she had retained her seat, in what was regarded as a 'safe' PLP constituency, by just eight votes. She was replaced by the erstwhile Minister of Works and Engineering, W. Alexander Scott. The Deputy Premier and Minister of Finance, C. Eugene Cox, also resigned, and was succeeded by Dr Ewart F. Brown, previously the Minister of Transport. Scott's first Cabinet, announced a few days after the elections, consisted of 12 members (including Cox, who retained the finance portfolio), despite the PLP's pre-election pledge to reduce its size to eight in order to reduce costs.

Following the death of C. Eugene Cox in January 2004, his daughter, Paula Ann Cox, the erstwhile Minister of Education and Justice and Attorney-General, was appointed as Minister of Finance. Senator Larry Mussenden became the new Minister of Justice and Attorney-General, while Terry Lister moved to the education ministry; Ashfield DeVent, hitherto Minister without Portfolio, was given responsibility for works, engineering and housing. Michael Scott, the Government's leader in the Senate and Minister of Legislative Affairs, succeeded Eugene Cox as the deputy for the Sandys North constituency after winning a by-election; the constituency has traditionally been regarded a PLP stronghold. In February the Prime Minister announced that his Government would initiate a 'relaxed public discussion' on the subject of achieving independence from the United Kingdom.

In July 2004 Scott effected a minor cabinet reshuffle after Maurine Webb resigned as Minister of Tourism, Telecommunications and E-Commerce. Responsibility for tourism transferred to Brown and Michael Scott, the erstwhile Minister of Legislative Affairs, was allocated Webb's other portfolios. The Ministry of Legislative Affairs was added to the responsibilities of Mussenden.

In December 2004 the Senate approved the Ombudsman Act, which created the post of Ombudsman to investigate complaints against the Government, although ministers would be exempt from the Ombudsman's remit; Arlene Brock was appointed to this post in May 2005. In January of that year the Bermuda Independence Commission was established, in co-operation with the UN Committee on Decolonization, in order to investigate and report on the implications of transition to self-rule. The British Government had recommended another referendum on such a decision, although opinion polls indicated that a majority of the public would vote against independence once again, and it was therefore believed that the Government would seek to avoid that method of resolution (and instead opt to incorporate the decision with the next general election). In February a Supreme Court judge ruled as discriminatory and unlawful the Government's policy of deferring the right to apply for parole for prisoners convicted of drugs-smuggling until they had served one-half of their sentence. Prisoners convicted of non-drugs, non-sexual or non-violent offences were allowed to apply for parole after serving one-third of their sentence. (In July parliament passed an amendment to legislation on the misuse of drugs, granting courts greater powers to fine and sentence those convicted; similar legislation was enacted with regard to the carrying of offensive weapons.)

In July 2003 Bermuda became an associate member of the Caribbean Community and Common Market (CARICOM).

# Economy

Despite the growth of financial services, it was estimated that tourism still accounted for almost one-half of foreign-exchange earnings and, directly and indirectly, 65% of employment in 1996; the service industries as a whole engaged 86.6% of the employed labour force in 2003. Visitor expenditure accounted for 10.2% of GDP in 2002. In 2003 12.6% of the employed labour force worked in restaurants and hotels. The tourism sector earned an estimated B $353.7m. in 2004. The majority of visitors come from the USA (some 77.0% of total arrivals by air in 2004), Canada and the United Kingdom. Although the number of cruise-ship passengers increased in 1998, 1999 and 2000, there was a large decrease (14.4%) in 2001, partly owing to the repercussions of the terrorist attacks on the USA on 11 September. The number of air arrivals fell consistently in 1998–2001. In late 2002, in an effort to halt the consequent decline in tourism receipts, the Bermuda Alliance for Tourism launched a rebranding of Bermuda as a luxury destination, in order to attract higher-spending visitors. The number of cruise-ship passengers and air arrivals increased in that year (by 11.5% and 3.1%, respectively); however, the sector suffered a further decline in 2003 when the number of arrivals by air fell by 9.5%. In terms of tourism receipts, which decreased by 9.6%, this decline was not significantly offset by a 13% rise in the number of cruise-ship passengers, who generally spend far less per-head than tourists arriving by air. In 2004 a total of 477,757 tourists visited Bermuda; the low level of expenditure in 2004 ($353.7m.) was attributed to the closure of a major hotel following damage caused by 'Hurricane Fabian' near the end of 2003. Between 1980 and 2000 no new hotels were built. In August 2000, however, the Smith Government granted tax concessions for hotel construction. The measure was intended to create a better climate for investment in the sector. The ongoing decline in the tourism sector was likely also to affect the retail sector, and to lead to an overall increase in unemployment.

Some 28.8% of the employed work-force were engaged directly in the financial, insurance, real-estate and business sectors in 2003. 'Offshore' commercial and financial services were a significant foreign-exchange earner. International business accounted for 20.6% of GDP in 2003, when the number of companies registered in Bermuda totalled 13,528 (compared with 13,305 in 2001). The sector's real GDP was reported to have increased by some 0.9% in 2002. The Bermudian insurance industry was estimated to have doubled during the last decade of the 20th century and by 2000 was the world's third largest. In March 2002 the US Department of the Treasury began an investigation into the reasons why many formerly US-based firms had relocated to the island. In late 2001 the industry came under huge pressure from claims made in connection with the September terrorist attacks in the USA, although increased demand for insurance and reinsurance services led to a number of new companies being formed at the same time.

Bermuda is very dependent upon food and energy imports, mainly from the USA, and has few export commodities. Therefore, it consistently records a large visible trade deficit (an estimated B $781m. in 2003). Receipts from the service industries normally ensure a surplus on the current account of the balance of payments (an estimated $172m. in 2004). The USA is the principal source of imports (80.0% of total imports in 2003) and the principal market for exports. Manufacturing and construction accounted for some 7.4% of GDP in 2003 and employed 10.7% of the work-force. Smaller industries include ship repair, small boat building and manufacture of paint, perfume, pharmaceuticals, mineral-water extracts and handicrafts.

In 2003 Bermuda's estimated gross domestic product (GDP) at current prices was B $3,966.3m. In 2002 GDP per head was equivalent to some $52,457, one of the highest levels in the world. In 2003 Bermuda recorded a provisional budgetary surplus of some $30.0m. The average annual rate of inflation was 2.1% in 1993–2003. The rate was estimated at 3.2% in 2003. The weakness of the US dollar posed a threat to the Bermudian rate of inflation, owing to the country's economic dependence upon imports and, in the longer term, the likelihood of labour shortages would also present a risk to the inflation rate. Some 7.8% of the labour force were registered unemployed in July 2002, although unofficial sources estimated that the rate might be as high as 15%.

# Statistical Survey

Source: Department of Statistics, POB HM 3015, Hamilton HM MX; tel. 297-7761; fax 295-8390; e-mail statistics@gov.bm; internet www.gov.bm.

## AREA AND POPULATION

**Area:** 53.3 sq km (20.59 sq miles).

**Population** (civilian, non-institutional): 58,460 at census of 20 May 1991; 62,059 (males 29,802, females 32,257) at census of 20 May 2000.

**Density** (census of 2000): 1,164.3 per sq km.

**Principal Towns** (population at 1991 census): St George's 1,648; Hamilton (capital) 1,100.

**Births, Marriages and Deaths** (2002): Live births 830 (birth rate 13.4 per 1,000); Marriage rate 15.1 per 1,000; Deaths 404 (death rate 6.5 per 1,000).

**Expectation of Life** (years at birth, official estimates in 2002): Males 75.2; Females 79.3. Source: Pan American Health Organization.

**Employment** (excluding unpaid family workers, 2003): Agriculture, forestry, fishing, mining and quarrying 638; Manufacturing 1,063; Electricity, gas and water 405; Construction 2,959; Wholesale and retail trade 5,015; Restaurants 1,779; Hotels 2,981; Transport and communications 2,861; Financial intermediation 2,821; Real estate 507; Business activities 3,756; Public administration 3,982; Education, health and social services 2,916; Other community, social and personal services 2,222; International business activity 3,781; Total employed 37,686 (males 19,597, females 18,089).

## HEALTH AND WELFARE

**Physicians** (per 1,000 head, 2003): 1.9.

**Hospital Beds** (per 1,000 head, 2003): 22.6.

**Health Expenditure** (% of GDP, 1997): 4.2.

**Health Expenditure** (public, % of total, 1995): 53.2.
Source: partly Pan American Health Organization.
For definitions, see explanatory note on p. vi.

## AGRICULTURE, ETC.

**Principal Crops** (FAO estimates, metric tons, 2003): Potatoes 700; Carrots 340; Vegetables and melons 2,855; Bananas 330.

**Livestock** (FAO estimates, 2003): Cattle 600; Horses 900; Pigs 600.

**Livestock Products** (FAO estimates, metric tons, 2003): Cows' milk 1,350; Hen eggs 280.

**Fishing** (metric tons, live weight, 2003): Groupers 33; Snappers and jobfishes 31; Wahoo 88; Yellowfin tuna 47; Carangids 41; Caribbean spiny lobster 31; Total catch (incl. others) 358.
Source: FAO.

## INDUSTRY

**Electric Energy** (consumption, million kWh): 590 in 2003.

BERMUDA                                                                                                      Directory

## FINANCE

**Currency and Exchange Rates:** 100 cents = 1 Bermuda dollar (B $). *Sterling, US Dollar and Euro Equivalents* (31 May 2005): £1 sterling = B $1.8181; US $1 = B $1.000; €1 = B $1.2331; B $100 = £55.00 = US $100.00 = €81.11. *Exchange Rate:* The Bermuda dollar is at par with the US dollar. Note: US and Canadian currencies are also accepted.

**Budget** (B $ million, 2002/03, provisional): *Total Revenue* 671.1 (Customs duties 185.0; Payroll tax 203.4; Hotel occupancy tax 10.9; Passenger tax 22.5; Land tax 40.5; International company tax 47.7; Stamp duties 34.9). *Total Expenditure* 641.4 (Salaries and wages 248.3; Other goods and services 151.3; Grants and contributions 140.2; Capital expenditure 70.8).

**Cost of Living** (Consumer Price Index; base: 1993 = 100): 119.6 in 2002; 123.4 in 2003; 127.8 in 2004.

**Expenditure on the Gross Domestic Product** (B $ million at current prices, year ending 31 March 2001): Government final consumption expenditure 367; Private final consumption expenditure 2,079; Gross fixed capital formation 679; *Total domestic expenditure* 3,125; Exports of goods and services 1,597; *Less* Imports of goods and services 1,325; *GDP in purchasers' values* 3,397. Source: UN, *National Accounts Statistics*.

**Gross Domestic Product by Economic Activity** (provisional, B $ million at current prices, 2003): Agriculture, forestry and fishing 31.3; Mining and quarrying 7.2; Manufacturing 82.2; Electricity, gas and water 90.2; Construction 213.0; Wholesale and retail trade 345.5; Restaurants and hotels 232.1; Transport and communications 258.6; Financial intermediation 478.9; Real-estate 467.6; Business services 341.0; Public administration 246.0; Education, health and social work 267.2; Other community, social and personal services 99.2; International business activity 819.3; *Sub-total* 3,979.3; *Less* Imputed bank service charges 209.3; Indirect taxes, less subsidies 196.3; *GDP in purchasers' values* 3,966.3.

**Balance of Payments** (B $ million, 2004, estimates): Gross receipts 2,241 (Tourism 354; Professional services 1,300); Gross payments –2,069 (Imports 965); *Overall balance* (net) 172. Source: Bermuda Monetary Authority.

## EXTERNAL TRADE

**Principal Commodities** (B $ million, 2004): *Imports:* Food, beverages and tobacco 175.0; Clothing 42.5; Fuels 105.4; Chemicals 108.4; Basic material and semi-manufacturing 144.4; Machinery 183.8; Transport equipment 64.6; Finished equipment 144.8; Miscellaneous 0.3; Total (incl. others) 969.1. *Exports* (2003, preliminary estimate): Total 52.0.

**Principal Trading Partners** (US $ million): *Imports* (1999): Canada 40.2; United Kingdom 34.5; USA 518.3; Total (incl. others) 712.1. *Exports* (1995): France 7.5; United Kingdom 3.9; USA 31.3; Total (incl. others) 62.9.

Source: partly UN, *International Trade Statistics Yearbook*.

## TRANSPORT

**Road Traffic** (vehicles in use, 2003): Private cars 20,976; Motorcycles 20,331; Buses, taxis and limousines 794; Trucks and tank wagons 3,818; Other 764; *Total* 46,683.

**Shipping:** *Ship Arrivals* (1999): Cruise ships 140; Cargo ships 171; Oil and gas tankers 17. *Merchant Fleet\** (registered at 31 December 2004): 122; Total displacement 6,166,162 grt. *International Freight Traffic†* ('000 metric tons, 1990): Goods loaded 130; Goods unloaded 470.
\*Source: Lloyd's Register-Fairplay, *World Fleet Statistics*.
†Source: UN, *Monthly Bulletin of Statistics*.

**Civil Aviation** (1999): Aircraft arrivals 6,024; Passengers 354,026; Air cargo 4,761,444 kg; Air mail 422,897 kg.

## TOURISM

**Visitor Arrivals:** 483,622 (arrivals by air 283,557, cruise-ship passengers 200,065) in 2002; 482,673 (arrivals by air 256,576, cruise-ship passengers 226,097) in 2003; 477,757 (arrivals by air 271,617, cruise-ship passengers 206,140) in 2004.

**Tourism Receipts** (B $ million, estimated): 378.8 in 2002; 342.5 in 2003; 353.7 in 2004.

## COMMUNICATIONS MEDIA

**Radio Receivers** (1997): 82,000 in use.
**Television Receivers** (1999): 70,000 in use.
**Telephones** (2001): 56,300 main lines in use.
**Mobile Cellular Telephones** (2001): 13,300 subscribers.
**Personal Computers** (2001): 32,000 in use.
**Internet Users** (1999): 25,000.
**Daily Newspapers** (2000): 1 (estimated circulation 17,700).
**Non-daily Newspapers** (2000): 2 (estimated circulation 11,600).
Sources: mainly UNESCO, *Statistical Yearbook*; UN, *Statistical Yearbook*; International Telecommunication Union.

## EDUCATION

**Pre-primary** (1999): 12 schools; 191 teachers; 429 pupils.
**Primary** (1999): 25 schools; 368 teachers; 4,980 pupils.
**Senior** (1999): 18 schools; 399 teachers; 2,335 pupils\*.
**Higher** (2002): 1 institution; 544 students.
\*Including six private schools.

**Adult Literacy Rate** (UNESCO estimates): 99% (males 98%; females 99%) in 1998 (Source: UNESCO, *Statistical Yearbook*).

# Directory

## The Constitution

The Constitution, introduced on 8 June 1968 and amended in 1973 and 1979, contains provisions relating to the protection of fundamental rights and freedoms of the individual; the powers and duties of the Governor; the composition, powers and procedure of the Legislature; the Cabinet; the judiciary; the public service and finance.

The British monarch is represented by an appointed Governor, who retains responsibility for external affairs, defence, internal security and the police.

The Legislature consists of the monarch, the Senate and the House of Assembly. Three members of the Senate are appointed at the Governor's discretion, five on the advice of the Government leader and three on the advice of the Opposition leader. The Senate elects a President and Vice-President. The House of Assembly, consisting of 36 members elected under universal adult franchise, elects a Speaker and a Deputy Speaker, and sits for a five-year term.

The Cabinet consists of the Premier and at least six other members of the Legislature. The Governor appoints the majority leader in the House of Assembly as Premier, who in turn nominates the other members of the Cabinet. They are assigned responsibilities for government departments and other business and, in some cases, are assisted by Permanent Cabinet Secretaries.

The Cabinet is presided over by the Premier. The Governor's Council enables the Governor to consult with the Premier and two other members of the Cabinet nominated by the Premier on matters for which the Governor has responsibility. The Secretary to the Cabinet, who heads the public service, acts as secretary to the Governor's Council.

Voters must be British subjects aged 18 years or over (lowered from 21 years in 1990), and, if not possessing Bermudian status, must have been registered as electors on 1 May 1976. Candidates for election must qualify as electors, and must possess Bermudian status.

Under the British Overseas Territories Act, which entered into effect in May 2002, Bermudian citizens have the right to United Kingdom citizenship and the right of abode in the United Kingdom. British citizens do not enjoy reciprocal rights.

## The Government

**Governor and Commander-in-Chief:** Sir JOHN VEREKER (took office 21 November 2001).
**Deputy Governor:** NICK CARTER.

### CABINET
(July 2005)

**Premier:** WILLIAM ALEXANDER SCOTT.

BERMUDA

*Directory*

**Deputy Premier, Minister of Tourism and Transport:** Dr EWART FREDERICK BROWN.
**Minister of Finance:** PAULA ANN COX.
**Minister of the Environment:** NELETHA I. BUTTERFIELD.
**Minister of Education and Development:** TERRY E. LISTER.
**Minister of Health and Family Services:** PATRICE K. MINORS.
**Minister of Labour, Home Affairs and Public Safety:** K. H. RANDOLPH HORTON.
**Minister of Telecommunications and E-Commerce:** MICHAEL SCOTT.
**Minister of Works, Engineering and Housing:** ASHFIELD DEVENT.
**Minister of Community Affairs and Sports:** DALE D. BUTLER.
**Attorney-General and Minister of Justice and Legislative Affairs:** Sen. LARRY D. MUSSENDEN.

### MINISTRIES

**Office of the Governor:** Government House, 11 Langton Hill, Pembroke HM 13; tel. 292-3600; fax 292-6831; e-mail depgov@ibl.bm; internet www.gov.bm.

**Office of the Premier:** Cabinet Office, Cabinet Bldg, 105 Front St, Hamilton HM 12; tel. 292-5501; fax 292-8397; e-mail ascott@gov.bm; internet www.gov.bm.

**Ministry of Community Affairs and Sports:** Old Fire Station Bldg, 81 Court St, Hamilton HM 12; tel. 295-0855; fax 295-6292.

**Ministry of Education and Development:** Dundonald Place, 14 Dundonald St, POB HM 1185, Hamilton HM EX; tel. 278-3300; fax 278-3348; internet www.moe.bm.

**Ministry of the Environment:** Government Administration Bldg, 30 Parliament St, Hamilton HM 12; tel. 297-7590; fax 292-2349; e-mail browlinson@gov.bm.

**Ministry of Finance:** Government Administration Bldg, 30 Parliament St, Hamilton HM 12; tel. 295-5151; fax 295-5727.

**Ministry of Health and Family Services:** 7 Point Finger Rd, Paget DV 04, POB HM 380, Hamilton HM BX; tel. 236-0224; fax 236-3971; e-mail ejjoell@gov.bm; internet www.healthandfamily.gov.bm.

**Ministry of Justice and Legislative Affairs and Attorney-General's Chambers:** Penthouse Floor, Global House, 43 Church St, Hamilton HM 12; tel. 292-2463; fax 292-3608; e-mail agc@gov.bm.

**Ministry of Labour, Home Affairs and Public Safety:** Global House, 43 Church St, Hamilton HM 12; tel. 292-5998; fax 295-5267; e-mail lhaps@gov.bm.

**Ministry of Telecommunications and E-Commerce:** POB HM 101, Hamilton HM AX; tel. 292-4595; fax 295-1462; e-mail gtelecom@gov.bm; internet www.mtec.bm.

**Ministry of Tourism and Transport:** Global House, 43 Church St, Hamilton HM 12; tel. 295-3130; fax 295-1013.

**Ministry of Works, Engineering and Housing:** 3rd Floor, General Post Office Bldg, 56 Church St, POB HM 525, Hamilton HM 12; tel. 297-7699; fax 295-0170; e-mail nfox@bdagov.bm.

## Legislature

### SENATE

**President:** ALFRED OUGHTON.
**Vice-President:** Dr IDWAL WYN (WALWYN) HUGHES.

There are 11 nominated members.

### HOUSE OF ASSEMBLY

**Speaker:** STANLEY W. LOWE.
**Deputy Speaker:** Dr JENNIFER M. SMITH.

Following a constutional change effected in mid-2003, there are 36 members of the House of Assembly, each of whom is elected by a single-member constituency. The House of Assembly meets on Fridays when in session, and on Mondays, Wednesdays and Fridays during the period that the annual budget is debated.

**Clerk to the Legislature:** Y. MURIEL ROACH (The Legislature, Hamilton), tel. 292-7408; fax 292-2006; e-mail mroach@gov.bm.

*General Election, 24 July 2003*

| Party | % of votes | Seats |
|---|---|---|
| Progressive Labour Party | 51.6 | 22 |
| United Bermuda Party | 48.0 | 14 |
| **Total** (incl. others) | 100.0 | 36 |

## Political Organizations

**National Liberal Party (NLP):** 53 Church St, Hamilton HM 12, POB HM 1794, Hamilton HM FX; tel. 292-8587; f. 1985; Leader DESSALINE WALDRON; Chair. GRAEME OUTERBRIDGE.

**Progressive Labour Party (PLP):** Alaska Hall, 16 Court St, POB 1367, Hamilton HM 12; tel. 292-2264; fax 295-7890; e-mail info@plp.bm; internet www.plp.bm; f. 1963; advocates the 'Bermudianization' of the economy, more equitable taxation, a more developed system of welfare and preparation for independence; Leader WILLIAM ALEXANDER SCOTT; Deputy Leader Dr EWART FREDERICK BROWN; Chair. RODERICK BURCHALL.

**United Bermuda Party (UBP):** Central Office, 3rd Floor, Bermudiana Arcade, 27 Queen St, Hamilton HM 12; tel. 295-0729; fax 292-7195; e-mail info@ubp.bm; internet www.ubp.bm; f. 1964; policy of participatory democracy, supporting system of free enterprise; Leader GRANT GIBBONS; Chair. AUSTIN B. WOODS.

## Judicial System

**Chief Justice:** RICHARD GROUND.
**President of the Court of Appeal:** Sir JAMES ASTWOOD.
**Registrar of Supreme Court and Court of Appeal:** MICHAEL J. MELLO.
**Attorney-General:** Sen. LARRY MUSSENDEN.
**Director of Public Prosecutions:** KULANDRA RATNESER (acting).
**Solicitor-General:** WILHELM C. BOURNE.

The Court of Appeal was established in 1964, with powers and jurisdiction of equivalent courts in other parts of the Commonwealth. The Supreme Court has jurisdiction over all serious criminal matters and has unlimited civil jurisdiction. The Court also hears civil and criminal appeals from the Magistrates' Courts. The three Magistrates' Courts have jurisdiction over all petty offences, and have a limited civil jurisdiction.

## Religion

### CHRISTIANITY

In 2000 it was estimated that 23% of the population were members of the Anglican Communion, 15% were Roman Catholics, 11% were African Methodist Episcopalians, 7% were Seventh-day Adventists and 4% were Wesleyan Methodists. The Presbyterian Church, the Baptist Church and the Pentecostal Church are also active in Bermuda.

#### The Anglican Communion

The Anglican Church of Bermuda consists of a single, extra-provincial diocese, directly under the metropolitan jurisdiction of the Archbishop of Canterbury, the Primate of All England. There are about 23,000 Anglicans and Episcopalians in Bermuda.

**Bishop of Bermuda:** Rt Rev. EWEN RATTERAY, Bishop's Lodge, 18 Ferrar's Lane, Pembroke HM 08, POB HM 769, Hamilton HM CX; tel. 292-6987; fax 292-5421; e-mail bishopratteray@ibl.bm; internet www.anglican.bm.

#### The Roman Catholic Church

Bermuda forms a single diocese, suffragan to the archdiocese of Kingston in Jamaica. At 31 December 2003 there were an estimated 9,275 adherents in the Territory. The Bishop participates in the Antilles Episcopal Conference (currently based in Port of Spain, Trinidad and Tobago).

**Bishop of Hamilton in Bermuda:** ROBERT JOSEPH KURTZ, 2 Astwood Rd, POB HM 1191, Hamilton HM EX; tel. 232-4414; fax 232-4447; e-mail rjkurtz@northrock.bm.

### Protestant Churches

**Baptist Church:** Emmanuel Baptist Church, 35 Dundonald St, Hamilton HM 10; tel. 295-6555; fax 296-4461; Pastor RONALD K. SMITH.

**Wesley Methodist Church:** 41 Church St, POB HM 346, Hamilton HM BX; tel. 292-0418; fax 295-9460; e-mail info@wesley.bm; internet www.wesley.bm; Rev. JEFF CHANT.

## The Press

**Bermuda Homes and Gardens Magazine:** Suite 16352, 48 Par-la-Ville Rd, Hamilton HM 11; tel. 295-5845; e-mail info@bdahomesandgardens.bm; monthly, free.

**Bermuda Magazine:** POB HM 283, Hamilton HM HX; tel. 295-0695; fax 295-8616; e-mail cbarclay@ibl.bm; f. 1990; quarterly; Editor-in-Chief CHARLES BARCLAY.

**Bermudian Business Online:** POB HM 283, Hamilton HM AX; tel. 295-0695; e-mail berpub@ibl.bm; internet www.bermudianbusiness.com; Publisher TINA STEVENSON.

**The Bermuda Sun:** 19 Elliott St, POB HM 1241, Hamilton HM FX; tel. 295-3902; fax 292-5597; e-mail bdasun@ibl.bm; internet www.bermudasun.bm; f. 1964; 2 a week; official government gazette; Editor TONY MCWILLIAM; circ. 12,500.

**The Bermudian:** 13 Addendum Lane, Pitt's Bay Rd, Pembroke HM07; POB HM 283, Hamilton HM AX; tel. 295-0695; fax 295-8616; e-mail berpub@ibl.bm; f. 1930; monthly; pictorial and resort magazine; Editor MEREDITH EBBIN; circ. 7,500.

**Cable TV Guide:** 41 Victoria St, Hamilton HM 12; tel. 295-3902; fax 295-5597.

**The Mid-Ocean News:** 2 Par-la-Ville Rd, POB HM 1025, Hamilton HM DX; tel. 295-5881; fax 295-1513; f. 1911; weekly with *TV Guide*; Editor TIM HODGSON; Gen. Man. KEITH JENSEN; circ. 14,500.

**The Royal Gazette:** 2 Par-la-Ville Rd, POB HM 1025, Hamilton HM DX; tel. 295-5881; fax 295-1513; internet www.theroyalgazette.com; f. 1828; morning daily; Editor WILLIAM J. ZUILL; Gen. Man. KEITH JENSEN; circ. 17,500.

**TV Week:** 2 Par-la-Ville Rd, Hamilton HM 08; tel. 295-5881.

**The Worker's Voice:** 49 Union Sq., Hamilton HM 12; tel. 292-0044; fax 295-7992; e-mail biu@ibl.bm; fortnightly; organ of the Bermuda Industrial Union; Editor Dr B. B. BALL.

## Publisher

**Bermudian Publishing Co:** POB HM 283, Hamilton HM AX; tel. 295-0695; fax 295-8616; e-mail berpub@ibl.bm; social sciences, sociology, sports; Editor KEVIN STEVENSON.

## Broadcasting and Communications

### TELECOMMUNICATIONS

**Bermuda Telephone Co (BTC):** 30 Victoria St, POB 1021, Hamilton HM DX; tel. 295-1001; fax 295-1192; internet www.btc.bm; f. 1987; Pres. FRANCIS R. MUSSENDEN.

**Bermuda Digital Communications:** 22 Reid St, Hamilton HM 11; tel. 296-4010; fax 296-4020; e-mail info@bdc.bm; internet www.bdc.bm; mobile cellular telephone operator; f. 1998; Chair. and CEO KURT EVE.

**Cable & Wireless (Bermuda) Ltd:** 1 Middle Rd, Smith's FL 03, POB HM 151, Hamilton HM AX; tel. 297-7000; fax 295-8995; e-mail helpdesk@bda.cwplc.com; internet www.cw.com/bermuda; Gen. Man. EDDIE SAINTS.

**TeleBermuda International (TBI) Ltd:** Bermuda Commercial Bank Bldg, 2nd Floor, 43 Victoria St, Hamilton HM 12; tel. 296-9000; fax 296-9010; e-mail save@telebermuda.com; internet www.telebermuda.com; f. 1997; a division of GlobeNet Communications, provides an international service; owns a fibre-optic network connecting Bermuda and the USA; Gen. Man. JAMES FITZGERALD.

### BROADCASTING

#### Radio

**Bermuda Broadcasting Company:** POB HM 452, Hamilton HM BX; tel. 295-2828; fax 295-4282; e-mail zbmzfb@bermudabroadcasting.com; internet www.bermudabroadcasting.com; f. 1982 as merger of ZBM (f. 1943) and ZFB (f. 1962); operates 4 radio stations; CEO ULRIC P. RICHARDSON; Operations Man. E. DELANO INGHAM.

**DeFontes Broadcasting Co Ltd (VSB):** POB HM 1450, Hamilton HM FX; tel. 292-0050; fax 295-1658; e-mail vsbnews@ibl.bm; internet www.vsb.bm; f. 1981 as St George's Broadcasting Co; commercial; 4 radio stations; Pres. KENNETH DEFONTES; Station Man. MIKE BISHOP.

#### Television

**Bermuda Broadcasting Company:** see Radio; operates 2 TV stations (Channels 7 and 9).

**Bermuda Cablevision Ltd:** 19 Laffan St, Hamilton; tel. 292-5544; fax 296-3023; e-mail info@cablevision.bm; internet www.cablevision.bm; f. 1988; 180 channels; Pres. DAVID LINES; Gen. Man. JEREMY ELMAS.

**DeFontes Broadcasting Co Ltd (VSB):** see Radio; operates 1 TV station.

## Finance

(cap. = capital; res = reserves; dep. = deposits; m. = million; brs = branches; amounts in Bermuda dollars)

### BANKING

#### Central Bank

**Bermuda Monetary Authority:** 31 Reid St, Hamilton, HM 12; tel. 295-5278; fax 292-7471; e-mail info@bma.bm; internet www.bma.bm; f. 1969; central issuing and monetary authority; cap. 10.6m., res 15.3m., total assets 117.0m. (Dec. 2002); Chair., CEO and Controller of Foreign Exchange CHERYL-ANN LISTER.

#### Commercial Banks

**Bank of Bermuda Ltd:** 6 Front St, POB HM 1020, Hamilton HM DX; tel. 295-4000; fax 295-7093; e-mail corpcom@ibl.bm; internet www.bankofbermuda.bm; f. 1889; acquired by HSBC in Feb. 2004; cap. 29.0m., res 390.4m., dep. 10,165.2m. (Dec. 2002); CEO PHILLIP BUTTERFIELD; COO ANDY GENT; 6 brs.

**Bank of N. T. Butterfield & Son Ltd:** 65 Front St, POB HM 195, Hamilton HM AX; tel. 295-1111; fax 292-4365; e-mail contact@bntb.bm; internet www.bankofbutterfield.com; f. 1858; inc. 1904; cap. 21.3m., res 143.0m., dep. 5,246.5m. (June 2002); Chair. Dr JAMES A. C. KING; Pres. and CEO ALAN R. THOMPSON; 4 brs.

**Bermuda Commercial Bank Ltd:** Bermuda Commercial Bank Bldg, 43 Victoria St, Hamilton HM 12; tel. 295-5678; fax 295-8091; e-mail enquiries@bcb.bm; internet www.bermuda-bcb.com; f. 1969; 48.06%-owned by First Curaçao International Bank NV; cap. 10.3m., res 10.8m., dep. 533.2m. (Sept. 2002); Pres. JOHN CHR. M. A. M. DEUSS; Pres. and COO TIMOTHY W. ULRICH.

### STOCK EXCHANGE

**Bermuda Stock Exchange:** 3rd Floor, Washington Mall, Church St, Hamilton HM FX; tel. 292-7212; fax 292-7619; e-mail info@bsx.com; internet www.bsx.com; f. 1971; 380 listed equities, funds, debt issues and depositary programmes; Chair. DAVID BROWN; Pres. and CEO GREG WOJCIECHOWSKI.

### INSURANCE

Bermuda had a total of some 1,600 registered insurance companies in 2002, the majority of which are subsidiaries of foreign insurance companies, or owned by foreign industrial or financial concerns. Many of them have offices on the island.

**Insurance Information Office:** Cedarpark Centre, 48 Cedar Ave, POB HM 2911, Hamilton HM LX; tel. 292-9829; fax 295-3532; e-mail biminfo@bii.bm; internet www.bermuda-insurance.org.

#### Major Companies

**ACE Bermuda:** ACE Bldg, 30 Woodbourne Ave, POB HM 1015, Hamilton HM DX; tel. 295-5200; fax 295-5221; e-mail info@mail.ace.bm; internet www.ace.bm; total revenues US $3,017.0m. (Dec. 1999); Pres. EVAN GREENBERG; Chair. and CEO BRIAN DUPERREAULT.

**Argus Insurance Co Ltd:** Argus Insurance Bldg, 12 Wesley St, POB HM 1064, Hamilton HM EX; tel. 295-2021; fax 292-6763; e-mail insurance@argus.bm; internet www.argus.bm; Pres. and CEO GERALD D. E. SIMONS.

**Bermuda Insurance Management Association (BIMA):** POB HM 1752, Hamilton HM GX; tel. 295-4864; fax 292-7375.

**Paumanock Insurance Co Ltd:** POB HM 2267, Hamilton HM JX; tel. 292-2404; fax 292-2648.

# BERMUDA

**X. L. Insurance Co Ltd:** 1 Bermudiana Rd, Hamilton HM 11; tel. 292-8515; fax 292-5226; e-mail info@xl.bm; internet www.xlinsurance.com; Pres. and Chief Exec. BRIAN O'HARA.

## Trade and Industry

### GOVERNMENT AGENCY

**Bermuda Registrar of Companies:** Government Administration Bldg, 30 Parliament St, Hamilton HM 12; tel. 297-7574; fax 292-6640; e-mail jfsmith@gov.bm; internet www.roc.gov.bm; Registrar of Companies STEPHEN LOWE.

### DEVELOPMENT ORGANIZATION

**Bermuda Small Business Development Corpn:** POB HM 637, Hamilton HM CX; tel. 292-5570; fax 295-1600; e-mail bdasmallbusiness@gov.bm; internet www.bsbdc.bm; f. 1980; funded jointly by the Government and private banks; guarantees loans to small businesses; assets $2m. (March 2004); Gen. Man. LUCRECIA MING.

### CHAMBER OF COMMERCE

**Bermuda Chamber of Commerce:** 1 Point Pleasant Rd, POB HM 655, Hamilton HM CX; tel. 295-4201; fax 292-5779; e-mail info@bermudacommerce.com; internet www.bermudacommerce.com; f. 1907; Pres. CHARLES GOSLING; Exec. Vice-Pres. DIANE GORDON; 750 mems.

### INDUSTRIAL AND TRADE ASSOCIATION

**Bermuda International Business Association (BIBA):** Ground Floor, 20 Victoria St, Hamilton HM 12; tel. 292-0632; fax 292-1797; e-mail lrawlins@biba.org; internet www.biba.org; Chair. GREG HAYCOCK; CEO DEBORAH MIDDLETON.

### EMPLOYERS' ASSOCIATIONS

**Bermuda Employers' Council:** Reid House, Ground Floor, 31 Church St, Hamilton HM 12; tel. 295-5070; fax 295-1966; e-mail emp.org@bec.bm; internet www.bec.bm; f. 1960; advisory body on employment and labour relations; Pres. GERALD SIMONS; Exec. Dir ANDREA MOWBRAY; 347 mems.

**Construction Association of Bermuda:** POB HM 238, Hamilton HM AX; tel. 236-5537; fax 236-5485; e-mail administrator@constructionbermuda.com; internet www.constructionbermuda.com; f. 1968; Pres. TOMAS SMITH; 65 mems.

**Hotel Employers of Bermuda:** c/o Bermuda Hotel Association, 'Carmel', 61 King St, Hamilton HM 19; tel. 295-2127; fax 292-6671; e-mail johnh@ibl.bm; f. 1968; Pres. FRANK STOCEK; CEO JOHN HARVEY; 8 mems.

### UTILITY

**BELCO Holdings Ltd:** 27 Serpentine Rd, POB HM 1026, Hamilton HM DX; tel. 295-5111; fax 292-8975; e-mail webmaster@belco.bhl.bm; internet www.belcoholdings.bm; f. 1908; holding co for Bermuda Electric Light Co Ltd, Bermuda Gas and Utility Co Ltd, and BELCO Energy Services Co; Chair. J. MICHAEL COLLIER; Pres. and CEO GARRY A. MADEIROS.

### TRADE UNIONS

In 1997 trade union membership was estimated at 8,859. There are nine registered trade unions, the principal ones being:

**Bermuda Federation of Musicians and Variety Artists:** Reid St, POB HM 6, Hamilton HM AX; tel. 291-0138; Sec.-Gen. LLOYD H. L. SIMMONS; 318 mems.

**Bermuda Industrial Union:** 49 Union Sq., Hamilton HM 12; tel. 292-0044; fax 295-7992; e-mail biu@biu.bm; f. 1946; Pres. DERRICK BURGESS; Gen. Sec. HELENA BURGESS; 5,202 mems.

**Bermuda Public Services Union:** POB HM 763, Hamilton HM CX; tel. 292-6985; fax 292-1149; e-mail info@bpsu.bm; re-formed 1961; Pres. NIGEL A. D. PEMBERTON; Gen. Sec. EDWARD G. BALL, Jr; 3,454 mems.

**Bermuda Union of Teachers:** POB HM 726, Hamilton HM CX; tel. 292-6515; fax 292-0697; e-mail butunion@ibl.bm; f. 1919; Pres. ANTHONY E. WOLFFE; Gen. Sec. MICHAEL A. CHARLES; 700 mems.

## Transport

### ROADS

There are some 225 km (140 miles) of public highways and 222 km (138 miles) of private roads, with almost 6 km (4 miles) reserved for cyclists and pedestrians. Each Bermudian household is permitted only one passenger vehicle, and visitors may only hire mopeds, to limit traffic congestion.

### SHIPPING

The chief port of Bermuda is Hamilton, with a secondary port at St George's. Both are used by freight and cruise ships. There is also a 'free' port, Freeport, on Ireland Island. In 2000 it was proposed to enlarge Hamilton docks in order to accommodate larger cruise ships. There remained, however, fears that such an enlargement would place excessive strain on the island's environment and infrastructure. Bermuda is a free-flag nation, and at December 2004 the shipping register comprised 122 vessels totalling 6,166,162 grt.

**Department of Marine and Ports Services:** POB HM 180, Hamilton HM AX; tel. 295-6575; fax 295-5523; e-mail marineports@bolagov.bm; Dir of Marine and Ports Services BARRY COUPLAND; Deputy Dir and Habour Master MICHAEL DOLDING.

**Department of Maritime Administration:** POB HM 1628, Hamilton HM GX; tel. 295-7251; fax 295-3718; e-mail maradros@gov.bm; Chief Surveyor DUNCAN CURRIE; Registrar of Shipping ANGELIQUE BURGESS.

#### Principal Shipping Companies

**A. M. Services Ltd:** Belvedere Bldg, 69 Pitts Bay Rd, Pembroke HM 08; tel. 295-0850; fax 292-3704; e-mail amsl@northrock.bm.

**Atlantic Marine Limited Partnership:** Richmond House, 12 Par-la-Ville Rd, POB HM 2089, Hamilton HM HX; tel. 295-0614; fax 292-1549; e-mail management@amlp.bm; internet www.amlp.bm; f. 1970; Pres. JENS ALERS.

**B & H Ocean Carriers Ltd:** 3rd Floor, Par-la-Ville Place, 14 Par-la-Ville Rd, POB HM 2257 HM JX, Hamilton; tel. 295-6875; fax 295-6796; e-mail info@bhcousa.com; internet www.bhocean.com; f. 1987; Chair. MICHAEL S. HUDNER.

**Benor Tankers Ltd:** Cedar House, 41 Cedar Ave, HM 12 Hamilton; Pres. CARL-ERIK HAAVALDSEN; Chair. HARRY RUTTEN.

**Bermuda International Shipping Ltd:** Waverley Bldg, 35 Church St, Hamilton HM 12; tel. 296-9798; fax 295-4556; e-mail meyerfreight@ibl.bm; internet www.meyer.bm; Dir J. HENRY HAYWARD.

**Container Ship Management Ltd:** 14 Par-la-Ville Rd, Hamilton HM 08; tel. 295-1624; fax 295-3781; e-mail csm@csm.bm; internet www.bcl.bm/csm; Pres. GEOFFREY FRITH.

**Gearbulk Holding Ltd:** Par-la-Ville Place, 14 Par-la-Ville Rd, HM JX Hamilton; tel. 295-2184; fax 295-2234; internet www.gearbulk.com; Pres. ARTHUR E. M. JONES.

**Golden Ocean Management:** Par-la-Ville Place, 14 Par-la-Ville Rd, POB HM 1593, Hamilton HM 08; tel. 295-6935; fax 295-3494; internet www.goldenocean.no; Man. Dir HERMAN BILLUNG.

**Norwegian Cruise Line:** 3rd Floor, Reid House, Church St, POB 1564, Hamilton; internet www.ncl.com; Chair. EINAR KLOSTER.

**Shell Bermuda (Overseas) Ltd:** Shell House, Ferry Reach, POB 2, St George's 1.

**Unicool Ltd:** POB HM 1179, HM EX Hamilton; tel. 295-2244; fax 292-8666; Pres. MATS JANSSON.

**Worldwide Shipping Managers Ltd:** Suite 402, 7 Reid St, POB HM 1862, HM 11 Hamilton; tel. 295-3770; fax 295-3801.

### CIVIL AVIATION

The former US Naval Air Station (the only airfield) was returned to the Government of Bermuda in September 1995, following the closure of the base and the withdrawal of US forces from the islands.

**Department of Civil Aviation:** POB GE 218, St George's GE BX; tel. 293-1640; fax 293-2417; e-mail info@dca.gov.bm; internet www.dca.gov.bm; responsible for all civil aviation matters; Dir of Civil Aviation IAN MACINTYRE.

**Bermuda International Airport:** 3 Cahow Way, St George's GE CX; tel. 293-2470; e-mail dao@gov.bm; internet www.bermudaairport.com; Gen. Man. JAMES G. HOWES.

**Delta Airlines:** Kindley Field; tel. 293-2050; internet www.delta.com; passenger and air cargo service.

## Tourism

Tourism is the principal industry of Bermuda and is government-sponsored. The great attractions of the islands are the climate, scenery, and facilities for outdoor entertainment of all types. In 2004 a total of 477,757 tourists (including 206,140 cruise-ship passengers) visited Bermuda. In the same year the industry earned $353.7m. In 2002 there were 6,523 hotel beds.

**Department of Tourism:** Global House, 43 Church St, Hamilton HM 12; tel. 292-0023; fax 292-7537; e-mail bermudatourism@hillsbalfour.com; internet www.bermudatourism.com; Dir of Tourism CHERIE WITTER.

**Bermuda Hotel Association:** 'Carmel', 61 King St, Hamilton HM 19; tel. 295-2127; fax 292-6671; e-mail johnh@ibl.bm; internet www.experiencebermuda.com; Chair. NORMAN MASTALIR; Exec. Dir JOHN HARVEY; Pres. MICHAEL WINFIELD; 37 mem. hotels.

## Defence

The local defence force is the Bermuda Regiment, with a strength of some 630 men and women in 1999. The Regiment employs selective conscription.

## Education

There is free compulsory education in government schools between the ages of five and 16 years, and a number of scholarships are awarded for higher education and teacher training. There are also seven private secondary schools, which charge fees. The Bermuda College, founded in 1972, accepts students over the age of 16, and is the only post-secondary educational institution. Extramural degree courses are available through Queen's University, Canada, and Indiana and Maryland Universities, USA. A major programme to upgrade the education system, involving the establishment of five new primary and two secondary schools, was implemented between 1996 and 2002.

# BOLIVIA

## Geography

### PHYSICAL FEATURES

The Republic of Bolivia lies at the heart of the South American continent, in the centre-west, set on the high Andes, but reaching down into the Amazon basin. The country is landlocked, and has been since 1884, when it lost the Atacama Desert region to Chile (with which it still maintains demands for a sovereign corridor to the Pacific). The border with Chile, 861 km (535 miles) in extent, is in the south-west, while the rest of the western border, further north, is with Peru (900 km in extent). Brazil (with which Bolivia has a long, 3,400-km border) is to the north and north-east, the other landlocked country of the continent, Paraguay (750 km), to the south-east, and Argentina (832 km) to the south. The country is the fifth largest in South America, the size of France and Spain together, and covers 1,098,581 sq km (424,164 sq miles).

Bolivia's maximum length, from north to south, is about 1,530 km and its maximum breadth 1,450 km. Known as a 'rooftop of the world' from the setting of its lofty plateau amid the high Andes (which account for about one-third of the territory), the country also reaches down the eastern slopes of the mountains into vast grassy plains threaded by rivers with densely forested banks (Oriente). The high plateau is known as the Altiplano and the lower slopes and valleys as the Yungas, while in the southeast the plains of the Amazonian–Chaco lowlands are known as the Llanos. The Andes form two main ranges in western Bolivia, the lower Cordillera Occidental along the border with Chile and, further inland, the great peaks of the broader Cordillera Oriental (Cordillera Real), also crossing from north to south, but in the centre-west of the country. Between the ranges is the main plateau, about 800 km long and 130 km wide, arid in the south, but in the north the country shares Lake Titicaca, the highest navigable lake in the world (at 3,805 m or 12,488 ft), with Peru. The region consists of snowy peaks and broad, windy plateaux over 4,000 m above sea level, often barren but for ichu (a coarse grass). The highest peak is Illimani (6,462 m). The Yungas consists of the lower, eastern slopes of the Cordillera Oriental, the fertile, forested and well watered, but steep, valleys, separating the plateau region and the plains. Then, east and northeast of the mountains, are the great Amazonian plains. South of them, beyond the low Chiquitos hills, is the Bolivian portion of the dry plains of the fought-over Chaco, the Llanos. Vast swathes of the plains become swampland during the rainy season, but the drier parts provide rich grazing for livestock. The plains are covered by great grasslands, although along the rivers, particularly in the north-east, there are stretches of dense tropical rainforest (forests cover about one-half of the country). Rivers drain either into the Amazon basin or into the system of the Río de la Plata (River Plate)—the lowest point in Bolivia, at 90 m above sea level, is the Paraguay river, part of the latter system, as it leaves the country. This vast, sparsely populated and rough countryside also enables Bolivia to have become the world's third largest cultivator of coca (after Colombia and Peru).

### CLIMATE

The country is situated entirely in the tropics, but its varied elevation gives it a wide range of climates. It is cold and dry in the mountainous south-west, but it is much warmer and wetter lower down. Being south of the Equator, winter is in the middle of the calendar year. The mean annual temperature in the Altiplano, where most people live, is 8°C (46°F) and in the Llanos 26°C (79°F). The Yungas is more subtropical than the higher slopes. The main rain-bearing winds are those that cross the Amazon basin, and the north-east, particularly, is prone to flooding during the wet season (December–January), although

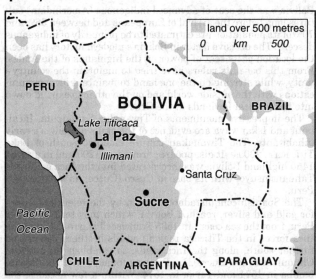

droughts here are equally possible. The Llanos gets drier towards the south.

### POPULATION

Most people are Amerindian, native American, and have only been involved in the activities of the state and the benefits of its economy since the 1950s, when greater social mobility was encouraged, to the detriment of the white ruling élite. About 56% of the population are reckoned to be Amerindian, or predominantly Amerindian, and 30% Mestizo (predominantly of Spanish descent), with the rest white, apart from a small number of those of African descent. Bolivia was, anciently, home to a flourishing Aymará civilization based around Titicaca, but was later conquered by the Quechua-speaking Incas from the north. The Quechua peoples alone are put at about one-quarter of the population, while the Aymará now fall short of one-fifth. Given that only 36% of Bolivians use Spanish as their principal language, it is no surprise to find that Quechua and Aymará have joined it as official languages. Guaraní is the main one of the other indigenous languages in use. Although other aspects of native tradition are powerful in Bolivian culture, some homogeneity was provided by the prevalence of at least nominal, if adapted, Roman Catholic Christianity (still over 80% of the population). There are also increasing numbers of Protestant denominations, particularly the more evangelical ones, while the Bahá'ís claim almost 3%.

The total population at mid-2004 was put at an estimated 9.2m., with most people living in the mountains (although the eastern lowlands have become more populated since the second half of the 20th century, owing to the exploitation of the hydrocarbons resources here), but less than two-fifths in rural areas. Bolivia is one of the least-densely populated Latin American countries. The administrative capital and largest city (with its suburbs) is La Paz, in the west, at the northern end of the Altiplano, and, at 3,640 m, the highest capital in the world. The judicial or constitutional capital is Sucre, in the centre-south. Between them, also in the highlands, are Cochabamba, at the centre of a fertile farming area, and the mining town of Oruro, more to the west. To the east, on the edge of the plains, is Santa Cruz de la Sierra, which is the only metropolis in Bolivia to have over 1m. people in the city proper. The country is divided into nine administrative districts.

# History
## Prof. PETER CALVERT

Bolivia was the seat of advanced indigenous (Amerindian) cultures long before the arrival of Europeans and between 56% and 70% of its population are estimated to be primarily of indigenous descent. The decisive factor in Bolivia's modern history has been the location of its seat of power on the highlands of the Andes. From this base, its rulers have tried to maintain the country's unity, while striving on the one hand to maintain its communications with the outside world and on the other to expand down into the tropical lowlands of the Oriente.

The impressive monuments of Tiwanaku, Samaipata, Incallajta and Iskanwaya are evidence of the skills of Bolivia's early inhabitants. The Tiwanakan culture emerged south of Lake Titicaca c. 200 BC. It disappeared around AD 1200 and in around 1450 highland Bolivia was incorporated into the Inca empire of Tahuantinsuyo, with its base at Cusco (Cuzco), in what is now Peru.

The Spanish conquistadores, driven by the relentless desire for gold and silver, reached Bolivia, which they termed Upper Peru, from the sea coast. In 1538 Francisco Pizarro defeated the Inca forces in the Titicaca region and his brothers penetrated further south, along the Inca roads, to establish the town of Chuquisaca (now Sucre). La Paz was founded in 1549, Cochabamba in 1571 and Tarija in 1574. Within a few decades, the mines of Upper Peru were being exploited directly to fund the imperial ambitions of Spain. At its height Potosí, with a population of more than 100,000, was the largest urban area in the Western hemisphere. Socially and politically, however, it remained a distant dependency of the Viceregal Court at Lima (Peru), while in 1776 the Charcas Valley in the south was separated from Peru when it became part of the new Viceroyalty of the Río de la Plata.

### INDEPENDENCE

The province of Charcas was represented at Tucumán in 1816, when independence was proclaimed by the United Provinces (now Argentina). The highlands of Upper Peru had to wait for the arrival of the Liberator, Simón Bolívar, in the early 1820s to throw off its allegiance both to Spain and to Peru. In 1825 its leader Gen. Antonio Sucre offered to name the country after Bolívar if he would recognize its independence and give a Constitution to what was initially to become one of the largest independent states in the Americas, and this he did.

However, from the beginning Bolivia's lack of unity condemned it both to internal political turbulence and to constant involvement in regional conflict. A short-lived attempt at confederation with Peru was successfully frustrated by Chile in 1841. In the 1850s silver mining revived, but workers from Chile were needed to develop a new source of wealth in the nitrate fields of Antofagasta. Fearing Chile's ambitions, Bolivia signed a secret defence treaty with Peru. However, in the War of the Pacific (1879–83) Chile won a decisive victory over both Bolivia and Peru, gaining control of the valuable nitrate fields and depriving Bolivia of its coastal territory, a loss which ever since has remained a source of friction between the two countries. It was not until 1904 that a peace treaty obliged Chile to build a railway from La Paz to Arica (Chile), which became Bolivia's main outlet to the outside world.

Towards the end of the 19th century silver production reached a peak. However, with the general adoption of the gold standard, prices collapsed. Nevertheless, Bolivia was also rich in tin, often found associated with silver, but largely ignored until the rise of demand from the canning industry. It was tin, not silver, that drew Bolivia into the modern world economy. The industry came to be dominated by three business empires, the Patiño, Aramayo and Hochschild interests. Labour was recruited from the Indian population of the Altiplano and railways were built, linking the mines to the Chilean ports of Antofagasta and Mollendo.

Political instability was replaced, between 1884 and 1920, by stable government under a two-party system of Conservatives and Liberals. Essentially, these represented the rural landowning oligarchy and the powerful mining interests, an élite often known as the Rosca (circle). However, from 1920 to 1932 consensus government began to break down, with the emergence of the Republican party, the growth of trade unions among miners and railwaymen and the increasing discontent among the peasantry, traders and artisans of the cities, especially in La Paz. The expansion of Bolivian forces eastwards into the sparsely populated territory of the Chaco Boreal was checked by Paraguay in 1929. In the Chaco War (1932–35) Bolivian forces suffered a disastrous defeat and Paraguayan forces came within sight of the southern Bolivian oilfield before peace could be secured.

### THE 'BOLIVIAN REVOLUTION'

Military governments dominated the years between 1938 and 1952, as former commanders sought to justify their existence. However, new political parties competed for the support of the powerful miners' union, the urban workers and the new middle class. The most important of these parties was the Movimiento Nacionalista Revolucionario (MNR—Nationalist Revolutionary Movement), founded in 1942. After a presidential election in May 1951 had given him the largest share of the popular vote, a military coup prevented the MNR's candidate, Víctor Paz Estenssoro, from assuming power. However, in April 1952 the miners fought their way into La Paz and brought about a true, if limited, social revolution.

The MNR took power under the presidency of Paz Estenssoro, with Hernán Siles Zuazo as Vice-President. The new Government introduced universal adult suffrage, nationalized the mining sector, implemented a land-reform programme and, for a time, dismantled the power of the Army. Trade unions were organized into a confederation (Central Obrera Boliviana—COB). Emphasis was given to economic development, with US aid. The national petroleum company, Yacimientos Petrolíferos Fiscales Bolivianos (YPFB), was reorganized. A new petroleum code invited exploration and development by foreign companies. As new roads were built, encouragement was given to colonization in the region of Santa Cruz. The MNR remained in power until 1964, with Paz Estenssoro and Siles Zuazo alternating as President. This period saw the MNR gradually retreating from its earlier policies of radical reform and introducing measures of monetary stabilization, but also slowly alienating both miners and peasants.

### MILITARY GOVERNMENTS OF 1964–82

The military coup which toppled Paz Estenssoro and ended the MNR regime in 1964 heralded the start of one of the most turbulent and complex periods in Bolivia's history. After two years of joint rule with Gen. Alfredo Ovando Candía, elections in 1966 returned Gen. René Barrientos Ortuño as sole President, the first in Bolivia's history to speak Quechua. It was during his term that the Argentine-born Cuban revolutionary, Ernesto ('Che') Guevara, led his ill-fated expedition to Bolivia. Distrustful peasants failed to support his small group. It was quickly rounded up, and Guevara himself was shot at Vallegrande on 9 October 1967. In April 1969, however, Gen. Barrientos Ortuño was killed in an air crash. Gen. Ovando Candía deposed his civilian successor, Luis Adolfo Siles Salinas, in September, and reassumed the presidency. A power struggle ensued between right- and left-wing army officers. In 1970 Gen. Juan José Torres González, supported by the leftist military faction, was installed as President and sought to implement a programme of radical reforms, including worker participation in management and agrarian reform. However, in August 1971 Col (later Gen.) Hugo Bánzer Suárez deposed the Torres Government in a brief battle, supported by right-wing members of the

Army, the right-wing Falange Socialista Boliviana and a section of the MNR.

The Bánzer dictatorship (1971–78) was the longest period of continuous rule in the country's history and proved to be a period of growth and relative stability. The price was a ban on all political and trade-union activity. Bánzer Suárez's coup had been supported by the Brazilian military regime in return for a promise to upgrade communications links to Brazil and allow Brazil access to the iron-ore deposits of the Oriente. Bolivia had joined the Andean Pact (Acuerdo de Cartagena—known as the Andean Community of Nations from March 1996) in May 1969, but Bánzer was keen to develop a direct link both to the Atlantic and to the Pacific, for which he hoped to persuade Chile to cede land. Not only was he unsuccessful, but the issue provoked the Bolivian military again to intervene in politics. In 1978 Gen. Juan Pereda Asbún led a successful coup against President Bánzer, although he himself was then ousted, in November of the same year, by Gen. David Padilla Aranciba, with the assistance of national left-wing groups.

Presidential and legislative elections were held in July 1979. The presidential contest resulted in almost equal support for two former presidents, Siles Zuazo and Paz Estenssoro, each leading a rival faction of the MNR. The Congreso Nacional (Congress), in turn, failed to give a majority to either candidate and an interim Government was formed under Walter Guevara Arce, President of the Cámara de Senadores (Senate). In November 1979 the temporary administration was overthrown by a right-wing army officer, Col Alberto Natusch Busch. He resigned 15 days later, after failing to gain the support of the Congreso Nacional. Lidia Gueiler Tejada, President of the Cámara de Diputados (Chamber of Deputies), became Bolivia's first (and Latin America's second) woman head of state.

### The García Meza 'Narco-Regime', 1980–81

In the June 1980 presidential election Siles Zuazo and Paz Estenssoro were again the main protagonists. Neither candidate obtained a clear majority. Before the Congreso Nacional could resolve the issue a military junta seized power, led by Gen. Luis García Meza Tejada. The regime of Gen. García Meza, characterized by corruption and oppression, was sponsored by the Argentine military Government and supported by Argentine special forces. Their aim was to pre-empt any return to democracy and secure a reliable ally in the event of war between Argentina and Chile. The dictatorship was supported by a section of the military leadership deeply involved in the illegal drugs trade. With the country in economic crisis, farmers readily turned to growing coca. The USA refused to recognize the new Government. International aid to Bolivia was suspended and Bolivia withdrew temporarily from the Andean Pact.

In August 1981 Gen. García Meza was ousted from power by a military junta. (He was returned to Bolivia in 1995, to serve a 30-year prison term to which he had been sentenced *in absentia* in 1993.) Elections were planned for April 1983 but by September 1982 Bolivia's economic situation had deteriorated to such an extent that the military regime had no option but to bring forward the transfer of power to a civilian government to enable the country to negotiate successfully with foreign governments and the international lending agencies. In October 1982 Siles Zuazo was installed as President for a four-year term, on the basis of the 1980 election results.

### THE RETURN TO ELECTED GOVERNMENT

Siles Zuazo's electoral alliance of his own Movimiento Nacionalista Revolucionario de Izquierda (MNR-I—Left-Wing Nationalist Revolutionary Movement), the Movimiento de la Izquierda Revolucionaria (MIR—Movement of the Revolutionary Left) and the Partido Comunista de Bolivia (Communist Party of Bolivia) was reflected in his first Cabinet. However, consensus within the alliance was often elusive and resulted in a series of government reorganizations. Other characteristics of Siles Zuazo's second period of government were persistent coup rumours and industrial unrest. The elections of 14 July 1985 gave a narrow lead to the Acción Democrática Nacionalista (ADN—Nationalist Democratic Action), under Bánzer, over Paz Estenssoro's Movimiento Nacionalista Revolucionario Histórico (MNR-H—Historic Nationalist Revolutionary Movement). In the absence of an overall majority, the issue was resolved by the Congreso Paz Estenssoro, with the support of the left, narrowly defeated Bánzer and again took office as President.

The new Government focused its attention on rescuing the country's economy from collapse. At the end of August 1985 President Paz Estenssoro announced his New Economic Policy, which introduced a series of austere domestic measures, including a wage 'freeze' in the public sector, the removal of subsidies, and the decentralization of state enterprises. The severity of the measures was a great shock to the Bolivian people and brought the Government into direct confrontation with the COB, prompting a state of siege to be declared in September. Within months, however, inflation had fallen from a record 16,000% per year to double figures.

The effect of the measures was greatly increased by the sudden collapse of the world tin market in late 1985, which exacerbated the long-standing crisis in the country's mining sector. By this time President Paz Estenssoro had secured majority support for his policies in the Congreso Nacional, by formally agreeing to an alliance with Bánzer and the ADN, entitled the Pact for Democracy (Pacto para la Democracia). By early 1987 favourable results were emerging: inflation had been further reduced; debts had been rescheduled; and the IMF had resumed the allocation of stand-by credits. There was much discontent with the Government's austerity measures but the Pact for Democracy remained in force.

The results of the presidential elections of May 1989 gave Gonzalo Sánchez de Lozada, the MNR candidate, 23% of the votes cast, Bánzer, the ADN candidate, 22% and Jaime Paz Zamora of the MIR 20%. As in 1985 no candidate obtained a majority, leaving responsibility for the choice of President with the Congreso Nacional. In an unexpected move, Bánzer chose to support his former rival, Paz Zamora. In return for its support, the ADN was given important positions in the coalition Government of 'national unity', under the Acuerdo Patriótico (Patriotic Accord).

Austerity measures maintained by the new Government caused unrest and provoked the reimposition of a state of siege in November 1989. The privatization programme proceeded slowly and permitted foreign participation in the Bolivian mining and petroleum industries. A major success was the signing of an agreement with Peru, in January 1992, granting Bolivia free-port status at the Peruvian-owned Pacific port of Ilo. In June a shift towards a new alignment was signalled when Paz Zamora attended a presidential summit of the Southern Common Market (Mercado Común del Sur—Mercosur), the new economic grouping of Argentina, Brazil, Paraguay and Uruguay. The burden of the external debt lessened under President Paz Zamora, but the cost to the Bolivian people was high. The standard of living, already one of the lowest in South America, declined precipitously.

In January 1992 a pact, signed between the COB and the Government, gave miners' unions consultative rights in the proposed mines-privatization programme. However, the COB's power was in decline. Urban and industrial expansion had seen the articulation of interests of other workers besides those of the miners. The peasantry found an effective voice only as a result of land reform, during which many peasant unions were hastily formed. Peasant leaders became an important pressure group, while indigenous Indian groups united to form their own party. At the same time a new agro-commercial élite emerged to eclipse the old rural oligarchy. The private mining sector grew in importance in the 1970s, and the expansion of the public sector and industrial growth created a new class of professionals, white-collar workers and business entrepreneurs. Meanwhile, a wealthy society emerged in the Oriente of eastern Bolivia, largely dependent on illicit funds. This new grouping had sufficient power, in terms of both patronage and force, to be a significant challenge to the Government. Now under pressure from the USA, the Government initiated a series of operations against the traffickers and their processing plants. Political opponents claimed that Bolivia's national sovereignty was being undermined. However, the inflow of illegal revenue from the drugs trade helped to stabilize the exchange rate and moderate the money supply, both essential contributory factors in a return to the international economic system. President Paz Zamora had estimated that 70% of Bolivia's real gross domestic product was drugs-related and that more than 50% of its exports were financed by the illegal trade. Hence, the balance of power and advantage was a delicate one.

In June 1993 Gonzalo 'Goni' Sánchez de Lozada, the architect of President Paz Estensorro's lauded New Economic Policy, was elected President. His candidate for the vice-presidency was Víctor Hugo Cárdenas, the Aymaran leader of the Movimiento Revolucionario Túpaj Katarí de Liberación (MRTKL—Túpaj Katarí Revolutionary Liberation Movement). The latter's appeal to peasant and Indian voters was crucial; as no candidate secured the requisite absolute majority on the first ballot, a congressional vote was scheduled for 4 August. On 9 June, however, Bánzer withdrew from the contest, leaving Sánchez de Lozada's candidacy unopposed. Cabinet portfolios in his Government were allocated to members of the Unión Cívica Solidaridad (UCS) and Movimiento Bolivia Libre (MBL), sealing an alliance which gave the Government 97 of the 157 seats in the Congreso Nacional, an alliance which lasted until September 1994.

Industrial action by teachers, supported by the COB, culminated in early 1995 in an indefinite strike. In response, the Government declared a state of siege for 90 days. Military units were deployed throughout the country and 370 union leaders (including the Secretary-General of the COB, Oscar Salas) were arrested and banished to remote areas. Meanwhile, however, more senior public officials were implicated in the illegal drugs trade. In September four members of the Fuerza Especial para la Lucha Contra los Narcóticos (FELCN—Special Force for the Fight Against Drugs), including the second-in-command, Col Fernando Tarifa, were dismissed after an investigation into their involvement with drugs-traffickers. Moreover, there was intense civil unrest in the Chapare valley, where, despite a voluntary coca-eradication programme, forces of the Unidad Móvil de Patrullaje Rural (UMOPAR—Mobile Rural Patrol Unit) had begun to occupy villages and to destroy coca plantations. Violent clashes between peasant farmers and UMOPAR personnel resulted in the deaths of several peasants and the arrest of almost 1,000 coca growers. As a consequence, negotiations between coca growers and the Government broke down. In 1996 there were protests from women coca growers against both the US-sponsored compulsory eradication and the alleged denial of basic human rights. A series of strikes in early 2001, in protest at the sale to a Chilean corporation of the Eastern Railway and the 'capitalization' of the state petroleum corporation, YPFB, culminated in a general strike on 25 March and clashes with police in which one demonstrator died. However, the Government negotiated directly with individual unions and, in mid-April, the COB was forced to accept the original public-sector pay agreement with no concessions on the privatization programme.

## THE RETURN OF BÁNZER SUÁREZ, 1997–2001

The acrimonious campaign for the presidential election of 1 June 1997 was marked by expressions of increasing discontent with the Government's economic-reform programme. Despite outspoken criticism of his human rights record while in power, the populist campaign of former dictator Hugo Bánzer Suárez secured the largest share of the votes cast, although not the requisite absolute majority. At legislative elections held concurrently the ADN won 46 congressional seats, the MIR won 31 and the MNR secured 29 seats. The UCS and Conciencia de Patria (Condepa—Conscience of the Fatherland) finished with 23 and 20 seats, respectively. The ADN subsequently concluded a pact with the MIR, the UCS and Condepa to secure a congressional majority. As a result, on 5 August Bánzer was elected to the presidency, with the support of 118 of the 157 deputies, becoming the first former dictator in South America since 1952 to return as a democratically elected President.

President Bánzer undertook to continue the work of the previous Government, enthusiastically adopting as his own both the 'Washington Consensus' on free-market economic-reform and the US-supported campaign to combat illicit coca production. In August 1997 the new Government signed a co-operation agreement with the USA which provided funds for the continuation of compensation payments to coca growers who agreed to the destruction of their crops. This policy had limited impact on levels of coca production in Bolivia, as farmers were encouraged to plant extra crops, solely to be surrendered, or used the payments to replant their crops in more remote areas. Nevertheless, in 1998 the Government announced that the scheme would be extended, with the aim of eradicating all illegal coca cultivation by 2002. In April and May there were violent clashes between farmers in the Cochabamba region and security forces engaged in crop eradication and in August more than 1,000 coca growers marched 800 km from Chapare to La Paz to demand an end to land confiscation, the incarceration of coca growers and reductions in compensation payments to farmers.

President Bánzer was a controversial figure, as domestic and international attention focused increasingly on the human-rights abuses committed during his dictatorship in the 1970s, and his coalition Government experienced instability and embarrassment. In July 1998 divisions within Condepa, which caused the resignation of Freddy Conde, the agriculture minister, prompted the President to expel that party from the governing coalition. The resignations, in May 1999, of labour minister Leopoldo López and of senior police commander Ivar Narváez, a confidant of President Bánzer, amid allegations of corruption, tainted the administration further. However, it was hoped that reforms to the judicial system, announced in 1999, would reduce corruption and increase the accountability of the police force, though there was little evidence of this. Bolivia's economic position remained critical. Nevertheless, in recognition of a decade of progress in economic stabilization, in 1998 Bolivia became the first country in Latin America to be designated eligible for debt-service relief under the World Bank-led heavily indebted poor countries (HIPC) initiative and from 1999 onwards became a pilot country for the Comprehensive Development Framework, co-sponsored by the IMF and the World Bank.

In October 1999 the Congreso Nacional approved a new penal code, incorporating, for the first time, customary Quechua and Aymará law. However, in the meanwhile a sharp increase in water charges in Cochabamba in early 2000 ignited violent demonstrations, which spread to other areas. In April a brief strike by the police forced the Government to withdraw the measure, which had been intended to help the newly privatized water authority fund improvements in supply. The Government was put under further pressure in September 2000 when demonstrations by striking teachers demanding higher salaries, and farmers protesting against the Government's plans to restrict the cultivation of coca, brought the country to a standstill. Violence between the protesters and riot police resulted in at least 10 fatalities. The protests ended in October when the Government agreed to offer the teachers bonus payments and, as a concession to the farmers, to stop the construction of three military bases in the Chapare region. However, at the same time, the Government (under pressure from the USA) made it clear that the coca-eradication programme was not negotiable, and claimed optimistically that in 2000 the area cultivated for coca had been successfully reduced by some 45%. In June 2001 the IMF and the World Bank agreed to reduce Bolivia's debt servicing by US $120m. per year over the next 10 years under the HIPC initiative. In July the 'Paris Club' of creditor governments agreed to write off some $262m. of Bolivia's national debt and, at a meeting in La Paz in early October, approved a grant of $1,360m. in aid to underpin further economic and political reform. Nevertheless, widespread demonstrations continued throughout the year. The protesters complained that the Government had not honoured the agreements made in October 2000. They demanded, *inter alia,* higher wages for public-sector workers, abandonment of water privatization, rescinding of the Agrarian Reform Law and an end to the destruction of coca plantations in Chapare. At the end of June the coca-eradication programme was temporarily suspended following clashes between the security forces and coca farmers in the Yungas. Similar clashes in the Chapare region in November resulted in four deaths and a further suspension of the eradication programme.

## CONTINUING UNREST

On 27 July 2001 President Bánzer, who had been embarrassed in January by the release of documents in Chile implicating him in 'Plan Condor' (a scheme among right-wing dictators in Argentina, Chile, Uruguay and Bolivia to kill political opponents living in exile during the 1970s), announced that he had been diagnosed with terminal lung and liver cancer and would resign on 6 August. The Vice-President, 41-year old Jorge Quiroga Ramírez, who had previously served as Minister of Finance, was sworn in as President the following day and appointed a radically different Cabinet in which no fewer than seven portfolios were given to young technocrats. Two members of the conservative ADN were, however, appointed to the Ministries of Interior and Defence. The stated aims of the new administration were to stabilize the economy, implement anti-corruption measures and continue the dialogue with the protesting coca farmers and indigenous groups.

In January 2002 it was reported that seven people were killed during demonstrations by coca farmers protesting against a presidential decree ordering the closure of established coca markets, and against the actions of the security forces in the Chapare region. In the following month the Government and representatives of the coca farmers reached an agreement that suspended the decree in return for a halt to the protests and roadblocks that were paralysing the Chapare region.

Presidential and legislative elections were held on 30 June 2002. The main contenders for the presidency were former head of state Goni Sánchez de Lozada, again representing the MNR, Manfred Reyes Villa of the centre-right Nueva Fuerza Republicana (NFR—New Republican Force) and Evo Morales of Movimiento al Socialismo—Instrumento Político por la Soberanía de los Pueblos (MAS). Sánchez de Lozada won the largest percentage of the votes cast (22.5%) and, unexpectedly, Morales, who strongly opposed free-market economic policies and the Government's drugs-eradication programme, achieved second place with 20.9% of the ballot. However, as neither candidate achieved the required absolute majority, the power to appoint the new President again passed to the newly elected Congreso Nacional. It was thought that a statement made by the US Ambassador to Bolivia before the election, warning that US financial aid could be withdrawn if Morales were elected, had actually benefited the MAS leader's campaign. In the legislative elections the MNR won the majority of seats in both the upper and lower chambers of the Congreso Nacional (11 and 36, respectively), while MAS won eight seats in the Cámara de Senadores and 27 in the Cámara de Diputados. On 4 August the Congreso Nacional voted, by 84 votes to 43, to appoint Sánchez de Lozada to the presidency. He was inaugurated on 6 August.

The new Cabinet was dominated by the MNR, although the MIR's Carlos Saavedra Bruno was appointed Minister of Foreign Affairs. In an important concession to the pro-indigenous groups, on 1 September 2002 it was announced that the Government planned to redistribute some 500,000 ha of agricultural and 700,000 ha of forestry land to 11,000 landless families. The programme would cost an estimated US $2,500m. However, the new Government, already unpopular owing to public suspicion about President Sánchez de Lozada's business dealings, faced unrest in various quarters. In early 2003 a crisis arose when the new Government presented budget proposals to raise taxes in order to comply with IMF requirements. The proposed introduction of an *impuestazo*, or 'tax shock', attracted widespread criticism and civil unrest. On 12 February the situation worsened when, following the Government's rejection of their 40% pay demand, the police joined demonstrators in the streets. The President called in more troops to keep order, but soldiers attempting to guard the presidential palace fired on the crowds besieging it and the dissident police officers fired back. While the President fled from the building in an ambulance, eight police officers, two soldiers and seven civilians were killed in a series of running battles. In all 32 died in the unrest and an unknown number were injured. On 19 February the entire Cabinet resigned in response to the protests, although seven ministers were reappointed to the new council of ministers that took office the next day.

Protests continued in 2003, with sporadic outbreaks of violence resulting in several fatalities. In February the Government announced a review of the nation's legal coca consumption, with a view to allowing a limited extension of lawful production. The suggestion was enough to lead to renewed threats from the US Administration to suspend US and IMF aid. Nevertheless, renewed government efforts to reform the economy led to the IMF disbursing some US $15m. in loans in early July, despite the Government's effective abandonment of the IMF-approved budget proposals earlier in the year. Hostility to the USA remained strong, nevertheless, because of that country's continuing pressure to reduce coca cultivation. Following the publication of a UN report in June 2005 stating that the area under coca cultivation in Bolivia had increased by 17% in 2004, in early July the Government reassured coca growers that eradication efforts would not be increased as a result.

In September 2003 a general strike was organized by peasant organizations in protest at Pacific LNG (liquefied natural gas) project, a plan to export liquefied natural gas via Chile to the USA and Mexico. Opponents of the project accused the Government of selling the country's natural resources to foreign countries, and were hostile to the pipeline's terminal being located in the Chilean port of Patillos, which had belonged to Bolivia until the late 19th century. The strike was the culmination of months of increasingly vociferous protests against the project. Another general strike was held at the end of September, and in early October disturbances in a suburb of La Paz led to the death of many protesters, after the army intervened to restore order. This in turn caused another protest, the next day, prompting military intervention and a further 13 deaths in what soon became known as the 'Gas War'. On 13 October the Government announced the suspension of the gas-export plan and on the same day Vice-President Carlos Diego Mesa Gisbert resigned, in protest at the army's suppression of the demonstrations. The civil unrest continued, nevertheless. Finally, on 17 October, after 27 demonstrators had been killed in one clash, Sánchez de Lozada resigned as President and left the country for exile in Miami, Florida, USA. The Congreso Nacional approved the appointment to the presidency of former Vice-President Mesa.

Mesa Gisbert was sworn in as President on 18 October 2003. He immediately appointed the country's first Minister of Ethnic Affairs and sought to calm unrest by publicly undertaking to meet three of the opposition's demands, the so-called 'October Agenda'. These were: to reform the hydrocarbons law in order to bring national resources back under state control; to submit the proposed gas export project to referendum; and, if the Congreso Nacional concurred, to hold early elections to a Constituent Assembly. The promised referendum was held on 18 July 2004. The majority of voters were in favour of the changes as proposed by the Government, which included the abrogation of the existing hydrocarbons legislation, but allowed for the export of the country's natural gas reserves, although with increased state control. (Several foreign hydrocarbon companies, notably Petrobrás and Repsol, subsequently warned they would oppose any measures they believed to be confiscatory and would consider legal action.) Voters also sanctioned the construction of a natural-gas pipeline via Chile as a means of regaining access to the Pacific coast. The results of the ballot were interpreted as an endorsement of the policies of President Mesa. However, this did not calm the unrest, and demonstrations and strikes continued. The new hydrocarbons law was eventually promulgated in May 2005.

Meanwhile, civil and business groups from Santa Cruz and Tarija, both oil- and gas-producing regions, campaigned vociferously against the proposals, fearing that increased Government intervention in the sector would deter foreign investment, while increasing corruption and inefficiency. This campaign was part of a wider movement, led by the Comité Cívico pro Santa Cruz (CCSC), for increased regional autonomy or even full secession. In furtherance of this aim, a widely supported general strike was held in the two departments in November 2004. On 28 January 2005 a mass rally was held in Santa Cruz, at which the CCSC declared the department to be autonomous and announced that it was to establish a provincial interim government. In response, President Mesa announced that provincial gubernatorial elections would be held on 12 June, although this date was later deferred to 12 August (and then, in July, postponed to December—see below). As departmental prefects were hitherto appointed by the President, these elections could not be held without a constitutional amendment; to this end, President Mesa announced that a constituent assembly would be established, with elections to it to be held later in the year. However, the secessionists in Santa Cruz insisted that a referendum on regional autonomy be held at the same time as the provincial elections, that is, before a constituent assembly could be established.

Early in 2005 the protests in the Altiplano increasingly focused on the unpopular privatization of the water and sewerage networks. In January, in response to the demonstrations, the Government cancelled the contract held by Aguas de Illimani (controlled by the French company Suez) to supply the town of El Alto, adjoining the capital. However, the contract was not immediately transferred to the municipal water company, and unrest continued. On 6 March President Mesa appeared on national television to announce that he was submitting his resignation to the Congreso Nacional the following day. He claimed that the protests, organized by Evo Morales, leader of the coca growers' movement and erstwhile presidential candidate, were unreasonable. It was clear that Mesa wanted to forestall a protest march from El Alto to La Paz planned for 8 March, at which Morales' followers had threatened to occupy the congressional building. Instead, early next morning, thousands of the President's middle-class supporters surrounded the building, and when the Congreso Nacional met it refused to accept Mesa's resignation, agreeing in principle to a four-point governability pact that would allow the approval of the hydrocarbons law and of an enabling law to establish a constituent assembly, provincial elections and a national referendum on autonomy. The agreement also would end the opposition-organized protests and roadblocks. Opposition groups, however, were quick to reject the deal as inadequate, and the situation remained extremely unstable.

Blockades, demonstrations and marches continued to be held throughout April and May 2005, causing great disruption, with sporadic outbreaks of violence. Shortages of essential goods were reported in many towns and cities, including La Paz.

President Mesa, nevertheless, refused to authorize the use of potentially lethal force by the police and military personnel against the protesters. Finally considering his position untenable, on 6 June President Mesa submitted to the Congreso Nacional his resignation, which was accepted following an emergency session convened in Sucre on 9 June. The outgoing President appealed for a halt to the disruption, warning that the country was on the brink of civil war. The speakers of the Cámara de Senadores and the Cámara de Diputados, Hormando Vaca Díez and Mario Cossío Cortez, respectively, both waived their constitutional right to assume the presidency, which was instead assumed the following day by Eduardo Rodríguez Veltzé, hitherto President of the Supreme Court. It was understood that Vaca Díez, a former business leader from Santa Cruz opposed to nationalization, had declined the presidency owing to fears that his investiture would provoke a civil conflict between his partisans and those groups who vociferously opposed him, including, notably, MAS.

A new presidential election was to be held within 150 days, as stipulated in the Constitution. In early July 2005 the Congreso Nacional reached agreement that presidential and congressional elections would be held concurrently, on 4 December. Opposition parties from both the left and right wing had originally demanded that elections to a constituent assembly should take precedence over other ballots; however, the vote on a constituent assembly was postponed until 2 July 2006. A referendum on autonomy in Santa Cruz would be held on the same day. The secessionists also acquiesced to an announcement by President Rodríguez that gubernatorial elections in that department would also be postponed until December 2005.

# Economy

## Prof. PETER CALVERT

Bolivia's economy is conditioned by its geography. It has substantial natural resources but is hindered economically by its landlocked position and the high altitude of much of the country. The latter limits the possibilities for agriculture and historically forced its dependence on the export of minerals. Declining world prices for minerals, however, have made these much less profitable than in the past. Since the beginning of the 21st century, as a result, Bolivia has suffered from endemic social and political unrest as its economy has stagnated. Bolivia has an area of 1,098,581 sq km (424,164 sq miles), ranking 28th in size in the world, but has a population of only 9.2m., according to 2004 midyear estimates, giving an average population density of some 8.5 people per sq km. Population growth per year averaged 2.4% in 1990–2003, but declined to 2.2% in 2004. In 2004 the crude birth rate was 23.76 per 1,000 and death rate 7.64 per 1,000; infant mortality was estimated to have fallen from 92.0 per 1,000 in 1991 to 53.11 per 1,000 in 2004. Life expectancy at birth in 2004, at 65.5 years (males 62.89; females 68.25), was steadily improving, but remained among the lowest in Latin America. Some 63% of the population were below the national poverty line. Gross national income (GNI), measured at average 2001–2003 prices, in 2003 was $7,722m., according to World Bank estimates, and GNI was only $890 per head (or $2,450 per head on an international purchasing-power parity basis), placing the country at the lower end of the World Bank's lower-middle-income countries. Annual growth in 2004, however, was put at 3.6%.

Some 70% of the population live on the high plateau known as the Altiplano, which occupies approximately one-half of the national territory. The Altiplano, at between 3,660 m and 3,800 m above sea-level, is flanked by Andean ranges: to the west by the Cordillera Occidental, and to the east by the Cordillera Real, which is rich in mineral deposits. The northern Altiplano has a more moist climate, receiving a mean annual rainfall of 600–700 mm, concentrated in the months from October to March. Temperatures are adequate for arable farming, though temperatures fall sharply at night (by up to 20°C or more). Occasional frosts, hail and droughts can threaten agricultural production. The more sparsely populated south-western Altiplano is dry and cold.

To the east and north-east of the high plateau lie the valleys and foothills, which form a geologically dissected and diverse region fringing the Cordillera Real. At intermediate altitudes they enjoy mild climates suitable for temperate-zone crops such as wheat and soya beans. The semi-tropical Yungas region to the north-east of (and in close proximity to) the capital, La Paz, supports the commercial cultivation of coffee, maize, cassava, bananas and oranges. The eastern plains of the Oriente can be subdivided into the rainforests, savannah grasslands and swamps of the Amazonian north-east, and the seasonally dry and hot scrublands and grasslands of the Gran Chaco in the east. Population growth in the low-lying Oriente, in particular in and around Santa Cruz, fuelled by the discovery of significant petroleum and natural-gas deposits, resulted in the region accounting for more than one-quarter of the population by the late 1980s. In 2001 62.4% of the population lived in urban areas.

### AGRICULTURE

In 2002 only 2.67m. ha (1.8%) of Bolivia's total land area of 108.4m. ha were classified as arable. A further 206,000 ha (0.2%) were given over to permanent crops and 33.8m. ha (31.2%) to permanent pasture. In 2001 some 53% was classified as forest and woodland and 21% as other terrain (mostly mountain). Agriculture, forestry and fishing accounted for some 28.5% of Bolivia's gross domestic product (GDP) in 1986. By 2004, however, the sector's contribution had fallen to a provisional 15.1%, despite employing an estimated 43.4% of the economically active population in 2003, according to the UN's Food and Agriculture Organization (FAO). Of the 1,588,000 people estimated to be employed in the agricultural sector in mid-2003, about two-thirds were engaged in subsistence agriculture, predominantly in the central highlands, accounting for many of the 63% of the population who lived below the national poverty line. The fertile tropical lowlands of the Oriente are dominated by more capital-intensive commercial farms, many of which attract significant foreign investment.

The performance of the agricultural sector is prone to marked fluctuations, dictated, largely, by weather conditions. The importance of the sector in terms of national subsistence as well as GDP makes these fluctuations significant for Bolivia's economic well-being. Regular floods in the north-east of the country in March or April are compounded by the torrential rains and lowland flooding typically brought by the climatic phenomenon known as El Niño (a warm current that periodically appears along the Pacific coast of South America). Agricultural GDP increased at an annual average rate of 2.8% in 1990–2003.

Maize, wheat, rice and potatoes are the principal foodstuffs grown for domestic consumption. However, grain imports had to supplement low domestic production from the late 1980s onwards. In 2002 Bolivia imported some 489,503 metric tons of cereals. According to the FAO, total cereal production in 2003 was 1.486.3m. tons. In that year wheat production, which fluctuated considerably also in the 1990s, was 88,674 tons, down from 118,149 tons the previous year. Maize production was steady at 707,738 tons and output of rice (paddy) well up at 424,454 tons. Cassava (manioc) production, most of which was for domestic consumption, in 2003 remained steady at 392,268 tons. Production of oilcrops in 2003 was 364,201 tons.

In the 1990s, in response to an increase in overseas demand, there was a substantial increase in the growth of soya (soybeans), cultivated chiefly in the Santa Cruz area, where the climate allows for two harvests each year. In 2003 the area under soybean cultivation was 653,220 ha, compared with just 175,000 ha in 1991, and production stood at 1,550,800 metric tons. Soya was Bolivia's most important legal agricultural export, although cotton, coffee and sugar were also grown. Soya even rivalled gas in its importance to the national economy. In 2003 Bolivia exported 115,229 tons of soya, with a value of US $35.46m. Cotton lint exports of 1,439 tons had a value of $1.87m. and exports of coffee (4,453 tons) were valued at $6.3m. In 2003 105,000 ha of land were used to cultivate sugar cane and 4.9m. tons were produced; 90,627 tons of refined sugar had been produced for export, with a value of $21.88m. Forestry was an important export earner, but had also been adversely affected by recession: in 2003 Bolivia produced 347,000 cu m and exported 42,600 cu m of sawnwood with a value of $224.1m., a useful increase on the 299,000 tons produced the previous year.

However, Bolivia's most valuable agricultural exports remained coca and its illegal derivatives. Coca has been grown from pre-Colombian times, although its modern economic significance dated only from the 1960s. Bolivia is the third largest cultivator of coca in the world and in recent years has produced approximately one-third of the world's cocaine, predominantly in the departments of Chapare and Yungas. Official sources estimated that around 80,000 peasant farmers were engaged directly in its production; however, a far greater number of Bolivians depended on the industry for their livelihood. Under strong pressure from the US Government, cultivation for non-traditional use was made illegal in 1988 and legal cultivation was supposed to be confined to 12,000 ha after 2002.

The importance of Bolivian coca in the international cocaine trade prompted considerable official US interest in the restriction of coca cultivation. Any US military and economic assistance was made dependent upon meeting eradication targets of 5,000–8,000 ha each year and programmes were instituted to compensate farmers for the destruction of their coca and to encourage the development of alternative crops. In 2000 the Bolivian Government claimed that the area illegally cultivated had been reduced by some 45%. The past record suggested this was unlikely, however, with large sections of the rural community dependent upon coca cultivation for their survival, and without the realistic prospect of substituting other crops, and in the event the claim was admitted to be untrue. In 2001 it was conceded that the Government's eradication efforts had brought disaster to rural communities and cost it some US $500m. per year in lost revenues. It also mobilized intense opposition among indigenous communities which repeatedly descended into violence. Furthermore, in mid-June 2005 the UN's Office on Drugs and Crime announced that coca production in Bolivia had risen by 17% during 2004, primarily owing to displacement of coca-growing operations from Colombia, where eradication programmes had successfully reduced cultivation by over 50% in 1999–2004.

Cattle stocks were estimated at 6.7m. head in 2003. There were a total of 8.6m. sheep and 1.9m. camelids (including llama, used as beasts of burden, and alpaca, valued for their wool). Some 75m. chickens, 2.9m. pigs and 425,000 cavies and other edible rodents were also kept, largely for domestic consumption. In 2003 the total fish catch was 6,974 metric tons, of which freshwater fish caught accounted for an estimated 5,615 tons.

## MINING

The importance of the mining sector for the Bolivian economy was belied by its relatively modest contribution to the country's GDP. In 2004 mining and quarrying contributed a provisional 10.4% of GDP and, in 2000, employed only about 1.7% of the working population. However, it was in Bolivia's trade balance that the sector's significance became clear. In 2002 the mining industry generated an estimated US $724m. in export revenue, accounting for almost one-half (44.6%) of the country's foreign exchange.

Rich reserves of high-content ores made Bolivia the world's second largest tin producer, after Malaysia, for much of the 20th century. A state-owned mining corporation, Corporación Minera de Bolivia (COMIBOL), was founded in 1952, following President Paz Estenssoro's nationalization of Bolivia's larger mines, and produced 27% of Bolivian non-ferrous metal output. However, by the early 1980s Bolivia's share of the world tin market was already in decline, as richer seams neared exhaustion and production inefficiencies limited competitiveness. Then, in October 1985, the world price of tin decreased dramatically, following the withdrawal of the International Tin Council (see section on Tin in Major Commodities of Latin America) from the market. COMIBOL was severely affected, as a result of its size, its relative inefficiency and its vulnerability to both external pressures and direct state intervention. Tin production fell to one-fifth of 1970s' production levels.

In 2005 Bolivian deep-mined tin remained uncompetitive compared with the low-cost alluvial tin produced by Malaysia, Indonesia and Thailand. Bolivia was also surpassed by neighbours Brazil and Peru in terms of the volume of tin-concentrate production. As early as 1993 the Bolivian tin industry was running at a loss, prompting a focus on the best, and therefore the cheapest, seams. Small-mine owners demanded government subsidies in order to increase prices which, along with subsidized inputs, would, they claimed, allow their mines to stay operative. However, while prices remained roughly constant, production again declined, stabilizing at 13,210 metric tons in 2002.

The mining industry diversified towards the end of the 20th century, however, and increased production of gold, silver and zinc, in particular, offset the decline in tin mining. Gold production increased from just 500 kg in 1990 to 11,269 kg in 2002 and output of silver rose from 310 tons to 461 tons in the same period. In 2002 output of zinc was an estimated 141,708 tons. Production of other industrial metals in 2002 included 9,268 tons of lead, 2,343 tons of antimony and 474 tons of tungsten. The discovery of huge deposits of silver, zinc and lead in the depressed San Cristóbal region of southern Bolivia in 1996 had reinvigorated the mining sector. Apex Silver (USA) planned to invest US $300m. to bring the mine into production, when it was expected to yield an average annual output of 132,700 tons of zinc, 39,500 tons of lead and 435 tons of silver, making it the world's fourth largest silver mine; however, low world prices continued to discourage development.

Other potentially viable concessions, such as the huge deposits of iron ore at Mutún, near Puerto Suárez, remained under-exploited, owing to the combination of an adverse political climate, the need for high levels of new investment and poor infrastructure. In 1991 legislation was passed that allowed foreign companies to mine within the 50-km frontier zone and to forge joint ventures with COMIBOL for the first time. In the same year the Government awarded the concession for the previously unexploited lithium deposits in the Salar de Uyuni, the world's largest salt pan, to the multinational mining concern, Lithco, raising serious environmental concerns. In mid-1995 COMIBOL was designated for capitalization, a variant of privatization in which private-sector companies would bid for 50% of the shares and full management control of the state company, the funds to be invested directly into the company's operations. COMIBOL also planned to capitalize properties grouped under its subsidiary company, Vinto; this included deposits and mines, as well as the tin and antimony smelters and a lead-silver refinery. However, the Vinto capitalization suffered lengthy delays, principally owing to industrial unrest from miners opposed to the private development of the sector, and this was typical of the faltering progress of the initiative. In early 2000 the Huanuni mine was offered for sale as a joint venture and the Vinto smelter outright. Both were bought by a US firm, and transferred to an Indian-British consortium, RBG Resources, which was subsequently discovered by the US Government to be undercapitalized and went bankrupt. In July 2002 protests by miners and other local residents forced the Government of Jorge Quiroga Ramírez to rescind the privatization of the Huanuni mine and Vinto smelter.

## ENERGY

Bolivia is effectively self-sufficient in energy. Domestic energy consumption remained low. According to the US Energy Information Administration (EIA), in 2002 petroleum accounted for 60% of energy consumed and gas for 25%, with the balance being supplied by traditional combustibles, and electricity production satisfying only a limited proportion of demand. In July 2004 proposals to increase government control over the hydrocarbon sector, while permitting the export of natural gas, were approved overwhelmingly at a referendum. However, considerable controversy surrounded the resulting legislation, which was finally promulgated in May 2005 following several amendments (see History).

In January 2004 total proven crude petroleum reserves were estimated to be equivalent to 441m. barrels, although official estimates were 462m. barrels. Production in 2004 was 39,000 b/d (barrels per day), although domestic demand was estimated at 53,000 b/d, with the shortfall met primarily by diesel imports. The 34 productive fields were concentrated in the east and south-east of the country. Camiri had remained an important production site since it began production in 1927, and the oilfields in the departments of Chuquisica, Santa Cruz and Tarija continued to produce crude petroleum in significant quantities. Production peaked at 49,000 b/d in 1973, but declined to about 19,000 b/d by 1987. Production and exports expanded in the 1990s, however, with petroleum and its products earning US $60.0m. in export revenues by 2001. Of the 11.3m. barrels produced in 2002, about one-half were produced by the capitalized sector. In the first 10 months of 2003 40,700 b/d were produced, 6,024 b/d of which were exported to Chile. Consumption in the same period was estimated at 53,000 b/d.

In 1937, the year in which Yacimientos Petrolíferos Fiscales Bolivianos (YPFB) was created, petroleum companies were nationalized. However, the sector suffered from a lack of capital input as successive Governments failed to reinvest profits. From the 1930s foreign exploration companies also operated intermittently in Bolivia. YPFB was capitalized in 1996.

In January 2004 proven reserves of natural gas were estimated at 27,600,001m. cu ft. Since the main fields were in the south and south-east of the country, and the country effectively had two pipeline networks, a problem remained as to how this bounty was to be exploited. Production in 2004 was 298,056m. cu ft.

In 1998 a major new gas deposit was found near Santa Cruz, with reserves estimated at 1,700,000m. cu ft, and in June 1999 came the discovery of a potentially huge field in Tarija, with reserves in the region of 10,000,000m. cu ft. In October 2004 TotalElfFina announced the discovery of a major field in Chuquiasca. In 2001 the value of natural and manufactured gas exports totalled US $241.9m., a substantial increase on the $127.2m. earned in the previous year. In September 1996 work began on a 3,150-km natural-gas pipeline from Río Grande, near Santa Cruz, to São Paulo state, in Brazil, the largest project of its kind in South America. The section from Santa Cruz to Guararema, Brazil, to connect with the existing Rio–São Paulo pipeline, came on-stream in July 1999 but only reached its 1,060m. cu ft of gas per day (cu ft/d) capacity when the fitting of compressors was completed in 2004.

Under the terms of its supply contract, Bolivia was scheduled to increase its exports to Brazil from about 180m. cu ft/d to 320m. cu ft/d by 2000, and then to more than 1,000m. cu ft/d by 2005, with that level to be maintained until 2019. In 2000 a new gas supply contract was signed with Brazil. Though Bolivia was supplying Brazil with some 310m. cu ft/d, the contract for 2000 was for 740m. cu ft/d and the target for 2004 was set at 3,000m. cu ft/d. In July 2003 Brazil offered to increase guaranteed purchases if the cost was lowered by 30%. Meanwhile, Bolivia resumed exports to Argentina in 2002 at a reported price of US $0.82 per million British Thermal Units (BTU), compared with the current price to Brazil of $1.94. In October 2004 Bolivia signed an agreement to supply Argentina with 700m. cu ft/d of gas from 2007, conditional upon the construction of a new cross-border pipeline.

Efforts were also made to increase capacity to produce liquefied natural gas (LNG) for export. From 2001 a consortium called Pacific LNG, comprising Repsol, British Gas and British Petroleum, was developing a US $5,800m. private-sector project to build a pipeline that would ship LNG from Bolivia, via either Peru or Chile, to North America. In 2003 it was announced that the consortium favoured Chile. However, given the long history of poor relations with Chile, the project led to a strong nationalist reaction. Accordingly, in October 2003 the Government announced its suspension, pending a referendum on the issue, which was duly held in July 2004. If the project proceeded, it was not expected to come on-stream until 2006 at the earliest (for further details, see History).

Electricity production was 4,100m. kWh in 2002, 72% of which was thermal and 28% hydroelectric. Consumption was 3,848m. kWh or 402.6 kWh per head. However, in the same year Bolivia exported 3m. kWh and imported 9m. kWh of electricity. Bolivia has enormous unexploited hydroelectric potential, estimated at 38,857 MW, with some 34,200 MW on the Amazonian margins of the Andes, which the Government remained keen to exploit.

## MANUFACTURING

Non-durable consumer goods accounted for approximately 60% of the manufacturing sector's output. The manufacture of durable consumer goods was modest. Processing of agricultural output and the manufacture of construction materials were potential growth areas. Demand for beer, aerated drinks, conserved beef, cement and tiles was high; growth was also strong in the frozen meat, textiles, flour-milling and baking sectors. The mining and energy sectors supported a number of ancillary operations. Bolivia also undertook some petroleum refining and had a small petrochemical industry. The manufacturing sector grew at an annual average rate of 3.1% in 1990–2003 and accounted for a provisional 13.9% of Bolivia's GDP in 2004.

Accurate assessment of the manufacturing sector was difficult, as many small enterprises lay outside the formal sector. These were frequently engaged in the production of textiles, handicrafts and food processing. Historically, the manufacturing sector was adversely affected by expensive credit, the high cost of imported inputs, foreign competition and contraband; demand was subdued by wage controls. Structurally, Bolivia's poor transport network further limited the efficiency of the domestic manufacturing sector, and the size and low purchasing power of the population restricted the potential market. According to provisional figures, in 2004 industry (including manufacturing) accounted for 29.7% of GDP and the services sector for 55.2%.

## TRANSPORT AND COMMUNICATIONS

Bolivia's trade regime was dependent on its access to seaports, which, historically, was limited and subject to disruption. Bolivia lost its coastal territory of Antofagasta (now Atacama) to Chile in the War of the Pacific (1879–83) and thereafter was dependent on Chile for its main access to the sea at Arica. The Treaty of Ilo, signed in January 1992, granted Bolivia use of that Pacific seaport in Peru and Bolivia also has free-port privileges in ports in Argentina, Brazil and Paraguay. In addition, an extensive and reliable transport system, essential for the successful exploitation of Bolivia's mineral and energy reserves, for the operation of the commercial farming sector, for national integration and for the effective delivery of social services, still did not exist in the early 21st century.

Bolivia's national rail network, totalling 3,608 km, almost all metre gauge, consisted of two distinct systems, eastern and western, both of which were owned and managed by the formerly state-owned railway company, Empresa Nacional de Ferrocarriles (ENFE—capitalized in 1995). The eastern system ran in the Santa Cruz lowlands and linked Bolivia with Argentina at La Quiaca and with Brazil at Corumbá, with connections to the Atlantic coast at São Paulo. Plans to lay track between Santa Cruz and Cochabamba, which would complete an Atlantic–Pacific rail connection, took a step forward with a feasibility study into a Santa Cruz–Arquile link.

The road system in 2004 covered an estimated 60,282 km, of which an estimated 3,979 km (6.6%) were paved. The country's major road artery links La Paz with Santa Cruz via Oruro and Cochabamba. The Pan-American Highway links Bolivia directly with Argentina and Peru. A paved road link to the Pacific coast at Arica (Chile) was completed in 1996. Mining operations in the east and south-east of the country were impeded by a poor road network, while seasonal rains in the Amazonian north-east periodically severed overland communications. There were 254,853 passenger cars in 2001.

In contrast to the aridity of the highlands, Bolivia has some 14,000 km of navigable waterways in the east of the country, along the Rivers Chapare, Guaporé and Mamoré, in the Amazonian north-east and on the River Paraguay. The main port is at Puerto Aguirre at the border with Brazil on the Paraguay–Paraná system. In 1994 plans were finalized to widen and deepen the River Paraguay, thus providing a waterway linking Bolivia with the Atlantic coast in Uruguay. The project was expected to take three years to complete and cost some US $7,000m., but work only began in 2003, principally owing to environmental concerns.

Bolivia had 30 major airports in 2005, of which only 16 had paved runways, including international terminals at La Paz (El Alto) and Santa Cruz. There were also more than 1,000 unpaved airfields. The national air carrier, Lloyd Aéreo Boliviano (LAB), was privatized in 1995, when a Brazilian airline, VASP (Viação Aérea de São Paulo), paid US $47.5m. for a 49% stake, becoming the principal shareholder. There were 28,900 aircraft departures in 2003.

The state telecommunications company, Empresa Nacional de Telecomunicaciones (ENTEL), was privatized in 1995. In 2003 there were 600,100 fixed lines in use and 1,401,500 mobile cellular telephone subscribers. There were 171 MW/LW radio stations and 73 FM stations in 1999, as well as 77 short-wave stations and 48 television stations broadcasting to some 960,000 receivers. In 2002 there were 270,000 internet users.

## TOURISM

There was a steady increase in both numbers of tourist arrivals and receipts from the 1990s. In 2002 382,200 foreign visitors arrived at Bolivian hotels. In 2003 revenue from tourism was US $176m.

## GOVERNMENT FINANCE AND INVESTMENT

Inflation began to rise after 1978, accompanied by frequent devaluations of the currency against the US dollar. Governments tried to cope with the effects by indexation of minimum wages, fixing of exchange rates and price controls. In the absence of foreign credits, the deficit had to be covered domestically. This resulted in rapid monetary growth and 'hyperinflation'. In August 1985, amid growing budget deficits, goods shortages and devaluations, Bolivia's annualized inflation rate reached 16,000%.

In that month President Víctor Paz Estenssoro introduced the New Economic Policy (NEP), an economic austerity programme that caused considerable social hardship and unrest but reduced inflation to double figures within a few months. In 1987 a new currency was adopted, the boliviano, equivalent to 1m. pesos. Average annual inflation in 1985–95 was 18.4%. The NEP reduced the wages and salaries and the number of public-sector employees; it also ended many trade restrictions, price controls and subsidies, thus further reducing state-sector expenditure.

Successive Governments maintained these austerity measures. Since a large proportion of public revenue came from taxes and royalties from the mining and energy sectors, sectors which were vulnerable to demand and price fluctuations, efforts were also made to broaden the tax base. Value-added tax (VAT) on consumer goods was introduced in 1986. However, the fiscal deficit continued to rise, and was equal to 3.9% of GDP in 2000. Despite debt relief under the enhanced heavily indebted poorer countries (HIPC II) initiative, the fiscal deficit rose further, to 9.3% of GDP in 2001 and to 8.6% in 2002, before falling slightly to 6.5% in 2003. Following the severe difficulties in implementing the 2003 budget, and its acknowledgement of the need to modernize the financial system, in April 2003 the Bolivian Government agreed a one-year stand-by arrangement with the IMF, worth some US $118m., the first tranche of which was released in July. It later requested further support under the IMF's Poverty Reduction and Growth Facility.

Bolivia's GDP increased, in real terms, by an annual average of 3.4% in 1990–2003. It increased by 2.8% in 2003 and by a further 3.6% in 2004. In 2004 the annual rate of inflation was 4.4% (compared with 0.9% in 2002 and 3.3% in 2003).

## FOREIGN TRADE AND BALANCE OF PAYMENTS

Export markets continued to diversify in the first years of the 21st century: in 1986 53.2% of exports went to Argentina, much of it natural gas, payment for which was often well in arrears. In 2004 the total value of exports was estimated at US $2,254.4m. (including re-exports). The principal exports in that year were gas, soybeans and their products, crude petroleum, zinc ore and tin and their main destinations were Brazil (31.7%), the USA (15.9%), Venezuela (10.8), Peru (6.1%) and Argentina (5.8%). In the same year total imports were valued at $1,887.8m., 25.8% of which came from Brazil, 15.6% from Argentina, 13.8% from the USA, 6.7% from Peru and 5.9% from Chile. They included petroleum products, plastics, paper, aircraft and aircraft parts, prepared foods, passenger cars, insecticides and soybeans.

Exports to the European Union (EU, known as the European Community until November 1993) increased significantly from the 1980s, although trade was hindered by distance and high transportation costs. Bolivian products (clothes and leather excepted) entered the USA free of duty, on a concessionary basis, under the Andean Trade Preference Act (ATPA, succeeded by the Andean Trade Promotion and Drug Eradication Act in 2002). Bolivia was a member of the Andean Community of Nations (Comunidad Andina de Naciones—CAN) from its foundation in 1969 (when it was known as the Andean Pact). However, during the mid-1990s Bolivia favoured closer links with the Southern Common Market (Mercado Común del Sur—Mercosur), which consisted of Argentina, Brazil, Paraguay and Uruguay. Bolivia became an associate member in December 1996, committed to the phasing out of trade tariffs with the Mercosur countries within eight years, and in 2000 Bolivia declared its intention to strengthen links with Mercosur, with the aim of eventually becoming a full member. Meanwhile, however, Mercosur instead accepted the other members of the CAN as four new associate members in 2003 and 2004. In December 2004 Bolivia was one of the founder signatories to the agreement signed in Cusco, Peru, creating the South American Community of Nations (Comunidad Sudamericana de Naciones), intended to promote greater regional economic integration, due to become operational by 2007.

In 2003 Bolivia recorded a trade surplus of US $75.1m. However, until that year the country's trade balance had persistently been in deficit (a negative balance of $340.0m. was recorded in 2002). In 2002 the current-account deficit on the balance of payments was $352.0m., although a surplus of $35.5m. was recorded in the following year.

After 1980 servicing of the debt accumulated during the latter half of the 1970s became difficult, and interest payments were suspended in 1984. The Government's austerity programme incorporated the basic features of an IMF structural-adjustment plan, in return for which official donors financed the 'buy-back' of Bolivia's commercial debt by agreement with the 'Paris Club' of Western creditor governments. Debt was exchanged on favourable terms with Argentina and Brazil and, under the Enterprise for the Americas Initiative, the USA cancelled government-held debt. By May 1993 Bolivia had all but cleared its debt with the international commercial banks, notwithstanding that the success of the Initiative largely depended on the recycling of US dollars illegally earned by the narcotics trade.

At the end of 2004 Bolivia's total external debt stood at US $5,900m. The cost of debt servicing was equivalent to 21% of the total value of exports of goods and services, compared with 37.1% in 2000. In recognition of the progress made over a decade in macroeconomic stabilization, in September 1998 the World Bank and the IMF approved a debt-relief package worth some $760m. Bolivia became the first country in Latin America to receive debt-service relief under the terms of the World Bank-led HIPC programme. It later qualified for HIPC II and since 1999 has been regarded as a pilot country for the World Bank and IMF-sponsored Comprehensive Development Framework, a programme which is intended, in part, to complement debt-reduction plans. In June 2001 the IMF and the World Bank agreed to reduce Bolivia's debt servicing by $120m. per year over the next 10 years under the HIPC initiative, and in July the 'Paris Club' agreed to write off some $262m. of the country's national debt. In June 2005, in advance of a summit meeting of the Group of Eight (G-8) industrialized nations in Gleneagles, Scotland, G-8 finance ministers announced that Bolivia was one of 18 countries to have satisfied all the criteria under the HIPC initiative to qualify for immediate debt relief. At the end of 2003 reserves, including gold, totalled $1,096.2m.

## OUTLOOK

In October 2003 President Gonzalo ('Goni') Sánchez de Lozada was forced to resign and flee to the USA, owing to public hostility to plans to increase gas exports. Since then, it has been estimated that the economic loss to the country from protests and demonstrations was US $3.8m. per day, by discouraging both private-sector investment and private-sector capital inflows, which had amounted to $676.6m. in 2002. There was also still significant need for expansion and improvement of the country's infrastructure. The outlook for metals was not encouraging, and the prospects for hydrocarbons remained uncertain, owing to the deep controversy over gas exports and further nationalization of the sector (see above). Moreover, the US-funded coca-eradication programme had sharply reduced the informal national income while, at the same time, encountering determined resistance from rural interests that threatened to escalate into outright insurrection. Although there was consistent economic growth, it was still insufficient to make a significant impact on poverty, which continued to present the major challenge to the Government. Urban unemployment was officially 12.9% in October 2002, but this figure clearly understated the extent of underemployment. In 2000 malnutrition affected 9% of children under the age of five years and 60% of the population did not have access to a safe water supply.

In 2002 Bolivia ranked 114th on the UN Development Programme's Human Development Index. The sustained austerity measures, wage controls and taxation increases of the 1990s stabilized the economy, but reinforced the position of Bolivia as mainland Latin America's poorest country; a position from which it was difficult to envisage an escape route, given the current economic and political climate.

# Statistical Survey

Sources (unless otherwise indicated): Instituto Nacional de Estadística, José Carrasco 1391, Casilla 6129, La Paz; tel. (2) 222-2333; fax (2) 222-693; internet www.ine.gov.bo; Banco Central de Bolivia, Ayacucho esq. Mercado, Casilla 3118, La Paz; tel. (2) 240-9090; fax (2) 240-6614; e-mail bancocentraldebolivia@bcb.gov.bo; internet www.bcb.gov.bo.

## Area and Population

### AREA, POPULATION AND DENSITY

| | |
|---|---:|
| Area (sq km) | |
|   Land | 1,084,391 |
|   Inland water | 14,190 |
|   Total | 1,098,581* |
| Population (census results)† | |
|   3 June 1992 | 6,420,792 |
|   5 September 2001 | |
|     Males | 4,123,850 |
|     Females | 4,150,475 |
|     Total | 8,274,325 |
| Population (official estimates at mid-year) | |
|   2003 | 9,024,922 |
|   2004 | 9,226,506 |
| Density (per sq km) at mid-2004 | 8.5‡ |

* 424,164 sq miles.
† Figures exclude adjustment for underenumeration. This was estimated at 6.92% in 1992.
‡ Land area only.

### DEPARTMENTS
(2001 census)

| | Area (sq km) | Population | Density (per sq km) | Capital |
|---|---:|---:|---:|---|
| Beni | 213,564 | 362,521 | 1.7 | Trinidad |
| Chuquisaca | 51,524 | 531,522 | 10.3 | Sucre |
| Cochabamba | 55,631 | 1,455,711 | 26.2 | Cochabamba |
| La Paz | 133,985 | 2,350,466 | 17.5 | La Paz |
| Oruro | 53,588 | 391,870 | 7.3 | Oruro |
| Pando | 63,827 | 52,525 | 0.8 | Cobija |
| Potosí | 118,218 | 709,013 | 6.0 | Potosí |
| Santa Cruz | 370,621 | 2,029,471 | 5.5 | Santa Cruz de la Sierra |
| Tarija | 37,623 | 391,226 | 10.4 | Tarija |
| **Total** | **1,098,581** | **8,274,325** | **7.5** | |

### PRINCIPAL TOWNS
(2001 census)

| | | | |
|---|---:|---|---:|
| Santa Cruz de la Sierra | 1,135,526 | Oruro | 215,660 |
| La Paz (administrative capital) | 793,293 | Tarija | 153,457 |
| El Alto | 649,958 | Potosí | 145,057 |
| Cochabamba | 517,024 | Sacaba | 117,100 |
| Sucre (legal capital) | 215,778 | Quillacollo | 104,206 |

# BOLIVIA

*Statistical Survey*

## BIRTHS AND DEATHS
(UN estimates, annual averages)

|  | 1985–90 | 1990–95 | 1995–2000 |
|---|---|---|---|
| Birth rate (per 1,000) | 36.8 | 35.8 | 32.6 |
| Death rate (per 1,000) | 11.4 | 10.0 | 8.9 |

Source: UN, *World Population Prospects: The 2004 Revision*.

**Expectation of life** (WHO estimates, years at birth): 65 (males 63; females 67) in 2003 (Source: WHO, *World Health Report*).

## ECONOMICALLY ACTIVE POPULATION
(labour force surveys, '000 persons aged 10 years and over, at November, urban areas only)

|  | 1999 | 2000 |
|---|---|---|
| Agriculture, hunting, forestry and fishing | 77.4 | 102.7 |
| Mining and quarrying | 17.2 | 35.3 |
| Manufacturing | 370.5 | 320.1 |
| Electricity, gas and water supply | 5.5 | 15.9 |
| Construction | 176.5 | 218.9 |
| Wholesale and retail trade; repair of motor vehicles, motorcycles and personal and household goods | 542.5 | 536.1 |
| Hotels and restaurants | 126.8 | 124.4 |
| Transport, storage and communications | 173.0 | 144.3 |
| Financial intermediation | 17.5 | 20.0 |
| Real estate, renting and business activities | 71.7 | 95.8 |
| Public administration and defence; compulsory social security | 78.7 | 72.7 |
| Education | 135.6 | 132.8 |
| Health and social work | 63.2 | 48.8 |
| Other community, social and personal service activities | 76.6 | 98.8 |
| Private households with employed persons | 83.0 | 126.8 |
| Extra-territorial organizations and bodies | 1.2 | 2.8 |
| **Total employed** | 2,017.0 | 2,096.0 |
| Unemployed | 156.7 | 167.5 |
| **Total labour force** | 2,173.7 | 2,263.5 |
| Males | 1,204.6 | 1,248.1 |
| Females | 969.1 | 1,015.4 |

Source: ILO.

**Mid-2003** (estimates in '000): Agriculture, etc. 1,588; Total labour force 3,662 (Source: FAO).

## Health and Welfare

### KEY INDICATORS

| | |
|---|---|
| Total fertility rate (children per woman, 2003) | 3.8 |
| Under-5 mortality rate (per 1,000 live births, 2003) | 66 |
| HIV/AIDS (% of persons aged 15–49, 2003) | 0.1 |
| Physicians (per 1,000 head, 2001) | 0.73 |
| Hospital beds (per 1,000 head, 1996) | 1.67 |
| Health expenditure (2002): US $ per head (PPP) | 179 |
| Health expenditure (2002): % of GDP | 7.0 |
| Health expenditure (2002): public (% of total) | 59.8 |
| Access to water (% of persons, 2002) | 85 |
| Access to sanitation (% of persons, 2002) | 45 |
| Human Development Index (2002): ranking | 114 |
| Human Development Index (2002): value | 0.681 |

For sources and definitions, see explanatory note on p. vi.

## Agriculture

### PRINCIPAL CROPS
('000 metric tons)

|  | 2001 | 2002 | 2003 |
|---|---|---|---|
| Wheat | 117 | 118 | 89 |
| Rice (paddy) | 287 | 202 | 424 |
| Barley | 64 | 62 | 60 |
| Maize | 678 | 673 | 708 |
| Sorghum | 105 | 172 | 176 |
| Potatoes | 902 | 794 | 787 |
| Cassava (Manioc) | 517 | 393 | 392 |
| Other roots and tubers* | 100 | 95 | 96 |
| Sugar cane | 3,859 | 4,735 | 4,872 |
| Brazil nuts* | 38 | 38 | 39 |
| Chestnuts* | 35 | 35 | 35 |
| Soybeans (Soya beans) | 834 | 1,298 | 1,550 |
| Sunflower seeds | 149 | 173 | 179 |
| Cottonseed | 48† | 48* | 48* |
| Tomatoes | 138 | 117 | 135 |
| Pumpkins, squash and gourds* | 110 | 111 | 112 |
| Onions (dry) | 49 | 49 | 49 |
| Peas (green) | 30 | 24 | 24 |
| Broad beans (green) | 66 | 60 | 59 |
| Carrots | 36 | 36 | 36 |
| Green corn* | 60 | 62 | 61 |
| Other vegetables* | 74 | 75 | 74 |
| Watermelons | 23 | 23 | 23 |
| Bananas | 688 | 624 | 627 |
| Plantains* | 187 | 188 | 185 |
| Oranges | 110 | 105 | 104 |
| Tangerines, mandarins clementines and satsumas | 62 | 58 | 58 |
| Lemons and limes | 64 | 64 | 64 |
| Grapefruit and pomelos | 29 | 29 | 29 |
| Peaches and nectarines* | 37 | 37 | 37 |
| Grapes | 29 | 30 | 30 |
| Pineapples | 60 | 58 | 58 |
| Papayas | 23 | 23 | 23 |
| Other fruits and berries | 51 | 51 | 52 |
| Coffee (green) | 25 | 25 | 25 |
| Cotton (lint) | 26† | 26* | 26* |

* FAO estimate(s).
† Unofficial figure.

Source: FAO.

### LIVESTOCK
('000 head, year ending September)

|  | 2001 | 2002 | 2003 |
|---|---|---|---|
| Horses* | 322 | 323 | 323 |
| Mules* | 81 | 82 | 82 |
| Asses* | 631 | 632 | 635 |
| Cattle | 6,457 | 6,576 | 6,680 |
| Pigs | 2,851 | 2,851 | 2,925 |
| Sheep | 8,902 | 8,902 | 8,596 |
| Goats* | 1,500 | 1,501 | 1,501 |
| Chickens* | 74,000 | 74,500 | 75,000 |
| Ducks* | 280 | 280 | 295 |
| Turkeys* | 150 | 150 | 155 |

* FAO estimates.

Source: FAO.

# BOLIVIA

## Statistical Survey

### LIVESTOCK PRODUCTS
('000 metric tons)

|  | 2001 | 2002 | 2003 |
|---|---|---|---|
| Beef and veal | 160.9 | 164.6 | 168.2 |
| Mutton and lamb | 16.3 | 16.9 | 17.6 |
| Goat meat* | 5.8 | 5.8 | 5.8 |
| Pig meat* | 97.0 | 100.6 | 104.2 |
| Poultry meat | 129.0 | 136.0 | 136.0 |
| Cows' milk | 170.0 | 291.0 | 240.0* |
| Sheep's milk* | 29.1 | 29.1 | 29.5 |
| Goats' milk* | 11.5 | 11.6 | 12.0 |
| Cheese* | 6.8 | 6.8 | 7.0 |
| Hen eggs | 38.7 | 38.7 | 38.7 |
| Wool: greasy* | 8.9 | 8.9 | 8.9 |
| Cattle hides (fresh)* | 19.6 | 19.9 | 20.4 |
| Sheepskins (fresh)* | 5.8 | 5.8 | 4.4 |

* FAO estimate(s).
Source: FAO.

## Forestry

### ROUNDWOOD REMOVALS
('000 cubic metres, excl. bark)

|  | 2001 | 2002 | 2003 |
|---|---|---|---|
| Sawlogs, veneer logs and logs for sleepers | 559 | 544 | 544 |
| Pulpwood | 7,923 | 7,062 | 7,062 |
| Other industrial wood | 404 | 448 | 448 |
| Fuel wood* | 2,163 | 2,184 | 2,206 |
| **Total** | 11,049 | 10,238 | 10,260 |

* FAO estimates.
Source: FAO.

### SAWNWOOD PRODUCTION
('000 cubic metres, incl. railway sleepers)

|  | 2001 | 2002 | 2003 |
|---|---|---|---|
| **Total** (all broadleaved) | 308 | 299 | 347 |

Source: FAO.

## Fishing
(metric tons, live weight)

|  | 2001 | 2002 | 2003 |
|---|---|---|---|
| Capture | 5,940 | 6,300* | 6,599 |
| Freshwater fishes | 4,900 | 5,303* | 5,615 |
| Rainbow trout | 280 | 197 | 124 |
| Silversides (sand smelts) | 760 | 800* | 860 |
| Aquaculture | 320 | 418 | 375 |
| Rainbow trout | 250 | 328 | 274 |
| **Total catch** | 6,260 | 6,718* | 6,974 |

* FAO estimate.
Note: Figures exclude crocodiles and alligators, recorded by number rather than by weight. The number of spectacled caimans caught was: 28,170 in 2001; 31,021 in 2002; 31,108 in 2003.
Source: FAO.

## Mining

(metric tons, unless otherwise indicated; figures for metallic minerals refer to the metal content of ores)

|  | 2002 | 2003* | 2004* |
|---|---|---|---|
| Crude petroleum ('000 barrels) | 11,338 | 12,223 | 14,192 |
| Natural gas (million cu feet, gross production) | 314,336 | 360,272 | 447,537 |
| Copper | 120 | 344 | 564 |
| Tin | 13,210 | 16,386 | 17,609 |
| Lead | 9,268 | 9,353 | 10,871 |
| Zinc | 141,708 | 145,490 | 148,149 |
| Tungsten (Wolfram) | 474 | 556 | 451 |
| Antimony | 2,343 | 2,432 | 3,119 |
| Silver | 461 | 466 | 414 |
| Gold (kg) | 11,269 | 9,361 | 6,159 |

* Preliminary.

## Industry

### SELECTED PRODUCTS
('000 metric tons, unless otherwise indicated)

|  | 1999 | 2000 | 2001* |
|---|---|---|---|
| Flour | 567 | 731 | 788 |
| Cement | 1,180 | 1,008 | 903 |
| Refined sugar | 294 | 311 | 333 |
| Carbonated drinks ('000 hectolitres) | 2,192 | 2,343 | 1,996 |
| Beer ('000 hectolitres) | 1,797 | 1,509 | 1,586 |
| Cigarettes (packets) | 67,332 | 67,211 | 75,373 |
| Alcohol ('000 litres) | 26,412 | 30,981 | 29,099 |
| Diesel oil ('000 barrels) | 3,063 | 2,727 | 2,923 |
| Motor spirit (petrol) ('000 barrels) | 4,176 | 3,852 | 3,439 |
| Electric energy (million kWh) | 3,899 | 3,952 | 3,972 |

* Provisional figures.
Source: partly UN, *Industrial Commodity Statistics Yearbook*.

**2002:** Electric energy 4,188 million kWh (Source: UN Commission for Latin America and the Caribbean, *Statistical Yearbook*).

## Finance

### CURRENCY AND EXCHANGE RATES

**Monetary Units**
100 centavos = 1 boliviano (B).

**Sterling, Dollar and Euro Equivalents** (31 May 2005)
£1 sterling = 14.708 bolivianos;
US $1 = 8.090 bolivianos;
€1 = 9.976 bolivianos;
1,000 bolivianos = £67.99 = $123.61 = €100.24.

**Average Exchange Rate** (bolivianos per US $)
2002   7.1700
2003   7.6592
2004   7.9363

### GENERAL BUDGET
(million bolivianos, preliminary)

| Revenue | 2002 | 2003 | 2004 |
|---|---|---|---|
| Current revenue | 12,264.7 | 13,096.0 | 16,645.0 |
| Tax revenue | 7,468.9 | 8,174.1 | 10,838.6 |
| Internal | 6,832.9 | 7,564.7 | 10,093.7 |
| Customs | 588.3 | 558.9 | 661.9 |
| Mining royalties | 47.8 | 50.4 | 83.0 |
| Duties on hydrocarbons | 2,610.5 | 2,831.2 | 3,479.9 |
| Royalties | 1,300.4 | 1,762.9 | 2,333.0 |
| Sale of public goods and services | 100.0 | 92.0 | 96.8 |
| Current transfers | 646.3 | 751.5 | 700.4 |
| Other current revenue | 1,439.0 | 1,247.2 | 1,529.3 |
| Capital revenue | 1,293.1 | 1,799.0 | 1,745.3 |
| **Total** | 13,557.9 | 14,894.9 | 18,390.3 |

# BOLIVIA

| Expenditure | 2002 | 2003 | 2004 |
|---|---|---|---|
| Current expenditure | 11,121.9 | 11,776.4 | 12,818.7 |
| Personal services | 5,341.8 | 5,868.6 | 6,258.5 |
| Insurance claims | 56.7 | 30.3 | 35.2 |
| Awards | 227.6 | 223.2 | 205.7 |
| Other remunerations | 5,057.5 | 5,615.1 | 6,017.6 |
| Goods and services | 1,670.2 | 1,584.0 | 1,426.2 |
| Interest on debt | 1,156.9 | 1,594.0 | 1,819.8 |
| External | 557.7 | 732.0 | 761.0 |
| Internal | 599.2 | 862.0 | 1,058.8 |
| Current transfers | 1,037.9 | 975.2 | 1,427.1 |
| Other current expenditure (incl. unidentified) | 1,914.9 | 1,754.7 | 1,887.2 |
| Capital expenditure | 4,700.9 | 4,963.2 | 6,291.4 |
| **Total** | 15,822.8 | 16,739.6 | 19,110.1 |

## INTERNATIONAL RESERVES
(US $ million at 31 December)

| | 2002 | 2003 | 2004 |
|---|---|---|---|
| Gold* | 316.4 | 379.4 | 399.4 |
| IMF special drawing rights | 27.3 | 27.1 | 26.6 |
| Reserve position in IMF | 8.9 | 8.9 | 8.9 |
| Foreign exchange | 531.2 | 663.3 | 817.3 |
| **Total** | 883.8 | 1,078.7 | 1,252.2 |

* National valuation.
Source: IMF, *International Financial Statistics*.

## MONEY SUPPLY
(million bolivianos at 31 December)

| | 2002 | 2003 | 2004 |
|---|---|---|---|
| Currency outside banks | 2,707 | 3,231 | 3,917 |
| Demand deposits at commercial banks | 1,271 | 1,385 | 1,427 |
| **Total money** (incl. others) | 4,725 | 5,636 | 6,686 |

Source: IMF, *International Financial Statistics*.

## COST OF LIVING
(Consumer Price Index for urban areas; base: 2000 = 100)

| | 2001 | 2002 | 2003 |
|---|---|---|---|
| Food and beverages | 100.6 | 99.7 | 103.2 |
| Fuel and light | 104.4 | 108.9 | 114.2 |
| Clothing and footwear | 102.9 | 105.9 | 109.0 |
| Rent | 100.2 | 101.1 | 109.0 |
| **All items** (incl. others) | 101.6 | 102.5 | 105.9 |

**2004:** All items 110.6.
Source: ILO.

## NATIONAL ACCOUNTS
**Expenditure on the Gross Domestic Product**
(million bolivianos at current prices)

| | 2002 | 2003 | 2004 |
|---|---|---|---|
| Government final consumption expenditure | 9,051 | 10,227 | 10,551 |
| Private final consumption expenditure | 41,802 | 43,987 | 47,458 |
| Increase in stocks | 497 | 299 | −179 |
| Gross fixed capital formation | 8,915 | 7,973 | 8,787 |
| **Total domestic expenditure** | 60,265 | 62,486 | 66,617 |
| Exports of goods and services | 12,263 | 15,796 | 21,373 |
| *Less* Imports of goods and services | 15,710 | 16,322 | 18,364 |
| **GDP at market prices** | 56,818 | 61,959 | 69,626 |
| **GDP at constant 1990 prices** | 23,286 | 23,934 | 24,792 |

Source: IMF, *International Financial Statistics*.

**Gross Domestic Product by Economic Activity**
(million bolivianos at current prices, provisional figures)

| | 2002 | 2003 | 2004 |
|---|---|---|---|
| Agriculture, hunting, forestry and fishing | 7,359.6 | 8,308.9 | 9,385.6 |
| Mining and quarrying | 3,654.5 | 4,661.1 | 6,493.4 |
| Manufacturing | 7,451.3 | 7,974.9 | 8,626.9 |
| Electricity, gas and water | 1,650.6 | 1,820.4 | 1,910.2 |
| Construction | 1,767.3 | 1,449.1 | 1,482.6 |
| Trade | 4,036.0 | 4,269.1 | 4,911.3 |
| Hotels and restaurants | 1,827.5 | 1,961.9 | 2,143.3 |
| Transport, storage and communications | 6,763.7 | 7,506.4 | 8,208.6 |
| Finance, insurance, real estate and business services | 6,855.1 | 6,829.6 | 6,954.4 |
| Government services | 7,073.4 | 7,815.3 | 8,507.0 |
| Other community, social and personal services | 3,123.8 | 3,364.6 | 3,655.5 |
| **Sub-total** | 51,562.7 | 55,961.2 | 62,278.8 |
| Value-added tax / Import duties | 7,394.7 | 7,993.1 | 9,294.0 |
| *Less* Imputed bank charge | 2,139.2 | 1,995.6 | 1,946.9 |
| **GDP in purchasers' values** | 56,818.2 | 61,958.7 | 69,625.9 |

## BALANCE OF PAYMENTS
(US $ million)

| | 2001 | 2002 | 2003 |
|---|---|---|---|
| Exports of goods f.o.b. | 1,284.8 | 1,298.7 | 1,573.4 |
| Imports of goods f.o.b. | −1,580.0 | −1,638.7 | −1,498.3 |
| **Trade balance** | −295.2 | −340.0 | 75.1 |
| Exports of services | 235.9 | 256.6 | 298.5 |
| Imports of services | −399.4 | −433.3 | −477.9 |
| **Balance on goods and services** | −458.7 | −516.7 | −104.3 |
| Other income received | 121.2 | 103.2 | 71.7 |
| Other income paid | −332.5 | −307.9 | −372.8 |
| **Balance on goods, services and income** | −670.1 | −721.4 | −405.5 |
| Current transfers received | 431.6 | 407.8 | 478.9 |
| Current transfers paid | −35.5 | −38.4 | −37.9 |
| **Current balance** | −274.0 | −352.0 | 35.5 |
| Direct investment abroad | −2.5 | −2.5 | −2.5 |
| Direct investment from abroad | 705.8 | 676.6 | 166.8 |
| Portfolio investment assets | −23.0 | −19.3 | −68.2 |
| Other investment assets | −166.7 | −193.4 | −463.3 |
| Other investment liabilities | −72.9 | 187.7 | 303.0 |
| Net errors and omissions | −202.7 | −639.8 | −33.0 |
| **Overall balance** | −36.0 | −342.7 | −61.6 |

Source: IMF, *International Financial Statistics*.

# External Trade

**PRINCIPAL COMMODITIES**
(distribution by SITC, US $ million)

| Imports c.i.f. | 2001 | 2002 | 2003 |
|---|---|---|---|
| Vegetable products | 145.9 | 144.6 | 130.2 |
| Prepared foodstuffs (incl. beverages and tobacco) | 78.8 | 71.1 | 78.7 |
| Mineral fuels, lubricants, etc. | 126.6 | 93.9 | 128.7 |
| Chemicals and related products | 238.9 | 223.1 | 241.0 |
| Plastic and articles thereof | 121.9 | 123.4 | 123.5 |
| Wood and wood products | 81.3 | 74.0 | 75.3 |
| Textiles and textile products | 100.3 | 79.4 | 78.9 |
| Basic manufactures | 157.9 | 269.4 | 161.9 |
| Machinery and mechanical appliances (incl. electrical equipment and parts thereof) | 409.5 | 447.0 | 350.6 |
| Transport equipment | 88.4 | 150.5 | 174.6 |
| **Total** (incl. others) | 1,708.3 | 1,832.0 | 1,684.6 |

# BOLIVIA

*Statistical Survey*

| Exports f.o.b. | 2001 | 2002 | 2003 |
|---|---|---|---|
| Vegetable products | 63.8 | 73.5 | 105.7 |
| Prepared foodstuffs (incl. beverages and tobacco) | 220.1 | 240.8 | 253.3 |
| Animal and vegetable oils, fats and waxes | 108.6 | 116.6 | 129.4 |
| Mineral fuels, lubricants, etc. | 492.5 | 545.0 | 723.7 |
| Wood and wood products | 41.0 | 41.1 | 42.6 |
| Textiles and textile products | 42.6 | 33.0 | 52.6 |
| Jewellery, goldsmiths' and silversmiths' wares, etc. | 188.2 | 214.1 | 185.1 |
| Basic manufactures | 53.5 | 51.9 | 62.4 |
| **Total*** (incl. others) | 1,265.6 | 1,377.2 | 1,621.7 |

* Including re-exports (US $ million): 123.7 in 2001; 54.4 in 2002; 79.6 in 2003.

## PRINCIPAL TRADING PARTNERS
(US $ million)

| Imports c.i.f. | 2002 | 2003 | 2004 |
|---|---|---|---|
| Argentina | 308.5 | 281.4 | 295.3 |
| Brazil | 392.5 | 343.4 | 486.7 |
| Chile | 126.1 | 122.0 | 111.4 |
| China, People's Republic | 95.9 | 84.6 | 107.9 |
| Colombia | 42.6 | 48.1 | 61.3 |
| France | 19.9 | 19.5 | 21.7 |
| Germany | 31.6 | 36.7 | 36.9 |
| Italy | 24.3 | 21.2 | 19.0 |
| Japan | 101.3 | 83.5 | 105.3 |
| Korea, Republic | 17.2 | 17.2 | 14.0 |
| Mexico | 32.9 | 34.7 | 36.3 |
| Paraguay | 18.1 | 16.7 | 19.2 |
| Peru | 95.0 | 104.4 | 127.3 |
| Spain | 24.3 | 23.6 | 27.4 |
| USA | 309.9 | 302.3 | 260.3 |
| **Total** (incl. others) | 1,832.0 | 1,684.6 | 1,887.8 |

| Exports | 2002 | 2003 | 2004 |
|---|---|---|---|
| Argentina | 27.9 | 56.3 | 131.5 |
| Belgium | 8.7 | 15.7 | 32.7 |
| Brazil | 336.6 | 495.7 | 713.6 |
| Chile | 32.8 | 43.5 | 51.2 |
| Colombia | 139.5 | 170.3 | 119.7 |
| Germany | 66.5 | 67.5 | 10.3 |
| Italy | 13.8 | 16.4 | 15.4 |
| Japan | 5.9 | 18.6 | 68.4 |
| Mexico | 19.9 | 21.0 | 27.0 |
| Peru | 74.1 | 83.7 | 137.8 |
| Switzerland | 214.9 | 165.9 | 52.5 |
| United Kingdom | 32.0 | 33.5 | 53.6 |
| USA | 193.0 | 235.3 | 358.6 |
| Venezuela | 175.4 | 155.4 | 244.5 |
| **Total*** (incl. others) | 1,376.9 | 1,650.7 | 2,254.4 |

* Including re-exports (US $ million): 54.4 in 2002; 79.6 in 2003; 68.1 in 2004.

# Transport

## RAILWAYS
(traffic)

| | 1999 | 2000 | 2001 |
|---|---|---|---|
| Passenger-kilometres (million) | 271 | 259 | 267 |
| Freight ton-kilometres (million) | 832 | 856 | 750 |

Source: UN, *Statistical Yearbook*.

## ROAD TRAFFIC
(motor vehicles in use at 31 December)

| | 1999* | 2000* | 2001 |
|---|---|---|---|
| Passenger cars | 168,611 | 181,409 | 254,853 |
| Buses | 19,581 | 21,183 | 21,808 |
| Lorries and vans | 213,455 | 234,746 | 133,686 |
| Tractors | 804 | 944 | n.a. |
| Motorcycles | 25,917 | 27,897 | 27,626 |

* Estimates.

Source: IRF, *World Road Statistics*.

## SHIPPING
**Merchant Fleet**
(registered at 31 December)

| | 2002 | 2003 | 2004 |
|---|---|---|---|
| Number of vessels | 96 | 136 | 85 |
| Total displacement ('000 grt) | 358.1 | 420.4 | 580.3 |

Source: Lloyd's Register-Fairplay, *World Fleet Statistics*.

## CIVIL AVIATION
(traffic on scheduled services)

| | 1999 | 2000 | 2001 |
|---|---|---|---|
| Kilometres flown (million) | 21 | 20 | 18 |
| Passengers carried ('000) | 1,873 | 1,757 | 1,557 |
| Passenger-km (million) | 1,851 | 1,809 | 1,567 |
| Freight ton-km (million) | 186 | 181 | 159 |

Source: UN, *Statistical Yearbook*.

# Tourism

## ARRIVALS AT HOTELS
(regional capitals only)

| Country of origin | 2001 | 2002 | 2003 |
|---|---|---|---|
| Argentina | 42,101 | 35,250 | 31,242 |
| Brazil | 24,015 | 26,265 | 23,810 |
| Canada | 7,832 | 8,205 | 7,429 |
| Chile | 23,199 | 21,911 | 17,152 |
| France | 24,533 | 26,672 | 24,356 |
| Germany | 23,415 | 21,781 | 19,056 |
| Israel | 9,250 | 10,569 | 12,003 |
| Italy | 7,561 | 7,871 | 7,631 |
| Japan | 7,198 | 6,964 | 6,379 |
| Netherlands | 12,935 | 12,281 | 10,444 |
| Peru | 52,619 | 59,736 | 62,164 |
| Spain | 9,462 | 10,850 | 10,964 |
| Switzerland | 10,365 | 9,225 | 8,613 |
| United Kingdom | 19,608 | 20,103 | 20,434 |
| USA | 37,902 | 38,515 | 36,801 |
| **Total** (incl. others) | 378,551 | 382,185 | 367,036 |

**Total tourism receipts** (US $ million, incl. passenger transport): 119 in 2001; 143 in 2002; 176 in 2003.

Source: World Tourism Organization.

## Communications Media

|  | 2001 | 2002 | 2003 |
|---|---|---|---|
| Telephones ('000 main lines in use) | 524.4 | 530.9 | 600.1 |
| Mobile cellular telephones ('000 subscribers) | 779.9 | 872.7 | 1,401.5 |
| Personal computers ('000 in use)* | 170 | 190 | n.a. |
| Internet users ('000)* | 180 | 270 | n.a. |

* Estimates.
Source: International Telecommunication Union.

**Television receivers** ('000): 990 in use in 2000 (Source: International Telecommunication Union).

**Radio receivers** ('000): 5,250 in use in 1997 (Source: UNESCO, *Statistical Yearbook*).

**Daily newspapers:** 18 in 1996 (average circulation 420,000 copies) (Source: UNESCO, *Statistical Yearbook*).

## Education

(2001)

|  | Institutions | Teachers | Students |
|---|---|---|---|
| Pre-primary | 2,294* | 4,133 | 198,640 |
| Primary | 12,639† | 66,339 | 1,666,150 |
| Secondary | n.a. | 16,507 | 398,360 |
| Higher: |  |  |  |
| state universities | n.a. | 7,738‡ | 212,594‡ |
| private universities | n.a. | 3,332§ | 48,817‖ |

* 1988.
† 1987.
‡ 2000.
§ 1998.
‖ Preliminary figure.
Source: partly UNESCO, *Statistical Yearbook*.

**Adult literacy rate** (census data): 86.7% (males 93.1%; females 80.7%) in 2002 (Source: UN Development Programme, *Human Development Report*).

# Directory

## The Constitution

Bolivia became an independent republic in 1825 and received its first Constitution in November 1826. Since that date a number of new Constitutions have been promulgated. Following the *coup d'état* of November 1964, the Constitution of 1947 was revived. Under its provisions, executive power is vested in the President, who chairs the Cabinet. According to the revised Constitution, the President is elected by direct suffrage for a five-year term (extended from four years in 1997) and is not eligible for immediate re-election. In the event of the President's death or failure to assume office, the Vice-President or, failing the Vice-President, the President of the Senate becomes interim Head of State.

The President has power to appoint members of the Cabinet and diplomatic representatives from a panel proposed by the Senate. The President is responsible for the conduct of foreign affairs and is also empowered to issue decrees, and initiate legislation by special messages to Congress.

The Congreso Nacional (Congress) consists of a 27-member Cámara de Senadores (Senate) and a 130-member Cámara de Diputados (Chamber of Deputies). The Congreso meets annually and its ordinary sessions last only 90 working days, which may be extended to 120. Each of the nine departments (La Paz, Chuquisaca, Oruro, Beni, Santa Cruz, Potosí, Tarija, Cochabamba and Pando), into which the country is divided for administrative purposes, elects three senators. Members of both houses are elected for five years.

The supreme administrative, political and military authority in each department is vested in a prefect appointed by the President. The sub-divisions of each department, known as provinces, are administered by sub-prefects. The provinces are further divided into cantons. There are 94 provinces and some 1,000 cantons. The capital of each department has its autonomous municipal council and controls its own revenue and expenditure.

Public order, education and roads are under national control.

A decree issued in July 1952 conferred the franchise on all persons who had reached the age of 21 years, whether literate or illiterate. Previously the franchise had been restricted to literate persons. (The voting age for married persons was lowered to 18 years at the 1989 elections.)

## The Government

### HEAD OF STATE

**President:** Eduardo Rodríguez Veltzé (took office 9 June 2005).

### THE CABINET
(August 2005)

**Minister of Foreign Affairs and Worship:** Armando Loayza Mariaca.
**Minister of the Interior:** Gustavo Avila Bustamante.
**Minister of National Defence:** Gonzalo Méndez Gutiérrez.
**Minister of Finance:** Waldo Gutiérrez.
**Minister of Sustainable Development and Planning:** Martha Bozo.
**Minister of the Presidency:** Iván Avilés Mantilla.
**Minister of Health and Sport:** Alvaro Muñoz Reyes Navarro.
**Minister of Labour:** Carlos Laguna Navarro.
**Minister of Education:** María Cristina Mejía Barragán.
**Minister of Agriculture, Livestock and Rural Development:** Guillermo Rivera Cuéllar.
**Minister of Economic Development:** Carlos Melchor Díaz Villavicencio.
**Minister of Hydrocarbons and Energy:** Jaime Eduardo Dunn Castellanos.
**Minister of Mines and Metallurgy:** Dionisio Garzón Martínez.
**Minister of Public Works and Services:** Mario Moreno Viruez.
**Minister without Portfolio responsible for Popular Participation:** Naya Ponce Fortún.
**Minister without Portfolio responsible for Indigenous and Ethnic Peoples' Affairs:** Pedro Ticona Cruz.
**Presidential Delegate for Political Affairs:** Jorge Lazarte Rojas.

### MINISTRIES

**Ministry of Agriculture, Livestock and Rural Development:** Avda Camacho 1471, La Paz.

**Ministry of Economic Development:** Edif. Palacio de Comunicaciones, Avda Mariscal Santa Cruz, La Paz; tel. (2) 237-7234; fax (2) 235-9955; e-mail contactos@desarrollo.gov.bo; internet www.desarrollo.gov.bo.

**Ministry of Education:** Casilla 6500, La Paz; tel. (2) 220-3576; fax (2) 220-3576; internet www.minedu.gov.bo.

**Ministry of Finance:** Edif. Palacio de Comunicaciones, Avda Mariscal Santa Cruz, La Paz; tel. (2) 237-7234; fax (2) 235-9955.

**Ministry of Foreign Affairs and Worship:** Calle Ingavi, esq. Junín, La Paz; tel. (2) 237-1150; fax (2) 237-1155; e-mail mreuno@rree.gov.bo; internet www.rree.gov.bo.

**Ministry of Health and Sport:** Plaza del Estudiante, La Paz; tel. (2) 237-1373; fax (2) 239-1590; e-mail minsalud@ceibo.entelnet.bo; internet www.sns.gov.bo.

**Ministry of Hydrocarbons and Energy:** La Paz.

**Ministry of the Interior:** Avda Arce 2409, esq. Belisario Salinas, La Paz; tel. (2) 237-0460; fax (2) 237-1334; e-mail mail@mingobierno.gov.bo; internet www.mingobierno.gov.bo.

**Ministry of Labour:** Calle Yanacocha, esq. Mercado, La Paz; tel. (2) 236-4164; fax (2) 237-1387; e-mail mintrabajo@unete.com; internet www.mintrabajo.gob.bo.

**Ministry of Mines and Metallurgy:** Avda Mariscal Santa Cruz, esq. Oruro, Edif., Palacio de Telecomunicaciones, 12°, La Paz; tel. (2) 237-4050; fax (2) 239-2758.

**Ministry of National Defence:** Plaza Avaroa, esq. Pedro Salazar y 20 de Octubre 2502, La Paz; tel. (2) 232-0225; fax (2) 243-3153; e-mail comunicaciones@mindef.gov.bo; internet www.mindef.gov.bo.

**Ministry of the Presidency:** Palacio de Gobierno, Plaza Murillo, La Paz; tel. (2) 237-1082; fax (2) 237-1388.

**Ministry of Public Works and Services:** La Paz.

**Ministry of Sustainable Development and Planning:** Arce 2147, 4°, Casilla 12814, La Paz; tel. (2) 231-0860; fax (2) 231-7320; e-mail sdnp@coord.rds.org.bo; internet www.rds.org.bo.

## President and Legislature

### PRESIDENT

At the presidential election that took place on 30 June 2002 the majority of votes were spread between five of the 11 candidates. Gonzalo ('Goni') Sánchez de Lozada of the Movimiento Nacionalista Revolucionario (MNR) obtained 22.49% of the votes cast, Evo Morales of the Movimiento al Socialismo (MAS) won 20.94%, Manfred Reyes Villa of the Nueva Fuerza Republicana (NFR) won 20.91%, Jaime Paz Zamora of the Movimiento de la Izquierda Revolucionaria (MIR) secured 16.31% and Felipe Quispe Huanca of Movimiento Indígena Pachakuti (MIP) won 6.09%. As no candidate obtained the requisite absolute majority, responsibility for the selection of the President passed to the new Congreso Nacional (National Congress), which, on 4 August 2002, voted, by 84 votes to 43, to appoint Goni Sánchez de Lozada to the presidency. On 18 October 2003, following Sánchez de Lozada's resignation, Carlos Diego Mesa Gisbert (Vice-President until he had resigned on 13 October) was sworn in as President. Mesa resigned on 9 June 2005 and was replaced by Eduardo Rodríguez Veltzé, hitherto President of the Supreme Court.

### CONGRESO NACIONAL

**President of the Cámara de Senadores:** HORMANDO VACA DÍEZ.
**President of the Cámara de Diputados:** MARIO COSSÍO CORTEZ.

**General Election, 30 June 2002**

| Party | Cámara de Diputados | Cámara de Senadores |
|---|---|---|
| Movimiento Nacionalista Revolucionario (Histórico) (MNR) | 36 | 11 |
| Movimiento al Socialismo (MAS) | 27 | 8 |
| Movimiento de la Izquierda Revolucionaria (MIR) | 26 | 5 |
| Nueva Fuerza Republicana (NFR) | 25 | 2 |
| Movimiento Indígena Pachakuti (MIP) | 7 | — |
| Unión Cívica Solidaridad (UCS) | 5 | — |
| Acción Democrática Nacionalista (ADN) | 4 | 1 |
| Partido Socialista (PS) | 1 | — |
| **Total** | **130** | **27** |

## Political Organizations

**Acción Democrática Nacionalista (ADN):** Calle Uruguay 454, La Paz; tel. (2) 242-3067; fax (2) 242-3412; e-mail mvberteros@aol.com; internet www.bolivian.com/adn; f. 1979; right-wing; Leader MAURO BERTERO GUTIÉRREZ; Nat. Exec. Sec. JORGE LANDÍVAR.

**Alianza de Renovación Boliviana (ARBOL):** La Paz; f. 1993; conservative; Leader CASIANO ACALLE CHOQUE.

**Alianza Social (AS):** Potosí; f. 2004; Leader RENÉ JOAQUINO.

**Bolivia Insurgente:** La Paz; f. 1996; populist party; Leader MÓNICA MEDINA.

**Comité Cívico pro Santa Cruz (CCSC):** Avda Cañada Strongest 70, CP 1801, Santa Cruz; tel. (3) 334-2777; fax (3) 334-1812; e-mail info@comiteprosantacruz.org; internet www.comiteprosantacruz.org; f. 1950; right-wing autonomist grouping; Pres. GERMÁN ANTELO; Vice-Pres. NINO GANDARILLA.

**Comité Cívico de Tarija (CCT):** Tarija; right-wing autonomist grouping; Pres. ROBERTO RUIZ.

**Federación de Juntas Vecinales (FEJUVE):** El Alto; left-wing grouping campaigning for civil rights and the nationalization of industries and utilities; Pres. ABEL MAMANI; Sec.-Gen. JORGE CHURA.

**Frente Revolucionario de Izquierda (FRI):** La Paz; left-wing; Leader OSCAR ZAMORA.

**Frente Unidad Nacional (FUN):** Cochabamba; f. 2003; left-wing; Leader SAMUEL DORIA MEDINA.

**Movimiento Bolivia Libre (MBL):** Edif. Camiri, Of. 601, Calle Comercio 972, esq. Yanacocha, Casilla 10382, La Paz; tel. (2) 234-0257; fax (2) 239-2242; f. 1985; left-wing; breakaway faction of MIR; 60,000 mems; Leader FRANK BARRIOS.

**Movimiento Indígena Pachakuti (MIP):** indigenous movement; f. 2002; Leader FELIPE QUISPE HUANCA.

**Movimiento de la Izquierda Revolucionaria (MIR):** Avda América 119, 2°, La Paz; e-mail mir@ceibo.entelnet.bo; internet www.cibergallo.com; f. 1971; split into several factions in 1985; left-wing; 150,000 mems; Leader JAIME PAZ ZAMORA; Sec.-Gen. OSCAR EID FRANCO.

**Movimiento sin Miedo:** La Paz; f. 1999; left-wing; Leader JUAN DEL GRANADO.

**Movimiento Nacionalista Revolucionario (Histórico) (MNR):** Calle Nicolás Acosta 574, La Paz; tel. (2) 249-0748; fax (2) 249-0009; formerly part of the Movimiento Nacionalista Revolucionario (MNR, f. 1942); centre-right; Leader GONZALO SÁNCHEZ DE LOZADA; Sec.-Gen. CARLOS SÁNCHEZ BERZAIN; 165,000 mems.

**Movimiento Revolucionario Túpac Katarí de Liberación (MRTKL):** Avda Baptista 939, Casilla 9133, La Paz; tel. (2) 235-4784; f. 1978; peasant party; Leader VÍCTOR HUGO CÁRDENAS CONDE; Sec.-Gen. NORBERTO PÉREZ HIDALGO; 80,000 mems.

**Movimiento al Socialismo—Instrumento Político por la Soberanía de los Pueblos (MAS):** La Paz; e-mail webmaster@masbolivia.org; internet www.masbolivia.org; f. 1987; left-wing; Leader EVO MORALES.

**Nueva Fuerza Republicana (NFR):** Cochabamba; internet www.bolivian.com/nfr; f. 1996; centre-right; Leader MANFRED REYES VILLA.

**Partido Comunista de Bolivia (PCB):** La Paz; f. 1950; Leader MARCOS DOMIC; First Sec. SIMÓN REYES RIVERA.

**Partido Demócrata Cristiano (PDC):** Casilla 4345, La Paz; f. 1954; Pres. BENJAMÍN MIGUEL HARB; Sec. ANTONIO CANELAS-GALATOIRE; 50,000 mems.

**Partido Obrero Revolucionario (POR):** Correo Central, La Paz; f. 1935; Trotskyist; Leader GUILLERMO LORA.

**Partido Revolucionario de la Izquierda Nacionalista (PRIN):** Calle Colón 693, La Paz; f. 1964; left-wing; Leader JUAN LECHIN OQUENDO.

**Partido Socialista (PS):** La Paz; f. 1987; Leader JERES JUSTINIANO.

**Partido de Vanguardia Obrera:** Plaza Venezuela 1452, La Paz; Leader FILEMÓN ESCOBAR.

**Plan Progreso:** La Paz; f. 2004, by fmr mems of the MIR; Leader JOSÉ LUIS PAREDES.

**Unión Cívica Solidaridad (UCS):** Calle Mercado 1064, 6°, La Paz; tel. (2) 236-0297; fax (2) 237-2200; f. 1989; populist; 102,000 mems; Leader JOHNNY FERNÁNDEZ.

## Diplomatic Representation

### EMBASSIES IN BOLIVIA

**Argentina:** Calle Aspiazú 497, CP 64, La Paz; tel. (2) 241-7737; fax (2) 242-2727; e-mail embarbol@adslmail.entelnet.bo; Ambassador HORACIO ANTONIO MACEDO.

**Belgium:** Calle 9, No 6, Achumani, Casilla 2433, La Paz; tel. (2) 277-1430; fax (2) 279-1219; e-mail lapaz@diplobel.be; internet www.diplomatie.be/lapazes; Ambassador FRANK VAN DE CRAEN.

**Brazil:** Edif. Multicentro, Torre B, Avda Arce s/n, esq. Rosendo Gutiérrez, Sopocachi, La Paz; tel. (2) 244-0202; fax (2) 244-0043; e-mail webmaster@brasil.org.bo; internet www.brasil.org.bo; Ambassador ANTONINO MENA-GONÇALVES.

**China, People's Republic:** Calle 1, Los Pinos 8532, Casilla 10005, La Paz; tel. (2) 279-3851; fax (2) 279-7121; e-mail emb-china@kolla.net; Ambassador ZHANG TUO.

**Colombia:** Calle 9, 7835 Calacoto; tel. (2) 271-2413; fax (2) 278-6510; e-mail elapaz@minrelext.gov.co; Ambassador GUILLERMO ANTONIO VANEGAS SIERRA.

**Costa Rica:** Edif. San Miguel Arcángel 1420, 1°, Of. 102, Avda Montenegro en Calacoto, La Paz; tel. and fax (2) 279-8930; e-mail embcrbo@entelnet.bo; Ambassador DIDIER CARRANZA RODRÍGUEZ.

# BOLIVIA

**Cuba:** Bajo Irpavi Avda Gobles 20, entre 13 y 14, La Paz; tel. (2) 272-1157; fax (2) 272-3419; e-mail embacuba@acelerate.com; Ambassador LUIS FELIPE VÁSQUEZ VÁSQUEZ.

**Denmark:** Edif. Fortaleza, Avda Arce 2799, esq. Cordero, 9°, Casilla 9860, La Paz; tel. (2) 243-2070; fax (2) 243-3150; e-mail lpbamb@um.dk; Ambassador MOGENS PEDERSEN.

**Ecuador:** Edif. Hermman, 14°, Plaza Venezuela 1440, Casilla 406, La Paz; tel. (2) 233-1588; fax (2) 231-9739; e-mail mecuabol@caoba.entelnet.bo; Ambassador Dr FERNANDO CÓRDOVA BOSSANO.

**Egypt:** Avda Ballivián 599, Casilla 2956, La Paz; tel. (2) 278-6511; fax (2) 278-4325; e-mail embegylp@ceibo.entelnet.bo; Ambassador NAGWA MOHAMED AFIFI.

**France:** Avda Hernando Silés 5390, esq. Calle 8, Obrajes, Casilla 717, La Paz; tel. (2) 214-9900; fax (2) 214-9904; e-mail information@ambafrance-bo.org; internet www.ambafrance-bo.org; Ambassador ANITA LIMIDO.

**Germany:** Avda Arce 2395, esq. Belisario Salinas, Casilla 5265, La Paz; tel. (2) 244-0066; fax (2) 244-1441; e-mail info@embajada-alemana-bolivia.org; internet www.embajada-alemana-bolivia.org; Ambassador Dr BERND SPROEDT.

**Holy See:** Avda Arce 2990, Casilla 136, La Paz; tel. (2) 243-1007; fax (2) 243-2120; e-mail nunapobol@acelerate.com; Apostolic Nuncio Most Rev. IVO SCAPOLO (Titular Archbishop of Tagaste).

**Italy:** Avda 6 de Agosto 2575, Casilla 626, La Paz; tel. (2) 243-4929; fax (2) 243-4975; e-mail ambitlap@ceibo.entelnet.bo; internet www.ambital.org.bo; Ambassador Dr MAURIZIO ZANINI.

**Japan:** Calle Rosendo Gutiérrez 497, esq. Sánchez Lima, Casilla 2725, La Paz; tel. (2) 241-9110; fax (2) 241-1919; e-mail coopjapon@acelerate.com; internet www.bo.emb-japan.go.jp; Ambassador MITSUNORI SHIRAKAWA.

**Korea, Republic:** La Paz; Ambassador CHUNG JIN-HO.

**Mexico:** Sánchez Bustamante 509, Casilla 430, La Paz; tel. (2) 277-2133; fax (2) 277-6085; e-mail embamex@acelerate.com; internet www.sre.gob.mx/bolivia; Ambassador JOSÉ ANTONIO ZABALGOITIA TREJO.

**Netherlands:** Edif. Hilda, 7°, Avda 6 de Agosto 2455, La Paz; tel. (2) 244-4040; fax (2) 244-3804; e-mail nllap@caoba.entelnet.bo; internet www.embholanda-bo.org; Ambassador RONALD C. J. MUYZERT.

**Panama:** Calle 10, No 7853 de Calacoto, Casilla 678, La Paz; tel. (2) 278-7334; fax (2) 279-7290; e-mail empanbol@ceibo.entelnet.bo; Ambassador AUGUSTO LUIS VILLARREAL AMARANTO.

**Paraguay:** Edif. Illimani II, 1°, Avda 6 de Agosto y Pedro Salazar, Casilla 882, La Paz; tel. (2) 243-3176; fax (2) 243-2201; e-mail embapar@acelerate.com; Ambassador NIMIA OVIEDO DE TORALES.

**Peru:** Calle F. Guachalla 300, Casilla 668, Sopocachi, La Paz; tel. (2) 244-1250; fax (2) 244-1240; e-mail embbol@caoba.entelnet.bo; Ambassador LUZMILA ZANABRIA.

**Russia:** Calacoto, Avda Walter Guevara Arze 8129, La Paz; tel. (2) 278-6419; fax (2) 278-6531; e-mail embrusia@ceibo.entelnet.bo; Ambassador VLADIMIR L. KULIKOV.

**Spain:** Avda 6 de Agosto 2827, Casilla 282, La Paz; tel. (2) 243-3518; fax (2) 211-3267; e-mail embespa@ceibo.entelnet.bo; Ambassador JUAN FRANCISCO MONTALBÁN CARRASCO.

**Switzerland:** Calle 13, esq. Avda 14 de Setiembre, Obrajes, Casilla 9356, La Paz; tel. (2) 275-1001; fax (2) 214-0885; e-mail vertretung@paz.rep.admin.ch; internet www.eda.admin.ch/lapaz; Chargé d'affaires a.i. JACQUES GREMAUD.

**United Kingdom:** Avda Sopocachi, Arce 2732, Casilla 694, La Paz; tel. (2) 243-3424; fax (2) 243-1073; e-mail ppa@megalink.com; internet www.britishembassy.gov.uk/bolivia; Ambassador PETER BATEMAN.

**USA:** Avda Arce 2780, Casilla 425, La Paz; tel. (2) 243-0251; fax (2) 243-3900; internet www.megalink.com/usemblapaz; Ambassador DAVID N. GREENLEE.

**Uruguay:** Edif. Monroy Velez, 7°, Calle 21 No 8350, Calacoto, La Paz, La Paz; tel. (2) 279-1482; fax (2) 212-9413; e-mail urulivia@acelerate.com; Ambassador JUAN A. PACHECO RAMÍREZ.

**Venezuela:** Edif. Illimani, 4°, Avda Arce, esq. Campos, Casilla 441, La Paz; tel. (2) 243-1365; fax (2) 243-2348; e-mail embvzla@acelerate.com; Chargé d'affaires AZAEL VALERO ANGULO.

# Judicial System

## SUPREME COURT

### Corte Suprema

Calle Pilinco 352, Sucre; tel. (4) 645-1883; fax (4) 646-2696; e-mail csuprema@poderjudicial.gov.bo; internet www.poderjudicial.gov.bo. Judicial power is vested in the Supreme Court. There are 12 members, appointed by Congress for a term of 10 years. The court is divided into four chambers of three justices each. Two chambers deal with civil cases, the third deals with criminal cases and the fourth deals with administrative, social and mining cases. The President of the Supreme Court presides over joint sessions of the courts and attends the joint sessions for cassation cases.

**President of the Supreme Court:** HÉCTOR SANDÓVAL PARADA (acting).

## DISTRICT COURTS

There is a District Court sitting in each Department, and additional provincial and local courts to try minor cases.

## ATTORNEY-GENERAL

In addition to the Attorney-General at Sucre (appointed by the President on the proposal of the Senate), there is a District Attorney in each Department as well as circuit judges.

**Attorney-General:** Dr OSCAR CRESPO SOLIZ.

# Religion

The majority of the population are Roman Catholics; there were an estimated 8.3m. adherents at 31 December 2003, equivalent to 84.8% of the population. Religious freedom is guaranteed. There are a number of Bahá'ís and a small Jewish community, as well as various Protestant denominations, in Bolivia.

## CHRISTIANITY

### The Roman Catholic Church

Bolivia comprises four archdioceses, six dioceses, two Territorial Prelatures and five Apostolic Vicariates.

#### Bishops' Conference

Conferencia Episcopal Boliviana, Calle Potosí 814, Casilla 2309, La Paz; tel. (2) 240-6855; fax (2) 240-6941; e-mail asc@scbbs-bo.com. f. 1972; Pres. Cardinal JULIO TERRAZAS SANDOVAL (Archbishop of Santa Cruz de la Sierra).

**Archbishop of Cochabamba:** Most Rev. TITO SOLARI, Avda Heroínas 152, esq. Zenteno Anaya, Casilla 129, Cochabamba; tel. (4) 425-6562; fax (4) 425-0522; e-mail arz_cbba@supernet.com.bo.

**Archbishop of La Paz:** Most Rev. EDMUNDO LUIS FLAVIO ABASTOFLOR MONTERO, Calle Ballivián 1277, Casilla 259, La Paz; tel. (2) 220-3690; fax (2) 220-3840; e-mail arzonslp@ceibo.entelnet.bo.

**Archbishop of Santa Cruz de la Sierra:** Cardinal JULIO TERRAZAS SANDOVAL, Calle Ingavi 49, Casilla 25, Santa Cruz; tel. (3) 332-4416; fax (3) 333-0181; e-mail asc@scbbs-bo.com.

**Archbishop of Sucre:** Most Rev. JESÚS GERVASIO PÉREZ RODRÍGUEZ, Calle Bolívar 702, Casilla 205, Sucre; tel. (4) 645-1587; fax (4) 646-0336; e-mail arzsucre@mara.scr.entelnet.bo.

### The Anglican Communion

Within the Iglesia Anglicana del Cono Sur de América (Anglican Church of the Southern Cone of America), Bolivia forms part of the diocese of Peru. The Bishop is resident in Lima, Peru.

### Protestant Churches

**Baptist Union of Bolivia:** Casilla 2199, La Paz; tel. (2) 222-9538; Pres. Rev. AUGUSTO CHUIJO.

**Convención Bautista Boliviana** (Baptist Convention of Bolivia): Casilla 3147, Santa Cruz; tel. (3) 334-0717; fax (3) 334-0717; f. 1947; Pres. EIRA SORUCO DE FLORES.

**Iglesia Evangélica Metodista en Bolivia** (Evangelical Methodist Church in Bolivia): Casillas 356 y 8347, La Paz; tel. (2) 249-1628; fax (2) 249-1624; autonomous since 1969; 10,000 mems; Bishop Rev. CARLOS INTIPAMPA.

BOLIVIA                                                                                                           *Directory*

## BAHÁ'Í FAITH

**National Spiritual Assembly of the Bahá'ís of Bolivia:** Casilla 1613, La Paz; tel. (2) 278-5058; fax (2) 278-2387; e-mail secretariat@bahai.org.bo; internet bahai.org.bo; mems resident in 5,161 localities.

# The Press

## DAILY NEWSPAPERS

### Cochabamba

**Opinión:** General Acha 252, Casilla 287, Cochabamba; tel. (4) 425-4400; fax (4) 441-5121; e-mail opinion@opinion.com.bo; internet www.opinion.com.bo; f. 1985; Dir EDWIN TAPIA FRONTANILLA; Man. GRACIELA MÉNDEZ DE ESCOBAR.

**Los Tiempos:** Edif. Los Tiempos, Plaza Quintanilla-Norte, Casilla 525, Cochabamba; tel. (4) 425-4562; fax (4) 425-4577; e-mail lostiempos@lostiempos-bolivia.com; internet www.lostiempos.com; f. 1943; morning; independent; Dir FERNANDO CANELAS; Man. Editor ALCIDES FLORES MONCADA; circ. 19,000.

### La Paz

**El Diario:** Calle Loayza 118, Casilla 5, La Paz; tel. (2) 239-0900; fax (2) 236-3846; e-mail contacto@eldiario.net; internet www.eldiario.net; f. 1904; morning; conservative; Dir JORGE CARRASCO JAHNSEN; Man. Editor MAURICIO CARRASCO; circ. 55,000.

**Jornada:** Edif. Almirante Grau 672, Zona San Pedro, Casilla 1628, La Paz; tel. (2) 248-8163; fax (2) 248-7487; e-mail cartas@jornadanet.com; internet www.jornadanet.com; f. 1964; evening; independent; Dir DAVID RÍOS ARANDA; circ. 11,500.

**La Razón:** Colinas de Santa Rita, Auquisamaña, Casilla 13100, La Paz; tel. (2) 227-1415; fax (2) 277-0908; e-mail larazon@la-razon.com; internet www.la-razon.com; f. 1990; Dir JUAN CARLOS ROCHA CHAVARRÍA; Gen. Man. GERARDO TÓRREZ OSSIO; circ. 35,000.

### Oruro

**El Expreso:** Potosí 4921, esq. Villarroel, Oruro; f. 1973; morning; independent; right-wing; Dir ALBERTO FRONTANILLA MORALES; circ. 1,000.

**La Patria:** Avda Camacho 1892, Casilla 48, Oruro; tel. (2) 525-0761; fax (2) 525-0781; f. 1919; morning; independent; Pres. MARCELO MIRRALLES BOVÁ; Dir ENRIQUE MIRALLES BONNECARRERE; circ. 6,000.

### Potosí

**El Siglo:** Calle Linares 99, Casilla 389, Potosí; f. 1975; morning; Dir WILSON MENDIETA PACHECO; circ. 1,500.

### Santa Cruz

**El Deber:** Avda El Trompillo 1144, Casilla 2144, Santa Cruz; tel. (3) 353-8000; fax (3) 353-9053; e-mail web@eldeber.com.bo; internet www.eldeber.com.bo; f. 1955; morning; independent; Dir Dr PEDRO RIVERO MERCADO; Man. Editor TUFFÍ ARÉ VÁZQUEZ; circ. 35,000.

**La Estrella del Oriente:** Calle Sucre 558, Casilla 736, Santa Cruz; tel. (3) 337-0707; fax (3) 337-0557; e-mail estrella@mitai.nrs.bolnet.bo; f. 1864; Pres. JORGE LANDÍVAR ROCA; Man. Editor TUFFI ARÉ.

**El Mundo:** Parque Industrial MZ-7, Casilla 1984, Santa Cruz; tel. (3) 346-4646; fax (3) 346-5057; e-mail redaccion@mail.elmundo.com.bo; internet www.elmundo.com.bo; f. 1979; morning; owned by Santa Cruz Industrialists' Asscn; Pres. WALTER PAREJAS MORENO; Dir JUAN JAVIER ZEBALLOS GUTIÉRREZ; circ. 15,000.

**El Nuevo Día:** Calle Independencia 470, Casilla 5344, Santa Cruz; tel. (3) 333-7474; fax (3) 336-0303; e-mail nuevodia@el-nuevodia.com; internet www.el-nuevodia.com; f. 1987; Dir NANCY EKLUND VDA DE GUTIÉRREZ; Man. Editor JORGE ORÍAS HERRERA.

### Sucre

**Correo del Sur:** Sucre; internet correodelsur.net; f. 1987; Pres. GONZALO CANELAS TARDÍO; Gen. Man. JULIO AUZA ANGLARILL.

### Trinidad

**La Razón:** Avda Bolívar 295, Casilla 166, Trinidad; tel. (3) 462-1377; f. 1972; Dir CARLOS VÉLEZ.

## PERIODICALS

**Actualidad Boliviana Confidencial (ABC):** Fernando Guachalla 969, Casilla 648, La Paz; f. 1966; weekly; Dir HUGO GONZÁLEZ RIOJA; circ. 6,000.

**Aquí:** Casilla 10937, La Paz; tel. (2) 234-3524; fax (2) 235-2455; f. 1979; weekly; circ. 10,000.

**Bolivia Libre:** Edif. Esperanza, 5°, Avda Mariscal Santa Cruz 2150, Casilla 6500, La Paz; fortnightly; govt organ.

**Carta Cruceña de Integración:** Casilla 3531, Santa Cruz de la Sierra; weekly; Dirs HERNÁN LLANOVARCED A., JOHNNY LAZARTE J.

**Comentarios Económicos de Actualidad (CEA):** Casilla 312097, La Paz; tel. (2) 242-4766; fax (2) 242-4772; e-mail veceba@caoba.entelnet.bo; f. 1983; fortnightly; articles and economic analyses; Editor GUIDO CESPEDES.

**Información Política y Económica (IPE):** Calle Comercio, Casilla 2484, La Paz; weekly; Dir GONZALO LÓPEZ MUÑOZ.

**Informe R:** La Paz; weekly; Editor SARA MONROY.

**Notas:** Edif. Mariscal de Ayacucho, 5°, Of. 501, Calle Loayza 233, Casilla 5782, La Paz; tel. (2) 233-5577; fax (2) 233-7607; e-mail anf@ceibo.entelnet.bo; internet www.agenciadenoticiasfides.net; f. 1963; weekly; political and economic analysis; Editor JOSÉ GRAMUNT DE MORAGAS.

**El Noticiero:** Sucre; weekly; Dir DAVID CABEZAS; circ. 1,500.

**Prensa Libre:** Sucre; tel. (4) 646-2447; fax (4) 646-2768; e-mail prelibre@mara.scr.entelnet.bo; f. 1989; weekly; Dir JULIO PEMINTEL A.

**Servicio de Información Confidencial (SIC):** Elías Sagárnaga 274, Casilla 5035, La Paz; weekly; publ. by Asociación Nacional de Prensa; Dir JOSÉ CARRANZA.

**Siglo XXI:** La Paz; weekly.

**Unión:** Sucre; weekly; Dir JAIME MERILES.

**Visión Boliviana:** Calle Loayza 420, Casilla 2870, La Paz; 6 a year.

## PRESS ASSOCIATIONS

**Asociación Nacional de la Prensa:** Avda 6 de Agosto 2170, Casilla 477, La Paz; tel. (2) 236-9916; Pres. Dr CARLOS SERRATE REICH.

**Asociación de Periodistas de La Paz:** Avda 6 de Agosto 2170, Casilla 477, La Paz; tel. (2) 236-9916; fax (2) 232-3701; f. 1929; Pres. MARIO MALDONADO VISCARRA; Vice-Pres. MARÍA EUGENIA VERASTEGUI A.

## NEWS AGENCIES

**Agencia de Noticias Fides (ANF):** Edif. Mariscal de Ayacucho, 5°, Of. 501, Calle Loayza, Casilla 5782, La Paz; tel. (2) 236-5152; fax (2) 236-5153; e-mail anf@ceibo.entelnet.bo; internet www.agenciadenoticiasfides.net; f. 1963; owned by Roman Catholic Church; Dir JOSÉ GRAMUNT DE MORAGAS; Man. Editor WALTER PATIÑO.

### Foreign Bureaux

**Agencia EFE** (Spain): Edif. Anibal Mz 01, Avda Sánchez Lima 2520, Casilla 7403, La Paz; tel. (2) 241-9222; fax (2) 241-8388; e-mail efebol@entelnet.bo; Bureau Chief ESTHER REBOLLO.

**Agenzia Nazionale Stampa Associata (ANSA)** (Italy): La Paz; tel. (2) 235-5521; fax (2) 236-8221; Correspondent RAÚL PENARANDA UNDURRAGA.

**Associated Press (AP)** (USA): Edif. Mariscal de Ayacucho, Of. 1209, Calle Loayza 273, Casilla 9569, La Paz; tel. (2) 220-1557; fax (2) 220-1558; e-mail varrington@ap.org; Correspondent VANESSA ARRINGTON.

**Deutsche Presse-Agentur (dpa)** (Germany): Edif. Esperanza, 9°, Of. 3, Av. Mariscal Santa Cruz 2150, Casilla 13885, La Paz; tel. (2) 235-2684; fax (2) 239-2488; Correspondent ROBERT BROCKMANN.

**Informatsionnoye Telegrafnoye Agentstvo Rossii—Telegrafnoye Agentstvo Suverennykh Stran (ITAR—TASS)** (Russia): Casilla 6839, San Miguel, Bloque 0–33, Casa 958, La Paz; tel. (2) 279-2108; Correspondent ELDAR ABDULLAEV.

**Inter Press Service (IPS)** (Italy): Edif. Esperanza, 6°, Of. 6, Casilla 4313, La Paz; tel. (2) 236-1227; Correspondent RONALD GREBE LÓPEZ.

**Prensa Latina** (Cuba): La Paz; tel. (2) 232-3479; Correspondent MANUEL ROBLES SOSA.

**Reuters** (United Kingdom): Edif. Loayza, 3°, Of. 301, Calle Loayza 349, Casilla 4057, La Paz; tel. (2) 235-1106; fax (2) 239-1366; Correspondent RENÉ VILLEGAS MONJE.

**Rossiyskoye Informatsionnoye Agentstvo—Novosti (RIA—Novosti)** (Russia): La Paz; tel. (2) 237-3857; Correspondent VLADIMIR RAMÍREZ.

Agence France-Presse and Telam (Argentina) are also represented.

# Publishers

**Ediciones Runa:** Calle España 459, Cochabamba; tel. (4) 523-389; e-mail edicionesruna@yahoo.com; internet www.geocities.com/edicionesruna/; f. 1968; juvenile, educational and scholarly.

**Editora Khana Cruz SRL:** Avda Camacho 1372, Casilla 5920, La Paz; tel. (2) 237-0263; Dir GLADIS ANDRADE.

**Editora Lux:** Edif. Esperanza, Avda Mariscal Santa Cruz, Casilla 1566, La Paz; tel. (2) 232-9102; fax (2) 234-3968; f. 1952; Dir FELICISIMO TARILONTE PÉREZ.

**Editorial los Amigos del Libro:** Avda Ayacucho 0-156, Casilla 450, Cochabamba; tel. (4) 450-4150; fax (4) 411-5128; e-mail amigol@amigol.bo.net; e-mail gutten@amigol.bo.net; internet www.librosbolivia.com; f. 1945; general; Man. Dir WERNER GUTTENTAG; Gen. Man. INGRID GUTTENTAG.

**Editorial Bruño:** Loayza 167, Casilla 4809, La Paz; tel. (2) 233-1254; fax (2) 233-5043; e-mail brunol@caoba.entelnet.bo; f. 1964; Dir IGNACIO LOMAS.

**Editorial Don Bosco:** Avda 16 de Julio 1899, Casilla 4458, La Paz; tel. (2) 237-1449; fax (2) 236-2822; f. 1896; social sciences and literature; Dir GRAMAGLIA MAGLIANO.

**Editorial Icthus:** La Paz; tel. (2) 235-4007; f. 1967; general and textbooks; Man. Dir DANIEL AQUIZE.

**Editorial Popular:** Plaza Pérez Velasco 787, Casilla 4171, La Paz; tel. (2) 235-0701; f. 1935; textbooks, postcards, tourist guides, etc; Man. Dir GERMÁN VILLAMOR.

**Editorial Puerta del Sol:** Edif. Litoral Sub Suelo, Avda Mariscal Santa Cruz, La Paz; tel. (2) 236-0746; f. 1965; Man. Dir OSCAR CRESPO.

**Empresa Editora Proinsa:** Avda Saavedra 2055, Casilla 7181, La Paz; tel. (2) 222-7781; fax (2) 222-6671; f. 1974; school books; Dirs FLOREN SANABRIA G., CARLOS SANABRIA C.

**Gisbert y Cía, SA:** Calle Comercio 1270, Casilla 195, La Paz; tel. (2) 220-2626; fax (2) 220-2911; e-mail libgis@entelnet.bo; f. 1907; textbooks, history, law and general; Pres. JAVIER GISBERT; Promotions Man. MARÍA DEL CARMEN SCHULCZEWSKI; Admin. Man. ANTONIO SCHULCZEWSKI.

**Ivar American:** Calle Potosí 1375, Casilla 6016, La Paz; tel. (2) 236-1519; Man. Dir HÉCTOR IBÁÑEZ.

**Librería Editorial Juventud:** Plaza Murillo 519, Casilla 1489, La Paz; tel. (2) 240-6248; f. 1946; textbooks and general; Dir GUSTAVO URQUIZO MENDOZA.

**Librería El Ateneo SRL:** Calle Ballivián 1275, Casilla 7917, La Paz; tel. (2) 236-9925; fax (2) 239-1513; Dirs JUAN CHIRVECHES D., MIRIAN C. DE CHIRVECHES.

**Librería Dismo Ltda:** Calle Comercio 806, Casilla 988, La Paz; tel. (2) 240-6411; fax (2) 31-6545; e-mail dismo@caoba.entelnet.bo; Dir TERESA GONZÁLEZ DE ALVAREZ.

**Librería La Paz:** Calle Campos y Villegas, Edif. Artemis, Casilla 539, La Paz; tel. (2) 243-4927; fax (2) 243-5004; e-mail liblapaz@ceibo.entelnet.bo; f. 1900; Dirs EDUARDO BURGOS R., CARLOS BURGOS M.

**Librería La Universal SRL:** Calle Ingavi 780, Casilla 2869, La Paz; tel. (2) 228-6634; f. 1958; Man. Dir ROLANDO CONDORI SALINAS.

**Librería San Pablo:** Calle Colón 627, Casilla 3152, La Paz; tel. (2) 232-6084; f. 1967; Man. Dir MARÍA DE JESÚS VALERIANO.

**Santillana de Ediciones SA:** Avda Arce 2333, La Paz; tel. (2) 441-122; fax (2) 442–208; internet www.santillanabo.com; Gen. Man. ANDRÉS CARDÓ.

### PUBLISHERS' ASSOCIATION

**Cámara Boliviana del Libro:** Calle Capitán Ravelo 2116, Casilla 682, La Paz; tel. (2) 244-4239; fax (2) 244-1523; e-mail cabolib@ceibo.entelnet.bo; f. 1947; Pres. AMPARO LINARES; Vice-Pres. ERNESTO MARTÍNEZ; Gen. Man. ANA PATRICIA NAVARRO.

# Broadcasting and Communications

### TELECOMMUNICATIONS

**Cámara Nacional de Medios de Comunicación:** Casilla 2431, La Paz.

**Empresa Nacional de Telecomunicaciones (ENTEL):** Calle Federico Zuazo 1771, Casilla 4450, La Paz; tel. (2) 231-3030; fax (2) 239-1789; tel. contacto@entelsa.entelnet.bo; internet www.entel.bo; f. 1965; privatized under the Govt's capitalization programme in 1995; owned by Telecom Italia; Exec. Pres. MARIO FUMI.

**Superintendencia de Telecomunicaciones:** Calle 13, No 8260, Calacoto, La Paz; tel. (2) 277-2266; fax (2) 277-2299; e-mail supertel@ceibo.entelnet.bo; internet www.sittel.gov.bo; f. 1995; govt-controlled broadcasting authority; Supt RENÉ BUSTILLO PORTOCARRERO.

### BROADCASTING

#### Radio

The majority of radio stations are commercial. Broadcasts are in Spanish, Aymará and Quechua. In 1999 there were 171 MW and LW radio stations and 73 FM stations, as well as 77 short-wave stations.

**Asociación Boliviana de Radiodifusoras (ASBORA):** Edif. Jazmín, 10°, Avda 20 de Octubre 2019, Casilla 5324, La Paz; tel. (2) 236-5154; fax (2) 236-3069; broadcasting authority; Pres. TERESA SANJINÉS L.; Vice-Pres. LUIS ANTONIO SERRANO.

**Educación Radiofónica de Bolivia (ERBOL):** Calle Ballivian 1323, 4°, Casilla 5946, La Paz; tel. (2) 235-4142; fax (2) 239-1985; asscn of 28 educational radio stations in Bolivia; Gen. Sec. RONALD GREBE LÓPEZ.

**Radio Fides:** La Paz; internet www.fides.com; network of 19 radio stations; Catholic; Dir EDUARDO PÉREZ IRIBARNE.

**Radio Illimani:** Avda Camacho 1485, 6°, La Paz; tel. (2) 220-0473; e-mail illimani@comunica.gov.bo; internet www.comunica.gov.bo/illimani; f. 1932 as Compañía Radio Boliviana; govt-owned.

**Radio Panamerica:** Edif. 16 de Julio, 9°, Of. 902, El Prado, La Paz; tel. (2) 233-4271; e-mail pana@panamericana-bolivia.com; internet www.panamericanabolivia.com; f. 1972; Dir DANIEL SÁNCHEZ.

#### Television

There were 48 television stations in 1999.

**ATB Red Nacional** (Canal 9): Avda Argentina 2057, Casilla 9285, La Paz; tel. and fax (2) 222-9922; internet www.atb.com.bo; f. 1985; privately owned television network.

**Bolivisión** (Canal 4): Santa Cruz; internet www.bolivisiontv.com; f. 1997; privately owned television network; Exec. Pres. Ing. ERNESTO ASBÚN GAZAUI; Gen. Man. Lic. JEANNETTE ARRÁZOLA RIVERO.

**Empresa Nacional de Televisión Boliviana** (Canal 7): Edif. La Urbana, 6° y 7°, Avda Camacho 1486, Casilla 900, La Paz; tel. (2) 237-6356; fax (2) 235-9753; f. 1969; govt network operating stations in La Paz, Oruro, Cochabamba, Potosí, Chuquisaca, Pando, Beni, Tarija and Santa Cruz; Gen. Man. MIGUEL N. MONTERO VACA.

**Televisión Universitaria** (Canal 13): Edif. 'Hoy', 12°–13°, Avda 6 de Agosto 2170, Casilla 13383, La Paz; tel. (2) 235-9297; fax (2) 235-9298; internet www.umsanet.edu.bo; f. 1980; educational programmes; stations in Oruro, Cochabamba, Potosí, Sucre, Tarija, Beni and Santa Cruz; Dir Lic. ROBERTO CUEVAS RAMÍREZ.

**Unitel** (Canal 9): Km 5, Carretera antigua a Cochabamba, Santa Cruz; tel. (3) 352-7686; fax (3) 352-7688; e-mail webmaster@unitel.com.bo; internet www.unitel.tv; f. 1997; privately owned television network; Gen. Man. Ing. JUAN ALBERTO ROJAS.

# Finance

(cap. = capital; res = reserves; dep. = deposits; m. = million; brs = branches; amounts are in bolivianos, unless otherwise stated)

### BANKING

#### Supervisory Authority

**Superintendencia de Bancos y Entidades Financieras:** Plaza Isabel la Católica 2507, Casilla 447, La Paz; tel. (2) 243-1919; fax (2) 243-0028; e-mail sbef@sbef.gov.bo; internet www.sbef.gov.bo; f. 1928; Supt Dr LUIS FERNANDO CALVO UNZUETA.

#### Central Bank

**Banco Central de Bolivia:** Avda Ayacucho, esq. Mercado, Casilla 3118, La Paz; tel. (2) 237-4151; fax (2) 239-2398; e-mail vmarquez@mail.bcb.gov.bo; internet www.bcb.gov.bo; f. 1911 as Banco de la Nación Boliviana, name changed as above in 1928; bank of issue; cap. 515.8m., res 4,438.6m., dep. 11,079.5m. (Dec. 2002); Pres. Dr JUAN ANTONIO MORALES ANAYA; Gen. Man. Lic. JAIME VALENCIA.

#### Commercial Banks

**Banco Bisa SA:** Avda 16 de Julio 1628, Casilla 1290, La Paz; tel. (2) 235-9471; fax (2) 239-2013; e-mail bancobisa@grupobisa.com; internet www.grupobisa.com; f. 1963; cap. 479.7m., res 165.3m., dep. 4,353.8m. (Dec. 2002); Pres. and CEO Ing. JULIÓ LEÓN PRADO.

**Banco de Crédito de Bolivia, SA:** Calle Colón, esq. Mercado 1308, Casilla 907, La Paz; tel. (2) 220-1154; fax (2) 223-2203; internet www.bancodecredito.com.bo; f. 1993 as Banco Popular, SA, name changed as above 1994; owned by Banco de Crédito del Perú; cap. 315.5m., res 109.1m., dep. 2,995.1m. (Dec. 2002); Chair. DIONISIO ROMERO SEMINARIO; Gen. Man. DAVID SAETTONE WATMOUGH; 6 brs.

**Banco Económico SA-SCZ:** Calle Ayacucho 166, Casilla 5603, Santa Cruz; tel. (3) 336-1177; fax (3) 336-1184; e-mail baneco@baneco.com.bo; f. 1990; dep. US $174.3m., cap. US $20.99m., total assets US $240.7m. (Dec. 2003); Pres. LUIS PERROGÓN TOLEDO; Gen. Man. Ing. JUSTO YEPEZ KAKUDA; 14 brs.

**Banco Ganadero SA-Santa Cruz:** Calle 24 de Setiembre 110, Casilla 4492, Santa Cruz; tel. (3) 336-1616; fax (3) 333-2567; e-mail hkrutzfeldt@bancoganadero.co.bo; internet www.bancoganadero.com.bo; f. 1994; cap. and res 108.1m., dep. 1,078.1m. (Dec. 2000); Pres. FERNANDO MONASTERIO NIEME; Gen. Man. RONALD GUTIÉRREZ LÓPEZ.

**Banco Mercantil, SA:** Calle Ayacucho, esq. Mercado 295, Casilla 423, La Paz; tel. (2) 240-9040; fax (2) 240-9158; e-mail bercant@bancomercantil.com.bo; internet www.bancomercantil.com.bo; f. 1905; cap. and res 57.7m., dep. 457m. (Dec. 2003); Pres. EDUARDO QUINTANILLA; Exec. Vice-Pres. EMILIO UNZUETA ZEGARRA; 41 brs.

**Banco Nacional de Bolivia:** Avda Camacho, esq. Colón 1312, Casilla 360, La Paz; tel. (2) 311-139; fax (2) 334-723; e-mail info@bnb.com.bo; internet www.bnb-bol.com; f. 1871; cap. 261.0m., res 84.9m., dep. 3,853.7m. (Dec. 2004); Pres. GONZALO ARGANDOÑA; Gen. Man. PABLO BEDOYA; 9 brs.

**Banco Santa Cruz, SA:** Calle Junín 154, Casilla 865, Santa Cruz; tel. (3) 336-9911; fax (3) 335-0114; e-mail ggeneral@mail.bsc.com.bo; internet www.bsc.com.bo; f. 1965; 96% owned by Santander Central Hispano (Spain); Gen. Man. LISARDO PELÁEZ ACERO; 19 brs.

**Banco Solidario, SA** (BancoSol): Calle Nicolás Acosta, esq. Cañada Strongest 289, Casilla 13176, La Paz; tel. (2) 248-6563; fax (2) 248-6533; e-mail info@bancosol.com.bo; internet www.bancosol.com.bo; f. 1992; Pres. HERBERT MÜLLER; Gen. Man. KURT KÖNIGSFEST.

**Banco Unión, SA:** Calle Libertad 156, Casilla 4057, Santa Cruz; tel. (3) 336-6869; fax (3) 334-0684; e-mail info@bancounion.com.bo; internet www.bancounion.com.bo; f. 1982; cap. US $40.9m., res US $3.1m., dep. US $307.1m. (Dec. 2002); Pres. Ing. ANDRÉS PETRICEVIC; Gen. Man. JULIAN SECO; 2 brs.

### Foreign Banks

**Banco ABN AMRO Real, SA** (Netherlands): Avda 16 de Julio 1642, Casilla 10008, La Paz; tel. (2) 233-4477; fax (2) 233-5588; f. 1978 as Banco Real, name changed as above 2000; dep. US $8.3m., total assets US $17.5m. (Dec. 1997); Operations Man. JOSÉ PORFIRIO VASCONCELOS.

**Banco do Brasil, SA:** Avda 16 de Julio 1642, El Prado, La Paz; tel. (2) 231-0909; fax (2) 231-1788; e-mail lapaz@bb.com.br; f. 1961; Gen. Man. JOSÉ ENEAS BUENO JUNIOR; 1 br.

**Banco de la Nación Argentina:** Junín 22, Plaza 24 de Setiembre, Santa Cruz; tel. (3) 334-3777; fax (2) 334-3729; e-mail eaira@bnasc.com.bo; internet www.bna.com.ar; f. 1891; Man. ESTEBAN SALVADOR AIRA; 1 br.

### Banking Association

**Asociación de Bancos Privados de Bolivia (ASOBAN):** Edif. Cámara Nacional de Comercio, 15°, Avda Mariscal Santa Cruz, esq. Colombia 1392, Casilla 5822, La Paz; tel. (2) 237-6164; fax (2) 239-1093; e-mail info@asoban.bo; internet www.asoban.bo; f. 1957; Pres. EMILO UNZUETA ZEGARRA; Vice-Pres PABLO BEDOYA, FRANCISCO MONASTERIOS; 18 mems.

## STOCK EXCHANGE
### Supervisory Authorities

**Superintendencia de Pensiones, Valores y Seguros:** Calle Reyes Ortiz, esq. Federico Zuazo, Edif. Torres Gundlach Este, 3°, Casilla 6118, La Paz; tel. (2) 233-1212; fax (2) 233-0001; e-mail spvs@spvs.gov.bo; internet www.spvs.gov.bo; Supt GUILLERMO APONTE; Operations Dir LUIS TEJADA PONCE.

**Bolsa Boliviana de Valores SA:** Edif. Zambrana P.B., Calle Montevideo 142, Casilla 12521, La Paz; tel. (2) 244-3232; fax (2) 244-2308; e-mail info@bolsa-valores-bolivia.com; internet www.bolsa-valores-bolivia.com; f. 1989; Gen. Man. Lic. ARMANDO ALVAREZ ARNAL.

## INSURANCE
### Supervisory Authority

**Superintendencia de Pensiones, Valores y Seguros:** see above.

### National Companies

**Adriatica Seguros y Reaseguros, SA:** Calle Libertad, esq. Cañoto 879, Casilla 1515, Santa Cruz; tel. (3) 336-6667; fax (3) 336-0600; e-mail elandivar@adriatica.com.bo; internet www.adriatica.com.bo; f. 1995; Pres. ANTONIO OLEA BAUDOIN.

**Alianza, Cía de Seguros y Reaseguros, SA:** Avda 20 de Octubre 2680, esq. Campos, Zona San Jorge, Casilla 1043, La Paz; tel. (2) 243-2121; fax (2) 243-2713; e-mail info@alianzaseguros.com; f. 1991; Pres. JUAN MANUEL PEÑA ROCA.

**Alianza Vida Seguros y Reaseguros, SA:** Avda Viedma 19, Casilla 7181, Santa Cruz; tel. (3) 337-5656; fax (3) 337-5666; e-mail alejandroy@alianzaseguros.com; f. 1999; Pres. RAÚL ADLER K; Gen. Man. ALEJANDRO YBARRA CARRASCO.

**Bisa Seguros y Reaseguros, SA:** Edif. San Pablo, 13°, Avda 16 de Julio 1479, Casilla 3669, La Paz; tel. (2) 235-2123; fax (2) 239-2500; e-mail bisaseguros@grupobisa.com; internet www.grupobisa.com; f. 1991; Pres. JULIO CESAR LEÓN PRADO.

**La Boliviana Ciacruz de Seguros y Reaseguros, SA:** Calle Colón, esq. Mercado 288, Casilla 628, La Paz; tel. (2) 220-3131; fax (2) 220-4087; e-mail rodrigo.bedoya@zurich.com; internet www.boliviana-ciacruz.com; f. 1946; owned by Zurich group; all classes; Pres. GONZALO BEDOYA HERRERA; Gen. Man. RODRIGO BEDOYA DIEZ DE MEDINA.

**Compañía de Seguros y Reaseguros Fortaleza, SA:** Avda Virgen de Cotoca 2080, Casilla 1366, Santa Cruz; tel. and fax (3) 348-7273; e-mail nhinojosa@grupofortaleza.com.bo; Pres. GUIDO HINOJOSA; Gen. Man. NELSON HINOJOSA.

**Credinform International SA de Seguros:** Edif. Credinform, Calle Potosí, esq. Ayacucho 1220, Casilla 1724, La Paz; tel. (2) 231-5566; fax (2) 220-3917; e-mail maicobarragan@credinformsa.com; f. 1962; all classes; Pres. Dr ROBÍN BARRAGÁN PELÁEZ; Gen. Man. MIGUEL ANGEL BARRAGÁN IBARGUEN.

**Nacional Vida Seguros de Personas, SA:** Edif. Sucumbé, Avda Gral. San Martín, esq. Ricardo Jamies Freire, Casilla 4387, Santa Cruz; tel. (3) 333-6262; fax (3) 333-7969; e-mail nv_jlcamacho@cotas.com.bo; f. 1999; Pres. CARMEN SEEGHERS DE ZWAHLEN; Gen. Man. JOSÉ LUIS CAMACHO.

**Seguros y Reaseguros Generales 24 de Septiembre, SA:** Avda Ejercito Nacional 487, Santa Cruz; tel. and fax (3) 354-8484; e-mail seguros@caoba.entelnet.com.bo; f. 2001; Pres. Col. ROBERTO FORONDA; Gen. Man. JOSEFINA SOLIZ DE FORONDA.

**Seguros Illimani, SA:** Edif. Mariscal de Ayacucho, 10°, Calle Loayza 233, Casilla 133, La Paz; tel. (2) 220-3040; fax (2) 239-1149; e-mail sisalp@entelnet.bo; internet www.segurosillimani.com; f. 1979; all classes; Pres. FERNANDO ARCE GRANDCHANT.

**Seguros Próvida, SA:** Edif. San José, Avda 20 de Octubre 2524, Casilla 133, La Paz; tel. (2) 243-0300; fax (2) 243-0318; e-mail provida@segurosprovida.com.bo; f. 1999; Exec. Pres. FERNANDO ARCE GRANDCHANT; Gen. Man. JUSTINO AVENDAÑO.

**La Vitalicia Seguros y Reaseguros de Vida, SA:** Edif. Hoy, Avda 6 de Agosto 2860, Casilla 8424, La Paz; tel. (2) 212-5355; fax (2) 211-3480; e-mail aibanez@grupobisa.com; internet www.grupobisa.com/somosvitalicia.html; f. 1988; part of Grupo Bisa; Pres. JULIO CESAR LEÓN PRADO; Exec. Vice-Pres. Lic. LUIS ALFONSO IBAÑEZ MONTES.

There are also four foreign-owned insurance companies operating in Bolivia: American Life Insurance Co, American Home Assurance Co, United States Fire Insurance Co and International Health Insurance Danmarck.

### Insurance Association

**Asociación Boliviana de Aseguradores:** Edif. Castilla, 5°, Of. 510, Calle Loayza, esq. Mercado 250, Casilla 4804, La Paz; tel. (2) 220-1014; fax (2) 220-1088; e-mail aba@ababolivia.org; internet www.ababolivia.org; f. 1962; Pres. LUIS ALFONSO IBÁNEZ MONTES; Gen. Man. CARLOS BAUDOIN DÁVALOS.

# Trade and Industry

## GOVERNMENT AGENCIES

**Cámara Nacional de Exportadores (CAMEX):** Avda Arce 2017, esq. Goitia, Casilla 12145, La Paz; tel. (2) 234-1220; fax (2) 236-1491; e-mail camex@caoba.entelnet.bo; internet www.camex-lpb.com; f. 1970; Pres. LUIS NEMTALA YAMIN; Gen. Man. JORGE ADRIAZOLA REIMERS.

**Instituto Nacional de Inversiones (INI):** Edif. Cristal, 10°, Calle Yanacocha, Casilla 4393, La Paz; tel. (2) 237-5730; fax (2) 236-7297; e-mail abeseg@kolla.net.bo; f. 1971; state institution for the promotion of new investments and the application of the Investment Law; Exec. Dir Ing. JOSÉ MARIO FERNÁNDEZ IRAHOLA.

**Sistema de Regulación Sectorial (SIRESE):** Edif. Capitán Ravelo, 8°, Casilla 9647, La Paz; tel. (2) 244-4545; fax (2) 244-4017; e-mail sg@sirese.gov.bo; internet www.sirese.gov.bo; f. 1994; regulatory body for the formerly state-owned companies and utilities; oversees the general co-ordination and growth of the regulatory system and the work of its Superintendencies of Electricity, Hydrocarbons, Telecommunications, Transport and Water; Supt-Gen. CLAUDE BESSE ARZE.

## DEVELOPMENT ORGANIZATIONS

**Centro de Estudios para el Desarrollo Laboral y Agrario (CEDLA):** Avda Jaimes Freyre 2940, esq. Muñoz Cornejo, Casilla 8630, La Paz; tel. (2) 241-3175; fax (2) 241-4625; e-mail cedla@cedla.org; internet www.cedla.org; f. 1985; agrarian and labour development; Exec. Dir CARLOS ARZE.

**Consejo Nacional de Acreditación y Medición de la Calidad Educativa (CONAMED):** La Paz; f. 1994; education quality board.

**Consejo Nacional de Planificación (CONEPLAN):** Edif. Banco Central de Bolivia, 26°, Calle Mercado, esq. Ayacucho, Casilla 3118, La Paz; tel. (2) 237-4151; fax (2) 235-3840; e-mail claves@mail.bcb.gov.bo; internet www.bcb.gov.bo; f. 1985.

**Corporación de las Fuerzas Armadas para el Desarrollo Nacional (COFADENA):** Avda 6 de Agosto 2649, Casilla 1015, La Paz; tel. (2) 237-7305; fax (2) 236-0900; f. 1972; industrial, agricultural and mining holding co and development org. owned by the Bolivian armed forces; Gen. Man. Col JUAN MANUEL ROSALES.

**Corporación Regional de Desarrollo de La Paz (CORDEPAZ):** Avda Arce 2529, Edif. Santa Isabel, Casilla 6102, La Paz; tel. (2) 243-0313; fax (2) 243-2152; f. 1972; decentralized government institution to foster the development of the La Paz area; Pres. Lic. RICARDO PAZ BALLIVIÁN; Gen. Man. Ing. JUAN G. CARRASCO R.

**Instituto para el Desarrollo de la Pequeña Unidad Productiva:** La Paz.

## CHAMBERS OF COMMERCE

**Cámara Nacional de Comercio:** Edif. Cámara Nacional de Comercio, Avda Mariscal Santa Cruz 1392, 1° y 2°, Casilla 7, La Paz; tel. (2) 237-8606; fax (2) 239-1004; e-mail cnc@boliviacomercio.org.bo; internet www.boliviacomercio.org.bo; f. 1890; 30 brs and special brs; Pres. GUILLERMO MORALES FERNÁNDEZ; Gen. Man. RODRIGO AGREDA GÓMEZ.

**Cámara de Comercio de Oruro:** Pasaje Guachalla s/n, Casilla 148, Oruro; tel. and fax (2) 525-0606; e-mail camacor@coteor.net.bo; f. 1895; Pres. ALVARO CORNEJO GAZCÓN; Gen. Man. LUIS CAMACHO VARGAS.

**Cámara Departamental de Industria y Comercio de Santa Cruz:** Calle Suárez de Figueroa 127, esq. Saavedra, 3°, Casilla 180, Santa Cruz; tel. (3) 333-4555; fax (3) 334-2353; e-mail cainco@cainco.org.bo; internet www.cainco.org.bo; f. 1915; Pres. GABRIEL DABDOUB; Gen. Man. Lic. OSCAR ORTIZ ANTELO.

**Cámara Departamental de Comercio de Cochabamba:** Calle Sucre E-0336, Casilla 493, Cochabamba; tel. (4) 425-7715; fax (4) 425-7717; e-mail camcom@pino.cbb.entelnet.bo; f. 1922; Pres. Ing. CARLOS OLMEDO Z.; Gen. Man. Lic. JUAN CARLOS AVILA S.

**Cámara Departamental de Comercio e Industria de Potosí:** Calle Matos 12, Casilla 159, Potosí; tel. (2) 622-2641; fax (2) 622-2641; Pres. JAVIER FLORES CASTRO; Gen. Man. WALTER ZAVALA AYLLON.

**Cámara Departamental de Industria y Comercio de Chuquisaca:** Calle España 64, Casilla 33, Sucre; tel. (4) 645-1194; fax (4) 645-1850; e-mail cicch@camara.scr.entelnet.bo; f. 1923; Pres. MARCO MIHAIC; Gen. Man. Lic. ALFREDO YÁÑEZ MERCADO.

**Cámara Departamental de Comercio e Industria de Pando—Cobija:** Plaza Germán Busch, Casilla 110, Cobija; tel. (3) 842-3139; fax (3) 842-2291; Pres. NEMESIO RAMÍREZ.

**Cámara Departamental de Industria y Comercio de Tarija:** Avda Bolívar 0413, esq. General Trigo, 1°, Casilla 74, Tarija; tel. and fax (6) 642-2737; e-mail metfess@olivo.tja.entelnet.bo; Pres. RENE SILBERMANN U.; Gen. Man. VÍCTOR FERNÁNDEZ ARAMAYO.

**Cámara Departamental de Comercio de Beni:** Casilla 96, Trinidad; tel. (3) 462-2365; fax (3) 462-1400; Pres. EDUARDO AVILA ALVERDI; Sec.-Gen. JOSÉ MAMERTO DURÁN.

**Cámara de Exportadores de Santa Cruz (CADEX):** Avda Velarde 131, Santa Cruz; tel. (3) 336-2030; fax (3) 332-1509; e-mail cadex@cadex.org; internet www.cadex.org; f. 1986; Gen. Man. JUAN MANUEL ARIAS CASTRO.

## INDUSTRIAL AND TRADE ASSOCIATIONS

**Asociación Nacional de Exportadores de Café (ANDEC):** Calle Nicaragua 1638, Casilla 9770, La Paz; tel. (2) 224-4290; fax (2) 224-4561; e-mail andec@caoba.entelnet.bo; internet www.boliviancoffee.bo; controls the export, quality and marketing of coffee producers; Exec. Pres. CARMEN DONOSO DE ARAMAYO.

**Cámara Agropecuaria del Oriente:** 3 anillo interno zona Oeste, Casilla 116, Santa Cruz; tel. (3) 352-2200; fax (3) 352-2621; e-mail caosrz@bibosi.scz.entelnet.bo; f. 1964; agriculture and livestock association for eastern Bolivia; Pres. RICARDO FRERKING ORTIZ; Gen. Man. WALTER NÚÑEZ RODRÍGUEZ.

**Cámara Agropecuaria de La Paz:** Avda 16 de Julio 1525, Casilla 12521, La Paz; tel. (2) 239-2911; fax (2) 235-2308; Pres. ALBERTO DE OLIVA MAYA; Gen. Man. JUAN CARLOS ZAMORANO.

**Cámara Boliviana de Hidrocarburos:** Radial 17 1/2 y Sexto Anillo, Santa Cruz; tel. (3) 353-8799; fax (3) 357-7868; e-mail cbh@cbh.org.bo; internet www.cbh.org.bo; f. 1986; Pres. RAÚL KIEFFER.

**Cámara Forestal de Bolivia:** Prolongación Manuel Ignacio Salvatierra 1055, Casilla 346, Santa Cruz; tel. (3) 333-2699; fax (3) 333-1456; e-mail camaraforestal@cfb.org.bo; internet www.cfb.org.bo; f. 1969; represents the interests of the Bolivian timber industry; Pres. JUAN ABUAWAD CHAHUÁN; Gen. Man. Lic. ARTURO BOWLES OLHAGARAY.

**Cámara Nacional de Industrias:** Edif. Cámara Nacional de Comercio, 14°, Avda Mariscal Santa Cruz 1392, Casilla 611, La Paz; tel. (2) 237-4477; fax (2) 236-2766; e-mail cni@entelnet.bo; internet www.bolivia-industry.com; f. 1931; Pres. ROBERTO MUSTAFÁ; Man. GERARDO VELASCO T.

**Cámara Nacional de Minería:** Pasaje Bernardo Trigo 429, Casilla 2022, La Paz; tel. (2) 235-0623; f. 1953; mining institute; Pres. Ing. LUIS PRADO BARRIENTOS; Sec.-Gen. GERMÁN GORDILLO S.

**Comité Boliviano de Productores de Antimonio:** Edif. El Condor, Batallón Colorados 1404, 14°, Casilla 14451, La Paz; tel. (2) 244-2140; fax (2) 244-1653; f. 1978; controls the marketing, pricing and promotion policies of the antimony industry; Pres. MARIO MARISCAL MORALES; Sec.-Gen. Dr ALCIDES RODRÍGUEZ J.

**Comité Boliviano del Café (COBOLCA):** Calle Nicaragua 1638, Casilla 9770, La Paz; tel. (2) 222-3883; fax (2) 224-4591; e-mail cobolca@ceibo.entelnet.bo; controls the export, quality, marketing and growing policies of the coffee industry; Gen. Man. MAURICIO VILLARROEL.

## EMPLOYERS' ASSOCIATIONS

**Asociación Nacional de Mineros Medianos:** Calle Pedro Salazar 600, esq. Presbítero Medina, Casilla 6190, La Paz; tel. and fax (2) 241-4123; e-mail anmm@caoba.entelnet.bo; f. 1939; asscn of 14 private medium-sized mining cos; Pres. Dr OSCAR BONIFAZ G.; Sec.-Gen. Dr HUGO URIONA.

**Confederación de Empresarios Privados de Bolivia (CEPB):** Calle Méndez Arcos 117, Plaza España, Zona Sopacachi, Casilla 4239, La Paz; tel. (2) 242-0999; fax (2) 242-1272; e-mail cepb@cepb.org.bo; internet www.cepb.org.bo; largest national employers' organization; Pres. ROBERTO MUSTAFÁ SCHNOR; Dir of Social and Institutional Affairs Lic. MAX GASTELÚ ZACONETA.

There are also employers' federations in Santa Cruz, Cochabamba, Oruro, Potosí, Beni and Tarija.

## MAJOR COMPANIES

**ADM-SAO, SA:** Parque Industrial Pl-9, Casilla 1295, Santa Cruz; tel. (3) 346-0888; fax (3) 346-3941; e-mail admsao_gg@admworld.com; internet www.admsao.com; f. 1976; edible vegetable oils and soya-bean products; Pres. ANDRÉS PETRICEVIC; Gen. Man. CARLOS E. KEMPFF; 560 employees.

**Air BP:** Edif. Centro Empresarial Equipetrol, 8°, Avda San Martín 1700, Santa Cruz; tel. (3) 343-7100; fax (3) 343-7200; internet www.bp.com; f. 1999; wholly owned subsidiary of BP (British Petroleum, UK); oil and gas exploration and production.

**Cervecería Boliviana Nacional, SA:** Avda Montes 400, Casilla 421, La Paz; tel. (2) 245-5455; fax (2) 245-5375; e-mail cbn@pacena.com; f. 1920; brewing; Pres. JOHNNY FERNÁNDEZ SAUCEDO; Gen. Man. JUAN MEDINACELLI VALENCIA; 275 employees.

**Cervecería Santa Cruz, SA:** Avda Busch, 3er Anillo Interno, Santa Cruz; tel. (3) 353-5000; fax (3) 353-7070; f. 1952; brewing; Pres. HERMANN WILLE; Gen. Man. RICARDO LÓPEZ ECHEVERRÍA; 220 employees.

**Cervecería Taquiña, SA:** Avda Centenario Final, Zona Taquiña, La Paz; tel. (2) 228-7500; fax (2) 229-6403; f. 1895; brewing; Pres. ERNESTO ASBÚN GIZAUI; Gen. Man. FAUSTINO ARIAS REY; 345 employees.

**Compañía Industrial Azucarera San Aurelio, SA:** Avda San Aurelio, esq. 4° anillo, Zona Sud, Casilla 94, Santa Cruz; tel. (3) 352-2882; fax (3) 352-1182; e-mail ciasa@unete.com; sugar refining and alcohol distillery; CEO EDUARDO GUTIÉRREZ SOSA; Gen. Man. MARIO E. TEJADA; 800 employees.

**Compañía Industrial Maderera Ltda** (CIMAL): Parque Industrial Pesado 10, Santa Cruz; tel. (3) 346-0404; fax (3) 346-1502; f. 1974; sawmill operations; Pres. CRISTÓBAL RODA DAZA; Gen. Man. LARRY HANSLER; 265 employees.

**Compañía Industrial de Tabacos, SA:** Avda Montes 515, Casilla 210, La Paz; tel. (2) 231-9090; fax (2) 235-0104; e-mail citsa@ceibo.entelnet.bo; f. 1934; cigarette manufacturers; Pres. RODOLFO KAVLIN; Gen. Man. JORGE H. PAREJA; 200 employees.

**Compañía Minera del Sur (COMSUR):** Edif. Multicentro Avda 11, Torre B, La Paz; tel. and fax (2) 244-4766; e-mail comsur@caoba.entelnet.bo; f. 1965; mining and processing of lead and zinc ores and precious metals; Pres. JAIME URJEL; 1,456 employees.

**Cooperativa Boliviana de Cemento Industrias y Servicios (COBOCE):** Avda San Martín 558, Cochabamba; tel. (2) 232-5366; fax (2) 422-2485; internet www.coboce.com; f. 1966; manufacture and distribution of cement; Pres. LUIS SÁINZ HINOJOSA; Gen. Man. JAIME MÉNDEZ QUIROGA; 455 employees.

**Corporación Minera de Bolivia (COMIBOL):** Avda Camacho, esq. Loayza, La Paz; tel. (2) 236-7681; fax (2) 236-7483; e-mail edmundo.zogbi@comibol.gov.bo; internet www.bolivia.com/empresas/comibol; f. 1952; state mining corpn; taken over by FSTMB (miners' union) in April 1983; owns both mines and processing plants; Pres. GONZALO MARTÍNEZ; 26,000 employees.

**Drogueria Inti, SA:** Calle Socabaya 242–6, Casilla 1421, La Paz; tel. (2) 240-8282; fax (2) 240-6450; e-mail drogueria@inti.com.bo; internet www.inti.com.bo; f. 1947; manufacture and distribution of pharmaceuticals; Pres. FRIEDRICH OHNES TANZER; Man. DIETER SCHILLING KRIETE; 365 employees.

**Empresa Metalúrgica Vinto (EMV):** Avda Villazón 1966, La Paz; tel. (2) 232-4209; fax (2) 236-3722; f. 1966; state co for the smelting of non-ferrous minerals and special alloys; Exec. Pres. FERNANDO FREUDENTHAL; Gen. Man. RENÉ CANDIA; 950 employees.

**Empresa Minera Inti Raymi, SA:** Corneta Mamani 1989, Edif. EMUSA, La Paz; tel. (2) 279-7676; fax (2) 279-7273; f. 1982; gold mining; owns the Kori Kollo mine; Gen. Man. ALVARO UGALDE; 600 employees.

**Empresa Minera Unificada, SA (EMUSA):** Avda 20 de Octubre 1963, La Paz; tel. (2) 232-1098; fax (2) 235-8462; f. 1946; mining and processing of metal ores; Pres. LUIS MERCADO; 416 employees.

**Empresa Petrolera Chaco SA:** Edif. Centro Empresarial Equipetrol, 6°, Avda San Martín 1700, Santa Cruz; tel. (3) 345-3700; fax (3) 345-3710; internet www.bp.com; f. 1999; wholly owned subsidiary of BP (British Petroleum, UK); oil and gas exploration and production.

**Fábrica Nacional de Cemento, SA (FANCESA):** Pasaje Armando Alba No 80, Sucre; tel. (4) 645-3882; fax (4) 644-1221; e-mail info@fancesa.com; internet www.fancesa.com; f. 1959; manufacturers of cement; Pres. SAMUEL DORIA MEDINA AUZA; Gen. Man. GONZALO ARCE ARANCIBIA; 300 employees.

**Ferrari Ghezzi Ltda:** Avda 6 de Octubre 4671, Lira y Sargento Flores, Casilla 371, Oruro; tel. (5) 524-1340; fax (5) 524-0110; e-mail ferrari@nogal.oru.entelnet.bo; internet www.oru.entelnet.bo/FGL; f. 1935; food production and processing; Gen. Man. CARLOS FERRARI; 500 employees.

**Gravetal Bolivia, SA:** Edif. Banco Nacional de Bolivia, 6°, René Moreno, Santa Cruz; tel. (3) 336-3601; fax (3) 332-4723; e-mail gravetal@gravetal.com; internet www.gravetal.com.bo; f. 1992; production of soyabean oil and soyabean meal; Pres. NOHEMÍ CALERO; Vice-Pres. Dr FELIPE RAFFO; Gen. Man. JORGE ARIAS LAZCANO.

**Industrias de Aceite, SA** (FINO): Carretera al Norte Km 6.5; Casilla 1759, Santa Cruz; tel. (3) 344-3000; fax (3) 344-3020; e-mail fino@fino.com.bo; internet www.fino.com.bo; f. 1944; owned by Grupo Romero of Peru; manufacture of edible vegetable oils; Pres. RONALD CAMPBELL GARCÍA; employees 1,080.

**Ingenio Azucarero Guabira, SA:** Guabira, Montero, Santa Cruz; tel. (92) 20225; fax (92) 20633; e-mail guabira@mail.cotas.com.bo; f. 1956; processing and refining of sugar cane and alcohol distillation; Pres. Ing. MARIANO AGUILERA; Gen. Man. Ing. RUDIGER TREPP; 575 employees.

**La Papelera, SA:** Calle Loayza 178, Casilla 4601, La Paz; tel. (2) 231-3023; fax (2) 231-2906; e-mail lapapelera@papelera.com; internet www.lapapelera.com; f. 1941; paper and plastics manufacturers; Pres. EMILIO VON BERGEN; Gen. Man. JUAN CARLOS ARNEZ; 150 employees.

**Manufactura Boliviana, SA (MANACO):** Casilla 513, Cochabamba; tel. (4) 426-2900; fax (4) 426-3013; e-mail manaco1@manaco.cnb.net; internet www.bata.com; f. 1940; manufacturers of footwear; part of the Bata group; Pres. PABLO DERMIZAKY PEREDO; Gen. Man. CARLOS BUSTAMENTE MORALES; 771 employees.

**Manufacturas Textiles Forno, SA:** Avda Chacaltaya 789, Casilla 881, La Paz; tel. (2) 235-2520; fax (2) 235-1600; production of textiles; Pres. Lic. JAVIER GISBERT; Gen. Man. CARLOS FORNO H.; 465 employees.

**Petrobrás Bolivia, SA:** Carretera Antígua a Cochabamba, Km 2, Calle los Troncos 5, Casilla 6866, Santa Cruz; tel. (3) 358-6030; fax (3) 358-6031; e-mail petrobrasbolivia@petrobras.com.bo; internet www.petrobras.com.br; subsidiary of Petrobrás (Brazil); oil and gas exploration and production; f. 1995; Gen. Man. DÉCIO FABRÍCIO ODDONE DA COSTA.

**Petroquim, SRL:** Zona Franca Winner Módulo, CP 3445, Santa Cruz; tel. (3) 348-8000; fax (3) 348-8200; e-mail info@petroquim.net; internet www.petroquim.net; petrochemicals.

**Pluspetrol Exploración y Producción, SA:** Avda Grigota, esq. Calle Las Palmas, Santa Cruz; tel. (3) 352-0606; fax (3) 354-8080; internet www.pluspetrol.net; oil and gas exploration and production; Chair. and Pres. LUIS ALBERTO REY.

**Prosegur Bolivia:** Calle Macario Pinilla 418, Casilla 6264, La Paz; tel. (2) 243-2920; fax (2) 239-2332; 100% stake sold by Prosegur (Spain) to local investors in June 2005; security products and services.

**Sagic, SA:** Avda Andrés Julio C. Patiño 425, Calacoto, La Paz; tel. (2) 279-4344; fax (2) 279-4347; e-mail sagic@caoba.entelnet.bo; f. 1925; producer and exporter of alcoholic beverages and fresh fruit; Gen. Mans CARLOS CALVO GALINDO, ERNESTO REINAGA.

**Servicio Nacional de Caminos (SNC):** Avda Mariscal Santa Cruz, Edif. Centro de Comunicaciones, La Paz; tel. (2) 235-7200; fax (2) 239-1764; internet www.snc.gov.bo; f. 1961; state co responsible for road construction and maintenance; Exec. Dir JOSÉ MARÍA BACKOVIC; Sec.-Gen. FREDDY VARGAS; 2,100 employees.

**Sociedad Boliviana de Cemento, SA (SOBOCE):** Calle Mercado 1075, 1°, Casilla 557, La Paz; tel. (2) 240-6040; fax (2) 240-7440; e-mail info@soboce.com; internet www.soboce.com; f. 1925; manufacturers of cement; Pres. SAMUEL DORIA MEDINA; Gen. Man. Lic. ARMANDO GUMUCIO; 710 employees.

**Sociedad Comercial e Industrial Hansa Ltda (HANSA):** Edif. Hansa, Calle Yanacocha, esq. Mercado 1004, Casilla 10800, La Paz; tel. (2) 231-4445; fax (2) 237-0397; e-mail cvasquez@hansa.com.bo; internet www.hansa.com.bo; f. 1954; import and trading of telecommunications equipment, hardware, industrial machinery, motor vehicles and mining equipment; Gen. Man. GEORGES PETIT; 370 employees.

## UTILITIES

### Electricity

**Superintendencia de Electricidad:** Avda 16 de Julio 1571, La Paz; tel. (2) 231-2401; fax (2) 231-2393; e-mail superele@superele.gov.bo; internet www.superele.gov.bo; f. 1994; regulates the electricity sector; Superintendent ALEJANDRO NOWOTNY VERA; Gen. Sec. ROLANDO LÓPEZ.

**Alternative Energy Systems Ltd** (Talleres AES): Pasaje Alexander 15, final oeste Avda América, CP 4082, Cochabamba; tel. and fax (4) 442-1777; e-mail aes@bo.net; internet aesbol.freeyellow.com; alternative energy products.

**Compañía Boliviana de Energía Eléctrica, SA (COBEE):** Avda Hernando Siles 5635, Casilla 353, La Paz; tel. (2) 278-2474; fax (2) 278-5920; e-mail cobee@cobee.com; internet www.cobee.com; f. 1925; largest private power producer and distributor, serving the areas of La Paz and Oruro; generated 27.2% of Bolivia's total electricity output in 2002; mainly hydroelectric; Gen. Man. JULIO LEMAITRE SOLARES.

**Compañía Eléctrica Central Bulo Bulo, SA** (CECBB): Avda San Martín 1700, Equipetrol Norte, 6°, Santa Cruz; tel. (3) 346-0314; fax (3) 349-7800; e-mail carlosg@ipolbolivia.com.bo; f. 1999; generator co; 101.2 MW capacity in 2002; Gen. Man. RAMÓN BASCOPE PARADA.

**Compañía Eléctrica Sucre, SA (CESSA):** Calle Ayacucho 254, Sucre; tel. (4) 645-3126; fax (4) 646-0292; e-mail cessa@mara.scr.entelnet.bo; electricity distributor; Gen. Man. Ing. JOSÉ ANAVE LEÓN.

**Cooperativa Rural de Electrificación Ltda (CRE):** Avda Busch, esq. Honduras, Santa Cruz; tel. (3) 336-7777; fax (3) 336-9391; e-mail cre@cre.com.bo; internet www.cre.com.bo; electricity distributor; Gen. Man. Ing. LUIS FERNANDO AÑEZ PEREIRA; Pres. JUAN CARLOS ANTELO SALMON.

**Electropaz:** Avda Illimani 1973, Miraflores, La Paz; tel. (2) 222-2200; fax (2) 222-3756; e-mail covideo@electropaz.com; internet www.electropaz.com; distributor serving La Paz area; Gen. Man. Ing. CARLOS PASTOR AREITI.

**Empresa Corani, SA:** Avda Oquendo 654, Edif. Las Torres Soler I, 9°, Cochabamba; tel. (4) 423-5700; fax (4) 425-9148; e-mail corani@corani.com; internet www.corani.com; generator co; 126 MW capacity in 2002; Pres. MICHAEL DULANEY.

# BOLIVIA

*Directory*

**Empresa Eléctrica Valle Hermoso, SA** (EVH): Calle Tarija 1425, esq. Adela Zamudio, Cala Cala, Cochabamba; tel. (4) 428-6600; fax (4) 428-6838; e-mail vhermoso@pino.cbb.entelnet.bo; generator co; 347.41 MW capacity in 2002; Gen. Man. Ing. Carlos Querejazu Ortiz.

**Empresa de Generación Guaracachi, SA** (EGSA): Avda Brasil y Tercer Anillo Interno s/n, Santa Cruz; tel. (3) 346-0314; fax (3) 346-5888; e-mail central@egsa-bol.com; internet www.egsa-bol.com; generator co; 347.41 MW capacity in 2002; Gen. Man. Ing. Mauricio Peró Diez de Medina.

**Empresa de Luz y Fuerza Eléctrica Cochabamba, SA (ELFEC):** Avda Heroínas 0686, Cochabamba; tel. (4) 425-9410; fax (4) 425-9427; e-mail sugerencias@elfec.com; internet www.elfec.com; electricty distributor; Gen. Man. Ing. José D. Troncoso Esparza.

**Empresa Nacional de Electricidad, SA** (ENDE): Avda Ballivián 503, esq. México, 7°, Casilla Correo 565, Cochabamba; tel. (4) 452-0322; fax (4) 452-0318; e-mail ende@pino.cbb.entelnet.bo; f. 1962; former state electricity company; privatized under the Govt's capitalization programme in 1995 and divided into three arms concerned with generation, transmission and distribution, respectively; Gen. Man. Dr Enrique Gómez D'Angelo.

**Hidroeléctrica Boliviana, SA:** Avda Fuerza Naval 22, La Paz; tel. (2) 277-0765; fax (2) 277-0933; e-mail hb@hidrobol.com; internet www.hidrobol.com; Gen. Man. Ing. Angel Zannier Claros.

## Gas

Numerous distributors of natural gas exist throughout the country, many of which are owned by the petroleum distributor, Yacimientos Petrolíferos Fiscales Bolivianos (YPFB).

**Yacimientos Petrolíferos Fiscales Bolivianos (YPFB):** Calle Bueno 185, Casilla 401, La Paz; tel. (2) 235-6540; fax (2) 239-2596; f. 1936; exploration, drilling, production, refining, transportation and distribution of petroleum; partially privatized in 1996; Pres. Carlos D'Arlach Lema (acting); 4,900 employees.

## Water

**Aguas de Illimani (Suez), SA:** Avda de las Américas 705, La Paz; tel. (2) 221-1181; fax (2) 221-2451; e-mail aguailli@datacom-bo.net; majority stake owned by Suez Group (France); supplies drinking water and sewerage in La Paz and El Alto; Govt commenced annulment of contract in Jan. 2005; the company's operations were due to end in mid-2005; Man. Roberto Bianchi.

## CO-OPERATIVE

**Instituto Nacional de Co-operativas (INALCO):** Edif. Lotería Nacional, 4°, Avda Mariscal Santa Cruz y Cochabamba, La Paz; tel. (2) 237-4366; fax (2) 237-2104; e-mail inalcolp@ceibo.entelnet.bo; f. 1974; Pres. David Ayaviri.

## TRADE UNIONS

**Central Obrera Boliviana (COB):** Edif. COB, Calle Pisagua 618, Casilla 6552, La Paz; tel. (2) 352-426; fax (2) 281-201; e-mail postmast@cob-bolivia.org; internet www.cob-bolivia.org; f. 1952; main union confederation; 800,000 mems; Exec. Sec. Jaime Solares Quintanilla; Pres. Pedro Cruz; Sec.-Gen. Luis Choqueticlla Véliz.

Affiliated unions:

**Central Obrera Departamental de La Paz:** Estación Central 284, La Paz; tel. (2) 235-2898; Exec. Sec. Genaro Torrico.

**Confederación Sindical Unica de los Trabajadores Campesinos de Bolivia (CSUTCB):** Calle Sucre, esq. Yanacocha, La Paz; tel. (2) 236-9433; f. 1979; peasant farmers' union; Sec.-Gen. Felipe Quispe Huanca.

**Federación de Empleados de Industria Fabril:** Edif. Fabril, 5°, Plaza de San Francisco, La Paz; tel. (2) 240-6799; fax (2) 240-7044; Exec. Sec. Alex Gálvez.

**Federación Sindical de Trabajadores Mineros de Bolivia (FSTMB):** Plaza Venezuela 1470, Casilla 14565, La Paz; tel. (2) 235-9656; fax (2) 248-4948; f. 1944; mineworkers' union; Leader Miguel Zubieta; Gen. Sec. Edgar Ramírez Santiestéban; 27,000 mems.

**Federación Sindical de Trabajadores Petroleros de Bolivia:** Calle México 1504, La Paz; tel. (2) 235-1748; Exec. Sec. Neftalymendoza Durán.

**Central Obrera Regional (COR):** El Alto; Exec. Sec. Edgar Patana.

**Confederación General de Trabajadores Fabriles de Bolivia (CGTFB):** Avda Armentia 452, Casilla 21590, La Paz; tel. (2) 237-1603; fax (2) 232-4302; e-mail dirabc@bo.net; f. 1951; manufacturing workers' union; Exec. Sec. Angel Asturizaga; Gen. Sec. Roberto Encinas.

# Transport

## RAILWAYS

**Empresa Nacional de Ferrocarriles (ENFE):** Estación Central de Ferrocarriles, Plaza Zalles, Casilla 428, La Paz; tel. (2) 232-7401; fax (2) 239-2677; f. 1964; privatized under the Government's capitalization programme in 1995; administers most of the railways in Bolivia; holding co for unauctioned former state assets; total networks: 3,608 km (1999); Andina network: 2,274 km; Oriental (Eastern) network: 1,424 km; Pres. J. L. Landívar.

**Empresa Ferrocarril Andino, SA:** Casilla 4350, La Paz; tel. and fax (2) 239-145; e-mail efasa@fca.com.bo; internet www.fca.com.bo; f. 1996; Gen. Man. Eduardo Maclean Avaroa.

**Empresa Ferroviaria Oriental, SA (FCOSA):** Avda Montes Final s/n, Casilla 108, Santa Cruz; tel. (3) 346-3900; fax (3) 346-3920; e-mail fcosa@fcosa.com; internet www.fcosa.com; f. 1996; Gen. Man. Jaime Valencia.

There are plans to construct a railway line with Brazilian assistance, to link Cochabamba and Santa Cruz. There were also plans for the construction of a rail link between Santa Cruz and Mutún on the border with Brazil.

## ROADS

In 2004 Bolivia had some 60,282 km of roads, of which an estimated 3,979 km (6.6%) were paved. Almost the entire road network is concentrated in the Altiplano region and the Andes valleys. A 560-km highway runs from Santa Cruz to Cochabamba, serving a colonization scheme on virgin lands around Santa Cruz. The Pan-American Highway, linking Argentina and Peru, crosses Bolivia from south to north-west. In 1997 the Government announced the construction of 1,844 km of new roads in the hope of improving Bolivia's connections with neighbouring countries.

## INLAND WATERWAYS

By agreement with Paraguay in 1938 (confirmed in 1939), Bolivia has an outlet on the River Paraguay. This arrangement, together with navigation rights on the Paraná, gives Bolivia access to the River Plate and the sea. The River Paraguay is navigable for vessels of 12-ft draught for 288 km beyond Asunción, in Paraguay, and for smaller boats another 960 km to Corumbá in Brazil. In late 1994 plans were finalized to widen and deepen the River Paraguay, providing a waterway from Bolivia to the Atlantic coast in Uruguay. However, work on the project was delayed, owing largely to environmental concerns.

In 1974 Bolivia was granted free duty access to the Brazilian coastal ports of Belém and Santos and the inland ports of Corumbá and Port Velho. In 1976 Argentina granted Bolivia free-port facilities at Rosario on the River Paraná. In 1992 an agreement was signed with Peru, granting Bolivia access to (and the use, without customs formalities, of) the Pacific port of Ilo. Most of Bolivia's foreign trade is handled through the ports of Matarani (Peru), Antofagasta and Arica (Chile), Rosario and Buenos Aires (Argentina) and Santos (Brazil). An agreement between Bolivia and Chile to reform Bolivia's access arrangements to the port of Arica came into effect in January 1996.

Bolivia has over 14,000 km of navigable rivers, which connect most of Bolivia with the Amazon basin.

**Bolivian River Navigation Company:** f. 1958; services from Puerto Suárez to Buenos Aires (Argentina).

## CIVIL AVIATION

Bolivia has 30 airports, including the two international airports at La Paz (El Alto) and Santa Cruz (Viru-Viru).

**Dirección General de Aeronáutica Civil:** Avda Mariscal Santa Cruz 1278, Casilla 9360; La Paz; tel. (2) 237-4142; e-mail dgacbol@ceibo-entelnet.bo; internet www.dgac.gov.bo; f. 1947; Dir-Gen. Orlando Montoya Koëster.

**AeroSur:** Calle Colón y Avda Irala 616, Casilla 3104, Santa Cruz; tel. (3) 336-4446; fax (3) 333-0666; e-mail aerosur@aerosur.com; internet www.aerosur.com; f. 1992 by merger of existing charter cos following deregulation; privately owned; Pres. Oscar Alcocer; Gen. Man. Fernando Prudencio.

**Lloyd Aéreo Boliviano, SAM (LAB):** Casilla 132, Aeropuerto 'Jorge Wilstermann', Cochabamba; tel. (4) 425-1270; fax (4) 425-0766; e-mail presidencia@labairlines.com.bo; internet www.labairlines.com; f. 1925; privatized under the Government's capitalization programme in 1995; jtly owned by Bolivian Govt (48%), and private interests (52%); operates a network of scheduled services to 12 cities within Bolivia and to 21 international destinations in South America, Central America and the USA; Pres. Ing. Ernesto Asbún; Gen. Man. José Rodríguez.

**Transportes Aéreos Bolivianos (TAB):** Casilla 12237, La Paz; tel. (2) 237-8325; fax (2) 235-9660; f. 1977; regional scheduled and charter cargo services; Gen. Man. LUIS GUERECA PADILLA; Chair. CARLO APARICIO.

**Transportes Aéreos Militares:** Avda Montes 738, La Paz; tel. (2) 237-9286; internal passenger and cargo services; Dir-Gen. REMBERTO DURÁN.

## Tourism

Bolivia's tourist attractions include Lake Titicaca, at 3,810 m (12,500 ft) above sea-level, pre-Incan ruins at Tiwanaku, Chacaltaya in the Andes mountains, which has the highest ski-run in the world, and the UNESCO World Cultural Heritage Sites of Potosí and Sucre. In 2002 382,185 foreign visitors arrived at hotels in Bolivian regional capitals. In 2001 receipts from tourism totalled US $156m. Tourists come mainly from South American countries, the USA and Europe.

**Asociación Boliviana de Agencias de Viajes y Turismo:** Edif. Litoral, Avda Mariscal Santa Cruz 1351, Casilla 3967, La Paz; f. 1984; Pres. EUGENIO MONROY VÉLEZ.

**Viceministerio de Turismo:** Avda Mariscal Santa Cruz, 16°, Casilla 1868, La Paz; tel. (2) 235-8312; fax (2) 237-4630; Vice-Minister of Tourism VÍCTOR HUGO ORDÓÑEZ.

## Defence

In August 2004 Bolivia's Armed Forces numbered 31,500 (there were plans to increase this to 35,000): Army 25,000 (including 18,000 conscripts), Navy 3,500, Air Force 3,000. Military service, lasting one year, is selective.

**Defence Expenditure:** Expenditure on defence by the central Government in 2004 was an estimated 1,000m. bolivianos.

**Commander-in-Chief of the Armed Forces:** Adm. MARCO ANTONIO JUSTINIANO ESCALANTE.

**Chief of Staff of the Armed Forces:** Gen. CARLOS DELFÍN MESA (acting).

**General Commander of the Army:** Gen. MARCELO ANTEZANA RUIZ (acting).

**General Commander of the Air Force:** Gen. ANDRÉS QUIROZ RICO (acting).

**General Commander of the Naval Forces:** Vice-Adm. JORGE BOTELHO MONJE (acting).

## Education

Education in Bolivia is free and, where possible, compulsory between the ages of six and 14 years. In 1990 the total enrolment at primary and secondary schools was equivalent to 77% of the school-age population (81% of boys; 73% of girls). In that year enrolment at primary schools included an estimated 91% of children in the relevant age-group (95% of boys; 87% of girls), while the comparable ratio for secondary enrolment was only 29% (32% of boys; 27% of girls). In 1991 the total enrolment at primary and secondary schools was equivalent to 79% of the school-age population. There are eight state universities and two private universities. Expenditure on education by the central Government in 2001 amounted to 19.9% of government expenditure, some 2,796.9m. bolivianos.

## Bibliography

For works on South America generally, see Select Bibliography (Books)

Crabtree, J. *Patterns of Protest: Politics and Social Movements in Bolivia*. London, Latin America Bureau, 2005.

Farcau, B. *The Chaco War: Bolivia and Paraguay, 1932–1935*. New York, NY, Praeger Publrs, 1996.

Healy, K. *Llamas, Weavings and Organic Chocolate: Multilateral Grassroots Development in the Andes and Amazon of Bolivia*. Notre Dame, IN, University of Notre Dame Press, 2000.

James, D. (Ed.). *The Complete Bolivian Diaries of Che Guevara*. New York, NY, Cooper Square Press, 2000.

Jemio, L. C. *Debt, Crisis and Reform in Bolivia: Biting the Bullet*. The Hague, Institute of Social Studies, 2001.

Klein, H. S. *Bolivia: The Evolution of a Multi-Ethnic Society*. New York, NY, Oxford University Press, 1982.

*A Concise History of Bolivia*. Cambridge, Cambridge University Press, 2003.

Lehman, K. D. *Bolivia and the United States: A Limited Partnership*. Athens, GA, University of Georgia Press, 1999.

López Levy, M. *Bolivia*. Oxford, Oxfam Publishing, 2001.

Menzel, S. H. *Fire in the Andes: US Foreign Policy and Cocaine Politics in Bolivia and Peru*. Lanham, MD, University Press of America, 1996.

Painter, J. *Bolivia and Coca: A Study in Dependency*. London, Lynne Rienner Publrs, 1994.

Rhyne, E. *Mainstreaming Microfinance: How Lending to the Poor Began, Grew and Came of Age in Bolivia*. Bloomfield, CT, Kumarian Press, 2001.

Saldana, R. *Fertile Ground: Che Guevara and Bolivia*. New York, NY, Pathfinder Press, 2001.

Siekmeier, J. F. *Aid, Nationalism and Inter-American Relations: Guatemala, Bolivia and the United States 1945–1961*. Lewiston, NY, Edwin Mellen Press, 1999.

Stephnson, M. *Gender and Modernity in Andean Bolivia*. Austin, TX, University of Texas Press, 1999.

Villegas, H. *Pombo, A Man of Che's Guerrilla: With Che Guevara in Bolivia 1966–68*. London, Pathfinder Press, 1997.

# BRAZIL
## Geography

### PHYSICAL FEATURES

The Federative Republic of Brazil is the largest country in Latin America, occupying much of the east of South America. Nearly one-half of the continent is in Brazil. Its longest border is with Bolivia (3,400 km or 2,111 miles), which lies to the south-west, then Paraguay (1,290 km of border). Brazil then thrusts southwards, between the Atlantic coast and an extension of Argentina (1,224 km) to the west, ending in a 985-km border with Uruguay to the south. In the north is the Guianan coast (from east to west, French Guiana—673 km, Suriname—597 km and Guyana—1,119 km) and then Venezuela (2,200 km). Colombia pushes south to form the northern part of the western border (1,643 km), in north-west Brazil, with Peru also lying to the west, beyond a 1,560-km frontier. Of all the South American territories, only Ecuador and Chile do not have borders with Brazil. Brazil has an uncontested territorial dispute with Uruguay over small river islands in the Quarai (Cuareim) and the Arroio Invernada (Arroyo de la Invernada). Somewhat smaller than the USA, Brazil covers an area of 8,547,404 sq km (3,300,170.9 sq miles), making it the fifth largest country in the world.

Brazil, which also includes a number of offshore islands and islets, has 7,491 km of coastline, formed where South America bulges eastwards and then begins to taper south. The Amazon enters the Atlantic on the north-eastern coast of South America, and the mouth of the river is complicated by many channels and islands (the largest is Marajó), as well as swamps, mangroves and flooding, features common until higher land begins in eastern Brazil. Here the north-eastern highlands make the coast more defined, smoother and drier, with stretches of dunes, although there are still occasional mangroves and lagoons beyond São Roque cape, where the eastern bulge of the continent turns south. To the south-west, beneath the south-eastern highlands, the shore is varied by sandy spits and beaches, as well as lagoons and marshes, but for 1,000 km beyond Rio de Janeiro the coastal plains are reduced to occasional patches, as the highlands often come sheer to the sea. Most of the country's territory, however, is defined by political rather than natural borders, although in the north the Guianan highlands help establish the line. These forested heights cover only 2% of the country and are generally considered to be part of the Amazon basin, but they also include Brazil's highest mountain, Pico da Neblina (3,014 m or 9,892 ft), which lies in the north-west, on the border with southern Venezuela. The Amazon basin itself, which accounts for about one-third of the country, spreads across the north of Brazil, pushing it west, deeply into the heart of the continent. Brazil also shares the basin of the much smaller River Plate (Río de la Plata) system, in the south, while coastal plains constitute the only other area of lowland in the country. In the midst of these other features, and in marked contrast to the dense jungles of the Amazonian lowlands, are the open Brazilian highlands, an eroded plateau of jumbled mountain ranges and river valleys, running from the easternmost end of the country towards the south-west, generally just inland from the coast. However, once human and economic geography is taken into account, Brazil is usually described as consisting of five regions: the north (most of the Amazon basin and the Guianan highlands—45% of the territory, but only 7% of the population); the north-east (essentially the eastern bulge, the north-eastern end of the Brazilian highlands—the area first settled by Europeans and their African slaves); the south-east (the other, higher end of the highlands—11% of the territory, but 43% of the population); the south (the smallest region, temperate in climate); and, finally, the landlocked centre-west, sparsely populated, but including the capital city, Brasília (this region is a transitional region, including the edges of the Amazonian plains to the north, the Brazilian highlands to the south and east, and the upper lowlands of the River Plate basin in the west). All these vast territories include surprisingly little fertile land and, although the range of crops produced is wide, relatively small amounts of land are cultivated. Grasslands are used extensively for pasture.

The main lowland area of Brazil is the Amazon basin, which is flat, or gently rolling, seldom exceeding 150 m above sea level and covered in the largest rainforest in the world. The Amazon and its tributaries are prone to seasonal flooding, inundating the level, swampy areas known as varzeas. Similarly, the headwaters of the Paraná and Paraguay can flood the important wetlands of the Pantanal, where the hills of the Brazilian highlands yield to the plains of the River Plate basin (the Chaco spreads through Paraguay to the south). The Pantanal forms the western end of the centre-west region, dividing the Amazonian north from the south-eastern highlands. Finally, there are the coastal plains, extending for thousands of kilometres from the north-east to the border with Uruguay. Up to 60 km in width in the north-east, the coastal plains are negligible south of Rio de Janeiro, where the Serra do Mar form a sharp edge along the shore. The plains only broaden again in the far south, as they widen towards the Pampas and, inland, the Chaco.

The Brazilian highlands are a huge block of geologically ancient rocks, falling away to the north-east and north. In the south-eastern region of the country the highlands consist of a complex mass of ridges and ranges, some dropping steeply into the sea, and generally with elevations of around 1,200 m, although the highest summits reach about 2,800 m. The main ranges include the Serra do Mar and the Serra da Mantiqueira. Inland from the coastal ranges is a broad plateau, hills lowering themselves into the centre-west and towards the Amazonian lowlands. Likewise, the highlands fall away to the north-east, as they parallel the coast and form the solid core of the eastern bulge of South America. Here are low, rolling hills, with the semi-arid interior known as the Sertão. As mentioned above, the north side of the Amazon basin is defined by the highlands separating Brazil from the Guiana coast and from the drainage area of the Orinoco. Out of these highlands, but mainly from the Andes to the west, flow the main rivers of Brazil, which can be grouped in eight systems, together carrying about one-fifth of the world's running water. The Amazon itself is the second longest river in the world, after the Nile in Africa, at 6,516 km (most of it flowing through Brazil, but still navigable into Peru). However, some of its tributaries are mighty rivers in themselves, notably the Tocantins, which joins the Amazon from the south, near its mouth. The second river system is that of the south-draining Paraná, which empties into the Plate (between Uruguay and Argentina), draining much of the south, south-east and centre-west of Brazil. The principal river of the eastern plateau region is the São Francisco, which flows north through the highlands until it turns east into the Atlantic.

The vegetation of this varied landscape is diverse, ranging from tropical and temperate woodland, through savannah and often swampy grasslands, to semi-arid scrub. Many of the species sheltering in these environments are still unknown, although already threatened, particularly by deforestation, and also by mining and industrial pollution. Wooded areas account for 58% of Brazil's total, but it is the great jungle of the Amazon, the largest rainforest (despite massive and continuing encroachments) in the world, covering two-fifths of Brazil's territory, which dominates. The luxuriant vegetation hosts a massively varied array of ecosystems and a good proportion of the many species found in Brazil. There are almost 400 species of mammal found in Brazil—such as endangered jaguars, rare bush dogs, anteaters, deer (for instance, the endangered Pantanal deer), monkeys, and tapirs—but, more impressively perhaps, the country has among the most diverse populations of birds (1,635 species) and amphibians (502 species), as well as 1,500 species of freshwater fish, of which more than two-thirds are found in the Amazon basin. There are over 100,000 invertebrate species, of which 70,000 are insects, although it should be noted that, given the scale, all these figures are estimates. Much of this wealth is threatened, the Amazon increasingly affected by deforestation since the 1970s (much of the eastern and southern uplands have already been denuded of their widest variety since 1500), although some 3.3m. sq km of rainforest remain. One tree species should be mentioned, as it gave the country its name—the pau brasil or brazilwood tree provided dyewood, the first commodity to be exported by Europeans from Brazil. Grasslands have also been economically exploited (pasture covers 22% of the country), with both savannah and rich wetlands used for ranching—in areas such as the Pantanal, the Sertão, the Cerrado (in the centre-west, where the rainforest yields to a more open and varied landscape of trees and bushes, as well as grassland) and the Campos of the far south.

## CLIMATE

Over such a large area, the climate is obviously extremely varied. All but the extreme north (the Equator passes across the mouth of the Amazon) and the far south lies within the Tropic of Capricorn (which passes through São Paulo). Most of the country has annual average temperatures of over 22°C (72°F), but in the far south and in the high country it occasionally falls lower than this, with seasonal variations also more pronounced. Northern Brazil largely has a tropical wet climate, with much rainfall and virtually no dry season. Temperatures average 25°C (77°F), varying more between night and day than by time of year, and average rainfall is about 2,200 mm per year. It is oppressively humid here in the Amazon. In central Brazil rainfall (1,600 mm annually) is more seasonal, typical for a savannah area (80% falls in summer, October–March). The interior north-east, or Sertão, is even more pronouncedly seasonal in its little rainfall (only 800 mm per year—almost all falling within only two or three months), although precipitation is very liable to fail completely, causing drought (temperatures are extremely hot, able to exceed 40°C—104°F). In the south-east the tropical climate is moderated by altitude, with winter temperatures averaging below 18°C (64°F) and annual rainfall at 1,400 mm (falling mainly in the summer). The south has a subtropical climate, verging on the temperate, with cool winters that can produce a few frosts and even some snow at higher elevations. Annual rainfall is 1,500 mm, fairly evenly spread throughout the year.

## POPULATION

Brazil is the only lusophone country in the Western hemisphere, owing to the 1494 Treaty of Tordesillas, which modified a papal arbitration between Portugal and Spain of the previous year. Most of the 'New World' was accorded to Spain, but moving the original Line of Demarcation further west ensured that Portugal gained territory here too. By the 1777 Treaty of Ildefonso (confirming principles established in 1750), Portugal also gained vast territories west again of the 1494 Line. As a result, modern Brazil is a unique blending of Portuguese settlers and their forcibly imported African labour with native Amerindians and later waves of immigration (usually from Europe). There is also a noticeably more relaxed attitude to race in Brazil than in many countries, with the Portuguese joined by many other European settlers and the African slaves taken not only from West Africa (as in much of the Caribbean), but also from Congo, Angola and Mozambique. About 55% of the population are classed as white (principally, 15% of mainly Portuguese descent, 11% Italian, 10% Spanish and 3% German), 22% of mixed white and black descent (mulatto), 12% Mestizo (mixed white and Amerindian) and 6% black. Amerindians now account for only about 1% of the population, with some other minority immigrant communities (such as those of Arab or Japanese descent) also enjoying a similar size. There is considerable regional variation in the distribution of the ethnic groups, with over 70% black or mulatto in the north-east, the old slave-worked region, while the urban south-east is 66% white and only 33% black, and the south is 82% white, being settled after the slave era by a variety of groups from Europe. There is some element of standing in racial definition, with the élite generally white, but those of mixed race particularly enjoy considerable social mobility. Portuguese, with some regional variation, is the most widely used as well as the official language, although German and Italian, for instance, are still used in parts of the south (English and French tend to be the main second languages of the educated). Local dialects have incorporated Amerindian and African words. There are still over 100 indigenous Amerindian languages, of which the main ones belong to the Tupí, Gê, Arawak, Carib (Garib) and Nambicuara groups. Caribs and Arawaks are the main peoples of the north, the Tupí-Guaraní of the east coast and the Amazon river valley, the Gê of eastern and southern Brazil and the Pano in the west. Most Amerindians survive in the north and west. Virtually all groups, excluding the more remote tribes of the Amazon and more recent immigrants from non-Christian backgrounds, tend to be at least nominally Roman Catholic (79%), making Brazil the largest Roman Catholic country in the world. Christian adherence is sometimes supplemented by parallel belief systems. Perhaps 6% of the population are classified as Protestant Christian. Some of the urban middle classes are also followers of Spiritualism (involving reincarnation, communication with the dead, etc.), while Afro-Brazilian blendings have produced religions such as Candomblé, Macumba and Umbanda (the identity of Christian saints with African deities, the distinctive use of music and a belief in spirit possession, etc.), relatively widespread in areas such as the north-east. There are also some Buddhists and Jews, as well as a number of continuing, if with few adherents, aboriginal belief systems.

The total population was estimated at 181.6m. in mid-2004, making Brazil the second most populous country in the Americas after the USA. The social legacy of a plantation society means there is still much inequality and poverty, especially in the countryside, although after massive urban migration the rural population accounts for only about one-fifth of the total. One-third of Brazilians live in cities of over 1m., the largest being São Paulo, the biggest city in South America and the country's main industrial centre (10.4m. at the 2000 census—results for the municipal area). The second most populous city is the old capital (1763–1960), Rio de Janeiro, which remains an important port and the commercial centre of the country (5.9m.). Both cities were founded on the scant coastal plains of the south-east region, Rio de Janeiro further east up the coast. The cities, with the Federal District, are the most densely populated parts of the country. Brasília, the federal capital since 1960, was purposely located away from the coast (four-fifths of Brazilians live within 350 km of the sea) and from the crowded south-east, nearer the centre of the country (2.3m. in 2004). Salvador (2.4m. in 2000) and Fortaleza (formerly Ceará—2.1m.), both in the north-east, and Belo Horizonte (2.2m.), in Minas Gerais, to the north of Rio de Janeiro, are all more populous than the capital. The other cities of over 1m. are Curitiba and Porto Alegre in the south, Recife (formerly Pernambuco) on the north-east coast, the

# History

Dr DAVID FLEISCHER

Based on an earlier text by Dr PAUL CAMMACK

Brazil was Portugal's only American colony, and survived as a single unit after independence to become Latin America's only Portuguese-speaking nation. During the period of Portuguese colonial rule, millions of Africans were forcibly transported to Brazil to work as slaves. As a result of the flight of the Portuguese royal family to the colony in 1808, Brazil attained independence in 1822 under Prince Pedro, who became Emperor Pedro I of Brazil. The country remained a monarchy until 1889, when a republic was declared, one year after the abolition of slavery. A federalist Constitution was adopted in 1891. This First Republic became a decentralized, federal regime and endured until it was overthrown in 1930, in a revolution that brought Dr Getúlio Vargas to power. Vargas oversaw the introduction of a new and more centralized Constitution in 1934, but established a military-backed dictatorship in 1937 rather than retire from the presidency following elections held in 1938. During this 15-year period, Vargas recentralized the political system, initiated state reforms, encouraged import substitution and developed the steel industry. The new regime (the 'Estado Novo') lasted until 1945, when the military withdrew its support and forced Vargas from power.

## THE RESTORATION OF DEMOCRACY

The restoration of democracy with the new Constitution of 1946 gave most Brazilians their first experience of political involvement and inaugurated nearly two decades of continuous but unstable party competition. The period was dominated by Vargas, now presenting himself, with some success, as a champion of the masses, and his heirs. They were grouped in the broadly conservative, rural-based Partido Social Democrático (PSD) and the leftist and increasingly influential, urban-based Partido Trabalhista Brasileiro (PTB), and opposed by the liberal União Democrática Nacional (UDN) and at times by the Partido Social Progressista. Vargas was elected to the presidency in 1950, but committed suicide in 1954, when the military demanded his resignation.

Over the next 10 years Brazil gradually declined into a state of acute political crisis. Industrialization proceeded rapidly under the presidency of Juscelino Kubitschek (1956–61), but the economic strains that were created, and the political tensions arising out of urbanization and swift social change proved too great for the fragile political system. As pressure mounted for social and structural reform, the UDN secured the presidency for the first time, through an independent, Jânio Quadros, at elections in October 1960. Within seven months of taking power in January 1961, Quadros resigned, alleging lack of support from the Congresso Nacional (National Congress), and the country was plunged into crisis. He was succeeded by the Vice-President, PTB leader João Goulart, after the military had forced the Congresso to change from a presidential to a parliamentary system, with Tancredo de Almeida Neves as Prime Minister.

Under pressure from the left to adopt a radical programme, Goulart at first hesitated, but, after regaining presidential powers in a referendum held in January 1963, and lacking a majority in the Congresso Nacional, he moved to respond to such demands by decree. Before the ensuing radicalization was far advanced, the military intervened, seizing power on 1 April 1964. The military coup brought an end to two decades of fragile democracy, marked by a refusal on the part of the privileged élites to countenance any degree of social reform, and a general failure on the part of political parties to establish themselves as independent actors, rather than as clientelistic groupings reliant upon the patronage powers of the state.

## MILITARY RULE AFTER 1964

The 21-year military regime was a curious hybrid, quite distinct from the military governments in Argentina, Chile, Peru and Uruguay. The armed forces concentrated power in their own hands, but kept the Congresso Nacional in session (except for an extended period in 1968–69 and briefly in 1977) while denying it autonomy, and held regular elections for the Congresso, state legislatures and local mayors and city councils. Successive purges removed all but the most moderate opponents of the regime. The five military Presidents were vested with power to govern by decree, and the parties existing in 1964 were replaced by a two-party system in 1966, with pro-Government forces congregating in the majority Aliança Renovadora Nacional (ARENA) and the remaining opposition members grouped in the Movimento Democrático Brasileiro (MDB).

The dictatorship was at its most harsh between 1968 and 1974, particularly under Gen. Emílio Garrastazú Médici. The already highly authoritarian Constitution approved in 1967 was heavily amended in 1969 to strengthen further the power of the military executive, and the elections of 1970, held in conditions which made meaningful competition impossible, gave emphatic majorities to the government party. Throughout this period, the retention of a system of political parties combined with a concentration of powers of decision in the military executive, pushed to extremes the tendency for the governing party to act as a clientelistic machine. In 1974 Gen. Ernesto Geisel (1974–79), relying on the appeal of limited liberalization and Brazil's burst of economic growth after a period of recession had ended in 1967, allowed more open elections. The electoral system protected the government majority, but the unexpected gains that were made by the MDB, particularly in the elections for one-third of the Senado Federal (Federal Senate), may be seen in retrospect as marking the beginning of the long retreat of the military from power. Following the 1973 petroleum crisis, Brazil's economy stagnated and Geisel's decision to allow more open elections in November 1974 provided political legitimacy to, and bolstered support for, his regime.

From 1974 onwards the military lacked a natural majority in the country, but persisted in holding elections on schedule, seeking to maintain their hold on power by a series of expedient measures such as the indirect election (to all intents and purposes the appointment) of one-third of the Senado Federal in 1978. This failed to conceal either their unpopularity or their waning self-confidence. Geisel's successor, Gen. João Baptista de Figueiredo (1979–85), was the beneficiary of a decision to prolong the presidential mandate by one year, but it fell to him to oversee the departure of the military from power.

An attempt to regain the initiative by dissolving the two-party system in 1979 in a bid to divide the opposition and to halt the advance of the reorganized and increasingly effective MDB, failed in its objective. ARENA was reconstituted, shorn of some of its moderate elements, as the Partido Democrático Social (PDS), while a number of new opposition parties appeared, led by the renamed Partido do Movimento Democrático Brasileiro (PMDB), which was reduced to half the strength of its predecessor, the MDB. Most prominent among the new parties were the Partido dos Trabalhadores (PT), led by Luiz Inácio ('Lula') da Silva, a labour-union organizer, and the Partido Democrático Trabalhista (PDT), led by Goulart's brother-in-law, Leonel de

Moura Brizola. The PDT was a splinter group from the reincarnation of the old PTB, led by Ivete Vargas, a niece of the former dictator. A centrist Partido Popular (PP) also briefly appeared, led by veteran politician Tancredo Neves, formerly of the pre-coup PSD. When the Government introduced legislation to prevent cross-party voting and coalitions in the 1982 elections, in a further attempt to stem the tide of opposition gains, the PP dissolved itself and the majority of its members joined the PMDB.

## THE WANING OF MILITARY AUTHORITY

By 1982 five years of social mobilization and protest, focused primarily on the factories and working-class communities of São Paulo and co-ordinated as much by the Roman Catholic Church as by the parties and labour unions, had put the military on the defensive. The 1982 elections gave the governorships of the 10 leading states to the opposition (with direct gubernatorial elections for the first time since 1965) and would have given the combined opposition forces a majority in the Câmara dos Deputados (Chamber of Deputies) had the PTB not formed a coalition with the PDS. In April 1984 a substantial vote by the Câmara in favour of an amendment to the Constitution, introducing direct elections, failed to gain the required two-thirds' majority. However, the military executive lost control of its own party, and the official nomination for its presidential candidate went to the civilian industrialist and financier, Paulo Salim Maluf, former Governor of São Paulo. His aggressive style provoked a division of the government party and led, eventually, to the formation of the Partido da Frente Liberal (PFL). This grouping gave its support to the PMDB candidate, Tancredo Neves, and secured the vice-presidential candidacy for José Sarney, erstwhile leader of the pro-Government PDS. As a result, the electoral college that was to elect the President became opposition-controlled. Lacking other options, the military accepted Neves' victory, achieved by a massive 300-vote majority in the 686-member college when voting took place on 15 January 1985. The transfer of power took place as scheduled on 15 March, but Neves, then 74 years old, required emergency surgical treatment on the eve of his accession. As a result, José Sarney was sworn in as acting President. He assumed full presidential powers after Neves' death in April

## THE RETURN TO DEMOCRACY, 1985

Brazil returned to competitive liberal democracy in challenging circumstances. Economic growth was faltering as inflation spiralled far beyond the levels it had reached when the military intervened in 1964, but somewhat alleviated by monthly monetary correction of contracts and salaries. Socially, the strains of rapid industrialization and urbanization over previous decades had been exacerbated by the sharp worsening of income distribution over the period of military rule, leading to growing malnutrition and absolute poverty in urban and rural areas alike. Amid a general recognition of the need for political and social reform and substantial economic redistribution, civilian politicians were under pressure to address a range of issues that had been neglected during the military period. Initial suspicion arising from Sarney's recent links with the armed forces limited his popular appeal, and his relations with the dominant PMDB proved to be difficult. However, he pledged to implement the programme that Neves had proposed, including the convocation of a National Constituent Assembly and the introduction of direct elections to the presidency, and he reached a peak of popularity in 1986, as a consequence of the temporary success of the Cruzado Plan, an anti-inflation, price-wage freeze programme announced in February. The election, in November, of the Congresso Nacional marked the first stage in the transition to the adoption of a new constitution and a full return to democracy; the election was also a zenith for the PMDB, which won 22 of the 23 state governorships and absolute majorities in both houses of the Congresso.

However, the apparent economic and political success of the transition to democracy proved short-lived. The key measures of the Cruzado Plan were abandoned immediately after the November 1986 elections, and it collapsed altogether in early 1987. Debates over the new constitution were dominated by rivalry between the President and the unicameral National Constituent Assembly, and conflict over the extent to which commitments to social reform should be written into the document. Part of the problem lay in the changed character of the once reformist PMDB. Since its establishment as the leading opposition force in the 1970s, it had attracted the support of conservatives who abandoned the PDS as its prospects faded, and embraced the PMDB. Thus, by 1986 the PMDB was no longer a party committed to genuine reform. A new Constitution was finally promulgated in October 1988. It provided for a five-year presidential term and adopted a conservative stance with regard to land reform in particular. The PMDB split in June 1988, with many of its founders moving into a new social democratic party, the Partido da Social Democracia Brasileira (PSDB), including the political exile and São Paulo senator, Fernando Henrique Cardoso, senator Mário Covas and federal deputy José Serra. Signs of a serious challenge from the left emerged in the municipal elections of November 1988, at which da Silva's hitherto small PT gained control of the city of São Paulo, as well as 37 other major towns and cities across the country, while the PDT was successful in Rio de Janeiro.

## THE RISE AND FALL OF COLLOR DE MELLO

The first direct presidential election since 1960 was held on 15 November 1989. In the first round of voting Fernando Collor de Mello, leader of the tiny Partido de Reconstrução Nacional (PRN) took 30.5% of the valid votes cast, compared with da Silva's 17.2%, with Brizola coming third with 16.5%. As the Constitution required an absolute majority, a second round of voting was held, just over one month later, at which Collor defeated da Silva, obtaining 53% of the votes cast compared with 47% for da Silva.

President Collor de Mello introduced a radical reform programme aimed at reducing public employment, lowering government expenditure and liberalizing the economy. At the same time, in 1991, Brazil, in conjunction with Argentina, Paraguay and Uruguay, began to establish the free trade zone known as Mercosul (Mercado Comum do Sul or, in Spanish, Mercado Común del Sur—Mercosur). However, the initial results of these reforms were disappointing, with a sharp recession in 1990 followed by a resumption of inflation in 1991. Collor de Mello replaced his entire economic team as a result, but found his popularity and congressional support dwindling, even after the poor performance of the left at legislative elections held in October 1990, where the PRN also achieved disappointing results.

President Collor de Mello's position deteriorated further in 1992 as he failed to persuade the PSDB and other congressional parties to support a new reform programme and he resorted to governing by decree. The Câmara dos Deputados subsequently voted to bring forward a referendum on changing to a parliamentary system of government. However, Collor intervened and in December 1991 the Senado defeated the measure. After a series of corruption scandals emerged in early 1992, causing further ministerial resignations, in May Collor de Mello's brother, Pedro, began a national campaign against the President's campaign manager, treasurer and confidant, Paulo César Farias. The ensuing succession of corruption scandals soon involved the President himself and ultimately led to his downfall. After a joint congressional investigating committee reported in favour of impeachment, the Câmara voted to commence impeachment proceedings and Collor de Mello was suspended from office on 28 September. The Senado later convicted Collor de Mello of 'political crimes' and he was impeached on 30 December. The Vice-President, a relatively unknown nationalist politician from Minas Gerais, former senator Itamar Franco, who had been appointed acting President in September, was confirmed as President on the same day.

## THE TRANSITION TO DEMOCRACY IN CRISIS, 1993–94

With the coming to power of Franco, the crisis surrounding the attempted transition to democracy deepened further. A number of serious problems, some with deep historical roots, made decisive action imperative, but Franco lacked both the necessary political experience and the organized political support that would have made effective government possible. Inflation continued to worsen and threatened to spiral entirely out of control. Economic growth, also, continued to falter, while levels of foreign investment dwindled as international confidence in the Brazilian economy declined further. On the political side, pressure was already mounting for reform of the 1988 Constitution, while public discontent with politicians in general, brought the future of the transition into question. Most seriously, Brazil's political party system, chronically weak and prone to fragmentation throughout the republican period, appeared once again to be in terminal decline. No party was able to elect a President, to provide majority support in the Congresso Nacional for the resulting administration, nor to exert sufficient authority over powerful élites to achieve either economic stability or social reform. These difficulties combined to make the situation of the Franco Government and the political system as a whole extremely precarious by the beginning of 1994.

Economic affairs were dominated by the effort to introduce a credible programme of economic adjustment backed by fiscal reforms. There was an urgent need to dramatically reduce inflation, restore growth, balance the budget and address the pressing problem of the steadily worsening distribution of income. The new Minister of Finance, Fernando Henrique Cardoso, therefore, sought repeatedly, but largely unsuccessfully, to introduce a series of wide-ranging structural reforms to the fiscal system. By early 1994 he had succeeded in introducing a limited fiscal reform package incorporating selected expenditure reductions and the establishment of a Social Emergency Fund. The compulsory transfer of large resources from the federal to the state and local governments was also reduced, but remained a serious, and constitutionally embedded, obstacle to the achievement of a balanced budget.

## THE PRESIDENTIAL ELECTION AND THE REAL PLAN, 1994

With the Government weak and the Congresso Nacional discredited, the immediate beneficiaries were the PT and da Silva, who emerged as the left-wing's leading contender in the presidential election that was scheduled for October 1994. A consensus eventually emerged in the Government, with business and army circles favouring the candidacy of centrist Minister of Finance Cardoso. The final stages of the Franco presidency were dominated by the long-awaited and much-postponed programme of economic reform, the centrepiece of which was the introduction, on 1 July 1994, of a new currency, the real, pegged to the US dollar. The initial impact of the measures, known as the Real Plan appeared broadly positive as inflation was dramatically reduced and the real incomes of poorer groups increased. This was a very powerful election instrument, and at the election, on 3 October, Cardoso secured the presidency without the need to proceed to a second ballot. PSDB governors were also elected in the important central states of São Paulo, Minas Gerais and Rio de Janeiro. The election of Cardoso and his allies appeared to demonstrate that Brazilian voters had opted for a path of continuity with moderate reform rather than radical change or conservative reaction.

## CARDOSO'S FIRST TERM, 1995–98

When President Cardoso assumed office on 1 January 1995 he had committed himself to a series of key constitutional reforms, aimed at accelerating the modernization of the economic and social fabric and overcoming the federal Government's fiscal crisis. The principal reforms envisaged were: a liberalization of the petroleum, electricity and telecommunications sectors; the allowance of foreign investment in mining and hydroelectric projects; a reform of the civil service; a major overhaul of the social security system; fundamental alterations to the federal Government's taxation and budgetary regimes; and an increased emphasis on achieving a more even pattern of landholding in rural areas. Cardoso achieved some initial success with his proposed reforms with the approval, in early 1995, of constitutional changes terminating state monopolies and permitting foreign investment in the sectors mentioned above. In addition, the programme of economic liberalization was given impetus by the full implementation of the Mercosul free trade area on 1 January 1995. Furthermore, in December Brazil signed an agreement to establish a Free Trade Area of the Americas (FTAA). However, the reform programme encountered a number of obstacles during 1996 and 1997. In particular, as had long been the case, the Government found it very difficult to exercise control over the budget deficit. The constitutionally mandated transfers of funds between the federal Government and the state and municipal governments led to a weakening of the federal Government's fiscal position as expenditures expanded. Slow progress in the reform of the civil service and the social security system meant that other major items of expenditure could not be reduced in compensation. The lack of progress in the reform of the landholding system resulted in a growing number of confrontations between landless peasants represented by the Movimento dos Sem Terra (MST) and landowners. The impasse in the reform programme was exacerbated by divisions between the two main parties of the coalition: President Cardoso's social democratic PSDB and the conservative PFL.

Although fiscal, social-security and administrative reforms remained delayed by the Congresso Nacional throughout 1996 and most of 1997, considerably more progress was made with the privatization and economic liberalization programmes. In 1996 steps were taken towards the sale of the Brazilian electricity network, Centrais Elétricas Brasileiras, SA (Eletrobras). The privatization of the federal rail network, Rede Ferroviária Federal, SA was completed in the following year. The huge mining group, Companhia Vale do Rio Doce, SA, was privatized in May 1997. From 1997 onwards a series of other major privatizations occurred, many in the energy sector. By far the most significant privatization to date, however, was implemented in the telecommunications sector with the sale of the subsidiaries of Telecomunicações Brasileiras, SA in July 1998. The Government also accelerated its attempts to privatize the state banking sector in the late 1990s. In August 1997 legislation enabling the liberalization of the petroleum sector became law. In a radical departure, the new regulatory framework for the sector allowed the participation of foreign enterprises in the exploration for petroleum within Brazil. This clearly eroded the monopoly status which the state-owned Petróleo Brasileiro, SA (Petrobras) had enjoyed since its foundation in 1953. With the new legislation in place, a series of exploration concessions were auctioned off to both domestic and foreign bidders between 1998 and 2001.

Despite the achievements in economic policy described above, lack of progress on constitutional reform left unaddressed a series of lingering macroeconomic problems. In particular, the Brazilian economy remained unable to escape from its tendency to accumulate heavy internal and external deficits. The persistence of these deficits became an increasing source of concern to international investors in 1997 and serious doubts were expressed in international financial markets as to the ability of the Brazilian Government to avoid a rapid, unplanned devaluation of the real. In order to maintain the valuation of the currency and avoid a resurgence of inflation, President Cardoso introduced a series of emergency measures intended to lower the budget deficit in November. The atmosphere of urgency surrounding the implementation of the measures, did, however, have some favourable political effects. In underlining the need for further progress on structural reform, the crisis induced notable advances in the passage of important legislation through the Câmara dos Deputados. After an extraordinary session of the Congresso Nacional in January 1998, crucial social security reforms were successfully enacted, while the Government's civil service reforms were finally approved in March, although amendments were enforced by the Congresso. By the beginning of August 1998, a second term in office for Cardoso's Government appeared inevitable. However, international financial events during that month caused the Government considerable unease and demonstrated once again the vulnerability of Brazil's economy and prompted renewed concern among investors over the scale of Brazil's external and fiscal deficits. As financial disruption increased world-wide, investors began to withdraw resources from Brazil in ever-increasing quantities, causing a sharp decline in international reserves and testing the Government's ability to defend the value of the real.

By September 1998 Brazil was experiencing a period of economic crisis in which the sustainability of the Real Plan seemed increasingly in doubt. The political effects of this crisis, however, ultimately proved far from unfavourable for the Government. Emphasizing the technical competence of his economic policy team, President Cardoso undertook an effective last-minute election campaign. At the presidential election held on 4 October Cardoso became the first President to be re-elected for a consecutive term in office, securing 53% of the valid votes cast, compared with 32% for his closest rival, da Silva. The former Minister of Finance and Partido Progressista Socialista (PPS) candidate, Ciro Gomes, came third with 11% of the votes. The results of the legislative and gubernatorial elections, however, were not as favourable for Cardoso's PSDB. Although the governing coalition in the Câmara dos Deputados secured 377 of the 513 seats, the PSDB's 99 seats did not constitute a significant increase in the party's overall legislative representation, but it was considerably more than the 60 seats won by Cardoso's party in 1994. The election of the populist, anti-Cardoso, state governors in Minas Gerais, Rio de Janeiro and Rio Grande do Sul, provided focal points for increasingly vocal regional opposition to the federal Government.

## CARDOSO'S SECOND TERM, 1998–2002

Despite the generally favourable domestic and international reaction to Cardoso's re-election, Brazil's economic situation continued to deteriorate in late 1998. In November a US $41,000m. agreement was concluded with the IMF, imposing stronger fiscal austerity on Brazil. On 2 December the Câmara dos Deputados voted to reject a significant government fiscal reform measure affecting public-employee pension contributions. The failure of this legislation increased concerns over the Government's ability to meet IMF targets and led to further outflows of foreign capital. By January 1999 it became apparent that the fixed exchange-rate policy pursued by the Government was becoming untenable. Moreover, in the same month, former President Itamar Franco, the newly elected Governor of Minas Gerais, declared that the state was defaulting on its debt to the federal Government and on a loan to a French bank consortium, thus indirectly precipitating the devaluation of the real. By the second week of January the drain of foreign exchange reserves had become significant. Faced with the imminent prospect of a complete depletion of reserves, on 13 January the Central Bank announced that the real was to float freely against the US dollar. Following the flotation, the real swiftly depreciated. With the Real Plan apparently in ruins, one of its key architects, the President of the Central Bank, Gustavo Franco, resigned.

While the devaluation of the real did not affect the economy as severely as had been predicted, it did have the effect of galvanizing congressional opinion in favour of accelerating the fiscal and structural reform programme. In an important measure designed to reduce the recurrent deficits of the social security system, the Congresso Nacional finally approved the Fator Previdenciário in November 1999, which created a greater correspondence between social security contributions and pension payments in the private sector. Furthermore, in April 2000 legislation on fiscal responsibility was approved by the Senado, establishing stricter regulations for the setting of state and municipal budgets with harsh penalties for deficit spending. Moreover, the Congresso approved both the budgets for 2000 and 2001 with minimal amendments. However, the Cardoso Government remained frustrated in its attempts to introduce comprehensive taxation reform.

Accelerated progress in the area of fiscal reform coincided with a number of political developments in early 2000. In mid-February Cardoso's PSDB announced a new congressional alliance with the centrist PTB. This alliance further increased the divisions between the PSDB and its leading coalition partner in the Congresso Nacional, the centre-right PFL. In August the PTB announced that it was to dissolve its alliance with the PSDB, thus re-establishing the PFL as the largest party in the Câmara dos Deputados.

Left-wing opposition parties performed well in the municipal elections of October 2000, in advance of the presidential election of 2002. Furthermore, throughout 2000 and 2001 the Government was embroiled in an extensive scandal involving allegations of embezzlement and corruption at senior levels. In early 2000 a former presidential adviser, Eduardo Jorge Caldas Pereira, a prominent senator, Luiz Estevão, and a federal labour court judge, Nicolau dos Santos Neto, were alleged to be among the participants in a scheme to divert some R $170m. in public funds destined for the construction of a new labour court building in São Paulo. As a result of the allegations, Estevão was expelled from the Senado while dos Santos Neto went into hiding (he was arrested in early 2001). Moreover, further evidence of high-level corruption had emerged in November 2000 following the publication of a report by another congressional investigating committee on organized drugs-trafficking and crime in Brazil, implicating a number of state deputies and mayors, as well as members of the Congresso Nacional, prosecutors, police, the judiciary and the armed forces.

Among its other effects, the eruption of the São Paulo labour court scandal added fresh impetus to the long-running campaign against sleaze in government led by Senado President Antônio Carlos Magalhães. In February 2001 Magalhães turned his attention to the election of the legislative speakers, alleging that his possible successor as Senado leader, the President of the PMDB, Jader Barbalho, was involved in corrupt activities. Following Barbalho's election, Magalhães increased his attacks on the Government. However, in July Barbalho relinquished his position for two months, after it emerged that he had been involved in fraudulent and corrupt activities while Governor of the state of Pará in the 1980s. In October Barbalho resigned from the Senado, in an attempt to prevent his own impeachment, were he to be found guilty of corruption. Meanwhile, in April the release of telephone transcripts secretly recorded by the police investigating the disappearance of some US $830m. from an Amazonian development company, SUDAM (Superintendência do Desenvolvimento da Amazônia), caused further embarrassment to the ruling coalition.

Despite his attempts to expose corruption in government, Magalhães himself became involved in allegations of impropriety. Following a congressional investigation in early 2001 it emerged that he may have improperly instigated the disclosure of confidential computerized voting records relating to the Senado's expulsion of Luiz Estevão. Magalhães resigned his senate seat in May. In the same month the Minister for National Integration, Fernando Bezerra, resigned following allegations of bribery; it was believed that he had received funds from a development agency, SUDENE (Superintendência do Desenvolvimento do Nordeste), in return for favours. (SUDENE and SUDAM were both dissolved by President Cardoso in early May.) It was subsequently discovered by federal prosecutors that some R $4,000m. had been embezzled from the two development agencies. In March 2002 accusations involving the SUDAM case re-emerged, discrediting the PFL's presidential candidate, Maranhão Governor Roseana Sarney.

The controversy surrounding allegations of high-level sleaze and the departure of Magalhães both acted to stall the Government's ambitious legislative programme and to taint the Government's hard-won reputation for competent economic management. In the first half of 2001 the real significantly weakened against the US dollar and other major currencies as a consequence of the economic crisis in Argentina, as well as the domestic political disarray. By mid-2001 the weakening of the real prompted the authorities to increase interest rates and lower growth forecasts. To add to the Government's difficulties, unusually low levels of rainfall had led to water in hydroelectric reservoirs falling to critically low levels. The authorities were forced to implement an emergency programme of energy conservation that further reduced economic growth and adversely affected the Government's popularity, and demonstrated that the privatization programme for the electric sector had been a failure. Thus, in the year preceding the 2002 elections, the Government faced the serious challenge of attempting to consolidate its considerable achievements against a background of political infighting and a deteriorating economy.

## THE 2002 ELECTIONS

By mid-2001 several potential opposition candidates for the October 2002 presidential election had emerged. Among the prominent were 'Lula' da Silva as the PT candidate, former

Minister of Finance Ciro Gomes for the PPS, and the Governor of Rio de Janeiro, Antônio Garotinho, representing the PSB. The PMDB had several potential candidates: Itamar Franco, Pedro Simon and Orestes Quércia, although none of these had succeeded in gaining much popularity. In August 2001 a potential candidate for the PFL, the Governor of Maranhão, Roseana Sarney, stepped into the vacuum created by the lack of an obvious candidate from the Partido Progressista Brasileiro (PPB) and the PSDB, and began an active and well-organized television campaign. She enjoyed increasing popularity in the opinion polls until, on 1 March 2002, the offices of her husband were raided by federal police as part of the investigation into the SUDAM scandal (see above). Two weeks later Sarney withdrew her presidential candidacy and, alleging that the raid had been instigated by allies of the recently declared PSDB candidate, José Serra, the PFL formally withdrew from Cardoso's governing coalition.

As a result, in the months preceding the 2002 elections President Cardoso's coalition found itself in severe difficulties. Its most cohesive party, the PFL, had declared itself independent, the support of the PMDB was doubtful, and the PTB had joined the Aliança Trabalhista with the PDT in support of Ciro Gomes's candidacy. Nevertheless, Cardoso attempted to mobilize support for his preferred successor, Serra. Furthermore, the PMDB was persuaded to enter into a formal alliance with the PSDB. In an attempt to strengthen the PSDB-PMDB alliance, in April the Supreme Court upheld a ruling by the Superior Election Court (Tribunal Superior Eleitoral—TSE) requiring party coalitions in gubernatorial and state elections to be the same as those formed for national presidential and legislative elections (known as the 'verticalization' of coalitions). By April 2002 Serra had emerged as second favourite in the presidential contest, behind da Silva, who was running in alliance with the Partido Liberal (PL) candidate, José Alencar. However, following press revelations in May regarding a number of scandals dating back to the mid- and late 1990s, the PSDB candidate's popularity decreased significantly. In the same month it also became apparent to international financial markets that Brazil's macroeconomic situation was deteriorating and analysts questioned Brazil's capacity to honour its large debt (equivalent to 55% of GDP) in 2003 and beyond. Amid insinuations that Brazil was in danger of suffering the economic crisis experienced by Argentina were an 'incompetent' President and economic team to be elected, da Silva's popularity rating also fell. The concerns by the international financial community regarding an imminent da Silva victory had considerable negative impact on Brazil in mid-2002. Banks sharply reduced short-term trade credits, Brazil's risk evaluation soared, the real devalued strongly against the US dollar, and international reserves dwindled. As in 1998, the Cardoso Government again sought IMF assistance and quickly concluded a new 15-month agreement, worth US $30,000m., in August 2002. Da Silva and the other opposition candidates accused the Government of practising 'economic terrorism'. Nevertheless, in order to allay fears, the PT's campaign platform took a sharp turn to the centre and attracted the support of several prominent business leaders.

The elections of October 2002 produced a decisive victory for the left in Brazil. In the presidential contest 'Lula' da Silva nearly achieved an absolute majority (46.4%) at the first round, held on 6 October. Ciro Gomes (PPS-PDT-PTB) and Antônio Garotinho (PSB) supported da Silva in the 27 October run-off, in which the PT candidate defeated José Serra (PSDB-PMDB) by an unprecedented margin, attracting 61.3% of the valid votes cast. In several states, the PMDB ignored the TSE coalition verticalization ruling and actively supported da Silva in the first and second rounds. Having no presidential candidate in 2002, the PFL and PPB were free to operate diverse coalitions in the states, and large segments of the PFL supported da Silva. The centre-right coalition parties elected 16 governors (the PSDB seven, the PMDB five, and the PFL four). The PT coalition parties elected 10 governors (the PT three, the PSB four, the PPS two, and the PDT one). In the Congresso Nacional, the pro-da Silva parties elected 218 deputies and 31 senators, a considerable increase over 1998, but still fewer than the respective 257- and 41-member absolute majorities. The PT returned the largest delegation in the Câmara dos Deputados, with 91 federal deputies.

## THE PT IN POWER, 2003–

'Lula' da Silva was sworn in as President on 2 January 2003, and the new Congresso Nacional took office on 1 February. Da Silva's Cabinet was recruited from his electoral coalition, with a heavy concentration of PT militants, especially from São Paulo and Rio Grande do Sul, and representatives from the PL, PDT, PSB, PTB, PPS, the Partido Comunista do Brasil and the Partido Verde. Many PT ministers were defeated gubernatorial and senatorial candidates, and several had been active guerrilla fighters in the 1968–74 period. However, two prominent business leaders were also appointed, to the Ministries of Development, Industry and Trade and of Agriculture, Fisheries and Food Supply. The choice of former BankBoston President Henrique Meirelles to head the Central Bank particularly irritated the more radical PT deputies and senators. Many had feared that the new da Silva Government's policy initiatives would be impeded in the Congresso Nacional by a lack of majorities. However, the new President's team (led by Cabinet Chief José Dirceu) adroitly used the same power mechanisms (federal appointments and disbursements) as Cardoso to consolidate absolute majorities and even constitutional quorums (60%). In a successful mutual support effort, José Sarney (PMDB) and José Paulo Cunha (PT) were elected Presidents of the Senado and Câmara, respectively.

The new Minister of Finance, Antônio Palocci, and his economic team unequivocally pursued a fiscal austerity programme even more rigorous than that of the previous Government. Such a policy won enthusiastic approval from the IMF, which endorsed loan disbursements scheduled for March and June 2003. In the first five months of 2003 the trade surplus, tax collections and foreign direct investment (FDI) all increased significantly, while the current-account deficit decreased. Brazil's risk evaluation improved significantly. Inflation declined in April, May and June, but the Central Bank maintained a high basic interest rate. As a result, GDP declined by 0.1% in the first quarter of 2003 and unemployment increased to over 20%. The Government claimed this was part of the negative economic situation inherited from the Cardoso period.

On assuming office, the PT Government faced the dilemma of high deficits in a stagnating economy with low inflation. Thus, the da Silva team opted for a quite courageous social policy reform. In addition to enhancing his majorities in the Congresso Nacional, da Silva struck a very effective alliance with the 27 state governors elected in October 2002, through several conclaves in Brasília during 2003. The new President garnered total support for his two very important 'showcase' initiatives: social security and tax reforms. Da Silva's proposals to reform Brazil's deficit-ridden social security system were deemed 'heresy' by PT radicals. The PT had shown strong opposition in Congresso to the reform proposals (especially social security) of the Cardoso Government the da Silva team was proposing even more wide-ranging changes. A first round vote on social security reform was held in the lower house in August. Although it was approved, the vote was only just quorate and the government coalition was able to muster only 293 votes in favour. The separate vote on the introduction of an 11% levy on public servants' pensions was even more of a trial for the Government, being approved only with votes from opposition deputies. The second round votes were easier, and in late September a more cohesive da Silva coalition was able to approve the tax reform in the lower house without help from the opposition.

Following a record five months of deliberations, in December 2003 the Senado finally approved the two reforms, but only after senators had demanded that changes be made to the legislation. The Government, desiring final approval of both reforms during the 2003 fiscal year, agreed to dividing up the tax reform proposals into three sections, only one of which was approved.

In May 2003 President da Silva had the extraordinary opportunity to appoint three new judges to the Supreme Federal Tribunal (supreme court), including the first Afro-Brazilian justice. The President made his fourth supreme court appointment in April 2004 and in 2005 would make his fifth nomination to the 11-member court. A sympathetic majority on the Supreme

Federal Tribunal was crucial to the successful passage of the Government's ambitious judicial reforms as the Government hoped to conclude a controversial judicial reform in 2004 (including an external control unit and the concept of 'tied jurisprudence') and planned to present proposals for changes in labour legislation (and norms for labour unions) in early 2005. This latter change would significantly reduce the social overhead costs for workers, thus reducing production costs and stimulating new jobs. The reforms to the judiciary were approved in early 2005 and took effect in June of that year. In August 2004 the supreme court ruled against the compulsory taxing of pensions.

In January 2004 President da Silva effected a cabinet reshuffle. Seven members of the Government either were dismissed or had their duties altered, and da Silva also brought in representatives of the PMDB in an attempt to shore up support for his Government in the Congresso Nacional. Nevertheless, in mid-February the President encountered several political reverses and defeats. A videotape exposed allegedly corrupt campaign finance dealings in May 2002 by his congressional relations chief, Waldomiro Diniz. This episode had a very negative impact on Diniz's superior, da Silva's Cabinet Chief, José Dirceu. As a result, da Silva issued a decree prohibiting bingo operations, but the Senado overruled him in early May 2004. Diniz then had been implicated in intermediation attempts in 2003 between a lottery contractor and the Caixa Econômica Federal. In mid-April 2004 the PL's three senators declared themselves no longer aligned with the the ruling coalition, thus reducing the Government's majority in the Senado to 42 (just one above the absolute majority). In May a constitutional amendment to allow the re-election of the presidents of the Câmara dos Deputados and Senado to a second consecutive two-year term in the same legislative session was rejected in the lower house by a six-vote margin. Da Silva was 'neutral' on this amendment, but the bill's failure incurred resentment from Senado President José Sarney. In June the da Silva Government encountered great difficulties in sustaining its decree increasing the minimum wage by only R $20 in successive votes in the Congresso. On 17 June the Senado rejected the Government's plan and voted to increase the minimum wage by a further R $15, to R $275 per month. Finally, the supreme court threatened to declare a major part of the 2003 social security reform unconstitutional.

In 2003 the da Silva Government resettled 36,000 families on farms as part of an agreement made to the MST. This was a significantly smaller number than the 60,000 the Government had promised to resettle in that year; however, officials blamed excessive bureaucracy for the shortfall. In May 2004 an unofficial MST truce, which had lasted since December 2003, came to an end and several farms were occupied in subsequent months. In 2004 a total of 327 farms were occupied. In May 2005 some 12,000 MST members participated in a march to Brasília to protest against the Government's comparative lack of progress in resettling landless families.

Although the da Silva Government got off to a faster and more effective start than most anticipated, by 2004 the political coalition had weakened. The results of the October 2004 municipal elections were not as favourable for PT candidates as they had been in October 2000. Although 411 PT mayors were elected (compared with 187 in 2002) and the party's total vote increased from 11.9m. to 16.3m., in 2004 the party lost São Paulo to the PSDB and Porto Alegre to the PPS; the PT had governed the latter city for 16 years. Overall, PT mayors were elected in 23 of the largest 96 cities in Brazil (compared with 25 in 2000). However, the party did increase its share of mayors in smaller cities, from 60 to 134 in cities with populations of between 10,000 and 50,000, and from 80 to 219 in towns with populations of less than 10,000. The PMDB had its candidates elected in 1,057 mayoral contests (200 fewer than in 2000), while the PSDB's mayoral representation declined from 990 to 871, and the PFL's from 1,028 to 790.

Owing to the long 'election recess' of the Congresso Nacional in 2004 (from July to November), the Government's reform agenda was delayed in that year. However, in early 2005 Congresso finally approved the judicial reforms that established external control councils for Brazil's judiciary and public prosecutors. A long-awaited new bankruptcy law, known as the 'business recovery law', was also approved and took effect in early June. The legislation enabled struggling businesses to negotiate debts and therefore to remain operational. Also, a proposed public-private partnership mechanism received legislative approval in early 2005, which, it was hoped, would facilitate new private investment in badly needed infrastructure projects in 2006 and 2007. Other reforms (political, labour legislation, new rules for regulatory agencies, plus additional stages of the tax and fiscal and social security reforms) remained stalled in mid-2005, however.

Although the da Silva Government made advances in some areas of its reform programme, it was weakened in early 2005 as the President's planned cabinet reform, scheduled for January or February, was postponed after the PT candidate was defeated in the election, in mid-February, to the presidency of the Câmara dos Deputados. Severino Cavalcanti of the PP defeated both the 'official' PT candidate, Eduardo Greenhalgh, and the 'dissident' PT candidate, Virgílio Guimarães. Although the PP was part of the pro-Government bloc in the Câmara, Cavalcanti charted a very independent strategy as head of the lower house, and, as a result, the Government's legislative agenda encountered obstacles. In late May, an opinion poll conducted by Datafolha showed that da Silva's approval rating had declined 10 points since December 2004, from 45% to 35%.

In May 2005 several cases of government corruption were revealed by federal police investigations, and a major scandal erupted in early June. A videotape emerged that allegedly showed proof that a bribes-for-contracts scheme was in operation at the state postal service, the Empresa Brasileira de Correios e Telégrafos. The recipient of the bribes, Maurício Marinho, a senior postal service employee, claimed on the tape that the scheme was operated by PTB appointees, organized by Roberto Jefferson Monteiro, the party's national president. A similar scheme was subsequently revealed to be in operation at the state-owned reinsurance institution, the IRB-Brasil Resseguros. Jefferson denied the allegation and, in turn, publicly accused the PT of operating a bribery scheme in 2003 and 2004, whereby PL and PP deputies received a monthly allowance (*mensalão*) of R $30,000 for voting for government-sponsored legislation. Jefferson also alleged that in 2004 the PT had promised the PTB some R $20m. in financial support for the municipal election campaign, but that only R $4m. had actually been transferred to party funds. Testifying before the Câmara dos Deputados' Conselho de Ética e Decoro Parlamentar (Ethics Council) on 14 June, Jefferson repeated his earlier accusation that President da Silva's Cabinet Chief, José Dirceu, had organized the bribery scheme. As a result, on 16 June Dirceu resigned his influential position in the da Silva Government. The PT's Secretary-General, President and Treasurer also stepped down from their posts in July after being implicated in the scandal. In late May the Congresso Nacional installed an investigating committee to examine the accusations relating to the postal service, and, following Jefferson's testimony, into the *mensalãos*. On 17 June Jefferson took a leave of absence from the PTB presidency while the investigations continued.

In late June 2005 President da Silva appointed Dilma Rousseff, hitherto Minister of Mines and Energy, as Dirceu's successor. A further reallocation of cabinet portfolios was carried out in early July. The number of PMDB ministers in the Government was increased, in order to ensure that party's support in the Congresso Nacional. However, the Minister of Social Security, Romero Jucá (of the PMDB), was replaced. Jucá was implicated in a separate corruption scandal involving irregular loans. Meanwhile, the Governor of the Central Bank, Henrique de Campos Meirelles, also faced an investigation into allegations of finanical irregularities.

The corruption allegations put the PT on the defensive and further eroded public confidence in the Government. In mid-June 2005 Datafolha ran another poll that indicated voter approval of the da Silva Government steady at 36%, and at 49% for President da Silva. Among those polled, 70% said corruption existed in the da Silva Government, but only 29% said the President was 'very responsible'. (Jefferson had made it clear that da Silva had not known of the supposed *mensalão* scheme.) In contrast, the approval rating for the Congresso Nacional was only 15%.

During the first half of his term in office, President da Silva was very active in foreign affairs and undertook many visits overseas. Brazil continued to pursue its ambition to gain access to one of the new permanent regional seats on the UN Security Council, a goal opposed by Argentina and Mexico. To this end, in 2004 Brazil accepted the command of the UN Stabilization Mission in Haiti and contributed 1,200 troops. In that year Brazil was also involved in the so-called 'G-3' (Brazil, India and South Africa) group of countries, and in 2005 the 'G-4' (Brazil, India, Japan and Germany) grouping was organized to seek this objective collectively. Brazil led the opposition by the 'G-20' emerging nations to pressure the G-8 grouping of industrialized countries to reduce their subsidies and protectionism, particularly subsidies on agricultural products. The strong stance adopted by Brazil on this issue at the fifth Ministerial Conference of the World Trade Organization (WTO) in Cancún, Mexico, in September 2003 effectively deadlocked negotiations towards the FTAA. Negotiations resumed in February 2005, but no advances were made. In June 2004 a complaint against US cotton subsidies brought by Brazil was upheld by the WTO, and in August a similar complaint against European Union (EU) sugar subsidies was sustained. Trade negotiations between Mercosul and the EU were promising in 2004 but became stalled in 2005. In 2004 President da Silva made two visits to sub-Saharan Africa, as well as trade promotion and investment missions to South Korea, Japan and the People's Republic of China. In January 2005 President da Silva repeated his 2003 agenda and made consecutive visits to the World Social Forum in Porto Alegre, RS, and the World Economic Forum in Davos, Switzerland. In May 2005 Brazil hosted the first Summit of (22) Arab and (12) South American nations in Brasília.

In December 2004 Brazil was one of 12 countries that were signatories to the agreement, signed in Cusco, Peru, creating the South American Community of Nations (Comunidade Sul-Americana de Nações), intended to promote greater regional economic integration and due to become operational by 2007. Brazil's development bank (Banco Nacional do Desenvolvimento Econômico e Social) provided loans for infrastructure projects in Venezuela, Peru, Ecuador, Colombia and Bolivia in 2004. In late 2004 inspectors of the International Atomic Energy Agency demanded full access to Brazil's new uranium enrichment plant in Resende, RJ, and in 2005 a compromise solution was achieved.

# Economy
## Dr GABRIEL PALMA

In the colonial period and in the first century of independence, Brazil experienced three successive export cycles. The first based on sugar, in the north-east around Recife and Salvador; the second based on gold and diamond mining, in what is now the central state of Minas Gerais; and the third, from the early 19th century onwards, based on coffee. Initially grown with the help of slave labour in the state of Rio de Janeiro, coffee production achieved its most dynamic expansion in the state of São Paulo after 1890, as European migrants, predominantly Italians, moved into the region in large numbers. It was only in the 1980s that coffee lost its place to soya beans as Brazil's leading agricultural export. In terms of the economic geography of the country, although there was spectacular expansion of rubber production in the Amazon region between 1890 and 1920 (brought to an abrupt halt by the establishment of plantations in South-East Asia, from Brazilian stock), the centre of the Brazilian economy shifted from the north-east to the centre-south of the country. This was marked by the transfer of the capital from Bahia (now Salvador) to Rio de Janeiro in 1761, and by the emergence in the 20th century of São Paulo as the country's dominant industrial, financial and commercial centre.

The country's economy was dominated by the 'industrial triangle' of São Paulo–Rio de Janeiro–Belo Horizonte. This was in spite of the establishment of nuclei of development in the north-east and in the south, and increasing flows of investment into Amazonia. The cities and rural areas of the centre-south attracted a persistent flow of migrants from the impoverished and overpopulated north-east, partly offset by a counter-flow from the south into Amazonia.

The remarkable collapse of the coffee prices during the depression of the 1930s was a determining factor in the political élite finally switching the engine of growth from the coffee-based export sector towards an already emerging and dynamic manufacturing sector. For six decades thereafter, governments of very different political ideologies pursued a fairly successful state-led industrial development through often unorthodox interventionist policies. As a result, Brazil emerged as a major industrial power in the developing world, and by the late 1980s had become the 10th largest economy in the world.

Despite bouts of economic nationalism, foreign capital was generally welcomed into the manufacturing sector, particularly since the 1950s. However, the state frequently insisted upon stringent requirements relating to the source of capital, the transfer of technology, and joint ventures (often with state, rather than private, capital). One of the explicit aims of economic reforms in the 1990s was to reverse this pattern of state-led development in favour of financial and trade liberalization, deregulation of the economy, and the integration with Argentina, Paraguay and Uruguay into a regional common market, Mercosul (Mercado Comum do Sul, or more familiarly in Spanish, Mercosur—Mercado Comun del Sur).

A striking feature of state-led development was a massive building of productive capacity, particularly in the areas of energy, heavy industry and capital goods. As a consequence of this, and of Brazil's large internal market and remarkably abundant natural resources, the country experienced rapid growth after the Second World War. Between 1947 and 1980, an average compound real rate of growth of output of 7% per year was achieved, placing Brazil in the East Asian rather than in the Latin American 'growth league' during this period. As a result, domestic output increased nearly 10-fold in this 33-year period. However, first the 1982 debt crisis and later the dogmatically implemented 'neo-liberal' reforms of the early 1990s not only brought this long period of rapid growth to an abrupt end, but also marked the return of Brazil to the Latin American 'growth league', characterized by poor and highly volatile economic performance.

After the 1982 crisis Brazil experienced a severe recession, made more acute by rising inflation and a heavy debt burden. From 1984 until the implementation of the 'Real Plan' in 1994 (see below), Brazil's real gross domestic product (GDP) fluctuated between small increases and contractions. During the first months of the Plan, however, GDP growth accelerated rapidly, but this growth soon proved unsustainable owing to both external and internal constraints. The main external factor was the so-called 'tequila effect', which followed the Mexican financial crisis of December 1994; this made the whole issue of Brazil's excessive reliance on foreign capital (the cornerstone of the new economic policies) much more uncertain. Internal constraints began to emerge (particularly in the current account) as a result of the excess expenditure of both the private and the public sectors. The former was the result of both the sudden disappearance of the 'inflation tax' and the new availability of foreign exchange (brought about by the massive increase in foreign inflows). The latter was a result of the implementation of the stabilization plan being characterized by a dramatic increase in public debt, mainly as a consequence of rigidities in public finances, the cost of 'sterilizing' foreign inflows and high interest rates. These rates not only became a permanent feature of the Plan, but also remained continuously (and unnecessarily) higher than international interest rates plus Brazil's 'country risk'. They also remained significantly higher than the growth of public revenues; this meant that the Government was forced to 'capitalize' a growing part of its interest payments (i.e., borrow in order to be able to pay for its existing obligations). Furthermore, the Real Plan added indirectly to the stock of public debt, as a result of its association with a series of costly banking crises both in the private and public sectors, the first of which occurred in 1995. Finally, high interest rates, the peculiarities of state politics and the weakness of the political coalition supporting President Fernando Henrique Cardoso also were responsible for increasing the public debt via continuous large rescue operations of state finances by the federal Government and Central Bank.

With mounting internal and external problems, the Government took a series of drastic measures in 1995 to ensure a sharp decline in the rate of growth of private expenditure, notably, a further increase in interest rates. These measures proved effective in the short term. However, perhaps as a result of the very effectiveness of these measures and the ease with which it could initially finance its growing public debt, the Government was deluded into believing that much needed (but politically controversial) fiscal reforms were no longer essential. Such an omission resulted thereafter in increased financial fragility, particularly in the banking system and the public sector.

Following the decline in output growth, the Government began to ease monetary and fiscal conditions and the economy began to recover again. However, first the East Asian crisis in July 1997 and then the Russian default in August 1998 revealed the remarkable internal and external fragility of the Brazilian economy, which returned to recession in the second half of 1998. The Government of President Fernando Henrique Cardoso responded to the difficult economic conditions by the only way it seemed to know, further tightening monetary policy. Finally, the general uncertainties prevailing both within the country and in international financial markets with regard to the Brazilian economy (in particular, the Government's ability to manage its domestic and external debts, the banking-sector fragility and the growing deficit in the balance of payments) were made unmanageable by the declaration by the Governor of Minas Gerais, former President Itamar Franco, that the state was defaulting on its debt to the federal Government. Although the amounts involved were low, it was feared that other state governors (many of them belonging to opposition parties and with large debts owing to the federal Government) would follow suit. Such fears led to a massive withdrawal of funds from the Central Bank (about one-half of its total reserves) and the Government was forced to float the real on 13 January 1999.

Most observers expected that, as a result of this devaluation and financial collapse, Brazil would enter (as Mexico had done in 1994, several East Asian countries in 1997 and Russia in 1998) into a period of acute recession, increased inflation and exchange-rate volatility. However, owing to a combination of factors, in particular the policies followed by Arminio Fraga, the newly appointed President of the Central Bank, who brought to an end his predecessor's obsession with severe monetary policy, GDP actually increased by 0.8% in 1999; this recovery accelerated in 2000, with the economy growing at an annual rate of 4.5%. Furthermore, consumer prices increased by just 4.9% in 1999 and by 7.1% in 2000. After the crisis, the real finally stabilized at a rate just under two reais per US dollar, a relative stability that continued into 2000.

The remarkable low inflationary effect of the devaluation was owing to several factors, not least the fact that the country was already in recession at the time of the devaluation of January 1999. Another factor was that the country's imports represented only about 10% of GDP. Furthermore, the IMF had already agreed before the January 1999 crisis a large rescue package for the country and, having learnt from its mistakes in the East Asian crisis, was prepared to be much more flexible in its terms. Finally, international financial markets did not panic as much as was initially feared.

Although the 1994 Real Plan ended badly, it at least succeeded in its initial aim, that of putting an end to the long-running hyperinflation. Prior to this, attempts to halt periodically rampant inflation since the 1982 debt crisis had experienced only brief success. With the rate of increase in consumer prices at over 200% per year when the military left power in 1985, it was then held at close to zero for just nine months in 1986, following the implementation of President José Sarney's 'Cruzado Plan'. Under this Plan the cruzeiro was replaced by a new currency, the cruzado, which was equivalent to 1,000 units of the old currency. Owing to the balance of payment problems brought about by the expansionary effect of the price-freeze of the Cruzado Plan, inflation accelerated rapidly again. In March 1990, with an annual rate of inflation threatening to reach 5,000%, the new President, Fernando Collor de Mello, introduced the 'New Brazil' programme, or 'Collor Plan', which restored the cruzeiro as the national currency and implemented a set of measures that succeeded in reducing inflation to 3,000% for the year as a whole, and to 473% in 1991. However, following the collapse of the Collor Government in late 1992, inflation rose again, reaching some 2,500% in 1993. Following the successful implementation of the Real Plan in mid-1994, the annual rate of inflation decreased sharply, reaching 3.2% in 1998.

Despite the difficult economic problems inherited from the long period of military rule, this phase saw the completion of a process which was already well under way before the 1964 coup: the emergence of Brazil as a major industrial world power. However, the legacy of the military regimes was a complex one, not only because of the high rates of inflation and huge levels of foreign debt. When the military departed from government it left what proved to be a weak and ineffective institutional setting (as has been the case in many other similar transitions in Latin America, not least in order to keep as much power and influence as possible in the new democratic climate), a culture of political and economic corruption, and an extraordinary level of inequality—if Brazil's economy had significantly improved during the military rule, the lot of the majority of its people certainly had not. All these, as well as the legacy of the debt crisis, rendered Cardoso's attempt at reform and economic and financial liberalization, at best, a particularly difficult process. However, the main obstacle for the success of these reforms was the extreme degree of dogmatism and rigidity with which they were implemented. In fact, the 'magnetic north' of the Brazilian neoliberal 'compass', as with most of Latin America, was simply a reversal of as many aspects as possible of the previous growth strategy. According to Gustavo Franco, President of the Brazilian Central Bank until the 1999 financial crisis, 'our real task is to undo 40 years of stupidity [besteira] ... The alternative in Brazil is to be neo-liberal or neo-idiotic.' (Veja, 15 November 1996). The fact that Brazil's previous development strategy delivered for most of those 40 years one of the fastest growth rates in the world was, according to Franco, a mere detail of history.

## DEREGULATION, PRIVATIZATION AND REGIONAL INTEGRATION

The 1982 debt crisis and the subsequent acute recession placed the model of state-led development under considerable strain throughout Latin America. In Brazil, this was compounded by the economic and political difficulties of a fragile transition to democracy. As had happened in the 1930s, a massive and continuous external shock that found Latin America in an extremely vulnerable economic and political position not only brought about the need for a very painful internal and external macroeconomic adjustment, but also laid the foundations for a radical and widespread change in economic thinking. The resulting ideological transformation eventually led to a generalized change in the economic paradigm of the region. In this case, it was characterized by an extreme move towards trade and financial liberalization, wholesale privatization and market deregulation, along the lines begun in Chile in 1973. In this way, a key element to understanding these reforms, particularly the 'fundamentalist' way in which they were implemented throughout the region, is that they were mostly carried out as a result of the perceived weaknesses of these economies. This fact helps to explain the different degrees of intensity and rigidity with which the reforms were implemented in Latin America, as opposed to in East Asia. Although initially these reforms were a desperate attempt to reverse capital flight, reduce hyperinflation and bring the economies out of recession, they permanently changed the direction of Latin American economic development. Brazil's economic reforms actually began in 1990, before the Real Plan, when the incoming President, Fernando

Collor de Mello, announced the removal of subsidies for exports, and phased reductions in tariffs, as part of his Collor Plan. This was followed in 1991 by a large-scale privatization programme, the deregulation of the fuel market, and the dissolution of the coffee and sugar trading boards. Despite the political weakness of Collor de Mello's short presidency and its unsure handling of macroeconomic policy, the initiatives that it launched succeeded in permanently changing the nature of the Brazilian economy.

Tariffs were set to halve in three years, with no tariff exceeding 35% by the end of that period. Of equal moment was the commitment to remove the cumbersome system of import licensing. The programme of privatization was initially intended to raise US $18,000m. from the disposal of 27 state companies. Privatization eventually began in October 1991, with the sale of the Usinas Siderúrgicas de Minas Gerais, SA steel mill. Over the following months further privatizations took place in smaller concerns, but receipts tended to fall far short of targets. From 1995 the Government of President Cardoso gave a further impetus to the privatization process.

A related policy initiative of the Collor Government was the creation, in March 1991, of Mercosul. Trade between Brazil and Argentina grew by 45% in the first year alone; in all, exports to Argentina grew more than four-fold between 1991 and 1998, and imports from Argentina five-fold. However, the growing economic problems of Brazil and Argentina raised doubts as to whether the harmonization of tariffs and economic integration could be sustained. In fact, intra-Mercosul trade, after having grown by 20% in 1997, fell by 3% with the worsening economic situation of 1998, and by 24% in 1999; although this trade recovered in 2000 (by about 16%), it decreased again after the December 2001 Argentine crisis. Nevertheless, the election in October 2002 of Luiz Inácio ('Lula') da Silva to the presidency, and the election in Argentina of President Néstor Kirchner in April 2003, both apparently committed to Mercosul, proved to be a positive incentive to further economic integration in the region.

Economic policy under President Collor de Mello's successor, Itamar Franco, continued very much unchanged in 1993, with its principal objectives being the liberalization of most prices, control over public expenditure and a strict monetary policy through high interest rates. In May, after several changes of finance minister, President Franco appointed Fernando Henrique Cardoso to the position. Together with a group of well-known economists, Cardoso devised the all-encompassing Real Plan (named after the new currency hereby introduced, the real), which began operations on 1 July 1994. The main characteristic of this new Plan was that, as opposed to most of its predecessors, it intended to avoid 'shock treatments', unsustainable price freezes or surprise announcements. Instead, it attempted to reduce prices by three interconnected means: first, by reducing inflationary expectations, principally through a relative fix of the real to the US dollar, initially at a rate of one real to the dollar (crucially, the real was not rigidly 'pegged' to the dollar, as in Argentina); second, by reducing inflationary 'inertia' by dismantling a complex system of indexation; and, third, through the progressive achievement of an internal and external macroeconomic equilibrium. The Plan acquired an overwhelming degree of consensus and public support. Its initial successes in mid-1994 helped Cardoso's campaign for the presidency, to which he was elected in October of that year.

Despite the huge initial success of the Real Plan in containing inflation, by the time President Cardoso finished his first term in office the economy was again on the brink of a major financial collapse, which occurred during the first month of his second term. Furthermore, when President Cardoso finished his second four-year term, in 2002, the economy was again in recession, the government coalition had collapsed, the domestic debt had more than doubled as a percentage of GDP (in comparison with the level before President Cardoso took office) and debt-servicing had reached more than 10% of GDP. In other words, President Cardoso reached the end of his second term in office in much the same way as he did his first, with an economy apparently heading for a major collapse and a Government that was virtually paralysed, just hoping for the crisis to be delayed until after the presidential election. In this Cardoso was successful, but the economic legacy fpr his successor, President da Silva, was a daunting one.

## AGRICULTURE

Agriculture (including hunting, forestry and fishing) only employed about 18% of the total labour force in 2004, compared with more than 30% in 1980. The sector accounted for 9.4% of GDP in 2003, compared with 12.3% in 1970 and 20.6% in 1960. In terms of exports, the sector (including processed foods, such as orange juice) contributed 28% of the total in 2004, compared with 71% in 1970. Sectoral growth increased an average annual rate of 3.4% between 1990 and 2004. Nevertheless, with the exception of soybeans, by 2005 the sector had failed to respond properly to the challenge of generating significant additional foreign exchange following the economic crisis and huge devaluation of January 1999. Furthermore, concern was expressed at the implications for Brazil of the Farm Security and Rural Investment Act of 2002, approved by the US Congress in mid-2002. This legislation increased subsidies to US farmers, and consequently reduced demand for a variety of export crops produced by Brazil (notably soya and cotton). Brazil responded with a strong challenge to this legislation through the World Trade Organization, and in June 2004 its complaint against US cotton subsidies was upheld.

The new pattern of export-led growth was supposed to recover the role of agriculture as a substantial source of foreign exchange. Emphasis was given to large-scale commercial farming and to export crops. However, not only did most of these exports perform disappointingly, but also food production for the internal market suffered, with many thousands of peasants and small farmers being forced from their land. Annual coffee production averaged some 3m. metric tons per year at the beginning of the 1990s. Output then fluctuated in the mid-1990s, before expanding at the end of the decade; in 2004 coffee production stood at 2.5m. tons, but less than 2m. tons was exported (generating US $1,750m., below 2% of total export revenues).

By 1990, partly as a result of dramatically decreasing coffee prices, earnings from soybeans were more than double those from coffee. The harvest rose from 15m. metric tons in 1980, to 24m. tons in 1989. The crop then fell, but recovered in the mid-1990s; by 2003 production stood at 51.5m. tons. The harvested area in that year was more than 10 times larger than that in 1970, and in 2004 exports of soya and related products reached nearly US $10,000m. There were also spectacular increases in sugar-cane harvests during the 1980s and 1990s, with production increasing from 103m. tons in 1976, when the alcohol programme (see Petroleum and Ethanol, below) was introduced, to 389.5m. tons in 2003. The harvested area that year was twice that of 1980.

Orange juice became Brazil's third most important agricultural export in the 1980s, as a result of increased production (which had doubled since 1979) and rising prices. Output of oranges stood at 16.9m. metric tons in 2003. Banana production grew steadily in the 1980s and more erratically in the 1990s, from 4.5m. tons in 1990 to 6.8m. tons in 2003.

The bias towards exports had adverse effects on domestic food supplies, as areas hitherto devoted to food crops diverted production. According to some estimates, food output per head actually declined in the period 1965–85, and stagnated thereafter. Production of paddy rice reached a peak of 9.6m. metric tons in 1976, a figure barely surpassed by 2003 (10.3m. tons); however, by then productivity had increased significantly as the latter output was obtained with a harvested area of about half that of the late 1970s. Production of beans (the other staple crop) was also only slightly higher than the levels of the early 1970s, standing at 3.3m. tons in 2003. Production of maize (corn) fluctuated in the 1990s, and stood at 48.0m. tons in 2003. Wheat, once heavily subsidized, recovered from a poor harvest of less than 2m. tons in 1984, but declined subsequently to a low of 1.5m. tons in 1995, obviously finding it difficult to compete with new Mercosul imports from Argentina. In 2003, however, wheat output jumped to nearly twice the levels of 2002, reaching 6.0m. tons. Again, as in rice production, the area harvested in 2003 was only about half that of 1980.

## MINING

Although mining (including quarrying) accounted overall for only a small proportion of Brazil's GDP (about 3.7% in 2003), it was important to exports and to industrial production. Between

1990 and 2003 the sector grew at an average of 4% per year. In 2003 Brazil produced iron ore with some 155.7m. metric tons of iron content and was the world's largest exporter. In 2003 the value of exports of iron ore reached US $3,500m. (and those of processed iron and steel $7,400m.). The bulk of iron-ore exports traditionally came from the 'iron quadrangle' of Minas Gerais, from the Brazilian mining groups Mineraçoes Brasileiras Reunidas, SA (MBR) and CVRD—Companhia Vale do Rio Doce, SA, the latter being the largest iron-ore producing and exporting company in the world. By the 1990s deposits in the hitherto inaccessible Amazon state of Pará were reached by road and rail and were supplied with electricity from the Tucuruí hydroelectric power station (see Power, below). By 2004 they provided about one-quarter of national production.

Brazil is also a significant producer of manganese, with output of about 2.5m. metric tons in 2003, and of tin, where rapid expansion in the 1980s took total output to a peak of some 50,000 tons in 1989, before decreasing again to nearly 10,761 tons in 2003. In addition, the country is a large exporter of gold, although production is difficult to estimate, as about two-thirds comes from independent prospectors (*garimpeiros*) in frontier areas, where conflict with native Indian (Amerindian) inhabitants is frequent and violent. Output was estimated at 44.4 tons in 2003. Brazil has 1,600m. tons of bauxite reserves, of between 40% and 50% purity, largely in Minas Gerais and the Amazon region. Annual output has stagnated at around 13.5m. tons.

## POWER

Brazil exploited only about one-quarter of its potential for the generation of hydroelectric power. Even so, power from this source met about 95% of the country's total electricity needs. The major increase in total capacity from the late 1980s stemmed from two significant developments in 1984: the entry into operation of the first turbines at the Itaipú hydroelectricity complex on the Paraguayan border (which was expected to produce over one-third of Brazil's total electricity requirements); and the opening of the Tucuruí plant, on the Tocantins river, in the Amazon region. The former plant provided additional generating capacity of about 12,000 MW to the national grid, to the benefit, primarily, of the centre-south of the country; the latter, capable of supplying some 2,000 MW, served the mining and processing operations centred on the Carajá region. In all, the production of electricity by hydroelectric generation in 2003 was well over 50 times that of 1980. Moreover, since 2000 there has been work further to expand the hydroelectric power station at Itaipú in order to increase its generating capacity to 14,000 MW. However, a drought and lack of finance for the expansion project reduced the production of this type of electricity in 2001; to compensate for this, and avoid rolling blackouts, in June 2001 the Government implemented a six-month rationing system in an attempt to reduce consumption by 20%. However, the drought ended later in the year and electricity production returned to normal levels, although consumption remained somewhat lower.

The nuclear power industry in Brazil has a troubled history and has been an extraordinarily expensive and unsuccessful experiment. The nuclear plant Angra I, which began production in 1985, had to be closed down almost immediately for safety reasons. It was again closed in 1987, because of generating difficulties, and has operated only intermittently thereafter, and at very low capacity (often below 10%). Financial constraints hindered the completion of Angra II, which became operational only in mid-2000. A further plant, Angra III, was scheduled to become operational in 2008. In 2003 more than 90% of energy was produced from hydroelectric power, about 4% from petroleum, 3% from coal and only 1% from nuclear power.

## PETROLEUM AND ETHANOL

In the early 1980s Brazil's petroleum sector underwent a transformation of momentous consequence. The first advance was a doubling of domestic production, as a result of the discovery and development of the Campos oilfield, off shore from the state of Rio de Janeiro. The second factor was a massive programme to reduce oil imports via its substitution by alcohol, through the production of ethanol derived from sugar cane. The combination of these two factors, as well as low petroleum prices during the 1990s (a period in which demand was stable or even declining), resulted in a decrease in the share of imports accounted for by petroleum (and lubricants), from over 50% in 1979 to less than 7.0% in 1999. As Brazil met about 80% of its domestic petroleum requirements, oil price increases thereafter had moderate effect on imports (in 2004 these imports represented 10.8% of total imports).

After years in which Petrobras (Petróleo Brasileiro, SA) had invested heavily in a largely fruitless search for petroleum, in 1974 the first strike was made in the Campos field. In 1981 production was 220,000 b/d (barrels per day) and, in the following year, the Campos field came on stream. In the early 1990s record levels of production, of over 700,000 b/d, were achieved, placing Brazil third among regional producers, after Mexico and Venezuela. Production continued to increase thereafter and in 1999, as part of plans to reorganize and modernize the company, it was announced that Petrobras would be transformed into a regional, rather than just a national, entity, with a new mixed-ownership structure (although the state would permanently retain a clear dominant control). One of the first components of the modernization project was a US $1,000m. scheme to improve safety following two major oil spills. As such, government plans for the petroleum industry seemed to differ from those for other industries where there was previous state involvement, and one in which outright privatization has been explicitly ruled out. In 2004 Petrobras was supplying more than 90% of Brazil's national petroleum consumption, and Brazil was expected soon to become self-sufficient. In addition to petroleum, Brazil produced 13,500m. cu m of natural gas in 2003. In June 2005 the Presidents of the Mercosul countries signed an agreement creating an 'energy ring' (*ariel energético*), which would facilitate the supply and storage of natural gas from the Camisea field in Peru.

The controversial ethanol programme dated from the founding of PROALCOOL in 1976. From that time thousands of hectares were planted with sugar cane, more than 400 processing plants were constructed and a capacity of 160m. hectolitres of ethanol production per year was achieved. The macroeconomic and environmental benefits of this programme have always been disputed. Soon after the beginning of the programme a 20% ethanol content in fuel oil was mandatory and, in the mid-1980s, the automobile and truck industries in Brazil changed the bulk of their production to engines that burnt either 'gasohol' or 100% ethanol. By 1986 more than 50% of fuel oil consumption was of ethanol and production was approaching 120m. hectolitres, but, as a result of rapidly falling petroleum prices, the production cost of hydrated alcohol reached a level twice that of imported petroleum. The removal of most subsidies and price controls in 1990 was followed by the closure of 67 of the units producing ethanol. Thereafter, production continued to decline as demand fell away, placing the future of the whole programme in doubt. However, the sharp increase in the price of petroleum at the end of the 1990s and again since 2003 brought a new lease of life to the industry.

## MANUFACTURING

In 2003 manufacturing employed 10.7% of the active labour force, almost one-half the level of 1980. In that year the sector contributed 22.9% of GDP, compared with 31% in 1982; however, according to the World Bank, in US dollar terms the decline of the share of manufacturing in domestic output was even more remarkable: from 35% of GDP in 1982 (at the time of the debt-crisis), to 25% in 1993 (the year before the Real Plan), to just 11.4% in 2003. According to many analysts, this rapid process of 'de-industrialization' was the main reason for the slowdown in the rate of economic growth of the Brazilian economy since 1980. However, despite the relative shrinkage of the manufacturing sector, the industrial structure of the country was still remarkably diverse, and this was fully reflected in the breadth of the manufacturing sector. Links with the primary sector played an important part in Brazil's advanced manufacturing base. As noted above, a substantial proportion of sugar cane was utilized for the production of ethanol. In the mining sector, a significant amount of iron ore was transformed into steel, with production totalling 32.9m. metric tons by 2004. Moreover, the country was largely self-sufficient in the areas of hydroelectricity and transmission systems, producing many of its own turbines, generators, transformers and reactors. Increasing competitiveness in deep-sea petroleum exploration and extraction won Petrobras substantial contracts to provide drilling platforms and other equipment for Middle Eastern and other oilfields. Industrialized inputs into agriculture came increasingly from the domestic chemicals sector.

A second source of stimulus to the sector was the demand coming from a large internal market. Here, the vehicle and components industry played a key role. Rapid expansion of the passenger motor vehicle sector in the 1960s and 1970s, following the entry of European and then US manufacturers, resulted in production levels in Brazil of about 1m. vehicles per year by the end of the 1970s. The recession that began in 1982 adversely affected this sector, reducing domestic demand by 40%. However, some of the large foreign investments in the country in the 1990s were intended to modernize and enlarge this sector. Production recovered to 2.1m. in 1997, before falling to 1.8m. in 2003.

Partial trade liberalization had a double-edged effect on the vehicle industry. Imports increased very rapidly, from 3% of apparent car consumption in 1991, to well over 33% in 1995. In terms of value, imports of passenger motor cars grew nearly nine-fold in only six years, from US $358m. in 1989 to over $3,000m. in 1995; however, there was a sharp decrease in this figure in 1996, to $1,600m., following an increase in tariffs for non-Mercosul imports and the introduction of stricter monetary policy. This figure increased again in the late 1990s, to about $2,500m., but then fell again in 2000, to just $1,200m., and continued to fluctuate thereafter. Exports also increased very rapidly, albeit erratically, mainly owing to the depressed state of the internal market, increased foreign investment, the development of Mercosul and some improvement in output quality (a long-standing problem for Brazilian exports). Exports of passenger vehicles increased by 136% in 1997 before declining by nearly 33% in 1999. A recovery was under way from 2000, and in 2004 passenger cars generated $3,351.5m. in export earnings. The rising proportion of exports from the manufacturing sector was an indicator of the strength and some degree of sophistication of industrial production. In addition to the exports of cars, trucks and buses, Brazil exported passenger jet aircraft (particularly the quiet and economical Bandeirante) and military aircraft (such as the trainer aeroplane, the Tucano). The state-owned Embraer (Empresa Brasileira de Aeronáutica, SA) developed a range of models for export and engaged in joint ventures, in the civilian and military fields, with a number of foreign governments and corporations.

The recession experienced by Brazil in the early 1980s severely affected the industrial sector. The crisis was exacerbated by sharp reductions in state investment, historically important in all areas and central to the development of the capital-goods industry. President Collor de Mello's reforms in 1990 provoked a further sharp recession. Output recovered in the mid-1990s, but difficult conditions at the end of the decade led to a contraction in the sector. Manufacturing output grew at an average annual rate of just 1.1% between 1990 and 2003; however, according to preliminary figures, the sector expanded by 7.7% in 2004, and was forecast to grow by 5.5% in 2005. Preliminary data for 2004 indicated that the recovery was led by consumer durables and capital goods, while non-durables and intermediate inputs grew at a much lower rate. Trade liberalization (though not as sharp as in most other Latin American countries), in a context of a remarkably strict monetary policy, financial deregulation, and an unstable domestic and international environment, proved to be a major challenge for Brazil's manufacturing industry, the outcome of which still remained uncertain.

## TRANSPORT INFRASTRUCTURE

One of the key characteristics of industrial policy in Brazil from the 1960s was its emphasis on road construction, as a stimulus to the domestic car and truck industry. This was the case even with regard to the opening up of the Amazon region and the agricultural frontier. In 2004 Brazil had more than 25m. passenger cars and there were about 1.7m. km of roads, less than 10% of which were paved. In addition, there were some 29,000 km of railway line in 2003, much of which served the mining areas of Minas Gerais and Carajá. Rail projects for lines between the mining areas and the ports of Tubarao and São Luis accounted for the bulk of recent investment in the railways.

There are more than 40 deep-water ports, the largest being Santos, Rio de Janeiro, Paranaguá, Recife and Vitória. Santos handles some 30m. metric tons of cargo per year, and important new facilities have been developed for 'roll on-roll off' containers there. Brazil has the largest merchant fleet in Latin America. The River Amazon is navigable for 3,700 km, as far as Iquitos, in Peru; and ocean-going ships can reach as far as Manaus, 1,600 km upstream. On account of Brazil's size and the obstacles presented to other means of transport, by jungles, rivers and mountains, air travel is very important. There are about 1,500 airports and airstrips, of which 22 handle international flights, although the bulk of international traffic goes through (the overstretched) São Paulo airport and (the much under-utilized) Rio de Janeiro airport.

## TOURISM

Tourism had the potential to become an important net source of foreign exchange for Brazil; however, earnings were not only tiny with respect to the extraordinary potential of a country such as Brazil, but were only a fraction of the expenditure in tourism by Brazilians abroad. It is obvious that with improved security for tourists (the rapidly growing crime rate in Brazil, particularly in Rio de Janeiro, proved to be the main obstacle for the tourism industry), more investment in infrastructure and better promotion, Brazil could significantly develop its tourism industry. In 2003 some 4.1m. tourists visited Brazil, down from 4.8m. in 2001, although most of the decline was owing to a fall in tourist numbers from Argentina. About one-third came from other Mercosul countries and Chile. Receipts from tourism stood at US $3,386m. in 2003. Further development of the industry seemed to be one of the most obvious foreign-exchange targets of government policy, in view both of the attractions of the country and of the structural changes taking place in the world market for tourism, with emphasis being placed upon more 'exotic' locations. However, it seemed inevitable that unless the authorities made substantial improvements in law and order that tourist numbers would not increase significantly in the near future.

## ENVIRONMENTAL ISSUES

Environmental damage and ecological destruction was a growing feature of Brazilian development from the mid-20th century, and of extreme significance from the 1970s. After 1964 the military obsession with 'occupying Amazonia' meshed with a 'developmentalism at any cost' mentality, hospitality to foreign capital and enthusiasm for large-scale projects. This resulted in a series of ill-advised road-building, farming and mining ventures, the impact of which on the environment was severe and destructive. The worst of these were the Transamazonian Highway, the highly corrupt, heavily subsidized and economically unproductive ranching projects of the 1970s, the controversial project of the Tucuruí dam, and the promotion of energy-intensive aluminium production. By the end of the 1980s much of the ranching had been abandoned, but mining had expanded considerably.

It was estimated that during the 1990s on average as much as 200,000 sq km of the Amazon rainforest were destroyed every year. In spite of the decline in large-scale ventures, the burning of rainforest for land clearance continued apace. This figure reached record levels at the beginning of 1998, owing to deliberate large-scale fires near the border with Venezuela (close to the Yanomani reserves) that became uncontrollable. Extensive logging for export (a large amount of which was illegal) and large-scale charcoal production, without which the Carajá iron-ore project would have been uneconomical, also continued into the 2000s. Furthermore, the conflict between an estimated 50,000 gold prospectors and 9,000 Yanomani Indians (the latter groups ravaged by new diseases brought by the miners) was accompanied by the equally insidious effects of the introduction into the Amazon river system of as much as 200 metric tons of mercury every year, from the process of gold production. These developments, along with the well-publicized conflicts between rubber-tappers and local warlords in the far north of the country, culminating in the death of union activist Chico Mendes in December 1988, led to the emergence of powerful local and international opposition movements. Throughout the 1980s and 1990s government policy in addressing environmental issues was ineffective. By mid-2005 President da Silva's Government had not proved any better in this respect. In fact, recent official data showed that in 2003 alone, an area of Amazon rainforest nearly the size of Belgium was destroyed, in part by large-scale soya-producing farms attempting to expand their activities. Although the Government invested in an expensive modern satellite-tracking system to monitor the situation, it did little to address the situation.

In spite of such spectacular initiatives as the satellite-tracking system, and the related discussions on linking debt reduction to the creation of forest reserves, it remained the case that Brazilian Governments so far have lacked the capacity and the will to rationalize the expansion of commercial farming and control the activities of the hundreds of thousands of individuals pushed into the Amazon by poverty. In the immediate future the ecological impact of Amazonian development probably depends more on international pressure on foreign investors and funding institutions, and the overall rhythm of economic activity, than on the efforts of the Brazilian Government.

## INVESTMENT AND FINANCE

Throughout the period of Brazil's rapid industrial expansion, public-sector investment consistently surpassed private investment. Therefore, the reductions in state investment from the mid-1980s had a significant effect on the rate of expansion of productive capacity and the economy as a whole. In terms of percentages of GDP, gross domestic investment fell from over 32% of GDP in 1980 to to 18.6% in 2004 (a figure almost identical to the Latin American average for the year). The composition of this investment between the public and private sector changed sharply, with public investment all but disappearing. According to official figures for 2003, while current expenditure by the federal Government reached 23.8% of GDP, public expenditure by the federal Government in capital formation represented just above 1% of GDP. The corresponding figure for the various states and municipalities was 16% and 1.4%, respectively. As public expenditure in education and health also fell as a percentage of GDP, total public expenditure in physical and human capital fell from (the already low levels of) 11.4% in 1993 to about 6% in 2003. The decline in investment was evident in areas such as electricity generation and public roads: the average rate of growth in electricity generation (in terms of megawatts per hour) between 1964 and 1980 was 10%, whereas this figure fell to just 3.5% between 1994 and 2003. Similarly, the respective rate of growth in public roads (measured in terms of kilometres) was 5.6% in 1964–80, compared with 0.7% in 1994–2003.

The finance of investment also changed significantly after 1980. In that year gross national savings financed 88% of investment, while foreign savings financed the remaining 12%. However, as foreign savings became more difficult to obtain following the 1982 debt crisis, until 1994 investment had to be fully financed with national savings. This picture changed abruptly with financial liberalization: the inflow of foreign savings increased from 0.3% of GDP in 1994 to 5.5% in 1998 (despite the collapse following the Russian default in August that year). Foreign savings have remained at about 4% of GDP since then. Useful as these foreign savings could be, in Brazil, as in the rest of Latin America, increased foreign savings became a substitute for, instead of a complement to, national savings (as in East Asia). In Brazil these decreased sharply from the mid-1990s.

One of the main aims of President Collor's New Brazil programme in 1990 was to transform the increasing federal deficit (of about 8% of GDP) into a surplus of 2% through a mixture of tax increases, expenditure reductions and privatization receipts. Although a budget surplus was achieved in 1990 (equivalent to 1.4% of GDP), the Government was not able to sustain this achievement. Following a change of economy minister in 1993, more orthodox policies, centred on high real interest rates, were implemented. However, these not only failed to reduce inflation, but also were unable to achieve a sustainable equilibrium in the public-sector accounts, let alone to prompt a sustained resumption of growth. One of the obvious reasons why Collor de Mello had trouble balancing the budget was that his own policy of high interest rates trebled the servicing of public debt in just one year (to 4.4% of GDP).

The economic situation did not change much until the implementation of the Real Plan in 1994, which aimed to attack the root causes of inflation, while at the same time attempting to achieve financial balance in the public sector. As far as monetary policy was concerned, instruments and rules were devised to guarantee price stability. Following implementation of the Real Plan, the monthly rate of inflation, which had reached a record of nearly 50% in June, fell immediately to approximately 2% in three months. However, the Plan was not equally successful as far as the public-sector deficit was concerned. Even though an initial surplus was achieved in 1994 (equivalent to 1.1% of GDP), by 1995 this had already turned into a substantial deficit of 4.9% of GDP. Partly as a result of the increase in interest rates that followed the East Asian financial crisis in 1997 and the Russian devaluation in 1998, and partly owing to further absorption of 'bad' debt from the private financial sector and state governments, by 1998 the deficit reached 7.4% of GDP. As a result, net public debt (that is, total public debt minus international reserves and public financial assets) increased at a very rapid pace. Brazil's public debt, although not excessively large as a share of GDP compared with other countries, became unmanageable, owing to continuous high levels of interest rates (by March 2005 increased deposit interest rates reached 19.2%, at a time in which the accumulated inflation rate in the previous 12 months stood at 6.6%). In 2004 the respective figures for the net debt of the federal Government and the Central Bank, state governments and municipalities, and public corporations were 32%, 19%, and 0.2% of GDP. Owing to the high levels of reserves (US $54,000m. in January 2005), about 80% of the total public-sector net debt was domestic.

As the 'primary' accounts of the federal Government and Central Bank were either in surplus or roughly in balance throughout the period of the Plan, the large increase in debt in the 1990s was caused by a number of other factors. The first of these was the large and continuous absorption of bad financial debt from the private sector and state governments. The key problem of the banking sector arose mainly because financial liberalization created a disparity between the requirements of its assets and liabilities; the former were domestic and could only perform if borrowers had to pay low interest rates, the latter had a substantial foreign-exchange exposure and required high interest rates to sustain the exchange rate. As interest rates could not be both low and high at the same time, continuous high interest rates made the banking sector increasingly fragile on its assets side. The second factor was that state governments were unable to keep their finances in order owing to high interest; this led to frequent rescue operations from the federal Government and Central Bank. The third factor was the cost of 'sterilization' of high foreign inflows. Last, the fourth main factor was that interest rates also had to stay high during the second half of the 1990s as a poor substitute for the lack of fiscal reforms and political stalemate during most of Cardoso's administration. All these led to a situation in which Brazil violated a crucial 'golden rule' of public finance: the interest rate paid on public debt was continuously higher than both the growth of public revenues and the returns on foreign-exchange reserves; as a result of this asymmetry Brazil had to borrow just to be able to service the existing debt.

As a result of the fiscal adjustment agreed with the IMF following the January 1999 devaluation, the public sector increased its primary surplus to 3.1% of GDP in 1999 (from 0.01% in 1998), but as its interest payments increased dramatically, to 13.3% of GDP (as a result of high interest rates and the large devaluation of the real), the total deficit of the public sector (in nominal terms) reached 10.2% of GDP. In 2002, at the end of Cardoso's Government, the primary surplus had improved to nearly 4.0% of GDP, but interest payments stood at 14.4% of GDP (and the overall public-sector deficit reached 10.5% of GDP). One of the main aims of da Silva's Government was to increase even further the primary surplus; in 2003 this reached 4% of GDP, and in 2004 4.6% of GDP.

Brazil's long-term failure to generate a trade surplus sufficiently large to pay for its existing foreign obligations, and the long periods it enjoyed of easy access to international finance, led to a rapid increase in its foreign debt. The first surge in foreign debt took place in the 1970s; this was a direct consequence of Brazil having made the same mistake as most Latin American countries during this period: it opened the capital account of the balance of payments at a time of high international liquidity and borrowed heavily at variable interest rates, without having had a clear programme of export expansion able to generate the foreign exchange required to service the resulting growing foreign debt. In 1987 the country declared a moratorium on payments on its medium- and long-term debt, by then worth some US $120,000m. Following several negotiations and the implementation of the 'Brady Plan', Brazil once more gained access to international finance and again began to borrow heavily abroad. Therefore, foreign debt continued to increase, reaching a peak of $260,000m. in 1998; thereafter it began to decline, reaching $221,000m. in 2004 (still equivalent to more than twice its level of exports of goods and services). The structure of this debt, in terms of the share owed by the public and private sectors, also changed substantially after the implementation of the Real Plan: in the early 1990s about 80% of the external debt was public; by 2004 this share had fallen by about one-half.

The reduction of international interest rates and the resumption of voluntary international capital movement into the country in the early 1990s were enormously helpful to Brazil in coping with its foreign-debt problem in the short-term. Private capital inflows into Brazil (in the form of foreign direct investment, bonds and portfolio equity) recorded a significant increase. Foreign direct investment flows from European countries into Brazil reached a total of US $3,300m. in 1987–90 (about one-half of the Latin American total), while those from the USA reached $5,400m. By 1996 the net inflow of foreign capital had increased dramatically, to $34,300m. (or 72% of exports). In 1997, mainly as a result of the East Asian crisis, these net inflows were reduced to $26,000m., and in 1998 there was a further decrease owing to the Russian default. Following the January 1999 crisis practically all non-foreign direct investment inflows into Brazil ceased, but foreign direct investment remained very high until 2002 when they fell by almost 50%, to $14,000m. In 2003 inflows totalled just $7,100m.

One of the major problems of the Real Plan was that it had to be implemented during a period in which three significant external shocks had severe implications for Brazil's external finances. In 1997 the upheaval in international financial markets reduced the net inflow into Brazil, although foreign direct investment continued to increase. The impact of the Russian devaluation in 1998 and, later, Brazil's own financial crisis in January 1999 resulted in further massive outflows. In terms of net transfer of resources (net capital inflows minus net interest payments and profit remittance), Brazil posted a positive figure of US $20,000m. in 1996; by 2003 this figure had been reversed to a negative outflow of $32,000m.

## FOREIGN TRADE AND THE BALANCE OF PAYMENTS

After the 1982 debt crisis Brazil's traditional trade deficit had to be rapidly transformed into a surplus in order to generate the foreign exchange needed for the financial requirements of the current account. This was achieved mainly through weak domestic demand, the domestic production of and substitution for petroleum, growing capital-goods self-sufficiency, and a policy of export promotion. The trade surpluses continued until the implementation of the Real Plan in 1994, when the rapid increase in imports brought about a significant deterioration in the trade surplus. Following the East Asian crisis in mid-1997 the Government introduced strict policies to reduce the mounting trade deficit, which was mainly the result of a rapid increase in the value of merchandise imports that followed the process of import liberalization and the overvaluation of the currency. Exports also increased, but not at the same rate. As a result of the January 1999 crisis, additional policies were implemented in an attempt to turn again the deficit into a surplus that would help finance the large obligations of the current account and to comply with the conditions of the IMF rescue package. Finally, in 2001, a trade surplus (of some US $2,650m.) was registered; thereafter further improvements in the trade balance of goods were registered, with surpluses of $24,801m. and $32,966m. in 2003 and 2004, respectively.

Heavy debt obligations produced deficits in the current account of the balance of payments in the years immediately preceding the implementation of the Real Plan. The Real Plan started in 1994 with a small current-account deficit of US $1,153m.; however, by 1995 this deficit had risen dramatically to $18,136m. The deterioration in the current account continued thereafter, with the deficit reaching $33,829m. in 1998 (equivalent to 4.5% of GDP and 70% of merchandise exports). In spite of the January 1999 devaluation and the improvement in the trade balance, the current-account deficit did not improve significantly until 2002, when a deficit of $7,637m. was registered. The huge trade surpluses of 2003 and 2004 brought the current account into surpluses of $4,016m. and $11,094m., respectively.

Substantial as the pre-2003 deficits were, for most of the time in the 1990s capital inflows into the country were even larger, massively increasing the level of foreign reserves (which reached US $58,322m. in 1996). By December 2001 these reserves had decreased to $35,729m., but by December 2004 they had recovered to $52,770m.

As in the aftermath of the Mexican peso crisis in 1995, most analysts were surprised by the speed of the return of foreign capital to Brazil following the 1997 East Asian crisis. This was clearly helped by the rapid reaction from the Brazilian authorities, who swiftly increased interest to an annualized level of 43%, raised some import tariffs significantly, and took measures to facilitate these inflows. However, this short-term success carried with it the danger of complacency and by mid-1998, particularly owing to the massive inflow of foreign capital in the first half of the year, the East Asian crisis appeared to have been totally forgotten. As it was, by then, and before the Russian default, Brazil already needed to borrow abroad at a premium, sell some of its export products at a discount, and cope with reduced demand for its exports from Asia and Argentina. Added to this was the urgent problem of the public accounts. However, in an election year little was done. This might have helped President Cardoso's re-election, but it did little to strengthen the Brazilian economy. As a result, the Russian crisis affected the Brazilian economy much more adversely than either of the two previous crises, and by January 1999 the Central Bank had little option but massively to devalue the currency. To contain the subsequent crisis, the Government not only had to increase deposit interest rates yet again, but also had to be more serious about its determination to decrease the public-sector deficit. The IMF made the latter a condition of its huge financial rescue programme.

The initial recovery from the crisis was impressive, but many of the problems that led to the crisis still persisted by the end of the Cardoso Government. Both the internal and external debts continued to tax heavily both the public finance and the balance of payments. In 1999 the Government still had to use about 30% of its revenues in debt service and pay US $9,000m. in interest on its foreign debt, and an additional $27,000m. for the amortization of this debt in the first six months of 2000 alone.

## THE GOVERNMENT OF 'LULA' DA SILVA

By the time of the October 2002 elections the Brazilian economy was facing a set of adverse shocks and President Cardoso's room for manoeuvre had long since disappeared. The government coalition had collapsed, no plans for serious fiscal reform had been agreed upon between the Government and the Congresso Nacional (National Congress), and the level of economic activity in the first half of 2002 showed an economy already in recession. The economic downturn in the USA and the continuing financial crisis in Argentina were part of the problem; however, the real problem rested with a Government that, towards the end of its second period in office, had run out of resources and ideas. Furthermore, President Cardoso's attempt to frighten the electorate regarding the economic consequences of a victory by the left-wing opposition candidate, 'Lula' da Silva, backfired; Cardoso's rhetoric succeeded only in frightening international financial markets, which responded by increasing Brazil's country risk premium to 16 percentage points (after Argentina, the highest in the world).

Immediately after taking office in January 2003, da Silva committed his Government to an extreme fiscal and monetary austerity, even stricter than that agreed by the previous Government with the IMF. Furthermore, by the end of the first quarter of 2003 he had succeeded in this not only by increasing interest rates to 26.5% (and by doing so bringing the economy again into recession), but also by increasing the primary surplus via a drastic reduction in (already particularly low) social expenditure (including health and education) of nearly two percentage points of GDP. The trade surplus continued to improve rapidly and the deficits in services and income also continued to narrow, bringing the current account into an unprecedented surplus. Da Silva also succeeded in winning the approval of the Congresso Nacional for some fiscal reforms.

However, on the political front, failure to deliver social change and President da Silva's opposition to a basic increase of the minimum wage (despite an election campaign pledge) at a time when the Government was indiscriminately and massively increasing (non-performance-related) subsidies to foreign and domestic corporations seriously eroded the President's popular support in 2004 and had negative consequences for the Government's ability to secure approval for the rest of its ambitious reforms in the Congresso. On the economic front, nevertheless, the Brazilian economy managed a substantial recovery in 2004: annual GDP growth increased to 5.2% (compared with 0.5% in 2003), with growth per head of 3.7%; average wages grew by 9%, and capacity utilization increased to 82.8%. In March 2005 the Government announced that the agreement with the IMF would not be renewed. Unless there was a major upheaval in the world economy in the second half of 2005, the positive economic outlook of 2004 was expected to be repeated in 2005, placing President da Silva in an ideal position for a re-election attempt in 2006.

Probably the crucial test for the remainder of President da Silva's term in office was not so much how to deal with the prevailing economic problems of inflation, and overvalued exchange rate, etc., but the capacity to find an imaginative solution to the acute problem of the huge domestic debt. As by 2005 both 'corner' solutions seemed impossible (full payment being economically unachievable and an Argentine-style unilateral default politically unacceptable, both at home and abroad), the real challenge for the Government in its remaining period in office would be to find a 'third way' solution, much as the Brady Plan did for Latin America's unpayable foreign debt at the end of the 1980s.

# Statistical Survey

Sources (unless otherwise stated): Economic Research Department, Banco Central do Brasil, SBS, Q 03, Bloco B, Brasília, DF; tel. (61) 414-1074; fax (61) 414-2036; e-mail coace.depec.@bcb.gov.br; internet www.bcb.gov.br; Instituto Brasileiro de Geografia e Estatística (IBGE), Centro de Documentação e Disseminação de Informações (CDDI), Rua Gen. Canabarro 706, 2° andar, Maracanã, 20271-201 Rio de Janeiro, RJ; tel. (21) 2142-4781; fax (21) 2142-4933; e-mail webmaster@ibge.bov.br; internet www.ibge.gov.br.

## Area and Population

### AREA, POPULATION AND DENSITY

| | |
|---|---:|
| Area (sq km) | 8,547,404* |
| Population (census results)† | |
| 1 August 1996 | 157,070,163 |
| 1 August 2000 | |
| Males | 83,576,015 |
| Females | 86,223,155 |
| Total | 169,799,170 |
| Population (official estimates at mid-year) | |
| 2002 | 176,391,015 |
| 2003 | 178,985,306 |
| 2004 | 181,586,030 |
| Density (per sq km) at mid-2004 | 21.2 |

* 3,300,170.9 sq miles.
† Excluding Indian jungle population (numbering 45,429 in 1950).

### ADMINISTRATIVE DIVISIONS
(population at census of 1 August 2000)

| State | Population | Capital |
|---|---:|---|
| Acre (AC) | 557,526 | Rio Branco |
| Alagoas (AL) | 2,822,621 | Maceió |
| Amapá (AP) | 477,032 | Macapá |
| Amazonas (AM) | 2,812,557 | Manaus |
| Bahia (BA) | 13,070,250 | Salvador |
| Ceará (CE) | 7,430,661 | Fortaleza |
| Espírito Santo (ES) | 3,097,232 | Vitória |
| Goiás (GO) | 5,033,228 | Goiânia |
| Maranhão (MA) | 5,651,475 | São Luís |
| Mato Grosso (MT) | 2,504,353 | Cuiabá |
| Mato Grosso do Sul (MS) | 2,078,001 | Campo Grande |
| Minas Gerais (MG) | 17,891,494 | Belo Horizonte |
| Pará (PA) | 6,192,307 | Belém |
| Paraíba (PB) | 3,443,825 | João Pessoa |
| Paraná (PR) | 9,563,458 | Curitiba |
| Pernambuco (PE) | 7,918,344 | Recife |
| Piauí (PI) | 2,843,278 | Teresina |
| Rio de Janeiro (RJ) | 14,391,282 | Rio de Janeiro |
| Rio Grande do Norte (RN) | 2,776,782 | Natal |
| Rio Grande do Sul (RS) | 10,187,798 | Porto Alegre |
| Rondônia (RO) | 1,379,787 | Porto Velho |
| Roraima (RR) | 324,397 | Boa Vista |
| Santa Catarina (SC) | 5,356,360 | Florianópolis |
| São Paulo (SP) | 37,032,403 | São Paulo |
| Sergipe (SE) | 1,784,475 | Aracaju |
| Tocantins (TO) | 1,157,098 | Palmas |
| Distrito Federal (DF) | 2,051,146 | Brasília |
| **Total** | **169,799,170** | — |

# BRAZIL

## PRINCIPAL TOWNS
(population at census of 1 August 2000)*

| | | | | |
|---|---|---|---|---|
| São Paulo | 10,434,252 | Santo André | 649,331 |
| Rio de Janeiro | 5,857,904 | João Pessoa | 597,934 |
| Salvador | 2,443,107 | Jaboatão | 581,556 |
| Belo Horizonte | 2,238,526 | São José dos Campos | 539,313 |
| Fortaleza | 2,141,402 | Contagem | 538,017 |
| Brasília (capital) | 2,051,146 | Ribeirão Preto | 504,923 |
| Curitiba | 1,587,315 | Uberlândia | 501,214 |
| Recife | 1,422,905 | Sorocaba | 493,468 |
| Manaus | 1,405,835 | Cuiabá | 483,346 |
| Porto Alegre | 1,360,590 | Feira de Santana | 480,949 |
| Belém | 1,280,614 | Aracaju | 461,534 |
| Goiânia | 1,093,607 | Niterói | 459,451 |
| Guarulhos | 1,072,717 | Juíz de Fora | 456,796 |
| Campinas | 969,396 | São João de Meriti | 449,476 |
| Nova Iguaçu | 920,599 | Londrina | 447,065 |
| São Gonçalo | 891,119 | Joinville | 429,604 |
| São Luís | 870,028 | Santos | 417,983 |
| Maceió | 797,759 | Campos dos Goytacazes | 406,989 |
| Duque de Caxias | 775,456 | Ananindeua | 393,392 |
| Teresina | 715,360 | Olinda | 367,902 |
| Natal | 712,317 | Mauá | 363,392 |
| São Bernardo do Campo | 703,177 | São José do Rio Preto | 358,523 |
| Campo Grande | 663,621 | Diadema | 356,389 |
| Osasco | 652,593 | | |

* Figures refer to *municípios*, which may contain rural districts.

**Brasília (Distrito Federal)** (official population estimates at mid-year): 2,180,279 in 2002; 2,231,101 in 2003; 2,282,049 in 2004.

## BIRTHS, MARRIAGES AND DEATHS
(official estimates based on annual registrations)

| | Birth rate (per 1,000) | Death rate (per 1,000) |
|---|---|---|
| 1997 | 21.9 | 5.7 |
| 1998 | 25.5 | 5.8 |
| 1999 | 25.1 | 5.8 |
| 2000 | 24.1 | 5.6 |
| 2001 | 21.7 | 5.5 |
| 2002 | 22.1 | 5.7 |

**Registered births** (incl. births registered but not occurring during that year): 3,576,890 in 1997 (1,231,816 not occurring in that year); 4,218,204 in 1998 (1,758,929 not occurring in that year); 4,209,768 in 1999 (1,552,155 not occurring in that year); 4,107,757 in 2000 (1,496,335 not occurring in that year); 3,743,651 in 2001 (1,234,297 not occurring in that year); 3,853,869 in 2002 (1,272,814 not occurring in that year); 3,649,996 in 2003 (827,534 not occurring in that year).

**Registered marriages:** 724,738 in 1997; 698,614 in 1998; 788,744 in 1999; 732,721 in 2000; 710,121 in 2001; 715,166 in 2002; 746,727 in 2003.

**Registered deaths** (incl. deaths registered but not occurring during that year): 932,858 in 1997 (23,541 not occurring in that year); 960,328 in 1998 (20,880 not occurring in that year); 966,010 in 1999 (22,486 not occurring in that year); 945,492 in 2000 (20,792 not occurring in that year); 953,519 in 2001 (25,174 not occurring in that year); 987,966 in 2002 (29,491 not occurring in that year); 996,729 in 2003 (29,012 not occurring in that year).

**Expectation of life** (WHO estimates, years at birth): 69 (males 66; females 73) in 2003 (Source: WHO, *World Health Report*).

## ECONOMICALLY ACTIVE POPULATION
('000 persons aged 10 years and over, labour force sample survey at September)*

| | 2002 | 2003 |
|---|---|---|
| Agriculture, hunting, forestry and fishing | 16,141.3 | 16,409.4 |
| Industry (excl. construction) | 11,131.5 | 11,387.0 |
| Manufacturing industries | 10,569.0 | 10,749.1 |
| Construction | 5,558.4 | 5,157.6 |
| Commerce and repair of motor vehicles and household goods | 13,414.1 | 14,047.5 |
| Hotels and restaurants | 2,901.5 | 2,858.3 |
| Transport, storage and communication | 3,654.3 | 3,680.6 |
| Public administration | 3,830.4 | 3,942.2 |
| Education, health and social services | 6,990.6 | 7,087.3 |
| Domestic services | 6,047.7 | 6,081.9 |
| Other community, social and personal services | 3,116.2 | 2,947.0 |
| Other activities | 5,193.5 | 5,455.6 |
| Activities not adequately defined | 200.1 | 196.2 |
| **Total employed** | **78,179.6** | **79,250.6** |
| Unemployed | 7,876.0 | 8,537.0 |
| **Total labour force** | **86,055.6** | **87,787.7** |

* Data coverage excludes rural areas of Acre, Amapá, Amazonas, Pará, Rondônia and Roraima.

**2001** ('000 persons aged 10 years and over, labour force sample survey at September): Total employed 75,458.2; Unemployed 7,785.1; Total labour force 83,243.2.

# Health and Welfare

## KEY INDICATORS

| | |
|---|---|
| Total fertility rate (children per woman, 2003) | 2.2 |
| Under-5 mortality rate (per 1,000 live births, 2003) | 35 |
| HIV/AIDS (% of persons aged 15–49, 2003) | 0.7 |
| Physicians (per 1,000 head, 2001) | 2.06 |
| Hospital beds (per 1,000 head, 1996) | 3.11 |
| Health expenditure (2002): US $ per head (PPP) | 611 |
| Health expenditure (2002): % of GDP | 7.9 |
| Health expenditure (2002): public (% of total) | 45.9 |
| Access to water (% of persons, 2002) | 89 |
| Access to sanitation (% of persons, 2002) | 75 |
| Human Development Index (2002): ranking | 72 |
| Human Development Index (2002): value | 0.775 |

For sources and definitions, see explanatory note on p. vi.

# Agriculture

## PRINCIPAL CROPS
('000 metric tons)

| | 2001 | 2002 | 2003 |
|---|---|---|---|
| Wheat | 3,365 | 3,106 | 6,029 |
| Rice (paddy) | 10,184 | 10,457 | 10,319 |
| Barley | 298 | 245 | 336 |
| Maize | 41,955 | 35,933 | 47,988 |
| Oats | 342 | 299 | 415 |
| Sorghum | 914 | 787 | 1,755 |
| Buckwheat* | 50 | 48 | 48 |
| Potatoes | 2,849 | 3,126 | 3,047 |
| Sweet potatoes | 485 | 498 | 495* |
| Cassava (Manioc) | 22,577 | 23,066 | 22,147 |
| Yams* | 235 | 230 | 230 |
| Sugar cane | 345,942 | 363,721 | 389,849 |
| Dry beans | 2,453 | 3,064 | 3,310 |
| Brazil nuts | 28 | 27 | 28* |
| Cashew nuts | 124 | 164 | 178 |
| Soybeans (Soya beans) | 37,881 | 42,125 | 51,482 |
| Groundnuts (in shell) | 202 | 195 | 177 |
| Coconuts | 2,131 | 2,892 | 2,851 |
| Oil palm fruit* | 460 | 450 | 516 |
| Castor beans | 100 | 171 | 78 |
| Sunflower seed† | 158 | 150 | 168 |
| Cottonseed† | 1,665 | 1,363 | 1,402 |
| Tomatoes | 3,103 | 3,653 | 3,694 |
| Onions (dry) | 1,050 | 1,222 | 1,194 |
| Garlic | 102 | 114 | 122 |

BRAZIL

*— continued*

| | 2001 | 2002 | 2003 |
|---|---|---|---|
| Other fresh vegetables* | 2,200 | 2,250 | 2,250 |
| Watermelons* | 600 | 620 | 620 |
| Cantaloupes and other melons* | 150 | 155 | 155 |
| Bananas | 6,177 | 6,423 | 6,775 |
| Oranges | 16,983 | 18,531 | 16,903 |
| Tangerines, mandarins, clementines and satsumas | 1,125 | 1,263 | 1,263* |
| Lemons and limes | 965 | 984 | 950* |
| Grapefruit and pomelos* | 66 | 67 | 67 |
| Apples | 716 | 857 | 835 |
| Peaches and nectarines | 223 | 218 | 215* |
| Grapes | 1,058 | 1,149 | 1,065 |
| Mangoes | 782 | 842 | 845* |
| Avocados | 154 | 174 | 173* |
| Pineapples | 1,430 | 1,433 | 1,406 |
| Persimmons* | 65 | 65 | 66 |
| Cashew-apple* | 1,550 | 1,600 | 1,603 |
| Papayas | 1,489 | 1,598 | 1,600* |
| Coffee beans (green)‡ | 1,820 | 2,650 | 1,997 |
| Cocoa beans | 186 | 175 | 170 |
| Mate | 646 | 513 | 550* |
| Sisal | 181 | 171 | 184 |
| Cotton (lint)† | 872 | 713 | 737 |
| Other fibre crops | 104 | 107 | 108* |
| Tobacco (leaves) | 568 | 670 | 656 |
| Natural rubber | 88† | 96† | 96* |

* FAO estimate(s).
† Unofficial figure(s).
‡ Official figures, reported in terms of dry cherries, have been converted into green coffee beans at 50%.

Source: FAO.

### LIVESTOCK
('000 head, year ending September)

| | 2001 | 2002 | 2003 |
|---|---|---|---|
| Cattle | 176,389 | 185,347 | 189,513 |
| Buffaloes | 1,119 | 1,115 | 1,200* |
| Horses | 5,801 | 5,900* | 5,900* |
| Asses | 1,239 | 1,250* | 1,250* |
| Mules | 1,346 | 1,350* | 1,350* |
| Pigs | 32,605 | 32,013 | 32,605 |
| Sheep | 14,639 | 14,287 | 14,182† |
| Goats | 9,537 | 9,429 | 9,087† |
| Chickens | 883 | 908 | 1,050* |
| Ducks* | 3,400 | 3,500 | 3,550 |
| Turkeys* | 12,600 | 13,000 | 13,500 |

* FAO estimate(s).
† Unofficial figure.

Source: FAO.

### LIVESTOCK PRODUCTS
('000 metric tons)

| | 2001 | 2002 | 2003 |
|---|---|---|---|
| Beef and veal | 6,823 | 7,139 | 7,230 |
| Mutton and lamb* | 71 | 69 | 68 |
| Goat meat* | 39 | 40 | 40 |
| Pig meat | 2,637 | 2,798 | 3,059 |
| Horse meat* | 21 | 21 | 21 |
| Poultry meat | 6,380 | 7,239 | 7,967 |
| Cows' milk | 21,146 | 22,315 | 23,315† |
| Goats' milk* | 138 | 138 | 138 |
| Butter and ghee† | 75 | 77 | 79 |
| Cheese* | 38 | 39 | 39 |
| Hen eggs | 1,539† | 1,547† | 1,550* |
| Other poultry eggs* | 65 | 59 | 59 |
| Honey | 22 | 24 | 24* |
| Wool: greasy | 12 | 11 | 11* |
| Wool: scoured* | 7 | 8 | 8 |
| Cattle hides (fresh)* | 670 | 715 | 715 |
| Goatskins (fresh)* | 5 | 5 | 5 |
| Sheepskins (fresh)* | 15 | 16 | 16 |

* FAO estimate(s).
† Unofficial figure(s).

Source: FAO.

# Forestry

### ROUNDWOOD REMOVALS
(FAO estimates, '000 cubic metres, excl. bark)

| | 2001 | 2002 | 2003 |
|---|---|---|---|
| Sawlogs, veneer logs and logs for sleepers | 49,290 | 49,290 | 49,290 |
| Pulpwood | 45,861 | 45,861 | 45,861 |
| Other industrial wood | 7,843 | 7,843 | 7,843 |
| Fuel wood | 133,428 | 134,473 | 135,542 |
| **Total** | 236,422 | 237,467 | 238,536 |

Source: FAO.

### SAWNWOOD PRODUCTION
(FAO estimates, '000 cubic metres, incl. railway sleepers)

| | 2000 | 2001 | 2002 |
|---|---|---|---|
| Coniferous (softwood) | 7,800 | 6,050 | 6,400 |
| Broadleaved (hardwood) | 15,300 | 14,800 | 14,800* |
| **Total** | 23,100 | 20,850 | 21,200 |

* FAO estimate.
**2003:** Figures assumed to be unchanged from 2002 (FAO estimates).

Source: FAO.

# Fishing

('000 metric tons, live weight)

| | 2001 | 2002 | 2003 |
|---|---|---|---|
| Capture | 830.4 | 855.6 | 808.9* |
| Characins | 90.2 | 95.6 | 90.0* |
| Freshwater siluroids | 70.4 | 79.3 | 75.0* |
| Weakfishes | 44.3 | 45.4 | 45.0* |
| Whitemouth croaker | 40.3 | 43.1 | 36.0* |
| Brazilian sardinella | 39.8 | 22.1 | 25.3 |
| Aquaculture | 203.7 | 242.6 | 277.6 |
| Common carp | 55.5 | 54.5 | 49.6 |
| Tilapias | 35.8 | 42.0 | 62.6 |
| **Total catch** | 1,034.1 | 1,098.2 | 1,086.5* |

* FAO estimate.
Note: Figures exclude aquatic mammals, recorded by number rather than by weight. The number of toothed whales caught was: 1,057 in 2000; 23 in 2001; 131 in 2002. Also excluded are crocodiles. The number of spectacled caimans caught was: 1,253 in 2001; 6,048 in 2002; 12,851 in 2003.

Source: FAO.

## Mining

('000 metric tons, unless otherwise indicated)

|  | 2001 | 2002 | 2003[1] |
|---|---|---|---|
| Hard coal[2] | 6,000 | 6,000 | 6,000 |
| Crude petroleum ('000 barrels) | 487,640 | 547,134 | 620,865 |
| Natural gas (million cu m) | 14,112 | 13,274 | 13,503 |
| Iron ore: | | | |
|   gross weight | 208,438 | 214,560 | 234,478 |
|   metal content | 133,713 | 142,468 | 155,693 |
| Copper (metric tons)[3] | 212,243 | 189,651 | 174,000 |
| Nickel ore (metric tons)[3,4] | 45,300 | 45,300 | 31,100 |
| Bauxite | 13,790 | 13,189 | 13,148 |
| Lead concentrates (metric tons)[4] | 10,725 | 9,253 | 10,652 |
| Zinc (metric tons)[3] | 197,037 | 249,434 | 250,000 |
| Tin concentrates (metric tons)[3,4] | 12,168 | 11,675 | 11,500 |
| Manganese ore[1] | 6,500 | 6,500 | 6,500 |
| Chromium ore (metric tons)[5] | 178,013 | 113,811 | 114,000 |
| Tungsten concentrates (metric tons)[4] | 22 | 24 | 24 |
| Ilmenite (metric tons) | 111,113 | 174,382 | 174,000 |
| Rutile (metric tons) | 1,791 | 2,645 | 2,650 |
| Zirconium concentrates (metric tons)[6] | 20,553 | 20,000 | 20,500 |
| Silver (kilograms)[7] | 46,046 | 33,000 | 35,000 |
| Gold (kilograms)[8] | 51,867 | 37,886 | 44,400 |
| Bentonite (beneficiated) | 160.4 | 174.9 | 175.0 |
| Kaolin (beneficiated) | 1,817.4 | 1,708.5 | 1,708.5 |
| Magnesite (beneficiated) | 265.7 | 269.2 | 269.0 |
| Phosphate rock[9] | 4,805 | 4,883 | 4,900 |
| Potash salts[10] | 318.6 | 337.3 | 337.3 |
| Fluorspar (Fluorite) (metric tons)[11] | 43,734 | 47,899 | 48,100 |
| Barite (Barytes) (beneficiated) (metric tons) | 54,790 | 54,895 | 55,000 |
| Quartz (natural crystals) (metric tons) | 4,350 | 4,300 | 4,300 |
| Salt (unrefined): | | | |
|   marine | 4,370 | 4,835 | 4,840 |
|   rock | 1,208 | 1,274 | 1,270 |
| Gypsum and anhydrite (crude) | 1,507 | 1,633 | 1,630 |
| Graphite (natural) (metric tons) | 70,091 | 60,922 | 61,000 |
| Asbestos (fibre) (metric tons) | 172,695 | 194,732 | 194,750 |
| Mica (metric tons)[1] | 4,000 | 4,000 | 4,000 |
| Vermiculite (beneficiated) (metric tons) | 21,464 | 22,577 | 22,600 |
| Talc (crude) | 370 | 390[1] | 395 |
| Pyrophyllite (crude) | 189 | 200[1] | 200 |
| Diamonds ('000 carats):[2] | | | |
|   gem | 700 | 500 | 500 |
|   industrial | 600 | 600 | 600 |

[1] Estimated production.
[2] Figures refer to marketable products.
[3] Figures from the Ministério de Minas e Energia.
[4] Figures refer to the metal content of ores and concentrates.
[5] Figures refer to the chromic oxide ($Cr_2O_3$) content.
[6] Including production of baddeleyite-caldasite.
[7] Figures refer to primary production only. The production of secondary silver (in kilograms) was: 50,000 per year in 1999–2001.
[8] Including official production by independent miners (garimpeiros): 9,055 kg in 1999; 10,395 kg in 2000; 5,866 kg in 2001.
[9] Figures refer to the gross weight of concentrates. The phosphoric acid ($P_2O_5$) content (in '000 metric tons) was: 1,560 in 1998; 1,528 in 1999.
[10] Figures refer to the potassium oxide ($K_2O$) content.
[11] Acid-grade and metallurgical-grade concentrates.
Source: US Geological Survey.

## Industry

**SELECTED PRODUCTS**

('000 metric tons, unless otherwise indicated)

|  | 2000 | 2001 | 2002 |
|---|---|---|---|
| Beer ('000 hl) | 66,954 | 67,905 | 64,576 |
| Soft drinks ('000 hl) | 59,651 | 62,261 | 63,385 |
| Wood pulp (sulphate and soda) | 6,668 | 6,789 | 6,789 |
| Newsprint | 266 | 230 | 230 |
| Caustic soda | 1,353 | 1,174 | 1,206 |
| Fertilizers* | n.a. | 7,597 | 8,071 |
| Electric energy (million kWh) | 348,909 | 328,509 | n.a. |
| Pig-iron | 27,723 | 27,391 | 29,694 |
| Crude steel | 27,843 | 26,691 | 29,572 |
| Cement† | 39,559 | 38,735 | 38,104 |
| Passenger cars (units) | 260,000 | 296,000 | 326,000 |
| Buses and Coaches (units) | 22,672 | 23,163 | 22,826 |
| Lorries (units) | 71,686 | 77,431 | 68,558 |
| Motorcycles (units) | 621,000 | 741,000 | 851,000 |

* Source: Ministério de Desenvolvimento, Indústria e Comércio Exterior, Brasília.

† Portland cement only.

Source: mostly UN, *Industrial Commodity Statistics Yearbook*.

**2003** ('000 metric tons unless otherwise indicated): Fertilizers 9,353; Paper 7,916; Crude steel 31,147; Aluminium 1,381; Cement 34,010; Cellulose 9,069; Motor vehicles ('000 units) 1,827 (Source: Ministério de Desenvolvimento, Indústria e Comércio Exterior, Brasília).

**2004** ('000 metric tons unless otherwise indicated): Fertilizers 9,784; Paper 8,221; Crude steel 32,913; Aluminium 1,457; Cement 34,402; Cellulose 9,529; Motor vehicles ('000 units) 2,210 (Source: Ministério de Desenvolvimento, Indústria e Comércio Exterior, Brasília).

## Finance

**CURRENCY AND EXCHANGE RATES**

**Monetary Units**
100 centavos = 1 real (plural: reais).

**Sterling, Dollar and Euro Equivalents** (31 May 2005)
£1 sterling = 4.3689 reais;
US $1 = 2.4030
€1 = 2.9631 reais;
100 reais = £22.89 = $41.61 = €33.75.

**Average Exchange Rates** (reais per US $)
2002   2.9208
2003   3.0771
2004   2.9251

Note: In March 1986 the cruzeiro (CR $) was replaced by a new currency unit, the cruzado (CZ $), equivalent to 1,000 cruzeiros. In January 1989 the cruzado was, in turn, replaced by the new cruzado (NCZ $), equivalent to CZ $1,000 and initially at par with the US dollar (US $). In March 1990 the new cruzado was replaced by the cruzeiro (CR $), at an exchange rate of one new cruzado for one cruzeiro. In August 1993 the cruzeiro was replaced by the cruzeiro real, equivalent to CR $1,000. On 1 March 1994, in preparation for the introduction of a new currency, a transitional accounting unit, the Unidade Real de Valor (at par with the US $), came into operation, alongside the cruzeiro real. On 1 July 1994 the cruzeiro real was replaced by the real (R $), also at par with the US $ and thus equivalent to 2,750 cruzeiros reais.

# BRAZIL

## BUDGET
(R $ million)*

| Revenue | 2002 | 2003 | 2004 |
|---|---|---|---|
| Tax revenue | 243,006 | 273,358 | 322,556 |
| Income tax | 85,803 | 93,016 | 102,820 |
| Tax on profits of legal entities | 13,364 | 16,750 | 19,647 |
| Value-added tax on industrial products | 19,799 | 19,674 | 22,830 |
| Tax on financial operations and export duty | 4,022 | 4,450 | 5,253 |
| Import duty | 7,972 | 8,143 | 9,200 |
| Other taxes and duties | 26,538 | 31,377 | 40,030 |
| Social security contributions | 52,267 | 59,565 | 76,890 |
| Contributions to Social Integration Programme and Financial Reserve Fund for Public Employees | 12,872 | 17,337 | 19,454 |
| Provisional contribution on financial transactions | 20,369 | 23,046 | 26,432 |
| Social welfare contributions | 71,027 | 80,732 | 93,765 |
| Other income | 7,809 | 3,801 | 6,129 |
| **Total** | **321,842** | **357,891** | **422,450** |

| Expenditure | 2002 | 2003 | 2004 |
|---|---|---|---|
| Transfers to state and local governments† | 56,138 | 60,226 | 67,559 |
| Personnel | 73,304 | 78,068 | 87,730 |
| Social service and welfare transfers | 88,029 | 107,135 | 125,751 |
| Other expenditures | 71,912 | 72,978 | 91,689 |
| **Total** | **289,383** | **318,407** | **372,730** |

\* Figures refer to cash operations of the National Treasury, including the collection and transfer of earmarked revenues for social expenditure purposes. The data exclude the transactions of other funds and accounts controlled by the Federal Government.
† Constitutionally mandated participation funds.

## CENTRAL BANK RESERVES
(US $ million at 31 December)

| | 2002 | 2003 | 2004 |
|---|---|---|---|
| Gold* | 153 | 186 | 195 |
| IMF special drawing rights | 275 | 2 | 4 |
| Foreign exchange | 37,409 | 49,108 | 52,736 |
| **Total** | **37,836** | **49,297** | **52,935** |

\* Valued at market-related prices.
Source: IMF, *International Financial Statistics*.

## MONEY SUPPLY
(R $ million at 31 December)

| | 2002 | 2003 | 2004 |
|---|---|---|---|
| Currency outside banks | 42,351 | 43,065 | 52,014 |
| Demand deposits at deposit money banks | 64,680 | 65,879 | 74,172 |
| **Total money** (incl. others) | **107,209** | **111,542** | **130,110** |

Source: IMF, *International Financial Statistics*.

## COST OF LIVING
(Consumer Price Index; base: 2000 = 100)

| | 2002 | 2003 | 2004 |
|---|---|---|---|
| All items | 115.9 | 132.9 | 141.7 |

Source: IMF, *International Financial Statistics*.

## NATIONAL ACCOUNTS
(R $ million at current prices)

### National Income and Product

| | 2001 | 2002 | 2003 |
|---|---|---|---|
| Compensation of employees | 444,067 | 486,457 | 554,149 |
| Net operating surplus | 490,327 | 564,323 | 668,926 |
| Net mixed income | 60,469 | 61,618 | 69,757 |
| **Gross domestic product (GDP) at factor cost** | **994,863** | **1,112,397** | **1,292,832** |
| Taxes on production and imports | 208,578 | 237,061 | 263,350 |
| *Less* Subsidies | 4,704 | 3,430 | |
| **GDP in market prices (purchasers' values)** | **1,198,736** | **1,346,028** | **1,556,182** |
| Primary incomes received from abroad | 8,185 | 10,434 | 10,902 |
| *Less* Primary incomes paid abroad | 53,689 | 62,706 | 66,384 |
| Statistical discrepancy | 220 | 328 | 332 |
| **Gross national income (GNI)** | **1,153,452** | **1,294,084** | **1,501,032** |
| Current transfers from abroad | 4,936 | 8,341 | 9,694 |
| *Less* Current transfers paid abroad | 1,069 | 1,074 | 941 |
| **Net national disposable income** | **1,157,318** | **1,301,351** | **1,509,785** |

### Expenditure on the Gross Domestic Product

| | 2002 | 2003 | 2004 |
|---|---|---|---|
| Final consumption expenditure | 1,052,139 | 1,192,613 | 1,310,323 |
| Households | | | |
| Non-profit institutions serving households | 781,174 | 882,983 | 977,991 |
| General government | 270,965 | 309,631 | 33,332 |
| Gross capital formation | 265,953 | 307,491 | 376,408 |
| Gross fixed capital formation | 246,606 | 276,741 | 346,258 |
| Changes in inventories | | | |
| Acquisitions, less disposals, of valuables | 19,348 | 30,750 | 30,151 |
| **Total domestic expenditure** | **1,318,092** | **1,500,104** | **1,686,731** |
| Exports of goods and services | 208,489 | 254,832 | 318,387 |
| *Less* Imports of goods and services | 180,554 | 198,754 | 235,917 |
| **GDP in purchasers' values (market prices)** | **1,346,028** | **1,556,182** | **1,769,202** |
| **GDP at constant 2004 prices** | **1,672,954** | **1,682,071** | **1,769,202** |

### Gross Domestic Product by Economic Activity

| | 2001 | 2002 | 2003 |
|---|---|---|---|
| Agriculture, hunting, forestry and fishing | 89,287 | 104,908 | 138,191 |
| Mining and quarrying | 30,538 | 40,725 | 54,888 |
| Manufacturing | 240,834 | 279,907 | 337,457 |
| Electricity, gas and water | 38,796 | 43,206 | 47,594 |
| Construction | 91,006 | 95,469 | 100,951 |
| Trade, restaurants and hotels | 79,431 | 92,190 | 107,501 |
| Transport | 28,696 | 30,912 | 34,186 |
| Communications | 28,780 | 32,570 | 44,151 |
| Finance, insurance, real estate and business services | 70,126 | 92,190 | 97,459 |
| Government services | 173,302 | 195,933 | 220,458 |
| Rents | 127,546 | 135,629 | 142,544 |
| Other community, social and personal services | 120,080 | 130,839 | 144,884 |
| **Sub-total** | **1,118,422** | **1,274,478** | **1,470,264** |
| *Less* Financial intermediation services indirectly measured | 54,653 | 75,332 | 74,661 |
| **Gross value added in basic prices** | **1,063,769** | **1,199,146** | **1,395,604** |
| Taxes, less subsidies, on products | 134,967 | 146,883 | 160,578 |
| **GDP in market prices** | **1,198,736** | **1,346,028** | **1,556,182** |

# BRAZIL

## BALANCE OF PAYMENTS
(US $ million)

|  | 2002 | 2003 | 2004 |
|---|---:|---:|---:|
| Exports of goods f.o.b. | 60,362 | 73,084 | 96,475 |
| Imports of goods f.o.b. | −47,240 | −48,290 | −62,782 |
| **Trade balance** | 13,121 | 24,794 | 33,693 |
| Exports of services | 9,551 | 10,447 | 12,442 |
| Imports of services | −14,509 | −15,378 | −17,215 |
| **Balance on goods and services** | 8,164 | 19,863 | 28,921 |
| Other income received | 3,295 | 3,339 | 3,199 |
| Other income paid | −21,486 | −21,891 | −23,719 |
| **Balance on goods, services and income** | −10,026 | 1,311 | 8,401 |
| Current transfers received | 2,627 | 3,132 | 3,582 |
| Current transfers paid | −237 | −265 | −314 |
| **Current balance** | −7,637 | 4,177 | 11,669 |
| Capital account (net) | 433 | 498 | 703 |
| Direct investment abroad | −2,482 | −249 | −9,471 |
| Direct investment from abroad | 16,590 | 10,144 | 18,166 |
| Portfolio investment assets | −321 | 179 | −755 |
| Portfolio investment liabilities | −4,797 | 5,129 | −3,996 |
| Financial derivatives assets | 933 | 683 | 467 |
| Financial derivatives liabilities | −1,289 | −834 | −1,145 |
| Other investment assets | −3,211 | −9,483 | −1,462 |
| Other investment liabilities | −9,331 | −5,724 | −5,456 |
| Net errors and omissions | −154 | −933 | −2,122 |
| **Overall balance** | −11,266 | 3,586 | 6,599 |

Source: IMF, *International Financial Statistics*.

## External Trade

### PRINCIPAL COMMODITIES
(distribution by SITC, US $ million)

| Imports f.o.b. | 1999 | 2000 | 2001 |
|---|---:|---:|---:|
| **Food and live animals** | 3,560.3 | 3,443.1 | 2,938.6 |
| Cereals and cereal preparations | 1,611.0 | 1,662.0 | 1,524.1 |
| **Crude materials (inedible) except fuels** | 1,680.9 | 1,978.6 | 1,532.3 |
| **Mineral fuels, lubricants, etc.** | 5,887.4 | 8,912.4 | 8,454.1 |
| Petroleum, petroleum products, etc. | 4,634.8 | 7,121.1 | 6,714.3 |
| Crude petroleum and bituminous oils | 2,277.9 | 3,304.9 | 3,320.9 |
| Refined petroleum products | 2,245.7 | 3,671.3 | 3,202.1 |
| Gasoline and other light oils | 1,203.9 | 1,963.2 | 1,431.0 |
| **Chemicals and related products** | 9,512.4 | 10,363.4 | 10,561.6 |
| Organic chemicals | 3,041.7 | 3,159.8 | 3,269.9 |
| Medicinal and pharmaceutical products | 1,950.5 | 1,804.3 | 1,910.2 |
| Artificial resins and plastic materials | 1,295.0 | 1,656.2 | 1,593.0 |
| **Basic manufactures** | 5,161.6 | 6,016.7 | 5,919.3 |
| **Machinery and transport equipment** | 22,197.2 | 24,211.7 | 25,136.2 |
| Power-generating machinery and equipment | 2,335.1 | 2,165.8 | 3,129.0 |
| Machinery specialized for particular industries | 2,383.6 | 2,196.4 | 2,205.7 |
| General industrial machinery equipment and parts | 2,888.7 | 2,944.7 | 3,323.4 |
| Office machines and automatic data-processing machines | 1,614.5 | 2,054.9 | 1,905.3 |

| Imports f.o.b.— *continued* | 1999 | 2000 | 2001 |
|---|---:|---:|---:|
| Telecommunications and sound equipment | 2,673.0 | 3,206.4 | 3,156.3 |
| Other electrical machinery apparatus, etc. | 4,604.8 | 6,041.1 | 5,888.7 |
| Thermionic, microcircuits, transistors and valves, etc. | 1,634.8 | 2,638.6 | 2,118.9 |
| Road vehicles and parts* | 3,583.0 | 3,861.6 | 3,894.7 |
| Other transport equipment* | 1,581.7 | 1,741.2 | 1,644.9 |
| **Miscellaneous manufactured articles** | 3,295.3 | 3,533.3 | 3,594.4 |
| **Total** (incl. others) | 51,747.4 | 58,931.2 | 58,509.9 |

* Excluding tyres, engines and electrical parts.

| Exports f.o.b. | 1999 | 2000 | 2001 |
|---|---:|---:|---:|
| **Food and live animals** | 10,384.8 | 9,230.9 | 11,647.3 |
| Meat and meat preparations | 1,929.1 | 1,927.2 | 2,883.0 |
| Vegetables and fruit | 1,690.9 | 1,525.0 | 1,288.3 |
| Sugar, sugar preparations and honey | 2,010.1 | 1,295.2 | 2,403.8 |
| Sugar and honey | 1,925.1 | 1,203.9 | 2,289.2 |
| Coffee, tea, cocoa and spices | 2,760.7 | 2,066.8 | 1,718.6 |
| Coffee and coffee substitutes | 2,463.9 | 1,784.6 | 1,417.1 |
| Unroasted coffee, coffee husks and skins | 2,330.8 | 1,559.6 | 1,207.7 |
| Feeding-stuff for animals (excl. unmilled cereals) | 1,587.1 | 1,716.0 | 2,166.8 |
| Oilcake and other residues of soya beans | 1,503.6 | 1,652.6 | 2,067.4 |
| **Beverages and tobacco** | 1,017.8 | 911.6 | 1,001.4 |
| Tobacco and tobacco manufactures | 961.2 | 841.5 | 944.3 |
| **Crude materials (inedible) except fuels** | 7,153.2 | 8,650.9 | 8,797.8 |
| Oil seeds and oleaginous fruit | 1,595.1 | 2,189.9 | 2,731.3 |
| Soya beans | 1,593.3 | 2,187.9 | 2,725.5 |
| Pulp and waste paper | 1,243.6 | 1,602.4 | 1,247.6 |
| Metalliferous ores and metal scrap | 3,168.2 | 3,536.1 | 3,389.7 |
| Iron ore and concentrates | 2,746.0 | 3,048.2 | 2,931.5 |
| Iron ore and concentrates, not agglomerated | 1,726.0 | 1,852.9 | 1,916.9 |
| **Chemicals and related products** | 2,980.4 | 3,502.5 | 3,149.6 |
| **Basic manufactures** | 9,879.8 | 11,219.3 | 10,351.1 |
| Iron and steel | 3,139.3 | 3,685.0 | 3,191.2 |
| Iron and steel in primary form | 1,295.0 | 1,635.0 | 1,221.9 |
| Non-ferrous metals | 1,499.3 | 1,772.0 | 1,352.6 |
| **Machinery and transport equipment** | 11,366.2 | 15,485.1 | 15,542.4 |
| Power-generating machinery and equipment | 1,384.3 | 1,514.5 | 1,695.5 |
| General industrial machinery equipment and parts | 1,442.0 | 1,598.0 | 1,468.0 |
| Telecommunications and sound equipment | 724.7 | 1,648.5 | 1,793.2 |
| Road vehicles and parts* | 3,499.6 | 4,367.8 | 4,325.6 |
| Passenger motor cars (excl. buses) | 1,138.5 | 1,768.5 | 1,951.4 |
| Other transport equipment* | 1,937.7 | 3,618.7 | 3,637.8 |
| **Miscellaneous manufactured articles** | 2,851.5 | 3,443.2 | 3,490.0 |
| **Total** (incl. others) | 48,011.4 | 55,282.5 | 58,222.6 |

* Excluding tyres, engines and electrical parts.
Source: UN, *International Trade Statistics Yearbook*.

# BRAZIL

**2002** (US $ million): *Imports:* Crude petroleum 3,247.3; Motors, generators and electrical equipment 1,685.8; Motor vehicle parts 1,357.7; Medications for human and veterinary medicine 1,352.5; Electronic equipment 1,250.2; Fuel oils 1,097.6; Total (incl. others) 47,240.5. *Exports:* Metalliferous ores and concentrates 3,048.9; Soya 3,032.0; Aeroplanes 2,335.5; Oil-cake and other residues from soya beans 2,198.9; Passenger cars 2,005.2; Radio transmitters and receivers and parts thereof 1,782.3; Total (incl. others) 60,361.8 (Source: Ministério do Desenvolvimento, Indústria e Comércio Exterior, Brasília).

**2003** (US $ million): *Imports:* Crude petroleum 3,780.5; Motor vehicle parts 1,500.8; Electronic equipment 1,470.5; Medications for human and veterinary medicine 1,397.3; Motors, generators and electrical equipment 996.9; Wheat grain 1,009.7; Total (incl. others) 48,291.0. *Exports:* Soya 4,290.4; Metalliferous ores and concentrates 3,455.9; Passenger cars 2,655.7; Oil-cake and other residues from soya beans 2,602.4; Petroleum derivatives 2,121.9; Aeroplanes 1,938.6; Total (incl. others) 73,084.1 (Source: Ministério do Desenvolvimento, Indústria e Comércio Exterior, Brasília).

**2004** (US $ million): *Imports:* Crude petroleum 6,758.1; Motor vehicle parts 2,041.1; Electronic equipment 2,035.8; Medications for human and veterinary medicine 1,630.5; Motors, generators and electrical equipment 1,586.9; Fertilizers 1,301.9; Total (incl. others) 62,781.8. *Exports:* Soya 5,394.9; Metalliferous ores and concentrates 4,758.9; Passenger cars 3,351.5; Oil-cake and other residues from soya beans 3,270.9; Aeroplanes 3,268.8; Petroleum derivatives 2,527.7; Total (incl. others) 96,475.2 (Source: Ministério do Desenvolvimento, Indústria e Comércio Exterior, Brasília).

## PRINCIPAL TRADING PARTNERS
(US $ million)*

| Imports f.o.b. | 2001 | 2002 | 2003 |
|---|---|---|---|
| Algeria | 1,096 | 1,098 | n.a. |
| Argentina | 6,207 | 4,743 | 4,673 |
| Belgium-Luxembourg | 585 | 546 | 515 |
| Canada | 927 | 740 | 749 |
| Chile | 862 | 649 | 798 |
| China, People's Republic | 1,328 | 1,554 | 2,148 |
| France | 2,083 | 1,777 | 1,768 |
| Germany | 4,812 | 4,419 | 4,205 |
| Italy | 2,185 | 1,762 | 1,757 |
| Japan | 3,064 | 2,348 | 2,521 |
| Korea, Republic | 1,574 | 1,067 | 1,079 |
| Mexico | 695 | 580 | 533 |
| Netherlands | 532 | 535 | 509 |
| Paraguay | n.a. | 383 | 475 |
| Spain | 1,226 | 975 | 974 |
| United Kingdom | 1,235 | 1,345 | 1,202 |
| USA | 12,894 | 10,438 | 9,725 |
| Uruguay | 503 | 485 | 538 |
| **Total** (incl. others) | 55,581 | 47,240 | 48,260 |

| Exports f.o.b. | 2001 | 2002 | 2003 |
|---|---|---|---|
| Argentina | 5,002 | 2,342 | 4,561 |
| Belgium-Luxembourg | 1,812 | 1,892 | 1,795 |
| Canada | 555 | 732 | 978 |
| Chile | 1,352 | 1,461 | 1,880 |
| China, People's Republic | 1,902 | 2,520 | 4,533 |
| France | 1,648 | 1,525 | 1,715 |
| Germany | 2,502 | 2,537 | 3,136 |
| Italy | 1,809 | 1,817 | 2,208 |
| Japan | 1,986 | 2,098 | 2,311 |
| Korea, Republic | 736 | 852 | 1,223 |
| Mexico | 1,868 | 2,342 | 2,741 |
| Netherlands | 2,863 | 3,182 | 4,246 |
| Paraguay | 720 | 558 | 707 |
| Spain | 1,042 | 1,120 | 1,552 |
| United Kingdom | 1,705 | 1,769 | 1,899 |
| USA | 14,190 | 15,355 | 16,900 |
| Uruguay | 641 | 410 | 404 |
| **Total** (incl. others) | 58,270 | 60,362 | 73,084 |

* Imports by country of purchase; exports by country of last consignment.

Source: Ministério do Desenvolvimento, Indústria e Comércio Exterior, Brasília.

# Transport

## RAILWAYS*

|  | 1998 | 1999 | 2000 |
|---|---|---|---|
| Passengers ('000) | 2,449 | 1,587 | 1,614 |
| Passenger-km ('000) | 430,637 | 437,500 | 601,908 |
| Freight ('000 metric tons) | 263,140 | 261,920 | 292,500 |
| Freight ton-km (million) | 143,170 | 140,600 | 155,590 |

* Including suburban and metro services.

Source: Secretaria de Transportes Terrestres, Ministério dos Transportes, Brasília.

**Freight transport:** 162,235m. ton-km in 2001; 320,992 metric tons and 170,177m. ton-km in 2002; 344,996 metric tons and 182,648m. ton-km in 2003.

Source: Agência Nacional de Transportes Terrestres, Ministério dos Transportes, Brasília.

## ROAD TRAFFIC
(motor vehicles in use at 31 December)

|  | 1998 | 1999 | 2000 |
|---|---|---|---|
| Passenger cars | 21,313,351 | 22,347,423 | 23,241,966 |
| Buses and coaches | 430,062 | 400,048 | 427,213 |
| Light goods vehicles | 3,313,774 | 3,193,058 | 3,469,927 |
| Heavy goods vehicles | 1,755,877 | 1,778,084 | 1,836,203 |
| Motorcycles and mopeds | 3,854,646 | 4,222,705 | 4,732,331 |

Source: Empresa Brasileira de Planejamento de Transportes (GEIPOT).

## SHIPPING

**Merchant Fleet**
(registered at 31 December)

|  | 2002 | 2003 | 2004 |
|---|---|---|---|
| Number of vessels | 476 | 482 | 494 |
| Total displacement ('000 grt) | 3,449 | 3,258 | 2,628 |

Source: Lloyd's Register-Fairplay, *World Fleet Statistics*.

**International Sea-borne Freight Traffic**
('000 metric tons)

|  | 2000 | 2001 | 2002 |
|---|---|---|---|
| Goods loaded | 288,202 | 315,135 | 342,675 |
| Goods unloaded | 191,072 | 191,072 | 186,330 |

Source: Ministério dos Transportes.

## CIVIL AVIATION
(embarked passengers, mail and cargo)

|  | 2001 | 2002 | 2003 |
|---|---|---|---|
| Number of passengers ('000) | 36,001 | 35,923 | 33,420 |
| Passenger-km (million) | 50,789 | 50,526 | n.a. |
| Freight ton-km ('000)* | 6,930,295 | 6,976,683 | 6,801,204 |

* Including mail.

Source: Departamento de Aviação Civil (DAC), Comando da Aeronáutica, Ministério da Defesa, Brasília.

## Tourism

**FOREIGN TOURIST ARRIVALS**

| Country of origin | 2001 | 2002 | 2003 |
|---|---|---|---|
| Argentina | 1,374,584 | 699,177 | 792,753 |
| Bolivia | 107,673 | 67,673 | 60,487 |
| Canada | 55,629 | 67,531 | 68,585 |
| Chile | 154,093 | 112,451 | 114,562 |
| France | 185,033 | 206,262 | 225,235 |
| Germany | 320,602 | 296,157 | 315,532 |
| Italy | 216,517 | 183,469 | 214,141 |
| Netherlands | 44,057 | 55,088 | 77,693 |
| Paraguay | 285,752 | 218,653 | 186,457 |
| Portugal | 165,908 | 168,329 | 228,153 |
| Spain | 126,973 | 110,177 | 120,234 |
| Switzerland | 71,562 | 56,175 | 62,829 |
| United Kingdom | 143,823 | 146,513 | 155,877 |
| USA | 594,309 | 636,063 | 670,863 |
| Uruguay | 305,084 | 222,410 | 239,885 |
| **Total** (incl. others) | 4,772,575 | 3,783,400 | 4,090,590 |

**Receipts from tourism** (US $ million): 3,700.9 in 2001; 3,120.1 in 2002; 3,386.0 in 2003.

Source: Instituto Brasileiro de Turismo—EMBRATUR, Brasília.

## Communications Media

| | 2001 | 2002 | 2003 |
|---|---|---|---|
| Personal computers ('000 in use) | 10,800 | 13,000 | n.a. |
| Internet users ('000) | 8,000 | 14,300 | n.a. |
| Telephones in use ('000 main lines) | 37,431 | 38,810 | 39,222 |
| Mobile cellular telephones ('000 subscribers) | 28,746 | 34,881 | 46,373 |

**Radio receivers** ('000 in use): 71,000 in 1997 (Source: UNESCO, *Statistical Yearbook*).

**Television receivers** ('000 in use): 58,283 in 2000 (Source: International Telecommunication Union (ITU)).

**Book production** ('000 titles): 21,689 in 1998 (Source: UNESCO Institute for Statistics).

**Daily newspapers:** 465 (average circulation, '000 copies: 7,883) in 2000 (Source: UNESCO Institute for Statistics).

**Non-daily newspapers:** 2,010 in 2000 (Source: UNESCO Institute for Statistics).

**Facsimile machines:** 500,000 in 1997 (Source: UN, *Statistical Yearbook*).

## Education

(2003, unless otherwise indicated)

| | Institutions | Teachers | Students |
|---|---|---|---|
| Pre-primary | 94,741 | 270,575 | 5,555,525* |
| Literacy classes (Classe de Alfabetização) | 27,670 | 37,508 | 598,589 |
| Primary | 169,075 | 1,603,851 | 34,012,434* |
| Secondary | 23,118 | 488,376 | 9,169,357* |
| Higher | 1,859 | 268,816 | 4,900,023 |

* 2004 figure.

**Adult literacy rate:** 87.6% in 2001.

Source: Ministério da Educação, Brasília.

# Directory

## The Constitution

A new Constitution was promulgated on 5 October 1988. The following is a summary of the main provisions:

The Federative Republic of Brazil, formed by the indissoluble union of the States, the Municipalities and the Federal District, is constituted as a democratic state. All power emanates from the people. The Federative Republic of Brazil seeks the economic, political, social and cultural integration of the peoples of Latin America.

All are equal before the law. The inviolability of the right to life, freedom, equality, security and property is guaranteed. No one shall be subjected to torture. Freedom of thought, conscience, religious belief and expression are guaranteed, as is privacy. The principles of habeas corpus and 'habeas data' (the latter giving citizens access to personal information held in government data banks) are granted. There is freedom of association, and the right to strike is guaranteed.

There is universal suffrage by direct secret ballot. Voting is compulsory for literate persons between 18 and 69 years of age, and optional for those who are illiterate, those over 70 years of age and those aged 16 and 17.

Brasília is the federal capital. The Union's competence includes maintaining relations with foreign states, and taking part in international organizations; declaring war and making peace; guaranteeing national defence; decreeing a state of siege; issuing currency; supervising credits, etc.; formulating and implementing plans for economic and social development; maintaining national services, including communications, energy, the judiciary and the police; legislating on civil, commercial, penal, procedural, electoral, agrarian, maritime, aeronautical, spatial and labour law, etc. The Union, States, Federal District and Municipalities must protect the Constitution, laws and democratic institutions, and preserve national heritage.

The States are responsible for electing their Governors by universal suffrage and direct secret ballot for a four-year term. The organization of the Municipalities, the Federal District and the Territories is regulated by law.

The Union may intervene in the States and in the Federal District only in certain circumstances, such as a threat to national security or public order, and then only after reference to the National Congress.

### LEGISLATIVE POWER

Legislative power is exercised by the Congresso Nacional (National Congress), which is composed of the Câmara dos Deputados (Chamber of Deputies) and the Senado Federal (Federal Senate). Elections for deputies and senators take place simultaneously throughout the country; candidates for the Congresso must be Brazilian by birth and have full exercise of their political rights. They must be at least 21 years of age in the case of deputies and at least 35 years of age in the case of senators. The Congresso meets twice a year in ordinary sessions, and extraordinary sessions may be convened by the President of the Republic, the Presidents of the Câmara and the Senado, or at the request of the majority of the members of either house.

The Câmara is made up of representatives of the people, elected by a system of proportional representation in each State, Territory and the Federal District for a period of four years. The total number of deputies representing the States and the Federal District will be established in proportion to the population; each Territory will elect four deputies.

The Senado is composed of representatives of the States and the Federal District, elected according to the principle of majority. Each State and the Federal District will elect three senators with a mandate of eight years, with elections after four years for one-third of the members and after another four years for the remaining two-thirds. Each Senator is elected with two substitutes. The Senado approves, by secret ballot, the choice of Magistrates, when required by the Constitution; of the Attorney-General of the Republic, of the Ministers of the Accounts Tribunal, of the Territorial Governors, of the president and directors of the central bank and of the permanent heads of diplomatic missions.

The Congresso is responsible for deciding on all matters within the competence of the Union, especially fiscal and budgetary arrangements, national, regional and local plans and programmes, the strength of the armed forces and territorial limits. It is also responsible for making definitive resolutions on international treaties, and for authorizing the President to declare war.

The powers of the Câmara include authorizing the instigation of legal proceedings against the President and Vice-President of the Republic and Ministers of State. The Senado may indict and impose sentence on the President and Vice-President of the Republic and Ministers of State.

Constitutional amendments may be proposed by at least one-third of the members of either house, by the President or by more than one-half of the legislative assemblies of the units of the Federation. Amendments must be ratified by three-fifths of the members of each house. The Constitution may not be amended during times of national emergency, such as a state of siege.

## EXECUTIVE POWER

Executive power is exercised by the President of the Republic, aided by the Ministers of State. Candidates for the Presidency and Vice-Presidency must be Brazilian-born, be in full exercise of their political rights and be over 35 years of age. The candidate who obtains an absolute majority of votes will be elected President. If no candidate attains an absolute majority, the two candidates who have received the most votes proceed to a second round of voting, at which the candidate obtaining the majority of valid votes will be elected President. The President holds office for a term of four years and (under an amendment adopted in 1997) is eligible for re-election.

The Ministers of State are chosen by the President and their duties include countersigning acts and decrees signed by the President, expediting instructions for the enactment of laws, decrees and regulations, and presentation to the President of an annual report of their activities.

The Council of the Republic is the higher consultative organ of the President of the Republic. It comprises the Vice-President of the Republic, the Presidents of the Câmara and Senado, the leaders of the majority and of the minority in each house, the Minister of Justice, two members appointed by the President of the Republic, two elected by the Senado and two elected by the Câmara, the latter six having a mandate of three years.

The National Defence Council advises the President on matters relating to national sovereignty and defence. It comprises the Vice-President of the Republic, the Presidents of the Câmara and Senado, the Minister of Justice, military Ministers and the Ministers of Foreign Affairs and of Planning.

## JUDICIAL POWER

Judicial power in the Union is exercised by the Supreme Federal Tribunal; the Higher Tribunal of Justice; the Regional Federal Tribunals and federal judges; Labour Tribunals and judges; Electoral Tribunals and judges; Military Tribunals and judges; and the States' Tribunals and judges. Judges are appointed for life; they may not undertake any other employment. The Tribunals elect their own controlling organs and organize their own internal structure.

The Supreme Federal Tribunal, situated in the Union capital, has jurisdiction over the whole national territory and is composed of 11 ministers. The ministers are nominated by the President after approval by the Senado, from Brazilian-born citizens, between the ages of 35 and 65 years, of proved judicial knowledge and experience.

# The Government

## HEAD OF STATE

**President:** LUIZ INÁCIO (LULA) DA SILVA (PT) (took office 2 January 2003).

**Vice-President:** JOSÉ ALENCAR GOMES DA SILVA (PL).

## THE CABINET
(July 2005)

The Cabinet is composed of members of the Partido dos Trabalhadores (PT), the Partido Comunista do Brasil (PC do B), the Partido Liberal (PL), the Partido Socialista Brasileiro (PSB), the Partido do Movimento Democrático Brasileiro (PMDB), the Partido Progressista (PP), the Partido Verde (PV) and Independents.

**Minister of Foreign Affairs:** CELSO AMORIM (Ind.).
**Minister of Justice:** MÁRCIO THOMAZ BASTOS (PT).
**Minister of Finance:** ANTÔNIO PALOCCI, Filho (PT).
**Minister of Defence:** JOSÉ ALENCAR GOMES DA SILVA (PL).
**Minister of Agriculture, Fisheries and Food Supply:** ROBERTO RODRIGUES (Ind.).
**Minister of Agrarian Development:** MIGUEL SOLDATELLI ROSSETTO (PT).
**Minister of Labour and Employment:** RICARDO BERZOINI (PT).
**Minister of Transport:** ALFREDO PEREIRA DO NASCIMENTO (PL).
**Minister of Cities:** MÁRCIO FORTES (PP).
**Minister of Planning, Budget and Administration:** PAULO BERNARDO SILVA (PT).
**Minister of Mines and Energy:** SILAS RONDEAU (PMDB).
**Minister of Culture:** GILBERTO PASSOS GIL MOREIRA (PV).
**Minister of the Environment:** MARIA OSMARINA MARINA DA SILVA VAZ DE LIMA (PT).
**Minister of Development, Industry and Trade:** LUIZ FERNANDO FURLÁN (Ind.).
**Minister of Education:** FERNANDO HADDAD (PT).
**Minister of Health:** JOSÉ SARAIVA FELIPE (PMDB).
**Minister of National Integration:** CIRO FERREIRA GOMES (Ind.).
**Minister of Social Security:** NÉLSON MACHADO (PT).
**Minister of Social Development and the Fight against Hunger:** PATRUS ANANÍAS (PT).
**Minister of Communications:** HÉLIO COSTA (PMDB).
**Minister of Science and Technology:** SÉRGIO REZENDE (PSB).
**Minister of Sport:** AGNELO SANTOS QUEIROZ, Filho (PC do B).
**Minister of Tourism:** WALFRIDO DO MARES GUIA (Ind.).
**Comptroller-General:** WALDIR PIRES (PT).
**Cabinet Chief:** DILMA VANA ROUSSEFF (Ind.).

There are also eight Secretaries of State

## MINISTRIES

**Office of the President:** Palácio do Planalto, Praça dos Três Poderes, 70150-900 Brasília, DF; tel. (61) 411-1573; fax (61) 323-1461; e-mail protocolo@planalto.gov.br; internet www.presidencia.gov.br.

**Office of the Civilian Cabinet:** Palácio do Planalto, 4° andar, Praça dos Três Poderes, 70150-900 Brasília, DF; tel. (61) 211-1034; fax (61) 321-5804; e-mail gabcivil@planalto.gov.br; internet www.presidencia.gov.br/casacivil.

**Ministry of Agrarian Development:** Esplanada dos Ministérios, Bloco A, Ala Norte, 70054-900 Brasília, DF; tel. (61) 223-8002; fax (61) 322-0492; e-mail miguel.rossetto@mda.gov.br; internet www.mda.gov.br.

**Ministry of Agriculture, Fisheries and Food Supply:** Esplanada dos Ministérios, Bloco D, 70043-900 Brasília, DF; tel. (61) 218-2828; e-mail cenagri@agricultura.gov.br; internet www.agricultura.gov.br.

**Ministry of Cities:** Esplanada dos Ministérios, Bloco A, Anexo 2, 70050-901 Brasília, DF; tel. (61) 2108-1000; fax (61) 226-2719; e-mail cidades@cidades.gov.br; internet www.cidades.gov.br.

**Ministry of Communications:** Esplanada dos Ministérios, Bloco R, 8° andar, 70044-900 Brasília, DF; tel. (61) 311-6079; fax (61) 311-6731; e-mail webmaster@mc.gov.br; internet www.mc.gov.br.

**Ministry of Culture:** Esplanada dos Ministérios, Bloco B, 3° andar, 70068-900 Brasília, DF; tel. (61) 316-2171; fax (61) 225-9162; e-mail cgm@minc.gov.br; internet www.cultura.gov.br.

**Ministry of Defence:** Esplanada dos Ministérios, Bloco Q, 70049-900 Brasília, DF; tel. (61) 312-4000; fax (61) 321-2477; e-mail faleconosco@defesa.gov.br; internet www.defesa.gov.br.

**Ministry of Development, Industry and Trade:** Esplanada dos Ministérios, Bloco J, 70053-900 Brasília, DF; tel. (61) 2109-7000; fax (61) 325-2063; e-mail ascom@desenvolvimento.gov.br; internet www.desenvolvimento.gov.br.

**Ministry of Education:** Esplanada dos Ministérios, Bloco L, 70047-900 Brasília, DF; tel. (61) 2104-8484; fax (61) 2104-9172; e-mail acordabr@acb.mec.gov.br; internet www.mec.gov.br.

**Ministry of the Environment:** Esplanada dos Ministérios, Bloco B, 5°–9° andar, 70068-900 Brasília, DF; tel. (61) 4009-1000; fax (61) 4009-1756; e-mail marina.silva@mma.gov.br; internet www.mma.gov.br.

**Ministry of Finance:** Esplanada dos Ministérios, Bloco P, 5° andar, 70048-900 Brasília, DF; tel. (61) 412-2000; fax (61) 226-9084; e-mail acs.df.gmf@fazenda.gov.br; internet www.fazenda.gov.br.

# BRAZIL

**Ministry of Foreign Affairs:** Palácio do Itamaraty, Esplanada dos Ministérios, Bloco H, 70170-900 Brasília, DF; tel. (61) 411-6161; fax (61) 225-1272; internet www.mre.gov.br.

**Ministry of Health:** Esplanada dos Ministérios, Bloco G, 70058-900 Brasília, DF; tel. (61) 315-2425; fax (61) 224-8747; internet www.saude.gov.br.

**Ministry of Justice:** Esplanada dos Ministérios, Bloco T, Edif. Sede, 70064-900 Brasília, DF; tel. (61) 429-3000; fax (61) 322-6817; e-mail acs@mj.gov.br; internet www.mj.gov.br.

**Ministry of Labour and Employment:** Esplanada dos Ministérios, Bloco F, 5° andar, 70059-900 Brasília, DF; tel. (61) 317-6000; fax (61) 226-3577; e-mail internacional@mte.gov.br; internet www.mte.gov.br.

**Ministry of Mines and Energy:** Esplanada dos Ministérios, Bloco U, 70065-900 Brasília, DF; tel. (61) 319-5555; fax (61) 225-5407; internet www.mme.gov.br.

**Ministry of National Integration:** Esplanada dos Ministérios, Bloco E, 70067-901 Brasília, DF; tel. (61) 414-5972; fax (61) 414-5483; e-mail impresa@integracao.gov.br; internet www.integracao.gov.br.

**Ministry of Planning, Budget and Administration:** Esplanada dos Ministérios, Bloco K, 70040-906 Brasília, DF; tel. (61) 429-4102; fax (61) 225-7287; e-mail ministro@planejamento.gov.br; internet www.planejamento.gov.br.

**Ministry of Science and Technology:** Esplanada dos Ministérios, Bloco E, 4° andar, 70067-900 Brasília, DF; tel. (61) 317-7600; fax (61) 225-1441; e-mail webgab@mct.gov.br; internet www.mct.gov.br.

**Ministry of Social Development and the Fight against Hunger:** Esplanada dos Ministérios, Bloco C, 5° andar, 70046-900 Brasília, DF; tel. (61) 313-1822; internet www.desenvolvimentosocial.gov.br.

**Ministry of Social Security:** Esplanada dos Ministérios, Bloco F, 70059-900 Brasília, DF; tel. (61) 317-5151; fax (61) 317-5407; internet www.mps.gov.br.

**Ministry of Sport:** Esplanada dos Ministérios, Bloco A, 70054-906 Brasília, DF; tel. (61) 217-1800; fax (61) 217-1707; internet www.esporte.gov.br.

**Ministry of Tourism:** Esplanada dos Ministérios, Bloco U, 3°–4° andares, 70065-900 Brasília, DF; tel. (61) 310-9491; e-mail sidney.acosta@turismo.gov.br; internet www.turismo.gov.br.

**Ministry of Transport:** Esplanada dos Ministérios, Bloco R, 70044-900 Brasília, DF; tel. (61) 311-7001; fax (61) 311-7876; e-mail alfredo.nascimento@transportes.gov.br; internet www.transportes.gov.br.

# President and Legislature

## PRESIDENT

**Election, First Round, 6 October 2002**

| Candidate | Valid votes cast | % valid votes cast |
|---|---|---|
| Luiz Inácio (Lula) da Silva (PT) | 39,443,876 | 46.4 |
| José Serra (PSDB) | 19,700,470 | 23.2 |
| Antônio Garotinho (PSB) | 15,175,776 | 17.9 |
| Ciro Gomes (PPS) | 10,167,650 | 12.0 |
| Others | 440,646 | 0.5 |
| **Total** | **84,928,418** | **100.0** |

**Election, Second Round, 27 October 2002**

| Candidate | Valid votes cast | % valid votes cast |
|---|---|---|
| Luiz Inácio (Lula) da Silva (PT) | 52,793,364 | 61.3 |
| José Serra (PSDB) | 33,370,739 | 38.7 |
| **Total** | **86,164,103** | **100.0** |

## CONGRESSO NACIONAL

### Câmara dos Deputados

**Chamber of Deputies:** Palácio do Congresso Nacional, Edif. Principal, Praça dos Três Poderes, 70160-900 Brasília, DF; tel. (61) 216-0000; internet www.camara.gov.br.

**President:** SEVERINO CAVALCANTI (PP).

The Chamber has 513 members who hold office for a four-year term.

**General Election, 6 October 2002**

| Party | Seats at election | Seats at August 2005 |
|---|---|---|
| Partido dos Trabalhadores (PT) | 91 | 90 |
| Partido da Frente Liberal (PFL) | 84 | 59 |
| Partido do Movimento Democrático Brasileiro (PMDB) | 75 | 86 |
| Partido da Social Democracia Brasileira (PSDB) | 70 | 50 |
| Partido Progressista Brasileiro (PPB)* | 49 | 55 |
| Partido Liberal (PL)† | 26 | 51‡ |
| Partido Trabalhista Brasileiro (PTB) | 26 | 45 |
| Partido Socialista Brasileiro (PSB) | 22 | 20 |
| Partido Democrático Trabalhista (PDT) | 21 | 14 |
| Partido Popular Socialista (PPS) | 15 | 15 |
| Partido Comunista do Brasil (PC do B) | 12 | 10 |
| Partido de Reedificação da Ordem Nacional (PRONA) | 6 | 2 |
| Partido Verde (PV) | 5 | 7 |
| Partido Social Cristão (PSC) | 1 | 2 |
| Partido Social Liberal (PSL)† | 1 | — |
| Partido Republicano Progressista (PRP) | — | 1 |
| Independents | 9 | 6 |
| **Total** | **513** | **513** |

* Known as the Partido Progressista (PP) from 2003.
† The PL and the PSL subsequently formed a coalition.
‡ Including seats belonging to the PSL.

### Senado Federal

**Federal Senate:** Palácio do Congresso Nacional, 70165-900 Brasília, DF; tel. (61) 311-4141; fax (61) 311-3190; e-mail webmaster.secs@senado.gov.br; internet www.senado.gov.br.

**President:** RENAN CALHEIROS (PMDB).

The 81 members of the Senate are elected by the 26 States and the Federal District (three Senators for each) according to the principle of majority. The Senate's term of office is eight years, with elections after four years for one-third of the members and after another four years for the remaining two-thirds.

In the elections of 6 October 2002 54 seats were contested. In August 2005 the PMDB was represented by 23 senators, the PFL by 15, the PT by 13, the PSDB by 12, the PDT by four, the PL, the PSB and the PTB by three each, the P-SOL by two, and the PP by one. There were two independents.

# Governors

## STATES

**Acre:** JORGE NEY VIANA (PT).
**Alagoas:** RONALDO AUGUSTO LESSA (PSB).
**Amapá:** ANTÔNIO WALDEZ GÓES (PDT).
**Amazonas:** EDUARDO BRAGA (PPS).
**Bahia:** PAULO SOUTO (PFL).
**Ceará:** LÚCIO ALCÂNTARA (PSDB).
**Espírito Santo:** PAULO HARTUNG (PSB).
**Goiás:** MARCONI FERREIRA PERILLO (PSDB).
**Maranhão:** JOSÉ REINALDO (PFL).
**Mato Grosso:** BLAIRO MAGGI (PPS).
**Mato Grosso do Sul:** JOSÉ ORCÍRIO ZECA (PT).
**Minas Gerais:** AÉCIO NEVES (PSDB).
**Pará:** SIMÃO ROBISON OLIVEIRA JATENE (PSDB).
**Paraíba:** CASSIO CUNHA LIMA (PSDB).
**Paraná:** ROBERTO REQUIÃO (PMDB).
**Pernambuco:** JARBAS DE ANDRADE VASCONCELOS (PMDB).
**Piauí:** WELLINGTON DIAS (PT).
**Rio de Janeiro:** ROSINHA MATHEUS GAROTINHO (PSB).
**Rio Grande do Norte:** VILMA DE FARIA (PSB).
**Rio Grande do Sul:** GERMANO RIGOTTO (PMDB).
**Rondônia:** IVO CASSOL (PSDB).
**Roraima:** FLAMARION POTELA (PSL).
**Santa Catarina:** LUIZ ENRIQUE DA SILVEIRA (PMDB).
**São Paulo:** GERALDO ALCKMIN (PSDB).

BRAZIL                                                                                                                                    *Directory*

**Sergipe:** JOÃO ALVES (PFL).
**Tocantins:** MARCELO MIRANDA (PFL).

### FEDERAL DISTRICT

**Brasília:** JOAQUIM DOMINGOS RORIZ (PMDB).

## Political Organizations

**Partido Comunista do Brasil (PC do B):** Av. Sarutaiá 185, Jardim Paulista, 01403-010 São Paulo, SP; tel. (11) 3054-1800; fax (11) 3051-7738; e-mail secretariageral@pcdob.org.br; internet www.vermelho.org.br/pcdob; f. 1922; Pres. RENATO REBELO; 185,000 mems.

**Partido Democrático Trabalhista (PDT):** Rua Marechal Câmara 160, 4°, 20050 Rio de Janeiro, RJ; fax (21) 2262-8834; e-mail pdtpage@uol.com.br; internet www.pdt.org.br; f. 1980; formerly the PTB (Partido Trabalhista Brasileiro), renamed 1980 when that name was awarded to a dissident group following controversial judicial proceedings; mem. of Socialist International; Pres. (vacant); Gen. Sec. MANOEL DIAS.

**Partido da Frente Liberal (PFL):** Senado Federal, Anexo 1, 26° andar, 70165-900 Brasília, DF; tel. (61) 311-4305; fax (61) 224-1912; e-mail pfl25@pfl.org.br; internet www.pfl.org.br; f. 1984 by moderate members of the PDS and PMDB; Pres. JORGE BORNHAUSEN; Gen. Sec. JOSÉ CARLOS ALELUIA.

**Partido Liberal (PL):** Câmara dos Deputados, Anexo 1, 26° andar, 70160-900 Brasília, DF; tel. (61) 3202-9922; e-mail pl@pl.org.br; internet www.pl.org.br; Pres. VALDEMAR COSTA NETO; Sec.-Gen. GUILHERME FARHAT FERRAZ.

**Partido do Movimento Democrático Brasileiro (PMDB):** Câmara dos Deputados, Edif. Principal, 70160-900 Brasília, DF; tel. (61) 318-5120; e-mail presidente@pmdb.org.br; internet www.pmdb.org.br; f. 1980; moderate elements of former MDB; merged with Partido Popular in February 1982; Pres. MICHEL TEMER; Sec.-Gen. SARIVA FELIPE; factions include the **Históricos** and the **Movimento da Unidade Progressiva (MUP).**

**Partido Popular Socialista (PPS):** SCS Quadra 7, Bloco A, Edif. Executive Tower, SL 826/828, Pátio Brasil Shopping, Setor Comerical Sul, 70307-000 Brasília, DF; tel. (61) 223-0623; fax (61) 323-3623; e-mail pps23@pps.org.br; internet www.pps.org.br; f. 1922; Pres. ROBERTO FREIRE; Sec.-Gen. RUBENS BUENO.

**Partido Progressista (PP):** Senado Federal, Anexo 1, 17° andar, 70165-900 Brasília, DF; tel. (61) 311-3041; fax (61) 226-8192; internet www.pp.org.br; f. 1995 as Partido Progressista Brasileiro (PPB) by merger of Partido Progressista Reformador (PPR), Partido Progressista (PP) and the Partido Republicano Progressista (PRP); changed name as above in 2003; right-wing; Pres. PEDRO CORRÊA; Sec.-Gen. BENITO DOMINGOS.

**Partido de Reedificação da Ordem Nacional (PRONA):** SCN, Quadra 01, Bloco E, 50, Sala 114, Edif. Central Park, Asa Norte, 70711-903 Brasília, DF; tel. (61) 3964-5656; internet www.prona.org.br; Pres. Dr ENÉAS FERREIRA CARNEIRO.

**Partido Social Cristão (PSC):** tel. (21) 2220-1919; e-mail psc@psc.org.br; internet www.psc.org.br; f. 1970 as Partido Democrático Republicano; Pres. VITOR JORGE ADBALA NÓSSEIS; Sec.-Gen. SÉRGIO BUENO.

**Partido da Social Democracia Brasileira (PSDB):** Av. L2 Sul, Quadra 607, Edif. Metrópolis, 70200-670 Brasília, DF; tel. (61) 424-0500; fax (61) 424-0519; e-mail tucano@psdb.org.br; internet www.psdb.org.br; f. 1988; centre-left; formed by dissident members of the PMDB (incl. Históricos), PFL, PDS, PDT, PSB and PTB; Pres. EDUARDO AZEREDO; Sec.-Gen. BISMARCK MAIA.

**Partido Social Liberal (PSL):** SCS, Quadra 01, Bloco E, Sala 1004, Edif. Ceará, 70303-900 Brasília, DF; tel. and fax (61) 322-1721; fax (61) 3032–6832; e-mail contato@pslnacional.org.br; internet www.psl.org.br; f. 1994; Pres. EMMANUEL MAYRINCK DE SOUZA GAYOSO; Sec.-Gen. RONALDO NÓBREGA MEDEIROS.

**Partido Socialismo e Liberdade (P-SOL):** SDS, CONIC, Edif. Venâncio V, Loja 28, 70300-000 Brasília, DF; tel. and fax (61) 225-8322; internet www.psol.org.br; f. 2004 by fmr PT mems; Pres. HELOISA HELENA.

**Partido Socialista Brasileiro (PSB):** Câmara dos Deputados, Anexo 2, Brasília, DF; tel. (61) 215-9650; fax (61) 215-9663; e-mail lid.psb@camara.gov.br; internet www.psb.org.br; f. 1947; Pres. MIGUEL ARRAES; Sec.-Gen. RENATO SOARES.

**Partido dos Trabalhadores (PT):** Rua Silveira Martins 132, Centro, 01019-000 São Paulo, SP; tel. (11) 3243-1313; fax (11) 3243-1345; e-mail presidencia@pt.org.br; internet www.pt.org.br; f. 1980; first independent labour party; associated with the *autêntico* branch of the trade union movement; 350,000 mems; Pres. TARSO GENRO; Nat. Sec.-Gen. RICARDO BERZOINI (acting).

**Partido Trabalhista Brasileiro (PTB):** SCLN 303, Bloco C, Sala 105, 70735-530 Brasília, DF; tel. (61) 2101-1414; fax (61) 2101-1400; e-mail ptb@ptb.org.br; internet www.ptb.org.br; f. 1980; Pres. FLÁVIO DE CASTRO MARTINEZ (acting—in June 2005 Roberto Jefferson Monteiro Francisco stepped down temporarily as President); Sec.-Gen. LUIS ANTÔNIO FLEURY, Filho.

**Partido Verde (PV):** Rua dos Pinheiros 812, Pinheiros, 05422-001 São Paulo, SP; tel. (11) 3083-1722; fax (11) 3083-1062; e-mail pvsp@pv.org.br; internet www.pv.org.br; Pres. JOSÉ LUIZ DE FRANÇA PENNA; Organizing Sec. ANTÔNIO JORGE MELO VIANA.

Other political parties include the Partido Republicana Progressista (PRP), the Partido Social-Democrático (PSD), the Partido Social Trabalhista (PST), the Partido da Mobilização Nacional (PMN; internet www.pmn.org.br) and the Partido Social Democrata Cristão (PSDC; internet www.psdc.org.br).

### OTHER ORGANIZATIONS

**Movimento dos Trabalhadores Rurais Sem Terra (MST):** Alameda Barão de Limeira, 1232 Campos Elíseos, 01202-002 São Paulo, SP; tel. (11) 3361-3866; e-mail semterra@mst.org.br; internet www.mst.org.br; landless peasant movt; Pres. JOÃO PERO STÉDILE; Nat. Coordinator GILMAR MAURO.

The Organização da Luto no Campo (OLC) is another rural peasant movement.

## Diplomatic Representation

### EMBASSIES IN BRAZIL

**Algeria:** SHIS, QI 09, Conj. 13, Casa 01, Lago Sul, 70472-900 Brasília, DF; tel. (61) 248-4039; fax (61) 248-4691; e-mail sanag277@bsb.terra.com.br; Ambassador LAHCÈNE MOUSSAOUI.

**Angola:** SHIS, QL 06, Conj. 5, Casa 01, 71620-055 Brasília, DF; tel. (61) 248-4489; fax (61) 248-1567; e-mail emb.angola@tecnolink.com.br; internet www.angola.org.br; Ambassador ALBERTO CORREIA NETO.

**Argentina:** SHIS, QL 02, Conj. 01, Casa 19, Lago Sul, 70442-900 Brasília, DF; tel. (61) 364-7600; fax (61) 364-7666; e-mail embarg@embarg.org.br; internet www.embarg.org.br; Ambassador JUAN PABLO LOHLÉ.

**Australia:** SES, Av. Das Nações, Quadra 801, Conj. K, Lote 7, 70200-010 Brasília, DF; tel. (61) 226-3111; fax (61) 226-1112; e-mail australemb@terra.com.br; internet www.brasil.embassy.gov.br; Ambassador JOHN WILLIAM SULLIVAN.

**Austria:** SES, Av. das Nações, Quadra 811, Lote 40, 70426-900 Brasília, DF; tel. (61) 443-3111; fax (61) 443-5233; e-mail brasilia-ob@bmaa.gv.at; Ambassador WERNER BRANDSTETTER.

**Belgium:** SES, Av. das Nações, Quadra 809, Lote 32, 70422-900 Brasília, DF; tel. (61) 443-1133; fax (61) 443-1219; e-mail brasilia@diplobel.org; internet www.belgica.org.br; Ambassador (vacant).

**Bolivia:** SHIS, QI 19, Conj. 13, Casa 19, 71655-130 Brasília, DF; tel. (61) 366-3432; fax (61) 366-3136; e-mail embolivia-brasilia@embolivia-brasil.org.br; internet www.embolivia-brasil.org.br; Ambassador EDGAR CAMACHO OMISTE.

**Bulgaria:** SEN, Av. das Nações, Quadra 801, Lote 08, 70432-900 Brasília, DF; tel. (61) 223-6193; fax (61) 323-3285; e-mail bulagria@abordo.com.br; Ambassador VENTZISLAV ANGUELOV IVANOV.

**Cameroon:** SHIS, QI 09, Conj. 07, Casa 01, 71625-070 Brasília, DF; tel. (61) 248-5403; fax (61) 248-0443; e-mail embcameroun@embcameroun.org.br; internet www.embcameroon.org.br; Ambassador MARTIN MBARGA NGUELE.

**Canada:** SES, Av. das Nações, Quadra 803, Lote 16, 70410-900 Brasília, DF; tel. (61) 424-5400; fax (61) 424-5490; e-mail brsla@international.gc.ca; internet www.dfait-maeci.gc.ca/brazil; Ambassador SUZANNE LAPORTE.

**Cape Verde:** SHIS, QL 08, Conj. 08, Casa 07, 71620-285 Brasília, DF; tel. (61) 248-0543; fax (61) 364-4059; e-mail embcaboverde@rudah.com.br; Ambassador LUÍS ANTÓNIO VALADARES DUPRET.

**Chile:** SES, Av. das Nações, Quadra 803, Lote 11, 70407-900 Brasília, DF; tel. (61) 2103-5151; fax (61) 322-0714; e-mail embchile@embchile.org.br; Ambassador (vacant).

**China, People's Republic:** SES, Av. das Nações, Quadra 813, Lote 51, 70443-900 Brasília, DF; tel. (61) 346-4436; fax (61) 346-3299; internet www.embchina.org.br; Ambassador JIANG YUANDE.

**Colombia:** SES, Av. das Nações, Quadra 803, Lote 10, 70444-900 Brasília, DF; tel. (61) 226-8997; fax (61) 224-4732; e-mail embjcol@embcol.org.br; Ambassador CLAUDIA RODRÍGUEZ CASTELLANOS.

# BRAZIL — Directory

**Congo, Democratic Republic:** SHIS, QI 06, Conj. 04, Casa 15, Lago Sul, CP 71620-045, Brasília, DF; tel. (61) 365-4822; fax (61) 365-4823; e-mail ambaredeco@ig.com.br; Chargé d'affaires a.i. Daniel Muanda Lendo.

**Costa Rica:** SRTV/N 701, Conj. C, Ala A, Salas 308/310, Edif. Centro Empresarial Norte, 70710-200 Brasília, DF; tel. (61) 328-2219; fax (61) 328-2243; e-mail embrica@solar.com.br; Ambassador Sara Faigenzicht de Gloobe.

**Côte d'Ivoire:** SEN, Av. das Nações, Lote 09, 70473-900, Brasília, DF; tel. (61) 321-7320; fax (61) 321-1306; e-mail cotedivoire@cotedivoire.org.br; internet www.cotedivoire.org.br; Ambassador Colette Gallie Lambin.

**Croatia:** SHIS, QI 09, Conj. 11, Casa 03, 71625-110 Brasília, DF; tel. and fax (61) 248-0610; e-mail embaixada.croacia@zaz.com.br; Ambassador (vacant).

**Cuba:** SHIS, QI 05, Conj. 18, Casa 01, 71615-180 Brasília, DF; tel. (61) 248-4710; fax (61) 248-6778; e-mail embacuba@uol.com.br; internet www.embaixadacuba.org.br; Ambassador Pedro Juan Núñez Mosquera.

**Czech Republic:** Via L-3, Quadra 805, Lote 21, 70414-900, CP 170 Brasília, DF; tel. (61) 242-7785; fax (61) 242-7833; e-mail brasilia@embassy.mzv.cz; internet www.mzv.cz/brasilia; Ambassador Václav Hubinger.

**Denmark:** SES, Av. das Nações, Quadra 807, Lote 26, 70416-900 Brasília, DF; tel. (61) 445-3443; fax (61) 445-3509; e-mail bsbamb@um.dk; internet www.denmark.org.br; Ambassador Christian Konigsfeldt.

**Dominican Republic:** SHIS, QL 14, Conj. 05, Casa 08/10, 71640-055 Brasília, DF; tel. (61) 248-1405; fax (61) 364-3214; e-mail embajadombrasil@nutecnet.com.br; Ambassador (vacant).

**Ecuador:** SHIS, QI 11, Conj. 09, Casa 24, 71625-290 Brasília, DF; tel. (61) 248-5560; fax (61) 248-1290; e-mail embeq@solar.com.br; Ambassador Diego Ribadeneira Espinosa.

**Egypt:** SEN, Av. das Nações, Lote 12, 70435-900 Brasília, DF; tel. (61) 323-8800; fax (61) 323-1039; e-mail embegito@opengate.com.br; internet www.opengate.com.br/embegito; Ambassador Muhammad Abd-al Fattah Abdallah.

**El Salvador:** SHIS, QI 10, Conj. 01, Casa 15, 71630-100 Brasília, DF; tel. (61) 364-4141; fax (61) 364-2459; e-mail brasemb@es.com.sv; internet www.elsalvador.hpg.com.br; Ambassador (vacant).

**Finland:** SES, Av. das Nações, Quadra 807, Lote 27, 70417-900 Brasília, DF; tel. (61) 443-7151; fax (61) 443-3315; e-mail brasilia@finlandia.org.br; internet www.finlandia.org.br; Ambassador Hannu Uusi-Videnoja.

**France:** SES, Av. das Nações, Quadra 801, Lote 04, 70404-900 Brasília, DF; tel. (61) 312-9100; fax (61) 312-9108; e-mail france@ambafrance.org.br; internet www.ambafrance.org.br; Ambassador Jean de Gliniasty.

**Gabon:** SHIS, QI 09, Conj. 11, Casa 24, 71625-110 Brasília, DF; tel. (61) 248-3536; fax (61) 248-2241; e-mail mgabao@terra.com.br; Ambassador Marcel Odongui-Bonnard.

**Germany:** SES, Av. das Nações, Quadra 807, Lote 25, 70415-900 Brasília, DF; tel. (61) 442-7000; fax (61) 443-7508; e-mail info.brasilia@alemanha.org.br; internet www.alemanha.org.br; Ambassador Friedrich Prot von Kunow.

**Ghana:** SHIS, QL 10, Conj. 08, Casa 02, 70466-900 Brasília, DF; tel. (61) 248-6047; fax (61) 248-7913; e-mail ghaembra@zaz.com.br; Ambassador Daniel Yaw Adjei.

**Greece:** SES, Av. das Nações, Quadra 805, Lote 22, 70480-900 Brasília, DF; tel. (61) 443-6573; fax (61) 443-6902; e-mail info@emb-grecia.org.br; internet www.emb-grecia.org.br; Ambassador Andonios Nicolaidis.

**Guatemala:** SHIS, QI 03, Conj. 10, Casa 01, Lago-Sul, 71605-300 Brasília, DF; tel. (61) 365-1908; fax (61) 365-1906; e-mail embaguate-brasil@minex.gob.gt; Ambassador Manuel Estuardo Roldán Barillas.

**Guyana:** SHIS, QI 05, Conj. 19, Casa 24, 71615-190 Brasília, DF; tel. (61) 248-0874; fax (61) 248-0886; e-mail embguyana@apis.com.br; Ambassador Marilyn Cheryl Miles.

**Haiti:** SHIS, QL 10, Conj. 06, Casa 16, Lago Sul, 71630-065 Brasília, DF; tel. (61) 248-6860; fax (61) 248-7472; e-mail embhaiti@zaz.com.br; Ambassador Antonio Fenelon.

**Holy See:** SES, Av. das Nações, Quadra 801, Lote 01, 70401-900 Brasília, DF; tel. (61) 223-0794; fax (61) 224-9365; e-mail nunapost@solar.com.br; Apostolic Nuncio Most Rev. Lorenzo Baldisseri (Titular Archbishop of Diocletiana).

**Honduras:** SHIS, QI 19, Conj. 07, Casa 34, Lago Sul, 71655-070 Brasília, DF; tel. (61) 366-4082; fax (61) 366-4618; e-mail embhonduras@ig.com.br; Ambassador Victor Manuel Lozano Urbina.

**Hungary:** SES, Av. das Nações, Quadra 805, Lote 19, 70413-900 Brasília, DF; tel. (61) 443-0836; fax (61) 443-3434; e-mail huembbrz@terra.com.br; internet www.hungria.org.br; Ambassador Dr József Németh.

**India:** SHIS, QL 08, Conj. 08, Casa 01, 71620-285 Brasília, DF; tel. (61) 248-4006; fax (61) 248-7849; e-mail indemb@indianembassy.org.br; internet www.indianembassy.org.br; Ambassador Hardeep Singh Puri.

**Indonesia:** SES, Av. das Nações, Quadra 805, Lote 20, 70479-900 Brasília, DF; tel. (61) 443-8800; fax (61) 443-6732; e-mail kbribrasilia@persocom.com.br; Ambassador Pieter Taruyu Vau.

**Iran:** SES, Av. das Nações, Quadra 809, Lote 31, 70421-900 Brasília, DF; tel. (61) 242-5733; fax (61) 224-9640; e-mail webiran@webiran.org.br; internet www.webiran.org.br; Ambassador Seyed Jafar Hashemi.

**Iraq:** SES, Av. das Nações, Quadra 815, Lote 64, 70430-900 Brasília, DF; tel. (61) 346-2822; fax (61) 346-7034; Chargé d'affaires Ala Alddin Abbas Hilmi.

**Ireland:** SHIS QL 12, Conj. 05, Casa 09, Lago Sul, Brasília, DF; tel. (61) 248-8800; fax (61) 248-8816; e-mail irishembassybrasilia@eircom.net; Ambassador Martin Greene.

**Israel:** SES, Av. das Nações, Quadra 809, Lote 38, 70424-900 Brasília, DF; tel. (61) 2105-0500; fax (61) 2105-0555; e-mail info@brasilia.mfa.gov.il; internet www.brasilia.mfa.gov.il; Ambassador Tzipora Rimon.

**Italy:** SES, Av. das Nações, Lote 30, 70420-900 Brasília, DF; tel. (61) 442-9900; fax (61) 443-1231; e-mail embitalia@embitalia.org.br; internet www.embitalia.org.br; Ambassador Michele Valensise.

**Japan:** SES, Av. das Nações, Quadra 811, Lote 39, 70425-900 Brasília, DF; tel. (61) 442-4200; fax (61) 242-0738; e-mail culturaljapao1@yawl.com.br; internet www.japao.org.br; Ambassador Takahiko Horimura.

**Jordan:** SHIS, QI 09, Conj. 18, Casa 14, 70408-900 Brasília, DF; tel. (61) 248-5407; fax (61) 248-1698; e-mail jordania@apis.com.br; Ambassador (vacant).

**Korea, Democratic People's Republic:** SHIS, QI 25, Conj. 10, Casa 11, Brasília, DF; tel. (61) 367-1940; fax (61) 367-3177; e-mail embcorea@hotmail.com; Ambassador (vacant).

**Korea, Republic:** SEN, Av. das Nações, Lote 14, 70436-900 Brasília, DF; tel. (61) 321-2500; fax (61) 321-2508; e-mail ebcoreia@terra.com.br; Ambassador Kim Kwang-dong.

**Kuwait:** SHIS, QI 05, Chácara 30, Lago Sul, 71600-580 Brasília, DF; tel. (61) 248-2323; fax (61) 248-0969; e-mail kuwait@opendf.com.br; Ambassador Hamood Y. M. Ar-Roudhan.

**Lebanon:** SES, Av. das Nações, Quadra 805, Lote 17, 70411-900 Brasília, DF; tel. (61) 443-5552; fax (61) 443-8574; e-mail embaixada@libano.org.br; internet www.libano.org.br; Ambassador Fouad el-Khoury Ghanem.

**Libya:** SHIS, QI 15, Chácara 26, 71600-750 Brasília, DF; tel. (61) 248-6710; fax (61) 248-0598; e-mail emblibia@terra.com.br; Ambassador Mohamed Heimeda Saad Matri.

**Malaysia:** SHIS, QI 05, Chácara 62, 70477-900 Brasília, DF; tel. (61) 248-5008; fax (61) 248-6472; e-mail mwbrasilia@persocom.com.br; Ambassador Tai Kat Meng.

**Mexico:** SES, Av. das Nações, Quadra 805, Lote 18, 70412-900 Brasília, DF; tel. (61) 244-1011; fax (61) 244-1755; e-mail embamexbra@cabonet.com.br; internet www.mexico.org.br; Ambassador Cecilia Soto González.

**Morocco:** SEN, Av. das Nações, Quadra 801, Lote 02, 70432-900 Brasília, DF; tel. (61) 321-4487; fax (61) 321-0745; e-mail sifamarbr@onix.com.br; Ambassador Ali Achour.

**Mozambique:** SHIS, QL 12, Conj. 07, Casa 09, 71630-275 Brasília, DF; tel. (61) 248-4222; fax (61) 248-3917; e-mail embamoc-bsb@uol.com; Ambassador (vacant).

**Myanmar:** SHIS, QI 07, Conj. 14, Casa 05, 71615-340 Brasília, DF; tel. (61) 248-3747; fax (61) 364-2747; e-mail mebrsl@brnet.com.br; Ambassador Hla Myint.

**Namibia:** SHIS QI 09, Conj. 08, Casa 11, Lago Sul, 71625-080 Brasília, DF; tel. (61) 248-6274; fax (61) 248-7135; e-mail namibianembassy-brazil@ibest.com.br; Ambassador Patrick Nandago.

**Netherlands:** SES, Av. das Nações, Quadra 801, Lote 05, 70405-900 Brasília, DF; tel. (61) 321-4769; fax (61) 321-1518; e-mail bra@minbuza.nl; internet www.embaixada-holanda.org.br; Ambassador (vacant).

**New Zealand:** SHIS, QI 09, Conj. 16, Casa 01, 71625-160 Brasília, DF; tel. (61) 248-9900; fax (61) 248-9916; e-mail zelandia@nwi.com.br; Ambassador DENISE JOAN ALMÃO.

**Nicaragua:** SHIS, QI 16, Conj. 04, Casa 15, 71640-245 Brasília, DF; tel. (61) 248-1115; fax (61) 248-1120; e-mail embanibra@zaz.com.br; Chargé d'affaires a.i. LIZA TUCKER.

**Nigeria:** SEN, Av. das Nações, Lote 05, 70459-900 Brasília, DF; tel. (61) 226-1717; fax (61) 226-5192; e-mail nigeria@persocom.com.br; internet www.nigerianembassy-brasilia.org.br; Ambassador JOSEF SOOKORE EGBUSON.

**Norway:** SES, Av. das Nações, Quadra 807, Lote 28, 70418-900 Brasília, DF; tel. (61) 443-8720; fax (61) 443-2942; e-mail embno@terra.com.br; internet www.noruega.org.br; Ambassador JAN GERHARD LASSEN.

**Pakistan:** SHIS, QL 12, Conj. 02, Casa 19, 71630-225 Brasília, DF; tel. (61) 364-1632; fax (61) 248-0246; e-mail parepbra@bruturbo.com; Ambassador KHALID KHATTAK.

**Panama:** SHIS, QI 07, Conj. 09, Casa 08, 71615-290 Brasília, DF; tel. (61) 248-7309; fax (61) 248-2834; e-mail empanama@nettur.com.br; Ambassador (vacant).

**Paraguay:** SES, Av. das Nações, Quadra 811, Lote 42, 70427-900 Brasília, DF; tel. (61) 242-3732; fax (61) 242-4605; e-mail embapar@yawl.com.br; Ambassador LUIS GONZÁLEZ ARIAS.

**Peru:** SES, Av. das Nações, Quadra 811, Lote 43, 70428-900 Brasília, DF; tel. (61) 242-9933; fax (61) 244-9344; e-mail embperu@embperu.org.br; internet www.embperu.org.br; Ambassador HERNÁN COUTURIER MARIÁTEGUI.

**Philippines:** SEN, Av. das Nações, Lote 01, 70431-900 Brasília, DF; tel. (61) 223-5143; fax (61) 226-7411; e-mail pg@persocom.com.br; internet www.yehey.com; Ambassador (vacant).

**Poland:** SES, Av. das Nações, Quadra 809, Lote 33, 70423-900 Brasília, DF; tel. (61) 443-3438; fax (61) 242-8543; e-mail embaixada@polonia.org.br; Ambassador KRZYSZTOF JACEK HINZ.

**Portugal:** SES, Av. das Nações, Quadra 801, Lote 02, 70402-900 Brasília, DF; tel. (61) 3032-9600; fax (61) 3032-9642; e-mail embaixadaportugal@embaixadaportugal.org.br; Ambassador FRANCISCO MANUEL SEIXAS DE COSTA.

**Romania:** SEN, Av. das Nações, Lote 06, 70456-900 Brasília, DF; tel. (61) 226-0746; fax (61) 226-6629; e-mail romenia@solar.com.br; Ambassador MONICA-MARIANA GRIGORESCU.

**Russia:** SES, Av. das Nações, Quadra 801, Lote A, 70476-900 Brasília, DF; tel. (61) 223-3094; fax (61) 226-7319; e-mail embr@embrus.brte.com.br; internet www.brazil.mid.ru; Ambassador VLADIMIR L. TYURDENEV.

**Saudi Arabia:** SHIS, QL 10, Conj. 09, Casa 20, 70471-900 Brasília, DF; tel. (61) 248-3523; fax (61) 284-2905; e-mail embsaud@tba.com.br; Ambassador (vacant).

**Senegal:** SEN, Av. das Nações, Lote 18, QL 12, Conj. 04, Casa 13, Lago Sul, 70800-400 Brasília, DF; tel. (61) 223-6110; fax (61) 322-7822; e-mail senebrasilia@bol.com.br; Ambassador MOUHAMADOU DOUDOU LO.

**Serbia and Montenegro:** SES, Av. das Nações, Quadra 803, Lote 15, 70409-900 Brasília, DF; tel. (61) 223-7272; fax (61) 223-8462; e-mail embiugos@nutecnet.com.br; Ambassador RADIVOJE LAZAREVIĆ.

**Slovakia:** Av. das Nações, Quadra 805, Lote 21, 70414-900 Brasília, DF; tel. (61) 443-1263; fax (61) 443-1267; e-mail eslovaca@brasil.mfa.sk; Ambassador MARIÁN MASARIK.

**South Africa:** SES, Av. das Nações, Quadra 801, Lote 06, 70406-900 Brasília, DF; tel. (61) 312-9500; fax (61) 312-8491; e-mail saemb@solar.com.br; internet www.africadosul.org.br; Ambassador LINDIWE DAPHNE ZULU.

**Spain:** SES, Av. das Nações, Quadra 811, Lote 44, 70429-900 Brasília, DF; tel. (61) 244-2776; fax (61) 242-1781; e-mail embespbr@correo.mae.es; Ambassador RICARDO PEIDRÓ CONDE.

**Sri Lanka:** SHIS, QI 09, Conj. 09, Casa 07, Lago Sul, 71625-090 Brasília DF; tel. (61) 248-2701; fax (61) 364-5430; e-mail lankaemb@yawl.com.br; Ambassador Gen. ROHAN DE SILVA DALUWATTE.

**Sudan:** SHIS, QI 11, Conj. 5, Casa 13, Lago Sul, Brasília, DF; tel. (61) 248-4835; fax (61) 248-4833; e-mail sdbrasilia@sudanbrasilia.org; internet www.sudanbrasilia.org; Ambassador RAHAMTALLA MOHAMED OSMAN.

**Suriname:** SHIS, QI 09, Conj. 08, Casa 24, 71625-080 Brasília, DF; tel. (61) 248-6705; fax (61) 248-3791; e-mail surinameemb@terra.com.br; Ambassador RADJENDRAKUMAR NIHALCHAND SONNY HIRA.

**Sweden:** SES, Av. das Nações, Quadra 807, Lote 29, 70419-900 Brasília, DF; tel. (61) 443-1444; fax (61) 443-1187; e-mail swebra@opengate.com.br; Ambassador MARGARETA WINBERG.

**Switzerland:** SES, Av. das Nações, Lote 41, 70448-900 Brasília, DF; tel. (61) 443-5500; fax (61) 443-5711; e-mail vertretung@bra.rep.admin.ch; internet www.dfae.admin.ch/brasilia; Ambassador RUDOLF BAERFUSS.

**Syria:** SEN, Av. das Nações, Lote 11, 70434-900 Brasília, DF; tel. (61) 226-0970; fax (61) 223-2595; e-mail embsiria@uol.com.br; Ambassador ALI DIAB.

**Thailand:** SEN, Av. das Nações, Lote 10, 70433-900 Brasília, DF; tel. (61) 224-6943; fax (61) 223-7502; e-mail thaiemb@linkexpress.com.br; Ambassador SIREE BUNNAG.

**Trinidad and Tobago:** SHIS, QL 02, Conj. 02, Casa 01, 71665-028 Brasília, DF; tel. (61) 365-1132; fax (61) 365-1733; e-mail trinbago@terra.com.br; Ambassador WINSTON CLYDE MOORE.

**Tunisia:** SHIS, QI 09, Conj. 16, Casa 20, 71625-160 Brasília, DF; tel. (61) 248-7277; fax (61) 248-7355; e-mail at.brasilia@terra.com.br; Ambassador HASSINE BOUZID.

**Turkey:** SES, Av. das Nações, Quadra 805, Lote 23, 70452-900 Brasília, DF; tel. (61) 242-1850; fax (61) 242-1448; e-mail turquia@conectanet.com.br; Ambassador AHMET GÜRKAN.

**Ukraine:** SHIS QI 05, Conj. 04, Casa 02, 71615-040 Brasília, DF; tel. (61) 365-3889; fax (61) 365-2127; e-mail brucremb@zaz.com.br; Ambassador YURII BOHAIEVSKYI.

**United Arab Emirates:** SHIS, QI 05, Chácara 54, 71600-580 Brasília, DF; tel. (61) 248-0717; fax (61) 248-7543; e-mail uae@uae.org.br; internet www.uae.org.br; Ambassador YOUSEF ALI AL-USAIMI.

**United Kingdom:** SES, Quadra 801, Conj. K, Lote 08, 70408-900 Brasília, DF; tel. (61) 329-2300; fax (61) 329-2369; e-mail contato@reinounido.org.br; internet www.reinounido.org.br; Ambassador PETER SALMON COLLECOTT.

**USA:** SES, Av. das Nações, Quadra 801, Lote 03, 70403-900 Brasília, DF; tel. (61) 321-7000; fax (61) 321-7241; internet www.embaixada-americana.org.br; Ambassador JOHN J. DANILOVICH.

**Uruguay:** SES, Av. das Nações, Lote 14, 70450-900 Brasília, DF; tel. (61) 322-1200; fax (61) 322-6534; e-mail urubras@emburuguai.org.br; internet www.emburuguai.org.br; Ambassador (vacant).

**Venezuela:** SES, Av. das Nações, Quadra 803, Lote 13, 70451-900 Brasília, DF; tel. (61) 322-1011; fax (61) 226-5633; e-mail embven1@rudah.com.br; Ambassador JULIO JOSÉ GARCÍA MONTOYA.

**Viet Nam:** SHIS, QI 09, Conj. 10, Casa 01, 71625-100 Brasília, DF; tel. (61) 364-7856; fax (61) 364-5836; e-mail embavina@uol.com.br; Ambassador NGUYEN VAN HUYNH.

**Zimbabwe:** SHIS, QI 01, Conj. 04, Casa 25, Brasília, DF; tel. (61) 365-4801; fax (61) 365-4803; e-mail zimbrasilia@uol.com.br; Ambassador THOMAS SUKUTAI BVUMA.

# Judicial System

The judiciary powers of the State are held by the following: the Supreme Federal Tribunal, the Higher Tribunal of Justice, the five Regional Federal Tribunals and Federal Judges, the Higher Labour Tribunal, the 24 Regional Labour Tribunals, the Conciliation and Judgment Councils and Labour Judges, the Higher Electoral Tribunal, the 27 Regional Electoral Tribunals, the Electoral Judges and Electoral Councils, the Higher Military Tribunal, the Military Tribunals and Military Judges, the Tribunals of the States and Judges of the States, the Tribunal of the Federal District and of the Territories and Judges of the Federal District and of the Territories.

  The Supreme Federal Tribunal comprises 11 ministers, nominated by the President and approved by the Senate. Its most important role is to rule on the final interpretation of the Constitution. The Supreme Federal Tribunal has the power to declare an act of Congress void if it is unconstitutional. It judges offences committed by persons such as the President, the Vice-President, members of the Congresso Nacional, Ministers of State, its own members, the Attorney General, judges of other higher courts, and heads of permanent diplomatic missions. It also judges cases of litigation between the Union and the States, between the States, or between foreign nations and the Union or the States; disputes as to jurisdiction between higher Tribunals, or between the latter and any other court, in cases involving the extradition of criminals, and others related to the writs of habeas corpus and habeas data, and in other cases.

  The Higher Tribunal of Justice comprises at least 33 members, appointed by the President and approved by the Senado. Its jurisdiction includes the judgment of offences committed by State Governors. The Regional Federal Tribunals comprise at least seven judges, recruited when possible in the respective region and appointed by the President of the Republic. The Higher Labour Tribunal comprises 27 members, appointed by the President and approved by the Senado. The judges of the Regional Labour Tribu-

# BRAZIL

nals are also appointed by the President. The Regional Electoral Tribunals are also composed of seven members. The Higher Military Tribunal comprises 15 life members, appointed by the President and approved by the Senate; three from the navy, four from the army, three from the air force and five civilian members. The States are responsible for the administration of their own justice, according to the principles established by the Constitution.

### SUPREME FEDERAL TRIBUNAL

**Supremo Tribunal Federal:** Praça dos Três Poderes, 70175-900 Brasília, DF; tel. (61) 217-3000; fax (61) 316-5483; e-mail webmaster@stf.gov.br; internet www.stf.gov.br.
**President:** NELSON DE ACEVEDO JOBIM, Filho.
**Vice-President:** ELLEN GRACIE.
**Justices:** JOSÉ PAULO SEPÚLVEDA PERTENCE, JOSÉ CELSO DE MELLO, CARLOS VELLOSO, MARCO AURÉLIO, GILMAR MENDES, CEZAR PELUSO, CARLOS BRITTO, JOAQUIM BARBOSA, EROS GRAU.
**Attorney-General:** CLÁUDIO LEMOS FONTELES.
**Director-General (Secretariat):** MIGUEL AUGUSTO FONSECA DE CAMPOS.

# Religion

According to the census of 2000, there were around 125m. Roman Catholics, 26m. Evangelical Christians, 1m. Jehovah's Witnesses, and 2m. Spiritualists. Other faiths include Islam (27,239), Buddhism (214,873) and Judaism (86,825). There are also followers of African and indigenous religions.

### CHRISTIANITY

**Conselho Nacional de Igrejas Cristãs do Brasil (CONIC)** (National Council of Christian Churches in Brazil): SCS, Quadra 01, Bloco E, Edif. Ceará, Sala 713, 70303-900 Brasília, DF; tel. and fax (61) 321-8341; e-mail conic.brasil@terra.com.br; internet www.conic.org.br; f. 1982; seven mem. churches; Pres. Bishop ADRIEL DE SOUZA MAIA; Exec. Sec. P. ERVINO SCHMIDT.

### The Roman Catholic Church

Brazil comprises 41 archdioceses, 219 dioceses (including one each for Catholics of the Maronite, Melkite and Ukrainian Rites), 13 territorial prelatures and one territorial abbacy. The Archbishop of São Sebastião do Rio de Janeiro is also the Ordinary for Catholics of other Oriental Rites in Brazil (estimated at 10,000 in 1994). The great majority of Brazil's population are adherents of the Roman Catholic Church. Adherents at 1 December 2003 represented an estimated 79.3% of the population.

**Bishops' Conference:** Conferência Nacional dos Bispos do Brasil, SE/Sul Quadra 801, Conj. B, CP 02067, 70259-970 Brasília, DF; tel. (61) 313-8300; fax (61) 313-8303; e-mail cnbb@cnbb.org.br; internet www.cnbb.org.br; f. 1980; statutes approved 2002; Pres. Cardinal GERALDO MAJELLA AGNELO (Archbishop of São Salvador da Bahia, BA); Sec.-Gen. ODILO PEDRO SCHERER.

#### Latin Rite

**Archbishop of São Salvador da Bahia, BA:** Cardinal GERALDO MAJELLA AGNELO, Primate of Brazil, Cúria Metropolitana, Rua Martin Afonso de Souza 270, 40100-050 Salvador, BA; tel. (71) 328-6699; fax (71) 328-0068; e-mail gma@atarde.com.br.
**Archbishop of Aparecida, SP:** RAYMUNDO DAMASCENO ASSIS.
**Archbishop of Aracaju, SE:** JOSÉ PALMEIRA LESSA.
**Archbishop of Belém do Pará, PA:** ORANI JOÃO TEMPESTA.
**Archbishop of Belo Horizonte, MG:** WALMOR OLIVEIRA DE AZEVEDO.
**Archbishop of Botucatu, SP:** ALOYSIO JOSÉ LEAL PENNA.
**Archbishop of Brasília, DF:** JOÃO BRAZ DE AVIZ.
**Archbishop of Campinas, SP:** BRUNO GAMBERINI.
**Archbishop of Campo Grande, MS:** VITÓRIO PAVANELLO.
**Archbishop of Cascavel, PR:** LÚCIO IGNÁCIO BAUMGAERTNER.
**Archbishop of Cuiabá, MT:** MILTON ANTÔNIO DOS SANTOS.
**Archbishop of Curitiba, PR:** MOACYR JOSÉ VITTI.
**Archbishop of Diamantina, MG:** PAULO LOPES DE FARIA.
**Archbishop of Feira de Santana, BA:** ITAMAR VIAN.
**Archbishop of Florianópolis, SC:** MURILO SEBASTIÃO RAMOS KRIEGER.
**Archbishop of Fortaleza, CE:** JOSÉ ANTÔNIO APARECIDO TOSI MARQUES.

**Archbishop of Goiânia, GO:** WASHINGTON CRUZ.
**Archbishop of Juiz de Fora, MG:** EURICO DOS SANTOS VELOSO.
**Archbishop of Londrina, PR:** ALBANO BORTOLETTO CAVALLIN.
**Archbishop of Maceió, AL:** JOSÉ CARLOS MELO.
**Archbishop of Manaus, AM:** LUIZ SOARES VIEIRA.
**Archbishop of Mariana, MG:** LUCIANO PEDRO MENDES DE ALMEIDA.
**Archbishop of Maringá, PR:** ANUAR BATTISTI.
**Archbishop of Montes Claros, MG:** GERALDO MAJELA DE CASTRO.
**Archbishop of Natal, RN:** MATIAS PATRÍCIO DE MACÊDO.
**Archbishop of Niterói, RJ:** ALANO MARIA PENA.
**Archbishop of Olinda e Recife, PE:** JOSÉ CARDOSO SOBRINHO.
**Archbishop of Palmas, PR:** ALBERTO TAVEIRA CORRÊA.
**Archbishop of Paraíba, PB:** ALDO DE CILLO PAGOTTO.
**Archbishop of Porto Alegre, RS:** DADEUS GRINGS.
**Archbishop of Porto Velho, RO:** MOACYR GRECHI.
**Archbishop of Pouso Alegre, MG:** RICARDO PEDRO CHAVES PINTO, Filho.
**Archbishop of Ribeirão Preto, SP:** ARNALDO RIBEIRO.
**Archbishop of São Luís do Maranhão, MA:** PAULO EDUARDO DE ANDRADE PONTE.
**Archbishop of São Sebastião do Rio de Janeiro, RJ:** Cardinal EUSÉBIO OSCAR SCHEID.
**Archbishop of Sorocaba, SP:** EDUARDO BENES DE SALES RODRIGUES.
**Archbishop of Teresina, PI:** CELSO JOSÉ PINTO DA SILVA.
**Archbishop of Uberaba, MG:** ALOÍSIO ROQUE OPPERMANN.
**Archbishop of Vitória, ES:** LUIZ MANCILHA VILELA.
**Archbishop of Vitória da Conquista, BA:** GERALDO LYRIO ROCHA.

#### Maronite Rite

**Bishop of Nossa Senhora do Líbano em São Paulo, SP:** JOSEPH MAHFOUZ.

#### Melkite Rite

**Bishop of Nossa Senhora do Paraíso em São Paulo, SP:** FARES MAAKAROUN.

#### Ukrainian Rite

**Bishop of São João Batista em Curitiba, PR:** EFRAIM BASÍLIO KREVEY.

### The Anglican Communion

Anglicans form the Episcopal Anglican Church of Brazil (Igreja Episcopal Anglicana do Brasil), comprising seven dioceses.

**Igreja Episcopal Anglicana do Brasil:** Av. Ludolfo Boehl 256, Teresópolis, 91720-150 Porto Alegre, RS; tel. and fax (51) 3318-6200; e-mail ieabl@ieab.org.br; internet www.ieab.org.br; f. 1890; 103,021 mems (1997); Primate Right Rev. ORLANDO SANTO DE OLIVEIRA; Gen. Sec. CHRISTINA TAKATSU WINNISCHOFER; e-mail cwinnischofer@ieab.org.br.

### Protestant Churches

**Igreja Cristã Reformada do Brasil:** Rua Domingos Rodrigues, 306/Lapa, 05075-000 São Paulo, SP; tel. (11) 260-7514; f. 1932; Pres. Rev. ANTÔNIO BONZOI.

**Igreja Evangélica de Confissão Luterana no Brasil (IECLB):** Rua Senhor dos Passos 202, 4° andar, CP 2876, 90020-180 Porto Alegre, RS; tel. (51) 3221-3433; fax (51) 3225-7244; e-mail secretariageral@ieclb.org.br; internet www.ieclb.org.br; f. 1949; 715,000 mems; Pres. Pastor WALTER ALTMANN; Sec.-Gen. Dr NESTOR PAULO FRIEDRICH.

**Igreja Evangélica Congregacional do Brasil:** CP 414, 98700 Ijuí, RS; tel. (55) 332-4656; f. 1942; 148,836 mems (2000); Pres. Rev. H. HARTMUT W. HACHTMANN.

**Igreja Evangélica Luterana do Brasil:** Rua Cel. Lucas de Oliveira 894, 90440-010 Porto Alegre, RS; tel. (51) 3332-2111; fax (51) 3332-8145; e-mail ielb@ielb.org.br; internet www.ielb.org.br; f. 1904; 223,588 mems; Pres. Rev. CARLOS WALTER WINTERLE; Sec. Rev. RONY RICARDO MARQUARDT.

**Igreja Maná do Brasil:** Rua Marie Satzke, 47, Jardim Marajoara, 04664-150, São Paulo, SP; tel. (11) 4033-5091; e-mail adm_brasil@igrejamana.com; internet www.igrejamana.com.

**Igreja Metodista do Brasil:** Av. Piassanguaba 3031, Planalto Paulista, 04060-004 São Paulo, SP; tel. (11) 5585-0032 ; fax (11) 5594-3328; e-mail isede.nacional@metodista.org.br; internet www

BRAZIL	Directory

.metodista.org.br; 340,963 mems (2000); Exec. Sec. Bishop JOÃO ALVES DE OLIVEIRA.

**Igreja Presbiteriana Unida do Brasil (IPU):** Av. Princesa Isabel 629, 1210/1, Edif. Vitória Center, 29010-360 Vitória, ES; tel. and fax (27) 3222-8024; e-mail ipu@terra.com.br; internet www.ipu.org.br; f. 1978; Moderator Rev. GERSON ANTÔNIO URBAN.

### BAHÁ'Í FAITH

**Bahá'í Community of Brazil:** SHIS, QL 08, Conj. 02, CP 7035, 71620-285 Brasília, DF; tel. (61) 364-3594; fax (61) 364-3470; e-mail info@bahai.org.br; internet www.bahai.org.br; f. 1965; Sec. CARLOS ALBERTO SILVA.

### BUDDHISM

**Sociedade Budista do Brasil** (Rio Buddhist Vihara): Dom Joaquim Mamede 45, Lagoinha, Santa Tereza, 20241-390 Rio de Janeiro, RJ; tel. (21) 2526-1411; e-mail sbbrj@yahoo.com; internet www.geocities.com/sbbrj; f. 1972; Principal Dr PUHULWELLE VIPASSI.

**Sociedade Taoísta do Brasil:** Rua Cosme Velho 355, Cosme Velho, 22241-090 Rio de Janeiro, RJ; tel. (21) 2225-2887; e-mail info@taoismo.com.br; internet www.taoismo.com.br; f. 1991.

## The Press

The most striking feature of the Brazilian press is the relatively small circulation of newspapers in comparison with the size of the population. The newspapers with the largest circulations are *O Dia* (250,000), *O Globo* (350,000), *Fôlha de São Paulo* (560,000), and *O Estado de São Paulo* (242,000). The low circulation is mainly owing to high costs resulting from distribution difficulties. In consequence there are no national newspapers. In 2000 a total of 465 daily newspaper titles, with an average circulation of 7,883,000, and 2,010 non-daily newspapers were published in Brazil.

### DAILY NEWSPAPERS

#### Belém, PA

**O Liberal:** Rua Gaspar Viana 253, 66020 Belém, PA; tel. (91) 222-3000; fax (91) 224-1906; e-mail redacao@orm.com.br; internet www.oliberal.com.br; f. 1946; Pres. LUCIDEA MAIORANA; circ. 20,000.

#### Belo Horizonte, MG

**Diário da Tarde:** Av. Getúlio Vargas 291, 30112-020 Belo Horizonte, MG; tel. (31) 3263-5229; internet www.estaminas.com.br/dt; f. 1931; evening; Editor FÁBIO PROENÇA DOYLE; total circ. 150,000.

**Diário do Comércio:** Av. Américo Vespúcio 1660, Nova Esperança, 31230-250 Belo Horizonte, MG; tel. (31) 3469-1011; fax (31) 3469-1080; e-mail redacaodc@diariodocomercio.com.br; internet www.diariodocomercio.com.br; f. 1932; Pres. LUIZ CARLOS MOTTA COSTA.

**Estado de Minas:** Av. Getúlio Vargas 291, 30112-020 Belo Horizonte, MG; tel. (31) 3263-5000; fax (31) 3263-5070; internet www.uai.com.br/em.html; f. 1928; morning; independent; Pres. BRITALDO SILVEIRA SOARES; circ. 65,000.

**Hoje em Dia:** Rua Padre Rolim 652, Santa Efigênia, 30130-090 Belo Horizonte, MG; tel. (31) 3236-8000; fax (31) 3236-8046; e-mail comercial@hojeemdia.com.br; internet www.hojeemdia.com.br; Man. Dir REINALDO GILLI.

#### Blumenau, SC

**Jornal de Santa Catarina:** Rua São Paulo 1120, 89010-000 Blumenau, SC; tel. (2147) 340-1400; e-mail redacao@santa.com.br; internet www.santa.com.br; f. 1971; Dir ALVARO IAHNIG; circ. 25,000.

#### Brasília, DF

**Correio Brasiliense:** SIG, Quadra 02, Lotes 340, 70610-901 Brasília, DF; tel. (61) 214-1100; fax (61) 214-1157; e-mail geral@correioweb.com.br ; internet www.correioweb.com.br; f. 1960; Pres. ÁLVARO TEIXEIRA DA COSTA; circ. 30,000.

**Jornal de Brasília:** SIG, Trecho 1, Lotes 585/645, 70610-400 Brasília, DF; tel. (61) 225-2515; internet www.jornaldebrasilia.com.br; f. 1972; Dir-Gen. FERNANDO CÔMA; circ. 25,000.

#### Campinas, SP

**Correio Popular:** Rua 7 de Setembro 189, Vila Industrial, 13035-350 Campinas, SP; tel. (19) 3736-3050; fax (19) 3234-8984; e-mail sygon@rac.com.br; internet www.cpopular.com.br; f. 1927; Editorial Dir NELSON HOMEM DE MELLO; circ. 40,000.

#### Curitiba, PR

**O Estado do Paraná:** Parque Gráfico e Administração, Rua João Tschannerl 800, Cidade da Comunicação, Jardim Mercês, 80820-010 Curitiba, PR; tel. (41) 331-5000; fax (41) 335-2838; e-mail oestado@parana-online.com.br; internet www.parana-online.com.br; f. 1951; Pres. PAULO CRUZ PIMENTEL; Dir. MUSSA JOSÉ ASSIS; circ. 15,000.

**Gazeta do Povo:** Rua Pedro Ivo 459, Centro, 80010-020, Curitiba, PR; tel. (41) 224-0522; fax (41) 225-6848; e-mail atendimento@tudoparana.com; internet tudoparana.globo.com/gazetadepovo; f. 1919; Pres. FRANCISCO CUNHA PEREIRA, Filho; circ. 40,000.

**Jornal Tribuna do Paraná:** Parque Gráfico e Administração, Rua João Tschannerl 800, Cidade da Comunicação, Jardim Mercês, 80820-010 Curitiba, PR; tel. (41) 331-5000; fax (41) 335-2838; e-mail tribuna@parana-online.com.br; internet www.parana-online.com.br; f. 1956; Pres. PAULO CRUZ PIMENTEL; circ. 15,000.

#### Florianópolis, SC

**O Estado:** Rodovia SC-401, Km 3, 88030-900 Florianópolis, SC; tel. and fax (482) 239-8888; internet www.oestado.com.br; f. 1915; Pres. JOSÉ MATUSALÉM DE CARVALHO COMELLI; Gen. Editor SANDRA ANNUSECK; circ. 20,000.

#### Fortaleza, CE

**Jornal O Povo:** Av. Aguanambi 282, 60055 Fortaleza, CE; tel. (85) 3211-9666; fax (85) 3231-5792; internet www.noolhar.com/opovo; f. 1928; evening; Pres. DEMÓCRITO ROCHA DUMMAR; Editor-in-Chief FÁTIMA SUDÁRIO; circ. 20,000.

**Tribuna do Ceará:** Av. Desembargador Moreira 2900, 60170-002 Fortaleza, CE; tel. (85) 3247-3066; fax (85) 3272-2799; e-mail tc@secrl.com.br; f. 1957; Dir JOSÉ A. SANCHO; circ. 12,000.

#### Goiânia, GO

**Diário da Manhã:** Av. Anhanguera 2833, Setor Leste Universitário, 74000 Goiânia, GO; tel. (62) 261-7371; internet www.dm.com.br; f. 1980; Editor BATISTA CUSTÓDIO; circ. 16,000.

**Jornal O Popular:** Rua Thómas Edson Q7, Setor Serrinha, 74835-130 Goiânia, GO; tel. (62) 250-1000; fax (62) 241-1018; f. 1938; Pres. JAIME CÂMARA JÚNIOR; circ. 65,000.

#### João Pessoa, PB

**Correio da Paraíba:** Av. Pedro II, Centro, João Pessoa, PB; tel. (83) 216-5000; fax (83) 216-5009; e-mail beatriz@portalcorreio.com.br; internet www.correiodaparaiba.com.br; Exec. Dir BEATRIZ RIBEIRO.

#### Londrina, PR

**Fôlha de Londrina:** Rua Piauí 241, 86010-909 Londrina, PR; tel. (43) 3374-2000; fax (43) 3321-1051; e-mail caf@folhadelondrina.com.br; internet www.bonde.com.br/folha; f. 1948; Pres. JOÃO MILANEZ; circ. 40,000.

#### Manaus, AM

**A Crítica:** Av. André Araújo, Km 3, 69060 Manaus, AM; tel. (92) 643-1233; fax (92) 643-1234; internet www.acritica.com.br; f. 1949; Dir RITA ARAÚJO CADERARO; circ. 19,000.

#### Natal, RN

**Diario de Natal:** Av. Deodoro 245, Petrópolis, 59012-600, Natal, RN; tel. (84) 220-0166; e-mail diretorger@diariodenatal.com.br; internet www.diariodenatal.com.br; Pres. GLADSTONE VIEIRA BELO; Dir ALBIMAR FURTADO.

#### Niterói, RJ

**O Fluminense:** Rua da Conceição 188, Lj 118, Niterói Shopping, Niterói, RJ; tel. (21) 2620-3311; internet www.ofluminense.com.br; f. 1878; Dirs NINA RITA TORRES, ALEXANDRE TORRES; circ. 80,000.

**A Tribuna:** Rua Barão do Amazonas 31, 2403-0111 Niterói, RJ; tel. (13) 2102-7000; fax (13) 3219-4466; e-mail icarai@urbi.com.br; f. 1936; daily; Dir-Supt GUSTAVO SANTANO AMÓRO; circ. 10,000.

#### Palmas, TO

**O Girassol:** Av. Teotônio Segurado 101 Sul, Conj. 01, Edif. Office Center, 77015-002 Palmas, TO; tel. and fax (63) 225-5456; e-mail ogirassol@uol.com.br; internet www.ogirassol.com.br; Pres. WILBERGSON ESTRELA GOMES; Exec. Editor SONIELSON LUCIANO DE SOUSA.

#### Porto Alegre, RS

**Zero Hora:** Av. Ipiranga 1075, Azenha, 90169-900 Porto Alegre, RS; tel. (51) 3218-4900; fax (51) 3218-4700; internet www.zerohora.com.br; f. 1964; Pres. NELSON SIROTSKY; circ. 165,000 Monday, 170,000 weekdays, 240,000 Sunday.

### Recife, PE

**Diário de Pernambuco:** Rua do Veiga 600, Santo Amaro, 50040-110 Recife, PE; tel. (81) 2122-7666; fax (81) 2122-7603; e-mail faleconosco@pernambuco.com; internet www.pernambuco.com/diario; f. 1825; morning; independent; Pres. JOEZEIL BARROS; circ. 31,000.

### Ribeirão Preto, SP

**Jornal Tribuna da Ribeirão Preto:** Rua São Sebastião 1380, Centro, 14015-040 Ribeirão Preto, SP; tel. and fax (16) 3632-2200; e-mail tribuna@tribunariberao.com.br; internet www.tribunaribeirao.com.br; Dir FRANCISCO JORGE ROSA, Filho.

### Rio de Janeiro, RJ

**O Dia:** Rua Riachuelo 359, 20235-900 Rio de Janeiro, RJ; fax (21) 2507-1228; internet odia.ig.com.br; f. 1951; morning; centrist labour; Pres. GIGI CARVALHO; Editor-in-Chief ALEXANDRE FREELAND; circ. 250,000 weekdays, 500,000 Sundays.

**O Globo:** Rua Irineu Marinho 35, CP 1090, 20233-900 Rio de Janeiro, RJ; tel. (21) 2534-5000; fax (21) 2534-5510; internet oglobo.globo.com; f. 1925; morning; Editor-in-Chief JOYCE JANE; circ. 350,000 weekdays, 600,000 Sundays.

**Jornal do Brasil:** Rua São José 90, Centro, 20010-020 Rio de Janeiro, RJ; tel. (21) 2323-1000; e-mail jb@jbonline.com.br; internet www.jb.com.br; f. 1891; morning; Catholic, liberal; Vice-Pres. RICARDO CARVALHO; circ. 200,000 weekdays, 325,000 Sundays.

**Jornal do Commercio:** Rua do Livramento 189, 20221-191 Rio de Janeiro, RJ; tel. and fax (21) 2223-8500; e-mail jornaldocommercio@jcom.com.br; internet www.jornaldocommercio.com.br; f. 1827; morning; Pres. MAURICIO DINEPI; circ. 31,000 weekdays.

**Jornal dos Sports:** Rua Pereira de Almeida 88, Praça de Bandeira, 20260-100 Rio de Janeiro, RJ; tel. (21) 2563-0363; e-mail falatorcedor@jsports.com.br; internet www.jsports.com.br; f. 1931; morning; sporting daily; Man. Dir ARMANDO G. COELHO; circ. 39,000.

### Salvador, BA

**Correio da Bahia:** Rua Aristides Novis 123, Fereração, 40210-630 Salvador, BA; tel. (71) 32721-2850; fax (71) 203-1811; e-mail comercial.correio@redebahia.com.br; internet www.correiodabahia.com.br; f. 1979; Pres. ARMANDO GONÇALVES.

**A Tarde:** Av. Tancredo Neves 1092, Caminho das Árvores, 41822-900 Salvador, BA; tel. (71) 340-8500; fax (71) 231-1064; e-mail presidencia@atarde.com.br; internet www.atarde.com.br; f. 1912; evening; Pres. REGINA SIMÕES DE MELLO LEITÃO; circ. 54,000.

### Santarém, PA

**O Impacto—O Jornal da Amazônia:** Av. Presidente Vargas, 3728, Caranaza, 68040-060 Santarém, PA; tel. (93) 523-3330; fax (93) 523-9131; e-mail oimpacto@oimpacto.com.br; internet www.oimpacto.com.br; Pres. ADMILTON ALMEIDA.

### Santo André, SP

**Diário do Grande ABC:** Rua Catequese 562, Bairro Jardim, 09090-900 Santo André, SP; tel. (11) 4435-8100; fax (11) 4434-8307; e-mail online@dgabc.com.br; internet www.dgabc.com.br; f. 1958; Pres. MAURY DE CAMPOS DOTTO; circ. 78,500.

### Santos, SP

**A Tribuna:** Rua General Câmara 90/94, 11010-903 Santos, SP; tel. (13) 3211-7000; fax (13) 3219-6783; e-mail digital@atribuna.com.br; internet www.atribuna.com.br; f. 1984; Dir ROBERTO M. SANTINI; circ. 40,000.

### São Luís, MA

**O Imparcial:** Empresa Pacotilha Ltda, Rua Assis Chateaubriand s/n, Renascença 2, 65075-670 São Luís, MA; tel. (98) 3212-2030; e-mail redacao@pacotilha.com.br; internet www.oimparcial.com.br; f. 1926; Dir-Gen. PEDRO BATISTA FREIRE.

### São Paulo, SP

**Diário Comércio, Indústria e Serviços:** Rua Bacaetava 191, 1°, 04705-010 São Paulo, SP; tel. (11) 5094-5200; e-mail redacao@dci.com.br; internet www.dci.com.br; f. 1933; morning; Man. Dir ANTÓNIO CARLOS RIOS CORRAL; circ. 50,000.

**Diário de São Paulo:** Rua Major Quedinho 90, Centro, São Paulo, SP; tel. (11) 3658-8000; internet www.diariosp.com.br; f. 1884; fmrly Diário Popular; evening; owned by O Globo; Exec. Dir RICARDO GANDOUR; circ. 90,000.

**O Estado de São Paulo:** Av. Celestino Bourroul 68, 1°, 02710-000 São Paulo, SP; tel. (11) 3856-2122; fax (11) 3266-2206; e-mail falecom@estado.com.br; internet www.estado.com.br; f. 1875; morning; independent; Exec. Editor LUIZ OCTÁVIO LIMA; circ. 242,000 weekdays, 460,000 Sundays.

**Fôlha de São Paulo:** Alameda Barão de Limeira 425, 6° andar, Campos Elíseos, 01202-900 São Paulo, SP; tel. (11) 3224-3678; fax (11) 3224-7550; e-mail falecomagente@folha.com.br; internet www.folha.com.br; f. 1921; morning; Exec. Dir ANA LUCIA BUSCH; circ. 557,650 weekdays, 1,401,178 Sundays.

**Gazeta Mercantil:** Vila Olimpia, Rua Ramos Batista 444, 11° andar, CP 04552-020 São Paulo, SP; tel. (11) 4501-1112; e-mail red@gazetamercantil.com.br; internet www.gazeta.com.br; f. 1920; business paper; Pres. LUIZ FERREIRA LEVY; circ. 80,000.

**Jornal da Tarde:** Av. Eng. Caetano Álvares 55, Limão, 02598-000 São Paulo, SP; tel. (11) 3856-2122; fax (11) 3856-2257; e-mail pergunta@jt.com.br; internet www.jt.com.br; f. 1966; evening; independent; Dir FERNÃO LARA MESQUITA; circ. 120,000, 180,000 Mondays.

### Vitória, ES

**A Gazeta:** Rua Charic Murad 902, 29050 Vitória, ES; tel. (27) 3321-8333; fax (27) 3321-8720; e-mail aleite@redegazeta.com.br; internet gazetaonline.globo.com; f. 1928; Editorial Dir ANTÓNIO CARLOS LEITE; circ. 19,000.

## PERIODICALS

### Rio de Janeiro, RJ

**Amiga:** Rua do Russel 766/804, 22214 Rio de Janeiro, RJ; tel. (21) 2285-0033; fax (21) 2205-9998; weekly; women's interest; Pres. ADOLPHO BLOCH; circ. 83,000.

**Antenna-Eletrônica Popular:** Av. Marechal Floriano 151, 20080-005 Rio de Janeiro, RJ; tel. (21) 2223-2442; fax (21) 2263-8840; e-mail antenna@anep.com.br; internet www.anep.com.br; f. 1926; monthly; telecommunications and electronics, radio, TV, hi-fi, amateur and CB radio; Dir MARIA BEATRIZ AFFONSO PENNA; circ. 15,000.

**Carinho:** Rua do Russel 766/804, 22214 Rio de Janeiro, RJ; tel. (21) 2285-0033; fax (21) 2205-9998; monthly; women's interest; Pres. ADOLPHO BLOCH; circ. 65,000.

**Conjuntura Econômica:** Praia de Botafogo 190, Sala 923, 22253-900 Rio de Janeiro, RJ; tel. (21) 2559-6040; fax (21) 2559-6039; e-mail conjunturaeconomica@fgv.br; internet www.fgv.br/conjuntura.htm; f. 1947; monthly; economics and finance; published by Fundação Getúlio Vargas; Editor-in-Chief ROBERTO FENDT; circ. 25,000.

**Desfile:** Rua do Russel 766/804, 22214 Rio de Janeiro, RJ; tel. (21) 2285-0033; fax (21) 2205-9998; f. 1969; monthly; women's interest; Dir ADOLPHO BLOCH; circ. 120,000.

**ECO21:** Av. Copacabana 2, No 301, 28010-122 Rio de Janeiro, RJ; tel. (21) 2275-1490; internet www.eco21.com.br; ecological issues; Dirs LÚCIA CHAYB, RENÉ CAPRILES.

**Ele Ela:** Rua do Russel 766/804, 22214 Rio de Janeiro, RJ; tel. (21) 2285-0033; fax (21) 2205-9998; f. 1969; monthly; men's interest; Dir ADOLPHO BLOCH; circ. 150,000.

**Manchete:** Rua do Russel 766/804, 20214 Rio de Janeiro, RJ; tel. (21) 285-0033; fax (21) 205-9998; f. 1952; weekly; general; Dir ADOLPHO BLOCH; circ. 110,000.

### São Paulo, SP

**Caros Amigos:** Rua Fidalga 162, Vila Madalena, São Paulo, SP; tel. (11) 3819-0130; internet www.carosamigos.com.br; f. 1997; monthly; political; Editor SÉRGIO DE SOUZA; circ. 50,000.

**Casa e Jardim:** Av. Jaguaré 1485, Jaguaré 05346-902 São Paulo, SP; tel. (11) 3797-7986; fax (11) 3797-7936; e-mail casaejardim@edglobo.com,br; internet revistacasaejardim.globo.com; f. 1953; monthly; homes and gardens, illustrated; Dir JUAN OCERIN; circ. 120,000.

**Claudia:** Editora Abril, Av. das Nações Unidas 7221, Pinheiros 05425-902, São Paulo, SP; tel. (11) 3037-2000; internet www.claudia.abril.uol.com.br; f. 1962; monthly; women's magazine; Dir ROBERTO CIVITA; circ. 460,000.

**Criativa:** Av. Jaguaré 1485, 05346-902 São Paulo, SP; tel. (11) 3767-7812; fax (11) 3767-7771; e-mail criativa@edglobo.com.br; internet revistacriativa.globo.com; monthly; women's interest; Dir JUAN OCERIN; circ. 121,000.

**Digesto Econômico:** Associação Comercial de São Paulo, Rua Boa Vista 51, 01014-911 São Paulo, SP; tel. (11) 3244-3092; fax (11) 3244-3355; e-mail admdiario@acsp.com.br; every 2 months; Pres. ELVIO ALIPRANDI; Chief Editor JOÃO DE SCANTIMBURGO.

**Disney Especial:** Av. das Nações Unidas 7221, 05477-000 São Paulo, SP; tel. (11) 3037-2000; fax (11) 3037-4124; every 2 months; children's magazine; Dir ROBERTO CIVITA; circ. 211,600.

**Elle:** Av. das Nações Unidas 7221, 05425-902 São Paulo, SP; tel. (11) 3037-5197; fax (11) 3037-5451; internet www.uol.com.br/elle; f. 1988; monthly; women's magazine; Editor Carlos Costa; circ. 100,000.

**Exame:** CP 11079, 05422-970 São Paulo, SP; fax (11) 3037-2027; e-mail portalexame@abril.com.br; internet portalexame.abril.com.br; f. 1967; 2 a week; business; Editorial Dir Clayton Netz; circ. 168,300.

**Isto É:** Rua William Speers 1000, 05067-900 São Paulo, SP; tel. (11) 3618-4566; fax (11) 3611-7211; e-mail atendimento@editora3.com.br; internet www.terra.com.br/istoe; politics and current affairs.

**Máquinas e Metais:** Alameda Olga 315, 01155-900, São Paulo, SP; tel. (11) 3824-5300; fax (11) 3662-0103; e-mail info@arandanet.com.br; internet www.arandanet.com.br; f. 1964; monthly; machine and metal industries; Editor José Roberto Gonçalves; circ. 15,000.

**Marie Claire:** Av. Jaguaré 1485, 05346-902 São Paulo, SP; tel. (11) 3767-7000; internet revistamarieclaire.globo.com; monthly; women's magazine; Editorial Dir Mônica de Albuquerque Lins Serino; circ. 273,000.

**Mickey:** Av. das Nações Unidas 7221, 05477-000 São Paulo, SP; tel. (11) 3037-2000; fax (11) 3037-4124; monthly; children's magazine; Dir Roberto Civita; circ. 76,000.

**Micromundo-Computerworld do Brasil:** Rua Caçapava 79, 01408 São Paulo, SP; tel. (11) 3289-1767; monthly; computers; Gen. Dir Eric Hippeau; circ. 38,000.

**Pato Donald:** Av. das Nações Unidas 7221, 05477-000 São Paulo, SP; tel. (11) 3037-2000; fax (11) 3037-4124; every 2 weeks; children's magazine; Dir Roberto Civita; circ. 120,000.

**Placar:** Av. das Nações Unidas 7221, 14° andar, 05477-000 São Paulo, SP; tel. (11) 3037-5816; fax (11) 3037-5597; e-mail placar.leitor@email.abril.com.br; f. 1970; monthly; soccer magazine; Dir Marcelo Durate; circ. 127,000.

**Quatro Rodas:** Av. das Nações Unidas 7221, 14°, 05425-902 São Paulo, SP; fax (11) 3037-5039; internet quatrorodas.abril.com.br; f. 1960; monthly; motoring; Dir Paulo Nogueira; circ. 250,000.

**Revista O Carreteiro:** Rua Palacete das Aguias 395, 04635-021 São Paulo, SP; tel. (11) 5031-8646; fax (11) 5031-8647; e-mail redacao@ocarreteiro.com.br; internet www.ocarreteiro.com.br; f. 1970; monthly; transport; Dirs João Alberto Antunes de Figueiredo, Edson Pereira Coelho; circ. 100,000.

**Saúde:** Av. das Nações Unidas 7221, 05477-000 São Paulo, SP; tel. (11) 3031-7675; fax (11) 3813-9115; internet saude.abril.com.br; monthly; health; Dir Angelo Rossi; circ. 180,000.

**Veja:** Av. das Nações Unidas 7221, CP 11079, 05422-970 São Paulo, SP; fax (11) 3037-5638; e-mail veja@abril.com.br; internet www.veja.com.br; f. 1968; news weekly; Dirs José Roberto Guzzo, Tales Alvarenga, Euripedes Alcântara; circ. 1,200,000.

### NEWS AGENCIES

**Editora Abril, SA:** Av. Otaviano Alves de Lima 4400, CP 2372, 02909-970 São Paulo, SP; tel. (11) 3977-1322; fax (11) 3977-1640; f. 1950; Pres. Roberto Civita.

**Agência ANDA:** Edif. Correio Brasiliense, Setor das Indústrias Gráficas 300/350, Brasília, DF; Dir Edilson Varela.

**Agência o Estado de São Paulo:** Av. Eng. Caetano Alvares 55, 02588-900 São Paulo, SP; tel. (11) 3856-2122; Rep. Samuel Dirceu F. Bueno.

**Agência Fôlha de São Paulo:** Alameda Barão de Limeira 425, 4° andar, 01290-900 São Paulo, SP; tel. (11) 224-3790; fax (11) 221-0675; Dir Marion Strecker.

**Agência Globo:** Rua Irineu Marinho 35, 2° andar, Centro, 20233-900 Rio de Janeiro, RJ; tel. and fax (21) 2292-2000; Dir Carlos Lemos.

**Agência Jornal do Brasil:** Av. Brasil 500, 6° andar, São Cristóvão, 20949-900 Rio de Janeiro, RJ; tel. (21) 2585-4453; fax (21) 2580-9944; f. 1966; Exec. Dir. Edgar Lisboa.

#### Foreign Bureaux

**Agence France-Presse (AFP)** (France): Av. Almirante Barroso 52, 1002, 20031-000 Rio de Janeiro, RJ; tel. (21) 2215-0222; fax (21) 2262-2399; e-mail afpriored@afp/com; Bureau Chief (Brazil) Alain Boebion.

**Agencia EFE** (Spain): Praia de Botafogo 228, Bloco B, Gr. 907, 22359-900 Rio de Janeiro, RJ; tel. (21) 2553-6355; fax (21) 2553-8823; e-mail efe@efebrasil.com.br; internet www.efebrasil.com.br; Bureau Chief Francisco R. Figuéroa.

**Agência Lusa de Informação** (Portugal): São Paulo, SP; Bureau Chief Gonçalo César de Sá.

**Agenzia Nazionale Stampa Associata (ANSA)** (Italy): Av. São Luís 258, 20° andar, Conj. 2004, 01046-000 São Paulo, SP; tel. (11) 3258-7022; fax (11) 3129-5543; e-mail ansasp@uol.com.br; internet www.eurosul.com; Dir Oliviero Pluviano.

**Associated Press (AP)** (USA): Praia de Botafogo 228, Sala 1105, Botafogo, 22250-906 Rio de Janeiro, RJ; tel. (21) 3288-5000; fax (21) 3288-5049; Correspondents Michael Astor, José Harold Olmos.

**Deutsche Presse-Agentur (dpa)** (Germany): Av. das Américas, 700, Bloco 2, Sala 3, Barra da Tijuca, 22640-102 Rio de Janeiro, RJ; tel. (21) 3803-7723; fax (21) 3803-7721; Bureau Chief Esteban Engel.

**Informatsionnoye Telegrafnoye Agentstvo Rossii—Telegrafnoye Agentstvo Suverennykh Stran (ITAR—TASS)** (Russia): Av. Gastão Senges 55, 2003, 22631-280 Rio de Janeiro, RJ; e-mail itartass@skydome.net.

**Inter Press Service (IPS)** (Italy): Av. Augusto Severo 156, 1106, 20021-040 Rio de Janeiro, RJ; tel. and fax (21) 2509-4842; Correspondent Ricardo Bittencourt.

**Jiji Tsushin-Sha (Jiji Press)** (Japan): Av. Paulista 854, 13° andar, Conj. 133, Bela Vista, 01310-913 São Paulo, SP; tel. (11) 3285-0025; fax (11) 3285-3816; e-mail jijisp@nethall.com.br; f. 1958; Chief Correspondent Mutsuhiro Takabayashi.

**Kyodo Tsushin** (Japan): Av. Oswaldo Cruz 149, 1206, Flamengo, 22250-060 Rio de Janeiro, RJ; tel. (21) 2533-5561; fax (21) 2553-0276; e-mail kyodonews@uol.com.br; Bureau Chief Takayoshi Makita.

**Prensa Latina** (Cuba): Marechal Mascarenhas de Moraís 121, Apto 602, Copacabana, 22030-040 Rio de Janeiro, RJ; tel. and fax (21) 2237-1766; Correspondent Francisco Forteza.

**Reuters** (United Kingdom): Av. das Nações Unidas 17891, 8° andar, 04795-100 São Paulo, SP; tel. (11) 5644-7720; fax (11) 5644-7878; e-mail saopaolo.newsroom@reuters.com; internet www.reuters.com.br; Bureau Chief (News and Television) Adrian Dickson; also represented in Rio de Janeiro and Brasília.

**Rossiskoye Informatsionnoye Agentstvo—Novosti (RIA—Novosti)** (Russia): Rua Sambaíba 176, 701, 22450-140 Rio de Janeiro, RJ; tel. and fax (21) 2512-4729; e-mail novosti@uol.com.br; Correspondent Vladimir Stepanov.

**Xinhua (New China) News Agency** (People's Republic of China): Rua do México 21, 701–02, 20031-144 Rio de Janeiro, RJ; tel. (21) 2275-6173; fax (21) 2524-6111; e-mail xinhuabr@hotmail.com; Chief Correspondent Wang Zhigen.

Central News Agency (Taiwan) is also represented in Brazil.

### PRESS ASSOCIATIONS

**Associação Brasileira de Imprensa:** Rua Araújo Porto Alegre 71, Centro, 20030-010 Rio de Janeiro, RJ; f. 1908; 4,000 mems; Pres. Maurício Azêdo; Sec. Fichel Davit Chargel.

**Associação Nacional de Editores de Revistas (ANER):** Rua Deputado Lacerda Franco 300, 15°, Conj. 55, 05418-000 São Paulo, SP; tel. (11) 3030-9392; fax (11) 3030-9393; e-mail mceliafurtado@aner.org.br; Pres. Angelo Rossi; Exec. Dir Maria Célia Furtado.

**Federação Nacional dos Jornalistas (FENAJ):** Higs 707, Bloco R, Casa 54, 70351-718 Brasília, DF; tel. (61) 244-0650; fax (61) 242-6616; e-mail fenaj@fenaj.org.br; internet www.fenaj.org.br; f. 1946; represents 30 regional unions; Pres. Elisabeth Villela da Costa.

## Publishers

### RIO DE JANEIRO, RJ

**Ao Livro Técnico Indústria e Comércio Ltda:** Rua Sá Freire 36/40, São Cristóvão, 20930-430 Rio de Janeiro, RJ; tel. (21) 2580-6230; fax (21) 2580-9955; internet www.editoraaolivrotécnico.com.br; f. 1933; textbooks, children's and teenagers' fiction and non-fiction, art books, dictionaries; Man. Dir Reynaldo Max Paul Bluhm.

**Ediouro Publicações, SA:** Rua Nova Jerusalém 345, CP 1880, Bonsucesso, 21042-230 Rio de Janeiro, RJ; tel. (21) 3882-8200; e-mail editoriallivros@ediouro.com.br; internet www.ediouro.com.br; f. 1939; general; Pres. Jorge Carneiro.

**Editora Campus:** Rua Sete de Setembro 111, 16° andar, 20050-002 Rio de Janeiro, RJ; tel. (21) 3970-9300; fax (21) 2507-1991; e-mail c.rothmuller@campus.com.br; internet www.campus.com.br; f. 1976; business, computing, non-fiction; Man. Dir Claudio Rothmuller.

**Editora Delta, SA:** Av. Nilo Peçanha 50, 2817, 20020-100 Rio de Janeiro, RJ; tel. (21) 2262-5243; internet www.delta.com.br; f. 1930; reference books; Pres. André Koogan Breitman.

**Editora Expressão e Cultura—Exped Ltda:** Estrada dos Bandeirantes 1700, Bloco H, 22710-113, Rio de Janeiro, RJ; tel. (21)

# BRAZIL

2444-0649; fax (21) 2444-0651; e-mail exped@ggh.com.br; internet www.exped.com.br; f. 1966; textbooks, literature, reference; Gen. Man. Ricardo Augusto Pamplona Vaz.

**Editora e Gráfica Miguel Couto, SA:** Rua da Passagem 78, Loja A, Botafogo, 22290-030 Rio de Janeiro, RJ; tel. (21) 541-5145; f. 1969; engineering; Dir Paulo Kobler Pinto Lopes Sampaio.

**Editora Nova Fronteira, SA:** Rua Bambina 25, Botafogo, 22251-050 Rio de Janeiro, RJ; tel. (21) 2131-1111; fax (21) 2286-6755; e-mail sac@novafronteira.com.br; internet www.novafronteira.com.br; f. 1965; fiction, psychology, history, politics, science fiction, poetry, leisure, reference; Pres. Carlos Augusto Lacerda.

**Editora Record, SA:** Rua Argentina 171, São Cristóvão, CP 884, 20001-970 Rio de Janeiro, RJ; tel. (21) 2585-2000; fax (21) 2585-2085; e-mail record@record.com.br; internet www.record.com.br; f. 1941; general fiction and non-fiction, education, textbooks, fine arts; Pres. Sérgio Machado.

**Editora Vozes, Ltda:** Rua Frei Luís 100, CP 90023, 25689-900 Petrópolis, RJ; tel. (24) 2233-9000; fax (24) 2231-4676; e-mail editorial@vozes.com.br; internet www.vozes.com.br; f. 1901; Catholic publishers; management, theology, philosophy, history, linguistics, science, psychology, fiction, education, etc.; Dir Antônio Moser.

**Livraria Francisco Alves Editora, SA:** Rua Uruguaiana, 94, 13° andar, centro, 20050-091 Rio de Janeiro, RJ; tel. (21) 2221-3198; fax (21) 2242-3438; f. 1854; textbooks, fiction, non-fiction; Pres. Carlos Leal.

**Livraria José Olympio Editora, SA:** Rua da Glória 344, 4° andar, Glória, 20241-180 Rio de Janeiro, RJ; tel. (21) 2509-6939; fax (21) 2242-0802; f. 1931; juvenile, science, history, philosophy, psychology, sociology, fiction; Dir Manoel Roberto Domingues.

**Pallas Editora:** Rua Frederico de Albuquerque 56, Higienópolis, 21050-840 Rio de Janeiro, RJ; tel. and fax (21) 2270-0186; e-mail pallas@alternex.com.br; internet www.pallaseditora.com.br; f. 1980; Afro-Brazilian culture.

## SÃO PAULO, SP

**Atual Editora, Ltda:** São Paulo, SP; tel. (11) 5071-2288; fax (11) 5071-3099; e-mail atendprof@atualeditora.com.br; internet www.atualeditora.com.br; f. 1973; school and children's books, literature; Dirs Gelson Iezzi, Osvaldo Dolce.

**Barsa Planeta Internacional:** Rua Rego Freitas 192, Vila Buarque, CP 299, 01059-970 São Paulo, SP; tel. (11) 3225-1900; fax (11) 3225-1960; e-mail atendimento@barsaplaneta.com.br; internet www.barsa.com; f. 1951; reference books.

**Ebid-Editora Páginas Amarelas Ltda:** Av. Liberdade 956, 5° andar, 01502-001 São Paulo, SP; tel. (11) 3208-6622; fax (11) 3209-8723; f. 1947; commercial directories.

**Editora Abril, SA:** Av. das Nações Unidas 7221, 10°, 05425-902 São Paulo, SP; tel. (11) 3037-2000; fax (11) 3037-2100; e-mail fmendia@abril.com.br; internet www.abril.com.br; f. 1950; magazines; Pres. and Editor Roberto Civita.

**Editora Atica, SA:** Rua Barão de Iguape 110, 01507-900 São Paulo, SP; tel. (11) 3346-3000; fax (11) 3277-4146; e-mail editora@atica.com.br; internet www.atica.com.br; f. 1965; textbooks, Brazilian and African literature; Pres. Vicente Paz Fernandez.

**Editora Atlas, SA:** Rua Conselheiro Nébias 1384, Campos Elíseos, 01203-904 São Paulo, SP; tel. (11) 3357-9144; e-mail diretoria@editora-atlas.com.br; internet www.editoraatlas.com.br; f. 1944; business administration, data-processing, economics, accounting, law, education, social sciences; Pres. Luiz Herrmann.

**Editora Brasiliense, SA:** Rua Airi 22, Tatuapé, 03310-010 São Paulo, SP; tel. and fax (11) 6198-1488; e-mail brasilienseedit@uol.com.br; internet www.editorabrasiliense.com.br; f. 1943; education, racism, gender studies, human rights, ecology, history, literature, social sciences; Man. Yolanda C. da Silva Prado; Vice-Pres. Maria Teresa B. de Lima.

**Editora do Brasil, SA:** Rua Conselheiro Nébias 887, Campos Elíseos, CP 4986, 01203-001 São Paulo, SP; tel. (11) 3226-0211; fax (11) 3226-9655; e-mail editora@editoradobrasil.com.br; internet www.editoradobrasil.com.br; f. 1943; education; Pres. Dr Carlos Costa.

**Editora FTD, SA:** Rua Rui Barbosa 156, Bairro Bela Vista, 01326-010 São Paulo, SP; tel. (11) 3253-5011; fax (11) 288-0132; internet www.ftd.com.br; f. 1902; textbooks; Pres. João Tissi.

**Editora Globo, SA:** Av. Jaguaré 1485/1487, 05346-902 São Paulo, SP; tel. (11) 3767-7400; fax (11) 3767-7870; e-mail globolivros@edglobo.com.br; internet globolivros.globo.com; f. 1957; fiction, engineering, agriculture, cookery, environmental studies; Gen. Man. Juan Ocerin.

**Editora Melhoramentos Ltda:** Rua Tito 479, 05051-000 São Paulo, SP; tel. (11) 3874-0854; fax (11) 3874-0855; e-mail blerner@melhoramentos.com.br; internet www.melhoramentos.com.br; f. 1890; general non-fiction; children's books, dictionaries; Dir Breno Lerner.

**Editora Michalany Ltda:** Rua Laura dos Anjos Ramos 420, Jardim Santa Cruz—Interlagos, 04455-350 São Paulo, SP; tel. (11) 5611-3414; fax (11) 5614-1592; e-mail editora@editoramichalany.com.br; internet www.editoramichalany.com.br; f. 1965; biographies, economics, textbooks, geography, history, religion, maps; Dir Douglas Michalany.

**Editora Moderna, Ltda:** Rua Padre Adelino 758, Belenzinho, 03303-904, São Paulo, SP; tel. (11) 6090-1316; fax (11) 6090-1369; e-mail moderna@moderna.com.br; internet www.moderna.com.br; Pres. Ricardo Arissa Feltre.

**Editora Revista dos Tribunais Ltda:** Rua do Bosque 820, 01136-000 São Paulo, SP; tel. (11) 3613-8400; fax (11) 3613-8474; e-mail gmarketing@rt.com.br; internet www.rt.com.br; f. 1912; law and jurisprudence books and periodicals; Dir Carlos Henrique de Carvalho, Filho.

**Editora Rideel Ltda:** Alameda Afonso Schmidt 879, Santa Terezinha, 02450-001 São Paulo, SP; tel. (11) 6977-8344; fax (11) 6976-7415; e-mail sac@rideel.com.br; internet www.rideel.com.br; f. 1971; general; Dir Italo Amadio.

**Editora Saraiva:** Av. Marquês de São Vicente 1697, CP 2362, 01139-904 São Paulo, SP; tel. (11) 3933-3366; fax (11) 861-3308; e-mail diretoria.editora@editorasaraiva.com.br; internet www.editorasaraiva.com.br; f. 1914; education, textbooks, law, economics; Pres. Jorge Eduardo Saraiva.

**Editora Scipione Ltda:** Praça Carlos Gomes 46, 01501-040 São Paulo, SP; tel. (11) 3241-2255; e-mail scipione@scipione.com.br; internet www.scipione.com.br; f. 1983; owned by Editora Abril, SA; school-books, literature, reference; Dir Luiz Esteves Sallum.

**Instituto Brasileiro de Edições Pedagógicas, Ltda** (Editoras IBEP Nacional): Avda Alexandre Mackenzie 619, Jaguaré, 05322-000 São Paulo, SP; tel. (11) 6099-7799; fax (11) 6694-5338; e-mail editoras@ibep-nacional.com.br; internet www.ibep-nacional.com.br; f. 1965; textbooks, foreign languages and reference books; Dirs Jorge Yunes, Paulo C. Marti.

**Lex Editora, SA:** Rua da Consolação 77, 01301-000 São Paulo, SP; tel. (11) 3124-3030; fax (11) 3124-3040; e-mail wsoares@aduaneiras.com.br; internet www.lex.com.br; f. 1937; legislation and jurisprudence; Dir Milton Nicolau Vitale Patara.

**Thomson Pioneira:** Rua Traipu 114, Perdizes, 01235-000 São Paulo, SP; tel. (11) 3665-9900; fax (11) 3665-9901; internet www.thomsonlearning.com.br; f. 1960 as Editora Pioneira; acquired by Thomson Learning in 2000; architecture, computers, political and social sciences, business studies, languages, children's books; Dirs Roberto Guazzelli, Liliana Guazzelli.

## BELO HORIZONTE, MG

**Editora Lê, SA:** Av. D. Pedro II 4550, Jardim Montanhês, CP 2585, 30730 Belo Horizonte, MG; internet www.le.com.br; tel. (31) 3413-1720; f. 1967; textbooks.

**Editora Lemi, SA:** Av. Nossa Senhora de Fátima 1945, CP 1890, 30000 Belo Horizonte, MG; tel. (31) 3201-8044; f. 1967; administration, accounting, law, ecology, economics, textbooks, children's books and reference books.

**Editora Vigília, Ltda:** Belo Horizonte, MG; e-mail lerg@planetarium.com.br; tel. (31) 3337-2744; fax (31) 3337-2834; f. 1960; general.

## CURITIBA, PR

**Editora Educacional Brasileira, SA:** Rua XV de Novembro 178, Salas 101/04, CP 7498, 80000 Curitiba, PR; tel. (41) 223-5012; f. 1963; biology, textbooks and reference books.

## PUBLISHERS' ASSOCIATIONS

**Associação Brasileira de Editores de Livros (Abrelivros):** Rua Turiassu 143, conj. 101/102, 05005-001 São Paulo, SP; tel. and fax (11) 3826-9071; e-mail abrelivros@abrelivros.org.br; internet www.abrelivros.org.br; 30 mem. cos; Pres. João Arinos dos Santos Soares.

**Associação Brasileira do Livro (ABL):** Av. 13 de Maio 23, 16°, Sala 1619/1620, 20031-000 Rio de Janeiro, RJ; tel. and fax (21) 2240-9115; e-mail abralivro@uol.com.br; Pres. Adenilson Jarbas Cabral.

**Câmara Brasileira do Livro:** Cristiano Viana 91, 05411-000 São Paulo, SP; tel. (11) 3069-1300; e-mail marketing@cbl.org.br; internet www.cbl.org.br; f. 1946; Pres. Oswaldo Siciliano.

**Sindicato Nacional dos Editores de Livros (SNEL):** Rua da Ajuda 35, 18° andar, Centro, 20040-000 Rio de Janeiro, RJ; tel. (21)

2533-0399; fax (21) 2533-0422; e-mail snel@snel.org.br; internet www.snel.org.br; 200 mems; Pres. PAULO ROBERTO ROCCO.
There are also regional publishers' associations.

# Broadcasting and Communications

## TELECOMMUNICATIONS

**Amazônia Telecom:** Rua Levindo Lopes 258, 4° andar, Savassi, 31040-170 Belo Horizonte, MG; internet www.amazoniacelular.com.br; mobile cellular provider in the Amazon region; 1.2m. customers.

**Brasil Telecom:** SIA Sul, Área dos Servicos Públicos, Lote D, Bloco B, 71215-000 Brasília, DF; tel. (61) 415-1128; fax (61) 415-1133; internet www.brasiltelecom.com.br; fixed line services in 10 states; also operate mobile cellular network, Brasil Telecom Celular; Chair. MODESTO SOUZA BARROS CARVALHOSO; Pres. HENRIQUE SUTTON DE SOUSA NEVES.

**Claro:** Rua Mena Barreto 42, Botafogo, 22271-100 Rio de Janeiro, RJ; internet www.claro.com.br; f. 2003 by mergers; mobile cellular provider; 14.3m. customers.

**CTBC Telecom:** Av. Afonso Pena 3928, Bairro Brasil, 38400-668 Uberlândia, MG; internet www4.ctbctelecom.com.br; mobile and fixed-line provider in central Brazil.

**Empresa Brasileira de Telecomunicações, SA (Embratel):** Av. Pres. Vargas 1012, CP 2586, 20179-900 Rio de Janeiro, RJ; tel. (21) 2519-8182; e-mail cmsocial@embratel.net.br; internet www.embratel.com.br; f. 1965; operates national and international telecommunications system; controlled by Telmex, Mexico; Chair. DANIEL CRAWFORD.

**Sercomtel, SA:** Rua João Cândido 555, 86010-000 Londrina, PR; tel. 0800 400 4343; e-mail casc@sercomtel.com.br; internet www.sercomtelcelular.com.br; f. 1998; fixed-line and mobile provider.

**Telefônica:** Rua Martiniano de Carvalho 851, Bela Vista, 01321-000 São Paulo, SP; tel. (11) 3549-7200; fax (11) 3549-7202; e-mail webmaster@telesp.com.br; internet www.telesp.com.br; formerly Telecomunicações de São Paulo (Telesp); subsidiary of Telefónica, SA (Spain); 41m. customers.

**Telemar:** Rua Lauro Müller 116, 22° andar, Botafogo, Rio de Janeiro, RJ; tel. (21) 2815-2921; fax (21) 2571-3050; internet www.telemar.com.br; f. 1998 as Tele Norte Leste; fixed line services in 16 states; operates mobile subsidiary, Oi; Chair. CARLOS JEREISSATI; Pres. MANOEL DE SILVA.

**Telemig Celular:** Rua Levindo Lopes 258, 4° andar, Savassi, 30140-170 Belo Horizonte, MG; internet www.telemigcelular.com.br; mobile cellular provider in Minas Gerais; 2.6m. customers.

**TIM:** Av. das Américas 3434, 5° andar, Barra da Tijuca, 22640-102 Rio de Janeiro, RJ; internet www.tim.com.br; f. 1998 in Brazil; part of Telecom Italia Mobile; mobile cellular provider.

**Vivo:** Av. Chucri Zaidan 2460, 5°, 04583-110 São Paulo, SP; tel. (11) 5105-1001; internet www.vivo.com.br; owned by Telefónica Móviles, SA of Spain (50%) and Portugal Telecom; mobile telephones; Pres. ROBERTO DE OLIVEIRA LIMA; 24.6m. customers.

### Regulatory Authority

**Agência Nacional de Telecomunicações (ANATEL):** SAUS Quadra 06, Blocos E e H, 70070-940 Brasília, DF; tel. (61) 312-2000; fax (61) 312-2002; e-mail biblioteca@anatel.gov.br; internet www.anatel.gov.br; f. 1998; regional office in each state; Pres. ELIFAS CHAVES GURGEL DO AMARAL.

## BROADCASTING

### Radio

In April 1992 there were 2,917 radio stations in Brazil, including 20 in Brasília, 38 in Rio de Janeiro, 32 in São Paulo, 24 in Curitiba, 24 in Porto Alegre and 23 in Belo Horizonte.

The main broadcasting stations in Rio de Janeiro are: Rádio Nacional, Rádio Globo, Rádio Eldorado, Rádio Jornal do Brasil, Rádio Tupi and Rádio Mundial. In São Paulo the main stations are Rádio Bandeirantes, Rádio Mulher, Rádio Eldorado, Rádio Gazeta and Rádio Excelsior; and in Brasília: Rádio Nacional, Rádio Alvorada, Rádio Planalto and Rádio Capital.

**Rádio Nacional do Brásil (RADIOBRÁS):** CP 070747, 70359-970 Brasília, DF; tel. 321-3949; fax 321-7602; e-mail maurilio@radiobras.gov.br; internet www.radiobras.gov.br; Pres. EUGENIO BUCCI.

### Television

The main television networks are:

**RBS TV-TV Gaúcha, SA:** Rua Rádio y TV Gaúcha 189, 90850-080 Porto Alegre, RS; tel. (51) 3218-5119; fax (51) 3218-5144; e-mail info@rbstv.com.br; internet www.rbs.com.br; Vice-Pres WALMOR BERGESCH.

**TV Bandeirantes:** Rádio e Televisão Bandeirantes Ltda, Rua Radiantes 13, Morumbi, 05699-900 São Paulo, SP; tel. (11) 3742-3011; fax (11) 3745-7622; e-mail cat@band.com.br; internet www.band.com.br; 65 TV stations and repeaters throughout Brazil; Pres. JOÃO JORGE SAAD.

**TV Record—Rede Record de Televisão—Radio Record, SA:** Rua de Várzea 240, Barra Funda, 01140-080 São Paulo, SP; tel. (11) 3660-4761; fax (11) 3660-4756; e-mail tvrecord@rederecord.com.br; internet www.tvrecord.com.br; Pres. JOÃO BATISTA R. SILVA; Exec. Vice-Pres. H. GONÇALVES.

**TV Rede Globo:** Rua Lopes Quintas 303, Jardim Botânico, 22460-010 Rio de Janeiro, RJ; tel. (21) 2540-2000; fax (21) 2294-2092; e-mail webmaster@redeglobo.com.br; internet www.redeglobo.com.br; f. 1965; 8 stations; national network; Dir ADILSON PONTES MALTA.

**TV SBT—Sistema Brasileira de Televisão—Canal 4 de São Paulo, SA:** Av. Das Comunicações 4, Vila Jaraguá, 06278-905 São Paulo, SP; tel. (11) 7087-3000; fax (11) 7087-3509; internet www.sbt.com.br; Vice-Pres. GUILHERME STOLIAR.

### Broadcasting Associations

**Associação Brasileira de Emissoras de Rádio e Televisão (ABERT):** Centro Empresarial Varig, SCN Quadra 04, Bloco B, Conj. 501, Pétala A, 70710-500 Brasília, DF; tel. (61) 327-4600; fax (61) 327-3660; e-mail abert@abert.org.br; internet www.abert.org.br; f. 1962; mems: 32 shortwave, 1,275 FM, 1,574 medium-wave and 80 tropical-wave radio stations and 258 television stations (1997); Pres. PAULO MACHADO DE CARVALHO NETO; Exec. Dir OSCAR PICONEZ.

There are regional associations for Bahia, Ceará, Goiás, Minas Gerais, Rio Grande do Sul, Santa Catarina, São Paulo, Amazonas, Distrito Federal, Mato Grosso and Mato Grosso do Sul (combined) and Sergipe.

# Finance

(cap. = capital; dep. = deposits; res = reserves; m. = million; brs = branches; amounts in reais, unless otherwise stated)

## BANKING

**Conselho Monetário Nacional (CMN):** Setor Bancário Sul, Quadra 03, Bloco B, Edif. Sede do Banco do Brasil, 21° andar, 70074-900 Brasília, DF; tel. (61) 414-1945; fax (61) 414-2528; e-mail cmn@bcb.gov.br; f. 1964 to formulate monetary policy and to supervise the banking system; Pres. ANTÔNIO PALOCCI, Filho (Minister of Finance).

### Central Bank

**Banco Central do Brasil:** Setor Bancário Sul, Quadra 03, Mezanino 01, Bloco B, 70074-900 Brasília, DF; tel. (61) 414-1955; fax (61) 223-1033; e-mail secre.surel@bcb.gov.br; internet www.bcb.gov.br; f. 1965 to execute the decisions of the Conselho Monetário Nacional; bank of issue; cap. 2,576.4m., res 4,467.7m., dep. 317,639.4m. (Dec. 2003); Gov. HENRIQUE DE CAMPOS MEIRELLES; 10 brs.

### State Commercial Banks

**Banco do Brasil, SA:** Setor Bancário Sul, Quadra 01, Bloco C, Lote 32, Edif. Sede III, 70073-901 Brasília, DF; tel. (61) 310-4500; fax (61) 310-2444; internet www.bb.com.br; f. 1808; cap. 8,366.2m., res 3,805.6m., dep. 151,713.6m. (Dec. 2003); Pres. CÁSSIO CASSEB LIMA; 3,241 brs.

**Banco do Estado do Ceará (BEC):** Rua Pedro Pereira 481, 3° andar, 60035-000 Fortaleza, CE; tel. (85) 3255-1818; fax (85) 3255-1933; e-mail bec@bec.com.br; internet www.bec.com.br; f. 1964; cap. 242.3m., res 37.7m., dep. 803.8m. (June 2003); Pres. JOÃO BATISTA SANTOS; 4 brs.

**Banco do Estado do Rio Grande do Sul, SA** (Banrisul): Rua Caldas Junior 108, 7° andar, 90018-900 Porto Alegre RS; tel. (51) 3215-2501; fax (51) 3215-1715; e-mail cambio_dg@banrisul.com.br; internet www.banrisul.com.br; f. 1928; cap. 600.0m., res 200.8m., dep. 9,515.7m. (Dec. 2003); Pres. ARIO ZIMMERMANN; 352 brs.

**Banco do Estado de Santa Catarina SA:** Rodovia SC 401, Km 5, 4600, Bairro Saco Grande II, 88032-000 Florianpólis, SC; tel. (48) 239-9000; fax (48) 239-9052; e-mail decam/dinte@besc.com.br; internet www.besc.com.br; f. 1962; cap. 1,319.0m., res -1,201.3m., dep. 1,690.8m. (Dec. 2003); Pres. VITOR FONTANA.

**Banco do Estado de São Paulo, SA (Banespa):** Rua Antonio Prado 6, Centro, 01010-010 São Paulo, SP; tel. (11) 5538-7450; fax (11) 5538-6680; e-mail relacoesmercado@santander.combr; internet www.banespa.com.br; f. 1909; owned by Banco Santander (Spain);

cap. 2,536.8m., res 2,219.9m., dep. 15,408.8m. (Dec. 2003); Pres. GABRIEL JARAMILLO SANINT; 1,404 brs.

**Banco do Nordeste do Brasil, SA:** Av. Paranjana 5700, Passaré, 60740-000 Fortaleza, CE; tel. (85) 3299-3333; fax (85) 3299-3878; e-mail info@banconordeste.gov.br; internet www.banconordeste.gov.br; f. 1952; cap. 1,162.0m., res 153.2m., dep. 2,946.6m. (Dec. 2003); Pres. and CEO BYRON COSTA DE QUEIROZ; 186 brs.

**Banco Nossa Caixa, SA:** São Paulo, SP; tel. (11) 3856-7770; fax (11) 3856-7771; internet www.nossacaixa.com.br; f. 1917 as Caixa Econômica do Estado de São Paulo; cap. 1,27.4m., res 171.2m., dep. 16,793.3m. (June 2002); Pres. VALDERY FROTA DE ALBUQERQUE; 498 brs.

Other state commercial banks are the Banco do Estado do Pará and Banestes—the Banco do Estado do Espirito Santo.

### Private Banks

**Banco ABC Brasil SA:** Av. Paulista 37, 01311-902 São Paulo, SP; tel. (11) 3170-2000; fax (11) 3170-2001; e-mail abcbrasil@abcbrasil.com.br; internet www.abcbrasil.com.br; f. 1989 as Banco ABC—Roma SA; cap. 222.1m., res 157.0m., dep. 998.2m. (Dec. 2003); Pres. and Gen. Man. TITO ENRIQUE DA SILVA NETO; 2 brs.

**Banco ABN AMRO SA:** Av. Paulista 1374, 3°, 01310-916 São Paulo, SP; tel. (11) 3174-9615; fax (11) 3174-7052; internet www.bancoreal.com.br; cap. 7,458.2m., res 646.4m., dep. 25,822.3m. (Dec. 2003); 847 brs.

**Banco Alfa de Investimento SA:** Almadena Santos 466, 4° andar, 01418-000 Paraíso, SP; tel. (11) 3175-5074; fax (11) 3171-2438; internet www.bancoalfa.com.br; f. 1998; cap. 226.0m., res 357.9m., dep. 3,293.6m. (Dec. 2003); Pres. PAULO GUIHERME MONTEIRO LOBATO RIBEIRO; 9 brs.

**Banco da Amazônia, SA:** Av. Presidente Vargas 800, 66017-000 Belém, PA; tel. (91) 4008-3252; fax (91) 223-5403; e-mail cambio@bancoamazionia.com.br; internet www.bancoamazonia.com.br; f. 1942; cap. 1,205.2m., res 221.4m., dep. 1,588.3m. (Dec. 2003); Pres. MÂNCIO LIMA CORDEIRO; 88 brs.

**Banco Banestado (BANESTADO):** Rua Monsenhor Celso 151, 11° andar, 81110-150 Curitiba, PR; tel. (11) 5019-8101; fax (11) 5019-8103; e-mail helphb@email.banestado.com.br; internet www.banestado.com.br; f. 1928; Pres. ROBERTO EGYDIO SETUBAL; 374 brs.

**Banco BBM SA:** Rua Miguel Calmon 57, 2° andar, Comercis, 40015-010 Salvador, BA; tel. (71) 243-7675; fax (71) 243-2987; internet www.bancobbm.com.br; f. 1858 as Banco de Bahia; cap. 162.1m., res 150.5m., dep. 2,910.6m. (Dec. 2003); Pres. PEDRO HENRIQUE MARIANI; 4 brs.

**Banco BCN SA:** Av. das Nações Unidas 12901, CENU—Torre Oeste, 15° andar, 04578-000 São Paulo, SP; tel. (11) 5509-2444; fax (11) 5509-2804; internet www.bcn.com.br; f. 1929 as Casa Bancaria Conde & Cia; cap. 2,270.1m., res 957.9m., dep. 14,797.6m. (Dec. 2003); Pres. JOSÉ LUIZ ACAR PEDRO; 222 brs.

**Banco BMC, SA:** Av. das Nações Unidas 12995, 24° andar, 04578-000 São Paulo, SP; tel. (11) 5503-7500; fax (11) 5503-7676; e-mail bancobmc@bmc.com.br; internet www.bmc.com.br; f. 1939; adopted current name in 1990; cap. 164.8m., res 53.6m., dep. 1,562.1m. (Dec. 2003); Chair. and CEO FRANCISCO JAIME NOGUEIRA PINHEIRO; 9 brs.

**Banco BMG, SA:** Av. Alvares Cabral 1707, Santo Agostinho, 30170-001 Belo Horizonte, MG; tel. (31) 3290-3700; fax (31) 3290-3168; e-mail bancobmg@bancobmg.com.br; internet www.bancobmg.com.br; f. 1930; cap. 201.8m., res 130.2m., dep. 1,498.4m. (Dec. 2003); Pres. FLÁVIO PENTAGNA GUIMARÃES; 10 brs.

**Banco Bemge, SA:** Rua Albita 131, 30310-160 Belo Horizonte, MG; tel. (31) 5019-8101; fax (31) 5019-8103; f. 1967; fmrly Banco do Estado de Minas Gerais SA; acquired by Banco Itaú, SA, in Sept. 1998; cap. 250.8m., res –157.6m., dep. 2,450.3m. (Dec. 1997); Pres. and CEO JOSÉ AFONSO BELTRÃO DA SILVA.

**Banco Boavista Interatlântico, SA:** Praça Pio X 118-11, 20091-040 Rio de Janeiro, RJ; tel. (21) 3849-5535; e-mail boavista@ibm.net; internet www.boavista.com.br; f. 1997; acquired by Banco Bradesco in 2000; cap. 4,401.9m., res 278.7m., dep. 4,241.4m. (Dec. 2002); Pres. JOSÉ LUIZ ACAR PEDRO.

**Banco Bradesco, SA:** Cidade de Deus, Vila Yara, 06029–900 Osasco, SP; tel. (11) 3681-4011; fax (11) 3684-4630; internet www.bradesco.com.br; f. 1943; fmrly Banco Brasileiro de Descontos; cap. 7,000.0m., res 6,546.9m., dep. 97,663.5m. (Dec. 2003); Chair. LÁZARO DE MELLO BRANDÃO; Vice-Chair. ANTÔNIO BORNIA; 3,054 brs.

**Banco Brascan SA:** Av. Almirante Barroso 52, 30° andar, Centro, 20031-000 Rio de Janeiro, RJ; tel. (21) 3231-3000; fax (21) 3231-3231; internet www.bancobrascan.com.br; cap. 167.9m., res 77.9m., dep. 621.7m. (Dec. 2003); Pres. ANTÔNIO PAULO DE AZEVEDO SODRE.

**Banco Dibens, SA:** Rua Boa Vista 162, 6° andar, Centro, 01014-902 São Paulo, SP; tel. (11) 3243-7533; fax (11) 3243-7534; internet www.dibens.com.br; f. 1989; jt owned by Unibanco and Grupo Verdi; cap. 110.7m., res 31.7m., dep. 1,993.3m. (Dec. 2001); Pres. MAURO SADDI; 23 brs.

**Banco Fibra:** Av. Brigadeiro Faria Lima 3064, Itaim Bibi, 01451-000 São Paulo, SP; tel. (11) 3847-6700; fax (11) 3847-6969; e-mail bancofibra@bancofibra.com.br; internet www.bancofibra.com.br; f. 1989; cap. 170.0m., res 244.7m., dep. 5,393.7m. (Dec. 2003); Pres. BENJAMIN STEINBRUCH; CEO JOÃO AYRES RABELLO, Filho.

**Banco Industrial do Brasil:** Av. Juscelino Kubitschek 1703, Itaim Bibi, 04543-000 São Paulo, SP; tel. (11) 3049-9700; fax (11) 3049-9810; internet www.bancoindustrial.com.br; cap. 103.0m., res 22.6m., dep. 562.8m. (Dec. 2003); Pres. CARLOS ALBERTO MANSUR.

**Banco Industrial e Comercial, SA (Bicbanco):** Av. Paulista 1048, Bela Vista, 01310-100 São Paulo, SP; tel. (11) 3179-9000; fax (11) 3179-9277; e-mail internationa@bicbanco.com.br; internet www.bicbanco.com.br; f. 1938; cap. 300.0m., res 61.1m., dep. 2,833.4m. (Dec. 2003); Pres. JOSÉ BEZERRA DE MENEZES; 36 brs.

**Banco Itaú, SA:** Praça Alfredo Egydio de Souza Aranha 100, Torre Itaúsa, 04344-902 São Paulo, SP; tel. (11) 5019-1549; fax (11) 5019-1133; e-mail investor.relations@itau.com.br; internet www.itau.com.br; f. 1944; cap. 5,525.0m., res 1,531.7m., dep. 55,326.2m. (Dec. 2003); Chair. OLAVO EGYDIO SETÚBAL; Pres. and CEO ROBERTO EGYDIO SETÚBAL; 1,677 brs.

**Banco Itau BBA, SA:** Av. Brigadeiro Faria Lima 3400, 4°, 04538-132 São Paulo, SP; tel. (11) 3708-8000; fax (11) 3708-8172; e-mail bancoitaubba@itaubba.com.br; internet www.itaubba.com.br; f. 1988; acquired by Banco Itaú in 2002; cap. 2,430.0m., res 685.6m., dep. 9,311.6m. (Dec. 2003); Pres. FERNÃO CARLOS BOTELHO BRACHER; 6 brs.

**Banco Mercantil do Brasil, SA:** Rua Rio de Janerio 654, 10°, Centro, 30160–912 Belo Horizonte, MG; tel. (31) 3239-6314; fax (31) 3239-6975; e-mail sac@mercantil.com.br; internet www.mercantil.com.br; cap. 173.2m., res 215.1m., dep. 3,324.4m. (Dec. 2003); Chair., Pres. and CEO MILTON DE ARAÚJO; 204 brs.

**Banco Mercantil de São Paulo:** CP 4077, Av. Paulista 1450, 01310-917 São Paulo, SP; tel. (11) 3145-2121; fax (11) 3145-3180; e-mail finasa@finasa.com.br; internet www.finasa.com.br; f. 1938; owned by Banco Bradesco; cap. 817.4m., res 189.6m., dep. 3,832.6m. (Dec. 1999); Pres. Dr GASTÃO EDUARDO DE BUENO VIDIGAL.

**Banco Pactual:** Av. República do Chile 230, 28° e 29° andares, 20031-170 Rio de Janeiro, RJ; tel. (21) 2514-9600; fax (21) 2514-8600; e-mail webmaster@pactual.com.br; internet www.pactual.com.br; f. 1983; cap. 208.1m., res 296.5m., dep. 1,261.6m. (Dec. 2002); Pres. EDUARDO PLASS; 4 brs.

**Banco de Pernambuco SA** (Bandepe): Edif. Bandepe 5°, Cais do Apolo 222, Bairro do Recife, 50030-230 Recife, PE; tel. (11) 3425-6385; fax (11) 0812-2097; internet www.bandepe.com.br; f. 1938 as Banco do Estado de Pernambuco; cap. 2,412.4m., res –684.2m., dep. 868.1m. (Dec. 2002); Pres. FLORIS GUSTAAF HENDKIK DECKERS; 52 brs.

**Banco Pine SA:** Alameda Santos 1940, 12-13°, Cerqueira Cesar, 01418-200 São Paulo, SP; tel. (11) 3372-5200; fax (11) 3372-5403; e-mail bancopine@uol.com.br; internet www.bancopine.com.br; f. 1997; cap. 113.3m., res 38.6m., dep. 631.1m. (Dec. 2003); Man. Dirs Dr NELSON NOGUEIRA PINHEIRO, NORBERTO NOGUEIRA PINHEIRO.

**Banco Rural SA:** Av. Presidente Wilson 927, 6°, 20065-900 Rio de Janeiro, RJ; tel. (21) 3210-1310; fax (21) 3220-0198; internet www.rural.com.br; f. 1964; cap. 177.7m., res 432.0m., dep. 3,478.8m. (Dec. 2003); Pres. SABINO CORRÊA RABELLO; 78 brs.

**Banco Safra, SA:** Av. Paulista 2100, 9° andar, 01310-930 São Paulo, SP; tel. (11) 3175-7575; fax (11) 3175-7211; e-mail safrapact@uol.com.br; internet www.safra.com.br; f. 1940; cap. 1,411.9m., res 1,646.0m., dep. 16,285.8m. (Dec. 2003); Pres. CARLOS ALBERTO VIEIRA; 77 brs.

**Banco Santander Brasil, SA:** Rua Amador Bueno 474, 04752-005 São Paulo, SP; tel. (11) 5538-6555; fax (11) 5538-7750; internet www.santander.com.br; f. 1942; fmrly Banco Geral do Comercio; cap. 1,248.2m., res 341.0m., dep. 15,839.0m. (Dec. 2003); Pres. GABRIEL JARAMILLO SANINT; 41 brs.

**Banco Santander Meridional, SA:** Rua Siqueira Campos 1125, Centro, 90010-001 Porto Alegre, RS; tel. (51) 3287-5700; fax (51) 3287-5672; internet www.meridional.com.br; f. 1973; fmrly Banco Sulbrasileiro, SA; taken over by the Govt in Aug. 1985; acquired by Banco Bozano, Simonsen in 1997; name changed in 2000 following acquisition by Banco Santander Brasil, SA; cap. 815.2m., res 149.7m., dep. 1,717.7m. (Dec. 2003); Pres. GABRIEL JARAMILLO SANINT; 227 brs.

**Banco Société Générale Brasil SA:** Av. Paulista 2300, 9° andar, Cerqueira Cesar, 01310-300 São Paulo, SP; tel. (11) 3217-8000; fax (11) 3217-8090; internet www.sgbrasil.com.br; f. 1981 as Banco Sogeral; cap. 125.1m., res –49.5m., dep. 206.6m. (Dec. 2003); Pres. FRANÇOIS DOSSA.

**Banco Sudameris Brasil, SA:** Av. Paulista 1000, 01310-912 São Paulo, SP; tel. (11) 3170-9899; fax (11) 3289-1239; internet www.sudameris.com.br; f. 1910; cap. 1,138.7m., res 200.8m., dep. 7,944.6m. (Dec. 2003); Exec. Dir. YVES L. J. LEJEUNE; 270 brs.

**Banco Votorantim, SA:** Av. Roque Petroni, Jr 999, 16° andar, 04707-910 São Paulo, SP; tel. (11) 5185-1700; fax (11) 5185-1900; internet www.bancovotorantim.com.br; cap. 1,500.2m., res 358.2m., dep. 12,610.1m. (June 2004); 4 brs.

**UNIBANCO** (União de Bancos Brasileiros, SA): Av. Eusébio Matoso 891, 05423-901 São Paulo, SP; tel. (11) 3097-1313; fax (11) 3813-6182; e-mail investor.relations@unibanco.com.br; internet www.unibanco.com.br; f. 1924; cap. 3,690.6m., res 3,465.3m., dep. 36,005.8m. (Dec. 2003); Chair. PEDRO MOREIRA SALLES; 809 brs.

### Development Banks

**Banco de Desenvolvimento de Minas Gerais, SA (BDMG):** Rua da Bahia 1600, CP 1026, 30160-907 Belo Horizonte, MG; tel. (31) 3219-8000; fax 3226-3292; e-mail contastos@bdmg.mg.gov.br; internet www.bdmg.mg.gov.br; f. 1962; long-term credit operations; cap. 464.1m., res −178.8m. (Dec. 2002); Pres. ROMEU SCARIOLI.

**Banco Nacional do Desenvolvimento Econômico e Social (BNDES):** Av. República do Chile 100, Centro, 20031-917 Rio de Janeiro, RJ; tel. (21) 2277-7447; fax (21) 3088-7447; internet www.bndes.gov.br; f. 1952 to act as main instrument for financing of development schemes sponsored by the Government and to support programmes for the development of the national economy; charged with supervision of privatization programme of the 1990s; cap. 12,500.1m., res 576.4m., dep. 8,617.0m. (June 2002); Pres. GIUDO MANTEGA; 1 br.

**Banco Regional de Desenvolvimento do Extremo Sul (BRDE):** Rua Uruguai 155, 4°, CP 139, 90010-140 Porto Alegre, RS; tel. (51) 3215-5000; fax (51) 3215-5050; e-mail brde@brde.com.br; internet www.brde.com.br; f. 1961; development bank for the states of Paraná, Rio Grande do Sul and Santa Catarina; finances small and medium-sized enterprises; cap. 15m. (Dec. 1993); Pres. CASILDO JOÃO MALDANER; 3 brs.

### Investment Banks

**Banco de Investimentos Credit Suisse First Boston:** Av. Brigadeiro Faria Lima 3064, 13° andar, Itaim Bibi, 01451-000 São Paulo, SP; tel. (11) 3842-9700; fax (11) 3842-9747; e-mail csam@csam.br; internet www.csfb.com.br; f. 1971; fmrly Banco de Investimentos Garantia; subsidiary of Credit Suisse, Switzerland; cap. 956.6m., res 107.2m., dep. 1,794.7m. (Dec. 1996); Dir MAURO BERGSTEIN; 1 br.

Other investment banks include Banco Finasa de Investimento SA, Banco Fininvest SA.

### State-owned Savings Bank

**Caixa Econômica Federal:** Brasília, DF; tel. (61) 321-9209; fax (61) 225-0215; internet www.caixa.gov.br; f. 1860; cap. 4,301.1m., res 564.4m., dep. 73,611.2m. (June 2002); Pres. EMÍLIO CARRAZAI; 2,026 brs.

### Foreign Banks

**American Express Bank Brasil** (USA): Av. Bridadeiro Faria Lima 1355, 16° andar, 01452-022 São Paulo, SP; tel. (11) 3030-3000; fax (11) 3030-3030; e-mail biamex@biamex.com.br; internet www.biamex.com.br; f. 1988 as Mantrust SRL; present name adopted Dec. 2003; owned by American Express Bank; cap. 102.3m., res 1.6m., dep. 389.2m. (Dec. 2003); Pres. MARK W. GROSS.

**Banco Barclays SA:** Praça Prof. José Lannes 40, 5°, 04157-100 São Paulo, SP; tel. (11) 5509-3200; fax (11) 5509-3340; e-mail barclays@uol.com.br; internet www.barclays.com; f. 1967 as Banco de Investimento; cap. 488.6m., res −320.2m., dep. 313.0m. (Dec. 2002); Pres. PETER ANDERSON.

**Banco de la Nación Argentina:** Av. Paulista 2319, Sobreloja, 01310 São Paulo, SP; tel. (11) 3083-1555; fax (11) 3081-4630; e-mail bnaspbb@dialdata.com.br; f. 1891; Dir-Gen. GERARDO LUIS PONCE; 2 brs.

**Banco JP Morgan SA:** Av. Brigadeiro Faria Lima 3729, 04538-905 São Paulo, SP; tel. (11) 3048-3700; fax (11) 3048-3500; internet www.jpmorgan.com; f. 1969; Pres. CHARLES WORTMAN.

**Banco Sumitomo Mitsui Brasileiro SA:** Av. Paulista 37, 12°, Paraiso, 01311-902 São Paulo, SP; tel. (11) 3178-8000; fax (11) 3289-1668; f. 1958; cap. 309.4m., res 12.8m., dep. 132.5m. (Dec. 2003); Pres. TAKEAKI MISUMI; 1 br.

**Banco de Tokyo-Mitsubishi SA:** Av. Paulista 1274, Bela Vista, 01310-925 São Paulo, SP; tel. (11) 3268-0211; fax (11) 3268-0453; internet www.btm.com.br; f. 1972 as Banco de Tokyo; cap. 186.9m., res 150.3m., dep. 446.8m. (Dec. 2003); Pres. YOSHIO NOZAKI; 2 brs.

**BankBoston SA:** Av. Doutor Chucri Zaidan 246, Villa Cordeiro, Morumbi, 04583-110 São Paulo, SP; tel. (11) 3249-5622; fax (11) 3249-5529; internet www.bankboston.com.br; cap. 1,610.0m., res 304.6m., dep. 8,427.9m. (Dec. 2003); Pres. GERALDO JOSÉ CARBONE; 59 brs.

**Deutsche Bank SA Banco Alemao:** Rua Alexandre Dumas 2200, 04717-910 São Paulo, SP; tel. (11) 5189-5000; fax (11) 5189-5100; internet www.deutschebank.com.br; f. 1911; cap. 138.9m., res 118.3m., dep. 2,506.6m. (Dec. 2003); Pres. ROGER I. KARAM.

**HSBC Bank Brasil SA-Banco Multiplo:** Travessa Oliveira Belo 34, 4° andar, Centro, 80020-030 Curitiba, PR; tel. (41) 321-6161; fax (41) 340-2660; internet www.hsbc.com.br; f. 1997; cap. 1,340.6m., res 296.8m., dep. 16,934.2m. (Dec. 2003); Pres. YOUSSEF ASSAAD NASR; 966 brs.

**ING Bank NV** (Netherlands): Av. Brigadeiro Faria Lima 3064, 10°, 01451-000 São Paulo, SP; tel. (11) 3847-6274; fax (11) 3847-8274; e-mail deiwes.rubira@ing-barings.com.

### Banking Associations

**Associação Nacional dos Bancos de Investimentos—ANBID:** Av. Brigadeiro Faria Lima 2179, 2° andar, 01451-001 São Paulo, SP; tel. (11) 3471-4200; fax (11) 3471-4230; e-mail anbid@anbid.com.br; internet www.anbid.com.br; investment banks; Pres. ALFREDO EGYDIO SETÚBAL; Supt LUIZ KAUFMAN.

**Federação Brasileira dos Bancos:** Rua Líbero Badaró 425, 17°, 01009-905 São Paulo, SP; tel. (11) 3244-9800; fax (11) 3107-8486; internet www.febraban.org.br; f. 1966; Dir-Gen. WILSON ROBERTO LEVORATO.

**Sindicato dos Bancos dos Estados do Rio de Janeiro e Espírito Santo:** Av. Rio Branco 81, 19° andar, Rio de Janeiro, RJ; Pres. THEÓPHILO DE AZEREDO SANTOS; Vice-Pres. Dr NELSON MUFARREJ.

**Sindicato dos Bancos dos Estados de São Paulo, Paraná, Mato Grosso e Mato Grosso do Sul:** Rua Líbero Badaró 293, 13° andar, 01905 São Paulo, SP; f. 1924; Pres. PAULO DE QUEIROZ.

There are other banking associations in Maceió, Salvador, Fortaleza, Belo Horizonte, João Pessoa, Recife and Porto Alegre.

## STOCK EXCHANGES

**Comissão de Valores Mobiliários (CVM):** Rua 7 de Setembro 111, 13°, 26° e 34°, Centro, 20159-900 Rio de Janeiro, RJ; tel. (21) 3233-0200; fax (21) 2221-6769; e-mail soi@cvm.gov.br; internet www.cvm.gov.br; f. 1977 to supervise the operations of the stock exchanges and develop the Brazilian securities market; regional offices in Brasília and São Paulo; Chair. JOSÉ LUIZ OSORIO DE ALMEIDA, Filho.

**Bolsa de Mercadorias e Futuros (BM&F)** (Brazilian Mercantile and Futures Exchange): Praça Antônio Prado 48, 01010-901 São Paulo, SP; tel. (11) 3199-2000; fax (11) 3107-9911; e-mail bmf@bmf.com.br; internet www.bmf.com.br; f. 1985; trading centre and clearing house for assets, securities and derivatives; offices in Brasília, Rio de Janeiro, Campo Grande, Santos, New York (USA) and Shanghai (China); Pres. MANOEL FELIX CINTRA NETO; CEO EDEMIR PINTO.

**Bolsa de Valores do Rio de Janeiro:** Praça XV de Novembro 20, 20010-010 Rio de Janeiro, RJ; tel. (21) 2514-1069; fax (21) 2514-11107; e-mail info@bvrj.com.br; internet www.bvrj.com.br; f. 1845; focuses on the trading of fixed income government bonds and foreign exchange; Chair. EDSON FIGUEIREDO MENEZES; Supt-Gen. SÉRGIO PÓVOA.

**Bolsa de Valores de São Paulo (BOVESPA):** Rua XV de Novembro 275, CP 3456, 01013-001 São Paulo, SP; tel. (11) 3233-2000; fax (11) 3233-3550; e-mail bovespa@bovespa.com.br; internet www.bovespa.com.br; f. 1890; offices in Rio de Janeiro and Porto Alegre; 550 companies listed in 1997; Pres. RAYMUNDO MAGLIANO, Filho; CEO GILBERTO MIFANO.

There are commodity exchanges at Paraná, Porto Alegre, Vitória, Recife, Santos and São Paulo.

## INSURANCE

### Supervisory Authorities

**Superintendência de Seguros Privados (SUSEP):** Rua Buenos Aires 256, 4°, 20061-000 Rio de Janeiro, RJ; tel. (21) 3086-9800; e-mail gabin@susep.gov.br; internet www.susep.gov.br; f. 1966; within Ministry of Finance; offices in Brasília, São Paulo and Porto Alegre; Pres. ANTÔNIO PALOCCI, Filho (Minister of Finance); Supt RENÊ DE OLIVEIRA GARCIA JÚNIOR.

**Conselho de Recursos do Sistema Nacional de Seguros Privados, de Previdência Abierta e de Capitalização (CRSNSP):** Rua Buenos Aires 256, 20061-000 Rio de Janeiro, RJ; tel. (21) 3806-

9815; f. 1966 as Conselho Nacional de Seguros Privados (CNSP); changed name in 1998; Sec. THERESA CHRISTINA CUNHA MARTINS.

**Federação Nacional dos Corretores de Seguros Privados, de Capitalização, de Previdência Privada e das Empresas Corretoras de Seguros (FENACOR):** Av. Rio Branco 147, 6°, 20040-006 Rio de Janeiro, RJ; tel. (21) 2507-0033; fax (21) 2507-0037; e-mail leoncio@fenacor.com.br; internet www.fenacor.com.br; Pres. ARMANDO VERGÍLIO DOS SANTOS JUNIOR.

**Federação Nacional das Empresas de Seguros Privados e de Capitalização (FENASEG):** Rua Senador Dantas 74, Centro, 20031-200 Rio de Janeiro, RJ; tel. (21) 2510-7777; e-mail fenaseg@fenaseg.org.br; internet www.fenaseg.org.br; f. 1951; Pres. JOÃO ELISIO FERRAZ DE CAMPOS.

**IRB-Brasil Resseguros:** Av. Marechal Câmara 171, Castelo, 20020-901 Rio de Janeiro, RJ; tel. (21) 2272-0200; fax (21) 2240-8775; e-mail info@irb-brasilre.com.br; internet www.irb-brasilre.com.br; f. 1939; fmrly Instituto de Resseguros do Brasil; offices in Brasília, São Paulo, Porto Alegre, New York (USA) and London (UK); reinsurance; Pres. MARCOS DE BARROS LISBOA.

### Principal Companies

The following is a list of the principal national insurance companies, selected on the basis of assets. The total assets of insurance companies operating in Brazil were US $ 21,465.7m. in December 2003.

**AGF Vida e Previdência, SA:** São Paulo, SP; separated from AGF Brasil Seguros and sold to Itau in 2003; total assets US $218.2m. (Dec. 2003).

**Bradesco Saúde, SA:** Rua Barão de Itapagipe 225, Rio Comprido, Rio de Janeiro, RJ 20261-901; tel. (21) 2503-1101; fax (21) 2293-9489; internet www.bradescosaude.com.br; health insurance; Pres. LUIZ CARLOS TRABUCO CAPPI.

**Bradesco Seguros e Previdência, SA:** Rua Barão de Itapagipe 225, 20269-900 Rio de Janeiro, RJ; tel. (21) 2503-1101; fax (21) 2293-9489; internet www.bradescoseguros.com.br; f. 1934; general; total assets US $354.7m. (Dec. 2003); Pres. LUIZ CARLOS TRABUCO CAPPI.

**Bradesco Vida e Previdência, SA:** Cidade de Deus s/n, Vila Yara, São Paulo, SP; tel. (11) 3684-2122; fax (11) 3684-5068; internet www.bradescoprevidencia.com.br; f. 2001; life insurance; total assets US $8,530.2m. (Dec. 2003); Pres. MARCO ANTÔNIO ROSSI.

**Brasilprev Seguros e Prevedência, SA:** Rua Verbo Divino 1711, Chácara Santo Antônio, 04719-002 São Paulo, SP; tel. (11) 5185-4240; e-mail atendimento@brasilprev.com.br; internet www.brasilprev.com.br; f. 1993; all classes; 50% owned by Banco do Brasil; total assets US $1,933.5m. (Dec. 2003); Pres. FUAD NOMAN.

**Caixa Seguros:** SCN Quadra 01, Bloco A, 15°–17° andares, Asa Norte, 70711-900 Brasília, DF; tel. (61) 429-2400; fax (61) 328-0600; internet www.caixaseguros.com.br; f. 1967; fmrly Sasse, Cia Nacional de Seguros; changed name to Sasse Caixa Seguros in 1998 and as above in 2000; general; total assets US $425.0m. (Dec. 2003); Pres. THIERRY MARC CLAUDE CLAUDON.

**Caixa Vida e Previdência, SA:** SCN Quadra 1, Bloco A, 15°, Asa Norte, 70711-900 Brasília, DF; tel. (61) 429-2400; fax (61) 328-0600; internet www.caixaprevidencia.com.br; part of Caixa Seguros group; total assets US $451.6m. (Dec. 2003).

**Cia de Seguros Aliança do Brasil, SA:** Rua Manuel da Nóbrega 1280, 9°, 04001-004 São Paulo, SP; tel. (11) 4689-5638; internet www.aliancadobrasil.com.br; f. 1996; total assets US $324.4m. (Dec. 2003); Pres. KHALID MOHAMMED RAOUF.

**HSBC Vida e Previdência(Brasil), SA:** Rua Ten. Francisco Ferreira de Souza 805, Bloco 1, Ala 4, Vila Hauer, 81570-340 Curitaba, PR; tel. (41) 217-4555; fax (41) 321-8800; e-mail spariz@hsbc.com.vr; internet www.hsbc.com.br; f. 1938; all classes; total assets US $499.3m. (Dec. 2004); Supt Dir VILSON ANDRADE; Pension Funds Dir SIDNEY PARIZ.

**Icatu Hartford Seguros, SA:** Av. Presidente Wilson 231, 10° andar, 20030-021 Rio de Janeiro, RJ; tel. 0800-285-3000; e-mail online@icatu.com.br; internet www.icatu-hartford.com.br; total assets US $329.2m. (Dec. 2003).

**Itaú Seguros, SA:** Praça Alfredo Egydio de Souza Aranha 100, Bloco A, 04344-920 São Paulo, SP; tel. (11) 5019-3322; fax (11) 5019-3530; e-mail itauseguros@itauseguros.com.br; internet www.itauseguros.com.br; f. 1921; all classes; total assets US $325.4m. (Dec. 2003); Pres. LUIZ DE CAMPOS SALLES.

**Itaú Vida e Previdência, SA:** internet www.itau.com.br; part of the Itaú group; total assets US $1,808.4m. (Dec. 2003).

**Liberty Paulista Seguros, SA:** Rua Dr Geraldo Campos Moreira 110, 04571-020 São Paulo, SP; tel. (11) 5503-4000; fax (11) 5505-2122; internet www.libertypaulista.com.br; f. 1906; general; total assets US $156.7m. (Dec. 2003); Pres. LUIS MAURETTE.

**Marítima Seguros, SA:** Rua Col. Xavier de Toledo 114, 10°, São Paulo, SP; tel. (11) 3156-1000; fax (11) 3156-1712; internet www.maritimaseguros.com.br; f. 1943; Pres. FRANCISCO CAIUBY VIDIGAL; Dir-Gen. MILTON BELLIZIA BELLIZIA, Filho.

**Porto Seguro Cia de Seguros Gerais:** Rua Guaianazes 1238, 12°, 01204-001 São Paulo, SP; tel. (11) 3366-5199; fax (11) 3366-5140; internet www.portoseguro.com.br; f. 1945; life, automotive and risk; total assets US $518.6m. (Dec. 2003); Pres. ROSA GARFINKEL.

**Real Seguros, SA:** Rua Samapiao Viana 44, 04004-902 Paraíso, SP; tel. (11) 3054-7000; internet www.realseguros.com.br; f. 1969; in 1999 became part of ABM AMRO; general; total assets US $165.8m. (Dec. 2003).

**Real Vida e Previdência, SA:** Rua Sampaio Viana 44, 04004-902 Paraíso, SP; tel. (11) 3054-7000; internet www.bancoreal.com.br; f. 1969; became part of the ABM AMRO group in 1999; total assets US $495.1m. (Dec. 2003).

**Santander Seguros, SA:** internet www.santander.com.br; part of the Banco Santander; total assets US $467.3m. (Dec. 2003).

**Sul América, Cia Nacional de Seguros:** Rua da Quitanda 86, 20091-000 Rio de Janeiro, RJ; tel. (21) 2506-8585; fax (21) 2506-8807; internet www.sulamerica.com.br; f. 1895; life and risk; total assets US $221.9m. (Dec. 2003); Pres. RONY CASTRO DE OLIVEIRA LYRIO.

**Sul América Seguros de Vida e Previdência, SA:** Rua Pedro Avacine 73, Morumbi, 05679-160 São Paulo, SP; tel. (11) 3779-7000; fax (11) 3758-8972; internet www.sulamerica.com.br; total assets US $348.3m. (Dec. 2003).

**Unibanco AIG Previdência, SA:** Av. Eusébio Matoso 1375, 0523-180 São Paulo, SP; tel. (11) 3039-4082; fax (11) 3039-4074; internet www.unibancoseguros.com.br; f. 2001 by merger of Unibanco Seguros and AIG Brasil; part of the Unibanco AIG group; Pres. JOSÉ CASTRO ARAÚJO RUDGE.

**Unibanco AIG Seguros:** Av. Eusébio Matoso 1375, 05423-180 São Paulo, SP; tel. (11) 3039-4082; fax (11) 3039-4074; internet www.unibancoseguros.com.br; f. 2001 by merger of Unibanco Seguros and AIG Brasil; part of the Unibanco AIG group; life; Pres. JOSÉ CASTRO ARAÚJO RUDGE.

# Trade and Industry

## GOVERNMENT AGENCIES

**Agência Nacional de Petróleo (ANP):** SGAM, Quadra 603, Módulo I, 3°, 70830-902 Brasília, DF; tel. (61) 226-0444; fax (61) 226-0699; internet www.anp.gov.br; f. 1998; regulatory body of the petroleum industry; Dir-Gen. VITOR MARTINS.

**Agência de Promoção de Exportações do Brasil:** SBN Quadra 1, Bloco B, 10° andar, Edif. CNC, 70041-902 Brasília, DF; tel. (61) 426-0202; fax (61) 426-0222; e-mail apex@apexbrasil.com.br; internet www.apexbrasil.com.br; f. 2003; promotes Brazilian exports; Pres. JUAN MANUEL QUIRÓS.

**Comissão de Fusão e Incorporação de Empresa (COFIE):** Ministério da Fazenda, Edif. Sede, Ala B, 1° andar, Esplanada dos Ministérios, Brasília, DF; tel. (61) 225-3405; mergers commission; Pres. SEBASTIÃO MARCOS VITAL; Exec. Sec. EDGAR BEZERRA LEITE, Filho.

**Conselho de Desenvolvimento Comercial (CDC):** Bloco R, Esplanada dos Ministérios, 70044 Brasília, DF; tel. (61) 223-0308; commercial development council; Exec. Sec. Dr RUY COUTINHO DO NASCIMENTO.

**Conselho de Desenvolvimento Econômico (CDE):** Bloco K, 7° andar, Esplanada dos Ministérios, 70063 Brasília, DF; tel. (61) 215-4100; f. 1974; economic development council; Gen. Sec. JOÃO BATISTA DE ABREU.

**Conselho de Desenvolvimento Social (CDS):** Bloco K, 3° andar, 382, Esplanada dos Ministérios, 70063 Brasília, DF; tel. (61) 215-4477; social development council; Exec. Sec. JOÃO A. TELES.

**Conselho Nacional do Comércio Exterior (CONCEX):** Fazenda, 5° andar, Gabinete do Ministro, Bloco 6, Esplanada dos Ministérios, 70048 Brasília, DF; tel. (61) 223-4856; f. 1966; responsible for foreign exchange and trade policies and for the control of export activities; Exec. Sec. NAMIR SALEK.

**Conselho Nacional de Desenvolvimento Científico e Tecnológico (CNPq):** SEPN 507, 3° andar, Brasília, DF; tel. (61) 348-9401; fax (61) 348-3941; e-mail presidencia@cnpq.br; internet www.cnpq.br; f. 1951; scientific and technological development council; Pres. ERNEY FELÍCIO PLESSMANN DE CAMARGO.

**Conselho Nacional de Desenvolvimento Rural Sustentável (CNDRS):** SCN, Quadra 01, Edif. Brasília Trade Centre, Sala 501, 70711-902 Brasília, DF; tel. (61) 328-5218; fax (61) 328-5175;

internet www.cndrs.org.br; f. 2000 to promote sustainable rural development.

**Conselho de Não-Ferrosos e de Siderurgia (CONSIDER):** Ministério da Indústria e Comércio, Esplanada dos Ministérios, Bloco 7, 7° andar, 70056-900 Brasília, DF; tel. (61) 224-6039; f. 1973; exercises a supervisory role over development policy in the non-ferrous and iron and steel industries; Exec. Sec. WILLIAM ROCHA CANTAL.

**Instituto Brasileiro de Geografia e Estatística (IBGE):** Centro de Documentação e Disseminação de Informações (CDDI), Rua Gen. Canabarro 706, 2° andar, Maracanã, 20271-201 Rio de Janeiro, RJ; tel. (21) 2142-4781; fax (21) 2142-4933; e-mail webmaster@ibge.gov.br; internet www.ibge.gov.br; f. 1936; produces and analyses statistical, geographical, cartographic, geodetic, demographic and socioeconomic information; Pres. (IBGE) EDUARDO PEREIRA NUNES; Supt (CDDI) DAVID WU TAI.

**Instituto Nacional de Colonização e Reforma Agraria (INCRA):** Edif. Palácio do Desenvolvimento, 70057-900 Brasília, DF; e-mail incra@incra.gov.br; internet www.incra.gov.br; f. 1970; land reform agency; Pres. RALF HACKBART.

**Instituto Nacional de Metrologia, Normalização e Qualidade Industrial (INMETRO):** Rua Santa Alexandrina 416, 5°, Rio Comprido, 20261-232 Rio de Janeiro, RJ; tel. (21) 2502-1009; fax (21) 2563-2970; e-mail gabin@inmetro.gov.br; internet www.inmetro.gov.br; f. 1973; in 1981 INMETRO absorbed the Instituto Nacional de Pesos e Medidas (INPM), the weights and measures institute; Pres. ARMANDO MARIANTE CARVALHO.

**Instituto de Pesquisa Econômica Aplicada (IPEA):** Av. Presidente António Carlos 51, 13° andar, CP 2672, 20020-010 Rio de Janeiro, RS; tel. (21) 3804-8000; fax (21) 2240-1920; e-mail editrj@ipea.gov.br; internet www.ipea.gov.br; also has an office in Brasília; f. 1970; economics and planning institute; Pres. GLAUCO ARBIX.

**Secretária de Desenvolvimento da Produção:** Esplanada dos Ministérios, Bloco J, 5° andar, Brasília, DF; tel. (61) 2109-7182; internet www.desenvolvimento.gov.br; developing commercial, industrial and service production; Sec. CARLOS GASTALDONI.

## REGIONAL DEVELOPMENT ORGANIZATIONS

**Agência de Desenvolvimento da Amazônia (ADA):** Av. Almirante Barroso 426, Marco, 66090-900 Belém, PA; tel. (91) 210-5440; fax (91) 266-0366; e-mail gabinete@ada.gov.br; internet www.ada.gov.br; f. 2001 to co-ordinate the development of resources in Amazonia; Dir-Gen. DJALMA BEZERRA MELLO.

**Agência de Desenvolvimento do Nordeste (ADENE):** Praça Ministro João Gonçalves de Souza s/n, Edif. Sudene, 50670-900 Recife, PE; tel. (81) 3416-2108; e-mail webmaster@adene.gov.br; internet www.adene.gov.br; f. 2001 to replace the defunct Superintendência de Desenvolvimiento do Nordeste (SUDENE); Dir-Gen. MANOEL BRANDÃO.

**Companhia de Desenvolvimento dos Vales do São Francisco e do Parnaíba (CODEVASF):** SGAN, Quadra 601, Lote 1, Edif. Manoel Novaes, 70830-901 Brasília, DF; tel. (61) 312-4747; fax (61) 311-7814; e-mail divulgacao@codevasf.gov.br; internet www.codevasf.gov.br; f. 1974; promotes integrated development of resources of São Francisco and Parnaíba Valley; Pres. LUIZ CARLOS EVERTON DE FARIAS.

**Superintendência da Zona Franca de Manaus (SUFRAMA):** Rua Ministro João Gonçalves de Souza s/n, Distrito Industrial, 69075-783 Manaus, AM; tel. (92) 237-1691; fax (92) 237-6549; e-mail gabin@suframa.gov.br; internet www.suframa.gov.br; to assist in the development of the Manaus Free Zone; Supt FLÁVIA SKROBOT BARBOSA GROSSO.

## AGRICULTURAL, INDUSTRIAL AND TRADE ORGANIZATIONS

**Associação Brasileira de Celulose e Papel—Bracelpa:** Rua Afonso de Freitas 409, Bairro Paraíso, 04006-900 São Paulo, SP; tel. (11) 3885-1845; fax (11) 3885-3689; e-mail faleconosco@bracelpa.org.br; internet www.bracelpa.org.br; f. 1932; pulp and paper association; Pres. OSMAR ELIAS ZOGBI.

**Associação Brasileira dos Produtores de Algodão—Abrapa:** Sector Bancário Norte, Quadra 1, Bloco F, Sala 401, Edif. Palácio da Agricultura, 70040-908 Brasília, DF; tel. and fax (61) 425-2762; e-mail abrapa@abrapa.com.br; internet www.abrapa.com.br; f. 1999; cotton producers' association; Pres. JOÃO LUIZ RIBAS PESSA.

**Associação de Comércio Exterior do Brasil (AEB):** Av. General Justo 335, 4° andar, 20021-130 Rio de Janeiro, RJ; tel. (21) 2544-0048; fax (21) 2544-0577; e-mail aebbras@aeb.org.br; internet www.aeb.org.br; exporters' association.

**Associação Comercial do Rio de Janeiro:** Rua da Calendária 9, 11°–12° andares, Centro, 20091-020 Rio de Janeiro, RJ; tel. (21) 2291-1229; fax (21) 2253-6236; e-mail acrj@acrj.org.br; internet www.acrj.org.br; f. 1820; Pres. MARCÍLIO MARQUES MOREIRA.

**Centro das Indústrias do Estado de São Paulo (CIESP):** Rua Cel. Santos Cardos 537, Jardim Santista, CE, 08730-110 Mogi das Cruzes, SP; tel. 4735-3447; fax 4722-7251; e-mail ciesp@ciesp.com.br; internet www.ciesp.com.br; f. 1928; asscn of small and medium-sized businesses; Pres. CLÁUDIO VAZ.

**Companhia de Pesquisa de Recursos Minerais (CPRM):** Serviço Geológico do Brasil, Av. Pasteur 404, Urca, 22290-240 Rio de Janeiro, RJ; tel. (21) 2295-5337; fax (21) 2542-3647; e-mail cprm@cprm.gov.br; internet www.cprm.gov.br; mining research, attached to the Ministry of Mines and Energy; Pres. UMBERTO RAIMUNDO COSTA.

**Confederação da Agricultura e Pecuária do Brasil (CNA):** SBN, Quadra 01, Bloco F, 3°-5°, Edif. Palácio da Agricultura, 70048-908 Brasília, DF; tel. (61) 424-1400; fax (61) 424-1490; e-mail cna@cna.org.br; internet www.cna.org.br; f. 1964; national agricultural confederation; Pres. ALYSSON PAULINELLI.

**Confederação Nacional do Comércio (CNC):** SCS, Edif. Presidente Dutra, 4° andar, Quadra 11, 70327 Brasília, DF; tel. (61) 223-0578; national confederation comprising 35 affiliated federations of commerce; Pres. ANTÔNIO JOSÉ DOMINGUES DE OLIVEIRA SANTOS.

**Confederação Nacional da Indústria (CNI):** Rua Mariz e Barros 678, 2°, Maracanã, 20270-002 Rio de Janeiro, RJ; tel. (21) 2204-9513; fax (21) 2204-9522; e-mail sac@cni.org.br; internet www.cni.org.br; f. 1938; national confederation of industry comprising 27 state industrial federations; Pres. ARMANDO DE QUEIROZ MONTEIRO METO; First Vice-Pres. CARLOS EDUARDO MOREIRA FERREIRA.

**Conselho dos Exportadores de Café Verde do Brasil (CECAFE):** Av. Nove de Julho 4865, Torre A, Conj. 61, Chácara Itaim, 01407-200 São Paulo, SP; tel. (11) 3079-3755; fax (11) 3167-4060; internet www.cecafe.com; f. 1999, through merger of Federação Brasileira dos Exportadores de Café and Associação Brasileira dos Exportadores de Café; council of green coffee exporters.

**Departamento Nacional da Produção Mineral (DNPM):** SAN, Quadra 1, Bloco B, 3° andar, 70040-200 Brasília, DF; tel. (61) 224-7097; fax (61) 225-8274; e-mail webmaster@dnpm.gov.br; internet www.dnpm.gov.br; f. 1934; responsible for geological studies and control of exploration of mineral resources; Dir-Gen. JOÃO R. PIMENTEL.

**Empresa Brasileira de Pesquisa Agropecuária (EMBRAPA):** SAIN, Parque Rural, W/3 Norte, CP 040315, 70770-901 Brasília, DF; tel. (61) 348-4433; fax (61) 347-1041; f. 1973; attached to the Ministry of Agriculture, Fisheries and Food Supply; agricultural research; Pres. ALBERTO DUQUE PORTUGAL.

**Federação das Indústrias do Estado do Rio de Janeiro (FIRJAN):** Centro Empresarial FIRJAN, Avda Graça Aranha 1, Rio de Janeiro, RJ; tel. (21) 2563-4389; e-mail centrodeatendimento@firjan.org.br; internet www.firjan.org; regional manufacturers' association; 102 affiliated sindicates.

**Federação das Indústrias do Estado de São Paulo (FIESP):** Av. Paulista 1313, 01311-923 São Paulo, SP; tel. (11) 3549-4499; internet www.fiesp.org.br; regional manufacturers' association; Pres. PAULO SKAF.

**Instituto Brasileiro do Meio Ambiente e Recursos Naturais Renováveis (IBAMA):** Edif. Sede IBAMA, Av. SAIN, L4 Norte, Bloco C, Subsolo, 70800-200 Brasília, DF; tel. (61) 316-1205; fax (61) 226-5094; e-mail cnia@sede.ibama.gov.br; internet www.ibama.gov.br; f. 1967; responsible for the annual formulation of national environmental plans; merged with SEMA (National Environmental Agency) in 1988 and replaced the IBDF in 1989; Pres. EDUARDO MAILIUS.

**Instituto Brasileiro do Mineração (IBRAM):** SCS, Quadra 2, Bloco D, Edif. Oscar Niemeyer, 11° andar, 70316-900 Brasília, DF; tel. (61) 226-9367; fax (61) 226-9580; e-mail ibram@ibram.org.br; internet ibram.org.br; f. 1976 to foster the development the mining industry; Pres. JOÃO SÉRGIO MARINHO NUNES.

**Instituto Nacional da Propriedade Industrial (INPI):** Praça Mauá 7, 18° andar, 20081-240 Rio de Janeiro, RJ; tel. (21) 2206-3000; fax (21) 2263-2539; e-mail inpipres@inpi.gov.br; internet www.inpi.gov.br; f. 1970; intellectual property, etc.; Pres. LUIZ OTÁVIO BEAKLINI.

**Instituto Nacional de Tecnologia (INT):** Av. Venezuela 82, 8° andar, 20081-310 Rio de Janeiro, RJ; tel. (21) 2123-1238; fax (21) 2123-1136; e-mail int@int.gov.br; internet www.int.gov.br; f. 1921; co-operates in national industrial development; Dir JOÃO LUIZ HANRIOT SELASCO.

**União Democrática Ruralista (UDR):** Av. Col. Marcondes 983, 6° andar, Sala 62, Centro, 19010-080 Presidente Prudente, SP; tel. (11) 3221-1082; fax (11) 3232-4622; e-mail udr.org@uol.com.br; internet www.udr.org.br; landowners' organization; Pres. LUIZ ANTÔNIO NABHAN GARCIA.

# MAJOR COMPANIES

### Chemicals, Petrochemicals and Petroleum

**Braskem, SA:** Av. das Nações Unidas 4777, 05477-000 São Paulo, SP; tel. (11) 3443-9999; fax (11) 3023-0415; internet www.braskem.com.br; f. 2002 by merger of petrochemical operations of Odebrecht with Copene and Grupo Mariani; 13 chemical plants; Chair. CARLOS MARIANI BITTENCOURT; CEO JOSÉ CARLOS GRUBISICH; 3,000 employees (2005).

**Companhia Petroquímica do Sul (COPESUL):** BR 386 Km 419 Rodovia Tabaí-Canoas, Pólo Petroquímico, 95853-000 Triunfo, RS; tel. (51) 457-6000; fax (51) 457-6050; e-mail comunicacao@copesul.com.br; internet www.copesul.com.br; f. 1976; manufacturers of industrial chemicals and petrochemicals; Pres. EDUARDO EUGÊNIO GOUVÊA VIEIRA; CEO LUIZ FERNANDO CIRNE LIMA; 920 employees.

**Empresas Petróleo Ipiranga, SA:** Rua Francisco Eugênio 329, 20948-900 Rio de Janeiro, RJ; tel. (21) 2574-5858; fax (21) 2569-8796; internet www.ipiranga.com.br; f. 1959; petroleum, petroleum products and natural gas; Pres. JOÃO PEDRO GOUVÊA VIEIRA; Vice-Pres. SÉRGIO SILVEIRA SARAIVA; 1,812 employees.

**Distribuidora de Productos de Petróleo Ipiranga SA:** Av. Dolores Alcaraz Caldas 90, Praia das Belas, 90110-180 Porto Alegre, RS; tel. (51) 3216-4411; fax (51) 3224-0403; internet www.ipiranga.com.br; f. 1957; distribution of petroleum derivatives; cap. US $103.4m., sales $830.2m. (2001); Pres. SÉRGIO SILVEIRA SARAIVA; Vice-Pres. CARLOS ALBERTO MARTINS BASTOS; 363 employees.

**Petróleo Brasileiro, SA (Petrobras):** Av. República do Chile 65, 20035-900 Rio de Janeiro, RJ; tel. (21) 2534-4477; fax (21) 2220-5052; internet www.petrobras.com.br; f. 1953; production of petroleum and petroleum products; owns 16 oil refineries; net profit R $17.9m. (2004); Pres. and CEO JOSÉ EDUARDO DE BARROS DUTRA; 34,520 employees.; subsidiary cos are Petrobras Transporte, SA (Transpetro), Petrobras Comercializadora de Energia, Ltda, Petrobras Negócios Eletrônicos, SA, Petrobras International Finance Company (PIFCO) and Downstream Participações, SA, and cos listed below.

**Petrobras Distribuidora, SA:** Rua General Canabarro 500, Maracanã, 20271-905 Rio de Janeiro, RJ; tel. (21) 3876-4477; fax (21) 3876-4977; internet www.br.com.br; f. 1971; distribution of all petroleum by-products; Pres. LUIS RODOLFO LANDIM MACHADO; 3,536 employees.

**Petrobras Química, SA (Petroquisa):** Rua Buenos Aires 40, 2° andar, 20070-020 Rio de Janeiro, RJ; tel. (21) 2297-3778; fax (21) 2262-4294; internet www.petroquisa.com.br; f. 1968; petrochemicals industry; controls 27 affiliated companies and four subsidiaries; Pres. KUNIUUKI TERABE.

### Metals and Mining

**Aços Villares, SA:** Av. Maria Coelho Aguiar 215—BLA, 5A, Jardim São Luis, 05840-900 São Paulo, SP; tel. (11) 3748-9000; fax (11) 3748-2212; f. 1944; steel producers; Chair. SABINO ARRIETA HERAS; Vice Chair. JOSÉ M. MONTERO OLIDEN; 4,307 employees.

**Alcan Alumínio do Brasil Ltda:** Av. das Nações Unidas 12551, 15°, 04578-000 São Paulo, SP; tel. (11) 5503-0722; fax (11) 5503-0791; internet www.alcan.com.br; f. 1940; extraction and processing of bauxite, packaging materials.

**Alcoa Alumínio, SA:** Av. Maria Coelho de Aguiar 215, 4°, Jd. São Luis, Santo Amaro, 05804-900 São Paulo, SP; tel. (11) 3741-5988; fax (11) 3741-8300; internet www.alcoa.com.br; f. 1965; subsidiary of Alcoa Inc., USA; extraction and processing of bauxite; Pres. FAUSTO PENNA MOREIRA, Filho; 6,000 employees.

**Companhia Brasileira de Metalurgia e Mineração (CBMM):** Córrego da Mata s/n, CP 8, 38180-970 Araxá, MG; tel. (34) 3669-3000; fax (34) 3669-3100; internet www.us.cbmm.com.br; f. 1955; extraction, processing of niobium, manufacturing of niobium products.

**Caemi Mineração e Metalurgia, SA (CAEMI):** Praia de Botafogo 300, 8°, 22250-040 Rio de Janeiro, RJ; tel. (21) 2536-4100; fax (21) 2552-2745; internet www.caemi.com.br; f. 1987; mining and processing of iron ore, bauxite, kaolin and chromite; 60% owned by CVRD; Pres. WANDERLEI VIÇOSO FAGUNDES; 2,400 employees.

**Companhia Aços Especiais Itabira (ACESITA):** Av. João Pinheiro 580, Centro, Belo Horizonte, MG 30130-180; tel. (31) 3235-4200; fax (31) 3235-4294; e-mail invest@acesita.com.br; internet www.acesita.com.br; f. 1944; privatized in 1992; iron and steel producers; Pres. JOAQUIM FERREIRA AMARO; 5,545 employees.

**Companhia Siderúrgica Belgo-Mineira:** Av. Carandaí 1115, 25° andar, 30130-915 Belo Horizonte, MG; tel. (31) 3219-1122; fax (31) 3273-2927; internet www.belgomineira.com.br; f. 1921; steel mill; CEO FRANÇOIS MOYEN; 2,986 employees.

**Companhia Siderúrgica Nacional (CSN):** San José 20, 1602, Centro, 20010-020 Rio de Janeiro, RJ; tel. (21) 2215-3021; fax (21) 2586-1400; internet www.csn.com.br; f. 1941; privatized 1993; steel; Chair. and CEO BANJAMIN STEINBRUCH; 8,000 employees.

**Companhia Siderúrgica Paulista (COSIPA):** Av. do Café 277, Torre B, 8°–9°, Vila Guarani, 04311-000 São Paulo, SP; tel. (11) 5070-8800; fax (11) 5070-8863; internet www.cosipa.com.br; steel; 49.8% owned by USIMINAS; Pres. ALDO NARCISI; CEO OMAR SILVA JÚNIOR.

**Companhia Siderúrgica de Tubarão (CST):** Av. Brigadeiro Eduardo Gomes 930, Jardim Limoeiro, 29163-970 Serra, ES; tel. (27) 3348-1333; fax (27) 3348-1520; internet www.cst.com.br; f. 1976; manufacturers of steel slabs; Dir Pres. JOSÉ ARMANDO CAMPOS; 3,960 employees (2004).

**Companhia Vale do Rio Doce, SA (CVRD):** Av. Graça Aranha 26, 6° andar, Bairro Castelo, 20030-000 Rio de Janeiro, RJ; tel. (21) 3814-4477; fax (21) 3814-4040; internet www.cvrd.com.br; f. 1942; fmr state-owned mining co, privatized in 1997; owns and operates two systems: in the north, the Carajás iron-ore mine and railway, and port of Ponta da Madeira; in the south, the Itabira iron-ore mine, the Vitória–Minas railway and the port of Tubarão; largest gold producer in Latin America; also involved in forestry and pulp production, aluminium and other minerals; also has extensive overseas operations; CEO ROGER AGNELLI; 15,500 employees.

**Gerdau:** Av. Farrapos 1811, 90220-005 Porto Alegre, RS; tel. (51) 3323-2000; fax (51) 3323-2222; internet www.gerdau.com.br; f. 1901; long steel; Pres. JORGE GERDAU JOHANNPETER.

**Gerdau Açominas, SA:** Ruas dos Inconfidentes 871, Funcionários 30140-120 Belo Horizonte, MG; tel. (31) 3269-4100; fax (31) 3269-4300; internet www.acominas.com.br; f. 1986; manufacturers of iron and steel products; privatized in 1993: 28.7% owned by Gerdau; Chair. JORGE GERDAU JOHANNPETER; CEO LUIZ ANDRÉ RICO VICENTE; 4,200 employees (2005).

**Minerações Brasileiras Reunidas, SA (MBR):** Praia de Botafogo 300, 8° andar, Rio de Janeiro, RJ 22259-900; tel. (21) 2536-4314; fax (21) 2552-2346; e-mail nsa@mbr.com.br; internet www.mbr.com.br; f. 1964; 85% controlled by Caemi Mineração e Metalurgia, SA; mining of iron ores; Pres. TITO MARTINS; 2,200 employees.

**Rio Tinto Brasil:** Rua Lauro Müller 116, 35° andar, Botafogo, 22290-160 Rio de Janeiro, RJ; tel. (21) 2541-9000; fax (21) 2542-1100; e-mail externo@riotinto.com.br; internet www.riotinto.com.br; f. 1971; mining and transport of minerals; subsidiary cos include Rio Tinto Desenvolvimentos Minerais, Ltda and Transbarge Navegación, SA.

**Mineracaõ Corumbaense Reunida, SA (Corumbá):** Rua do Cabral 1555, 79332-030 Corumbá, MS; tel. (67) 231-2333; fax (67) 231-2236; e-mail mcr@mcr.riotinto.com.br; internet www.riotinto.com.br; owned by Rio Tinto; iron ore production.

**SMS Demag, Ltda:** Av. das Nações Unidas 999, 33200-000 Vespasiano, MG; e-mail vendas@smsdemag.com.br; internet www.smsdemag.com.br; f. 1977; subsidiary of SMS Demag, Germany; steel mills and hot metal production.

**Usinas Siderúrgicas de Minas Gerais, SA (USIMINAS):** Rua Prof. José Vieira de Mendonça 3011, 31310-260 Belo Horizonte, MG; tel. (31) 3499-8000; fax (31) 3499-8899; e-mail pgn@usiminas.com.br; internet www.usiminas.com.br; f. 1956; steel mill; privatized in 1991; Pres. ADEMAR DE CARVALHO BARBOSA; 8,448 employees.

**Votorantim Metais:** Praça Ramos de Azevedo 254, 6° andar, 01037-912 São Paulo, SP; tel. (11) 3225-3100; fax (11) 3361-3628; internet www.vmetais.com.br; aluminium and zinc extraction and processing, long steel; Man. Dir JOÃO BOSCO SILVA; 3,500 employees; subsidiary cos include Companhia Mineira do Metais, Companhia Niquel Tocatins and Companhia Paraibuna de Metais.

**Siderúrgica Barra Mansa:** Av. Homero Leite 1051, Saudade, 27355-501 Barra Mansa, RJ; tel. (24) 3324-9655; fax (24) 3324-9600; e-mail sbm@votocaco.com.br; f. 1937; steel.

**White Martins:** Rua Mayrink Veiga 9, 20090-050 Rio de Janeiro, RJ; tel. (21) 2588-6622; fax (21) 2588-6683; internet www.praxair.com; f. 1912; almost 100% owned by Praxair; manufacturers and distributors of industrial gases, welding equipment and seamless cylinders; CEO RICARDO S. MALFITANO; 10,000 employees.

### Motor Vehicles and Aircraft

**Companhia Fabricadora de Peças (COFAP):** Av. Alexandre de Gusmão 1395, Capuava, 09110-901 Santo André, SP; tel. (11) 411-8211; fax (11) 411-7979; internet www.cofap.com.br; f. 1951; manufacturers of motor-vehicle components; Pres. CLEDORVINO BELINI; 7,100 employees.

**DaimlerChrysler do Brasil, SA:** Av. Mercedes Benz 679, Distrito Industrial, 13054-750 Campinas, SP; tel. 0800 909090; fax (19) 3725-3635; internet www.daimlerchrysler.com.br; f. 1953 as Mercedes

# BRAZIL
*Directory*

**Benz do Brasil, SA;** subsidiary of DaimlerChrysler AG of Germany; motor-car, truck and bus-chassis production; Pres. BEN VAN SCHAIK; 13,041 mployees (2004).

**Empresa Brasileira de Aeronáutica, SA (Embraer):** Av. Brig. Faria Lima 2170, 12227-901 São José dos Campos, SP; tel. (12) 3927-1000; fax (12) 3921-2394; e-mail mercapit@embraer.com.br; internet www.embraer.com; f. 1969; aeronautics industry; Chair. CARLYLE WILSON; Pres. and CEO MAURÍCIO NOVIS BOTELHO; 14,500 employees in Brazil and abroad.

**Fiat do Brasil, SA:** Rodovia Fernão Dias s/n, BR 381 km 429, 32560-460 Betim, MG; internet www.fiat.com.br; tel. (31) 3589-4000; 18,500 employees.

**Ford Brasil, Ltda:** Av. Taboão 899, predio 6, 09655-900 São Bernardo do Campo, SP; tel. 0800 7033673; fax (11) 848-9057; internet www.ford.com.br; f. 1987; subsidiary of Ford Motor Co of the USA; motor vehicles; Pres. JAMES PADILLA; 3,700 employees.

**General Motors do Brasil:** Rodovia BR 290, Km 67, Gravataí, RS; tel. (51) 3430-1718; internet gmb.chevrolet.com.br; f. 1925; Pres. RAY G. YOUNG.

**Iochpe-Maxion, SA:** Rua Luigi Galvani 146, 13° andar, 04575-020 São Paulo, SP; tel. (11) 5508-3800; fax (11) 5506-7353; internet www.iochpe-maxion.com.br; f. 1918; motor-vehicle manufacturers; Pres. IVONCY BROCHMANN IOSCHPE; CEO DANIEL IOSCHPE; 3,919 employees.

**Volkswagen do Brasil, SA:** Rua Volkswagen 291, POB 8890, 04344-900 São Paulo, SP; tel. (11) 5582-5030; fax (11) 578-0947; e-mail info@volkswagen.com.br; internet www.volkswagen.com.br; f. 1953; subsidiary of Volkswagen AG of Germany; manufacture of trucks and passenger commercial vehicles; Pres. HANS CHRISTIAN MAERGNER; 47,000 employees.

## Rubber, Textiles and Paper

**Aracruz Celulose, SA:** Rua Lauro Müller 116, 40° andar, 22299-900 Rio de Janeiro, RJ; tel. (21) 3820-8111; fax (21) 3820-2802; e-mail aracruz@aracruz.com.br; internet www.aracruz.com.br; f. 1972; wood and bleached eucalyptus pulp; Chair. CARLOS ALBERTO VIEIRA; Pres. and CEO CARLOS AUGUSTO LIRA AGUIRA; 2,281 employees.

**Celulose Nipo-Brasileira—Cenibra:** Rua Bernardo Guimarães, 8 e 9°, Bairro Funcionários, 30140-080 Belo Horizonte, MG; tel. (31) 3274-7355; fax (31) 3273-2787; internet www.cenibra.com.br; f. 1973; eucalyptus pulp paper; Chair. FERNANDO HENRIQUE DA FONSECA; 1,600 employees.

**Companhia Suzano de Papel e Celulose:** Av. Brigadeiro Faria Lima 1355, 10° andar, Pinheiro, 01452-919 São Paulo, SP; tel. (11) 3037-9326; fax (11) 3037-9313; e-mail suzano@suzano.com.br; internet www.suzano.com.br; f. 1923; makes and distributes eucalyptus pulp and paper products; Pres. DAVID FEFFER; 3,562 employees.

**Fibra, SA:** Av. São Jerônimo 4600, Bairro São Jerônimo, 13470-900 Americana, SP; tel. (19) 3471-2000; fax (19) 3471-2008; e-mail endasfilamentos@fibra.com.br; internet www.fibra.com.br; f. 1949; synthetic fibres and filaments.

**Indústrias Klabin de Papel e Celulose, SA (IKPC):** Rua Formosa 367, 18° andar, 01049-000 São Paulo, SP; tel. (11) 250-4000; fax (11) 250-4067; e-mail klabin@klabin.com.br; internet www.klabin.com.br; f. 1934; manufacturers of paper and paper products; Chair. DANIEL MIGUEL KLABIN; CEO JOSMAR VERILLO; 7,324 employees.

**Petroflex:** Rua Marambi 600, Campos Elíseos, 54510-900 Cabo, PE; tel. (81) 3521-7300; fax (81) 3521-1063; e-mail cliente@petroflex.com.br; internet www.petroflex.com.br; f. 1977 as a subsidiary of Petroquisa; sold in 1992; synthetic rubber; CEO SERGIO VAN KLAVERN; 550 employees.

**Pirelli Pneus, SA:** Av. Giovanni Battista Pirelli 871, Vila Homero Thon, 09111-340 Santo André, SP; tel. 0800-787638; fax (11) 4998-5077; e-mail webpneus@pirelli.com.br; internet www.pirelli.com.br; f. 1988; owned by Pirelli of Italy; makers of rubber inner tubes and tyres; Pres. GIORGIO DELLA SETA FERRARI CORBELLI GRECO; 4,864 employees.

**Ripasa SA Celulose e Papel:** Rua Clodomiro Amazonas 249, 10° andar, 04537-010 São Paulo, SP; tel. (11) 3491-5000; fax (11) 3491-5012; internet www.ripasa.com.br; f. 1959; production of pulp, paper and paper products; Chair. WALTER ZARZUR DERANI; Pres. OSMAR ELIAS ZOGBI; 2,783 employees (2003).

**Tecelagem Kuehnrich, SA (TEKA):** Rua Paulo Kuehnrich 68, Bairro Itoupava Norte, 89052-900 Blumenau, SC; tel. (47) 321-5000; fax (47) 321-5050; e-mail sac@teka.com.br; internet www.teka.com.br; f. 1935; textile manufacturers; Pres. ROLF KUEHNRICH; 4,500 employees.

**Votorantim Cellulose e Papel, SA:** Alameda Santos 1357, 8°, 01419-908 São Paulo, SP; tel. (11) 3269-4000; fax (11) 3269-4065; internet www.vcp.com.br; pulp and paper products; part of Grupo Votorantim; Dir-Pres. JOSÉ LUCIANO PENIDO.

## Construction

**Andrade Gutierrez Construção Pesado Brasil, SA:** Rua Dr Geraldo Campos Moreira 375, 10°, 04571-020 São Paulo, SP; tel. (11) 5502-2000; internet www.agsa.com.br; f. 1948; heavy construction and civil engineering; CEO ROGÉRIO NORA DE SÁ; 14,200 employees.

**Construções e Comércio Camargo Corrêa, SA:** Rua Funchal 160, Vila Olímpia, 04551-903 São Paulo, SP; tel. (11) 3841-5511; fax (11) 3841-5109; e-mail camargo@camargocorrea.com.br; internet www.cccc.camargocorrea.com.br; f. 1946; heavy construction; Pres. CARLOS NOGUEIRA ROSA; 6,470 employees.

**Construtora Norberto Odebrecht, SA:** Praia de Botafogo 300, 4° e 11°, Botafogo, 22250-040 Rio de Janeiro, RJ; tel. (21) 2559-3000; fax (21) 2552-4448; e-mail info@odebrecht.com.br; internet www.odebrecht.com; f. 1944; Pres. MARCELO ODEBRECHT.

**Construtora Queiroz Galvão, SA:** Av. Rio Branco 156, 20043-900 Rio de Janeiro, RJ; tel. (21) 2292-3993; fax (21) 2240-9367; e-mail suporte@ggalvao.com.br; internet www.qgalvao.com; f. 1953; civil-engineering and construction projects; Pres. ANTÔNIO DE QUEIROZ GALVÃO; Man. Dir JOÃO ANTÔNIO DE QUEIROZ GALVÃO; 6,853 employees.

## Food Products, Food Processing and Retail

**Bompreço, SA, Supermercados do Nordeste:** Av. Caxangá 3841, Iputinga, 50670-902 Recife, PE; tel. (81) 271-7339; fax (81) 271-7337; e-mail ccollier@bompreco.com.br; internet www.bompreco.com.br; f. 1935; owned by Walmart (USA); retail trade: 68 supermarkets, 27 hypermarkets and 19 other stores; Pres. JOÃO CARLOS PAES MENDONÇA; 17,633 employees.

**Bunge Alimentos, SA:** Rodovia Jorge Lacerda, Km 20, Poço Grande, CP 45, 89110-000 Gaspar, SC; tel. (47) 331-2222; fax (47) 331-2005; internet www.bungealimentos.com.br; f. 2000 by merger of Santista Alimentos and Ceval Alimentos; agricultural processing; 7,000 employees.

**Carrefour Brasil:** internet www.carrefour.com.br; 86 hypermarkets, 98 supermarkets, 61 discount stores.

**Companhia Brasileira de Distribuição/Grupo Pão de Açúcar:** Av. Brigadeiro Luis Antonio 3, 126, 01402-901 São Paulo, SP; tel. (11) 3886-0421; fax (11) 3884-2677; internet www.grupopaodeacucar.com.br; f. 1948; supermarkets, hypermarkets and electrical stores; Chair. and CEO ABILIO DOS SANTOS DINIZ; 70,000 employees.

**InBev:** CP 60550, 05804-970 São Paulo, SP; f. 2004 following purchase of AmBev (Companhia de Bebidas das Américas) by Interbrew (Belgium); beer and soft drinks maker; CEO CARLOS ALVO BRITO.

**Perdigão Agroindustrial, SA:** Av. Escola Politécnica 760, 05350-901 São Paulo, SP; tel. (11) 3718-5300; fax (11) 3768-8070; e-mail perdigao@perdigao.com.br; internet www.perdigao.com.br; f. 1934; meat processing and packaging; Pres. EGGON JOÃO DA SILVA; 27,000 employees.

**Sadia, SA:** Alameda Tocantins 525, 06455-921 Barueri, SP; tel. (11) 7296-4432; fax (11) 7296-4510; internet www.sadia.com.br; f. 1944; refrigeration, meat packing, animal feeds; Pres. WALTER FONTANA, Filho; 22,331 employees.

**Sonae Distribuição Brasil, SA:** Av. Sertorio 6600, sobreloja, Sarandi, 91110-580 Porto Alegre, RS; tel. (51) 3349-4444; fax (51) 3349-4545; internet www.sonae.com.br; f. 1998; 93 hypermarkets, 45 hypermarkets, 15 specialized stores; 22,000 employees.

## Pharmaceuticals

**Aché Laboratórios Farmacêuticos, SA:** Rodovia Presidente Dutra, Km 222,2, Porto da Igreja, 07034-904 Guarulhos, SP; tel. (11) 0800-12-5005; e-mail cdebora@ache.com.br; internet www.ache.com.br; f. 1966 as Prodoctor Produtos Farmacêuticos; Dir-Gen. JOSÉ EDUARDO BANDEIRO DE MELHO; 2,600 employees.

**Biosintética:** Av. das Nações Unidas 22428, Jurubatuba, 04795-916 São Paulo, SP; tel. (11) 5546-6822; fax (11) 5546-6800; internet www.biosintetica.com.br; cardiovascular medication and other products.

**EMS Indústria Farmacêutica, Ltda:** Rodovia SP 101, Km 08, 13186-481 Hortolândia, SP; tel. (19) 3887-9800; fax (19) 3887-9515; e-mail sac@ems.com.br; internet www.ems.com.br; f. 1964.

**Eurofarma:** tel. 0800-704-3876; e-mail euroatende@eurofarma.com.br; internet www.eurofarma.com.br; f. 1972 as Billi Farmacêutica; products include prescription and oncological medications.

**JP Indústria Farmacêutica, SA:** Av. Pres. Castelo Branco 999, Lagoinha, 14095-000 Ribeirão Preto, SP; tel. (16) 3629-5100; fax (16) 3629-7171; e-mail sac@jpfarma.com.br; internet www.jpfarma.com.br; f. 1966; hospital supplies and medications.

**Libbs:** Rua Josef Kryss 250, Parque Industrial Thomas Edson, 01140-050 São Paulo, SP; tel. (11) 2879-2500; fax (11) 3879-0957; e-mail bd@libbs.com.br; internet www.libbs.com.br; f. 1958; general pharmaceutical products; 820 employees.

# BRAZIL

**Medley, SA, Indústria Farmacêutica:** Rua Macedo Costa 55, Jardim Santa Genebra, 13080-180 Campinas, SP; tel. (19) 3708-8222; fax (19) 3708-8227; internet www.medley.com.br; brand and generic products.

### Other

**Duratex, SA:** Av. Paulista 1938, Bela Vista, 01310-942 São Paulo, SP; tel. (11) 3179-7733; fax (11) 3179-7355; internet www.duratex.com; f. 1951; manufacturers of hardboard and plywood, ceramic products and bathroom fixtures; Chair. EUDORO VILLELA; Pres. PAULO SETUBAL; 6,288 employees.

**Electrolux do Brasil, SA:** Rua Verbo Divino 1488, 7°, 72B, 04719-904 São Paulo, SP; tel. (11) 5188-1155; fax (11) 5188-1281; e-mail eluxfct@electrolux.com.br; internet www.electrolux.com.br; f. 1926; makers of refrigerators, freezers and vacuum cleaners; CEO RUY HIRSCHHEIMER; 431 employees.

**Empresa Brasileira de Correios e Telégrafos (ECT):** SBN Lote 3, Edif. Sede, 17° andar, Conj. 3, Bloco A, 70002-900 Brasília, DF; tel. (61) 426-2450; fax (61) 426-2310; e-mail robervalcorrea@correios.com.br; internet www.correios.com.br; f. 1969; posts and telegraph; Pres. HASSAN GEBRIM; 82,253 employees.

**Globex Utilidades, SA:** Av. Tenente Rebelo 675, 21241-460 Irajá, RJ; tel. (21) 2472-8509; fax (21) 3372-7019; internet www.pontofrio.com.br; f. 1946; retail of household goods; Chair. and CEO SIMON M. ALOUAN.

**Indústrias Villares, SA:** Av. Interlagos 4455, 04669-900 São Paulo, SP; tel. (11) 5525-3222; fax (11) 5548-2212; f. 1918; produces and maintains lifts and escalators; Pres. PAULO DIEDERICHSEN VILLARES; Vice-Pres. WILSON NÉLIO BRUMER; 3,856 employees.

**Lojas Americanas, SA:** Rua Sacadura Cabral 102, Saúde, 20081-260 Rio de Janeiro, RJ; tel. (21) 2206-6505; fax (21) 2206-6898; e-mail gisomar.marinho@lasa.com.br; internet www.lasa.com.br; f. 1929; retail traders; sold to Comptoirs Modernes, SA, in November 1998; Pres. CARLOS ALBERTO DA VEIGA SICUPIRA; 11,669 employees.

**Multibrás SA Eletrodomésticos:** Edif. Plaza Centenário, Av. das Nações Unidas 12995, 04578-000 São Paulo, SP; tel. (11) 5586-6100; fax (11) 5586-6388; internet www.multibras.com.br; f. 1945 as car distributor; household appliances.

**Ryder do Brasil Ltda:** Rua Laguna 276, Chácara Santo Antonio, 04728-000 São Paulo, SP; tel. (11) 5644-9644; fax (11) 5641-5780; internet www.ryder.com/south_america_br; logistics; subsidiary of Ryder, USA.

**Saint-Gobain Vidros, SA:** Av. Santa Marina 482, Agua Blanca, 05036-903 São Paulo, SP; tel. (11) 3874-7988; fax (11) 3611-0299; internet www.saint-gobain-vidros.com.br; f. 1896 as Companhia Vidraria Santa Marina; subsidiary of Groupe Saint-Gobain, France; glass manufacturers; Chair. JEAN JACQUES FAUST; 3,900 employees.

**Souza Cruz, SA:** Av. Presidente Vargas 3131, 7°, Col. Centro, 20210-030 Rio de Janeiro, RJ; tel. 0800-888-2223; fax (21) 2503-5154; e-mail sac@scruz.com.br; internet www.souzacruz.com.br; f. 1903; manufacturers of cigarettes and tobacco; 75% owned by British American Tobacco; Pres. NICANDRO DURANTE; 4,500 employees.

**Votorantim Participações:** Rua Amauri 255, 01448-000 São Paulo, SP; tel. (11) 3704-3300; internet (11) 3167-1550; f. 1918; holding co with interests in cement, paper, metals, chemicals and financial services; 22,000 employees.

**Xerox Brasil:** Av. Rodriques Alves 261, 20220-360 Rio de Janeiro, RJ; tel. (21) 2271-1212; fax (21) 2271-1649; internet www.xerox.com.br; office equipment, technology; Pres. PEDRO FÁBREGA.

### UTILITIES

#### Regulatory Agencies

**Agência Nacional de Energia Elétrica (ANEEL):** SGAN 603, Módulo J, 70830-030 Brasília, DF; e-mail aneel@aneel.gov.br; internet www.aneel.gov.br; f. 1939 as Conselho Nacional de Águas e Energia Elétrica, present name adopted 1996; Pres. JERSON KELMAN; Dirs ISAAC PINTO AVERBUCH, JACONIAS DE AGUIAR.

**Comissão Nacional de Energia Nuclear (CNEN):** Rua General Severiano 90, Botafogo, 22290-901 Rio de Janeiro, RJ; tel. (21) 2295-9596; fax (21) 2295-86962; e-mail corin@cnen.gov.br; internet www.cnen.gov.br; f. 1956; state organization responsible for management of nuclear power programme; Pres. ODAIR DIAS GONCALVES.

#### Electricity

The auctioning of electricity supply contracts began in December 2004.

**Centrais Elétricas Brasileiras, SA (Eletrobrás):** Edif. Petrobrás, Rua Dois, Setor de Autarquias Norte, 70040-903 Brasília, DF; tel. (61) 223-5050; fax (61) 225-5502; e-mail maryann@eletrobras.gov.br; internet www.eletrobras.gov.br; f. 1962; govt holding company responsible for planning, financing and managing Brazil's electrical energy programme; 52% govt-owned; scheduled for division into eight generating cos and privatization; Pres. SILAS RONDEAU CAVALCANTE SILVA.

**Centrais Elétricas do Norte do Brasil, SA (Eletronorte):** SCN, Quadra 6, Conj. A, Blocos B/C, Sala 602, Super Center Venâncio 3000, 70718-500 Brasília, DF; tel. (61) 429-5151; fax (61) 328-1566; e-mail elnweb@eln.gov.br; internet www.eln.gov.br; f. 1973; Pres. BENEDITO APARECIDO CARRARO.

**Boa Vista Energia, SA:** Av. Capitão Ene Garcêz 691, Centro, Boa Vista, RR; tel. (95) 2621-1400; internet www.boavistaenergia.gov.br; f. 1997; subsidiary of Eletronorte; electricity distribution; Pres. CARLOS AUGUSTO ANDRADE SILVA.

**Manaus Energia, SA:** Manaus, AM; internet www.manausenergia.com.br; f. 1895; became subsidiary of Eletronorte in 1997; electricity distributor; Pres. Dr WILLAMY MOREIRA FROTA.

**Centrais Elétricas do Sul do Brasil, SA (Eletrosul):** Rua Deputado Antônio Edu Vieira 999, Pantanal, 88040-901 Florianópolis, SC; tel. (48) 231-7000; fax (48) 234-3434; internet www.eletrosul.gov.br/home; Gerasul responsible for generating capacity; f. 1968; Pres. MILTON MENDES DE OLIVEIRA.

**Companhia de Geração Térmica de Energia Elétrica (CGTEE):** Rua Sete de Setembro 539, 90010-190 Porto Alegre, RS; tel. (51) 3287-1500; fax (51) 3287-1566; internet www.cgtee.gov.br; f. 1997; became part of Eletrobrás in 2000; Pres. VALTER LUIZ CARDEAL DE SOUZA; Exec. Dir JÚLIO CÉSAR RIEMENSCHNEIDER DE QUADROS.

**Companhia Hidro Elétrica do São Francisco (Chesf):** 333 Edif. André Falcão, Bloco A, Sala 313 Bongi, Rua Delmiro Golveia, 50761-901 Recife, PE; tel. (81) 229-2000; fax (81) 229-2390; e-mail chesf@chesf.com.br; internet www.chesf.com.br; f. 1948; Exec. Dir DILTON DA CONTI OLIVEIRA.

**Eletrobrás Termonuclear, SA (Eletronuclear):** Rua da Candelária 65, Centro, 20091-020 Rio de Janeiro, RJ; tel. (21) 2588-7000; fax (21) 2588-7200; internet www.eletronuclear.gov.br; f. 1997 by fusion of the nuclear branch of Furnas with Nuclebrás Engenharia (NUCLEN); operates two nuclear facilities, Angra I and II; Angra III under construction and scheduled to come into operation in 2008; Pres. JOSÉ DRUMMOND SARAIVA; Exec. Dir PAULO FIGUEIREDO.

**Furnas Centrais Elétricas, SA:** Rua Real Grandeza 219, Bloco A, 16° andar, Botafogo, 22281-031 Rio de Janeiro, RJ; tel. (21) 2528-3970; fax (21) 2528-4480; e-mail presiden@furnas.com.br; internet www.furnas.com.br; f. 1957; Pres. JOSÉ PEDRO RODRIQUES DE OLIVEIRA.

**Centrais Elétricas de Rondônia, SA (CERON):** Rua José de Alencar 2.613, Centro Porto Velho, RO; tel. (69) 216-4000; internet www.ceron.com.br; f. 1968; Pres. EURÍPEDES MIRANDA BOTELHO.

**Centrais Elétricas de Santa Catarina, SA (CELESC):** Rodovia SC 404, Km 3, Itacorubi, 88034-900 Florianópolis, SC; tel. (48) 231-5000; fax (48) 231-6530; e-mail celesc@celesc.com.br; internet www.celesc.com.br; production and distribution of electricity throughout state of Santa Catarina; Pres. FRANCISCO DE ASSIS KÜSTER.

**Companhia de Eletricidade do Acre (ELETROACRE):** Rua Valério Magalhães 226, Bairro do Bosque, 69909-710 Rio Branco, AC; tel. (68) 212-5700; fax (68) 223-1142; internet www.eletroacre.com.br; f. 1965; Pres. EDILSON SOMÕES CADAXO SOBRINHO.

**Companhia de Eletricidade do Estado da Bahia (COELBA):** Av. Edgard Santos 300, Cabula IV, 41186-900 Salvador, BA; tel. (71) 370-5130; fax (71) 370-5132; Pres. EDUARDO LÓPEZ ARANGUREN MARCOS.

**Companhia de Eletricidade do Estado do Rio de Janeiro (CERJ):** Rua Visconde do Rio Branco 429, Centro, 24020-003 Niterói, RJ; tel. (21) 2613-7120; fax (21) 2613-7196; e-mail cerj@cerj.com.br; internet www.cerj.com.br; f. 1907; privatized in 1996; Pres. ALEJANDRO DANÚS CHIRIGHIN.

**Companhia Energética de Alagoas (CEAL):** Av. Fernandes Lima 3496, Gruta de Lourdes, 57057-900 Maceió, AL; tel. (82) 218-9300; internet www.ceal.com.br; f. 1961; Man. Dir JOAQUIM BRITO.

**Companhia Energética do Amazonas (CEAM):** Manaus, AM; internet www.ceam-am.com.br; electricity generating co; Pres. Dr WILLAMY MOREIRA FROTA.

**Companhia Energética de Brasília (CEB):** SCRS, Quadro 503, Bloco B, Lotes 13/14, Brasília, DF; tel. 0800 610196; e-mail info@ceb.com.br; internet www.ceb.com.br; services in DF; also operates gas distribution co CEBGAS.

**Companhia Energética do Ceará (COELCE):** Av. Barão de Studart 2917, 60120-002 Fortaleza, CE; tel. (85) 3247-1444; fax (85)

# BRAZIL

3272-4711; internet www.coelce.com.br; f. 1971; Pres. CARLOS EDUARDO CARVALHO ALVES.

**Companhia Energética do Maranhão (CEMAR):** Av. Colares Moreira 477, Renascença II, São Luis, MA ; internet www.cemar-ma.com.br; f. 1958 as Centrais Elétricas do Maranhão; changed name as above in 1984; owned by PPL Global, Inc, USA; Tech. Dir MARCELINO MACHADO DA CUNHA NETO.

**Companhia Energética de Minas Gerais (CEMIG):** Av. Barbacena 1200, 30190-131 Belo Horizonte, MG; tel. (31) 3299-4900; fax (31) 3299-3700; e-mail atendimento@cemig.com.br; internet www.cemig.com.br; fmrly state-owned, sold to a Brazilian-US consortium in May 1997; Pres. DJALMA BASTOS DE MORAIS.

**Companhia Energética de Pernambuco (CELPE):** Av. João de Barros 111, Sala 301, 50050-902 Recife, PE; tel. (81) 3217-5168; e-mail celpe@celpe.com.br; internet www.celpe.com.br; state distributor of electricity; CEO JOÃO BOSCO DE ALMEIDA.

**Companhia Energética do Piauí (CEPISA):** Av. Maranhão 759, Sul, 64.001-010 Teresina, PI; tel. (86) 221-8000; internet www.cepisa.com.br; f. 1962; 99% of shares bought by Eletrobrás in 1997; distributor of electricity in state of Piauí; Pres. CARLOS EVANDRO DE OLIVEIRA; Exec. Dir JORGE TARGA JUNI.

**Companhia Energética de São Paulo (CESP):** Av. Nossa Senhora do Sabará 5312, Bairro Pedreira, 04447-011 São Paulo, SP; tel. (11) 5613-2100; fax (11) 3262-5545; e-mail inform@cesp.com.br; internet www.cesp.com.br; f. 1966; Pres. MAURO GUILHERME JARDIM ARCE.

**Companhia Força e Luz Cataguazes-Leopoldina:** Praça Rui Barbosa 80, 36770-000 Cataguases, MG; tel. (32) 3429-6000; fax (32) 3421-4240; internet www.cataguazes.com.br; f. 1905; Pres. MANOEL OTONI NEIVA.

**Companhia Paranaense de Energia (COPEL):** Rua Coronel Dulcídio 800, 80420-170 Curitiba, PR; tel. (41) 322-3535; fax (41) 331-4145; e-mail copel@mail.copel.com.br; internet www.copel.com.br; f. 1954; state distributor of electricity and gas; Pres. JOÃO BONIFÁCIO CABRAL JÚNIOR; Exec. Dir RUBENS GHILARDI.

**Companhia Paulista de Força e Luz (CPFL):** Rodovia Campinas Mogi-Mirim Km 2.5, 10388-900 Campinas, SP; tel. (19) 3253-8704; fax (19) 3252-7644; internet www.cpfl.com.br; provides electricity through govt concessions; Pres. WILSON PINTO FERREIRA JÚNIOR.

**Eletricidade de São Paulo, SA (ELETROPAULO):** Av. Alfredo Egidio de Souza Aranha 100, 04791-900 São Paulo, SP; tel. (11) 5546-1467; fax (11) 3241-1387; e-mail administracao@eletropaulo.com.br; internet www.eletropaulo.com.br; f. 1899; acquired by AES in 2001; Pres. MARC ANDRÉ PERREIRA.

**Espírito Santo Centrais Elétricas, SA (ESCELSA):** Rua Sete de Setembro 362, Centro, CP 01-0452, 29015-000 Vitória, ES; tel. (27) 3321-9000; fax (27) 3322-0378; internet www.escelsa.com.br; f. 1968; Pres. JOSÉ GUSTAVO DE SOUZA.

**Indústrias Nucleares do Brasil, SA (INB):** Rua Mena Barreto 161, Botafogo, 22271-100 Rio de Janeiro, RJ; tel. (21) 2536-1600; fax (21) 2537-9391; e-mail inbrio@inb.gov.br; internet www.inb.gov.br; Pres. ROBERTO NOGUEIRA DA FRANCA.

**Itaipú Binacional:** Av. Tancredo Neves 6731, 85856-970 Foz de Iguaçu, PR; tel. (45) 520-5252; e-mail itaipu@itaipu.gov.br; internet www.itaipu.gov.br; f. 1974; jtly owned by Brazil and Paraguay; hydroelectric power-station on Brazilian-Paraguayan border; 1.3m. GWh of electricity produced in 2004; Dir-Gen. (Brazil) JORGE MIGUEL SAMEK.

**LIGHT—Serviços de Eletricidade, SA:** Av. Marechal Floriano 168, CP 0571, 20080-002 Rio de Janeiro, RJ; tel. (21) 2211-7171; fax (21) 2233-1249; e-mail light@lightrio.com.br; internet www.lightrio.com.br; f. 1905; electricity generation and distribution in Rio de Janeiro; formerly state-owned, sold to a Brazilian-French-US consortium in 1996; Pres. LUIZ DAVID TRAVESSO.

## Gas

**Companhia Distribuidora de Gás do Rio de Janeiro (CEG):** Av. Pedro II 68, São Cristóvão, 20941-070 Rio de Janeiro, RJ; tel. (21) 2585-7575; fax (21) 2585-7070; internet www.ceg.com.br; f. 1969; gas distribution in the Rio de Janeiro region; privatized in July 1997.

**Companhia de Gás de Alagoas, SA (ALGÁS):** Rua Comendador Palmeria 129, Farol, 57051-150 Maceió, AL; tel. (82) 216-3600; fax (82) 216-3628; e-mail algas@algas.com.br; internet www.algas.com.br; 51% state-owned; Dir and Pres. Dr GERSON DA FONSECA.

**Companhia de Gás de Bahia (BAHIAGÁS):** Avda Tancredo Neves 450, Sala 1801, Edif. Suarez Trade, 41820-020 Salvador, BA; tel. (71) 340-9000; fax (71) 341-9001; e-mail bahiagas@bahiagas.com.br; internet www.bahiagas.com.br; 51% state-owned; Pres. Dr LUIZ FERNANDO MUELLER KOSER.

*Directory*

**Companhia de Gás do Ceará (CEGÁS):** Av. Santos Dumont 7700, 5°-6° andares, 20941-070 Fortaleza, CE; tel. (85) 3265-1144; fax (85) 3265-2026; e-mail cegas@secrel.com.br; internet www.cegas.com.br; 51% owned by the state of Amazonas; Pres. Dr JOSÉ REGO, Filho.

**Companhia de Gás de Minas Gerais (GASMIG):** Av. Álvares Cabral 1740, 7° andar, 30170-001 Belo Horizonte, MG; tel. (31) 3291-2001; e-mail gasmig@gasmig.com.br; internet www.gasmig.com.br; Pres. DJALMA BASTOS DE MORAIS.

**Companhia de Gás de Pernambuco (COPERGÁS):** Av. Eng. Domingo Ferreira 4060, 15° andar, 51021-040 Recife, PE; tel. (81) 3463-2000; e-mail copergas@copergas.com.br; internet www.copergas.com.br; 51% state-owned; Pres. Dr ROMERO DE OLIVEIRA E SILVA.

**Companhia de Gás do Rio Grande do Sul (SULGÁS):** Travessa Francisco Leonardo Truda 40, 90010-050 Porto Alegre, RS; tel. (51) 3227-4111; internet www.sulgas.rs.gov.br; 51% state-owned; Pres. Dr GILES CARRECONDE AZEVEDO.

**Companhia de Gás de Santa Catarina (SCGÁS):** Rua Antônia Luz 255, Centro Empresarial Hoepcke, 88010-410 Florianápolis, SC; tel. (48) 229-1200; fax (48) 229-1230; e-mail scgasonline.com.br; internet www.scgas.com.br; 51% state-owned; Pres. Dr LUIZ GOMES.

**Companhia de Gás de São Paulo (COMGÁS):** Rua Augusta 1600, 9° andar, 01304-901 São Paulo, SP; tel. (11) 3177-5000; fax (11) 3177-5359; e-mail comgas@comgas.com.br; internet www.comgas.com.br; f. 1978; distribution in São Paulo of gas; sold in April 1999 to consortium including British Gas PLC and Royal Dutch/Shell Group; Pres. LUIS DOMENECH.

**Companhia Paraibana de Gás (PBGÁS):** Av. Epitácio Pessoa 4840, Sala 210, 1° andar, Tambaú, 58030-001 João Pessoa, PB; tel. (83) 247-2244; e-mail cicero@pbgas.com.br; internet www.pbgas.com.br; 51% state-owned; Pres. Dr CÍCERO ERNESTO LEITE DE SOUZA.

**Companhia Paranaense de Gás (COMPAGÁS):** Rua Pasteur 463, 7° andar, Batel, 80250-080 Curitiba, PR; tel. (41) 312-1900; fax (41) 222-6633; e-mail compagas@mail.copel.br; internet www.compagas.com.br; Pres. Dr ANTÔNIO FERNANDO KREMPEL.

**Companhia Potiguar de Gás (POTIGÁS):** Rua Dão Silveira 3672, Candelária, 59066-180 Natal, RN; tel. (84) 217-3322; fax (84) 217-3309; e-mail ismael@potigas.com.br; internet www.potigas.com.br; 17% state-owned; Pres. Dr ISMAEL WANDERLEY GOMES, Filho.

**Companhia Rondoniense de Gás, SA (RONGÁS):** Av. Carlos Gomes 1223, Sala 403, Centro, 78903-000 Porto Velho, RO; tel. and fax (69) 229-0333; e-mail rongas@enter-net.com.br; internet www.rongas.com.br; f. 1998; 17% state-owned; Pres. GERSON ACURSI.

**Empresa Sergipana de Gás, SA (EMSERGÁS):** Rua Dom Bosco 1223 B, Suíça, 49050-220 Aracaju, SE; tel. and fax (79) 211-5213; e-mail emsergas@infonet.com.br; internet www.emsergas.com.br; Pres. Dr LUIZ MACHADO MENDONÇA.

## Water

**Agua e Esgotos do Piauí (AGESPISA):** Av. Mal Castelo Branco 101, Cabral, 64000 Teresina, PI; tel. (86) 2239-300; f. 1962; state-owned; water and waste management; Pres. OLIVÃO GOMES DA SOUSA.

**Companhia de Agua e Esgosto de Ceará (CAGECE):** Rua Lauro Vieira Chaves 1030, Fortaleza, CE; tel. (85) 3247-2422; internet www.cagece.com.br; state-owned; water and sewerage services; Gen. Man. JOSÉ DE RIBAMAR DA SILVA.

**Companhia Algoas Industrial (CINAL):** Rodovia Divaldo Suruagy, Km 12, 57160-000 Marechal Deodoro, AL; tel. (82) 269-1100; fax (82) 269-1199; internet www.cinal.com.br; f. 1982; management of steam and treated water; Dir PAULO ALBERQUERQUE MARANHÃO.

**Companhia Espírito Santense de Saneamento (CESAN):** Av. Governador Bley, 186, Edif. BEMGE, 29010-150 Vitória ES; tel. (27) 3322-8399; fax (27) 3322-4551; internet www.cesan.com.br; f. 1968; state-owned; construction, maintenance and operation of water supply and sewerage systems; Pres. CLÁUDIO DE MORAES MACHADO.

**Companhia Estadual de Aguas e Esgotos (CEDAE):** Rua Sacadura Cabral 103, 9° andar, 20081-260 Rio de Janeiro, RJ; tel. (21) 2296-0025; fax (21) 2296-0416; internet www.cedae.rj.gov.br; state-owned; water supply and sewerage treatment; Pres. ALBERTO JOSÉ MENDES GOMES.

**Companhia Pernambucana Saneamento (COMPESA):** Av. Cruz Cabugá 1387, Bairro Santo Amaro, 50040-905 Recife, PE; tel. (81) 421-1711; fax (81) 421-2712; state-owned; management and operation of regional water supply in the state of Pernambuco; Pres. GUSTAVO DE MATTO PONTUAL SAMPAIO.

**Companhia Riograndense de Saneamento (CORSAN):** Rua Caldas Júnior 120, 18° andar, 90010-260 Porto Alegre, RS; tel. (51) 3215-5691; e-mail ascom@corsan.com.br; internet www.corsan.com.br; f. 1965; state-owned; management and operation of regional water supply and sanitation programmes; Dir DIETER WARTCHOW.

**Companhia de Saneamento Básico do Estado de São Paulo (SABESP):** Rua Costa Carvalho 300, 05429-000 São Paulo, SP; tel. (11) 3030-4000; internet www.sabesp.com.br; f. 1973; state-owned; supplies basic sanitation services for the state of São Paulo, including water treatment and supply; Pres. ARIOVALDO CARMIGNANI.

## TRADE UNIONS

**Central Unica dos Trabalhadores (CUT):** Rua Caetano Pinto 575, Brás, 03041-000 São Paulo, SP; tel. (11) 3272-9411; e-mail duvaier@cut.org.br; internet www.cut.org.br; f. 1983; central union confederation; left-wing; Pres. LUIZ MARINHO; Gen. Sec. GILMAR CARNEIRO.

**Confederação Geral dos Trabalhadores (CGT):** Rua Tomaz Gonzaga 50, 2° andar, Liberdade, 01506-020 São Paulo, SP; tel. (11) 3209-6577; e-mail cgt@cgt.org.br; internet www.cgt.org.br; f. 1986; fmrly Coordenação Nacional das Classes Trabalhadoras; represents 1,012 labour organizations linked to PMDB, containing 6.3m workers; Pres. ANTÔNIO CARLOS DOS REIS MEDEIROS; Sec.-Gen. FRANCISCO CANÍNDE PEGADO DO NASCIMENTO.

**Confederação Nacional dos Metalúrgicos** (Metal Workers): e-mail imprensa@cnmcut.org.br; internet www.cnmcut.org.br; f. 1985; Pres. HEIGUIBERTO GUIBA DELLA BELLA NAVARRO; Gen. Sec. FERNANDO AUGUSTO MOREIRA LOPES.

**Confederação Nacional das Profissões Liberais (CNPL)** (Liberal Professions): SAU/SUL, Edif. Belvedere Gr. 202, 70070-000 Brasília, DF; tel. (61) 223-1683; fax (61) 223-1944; e-mail cnpl@cnpl.org.br; internet www.cnpl.org.br; f. 1953; Pres. LUÍS EDUARDO GAUTÉRIO GALLO; Exec. Sec. JOSÉ ANTÔNIO BRITO ANDRADE.

**Confederação Nacional dos Trabalhadores na Indústria (CNTI)** (Industrial Workers): Av. W/3 Norte, Quadra 505, Lote 01, 70730-517 Brasília, DF; tel. (61) 274-4150; fax (61) 274-7001; f. 1946; Pres. JOSÉ CALIXTO RAMOS.

**Confederação Nacional dos Trabalhadores no Comércio (CNTC)** (Commercial Workers): Av. W/5 Sul, SGAS Quadra 902, Bloco C, 70390-020 Brasília, DF; tel. (61) 217-7100; fax (61) 217-7122; e-mail secretaria@cntc.org.br; internet www.cntc.org.br; f. 1946; Pres. ANTÔNIO DE OLIVEIRA SANTOS.

**Confederação Nacional dos Trabalhadores em Transportes Marítimos, Fluviais e Aéreos (CONTTMAF)** (Maritime, River and Air Transport Workers): Av. Pres. Vargas 446, gr. 2205, 20071 Rio de Janeiro, RJ; tel. (21) 2233-8329; f. 1957; Pres. MAURÍCIO MONTEIRO SANT'ANNA.

**Confederação Nacional dos Trabalhadores em Comunicações e Publicidade (CONTCOP)** (Communications and Advertising Workers): SCS, Edif. Serra Dourada, 7° andar, No 705/709, Q 11, 70315 Brasília, DF; tel. (61) 224-7926; fax (61) 224-5686; f. 1964; 350,000 mems; Pres. ANTÔNIO MARIA THAUMATURGO CORTIZO.

**Confederação Nacional dos Trabalhadores nas Empresas de Crédito (CONTEC)** (Workers in Credit Institutions): SEP-SUL, Av. W4, EQ 707/907 Lote E, 70351 Brasília, DF; tel. (61) 244-5833; e-mail contec@yawl.com.br; internet www.contec.org.br; f. 1958; 814,532 mems (1988); Pres. LOURENÇO FERREIRA DO PRADO.

**Confederação Nacional dos Trabalhadores em Estabelecimentos de Educação e Cultura (CNTEEC)** (Workers in Education and Culture): SAS, Quadra 4, Bloco B, 70070-908 Brasília, DF; tel. (61) 321-4140; fax (61) 321-2704; internet www.cnteec.org.br; f. 1966; Pres. MIGUEL ABRÃO NETO.

**Confederação Nacional dos Trabalhadores na Agricultura (CONTAG)** (Agricultural Workers): SDS, Edif. Venâncio VI, 1° andar, 70393-900 Brasília, DF; tel. (61) 321-2288; fax (61) 321-3229; internet www.contag.org.br; f. 1964; represents 25 state federations and 3,630 syndicates, 15m. mems; Pres. MANOEL JOSÉ DOS SANTOS.

**Força Sindical (FS):** Rua Galvão Bueno 782, Liberdade, São Paulo, SP; tel. (11) 3277-5877; e-mail secgeral@fsindical.org.br; internet www.fsindical.org.br; f. 1991; 6m. mems (1991); Pres. PAULO PEREIRA DA SILVA.

## Transport

**Ministério dos Transportes:** see section on the Government (Ministries).

### RAILWAYS

In 2003 there were 29,000 km of railway lines. There are also railways owned by state governments and several privately owned railways.

**Companhia Brasileira de Trens Urbanos (CBTU):** Estrada Velha da Tijuca 77, Usina, 20531-080 Rio de Janeiro, RJ; tel. (21) 2575-3399; fax (21) 2571-6149; internet www.cbtu.gov.br; f. 1984; fmrly responsible for surburban networks and metro systems throughout Brazil; 252 km in 1998; the transfer of each city network to its respective local government was under way; Pres. LUIZ OTÁVIO MOTA VALADARES.

**Gerência de Trens Urbanos de João Pessoa (GTU/JP):** Praça Napoleão Laureano 1, 58010-040 João Pessoa, PB; tel. (83) 241-4240; fax (83) 241-6388; 30 km.

**Gerência de Trens Urbanos de Maceió (GTU/MAC):** Rua Barão de Anadiva 121, 57020-630 Maceió, AL; tel. (82) 221-1839; fax (82) 223-4024; 32 km.

**Gerência de Trens Urbanos de Natal:** Praça Augusto Severo 302, 59012-380 Natal, RN; tel. (84) 221-3355; fax (84) 221-0181; 56 km.

**Superintendência de Trens Urbanos de Belo Horizonte (STU/BH-Demetrô):** Av. Afonso Pena 1500, 11°, 30130-005 Belo Horizonte, MG; tel. (31) 3250-4021; fax (31) 3250-4053; e-mail metrobh@gold.horizontes.com.br; f. 1986; 21.3 km open in 2002; Gen. Man. M. L. L. SIQUEIRA.

**Superintêndencia de Trens Urbanos de Recife (STU/REC):** Rua José Natário 478, Areias, 50900-000 Recife, PE; tel. (81) 3455-4655; fax (81) 3455-4422; f. 1985; 53 km open in 2002; Supt FERNANDO ANTÔNIO C. DUEIRE.

**Superintendência de Trens Urbanos de Salvador (STU/SAL):** Praça Onze de Decembro, s/n, Bairro Calçada, 40410-360 Salvador, BA; tel. (71) 313-9512; fax (71) 313-8760; 14 km.

**América Latina Logística do Brasil, SA:** Av. Setembro 2645, 80230-010 Curitaba, PR; tel. (41) 321-7459; e-mail call@all-logistica.com; internet www.all-logistica.com; f. 1997; 6,586 km in 2003; Pres. ALEXANDRE BEHRING COSTA.

**Companhia Cearense de Transportes Metropolitanos, SA (Metrofor):** Rua 24 de Maio 60, 60020-001 Fortaleza, CE; tel. (85) 3455-7000; fax (85) 3212-4848; e-mail metrofor@metrofor.ce.gov.br; internet www.metrofor.ce.gov.br; f. 1997; 46 km.

**Companhia Ferroviária do Nordeste:** Av. Francisco de Sá 4829, Bairro Carlito Pamplona, 60310-020 Fortaleza, CE; tel. (85) 3286-2525; fax (85) 3286-6156; e-mail kerley@cfn.com.br; internet www.cfn.com.br; 4,534 km in 2003; Dir MARTINIANO DIAS.

**Companhia do Metropolitano de São Paulo:** Rua Augusta 1626, 01304-902 São Paulo, SP; tel. (11) 3371-7411; fax (11) 3283-5228; internet www.metro.sp.gov.br; f. 1974; 4-line metro system, 57.6 km open in 2002; Pres. LUIZ CARLOS FRAYZE DAVID.

**Companhia Paulista de Trens Metropolitanos (CPTM):** Av. Paulista 402, 5° andar, 01310-000 São Paulo, SP; tel. (11) 3371-1530; fax (11) 3285-0323; e-mail ctpm@ctpm.sp.gov.br; internet www.ctpm.sp.gov.br; f. 1992 to incorporate suburban lines fmrly operated by the CBTU and FEPASA; 286 km; Dir and Pres. MÁRIO MANUEL SEABRA RODRIGUES BANDEIRA.

**Departamento Metropolitano de Transportes Urbanos:** SES, Quadra 4, Lote 6, Brasília, DF; tel. (61) 317-4090; fax (61) 226-9546; internet www.dmtu.df.gov.br; the first section of the Brasília metro, linking the capital with the western suburb of Samambaia, was inaugurated in 1994; 38.5 km open in 1997; Dir LEONARDO DE FARIA E SILVA.

**Empresa de Trens Urbanos de Porto Alegre, SA:** Av. Ernesto Neugebauer 1985, 90250-140 Porto Alegre, RS; tel. (51) 3371-5000; fax (51) 3371-1219; e-mail secos@trensurb.com.br; internet www.trensurb.gov.br; f. 1985; 31 km open in 1998; Pres. MARCO MAIA.

**Estrada de Ferro do Amapá (EFA):** Av. Santana 429, Porto de Santana, 68925-000 Santana, AP; tel. (96) 231-1719; fax (96) 281-1175; f. 1957; operated by Indústria e Comércio de Minérios, SA; 194 km open in 1998; Dir Supt JOSÉ LUIZ ORTIZ VERGULINO.

**Estrada de Ferro Campos do Jordão:** Rua Martin Cabral 87, CP 11, 12400-000 Pindamonhangaba, SP; tel. (22) 3644-7408; fax (22) 3643-2951; internet www.efcj.com.br; f. 1914; operated by the Tourism Secretariat of the State of São Paulo; 47 km open in 1998; Dir ARTHUR FERREIRA DOS SANTOS.

**Estrada de Ferro Carajás:** Av. dos Portugueses s/n, 65085-580 São Luís, MA; tel. (98) 3218-4000; fax (98) 3218-4530; f. 1985 for movement of minerals from the Serra do Carajás to the port at Ponta da Madeira; operated by the Companhia Vale do Rio Doce; 892 km open in 2002; Supt JUARES SALIBRA.

**Estrada de Ferro do Jari:** Vila Mongouba s/n, Monte Dourado, 68230-000 Pará, PA; tel. (91) 3736-6526; fax (91) 3736-6490; transportation of timber and bauxite; 68 km open; Dir ARMINDO LUIZ BARETTA.

**Estrada de Ferro Mineração Rio do Norte, SA:** Praia do Flamengo 200, 5° e 6° andares, 22210-030 Rio de Janeiro, RJ; tel. (21) 2205-9112; fax (21) 2545-5717; 35 km open in 1998; Pres. ANTÔNIO JOÃO TORRES.

**Estrada de Ferro Paraná-Oeste, SA (Ferroeste):** Av. Iguaçú 420-7°, 80230-902 Curitiba, PR; tel. (41) 322-1811; fax (41) 233-2147; internet www.pr.gov.br/ferroeste; f. 1988 to serve the grain-producing regions in Paraná and Mato Grosso do Sul; 249 km in 2003; privatized in late 1996, Brazilian company, Ferropar, appointed as administrator; Pres. JOSÉ HERALDO CARNEIRO LOBO.

**Estrada de Ferros Trombetas:** Porto de Trombetas, 68275-000 Oriximiná, PA; tel. (91) 549-7010; fax (91) 549-1482; e-mail jcarlos@mrn.com.br; 30 km; bauxite transport to Trombetas port; operated by Mineração Rio do Norte; Pres. JOSÉ CARLOS SOARES.

**Estrada de Ferro Vitória-Minas:** Av. Carandaí 1115, 13°, Funcionários, 30130-915 Belo Horizonte, MG; tel. (31) 3279-4545; fax (31) 3279-4676; f. 1942; operated by Companhia Vale de Rio Doce; transport of iron ore, general cargo and passengers; 905 km open in 2003; Dir JOSÉ FRANCISCO MARTINS VIVEIROS.

**Ferrovia Bandeirante, SA (Ferroban):** Rua Dr Sales de Oliveira 1380, Vila Industrial, 13035-270 Campinas, SP; tel. (19) 3735-3100; fax (19) 3735-3196; e-mail ferroban@ferroban.com.br; internet www.ferroban.com.br; f. 1971 by merger of five railways operated by São Paulo State; transferred to private ownership in Nov. 1998; fmrly Ferrovia Paulista; 4,236 km open in 2003; Dir JOÃO GOUVEIA FERRÃO NETO.

**Ferrovia Centro Atlântica, SA:** Rua Sapucaí 383, Floresta 30150-904, Belo Horizonte, MG; tel. (31) 3279-5520; fax (31) 3279-5709; e-mail thiers@centro-atlantica.com.br; internet www.fcasa.com.br; f. 1996 following the privatization of Rede Ferroviária Federal, SA; industrial freight; 7,080 km; Man. Dir THIERS MANZANO BARSOTTI.

**Ferrovia Norte-Sul:** Avda Marechal Floriano 45, Centro, 20080-003 Rio de Janeiro, RJ; tel. (21) 2253-9659; fax (21) 2263-9119; e-mail valecascom@ferrovianortesul.com.br; internet www.ferrovianortesul.com.br; 2,066 km from Belém to Goiânia; Man. Dir LUIZ RAIMUNDO CARNEIRO DE AZEVEDO.

**Ferrovia Novoeste, SA:** Rua do Rócio 351, 3°, 04552-905 São Paulo, SP; tel. (11) 3845-4966; fax (11) 3841-9252; e-mail silviam@uol.com.br; 1,622 km in 2003; Man. Dir NÉLSON DE SAMPAIO BASTOS.

**Ferrovia Tereza Cristina, SA:** Rua dos Ferroviários 100, Oficinas, 88702-230 Tubarão, SC; tel. (48) 621-7700; fax (48) 621-7747; e-mail ftc@ftc.com.br; internet www.ftc.com.br; 164 km in 2003; Man. Dir BENONY SCHMIDT, Filho.

**Ferrovias Norte do Brasil, SA (FERRONORTE):** Rua do Rócio 351, 3° andar, Vila Olímpia, 04552-905 São Paulo, SP; tel. (11) 3845-4966; fax (11) 3841-9252; e-mail jhomero@novoeste.com.br; internet www.ferronorte.com.br; f. 1988; 403 km in 2003; affiliated with Ferrovia Novoeste and Ferroban; Man. Dir NELSON DE SAMPAIO BASTOS.

**Metrô Rio:** Av. Presidente Vargas 2000, Col. Centro, 20210-031 Rio de Janeiro, RJ; tel. (21) 3211-6300; e-mail sac@metrorio.com.br; internet www.metrorio.com.br; 2-line metro system, 42 km open in 1997; operated by Opportans Concessão Metroviária, SA; Pres. ALVARO J. M. SANTOS.

**MRS Logística, SA:** Praia de Botafogo 228, Sala 1201E, Ala B, Botafogo, 22359-900, Rio de Janeiro, RJ; tel. (21) 2559-4610; internet www.mrs.com.br; f. 1996; 1,974 km in 2003; CEO JULIO FONTANA NETO.

## ROADS

In 2003 there were an estimated 1.7m. km of roads in Brazil, of which less than 10% were paved. Brasília has been a focal point for inter-regional development, and paved roads link the capital with every region of Brazil. The building of completely new roads has taken place predominantly in the north. Roads are the principal mode of transport, accounting for 63% of freight and 96% of passenger traffic, including long-distance bus services, in 1998. Major projects include the 5,000-km Transamazonian Highway, running from Recife and Cabedelo to the Peruvian border, the 4,138-km Cuibá–Santarém highway, which will run in a north–south direction, and the 3,555-km Trans-Brasiliana project, which will link Marabá, on the Trans-Amazonian highway, with Aceguá, on the Uruguayan frontier. A 20-year plan to construct a highway linking São Paulo with the Argentine and Chilean capitals was endorsed in 1992 within the context of the development of the Southern Cone Common Market (Mercosul). In 2004 an agreement was reached to construct a highway linking the Brazilian state of Acre with the coast of Peru.

**Departamento Nacional de Estradas de Rodagem (DNER)** (National Roads Development): SAN, Quadra 3, Blocos N/O, 4° andar, Edif. Núcleo dos Transportes, 70040-902 Brasília, DF; tel. (61) 315-4100; fax (61) 315-4050; e-mail diretoria.geral@dner.gov.br; internet www.dner.gov.br; f. 1945 to plan and execute federal road policy and to supervise state and municipal roads in order to integrate them into the national network; Exec. Dir ROGÉRIO GONZALES ALVES.

## INLAND WATERWAYS

River transport plays only a minor part in the movement of goods. There are three major river systems, the Amazon, Paraná and the São Francisco. The Amazon is navigable for 3,680 km, as far as Iquitos in Peru, and ocean-going ships can reach Manaus, 1,600 km upstream. Plans have been drawn up to improve the inland waterway system and one plan is to link the Amazon and Upper Paraná to provide a navigable waterway across the centre of the country. In 1993 the member governments of Mercosul, together with Bolivia, reaffirmed their commitment to a 10-year development programme (initiated in 1992) for the extension of the Tietê Paraná river network along the Paraguay and Paraná rivers as far as Buenos Aires, improving access to Atlantic ports and creating a 3,442 km waterway system, navigable throughout the year.

**Agência Nacional de Transportes Aquaviários:** Ministério dos Transportes, SAN, Quadra 3, Blocos N/O, 70040-902 Brasília, DF; tel. (61) 315-8102; Sec. WILDJAN DA FONSECA MAGNO.

**Administração da Hidrovia do Nordeste (AHINOR):** Rua da Paz 561, 65020-450 São Luiz, MA.

**Administração da Hidrovia do Paraguai (AHIPAR):** Rua Treze de Junho 960, Corumbá, MS; tel. (67) 231-2841; fax (67) 231-2661; internet www.ahiper.gov.br; Supt PAULO CÉSAR C. GOMES DA SILVA.

**Administração da Hidrovia do Paraná (AHRANA):** Av. Brigadeiro Faria Lima 1884, 6°, 01451-000 São Paulo, SP; tel. (11) 3034-0270; fax (11) 3815-5435; e-mail ahrana@ahrana.gov.br; internet www.ahrana.gov.br; Supt LUIZ EDUARDO GARCIA.

**Administração da Hidrovia do São Francisco (AHSFRA):** Praça do Porto 70, Distrito Industrial, 39270-000 Pirapora, MG; tel. (38) 3741-2555; fax (38) 3741-2510; internet www.ahsfra.gov.br; Supt JOSÉ H. BORATO JABUR JÚNIOR.

**Administração das Hidrovias da Amazônia Oriental (AHIMOR):** Rua Joaquim Nabuco 8, Nazaré, 66055-300 Belém, PA; tel. (91) 241-3995; fax (91) 241-4589; e-mail ahimor@ahimor.gov.br; internet www.ahimor.gov.br.

**Administração das Hidrovias do Sul (AHSUL):** Praça Oswaldo Cruz 15, 3° andar, 90030-160 Porto Alegre, RS; tel. (51) 3228-3677; fax (51) 3226-9068; Supt JOSÉ LUIZ F. DE AZAMBUJA.

**Administração das Hidrovias do Tocantins e Araguaia (AHITAR):** Rua 85 971, Setor Sul, 74080-010 Goiânia, GO; tel. (62) 225-1744; e-mail ahitar@terra.com.br; internet www.ahitar.com.br.

**Empresa de Navegação da Amazônia, SA (ENASA):** Belém, PA; tel. (91) 223-3878; fax (91) 224-0528; f. 1967; cargo and passenger services on the Amazon river and its principal tributaries, connecting the port of Belém with all major river ports; Pres. ANTÔNIO DE SOUZA MENDONÇA; 48 vessels.

## SHIPPING

There are more than 40 deep-water ports in Brazil, all but one of which (Imbituba) are directly or indirectly administered by the Government. The majority of ports are operated by state-owned concerns (Cia Docas do Pará, Estado de Ceará, Estado do Rio Grande do Norte, Bahia, Paraíba, Espírito Santo, Rio de Janeiro and Estado de São Paulo), while a smaller number (including Suape, Cabedelo, São Sebastião, Paranaguá, Antonina, São Francisco do Sul, Porto Alegre, Itajaí, Pelotas and Rio Grande) are administered by state governments.

The ports of Santos, Rio de Janeiro and Rio Grande have specialized container terminals handling more than 1,200,000 TEUs (20-ft equivalent units of containerized cargo) per year. Santos is the major container port in Brazil, accounting for 800,000 TEUs annually. The ports of Paranaguá, Itajaí, São Francisco do Sul, Salvador, Vitória and Imbituba cater for containerized cargo to a lesser extent.

Total cargo handled by Brazilian ports in 2002 amounted to 529m. tons, compared with 506m. tons in 2001. Some 43,000 vessels used Brazil's ports in 1998.

Brazil's merchant fleet comprised 494 vessels, with a combined aggregate displacement of 2,628,338 grt, in December 2004.

**Departamento de Marinha Mercante:** Coordenação Geral de Transporte Maritimo, Av. Rio Branco 103, 6° e 8° andar, 20040-004 Rio de Janeiro, RJ; tel. (21) 2221-4014; fax (21) 2221-5929; Dir PAULO OCTÁVIO DE PAIVA ALMEIDA.

### Port Authorities

**Departamento de Portos:** SAN, Quadra 3, Blocos N/O, CEP 70040-902 Brasília, DF; Dir PAULO ROBERTO K. TANNENBAUM.

**Administração do Porto de Manaus (SNPH):** Rua Governador Vitório 121, Centro, 69000 Manaus, AM; tel. (92) 621-4300; fax (92) 635-9464; internet faleconosco@portodemanaus.com.br; internet www.portodemanaus.com.br; private; operates the port of Manaus.

**Administração do Porto de São Francisco do Sul (APSFS):** Av. Eng. Leite Ribeiro 782, CP 71, 89240-000 São Francisco do Sul, SC;

# BRAZIL

tel. (47) 471-1200; fax (47) 471-1211; e-mail porto@apsfs.sc.gov.br; internet www.apsfs.sc.gov.br; Dir-Gen. ARNALDO S. THIAGO.

**Administração dos Portos de Paranaguá e Antonina (APPA):** Rua Antonio Pereira 161, 83221-030 Paranaguá, PR; tel. (41) 420-1100; fax (41) 423-4252; internet www.pr.gov.br/portos; Port Admin. Eng. OSIRIS STENGHEL GUIMARÃES.

**Companhia Docas do Espírito Santo (CODESA):** Av. Getúlio Vargas 556, Centro, 29020-030 Vitória, ES; tel. (27) 3132-7360; fax (27) 3132-7311; e-mail assecs@codesa.com.br; internet www.codesa.com.br; f. 1983; Dir JOÃO LUIZ ZAGANELLI.

**Companhia das Docas do Estado de Bahia:** Av. da França 1551, 40010-000 Salvador, BA; tel. (71) 243-5066; fax (71) 241-6712; internet www.codeba.com.br; administers the ports of Aracaju, Salvador, Aratu, Ilhéus and Pirapora, and the São Francisco waterway (AHSFRA); Pres. JORGE FRANCISCO MEDAUAR.

**Companhia Docas do Estado de Ceará (CDC):** Praça Amigos da Marinha s/n, 60182-640 Fortaleza, CE; tel. (85) 3263-1551; fax (85) 3263-2433; internet www.docasdoceara.com.br; administers the port of Fortaleza; Dir MARCELO MOTA TEIXEIRA.

**Companhia Docas do Estado de São Paulo (CODESP):** Av. Conselheiro Rodrigues Alves s/n, Macuco, 11015-900 Santos, SP; tel. (13) 3233-6565; fax (13) 3222-3068; e-mail codesp@carrier.com.br; internet www.portodesantos.com; administers the ports of Santos, Charqueadas, Estrela, Cáceres, Corumbá/Ladário, and the waterways of Paraná (AHRANA), Paraguai (AHIPAR) and the South (AHSUL); Pres. JOSÉ CARLOS MELLO REGO.

**Companhia Docas de Imbituba (CDI):** Porto de Imbituba, Av. Presidente Vargas 100, 88780-000 Imbituba, SC; tel. (48) 255-0080; fax (48) 255-0701; e-mail docas@cdiport.com.br; internet www.cdiport.com.br; private-sector concession to administer the port of Imbituba; Man. Dir NILTON GARCIA DE ARAUJO.

**Companhia Docas do Pará (CDP):** Av. Pres. Vargas 41, 2° andar, 66010-000 Belém, PA; tel. (91) 216-2011; fax (91) 241-1741; internet www.cdp.com.br; f. 1967; administers the ports of Belém, Macapá, Santarém and Vila do Conde, and the waterways of the Eastern Amazon (AHIMOR) and Tocantins and Araguaia (AHITAR); Dir-Pres. CARLOS ACATAUSSÚ NUNES.

**Companhia Docas da Paraíba (DOCAS-PB):** Rua Presidente João Pessoa, s/n, 58310-000 Cabedelo, PB; internet www.pbnet.com.br/zaitek/porto-pb/index.htm; administers the port of Cabedelo.

**Companhia Docas do Rio de Janeiro (CDRJ):** Rua do Acre 21, 20081-000 Rio de Janeiro, RJ; tel. (21) 2219-8544; fax (21) 2253-0528; e-mail cdrj@portosrio.gov.br; internet www.portosrio.gov.br; administers the ports of Rio de Janeiro, Niterói, Sepetiba and Angra dos Reis; Pres. ANTÔNIO CARLOS SOARES LIMA.

**Companhia Docas do Rio Grande do Norte (CODERN):** Av. Hildebrando de Góis 220, Ribeira, 59010-700 Natal, RN; tel. (84) 211-5311; fax (84) 221-6072; administers the ports of Areia Branca, Natal, Recife and Maceió; Dir JOSÉ WALTER DE CARVALHO.

**Empresa Maranhense de Administração Portuária (EMAP):** Porto do Itaquí, s/n, 65085-370 São Luís, MA; tel. (98) 216-6000; fax (98) 216-6060; internet www.portodoitaqui.com.br; f. 2001 to administer ports of Itaquí as concession from the state of Maranhão; Pres. FERNANDO ANTONIO BRITO, Filho.

**Sociedade de Portos e Hidrovias do Estado de Rondônia (SOPH):** Terminal dos Milagres 400, Caiari; tel. (69) 229-2134; fax (69) 229-3943; operates the port of Porto Velho; Dir OBEDES OLIVEIRA DE QUEIROZ.

**SUAPE—Complexo Industrial Portuário Governador Eraldo Gueiros:** Rodovia PE-060, Km 10, Engenho Massangana, 55590-972 Município de Ipojuca, PE; tel. (81) 3527-5000; fax (81) 3527-5064; e-mail presidênciasuape.pe.gov.br; administers the port of Suape.

**Superintendência do Porto de Itajaí:** Rua Blumenau 5, Col. Centro, 88305-101 Itajaí, SC; tel. (47) 341-8000; e-mail atendimento@portoitajai.com.br; internet www.portoitajai.com.br.

**Superintendência do Porto de Rio Grande (SUPRG):** Av. Honório Bicalho, s/n, CP 198, 96201-020 Rio Grande do Sul, RS; tel. (53) 231-1366; fax (53) 231-1857; internet www.portoriogrande.com.br; f. 1996; Port Dir LUIZ FRANCISCO SPOTORNO.

**Superintendência do Porto de Tubarão:** Porto de Tubarão, 29072-970 Vitória, ES; tel. (27) 3228-1053; fax (27) 3228-1682; operated by the Companhia Vale do Rio Doce mining co; Port Dir CANDIDO COTTA PACHECO.

**Superintendência de Portos e Hidrovias do Estado do Rio Grande do Sul (SPH):** Av. Mauá 1050, 4°, 90010-110 Porto Alegre, RS; tel. (51) 3211-4849; fax (51) 3225-8954; e-mail executiva@sph.rs.gov.br; internet www.sph.rs.gov.br; administers the ports of Porto Alegre, Pelotas and Cachoeira do Sul, the São Gonçalo canal and other waterways.

## State-owned Company

**Companhia de Navegação do Estado de Rio de Janeiro:** Praça 15 de Novembro 21, 20010-010 Rio de Janeiro, RJ; tel. (21) 2533-6661; fax (21) 2252-0524; Pres. MARCOS TEIXEIRA.

## Private Companies

**Aliança Navegação e Logística Ltda:** Rua Verbo Divino 1547, Bairro Chácara Santo Antônio, 04719-002 São Paulo, SP; tel. (11) 5185-5600; fax (11) 5185-5624; e-mail alianca@sao.alianca.com.br; internet www.alianca.com.br; f. 1950; cargo services to Argentina, Uruguay, Europe, Baltic, Atlantic and North Sea ports; Pres. ARSÊNIO CARLOS NÓBREGA.

**Companhia de Navegação do Norte (CONAN):** Av. Rio Branco 23, 25° andar, 20090-003 Rio de Janeiro, RJ; tel. (21) 2223-4155; fax (21) 2253-7128; f. 1965; services to Brazil, Argentina, Uruguay and inland waterways; Chair. J. R. RIBEIRO SALOMÃO.

**Companhia de Navegação do São Francisco:** Av. São Francisco 1517, 39270-000 Pirapora, MG; tel. (38) 3741-1444; fax (38) 3741-1164; Pres. JOSÉ HUMBERTO BARATA JABUR.

**Companhia Interamericana de Navegação e Comercio—CINCO:** Av. 14 de Março 1700, 79370-000 Ladário, MS; tel. (67) 226-1010; fax (67) 226-1718; e-mail luiz.assy@cinconav.com.br; internet www.cinconav.com.br; f. 1989; Paraguay and Parana Rivers water transportation; Dirs LUIZ ALBERTO DO AMARAL ASSY, MICHEL CHAIM.

**Companhia Libra de Navegação:** Rua São Bento 8, 8°, Centro, 20090-010 Rio de Janeiro, RJ; tel. (21) 2203-5000; fax (21) 2283-3001; internet www.libra.com.br.

**Frota Oceânica e Amazonica, SA (FOASA):** Av. Venezuela 110, CP 21-020, 20081-310 Rio de Janeiro, RJ; tel. (21) 2203-3838; fax (21) 2253-6363; e-mail foasa@pamar.com.br; f. 1947; Pres. JOSÉ CARLOS FRAGOSO PIRES; Vice-Pres. LUIZ J. C. ALHANATI.

**Petrobrás Transporte, SA (TRANSPETRO):** Edif. Visconde de Itaboraí, Av. Presidente Vargas 328, Rio de Janeiro, RJ; internet www.transpetro.com.br; f. 1998; absorbed the Frota Nacional de Petroleiros (FRONAPE) in 1999; transport of petroleum and related products; 55 vessels; Pres. MAURO OROFINO CAMPOS.

**Vale do Rio Doce Navegação, SA (DOCENAVE):** Av. Graça Aranha 26, 8°–9° andar, 20005-900 Rio de Janeiro, RJ; fax (21) 3814-4971; internet www.docenave.com.br; bulk carrier to Japan, Arabian Gulf, Europe, North America and Argentina; Pres. ALVARO DE OLIVEIRA, Filho.

**Wilson Sons Agência Marítima:** Rua Jardim Botânico 518, 3° andar, 22461-000 Rio de Janeiro, RJ; tel. (21) 2126-4222; fax (21) 2126-4190; e-mail box@wilsonsons.com.br; internet www.wilsonsons.com.br; f. 1837; shipping agency, port operations, towage, small shipyard.

## CIVIL AVIATION

There are about 2,014 airports and airstrips and 417 helipads. Of the 67 principal airports 22 are international, although most international traffic is handled by the two airports at Rio de Janeiro and two at São Paulo. There were 16,454 aircraft registered in Brazil in 2003.

**Departamento de Aviação Civil:** Rua Santa Luiza 651, Castelo, 20030-040 Rio de Janeiro, RJ; tel. (21) 3814-6700; internet www.dac.gov.br; studies, plans and coordinates develops activities of public and private civil aviation; Dir-Gen. Maj.-Brig. CARLOS DE CAMPOS MACHADO.

**Empresa Brasileira de Infra-Estrutura Aeroportuária (INFRAERO):** SCS, Quadra 04, NR 58, Edif. Infraero, 6° andar, 70304-902 Brasília, DF; tel. (61) 312-3170; fax (61) 312-3105; e-mail fernandalima@infraero.gov.br; internet www.infraero.gov.br; Pres. EDUARDO BOGALHO PETTENGILL.

### Principal Airlines

**GOL Transportes Aéreos, SA:** Rua Tamios 246, Jardim Aeropuerto, 04630-000 São Paulo, SP; tel. (11) 5033-4200; internet www.voegol.com.br; 2001; low-cost airline; Man. Dir CONSTANTINO OLIVEIRA JÚNIOR.

**Líder Táxi Aéreo, SA:** Av. Santa Rosa 123, 30535-630 Belo Horizonte, MG; tel. (31) 3490-4500; fax (31) 3490-4600; internet www.lideraviacao.com.br; helicopters and small jets; f. 1958; Pres. JOSÉ AFONSO ASSUMPÇÃO.

**Oceanair Linhas Aéreas Ltda:** Av. Marechal Câmara 160, Sala 1532, Centro, 20020-080 Rio de Janeiro, RJ; tel. (21) 2544-2181; fax (21) 2215-7181; internet www.oceanair.com.br; f. 1998; domestic services; Pres. GERMAN EFROMOVICH.

**Pantanal Linhas Aéreas Sul-Matogrossenses, SA:** Av. das Nações Unidas 10989, 8° andar, São Paulo, SP; tel. (11) 3040-3900; fax (11) 3846-3424; e-mail pantanal@uninet.com.br; internet www

.pantanal-airlines.com.br; f. 1993; regional services; Pres. MARCOS FERREIRA SAMPAIO.

**TAM Linhas Aéreas, SA** (Transportes Aéreos Regionais—TAM): Av. Jurandir 856, Lote 4, 1°, Jardim Ceci, 04072-000 São Paulo, SP; tel. (11) 5582-8811; fax (11) 578-5946; e-mail tamimprensa@tam.com.br; internet www.tam.com.br; f. 1976; scheduled passenger and cargo services from São Paulo to destinations throughout Brazil; Pres. DANIEL MANDELLI MARTIN.

**VARIG, SA** (Viação Aérea Rio-Grandense): Av. Almirante Silvio de Noronha 365, Bloco C, 4°, 20021-010 Rio de Janeiro, RJ; tel. (21) 3814-5869; fax (21) 3814-5703; internet www.varig.com.br; f. 1927; granted bankruptcy protection mid-2005; international services throughout North, Central and South America, Africa, Western Europe and Asia; domestic services to major Brazilian cities; cargo services; also operates the domestic regional subsidiary airlines Nordeste and Rio-Sul; Chair. DAVID ZYLBERSZTAJN; Man. Dir MANOEL GUEDES.

**VASP, SA** (Viação Aérea São Paulo): Praça Comte-Lineu Gomes s/n, Aeroporto Congonhas, 04626-910 São Paulo, SP; tel. (11) 5532-3000; fax (11) 5542-0880; internet www.vasp.com.br; f. 1933; privatized in Sept. 1990; domestic services throughout Brazil; international services to Argentina, Belgium, the Caribbean, South Korea and the USA; Pres. WAGNER CANHEDO AZEVEDO.

## Tourism

In 2003 some 4.1m. tourists visited Brazil. In that year receipts from tourism totalled US $3,386m. Rio de Janeiro, with its famous beaches, is the centre of the tourist trade. Like Salvador, Recife and other towns, it has excellent examples of Portuguese colonial and modern architecture. The modern capital, Brasília, incorporates a new concept of city planning and is the nation's show-piece. Other attractions are the Iguaçu Falls, the seventh largest (by volume) in the world, the tropical forests of the Amazon basin and the wildlife of the Pantanal.

**Associação Brasileira da Indústria de Hotéis (ABIH):** Avda Nilo Peçanha 12, Grups 1005, 20020-100 Rio de Janeiro, RJ; tel. (21) 2533-5768; fax (21) 2533-7632; e-mail abihnacional@abih.com.br; internet www.abih.com.br; f. 1936; hoteliers' association; Pres. LUIZ CARLOS NUNES; Sec. MARLUCE MARGALHÃES.

**Instituto Brasileiro de Turismo (EMBRATUR):** SCN, Quadra 02, Bloco G, 3° andar, 70710-500 Brasília, DF; tel. (61) 224-9100; fax (61) 223-9889; internet www.embratur.gov.br; f. 1966; Pres. CAIO LUIZ DE CARVALHO.

**Seção de Feiras e Turismo/Departamento de Promoção Comercial:** Ministério das Relações Exteriores, Esplanada dos Ministérios, 5° andar, Sala 523, 70170-900 Brasília, DF; tel. (61) 411-6394; fax (61) 322-0833; e-mail docstt@mre.gov.br; internet www.braziltradenet.gov.br; f. 1977; organizes Brazil's participation in trade fairs and commercial exhibitions abroad; Principal Officer ANTÓNIO J. M. DE SOUZA E SILVA.

## Defence

In August 2004 Brazil's Armed Forces numbered 302,909 men: Army 189,000 (including 40,000 conscripts); Navy 48,600 (3,200 are conscripts; also including 1,150 in the naval air force and 14,600 marines); and Air Force 65,309 (including 2,507 conscripts). Reserves number 1,115,000 and there were some 385,600 in the paramilitary Public Security Forces, state militias under Army control. Military service lasts for 12 months and is compulsory for men between 18 and 45 years of age.

**Defence Budget:** R $27,900m. in 2004.

**Chief of Staff of the Air Forces:** Gen. LUIS CARLOS DA SILVA BUENO.

**Chief of Staff of the Army:** Gen. ROBERTO DE ALBUQUERQUE.

**Chief of Staff of the Navy:** Adm. ROBERTO DE GUIMARÃES CARVALHO.

## Education

Education is free in Brazil in official schools at primary and secondary level. Primary education is compulsory between the ages of seven and 14 years and lasts for eight years. Secondary education begins at 15 years of age and lasts for three years. The federal Government is responsible for higher education, and in 2002 there were 128 universities, of which about one-half were state administered. In 2000–01, according to UNESCO, 97% of children in the relevant age-group were enrolled at primary schools, and 71% of those aged 15 to 17 were enrolled at secondary schools. In 2003, according to official figures, 34,438,749 children were enrolled at 169,075 primary schools, while 9,072,942 were enrolled at 23,118 secondary schools. There were 94,741 pre-primary schools with 5,155,676 pupils. There is a large number of private institutions at all levels of education. Central government expenditure on education was R $15,314m. in 2003.

# Bibliography

For works on South America generally, see Select Bibliography (Books)

Andersen, L., Granger, C. W. J., Reis, J. E., Weinhold, D., and Wunder, S. *The Dynamic of Deforestation and Economic Growth in the Brazilian Amazon*. Cambridge, Cambridge University Press, 2002.

Arruda, M. *External Debt (Brazil and the International Financial Crisis)*. London, Pluto Press Ltd, 2000.

Baer, W. *The Brazilian Economy (Growth and Development)*. New York, NY, Praeger Publrs, 2001.

Baiocchi, G. *Militants and Citizens: The Politics of Participatory Democracy in Porto Alegre*. Stanford, CA, Stanford University Press, 2005.

Baumann, R. *Brazil in the 1990s*. Basingstoke, Palgrave Macmillan, 2001.

Branford, S., and Kucinski, B. *Brazil: Carnival of the Oppressed*. London, Latin America Bureau, 1995.

*Lula and the Workers Party in Brazil*. New York, The New Press, 2005.

Cardoso, F. E., and Font, M. (Ed.) *Charting a New Course*. Lanham, MD, Rowman & Littlefield Publrs, 2001.

Castro, P. F. *Fronteras Abiertas: Expansionismo y Geopolítica en el Brasil Contemporáneo*. Madrid, Editores Siglo XXI, 2002.

Conrad, R. E. (Ed.). *Children of God's Fire: A Documentary History of Black Slavery in Brazil*. Philadelphia, PA, Pennsylvania State University Press, 1994.

Dillon Soares, G. A. *A Democracia Interrompida*. Rio de Janeiro, RJ, Editora FGV, 2001.

Font, M. A., Spanakos, A., and Bordin, C. *Reforming Brazil (Western Hemisphere Studies)*. Lanham, MD, Lexington Books, 2004.

Foweraker, J. *The Struggle for Land: A Political Economy of the Pioneer Frontier in Brazil from 1930 to the Present Day*. Cambridge, Cambridge University Press, 2002.

Freyre, G. *The Masters and the Slaves: A Study in the Development of Brazilian Civilization*. New York, NY, Alfred A. Knopf, 1946.

Goertzel, T. G. *Fernando Henrique Cardoso: Reinventing Democracy in Brazil*. Boulder, CO, Lynne Rienner Publrs, 1999.

Gordon, L. *Brazil's Second Chance: En Route toward the First World*. Washington, DC, The Brookings Institution, 2001.

Guilhoto, J. M. J., and Hewings, G. J. D. (Eds). *Structure and Structural Change in the Brazilian Economy*. Aldershot, Ashgate Publishing Ltd, 2001.

Hirst, M. *The United States and Brazil: A Long Road of Unmet Expectations*. London, Routledge, 2005.

Johnson, O. A., III. *Brazilian Party Politics and the Coup of 1964*. Gainesville, FL, University Press of Florida, 2001.

Loureiro, M. R. *Os Economistas no Governo: Gestão Econômica e Democracia*. Rio de Janeiro, RJ, Editora FGV, 1997.

Mendes, C. *Fight for the Forest: Chico Mendes in his own Words*. London, Latin America Bureau, 1989.

Navarro, Z. 'Democracy, citizenship and representation: rural social movements in southern Brazil, 1978–1990' in *Bulletin of Latin American Research*, Vol. 13, No. 2 (May), pp 129–154. 1994.

Newitt, M. (Ed.). *The First Portuguese Colonial Empire*. Exeter, University of Exeter Press, 2002.

Revkin, A. *The Burning Season: The Murder of Chico Mendes and the Fight for the Amazon Rain Forest.* Washington, DC, Shearwater Books, 2004.

Ribeiro, D. *The Brazilian People (The Formation and Meaning of Brazil).* Gainesville, FL, University Press of Florida, 2000.

Rocha, S. *Pobreza no Brasil: Afinal de que se trata?* Rio de Janeiro, RJ, Editora, FGV, 2003.

Rosenn, K. S., and Downes, R. *Corruption and Political Reform in Brazil (The Aftermath of Fernando Collor De Mello).* Boulder, CO, Lynne Rienner Publrs, 1999.

Schneider, R. M. *Brazil: Culture and Politics in a New Industrial Powerhouse.* Boulder, CO, Westview Press, 1996.

Skidmore, T. E. *The Politics of Military Rule in Brazil, 1964–85.* Oxford, Oxford University Press, 1990.

Smith, J. *A History of Brazil.* Harlow, Longman, 2002.

# THE BRITISH VIRGIN ISLANDS

## Geography

### PHYSICAL FEATURES

The British Virgin Islands is an Overseas Territory of the United Kingdom in the West Indies. The Virgin Islands lie at the north-western end of the Lesser Antilles, the chain that defines the edge of the Caribbean Sea, north and east of which is the Atlantic Ocean. To the east of the British Virgin Islands, beyond the shipping lane known as the Anegada Passage, Anguilla (another British dependency) and the other Leeward Islands continue the arc of the Lesser Antilles south-eastwards, while to the west of the US Virgin Islands (formerly the Danish West Indies) is Puerto Rico and the other Greater Antilles. The Virgin Islands themselves are divided between two sovereignties, the smaller, eastern group being British, the rest constituting a territory of the USA. South-west from Tortola, the main island of the British Virgin Islands, across a narrow sea channel, is the US Virgin Island of St John. The main island of Puerto Rico (another US territory) is almost 100 km (60 miles) to the west. The British Virgin Islands has a total area of 153 sq km (59 sq miles).

The British Virgin Islands consists of between 40 and 60 islands, islets and cays (with only about 80 km of coastline between them) strewn over almost 3,450 sq km of sea. Of the main islands, 16 are inhabited and 20 uninhabited. The largest island is Tortola (54 sq km), where the capital is located. The next in size are Anegada (39 sq km), Virgin Gorda (21 sq km) and Jost Van Dyke (9 sq km). The last is west of Tortola, while Virgin Gorda is to the east (beyond the small clump of the Dog Islands), and the more isolated Anegada is north of Virgin Gorda. At the centre of the archipelago, like an 'inland sea', is the 30-km Sir Francis Drake Channel, which runs north-eastwards from St John to Virgin Gorda, flanked by Tortola (on the northern side) and by a string of islands including Norman, Peter, Salt, Cooper and Ginger (to the south). Most of the main islands are hilly and steep (the highest point is Mt Sage on Tortola, at 521 m or 1,710 ft), the result of long-past volcanic activity, which has made the islands fertile and lush with tropical greenery. However, there are also extensive coral reefs, themselves adding to the land area, most notably in the northern island of Anegada, which is flat and coralline. South of Anegada is the 18-km Horseshoe Reef, one of the largest reefs in the world—some 300 ships are believed to have been wrecked on and around the island. Only on Tortola are there open streams, and these are seasonal. The complicated island geography has, historically, attracted pirates, but it is now tourists, particularly yachters, who are drawn to the scattered islands, sometimes perilous reefs, hidden coves and sandy beaches. Some tree species can only be found on the Virgin Islands, as can the smallest lizard in the world, the cotton ginner or dwarf gecko, while Anegada, by contrast, shelters the last survivors of the indigenous variety of rock iguana, which can grow to over 1.5 m.

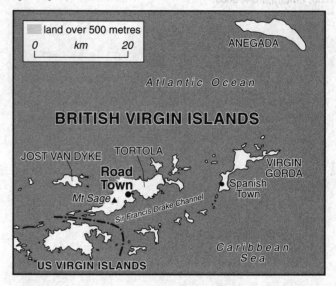

### CLIMATE

The climate is subtropical and humid, moderated by the trade winds off the Atlantic. The hilly terrain helps capture some of the 1,350 mm (53 ins) of rainfall received in an average year. However, the British Virgin Islands are in the hurricane belt, which can strike to devastating effect. Temperature variations are relatively mild, ranging between 22°–28°C (72°–82°F) in winter (December–March) and 26°–32°C (79°–90°F) in summer.

### POPULATION

Most of the population is black (83%) and Christian (33% Methodist, 21% from other Protestant denominations, 17% Anglican—Episcopal and 10% Roman Catholic). English is the official and most widely spoken language. Links with the neighbouring US Virgin Islands are strong.

The total population was estimated at some 21,689 in mid-2004. It is reckoned to have almost doubled in the last 20 years of the 20th century, with only about one-half of the present population reckoned to be of British Virgin Islands origin, the rest being drawn by opportunities in tourism and construction. The capital is Road Town, on the south coast of Tortola, the most populous (82% of the total) as well as the largest island. 15 other islands are also inhabited, although most of the rest of the population is on Virgin Gorda (itself once the centre of population), the chief settlement of which is Spanish Town.

## History

### MARK WILSON

The British Virgin Islands is a United Kingdom Overseas Territory. A Governor, who is the representative of the British monarch, has important reserve powers, including responsibility for national security and defence, the civil service, the judiciary and certain financial matters. A Chief Minister is responsible to the Legislative Council, similar in function to a parliament, which contains a majority of elected members; nine represent constituencies, while four 'at large' members are elected by an overall vote. The Governor—currently Thomas Macan—presides over an Executive Council, similar in function to a cabinet, which includes the Chief Minister and the other ministers. A constitutional review was submitted to the Chief

Minister and the Governor for consideration in April 2005, with a total of 116 recommendations, but with the overall conclusion that the people of the Territory preferred the sharing of responsibilities between the Government of the Virgin Islands and the Government of the United Kingdom to be continued.

Few traces remain of the original Amerindian inhabitants. The Virgin Islands were named by Columbus in 1493. The first European settlers were from the Netherlands, but the islands were British from 1666. There was an elected Assembly from 1773. From 1871 to 1956 the British Virgin Islands formed part of the Leeward Islands Federation. From 1951, the local Legislative Council was given four elected representatives, equal in number to the nominated members, with the franchise extended to adults passing a literacy test. The islands did not join the Federation of the West Indies in 1958, possibly because islanders saw their future as more closely linked to the neighbouring US Virgin Islands. Instead they were from 1960 a separate British Dependent Territory, first under an Administrator, then under a Governor. From February 1998, along with the United Kingdom's other remaining dependencies, they were designated a United Kingdom Overseas Territory.

In 1967 a new Constitution established the office of Chief Minister. For most of the succeeding period, the Virgin Islands Party (VIP) has held office, led by H. Lavity Stoutt until his death in May 1995, and then by Ralph T. O'Neal. The United Party (UP) formed a coalition during 1983–86, with an independent member, Cyril Romney, as Chief Minister.

In 1986 the VIP then returned to office, winning five out of the nine elected seats in that year, and six seats in 1989, but losing its absolute majority in a general election on 20 February 1995. Four 'at large' members had been added to the nine constituency representatives, against the wishes of the VIP; with the composition of the Legislative Council altered in this way, the VIP took six seats, with two for the UP, two for the Concerned Citizens' Movement (CCM), and three independents. However, one of the independents joined the VIP, which was thus able to form an administration with a majority of a single seat. At the general election of 17 May 1999 the VIP retained its single-seat majority, winning seven seats, in spite of a strong challenge from the National Democratic Party (NDP), which had been formed only in 1998 and which took five seats. The remaining seat was held by the CCM.

A brief political crisis began on 20 July 2000, when the appointment of Eileene Parsons as Minister for Health, Education, Culture and Welfare was revoked. Parsons then joined the NDP, a move which would have deprived the VIP of its majority, had not the single CCM member, Ethlyn E. Smith, crossed the floor to join the Government as Minister for Health and Welfare. O'Neal's support was increased from seven to eight seats in February 2001 when another opposition member, Mark Vanterpool, joined the VIP.

Reported corruption on an airport-improvement project precipitated a serious political crisis in 2001–02. Following a report presented to the Governor on 9 November 2001, several people were arrested, including the Government's financial secretary Allen Wheatley, the budget co-ordinator and the former head of the telephone services management unit. In March 2002 they were charged in court with conspiracy to defraud the Government and related offences; they received sentences of from six to nine months in January 2004. The Attorney-General appealed in March 2004 against the leniency of the sentences. A report by the Government's chief auditor in September 2003 also expressed concern over the neglect of established procedures for the award of government contracts. An opposition-proposed motion of 'no confidence' in the Legislative Council was defeated in May, with O'Neal expressing a lack of concern about the issue. The standing of the Government was severely weakened and, with a constitutional review in progress, strong pressure for stricter financial management could be expected from the British authorities. However, the Chief Minister has argued for greater local autonomy and a reduction in the Governor's reserve powers, while a report by the United Kingdom's Centre for Management and Policy Studies commented on the 'almost total breakdown' in the relationship between elected ministers and senior civil servants. In addition to corruption, drugs-related crime is also a serious problem. Several murders in 2003–05 appeared to be drugs-related, and, with a small population, resulted in a high per-head murder rate.

In May 2002 the British Overseas Territories Act, having received royal assent in the United Kingdom in February, came into force and granted British citizenship to the people of its Overseas Territories, including the British Virgin Islands. Under the new law, British Virgin Islanders would be able to hold British passports and work in the United Kingdom and anywhere else in the European Union. Meanwhile, in the same month, Julian Frazer was appointed Deputy Chief Minister and Minister of Communications and Works in succession to J. Alvin Christopher. Although O'Neal denied the claims, it was reported that Christopher had been removed from office because of his support for corruption allegations against the Government in connection with the airport development project. Reeial George replaced Frazer as Minister for Natural Resources and Labour.

After 17 years in office, the VIP relinquished control of the legislature after a general election, which was held on 16 June 2003. The NDP secured eight seats compared with the VIP's five after a campaign that was dominated by the issues of alleged corruption, management of public-sector capital projects and relations with the United Kingdom. The VIP, which claimed returning officers failed to account for all ballots issued and votes cast, demanded a recount in two constituencies; however, the complaint was rejected by the High Court. The new Chief Minister, Orlando Smith, appointed Ronnie Skelton as Minister of Finance, Health and Welfare, Paul Wattley as Minister of Communications and Works, Lloyd Black as Minister of Education and Culture, and J. Alvin Christopher as Minister of Natural Resources and Labour. In an effort to improve strategic planning for public-sector capital spending, the new Government decided to reappraise a controversial $77m. hospital project, and ended a dispute with the Caribbean Development Bank over the appointment of contractors to complete the airport project; as a result, work on the runway started in October 2003, after a delay of 15 months, and was completed in May 2004. Anti-corruption legislation targeted at politicians and public figures was debated by the Legislative Council in December 2003, while a constitutional review commission reported in April 2005; consultations arising from the report were being conducted at mid-2005. The Minister of Communications and Works, Paul Wattley, died in July of that year.

# Economy

## MARK WILSON

The British Virgin Islands is situated in the eastern Caribbean and, in 2004, had an estimated population of 21,689 on 153 sq km. Immigration resulting from economic prosperity resulted in an average annual population growth of 2.1% in 1992–2002. At the 1991 census, 81.5% of the population lived on the island of Tortola, but Virgin Gorda is developing rapidly, and doubled its population in 1991–97.

The British Virgin Islands is an associate member of the Caribbean Community and Common Market, or CARICOM, which, in principle, was attempting to develop a single market

# THE BRITISH VIRGIN ISLANDS

and economy by 2006. It is also a member of the Organisation of Eastern Caribbean States, which links nine of the smaller Caribbean territories. However, the territory does not participate in the Eastern Caribbean Central Bank and has no separate central banking arrangements, using US currency for all purposes.

The islands have an extremely prosperous economy, with a per-head gross domestic product (GDP) of $38,643 in 2003, and an average annual growth rate of close to 4% in 1995–99.

After several years of buoyancy, the economy slowed sharply in 2000–03, largely because of an international downturn in tourism. Revenue in 2002 was almost 10% below the original budget projection, forcing the Government to draw heavily on reserves. Owing to the consequent difficulties in financial planning, the budget for 2003, which would normally have been agreed in December of the previous year, was presented to the Legislative Council only in March, and passed on 6 April. None the less, the 2004 budget was presented in December 2003, according to the normal timetable, and projected a current-account surplus of 1.5% of GDP, while raising the income-tax threshold from US $3,000 to $7,500. Economic growth in 2004 was estimated at 2.5%. The 2005 budget projected a current fiscal surplus of $27.9m., equivalent to 3.5% of GDP. Of this amount, $24.1m. was to be transferred to capital budget, with $1m. to an emergency and disaster fund, and $2m. to a reserve fund to provide a cushion against fiscal shocks. The balance of the capital budget was to be financed from new borrowing. The national debt was $90.0m. at the end of 2004, 1.1% greater than one year earlier. Of this amount, $50.0m. was foreign debt and $24.75m. represented the liabilities of statutory boards and public corporations.

Financial and business services made up 35.9% of GDP in 2003. The very large 'offshore' financial sector, administered by an independent Financial Services Commission, specializes in the registration of International Business Companies, of which there was a cumulative total of 544,000 at the end of 2004, with new registrations running at more than 50,000 per year. There were 312 local and captive 'offshore' insurance companies. Management of the 2,023 licensed mutual funds is another area of activity, but 'offshore' banking is not well developed, a deliberate choice by the authorities who are hesitant to take on the regulatory problems involved. Legislation on trusts introduced in 2003 was designed to broaden the basis of the 'offshore' sector, and to control possible abuse of existing regulations. Fees from the 'offshore' sector totalled US $100m. in 2002, making up over one-half of government revenue, which totalled $183m. Following international pressure, in particular from the Organisation for Economic Co-operation and Development (OECD), which, in June 2000, had included the British Virgin Islands on its list of tax 'havens', legislation to improve transparency in the financial sector was introduced. In early 2002 the Government committed itself to improving the islands' financial sector to meet OECD guide-lines by 2005. A Tax Information Exchange Treaty with the USA was signed in April 2002. In 2004 further legislation was introduced, eliminating the distinction between 'onshore' and 'offshore' taxation regimes, specifically in order to correspond to OECD and European Union standards; the transition was to be completed by 2007.

Tourism is the other mainstay of the economy, with hotels and restaurants accounting for 16% of GDP in 2003. The main attractions are the tranquillity of the islands, the clear surrounding waters, low rainfall, and white sandy beaches. On the smaller islands in the group, there is the added benefit of near-complete privacy. The number of stop-over tourists increased by an average of 4.9% per year in 1997–2001, reaching 295,625 in 2001, before decreasing slightly to some 281,700 in 2002 and further, to 270,700 in 2003. However, stop-over numbers rebounded to a record figure of 331,815 in 2004. Yachting is an important segment of the tourism industry, with 56% of stop-over tourists in 2001 staying on charter boats. Most tourists stay in luxury accommodation; their high spending power per head is of further economic benefit. However, there were complaints that cruise-ship traffic was diluting the islands' 'exclusive' image (there were some 466,601 cruise-ship passengers in 2004). Visitor expenditure totalled US $368.0m. in 2003.

Residential, commercial, public-sector and tourism-related investment have resulted in a high level of construction activity. Mismanagement and alleged corruption in public-sector capital projects has been a major concern; the incoming National Democratic Party Government attempted to address this issue from its election in June 2003. Consequently, a US $77m. hospital project was reappraised, while agreement reached with the Caribbean Development Bank allowed a long-delayed airport runway improvement to be completed.

Agriculture comprised only 0.4% of GDP in 2001, with a few small farmers keeping livestock and growing food crops. There is also a small-scale fishing industry, contributing 1.3% of GDP in the same year. Manufacturing (3.3% of GDP in 2003) is limited to small-scale activities such as printing and the blending and bottling of rum. Electric power supply has been a problem, in spite of the addition of new generating capacity.

The territory is in the heart of the hurricane belt, and has been damaged by several storms in recent years. There is, however, no volcanic risk.

# Statistical Survey

Source: Development Planning Unit, Central Administrative Complex, Road Town, Tortola; tel. 494-3701; fax 494-3947; e-mail dpu@dpu.org; internet www.dpu.gov.vg.

## AREA AND POPULATION

**Area:** 153 sq km (59 sq miles). *Principal Islands* (sq km): Tortola 54.4; Anegada 38.8; Virgin Gorda 21.4; Jost Van Dyke 9.1.

**Population:** 16,644 (males 8,570, females 8,074) at census of 12 May 1991; 19,864 (males 10,234, females 9,630) in 1999; 21,689 (estimate) at mid-2004; *By Island* (1980): Tortola 9,119; Virgin Gorda 1,412; Anegada 164; Jost Van Dyke 134; Other islands 156; (1991) Tortola 13,568. Source: partly Caribbean Development Bank, *Social and Economic Indicators*.

**Density** (mid-2004): 141.7 per sq km.

**Principal Town:** Road Town (capital), population 12,000 (UN estimate, incl. suburbs, mid-2003). Source: UN, *World Urbanization Prospects: The 2003 Revision*.

**Births, Marriages and Deaths** (registrations, 2004): 316 live births (birth rate 14.7 per 1,000); 426 marriages (marriage rate 19.6 per 1,000); 120 deaths (death rate 5.5 per 1,000).

**Expectation of Life** (years at birth, estimates): 73.8 (males 69.9; females 78.5) in 2004.

**Employment** (2003): Agriculture, hunting and forestry 43; Fishing 13; Mining and quarrying 80; Manufacturing 283; Electricity, gas and water supply 196; Construction 1,208; Wholesale and retail trade 1,359; Hotels and restaurants 2,486; Transport, storage and communications 613; Financial 712; Real estate, renting and business activities 1,142; Public administration and social security 4,770; Education 981; Health and social work 124; Other community, social and personal service activities 416; Private households with employed persons 384; Not classifiable by economic activity 5; Total 14,815.

## HEALTH AND WELFARE

**Physicians** (per 1,000 head, 1999): 1.15.

**Hospital Beds** (per 1,000 head, 2003): 2.0.

**Health Expenditure** (% of GDP, 1995): 3.9. *2003* (public expenditure only): 3.1.

**Health Expenditure** (public, % of total, 1995): 36.5.
Source: Pan American Health Organization.
For definitions, see explanatory note on p. vi.

# THE BRITISH VIRGIN ISLANDS

## AGRICULTURE, ETC.

**Livestock** (FAO estimates, '000 head, 2003): Cattle 2.4; Sheep 6.1; Goats 10.0; Pigs 1.5.

**Fishing** (FAO estimates, metric tons, live weight, 2003): Swordfish 2, Caribbean spiny lobster 3, Stromboid conchs 6; Total catch (incl. others) 50.
Source: FAO.

## INDUSTRY

**Electric Energy** (production, million kWh): 43 in 1988; 44 in 1989; 45 in 1990. 1991–2001 annual production as in 1990 (UN estimates).
Source: UN, *Industrial Commodity Statistics Yearbook*.

## FINANCE

**Currency and Exchange Rate:** United States currency is used: 100 cents = 1 US dollar ($). *Sterling and Euro Equivalents* (31 May 2005): £1 sterling = US $1.8181; €1 = US $1.2331; US $100 = £55.03 = €81.10.

**Budget** (estimates, $ million, 2003/04): Recurrent revenue 190.0 (Tax revenue 66.0 *of which:* import duties 21.0, income and property tax 40.0, passenger and hotel tax 5.0; Non-tax revenue 124.0 *of which:* financial services sector 102.2); Recurrent expenditure 178.0 (Goods and services 75.1; Wages and salaries 73.5; Subsidies and transfers 28.3); Capital expenditure 38.2.

**Cost of Living** (Consumer Price Index; base: 1995 = 100): 127.2 in 2002; 131.8 in 2003; 132.7 in 2004.

**Gross Domestic Product** ($ million, at market prices): 682.9 in 2000; 824.6 in 2001; 840.2 in 2002; 824.4 in 2003. Source: Caribbean Development Bank, *Social and Economic Indicators*.

**Expenditure on the Gross Domestic Product** ($ million at current prices, 1999): Government final consumption expenditure 73; Private final consumption expenditure 269; Gross fixed capital formation 158; *Total domestic expenditure* 500; Exports of goods and services 691; *Less* Imports of goods and services 530; *GDP in purchasers' values* 662. Source: UN, *National Accounts Statistics*.

**Gross Domestic Product by Economic Activity** (rounded figures in current prices, $ million, 1999): Agriculture, hunting and forestry 4.1; Fishing 5.4; Mining and quarrying 0.4; Manufacturing 27.0; Electricity, gas and water 11.1; Construction 49.0; Wholesale and retail trade 90.0; Hotels and restaurants 98.0; Transport, storage and communications 79.0; Financial intermediation 43.0; Real estate, renting and business services 188.0; Public administration 35.0; Education 13.6; Health and social work 9.5; Other community, social and personal services 15.8; Private households with employed persons 6.0; *Sub-total* 674.0; *Less* Imputed bank service charges 35.0; Indirect taxes, less subsidies 23.0; *GDP in purchasers' values* 662.0. Source: UN, *National Accounts Statistics*.

## EXTERNAL TRADE

**Principal Commodities** ($ '000): *Imports c.i.f.* (1997): Food and live animals 31,515; Beverages and tobacco 8,797; Crude materials (inedible) except fuels 1,168; Mineral fuels, lubricants, etc. 9,847; Chemicals 8,816; Basic manufactures 31,715; Machinery and transport equipment 47,019; Total (incl. others) 116,379. *Exports f.o.b.* (1996): Food and live animals 368; Beverages and tobacco 3,967; Crude materials (inedible) except fuels 1,334; Total (incl. others) 5,862. *2001* (exports, $ million) Animals 0.1; Fresh fish 0.7; Gravel and sand 1.4; Rum 3.6; Total 28.13.

**Principal Trading Partners** ($ '000): *Imports c.i.f.* (1997): Antigua and Barbuda 1,807; Trinidad and Tobago 2,555; United Kingdom 406; USA 94,651; Total (incl. others) 166,379. *Exports f.o.b.* (1996): USA and Puerto Rico 1,077; US Virgin Islands 2,001; Total (incl. others) 5,862. *1999:* Imports 208,419; Exports 2,081.
Source: mainly UN, *International Trade Statistics Yearbook*.

## TRANSPORT

**Road Traffic** (motor vehicles in use): 6,900 in 1992; 6,700 in 1993; 7,000 in 1994. *2001* (motor vehicles registered and licensed): 11,313 (private cars 3,490). Source: mainly UN, *Statistical Yearbook*.

**Shipping:** *International Freight Traffic* ('000 metric tons, 2002): Goods unloaded 145.6. *Cargo Ship Arrivals* (2002): 2,027. *Merchant Fleet* (vessels registered, at 31 December 2004): 14; Total displacement 2,520 grt (Sources: British Virgin Islands Port Authority, and Lloyd's Register-Fairplay, *World Fleet Statistics*).

**Civil Aviation** (passenger arrivals): 145,929 in 2000; 144,914 in 2001; 134,690 in 2002.

## TOURISM

**Visitor Arrivals** ('000): 281.7 stop-over visitors, 230.1 cruise-ship passengers in 2002; 270.7 stop-over visitors, 299.2 cruise-ship passengers in 2003; 331.8 stop-over visitors, 466.6 cruise-ship passengers in 2004.

**Visitor Expenditure** (US $ million, estimates): 373.8 in 2001; 356.5 in 2002; 368.0 in 2003.
Source: Caribbean Development Bank, *Social and Economic Indicators*.

## COMMUNICATIONS MEDIA

**Radio Receivers** (1997): 9,000 in use.

**Television Receivers** (1999): 4,000 in use.

**Telephones** (1996): 10,000 main lines in use.

**Facsimile Machines** (1996): 1,200 in use.

**Non-daily Newspapers** (1996): 2 (estimated circulation 4,000).
Sources: UNESCO, *Statistical Yearbook*; UN, *Statistical Yearbook*.

## EDUCATION

**Pre-primary:** 5 schools (1994/95); 46 teachers (2001/02); 628 pupils (2001/02).

**Primary:** 20 schools (1993/94); 168 teachers (2001/02); 2,735 pupils (2003).

**Secondary:** 4 schools (1988); 166 teachers (UNESCO estimate, 2001/02); 1,633 pupils (2003).

**Tertiary:** 1,916 pupils enrolled (2003).
Source: UNESCO Institute for Statistics; Caribbean Development Bank, *Social and Economic Indicators*.

# Directory

## The Constitution

The British Virgin Islands have had a representative assembly since 1774. The present Constitution took effect from June 1977. Under its terms, the Governor is responsible for defence and internal security, external affairs, terms and conditions of service of public officers, and the administration of the Courts. The Governor also possesses reserved legislative powers in respect of legislation necessary in the interests of his special responsibilities. There is an Executive Council, with the Governor as Chairman, one *ex-officio* member (the Attorney-General), the Chief Minister (appointed by the Governor from among the elected members of the Legislative Council) who has responsibility for finance, and three other ministers (appointed by the Governor on the advice of the Chief Minister); and a Legislative Council consisting of a Speaker, chosen from outside the Council, one *ex-officio* member (the Attorney-General) and 13 elected members (nine members from one-member electoral districts and four members representing the Territory 'at large').

The division of the islands into nine electoral districts, instead of seven, came into effect at the November 1979 general election. The four 'at large' seats were introduced at the February 1995 general election. The minimum voting age was lowered from 21 years to 18 years.

Under the British Overseas Territories Act, which entered into effect in May 2002, British Virgin Islanders have the right to United Kingdom citizenship and the right of abode in the United Kingdom. British citizens do not enjoy reciprocal rights.

## The Government

**Governor:** THOMAS MACAN (assumed office October 2002).

# THE BRITISH VIRGIN ISLANDS

**Deputy Governor:** DANCIA PENN.

### EXECUTIVE COUNCIL
(July 2005)

**Chairman:** TOM MACAN (The Governor).

**Chief Minister:** Dr ORLANDO SMITH.

**Deputy Chief Minister and Minister of Finance, Health and Welfare:** RONNIE SKELTON.

**Minister of Natural Resources and Labour:** J. ALVIN CHRISTOPHER.

**Minister of Education and Culture:** LLOYD BLACK.

**Minister of Communications and Works:** (vacant).

**Attorney-General:** CHERNO JALLOW.

### MINISTRIES

**Office of the Governor:** Government House, POB 702, Road Town, Tortola; tel. 494-2345; fax 494-5790; e-mail bvigovernor@gov.vg; internet www.bvi.gov.vg.

**Office of the Deputy Governor:** Central Administration Bldg, Road Town, Tortola; tel. 468-3701; fax 494-6481; e-mail dpenn@gov.vg; internet www.bvi.gov.vg.

**Office of the Chief Minister:** Road Town, Tortola; tel. 494-3701; fax 494-4435; e-mail pcsmo@bvigovernment.org.

**Ministry of Communications and Works:** Road Town, Tortola; tel. 494-3701; e-mail mcw@bvigovernment.org.

**Ministry of Education and Culture:** Road Town, Tortola; tel. 494-3887.

**Ministry of Finance, Health and Welfare:** Pasea Estate, Road Town, Tortola; tel. 494-3701; fax 494-6180; e-mail finance@mail.caribsurf.com.

**Ministry of Natural Resources and Labour:** Road Town, Tortola; tel. 494-3614.

All ministries are based in Road Town, Tortola, mainly at the Central Administration Building (fax 494-4435; e-mail tsm@bvigovernment.org; internet www.bvigovernment.org).

### LEGISLATIVE COUNCIL

**Speaker:** V. INEZ ARCHIBALD.

**Clerk:** JULIA LEONARD-MASSICOTT.

**General Election, 16 June 2003**

| Party | % of vote | Seats |
| --- | --- | --- |
| National Democratic Party | 52.4 | 8 |
| Virgin Islands Party | 42.2 | 5 |
| Other | 5.4 | — |
| Total | 100.0 | 13 |

## Political Organizations

**Concerned Citizens' Movement (CCM):** Road Town, Tortola; f. 1994 as successor to Independent People's Movement; Leader ETHLYN SMITH.

**National Democratic Party (NDP):** Road Town, Tortola; f. 1998; Chair. RUSSELL HARRIGAN; Leader ORLANDO SMITH.

**United Party (UP):** POB 3348, Road Town, Tortola; tel. 495-2656; fax 494-1808; e-mail liberatebvi@msn.com; f. 1967; Chair. ULRIC SCATLIFFE; Pres. CONRAD MADURO.

**Virgin Islands Party (VIP):** Road Town, Tortola; Leader RALPH T. O'NEAL.

## Judicial System

Justice is administered by the Eastern Caribbean Supreme Court, based in Saint Lucia, which consists of two divisions: The High Court of Justice and the Court of Appeal. There are two resident High Court Judges, and a visiting Court of Appeal which is comprised of the Chief Justice and two Judges of Appeal and which sits twice a year in the British Virgin Islands. There is also a Magistrates' Court, which hears prescribed civil and criminal cases. The final Court of Appeal is the Privy Council in the United Kingdom.

**Resident Judges:** STANLEY MOORE, KENNETH BENJAMIN.

**Magistrate:** DORIEN TAYLOR.

**Registrar:** GAIL CHARLES.

**Magistrate's Office:** Road Town, Tortola; tel. 494-3460; fax 494-2499.

## Religion

### CHRISTIANITY

#### The Roman Catholic Church

The diocese of St John's-Basseterre, suffragan to the archdiocese of Castries (Saint Lucia), includes Anguilla, Antigua and Barbuda, the British Virgin Islands, Montserrat and Saint Christopher and Nevis. The Bishop is resident in St John's, Antigua.

#### The Anglican Communion

The British and US Virgin Islands form a single, missionary diocese of the Episcopal Church of the United States of America. The Bishop of the Virgin Islands is resident on St Thomas in the US Virgin Islands.

#### Protestant Churches

Various Protestant denominations are represented, principally the Methodist Church. Others include the Seventh-day Adventist, Church of God and Baptist Churches.

## The Press

**The BVI Beacon:** POB 3030, 10 Russell Hill Rd, Road Town, Tortola; tel. 494-3434; fax 494-6267; e-mail bvibeacn@surfbvi.com; internet www.bvibeacon.com; f. 1984; Thursdays; covers local and international news; Editor LINNELL M. ABBOTT; circ. 3,400.

**The Island Sun:** POB 21, 112 Main St, Road Town, Tortola; tel. 494-2476; fax 494-5854; e-mail issun@candwbvi.net; internet www.islandsun.com; f. 1962; Fridays; Editor VERNON PICKERING; circ. 2,850.

**The BVI StandPoint:** POB 4311, Wickhams Cay, Road Town, Tortola; tel. 494-8106; fax 494-8647; e-mail bvistandpoint@surfbvi.com; internet www.bvistandpoint.net; fmrly BVI PennySaver, adopted current name in 2001; Tuesdays; covers local and international news; Editor SUSAN HENIGHAN; circ. 18,000.

**The Welcome:** POB 133, Road Town, Tortola; tel. 494-2413; fax 494-4413; e-mail info@bviwelcome.com; internet www.bviwelcome.com; f. 1971; every 2 months; general, tourist information; Publr PAUL BACKSHALL; Editor CLAUDIA COLLI; annual circ. 172,000.

## Publishers

**Caribbean Publishing Co (BVI) Ltd:** POB 3403, Road Town, Tortola; tel. 494-2060; fax 494-3060; e-mail bvi-sales@caribpub.com; Sales Man. ARTHUR RUBAINE.

**Island Publishing Services Ltd:** POB 133, Road Town, Tortola; tel. 494-2413; fax 494-658; e-mail cpcips@surfbvi.com; internet www.bviwelcome.com; publishes *The British Virgin Islands Welcome Tourist Guide*.

## Broadcasting and Communications

### TELECOMMUNICATIONS

Plans to liberalize the telecommunications sector were under way in early 2005.

**Telephone Services Management:** Central Administration Building, Road Town, Tortola; tel. 494-4728; fax 494-6551; e-mail tsmu@caribsurf.com; government agency.

**Cable & Wireless (WI) Ltd:** Cutlass Bldg, Wickhams Cay 1, POB 440, Road Town, Tortola; tel. 494-4444; fax 494-2506; e-mail candw@candwbvi.net; internet www.candw.vg; f. 1967; CEO VANCE LEWIS.

**CCT Boatphone:** Geneva Place, Road Town, Tortola; tel. 444-4444; e-mail info@bvicellular.com; internet www.bvicellular.com; mobile cellular telephone operator.

### BROADCASTING

#### Radio

**Caribbean Broadcasting System:** POB 3049, Road Town, Tortola; tel. 494-4990; commercial; Gen. Man. ALVIN KORNGOLD.

# THE BRITISH VIRGIN ISLANDS

**Virgin Islands Broadcasting Ltd—Radio ZBVI:** Baughers Bay, POB 78, Road Town, Tortola; tel. 494-2250; fax 494-1139; e-mail zbvi@caribsurf.com; internet www.zbvi.vi; f. 1965; commercial; Gen. Man. HARVEY HERBERT; Ops Man. SANDRA POTTER WARRICAN.

### Television

**BVI Cable TV:** Fishlock Rd, POB 694, Road Town, Tortola; tel. 494-3205; fax 494-2952; programmes from US Virgin Islands and Puerto Rico; 53 channels; Man. Dir TODD KLINDWORTH.

**Television West Indies Ltd (ZBTV):** Broadcast Peak, Chawell, POB 34, Tortola; tel. 494-3332; commercial.

# Finance

## BANKING

### Regulatory Authority

**Financial Services Commission:** internet www.bvifsc.vg; f. 2002; independent financial services regulator; Chair. MICHAEL RIEGELS; CEO ROBERT MATHAVIOUS.

### Commercial Banks

**Ansbacher (BVI) Ltd:** POB 659, International Trust Bldg, Road Town, Tortola; tel. 494-3215; fax 494-3216.

**Banco Popular de Puerto Rico:** POB 67, Road Town, Tortola; tel. 494-2117; fax 494-5294; internet www.bancopopular.com; Man. SANDRA SCATLIFFE.

**Bank of East Asia (BVI) Ltd:** POB 901, Road Town, Tortola; tel. 495-5588; fax 494-4513; Man. ELIZABETH WILKINSON.

**Bank of Nova Scotia** (Canada): Wickhams Cay 1, POB 434, Road Town, Tortola; tel. 494-2526; fax 494-4657; e-mail Michael.Rolle@scotiabank.com; internet www.scotiabank.com; f. 1967; Man. Dir MICHAEL ROLLES.

**First Bank Virgin Islands:** Wickham's Cay, 1 Road Town, Tortola; tel. 494-2662; fax 494-5106; internet www.firstbankvi.com.

**Citco Bank (BVI) Ltd:** Citco Bldg, Wickhams Cay, POB 662, Road Town, Tortola; tel. 494-2218; fax 494-3917; e-mail bvi-trust@citco.com; internet www.citco.com; f. 1978; Man. RENE ROMER.

**DISA Bank (BVI) Ltd:** POB 985, Road Town, Tortola; tel. 494-6036; fax 494-4980; Man. ROSA RESTREPO.

**FirstCaribbean International Bank Ltd:** Wickhams Cay 1, POB 70, Road Town, Tortola; tel. 494-2171; fax 494-4315; e-mail care@firstcaribbeanbank.com; internet www.firstcaribbeanbank.com; f. 2003 following merger of Caribbean operations of Barclays Bank PLC and CIBC; CEO CHARLES PINK; Exec. Dir SHARON BROWN.

**HSBC Guyerzeller Bank (BVI) Ltd:** POB 3162, Road Town, Tortola; tel. 494-5416; fax 494-5417; e-mail rhbvi@surfbvi.com; Dir KENNETH W. MORGAN.

**Rathbone Bank (BVI) Ltd:** POB 986, Road Town, Tortola; tel. 494-6544; fax 494-6532; e-mail rathbone@surfbvi.com; Man. CORNEL BAPTISTE.

**VP Bank (BVI) Ltd:** POB 3463, Road Town, Tortola; tel. 494-1100; fax 494-1199; e-mail vpbank@surfbvi.com; internet www.vpbank.com; Gen-Man. PETER REICHENSTEIN.

### Development Bank

**Development Bank of the British Virgin Islands:** Wickhams Cay 1, POB 275, Road Town, Tortola; tel. 494-3737; fax 494-3119; state-owned; Chair. MEADE MALONE.

## TRUST COMPANIES

**Abacus Trust and Management Services Ltd:** POB 3339, Road Town, Tortola; tel. 494-4388; fax 494-3088; e-mail mwmabacus@surfbvi.com; Man. MEADE MALONE.

**Aleman, Cordero, Galindo and Lee Trust (BVI) Ltd:** POB 3175, Road Town, Tortola; tel. 494-4666; fax 494-4679; e-mail alcogalbvi@alcogal.com; Man. GABRIELLA CONTE.

**AMS Trustees ltd:** POB 116, Road Town, Tortola; tel. 494-3399; fax 494-3041; e-mail ams@amsbvi.com; Man. ROGER DAWES.

**Belmont Trust Ltd:** POB 3443, Road Town, Tortola; tel. 494-5800; fax 494-2545; e-mail belmont@kpmgbvi.net; Man. ANDREA DOUGLAS.

**CCP Financial Consultants Ltd:** POB 681, Road Town, Tortola; tel. 494-6777; fax 494-6787; e-mail mail@ccpbvi.com; internet www.ccpbvi.com; Man. JOSEPH ROBERTS.

**Citco BVI Ltd:** POB 662, Road Town, Tortola; tel. 494-2217; fax 494-3917; e-mail bvi-trust@citco.com; internet www.citco.com; Man. REINIER TEIXEIRA DE MATTOS.

**Guardian Trust & Securities Ltd:** POB 438, Road Town, Tortola; tel. 494-2616; fax 494-2704; e-mail infovg@equitytrust.com; Man. LINDA ROMNEY-LEUE.

**HSBC International Trustee (BVI) Ltd:** POB 916, Road Town, Tortola; tel. 494-5414; fax 494-5417; e-mail kenneth.morgan@rawlinson-hunter.vg; Man. K. W. MORGAN.

**Hunte & Co Services Ltd:** POB 3504, Road Town, Tortola; tel. 495-0232; fax 495-0229; internet www.hunteandco.com; Man. LEWIS HUNTE.

**Maples Finance BVI Ltd:** POB 173, Road Town, Tortola; tel. 494-3384; fax 494-4643; internet www.maplesandcalder.com; Man. ANTHONY LYNTON.

**Moore Stephens International Services (BVI) Ltd:** POB 3186, Road Town, Tortola; tel. 494-3503; fax 494-3592; e-mail moorestephens@surfbvi.com; internet www.moorestephens.com; Man. CRAIG MURPHY-PAIGE.

**Midocean Management and Trust Services (BVI) Ltd:** POB 805, Road Town, Tortola; tel. 494-4567; fax 494-4568; e-mail midocean@maitlandbvi.com; Man. ELIZABETH WILKINSON.

**Totalserve Trust Company Ltd:** POB 3200, Road Town, Tortola; tel. 494-7295; fax 494-7296; e-mail bvi@jerseytrustco.com; internet www.jerseytrustco.com; Man. MICHELLE D. CELESTINE.

**TMF (BVI) Ltd:** POB 964, Road Town, Tortola; tel. 494-4997; fax 494-4999; e-mail tmfbvi@surfbvi.com; Man. GRAHAM COOK.

**Tricor Services (BVI) Ltd:** POB 3340, Road Town, Tortola; tel. 494-6004; fax 494-6404; e-mail eybvi@surfbvi.com; Man. PATRICK A. NICHOLAS.

**Trident Corporate Services (BVI) Ltd:** POB 659, Road Town, Tortola; tel. 494-3215; fax 494-3216; e-mail tcbvi@tridenttrust.com; internet www.tridenttrust.com; Man. BARRY R. GOODMAN.

## INSURANCE

**AMS Insurance Management Services Ltd:** POB 116, Road Town, Tortola; tel. 494-4078; fax 494-2519; e-mail amsins@amsbvi.com; Man. GREGORY M. TAYLOR.

**Atlas Insurance Management (BVI) Ltd:** POB 129, Road Town, Tortola; tel. 494-2728; fax 494-4393; e-mail cil@surfbvi.com; Man. JOHN WILLIAMS.

**ATU Insurance Management (BVI) Ltd:** POB 3463, Road Town, Tortola; tel. 494-1100; fax 494-1199; e-mail atu@atubvi.com; internet www.atubvi.com; Man. PETER REICHENSTEIN.

**Belmont Insurance Management Ltd:** POB 3443, Road Town, Tortola; tel. 494-5800; fax 494-6565; e-mail kpmginsurance@surfbvi.com; Man. PAUL MARTIN.

**Captiva Managers Ltd:** POB 765, Road Town, Tortola; tel. 494-0998; fax 494-9474; internet www.captivamanagers.com; Man. HARRY THOMPSON.

**Caribbean Insurance Management:** POB 129, Road Town, Tortola; tel. 494-8239; fax 494-4393; e-mail cil@surfbvi.com; Man. JOHN WILLIAMS.

**Caribbean Insurers Ltd:** POB 129, Road Town, Tortola; tel. 494-2728; fax 494-4393; e-mail cil@surfbvi.com; Man. JOHN WILLIAMS.

**Codan Management (BVI) Ltd:** POB 3140, Road Town, Tortola; tel. 494-2065; fax 494-4929; e-mail bvi@cdp.bm; Man. A. GUY ELDRIDGE.

**Euro-American Insurance Management Ltd:** POB 3161, Road Town, Tortola; tel. 494-4692; fax 494-4695; e-mail eamsbvi@surfbvi.com; Man. KAY REDDY.

**HWR Insurance Management Services Ltd:** POB 71, Road Town, Tortola; tel. 494-2233; fax 494-3547; e-mail mail@harneys.com; Man. CONNOR JENNINGS.

**Marine Insurance Office (BVI) Ltd:** POB 874, Road Town, Tortola; tel. 494-3795; fax 494-4540; e-mail marinein@surfbvi.com; Man. WESLEY WOODHOUSE.

**Marsh Management Services (BVI) Ltd:** POB 3140, Road Town, Tortola; tel. 494-4850; fax 494-7467; Man. A. GUY ELDRIDGE.

**Osiris Insurance Management Ltd:** POB 2221, Road Town, Tortola; tel. 494-8920; fax 494-6934; e-mail osiristrust@surfbvi.com; Man. TERRANCE M. ROLLINS.

**R & H Insurance Services Ltd:** POB 3162, Road Town, Tortola; tel. 494-5414; fax 494-5417; e-mail edith.steel@rawlinson-hunter.vg; Man. EDITH STEEL.

# THE BRITISH VIRGIN ISLANDS

**Trafford Insurance Management Services Ltd:** POB 116, Road Town, Tortola; tel. 494-4078; fax 494-2519; e-mail amsims@amsbvi.com; Man. GREGORY M. TAYLOR.

**Trident Insurance Management (BVI) Ltd:** POB 146, Road Town, Tortola; tel. 494-2434; fax 494-3754; e-mail trident@surfbvi.com; Man. GREGORY M. TAYLOR.

**TSA Insurance Management Ltd:** POB 3443, Road Town, Tortola; tel. 494-5800; fax 494-6563; e-mail kpmginsurance@surfbvi.com; Man. PAUL MARTIN.

**USA Risk Group (BVI) Inc.:** POB 3140, Road Town, Tortola; tel. 494-2065; fax 494-4929; Man. A. GUY ELDRIDGE.

Several US and other foreign companies have agents in the British Virgin Islands.

## Trade and Industry

### GOVERNMENT AGENCY

**Trade and Investment Promotion Department:** Office of the Chief Minister, Road Town, Tortola; tel. 494-3701; fax 494-6413; internet www.trade.gov.vg.

### CHAMBER OF COMMERCE

**British Virgin Islands Chamber of Commerce and Hotel Association:** James Frett Bldg, Wickhams Cay 1, POB 376, Road Town, Tortola; tel. 494-3514; fax 494-6179; e-mail bviccha@surfbvi.com; internet www.bvihotels.org; f. 1986; Chair. NELDA FARRINGTON-BRYDON; Pres. (Business and Commerce) FRED COLVILLE; Pres. (Hotels and Tourism) LOUIS SCHWARTZ.

### UTILITIES

#### Electricity

**British Virgin Islands Electricity Corpn (BVIEC):** Long Bush, POB 268, Road Town, Tortola; tel. 494-3911; fax 494-4291; e-mail bviecce@caribsurf.com; f. 1979; state-owned, privatization pending; Chair. MARGARET ALMYRA PENN.

#### Water

**Water and Sewerage Dept:** Baughers Bay, Road Town, Tortola; tel. 468-3701; fax 494-6746.

## Transport

### ROADS

In 2002 there were 82 miles of access roads, 48 miles of primary roads, 23 miles of secondary roads and 56 miles of tertiary roads. In 2001 11,313 vehicles were licensed, 3,490 of which were private vehicles.

**Public Works Department:** Baughers Bay, Tortola; tel. 468-2722; fax 468-4740; e-mail pwd@mailbvigovernment.com; responsible for road maintenance.

### SHIPPING

There are two direct steamship services, one from the United Kingdom and one from the USA. Motor launches maintain daily mail and passenger services with St Thomas and St John, US Virgin Islands. A new cruise-ship pier, built at a cost of US $6.9m. with assistance from the Caribbean Development Bank, was opened in Road Town in 1994 and was later expanded.

**British Virgin Islands Port Authority:** Port Purcell, POB 4, Road Town, Tortola; tel. 494-3435; fax 494-2642; e-mail bviports@candwbvi.net; internet www.bviports.org; f. 1991; Man. Dir GENE E. CREQUE (acting).

**Tropical Shipping:** POB 250, Pasea Estate, Road Town, Tortola; tel. 494-2674; fax 494-3505; e-mail lmoses@tropical.com; internet www.tropical.com; Man. LEROY MOSES.

### CIVIL AVIATION

Beef Island Airport, about 16 km (10 miles) from Road Town, has a runway with a length of 1,100 m (3,700 ft). A new US $65m. airport terminal was opened at the airport in March 2002. After a long delay (see History), in October 2003 work began to lengthen the runway to 1,500 m (4,700 ft). Upon completion of the renovation in May 2004 the airport was renamed Terrence B. Lettsome Airport. Captain Auguste George Airport on Anegada has been designated an international point of entry and was resurfaced in the late 1990s. The airport runway on Virgin Gorda was to be extended to 1,200 m (4,000 ft) to allow larger aircraft to land.

**Director of Civil Aviation:** MILTON CREQUE, tel. 494-3701; fax 494-3437.

## Tourism

The main attraction of the islands is their tranquillity and clear waters, which provide excellent facilities for sailing, fishing, diving and other water sports. In 1998 there 1,231 hotel rooms in the islands and 365 rooms in guest-houses and rented apartments. There are also many charter yachts offering overnight accommodation. There were some 331,815 stop-over visitors and 466,601 cruise-ship passengers in 2004. The majority of tourists are from the USA. Receipts from tourism totalled US $368.0m. in 2003.

**British Virgin Islands Tourist Board:** Social Security Bldg, Waterfront Dr., POB 134, Road Town, Tortola; tel. 494-3134; fax 494-3866; e-mail bvitor@bvitouristboard.com; internet www.bvitouristboard.com; Dir KEDRICK MALONE.

**British Virgin Islands Chamber of Commerce and Hotel Association:** see Chamber of Commerce.

## Defence

The United Kingdom is responsible for the defence of the islands. In September 1999 it was announced that Royal Air Force patrols were to be undertaken from the islands in order to intercept drugs-traffickers. Central government expenditure on defence totalled US $0.4m. in 1997.

## Education

Primary education is free, universal and compulsory between the ages of five and 11. Secondary education is also free and lasts from 12 to 16 years of age. In 1997 some 206 pupils were enrolled in pre-primary schools and in 2003 there were 2,735 pupils attending primary schools. In 2003 some 1,633 pupils were enrolled in secondary education. Higher education is available at the University of the Virgin Islands (St Thomas, US Virgin Islands) and elsewhere in the Caribbean, in North America and in the United Kingdom. In 1990 only 0.7% of the adult population had received no schooling. Central government expenditure on education in 1997 was US $17.1m.

# THE CAYMAN ISLANDS

## Geography

### PHYSICAL FEATURES

The Cayman Islands is a British Overseas Territory in the Caribbean Sea, mid-way between Cuba and Honduras. The islands are located about 240 km (150 miles) south of Cuba and some 290 km north-west of Jamaica, to which the territory was once dependent (before Jamaican independence). The Caymans are separated from Jamaica by the Cayman Trench, the deepest part of the Caribbean. The territory covers an area of only 262 sq km (102 sq miles).

The three islands are low, limestone-and-coral formations, largely surrounded by coral reef and with shorelines ill defined by mangrove swamps. The largest island is Grand Cayman (about 35 km—22 miles—long, with an average width of some 6 km), which constitutes three-quarters of the territory's land area, although about one-half of it is wetland. The island is aligned along an east–west axis, tapering in the south-west before a northward-extending spit defines the western shore of the 110-sq-km shallow lagoon, the North Sound, as well as the sheltered west coast of the whole island (notably the West Bay and its Seven-Mile Beach). The other two islands are long and narrow, aligned more to the north-east, end to end and separated by an 8-km channel. They lie 128 km to the east of Grand Cayman, and a little north, slightly closer to both Cuba and Jamaica than the largest island. Low-lying and swampy Little Cayman (26 sq km) is about 16 km long and seldom more than 1.5 km wide. Cayman Brac (39 sq km) is the most easterly of the islands. It is about 19 km long and about 2 km wide, but reaches the highest point in the territory, where a huge limestone outcrop, called The Bluff (43 m—a Gaelic word for a bluff is brac), rises along the centre of the island. The islands' land and marine environments are extensively protected, in an effort to protect the bird and animal life, some of it unique (a species of orchid on Grand Cayman, for instance, trees on Cayman Brac and Cayman parrots). The islands were originally named, and, indeed, settled for the large turtle population, although they have long been named with the Carib word for the marine crocodile that also lived here.

### CLIMATE

The climate is subtropical marine, with warm, wet summers between May and October (in the latter part of this season hurricanes occasionally occur) and milder, dryer winters. Rain is

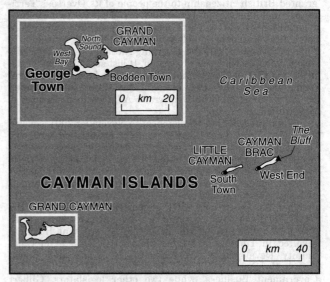

essential to the freshwater supply. Average annual rainfall is about 1,290 mm (50 ins), the average annual temperature 28.5°C (83.3°F).

### POPULATION

A mixed-race population (40%) is balanced by roughly equal black and white communities (each about 20%), as well as expatriates of various ethnic groups. About one-third of the population are foreign workers. The islanders are predominantly Christian and all English-speaking.

The population in December 2003 was officially estimated at 44,144, 95% of whom lived on Grand Cayman (with almost 2,000 on Cayman Brac and just over 100 on Little Cayman). The capital is George Town in the south-west of Grand Cayman, which had an estimated population of 23,940 at mid-2003. West Bay (to the north of the capital) and Bodden Town (on the central southern coast) are the other main centres on Grand Cayman. The main settlements on the other two islands are West End on Cayman Brac and South Town on Little Cayman. The territory is divided into eight administrative districts.

## History

The Cayman Islands constitute a United Kingdom Overseas Territory, with a Cabinet (known as the Executive Council until 2003) headed by the Governor, who is the representative of the British monarch.

Grand Cayman became a British colony in 1670, with the two smaller islands settled only in 1833. Until 1959 all three islands were a dependency of Jamaica. In 1962 a separate Administrator for the islands was appointed (redesignated Governor in 1971). The Constitution was revised in 1972, 1992 and 1994. The Governor is Chairman of the Cabinet (comprising four other official members and five members elected by the Legislative Assembly from among their own number). The Legislative Assembly comprises three members appointed by the Governor and 15 elected members.

In the absence of formal political parties, elections to the Assembly were contested every four years by independents and by individuals standing as 'teams'. In August 1991, however, the territory's first formal political organization since the 1960s was formed, the Progressive Democratic Party. It opposed various provisions of the proposed constitutional reforms. It developed into a broad coalition and was renamed the National Team, and then won 12 of the 15 elective seats at the general election of November 1992. The new Government did not seek the introduction of a post of chief minister. At a general election on 20 November 1996 the National Team remained in power, winning nine seats in the Legislative Assembly.

A serious problem for the Cayman Islands in recent years has been drugs-related crime. In the 1990s an estimated 75% of all thefts and burglaries in the islands were attributed, directly or indirectly, to the drugs trade. The Mutual Legal Assistance Treaty (signed in 1986, and ratified by the US Senate in 1990) between the Cayman Islands and the USA provides for the mutual exchange of information for use in combating crime (particularly drugs-trafficking and money-laundering). Further legislation relating to the abuse of the financial sector by criminal organizations was approved in November 1996.

In March 1999 the British Government published draft legislation confirming that its Dependencies were to be referred to as United Kingdom Overseas Territories; the document stated that all such territories would be required to comply with European standards on human rights and financial regulation. In May 2002 the British Overseas Territories Act, having received royal assent in the United Kingdom in February, came into force and granted British citizenship to the people of its Overseas Territories, including the Cayman Islands. Under the new law, Caymanians would be able to hold British passports and work in the United Kingdom and elsewhere in the European Union.

On 8 November 2000 the governing National Team suffered a heavy defeat in the general election. It lost six of its nine seats, including that held by Truman Bodden, the Leader of Government Business. The newly elected Legislative Assembly voted for Kurt Tibbetts of the Democratic Alliance to be Leader of Government Business.

In an unprecedented development in November 2001, several members of the Legislative Assembly, dissatisfied with the Government's leadership during the economic slowdown (see Economy), formed the United Democratic Party (UDP). McKeeva Bush, the leader of the new party and Deputy Leader of Government Business and Minister of Tourism, Environment and Transport, claimed that at least 10 members of the 15-member Assembly were UDP supporters. Subsequently, the passing of a motion of 'no confidence' (by nine votes to five, with one abstention) against the Leader of Government Business resulted in Tibbetts and Edna Moyle, the Minister of Community Development, Women's Affairs, Youth and Sport, leaving the Executive Council. Bush became the new Leader of Government Business, while two other legislators, Gilbert McLean and Frank McField, both members of the UDP, joined the Cabinet. In May 2002, at the UDP's inaugural convention, Bush was formally elected leader of the party. In the same month it was announced that the five opposition members of the Legislative Assembly had formed a new political party, the People's Progressive Movement (PPM), led by Tibbetts.

In March 2002 a three-member Constitutional Review Commission, appointed by the Governor in May 2001, submitted a new draft constitution. The proposed document had to be debated by the Legislative Assembly and then approved by the British Parliament before being formally adopted. The draft constitution, made public in April, included the creation of the office of Chief Minister, proposed a full ministerial government, and incorporated a bill of rights. The opposition demanded a referendum on the recommendations, on the grounds that the UDP had rejected several of the proposals, despite strong public support for the changes, but Bush forwarded the proposals to the British Foreign and Commonwealth Office (FCO), via the Governor's office. In December Bush and the leader of the opposition, Tibbets, travelled to the United Kingdom in order to review the proposed constitution with FCO officials. Consensus, however, proved impossible to achieve, and the FCO was expected to produce a revised draft constitution for public discussion and debate in 2004. Furthermore, the UDP announced in February 2004 that it would not participate in the constitutional review process before the next elections, due by November of that year. For their part, the PPM called for a public referendum on the issue.

In May 2002 it was announced that the Cayman Islands had become an associate member of the Caribbean Community and Common Market (CARICOM). In the same month Bruce Dinwiddy succeeded Peter Smith as Governor.

In January 2003 Bush accused the United Kingdom of undermining the course of Cayman Islands' justice, after a routine money-laundering case was dismissed amid allegations of espionage and obstruction of justice by British intelligence agents. The trial collapsed after it was alleged that the Director of the Cayman Islands Financial Reporting Unit and a key witness in the trial had passed information about the case to an unnamed agency of the British Government, understood to be the Secret Intelligence Service (MI6). It was claimed that MI6 wished to protect the names of its sources within the Caribbean 'offshore' banking community. Bush demanded that the British Government pay for the failed trial, estimated to have cost some US $5m., and for any negative repercussions the affair might have for the reputation of the territory's banking sector. The United Kingdom, however, refused to compensate the Cayman Islands and maintained that, although its intelligence agencies might have helped in the investigation, they had never interfered in the case. In March the Attorney-General, David Ballantyne, resigned amid accusations that he was aware that British intelligence agents were working covertly in the Cayman Islands; in July Samuel Bulgin, the Solicitor-General, was appointed as Ballantyne's permanent successor.

In June 2003 the Government appointed Orrett Connor to a new post of Cabinet Secretary. Connor, who joined the Executive Council, thereby increasing its membership to 10, was to be responsible for providing neutral advice and technical support to the policy-making process. In October Linford Pierson, the Deputy Leader of Government Business and Minister of Planning, Communications, Works and Information Technology, resigned his position in the Cabinet (called the Executive Council until 2003), reportedly owing to ill health. It was announced that he would swap roles with the Speaker of the Legislative Assembly, Julianna O'Connor-Connolly. The change led to a small realignment of ministerial responsibilities, with O'Connor-Connolly surrendering the Ministry of Works to Gilbert McLean in exchange for the Ministry of District Administration. It was also announced that McLean would succeed Pierson as Deputy Leader of Government Business.

The legislative election scheduled to take place in November 2004 was postponed by six months following the devastation caused by 'Hurricane Ivan' in September of that year. In response to the hurricane Governor Dinwiddy enacted the Emergency Powers Act, allowing him to extend the remit of his authority and impose a curfew until the restoration of full electricity supplies, as a deterrent to looting. In early 2005 Linford Pierson left the UDP and established the People's Democratic Alliance, in advance of the rescheduled election. The opposition PPM secured nine of the 15 seats available in the ballot that was held on 11 May; the UDP won only five mandates and the remaining seat was occupied by an independent candidate. The outgoing UDP Government had been criticized for its management of the aftermath of the hurricane and also for the decision to grant Caymanian 'belonger' status to 3,000 residents in 2003, which was interpreted by many as an attempt to increase in the party's favour, those eligible to vote, prior to the election. The leader of the PPM, Kurt Tibbetts, once again became the Leader of Government Business and committed the administration to holding a referendum on increased autonomy in the Territory. On 18 May the Legislative Assembly elected Anthony Eden, Alden McLaughlin, Arden McLean and Charles Clifford to the Cabinet. Stuart Jack was appointed by the British Government to succeed Bruce Dinwiddy as Governor from November 2005.

# Economy

Since the introduction of secrecy laws in respect of bank accounts and other professional information in the late 1960s, and the easing of foreign-exchange regulations in the 1970s (finally abandoned entirely in 1980), the islands have developed as one of the world's major 'offshore' financial markets. In 1999 the sector contributed some 36% of gross domestic product (GDP) and employed more than 10% of the labour force. In 2005 there were 438 banks and trust companies and 6,268 mutual funds on the islands. The absence of any form of direct taxation also made the islands notorious as a tax 'haven'.

In 1986 a treaty of mutual assistance was signed with the USA, providing access for US law-enforcement agencies to the financial records of Cayman Islands banks, in cases where serious criminal activity is suspected. In November 1996 the powers of the authorities to investigate such cases were augmented. The Cayman Islands Monetary Authority (CIMA), responsible for managing the islands' currency and reserves and for regulating the financial sector, began operations in early 1997.

In June 2000 the Cayman Islands were included on a list, compiled by the Financial Action Task Force on Money Laundering (FATF, based in Paris, France), of those jurisdictions considered to be 'non-co-operative' in international efforts to combat money-laundering. The Government's prompt response in addressing the FATF's concerns resulted in the Cayman Islands' removal from the list in 2001. The CIMA was given increased powers, and efforts were made to eliminate so-called 'shell banks' on the islands (those banks registered in the jurisdiction but with no staff or offices).

In anticipation of the publication by the Organisation for Economic Co-operation and Development (OECD) of a blacklist of 'unco-operative tax havens', in early 2000 the Government of the Cayman Islands made a commitment to the elimination of harmful tax practices. In November 2001, in a further move to increase international confidence in its financial regulation, the Government signed a tax information exchange agreement with the USA, designed to reduce the potential for abuse of the tax system. (The agreement took effect with regard to criminal tax evasion in January 2004 and would do so for further aspects from 2006.) The Government also agreed to respond to any US requests for assistance in investigations into the collapse of the major US energy company Enron, which was suspected of financial malpractice. (Reports presented to the US Securities and Exchange Commission in 2002 claimed that Enron used 692 companies in the Cayman Islands to avoid paying taxes in the USA.) However, fears remained that reforms to the financial services sector would reduce the islands' ability to attract international business. Indeed, in October 2001 the Chamber of Commerce linked an economic downturn to the negative impact of OECD pressure on the financial services sector. Moreover, the Government's decision to increase the annual licensing fee charged to 'offshore' financial institutions in the 2002 budget was widely criticized as being potentially damaging to the industry.

From late 2002 the financial sector faced further disruption when the United Kingdom, under pressure from a European Union (EU) investigation into tax evasion, demanded that the Cayman Islands disclose the identities and account details of Europeans holding private savings accounts on the islands. The Cayman Islands, along with some other British Overseas Territories facing similar demands, claimed it was being treated unfairly compared with more powerful European countries, such as Switzerland and Luxembourg, and refused to make any concessions. The British Paymaster-General demanded that the Cayman Islands enact the necessary legislation to implement the EU's Savings Tax Directive by 30 June 2004; the Legislative Assembly, after the British Government pledged to safeguard the territory's interests, voted to accept British/EU demands by 1 January 2005. (As a result of the agreement, recognition was granted to the Cayman Islands Stock Exchange by the United Kingdom's Inland Revenue body.) However, the implementation of the regulations was delayed until July 2005. Meanwhile, the investigation into the collapse of Parmalat, an Italian food company, which had 10 subsidiary companies in the Cayman Islands, focused further attention on the regulatory apparatus of the territory's financial sector in 2003–04.

In addition to 'offshore' financial services, the tourism sector contributes strongly to the economy. Both of these sectors benefit from the Cayman Islands' political stability, good infrastructure and extensive development. Although the industry suffered a decline owing to the global economic slowdown and the adverse effects of the September 2001 terrorist attacks in the USA, tourism was still estimated to account for almost one-quarter of GDP and, directly or indirectly, 50% of employment. Visitor expenditure earned an estimated US $607m. in 2002. Most tourists are from the USA (71% of stop-over arrivals in 1999). The industry is marketed toward the wealthier visitor and, since 1987, the Government has limited the number of cruise-ship passengers. The Cayman Islands also maintains the largest registry of luxury yachts of over 100 ft in length in the world. The tourism sector was adversely affected by the destruction wreaked by 'Hurricane Ivan' in September 2004 and the ensuing government-imposed curfew and travel restrictions (although these had ceased by the end of the year). Many of the hotels that had suffered damage had reopened by early 2005, but the new Ritz-Carlton hotel was forced to postpone its opening until October. Since the late 1990s the Government has attempted to limit tourist numbers in order to minimize damage to the environment. There is also an active construction sector (contributing 8.9% of GDP in 1991), although the agricultural and light industrial sectors are small, and more than 90% of the islands' food needs are imported. In particular, agriculture is limited by infertile soil, low rainfall and high labour costs. Traditional fishing activities, chiefly of turtles, declined from 1970, particularly after the USA imposed an import ban on turtle products in 1979 (the islands possess the world's only commercial turtle farm). The economic growth rate was 1.0% in 2000 and the economy expanded by just 0.6% and 1.7% in 2001 and 2002, respectively. Growth remained stable, at 2.0%, in 2003. Growth of some 3%–4% had been forecast in 2004, on the basis of economic performance in the first eight months of the year; however, the occurrence of 'Hurricane Ivan' caused a severe setback, and a potentially negative overall rate of growth for the year. An estimated 90% of buildings sustained damage and the cost of the disaster to the country was some CI $2,900m. (US $3,500m.).

# Statistical Survey

Sources: Government Information Services, Cricket Sq., Elgin Ave, George Town, Grand Cayman; tel. 949-8092; fax 949-5936; The Information Centre, Economic and Statistics Office, Government Administration Bldg, Grand Cayman; tel. 949-0940; fax 949-8782; e-mail infostats@gov.ky; internet www.eso.ky.

## AREA AND POPULATION

**Area:** 262 sq km (102 sq miles). The main island of Grand Cayman is about 197 sq km (76 sq miles), about one-half of which is swamp. Cayman Brac is 39 sq km (15 sq miles); Little Cayman is 26 sq km (11 sq miles).

**Population:** 39,410 (males 19,311, females 20,099) at census of 10 October 1999 (Grand Cayman 37,473, Cayman Brac 1,822, Little Cayman 115). *2003* (estimated population at 31 December): 44,144.

**Density** (31 December 2003): 168.5 per sq km.

**Principal Towns** (population at 1999 census): George Town (capital) 20,626; West Bay 8,243; Bodden Town 5,764. *Mid-2003* (UN estimate, incl. suburbs): George Town 23,940 (Source: UN, *World Urbanization Prospects: The 2003 Revision*).

**Births, Marriages and Deaths** (estimates, 2001): Live births 622 (birth rate 15.0 per 1,000); Marriage rate 9.7 per 1,000 (2000); Deaths 132 (death rate 3.2 per 1,000).

**Expectation of Life** (years at birth): 79.8 in 2004. Source: Pan-American Health Organization.

**Economically Active Population** (sample survey, persons aged 15 years and over, October 1995): Agriculture, hunting, forestry and fishing 270; Manufacturing 320; Electricity, gas and water 245; Construction 1,805; Trade, restaurants and hotels 5,555; Transport, storage and communications 1,785; Financing, insurance, real estate and business services 3,570; Community, social and personal services 5,295; Total employment 18,845 (males 8,930, females 9,910). Figures exclude persons seeking work for the first time, totalling 100 (males 60, females 40), and other unemployed persons, totalling 900 (males 455, females 445) (Source: ILO). *2004* (sample survey of 20–27 November, Grand Cayman): Total employed 22,420; Unemployed 1,034 (males 368, females 666); Total labour force

# THE CAYMAN ISLANDS

23,453 (males 12,432, females 11,021) (Note: Totals may not be equal to the sum of their constituent parts, as figures are approximations based on a reduced sample.).

### HEALTH AND WELFARE

**Physicians** (per 1,000 head, 2003): 2.2.

**Hospital Beds** (per 1,000 head, 2003):

**Health Expenditure:** % of GDP (1997): 4.2.

**Health Expenditure:** public (% of total, 1997): 53.2.
Source: mainly Pan American Health Organization.
For definitions, see explanatory note on p. vi.

### AGRICULTURE, ETC.

**Livestock** (FAO estimates, '000 head, 2003): Cattle 1.3; Goats 0.3; Pigs 0.4; Chickens 6.

**Fishing** (metric tons, live weight, 2003): Total catch 125 (all marine fishes).
Source: FAO.

### INDUSTRY

**Electric Energy** (consumption, million kWh): 414.6 in 2002; 429.3 in 2003; 450.3 in 2004. Source: Caribbean Utilities Company Ltd.

### FINANCE

**Currency and Exchange Rates:** 100 cents = 1 Cayman Islands dollar (CI $). *Sterling, US Dollar and Euro Equivalents* (31 May 2005): £1 sterling = CI $1.515; US $1 = 83.3 CI cents; €1 = CI $1.028; CI $100 = £66.03 = US $120.00 = €81.10. *Exchange rate:* Fixed at CI $1 = US $1.20.

**Budget** (CI $ million, 2003): *Revenue:* Taxes on international trade and transactions 117.6; Taxes on other domestic goods and services 153.4; Taxes on property 17.7; Other tax revenue 5.4; Other current revenue 32.1; Total 326.2. *Expenditure:* Current expenditure 283.7 (Personnel costs 138.9, Supplies and consumable goods 61.3, Subsidies 58.8, Transfer payments 18.8, Interest payments 5.9); Capital expenditure and net lending 21.5; Total 305.2.

**Cost of Living** (Consumer Price Index; base: 2000 = 100): 101.1 in 2001; 103.6 in 2002; 104.3 in 2003. Source: ILO.

**Gross Domestic Product** (estimates, CI $ million in current prices): 1,482.3 in 2001; 1,546.0 in 2002; 1,603.2 in 2003.

**Gross Domestic Product by Economic Activity** (CI $ million in current prices, 1991): Primary industries 5; Manufacturing 9; Electricity, gas and water 19; Construction 54; Trade, restaurants and hotels 138; Transport, storage and communications 65; Finance, insurance, real estate and business services 210; Community, social and personal services 42; Government services 63; Statistical discrepancy 1; *Sub-total* 606; Import duties less imputed bank service charge 11; *GDP in purchasers' values* 617.

### EXTERNAL TRADE

**Principal Commodities** (US $ million, 1996): *Imports c.i.f.:* Food and live animals 70.3; Beverages and tobacco 15.2; Mineral fuels, lubricants, etc. 41.2 (Refined petroleum products 39.8); Chemicals 22.2; Basic manufactures 51.3; Machinery and transport equipment 98.4; Miscellaneous manufactured articles 69.9; Total (incl. others) 377.9. *Exports f.o.b.:* Total 3.96.

**Principal Trading Partners** (US $ million): *Imports c.i.f.* (1996): Japan 8.2; Netherlands Antilles 40.1; United Kingdom 6.8; USA 289.7; Total (incl. others) 377.9. *Exports f.o.b.* (1994): USA 0.9; Total 2.6. *2003* (CI $ million): Total imports 553.5 (Fuel 43.2); Total exports 4.3.
Source: mainly UN, *International Trade Statistics Yearbook*.

### TRANSPORT

**Road Traffic** (2000): Motor vehicles in use 24,791.

**Shipping:** *International Freight Traffic* ('000 metric tons): Goods loaded 735 (1990); Goods unloaded 239,138 (2000). *Cargo Vessels* (1995): Vessels 15, Calls at port 266. *Merchant Fleet* (vessels registered at 31 December 2004): 156; Total displacement 2,608,796 grt. (Source: Lloyd's Register-Fairplay, *World Fleet Statistics*).

### TOURISM

**Visitor Arrivals** ('000): 1,548.9 (arrivals by air 334.1, cruise-ship passengers 1,214.8) in 2001; 1,877.6 (arrivals by air 302.7, cruise-ship passengers 1,574.8) in 2002; 2,112.5 (arrivals by air 293.5, cruise-ship passengers 1,819.0) in 2003.

**Visitor Expenditure** (estimates, US $ million): 559.2 in 2000; 585.1 in 2001; 607.0 in 2002.
Source: Caribbean Development Bank, *Social and Economic Indicators*.

### COMMUNICATIONS MEDIA

**Radio Receivers:** 36,000 in use in 1997.

**Television Receivers:** 23,239 in use in 1999.

**Telephones:** 32,967 main lines in use in 2003.

**Facsimile Machines:** 116 in use in 1993.

**Mobile Cellular Telephones** (subscribers): 11,370 in 2000.

**Internet Connections:** 9,909 in 2003.

**Daily Newspapers:** 1 (circulation 10,500) in 2000.

### EDUCATION

**Institutions** (2001): 10 state primary schools (with 2,246 pupils); 9 private primary and secondary schools (2,238 pupils); 3 state high schools (1,750 pupils); 1 community college; 1 private college (519 pupils).

# Directory

## The Constitution

The Constitution of 1959 was revised in 1972, 1992 and 1994. Under its terms, the Governor, who is appointed for four years, is responsible for defence and internal security, external affairs, and the public service. The Cabinet (known as the Executive Council until 2003) comprises the Chairman (the Governor), the Chief Secretary, the Financial Secretary, the Attorney-General, the Cabinet Secretary (all four of whom are appointed by the Governor) and five other Ministers elected by the Legislative Assembly from their own number. The Governor assigns ministerial portfolios to the elected members of the Cabinet. There are 15 elected members of the Legislative Assembly (elected by direct, universal adult suffrage for a term of four years) and three official members appointed by the Governor. The Speaker presides over the Assembly. The United Kingdom retains full control over foreign affairs. In May 2001 the Governor appointed a Constitutional Review Commission to make recommendations on changes to the Cayman Islands' political structure and processes.

## The Government

**Governor:** BRUCE H. DINWIDDY (assumed office May 2002).

**Governor-designate:** STUART JACK (from November 2005).

### CABINET
(July 2005)

**Chairman:** BRUCE H. DINWIDDY (The Governor).

**Chief Secretary and Minister of Internal and External Affairs\*:** GEORGE A. MCCARTHY.

**Attorney-General and Minister of Legal Affairs\*:** SAMUEL W. BULGIN.

**Financial Secretary\*:** KEN JEFFERSON.

**Cabinet Secretary\*:** ORRETT CONNOR.

**Leader of Government Business and Minister for District Administration, Planning, Agriculture and Housing:** D. KURT TIBBETTS.

**Minister for Health and Human Services:** ANTHONY S. EDEN.

**Minister for Education, Employment Relations, Youth, Sports and Culture:** ALDEN MCLAUGHLIN.

**Minister for Communications, Works and Infrastructure:** V. ARDEN MCLEAN.

**Minister for Tourism, Environment, Investment and Commerce:** CHARLES CLIFFORD.

A District Commissioner, Kenny Ryan, represents the Governor on Cayman Brac and Little Cayman.

\* Appointed by the Governor.

# THE CAYMAN ISLANDS

## LEGISLATIVE ASSEMBLY

**Members:** The Chief Secretary, the Financial Secretary, the Attorney-General, and 15 elected members. The most recent election to the Assembly was on 11 May 2005. In February 1991 a Speaker was elected to preside over the Assembly (despite provision for such a post in the Constitution, the functions of the Speaker had hitherto been assumed by the Governor).

**Speaker:** EDNA MOYLE.

**Leader of Government Business:** D. KURT TIBBETTS.

## GOVERNMENT OFFICES

**Office of the Governor:** Government Administration Bldg, Elgin Ave, George Town, Grand Cayman; tel. 244-2401; fax 945-4131; e-mail staffoff@candw.ky; internet www.gov.ky.

All official government offices and ministries are located in the Government Administration Bldg, Elgin Ave, George Town, Grand Cayman.

## Political Organizations

**People's Democratic Alliance (PDA):** George Town, Grand Cayman; internet www.pda.ky; f. 2005; Leader LINFORD PIERSON.

**People's Progressive Movement (PPM):** POB 10526 APO, Grand Cayman; tel. 945-1776; e-mail ppm@governmentyoucantrust.org; internet www.governmentyoucantrust.org; f. 2002; Leader KURT TIBBETTS; Chair. ANTONY DUCKWORTH.

**United Democratic Party (UDP):** POB 10009, Grand Cayman; e-mail info@udp.ky; internet www.udp.ky; f. 2001; Leader W. MCKEEVA BUSH; Chair. BILLY REID.

## Judicial System

There is a Grand Court of the Islands (with Supreme Court status), a Summary Court, a Youth Court and a Coroner's Court. The Grand Court has jurisdiction in all civil matters, admiralty matters, and in trials on indictment. Appeals lie to the Court of Appeal of the Cayman Islands and beyond that to the Privy Council in the United Kingdom. The Summary Courts deal with criminal and civil matters (up to a certain limit defined by law) and appeals lie to the Grand Court.

**Chief Justice:** ANTHONY SMELLIE.

**President of the Court of Appeal:** EDWARD ZACCA.

**Solicitor-General:** CHERYLL RICHARDS.

**Registrar of the Grand Court of the Islands:** DELENE M. BODDEN, Court's Office, George Town, Grand Cayman; tel. 949-4296; fax 949-9856.

## Religion

### CHRISTIANITY

The oldest-established denominations are (on Grand Cayman) the United Church of Jamaica and Grand Cayman (Presbyterian), and (on Cayman Brac) the Baptist Church. Anglicans are adherents of the Church in the Province of the West Indies (Grand Cayman forms part of the diocese of Jamaica). Within the Roman Catholic Church, the Cayman Islands forms part of the archdiocese of Kingston in Jamaica. Other denominations include the Church of God, Church of God (Full Gospel), Church of Christ, Seventh-day Adventist, Wesleyan Holiness, Jehovah's Witnesses, Church of the Latter Day Saints, Bahá'í and Church of God (Universal). In 1999 there were an estimated 90 churches in the Cayman Islands, including seven churches on Cayman Brac, and a Baptist Church on Little Cayman.

## The Press

**Cayman Islands Journal:** The Compass Centre, Shedden Rd, POB 1365, George Town, Grand Cayman; tel. 949-5111; fax 949-7675; monthly; broadsheet business newspaper; Publr BRIAN UZZELL.

**Cayman Net News:** 85 North Sound Rd, Alissta Towers, POB 10707, Grand Cayman; tel. 946-6060; fax 949-0679; e-mail caymanet@candw.ky; internet www.caymannetnews.com; internet news service; Publr DESMOND SEALES.

**Caymanian Compass:** The Compass Centre, Shedden Rd, POB 1365, George Town, Grand Cayman; tel. 949-5111; fax 949-7675; internet www.caycompass.com; f. 1965; 5 a week; Publr BRIAN UZZELL; circ. 10,000.

**Chamber in Action:** POB 1000, George Town, Grand Cayman; tel. 949-8090; fax 949-0220; e-mail info@caymanchamber.ky; internet www.caymanchamber.ky; f. 1965; monthly; newsletter of the Cayman Islands Chamber of Commerce; Editor WIL PINEAU; circ. 5,000.

**Key to Cayman:** The Compass Centre, Shedden Rd, POB 1365GT; tel. 949-5111; fax 949-2698; e-mail info@cfp.ky; internet www.caymanfreepress.com; 2 a year; free tourist magazine.

**The Executive:** Crewe Rd, George Town, POB 173 GT; tel. 949-5111; fax 949-7033; e-mail cfp@candw.ky; quarterly; circ. 7,500.

**The New Caymanian:** Grand Cayman; tel. 949-7414; fax 949-0036; weekly; Publr and Editor-in-Chief PETER JACKSON.

## Publishers

**Caribbean Publishing Co (Cayman) Ltd:** 1 Paddington Pl., Suite 306, North Sound Way, POB 688, George Town, Grand Cayman; tel. 949-7027; fax 949-8366; internet www.caribbeanwhitepages.com; f. 1978.

**Cayman Free Press Ltd:** The Compass Centre, Shedden Rd, POB 1365, George Town, Grand Cayman; tel. 949-5111; fax 949-7675; e-mail info@cfp.ky; internet www.caymanfreepress.com; f. 1965.

**Progressive Publications Ltd:** Economy Printers Bldg, POB 764, George Town, Grand Cayman; tel. 949-5780; fax 949-7674.

**Tower Marketing and Publishing:** tel. 946-6000; fax 946-6001; e-mail info@tower.com.ky; internet www.tower.com.ky.

## Broadcasting and Communications

### TELECOMMUNICATIONS

**Cable & Wireless (Cayman Islands) Ltd:** Anderson Sq., POB 293, George Town, Grand Cayman; tel. 949-7800; fax 949-7962; e-mail cs@candw.ky; internet www.candw.ky; f. 1966; Cable & Wireless' monopoly over the telecommunications market ended in 2004; Man. Dir TONEY HEART; Gen. Man. TIMOTHY ADAM.

**Digicel:** POB 700, George Town, Grand Cayman; tel. 345-9433; fax 945-1351; e-mail caycustomercare@digicelgroup.com; internet www.digicelcayman.com; f. 2003; owned by an Irish consortium; acquired the operations of Cingular Wireless (formerly those of AT & T Wireless) in the country in 2005 (www.cingular.ky).

**TeleCayman:** 4th Floor, Cayman Corporate Centre, POB 704 GT, Grand Cayman; e-mail gilbert.chalifoux@telecaymen.com; internet www.telecayman.com; Man. GILBERT CHALIFOUX.

### BROADCASTING

#### Radio

**Radio Cayman:** Elgin Ave, POB 1110 GT, George Town, Grand Cayman; tel. 949-7799; fax 949-6536; e-mail radiocym@candw.ky; internet www.gov.ky/radiocayman; started full-time broadcasting 1976; govt-owned commercial radio station; service in English; Dir LOXLEY E. M. BANKS.

**Radio Heaven 97 FM:** POB 31481 SMB, Industrial Park, George Town, Grand Cayman; tel. 945-2797; fax 945-2707; e-mail heaven97@candw.ky; internet www.heaven97.com; f. 1997; Christian broadcasting, music and news; commercial station.

**Radio ICCI-FM:** International College of the Cayman Islands, Newlands, Grand Cayman; tel. 947-1100; fax 947-1210; e-mail icci@candw.ky; f. 1973; educational and cultural; Pres. Dr ELSA M. CUMMINGS.

**Radio Z99.9 FM:** 256 Crewe Rd, Suite 201, Crighton Bldg, Seven Mile Beach, POB 30110, Grand Cayman; tel. 945-1166; fax 945-1006; e-mail info@z99.ky; internet www.z99.ky; Gen. Man. RANDY MERREN.

#### Television

**Cayman Adventist Television Network (CATN/TV):** George Town, Grand Cayman; tel. 949-2739; internet www.tagnet.org/cayman/tv.html; f. 1996; local and international programmes, mainly religious.

**Cayman Christian TV Ltd:** POB 30213, George Town, Grand Cayman; Vice-Pres. FRED RUTTY; relays Christian broadcasting from the Trinity Broadcasting Network (USA).

# THE CAYMAN ISLANDS

**CITN Cayman 27:** POB 55G, Sound Way, George Town, Grand Cayman; tel. 945-2739; fax 945-1373; e-mail citn@Cayman27.com.ky; internet www.cayman27.com.ky; f. 1992 as Cayman International Television Network; 24 hrs daily; local and international news and US entertainment; 10-channel cable service of international programmes by subscription; Mans COLIN WILSON, JOANNE WILSON.

**Weststar TV Ltd:** POB 30563 SMB, Grand Cayman; e-mail weststar@candw.ky; internet www.weststartv.net.ky; Man. ROD HANSEN.

## Finance

(cap. = capital; res = reserves; dep. = deposits; m. = million; brs = branches)

Banking facilities are provided by commercial banks. The islands have become an important centre for 'offshore' companies and trusts. At the end of 2003 there were 68,078 companies registered in the Cayman Islands. In 2005 there were also 438 licensed banks and trusts and some 6,268 registered mutual funds. The islands were well-known as a tax 'haven' because of the absence of any form of direct taxation. At the end of 2003 assets held by banks registered in the Cayman Islands totalled US $1,058,900m.

**Cayman Islands Monetary Authority:** 80 Shedden Rd, POB 10052 APO, Elizabethan Sq., George Town, Grand Cayman; tel. 949-7089; fax 945-2532; e-mail admin@cimoney.com.ky; internet www.cimoney.com.ky; f. 1997; responsible for managing the Territory's currency and reserves and for regulating the financial services sector; cap. CI $7.1m., res CI $8.1m., dep. CI $51.6m. (Dec. 2002); Chair. TIMOTHY RIDLEY; Man. Dir CINDY SCOTLAND.

### PRINCIPAL BANKS AND TRUST COMPANIES

**AALL Trust and Banking Corpn Ltd:** AALL Bldg, POB 1166, George Town, Grand Cayman; tel. 949-5588; fax 949-8265; Chair. ERIC MONSEN; Man. Dir KEVIN DOYLE.

**Ansbacher (Cayman) Ltd:** POB 887, George Town, Grand Cayman; tel. 949-8655; fax 949-7946; e-mail info@ansbacher.com.ky; internet www.ansbacher.com; f. 1971; cap. US $10.0m., res US $3.8m., dep. US $198.3m. (June 2002); Regional Man. Dir and Chair. MICHAEL L. HODGSON.

**Atlantic Security Bank:** POB 10340, George Town, Grand Cayman; f. 1981 as Banco de Crédito del Peru International, name changed as above 1986; Chair. DIONISIO ROMERO; Pres. CARLOS MUÑOZ.

**Julius Baer Bank and Trust Co Ltd:** Windward Bldg 3, Safe Haven Corporate Centre, West Bay Rd, POB 1100, George Town, Grand Cayman; tel. 949-7212; fax 949-0993; internet www.juliusbaer.ch; f. 1974; Man. Dir CHARLES FARRINGTON.

**Banca Unione di Credito (Cayman) Ltd:** POB 10182 APO, Anderson Sq. Bldg, George Town, Grand Cayman; tel. 949-7129; fax 949-2168; internet www.buc.ch; cap. Sw Fr 10.0m., dep. Sw Fr 704.4m., total assets Sw Fr 733.1m. (Dec. 2002); Gen. Man. URS FREI.

**Banco Português do Atlântico:** POB 30124, Grand Cayman; tel. 949-8322; fax 949-7743; e-mail bcpjvic@candw.ky; Gen. Man. HELENA SOARES CARNEIRO.

**Banco Safra (Cayman Islands) Ltd:** c/o Bank of Nova Scotia, POB 501, George Town, Grand Cayman; tel. 949-2001; fax 949-7097; f. 1993; cap. US $60.0, res US $124.5m., dep. US $254.3m. (Dec. 2002).

**BANIF-Banco Internaçional do Funchal (Cayman) Ltd:** Genesis Bldg, 3rd Floor, POB 32338 SMB, George Town, Grand Cayman; tel. 945-8060; fax 945-8069; e-mail banifcay@candw.ky; internet www.banif.pt; Chair. and CEO Dr JOAQUIM FILIPE MARQUES DOS SANTOS; Gen. Man. VALDEMAR B. LOPES.

**Bank of America Trust and Banking Corpn (Cayman) Ltd:** Fort St, POB 1092, George Town, Grand Cayman; tel. 949-7888; fax 949-7883; f. 1999; Man. Dir CHARLES FARRINGTON.

**Bank of Bermuda (Cayman) Ltd:** POB 513, 3rd Floor, British American Tower, George Town, Grand Cayman; tel. 949-9898; fax 949-7959; internet www.bankofbermuda.bm; f. 1968 as a trust; converted to a bank in 1988; total assets US $1,041m. (July 2001); Chair. HENRY B. SMITH; Man. Dir ALLEN BERNARDO.

**Bank of Butterfield International (Cayman) Ltd:** 1 Butterfield Place, 7 Main St, George Town, Grand Cayman; tel. 949-7055; fax 949-7761; e-mail info@butterfieldbank.ky; internet www.butterfieldbank.ky; f. 1967; subsidiary of N. T. Butterfield & Son Ltd, Bermuda; cap. US $16.5m., res US $35.0m., dep. US $1,092.7m. (June 2002); Chair. ALAN THOMPSON; Man. Dir CONOR J. O'DEA; 3 brs.

**Bank of Nova Scotia:** 6 Cardinal Ave, POB 689, George Town, Grand Cayman; tel. 949-7666; fax 949-0020; e-mail scotiaci@candw.ky; internet www.scotiabank.com; also runs trust company; Man. Dir FARRIED SULLIMAN.

**Bank of Novia Scotia Trust Company (Cayman) Ltd:** Scotiabank Bldg, 6 Cardinal Ave, POB 501 GT, George Town, Grand Cayman; tel. 949-2001; fax 949-7097; e-mail cayman@scotiatrust.com; internet www.scotiabank.com; Man. JOHN FLETCHER.

**BankBoston Trust Co (Cayman Islands) Ltd:** The Bank of Nova Scotia Trust Company (Cayman) Ltd, Scotiabank Bldg, 6 Cardinal Ave, POB 501, George Town, Grand Cayman; tel. 949-8066; fax 949-8080; e-mail info@maples.candw.ky; internet www.bankbostoninternational.com; f. 1997.

**Bermuda Trust (Cayman) Ltd:** 5th Floor, Bermuda House, POB 513, George Town, Grand Cayman; tel. 949-9898; fax 949-7959; internet www.bankofbermuda.bm; f. 1968 as Arawak Trust Co; became subsidiary of Bank of Bermuda in 1988; bank and trust services; Chair. JOSEPH JOHNSON; Man. Dir KENNETH GIBBS.

**BFC Bank (Cayman) Ltd:** Trafalgar Pl., POB 1765, George Town, Grand Cayman; tel. 949-8748; fax 949-8749; e-mail bfc@candw.ky; f. 1985; FC Financière de la Cité, Geneva, 99.9%; cap. US $0.8m., res US $3.7m., dep. US $33.5m. (March 2001); Chair. SIMON C. TAY; Resident Man. CHERRYLEE BUSH.

**Caledonian Bank and Trust Ltd:** POB 1043, George Town, Grand Cayman; tel. 949-0050; fax 949-8062; e-mail info@caledonian.com; internet www.caledonian.com; f. 1970; Chair. WILLIAM S. WALKER; Man. Dir DAVID S. SARGISON.

**Cayman National Bank Ltd:** Cayman National Bank Bldg, 4th Floor, 200 Elgin Ave, POB 1097, George Town, Grand Cayman; tel. 949-4655; fax 949-7506; e-mail cnb@caymannational.com; internet www.caymannational.com; f. 1974; subsidiary of Cayman National Corpn; cap. CI $2.4m., res CI $41.4m., dep. CI $466.0m. (Sept. 2002); Chair. ERIC J. CRUTCHLEY; Pres. DAVID J. MCCONNEY; 6 brs.

**Coutts (Cayman) Ltd:** Coutts House, 1446 West Bay Rd, POB 707, George Town, Grand Cayman; tel. 945-4777; fax 945-4799; internet www.coutts.com; f. 1967; fmrly NatWest International Trust Corpn (Cayman) Ltd; Chair. GERALD C. WILLIAMS; Man. Dir ANDREW GALLOWAY.

**Deutsche Bank (Cayman) Ltd:** Elizabethan Sq., POB 1984, George Town, Grand Cayman; tel. 949-8244; fax 949-8178; e-mail dmg-cay@candw.ky; internet www.dboffshore.com; f. 1983 as Morgan Grenfell (Cayman) Ltd; name changed to Deutsche Morgan Grenfell (Cayman) Ltd in 1996; name changed as above in 1998; cap. US $5.0m., res US $20.3m., dep. US $129.3m. (Dec. 1998); Chair. HANS JUERGEN KOCH; Branch Man. JANET HISLOP.

**Deutsche Bank International Trust Co (Cayman) Ltd:** POB 1967, George Town, Grand Cayman; tel. 949-8244; fax 949-7866; f. 1999.

**Deutsche Girozentrale Overseas Ltd:** POB 852, George Town, Grand Cayman; tel. 914-1066; fax 914-4060.

**Fidelity Bank (Cayman) Ltd:** POB 914, George Town, Grand Cayman; tel. 949-7822; fax 949-6064; Pres. and CEO BRETT HILL.

**FirstCaribbean International Bank Ltd:** POB 68, George Town, Grand Cayman; tel. 949-7300; fax 815-2292; internet www.firstcaribbeanbank.com; f. 2002 following merger of Caribbean operations of Barclays Bank PLC and CIBC; Exec. Chair. MICHAEL MANSOOR; CEO CHARLES PINK.

**Fortis Bank (Cayman) Ltd:** POB 2003, George Town, Grand Cayman; tel. 949-7942; fax 949-8340; e-mail phil.brown@ky.fortisbank.com; internet www.fortis.com; f. 1984 as Pierson Heldring & Pierson (Cayman) Ltd; name changed to Mees Pierson (Cayman) Ltd in 1993; present name adopted in June 2000; Man. Dir ROGER HANSON.

**HSBC Financial Services (Cayman) Ltd:** Strathvale House, 2nd Floor, 90 North Church St, POB 1109, George Town, Grand Cayman; tel. 949-7755; fax 949-7634; e-mail hfsc.info@ky.hsbc.com; internet www.hsbc.ky; f. 1982; Dirs TOM CLARK, DAVID A. WHITEFIELD.

**IBJ Whitehall Bank and Trust Co:** West Wind Bldg, POB 1040, George Town, Grand Cayman; tel. 949-2849; fax 949-5409; Man. ROGER HEALY.

**Intesa Bank Overseas Ltd:** c/o Coutts (Cayman) Ltd, Coutts House, West Bay Rd, POB 707, George Town, Grand Cayman; tel. 945-4777; fax 945-4799; f. 1994 as Ambroveneto International Bank; name changed as above in March 1999; cap. US $10.0m., total assets US $1,339.8m. (Dec. 1999); Chair. FRANCESCO DE VECCHI; Man. Dirs RICHARD AUSTIN, ANDREW GALLOWAY.

**LGT Bank in Liechtenstein (Cayman) Ltd:** UBS House, POB 852, George Town, Grand Cayman; tel. 949-7676; fax 949-8512; internet www.lgt-bank-in-liechtenstein.com.

**Lloyds TSB Bank and Trust (Cayman) Ltd:** Grand Cayman; tel. 949-7854; fax 949-0090; Man. ROGER C. BARKER.

**Mercury Bank and Trust Ltd:** POB 2424, George Town, Grand Cayman; tel. 949-0800; fax 949-0295; Man. VOLKER MERGENTHALER.

# THE CAYMAN ISLANDS

**Merrill Lynch Bank and Trust Co (Cayman) Ltd:** POB 1164, George Town, Grand Cayman; tel. 949-8206; fax 949-8895; internet www.ml.com.

**Royal Bank of Canada:** 24 Shedden Rd, POB 245, Grand Cayman; tel. 949-4600; fax 949-7396; internet www.royalbank.com; Man. HARRY C. CHISHOLM.

**Royal Bank of Canada Trust Co (Cayman) Ltd:** 24 Shedden Rd, POB 1586 GT, Grand Cayman; tel. 949-9107; fax 949-5777; internet www.rbcprivatebanking.com/cayman-islands.html; Man. Dir RALPH AWREY.

**UBS (Cayman Islands) Ltd:** UBS House, 227 Elgin Ave, POB 852, George Town, Grand Cayman; tel. 914-1060; fax 914-4060; internet www.ubs.com/cayman-funds; CEO WALTER EGGENSCHWILER.

### Development Bank

**Cayman Islands Development Bank:** Cayman Financial Centre, 36B Dr Roy's Dr., POB 2576, George Town; tel. 949-7511; fax 949-6168; e-mail cidb@gov.ky; f. 2002; replaced the Housing Devt Corpn and the Agricultural and Industrial Devt Bd; Devt Finance Institution Gen. Man. ANGELA J. MILLER.

### Banking Association

**Cayman Islands Bankers' Association:** Macdonald Sq., Fort St, POB 676, George Town, Grand Cayman; tel. 949-0330; fax 945-1448; e-mail ciba@candw.ky; Pres. TIMOTHY GODBER.

### STOCK EXCHANGE

**Cayman Islands Stock Exchange (CSX):** 4th Floor, Elizabethan Sq., POB 2408, George Town, Grand Cayman; tel. 945-6060; fax 945-6061; e-mail csx@csx.com.ky; internet www.csx.com.ky; f. 1996; 872 companies listed, including 788 mutual funds (Feb. 2005); CEO VALIA THEODORAKI.

### INSURANCE

Several foreign companies have agents in the islands. A total of 672 captive insurance companies were registered at the end of 2003. In particular, the islands are a leading international market for health insurance. Local companies include the following:

**British Caymanian Insurance Agency Ltd:** Elizabethan Sq., POB 74 GT, Grand Cayman; tel. 949-8699; fax 949-8411.

**Caribbean Home Insurance:** POB 931, Commerce House 7, Genesis Close, George Town, Grand Cayman; tel. 949-7788; fax 949-8422.

**Cayman General Insurance Co Ltd:** Cayman National Bank Bldg, 200 Elgin Ave, POB 2171, George Town, Grand Cayman; tel. 949-7028; fax 949-7457; e-mail cgi@caymannational.com; internet www.caymannational.com; Chair. BENSON O. EBANKS; Pres. DANNY A. SCOTT.

**Cayman Insurance Centre:** POB 10056, Cayman Business Park; tel. 948-1382; internet www.cic.com.ky; Pres. LINDA CHAPMAN-KY.

**Cayman Islands National Insurance Co (CINICO):** Elizabethan Sq., Phase 3, 1st Floor, POB 512 GT, Grand Cayman; tel. 949-8101; fax 949-8226; internet www.cinico.ky; CEO and Pres. RON SULISZ.

**Island Heritage Insurance Co Ltd:** POB GT, Grand Cayman; tel. 949-7280; fax 945-6765; e-mail info@islandheritage.com.ky; internet www.island-heritage.com; Chair. ROBERT CLEMENTS.

**Global Life Assurance Co Ltd:** Global House, POB 1087, North Church St, Grand Cayman; tel. 949-8211; fax 949-8262; f. 1992; Man. WINSOME RUDDOCK.

**Sagicor Life of the Cayman Islands Ltd:** 198 North Church St, George Town, Grand Cayman; tel. 949-8211; fax 949-8262; e-mail global@candw.ky; internet www.themutual.com; f. in 2004 by merger between Global Life and Capital Life; Man. MICHEL TRUMBACH.

## Trade and Industry

### CHAMBER OF COMMERCE

**Cayman Islands Chamber of Commerce:** Macdonald Sq., Fort St, POB 1000, George Town, Grand Cayman; tel. 949-8090; fax 949-0220; e-mail info@caymanchamber.ky; internet www.caymanchamber.ky; f. 1965; Pres. JOSEPH HEW; Chief Exec. WIL PINEAU; 730 local mems.

### TRADE ASSOCIATION

**Cayman Islands Financial Services Association (CIFSA):** POB 11048, Grand Cayman; tel. 946-6000; fax 946-6001; e-mail info@caymanfinances.com; internet www.caymanfinances.com; f. 2003; Dir EDUARDO D'ANGELO P. SILVA.

### EMPLOYERS' ORGANIZATION

**Labour Office:** 4th Floor, Tower Bldg; tel. 949-0941; fax 949-6057; Dir DALE M. BANKS.

The Cayman Islands have had a labour law since 1942, but only three trade unions have been registered.

### UTILITIES

#### Electricity

**Caribbean Utilities Co. Ltd (CUC):** Corporate HQ & Plant, North Sound Rd, POB 38, George Town, Grand Cayman; tel. 949-5200; fax 949-4621; e-mail info@cuc.ky; internet www.cuc-cayman.com; Pres. and CEO PETER A. THOMSON; Chair. DAVID RITCH.

**Cayman Brac Power and Light Co. Ltd:** Stake Bay Point, POB 95, Stake Bay, Cayman Brac; tel. 948-2224; fax 948-2204.

**West Indies Power Corpn Ltd:** CUC Corporate Centre, North Sound Rd, POB 38, George Town, Grand Cayman; tel. 949-2250.

#### Water

**Cayman Islands Water Authority:** 13G Red Gate Rd, POB 1104, George Town, Grand Cayman; tel. 949-6352; fax 949-0094; e-mail wac@candw.ky.

**Consolidated Water Co Ltd (CWCO):** Regatta Business Park, Windward 3, 4th Fl., POB 1114, George Town, Grand Cayman; tel. 945-4277; fax 949-2957; e-mail info@cwco.com; internet www.cwco.com; Pres. and CEO FREDERICK W. MCTAGGART; Chair. JEFFREY M. PARKER.

## Transport

### ROADS

There are some 406 km (252 miles) of motorable roads, of which 304 km (189 miles) are surfaced with tarmac. The road network connects all districts on Grand Cayman and Cayman Brac (which has 76 km (47 miles) of motorable road), and there are 27 miles of motorable road on Little Cayman (of which about 11 miles are paved).

### SHIPPING

George Town is the principal port and a new port facility was opened in July 1977. Cruise liners, container ships and smaller cargo vessels ply between the Cayman Islands, Florida, Jamaica and Costa Rica. There is no cruise-ship dock in the Cayman Islands. Ships anchor off George Town and ferry passengers ashore to the North or South Dock Terminals in George Town. In February 2002 the Florida Caribbean Cruise Ship Association agreed to provide up to US $10m. for improvements to the port's cruise-ship facilities. In 1993 the Government limited the number of cruise-ship passengers to 6,000 per day. The port of Cayman Brac is Creek; there are limited facilities on Little Cayman. In December 2004 the shipping register comprised 156 vessels totalling 2,608,796 grt.

**Port Authority of the Cayman Islands:** Harbour Dr., POB 1358, George Town, Grand Cayman; tel. 949-2055; fax 949-5820; e-mail info@caymanport.com; internet www.caymanport.com; Port Dir PAUL HURLSTON.

**Cayman Islands Shipping Registry:** 3rd Floor, Kirk House, 22 Albert Panton St, POB 2256, George Town, Grand Cayman; tel. 949-8831; fax 949-8849; e-mail cisr@candw.ky; internet www.caymarad.org; Dir JOEL WALTON.

**Cayman Freight Shipping Ltd:** Mirco Commercial Centre, 2nd Floor, Industrial Park, POB 1372, George Town, Grand Cayman; tel. 949-4977; fax 949-8402; e-mail cfssl@candw.ky; internet www.seaboardmarinecayman.ky; Man. Dir ROBERT FOSTER.

**Thompson Shipping Co Ltd:** POB 188, Terminal Eastern Ave, George Town, Grand Cayman; tel. 949-8044; fax 949-8349; f. 1977.

### CIVIL AVIATION

There are two international airports in the Territory: Owen Roberts International Airport, 3.5 km (2 miles) from George Town, and Gerrard Smith International Airport on Cayman Brac. Both are capable of handling jet-engined aircraft. Edward Bodden Airport on Little Cayman can cater for light aircraft. Several scheduled carriers serve the islands.

**Civil Aviation Authority of the Cayman Islands:** Unit 4, Cayman Grand Harbour, POB 10277 APO, George Town, Grand Cayman; tel. 949-7811; fax 949-0761; e-mail civil.aviation@

caacayman.com; internet www.caacayman.com; f. 1987; Dir-Gen. RICHARD SMITH.

**Cayman Airways Ltd:** 233 Owen Roberts Dr., POB 10092 APO, Grand Cayman; tel. 949-8200; fax 949-7607; e-mail customerrelations@caymanairways.net; internet www.caymanairways.com; f. 1968; wholly govt-owned since 1977; operates local services and scheduled flights to Jamaica, Honduras and the USA; Chair. ROY MCTAGGART; CEO MIKE ADAM.

**Island Air:** POB 2433, Airport Rd, George Town, Grand Cayman; tel. 949-5252; fax 949-7044; e-mail iair@candw.ky; operates daily scheduled services between Grand Cayman, Cayman Brac and Little Cayman.

## Tourism

The Cayman Islands are a major tourist destination, the majority of visitors coming from North America. The tourism industry was badly affected by the damage caused by 'Hurricane Ivan' in September 2004, and a major reconstruction effort was in progress in early 2005. The beaches and opportunities for diving in the offshore reefs form the main attraction for most tourists. Major celebrations include Pirates' Week in October and the costume festivals on Grand Cayman (Batabano), at the end of April, and, one week later, on Cayman Brac (Brachanal). In 1995 there were an estimated 7,648 hotel beds. In 2003 there were 293,500 arrivals by air and 1,819,000 cruise visitors. In 2002 the tourism industry earned an estimated US $607.0m.

**Cayman Islands Department of Tourism:** Cricket Sq., POB 67, George Town, Grand Cayman; tel. 949-0623; fax 949-4053; internet www.caymanislands.ky; f. 1965; Dir PILAR BUSH.

**Cayman Islands Tourism Association (CITA):** 73 Lawrence Blvd, Islander Complex, POB 31086 SMB, Grand Cayman; tel. and fax 949-8522; fax 946-8522; e-mail info@cita.ky; internet www.cita.ky; f. 2001 as a result of the amalgamation of the Cayman Tourism Alliance and the Cayman Islands Hotel and Condominium Asscn; Pres. MARK BASTIS.

**Sister Islands Tourism Association:** Stake Bay, POB 187, Cayman Brac; tel. and fax 948-1345; e-mail sita@candw.ky; internet www.sisterislands.com; Pres. MAX HILLIER.

## Defence

The United Kingdom is responsible for the defence of the Cayman Islands.

## Education

Schooling is compulsory for children between the ages of five and 15 years. It is provided free in 10 government-run primary schools, and there are also three state secondary schools, as well as six church-sponsored schools (five of which offer secondary as well as primary education). Primary education, from five years of age, lasts for six years. Secondary education is for seven years. Government expenditure on education in 1994 was CI $31.5m. (20.9% of total spending). Some CI $17.7m. (9.9% of total expenditure) was allocated for education in 1995.

# CHILE
## Geography

### PHYSICAL FEATURES

The Republic of Chile occupies the narrow strip of territory between the Andes and the Pacific in the southern part of South America. The country is 4,329 km (2,688 miles) in length, but never more than 180 km in width, and its border along the mountainous Continental Divide is primarily with Argentina (5,150 km in extent). The north of the country (Atacama was gained from Bolivia in the 1880s, although that country still pursues a claim for a sovereign corridor to the Pacific through Chile) has a western frontier with land-locked Bolivia (861 km of border) and a short northern one with Peru (160 km). Chile has an issue with Peru over maritime economic boundaries, otherwise its only other territorial dispute concerns its claim in Antarctica (which conflicts with Argentine and British claims). Chile has a total area of 756,096 sq km (291,930 sq miles—excluding the 1.25m. sq km of the Antarctic claim).

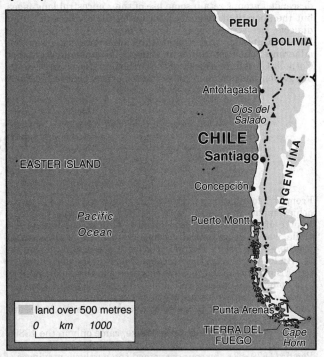

Chile includes many of the clustered islands into which the continent disintegrates in the south-west and in the eastward flick of its tail, as well as an often deeply indented shore, so that it has a total coastline of 6,435 km. The country also includes a number of islands in the Pacific proper—notably Isla de Pascua (Easter Island or Rapa Nui—the most distant, at 3,790 km west of the north-central mainland), but also Sala y Gómez, Islas de los Desventurados and Archipiélago Juan Fernández. The first is famous for the isolated Polynesian civilization that flourished here long before the arrival of Europeans and erected the great stone statues (moai), and the last includes one island named for a shipwrecked sailor (Alexander Selkirk) and another for the fictional character he inspired (Robinson Crusoe).

Chile itself consists of the narrow lands west of the Andes, from the Atacama Desert in the north to the Patagonian icefields. It is a geologically unstable area, with earthquakes common and volcanoes active in the mountains, and the coasts are sometimes struck by tsunamis. The arid north is dominated by some of Chile's highest mountains, where the main Andean cordillera is at its widest and includes broad plateaux and soaring peaks (including the country's highest, Ojos del Salado—6,880 m or 22,580 ft—on the border with Argentina). Here the country forms the Atacama Desert, reputedly the driest in the world, but also rich in minerals and, especially, nitrates. Central Chile, with its more Mediterranean-type climate, sees a clear definition between the main line of the Andes and the lower coastal range, with an intervening Central Valley. The Andes are lower and narrower here, pierced by the best passes to the east, and the coast has many good harbours. Between is the Central Valley, some 40–80 km in width, with fertile, deep alluvial soils, particularly between the Aconcagua and Bíobío rivers (from just north of Santiago, the capital, south to Concepción). South of Puerto Montt the interior valley or plateau between the Andean and coastal ranges disappears beneath the sea, the coastal mountains continuing as a mass of islands off shore, itself untidy with fjords and inlets. Archipelagic Chile extends from Chiloé all the way down to Tierra del Fuego (of which Argentina has the eastern part) and the Isla Hornos, with its famous Cabo de Hornos (Cape Horn). On the mainland, the Andes are lower than in the north (seldom more than 1,800 m), the lowlands forming the desolate, undulating plains of Patagonia that sweep down to the tip of the continent and over into Argentina. Apart from areas of permanent ice (the Patagonian ice cap is reckoned to be the largest after those of Antarctica and Greenland), this is a steppe country of coarse grasses and shrubs, while the north of Chile is a true desert, with very little vegetation at all. The vegetation of the central regions ranges, however, from cacti and scrub in the north to some dense rainforest south of Valdivia. Flora and fauna is not as rich in Chile as it is beyond the high Andes, although the coastal waters are teeming with life. The rivers, all short and steep (vital to irrigation and electricity generation), have few fish (mainly the introduced trout) and few of the larger birds common on the continent are found here. However, there is a Chilean pine (with edible nuts) and animal species include the llama, the alpaca, the chinchilla, the Andean wolf, the puma, the huemal (a large deer) and the pudu (a small deer).

### CLIMATE

The climate is very varied across such a long country, but is generally moderated by oceanic influences. The northern region is one of the driest areas of the world, although the temperatures are moderated by the cold offshore Humboldt or Peru Current. Average temperatures at Antofagasta range from 12°–16°C (54°–61°F) in July up to 18°–23°C (64°–73°F) in January, but rainfall is only 2 mm (barely 0.1 ins) per year. Rainfall increases southwards, reaching 360 mm (14 ins) at Santiago in the Central Valley, where it falls mainly in the winter months of May–August. The winters, apart from being fairly wet, are mild, and the summers are cool. The average temperatures range from 3°–15°C (37°–59°F) in July to 12°–29°C (54°–84°F) in January. In the south it is a temperate marine climate, cooler and still wetter, with rain falling fairly evenly throughout the year and reaching levels of, at most, about 5,000 mm (much falling as snow) near the Straits of Magellan. Strong winds and cyclonic storms are common. The annual average temperature at Punta Arenas is 7°C (45°F).

### POPULATION

The early, mainly Spanish, settlers from Europe intermarried with the local Amerindians (predominantly the Araucanian Mapuche tribes, who boast of never having succumbed to the Quechua-speaking Incas) and mixed-race Mestizos now comprise well over 90% of the population. Subsequent European immigration was not as important to Chile as to many other South American countries, although immigration by Germans to the Valdivia–Puerto Montt area was influential, and other groups to make a significant contribution were from the United Kingdom, France, Italy, Switzerland, Austria and the former Yugoslavia. Only about 2% of the population is now unmixed

European and there is somewhere between 3% and 6% unmixed Amerindians. The latter survive mainly in the south, the Mapuche, with some Aymará in the north (5%) and the Polynesian-descended Rapa Nui (2%) on Easter Island. There are a few other groups, including the aboriginal peoples of Patagonia and Tierra del Fuego. The official language, however, and that in almost total use is Spanish. Some of the other European-descended groups retain some use of their ancestral languages, but the aboriginal languages, including the seven Araucanian dialects, are struggling to survive. Most of the population is Roman Catholic, with about three-quarters acknowledging adherence, but Protestants (in Chile defined as all non-Roman Catholic Christians except for the Orthodox, Mormons, Seventh-day Adventists and Jehovah's Witnesses) now claim 12%. There is a relatively high figure of 6% given for those expressing no religious affiliation.

According to official estimates, the total population was 16.1m. in mid-2004. Some 90% of the population live in the central region, with 86% defined as urbanized (1992 figures). About one-third of the total population lived in and around the capital city alone. Greater Santiago had a population of 4.7m. in 2002, with the next largest city being just to the south-west of the capital proper—Puente Alto (0.5m.). Other major population centres are Antofagasta, in the north, the neighbouring cities of Viña del Mar and Valparaíso, on the coast to the north-west of Santiago, and Talcahuano and Concepción, also close to each other, mid-way down the central region. The country is divided into 13 regions.

# History

## PHILIP J. O'BRIEN

From the foundation of its capital, Santiago, in 1541, to the achievement of independence in 1818, Chile was the most remote of Spain's American colonies. Throughout that period Chile was a geographically compact territory with 'natural' frontiers—the Pacific Ocean to the west, the mountain chain of the Andes to the east, the Atacama Desert to the north and the River Bíobío to the south. Chile was then one-third of its present size, some 1,000 km long and at no point more than 160 km wide. It was a rich, agricultural region with a Mediterranean climate and natural irrigation provided by rivers rising in the Andes and flowing to the Pacific.

Chile's expansion, to its present size, came only in the 1880s. Following its victory over Bolivia and Peru, in the War of the Pacific (1879–83), the country gained the northern provinces of Antofagasta (from Bolivia) and Tarapacá (from Peru). Its territory further expanded with the subjugation of the Araucanian Indians (Amerindians) south of the Bíobío, which had been a military frontier throughout the colonial period.

The European minority dominated society and the economy as landowners, mine-owners and traders, while the largely illiterate mestizo masses obeyed them. Independence from Spain meant little change socially and economically. However, the compact nature of the colony and the solidarity of its upper class gave Chile an early sense of nationality and, after a short period of turbulence, constitutional and political stability. The Constitution of 1833 (which lasted until 1925) conferred extensive powers on the President; the military played little part in politics and a multi-party system emerged, in which power changed hands without recourse to arms. Only three attempts at revolution occurred in the 19th century, and only one of these, in 1891, succeeded.

### THE DEVELOPMENT OF PARTY POLITICS

However, evolutionary progress was fitful. The population grew and cities expanded, while politics remained a largely aristocratic pursuit. This caused new social pressures from the lower classes which began to challenge the traditional order. Moreover, the benefits of economic growth in the 19th and 20th centuries, based primarily on exports of wheat, silver, copper and, between 1880 and 1920, nitrates, were unevenly distributed. Trade unions were established and new political parties emerged, alongside the older Conservative, Liberal and Radical parties. The first important left-wing party was founded in 1912, becoming the Communist Party in 1922. Government repression of strikes, particularly in the cities and the nitrate works of the north, by the 1920s created a divided nation, ruled by an indifferent political class.

In 1920 Arturo Alessandri Palma campaigned for the presidency, on a programme of economic and social reform, winning the election narrowly. However, his programme was obstructed by congressional opposition. The military, unpaid for months like all other public servants, intervened to break Chile's legalistic tradition in 1924. Alessandri left the country and a start was made to secure better administration, with the Congreso (Congress) being forced to enact some reforms. A second military coup in 1925, organized by junior officers who suspected the high command of delaying the programme of reforms, resulted in Alessandri's return to power. His second period of government lasted only a few months, during which time, however, a new Constitution was introduced. Alessandri was ousted by the Minister of War, Col Carlos Ibáñez del Campo, who became President in 1927.

During Ibáñez's four years of government there was a sizeable expansion of public works, funded by foreign loans and administrative reform. However, the Great Depression of 1929–31 precipitated his downfall. Several short-lived, contrasting Governments, including a 'Socialist Republic', followed rapidly until in 1932 new elections returned Alessandri to the presidency. This six-year term saw an austere but effective economic recovery and the return to the constitutional tradition.

The Great Depression, however, had created other parties, notably the Socialist Party in 1933 and a Falange Party, which later became the Partido Demócrata Cristiano (PDC—Christian Democratic Party). Thus, an increasing number of parties competed for votes and rarely could one party rule alone, thereby necessitating compromise, at times at the cost of efficiency. Moreover, fundamental social problems remained. The previous system had some achievements, particularly in education. However, as late as the 1940s, 80% of the arable land was owned by 3% of landholders, and the agrarian population, which comprised about one-third of the total, remained unorganized until the late 1960s.

The middle-class radical Frente Popular dominated government from 1938 to 1952, firstly in coalition with the left-wing parties and then with a variety of others. State intervention in the economy grew markedly in this period, though the shifting alliances or coalitions in government, culminating in the outlawing of the Communist Party in 1948, created public disillusion with politics. In 1952 Ibáñez was re-elected to the presidency. However, he failed to achieve stability in government or address the increasing economic problems. In 1958 Arturo Alessandri's son, Jorge, a prominent right-wing businessman, was elected President, beating the Marxist Salvador Allende Gossens by a narrow margin. Alessandri governed with a pragmatically formed series of Cabinets. Meanwhile, a new central political force had been growing, to challenge the dominance of the Radical party. The PDC, led by Eduardo Frei Montalva, had an extensive reform programme, which held increasing appeal. In 1964 Frei won the presidency with over 50% of the votes cast, the first clear public mandate for decades. One year later, the Christian Democrats gained control of the lower house of the legislature, in congressional elections, the first single party to do so in over 100 years. However, they did not have a majority in the Senado (Senate), which prevented much of Frei's reformist legislation. Nevertheless, in housing, health and education reform, much was achieved; an agrarian reform law was finally enacted and the Government secured a 50% stake in the largely US-owned major copper mines.

## THE ELECTION OF THE ALLENDE GOVERNMENT

Obstructed in the Senado and hindered by a drought in the late 1960s, Frei's Government could not fulfil its electoral promises. On the left, the Socialist and Communist parties united with the Radicals in a coalition (Popular Unity), with Salvador Allende as its presidential candidate for the 1970 election. The Christian Democrats chose Radomiro Tomic, on the left of the party, while the right-wing National Party backed Jorge Alessandri.

Allende won a close election, beating Alessandri into second place by only 39,000 of the votes cast, in an electorate of 3.5m. Like Frei's 'revolution in liberty', Allende's 'Chilean road to socialism' was intended to preserve the constitutional, democratic system while executing a more radical programme of economic and social change than any hitherto attempted.

An extensive programme of nationalization of monopolies, such as the copper industry, the domestic banking system and other sectors, as well as exchange and commodity controls, was soon implemented. The Government extended its relations with socialist (Communist) states, inviting the Cuban leader, Fidel Castro Ruz, to visit Chile in November 1971. Using both a long-forgotten law of the 'Socialist Republic' of 1932, which gave the state power to take over industries deemed to have 'failed to supply the people', and the state development corporation's (Corporación de Fomento de la Producción—CORFO) purchase of private shares, many banks and other businesses passed into state control. In addition, revolutionary groups pressured the Government to introduce more radical policies by inciting workers to seize factories and landed estates, thereby disrupting production. While Allende condemned such activities, the opposition accused the Government of doing nothing to stop them. Political polarization developed rapidly and violence on the streets increased.

The economy declined dramatically, primarily because of the internal conflicts, but also because of the reduction in foreign aid and the 'informal blockade' imposed by the USA. Inflation had increased rapidly, and there were widespread shortages. The economy was in crisis. Allende's attempt to pacify the opposition by bringing members of the military into his Government failed. The country moved towards a political crisis.

## MILITARY DICTATORSHIP AND PINOCHET

Chile's internal disarray was resolved on 11 September 1973 by the intervention of the armed services, including the paramilitary police, in a combined operation to overthrow the Government. President Allende died in an attack on the presidential palace. Many of his supporters were killed, arrested or exiled. All political activity was suspended. In addition, the Congreso was dissolved and censorship introduced. The constitutional, democratic system of government, which had lasted for so long, was replaced by a military regime, dominated by the head of the Army, Gen. Augusto Pinochet Ugarte. His policy was to extirpate Marxism from Chile and to allow only a slow return to a competitive party system. He also aimed not only to repair the economy but, indeed, to reverse the patterns of the past, by severely reducing the role of the state in economic management.

The regime was authoritarian and brutal in its repression of discontent. It maintained a considerable measure of support from sectors of the population that preferred strict order to the threat of chaos. Restrictions on the non-Marxist parties meant that the most important, but cautious, critic of the Government, particularly in the area of human rights, became the Roman Catholic Church. Institutionally, Pinochet's plan was contained in a new Constitution, implemented in 1981, which had been endorsed by two-thirds of the electorate in a referendum held in the previous year and which gave the President wide-reaching powers.

## ECONOMIC RECESSION AND POLITICAL OPPOSITION

The military regime followed an economic stabilization plan, which, however, had substantial social costs, in terms of jobs and the living standards of the majority of the population. The plan reduced inflation and import tariffs, and diversified Chile's exports. By 1981 Chile was one of the most free-enterprise, open economies in the world. However, falling copper prices in a world recession, a heavy concentration of economic power with too little supervision, the massive growth of foreign debt and a policy of maintaining a fixed exchange rate led to a recession in Chile, in 1982 and 1983.

Internationally, the Pinochet regime was always isolated because of its record on human rights. However, in 1983 Chile did succeed, with the return of democracy to Argentina, in resolving a protracted territorial dispute over three islands in the Beagle Channel. The dispute, which had continued for a century, almost led the countries to war in 1978, but, with the assistance of papal mediation, an agreement was reached in October 1984.

The reduction in economic prosperity, coupled with social and political discontent, led to political protests. Under the leadership of the Confederación de Trabajadores del Cobre (Federation of Copper Workers), the trade unions formed the Comando Nacional de Trabajadores (National Workers' Command), and organized national protest days, with the aim of forcing Pinochet to resign. The initial protests, bringing together trade unions, students and shanty-town dwellers, as well as the opposition political parties, surpassed expectations. In the August 1983 protest Pinochet declared a state of siege and ordered 18,000 troops on to the streets of Santiago. The soldiers killed 32 protesters and arrested 1,200. The protest forced Pinochet to appoint a civilian as Minister of the Interior, to negotiate with the opposition forces. The opposition, however, was divided, with the centre parties in a new coalition, the Alianza Democrática (Democratic Alliance), while the left-wing parties formed the Movimiento Democrático Popular (Popular Democratic Movement). The opposition failed to convert general public discontent into effective political support. Pinochet was determined to repress the protests and not concede to the opposition's demands for congressional elections to be held before 1989.

Despite the Government's attempts, including assassinations, to eradicate internal opposition, the campaign of public protests continued. In September 1986, after an attempt to assassinate Pinochet by the Frente Patriótico Manuel Rodríguez (FPMR), the guerrilla group of the banned Communist Party, a highly repressive state of siege was reimposed. However, in January 1987, in anticipation of a visit by the head of the Roman Catholic Church, Pope John Paul II, Pinochet ended the state of siege and began to allow some political exiles to return home.

## THE DEFEAT OF PINOCHET

In March 1987 the Government promulgated a law whereby non-Marxist political parties were allowed to register officially to organize for the plebiscite to be held on 5 October 1988. The plebiscite gave voters a simple option: a 'yes' vote would confirm as President the sole candidate nominated by the Pinochet regime; a 'no' vote would oblige the regime to hold open elections within one year. Initially, 13 parties and political groups campaigned for a 'no' vote and in June 1988 the Communist Party joined this campaign. Following US pressure, the Government lifted the state of emergency in preparation for the plebiscite. Despite internal disagreements over Pinochet's candidacy, the junta nominated him in August as the regime's candidate.

The plebiscite resulted in a defeat for Pinochet: 55% voted 'no', 43% 'yes'. Under the 1981 Constitution, Pinochet was obliged to hold elections for the presidency and for the two houses of the Congreso Nacional (though part of the Senado was nominated by the President). The elections were announced for December 1989, with the newly elected President taking office on 11 March 1990.

After negotiations with Concertación de Partidos por la Democracia (CPD—Co-ordination of the Parties for Democracy), a 17-party grouping of the opposition and the right-wing Partido Renovación Nacional (RN—National Renovation), Pinochet agreed to 54 reforms to the 1981 Constitution, which were approved in a further national plebiscite. These included the reduction of the next President's term to four years, the redrafting of Article Eight (which banned Marxist parties), the introduction of military-civilian parity on a National Security Council (with its veto powers curtailed), an increase in the

## THE GOVERNMENT OF PRESIDENT AYLWIN

In the elections of 15 December 1989, Patricio Aylwin Azócar, the leader of the Christian Democrats and head of the CPD coalition, polled 55% of the votes cast, Hernán Büchi, the right-wing candidate, 29% and Francisco Javier Errázuriz, an independent, 15%. The PDC emerged as the main party, winning 13 of the 38 elected senatorial seats and 38 of the 120 seats in the lower chamber. The CPD alliance, dominated by the Christian Democrats and the Socialists, as a whole won a clear majority in the Cámara de Diputados (72 seats out of 120), but faced a potential opposition majority in the Senado, where it held 22 seats, as opposed to the right wing's 16 seats, which could combine with the nine (later reduced to eight) non-elected senators.

The new Government had three main challenges. The first was what to do about past human rights abuses. The second was how to deal with Pinochet and the Army. According to the Constitution implemented in 1981, Gen. Pinochet was allowed to remain in office for eight years as Commander-in-Chief of the Army after he relinquished the presidency in 1989. The third was how to consolidate democracy and maintain macroeconomic balances, while responding to the demands of the poorer sectors of society.

## THE RETTIG REPORT

On April 24 1990 President Aylwin signed a decree creating the Commission for Truth and Reconciliation (Comisión por la Verdad y la Reconciliación), chaired by former Senator Raul Rettig, to investigate the most serious allegations of human rights violations between September 1973 and March 1990. The Rettig Report, on human rights abuses during the Pinochet regime, was published on 5 March 1991. It documented the deaths of 2,279 people, nearly all caused by agents of the state. The Report concluded that the Pinochet regime had executed a 'systematic policy of extermination', that the responsibility reached as high as Gen. Pinochet and that the judiciary not only did not attempt to prevent human rights abuses, but allowed military courts to pass death sentences. This added to pressure for Pinochet's resignation, although, following the military's declaration of loyalty to him, Pinochet rejected the Report's conclusions.

On 1 April 1991 Senator Jaime Guzmán, a leading supporter of the Pinochet regime, was murdered. Guzmán's assassination prevented any follow up to the Rettig Report. Aylwin declared that Chileans must not think about justice for those guilty of human rights abuses, because that could destabilize democracy. Compensation was given to families of the victims, but little effort was made to bring the guilty to justice, with the exception of those responsible for the murder in Washington, DC, USA, of the former Chilean ambassador to the USA, Orlando Letelier, and his US associate, Ronnie Moffat, and the sentencing of three police-officers to life imprisonment for the murder of three teachers in 1985.

## CONSOLIDATING DEMOCRACY

In 1992 the Government declared its two main aims to be the 'consolidation of democracy' and 'growth with equity'. A fundamental step towards the achievement of the first objective was the municipal elections on 28 June, which represented the first opportunity for Chileans to vote for local authorities since 1971. The government parties won over one-half of the votes cast, demonstrating a clear endorsement of the ruling coalition. However, constitutional amendments envisaged by President Aylwin (including plans to restore presidential power to remove armed forces Commanders-in-Chief, to counter right-wing bias in the electoral system, to balance politically the composition of the constitutional tribunal and to abolish government-appointed senators) failed to secure a majority in the Senado. In December 1993, after much deliberation, a proposal to extend the presidential mandate to a six-year, non-renewable term was approved by both houses.

## THE GOVERNMENT OF PRESIDENT FREI

The presidential election of 11 December 1993 resulted in an overwhelming victory for the CPD candidate, Eduardo Frei Ruiz-Tagle, a Christian Democrat senator and son of the late Eduardo Frei Montalva, the former President, with 58% of the votes cast. Arturo Alessandri Besa (also the son of a former President), the candidate of the right-wing coalition, the Unión para el Progreso de Chile (Union for the Progress of Chile—UPC), came second with 24% of the votes cast. However, the ruling coalition failed to make significant gains at concurrently conducted congressional elections. Without a two-thirds' majority in the Congreso Nacional, President Frei was unable to amend what remained of Pinochet's 1981 Constitution, a task he had indicated would be one of his priorities.

This failure to change the Constitution bedevilled the Government. Furthermore, there was a series of conflicts with the military over such matters as early retirements and imprisonments over human rights issues. The spectacular escape from prison of the leaders of the FPMR guerrilla group strengthened the position of the military, as President Frei was obliged to create a new intelligence system. Throughout, military leaders remained highly resentful of any attempt either to accuse them of crimes committed under Pinochet or to question their authority or budgets, said to be the largest in Latin America. When Pinochet retired as head of the armed forces in March 1998 he assumed his seat-for-life in the Senado, which guaranteed immunity from prosecution. His successor, Ricardo Izurieta, promised a new era in military-civilian relations.

Congressional elections to renew all 120 seats in the lower house and 20 of the elective seats in the upper house were held on 11 December 1997. With about 40% of the potential electorate abstaining, the governing coalition won just over 50% of the vote, while the right-wing opposition won 36.2% of the ballot, an increase of 3% over 1993. This meant that the Government's majority was insufficient to vote through major constitutional changes, including the abolition of the 'designated' senators.

## THE PINOCHET AFFAIR

In spite of these tensions, by late 1998 Chile's progress towards full democracy, although slow, looked secure. Then, on 16 October, Gen. Pinochet was arrested during a visit to London, United Kingdom, in response to a preliminary extradition request issued by a Spanish judge, Baltasar Garzón, regarding charges of crimes of torture and murder committed against some 4,000 people, including Spanish nationals, by his administration during 1973–90. The arrest led to lengthy legal wrangles in the United Kingdom and threatened Chile's delicate political and military balance. The Chilean Government formally requested that Pinochet be allowed to return to Chile on the grounds of ill health. Following several legal rulings in 1999 denying Pinochet immunity to prosecution, in January 2000 a group of independent medical specialists examined Pinochet and, upon receiving their report that Pinochet was medically unfit to stand trial in Spain, on 2 March the British Government announced that Pinochet was to be released; he returned to Chile the following day.

Within Chile the repercussions of Pinochet's detention in the United Kingdom were enormous. The Chilean Government consistently argued that Pinochet's arrest was an unwarranted interference in the country's internal affairs, and that Pinochet should be returned to Chile where he could be tried. Chilean society was divided over the affair. There were demonstrations and counter-demonstrations, and the armed forces, increasingly worried about judicial proceedings against former generals in the wake of Pinochet's arrest, expressed their disquiet. However, in spite of the tensions, the presidential election of late 1999 and early 2000 proceeded peacefully.

## THE GOVERNMENT OF PRESIDENT LAGOS

The detention of Pinochet, coupled with an economic downturn, accentuated the problems within the governing coalition. After a bitter primary campaign, Ricardo Lagos Escobar won the left's candidacy for the presidential election, but leaving the PDC in some disarray. In contrast, the right wing united around the populist Joaquín Lavín Infante of the Partido Unión Demócrata

Independiente (UDI). In the first round, held on 12 December 1999, Lagos received 47.96% of the total votes cast, with Lavín obtaining 47.52%. At the second round of voting, on 16 January 2000, Lagos emerged victorious, with 51.31% of the total votes, and thus became the first Socialist President in Chile since Salvador Allende.

President Lagos inherited the same problem that had beset his two predecessors—how to secure constitutional change that would remove 'the authoritarian enclaves' remaining from the Pinochet Constitution while lacking a sufficient majority in both houses of the Congreso. Moreover, even within the CPD, the PS had far fewer seats than the Christian Democrats in both chambers. In selecting his Cabinet, Lagos needed to keep the CPD united by carefully ensuring a balance between the constituent parties.

The decline of the CPD was confirmed in the legislative elections of 16 December 2001, when the right-wing Alianza por Chile (which included the UDI and was led by Lavín) made significant gains. In the lower chamber the CPD won 62 seats, a substantially reduced majority, while the Alianza por Chile coalition secured 57 seats. In the Senado, where 18 of the 38 elective seats were contested, the CPD won an overall 20 seats, but lost its one-seat majority, while the Alianza por Chile increased its representation to 16 seats (the UDI's total increased to nine seats). The main winner was the UDI, which saw its vote increase from 17% to 25%, and which replaced the PDC as Chile's largest political party.

A number of issues confronted the new President: the continuing tensions over human rights trials, including possibly that of Pinochet, and how to secure constitutional change; Mapuche militancy against timber and other companies taking over and damaging the environment of lands they claimed under ancestral right (see below); corruption; ongoing economic problems; international relations; and the rising popularity of the right-wing and former rival candidate, Lavín.

**Human Rights and Constitutional Reform**

The return of Pinochet to Chile in March 2000, marked by a public welcome by the heads of the armed forces, was a clear warning that any attempt to put the former dictator on trial in Chile would be met with obstacles. After many legal delays, in August 2000 the Supreme Court voted 14 to six to confirm the Santiago Appeal Court's decision to revoke Pinochet's immunity. With a total of 240 lawsuits filed against Pinochet, in December 2000 Judge Juan Guzmán indicted Pinochet on charges of aggravated kidnapping and murder. However, in Chile all persons over 70 years of age must be subject to medical examination to determine whether they are mentally fit to stand trial. The Supreme Court finally ruled that Pinochet must undergo the medical tests.

After appeals and counter-appeals, in March 2001 the Santiago Court of Appeal reduced the charges against Pinochet from being the intellectual author of murder and kidnappings to acting as an accessory after the fact. On 1 July 2002 the Supreme Court voted (by four to one) to close permanently the case against Pinochet on the grounds that he was mentally unfit to stand trial. Just four days later, following clandestine talks, Pinochet relinquished his post as senator-for-life. The negotiated agreement left the former dictator with his pension, immunity from prosecution and the status of a former President. Thousands protested, and there were violent scenes in the Senado during the ratification process, but the agreement was approved.

However, this apparent conclusion of the Pinochet affair did not end controversies over human rights. In February 2003 five senior officers were charged with the murder of the former head of the Chilean armed forces, Gen. Carlos Prats, and his wife in 1974, and there were still many more cases pending. Then, in May 2004, accusations against Pinochet resurfaced. The Santiago Court of Appeal voted by 14 to nine to remove Pinochet's immunity from prosecution, this time over Pinochet's role in 'Plan Condor' (an intelligence operation involving the kidnapping and murder of left-wing dissidents in Brazil, Argentina, Paraguay, Chile, and Uruguay in the 1970s). The publication of the ruling by the Court in early July 2004 affirmed Pinochet could not have been unaware of the criminal actions of his subordinates. The subsequent claim by a US Senate report that Pinochet had deposited between US $4m. and $8m. in secret accounts with the US-based Riggs bank (a later US Senate report in March 2005 put the figure at $15m.) prompted the Congreso Nacional to immediately approve an investigation into Pinochet's finances. The legislature also approved the declassification of secret laws passed under Pinochet. In October 2004 the Government published a report detailing the political imprisonment, torture or murder of some 27,000 detainees under Pinochet. (Details of abuses against 1,200 further persons, including minors, were included in June 2005.) In these circumstances, defending Pinochet or any of the laws made by his regime became politically difficult. In November 2004 the Supreme Court overturned the 1978 Amnesty Law, which protected the armed forces against accusation of abuses of human rights. In a short space of time 384 military officers were facing legal actions. In May 2005 Gen. (retd) Manuel Contreras, the former head of Pinochet's secret police, gave an account of what had happened to some of the disappeared, holding Pinochet ultimately responsible for the crimes. In June, however, the Santiago Court of Appeals revoked charges against Pinochet relating to 'Plan Condor', owing to his deteriorating health. Nevertheless, in a separate ruling, the Court annulled Pinochet's immunity from prosecution for his alleged financial improprieties. In July the Court annulled, by a vote of 11 to 10, Pinochet's immunity from prosecution on charges relating to 'Plan Colombo', in which 15 left-wing activists 'disappeared' in 1975.

In mid-July 2005 the Senado approved constitutional reforms (passed by the Cámara de Diputados in October 2004), that, *inter alia*, ended the system of appointed senators and senators-for-life, restored the presidential prerogative to dismiss the heads of the armed forces, and permitted Chilean citizenship to be granted to children born to Chilean expatriates. President Lagos claimed the reforms concluded the transition in Chile from dictatorship to democracy.

**Indigenous Groups and the Environment**

As a consequence of economic and industrial growth, a new problem emerged in Chile in recent years: environmental degradation. Chile's rapid economic growth depended on the exploitation of the country's vast natural resources, many of which showed signs of severe environmental stress. For example, salmon and trout farming were causing serious damage to lakes and coastal waters in the south of Chile and more than 62% of Chilean territory was affected by creeping desertification. Furthermore, Chile was rapidly losing its native tree species. In addition, mining, dam and forestry projects were causing violent protests, particularly by the Mapuche (or Araucano) indigenous peoples campaigning for the restitution of lands they claimed under ancestral rights. Following violent protests in Temuco against a court decision to allow the Spanish-owned national electricity company, ENDESA (Empresa Nacional de Electricidad, SA), to resume its controversial Ralco dam project, President Lagos announced a Historical Truth and New Deal Commission to consider the demands and needs of the Mapuche communities. The report recommended a change to Chile's Constitution, giving greater rights to Chile's indigenous population. However, in May 2003 the Senado rejected such changes. This, as well as the continuation of work on the Ralco dam, led to a resumption of violent demonstrations by Mapuche groups. Several Mapuche activists opposed to the activities of the timber industry were arrested under anti-terrorism legislation and accused of arson or conspiracy to other 'terrorist acts'. The invocation of anti-terrorism legislation passed under the Pinochet Government was widely criticized as disproportionate. In November 2004 seven Mapuche activists were acquitted of terrorism charges. Meanwhile, following years of legal dispute between ENDESA and ecological and Pehuenche activist groups, the Ralco dam was finally inaugurated in September. It was thought the dam would supply 3,100 GWh annually, equivalent to some 9% of Chile's electricity requirements. Despite promising stronger measures to protect the environment, President Lagos has consistently eased environmental regulations, leading to increasing protests from environmental groups.

**International Relations**

Following a number of differences with its neighbours, in 2004 Chile found itself more isolated within Latin America than it had been since the days of the Pinochet dictatorship. Tensions over Bolivia's long-standing demand for access to the Pacific, which it had lost to Chile in 1879 at the end of the War of the Pacific, increased markedly during a number of Latin American summits in 2003–04. To Chile's immense annoyance, Bolivia received outspoken support from President Hugo Rafael Chávez Frías of Venezuela. Although most Chileans were opposed to ceding Bolivia any Chilean territory, even the USA supported a process to solve Bolivia's maritime claim. In March 2004 Peru caused further controversy by restating its long-standing claim to move its maritime boundaries south. Relations with Peru continued to be tense in early 2005, when it was revealed that Chile had sold arms to Ecuador during its 1995 conflict with Peru. Although both Governments declared in May that the dispute over the arms sales had been resolved, tensions arose again in that month following the announcement that the Chilean Government was to purchase armaments, including F-16 fighter planes, worth an estimated US $500m.

In May 2004 the Argentine Government unilaterally reduced its supply of natural gas to Chile in order to meet a domestic energy shortfall. The Chilean Government reacted angrily, accusing Argentina of breaching a treaty between the two countries. In the following month Bolivia consented to export emergency gas supplies to Argentina, provided Argentina did not re-export any of this to a third country, which prompted Chile to make a formal complaint to Bolivia. In mid-June 2005, however, President Kirchner announced at a meeting with President Lagos in Buenos Aires that Argentina was committed to maintaining the supply of gas to Chile in the future and that a new gas pipeline linking Peru, Chile and Argentina was under consideration.

Traditionally, Chile has good relations with Brazil, but Chile's decision unilaterally to negotiate a free trade agreement with the USA in 2003 irked the Brazilian President, Luiz Inácio ('Lula') da Silva, who was keen to advance the Mercado Común del Sur (Mercosur), of which Chile was an associate member. In response, the Brazilian Government began to favour closer relations with Argentina (at Chile's expense). However, Chile restored its traditional friendships in South America when the whole continent supported the ultimately successful candidature of the Chilean Minister of the Interior, José Miguel Insulza, against the US-supported candidates to become Secretary-General of the Organization of American States.

## TOWARDS THE 2005 ELECTIONS

For a considerable period before the presidential election scheduled to be held in December 2005, Joaquín Lavín Infante, the likely candidate of the right-wing UDI, commanded an impressive lead in the opinion polls. However, in October 2003 an RN legislator, María Pía Guzmán, accused two UDI senators of attending parties at which the sexual abuse of children allegedly occurred. The UDI accused Guzmán of trying to score political points, triggering a political feud between the UDI and the RN, both members of the Alianza por Chile. Although the case against them collapsed, the controversy adversely affected Lavín's ratings in the opinion polls. At the same time, with the economy improving, President Lagos's approval ratings rose sharply, from 47% in December 2003 to 61% in May 2004. Also, the mounting evidence against Pinochet further weakened support for the right-wing opposition, despite its increasing efforts to distance itself from the former dictator and his legacy. In the October 2004 municipal elections the UDI lost its position as the largest party in Chile; subsequently Lavín, who was confirmed as the UDI's nominee in June 2005, consistently trailed behind the CPD's candidate, former Minister of National Defence Michelle Bachelet Jeria. The contest to secure the presidential candidacy of the ruling Concertación grouping became a race between two women: Bachelet of the PS and the former Minister of Foreign Affairs, María Soledad Alvear Valenzuela of the PDC, both of whom had resigned their ministerial positions to run for the candidacy. Each was widely regarded as having performed well in her respective portfolio, and each, according to opinion polls, enjoyed a much more positive image than Lavín. In May 2005, however, Alvear withdrew from the contest to support Bachelet in order not to divide the left-wing vote, following the announcement that Sebastián Piñera, a wealthy businessman, was to stand as presidential candidate of the moderately rightwing RN. At that time Bachelet remained a firm favourite in the opinion polls.

# Economy

## PHILIP J. O'BRIEN

With a length of some 4,300 km and a maximum width of 180 km, Chile has a wider variety of climate and topography than any other country of comparable area. This ranges from the desert north, rich in minerals, through the fertile Central Valley, the agricultural heart of the country, to the southern, wetter forest region, source of the country's wood products. Further south still are the open grasslands of Chilean Patagonia, the chief sheep-rearing region and, offshore, the country's major source of petroleum. Since Chile's entire western frontier is the Pacific shoreline, its resources of fish products are immense.

Like other Latin American countries, Chile's economic history was dominated by the export of primary products: wheat in the colonial period; copper and nitrates in the 19th and early 20th centuries; and, even now, primarily copper. From 1880 to 1920 export taxes on nitrates accounted for roughly one-half of government income, until synthetics destroyed the world market in the natural product. Thereafter, with heavy foreign capital and technological investment, copper dominated the economy. Chile possessed an estimated one-quarter of the world's proven reserves of copper and, in 1982, became the world's leading producer. However, this dependence on primary exports, subject to fluctuations in world markets, proved disadvantageous at times. During the Great Depression Chile was affected more severely than any other country, since 80% of government revenue came from copper and nitrates.

The reaction to this Depression, in the 1930s and onwards, was to pursue import-substitution policies in industry, under the protection of high tariffs. Moreover, a tradition of state intervention in the economy was accentuated, notably by the establishment, in 1939, of the National Development Corporation (Corporación de Fomento de la Producción—CORFO), under the aegis of which a large number of state-owned enterprises prospered. By the end of the 1960s the results of increased state control were not impressive: per-head economic growth rates were low; and unemployment and inflation high. By 1970 copper accounted for 75% of exports by value, indicating an unbalanced economy.

The Socialist Government (1970–73), under Salvador Allende Gossens, accentuated state control, nationalized the copper industry, the banking sector and many private enterprises, and redistributed land to peasants. The economic upheaval that followed facilitated the military intervention of September 1973. The military regime, led by Gen. Augusto Pinochet Ugarte, initiated policies to reverse the trends of decades, with the assistance of successive economic teams. The Government's principal objectives were: to reduce inflation; to reduce overstaffing in highly protected, inefficient industries; to diversify the export pattern, thus reducing dependence on copper and to expose Chilean industry to foreign competition by reducing import tariffs. In short, the aim was to enable Chile to realize better the potential of its enormous natural resources.

To achieve these aims: public expenditure was reduced; tariff barriers fell to the lowest in the world (averaging 10%, apart from the automobile industry, where a more 'gradualist' policy was followed); denationalization of banks and credit was introduced; and foreign investment was encouraged, under liberal laws on taxation and remittance of profits. By 1980 Chile was one of the most free-market economies in the world. Although the social costs were high (unemployment reached over 20% in several years), the economic results were impressive. Annual average inflation was reduced from over 500% in 1973 to 9.9% in 1982. Dependence on copper for export revenue fell from 75% in 1970, to under 50% in 1980, as 'non-traditional' exports such as wine, fruit, wood and fish products were encouraged. Foreign capital, invested chiefly in mining, increased to over US $4,000m. by 1980.

These results were reflected in sustained economic growth from 1976 until 1981. However, both gross domestic product (GDP) and gross national product (GNP) were reduced substantially from 1981 as the economy went into recession. This was caused by a combination of circumstances. The impact of the growing world recession reduced copper prices, lax financial management led to a banking crisis, and the maintenance of a fixed exchange rate for years kept inflation down but at the same time encouraged imports and discouraged exports, leading to a huge current-account deficit. In general, moreover, the dogmatic application of the free-market model, was also to blame. Bankruptcies proliferated and unemployment increased dramatically, to 17.8% in mid-1983.

In 1983 the banking system in Chile virtually collapsed, and had to be rescued by massive state intervention. Foreign confidence disappeared, and the Chilean peso lost 90% of its value in 12 months. GDP decreased by 14.1% and urban unemployment became the highest in Latin America, resulting in considerable social unrest. Imports declined by almost one-half and the foreign debt reached US $17,200m., or 71.1% of GNP. Almost 80% of export earnings were needed for debt-servicing. Rescheduling the foreign debt, under the auspices of the IMF, became imperative.

From 1984, however, a free-market, export-led, 'crawling peg' exchange-rate policy was pursued, which initiated a period of sustained economic recovery, averaging over 5.5% annual GDP growth, for the rest of the 1980s. Chile's economic results from 1984 onwards stand out from those in other Latin American countries, registering the best growth rate in the region during the 1980s.

The Government of President Patricio Aylwin Azócar, which assumed power in March 1990, continued the expansionary export-led growth policies of its predecessor. In the four years of President Aylwin's administration Chile's GDP grew, on average, by 6.2% per year, by far the best growth under any Chilean Government, and some three times faster than that of the rest of Latin America. Investment also expanded rapidly, almost doubling in the four years of Aylwin's Government.

Eduardo Frei Ruiz-Tagle, who assumed the presidency in March 1994, inherited a sound economy. He, too, continued to reaffirm Chile's commitment to low inflation and export-led growth. GDP continued to expand rapidly, with average annual growth at 3.0% between 1981 and 1990 and at 6.6% from 1990 to 2000. In the early 1990s domestic savings and investment were at a high level, ensuring that Chile avoided the recessions that occurred elsewhere in the continent following the Mexican economic crisis of 1994. Unsurprisingly, the country's economic performance earned praise from foreign banks, attracting good rates of investment. With inflation under control, rapidly expanding exports, high levels of liquidity and sustained economic growth, Chile was even favourably compared to the successful 'tiger' economies of Asia. However, economic problems began to accumulate in 1998, and the economy contracted by 1.0% in 1999. Nevertheless, GDP growth recovered to stand at 4.4% in 2000, the year that the Government of Ricardo Lagos Escobar took office; growth slowed in 2001 and 2002, when GDP increased by 3.4% and 2.2%, respectively, reflecting a general downturn in the world economy, particularly that of the USA. However, the economy increased by 3.7% in 2003, by a provisional 6.1% in 2004, and growth of some 5%–6% was forecast for 2005; these high growth rates could be attributed primarily to the dramatic increase in copper prices.

In spite of the expanding economy, the rate of unemployment remained high, however. In 2002 the rate reached 10.0%, with unemployment heavily concentrated in regions such as Valparaíso (12.0%). The rate of unemployment among the young was persistently high, at around 22%. Beginning in 2001 the Government allocated special funds to reduce unemployment, and in 2003 unemployment fell to 9%. However, in 2004 the rate of unemployment rose slightly to around 9.7% and still remained high, at 7.6% in mid-2005.

In contrast, inflation remained well under control. From 26.0% in 1991, the annual rate decreased during the 1990s to 2.3% in 1999. It increased to 3.9% in 2000, before falling to 3.5% in 2001 and 2.4% in 2002. Following an increase to 2.8% in 2003, in 2004 the annual inflation rate was just 1.1%. These favourable results were achieved by maintaining strict control over external accounts and finances, and encouraging investments, particularly in exports.

The Pinochet Government promoted private investment, through a controversial policy of denationalizing state assets, and the encouragement of exports through selective tariffs, technical aid and supported prices. At the same time, encouraged by strict fiscal policy, in accordance with the IMF's requirements, foreign investment also increased, after the years of recession. Its role in the economy was enhanced by the Government's policy of capitalizing external debt. By aggressively using 'debt-swap' mechanisms, under which external creditors were able to sell or exchange their debt for Chilean assets, the country slightly decreased its overall debt between 1985 and 1990, from US $20,400m. to $19,114m. The foreign debt increased to $43,231m. at the end of 2003. However, since very little (17%) of this was short-term debt, Chile's debt rating remained unchanged, with an outlook assessed as positive.

Crucially, during the mid-1990s Chile managed to maintain a steady flow of foreign capital inflows without being destabilized by them. It achieved this mainly by encouraging direct foreign investment rather than short-term speculative capital inflows, thus making the balance of payments less sensitive to interest-rate changes and financial-market perceptions. Furthermore, the private pension funds and private health care schemes, established in the 1980s, contributed to the development of a local capital market and played a key role in the high rates of savings and investment. However, from 1999 foreign investment began to fall, and in 2000 was US $3,700m., the lowest level since 1995. From 2000 to 2002 foreign investment rose, particularly in 2002, owing to the sale of La Disputada mine. However, in 2003 foreign investment fell by 35.5%, to $2,460m., the lowest level since 1993, although rates improved again in 2004 and 2005.

Declines in world petroleum prices (Chile imports some 50% of its domestic requirements), a prosperous export sector and reductions in international interest rates were also important factors in the country's rapid growth. Prices for exports improved fairly consistently; the value of merchandise exports grew from US $2,500m. in 1978 to $16,663m. in 1997. Following the Asian financial crisis, merchandise exports fell to $14,842m. in 1998, but soon recovered, and stood at $19,807m. in 2003, owing mainly to the increase in the copper price. There was also a rapid diversification of export markets, with notable expansion in Asia. Framework agreements signed with the European Union (EU), the establishment of the Mercado Común del Sur (Mercosur—comprising Argentina, Brazil, Paraguay and Uruguay), and the bilateral trade agreements with Canada, Mexico and the USA would help to secure Chile's high level of exports in the future. In 2004 Chile finally secured a free trade agreement with the Republic of Korea (South Korea).

Although the economy grew from the mid-1980s, the social cost was high for a long time. The minimum monthly wage, earned by one-fifth of the country's 5m.-strong work-force, was considerably lower in 1990 than 10 years earlier. Moreover, wages' share of GDP fell from 34.8% in 1985/86 to 33.4% in 1992/93, while profits' share increased from 38.3% to 44.0% in the same period. Improvements have been made in decreasing poverty: official government statistics showed that in 1990 40% of the population were still classified as extremely poor; in 1997 the figure was 25%, with 6.5% defined as 'hard-core' poor. This meant that some 3.2m. Chileans lived below the official poverty line, with urban poverty accounting for over 80%. In mid-2005, with unemployment remaining high, there was still a persistent, although declining, high level of poverty, despite attempts by the Lagos Government to reduce the poverty rate to below 22%.

## AGRICULTURE, FORESTRY AND FISHING

Agriculture had been the mainstay of Chile's colonial economy, with wheat and cattle (primarily for hides and tallow) the main products. The country's Central Valley is very fertile and suitable for a wide variety of crops and fruit. Commercialization of the agrarian sector began under the Pinochet regime, with a consequent decrease in the rural population from 27% of the total population in 1975 to 13% by 2002.

Agriculture, forestry and fishing as a percentage of GDP was consistently high during the 1980s, reaching 9.5% in 1988. Thereafter the sector's share fell, and stood at a provisional 5.8% in 2004. The main food crops are wheat, maize, barley, oats, rice, rye, potatoes and other vegetables, with levels of production fluctuating, depending on areas sown and climatic factors. Traditional crops in the 1990s had such low profitability that many farmers changed to the cultivation of forestry and fruit. The future for some of Chile's traditional food crops such as wheat, rice and maize were increasingly bleak. Agriculture in general was badly affected in 1997 by the strong effects of El Niño, the warm current that periodically appears along the Pacific coast (displacing the more usual cold current). Then, in 1998 and 1999 Chile experienced its worst drought for 50 years, and all areas of agriculture had substantial losses. Nevertheless, in 2000 and 2001 agricultural GDP recovered dramatically, growing by 6.7% and 6.5%, respectively. Although the sector contracted in 2002, it grew by 4.1% in 2003 and continued to grow in 2004 and the first half of 2005.

By the end of the 1980s fruit production had become second only to copper as an export earner. At this time the fruit industry employed some 250,000 people, and generated 12% of the country's export income. The rapid increases in fruit exports came from plums, apples, pears, grapes, berries and kiwi fruit. Total fruit exports earned US $1,221m. in 2001 (equivalent to 7% of export earnings). In 1997 El Niño severely affected several fruit crops, and the severe drought in 1998/99 continued to affect adversely many crops; however, there were signs of a significant recovery in 2000–05.

Wine was one of Chile's best-known products for decades, although until the 1980s relatively little was exported. During the 1990s Chilean wines had earned popularity abroad, particularly within its main export markets of Canada, the United Kingdom and the USA. In 2003 production was an estimated 668,222 metric tons, and in 2001 wine exports accounted for US $645m. Following drought in 1999, production and sales of wine recovered from 2000.

Forestry products were one of the most dynamic areas in the Chilean economy in the 1980s, principally as a result of the superb growing conditions in the south, where pine grows faster than anywhere else in the world, and of the country's generous reforestation laws. Chile has more than one-half of the world's temperate rainforests, although at the beginning of the 21st century this was fast disappearing. Up to 80% of Chile's natural forests were damaged, with 120,000 ha per year being lost. Over 60% of this loss can be attributed to the native trees being replaced by fast-growing exotic varieties. Forestry employs some 60,000 workers directly and another 60,000 indirectly. Over 50 countries import Chilean pulp, sawn-wood, logs, newsprint and other forestry products. Some 72,900 ha of land were reforested in 1988, entirely by private companies. Forestry products accounted for only 2.8% of export revenues in 1973, but by 1993 this figure had increased to 13.1%, becoming Chile's second largest foreign-exchange earner. This sector continued to expand, increasingly producing processed articles, rather than primary products. A fall in world prices for forestry products led to a decrease in the value of exports in the late 1990s. Prices recovered in the early 2000s and in 2001 exports of cork and wood stood at US $780.3m. Investment in the sector continued to be high, even though many forestry projects caused conflict with local Mapuche communities who viewed some of them as encroachments on their ancestral lands. In 2002 the forestry company Arauco, Celulosa y Constitución, SA began its massive $1,000m. investment programme, mainly in a new cellulose plant in southern Chile. In 2005 the forestry sector continued to be a major exporter.

The fishing catch stood at 4.2m. metric tons in 2003. Canning was carried out in plants, situated mostly in the north, with an installed capacity to process some 1,300 tons of raw materials per day. Over 75% of the fishing catch was processed into fishmeal. Fishery exports, including processed products, increased from US $62m. in 1985 to $1,799.2m. in 2003. Exports of frozen fish also increased rapidly. A new development was the growth of fish farms in the south of Chile. In 1995 Chile became the world's second largest salmon exporter after Norway, exporting about 145,000 tons of salmon, worth about $614m. in 1997. However, the swift and spectacular penetration of the US market (accounting for about 30% of Chilean salmon exports) led the USA to impose additional tariffs against two salmon-exporting firms on the grounds of 'dumping' (the anti-competitive practice of maintaining the price of export commodities at artificially low levels). The Chilean salmon industry has spent some $22m. on contesting the US allegations. By mid-2003 17 Chilean salmon companies had been cleared of the dumping charge. Despite these obstacles, the value of exports of Chilean salmon reached $1,000m. in 2000, an increase of 10% on 1999. In 2002 a harmful algae bloom caused immense damage to the small-scale fishing industry along the southern coast of Chile. During 2003, in spite of an increase in salmon prices, Chilean production fell by 9.2% and had not returned to its previous levels by mid-2005.

## MINING AND POWER

Mining was always of crucial importance to the Chilean economy. The country has an impressive share of world mining reserves: 22% of the world's copper reserves; 23% of selenium; 20% of molybdenum; 12% of iodine; and 5% of gold. Chile also has most of the world's nitrate reserves. Although it employs only some 1.3% of the work-force, the mining sector is, as it has been historically, a critical export sector, contributing a provisional 14.1% of GDP in 2004.

In 1982 Chile became the world's leading copper producer, with production of 1.26m. metric tons (copper content of ores). Production continued to increase thereafter, and stood at 4.6m. tons in 2002. Large investments were made in mining in the mid-1990s, mainly in the development of new copper mines. From 1987 the copper price rose steadily to reach its highest point of US $1.34 per lb in 1995. However, the copper-trading scandal of mid-1996, involving a Japanese company, Sumitomo Corporation, reduced the price, causing both a fall in Chile's copper revenues and the postponement of some investment plans. Furthermore, partly as a result of the Asian financial crisis in 1998, copper prices decreased to their lowest value, in real terms, of the 20th century, at $0.61 per lb in February 1999. Thus, although in 1998 copper sales increased by some 12%, their value declined by some 24%. After February 1999 copper prices began to increase steadily. However, copper prices fell considerably in 2001 and remained low in 2002. They did, however, increase by 13% in 2003, and in 2004 began to rise very sharply, surpassing the high price of 1995. In June 2005 the copper price reached a record $1.60 per lb. The high copper price can be attributed to an improvement in the US economy and to demand from the People's Republic of China, which imported 35% more copper in January 2004 than in January 2003. It was predicted that Chinese demand for copper could grow by as much as 8%–10% per year until 2010.

As a result of such price fluctuations, copper's share of total export revenue did not increase in line with output: in the 1970s export earnings from copper contributed more than 70% of total export earnings; by 1992 this figure fell to 38.5% of total exports. Thereafter, however, mining increased in importance, and in 2001 accounted for almost one-half of all export earnings, 32.7% of which were from copper. Other minerals, such as molybdenum, iron ore, nitrate and other salts, zinc, lead, coal, gold and silver, are also important foreign-exchange earners.

In 2004 the Corporación Nacional del Cobre de Chile (CODELCO) made record profits of US $1,530m., an increase of 44% compared with 2003. Production from the corporation stood at 1.8m. metric tons in 2004. However, the long-term future of Chilean copper was far from secure: reserves could be exhausted within 60 years, with the large private-sector developments which began in 1988 surviving for just 20 years.

The large-scale copper industry, concentrated at Chuquicamata, Andina and Salvador, in the north, and at El Teniente, in the Central Valley, was nationalized by the Allende Government. Copper production was dominated by CODELCO. Following the loss of over US $200m. in copper futures trading in 1993 there were demands to privatize CODELCO. The incident, usually attributed to trading errors (although there were allegations of fraud), led to a modernization of the corporation, including the decentralization of the four main mines. In mid-2005 the Government also intended to curtail the transfer of 10% of CODELCO's earnings to the military, but this proposal met with vociferous opposition from military leaders. In 2004 CODELCO doubled its export revenues. In that year the People's Republic of China became the leading customer of CODELCO.

The increasingly important role of foreign investment in mining was not without controversy. The La Escondida copper-mining project, which was the largest mine in the country, was entirely owned by foreign multinationals: Utah International owns 60%, Rio Tinto Zinc 30%, and Mitsubishi Corporation and Nippon Mining Co 10%. The La Escondida deal effectively ended the Chilean state's domination of copper production and, arguably, the country's direct benefits from the enterprise were thus reduced. Subsequently, major private mines were opened at La Disputada, La Candelaria, and Zaldivar. Other developments included the acquisition of a 12% stake in the Collahuasi copper project, expected to be one of the world's biggest and cheapest mines, by a Japanese consortium of Mitsui and Nippon Mining. The mine was controlled by Minorco, a subsidiary of Anglo-American Corporation of South Africa and Falconbridge of Canada. In 2003 CODELCO came under increasing public scrutiny amid claims that it was involved in irregular awarding of contracts, and there was pressure put on the company to be audited by the state auditor. CODELCO claimed that, although it was a state concern, its statutes allowed it to operate as a private company. In 2004 the Government proposed a modest royalty of 3% on the annual sales of metallic products, but the opposition parties blocked the proposal.

Chile's energy resources are based on hydroelectric capacity and petroleum, both of which were expanded in the 1980s. Given the physical configuration of a large number of fast-flowing rivers, fed by the melting snows of the Andes and running east-to-west to the Pacific, Chile has the highest hydroelectric potential in the world. Total potential generating capacity has been estimated at 18,700 MW. Hydroelectric power, according to the US Energy Information Administration (EIA), provided some 52% of electricity generated in 2002. The Colbún-Machicura station, on the River Maule, began operating in 1985, producing a further 3,060m. kWh per year, about 25% of Chile's total generating capacity. Six further plants also began operations in the late 1980s. In the early 1990s construction of the Pehuenche plant (capacity 500 MW) was completed, with the help of a World Bank loan of US $95m. The project met with much opposition from environmental and indigenous groups.

Following the privatization, at a very low price, of the state energy company, Empresa Nacional de Electricidad (ENDESA), in 1988, there was much criticism of the management of the energy sector. In 1997 ENERSIS of Spain acquired virtual control of ENDESA, and thus about 40% control of Chilean energy. The deal aroused accusations of 'insider dealing' (benefiting from information not publicly known) and provoked government criticism. Disquiet with the privatization process reached a crisis in 1999 when, partly owing to the drought, there were severe energy shortages. The issue became highly political, particularly given the virtual monopoly of the Chilean electricity system by ENDESA. President Frei ordered the private utilities to increase their generating capacity or face severe fines. Legislation was introduced to increase potential penalties from US $26,800 to $6m. and making the utilities liable to pay compensation to users for non-compliance with the terms of their concessions. ENDESA recovered in 2000, when it made a profit of $189.8m., in contrast with the substantial losses of previous years. In 2003 ENERSIS was allowed to increase its stake in ENDESA to 85% by swapping a $1,400m. loan owed by ENDESA for shares.

In 2002 natural gas provided some 23% of the country's electrical energy needs, according to the EIA. It was expected that this figure would increase to 25% by 2005; there were plans for the construction of three gas-fired power-stations in Chile, with a total capacity in excess of 1,100 MW. Total investments in gas projects amounted to some US $3,000m.

However, problems with Chile's energy supplies were acute in the first half of 2004 because Argentina, itself suffering a severe energy shortage, twice cut its gas supplies to Chile. Argentina, after the 1995 protocol, supplied 90% of the natural gas of Chile's power generation. The suspensions precipitated a major diplomatic rift (subsequently resolved—see History), but did not, by mid-2005, seem to have significantly dented Chile's economic prospects.

Petroleum was exploited from the 1940s, when fields in Tierra del Fuego and Magallanes began operations. Following extensive exploration and exploitation in the early 1980s by the Empresa Nacional de Petróleo (ENAP), 45% of the country's requirements were met domestically. However, since this time petroleum production has steadily declined, from 49,000 b/d in 1983 to 18,500 b/d in 2003, according to the EIA. With demand at 235,000 b/d in 2003, the large shortfall was met by imports, primarily form Argentina. In June 2004 plans were announced to construct Chile's fourth refinery, in the Bíobío region.

## INDUSTRY

Industry was the sector that experienced the widest fluctuations in output in recent years. Chilean industry developed from the 1930s with a high degree of tariff protection in a policy of import substitution until 1973. The Pinochet Government's 'shock treatment' of dismantling tariff barriers, thus encouraging competitive foreign imports, was a shock to many industries, notably textiles. The effects of the free-market policy pursued from the mid-1970s to the early 1980s reduced the contribution to GDP of industry as a whole from 25% in 1975 to only 17.7% in 1992. The sector recovered thereafter, however; during 1990–2003 industrial GDP increased by an annual average of 5.2%. In 2004 industry as a percentage of GDP stood at a provisional 43.2%.

The monetarist model adopted caused an increased concentration of wealth and economic power in fewer hands, notably banking and financial groups, rather than industrialists. Competition from abroad and low levels of domestic investment contributed to the decline of the industrial sector, which only began its recovery in the 1990s. The plastics and transport-equipment sectors, in particular, grew rapidly. Manufacturing as a percentage of total GDP stood at a provisional 18.1% in 2004. In 2003 growth in Chilean manufacturing was modest, at only 2.4% in 2003, although it growth increased slightly in 2004. The key sectors within manufacturing were textiles, motor cars, chemicals, rubber products, cement and consumer goods. Manufacturing exports have not grown as much as was hoped, mainly because of the slow growth rates of Chile's main trading partners in Mercosur and the economic crisis in Argentina in 2001–02.

The steel industry, dating effectively from 1950 and based at Huachipato near Talcahuano, expanded during the late 1990s, following a decrease in exports and production from the late 1980s. Construction, too, was a sector much affected by fluctuations in the Chilean economy: in 2002 the sector's share of GDP stood at 19.5%, but fell to a provisional 8.0% in 2004.

For many years foreign investment in Chilean industry was low, despite the free-market policies of the Pinochet Government. However, having increased by 140% between 1986 and 1987, foreign investment continued to grow, owing to the 'debt-for-equity' conversions and the extensive privatization process. Although the majority of foreign capital was invested in mining, industry attracted an increasing share. US investments remained preponderant in Chile, although British interests were also traditionally strong. Chile's good economic performance encouraged foreign bankers and businessmen in general to invest in the country's stock market and to take advantage of the debt-conversion schemes. The Frei Government maintained the pace of privatization, selling the state's shares in a number of companies, including the airline carrier Línea Aérea Nacional de Chile (LAN-Chile), the electricity supplier Empresa Eléctrica del Norte (EDELNOR), the shipping company Empresa Marítima (Empremar Chile), and, in June 1999, Empresa Metropolitana de Obras Sanitarias (EMOS), Chile's biggest water company. The economic recession in the late 1990s badly affected the manufacturing sector, particularly the capital goods sector, and both foreign and domestic investment declined. In 2001 foreign investment increased, but then slowed down in 2002 and 2003.

## TRANSPORT AND COMMUNICATIONS

Internal transport in Chile is well developed. Railway lines, which were being increasingly electrified, linked Iquique, in the desert north, to Puerto Montt, in the south, with feeder lines running laterally from this central line to important towns and cities. Four international railways extended to Argentina, Bolivia and Peru. The Frei Government began the process of privatizing the Chilean railways. Freight services were sold in 1995, and the rest of the rail service was to be privatized in stages.

The road network in Chile in 2001 covered 79,604 km (of which 16,409 km was paved), including 3,455 km of Chile's section of the Pan-American Highway. The national bus service, linking main towns from north to south of the Central Valley, is considered excellent, as is Santiago's underground railway system.

Shipping facilities are also well developed, essential for a country with such a long coastline. In the early 2000s about a dozen companies were engaged in the coastal and international trade. The state controlled the port authority and roughly 40% of the merchant marine. The main ports are Valparaíso, Talcahuano, San Antonio and Antofagasta. Following a severe earthquake in March 1985, more than US $1,600m. was spent on the reconstruction of San Antonio, and its port was upgraded to handle 3m. metric tons of imports annually, compared with the previous 1m. tons.

Owing to its topography, Chile needs a large number of airports to facilitate communications. The international airport, Santiago, is served by some 18 international airlines and two national ones. Chile's other main airport is Chacalluta, 14 km north-east of Arica. The national carrier, LAN-Chile, serves major cities within the country and is also an international line. Internally, it is supplemented by Línea Aérea del Cobre (LADECO), the airline of CODELCO.

Chileans responded well to the new communication industries. Following its privatization in the late 1980s, the telecommunications industry grew rapidly. In 2003 there were 7.6m. mobile telephones in use and 4.0m. internet users. The state telephone company was privatized in 1988 and subsequently controlled by Telefónica of Spain. Deregulation of the market produced fierce competition, resulting in a reduction in costs to the consumer, and a massive increase in usage.

## TOURISM

The tourism industry grew rapidly in the 1990s. This growth was attributed to Chile's wide range of scenic beauty as well as government efforts to promote tourism, both internally and externally, through the Servicio Nacional de Turismo. Much private investment went into tourist centres, for example Marbella, to enable Chile to compete with resorts in the region such as Punta del Este in Uruguay. Efforts were made to enhance the attraction of major cities, notably Santiago, although the capital remained one of the most polluted cities in the world. In 2003 tourism receipts (including passenger transport) stood at US $1,362m. The number of foreign arrivals decreased from 1,613,523 in 2003 to 1,421,076 in 2004.

## FOREIGN TRADE, FOREIGN INVESTMENT AND REGIONAL BLOCS

Owing, in part, to favourable prices, merchandise exports increased steadily from the mid-1980s and stood at US $19,807.3m. in 2003. Chile was one of the first countries to explore the possibilities of establishing a free trade zone with the USA, following US President George Bush's (1989–93) 'Initiative for the Americas'. The country wanted to participate in the North American Free Trade Agreement (NAFTA), which formally came into effect at the beginning of 1994. However, Chile failed to secure 'fast track' status for its bid to join the grouping, and the Government then decided actively to pursue other trade negotiations, securing bilateral free trade agreements with Canada and Mexico. However, the key for Chile remained a free trade agreement with the USA. Finally, in January 2004, after 14 rounds of negotiations, such an agreement came into effect. While actively pursuing a bilateral free trade strategy, Chile has also been keen to have strong ties with Mercosur. Chile and Mercosur agreed an association accord whereby Chile would become a member of the Mercosur free trade zone but not the customs union, as Chile's flat rate 11% external tariff was at variance with the differential common external tariff of the four Mercosur countries. Special agreements were made on sensitive items such as Chilean wheat, and some sectors, for example vehicle-manufacturing and services, were excluded from the agreement altogether. Given that almost one-half of Chile's investments abroad were in Mercosur countries, particularly Argentina, the agreement significantly improved the country's foreign economic links.

Chile was admitted to Mercosur as an associate member in October 1996. However, the early optimism about the country becoming a full member did not last. Moves towards a common tariff could not be maintained as the Argentine economic crisis and problems with the Brazilian currency made convergence difficult. The Brazilian Government wanted Mercosur countries to negotiate as a unified bloc with other bodies, and would have preferred Chile to be less bilateral. Meanwhile, within Chile environmentalists and labour leaders voiced concerns that the free trade agreement with the USA would weaken the South American trade grouping. Also in 2003, an association agreement with the EU came into effect, which would facilitate trade liberalization with this key trading partner. This followed the May 2002 bilateral free trade agreement signed with the EU at the EU-Latin America Summit held in Madrid, Spain. According to the agreement, the EU would open 95% of its market to Chile within a period of three years. In return, Chile would open its manufacturing market to the EU by 2007 and had 10 years to dismantle trade barriers that protected its agricultural sector. It was also agreed that EU member states could wholly own fishing companies in Chile, ending the maximum 49% ownership hitherto allowed. In June 2003 a further free trade agreement was reached with the European Free Trade Association, comprising non-EU European countries. Chile joined Asia-Pacific Economic Co-operation (APEC) in 1994, emphasizing the importance of Pacific trade. In early 2004, in spite of fierce protests by South Korean farmers, Chile and South Korea signed a free trade agreement. In that year the Chilean Government began negotiations towards free trade agreements with New Zealand and Singapore, and with the People's Republic of China. Bilateral trade with that country increased by 35% in 2003, compared with the previous year. In 2004 Chile's international reserves stood at US $15,996.8m.

## CONCLUSION

Between 1990 and 2003 Chile achieved a remarkable annual average of 5.6% in GDP growth. It achieved this sustained growth by having the most open, stable and liberalized economy in Latin America. Chile underwent a profound economic, social and political change, with productivity gains, high growth, strong external position, low single-figure inflation and improvements in nearly all social indicators, including overall literacy, malnutrition, education standards, infant mortality and life expectancy.

Following a period of long and stable economic growth, at the end of the 1990s Chile returned to its more familiar fluctuating economic cycle. The full opening of the Chilean economy to the international economy always made it vulnerable to significant changes. So it proved when a major crisis affected the East Asian economies at the end of 1997. Subsequently, financial crisis in Argentina, a lack of confidence in the Brazilian economy and slow-downs in the US economy resulted in lower Chilean growth in 2001 and 2002 (of 2.8% and 2.2%, respectively). However, in 2003 growth recovered to 3.7%, mainly because of increased copper production. In 2004 and 2005 growth continued to improve, owing, in part, to a particularly high copper price. However, the success of various government measures to stimulate the Chilean economy, as always, depended upon the state of the world economy in general and of the economy of the USA—and, from the early 2000s, of the People's Republic of

China—in particular. In 2004 the World Trade Organization warned that Chile's trade, particularly in commodities, accounted for an ever increasing share of GDP (from 56% in 1997 to 66% in 2002), thus making Chile particularly vulnerable to any downturn in the global economy. However, with demand for copper, in particular from the People's Republic of China, remaining high, the prospects for the Chilean economy in both the short and medium term looked encouraging.

# Statistical Survey

Sources (unless otherwise stated): Instituto Nacional de Estadísticas, Avda Bulnes 418, Casilla 498-3, Correo 3, Santiago; tel. (2) 366-7777; fax (2) 671-2169; e-mail inesdadm@reuna.cl; internet www.ine.cl; Banco Central de Chile, Agustinas 1180, Santiago; tel. (2) 696-2281; fax (2) 698-4847; e-mail bcch@bcentral.cl; internet www.bcentral.cl.

## Area and Population

### AREA, POPULATION AND DENSITY*

| | |
|---|---:|
| Area (sq km) | 756,096† |
| Population (census results)‡ | |
| 22 April 1992 | 13,348,401 |
| 24 April 2002 | |
|   Males | 7,447,695 |
|   Females | 7,668,740 |
|   Total | 15,116,435 |
| Population (official estimates at mid-year) | |
| 2003 | 15,773,504 |
| 2004 | 15,955,631 |
| 2005 | 16,136,137 |
| Density (per sq km) at mid-2005 | 21.3 |

* Excluding Chilean Antarctic Territory (approximately 1,250,000 sq km).
† 291,930 sq miles.
‡ Excluding adjustment for underenumeration.

### REGIONS
(2002 census)

| | Area (sq km) | Population ('000) | Density (per sq km) | Capital |
|---|---:|---:|---:|---|
| I De Tarapacá | 59,099.1 | 428.6 | 7.3 | Iquique |
| II De Antofagasta | 126,049.1 | 494.0 | 3.9 | Antofagasta |
| III De Atacama | 75,176.2 | 254.3 | 3.4 | Copiapó |
| IV De Coquimbo | 40,579.9 | 603.2 | 14.9 | La Serena |
| V De Valparaíso | 16,396.1 | 1,539.9 | 93.9 | Valparaíso |
| VI Del Libertador Gen. Bernardo O'Higgins | 16,387.0 | 780.6 | 47.6 | Rancagua |
| VII Del Maule | 30,296.1 | 908.1 | 30.0 | Talca |
| VIII Del Bíobío | 37,062.6 | 1,861.6 | 50.2 | Concepción |
| IX De la Araucanía | 31,842.3 | 869.5 | 27.3 | Temuco |
| X De Los Lagos | 67,013.1 | 1,073.1 | 16.0 | Puerto Montt |
| XI Aisén del Gen. Carlos Ibáñez del Campo | 108,494.4 | 91.5 | 0.8 | Coihaique |
| XII De Magallanes y Antártica Chilena | 132,297.2 | 150.8 | 1.1 | Punta Arenas |
| Metropolitan Region (Santiago) | 15,403.2 | 6,061.2 | 393.5 | — |
| **Total** | 756,096.3 | 15,116.4 | 19.9 | — |

### PRINCIPAL TOWNS
(2002 census)

| | | | |
|---|---:|---|---:|
| Gran Santiago (capital) | 4,668,473 | Talca | 201,797 |
| Puente Alto | 492,915 | Arica | 185,268 |
| Viña del Mar | 286,931 | Puerto Montt | 175,938 |
| Antofagasta | 296,905 | Los Angeles | 166,556 |
| Valparaíso | 275,982 | Coquimbo | 163,036 |
| Talcahuano | 250,348 | Chillán | 161,953 |
| Temuco | 245,347 | La Serena | 160,148 |
| San Bernardo | 246,762 | Osorno | 145,475 |
| Concepción | 216,061 | Valdivia | 140,559 |
| Iquique | 216,419 | Calama | 138,402 |
| Rancagua | 214,344 | | |

## BIRTHS, MARRIAGES AND DEATHS

| | Registered live births* | | Registered marriages | | Registered deaths | |
|---|---:|---:|---:|---:|---:|---:|
| | Number | Rate (per 1,000) | Number | Rate (per 1,000) | Number | Rate (per 1,000) |
| 1994 | 288,175 | 20.6 | 91,555 | 6.5 | 75,445 | 5.4 |
| 1995 | 279,928 | 19.7 | 87,205 | 6.1 | 78,531 | 5.5 |
| 1996 | 278,729 | 19.3 | 83,547 | 5.8 | 79,123 | 5.5 |
| 1997 | 273,641 | 18.7 | 78,077 | 5.3 | 78,472 | 5.4 |
| 1998 | 270,637 | 18.3 | 73,456 | 5.0 | 80,257 | 5.4 |
| 1999 | 263,867 | 17.6 | 69,765 | 4.6 | 81,984 | 5.5 |
| 2000 | 261,993 | 17.2 | 66,607 | 4.4 | 78,814 | 5.2 |
| 2001 | 259,069 | 16.8 | 64,088 | 4.2 | 81,873 | 5.3 |

* Adjusted for underenumeration.
Source: partly UN, *Demographic Yearbook*.

**Expectation of life** (WHO estimates, years at birth): 77 (males 74; females 80) in 2003 (Source: WHO, *World Health Report*).

### ECONOMICALLY ACTIVE POPULATION*
('000 persons aged 15 years and over, October–December)

| | 2002 | 2003† | 2004† |
|---|---:|---:|---:|
| Agriculture, hunting, forestry and fishing | 746.6 | 771.8 | 783.2 |
| Mining and quarrying | 69.8 | 71.9 | 73.8 |
| Manufacturing | 780.4 | 797.2 | 805.1 |
| Electricity, gas and water | 29.5 | 31.2 | 31.9 |
| Construction | 439.9 | 427.4 | 473.0 |
| Trade, restaurants and hotels | 1,073.1 | 1,066.8 | 1,127.5 |
| Transport, storage and communications | 444.8 | 483.3 | 461.2 |
| Financing, insurance, real estate and business services | 430.9 | 453.5 | 472.6 |
| Community, social and personal services | 1,516.2 | 1,572.1 | 1,634.7 |
| **Total employed** | 5,531.3 | 5,675.1 | 5,862.9 |
| Unemployed | 468.7 | 453.1 | 494.7 |
| **Total labour force** | 5,999.9 | 6,128.2 | 6,357.6 |

* Figures are based on sample surveys, covering 36,000 households, and exclude members of the armed forces. Estimates are made independently, therefore totals are not always the sum of the component parts.
† Preliminary figures.

## Health and Welfare

### KEY INDICATORS

| | |
|---|---:|
| Total fertility rate (children per woman, 2003) | 2.3 |
| Under-5 mortality rate (per 1,000 live births, 2003) | 9 |
| HIV/AIDS (% of persons aged 15–49, 2003) | 0.3 |
| Physicians (per 1,000 head, 1998) | 1.15 |
| Hospital beds (per 1,000 head, 1996) | 2.67 |
| Health expenditure (2002): US $ per head (PPP) | 642 |
| Health expenditure (2002): % of GDP | 5.8 |
| Health expenditure (2002): public (% of total) | 45.1 |
| Access to water (% of persons, 2002) | 95 |
| Access to sanitation (% of persons, 2002) | 92 |
| Human Development Index (2002): ranking | 43 |
| Human Development Index (2002): value | 0.839 |

For sources and definitions, see explanatory note on p. vi.

## Agriculture

**PRINCIPAL CROPS**
('000 metric tons)

|  | 2001 | 2002 | 2003 |
|---|---|---|---|
| Wheat | 1,780 | 1,820 | 1,780* |
| Rice (paddy) | 143 | 142 | 141 |
| Barley | 65 | 77 | 50* |
| Maize | 778 | 924 | 1,190 |
| Oats | 345 | 416 | 420* |
| Potatoes | 1,210 | 1,303 | 1,094 |
| Sugar beet | 3,232 | 3,555 | 2,100 |
| Dry beans | 60 | 45 | 48 |
| Rapeseed | 59 | 2 | 13* |
| Cabbages† | 68 | 68 | 70 |
| Lettuce† | 85 | 86 | 87 |
| Tomatoes | 1,263* | 1,287* | 1,300† |
| Pumpkins, squash and gourds† | 105 | 106 | 105 |
| Green chillies and peppers† | 60 | 62 | 63 |
| Dry onions† | 285 | 287 | 290 |
| Carrots† | 98 | 99 | 99 |
| Green corn† | 240 | 245 | 250 |
| Other vegetables† | 335 | 342 | 338 |
| Watermelons† | 80 | 83 | 83 |
| Cantaloupes and other melons† | 63 | 64 | 65 |
| Oranges | 101 | 114 | 118 |
| Lemons and limes | 132 | 140 | 150 |
| Apples | 1,135 | 1,050 | 1,100 |
| Pears | 205 | 202 | 205 |
| Peaches and nectarines | 290 | 274 | 275† |
| Plums | 211 | 230 | 240 |
| Grapes | 1,785* | 1,720 | 1,750† |
| Avocados | 110 | 130 | 135 |
| Kiwi fruit | 135† | 128 | 125 |

* Unofficial figure.
† FAO estimate(s).

Source: FAO.

**LIVESTOCK**
('000 head, year ending September)

|  | 2001 | 2002 | 2003 |
|---|---|---|---|
| Horses* | 650 | 660 | 670 |
| Cattle† | 3,980 | 3,927 | 3,932 |
| Pigs* | 2,750 | 3,100 | 3,200 |
| Sheep* | 4,090 | 4,050 | 4,100 |
| Goats* | 750 | 880 | 990 |
| Chickens* | 82,000 | 74,700 | 80,000 |
| Turkeys* | 24,900 | 23,100 | 18,500 |

* FAO estimates.
† Unofficial figures.

Source: FAO.

**LIVESTOCK PRODUCTS**
('000 metric tons)

|  | 2001 | 2002 | 2003 |
|---|---|---|---|
| Beef and veal | 217.6 | 200.0 | 190.0 |
| Mutton and lamb | 10.9 | 9.9 | 10.8 |
| Pig meat | 303.0 | 350.7 | 385.8 |
| Horse meat | 10.8 | 11.1 | 11.2* |
| Poultry meat* | 487.8 | 452.6 | 457.6 |
| Cows' milk | 2,190 | 2,170 | 2,170 |
| Goats' milk* | 10.4 | 10.5 | 10.5 |
| Butter | 11.8 | 11.6 | 11.6* |
| Cheese | 57.2 | 59.7 | 59.7* |
| Hen eggs | 115.6† | 113.8† | 116.5 |
| Wool: greasy* | 16.2 | 16.0 | 16.0 |
| Cattle hides* | 31.3 | 28.8 | 27.0 |

* FAO estimate(s).
† Unofficial figure.

Source: FAO.

## Forestry

**ROUNDWOOD REMOVALS**
('000 cubic metres, excluding bark)

|  | 2001 | 2002 | 2003 |
|---|---|---|---|
| Sawlogs, veneer logs and logs for sleepers | 13,107 | 13,962 | 15,087 |
| Pulpwood | 12,386 | 11,321 | 12,221 |
| Other industrial wood | 189 | 208 | 183 |
| Fuel wood | 12,108 | 12,326* | 12,712* |
| **Total** | 37,790 | 37,817 | 40,203 |

* FAO estimate.
Source: FAO.

**SAWNWOOD PRODUCTION**
('000 cubic metres, including railway sleepers)

|  | 2001 | 2002 | 2003 |
|---|---|---|---|
| Coniferous (softwood) | 5,581 | 6,193 | 6,758 |
| Broadleaved (hardwood) | 291 | 246 | 246 |
| **Total** | 5,872 | 6,439 | 7,004 |

Source: FAO.

## Fishing

('000 metric tons, live weight)

|  | 2001 | 2002 | 2003 |
|---|---|---|---|
| Capture | 3,797.1 | 4,271.5 | 3,621.8 |
| Patagonian grenadier | 162.1 | 133.4 | 85.9 |
| Araucanian herring | 324.6 | 347.4 | 303.4 |
| Anchoveta (Peruvian anchovy) | 852.8 | 1,526.9 | 821.9 |
| Chilean jack mackerel | 1,649.9 | 1,519.0 | 1,420.9 |
| Chub mackerel | 365.0 | 343.4 | 571.5 |
| Aquaculture | 566.1 | 545.7 | 563.4 |
| Atlantic salmon | 253.9 | 265.7 | 280.5 |
| Coho (silver) salmon | 136.9 | 102.5 | 85.3 |
| **Total catch** | 4,363.2 | 4,817.1 | 4,185.2 |

Note: Figures exclude aquatic plants ('000 metric tons): 299.8 (capture 234.3, aquaculture 65.5) in 2001; 315.7 in 2002 (capture 244.0, aquaculture 71.6); 378.3 (capture 308.6, aquaculture 69.7) in 2003.

Source: FAO.

## Mining

('000 metric tons, unless otherwise indicated)

|  | 2000 | 2001 | 2002 |
|---|---|---|---|
| Copper (metal content) | 4,602.0 | 4,739.0 | 4,580.6 |
| Coal | 503.4 | 568.1 | 451.6 |
| Iron ore* | 8,728.9 | 5,437.1 | 7,268.8 |
| Calcium carbonate | 5,395.2 | 5,563.2 | 5,887.7 |
| Sodium sulphate (metric tons) | 56,501 | 67,760 | 70,776 |
| Zinc—metal content (metric tons) | 31,403 | 32,762 | 36,161 |
| Molybdenum—metal content (metric tons) | 33,639 | 33,492 | 29,467 |
| Manganese (metric tons)† | 41,716 | 31,320 | 12,195 |
| Gold (kilograms) | 54,143 | 42,673 | 38,688 |
| Silver (kilograms) | 1,242.2 | 1,348.7 | 1,210.5 |
| Petroleum ('000 cubic metres) | 325.8 | 309.8 | 336.4 |

* Gross weight. The estimated iron content is 61%.
† Gross weight. The estimated metal content is 32%.

Sources: Servicio Nacional de Geología y Minería; CODELCO.

## Industry

**SELECTED PRODUCTS**
('000 metric tons, unless otherwise indicated)

|  | 1999 | 2000 | 2001 |
|---|---:|---:|---:|
| Refined sugar | 434 | 432 | 476 |
| Beer (million litres) | 334 | 322 | 337 |
| Wine* | 480.7 | 667.4 | 565.2 |
| Soft drinks (million litres) | 1,154 | 1,225 | 2,212 |
| Cigarettes (million) | 13,271 | 13,796 | 13,305 |
| Non-rubber footwear ('000 pairs) | 6,237 | 5,735 | 5,251 |
| Particle board ('000 cu metres) | 301 | 366 | 360 |
| Mattresses ('000) | 1,082 | 1,132 | 1,016 |
| Sulphuric acid | 2,436 | 2,363 | 2,736 |
| Jet fuel | 603 | 569 | 693 |
| Motor spirit (petrol) | 2,129 | 2,170 | 2,035 |
| Kerosene | 227 | 147 | 168 |
| Distillate fuel oils | 3,436 | 3,716 | 3,797 |
| Residual fuel oils | 1,425 | 1,592 | 1,574 |
| Cement | 2,508 | 2,686 | 3,145 |
| Tyres ('000) | 2,551 | 3,084 | 3,246 |
| Glass sheets ('000 sq metres) | 21,523 | 23,180 | 20,714 |
| Blister copper | 2,835 | n.a. | n.a. |
| Refined copper, unwrought | 2,666 | 2,668 | 2,882 |
| Copper wire | 5.8 | 5.9 | 5.7 |
| Electric energy (million kWh) | 38,389 | 41,268 | 43,918 |

**2002:** Wine 575.2*; Particle board 448; Refined copper, unwrought 2,850.
**2003:** Wine 668.2*.

* Source: FAO.

Source (unless otherwise indicated): UN, *Industrial Commodity Statistics Yearbook*.

## Finance

**CURRENCY AND EXCHANGE RATES**

**Monetary Units**
  100 centavos = 1 Chilean peso.

**Sterling, Dollar and Euro Equivalents** (31 May 2005)
  £1 sterling = 1,054.86 pesos;
  US $1 = 580.20 pesos;
  €1 = 715.44 pesos;
  10,000 Chilean pesos = £9.48 = $17.24 = €13.98.

**Average Exchange Rate** (pesos per US $)
  2002   688.936
  2003   691.433
  2004   609.369

**BUDGET**
('000 million pesos)

| Revenue | 2002 | 2003* | 2004* |
|---|---:|---:|---:|
| Current revenue | 9,917.23 | 10,845.52 | 11,307.07 |
|   Operating revenue | 665.12 | 696.99 | 699.63 |
|   Social security contributions | 678.90 | 718.96 | 817.48 |
|   Net tax revenues | 7,709.00 | 8,336.28 | 8,899.21 |
|   Net copper revenue | 345.59 | 447.18 | 496.72 |
|   Transfers | 85.72 | 90.25 | 38.93 |
|   Other revenue | 432.89 | 528.46 | 326.64 |
|   Operating revenue of previous years | — | 27.39 | 28.46 |
| Capital revenue | 211.30 | 201.12 | 177.48 |
|   Sale of assets | 36.90 | 30.34 | 25.93 |
|   Repayment of loans | 174.40 | 170.78 | 151.55 |
| **Total revenue** | 10,128.52 | 11,046.64 | 11,484.55 |

* Projections.

| Expenditure* | 2000 | 2001 | 2002 |
|---|---:|---:|---:|
| Government | 332.79 | 365.24 | 415.01 |
| Defence | 708.78 | 746.25 | 780.70 |
| Public order and safety | 508.12 | 569.80 | 627.57 |
| Education | 1,580.11 | 1,774.41 | 1,958.54 |
| Health | 1,099.11 | 1,232.84 | 1,326.35 |
| Social security and welfare | 2,599.64 | 2,830.24 | 2,916.89 |
| Housing | 355.54 | 382.01 | 409.62 |
| Subsidies | 246.96 | 256.32 | 272.50 |
| Other social services | 411.11 | 441.20 | 446.29 |
| Economic services | 1,035.62 | 1,107.09 | 1,204.51 |
| Interest payments | 180.34 | 202.76 | 135.09 |
| **Sub-total** | 9,058.09 | 9,908.15 | 10,493.06 |
| *Less* Lending included in expenditure | 204.76 | 174.35 | 178.03 |
| **Total expenditure** | 8,853.33 | 9,733.81 | 10,315.03 |
| Current | 7,555.88 | 8,269.76 | 8,713.39 |
| Capital | 1,297.45 | 1,464.05 | 1,601.64 |

* Excluding net lending.

**2003** ('000 million pesos, excluding net lending, projections): Current expenditure 9,306.92; Capital expenditure 1,818.50; Total expenditure 11,125.42.

**2004** ('000 million pesos, excluding net lending, projections): Current expenditure 9,872.33; Capital expenditure 1,847.97; Total expenditure 11,720.30.

Source: Dirección de Presupuestos, Santiago.

**INTERNATIONAL RESERVES**
(US $ million at 31 December)

|  | 2002 | 2003 | 2004 |
|---|---:|---:|---:|
| Gold* | 2.3 | 2.7 | 3.0 |
| IMF special drawing rights | 36.5 | 45.7 | 52.6 |
| Reserve position in IMF | 490.7 | 582.9 | 445.8 |
| Foreign exchange | 14,813.9 | 15,211.0 | 15,495.4 |
| **Total** | 15,343.4 | 15,842.3 | 15,996.8 |

* National valuation.

Source: IMF, *International Financial Statistics*.

**MONEY SUPPLY**
('000 million pesos at 31 December)

|  | 2002 | 2003 | 2004 |
|---|---:|---:|---:|
| Currency outside banks | 1,301.5 | 1,379.7 | 1,601.7 |
| Demand deposits at commercial banks | 3,129.3 | 3,551.0 | 4,144.9 |
| **Total money** | 4,430.8 | 4,930.7 | 5,746.6 |

Source: IMF, *International Financial Statistics*.

**COST OF LIVING**
(Consumer Price Index for Santiago; base: 2000 = 100)

|  | 2001 | 2002 | 2003 |
|---|---:|---:|---:|
| Food (incl. beverages) | 100.8 | 102.9 | 105.8 |
| Rent, fuel and light | 106.1 | 109.6 | 115.0 |
| Clothing (incl. footwear) | 94.3 | 90.6 | 86.2 |
| **All items** (incl. others) | 103.6 | 106.1 | 109.1 |

Source: ILO.

## NATIONAL ACCOUNTS
('000 million pesos at current prices)

### Expenditure on the Gross Domestic Product

|  | 2002 | 2003* | 2004† |
|---|---|---|---|
| Government final consumption expenditure | 5,943.9 | 6,313.7 | 6,656.3 |
| Private final consumption expenditure | 29,241.8 | 31,229.5 | 33,097.8 |
| Increase in stocks | 157.0 | 393.8 | 617.0 |
| Gross fixed capital formation | 9,879.1 | 10,769.3 | 11,822.2 |
| **Total domestic expenditure** | 45,221.8 | 48,706.3 | 52,193.3 |
| Exports of goods and services | 15,773.5 | 18,553.3 | 23,487.9 |
| Less Imports of goods and services | 14,653.5 | 16,528.9 | 18,324.2 |
| **GDP in purchasers' values** | 46,341.8 | 50,730.7 | 57,357.0 |
| **GDP at constant 1996 prices** | 37,655.1 | 39,060.1 | 41,427.3 |

* Provisional figures.
† Preliminary data.

### Gross Domestic Product by Economic Activity

|  | 2002 | 2003* | 2004† |
|---|---|---|---|
| Agriculture and forestry | 1,819.8 | 1,961.6 | 2,002.0 |
| Fishing | 584.9 | 756.6 | 1,174.3 |
| Mining and quarrying | 3,202.4 | 4,301.2 | 7,658.0 |
| Manufacturing | 8,507.1 | 9,103.9 | 9,825.3 |
| Electricity, gas and water | 1,437.7 | 1,549.0 | 1,641.0 |
| Construction | 3,429.2 | 3,987.4 | 4,333.8 |
| Trade, restaurants and hotels | 4,360.2 | 4,718.7 | 4,815.3 |
| Transport, storage and communications | 3,279.0 | 3,605.5 | 3,874.5 |
| Financial services‡ | 6,380.8 | 6,754.7 | 7,090.2 |
| Sale of real estate | 2,549.6 | 2,649.0 | 2,716.5 |
| Personal services§ | 6,066.3 | 6,541.0 | 6,993.9 |
| Public administration | 2,010.9 | 2,133.8 | 2,238.4 |
| **Sub-total** | 43,627.8 | 48,062.4 | 54,363.2 |
| Value-added tax | 3,716.6 | 3,899.4 | 4,262.2 |
| Import duties | 664.1 | 523.0 | 482.0 |
| Less Imputed bank service charge | 1,666.6 | 1,754.2 | 1,750.4 |
| **GDP in purchasers' values** | 46,341.8 | 50,730.7 | 57,357.0 |

* Provisional figures.
† Preliminary data.
‡ Including insurance, renting of property and business loans.
§ Including education.

## BALANCE OF PAYMENTS
(US $ million)

|  | 2002 | 2003 | 2004 |
|---|---|---|---|
| Exports of goods f.o.b. | 18,179.8 | 21,523.6 | 32,024.9 |
| Imports of goods f.o.b. | −15,794.2 | −18,001.7 | −23,005.7 |
| **Trade balance** | 2,385.6 | 3,521.9 | 9,019.2 |
| Exports of services | 4,385.5 | 4,949.5 | 5,956.4 |
| Imports of services | −5,087.0 | −5,566.7 | −6,536.7 |
| **Balance on goods and services** | 1,684.1 | 2,904.7 | 8,438.9 |
| Other income received | 1,113.7 | 1,435.5 | 1,509.6 |
| Other income paid | −3,960.4 | −6,041.8 | −9,610.3 |
| **Balance on goods, services and income** | −1,162.6 | −1,701.5 | 338.3 |
| Current transfers received | 954.3 | 928.5 | 1,395.2 |
| Current transfers paid | −371.7 | −329.1 | −343.8 |
| **Current balance** | −580.1 | −1,102.1 | 1,389.7 |
| Capital account (net) | 83.0 | — | 5.1 |
| Direct investment abroad | −343.2 | −1,884.1 | −943.1 |
| Direct investment from abroad | 2,549.9 | 4,385.4 | 7,602.8 |
| Portfolio investment assets | −3,315.8 | −4,171.7 | −4,557.0 |
| Portfolio investment liabilities | 998.5 | 2,053.7 | 1,123.2 |
| Financial derivatives liabilities | 1,140.6 | −384.4 | −2,965.9 |
| Other investment assets | 727.9 | 1,544.5 | −700.2 |
| Other investment liabilities | −123.7 | 117.8 | −84.0 |
| Net errors and omissions | −952.2 | −916.3 | −1,062.0 |
| **Overall balance** | 185.0 | −357.2 | −191.5 |

Source: IMF, *International Financial Statistics*.

# External Trade

**PRINCIPAL COMMODITIES**
(distribution by SITC, US $ million)

| Imports c.i.f. | 1999 | 2000 | 2001 |
|---|---|---|---|
| **Food and live animals** | 999.8 | 1,080.7 | 1,053.8 |
| **Mineral fuels, lubricants, etc.** | 1,869.3 | 3,022.8 | 2,738.9 |
| Petroleum, petroleum products, etc. | 1,387.4 | 2,428.8 | 2,111.9 |
| Crude petroleum oils, etc. | 1,099.9 | 1,982.6 | 1,726.6 |
| Gas, natural and manufactured | 299.5 | 418.3 | 486.1 |
| **Chemicals and related products** | 1,865.4 | 2,044.4 | 2,141.3 |
| **Basic manufactures** | 1,969.5 | 2,342.9 | 2,423.3 |
| Iron and steel | 284.2 | 384.0 | 406.1 |
| **Machinery and transport equipment** | 5,081.4 | 5,791.0 | 5,598.9 |
| Machinery specialized for particular industries | 550.4 | 681.4 | 733.9 |
| General industrial machinery equipment and parts | 909.0 | 981.6 | 1,032.5 |
| Office machines and automatic data-processing equipment | 614.2 | 654.4 | 573.3 |
| Automatic data-processing equipment | 435.3 | 469.2 | 414.7 |
| Telecommunications and sound equipment | 891.3 | 960.4 | 817.6 |
| Other electrical machinery apparatus, etc. | 621.4 | 656.1 | 702.7 |
| Road vehicles and parts* | 982.6 | 1,529.1 | 1,343.0 |
| Passenger motor cars (excl. buses) | 415.2 | 629.4 | 535.0 |
| Lorries and trucks | 270.3 | 548.1 | 467.2 |
| **Miscellaneous manufactured articles** | 1,598.5 | 1,808.2 | 1,761.1 |
| **Total** (incl. others) | 13,891.5 | 16,619.7 | 16,136.2 |

* Data on parts exclude tyres, engines and electrical parts.

# CHILE

## Statistical Survey

| Exports f.o.b. | 1999 | 2000 | 2001 |
|---|---|---|---|
| **Food and live animals** | 3,684.0 | 3,788.2 | 4,197.0 |
| Fish, crustaceans and molluscs and preparations thereof | 1,404.3 | 1,546.2 | 1,629.9 |
| Fish, fresh (live or dead), chilled or frozen | 1,060.4 | 1,190.1 | 1,259.3 |
| Frozen fish | 509.8 | 537.2 | 547.3 |
| Vegetables and fruit | 1,533.4 | 1,536.7 | 1,613.2 |
| Fruit and nuts (excl. oil nuts) fresh or dried | 1,109.8 | 1,191.7 | 1,221.0 |
| Feeding-stuff for animals (excl. unmilled cereals) | 303.6 | 255.0 | 328.3 |
| **Beverages and tobacco** | 569.3 | 607.2 | 679.0 |
| Beverages | 556.3 | 593.8 | 661.5 |
| Wine of fresh grapes | 523.7 | 576.8 | 645.0 |
| **Crude materials (inedible) except fuels** | 3,877.3 | 4,929.3 | 4,774.0 |
| Cork and wood | 649.4 | 643.4 | 780.3 |
| Pulp and waste paper | 768.3 | 1,113.8 | 1,068.2 |
| Chemical wood pulp | 766.7 | 1,110.4 | 1,066.2 |
| Chemical wood pulp, bleached or semi-bleached | 641.3 | 955.6 | 898.5 |
| Metalliferous ores and metal scrap | 2,164.9 | 2,877.7 | 2,622.9 |
| Copper ores and concentrates (excl. matte) | 1,735.1 | 2,393.7 | 2,162.8 |
| **Chemicals and related products** | 750.6 | 1,015.3 | 1,158.6 |
| **Basic manufactures** | 5,290.7 | 6,162.8 | 6,088.6 |
| Non-ferrous metals | 4,383.2 | 5,164.6 | 4,921.3 |
| Copper | 4,249.8 | 5,063.8 | 4,816.2 |
| Copper and copper alloys refined or not, unwrought | 4,154.5 | 4,954.5 | 4,689.4 |
| Refined copper (incl. copper alloys other than master alloys), unwrought | 3,909.6 | 4,661.5 | 4,396.4 |
| **Machinery and transport equipment** | 495.0 | 494.6 | 522.6 |
| **Total (incl. others)** | 15,619.2 | 18,214.5 | 18,745.4 |

Source: UN, *International Trade Statistics Yearbook*.

**2002** (US $ million): *Imports c.i.f.* (provisional): Consumer goods 2,836.2 (Durables 1,001.7; Semi-durables 1,044.3); Intermediate goods 9,634.2 (Mineral fuels and lubricants 2,462.7, *of which* Petroleum 1,615.1); Capital goods 3,470.2; Franc zone 986.6; Total 16,927.2. *Exports f.o.b.:* Food and live animals 4,545.3 (Fish, crustaceans and molluscs and preparations thereof 1,657.2; Vegetables and fruit 1,955.7); Beverages and tobacco 627.0; Crude materials (inedible) except fuels 4,063.4 (Cork and wood 749.2; Pulp and waste paper 813.3; Metalliferous ores and metal scrap 2,235.3); Chemicals and related products 1,123.8; Basic manufactures 5,758.6 (Non-ferrous metals 4,726.2); Total (incl. others) 17,051.4.

**2003** (US $ million): *Imports c.i.f.* (provisional): Consumer goods 3,166.6 (Durables 1,186.1; Semi-durables 1,080.5); Intermediate goods 10,956.2 (Mineral fuels and lubricants 3,100.2, *of which* Petroleum 2,125.6); Capital goods 3,667.9; Franc zone 1,291.6; Total 19,082.4. *Exports f.o.b.:* Food and live animals 5,094.3 (Fish, crustaceans and molluscs and preparations thereof 1,799.2; Vegetables and fruit 2,201.9); Beverages and tobacco 701.6; Crude materials (inedible) except fuels 5,120.4 (Cork and wood 813.6; Pulp and waste paper 890.6; Metalliferous ores and metal scrap 3,094.8); Chemicals and related products 1,327.8; Basic manufactures 6,453.4 (Non-ferrous metals 5,275.2); Total (incl. others) 19,807.3.

## PRINCIPAL TRADING PARTNERS
(US $ million)

| Imports c.i.f. | 1999 | 2000 | 2001 |
|---|---|---|---|
| Argentina | 2,014.7 | 2,866.4 | 3,074.1 |
| Brazil | 965.7 | 1,329.8 | 1,490.8 |
| Canada | 406.1 | 508.4 | 423.3 |
| China, People's Republic | 658.4 | 949.5 | 1,013.1 |
| Colombia | 165.9 | 204.9 | 188.8 |
| Ecuador | 227.4 | 254.2 | 123.3 |
| France | 411.2 | 442.4 | 572.9 |
| Germany | 615.1 | 599.8 | 684.1 |
| Italy | 513.1 | 417.6 | 435.0 |
| Japan | 631.6 | 701.8 | 548.5 |
| Korea, Republic | 404.7 | 534.6 | 538.5 |
| Mexico | 578.4 | 615.1 | 532.1 |
| Nigeria | 131.4 | 313.0 | 122.6 |
| Peru | 168.4 | 255.4 | 285.3 |
| Spain | 408.7 | 426.3 | 464.1 |
| Sweden | 270.1 | 281.6 | 180.2 |
| Switzerland | 153.8 | 123.1 | 109.1 |
| United Kingdom | 181.2 | 176.3 | 193.4 |
| USA | 2,986.1 | 3,273.2 | 2,976.1 |
| Venezuela | 210.1 | 236.7 | 184.8 |
| **Total (incl. others)** | 13,891.5 | 16,619.7 | 16,136.2 |

| Exports f.o.b. | 1999 | 2000 | 2001 |
|---|---|---|---|
| Argentina | 726.3 | 639.0 | 618.7 |
| Belgium | 287.6 | 374.6 | 244.0 |
| Bolivia | 190.3 | 164.0 | 165.2 |
| Brazil | 701.3 | 969.4 | 909.6 |
| Canada | 175.5 | 244.1 | 285.4 |
| China, People's Republic | 357.3 | 901.8 | 1,065.0 |
| Colombia | 205.8 | 236.3 | 286.5 |
| Ecuador | 109.6 | 158.7 | 260.9 |
| France | 492.1 | 631.7 | 621.1 |
| Germany | 532.7 | 458.7 | 547.0 |
| Italy | 638.9 | 822.5 | 830.2 |
| Japan | 2,279.7 | 2,546.6 | 2,311.9 |
| Korea, Republic | 693.5 | 809.1 | 596.0 |
| Mexico | 621.0 | 815.6 | 917.0 |
| Netherlands | 506.6 | 446.5 | 551.6 |
| Peru | 356.6 | 439.3 | 586.7 |
| Spain | 312.5 | 377.0 | 354.1 |
| United Kingdom | 1,063.1 | 1,064.6 | 1,243.2 |
| USA | 2,811.1 | 3,007.8 | 3,483.6 |
| Venezuela | 195.2 | 228.4 | 306.8 |
| **Total (incl. others)** | 15,619.2 | 18,214.5 | 18,745.4 |

Source: UN, *International Trade Statistics Yearbook*.

# Transport

## PRINCIPAL RAILWAYS

| | 2001 | 2002 | 2003 |
|---|---|---|---|
| Passenger journeys ('000) | 16,095 | 14,052 | 14,444 |
| Passenger-kilometres ('000) | 870,836 | 770,392 | 829,338 |
| Freight ('000 metric tons) | 22,514 | 20,495 | 22,778 |
| Freight ton-kilometres (million) | 3,318 | 3,338 | 3,575 |

## ROAD TRAFFIC
(motor vehicles in use)

| | 2001 | 2002 | 2003 |
|---|---|---|---|
| Passenger cars (excl. taxis) | 1,247,985 | 1,270,516 | 1,302,602 |
| Buses and coaches (incl. taxis) | 176,275 | 177,001 | 175,504 |
| Lorries and vans | 656,209 | 675,518 | 678,338 |
| Specialized vehicles (incl. tractors) | 16,204 | 16,426 | 14,830 |
| Motorcycles and mopeds | 26,318 | 24,761 | 24,315 |

# CHILE

## SHIPPING

**Merchant Fleet**
(registered at 31 December)

|  | 2002 | 2003 | 2004 |
|---|---|---|---|
| Number of vessels | 525 | 532 | 539 |
| Total displacement ('000 grt) | 879.6 | 964.0 | 947.1 |

Source: Lloyd's Register-Fairplay, *World Fleet Statistics*.

**International Sea-borne Shipping**
(freight traffic, '000 metric tons)

|  | 2001 | 2002 | 2003 |
|---|---|---|---|
| Goods loaded | 37,113 | 36,109 | 39,712 |
| Goods unloaded | 15,257 | 17,646 | 18,026 |

## CIVIL AVIATION
(traffic on scheduled services)

|  | 1999 | 2000 | 2001 |
|---|---|---|---|
| Kilometres flown (million) | 107 | 108 | 110 |
| Passengers ('000) | 5,188 | 5,175 | 5,316 |
| Passenger-km (million) | 10,650 | 10,859 | 11,520 |
| Freight (million ton-km) | 2,107 | 2,296 | 2,329 |

Source: UN, *Statistical Yearbook*.

## Tourism

**ARRIVALS BY NATIONALITY**

|  | 2002 | 2003 | 2004 |
|---|---|---|---|
| Argentina | 514,711 | 536,010 | 473,003 |
| Bolivia | 107,703 | 132,312 | 108,911 |
| Brazil | 79,198 | 100,341 | 99,324 |
| France | 38,241 | 42,644 | 37,253 |
| Germany | 41,598 | 54,402 | 42,095 |
| Peru | 154,842 | 163,718 | 153,215 |
| Spain | 34,655 | 42,841 | 36,668 |
| United Kingdom | 37,269 | 50,984 | 39,752 |
| USA | 130,568 | 147,321 | 129,252 |
| **Total** (incl. others) | 1,412,315 | 1,613,523 | 1,421,076 |

**Tourism receipts** (US $ million, incl. passenger transport): 1,184 in 2001; 1,221 in 2002; 1,362 in 2003.

Sources: World Tourism Organization; Servicio Nacional de Turismo.

## Communications Media

|  | 2001 | 2002 | 2003 |
|---|---|---|---|
| Television receivers ('000 in use) | 4,400 | n.a. | n.a. |
| Telephones ('000 main lines in use) | 3,478.5 | 3,467.0 | 3,250.9 |
| Mobile cellular telephones ('000 subscribers) | 5,271.6 | 6,445.7 | 7,520.3 |
| Personal computers ('000 in use) | 1,640 | 1,796 | n.a. |
| Internet users ('000) | 3,102 | 3,575 | 4,000 |

**Radio receivers** ('000 in use): 5,180 in 1997.

**Facsimile machines:** 40,000 in use in 1997.

**Daily newspapers:** 52 in 1996.

Sources: mainly UNESCO, *Statistical Yearbook*; UN, *Statistical Yearbook*; International Telecommunication Union.

## Education

(2003)

|  | Institutions | Teachers | Students |
|---|---|---|---|
| Pre-primary |  | 9,997 | 286,381 |
| Special primary |  | 4,865 | 82,999 |
| Primary | n.a.* | 71,031 | 2,312,274 |
| Secondary |  | 32,444 | 947,057 |
| Adult |  | 1,897 | 152,453 |
| Higher (incl. universities) | 226 | n.a. | 567,114 |

* Many schools offer more than one level of education. A detailed breakdown is given below.

**Schools** (2003): Pre-primary: 662; Special 690; Primary 3,743; Secondary 516; Adult 279; Pre-primary and special 7; Pre-primary and primary 3,165; Pre-primary and secondary 1; Pre-primary and adult 1; Special and primary 17; Special and adult 4; Primary and secondary 377; Primary and adult 96; Secondary and adult 158; Pre-primary, special and primary 52; Pre-primary, primary and secondary 1,021; Pre-primary, primary and adult 278; Special, primary and secondary 2; Primary, secondary and adult 45; Pre-primary, special, primary and secondary 9; Pre-primary, special, primary and adult 11; Pre-primary, primary, secondary and adult 85; Pre-primary, special, primary, secondary and adult 4; Total 11,223.

**Adult literacy rate** (UNESCO estimates): 95.7% (males 95.8%; females 95.6%) in 2002 (Source: UN Development Programme, *Human Development Report*).

# Directory

## The Constitution

The 1981 Constitution, described as a 'transition to democracy', separated the presidency from the Junta and provided for presidential elections and for the re-establishment of the bicameral legislature, consisting of an upper chamber (Senado) of both elected and appointed senators, who are to serve an eight-year term, and a lower chamber (Cámara de Diputados) of 120 deputies elected for a four-year term. All former Presidents are to be senators for life. There is a National Security Council consisting of the President of the Republic, the heads of the armed forces and the police, and the Presidents of the Supreme Court and the Senado.

In July 1989 a national referendum approved 54 reforms to the Constitution, including 47 proposed by the Government and seven by the Military Junta. Among provisions made within the articles were an increase in the number of directly elected senators from 26 to 38, the abolition of the need for the approval of two successive Congresos for constitutional amendments (the support of two-thirds of the Cámara de Diputados and the Senado being sufficient), the reduction in term of office for the President to be elected in 1989 from eight to four years, with no immediate re-election possible, and the redrafting of the provision that outlawed Marxist groups so as to ensure 'true and responsible political pluralism'. The President's right to dismiss the Congreso and sentence to internal exile were eliminated.

In November 1991 the Congreso approved constitutional changes to local government. The amendments provided for the replacement of centrally appointed local officials with directly elected representatives.

In February 1994 an amendment to the Constitution was approved whereby the length of the presidential term was reduced from eight to six years.

## The Government

### HEAD OF STATE

**President:** RICARDO LAGOS ESCOBAR (took office 10 March 2000).

### THE CABINET
(July 2005)

A coalition of parties represented in the Concertación de los Partidos de la Democracia (CPD) (including the Partido Demócrata Cristiano—PDC, the Partido Socialista de Chile—PS, the Partido por la Democracia—PPD, and the Partido Radical Socialdemócrata—PRSD) and five Independents (Ind.).

**Minister of the Interior:** FRANCISCO VIDAL SALINAS (PPD).

**Minister of Foreign Affairs:** IGNACIO WALKER PRIETO (PDC).

# CHILE

**Minister of National Defence:** JAIME RAVINET DE LA FUENTE (PDC).
**Minister of Finance:** NICOLÁS EYZAGUIRRE GUZMÁN (PPD).
**Minister, Secretary-General of the Presidency:** EDUARDO DOCKENDORFF VALLEJOS (PDC).
**Minister, Secretary-General of the Government:** OSVALDO PUCCIO (PS).
**Minister of the Economy and Energy:** JORGE RODRÍGUEZ GROSSI (PDC).
**Minister of Mining:** ALFONSO DULANTO RENCORET (Ind.).
**Minister of Planning and Co-operation:** YASNA PROVOSTE (Ind.).
**Minister of Education:** SERGIO BITAR CHACRA (PPD).
**Minister of Justice:** LUIS BATES HIDALGO (Ind.).
**Minister of Labour and Social Security:** RICARDO SOLARI SAAVEDRA (PS).
**Minister of Public Works, Transport and Telecommunications:** JAVIER ETCHEVERRY CELHAY (PPD).
**Minister of Health:** PEDRO GARCÍA ASPILLAGA (Ind.).
**Minister of Housing and Urban Planning:** SONIA TSCHORNE (PS).
**Minister of Agriculture:** JAIME CAMPOS QUIROGA (PRSD).
**Minister of the National Women's Service (Sernam):** CECILIA PÉREZ DÍAZ (Ind.).

## MINISTRIES

**Ministry of Agriculture:** Teatinos 40, Santiago; tel. (2) 393-5000; fax (2) 393-5050; e-mail contacto@minagri.gob.cl; internet www.minagri.gob.cl.

**Ministry of the Economy and Energy:** Teatinos 120, 10°, Santiago; tel. (2) 672-5522; fax (2) 696-6305; e-mail economia@minecon.cl; internet www.economia.cl.

**Ministry of Education:** Alameda 1371, 7°, Santiago; tel. (2) 390-4000; fax (2) 380-0317; internet www.mineduc.cl.

**Ministry of Finance:** Teatinos 120, 12°, Santiago; tel. (2) 675-5800; fax (2) 671-8064; e-mail webmaster@minhda.cl; internet www.minhda.cl.

**Ministry of Foreign Affairs:** Catedral 1158, Santiago; tel. (2) 679-4200; fax (2) 699-4202; e-mail info@minrel.cl; internet www.minrel.cl.

**Ministry of Health:** Enrique MacIver 541, 3°, Santiago; tel. (2) 639-4001; fax (2) 633-5875; e-mail consulta@minsal.cl; internet www.minsal.cl.

**Ministry of Housing and Urban Development:** Alameda 924, Santiago; tel. (2) 638-0801; fax (2) 633-3892; e-mail contactenos@minvu.cl; internet www.minvu.cl.

**Ministry of the Interior:** Palacio de la Moneda, Santiago; tel. (2) 690-4000; fax (2) 699-2165; internet www.interior.cl.

**Ministry of Justice:** Morandé 107, Santiago; tel. (2) 674-3100; fax (2) 698-7098; internet www.minjusticia.cl.

**Ministry of Labour and Social Security:** Huérfanos 1273, 6°, Santiago; tel. (2) 695-5133; fax (2) 698-8473; e-mail mintrab@mintrab.gob.cl; internet www.mintrab.gob.cl.

**Ministry of Mining:** Teatinos 120, Santiago; tel. (2) 671-4373; fax (2) 698-9262; internet www.minmineria.cl.

**Ministry of National Defence:** Villavicencio 364, 22°, Edif. Diego Portales, Santiago; tel. (2) 222-1202; fax (2) 633-0568; internet www.emdn.cl.

**Ministry of the National Women's Service (Sernam):** Teatinos 950, 5°, Santiago; tel. (2) 549-6100; fax (2) 549-6148; e-mail mlobos@sernam.gov.cl; internet www.sernam.gov.cl.

**Ministry of Planning and Co-operation (MIDEPLAN):** Ahumada 48, 7°, Santiago; tel. (2) 675-1400; fax (2) 672-1879; internet www.mideplan.cl.

**Ministry of Public Works:** Morandé 59, Of. 545, Santiago; tel. (2) 361-2641; fax (2) 361-2700; internet www.mop.cl.

**Ministry of Transport:** Amunátegui 139, 3°, Santiago; tel. (2) 421-3000; fax (2) 421-3552; e-mail mtt@mtt.cl; internet www.mtt.cl.

**Office of the Minister, Secretary-General of the Government:** Palacio de la Moneda, Santiago; tel. (2) 690-4160; fax (2) 697-1756; e-mail cmladini@segegob.cl; internet www.segegob.cl.

**Office of the Minister, Secretary-General of the Presidency:** Palacio de la Moneda, Santiago; tel. (2) 690-4218; fax (2) 690-4329.

# President and Legislature

## PRESIDENT

Election, 12 December 1999 and 16 January 2000

|  | % of votes cast, 12 Dec. 1999 | % of votes cast, 16 Jan. 2000 |
|---|---|---|
| Ricardo Lagos Escobar (CPD) | 48.0 | 51.3 |
| Joaquín Lavín Infante (Alianza por Chile) | 47.5 | 48.7 |
| Gladys Marín Millie (PCCh) | 3.2 | — |
| Tomás Hirsch Goldschmidt (PH) | 0.5 | — |
| Sara Larraín Ruiz-Tagle | 0.4 | — |
| Arturo Frei Bolívar (UCCP) | 0.4 | — |
| **Total** | **100.0** | **100.0** |

## CONGRESO NACIONAL

### Senado*
(Senate)

**President:** ANDRÉS ZALDÍVAR LARRAÍN (PDC).

General Election, 16 December 2001*

|  | % of valid votes | Seats |
|---|---|---|
| Concertación de Partidos por la Democracia (CPD)† | 51.3 | 20 |
| Alianza por Chile‡ | 44.0 | 16 |
| Partido Comunista de Chile (PCCh) | 2.6 | — |
| Independents | 1.6 | 2 |
| Partido Humanista (PH) | 0.4 | — |
| Partido Liberal (PL) | 0.1 | — |
| **Total** | **100.0** | **38** |

1,716,942 valid votes were cast. In addition, there were 71,522 blank and 68,802 spoiled votes.

* Results of elections to renew 18 of the 38 elective seats in the Senado. In addition, there are nine designated senators, and a constitutional provision for former Presidents to assume a seat for life, in an ex-officio capacity. In July 2002 former President Pinochet resigned his senatorial seat, bringing the total number of senators to 48.
† Including the Partido Demócrata Cristiano (PDC), which won 12 seats, the Partido Socialista de Chile (PS), with five seats, and the Partido por la Democracia (PPD), with three seats.
‡ Including the Partido Unión Demócrata Independiente (UDI), which won nine seats, and the Partido Renovación Nacional (RN), which won seven seats.

### Cámara de Diputados
(Chamber of Deputies)

**President:** ISABEL ALLENDE (PPD).

General Election, 16 December 2001

|  | Valid votes | % of valid votes | Seats |
|---|---|---|---|
| Concertación de Partidos por la Democracia (CPD)* | 2,925,800 | 47.9 | 62 |
| Alianza por Chile† | 2,703,701 | 44.3 | 57 |
| Partido Comunista de Chile (PCCh) | 318,638 | 5.2 | — |
| Independents | 86,283 | 1.4 | 1 |
| Partido Humanista (PH) | 69,265 | 1.1 | — |
| Partido Liberal (PL) | 3,453 | 0.1 | — |
| **Total** | **6,107,140** | **100.0** | **120** |

In addition, there were 236,132 blank and 648,232 spoiled votes.

* Including the Partido Demócrata Cristiano (PDC), which won 24 seats, the Partido por la Democracia (PPD), which won 21 seats, the Partido Socialista de Chile (PS), with 11 seats, and the Partido Radical Socialdemócrata (PRSD), with six seats.
† Including the Partido Unión Demócrata Independiente (UDI), which won 35 seats, and the Partido Renovación Nacional (RN), which won 22 seats.

CHILE

## Political Organizations

**Alianza por Chile:** Santiago; f. 1996 as the Unión por Chile; name changed to above in 1999; right-wing alliance; Leader JOAQUÍN LAVÍN INFANTE.

   **Partido Renovación Nacional (RN):** Antonio Varas 454, Providencia, Santiago; tel. (2) 373-8740; fax (2) 373-8704; e-mail cmonckeberg@rn.cl; internet www.rn.cl; f. 1987; right-wing; Pres. SERGIO DIEZ URZÚA; Sec.-Gen. CRISTIAN MONCKEBERG BRUNETT.

   **Partido Unión Demócrata Independiente (UDI):** Suecia 286, Providencia, Santiago; tel. (2) 241-4200; fax (2) 233-6189; e-mail udi@caudi.cl; internet www.udi.cl; f. 1989; right-wing; Pres. PABLO LONGUEIRA MONTES; Sec.-Gen. PATRICIO MELERO ABAROA.

**Concertación de Partidos por la Democracia (CPD):** Londres 57, Santiago; tel. and fax (2) 639-7170; fax (2) 639-7449; e-mail concert@ctcreuna.cl; f. 1988 as the Comando por el No, an opposition front to campaign against the military regime in the plebiscite of 5 October 1988; name changed to above following plebiscite; Leader RICARDO LAGOS ESCOBAR.

   **Partido Demócrata Cristiano (PDC):** Alameda B. O'Higgins 1460, 2°, Santiago; tel. (2) 757-4400; fax (2) 757-4400; e-mail info@pdc.cl; internet www.pdc.cl; f. 1957; member of CPD; Pres. ADOLFO ZALDÍVAR; Sec.-Gen. JAIME MULET MARTÍNEZ.

   **Partido Liberal (PL):** Eduardo de la Barra 1384, Of. 404, Providencia, Santiago; tel. (2) 335-5233; fax (2) 233-2750; e-mail info@partidoliberal.cl; internet www.partidoliberal.cl; liberal party; f. 1998 by dissident centrist politicians; Pres. JOSÉ DUCCI CLARO; Sec.-Gen. CELSO HORMAZÁBAL SUAZO.

   **Partido por la Democracia (PPD):** Erasmo Escala 2154, Santiago; tel. (2) 735-2824; fax (2) 735-1692; internet www.ppd.cl; f. 1989; Pres. VÍCTOR BARRUETO; Sec.-Gen. ESTEBAN VALENZUELA VAN TREEK.

   **Partido Radical Socialdemócrata (PRSD):** Miraflores 495, Santiago; tel. (2) 632-2161; fax (2) 632-2161; centre-left; allied to CPD; Sec.-Gen. ISIDRO SOLIS PALMA.

   **Partido Socialista de Chile (PS):** Paris 873, Santiago; tel. (2) 630-6900; fax (2) 672-0507; e-mail pschile@pschile.cl; internet www.pschile.cl; f. 1933; left-wing; mem. of Socialist International; Pres. RICARDO NÚÑEZ; Sec.-Gen. ARTURO BARRIOS OTEÍZA.

**Partido Comunista de Chile (PCCh):** Avda Vicuña Mackenna 31, Santiago; tel. and fax (2) 695-4791; fax (2) 695-1150; e-mail www@pcchile.cl; internet www.pcchile.cl; f. 1912; achieved legal status in October 1990; Sec-Gen. GUILLERMO TELLIER.

**Partido Humanista (PH):** Quebec 564, Providencia, Santiago; tel. (2) 269-6116; internet www.partidohumanista.cl; e-mail secretario@partidohumanista.cl; Pres. EFREN OSORIO; Sec.-Gen. MARILEN CABRERA.

**Partido Izquierda Cristiana (PIC):** Compañia 2404, Santiago; tel. (2) 671-7681; fax (2) 671-7837; e-mail hmario@ctcinternet.cl; Pres. CARLOS DONOSO PACHECO; Sec.-Gen. PATRICIO VÉJAR MERCADO.

## Diplomatic Representation

### EMBASSIES IN CHILE

**Argentina:** Miraflores 285, Santiago; tel. (2) 582-2500; fax (2) 639-3321; e-mail embajador@embargentina.cl; internet www.embargentina.cl; Ambassador CARLOS ENRIQUE ABIHAGGLE.

**Australia:** Isidora Goyenechea 3621, 12° y 13°, Casilla 33, Correo 10 Las Condes, Santiago; tel. (2) 550-3500; fax (2) 331-5960; e-mail consular.santiago@dfat.gov.au; internet www.chile.embassy.gov.au; Ambassador ELIZABETH SCHICK.

**Austria:** Isidora Goyenechea 2934, Of. 601, Las Condes, Santiago; tel. (2) 233-0557; fax (2) 233-6971; e-mail santiagodechile@bmaa.gv.at; Ambassador WALTER HOWADT.

**Belgium:** Edif. Forum, Providencia 2653, 11°, Of. 1103, Santiago; tel. (2) 232-1070; fax (2) 232-1073; e-mail santiago@diplobel.org; internet www.diplomatie.be/santiago; Ambassador FRANCIS DE SUTTER.

**Brazil:** Alonso Ovalle 1665, Casilla 1497, Santiago; tel. (2)698-2486; fax (2) 671-5961; internet www.brasembsantiago.cl; e-mail embrasil@brasembsantiago.cl; Ambassador GELSON FONSECA JÚNIOR.

**Bulgaria:** Rodolfo Bentjerodt 4895, Vitacura, Santiago; tel. (2) 228-3110; fax (2) 208-0404; e-mail embul@entelchile.net; Chargé d'affaires PETER D. ATANASSOV.

**Canada:** Edif. World Trade Center, Torre Norte, 12°, Nueva Tajamar 481, Santiago; tel. (2) 362-9660; fax (2) 362-9663; e-mail stago@international.gc.ca; internet www.dfait-maeci.gc.ca/santiago; Ambassador BERNARD GIROUX.

**China, People's Republic:** Pedro de Valdivia 550, Santiago; tel. (2) 233-9880; fax (2) 234-1129; e-mail embajadachina@entelchile.net; Ambassador LI CHANGHUA.

**Colombia:** Presidente Errázuriz 3943, Las Condes, Santiago; tel. (2) 206-1314; fax (2) 208-0712; e-mail esantiag@minrelext.gov.co; Chargé d'affaires a.i. SALVADOR ARAÑA SUS.

**Costa Rica:** Calle Zurich 255, Dpto 85, Las Condes, Santiago; tel. (2) 334-9486; fax (2) 334-9490; e-mail embacostarica@adsl.tie.cr; Ambassador GUSTAVO CAMPOS FALLAS.

**Croatia:** Ezequias Alliende 2370, Providencia, Santiago; tel. (2) 269-6141; fax (2) 269-6092; e-mail embajada@croacia.cl; Ambassador BORIS MARUNA.

**Cuba:** Avda Los Leones 1346, Providencia, Santiago; tel. (2) 494-1485; fax (2) 494-1495; e-mail afragap@vtr.net; Ambassador ALFONSO FRAGA PÉREZ.

**Czech Republic:** Avda El Golf 254, Santiago; tel. (2) 232–1066; fax (2) 232-0707; e-mail santiago@embassy.mzv.cz; internet www.mfa.cz/santiago; Ambassador LUBOMÍR HLADÍK.

**Denmark:** Jacques Cazotte 5531, Casilla 13430, Vitacura, Santiago; tel. (2) 941-5100; fax (2) 218-1736; e-mail sclamb@um.dk; internet www.ambsantiago.um.dk/la; Ambassador ANITA HUGAU.

**Dominican Republic:** Augusto Leguia Sur 79, Of. 1802, Las Condes, Santiago; tel. (2) 245-0667; fax (2) 245-1648; e-mail embajada@rd.tie.cl; Ambassador AMABLE PADILLA GUERRERO.

**Ecuador:** Avda Providencia 1979 y Pedro Valdivia, 5°, Santiago; tel. (2) 231-5073; fax (2) 232-5833; e-mail embajadaecuador@adsl.tie.cl; Ambassador GONZALO SALVADOR HOLGUÍN.

**Egypt:** Roberto del Río 1871, Providencia, Santiago; tel. (2) 274-8881; fax (2) 274-6334; e-mail egipto@ctcinternet.cl; Ambassador ABD-EL MOHSEN OMAR MAKHION.

**El Salvador:** Coronel 2330, 5°, Of. 51, Santiago; tel. (2) 233-8324; fax (2) 231-0960; e-mail embajada.deels001@chilnet.cl; Ambassador AIDA ELENA MINERO REYES.

**Finland:** Alcántara 200, Of. 201, Las Condes, Santiago; tel. (2) 263-4917; fax (2) 263-4701; e-mail sanomat.snt@formin.fi; Ambassador PEKKA J. KORVENHEIMO.

**France:** Condell 65, Casilla 38D, Providencia, Santiago; tel. (2) 470-8000; fax (2) 470-8050; e-mail ambafran@ia.cl; internet www.france.cl; Ambassador ELISABETH BETON-DÉLÈGUE.

**Germany:** Las Hualtatas 5677, Vitacura, Santiago; tel. (2) 463-2500; fax (2) 463-2525; e-mail central@embajadadealemania.cl; internet www.embajadadealemania.cl; Ambassador JOACHIM SCHMILLEN.

**Greece:** Jorge Sexto 306, Las Condes, Santiago; tel. (2) 212-7900; fax (2) 212-8048; e-mail embassygr@tie.cl; internet www.grecia.cl; Ambassador HARIS DIMITRIOU.

**Guatemala:** Séptimo de Línea 1262, Providencia, Santiago; tel. (2) 264-0525; fax (2) 264-1146; e-mail embchile@minex.gob.gt; Ambassador ANTONIO R. CASTELLANOS.

**Haiti:** Zurich 255, Of. 21, Las Condes, Santiago; tel. (2) 650-8180; fax (2) 334-0384; e-mail embhai@terra.cl; Ambassador RONALD PIERRE.

**Holy See:** Calle Nuncio Sótero Sanz 200, Casilla 16.836, Correo 9, Santiago (Apostolic Nunciature); tel. (2) 231-2020; fax (2) 231-0868; e-mail nunciatu@entelchile.net; Nuncio Most Rev. ALDO CAVALLI (Titular Archbishop of Vibo Valentia).

**Honduras:** Zurich 255, Dpto 51, Las Condes, Santiago; tel. (2) 234-4069; fax (2) 334-7946; e-mail honduras@entelchile.net; Ambassador EDUARDO KAWAS GATTAS.

**Hungary:** Avda Los Leones 2279, Providencia, Santiago; tel. (2) 204-2210; fax (2) 234-1227; e-mail huembstg@entelchile-net; Ambassador GYULA BARCSI.

**India:** Triana 871, Casilla 10433, Santiago; tel. (2) 235-2005; fax (2) 235-9607; e-mail embindia@entelchile.net; Ambassador SUSMITA GONGULEE THOMAS.

**Indonesia:** Nueva Costanera 3318, Vitacura, Santiago; tel. (2) 207-6266; fax (2) 207-9901; e-mail kbristgo@mi-mail.cl; Ambassador SUWARNO ATMOPRAWIRO.

**Israel:** San Sebastián 2812, 5°, Las Condes, Santiago; tel. (2) 750-0500; fax (2) 750-0555; e-mail info@santiago.mfa.gov.il; internet santiago.mfa.gov.il; Ambassador JOSEF REGEV.

**Italy:** Clemente Fabres 1050, Providencia, Santiago; tel. (2) 470-8400; fax (2) 223-2467; e-mail info@embitalia.cl; internet www.embitalia.cl; Ambassador GIOVANNI FERRERO.

**Japan:** Avda Ricardo Lyon 520, Santiago; tel. (2) 232-1807; fax (2) 232-1812; internet www.cl.emb-japan.go.jp; e-mail embajada.dejap001@chilnet.cl; Ambassador Hajime Ogawa.

**Jordan:** San Pascual 446, Las Condes, Casilla 00562, Santiago; tel. (2) 325-7748; fax (2) 325-7754; e-mail jordanemb@vtr.net; Ambassador (vacant).

**Korea, Democratic People's Republic:** Ambassador Yu Chang Un (resident in Peru).

**Korea, Republic:** Alcántara 74, Casilla 1301, Santiago; tel. (2) 228-4214; fax (2) 206-2355; e-mail corembad@tie.cl; Ambassador Shin Jang-bum.

**Lebanon:** Alianza 1728, Casilla 1950, Santiago; tel. (2) 219-9724; fax (2) 219-3502; e-mail libano@netline.cl; Ambassador Mourad Jammal.

**Malaysia:** Tajamar 183, 10° y 11°, Of. 1002, Correo 35, Las Condes, Santiago; tel. (2) 233-6698; fax (2) 234-3853; e-mail mwstg@embdemalasia.cl; Ambassador A. Ganapathy.

**Mexico:** Félix de Amesti 128, Las Condes, Santiago; tel. (2) 206-6133; fax (2) 206-6147; e-mail embamex@mi.cl; Ambassador Ricardo Villanueva Hallal.

**Morocco:** Avda Luis Pasteur 5850, Of. 203, Vitacura, Santiago; tel. (2) 218-0311; fax (2) 219-4280; e-mail ambmarch@terra.cl; internet www.marruecos.cl; Ambassador Abdelhadi Boucetta.

**Netherlands:** Las Violetas 2368, Casilla 56-D, Santiago; tel. (2) 756-9200; fax (2) 756-9226; e-mail stg@minbuza.nl; internet www.holanda-paisesbajos.cl; Ambassador Hinkinus Nijenhuis.

**New Zealand:** El Golf 99, Of. 703, Las Condes, Santiago; tel. (2) 290-9800; fax (2) 458-0940; e-mail embajada@nzembassy.cl; internet www.nzembassy.com/home.cfm?c=16; Ambassador Nigel Fyfe.

**Nicaragua:** Zurich 255, Of. 111, Las Condes, Santiago; tel. (2) 234-1808; fax (2) 234-5071; e-mail embanic@embajadadenicaragua.tie.cl; Ambassador Edgard Manuel Escobar Fornos.

**Norway:** San Sebastián 2839, Of. 509, Casilla 2431, Santiago; tel. (2) 234-2888; fax (2) 234-2201; e-mail emb.santiago@mfa.no; internet www.noruega.cl; Ambassador Pål Moe.

**Panama:** La Reconquista 640, Las Condes, Santiago; tel. (2) 202-6318; fax (2) 202-5439; e-mail embajada@panamachile.tie.cl; Ambassador Alejandro Young Downey.

**Paraguay:** Huérfanos 886, 5°, Ofs 514–515, Santiago; tel. (2) 639-4640; fax (2) 633-4426; e-mail epychemb@entelchile.net; Ambassador Juan Andrés Cardozo Domínguez.

**Peru:** Avda Andrés Bello 1751, Casilla 16277, Providencia, Santiago; tel. (2) 235-2356; fax (2) 235-8139; e-mail embstgo@entelchile.net; Ambassador José Antonio Meier Espinoza.

**Philippines:** Félix de Amesti 367, Santiago; tel. (2) 208-1313; fax (2) 208-1400; e-mail embajada.defil001@chilnet.cl; Ambassador Hermenegildo C. Cruz.

**Poland:** Mar del Plata 2055, Santiago; tel. (2) 204-1213; fax (2) 204-9332; internet www.polonia.cl; e-mail embchile@entelchile.net; Ambassador Jarosław Spyra.

**Portugal:** Nueva Tajamar 555, Torre Norte 16°, Las Condes, Santiago; tel. (2) 203-0542; fax (2) 203-0605; e-mail embaixada.portugal@entelchile.net; Ambassador António Machado de Faria e Maya.

**Romania:** Benjamín 2955, Las Condes, Santiago; tel. (2) 231-1893; fax (2) 232-3441; e-mail embajada@rumania.tie.cl; internet www.rumania.cl; Chargé d'affaires Cristian Lazarescu.

**Russia:** Cristobal Colón 4152, Las Condes, Santiago; tel. (2) 208-6254; fax (2) 206-8892; e-mail embajada@rusia.tie.cl; internet www.chile.mid.ru/2005/bull_006.html; Ambassador Vladimir V. Chkhikvadze.

**South Africa:** Avda 11 de Septiembre 2353, 16°, Torre San Ramón, Santiago; tel. (2) 231-2860; fax (2) 231-3185; e-mail embsachi@interaccess.cl; internet www.embajada-sudafrica.cl; Ambassador Victor Zazeraj.

**Spain:** Avda Andrés Bello 1895, Casilla 16456, Providencia, Santiago; tel. (2) 235-2755; fax (2) 235-1049; e-mail embespcl@correo.mae.es; Ambassador José Antonio Martínez de Villareal y Boena.

**Sweden:** CP 16639, Santiago; tel. (2) 940-1700; fax (2) 940-1730; e-mail ambassaden.santiago-de-chile@foreign.ministry.se; internet www.embajadasuecia.cl; Ambassador Arne Rodin.

**Switzerland:** Avda Américo Vespucio Sur 100, 14°, Las Condes, Santiago; tel. (2) 263-4211; fax (2) 263-4094; e-mail vertretung@san.rep.admin.ch; internet www.eda.admin.ch/santiago_emb/s/home.html; Ambassador André Regli.

**Syria:** Carmencita 111, Casilla 12, Correo 10, Santiago; tel. (2) 232-7471; Ambassador Hisham Hallaj.

**Thailand:** Avda Américo Vespucio 100, 15°, Las Condes, Santiago; tel. (2) 263-0710; fax (2) 263-0803; e-mail thaichil@ctcreuna.cl; internet www.rte-chile.thaiembdc.org; Ambassador Pithaya Pookaman.

**Turkey:** Edif. Montolin, Of. 71, Monseñor Sotero Sanz 55, Providencia, Santiago; tel. (2) 231-8952; fax (2) 231-7762; e-mail turquia@manquehue.net; Ambassador Aysenur Alpaslan.

**United Kingdom:** Avda el Bosque Norte 0125, Casilla 72-D, Santiago; tel. (2) 370-4100; fax (2) 370-4180; e-mail chancery.santiago@fco.gov.uk; internet www.britemb.cl; Ambassador Howard Drake.

**USA:** Avda Andrés Bello 2800, Las Condes, Santiago; tel. (2) 232-2600; fax (2) 330-3710; internet www.usembassy.cl; Ambassador Craig Kelly.

**Uruguay:** Avda Pedro de Valdivia 711, Santiago; tel. (2) 204-7988; fax (2) 274-4066; e-mail urusgo@uruguay.cl; internet www.uruguay.cl; Ambassador Alejandro Antonio Lorenzo y Losada Aldunate.

**Venezuela:** Bustos 2021, Providencia, Santiago; tel. (2) 225-0021; fax (2) 223-1170; e-mail embajada@embavenez.cl; internet www.embavenez.cl; Ambassador Víctor Eloy Delgado Monsalve.

# Judicial System

The Supreme Courts consist of 21 members.

There are Courts of Appeal (in the cities or departments of Arica, Iquique, Antofagasta, Copiapó, La Serena, Valparaíso, Santiago, San Miguel, Rancagua, Talca, Chillán, Concepción, Temuco, Valdivia, Puerto Montt, Coyhaique and Punta Arenas) whose members are appointed from a list submitted to the President of the Republic by the Supreme Court. The number of members of each court varies. Judges of the lower courts are appointed in a similar manner from lists submitted by the Court of Appeal of the district in which the vacancy arises. Judges and Ministers of the Supreme Court do not continue in office beyond the age of 75 years.

In March 1998 a major reform of the judiciary was implemented, including an increase, from 17 to 21, in the number of Ministers of the Supreme Court.

### Corte Suprema
Bandera 344, 2°, Santiago; tel. (2) 873-5258; fax (2) 873-5276; e-mail mgonzalezp@poderjudicial.cl; internet www.poderjudicial.cl.

**President of the Supreme Court:** Marcos Libedinsky Tschorne.

**Ministers of the Supreme Court:** Hernán Alvarez García, Eleodoro Ortiz Sepúlveda, José Benquis Camhi, Enrique Tapia Witting, Ricardo Gálvez Blanco, Alberto Chaigneau del Campo, Jorge Rodríguez Ariztía, Enrique Cury Urzúa, José Luis Pérez Zañartu, Orlando Alvarez Hernández, Urbano Marín Vallejo, Domingo Yurac Soto, Humberto Espejo Zúñiga, Jorge Medina Cuevas, Domingo Alfonso Mourgues, Milton Ivan Arancibia, Nibaldo Segura Peña, Maria Morales Villagran, Adalis Oyarzun Miranda, Jaime del Carmen Rodríguez Espoz.

**Attorney-General:** Monica Maldonado Croquevielle.

**Secretary of the Court:** Carlos A. Meneses Pizarro.

# Religion

## CHRISTIANITY

### The Roman Catholic Church

Some 76% of the population are Roman Catholics; there were an estimated 11.0m. adherents at 31 December 2003. Chile comprises five archdioceses, 18 dioceses, two territorial prelatures and one apostolic vicariate.

### Bishops' Conference
Conferencia Episcopal de Chile, Echaurren 4, 6°, Casilla 517-V, Correo 21, Santiago; tel. (2) 671-7733; fax (2) 698-1416; e-mail secretariageneral@episcopado.cl; internet www.iglesia.cl.

f. 1955 (statutes approved 2000); Pres. Mgr Alejandro Goic Karmelic O. Rancagua (Archbishop of Santiago de Chile).

**Archbishop of Antofagasta:** Pablo Lizama Riquelme, San Martín 2628, Casilla E, Antofagasta; tel. and fax (55) 26-8856; e-mail antofagasta@episcopado.cl.

**Archbishop of Concepción:** Antonio Moreno Casamitjana, Calle Barros Arana 544, Casilla 65-C, Concepción; e-mail amoreno@episcopado.cl; tel. (41) 22-8173; fax (41) 23-2844.

**Archbishop of La Serena:** Manuel Gerardo Donoso Donoso, Los Carrera 450, Casilla 613, La Serena; tel. (51) 21-2325; fax (51) 22-5886; e-mail laserena@episcopado.cl.

**Archbishop of Puerto Montt:** CRISTIÁN CARO CORDERO, Calle Benavente 385, Casilla 17, Puerto Montt; tel. (65) 25-2215; fax (65) 27-1861; e-mail puertomontt@episcopado.cl.

**Archbishop of Santiago de Chile:** Cardinal FRANCISCO JAVIER ERRÁZURIZ OSSA, Erasmo Escala 1884, Casilla 30-D, Santiago; tel. (2) 696-3275; fax (2) 671-2042; e-mail curiasantiago@arzobispado.tie.cl.

### The Anglican Communion

Anglicans in Chile come within the Diocese of Chile, which forms part of the Anglican Church of the Southern Cone of America, covering Argentina, Bolivia, Chile, Paraguay, Peru and Uruguay.

**Bishop of Chile:** Rt Rev. H. F. ZAVALA M, Iglesia Anglicana, José Miguel de la Barra 480, Of. 205, Casilla 50675, Correo Central, Santiago, Santiago; tel. (2) 639-1509; fax (2) 639-4581; e-mail fzavala@evangel.cl; internet www.iglesiaanglicana.cl.

### Other Christian Churches

**Baptist Evangelical Convention:** Casilla 41-22, Santiago; tel. (2) 222-4085; fax (2) 635-4104; f. 1908; Pres. MOISÉS PINTO; Gen. Sec. VÍCTOR OLIVARES.

**Evangelical Lutheran Church:** Pedro de Valdivia 3420-H, Dpto 33, Nuñoa, Casilla 15167, Santiago; tel. (2) 225-0091; fax (2) 205-2193; e-mail iglesia_evangelica_ielch@adsl.tie.cl; f. 1937 as German Evangelical Church in Chile; present name adopted in 1959; Pres. GLORIA ROJAS; 3,000 mems.

**Jehovah's Witnesses:** Avda Concha y Toro 3456, Puente Alto; tel. (2) 288-1264; fax (2) 288-1257; Dir PEDRO J. LOVATO GROSSO.

**Methodist Church:** Sargento Aldea 1041, Casilla 67, Santiago; tel. (2) 556-6074; fax (2) 554-1763; autonomous since 1969; 7,317 mems; Bishop NEFTALÍ ARAVENA BRAVO.

**Orthodox Church of the Patriarch of Antioch:** Avda Perú 502, Recoleta, Santiago; tel. and fax (2) 737-4697; Archbishop Mgr. SERGIO ABAD.

**Pentecostal Church:** Calle Pena 1103, Casilla de Correo 2, Curicó; tel. (75) 1035; f. 1945; 90,000 mems; Bishop ENRIQUE CHÁVEZ CAMPOS.

**Pentecostal Church Mission:** Calle Passy 32, Santiago; tel. (2) 634-6785; fax (2) 634-6786; f. 1952; Sec. Rev. DANIEL GODOY FERNÁNDEZ; Pres. Rev. ERASMO FARFÁN FIGUEROA; 12,000 mems.

### JUDAISM

**Comité Representativo de las Entidades Judías en Chile (CREJ):** Avda Ricardo Lyon 812, Providencia, Santiago; tel. (2) 274-7101; fax (2) 269-7005; Pres. ELIMAT Y. JASON.

**Comunidad Israelita Sefardi de Chile:** Avda Ricardo Lyon 812, Providencia, Santiago; tel. (2) 209-8086; fax (2) 204-7382; Pres. SALOMON CAMHI AVAYU, Rabbi IOSEF GABAY.

### ISLAM

**Sociedad Unión Musulmana:** Mezquita As-Salam, Campoamor 2975, esq. Chile-España, Ñuñoa, Santiago; tel. (2) 343-1376; fax (2) 343-11378; Pres. OUSAMA ABUGHAZALÉ.

### BAHÁ'Í FAITH

**National Spiritual Assembly:** Manuel de Salas 356, Casilla 3731, Ñuñoa, Santiago; tel. (2) 269-2005; fax (2) 225-8276; e-mail secretaria@bahai.cl; internet www.bahai.cl; Co-ordinator REED CHANDLER REED; Sec. FARBOD YOUSSEFI.

## The Press

Most newspapers of nation-wide circulation in Chile are published in Santiago.

### DAILIES

Circulation figures listed below are supplied mainly by the Asociación Nacional de la Prensa. Other sources give much lower figures.

#### Santiago

**La Cuarta:** Diagonal Vicuña Mackenna 1842, Casilla 2795, Santiago; tel. (2) 555-0034; fax (2) 556-1017; e-mail lacuarta@copesa.cl; internet www.lacuarta.cl; morning; Gen. Man. JUAN CARLOS LARRAÍN WORMALD.

**El Diario Financiero:** San Crescente 81, 3°, Las Condes, Santiago; tel. (2) 339-1000; fax (2) 231-3340; e-mail buzon@eldiario; internet www.eldiario.cl; f. 1988; morning; Dir GUILLERMO TURNER OLEA; Gen. Man. JOSÉ MIGUEL RESPALDIZA CHICHARRO; circ. 20,000.

**Diario Oficial de la República de Chile:** Casilla 81-D, Agustinas 1269, Santiago; tel. (2) 698-3969; fax (2) 698-1059; internet www.diarioficial.cl; f. 1877; Dir FLORENCIO CEBALLOS BUSTOS; circ. 10,000.

**Estrategia:** Luis Carrera 1298, Vitacura, Santiago; tel. (2) 655-6228; fax (2) 655-6256; e-mail estrategia@edgestion.cl; internet www.estrategia.cl; f. 1978; morning; Editor MÓNICA HABERLAND.

**El Mercurio:** Avda Santa María 5542, Casilla 13-D, Santiago; tel. (2) 330-1111; fax (2) 242-1131; e-mail jpillanes@mercurio.cl; internet www.elmercurio.cl; f. 1827; morning; conservative; Gen. Man. FERNANDO CISTERNAS BRAVO; circ. 120,000 (weekdays), 280,000 (Sun.).

**La Nación:** Agustinas 1269, Casilla 81-D, Santiago; tel. (2) 787-0100; fax (2) 698-1059; internet www.lanacion.cl; f. 1917 to replace govt-subsidized *El Cronista*; morning; financial; Propr Soc. Periodística La Nación; Dir JUAN WALKER EDWARDS; Gen. Man. FRANCISCO FERES NAZARALA; circ. 45,000.

**Santiago Times:** tel. (2) 777-5376; e-mail info@santiagotimes.cl; internet www.santiagotimes.cl; daily; national news in English; Editor STEPHEN ANDERSON; 10,000 subscribers.

**La Segunda:** Avda Santa María 5542, Casilla 13-D, Santiago; tel. (2) 330-1111; fax (2) 228-9289; e-mail redaccion@lasegunda.cl; internet www.lasegunda.com; f. 1931; owned by proprs of *El Mercurio*; evening; Dir CRISTIÁN ZEGERS ARIZTÍA; circ. 40,000.

**La Tercera:** Avda Vicuña Mackenna 1962, Ñuñoa, Santiago; tel. (2) 550-7000; fax (2) 550-7999; e-mail latercera@copesa.cl; internet www.latercera.cl; f. 1950; morning; Dir CRISTIÁN BOFILL RODRÍGUEZ; circ. 200,000.

**Las Ultimas Noticias:** Bellavista 0112, Providencia, Santiago; tel. (2) 730-3000; fax (2) 730-3331; e-mail rodolfo.gambetti@lun.cl; internet www.lun.cl; f. 1902; morning; Dir AGUSTÍN EDWARDS DEL RÍO; Gen. Man. JUAN ENRIQUE CANALES BESA; owned by the proprs of *El Mercurio*; circ. 150,000 (except Sat. and Sun.).

#### Antofagasta

**La Estrella del Norte:** Manuel Antonio Matta 2112, Antofagasta; tel. (55) 26-4835; f. 1966; evening; Dir CAUPOLICÁN MÁRQUEZ VERGARA; circ. 5,000.

**El Mercurio:** Manuel Antonio Matta 2112, Antofagasta; tel. (55) 45-3600; fax (55) 425-3612; e-mail cronicafta@mercurio.cl; internet www.mercurioantofagasta.cl; f. 1906; morning; conservative independent; Proprs Soc. Chilena de Publicaciones; Dir ARTURO ROMÁ HERRERA; circ. 9,000.

#### Arica

**La Estrella de Arica:** San Marcos 580, Arica; tel. (58) 22-5024; fax (58) 25-2890; f. 1976; Dir REINALDO NEIRA RUIZ; circ. 10,000.

#### Atacama

**Chañarcillo:** Los Carrera 801, Chañaral, Atacama; tel. (52) 21-9044; f. 1992; morning; Dir LUIS CERPA HIDALGO.

#### Calama

**El Mercurio:** Abaroa 2051, Calama; tel. (55) 34-1604; fax (55) 36-4255; e-mail cronicacalama@mercurio.cl; internet www.mercuriocalama.cl; f. 1968; propr Soc. Chilena de Publicaciones; Dir ROBERTO GAETE; circ. 4,500 (weekdays), 7,000 (Sun.).

#### Chillán

**La Discusión de Chillán, SA:** Calle 18 de Septiembre 721, Casilla 479, Chillán; tel. (42) 21-2650; fax (42) 21-3578; f. 1870; morning; independent; Dir TITO CASTILLO PERALTA; circ. 5,000.

#### Concepción

**El Sur:** Calle Freire 799, Casilla 8-C, Concepción; tel. (41) 23-5825; f. 1882; morning; independent; Dir RAFAEL MAIRA LAMAS; circ. 28,000 (weekdays), 45,000 (Sun.).

#### Copiapó

**Atacama:** Manuel Rodríguez 740, Copiapó; tel. (52) 2255; morning; independent; Dir SAMUEL SALGADO; circ. 6,500.

#### Coyhaique

**El Diario de Aisén:** 21 de Mayo 410, Coyhaique; tel. (67) 234-850; fax (67) 232-318; Dir ALDO MARCHESSE COMPODÓNICO.

#### Curicó

**La Prensa:** Merced 373, Curicó; tel. (75) 31-0453; fax (75) 31-1924; e-mail laprensa@entelchile.net; internet diariolaprensa.cl; f. 1898; morning; right-wing; Man. Dir MANUEL MASSA MAUTINO; Editor CARLOS POZO CARVACHO; Pres. CARLOS LAZCANO ALFONSO; circ. 6,000.

CHILE                                                                                                                    *Directory*

### Iquique

**La Estrella de Iquique:** Luis Uribe 452, Iquique; tel. (57) 42-2805; fax (57) 42-7975; f. 1966; evening; Dir Sergio Montivera Bruna; circ. 10,000.

**El Nortino:** Baquedano 1470, Iquique; tel. (57) 41-6666; fax (57) 41-2997; f. 1992; morning; Dir-Gen. Reynaldo Berríos González.

### La Serena

**El Día:** Brasil 431, La Serena; tel. (51) 22-2863; fax (51) 22-2844; f. 1944; morning; Dir Antonio Puga Rodríguez; circ. 10,800.

### Los Angeles

**La Tribuna:** Calle Colo Colo 464, Casilla 15-D, Los Angeles; tel. (43) 31-3315; fax (43) 31-1040; independent; Dir Cirilo Guzmán de la Fuente; circ. 4,500.

### Osorno

**El Diario Austral:** Yungay 499, Valdivia; tel. (64) 23-5191; fax (64) 23-5192; e-mail vpineda@australvaldivia.cl; internet www.australvaldivia.cl; f. 1982; Dir Verónica Moreno Aguilera; circ. 6,500 (weekdays), 7,300 (Sun.).

### Ovalle

**El Ovallino:** Victoria 323-B, Ovalle; tel. and fax (53) 627-557; Dir Jorge Contador Araya.

### Puerto Montt

**El Llanquíhue:** Antonio Varas 167, Puerto Montt; tel. (65) 25-5115; fax (65) 432-401; f. 1885; Dir Ernesto Montalba; circ. 4,800 (weekdays), 5,700 (Sun.).

### Punta Arenas

**La Prensa Austral:** Waldo Seguel 636, Casilla 9-D, Punta Arenas; tel. (61) 20-4000; fax (61) 24-7406; e-mail direccion@laprensaaustral.cl; internet www.laprensaaustral.cl; f. 1941; morning; independent; Dir Manuel González Araya; circ. 10,000, Sunday (*El Magallanes*; f. 1894) 12,000.

### Quillota

**El Observador:** La Concepción 277, Casilla 1-D, Quillota; tel. (33) 312-096; fax (33) 311-417; e-mail elobser@entelchile.net; internet www.diarioelobservador.cl; f. 1970; Tuesdays and Fridays; Man. Dir Roberto Silva Bijit.

### Rancagua

**El Rancagüino:** O'Carroll 518, Casilla 50, Rancagua; tel. (72) 23-0358; fax (72) 22-1483; e-mail prensaelrancaguino@adsl.tie.cl; internet www.elrancaguino.cl; f. 1915; independent; Dir Alejandro González; Gen. Man. Fernando Reyes; circ. 10,000.

### Talca

**El Centro:** Tres Oriente 798, Talca; tel. (71) 22-0946; fax (71) 22-0924; f. 1989; Gen. Man. Hugo Saavedra Oteiza.

### Temuco

**El Diario Austral:** Antonio Varas 945, Casilla 1-D, Temuco; tel. (45) 21-2575; fax (45) 23-9189; f. 1916; morning; commercial, industrial and agricultural interests; Dir Marco Antonio Pinto Zepeda; propr Soc. Periodística Araucanía, SA; circ. 15,100 (weekdays), 23,500 (Sun.).

### Tocopilla

**La Prensa de Tocopilla:** Bolívar 1244, Tocopilla; tel. (83) 81-3036; f. 1924; morning; independent; Gen. Man. Jorge Leiva Concha; circ. 3,000.

### Valdivia

**El Diario Austral:** Yungay 499, Valdivia; tel. (63) 24-2200; fax (63) 24-2209; internet www.australvaldivia.cl; f. 1982; Dir Gustavo Serrano Cotapos; circ. 5,600.

### Valparaíso

**La Estrella:** Esmeralda 1002, Casilla 57-V, Valparaíso; tel. (32) 26-4264; fax (32) 26-4241; internet www.mercuriovalpo.cl; e-mail purzua@mercurio.cl; f. 1921; evening; independent; Dir Alfonso Castagneto; owned by the proprs of *El Mercurio*; circ. 28,000 (weekdays), 35,000 (Sat.).

**El Mercurio:** Esmeralda 1002, Casilla 57-V, Valparaíso; tel. (32) 26-4264; fax (32) 26-4138; internet www.elmercuriovalpo.cl; f. 1827; morning; Dir Marco Antonio Pinto Zepada; owned by the proprs of *El Mercurio* in Santiago; circ. 65,000.

## PERIODICALS

### Santiago

**América Economía:** Apoquindo 4499, 10°, Las Condes, Santiago; tel. (2) 290-9400; fax (2) 206-6005; e-mail icolodro@aeconomia.cl; internet www.americaeconomia.com; e-mail rferro@aeconomia.cl; f. 1986; monthly; business; CEO Elías Selman; Publr Nils Strandberg; Editorial Dir Raúl Ferro Isakaz.

**Apsi:** Santiago; tel. (2) 77-5450; f. 1976; fortnightly; current affairs; Dir Marcelo Contreras Nieto; circ. 15,000.

**La Bicicleta:** José Fagnano 614, Santiago; tel. (2) 222-3969; satirical; Dir Antonio de la Fuente.

**CA (Ciudad/Arquitectura) Revista Oficial del Colegio de Arquitectos de Chile AG:** Manuel Montt 515, Santiago; tel. (2) 235-3368; fax (2) 235-8403; f. 1964; 4 a year; architects' magazine; Editor Arq. Jaime Márquez Rojas; circ. 3,500.

**Caras:** Reyes Lavalle 3194, Las Condes, Santiago; tel. (2) 399-6399; fax 399-6299; internet www.caras.cl; e-mail revista@caras.cl; women's interest; Dir Patricia Guzmán.

**Carola:** San Francisco 116, Casilla 1858, Santiago; tel. (2) 33-6433; fortnightly; women's magazine; published by Editorial Antártica, SA; Dir Isabel Margarita Aguirre de Maino.

**Cauce:** Huérfanos 713, Of. 604–60, Santiago; tel. (2) 38-2304; fortnightly; political, economic and cultural affairs; Dir Angel Flisfich; circ. 10,000.

**Chile Agrícola:** Teresa Vial 1170, Casilla 2, Correo 13, Santiago; tel. and fax (2) 522-2627; e-mail chileagricola@hotmail.com; f. 1975; 6 a year; organic farming; Dir Ing. Agr. Raúl González Valenzuela; circ. 7,000.

**Chile Forestal:** Avda Bulnes 285, Of. 601, Santiago; tel. (2) 390-0213; fax (2) 696-6724; e-mail mespejo@conaf.cl; internet www.conaf.cl; f. 1974; 10 a year; state-owned; technical information and features on forestry sector; Editor Ricardo San Martín; Dir Mariela Espejo Suazo; circ. 4,000.

**Cinegrama:** Avda Holanda 279, Providencia, Santiago; tel. (2) 422-8500; fax (2) 422-8570; internet www.cinegrama.cl; e-mail cinegrama@holanda.cl; monthly; cinema; Dir Jaime Godoy; Editor Leyla López.

**The Clinic:** Santo Domingo 550, Santiago; tel. (2) 633-9584; fax (2) 639-6584; fortnightly; political and social satire; Dir Patricio Fernández.

**Conozca Más:** Reyes Lavalle 3193, Las Condes, Santiago; tel. (2) 366-7100; fax (2) 246-2810; e-mail viamail@conozcamas.cl; internet www.conozcamas.cl; monthly; science; Dir Paula Aviles; circ. 90,000.

**Cosas:** Almirante Pastene 329, Providencia, Santiago; tel. (2) 364-5100; fax (2) 235-8331; f. 1976; fortnightly; entertainment and lifestyle; Dir Mónica Comandari Kaiser; circ. 40,000.

**Creces:** Luis Uribe 2610, Ñuñoa, Santiago; tel. and fax (2) 341-5829; internet www.creces.cl; e-mail crecesawebmaster@entelchile.net; monthly; science and technology; Dir Sergio Prenafeta; circ. 12,000.

**Deporte Total:** Santiago; tel. (2) 251-6236; fax (2) 204-7420; f. 1981; weekly; sport, illustrated; Dir Juan Ignacio Oto Larios; circ. 25,000.

**Ercilla:** Avda Holanda 279, Providencia, Santiago; tel. (2) 422-8500; fax (2) 422-8570; e-mail ercilla@holanda.cl; internet www.ercilla.cl; f. 1936; weekly; general interest; conservative; Dir Eugenio González; circ. 28,000.

**Gestión:** Luis Carrera 1289, Vitacura, Santiago; tel. (2) 655-6100; fax (2) 655-6439; f. 1975; monthly; business matters; Dir Maira Jordan; circ. 38,000.

**Internet:** Avda Carlos Valdovinos 251; tel. (2) 552-5599; monthly; internet and new technology; Dir Florencio Uteras.

**Mensaje:** Almirante Barroso 24, Casilla 10445, Santiago; tel. (2) 696-0653; fax (2) 671-7030; e-mail rrpp@mensaje.cl; internet www.mensaje.cl; f. 1951; monthly; national, church and international affairs; Dir Antonio Delfau; circ. 6,000.

**Mujer a Mujer:** Vicuña McKenna 1870, Ñuñoa, Santiago; tel. (2) 550-7000; fax (2) 550-7379; internet www.mujeramujer.cl; e-mail mujer@latercera.cl; weekly; women's interest; Dir Jackeline Otey.

**News Review:** Casilla 151/9, Santiago; tel. (2) 236-9511; fax (2) 236-0887; f. 1991; weekly; English language news; Dir Graham A. Wigg.

**Paula:** Avda Santa María 0120, Providencia, Santiago; tel. (2) 200-0407; fax (2) 200-0490; e-mail paula@paulacom.cl; internet www.paula.cl; f. 1967; monthly; women's interest; Dir Paula Recart; circ. 85,000.

CHILE    *Directory*

**Punto Final:** San Diego 31, Of. 606, Casilla 13954, Correo 21, Santiago; tel. and fax (2) 697-0615; e-mail punto@interaccess.cl; internet www.puntofinal.cl; f. 1965; fortnightly; politics; left-wing; Dir MANUEL CABIESES DONOSO; circ. 15,000.

**¿Qué Pasa?:** Vicuña Mackenna 1870, Ñuñoa, Santiago; tel. (2) 550-7523; fax (2) 550-7529; internet www.quepasa.cl; e-mail quepasa@copesa.cl; f. 1971; weekly; general interest; Dir BERNADITA DEL SOLAR VERA; circ. 30,000.

**Semanario Datos Sur:** Avda Urmeneta 231, Puerto Montt; tel. (65) 26-6934; e-mail datossur@telsur.cl; internet www.datossur.cl; weekly; regional, national and business news; Dir ALEX BERKHOFF ALCARRAZ.

**Semanario El Siglo:** Diagonal Paraguay 458, Of. 1, Casilla 13479, Correo 21, Santiago; tel. and fax (2) 633-0074; e-mail elsiglo@elsiglo.cl; internet www.elsiglo.cl; f. 1940; fortnightly; published by the Communist Party of Chile (PCCh); Dir CLAUDIO DE NEGRI QUINTANA; circ. 15,000.

**Vea:** Avda Holanda 279, Providencia, Santiago; tel. (2) 422-8500; fax (2) 422-8571; e-mail revistavea@holanda.cl; internet www.vea.cl; f. 1939; weekly; general interest, illustrated; Dir JAIME GODOY CARTES; circ. 150,000.

## PRESS ASSOCIATION

**Asociación Nacional de la Prensa:** Carlos Antúnez 2048, Santiago; tel. 232-1004; fax 232-1006; e-mail info@anp.cl; internet www.anp.cl; f. 1951; Pres. JUAN LUIS SOMMERS COMANDARI; Sec. DANIELA ROJAS.

## NEWS AGENCIES

**Agencia Chile Noticias (ACN):** Carlos Antúnez 1884, Of. 104, Providencia, Santiago; tel. and fax (2) 223 0205; e-mail prensa@chilenoticias.cl; internet www.chilenoticias.cl; f. 1993; Dir JEANETE FRANCO N.

**Agencia Orbe:** Avda Phillips 56, Of. 66, Santiago; tel. (2) 639-4774; fax (2) 639-6826; Bureau Chief PATRICIA ESCALONA CÁCERES.

**Agencia UPI:** Avda Nataniel 47, 9°, Santiago; tel. (2) 696-0162; fax (2) 698-6605; Bureau Chief FRANCISCO JARA ARANCIBIA.

**Business News Americas:** Carmencita 106, Las Condes, Santiago; tel. (2) 232-0302; fax (2) 232-9376; e-mail info@bnamericas.com; internet www.bnamericas.com; internet-based business information; CEO GREGORY BARTON.

**Chile Information Project (CHIP):** Avda Santa María 227, Of. 12, Recoleta, Santiago; tel. (2) 738-0150; fax (2) 735-2267; e-mail anderson@chip.mic.cl; internet www.santiagotimes.cl; English language; Dir STEPHEN J. ANDERSON.

**Europa Press:** Biarritz 1913, Providencia, Santiago; tel. and fax (2) 274-3552; e-mail europapress@rdc.cl; Dir JOSÉ RÍOS VIAL.

### Foreign Bureaux

**Agence France-Presse** (France): Avda B. O'Higgins 1316, 9°, Apdo 92, Santiago; tel. (2) 696-0559; fax (2) 695-5036; Correspondent HUMBERTO ZUMARÁN ARAYA.

**Agencia EFE** (Spain): Coronel Santiago Bueras 188, Santiago; tel. (2) 638-0179; fax (2) 633-6130; e-mail direccion@agenciaefe.tie.cl; internet www.efe.es; f. 1966; Bureau Chief Man. MANUEL FUENTES GARCÍA.

**Agenzia Nazionale Stampa Associata (ANSA)** (Italy): Moneda 1040, Of. 702, Santiago; tel. (2) 698-5811; fax (2) 698-3447; e-mail ansacile@entelchile.net; f. 1945; Bureau Chief GIORGIO BAGONI BETTOLLINI.

**Associated Press (AP)** (USA): Tenderini 85, 10°, Of. 100, Casilla 2653, Santiago; tel. (2) 633-5015; fax 633-8368; Bureau Chief EDUARDO GALLARDO.

**Bloomberg News** (USA): Miraflores 222, Santiago; tel. (2) 638-6820; fax (2) 698-3447; Dir MIKE SMITH.

**Deutsche Presse-Agentur (dpa)** (Germany): San Antonio 427, Of. 306, Santiago; tel. (2) 639-3633; Correspondent CARLOS DORAT.

**Inter Press Service (IPS)** (Italy): Santiago; tel. (2) 39-7091; Dir and Correspondent GUSTAVO GONZÁLEZ RODRÍGUEZ.

**Prensa Latina** (Cuba): Bombero Ossa 1010, Of. 1104, Santiago; tel. (2) 671-8222; fax (2) 695-8605; Correspondent LIDIA SEÑARIS CEJAS.

**Reuters** (United Kingdom): Nueva York 33, 10°, Casilla 4248, Santiago; tel. (2) 672-8800; fax (2) 696-0161; Correspondent RAÚL CUEVAS.

**Xinhua (New China) News Agency** (People's Republic of China): Biarritz 1981, Providencia, Santiago; tel. (2) 25-5033; Correspondent SUN KUOGUOWEIN.

### Association

**Asociación de Corresponsales de la Prensa Extranjera en Chile:** Coronel Santiago Bueras 188, Santiago; tel. (2) 688-0424; fax (2) 633-6130; Pres. OMAR RUZ.

## Publishers

**Arrayán Editores:** Bernarda Morin 435, Providencia, Santiago; tel. (2) 431-4200; fax (2) 274-1041; internet www.arrayan.cl; f. 1982; general; Gen. Man. PABLO MARINKOVIC.

**Distribuidora Molino, SA:** Abtao 574, Quinta Normal, Santiago; tel. (2) 786-3000; fax (2) 776-6425; e-mail molino@molinoxxi.cl; Admin. Man. JORGE VARGAS ARAYA.

**Ediciones y Comunicaciones Ltda:** Luis Thayer Ojeda 0115, Of. 402, Providencia, Santiago; tel. (2) 232-1241; fax (2) 234-9467; e-mail edicom@chilnet.cl; internet www.chilnet.cl/edicom/; publs include *Anuario Farmacológico*; Gen. Man. MARIO SILVA MARTÍNEZ.

**Ediciones San Pablo:** Vicuña Mackenna 10777, Casilla 3746, Santiago; tel. (2) 288-2025; fax (2) 288-2026; e-mail dgraledi@cnet.net; Catholic texts; Dir-Gen. P. LUIS NEIRA RAMÍREZ.

**Ediciones Técnicas Ltda:** El Condor 844, Of. 203, Huechuraba, Santiago; tel. (2) 757-4200; fax (2) 757-4201; e-mail editec@editec.cl; internet www.editec.cl; Pres. RICARDO CORTES DONOSO; Gen. Man. ROLY SOLIS SEPÚLVEDA.

**Ediciones Universitarias de Valparaíso:** Universidad Católica de Valparaíso, 12 de Febrero 187, Casilla 1415, Valparaíso; tel. (32) 27-3087; fax (32) 27-3429; e-mail euvsa@ucv.cl; internet www.euv.cl; f. 1970; literature, social and general sciences, engineering, education, music, arts, textbooks; Gen. Man. ALEJANDRO DAMIÁN VILARREAL.

**Ediciones Urano:** Avda Francisco Bilbao 2809, Providencia, Santiago; tel. (2) 341-6731; fax (2) 225-3896; e-mail chile@edicionesurano.com; internet www.edicionesurano.com; f. 1983; self-help, mystical and scholarly.

**Editorial Andrés Bello/Jurídica de Chile:** Avda Ricardo Lyon 946, Casilla 4256, Providencia, Santiago; tel. (2) 461-9500; fax (2) 225-3600; e-mail julio_serrano@entelchile.net; internet www.editorialandresbello.com; f. 1947; history, arts, literature, politics, economics, textbooks, law and social science; Gen. Man. JULIO SERRANO LAMAS.

**Editorial Antártica, SA:** San Francisco 116, Santiago; tel. (2) 639-3476; fax (2) 633-3402; e-mail plaborde@antartica; internet www.antartica.cl; f. 1978; Gen. Man. PAUL LABORDE U.

**Editorial Borlando:** Avda Victoria 151, Santiago; tel. (2) 555-9566; fax (2) 555-9564; internet www.editorialborlando.cl; scholarly, juvenile, educational and reference; f. 1984.

**Editorial Cuatro Vientos Ltda:** Jaime Guzmán Errázuriz 3293, Ñuñoa, Santiago; tel. (2) 225-8381; fax (2) 341-3107; e-mail 4vientos@netline.cl; internet www.cuatrovientos.net; f. 1980; Man. Editor JUAN FRANCISCO HUNEEUS COX.

**Editorial El Sembrador:** Sargento Aldea 1041, Casilla 2037, Santiago; tel. (2) 556-9454; Dir ISAÍAS GUTIÉRREZ.

**Editorial y Distribuidora Lenguaje y Pensamiento Ltda:** Avda Apoquindo 6275, Of. 36, Las Condes, Santiago; tel. (2) 245-2909; fax (2) 202-8263; internet www.juegosdidacticos.cl.

**Editorial Evolución, SA:** General del Canto 105, Of. 707, Santiago; tel. (2) 236-4789; fax (2) 236-4796; e-mail jbravo@evolucion.cl; internet www.evolucion.cl; business and management.

**Editorial Fondo de Cultura Económica Chile SA:** Paseo Bulnes 152, Santiago; tel. (2) 699-0189; fax (2) 696-2329.

**Editorial Renacimiento:** Huérfanos 623, Santiago; tel. (2) 632-7334; fax (2) 633-9374; internet www.feriachilenadellibro.cl; Gen. Man. MANUEL VILCHES.

**Editorial Terra Chile:** Pabellón 11, Santiago; tel. (2) 737-4455; fax (2) 738-0445; Gen. Man. ORLANDO MILESI.

**Editorial Texido:** Einstein 921, Recoleta, Santiago; tel. (2) 622-4652; fax (2) 622-4660; f. 1969; Gen. Man. ELSA ZLATER.

**Editorial Tiempo Presente Ltda:** Almirante Pastene 329, Casilla 303, Correo 22, Providencia, Santiago; tel. (2) 364-5100; fax (2) 235-8331; e-mail cosas@cosas.com; internet www.cosas.com; Gen. Man. JUAN LUIS SOMMERS.

**Editorial Trineo, SA:** Los Olmos 3685, Macul, Santiago; tel. (2) 750-1000; fax (2) 750-1001; e-mail trineo@trineo.cl; internet www.trineo.cl; Gen. Man. CARLOS JÉREZ HERNÁNDEZ.

**Editorial Universitaria, SA:** María Luisa Santander 0447, Casilla 10220, Providencia, Santiago; tel. (2) 487-0700; fax (2) 487-0702; e-mail comunicaciones@universitaria.cl; internet www

CHILE
*Directory*

.universitaria.cl; f. 1947; general literature, social science, technical, textbooks; Man. Dir RODRIGO CASTRO.

**Empresa Editora Zig-Zag SA:** Los Conquistadores 1700, 17°, Of. B, Providencia, Santiago; tel. (2) 335-7477; fax (2) 335-7445; e-mail zigzag@zigzag.cl; internet www.zigzag.cl; f. 1934; general publishers of literary works, reference books and magazines; Pres. GONZALO VIAL C; Gen. Man. FRANCISCO PÉREZ FRUGONE.

**Lexis-Nexis Chile:** Miraflores 383, 11°, Santiago; tel. (2) 510-5000; fax (2) 510-5110; internet www.lexisnexis.cl; legal, business, government and academic; f. 1973.

**McGraw-Hill/Interamericana de Chile Ltda:** Pocuro 2151, Providencia, Casilla 150, Correo 29, Santiago; tel. (2) 373-3000; fax (2) 635-4467; e-mail gerardo_aguayo@mcgraw-hill.com.cl; internet www.mcgraw-hill.cl.

**Norma de Chile, SA:** Avda Providencia 1760, Of. 502, Providencia, Santiago; tel. (2) 236-3355; fax (2) 236-3362; e-mail ventasnorma@carvajal.cl; internet www.norma.com; f. 1960; part of Editorial Norma of Colombia; Gen. Man. ELSY SALAZAR CAMPO.

**Pehuen Editores Ltda:** Antonio Varas 2043, Santiago; tel. (2) 204-9399; fax (2) 204-9399; e-mail pehuen@cmet.net; f. 1983; literature and sociology; Gen. Man. ALICIA Z. CERDA.

**Publicaciones Técnicas, SA (PUBLITECSA):** Miraflores 383, 11°, Santiago; tel. (2) 510-5000; fax (2) 365-8010; e-mail acliente@publitecsa.cl; internet www.publitecsa.cl; official information; Commercial Man. CARLOS MUNIZAGA.

**RIL Editores** (fmrly Red Internacional del Libro Ltda): Alférez Real 1464, Providencia, CP 750-0960, Santiago; tel. (2) 223-8100; fax (2) 225-4269; e-mail ril@rileditores.com; internet www.rileditores.com; literature, poetry, scholarly and political; f. 1991; Dir ELEONORA FINKELSTEIN; Dir of Publications DANIEL CALABRESE.

### PUBLISHERS' ASSOCIATION

**Cámara Chilena del Libro AG:** Avda B. O'Higgins 1370, Of. 501, Casilla 13526, Santiago; tel. (2) 698-9519; fax (2) 687-4088; e-mail camlibro@terra.cl; internet www.camlibro.cl; Pres. EDUARDO CASTILLO GARCÍA; Exec. Sec. RAQUEL TORNERO GÓMEZ.

## Broadcasting and Communications

### TELECOMMUNICATIONS

#### Regulatory Authority

**Subsecretaría de Telecomunicaciones** (Department of Telecommunications, Ministry of Public Works): Amunátegui 139, 5°, Casilla 120, Correo 21, Santiago; tel. (2) 421-3000; fax (2) 421-3553; e-mail subtel@subtel.cl; Under-Sec. CHRISTIAN NICOLAI ORELLANA.

#### Major Operators

**AT&T Chile:** Rinconada El Salto 202, Huechuraba, Santiago; tel. (2) 582 5000; fax (2) 585-5079; e-mail info@firstcom.cl; internet www.attla.cl; Pres. ALEJANDRO ROJAS; Dir-Gen. JAIME CHICO PARDO.

**Chilesat:** Rinconada El Salto 202, Huechuraba, Casilla 12, Santiago; tel. (2) 380-0171; fax (2) 382-5142; e-mail tlchile@chilesat.net; internet www.chilesat.net; Pres. JUAN EDUARDO IBÁÑEZ WALKER; Gen. Man. RAMÓN VALDIVIESO.

**CMET Compañía de Telefónos:** Avda Los Leones 1412, Providencia, Santiago; tel. (2) 251-3333; fax (2) 274-9573; Pres. RAFAEL BARRA CARMONA; Gen. Man. AGUSTÍN CASTELLÓN RAUCH.

**Empresa Nacional de Telecomunicaciones, SA—ENTEL Chile, SA:** Andrés Bello 2687, 14°, Casilla 4254, Las Condes, Santiago; tel. (2) 360-0123; fax (2) 360-3424; internet www.entel.cl; f. 1964; operates the Chilean land satellite stations of Longovilo, Punta Arenas and Coihaique, linked to INTELSAT system; 52% owned by Telecom Italia; Pres. JUAN HURTADO VICUÑA.

**Grupo GTD:** Santiago; e-mail soporte@gtdinternet.com; internet www.grupogtd.com; internet and telephone service provider.

**Smartcom:** Santiago; internet www.smartcom.cl; acquired in Aug. 2005 by América Móvil, SA de CV (Mexico); Pres. JORGE ROSENBLUT.

**Telefónica Chile:** Providencia 111, Santiago; tel. (2) 691-2020; fax (2) 691-2018; internet www.ctc.cl; fmrly Compañía de Telecomunicaciones de Chile, SA; privatized in 1988; owned by Telefónica, SA (Spain); Pres. BRUNIO PHILIPPI.

   **Telefónica Móvil de Chile:** Miraflores 130, Santiago; tel. (2) 661-6000; internet www.telefonicamovil.cl; f. 1996; mobile telephone services; Gen. Man. OLIVER FLÖGEL.

**TELEX-CHILE, SA:** Rinconada El Salto 202, Huechuraba, Santiago; tel. (2) 382-5786; fax (2) 382-5142; e-mail tlchile@chilesat.net; internet www.telex.cl; Pres. JORGE AWAD MEHECH; Gen. Man. RAMÓN VALDIVIESO RÍOS.

**VTR GlobalCom:** Reyes Lavalle 3340, 9°, Las Condes, Santiago; tel. (2) 310-1000; fax (2) 310-1560; internet www.vtr.cl; Pres. BLAS TOMIC; Gen. Man. PEDRO GUTIÉRREZ SÁNCHEZ.

### BROADCASTING

#### Regulatory Authority

**Asociación de Radiodifusores de Chile (ARCHI):** Pasaje Matte 956, Of. 801, Casilla 10476, Santiago; tel. (2) 639-8755; fax (2) 639-4205; e-mail archi@archiradios.cl; internet www.archiradios.cl; f. 1936; 455 broadcasting stations; Pres. JAIME BELLOLIO.

#### Radio

In December 2000 there were 1,153 radio stations (981 FM and 172 AM) transmitting in Chile.

**Agricultura (AM y FM):** Avda Manuel Rodríguez 15, Santiago; tel. (2) 695-3088; fax (2) 672-2749; owned by Sociedad Nacional de Agricultura; Pres. MANUEL VALDÉS VALDÉS; Gen. Man. GUIDO ERRÁZURIZ MORENO.

**Aurora FM:** Eliodoro Yáñez 1783, Providencia, Santiago; tel. (2) 390-2112; fax (2) 632-5860; f. 1982; part of Ibero American media group; Pres. ERNESTO CORONA BOZZO; Gen. Man. JUAN CARRASCO HERNÁNDEZ.

**Beethoven FM:** Garibaldi 1620, Ñuñoa, Santiago; tel. (2) 225-0222; fax (2) 274-3323; e-mail director@redfm.cl; internet www.beethovenfm.cl; f. 1981; mainly classical music; affiliate stations in Viña del Mar and Temuco; Exec. Dir RICARDO GUTIÉRREZ GATICA.

**Belén AM:** Benavente 385, 3°, Casilla 17, Puerto Montt; tel. (65) 25-8048; fax (65) 25-8097; e-mail radiobel.en001@chilnet.cl; f. 1990; owned by Archbishopric of Puerto Montt; Dir NELSON GONZÁLEZ ANDRADE; Gen. Man. CARLOS WAGNER CATALÁN.

**Bío Bío La Radio:** Avda Libertador Bernardo O'Higgins 680, Concepción; tel. (41) 225-660; fax (41) 226-742; e-mail pandrade@laradio.cl; internet www.radiobiobio.cl; affiliate stations in Concepción, Los Angeles, Temuco, Ancud, Castro, Osorno, Puerto Montt, Santiago and Valdivia; Man. PATRICIO ANDRADE.

**Compañía Radio Chilena:** Phillips 40, 2°, Casilla 10277, Santiago; tel. (2) 463-5000; fax (2) 463-5050; e-mail radio@radiochilena.cl; internet www.radiochilena.com; f. 1922; news; FM, AM and satellite; Exec. Dir JUAN LUIS SILVA DIBARRART.

**Radio El Conquistador FM:** El Conquistador del Monte 4644, Huechuraba, Santiago; tel. (2) 740-0000; fax (2) 740-0259; e-mail rconquis@entelchile.net; internet www.elconquistadorfm.cl; f. 1962; affiliate stations in Santiago, Iquique, Antofagasta, La Serena, Viña del Mar, Rancagua, Talca, Chillán, Concepción, Talcahuano, Pucón, Temuco, Villarrica, Lago Llanquihue, Osorno, Puerto Montt, Puerto Varas, Valdivia and Punta Arenas; Pres. JOAQUÍN MOLFINO.

**Radio Cooperativa (AM y FM):** Antonio Bellet 353, Casilla 16367, Correo 9, Santiago; tel. (2) 364-8000; fax (2) 236-0535; e-mail info@cooperativa.cl; internet www.cooperativa.cl; f. 1936; affiliate stations in Copiapó, Arica, Coquimbo, La Serena, Valparaíso, Concepción, Calama, Temuco and Castro; Pres. LUIS AJENJO ISASI; Gen. Man. SERGIO PARRA GODOY.

**La Clave FM:** Monjitas 454, Of. 406, Santiago; tel. (2) 633-1621; fax (2) 639-2914; f. 1980; Pres. MIGUEL NASUR ALLEL; Gen. Man. VÍCTOR IBARRA NEGRETE.

**Duna FM:** Eliodoro Yáñez 1804, Providencia, Santiago; tel. (2) 225-5494; fax (2) 225-6013; e-mail aholuigue@duna.cl; internet www.duna.cl; affiliate stations in Viña del Mar and Concepción; Pres. FELIPE LAMARCA CLARO; Dir A. HOLUIGUE.

**Estrella del Mar AM:** Eleuterio Ramírez 207, Ancud-Isla de Chiloé; tel. (65) 63-2900; fax (65) 63-2900; f. 1982; affiliate stations in Castro and Quellón; Dir MIGUEL ANGEL MILLAR SILVA.

**Festival AM:** Quinta 124 A, Viña del Mar; tel. (32) 68-9328; fax (32) 68-0266; e-mail schiesa@festival.cl; internet www.festival.cl; f. 1976; Pres. LUIS MUÑOZ AHUMADA; Dir-Gen. SANTIAGO CHIESA HOWARD.

**Finísima FM:** Luis Thayer Ojeda 1145, Santiago; tel. (2) 233-5771; fax (2) 231-0611; affiliate stations in Santiago, Arica, Iquique, Calama, Copiapó, La Serena, Ovalle, Isla de Pascua, Quilpe, San Antonio, San Felipe, Villa Alemana, Viña del Mar, Rancagua, Talca, Chillán, Concepción, Los Angeles, Temuco, Puerto Montt, Coihayque, Puerto Aysen and Punto Arenas; Gen. Man. CRISTIÁN WAGNER MUÑOZ.

**FM-Hit:** Eliodoro Yáñez 1783, Providencia, Santiago; tel. (2) 274-6737; fax (2) 274-8928; internet www.concierto.cl; f. 1999; part of Ibero American media group; affiliate stations in Santiago, Iquique, Antofagasta, San Antonio, La Serena, Viña del Mar, Concepción, Temuco, Osorno and Puerto Montt; Gen. Man. JAIME VEGA DE KUYPER.

CHILE                                                                                                                   *Directory*

**Horizonte:** Avda Los Leones 1625, Providencia, Santiago; tel. (2) 274-6737; fax (2) 410-5400; internet www.horizonte.cl; f. 1985; affiliate stations in Arica, Antofagasta, Iquique, La Serena, Viña del Mar, Concepción, San Antonio, Temuco, Villarrica, Puerto Montt, Punta Arenas and Osorno.

**Infinita FM:** Avda Los Leones 1285, Providencia, Santiago; tel. (2) 754-4400; fax (2) 341-6727; internet www.infinita.cl; f. 1977; affiliate stations in Santiago, Viña del Mar, Concepción and Valdivia; Gen. Man. CARLOS ALBERTO PEÑAFIEL GUARACHI.

**Radio Nacional de Chile:** Argomedo 369, Santiago; tel. (2) 638-1348; fax (2) 632-1065; affiliate stations in Arica and Punta Arenas; Gen. Man. SANTIAGO AGLIATI.

**Para Ti FM:** El Conquistador del Monte 4644, Huechuraba, Santiago; tel. (2) 740-9393; fax (2) 740-0405; internet www.fmparati.cl; 16 affiliate stations throughout Chile; Gen. Man. FELIPE MOLFINO BURKERT.

**Radio Polar:** Bories 871, 2°, Punta Arenas; tel. (61) 24-1909; fax (61) 22-8344; f. 1940; Pres. RENÉ VENEGAS OLMEDO.

**Pudahuel FM:** Eliodoro Yáñez 1783, Providencia, Santiago; tel. (2) 223-0704; fax (2) 223-7589; e-mail radio@pudahuel.cl; internet www.pudahuel.cl; f. 1966; part of Ibero American media group; affiliate stations in Arica, Iquique, Antofagasta, Calama, Copiapó, Coquimbo, La Serena, Ovalle, San Felipe, Valparaíso, Viña del Mar, Rancagua, Curico, Linares, Talca, Chillán, Concepción, Los Angeles, Talcahuano, Pucón, Temuco, Villarrica, Ancud-Castro, Osorno, Puerto Montt, Valdivia and Punta Arenas; Pres. SUSANA MUTINELLI ANCHUBIDART; Gen. Man. JOAQUÍN BLAYA BARRIOS.

**Santa María de Guadalupe:** Miguel Claro 161, Casilla 2626, Santiago; tel. (2) 235-7996; fax (2) 235-8527; e-mail radio@santamariadeguadalupe.com; internet www.santamariadeguadalupe.com; religious broadcasting; affiliate stations in Arica, Iquique, Antofagasta, La Serena, Viña del Mar, Temuco, Puerto Varas, Coihayque and Punta Arenas; Dir ALFONSO CHADWICK.

**Superandina FM:** Santa Rosa 441, Of. 34-36, Casilla 401, Los Andes; tel. (34) 42-2515; fax (34) 90-4091; e-mail radio@superandina.cl; internet www.superandina.cl; f. 1987; Dir JOSÉ ANDRÉS GÁLVEZ.

**Universo FM:** Félix de Amesti 124, 8°, Las Condes, Santiago; tel. (2) 206-6065; fax (2) 206-6049; affiliate stations in 18 cities, including Iquique, Copiapó, La Serena, Ovalle, Concepción, Temuco, Puerto Montt, Coihayque and Punta Arenas; Pres. ÁLVARO LARRAÍN.

### Television

**Corporación de Televisión de la Universidad Católica de Chile—Canal 13:** Inés Matte Urrejola 0848, Casilla 14600, Providencia, Santiago; tel. (2) 251-4000; fax (2) 630-2040; e-mail dasein@reuna.cl; internet www.canal13.cl; f. 1959; non-commercial; Exec. Dir ELEODORO RODRÍGUEZ MATTE; Gen. Man. JAIME BELLOLIO RODRÍGUEZ.

**La Red Televisión, S.A./TV Azteca Chile, S.A.:** Manquehue Sur 1201, Las Condes, Santiago; tel. (2) 385-4000; fax (2) 385-4020; internet www.redtv.cl; e-mail administracion@lared.cl; f. 1991; Pres. JUAN CARLOS LATORRE; Gen. Man. JOSÉ MANUEL LARRAÍN.

**Megavisión, S.A.—Canal 9:** Avda Vicuña Mackenna 1348, Ñuñoa, Santiago; tel. (2) 810-8000; fax (2) 551-8369; e-mail mega@mcl.cl; internet www.mega.cl; f. 1990; Pres. RICARDO CLARO VALDÉS; Gen. Man. CRISTÓBAL BULNES SERRANO.

**Red de Televisión SA/Chilevisión—Canal 11:** Inés Matte Urrejola 0825, Casilla 16547, Correo 9, Providencia, Santiago; tel. (2) 461-5100; fax (2) 461-5371; e-mail info@chilevision.cl; internet www.chilevision.cl; part of Ibero American media group; News Dir FELIPE POZO.

**Televisión Nacional de Chile—Canal 7:** Bellavista 0990, Casilla 16104, Providencia, Santiago; tel. (2) 707-7777; fax (2) 707-7766; e-mail rrpp@tvn.cl; internet www.tvn.cl; government network of 140 stations and an international satellite signal; Chair. LUIS ORTIZ QUIROGA; Exec. Dir PABLO PIÑEIRA ECHENIQUE.

**Corporación de Televisión de la Universidad Católica de Valparaíso:** Agua Santa Alta 2455, Casilla 247, Viña del Mar; tel. (32) 616-000; fax (32) 610-505; e-mail tv@ucv.cl; f. 1957; Dir JORGE A. BORNSCHEUER.

# Finance

(cap. = capital; res = reserves; dep. = deposits; m. = million; brs = branches; amounts in pesos unless otherwise specified)

## BANKING

### Supervisory Authority

**Superintendencia de Bancos e Instituciones Financieras:** Moneda 1123, 6°, Casilla 15-D, Santiago; tel. (2) 442-6200; fax (2) 441-0914; e-mail superintendente@sbif.cl; internet www.sbif.cl; f. 1925; affiliated to Ministry of Finance; Supt ENRIQUE MARSHALL RIVERA.

### Central Bank

**Banco Central de Chile:** Agustinas 1180, Santiago; tel. (2) 670-2000; fax (2) 670-2099; e-mail bcch@bcentral.cl; internet www.bcentral.cl; f. 1926; under Ministry of Finance until Dec. 1989, when autonomy was granted; bank of issue; cap. 257,604.6m., total assets 17,985,161.2m. (Dec. 2002); Pres. VITTORO CORBO; Gen. Man. CAMILO CARRASCO.

### State Bank

**Banco del Estado de Chile:** Avda B. O'Higgins 1111, Santiago; tel. (2) 670-7000; fax (2) 670-5711; internet www.bancoestado.cl; f. 1953; state bank; cap. and res 328,551m., dep. 3,492,359m. (May 2002); Chair. JAIME ESTÉVEZ; Gen. Man. JOSÉ MANUEL MENA; 294 brs.

### Commercial Banks

**Banco de A. Edwards:** Huérfanos 740, Santiago; tel. (2) 388-3000; fax (2) 388-4428; internet www.banedwards.cl; f. 1851; acquired by Banco de Chile in Jan. 2002.

**Banco BICE:** Teatinos 220, Santiago; tel. (2) 692-2000; fax (2) 696-5324; e-mail webmaster@bice.cl; internet www.bice.cl; f. 1979 as Banco Industrial y de Comercio Exterior; name changed as above in 1988; cap. and res 94,987m. (Dec. 2003); Pres. BERNARDO MATTE LARRAÍN; Gen. Man. RENÉ LÉHUEDÉ FUENZALIDA; 9 brs.

**Banco de Chile:** Ahumada 251, Casilla 151-D, Santiago; tel. (2) 637-1111; fax (2) 637-3434; internet www.bancochile.cl; f. 1894; 42% owned by SAOS, SA; cap. 482,504m., res 82,619m., dep. 7,265,620m. (Dec. 2003); Chair. SEGISMUNDO SCHULIN-ZEUTHEN SERRANO; CEO PABLO GRANIFO LAVÍN; 243 brs.

**Banco Conosur:** Avda B. O'Higgins 1980, 7°, Santiago; tel. (2) 697-1491; fax (2) 696-3133; internet www.bancoconosur.cl; Pres. LIONEL OLAVARRÍA LEYTON; Vice-Pres. ADOLFO GARCÍA-HUIDOBRO OCHAGAVÍA.

**Banco de Crédito e Inversiones (Bci):** Huérfanos 1134, Casilla 136-D, Santiago; tel. (2) 692-7000; fax (2) 695-3775; e-mail webmaster@bci.cl; internet www.bci.cl; f. 1937; cap. and res 287,854m., dep. 4,126,771m. (Dec. 2003); Pres. LUIS ENRIQUE YARUR REY; Gen. Man. LIONEL OLAVARRÍA; 112 brs.

**Banco del Desarrollo:** Avda B. O'Higgins 949, 3°, Casilla 320-V, Casilla 1, Santiago; tel. (2) 674-5000; fax (2) 671-5547; e-mail bdd@bandes.cl; internet www.bdd.cl; f. 1983; cap. 117,417m., res 209.6m., dep. 1,190,039m. (Dec. 2003); Chair. VICENTE CARUZ MIDDLETON; Gen. Man. HUGO TRIVELLI; 83 brs.

**Banco Internacional:** Moneda 818, Casilla 135-D, Santiago; tel. (2) 369-7000; fax (2) 369-7367; e-mail banco@binter.cl; internet www.bancointernacional.cl; f. 1944; cap. and res 14,489m., dep. 153,025m. (Dec. 2002); Pres. and Chair. ALEJANDRO L. FURMAN SIHMAN; Gen. Man. ALVARO ACHONDO GONZÁLEZ; 11 brs.

**Banco Santander Chile:** Bandera 140, 13°, Casilla 57-D, Santiago; tel. (2) 320-2000; fax (2) 320-8877; e-mail webmaster@santander.cl; internet www.santandersantiago.cl; f. 1926; cap. and res 810,420m., dep. 8,750,867m. (Dec. 2003); subsidiary of Banco de Santander (Spain); incorporated Banco Osorno y La Unión in 1996; Chair. MAURICIO LARRAÍN GARCES; CEO FERNANDO CAÑAS BERKOWITZ; 72 brs.

**Banco Security:** Apoquindo 3100, Las Condes, Santiago; tel. (2) 584-4000; fax (2) 270-4001; e-mail bancosecurity@security.cl; internet www.security.cl; f. 1981; fmrly Banco Urquijo de Chile; cap. and res 80,606m., dep. 970,284m. (Dec. 2003); Pres. and Chair. FRANCISCO SILVA S.; Gen. Man. RAMÓN ELUCHANS O.; 11 brs.

**BBVA Banco BHIF:** Huérfanos 1234, Casilla 517, Santiago; tel. (2) 679-1000; fax (2) 698-5640; e-mail bhif@bhif.cl; internet www.bhif.cl; f. 1883; was merged with Banco Nacional in 1989; acquired Banesto Chile Bank in Feb. 1995; controlling interest acquired by Banco Bilbao Vizcaya (Spain) in Sept. 1998; name changed as above in 1999; cap. and res 228,163m., dep. 1,349,303m. (May 2002); Pres. JOSÉ SAID SAFFIE; CEO RAMÓN MONELL V.; 78 brs.

**Corpbanca:** Huérfanos 1072, Casilla 80-D, Santiago; tel. (2) 687-8000; fax (2) 696-5763; e-mail ggeneral@corpbanca.cl; internet www.corpbanca.cl; f. 1871 as Banco de Concepción, current name adopted

CHILE  *Directory*

in 1997; cap. 270,002m., res 18,895m., dep. 2,144,606m. (Dec. 2003); Chair. CARLOS ABUMOHOR TOUMA; CEO CHRISTIAN SAMSING S.; 64 brs.

**Dresdner Bank Lateinamerika AG:** Huérfanos 1219, Casilla 10492, Santiago; tel. (2) 731-4444; fax (2) 671-3307; e-mail info@dbla.com; internet www.dbla.cl; f. 1958 as Banco Continental; bought by Crédit Lyonnais in Sept. 1987, subsequently by Dresdner BNP; current name adopted in 2002; cap. 20,131m., res 6,139m., dep. 149,107m. (Dec. 2003); Pres. EWALD DOERNER; Gen. Man. ROLAND JACOB; 4 brs.

**Scotiabank Sud Americano:** Morandé 226, Casilla 90-D, Santiago; tel. (2) 692-6000; fax (2) 698-6008; e-mail scotiabank@scotiabank.cl; internet www.scotiabank.cl; f. 1944; cap. 65,366m., res 32,433m., dep. 1,407,886m. (Dec. 2002); Chair. PETER C. CARDINAL; CEO LUIS FERNANDO TOBÓN; 49 brs.

### Finance Corporations

**Financiera Atlas, SA:** Nueva de Lyon 72, 7°, Santiago; tel. (2) 233-3151; fax (2) 233-3152; Gen. Man. NEIL A. DENTON FEILMANN.

### Banking Association

**Asociación de Bancos e Instituciones Financieras de Chile AG:** Ahumada 179, 12°, Santiago; tel. (2) 636-7100; fax (2) 698-8945; e-mail general@abif.cl; internet www.abif.cl; f. 1945; Pres. HERNÁN SOMERVILLE SENN; Gen. Man. ALEJANDRO ALARCÓN PÉREZ.

### Other Financial Supervisory Bodies

**Superintendencia de Administradoras de Fondos de Pensiones (SAFP)** (Superintendency of Pension Funds): Huérfanos 1273, 9°, Casilla 3955, Santiago; tel. (2) 753-0100; fax (2) 753-0122; internet www.safp.cl; f. 1981; Supt GUILLERMO LARRAÍN RÍOS.

**Superintendencia de Seguridad Social** (Superintendency of Social Security): Huérfanos 1376, 5°, Santiago; tel. (2) 620-4500; fax (2) 696-4672; e-mail secgral@suseso.cl; internet www.suseso.gov.cl; f. 1927; Supt XIMENA C. RINCÓN GONZÁLEZ.

## STOCK EXCHANGES

**Bolsa de Comercio de Santiago:** La Bolsa 64, Casilla 123-D, Santiago; tel. (2) 399-3000; fax (2) 318-1961; e-mail chathaway@bolsadesantiago.com; internet www.bolsadesantiago.com; f. 1893; 34 mems; Pres. PABLO YRARRÁZAVAL VALDÉS; Gen. Man. JOSÉ ANTONIO MARTÍNEZ ZUGARRAMURDI.

**Bolsa de Corredores—Valores de Valparaíso:** Prat 798, Casilla 218-V, Valparaíso; tel. (32) 25-0677; fax (32) 21-2764; e-mail bolsadec.orred001@chilnet.cl; internet www.bovalpo.com; f. 1905; Pres. CARLOS F. MARÍN ORREGO; Man. ARIE JOEL GELFENSTEIN FREUNDLICH.

**Bolsa Electrónica de Chile:** Huérfanos 770, 14°, Santiago; tel. (2) 639-4699; fax (2) 633-4174; e-mail info@bolchile.cl; internet www.bolchile.cl; Gen. Man. JUAN CARLOS SPENCER OSSA.

## INSURANCE

In 2002 there were 55 general, life and reinsurance companies operating in Chile.

### Supervisory Authority

**Superintendencia de Valores y Seguros:** Avda Libertador Bernardo O'Higgins 1449, Santiago 834-0518; tel. (2) 473-4000; fax (2) 473-4101; internet www.svs.cl; f. 1931; under Ministry of Finance; Supt ALEJANDRO FERREIRO YAZIGI.

### Principal Companies

**Aetna Chile Seguros de Vida, SA:** Suecia 211, 7°, Santiago; tel. (2) 364-2000; fax (2) 364-2010; e-mail jdupre@aetna.cl; internet www.aetna.cl; f. 1981; life; Pres. SERGIO BAEZA VALDÉS; Gen. Man. FERNANDO HASENBERG NATOLI.

**Aseguradora Magallanes, SA:** Agustinas 1022, Of. 722; tel. (2) 365-4848; fax (2) 365-4860; e-mail fvarela@magallanes.cl; internet www.magallanes.cl; f. 1957; general; Pres. SERGIO LARRAÍN PRIETO; Gen. Man. FERNANDO VARELA.

**Axa Seguros Generales, SA:** Huérfanos 1189, 2°, 3° y 4°, Casilla 429-V, Santiago; tel. (2) 679-9200; fax (2) 679-9300; internet www.axa.cl; f. 1936; general; Gen. Man. BERNARDO SERRANO LÓPEZ.

**Chilena Consolidada, SA:** Pedro de Valdivia 195, Casilla 16587, Correo 9, Providencia, Santiago; tel. (2) 200-7000; fax (2) 274-9933; internet www.chilena.cl; f. 1853; owned by Zurich group; general and life; Gen. Man. IGNACIO BARRIGA UGARTE.

**Chubb de Chile, SA:** Gertrudis Echeñique 30, 4°, Santiago; tel. (2) 206-2191; fax (2) 206-2735; internet www.chubb.com/chile; f. 1992; general; Gen. Man. CLAUDIO M. ROSSI.

**Cía de Seguros de Crédito Continental, SA:** Avda Isidora Goyenechea 3162, 6°, Edif. Parque 1 Golf, Santiago; tel. (2) 636-4000; fax (2) 636-4001; e-mail comer@continental.cl; f. 1990; general; Gen. Man. FRANCISCO ARTIGAS CELIS.

**Cía de Seguros Generales Aetna Chile, SA:** Suecia 211, Santiago; tel. (2) 364-2000; fax (2) 364-2060; internet www.aetna.cl; f. 1899; general; Pres. SERGIO BAEZA VALDÉS; Gen. Man. MÁXIMO ERRÁZURIZ DE SOLMINIHAC.

**Cía de Seguros Generales Consorcio Nacional de Seguros, SA:** El Bosque Sur 130, Santiago; tel. (2) 230-4000; fax (2) 250-2525; f. 1992; general; Gen. Man. MARCOS BÜCHI BUC.

**Cía de Seguros Generales Cruz del Sur, SA:** Avda El Golf 150, Santiago; tel. (2) 461-8000; fax (2) 461-8715; internet www.cruzdelsur.cl; f. 1974; general; Pres. JOSÉ TOMÁS GUZMÁN DUMAS; Gen. Man. MIKEL URIARTE PLAZAOLA.

**Cía de Seguros Generales Euroamérica, SA:** Agustinas 1127, 2°, Casilla 180-D, Santiago; tel. (2) 672-7242; fax (2) 696-4086; internet www.euroamerica.cl; f. 1986; general; Gen. Man. PATRICIA JAIME VÉLIZ.

**Cía de Seguros de Vida Consorcio Nacional de Seguros, SA:** Avda El Bosque Sur 180, 3°, Casilla 232, Correo 35, Providencia, Santiago; tel. (2) 230-4000; fax (2) 230-4050; f. 1916; life; Pres. JUAN BILBAO HORMAECHE; Gen. Man. MARCOS BÜCHI BUC.

**Cía de Seguros de Vida La Construcción, SA:** Avda Providencia 1806, 11°–18°, Providencia, Santiago; tel. (2) 340-3000; fax (2) 340-3024; internet www.laconstruccion.sa.cl; f. 1985; life; Pres. VÍCTOR MANUEL JARPA RIVEROS; Gen. Man. MANUEL ZEGERS IRARRÁZAVAL.

**Cía de Seguros de Vida Cruz del Sur, SA:** El Golf 150, Santiago; tel. (2) 461-8000; fax (2) 461-8715; internet www.cruzdelsur.cl; f. 1992; life; Pres. ROBERTO ANGELINI; Gen. Man. MIKEL URIARTE PLAZAOLA.

**Cía de Seguros de Vida Euroamérica, SA:** Agustinas 1127, 3°, Casilla 21-D, Santiago; tel. (2) 782-7000; fax (2) 699-0732; e-mail deptoservicio@eurovida.cl; internet www.eurovida.cl; f. 1962; life; Pres. BENJAMIN DAVIS CLARKE; Gen. Man. PATRICIA JAIME VÉLIZ.

**Cía de Seguros de Vida Santander, SA:** Bandera 150, Santiago; tel. (2) 640-1177; fax (2) 640-1377; e-mail servicio@santanderseg.cl; internet www.netra.santanderseg.cl/index; f. 1989; life; Pres. FRANCISCO MARTÍN LÓPEZ-QUESADA.

**ING, Seguros de Vida, SA:** Avda Suecia 211, Providencia, Casilla 13224, Santiago; tel. (2) 252-1464; fax (2) 364-2060; internet www.ing.cl; f. 1989; life; Gen. Man. ÁLVARO JIMÉNEZ URRUTIA.

**La Interamericana Compañía de Seguros de Vida:** Agustinas 640, 9°, Casilla 111, Correo Central, Santiago; tel. (2) 630-3000; fax (2) 639-5859; internet www.interamericana.cl; f. 1980; life; Pres. RICARDO PERALTA VALENZUELA; Gen. Man. RICARDO GARCÍA HOLTZ.

**Le Mans—ISE Compañía Seguros Generales, SA:** Encomenderos 113, Casilla 185-D, Centro 192, Las Condes, Santiago; tel. (2) 422-9000; fax (2) 232-0471; e-mail lemans@lemans.cl; internet www.lemansise.cl; f. 1888; general; Pres. IGNACIO WALKER CONCHA; Gen. Man. MARC GARÇON.

**Mapfre Garantías y Crédito, SA:** Teatinos 280, 5°, Santiago; tel. (2) 870-1500; fax (2) 870-1501; internet www.mapfregc.cl; f. 1991; general; Gen. Man. RODRIGO CAMPERO PETERS.

**Renta Nacional Compañía de Seguros de Vida, SA:** Amunátegui 178, 2°, Santiago; tel. (2) 670-0200; fax (2) 670-0045; f. 1982; life; Gen. Man. GENARO LAYMUNS HEILMAIER.

**Seguros Vida Security Previsión, SA:** Hendaya 60, 7°, Las Condes, Santiago; tel. (2) 750-2300; fax (2) 750-2440; internet www.vidasecurity.cl; f. 2002 through merger of Seguros Security and Seguros Previsión Vida; Gen. Man. ALEJANDRO ALZÉRRECA LUNA.

### Reinsurance

**Caja Reaseguradora de Chile, SA:** Apoquindo 4449, 8°, Santiago; tel. (2) 338-1200; fax (2) 206-4063; f. 1934; life; Pres. ANDRÉS JIMÉNES.

### Insurance Association

**Asociación de Aseguradores de Chile, AG:** La Concepción 322, Of. 501, Providencia, Santiago; tel. (2) 236-2596; fax (2) 235-1502; e-mail seguros@aach.cl; internet www.aach.cl; f. 1931; Pres. MIKEL URIARTE PLAZAOLA; Gen. Man. JORGE CLAUDE BOURDEL.

# Trade and Industry

## GOVERNMENT AGENCIES

**Comisión Nacional del Medio Ambiente (CONAMA):** Teatinos 254-258, Santiago; tel. (2) 240-5600; fax (2) 244-1262; e-mail informacion@conama.cl; internet www.conama.cl; f. 1994; environmental regulatory body; Exec. Dir PAULINA SABALL.

**Corporación de Fomento de la Producción (CORFO):** Moneda 921, Casilla 3886, Santiago; tel. (2) 631-8200; e-mail info@corfo.cl; internet www.corfo.cl; f. 1939; holding group of principal state enterprises; grants loans and guarantees to private sector; responsible for sale of non-strategic state enterprises; promotes entrepreneurship; CEO and Exec. Vice-Pres. OSCAR LANDERRETCHE GACITÚA; Gen. Man. BERNARDO ESPINOZA; 13 brs.

**PROCHILE** (Dirección General de Relaciones Económicas Internacionales): Alameda 1315, 2°, Casilla 14087, Correo 21, Santiago; tel. (2) 565-9000; fax (2) 696-0639; e-mail info@prochile.cl; internet www.prochile.cl; f. 1974; bureau of international economic affairs; Dir HUGO LAVADES M.

**Servicio Nacional de Capacitación y Empleo** (National Training and Employment Service): Teatinos 333, 8°, Santiago; tel. (2) 870-6222; fax (2) 696-7103; internet www.sence.cl; attached to Ministry of Labour and Social Security; Dir JOSSIÉ ESCÁRATE MÜLLER.

### STATE CORPORATIONS

**Corporación Nacional del Cobre de Chile (CODELCO-Chile):** Huérfanos 1270, Casilla 150-D, Santiago; tel. (2) 690-3000; fax (2) 690-3059; e-mail comunica@codelco.cl; internet www.codelco.com; f. 1976 as a state-owned enterprise with copper-producing operational divisions at Chuquicamata, Radomiro Tomić, Salvador, Andina, Salvador, Talleres Rancagua and El Teniente; attached to Ministry of Mining; Pres. ALFONSO DULANTO RENCORET (Minister of Mining); Exec. Pres. JUAN VILLARZÚ RHODE; 18,496 employees.

**Empresa Nacional de Petróleo (ENAP):** Vitacura 2736, 10°, Las Condes, Santiago; tel. (2) 280-3000; fax (2) 280-3199; e-mail webenap@enap.cl; internet www.enap.cl; f. 1950; state-owned petroleum and gas exploration and production corporation; Pres. ALFONSO DULANTO RENCORET (Minister of Mining); Gen. Man. ENRIQUE DÁVILA ALVEAL; Sec.-Gen. PAULA HIDALGO MANDUJANO; 3,286 employees.

### DEVELOPMENT ORGANIZATIONS

**Comisión Chilena de Energía Nuclear:** Amunátegui 95, Casilla 188-D, Santiago; tel. (2) 699-0070; fax (2) 699-1618; e-mail lvillanu@CCHEN.cl; internet www.cchen.cl; f. 1965; government body to develop peaceful uses of atomic energy; concentrates, regulates and controls all matters related to nuclear energy; Pres. Dr ROBERTO HOJMAN G.; Exec. Dir LORETO VILLANUEVA Z.

**Corporación Nacional de Desarrollo Indígena (Conadi):** Aldunate 620, 8°, Temuco, Chile; tel. (45) 641-500; fax (45) 641-520; e-mail ctranamil@conadi.gov.cl; internet www.conadi.cl; promotes the economic and social development of indigenous communities; Nat. Dir AROLDO CAYUN ANTICURA.

**Corporación Nacional Forestal (CONAF):** Región Metropolitana, Valenzuela Castillo 1868, Santiago; tel. (2) 225-0428; fax (2) 225-00641; e-mail consulta@conaf.cl; internet www.conaf.cl; f. 1970 to promote forestry activities, to enforce forestry law, to promote afforestation, to administer subsidies for afforestation projects and to increase and preserve forest resources; manages 13.97m. ha designated as National Parks, Natural Monuments and National Reserves; under Ministry of Agriculture; Exec. Dir Ing. CARLOS WEBER BONTE.

**Empresa Nacional de Minería (ENAMI):** MacIver 459, 2°, Casilla 100-D, Santiago; tel. (2) 637-5278; fax (2) 637-5452; e-mail eiturra@enami.cl; internet www.enami.cl; promotes the development of small and medium-sized mines; attached to Ministry of Mining; partially privatized; Exec. Vice-Pres. JAIME PÉREZ DE ARCE ARAYA.

### CHAMBERS OF COMMERCE

There are chambers of commerce in all major towns.

**Cámara de Comercio de Santiago de Chile, AG:** Edif. Del Comercio, Monjitas 392, Santiago; tel. (2) 360-7000; fax (2) 633-3595; e-mail cpn@ccs.cl; internet www.ccs.cl; f. 1919; 1,300 mems; Pres. CARLOS E. JORQUIERA M.; Gen. Man. CLAUDIO ORTIZ TELLO.

**Cámara de la Producción y del Comercio de Concepción:** Cauplicán 567, 2°, Concepción; tel. (41) 241-121; fax (41) 227-903; e-mail lmandiola@cpcc.cl; internet www.cpcc.cl; Pres. PEDRO SCHLACK HARNECKER; First Vice-Pres. GUSTAVO VALENZUELA G.; Gen. Man. LEONCIO TORO ARAYA.

**Cámara Nacional de Comercio, Servicios y Turismo de Chile:** Merced 230, Santiago; tel. (2) 365-4000; fax (2) 365-4001; internet www.cnc.cl; f. 1858; Pres. PEDRO CORONA BOZZO; Vice-Pres. HÉCTOR VALENCIA BRINGAS; 120 mems.

**Cámara de Comercio de Antofagasta:** Latorre 2580, Dpto 21, Antofagasta; tel. (55) 225-125; fax (55) 222-053; e-mail info@comercioantofagasta.cl; internet www.comercioantofagasta.cl; f. 1924; Man. JORGE MOZÓ GREZ.

**Cámara de Comercio, Servicios y Turismo de Temuco, A.G.:** Vicuña Mackenna 396, Temuco; tel. (45) 210-556; fax (45) 237-047; internet www.camaratemuco.cl; Pres. HANS KÜHN BEISSER; Vice-Pres. MAX HENZI IBARRA.

### INDUSTRIAL AND TRADE ASSOCIATIONS

**Servicio Agrícola y Ganadero (SAG):** Avda Bulnes 140, Santiago; tel. (2) 345-1100; fax (2) 345-1102; e-mail dirnac@sag.gob.cl; internet www.sag.cl; under Ministry of Agriculture; responsible for the protection and development of safe practice in the sector; Nat. Dir DIONISIO FAULBAUM.

**Servicio Nacional de Pesca (SERNAPESCA):** Victoria 2832, Valparaíso; tel. (32) 81-9100; fax (32) 25-6311; e-mail informaciones@sernapesca.cl; internet www.sernapesca.cl; f. 1978; government regulator of the fishing industry.

**Sociedad Agrícola y Servicios Isla de Pascua (SASIPA):** Alfredo Lecannelier 1940, Providencia, Santiago; tel. (2) 232-7497; fax (2) 232-7497; administers agriculture and public services on Easter Island; Gen. Man. GERARDO VELASCO.

### EMPLOYERS' ORGANIZATIONS

**Confederación de la Producción y del Comercio:** Monseñor Sótero Sanz 182, Providencia, Santiago; tel. (2) 231-9764; fax (2) 231-9808; e-mail procomer@entelchile.net; internet www.cpc.cl; f. 1936; Pres. HERNÁN GUILLERMO SOMERVILLE SENN.

Affiliated organizations:

**Asociación de Bancos e Instituciones Financieras de Chile AG:** see Finance (Banking Association).

**Cámara Nacional de Comercio, Servicios y Turismo:** see above.

**Cámara Chilena de la Construcción:** Marchant Pereira 10, 3°, Providencia, Código Postal 6640721, Santiago; tel. (2) 376-3300; fax (2) 371-3430; internet www.camaraconstruccion.cl; f. 1951; Pres. OTTO KUNZ SOMMER; Vice-Pres. JUAN MACKENNA IÑIGUEZ; 17,442 mems.

**Sociedad de Fomento Fabril, FG (SOFOFA):** Avda Andrés Bello 2777, 3°, Las Condes, Santiago; tel. (2) 391-3100; fax (2) 391-3200; internet www.sofofa.cl; f. 1883; largest employers' organization; Pres. JUAN CLARO; Sec.-Gen. ANDRÉS CONCHA; 2,500 mems.

**Sociedad Nacional de Agricultura—Federación Gremial (SNA):** Tenderini 187, 2°, CP 6500978, Santiago; tel. (2) 639-6710; fax (2) 633-7771; e-mail info@sna.cl; internet www.sna.cl; f. 1838; landowners' association; controls Radio Stations CB 57 and XQB8 (FM) in Santiago, CB-97 in Valparaíso, CD-120 in Los Angeles, CA-144 in La Serena, CD-127 in Temuco; Pres. ANDRÉS SANTA CRUZ; Dir JUAN MIGUEL OVALLE.

**Sociedad Nacional de Minería (SONAMI):** Avda Apoquindo 3000, 5°, Santiago; tel. (2) 335-9300; fax (2) 334-9700; e-mail monica.cavallini@sonami.cl; internet www.sonami.cl; f. 1883; Pres. ALFREDO OVALLE RODRÍGUEZ; Gen. Man. JAIME ALÉ YARAD; 48 mem. cos.

**Confederación de Asociaciones Gremiales y Federaciones de Agricultores de Chile:** Lautaro 218, Los Angeles; regd 1981; Pres. DOMINGO DURÁN NEUMANN; Gen. Sec. ADOLFO LARRAÍN V.

**Confederación del Comercio Detallista y Turismo de Chile, AG:** Merced 380, 8°, Of. 74, Santiago; tel. (2) 639-1264; fax (2) 638-0338; e-mail comerciodetallista@confedech.cl; internet www.comerciodetallista.cl; f. 1938; retail trade; Nat. Pres. RAFAEL CUMSILLE ZAPAPA; Sec.-Gen. ROBERTO ZUÑIGA BELAUZARÁN.

**Confederación Gremial Nacional Unida de la Mediana y Pequeña Industria, Servicios y Artesanado (CONUPIA):** General Parra 703, Providencia, Santiago; tel. (2) 235-8022; internet www.conupia.cl; e-mail gerencia@conupia.cl; small and medium-sized industries and crafts; Pres. GERMÁN DASTRES GONZÁLEZ; Vice-Pres. CHAQUIB SUFÁN AIDAR; Sec.-Gen. MARIO PONCE DÍAZ.

There are many federations of private industrialists, organized by industry and region.

### MAJOR COMPANIES

#### Petroleum and Mining

**Compañía de Petróleos de Chile, SA (COPEC):** Agustinas 1382, 1°–7°, Casilla 9391, Santiago 6500586; tel. (2) 690-7000; fax (2) 672-5119; e-mail icontact@copec.cl; internet www.copec.cl; f. 1934; manufacturers of petroleum products; owned by Angelini group following privatization in 1986; Pres. ROBERTO ANGELINI ROSSI; Gen. Man. EDUARDO NAVARRRO BELTRÁN; 7,800 employees.

# CHILE

**Minera Escondida Ltda:** Avda Américo Vespucio Sur 100, 7°, Las Condes, Santiago; tel. (2) 330-5000; fax (2) 207-6520; internet www.escondida.cl; f. 1985; 57.5% owned by BHP of Australia, 30% owned by Rio Tinto; copper mining and cathodes production; Pres. BRUCE L. TURNER; 2,189 employees.

**Sociedad Punta del Cobre, SA:** Avda El Bosque Sur 130, 14°, Santiago; tel. (2) 379-4560; fax (2) 379-4570; f. 1989; copper processing; Chair. FERNANDO HARAMBILLET ALONSO; Gen. Man. GONZALO CASTILLO OLIVARES; 360 employees.

**Sociedad Química y Minera de Chile, SA (SQM, SA):** El Trovador 4285, Las Condes, Santiago; tel. (2) 425-2000; fax (2) 425-2493; e-mail admin_web@sqm.cl; internet www.sqm.cl; f. 1968; mining co; nitrates etc.; Pres. JULIO PONCE LEROU; Gen. Man. PATRICIO CONTESSE GONZÁLEZ; 2,745 employees.

**Soprocal, Calerías e Industrias, SA:** Avda Pedro de Valdivia 0193, Of. 31, Providencia, Santiago; tel. (2) 231-8874; fax (2) 233-3396; e-mail info@soprocal.cl; internet www.soprocal.cl; f. 1940; producers of lime; Exec. Chair. ALFONSO ROZAS OSSA; CEO ISMAEL CUEVAS ZANARTE; 72 employees.

## Food and Beverages

**Agrícola Nacional, S.A.C.E.I.:** 300 Almirante Pastene, Santiago; tel. (2) 470-6800; e-mail info@anasac.cl; internet www.anasac.cl; f. 1948; food, agricultural chemicals and pest control; Chair. FERNANDO MARTÍNEZ PEREZ-CANTO; 553 employees; CEO EUGENIO DE MARCHENA GÚZMAN.

**Bodegas y Viñedos Santa Emiliana, SA:** Edif. World Trade Center, Avda Nueva Tajamar 481, Las Condes, Santiago; tel. (2) 353-9130; fax (2) 203-6227; e-mail info@emiliana.cl; internet www.vinedosemiliana.cl; f. 1986; wines and spirits producers; Chair. MARIANO FONTECILLA DE SANTIAGO CONCHA; Gen. Man RAFAEL GIULISASTI GANA; 765 employees.

**Coca-Cola Embonor, SA:** Santa Maria 2652, Arica; tel. (5) 824-1530; internet www.embonor.cl; bottling company for Coca-Cola; Chair. HERNÁN VICUÑA REYES; 4,193 employees.

**Compañía Cervecerías Unidas, SA (CCU):** Bandera 84, 6°–11°, Casilla 1977, Santiago; tel. (2) 427-3000; fax (2) 427-3222; e-mail ccuir@ccu-sa.com; internet www.ccu-sa.com; f. 1902; part of Quiñenco conglomerate; 20% owned by Anheuser Busch of the USA; beverages; Pres. GUILLERMO LUKSIC CRAIG; Gen. Man. PATRICIO JOTTAR NASRALLAH; 3,892 employees.

**Concha y Toro, SA:** Virginia Subercaseaux 210, Pirque, Santiago; tel. (2) 476-5000; fax (2) 853-1034; e-mail conchaytoro@banfivintners.com; internet www.conchaytoro.com; f. 1883; vintners; Chair. ALFONSO LARRAÍN; Vice-Chair. RAFAEL GUILISASTI GANA.

**Copefrut, SA:** Panamericana Sur, Km 185, Curico; tel. (2) 209-220; fax (2) 380-905; e-mail copefrut@copefrut.cl; internet www.copefrut.cl; f. 1955; fruit producers and exporters; Chair. JOSÉ SOLER MALLAFRÉ; Gen. Man. MARCELO CARDOEN QUINTANA; 309 employees.

**Dos en Uno, SA:** Arauco 1050, Santiago; tel. (2) 556-1078; internet www.arcor.cl; f. 1989; part of Arcor and Danone; makers of biscuits, cakes, etc.; CEO RAÚL QUEMADA LERIA; 1,700 employees.

**Embotelladora Andina, SA:** Avda Andres Bello 2687, Santiago; tel. (2) 338-0520; e-mail inv.rel@koandina.com; internet www.koandina.com; f. 1946; bottling company for Coca-Cola; also operates in Brazil and Argentina; Chair. JUAN CLARO GONZÁLEZ; CEO JAIME GARCÍA; 4,284 employees.

**Empresas Iansa, SA:** Bustamente 26, Casilla 189, Correo 22, Santiago; tel. (2) 565-5500; fax (2) 565-5525; e-mail empresas@iansa.cl; internet www.empresasiansa.cl; f. 1953; sugar production, frozen fruit and vegetables, fruit juices, animal feed; fmrly known as Industria Azucarera Nacional, SA; Chair. OSCAR GUILLERMO GARRETÓN; CEO NICOLÁS BAUTISTA; 1,255 employees (Dec. 2004).

**Iansagro, SA:** Km 385 Panamericana Sur, San Carlos, Chillán; tel. (42) 273-646; fax (2) 271-795; internet www.iansagro.cl; agricultural produce and animal feedstuffs; owned by Empresas Iansa, SA; Gen. Man. JUAN FABRI; 2,173 employees.

**Empresa Pesquera Eperva, SA:** Huérfanos 863, 3°, Casilla 4179, Santiago; tel. (2) 633-1155; fax (2) 639-3436; f. 1955; fish oil and flour producers; CEO CLAUDIO ELGUETA VERA; Chair. FELIPE ZALDÍVAR LARRAÍN; 600 employees.

**Empresas Santa Carolina, SA:** El Bosque Norte 0177, Las Condes, Santiago; tel. (2) 332-0150; fax (2) 238-0307; internet www.santacarolina.com; f. 1874; wine producers; part of Grupo CB conglomerate; Gen. Man. FELIPE DE LA JARA; 1,117 employees.

**Fruticola Viconto, SA:** Apoquindo 4775, 16°, Las Condes, Santiago; tel. (2) 707-4200; fax (2) 707-4250; e-mail viconto@viconto.cl; internet www.viconto.cl; f. 1986; wholesale fruit exporters; Pres. PABLO GUILISASTI GANA; Gen. Man. ENRIQUE TALA SAPAG; 200 employees.

**Industrias Alimentícias Carozzi, SA:** Camino Longitudinal Sur 5201, San Bernardo, Santiago; tel. (2) 377-6400; fax (2) 857-2579; internet www.carozzi.cl; f. 1898; food-processing; Pres. GONZALO BOFILL DE CASO; Gen. Man. JOSÉ JUAN LLUGANY RIGO-RIGHY; 5,584 employees.

**Pesquera Coloso, SA:** Avda El Bosque Norte 0440, Santiago 134 C 35, Santiago; tel. (2) 203-5300; fax (2) 203-5301; e-mail caracena@coloso.cl; internet www.coloso.cl; Commercial Vice-Pres. JUAN CARLOS FERRER; processed fish exporters; owners of Pesquera San José, SA; 731 employees.

**Pesquera Itata, SA:** Presidente Riesco 5711, Of. 1201, Las Condes, Santiago; tel. (2) 782-5400; fax (2) 231-0973; e-mail info@itata.com; internet www.itata.com; f. 1948; fish and fish products; Gen. Man. GERARDO BALBONTÍN; 307 employees.

**Viña Errázuriz, SA:** Avda Nueva Tajamar 481, Torre Sur, Of. 503, Providencia, Santiago; tel. (2) 203-6688; fax (2) 203-6690; e-mail wine.report@errazuriz.cl; internet www.errazuriz.com; f. 1870; wine producers; Pres. EDUARDO CHADWICK; Man. Dir RICARDO URRESTI; 180 employees.

**Viña San Pedro, SA:** Vitacura 4380, 6°, Vitacura, Santiago; tel. (2) 477-5339; fax (2) 477-5309; e-mail info@sanpedro.cl; internet www.sanpedro.cl; f. 1865; wine producers; part of Quiñenco group; Pres. PATRICIO JOTTAR NASRALLAH; Gen. Man. PABLO TURNER GONZÁLEZ; 347 employees.

**Viña Undurraga, SA:** Avda Vitacura 2939, 21°, Las Condes, Santiago; tel. (2) 372-2900; fax (2) 372-2901; e-mail info@undurraga.cl; internet www.undurraga.cl; f. 1885; wine producers; Chair. JORGE VIAL SUBERCASEAUX; CEO ALFONSO UNDURRAGA MACKENNA; 497 employees.

## Wood, Pulp and Paper

**Arauco, Celulosa y Constitución, SA:** El Golf 150, 14°, Santiago; tel. (2) 461-7200; fax (2) 698-5987; internet www.arauco.cl; f. 1979; wood pulp and timber group; Chair. JOSÉ TOMÁS GUZMÁN; CEO ALEJANDRO PÉREZ; 2,400 employees.

**Empresas CMPC, SA:** Agustinas 1343, 8°, Santiago; tel. (2) 441-2000; fax (2) 671-1957; internet www.cmpc.cl; f. 1920; paper and packaging manufacturers, cellulose and wood pulp; Chair. ELIODORO MATTE L.; CEO ARTURO MACKENNA I.; 8,146 employees.

**Industrias Forestales, SA:** 1357 Agustinas, 9°, Santiago; tel. (2) 441-2050; fax (2) 695-7809 ; internet www.inforsa.cl; f. 1956; paper producers; Pres. ANTONIO ALBARRÁN RUIZ-CLAVIJO; Gen. Man. ANDRÉS LARRAÍN MARCHANT; 372 employees.

## Construction Materials

**Besalco, SA:** Ebro 2705, Las Condes, Santiago; tel. (3) 334-4000; fax (3) 334-4031; e-mail besalco@besalco.cl; internet www.besalco.cl; f. 1944; civil engineering and construction; Pres. VICTOR BEZANILLA SAAVEDRA; Gen. Man. PAULO BEZANILLA SAAVEDRA; 7,200 employees.

**Cemento Polpaico, SA:** Avda El Bosque Norte 0177, Las Condes, Santiago; tel. (2) 337-6456; fax (2) 337-6324; e-mail ventaspolpaico@polpaico.cl; internet www.polpaico.cl; f. 1948; jtly owned by GASCO and Holcim of Switzerland; Pres. JUAN ANTONIO GUZMÁN MOLINARI; Gen. Man. EDUARDO KRETSCHMER CASTAÑEDA; 890 employees.

**Cementos Bío-Bío, SA:** Barros Errazuriz 1968, 4°, Santiago; tel. (2) 560-7000; fax (2) 560-7001; internet www.cbb.cl; f. 1957; cement, forestry and raw materials; Pres. HERNÁN BRIONES GOROSTIAGA; CEO ULISES PALLI BAVESTRELLO; 1,128 employees.

**Cerámicas Cordillera, SA:** Circunvalación Americo Vespucio 1001, Quilicura, Santiago; tel. (2) 387-4200; fax (2) 739-0054; e-mail cemento@melon.lafarge.cl; internet www.cordillera.cl; f. 1984; tiles, building materials; owned by Empresas Pizarreño SA; CEO CHRISTIAN PÉREZ MOORE; 243 employees.

**Compañía Industrial El Volcán, SA:** Agustinas 1357, 10°, Santiago; tel. (2) 380-9700; fax (2) 380-9710; f. 1916; makers of insulation and gypsum products; Pres. BERNARDO MATTE LARRAÍN; 287 employees.

**Edelpa, SA (Envases del Pacífico, SA):** Camino a Melipilla 13320, Maipú, Santiago 45; tel. (2) 385-4500; fax (2) 385-4600; e-mail comercial@edelpa.cl; internet www.edelpa.cl; f. 1984; plastic packaging and bottling producers; owners of Italprint, SA; Chair. FERNANDO AGÜERO GARCÉS; CEO LUIS ALBERTO DOMÍNGUEZ TIHISTA; 506 employees.

**Empresas Melón, SA:** Vitacura 2939, Las Condes, Santiago; tel. (2) 280-0000; fax (2) 280-0412; internet www.melon.cl; f. 1908; owned by Lafarge of France; manufacturers of cement; Gen. Man. DENIS BERTHON; 77,000 employees (group).

**Empresas Pizarreño, SA:** Avda Andres Bello 2777, 22°, Santiago; tel. (2) 203-3392; fax (2) 203-3399; internet www.pizarreno.cl; f. 1935; owned by Etex of Belgium; plastic building materials; controls

Cerámica Cordillera, SA; Chair. Canio Corbo Lioi; Gen. Man. Jorge Bennett Urrutia; 2,553 employees.

**Norte Grande, SA:** 331 Paulino Alfonso, Santiago; tel. (2) 633-5007; fax (2) 632-4872; tiles and building materials; CEO Pedro Díaz Guajardo.

### Metals and Chemicals

**Aceros Chile, SA:** Avda Diego Portales 3499-A, San Bernardo, Santiago; tel. (2) 857-3199; fax (2) 857-1579; steelmakers; Gen. Man. Hugo Gaido Gómez; 188 employees.

**CAP, SA:** Huérfanos 669, 8°, Santiago; tel. (2) 520-2000; fax (2) 633-7082; e-mail webmaster@cap.cl; internet www.cap.cl; f. 1946; steel producer and exporter; Chair. Jaime Arbildua Aramburu; Pres. Mario Seguel Santana; 4,746 employees.

**Compañía Electrometalúrgica, SA (ELECMETAL):** Avda Vicuña Mackenna 1570, Santiago; tel. (2) 361-4020; fax (2) 361-4021; e-mail gerencia@elecmetal.cl; internet www.elecmetal.cl; f. 1917; metal foundry, manufactures parts for heavy machinery; part of Cristalchile group; Chair. Ricardo Claro Valdés; Gen. Man. Rolando Medeiros Soux; 300 employees.

**Enaex, SA:** 3859 Renato Sánchez, Las Condes, Santiago; tel. (2) 210-6600; fax (2) 206-67; e-mail enaex@enaex.cl; internet www.enaex.cl; f. 1920; industrial chemicals, incl. explosives; Chair. Ramón Aboitiz Musatadi; 863 employees.

**Industrias Tricolor, SA:** Avda Claudio Arrau 9440, Pudahuel, Santiago; tel. (2) 290-8700; fax (2) 601-0055; e-mail contactenos@tricolor.cl; internet www.tricolor.cl; part of Grupo CB conglomerate; Gen. Man. Santiago Larraín Cruzat; 606 employees.

**Instituto Sanitas, SA:** Avda Américo Vespucio 01260, Quilicura, Santiago; tel. (2) 444-6600; fax (2) 444-6651; e-mail sanitas@sanitas.cl; internet www.sanitas.cl; f. 1920; medicines and vaccines; Chair. Joaquín Barros Fontaine; 171 employees.

**Laboratorio Chile, SA:** Avda Maratón 1315, Ñuñoa, Santiago; tel. (2) 365-5000; fax (2) 365-5100; e-mail pilar.rodriguez@labchile.cl; internet www.labchile.cl; f. 1896; pharmaceutical co; owned by Ivax Corpn of the USA; CEO Hernán Pfeifer; 1,000 employees.

**Madeco, SA:** Ureta Cox 930, Casilla 116-D-2379–10508, Santiago; tel. (2) 520-1000; fax (2) 520-1140; e-mail cgt@madeco.cl; internet www.madeco.cl; f. 1944; metallurgy and packaging; part of Quiñenco conglomerate; Chair. Guillermo Luksic Craig; Gen. Man. Tiberio Dall'Olio; 2,949 employees.

**Molymet, SA (Molibdenos y Metales, SA):** Huérfanos 812, 6°, Santiago; tel. (2) 368-3600; fax (2) 368-3653; e-mail info@molymet.cl; internet www.molymet.cl; f. 1975; producers of industrial chemicals and ferroalloy ores; Chair. Carlos Hurtado; CEO John Graell Moore; 602 employees.

### Textiles

**Bata, SA:** Camino Melipilla 9460, Maipú, Santiago; tel. (02) 560-4200; fax (02) 533-2931; e-mail bsochile@bata.cl; internet www.bata.com; manufacturers and distributors of sportswear and shoes; Chair. Fernando Rivera Jiménez; Gen. Man. Atty Marcelo Villagran Bravo; 4,156 employees.

### Retail

**D y S, SA (Distribución y Servicio, SA):** Avda Presidente Eduardo Frei Montalva 8301, Quilicura, Santiago 7490562; tel. (2) 200-5201; fax (2) 624-2401; f. 1893; supermarket group, owns Ekono, Almac and Líder chains; Chair. Felipe Ibañez Scott; CEO Nicolás Ibañez Scott; 15,998 employees.

**Empresas Almacenes Paris, SA:** Ricardo Lyon 222 Of. 802, Providencia, Santiago; tel. (2) 336-1712; fax (2) 336-7210; e-mail contacto@almacenes-paris.cl; internet www.almacenes-paris.cl; f. 1973; dept stores, furniture and shopping malls; Chair. Juan Galmez Couso; CEO José Miguel Galmez Puig; 6,165 employees.

**Falabella, S.A.C.I.:** Rosas 1665, Santiago; tel. (2) 620-2000; fax (2) 620-2000; internet www.falabella.com; f. 1889; dept stores and textiles; Pres. Reinaldo Solari Magnasco; CEO Pablo Turner Gonzales; 13,935 employees.

**Santa Isabel, SA:** Avda Kennedy 9001, 4°, Las Condes, Santiago; tel. (2) 959-0400; fax (2) 959-0490; internet www.santaisabel.cl; f. 1976; supermarket chain; partly owned by Disco SA of Argentina; Pres. Horst Paulmann; CEO Laurence Golborne; 12,720 employees.

**Sodimac, SA:** Avda Presidente Eduardo Frei 3092, Renca, Casilla 3110, Santiago; tel. (2) 738-1000; fax (2) 641-8271; internet www.sodimac.com; f. 1982; home improvement products retailers; Chair. Juan Pablo del Río Goudie; Gen. Man. Guillermo Aguero Piwonka; 4,100 employees.

### Information Technology

**Adexus, SA:** Miraflores 383, 22°, Santiago; tel. (2) 686-1000 ; fax (2) 686-1201; e-mail adexus@adexus.cl; internet www.adexus.com; information technology systems contractors; f. 1990 as Tandem Chile, SA; name changed 1998; Chair. Patricio del Sante Scroggie; CEO Carlos Busso Vyhmeister; 500 employees.

**Sonda, SA:** Teatinos 500, Santiago; tel. (2) 560-5000; fax (2) 560-5410; e-mail info@sonda.com; internet www.sonda.com; f. 1974; information technology services; Pres. Andrés Navarro; Gen. Man Mario Pavón; 2,400 employees.

### Miscellaneous

**Compañía Chilena de Fósforos, SA:** Los Conquistadores 1700, 15°, Providencia, Santiago; tel. (2) 232-4990; fax (2) 231-5072; internet www.fosforos.cl; f. 1913; makers of matches and producers of wine; Gen. Man. José Luis Vender; 1,280 employees.

**Compañía Chilena de Tabacos, SA (Chiletabacos):** El Bosque Norte 0125, Las Condes, Santiago; tel. (2) 464-6000; fax (2) 464-6241; internet www.chiletabacos.cl; f. 1909; subsidiary of British American Tobacco Co Ltd, United Kingdom; tobacco co; Gen. Man. Michael Hardy Tudor; 752 employees.

**Compañía Tecno Industrial, SA (CTI, SA):** Alberto Llona 777, Maipú, Santiago; tel. (2) 530-6100; fax (2) 530-6165; e-mail wadm@cti.cl; internet www.cti.cl; f. 1905; manufacturers of domestic electrical appliances; Chair. Ramón Aboitiz Musatadi; CEO Cirilo Pablo de Cordova; 840 employees.

**Compañias CIC, SA:** Avda Esquina Blanca 960, Maipú, Santiago; tel. (2) 530-4000; fax (2) 557-4362; e-mail cic@cic.cl; internet www.cic.cl; f. 1912; exporters of wooden furniture, mattresses, etc.; Chair. Leonidas E. Vial Echeverría; Gen. Man. Alejandro Bordeu Schwarz; 1,100 employees.

**Cristalerías de Chile, SA (Cristalchile):** Hendaya 60, Of. 201, Las Condes, Santiago; tel. (2) 787-8888; fax (2) 787-8800; e-mail gerencia@cristalchile.cl; internet www.cristalchile.cl; f. 1904; bottle and packaging producers; the group also controls winemakers Santa Rita, the *Diario Financiero* newspaper and the Megavisión television network; Chair. Ricardo Claro Valdés; CEO Cirilo Elton González; 700 employees.

**Hoteles Carrera, SA:** Teatinos 180, Santiago; tel. (2) 698-2011; fax (2) 672-1083; e-mail hotelcarrera@carrera.cl; internet www.carrera.cl; part of Quiñenco conglomerate; f. 1981; Chair. Jorge Olavarría Romussi; Gen. Man. Gustavo Yurjevic Marshall; 360 employees.

**Sociedad El Tattersall, SA:** Isidora Goyenechea 3250, 2°, Las Condes, Santiago; tel. (2) 362-3005; fax (2) 362-3002; e-mail sociedad@tattersall.cl; internet www.tattersall.cl; f. 1913; distributors of agricultural machinery; also represent Budget car rentals in Chile; Chair. Carlos Eugenio Jorquiera Malschafsky; 501 employees.

**Sociedad Hipódromo Chile, SA:** Avda Hipódromo Chile 1715, Santiago; tel. (2) 270-9237; fax (2) 777-2089; internet www.hipodromo.cl; f. 1904; sports club; Chair. Juan Cuneo Solari; Gen. Man. Luis Ignacio Salas Maturana; 584 employees.

**Somela, SA:** A. Escobar Williams 600, Cerillos, Santiago; tel. (2) 557-4225; fax (2) 557-5667; e-mail somela@somela.cl; internet www.somela.cl; f. 1950; manufacturers of domestic appliances; Chair. Ramón Aboitiz Musatadi; Gen. Man. Wayhí Yousef Allel; 263 employees.

## UTILITIES

**Comisión Nacional de Energía:** Teatinos 120, 7°, Clasificador 14, Correo 21, Santiago; tel. (2) 460-6800; fax (2) 365-6800; e-mail energia@cne.cl; internet www.cne.cl; Minister-Pres. Jorge Rodríguez Grossi; Exec. Sec. Luis Sánchez Castellón.

**Superintendencia de Electricidad y Combustibles (SEC):** Amunátegui 58, Santiago; internet www.sec.cl; Supt Sergio Espejo Yaksic; Gen. Sec. Marta Cabeza Vargas.

### General

**Colbún, SA:** Avda 11 de Septiembre 2355; tel. (2) 460-4000; fax (2) 460-4005; e-mail contacto@colbun.cl; internet www.colbun.cl; f. 1986; hydroelectricity producer; fmrly state-owned, acquired by Tracetebel SA of Belgium; Pres. Emilio Pellegrini Ripamonti; CEO Francisco J. Courbis Grez; 226 employees.

### Electricity

The four national power grids are SING (operated by CODELCO and EDELNOR), SIC, Aysén and Magallanes.

**AES Gener, SA:** Mariano Sánchez Fontecilla 310, 3°, Santiago; tel. (2) 686-8900; fax (2) 686-8991; e-mail gener@gener.cl; internet www.gener.cl; f. 1981 as Chilectra Generación, SA, following the restructuring of Compañía Chilena de Electricidad, SA; privatized in 1988 and Chilgener, SA, adopted in 1989; current name adopted in 1998; owned by AES Corpn (USA); responsible for operation of power plants Renca, Ventanas, Laguna Verde, El Indio, Altalfal, Maitenes, Queltehues and Volcán; total output 10,169.0 GWh (Dec. 2002); also operates subsidiaries in Argentina and Colombia; Pres. JOSEPH BRANDT; Gen. Man. FELIPE CERÓN; 1,121 employees (group).

**Empresa Eléctrica Guacolda, SA:** Miraflores 222, 16°, Santiago; tel. (2) 362-4000; fax (2) 360-1675; internet www.guacolda.cl; operates a thermoelectric power-station in Huasco; installed capacity of 304 MW; Gen. Man. JOSÉ FLORENCIO GUZMÁN.

**Empresa Eléctrica Santiago:** Jorge Hirmas 2964, Renca, Santiago; tel. (2) 680-4760; fax (2) 680-4743; operates the Nueva Renca thermoelectric plant in Santiago; installed capacity of 379 MW; Gen. Man. HÉCTOR ROJAS.

**Energía Verde:** O'Higgins 940, Of. 90, Concepción; tel. (43) 431-363 ; fax (43) 431-357; operates two co-generation power-stations at Constitución and Laja and a steam plant at Nacimiento; supplies the Cabrero industrial plant; CEO JAIME ZUAZAGOITÍA.

**Norgener, SA:** Jorge Hirmas 2964, Renca, Santiago; tel. (2) 680-4870; fax (2) 680-4895; northern subsidiary supplying the mining industry; operates power plants with installed capacity of 276 MW; Gen. Man. CARLOS AGUIRRE.

**Arauco Generación:** Vitacura 2771, 9°, Las Condes, Santiago; tel. (2) 560-6700; fax (2) 236-5090; e-mail gic@arauco.cl; f. 1994 to commercialize surplus power from pulp processing facility; Pres. CARLOS CROXATTO S.; Gen. Man. HERNÁN ARRIAGADA.

**BMV Industrias Eléctricas:** Avda Vicuña Mackenna 1540, Ñuñoa, Santiago; tel. (2) 555-8806; fax (2) 555-8807; e-mail bmv@bmv.cl; internet www.bmv.cl; Gen. Man. ANGÉLICA PADOVANI.

**Chilquinta Energía, SA:** General Cruz 222, Valparaíso; tel. (32) 502-000; fax (32) 231-171; e-mail contactoweb@chiquinta.cl; internet www.chilquinta.cl; f. 1995; owned by Sempra and PSEG of the USA; Pres. HÉCTOR MADARIAGA; Gen. Man. CRISTIAN ARNOLDS.

**Compañía Eléctrica del Litoral, SA:** San Sebastián 2952, Of. 202, Las Condes, Santiago; tel. (2) 481-195; fax (2) 483-313; internet www.litoral.cl; e-mail fmartine@litoral.cl; f. 1949; Gen. Man. MARCELO LUENGO.

**Compañía General de Electricidad, SA (CGE):** Teatinos 280, Santiago; tel. (2) 624-3243; fax (2) 680-7104; e-mail cge@cge.cl; internet www.cge.cl; installed capacity of 662 MW; Pres. GABRIEL DEL REAL; Gen. Man. GUILLERMO MATTA FUENZALIDA.

**Empresa Eléctrica Emec, SA:** Los Talleres 1831, Barrio Industrial, Coquimbo; tel. (51) 20-1000; fax (51) 20-1188; internet www.emec.cl; e-mail info@emec.cl; f. 1980; Pres. JOSÉ HORNAUER LÓPEZ; Gen. Man. SERGI JORDANA DE BUEN.

**Empresa Eléctrica de Aysén, SA:** Francisco Bilbao 412, Casilla 280, Coyhaique; tel. (67) 23-3105; e-mail ggral@entelchile.net; f. 1983; Gen. Man. JORGE ILLANES SENN.

**Empresa Eléctrica de Magallanes, SA (Edelmag, SA):** Croacia 444, Punta Arenas; tel. (71) 40-00; fax (71) 40-77; e-mail edelmag@edelmag.cl; internet www.edelmag.cl; f. 1981; Pres. JORGE JORDAN FRANULIC; Gen. Man. CARLOS YÁÑEZ ANTONUCCI; 111 employees.

**Empresa Eléctrica del Norte Grande (EDELNOR):** Avda Apoquindo 3721, Santiago; tel. (2) 353-3200; fax (2) 206-2154; f. 1981; acquired by Codelco and Tractebel of Belgium in Dec. 2002; Gen. Man. JUAN CLAVERÍA A.; 365 employees.

**Empresas Emel, SA:** Avda Libertador Bernardo O'Higgins 886, Central Post Office, 5° y 6°, Santiago; tel. (2) 376-6500; fax (2) 633-3849; holding co for the Emel group of electricity cos, owned by PPL Corpn of the USA; output in 2000 totalled 18,6124 MWh; CEO ALFONSO G. TORO; Emel group includes:

**ELECDA** (Empresa Eléctrica de Antofagasta, SA): José Miguel Carrera 1587, Antofagasta 1250; tel. (55) 649-100; Dir EDUARDO ARTURO APABLAZA DAU.

**ELIQSA** (Empresa Eléctrica de Iquique, SA): Zegeres 469, Iquique; tel. (57) 40-5400; fax (57) 42-7181; e-mail eliqsa@eliqsa.cl; CEO AQUILES IVAN MELÉNDEZ VARGAS.

**EMELAT** (Empresa Eléctrica Atacama, SA): Circunvalación Ignacio Carrera Pinto 51, Copiapó; tel. (52) 205-100; fax (52) 205-103; f. 1981; distribution company; Pres. MICHAEL FRIEDLANDER; Gen. Man. EDUARDO ARTURO APABLAZA DAU.

**EMELARI** (Empresa Eléctrica de Arica, SA): Baquedano 731, Arica; tel. (58) 201-100; fax (58) 23-1105; Dir JOHN FOGARTY.

**ENERSIS, SA:** Avda Kennedy 5454, Casilla 1557, Vitacura, Santiago; tel. (2) 353-4400; fax (2) 378-4768; e-mail comunicacion@e.enersis.cl; internet www.enersis.com; f. 1981; holding co for Spanish group generating and distributing electricity through its subsidiaries throughout South America; 60.62% owned by Endesa S.A.; Gen. Man. MARIO VALCARCE; 10,957 employees.

**Chilectra, SA:** Santo Domingo 789, Casilla 1557, Santiago; tel. (2) 632-2000; fax (2) 639-3280; e-mail rrpp@chilectra.cl; internet www.chilectra.cl; f. 1921; transmission and distribution arm of ENERSIS (see above); supplies distribution cos, including the Empresa Eléctrica Municipal de Lo Barnechea, Empresa Municipal de Til-Til, and the Empresa Eléctrica de Colina, SA; holds overseas distribution concessions in Argentina, Peru and Brazil; acquired by ENERSIS of Spain in 1999; Pres. JORGE ROSENBLUT; Gen. Man. JULIO VALENZUELA.

**Empresa Eléctrica Pehuenche, SA (EEP):** Santa Rosa 76, 14°, Santiago; tel. (2) 630-9000; fax (2) 696-5568; f. 1986; part of ENERSIS group; Gen. Man. LUCIO CASTRO MÁRQUEZ.

**Empresa Nacional de Electricidad, SA (ENDESA):** Santa Rosa 76, Casilla 1392, Santiago; tel. (2) 630-9000; fax (2) 635-4720; e-mail comunicacion@endesa.cl; internet www.endesa.cl; f. 1943; installed capacity 4,035 MW (Feb. 2002); ENERSIS obtained majority control of ENDESA Chile in April 1999; Pres. LUIS RIVERA NOVO; Gen. Man. HECTOR LÓPEZ VILASECO.

**SAESA** (Sociedad Austral de Electricidad, SA): Manuel Bulnes 441, Osorno; tel. (64) 20-6200; fax (64) 20-6209; e-mail saesa@saesa.cl; internet www.saesa.cl; owned by PSEG Corpn of the USA; Gen. Man. JORGE BRAHM BARRIL.

## Gas

**Abastecedora de Combustible (Abastible, SA):** Avda Vicuña Mackenna 55, Providencia, Santiago; tel. (2) 693-0000; fax (2) 693-9304; internet www.abastible.cl; owned by COPEC; Pres. FELIPE LAMARCA CLARO; Gen. Man. JOSÉ ODONE.

**AGA Chile, SA:** Juan Bautista Pistene 2344, Santiago; tel. (2) 907-6888; f. 1920; natural and industrial gases utility; owned by Linde Gas Corpn of Germany.

**Compañía de Consumidores de Gas de Santiago (GASCO, SA):** 1061 Santo Domingo, Casilla 8-D, Santiago; tel. (2) 694-4444; fax (2) 694-4370; e-mail info@gasco.cl; internet www.gasco.cl; natural gas utility; supplies Santiago and Punta Arenas regions; owned by CGE; Pres. MATÍAS PÉREZ CRUZ; Gen. Man. JOSÉ LUIS HORNAUER HERRMANN.

**Electrogas:** Evaristo Lillo 78, Of. 41, Las Condes, Santiago; tel. (2) 377-1458; fax (2) 233-4931; owned by ENDESA (q.v.); Gen. Man. CARLOS ANDREANI.

**Gas Valpo:** Camino Internacional 1420, Viña del Mar; tel. (32) 27-7000; fax (32) 21-3092; e-mail info@gasvalpo.cl; internet www.gasvalpo.cl; f. 1853; owned by AGL of Australia.

**GasAndes:** Avda. Isidora Goyenechea 3600, Las Condes, Santiago; tel. (2) 362-4200; distributes gas transported from the Argentine province of Mendoza via a 463-km pipeline.

**GasAtacama:** natural-gas producer and transporter; CEO RUDOLF ARANEDA.

**Lipigas:** Las Urbinas 53, 13°, Of. 131, Providencia, Santiago; tel. (32) 656-500; fax (32) 656-595; e-mail info@lipigas.cl; internet www.lipigas.cl; Pres. JAIME SANTA CRUZ; Gen. Man. MARIO FERNÁNDEZ.

**Agrogas:** Lo Ovalle 1321, Santiago; tel. (2) 511-0904; owned by Empresas Lipigas, SA; supplier of liquid gas.

**Industrias Codigas:** Las Urbinas 53, 9°, Providencia, Santiago; tel. (2) 520-4700; fax (2) 520-4733; e-mail info@codigas.cl; internet www.codigas.cl; f. 1959; owned by Empresas Lipigas, SA; Pres. HUGO YACONI MERINO; Gen. Man. JOSÉ LUIS MEIER.

**Metrogas:** El Bosque Norte 0177, 11°, Las Condes, Santiago; tel. (2) 337-8000; fax (2) 337-8173; internet www.metrogas.cl; natural gas utility; owned by GASCO and COPEC; Pres. MATÍAS PÉREZ CRUZ; Gen. Man. EDUARDO MORANDÉ.

## Water

**Aguas Andinas, SA:** Avda Presidente Balmaceda 1398, Santiago; tel. (2) 688-1000; fax (2) 698-5871; e-mail info@aguasandinas.cl; internet www.aguasandinas.cl; water supply and sanitation services to Santiago and the surrounding area; sold to a French-Spanish consortium in June 1999; Gen. Man. ANGEL SIMÓN GRIMALDOS.

**Empresa de Obras Sanitarias de Valparaíso, SA (Esval):** Cochrane 751, Valparaíso; tel. (32) 209-000; fax (32) 209-502; e-mail infoesval@entelchile.net; internet www.esval.cl; f. 1989; sanitation and irrigation co serving Valparaíso; Chair. EDMUNDO DUPRÉ ECHEVERRÍA; Gen. Man. GUSTAVO GONZÁLEZ DOORMAN; 377 employees.

**Sigsig Ltda (Tecnagent):** Presidente Errázuriz 3262, Casilla 7550295, Las Condes, Santiago; tel. (2) 335-2001; fax (2) 334-8466; e-mail tecnagent@tecnagent.cl; internet www.tecnagent.cl; Pres. RAÚL SIGREN BINDHOFF; Gen. Man. RAÚL A. SIGREN ORFILA.

CHILE

*Directory*

## TRADE UNIONS

There are more than 50 national labour federations and unions.

### Central Unions

**Central Autónoma de Trabajadores (CAT):** Sazié 1761, Casilla 6510480, Santiago; tel. and fax (2) 695-3388; e-mail catchile@entelchile.net; Pres. OSVALSO ERBACH ALVAREZ; Sec. Gen. PEDRO SAAVEDRA.

**Central Unitaria de Trabajadores de Chile (CUT):** Alameda 1346, Santiago; tel. (2) 361-9433; fax (2) 361-9452; e-mail presidencia.cut@entelchile.net; internet www.cutchile.cl; f. 1988; 12 associations, 32 confederations, 49 federations; 36 regional headquarters, 16 trade unions; Pres. ARTURO MARTÍNEZ MOLINA; 411,000 mems; Gen. Sec. JOSÉ ORTIZ ARCOS.

**Movimiento Unitario Campesino y Etnias de Chile (MUCECH):** Portugal 623, Of. 1-A, Santiago; tel. (2) 222-6572; fax (2) 635-1518; e-mail mucech@ia.cl; Pres. EUGENIO LEÓN GAJARDO; Nat. Sec. RIGOBERTO TURRA PAREDES.

### Union Confederations

There are 37 union confederations, of which the following are among the most important:

**Agrupación Nacional de Empleados Fiscales (ANEF):** Edif. Tucapel Jiménez, Alameda 1603, Santiago; tel. (2) 696-2957; fax 699-3806; internet www.anef.cl; f. 1943; affiliated to CUT; public-service workers; Pres. RAÚL DE LA PUENTE PEÑA; Sec.-Gen. JEANETTE SOTO FUENTES.

**Confederación Bancaria:** Santiago; tel. (2) 699-5597; affiliated to CUT; Pres. DIEGO OLIVARES ARAVENA; Sec.-Gen. RAÚL REQUENA MARTÍNEZ.

**Confederación de Empleados Particulares de Chile (CEPCH):** Santiago; tel. (2) 72-2093; trade union for workers in private sector; affiliated to CUT; Pres. ANGÉLICA CARVALLO PRENAFETA; Sec.-Gen. ANDRÉS BUSTOS GONZÁLEZ.

**Confederación General de Trabajadores del Transporte Terrestre (CGTT):** Almirante Latorre 355, 2°, Of. 3, Santiago; tel. and fax (2) 695-9551; affiliated to CUT; road transport workers; Pres. ULISES MARTÍNEZ SEPÚLVEDA; Sec.-Gen. RODOLFO DOSSETTO UGALDE.

**Confederación Nacional Campesina:** Gorbea 1769, Santiago; tel. and fax (2) 695-2017; affiliated to CUT; Pres. EUGENIO LEÓN GAJARDO; Sec.-Gen. RENÉ ASTUDILLO R.

**Confederación Nacional de Federaciones y Sindicatos de Empresas e Interempresas de Trabajadores del Transporte Terrestre y Afines (CONATRACH):** Concha y Toro 2A, 2°, Santiago; tel. and fax (2) 698-0810; Pres. PEDRO MONSALVE FUENTES; Sec.-Gen. PEDRO JARA ESPINOZA.

**Confederación Nacional de Federaciones y Sindicatos de Trabajadores Textiles y Ramos Similares (CONTEXTIL):** Avda Expaña 347, Santiago; tel. and fax (2) 696-8098; affiliated to CUT; Pres. PATRICIA COÑOMÁN CARRILLO; Sec.-Gen. MARÍA FELISA GARAY ASTUDILLO.

**Confederación Nacional de Gente de Mar, Marítimos, Portuarios y Pesqueros (CONGEMAR):** Tomás Ramos 158–172, Valparaíso; tel. (32) 255-430; fax (32) 257-580; affiliated to CUT; Pres. WALTER ASTORGA LOBOS; Sec.-Gen. JUAN GUZMÁN CARRASCO.

**Confederación Nacional de Sindicatos Agrícolas—Unidad Obrero Campesina (UOC):** Eleuterio Ramírez 1463, Santiago; tel. and fax (2) 696-6342; e-mail confe.uocchile@uocchile.cl; affiliated to CUT; Pres. SEGUNDO JUAN CORVALÁN HUERTA; Sec.-Gen. PEDRO REYES BAHAMONDEZ.

**Confederación Nacional de Sindicatos de Trabajadores de la Construcción, Maderas, Materiales de Edificación y Actividades Conexas:** Almirante Hurtado 2069, Santiago; tel. (2) 695-3908; fax (2) 696-4536; affiliated to CUT; Pres. MIGUEL ANGEL SOLÍS VIERA; Sec.-Gen. ADRIAN FUENTES HERMOSILLA.

**Confederación Nacional de Trabajadores Metalúrgicos (CONSTRAMET):** Santa Rosa 101, esq. Alonso Ovalle, Santiago; tel. (2) 664-8581; fax (2) 638-3694; e-mail contrame@ctcinternet.cl; affiliated to CUT and the International Metalworkers' Federation; Pres. MIGUEL SOTO ROA; Sec.-Gen. MIGUEL CHÁVEZ SOAZO.

**Confederación Nacional de Suplementeros de Chile (CONASUCH):** Tucapel Jiménez 26, Santiago; tel. (2) 695-7639; fax (2) 699-1646; f. 1942; Pres. IVAN ENCINA CARO; Sec. JOSÉ CANALES ORTIZ.

**Confederación Nacional de Trabajadores de la Alimentación y Afines (CONTALCH):** Liszt 3082, San Joaquín, Santiago; tel. and fax (2) 553-2193; affiliated to CUT; Pres. CIJIFREDO VERA VERA.

**Confederación Nacional de Trabajadores de la Industria del Pan (CONAPAN):** Tucapel Jiménez 32, 2°, Santiago; tel. and fax (2) 672-1622; affiliated to CUT; Pres. LUIS ALEGRÍA ALEGRÍA; Sec. LUIS PALACIOS CAMPOS.

**Confederación Nacional de Trabajadores de la Industria Textil (CONTEVECH):** Agustinas 2349, Santiago; tel. (2) 699-3442; fax (2) 687-3269; affiliated to CUT; Pres. MIGUEL VEGA FUENTES; Sec.-Gen. OSCAR CÁCERES YÁÑEZ.

**Confederación Nacional de Trabajadores del Comercio (CONATRADECO):** Santiago; tel. and fax (2) 638-6718; Pres. EDMUNDO LILLO ARAVENA; Sec.-Gen. FEDERICO MUJICA CANALES.

**Confederación Nacional de Trabajadores del Comercio (CONSFECOVE):** Santiago; tel. (2) 632-2950; fax (2) 632-2884; affiliated to CUT; Pres. JOSÉ LUIS ORTEGA; Sec.-Gen. SUSANA ROSAS VALDEBENITO.

**Confederación Nacional de Trabajadores del Cuero y Calzado (FONACC):** Arturo Prat 1490, Santiago; tel. (2) 556-9602; affiliated to CUT; Pres. MANUEL JIMÉNEZ TORRES; Sec.-Gen. VÍCTOR LABBÉ SILVA.

**Confederación Nacional de Trabajadores Electrometalúrgicos, Mineros y Automotrices de Chile (CONSFETEMA):** Vicuña Mackenna 3101, San Joaquín, Santiago; tel. (2) 238-1732; fax (2) 553-6494; e-mail consfetema@123mail.cl; Pres. LUIS SEPÚLVEDA DEL RÍO.

**Confederación Nacional de Trabajadores Forestales (CTF):** Rengo 884, Casilla 2717, Concepción; tel. and fax (41) 220-0407; Pres. JORGE GONZÁLEZ CASTILLO; Sec.-Gen. GUSTAVO CARRASO SALAZAR.

**Confederación Nacional de Trabajadores Molineros:** Bascuñan Guerrero 1739, Casilla 703, Correo 21, Santiago; tel. and fax (2) 683-8882; Pres. JOSÉ VÁSQUEZ ALIAGA.

**Confederación Nacional Minera:** Príncipe de Gales 88, Casilla 10361, Correo Central, Santiago; tel. and fax (2) 696-6945; Pres. MOISÉS LABRAÑA MENA; Sec.-Gen. JOSÉ CARRILLO BERMEDO.

**Confederación Nacional Sindical Campesina y del Agro 'El Surco':** Chacabuco 625, Santiago; tel. and fax (2) 681-1032; e-mail asurco@entelchile.net; affiliated to CUT; Pres. FERNANDO VALÁSQUEZ SERRANO; Sec.-Gen. SERGIO DÍAZ TAPIA.

There are also 45 union federations and over 100 individual unions.

## Transport

### RAILWAYS

#### State Railways

**Empresa de los Ferrocarriles del Estado:** Avda B. O'Higgins 3170, Santiago; tel. (2) 376-8500; fax (2) 776-2609; e-mail contacto@efe.cl; internet www.efe.cl; f. 1851; 3,977 km of track (2000); the State Railways are divided between the Ferrocarril Arica–La Paz, La Calera Puerto Montt, and Metro Regional de Valparaíso (passenger service only); several lines scheduled for privatization; Pres. LUIS AJENJO ISASI; Gen. Man. EDUARDO CASTILLO AGUIRRE.

#### Parastatal Railways

**Ferrocarril del Pacífico, SA (FEPASA):** Málaga 120, 5°, Las Condes, Santiago; tel. (2) 412-1000; fax (2) 412-1040; e-mail oguevara@fepasa.cl; internet www.fepasa.cl; f. 1993; privatized freight services; Pres. RAMÓN ABOITIZ MUSATADI; Vice-Pres. JUAN EDUARDO ERRÁZURIZ OSSA; Gen. Man. GAMALIEL VILLALOBOS ARANDA.

**Metro de Santiago:** Empresa de Transporte de Pasajeros Metro, SA, Avda B. O'Higgins 1414, Santiago; tel. (2) 250-3000; fax (2) 699-2475; internet www.metrosantiago.cl; started operations 1975; 40.2 km (2003); 3 lines; Pres. FERNANDO BUSTAMENTE HUENTA; Gen. Man. R. AZÓCAR HIDALGO.

#### Private Railways

**Empresa de Transporte Ferroviario, SA (Ferronor):** Huérfanos 587, Ofs 301 y 302, Santiago; tel. (2) 638-0430; e-mail ferronor@ferronor.cl; internet www.ferronor.cl; 2,300 km of track (2002); established as a public/private concern, following the transfer of the Ferrocarril Regional del Norte de Chile to the then Ministry of Production Development (CORFO) as a *Sociedad Anónima* in 1989; operates cargo services only; Commercial Man. PABLO ARRANZ; Pres. ROBERTO PIRAZZOLI.

**Ferrocarril Codelco-Chile:** Barquito, Region III, Atacama; tel. (52) 48-8521; fax (52) 48-8522; Gen. Man. B. BEHN THEUNE.

**Diego de Almagro a Potrerillos:** transport of forest products, minerals and manufactures; 99 km.

**Ferrocarril Rancagua–Teniente:** transport of forest products, livestock, minerals and manufactures; 68 km.

**Ferrocarril de Antofagasta a Bolivia (FCAB):** Bolívar 255, Casillas ST, Antofagasta; tel. (55) 20-6700; fax (55) 20-6220; e-mail webmaster@fcab.cl; internet www.fcab.cl; f. 1888; owned by Luksic

# CHILE
*Directory*

conglomerate (Quiñenco); operates an international railway to Bolivia and Argentina; cargo forwarding services; total track length 934 km; Chair. ANDRÓNICO LUKSIC ABAROA; Gen. Man. M. V. SEPÚLVEDA.

**Ferrocarril Tocopilla–Toco:** Calle Arturo Prat 1060, Casilla 2098, Tocopilla; tel. (55) 81-2139; fax (55) 81-2650; owned by Sociedad Química y Minera de Chile, SA; 117 km (1995); Gen. Man. SEGISFREDO HURTADO GUERRERO.

## Association

**Asociación Chilena de Conservación de Patrimonio Ferroviario** (ACCPF—Chilean Railway Society): Concha y Toro 10, Barrio Brasil, Santiago; tel. (2) 699-4607; fax (2) 280-0252; e-mail rsandoval@ucinf.cl; internet www.accpf.cl; Pres. CARLOS ROA VALANZUELA; Vice-Pres. SERGIO GONZÁLEZ RODRÍGUEZ.

## ROADS

The total length of roads in Chile in 2001 was an estimated 79,605 km, of which some 6,279 km were highways and some 16,410 km were secondary roads. The road system includes the entirely paved Pan-American Highway, extending 3,455 km from north to south. Toll gates exist on major motorways. The 1,200-km Carretera Austral (Southern Highway), linking Puerto Montt and Puerto Yungay, was completed in 1996, at an estimated total cost of US $200m.

## SHIPPING

As a consequence of Chile's difficult topography, maritime transport is of particular importance. In 1997 90% of the country's foreign trade was carried by sea (51m. metric tons). The principal ports are Valparaíso, Talcahuano, Antofagasta, San Antonio, Arica, Iquique, Coquimbo, San Vicente, Puerto Montt and Punta Arenas. Most port operations were privatized in the late 1990s.

Chile's merchant fleet amounted to 963,966 grt (comprising 532 vessels) at December 2003.

### Supervisory Authorities

**Asociación Nacional de Armadores:** Blanco 869, 3°, Valparaíso; tel. (32) 21-2057; fax (32) 21-2017; e-mail armadore@entelchile.net; f. 1931; shipowners' association; Pres. ERICH STRELOW CASTILLO; Gen. Man. ARTURO SIERRA MERINO.

**Cámara Marítima y Portuaria de Chile, AG:** Blanco 869, 2°, Valparaíso; tel. (32) 25-3443; fax (32) 25-0231; e-mail camarit@entelchile.net; Pres. BELTRÁN SAÉZ M.; Vice-Pres. RODOLFO GARCÍA SÁNCHEZ.

**Dirección General de Territorio Marítimo y Marina Mercante:** Errázuriz 537, 4°, Valparaíso; tel. (32) 20-8000; fax (32) 25-2539; e-mail webmaster@directemar.cl; internet www.directemar.cl; maritime admin. of the coast and national waters, control of the merchant navy; ship registry; Dir Rear Adm. FERNANDO LAZCANO.

### Cargo Handling Companies

**Empresa Portuaria Antofagasta:** Grecia s/n, Antofagasta; tel. (55) 25-1737; fax (55) 22-3171; e-mail afernandez@puertoantofagasta.cl; internet www.puertoantofagasta.cl; Pres. BLAS ENRIQUE ESPINOZA SEPÚLVEDA; Gen. Man. ALVARO FERNÁNDEZ SLATER.

**Empresa Portuaria Arica:** Máximo Lira 389, Arica; tel. (58) 25-5078; fax (58) 23-2284; e-mail puertoarica@puertoarica.cl; internet www.puertoarica.cl; Pres. CARLOS EDUARDO MENA KEYMER; Gen. Man. PATRICIO CAMPAÑA CUELLO.

**Empresa Portuaria Austral:** B. O'Higgins 1385, Punta Arenas; tel. (61) 24-1111; fax (61) 24-1111; e-mail portspuq@epa.co.cl; internet www.epa.co.cl; Pres. LAUTARO HERNÁN POBLETE KNUDTZON-TRAMPE; Dir FERNANDO ARTURO JOFRÉ WEISS.

**Empresa Portuaria Chacabuco:** B. O'Higgins s/n, Puerto Chacabuco; tel. (67) 35-1198; fax (67) 35-1174; e-mail ptochb@entelchile.net; Pres. LUIS MUSALEM MUSALEM; Dir RAIMUNDO CRISTI SAAVEDRA.

**Empresa Portuaria Coquimbo:** Melgareja 676, Coquimbo; tel. (51) 31-3606; fax (51) 32-6146; e-mail ptoqq@entelchile.net; Pres. ARMANDO ARANCIBIA CALDERÓN; Gen. Man. MIGUEL ZUVIC MUJICA.

**Empresa Portuaria Iquique:** Jorge Barrera 62, Iquique; tel. (57) 40-0100; fax (57) 41-3176; e-mail epi@epi.cl; internet www.port-iquique.cl; f. 1998; Pres. PATRICIO ARRAU PONS; Gen. Man. PEDRO DÁVILA PINO.

**Empresa Portuaria Puerto Montt:** Angelmó 1673, Puerto Montt; tel. (65) 25-2247; e-mail info@empormontt.cl; internet www.empormontt.cl; Pres. JOSÉ DANIEL BARRETA SÁEZ; Gen. Man. LUÍS RIVAS APABLAZA.

**Empresa Portuaria San Antonio:** Alan Macowan 0245, San Antonio; tel. (35) 58-6000; fax (35) 58-6015; e-mail correo@saiport.cl; internet www.saiport.cl; f. 1998; Pres. PATRICIO ARRAU PONS (designate); Gen. Man. RUBEN ALVARADO VIGAR.

**Empresa Portuaria Talcahuano-San Vicente:** Avda Blanco Encalada 547, Talcahuano; tel. (41) 79-7600; fax (41) 79-7627; e-mail eportuaria@puertotalcahuano.cl; Pres. JUAN ENRIQUE COEYMANS AVARIA; Gen. Man. LUIS ALBERTO ROSENBERG NESBET.

**Empresa Portuaria Valparaíso:** Errázuriz 25, 4°, Of. 1, Valparaíso; tel. (32) 44-8800; fax (32) 22-4190; e-mail gcomercial@portvalparaiso.cl; internet www.portvalparaiso.cl; Pres. GABRIEL ALDONEY V.; Gen. Man. HARALD JAEGER KARL.

### Principal Shipping Companies

#### Santiago

**Cía Chilena de Navegación Interoceánica, SA:** Plaza de la Justicia 59, Valparaíso; tel. (32) 275-500; fax (32) 255-949; e-mail info@ccni.cl; internet www.ccni.cl; f. 1930; regular sailings to Japan, Republic of Korea, Taiwan, Hong Kong, USA, Mexico, South Pacific, South Africa and Europe; bulk and dry cargo services; owned by Empresas Navieras, SA; Pres. CARLOS ALLIMANT; Gen. Man. ALEJANDRO PATTILLO MOREIRA; 116 employees.

**Naviera Magallanes, SA (NAVIMAG):** Avda El Bosque, Norte 0440, 11°, Of. 1103/1104, Las Condes, Santiago; tel. (2) 442-3150; fax (2) 442-3156; f. 1979; Chair. PEDRO LECAROS MENÉNDEZ; Gen. Man. HÉCTOR HENRÍQUEZ NEGRÓN.

**Nisa Navegación, SA:** Avda El Bosque Norte 0440, 11°, Casilla 2829, Santiago; tel. (2) 203-5180; fax (2) 203-5190; Chair. PEDRO LECAROS MENÉNDEZ; Fleet Man. C. SALINAS.

**Sociedad Anónima de Navegación Petrolera (SONAP):** Moneda 970, 20°, Casilla 13-D, Santiago; tel. (2) 630-1009; fax (2) 630-1041; e-mail valsonap@sonap.cl; f. 1954; tanker services; Chair. FELIPE VIAL C.; Gen. Man. JOSÉ THOMSEN Q.

#### Valparaíso

**Agencias Universales, SA (AGUNSA):** Avda Andrés Bello 2687, 15°, Las Condes, Santiago; tel. (2) 203-9000; fax (2) 203-9009; internet www.agunsa.com; f. 1960; maritime transportation and shipping, port and docking services; owned by Empresas Navieras, SA; Chair. JOSÉ MANUEL URENDA SALAMANCA; Gen. Man. FRANCO MONTALBETTI MOLTEDO.

**A. J. Broom y Cía, SAC:** Blanco 951, Casilla 910, Valparaíso and MacIver 225, 10°, Casilla 448, Santiago; tel. (32) 26-8200; fax (32) 21-3308; e-mail genmanager@ajbroom.cl; internet www.broomgroup.com; f. 1920; ship owners and brokers; Exec. Vice-Pres. LARS SORENSEN; Man. Dir JAMES C. WELLS M.

**Cía Sud-Americana de Vapores:** Plaza Sotomayor 50, Casilla 49-V, Valparaíso; also Hendaya 60, 12°, Santiago; tel. (32) 20-3000; tel. (2) 330-7000; fax (32) 20-3333; e-mail info@csav.com; internet www.csav.com; f. 1872; regular service between South America and US/Canadian ports, US Gulf ports, North European, Mediterranean, Scandinavian and Far East ports; bulk carriers, tramp and reefer services; part of Cristalchile group; Chair. RICARDO CLARO VALDÉS; Gen. Man. FRANCISCO SILVA DONOSO.

**Empresa Marítima, SA (Empremar Chile):** Almirante Gómez Carreño 49, Casilla 105-V, Valparaíso; tel. (32) 25-8061; fax (32) 21-3904; f. 1953; international and coastal services; Chair. LORENZO CAGLEVIC; Gen. Man. E. ESPINOZA.

**Naviera Chilena del Pacífico, SA (Nachipa):** Almirante Señoret 70, 6°, Casilla 370, Valparaíso; also Serrano 14, Of. 502, Casilla 2290, Santiago; tel. (32) 25-1253; fax (32) 25-3869; e-mail valparaiso@nachipa.com; cargo; Pres. ARTURO FERNÁNDEZ ZEGERS; Gen. Man. PABLO SIMIAN ZAMORANO.

**Sudamericana Agencias Aéreas y Marítimas, SA. (SAAM):** Blanco 895, Valparaíso; tel. (32) 20-1000; fax (32) 20-1481; e-mail gerenciavap@saamsa.com; internet www.saamsa.com; f. 1961; cargo services; Pres. RICARDO DE TEZANOS PINTO D.; Gen. Man. ALEJANDRO GARCÍA-HUIDOBRO O.

**Transmares Naviera Chilena Ltda:** Moneda 970, 20°, Edif. Eurocentro, Casilla 193-D, Santiago; also Cochrane 813, 8°, Casilla 52-V, Valparaíso; tel. (32) 20-2000; fax (32) 25-6607; e-mail transmares@transmares.cl; f. 1969; dry cargo service Chile–Uruguay–Brazil; Chair. WOLF VON APPEN; Gen. Man. C. KUHLENTHAL.

Several foreign shipping companies operate services to Valparaíso.

#### Punta Arenas

**Cía Marítima de Punta Arenas, SA:** Avda Independencia 830, Casilla 337, Punta Arenas; also Casilla 2829, Santiago; tel. (61) 22-1871; tel. (2) 203-5180; fax (61) 22-7514; fax (2) 203-5191; f. 1949; shipping agents and owners operating in the Magellan Straits; Pres. PEDRO LECAROS MENÉNDEZ; Gen. Man. ARTURO STORAKER MOLINA.

### Puerto Montt

**Transporte Marítimo Chiloé-Aysén, SA:** Angelmo 2187, Puerto Montt; tel. (65) 27-0419; Deputy Man. PEDRO HERNÁNDEZ LEHMAN.

### CIVIL AVIATION

There are 325 airfields in the country, of which eight have long runways. Arturo Merino Benítez, 20 km north-east of Santiago, and Chacalluta, 14 km north-east of Arica, are the principal international airports.

**Aero Continente:** Marchant Pereira 381, Providencia, Santiago; tel. (2) 242-4242; fax (2) 205-2575; e-mail n6clreservas@aerocontinente.net; internet www.aerocontinente.com; operations suspended in 2002; owned by Nordic Air of Finland; Pres. LUPE ZEVALLOS.

**Aerocardal:** Aeropuerto CAMB, Pudahuel, Casilla 64, Santiago; tel. (2) 377-7400; fax (2) 377-7406; e-mail aerocard@aerocardal.cl; f. 1989; charter services; Chair. ALEX CASASEMPERE.

**Aerovías DAP:** Avda B. O'Higgins 891, Casilla 406, Punta Arenas; tel. (61) 22-3340; fax (61) 22-1693; e-mail ventas@aeroviasdap.cl; internet www.aeroviasdap.cl; f. 1980; domestic services; CEO ÁLEX PISCEVIC.

**Línea Aérea Nacional de Chile (LAN-Chile):** Américo Vespucio 901, Renca, Santiago; tel. (2) 565-2525; fax (2) 565-1729; internet www.lanchile.com; f. 1929; operates scheduled domestic passenger and cargo services, also Santiago–Easter Island; international services to French Polynesia, Spain, and throughout North and South America; under the Govt's privatization programme, 99% of LAN-Chile shares have been sold to private interests since 1989; Pres. JORGE AWAD MEHECH; Exec. Vice-Pres. ENRIQUE CUETO.

## Tourism

Chile has a wide variety of attractions for the tourist, including fine beaches, ski resorts in the Andes, lakes, rivers and desert scenery. There are many opportunities for hunting and fishing in the southern archipelago, where there are plans to make an integrated tourist area with Argentina, requiring investment of US $120m. Isla de Pascua (Easter Island) may also be visited by tourists. In 2004 there were an estimated 1,421,076 tourist arrivals. In 2003 receipts from tourism totalled $1,362m.

**Servicio Nacional de Turismo (SERNATUR):** Avda Providencia 1550, 2°, Santiago; tel. (2) 731-8419; fax (2) 236-1417; e-mail pcasanova@sernatur.cl; internet www.sernatur.cl; f. 1975; Dir Dr OSCAR SANTELICES ALTAMIRANO.

**Asociación Chilena de Empresas de Turismo (ACHET):** Moneda 973, Of. 647, Santiago; tel. (2) 699-2140; fax (2) 699-4245; e-mail achet@achet.cl; internet www.achet.cl; f. 1945; 155 mems; Pres. GUILLERMO CORREA SANFUENTES; Dir GUIDO FERNÁNDEZ SALINAS.

## Defence

At 1 August 2004 Chile's Armed Forces numbered 77,700, including 22,400 conscripts: the Army 47,700 (20,700 conscripts); the Navy 19,000 (including naval air force and marines; 1,000 conscripts); and the Air Force 11,000 (700 conscripts). There were also paramilitary forces of 38,000 carabineros. Military service in the Navy or the Air Force is compulsory for men at 19 years of age and lasts for 22 months. Military service in the Army lasts for 12 months.

**Defence Expenditure:** Expenditure was budgeted at 947,000m. pesos in 2004.

**Commander-in-Chief of the Army:** Gen. JUAN EMILIO CHEYRE ESPINOSA.

**Commander-in-Chief of the Navy:** Adm. MIGUEL ANGEL VERGARA VILLALOBOS.

**Commander-in-Chief of the Air Force:** Gen. OSVALDO SARABIA VILCHES.

## Education

Primary education in Chile is free and compulsory for eight years, beginning at six or seven years of age. It is divided into two cycles: the first lasts for four years and provides a general education; the second cycle offers a more specialized schooling. Secondary education is divided into the humanities-science programme (lasting four years), with the emphasis on general education and possible entrance to university, and the technical-professional programme (lasting for up to six years), designed to fulfil the requirements of specialist training. In 2002 298,419 pupils attended kindergarten, 2,341,519 pupils attended primary school and 896,470 attended secondary school. There are three types of higher education institution: universities, professional institutes and centres of technical information. In 2002 there were 521,609 students in higher education. The provision for education in the 2002 central government budget was 1,958.54m. pesos (19.0% of total expenditure).

## Bibliography

For works on South America generally, see Select Bibliography (Books)

Adler Lomnitz, L., and Melnick, A. *Chile's Political Culture and Parties.* Notre Dame, IN, University of Notre Dame Press, 2000.

Arceneaux, C. L. *Bounded Missions: Military Regimes and Democratization in the Southern Cone and Brazil.* University Park, PA, Penn State University Press, 2001.

Barr-Melej, P. *Reforming Chile: Cultural Politics, Nationalism and the Rise of the Middle Class.* Chapel Hill, NC, University of North Carolina Press, 2001.

Beckett, A. *Pinochet in Piccadilly.* London, Faber and Faber, 2002.

Bethell, L. (Ed.). *Chile since Independence.* Cambridge, Cambridge University Press, 1993.

Borzutzky, S. *Vital Connections: Politics, Social Security, and Inequality in Chile.* Notre Dame, IN, University of Notre Dame Press, 2002.

Bouvier, V. *Alliance or Compliance, Implications of the Chilean Experience for the Catholic Church in Latin America.* Syracuse, NY, Syracuse University Press, 1983.

Collier, S., and Sater, W. *A History of Chile, 1808–1994.* Cambridge University Press, Cambridge Latin American Studies, 1996.

*A History of Chile, 1808–2002.* Cambridge University Press, Cambridge Latin American Studies, 2004 (2nd edn, Ed. Alan Knight).

French-Davis, R. *Economic Reforms in Chile: From Dictatorship to Democracy.* Michigan, MI, University of Michigan Press, 2001.

Hawkins, D. G. *International Human Rights and Authoritarian Rule in Chile.* Lincoln, NE, University of Nebraska Press, 2002.

Hite, K. *When the Romance Ended: Leaders of the Chilean Left, 1968–1998.* New York, NY, Columbia University Press, 2000.

Londregan, J. *Legislative Institutions and Ideology in Chile: Political Economy of Institutions and Decisions.* Cambridge, Cambridge University Press, 2000.

Loveman, B. *Chile: The Legacy of Hispanic Capitalism.* New York, NY, Oxford University Press Inc, 2001.

Mamalakis, M. J. *The Growth and Structure of the Chilean Economy.* New Haven, CT, and London, Yale University Press, 1996.

Meller, P. *The Unidad Popular and the Pinochet Dictatorship: A Political Analysis.* New York, NY, St Martins Press, 2000.

Mount, G. *Chile and the Nazis: From Hitler to Pinochet.* Montréal, QC, Black Rose Books, 2001.

O'Brien, P., and Roddick, J. *Chile, the Pinochet Decade.* London, Macmillan, 1983.

Oppenheim, L. H. *Politics in Chile: Democracy, Authoritarianism and the Search for Development.* Boulder, CO, Westview Press, 1998.

O'Shaughnessy, H. *Pinochet: The Politics of Torture.* London, Latin American Bureau, 1999.

Paley, J. *Marketing Democracy: Power and Social Movements in Post-Dictatorship Chile.* Los Angeles, CA, University of California Press, 2001.

Pollack, M. *The New Right in Chile 1973–1977.* Basingstoke, Macmillan, 2000.

Power, M. *Right-Wing Women in Chile: Feminine Power and the Struggle Against Allende.* Pennsylvania, PA, Penn State University Press, 2002.

*Report of the National Commission on Political Imprisonment and Torture.* Santiago, 2004.

Roberts, K. M. *Deepening Democracy? The Modern Left and Social Movements in Chile and Peru*. Stanford, CA, Stanford University Press, 1999.

Rosemblatt, K. A. *Gendered Compromises: Political Cultures and the State in Chile, 1920–1950*. Chapel Hill, NC, University of North Carolina Press, 2000.

Siavelis, P. *The President and Congress in Post-Authoritarian Chile: Institutional Constraints to Democratic Consolidation*. University Park, PA, Penn State University Press, 1999.

Sigmund, P. E. *The United States and Democracy in Chile*. Baltimore, MD, Johns Hopkins for Twentieth Century Fund, 1993.

Spooner, M. H. *Soldiers in a Narrow Land: The Pinochet Regime in Chile*. Los Angeles, CA, University of California Press, 1999.

Tinsman, H. *Partners in Conflict: The Politics of Gender, Sexuality and Labor in the Chilean Agrarian Reform, 1950–1973*. Durham, NC, Duke University Press, 2002.

Valdés, J. G. *Pinochet's Economists*. Cambridge, Cambridge University Press, 1995.

Verdugo, P. *Chile, Pinochet and the Caravan of Death*. Boulder, CO, Lynne Rienner Publrs, 2001.

# COLOMBIA

## Geography

### PHYSICAL FEATURES

The Republic of Colombia is in north-western South America, the only country on that continent to have coastlines on both the Pacific Ocean and the Caribbean Sea, separated mid-way by the westward-heading Isthmus of Panama. The country's shortest land border (225 km or 140 miles), therefore, is with Panama, to the west, across the start of the Central American land bridge (until 1903 Panama was a province of Colombia). To the south lie Ecuador and Peru, the former on the coast beyond a 590-km frontier and the latter inland, the border extending for 1,496 km onto the Amazonian plains. Here Colombia also meets Brazil, which lies east and south of a 1,643-km border in the south-east of the country. The longest border, however, is with Venezuela, which lies to the east. Central Colombia thrusts further east than the rest of the country, while in the north Venezuela encroaches into the west. However, Colombia still has 1,760 km of north-west-facing shores on the Caribbean (1,448 km on the Pacific), giving it the right to maintain its possession of a number of islands and islets off shore. These are grouped in a single administrative unit, the smallest department in the country (44 sq km or 17 sq miles), San Andrés and Providencia, which also has jurisdiction over Roncador Cay and the Quita Sueño, Serrana and Serranilla Banks. However, Colombian possession of this territory impelled Nicaragua in 2001 to pursue a claim with the International Court of Justice (based in The Hague, Netherlands) involving 50,000 sq km of territorial waters—San Andrés is only 180 km east of Nicaragua, whereas it is some 700 km north of the Colombian mainland. Providencia is a further 80 km north. The still more distant Serranilla Bank is claimed by the USA (which recognized Colombian possession of Roncador, Serrana and Quita Sueño in 1981, when a 1972 treaty took effect) and, on occasion, by Honduras. Colombia also has a dispute over maritime boundaries in the Gulf of Venezuela, with Venezuela. The Pacific coast involves fewer formal international problems—Isla de Malpelo is the only Pacific island to be included in Colombian territory, although it adds little area to the overall national territory of 1,141,748 sq km (making the country a little bigger than Bolivia).

The western two-fifths of Colombia is dominated by the Andes and the coastal lowlands, with the east and south dominated, respectively, by the Llanos (the grassland plains of the Orinoco basin) and by the Selvas (the flat rainforest region typical of the Amazon basin). These torrid lowlands of the east are sparsely inhabited and little explored, watered by rivers that drain into the Atlantic—the Llanos by the Meta and other tributaries of the Orinoco, and the Selvas by the Caquetá and other Amazon tributaries. The most important river of Colombia, however, is the Magdalena, which cuts north through the Andes for about 1,540 km, through the most settled parts of the country, to empty into the Caribbean. The great Andean chain enters the country in the south-west, then splitting into three cordilleras, the western, the lowest, the central, the highest, and the eastern. Like the coast, the volcanic ranges run slightly north of a south-west to north-east course. The region consists of soaring ranges separated by high plateaux, broad upland basins and deep, fertile valleys carrying powerful rivers. The Cordillera Occidental is a sheer wall of barren peaks rising to some 3,700 m (over 12,000 ft). The Cordillera Central has peaks over 5,500 m, the Cordillera Oriental some that are not much less, and both are under permanent snow at their summits. This, and a timberline at about 3,000 m, contrasts dramatically with the swampy tropical jungle the mountains descend to some 240 km short of the Caribbean coast. The Cordillera Oriental is distinguished by its densely populated plateaux and basins, usually between 2,400 m and 2,700 m, in one of which is the capital city, although there are also large centres in the Cordillera Central. Between the two ranges clefts the mighty Magdalena, serving as a transport conduit to the Atlantic (Caribbean) coast. On the other

side of the Cordillera Central, to the west, is the Cauca, a tributary of the Magdalena, joining it some 320 km before it reaches the sea. These rivers link the highlands and the Atlantic lowlands, which are often marshy, but long settled. These lowlands are separated from Venezuela by a northward extension of the Cordillera Oriental, the Sierra de Perijá, and, on the north-western coast, the flat, semi-arid Gujaira peninsula (which forms the western bluff of the Gulf of Venezuela) to the east and the isolated mountain mass of the Sierra Nevada de Santa Marta to the west. This range on the Caribbean includes the country's highest point, Pico Cristóbal Colón (5,776 m or 18,957 ft)—named, like the country, for Christopher Columbus, the Genoese (Italian) navigator who claimed much of the continent for Spain. The nearby Pico Simón Bolívar, named for the great liberator of South America, has a similar elevation. In the south the Atlantic plains narrow, and the densely forested region on the border with Panama leads onto the Pacific coast, first the Serrania de Baudo and then the jungles and swamps of the coastal plains, watered by relatively short Andean rivers, such as the Patía in the south. This varied terrain gives Colombia a biodiversity reckoned to be second only to Brazil, sheltering in forests that cover almost one-half of the country and on pastureland that covers about two-fifths (although some of this is on the bleak high moors, páramo, between the mountain basins). The forest is densest in the tropical east, but deforestation is probably a greater threat in the north and west. There is also a problem with illegal smuggling of animals, which can have a severe effect on endangered populations—particularly threatened species include the yellow-eared parrot, the condor, the giant armadillo, the cotton-top marmoset, the white-footed tamarin, tapir and some alligators. Other fauna that flourish in the natural conditions of Colombia include hummingbirds, toucans, storks, pumas, jaguars, red deer, sloths and monkeys, or, of flora, from coconut and mangrove, through mahogany, oak, pine, balsam, rubber, ginger, tonka beans, etc., to the extensive (illegal) cultivation of coca (Colombia is the world's leading producer), opium poppies and cannabis.

## CLIMATE

The Equator passes through the far south-east of Colombia, so most of the country lies within the Tropic of Cancer, but elevation makes a dramatic difference to the climate. The coastal lowlands and the deep Magdalena and Patía valleys, for instance, are very hot, with average annual temperatures of 24°–27°C (75°–81°F). From about 500 m the climate becomes subtropical, and then from about 2,300 m temperate (many people live at this level). It is only cold above 3,000 m (average temperatures ranging from –18°C to 13°C, 0°–13°F). Seasonal variation, however, is slight—the capital, Bogotá, has average high temperatures of 19°C (66°F) in July and of 20°C (68°F) in January. The main seasons are the wet and dry seasons, the two periods of rain being in March–May and September–November, except on the Atlantic coast, where there is one long wet season, in May–October. Rain is heaviest on the Pacific coast, and it is drier in the north and on the slopes of the Cordillera Oriental. At Bogotá the average annual rainfall is 1,050 mm (41 ins), but at Barranquilla, on the Caribbean, it is 800 mm.

## POPULATION

Before the advent of Europeans, there was a large Amerindian population in what is now Colombia, notably of the Chibcha (Muisca) people. There remain about 60 tribes scattered throughout the country, but unmixed Amerindians account for only about 1% of the population. The Spanish settlement and long years of colonial rule, as an imperial centre moreover, engrained a socially rigid class stratification, which is still strong. Family lineage, inherited wealth and racial background remain extremely important. The extremes of poverty in the country have not helped with widespread problems of social and political violence. Despite the pre-eminence of the old élite, most of the population (58%) are actually Mestizo, of mixed Spanish and Amerindian ancestry. However, 20% are of unmixed European ancestry, with 14% mulatto (black-white), 4% black and 3% black-Amerindian. The official language is Spanish, with Colombian Spanish said to be the purest in Latin America, but some native languages are still spoken by remoter groups and now have recognition under the Constitution. The Roman Catholic Church (which claims the nominal adherence of almost 90% of the population) also enjoys some official sanction, although it is not, formally, the state religion. There are small Protestant and Jewish minorities, with some even smaller Arab communities (in which there are some Muslims).

Estimates put the total population at 45.3m. in mid-2004. Almost three-quarters live in urban centres, most above the courses of the Magdalena and the Cauca and on the Atlantic coast. The national capital is Bogotá. It is located in the centre of the country, towards the southern end of the Cordillera Oriental, and is the largest city in Colombia (6.3m. at mid-1999). The cities of Calí (2.1m.), in the southern Cauca valley, and Medellín (1.9m.), in the north of the Cordillera Central, are next in size, followed by the Caribbean cities of Barranquilla, at the mouth of the Magdalena, and Cartagena, south-west along the coast. Colombia is a unitary republic consisting of 32 departments and one capital district (distrito capital—Bogotá).

# History

## Sir KEITH MORRIS

Colombia shares many features with the other Latin American countries and particularly with its Andean neighbours. However, its geography, pre-Columbian and colonial history gave the country distinctive characteristics that were accentuated following independence and became increasingly marked in the 20th century. The Andes mountain range divides into three cordilleras when it enters Colombia. The Pacific coast is largely jungle and mangrove swamps. The 60% of the country to the east of the Andes is divided between the Llanos (savannah, much of which is flooded for nine months of the year) and Amazonian jungle. Many places are only accessible by air. With its capital at Bogotá (500 miles from the Caribbean ports of entry and 8,600 ft high in the eastern cordillera), the country was inevitably inward-looking and regional.

The regionalism was reinforced by the country's Amerindian heritage. Although Colombia had many different civilizations before the Spanish conquest in the 16th century (they reached a high level of sophistication, producing the finest gold work in the Americas), they were never united in a large state like the Inca or Aztec empires. Few of them have survived as distinct groups and most were hispanicized, unlike those in Ecuador, Peru and Bolivia to the south. By the end of the colonial period the majority of Colombians were Mestizos (of mixed European and Amerindian descent) with significant European and mulatto minorities—the latter descended from the African slaves imported to work in the gold mines. The resulting lack of communal identity added individualism to the regionalism and localism, which geography and pre-colonial history had encouraged.

### INDEPENDENCE AND THE 19TH CENTURY

Nueva Granada became a Viceroyalty in 1739. Santa Fe de Bogotá, as it was called in colonial times, inevitably had a great concentration of the region's lawyers and administrators. Simón Bolívar made it the target of his great independence campaign of 1819 and there established the capital of Gran Colombia, comprising present-day Colombia, Venezuela, Ecuador and Panama. However, Gran Colombia broke up amid much bitterness in 1830, leaving the Colombians with a strong preference for civilian government after their experience with Bolívar and his largely Venezuelan generals.

Following independence, Colombia's politics in the 19th century bore a great resemblance to those of its neighbours. It was a turbulent period with nine civil wars. These were essentially struggles for power between the two main currents of national political life that had, by the middle of the century, emerged as the Liberal and Conservative Parties (Partido Liberal Colombiano—PL and Partido Conservador Colombiano—PCC). The only issue that consistently divided them was the greater or lesser role of the Roman Catholic Church; the PL contained anti-clerical elements. There was little dispute over economic policy and both parties were at times federalist, at times centralist, though the Liberals inclined more towards the former. However, party allegiance was often decided as much by family and locality as by doctrine.

The collapse of Gran Colombia had other lasting effects. It left Colombia with the largest share of the Gran Colombian debt. The Colombian State's finances were therefore poor from the beginning. It remained poor for the rest of the century, first because of the lack of large commodity discoveries and, second, owing to a weak external sector (the only effective form of taxation in 19th century Latin America was customs duties, which depended on foreign trade). As a result, the state was chronically weak with an army of only 2,000–3,000 men, which was frequently incapable of maintaining public order. No Colombian President could exercise the sort of authority that later enabled the Venezuelan leader, Gen. Juan Vicente Gómez (1908–35), effectively to disarm the Venezuelan population in the early 1900s. The Colombian Constitution alternated between extreme federalism (1863) and excessive centralism (1886). The latter was confirmed by the War of the Thousand Days (1899–1902), but the centralism was more policy than practice. Governments had to pay due respect to regional and local feeling.

## EARLY 20th CENTURY

Colombia's story in the 20th century was to diverge greatly from that of its neighbours. It was to have much greater constitutional stability (only one four-year military regime) and steadier economic development, but, paradoxically, more violence.

A consequence of the difficult geography and the poverty of the state, which was to plague Colombia throughout the 20th century, was the development of a frontier tradition. Colombia became a land of many internal frontiers as colonos (colonists) cleared the river valleys and advanced ever higher into the sierras, as well as opening up the Llanos and the jungle. As the state was absent in most of these areas, traditions of private justice prevailed. The ability to defend oneself became much admired, which may explain some of the tolerance shown to guerrillas to this day. The rural conflicts that afflicted Colombia in the late 20th and early 21st centuries led to many colonos taking their weapons and their frontier customs to the cities.

From the end of the 'War of the Thousand Days' (1899–1902) until the mid-1940s, Colombia enjoyed relative tranquillity. The Conservatives remained in power until 1930. The coffee and textile industries developed greatly, mainly in Medellín. In 1930 the Conservatives divided into factions, which allowed a moderate Liberal, Enrique Olaya Herrera, to govern in coalition with Conservatives. An attack by Peru on Colombia's Amazonian territories in 1931 ensured wide support for the new Government. President Alfonso López Pumarejo, who succeeded Olaya in 1934, introduced 'New Deal' type reforms, consolidating the Liberal's popular support. His successor, Eduardo Santos Montejo (1938–42), slowed the pace of reform.

Divisions within the PL led to a Conservative victory in 1946. However, by 1948 the Liberals had reunited behind the popular figure of Jorge Eliécer Gaitán, the dissident Liberal candidate in the 1946 elections. The assassination of Gaitán in Bogotá on 9 April 1948 led to an outbreak of civil unrest, known as the Bogotazo, with days of rioting, leaving several thousand dead. Gaitán, whose fiery oratory had won him a strong personal following, had been expected not only to regain the presidency for the Liberals in the 1950 elections, but also to introduce significant social change. The Government managed to restore order in the cities, but the conflict spread to the rural areas. 'La Violencia', as the period became known, continued until 1958 and may have claimed the lives of as many as 200,000 people.

## FRENTE NACIONAL, 1958–74

The military, led by Gen. Gustavo Rojas Pinilla, took power in 1953. The coup, the only one in the 20th century, initially enjoyed popular support; this, however, waned as it became clear that Rojas did not intend to restore constitutional government. A military junta removed Rojas in 1957 and, in the following year, power was transferred to a Frente Nacional (National Front). This power-sharing agreement between the two traditional political parties provided for them to alternate in the presidency for four terms and to have an equal number of seats in the Cabinet and the Congreso (Congress). This was less undemocratic than it sounds as under the Colombian system anyone could claim to be a Liberal or Conservative and the seats on both sides were strongly contested, with Communists winning some Liberal seats and Rojas's movement, the Alianza Nacional Popular (ANAPO), well represented on both sides.

Violence declined under the Frente Nacional as most of the remaining armed groups relinquished violence or were suppressed. However, the success of the Cuban revolution in 1959 gave fresh impetus to guerrilla activity. One of the surviving groups of Liberal guerrillas relaunched itself in the mid-1960s as the Fuerzas Armadas Revolucionarias de Colombia (FARC), the military wing of the pro-Soviet Communist Party, with strong support in some rural areas. The Ejército de Liberacíón Nacional (ELN), a Cuban-orientated movement, whose members were originally middle-class students and included several Roman Catholic priests, was founded at the same time. The Ejército Popular de Liberación (EPL), a smaller, Maoist guerrilla movement, followed in 1969.

Generally, the Frente Nacional's period of rule, which formally ended in 1974, was one of good economic growth and social progress, especially under Carlos Lleras Restrepo (1966–70), who gave much impetus to agrarian and administrative reform.

In the 1970 presidential election, the narrow victory of the official Conservative candidate, Misael Pastrana, was challenged by the second-placed candidate, Gen. Rojas, representing ANAPO. When Pastrana's victory was confirmed there were mass protests, as the result reinforced the popularly held view that the system was unfair and could not produce change peacefully. One consequence was the founding by some ANAPO supporters in 1974 of the Movimiento 19 de Abril (M-19), a non-Marxist guerrilla group, which, unlike the others, was initially city-based.

## RETURN TO LIBERAL GOVERNMENT, 1974–82

In the presidential election of 1974, the Liberal candidate, Alfonso López Michelsen, won a decisive victory over the Conservative Álvaro Gómez and the ANAPO contender, Maria Eugenia Rojas de Moreno Díaz. Curiously, the fathers of all three were former Presidents. The expectations aroused by López's victory were great. He was the first Liberal to win a fully competitive election since his father, whose name still symbolized progressive liberalism. However, he was committed to continue to govern in coalition with the Conservatives and any attempt at constitutional reform faced formidable opposition in the Congreso and the Supreme Court. The world economy was also in recession following the 1973 oil crisis. In fact, the López Michelson administration's most lasting achievement was probably the introduction of association contracts for oil exploration, at a time when other Latin American countries were nationalizing their oil industries. This would lead to the great discoveries at Caño Limón in 1982 and at Cusiana in 1991.

President López Michelsén's successor was Julio Cesar Turbay Ayala. Turbay took a firm and unpopular stand against the Argentine invasion of the Falkland Islands (Islas Malvinas) in 1982. He also sought to solve the problems of urban terrorism and drugs-trafficking. His efforts met with some success, although his counter-insurgency campaign against guerrillas in 1982 provoked many allegations of human rights abuses by the Armed Forces.

## THE DRUGS TRADE

The illegal drugs trade became the key factor in Colombia from the late 20th century. It began quietly in the 1970s with the cultivation and export of marijuana. Then, some Colombians saw the opportunity to gain a dominant role in the cocaine business. The coca paste was produced largely in Peru and Bolivia and flown to Colombia, which was strategically placed to process it into cocaine and ship it to the USA. By the early 1980s two groups in Medellín and Cali controlled most of the trade. Their activities were already on a large scale before the Government or society realized the extent of the threat. When challenged by the Government they retaliated and unleashed a cycle of violence that has continued to the present. In the case of the Medellín cartel under Pablo Escobar, violence escalated into 'narco-terrorism' (a direct assault on the state to force it to abandon the policy of extradition to the USA). The traffickers also hired paramilitaries to defend their newly acquired ranches from attack by the guerrillas. These paramilitary groups, often originally formed by the Army, have gone increasingly on the offensive and have killed many civilians in their counter-guerrilla war. Ironically, the traffickers have often financed the very guerrillas that their paramilitaries have been fighting. Many cocaine laboratories and much of the coca cultivation were situated in the jungle in south-east Colombia where the FARC was strong. The drugs cartels paid the FARC 'protection' money, which rapidly made the FARC into the world's richest guerrilla group. The advent of opium-poppy cultivation in the early 1990s and the increase in coca cultivation from 1995 made the FARC and their paramilitary enemies in the Autodefensas Unidas de Colombia (AUC) even richer. From 2001 'Plan Colombia' (see below) led to a significant reduction in coca production, but without to date making much impact on illicit revenues.

The impact of the drugs trade on Colombia went much further than the direct effects described above. It diverted and weakened the judicial system and security forces, allowing common criminality greater impunity and creating a culture of violence and contempt for any legal or moral restraints.

## REFORM, PEACE AND NARCO-TERRORISM, 1982–90

In May 1982 the Conservative candidate, Belisario Betancur Cuartas, was elected to the presidency, mainly owing to divisions within the PL. Betancur had moved from the right of the party to its far left. However, with a Liberal majority in the Congreso he had to continue the tradition of coalition government. Like his predecessors, he followed a prudent economic policy and encouraged foreign investment. His innovations focused on foreign policy and peace issues. Under his leadership, Colombia, traditionally a loyal US ally, became a member of the Non-Aligned Movement as well as the Contadora Group, which assisted efforts to find a peaceful solution to the conflicts in Central America.

Domestically, Betancur attempted to resolve Colombia's internal conflict by agreement. He granted an amnesty to guerrilla prisoners and concluded cease-fires with the FARC (which, in 1982, added the name Ejército del Pueblo to its official title, to become FARC—EP), M-19 and the EPL. The FARC—EP founded a political party, the Unión Patriótica (UP), which contested the 1986 elections. However, the cease-fires with both M-19 and EPL broke down and in November 1985 the M-19 seized the Palace of Justice. In the ensuing recapture of the building by the Army about 100 people were killed, including 11 judges, leading to strong public criticism of both the Government and the Army. Many observers concluded that the cease-fires had benefited the guerrillas, particularly the FARC—EP, which had used the time to build up its forces. Betancur also faced the beginnings of narco-terrorism when, in 1984, drugs-traffickers from Medellín assassinated the justice minister, Rodrigo Lara Bonilla, who had taken the first serious measures to combat their activities. Betancur concluded that extradition to the USA was the only effective means of addressing the problem.

The drugs-trafficking problem was to dominate the presidency of the Liberal Virgilio Barco Vargas, elected in 1986 by a decisive majority. His offer to the Conservatives of a limited participation in government was refused, which resulted in the first single-party Government since 1953. Barco shared Betancur's belief in extradition, but the Supreme Court twice ruled that such a treaty with the USA was unconstitutional. Barco, however, used emergency decrees to proceed with extraditions. The Medellín drugs cartel began a campaign of terror to force the Government to abandon this policy. In August 1989 it assassinated Luis Carlos Galán, the favourite to win the PL's presidential nomination in 1990. The M-19 and UP presidential candidates were also assassinated in early 1990. An AVIANCA aeroplane was blown up, as were government offices, and in the first seven months of 1990 over 200 policemen were killed in Medellín. Barco refused to be intimidated and called successfully for international support to counter the cartel's threat.

Barco also continued the peace process with the guerrillas. The cease-fire with the FARC—EP broke down in 1987, but in 1989 a settlement was reached with the M-19. They regrouped as the Alianza Democrática—M-19 (AD—M-19) and participated in the 1990 elections. Their presidential candidate, Carlos Pizarro, was assassinated by paramilitaries. Successful negotiations with the EPL, the Partido Revolucionario de Trabajadores (PRT) and the Comando Quintín Lame were also concluded in 1990. Sadly, hopes that the FARC—EP and the ELN might also enter into peaceful dialogue were reduced by the killing of over 2,000 members of the UP, largely by paramilitaries linked to the Medellín cartel, although the FARC—EP and ELN's strong financial position, despite the loss of support from the communist states, was probably the decisive factor. Many of the paramilitary groups had been set up by the Army, but as they fell increasingly under the control of drugs cartels, they had been declared illegal in 1989.

## CÉSAR GAVIRIA TRUJILLO: THE REFORM PROJECT, 1990–94

César Gaviria Trujillo, the Liberal candidate elected President in May 1990, was determined to accelerate political reform and the liberalization of the economy, a policy known as *apertura* (opening). After decades in which the Congreso and the Supreme Court had opposed almost all constitutional change, an informal referendum, held at the time of the presidential election in 1990, produced a huge majority in favour of the election of a Constituent Assembly. It was elected in December and the AD—M-19 and the Liberals received the largest share of the vote (27% each). Seats were also allocated to the EPL, the PRT and the Comando Quintín Lame as part of those groupings' peace settlements.

The Constituent Assembly drafted a new Constitution in 1991. It guaranteed every conceivable human right and took decentralization further, through the election of governors and the transfer of functions and central funds to departments and municipalities. It weakened the presidency by limiting emergency powers and providing for censure of ministers. A Constitutional Court was created, as was a prosecution service. Citizens were given the right to challenge almost any measure through an injunction (*tutela*) and extradition was prohibited. The Medellín cartel had halted its mass terrorists attacks when Gaviria took office, but had kidnapped several prominent figures. Gaviria offered the cartel the possibility of avoiding extradition if they released their hostages and surrendered, an offer that several prominent cartel members, including its leader, Pablo Escobar, accepted.

In his first year in office Gaviria liberalized labour markets, removed price controls and improved terms for foreign investment. He also abolished import licences and drastically reduced tariffs. In an attempt to ensure that Colombian industry benefited from the sudden advent of international competition, Gaviria promoted rapid integration within the Andean community, especially with Venezuela and Ecuador. However the liberalization of the political system was not completed by a peace settlement with the FARC—EP and the ELN. Negotiations with both groups failed in Caracas, Venezuela, in 1991 and in Tlaxcala, Mexico, in 1992. This was followed in April 1992 by a drought, which, in a country that was 80% dependent on hydropower, resulted in 13 months of power cuts. In July the Government was humiliated when Pablo Escobar escaped from his luxurious prison outside Medellín and returned to narco-terrorism. However, his organization was gradually dismantled and he was killed by the police in Medellín in December 1993. The economy began to recover from the 1992 drought, and grew at over 5% in 1993 and 1994. Gaviria's determination to persist with his reforms, in spite of reverses, was admired and when he left office in 1994 his popularity was higher than that of any previous retiring President. He was immediately elected Secretary-General of the Organization of American States, in which post he served two terms, retiring in 2004.

## ERNESTO SAMPER PIZANO: SOCIAL REFORM AND POLITICAL CRISIS

In June 1994 the PL's Ernesto Samper was elected by a narrow margin in Colombia's first two-round presidential election. Two days later his defeated Conservative opponent, Andrés Pastrana Arango, disclosed the existence of taped conversations suggesting that the Cali drugs cartel had partly financed Samper's campaign, an accusation that cast a shadow over his whole presidency. An initial investigation cleared Samper, but one year later the case was reopened when his treasurer, Santiago Medina, and then his campaign manager (at the time his Minister of Defence) accused him of personal involvement. The Cámara de Representantes (House of Representatives) finally voted to clear him of any wrongdoing in June 1996, but this was perceived by many as a political, rather than a legal, verdict. This long-running political crisis, which was exacerbated by US policy (see below), made it difficult for Samper to carry out the social-reform programme on which he had been elected. Always on the social-democratic wing of the party, Samper had made clear his reservations about the rapid pace of *apertura* pursued by Gaviria (under whom he had served as Minister of Economic Development until late 1991). As President he aimed not to reverse this policy, but to moderate it. He also promised to increase social spending to alleviate poverty, a problem hitherto neglected.

Samper's domestic problems were exacerbated by the response of the USA. Initially, the US Government stated that Samper, despite the allegations, would be judged on the results of his anti-narcotics policy. This proved to be quite successful, with the leaders of the Cali cartel captured in 1995. However, when Medina made his accusations against Samper in mid-1995, US policy shifted and the Administration of President Bill Clinton (1993–2001) openly expressed its lack of confidence in Samper and demanded that further anti-narcotics legislation be passed (seizure of assets, stricter penalties and even the reintroduction of extradition). Samper eventually succeeded in getting these measures approved by the Congreso. Colombia was, meanwhile, refused certification for its anti-narcotic efforts in both 1996 and 1997 and only received a conditional certification in 1998. The consequences of decertification were severe: Colombia received no US export credits and the USA voted against loans to Colombia from multilateral banks. The confidence of both domestic and foreign investors inevitably declined. The fiscal deficit and foreign debt rose and economic growth slowed. The FARC—EP and the ELN were correspondingly encouraged and saw no reason to negotiate with a President whom the USA considered corrupt. Foreign pressure, although it damaged his Government and the country, undoubtedly helped Samper to maintain a considerable level of popular support, as many Colombians resented such blatant US interference.

## ANDRÉS PASTRANA ARANGO: THE PEACE PROCESS AND THE 'PLAN COLOMBIA'

The 1998 presidential contest was one of the closest fought in Colombian history. In the first round the Liberal candidate, Horacio Serpa, was less than one percentage point ahead of Pastrana, the defeated 1994 Conservative candidate. In the second round, however, Pastrana won by 500,000 votes. Many leading Liberals had supported Pastrana, believing him better placed to end Colombia's isolation, restore confidence in the economy and restart the peace process.

Unfortunately, the means to achieving the last two goals were often in conflict. Restoring economic confidence meant reducing the fiscal deficit sharply by cutting public expenditure and raising taxes (see Economy, below). In the short term this worsened the recession Pastrana had inherited from the previous administration. It made it very difficult to maintain popular support for the peace process, which ideally required increased military spending to protect the population and to put pressure on the guerrillas and increased social spending to mitigate rising unemployment.

Pastrana took the initiative on peace, by meeting Manuel Marulanda Vélez, leader of the FARC—EP, in his jungle hideout. Such a dramatic step launched the peace process. The new President's cabinet appointments reassured the markets that Colombia was returning to its prudent tradition of orthodox financial management. Relations with the USA immediately improved. Pastrana made a state visit to the USA in October 1998, the first in 23 years. The USA granted Colombia full certification in March 1999 and in 2000 gave US $1,300m. over two years to help restore peace and stability and reduce the drugs trade. This funding formed the US portion of Plan Colombia (see below). To persuade the rebels to negotiate, Pastrana had to cede them, temporarily, 41,000 sq km in southeast Colombia, from which all government troops were withdrawn. The FARC—EP, having received this territory, successfully insisted on the removal of any restrictions before they would agree an agenda for the talks. Public support soon decreased sharply when it became clear the FARC—EP was using the demilitarized zone as a safe haven and was keeping military prisoners and kidnapped civilians there. Public discontent increased when both the FARC—EP and the ELN began to kidnap more indiscriminately, taking many middle-class and child victims.

President Pastrana's support for firmer military action against the FARC—EP, while negotiating, and his success in winning substantially increased US aid for the Armed Forces ensured that military discontent with the peace process was contained. From 1998 the Armed Forces, the combined operations and intelligence capacity of which greatly improved, won all major engagements against the FARC—EP. This contrasted with the last two years of the Samper administration when the Army suffered several humiliating defeats. However, the FARC—EP's capacity to launch guerrilla attacks was not affected and the number of kidnappings close to large cities increased. This led to increased support for the paramilitary AUC, whose numbers rose faster than those of the FARC—EP. The AUC put the ELN under great pressure. The latter, which from 2000 was engaged sporadically in talks about talks with the Government, in turn exerted pressure with periodic mass kidnappings.

Central to Pastrana's strategy from 2000 was Plan Colombia. Its aims included increasing the efficiency of both the security forces and the judicial system, eliminating drugs production through both eradication and crop substitution, and reducing unemployment. The international community was to fund almost 50% of the US $7,500m. Plan. The initial US contribution of $1,300m. included a military component of $1,000m., but also significant sums for judicial reform and human rights education. In 2001 the USA committed itself to providing a further $882m. under the new Andean Region Initiative, introduced following criticism of the Plan by worried neighbouring countries. The member states of the European Union (EU) also agreed to provide $300m. This was all to be devoted to economic and social projects and the EU made it clear that its aid was not linked to Plan Colombia, with the military emphasis of which the EU differed.

However, the slow progress of peace negotiations, halted several times by the FARC—EP, who made no concessions of substance, frustrated the Colombian public. President Pastrana visited Marulanda several times to keep the process going, but each time the FARC—EP stalled the talks and committed another terrorist act. Some degree of international involvement from February 2001 raised hopes that the peace process would succeed, but, following the hijacking of an aeroplane and the kidnapping of a prominent senator by the FARC—EP, an utterly disillusioned Pastrana ended the peace process on 20 February 2002 and ordered the Armed Forces to regain control of the demilitarized zone. The FARC—EP responded by resorting to urban terrorism in an attempt to intimidate the public into calling for peace again. The result was the election of Alvaro Uribe Vélez to the presidency on 26 May, with 54.0% of the votes cast. The other main contender was Liberal Horacio Serpa, who won 32.3% of the votes. The Conservatives did not even field an official presidential candidate, preferring to support Uribe. Uribe was a dissident Liberal candidate who had been a strong critic of Pastrana's peace process and he proposed a dramatic increase in the security forces. In legislative elections, held earlier, on 10 March, the most significant gains were made by independents and supporters of Uribe. The official PL remained the largest single party in the Senado (Senate) and the Cámara de Representantes, with 29 and 54 seats, respectively. However, its representation was substantially reduced, while the representation of the official Conservatives diminished to 13 seats in the Senado and 21 lower-house seats.

The failure of the peace process to resolve or even reduce the internal conflict eroded public support for President Pastrana. The austere economic measures his administration had to implement to restore Colombia's finances, while increasing spending on security, lost him even more support. Pastrana approached the end of his term as a deeply unpopular President. History may be kinder. Without his extraordinarily generous peace process it was unlikely that the international community would have given Colombia so much financial support or that Colombian public opinion would have decided that much greater sacrifices would be required to restore the authority of the state in order to end the internal conflict.

## ALVARO URIBE VÉLEZ: TOWARDS DEMOCRATIC SECURITY, 2002–

Colombian voters entered new territory in electing Alvaro Uribe Vélez to the presidency in 2002. For the first time they elected someone who was not the official candidate of one of the two traditional parties. A dissident Liberal, Uribe Vélez ran as an independent, without even the support of a faction within the PL. It was not until it became clear that he had a real chance of winning that many Liberal politicians and the official PCC gave him their backing. He was also the first President committed to ensuring the state's control of Colombia's entire territory, and protecting the lives of all Colombians became the central tenet of his election manifesto. Launched as the Democratic Security Policy, it was an immense task. The illegal groups had grown to unprecedented levels, and the FARC—EP saw him as a serious threat. Consequently, they fired mortar bombs at the presidential palace during his inauguration on 7 August, although the rounds fell short and killed 20 of the capital's poorest inhabitants. The incident signalled the beginning of a sporadic urban terrorist campaign. In order to defeat the illegal groups or convince them to negotiate seriously, President Uribe announced plans to double the number of professional soldiers to 100,000 and to recruit a further 100,000 police-officers. However, the economic restraints on these proposals were considerable. The economy was recovering very slowly from the 1999 recession, and the public-sector deficit had to be reduced as Colombia's foreign debt neared the limits of sustainability. Upon taking office in August 2002 the new administration found was that its predecessor had left a larger deficit than had been officially declared. Nevertheless, the new President had two factors in his favour. Firstly, the security forces had been modernized under Pastrana and the military equipment and training provided by the USA under Plan Colombia were at last starting to show results by 2002. Secondly, the Uribe Government enjoyed, and has continued to enjoy, exceptionally high levels of public support. The public had decided that the internal conflict was simply no longer tolerable and felt that under Uribe progress was being made in addressing it. Constant activity has been the characteristic of his administration thus far. The President's community meetings every Saturday in a different provincial town, often lasting 14 or 15 hours, exemplified his personal commitment to resolve the country's problems.

Within days of taking office in August 2002 President Uribe declared a state of emergency under which he levied a wealth tax to finance increased security. The move sent a message to Colombia and to the international community. The expansion of the Army and the police force was reinforced by the recruitment of 'peasant soldiers', who would serve in their own districts. A network of 'informers' was also established to help protect the roads. Escorted caravans at holiday weekends were another method of helping Colombians reclaim control of their country. The response was enthusiastic. Within months the main roads were largely safe during daytime. Police began moving back to the 160 or so municipalities whence they had been withdrawn following guerrilla pressure. By the end of 2003 all had a police presence again. The coca-eradication programme was stepped up, and in 2002 and 2003 the area under coca cultivation was reduced by one-half. The FARC—EP responded to the Government's measures by increasing their campaign of urban terrorism and attacks on the country's infrastructure; however, the group was increasingly on the defensive, and under strong pressure from the security forces they withdrew from the area around Bogotá back towards their bases in the jungles of the south-east. The FARC—EP rejected the President's terms for negotiations, which included discussions under UN auspices, on condition of a cease-fire, and demanded a much larger demilitarized zone than the one that had operated under Pastrana. The ELN seemed to be falling increasingly under FARC—EP influence as that group moved to protect them from the AUC. However, in mid-2004 the ELN expressed interest in entering negotiations with the Government under Mexican mediation, but withdrew from talks in early 2005, supposedly in protest at Mexican criticism of Cuba. Meanwhile, the majority of the AUC (whose leader, Carlos Castaño, disappeared in early 2004) in July 2003 committed themselves to a cease-fire, to be followed by negotiations and a phased demobilization. After many breaches of the cease-fire, in June 2004 the long-postponed negotiations began and were ongoing in mid-2005. Several thousand paramilitaries had already been demobilized, but before concluding the negotiations the leaders were awaiting the successful approval by the legislature of a 'Justice and Peace' act setting the terms for their surrender; the legislation had been approved by both the Senado and the Cámara de Representantes in June and at this time awaited final revision by the 'Reconciliation Committee' before its implementation.

In December 2002 Uribe had succeeded in gaining congressional approval for unpopular tax, pension and labour reforms. These did much to ensure that he could increase security expenditure while reducing the public-sector deficit, in line with a new IMF agreement. However more reforms were needed, particularly in the pensions sector, to secure the fiscal position in the second half of his term. Uribe decided to achieve these reforms, together with political measures such as a reduction in the size of the Congreso, in a referendum in October 2003. The measures were approved by a large majority of those who voted, but the referendum failed because the turn-out was just below the necessary 25%. This was not altogether surprising, as voters had to cast their ballots in local elections the next day, and many were being asked to vote to cut their own pensions or freeze their salaries. The election of left-wing mayors in Bogotá and Medellín and a left-wing governor in the Valle del Cauca (Cali) was seen as a second defeat for Uribe, although it could also be interpreted as a sign that Colombian democracy was much more authentic than critics had claimed. President Uribe had to return to the Congreso to obtain approval for further tax increases to meet the shortfall in finances forecast for 2005 and 2006. Despite the electoral set-backs and the unpopularity of extra taxes, in 2005 Uribe's approval rating remained above 70%, owing to steadily improving security, with rates of murders, kidnappings, massacres and attacks on small towns all lower than in previous years and the economy recovering faster than expected. In both 2003 and 2004 the economy increased, with private investment recovering sharply and the unemployment rate falling. Economic growth of at least 4% was forecast for 2005. Against this background, a proposal to amend the Constitution to allow the re-election of a President, a proposal that had failed in the Congreso in 2003, was revived in 2004 and approved by the Congreso at the end of the year. In mid-2005 the constitutional amendment still awaited the approval of the Constitutional Court. The measure had been opposed strongly by the official Liberals and by the recently formed left-wing party, the Polo Democrático Independiente, but enjoyed majority support in the country.

## INTERNATIONAL POLICY

Given Colombia's association in the popular mind with violence, it is worth stressing that the country has an impeccable record in international matters. Colombia has never attacked another country and lost Panama through US intervention in 1903. Colombia has been a consistent opponent of the use of force to settle disputes and voted to condemn the Argentine invasion of the Falkland Islands in 1982, despite the pressure of most other Latin American countries. The country has contributed to UN peace-keeping operations since their inception.

A founder of the Andean Pact in 1969, Colombia under President Gaviria played a leading role in the 1990s in making economic integration a reality, initially with Venezuela and Ecuador. This was followed by the formation of a free trade area with Venezuela and Mexico in 1994. Colombia has since concluded free trade agreements with Chile and, in 2004, with the Mercado Común del Sur (Mercosur). In June 2004 Colombia, together with Ecuador and Peru, began negotiations for a free trade agreement with the USA. The absence of Venezuela from these negotiations reflected the strained relations of both the USA and Colombia with Venezuela under the presidency of Lt-Col (retd) Hugo Rafael Chávez Frías. There have been repeated allegations of Chávez's support for the FARC—EP, although none have been confirmed and the Colombian Government has attempted with some success to avoid confrontation or involvement in Venezuela's own internal disputes.

Traditionally, Colombia has been a loyal US ally, although its role in the Non-Aligned Movement from the 1980s led to some more independent stands. Close collaboration with the USA on anti-narcotics policy was established under President Barco and continued thereafter, albeit with difficulties under President Samper (see above). Under the Uribe administration the relationship has become even closer, and Colombia supported the US-led coalition against the regime of Saddam Hussain in Iraq in 2003. Colombia has long ceased to be the introverted Andean republic that President López Michelsen referred to as an 'Andean Tibet'. It has become an active and respected player, but one which, at the beginning of the 21st century, needed international support to help cope with its internal problems.

## OUTLOOK

President Uribe achieved much more in his first three years in office than anyone could have imagined. He restored Colombia's belief in itself. He made much progress in providing the democratic security that he had promised, although the methods used were strongly criticized by non-governmental organizations (NGOs). He returned the economy to its traditional steady growth path. However, illegal groups remained large and well-funded and capable of doing immense damage. The FARC—EP's repeated attempts to assassinate the President were a constant reminder of the precariousness of the situation. Uribe's presidency was a very personal one—excessively so, even in the opinion of his friends. The ongoing peace process with the AUC was far from certain to end successfully. There was much criticism, both domestic and international, of the 'Justice and Peace' legislation (see above). Critics argued that the legislation did not provide for sufficiently severe punishments for those demobilized combatants found guilty of serious crimes. Meanwhile, government attempts to negotiate a peace settlement with the ELN had reached an impasse in mid-2005. There was still a long way to go before the FARC—EP might decide that it should negotiate a settlement rather than risk losing everything. From early 2005 the grouping started to make bold attacks on military units, especially in south-west Colombia, probably having calculated that a campaign of this kind, if maintained into the months preceding the legislative and presidential elections that were scheduled for March and May 2006, respectively, would undermine popular resolution and lead to renewed calls for peace talks on its own terms. Moreover, while the economic recovery in 2003–04 had been stronger than expected, and growth remained robust in mid-2005, some of the measures necessary to secure the fiscal situation in the medium term, especially a broadening of the tax base, had still to be undertaken, and the benign climate for developing countries in the financial markets was unlikely to continue as US interest rates began to rise. However, the coffee industry had begun to recover and Colombia's exporters had managed to increase exports by a record 25%, despite an 11% revaluation in 2004. An element of uncertainty on all fronts had been added by the proposed change to the Constitution allowing presidential re-election. It was passed by the Congreso at the end of 2004, but at mid-2005 still awaited approval from the Constitutional Court. Despite popular support, it was quite uncertain which way the Court would vote. In early July the Attorney-General, Edgardo José Maya Villazón, advised the Constitutional Court not to approve the amendment, citing irregularities in its passage through the Congreso. Were the amendment to be approved, President Uribe's chances of winning another term in office looked very good indeed. This prospect would encourage investment and might even persuade the FARC—EP to negotiate sooner rather than later. To many in the opposition these advantages were far outweighed by the risks they saw from an irresistible shift of power towards the executive. Whatever the outcome, continued international support for Colombia seemed assured. US military aid (usually worth some US $500m. per year) was set to continue, even though Plan Colombia was officially scheduled to end in 2005. The EU seemed prepared to continue its economic aid and political support, despite a vigorous campaign by European NGOs against the Uribe administration's human rights record and, in particular, the 'Justice and Peace' legislation. In 2004 the EU added the ELN to its list of terrorist organizations, on which the FARC—EP and the AUC already figured.

# Economy

## Sir KEITH MORRIS

### INTRODUCTION

Throughout most of its history as both a Spanish colony and an independent state, Colombia has been impoverished. Neither gold-mining, which generated most of the money in the colonial era, nor agriculture, which provided most of the employment, proved to be a basis for sustained economic development. Connections and trade with the rest of the world were hampered by geography, as Colombia's main centres of population are in the highlands, with difficult access to the outside world. The lack of a major export commodity during most of the 19th century meant that, in a time when customs dues were the only effective means of raising revenue, the Colombian state lacked the resources either to maintain law and order or to build infrastructure. These factors and the small size of the market deterred both foreign investors and also immigrants. As a result, most of the economy was in Colombian hands; coffee, the country's main export from the late 19th century, remained firmly a Colombian concern. During the 19th century Medellín became the country's economic stronghold, dominating coffee, gold-mining and the nascent textile industry. By the 1920s Colombia had become the world's leading exporter of mild coffee, but a brief coffee-led period of prosperity was ended by the Great Depression in the 1930s. This led to a policy of import-substitution, with the aim of both promoting local industry and insulating the country against the world economy. With this policy in place, Colombia grew steadily until the 1990s, becoming a rare model of economic and financial stability in Latin America, especially during the chaotic 1970s and 1980s. It was the Government of President César Gaviria Trujillo (1990–94) that changed the direction of Colombia's economy, opening it up to global competition by reducing tariffs, abolishing many import restrictions, inviting foreign investment, embarking on a privatization programme and allowing the Colombian peso to float relatively freely.

However, the economy weakened in the second half of the 1990s. Although the administration of Ernesto Samper Pizano (1994–98) largely maintained the free-market policies implemented by Gaviria, the economic downturn was triggered by a steep increase in central government spending. It was further aggravated by a clause in the new Constitution of 1991 that stipulated an increase in the amount of central government funds be equalled in regional authority expenditure. The increasing cost of the civil conflict during Samper's presidency also undermined economic performance. The civil conflict and related violence cost the equivalent of an estimated 4% of gross domestic product (GDP). These factors combined to produce a sharp increase in the fiscal deficit, generating inflationary pressure, which led to interest- and exchange-rate instability, but failed to generate significant economic growth. Colombia's overall GDP increased by an annual average of 2.4% in 1990–2003.

The Government of Andrés Pastrana Arango (1998–2002) inherited an incipient recession. The downturn was exacerbated by a sharp decline in international commodity prices and prevailing instability in emerging markets. In 1999 the economy registered a 4.2% contraction, the country's worst economic performance since records began at the start of the 20th century. The economic decline was in sharp contrast to the average 4% annual growth witnessed throughout the 1970s, 1980s and most of the 1990s. The depth of the recession in 1999 reduced inflation to a 30-year low of 11.2%, compared with 20.4% in 1998, but unemployment increased to 19.4%.

The Government's fiscally conservative, orthodox economic programme for recovery was based on a three-year, US $3,000m., extended fund facility agreement approved by the IMF in December 1999. Conditions of the agreement included a reduction in the fiscal deficit to 3.6% of GDP from the unusually high 5.4% in 1999, growth of 3% and an inflation target of 10%. To meet the targets the Government reduced sharply social spending, increased taxes and limited automatic transfers to local authorities. However, its pension reforms were only partly completed and its privatization programme was much smaller than intended, owing to market conditions. Colombia lost its investment-grade sovereign debt rating in 1999 and was downgraded further in 2000.

Despite these negative signals, the economy began to recover in 2000, when GDP growth of 2.9% was recorded. The IMF targets for the reductions in the public-sector deficit and inflation were exceeded, with rates of 3.4% and 9.5%, respectively. The current account of the balance of payments went into surplus, helped by an increase in exports of 13%. However, the Pastrana Government's hopes that growth would increase at a faster rate in 2001 were disappointed. The US recession, exacerbated by the terrorist attacks of 11 September 2001, made it more difficult to export Colombian goods. Falling coffee prices and declining petroleum prices and production did not help. The economy grew by 1.5% in 2001. However, annual inflation was reduced to 8.6% and interest rates remained historically low.

In 2002 growth was still sluggish, at 1.9%, and owed much to a recovery in construction, which was boosted by an increase in government-subsidized social housing. However, despite further cuts in spending in early 2002, the fiscal deficit rose to 3.7%, against an IMF target of 2.6% of GDP. The state of the public finances was an unpleasant surprise to the incoming administration of President Alvaro Uribe Vélez, elected in May 2002, which found that it would have to raise an extra 1% of GDP in addition to the 2% it was already committed to spending on increased security. A wealth tax of 1% for this purpose was immediately introduced under a state of emergency decreed by President Uribe in August. By the end of 2002 the legislature had approved significant tax, pension and labour reforms, and in January 2003 the IMF agreed a new, US $2,100m., three-year financing programme. The new Government's economic measures, combined with its success in improving the security situation, did much to restore confidence and investment recovered. Gross domestic investment increased by 17.6% in 2003 and by a further 19.1% in 2004. In 2003 annual GDP growth was 3.7% (with industrial GDP rising by 4.2%) and the fiscal deficit was reduced to 2.7%, against the IMF target of 2.8%. Growth in 2004 was 4.0% and the fiscal deficit was reduced to 1.3% of GDP. Growth was forecast to stay at 4.0% in 2005. Manufacturing expanded by 5.9% in 2004, led by demand from exports. Non-traditional exports, in which manufactures figured prominently, grew by 25% during 2004. Manufacturing continued to perform well in early 2005, with production up by 7.9% in the first four months of the year, compared with the same period in 2004. Public-sector debt was reduced from 60% of GDP in 2002 to 53% in 2004. Unemployment fell from a high of 20% in 2000 to 12% in 2004. Poverty was reduced from a high of 60% in 1999 to 52% in 2004.

## AGRICULTURE

Until the 1990s Colombia relied heavily on agriculture, a sector of the economy that was immensely diverse, owing to the country's varied topography. In 1960 agriculture (including forestry, fishing and hunting) employed over one-half of Colombia's total work-force and, as recently as 1976, it accounted for 26% of GDP. In 2004 agricultural employment had fallen to 21% and the sector contributed a provisional 12.1% of GDP. The long-term decline in agricultural employment was sharply accelerated during the economic liberalization of the early 1990s, when an estimated 100,000 jobs were lost in agriculture, as reduced tariffs made imports affordable. After 1994 the sector became more stable, despite the continuing violence in much of Colombia's rural areas. However, since 1999 production has been increasing steadily. Production increased by 3% in 2003, by 2% in 2004, and was forecast to grow by 3% in 2005.

### Coffee

In 2004 coffee remained Colombia's leading legal cash and export crop, and Colombian coffee still enjoyed an excellent reputation around the world, even though Colombia had lost its position as the world's second largest producer after Brazil to Viet Nam, the newest entrant in the coffee market. Coffee had been overtaken in the 1990s by petroleum and its derivatives as the single most important export commodity. In the early 1980s coffee accounted for roughly one-half of total export earnings, but this figure had fallen to 5.7% by 2004. Coffee still accounted for about one-third of employment in agriculture, with small and medium-sized holdings still responsible for a considerable percentage of the crop. There were about 400,000 coffee farms, which created a rural middle class unusual in Latin America.

The coffee sector was well regulated. Policy was set by the semi-official Federación Nacional de Cafeteros de Colombia (FNC—National Federation of Coffee Growers). The Fondo Nacional del Café (National Coffee Fund) was established to help producers overcome sharp fluctuations in world prices. However, not even the Fund could cope with the fall in the coffee price following the collapse of the International Coffee Agreement in 1989. Colombia's response was to increase production, relying on its reputation as a high-quality producer. Production reached a high of 1.1m. metric tons in 1992. However, prices, which fluctuated during much of the 1990s, fell to a 100-year low, in real terms, by the turn of the century. This was owing, in part, to the dramatic entry of Vietnamese coffee (predominantly of the relatively low-quality robusta varieties, rather than the high-quality arabicas more commonly grown in Colombia), which rapidly overtook Colombia's own production capacity. Colombia's small farmers could not compete on cost with the large-scale Brazilian producers or the Vietnamese with their extremely low labour costs. Coffee exports in 2004 were worth US $949m., an increase on the previous year's figure of $806m., but a significant fall from the $2,895m. recorded in 1997. In 2002 the sector launched a series of high-quality specialist brands under the still popular 'Café de Colombia' name. By the end of 2004 the coffee price had reached $1.05/lb, more than enough to cover costs, and by April 2005 it was up to $1.25. There were clear signs of recovery in the industry and in the coffee zone, once Colombia's most prosperous and peaceful, which had suffered greatly in recent years with coca cultivation appearing and the inevitable subsequent arrival of guerrilla and paramilitary organizations.

### Bananas

Bananas have been Colombia's most contentious legal export, the subject of long-standing trade disputes. Although the banana is a traditional Colombian crop, it became part of the country's drive towards agricultural diversification designed to protect the economy from the fluctuations of the coffee market. Banana exports peaked in 1998 at US $559m. Exports were worth $426m. in 2003 and $431m. in 2004. Most of the bananas exported were grown in Uraba, in the north-west of the country, and Santa Marta in the central north. The long-running dispute between the USA and the European Union (EU) about the latter's banana quotas was resolved by an agreement in 2001 under which quotas would be phased out by 2006. However, the EU's proposed tariff of €230 per metric ton for Latin American banana imports was vociferously contested by Colombia (together with Brazil, Ecuador, Venezuela, Honduras, Costa Rica, Panama and Guatemala) in late March 2005 and was scheduled to be submitted to the arbitration of the World Trade Organization.

### Flowers

The cut-flower industry was the biggest success story of Colombia's agricultural diversification. It was launched in the late 1960s, and by the 1990s Colombia had become the world's second largest exporter. The industry is largely concentrated on the plain surrounding Bogotá, a choice influenced both by the climate and the presence of the international airport, vital for rapid movement of the flowers, which are mostly bound for the USA. The industry successfully coped with the overvalued peso of the mid-1990s and with increasing international competition. In 2004 cut-flower exports earned Colombia US $701m., the highest annual figure yet recorded.

### Palm Oil

In 2003 Colombia, with 526,610 metric tons, was the world's fifth largest producer of palm oil, although its market share was very small when compared with Malaysia and Indonesia. Exports totalled 110,000 tons in the same year. The sector was encouraged by successive Colombian Governments, with a flexible credit system introduced in the late 1990s. The industry had considerable potential because the yields per hectare were some of the highest in the world and only a small proportion of the suitable land has yet been utilized. High costs, particularly of transportation from the interior, have been the brake on expansion. There is also considerable potential for the export of the exotic fruits native to Colombia, such as star fruit and guanabana, although volumes will inevitably remain modest.

### Staples

The traditional elements of the Colombian diet—apart from meat—are potatoes, maize, beans, rice, plaintains, cassava and citrus. Domestic production of these crops was inevitably affected by the import liberalization of the early 1990s. Production stabilized thereafter, with a slight increase generally recorded. Potato output grew from 2.4m.–2.6m. metric tons in the early 1990s to 2.7m.–2.9m. tons in 2000–03. In 2004 output was 2.8m. tons. Rice production, which reached a low of 1.6m. tons in 1993, had recovered to 2.5m. by 2003, but fell slightly, to 2.4m. tons in 2004. Sugar production rose from 1.6m. tons in 1990 to reach an 2.5m. tons in 2004.

### Livestock

Government initiatives to promote the supply of meat and dairy products through the use of subsidies led to an expansion in the amount of arable land dedicated to livestock. Almost 80% of the cattle population was based in the north-eastern pasture land of the llanos. Production of beef and veal rose from 3.5m. head slaughtered in 1995 to 3.8m. in 2000, but declined between 2001 and 2003, before recovering to 3.6m. in 2004. A problem facing many ranchers was endemic guerrilla violence in cattle-raising areas. Wealthy ranch owners, particularly those with connections to the drugs trade, could afford to defend their landholdings, but, for the increasingly impoverished peasant class, the general level of insecurity was indicated by a fall of 600,000 ha in the area sown with arable crops between 1990 and 1995. In 2004 the threat of guerrilla and paramilitary violence continued to depress levels of rural investment and discourage foreign involvement in the sector.

### Agrarian Reform

Successive Governments attempted to improve economic and social infrastructure in rural areas. President Samper pledged to increase such expenditure, but was impeded by the economic slowdown during his period of office. Despite the introduction of the first agrarian reform law in 1961, Colombia's record in this field has been disappointing. Considerable progress was made under the administration of President Carlos Lleras Restrepo (1966–70), but redistribution of land slowed thereafter. However, in 2003 the seizure of many ranches owned by drugs-traffickers promised a new opportunity for agrarian reform. The pattern of landholding varies greatly from region to region. The highlands of the Cordillera Oriental are mainly cultivated by smallholders, the majority of whom are often subsistence farmers. The coffee zone in the Cordillera Central is farmed in family-size commercial farms. The country's large cattle ranches are situated on the Caribbean coast and the llanos. A continuing obstacle to reforming the agricultural sector has been the question of legal title, as many internal migrants, or colonos, occupied land on Colombia's many internal frontiers. The legal position of these holdings has often remained obscure.

### Drugs

The illicit drugs trade has undoubtedly both contributed to Colombia's economic growth and hindered it. Marijuana and coca have long been grown in the country, but the drugs trade really took off with the processing of cocaine from the late 1970s. Until the mid-1990s most of the coca paste used came from Peru and Bolivia, but when the supply, largely by light aircraft, was disrupted, coca cultivation in Colombia increased five-fold to replace it. Increased aerial spraying under 'Plan Colombia' reduced the overall acreage under cultivation by over one-half between 2001 and 2004, despite considerable replanting. Nevertheless, Colombia remained the world's largest producer, accounting for some 56% of production, according to a report in mid-2005 by the UN Office on Drugs and Crime. Moreover, in the early 1990s there was an expansion, mainly in the south-west of the country, in the cultivation of the opium poppy, used to make heroin. The benefits or otherwise of this hidden input were disputed, as, by its nature, the trade resisted quantification and many conflicting and extravagant estimates have been made. One reputable study estimated that Colombian traffickers' net profits averaged between US $1,500m. and $2,500m. from 1982 to 1998. and that two thirds of this was brought back to Colombia. This would represent an average of 3.8% of GDP per year.

Government estimates have put the figure at closer to $1,000m a year, representing an average of 1.5%–2.0% of GDP, which seemed more likely, given that, for obvious reasons, a high proportion of drugs money is kept outside the country. The Government also regularly argued that the cost in terms of extra security expenditure and lost foreign investment was much higher than any gains. Certainly, the extra drugs-related crime and violence in rural areas, along with associated environmental damage, had a major negative impact. Nevertheless, while some rural areas remained effectively isolated from markets for legal agricultural produce, the attraction of growing illegal drugs crops, which would be collected by traffickers and which commanded a higher price, would persist. However, it was worth noting that many of the growers went to such areas specifically to grow such crops, often sent by the traffickers.

## MINING AND ENERGY

Despite a long history of oil production, the largest coal reserves in Latin America, the world's largest emerald deposits and much gold, as well as great hydraulic resources, mining and energy only accounted for 1% of GDP in 1980. Major developments saw this figure reach 8.8% by 1999; declining oil production saw the sector's share of GDP fall to 5.4% in 2001, but it had recovered to 7.0% in 2004, according to provisional figures.

### Coal

Colombia had the largest coal reserves in Latin America, estimated by the US Energy Information Administration (EIA) at 7,287m. metric tons in 2003. It is now the world's fourth largest producer and exporter, with over 90% of domestic production exported. Production in 2004 was 50.8m. tons. The leading coal producer was the El Cerrejón open-cast mine in the Guajira, one of the largest mines in the world. It was jointly developed by the state company, Carbones de Colombia (CARBOCOL), and International Resources Corporation (Intercor), a subsidiary of Exxon. Both CARBOCOL and Intercor were sold to an international consortium of BHP Billiton PLC, Anglo-American PLC and Glencore International AG in 2000 and 2002, respectively. El Cerrejón's production was 21.6m. tons in 2004, with plans to expand ongoing. The second largest producer was the US company Drummond Ltd at its mine at La Loma in the César department, with production of 20m. tons per year. Coal was Colombia's third most valuable export. Exports rose from US $595m. in 1995 to $1,841m. in 2004, equivalent to 11.1% of total exports.

### Petroleum and Gas

Colombia was a modest oil producer, usually producing a small surplus for export, from early in the 20th century until the major discoveries in the 1980s and 1990s. These followed the introduction of association contracts for oil exploration in the mid-1970s that kept Colombia open to foreign companies just when most other developing countries were nationalizing their oil industries. The first important discovery was the Caño Limón field of an estimated 1,000m. barrels in Arauca in the early 1980s by Occidental. In 1991 a consortium led by BP discovered the Cusiana field in Casanare. This and the neighbouring Cupiagua field were estimated to hold between 1,500m. and 2,000m. barrels. Production at this field has fallen from 434,000 barrels per day (b/d) in 1998, before falling to 193,000 b/d in 2003. Total Colombian petroleum production in 2004 has been estimated at 530,000 b/d by the EIA. Exports reached a record 617,000 b/d at the end of 1998, but had declined to 171,000 b/d by the end of 2004. Proven crude oil reserves in 2004 stood at 1,478m. barrels. A tightening of the terms of the association contracts in 1989 led to a sharp drop in new exploration in the 1990s, which threatened to make Colombia a net importer by the mid-2000s. Much improved terms introduced in 1999 prompted a renewal of interest. Two medium-sized fields were discovered in 2001, and in March 2003 the discovery of large reserves (initially estimated at over 200m. barrels) at the Gibraltar-1 field was announced (a subsequent survey in 2004, however, greatly reduced the estimate of potential reserves, to 15m. barrels). The creation in 2003 of the Agencia Nacional de Hidrocarburos (ANH—National Hydrocarbons Agency) to take over the regulatory side of the activities of the state-run Empresa Colombiana de Petróleos (ECOPETROL) and the new, improved terms that the ANH introduced in June 2004, the most attractive in Latin America, were warmly welcomed by the international oil industry. Furthermore, the exploration of blocs off the Caribbean coast had already attracted interest from leading companies. The ANH signed 28 new contracts in the first half of 2005 alone.

Colombia's gas potential has never received the same investment as oil, owing largely to low local demand. However, there were important stand-alone natural-gas fields off the Caribbean coast and a great deal of associated gas in the main oilfields. Gas reserves were estimated at 4,187,000m. cu ft at the end of 2004, while production in 2003 was estimated by the EIA at 215,000m. cu ft. The Empresa Colombiana de Gas (Ecogás) was established in 1997 to develop the distribution network, in which it invested US $1,000m. Ecogás was scheduled for privatization during 2005. The freeing of the price of associated gas in 2002 made it feasible to start building, in 2003, a plant to process Cusiana-Cupiagua gas to boost supplies to the network.

### Metals

Despite Colombia's position as a leading producer of gold, the country's mines were not only small-scale and technologically unsophisticated, but also located in remote areas heavily affected by guerrilla activity in the late 1990s. As a result, gold exports, which reached US $205m. in 1996, fell to $68m. in 2001, before recovering to $103m. in 2002. With a higher gold price they reached $592m. in 2003 and $571m. in 2004. In contrast the nickel industry, concentrated at the Cerromatoso plant, expanded steadily in the 1990s. Exports rose from $154m. in 1999, to $414m. in 2003 and $626m in 2004. Colombia was the world's fifth largest exporter of nickel.

### Emeralds

It was estimated that Colombia produced about 60% of the world's emeralds; however, large-scale smuggling made measurement exceptionally difficult. Official figures for exports were US $452m. in 1995, making them Colombia's fourth most valuable export commodity, although exports declined steadily, reaching $74m. in 2004.

### Power

Colombia made a massive investment in large hydroelectric projects in the 1970s and 1980s and, by the early 2000s, this sector represented 77% of generating capacity. The El Niño extreme weather pattern of 1992–93 caused a severe drought that exposed the economy's vulnerability to this over-reliance on hydroelectric power. A rapid expansion of the thermal power-stations was then undertaken both to reduce this reliance and to meet increasing demand. In the event, demand slowed with the 1999 recession and between 1995 and 2001 production only rose from 41,573 gWh to 43,173 gWh. Production reached 44,900 gWh in 2002, according to the EIA, with domestic consumption of 41,100 gWh. The thermal sector's capacity had risen to about 30% by 1998. The mid-1990s saw a major privatization programme, with the sale of eight companies for more than US $5,000m., largely to Spanish electricity concerns.

## MANUFACTURING

Manufacturing, which accounted for 15.1% of GDP in 2004, grew steadily under import-substitution policies from the 1930s to 1990. The drastic reduction in tariffs and removal of import controls in 1991 under President Gaviria's policy of economic liberalization, known as the *apertura* (opening), led to an annual average contraction in the sector of 0.6% in 1990–2003, compared with average growth of 5% per year in the 1980s. As well as facing increased international competition, manufacturing had to cope with an overvalued currency for most of the decade and large-scale smuggling that drugs-traffickers often used for money-laundering. Production fell by 8.6% in the 1999 recession, but increased by 9.7% in 2000 as exports rose sharply. It fell again, by 1.3%, in 2001, but grew by 1.1% in 2002. In 2003 production rose by over 4.4% and, it was estimated, by almost 6% in 2004. Food and beverages, the largest sector, was much affected by increased import penetration and accounted for 3.5% of GDP in 2004. The chemicals industry, in contrast, prospered, owing to major investment by multinational companies. The value of chemical exports increased from US $235m. in 1990 to $1,692m. in 2000, falling to $1,371m. in 2003, but recovering to $1,623m. in 2004, representing about 10% of total export revenues. The automobile industry was also in the hands of foreign multinationals. Three car-assembly plants produced a record 80,000 units in 1997, but following the recession of 1999 annual production fell to 32,000. It recovered to 64,000 units in 2003 and reached a new high of 91,000 units in 2004. The textiles sector, based largely in Medellín, drove Colombia's industrialization from the beginning of the 20th century. It was particularly adversely affected by the *apertura* and the overvalued currency in the 1990s. However, clothing manufacturers, in particular, became successful exporters.

## CONSTRUCTION

From the mid-1990s the highly cyclical construction industry suffered seriously from the economic slowdown and then the recession. The value of the construction sector declined by 27.0% and 2.9% in 1999 and 2000, respectively. A reduction in the amount of drugs money directed at speculative high-cost housing was one element in this decline, as was the increasingly rapid emigration of middle-class professionals. The Government attempted to revive the sector and employment by increasing expenditure on low-cost housing, but was constrained in the short term by budgetary restrictions and an unhelpful Constitutional Court ruling on mortgage rates. However, a recovery began in late 2000, which strengthened in 2001 (construction GDP increased by 3.9% in the year) and continued in 2002 (when the sector increased by a provisional 12.5%). Further provisional growth figures of 13.4% and 9.7% were recorded in 2003 and 2004, respectively. The sector's proportion of overall GDP, which had fallen to 4.2% in 2000, recovered to 5.4% in 2004.

## TELECOMMUNICATIONS

Telecommunications, formerly the exclusive monopoly of the state communications company, were deregulated in the 1990s. Mobile licences were opened to international competition in the mid-1990s and competition to the state concern for long-distance calls, the Empresa Nacional de Telecomunicaciones (TELECOM), was introduced in the late 1990s. In 2003 TELECOM itself, which was heavily indebted, was dissolved by the Government and reconstituted as Colombia Telecomunicaciones. The industry as a whole expanded rapidly, outpacing the rest of the official economy, and 2004 represented 2.8% of GDP. This rapid growth was expected to continue; in terms of access to fixed-line telecommunications, Colombia ranked fifth in Latin America in 2003, with 19.7 lines per 100 inhabitants.

## FINANCIAL SERVICES

Colombia's financial services sector remained well developed by regional standards, particularly in banking. Banking supervision had been more rigorously enforced after a crisis in 1982, when the Government was forced to renationalize 70% of the banking sector. In the 1990s most of the sector returned to private ownership, with foreign investors, principally from Spain, a significant presence. The banking sector prospered in the early 1990s, in large part owing to the privatization and liberalization process. However, as elsewhere in Latin America, the product range of banks was very limited and the banks were, therefore, over-dependent on loan income. As a result, they were not always prudent in their lending and, when the economy entered recession in 1998, many of the institutions found themselves in grave difficulty with large, unrecoverable debts. The problem was aggravated when interest rates were increased to record levels in late 1998, in order to reduce the public-sector deficit. In 1999 the Government was forced to carry out a major bank rescue, estimated to have cost 7% of GDP. By mid-2001 most banks had returned to profitability and foreign banks accounted for 31% of the system's assets. In 2004 the banking sector made profits of US $1,470m., and profits increased by a further 20% in the first half of 2005, compared to the same period of the previous year. Despite the increase in foreign ownership, Colombia was unique in Latin America in that a majority of its banks were locally owned. The stock exchanges in Bogotá, Cali and Medellín also grew rapidly as a result of the privatization and deregulation measures undertaken. None the less, market capitalization remained small by international standards.

## PRIVATIZATION AND INFRASTRUCTURE

The Gaviria administration made a modest start to privatizing Colombia's inadequate infrastructure in the early 1990s by transferring ports, railways and some power-stations to the private sector. Concessions were let for construction or modernization of airports, roads and water utilities, and several banks were also sold. The Samper administration privatized a major part of the power-generation industry, as well as several other banks. The Pastrana administration was forced to cancel its plans to sell the remainder of the power sector, however, owing to changed market conditions, but the state coal company, CARBOCOL, was sold in late 2000. In mid-2004 the Uribe administration announced plans to privatize US $10,000m. of state assets over a five-year period. Those likely to be sold first were two banks, several regional electricity distributors and the state's stake in the electricity transmission company. ECOGAS was also to be sold off during 2005. However, the prospects for disposing of the other assets quickly or well remained unpromising.

## FOREIGN TRADE AND BALANCE OF PAYMENTS

Colombia's export earnings were greatly dependent first on coffee and then on petroleum, both of which were vulnerable to highly volatile international prices. Petroleum exports reached US $4,569m. in 2000 (35% of total exports), but fell to $3,383m. (26%) in 2003, before recovering to $4,179m. (25%) in 2004. Coal and nickel were increasingly important exports. Non-traditional exports, in particular manufactures, bananas and cut flowers, which benefited from privileged access to the EU and US markets, were increasingly important, although held back by the currency overvaluation in the mid-1990s. However, these products benefited from the devaluations at the end of the decade and increased from 48% of total exports in 1994 to 53% in 2002. The *apertura* of the early 1990s saw Colombia's imports—which had traditionally been fairly modest—grow rapidly, reaching $14,635m. in 1998. This led to increasingly large trade and current-account deficits. However, the recession caused imports to drop sharply and, with a subsequent growth in exports, Colombia was able to report positive trade balances from 1999. The surplus stood at $1,134m. in 2004. With renewed economic growth, imports recovered from $13,258m in 2003 to a record $15,878m in 2004. Export growth was even stronger: from $13,782m. in 2003 to $17,011m. in 2004. The current account was in surplus in 1999 and 2000, but returned to a deficit in 2001–04, totalling $1,111m. in 2004.

## INVESTMENT AND INDEBTEDNESS

Foreign direct investment (FDI) increased greatly during the 1990s, after the Andean Pact's abolition of its rule limiting foreign holdings to 49% in 1990. Much European money went into the petroleum industry and then into banks and power generation when these were privatized. FDI reached US $5,639m. in 1997. Following the end of the privatization programme, inflows declined to $2,237m. in 2000. In 2004 renewed interest by foreign investors was reflected in an increase to $2,739m. in FDI, compared with $1,793m. in the previous year. Colombia was the only Latin American country that avoided major debt rescheduling in the 1980s. External debt grew slowly; from $17,000m. in 1986, it reached $24,912m. in 1995, although this represented a real reduction, from 46% to 27% of GDP. Borrowing increased rapidly thereafter and debt reached $39,800m.—or 47% of GDP—at the end of 2001. Debt fell back to $37,232m. at the end of 2002, but increased to $38,226m. in 2003, when it represented 49.3% of GDP. In 2004, however, owing to a sharp revaluation of the peso, it stood at $39,561m. (equivalent to 37% of GDP). Of this, $25,779m. was public debt and $13,782m. was private debt.

## OUTLOOK

In 2005 the Colombian economy gave every sign of having returned to its traditional steady growth. After the expansion in GDP of 3.7% in 2003, and of the 4.0% in 2004, the figure for 2005 looked likely to be just as good or better, not least because the recovery was broadly based, with a remarkable 25% growth in exports (despite an 11% revaluation of the peso during 2004) and domestic investment up to 19% of GDP, of which a significant amount came from Colombians returning home. An increase in foreign direct investment in 2004 (see above) also held promise of further growth. There remained many uncertainties, but President Uribe had managed, in his first three years in office, to restore confidence to an impressive extent. The great majority of Colombians supported his determination to effect a root-and-branch solution to the country's problems. The significant progress made in strengthening the security forces, regaining territory from illegal groups, reducing drugs cultivation, murders, kidnapping and extortion also cheered the business community. The tax, pension and labour reforms made in the administration's first year also raised investors' expectations, as did several decisive measures to restructure state agencies such as the Instituto de Seguros Sociales, TELECOM and ECOPETROL. Historically low rates of both inflation (of around 5%) and interest, combined with a sharp, if insufficient, fall in both unemployment and poverty, further indicated a return to strong economic fundamentals. The international financial institutions were also clearly impressed by the way the Government coped with the unexpectedly large deficit left by its predecessor. They made explicit their continuing support for Colombia's efforts with an IMF agreement signed in January 2003 and substantial loans from the development banks. The US Administration also maintained its considerable aid to the country. Nevertheless, considerable challenges remained, notably the need for more far-reaching fiscal reform to maintain the increased spending on security, to restore public investment and social spending and still to reduce the deficit under the IMF programme. The success in reducing the fiscal deficit to 1.3% in 2004 owed much to a revaluation of the peso, high oil prices and a sharp downward reduction in local-government spending. Such a combination of factors was unlikely to recur. At the same time, the tax burden needed to be spread more fairly in order to encourage growth. The Uribe administration had hoped to gain approval for much of this in a referendum in October 2003. However, this plan proved unsuccessful, and the substitute tax reform passed by the Congreso in December was too limited. Further reforms, including the completion of earlier pension reforms, were put to the Congreso in the second half of 2004, but encountered much opposition, not least because controversial change to the Constitution to allow re-election of a President occupied both congressional time and presidential political capital. A new pension reform aimed at significantly reducing the cost of some public-sector workers' pensions was finally approved in mid-2005, although the final legislation was less radical than the Government had planned. The Government's room for manoeuvre was narrow, both domestically, where it would need to avoid raising the tax burden to a point where it would discourage renewed growth, and in the margin for extra borrowing allowed within the IMF deficit targets, or indeed within the sustainable limits of foreign indebtedness, although revaluation of the peso eased this problem in the short term. Fortunately, the coffee industry, once the solid core of the economy, was at last emerging from its deep crisis. Furthermore, Colombian exporters, who in 2003 had managed in other markets to more than compensate for the loss of exports to Venezuela—traditionally Colombia's second most important market—in 2004 continued to increase sales to the rest of the world while recovering the Venezuelan market. However, oil production continued to fall and, without a major discovery of the kind BP had hoped to make at Niscota, Colombia risked becoming a net importer by the end of the decade. There was, nevertheless, widespread confidence that these problems could be overcome by an administration unafraid of taking on vested interests and enjoying unprecedented levels of public support.

# Statistical Survey

Sources (unless otherwise stated): Departamento Administrativo Nacional de Estadística (DANE), Transversal 45 No 26-70, Interior I-CAN, Bogotá, DC; tel. (1) 597-8300; fax (1) 597-8399; e-mail dane@dane.gov.co; internet www.dane.gov.co; Banco de la República, Carrera 7A, No 14-78, 5°, Apdo Aéreo 3531, Bogotá, DC; tel. (1) 343-1111; fax (1) 286-1686; e-mail wbanco@banrep.gov.co; internet www.banrep.gov.co.

## Area and Population

### AREA, POPULATION AND DENSITY

| | |
|---|---:|
| Area (sq km) | 1,141,748* |
| **Population (census results)†** | |
| 15 October 1985 | |
| Males | 14,642,835 |
| Females | 14,838,160 |
| Total | 29,480,995 |
| 24 October 1993 | 37,664,711 |
| **Population (official estimates at mid-year)** | |
| 2002 | 43,834,117 |
| 2003 | 44,583,575 |
| 2004 | 45,325,260 |
| Density (per sq km) at mid-2004 | 39.7 |

* 440,831 sq miles.
† Revised figures, including adjustment for underenumeration. The enumerated total was 27,853,436 (males 13,785,523, females 14,067,913) in 1985 and 33,109,840 (males 16,296,539, females 16,813,301) in 1993.

### DEPARTMENTS
(census of 24 October 1993)*

| Department | Area (sq km) | Population | Capital (with population) |
|---|---:|---:|---|
| Amazonas | 109,665 | 56,399 | Leticia (30,045) |
| Antioquia | 63,612 | 4,919,619 | Medellín (1,834,881) |
| Arauca | 23,818 | 185,882 | Arauca (59,805) |
| Atlántico | 3,388 | 1,837,468 | Barranquilla (1,090,618) |
| Bolívar | 25,978 | 1,702,188 | Cartagena (747,390) |
| Boyacá | 23,189 | 1,315,579 | Tunja (112,807) |
| Caldas | 7,888 | 1,055,577 | Manizales (345,539) |
| Caquetá | 88,965 | 367,898 | Florencia (107,620) |
| Casanare | 44,640 | 211,329 | Yopal (57,279) |
| Cauca | 29,308 | 1,127,678 | Popayán (207,700) |
| César | 22,905 | 827,219 | Valledupar (278,216) |
| Chocó | 46,530 | 406,199 | Quibdó (122,371) |
| Córdoba | 25,020 | 1,275,623 | Montería (308,506) |
| Cundinamarca | 22,623 | 1,875,337 | Bogotá† |
| Guainía | 72,238 | 28,478 | Puerto Inírida (18,270) |
| La Guajira | 20,848 | 433,361 | Riohacha (109,474) |
| Guaviare | 42,327 | 97,602 | San José del Guaviare (48,237) |
| Huila | 19,890 | 843,798 | Neiva (278,350) |
| Magdalena | 23,188 | 1,127,691 | Santa Marta (313,072) |
| Meta | 85,635 | 618,427 | Villavicencio (272,118) |
| Nariño | 33,268 | 1,443,671 | Pasto (331,866) |
| Norte de Santander | 21,658 | 1,162,474 | Cúcuta (538,126) |
| Putumayo | 24,885 | 264,291 | Mocoa (25,910) |
| Quindío | 1,845 | 495,212 | Armenia (258,990) |
| Risaralda | 4,140 | 844,184 | Pereira (401,909) |
| San Andrés y Providencia Islands | 44 | 61,040 | San Andrés (56,361) |
| Santander del Sur | 30,537 | 1,811,740 | Bucaramanga (472,461) |
| Sucre | 10,917 | 701,105 | Sincelejo (194,962) |
| Tolima | 23,562 | 1,286,078 | Ibagué (399,838) |
| Valle del Cauca | 22,140 | 3,736,090 | Cali (1,847,176) |
| Vaupés | 65,268 | 24,671 | Mitú (13,177) |
| Vichada | 100,242 | 62,073 | Puerto Carreño (11,452) |
| **Capital District** | | | |
| Bogotá, DC | 1,587 | 5,484,244 | — |
| **Total** | 1,141,748 | 37,664,711 | — |

* Adjusted for underenumeration.
† The capital city, Bogotá, exists as the capital of a department as well as the Capital District. The city's population is included only in Bogotá, DC.

### PRINCIPAL TOWNS
(estimated population at mid-1999)

| | | | |
|---|---:|---|---:|
| Bogotá, DC (capital) | 6,260,862 | Neiva | 300,052 |
| Cali | 2,077,386 | Soledad | 295,058 |
| Medellín | 1,861,265 | Armenia | 281,422 |
| Barranquilla | 1,223,260 | Villavicencio | 273,140 |
| Cartagena | 805,757 | Soacha | 272,058 |
| Cúcuta | 606,932 | Valledupar | 263,247 |
| Bucaramanga | 515,555 | Montería | 248,245 |
| Ibagué | 393,664 | Itagüí | 228,985 |
| Pereira | 381,725 | Palmira | 226,509 |
| Santa Marta | 359,147 | Buenaventura | 224,336 |
| Manizales | 337,580 | Floridablanca | 221,913 |
| Bello | 333,470 | Sincelejo | 220,704 |
| Pasto | 332,396 | Popayán | 200,719 |

**Mid-2003** (UN estimate, incl. suburbs): Bogotá 7,289,646 (Source: UN, *World Urbanization Prospects: The 2003 Revision*).

### BIRTHS, MARRIAGES AND DEATHS*

| | Registered live births | Registered deaths |
|---|---:|---:|
| 1996 | n.a. | 173,506 |
| 1997 | n.a. | 170,753 |
| 1998 | 720,984 | 175,363 |
| 1999 | 746,194 | 183,553 |
| 2000 | 752,834 | 187,432 |
| 2001 | 724,319 | 191,513 |
| 2002 | 700,455 | 192,262 |
| 2003† | 697,029 | 189,073 |

* Data are tabulated by year of registration rather than by year of occurrence, although registration is incomplete. According to UN estimates, the average annual rates in 1985–90 were births 27.8 per 1,000; deaths 6.5 per 1,000; in 1990–95: births 27.0 per 1,000; deaths 6.4 per 1,000; and in 1995–2000: births 24.5 per 1,000; deaths 5.7 per 1,000 (Source: UN, *World Population Prospects: The 2004 Revision*).
† Preliminary figures.

**Registered marriages:** 102,448 in 1980; 95,845 in 1981; 70,350 in 1986.

**Expectation of life** (WHO estimates, years at birth): 72 (males 68; females 77) in 2003 (Source: WHO, *World Health Report*).

### ECONOMICALLY ACTIVE POPULATION
(household survey, '000 persons, September)

| | 2001 | 2002 | 2003 |
|---|---:|---:|---:|
| Agriculture, hunting, forestry and fishing | 3,661 | 3,492 | 3,769 |
| Mining and quarrying | 192 | 181 | 177 |
| Manufacturing | 2,119 | 2,200 | 2,328 |
| Electricity, gas and water | 88 | 74 | 62 |
| Construction | 643 | 760 | 767 |
| Trade, restaurants and hotels | 4,235 | 4,193 | 4,375 |
| Transport, storage and communications | 1,069 | 1,069 | 1,133 |
| Finance, insurance, real estate and business services | 843 | 888 | 935 |
| Community, social and personal services | 3,645 | 3,745 | 3,917 |
| Activities not adequately described | 5 | 18 | 5 |
| **Total labour force** | 16,498 | 16,619 | 17,467 |

Source: ILO.

COLOMBIA  
*Statistical Survey*

## Health and Welfare

**KEY INDICATORS**

| | |
|---|---:|
| Total fertility rate (children per woman, 2003) | 2.6 |
| Under-5 mortality rate (per 1,000 live births, 2003) | 21 |
| HIV/AIDS (% of persons aged 15–49, 2003) | 0.7 |
| Physicians (per 1,000 head, 2002) | 1.35 |
| Hospital beds (per 1,000 head, 1996) | 1.46 |
| Health expenditure (2002): US $ per head (PPP) | 536 |
| Health expenditure (2002): % of GDP | 8.1 |
| Health expenditure (2002): public (% of total) | 82.9 |
| Access to water (% of persons, 2002) | 92 |
| Access to sanitation (% of persons, 2002) | 86 |
| Human Development Index (2002): ranking | 73 |
| Human Development Index (2002): value | 0.773 |

For sources and definitions, see explanatory note on p. vi.

## Agriculture

**PRINCIPAL CROPS**  
('000 metric tons)

| | 2001 | 2002 | 2003 |
|---|---:|---:|---:|
| Rice (paddy) | 2,314 | 2,348 | 2,543 |
| Maize | 1,197 | 1,174 | 1,289 |
| Sorghum | 212 | 223 | 260 |
| Potatoes | 2,874 | 2,835 | 2,872 |
| Cassava (Manioc) | 1,980 | 1,779 | 1,841 |
| Yams | 255 | 237 | 283 |
| Other roots and tubers* | 73 | 64 | 75 |
| Sugar cane* | 33,400 | 35,800 | 37,000 |
| Dry beans | 124 | 120 | 136 |
| Soybeans (Soya beans) | 56 | 57 | 58 |
| Coconuts | 99 | 96 | 105 |
| Oil palm fruit* | 2,600 | 2,600 | 2,633 |
| Cottonseed* | 70 | 60 | 88 |
| Cabbages* | 295 | 300 | 305 |
| Tomatoes | 398 | 423 | 369 |
| Green chillies and peppers | 46 | 53 | 27 |
| Dry onions | 412 | 419 | 583 |
| Carrots | 190 | 152 | 180 |
| Other vegetables* | 200 | 201 | 205 |
| Watermelons | 64 | 70 | 66 |
| Bananas | 1,375 | 1,424 | 1,511 |
| Plantains | 2,928 | 2,921 | 2,911 |
| Oranges | 238 | 298 | 300* |
| Mangoes | 134 | 141 | 167 |
| Avocados | 137 | 142 | 162 |
| Pineapples | 314 | 353 | 406 |
| Papayas | 111 | 86 | 88 |
| Other fruit* | 1,270 | 1,325 | 1,352 |
| Coffee (green) | 656 | 691 | 703 |

* FAO estimate(s).  
Source: FAO.

**LIVESTOCK**  
('000 head, year ending September)

| | 2001 | 2002 | 2003 |
|---|---:|---:|---:|
| Horses* | 2,600 | 2,650 | 2,700 |
| Mules* | 605 | 610 | 615 |
| Asses* | 720 | 723 | 725 |
| Cattle | 24,510 | 24,765 | 25,000* |
| Pigs | 2,198 | 2,234 | 2,300* |
| Sheep | 2,256 | 2,045 | 2,100* |
| Goats | 1,136 | 1,105 | 1,150* |
| Poultry* | 110,000 | 115,000 | 118,000 |

* FAO estimate(s).  
Source: FAO.

**LIVESTOCK PRODUCTS**  
(FAO estimates, '000 metric tons)

| | 2001 | 2002 | 2003 |
|---|---:|---:|---:|
| Beef and veal | 700 | 676 | 680 |
| Mutton and lamb | 6.9 | 6.2 | 6.5 |
| Goat meat | 6.4 | 6.3 | 6.5 |
| Pig meat | 98.4 | 110.0 | 110.0 |
| Horse meat | 5.5 | 5.6 | 5.7 |
| Poultry meat | 581 | 629 | 630 |
| Cows' milk | 5,877 | 6,021 | 5,920 |
| Cheese | 53 | 54.0 | 54.8 |
| Butter and ghee | 18.8 | 19.2 | 19.6 |
| Hen eggs | 354.9 | 314.4 | 325.0 |
| Cattle hides | 77.9 | 75.3 | 75.9 |

Source: FAO.

## Forestry

**ROUNDWOOD REMOVALS**  
('000 cu metres, excl. bark)

| | 2001 | 2002 | 2003 |
|---|---:|---:|---:|
| Sawlogs, veneer logs and logs for sleepers | 1,190 | 1,161 | 1,148 |
| Pulpwood | 500 | 778 | 836 |
| Other industrial wood | 51 | 73 | 84 |
| Fuel wood | 10,760 | 9,598 | 7,891 |
| **Total** | **12,501** | **11,610** | **9,959** |

Source: FAO.

**SAWNWOOD PRODUCTION**  
('000 cu metres, incl. railway sleepers)

| | 2001 | 2002 | 2003 |
|---|---:|---:|---:|
| Coniferous (softwood) | 18 | 18 | 144 |
| Broadleaved (hardwood) | 521 | 509 | 455 |
| **Total** | **539** | **527** | **599** |

Source: FAO.

## Fishing

('000 metric tons, live weight)

| | 2001 | 2002 | 2003 |
|---|---:|---:|---:|
| Capture | 137.4* | 148.0* | 157.8 |
| Characins | 10.0* | 15.0* | 21.4 |
| Freshwater siluroids | 9.8 | 9.2* | 8.6 |
| Other freshwater fishes | 15.2* | 20.8* | 30.5 |
| Pacific anchoveta | 25.1 | 25.1* | 25.3 |
| Skipjack tuna | 1.9 | 2.3 | 12.6 |
| Yellowfin tuna | 25.6 | 31.0 | 48.5 |
| Aquaculture | 57.7* | 57.2* | 60.9 |
| Tilapias | 18.0* | 17.0* | 15.6 |
| Pirapatinga | 10.0* | 5.0* | 1.3 |
| Rainbow trout | 7.0* | 5.0* | 4.2 |
| Whiteleg shrimp | 12.0* | 14.0* | 16.5 |
| **Total catch** | **195.0*** | **205.2*** | **218.7** |

* FAO estimate.  
Note: Figures exclude crocodiles, recorded by number rather than by weight. The number of spectacled caimans caught was: 704,313 in 2001; 585,000 in 2002; 555,219 in 2003.  
Source: FAO.

## Mining

('000 metric tons, unless otherwise indicated)

|  | 2001 | 2002 | 2003 |
|---|---|---|---|
| Gold (kilograms) | 21,813 | 20,799 | 46,515 |
| Silver (kilograms) | 7,242 | 6,986 | 9,511 |
| Salt | 395 | 527 | 447 |
| Hard coal | 43,910 | 39,500 | 49,300 |
| Iron ore* | 637 | 688 | 625 |
| Crude petroleum ('000 barrels) | 220,460 | 211,007 | 197,586 |

* Figures refer to the gross weight of ore. The estimated iron content is 46%.
Source: US Geological Survey.

## Industry

**SELECTED PRODUCTS**
('000 metric tons, unless otherwise indicated)

|  | 2000 | 2001 | 2002 |
|---|---|---|---|
| Sugar | 2,391 | 2,242 | 2,531 |
| Cement | 7,131 | 6,776 | 6,633 |
| Crude steel ingots (incl. steel for casting)* | 659.9 | 638.3 | 663.7 |
| Semi-manufactures of iron and steel (hot-rolled)* | 650 | 552 | 575 |
| Gas-diesel (distillate fuel) oils | 3,038 | 3,312 | n.a. |
| Residual fuel oils | 2,922 | 2,986 | n.a. |
| Motor spirit (petrol) | 4,900 | 5,092 | n.a. |

* Source: US Geological Survey.
Source: mostly UN, *Industrial Commodity Statistics Yearbook*.

**2003** ('000 metric tons): Crude steel ingots (incl. steel for casting) 668.4; Semi-manufactures of iron and steel (hot-rolled) 550 (Source: US Geological Survey).

## Finance

**CURRENCY AND EXCHANGE RATES**

**Monetary Units**
100 centavos = 1 Colombian peso.

**Sterling, Dollar and Euro Equivalents** (31 May 2005)
£1 sterling = 4,241.24 pesos;
US $1 = 2,332.79 pesos;
€1 = 2,876.56 pesos;
10,000 Colombian pesos = £2.36 = $4.29 = €3.48.

**Average Exchange Rate** (pesos per US $)
2002   2,504.24
2003   2,877.65
2004   2,628.61

**CENTRAL GOVERNMENT BUDGET**
('000 million pesos)

| Revenue | 2001 | 2002 | 2003 |
|---|---|---|---|
| Tax revenue | 24,802.3 | 27,086.7 | 31,372.9 |
| Tax on earnings | 10,022.5 | 10,626.4 | 12,218.1 |
| Value-added tax (domestic) | 6,931.5 | 7,363.7 | 8,887.9 |
| Value-added tax (external) | 3,079.4 | 3,264.8 | 4,183.5 |
| Levies | 2,150.1 | 2,083.5 | 2,158.9 |
| Tax on petrol | 1,106.4 | 976.6 | 1,025.0 |
| Tax on financial transactions | 1,421.3 | 1,443.2 | 1,621.5 |
| Other taxes* | 91.0 | 1,328.6 | 1,277.9 |
| Non-tax revenue | 176.5 | 162.8 | 93.8 |
| Special funds | 240.5 | 316.0 | 321.0 |
| Capital revenue | 2,071.4 | 2,523.6 | 2,419.0 |
| **Total** | **27,290.6** | **30,089.1** | **34,206.6** |

| Expenditure† | 2001 | 2002 | 2003 |
|---|---|---|---|
| Interest payments | 8,340.1 | 8,751.4 | 10,309.2 |
| Foreign | 3,017.5 | 3,575.7 | 4,355.3 |
| Domestic | 5,322.6 | 5,175.7 | 5,953.8 |
| Operational expenditures | 28,533.4 | 29,848.1 | 33,007.7 |
| Personal services | 6,546.5 | 6,777.9 | 7,490.3 |
| General expenditure | 1,774.3 | 1,982.1 | 2,162.7 |
| Transfers | 20,212.6 | 21,088.1 | 23,354.7 |
| Investment | 2,904.0 | 2,955.0 | 2,882.8 |
| **Total** | **39,777.6** | **41,554.6** | **46,199.6** |

* Beginning in 2002 includes tax on wealth.
† Excluding net lending ('000 million pesos): 1,470.6 in 2001; 1,258.7 in 2002; 864.8 in 2003.

**Budgetary central government** ('000 million pesos): *Revenue:* 40,884.4 in 2004. *Expenditure:* 51,795.9 (excl. lending minus repayments 151.7) in 2004 (Source: IMF, *International Financial Statistics*).

**INTERNATIONAL RESERVES**
(US $ million at 31 December)

|  | 2002 | 2003 | 2004 |
|---|---|---|---|
| Gold* | 112 | 136 | 143 |
| IMF special drawing rights | 154 | 171 | 181 |
| Reserve position in IMF | 389 | 425 | 444 |
| Foreign exchange | 10,190 | 10,188 | 12,769 |
| **Total** | **10,844** | **10,920** | **13,537** |

* National valuation.
Source: IMF, *International Financial Statistics*.

**MONEY SUPPLY**
('000 million pesos at 31 December)

|  | 2002 | 2003 | 2004 |
|---|---|---|---|
| Currency outside banks | 10,188.5 | 12,196.6 | 13,948.7 |
| Demand deposits at commercial banks | 11,256.3 | 12,545.1 | 14,769.6 |
| **Total money** (incl. others) | **21,576.4** | **24,916.9** | **28,904.8** |

Source: IMF, *International Financial Statistics*.

**COST OF LIVING**
(Consumer Price Index for low-income families; base: 2000 = 100)

|  | 2002 | 2003 | 2004 |
|---|---|---|---|
| Food and beverages | 118.4 | 127.2 | 134.6 |
| Clothing and footwear | 105.8 | 107.2 | 108.9 |
| Rent, fuel and light* | 105.1 | 108.3 | 112.7 |
| **All items** (incl. others) | **116.5** | **125.0** | **132.5** |

* Including certain household equipment.
Source: ILO.

**NATIONAL ACCOUNTS**
('000 million pesos at current prices, rounded figures)

**Composition of the Gross National Product**

|  | 1999 | 2000 | 2001* |
|---|---|---|---|
| Compensation of employees | 56,482 | 62,118 | 66,289 |
| Operating surplus | 81,671 | 95,743 | 102,222 |
| Consumption of fixed capital |  |  |  |
| **Gross domestic product (GDP) at factor cost** | 138,153 | 157,861 | 168,512 |
| Indirect taxes | 14,242 | 18,048 | 20,590 |
| *Less* Subsidies | 831 | 1,013 | 1,166 |
| **GDP in purchasers' values** | 151,565 | 174,896 | 187,936 |
| Net factor income from abroad | −2,663 | −4,835 | −5,963 |
| **Gross national product (GNP)** | 148,902 | 170,061 | 181,973 |

* Provisional figures.

# COLOMBIA

## Statistical Survey

**Expenditure on the Gross Domestic Product**
(provisional figures)

|  | 2002 | 2003 | 2004 |
|---|---|---|---|
| Final consumption expenditure | 176,342 | 193,652 | 212,234 |
|   Households* | 132,937 | 146,066 | 159,811 |
|   General government | 43,405 | 47,586 | 52,423 |
| Gross capital formation | 31,517 | 42,029 | 48,258 |
| **Total domestic expenditure** | 207,859 | 235,681 | 260,492 |
| Exports of goods and services | 38,847 | 46,997 | 53,527 |
| *Less* Imports of goods and services | 42,176 | 52,211 | 57,156 |
| **GDP in market prices** | 204,530 | 230,467 | 256,862 |

* Including non-profit institutions serving households.

**Gross Domestic Product by Economic Activity**
(provisional figures)

|  | 2002 | 2003 | 2004 |
|---|---|---|---|
| Agriculture, hunting, forestry and fishing | 25,093 | 27,658 | 29,545 |
| Mining and quarrying | 10,207 | 14,119 | 17,052 |
| Manufacturing | 29,395 | 32,967 | 37,012 |
| Electricity, gas and water | 8,712 | 10,401 | 11,671 |
| Construction | 8,372 | 10,030 | 13,226 |
| Wholesale and retail trade; repair of motor vehicles, motorcycles, and personal and household goods | 17,586 | 20,093 | 22,801 |
| Hotels and restaurants | 4,266 | 4,757 | 5,255 |
| Transport, storage and communications | 16,419 | 17,609 | 18,559 |
| Financial intermediation | 9,818 | 11,630 | 13,460 |
| Real estate, renting and business activities | 20,974 | 22,359 | 23,880 |
| Government services | 30,641 | 33,492 | 36,833 |
| Other community, social and personal service activities | 14,065 | 14,852 | 15,745 |
| **Sub-total** | 195,547 | 219,967 | 245,038 |
| *Less* Financial intermediation services indirectly measured | 8,294 | 10,136 | 12,011 |
| **Gross value added in basic prices** | 187,253 | 209,831 | 233,028 |
| Taxes on products | 17,507 | 20,924 | 24,165 |
| *Less* Subsidies on products | 230 | 288 | 331 |
| **GDP in market prices** | 204,530 | 230,467 | 256,862 |

**BALANCE OF PAYMENTS**
(US $ million)

|  | 2002 | 2003 | 2004 |
|---|---|---|---|
| Exports of goods f.o.b. | 12,316 | 13,782 | 17,011 |
| Imports of goods f.o.b. | −12,077 | −13,258 | −15,878 |
| **Trade balance** | 239 | 524 | 1,134 |
| Exports of services | 1,867 | 1,900 | 2,134 |
| Imports of services | −3,302 | −3,332 | −3,860 |
| **Balance on goods and services** | −1,196 | −908 | −592 |
| Other income received | 711 | 548 | 664 |
| Other income paid | −3,559 | −3,994 | −4,849 |
| **Balance on goods, services and income** | −4,045 | −4,355 | −4,777 |
| Current transfers received | 3,008 | 3,568 | 3,935 |
| Current transfers paid | −304 | −234 | −269 |
| **Current balance** | −1,340 | −1,191 | −1,111 |
| Direct investment abroad | −857 | −938 | −142 |
| Direct investment from abroad | 2,115 | 1,793 | 2,739 |
| Portfolio investment assets | 2,030 | −1,753 | −1,650 |
| Portfolio investment liabilities | −933 | 130 | 1,136 |
| Financial derivatives liabilities | −111 | −101 | −190 |
| Other investment assets | 282 | 1,642 | 980 |
| Other investment liabilities | −1,246 | −39 | 442 |
| Net errors and omissions | 200 | 99 | 266 |
| **Overall balance** | 139 | −188 | 2,469 |

Source: IMF, *International Financial Statistics*.

## External Trade

**PRINCIPAL COMMODITIES**
(US $ million)

| Imports c.i.f. | 2001 | 2002 | 2003 |
|---|---|---|---|
| Vegetables and vegetable products | 724 | 836 | 855 |
| Prepared foodstuffs, beverages and tobacco | 852 | 788 | 799 |
| Textiles and leather products | 745 | 702 | 741 |
| Paper and paper products | 435 | 436 | 454 |
| Chemical products | 2,725 | 2,770 | 3,757 |
| Petroleum and its derivatives | 219 | 239 | 292 |
| Metals | 574 | 562 | 678 |
| Mechanical, electrical and transport equipment | 5,558 | 5,441 | 6,157 |
| **Total** (incl. others) | 12,821 | 12,691 | 13,881 |

| Exports f.o.b. | 2001 | 2002 | 2003 |
|---|---|---|---|
| Vegetables and vegetable products | 1,108 | 1,193 | 1,188 |
| Coffee | 764 | 772 | 806 |
| Coal | 1,197 | 991 | 1,422 |
| Petroleum and its derivatives | 3,285 | 3,275 | 3,383 |
| Prepared foodstuffs, beverages and tobacco | 980 | 944 | 1,000 |
| Textiles and leather products | 1,015 | 880 | 1,017 |
| Paper and publishing | 399 | 373 | 398 |
| Chemicals | 1,522 | 1,509 | 1,371 |
| Mechanical, electrical and transport equipment | 1,093 | 931 | 665 |
| **Total** (incl. others) | 12,330 | 11,975 | 13,092 |

Source: Dirección de Impuestos y Aduanas Nacionales.

**PRINCIPAL TRADING PARTNERS**
(US $ million)

| Imports c.i.f. | 2001 | 2002 | 2003 |
|---|---|---|---|
| Argentina | 202.3 | 211.1 | 257.1 |
| Bolivia | 131.2 | 139.9 | 198.3 |
| Brazil | 582.1 | 642.8 | 619.4 |
| Canada | 344.2 | 313.5 | 274.1 |
| Chile | 262.5 | 299.9 | 280.8 |
| China, People's Rep. | 475.4 | 532.8 | 688.7 |
| Ecuador | 318.2 | 367.2 | 409.7 |
| France | 350.0 | 236.1 | 381.3 |
| Germany | 560.1 | 508.3 | 611.7 |
| Italy | 401.0 | 275.0 | 289.5 |
| Japan | 559.0 | 619.4 | 642.9 |
| Korea, Rep. | 266.3 | 311.5 | 337.8 |
| Mexico | 596.2 | 678.0 | 744.4 |
| Peru | 159.2 | 160.3 | 193.4 |
| Spain | 246.7 | 260.4 | 255.5 |
| Switzerland | 158.8 | 147.7 | 171.0 |
| Taiwan | 166.4 | 161.8 | 164.5 |
| United Kingdom | 179.6 | 161.8 | 164.3 |
| USA | 4,413.9 | 4,020.1 | 4,081.2 |
| Venezuela | 792.7 | 788.1 | 727.4 |
| **Total** (incl. others) | 12,820.6 | 12,690.4 | 13,880.6 |

# COLOMBIA

*Statistical Survey*

| Exports f.o.b. | 2001 | 2002 | 2003 |
|---|---|---|---|
| Belgium-Luxembourg | 210.4 | 241.8 | 227.7 |
| Canada | 142.5 | 164.2 | 176.3 |
| Chile | 168.0 | 174.1 | 188.0 |
| Costa Rica | 149.8 | 192.5 | 153.9 |
| Ecuador | 699.9 | 814.0 | 778.6 |
| France | 134.0 | 138.3 | 145.2 |
| Germany | 416.9 | 332.3 | 264.1 |
| Italy | 196.1 | 222.4 | 279.9 |
| Japan | 164.7 | 193.8 | 198.7 |
| Mexico | 261.3 | 310.4 | 356.6 |
| Netherlands | 114.3 | 129.2 | 301.1 |
| Panama | 265.2 | 216.1 | 170.1 |
| Peru | 277.5 | 348.5 | 394.5 |
| Puerto Rico | 90.1 | 178.4 | 361.9 |
| Spain | 140.5 | 206.1 | 195.4 |
| United Kingdom | 287.3 | 159.3 | 182.7 |
| USA | 5,254.3 | 5,159.7 | 5,758.2 |
| Venezuela | 1,737.4 | 1,123.3 | 693.5 |
| **Total** (incl. others) | **12,301.5** | **11,938.9** | **13,010.1** |

## Transport

### RAILWAYS
(traffic)

| | 1996 | 1997 | 1998 |
|---|---|---|---|
| Freight ('000 metric tons) | 321 | 348 | 281 |
| Freight ton-km ('000) | 746,544 | 736,427 | 657,585 |

Source: Sociedad de Transporte Ferroviario, SA.

### ROAD TRAFFIC
(motor vehicles in use)

| | 1997 | 1998 | 1999 |
|---|---|---|---|
| Passenger cars | 1,694,323 | 1,776,100 | 1,803,201 |
| Buses | 126,362 | 131,987 | 134,799 |
| Goods vehicles | 179,530 | 183,335 | 184,495 |
| Motorcycles | 385,378 | 450,283 | 479,073 |

Source: IRF, *World Road Statistics*.

### SHIPPING
**Merchant Fleet**
(registered at 31 December)

| | 2002 | 2003 | 2004 |
|---|---|---|---|
| Number of vessels | 109 | 121 | 126 |
| Total displacement ('000 grt) | 67.8 | 71.1 | 75.1 |

Source: Lloyd's Register-Fairplay, *World Fleet Statistics*.

**Domestic Sea-borne Freight Traffic**
('000 metric tons)

| | 1987 | 1988 | 1989 |
|---|---|---|---|
| Goods loaded and unloaded | 772.1 | 944.8 | 464.6 |

**International Sea-borne Freight Traffic**
('000 metric tons)

| | 1999 | 2000 | 2001 |
|---|---|---|---|
| Goods loaded | 4,111 | 3,543 | 3,832 |
| Goods unloaded | 1,274 | 1,114 | 1,399 |

### CIVIL AVIATION
(traffic)

| | 2001 | 2002 | 2003 |
|---|---|---|---|
| **Domestic** | | | |
| Passengers carried ('000) | 7,560 | 7,732 | 7,438 |
| Freight carried (metric tons) | 103,826 | 121,636 | 131,187 |
| **International** | | | |
| Passengers ('000): | | | |
|   arrivals | 1,446 | 1,429 | 1,460 |
|   departures | 1,605 | 1,455 | 1,484 |
| Freight (metric tons): | | | |
|   loaded | 129,557 | 127,688 | 146,142 |
|   unloaded | 242,780 | 257,293 | 299,463 |

Source: Departamento Administrativo de Aeronáutica Civil.

## Tourism

**TOURIST ARRIVALS**

| Country of origin | 2001 | 2002 | 2003 |
|---|---|---|---|
| Argentina | 17,759 | 13,667 | 16,028 |
| Brazil | 14,612 | 14,040 | 16,938 |
| Canada | 20,560 | 15,944 | 16,292 |
| Costa Rica | 12,516 | 11,380 | 18,410 |
| Ecuador | 59,476 | 55,701 | 64,431 |
| France | 16,265 | 16,951 | 16,545 |
| Germany | 17,206 | 13,921 | 14,163 |
| Italy | 20,000 | 16,303 | 14,899 |
| Mexico | 24,476 | 25,092 | 26,998 |
| Panama | 16,938 | 16,614 | 18,764 |
| Peru | 23,889 | 23,216 | 25,732 |
| Spain | 29,030 | 30,109 | 32,525 |
| USA | 156,640 | 155,377 | 171,906 |
| Venezuela | 85,263 | 65,781 | 73,567 |
| **Total** (incl. others) | **615,623** | **566,759** | **624,909** |

**Tourism receipts** (US $ million, incl. passenger transport): 1,483 in 2001; 1,237 in 2002; 1,114 in 2003.

Source: World Tourism Organization.

## Communications Media

| | 2001 | 2002 | 2003 |
|---|---|---|---|
| Telephones ('000 main lines in use) | 7,371.5 | 7,766.0 | 8,768.1 |
| Mobile cellular telephones ('000 subscribers) | 3,265.3 | 4,596.6 | 6,186.2 |
| Personal computers ('000 in use) | 1,800 | 2,133 | n.a. |
| Internet users ('000) | 1,154.0 | 2,000.0 | 2,732.2 |

**1996:** 37 daily newspapers.
**1997** ('000 in use): 21,000 radio receivers; 173 facsimile machines.
**2000** ('000 in use): 11,936 television receivers.
**Book production:** 1,481 titles in 1991.

Sources: UN, *Statistical Yearbook*; UNESCO, *Statistical Yearbook*; International Telecommunication Union.

## Education

(2002)

| | Institutions | Teachers | Pupils |
|---|---|---|---|
| Nursery | 32,432 | 50,713 | 1,080,555 |
| Primary | 55,869 | 193,606 | 5,205,489 |
| Secondary | 12,921 | 177,940 | 3,515,716 |
| Higher (incl. universities) | 321 | 97,522 | 981,458 |

Source: partly Ministerio de Educación Nacional.

**2003** (number of pupils): Nursery 1,143,291; Primary 5,207,772; Secondary 3,603,949.

**Adult literacy rate** (UNESCO estimates): 92.1% (males 92.1%; females 92.2%) in 2002 (Source: UN Development Programme, *Human Development Report*).

# Directory

## The Constitution

A new, 380-article Constitution, drafted by a 74-member National Constituent Assembly, took effect from 6 July 1991. The new Constitution retained the institutional framework of a directly elected President with a non-renewable four-year term of office, together with a bicameral legislature composed of an upper house or Senate (with 102 directly elected members) and a lower house or House of Representatives (with 161 members, to include at least two representatives of each national department). On 14 December 2004 both chambers of the Congreso approved a constitutional amendment to allow re-election of the President. The reform was scheduled to be implemented before the presidential election that was due in May 2006. A Vice-President is elected at the same time as the President, and also holds office for a term of four years.

The new Constitution also contained comprehensive provisions for the recognition and protection of civil rights, and for the reform of the structures and procedures of political participation and of the judiciary.

The fundamental principles upon which the new Constitution is based are embodied in articles 1–10.

Article 1: Colombia is a lawful state, organized as a single Republic, decentralized, with autonomous territorial entities, democratic, participatory and pluralist, founded on respect for human dignity, on the labour and solidarity of its people and on the prevalence of the general interest.

Article 2: The essential aims of the State are: to serve the community, to promote general prosperity and to guarantee the effectiveness of the principles, rights and obligations embodied in the Constitution, to facilitate the participation of all in the decisions which affect them and in the economic, political, administrative and cultural life of the nation; to defend national independence, to maintain territorial integrity and to ensure peaceful coexistence and the validity of the law.

The authorities of the Republic are instituted to protect the residents of Colombia, in regard to their life, honour, goods, beliefs and other rights and liberties, and to ensure the fulfilment of the obligations of the State and of the individual.

Article 3: Sovereignty rests exclusively with the people, from whom public power emanates. The people exercise power directly or through their representatives in the manner established by the Constitution.

Article 4: The Constitution is the highest authority. In all cases of incompatability between the Constitution and the law or other juridical rules, constitutional dispositions will apply.

It is the duty of nationals and foreigners in Colombia to observe the Constitution and the law, and to respect and obey the authorities.

Article 5: The State recognizes, without discrimination, the primacy of the inalienable rights of the individual and protects the family as the basic institution of society.

Article 6: Individuals are solely responsible to the authorities for infringements of the Constitution and of the law. Public servants are equally accountable and are responsible to the authorities for failure to fulfil their function or abuse of their position.

Article 7: The State recognizes and protects the ethnic diversity of the Colombian nation.

Article 8: It is an obligation of the State and of the people to protect the cultural and natural riches of the nation.

Article 9: The foreign relations of the State are based on national sovereignty, with respect for self-determination of people and with recognition of the principles of international law accepted by Colombia.

Similarly, Colombia's external politics will be directed towards Caribbean and Latin American integration.

Article 10: Spanish (Castellano) is the official language of Colombia. The languages and dialects of ethnic groups are officially recognized within their territories. Education in communities with their own linguistic traditions will be bilingual.

## The Government

### HEAD OF STATE

**President:** ALVARO URIBE VÉLEZ (took office 7 August 2002).
**Vice-President:** FRANCISCO SANTOS CALDERÓN (took office 7 August 2002).

### CABINET
(July 2005)

A coalition of the Partido Conservador Colombiano, the Partido Liberal Colombiano and independents.

**Minister of the Interior and Justice:** SABAS PRETELT DE LA VEGA.
**Minister of Foreign Affairs:** CAROLINA BARCO.
**Minister of Finance and Public Credit:** ALBERTO CARRASQUILLA BARRERA.
**Minister of National Defence:** CAMILO OSPINA BERNAL.
**Minister of Agriculture and Rural Development:** CARLOS GUSTAVO CANO SANZ.
**Minister of Labour, Social Security and Health:** DIEGO PALACIO BETANCOURT.
**Minister of Mines and Energy:** LUIS ERNESTO MEJÍA CASTRO.
**Minister of Trade, Industry and Tourism:** JORGE HUMBERTO BOTERO.
**Minister of National Education:** CECILIA MARÍA VÉLEZ WHITE.
**Minister of the Environment, Housing and Territorial Development:** SANDRA SUÁREZ PÉREZ.
**Minister of Communications:** MARTHA ELENA PINTO DE HART.
**Minister of Transport:** ANDRÉS URIEL GALLEGO HENAO.
**Minister of Culture:** MARÍA CONSUELO ARAÚJO CASTRO.

### MINISTRIES

**Office of the President:** Palacio de Nariño, Carrera 8A, No 7-26, Bogotá, DC; tel. (1) 562-9300; fax (1) 286-8063.

**Ministry of Agriculture and Rural Development:** Avda Jiménez, No 7-65, Bogotá, DC; tel. (1) 334-1199; fax (1) 284-1775; e-mail minagric@colomsat.net.co; internet www.minagricultura.gov.co.

**Ministry of Communications:** Edif. Murillo Toro, Carrera 7 y 8, Calle 12 y 13, Apdo Aéreo 14515, Bogotá, DC; tel. (1) 286-6911; fax (1) 344-3434; internet www.mincomunicaciones.gov.co.

**Ministry of Culture:** Calle 8, No 6-97, Bogotá, DC; tel. (1) 342-4100; fax (1) 342-1721; e-mail fvasquez@mincultura.gov.co; internet www.mincultura.gov.co.

**Ministry of the Environment, Housing and Territorial Development:** Calle 37, No 8-40, Bogotá, DC; tel. (1) 288-6877; fax (1) 288-9788; internet www.minambiente.gov.co.

**Ministry of Finance and Public Credit:** Carrera 8A, No 6-64, Of. 305, Bogotá, DC; tel. (1) 381-1700; fax (1) 350-9344; internet www.minhacienda.gov.co.

**Ministry of Foreign Affairs:** Palacio de San Carlos, Calle 10A, No 5-51, Bogotá, DC; tel. (1) 282-7811; fax (1) 341-6777; internet www.minrelext.gov.co.

**Ministry of the Interior and Justice:** Palacio Echeverry, Carrera 8A, No 8-09, Bogotá, DC; tel. (1) 334-0630; fax (1) 341-9583; internet www.presidencia.gov.co/ministerios.

**Ministry of Labour, Social Security and Health:** Carrera 7A, No 34-50, Bogotá, DC; tel. (1) 287-3434; fax (1) 285-7091; e-mail oaai@tutopia.com.

**Ministry of Mines and Energy:** Centro Administrativo Nacional (CAN), Avda El Dorado, Bogotá, DC; tel. (1) 222-4555; fax (1) 222-3651; internet www.minminas.gov.co.

**Ministry of National Defence:** Centro Administrativo Nacional (CAN), 2°, Avda El Dorado, Bogotá, DC; tel. (1) 220-4999; fax (1) 222-1874; internet www.mindefensa.gov.co.

**Ministry of National Education:** Centro Administrativo Nacional (CAN), Of. 501, Avda El Dorado, Bogotá, DC; tel. (1) 222-2800; fax (1) 222-4578; e-mail dci@mineducacion.gov.co; internet www.mineducacion.gov.co.

**Ministry of Trade, Industry and Tourism:** Calle 28, No 13A-15, 18°, Edif. Centro de Comercio Internacional, Bogotá, DC; tel. (1) 382-1307; fax (1) 352-2101; internet www.mincomercio.gov.co.

**Ministry of Transport:** Centro Administrativo Nacional (CAN), Of. 409, Avda El Dorado, Bogotá, DC; tel. (1) 222-4411; fax (1) 222-1647; internet www.mintransporte.gov.co.

COLOMBIA

# President and Legislature

### PRESIDENT
Presidential Election, 26 May 2002

|  | Votes | % of votes cast |
|---|---|---|
| Alvaro Uribe Vélez | 5,862,655 | 54.0 |
| Horacio Serpa Uribe | 3,514,779 | 32.3 |
| Luis Eduardo Garzón | 680,245 | 6.3 |
| Noemí Sanín Posada | 641,884 | 5.9 |
| Others | 155,966 | 1.5 |
| Total* | 10,855,529 | 100.0 |

*Excluding 196,116 blank ballots.

### CONGRESO

#### Senado
(Senate)

**President:** GERMÁN VARGAS LLERAS.

General Election, 10 March 2002

|  | Seats | % of votes cast |
|---|---|---|
| Independent groups | 52 | 51.5 |
| Partido Liberal Colombiano (PL) | 29 | 30.6 |
| Partido Conservador Colombiano (PCC) | 13 | 9.9 |
| Coalitions | 6 | 6.3 |
| Indigenous groups* | 2 | 1.7 |
| Total | 102 | 100.0 |

*Under the reforms of the Constitution in 1991, at least two Senate seats are reserved for indigenous groups.

#### Cámara de Representantes
(House of Representatives)

**President:** ALONSO ACOSTA OSIO.

General Election, 10 March 2002

|  | Seats | % of votes cast |
|---|---|---|
| Independent groups | 69 | 52.6 |
| Partido Liberal Colombiano (PL) | 54 | 32.0 |
| Partido Conservador Colombiano (PCC) | 21 | 7.3 |
| Coalitions | 17 | 8.1 |
| Total | 161 | 100.0 |

# Political Organizations

**Alianza Democrática—M-19 (AD—M-19):** Transversal 28, No 37-78, Bogotá, DC; tel. (1) 368-9436; f. 1990; alliance of centre-left groups (including factions of Unión Patriótica, Colombia Unida, Frente Popular and Socialismo Democrático) which supported the M-19 campaign for elections to the National Constituent Assembly in December 1990; Leader DIEGO MONTAÑA CUÉLLAR.

**Alianza Nacional Popular (ANAPO):** Carrera 18, No 33-95, Bogotá, DC; tel. (1) 287-7050; fax (1) 245-3138; f. 1971 by supporters of Gen. Gustavo Rojas Pinilla; populist party; Leader MARÍA EUGENIA ROJAS DE MORENO DÍAZ.

**Democracia Cristiana:** Avda 42, No 18-08, Apdo 25867, Bogotá, DC; tel. (1) 285-6639; f. 1964; Christian Democrat party; 10,000 mems; Pres. JUAN A. POLO FIGUEROA; Sec.-Gen. DIEGO ARANGO OSORIO.

**Frente Social y Político:** f. 2001; left-wing; Presidential Candidate LUIS EDUARDO GARZÓN.

**Frente por la Unidad del Pueblo (FUP):** Bogotá, DC; extreme left-wing front comprising socialists and Maoists.

**Movimiento 19 de Abril (M-19):** Calle 26, No 13B-09, Of. 1401, Bogotá, DC; tel. (1) 282-7891; fax (1) 282-8129; f. 1970 by followers of Gen. Gustavo Rojas Pinilla and dissident factions from the FARC—EP (see below); left-wing urban guerrilla group, until formally constituted as a political party in Oct. 1989; mem. of the Coordinadora Nacional Guerrilla Simón Bolívar (CNGSB); Leader OTTY PATIÑO.

**Movimiento Colombia Unida (CU):** Bogotá, DC; left-wing group allied to the UP; Leader ADALBERTO CARVAJAL.

**Movimiento Nacional Conservador (MNC):** Carrera 16, No 33-24, Bogotá, DC; tel. (1) 245-4418; fax (1) 284-8529; Sec.-Gen. JUAN PABLO CEPERA MÁRQUEZ.

**Movimiento Nacional Progresista (MNP):** Carrera 10, No 19-45, Of. 708, Bogotá, DC; tel. (1) 286-7517; fax (1) 341-9368; Sec.-Gen. EDUARDO AISAMAK LEÓN BELTRÁN.

**Movimiento Obrero Independiente Revolucionario (MOIR):** Avda Caracas 33–39, Bogotá, DC; tel. (1) 287-0732; e-mail moir@moir.org.co; internet www.moir.org.co; left-wing workers' movement; Maoist; Leader HÉCTOR VALENGA.

**Movimiento de Salvación Nacional (MSN):** Carrera 7A, No 58-00, Bogotá, DC; tel. (1) 249-0209; fax (1) 310-1991; f. 1990; split from the Partido Conservador Colombiano.

**Mujeres para la Democracia:** Bogotá, DC; f. 1991; women's party; Leader ANGELA CUEVAS DE DOLMETSCH.

**Organización Nacional Indígena de Colombia (ONIC):** Calle 13, No 4–38, Bogotá; tel. (1) 284-2168; fax (1) 284-3465; e-mail onic@onic.org.co; internet www.onic.org.co; f. 1982; asscn of indigenous political and development groups.

**Partido Conservador Colombiano (PCC):** Avda 22, No 37-09, Bogotá, DC; tel. (1) 369-3923; fax (1) 369-0053; e-mail presidencia@partidoconservador.com; internet www.partidoconservador.com; f. 1849; 2.9m. mems; Pres. CARLOS HOLGUÍN SARDI; Sec.-Gen. ALVARO GUILLERMO RENDÓN LÓPEZ.

**Partido de Esperanza, Paz y Libertad (EPL):** f. 1969 as Maoist guerrilla movement Ejército Popular de Liberación; splinter group from Communist Party; abandoned armed struggle in March 1991; Leader FRANCISCO CARABALLO.

**Partido Liberal Colombiano (PL):** Avda Caracas, No 36-01, Bogotá, DC; tel. (1) 288-1138; fax (1) 288-4200; e-mail medios@partidoliberal.org.co; internet www.partidoliberal.org.co; f. 1848; divided into two factions: the official group (HERNANDO DURÁN LUSSÁN, MIGUEL PINEDO) and the independent group, Nuevo Liberalismo (New Liberalism, led by Dr ALBERTO SANTOFIMIO BOTERO, ERNESTO SAMPER PIZANO, EDUARDO MESTRE); Pres. CÉSAR GAVIRIA TRUJILLO; Sec.-Gen. JUAN CARLOS POSADA GARCÍA-PEÑA.

**Partido Nacional Cristiano (PNC):** Calle 22C, No 31-01, Bogotá, DC; tel. (1) 337-9211; fax (1) 269-3621; e-mail mision2@latino.net.co; Pres. LIÑO LEAL COLLAZOS.

**Polo Democrático Independiente (PDI):** Carrera 17A, No 37-27, Bogotá; tel. (1) 288-6188; internet www.polodemocratico.net; f. 2002 as electoral alliance, constituted as a political party in July 2003; founded by fmr mems of the Movimiento 19 de Abril (q.v.); left-wing; Leader SAMUEL MORENO ROJAS.

**Unidad Democrática de la Izquierda** (Democratic Unity of the Left): Bogotá, DC; f. 1982; left-wing coalition incorporating the following parties:

**Firmes:** Bogotá, DC; democratic party.

**Partido Comunista Colombiano (PC):** Calle 18A, No 14-56, Apdo Aéreo 2523, Bogotá, DC; tel. (1) 334-1947; fax (1) 281-8259; f. 1930; Marxist-Leninist party; Sec.-Gen. ALVARO VÁSQUEZ DEL REAL.

**Partido Socialista de los Trabajadores (PST):** Bogotá, DC; workers' socialist party; Leader María SOCORRO RAMÍREZ.

**Unión Patriótica (UP):** Carrera 13A, No 38-32, Of. 204, Bogota, DC; fax (1) 570-4400; f. 1985; Marxist party formed by the FARC—EP (see below); obtained legal status in 1986; Pres. ERNÁN PASTRANA; Exec. Sec. OVIDIO SALINAS.

The following guerrilla groups and illegal organizations were active:

**Autodefensas Campesinas de Córdoba y Urabá (ACCU):** rightwing paramilitary org.; Leaders CARLOS CASTAÑO, SALVATORE MANCUSO.

**Autodefensas Unidas de Colombia (AUC):** e-mail auc_inspeccion@yahoo.es; internet www.colombialibre.org; rightwing paramilitary org.; announced cease-fire in Dec. 2002; demobilization began in Nov. 2004; scheduled to be completed by end of 2005; c. 10,000 mems; Leader JULIÁN BOLÍVAR; Inspector-Gen. ADOLFO PAZ; Commdr-Gen. RAMÓN ISAZA; Political Dir ERNESTO BÁEZ.

**Ejército de Liberación Nacional (ELN):** internet www.eln-voces.com; Castroite guerrilla movement; f. 1964; 3,500 mems; political status recognized by the Govt in 1998; mem. of the Coordinadora Nacional Guerrilla Simón Bolívar (CNGSB); Leader NICOLÁS ROGRÍGUEZ BAUTISTA.

**Frente Popular de Liberación Nacional (FPLN):** f. 1994 by dissident mems of the ELN and the EPL.

**Fuerzas Armadas Revolucionarias de Colombia—Ejército del Pueblo (FARC—EP):** internet www.farcep.org; f. 1964, although mems active from 1949; name changed from Fuerzas Armadas Revolucionarias de Colombia to the above in 1982; fmrly military wing of the Communist Party; composed of 39 armed fronts and about 17,000 mems; political status recognized by the Govt in 1998; mem. of the Coordinadora Nacional Guerrilla Simón Bolívar (CNGSB); C-in-C Pedro Antonio Marín (alias Manuel Marulanda Vélez or 'Tirofijo') (reported dead in mid-2004); Leader of the Bolivarian Movt Guillermo León Saenz Vargas (alias Alfonso Cano).

**Movimiento de Autodefensa Obrera (MAO):** workers' self-defence movement; Trotskyite; Leader Adelaida Abadia Rey.

**Movimiento de Restauración Nacional (MORENA):** right-wing; Leader Armando Valenzuela Ruiz.

**Muerte a Secuestradores (MAS)** (Death to Kidnappers): right-wing paramilitary org.; funded by drugs-traffickers.

**Nuevo Frente Revolucionario del Pueblo:** f. 1986; faction of M-19; active in Cundinamarca region.

**Patria Libre:** f. 1985; left-wing guerrilla movement.

## Diplomatic Representation

### EMBASSIES IN COLOMBIA

**Algeria:** Carrera 11, No 93-53, Of. 302, Bogotá, DC; tel. (1) 635-0520; fax (1) 635-0531; e-mail ambalgbg@cable.net.co; Ambassador Omar Benchehida.

**Argentina:** Avda 40a, No 13-09, 16°, Apdo Aéreo 53013, Bogotá, DC; tel. (1) 288-0900; fax (1) 288-8868; e-mail ecolo@mrecic.gov.ar; Ambassador Martín Antonio Balza.

**Austria:** Edif. Fiducafe, 4°, Carrera 9, No 73-44, Bogotá, DC; tel. (1) 326-2680; fax (1) 317-7639; e-mail eaustria@cable.net.co; internet www.austria.cjb.net; Ambassador Hans Peter Glanzer.

**Belgium:** Calle 26, No 4a-45, 7°, Apdo Aéreo 3564, Bogotá, DC; tel. (1) 380-0370; fax (1) 380-0340; e-mail bogota@diplobel.org; internet www.diplobel.org/colombia; Ambassador Jean-Luc Bodson.

**Bolivia:** Transversal 14a, No 118a-26, Apdo Aéreo 96219, Santa Barbara, Bogotá, DC; tel. (1) 629-8237; fax (1) 619-4940; e-mail embolivia-bogota@rree.gov.bo; Ambassador Jesús Herman Antelo Laughlin.

**Brazil:** Calle 93, No 14-20, 8°, Bogotá, DC; tel. (1) 218-0800; fax (1) 218-8393; internet www.brasil.org.co; Ambassador María Celina de Acevedo Rodrigues.

**Canada:** Carretera 7, No 115-33, 14°, Apdo Aéreo 110067, Bogotá, DC; tel. (1) 657-9800; fax (1) 657-9912; e-mail bgota@international.gc.ca; internet www.dfait-maeci.gc.ca/colombia; Ambassador Jean-Marc Duval.

**Chile:** Calle 100, No 11b-44, Bogotá, DC; tel. (1) 214-7990; fax (1) 619-3863; e-mail echileco@colomsat.net.co; internet www.embajadadechile.com.co; Ambassador Augusto Bermudez Arancibia.

**China, People's Republic:** Carrera 16, No 98-30, Bogotá, DC; tel. (1) 622-3215; fax (1) 622-3114; e-mail chinaemb_co@mfa.gov.cn; Ambassador Wu Changsheng.

**Costa Rica:** Carrera 8, No 95-48, Bogotá, DC; tel. (1) 623-0205; fax (1) 691-8558; e-mail embacosta@andinet.com; Ambassador Melvin Alfredo Sáenz Biolley.

**Cuba:** Carrera 9, No 92-54, Bogotá, DC; tel. (1) 621-7054; fax (1) 611-4382; e-mail embacuba@cable.net.co; Ambassador Luis Hernández Ojeda.

**Czech Republic:** Carrera 7, No 115-33, Ofs 603 y 604, 6°, Bogotá, DC; tel. (1) 640-0600; fax (1) 640-0599; e-mail bogota@embassy.msv.cz; internet www.mzv.cz/bogota; Ambassador Dr Josef Rychtar.

**Dominican Republic:** Carrera 7, No 115-33, Of. 403, Torre ABN AMRO, Bogotá, DC; tel. (1) 640-0560; fax (1) 522-0102; e-mail embajado@cable.net.co; Ambassador Raúl Fernando Barrientos-Lara.

**Ecuador:** Edif. Fernando Mazuera, 7°, Calle 72, No 6-30, Bogotá, DC; tel. (1) 212-6512; fax (1) 212-6536; e-mail mecucol@cable.net.co; internet www.embajadadaecuacol.net; Ambassador Rodrigo Riofrío Machuca.

**Egypt:** Transversal 19a 101–10, Bogotá, DC; tel. (1) 256-2940; fax (1) 256-9255; e-mail embegyptbta@unete.com; Ambassador Atef Anwar Ali Hassanein.

**El Salvador:** Edif. El Nogal, Of. 503, Carrera 9a, No 80-15, Bogotá, DC; tel. (1) 349-6765; fax (1) 349-6670; e-mail elsalvadorcolombia@cable.net.co; Ambassador Joaquín Alexander Maza Martelli.

**France:** Carrera 11, No 93-12, Bogotá, DC; tel. (1) 638-1400; fax (1) 638-1430; internet www.ambafrance-co.org; Ambassador Camille Rohou.

**Germany:** Edif. World Business Port, Carrera 69, No 43b-44, 7°, Bogotá, DC; tel. (1) 423-2600; fax (1) 429-3145; e-mail embajalemana@andinet.com; internet www.embajada-alemana-bogota.de; Ambassador Matei Ion Hoffmann.

**Guatemala:** Diagonal 145a, No 32-37, Bogotá, DC; tel. (1) 636-1724; fax (1) 274-1196; e-mail embcolombia@minex.gob.gt; Ambassador Fernando Sesenna Olivero.

**Holy See:** Carrera 15, No 36-33, Apdo Aéreo 3740, Bogotá, DC (Apostolic Nunciature); tel. (1) 320-0289; fax (1) 285-1817; e-mail nunciatura@cable.net.co; Apostolic Nuncio Most Rev. Guillermo León Escobar Herrán.

**Honduras:** Calle 121, No 13a-59, Bogotá, DC; tel. (1) 215-4259; fax (1) 637-0686; e-mail emhoncol@andinet.com; Ambassador Vicente Machado Valle.

**Hungary:** Carrera 6a, No 77-46, Bogotá, DC; tel. (1) 347-1467; fax (1) 347-1469; e-mail admon@embajadahungria.org.co; internet www.embajadahungria.org.co; Ambassador Béla Bardócz.

**India:** Edif. Bancafe, Torre B, Carrera 7, No 71-21, Of. 1001, Bogotá, DC; tel. (1) 317-4865; fax (1) 317-4976; e-mail indembog@cable.net.co; internet www.embajadaindia.org; Ambassador Nilima Mitra.

**Indonesia:** Carrera 9, No 76-27, Bogotá, DC; tel. (1) 217-6738; fax (1) 326-2165; e-mail info@indonesiabogota.org.co; internet www.indonesiabogota.org.co; Ambassador Setijanto Poedjowarsito.

**Iran:** Calle 96, No 11a-16/20, Bogotá, DC; tel. (1) 218-6205; fax (1) 610-2556; e-mail embajadairan@andinet.com.co; Ambassador Abdulrahim Sadatifar.

**Israel:** Calle 35, No 7-25, 14°, Bogotá, DC; tel. (1) 288-4637; fax (1) 287-7783; e-mail embisrae@cable.net.co; Ambassador Yair Recanati.

**Italy:** Calle 93b, No 9-92, Apdo Aéreo 50901, Bogotá, DC; tel. (1) 218-6604; fax (1) 610-5886; e-mail ambasciata@ambitaliabogota.org; internet www.ambitaliabogota.org; Ambassador Francesco Camillo Peano.

**Jamaica:** Avda 19, No 106a-83, Of. 304, Apdo Aéreo 102428, Bogotá, DC; tel. (1) 612-3389; fax (1) 612-3479; e-mail emjacol@cable.net.com; Chargé d'affaires a.i. Elaine Townsend de Sánchez.

**Japan:** Carrera 7a, No 71-21, 11°, Torre B, Bogotá, DC; tel. (1) 317-5001; fax (1) 317-5007; e-mail emb.japon@etb.net.co; internet www.colombia.emb-japan.go.jp; Ambassador Wataru Hayashi.

**Korea, Republic:** Calle 94, No 9-39, Bogotá, DC; tel. (1) 616-7200; fax (1) 610-0338; e-mail embcorea@mofat.go.kr; Ambassador Park Sang-Kyoon.

**Lebanon:** Calle 74, No 12-44, Bogotá, DC; tel. (1) 212-8360; fax (1) 347-9106; e-mail emblibanco@hotmail.com; Ambassador Mounir Khreich.

**Mexico:** Edif. Teleport Business Park, Calle 114, No 9-01, Of. 204, Torre A, Bogotá, DC; tel. (1) 629-4989; fax (1) 629-5121; e-mail emcolmex@007mundo.com; internet www.sre.gob.mx/colombia; Ambassador Mario Chacón Carillo.

**Morocco:** Carrera 10, No 93-16, Bogotá, DC; tel. (1) 530-9300; fax (1) 530-8979; e-mail sifamabogot@aldato.com.co; internet www.embamarruecos.org.co; Ambassador Mohamed Maoulainine.

**Netherlands:** Carrera 13, No 93-40, 5°, Apdo Aéreo 43585, Bogotá, DC; tel. (1) 638-4200; fax (1) 623-3020; e-mail nlgovbog@etb.net.co; internet www.embajadadeholanda.org.co; Ambassador Frans B.A.M. van Haren.

**Nicaragua:** Calle 108a, No 25-42, Bogotá, DC; tel. (1) 619-8934; fax (1) 612-6050; e-mail embnicaragua@007mundo.com; Ambassador Donald Castillo Rivas.

**Norway:** Carrera 9, No 73-44, 8°, Bogotá, DC; tel. (1) 317-7851; fax (1) 317-7851; e-mail emb.bogota@mfa.no; internet www.noruega.org.co; Chargé d'affaires a.i. Tom Tyrihjell.

**Panama:** Calle 92, No 7a-40, Bogotá, DC; tel. (1) 257-5058; fax (1) 257-5067; e-mail embpacol@cable.net.co; Ambassador Carlos Ozores Typaldos.

**Paraguay:** Carrera 7, No 72-28, Of. 302, Bogotá, DC; tel. (1) 347-0322; e-mail embaparcolombia@yagua.com.py; Ambassador Felipe Robertti.

**Peru:** Calle 80a, No 6-50, Bogotá, DC; tel. (1) 257-0505; fax (1) 249-8581; e-mail lbogota@cable.net.co; Ambassador José Luis Pérez Sánchez-Cerro.

**Poland:** Calle 104a, No 23-48, Bogotá, DC; tel. (1) 214-0400; fax (1) 214-0854; e-mail polemb@cable.net.co; Ambassador Henrik Kobierowski.

**Portugal:** Carrera 12, No 93-37, Of. 302, Bogotá, DC; tel. (1) 622-1334; fax (1) 622-1134; e-mail embporbo@andinet.com; Ambassador Joaquim José Ferreira da Fonseca.

**Romania:** Carrera 7a, No 92-58, Bogotá, DC; tel. (1) 256-6438; fax (1) 256-6158; e-mail emrubog@colomsat.net.co; Ambassador Radu Urzica.

# COLOMBIA

**Russia:** Carrera 4, No 75-02, Apdo Aéreo 90600, Bogotá, DC; tel. (1) 212-1881; fax (1) 210-4694; e-mail embajadarusia@cable.net.co; internet www.colombia.mid.ru; Ambassador VLADIMIR TRUJANOVSKI.

**Spain:** Calle 92, No 12-68, Bogotá, DC; tel. (1) 622-0090; fax (1) 621-0809; e-mail embespco@correo.mae.es; Ambassador CARLOS GÓMEZ-MÚGICA SANZ.

**Sweden:** Calle 72-bis, No 5-83, 8°, Bogotá, DC; tel. (1) 325-2165; fax (1) 325-2166; e-mail embsueca@cable.net.co; internet www.swedenabroad.com/bogota; Chargé d'affaires INGEMAR CEDERBERG.

**Switzerland:** Carrera 9, No 74-08/1101, 11°, Bogotá, DC; tel. (1) 349-7230; fax (1) 349-7195; e-mail vertretung@bog.rep.admin.ch; internet www.eda.admin.ch/bogota; Ambassador THOMAS KUPFER.

**United Kingdom:** Edif. ING Barings, Carrera 9, No 76-49, 9°, Bogotá, DC; tel. (1) 317-6690; fax (1) 317-6265; e-mail britain@cable.net.co; internet www.britain.gov.co; Ambassador HAYDON WARREN-GASH.

**USA:** Calle 22D-bis, No 47-51, Apdo Aéreo 3831, Bogotá, DC; tel. (1) 315-0811; fax (1) 315-2197; internet bogota.usembassy.gov; Ambassador WILLIAM BRAUCHER WOOD.

**Uruguay:** Carrera 9A, No 80-15, 11°, Apdo Aéreo 101466, Bogotá, DC; tel. (1) 235-2748; fax (1) 248-3734; e-mail urucolom@007mundo.com; Ambassador EDUARDO CÉSAR AÑÓN NOCETI.

**Venezuela:** Carrera 11, No 87-51, 5°, Bogotá, DC; tel. (1) 640-1213; fax (1) 640-1242; e-mail embajada@embaven.org.co; internet www.embaven.org.co; Ambassador CARLOS RODOLFO SANTIAGO RAMÍREZ.

## Judicial System

The constitutional integrity of the State is ensured by the Constitutional Court. The Constitutional Court is composed of nine judges who are elected by the Senate for eight years. Judges of the Constitutional Court are not eligible for re-election.

**President of the Constitutional Court:** Dr CLARA INÉS VARGAS HERNÁNDEZ.

**Judges of the Constitutional Court:** JAIME ARAUJO RENTERÍA, ALFREDO BELTRÁN SIERRA, MANUEL JOSÉ CEPEDA ESPIÑOSA, JAIME CÓRDOBA TRIVIÑO, RODRIGO ESCOBAR GIL, MARCO GERARDO MONROY CABRA, EDUARDO MONTEALEGRE LYNETT, ALVARO TAFUR GAVIS.

The ordinary judicial integrity of the State is ensured by the Supreme Court of Justice. The Supreme Court of Justice is composed of the Courts of Civil and Agrarian, Penal and Laboral Cassation. Judges of the Supreme Court of Justice are selected from the nominees of the Higher Council of Justice and serve an eight-year term of office which is not renewable.

**Prosecutor-General:** MARIO IGUARÁN ARANA.

**Attorney-General:** EDGARDO JOSÉ MAYA VILLAZÓN.

### SUPREME COURT OF JUSTICE

#### Supreme Court of Justice

Carrera 7A, No 27-18, Bogotá, DC; fax (1) 334-8745; internet www.ramajudicial.gov.co.

**President:** CARLOS ISAAC NADER.

**Court of Civil and Agrarian Cassation** (seven judges): President JORGE ANTONIO CASTILLO RUGELES.

**Court of Penal Cassation** (nine judges): President YESID RAMIREZ BASTIDAS.

**Court of Laboral Cassation** (seven judges): President FERNANDO VÁSQUEZ BOTERO.

## Religion

### CHRISTIANITY

#### The Roman Catholic Church

At 31 December 2003 Roman Catholicism had 38.1m. adherents in Colombia, equivalent to 86.3% of the population. Colombia comprises 13 archdioceses, 51 dioceses and 10 apostolic vicariates.

#### Bishops' Conference

Conferencia Episcopal de Colombia, Carrera 47, No 84-85, Apdo Aéreo 7448, Bogotá, DC; tel. (1) 311-4277; fax (1) 311-5575; e-mail colcec@cable.net.co.

*Directory*

f. 1978; statutes approved 1996; Pres. Cardinal PEDRO RUBIANO SÁENZ (Archbishop of Santafé de Bogotá).

**Archbishop of Barranquilla:** RUBÉN SALAZAR GÓMEZ, Carrera 45, No 53-122, Apdo Aéreo 1160, Barranquilla 4, Atlántico; tel. (5) 349-1145; fax (5) 349-1530; e-mail arquidio@arquidiocesibaq.org.co.

**Archbishop of Bucaramanga:** VÍCTOR MANUEL LÓPEZ FORERO, Calle 33, No 21-18, Bucaramanga, Santander; tel. (7) 642-4387; fax (7) 642-1361; e-mail sarqdbu@col1.telecom.com.

**Archbishop of Cali:** JUAN FRANCISCO SARASTI JARAMILLO, Carrera 4, No 7-17, Apdo Aéreo 8924, Cali, Valle del Cauca; tel. (2) 889-0562; fax (2) 83-7980; e-mail jsarasti@andinet.com.

**Archbishop of Cartagena:** CARLOS JOSÉ RUISECO VIEIRA, Apdo Aéreo 400, Cartagena; tel. (5) 664-5308; fax (5) 664-4974; e-mail arzoctg@telecartagena.com.

**Archbishop of Ibagué:** FLAVIO CALLE ZAPATA, Calle 10, No 2-58, Ibagué, Tolima; tel. (82) 61-1680; fax (82) 63-2681; e-mail arguibague@hotmail.com.

**Archbishop of Manizales:** FABIO BETANCUR TIRADO, Carrera 23, No 19-22, Manizales, Caldas; tel. (68) 84-0114; fax (68) 82-1853; e-mail arquiman@epm.net.co.

**Archbishop of Medellín:** ALBERTO GIRALDO JARAMILLO, Calle 57, No 49-44, 3°, Medellín; tel. (4) 251-7700; fax (4) 251-9395; e-mail arquidiomed@epm.net.co.

**Archbishop of Nueva Pamplona:** GUSTAVO MARTÍNEZ FRÍAS, Carrera 5, No 4-87, Nueva Pamplona; tel. (75) 68-1329; fax (75) 68-4540; e-mail gumafri@hotmail.com.

**Archbishop of Popayán:** IVÁN ANTONIO MARÍN-LÓPEZ, Calle 5, No 6-71, Apdo Aéreo 593, Popayán; tel. (928) 24-1710; fax (928) 24-0101; e-mail ivanarzo@emtel.net.co.

**Archbishop of Santa Fe de Antioquia:** IGNACIO GÓMEZ ARISTIZÁBAL, Plazuela Martínez Pardo, No 12-11, Santa Fe de Antioquia; tel. (94) 853-1155; fax (94) 853-1596; e-mail arquistafe@edatel.net.co.

**Archbishop of Santafé de Bogotá:** Cardinal PEDRO RUBIANO SÁENZ, Carrera 7A, No 10-20, Bogotá, DC; tel. (1) 350-5511; fax (1) 350-7290; e-mail cancilleria@arquidiocesisbogota.org.co.

**Archbishop of Tunja:** LUIS AUGUSTO CASTRO QUIROGA, Calle 17, No 9-85, Apdo Aéreo 1019, Tunja, Boyacá; tel. (987) 742-2093; fax (987) 743-3130; e-mail arquidio@telecom.com.co.

**Archbishop of Villavicencio:** JOSÉ OCTAVIO RUIZ ARENAS, Carrera 39, No 34-19, Apdo Aéreo 2401, Villavicencio, Meta; tel. (8) 663-0337; fax (8) 665-3200; e-mail diocesisvillavicencio@andinet.com.

#### The Anglican Communion

Anglicans in Colombia are members of the Episcopal Church in the USA.

**Bishop of Colombia:** Rt Rev. BERNARDO MERINO BOTERO, Carrera 6, No 49-85, Apdo Aéreo 52964, Bogotá, DC; tel. (1) 288-3167; fax (1) 288-3248; there are 3,500 baptized mems, 2,000 communicant mems, 29 parishes, missions and preaching stations; 5 schools and 1 orphanage; 8 clergy.

#### Protestant Churches

**Baptist Convention:** Medellín; tel. (4) 38-9623; Pres. RAMÓN MEDINA IBÁÑEZ; Exec. Sec. Rev. RAMIRO PÉREZ HOYOS.

**Iglesia Evangélica Luterana de Colombia:** Calle 75, No 20-54, Apdo Aéreo 51538, Bogotá, DC; tel. (1) 212-5735; fax (1) 212-5714; e-mail ofcentral@ielco.org; 3,000 mems; Pres. Bishop SIGIFRIDO DANIEL BUITRAGO PACHÓN.

### BAHÁ'Í FAITH

**National Spiritual Assembly of the Bahá'ís of Colombia:** Apdo Aéreo 51387, Bogotá, DC; tel. (1) 268-1658; fax (1) 268-1665; e-mail bahaicol@colombianet.net; adherents in 1,013 localities.

### JUDAISM

There is a community of about 25,000 with 66 synagogues.

## The Press

### DAILIES

#### Bogotá, DC

**El Espacio:** Carrera 61, No 45-35, Apdo Aéreo 80111, Avda El Dorado, Bogotá, DC; tel. (1) 425-1570; fax (1) 410-4595; internet

COLOMBIA                                                                                      *Directory*

www.elespacio.com.co; f. 1965; evening; Dir Jaime Ardila Casamitjana; circ. 159,000.

**El Nuevo Siglo:** Calle 45A, No 102-02, Apdo Aéreo 5452, Bogotá, DC; tel. (1) 413-9200; fax (1) 413-8547; internet www.elnuevosiglo.com.co; f. 1925; Conservative; Dirs Juan Pablo Uribe, Juan Gabriel Uribe; circ. 68,000.

**La República:** Calle 46, No 103-59, Bogotá, DC; tel. (1) 413-5077; fax (1) 413-3725; e-mail diario@larepublica.com.co; internet www.la-republica.com.co; f. 1953; morning; finance and economics; Dir Rodrigo Ospina Hernández; Editor Jorge Emilio Sierra M.; circ. 55,000.

**El Tiempo:** Avda El Dorado, No 59-70, Apdo Aéreo 3633, Bogotá, DC; tel. (1) 294-0100; fax (1) 410-5088; internet www.eltiempo.com; f. 1911; morning; Liberal; Dir Enrique Santos Calderón; Editor Francisco Santos; circ. 265,118 (weekdays), 536,377 (Sundays).

### Barranquilla, Atlántico

**El Heraldo:** Calle 53B, No 46-25, Barranquilla, Atlántico; tel. (5) 371-5000; fax (5) 371-5091; internet www.elheraldo.com.co; f. 1933; morning; Liberal; Gen. Editor Hugo Penton; circ. 70,000.

**La Libertad:** Carrera 53, No 55-166, Barranquilla, Atlántico; tel. (5) 349-1175; e-mail libertad@metrotel.net.co; internet www.lalibertad.com.co; Liberal; Dir Roberto Esper Rebaje; circ. 25,000.

**El Tiempo Caribe:** Carrera 50B, No 41-18, Barranquilla, Atlántico; tel. (5) 379-1510; fax (5) 341-7715; e-mail orlgam@eltiempo.com.co; internet www.eltiempo.com; f. 1956; daily; Liberal; Dir Orlanda Gamboa; circ. 45,000.

### Bucaramanga, Santander del Sur

**El Frente:** Calle 35, No 12-22, Apdo Aéreo 665, Bucaramanga, Santander del Sur; tel. (7) 42-5369; fax (7) 33-4541; f. 1942; morning; Conservative; Dir Rafael Serrano Prada; circ. 10,000.

**Vanguardia Liberal:** Calle 34, No 13-42, Bucaramanga, Santander del Sur; tel. (7) 680-0700; fax (7) 630-2443; e-mail vanglibe@colomsat.net.co; internet www.vanguardia.com; f. 1919; morning; Liberal; Sunday illustrated literary supplement and women's supplement; Dir Eduardo Muñoz Serpa; Editor Marco Antonio Ibarra Penalosa; circ. 48,000.

### Cali, Valle del Cauca

**Occidente:** Calle 12, No 5-22, Cali, Valle del Cauca; tel. (2) 895-9756; fax (2) 884-6572; e-mail occidente@cali.cercol.net.co; f. 1961; morning; Conservative; Dir Jorge Enrique Almario; circ. 25,000.

**El País:** Carrera 2A, No 24-46, Apdo Aéreo 4766, Cali, Valle del Cauca; tel. (2) 883-1183; fax (2) 883-5014; e-mail diario@elpais-cali.com; internet www.elpais-cali.com; f. 1950; Conservative; Dir Eduardo Fernández de Soto; Gen. Man. María Elvira Domínguez; circ. 60,000 (weekdays), 120,000 (Saturdays), 108,304 (Sundays).

**El Pueblo:** Avda 3A, Norte 35-N-10, Cali, Valle del Cauca; tel. (2) 68-8110; morning; Liberal; Dir Luis Fernando Londoño Capurro; circ. 50,000.

### Cartagena, Bolívar

**El Universal:** Pie del Cerro Calle 30, No 17-36, Cartagena, Bolívar; tel. (5) 664-7310; fax (5) 666-1964; e-mail director@eluniversal.com.co; internet www.eluniversal.com.co; f. 1948; daily; Liberal; Editor Sonia Gedeon; Dir Pedro Luis Mogolón; circ. 167,000.

### Cúcuta, Norte de Santander

**La Opinión:** Avda 4, No 16-12, Cúcuta, Norte de Santander; tel. (75) 71-9999; fax (75) 71-7869; e-mail gerencialaopinion@coll.telecom.com.co; f. 1960; morning; Liberal; Dir Dr José Eustorgio Colmenares Ossa; circ. 27,000.

### Manizales, Caldas

**La Patria:** Carrera 20, No 21-51, Apdo Aéreo 70, Manizales, Caldas; tel. (68) 84-2460; fax (68) 84-7158; e-mail lapatria@lapatria.com; f. 1921; morning; Independent; Dir Dr Luis José Restrepo Restrepo; circ. 22,000.

### Medellín, Antioquia

**El Colombiano:** Carrera 48, No 30 sur-119, Apdo Aéreo 80636, Medellín, Antioquia; tel. (4) 331-5252; fax (4) 331-4858; e-mail elcolombiano@elcolombiano.com.co; internet www.elcolombiano.com.co; f. 1912; morning; Conservative; Dir Ana Mercedes Gómez Martínez; circ. 90,000.

**El Mundo:** Calle 53, No 74-50, Apdo Aéreo 53874, Medellín, Antioquia; tel. (4) 264-2800; fax (4) 264-3729; e-mail elmundo@elmundo.com; internet www.elmundo.com; f. 1979; Dir Guillermo Gaviria Echeverri; Man. Carlos Alberto Gil Valencia; circ. 20,000.

### Montería, Córdoba

**El Meridiano de Córdoba:** Avda Circunvalar, No 38-30, Montería, Córdoba; tel. (47) 82-6888; fax (47) 82-6996; e-mail publicidad@elmeridianodecordoba.com; internet www.elmeridianodecordoba.com; f. 1995; morning; Dir William Enrique Salleg Taboada; circ. 18,000.

### Neiva

**Diario del Huila:** Calle 8A, No 6-30, Neiva; tel. (8) 871-2542; fax (8) 871-2543; internet www.diariodelhuila.com; f. 1966; Gen. Editor Alfredo Rubio; circ. 12,000.

### Pasto, Nariño

**El Derecho:** Calle 20, No 26-20, Pasto, Nariño; tel. (277) 2170; f. 1928; Conservative; Pres. Dr José Elías del Hierro; Dir Eduardo F. Mazuera; circ. 12,000.

### Pereira, Risaralda

**Diario del Otún:** Carrera 8A, No 22-75, Apdo Aéreo 2533, Pereira, Risaralda; tel. (63) 51313; fax (63) 324-1900; e-mail eldiario@interco.net.co; internet www.eldiario.com.co; f. 1982; Financial Dir Javier Ignacio Ramírez Múnera; circ. 30,000.

**El Imparcial:** km 11 vía Pereira-Armenia, El Jordán, Pereira, Risaralda; tel. (63) 25-9935; fax (63) 25-9934; f. 1948; morning; Dir Zahur Klemath Zapata; circ. 15,000.

**La Tarde:** Carrera 9A, No 20-54, Pereira, Risaralda; tel. (63) 335-4666; fax (63) 335-4832; internet www.latarde.com; f. 1975; evening; Dir Sonia Díaz Mantilla; circ. 30,000.

### Popayán, Cauca

**El Liberal:** Carrera 3, No 2-60, Apdo Aéreo 538, Popayán, Cauca; tel. (28) 24-2418; fax (28) 23-3888; f. 1938; Man. Carlos Alberto Cabal Jiménez; circ. 6,500.

### Santa Marta, Magdalena

**El Informador:** Calle 21, No 5-06, Santa Marta, Magdalena; f. 1921; Liberal; Dir José B. Vives; circ. 9,000.

### Villavicencio, Meta

**Clarín del Llano:** Villavicencio, Meta; tel. (866) 23207; Conservative; Dir Elías Matus Torres; circ. 5,000.

## PERIODICALS

### Bogotá, DC

**Antena:** Bogotá, DC; television, cinema and show business; circ. 10,000.

**Arco:** Carrera 6, No 35-39, Apdo Aéreo 8624, Bogotá, DC; tel. (1) 285-1500; f. 1959; monthly; history, philosophy, literature and humanities; Dir Alvaro Valencia Tovar; circ. 10,000.

**ART NEXUS/Arte en Colombia:** Carrera 5, No 67-19, Apdo Aéreo 90193, Bogotá, DC; tel. (1) 312-9332; fax (1) 312-9252; e-mail artnexus@artnexus.com; f. 1976; quarterly; Latin American art, photography, visual arts; editions in English and Spanish; Dir Celia Sredni de Birbragher; Exec. Editor Ivonne Pini; circ. 26,000.

**El Campesino:** Carrera 39A, No 15-11, Bogotá, DC; f. 1958; weekly; cultural; Dir Joaquín Gutiérrez Macías; circ. 70,000.

**Consigna:** Diagonal 34, No 5-11, Bogotá, DC; tel. (1) 287-1157; fortnightly; Turbayista; Dir (vacant); circ. 10,000.

**Coyuntura Económica:** Calle 78, No 9-91, Apdo Aéreo 75074, Bogotá, DC; tel. (1) 312-5300; fax (1) 212-6073; e-mail bibliote@fedesarrollo.org.co; f. 1970; quarterly; economics; published by Fundación para Educación Superior y el Desarrollo (FEDESARROLLO); Editor María Angélica Arbelaez; circ. 1,500.

**Cromos Magazine:** Calle 70A, No 7-81, Apdo Aéreo 59317, Bogotá, DC; f. 1916; weekly; illustrated; general news; Dir Alberto Zalamea; circ. 102,000.

**As Deportes:** Calle 20, No 4-55, Bogotá, DC; f. 1978; sports; circ. 25,000.

**Economía Colombiana:** Dirección Economía y Finanzas, San Agustín 6-45, Of. 126A, Bogotá, DC; tel. (1) 282-4597; fax (1) 282-3737; f. 1984; published by Contraloría General de la República; 6 a year; economics.

**El Espectador:** Carrera 68, No 23-71, Apdo Aéreo 3441, Bogotá, DC; tel. (1) 294-5555; fax (1) 260-2323; e-mail redactor@elespectador.com; internet www.elespectador.com; f. 1887 as a daily paper, published weekly from 2001; Editor Fidel Cano; circ. 200,000.

**Estrategia:** Carrera 4A, 25A-12B, Bogotá, DC; monthly; economics; Dir Rodrigo Otero.

# COLOMBIA

**Guión:** Carrera 16, No 36-89, Apdo Aéreo 19857; Bogotá, DC; tel. (1) 232-2660; f. 1977; weekly; general; Conservative; Dir JUAN CARLOS PASTRANA; circ. 35,000.

**Hit:** Calle 20, No 4-55, Bogotá, DC; cinema and show business; circ. 20,000.

**Informe Financiero:** Dirección Economía y Finanzas, San Agustín 6-45, Of. 126A, Bogotá, DC; tel. (1) 282-4597; fax (1) 282-3737; published by Contraloría General de la República; monthly; economics.

**Insurrección:** internet www.eln-voces.com; f. 1998; fortnightly; fmrly Correo de Magdalena; organ of the Ejército de Liberación Nacional (ELN).

**Menorah:** Apdo Aéreo 9081, Bogotá, DC; tel. (1) 611-2014; f. 1950; independent monthly review for the Jewish community; Dir ELIÉCER CELNIK; circ. 10,000.

**Nueva Frontera:** Carrera 7A, No 17-01, 5°, Bogotá, DC; tel. (1) 334-3763; f. 1974; weekly; politics, society, arts and culture; Liberal; Dir CARLOS LLERAS RESTREPO; circ. 23,000.

**Pluma:** Apdo Aéreo 12190, Bogotá, DC; monthly; art and literature; Dir (vacant); circ. 70,000.

**Que Hubo:** Bogotá, DC; weekly; general; Editor CONSUELO MONTEJO; circ. 15,000.

**Resistencia:** e-mail elbarcinocolombia@yahoo.com; internet www.farcep.org/resistencia; f. 1993; quarterly; organ of the Fuerzas Armadas Revolucionarias de Colombia—Ejército del Pueblo (FARC—EP); Dir RAÚL REYES.

**Revista Escala:** Calle 30, No 17-752, Bogotá, DC; tel. (1) 287-8200; fax (1) 285-9882; e-mail escala@col-online.com; f. 1962; fortnightly; architecture; Dir DAVID SERNA CÁRDENAS; circ. 18,000.

**Revista Diners:** Calle 85, No 18-32, 6°, Bogotá, DC; tel. (1) 636-0508; fax (1) 623-1762; e-mail diners@cable.net.co; f. 1963; monthly; Dir GERMÁN SANTAMARÍA; circ. 110,000.

**Semana:** Calle 93B, No 13-47, Bogotá, DC; tel. (1) 622-2277; fax (1) 621-0475; general; Pres. FELIPE LÓPEZ CABALLERO.

**Síntesis Económica:** Calle 70A, No 10-52, Bogotá, DC; tel. (1) 212-5121; fax (1) 212-8365; f. 1975; weekly; economics; Dir FÉLIX LAFAURIE RIVERA; circ. 16,000.

**Tribuna Médica:** Calle 8B, No 68A-41, y Calle 123, No 8-20, Bogotá, DC; tel. (1) 262-6085; fax (1) 262-4459; f. 1961; monthly; medical and scientific; Editor JACK ALBERTO GRIMBERG; circ. 50,000.

**Tribuna Roja:** Bogotá, DC; tel. (1) 243-0371; e-mail tribojar@moir.org.co; internet www.moir.org.co/tribuna/tribuna.htm; f. 1971; quarterly; organ of the MOIR (pro-Maoist Communist party); Dir CARLOS NARANJO; circ. 300,000.

**Vea:** Calle 20, No 4-55, Bogotá, DC; weekly; popular; circ. 90,000.

**Voz La Verdad del Pueblo:** Carrera 8, No 19-34, Ofs 310–311, Bogotá, DC; tel. (1) 284-5209; fax (1) 342-5041; weekly; left-wing; Dir CARLOS A. LOZANO G.; circ. 45,000.

## NEWS AGENCIES

**Ciep—El País:** Carrera 16, No 36-35, Bogotá, DC; tel. (1) 232-6816; fax (1) 288-0236; Dir JORGE TÉLLEZ.

**Colprensa:** Diagonal 34, No 5-63, Apdo Aéreo 20333, Bogotá, DC; tel. (1) 287-2200; fax (1) 285-5915; e-mail colpre@elsitio.net.co; f. 1980; Dir ROBERTO VARGAS GALVIS.

### Foreign Bureaux

**Agence France-Presse (AFP):** Carrera 5, No 16-14, Of. 807, Apdo Aéreo 4654, Bogotá, DC; tel. (1) 281-8613; Dir MARIE SANZ.

**Agencia EFE** (Spain): Carrera 16, No 39A-69, Apdo Aéreo 16038, Bogotá, DC; tel. (1) 285-1576; fax (1) 285-1598; e-mail efecol@efebogota.com.co; Bureau Chief ENRIQUE IBAÑEZ.

**Agencia de Noticias Nueva Colombia (ANNCOL):** Vattenledningsvägen 47, 126 34 Hägersten, Stockholm, Sweden; e-mail news@anncol.com; internet www.anncol.org; f. 1996; extreme left-wing news agency reporting on Colombian political and military affairs; founded in Sweden by expatriate dissident Colombians and sympathizers of the Fuerzas Armadas Revolucionarias de Colombia—Ejército del Pueblo (FARC—EP); Pres. of Editorial Bd LEIF LARSEN; Sr Editor ROBERTO GUTIÉRREZ.

**Agenzia Nazionale Stampa Associata (ANSA)** (Italy): Carrera 4, No 67-30, Apdo Aéreo 16077, Bogotá, DC; tel. (1) 211-9617; fax (1) 212-5409; Bureau Chief ALBERTO ROJAS MORALES.

**Associated Press (AP)** (USA): Transversal 14, No 122-36, Apdo Aéreo 093643, Bogotá, DC; tel. (1) 619-3487; fax (1) 213-8467; e-mail apbogota@bigfoot.com; Bureau Chief FRANK BAJAK.

**Central News Agency Inc.** (Taiwan): Carrera 13A, No 98-34, Bogotá, DC; tel. (1) 25-6342; Correspondent CHRISTINA CHOW.

**Deutsche Presse-Agentur (dpa)** (Germany): Carrera 15, No 76-60, Of. 302, Bogotá, DC; tel. (1) 618-3788; fax (1) 635-2516; e-mail presse@hbg.dpa.de; internet www.dpa.de; Correspondent RODRIGO RUIZ TOVAR.

**Informatsionnoye Telegrafnoye Agentstvo Rossii—Telegrafnoye Agentstvo Suverennykh Stran (ITAR—TASS)** (Russia): Calle 20, No 7-17, Of. 901, Bogotá, DC; tel. (1) 243-6720; Correspondent GENNADII KOCHUK.

**Inter Press Service (IPS)** (Italy): Calle 19, No 3-50, Of. 602, Apdo Aéreo 7739, Bogotá, DC; tel. (1) 341-8841; fax (1) 334-2249; Correspondent MARÍA ISABEL GARCÍA NAVARRETE.

**Prensa Latina:** Carrera 3, No 21-46, Apdo Aéreo 30372, Bogotá, DC; tel. (1) 282-4527; fax (1) 281-7286; Bureau Chief FAUSTO TRIANA.

**Reuters** (United Kingdom): Calle 94A, No 13-34, 4°, Apdo Aéreo 29848, Bogotá, DC; tel. (1) 634-4090; fax (1) 610-7733; e-mail bogota.newsroom@reuters.com; Bureau Chief JASON WEBB.

**Xinhua (New China) News Agency** (People's Republic of China): Calle 74, No 4-26, Apdo Aéreo 501, Bogotá, DC; tel. (1) 211-5347; Dir HOU YAOQI.

### PRESS ASSOCIATIONS

**Asociación Colombiana de Periodistas:** Avda Jiménez, No 8-74, Of. 510, Bogotá, DC; tel. (1) 243-6056.

**Asociación Nacional de Diarios Colombianos (ANDIARIOS):** Calle 61, No 5-20, Apdo Aéreo 13663, Bogotá, DC; tel. (1) 212-8694; fax (1) 212-7894; f. 1962; 30 affiliated newspapers; Pres. LUIS MIGUEL DE BEDOUT; Vice-Pres. LUIS FERNANDO BAENA.

**Asociación de la Prensa Extranjera:** Pedro Meléndez, No 87-93, Bogotá, DC; tel. (1) 288-3011.

**Círculo de Periodistas de Bogotá, DC (CPB):** Calle 26, No 13A-23, 23°, Bogotá, DC; tel. (1) 282-4217; Pres. MARÍA TERESA HERRÁN.

# Publishers

## BOGOTÁ, DC

**Comunicadores Técnicos Ltda:** Carrera 18, No 46-58, Apdo Aéreo 28797, Bogotá, DC; technical; Dir PEDRO P. MORCILLO.

**Ediciones Aula XXI:** Calle 19, No 44-10, Bogotá, DC; tel. (1) 574-3990; fax (1) 244-6129; e-mail gerencia@edicionesaulaxxi.com; internet www.edicionesaulaxxi.com; general interest, reference, fiction and educational; Legal Rep. CARLOS MARIO ESCOBAR BRAVO.

**Ediciones Cultural Colombiana Ltda:** Calle 72, No 16-15 y 16-21, Apdo Aéreo 6307, Bogotá, DC; tel. (1) 217-6529; fax (1) 217-6570; f. 1951; textbooks; Dir JOSÉ PORTO VÁSQUEZ.

**Ediciones Gaviota:** Transversal 43, No 99-13, Bogotá, DC; tel. (1) 613-6650; fax (1) 613-9117; e-mail gaviotalibros@edicionesgaviota.com.co.

**Ediciones Lerner Ltda:** Calle 8A, No 68A-41, Apdo Aéreo 8304, Bogotá, DC; tel. (1) 420-0650; fax (1) 262-4459; f. 1959; general; Commercial Man. FABIO CAICEDO GÓMEZ.

**Ediciones Modernas:** Carrera 41A, No 22F, Bogotá, DC; tel. (1) 269-0072; fax (1) 244-0706; e-mail edimodernas@edimodernas.com.co; internet www.empresario.com.co/edimodernas; f. 1991; juvenile.

**Editora Cinco, SA:** Calle 61, No 13-23, 7°, Apdo Aéreo 15188, Bogotá, DC; tel. (1) 285-6200; recreation, culture, textbooks, general; Man. PEDRO VARGAS G.

**Editorial Cypres Ltda:** Carrera 15, No 80-36, Of. 302, Bogotá, DC; tel. (1) 691-0578; fax (1) 636-3824; e-mail cypres@etb.net.co; general interest and educational.

**Editorial El Globo, SA:** Calle 16, No 4-96, Apdo Aéreo 6806, Bogotá, DC.

**Editorial Hispanoámerica:** Carrera 56B, No 45-27, Bogotá, DC; tel. (1) 221-3020; fax (1) 315-5587; e-mail info@hispanoamerica.com.co; internet www.hispanoamerica.com.co; f. 1984; materials for primary education; Man. GABY TERESA CORTÉS; Editor ALVARO PINZÓN.

**Editorial Paulinas:** Calle 161A, No 31-50, Bogotá, DC; tel. (1) 522-0828; fax (1) 671-0992; e-mail ventasp@paulinas.org.co; internet www.paulinas.org.co; Christian and self-help.

**Editorial Presencia, Ltda:** Calle 23, No 24-20, Apdo Aéreo 41500, Bogotá, DC; tel. (1) 269-2188; fax (1) 269-6830; textbooks, tradebooks; Gen. Man. MARÍA UMAÑA DE TANCO.

**Editorial San Pablo:** Carrera 46, No 22A–90, Quintaparedes, Apdo Aéreo 080152, Bogotá, DC; tel. (1) 368-2099; fax (1) 244-4383; e-mail

# COLOMBIA
*Directory*

editorial@sanpablo.com.co; internet www.sanpablo.com.co; f. 1914; religion (Catholic); Editorial Dir P. VICENTE MIOTTO.

**Editorial Temis SA:** Calle 17, No 68D, 46 PBX, Bogotá, DC; tel. (1) 424-7855; fax (1) 292-5801; e-mail editorial@editorialtemis.com; internet www.editorialtemis.com; f. 1951; law, sociology, politics; Man. Dir JORGE GUERRERO.

**Editorial Voluntad, SA:** Carrera 7A, No 24-89, 24°, Bogotá, DC; tel. (1) 286-0666; fax (1) 286-5540; e-mail voluntad@colomsat.net.co; f. 1930; school books; Pres. GASTÓN DE BEDOUT.

**Fondo de Cultura Económica:** Calle 16, No 80-18, Bogotá, DC; tel. (1) 531-2288; fax (1) 531-1322; internet www.fce.com.co; f. 1934; academic; Dir CONSUELO SÁIZAR GUERRERO.

**Fundación Centro de Investigación y Educación Popular (CINEP):** Carrera 5A, No 33A-08, Apdo Aéreo 25916, Bogotá, DC; tel. (1) 285-8977; fax (1) 287-9089; f. 1977; education and social sciences; Man. Dir FRANCISCO DE ROUX.

**Instituto Caro y Cuervo:** Carrera 11, No 64-37, PBX 3456004, Apdo Aéreo 51502, Bogotá, DC; fax (1) 217-0243; e-mail direcciongeneral@caroycuervo.gov.co; internet www.caroycuervo .gov.co; f. 1942; philology, general linguistics and reference; Man. Dir HERNANDO CABARCAS ANTEQUERA; Gen. Sec. LILIANA RIVERA ORJUELA.

**Inversiones Cromos SA:** Calle 70A, No 7-81, Apdo Aéreo 59317, Bogotá, DC; tel. (1) 217-1754; fax (1) 211-2642; f. 1916; Dir ALBERTO ZALAMEA; Gen. Man. JORGE EDUARDO CORREA ROBLEDO.

**Legis, SA:** Avda El Dorado, No 81-10, Apdo Aéreo 98888, Bogotá, DC; tel. (1) 263-4100; fax (1) 295-2650; e-mail jcastrof@legis.com.co; internet www.legis.com.co; f. 1952; economics, law, general; Man. JUAN ALBERTO CASTRO.

**McGraw Hill Interamericana, SA:** Avda Américas, No 46-41, Apdo Aéreo 81078, Bogotá, DC; tel. (1) 406-9000; fax (1) 245-4786; university textbooks; Dir-Gen. CARLOS G. MÁRQUEZ.

**Publicar, SA:** Avda 68, No 75A-50, 4°, Centro Comercial Metrópolis, Apdo Aéreo 8010, Bogotá, DC; tel. (1) 225-5555; fax (1) 225-4015; e-mail m-navia@publicar.com; internet www.publicar.com; f. 1954; directories; CEO MARÍA SOL NAVIA.

**Siglo del Hombre Editores Ltda:** Carrera 32, No 25-46, Bogotá, DC; tel. (1) 337-7700; fax (1) 337-7665; e-mail siglodelhombre@sky .net.co; f. 1992; arts, politics, anthropology, history, fiction, etc.; Gen. Man. EMILIA FRANCO DE ARCILA.

**Tercer Mundo Editores SA:** Transversal 2A, No 67-27, Apdo Aéreo 4817, Bogotá, DC; tel. (1) 255-1539; fax (1) 212-5976; e-mail tmundoed@polcola.com.co; f. 1963; social sciences; Pres. SANTIAGO POMBO VEJARANO.

**Thomson PLM:** Calle 98, No 19A-21, Apdo Aéreo 52998, Bogotá, DC; tel. (1) 257-4400; fax (1) 616-7620; internet www.plmlatina.com; medical; Commerical Dir DANILO SÁNCHEZ.

**VillegasEditores:** Avda 82, No 11-50, Interior 3, Bogotá, DC; tel. (1) 616-1788; fax (1) 616-0020; internet www.villegaseditor.com; f. 1985; illustrated and scholarly.

## ASSOCIATIONS

**Cámara Colombiana del Libro:** Carrera 17A, No 37-27, Apdo Aéreo 8998, Bogotá, DC; tel. (1) 288-6188; fax (1) 287-3320; e-mail camlibro@camlibro.com.co; internet www.camlibro.com.co; f. 1951; Pres. ENRIQUE GONZÁLEZ VILLA; Exec. Dir JULIANA CÁLAD; 120 mems.

**Colcultura:** Biblioteca Nacional de Colombia, Calle 24, No 5-60, Apdo Aéreo 27600, Bogotá, DC; tel. (1) 282-8656; fax (1) 341-4028; e-mail mgiraldo@mincultura.go.co; Dir ISADORA DE NORDEN.

**Fundalectura:** Avda 40, No 16-46, Bogotá, DC; tel. (1) 320-1511; fax (1) 287-7071; e-mail contactenos@fundalectura.org.co; internet www.fundalectura.org.co; Exec. Dir CARMEN BARVO.

# Broadcasting and Communications

## GOVERNMENT AGENCIES

**Comisión de Regulación de Telecomunicaciones (CRT):** Bogotá, DC; internet www.crt.gov.co; f. 2000; regulatory body.

**Instituto Nacional de Radio y Televisión (INRAVISION):** Centro Administrativo Nacional (CAN), Avda El Dorado, Bogotá, DC; tel. (1) 222-0700; fax (1) 222-0080; internet www.inravision.gov .co; f. 1954; govt-run TV and radio broadcasting network; educational and commercial broadcasting; Dir GILBERTO SAMPER RAMIREZ VALBUENA.

**Ministerio de Comunicaciones, Dirección de Telecomunicaciones:** Edif. Murillo Toro, Carreras 7A y 8A, entre Calle 12A y 13, Apdo Aéreo 14515, Bogotá, DC; tel. (1) 344-3460; fax (1) 286-1185; internet www.mincomunicaciones.gov.co; broadcasting authority; Dir MARTHA ELENA PINTO DE HART (Minister of Communications).

## TELECOMMUNICATIONS

**BellSouth Colombia, SA:** Calle 100, No 7-33, Edif. Capital Tower, Bogotá, DC; tel. (1) 650-0000; internet www.bellsouth.com.co; f. 1994; owned by Telefónica Móviles, SA (Spain); mobile telephone services.

**Celumóvil SA:** Calle 71A, No 6-30, 18°, Bogotá, DC; tel. (1) 346-1666; fax (1) 211-2031; Sec.-Gen. CARLOS BERNARDO CARREÑO R.

**Colombia Telecomunicaciones, SA (TELECOM):** Calle 23, No 13-49, Bogotá, DC; tel. (1) 286-0077; fax (1) 282-8768; internet www .telecom.com.co; f. June 2003 following dissolution of Empresa Nacional de Telecomunicaciones (f. 1947); national telecommunications enterprise; Pres. ALFONSO GÓMEZ PALACIO.

**Empresa de Telecomunicaciones de Bogotá, SA (ETB):** Carrera 8, No 20-56, 3°–9°, Bogotá, DC; tel. (1) 242-3483; fax (1) 242-2127; e-mail adrimara@etb.com.co; internet www.etb.com.co; Bogotá telephone co; partially privatized in May 2003; Pres. RAFAEL ANTONIO ORDUZ MEDINA; Gen. Man. ADRIANA MARTÍNEZ MENDIETA.

## BROADCASTING

### Radio

The principal radio networks are as follows:

**Cadena Melodía de Colombia:** Calle 45, No 13-70, Bogotá, DC; tel. (1) 323-1500; fax (1) 288-4020; Pres. EFRAÍN PÁEZ ESPITIA.

**Cadena Radial Auténtica:** Calle 32, No 16-12, Apdo Aéreo 18350, Bogotá, DC; tel. (1) 285-3360; fax (1) 285-2505; f. 1983; stations include Radio Auténtica and Radio Mundial; Pres. JORGE ENRIQUE GÓMEZ MONTEALEGRE.

**Cadena Radial La Libertad Ltda:** Carrera 53, No 55-166, Apdo Aéreo 3143, Barranquilla; tel. (5) 31-1517; fax (5) 32-1279; news and music programmes for Barranquilla, Cartagena and Santa Marta; stations include Emisora Ondas del Caribe (youth programmes), Radio Libertad (classical music programmes) and Emisora Fuentes.

**Cadena Super:** Calle 16A, No 86A-78, Bogotá, DC; tel. (1) 618-1371; fax (1) 618-1360; internet www.889.com.co; f. 1971; stations include Radio Super and Super Stereo FM; Pres. JAIME PAVA NAVARRO.

**CARACOL, SA** (Primera Cadena Radial Colombiana, SA): Edif. Caracol RadioCalle 67, No 7-37, Bogotá, DC; tel. (1) 348-7600; fax (1) 337-7126; internet www.caracol.com.co; f. 1948; 107 stations; Pres. JOSÉ MANUEL RESTREPO FERNÁNDEZ DE SOTO.

**Circuito Todelar de Colombia:** Avda 13, No 84-42, Apdo Aéreo 27344, Bogotá, DC; tel. (1) 616-1011; fax (1) 616-0056; f. 1953; 74 stations; Pres. BERNARDO TOBÓN DE LA ROCHE.

**Colmundo Radio, SA** ('La Cadena de la Paz'): Diagonal 58, 26A-29, Apdo Aéreo 36750, Bogotá, DC; tel. (1) 217-8911; fax (1) 217-9358; f. 1989; Pres. Dr NÉSTOR CHAMORRO P.

**Organización Radial Olímpica, SA (ORO, SA):** Calle 72, No 48-37, 2°, Apdo Aéreo 51266, Barranquilla; tel. (5) 358-0500; fax (5) 345-9080; programmes for the Antioquia and Atlantic coast regions.

**Radio Cadena Nacional, SA (RCN Radio):** Carrera 13A, No 37-32, Bogotá, DC; tel. (1) 314-7070; fax (1) 288-6130; e-mail rcn@ impsat.net.co; internet www.rcn.com.co; 116 stations; official network; Pres. RICARDO LONDOÑO LONDOÑO.

**Radiodifusora Nacional de Colombia:** Centro Administrativo Nacional (CAN), Avda El Dorado, Bogotá, DC; tel. (1) 222-0415; fax (1) 222-0409; e-mail radiodifusora@hotmail.com; f. 1940; national public radio; Dir ATHALA MORRIS.

**Radiodifusores Unidos, SA (RAU):** Carrera 13, No 85-51, Of. 705, Bogotá, DC; tel. (1) 617-0584; commercial network of independent local and regional stations throughout the country.

### Television

Television services began in 1954, and the NTSC colour television system was adopted in 1979. The government-run broadcasting network, INRAVISION, controls three national stations. Two national, privately-run stations began broadcasts in mid-1988. There are also two regional stations and one local, non-profit station. Broadcasting time is distributed among competing programmers through a public tender.

**Cadena Uno:** Centro Administrativo Nacional (CAN), Avda El Dorado, Bogotá, DC; tel. (1) 342-3777; fax (1) 341-6198; e-mail cadena1@latino.net.co; internet www.cadena1.com.co; f. 1992; Dir FERNANDO BARRERO CHÁVEZ.

**Canal 3:** Centro Administrativo Nacional (CAN), Avda El Dorado, Bogotá, DC; tel. (1) 222-1640; fax (1) 222-1514; f. 1970; Exec. Dir RODRIGO ANTONIO DURÁN BUSTOS.

# COLOMBIA
*Directory*

**Canal A:** Calle 35, No 7-51, CENPRO, Bogotá, DC; tel. (1) 232-3196; fax (1) 245-7526; f. 1966; Gen. Man. Rocío Fernández del Castillo.

**Caracol Televisión, SA:** Calle 76, No 11-35, Apdo Aéreo 26484, Bogotá, DC; tel. (1) 319-0860; fax (1) 321-1720; internet www.canalcaracol.com; f. 1969; Pres. Ricardo Alarcón Gaviria.

**Teleantioquia:** Carrera 41, No 52-28, Edif. EDA, 3°, Apdo Aéreo 8183, Medellín, Antioquia; tel. (4) 261-2222; fax (4) 262-0832; e-mail comunicaciones@teleantioquia.com.co; internet www.teleantioquia.com.co; f. 1985.

**Telecafé:** Carrera 19A, Calle 43, Sacatín contiguo Universidad Autónoma, Manizales, Caldas; tel. (68) 86-2949; fax (68) 86-3009; e-mail telecafe2004@epm.net.co; f. 1986; Gen. Man. Alberto López Marín.

**Telecaribe:** Carrera 54, No 72-142, 11°, Barranquilla, Atlántico; tel. (5) 358-2297; fax (5) 356-0924; e-mail ealviz@canal.telecaribe.com.co; internet www.telecaribe.com.co; f. 1986; Gen. Man. Iván Ovalle Poveda.

**Telepacífico:** Calle 5A, No 38A-14, 3°, esq. Centro Comercial Imbanaco, Cali, Valle del Cauca; tel. (2) 589-933; fax (2) 588-281; e-mail telepacifico@emcali.net.co; Gen. Man. Luis Guillermo Restrepo.

**TV Cúcuta:** tel. (75) 74-7874; fax (75) 75-2922; f. 1992; Pres. José A. Armella.

## ASSOCIATIONS

**Asociación Nacional de Medios de Comunicación (ASOMEDIOS):** Carrera 22, No 85-72, Bogotá, DC; tel. (1) 611-1300; fax (1) 621-6292; e-mail asomedio@cable.net.co; internet www.asomedios.com; f. 1978; merged with ANRADIO (Asociación Nacional de Radio, Televisión y Cine de Colombia) in 1980; Pres. Sergio Arboleda Casas.

# Finance

(cap. = capital; res = reserves; dep. = deposits; m. = million; amounts in pesos, unless otherwise indicated)

**Contraloría General de la República:** Carrera 10, No 17-18, Torre Colseguros, 27°, Bogotá, DC; tel. (1) 353-7700; fax (1) 353-7616; e-mail ahernandez@contraloriagen.gov.co; internet www.contraloriagen.gov.co; Controller-Gen. Dr Antonio Hernández Gamarra.

## BANKING

### Supervisory Authority

**Superintendencia Bancaria:** Calle 7, No 4-49, 11°, Apdo Aéreo 3460, Bogotá, DC; tel. (1) 594-0200; fax (1) 350-7999; e-mail super@superbancaria.gov.co; internet www.superbancaria.gov.co; Banking Supt Jorge Pinzón Sánchez.

### Central Bank

**Banco de la República:** Carrera 7A, No 14-78, 5°, Apdo Aéreo 3531, Bogotá, DC; tel. (1) 343-1111; fax (1) 286-1686; e-mail wbanco@banrep.gov.co; internet www.banrep.gov.co; f. 1923; sole bank of issue; cap. 12.7m., res 20,900.0m., dep. 2,264.0m. (Dec. 2002); Man. Dir José Darío Uribe; 28 brs.

### Commercial Banks

Bogotá, DC

**ABN AMRO Bank (Colombia), SA** (fmrly Banco Real de Colombia): KRA7, No 115-33, 17°, Bogotá, DC; tel. (1) 521-9100; fax (1) 640-0675; f. 1975; Pres. Carlos Eduardo Arruda Penteado; 17 brs.

**Bancafe:** Calle 28, No 13A-15, 6°, Apdo Aéreo 240332, Bogotá, DC; tel. (1) 606-7614; fax (1) 606-7727; e-mail c.gaona@bancafe.com.co; internet www.bancafe.com; f. 1953; fmrly Banco Cafetero, SA; 99% owned by Fondo de Garantías de Instituciones Financieras de Colombia; cap. 108,794.0m., res 75,750.2m., dep. 5,015,437.5m. (Dec. 2003); 283 brs; Pres. Pedro Nel Ospina Santa María.

**Banco Agrario de Colombia:** Carrera 8, No 15-43, Bogotá, DC; tel. (1) 212-3404; fax (1) 345-2279; internet www.bancoagrario.gov.co; f. 1999; state-owned; Pres. Jorge Restrepo Palacios.

**Banco America Colombia** (fmrly Banco Colombo-Americano): Carrera 7, No 71-52, Torre B, 4°, Apdo Aéreo 12327, Bogotá, DC; tel. (1) 312-2020; fax (1) 312-1645; cap. 3,268m., res 6,205m., dep. 14,475m. (Dec. 1993); wholly-owned subsidiary of Bank of America; Pres. Eduardo Romero Jaramillo; 1 br.

**Banco de Bogotá:** Calle 36, No 7-47, 15°, Apdo Aéreo 3436, Bogotá, DC; tel. (1) 332-0032; fax (1) 338-3302; internet www.bancodebogota.com.co; f. 1870; acquired BANCO del Comercio in 1992; cap. 2,253.8m., res 858,248.1m., dep. 4,697,327.6m. (Dec. 2002); Pres. Dr Alejandro Figueroa Jaramillo; 274 brs.

**Banco Caja Social:** Carrera 7, No 77-65, 11°, Bogotá, DC; tel. (1) 313-8000; fax (1) 321-6912; e-mail ep_manrique@fundacion-social.com.co; internet www.bancocajasocial.com; f. 1911; savings bank; cap. 89,999m., res 74,527m., dep. 1,360,862m. (Dec. 2004); Pres. Eulalia Arboleda de Montes; 135 brs.

**Banco Colpatria, SA:** Carrera 7A, No 24-89, 12°, Apdo Aéreo 30241, Bogotá, DC; tel. (1) 286-8277; fax (1) 334-0867; internet www.colpatria.com.co; f. 1955; cap. 1,002m., res 15,575m., dep. 197,390m. (Dec. 1994); Pres. Carlos Escobar Barco; 26 brs.

**Banco de Comercio Exterior de Colombia, SA (BANCOLDEX):** Calle 28, No 13A-15, 38°, Apdo Aéreo 240092, Bogotá, DC; tel. (1) 382-1515; fax (1) 282-5071; internet www.bancoldex.com; f. 1992; provides financing alternatives for Colombian exporters; affiliate trust company FIDUCOLDEX, SA, manages PROEXPORT (Export Promotion Trust); Pres. Miguel Gómez Martínez.

**Banco de Crédito:** Calle 27, No 6-48, Bogotá, DC; tel. (1) 286-8400; fax (1) 251-8381; internet www.bancodecredito.com; f. 1963; cap. 36,952.8m., res 73,409.6m., dep. 944,129.3m. (Dec. 1999); Man. Luis Fernando Mesa Prieto; 24 brs.

**Banco Granahorrar:** Carrera 6, No 15-32, 11°, Apdo Aéreo 3637, Bogotá, DC; tel. (1) 382-2090; fax (1) 282-2802; internet www.granahorrar.com.co; f. 1932; fmrly Banco Central Hipotecario; assets 4,072,000m. (June 2003); Pres. Dr Alberto Montoya Puyana; 132 brs.

**Banco Mercantil de Colombia, SA** (fmrly Banco de los Trabajadores): Avda 82, No 12-18, 8°, Bogotá, DC; tel. (1) 635-0017; fax (1) 623-7770; e-mail amartinez@bancomercantil.com.co; internet www.bancomercantil.com; f. 1974; wholly-owned subsidiary of Banco Mercantil (Venezuela); Pres. Gustavo Marturet; 10 brs.

**Banco del Pacífico, SA:** Calle 72, 10-07, 2°, Bogotá, DC; tel. (1) 345-0111; fax (1) 217-9436; e-mail jcbernal@banpacifico.com.co; internet www.bp.fin.ec; f. 1994; wholly-owned by Banco del Pacífico, SA, Guayaquil (Ecuador); Exec. Pres. Marcel Laniado; 1 br.

**Banco Popular, SA:** Calle 17, No 7-43, 3°, Apdo Aéreo 6796, Bogotá, DC; tel. (1) 339-5449; fax (1) 281-9448; e-mail vpinternacional@bancopopular.com.co; internet www.bancopopular.com.co; f. 1951; res 155.8m., dep. 3,140.2m. (June 2004); Int. Man. Hernando Vivas V.; 158 brs.

**Banco Standard Chartered:** Calle 74, No 6-65, Bogotá, DC; tel. (1) 217-7200; fax (1) 212-5786; f. 1982; cap. 10,312m., res 5,574m., dep. 84,200m. (Dec. 1996); Pres. Hans Juerguen Heilkuhl Ochoa; 11 brs.

**Banco Sudameris Colombia** (fmrly Banco Francés e Italiano): Carrera 11, No 94A-03, 5°, Apdo Aéreo 3440, Bogotá, DC; tel. (1) 636-8785; fax (1) 636-7702; internet www.sudameris.com.co; affiliate of Banque Sudameris, SA (France); Pres. and Chair. Jorge Ramírez Ocampo; 6 brs.

**Banco Tequendama, SA:** Diagonal 27, No 6-70, 2°, Bogotá, DC; tel. (1) 343-3900; fax (1) 346-7127; f. 1976; wholly-owned subsidiary of Banco Construcción (Venezuela); Pres. Rubén Loaiza Negreiros; 11 brs.

**Banco Unión Colombiano** (fmrly Banco Royal Colombiano): Torre Banco Unión Colombiano, Carrera 7A, No 71-52, 2°, Apdo Aéreo 3438, Bogotá, DC; tel. (1) 312-0411; fax (1) 312-0843; internet www.bancounion.com.co; f. 1925; cap. 33,708.1m. res 18,818.4m., dep. 354,047.1m. (Dec. 2001); Pres. Juan Fernando Posada Corpas; 20 brs.

**BBVA Colombia:** Carrera 9A, No 72-21, 11°, Apdo Aéreo 53859, Bogotá, DC; tel. (1) 312-4666; fax (1) 347-1600; internet www.bbvaganadero.com; f. 1956 as Banco Ganadero; name changed as above 2003; 73.7% owned by Banco Bilbao Vizcaya Argentaria, SA (Spain); cap. 879,467.8m., dep. 2,989,659.8m. (Dec. 1998); fmrly BBV-Banco Ganadero; Exec. Pres. Luis B Juango Fitero; 279 brs.

**Citibank Colombia, SA:** Carrera 9A, No 99-02, 3°, Bogotá, DC; tel. (1) 618-4455; fax (1) 621-0259; internet www.latam.citibank.com/colombia; wholly-owned subsidiary of Citibank (USA); Pres. Antonio Uribe; 23 brs.

**Lloyds TSB Bank, SA** (fmrly Banco Anglo Colombiano): Carrera 7, No 71-21, Torre B, 16°, Apdo Aéreo 3532, Bogotá, DC; tel. (1) 286-3155; fax (1) 281-8646; e-mail contactenos@lloydstsbbank.com.co; internet www.grupolloyds.com.co; f. 1976; cap. 14,990.7m., res 24,983.8m., dep. 229,629.3m. (Dec. 1996); Pres. Nigel Luson; 52 brs.

Cali

**Banco de Occidente:** Carrera 4, No 7-61, 12°, Apdo Aéreo 7607, Cali, Valle del Cauca; tel. (2) 886-1111; fax (2) 886-1298; e-mail dinternacional@bancodeoccidente.com.co; internet www.bancodeoccidente.com.co; cap. 3,821.1m., res 467,898.1m., dep. 4,057,732.6m. (Dec. 2003); 78.2% owned by Grupo Aval Acciones y Valores; Pres. Efrain Otero Álvarez; 122 brs.

COLOMBIA                                                                                                                                                              *Directory*

### Medellín

**Bancolombia, SA** (fmrly Banco Industrial de Colombia): Carrera 50, No 51-66, Medellín, Antioquia; tel. (4) 576-6060; fax (4) 513-4827; e-mail mfranco@bancolombia.com.co; internet www.bancolombia.com.co; f. 1945; renamed 1998; following merger with Banco de Colombia; cap. 288,348m., res 942,515m., dep. 7,771,159m. (Dec. 2003); Pres. JORGE LONDOÑO SALDARRIAGA; Chair. JUAN CAMILO OCHOA; 361 brs.

**Banco Santander:** Edif. Bavaria, 16°, Carrera 46, No 52-36, Medellín, Antioquia; tel. (1) 284-3100; fax (1) 281-0311; internet www.bancosantander.com.co; f. 1961; fmrly Banco Comercial Antioqueño, SA; subsidiary of Banco Santander (Spain); Pres. MONICA INÉS MARÍA APARICIO SMITH; 30 brs.

### Banking Associations

**Asociación Bancaria y de Entidades Financieras de Colombia:** Carrera 9A, No 74-08, 9°, Bogotá, DC; tel. (1) 249-6411; fax (1) 211-9915; e-mail info@asobancaria.com; internet www.asobancaria.com; f. 1936; 56 mem. banks; Pres. PATRICIA CÁRDENAS SANTA MARÍA.

**Asociación Nacional de Instituciones Financieras (ANIF):** Calle 70A, No 7-86, Bogotá, DC; tel. (1) 310-1500; fax (1) 235-5947; internet www.anif.org; Pres. Dr FABIO VILLEGAS RAMÍREZ.

### STOCK EXCHANGES

**Superintendencia de Valores:** Avda El Dorado, Calle 26, No 68-85, Torre Suramericana, 2° y 3°, Bogotá, DC; tel. (1) 427-0222; fax (1) 427-0871; internet www.supervalores.gov.co; f. 1979 to regulate the securities market; Supt CLEMENTE DEL VALLE BORRÁEZ.

**Bolsa de Bogotá:** Carrera 8A, No 13-82, 4°–8°, Apdo Aéreo 3584, Bogotá, DC; tel. (1) 243-6501; fax (1) 243-7327; internet www.bolsabogota.com.co; f. 1928; Pres. AUGUSTO ACOSTA TORRES; Sec.-Gen. MARÍA FERNANDA TORRES.

### INSURANCE
#### Principal National Companies

**ACE Seguros, SA:** Calle 72, No 10-51, 6°, 7° y 8°, Apdo Aéreo 29782, Bogotá, DC; tel. (1) 319-0300; fax (1) 319-0304; internet www.ace-ina.com; fmrly Cigna Seguros de Colombia, SA; Pres. ALVARO A. ROZO PALOU.

**Aseguradora Colseguros, SA:** Carrera 13A, No 29-24, Parque Central Bavaria, Apdo Aéreo 3537, Bogotá, DC; tel. (1) 560-0600; fax (1) 561-6695; internet www.colseguros.com; f. 1874; Pres. MAX THIERMANN.

**Aseguradora El Libertador, SA:** Carrera 13, No 26-45, 9°, Apdo Aéreo 10285, Bogotá, DC; tel. (1) 281-2427; fax (1) 286-0662; e-mail aselib@impsat.net.co; Pres. FERNANDO ROJAS CÁRDENAS.

**Aseguradora Solidaria de Colombia:** Carrera 12, No 93-30, Apdo Aéreo 252030, Bogotá, DC; tel. (1) 621-4330; fax (1) 621-4321; e-mail eguzman@solidaria.com.co; Pres. CARLOS GUZMÁN.

**Chubb de Colombia Cía de Seguros, SA:** Carrera 7A, No 71-52, Torre B, 10°, Bogotá, DC; tel. (1) 326-6200; fax (1) 326-6210; e-mail esaavedra@chubb.com.co; internet www.chubb.com.co; f. 1972; Pres. MANUEL OBREGÓN; 4 brs.

**Cía Agrícola de Seguros, SA:** Carrera 11, No 93-46, Apdo Aéreo 7212, Bogotá, DC; tel. (1) 635-5827; fax (1) 635-5876; f. 1952; Pres. Dr JOSÉ F. JARAMILLO HOYOS.

**Cía Aseguradora de Fianzas, SA (Confianza):** Calle 82, No 11-37, 7°, Apdo Aéreo 056965, Bogotá, DC; tel. (1) 617-0899; fax (1) 610-8866; e-mail ccorreos@confianza.com.co; internet www.confianza.com.co; Pres. RODRIGO JARAMILLO ARANGO.

**Cía Central de Seguros, SA:** Carrera 7A, No 76-07, 9°, Apdo Aéreo 5764, Bogotá, DC; tel. (1) 319-0700; fax (1) 640-5553; e-mail recursos@centralseguros.com.co; internet www.centralseguros.com.co; f. 1956; Pres. SYLVIA LUZ RINCÓN LEMA.

**Cía de Seguros Atlas, SA:** Calle 21, No 23-22, 3°–11°, Apdo Aéreo 413, Manizales, Caldas; tel. (68) 84-1500; fax (168) 84-1447; e-mail satlas@andi.org.co; internet www.andi.org.co/seguros_atlas.htm; Pres. JORGE HOYOS MAYA.

**Cía de Seguros Bolívar, SA:** Carrera 10A, No 16-39, Apdo Aéreo 4421, Bogotá, DC; tel. (1) 341-0077; fax (1) 281-8262; internet www.segurosbolivar.com.co; f. 1939; Pres. (vacant).

**Cía de Seguros Colmena, SA:** Calle 72, No 10-07, 7° y 8°, Apdo Aéreo 6774, Bogotá, DC; tel. (1) 211-9111; fax (1) 211-4952; Pres. JUAN MANUEL DÍAZ-GRANADOS.

**Cía de Seguros Generales Aurora, SA:** Edif. Seguros Aurora, 1°, 2° y 3°, Carrera 7, No 74-21, Apdo Aéreo 8006, Bogotá, DC; tel. (1) 212-2800; fax (1) 212-2138; Pres. GERMÁN ESPINOSA.

**Cía Mundial de Seguros, SA:** Calle 33, No 6-94, 2° y 3°, Bogotá, DC; tel. (1) 285-5600; fax (1) 285-1220; Dir-Gen. CAMILO FERNÁNDEZ ESCOVAR.

**Cía Suramericana de Seguros, SA:** Centro Suramericana, Carrera 64B, No 49A-30, Apdo Aéreo 780, Medellín, Antioquia; tel. (4) 260-2100; fax (4) 260-3194; e-mail contactenos@suramericana.co; internet www.suramericana.com.co; f. 1944; Pres. Dr NICANOR RESTREPO SANTAMARÍA.

**Condor, SA, Cía de Seguros Generales:** Calle 119, No 16-59, Apdo Aéreo 57018, Bogotá, DC; tel. (1) 612-0666; fax (1) 215-6121; Pres. EUDORO CARVAJAL IBÁÑEZ.

**Generali Colombia—Seguros Generales, SA:** Carrera 7A, No 72-13, 1°, 7° y 8°, Apdo Aéreo 076478, Bogotá, DC; tel. (1) 217-8411; fax (1) 255-1164; Pres. MARCO PAPINI.

**La Ganadera Cía de Seguros, SA:** Carrera 7, No 71-52, Torre B, 11°, Apdo Aéreo 052347, Bogotá, DC; tel. (1) 312-2630; fax (1) 312-2599; e-mail ganadera@impsat.net.co; internet www.laganadera.com.co; Pres. CARLOS VERGARA GÓMEZ.

**La Interamericana Cía de Seguros Generales, SA:** Calle 78, No 9-57, 4° y 5°, Apdo Aéreo 92381, Bogotá, DC; tel. (1) 210-2200; fax (1) 210-2021; Pres. DIDIER SERRANO.

**La Previsora, SA, Cía de Seguros:** Calle 57, No 8-93, Apdo Aéreo 52946, Bogotá, DC; tel. (1) 211-2880; fax (1) 211-8717; Pres. ALVARO ESCALLÓN EMILIANI.

**Liberty Seguros, SA:** Calle 71A, No 6-30, 2°, 3°, 4° y 14°, Apdo Aéreo 100327, Bogotá, DC; tel. (1) 212-4900; fax (1) 212-7706; e-mail lhernandez@impsat.net.co; f. 1954; fmrly Latinoamericana de Seguros, SA; Pres. MAURICIO GARCÍA ORTIZ.

**Mapfre Seguros Generales de Colombia, SA:** Carrera 7A, No 74-36, 2°, Apdo Aéreo 28525, Bogotá, DC; tel. (1) 346-8702; fax (1) 346-8793; Pres. JOSÉ MANUEL INCHAUSTI.

**Pan American de Colombia Cía de Seguros de Vida, SA:** Carrera 7A, No 75-09, Apdo Aéreo 76000, Bogotá, DC; tel. (1) 212-1300; fax (1) 217-8799; Pres. ALFONSO PONTÓN.

**Real Seguros, SA:** Carrera 7A, No 115-33, 10°, Apdo Aéreo 7412, Bogotá, DC; tel. (1) 523-1400; fax (1) 523-4010; Gen. Man. JOSÉ LUIZ TOMAZINI.

**Royal and Sun Alliance Seguros (Colombia), SA:** Carrera 7A, No 32-33, 6°, 7°, 11° y 12°, Bogotá, DC; tel. (1) 561-0380; fax (1) 320-3726; internet www.royalsunalliance.com.co; fmrly Seguros Fénix, SA; Pres. DINAND BLOM.

**Segurexpo de Colombia, SA:** Calle 72, No 6-44, 12°, Apdo Aéreo 75140, Bogotá, DC; tel. (1) 217-0900; fax (1) 211-0218; Pres. JUAN PABLO LUQUE LUQUE.

**Seguros Alfa, SA:** Carrera 13, No 27-47, 22° y 23°, Apdo Aéreo 27718, Bogotá, DC; tel. (1) 344-4720; fax (1) 344-6770; Pres. JESÚS HERNANDO GÓMEZ.

**Seguros Colpatria, SA:** Carrera 7A, No 24-89, 9°, Apdo Aéreo 7762, Bogotá, DC; tel. (1) 616-6655; fax (1) 281-5053; Pres. Dr FERNANDO QUINTERO.

**Seguros La Equidad, OC:** Calle 19, No 6-68, 10°, 11° y 12°, Apdo Aéreo 30261, Bogotá, DC; tel. (1) 284-1910; fax (1) 286-5124; Pres. Dr JULIO ENRIQUE MEDRANO LEÓN.

**Seguros del Estado, SA:** Carrera 11, No 90-20, Apdo Aéreo 6810, Bogotá, DC; tel. (1) 218-6977; fax (1) 218-0971; Pres. Dr JORGE MORA SÁNCHEZ.

**Skandia Seguros de Vida, SA:** Avda 19, No 113-30, Apdo Aéreo 100327, Bogotá, DC; tel. (1) 214-1200; fax (1) 214-0038; Pres. RAFAEL JARAMILLO SAMPER.

#### Insurance Association

**Federación de Aseguradores Colombianos (FASECOLDA):** Carrera 7A, No 26-20, 11° y 12°, Apdo Aéreo 5233, Bogotá, DC; tel. (1) 210-8080; fax (1) 210-7090; e-mail fasecolda@fasecolda.com; internet www.fasecolda.com; f. 1976; 33 mems; Pres. Dr WILLIAM R. FADUL VERGARA.

## Trade and Industry

### GOVERNMENT AGENCIES

**Agencia Nacional de Hidrocarburos (ANH):** Edif. Guadalupe, Calle 37, 7-43, 5°, Bogotá, DC; tel. (1) 234-4519; fax (1) 234-4671; e-mail info@anh.gov.co; internet www.anh.gov.co; f. 2003; govt agency responsible for regulation of the petroleum industry; Dir-Gen. JOSÉ ARMANDO ZAMORA REYES.

**Departamento Nacional de Planeación:** Calle 26, No 13-19, 14°, Bogotá, DC; tel. (1) 336-1600; fax (1) 281-3348; e-mail vgonzalez@

dnp.gov.co; internet www.dnp.gov.co; f. 1958; supervises and administers devt projects; approves foreign investments; Dir-Gen. Dr SANTIAGO JAVIER MONTENEGRO TRUJILLO.

**Instituto Colombiano de la Reforma Agraria (INCORA):** Centro Administrativo Nacional (CAN), Avda El Dorado, Apdo Aéreo 151046, Bogotá, DC; tel. (1) 222-0963; fax (1) 222-1536; e-mail administra@incora.gov.co; internet www.incora.gov.co; f. 1962; a public institution which, on behalf of the govt, administers public lands and those it acquires; reclaims land by irrigation and drainage facilities, roads, etc., to increase productivity in agriculture and stock-breeding; provides technical assistance and loans; supervises the distribution of land throughout the country; Dir Dr LUIS CARLOS OCHOA CADAVID.

**Superintendencia de Industria y Comercio (SUPERINDUSTRIA):** Carrera 13, No 27-00, 5°, Bogotá, DC; tel. (1) 382-0840; fax (1) 382-2696; e-mail info@sic.gov.co; internet www.sic.gov.co; supervises chambers of commerce; controls standards and prices; Supt JAIRO RUBIO ESCOBAR.

**Superintendencia de Sociedades (SUPERSOCIEDADES):** Avda El Dorado, No 46-80, Apdo Aéreo 4188, Bogotá, DC; tel. (1) 324-5777; fax (1) 324-5000; e-mail webmaster@supersociedades.gov.co; internet www.supersociedades.gov.co; f. 1931; oversees activities of local and foreign corpns; Supt RODOLFO DANÍES LACOUTURE.

## DEVELOPMENT AGENCIES

**Agencia Colombiana de Cooperación Social (ACCI):** Calle 26 No 13-19, 34°, Bogotá, DC; tel. (1) 334-0855; fax (1) 351-9672; e-mail acci@dnp.gov.co; internet www.acci.gov.co; social devt org. funded through Plan Colombia.

**Asociación Colombiana de Ingeniería Sanitaria y Ambiental (ACODAL):** Calle 39 No 14–75, Bogotá, DC; tel. (1) 245-9539; fax (1) 323-1407; e-mail ; internet www.acodal.org.co; f. 1956 as Asociación Colombiana de Acueductos y Alcantarillados; asscn promoting sanitary and environmental engineering projects; Pres. JORGE ENRIQUE ANGEL GÓMEZ; Man. JOSÉ FERNANDO CÁRDENAS.

**Asociación Colombiana de Medianas y Pequeñas Industrias (ACOPI):** Carrera 15, No 36-70, Bogotá, DC; tel. and fax (1) 320-4783; e-mail comunicaciones@acopi.org.co; internet www.acopi.org.co; f. 1951; promotes small and medium-sized industries; Pres. JUAN ALFREDO PINTO SAAVEDRA.

**Centro Internacional de Educación y Desarrollo Humano:** Edif. Las Tres Carabelas, El Laguito, Cartagena; tel. (5) 665-5100; fax (5) 665-0319; e-mail cinde.org.co; internet www.cinde.org.co; education and social devt; f. 1977; Pres. JAIME OLARTE RESTREPO.

**Corporación para la Investigación Socioeconómica y Tecnológica de Colombia (CINSET):** Carrera 32 No 90-76, La Castellana, Bogotá, DC; tel. (1) 256-0961; fax (1) 218-6416; e-mail cinset@cinset.org.co; internet www.cinset.org.co; f. 1987; social, economic and technical devt projects.

**Corporación Región:** Calle 46 No 41-10, Medellín; tel. (4) 216-6822; fax (4) 239-5544; e-mail coregion@region.org.co; internet www.region.org.co; environmental, political and social devt; f. 1989; Pres. RUBÉN HERNANDO FERNÁNDEZ ANDRADE.

**Fondo Nacional de Proyectos de Desarrollo (FONADE):** Calle 26, No 13-19, 19°–22°, Apdo Aéreo 24110, Bogotá, DC; tel. (1) 594-0407; fax (1) 282-6018; e-mail fonade@colomsat.net.co; internet www.fonade.gov.co; f. 1968; responsible for channelling loans towards economic devt projects; administered by a cttee under the head of the Departamento Nacional de Planeación; FONADE works in close association with other official planning orgs; Gen. Man. Dr ELVIRA FORERO HERNÁNDEZ.

**Fundación para el Desarrollo Integral del Valle del Cauca (FDI):** Calle 8, No 3-14, 17°, Apdo Aéreo 7482, Cali, Valle del Cauca; tel. (2) 80-6660; fax (2) 82-4627; f. 1969; industrial devt org.; Pres. GUNNAR LINDAHL HELLBERG; Exec. Pres. FABIO RODRÍGUEZ GONZALEZ.

## CHAMBERS OF COMMERCE

**Confederación Colombiana de Cámaras de Comercio (CONFECAMARAS):** Carrera 13, No 27-47, Of. 502, Apdo Aéreo 29750, Bogotá, DC; tel. (1) 346-7055; fax (1) 346-7026; e-mail confecamaras@confecamaras.org.co; internet www.confecamaras.org.co; f. 1969; 56 mem. orgs; Exec. Pres. EUGENIO MARULANDA GÓMEZ.

**Cámara Colombo Venezolana:** Carrera 20, No 82–40, Of. 301, Bogotá, DC; tel. (1) 530-0626; fax (1) 530-0628; e-mail info@comvenezuela.com; internet www.comvenezuela.com; f. 1977; Colombian–Venezuelan trade asscn; 19 mem. cos; Pres. MARÍA LUISA CHIAPPE.

**Cámara de Comercio de Bogotá:** Carrera 9A, No 16-21, Bogotá, DC; tel. (1) 381-0328; fax (1) 284-7735; internet www.ccb.org.co; f. 1878; 3,650 mem. orgs; Dir ARIEL JARAMILLO JARAMILLO; Exec.-Pres. Dr MARÍA FERNANDA CAMPO SAAVEDRA.

**Cámara de Comercio Colombo Americano** (Colombian-American Chamber of Commerce): Of. 1209, Calle 98, No 22-64, Bogotá, DC; tel. (1) 623-7088; fax (1) 621-6838; e-mail website@amchamcolombia.com.co; internet www.amchamcolombia.com.co.

**Cámara de Comercio Colombo Británica** (British-Colombian Chamber of Commerce): Of. 403, Carrera 12A, No 77A-52, Bogotá, DC; tel. (1) 321-7077; fax (1) 321-7964; e-mail britanica@cable.net.co; internet www.colombobritanica.com.

There are also local Chambers of Commerce in the capital towns of all the Departments and in many of the other trading centres.

## INDUSTRIAL AND TRADE ASSOCIATIONS

**Colombiana de Minería (COLMINAS):** Bogotá, DC; state mining concern; Man. ALFONSO RODRÍGUEZ KILBER.

**Corporación de la Industria Aeronáutica Colombiana, SA (CIAC SA):** Aeropuerto Internacional El Dorado, Entrada 1 y 2, Apdo Aéreo 14446, Bogotá, DC; tel. (1) 413-9735; fax (1) 413-8673; Gen. Man. ALBERTO MELÉNDEZ.

**Empresa Colombia de Niquel (ECONIQUEL):** Bogotá, DC; tel. (1) 232-3839; administers state nickel resources; Dir JAVIER RESTREPO TORO.

**Empresa Colombiana de Uranio (COLURANIO):** Centro Administrativo Nacional (CAN), 4°, Ministerio de Minas y Energía, Bogotá, DC; tel. (1) 244-5440; f. 1977 to further the exploration, processing and marketing of radio-active minerals; initial cap. US $750,000; Dir JAIME GARCÍA.

**Empresa de Comercialización de Productos Perecederos (EMCOPER):** Bogotá, DC; tel. (1) 235-5507; attached to Ministry of Agriculture and Rural Development; Dir LUIS FERNANDO LONDOÑO RUIZ.

**Industria Militar (INDUMIL):** Diagonal 40, No 47-75, Apdo Aéreo 7272, Bogotá, DC; tel. (1) 222-3001; fax (1) 222-4889; attached to Ministry of National Defence; Man. Adm. (retd) MANUEL F. AVENDAÑO.

**Instituto Colombiano Agropecuario (ICA):** Calle 37, No 8-43, 4° y 5°, Bogotá, DC; tel. (1) 285-5520; fax (1) 232-4689; e-mail info@ica.gov.co; internet www.ica.gov.co; f. 1962; institute for promotion, co-ordination and implementation of research into and teaching and devt of agriculture and animal husbandry; Dir Dr ALVARO ABISAMBRA ABISAMBRA.

**Instituto de Fomento Industrial (IFI):** Calle 16, No 6-66, Edif. Avianca, 8° y 11°, Apdo Aéreo 4222, Bogotá, DC; tel. (1) 444-2200; fax (1) 286-4166; internet www.ifi.gov.co; f. 1940; state finance corpn for enterprise devt; cap. 469,808m. pesos, total assets 2,246,480m. pesos (Dec. 1998); Pres. ENRIQUE CAMACHO MATAMOROS.

**Instituto de Hidrología, Meteorología y Estudios Ambientales (IDEAM):** Diagonal 97, No 17-60, 1°, 2°, 3° y 7°, Bogotá, DC; tel. (1) 283-6927; fax (1) 635-6218; f. 1995; responsible for irrigation, flood control, drainage, hydrology and meteorology; Dir PABLO LEYVA.

**Instituto de Investigación e Información Geocientifica, Mineroambiental y Nuclear (INGEOMINAS):** Diagonal 53, No 34-53, Apdo Aéreo 4865, Bogotá, DC; tel. (1) 222-1811; fax (1) 220-0582; e-mail henciso@ingeominas.gov.co; internet www.ingeominas.gov.co; f. 1968; responsible for mineral research, geological mapping and research including hydrogeology, remote sensing, geochemistry, geophysics and geological hazards; Dir Dr ADOLFO ALARCÓN GUZMÁN.

**Instituto de Mercadeo Agropecuario (IDEMA):** Carrera 10A, No 16-82, Of. 1006, Bogotá, DC; tel. (1) 342-2596; fax (1) 283-1838; state enterprise for the marketing of agricultural products; Dir-Gen. ENRIQUE CARLOS RUIZ RAAD.

**Instituto Nacional de Fomento Municipal (INSFOPAL):** Centro Administrativo Nacional (CAN), Avda El Dorado, Bogotá, DC; tel. (1) 222-3177; Gen. Man. JAIME MARIO SALAZAR VELÁSQUEZ.

**Instituto Nacional de los Recursos Naturales Renovables y del Ambiente (INDERENA):** Diagonal 34, No 5-18, 3°, Bogotá, DC; tel. (1) 285-4417; f. 1968; govt agency regulating the devt of natural resources.

**Minerales de Colombia, SA (MINERALCO):** Calle 32, No 13-07, Apdo Aéreo 17878, Bogotá, DC; tel. (1) 287-7136; fax (1) 87-4606; administers state resources of emerald, copper, gold, sulphur, gypsum, phosphate rock and other minerals except coal, petroleum and uranium; Gen. Man. ORLANDO ALVAREZ PÉREZ.

**Sociedad Minera del Guainía (SMG):** Bogotá, DC; f. 1987; state enterprise for exploration, mining and marketing of gold; Pres. Dr JORGE BENDECK OLIVELLA.

There are several other agricultural and regional development organizations.

COLOMBIA                                                                                                                        *Directory*

## EMPLOYERS' AND PRODUCERS' ORGANIZATIONS

**Asociación Colombiana de Cooperativos (ASCOOP):** Transversal 29 No 35-29, Bogotá, DC; tel. (1) 368-3500; fax (1) 268-4230; e-mail comunicaciones@ascoop.coop; internet www.ascoop.coop; promotes co-operatives.

**Asociación de Cultivadores de Caña de Azúcar de Colombia (ASOCAÑA):** Calle 58N, No 3N-15, Apdo Aéreo 4448, Cali, Valle del Cauca; tel. (2) 64-7902; fax (2) 64-5888; f. 1959; sugar planters' asscn; Pres. Dr RICARDO VILLAVECES PARDO.

**Asociación Nacional de Empresarios (ANDI)** (National Asscn of Manufacturers): Calle 52, No 47-48, Apdo Aéreo 997, Medellín, Antioquia; tel. (4) 511-1177; fax (4) 511-7575; e-mail comercial@andi.com.co; internet www.andi.com.co; f. 1944; Pres. LUIS CARLOS VILLEGAS ECHEVERRI; 9 brs; 756 mems.

**Asociación Nacional de Exportadores (ANALDEX):** Carrera 10, No 27, Int. 137, Of. 902, Apdo Aéreo 29812, Bogotá, DC; tel. (1) 342-0788; fax (1) 284-6911; internet www.analdex.org; exporters' asscn; Pres. JORGE RAMÍREZ OCAMPO.

**Asociación Nacional de Exportadores de Café de Colombia:** Calle 72, No 10-07, Of. 1101, Bogotá, DC; tel. (1) 347-8419; fax (1) 347-9523; e-mail asoexport@asoexport.org; internet www.asoexport.org; f. 1938; private asscn of coffee exporters; Pres. JORGE E. LOZANO MANCERA.

**Expocafé Ltda:** Edif. Seguros Caribe, Carrera 7A, No 74-36, 3°, Apdo Aéreo 41244, Bogotá, DC; tel. (1) 217-8900; fax (1) 217-3554; f. 1985; coffee exporting org.; Gen. Man. LUIS JOSÉ ALVAREZ L.

**Federación Colombiana de Ganaderos (FEDEGAN):** Carrera 14, No 36-65, Apdo Aéreo 9709, Bogotá, DC; tel. (1) 245-3041; fax (1) 232-7153; e-mail fedegan@fedegan.org.co; internet www.fedegan.org.co; f. 1963; cattle raisers' asscn; about 350,000 affiliates; Pres. JORGE VISBAL MARTELO.

**Federación Nacional de Cacaoteros:** Carrera 17, No 30-39, Apdo Aéreo 17736, Bogotá, DC; tel. (1) 288-7188; fax (1) 288-4424; e-mail f_cacaoteros@hotmail.com; fed. of cocoa growers; Gen. Man. Dr JOSÉ OMAR PINZÓN USECHE.

**Federación Nacional de Cafeteros de Colombia (FEDERACAFE)** (National Federation of Coffee Growers): Calle 73, No 8-13, Apdo Aéreo 57534, Bogotá, DC; tel. (1) 217-0600; fax (1) 217-1021; internet www.cafedecolombia.com; f. 1927; totally responsible for fostering and regulating the coffee economy; Gen. Man. GABRIEL SILVA LUJÁN; 203,000 mems.

**Federación Nacional de Comerciantes (FENALCO):** Carrera 4, No 19-85, 7°, Bogotá, DC; tel. (1) 350-0600; fax (1) 350-9424; e-mail fenalco@fenalco.com.co; internet www.fenalco.com.co; fed. of businessmen; Pres. GUILLERMO BOTERO NIETO.

**Federación Nacional de Cultivadores de Cereales y Leguminosas (FENALCE):** Carrera 14, No 97-62, Apdo Aéreo 8694, Bogotá, DC; tel. (1) 218-2114; fax (1) 218-9463; e-mail fenalce@cable.net.co; f. 1960; fed. of grain growers; Gen. Man. JOSÉ ADEL CANCELADO PERRY; 30,000 mems.

**Sociedad de Agricultores de Colombia (SAC)** (Colombian Farmers' Society): Carrera 7A, No 24-89, 44°, Apdo Aéreo 3638, Bogotá, DC; tel. (1) 281-0263; fax (1) 284-4572; e-mail socdeagr@impsat.net.co; f. 1871; Pres. JUAN MANUEL OSPINA RESTREPO; Sec.-Gen. Dr GABRIEL MARTÍNEZ TELÁEZ.

There are several other organizations, including those for rice growers, engineers and financiers.

## MAJOR COMPANIES

The following are some of the leading industrial and commercial companies operating in Colombia:

**Acerías Paz del Río, SA:** Carrera 8, No 13-31, 7°, Bogotá, DC; tel. (1) 382-1768; fax (1) 382-1773; e-mail presidenciaapdr@hotmail.com; f. 1948; mining and processing of iron ores; Pres. RODRIGO MESA CADAVID; 2,450 employees.

**Almacenes Exito, SA:** Carrera 48, 32B Sur, Apdo Aéreo 3479, Medellín; tel. (4) 339-6565; fax (4) 339-6390; e-mail almacenes.exito@exito.com.co; internet www.exito.com.co; f. 1972; wholesaling and retailing; Pres. GONZALO RESTREPO LÓPEZ; 8,500 employees.

**Alpina Productos Alimenticios, SA:** Carrera 63, No 15-61, Bogotá, DC; tel. (1) 423-8600; fax (1) 414-1480; e-mail alpina@alpina.com; internet www.alpina.com.co; f. 1984; food and food processing; Pres. HERNÁN MÉNDEZ; 2,800 employees.

**Aluminio Reynolds Santo Domingo, SA:** Calle 79, No 40-362, Via 40, Baranquilla; tel. (5) 353-6610; fax (5) 353-6470; e-mail ventas@aluminioreynolds.com.co; internet www.aluminioreynolds.com.co; aluminium and metal products; f. 1955; Pres. EDUARDO GAITAN PARRA; 525 employees.

**Anglo-American PLC:** internet www.angloamerican.co.uk; f. 1917; mining and natural resources; part of consortium that bought Carbones de Colombia (CARBOCOL) in 2000 and Intercor (q.v.) in 2002; owns 50% of Cerrejón coal-mine; CEO TONY TRAHAR.

**Bavaria, SA:** Calle 94, No 7A-47, Bogotá, DC; tel. (1) 638-9000; fax (1) 610-0270; e-mail bavaria@bavaria.com.co; internet www.bavaria.com.co; f. 1889; holding co with principal interests in brewing and the manufacture of soft drinks; also transport, telecommunications, construction, forestry and fishing; Pres. RICARDO OBREGÓN TRUJILLO; Sec.-Gen. JUAN MANUEL ARBOLEDA PERDOMO; 18,895 employees.

**BHP Billiton:** internet www.bhpbilliton.com; f. 2001; following merger of BHP of Australia and British-South African co Billiton; mining and natural resources; part of consortium that bought Carbones de Colombia (CARBOCOL) in 2000 and Intercor (q.v.) in 2002.

**BP Exploration Company (Colombia) Ltd:** Carrera 9A, No 99-02, 4°, Bogotá, DC; tel. (1) 222-8855; fax (1) 218-3108; f. 1972; subsidiary of British Petroleum; exploration for hydrocarbons reserves; Legal Rep. ALVARO CAMARGO PATIÑO; 1,020 employees.

**Carvajal, SA:** Calle 29 Norte, No 6A-40, Apdo Aéreo 46, Cali; tel. (2) 667-5011; fax (2) 668-7644; internet www.carvajal.com.co; f. 1941; holding co with principal interests in printing and publishing; also construction, electronic components, telecommunications, trade, personal credit and the manufacture of office furniture; Pres. ALFREDO CARVAJAL SINISTERRA; 10,000 employees.

**Cerro Matoso, SA:** Calle 114, No 9-01, Torre A, Edif. Teleport Park, Montelíbano, Córdoba; tel. (5) 629-1570; fax (5) 629-1593; f. 1979; mining of ferrous ores; owned by BHP Billiton; Pres. WAYNE W. CLOWERY; 700 employees.

**Cerveca Aguila, SA:** Calle 10, No 38-280, Barranquilla; tel. (5) 341-1900; fax (5) 332-5366; internet www.cervezaaguila.com; f. 1967; brewery; parent co is Bavaria, SA; Pres. ALVARO PUPO PUPO; 980 employees.

**Cervecería Leona, SA:** Carretera Central del Norte Km 30, Vía Tunja, Tocancipa; tel. (1) 857-4425; fax (1) 857-4355; internet www.leona.com.co; f. 1992; brewery; Legal Rep. RICARDO HUMBERTO RESTREPO; 400 employees.

**Cervecería Unión, SA:** Carrera 50A, No 38-39, Itaguï, Medellín; tel. (4) 372-2400; fax (4) 372-3488; internet www.cerveceriaunion.com.co; f. 1931; brewery; parent company is Bavaria, SA; Dir-Gen. JORGE BONNELS GALINDO; 900 employees.

**Compañía Colombiana Automotriz, SA (CCA):** Carretera 11, No 94-02, Bogotá, DC; tel. (1) 218-4111; fax (1) 257-2410; internet www.mimazda.com; f. 1973; automobile manufacturers; majority of shares held by Mazda Ltd and Mitsubishi Corpn of Japan; 1,169 employees.

**Compañía Colombiana de Tejidos, SA (COLTEJER):** Calle 62, No 44-103, Itaguï, Apdo Aéreo 636, Medellín; tel. (4) 373-1133; fax (4) 372-8585; e-mail coltejer@coltejer.com.co; internet www.coltejer.com.co; f. 1907; textile manufacturers; Pres. Dr CARLOS ALBERTO BELTRÁN ARDILA; 5,922 employees.

**Comunicación Celular, SA (COMCEL):** Calle 90, No 14-37, Bogotá, DC; tel. (1) 616-9797; fax (1) 623-1287; internet www.comcel.com.co; f. 1992; mobile telecommunications; Pres. ADRIÁN EFRÉN HERNÁNDEZ URUETA; 1,203 employees.

**Cristalería Peldar, SA:** Calle 39s, No 48-180, Apdo Aéreo 215, Envigado; tel. (4) 378-8000; fax (4) 270-4225; e-mail peldar@peldar.com; internet www.peldar.com; f. 1962; manufacture of glass products; Pres. GILBERTO E. RESTREPO VÁSQUEZ; 1,800 employees.

**Empresa Colombiana de Petróleos (ECOPETROL):** Edif. Ecopetrol, Carrera 13, No 36-24, 8°, Apdo Aéreo 5938, Bogotá, DC; tel. (1) 234-4000; fax (1) 234-4099; e-mail webmaster@ecopetrol.com.co; internet www.ecopetrol.com.co; f. 1951; state-owned co for the exploration, production, refining and transportation of petroleum; Pres. ISAAC YANOVICH FARBAIARZ; 12,323 employees.

**Enka de Colombia, SA:** Carrera 63, No 49A-31, Medellín; tel. (4) 405-5055; fax (4) 405-5140; e-mail administrativa@enka.com.co; internet www.enka.com.co; f. 1964; manufacture of synthetic fibres; Pres. NELSON MEJÍA; 1,450 employees.

**ExxonMobil de Colombia, SA:** Bogotá, DC; f. 2001 following the merger of Esso Colombiana with ExxonMobil; petroleum and gas exploration and extraction.

**Fábrica Colombiana de Automotores, SA (COLMOTORES):** Avda Boyacá, No 36A-03 Sur, Bogotá, DC; tel. (1) 710-1111; fax (1) 270-8382; subsidiary of General Motors Corpn, USA; producers of passenger and commercial vehicles, spare parts and accessories; Pres. VÍCTOR HUGO COELLO; 1,254 employees.

**Fabricato Tejicondor:** Calle 50, No 38-320 Medellín; tel. (4) 454-2424; fax (4) 454-3407; e-mail info@fabricato.com.co; internet www.fabricato.com.co; f. 1923; manufacture and export of cotton, textiles and synthetic fibre goods; Pres. Dr LUIS MARIANO SANIN ECHEVERRI; 4,325 employees.

**Glencore International AG:** e-mail info@glencore.com; internet www.glencore.com; Swiss-owned co, mining and natural resources; part of consortium that bought Carbones de Colombia (CARBOCOL) in 2000 and Intercor (q.v.) in 2002.

**Industrias Alimenticias Noel, SA:** Carrera 52, No 2-38, Apdo 897, Medellín; tel. (4) 365-9999; fax (4) 285-3553; e-mail webmaster@noel.com.co; internet www.noel.com.co; f. 1933; food production and processing including meat products, confectionery, powdered soft drinks and vegetable protein; Pres. RAFAEL MARIO VILLA MORENO; 2,000 employees.

**Incauca, SA:** Carrera 9, No 28-103, Cali; tel. (2) 418-3000; fax (2) 438-4909; e-mail incauca@incauca.com; internet www.incauca.com; f. 1963; cultivation and processing of sugar cane; Pres. Bd of Dirs CARLOS ARDILLA LÜLLE; 5,109 employees.

**Ingenio Providencia, SA:** Carrera 28, No 28-103, Palmira; tel. (2) 438-4865; fax (2) 438-4929; e-mail ingprovidencia@ingprovidencia.com; internet www.ingprovidencia.com; f. 1926; cultivation and wholesale of sugar cane; Pres. ALCARDO MOLINA ABADIA; 2,700 employees.

**International Resources Corporation (Intercor):** Carrera 54, No 72-80, Bogotá, DC; tel. (1) 285-2080; f. 1975; mining; acquired by a consortium comprising Anglo-American PLC, BHP-Billiton and Glencore International AG (q.v.) in 2002; Pres. RAMÓN DE LA TORRE LAGO; 2,200 employees.

**Leonisa Internacional:** Carrera 51, No 13-158, Medellín; tel. (4) 265-4000; fax (4) 265-0617; e-mail info@leonisa.com; internet www.leonisa.com; f. 1956; manufacturers of men's and women's clothing; Pres. OSCAR ECHEVERRI RESTREPO; 2,000 employees.

**Occidental Petroleum Corporation:** Calle 77A, No 11-32, Apdo 92171, Bogotá, DC; tel. (1) 346-0111; fax (1) 211-6820; internet www.oxy.com; f. 1977; principal shareholder Occidental Petroleum Corpn of the USA; petroleum and gas exploration and production; Man. JUAN CARLOS UCROS RODRÍGUEZ; 681 employees.

**Petrobras Colombia (BPCOL):** Edificio Bancafe, Torre B, Carretera 7A 7121, Bogotá DC; tel. (1) 313-5000; fax (1) 313-5070; internet www.petrobras.com.br; f. 1972; oil exploration and production; Gen. Man. PAULO CÉZAR AMARO AQUINO.

**Productora de Papeles, SA (PROPAL):** Antigua Carrera Cali–Yumbo Km 12, Yumbo, Valle; tel. (2) 669-8859; fax (2) 669-9244; e-mail funpropal@colnet.com.co; internet www.propal.com.co; f. 1957; owners of two paper mills manufacturing paper products; Gen. Man. HENRY SÁNCHEZ; 1,100 employees.

**Promigas:** Calle 66, No 67-123, Barranquilla; tel. (95) 371-3444; fax (95) 368-0515; internet www.promigas.com; f. 1974; natural-gas distributor, network covers 60% of the country; Pres. ANTONIO CELIA MARTÍNEZ-APARICIO.

**Smurfit Cartón de Colombia, SA:** Avda Americas No 56-41, Apdo Aéreo 4584, Bogotá; tel. (2) 425-4500; fax (2) 262-5569; internet www.smurfit.com.co; manufacturers of paper and packaging materials; subsidiary of Jefferson Smurfit Group of Ireland; Pres. NICANOR RESTREPO S.; 1,500 employees.

**Sociedad de Fabricación de Automotores, SA (Sofasa Renault):** Carretera Central del Norte, Km 17, Chía, Cundinamarca; tel. (1) 676-0108; e-mail staffscl@sofasa.com.co; internet www.sofasa.com.co; f. 1973; manufacture of motor vehicles and spare parts; Legal Rep. JUAN MANUEL CUNILL; 873 employees.

**Supertiendas y Droguerías Olimpica, SA:** Carrera 36, No 38-03, Barranquilla; tel. (53) 41-5912; fax (53) 41-1516; internet www.olimpica.com.co; f. 1953; retailing; Pres. GUSTAVO ENRIQUE VISBAL GALOFRE; 4,600 employees.

**Tecnoquímicas, SA:** Calle 23, No 7-39, Cali; tel. (2) 882-5555; fax (2) 883-8859; e-mail amalvarez@tecnoquimicas.com.co; internet www.tecnoquimicas.com.co; f. 1934; manufacture of pharmaceuticals; Pres. FRANCISCO BARBERI ZAMORANO; 1,670 employees.

**Tejidos El Condor, SA (TEJICONDOR):** Carrera 65, No 45-85, Apdo 815, Medellín; tel. (4) 454-2424; fax (4) 454-2466; f. 1934; manufacture of textiles, incl. cotton weaving; Chair. L. M. SANIN ECHEVERRI; 1,515 employees.

**Triton Colombia, Inc:** Carrera 9A, No 99-02, Of. 407, Bogotá, DC; tel. (1) 618-2411; fax (1) 618-2553; f. 1982; subsidiary of Triton Corpn of the USA; petroleum exploration and production; Pres. MICHAEL MURPHY; 19 employees.

## UTILITIES

### Electricity

**Corporación Eléctrica de la Costa Atlántica (Corelca):** Calle 55, No 72-109, 9°, Barranquilla, Atlántico; tel. (5) 356-0200; fax (5) 356-2370; responsible for supplying electricity to the Atlantic departments; generates more than 2,000m. kWh annually from thermal power-stations; Man. Dir HERNÁN CORREA NOGUERA.

**Empresa de Energía Eléctrica de Bogotá, SA (EEB):** Avda El Dorado, No 55-51, Bogotá, DC; tel. (1) 221-1665; fax (1) 221-6858; internet www.eeb.com.co; provides electricity for Bogotá area by generating capacity of 680 MW, mainly hydroelectric; Man. Dir LUIS EDUARDO GARZÓN.

**Instituto Colombiano de Energía Eléctrica (ICEL):** Carrera 13, No 27-00, 3°, Apdo Aéreo 16243, Bogotá, DC; tel. (1) 342-0181; fax (1) 286-2934; formulates policy for the devt of electrical energy; constructs systems for the generation, transmission and distribution of electrical energy; Man. DOUGLAS VELÁSQUEZ JACOME; Sec.-Gen. PATRICIA OLIVEROS LAVERDE.

**Interconexión Eléctrica, SA (ISA):** Calle 12 Sur, No 18-168, El Poblado, Apdo Aéreo 8915, Medellín, Antioquia; tel. (4) 325-2270; fax (4) 317-0848; e-mail isa@isa.com.co; internet www.isa.com.co; f. 1967; created by Colombia's principal electricity production and distribution cos to form a national network; installed capacity of 2,641m. kWh; operates major power-stations at Chivor and San Carlos; public co; Man. Dir JAVIER GUTIÉRREZ PEMBERTHY.

**Isagen:** Medellín, Antioquia; e-mail isagen@isagen.com.co; internet www.isagen.com.co; f. 1995 following division of ISA (q.v.); state-owned, scheduled for privatization; generates electricity from three hydraulic and two thermal power plants; Gen. Man. Ing. FERNANDO RICO PINZÓN.

### Gas

**Empresa Colombiana de Gas (Ecogás):** Carrera 34, No 41-51, Bucaramanga, Santander; tel. (7) 632-0002; fax (7) 632-5525; e-mail jramirez@ecogas.com.co; internet www.ecogas.com.co; f. 1997; operation and maintenance of gas-distribution network; Pres. CARLOS ALBERTO GÓMEZ GÓMEZ; Sec.-Gen. DANIELA GALVIS VILLAREAL; Administrative Vice-Pres. Dr JOSÉ GREGORIO RAMÍREZ AMAYA.

**Gas Natural ESP:** Avda 40A, No 13-09, 9°, Bogotá, DC; tel. (1) 338-1199; fax (1) 288-0807; f. 1987; private gas corpn; Pres. ANTONI PERIS MINGOT.

**Gasoriente:** distributes gas to eight municipalities in north-eastern Colombia.

## TRADE UNIONS

According to official figures, an estimated 900 of Colombia's 2,000 trade unions are independent.

**Central Unitaria de Trabajadores (CUT):** Calle 35, No 7-25, 9°, Apdo Aéreo 221, Bogotá, DC; tel. (1) 323-7550; fax (1) 323-7550; e-mail cut@cut.org.co; f. 1986; comprises 50 feds and 80% of all trade-union members; Pres. CARLOS RODRÍGUEZ DÍAZ; Sec.-Gen. BORIS MONTES DE OCA ANAYA.

  **Unión Sindical Obrera de la Indústria del Petróleo (USO):** Avda del Ferrocarril, 28-43 Galán, Barrancabermeja, Santander; tel. (7) 622-7856; fax (7) 622-9150; e-mail usocol@1.telecom.com; internet www.nodo50.org/usocolombia; f. 1922; petroleum workers' union; affiliated to CUT; Sec.-Gen. (vacant); 3,200 mems.

  **Federación Nacional Sindical Unitaria Agropecuaria (FENSUAGRO-CUT):** Bogotá, DC; f. 1976 as Federación Nacional Sindical Agropecuaria (FENSA); comprises 37 unions, 7 peasant asscns, with 80,000 mems; Leader HERNANDO HERNANDEZ TAPASCO (detained by the authorities June 2005).

**Frente Sindical Democrática (FSD):** f. 1984; centre-right trade-union alliance; comprises:

  **Confederación de Trabajadores de Colombia (CTC)** (Colombian Confederation of Workers): Calle 39, No 26A-23, 5°, Apdo Aéreo 4780, Bogotá, DC; tel. (1) 269-7119; f. 1934; mainly Liberal; 600 affiliates, including 6 national orgs and 20 regional feds; admitted to ICFTU; Pres. ALVIS FERNÁNDEZ; 400,000 mems.

  **Confederación de Trabajadores Democráticos de Colombia (CTDC):** Carrera 13, No 59-52, Of. 303, Bogotá, DC; tel. (1) 255-3146; fax (1) 484-581; f. 1988; comprises 23 industrial feds and 22 national unions; Pres. MARIO DE J. VALDERRAMA.

  **Confederación General de Trabajadores Democráticos (CGTD):** Calle 39A, No 14-48, Apdo Aéreo 5415, Bogotá, DC; tel. (1) 288-1560; fax (1) 573-4021; e-mail cgt@etb.net.co; internet www.cgtcolombia.org; Sec.-Gen. JULIO ROBERTO GÓMEZ ESGUERRA.

# Transport

Land transport in Colombia is rendered difficult by high mountains, so the principal means of long-distance transport is by air. As a result of the development of the El Cerrejón coalfield, Colombia's first deep-water port was constructed at Bahía de Portete and a 150-

km rail link between El Cerrejón and the port became operational in 1989.

**Instituto Nacional del Transporte (INTRA):** Edif. Minobras (CAN), 6°, Apdo Aéreo 24990, Bogotá, DC; tel. (1) 222-4100; govt body; Dir Dr GUILLERMO ANZOLA LIZARAZO.

### RAILWAYS

In 2000 there were 3,304 km of track.

**Dirección General de Transporte Férreo y Masivo:** Ministerio de Transporte, Centro Administrativo Nacional (CAN), Avda El Dorado, Bogotá, DC; e-mail dgtuf@mintransporte.gov.co; internet www.mintransporte.gov.co/Ministerio/DGTFM/dgtferreo/html; part of the Ministry of Transport; formulates railway transport policies; Dir JUAN GONZALO JARAMILLO.

**Empresa Colombiana de Vías Férreas (Ferrovías):** Calle 16, No 96-64, 8°, Bogotá, DC; tel. (1) 287-9888; fax (1) 287-2515; responsible for the maintenance and devt of the national rail network; Dir JUAN GONZALO JARAMILLO.

**Ferroviario Atlántico, SA:** Calle 72, No 13-23, 2°, Bogotá, DC; tel. (1) 255-8684; fax (1) 255-8704; links port of Santa Marta with Bogotá, DC; operated on a 30-year concession, awarded in 1998 to Asociación Futura Ferrocarriles de la Paz (Fepaz); 1,490 km (1993).

**Fondo de Pasivo Social de Ferrocarriles Nacionales de Colombia:** Bogotá, DC; administers welfare services for existing and former employees of the FNC.

**El Cerrejón Mine Railway:** International Colombia Resources Corpn, Carrera 54, No 72-80, Apdo Aéreo 52499, Barranquilla, Atlántico; tel. (5) 350-5389; fax (5) 350-2249; f. 1989 to link the mine and the port at Bahía de Portete; 150 km (1996); Supt M. MENDOZA.

**Metro de Medellín Ltda:** Calle 44, No 46-001, Apdo Aéreo 9128, Medellín, Antioquia; tel. (4) 452-6000; fax (4) 452-4450; e-mail emetro@col3.telecom.com.co; two-line metro with 25 stations opened in stages in 1995–96; 29 km; Gen. Man. LUIS GUILLERMO GÓMEZ A.

**Sociedad de Transporte Férreo de Occidente, SA:** Avda Vásquez Cobo, Estación Ferrocarril, 2°, Cali; tel. (2) 660-3314; fax (2) 660-3320; runs freight services between Cali and the port of Buenaventura; Gen. Man. R. A. GUZMAN.

### ROADS

In 2002 there were 110,000 km of roads, of which 26,000 km was paved. The country's main highways are the Caribbean Trunk Highway, the Eastern and Western Trunk Highways, the Central Trunk Highway and there are also roads into the interior. There are plans to construct a Jungle Edge highway to give access to the interior, a link road between Turbo, Bahía Solano and Medellín, a highway between Bogotá and Villavicencio and to complete the short section of the Pan-American Highway between Panama and Colombia.

There are a number of national bus companies and road haulage companies.

**Dirección General de Carreteras:** Ministerio de Transporte, Centro Administrativo Nacional (CAN), Of. 508, Avda El Dorado, Bogotá, DC; tel. (1) 222-7308; fax (2) 428-6749; e-mail dcarreteras@mintransporte.gov.co; internet www.mintransporte.gov.co/Ministerio/DGTCarreteras/; part of the Ministry of Transport; formulates road transport policies; Dir MANUEL ARIAS MOLANO.

**Instituto Nacional de Vías:** Transversal 45, Entrada 2, Bogotá, DC; tel. (1) 428-0400; fax (1) 315-6713; e-mail director@latino.net.co; f. 1966; reorganized 1994; wholly state-owned; responsible to the Ministry of Transport; maintenance and construction of national road network; Gen. Man. LUIS E. TOBON CARDONA.

### INLAND WATERWAYS

The Magdalena–Cauca river system is the centre of river traffic and is navigable for 1,500 km, while the Atrato is navigable for 687 km. The Orinoco system has more than five navigable rivers, which total more than 4,000 km of potential navigation (mainly through Venezuela); the Amazon system has four main rivers, which total 3,000 navigable km (mainly through Brazil). There are plans to connect the Arauca with the Meta, and the Putumayo with the Amazon, and also to construct an Atrato–Truandó inter-oceanic canal.

**Dirección General de Transporte Fluvial:** Ministerio de Transporte, Centro Administrativo Nacional (CAN), Avda El Dorado, Bogotá, DC; tel. (1) 428-7332; fax (2) 428-6233; e-mail omedina@mintransporte.gov.co; internet www.mintransporte.gov.co/Ministerio/DGTFLUVIAL/dgtfluvial.html; part of the Ministry of Transport; formulates river transport policies; Dir LUIS ALEJANDRO ESPINOSA FIERRO.

**Dirección de Navegación y Puertos:** Edif. Minobras (CAN), Of. 562, Bogotá, DC; tel. (1) 222-1248; responsible for river works and transport; the waterways system is divided into four sectors: Magdalena, Atrato, Orinoquia, and Amazonia; Dir ALBERTO RODRÍGUEZ ROJAS.

### SHIPPING

The four most important ocean terminals are Buenaventura on the Pacific coast and Santa Marta, Barranquilla and Cartagena on the Atlantic coast. The port of Tumaco on the Pacific coast is gaining in importance and there are plans for construction of a deep-water port at Bahía Solano. In 2003 Colombia's merchant fleet totalled 71,131 grt.

**Dirección General de Transporte Marítimo y Puertos:** Ministerio de Transporte, Centro Administrativo Nacional (CAN), Avda El Dorado, Bogotá, DC; tel. (1) 428-7332; fax (2) 428-6233; e-mail omedina@mintransporte.gov.co; internet www.mintransporte.gov.co/Ministerio/DGTMARITIMO/dgtmar.html; part of the Ministry of Transport; formulates shipping policies; Dir OSCAR HUMBERTO MEDINA MORA.

#### Port Authorities

**Port of Barranquilla:** Sociedad Portuaria Regional de Barranquilla, Carrera 38, Calle 1A, Barranquilla, Atlántico; tel. (5) 379-9555; fax (5) 379-9557; e-mail sprbbaq@latino.net.co; internet www.sprb.com.co; privatized in 1993; Port Man. ANÍBAL DAU.

**Port of Buenaventura:** Empresa Puertos de Colombia, Edif. El Café, Of. 1, Buenaventura; tel. (224) 22543; fax (224) 34447; Port Man. VÍCTOR GONZÁLEZ.

**Port of Cartagena:** Sociedad Portuaria Regional de Cartagena, SA, Manga, Terminal Marítimo, Cartagena, Bolívar; tel. (5) 660-8071; fax (5) 650-2239; e-mail comercial@sprc.com.co; internet www.sprc.com.co; f. 1959; Port Man. ALFONSO SALAS TRUJILLO; Harbour Master Capt. GONZALO PARRA.

**Port of Santa Marta:** Empresa Puertos de Colombia, Calle 15, No 3-25, 11°, Santa Marta, Magdalena; tel. (54) 210739; fax (54) 210711; e-mail spsm@spsm.com.co; internet www.spsm.com.co; Port Man. JULIÁN PALACIOS.

#### Principal Shipping Companies

**Colombiana Internacional de Vapores, Ltda (Colvapores):** Avda Caracas, No 35-02, Apdo Aéreo 17227, Bogotá, DC; cargo services mainly to the USA.

**Flota Mercante Grancolombiana, SA:** Edif. Grancolombiana, Carrera 13, No 27-75, Apdo Aéreo 4482, Bogotá, DC; tel. (1) 286-0200; fax (1) 286-9028; f. 1946; owned by the Colombian Coffee Growers' Federation (80%) and Ecuador Development Bank (20%); f. 1946 one of Latin America's leading cargo carriers serving 45 countries world-wide; Pres. LUIS FERNANDO ALARCÓN MANTILLA.

**Líneas Agromar, Ltda:** Calle 73, Vía 40-350, Apdo Aéreo 3256, Barranquilla, Atlántico; tel. (5) 345-1874; fax (5) 345-9634; Pres. MANUEL DEL DAGO FERNÁNDEZ.

**Petromar Ltda:** Bosque, Diagonal 23, No 56-152, Apdo Aéreo 505, Cartagena, Bolívar; tel. (5) 662-7208; fax (5) 662-7592; Chair. SAVERIO MINERVINI S.

**Transportadora Colombiana de Graneles, SA (NAVESCO, SA):** Avda 19, No 118-95, Of. 214-301, Bogotá, DC; tel. (1) 620-9035; fax (1) 620-8801; e-mail navesco@colomsat.net.co; Gen. Man. GUILLERMO SOLANO VARELA.

Several foreign shipping lines call at Colombian ports.

### CIVIL AVIATION

Colombia has more than 100 airports, including 11 international airports: Bogotá, DC (El Dorado International Airport), Medellín, Cali, Barranquilla, Bucaramanga, Cartagena, Cúcuta, Leticia, Pereira, San Andrés and Santa Marta.

**Dirección General de Transporte Aéreo:** Ministerio de Transporte, Centro Administrativo Nacional (CAN), Of. 304, Avda El Dorado, Bogotá, DC; tel. (1) 428-7397; fax (2) 428-7034; e-mail aereo@mintransporte.gov.co; internet www.mintransporte.gov.co/Ministerio/DGTAEREO/dgtaereo.html; part of the Ministry of Transport; formulates air transport policies; Dir JUAN CARLOS SALAZAR GÓMEZ.

#### Airports Authority

**Unidad Administrativa Especial de Aeronáutica Civil:** Aeropuerto Internacional El Dorado, 4°, Bogotá, DC; tel. (1) 413-9500; fax (1) 413-9878; f. 1967 as Departamento Administrativo de Aeronáutica Civil, reorganized 1993; wholly state-owned; Dir ERNESTO HUERTAS ESACALLÓN.

### National Airlines

**Aerotaca, SA (Aerotransportes Casanare):** Avda El Dorado, Entrada 1, Interior 20, Bogotá, DC; tel. (1) 413-9884; fax (1) 413-5256; f. 1965; scheduled regional and domestic passenger services; Gen. Man. RAFAEL URDANETA.

**AVIANCA** (Aerovías Nacionales de Colombia, SA): Avda El Dorado, No 93-30, 5°, Bogotá, DC; tel. (1) 413-9511; fax (1) 413-8716; internet www.avianca.com; f. 1919; operates domestic services to all cities in Colombia and international services to the USA, France, Spain and throughout Central and Southern America; merged with Aerolíneas Centrales de Colombia, SA (ACES) and Sociedad Aeronáutica de Medellín Consolidada, SA (SAM) in 2002; Chair. ANDRÉS OBREGÓN SANTO DOMINGO; Pres. JUAN EMILIO POSADA.

**Intercontinental de Aviación:** Avda El Dorado, Entrada 2, Interior 6, Bogotá, DC; tel. (1) 413-9700; fax (1) 413-8458; internet www.insite-network.com/inter/; f. 1965 as Aeropesca Colombia (Aerovías de Pesca y Colonización del Suroeste Colombiano); operates scheduled domestic, regional and international passenger and cargo services; Pres. Capt. LUIS HERNÁNDEZ ZIA.

**Servicio de Aeronavegación a Territorios Nacionales (Satena):** Avda El Dorado, Entrada 1, Interior 11, Apdo Aéreo 11163, Bogotá, DC; tel. (1) 413-7100; fax (1) 413-8178; internet www.satena.com; f. 1962; commercial enterprise attached to the Ministry of National Defence; internal services; CEO and Gen. Man. Maj.-Gen. HECTOR CAMPO.

**Tampa Cargo SA:** Terminal de Carga, Aeropuerto Internacional José María Córdova, Apdo Aéreo 494, Ríonegro, Antioquia; tel. (574) 569-9200; fax (4) 569-9247; e-mail amartinez@tampacargo.com; internet www.tampacargo.com.co; f. 1973; operates international cargo services to destinations throughout the Americas; Pres. FREDERICK JACOBSEN; Head of Corp. Communications ISABEL MARTÍNEZ.

In addition, the following airlines operate international and domestic charter cargo services: Aerosucre Colombia, Aero Transcolombiana de Carga (ATC), Aires Colombia and Líneas Aéreas Suramericanas (LAS).

## Tourism

The principal tourist attractions are the Caribbean coast (including the island of San Andrés), the 16th-century walled city of Cartagena, the Amazonian town of Leticia, the Andes mountains rising to 5,700m above sea-level, the extensive forests and jungles, pre-Columbian relics and monuments of colonial art. In 2002 there were 541,372 visitors (compared with 615,719 in 2001), most of whom came from the USA, Ecuador and Venezuela. In 2003 tourism receipts were US $1,114m.

**Ministerio de Comercio, Industria y Turismo:** Calle 28, No 13A-15, 18°, Edif. Centro de Comercio Internacional, Bogotá, DC; tel. (1) 382-1307; fax (1) 352-2101; internet www.mincomercio.gov.co; Dir of Tourism ADOLFO TORO VELÁSQUEZ.

**Asociación Colombiana de Agencias de Viajes y Turismo (ANATO):** Carrera 21, No 83-63, Apdo Aereo 7088, Bogotá, DC; tel. (1) 610-7099; fax (1) 236-2424; e-mail direccionejecutiva@anato.org; internet www.anato.com.co; f. 1949; Pres. Dr OSCAR RUEDA GARCÍA; Exec. Dir LUIS BETANCUR.

## Defence

In August 2004 Colombia's Armed Forces numbered 207,000: Army 178,000 (including 63,800 conscripts), Navy 22,000 (including 14,000 marines), Air Force 7,000 (including some 3,900 conscripts). There was also a paramilitary National Police Force numbering about 129,000. Military service is compulsory for men (except for students) and lasts for 12–24 months.

**Defence Expenditure:** an estimated 7,400,000m. pesos in 2004.

**Chief of Staff of the Armed Forces:** Gen. CARLOS ALBERTO OSPINA OVALLE.

**Commander of the Army:** Gen. MARTÍN ORLANDO CARREÑO SANDOVAL.

**Commander of the Navy:** Adm. MAURICIO SOTO GÓMEZ.

**Commander of the Air Force:** Gen. EDGAR ALFONSO LESMEZ ABAD.

## Education

Education in Colombia commences at nursery level for children under six years of age. Primary education is free and compulsory for five years. Admission to secondary school is conditional upon the successful completion of these five years. Secondary education is for four years. Following completion of this period, pupils may pursue a further two years of vocational study, leading to the Bachiller examination. In 1995 the total enrolment at primary and secondary schools was equivalent to 85% and 50% of the school-age population, respectively. In 2002 there were an estimated 55,869 primary schools and 12,291 secondary schools. In that year there were an estimated 321 higher education institutes (including universities) in Colombia. There are plans to construct an Open University to meet the increasing demand for higher education. Expenditure on education by the central Government in 1999 was 5,789,000m. pesos, representing 19.7% of total spending.

# Bibliography

For works on South America generally, see Select Bibliography (Books)

Ardila Galvis, C. *The Heart of the War in Colombia*. London, Latin American Bureau, 2000.

Bergquist, C. W. *Violence in Colombia, 1990–2000: Waging War and Negotiating Peace*. Wilmington, DC, Scholarly Resources Inc, 2001.

Betancourt, I. *Until Death Do Us Part: My Struggle to Reclaim Colombia*. London, Ecco Press, 2001.

Croce, E. *Programación Financiera: Métodos y Aplicación al caso de Colombia*. Washington, DC, IMF Publications, 2002.

Dudley, S. *Walking Ghosts: Murder and Guerrilla Politics in Colombia*. London, Routledge, 2004.

Drexler, R. W. *Colombia and the United States: Narcotics Traffic and a Failed Foreign Policy*. Jefferson, NC, McFarland & Co, 1997.

Earle, R. A. *Spain and the Independence of Colombia, 1808–1825*. Exeter, University of Exeter Press, 2000.

Giraldo, J. *Colombia: The Genocidal Democracy*. Monroe, ME, Common Courage Press, 1996.

Henderson, J. *Modernization in Colombia: The Laureano Gómez Years, 1889–1965*. Gainesville, FL, University Press of Florida, 2001.

Hernández Gamarra, A. *A Monetary History of Colombia*. Bogotá, DC, Villegas Editores, 2002.

Jaramillo, F. *Liberalization and Crisis in Colombian Agriculture*. Boulder, CO, Westview Press, 1998.

Kline, H. F. *State Building and Conflict Resolution in Colombia, 1986–1994*. Tuscaloosa, AL, University of Alabama Press, 1999.

Londoño-Vega, P. *Religion, Society and Culture in Colombia: Antioquia and Medellín, 1850–1930*. Oxford, Clarendon Press, 2002.

McFarlane, A. *Colombia Before Independence*. Cambridge, Cambridge University Press, 2002.

Rausch, J. *Colombia: Territorial Rule and the Llanos Frontier*. Gainesville, FL, University Press of Florida, 1999.

Richani, N. *Systems of Violence: The Political Economy of War and Peace in Colombia*. Albany, NY, State University of New York Press, 2002.

Ruiz, B. *The Colombian Civil War*. Jefferson, NC, McFarland and Co, 2001.

Safford, F., and Palacios, M. *Colombia: Fragmented Land, Divided Society*. Oxford, Oxford University Press, 2001.

Sanchez, G., and Meertens, D. *Bandits, Peasants and Politics: The Case of 'La Violencia' in Colombia*. Austin, TX, University of Texas Press, 2001.

# COSTA RICA

## Geography

### PHYSICAL FEATURES

The Republic of Costa Rica is the southernmost of the Central American countries, with Panama beyond the southern frontier (330 km or 205 miles in length). Nicaragua lies to the north (309 km). Although the border with Panama is across a narrower part of the isthmus, it is more convoluted and, therefore, longer. The country has coasts on both the Pacific (1,015 km) and the Caribbean (usually referred to as the Atlantic coast—212 km). The territory of the country includes the rugged Isla del Coco (Cocos Island), some 480 km south-west of continental Costa Rica. The country has a dispute with Nicaragua over navigational rights on the San Juan river on the border, exacerbated by the presence of many illegal Nicaraguan immigrants in the country. Costa Rica has a total area of 51,060 sq km (19,730 sq miles).

The San Juan, the outflow of Lake Nicaragua, flows into the Pacific and forms much of the north-eastern border. In the north-west the border skirts the south-western edges of Lake Nicaragua, to the east of the Cordillera de Guanacaste, on the Pacific side of which is the Nicoya peninsula. These mountains head inland eventually to join the higher, central ranges, culminating in the Cordillera de Talamanca, which thrusts up the spine of the country from the south. The region is also volcanic, with four volcanoes near San José, the capital, two of them active—Irazu last erupted destructively in the mid-1960s. The highest point is south of here at Cerro Chirripo (3,810 m or 1,2504 ft), in the rugged Talamanca range. The capital city is located in a fertile, upland basin, the Meseta Central Valley, at an altitude of about 1,170 m. On either side of the mountains are coastal plains, the Pacific coast being more irregular in outline and the Atlantic coast lower, swampier and heavily forested (almost one-third of the country is wooded). Costa Rica is about 460 km at its maximum length, its axis running from south-east to north-west, with the northern border the widest part of the country (260 km). The country is very fertile and rich in biodiversity, among the most intense in the world, with a massive range of flora and fauna, notably bird life, flourishing in a huge range of ecosystems. Thirty nature reserves cover some 11% of the territory, with about twice as much again also gaining some form of protection, sheltering more than 200 species of mammals (including six species of wild cat), over 850 species of birds (including endemic species, such as two types of hummingbird and a tanager), 1,000 butterfly types and almost 200 amphibians and 220 reptiles. The country is reckoned to contain up to 13,000 varieties of flowering plant and 10% of the world's birds and butterflies. The isolation of the Isla del Coco has created another unspoilt natural haven for both land and marine life (here alone are three endemic bird species: the Cocos finch, flycatcher and cuckoo).

### CLIMATE

The climate is tropical and subtropical, varied by the highlands and the competing weather systems of the Pacific and the Caribbean. The dry season is December–April. The onset of the rainy (or green) season can bring flooding in the coastal lowlands and, later, landslides in the mountainous interior, in a

topography complicated by earthquakes and active volcanoes. Hurricanes are likely along the Caribbean coast, where rainfall is greater than on the Pacific coast. Rainfall is also greater in the mountains. Annual precipitation varies enormously according to location, some places recording a remarkable 6,000 mm (234 ins), others as relatively little as 1,500 mm. The average for the whole country is 3,300 mm. Temperatures vary mainly with altitude, the coast experiencing thermometer readings between about 27°C and 32°C (81°–90°F), the Central Valley around 22°C (72°F) and the mountains being cooler still.

### POPULATION

The white and Mestizo or mixed-race population, mainly of Spanish descent, accounts for 94% of the total, with blacks at 3%, and Amerindians and Chinese at 1% each. Traditionally a Roman Catholic country, the denomination still enjoys the adherence of 77% of the population; however, evangelical Christian groups have gained many new followers and non-Catholic Christians now represent 18% of the population. Spanish is the official language, but some English is spoken around Puerto Limón, which is the heart of black Creole culture, on the Caribbean. The Amerindian languages spoken in Costa Rica are all of the Chibcha group. Costa Ricans refer to themselves informally as 'Ticos'.

About one-half of the total population of 4.0m. (mid-2004 estimate) live in the central plateau (Meseta Central) around San José, the capital. There are several other large population centres in that region, but the principal cities of the Pacific coast are Puntarenas, on the Gulf of Nicoya, in the north-west, and further north, inland, Liberia. The main city of the Atlantic coast is Limón. Another important demographic statistic is the 1m. or so tourists who visit annually. The country consists of seven provinces.

# History

**DIEGO SÁNCHEZ-ANCOCHEA**

Based on an earlier essay by Prof. JENNY PEARCE

Rumours of gold and treasure led the navigator Christopher Columbus to bestow the name of Costa Rica (rich coast) on the country's Caribbean shore, where he landed in 1502 on his final voyage from Europe to the New World. Ironically, there was no gold and few other resources. This fact influenced Costa Rican history profoundly and it evolved distinctly from the rest of the Central American isthmus. The country's indigenous Indian (Amerindian) inhabitants were few and mainly nomadic but still managed to resist Spanish incursion until 1522, when settlers moved from the coast and colonized the central plateau, the Meseta Central. Diseases the colonizers brought with them decimated the indigenous population. Only some small Mayan tribes in the northern and southern highlands survived. According to the 2000 census, the indigenous population made up just 1% of the total population. Without reserves of Indian labour to draw on or minerals to exploit, the Spanish settlers became poor subsistence farmers. This development was aided by the region's distance and physical isolation from the colonial Government, the Captaincy-General of Guatemala, based in Antigua, Guatemala, which allowed a certain amount of autonomy. An individualistic and agrarian society of small landowners developed, more along European than colonial lines. Cartago was founded in 1563, but there was no expansion of settlement until the beginning of the 18th century, when small groups left the Meseta Central to establish other cities. San José, now the capital of Costa Rica, was founded in 1737.

## INDEPENDENCE

Costa Rica joined the other countries of the region in their declaration of independence from Spain in 1821 (following the Spanish Revolution of 1820). It became part of the newly formed Mexican Empire, which lasted for two years until 1823, when Costa Rica joined the United Provinces of Central America. When the federation collapsed in 1838 Costa Rica became an independent republic. Costa Rica's early governments were civilian, in contrast to the rest of the isthmus.

Coffee, introduced from Cuba in 1808, became the motor of Costa Rica's development. The Government offered free land and trees to coffee growers, encouraging the development of a large landowning peasantry, with a small aristocracy. The opening of the European market in 1845 and the establishment of a cart road to the port of Puntarenas on the Pacific the following year hastened development. In 1890 a railway opened between San José and Puerto Limón on the Caribbean, along the Reventazón valley. This facilitated the foundation of coffee plantations outside the Meseta Central towards the east. Bananas were introduced around Puerto Limón in 1878. Jamaican labourers were brought to Costa Rica to clear large areas of land along the Caribbean coast; when production reached its peak, in 1913, interest was then shown in the Pacific coast.

## THE MOVE TOWARDS STABILITY

For half the period between 1835 and 1899 Costa Rica was under military rule as Conservative and Liberal élites contested power. However, in the early 20th century rising literacy rates increased political participation dramatically, laying the foundation for one of Latin America's most successful democratic systems. Progress was reversed briefly in 1917, when Federico Tinoco ousted the elected President, Alfredo González, who had introduced new taxes and reforms damaging to the coffee aristocracy. Two years later Tinoco's military dictatorship was ousted by a combination of popular revolt and guerrilla activity by Costa Rican exiles in Nicaragua. Julio Acosta's 1919 election victory restored democracy and constitutional order until 1948. The Communist party had grown powerful through the recession of the 1930s, taking especially strong hold in the union movement based in the banana plantations. However, it abandoned revolutionary policies in favour of social welfare, working with the Government of Dr Rafael Angel Calderón Guardia in the 1940s to introduce labour safeguards and a social security system. Calderón Guardia also formed an unlikely alliance with the Roman Catholic Church. The presidential election campaign of 1948 was characterized by violent protests, including a 15-day general strike. When the opposition candidate, Otilio Ulate Blanco, won, the government candidate, Calderón Guardia, contested the result and it was nullified. In March 1948 José Figueres Ferrer, a coffee farmer, led an uprising in support of Ulate Blanco. A month of fighting ensued until a truce was agreed, and Santos León Herrera installed as the interim President. In May the Constitution was abrogated and Figueres Ferrer and his junta took over government. A constitutional assembly convened in January 1949, formulating a new Constitution, which abolished the army, assigning its budget to education. During its months in office, the junta also nationalized the banking system and created a bureaucracy, which would become a strong political force during the second half of the 20th century. In November the junta resigned and Ulate Blanco, President-elect, was inaugurated.

## THE EMERGENCE OF THE PLN

Figueres, a social democrat, found Ulate too moderate, and withdrew his support for his presidency in 1952, forming the Partido de Liberación Nacional (PLN). He dominated Costa Rican politics for the next two decades, serving as President between 1953 and 1958 and, again, from 1970 to 1974. The PLN developed the so-called 'tico' model of heavy state intervention. The Figueres Governments and that of his PLN successor between 1974 and 1978, Daniel Oduber Quirós, nationalized large areas of the economy and extended the welfare state. Between 1958 and 1980 public social spending as a proportion of GDP increased from 8.7% to 23.6%. Intervening conservative administrations tended to favour private enterprise but were prevented from reducing state benefits by public opinion. Conservative government returned between 1978 and 1982, under Rodrigo Carazo Odio.

Elections in February 1982 gave a decisive victory to the PLN's presidential candidate, Luis Alberto Monge Alvarez, with 58% of the votes cast. The PLN also won a clear working majority in the Asamblea Legislativa (Legislative Assembly). The Monge administration was confronted by two big challenges: the domestic economic crisis and the external problem of conflict in the region, focusing on Costa Rica's northern neighbour, Nicaragua.

## THE ARIAS PRESIDENCY

Although the Monge Government achieved only limited economic success, the PLN won a decisive, if surprising, victory at the presidential and legislative elections in February 1986. The PLN's victory was attributed to the dynamism and youth of its leader, Oscar Arias Sánchez, who, at the age of 44, was the youngest President in the country's history. It was also a reaction against the extremism of his main opponent, the right-wing Rafael Angel Calderón Fournier, the son of the former President. On taking office, President Arias announced his desire for the country to pursue a more independent policy, while at the same time recognizing the necessity of maintaining good relations with the USA to ensure continued foreign aid. The Government was keen to emphasize its commitment to the welfare state, promising the creation of 25,000 jobs and 20,000 dwellings each year.

However, its principal aim was reform of the economy through the promotion of a market-friendly model that concentrated on the promotion of exports. This set the political pattern for the following administrations, all of which promoted neoliberal reforms in the face of protests from a citizenry nourished on 40 years of state-sponsored welfare. The Government renegotiated its huge public debt after its 1982 default. However, it had to agree to a number of structural adjustment policies with the IMF. Public-sector strikes in 1988 were followed by protests by farmers, who were unhappy with the Government's policy of promoting the cultivation of cash crops to appease the IMF. Labour unrest increased in 1989 as trade unions, professional bodies and civic groups united to demonstrate against the Government. However, President Arias could claim substantial economic success. As well as the renegotiation of the country's foreign debt, during his term of office gross domestic product (GDP) grew by 4% each year, the annual inflation rate was reduced to less than 10% and unemployment declined to less than 4% of the total labour force.

In September 1989 a report by the Asamblea Legislativa's commission of inquiry into the extent of drugs-trafficking and related activities was published. The report was a blow to the PLN, as it implicated many political and business figures of involvement in illegal activities, including the former PLN President, Daniel Oduber Quirós, and Leonel Villalobos, a PLN deputy. They were both forced to resign from all public positions. This scandal, combined with a general desire for change, led to the victory of Calderón Fournier, of the Partido Unidad Social Cristiana (PUSC), in the presidential election of February 1990. The PUSC also secured a majority in the Asamblea Legislativa.

## FROM CALDERÓN TO RODRÍGUEZ

The battle over the future of Costa Rica's welfare state and public-sector utilities continued to dominate politics in the 1990s. It also divided the country deeply and led to a gradual erosion of confidence in politicians and the political system. Costa Ricans were justifiably protective of the exceptional social harmony that the country enjoyed compared to others in the region. Nevertheless, economic adjustment measures adopted in order to qualify for further IMF credits and loans threatened the 'welfarism' that underpinned that harmony.

The presidency of Calderón Fournier (1990–94) was dominated by this issue. Overall economic performance during his administration was good, but the Government's privatization programme and reductions in expenditure on education and public health increased social and political tensions. The opposition PLN prevented the realization of the third phase of the structural adjustment programme in 1993 by voting against its approval in the Asamblea Legislativa. A total of US $280m. in multilateral funding for this programme depended on congressional approval of legislation to deregulate and liberalize the economy, including the privatization of the telecommunications and health sectors and of the state petroleum company, Refinadora Costarricense de Petróleo (Recope). The PLN strongly opposed privatization and proposals to reduce the public sector work-force by 25,000.

In 1994 José María Figueres Olsen, son of José Figueres Ferrer, the former President, stood as the PLN's presidential candidate, promising to improve housing, education, health and other social programmes, rather than implement the public-sector reforms agreed by President Calderón Fournier with the World Bank. With this manifesto Figueres obtained a narrow victory in the elections of 6 February. In simultaneous legislative elections, the PLN obtained 28 seats in the 57-seat Asamblea Legislativa, one short of an overall majority. The PUSC won 24 seats, with the remaining five seats being secured by independent candidates (most of whom would often vote with the PLN).

As President, however, Figueres rapidly converted to free market economics: during 1995 he restored relations with the IMF and the World Bank by agreeing to a new structural-adjustment programme. In October the Government signed a letter of intent with the IMF, which led to the release of loans from the Inter-American Development Bank (IDB) and the World Bank. These loans had been delayed by the Government's failure to meet economic performance targets in 1994, partly because of PLN opposition. Figueres sought to raise taxes to reduce the large and persistent fiscal deficit. In April the Government made an agreement with the opposition PUSC to ensure the implementation of tax increases and austerity measures agreed with the IMF. The PUSC was divided over this support for Figueres, while the President succeeded in consolidating his position within his own party, in spite of internal opposition to his shift to neoliberal policies.

However, within the country as a whole social unrest increased. In 1996 protests and strikes led to the suspension of plans to privatize the energy and telecommunications monopoly, the Instituto Costarricense de Electricidad (ICE), in favour of a restructuring plan. President Figueres's popularity decreased dramatically in 1995–96 as living standards fell, taxes increased, and the social spending that protected many Costa Ricans from poverty was drastically reduced.

A presidential election on 1 February 1998 resulted in victory for Miguel Angel Rodríguez Echeverría of the PUSC, with 47% of the votes cast, compared with the 44% obtained by the PLN's presidential candidate, José Miguel Corrales. The PUSC obtained 27 seats in the 57-seat Asamblea Legislativa, compared with the PLN's 23 seats. There was a 28% abstention rate, compared with 18.9% in 1994, demonstrating the level of public disillusionment with the established parties. Rodríguez received fewer votes in winning the presidency than he had four years previously when he lost.

## THE RODRÍGUEZ ADMINISTRATION AND THE ACCELERATION OF SOCIAL CONFLICT

President Rodríguez, a neoliberal economist, sought to further extend the process of liberalization and deregulation. He attracted foreign investment, signed free trade agreements and lowered tariff barriers. Annual GDP growth averaged more than 4% during his administration, but was confined to the new export-orientated sectors such as information technology, located in the free trade zones, and tourism. Traditional areas such as farming, food processing and manufacturing suffered and most Costa Ricans did not feel the benefit of the economic expansion. Moreover, Rodríguez suffered a serious reverse in April 2000, when the largest popular protests in 30 years forced the Government to withdraw legislation that would have opened the energy and telecommunications industries to private investment. The protests reflected the popularity of the ICE among most Costa Ricans, but also the opposition of a majority of the population to the Government's neoliberal policies. According to a survey published in 2000, 84.1% of the Costa Ricans believed that the protests were 'part of a general discomfort' with the Government and the political élite. The Supreme Court subsequently rejected the proposed law on procedural grounds. A special commission, established in May, composed of representatives of government, the private sector and trade unions also failed to agree on the issue.

This privatization debate encapsulated the difficulty in implementing unpopular economic reforms aimed at reducing the level of state intervention in the economy. This was the most contentious issue in Costa Rican politics in the 1990s. President Rodríguez was committed to deepening the reforms adopted during the Arias, Calderón and Figueres administrations, privatizing the Banco de Costa Rica and adopting other market-friendly policies. However, popular resistance led by powerful trade unions made progress extremely slow.

The Government also faced opposition from farmers, in the form of violent demonstrations, over its policy of increasing tariffs on food imports, and from industrial and agricultural workers over its apparent favouring of employers' interests and its perceived hostility to independent trade unions. Several shipments of sugar from the USA were blocked at Costa Rican ports. The ratification of a free trade deal with Canada was delayed by potato farmers who lobbied legislators to end the importation of Canadian potatoes. In May 1999 the International Confederation of Free Trade Unions filed a complaint at the International Labour Organization over alleged mistreatment of trade-union members, and within the banana sector there were persistent reports of harassment and attacks on trade unionists.

As President Rodríguez's unpopularity reached new depths in his last year in office, he sought to focus on the social issues that most preoccupied voters. As a result, his approval ratings improved dramatically. He was aided by a division in the PLN in 2000. First, Oscar Arias Sánchez, the former President, sought to amend the Constitution to allow him to run for office again, a move opposed by the PLN's new generation of leaders, who wanted their turn in power. The leadership of the conservative PUSC initially supported his ambition as it meant that former PUSC President Calderón Fournier would also benefit. However, deputies in the Asamblea Legislativa voted against the change. Arias took the issue to the Constitutional Court, which ruled against him in September, temporarily ending his hopes of regaining the presidency.

In December 2000 Ottón Solís, a former PLN deputy and a former Minister of Planning, formed a breakaway party, the Partido de Acción Ciudadana (PAC). He sought to take advantage of social discontent with the economic reforms of the 1990s and the electorate's disillusionment with the two traditional parties. He attracted many members of the PLN and was able to use their experience to develop a national structure with surprising speed. He also attracted non-voters and some PUSC supporters.

The PAC campaigned for the February 2002 presidential and legislative elections on an anti-corruption platform and also pledged to halt privatization and free trade agreements. It won over many young, well-educated people, as well as rural voters. Rodolfo Araya Monge, who emerged as the PLN's presidential candidate after a fractious primary campaign, also promised assistance to farmers and other groups affected by economic liberalization. However, Abel Pacheco de la Espriella, the PUSC's nominee, surprisingly managed best to address the country's unease, and won the presidential election. The psychiatrist and television host distanced himself from the party leadership and emphasized his desire for unity in the country. In the first round of voting, on 3 February 2002, he attracted 38.5% of the ballot, compared with Araya's 30.9%. Solís' strong performance in the first round of voting, when he was third placed with 26.3% of the vote, meant that a second round of voting was held on 7 April. Pacheco won this, with 58% of the votes cast, against 42% polled by Araya. It was the first time that the PLN had lost successive elections and the result left it badly weakened.

In elections to the 57-seat Asamblea Legislativa, also held on 3 February 2002, the PUSC won 19 seats, the PLN 17 and the PAC 14. The success of the PAC together with a strong showing by the neoliberal Movimiento Libertario, which secured six seats, signalled a transformation of the Costa Rican political system from a two-party to a multiparty system. Nevertheless, the fracture of the PAC in February 2003, when eight of its 14 congressmen left the party following a dispute with Solís over its ethical code, brought into question the long-term survival of the alternatives to the two dominant parties.

## THE PACHECO ADMINISTRATION: SOCIAL STALEMATE AND CORRUPTION SCANDALS

President Pacheco's challenge in dealing with a disaffected public and a divided legislature was enormous. At his inauguration on 8 May 2002 he promised to be conciliatory while continuing to pursue moderate reform. His first challenge was to reduce the fiscal deficit to avoid further debt increases. The month before, a cross-party commission of former finance ministers had presented its long-delayed report, which concluded that higher taxes and reduced spending were needed to avoid severe financial and social instability by 2006, when the fiscal deficit would be 12.6% of GDP and debt 56.3% of GDP. The main proposals were the replacement of sales tax with value-added tax and an extension of profits tax to the export zones. Along with a 3% spending cut, the moves would eliminate the fiscal deficit. However, the measures were unpopular. While the Government won legislative approval in December 2002 for a series of temporary tax rises and spending cuts to limit the fiscal deficit to 3% of GDP in 2003, it had still to secure approval of a long-term tax reform in the Asamblea Legislativa by May 2005.

Despite his failure to implement an enduring solution to the fiscal problem, President Pacheco's approval ratings, which had reached 66% in December 2003, remained high in the first two years in office. This was mostly owing to renewed economic growth as the US economy recovered from its brief recession. Nevertheless, tentative steps towards introducing the free market into traditionally publicly run areas proved politically costly and the Government faced continuous opposition. In October 2002 plans for a private US company to supervise the construction and staffing of a new maximum-security prison met with controversy. Following his objections to private-sector involvement in the scheme, the Minister of Justice, José Miguel Villalobos, was dismissed. In November the Government angered directors of the ICE when the Comptroller-General ordered them to annul contracts awarded to an equipment supplier whose charges the Government considered to be too high. The Government also announced that it would investigate a decision by the ICE to rescind a contract given to an Israeli company for the sale of telecommunications equipment. In early 2003 electrical workers failed to receive salary bonuses and further plans by the Government to reduce the ICE budget were criticized by unions and the institute's directors. ICE employees began strike action in mid-May. At the same time, teachers also began industrial action in protest at their salary and pension arrangements. The disputes resulted in the resignation of the Ministers of Finance, Presidential Affairs and Education before the Government capitulated to the ICE's demands and agreed to make a bond issue on the international markets to improve the finances of the power supplier in June. In September 2004 protests by public sector workers, with the support of other civil society groups with various demands, led to an increase in public sector wages. The Minister of Finance's opposition to this measure led to his resignation, which in turn provoked the departure of seven senior government officials. This brought to 14 the number of cabinet members who had resigned or been dismissed during Pacheco's administration.

Political uncertainty increased even further with the disclosure of two major corruption scandals in October and November 2004. The first involved the alleged payment of commissions to senior managers of the Costa Rican social security institute, the Caja Costarricense de Seguro Social (CCSS), a member of its board of directors and the former President of the Republic, Rafael Angel Calderón Fournier, for the purchase of medical equipment from a Finnish company represented in Costa Rica by Corporación Fischel. Another finding of illegal commissions, this time in the ICE, followed this first scandal: the French telecommunication transnational corporation Alcatel allegedly made payments of several million US dollars to another former President, Miguel Angel Rodríguez Echeverría, as well as to senior ICE officials in order to secure a contract to supply 400,000 cellular mobile telephones. The corruption scandals resulted in the preventive imprisonment of both former Presidents (as well as Rodríguez's resignation as Secretary-General of the Organization of American States) and also affected the popular standing of the PUSC and of the PLN, after it was revealed that a third former President, José María Figueres Olsen, also received payments of US $1m. from Alcatel for consultancies carried out after his term in office.

## FOREIGN POLICY: FROM REGIONAL CONFLICT TO TRADE CONFLICTS

Costa Rican foreign policy was dominated in the 1980s by the regional conflicts in Central America. In 1983 an official proclamation of neutrality was made, stating that Costa Rica would not be used as a base for attack against its neighbours. However, in these years Costa Rica collaborated closely with US policy towards the region. In 1984 a Nicaraguan counter-revolutionary ('Contra') base was established in the north of the country. Moreover, the decision in 1985 to create an anti-guerrilla battalion, to be trained by US military advisers, fuelled public scepticism about the Government's commitment to Costa Rican neutrality.

The Government of President Arias, which assumed office in 1986, changed tack and distanced itself from the USA. It fully restored diplomatic relations with Nicaragua and a peace accord based on proposals by Arias was signed by the Presidents of the five Latin Central American countries (Costa Rica, El Salvador, Guatemala, Honduras and Nicaragua) at a meeting in Esquípulas, Guatemala, in 1987. Acceptance, albeit provisionally, of the plan was regarded as a personal triumph for President Arias, who was awarded the Nobel Peace Prize in October. Despite continuing US pressure Arias maintained his neutrality. In January 1988 he organized the first meeting between Nicaraguan government officials and Contra leaders, to discuss terms for a cease-fire. The Arias initiatives also led towards the UN-supervised peace settlement in El Salvador in 1992.

Following the restoration of relative peace to the Central American region in the early 1990s, trade issues subsequently dominated Costa Rican foreign policy. Costa Rica, together with Chile and Mexico, has been one of the most active negotiators of free trade agreement in Latin America. The country sought to join the North American Free Trade Agreement (NAFTA), indicating the priority of trade relations with the USA and Mexico over those with its neighbours in the Central American Common Market. Costa Rica signed a free trade agreement with Mexico in 1994, giving some 86% of Costa Rican exports duty-free access to the Mexican market. The free trade agreement led to an increase in Mexican investment in Costa Rica, but only a small number of transnational corporations based in Costa Rica were able to take advantage of the new export opportunities.

In November 1999 Costa Rica signed a free trade agreement with Chile (ratified in December 2000) and in April 2001 the country reached a similar accord with Canada. At the end of the year Costa Rica and its four Central American partners reached the basis of a deal with Panama, but disputes over telecommunications and sensitive agricultural products obstructed progress towards a final treaty for all except El Salvador, which signed a bilateral deal. The Central American countries also reached an agreement with the Dominican Republic, which contributed to expand Costa Rica exports to that markets significantly.

In March 2002 US President George W. Bush (2001–) visited El Salvador to make a formal offer of a US trade agreement with the five Central American countries, to be known as the Central American Free Trade Agreement (CAFTA). Negotiations began in January 2003 and proceeded swiftly. Costa Rica was reluctant to open its domestic state-run monopolies to competition and continued to hold out after its four partners concluded talks in December. Government representatives finally agreed terms for CAFTA in January 2004 and signed the treaty in May. As a result of the negotiations, trade in potatoes and onions, two sensitive products, were excluded from the Agreement, while the Costa Rican delegation ceded ground over privatization: the Government agreed progressively to open three sectors of the telecommunications market, namely private network services, internet services and wireless services, and agreed to establish a regulatory framework with the aim of promoting effective market access. Costa Rica also engaged fully to liberalize the insurance sector by 1 January 2011. The Government hoped the accord would provide Costa Rican industry with wider access to materials, equipment and technology at lower prices. Nevertheless, by May 2005 CAFTA had still not been debated in the Costa Rican Asamblea Legislativa and was facing much social resistance. Farmers and trade unions were set to protest, the PAC said it would oppose ratification of the accord and the PLN's position remained uncertain. Moreover, the survival of CAFTA also depended on its approval by both houses of the US Congress; however, in late July the US lower house approved the Agreement.

Another key component of Costa Rica's foreign policy at the beginning of the 21st century was its bilateral relations with Nicaragua. In early 1995 relations with Nicaragua became strained when it was alleged that a group of illegal Nicaraguan immigrants had been expelled from Costa Rica in a violent manner. A tightening of immigration policy in Costa Rica resulted in further expulsions, although there was a general amnesty after Hurricane Mitch in 1998. In July 1998 further antagonism developed as Nicaragua reasserted its sovereignty over the San Juan river, which formed the border between the two countries, by prohibiting Costa Rican civil guards from carrying arms while navigating the river. This was settled in June 2000 but other disputes arose periodically, precipitated by both sides. In July 2001 the symbolism of a wall built on the frontier between Costa Rica and Nicaragua to control border traffic led to renewed tensions. In 2002 the new Governments in both countries resolved to settle the border dispute within three years. This resolve was tested by continuing controversy over the treatment of Nicaraguan immigrants, with around 150 refused entry at the border each day. There were also tensions over maritime rights. In October 2000 Costa Rica ratified a maritime boundaries treaty with Colombia, recognizing the latter's control over 500,000 sq km of the Pacific Ocean.

## OUTLOOK

With presidential elections approaching in 2006, Costa Rican politics were mired in uncertainty in 2005. Over the previous two decades the country had maintained intact its democratic institutions in the midst of military tensions in the region. Costa Rica also succeeded in attracting new foreign investment and generating high—though fluctuating—economic growth. However, Costa Ricans became increasingly disillusioned with the direction of the country and with the quality of its political class. Recent corruption scandals, while signalling the institutional strength of the judiciary, further contributed to the disenchantment. The country needed to reinvigorate its democracy and find new ways to build political and social consensus. A peaceful discussion of the benefits and costs of CAFTA and the resolution of the nation's fiscal problems were also paramount. The possible return of Oscar Arias Sánchez to the presidency (since the Constitutional Court voted to remove the bar on the re-election of former heads of state in April 2003) could be an important step in this direction or it could increase social tensions even further.

# Economy

## DIEGO SÁNCHEZ-ANCOCHEA

Based on an earlier article by ANDREW BOUNDS

Costa Rica has one of the highest levels of human development in Latin America. In 2002 it was ranked 45 in the Human Development Index, below only Argentina and Chile; life expectancy at birth was 78.0 years, infant mortality was just 9 per 1,000 live births and the literacy rate was 95.8% of the population. In 2003, according to World Bank estimates, gross national income per head was US $4,280, compared to the Latin American average of $3,280. An historical commitment to social stability, political accountability and public spending in health and education historically supported Costa Rica's unique economic model. Costa Rica still has one of the highest levels of social spending per head in Latin America and the Caribbean; in 2000/01 social expenditure was $689 per head (a total spending equivalent to 18.4% of gross domestic product—GDP).

## STAGES OF ECONOMIC DEVELOPMENT

Belying its Spanish colonial name, Costa Rica, or rich coast, the country yielded few resources for its conquistadores to exploit. There were also few indigenous inhabitants to press into slave labour, and those that were there did not long survive the European invasion. Thus, subsistence agriculture was the norm until the introduction of first coffee and then bananas from the early 19th century. From then on, Costa Rica's development outpaced the rest of Central America. It was first to export coffee (in 1832), to establish a commercial bank (in 1864) and to build a railway (in 1890). After the 1948 civil war it departed even further from its Central American peers. Large sectors of the economy were nationalized and considerable state support given to smallholders, who made up the backbone of the agricultural economy and underpinned its commitment to democracy. A welfare state was established and public utilities expanded quickly under state ownership. In the 1960s the Government sought to diversify the economy and promote new exports through import substitution within the Central American Common Market (CACM), attempting to build a domestic manufacturing industry through protectionist tariffs and export subsidies. However, it was not until the 1990s when tourism blossomed and vast foreign investment fostered first a textile then a high-technology sector that Costa Rica reduced its reliance on coffee and bananas. It sought to transform itself into a diversified exporter with world-class companies through a mixture of government incentives and network of free trade agreements. A report by the UN's Economic Commission for Latin America and the Caribbean in 2003 singled out Costa Rica's achievement as an 'example of how to achieve better conditions via *maquila*'.

However, success in the export sector was accompanied by problems in other spheres. By the 1980s high debt threatened to swamp the country, which was also negatively impacted by the crisis of the CACM and the military crisis in the region. The Government was forced to co-operate on an unpopular structural adjustment programme with multilateral lending institutions. This proved difficult owing to the inflexible nature and independence of Costa Rica's state-owned companies and political system. Successive Governments also sought to develop non-traditional exports with incentive schemes and reined in government spending. Nevertheless, large deficits had to be funded by heavy domestic and overseas borrowing and in 1988 the country was forced to apply for IMF and World Bank help to reschedule its debts. While it gained temporary relief then, progress over the subsequent 15 years was painstakingly slow because of the public's reluctance either to scale back the welfare state or to pay for it. Tax collection in 2002 was just 13.2% of gross domestic product (GDP), levels comparable to the rest of Central America. While the debt restructuring of 1988–89 reduced external debt, which stood at an estimated 21.1% of GDP in 2004, domestic public debt steadily increased to 38.7% of GDP in that year. Interest payments on the internal and external debt represented 21.4% and 6.6% of total public spending, respectively, in 2004.

During the 1990s Costa Rica followed a conflictual path towards a neoliberal policy model. Significant reforms were introduced, including changes to the pensions and tax systems, and measures to promote foreign investment and to facilitate private-sector participation in activities, such as banking and power generation, formerly confined to the state sector. At the same time, however, other reforms such as the deregulation of the telecommunications industry failed, owing to the active opposition of a majority of the population. Export and growth prospects were transformed by the establishment of two microprocessor plants by the US manufacturer Intel in March 1998, which added 3%–4% to GDP growth in 1999, although a decline in microchip exports in the following year caused a decrease of 2.3%. Costa Rica's economy expanded at an average annual rate of 4.8% between 1990 and 2004, one of the highest rates in Latin America. However, the Costa Rican export-driven model created a dual economy and questioned the survival of the old institutional structure. While export sectors performed well, traditional producers faced the loss of subsidies and protective measures. This led to wage stagnation, job losses and an erosion of social benefits. The rate of unemployment in the early 2000s was relatively high by historical standards. In 2004 Costa Rica had a rate of unemployment of 6.5% and a rate of underemployment of 7.9%. Meanwhile, the administration of President Abel Pacheco de la Espriella, which took office in 2002, initially managed to reduce the level of poverty from 20.6% in 2002 to 18.5% in 2003, although in 2004 it increased again to an estimated 21.7%.

# AGRICULTURE

Agriculture, hunting, forestry and fishing contributed an estimated 9.9% of real GDP and employed 14.8% of the employed population in 2004. Since 1990 the agriculture sector, with an annual average rate of real growth of 3.4%, has consistently grown less than the rest of the economy. In 2004 primary production grew by 1.6%, with 4.2% for the economy as a whole. The major crops were coffee and bananas for export, and rice for domestic consumption. The Government's attempt to encourage non-traditional export crops meant that the country was no longer self-sufficient in staples and imported large quantities of beans, rice and maize. As a result of falling prices, agricultural export earnings, which were around US $1,800m. per year in 1996–98, fell to $1,422.2m. in 1999 and to $1,222m. in 2002. The recuperation of coffee prices together with the expansion of the internal demand for pineapple contributed to the expansion of agricultural exports in 2004 to $1,469.5m.

**Bananas**

Apart from a brief period in the mid-1980s, bananas were Costa Rica's main export commodity. In 1985 the closure of the Pacific-coast operations of Chiquita brought widespread economic depression to the region, since many of its towns were wholly dependent on the company's fortunes. Consequently, the Government purchased 1,700 ha of the abandoned plantations (which totalled 2,300 ha) and converted them to the cultivation of cocoa. Elsewhere also, bananas were being replaced by more profitable crops, such as African palm, sugar cane and exotic fruits.

The Costa Rican banana industry recovered gradually with incentives from the Government, but then faced a new threat from the European Union (EU) quota system from 1993. The quota system, designed to protect banana production in the former European colonies, placed an annual limit of 2.0m. metric tons on banana imports from Latin American countries (compared with actual imports of 2.6m. tons in 1992). This led to the flooding of non-EU markets and a significant decrease in banana prices in Costa Rica. Despite this, in 1995 total production was some 2.3m. tons and exports accounted for 25.7% of total export revenues. However, Costa Rican producers were then operating under severe pressure caused by falling prices, continuing market-access problems and climatic adversity. There was also increasing competition from Ecuador, which rapidly became a bigger banana producer than Costa Rica. While Costa Rica had the world's highest productivity levels, it also had higher costs: the average worker earned US $18 per day, compared with $2 in Ecuador. In April 2001 the EU and the USA settled the banana dispute. Under the terms of the deal, negotiated by the World Trade Organization, the EU was to introduce a transitional system, issuing licences according to historical trade patterns, in preparation for a tariff-only system from 1 January 2006. Nevertheless, a world banana surfeit continued, forcing down prices to producers from $5.60 per 40-lb box in June 1999 to $5.20 one year later. The surfeit led to the closure of four plantations, resulting in the loss of some 1,200 jobs, and the cancellation of contracts in Costa Rica by the three principal exporters, Chiquita Brands, the Banana Development Corporation and the Standard Fruit Company. This adversely affected independent farmers, who accounted for more than one-half of total production. The value of banana exports fell from $623.2m. in 1999 to $478.0m. (9.1% of total exports) in 2002. In April 2001, despite resistance from the multinationals, the Government fixed the price of a 40-lb box at $5.25. In May 2002 heavy rains destroyed $30m.-worth of bananas on the Atlantic (i.e. Caribbean coast), reducing output to 87m. boxes, the lowest for 11 years. Production in 2002 was an estimated 1.6m. tons. In 2003–04 both production and export earnings recovered. In the latter year Costa Rica produced 2m. tons of bananas, which generated $558.7m. in export earnings.

**Coffee**

Costa Rica grows only arabica highland coffee, which commands a premium on the world market, and uses technology to achieve some of the highest yields in the world. Output expanded steadily from the mid-1980s, to a peak of 168,000 metric tons by 1992. However, sharp falls in the international price of coffee, which caused bankruptcies and led many coffee growers to diversify into other crops, resulted in a decline in export earnings to 12% of total export revenue in that year. As part of an agreement with other producers, aimed at forcing increases in prices, Costa Rica held back about one-fifth of its 1993 coffee output from the world market and by 1998 output had regained its 1992 level. However, international prices reached a 30-year low in early 2001 and coffee production suffered, amounting to only 150,289 tons in that year. Export receipts fell to $161.9m., from $277.8m. in 2000. This contrasted with an annual average of $404m. in 1995–98. Central American producers, along with Mexico, Colombia and Brazil, agreed to retain up to 10% of their stocks in order to allow the price to rise; however, the arrangement proved difficult to enforce, especially in the face of increased production by Asian countries, particularly Viet Nam, which opposed the retention scheme. From 2002 Costa Rica implemented a plan to take the poorest quality 5% of coffee off the market, with some success. Prices fell by two-thirds between 1998 and 2002, but did recover in 2003 and, more notably, in 2004. Some 40% of coffee in the 2003–04 harvest was classed as higher quality, thus commanding a premium price. This contributed to the gradual recovery of export receipts, which amounted to $193.6m. in 2003 and $213.8m in 2004.

### Sugar

Owing to a decline in world sugar prices, by the mid-1980s the contribution of sugar to Costa Rican export revenues had fallen to about 1%, accompanied by significant reductions in total output. However, sugar production began to increase once again at the end of the 1990s, in spite of a reduction in the annual US sugar quota. Production continued to increase until 1996, when it reached 348,790 metric tons, and hovered around that mark thereafter, although in 2003 output increased to an estimated 380,000 tons. In 2002 export earnings from sugar totalled US $27.1m. However, export earnings had recovered by 2004; that year Costa Rica exported an estimated 161,000 tons of sugar, receiving $38.1m. in foreign exchange. The sector should also receive a boost from the advent of the Central America Free Trade Agreement (CAFTA), under which the USA has granted Costa Rica a tariff-free import quota.

### Other Crops

Crops for domestic consumption included maize, beans and rice, which were grown mainly on small farms with low yields, although advances were being made in rice cultivation, and the size of the units of production was also increasing. The Government gave priority for export incentives to non-traditional crops such as cocoa, African oil palm, cotton, vegetables, cut flowers, macadamia nuts, coconut and tropical fruit. Exports of these crops grew at an average rate of 11% per year in 1991–95. Pineapples, in particular, flourished, and in 2003 exports were worth US $207.6m.—meaning the fruit replaced coffee as the second biggest agricultural earner after bananas. Exports improved further in 2004, to $256.1m. Although the EU imposed an 8.8% tariff on most of Costa Rica's fruit and vegetable exports in May 2004 after ruling that it was too wealthy to continue to benefit from the General System of Preferences, non-traditional primary exports grew by 5.1% in that year. The extraordinary success of pineapple exports was behind this strong performance.

### Forestry and Fishing

Costa Rica had considerable forestry resources, but the Government imposed strict controls on their exploitation because of high historical rates of deforestation. A US $275m. programme to maintain and develop the forestry resource over the following 20 years was announced in 1990. A 1996 forestry law introduced incentives for reforestation. Some 25% of the country lies in protected national parks. However, demand for sustainable wood continues to grow among rich, environmentally sensitive consumers. Some Costa Rican companies are exploiting this demand and others could follow, owing to the country's reputation for conservation.

The country has also been active in the selling of carbon credit offsets for the clean air its forests produce. However, the effective US withdrawal (by its non-signature) from the Kyoto Protocol (an agreement, signed by 38 nations in December 1997, endorsing the mandatory reduction of harmful gas emissions between 2008 and 2012) has dampened hopes for a fully developed carbon trading market in the near future. The absence of any reduction target after 2012 is likely to all but end the market in 2006, according to the World Bank.

There is also an expanding fishing industry, mainly for shrimps, sardines, tuna and tilapia. The amount of tilapia farmed more than doubled from 6.6 metric tons in 1999 to 14.9 tons in 2003. In 2003 the total catch was 49,873 tons (live weight).

### MINING AND POWER

Costa Rica had deposits of iron ore (400m. metric tons), bauxite (150m. tons), sulphur (11m. tons), manganese, mercury, gold and silver. However, by the beginning of the 21st century only the last two were mined. Canadian companies were most active but mining remained a tiny industry because of strong environmental protection; in 2004 it accounted for just 0.1% of real GDP. A mining code adopted in 2001 reduced taxes and improved the legal environment for mining operations. However, a law banning open-cast operations was passed by the Pacheco administration in 2002. The two biggest operating mines, Bellavista and Las Crucitas, near the Nicaraguan border, were ruled exempt by the Supreme Court.

The country also had substantial reserves of petroleum, but they remained largely unexploited. In 1998 Costa Rica offered exploration concessions in the Caribbean to international companies. Harken Energy of the USA found significant reserves in a protected area. In May 2002 the state environmental regulator refused Harken a licence to prospect, on environmental grounds and in November of the same year the Government also refused an application from Mallon Oil Company. President Pacheco made it clear that he would not permit any drilling and plans to exploit coal reserves were put on hold as a result of new mining legislation. As a result of the state's refusal to allow the exploitation of reserves, Costa Rica was compelled to import all its oil, mostly from Venezuela, under the Acuerdo de Caracas signed in October 2000, which provided concessional financing for some of the cost. The petroleum refinery at Puerto Limón processed up to 25,000 barrels per day (b/d). A 320-km pipeline was to be built, linking the Caribbean and Pacific coasts, with a capacity of 1m. b/d. Petrotrin, Trinidad and Tobago's state oil company, was to help upgrade the state petroleum refinery, Refinadora Costarricense de Petróleo (Recope).

By the mid-1980s Costa Rica had virtually eliminated the need for petroleum products for electricity generation through its development of hydroelectric power resources and the use of fuel wood, bagasse (vegetable waste) and sugar-cane alcohol. Geothermal energy from volcanoes was also developed and investment in the national grid ensured 95% of the population was covered by 2002. In 2003 an estimated 79.6% of all the annual electricity generated was hydroelectric, 15.1% was geothermal, 3.0% was wind power and 2.2% was thermal. Although the state electricity company, the ICE (Instituto Costarricense de Electricidad), generates the bulk of power, private producers have existed since 1990. In 2003 private plants generated 14.6% of the total. In 2003 imports of mineral products accounted for an estimated 8.9% of the total import value.

In 2000 99.5% of electricity was generated from renewable sources. In that year the ICE announced plans to build a 1,250-MW hydroelectric plant at Boruca, south of San José, by 2011. A single grid and single electricity market for Central America and Panama was expected to be established by 2006, allowing Costa Rica to export more of its surplus capacity. To help pay for investment plans the Government authorized a 30% increase in tariffs in April 2001. In the 1990s repeated government attempts to pass legislation necessary to privatize the ICE and Recope or, at least, to open the energy sector to private-sector investment, were defeated by the Asamblea Legislativa (Legislative Assembly) or by the public-sector unions. In April 2000 President Miguel Angel Rodríguez Echeverría was forced to abandon these plans, following widespread street protests.

### MANUFACTURING

With the manufacturing sector generating 21.7% of GDP in 2004, Costa Rica was the most industrialized country in Central America. Fast growth in this sector during the 1960s and 1970s owing to the creation of the CACM resulted in a high level of diversification. In the 1980s a contraction in demand caused a manufacturing recession, but growth returned in the 1990s, with industrial GDP increasing, in real terms, by an average annual rate of 5.0% between 1990 and 2004. However, in the early 2000s manufacturing production was very irregular, falling significantly in 2000 (by 2.9%) and 2001 (9.1%), before recovering in 2002 (by 3.4%) and 2003 (8.7%). In 2004 real manufacturing GDP grew by only 1.5%, well below the aggregate rate of economic growth. In the mid-1990s over two-thirds of industry was involved in the manufacture of non-durable consumer goods, mainly food, beverages and tobacco, but chemicals, plastics and tyres were also produced. Free trade zones have also become increasingly important since the creation of the first privately owned zone in 1986.

Exports from the free trade zones increased from US $7m. in 1986 to $891.3m. in 1997 but, owing to their high import propensity, generated lower levels of foreign exchange (only $398.6m. in 1997). In 1998 Intel began production at an assembly plant in the free trade zone at Rivera de Belén, west of San José, the first of four facilities in a projected $500m. investment programme. A second plant opened in June 2004. Although only employing some 2,200 people, Intel galvanized Costa Rica's export and growth figures. In 1999 electronic circuitry accounted for almost 40% of all exports, eclipsing traditional mainstays such as coffee and bananas (although this figure fell to about 28% in 2000). Total exports from free trade zones increased from $2,000m. in 1998 to $3,558m. in 1999, accounting for 56% of all exports. In that year Costa Rica achieved a trade surplus for the first time in four decades, of $307.8m., although this reverted to a deficit of $538.8m. in 2000. Intel was affected by the global slowdown in 2001 and, as a result, exports from the free trade zones fell to $2,378m. Between 2001 and 2004 exports from the free trade zones increased by an estimated 37%, mainly owing to a significant expansion in 2003. The recovery of the export sector was partly owing to Intel, but also to the arrival of new multinational corporations such as Procter & Gamble, which created a global business centre, and Abbot Laboratories of the USA, which established a medical equipment plant. The indigenous computer software sector flourished and had exports of an estimated $60m. in 2002. Intel took a stake in ArtinSoft, a Costa Rican producer that had a contract with Microsoft, in 2001. However, older established textile plants, unable to withstand international competition, began closing in 2003. Costa Rica's high wages relative to its neighbours, and relative to the People's Republic of China, suggested that the move to higher value products would need to continue apace. It was unlikely that even the approval of CAFTA would revive the Costa Rican apparel sector.

## TOURISM

Costa Rica was one of the world's fastest-growing destinations during the 1990s, when investment in tourism expanded significantly. As a result, tourism became Costa Rica's largest single source of foreign-exchange earnings in that decade, with revenue increasing from US $136.2m. in 1987 to $679.2m. in 1994. In spite of a drastic decrease in visitors following the terrorist attacks on the USA of 11 September 2001, the overall visitor total for the year increased by 4%, but fell slightly (by 1.6%) in 2002. In 2003 the total number of tourists was 1,238,692 (an increase of 11.2% over the previous year) and the sector generated a total of $1,199.4m. The country's reputation for political stability and relatively low crime attracts tourists to its fine beaches and extensive system of national parks and protected areas. The rainforest is accessible and birdwatching and trekking is popular. Costa Rica is home to an incredible variety of flora and fauna and is estimated to have 5% of the world's biodiversity.

The Government aims to diversify from its dependence on North American visitors, who accounted for nearly 50% of tourists in 2003. It also encouraged a shift from small operators catering to independent tourists to large resorts, built around San José and the northern Pacific coast. This may in the long term erode tourism's vital contribuion to the Costa Rican economy, as more money would be repatriated by international owners of resorts. In the late 1990s some 48% of the money spent by tourists remained in the country.

## INFRASTRUCTURE

In 2000 Costa Rica had an estimated 35,892 km of roads, of which about 22% were paved. The main road was the Pan-American Highway that runs north–south (fully paved). From 1998 the Government began offering road improvement concessions to private contractors, but the programme was delayed by contractual disputes and a lack of money on the part of some of the contractors. However, an extensive maintenance programme improved the quality of some existing roads; in 2003 public investment in roads increased by more than 44%, in real terms, while total public investment in infrastructure increased by 37%.

A new port on the Gulf of Nicoya, at Caldera, replaced Puntarenas as the principal Pacific port in the 1980s, although facilities at the latter were improved in the 1990s, as part of a general upgrading of the transport network. Nevertheless, the poor quality of port infrastructure was beginning to affect the competitiveness of exports by the end of the decade. The Pacific ports were leased to a private operator at the end of 2001. Trade unions have prevented such a move on the Atlantic (Caribbean) side.

There were about 950 km of railway, of which a sizeable proportion were plantation lines. These were the only ones working after 1995 when the state railway company, Instituto Costarricense de Ferrocarriles (INCOFER), suspended operations to the public indefinitely, pending privatization, although the transport of cargo continued. In 2000 a feasibility study concluded that the railway could be run profitably; before its closure INCOFER transported some 750,000 metric tons of cargo per year. However, a US group interested in the railway withdrew because of a contractual dispute and there was no immediate prospect of privatization.

The main airport, Juan Santamaría, at El Coco, near San José, was one of the busiest in Central America, transporting an average of 1.5m. passengers per year. Its modernization was vital to the development of the tourism industry, not least because of safety concerns raised by the US Federal Aviation Administration. In 2000 a consortium, led by Bechtel of the USA, was given a 20-year management contract of the airport. It pledged to invest US $161m.

The ICE was also responsible for telecommunications and by the 1990s had built an impressive land network. There were 26 telephone lines per 100 inhabitants in 2002, compared with about 11 in the rest of Central America. However, the ICE lacked the money to invest in advanced cellular and internet technology. There were substantial delays connecting to the mobile cellular network and obtaining a new fixed line. Internet capacity could not meet demand, until May 2001 when the country was connected to the Maya fibre-optic cable. Radiográfica Costarricense (RACSA), a subsidiary of the ICE, was the only internet service provider. In 2001 the Government established internet connections in schools and post offices and, as a result, internet users more than doubled in 2002 to 800,000 people, from 384,000 internet users in 2001. The ICE spent US $113.1m. in 2003 on new fibre-optic lines and measures to improve internet access. It also bought a new cellular network with capacity for 600,000 lines for $120m. An attempt to deregulate the telecommunications sector failed in 2000 owing to intense opposition from a large share of the population. According to the provisions of CAFTA, however, the Government must allow private participation in private network services, internet services and wireless services over time. This might constitute one of the most fundamental obstacles to the approval of the free trade agreement in the Asamblea Legislativa.

## FINANCE AND INVESTMENT

Adverse economic conditions and a lack of fiscal reform precipitated a crisis in Costa Rica's public finances in the early 1980s, necessitating the implementation of a stabilization plan. Following a brief recovery, at the end of the decade the Government's failure to reach economic targets led to the withdrawal of proposed loans by the IMF and the World Bank. The

Government of President Rafael Angel Calderón Fournier (1990–94) was frustrated in its tax-reform efforts by an unco-operative legislature and focused on reducing inflation through strict monetary control. In this aim it was largely successful, with annual inflation falling from 28.7% in 1991 to 9.8% in 1993. GDP growth produced higher tax revenues and the public-sector deficit declined, also helped by lower total interest payments. However, the budget deficit subsequently increased once again, as did the annual inflation, which reached 23.2% in 1995. From 1996 the annual rate of inflation fluctuated between 10% and 14%; in 2004 the rate stood at 11.0%, increasing slightly in the first few months of 2005.

Although the fiscal deficit continued to be a major preoccupation of successive Governments, budget reductions and tax measures were politically very difficult to pursue. President José María Figueres Olsen (1994–98) agreed to IMF demands that the budget deficit be reduced to 0.5% of GDP by the end of 1996, but forceful resistance from labour organizations and from within the legislature prevented significant progress in government-expenditure reductions, tax increases and the divestment of public assets. The deficit instead widened, reaching an estimated 5.2% of GDP in 1996, before decreasing to 4.2% in the following year. The Government of President Rodríguez (1998–2002) attempted to overcome the political obstacles to modernization by initiating a consensus-building approach. However, in April 1999 the legislature's unwillingness to approve a package of constitutional amendments, designed to end state monopolies in energy, telecommunications and insurance, signalled the failure of this effort. In 1999 the creation of 3,000 public-sector jobs in the priority areas of education and the police, coupled with lower-than-expected tax revenues, led to renewed anxiety about the budget deficit, which reached 3.1% of GDP in 2000, owing to rising interest payments on the internal debt and lower tax receipts from exports.

In July 2001 the Asamblea Legislativa approved tax reform legislation that would decrease tax on basic goods, while increasing tax on some luxury goods and services. This law, combined with austerity measures, resulted in the fiscal deficit falling to 2.9% of GDP in 2001. In April 2002 a cross-party commission of former finance ministers produced a long-delayed report on how to eliminate the fiscal deficit by 2006. It concluded that a rise equivalent to 2.6% of GDP in tax collection was necessary, as was a 3% spending cut. However, this would only be possible if the Government was given more flexibility over spending. Without reform, it was estimated that the fiscal deficit would reach 11.6% of GDP in 2006. In 2002 the central government deficit reached 4.3% of GDP. In December of that year the Government accordingly approved further tax reforms aimed at reducing the fiscal deficit to 3% of GDP by the end of 2003. These included changes to the tax code and the levying of value-added tax (VAT) on a wider range of goods and services. The measures were to be accompanied by a 5.9% limit on any increases in government spending and the reduction in the budgets of the ICE, social security departments and Recope. These temporary measures were successful and in 2003 the budgetary deficit declined to 2.8% of GDP. Nevertheless, the failure to reach an agreement on tax reform, combined with interest payments and pressure to increase public sector salaries were likely to result in continued deficits in the near future.

There was an increase in foreign direct investment (FDI) in the 1990s, as Costa Rica moved from being predominantly an exporter of coffee and bananas, to an advanced technology and *maquila* (parts assembly) exporter, with a successful tourism industry. Total annual FDI increased from US $172m. in 1991 to a record $662m. in 2002 and in the following two years, FDI remained high both in absolute terms ($597m. in 2004) and as a percentage of GDP (3.2% in 2004). Slow progress on privatization meant that most of this investment was directed at establishing or upgrading facilities, rather than the acquisition of state-owned enterprises. In 2001 US companies accounted for 59% of incoming investment; some 52% of investment was channelled into manufacturing and 27% went into tourism.

## BALANCE OF PAYMENTS AND THE EXTERNAL DEBT

Exports continued to perform well in the 1990s, but further strong economic growth, and the reduction of import tariffs in 1993 to a maximum of 20%, contributed to a dramatic increase in the trade deficit. While the trade deficits were partially offset by increased tourism revenues and the operations of the free trade zones, they necessitated high levels of external borrowing. However, export performance improved from 1998, owing mainly to output by Intel, and in 1999 the trade balance went into surplus for the first time in four decades. However, while there was a trade surplus, the impact on the balance of payments was much more limited, since a large part of that which entered the country in export receipts left as profit repatriation. Moreover, the sharp growth in microprocessor exports masked a sluggish performance in traditional export sectors. In 2000 the trade deficit returned (totalling US $538.8m.), and by 2002 it had risen to $1,267.2m., owing to the economic slowdown in the USA. The deficit fell slightly to $1,169.6m. in 2003, owing to the moderate growth in exports.

Following the approval of agreements on debt relief in 1989 with the 'Paris Club' of Western creditor nations and under the terms of the US Brady Initiative, by 1995 Costa Rica's external debt as a percentage of GDP had been reduced to 13.7%. A new agreement with the IMF was signed in that year, which led to the release of further credits from the World Bank and the IDB. Debt-servicing accounted for over one-quarter of government revenues by 2002. Costa Rica's total external debt at the end of 2003 was $5,424m., and the cost of debt-servicing was equivalent to 9.7% of the value of exports of goods and services.

While external debt was the problem of the 1980s and 1990s, internal debt was that of the 2000s. Internal debt reached US $4,374m. in 2001 and total debt was $6,109m., equivalent to 38.8% of GDP. From 1998 the Government sought to exchange some of this internal debt for foreign debt at lower interest rates, by issuing bonds. In 1998–99 $500m. was placed. A five-year plan to issue a further $1,450m. authorized by the legislature in 2000 was completed in January 2004, with a $250m., 10-year issue. Debt reduction was to remain a priority if resources were to be released for productive investment.

## OUTLOOK

The Costa Rican illustrated as well as any other country in Latin America and the Caribbean both the opportunities and the threats that the current process of market-friendly globalization could bring. On the positive side, Costa Rica succeeded in increasing total exports and diversifying its export base. The arrival of Intel and other multinational corporations in the high-technology and service sectors led to the creation of new, high-paying jobs, demonstrating the long-term importance of public investment in education. At the same time, however, the country has also struggled to adapt to the new global conditions. Repeated efforts by different administrations to deepen the process of neoliberal reforms were met with significant public opposition. Dissatisfaction with an economic model that has resulted in increasing inequality, and the consolidation of a dual economy was increasing, but by 2005 no real alternative had been proposed. Meanwhile, recent episodes of corruption involving three former Presidents further reduced public faith in the political parties. Costa Rica was thus at a crossroads. Building on its export success, the country needed to redefine the role of the state, expand public revenues and eliminate the exclusionary features of the current growth model. It was not an easy task, but it was one that could not be postponed and that might be facilitated by the social debate around the advent of CAFTA in 2005.

# Statistical Survey

Sources (unless otherwise stated): Instituto Nacional de Estadística y Censos, de la Rotonda de la Bandera 450 metros oeste, sobre Calle Los Negritos, Edif. Ana Lorena, Mercedes de Montes de Oca, San José; tel. 280-9280; fax 224-2221; e-mail informacion@inec.go.cr; internet www.inec.go.cr; Banco Central de Costa Rica, Avdas Central y Primera, Calles 2 y 4, Apdo 10.058, 1000 San José; tel. 233-4233; fax 223-4658; internet www.bccr.fi.cr.

## Area and Population

### AREA, POPULATION AND DENSITY

| | |
|---|---:|
| Area (sq km) | |
| Land | 51,060 |
| Inland water | 40 |
| Total | 51,100* |
| Population (census results) | |
| 11 June 1984† | 2,416,809 |
| 28 June 2000 | |
| Males | 1,996,350 |
| Females | 1,928,981 |
| Total | 3,925,331 |
| Population (official estimate at mid-year) | |
| 2002 | 4,045,837 |
| 2003 | 4,103,116 |
| 2004 | 4,159,757 |
| Density (per sq km) at mid-2004 | 81.5 |

* 19,730 sq miles.
† Excluding adjustment for underenumeration.

### PROVINCES
(official estimates, 1 July 2004)

| | Area (sq km) | Population (estimates) | Density (per sq km) | Capital (with population) |
|---|---:|---:|---:|---|
| Alajuela | 9,757.5 | 783,116 | 80.3 | Alajuela (242,769) |
| Cartago | 3,124.7 | 469,982 | 150.4 | Cartago (143,274) |
| Guanacaste | 10,140.7 | 288,448 | 28.4 | Liberia (51,779) |
| Heredia | 2,657.0 | 386,259 | 145.4 | Heredia (113,189) |
| Limón | 9,188.5 | 376,209 | 40.9 | Limón (99,723) |
| Puntarenas | 11,265.7 | 393,226 | 34.9 | Puntarenas (112,232) |
| San José | 4,965.9 | 1,462,517 | 290.5 | San José (336,829) |
| **Total** | 51,100.0 | 4,159,757 | 83.8 | — |

### PRINCIPAL TOWNS
(estimates, 28 June 2000)

| | | | | |
|---|---:|---|---:|
| San José | 309,672 | San Francisco | 40,840 |
| Limón | 56,719 | Cartago | 38,363 |
| Ipís | 52,922 | Cinco Esquinas | 36,627 |
| Desamparados | 52,283 | Liberia | 34,469 |
| Alajuela | 42,889 | Puntarenas | 32,460 |

Source: Thomas Brinkhoff, *City Population* (internet www.citypopulation.de).

### BIRTHS, MARRIAGES AND DEATHS

| | Registered live births | | Registered marriages | | Registered deaths | |
|---|---:|---:|---:|---:|---:|---:|
| | Number | Rate (per 1,000) | Number | Rate (per 1,000) | Number | Rate (per 1,000) |
| 1995 | 80,306 | 24.1 | 24,274 | 7.3 | 14,061 | 4.2 |
| 1996 | 79,203 | 23.3 | 23,574 | 6.9 | 13,993 | 4.1 |
| 1997 | 78,018 | 22.5 | 24,300 | 7.0 | 14,260 | 4.1 |
| 1998 | 76,982 | 21.8 | 24,831 | 7.0 | 14,708 | 4.2 |
| 1999 | 78,526 | 21.9 | 25,613 | 7.1 | 15,052 | 4.2 |
| 2000 | 78,178 | 20.5 | 24,436 | 6.2 | 14,944 | 3.9 |
| 2001 | 76,401 | 19.2 | 23,790* | 6.0* | 15,609 | 4.0 |
| 2002 | 71,144 | 17.6 | n.a. | n.a. | 15,004 | 3.7 |

* Provisional figure.

**Expectation of life** (WHO estimates, years at birth): 77 (males 75; females 80) in 2003 (Source: WHO, *World Health Report*).

### ECONOMICALLY ACTIVE POPULATION*
('000 persons aged 12 years and over, household survey, July)

| | 2002 | 2003 | 2004 |
|---|---:|---:|---:|
| Agriculture, hunting and forestry | 242.74 | 239.81 | 237.26 |
| Fishing | 8.77 | 8.57 | 8.07 |
| Mining and quarrying | 2.31 | 2.24 | 3.56 |
| Manufacturing | 226.28 | 230.06 | 229.48 |
| Electricity, gas and water supply | 21.86 | 22.09 | 23.56 |
| Construction | 106.58 | 109.62 | 107.29 |
| Wholesale and retail trade | 303.36 | 322.39 | 329.92 |
| Hotels and restaurants | 82.46 | 89.93 | 91.42 |
| Transport, storage and communications | 90.24 | 94.03 | 96.30 |
| Financial intermediation | 32.02 | 35.54 | 36.69 |
| Real estate, renting and business activities | 103.19 | 101.23 | 101.98 |
| Public administration activities | 71.87 | 76.26 | 78.50 |
| Education | 91.90 | 98.38 | 95.94 |
| Health and social work | 53.00 | 49.25 | 51.29 |
| Other community, social and personal service activities | 58.53 | 72.10 | 62.92 |
| Private households with employed persons | 83.41 | 79.28 | 90.79 |
| Extra-territorial organizations and bodies | 2.55 | 2.38 | 3.89 |
| Not classifiable by economic activity | 5.46 | 7.23 | 5.05 |
| **Total employed** | 1,586.49 | 1,640.39 | 1,653.88 |
| Unemployed | 108.53 | 117.19 | 114.88 |
| **Total labour force** | 1,695.02 | 1,757.58 | 1,768.76 |

* Figures for activities are rounded to the nearest 10 persons, and totals may not be equivalent to the sum of component parts as a result.

## Health and Welfare

### KEY INDICATORS

| | |
|---|---:|
| Total fertility rate (children per woman, 2003) | 2.3 |
| Under-5 mortality rate (per 1,000 live births, 2003) | 10 |
| HIV/AIDS (% of persons aged 15–49, 2003) | 0.6 |
| Physicians (per 1,000 head, 1998) | 1.7 |
| Hospital beds (per 1,000 head, 1998) | 1.68 |
| Health expenditure (2002): US $ per head (PPP) | 743 |
| Health expenditure (2002): % of GDP | 9.3 |
| Health expenditure (2002): public (% of total) | 65.4 |
| Access to water (% of persons, 2002) | 97 |
| Access to sanitation (% of persons, 2002) | 92 |
| Human Development Index (2002): ranking | 45 |
| Human Development Index (2002): value | 0.834 |

For sources and definitions, see explanatory note on p. vi.

# Agriculture

**PRINCIPAL CROPS**
('000 metric tons)

|  | 2001 | 2002 | 2003 |
|---|---|---|---|
| Rice (paddy) | 211.6 | 190.3 | 180.0 |
| Potatoes | 89.2 | 86.8 | 81.7 |
| Cassava (Manioc) | 106.3 | 94.2 | 94.2 |
| Sugar cane | 3,670.0 | 3,462.3 | 3,923.9 |
| Watermelons* | 77.0 | 77.0 | 77.0 |
| Cantaloupes and other melons* | 208.0 | 210.0 | 215.0 |
| Oil palm fruit | 666.1 | 692.4 | 700.0* |
| Bananas | 1,739.3 | 1,612.0 | 1,863.0 |
| Plantains* | 65.0 | 67.0 | 70.0 |
| Oranges | 436.6 | 367.0 | 367.0 |
| Pineapples | 950.4 | 992.0 | 725.2 |
| Coffee (green) | 837.5 | 787.5 | 731.1 |

* FAO estimate(s).
Source: FAO.

**LIVESTOCK**
('000 head, year ending September)

|  | 2001 | 2002 | 2003 |
|---|---|---|---|
| Horses* | 115 | 115 | 115 |
| Mules* | 5 | 5 | 5 |
| Asses* | 8 | 8 | 8 |
| Cattle | 1,289 | 1,220 | 1,150 |
| Pigs* | 525 | 535 | 500 |
| Sheep* | 3 | 3 | 3 |
| Goats* | 3 | 4 | 4 |
| Chickens* | 16,900 | 18,200 | 18,500 |

* FAO estimates.
Source: FAO.

**LIVESTOCK PRODUCTS**
('000 metric tons)

|  | 2001 | 2002 | 2003 |
|---|---|---|---|
| Beef and veal | 76.0 | 68.3 | 74.1 |
| Pig meat | 35.7 | 36.0 | 35.8 |
| Poultry meat | 77.7 | 76.7 | 71.8 |
| Cows' milk | 737.2 | 761.9 | 788.6 |
| Cheese* | 7.3 | 9.1 | 9.1 |
| Butter* | 4.8 | 5.0 | 5.0 |
| Hen eggs | 46.5 | 47.7 | 47.1 |
| Honey* | 1.3 | 1.3 | 1.3 |
| Cattle hides (fresh)* | 11.0 | 9.6 | 9.4 |

* FAO estimates.
Source: FAO.

# Forestry

**ROUNDWOOD REMOVALS**
(FAO estimates, '000 cubic metres, excl. bark)

|  | 2001 | 2002 | 2003 |
|---|---|---|---|
| Sawlogs, veneer logs and logs for sleepers | 1,441 | 1,441 | 1,441 |
| Other industrial wood | 246 | 246 | 246 |
| Fuel wood | 3,474 | 3,463 | 3,454 |
| Total | 5,161 | 5,150 | 5,141 |

Source: FAO.

**SAWNWOOD PRODUCTION**
(FAO estimates, '000 cubic metres, incl. railway sleepers)

|  | 1998 | 1999 | 2000 |
|---|---|---|---|
| Coniferous (softwood)* | 12 | 12 | 12 |
| Broadleaved (hardwood) | 768* | 768* | 800 |
| Total* | 780 | 780 | 812 |

* FAO estimate(s).

**2001–03:** Annual production as in 2000 (FAO estimates).

Source: FAO.

# Fishing

('000 metric tons, live weight)

|  | 2001 | 2002 | 2003 |
|---|---|---|---|
| Capture* | 34.7 | 32.9 | 29.3 |
| Clupeoids | 2.2 | 4.2 | 2.6 |
| Marlins, sailfishes, etc. | 2.2 | 2.3 | 2.0 |
| Tuna-like fishes | 1.2 | 1.6 | 1.4 |
| Common dolphinfish | 11.2 | 7.8 | 1.4 |
| Sharks, rays, skates, etc. | 8.6 | 7.5 | 10.2 |
| Other marine fishes | 4.1 | 3.9 | 3.7 |
| Aquaculture | 10.5 | 17.9 | 20.5 |
| Tilapias | 8.5 | 13.2 | 14.9 |
| Whiteleg shrimp | 1.8 | 4.1 | 5.1 |
| Total catch* | 45.3 | 50.8 | 49.9 |

* FAO estimates.
Source: FAO.

# Industry

**SELECTED PRODUCTS**
('000 metric tons, unless otherwise indicated)

|  | 1999 | 2000 | 2001 |
|---|---|---|---|
| Non-cellulosic continuous fibres | 7.8 | 8.2 | 8.7 |
| Raw sugar | 366 | 369 | 365 |
| Kerosene | 2 | 3 | 7 |
| Distillate fuel oils | — | 2 | 89 |
| Residual fuel oils | — | 1 | 148 |
| Bitumen | 12 | 12 | — |
| Cement | 1,100 | 1,150 | 1,100 |
| Electric energy (million kWh) | 6,438 | 7,227 | 6,941 |

**2002** ('000 metric tons): Raw sugar 361; Cement 1,100.

Source: UN, *Industrial Commodity Statistics Yearbook*.

# Finance

**CURRENCY AND EXCHANGE RATES**

**Monetary Units**
100 céntimos =1 Costa Rican colón.

**Sterling, Dollar and Euro Equivalents** (31 May 2005)
£1 sterling = 862.706 colones
US $1 = 474.510 colones
€1 = 585.118 colones
1,000 Costa Rican colones = £1.16 = $2.11 = €1.71

**Average Exchange Rate** (colones per US $)
2002    359.8170
2003    398.6630
2004    437.9110

# COSTA RICA

*Statistical Survey*

## GENERAL BUDGET
(million colones)

| Revenue | 2000 | 2001 | 2002 |
|---|---|---|---|
| Taxation | 603,748 | 712,944 | 801,424.9 |
|   Income tax | 133,090 | 163,059 | 185,600.3 |
|   Social security contributions | 18,677 | 21,151 | 23,090.8 |
|   Taxes on property | 15,374 | 17,397 | 22,054.3 |
|   Taxes on goods and services | 389,381 | 459,234 | 513,751.0 |
|   Taxes on international trade | 47,226 | 52,104 | 56,928.5 |
| Other current revenue | 2,755 | 6,875 | 4,636.2 |
| Current transfers | 2,035 | 2,273 | 2,689.6 |
| Capital transfers | 1,600 | 3,472 | 51.3 |
| **Total** | 610,138 | 725,563 | 808,802.0 |

| Expenditure | 2000 | 2001 | 2002 |
|---|---|---|---|
| Current expenditure | 685,955 | 809,055 | 977,982.8 |
|   Wages and salaries | 225,873 | 267,037 | 314,748.7 |
|   Social security contributions | 29,480 | 38,954 | 46,303.7 |
|   Other purchases of goods and services | 22,837 | 26,625 | 36,146.0 |
|   Interest payments | 175,653 | 213,866 | 259,416.6 |
|     Internal | 149,549 | 174,019 | 211,223.4 |
|     External | 26,104 | 39,847 | 48,193.2 |
|   Current transfers | 232,111 | 262,573 | 321,367.8 |
| Capital expenditure | 75,351 | 73,217 | 90,130.5 |
|   Investment | 23,414 | 16,100 | 21,729.2 |
|   Capital transfers | 51,937 | 57,117 | 68,335.8 |
| **Total** | 761,306 | 882,272 | 1,068,113.5 |

## INTERNATIONAL RESERVES
(US $ million at 31 December)

| | 2002 | 2003 | 2004 |
|---|---|---|---|
| Gold* | 0.02 | 0.02 | 0.02 |
| IMF special drawing rights | 0.09 | 0.05 | 0.13 |
| Reserve position in IMF | 27.19 | 29.72 | 31.06 |
| Foreign exchange | 1,469.27 | 1,806.49 | 1,886.71 |
| **Total** | 1,496.57 | 1,836.29 | 1,917.92 |

* National valuation.

Source: IMF, *International Financial Statistics*.

## MONEY SUPPLY
('000 million colones at 31 December)

| | 2002 | 2003 | 2004 |
|---|---|---|---|
| Currency outside banks | 169.7 | 186.9 | 205.6 |
| Demand deposits at commercial banks | 662.6 | 779.0 | 837.6 |
| **Total money** (incl. others) | 843.1 | 970.1 | 1,046.7 |

Source: IMF, *International Financial Statistics*.

## COST OF LIVING
(Consumer Price Index; base: January 1995 = 100)

| | 2002 | 2003 | 2004 |
|---|---|---|---|
| Food, beverages and tobacco | 235.5 | 255.9 | 287.0 |
| Clothing and footwear | 162.1 | 170.0 | 176.7 |
| Housing | 220.3 | 236.9 | 257.1 |
| Medical care | 275.5 | 307.6 | 349.2 |
| Transport | 259.1 | 294.8 | 338.5 |
| Leisure and education | 201.5 | 220.5 | 244.5 |
| **All items** (incl. others) | 228.0 | 248.9 | 276.4 |

## NATIONAL ACCOUNTS
(million colones at current prices)

### Expenditure on the Gross Domestic Product

| | 2001 | 2002 | 2003 |
|---|---|---|---|
| Government final consumption expenditure | 772,575 | 900,615 | 1,013,482 |
| Private final consumption expenditure | 3,689,866 | 4,106,884 | 4,652,840 |
| Increase in stocks | 106,985 | 221,601 | 96,130 |
| Gross fixed capital formation | 987,279 | 1,143,140 | 1,332,290 |
| **Total domestic expenditure** | 5,556,705 | 6,372,240 | 7,094,741 |
| Exports of goods and services | 2,236,332 | 2,569,619 | 3,260,846 |
| *Less* Imports of goods and services | 2,398,442 | 2,882,965 | 3,384,774 |
| **GDP in purchasers' values** | 5,394,595 | 6,058,895 | 6,970,815 |
| **GDP at constant 1991 prices** | 1,438,695 | 1,480,666 | 1,577,362 |

### Gross Domestic Product by Economic Activity

| | 2001 | 2002 | 2003 |
|---|---|---|---|
| Agriculture, hunting, forestry and fishing | 429,098 | 467,695 | 551,239 |
| Mining and quarrying | 8,458 | 8,846 | 9,510 |
| Manufacturing | 1,066,499 | 1,179,164 | 1,344,992 |
| Electricity, gas and water | 142,503 | 149,642 | 170,343 |
| Construction | 232,998 | 256,406 | 292,502 |
| Trade, restaurants and hotels | 963,027 | 1,050,096 | 1,205,774 |
| Transport, storage and communications | 413,057 | 494,186 | 595,209 |
| Finance and insurance | 254,554 | 308,798 | 374,685 |
| Real estate | 222,968 | 244,942 | 267,666 |
| Other business services | 199,709 | 236,773 | 271,208 |
| Public administration | 201,267 | 239,941 | 269,464 |
| Other community, social and personal services | 927,254 | 1,079,933 | 1,243,560 |
| **Sub-total** | 5,061,392 | 5,716,422 | 6,596,151 |
| *Less* Imputed bank service charge | 185,670 | 239,214 | 274,699 |
| **GDP at basic prices** | 4,875,722 | 5,477,206 | 6,321,453 |
| Taxes, *less* subsidies, on products | 518,874 | 581,689 | 649,362 |
| **GDP in purchasers' values** | 5,394,595 | 6,058,895 | 6,970,815 |

## BALANCE OF PAYMENTS
(US $ million)

| | 2001 | 2002 | 2003 |
|---|---|---|---|
| Exports of goods f.o.b. | 4,923.2 | 5,269.9 | 6,124.7 |
| Imports of goods f.o.b. | −5,743.3 | −6,537.1 | −7,294.4 |
| **Trade balance** | −820.1 | −1,267.2 | −1,169.6 |
| Exports of services | 1,900.6 | 1,869.9 | 2,027.0 |
| Imports of services | −1,168.9 | −1,182.1 | −1,188.4 |
| **Balance on goods and services** | −88.4 | −579.5 | −331.0 |
| Other income received | 196.0 | 317.9 | 213.3 |
| Other income paid | −975.6 | −835.2 | −1,061.9 |
| **Balance on goods, services and income** | −867.9 | −1,096.8 | −1,179.9 |
| Current transfers received | 266.4 | 296.9 | 368.6 |
| Current transfers paid | −111.1 | −116.2 | −156.0 |
| **Current balance** | −712.7 | −916.1 | −967.0 |
| Capital account (net) | 12.4 | 5.7 | 26.1 |
| Direct investment abroad | −11.1 | −34.1 | −26.9 |
| Direct investment from abroad | 453.6 | 661.9 | 576.7 |
| Portfolio investment assets | −81.2 | 28.4 | −91.6 |
| Portfolio investment liabilities | −57.9 | −125.8 | −304.5 |
| Other investment assets | 106.2 | 217.5 | 170.6 |
| Other investment liabilities | −89.2 | 96.9 | 329.1 |
| Net errors and omissions | 243.9 | 28.2 | 68.0 |
| **Overall balance** | −136.0 | −37.4 | −219.5 |

Source: IMF, *International Financial Statistics*.

# External Trade

**PRINCIPAL COMMODITIES**
(US $ million)

| Imports c.i.f. | 2001 | 2002 | 2003 |
|---|---|---|---|
| Food and live animals | 276.0 | 342.4 | 355.5 |
| Mineral products | 484.8 | 620.5 | 700.1 |
| Basic manufactures | 194.4 | 231.9 | 240.6 |
| Chemicals and related products | 724.9 | 777.0 | 845.6 |
| Plastic materials and manufactures | 452.3 | 468.4 | 487.4 |
| Leather, hides and furs | 23.6 | 26.8 | 44.7 |
| Paper, paperboard and manufactures | 26.4 | 24.2 | 26.7 |
| Wood pulp and other fibrous materials | 325.7 | 319.6 | 351.8 |
| Silk, cotton and textile fibres | 691.9 | 657.3 | 600.5 |
| Footwear, hats, umbrellas, etc. | 46.8 | 49.2 | 51.7 |
| Stone manufactures, etc. | 82.4 | 89.3 | 89.7 |
| Natural and cultured pearls | 27.1 | 38.1 | 30.5 |
| Common metals and manufactures | 397.2 | 444.2 | 492.8 |
| Machinery and electrical equipment | 2,149.6 | 2,478.7 | 2,756.3 |
| Transport equipment | 393.2 | 447.0 | 403.6 |
| Optical and topographical apparatus and instruments etc. | 128.8 | 184.9 | 306.1 |
| **Total** (incl. others) | 6,552.5 | 7,314.6 | 7,906.0 |

| Exports f.o.b. | 2001 | 2002 | 2003 |
|---|---|---|---|
| Food and live animals | 1,343.1 | 1,415.3 | 1,535.9 |
| Mineral products | 54.9 | 65.8 | 43.0 |
| Basic manufactures | 357.2 | 411.6 | 433.3 |
| Chemicals and related products | 289.6 | 329.5 | 364.4 |
| Plastic materials and manufactures | 223.3 | 235.7 | 235.6 |
| Leather, hides and furs | 44.9 | 39.9 | 51.9 |
| Paper, paperboard and manufactures | 31.7 | 33.5 | 30.1 |
| Wood pulp and other fibrous materials | 81.7 | 103.5 | 82.0 |
| Silk, cotton and textile fibres | 676.4 | 692.0 | 584.7 |
| Footwear, hats, umbrellas, etc. | 4.5 | 1.0 | 1.0 |
| Stone manufactures, etc. | 66.2 | 72.8 | 67.4 |
| Natural and cultured pearls | 27.3 | 35.4 | 31.3 |
| Common metals and manufactures | 116.9 | 126.8 | 134.6 |
| Machinery and electrical equipment | 1,315.5 | 1,423.9 | 1,955.8 |
| Transport equipment | 34.1 | 19.4 | 28.8 |
| Optical and topographical apparatus and instruments, etc.. | 333.6 | 404.1 | 533.3 |
| **Total** (incl. others) | 5,059.5 | 5,466.6 | 6,164.0 |

**PRINCIPAL TRADING PARTNERS**
(US $ million)

| Imports c.i.f. | 2001 | 2002 | 2003 |
|---|---|---|---|
| Brazil | 110.8 | 163.4 | 269.1 |
| Canada | 73.5 | 70.8 | 73.3 |
| China, People's Republic | 100.5 | 121.7 | 169.7 |
| Colombia | 154.2 | 164.9 | 213.3 |
| El Salvador | 92.0 | 108.1 | 101.9 |
| France | 76.3 | 66.7 | 195.5 |
| Germany | 137.1 | 162.7 | 170.6 |
| Guatemala | 140.9 | 149.7 | 153.5 |
| Italy | 72.6 | 85.6 | 112.5 |
| Japan | 230.1 | 378.7 | 325.3 |
| Korea, Republic | 138.6 | 139.8 | 103.4 |
| Mexico | 381.7 | 369.0 | 380.4 |
| Netherlands | 128.7 | 130.9 | 152.7 |
| Panama | 110.7 | 122.9 | 114.3 |
| Spain | 104.6 | 136.5 | 131.8 |
| USA | 3,504.1 | 3,598.9 | 4,002.0 |
| Venezuela | 298.5 | 379.3 | 280.7 |
| **Total** (incl. others) | 6,546.3 | 7,303.6 | 7,906.0 |

| Exports f.o.b. | 2001 | 2002 | 2003 |
|---|---|---|---|
| Belgium* | 90.9 | 85.7 | 98.4 |
| El Salvador | 151.8 | 135.4 | 160.9 |
| Germany | 117.5 | 169.8 | 208.2 |
| Guatemala | 211.4 | 226.7 | 232.7 |
| Honduras | 123.3 | 147.5 | 147.1 |
| Italy | 94.4 | 89.7 | 89.2 |
| Malaysia | 155.0 | 122.9 | 215.6 |
| Mexico | 87.3 | 129.5 | 150.4 |
| Netherlands | 275.2 | 319.4 | 375.7 |
| Nicaragua | 163.3 | 174.9 | 188.1 |
| Panama | 142.7 | 131.8 | 141.6 |
| Puerto Rico | 156.7 | 167.1 | 166.0 |
| Sweden | 46.1 | 59.1 | 72.8 |
| United Kingdom | 128.1 | 110.7 | 162.3 |
| USA | 2,504.8 | 2,726.7 | 2,977.2 |
| **Total** (incl. others) | 5,042.7 | 5,466.6 | 6,333.4 |

* Includes Luxembourg.

# Transport

**ROAD TRAFFIC**
(motor vehicles in use at 31 December)

| | 2000 | 2001 | 2002 |
|---|---|---|---|
| Private cars | 341,990 | 354,394 | 367,832 |
| Buses and coaches | 11,983 | 12,419 | 12,891 |
| Goods vehicles | 177,875 | 184,326 | 191,315 |
| Road tractors | 24,027 | 24,898 | 25,842 |
| Motorcycles and mopeds | 85,427 | 88,526 | 91,883 |

Source: IRF, *World Road Statistics*.

**SHIPPING**

**Merchant Fleet**
(registered at 31 December)

| | 2002 | 2003 | 2004 |
|---|---|---|---|
| Number of vessels | 13 | 14 | 14 |
| Total displacement ('000 grt) | 4.0 | 4.6 | 4.6 |

Source: Lloyd's Register-Fairplay, *World Fleet Statistics*.

**International Sea-borne Freight Traffic**
('000 metric tons)

| | 1996 | 1997 | 1998 |
|---|---|---|---|
| Goods loaded | 3,017 | 3,421 | 3,721 |
| Goods unloaded | 3,972 | 4,522 | 5,188 |

Source: Ministry of Public Works and Transport.

**CIVIL AVIATION**
(scheduled services)

| | 1999 | 2000 | 2001 |
|---|---|---|---|
| Kilometres flown (million) | 27 | 20 | 24 |
| Passengers carried ('000) | 1,055 | 878 | 738 |
| Passenger-km (million) | 2,145 | 2,358 | 2,152 |
| Total ton-km (million) | 245 | 252 | 179 |

Source: UN, *Statistical Yearbook*.

## Tourism

**FOREIGN TOURIST ARRIVALS BY COUNTRY OF ORIGIN**

|  | 2001 | 2002 | 2003 |
|---|---:|---:|---:|
| Canada | 52,661 | 49,168 | 54,656 |
| Colombia | 47,547 | 35,220 | 26,645 |
| El Salvador | 35,054 | 33,531 | 33,892 |
| France | 15,558 | 18,309 | 23,606 |
| Germany | 23,995 | 23,848 | 29,151 |
| Guatemala | 32,574 | 33,150 | 35,174 |
| Honduras | 27,174 | 23,705 | 23,004 |
| Italy | 16,479 | 15,985 | 18,361 |
| Mexico | 36,841 | 37,870 | 46,113 |
| Netherlands | 18,922 | 19,938 | 24,665 |
| Nicaragua | 171,583 | 174,455 | 163,632 |
| Panama | 53,892 | 55,774 | 56,490 |
| Spain | 26,916 | 29,874 | 34,442 |
| United Kingdom | 18,922 | 19,037 | 23,019 |
| USA | 429,093 | 422,215 | 510,751 |
| **Total** (incl. others) | 1,131,406 | 1,113,359 | 1,238,692 |

**Tourism receipts** (US $ million): 1,278 in 2001; 1,078 in 2002; 1,199 in 2003.

**Hotels:** 370 in 2001; 371 in 2002; 379 in 2003.

## Communications Media

|  | 2000 | 2001 | 2002 |
|---|---:|---:|---:|
| Telephones ('000 main lines in use) | 1,003.4 | 945.0 | 1,038.0 |
| Mobile cellular telephones ('000 subscribers) | 209.1 | 311.3 | 459.8 |
| Personal computers ('000 in use) | n.a. | 700.0 | 817.0 |
| Internet users ('000) | 228.0 | 384.0 | 800.0 |

**Radio receivers** ('000 in use): 3,045 in 1999.

**Television receivers** ('000 in use): 930 in 2000.

**Facsimile machines** (number in use): 8,500 in 1997.

**Daily newspapers:** 8 in 2000.

**Non-daily newspapers:** 27 in 2000.

**Book production:** 1,464 titles (excluding pamphlets) in 1998.

Sources: UNESCO, *Statistical Yearbook*, UN, *Statistical Yearbook*, International Telecommunication Union.

## Education

(2001/02, unless otherwise indicated)

|  | Institutions* | Teachers* | Males | Females | Total |
|---|---:|---:|---:|---:|---:|
| Pre-primary | 1,821 | 3,604 | 47,804 | 45,929 | 93,733 |
| Primary | 3,768 | 20,185 | 287,197 | 265,105 | 552,302 |
| Secondary | 468 | 11,891 | 146,528 | 140,781 | 287,309 |
| General | 386 | 8,908 | 111,683 | 116,242 | 227,925 |
| Vocational | 82 | 2,983 | 33,849 | 25,535 | 59,384 |
| Tertiary | 52 | n.a. | 37,460 | 41,722 | 79,182 |

* 1999.

**Adult literacy rate** (UNESCO estimates): 95.8% (males 95.7%; females 95.9%) in 2002 (Source: UN Development Programme, *Human Development Report*).

# Directory

## The Constitution

The present Constitution of Costa Rica was promulgated in November 1949. Its main provisions are summarized below:

### GOVERNMENT

The government is unitary: provincial and local bodies derive their authority from the national Government. The country is divided into seven Provinces, each administered by a Governor who is appointed by the President. The Provinces are divided into Cantons, and each Canton into Districts. There is an elected Municipal Council in the chief city of each Canton, the number of its members being related to the population of the Canton. The Municipal Council supervises the affairs of the Canton. Municipal government is closely regulated by national law, particularly in matters of finance.

### LEGISLATURE

The government consists of three branches: legislative, executive and judicial. Legislative power is vested in a single chamber, the Legislative Assembly (Asamblea Legislativa), which meets in regular session twice a year—from 1 May to 31 July, and from 1 September to 30 November. Special sessions may be convoked by the President to consider specified business. The Assembly is composed of 57 deputies elected for four years. The chief powers of the Assembly are to enact laws, levy taxes, authorize declarations of war and, by a two-thirds' majority, suspend, in cases of civil disorder, certain civil liberties guaranteed in the Constitution.

Bills may be initiated by the Assembly or by the Executive and must have three readings, in at least two different legislative periods, before they become law. The Assembly may override the presidential vote by a two-thirds' majority.

### EXECUTIVE

The executive branch is headed by the President, who is assisted by the Cabinet. If the President should resign or be incapacitated, the executive power is entrusted to the First Vice-President; next in line to succeed to executive power are the Second Vice-President and the President of the Legislative Assembly.

The President sees that the laws and the provisions of the Constitution are carried out, and maintains order; has power to appoint and remove cabinet ministers and diplomatic representatives, and to negotiate treaties with foreign nations (which are, however, subject to ratification by the Legislative Assembly). The President is assisted in these duties by a Cabinet, each member of which is head of an executive department.

### ELECTORATE

Suffrage is universal, compulsory and secret for persons over the age of 18 years.

### DEFENCE

The Costa Rican Constitution has a clause outlawing a national army. Only by a continental convention or for the purpose of national defence may a military force be organized.

## The Government

### HEAD OF STATE

**President:** ABEL PACHECO DE LA ESPRIELLA (took office 8 May 2002).

**First Vice-President:** LINETH SABORÍO.

**Second Vice-President:** (vacant).

## THE CABINET
(July 2005)

**Minister of Finance:** Federico Carrillo Zürcher.
**Minister of the Economy, Industry and Commerce:** Gilberto Barrantes.
**Minister of the Presidency:** Lineth Saborío.
**Minister of Planning and Economic Policy:** Jorge Polinaris.
**Minister of Foreign Relations:** Dr Roberto Tovar Faja.
**Minister of Foreign Commerce:** Manuel Antonio González Sanz.
**Minister of the Interior, Police and Public Security:** Rogelio Ramos Martínez.
**Minister of Agriculture and Livestock:** Rodolfo Coto Pacheco.
**Minister of the Environment and Energy:** Carlos Manuel Rodríguez Echandi.
**Minister of Justice:** Patricia Vega Herrera.
**Minister of Labour and Social Security:** Fernando Trejos Ballestero.
**Minister of Public Education:** Manuel Antonio Bolaños Salas.
**Minister of Public Health:** María del Rocío Sáenz.
**Minister of Housing:** Helio Fallas Venegas.
**Minister of Public Works and Transport:** Randall Quirós Bustamente.
**Minister of Science and Technology:** Fernando Gutiérrez Ortiz.
**Minister of Culture, Youth and Sports:** Guido Sáenz González.
**Minister of Tourism:** Rodrigo Castro.
**Minister of Women's Affairs:** Georgina Vargas Pagán.
**Secretary-General of the Cabinet:** Hermes Navarro del Valle.

### MINISTRIES

**Ministry of Agriculture and Livestock:** Antigüo Colegio La Salle, Sabana Sur, Apdo 10094, 1000 San José; tel. 231-2344; fax 232-2103; e-mail magweb@mag.go.cr; internet www.mag.go.cr.

**Ministry of Culture, Youth and Sports:** Avdas 3 y 7, Calles 11 y 15, frente al parque España, San José; tel. 255-3188; e-mail mincjd@costarricense.cr; internet www.mcjdcr.go.cr.

**Ministry of the Economy, Industry and Commerce:** Edif. IFAM, Moravia del Colegio Lincoln, San José; tel. 235-2700; fax 236-7281; e-mail info@meic.go.cr; internet www.meic.go.cr.

**Ministry of the Environment and Energy:** Avdas 8 y 10, Calle 25, Apdo 10.104, 1000 San José; tel. 233-4533; fax 257-0697; e-mail emiliarg@minae.go.cr; internet www.minae.go.cr.

**Ministry of Finance:** Edif. Antigüo Banco Anglo, Avda 2a, Calle 3a, San José; tel. 257-9333; fax 255-4874; e-mail WebMaster1@hacienda.go.cr; internet www.hacienda.go.cr.

**Ministry of Foreign Commerce:** Montes de Oca, Apdo 96, 2050 San José; tel. 256-7111; fax 255-3281; e-mail info@comex.go.cr; internet www.comex.go.cr.

**Ministry of Foreign Relations:** Avda 7 y 9, Calle 11 y 13, Apdo 10027, 1000 San José; tel. 223-7555; fax 257-6597; e-mail despacho.ministro@rree.go.cr; internet www.rree.go.cr.

**Ministry of Housing:** Paseo de los Estudiantes, Apdo 222, 1002 San José; tel. 221-4411; fax 255-1976; internet www.mivah.go.cr.

**Ministry of the Interior, Police and Public Security:** Apdo 55, 4874 San José; tel. 227-4866; fax 226-6581; internet www.msp.go.cr.

**Ministry of Justice:** Zapote, frente a Price Mart, Registro Nacional, Edif. Administrativo, 4°, 1000 San José; tel. 280-9054; fax 234-7959; e-mail justicia@gobnet.go.cr; internet www.mj.go.cr.

**Ministry of Labour and Social Security:** Edif. Benjamín Núñez, 4°, Barrio Tournón, Apdo 10133, 1000 San José; tel. 257-8211; internet www.ministrabajo.go.cr.

**Ministry of Planning and Economic Policy:** Avdas 3 y 5, Calle 4, Apdo 10.127, 1000 San José; tel. 281-2700; fax 253-6243; e-mail pnd@ns.mideplan.go.cr; internet www.mideplan.go.cr.

**Ministry of Public Education:** Edif. Antigua Embajada, Apdo 10087, 1000 San José; tel. 233-9050; fax 233-0390; e-mail minieduc@sol.racsa.co.cr; internet www.mep.go.cr.

**Ministry of Public Health:** Calle 16, Avda 6 y 8, Apdo 10123, 1000 San José; tel. 223-0333; fax 255-4997; e-mail minsalud@netsalud.sa.cr; internet www.ministeriodesalud.go.cr.

**Ministry of Public Works and Transport:** Plaza González Víquez, C 9, Avda 20 y 22, Apdo 10176, 1000 San José; tel. 253-2000; fax 255-0242; internet www.mopt.go.cr.

**Ministry of Science and Technology:** 1.3 km al norte de la Embajada Americana, Apdo 5589, 1000 San José; tel. 290-1790; fax 290-4967; e-mail micit@micit.go.cr; internet www.micit.go.cr.

**Ministry of Women's Affairs:** San José.

## President and Legislature

### PRESIDENT
Elections, 3 February and 7 April 2002

| Candidate | First round % of votes | Second round % of votes |
|---|---|---|
| Abel Pacheco de la Espriella (PUSC) | 38.5 | 58.0 |
| Rolando Araya Monge (PLN) | 30.9 | 42.0 |
| Ottón Solís (PAC) | 26.3 | — |
| Otto Guevara Guth (ML) | 1.7 | — |
| **Total** (incl. others) | 100.0 | 100.0 |

### ASAMBLEA LEGISLATIVA
General Election, 3 February 2002

| Party | Seats |
|---|---|
| Partido Unidad Social Cristiana (PUSC) | 19 |
| Partido de Liberación Nacional (PLN) | 17 |
| Partido Acción Ciudadana (PAC) | 14 |
| Movimiento Libertario (ML) | 6 |
| Partido Renovación Costarricense (PRC) | 1 |
| **Total** | 57 |

## Political Organizations

**Alianza Nacional Cristiana (ANC):** Calle Vargas Araya, Condominio UNI, 50 m norte Bazar Tere, Apdo 353, 2050 San José; tel. 253-2772; f. 1981; national party; Pres. Víctor Hugo González Montero; Sec. Salvador Estaban Beatriz Porras.

**Bloque Patriótico Parlamentario:** San José; f. 2003 by dissident faction of the PAC.

**Cambio 2000:** 200 m al sur del Banco Popular de San José, Edif. esquinero, frente a la Sociedad de Seguros de Vida del Magisterio Nacional, San José; tel. 221-0694; f. 2000; Pres. Walter Coto Molina; Sec. Rosa María Artavia Rodríguez; Coalition comprising:

   **Acción Democrática Alajuelense:** Frente al Liceo de San Carlos, Barrio San Roque, Ciudad Quesada, Alajuela; tel. 460-0075; f. 1978; provincial party; Pres. Mariano Barquero Fonseca; Sec. Jorge Arroyo Castro.

   **Partido Pueblo Unido:** 11 Calle Fallas, Ciudadela Cucubres de la plaza de deportes, Apdo 4.665, 1000 San José; tel. 224-1904; fax 224-2364; e-mail upvargas@sol.racsa.co.cr; f. 1995; national party; Pres. Trino Barrantes Araya; Sec. Humberto Vargas Carbonell.

**Independiente Obrero:** Edif. Multifamiliares, Hatillo 5, Apdo 179, 1000 San José; tel. 254-5450; fax 283-7752; e-mail cuberopio@racsa.co.cr; f. 1971; national party; Pres. José Alberto Cubero Carmona; Sec. Luis Fernando Salazar Villegas.

**Movimiento Libertario (ML):** Of. de Cabinas San Isidro, Barrio Los Yoses Sur, Apdo 4.674, 1000 San José; tel. 281-3767 ; fax 224-169222; e-mail info@libertario.org; internet www.libertario.org; f. 1994; national party; Pres. Otto Guevara Guth; Sec.-Gen. Raúl Costales Domínguez.

**Nuevo Partido Democrático:** Central Barrio Naciones Unidas, detrás del Centro Comercial del Sur, Apdo 528, 2100 San José; tel. 227-1422; fax 283-4857; f. 1996; national party; Pres. Rodrigo Gutiérrez Schwanhauser; Sec. Rosa María Zeledon Gómez.

**Partido Acción Ciudadana (PAC):** 25 San Pedro, 425 metros sur del Templo Parroquial, San José; tel. 281-2727; fax 280-6640; e-mail pac2002@racsa.co.cr; internet www.pac.or.cr; f. 2000; centre party; Pres. Ottón Solís; Sec. Sadie Bravo Pérez.

**Partido Acción Laborista Agrícola (PALA):** 700m este y 50 norte del Palide Barva de Heredia, Apdo 698–1100 Tibás; tel. 236-6404; fax 240-6536; e-mail fmurillo66@hotmail.com; f. 1987; provincial party; Pres. Freddy Jesús Murillo Espinoza; Sec. Iris María Rodríguez Vargas.

**Partido Agrario Nacional:** Frente al Banco Nacional, Guácimo, Apdo 14, 7210 Limón; tel. 710-0493; fax 710-0493; f. 1988; provincial party; Pres. LUIS FRANCISCO SÁNCHEZ MOREIR; Sec. PEDRO CAMPOS PICADO.

**Partido Auténtico Limonense:** 70 m Oeste de la Escuela Rafael Yglesias, Apdo 1046, Limón; tel. 758-3563; f. 1976; provincial party; Pres. MARVIN WRIGHT LINDO; Sec. DELROY SEINOR GRANT.

**Partido Cambio Ya:** Farmacia de la Cruz, 2°, Ciudad Quesada, San Carlos, Alajuela; tel. 460-2501; fax 460-3855; f. 1995; provincial party; Pres. ANA PATRICIA GUILLÉN CAMPOS; Sec. MARÍA AURELIA ROJAS HERRERA.

**Partido Convergencia Nacional:** Del Restaurant Versalles, 50 m Oeste, Barrio Los Angeles, Cartago; tel. 551-4361; fax 552-3360; f. 1992; provincial party; Pres. EDUARDO CANTILLO ARIAS; Sec. SERGIO IZAGUIRRE CERDA.

**Partido Demócrata (PD):** Frente a las Oficinas del INVU, barrio Amón, contiguo a Restaurante La Criollita, Apdo 121, San José; tel. 256-4168; fax 256-0350; f. 1996; national party; Pres. ALVARO GONZÁLEZ ESPINOZA; Sec. ANA MARÍA PÉREZ GRANADOS.

**Partido Fuerza Agraria de los Cartagineses:** Comisión Campesina Asamblea Legislativa, Edif. Antiguo Sión, Cartago; tel. 551-8884; f. 1996; provincial party; Pres. ALICIA SOLANO BRAVO; Sec. JORGE ANGULO SOLANO.

**Partido Fuerza Democrática (FD):** Edif. Colón, Apdo 1129-1007, San José; tel. 258-7207; fax 258-7204; f. 1992 as coalition; later became national party; Pres. JUAN CARLOS CHAVES MORA; Vice-Pres. MARJORIE SANTAMARÍA MONGE.

**Partido Guanacaste Independiente:** Costado Norte de la Plaza de Fútbol, las Juntas, Guanacaste; tel. 662-0141; f. 1992; provincial party; Pres. JOSÉ ANGEL JARA CHAVARRÍA; Sec. TERESITA SALAS VINDAS.

**Partido Integración Nacional (PIN):** Apartamentos San Antonio, Contiguo a Banco Catay, San Pedro de Montes de Oca; tel. 225-5067; fax 280-7237; e-mail waltermunoz@costarricense.cr; f. 1996; national party; Pres. Dr WALTER MUÑOZ CÉSPEDES; Sec.-Gen. ANA LOURDES GÓLCHER GONZÁLEZ.

**Partido Liberación Nacional (PLN):** Mata Redonda, 125 m oeste del Ministerio de Agricultura y Ganadería, Casa Liberacionista José Figueres Ferrer, Apdo 10.051, 1000 San José; tel. 232-5133; fax 231-4097; e-mail palina@sol.racsa.co.cr; internet www.pln.or.cr; f. 1952; national social democratic party; affiliated to the Socialist International; 500,000 mems; Pres. FRANCISCO ANTONIO PACHECO FERNÁNDEZ; Sec.-Gen. OSCAR NÚÑEZ CALVO (acting).

**Partido Patriótico Nacional:** Curridabat, Residencial Hacienda Vieja, de la entrada principal 200 m al sur y 200 m al este, Casa esquinera, San José; tel. 272-0835; fax 253-5868; f. 1971; national party; Pres. DANIEL ENRIQUE REYNOLDS VARGAS; Sec. ERICK DELGADO LEÓN.

**Partido Renovación Costarricense (PRC):** Centro Educativo Instituto de Desarrollo de Inteligencia, Hatillo 1, Avda Villanea, Apdo 31, 1300 San José; tel. 254-3651; fax 252-3270; f. 1995; national party; Pres. GERARDO JUSTO OROZCO ALVAREZ; First Vice-Pres. RAFAEL MATAMORO MESÉN.

**Partido Rescate Nacional:** De la Iglesia Católica de San Pedro de Montes de Oca 150 m al oeste, contiguo al restaurante El Farolito, San José; tel. 234-9569; fax 225-0931; f. 1996; national party; Pres. CARLOS VARGAS SOLANO; Sec. BUENAVENTURA CARLOS VILLALOBOS BRENES.

**Partido Unidad Social Cristiana (PUSC):** Del Restaurante Kentucky Fried Chicken 75 metros al sur, frente a la Embajada de España, Paseo Colón, Apdo 10.095, 1000 San José; tel. 248-2470; fax 248-2179; f. 1983; national party; Pres. LORENA VÁSQUEZ BADILLA; Sec. JORGE EDUARDO SÁNCHEZ SIBAJA.

**Partido Unión Agrícola Cartagines:** Costado Este de la Iglesia, Cervantes, Frente al Estanco del CNP Alvarado, Apdo 534-4024, Cartago; tel. 534-4024; fax 534-8305; f. 1969; provincial party; Pres. JUAN GUILLERMO BRENES CASTILLO; Sec. CAROLINA MÉNDEZ ZAMORA.

**Partido Unión General (PUGEN):** Casa 256, Centro de Pérez Zeledón, Frente al Edif. de la Cámara de Comercio, Apdo 440, 8000, Pérez Zeledón; tel. 771-0524; fax 771-0737; e-mail pugen@apc.c.co.cr; f. 1980; national party; Pres. Dr CARLOS A. FERNÁNDEZ VEGA; Sec. MARÍA LOURDES RODRÍGUEZ MORALES.

## Diplomatic Representation

### EMBASSIES IN COSTA RICA

**Argentina:** Curridabat, Apdo 1963, 1000 San José; tel. 234-6520; fax 283-9983; e-mail embarg@racsa.co.cr; Ambassador JUAN JOSÉ ARCURI.

**Belgium:** Los Yoses, 4a entrada, 25 m sur, Apdo 3725, 1000 San José; tel. 225-6633; fax 225-0351; e-mail sanjose@diplobel.be; internet www.diplobel.org/costarica; Ambassador ROBERT VANREUSEL.

**Bolivia:** Barrio Rohrmoser 669, Apdo 84810, 1000 San José; tel. 296-3747; fax 232-7292; e-mail embocr@racsa.co.cr; internet www.embajada-bolivia-costarica.com; Ambassador SUSANA PEÑARANDA DE DEL GRANADO.

**Brazil:** Edif. Torre Mercedes, 6°, Apdo 10.132, 1000 San José; tel. 295-6875; fax 295-6874; e-mail embbrsjo@yahoo.com; Ambassador FRANCISCO SOARES ALVIN NETODE SOUZA DE GOMES.

**Canada:** Oficentro Ejecutivo La Sabana, Edif. 5, 3°, detrás de la Contraloría, Centro Colón, Apdo 351, 1007 San José; tel. 242-4400; fax 242-4410; e-mail sjcra@dfait.maeci.gc.ca; internet www.dfait-maeci.gc.ca/sanjose; Ambassador MARIO LAGUË.

**Chile:** De Autos Subarú en Los Yoses, 200 m norte, Barrio Dent, San Pedro de Montes de Oca, Apdo 10102-1000 San José; tel. 280-0037; fax 253-7016; e-mail politica@embchile.co.cr; Ambassador GERMÁN GUERRERO PAREZ.

**China (Taiwan):** 300 m al norte y 150 m al este de la Iglesia Sta Teresita, Barrio Escalante, Apdo 676, 2010 San José; tel. 224-8180; fax 253-8333; e-mail embajroc@racsa.co.cr; Ambassador TZU-DAN WU.

**Colombia:** Barrio Dent de Taco Bell, San Pedro, Apdo 3.154, 1000 San José; tel. 283-6871; fax 283-6818; e-mail emsanjose@minrelext.gov.co; Ambassador JULIO ANIBAL RIAÑO VELANDIA.

**Czech Republic:** 75 m oeste de la entrada principal del Colegio Humboldt, Apdo 12.041, 1000 San José; tel. 296-5671; fax 296-5595; e-mail sanjose@embassy.mzv.cz; internet www.mzv.cz; Ambassador (vacant).

**Dominican Republic:** McDonald's de Curridabat 400 sur, 90 m este, Apdo 4746-1000 San José; tel. 283-8103; fax 280-7604; e-mail embdominicanacr@racsa.co.cr; Ambassador ADONAIDA MEDINA RODRÍGUEZ.

**Ecuador:** Edif. de la esquina sureste del Museo Nacional, 125 m al este, Avda 2, Calles 19 y 21, Apdo 1.374, 1000 San José; tel. 232-1503; fax 232-2086; e-mail embecuar@sol.racsa.co.cr; Ambassador Dr JUAN LEORO ALMEIDA.

**El Salvador:** Paseo Colón, Calle 30, Avda 1, No 53, Apdo 1.378, 1000 San José; tel. 257-7855; fax 257-7683; e-mail embasacr@sol.racsa.co.cr; Ambassador HUGO ROBERTO CARRILLO CORLETO.

**France:** Carretera a Curridabat, del Indoor Club 200 m sur y 25 m oeste, Apdo 10.177, 1000 San José; tel. 234-4167; fax 234-4195; e-mail sjfrance@sol.racsa.co.cr; internet www.ambafrance-cr.org; Ambassador JEAN-PAUL MONCHAU.

**Germany:** Barrio Rohrmoser, de la Casa de Oscar Arias 200 m norte, 75 m este, Apdo 4.017, 1000 San José; tel. 232-5533; fax 231-6403; e-mail info@embajada-alemana-costarica.org; internet www.embajada-alemana-costarica.org; Ambassador VOLKER FINK.

**Guatemala:** De Pops Curridabat 500 m sur y 30 m este, 2a Casa Izquierda, Apdo 328, 1000 San José; tel. 283-2290; fax 283-2290; e-mail embcostarica@minex.gob.gt; Ambassador JORGE MARIO GARCÍA LAGUARDIA.

**Holy See:** Urbanización Rohrmoser, Sabana Oeste, Centro Colón, Apdo 992, 1007 San José (Apostolic Nunciature); tel. 232-2128; fax 231-2557; e-mail nuapcr@racsa.co.cr; Apostolic Nuncio Most Rev. OSVALDO PADILLA (Titular Archbishop of Pia).

**Honduras:** Los Yoses sur, del ITAN hacia la Presidencia la primera entrada a la izquierda, 200 m norte y 100 m este, Apdo 2.239, 1000 San José; tel. 234-9502; fax 253-2209; e-mail emhondcr@sol.racsa.co.cr; Ambassador ARISTIDES MEJÍA CASTRO.

**Israel:** Edif. Centro Colón, 11°, Calle 38 Paseo Colón, Apdo 5147, 1000 San José; tel. 221-6011; fax 257-0867; e-mail ambassador.sec@sanjose.mfa.gov.il; internet sanjose.mfa.gov.il; Ambassador ALEX BEN ZVI.

**Italy:** Los Yoses, 5a entrada, Apdo 1.729, 1000 San José; tel. 224-6574; fax 225-8200; e-mail ambiter@sol.racsa.co.cr; internet www.ambitcr.com; Ambassador GIOACCHINO CARLO TRIZZINO.

**Japan:** Oficentro Ejecutivo La Sabana, Edif. 7, 3°, detrás de la Contraloría, Sabana Sur, Apdo 501, 1000 San José; tel. 232-1255; fax 231-3140; e-mail embjapon@sol.racsa.co.cr; internet www.cr.emb-japan.go.jp; Ambassador TADANORI INOMATA.

**Korea, Republic:** Oficentro Ejecutivo La Sabana, Edif. 2, 3°, Sabana Sur, Apdo 838, 1007 San José; tel. 220-3160; fax 220-3168; e-mail koreasec@sol.racsa.co.cr; Ambassador IM CHANG-SOON.

**Mexico:** Avda 7, No 1371, Apdo 10.107, 1000 San José; tel. 257-0633; fax 258-2437; e-mail embamex@racsa.co.cr; internet www.embajadademexico.com; Ambassador MARÍA CARMEN OÑATE MUÑOZ.

**Netherlands:** Los Yoses, Avda 8, Calles 35 y 37, Apdo 10.285, 1000 San José; tel. 296-1490; fax 296-2933; e-mail nethemb@racsa.co.cr; internet www.nethemb.or.cr; Ambassador W. G. J. M. WESSELS.

# COSTA RICA

**Nicaragua:** Edif. Trianón, Avda Central 250, Barrio la California, Apdo 1.382, 1000 San José; tel. 222-2373; fax 221-5481; e-mail embanic@sol.racsa.co.cr; Ambassador Dr FIALLOS NAVARRO.

**Panama:** Del San Pedro de Montes de Oca, Apdo 103-2050, San José; tel. 257-3241; fax 257-4864; e-mail panaembacr@racsa.co.cr; Ambassador LUIS E. VERGARA I.

**Peru:** Del Colegio de Igenieros y Arquitctos, 350m al norte, Urb. Freses, Curridabat, Apdo 4248, 1000 San José; tel. 225-9145; fax 253-0457; e-mail embaperu@amnet.co.cr; internet www.embaperu-costarica.rree.gob.pe; Ambassador ALBERTO GUTIÉRREZ LA MADRID.

**Poland:** De la Iglesia Santa Teresita 300 m este, 3307, Barrio Escalante, Apdo 664, 2010 San José; tel. 225-1481; fax 225-1592; e-mail embajpolonia1@racsa.co.cr; internet www.polonia-emb-cr.com/indes.php; Ambassador RYSZARD SCHNEPF.

**Russia:** Curridabat, Lomas de Ayarco Sur, de la carretera a Cartago, 1a entrada, 100 m sur, Apdo 6.340, 1000 San José; tel. 272-1021; fax 272-0142; e-mail emrusa@sol.racsa.co.cr; Ambassador VLADIMIR N. KAZIMIROV.

**Spain:** Calle 32, Paseo Colón, Avda 2, Apdo 10.150, 1000 San José; tel. 222-1933; fax 222-4180; e-mail embespcr@correo.mae.es; Ambassador JUAN JOSÉ URTASUN ERRO.

**Switzerland:** Paseo Colón, Centro Colón, Apdo 895, 1007 San José; tel. 221-4829; fax 255-2831; e-mail vertretung@sjc.rep.admin.ch; Ambassador GABRIELA NÜTZI.

**United Kingdom:** Edif. Centro Colón, 11°, Apdo 815, 1007 San José; tel. 258-2025; fax 233-9938; e-mail britemb@racsa.co.cr; internet www.britishembassycr.com; Ambassador GEORGINA BUTLER.

**Uruguay:** Avda 14, Calles 35 y 37, Apdo 3.448, 1000 San José; tel. 253-2755; fax 234-9909; e-mail embajrou@sol.racsa.co.cr; Ambassador ANTONIO RICARDO MORELL BORDOLI.

**USA:** Calle 120 Avda 0, Pavas, Apdo 920, 1200 San José; tel. 519-2000; fax 220-2305; e-mail info@usembassy.or.cr; internet sanjose.usembassy.gov; Chargé d'affaires a.i. RUSSELL FRISBIE.

**Venezuela:** Avda Central, Los Yoses, 5a entrada, Apdo 10.230, 1000 San José; tel. 225-8810; fax 253-1335; e-mail embaven@racsa.co.cr; Ambassador NORA URIBE TRUJILLO.

# Judicial System

Ultimate judicial power is vested in the Supreme Court, the justices of which are elected by the Assembly for a term of eight years, and are automatically re-elected for an equal period, unless the Assembly decides to the contrary by a two-thirds vote. The Supreme Court justices sit in four courts, the First Court (civil, administrative, agrarian and commercial matters), the Second Court (employment and family), the Third Court (penal) and the Constitutional Court.

There are, in addition, appellate courts, criminal courts, civil courts and special courts. The jury system is not used. Judges of the lower courts are appointed by the Supreme Court's administrative body, the Supreme Council. The Supreme Council's five members are elected by the Supreme Court.

### The Supreme Court
Sala Constitucional de la Corte Suprema de Justicia, Apdo 5, 1003 San José; tel. 295-3000; fax 257-0801; e-mail sala4-informacion@poder-judicial.go.cr.

**President of the Supreme Court:** LUIS PAULINO MORA MORA.

### Supreme Council
Members: LUIS PAULINO MORA MORA, ALFONSO CHAVES RAMÍREZ, MIRIAM ANCHÍA PANIAGUA, MILENA CONEJO AGUILAR, LUPITA CHAVES CERVANTES, SILVIA NAVARRO ROMANINI, ALFREDO JONES LEÓN.

**Justices of the First Court:** Dr ANABELLE LEÓN FEOLI, Dr OSCAR GONZÁLEZ CAMACHO, Dr ROMÁN SOLÍS ZELAYA, CARMENMARÍA ESCOTO FERNÁNDEZ, LUIS GUILLERMO RIVAS LOÁICIGA.

**Justices of the Second Court:** ORLANDO AGUIRRE GÓMEZ, Dr BERNARDO VAN DER LAAT ECHEVERRÍA, ZARELA VILLANUEVA MONGE, JULIA VARELA ARAYA, Dr ROLANDO VEGA ROBERT.

**Justices of the Third Court:** Dr DANIEL GONZÁLEZ ALVAREZ, JOSÉ MANUEL ARROYO GUTIÉRREZ, JESÚS RAMÍREZ QUIRÓS, RODRIGO CASTRO MONGE, ALFONSO CHAVES RAMÍREZ.

**Justices of the Constitutional Court:** Dr LUIS FERNANDO SOLANO CARRERA, ANA VIRGINIA CALZADA MIRANDA, CARLOS ARGUEDAS RAMÍREZ, Dr GILBERT ARMIJO SANCHOL, ADRIÁN VARGAS BENAVIDES, LUIS PAULINO MORA MORA, ERNESTO JINESTA LOBO.

# Religion

Under the Constitution, all forms of worship are tolerated. Roman Catholicism is the official religion of the country. Various Protestant churches are also represented.

## CHRISTIANITY

### The Roman Catholic Church
Costa Rica comprises one archdiocese and six dioceses. At 31 December 2003 Roman Catholics represented some 77% of the total population.

### Bishops' Conference
Conferencia Episcopal de Costa Rica, Apdo 7288, 1000 San José; tel. 258-3053; fax 221-66-62; e-mail seccecor@racsa.co.cr.

f. 1977; Pres. Most Rev. JOSÉ FRANCISCO ULLOA ROJAS.

**Archbishop of San José de Costa Rica:** Most Rev. HUGO BARRANTES UREÑA, Arzobispado, Apdo 497, 1000 San José; tel. 258-1015; fax 221-2427; e-mail curiam@racsa.co.cr.

### The Anglican Communion
Costa Rica comprises one of the five dioceses of the Iglesia Anglicana de la Región Central de América.

**Bishop of Costa Rica:** Rt Rev. HECTOR MONTERROSO, Apdo 10520, 1000 San José; tel. 225-0790; fax 253-8331; e-mail iarca@amnet.co.cr.

### Other Churches
**Federación de Asociaciones Bautistas de Costa Rica:** Apdo 1.631, 2100 Guadalupe; tel. 253-5820; fax 253-4723; e-mail coven@racsa.co.cr; internet www.fabcr.org; f. 1946; represents Baptist churches; Pres. JOSÉ SOTO VILLEGAS.

**Iglesia Evangélica Luterana de Costa Rica** (Evangelical Lutheran Church of Costa Rica): Apdo 1512, Pavas, 1200 San José; tel. and fax 231-3345; e-mail evkirche@racsa.co.cr; internet www.ielcor.org; f. 1955; 600 mems; Pres. Rev. RENÉ LAMMER.

**Iglesia Evangélica Metodista de Costa Rica** (Evangelical Methodist Church of Costa Rica): Apdo 5481, 1000 San José; tel. 236-2171; fax 236-5921; e-mail semlfp@racsa.co.cr; autonomous since 1973; affiliated to the United Methodist Church; 6,000 mems; Pres. Bishop LUIS F. PALOMO.

## BAHÁ'Í FAITH

**Bahá'í Information Centre:** Apdo 553, 1150 San José; tel. 231-0647; fax 296-1033; adherents resident in 242 localities.

**National Spiritual Assembly of the Bahá'ís of Costa Rica:** Apdo 553, 1150 La Uruca; tel. 231-0647; fax 296-1033; e-mail bahaiscr@sol.racsa.co.cr.

# The Press

## DAILIES

**Al Día:** Llorente de Tibás, Apdo 10.138, 1000, San José; tel. 247-4640; fax 247-4665; e-mail aldia@nacion.co.cr; internet www.aldia.co.cr; f. 1992; morning; independent; Dir EDGAR FONSECA; Editor MÓNICA GÓMEZ; circ. 60,000.

**Boletín Judicial:** La Uruca, Apdo 5.024, San José; tel. 231-5222; internet www.boletinjudicial.go.cr; f. 1878; journal of the judiciary; Dir BIENVENIDO VENEGAS PORRAS; circ. 2,500.

**Diario Extra:** Calle 4, Avda 4, Apdo 177, 1009 San José; tel. 223-9505; fax 223-5921; internet www.diarioextra.com; f. 1978; morning; independent; Dir WILLIAM GÓMEZ VARGAS; circ. 120,000.

**La Gaceta:** La Uruca, Apdo 5.024, San José; tel. 231-5222; internet www.gaceta.go.cr; f. 1878; official gazette; Dir BIENVENIDO VENEGAS; circ. 5,300.

**El Heraldo:** 400 m al este de las oficinas centrales, Apdo 1.500, San José; tel. 222-6665; fax 222-3039; e-mail info@elheraldo.net; internet www.elheraldo.net; f. 1994; morning; independent; Chief Editor VANESSA ESQUIVEL S.; Gen. Man. RODRIGO BARBOZA S.; circ. 30,000.

**La Nación:** Llorente de Tibás, Apdo 10.138, 1000 San José; tel. 247-4747; fax 247-5002; e-mail cacortes@nacion.com; internet www.nacion.com; f. 1946; morning; independent; Pres. MANUEL F. JIMÉNEZ ECHEVERRÍA; circ. 120,000.

**La Prensa Libre:** Calle 4, Avda 4, Apdo 177.1009, San José; tel. 223-6666; fax 233-6831; e-mail plibre@prensalibre.co.cr; internet www.prensalibre.co.cr; f. 1889; evening; independent; Dir ANDRÉS BORRASÉ SANOU; circ. 56,000.

# COSTA RICA

**La República:** Barrio Tournón, Guadalupe, Apdo 2130.1000, San José; tel. 223-0266; fax 257-0401; e-mail info@larepublica.co.cr; internet www.larepublica.co.cr; f. 1950; reorganized 1967; morning; independent; Dir ALBERTO MUÑOZ; Pres. FRED BLASER; circ. 61,000.

### PERIODICALS

**Abanico:** Calle 4, Avda 4, Apdo 177.1009, San José; tel. 223-6666; fax 223-4671; internet www.prensalibre.co.cr; weekly supplement of *La Prensa Libre*; women's interests; circ. 50,000.

**Acta Médica Costarricense:** Colegio de Médicos y Cirujanos de Costa Rica, Sabana Sur, Apdo 548-1000, San José; tel. 232-3433; fax 232-2406; e-mail medicos@racsa.co.cr; internet www.medicos.sa.cr; f. 1957; journal of the Colegio de Médicos; 4 issues per year; Publr Dr MARÍA PAZ LEÓN BRATTI; Editor Dr CARLOS SALAZAR VARGAS; circ. 5,000.

**El Cafetalero:** Calle 1, Avdas 18 y 20, Apdo 37, 1000 San José; tel. 222-6411; fax 223-6025; e-mail evilla@icafe.go.cr; f. 1964 as Noticiero del Café; changed name in 2000; bi-monthly; coffee journal; owned by the Instituto del Café de Costa Rica; Editor ERNESTO VILLALOBOS; circ. 10,000.

**Contrapunto:** La Uruca, Apdo 7, 1980 San José; tel. 231-3333; f. 1978; fortnightly; publication of Sistema Nacional de Radio y Televisión; Dir-Gen. BELISARI SOLANO; circ. 10,000.

**Eco Católico:** Calle 22, Avdas 3 y 5, Apdo 1.064, San José; tel. 222-6156; fax 256-0407; e-mail ecocatolico@rasca.co.cr; f. 1931; Catholic weekly; Dir ARMANDO ALFARO; circ. 20,000.

**Perfil:** Llorente de Tibás, Apdo 1.517, 1100 San José; tel. 247-4345; fax 247-5110; e-mail perfil@nacion.co.cr; f. 1984; fortnightly; women's interest; Dir CAROLINA CARAZO BARRANTES; circ. 16,000.

**Polémica:** San José; tel. 233-3964; f. 1981; every 4 months; leftwing; Dir GABRIEL AGUILERA PERALTA.

**Rumbo:** Llorente de Tibás, Apdo 10.138, 1000 San José; tel. 240-4848; fax 240-6480; f. 1984; weekly; general; Dir ROXANA ZÚÑIGA; circ. 15,000.

**San José News:** Apdo 7, 2730 San José; 2 a week; Dir CHRISTIAN RODRÍGUEZ.

**Semanario Libertad:** Calle 4, Avdas 8 y 10, Apdo 6.613, 1000 San José; tel. 225-9024; fax 253-2628; f. 1962; weekly; organ of the Partido del Pueblo Costarricense; Dir RODOLFO ULLOA B.; Editor JOSÉ A. ZÚÑIGA; circ. 10,000.

**Semanario Universidad:** San José; tel. 207-5355; fax 207-4774; e-mail semana@cariari.ucr.ac.cr; internet cariari.ucr.ac.cr; f. 1970; weekly; general; Dir THAIS AGUILAR; circ. 15,000.

**The Tico Times:** Calle 15, Avda 8, Apdo 4.632, 1000 San José; tel. 258-1558; fax 223-6378; e-mail info@ticotimes.net; internet www.ticotimes.net; f. 1956; weekly; in English; Dir DERY DYER; circ. 15,210.

**Tiempos de Costa Rica:** 100m sur de Ferretería El Mar, San Pedro, San José; tel. 280-2332; fax 280-6840; e-mail admin@tdm.com; internet www.tdm.com; f. 1996; Costa Rican edition of the international *Tiempos de Mundo*; Dir WILLIAM COOK; Editor JOSÉ A. PASTOR.

### PRESS ASSOCIATIONS

**Colegio de Periodistas de Costa Rica:** Sabana Este, Calle 42, Avda 4, Apdo 5.416, San José; tel. 233-5850; fax 223-8669; e-mail ejecutiva@colegiodeperiodistas.org; internet www.colper.or.cr; f. 1969; 1,447 mems; Pres. RAMSÉS ROMÁN.

**Sindicato Nacional de Periodistas:** Sabana Este, Calle 42, Avda 4, Apdo 5.416, San José; tel. 222-7589; fax 258-3229; e-mail sindicato@colper.or.cr; f. 1970; 200 mems; Sec.-Gen. SERGIO FERNANDEZ.

### FOREIGN NEWS BUREAUX

**ACAN-EFE** (Central America): Costado Sur, Casa Matute Gómez, Casa 1912, Apdo 84.930, San José; tel. 222-6785; fax 233-7681; e-mail acanefe@sol.racsa.co.cr; Correspondent LORNA CHACÓN.

**Agence France-Presse** (France): Barrio Escalante, Del Farolito 175m al Este, Casa 3361, Apdo 5.276, San José; tel. 280-1773; fax 280-1755; Correspondent DOMINIQUE PETTIT.

**Agenzia Nazionale Stampa Associata (ANSA)** (Italy): c/o Diario La República, Barrio Tournón, Guadalupe, Apdo 545, 1200 San José; tel. 231-1140; fax 231-1140; internet www.ansa.it; Correspondent LUIS CARTÍN S.

**Deutsche Presse-Agentur (dpa)** (Germany): Edif. 152, 3°, Calle 11, Avdas 1 y 3, Apdo 7.156, San José; tel. 233-0604; fax 233-0604; Correspondent ERNESTO RAMÍREZ.

**Informatsionnoye Telegrafnoye Agentstvo Rossii—Telegrafnoye Agentstvo Suverennykh Stran (ITAR—TASS)** (Russia): De la Casa Italia 1,000 m este, 50 m norte, Casa 675, Apdo 1.011, San José; tel. 224-1560; Correspondent ENRIQUE MORA.

**Rossiyskoye Informatsionnoye Agentstvo—Novosti (RIA—Novosti)** (Russia): De la Casa Italiana 100 m este, 50 m norte, San José; tel. 224-1560; internet www.rian.ru.

**Xinhua (New China) News Agency** (People's Republic of China): Apdo 4.774, San José; tel. 231-3497; Correspondent XU BIHUA.

## Publishers

**Alef Editores:** Apdo 146, 1017 San José 2000; tel. 255-0202; fax 222-7878; e-mail alefreading@racsa.co.cr; Dir JOSÉ SUCCAR.

**Ediciones Farben, SA:** De la Sylvania 500 oeste, Condominio Industrial Pavas 19, Apdo 592-1200 Pavas; tel. 290-7125; fax 290-0061; e-mail fvanderlaat@farben.co.cr; internet www.farben.co.cr; Dir FERNANDO VANDERLAAT.

**Editorial Caribe:** Apdo 1.307, San José; tel. 222-7244; e-mail info@editorialcaribe.com; internet www.editorialcaribe.com; f. 1949; division of Thomas Nelson Publrs; religious textbooks; Dir JOHN STROWEL.

**Editorial Costa Rica:** Costado oeste del cementerio, Guadalupe de Goicoechea, Apdo 10.010, San José; tel. 255-2323; fax 253-5091; e-mail editocr@racsa.co.cr; internet www.editorialcostarica.com; f. 1959; government-owned; cultural; Gen. Man. HABIB SUCCAR GUZMÁN.

**Editorial Fernández Arce:** 50 este de Sterling Products, La Paulina de Montes de Oca, Apdo 2.410, 1000 San José; tel. 224-5201; fax 225-6109; f. 1967; textbooks for primary, secondary and university education; Dir Dr MARIO FERNÁNDEZ LOBO.

**Editorial de la Universidad Autónoma de Centro América (UACA):** Apdo 7.637, 1000 San José; tel. 234-0701; fax 224-0391; e-mail info@uaca.ac.cr; f. 1981; Editor GUILLERMO MALAVASSI.

**Editorial de la Universidad Estatal a Distancia (EUNED):** Apdo 474, 2050 San Pedro; tel. 253-2121; fax 234-9138; e-mail editoria@uned.ac.cr; internet www.uned.ac.cr; f. 1979; Dir RENÉ MUIÑOS.

**Editorial Universitaria Centroamericana (EDUCA):** Ciudad Universitaria Rodrigo Facio, San Pedro, Montes de Oca, Apdo 64, 2060 San Pedro; tel. 224-3727; fax 253-9141; e-mail educa@sp.cusa.ac.cr; f. 1969; organ of the CSUCA; science, literature, philosophy; Dir ANITA DE FORMOSO.

**Imprenta Nacional:** San José; tel. 296-9570; Dir-Gen. BIENVENIDO VENEGAS PORRAS.

**Librería Lehmann, Imprenta y Litografía, Ltda:** Calles 1 y 3, Avda Central, Apdo 10011, San José; tel. 524-0000; fax 233-0713; e-mail servicio@librerialehman.com; internet www.librerialehmann.com; f. 1896; general fiction, educational, textbooks; Man. Dir ANTONIO LEHMANN STRUVE.

**Trejos Hermanos Sucesores, SA:** Curridabat, Apdo 10.096, San José; tel. 224-2411; e-mail henry@trejoshnos.com; f. 1912; general and reference; Man. HENRY CHAMBERLAIN.

### PUBLISHING ASSOCIATION

**Cámara Costarricense del Libro:** Paseo de los Estudiantes, Apdo 1571, 1002 San José; tel. 225-1363; fax 252-4297; e-mail ccl@libroscr.com; internet www.libroscr.com; f. 1978; Pres. OSCAR CASTILLO.

## Broadcasting and Communications

### TELECOMMUNICATIONS

In 2002 there were plans to expand and modernize the telecommunications sector. The French company Alcatel was to oversee the project.

**Cámara Costarricense de Telecomunicaciones (CCTEL):** Edif. Centro Colón, Apdo 591, 1007 San José; tel. and fax 255-3422; e-mail cctel@cctel.org; internet cctel.org; Pres. MIGUEL LEÓN S.

**Cámara Nacional de Medios de Comunicación Colectiva (CANAMECC):** San José; tel. 222-4820; f. 1954; Pres. ANDRÉS QUINTANA CAVALLINI.

**Instituto Costarricense de Electricidad (ICE):** govt agency for power and telecommunications (see Trade and Industry: Utilities).

    **Radiográfica Costarricense, SA (RACSA):** Avda 5, Calle 1, Frente al Edif. Numar, Apdo 54, 1000 San José; tel. 287-0087; fax 287-0379; e-mail mcruz@sol.racsa.co.cr; internet www.racsa.co.cr; f. 1921; state telecommunications co., owned by ICE; Dir-Gen. MARCO A. CRUZ MIRANDA.

# COSTA RICA

## RADIO

**Asociación Costarricense de Información y Cultura (ACIC):** Apdo 365, 1009 San José; f. 1983; independent body; controls private radio stations; Pres. JUAN FEDERICO MONTEALEGRE MARTÍN.

**Cámara Nacional de Radio (CANARA):** Paseo de los Estudiantes, Apdo 1.583, 1002 San José; tel. 256-2338; fax 255-4483; e-mail info@canara.org; internet www.canara.org; f. 1947; Exec. Dir LAURA ARRIETA VARGAS.

**Control Nacional de Radio (CNR):** Dirección Nacional de Comunicaciones, Ministerio de Gobernación y Policía, Apdo 10.006, 1000 San José; tel. 221-0992; fax 283-0741; e-mail cnradio@racsa.co.cr; f. 1954; governmental supervisory department; Dir MELVIN MURILLO ALVAREZ.

### Non-commercial

**Faro del Caribe:** Apdo 2.710, 1000 San José; tel. 226-2573; fax 227-1725; e-mail juntadirectiva@farodelcaribe.org; internet www.farodelcaribe.org; f. 1948; religious and cultural programmes in Spanish and English; Man. GEOVANNY CALDERÓN CASTRO.

**Radio FCN Sonora** (Family Christian Network): Apdo: 60-2020 Zapote, San José; tel. 293-6912; fax 293-7993; e-mail info@fcnradio.com; internet www.fcnradio.com; Dir Dr DECAROL WILLIAMSON.

**Radio Fides:** Avda 4, Curia Metropolitana, Apdo 5.079, 1000 San José; tel. 258-1415; fax 233-2387; e-mail emiliom@radiofides.co.cr; internet www.radiofides.co.cr; f. 1952; Roman Catholic station; Dir EMILIO OTÁROLA.

**Radio Nacional:** 1 km oeste del Parque Nacional de Diversiones, La Uruca, Apdo 7, 1980 San José; tel. 231-7983; fax 220-0070; e-mail sinart@sol.racsa.co.cr; f. 1978; Dir RODOLFO RODRÍGUEZ.

**Radio Santa Clara:** Santa Clara, San Carlos, Apdo 221, Ciudad Quesada, Alajuela; tel. and fax 460-6666; e-mail radio@radiosantaclara.org; internet www.radiosantaclara.org; f. 1984; Roman Catholic station; Dir Rev. MARCO A. SOLÍS V.

**Radio Universidad:** Ciudad Universitaria Rodrigo Facio, San Pedro, Montes de Oca, Apdo 2.060, 1000 San José; tel. 207-5356; fax 207-5459; e-mail radioucr@cariari.ucr.ac.cr; internet cariari.ucr.ac.cr/~radioucr; f. 1949; classical music; Dir CARLOS MORALES.

### Commercial

There are about 80 commercial radio stations, including:

**Cadena de Emisoras Columbia:** Apdo 1.708, San José; tel. 224-0808; fax 225-9275; e-mail columbia@columbia.co.cr; internet www.columbia.co.cr; operates Radio Columbia, Radio Dos, Radio 955 Jazz; Dir C. ARNOLDO ALFARO CHAVARRA.

**Cadena Musical:** Apdo 854, 1000, San José; tel. 257-2789; fax 233-9975; internet www.radiomusical.com; f. 1954; operates Radio Musical, Radio Bomba; Gen. Man. JORGE JAVIER CASTRO.

**Grupo Centro:** Apdo 6.133, San José; tel. 240-7591; fax 236-3672; f. 1971; operates Radio Centro 96.3 FM, Radio 820 AM, Televisora Guanacasteca Channels 16 and 28; Dir ROBERTO HERNÁNDEZ RAMÍREZ.

**Radio Chorotega:** Conferencia Episcopal de Costa Rica, Casa Cural de Santa Cruz, Apdo 92, 5175 Guanacaste; tel. and fax 680-0447; f. 1983; Roman Catholic station; Dir Rev. HUGO BRENES VILLALOBOS.

**Radio Eco:** Apdo 585-1007, Centro Colón, San José; tel. 220-1001; fax 290-0970; e-mail info@radioeco.com; internet www.radioeco.com; Dir RICARDO ZAMORA; Gen. Man. LUIS ENRIQUE ORTIZ VAGLIO.

**Radio Emaús:** San Vito, Coto Brus; tel. and fax 773-3101; fax 773-4035; e-mail radioemaus@racsa.co.cr; f. 1962; Roman Catholic station; religious programmes; Dir Rev. MIGUEL ANGEL BERGANZA.

**Radio Monumental:** Avda Central y 2, Calle 2, Apdo 800, 1000 San José; tel. 296-6093; fax 231-1210; e-mail ventas@monumental.co.cr; internet www.monumental.co.cr; f. 1929; all news station; Gen. Man. TERESA MARÍA CHÁVES ZAMORA.

**Radio Musical:** Apdo 854, 1000 San José; tel. 257-2789; fax 233-9975; e-mail info@radiomusical.com; internet www.radiomusical.com; f. 1951.

**Radio Sendas de Vida:** San José; tel. 248-1148; fax 233-1259; e-mail info@radiosendas.com; internet www.radiosendas.com; f. 1982; Christian station.

## TELEVISION

### Government-owned

**Sistema Nacional de Radio y Televisión Cultural (SINART):** 1 km al oeste del Parque Nacional de Diversiones La Uruca, Apdo 7, 1980 San José; tel. 231-3333; fax 231-6604; e-mail sinart@racsa.co.cr; f. 1977; cultural; Dir-Gen. BELISARI SOLANO.

### Commercial

**Alphavisión (Canal 19):** Detrás de la Iglesia de Santa María y Griega, Carretera a Desamparados, Apdo 1.490, San José; tel. 226-9333; fax 226-9095; f. 1987; Gen. Man. CECILIA RAMÍREZ.

**Canal 54:** Detrás de la Iglesia Santa Marta, Carretera a Desamparados, San José; tel. 286-3344; fax 231-3408; e-mail canalcr@sol.racsa.co.cr; f. 1996; Pres. ANTONIO ALEXANDRE GARCÍA.

**Multivisión de Costa Rica, Ltda (Canal 9):** 150 m oeste del Centro Comercial de Guadalupe, Apdo 4.666, 1000 San José; tel. 233-4444; fax 221-1734; f. 1961; operates Radio Sistema Universal A.M. (f. 1956), Channel 9 (f. 1962), and FM (f. 1980); sold Channel 4 (f. 1964) to Repretel in 2000; Gen. Man. ARNOLD VARGAS.

**Televisora de Costa Rica (Canal 7), SA (Teletica):** Costado oeste Estadio Nacional, Apdo 3.876, San José; tel. 232-2222; fax 231-6258; e-mail info@teletica.com; internet www.teletica.com; f. 1960; operates Channel 7; Pres. OLGA COZZA DE PICADO; Gen. Man. RENÉ PICADO COZZA.

**Repretel (Canales 4, 6 y 11):** Apdo 2860, 1000 San José; tel. 290-6665; fax 232-4203; e-mail info@repretel.com; internet www.repretel.com; News Dir GRETTEL ALFARO.

# Finance

(cap. = capital; p.u. = paid up; res = reserves; dep. = deposits; m. = million; brs = branches; amounts in colones, unless otherwise indicated)

## BANKING

**Banco Central de Costa Rica:** Avdas Central y Primera, Calles 2 y 4, Apdo 10.058, 1000 San José; tel. 243-3333; fax 243-4559; internet www.bccr.fi.cr; f. 1950; total assets 944,168.9m. (2003); scheduled for privatization; Pres. Dr FRANCISCO DE PAULA GUTIÉRREZ GUTIÉRREZ; Man. ROY GONZÁLEZ.

### State-owned Banks

**Banco de Costa Rica:** Avdas Central y 2da, Calles 4 y 6, Apdo 10035, 1000 San José; tel. 287-9000; fax 255-0221; e-mail bancobcr@bancobcr.com; internet www.bancobcr.com; f. 1877; responsible for industry; cap. 32,537.4m., res 57,679.8m., dep.554,813.3m. (2003); Pres. MARIANO GUARDIA CAÑAS; Gen. Man. MARIO BARRENECHEA; 94 brs.

**Banco Nacional de Costa Rica:** Avda 1ra, Calle 2A, Apdo 10015, 1000 San José; tel. 221-2223; fax 2223-8054; e-mail bncr@bncr.fi.cr; internet www.bncr.fi.cr; f. 1914; responsible for the agricultural sector; cap. 30,042.7m., res 33,237.9m., dep. 873,049.6m. (2003); Pres. RODOLFO BRENES GÓMEZ; Gen. Man. GUILLERMO HEYDEN QUINTERO; 140 brs and agencies.

**Banco Popular y de Desarrollo Comunal:** Calle 1, Avda 2, Apdo 10190, San José; tel. 257-5797; fax 255-1966; e-mail relacionespublicas@bp.fi.cr; internet www.bancopopular.fi.cr; f. 1969; cap. 24,991.2m. (2003); Pres. FLORIBETH LÓPEZ UGALDE; Vice-Pres. HILDA VALVERDE AVALOS.

### Private Banks

**Banco Banex SA:** Barrio Tournón, Diagonal a Ulacit, Apdo 7983, 1000 San José; tel. 257-0522; fax 257-5967; e-mail interna@banex.co.cr; internet www.banex.co.cr; f. 1981 as Banco Agroindustrial y de Exportaciones, SA; adopted present name 1987; incorporated Banco Metropolitano in May 2001 and Banco Bancrecen in December 2002; cap. 125,892.9m., res 2,640.7m., dep. 190,808.9m. (Dec. 2003); Pres. ALBERTO VALLARINO; Gen. Man. FEDERICO ALBERT; 36 brs.

**Banco Bantec, SA:** Calle Principalla Uruca, Antigua Cana, Apdo 1164, 1000 San José; tel. 290-8585; fax 290-2939; e-mail consorcio@bantec.com.

**Banco BCT, SA:** 160 Calle Central, Apdo 7698, San José; tel. 257-6010; fax 222-3706; e-mail banco@bct.fi.cr; f. 1984; total assets 13,794m. (1999); merged with Banco del Comercio, SA in 2000; Pres. ANTONIO BURGUÉS; Gen. Man. LEONEL BARUCH.

**Banco de Crédito Centroamericano, SA (Bancentro):** De la Rotonda la Bandera, Barrio Escalante, 50 m oeste, Apdo 5099, 1000 San José; tel. 280-5555; fax 280-5090; e-mail info@lafise.fi.cr; f. 1974; owned by Grupo Lafise; cap. 3,403.6m. res 258.0m. total assets 3,983.1m. (Dec. 2004); Pres. ROBERTO J. ZAMORA LLANES; Gen. Man GILBERTO SERRANO.

**Banco Cuscatlan de Costa Rica, SA:** De La Rotonda Juan Pablo II, 150 m norte, Canton Central, La Uruca, Apdo 6531, 1000 San José; tel. 299-0299; fax 220-0297; e-mail cuscatlan@cuscatlancr.com; internet www.bancocuscatlan.com; f. 1984 as Banco de Fomento Agrícola; changed name to Banco BFA in 1994; changed name to above in 2000; cap. 3,865.7m., res 750.8m., dep. 86,671.3m. (Dec.

COSTA RICA                                                                                                                                                                 *Directory*

2003); Pres. ERNESTO ROHRMOSER; Gen. Man. ALVARO ENRIQUE SABORIO LEGERS.

**Banco Elca, SA:** Del Centro Colón, 200 m norte, Paseo Colón, Apdo 1112, 1000 San José; tel. 258-3355; fax 233-8383; e-mail banelca@sol.racsa.co.cr; internet www.bancoelca.fi.cr; f. 1985 as Banco de la Industria; changed name to above in 2000; Pres. ROBERTO ALVARADO MOYA; Gen. Man. CARLOS ALVARADO MOYA.

**Banco Improsa, SA:** 2985 Calle 29 y 31, Avda 5, Carmen, San José; tel. 257-0689; fax 223-7319; e-mail banimpro@sol.racsa.co.cr.

**Banco Interfin, SA:** Calle 3, Avdas 2 y 4, Apdo 6899, San José; tel. 210-4000; fax 210-4510; e-mail webmaster@interfin.fi.cr; internet www.interfin.fi.cr; f. 1982; dep. 15,450.0m., res 2,312.1m. total assets 23,786.3m. (Dec. 2004); Pres. Ing. LUIS LUKOWIECKI; Gen. Man. Dr LUIS LIBERMAN.

**Banco Promérica, SA:** Edif. Promérica, Sabana oeste canal 7, Mata Redonda, Apdo 1.289, 1200 San José; tel. 220-3090; fax 232-5727; e-mail solucion@promerica.fi.cr; internet www.promerica.fi.cr; 21 brs.

**Banco BAC San José, SA:** Calle Central, Avdas 3 y 5, Apdo 5.445, 1000 San José; tel. 295-9595; fax 222-7103; e-mail info@bancosanjose.fi.cr; internet www.bacsanjose.com; f. 1968; fmrly Bank of America, SA; total assets US $505m. (2003); Pres. ERNESTO CASTEGNARO ODIO; Vice-Pres. GERARDO CORRALES BRENES.

**Banco Uno, SA:** Calle 28, Paseo Colón, Merced, Apdo 5884, 1000 San José; tel. 257-1344; fax 257-2215; e-mail bunocr@grupo-uno.com; internet www.bancouno.fi.cr; cap. 2,673.9m., res 171.3m. (Dec. 2003); Pres. Dr ERNESTO FERNÁNDEZ HOLMANN.

**Citibank (Costa Rica), SA:** Edif. 3, Blvd Camino Real, Apdo 10277, San José; tel. 201-0800; fax 296-1496; e-mail citibab@sol.racsa.co.cr.

**Scotiabank Costa Rica:** Avda 1ra, Calles Central y 2, Apdo 5395, 1000, San José; tel. 287-8700; fax 255-3076; e-mail scotiacr@scotiabank.com; f. 1995; Gen. Man. BRIAN W. BRADY; 13 brs.

### Banking Association

**Asociación Bancaria Costarricense:** San José; tel. 253-2889; fax 225-0987; e-mail abc@abc.fi.cr; internet www.abc.fi.cr; Pres. JORGE MONGE.

### STOCK EXCHANGE

**Bolsa Nacional de Valores, SA:** Edif. Cartagena, 4°, Calle Central, Avda Primera, Santa Ana, Apdo 6.155, 1000 San José; tel. 204-4848; fax 204-4802; e-mail bnv@bnv.co.cr; internet www.bnv.co.cr; f. 1976; Pres. Dr LUIS LIBERMAN GINSBURG; Gen. Man. JOSÉ RAFAEL BRENES.

### INSURANCE

In 1998 the Legislative Assembly approved legislative reform effectively terminating the state monopoly of all insurance activities.

**Caja Costarricense de Seguro Social:** Apdo 10.105, San José; tel. 257-9122; fax 222-1217; internet www.info.ccss.sa.cr; accident and health insurance, state-owned; Exec. Pres. Dr ALBERTO SÁENZ PACHECO.

**Instituto Nacional de Seguros:** Calles 9 y 9 bis, Avda 7, Apdo 10.061, 1000 San José; tel. 223-5800; fax 255-3381; internet www.ins.go.cr; f. 1924; administers the state monopoly of insurance; services of foreign insurance companies may be used only by authorization of the Ministry of the Economy, Industry and Commerce, and only after the Instituto has certified that it will not accept the risk; Exec. Pres. LUIS JAVIER GUIER ALFARO; Gen. Man. MARIANO CAMPOS SALAS.

# Trade and Industry

## GOVERNMENT AGENCIES

**Instituto Nacional de Vivienda y Urbanismo (INVU):** Apdo 2.534, San José; tel. 221-5266; fax 223-4006; internet www.invu.go.cr; housing and town planning institute; Exec. Pres. ÁNGELO ALTAMURA CARRIERO; Gen. Man. PEDRO HERNÁNDEZ RUIZ.

**Ministry of Planning and Economic Policy:** See The Government (Ministries).

**Promotora del Comercio Exterior de Costa Rica (PROCOMER):** Calle 40, Avdas Central y 3, Centro Colón, Apdo 1.278, 1007 San José; tel. 256-7111; fax 233-4655; e-mail info@procomer.com; internet www.procomer.com; f. 1968 to improve international competitiveness by providing services aimed at increasing, diversifying and expediting international trade.

## DEVELOPMENT ORGANIZATIONS

**Cámara de Azucareros:** Calle 3, Avda Fernández Güell, Apdo 1.577, 1000 San José; tel. 221-2103; fax 222-1358; e-mail crazucar@racsa.co.cr; f. 1949; sugar growers; Pres. RODRIGO ARIAS SÁNCHEZ; 16 mems.

**Cámara Nacional de Bananeros:** Edif. Urcha, 3°, Calle 11, Avda 6, Apdo 10.273, 1000 San José; tel. 222-7891; fax 233-1268; e-mail canaba@racsa.co.cr; f. 1967; banana growers; Pres. DIANA GUZMÁN CALZADA; Exec. Dir MARÍA DE LOS ANGELES VINDAS.

**Cámara Nacional de Cafetaleros:** Condominio Oroki 4D, La Uruca, Apdo 1.310, San José; tel. and fax 296-8334; e-mail camcafe@racsa.co.cr; f. 1948; 70 mems; coffee millers and growers; Pres. RONALD PETER SEEVERS; Exec. Dir GABRIELA LOBO H.

**Cámara Nacional de Artesanía y Pequeña Industria de Costa Rica (CANAPI):** Calle 17, Avda 10, detrás estatua de San Martín, Apdo 1.783 Goicoechea, 2100 San José; tel. 223-2763; fax 255-4873; e-mail canapi@sol.racsa.co.cr; f. 1963; development, marketing and export of small-scale industries and handicrafts.

**Corporación Bananera Nacional:** Apdo 6504, 1000 San José; tel. 202-4700; fax 234-9421; e-mail corbana@racsa.co.cr; internet www.corbana.co.cr.

**Costa Rican Investment and Development Board (CINDE):** Edif. Los Balcones, Plaza Roble, 4°, Guachipelin, Ezcazú; tel. 201-2800; fax 201-2867; e-mail invest@cinde.org; internet www.cinde.or.cr; f. 1983; coalition for development of initiatives to attract foreign investment for production and export of new products; Chair. EMILIO BRUCE; CEO ENRIQUE EGLOFF.

**Instituto del Café de Costa Rica:** Calle 1, Avdas 18 y 20, Apdo 37, 1000 San José; tel. 243-7812; fax 222-2838; internet www.icafe.go.cr; e-mail achacon@icafe.go.cr; f. 1933 to develop the coffee industry, to control production and to regulate marketing; Dir. JUAN MAYO.

**Sistema de Información del Sector Agropecuario:** San José; tel. 296-2579; fax 296-1652; e-mail infoagro@mag.go.cr; internet www.infoagro.go.cr.

## CHAMBERS OF COMMERCE

**Cámara de Comercio de Costa Rica:** Urbanización Tournón, 150 m noroeste del parqueo del Centro Comercial El Pueblo, Apdo 1.114, 1000 San José; tel. 221-0005; fax 223-157; e-mail servicio@camara-comercio.com; internet www.camara-comercio.com; f. 1915; 900 mems; Pres. EVITA ARGUEDAS MAKLOUF; Dirs JOSÉ SEGOVIA ATENCIA, DIETER FIEBERG SILAKOWSKI.

**Cámara de Industrias de Costa Rica:** 350 m sur de la Fuente de la Hispanidad, San Pedro de Montes de Oca, Apdo 10.003, San José; tel. 281-0006; fax 234-6163; e-mail cicr@cicr.com; internet www.cicr.com; Pres. JACK LIBERMAN GINSBURG; Vice-Pres. CHESTER ZELAYA GOEBEL.

**Unión Costarricense de Cámaras y Asociaciones de la Empresa Privada (UCCAEP):** De McDonalds en Sabana Sur, 400 m al sur, 100 m al este, 25 m al sur, San José; tel. 290-5595; fax 290-5596; internet www.uccaep.or.cr; f. 1974; business federation; Pres. Ing. SAMUEL YANKELEWITZ BERGER; Exec. Dir ALVARO RAMÍREZ BOGANTES.

## INDUSTRIAL AND TRADE ASSOCIATIONS

**Asociación de Empresas de Zonas Francas (AZOFRAS):** La Uruca San José, Apdo 162, 3006 Barreal de Heredia; tel. 293-7073; fax 293-7094; e-mail azofras@racsa.co.cr; internet www.azofras.com; Pres. JORGE BRENES; Exec. Dir TIMOTHY SCOTT HALL.

**Cámara Nacional de Agricultura y Agroindustria:** Avda 10-10 bis, Cv. 23, Apdo 1.671, 1000 San José; tel. 225-8245; fax 280-0969; e-mail cnaacr@sol.racsa.co.cr; internet www.cnaacr.com; f. 1947; Pres. JOSÉ ANTONIO MADRIZ CARRILLO; Exec. Dir MÓNICA NAVARRO DEL VALLE.

**Consejo Nacional de Producción:** 125 m al sur de Yamuni La Sabana en Avda 10, Apdo 2.205, San José; tel. 255-0056; fax 256-9625; e-mail sim@cnp.go.cr; internet www.mercanet.cnp.go.cr; f. 1948 to encourage agricultural and fish production and to regulate production and distribution of basic commodities; Pres. JOSÉ JOAQUÍN ACUÑA; Gen. Man. CARLOS CRUZ CHANG.

**Instituto de Desarrollo Agrario (IDA):** Apdo 5.054, 1000 San José; tel. 224-6066; internet www.ida.go.cr; Exec. Pres. WALTER CÉSPEDES SALAZAR.

**Instituto Mixto de Ayuda Social (IMAS):** Calle 29, Avdas 2 y 4, Apdo 6.213, San José; tel. 225-5555; fax 224-8783; internet www.imas.go.cr; Pres. MARTA LORA.

**Instituto Nacional de Fomento Cooperativo:** Apdo 10.103, 1000 San José; tel. 256-2944; fax 255-3835; e-mail info@infocoop.go.cr; internet www.infocoop.go.cr; f. 1973 to encourage the establishment of co-operatives and to provide technical assistance and credit facilities; Pres. Rafael Angel Rojas Jiménez; Exec. Dir Luis Antonio Monge Román.

## MAJOR COMPANIES

**Abbot Laboratories:** San José; tel. 209-5000; fax 209-5308; f. 2001; medical equipment manufacturers.

**Abonos Superior SA:** La Lima de la Bomba Shell, 700 m Sur, 300 m Oeste y 300 m Norte, Cartago; tel. 573-7041; fax 573-8116; internet www.superior.co.cr; holding co with interests in the fertilizer and utilities sectors; Chair. Fourneir Vargas; Man. Carlos Arguedas.

**ArtinSoft:** Edif. Artinsoft, Calle Blancos, San José; tel. 247-3000; fax 241-1520; internet www.artinsoft.com; producer of electronic software; partially owned by Intel (USA); Pres. and CEO Carlos Araya.

**AS de Oros SA:** Apdo 428, 4005 Belén-Heredia; tel. 298-1895; fax 239-1322; e-mail asdeoros@sol.racsa.co.cr; f. 1981; animal feed manufacturers; owned by Rica Foods Inc, USA; Dir.-Gen. José Manuel Quesada Chaves; 612 employees.

**Atlas Eléctrica, SA:** Carretera a Heredia, Apdo 2.166, 1000 San José; tel. 277-2000; fax 260-3930; e-mail atlasele@sol.racsa.co.cr; internet www.atlas.co.cr; f. 1961; manufacturers of domestic cooking and refrigeration appliances; Chair. Walter Kissling; Man. Diego Artiñano; 750 employees.

**BASF de Costa Rica, SA:** Edif. Irma, 2°, Avda Central, Calles 29-33, Apdo 4.471, 1000 San José; tel. 253-2217; fax 224-7225; internet www.basf-costa-rica.com; chemicals; owned by BASF, AG (Germany); Man. Juan Carlos Cruz.

**Baxter Productos Médicos, Ltd:** San José; tel. 573-7811; fax 573-7433; e-mail zelayac@baxter.com; Pres. Chester Zelaya.

**Bayer Costa Rica, SA:** San Gabriel de Calle Blancos, Apdo 5.103, 1000 San José; tel. 223-6166; fax 255-0693; e-mail bayer.costarica.bc@bayer-ca.com; f. 1978; chemicals; Pres. Manfred Loeseken.

**BDF Costa Rica, SA:** Santa Teresita 2988, Apdo 416, 2010 Zapote; tel. and fax 253-5693; manufacturer of pharmaceuticals and toiletries; owned by Beiersdorf AG (Germany); Dir.-Gen. Luis Emilio Wong.

**Bticino Costa Rica, SA:** Frente al Cenada, Barreal de Heredia, Apdo 6.563, 1000 San José; tel. 298-0101; fax 293-0108; e-mail bticino@racsa.co.cr; manufacturer of electrical apparatus; Gen. Man. Soorrado Franché.

**CAMtronics, SA:** Edif. 50, Parque Industrial Zona Franca Cartago, Cartago; tel. 573-7366; fax 573-7225; e-mail marketing@camtronicscr.com; internet www.camtronicscr.com; manufacturers of electronic equipment; Pres. and CEO Enrique Ortiz Carazo.

**Cemex Costa Rica:** Apdo 6.558, 1000 San José; tel. and fax 201-8200; internet www.cemexcostarica.com; cement manufacturers; Pres. Lorenzo H. Zambrano.

**Cibertec International, SA:** 25 oeste de la Escuela St Joseph, Barrio la Guaria, Moravia, POB 149-2300, San José; tel. 297-0933; fax 240-3266; e-mail int.sales@cibertec.com; telecommunications equipment; CEO Mario Caravajal.

**Colgate Palmolive (Costa Rica), SA:** Calles 26 y 28 Avda 3, Apdo 10.040, 1000 San José; tel. 298-4600; fax 293-7171; e-mail juanita_espinoza@colpal.com; manufacturers of toiletries; Dir. Gen. Carl Mark.

**Componentes Intel de Costa Rica, SA:** Calle 129 La Ribera, Heredia; tel. 298-6000; fax 298-7206; internet www.intel.com; manufacturers of computer components; Gen. Man. Pat Raburn.

**Compañía Costarricense de Café, SA (CAFESA):** Frente al Hospital México, Apdo 4.588, 1000 San José; tel. 232-2255; fax 231-3640; e-mail cafesa@racsa.co.cr; manufacturer of agrochemicals, fertilizers; Pres. Carlos Abreu McDonough; Gen. Man Daniel Valerio Bolaños.

**Cooperativa Agrícola Industrial Victoria, RL (CoopeVictoria):** Apdo 176, San Isidro, Grecia, 41000 Alajuela; tel. 494-1866; fax 444-6346; e-mail victoria@coopevictoria.com; internet www.coopevictoria.com; f. 1949; local coffee and sugar growers and processors; Dir. José Eduardo Hernández Chaverri; Gen. Man. Dagoberto Rodríguez; 345 employees.

**Cooperativa de Productores de Leche Dos Pinos, RL:** Calles 21–23, Avdas 10–12, Apdo 605, 1000 San José; tel. 437-3000; fax 437-3010; internet www.dospinos.com; f. 1948; manufacturers of dairy products and fruit juices; Pres. Carlos Vargas Alfaro; 2,526 employees.

**Corporación Bananera Nacional, SA (CORBANA):** Zapote frente casa Presidencial, Apdo 6.504, 1000 San José; tel. 283-4114; fax 253-9117; e-mail corbana@racsa.co.cr; internet www.corbana.co.cr; f. 1971; cultivation and wholesale of agricultural produce, incl. bananas; Man. Jorge Sauma; 1,700 employees.

**Corporación de Desarrollo Pinero de Costa Rica, SA** (Pineapple Development Corporation—PINDECO): Apdo 4.084, 1000 San José; tel. 222-9211; fax 233-7808; f. 1978; subsidiary of Del Monte Fresh; cultivation and wholesale of pineapples; Dir-Gen. Rodrigo Jiménez; 4,800 employees.

**Corporación Fischel:** Apdo 434, 1000 San José; tel. 295-5700; fax 221-2574; internet www.fischel.co.cr; subsidiary of Instrumentarium Medko Medical Corpn; pharmaceutical manufacturers; Pres. Emilio Bruce Jiménez.

**Corporación Pipasa, SA:** 1.5 km al oeste de la Firestone, La Ribera de Belén, Apdo 22, 4005 San Antonio Belén; tel. 298-1800; fax 293-3492; e-mail pipasa@sol.racsa.co.cr; internet www.crica.com/biz/pipcorp.html; f. 1969; owned by Rica Foods Inc, USA; breeding and wholesale of poultry products; Dir-Gen. Jorge Manuel Quesada Chaves; 1,670 employees.

**Corrugados Belén, SA (CORBEL):** San Antonio de Belén, Apdo 100, 40005 Belén; tel. 239-0122; fax 239-1023; e-mail maep@corbelbox.com; internet www.corbelbox.com; manufacturers of corrugated cardboard boxes; Pres. Alvaro Esquivel; 2,131 employees.

**DEMASA** (Derivados de Maiz Alimenticios, SA): San Gabriel de Calle Blancos, Apdo 7.299, 1000 San José; tel. 232-9744; fax 231-1935; e-mail demasa@sol.racsa.co.cr; internet www.demasa.net; f. 1986; subsidiary of Gruma, Mexico; food processing; Pres. Hans J. Bucher; 2,500 employees.

**Dole:** Calle 5–7, 1° Avda, Fte Escorial, 1000 San José; tel. 287-3000; fax 256-2466; e-mail pgilmore@dla.co.cr; fmrly Standard Fruit Company; Dir Jerry Vriesenga; 3,250 employees.

**Durman Esquivel:** San José; tel. 212-5700; fax 256-7176; e-mail servicio@durman.com; internet www.durman.com; manufacturer of plastic products; Pres. Francis Durman.

**Fertilizantes de Centroamérica, SA (FERTICA):** Edif. Santiagomillas, Barrio Dent, Apdo 5.350, 1000 San José; tel. 224-3344; fax 224-9147; e-mail fertica@sol.racsa.co.cr; internet www.ferticacr.com; f. 1961; manufacturers of chemical fertilizers; Pres. Jaime Carey Tagle; 360 employees.

**Florida Ice and Farm Company, SA:** Calles 12, Avdas 4 y 6, Apdo 10.021, 1000 San José; tel. 221-3722; fax 223-7830; e-mail info@florida.co.cr; internet www.florida.co.cr; f. 1966; 3 main subsidiaries: Florida Inmobiliaria SA, Florida Capitales SA and Florida Bebidas SA; Pres. Rodolfo Jiménez; 1,465 employees.

**Holcim (Costa Rica), SA:** Barrio Tournón, Frente Periódico La República, Apdo 4.009, 1000 San José; tel. 552-8922; fax 552-8384; e-mail cemincsa@sol.racsa.co.cr; internet www.holcim.com; f. 1960 as Industria Nacional de Cemento (INCSA); owned by Holcim Ltd (Switzerland); manufacturers of cement; Pres. Bruno Stagno; 924 employees.

**Hules Técnicos, SA:** 150 m norte de La Cañada, Apdo 84.140, 1000 San José; tel. 231-2911; fax 220-4502; e-mail hultec@sol.racsa.co.cr; f. 1976; manufacturers rubber seals for PVC tubes; Pres. Samuel Guzoski Rose; 673 employees.

**Industria de Cerámica Costarricense, SA:** La Uruca, 400 m norte de Almeda, Apdo 4.120, 1000 San José; tel. 232-5266; fax 220-0044; f. 1957; manufacture of ceramic plumbing fixtures; Pres. Carlos Araya Lizano; 876 employees.

**Instituto Costarricense de Acueductos y Alcantarillados:** Calle 5, Avda Central, Apdo 5.120, 1000 San José; tel. 255-1125; fax 256-5642; e-mail aya@netsalud.sa.cr; internet www.aya.go.cr; f. 1961; construction and operation of water and sewerage services; Pres. Cristobal H. Zawadzki; 2,975 employees.

**Intel:** Heredia, San José; internet www.intel.com; f. 1998; microprocessor manufacture and assembly; operates two factories; Vice-Pres. Gulzar Mohd Ali; 2,100 employees.

**Merck Sharp & Dohme:** Apdo 10.135, 1000 San José; tel. 210-0210; fax 232-2384; e-mail mercksha@sol.racsa.co.cr; manufacturers of medical and pharmaceutical products; Dir-Gen. Clemens Caicendo.

**Motorola Costa Rica, SA:** Oficentro Plaza Mayor, 3°, Blvd Rohrmoser, San José; tel. 296-5385; fax 296-5387; internet www.motorola.com/cr; Corporate Regional Dir Juan Carlos Perdomo.

**Nestlé (Costa Rica), SA:** Apdo 1.349, 1000 San José; tel. 293-0729; fax 239-0527; e-mail nestlecr@sol.racsa.co.cr; internet www.nestle.co.cr; chocolate producers; Gen. Man. Nelson Salgado.

**Productos Roche, SA:** Apdo 3438-1000 San José; tel. 298-1500; fax 298-1607; e-mail rochesa@sol.racsa.co.cr; medication and diagnostic instruments; Man. Mario Granados.

**Refinadora Costarricense de Petróleo (Recope):** San Guadalupe de Iglesia 200, Apdo 4.351, 1000 San José; tel. 257-6544; fax

255-4993; internet www.recope.go.cr; f. 1961; state petroleum co; Pres. Littleton Bolton; 1,100 employees.

**Siemens, SA:** La Uruca, de la Plaza de Deportes 200 m al este, Apdo 1.002, 1000 San José; tel. 287-5070; fax 257-5050; e-mail siemens@sol.racsa.co.cr; internet www.siemens.de; electronic systems, telecommunications equipment; Dir-Gen. Mario Chaves Ortiz; Pres. Walter Beutel Streitberger.

**Textiles Industriales de Centroamerica, SA (TICATEX):** 400 Noreste del Cruce de San Antonio de Belén, Heredia, Apdo 10.077, 1000 San José; tel. 239-0011; fax 293-4235; e-mail ticatex@sol.racsa.co.cr; f. 1965; subsidiary of Toyobo Co Ltd, Japan; manufacturers of textiles; Pres. Nobuya Ishii; 400 employees.

**Vidrieria Centroamericana, SA (VICESA):** San Nicolás de Cartago, Apdo 355, 7050 Cartago; tel. 550-3200; fax 552-5113; e-mail vicesa@racsa.co.cr; f. 1973; manufacturers of glass; Pres. Willhelm Steinvorth Herrera; 500 employees.

## UTILITIES
### Electricity

**Cía Nacional de Fuerza y Luz, SA:** Calle Central y Primera, Avda 5, Apdo 10.026, 1000 San José; tel. 296-4608; fax 296-3950; internet www.cnfl.go.cr; f. 1941; electricity company; mem. of ICE Group; Pres. Pablo Cob.

**Instituto Costarricense de Electricidad (ICE)** (Costa Rican Electricity Institute): Apdo 10.032, 1000 San José; tel. 220-7720; fax 220-1555; e-mail ice-si@ice.co.cr; internet www.ice.co.cr; f. 1949; govt agency for power and telecommunications; Exec. Pres. Ing. Pablo Cob Saborío.

**JASEC** (Junta Administrativa del Servicio Eléctrico Municipal de Cartago): Apdo 179-7050, Cartago; tel. 550-6800; fax 551-2115; e-mail agomez@jasec.co.cr; internet www.jasec.co.cr; f. 1964; Pres. Elías Chavez Brenes.

**Servicio Nacional de Electricidad:** Apdo 936, 1000 San José; tel. 220-0102; fax 220-0374; co-ordinates the development of the electricity industry; Chair. Leonel Fonseca.

### Water

**Instituto Costarricense de Acueductos y Alcantarillados:** Avda Central, Calle 2, Apdo 5.120, 1000 San José; tel. 257-8822; fax 222-2259; e-mail administrador@aya.go.cr; internet www.aya.go.cr; water and sewerage; Pres. Everardo Rodríguez Bastos.

## TRADE UNIONS

**Asociación Nacional de Empleados Públicos (ANEP):** Apdo 5152-1000, San José; tel. 257-8233; fax 257-8859; e-mail info@anep.or.cr; internet www.anep.or.cr; f. 1958; Sec.-Gen. Albino Vargas.

**Central del Movimiento de Trabajadores Costarricenses (CMTC)** (Costa Rican Workers' Union): Calle 20, 75 este de la Cinta Amarilla, Apdo 4.137, 1000 San José; tel. 222-5893; fax 221-3353; e-mail cmtccr@solracsa.co.cr; internet cmtcr.org; Pres. Dennis Cabezas Badilla.

**Confederación Auténtica de Trabajadores Democráticos** (Democratic Workers' Union): Calle 13, Avdas 10 y 12, Solera; tel. 253-2971; Pres. Luis Armando Gutiérrez; Sec.-Gen. Prof. Carlos Vargas.

**Confederación Costarricense de Trabajadores Democráticos** (Costa Rican Confederation of Democratic Workers): Oficinas Centrales del Banco Nacional, 13°, Apdo 2.167, San José; tel. 223-7903; f. 1966; mem. ICFTU and ORIT; Sec.-Gen. Olger Chaves; 50,000 mems.

**Confederación Unitaria de Trabajadores (CUT):** Calles 1 y 3, Avda 12, Casa 142, Apdo 186, 1009 San José; tel. 233-4188; e-mail mcalde@racsa.co.cr; f. 1980 from a merger of the Federación Nacional de Trabajadores Públicos and the Confederación General de Trabajadores; 53 affiliated unions; Sec.-Gen. Miguel Marín Calderón; c. 75,000 mems.

**Federación Sindical Agraria Nacional (FESIAN)** (National Agrarian Confederation): Apdo 2.167, 1000 San José; tel. 233-5897; 20,000 member families; Sec.-Gen. Juan Mejía Villalobos.

The **Consejo Permanente de los Trabajadores**, formed in 1986, comprises six union organizations and two teachers' unions.

# Transport

**Ministry of Public Works and Transport:** See The Government (Ministries); the Ministry is responsible for setting tariffs, allocating funds, maintaining existing systems and constructing new ones.

**Cámara Nacional de Transportes:** San José; national chamber of transport.

### RAILWAYS

**Instituto Costarricense de Ferrocarriles (INCOFER):** Calle Central, Avda 22 y 24, Apdo 1, 1009 San José; tel. 222-8857; fax 222-6998; e-mail incofer@sol.racsa.co.cr; f. 1985; government-owned; Exec. Pres. Littleton Bolton Jones; 471 km, of which 388 km are electrified.

INCOFER comprised:

**División I:** Atlantic sector running between Limón, Río Frío, Valle la Estrella and Siquirres. Main line of 109 km, with additional 120 km of branch lines, for tourists and the transport of bananas; services resumed in 1999.

**División II:** Pacific sector running from San José to Puntarenas and Caldera; 116 km of track, principally for transport of grain, iron and stone.

Note: In 1995 INCOFER suspended most operations, pending privatization, although some cargo transport continued.

### ROADS

In 2000 there were 35,892 km of roads, of which 7,437 km were main roads and 28,455 km were secondary roads. An estimated 22% of the total road network was paved in that year. In 2001 the construction of four major roads across the country began; the first road to be built was 74 km long, running from San José to San Ramón.

### SHIPPING

Local services operate between the Costa Rican ports of Puntarenas and Limón and those of Colón and Cristóbal in Panama and other Central American ports. The multi-million dollar project at Caldera on the Gulf of Nicoya is now in operation as the main Pacific port; Puntarenas is being used as the second port. The Caribbean coast is served by the port complex of Limón/Moín. International services are operated by various foreign shipping lines. The Caldera and Puntarenas ports were opened up to private investment and operation from 2002.

**Junta de Administración Portuaria y de Desarrollo Económico de la Vertiente Atlántica (JAPDEVA):** Calle 17, Avda 7, Apdo 5.330, 1000 San José; tel. 233-5301; state agency for the development of Atlantic ports; Exec. Pres. Alberto Amador.

**Instituto Costarricense de Puertos del Pacífico (INCOP):** Calle 36, Avda 3, Apdo 543, 1000 San José; tel. 223-7111; fax 223-4348; e-mail incoppe2@racsa.co.cr; f. 1972; state agency for the development of Pacific ports; Exec. Pres. Enrique Montelalegre Martí.

### CIVIL AVIATION

Costa Rica's main international airport is the Juan Santamaría Airport, 16 km from San José at El Coco. Following a report by the US Federal Aviation Administration, the aviation sector was to undergo expansion and modernization from 2001. There is a second international airport, the Daniel Oduber Quirós Airport, at Liberia and there are regional airports at Limón and Pavas (Tobías Bolaños Airport).

**Nature Air:** San José; tel. 535-8832; e-mail NatureAir@centralamerica.com; flights from San José to 16 domestic destinations.

**Líneas Aéreas Costarricenses, SA (LACSA)** (Costa Rican Airlines): Edif. Lacsa, La Uruca, Apdo 1.531, San José; tel. 290-2727; fax 232-4178; internet www.flylatinamerica.com; f. 1945; operates international services within Latin America and to North America; Gen. Man. Richard Krug.

**Servicios Aéreos Nacionales, SA (SANSA):** Paseo Colón, Centro Colón, Apdo 999, 1007 San José; tel. 233-2714; fax 255-2176; e-mail info@flysansa.com; subsidiary of LACSA; international, regional and domestic scheduled passenger and cargo services; Man. Dir Carlos Manuel Delgado Aguilar.

**Servicios de Carga Aérea (SERCA):** Aeropuerto Internacional Juan Santamaría, Apdo 6.855, San José; f. 1982; operates cargo service from San José.

# Tourism

Costa Rica boasts a system of nature reserves and national parks unique in the world, which cover one-third of the country. The main tourist features are the Irazú and Poás volcanoes, the Orosí valley and the ruins of the colonial church at Ujarras. Tourists also visit San José, the capital, the Pacific beaches of Guanacaste and Puntarenas, and the Caribbean beaches of Limón. In 2001 there were numerous hotel and resort construction projects under way in the Guanacaste province. The projects were being undertaken by the Instituto Costarricense de Turismo. In May 2001 one of the country's principal tourist sites, La Casona de Santa Rosa, was almost completely destroyed by fire. Some 1,238,692 tourists visited Costa Rica in 2003, when tourism receipts totalled US $1,199m. Most visitors came from the USA (41.2%). There were 379 hotel in Costa Rica in 2003.

**Cámara Nacional de Turismo (Canatur):** tel. 234-6222; fax 253-8102; e-mail info@tourism.co.cr; internet www.costarica.tourism.co.cr.

**Instituto Costarricense de Turismo (ICT):** La Uruca, Costado Este del Puente Juan Pablo II, Apdo 777, 1000 San José; tel. 299-5880; fax 291-5675; internet www.visitcostarica.com; f. 1955; Exec. Pres. RODRIGO CASTRO FONSECA; Gen. Man. GUILLERMO ALVARADO HERRERA.

## Defence

Costa Rica has had no armed forces since 1948. In August 2004 Rural and Civil Guards totalled 2,000 and 4,400 men, respectively. In addition, there were 2,000 Border Security Police.

**Defence Budget:** an estimated 46,000m. colones (US $106m.) in 2004.

**Minister of the Interior, Police and Public Security:** ROGELIO RAMOS MARTÍNEZ.

## Education

Education in Costa Rica is free, and is compulsory between six and 15 years of age. Primary education begins at the age of six and lasts for six years. Official secondary education consists of a three-year basic course, followed by a more highly specialized course of two years. In 2001 an estimated 90.6% (males 89.9%; females 91.4%) of children aged six to 11 were enrolled at primary schools, while 50.7% (males 48.2%; females 53.4%) of those aged 12 to 16 received secondary education. In 1999 there were 3,768 primary schools and 468 secondary schools. In 2002, according to estimates by the UN Development Programme, Costa Rica's adult illiteracy rate was 4.2% (males 4.3%; females 4.1%). In 2001 government expenditure on the education system was about 21.1% of total spending.

## Bibliography

For works on Central America generally, see Select Bibliography (Books)

Biesanz, M., et al. *The Ticos: Culture and Social Change in Costa Rica*. Boulder, CO, Lynne Rienner Publrs, 1998.

Booth, J. A. *Costa Rica: Quest for Democracy*. Boulder, CO, Westview Press, 1999.

Calderon, G. *The Life of Costa Rica*. Santafé de Bogotá, DC, Villegas Editores, 2001.

Costa Rica Research Group. *Executive Report on Strategies in Costa Rica (Strategic Planning Series)*. San Diego, CA, Icon Group International, 2000.

Cruz, C. *Political Culture and Institutional Development in Costa Rica and Nicaragua: World-Making in the Tropics*. Cambridge, Cambridge University Press, 2005.

Edelman, M. *Peasants Against Globalisation: Rural Socialist Movements in Costa Rica*. Stanford, CA, Stanford University Press, 1999.

Evans, S. *The Green Republic: A Conservation History of Costa Rica*. Austin, TX, University of Texas Press, 1999.

*Foreign Investment in Latin America and the Caribbean*. Santiago, Economic Commission for Latin America and the Caribbean, 2004.

Harpelle, R. N. *The West Indians of Costa Rica*. Montréal, QC, McGill-Queen's University Press, 2001.

Helmuth, C. *Culture and Customs of Costa Rica*. Westport, CT, Greenwood Publishing Group, 2000.

Honey, M. *Hostile Acts: US Policy in Costa Rica in the 1980s*. Gainesville, FL, University Press of Florida, 1994.

Longley, K. *The Sparrow and the Hawk: Costa Rica and the United States During the Rise of José Figueres*. Tuscaloosa, AL, University of Alabama Press, 1997.

Paus, E. *Foreign Investment, Development, And Globalization: Can Costa Rica Become Ireland?* Basingstoke, Palgrave Macmillan, 2005.

Sandoval-Garcia, C. *Threatening Others: Nicaraguans and the Formation of National Identities in Costa Rica (Latin America S.)*. Athens, OH, Ohio University Press, 2004.

Steigenga, T. J. *Politics of the Spirit: The Political Implications of Pentecostalized Religion in Costa Rica and Guatemala*. Moscow, ID, Lexington Books, 2001.

Wilson, B. M. *Costa Rica: Politics, Economics and Democracy*. Boulder, CO, Lynne Rienner Publrs, 1998.

Yashar, D. J. *Demanding Democracy: Reform and Reaction in Costa Rica and Guatemala, 1870s–1950s*. Stanford, CA, Stanford University Press, 1997.

# CUBA
## Geography

### PHYSICAL FEATURES

The Republic of Cuba consists of the island of Cuba (the largest and westernmost of the Greater Antilles), the Isla de la Juventud (Isle of Youth, until 1978 the Isle of Pines—Isla de Pinos) and 1,600 small offshore islands. Cuba is the largest island in the West Indies, lying at the entrance to the Gulf of Mexico (to the north-west), and is washed by the Caribbean Sea to the south and west and by the Atlantic Ocean to the north-east. Cuba lies only 145 km (90 miles) north of Jamaica and, at the other, north-western end of the island, a similar distance south of Key West in Florida, USA. However, some of the more remote cays of the Bahamas lie much closer, and that country stretches across the north-western approaches to Cuba. Haiti is only 80 km to the east. The Yucatán peninsula of Mexico lies about 210 km to the west and the British dependency of the Cayman Islands some 240 km to the south. Cuba also has a 29-km land border in the south-west, where the USA has a lease on the area around its naval base on Guantánamo Bay, but Cuba retains sovereignty. Cuba, the largest country of the insular Caribbean, in terms of both extent and population, covers an area of 110,860 sq km (42,803 sq miles) or about 45% of the total surface area of the Antilles.

The island of Cuba, which is roughly the same size as the North Island of New Zealand or Newfoundland (Canada), is about 1,250 km long and between 32 km and 191 km wide. From a flattened, southern head of mountainous terrain, the island of Cuba extends back north-westwards into a narrowing tail that begins to turn towards the south-west when the island ends. This tail is attempting to encompass the 2,200-sq km Isla de la Juventud (Isla de Pinos), which lies about 100 km to the south (north of a large, bisecting swamp, this island, famed for its citrus fruits, is generally dry and flat, although there are some hills). The Isla de la Juventud is by far the largest of the myriad offshore islands and cays that, together with extensive coral reefs, further complicate the heavily indented, 3,735-km coast-line. In the north old coral and limestone has lifted into a steep shore of cliffs and bluffs, sheltering some fine harbours, while the south subsides into a low and often marshy littoral. Much of the terrain of this predominantly limestone island is flat or undulating plain, with wide and fertile valleys, and highlands of any significance only in the south-east. About one-quarter of Cuba's territory is mountainous, with three main ranges: the Cordillera de Guaniguanico (including the distinctive steep-sided, flat-topped mountains of the Sierra de los Organos and the more easterly Sierra del Rosario) at the north-western end of the island; the centre-west highlands, such as the Escambray; and the geologically more recent Sierra Maestra (largely volcanic in origin) in the far south-east, where Pico Turquino reaches 2,005 m (6,580 ft). Rivers tend to be short and fall steeply. The longest is the Cauto, in the east, at only 240 km, and the Toa (110 km long) is the widest river in Cuba.

Rivers and plentiful rainfall water an island still rich in biodiversity, although the almost entirely wooded island that the Spanish first settled on is now three-quarters savannah or plains (there are two main areas of savannah). About 4% of the island is swampy wetland and a reforestation programme aims to increase the area under trees to 27% of the total. Meanwhile, Cuba is home to an extraordinary range of terrains and vegetation types, with the eastern end of the south-east being particularly rich in biodiversity (it claims to be the most diverse in the Caribbean, with almost one-third of the endemic species of the island). Over 7,000 species of flora have been identified on Cuba, of which about 3,000 are endemic. However, almost 1,000 have been made extinct, rare or endangered since the 17th century. Examples of native plants include the world's only carnivorous epiphyte (plants that live on other plants, but are not parasitic, like moss or orchids), about 100 types of palm (90 of which are endemic) and one of nature's largest flowers. Cuba might be more generally associated with claims of the world's smallest products of nature, however, particularly in the animal kingdom: the smallest frog, the Cuban pygmy frog, which is 12 mm (0.5 ins); the smallest mammal, the 55-mm butterfly bat (*mariposa*); the smallest scorpion, the 10-mm dwarf scorpion; and the smallest bird, the 63-mm bee hummingbird (*zunzuncito*). There is also a pygmy owl, tiny salamanders and the insect-eating rodent, the *almiquí*, which is a long-surviving species, a 'living fossil' (like the cork palm, the Cuban or rhombifer crocodile and the Cuban alligator). In all, there are almost 14,000 species of fauna, although about 1,500 are said to be near extinction. Many of the species have diversified into distinctly Cuban varieties—40% of mammal species (including some of the larger ones, such as manatees or sea cows and hutias) are said to be endemic, 85% of reptiles, 90% of amphibians and 96% of molluscs. Protected areas cover 30% of Cuba, including its marine platform and four UNESCO biosphere reserves, although enforcement sometimes lacks infrastructure.

### CLIMATE

The climate is subtropical, warm and humid, with the Atlantic trade winds alleviating extremes. A rainy season falls from May to October. During this time hurricanes occasionally trouble the eastern coasts. Drought is a more usual and general problem. Average annual rainfall in Havana (La Habana—the capital, in the north-west) is 1,730 mm (67 ins), but 860 mm in the east. The average annual temperature in Cuba is 25°C (77°F), but readings can rise to 33°C (91°F) in the shade even in Havana (the south-eastern city of Santiago de Cuba is hotter and drier) during the summer. The drier, winter months of December–April see temperatures nearer 20°C (68°F), but they can occasionally fall to as low as 8°C (46°F) in a north wind.

### POPULATION

Just over one-half of the population (51%) is of mulatto or mixed-race ethnicity (mainly of Spanish and African descent), with a further 37% defined as white, 11% as black and 1% as Chinese. Numerous other influxes have influenced Cuba and its culture, notably in the south-east, where the French fled from revolution in Haiti in the early 19th century or, more recently, Jamaicans have settled. The oldest influences to modify Spanish colonial settlers, however, came from West Africa with the slaves brought to work the plantations. The most numerous group was of Yoruba speakers from south-western Nigeria, Benin (formerly Dahomey) and Togo. They are known in Cuba as *Lucumí*. The descendants of the Bantu-speaking peoples brought from the Congo basin are known as the *Congos*, while those from the Calabar region of southern Nigeria and Cameroon are known as the *Carabalí*. The Yoruba pantheon and legends were syncretized by the early *Lucumí* with the saints of the Christian religion they were formally required to adopt, forming the basis of the Regla de Ocha cult, or *Santería*. It is a religion without sects or missionaries and can co-exist without complication with both Christianity or the similarly syncretic cult of the *Congos*, the Regla Conga or Palo Monte. Some practise all three belief systems at once. The *Carabalí* originated not a cult, but a sort of masonic closed sect, the Abakuá Secret Society, which is open to men only and upholds values traditionally associated with masculinity (it is also known as *ñañiguismo*). Ostensibly, however, the predominant religion remains Roman Catholic Christianity and, prior to the advent of the current Communist regime and the official disapproval of religion (although this was eased from 1992), some 85% of the population were at least nominally Roman Catholic. In 2003 some 51% of the population were open adherents of Roman Catholicism (with another 10% belonging to other Christian denominations). There are also some Jews and a few that practise other faiths. Spanish is the official language.

About three-quarters of the total population of 11.2m. (as estimated in mid-2003) are urbanized, and some 2.2m. live in Havana, on the north coast. The capital is one of 14 provinces, while the Isla de la Juventud forms a special municipality and completes the administrative organization of the Republic. The second city is Santiago de Cuba (with a population of over 0.4m. in 1999), in the south-east, followed by Camagüey, Holguín, Santa Clara and Guantánamo (the last being east of Santiago, the others along the main route down the centre of the island). There is a high level of illicit emigration from Cuba, particularly to the USA.

# History

## MELANIE JONES

Cuban archaeological and anthropological studies identified a number of ethnic groups that lived on the island prior to the Spanish invasion of 1511. Christopher Columbus, the Genoese navigator in the service of the Spanish Crown, who landed on Cuba in 1492, made reports of a relatively sophisticated agricultural society in parts of the island. Researchers identified this group as the Taínos. There were Siboneys and Guanahuatabeys also in evidence prior to the Conquest. The indigenous population is calculated to have reached a peak of some 100,000 by 1511, before falling drastically to a mere 4,000 by 1550.

In 1500 a Spanish cartographer, Juan de la Cosa, drew a map of Cuba very similar to the one used to this day, in spite of Columbus' insistence that the island formed part of the continental mainland. However, Cuba was largely forgotten by Spain until the early 16th century. It was only in 1508 that another Spaniard, Sebastian de Ocampo, completed a circumnavigation of Cuba, thereby conclusively establishing its insular nature. The information gathered on this expedition was later to be used in the Spanish occupation of the island.

The Spanish conquest of Cuba took from 1511 until 1515. Indigenous resistance was destroyed, and with the execution of the political leader, Hatuey, Spanish dominance was secured. The first seven townships were founded during this period: Baracoa (1512); Bayamo (1513); Trinidad, Sancti Spíritus and Havana (1514); and Puerto Príncipe and Santiago de Cuba (1515). The last became the first capital of Cuba, remaining so for some decades.

### COLONIAL RULE AND INDEPENDENCE WARS

The colonial administrative apparatus was installed in the 16th century: *cabildos* (municipal councils) had authority over the townships, with a governor representing the Spanish Crown. The trade department, based in Seville, controlled colonial commerce, the other major power being the Roman Catholic Church. Efforts were at first focused on developing the mining industry. However, as deposits were exhausted and the indigenous population exterminated, attention was turned to agriculture, and a new, enslaved labour force was introduced from Africa. The first sugar mills appeared in the late 16th century and tobacco cultivation spread throughout the 17th century. However, the threat of piracy grew ever greater. When a French privateer, Jacques de Sores, occupied Havana (by now the capital) in 1555, causing the governor to flee, the Spanish authorities were forced to reconsider the country's strategic importance. Fortifications were built around the port, which had become a port of call for fleets *en route* to Spain with cargoes of gold and silver from the Americas.

The early centuries of poor economic development were also characterized by totalitarian domination on the part of Spain and the population of the island frequently resorted to dependence on contraband. Spain itself was weakened by constant wars with other European powers, and although the reign of Philip V (1700–46) resulted in some economic progress, the central government's over-zealous control of the tobacco industry led to a number of protests, which were violently crushed.

The British occupied Havana for a period of 11 months during 1762–63, eventually returning it to Spain in exchange for Florida. However, this short period created a number of trading possibilities for Cuba, and a certain divergence between the native Creoles and the Spanish emerged. The rapid expansion of the sugar trade defined the development of the Cuban economy at the end of the 18th century. Cuba was to become the world leader in sugar production, a situation that was to have a profound influence on political events even into the 20th century and beyond.

At a time when the slave trade was experiencing a general decline, Cuba saw the influx of some 500,000 slaves over a period of 50 years. Owing to the violent slave rebellion in the nearby French territory of Santo Domingo (now Haiti) in 1791, and the fear inspired by the growing numbers of blacks, the Creoles delayed any thoughts of pursuing independence from Spain. The slave trade was made illegal in 1817, but 'trafficking' was widespread. Rebellions were dealt with swiftly and brutally, and slave resistance tended to be manifested in the form of 'runaways' (*cimarrones*), who established communities in remote mountain areas known as *palenques*. The effective abolition of slavery did not take place until 1886.

Cuba was the last Spanish colony in the Americas, and only broke free from Spanish rule after three wars of independence. The first, known as the Ten-Year War or Yara Revolution, commenced in October 1868, when a landowner, Carlos Manuel de Céspedes (known as the 'Father of the Nation'), freed his slaves from the Demajagua sugar mill. Weaknesses were evident on both sides, the Spanish impeded by the chaos that followed the fall of their monarch, Isabella II, and the Cuban forces divided by an internal struggle between the reformists, those seeking independence and those who saw their future in annexation by the USA. The war finally ended in 1878, when the Zanjón Pact was signed, introducing freedom for rebel slaves and the extension of political rights to Cuba. However, attempts at reconciliation in the country were frustrated by the Baraguá protest, in which a new, mixed-race leader, Antonio Maceo, called for the full abolition of slavery and complete independence. Thus began the second conflict, known as the Little War, which lasted for only two years. Then, in 1894 Spain terminated the existing trade agreement between Cuba and the USA, causing an economic crisis. The final campaign began in February 1895, after the radical and popular elements had gained widespread support. José Martí, the ideologue of the independence forces, who had long warned of the dangers of US interference, was killed in an early skirmish. The war was long and bloody, eventually prompting the USA to intervene.

## US INTERVENTION AND THE 'PUPPET' REPUBLIC

From as early as 1805 US agents were instructed to monitor events in Cuba. On more than one occasion the USA attempted to enter into negotiations with Spain over Cuba, and prior to the US Civil War (1861–65) many slave owners harboured the hope that annexation would protect their interests. However, the USA never officially recognized the struggle of the *mambises*, as the independence forces were known, for both commercial and strategic reasons. With an end to the Cuban war for independence approaching, the mysterious explosion in February 1898 of the USS *Maine*, anchored in Havana port to provide protection for US citizens, presented the USA with the pretext to intervene against Spain. In a brief campaign US forces captured Cuba, Puerto Rico and the Philippines, so bringing an end to the war and to much of what remained of the Spanish colonial empire.

The Treaty of Paris, signed by the USA and Spain in December 1898, marked an official end to the hostilities and a formal renunciation of Spanish sovereignty in Cuba. There followed three years of US military occupation until, on 20 May 1902, Cuba was granted independence. Tomás Estrada Palma, a pragmatic proponent of annexation to the USA, became the first President of the new Republic. Estrada Palma had signed the Platt Amendment in the previous year, by virtue of which the USA reserved the right to intervene in Cuba and to maintain a military installation on the island. Indeed, a second period of US intervention lasted from 1906 to 1909, following a rebellion by disgruntled liberals who disputed the re-election of Estrada Palma to the presidency. Thereafter the country suffered successive political and economic crises, a situation compounded by widespread corruption.

In 1924–33 Gerardo Machado occupied the presidency in what was known as the 'bloody decade'. This period was characterized by a succession of scandals concerning political corruption, by an incipient trade-union movement and by the first signs of organized crime. A number of student and worker groups emerged, including the Federation of University Students (FEU), the Anti-Imperialist League and the José Martí People's University for Workers. The Communist Party was formed in 1925 by Carlos Baliño, a close collaborator of Martí, and Julio Antonio Mella, who was assassinated in Mexico in 1929. The parties were illegal and their leaders persecuted. In 1929 the Great Depression began and sugar prices plummeted, causing student and labour opposition groups to join forces in an attempt to bring about change. In March 1930 the country experienced an abortive general strike and in September a student radical was assassinated, unleashing a wave of public protest. By 1933 the country was in the midst of violent unrest, and in July the USA sent a diplomat, Sumner Welles, to mediate. The following month a general strike led the army to abandon support for Machado and he was overthrown by a military coup. Carlos Manuel de Céspedes (the younger) was installed as President. Machado fled to the Bahamas and later took up residence in the USA. In September the Sergeants' Revolt, led by Sgt (later Gen.) Fulgencio Batista Zaldivar, in turn overthrew Céspedes, who was replaced by Ramón Grau San Martín, a professor from the University of Havana.

Grau's 100-day Government comprised representatives from the entire political spectrum, with student leader Antonio Guiteras appointed Minister of the Interior. The Government introduced a number of social benefits, improving working conditions and land distribution, and gave women the right to vote. The USA, however, refused to recognize the Government and Grau resigned in 1934, leaving the way clear for Batista, who was to dominate Cuban politics for the next 25 years. Batista was the guiding force behind a series of 'puppet' presidents. In 1934 the Platt Amendment was abrogated, and a new Reciprocity Treaty enacted in its place. The workers' movement evolved with the establishment of the Confederation of Cuban Workers (CTC). In 1940 a new Constitution, aspiring to democracy and social justice, was completed, and Batista was proclaimed the constitutionally elected President, with the support of US investors.

Grau assumed the presidency once again in 1944 at the head of the Authentic Party, but this was a shadow of the Government of 1933. Carlos Prío Socarrás succeeded him in 1948, but his term was brought to an end in 1952 when Batista seized power in a bloodless coup. Eduardo Chibás left the Authentic Party in 1947 to form the more radical Orthodox Party. A strong candidate for the presidency, Chibás publicly committed suicide during a radio broadcast in 1951, in a personal protest against the social injustice and political corruption that he deemed to be rife in Cuba.

On assuming power Batista revoked the Constitution of 1940, annulling the legislative functions of government and any semblance of democracy. In the following year the first signs of the pending revolution were manifested. On 26 July 1953 a radical opposition group, led by Fidel Castro Ruz, staged an assault on the Moncada garrison in Santiago de Cuba. The attack failed, and Castro was imprisoned with those companions who were not killed in the attempt. The speech made by Castro in his own defence, recorded under the title of 'History Will Absolve Me', became the doctrine of the new revolutionary forces. Castro was sentenced to 15 years' imprisonment. In 1954, unopposed, Batista was elected to another four-year term in government. In an effort to court public opinion, President Batista released Castro into exile in 1955 under a general amnesty. He departed for Mexico as the leader of the newly formed '26 July Movement' to organize resistance forces against Batista. The future guerrillas, among them a young Argentine doctor, Ernesto ('Che') Guevara, followed a rigorous theoretical-military training programme.

## FIDEL CASTRO'S REVOLUTION

In December 1956 Castro and 81 followers landed in Cuba aboard the yacht *Granma*, miles from the planned location and missing the pre-arranged date of 30 November. The expedition was widely believed to have been a failure, since all but 12 revolutionaries were soon captured or killed. This nucleus, however, was sufficient to initiate a new guerrilla movement in the Sierra Maestra mountains of south-eastern Cuba. In March 1957 the University Directorate (Directorio Universitario), a group of intellectuals organized by José Antonio Echeverría, attacked the presidential palace in an attempt to assassinate Batista. The attack failed and Echeverría was killed, after having announced Batista's demise on a radio broadcast. However, as a result of the attempt President Batista was seen to be vulnerable, and his sporadic acts of retaliation further eroded his popularity.

The guerrilla movement expanded throughout the countryside and was paralleled by growing insurgency in the cities, which combined acts of sabotage against economic targets with an increasing number of political assassinations and kidnappings. An attempt to overthrow President Batista by means of a general strike in April 1958 failed. The Government planned a retaliatory offensive, but this was thwarted when Raúl Castro Ruz, Fidel Castro's brother, took several prominent foreigners, mostly North Americans, as hostages, thus provoking diplomatic pressure on Batista by the US Government, which imposed an arms embargo, tantamount to a withdrawal of support. This had a devastating effect on the morale of the armed forces, and the election, later that year, of the official candidate provoked further consternation among those who still hoped for a political solution. Meanwhile, revolutionary activity was gaining momentum. In Santa Clara, Guevara's forces received ample assistance from the local population and widespread support for the revolutionary forces was evident. On 1 January 1959 the army seized power. With Batista's subsequent flight from Cuba and the collapse of his regime, Castro's rebel forces occupied Havana.

Castro first unified all the remaining military groups under his control, incorporating the remnants of the University Directorate in order to avoid a repetition of the divisions in leadership that had brought an end to the revolution of 1933. In January 1959 a new Constitution, the Fundamental Law, was introduced. Manuel Urrutia Lléo was appointed President and in March Fidel Castro took the post of Prime Minister, with his brother, Raúl, as his deputy. Once established, the first government policies were swiftly effected, with the expropriation and nationalization of all major enterprises, including foreign sugar estates, and agrarian reform. Internal conflict and opposition from certain sectors of the population led Castro to consider resignation, but public opinion generally seemed to support the guerrilla leader, and so it was Urrutia who resigned. Osvaldo Dorticós Torrado was named President in his place. As measures grew more radical, acts of violence, conspiracy and desertion escalated. Approximately 700,000 mainly white, middle-class Cubans emigrated, their places, notably in the administration, filled by Communist militants. Prominent among them was Che Guevara, who was, for a time, the ideological guiding force of the revolution.

In April and May 1959 Castro visited the US cities of New York and Washington, DC. He was well received, in spite of adverse publicity over the execution of Batista supporters, but he failed to establish the desired links with the US Government. By May 1960, as Cuba and the USSR re-established diplomatic relations, those with the USA were steadily deteriorating. In July the sugar quota was suspended and in October all US business interests in Cuba were expropriated, without compensation. A full economic embargo was imposed by the USA in the same month. In January 1961 the USA severed its diplomatic links with Cuba, and pressurized other Latin American countries to do likewise. In 1962 Cuba was suspended from the Organization of American States (OAS), Mexico alone maintaining its traditional links with the country. Cuba became an essential part of the US external and internal political agenda, the Cuban exile community having become a powerful political lobby within the USA. In April 1961, with the support of the US Administration, the exiled forces launched an unsuccessful attack on the Bay of Pigs (Bahía de Cochinos). The failure of the invasion led Castro to proclaim Cuba a Socialist state 'of the poor, by the poor and for the poor' and all remaining internal dissent was suppressed. Some 120,000 exiles were taken prisoner and their trials broadcast throughout the country.

The new Government favoured the Soviet model of economic planning, 'democratic socialism'. Collectivization and industrialization were the main ingredients of the Cuban experiment, with a brief incursion into the Marxist policies of the Chinese Revolution (favouring agriculture) in the late 1960s. However, the attempt at industrialization was, to a large extent, unsuccessful, since although the country had mineral deposits, it had no known petroleum reserves. In addition, Cuba had no history of industrialization, the economy having been dominated by sugar production, and Soviet planning and techniques were inadequate for Cuba's needs. However, in July 1972 Cuba's links with the USSR were increased when the country joined the Council for Mutual Economic Assistance (CMEA), an organization linking the USSR and other Communist states. As a result of its membership, Cuba received preferential trading rights and technical advice from the USSR and other Eastern European countries. Between 1959 and 1984, in spite of Cuba's continued status as an underdeveloped country, substantial advances were made in the sectors of education and health, and a thriving cultural programme emerged. Illiteracy was virtually eradicated and infant mortality brought into line with developed countries.

At the political level, in 1961 the numerous pro-Government groups that had evolved were merged under one over-arching group, the Organizaciones Revolucionarias Integradas (ORI—Integrated Revolutionary Organizations). Highly criticized for its incompetence, the ORI was renamed the Partido Unido de la Revolución Socialista Cubana (PURSC—United Party of the Cuban Socialist Revolution) in 1962, Castro having declared himself a Marxist-Leninist, and in 1965 this in turn became the Partido Comunista de Cuba (PCC—Communist Party of Cuba).

In October 1962 the friction between Cuba and the USA ignited into an international incident known as the Cuban Missile Crisis, after US reconnaissance aircraft detected the presence of Soviet nuclear warheads on the island. As the USA established a naval blockade, the possibility of a superpower confrontation over Cuba seemed imminent. The crisis was eventually resolved between the USSR and the USA, and the missile bases were dismantled, but Cuba's position was, humiliatingly, ignored.

In 1965 Guevara left Cuba to continue the revolutionary struggle abroad. The abject failure of the Cuban style of guerrilla warfare (foquismo) and Guevara's subsequent death in Bolivia in 1967 dealt a bitter blow to Cuba's internationalist aspirations. Nevertheless, Che Guevara was seen as a martyr of the revolution, widely respected in Cuba and a revolutionary icon world-wide. The return of his body from Bolivia in 1997, for burial in Cuba, was a national event.

In 1970 Cuba suffered another reverse in its highly publicized campaign to produce a sugar harvest of 10m. metric tons, by means of which the country aspired to emerge from its condition of economic underdevelopment. This was to be the decisive test for the revolution and vast amounts of resources were diverted from other sectors. The failure, ultimately, to reach this target only served to emphasize the country's economic dependence on sugar and tobacco. So began a period of gentle liberalization, described as a 'retreat into socialism'. The 1971–85 period introduced decentralization and material incentives, but was accompanied by the expansion of the 'black' (parallel, illegal) market.

In June 1974 Cuba's first elections since the Revolution were held for municipal offices in Matanzas province. The following year marked the beginning of Cuban participation in the independence struggles of African nations such as Angola and Ethiopia, which lasted until the early 1990s. It was also the year of the First Congress of the PCC and the proclamation of a new, Socialist Constitution, under which Fidel Castro added Head of State to his former title of Head of Government in 1976. In October elections to municipal assemblies were held. The assemblies subsequently elected delegates to provincial assemblies and deputies to the Asamblea Nacional del Poder Popular (National Assembly of People's Power), inaugurated as 'the supreme organ of state' in December 1976. The Asamblea Nacional in turn selected the members of a new Council of State, with Castro as President. The Second Congress of the PCC was held in April 1980.

In April 1980 restrictions on emigration were temporarily lifted after 10,000 people occupied the grounds of the Peruvian embassy. The result was the emigration of some 125,000 Cubans to the USA (agreement was later reached on the return of some 2,500 emigrants deemed 'excludable', after they had served sentences in US prisons). In April 1982 tourism and investment in Cuba by US nationals was prohibited by the US authorities, although there was a trend towards improved links between Cuba and other Latin American countries at this time. A brief clash between US troops and Cuban construction workers during the 1983 US invasion of Grenada resulted in a renewed deterioration in relations between the two countries. In December 1984 agreement was reached on the resumption of Cuban immigration into the USA. However, when the Voice of America broadcasting network founded Radio Martí, a station transmitting news and other programmes from Florida, USA, to Cuba, relations became further strained, and the immigration accord was suspended until October 1997.

From 1986 there was a shift away from the liberal policies of previous years, which were replaced by a process of 'rectification of errors and correction of deviant trends'. The Third Congress of the PCC, held in February, resulted in the retirement of many veterans of the Revolution, and the promotion of women and blacks. In the same year contact with the Roman Catholic Church was restored. In a renewed effort to increase diplomatic links, Cuba honoured an agreement to withdraw all troops from Angola within a period of 30 months; this withdrawal was completed by 1991. In September 1988 Cuba established diplomatic links with the European Community (from 1993 known as the European Union—EU).

## THE SPECIAL PERIOD, 1990–2005

Until the demise of the USSR Cuba had received critical economic and military assistance from the Communist superpower. The final dissolution of the Soviet Federation in December 1991 left Cuba effectively isolated on both economic and ideological fronts. With the 1990 electoral defeat of the Sandinista regime in Nicaragua, Cuba lost the last of its allies in the region.

The declaration of a 'special period in time of peace' in late 1990 was marked by increasing shortages; Cuba's petroleum supplies having been severely curtailed, ox-carts and Chinese bicycles became the chief modes of transport. The crisis reached its peak in 1993. Many state enterprises were closed owing to a lack of fuel and raw materials, and key sectors, such as education and public health, were seriously affected. The adoption by the USA of the Cuban Democracy Act, also known as the Torricelli Act, in October 1992, tightened the embargo, further compounding the country's economic problems.

A meeting of the Asamblea Nacional, convened in July 1992, granted Castro, as Head of State, new emergency powers, introduced direct elections to the Asamblea Nacional, outlawed discrimination on the grounds of religious beliefs, and permitted foreign investment. Members of the 'old guard' were replaced by new figures. In the same year Castro offered his abdication in exchange for the suspension of the US trade embargo by the US President-elect, Bill Clinton (1993–2001).

While indicating his distaste for economic reform, Castro acknowledged its necessity. The reforms introduced were designed to allow space for the functioning of market mechanisms, under extensive state regulation, while safeguarding the standards of social welfare. In July 1993 a 30-year ban on the possession of foreign currency was lifted in an attempt to recoup the 'hard' or convertible currency, principally US dollars, circulating in the black market, and to increase remittances from Cuban exiles. This move introduced, in effect, a two-tier economy in which pesos would buy basic foodstuffs and US dollars would purchase most other goods. State farms were decentralized, with the formation of the semi-autonomous Unidades Básicas de Producción Cooperativa (UBPCs—Units of Basic Co-operative Production). Other reforms included the growing list of enterprises open to self-employment, such as the *paladares* (restaurants in private homes).

In April 1994 four new government ministries (Economic Co-operation, Economy and Planning, Finance and Prices, and Foreign Investment) were created, reflecting a significant change in the country's economic management. However, the economic restructuring came too late for some, and in July and August disturbances broke out throughout Havana. Castro made a television appearance announcing that if the US Government failed to halt illegal emigration, Cuba would suspend its own travel restrictions. Thousands of Cubans put out to sea in makeshift craft in the 'rafters' exodus, only to be returned to Cuba and held at the US naval base at Guantánamo, as President Clinton revoked the 1966 Cuban Adjustment Act, which conferred automatic refugee status on Cubans, until then one of the most firmly entrenched parts of US policy. Border restrictions were eventually reinstated in September, as part of a bilateral immigration accord, in which the USA resolved to admit a minimum of 20,000 Cubans to the USA each year. In 1995 the USA made a commitment to repatriate all Cubans reaching the US coasts illegally.

US-Cuban relations were once again strained in February 1996 when two light aircraft belonging to the US-based exile group, Brothers to the Rescue, were shot down after having allegedly entered Cuban airspace. As a direct response, in March President Clinton signed the punitive legislation known as the Cuban Liberty and Solidarity Act (commonly referred to as the Helms-Burton Act, after its instigators), which threatened to impose sanctions on countries trading with or investing in Cuba. Canada and Mexico sought to dispute the extraterritorial nature of the Act under the provisions of the North American Free Trade Agreement (NAFTA) and the EU requested a disputes panel with the World Trade Organization (WTO). The complaint was subsequently dropped in return for a US commitment to waive sanctions against member countries doing business in Cuba.

The efforts of an increasing number of US religious, humanitarian and business organizations to lift the economic restrictions on Cuba gained momentum throughout the 1990s because of mounting opinion that the sanctions had failed to foster change on the island. In January 1999 the US Government announced the introduction of measures to broaden remittances and humanitarian flights from the USA to Cuba, to increase co-operation in areas of mutual interest, such as drugs-trafficking and migration issues, and to encourage people-to-people contacts. US sources stressed that the aim of the measures was to facilitate a peaceful transition of power without providing any direct assistance to the Cuban Government.

In late 2000 the US Congress requested a detailed study from the International Trade Administration on the economic impact of the sanctions. A report from an independent 'think tank' in the USA, the Council of Foreign Relations, advocated an easing of restrictions in relations with Cuba, while stopping short of suggesting an end to all sanctions. It emphasized that current policy served to alienate sectors of Cuban society that were important to the US Administration, such as the Roman Catholic Church and certain dissident groups. US legislation was finally passed in October 2000 ending restrictions on the sale of food and medicines. However, in the weeks preceding the US presidential election in November, the Cuban-American lobby was successful in neutralizing the effects of this measure, securing assurances that there would be no access to US credit to make such purchases, that the travel ban on US citizens would be written into law and that sales of Cuban goods in the USA would still be prohibited.

Three days after 'Hurricane Michelle' struck Cuba in November 2001, causing widespread damage to the island's infrastructure and agriculture, the US Administration of George W. Bush (2001–) made an offer of humanitarian assistance. Although this offer was refused, Cuba agreed to purchase US food supplies to replenish dwindling emergency stocks. The first exports of US food in more than 40 years reached Cuban ports the following month. Both parties were anxious to emphasize that this was an isolated consignment, resulting from exceptional circumstances, but by 2003 Cuba had become the fourth largest importer of US food in the Caribbean. In October 2004 the UN General Assembly passed a resolution, by an overwhelming majority, condemning the US trade embargo on Cuba for the 13th time: the USA, Israel, the Marshall Islands and Palau voted against the resolution, while the Federated States of Micronesia abstained. In July 2005 Castro refused an offer of humanitarian aid from the USA and the EU after 'Hurricane Dennis' struck the island, killing at least 16 people and causing an estimated US $1,400m. in damage.

At the beginning of 2003 the PCC hailed a victory for socialism, announcing that more than 97% of those eligible to vote had taken part in Cuba's one-party general election, held in January. More than 8m. Cubans cast ballots for representatives to the Asamblea Nacional, electing 609 unopposed pro-Government candidates, including internationally known arts and sports figures, and Juan Miguel González, father of Elián (the boy at the centre of the child custody battle with his Miami relatives in 2000—see below). Fidel Castro was unanimously re-elected as President for a sixth consecutive term, having outlasted nine US Presidents. His tenure as the longest-serving head of government in the world has been marked by many achievements. Cuba now has the highest rate of life expectancy in Latin America and one of the lowest infant mortality rates in the world, the result of its citizens having access to one of the world's most extensive free public-health systems. It also has one of the highest adult literacy rates in the world, some 97% of the population in 2002.

## INTERNATIONAL RELATIONS

Throughout the 1990s Cuba continued to strengthen its links throughout Latin America and the Caribbean and was admitted as a full member of the Latin American Integration Association (LAIA, known as the Asociación Latinoamericana de Integración—ALADI, in Spanish) in November 1998. Cuba's normally solid relations with Mexico deteriorated in 2000 upon the election to the presidency of Vicente Fox Quesada, who secured Cuba's exclusion from the San José Agreement (which commits the region's petroleum producers to selling subsidized supplies, on favourable terms, to developing countries of the region). However, in February 2002 President Fox visited Cuba, where he and Castro stated their intention to build on their traditionally strong trade ties. In November 2000 Castro's friend and ally, Venezuelan President Lt-Col (retd) Hugo Chávez Frías, agreed to supply one-third of the country's petroleum requirements, on preferential terms. In early 2002, amid reports that Cuba had been delaying payments on deliveries of oil, the Venezuelan opposition called for a review of the agreement. The failed *coup d'état* in Venezuela in April increased pressure on Chávez to renegotiate Cuba's growing oil debt to Venezuela. However, subsequent agreements further increased commerce between the two countries, guaranteeing mutual preferential trade in over 1,000 products. Venezuela now supplies one-half of Cuba's oil at highly favourable rates. In exchange, Cuba shares its expertise in education and health with Venezuela. The two heads of state proposed the expansion of this model across the region as an alternative to the US-backed Free Trade Area of the

Americas (FTAA). Cuba also consolidated links with Colombia by acting as broker in peace talks between that Government and the dissident guerrilla group, the Ejército de Liberación Nacional (ELN). In January 2002 Honduras re-established diplomatic relations with Cuba, leaving Costa Rica and El Salvador as the only Latin American countries without diplomatic links with Castro's Government. The election of the candidate of the left-wing Partido dos Trabalhadores (PT), Luiz Inácio 'Lula' da Silva, to the presidency of Brazil in October 2002 saw another long-time friend of Castro enter government. However, the Brazilian leader was also anxious to improve relations with the USA. During a potentially controversial visit to Cuba in September 2003 he avoided discussion of politics, instead focusing on economic relations and co-operation agreements. In March 2005 Uruguay's first left-wing President, Tabaré Vázquez Rosas, was inaugurated. President Vázquez restored full diplomatic relations with Cuba immediately after being sworn in: Uruguay had broken off diplomatic ties with Cuba three years earlier, when outgoing President Jorge Batlle Ibáñez supported US efforts to condemn Cuba's human rights record in a vote at the UN Commission on Human Rights (UNCHR). In August 2004 Cuba suspended diplomatic relations with Panama following that country's pardon of four Cuban exiles (including Luis Posada Carriles, see below) suspected of involvement in an attempt to assassinate Castro in 2000. Diplomatic links between the two countries were restored after the new Panamanian President, Martín Torrijos Espino, denounced the pardons.

Trade links with many individual European nations continued to be strengthened, particularly as the US embargo left the Cuban market free from US competition. The EU opened a legation office in Havana in March 2003, and relations initially appeared likely to be upgraded, with the expansion of bilateral political dialogue and improved trade and investment ties. During a four-day visit in the same month, Poul Nielson, the European Development Commissioner, discussed the possibility of Cuba becoming a full signatory to the Cotonou Agreement, a preferential trade and aid pact that included 78 African, Caribbean and Pacific nations. Upon departure, Nielson professed satisfaction with Cuba's progress and recommended its inclusion. Cuba had previously withdrawn its request to join the Agreement after criticism of its human rights record had prompted speculation that the EU would impose discriminatory conditions.

In December 2000 the first visit to Cuba by a Russian head of state took place since the dissolution of the USSR effectively ruptured relations. The Russian President, Vladimir Putin, and Castro signed a number of accords aimed at increasing bilateral links, and discussed a possible reduction in Cuba's Soviet era debt. This new climate of friendship was soon tested by the Russian decision, in October 2001, to close the Lourdes Surveillance Base close to Havana. Putin stressed the decision was purely for economic reasons, insisting that relations with Cuba were not being scaled down. Cuba, however, described this act as a 'special present' to US President Bush, presaging a disarmament deal between the two leaders (a claim borne out by the US-Russian signing of a nuclear-armament reduction treaty in May 2002). During a visit to the island in November 2004, Chinese President Hu Jintao agreed to invest US $500m. in Cuba's nickel industry. The agreement confirmed the close links between the two countries. By 2005 the People's Republic of China was Cuba's third most important trade supplier after Venezuela and Spain.

### THE CUBAN DIASPORA

Throughout the 1990s there were indications of a tendency among the Cuban exile community, mainly in the USA, to seek contact with Cuba and to eschew violence as an option. Although those favouring a hard-line stance remained a potent force, with strong financial support, many had become disaffected with recent US policy and a generation gap emerged.

The case of Elián González, a five-year-old boy found floating on an inner tube in the Atlantic Ocean off the coast of Florida in November 1999, highlighted the growing divisions within the Cuban emigrant community in the USA. Elián was one of just three survivors of a shipwreck in which his mother and 10 others died. The Cuban exile community, with strong links to the Cuban-American National Foundation (CANF), attempted to initiate legislation to have him granted US citizenship as a political refugee, although he was a minor. A bitterly fought custody battle ensued between his Miami-based relatives and his father, who wished to return with him to Cuba. In April 2000 armed US Federal Bureau of Investigation (FBI) agents seized Elián from his Miami relatives and, following the US Supreme Court's refusal to hear further appeals, he finally returned to Cuba in June.

In Cuba the case was presented as an abduction by the USA, provoking public demonstrations and daily televised *tribunas abiertas* (open courts), condemning the boy's detention in the USA. The case cemented a sense of national unity among Cubans disaffected by isolation and deprivation, while, for once, Fidel Castro found himself on the same side as the US Administration and public opinion. The Government afterwards endeavoured to maintain the momentum of popular participation and mobilization, with the *tribunas abiertas* formalized to discuss concerns such as US sanctions, an attempt on Castro's life and the discriminatory migration policies of the USA, which were seen to stimulate dangerous illegal immigration bids. In 2002 the USA extended the 1966 Cuban Adjustment Act to include those arriving from third countries using false passports. The momentum of popular participation increased in late 2001 following the imprisonment in the USA of five Cubans on charges of spying and conspiracy. The Cuban Government maintained that they were carrying out legitimate acts against Cuban-American networks in the USA engaged in activities to undermine the security of Cuba. Cuba has since demanded their retrial, claiming that the FBI has documents proving that the information transmitted by the defendants was not sensitive and did not compromise US national security.

In recent years Castro has endeavoured to reach out to moderate Cubans abroad. In May 2004 hundreds of Cuban *émigrés* attended a conference in Havana hosted by the Cuban Government. As a gesture of reconciliation, seven men who took part in the 1961 Bay of Pigs invasion were reinstated as Cuban citizens. All had publicly renounced their hostility to the Cuban Government. The Administration of US President George W. Bush continued to cultivate links with opposition groups both in Cuba and in exile. In May 2001 controversial legislation was introduced to the US Senate for a Cuban Solidarity Pact, offering material support to dissidents working for change inside Cuba. The votes of the Cuban-American exiles in Florida were considered instrumental in the election of Bush to the presidency, while Democrats were punished for former President Clinton's association with the Elián González affair. This powerful interest group expected to have its support repaid by the adoption of a tougher position on Cuba.

### CUBAN-US RELATIONS

The context of the USA's global 'war on terror' following the 11 September 2001 terrorist attacks on New York and Washington, DC, elicited some efforts at compromise from the normally unyielding Castro Government. Although Cuba contested the subsequent US-led bombing campaign against the al-Qa'ida (Base) organization and its Taliban allies in Afghanistan, the anti-terrorism effort was taken as a means to focus attention on the activities of extremist groups in Miami. In December the Cuban Government introduced a new anti-terrorist law, with a wide definition of terrorism, including the perpetration or advocacy of actions designed to provoke alarm or fear. In addition, the Government adopted all 12 of the UN's anti-terrorism measures (only the third country to do so). Initially, Cuba did not oppose the internment of Taliban and al-Qa'ida prisoners in the long-contested territory occupied by the US naval base in Guantánamo. Instead, the Cuban authorities were quick to offer medical and logistical support, thus demonstrating their commitment to international efforts and attempting to ward off attempts by exile groups to have Cuba included on the list of states targeted by the USA.

In May 2002 former US President Jimmy Carter made a landmark visit to Cuba, the first by any former or acting US President in four decades. In an unprecedented move, Carter spoke uncensored to the Cuban people in a live television and radio broadcast. His speech offered a vision in which Cuba was fully integrated into the democratic hemisphere, participating in the FTAA. He stated that, as the most powerful nation, the USA should take the first step, since the embargo only served to perpetuate the existing impasse. However, he urged Cuba to allow inspection by the UN's High Commissioner for Human Rights and to permit a referendum on civil liberties, as demanded in a petition bearing some 11,000 signatures presented to the Asamblea Nacional by a dissident initiative known as the Varela Project. In the same year the leader of the initiative, Oswaldo Payá, was awarded the Sakharov Human Rights Award by the European Parliament for what was perceived to be his peaceful attempt to restore democracy in the country. Although committed to change, Payá insisted that it must take place from within Cuba itself. In October 2003 a further 14,000 signatures were delivered in a second petition.

Carter's visit was treated with antipathy by the Bush Administration. Days earlier Cuba, Libya and Syria were added to President Bush's 'axis of evil' (hitherto comprising just Iran, Iraq and North Korea), nations which the USA considered to be actively attempting to develop weapons of mass destruction and trading with 'rogue states'. This move anticipated a renewal of sanctions later in the month with only a few minor concessions, such as the resumption of a direct postal service between the two countries. President Bush took the opportunity to repeat that the embargo would only be lifted should Cuba hold 'certifiably free and fair elections and reform its economic system'. Castro's response was to formulate an amendment to the Cuban Constitution, declaring the socialist system to be 'untouchable' and to 'ratify that economic, diplomatic and political relations with any other state can never be negotiated in the face of aggression, threat or pressure from a foreign power'. The amendment was approved by the Asamblea Nacional on 27 June 2003.

## CONCLUSION

After a period of relative tolerance, in April 2003 there was a harsh clamp-down on political dissent on the island as some 75 independent journalists, pro-democracy activists and leaders of opposition political parties were detained and, following summary trials, sentenced to prison terms of up to 28 years. The detainees were accused of being mercenaries who were conspiring with the head of the US Interests Section in Cuba, James Cason, to bring down the Government. For almost a year Cason had openly courted dissidents, holding high-profile meetings with them at his home and lauding them as the future political leaders of the country. Speaking of the arrests, Castro claimed that Bush, following the success of the US-led campaign to overthrow the regime of Saddam Hussain in Iraq in early 2003, sought to foment social unrest in Cuba, thus providing a pretext for direct intervention. On 2 April a hijacked ferry carrying 50 passengers ran out of fuel on its way to Florida. Three of the captured hijackers were executed nine days later, ending a three-year moratorium on capital punishment in Cuba. The ferry incident was one of a series of similar plots stimulated, argued the Cuban Government, by a virtual guarantee on the part of the USA that any Cuban reaching US soil would be granted residency. Cuban officials stated that such severe measures had been necessary to halt a mass illegal migration that was already in full development. The actions provoked widespread international condemnation; the EU downgraded diplomatic contacts and announced that it would postpone Cuba's application to join the Cotonou Agreement in the light of the country's recent human rights record. (In May the Cuban Government announced the formal withdrawal of its application.)

In January 2005 the EU announced its intention temporarily to lift sanctions on Cuba in order to assess whether constructive dialogue would be a more effective policy to achieve the desired changes. It was stressed that enhanced relations would also be sought with peaceful opposition groups. European foreign ministers agreed to a Spanish proposal aimed at ending the 'cocktail wars', which broke out when EU officials invited opponents of the Cuban Government to EU embassy parties in Havana in protest at the 2003 crackdown on dissent. This improvement in relations followed the release of a number of those 75 dissidents, all of them on health grounds. The policy was to be reviewed after six months, at which time the EU would judge whether there had been an improvement in Cuba's human rights record.

In May 2005 about 200 dissidents held a public meeting to debate their pro-democracy projects, said by organizers to be the first such gathering since the 1959 revolution. However, a number of EU politicians were prevented from attending the conference; a Czech senator and a German legislator were expelled from the country prior to the event, while two Polish members of the European Parliament were denied entry into Cuba.

In April 2005 the UNCHR approved a motion requesting that the mandate of the UN's Special Rapporteur on Human Rights in Cuba be extended. Cuba has refused such access for the past three years. Similar resolutions, condemning Cuba's human rights record, were approved by the General Assembly in 2002–04. The issue was a highly charged one, with many developing countries convinced that such motions were a tactic used by the rich and powerful against countries they did not favour. Cuba responded to the 2005 resolution by calling on the Commission to examine alleged US human rights abuses. It recommended an independent inspection of conditions among prisoners in the Guantánamo naval base, which the Cuban Government described as 'a concentration camp'. The US naval base at Guantánamo Bay was leased under an agreement signed before 1959. Castro consistently described the agreement as illegal and pointedly refused to cash the cheques received for rent of the facility.

By 2005 Cuba's moves towards creating a new climate of openness appeared to have stalled, with the country's tentative progress towards a more market-orientated economy faltering. In the months preceding the presidential election in the USA in 2004, both candidates made efforts to secure the crucial vote of the Cuban-American exile community in Florida. This community has tended to vote as a bloc, in a state otherwise fairly evenly matched between Democrat and Republican, so candidates have needed to prove their willingness to adopt a stringent approach towards Cuba. In May 2004 President George W. Bush demonstrated such a commitment by ordering a series of measures intended to hasten the fall of Fidel Castro. These included: the imposition of new limits on family visits by Cuban-Americans; stricter controls on the level of remittances sent home to relatives by Cuban expatriates, a key source of revenue for Cuba; and the use of US military aircraft to broadcast pro-democracy radio and television programmes into Cuba, circumventing the efforts of the Cuban authorities to block such signals. President Bush stressed that the new strategy would encourage expenditure that would help organizations to protect dissidents and to promote human rights.

The response from Cuba to the intensification of the embargo was swift. In October 2004 Castro announced a ban on transactions in US dollars, and decreed that all conversions between the US dollar and Cuba's 'convertible' peso would be subject to a 10% tax. This move affected Cuban citizens receiving money from family members overseas, as well as impacting upon foreign visitors. The Cuban Government stressed that the tax would apply exclusively to the US dollar and that it had no option in view of the latest attempts by the US Government to suffocate the country. In May Castro led an estimated 1m. people on a protest march through Havana; in an address to the demonstrators, he likened himself to a Roman gladiator prepared to fight to the death to prevent Cuba becoming a US 'neo-colony'. Leading dissident Oswaldo Payá was among those who criticized the sanctions, insisting that it was inappropriate for any forces outside Cuba to try to influence the process of transition on the island.

In May 2005 the US Administration was faced with a new dilemma as declassified US government documents revealed that a man suspected of involvement in terrorist activities had previously worked for its Central Intelligence Agency (CIA). Luis Posada Carriles, a Cuban *émigré* resident in Venezuela, was detained in Miami by immigration agents after entering the USA illegally. In 1985 he had escaped from a Caracas prison while awaiting trial for the bombing of a Cuban passenger aeroplane near Barbados in 1976 that had killed 73 people; he was also implicated in a series of hotel bombings in Havana in 1997. In 2000 he was arrested in Panama and charged with involvement in a plot to kill President Castro. In April 2004 Posada Carriles was sentenced on a lesser charge to an eight-year prison term, but in August of that year, having been pardoned by the outgoing Panamanian President (which caused Cuba to break off diplomatic relations, see above), he fled the country.

While the US Administration was anxious to appease the politically influential Cuban-American community in the USA, it was equally important that it be seen to be taking a strict stance on suspected terrorists in the wake of its declaration of a 'war on terror'. The Venezuelan Government sought to extradite Posada Carriles, insisting that he would not be handed over to Cuba, while Castro stated that he would be happy to see him put on trial in Venezuela; the alternative feared by Castro was that he might be dispatched to a country such as El Salvador, where he faced fraud charges, where he could expect to be treated leniently (El Salvador dropped its extradition warrant for Posada Carriles in July 2005). The US authorities had until the end of August 2005 to decide whether to extradite Posada Carriles or to bring charges against him in the USA.

President Castro's dramatic fall before a crowd of 30,000 in the Cuban city of Santa Clara in October 2004 prompted renewed speculation about the leader's mortality. Castro, however, was swift to turn the situation to his advantage, assuring the crowd that he was still very much in control. At the age of 79 Castro has insisted that, for revolutionaries such as himself, retirement simply is not an option.

# Economy

## LILA HAINES

### ECONOMIC POLICY

For most of the period after the 1959 Revolution economic policy in Cuba was guided by a commitment to collective ownership of the means of production and, in particular following full membership (in 1972) of the Council for Mutual Economic Assistance (CMEA), to central planning. In the early 1990s the recession in Cuba's economy after the collapse of the Eastern European socialist (Communist) bloc led to the introduction of austerity measures and a cautious reassessment of economic policy. The new approach focused on seeking inward investment and developing tourism to obtain convertible ('hard') currency in an attempt to alleviate the social effects of a precipitous drop in foreign trade, including the loss of vital food and fuel imports from the socialist bloc, while remaining faithful to the basic principles of socialism. The Government initially reduced domestic spending and non-essential imports and declared a 'special period in time of peace', an austerity programme that was marked by food shortages, less expenditure on power and public transport, and attempts to improve the efficiency of state enterprises. By 1993, however, the system was clearly in danger of collapse: the money supply was increasing rapidly; subsidies to loss-making industries were soaring; and the 'black market' (parallel, illegal economy) was expanding, with the state often unable to supply even basic rations. In July 1993 President Fidel Castro Ruz announced the legalization of the use of hard currency by Cubans, one of several cautious, although significant, economic reforms arising from policy decisions taken at the Fourth Congress of the Partido Comunista de Cuba (PCC—Communist Party of Cuba), held in 1991. Constitutional amendments passed in 1992 marginally liberalized the concept of property and the state's economic role, promised protection to foreign investors and granted limited recognition to private enterprise, thus preparing for new legislation and the use of new policy instruments from 1993.

Thereafter, a shift away from direct control of production was discernible and state enterprises were given greater operational freedom. However, the state retained ownership of the resource base and the Government continued to set economic priorities, to control the flow of essential commodities and to monitor carefully the use of hard-currency finance. A law passed in 1998 aimed to raise managerial standards and to make enterprises self-financing, but also reaffirmed the state enterprise as the basic business unit in the national economy. This period also saw slow, but clear, changes in social attitudes and the business culture, resulting from overseas investment, a greater Western presence in foreign trade and the opening of new employment opportunities (mainly in tourism). Moves towards recentralizing economic management and tightening control over use of the US dollar in 2004–05 led to the emergence of concerns about the direction of policy. The concurrent increases in peso-denominated basic pay and social security rates and exchange-rate revaluation suggested that the moves were expedients to deal with current problems, part of the pattern of pragmatic economic measures within a policy framework designed to protect the social system while attracting inward investment and expanding trade. Alternatively, these moves might have signalled a decision to return to a full command economy, if viewed as a new departure marked by Head of State Fidel Castro Ruz's announcement in March 2005 that the 'special period' of economic hardship, in force since the early 1990s, was over.

In 1994 the Government also began to use monetary and fiscal instruments of economic policy, which subsequently gained in importance. As part of this policy it introduced new taxes aimed at eradicating surplus liquidity, and reintroduced taxation on income from self-employment and hard-currency earnings in 1995 and 1996, respectively. It also reduced subsidies to state enterprises and began to implement a system of profit taxation. Official data showed that these measures fulfilled the aims of increasing government revenues and curbing inflation. However, the effect on the money supply was difficult to measure owing to a number of factors, including the coexistence of several markets and currencies. For over three decades the state controlled the prices of all officially traded goods and provided a heavily subsidized 'basket' of essential products to all citizens. In 1994, when it became clear that it could no longer directly ensure minimum consumer supplies, the Government authorized the opening of non-regulated farm-produce markets and other retail outlets where prices responded to supply and demand. Together with the division of most state farms into smaller workers' co-operatives called Unidades Básicas de Producción Cooperativa (UBPC—Units of Basic Co-operative Production) from 1993, and the leasing of smallholdings to individuals, this measure helped to ease food shortages. However, agricultural output did not increase as much as had been hoped, owing, in large part, to the failure to address the inefficiencies in the sugar industry. In June 2002 plans were announced to restructure the sugar industry, thus finally recognizing the impracticality of trying to maintain agricultural and manufacturing practices left over from the era of secure socialist bloc markets. Much of the land used to cultivate sugar cane was to be turned over to other crops.

Cuba moved into the 21st century with its economy recovering from the collapse suffered a decade earlier, but still fragile and vulnerable to fluctuations in commodity prices on the world markets, particularly for sugar, oil and nickel, and to the repercussions of natural disasters such as hurricanes. Its biggest economic success, the development of the tourism sector, also brought with it an additional element of instability, as demonstrated by the decline in tourism in the region following the 11 September 2001 terrorist attacks on the USA. Growth of gross domestic product (GDP) slowed from 6.1% in 2000 to 3.0% in 2001. According to official government figures, in 2002 GDP contracted by 6.3%, before recovering in 2003, when it increased by 2.6%. However, according to figures from the UN Economic Commission for Latin America and the Caribbean (ECLAC), in 2002 the economy expanded, albeit slightly, by 1.1%. In 1994 Cuba began to use GDP in place of the Soviet system of Gross Material Product (GMP), which excluded the contribution of services (such as health and education). All data since then have been best estimates, using the 1981 GMP figures as the base year for the period 1994–2000. In 2001 the base year was adjusted to 1997, changing both the value and the sectoral composition of GDP in the years for which it is available; however, in 2005 it remained unclear what difference this would make to analysis of earlier years.

## THE FINANCIAL SECTOR

In 1994, in a radical departure from existing policy, foreign banks were authorized to open representative offices, but not branches, in Cuba. This move signalled the start of the reform of the domestic financial system. Banking had become a state monopoly in 1960 and, until the 1990s, the Banco Nacional de Cuba (BNC) and Banco Popular de Ahorro, the national savings bank, were the only domestic banking institutions. In 1995 a chain of exchange bureaux began to operate, trading in US dollars and both convertible and old pesos (see below). New banks were also established to provide commercial and private banking services and development financing. In 1997 the BNC's central banking functions passed to a new central bank, the Banco Central de Cuba (BCC); banking operations were also modernized. In addition, the insurance sector expanded and began to form joint ventures with overseas insurance companies. A number of overseas-based investment funds opened offices in Havana. In 1999 a monetary policy committee was established, to set domestic interest rates.

In the immediate aftermath of hard-currency liberalization in 1993, as the economy became more unbalanced and political tensions rose, the Cuban peso continued to fall rapidly in value against the US dollar in domestic transactions. The peso recovered, however, from its lowest rate of 120 pesos per dollar in mid-1994 to an average of 19–23 pesos per dollar from 1996 until late 2001, when it depreciated to approximately 26 pesos per dollar, largely owing to the negative effect on tourism of the 11 September terrorist attacks on the USA. A 'convertible peso', at par with the US dollar, was introduced in December 1994 and by the late 1990s the US dollar circulated freely alongside the traditional and the convertible peso. Thus, there were in effect three currencies: the Cuban peso and the convertible peso, neither of which could be exchanged directly for foreign currency, and the US dollar. There were also three exchange rates. In April 2005 the convertible peso was devalued, apparently in response to the weakening of the US dollar in world markets; in policy terms it may also have reflected a move towards an adjustable peg. A fourth official currency was introduced in 2002, when the euro became legal tender for hard-currency transactions within the main tourist resorts. The move was motivated both by politics and by a wish to lower transaction costs caused by the dependence on the US dollar. Full currency convertibility remained unlikely while Cuba's foreign-debt problem persisted. The budget deficit was reduced from an estimated 6.7% of GDP at current prices in 1994, to under 2.5% of GDP in 2001.

## US POLICY AND REFORM

Overall, US policy (notably the trade embargo) negatively affected the Cuban economy, mainly through higher import and other transaction costs, the loss of potential tourism and other services revenue, and a high-risk investment and trade climate. However, US trade sanctions, implemented with varying degrees of severity in response to political events and retained mainly as a result of pressure from a powerful Cuban exiles' lobby, failed to achieve their aim of bringing down the Castro Government. In October 2000, following an acrimonious passage through the US Congress, legislation was approved that allowed US food and medicine sales, but attached conditions (see below) that left most parties dissatisfied, including the then US President, Bill Clinton (1993–2001), who saw his power to ease or tighten travel restrictions removed. His successor, President George W. Bush (2001–), has demonstrated his intention to retain and implement trade sanctions (see below), despite growing opposition from mainstream US figures, including former President Jimmy Carter, who visited Cuba in May 2002 and advocated the end of the embargo.

## INFRASTRUCTURE

About one-half of Cuba's estimated 10,990 km of paved roads were built after 1960. A central highway links Pinar del Río in the west with Santiago de Cuba in the east. Causeways join the mainland to keys off the north coast, which, since 1990, have been developed as tourist resorts. From 1996 some maintenance and infrastructure work was recommenced, but roads were neglected again in 2002, when earnings from tourism and commodity exports fell and petroleum prices rose. More than one-half of the 14,000 km rail network serves the sugar industry, although it is unclear to what extent the decline of the latter has reduced such usage, while the rest offers passenger and other freight services. This too suffered seriously from neglect during the 1990s, although the railways began to attract some investment later in that decade. At the start of the new millennium private car ownership remained low and the principal mode of transport in Cuba remained the bus, which had begun to offer improved services after 1998 following a period of severe curtailment post-1990.

Havana handled more than one-half of the country's maritime cargo, making it by far the most important of 16 commercial seaports and 23 minor ports. The petroleum industry operated 11 sea terminals and 17 land storage facilities designed for domestic trade. New pipelines and tanker bases were constructed to support the growing nickel sector and, a little later, the expansion in the production and use of domestic crude oil.

Civil aviation enjoyed a dramatic, tourism-led expansion during the 1990s. The main carrier was the national airline, Empresa Consolidada Cubana de Aviación (Cubana), which flew to over 40 countries. Cubana, the smaller local air companies and the national air services enterprise were brought under the umbrella of the newly formed Corporación de la Aviación Cubana in 1997. State and overseas investment modernized the island's airports, the newest of which, situated on the Cayo Coco offshore tourist resort, was completed in 2002, bringing the total number of civilian airports in Cuba to 21, although just nine of them handled significant numbers of international passengers.

Modernization of the telephone system began after June 1994, when the Government approved the formation of a joint venture, Empresa de Telecomunicaciones de Cuba (ETECSA), 49% of which was owned by a Mexican company, Grupo Domos. Grupo Domos later withdrew and was replaced by Stet, an Italian company. Two small telecommunications companies operated in Cuba for some years until they were acquired by ETECSA in 2003. ETECSA continued to invest in modernization and reported that over 85% of users had access to the digital system by the end of 2004. It was reported that ETECSA also planned to extend the mobile cellular telephone network to remote areas, a move that would require arrangements for payments in local currency and would thus bring mobile cellular telephones officially into the peso economy for the first time. Internet services had been installed in post offices and this, together with a network of computer clubs and workplace use, gave almost 500,000 Cubans e-mail access. Recent modest improvements in government income (mainly from nickel, tourism and health products) were used in part to improve water services in the eastern provinces, which suffered regularly from drought.

## HUMAN RESOURCES AND EMPLOYMENT

The economic reforms of the 1990s initiated a shift away from the state domination of employment that had resulted from the nationalization programmes of the 1960s. The state sector accounted for 77.5% of civilian employment at the end of 2000, compared with 95.4% in 1989. The remainder were mainly private or co-operative farmers or were self-employed, with under 1% occupied in the new foreign sector. The unemployment rate in Cuba at the end of 2004 was, according to preliminary figures, 2.0%, and the Government claimed to have created over 92,000 new jobs in 2001–04, mainly in education, health, information technology and the media. However, it was probable that under-employment and low productivity were still serious problems. In the past, ECLAC estimated that the real unemployment figure was much higher than the official figure. It was thought that the discrepancy could have resulted, at least in part, from the retention of former staff at government ministries and other agencies and state enterprises on the payroll until they found alternative employment. This approach was adopted in the process of restructuring the sugar industry, when 200,000 redundant workers were guaranteed redeployment or retraining.

In contrast to the moral stimulus approach that characterized the Ernesto ('Che') Guevara era of 'socialist economics', in the 1990s the granting of material incentives to employees became an important strand of policies aimed at stimulating greater labour productivity. Bonuses, in the form of scarce goods or hard currency, were introduced, firstly in the sectors considered to be of greatest importance for economic recovery, such as tourism and mining, and then gradually in other sectors. Salaries in joint ventures involving foreign capital were 15%–30% above sectoral averages and both official and unofficial bonuses were often relatively generous. Salaries in the public services, in contrast, were largely frozen for a decade, until 1999 when employees in the education and public-health sectors received an average 30% pay rise, followed by those in other sectors such as the police. Increases in social security and minimum wage rates were announced in 2005 in order to raise the income of the poorest sectors of society and thus reduce inequality, which had worsened during the 1990s.

In 2002 agriculture (including hunting, forestry and fishing) employed around 26% of the economically active population and manufacturing industry 14%, compared with 40% and 13%, respectively, in 1960. The service sector employed about 53% of the work-force. In 1995 women formed 37.6% of the labour force, rising to 42.0% in the state sector and to 73.2% in health, a sector employing over 328,500 people in that year. Investment in universal education after 1959 produced a well-educated work-force, which was perceived as an asset in attracting foreign investment.

## FOREIGN TRADE

Prior to 1959 Cuba's main trading partner was the USA, which had consolidated its dominance after Cuban independence from Spain in 1898. In 1958, the year preceding the overthrow of Fulgencio Batista Zaldivar by the Castro-led revolutionaries, 68% of foreign trade was with the USA and Cuban sugar enjoyed preferential entry to the US market. This position changed radically in the early 1960s, when the new Government tried to implement a programme that it believed reflected its commitment to social justice.

After the Cuban Government had nationalized assets belonging to US companies, valued at over US $1,000m., the USA severed diplomatic relations in January 1961, supported the unsuccessful Bay of Pigs invasion by anti-Castro exiles in April and, in March 1962, extended to all goods the partial trade embargo that it had imposed on Cuba in 1960. The so-called Cuban Missile Crisis of October 1962, following the attempt by the Soviet Government to install nuclear weapons on the island, demonstrated Cuba's peripheral status in relation to the cold war 'superpowers' of the USA and the USSR. The country was left with little choice but to depend on the USSR for its trade revenue, as well as for aid to implement an ambitious social and economic development programme.

After 1962 foreign trade was conducted increasingly with the Communist bloc, although the People's Republic of China, Japan and some countries of Western Europe each took a small, but significant, share at various times. In 1989 over 80% of foreign trade was with Eastern Europe. As a result of the collapse of trade with the Soviet bloc from 1990, the value of merchandise trade plummeted from US $13,500m. in 1989 to less than $3,170m. in 1993. Recovery was slow and included a diversification of trading partners and greater integration with the world market. In 2003 Western Europe was the leading destination for Cuba's exports and main source of its imports, with Russia taking slightly less than 10% of exports and Canada over 15%. The People's Republic of China's share of both imports and exports was growing. The country was Cuba's third largest supplier of imports after Venezuela and Spain. A significant change in the profile of foreign trade was brought about by the re-entry of the USA in 2001, when an amendment to sanctions regulations allowed food to be exported to Cuba. The USA exported approximately $400m. of food to Cuba in 2004, accounting for about one-half of the island's food imports. However, the US Administration tightened restrictions on exports to Cuba during the US presidential election year of 2004, which appeared to lead to a fall in the US share of Cuban imports. Another significant recent change in the directon of trade resulted from agreements with Venezuela, primarily to exchange oil for Cuban health and education services.

The change in the composition of trade after 1990 was significant. Services comprised a growing share of overall foreign trade, mainly owing to increases in the tourism sector, but also as a result of the greatly diminished role of the sugar sector, whose exports were valued at just US $243m. in 2003, a year in which nickel exports reached US $634m. The other main exports were tobacco products and seafood. Exports of health products, though still small, were growing, particularly to Latin America and the People's Republic of China, although this sector appeared not to have lived up to the Government's high expectations. Total spending on imports of goods and services decreased to US $2,320m. in 1993, but increased thereafter. Imports diversified, but a shortage of purchasing capacity and credits constrained recovery. Under 1996 legislation the Government opened four free trade zones to promote exports of goods and services and to help resuscitate dormant domestic industry by attracting more foreign investment and spreading its benefits to new geographic and economic areas. However, they did not figure noticeably in exports, although it was believed that they might have attracted smaller companies that would not have invested where there was a greater risk of claims for future compensation from US-based individuals or corporations.

## THE US TRADE EMBARGO

The US trade embargo, implemented with varying degrees of severity since first imposed by President John F. Kennedy in 1960, was strengthened in 1992 when the Cuban Democracy Act (also known as the Torricelli Act) banned trade with Cuba by overseas subsidiaries of US companies. A further dimension was added in March 1996 when President Clinton signed into law the Cuban Liberty and Solidarity Act (commonly known as the Helms-Burton Act), which aimed to halt foreign investment in Cuba. It provoked exceptionally strong protests by other Western governments concerned by its attempted extraterritorial reach.

Contrary to international norms, this legislation potentially opened the US courts to claimants who had obtained US citizenship after their property in Cuba was nationalized. However, President Clinton repeatedly exercised his right to postpone the implementation of Title III, the section of the Act allowing US nationals to claim damages in US federal courts from overseas companies believed to be 'trafficking' in confiscated Cuban property. (Claims by former owners of such property resident in countries other than the USA had been settled.) Clinton's successor, George W. Bush, also suspended the implementation of Title III, despite his heightened anti-Cuban rhetoric. Title IV, which made executives of companies investing in Cuba (and their dependants) liable to exclusion from the USA, was implemented selectively.

In 1998 and 1999 President Clinton announced plans to ease restrictions relating to travel by Cuban Americans for family visits, cash remittances and the export of food and medicines to Cuba; as a result, additional flights were authorized. In 1999 Western Union was allowed to begin cash transfers to Cuba, in partnership with a Cuban enterprise, and the US Department of the Treasury licensed the pharmaceuticals company SmithKline Beecham (later GlaxoSmithKline) to market a Cuban meningitis vaccine. In October 2000 legislation was passed that allowed US food and medicine sales to Cuba. However, the potential for such trade was restricted by the conditions attached, such as a ban on financing by US banks or official US credits. Nevertheless, US exports to Cuba grew, totalling over US $400m. in 2004, about one-half of Cuba's food imports, though they appeared to be less buoyant during the first six months of 2005 as a result of further restrictions, including a requirement for payment in case, introduced by the Bush Administration. Imports of most Cuban goods remained illegal. Other sanctions legislation, such as the Torricelli and Helms-Burton Acts and the 'Trading with the Enemy' Act, remained operational. These included a clause of the Torricelli Act that forbade ships to enter US ports within six months of entering a Cuban port for the purpose of trade. However, in February 2001 the USA granted the first licence to run a scheduled route to Cuba to the shipping company Crowley Liner Services. In July the US Department of Commerce announced new regulations regarding the US sales to Cuba covered in the October 2000 legislation, although public and private financing to fund the sales remained prohibited. In July 2001 the US House of Representatives approved a proposal to lift restrictions on most travel to Cuba. A similar motion was approved in 2002, and again in 2003. However, each time the proposal to implement such an easing of travel was removed from the budget bill to which it was attached at the committee stage by Republican business managers. This was consistent with the more inflexible approach to Cuba adopted by the Bush Administration, which implemented sanctions more strictly than hitherto, increasing prosecutions against US residents for non-permitted travel and against third country nationals for infringing trade restrictions. In 2004 the Cuban authorities announced that they would waive the requirement that Cubans living in the USA apply for Cuban visas to visit the island. However, the US Government then introduced further travel restrictions, causing a fall in the number of US visitors to Cuba.

## FOREIGN INVESTMENT

Legal from 1982, foreign investment only became a priority for the Cuban Government following the collapse of the Communist bloc at the end of the 1980s. In September 1995 foreign investment (up to 100%) was legalized in all sectors of the economy except defence, public health and education, and in 1996 a law was passed allowing the establishment of free trade zones and industrial parks with a view to attracting foreign investment in such areas. Foreign companies were normally expected to invest in the new type of nominally autonomous limited company (Sociedad Anónima), in which the state (or its nominee) was the major or only shareholder. US companies were prevented by their own laws from investing in Cuba and the US Helms-Burton Act rendered negotiations with companies from third countries more difficult, although it did not halt the flow of overseas investment. At the end of 2001 cumulative investments and commitments totalled US $5,400m., according to official data, and almost 400 joint ventures and other forms of economic association with overseas companies were operating, with an estimated annual turnover of over $1,200m. However, at the end of 2003 there were just 342 joint ventures, a drop attributed to the rigorous administrative procedures required by the Cuban Government. In 2000 the Government halted foreign involvement in real estate, one of the most sought-after investment opportunities, and the Cuban partners in property ventures bought back the units that had already been built. This was not surprising, as political opposition had prevented a planned real estate law from being enacted. There was also a growing preference on the part of the Cuban authorities for larger projects in strategic sectors, such as energy and infrastructure, and for partners who could offer loan finance. The purchase in 2000 of 50% of the state tobacco marketing company by a French-Spanish company, Altadis, brought in $500m. In another effort to improve access to foreign capital, in March 2002 the first joint venture in the sugar industry was announced.

## FOREIGN DEBT

Cuba's central bank estimated that the hard-currency foreign debt amounted to US $12,000m. at the end of 2004, having risen from $11,300m. one year earlier. As Cuba has no access to the World Bank or IMF, most of its official foreign debt is bilateral. Cuba suspended the servicing of bank and bilateral debt with Western governments in 1986, and in 1991 it was granted a moratorium on Soviet debt estimated at 15,000m. roubles. From the mid-1990s Cuba recommenced talks with its creditors and reached several debt-rescheduling agreements. Talks with the 'Paris Club' of Western creditors took place in 2000, but progress was prevented, in part, by indecision regarding the inclusion of the debt owed to Russia. In 2002 Cuba announced that it was unable to meet the terms of its 1998 agreement to service $750m. of Japanese commercial debt. This was attributed to trade difficulties and seemed to be indicative of a change in the priorities of the Central Bank, forced upon it by the island's worsening trade balance. However, at the end of 2004 the Central Bank reported a stronger balance of payments, mainly as a result of strong growth in exports of both goods and services.

## AGRICULTURE, FORESTRY AND FISHING

Agriculture has been one of the most unpredictable sectors of the Cuban economy, suffering from frequent droughts, low investment, poor productivity in many crops and, crucially, the central role of sugar long after it had become economically unviable for such reliance to continue. Agriculture and fisheries accounted for 6.9% of GDP, measured at constant prices, in 2003. Sugar output and its economic contribution diminished dramatically during the 1990s. The fall in the agricultural sector's contribution to national income reflected the effects of the post-Soviet economic crisis and endemic problems, some of which were broached, though not adequately addressed, in the mid-1990s. Foreign financing helped to overcome input shortages in sugar cane, tobacco and citrus fruits, although relief was short-lived in the case of sugar. The 1994 law legalizing free-market sales of agricultural produce (with some exclusions) helped to increase food availability, but production was still below national needs. Although legal, there was little foreign investment in agriculture—the sale of land was not allowed.

A 1993 law divided most state farms into over 4,000 UBPCs. Small areas of land were leased to individuals for private cultivation, especially of coffee and tobacco, which helped stimulate output, although the number returning to the land was not significant. This reduced the land owned directly by the state from over 82% in 1989 to 33% in 1998. The rest was held by private farmers (16%), traditional co-operative farms (9%) and the UBPCs (41%), although the land on which the latter operated remained state property. Most UBPCs claimed to be penalized by the low prices paid by the state, to which they were obliged to sell 80% of their produce, as well as by overpriced inputs and services available only from state enterprises and by onerous repayment terms on the capital stock bought from state farms on formation. Indeed, in practice, according to some observers, many UBPCs were little more than subsidiaries of the enterprises from which they were formed, particularly in the sugar industry. From 1998 central government subsidies were gradually eliminated and UBPCs were required to take up bank loans instead. In 2002 a new law reduced profit tax for small farmers by allowing them to retain 75% of their profits rather than the 50% previously permitted.

The restructuring of the sugar industry, formerly Cuba's leading industry, which commenced in 2002, constituted the country's most significant economic policy reform since the early 1990s. The cumulative effect of shortages of inputs, spare parts and fuel had caused sugar production to fall significantly in the 1990s. The implementation by the USA of the Helms-Burton Act in 1996 led overseas banks and traders who had financed the industry to review their commitments. From 1997 the Government attempted to achieve greater efficiency and discipline, closing the least efficient sugar mills and ending the practice of cutting young cane. However, such measures did not prevent the harvest from falling to a 50-year low in 1997/98, or from remaining below 4m. metric tons in every year thereafter. This occurred at a time when export earnings were adversely affected by low sugar prices. In 2002, during the first phase of restructuring, the amount of land planted with sugar cane was reduced by 30%, but, despite the concentration of production on better land, output did not appear to improve and the 2004/05 sugar harvest produced an estimated 1.3m. tons, the lowest for a century.

The recovery of the tobacco industry, following a rapid decline during the early 1990s, was one of the clearest successes of economic reform. From an annual average of 43,000 metric tons in 1986–90, production fell rapidly, to 17,000 tons in 1993, before recovering in the late 1990s, owing to financing obtained from the main French and Spanish buyers and incentive schemes for producers. Output was further expanded by allocating more land to private farmers, who produced over 80% of tobacco. This allowed significant expansion in the production of cigars, to respond to growing demand on the world market. The sector opted for a high-quality image with premium prices on the international market; this approach was reinforced when Altadis purchased 50% of the state marketing board in 2000. Production stood at an estimated 34,500 tons in 2003.

Joint ventures provided inputs for a modern, foreign-financed citrus-juicing industry in the 1990s, which also boosted export income. As a result, citrus production increased during the 1990s and reached 961,301 metric tons in 2000. However, yields declined to 482,501 tons in 2002, a season in which most crops were severely damaged by hurricanes. By 2003 citrus production recovered to 797,600 tons in 2003. Other crops displayed an erratic production pattern, although the trend was mainly one of recovery. New sectors undergoing expansion included those of organic production, particularly of crops such as coffee, which could obtain premium export prices, and urban agriculture, which reportedly created thousands of new jobs while simultaneously improving domestic food supplies.

The dairy industry, which, with poultry, was one of the agricultural successes of the revolutionary period, was severely affected by the collapse of foreign trade relations with the Soviet bloc. The national herd was bred to produce high yields from a feedlot system (according to which cattle were kept indoors and fed automatically) and adapted badly to grazing when cheap Soviet feed imports were curtailed. Milk production was an estimated 589,700 metric tons in 2002, compared with its pre-1990 level of just over 1m. tons. Overall, private farmers recorded the most notable output improvements, a fact attributed largely to the incentive offered by the opening of deregulated farmers' markets. Irregularity characterized most food sectors, exacerbated since 1990 by the state of the economy but also by the susceptibility of Cuban agriculture to hurricane and other weather damage.

Fish farming was developed in the 1990s, and was the main source of fish for domestic consumers. Lobster and other shellfish were exported, as was the catch from Cuba's deep-sea fishing fleet. An ageing fleet, combined with fuel and other shortages, produced a 1994 catch of less than one-half that of 1989, although increases were recorded from 1995 onwards. Lobster-processing plants were modernized to respond to demand from exports and tourism. A scheme introduced in 1998, whereby fishing boats were leased as co-operatives to their crews, was reported to have raised the inshore catch.

Following a reforestation programme, by the mid-1990s forestry covered 21% of Cuban territory, up from 18% in 1987. However, like most of the economy, the sector suffered from resource shortages in the 1990s and by the end of the decade was seeking foreign investment. Part of the land taken out of sugar cane cultivation was planted with trees.

## MINING AND ENERGY

Gold was the first economic resource developed by the Spaniards in Cuba, in 1515–38. Copper was mined from 1530 onwards. However, mining was thereafter of little economic importance in Cuba. One study valued mining output at US $467m. between 1902 and 1950, approximately equivalent to a single post-Second World War sugar harvest.

The island developed production of industrial minerals such as zeolite, and in the 1990s copper, gold, silver and other metals received attention from foreign prospectors. However, deposits of such minerals were insignificant compared with Cuba's proven and estimated reserves of nickel and cobalt, which were among the world's largest. Nickel output increased from 26,900 metric tons in 1994 to 71,700 tons in 2003. The increase was mainly owing to significant joint-venture agreements, particularly with Sherritt International, a Canadian corporation, which included a 50% Cuban share in a Canadian refinery. Nickel became the largest commodity export and Cuba became the world's fifth largest nickel producer. Nickel exports helped Cuba offset recessionary factors. Further growth, to a predicted 120,000 metric tons by 2008, was predicted following investments announced in 2004 and 2005 by Sherritt International, the Chinese Government and the Cuban state.

Energy was one of the Cuban economy's weakest sectors, and the country depended on imported petroleum to generate over 80% of its electricity until the collapse of Soviet economic support forced Cuba to maximize its use of domestic sources. Some 13m. metric tons of petroleum were formerly imported annually from the USSR, on a barter basis, with around 2m. tons being re-exported. When Russian trade was established on a market-prices basis, Cuban petroleum imports decreased sharply. Cuba held international bidding rounds for petroleum exploration blocks in 1993, 1996 and 2000. Canadian and European companies began prospecting both onshore and offshore. While there were no major petroleum discoveries and some foreign companies withdrew from the project, small but significant discoveries were made. Domestic petroleum extraction rose from 0.7m. metric tons in 1990 to some 3.7m. tons in 2003, and combined petroleum and gas output in 2002 was 4.1m. tons (oil equivalent). The Spanish company Repsol discovered petroleum, although not in commercial quantities, in Cuba's deep-water territory in the Gulf of Mexico in 2004 and planned to continue exploration. Electricity generating plants were converted to use domestic fuel and Canadian investment allowed natural gas to be harnessed for energy generation. Ensuring reliable and affordable sources for the remainder of Cuba's petroleum requirements has proved problematic. Venezuela supplied oil to Cuba and Petróleos de Venezuela (PDVSA), the Venezuelan state oil company, was engaged in negotiations to modernize a Soviet-built refinery.

## INDUSTRY

From 1962 Cuba invested heavily in developing and diversifying an industrial base that had previously been dominated by sugar. In 1965–72 the main focus was on rehabilitating sugar mills (which were nationalized with most existing industry in the early 1960s) and building up the production of spare parts, agricultural equipment, cement and fertilizers. Investment spending accelerated in 1976–90, so that by the end of the 1980s Cuba had developed a varied industrial base, ranging from food processing and light industry to construction materials, chemicals, machine tools, paper and glass. However, there were several major weaknesses, which became more pronounced as industry came to a near standstill after Cuba's Communist bloc markets disappeared. A steep fall in crucial petroleum imports led to the closure of most plants. This was exacerbated by the fuel inefficiency of many factories constructed with Soviet or other Eastern European aid, which were often too large for Cuba's needs. Other by-products of the close integration with the CMEA included the unsuitability of many Cuban products for Western markets. There was also heavy dependence on imported raw materials for many sectors, perhaps most marked in the chemicals industry which, according to one leading Cuban analyst, was at least 20 years behind its counterparts in most large and medium-sized Latin American countries.

In the 1990s the Government attempted to make industry more efficient through measures such as dividing large enterprises into smaller and more manageable entities, using only the most efficient lines and reducing staff. Such measures were believed to be essential for attracting inward investment to manufacturing and some sectors recorded higher output after 1994. Results were mixed, however, and growth was often linked to tourism, in cases such as construction materials and furniture, or to inward investment, as in nickel-processing and the beverages and tobacco sectors. It seemed, however, that many plants were still producing at a fraction of installed capacity, although official data showed that manufacturing output was 20% higher in 2000 than it had been in 1995. Manufacturing, like the overall economy, was restrained by the sugar industry's disappointing results and by low levels of finance availability. Subsidies were cut and finance from the new banking sector was both relatively expensive and dependent on acceptable projections of future returns. By the early 21st century some satisfaction was being expressed about the progress that had been made in heavy industry, but light industry continued to disappoint.

## PHARMACEUTICALS AND HEALTH CARE

One area that continued to receive significant support from the Government throughout the 1990s crisis period was the indigenous pharmaceuticals and medical-goods sector. The Government justified its decision to continue investing by reference to the need to protect the national health-care service, and to recover the state's earlier spending on training and infrastructure. Between 1989 and 1995 over US $345m. in hard currency was invested in the medical-pharmaceutical industry. Existing laboratories and production lines were improved, the biotechnology sector's production capacity expanded and integrated research and production centres developed. It was claimed that, by 2002, the sector produced 90% of medicines consumed in Cuba.

A network of science parks organized the contribution of some 200 life-science institutions to the national economy and leading research and development centres began to form joint-venture agreements. No reliable recent data are available for biotechnology exports, but estimates suggest that production rose 12-fold after 1990. Earlier figures indicated that exports of medical products reached an estimated US $135m. in 1997. These included new vaccines, interferons, monoclonal antibodies and technologically advanced products developed by leading institutions such as the biotechnology and genetic-engineering centre in Havana. The sector received a potential boost in July 1999 when a joint venture was formed with the pharmaceutical manufacturer SmithKline Beecham (later GlaxoSmithKline) to market a Cuban meningitis vaccine, raising hopes that Cuban medical products might finally start to penetrate more lucrative Western markets. Indeed, vaccines and other medical products have been exported to about 50 countries, and the World Health Organization has licensed Cuba to provide a vaccine for the hepatitis B disease to the UN Children's Fund (UNICEF).

## TOURISM

In the 1990s the international tourism sector in Cuba experienced rapid growth, having virtually disappeared after the USA banned travel to the island by its citizens in 1961. Gross tourism revenue increased from US $243.4m. in 1990 to an estimated $1,846m. in 2003, and tourism surpassed sugar as the principal source of foreign currency earnings. Arrivals, mainly from Europe and Canada, rose from some 340,300 in 1990 to 1,774,541 in 2001. The sector, like the rest of the world's tourism market, suffered following the 11 September 2001 terrorist attacks on the USA, with the number of arrivals decreasing to 1,686,162 in 2002; however, it recovered to 1,905,682. arrivals in 2003, when revenue rose by 16%, compared with the previous year. One of the most significant developments at the end of the 1990s was the rise in the number of US visitors, from 34,956 in 1997 to 84,529 legal visitors from the USA in 2003; a further estimated 40,000 US citizens entered Cuba without US permission. The Cuban Government's announcement that it would allow Cubans resident in the USA to visit the island without having to apply for entry visas was expected to boost travel, although it was also perceived as a political gesture in a climate where the US Administration was strictly enforcing travel restrictions, thus severely hampering the growth potential of the Cuban tourism sector.

Having been orientated primarily towards domestic tourism for nearly 30 years, in the 1980s the tourism industry was identified as a potentially rich source of convertible currency. Overseas investment to help develop the industry was sought, and the first agreement to build a joint-venture hotel was signed in 1989 with a leading Spanish hotel group. Expansion plans for the period to 2000 required US $1,334m. in convertible currency, which Cuba sought from domestic and foreign sources, including joint ventures with foreign partners, conventional loans, mortgage-backed securities and offshore public companies. However, official sources claimed that foreign capital remained of less importance than that from national sources and, although the number and origin of foreign participants in the industry increased, most overseas companies managed, rather than owned, the hotels they operated. In 2002 the supply of hotel rooms for international tourism was reported to be 40,000, compared with 13,000 in 1990; there were, in addition, 6,000 rooms in private homes, although the latter source was estimated to have since contracted as a result of stricter regulation. The industry planned to have 47,500 rooms by 2005. It proved difficult to develop the cruise-ship circuit despite attempts by several cruise-ship operators, but some 100,000 cruise-ship passengers visited the island in 2003. The importance of US ports and visitors in the sector, the US ban on travel to Cuba and the uncertainty aroused by the shipping clause of the Torricelli Act (see above) were among the negative factors limiting expansion.

## ECONOMIC OUTLOOK

In the early 2000s there were indications that Cuba was beginning to recover from possibly the worst decline suffered by any Communist bloc economy. Real GDP continued to record positive growth, reflecting both some recovery in important sectors and the expansion of tourism. However, GDP contracted in 2002, largely owing to international factors and, despite steady recovery in some sectors, the economy was operating considerably below capacity. The restructuring of the sugar industry, which began in 2002, was expected to reduce economic dependence on this vulnerable sector, but this would happen slowly. The legislation introduced in the USA allowing exports of food and medicines was not considered likely to lead to a major improvement in US-Cuban trade owing to the limited range of goods that could be exported from the USA and the continuing ban on imports from Cuba. However, the US $400m. of US imports during 2004 provided an indication of the potential for growth in what had so far been one-way trade. Cubans appeared to want to expand this trade on purely economic grounds because of the lower costs that were achievable. The opposition of US President George W. Bush to any further relaxation of sanctions, together with heightened political tensions within Cuba, would hinder significant activity by the anti-sanctions lobby within the USA. An early and definitive solution to the problem of the hard-currency foreign debt was not expected, but an improved balance of trade and level of foreign currency reserves appeared to suggest there would be greater scope for both debt management and economic expansion. Cuba was not a member of the World Bank or the IMF and had little hope of an agreement that would give it access to cheaper new credits. The inflow of overseas investment was expected to continue, although more slowly.

The expansion of the tourism sector was likely to continue at a slower pace. The mining sector and several smaller export earners, such as the beverages and tobacco industries, were also expected to expand, but increasing revenue from these sources was likely to be partially offset by high-cost imports; however, the oil-for-services deal with Venezuela was likely to ease the latter problem, unless political instability in Venezuela worsened. The rise in world petroleum prices was expected to have a negative impact, although not as severe as was the case in the 1990s prior to the development of the domestic petroleum and gas sector. The net effect was thought likely to be slow but steady growth, a very gradual improvement in living standards, and some minor adjustment of policy instruments. It was probable that the decision radically to reduce sugar production, while having a negative effect in the short term on employment, would be economically positive in the longer term, freeing physical and human resources to concentrate on potential growth sectors. The Government was expected to continue with the same cautious approach to economic reform (with the exception of the drastic restructuring of the sugar industry), and to remain committed to a planned economy. The state was expected to face continued investment difficulties and to prioritize spending on human capital and social services. No significant departure from strict fiscal control and central planning was anticipated, but neither was it thought that the decentralization that had been implemented would be radically reversed. Cuba's relations with most Latin American countries were expected to further improve, with concomitant growth in trade and investment. GDP was likely to continue to grow, albeit slowly, and to depend heavily on tourism, nickel and energy, the new mainstays of the economy. Fidel Castro's announcement of an end of the 'special period' of economic hardship that characterized the 1990s, accompanied as it was by increased investment in social infrastructure and services and rises in basic pay and social security, was seen as a signal by the authorities that continued economic growth could be expected.

# Statistical Survey

Sources (unless otherwise stated): Cámara de Comercio de la República de Cuba, Calle 21, No 661/701, esq. Calle A, Apdo 4237, Vedado, Havana; tel. (7) 55-1321; fax (7) 33-3042; e-mail pdcia@camara.com.cu; internet www.camaracuba.cubaweb.cu; Oficina Nacional de Estadísticas, Calle Paseo 60, entre 3 y 5, Vedado, Havana; tel. (7) 30-0005; fax (7) 33-3083; internet www.cubagob.cu/otras_info/estadisticas.htm.

## Area and Population

### AREA, POPULATION AND DENSITY

| | |
|---|---:|
| Area (sq km) | 110,860* |
| Population (census results) | |
| 6 September 1970 | 8,569,121 |
| 11 September 1981 | |
|   Males | 4,914,873 |
|   Females | 4,808,732 |
|   Total | 9,723,605 |
| Population (official estimates at 31 December) | |
| 2001 | 11,172,300† |
| 2002 | 11,200,400† |
| 2003 | 11,230,076 |
| Density (per sq km) at 31 December 2003 | 101.3 |

* 42,803 sq miles.
† Figures rounded to nearest 100 persons.

### PROVINCES
(31 December 2003)*

| | Population (estimates) |
|---|---:|
| Camagüey | 786,400 |
| Ciego de Avila | 415,300 |
| Cienfuegos | 398,000 |
| Ciudad de la Habana | 2,200,400 |
| Granma | 828,100 |
| Guantánamo | 510,300 |
| La Habana | 719,900 |
| Holguín | 1,027,800 |
| Isla de la Juventud | 86,600 |
| Matanzas | 674,600 |
| Pinar del Rio | 730,400 |
| Sancti Spíritus | 462,800 |
| Santiago de Cuba | 1,042,100 |
| Las Tunas | 529,200 |
| Villa Clara | 818,200 |
| **Total** | **11,230,100** |

* Figures are rounded to nearest 100 persons.

### PRINCIPAL TOWNS
(population in 1999)

| | | | |
|---|---:|---|---:|
| La Habana (Havana, the capital) | 2,189,716 | Pinar del Rio | 148,500 |
| Santiago de Cuba | 441,524 | Bayamo | 143,600 |
| Camagüey | 306,049 | Cienfuegos | 137,513 |
| Holguín | 259,300 | Las Tunas | 137,331 |
| Santa Clara | 210,100 | Matanzas | 124,754 |
| Guantánamo | 208,030 | Sancti Spíritus* | 103,591 |

* 1998 figure.
Source: UN Citydata.

**Mid-2003** (UN estimate, incl. suburbs): La Habana (Havana) 2,188,919 (Source: UN, *World Urbanization Prospects: The 2003 Revision*).

### BIRTHS, MARRIAGES AND DEATHS*

| | Registered live births† | | Registered marriages‡ | | Registered deaths | |
|---|---:|---:|---:|---:|---:|---:|
| | Number | Rate (per 1,000) | Number | Rate (per 1,000) | Number | Rate (per 1,000) |
| 1996 | 140,276 | 12.7 | 65,009 | 5.9 | 79,662 | 7.2 |
| 1997 | 152,681 | 13.8 | 60,900 | 5.5 | 77,316 | 7.0 |
| 1998 | 151,080 | 13.6 | 64,900 | 5.8 | 77,565 | 7.0 |
| 1999 | 150,785 | 13.5 | 57,300 | 5.1 | 79,499 | 7.1 |
| 2000 | 143,528 | 12.8 | 57,000 | 5.1 | 76,448 | 6.8 |
| 2001 | 138,700 | 12.4 | 54,300 | 4.8 | 79,400 | 7.1 |
| 2002 | 141,200 | 12.6 | 56,900 | 5.1 | 73,900 | 6.6 |
| 2003 | 136,800 | 12.2 | 54,700 | 4.9 | 78,400 | 7.0 |

* Data are tabulated by year of registration rather than by year of occurrence.
† Births registered in the National Consumers Register, established on 31 December 1964.
‡ Including consensual unions formalized in response to special legislation.

**Expectation of life** (WHO estimates, years at birth): 77 (males 75; females 79) in 2003 (Source: WHO, *World Health Report*).

### ECONOMICALLY ACTIVE POPULATION
(official estimates, '000 persons aged 15 years and over)

| | 2000 | 2001 | 2002 |
|---|---:|---:|---:|
| Agriculture, hunting, forestry and fishing | 937.9 | 974.1 | 1,064.6 |
| Mining and quarrying | 50.5 | 33.1 | 28.3 |
| Manufacturing | 615.0 | 604.9 | 564.1 |
| Electricity, gas and water | 53.2 | 49.6 | 51.2 |
| Construction | 204.5 | 192.8 | 176.4 |
| Trade, restaurants and hotels | 473.6 | 470.6 | 507.9 |
| Transport, storage and communications | 194.9 | 207.1 | 201.7 |
| Financing, insurance, real estate and business services | 55.0 | 58.9 | 54.2 |
| Community, social and personal services | 1,258.4 | 1,377.8 | 1,375.7 |
| **Total labour force** (incl. others) | **3,843.0** | **3,968.9** | **4,024.1** |

Source: ILO.

**Mid-2003** (estimates in '000): Agriculture, etc. 742; Total labour force 5,654 (Source: FAO).

### CIVILIAN EMPLOYMENT IN THE STATE SECTOR
('000 persons)

| | 1998 | 1999 | 2000 |
|---|---:|---:|---:|
| Agriculture, hunting, forestry and fishing | 733.1 | 714.4 | 714.2 |
| Mining and quarrying | 47.3 | 20.8 | 20.8 |
| Manufacturing | 458.9 | 512.7 | 512.6 |
| Electricity, gas and water | 46.0 | 51.0 | 51.0 |
| Construction | 178.9 | 167.1 | 167.8 |
| Trade, restaurants and hotels | 355.2 | 375.2 | 375.2 |
| Transport, storage and communications | 175.2 | 157.4 | 157.4 |
| Financing, insurance, real estate and business services | 47.6 | 54.3 | 54.3 |
| Community, social and personal services | 964.5 | 952.1 | 951.8 |
| **Total** | **3,006.7** | **3,005.0** | **3,005.1** |

CUBA

## Health and Welfare

### KEY INDICATORS

| | |
|---|---|
| Total fertility rate (children per woman, 2003) | 1.6 |
| Under-5 mortality rate (per 1,000 live births, 2003) | 8 |
| HIV/AIDS (% of persons aged 15–49, 2003) | 0.10 |
| Physicians (per 1,000 head, 2002) | 5.91 |
| Hospital beds (per 1,000 head, 1996) | 5.13 |
| Health expenditure (2002): US $ per head (PPP) | 236 |
| Health expenditure (2002): % of GDP | 7.5 |
| Health expenditure (2002): public (% of total) | 86.5 |
| Access to water (% of persons, 2002) | 91 |
| Access to sanitation (% of persons, 2002) | 98 |
| Human Development Index (2002): ranking | 52 |
| Human Development Index (2002): value | 0.809 |

For sources and definitions, see explanatory note on p. vi.

## Agriculture

### PRINCIPAL CROPS
('000 metric tons)

| | 2001 | 2002 | 2003 |
|---|---|---|---|
| Rice (paddy) | 601.0 | 692.0 | 715.8 |
| Maize | 298.9 | 309.0 | 360.0 |
| Potatoes | 345.4 | 329.7 | 304.6 |
| Sweet potatoes | 360.3 | 392.4 | 503.4 |
| Cassava (Manioc) | 512.9 | 532.0 | 682.5 |
| Yautia (Cocoyam) | 102.2 | 122.5 | 139.6 |
| Sugar cane† | 3,210 | 3,470 | 2,290 |
| Dry beans | 99.1 | 107.3 | 127.0 |
| Groundnuts (in shell)* | 10 | 10 | 10 |
| Coconuts | 88.7 | 108.6 | 110.6 |
| Cabbages | 138.8 | 159.3 | 184.3 |
| Tomatoes | 452.7 | 496.0 | 643.7 |
| Pumpkins, squash and gourds | 351.8 | 416.6 | 481.8 |
| Cucumbers and gherkins | 248.0 | 300.2 | 347.2 |
| Chillies and peppers, green | 59.3 | 61.4 | 68.0 |
| Dry onions | 91.4 | 91.9 | 101.7 |
| Garlic | 40.3 | 31.7 | 36.6 |
| Other vegetables | 1,228.1 | 1,723.4 | 1,930.9 |
| Watermelons | 66.0 | 64.5 | 74.6 |
| Cantaloupes and other melons* | 25 | 24 | 25 |
| Bananas | 318.1 | 206.9 | 315.4 |
| Plantains | 649.9 | 523.0 | 797.2 |
| Oranges | 608.2 | 296.6 | 492.2 |
| Tangerines, mandarins, clementines and satsumas | 28.2 | 27.7 | 46.0 |
| Lemons and limes | 18.2 | 16.1 | 26.7 |
| Grapefruit and pomelos | 302.5 | 137.3 | 227.8 |
| Mangoes | 211.8 | 207.8 | 232.9 |
| Pineapples | 29.3 | 36.2 | 40.5 |
| Papayas | 135.1 | 107.2 | 120.1 |
| Other fruit* | 281.4 | 322.1 | 365.9 |
| Coffee (green)† | 15.7 | 14.7 | 15.0 |
| Fibre crops* | 27.5 | 27.5 | 27.5 |
| Tobacco (leaves) | 31.8 | 34.5 | 34.5* |

* FAO estimate(s).
† Unofficial figures.
Source: FAO.

### LIVESTOCK
('000 head, year ending September)

| | 2001 | 2002 | 2003 |
|---|---|---|---|
| Cattle | 4,038.4 | 3,972.3 | 4,025.4 |
| Horses* | 400 | 400 | 400 |
| Mules* | 23 | 23 | 23 |
| Pigs | 1,313.2 | 1,554.5 | 1,683.6 |
| Sheep† | 2,991.1 | 3,080.3 | 3,120.7 |
| Goats | 291 | 400† | 410† |
| Poultry | 26,168 | 16,700* | 23,210 |

* FAO estimate(s).
† Unofficial figure(s).
Source: FAO.

### LIVESTOCK PRODUCTS
('000 metric tons)

| | 2001 | 2002 | 2003 |
|---|---|---|---|
| Beef and veal | 75.0 | 65.8 | 56.2 |
| Pig meat | 75.9 | 89.8 | 94.0 |
| Chicken meat | 54.8 | 35.1 | 33.7 |
| Cows' milk | 620.7 | 589.7 | 607.5 |
| Cheese* | 14.7 | 14.2 | 14.2 |
| Butter* | 7.1 | 7.0 | 7.0 |
| Hen eggs | 67.1 | 78.2 | 78.5 |
| Honey | 6.6 | 5.6 | 7.2 |
| Hides and skins* | 13.2 | 12.4 | 10.4 |

* FAO estimates.
Source: FAO.

## Forestry

### ROUNDWOOD REMOVALS
('000 cubic metres, excl. bark)

| | 2001 | 2002 | 2003 |
|---|---|---|---|
| Sawlogs, veneer logs and logs for sleepers* | 400 | 400 | 400 |
| Other industrial wood | 408 | 408 | 408 |
| Fuel wood | 888 | 1,970 | 1,828 |
| **Total** | **1,696** | **2,778** | **2,636** |

* FAO estimates.
Source: FAO.

### SAWNWOOD PRODUCTION
('000 cubic metres, incl. railway sleepers)

| | 2001 | 2002* | 2003* |
|---|---|---|---|
| Coniferous (softwood) | 107 | 88 | 109 |
| Broadleaved (hardwood) | 83 | 59 | 72 |
| **Total** | **190** | **147** | **181** |

* FAO estimates.
Source: FAO.

## Fishing

('000 metric tons, live weight)

| | 2001 | 2002 | 2003 |
|---|---|---|---|
| Capture | 56.2 | 33.8 | 41.5 |
| Blue tilapia | 1.9 | 2.0 | 2.7 |
| Lane snapper | 1.9 | 1.8 | 1.1 |
| Caribbean spiny lobster | 6.8 | 8.0 | 5.3 |
| Aquaculture* | 25.6 | 27.0 | 26.9 |
| Silver carp | 13.0 | 12.8 | 13.1 |
| **Total catch*** | **81.7** | **60.9** | **68.4** |

* FAO estimates.
Note: Figures exclude sponges (metric tons): 51 in 2001; 54 in 2002; 57 in 2003.
Source: FAO.

## Mining

('000 metric tons, unless otherwise indicated)

| | 2001 | 2002 | 2003 |
|---|---|---|---|
| Crude petroleum | 2,773.4 | 3,533.4 | 3,691.0 |
| Natural gas (million cu metres) | 594.6 | 584.7 | 653.0 |
| Nickel and cobalt (metal content) | 76.5 | 75.1 | 71.7 |
| Crushed stone ('000 cu metres)* | 3,300.0 | n.a. | n.a. |

* Source: US Geological Survey.

# Industry

**SELECTED PRODUCTS**
('000 metric tons, unless otherwise indicated)

|  | 2001 | 2002 | 2003 |
|---|---|---|---|
| Crude steel | 269.6 | 264.1 | n.a. |
| Grey cement | 1,324.1 | 1,326.9 | 1,345.5 |
| Corrugated asbestos-cement tiles | 2,023.3 | 5,954.2 | 4,400.0 |
| Colour television sets ('000) | 79.4 | 105.0 | 350.0 |
| Fuel oil | 903.1 | 773.9 | 1,105.0 |
| Recapped tyres ('000) | 140.1 | 107.8 | n.a. |
| Woven textile fabrics (million sq metres) | 46.9 | 29.7 | 27.8 |
| Cigarettes ('000 million) | 11.8 | 12.6 | 14.3 |
| Cigars (million) | 339.2 | 327.2 | n.a. |
| Alcoholic beverages (excl. wines, '000 litres) | 664.4 | 714.5 | 777.7 |
| Beer ('000 hectolitres) | 2,197.1 | 2,331.0 | 2,409.2 |
| Soft drinks ('000 hectolitres) | 2,873.4 | 3,090.6 | n.a. |
| Electric energy (million kWh) | 15,301.3 | 15,699.8 | 15,810.5 |

# Finance

**CURRENCY AND EXCHANGE RATES**

**Monetary Units**
100 centavos = 1 Cuban peso.
1 Cuban peso = 1 convertible peso (official rate).

**Sterling, Dollar and Euro Equivalents** (31 May 2005)
£1 sterling = 1.683 convertible pesos;
US $1 = 0.926 convertible pesos;
€1 = 1.142 convertible pesos;
100 convertible pesos = £59.40 = $108.00 = €87.59.

Note: The foregoing information relates to official exchange rates. For the purposes of foreign trade, the peso was at par with the US dollar during each of the 10 years 1987–96. In addition, a 'convertible peso' was introduced in December 1994. Although officially at par with the Cuban peso, in March 2005 the 'unofficial' exchange rate prevailing in domestic exchange houses was adjusted to 24 pesos per convertible peso.

**STATE BUDGET**
(million pesos)

|  | 2001 | 2002 | 2003 |
|---|---|---|---|
| Total revenue | 15,033.5 | 16,196.7 | 17,250.0 |
| Road tax | 5,721.6 | 6,619.8 | 7,228.0 |
| Social security contributions | 1,247.8 | 1,336.3 | 1,405.0 |
| Total expenditure | 15,771.0 | 17,193.2 | 18,324.0 |
| Entrepreneurial activity | 2,622.2 | 3,285.8 | 3,473.0 |
| Education | 2,368.6 | 2,751.6 | 3,208.0 |
| Health care | 1,796.6 | 1,923.0 | 2,030.0 |
| Social security | 1,870.3 | 1,984.8 | 2,101.0 |
| Management | 565.2 | 611.1 | 602.9 |

Source: Ministry of Finance and Prices.

**INTERNATIONAL RESERVES**
(million pesos at 31 December)

|  | 1987 | 1988 |
|---|---|---|
| Gold and other precious metals | 17.5 | 19.5 |
| Cash and deposits in foreign banks (convertible currency) | 36.5 | 78.0 |
| **Sub-total** | 54.0 | 97.5 |
| Deposits in foreign banks (in transferable roubles) | 142.5 | 137.0 |
| **Total** | 196.5 | 234.5 |

**MONEY SUPPLY**
(million pesos)

|  | 2001 | 2002 | 2003 |
|---|---|---|---|
| Currency in circulation | 6,403.3 | 6,941.2 | 6,650.5 |
| Savings | 5,934.3 | 6,675.6 | 6,840.3 |
| **Total** | 12,337.6 | 13,616.8 | 13,490.8 |

# National Accounts

**Composition of Gross National Product**
(million pesos at current prices)

|  | 1998 | 1999 | 2000 |
|---|---|---|---|
| Compensation of employees | 10,328.3 | 11,146.9 | 11,965.8 |
| Operating surplus | 6,631.8 | 8,570.3 | 9,538.2 |
| Consumption of fixed capital | | | |
| **Gross domestic product (GDP) at factor cost** | 16,960.1 | 19,717.2 | 21,504.0 |
| Indirect taxes *less* subsidies | 6,940.7 | 5,786.4 | 6,130.7 |
| **GDP in purchasers' values** | 23,900.8 | 25,503.6 | 27,634.7 |
| *Less* Factor income paid abroad (net) | 599.2 | 514.1 | 693.0 |
| **Gross national product** | 23,301.6 | 24,989.5 | 26,941.7 |

Source: UN Economic Commission for Latin America and the Caribbean, *Statistical Yearbook*.

**Expenditure on the Gross Domestic Product**
(million pesos at current prices)

|  | 2000 | 2001 |
|---|---|---|
| Government final consumption expenditure | 6,686.9 | 7,731.0 |
| Private final consumption expenditure | 18,558.1 | 19,180.1 |
| Increase in stocks | 191.5 | 136.2 |
| Gross fixed capital formation | 3,630.2 | 3,225.8 |
| **Total domestic expenditure** | 29,066.7 | 30,273.1 |
| Exports of goods and services | 4,318.9 | 4,232.0 |
| *Less* Imports of goods and services | 5,179.6 | 5,103.3 |
| Statistical discrepancy | — | — |
| **GDP in purchasers' values** | 28,206.0 | 29,401.8 |
| **GDP at constant 1997 prices** | 26,482.1 | 27,273.7 |

Source: UN Economic Commission for Latin America and the Caribbean, *Statistical Yearbook*.

**Gross Domestic Product by Economic Activity**
(million pesos at constant 1997 prices)

|  | 2001 | 2002 | 2003 |
|---|---|---|---|
| Agriculture, hunting, forestry and fishing | 2,021.2 | 1,875.7 | 1,923.9 |
| Mining and quarrying | 434.9 | 463.6 | 478.0 |
| Manufacturing | 5,091.2 | 4,787.8 | 4,671.8 |
| Electricity, gas and water | 629.7 | 591.9 | 611.5 |
| Construction | 1,721.2 | 1,618.7 | 1,636.7 |
| Wholesale and retail trade, restaurants and hotels | 8,053.7 | 7,788.7 | 8,163.8 |
| Transport, storage and communications | 2,951.2 | 2,716.6 | 2,712.4 |
| Finance, insurance, real estate and business services | 1,961.6 | 2,101.2 | 2,186.0 |
| Community, social and personal services | 6,289.3 | 5,403.4 | 5,625.4 |
| **Sub total** | 29,154.0 | 27,347.6 | 28,009.5 |
| Import duties | 403.0 | 338.6 | 383.6 |
| **Total** | 29,557.0 | 27,686.2 | 28,393.1 |

# CUBA

*Statistical Survey*

## BALANCE OF PAYMENTS
(million pesos)

|  | 1996 | 1997 | 1998 |
|---|---|---|---|
| Exports of goods | 1,866.2 | 1,823.1 | 1,444.4 |
| Imports of goods | −3,656.5 | −4,087.6 | −4,229.7 |
| **Trade balance** | −1,790.3 | −2,264.5 | −2,785.3 |
| Services (net) | 1,372.4 | 1,519.0 | 2,168.2 |
| **Balance on goods and services** | −417.9 | −745.5 | −617.1 |
| Other income (net) | −492.6 | −482.9 | −599.2 |
| **Balance on goods, services and income** | −910.5 | −1,228.4 | −1,216.3 |
| Current transfers (net) | 743.7 | 791.7 | 820.0 |
| **Current balance** | −166.8 | −436.7 | −396.3 |
| Direct investment (net) | 82.1 | 442.0 | 206.6 |
| Other long-term capital (net) | 225.8 | 344.9 | 426.1 |
| Other capital (net) | −133.5 | −329.5 | −219.4 |
| **Overall balance** | 7.6 | 20.7 | 17.0 |

# External Trade

## PRINCIPAL COMMODITIES
('000 pesos)

| Imports c.i.f. | 1998 | 1999 | 2000 |
|---|---|---|---|
| Food and live animals | 704,200 | 722,396 | 671,801 |
| Cereals and cereal preparations | 348,260 | 313,755 | 285,541 |
| Wheat and meslin (unmilled) | 178,590 | 122,979 | 112,187 |
| Rice | 99,382 | 141,163 | 100,772 |
| Mineral fuels, lubricants, etc. | 687,030 | 730,763 | 1,158,071 |
| Petroleum and petroleum products | 664,844 | 710,881 | 1,137,418 |
| Crude petroleum oils, etc. | 65,358 | 88,087 | 279,901 |
| Gas oils | 151,819 | 198,056 | 301,551 |
| Chemicals and related products | 419,720 | 428,938 | 418,765 |
| Basic manufactures | 628,672 | 687,889 | 673,195 |
| Iron and steel | 117,098 | 150,889 | 111,940 |
| Manufactures of metal | 141,184 | 152,491 | 162,428 |
| Machinery and transport equipment | 1,130,414 | 1,144,441 | 1,202,330 |
| Power-generating machinery and equipment | 208,992 | 129,110 | 131,725 |
| General industrial machinery and equipment and machine parts | 249,889 | 269,441 | 259,300 |
| **Total** (incl. others) | 4,181,192 | 4,349,090 | 4,829,050 |

| Exports f.o.b. | 1998 | 1999 | 2000 |
|---|---|---|---|
| Food and live animals | 776,262 | 659,992 | 657,305 |
| Fish, crustaceans and molluscs and preparations thereof | 102,786 | 96,055 | 87,830 |
| Fresh, chilled or frozen fish | 102,661 | 95,267 | 87,176 |
| Fruit and vegetables | 35,201 | 74,980 | 89,618 |
| Sugar, sugar preparations and honey | 605,494 | 470,680 | 458,427 |
| Raw beet and cane sugars (solid) | 593,694 | 458,210 | 447,677 |
| Beverages and tobacco | 202,562 | 218,471 | 179,719 |
| Manufactured tobacco | 162,750 | 173,477 | 140,346 |
| Crude materials (inedible) except fuels | 358,781 | 436,808 | 623,089 |
| Metalliferous ores and metal scrap | 351,926 | 430,393 | 617,787 |
| Nickel ores, concentrates, etc. | 168,017 | 201,649 | 318,706 |
| Petroleum, petroleum products and related materials | 445 | 18,934 | 52,388 |
| Chemicals and related products | 45,566 | 39,662 | 41,235 |
| Basic manufactures | 97,202 | 94,268 | 99,230 |
| Iron and steel | 45,985 | 43,430 | 61,484 |
| **Total** (incl. others) | 1,512,197 | 1,495,783 | 1,675,868 |

## PRINCIPAL TRADING PARTNERS
('000 pesos)

| Imports c.i.f. | 1998 | 1999 | 2000 |
|---|---|---|---|
| Argentina | 108,827 | 129,938 | 106,424 |
| Brazil | 63,630 | 75,761 | 130,129 |
| Canada | 321,046 | 339,642 | 311,074 |
| China, People's Republic | 336,496 | 432,241 | 445,769 |
| Colombia | 30,312 | 36,978 | 85,156 |
| France | 318,381 | 276,337 | 289,876 |
| Germany | 78,140 | 75,967 | 77,841 |
| Italy | 253,203 | 264,063 | 296,602 |
| Mexico | 342,796 | 321,772 | 298,527 |
| Netherlands Antilles | 74,397 | 84,312 | 25,788 |
| Russia | 134,881 | 124,545 | 111,300 |
| Spain | 608,210 | 721,771 | 743,589 |
| United Kingdom | 54,022 | 41,193 | 56,864 |
| Venezuela | 385,570 | 451,451 | 898,393 |
| **Total** (incl. others) | 4,181,192 | 4,349,090 | 4,829,050 |

| Exports f.o.b. | 1998 | 1999 | 2000 |
|---|---|---|---|
| Belgium | 41,407 | 43,978 | 48,289 |
| Canada | 230,169 | 229,185 | 277,728 |
| China, People's Republic | 81,855 | 49,491 | 80,532 |
| Colombia | 20,610 | 5,853 | 18,941 |
| Dominican Republic | 31,272 | 44,580 | 19,823 |
| France | 43,110 | 50,517 | 42,411 |
| Germany | 67,859 | 72,596 | 120,882 |
| Iran | 14,442 | 19,620 | 14,179 |
| Italy | 27,979 | 39,283 | 34,198 |
| Japan | 19,478 | 49,506 | 29,374 |
| Korea, Democratic People's Republic | 24,609 | 35,157 | 43 |
| Mexico | 45,134 | 25,049 | 39,285 |
| Netherlands | 53,074 | 93,037 | 116,842 |
| Portugal | 14,562 | 26,424 | 17,844 |
| Russia | 355,254 | 302,776 | 324,577 |
| Spain | 141,069 | 159,409 | 149,656 |
| Sweden | 31,346 | 35,240 | 50,426 |
| Switzerland | 21,466 | 24,329 | 22,602 |
| Taiwan | 23,343 | 2,422 | 11,210 |
| Ukraine | 22,614 | 1,199 | 478 |
| United Kingdom | 24,687 | 30,206 | 38,101 |
| **Total** (incl. others) | 1,512,197 | 1,495,783 | 1,675,868 |

**Total imports** ('000 pesos): 4,793,300 in 2001; 4,129,400 in 2002.
**Total exports** ('000 pesos): 1,621,800 in 2001; 1,402,300 in 2002.

# Transport

## RAILWAYS

|  | 2000 | 2001 | 2002 |
|---|---|---|---|
| Passenger-kilometres (million) | 1,737 | 1,758 | 1,682 |
| Freight ton-kilometres (million) | 808 | 842 | 814 |

Source: UN, *Statistical Yearbook*.

## ROAD TRAFFIC
(motor vehicles in use at 31 December)

|  | 1996 | 1997 |
|---|---|---|
| Passenger cars | 216,575 | 172,574 |
| Buses and coaches | 28,089 | 28,861 |
| Lorries and vans | 246,105 | 156,634 |

Source: International Road Federation, *World Road Statistics*.

## SHIPPING

**Merchant Fleet**
(registered at 31 December)

|  | 2002 | 2003 | 2004 |
|---|---|---|---|
| Number of vessels | 91 | 90 | 96 |
| Total displacement ('000 grt) | 103 | 90 | 126 |

Source: Lloyd's Register-Fairplay, *World Fleet Statistics*.

## International Sea-borne Freight Traffic
('000 metric tons)

|  | 1988 | 1989 | 1990 |
|---|---|---|---|
| Goods loaded | 8,600 | 8,517 | 8,092 |
| Goods unloaded | 15,500 | 15,595 | 15,440 |

Source: UN, *Monthly Bulletin of Statistics*.

### CIVIL AVIATION
(traffic on scheduled services)

|  | 1999 | 2000 | 2001 |
|---|---|---|---|
| Kilometres flown (million) | 26 | 20 | 19 |
| Passengers carried ('000) | 1,259 | 1,007 | 882 |
| Passenger-kilometres (million) | 3,712 | 2,964 | 3,171 |
| Total ton-kilometres (million) | 421 | 335 | 361 |

Source: UN, *Statistical Yearbook*.

## Tourism

### ARRIVALS BY COUNTRY OF RESIDENCE*

|  | 2001 | 2002 | 2003 |
|---|---|---|---|
| Canada | 350,426 | 348,468 | 452,438 |
| France | 138,765 | 129,907 | 144,548 |
| Germany | 171,851 | 152,662 | 157,721 |
| Italy | 159,423 | 147,750 | 177,627 |
| Mexico | 98,495 | 87,589 | 88,787 |
| Spain | 140,125 | 138,609 | 127,666 |
| United Kingdom | 94,794 | 103,741 | 120,866 |
| USA | 78,789 | 77,646 | 84,529 |
| **Total** (incl. others) | 1,774,541 | 1,686,162 | 1,905,682 |

* Figures include same-day visitors (excursionists).

**Tourism receipts** (US $ million, incl. passenger transport): 1,692 in 2001; 1,633 in 2002; 1,846 in 2003.

Source: World Tourism Organization.

## Communications Media

|  | 2001 | 2002 | 2003 |
|---|---|---|---|
| Telephones ('000 main lines in use) | 572.6 | 665.6 | 724.3 |
| Mobile cellular telephones ('000 subscribers) | 8.1 | 17.9 | 35.4 |
| Personal computers ('000 in use) | 220 | 250 | 270 |
| Internet users ('000) | 120.0 | 160.0 | 98.0 |
| Book production (titles) | 1,004 | n.a. | n.a. |

**Radio receivers** ('000 in use): 3,900 in 1997.

**Television receivers** ('000 in use): 2,800 in 2000.

**Facsimile machines** (number in use): 392 in 1992.

**Daily newspapers:** 2 in 2000 (average estimated circulation 600,000 copies).

Sources: UNESCO, *Statistical Yearbook*; UN, *Statistical Yearbook*; International Telecommunication Union.

## Education

(2002/03)

|  | Institutions | Teachers | Students |
|---|---|---|---|
| Pre-primary | 1,115 | 18,600 | 147,600 |
| Primary | 9,397 | 93,000 | 873,700 |
| Secondary* | 2,032 | 85,600 | 992,000 |
| Universities | 64 | 23,700 | 192,000 |

* According to results of 1981 population and housing census.

**2003/04:** Primary: 79,600 teachers; 856,500 students. Universities: 273,100 students.

**Adult literacy rate** (UNESCO estimates): 96.9% (males 97.0%; females 96.8%) in 2002 (Source: UN Development Programme, *Human Development Report*).

# Directory

## The Constitution

Following the assumption of power by the Castro regime, on 1 January 1959, the Constitution was suspended and a Fundamental Law of the Republic was instituted, with effect from 7 February 1959. In February 1976 Cuba's first socialist Constitution came into force after being submitted to the first Congress of the Communist Party of Cuba, in December 1975, and to popular referendum, in February 1976; it was amended in July 1992. The main provisions of the Constitution, as amended, are summarized below.

Note: On 27 July 2002 the Constitution was further amended to enshrine the socialist system as irrevocable and to ratify that economic, diplomatic and political relations with another state cannot be negotiated in the face of aggression, threat or pressure from a foreign power. A clause was also introduced making it impossible to remove these amendments from the Constitution.

### POLITICAL, SOCIAL AND ECONOMIC PRINCIPLES

The Republic of Cuba is a socialist, independent, and sovereign state, organized with all and for the sake of all as a unitary and democratic republic for the enjoyment of political freedom, social justice, collective and individual well-being and human solidarity. Sovereignty rests with the people, from whom originates the power of the State. The Communist Party of Cuba is the leading force of society and the State. The State recognizes, respects and guarantees freedom of religion. Religious institutions are separate from the State. The socialist State carries out the will of the working people and guarantees work, medical care, education, food, clothing and housing. The Republic of Cuba bases its relations with other socialist countries on socialist internationalism, friendship, co-operation and mutual assistance. It reaffirms its willingness to integrate with and co-operate with the countries of Latin America and the Caribbean.

The State organizes and directs the economic life of the nation in accordance with a central social and economic development plan.

The State directs and controls foreign trade. The State recognizes the right of small farmers to own their lands and other means of production and to sell that land. The State guarantees the right of citizens to ownership of personal property in the form of earnings, savings, place of residence and other possessions and objects which serve to satisfy their material and cultural needs. The State also guarantees the right of inheritance.

Cuban citizenship is acquired by birth or through naturalization.

The State protects the family, motherhood and matrimony.

The State directs and encourages all aspects of education, culture and science.

All citizens have equal rights and are subject to equal duties.

The State guarantees the right to medical care, education, freedom of speech and press, assembly, demonstration, association and privacy. In the socialist society work is the right and duty, and a source of pride for every citizen.

### GOVERNMENT

#### National Assembly of People's Power

The National Assembly of People's Power (Asamblea Nacional del Poder Popular) is the supreme organ of the State and is the only organ with constituent and legislative authority. It is composed of deputies, over the age of 18, elected by free, direct and secret ballot, for a period of five years. All Cuban citizens aged 16 years or more, except those who are mentally incapacitated or who have committed a crime, are eligible to vote. The National Assembly of People's Power holds two ordinary sessions a year and a special session when requested by one-third of the deputies or by the Council of State. More than one-half of the total number of deputies must be present for a session to be held.

All decisions made by the Assembly, except those relating to constitutional reforms, are adopted by a simple majority of votes. The deputies may be recalled by their electors at any time.

The National Assembly of People's Power has the following functions:

- to reform the Constitution;
- to approve, modify and annul laws;
- to supervise all organs of the State and government;
- to decide on the constitutionality of laws and decrees;
- to revoke decree-laws issued by the Council of State and the Council of Ministers;
- to discuss and approve economic and social development plans, the state budget, monetary and credit systems;
- to approve the general outlines of foreign and domestic policy, to ratify and annul international treaties, to declare war and approve peace treaties;
- to approve the administrative division of the country;
- to elect the President, First Vice-President, the Vice-Presidents and other members of the Council of State;
- to elect the President, Vice-President and Secretary of the National Assembly;
- to appoint the members of the Council of Ministers on the proposal of the President of the Council of State;
- to elect the President, Vice-President and other judges of the People's Supreme Court;
- to elect the Attorney-General and the Deputy Attorney-Generals;
- to grant amnesty;
- to call referendums.

The President of the National Assembly presides over sessions of the Assembly, calls ordinary sessions, proposes the draft agenda, signs the Official Gazette, organizes the work of the commissions appointed by the Assembly and attends the meetings of the Council of State.

### Council of State

The Council of State is elected from the members of the National Assembly and represents that Assembly in the period between sessions. It comprises a President, one First Vice-President, five Vice-Presidents, one Secretary and 23 other members. Its mandate ends when a new Assembly meets. All decisions are adopted by a simple majority of votes. It is accountable for its actions to the National Assembly.

The Council of State has the following functions:

- to call special sessions of the National Assembly;
- to set the date for the election of a new Assembly;
- to issue decree-laws in the period between the sessions of the National Assembly;
- to decree mobilization in the event of war and to approve peace treaties when the Assembly is in recess;
- to issue instructions to the courts and the Office of the Attorney-General of the Republic;
- to appoint and remove ambassadors of Cuba abroad on the proposal of its President, to grant or refuse recognition to diplomatic representatives of other countries to Cuba;
- to suspend those provisions of the Council of Ministers that are not in accordance with the Constitution;
- to revoke the resolutions of the Executive Committee of the local organs of People's Power which are contrary to the Constitution or laws and decrees formulated by other higher organs.

For all purposes the Council of State is the highest representative of the Cuban state.

### Head of State

The President of the Council of State is the Head of State and the Head of Government and has the following powers:

- to represent the State and Government and conduct general policy;
- to convene and preside over the sessions of the Council of State and the Council of Ministers;
- to supervise the ministries and other administrative bodies;
- to propose the members of the Council of Ministers to the National Assembly of People's Power;
- to receive the credentials of the heads of foreign diplomatic missions;
- to sign the decree-laws and other resolutions of the Council of State;
- to exercise the Supreme Command of all armed institutions and determine their general organization;
- to preside over the National Defence Council;
- to declare a state of emergency in the cases outlined in the Constitution.

In the case of absence, illness or death of the President of the Council of State, the First Vice-President assumes the President's duties.

### The Council of Ministers

The Council of Ministers is the highest-ranking executive and administrative organ. It is composed of the Head of State and Government, as its President, the First Vice-President, the Vice-Presidents, the Ministers, the Secretary and other members determined by law. Its Executive Committee is composed of the President, the First Vice-President, the Vice-Presidents and other members of the Council of Ministers determined by the President.

The Council of Ministers has the following powers:

- to conduct political, economic, cultural, scientific, social and defence policy as outlined by the National Assembly;
- to approve international treaties;
- to propose projects for the general development plan and, if they are approved by the National Assembly, to supervise their implementation;
- to conduct foreign policy and trade;
- to draw up bills and submit them to the National Assembly;
- to draw up the draft state budget;
- to conduct general administration, implement laws, issue decrees and supervise defence and national security.

The Council of Ministers is accountable to the National Assembly of People's Power.

### LOCAL GOVERNMENT

The country is divided into 14 provinces and 169 municipalities. The provinces are: Pinar del Río, Habana, Ciudad de la Habana, Matanzas, Villa Clara, Cienfuegos, Sancti Spíritus, Ciego de Avila, Camagüey, Las Tunas, Holguín, Granma, Santiago de Cuba and Guantánamo.

Voting for delegates to the municipal assemblies is direct, secret and voluntary. All citizens over 16 years of age are eligible to vote. The number of delegates to each assembly is proportionate to the number of people living in that area. A delegate must obtain more than one-half of the total number of votes cast in the constituency in order to be elected. The Municipal and Provincial Assemblies of People's Power are elected by free, direct and secret ballot. Nominations for Municipal and Provincial Executive Committees of People's Power are submitted to the relevant assembly by a commission presided over by a representative of the Communist Party's leading organ and consisting of representatives of youth, workers', farmers', revolutionary and women's organizations. The President and Secretary of each of the regional and the provincial assemblies are the only full-time members, the other delegates carrying out their functions in addition to their normal employment.

The regular and extraordinary sessions of the local Assemblies of People's Power are public. More than one-half of the total number of members must be present in order for agreements made to be valid. Agreements are adopted by simple majority.

### JUDICIARY

Judicial power is exercised by the People's Supreme Court and all other competent tribunals and courts. The People's Supreme Court is the supreme judicial authority and is accountable only to the National Assembly of People's Power. It can propose laws and issue regulations through its Council of Government. Judges are independent but the courts must inform the electorate of their activities at least once a year. Every accused person has the right to a defence and can be tried only by a tribunal.

The Office of the Attorney-General is subordinate only to the National Assembly and the Council of State and is responsible for ensuring that the law is properly obeyed.

The Constitution may be totally or partially modified only by a two-thirds' majority vote in the National Assembly of People's Power. If the modification is total, or if it concerns the composition and powers of the National Assembly of People's Power or the Council of State, or the rights and duties contained in the Constitution, it also requires a positive vote by referendum.

## The Government

**Head of State:** Dr FIDEL CASTRO RUZ (took office 2 December 1976; re-elected December 1981, December 1986, March 1993, February 1998 and March 2003).

### COUNCIL OF STATE
(July 2005)

**President:** Dr FIDEL CASTRO RUZ.
**First Vice-President:** Gen. RAÚL CASTRO RUZ.

CUBA                                                                                             *Directory*

**Vice-Presidents:** Juan Almeida Bosque, Gen. Abelardo Colomé Ibarra, Carlos Lage Dávila, Juan Esteban Lazo Hernández, José Ramón Machado Ventura.

**Secretary:** Dr José M. Miyar Barruecos.

**Members:** José Ramón Balaguer Cabrera, Vilma Espín Guillois de Castro, Dr Armando Hart Dávalos, Orlando Lugo Fonte, Nidia Diana Martínez Piti, María T. Ferrer Madrazo, Marta Hernández Romero, Gen. Julio Casas Regueiro, Pedro Sáez Montejo, Gen. Roberto Ignacio González Planas, Francisco Soberón Valdés, Pedro Miret Prieto, Ramiro Valdés Menéndez, Julio Cristhian Jiménez Molina, Roberto Fernández Retamar, Felipe Ramón Pérez Roque, Marcos J. Portal León, Luis S. Herrera Martínez, Iris Betancourt Téllez, Pedro Ross Leal, Otto Rivero Torres, Carlos Manuel Valenciaga Díaz; 1 vacancy.

### COUNCIL OF MINISTERS

**President:** Dr Fidel Castro Ruz.
**First Vice-President:** Gen. Raúl Castro Ruz.
**Secretary:** Carlos Lage Dávila.
**Vice-Presidents:** Osmany Cienfuegos Gorriarán, Pedro Miret Prieto, José Luis Rodríguez García, José Ramón Fernández Alvarez.
**Minister of Agriculture:** Alfredo Jordán Morales.
**Minister of Foreign Trade:** Raúl de la Nuez Ramírez.
**Minister of Domestic Trade:** Barbara Castillo Cuesta.
**Minister of Computer Science and Communications:** Gen. Roberto Ignacio González Planas.
**Minister of Construction:** Fidel Fernando Figueroa de la Paz.
**Minister of Culture:** Abel Enrique Prieto Jiménez.
**Minister of Economy and Planning:** José Luis Rodríguez García.
**Minister of Education:** Luis Ignacio Gómez Gutiérrez.
**Minister of Higher Education:** Fernando Vecino Alegret.
**Minister of the Revolutionary Armed Forces:** Gen. Raúl Castro Ruz.
**Minister of Finance and Prices:** Georgina Barreiro Fajardo.
**Minister of the Food Industry:** Alejandro Roca Iglesias.
**Minister of Foreign Investment and Economic Co-operation:** Martha Lomas Morales.
**Minister of the Sugar Industry:** Gen. Ulises Rosales del Toro.
**Minister of Light Industry:** Estela Martha Dominguez Ariosa.
**Minister of the Fishing Industry:** Alfredo López Valdés.
**Minister of the Iron, Steel and Engineering Industries:** Fernando Acosta Santana.
**Minister of Basic Industry:** Yadirá García Vera.
**Minister of the Interior:** Gen. Abelardo Colomé Ibarra.
**Minister of Justice:** Roberto T. Díaz Sotolongo.
**Minister of Foreign Relations:** Felipe Ramón Pérez Roque.
**Minister of Labour and Social Security:** Alfredo Morales Cartaya.
**Minister of Public Health:** José Ramón Balaguer Cabrera.
**Minister of Science, Technology and the Environment:** Fernando Mario González Bermúdez (acting).
**Minister of Transportation:** Carlos Manuel Pazo Torrado.
**Minister of Tourism:** Manuel Marrero Cruz.
**Minister of Auditing and Control:** Lina Olinda Pedraza Rodríguez.
**Minister, President of the Banco Central de Cuba:** Francisco Soberón Valdés.
**Ministers without Portfolio:** Ricardo Cabrisas Ruiz, Wilfredo López Rodríguez.

### MINISTRIES

**Ministry of Agriculture:** Edif. MINAG, Avda Boyeros y Conill, Havana 10600; tel. (7) 884-5427; fax (7) 881-2837; e-mail gestion@agrinfor.cu.

**Ministry of Auditing and Control:** Havana; replaced the National Auditing Office in 2001.

**Ministry of Basic Industry:** Avda Salvador Allende 666, Havana; tel. (7) 70-7711; e-mail webmaster@one.gov.cu.

**Ministry of Construction:** Avda Carlos M. de Céspedes y Calle 35, Plaza de la Revolución, Havana; tel. (7) 81-8385; fax (7) 55-5303; e-mail despacho@micons.cu; internet www.micons.cu.

**Ministry of Computer Science and Communications:** Avdas Independencia y 19 de Mayo, Plaza de la Revolución, Havana; tel. (7) 66-8000; e-mail relaciones@mic.cu; internet www.mic.gov.cu.

**Ministry of Culture:** Calle 2, No 258, entre 11 y 13, Plaza de la Revolución, Vedado, CP 10400, Havana; tel. (7) 55-2260; fax (7) 66-2053; e-mail atencion@min.cult.cu; internet www.min.cult.cu.

**Ministry of Domestic Trade:** Calle Habana 258, Havana; tel. (7) 62-5790.

**Ministry of Economy and Planning:** 20 de Mayo, entre Territorial y Ayestarán, Plaza de la Revolución, Havana; tel. (7) 881-8789; fax (7) 33-3387; e-mail mep@ceniai.inf.cu.

**Ministry of Education:** Obispo 160, Havana; tel. (7) 61-4888; internet www.rimed.cu.

**Ministry of Finance and Prices:** Calle Obispo 211, esq. Cuba, Havana; tel. (7) 867-1920; fax (7) 33-8050; e-mail bhcifip@mfp.gov.cu; internet www.mfp.cu.

**Ministry of the Fishing Industry:** Avda 5, entre 246 y 248, No 24606, Barlovento, Playa, Havana; tel. (7) 209-7930; fax (7) 204-9168; e-mail info@fishnavy.inf.cu; internet www1.cubamar.cu.

**Ministry of the Food Industry:** Avda 41, No 4455, entre 48 y 50, Playa, Havana; tel. (7) 203-6801; fax (7) 204-0517; e-mail minal@minal.get.cma.net; internet www.minal.cubaindustria.cu.

**Ministry of Foreign Investment and Economic Co-operation:** Calle 1ra y 18, No 1803, Miramar, Havana; tel. (7) 203-7035; fax (7) 204-3496; e-mail ministro@minvec.cu; internet www.minvec.cu.

**Ministry of Foreign Relations:** Calzada 360, esq. G, Vedado, Havana; tel. (7) 55-3537; fax (7) 33-3460; e-mail cubaminrex@minrex.gov.cu; internet www.cubaminrex.cu.

**Ministry of Foreign Trade:** Infanta 16, esq. 23, Vedado, Havana; tel. (7) 55-0428; fax (7) 55-0376; e-mail opinion@mincex.cu; internet www.mincex.cu.

**Ministry of Higher Education:** Calle 23, No 565, esq. a F, Vedado, Havana; tel. (7) 55-2314; fax (7) 33-3295; e-mail develop@reduniv.edu.cu; internet www.mes.edu.cu.

**Ministry of the Interior:** Plaza de la Revolución, Havana; tel. (7) 30-1566; fax (7) 33-5261.

**Ministry of the Iron, Steel and Engineering Industries:** Avda Rancho Boyeros y Calle 100, Havana; tel. (7) 45-3911; e-mail sime@sime.co.cu; internet www.disaic.cu/sime.

**Ministry of Justice:** Calle O, No 216, entre 23 y Humboldt, Vedado, Apdo 10400, Havana 4; tel. (7) 32-6319; e-mail minjus@minjus.cu; internet www.minjus.cu.

**Ministry of Labour and Social Security:** Calle 23, esq. Calles O y P, Vedado, Havana; tel. (7) 55-0071; fax (7) 33-5816; e-mail mtssmin@ceniai.inf.cu.

**Ministry of Light Industry:** Empedrado 302, Havana; tel. (7) 67-0387; fax (7) 67-0329; e-mail ministro@minil.org.cu; internet www.ligera.cu.

**Ministry of Public Health:** Calle 23, No 301, Vedado, Havana; tel. (7) 32-2561; internet www.dne.sld.cu/minsap.

**Ministry of the Revolutionary Armed Forces:** Plaza de la Revolución, Havana; internet www.cubagob.cu/otras_info/minfar/far/minfar.htm.

**Ministry of Science, Technology and the Environment:** Industria y San José, Habana Vieja, Havana; tel. (7) 867-0618; fax (7) 33-8654; e-mail citma@ceniai.inf.cu; internet www.cuba.cu/ciencia/citma/.

**Ministry of Sugar:** Calle 23, No 171, Vedado, Havana; tel. (7) 55-3194; fax (7) 55-3260; e-mail relint@ocentral.minaz.cu.

**Ministry of Tourism:** Calle 19, No 710, entre Paseo y A, Vedado, Havana; tel. (7) 33-0545; fax (7) 33-4086; e-mail promo@mintur.mit.cma.net; internet www.cubatravel.cu.

**Ministry of Transportation:** Avda Boyeros, esq. Tulipán, Plaza de la Revolución, Havana; tel. (7) 55-5030; e-mail lupe.cic@mitrans.transnet.cu; internet www.cubagob.cu/des_eco/mitrans.

# Legislature

### ASAMBLEA NACIONAL DEL PODER POPULAR

The National Assembly of People's Power was constituted on 2 December 1976. In July 1992 the National Assembly adopted a constitutional amendment providing for legislative elections by direct vote. Only candidates nominated by the Partido Comunista de Cuba (PCC) were permitted to contest the elections. At elections to the National Assembly conducted on 19 January 2003 all 609 candidates were elected. Of the 8.2m. registered voters, 97.61%

CUBA — Directory

participated in the elections. Only 3.86% of votes cast were blank or spoilt.

**President:** RICARDO ALARCÓN DE QUESADA.
**Vice-President:** JAIME CROMBET HERNÁNDEZ MAURELL.
**Secretary:** Dr ERNESTO SUÁREZ MÉNDEZ.

## Political Organizations

**Partido Comunista de Cuba (PCC)** (Communist Party of Cuba): Havana; e-mail root@epol.cipcc.inf.cu; internet www.pcc.cu; f. 1961 as the Organizaciones Revolucionarias Integradas (ORI) from a fusion of the Partido Socialista Popular (Communist), Fidel Castro's Movimiento 26 de Julio and the Directorio Revolucionario 13 de Marzo; became the Partido Unido de la Revolución Socialista Cubana (PURSC) in 1962; renamed as the Partido Comunista de Cuba in 1965; 150-member Central Committee, Political Bureau (23 mems in 2005), and five Commissions; 706,132 mems (1994).

### Political Bureau

Dr FIDEL CASTRO RUZ Gen. RAÚL CASTRO RUZ; JUAN ALMEIDA BOSQUE; JOSÉ RAMÓN MACHADO VENTURA; JUAN ESTEBAN LAZO HERNÁNDEZ; Gen. ABELARDO COLOMÉ IBARRA; PEDRO ROSS LEAL; CARLOS LAGE DÁVILA; ALFREDO JORDÁN MORALES; Gen. ULISES ROSALES DEL TORO; CONCEPCIÓN CAMPA HUERGO; YADIRA GARCÍA VERA; ABEL ENRIQUE PRIETO JIMÉNEZ; Gen. JULIO CASAS REGUEIRO; Gen. LEOPOLDO CINTRA FRÍAS; RICARDO ALARCÓN DE QUESADA; JOSÉ RAMÓN BALAGUER CABRERA; MISAEL ENAMORADO DAGER; Gen. RAMÓN ESPINOSA MARTÍN; MARCOS JAVIER PORTAL LEÓN; JUAN CARLOS ROBINSON AGRAMONTE; PEDRO SÁEZ MONTEJO; JORGE LUIS SIERRA LÓPEZ

There are a number of dissident groups operating in Cuba. These include:

**Arco Progresista:** f. 2003; alliance of social-democratic groups in and outside Cuba.

**Concertación Democrática Cubana (CDC):** f. 1991; alliance of 11 dissident organizations campaigning for political pluralism and economic reform; Leader ELIZARDO SÁNCHEZ SANTA CRUZ.

**Cuban Democratic Platform:** f. 1990; alliance comprising three dissident organizations:

**Coordinadora Social Demócrata:** Havana; e-mail morm21@yahoo.com; internet www.cosodecu.org; Pres. BYRON MIGUEL.

**Partido Demócrata Cristiano de Cuba (PDC):** POB 558987, Miami, FL 33155, USA; tel. (305) 264-9411; fax (954) 489-1572; e-mail miyares@pdc-cuba.org; internet www.pdc-cuba.org; f. 1959 as Movimiento Demócrata Cristiano; name changed as above in 1991; Pres. MARCELINO MIYARES; Sec.-Gen. JOSÉ VAZQUEZ.

**Unión Liberal Cubana:** Paseo de la Retama 97, 29600 Marbella, Spain ; tel. (91) 4340201; fax (91) 5011342; e-mail cubaliberal@mercuryin.es; internet www.cubaliberal.org; mem. of Liberal International; Founder and Chair. CARLOS ALBERTO MONTANER.

**Partido Cubano Ortodoxo:** f. 1999; Leader NELSON AGUIAR.

**Partido Liberal Democrático de Cuba:** Pres. JULIA CECILIA DELGADO GLEZ (acting).

**Partido pro-Derechos Humanos:** f. 1988 to defend human rights in Cuba; Pres. HIRAM ABI COBAS; Sec.-Gen. TANIA DÍAZ.

**Partido Socialdemócrata de Cuba (PSC):** Calle 36, No 105, Nuevo Vedado, Havana 10600; tel. (7) 881-8203; e-mail infopsc@pscuba.org; internet pscuba.org; Pres. VLADIMIRO ROCA ANTÚNEZ; Sec.-Gen. ANTONIO SANTIAGO RUIZ.

**Partido Social Revolucionario Democrático de Cuba:** POB 351081, Miami, FL 33135, USA; tel. and fax (305) 649-2886; e-mail psrdc@psrdc.org; internet www.psrdc.org; f. 1992; Pres. JORGE VALLS.

**Partido Solidaridad Democrática (PSD):** POB 310063, Miami, FL 33131, USA; tel. (305) 408-2659; e-mail gladyperez@aol.com; internet www.ccsi.com/~ams/psd/psd.htm; Pres. FERNANDO SÁNCHEZ LÓPEZ.

**Solidaridad Cubana:** Leader FERNANDO SÁNCHEZ LÓPEZ.

## Diplomatic Representation

### EMBASSIES IN CUBA

**Algeria:** Avda 5, No 2802, esq. 28, Miramar, Havana; tel. (7) 204-2835; fax (7) 204-2702; Ambassador AHCENE KERMAN.

**Angola:** Avda 5, No 1012, entre 10 y 12, Miramar, Havana; tel. (7) 204-2474; fax (7) 204-0487; e-mail embangol@ceniai.inf.cu; Ambassador ANTONIO J. CONDESSE D. CARVAHLO.

**Argentina:** Calle 36, No 511, entre 5 y 7, Miramar, Havana; tel. (7) 204-2565; fax (7) 204-2140; e-mail ecuba@enet.cu; Ambassador DARÍO ALESSANDRO.

**Austria:** Calle 4, No 101, entre 1 y 3, Miramar, Havana; tel. (7) 204-2394; fax (7) 204-1235; e-mail havanna-ob@bmaa.gv.at; Ambassador Dr HELGA KONRAD.

**Belarus:** Avda 5, No 3802, entre 38 y 40, Miramar, Havana; tel. (7) 204-7330; fax (7) 204-7332; e-mail cuba@belembassy.org; Ambassador POLUYAN IGOR IVANOVICH.

**Belgium:** Calle 8, No 309, entre 3 y 5, Miramar, Havana; tel. (7) 204-2410; fax (7) 204-1318; e-mail havana@diplobel.org; internet www.diplomatie.be/havana; Ambassador PIERRE-EMMANUEL DE BAUW.

**Belize:** Avda 5, No 3608, entre 36 y 36A, Miramar, Havana; tel. (7) 204-3504; fax (7) 204-3506; e-mail belize.embassy@ip.etecsa.cu; Ambassador ASSAH SHOMAN.

**Benin:** Calle 20, No 119, entre 1 y 3, Miramar, Havana; tel. (7) 204-2179; fax (7) 204-2334; e-mail ambencub@ceniai.inf.cu; Ambassador GRÉGOIRE LAITIAN HOUDÉ.

**Bolivia:** Calle 26, No 113, entre 1 y 3, Miramar, Havana; tel. (7) 204-2426; fax (7) 204-2127; e-mail emboliviahabana@cubacel.net; Ambassador ELSA GUEVARA AGUIRRE.

**Brazil:** Calle Lamparilla, No 2, 4°K, Habana Vieja, Havana; tel. (7) 66-9052; fax (7) 66-2912; e-mail brasil@ceniai.inf.cu; Ambassador TILDEN JOSÉ SANTIAGO.

**Bulgaria:** Calle B, No 252, entre 11 y 13, Vedado, Havana; tel. (7) 33-3125; fax (7) 33-3297; e-mail embulhav@ceniai.inf.cu; Chargé d'affaires a.i. LUBOMIR ILIEV IVANOV.

**Burkina Faso:** Calle 7, No 8401, Miramar, Havana; tel. (7) 204-2895; fax (7) 204-1942; e-mail ambfaso@ceniai.inf.cu; Ambassador SALIF NÉBIÉ.

**Cambodia:** Avda 5, No 7001, entre 70 y 72, Miramar, Havana; tel. (7) 204-1496; fax (7) 204-6400; e-mail cambodia@ceniai.inf.cu; Ambassador CHIM PRORNG.

**Canada:** Calle 30, No 518, entre 5 y 7, Miramar, Havana; tel. (7) 204-2516; fax (7) 204-2044; e-mail havan@dfait-maeci.gc.ca; internet www.dfait-maeci.gc.ca/cuba; Ambassador ALEXANDRA BUGAILISKIS.

**Cape Verde:** Calle 20, No 2001, esq. 7, Miramar, Havana; tel. (7) 204-2979; fax (7) 204-1072; e-mail ecvc@ceniai.inf.cu; Ambassador CRISPINA ALMEIDA GOMES.

**Chile:** Avda 33, No 1423, entre 14 y 18, Miramar, Havana; tel. (7) 204-1222; fax (7) 204-1694; e-mail echilecu@ceniai.inf.cu; internet www.conchile-lahabana.cu; Ambassador MORENO LAVAL CELSO.

**China, People's Republic:** Calle 13, No 551, entre C y D, Vedado, Havana; tel. (7) 33-3005; fax (7) 33-3092; e-mail chinaemb_cu@mfa.gov.cn; Ambassador LI LIANTU.

**Colombia:** Calle 14, No 515, entre 5 y 7, Miramar, Havana; tel. (7) 204-1246; fax (7) 204-1249; e-mail elahabana@minrelext.gov.co; Ambassador LUIS HERNANDEZ OJEDA.

**Congo, Republic:** Avda 5, No 1003, Miramar, Havana; tel. and fax (7) 204-9055; Ambassador PASCAL ONGUEMBY.

**Czech Republic:** Avda Kohly, No 259, entre 41 y 43, Nuevo Vedado, Havana; tel. (7) 883-3201; fax (7) 883-3596; e-mail havana@embassy.mzv.cz; internet www.czechembassy.org/wwwo/?zu=havanna; Chargé d'affaires a.i. PETR STIEGLER.

**Dominican Republic:** Avda 5, No 9202, entre 92 y 94, Miramar, Havana; tel. (7) 204-8429; fax (7) 204-8431; Ambassador RAFAEL A. BÁEZ PÉREZ.

**Ecuador:** Avda 5A, No 4407, entre 44 y 46, Miramar, Havana; tel. (7) 204-2034; fax (7) 204-2868; e-mail embecuad@ceniai.inf.cu; Ambassador ALBA COELLO MATUTE.

**Egypt:** Avda 5, No 1801, esq. 18, Miramar, Havana; tel. (7) 204-2441; fax (7) 204-0905; e-mail emegipto@enet.cu; Ambassador HAMDI MOHAMED ABDEL ELMONEIM EL SHAZLY.

**France:** Calle 14, No 312, entre 3 y 5, Miramar, Havana; tel. (7) 201-3131; fax (7) 204-1439; e-mail la-havane.amba@diplomatie.gouv.fr; internet www.ambafrance-cu.org; Ambassador MARIE-FRANCE PAGNIER.

**Gambia:** Calle 24, No 307, entre 3 y 5, Miramar, Havana; tel. (7) 204-5315; fax (7) 204-5316; Ambassador BALA GARBA-JAHUMPA.

**Germany:** Calle 13, No 652, esq. B, Vedado, Havana; tel. (7) 833-2539; fax (7) 833-1586; e-mail alemania@enet.cu; internet www.deutschebotschaft-havanna.cu; Ambassador (vacant).

**Ghana:** Avda 5, No 1808, esq. 20, Miramar, Havana; tel. (7) 204-2153; fax (7) 204-2317; e-mail eghana@ceniai.inf.cu; internet www.ghanaembassy.cu; Ambassador Dr ISAAC ANTWI-OMANE.

**Greece:** Avda 5, No 7802, esq. 78, Miramar, Havana; tel. (7) 204-2995; fax (7) 204-1784; e-mail gremb@enet.cu; Ambassador PETROS AVIERINOS.

**Guatemala:** Calle 16, No 505, entre 3 y 5, Miramar, Havana; tel. (7) 204-3417; fax (7) 204-3200; e-mail embagucu@ceniai.inf.cu; Ambassador HUGO RENÉ GUZMÁN MALDONADO.

**Guinea:** Calle 20, No 504, entre 5 y 7, Miramar, Havana; tel. (7) 204-2003; fax (7) 204-2380; Chargé d'affaires a.i. ALMAMY IBRAHIMA SECK.

**Guinea-Bissau:** Calle 14, No 313, entre 3 y 5, Miramar, Havana; tel. (7) 204-2689; fax (7) 204-2794; Chargé d'affaires a.i. TOMÁS SANCAS.

**Guyana:** Calle 18, No 506, entre 5 y 7, Miramar, Havana; tel. (7) 204-2249; fax (7) 204-2867; e-mail embguyana@ip.etecsa.cu; Ambassador Dr TIMOTHY N. CRICHLOW.

**Haiti:** Avda 7, No 4402, esq. 44, Miramar, Havana; tel. (7) 204-5421; fax (7) 204-5423; e-mail embhaiti@enet.cu; internet www.embhaiti.cu; Ambassador Dra MARIE CARMELE ANDRINE CONSTANT.

**Holy See:** Calle 12, No 514, entre 5 y 7, Miramar, Havana (Apostolic Nunciature); tel. (7) 204-2700; fax (7) 204-2257; e-mail csa@pcn.net; Apostolic Nuncio Most Rev. LUIGI BONAZZI (Titular Archbishop of Atella).

**Honduras:** Interests Section: Casa Palhero 112, Planta Baja, Calle 30, entre 1 y 3, Miramar, Havana; tel. (7) 204-5497; fax (7) 204-5496; e-mail embhocu@ip.etecsa.cu; Chargé d'affaires a.i. REYNIERI DAVID AMADOR.

**Hungary:** Calle G, No 458, entre 19 y 21, Vedado, Havana; tel. (7) 833-3365; fax (7) 833-3286; e-mail embhuncu@ceniai.inf.cu; Ambassador JÓZSEF BALÁZS.

**India:** Calle 21, No 202, esq. a K, Vedado, Havana; tel. (7) 833-3777; fax (7) 833-3287; e-mail eoihav@ceniai.inf.cu; internet www.indembassyhavana.cu; Ambassador BHASKAR BALAKRISHNAN.

**Indonesia:** Avda 5, No 1607, esq. 18, Miramar, Havana; tel. (7) 204-9618; fax (7) 204-9617; e-mail indonhav@ceniai.inf.cu; internet www.indohav.cu; Ambassador INDRA TJAHJA SEMPURNAJAYA.

**Iran:** Avda 5, No 3002, esq. 30, Miramar, Havana; tel. (7) 204-2675; fax (7) 204-2770; e-mail embairan@ip.etecsa.cu; Ambassador AHMAD EDRISIAN.

**Iraq:** Avda 5, No 8201, entre 82 y 84, Miramar, Havana; tel. (7) 204-1607; fax (7) 204-2157; Chargé d'affaires a.i. MAZIN A. THIAB.

**Italy:** No 5, Avda 402, Calle 4, Miramar, Havana; tel. (7) 204-5615; fax (7) 204-5659; e-mail ambasciata.avana@esteri.it; internet sedi.esteri.it/avana/; Ambassador ELIO MENZIONE.

**Jamaica:** Avda 5, No 3608, entre 36 y 36A, Miramar, Havana; tel. (7) 204-2908; fax (7) 204-2531; e-mail embjmcub@enet.cu; Ambassador ELINOR FELIX-SHERLOCK.

**Japan:** Centro de Negocios Miramar, Avda 3, No 1, 5°, esq. 80, Miramar, Havana; tel. (7) 204-3355; fax (7) 204-8902; Ambassador TATSUAKI IWATA.

**Korea, Democratic People's Republic:** Calle 17, No 752, Vedado, Havana; tel. (7) 66-2313; fax (7) 33-3073; Ambassador PAK DONG CHUN.

**Laos:** Avda 5, No 2808, esq. 30, Miramar, Havana; tel. (7) 204-1056; fax (7) 204-9622; e-mail embalao@ip.etecsa.cu; Ambassador PHOUANGKEO LANGSY.

**Lebanon:** Calle 17A, No 16403, entre 164 y 174, Siboney, Havana; tel. (7) 208-6220; fax (7) 208-6432; e-mail lbcunet@ceniai.inf.cu; Ambassador SLEIMAN C. RASSI.

**Libya:** Avda 7, No 1402, esq. 14, Miramar, Havana; tel. (7) 204-2192; fax (7) 204-2991; e-mail oficinalibia@ip.etecsa.cu; Ambassador RAJAB JABA-ALAH ALMOSHAITI.

**Malaysia:** Avda 5 y 68, No 6612, Miramar, Havana; tel. (7) 204-8883; fax (7) 204-6888; e-mail mwhavana@kln.gov.my; Ambassador Dato' MOHD KAMAL BIN YAN YAHAYA.

**Mali:** Calle 36A, No 704, entre 7 y 9, Miramar, Havana; tel. (7) 204-5321; fax (7) 204-5320; e-mail ambamali@ceniai.inf.cu; Ambassador FIDELE DIARRA.

**Mexico:** Calle 12, No 518, Miramar, Playa, Havana; tel. (7) 204-2553; fax (7) 204-2717; e-mail embamex@ip.etecsa.cu; Ambassador MELBA PRIA.

**Mongolia:** Calle 66, No 505, esq. 5, Miramar, Havana; tel. (7) 204-2763; fax (7) 204-0639; e-mail monelch@ceniai.inf.cu; Ambassador GURJAV ERDENE.

**Mozambique:** Avda 7, No 2203, entre 22 y 24, Miramar, Havana; tel. (7) 204-2443; fax (7) 204-2232; e-mail embamoc@ceniai.inf.cu; Ambassador JERÓNIMO ROSA JOÃO CHIVAVI.

**Namibia:** Avda 5, No 4406, entre 44 y 46, Miramar, Havana; tel. (7) 204-1430; fax (7) 204-1431; e-mail embnamib@ceniai.inf.cu; Ambassador CLAUDIA GRACE UUSHONA.

**Netherlands:** Calle 8, No 307, entre 3 y 5, Miramar, Havana; tel. (7) 204-2511; fax (7) 204-2059; e-mail hav@minbuza.nl; Ambassador W. W. WILDEBOER.

**Nicaragua:** Calle 20, No 709, entre 7 y 9, Miramar, Havana; tel. (7) 204-1025; fax (7) 204-6323; e-mail embanicc@enet.cu; Chargé d'affaires a.i. CARLOS JOSÉ PÉREZ ROMAN.

**Nigeria:** Avda 5, No 1401, entre 14 y 16, Miramar, Havana; tel. (7) 204-2898; fax (7) 204-2202; e-mail enigera@ceniai.inf.cu; Ambassador NGAM NWACHUKWU.

**Panama:** Calle 26, No 109, entre 1 y 3, Miramar, Havana; tel. (7) 204-0858; fax (7) 204-1674; e-mail panembacuba@ip.etecsa.cu; Ambassador ABRAHAM BARCENAS ROSALES.

**Paraguay:** Calle 34, No 503, entre 5 y 7, Miramar, Havana; tel. (7) 204-0884; fax (7) 204-0883; e-mail cgphav@enet.cu; Chargé d'affaires a.i. AUGUSTO OCAMPOS CABALLERO.

**Peru:** Calle 30, No 107, entre 1 y 3, Miramar, Havana; tel. (7) 204-3570; fax (7) 204-2636; e-mail embaperu@embaperu.cu; Chargé d'affaires a.i. FÉLIX VÁSQUEZ SOLÍS.

**Philippines:** Avda 5, No 2207, esq. 24, Miramar, Havana; tel. (7) 204-1372; fax (7) 204-2915; e-mail philhavpe@enet.cu; Ambassador REGINA IRENE P. SARMIENTO.

**Poland:** Calle G, No 452, esq. 19, Vedado, Havana; tel. (7) 833-2439; fax (7) 833-2442; e-mail ambhawpl@ceniai.inf.cu; internet www.embajadapolonia.cu; Ambassador TOMASZ TUROWSKI.

**Portugal:** Avda 7, No 2207, esq. 24, Miramar, Havana; tel. (7) 204-0149; fax (7) 204-2593; e-mail embport@enet.cu; Ambassador MARIO GODINHO DE MATOS.

**Qatar:** Hotel Nacional, Havana 211; tel. (7) 33-3564; fax (7) 73-4700; Ambassador ALI BIN SAAD AL-KHARJI.

**Romania:** Calle 21, No 307, Vedado, Havana; tel. (7) 33-3325; fax (7) 33-3324; e-mail erumania@ceniai.inf.cu; Ambassador CONSTANTIN SIMIRAD.

**Russia:** Avda 5, No 6402, entre 62 y 66, Miramar, Havana; tel. (7) 204-2686; fax (7) 204-1038; e-mail embrusia@ceniai.inf.cu; Ambassador ANDREI VIKTOROVICH DIMITRIEV.

**Saudi Arabia:** Avda 5, No 8206, entre 82 y 84, Miramar, Havana; tel. (7) 204-1045; fax (7) 204-6401; e-mail erasdcu@ceniai.inf.cu; Ambassador MOHAMED MUSTAFA TLEIMIDI.

**Serbia and Montenegro:** Calle 42, No 115, entre 1 y 3, Miramar, Havana; tel. (7) 204-2488; fax (7) 204-2982; e-mail embyuhav@ceniai.inf.cu; Ambassador MILENA LUKOVIC JOVANOVIC.

**Slovakia:** Calle 66, No 521, entre 5B y 7, Miramar, Havana; tel. (7) 204-1884; fax (7) 204-1883; e-mail embeslovaca@enet.cu; Ambassador IVO HLAVACEK.

**South Africa:** Avda 5, No 4201, esq. 42, Miramar, Havana; tel. (7) 204-9671; fax (7) 204-1101; e-mail rsacuba@ceniai.inf.cu; internet www.sudafrica.cu; Ambassador THENJIWE ETHEL MTINTSO.

**Spain:** Cárcel No 51, esq. Zulueta, Havana; tel. (7) 866-8025; fax (7) 866-8006; e-mail embespcu@correo.mae.es; Ambassador CARLOS ALONSO ZALDIVAR.

**Sri Lanka:** Calle 32, No 307, entre 3 y 5, Miramar, Havana; tel. (7) 204-2562; fax (7) 204-2183; e-mail sri.lanka@ip.etecsa.cu; Ambassador PIYASIRI PREMACHANDRA VIJAYASEKERE.

**Sweden:** Calle 34, No 510, entre 5 y 7, Miramar, Havana; tel. (7) 204-2831; fax (7) 204-1194; e-mail ambassaden.havanna@foreign.ministry.se; internet www.swedenabroad/havana; Ambassador CHRISTER ELM.

**Switzerland:** Avda 5, No 2005, entre 20 y 22, Miramar, Havana; tel. (7) 204-2611; fax (7) 204-1148; e-mail swissem@enet.cu; Ambassador JEAN-CLAUDE RICHARD.

**Syria:** Avda 5, No 7402, entre 74 y 76, Miramar, Havana; tel. (7) 204-2266; fax (7) 204-2829; e-mail embsiria@ceniai.inf.cu; Ambassador CLOVIS KHOURY.

**Turkey:** Avda 5, No 3805, entre 36 y 40, Miramar, Havana; tel. (7) 204-1205; fax (7) 204-2899; e-mail turkemb@ip.etecsa.cu; Ambassador VEFAHAN OCAK.

**Ukraine:** Avda 5, No 4405, entre 44 y 46, Miramar, Havana; tel. (7) 204-2586; fax (7) 204-2341; e-mail cubukrem@ceniai.inf.cu; Ambassador VICTOR PASCHUK.

**United Kingdom:** Calle 34, No 702/4, esq. 7, Miramar, Havana; tel. (7) 204-1771; fax (7) 204-8104; e-mail embrit@ceniai.inf.cu; internet www.britishembassy.gov.uk/cuba; Ambassador JOHN ANTHONY DEW.

CUBA                                                                                              *Directory*

**USA** (Relations severed in 1961): Interests Section: Calzada, entre L y M, Vedado, Havana; tel. (7) 833-3551; fax (7) 833-1084; internet havana.usinterestsection.gov; Principal Officer JAMES C. CASON.

**Uruguay:** Calle 14, No 506, entre 5 y 7, Miramar, Havana; tel. (7) 204-2311; fax (7) 204-2246; e-mail urucub@ceniai.inf.cu; Ambassador JORGE ERNESTO MAZZAROVICH SEVERI.

**Venezuela:** Avda 1601, No 5, entre 16 y 18, Miramar, Havana; tel. (7) 204-2612; fax (7) 204-2773; e-mail vencuba@enet.cu; Ambassador ADÁN CHÁVEZ FRÍAS.

**Viet Nam:** Avda 5, No 1802, esq. 18, Miramar, Havana; tel. (7) 204-1502; fax (7) 204-1041; e-mail embaviet@ceniai.inf.cu; internet www.vietnamembassy.cu; Ambassador PHAM TIEN TU.

**Yemen:** Calle 16, No 503, entre 5 y 7, Miramar, Havana; tel. (7) 204-1506; fax (7) 204-1131; e-mail gamdan-hav@enet.cu; Ambassador AHMED AMIN MOHAMED ZAIDAN.

**Zimbabwe:** Avda 3, No 1001, esq. a 10, Miramar, Havana; tel. (7) 204-2857; fax (7) 204-2720; e-mail zimhavan@ip.etecsa.cu; Ambassador JEVANA BEN MASEKO.

## Judicial System

The judicial system comprises the People's Supreme Court, the People's Provincial Courts and the People's Municipal Courts. The People's Supreme Court exercises the highest judicial authority.

### PEOPLE'S SUPREME COURT

The People's Supreme Court comprises the Plenum, the six Courts of Justice in joint session and the Council of Government. When the Courts of Justice are in joint session they comprise all the professional and lay judges, the Attorney-General and the Minister of Justice. The Council of Government comprises the President and Vice-Presidents of the People's Supreme Court, the Presidents of each Court of Justice and the Attorney-General of the Republic. The Minister of Justice may participate in its meetings.

**President:** Dr RUBÉN REMIGIO FERRO.

**Vice-Presidents:** OSVALDO SÁNCHEZ MARTÍN, EMILIA GONZÁLEZ PÉREZ, EDUARDO RODRÍGUEZ GONZÁLEZ, PEDRO LUIS GONZÁLEZ CHÁVEZ.

#### Criminal Court
**President:** Dr JORGE L. BODES TORRES.

#### Civil and Administrative Court
**President:** CARLOS DÍAZ TENRREIRO.

#### Labour Court
**President:** Dr ANTONIO R. MARTÍN SÁNCHEZ.

#### Court for State Security
**President:** Dr GUILLERMO HERNÁNDEZ INFANTE.

#### Economic Court
**President:** NARCISO COBO ROURA.

#### Military Court
**President:** Col JUAN MARINO FUENTES CALZADO.

**Attorney-General:** JUAN ESCALONA REGUERA.

## Religion

There is no established Church, and all religions are permitted, though Roman Catholicism predominates. The Afro-Cuban religions of Regla de Ocha (Santéria) and Regla Conga (Palo Monte) also have numerous adherents.

### CHRISTIANITY

**Consejo Ecuménico de Cuba** (Ecumenical Council of Cuba): Calle 14, No 304, entre 3 y 5, Miramar, Playa, Havana; tel. (7) 33-1792; fax (7) 33-178820; f. 1941; 11 mem. churches; Pres. Rev. ORESTES GONZÁLEZ; Exec. Sec. Rev. JOSÉ LÓPEZ.

#### The Roman Catholic Church

Cuba comprises three archdioceses and eight dioceses. At 31 December 2003 there were 6,331,250 adherents, representing 50.5% of the total population.

**Conferencia de Obispos Católicos de Cuba—COCC (Bishops' Conference)**
Calle 26, No 314, entre 3 y 5, Miramar, Apdo 635, 11300 Havana; tel. (7) 22–3868; fax (7) 33–2168; e-mail adjunto@cocc.co.cu; internet www.celam.org/sitios/ce_cuba/COCC.htm.
f. 1983; Pres. Cardinal JAIME LUCAS ORTEGA Y ALAMINO (Archbishop of San Cristóbal de la Habana).

**Archbishop of Camagüey:** JUAN GARCÍA RODRÍGUEZ, Calle Luaces, No 55, Apdo 105, Camagüey 70100; tel. (322) 92268; fax (322) 87143; e-mail dei@cocc.co.cu.

**Archbishop of San Cristóbal de la Habana:** Cardinal JAIME LUCAS ORTEGA Y ALAMINO, Calle Habana No 152, esq. a Chacón, Apdo 594, Havana 10100; tel. (7) 62-4000; fax (7) 33-8109; e-mail cocc@brigadoo.com.

**Archbishop of Santiago de Cuba:** PEDRO CLARO MEURICE ESTÍU, Sánchez Hechevarría No 607, Apdo 26, Santiago de Cuba 90100; tel. (226) 25480; fax (226) 86186.

#### The Anglican Communion

Anglicans are adherents of the Iglesia Episcopal de Cuba (Episcopal Church of Cuba).

**Bishop of Cuba:** Rt Rev. JORGE PERERA HURTADO, Calle 6, No 273, Vedado, Havana 10400; tel. (7) 832-1120; fax (7) 33-3293; e-mail episcopal@ip.etecsa.cu; internet www.cuba.anglican.org.

#### Protestant Churches

**Convención Bautista de Cuba Oriental** (Baptist Convention of Eastern Cuba): San Jerónimo, No 467, entre Calvario y Carnicería, Santiago de Cuba 90100; tel. (226) 62-0173; fax (226) 68-7010; e-mail ecreyes@cbcor.com; internet www.cbcor.com; f. 1905; Pres. Rev. Dr ROY ACOSTA GARCÍA; Sec. Rev. IDALDO MATOS DELGADO.

**Iglesia Metodista en Cuba** (Methodist Church in Cuba): Calle K, No 502, 25 y 27, Vedado, Apdo 10400, Havana; tel. (7) 832-2991; fax (7) 832-0770; e-mail imecu@enet.cu; autonomous since 1968; 6,000 mems; Bishop RICARDO PEREIRA DÍAZ.

**Iglesia Presbiteriana Reformada en Cuba** (Presbyterian-Reformed Church in Cuba): Salud 222, entre Lealtad y Campanario, Havana; tel. (7) 862-1219; fax (7) 833-8819; e-mail presbit@enet.cu; internet www.prccuba.org; f. 1890; 8,000 mems; Moderator Rev. Dr HÉCTOR MÉNDEZ.

Other denominations active in Cuba include the Apostolic Church of Jesus Christ, the Bethel Evangelical Church, the Christian Pentecostal Church, the Church of God, the Church of the Nazarene, the Free Baptist Convention, the Holy Pentecost Church, the Pentecostal Congregational Church and the Salvation Army.

## The Press

### DAILY

In October 1990 President Castro announced that, in accordance with other wide-ranging economic austerity measures, only one newspaper, *Granma*, would henceforth be published as a nationwide daily. The other national dailies were to become weeklies or were to cease publication.

**Granma:** Avda Gen. Suárez y Territorial, Plaza de la Revolución, Apdo 6187, Havana; tel. (7) 881-3333; fax (7) 881-9854; e-mail correo@granma.cip.cu; internet www.granma.cubaweb.cu; f. 1965 to replace *Hoy* and *Revolución*; official Communist Party organ; Dir FRANK AGÜERO GÓMEZ; circ. 400,000.

### PERIODICALS

**Adelante:** Avda A, Rpto Jayamá, Camagüey; e-mail adelante@caonao.cmw.inf.cu; internet www.adelante.cu; f. 1959; Dir ZARITMA CARDOSO MORENO; circ. 42,000.

**Ahora:** Salida a San Germán y Circunvalación, Holguín; e-mail ahoraweb@ahora.cu; internet www.ahora.cu; f. 1962; Dir RADOBALDO MARTÍNEZ PÉREZ; circ. 50,000.

**Alma Mater:** Prado 553, esq. Teniente Ray, Habana Vieja, Havana; e-mail almamater@eabril.jovenclub.cu; internet www.almamater.cu; f. 1922; organ of the universities; Dir MARIO VIZCAÍNO SERRAT.

**ANAP:** Obispo 527, Apdo 605, Havana; f. 1961; monthly; information for small farmers; Editor RICARDO MACHADO; circ. 90,000.

**Bastión:** Territorial esq. a Gen. Suárez, Plaza de la Revolución, Havana; tel. (7) 79-3361; organ of the Revolutionary Armed Forces; Dir FRANK AGÜERO GÓMEZ; circ. 65,000.

**Bohemia:** Avda Independencia y San Pedro, Apdo 6000, Havana; tel. (7) 81-9213; fax (7) 33-5511; e-mail bohemia@bohemia.co.cu;

CUBA — *Directory*

internet www.bohemia.cubaweb.cu; f. 1908; weekly; politics; Dir José Fernández Vega; circ. 100,000.

**Boletín Alimentaria de Cuba:** Amargura 103, 10100 Havana; tel. (7) 62-9245; f. 1996; quarterly; food industry; Dir Antonio Campos; circ. 10,000.

**El Caimán Barbudo:** Paseo 613, Vedado, Havana; e-mail caiman@eabril.jovenclub.cu; internet www.caimanbarbudo.cu; f. 1966; monthly; cultural; Dir Fidel Díaz Castro; circ. 47,000.

**Casa de las Américas:** 3 y G Vedado, 10400 Havana; tel. (7) 55-2706; fax (7) 834-4554; e-mail webmaster@casa.cult.cu; internet www.casadelasamericas.com; f. 1959; 6 a year; cultural; Dir Pablo Armando Fernández.

**Cinco de Septiembre:** Calle 63 y Circunvalación, 2°, Cienfuegos; tel. (7) 52-2636; e-mail cip219@cip.enet.cu; internet www.5septiembre.cu; f. 1980; Dir Alina Rosell Chong; circ. 18,000.

**Cómicos:** Calle 28, No 112, entre 1 y 3, Miramar, Havana; tel. (7) 22-5892; monthly; humorous; circ. 70,000.

**Cuba Internacional:** Calle 21, No 406, entre F y G, Vedado, Havana; tel. (7) 832-3578; fax (7) 832-3268; e-mail cubainternac@pubs.prensa-latina.cu; internet www.prensa_latina.cu/NewsSection.asp?Section=PUB_CUBA; f. 1959; every 2 months; political; Editorial Dir Ana María Ruiz; circ. 30,000.

**Dedeté:** Territorial y Gen. Suárez, Plaza de la Revolución, Apdo 6344, Havana; tel. (7) 82-0134; fax (7) 81-8621; e-mail ddt@jrebelde.cip.cu; internet www.dedete.cubaweb.cu; f. 1969; monthly; humorous supplementary publication of Juventud Rebelde; Dir Alen Lauzán; circ. 70,000.

**La Demajagua:** Amado Estévez, esq. Calle 10, Rpto R. Reyes, Bayamo; tel. (23) 42-4221; e-mail cip225@cip.enet.cu; internet www.lademajagua.co.cu; f. 1977; Dir Luis Carlos Frómeta Agüero; circ. 21,000.

**El Deporte, Derecho del Pueblo:** Vía Blanca y Boyeros, Havana; tel. (7) 40-6838; f. 1968; monthly; sport; Dir Manuel Vaillant Carpente; circ. 15,000.

**El Economista:** Asociación Nacional de Economistas de Cuba, Calle 22, No 901 esq. a 9, Miramar, Havana; tel. (7) 209-3303; fax (7) 202-3456; e-mail anec@info.get.tur.cu; internet www.eleconomista.cubaweb.cu; monthly; business; Dir-Gen. Roberto Verrier Castro.

**Escambray:** Adolfo del Castillo 10, Sancti Spíritus; tel. (41) 23003; e-mail cip220@cip.enet.cu; internet www.escambray.islagrande.cu; f. 1979; Dir Juan Antonio Borrego Díaz; circ. 21,000.

**Girón:** Avda Camilo Cienfuegos No 10505, P. Nuero, Matanzas; e-mail giron@ma.cc.cu; internet www.giron.co.cu; f. 1960; Dir Clovis Ortega Castañeda; circ. 25,000.

**Guerrillero:** Colón esq. Delicias y Adela Azcuy, Pinar del Río; e-mail cip216@cip.enet.cu; internet www.guerrillero.co.cu; f. 1969; Dir Ernesto Osorio Roque; circ. 33,000.

**El Habanero:** Gen. Suárez y Territorial, Plaza de la Revolución, Apdo 6269, Havana; tel. (7) 6160; e-mail internet@habanero.cip.cu; internet www.elhabanero.cubaweb.cu; f. 1987; Dir Andrés Hernández Rivero; circ. 21,000.

**Industria Alimentaria:** Obispo 527, Apdo 605, Havana; tel. (7) 61-8453; f. 1968; quarterly; food industry; Editor Cesar Valdivia; circ. 10,000.

**Invasor:** Avda de los Deportes s/n , Ciego de Avila; internet www.invasor.islagrande.cu; f. 1979; Dir Migdalia Utrera Peña; circ. 10,500.

**Juventud Rebelde:** Avda Territorial y Gen. Suárez, Plaza de la Revolución, Apdo 6344, Havana; tel. (7) 882-0155; fax (7) 33-8959; e-mail cida@jrebelde.cip.cu; internet www.jrebelde.cubaweb.cu; f. 1965; organ of the Young Communist League; Dir Rogelio Polanco Fuentes; circ. 250,000.

**Juventud Técnica:** Prado 553, esq. Teniente Rey, Habana Vieja, Havana; tel. (7) 62-4330; e-mail jtecnica@eabril.jovenclub.cu; internet www.juventudtecnica.cu; f. 1965; every 2 months; scientific-technical; Dir Miriam Zito Valdés; circ. 20,000.

**Mar y Pesca:** San Ignacio 303, entre Amargura y Teniente Rey, Habana Vieja, Havana; tel. (7) 861-5518; fax (7) 861-6280; e-mail marpesca@fishnavy.inf.cu; f. 1965; quarterly; fishing; Dir Mario Guillot Vega; circ. 20,000.

**Moncada:** Ministry of the Interior, Plaza de la Revolución, Havana; tel. (7) 830-1566; fax (7) 33-5261; f. 1966; monthly; government publication; Dir Ricardo Martínez; circ. 70,000.

**Muchacha:** Galiano 264, esq. Neptuno, Havana; tel. (7) 861-5919; f. 1980; monthly; young women's magazine; Dir Silvia Martínez; circ. 120,000.

**Mujeres:** Galiano 264, entre Neptuno y Concordia, Havana 10200; tel. (7) 861-5919; e-mail mujeres@enet.cu; internet www.mujeres.cubaweb.cu; f. 1961; monthly; women's magazine; Dir-Gen. Carolina Aguilar Ayerra; circ. 270,000.

**El Muñe:** Calle 28, No 112, entre 1 y 3, Miramar, Havana; tel. (7) 202-5892; weekly; circ. 50,000.

**El Nuevo Fenix:** Independencia 52, esq. a Honorato del Castillo, Sancti Spíritus; tel. (41) 27902; e-mail plss@ip.etecsa.cu; internet www.fenix.islagrande.cu; f. 1999; Dir Yolanda Brito.

**Opciones:** Territorial esq. Gen. Suárez, Plaza de la Revolucíon, Havana; tel. (7) 881-8934; fax (7) 881-8621; e-mail opciones@jrebelde.cip.cu; internet www.opciones.cubaweb.cu; f. 1994; weekly; finance, commerce and tourism; Dir Rogelio Polanco Fuentes.

**Opina:** Edif. Focsa, M entre 17 y 19, Havana; f. 1979; 2 a month; consumer-orientated; published by Institute of Internal Demand; Dir Eugenio Rodríguez Balari; circ. 250,000.

**Pablo:** Calle 28, No 112, entre 1 y 3, Miramar, Havana; tel. (7) 22-5892; 16 a year; circ. 53,000.

**Palante:** Calle 21, No 954, entre 8 y 10, Vedado, Havana; tel. (7) 3-5098; e-mail palante@palan.cipcc.get.cma.net; f. 1961; weekly; humorous; Dir Rosendo Gutiérrez Román; circ. 235,000.

**Periódico 26:** Avda Carlos J. Finlay, Las Tunas; e-mail cip224@cip.enet.cu; internet www.periodico26.cu; Dir Ramiro Segura García.

**Pionero:** Calle 17, No 354, Havana; tel. (7) 32-4571; e-mail pionero@eabril.jovenclub.cu; internet www.pionero.cu; f. 1961; weekly; children's magazine; Editor Pilar Digat Landrove; circ. 210,000.

**Prisma:** Calle 21 y Avda G, No 406, Vedado, Havana; tel. (7) 832-3578; e-mail prisma@pubs.prensa-latina.cu; f. 1979; bimonthly; tourism; Man. Dir Luis Manuel Arce; circ. 15,000 (Spanish), 10,000 (English).

**RIL:** O'Reilly 358, Havana; tel. (7) 62-0777; f. 1972; 2 a month; technical; Dir Exec. Council of Publicity Dept, Ministry of Light Industry; Chief Officer Mireya Crespo; circ. 8,000.

**Sierra Maestra:** Avda de Los Desfiles, Santiago de Cuba; tel. (7) 2-2813; e-mail cip226@cip.enet.cu; internet www.sierramaestra.cu; f. 1957; weekly; Dir Arnaldo Clavel Carmenaty; circ. 45,000.

**Sol de Cuba:** Calle 19, No 60, entre M y N, Vedado, Havana 4; tel. (7) 832-9881; f. 1983; every 3 months; Spanish, English and French editions; Gen. Dir Alcides Giro Mitjans; Editorial Dir Doris Vélez; circ. 200,000.

**Somos Jóvenes:** Calle 17, No 354, esq. H, Vedado, Havana; tel. (7) 32-4571; e-mail eabril@jcce.org.cu; internet www.somosjovenes.cu; f. 1977; monthly; Dir Robin Marín Sánachez; circ. 200,000.

**Trabajadores:** Territorial esq. Gen. Suárez, Plaza de la Revolución, Havana; tel. (7) 79-0819; fax (7) 55-5927; e-mail digital@trabaja.cip.cu; internet www.trabajadores.cubaweb.cu; f. 1970; organ of the trade-union movement; Dir Jorge Luis Canela Ciurana; circ. 150,000.

**Tribuna de la Habana:** Territorial esq. Gen. Suárez, Plaza de la Revolución, Havana; tel. (7) 881-8021; e-mail redac@tribuna.cip.cu; internet www.tribuna.islagrande.cu; f. 1980; weekly; Dir Jesús Álvarez Ferrer; circ. 90,000.

**Vanguardia:** Calle Céspedes 5, esq. a Plácido, Santa Clara, Matanzas; e-mail cip218@cip.enet.cu; internet www.vanguardia.co.cu; f. 1962; Dir F. A. Chang L.; circ. 24,000.

**Venceremos:** Avda Ernesto Che Guevara, Km 1½, Guantánamo; tel. (7) 32-5402; e-mail cip227@cip.enet.cu; internet www.venceremos.co.cu; f. 1962; economic, political and social publication for Guantánamo province; Dir Elizabeth Santiesteban Pérez; circ. 33,500.

**Ventiseis:** Avda Carlos J. Finley, Las Tunas; f. 1977; Dir José Infantes Reyes; circ. 21,000.

**Verde Olivo:** Avda Independencia y San Pedro, Havana; tel. (7) 79-8373; f. 1959; monthly; organ of the Revolutionary Armed Forces; Dir Armando Diéguez Suárez; circ. 100,000.

**Victoria:** Carretera de la Fe, Km 1½, Plaza de la Revolución, Nueva Gerona, Isla de la Juventud; tel. (46) 32-4210; e-mail periodic@gerona.inf.cu; internet www.victoria.co.cu; f. 1967; Dir Sergio Rivero Carrasco; circ. 9,200.

**Zunzún:** Obispo 527, Apdo 605, Havana; e-mail zunzun@eabril.jovenclub.cu; internet www.zunzun.cu; f. 1990; children's magazine; Dir Martha Ríos.

## PRESS ASSOCIATIONS

**Unión de Periodistas de Cuba:** Avda 23, No 452, esq. a I, Vedado, 10400 Havana; tel. (7) 32-4550; fax (7) 33-3079; e-mail cip301@cip.enet.cu; internet www.cubaperiodistas.cu; f. 1963; Pres. Tubal Páez Hernández.

**Unión de Escritores y Artistas de Cuba (UNEAC):** Calle 17, No 354, entre G y H, Vedado, Havana; tel. (7) 53-5081; fax (7) 33-3158;

e-mail promeven@uneac.co.cu; internet www.uneac.com; Pres. CARLOS MARTÍ; Exec. Vice-Pres. MIGUEL BARNET.

### NEWS AGENCIES

**Agencia de Información Nacional (AIN):** Calle 23, No 358, esq. a J, Vedado, Havana; tel. (7) 32-5541; fax (7) 66-2049; e-mail ainnews@ain.cu; internet www.ain.cubaweb.cu; f. 1974; national news agency; Gen. Dir ESTEBAN RAMÍREZ ALONSO.

**Prensa Latina (Agencia Informativa Latinoamericana, SA):** Calle 23, No 201, esq. a N, Vedado, Havana; tel. (7) 55-3496; fax (7) 33-3068; e-mail difusion@prensa-latina.cu; internet www.prensa-latina.cu; f. 1959; Dir PEDRO MARGOLLES VILLANUEVA.

#### Foreign Bureaux

**Agence France-Presse (AFP):** Calle 17, No 4, 13°, entre N y O, Vedado, Havana; tel. (7) 33-3503; fax (7) 33-3034; e-mail mlsanz@ip.etecsa.cu; Bureau Chief MARIE SANZ.

**Agencia EFE** (Spain): Calle 36, No 110, entre 1 y 3, Miramar, Apdo 5, Havana; tel. (7) 204-2293; fax (7) 204-2272; e-mail efecuba@ip.etecsa.cu; Bureau Chief SOLEDAD MARÍN MARTÍN.

**Agenzia Nazionale Stampa Associata (ANSA)** (Italy): Edif. Someillán, Línea 5, Dpt 12, Vedado, Havana; tel. (8) 33-3542; e-mail ansacuba@enet.cu; internet www.ansa.it; Correspondent SILVINA SANTILLÁN.

**Associated Press (AP)** (USA): Lonja del Comercio, planta baja, Lamparilla 2, Local B, Habana Vieja, Havana; tel. (7) 33-0370; Correspondent ANITA SNOW.

**Bulgarska Telegrafna Agentsia (BTA)** (Bulgaria): Edif. Focsa, Calle 17, esq. M, Vedado, Apdo 22E, Havana; tel. (7) 32-4779; Bureau Chief VASIL MIKOULACH.

**Česká tisková kancelář (ČTK)** (Czech Republic): Edif. Fajardo, Calle 17 y M, Vedado, Apdo 3A, Vedado, Havana; tel. (7) 32-6101; Bureau Chief PAVEL ZOVADIL.

**Deutsche Presse-Agentur (dpa)** (Germany): Edif. Focsa, Calle 17 y M, Vedado, Apdo 2K, Havana; tel. (7) 33-3501; Bureau Chief VICTORIO COPA.

**Informatsionnoye Telegrafnoye Agentstvo Rossii—Telegrafnoye Agentstvo Suverennykh Stran (ITAR—TASS)** (Russia): Calle 96, No 317, entre 3 y 5, Miramar, Havana 4; tel. (7) 29-2528; Bureau Chief ALEKSANDR KANICHEV.

**Inter Press Service (IPS)** (Italy): Calle 36A, No 121 Bajos, esq. a 3, Miramar, Apdo 1, Havana; tel. (7) 22-1981; Bureau Chief CLAUDE JOSEPH HACKIN; Correspondent CARLOS BASTISTA MORENO.

**Korean Central News Agency** (Democratic People's Republic of Korea): Calle 10, No 613, esq. 25, Vedado, Apdo 6, Havana; tel. (7) 31-4201; Bureau Chief CHANG YON CHOL.

**Magyar Távirati Iroda (MTI)** (Hungary): Edif. Fajardo, Calle 17 y M, Apdo 2C, Havana; tel. (7) 32-8353; Bureau Chief ZOLTÁN TAKACS; Correspondent TIBOR CSÁSZÁR.

**Novinska Agencija Tanjug** (Serbia and Montenegro): Calle 5F, No 9801, esq. 98, Miramar, Havana; tel. (7) 22-7671; Bureau Chief DUŠAN DAKOVIĆ.

**Polska Agencja Prasowa (PAP)** (Poland): Calle 6, No 702, Apdo 5, entre 7 y 9, Miramar; Havana; tel. (7) 20-7067; Bureau Chief PIOTR SOMMERFED.

**Reuters** (United Kingdom): Edif. Someillán, Linea 5, 9°, Vedado, Havana; tel. (7) 33-3145; e-mail anthony.boadle@reuters.com; Bureau Chief ANTHONY BOADLE.

**Rossiyskoye Informatsionnoye Agentstvo—Novosti (RIA—Novosti)** (Russia): Calle 28, No 510, entre 5 y 7, Miramar, Havana; tel. (7) 22-4129; Bureau Chief YURII GOLOVIATENKO.

**Viet Nam Agency (VNA):** Calle 16, No 514, 1°, entre 5 y 7, Miramar, Havana; tel. (7) 32-4455; Bureau Chief PHAM DINH LOI.

**Xinhua (New China) News Agency** (People's Republic of China): Calle G, No 259, esq. 13, Vedado, Havana; tel. (7) 32-4616; Bureau Chief GAO YONGHUA.

## Publishers

**Artecubano Ediciones:** Calle 3, No 1205, entre 12 y 14, Playa, Havana; tel. (7) 203-8581; fax (7) 204-2744; e-mail cnap@cubarte.cult.cu; Dir RAFAEL ACOSTA DE ARRIBA.

**Casa de las Américas:** Calle 3 y Avda G, Plaza de la Revolución, Vedado, 10400 Havana; tel. (7) 55-2716; fax (7) 832-7272; e-mail prensa@casa.cult.cu; internet www.casadelasamericas.com; f. 1959; Latin American literature and social sciences; Dir PABLO ARMANDO FERNÁNDEZ.

**Ediciones Creart:** Calle 4, No 205, entre Línea y 11, Plaza de la Revolución, Vedado, Havana; tel. (7) 832-9691; fax (7) 66-2582; e-mail creart@cubarte.cult.cu; Dir TOMÁS VALDÉS BECERRA.

**Ediciones Unión:** Calle 17, No 354, entre G y H, Plaza de la Revolución, Vedado, 10400 Havana; tel. (7) 55-3112; fax (7) 33-3158; e-mail editora@uneac.co.cu; f. 1962; publishing arm of the Unión de Escritores y Artistas de Cuba; Cuban literature, art; Dir OLGA MARTA CABRERA.

**Editora Abril:** Prado 553, esq. Teniente Rey y Dragones, Habana Vieja, Havana; tel. (7) 862-7871; fax (7) 862-4330; e-mail eabril@jcce.org.cu; f. 1980; attached to the Union of Young Communists; children's literature; Dir NIURKA DUMÉNIGO.

**Editora Política:** Belascoaín No 864, esq. a Desagüe y Peñalver, Havana; tel. (7) 879-8553; fax (7) 881-1024; e-mail editora@epol.cipcc.get.cma.net; f. 1963; publishing institution of the Communist Party of Cuba; Dir SANTIAGO DÓRQUEZ PÉREZ.

**Editorial Academia:** Industria y Barcelona, Capitolio Nacional, 4°, Habana Vieja, 10200 Havana; tel. and fax (7) 863-0315; e-mail editorial@gecyt.cu; f. 1963; attached to the Ministry of Science, Technology and the Environment; scientific and technical; Dir GLADYS HERNÁNDEZ HERRERA.

**Editorial Arte y Literatura:** Calle O' Reilly, No 4, esq. Tacón, Habana Vieja, Havana; tel. (7) 862-4326; fax (7) 833-8187; e-mail publicaciones@icl.cult.cu; f. 1967; traditional Cuban literature and arts; Dir ELIZABETH DÍAZ GONZÁLEZ.

**Editorial Ciencias Médicas:** Línea esq. 1, 10°, Vedado, 10400 Havana; tel. (7) 832-3863; fax (7) 33-3063; e-mail cnicm@infomed.sld.cu; attached to the Ministry of Public Health; books and magazines specializing in the medical sciences; Dir GUILLERMO PADRÓN GONZÁLEZ.

**Editorial Ciencias Sociales:** Calle 14, No 4104, entre 41 y 43, Miramar, Playa, Havana; tel. (7) 203-3959; fax (7) 33-3441; e-mail nuevomil@icl.cult.cu; internet www.cubaliteraria.com/ciencias_sociales/editorial/catalogo_cs1.asp; f. 1967; attached to the Cuban Book Institute; social and political literature, history, philosophy, juridical sciences and economics; Dir ERNESTO ESCOBAR SOTO.

**Editorial Científico-Técnica:** Calle 2, No 58, entre 3 y 5, Vedado, Havana; tel. (7) 3-9417; internet www.cubaliteraria.com/ciencias_sociales/editorial/catalogo_ct.asp; f. 1967; attached to the Ministry of Culture; technical and scientific literature; Dir ISIDRO FERNÁNDEZ RODRÍGUEZ.

**Editorial Félix Varela:** San Miguel No 1011, entre Mazón y Basarrate, Plaza de la Revolución, Vedado, 10400 Havana; tel. (7) 877-5617; fax (7) 73-5419; e-mail elsa@enpses.co.cu; Dir ELSA RODRÍGUEZ.

**Editorial Gente Nueva:** Calle 2, No 58, entre 3 y 5, Plaza de la Revolución, Vedado, 10400 Havana; tel. (7) 830-6548; fax (7) 33-8187; e-mail gentenueva@icl.cult.cu; internet www.cubaliteraria.com/icl/editoriales/gente_nueva; f. 1967; books for children; Dir MIRTA GONZÁLEZ GUTIÉRREZ.

**Editorial José Martí:** Calzada 259, entre I y J, Apdo 4208, Plaza de la Revolución, 10400 Havana; tel. (7) 832-9838; fax (7) 33-3441; e-mail editjmal@icl.cult.cu; f. 1983; attached to the Ministry of Culture; foreign-language publishing; Dir LOURDES GONZÁLEZ.

**Editorial Letras Cubanas:** Calle O'Reilly, No 4, esq. Tacón, Habana Vieja, 10100 Havana; tel. (7) 862-4378; fax (7) 33-8187; e-mail elc@icl.cult.cu; internet www.cubaliteraria.com/icl/editoriales/letras_cubanas; f. 1977; attached to the Ministry of Culture; general, particularly classic and contemporary Cuban literature and arts; Dir DANIEL GARCÍA SANTOS.

**Editorial Oriente:** Santa Lucía 356, Santiago de Cuba; tel. (226) 22496; fax (226) 42387; e-mail edoriente@cultstgo.cult.cu; f. 1971; publishes works from the Eastern provinces; fiction, history, female literature and studies, art and culture, practical books and books for children; Dir AIDA BAHR.

**Editorial Pablo de la Torriente Brau:** Calle 11, No 160, entre K y L, Plaza de la Revolución, Vedado, 10400 Havana; tel. (7) 832-7581; e-mail edpablo@eventos.cip.cu; f. 1985; publishing arm of the Unión de Periodistas de Cuba; Dir IRMA DE ARMAS FONSECA.

**Editorial Pueblo y Educación:** Avda 3A, No 4601, entre 46 y 60, Playa, Havana; tel. (7) 202-1490; fax (7) 204-0844; e-mail epe@ceniai.inf.cu; f. 1971; textbooks and educational publications; publishes Revista Educación 3 times a year (circ. 2,200); Dir CATALINA LAJUD HERRERO.

**Editorial San Lope:** : Calle Gonzalo de Quesada, No 121, entre Lico Cruz y Lucas Ortiz, Las Tunas; tel. (31) 48191; fax (31) 47380; e-mail librolt@tunet.cult.cu; attached to the Ministry of Culture; Dir ELAINE GONZÁLEZ URGELLÉS.

**Editorial Si-Mar:** Calle 47, No 1210, entre 36 y Lindero, Plaza de la Revolución, Vedado, 10600 Havana; tel. (7) 881-8168; fax (7) 881-8523; e-mail edicion@simar.cu; Dir ELIO VILLAREAL ACEVEDO.

CUBA
*Directory*

## GOVERNMENT PUBLISHING HOUSES

**Instituto Cubano del Libro:** Palacio del Segundo Cabo, Calle O'Reilly, No 4, esq. a Tacón, Havana; tel. (7) 62-4789; fax (7) 33-8187; e-mail cclfilh@artsoft.cult.cu; internet www.cubaliteraria.com/icl/index.htm; f. 1967; printing and publishing organization attached to the Ministry of Culture which combines several publishing houses and has direct links with others; presides over the National Editorial Council (CEN); Pres. OMAR GONZÁLEZ JIMÉNEZ.

**Oficina Publicaciones del Consejo de Estado:** Calle 17, No 552, esq. a D, Plaza de la Revolución, Vedado, 10400 Havana; tel. (7) 55-1406; fax (7) 57-4578; e-mail palvarez@enet.cu; attached to the Council of State; books, pamphlets and other printed media on historical and political matters; Dir PEDRO ALVAREZ TABÍO.

## Broadcasting and Communications

### TELECOMMUNICATIONS

**Empresa de Radiocomunicación y Radiodifusión (RADIO-CUBA):** Edif. Western Union, No 406, entre Obispo y Obrapía, Habana Vieja, Havana; tel. (7) 860-3142; fax (7) 33-8301; e-mail radiocuba@radiocuba.cu; internet www.radiocuba.cu; f. 1967; voice, mobile satellite and television radiocommunication services; Dir-Gen. JULIO ANTONIO GONZÁLEZ GARCÍA.

**Empresa de Telecomunicaciones de Cuba, SA (ETECSA):** Avda 5, entre 76 y 78, Edif. Barcelona, Miramar, Havana; tel. (7) 266-6203; fax (7) 860-5144; e-mail presidencia@etecsa.cu; internet www.etecsa.cu; f. 1991; 27% owned by Telecom Italia International, SpA; merged with Empresa de Telecomunicaciones Celulares del Caribe, SA (C-Com) and Teléfonos Celulares de Cuba, SA (CUBACEL) in 2003; Exec. Pres. JOSÉ ANTONIO FERNÁNDEZ.

**Instituto de Investigación y Desarrollo de las Telecomunicaciones (LACETEL):** Avda Independencia, Km 14½, 1° de Mayo, Rancho Boyeros, Havana; tel. (7) 57-9265; fax (7) 33-5828; e-mail lacetel@lacetel.cu; internet www.lacetel.cu; Dir-Gen. ORLANDO PIÑEIRA OLIVA.

**Ministerio de la Informática y las Comunicaciones (Dirección General de Telecomunicaciones):** Avda Independencia y 19 de Mayo, Plaza de la Revolución, Havana; tel. (7) 81-7654; e-mail infosoc@mic.cu; internet www.mic.gov.cu; Dir CARLOS MARTÍNEZ ALBUERNE.

**Telecomunicaciones Móviles, SA (MOVITEL):** Avda 47, No. 3405, Reparto Kohly, Havana; tel. (7) 204-8400; fax (7) 204-4264; e-mail movitel@movitel.co.cu; internet www.movitel.co.cu; mobile telecommunications; Dir-Gen. ASELA FERNÁNDEZ LORENZO.

### BROADCASTING

**Ministerio de la Informática y las Comunicaciones (Dirección de Frecuencias Radioeléctricas):** Avda Independencia y 19 de Mayo, Plaza de la Revolución, Havana; tel. (7) 81-7654; internet www.mic.gov.cu; Dir CARLOS MARTÍNEZ ALBUERNE.

**Instituto Cubano de Radio y Televisión (ICRT):** Edif. Radiocentro, Avda 23, No 258, entre L y M, Vedado, Havana 4; tel. (7) 32-1568; fax (7) 33-3107; e-mail icrt@cecm.get.tur.cu; f. 1962; Pres. ERNESTO LÓPEZ DOMÍNGUEZ.

**Radio y Televisión Comercial:** Calle 26, No 301, esq. 21, Vedado, Havana; tel. (7) 66-2719; fax (7) 33-3939; e-mail g.general@rtvc.com.cu; Dir RENÉ DUQUESNE LÓPEZ.

#### Radio

In 1997 there were five national networks and one international network, 14 provincial radio stations and 31 municipal radio stations, with a total of some 170 transmitters.

**Radio Cadena Agramonte:** Calle Cisneros, No 310, entre Ignacio Agramonte y General Gómez, Camagüey; tel. (322) 29-1195; e-mail cip240@cip.etecsa.cu; internet www.cadenagramonte.cubaweb.cu; digital radio; serves Camagüey; Dir GUILLERMO PAVÓN PACHECO.

**Radio Enciclopedia:** Calle N, No 266, entre 21 y 23, Vedado, Havana; tel. (7) 81-2809; e-mail enciclop@ceniai.inf.cu; internet www.radioenciclopedia.co.cu; f. 1962; national network; instrumental music programmes; 24 hours daily; Dir EDELSA PALACIOS GORDO.

**Radio Habana Cuba:** Infanta 105, Apdo 6240, Havana; tel. (7) 877-6628; fax (7) 881-2927; e-mail radiohc@enet.cu; internet www.radiohc.org; f. 1961; shortwave station; broadcasts in Spanish, English, French, Portuguese, Arabic, Esperanto, Quechua, Guaraní and Creole; Dir-Gen. LUIS LÓPEZ.

**CMBF** (Radio Musical Nacional): Calle N, No 266, entre 21 y 23, Vedado, Plaza de la Revolución, Havana; tel. (7) 32-0085; e-mail musical@ceniai.inf.cu; f. 1948; national network; classical music programmes; 17 hours daily; Dir LUIZ LÓPEZ-QUINTANA.

**Radio Progreso:** Infanta 105, esq. a 25, 6°, Apdo 3042, Havana; tel. (7) 877-5519; e-mail progreso@ceniai.inf.cu; internet www.radioprogreso.cu; f. 1929; national network; mainly entertainment and music; 24 hours daily; Dir-Gen. MANUEL E. ANDRÉS MAZORRA.

**Radio Rebelde:** Edif. ICRT, Avda 23, No 258, entre L y M, Vedado, Apdo 6277, Havana; tel. (7) 831-3514; fax (7) 33-4270; e-mail webrebelde@rrebelde.icrt.cu; internet www.radiorebelde.com.cu; f. 1958; merged with Radio Liberación in 1984; national network; 24-hour news and cultural programmes, music and sports; Dir-Gen. GERARDO CALDERÍN GAÍNZA.

**Radio Reloj:** Edif. Radiocentro, Avda 23, No 258, entre L y M, Vedado, Havana; tel. (7) 55-4185; fax (7) 31-2706; e-mail radioreloj@rreloj.icrt.cu; internet www.radioreloj.cu; f. 1947; national network; 24-hour news service; Dir ISIDRO BETANCOURT SILVA.

**Radio Taino:** Edif. Radiocentro, Avda 23, No 258, entre L y M, Vedado, Havana; tel. (7) 55-4181; fax (7) 55-4490; e-mail radiotaino@89.1fm.cu; internet radiotaino.cubasi.cu; f. 1985; broadcasts in English and Spanish.

#### Television

The Cuban Government holds a 19% stake in the regional television channel Telesur (q.v.), which began operations in May 2005 and is based in Caracas, Venezuela.

**Instituto Cubano de Radiodifusión (Televisión Nacional):** Calle 23, No 258, Vedado, Havana; tel. (7) 55-4059; fax (7) 33-3107; f. 1950; broadcasts through four channels; Pres. ERNESTO LÓPEZ DOMÍNGUEZ.

**Canal Educativo:** Calle 23, No 258, Vedado, Havana; tel. (7) 55-4059; fax (7) 33-3107; f. 2002; broadcasts on channel 13; educational; Dir IVÁN BARRETO.

**Cubavisión:** Calle M, No 313, Vedado, Havana; e-mail info@cubavision.icrt.cu; internet www.cubavision.cubaweb.cu; broadcasts on channel 6.

**Tele Rebelde:** Mazón, No 52, Vedado, Havana; tel. (7) 32-3369; broadcasts on channel 2; Vice-Pres. GARY GONZÁLEZ.

**CHTV:** Habana Libre Hotel, Havana; f. 1990; subsidiary station of Tele-Rebelde; Dir ROSA MARÍA FERNÁNDEZ SOFÍA.

## Finance

(cap. = capital; p.u. = paid up; res = reserves; dep. = deposits; m. = million; brs = branches)

### BANKING

All banks were nationalized in 1960. Legislation establishing the national banking system was approved by the Council of State in 1984. A restructuring of the banking system, initiated in 1995, to accommodate Cuba's transformation to a more market-orientated economy, was proceeding in 2004. A new central bank, the Banco Central de Cuba (BCC), was created in 1997 to supersede the Banco Nacional de Cuba (BNC). The BCC was to be responsible for issuing currency, proposing and implementing monetary policy and the regulation of financial institutions. The BNC was to continue functioning as a commercial bank and servicing the country's foreign debt. The restructuring of the banking system also allowed for the creation of an investment bank, the Banco de Inversiones, to provide medium- and long-term financing for investment, and the Banco Financiero Internacional, SA, to offer short-term financing. A new agro-industrial and commercial bank was also to be created to provide services for farmers and co-operatives. The new banking system is under the control of Grupo Nueva Banca, which holds a majority share in each institution. In 2002 there were eight commercial banks, 18 non-banking financial institutions, 13 representative offices of foreign banks and four representative offices of non-banking financial institutions operating in Cuba.

#### Central Bank

**Banco Central de Cuba (BCC):** Calle Cuba, No 402, Aguiar 411, Habana Vieja, Havana; tel. (7) 866-8003; fax (7) 866-6601; e-mail plascncia@bc.gov.cu; internet www.bc.gov.cu; f. 1997; sole bank of issue; Pres. FRANCISCO SOBERÓN VALDEZ.

#### Commercial Banks

**Banco de Crédito y Comercio (BANDEC):** Amargura 158, entre Cuba y Aguiar, Habana Vieja, Havana; tel. (7) 861-4533; fax (7) 866-8968; e-mail ileana@oc.bandec.cu; f. 1997; Pres. ILEANA ESTÉVEZ.

**Banco Exterior de Cuba:** Calle 23, No 55, esq. a P, Vedado, Municipio Plaza, Havana; tel. (7) 55-0795; fax (7) 55-0794; e-mail

CUBA
*Directory*

bec@bec.co.cu; f. 1999; cap. 450m. pesos, res 2.8m. pesos, dep. 65.5m. pesos (Dec. 2002); Pres. Jacobo Peison Weiner.

**Banco Financiero Internacional, SA:** Avda 5, No 9009, esq. 92, Miramar, Municipio Playa, Havana; tel. (7) 267-5000; fax (7) 267-5002; e-mail bfi@bfi.com.cu; f. 1984; autonomous; finances Cuba's foreign trade; Pres. Ernesto Medina; Gen. Man. (Int.) Nivaldo Puldón.

**Banco Internacional de Comercio, SA:** 20 de Mayo y Ayestarán, Apdo 6113, Havana 10600; tel. (7) 883-6038; fax (7) 883-6028; e-mail bicsa@bicsa.colombus.cu; f. 1993; cap. US $95.0m., res $11.3m., dep. $383.8m. (Dec. 2002); Chair. and Pres. Marcos Díaz.

**Banco Metropolitano:** Avda 5 y Calle 112, Playa, Havana 11600; tel. (7) 204-3869; fax (7) 204-9193; e-mail bm@banco-metropolitano.com; internet www.banco-metropolitano.cu; f. 1996; offers foreign currency and deposit account facilities; Pres. Manuel Vale; Dir-Gen. Pedro de la Rosa González.

**Banco Nacional de Cuba (BNC):** Aguiar 456, entre Amargura y Lamparilla, Habana Vieja, Havana; tel. (7) 862-8896; fax (7) 866-9390; e-mail bancuba@bnc.cu; f. 1950; reorganized 1997; Chair. Diana Amelia Fernández Vila.

### Foreign Banks

There are 13 foreign banks represented in Cuba, including Banco Bilbao Vizcaya Argentaria (Spain), Banco de Comercio Exterior de México, Banco Exterior de España, Banco Sabadell (Spain), ING Bank (Netherlands) and Société General de France.

### Savings Bank

**Banco Popular de Ahorro:** Calle 16, No 306, entre 3 y 5, Miramar, Playa, Havana; tel. (7) 202-2545; fax (7) 204-1180; internet www.bancopopulardeahorro.com; f. 1983; savings bank; cap. 30m. pesos, dep. 5,363.7m. pesos; Pres. Manuel Vale Marrero; Sec. Lourdes Pérez Soler; 520 brs.

### Investment Bank

**Banco de Inversiones, SA:** Avda 5, No 6802 esq. a 68, Miramar, Havana; ; tel. (7) 204-3374; fax (7) 204-3377; e-mail inversiones@bdi.cu; internet www.bdi.cu; Pres. Raúl Rangel.

## INSURANCE

### State Organizations

**Empresa del Seguro Estatal Nacional (ESEN):** Calle 5, No 306, entre C y D, Vedado, Havana; tel. (7) 32-2500; fax (7) 33-8717; e-mail esen@esen.com.cu; internet www.esen.cu; f. 1978; motor and agricultural insurance; Dir-Gen. Rafael J. Gonzalez Pérez.

**Seguros Internacionales de Cuba, SA (Esicuba):** Cuba No 314, entre Obispo y Obrapía, Habana Vieja, Havana; tel. (7) 33-8400; fax (7) 33-8038; e-mail esicuba@esicuba.cu; f. 1963; reorganized 1986; all classes of insurance except life; Pres. Ramón Martínez Carrera.

# Trade and Industry

## GOVERNMENT AGENCIES

**Ministry of Foreign Investment and Economic Co-operation:** Calle 1ra y 18, No 1803, Miramar, Havana; tel. (7) 203-7035; fax (7) 204-3496; e-mail ministro@minvec.cu; internet www.minvec.cu.

**Free-Trade Zones National Office:** Calle 22, No 528, entre 3 y 5, Miramar, Havana; tel. (7) 204-7636; fax (7) 204-7637; created and regulated by the Ministry of Foreign Investment and Economic Co-operation.

## CHAMBER OF COMMERCE

**Cámara de Comercio de la República de Cuba:** Calle 21, No 661/701, esq. Calle A, Apdo 4237, Vedado, Havana; tel. (7) 830-4436; fax (7) 833-3042; e-mail pdcia@camara.com.cu; internet www.camaracuba.cu; f. 1963; mems include all Cuban foreign trade enterprises and the most important agricultural and industrial enterprises; Pres. Bertha Delgado Guanche García; Sec.-Gen. Frank Abel Portela.

## AGRICULTURAL ORGANIZATION

**Asociación Nacional de Agricultores Pequeños (ANAP)** (National Association of Small Farmers): Calle I, No 206, entre Linea y 13, Vedado, Havana; tel. (7) 32-4541; fax (7) 33-4244; f. 1961; 220,000 mems; Pres. Orlando Lugo Fonte; Vice-Pres. Evelio Pausa Bello.

## STATE IMPORT-EXPORT BOARDS

**Alimport** (Empresa Cubana Importadora de Alimentos): Infanta 16, 3°, Apdo 7006, Havana; tel. (7) 54-2501; fax (7) 33-3151; e-mail precios@alimport.com.cu; f. 1962; controls import of foodstuffs and liquors; Man. Dir Pedro Alvarez Borrego.

**Autoimport** (Empresa Central de Abastecimiento y Venta de Equipos de Transporte Ligero): Galiano 213, entre Concordia y Virtudes, Havana; tel. (7) 61-5322; fax (7) 66-6549; e-mail eric@autoimport.com.cu; imports cars, light vehicles, motor cycles and spare parts; Dir José Arañaburu.

**Aviaimport** (Empresa Cubana Importadora y Exportadora de Aviación): Calle 182, No 126, entre 1 y 5, Rpto Flores, Playa, Havana; tel. (7) 33-0142; fax (7) 33-6234; e-mail aviaimport@avianet.cu; import and export of aircraft and components; Man. Dir Marcos Lago Martínez.

**Caribex** (Empresa Exportadora del Caribe): Aparthotel Las Brisas, Apdo 3c23, Villa Panamericana, Havana; tel. (7) 95-1121; fax (7) 95-1120; e-mail acepex@acepex.telemar.cu; internet www1.cubamar.cu; export of seafood and marine products; Dir Jacinto Fierro Barefoot.

**Catec** (Empresa Cuban Importadora, Exportadora y Comercializadora de Productos de la Ciencia y la Técnica Agropecuaria): Calle 148, No 905, esq. 9, Miramar, Havana; tel. (7) 208-2164; fax (7) 204-6071; e-mail alina@catec.co.cu; exports, imports and markets scientific and technical products relating to the farming and forestry industries; Dir-Gen. Osvaldo Carvejal Gabela.

**Construimport** (Empresa Central de Abastecimiento y Venta de Equipos de Construcción y sus Piezas): Carretera de Varona, Km 1½, Capdevila, Havana; tel. (7) 45-2567; fax (7) 66-6180; e-mail construimport@colombus.cu; f. 1969; controls the import and export of construction machinery and equipment; Man. Dir Jesús Serrano Rodríguez.

**Consumimport** (Empresa Cubana Importadora de Artículos de Consumo General): Calle 23, No 55, 9°, Apdo 6427, Vedado, Havana; tel. (7) 54-3110; fax (7) 54-2142; e-mail comer@consumimport.infocex.cu; f. 1962; imports and exports general consumer goods; Dir Mercedes Rey Hechavarría.

**Copextel** (Corporación Productora y Exportadora de Tecnología Electrónica): Calle 194 y 7a, Siboney, Havana; tel. (7) 21-8400; fax (7) 33-1414; e-mail copextel@copextel.com.cu; internet www.copextel.com.cu; f. 1985; exports LTEL personal computers and microcomputer software; Dir Norma M. García Bruzón.

**Coprefil** (Empresa Comercial y de Producciones Filatélicas): Avda 49, No 2831, Reparto Kohly, Havana; tel. (7) 204-9668; fax (7) 204-5077; e-mail coprefil@coprefil.cu; imports and exports postage stamps, postcards, calendars, handicrafts, communications equipment, electronics, watches, etc.; Dir Nelson Iglesias Fernández.

**Cubaelectrónica** (Empresa Importadora y Exportadora de Productos de la Electrónica): Calle 22, No 510, entre 5 y 7, Miramar, Havana; tel. (7) 204-0178; fax (7) 204-1233; e-mail mariaisabel@columbus.cu; f. 1986; imports and exports electronic equipment and devices; Pres. María Isabel Morejón Pérez.

**Cubaexport** (Empresa Cubana Exportadora de Alimentos y Productos Varios): Calle 23, No 55, entre Infanta y P, 8°, Vedado, Apdo 6719, Havana; tel. (7) 54-3130; fax (7) 33-3587; e-mail cexport@infocex.cu; export of foodstuffs and industrial products; Man. Dir Milda Picos Rivers.

**Cubafrutas** (Empresa Cubana Exportadora de Frutas Tropicales): Calle 23, No 55, Apdo 6683, Vedado, Havana; tel. and fax (7) 79-5653; f. 1979; controls export of fruits, vegetables and canned foodstuffs; Dir Jorge Amaro Morejón.

**Cubahidráulica** (Empresa Central de Equipos Hidráulicos): Carretera Vieja de Guanabacoa y Linea de Ferrocarril, Reparto Mañana, Guanabacoa, Havana; tel. (7) 97-0821; fax (7) 97-1627; e-mail cubahidraulica@enet.cu; internet www.cubahidraulica.cu; imports and exports hydraulic and mechanical equipment, parts and accessories; Dir-Gen. José Marrero Carnacho.

**Cubalse** (Empresa para Prestación de Servicios al Cuerpo Diplomático): Avda 3 y Final, Miramar, Havana; tel. (7) 204-2284; fax (7) 204-2282; e-mail cubalse@cm.cubalse.cma.net; internet www.cubalse.cu; f. 1974; imports consumer goods for the diplomatic corps and foreign technicians residing in Cuba; exports beverages and tobacco, leather goods and foodstuffs; other operations include real estate, retail trade, restaurants, clubs, automobile business, state-of-the-art equipment and household appliances, construction, investments, wholesale, road transport, freight transit, shipping, publicity, photography and video, financing, legal matters; Pres. Reidal Roncourt Font.

**Cubametales** (Empresa Cubana Importadora de Metales, Combustibles y Lubricantes): Infanta 16, 4°, Apdo 6917, Vedado, Havana; tel. (7) 70-4225; fax (7) 33-3477; e-mail pedro@cubametal.infocex.cu;

controls import of metals (ferrous and non-ferrous), crude petroleum and petroleum products; also engaged in the export of petroleum products and ferrous and non-ferrous scrap; Dir PEDRO PEREZ RODRÍGUEZ.

**Cubaniquel** (Empresa Cubana Exportadora de Minerales y Metales): Calle 23, No 55, 8°, Apdo 6128, Havana; tel. (7) 33-5334; fax (7) 33-3332; e-mail bcorrea@moa.minbas.cu; f. 1961; sole exporter of minerals and metals; Man. Dir RICARDO GONZÁLEZ.

**Cubatabaco** (Empresa Cubana del Tabaco): Calle O'Reilly, No 104, Apdo 6557, Havana; tel. (7) 861-5775; fax (7) 33-8214; e-mail juan@cubatabaco.cu; f. 1962; controls export of leaf tobacco, cigars and cigarettes to France; Dir JUAN MANUEL DÍAZ TENORIO.

**Cubatécnica** (Empresa de Contratación de Asistencia Técnica): Calle 12, No 513, entre 5 y 7, Miramar, Havana; tel. (7) 202-3270; fax (7) 204-0923; e-mail comercial@cubatecnica.cu; internet www.cubatecnica.cu; f. 1976; controls export and import of technical assistance; Dir FÉLIX GONZÁLEZ NAVERÁN.

**Cubatex** (Empresa Cubana Importadora de Fibras, Tejidos, Cueros y sus Productos): Calle 23, No 55, Apdo 7115, Vedado, Havana; tel. (7) 70-2531; fax (7) 33-3321; controls import of fibres, textiles, hides and by-products and export of fabric and clothing; Dir LUISA AMPARO SESÍN VIDAL.

**Cubazúcar** (Empresa Cubana Exportadora de Azúcar y sus Derivados): Calle 23, No 55, 7°, Vedado, Apdo 6647, Havana; tel. (7) 54-2175; fax (7) 33-3482; e-mail producer@cubazucar.com; internet www.cubazucar.com; f. 1962; controls export of sugar, molasses and alcohol; Pres. JOSÉ LÓPEZ SILVERO.

**Ecimact** (Empresa Comercial de Industrias de Materiales, Construcción y Turismo): Calle 1c, entre 152 y 154, Miramar, Havana; tel. (7) 21-9783; controls import and export of engineering services and plant for industrial construction and tourist complexes; Dir OCTAVIO CASTILLA CANGAS.

**Ecimetal** (Empresa Importadora y Exportadora de Objetivos Industriales): Calle 23, No 55, esq. Plaza, Vedado, Havana; tel. (7) 55-0548; fax (7) 33-4737; e-mail ecimetal@infocex.cu; f. 1977; controls import and export of plant, equipment and raw materials for all major industrial sectors; Dir CONCEPCIÓN BUENO.

**Ediciones Cubanas** (Empresa de Comercio Exterior de Publicaciones): Obispo 527, esq. Bernaza, Apdo 47, Havana; tel. (7) 863-1989; fax (7) 33-8943; e-mail edicuba@cubarte.cult.cu; controls import and export of books and periodicals; Dir NANCY MATOS LACOSTA.

**Egrem** (Estudios de Grabaciones y Ediciones Musicales): Avda 3, No 1008, entre 10 y 12, Miramar, Playa, Havana; tel. (7) 204-1925; fax (7) 204-2519; e-mail relaciones@egrem.cult.cu; f. 1964; controls the import and export of records, tapes, printed music and musical instruments; Dir Gen. JULIO BALLESTER GUZMÁN.

**Emexcon** (Empresa Importadora y Exportadora de la Construcción): Calle 25, No 2602, esq. a 26, Playa, Havana; tel. (7) 204-2263; fax (7) 204-1862; e-mail enrique@emexcon.com.cu; f. 1978; consulting engineer services, contracting, import and export of building materials and equipment; Dir ENRIQUE MARTÍNEZ DE LA FÉ.

**Emiat** (Empresa Importadora y Exportadora de Suministros Técnicos): Avda 47, No 2828, entre 28 y 34, Rpto Kohly, Havana; tel. (7) 203-0345; fax (7) 204-9353; e-mail emiat@enet.cu; f. 1983; imports technical materials, equipment and special products; exports furniture, kitchen utensils and accessories; Man. FIDEL GARCÍA HERNÁNDEZ.

**Emidict** (Empresa Especializada Importadora, Exportadora y Distribuidora para la Ciencia y la Técnica): Calle 16, No 102, esq. Avda 1, Miramar, Playa, 13000 Havana; tel. (7) 203-5316; fax (7) 204-1768; e-mail emidict@ceniai.inf.cu; internet www.emidict.com.cu; f. 1982; controls import and export of scientific and technical products and equipment, live animals; scientific information; Dir-Gen. CARLOS CANALES ENRÍQUEZ.

**Energoimport** (Empresa Importadora de Objetivos Electro-energéticos): Amenidad No 124, entre Nueva y 20 de Mayo, Municipio Cerro, 10600 Havana; tel. (7) 70-2501; fax (7) 66-6079; e-mail energoimport@energonet.com.cu; internet www.energonet.com.cu; f. 1977; controls import of equipment for electricity generation; Man. ANDRÉS MONTEZ PERRERA.

**Eprob** (Empresa de Proyectos para las Industrias de la Básica): Avda 31A, No 1805, entre 18 y 20, Edif. Las Ursulinas, Miramar, Playa, Apdo 12100, Havana; tel. (7) 202-5562; fax (7) 204-2146; f. 1967; exports consulting services and processing of engineering construction projects, consulting services and supplies of complete industrial plants and turn-key projects; Man. Dir GLORIA EXPÓSITO DÍAZ.

**Eproyiv** (Empresa de Proyectos para Industrias Varias): Calle 33, No 1815, entre 18 y 20, Playa, Havana; tel. (7) 24-2149; e-mail eproyiv@ceniai.inf.cu; f. 1967; consulting services, feasibility studies, development of basic and detailed engineering models, project management and turn-key projects; Dir MARTA ELENA HERNÁNDEZ DÍAZ.

**Esi** (Empresa de Suministros Industriales): Calle Aguiar, No 556, entre Teniente Rey y Muralla, Havana; tel. (7) 62-0696; fax (7) 33-8951; f. 1985; imports machinery, equipment and components for industrial plants; Dir-Gen. FRANCISCO DÍAZ CABRERA.

**Fecuimport** (Empresa Cubana Importadora y Exportadora de Ferrocarriles): Avda 7A, No 6209, entre 62 y 66, Apdo 6003, Miramar, Havana; tel. (7) 203-3764; f. 1968; imports and exports railway equipment; Pres. DOMINGOS HERRERA.

**Ferrimport** (Empresa Cubana Importadora de Artículos de Ferretería): Calle 23, No 55, 2°, Vedado, Apdo 6258, Havana; tel. (7) 870-6678; fax (7) 879-4417; f. 1965; importers of industrial hardware; Dir-Gen. ALEJANDRO MUSTELIER.

**Fondo Cubano de Bienes Culturales:** Calle 36, No 4702, esq. Avda 47, Reparto Kohly, Playa, Havana; tel. (7) 204-8005; fax (7) 204-0391; e-mail fcbc@fcbc.cult.cu; f. 1978; controls export of fine handicraft and works of art; Dir-Gen. JOSÉ GONZÁLEZ FERNÁNDEZ-LARREA.

**Habanos, SA:** Avda 3, No 2006, entre 20 y 22, Miramar, Havana; tel. (7) 204-0510; fax (7) 204-0511; e-mail habanos@habanos.cu; internet www.habanos.net; f. 1994; controls export of leaf and pipe tobacco, cigars and cigarettes to all markets; jt venture with Altadis, SA (Spain).

**ICAIC** (Instituto Cubano del Arte e Industria Cinematográficos): Calle 23, No 1155, Vedado, Havana 4; tel. (7) 55-3128; fax (7) 33-3032; f. 1959; production, import and export of films and newsreel; Dir ANTONIO RODRÍGUEZ RODRÍGUEZ.

**Imexin** (Empresa Importadora y Exportadora de Infraestructura): Avda 5, No 1007, esq. a 12, Miramar, Havana; tel. (7) 204-0658; fax (7) 204-0622; e-mail imexinsa@ceniai.inf.cu; f. 1977; controls import and export of infrastructure; Man. Dir RAÚL BENCE VIJANDE.

**Imexpal** (Empresa Importadora y Exportadora de Plantas Alimentarias, sus Complementos y Derivados): Calle 22, No 313, entre 3 y 5, Miramar, Havana; tel. (7) 29-1671; controls import and export of food-processing plants and related items; Man. Dir Ing. CONCEPCIÓN BUENO CAMPOS.

**Maprinter** (Empresa Cubana Importadora y Exportadora de Materias Primas y Productos Intermedios): Edif. MINCEX, Calle 23 y 55, entre P y Infanta, 8°, Plaza de la Revolución, Havana; tel. (7) 878-0711; fax (7) 833-3535; e-mail direccion@maprinter.mincex.cu; internet www.maprinter.cu; f. 1962; controls import and export of raw materials and intermediate products; Dir-Gen. ODALYS ÁLDAMA VALDÉS.

**Maquimport** (Empresa Cubana Importadora de Maquinarias y Equipos): Calle 23, No 55, 6°, entre P y Infanta, Vedado, Apdo 6052, Havana; tel. (7) 55-0632; fax (7) 66-2217; e-mail direccion@maquimport.mincex.cu; internet www.maquimport-cuba.com; imports industrial goods and equipment; Dir ROBERTO E. TORRES.

**Marpesca** (Empresa Cubana Importadora y Exportadora de Buques Mercantes y de Pesca): Conill No 580, esq. Avda 26, Nuevo Vedado, Havana; tel. (7) 881-1846; fax (7) 879-1010; f. 1978; imports and exports ships and port and fishing equipment; Pres. JOSÉ CEREIJO CASAS.

**Medicuba** (Empresa Cubana Importadora y Exportadora de Productos Médicos): Máximo Gómez 1, esq. a Egido, Havana; tel. (7) 62-4061; fax (7) 33-8516; e-mail alfonso@medicuba.sld.cu; enterprise for the export and import of medical and pharmaceutical products; Dir ALFONSO SÁNCHEZ DÍAZ.

**Produimport** (Empresa Central de Abastecimiento y Venta de Productos Químicos y de la Goma): Calle Consulado 262, entre Animas y Virtudes, Havana; tel. (7) 62-0581; fax (7) 62-9588; f. 1977; imports and exports spare parts for motor vehicles; Dir JOSÉ GUERRA MATOS.

**Propes** (Empresa Importadoro y Proveedora de Productos para la Pesca): Calle 22, No 2, esq. Calzada, Vedado, Havana; tel. (7) 830-3770; fax (7) 55-1729; e-mail pesmar@apropes.fishnavy.inf.cu; importer and distributor of a wide variety of equipment and accessories pertaining to the fishing industry; Dir-Gen. PEDRO BLAS ARTEAGA.

**Quimimport** (Empresa Cubana Importadora y Exportadora de Productos Químicos): Calle 23, No 55, entre Infanta y P, Apdo 6088, Vedado, Havana; tel. (7) 33-3394; fax (7) 33-3190; e-mail global@quimimport.infocex.cu; controls import and export of chemical products; Dir ARMANDO BARRERA MARTÍNEZ.

**Suchel** (Empresa de Jabonería y Perfumería): Calzada de Buenos Aires 353, esq. a Durege, Apdo 6359, Havana; tel. (7) 649-8008; fax (7) 649-5311; e-mail direccion@suchel.co.cu; internet www.suchel.cu; f. 1977; imports materials for the detergent, perfumery and cosmetics industry, exports cosmetics, perfumes, hotel amenities and household products; Dir JOSÉ GARCÍA DÍAZ.

CUBA
*Directory*

**Tecnoazúcar** (Empresa de Servicios Técnicos e Ingeniería para la Agro-industria Azucarera): Calle 12, No 310, entre 3 y 5, Miramar, Playa, Havana; tel. (7) 29-5441; fax (7) 33-1218; e-mail promocion@tecnoazucar.cu; imports machinery and equipment for the sugar industry, provides technical and engineering assistance for the sugar industry; exports sugar-machinery equipment and spare parts; provides engineering and technical assistance services for sugar-cane by-product industry; Gen. Man. VICTOR R. HERNÁNDEZ MARTÍNEZ.

**Tecnoimport** (Empresa Importadora y Exportadora de Productos Técnicos): Edif. La Marina, Avda del Puerto 102, entre Justiz y Obrapía, Habana Vieja, Havana; tel. (7) 861-5552; fax (7) 66-9777; e-mail celeste@ti.gae.com.cu; f. 1968; imports technical products; Dir ADEL IZQUIERDO RODRÍGUEZ.

**Tecnotex** (Empresa Cubana Exportadora e Importadora de Servicios, Artículos y Productos Técnicos Especializados): Avda 47, No 3419, Playa, Havana; tel. (7) 861-3536; fax (7) 66-6270; e-mail ailede@tecnotex.qae.com.cu; f. 1983; imports specialized technical and radiocommunications equipment, exports outdoor equipment and geodetic networks; Dir RENÉ ROJAS RODRÍGUEZ.

**Tractoimport** (Empresa Central de Abastecimiento y Venta de Maquinaria Agrícola y sus Piezas de Repuesto): Avda Rancho Boyeros y Calle 100, Apdo 7007, Havana; tel. (7) 45-2166; fax (7) 267-0786; e-mail direccion@tractoimport.co.cu; f. 1962 for the import of tractors and agricultural equipment; also exports pumps and agricultural implements; Dir-Gen. ABDEL GARCÍA GONZÁLEZ.

**Transimport** (Empresa Central de Abastecimiento y Venta de Equipos de Transporte Pesados y sus Piezas): Calle 102 y Avda 63, Marianao, Apdo 6665, 11500 Havana; tel. (7) 260-0329; fax (7) 267-9050; e-mail direccion@transimport.co.cu; f. 1968; controls import and export of vehicles and transportation equipment; Dir-Gen. JUAN CARLOS TASSÉ BELLOT.

### OTHER MAJOR COMPANIES

**BrasCuba Cigarrilos, SA:** Calle Princesa, No 202, entre Reyes y San José, Luyanó, Havana; tel. (7) 55-7510; fax (7) 66-9306; e-mail brascuba@ceniai.inf.cu; f. 1995; manufactures and markets cigarettes; jt venture between TabaCuba (Grupo Empresarial del Tobaco) and Souza Cruz, SA (Brazil); Pres. FLAVIO DE ANDRADE.

**CariFin** (Caribbean Finance Investments Ltd): Calle 22, Nos 311 y 313, entre 3 y 5, Miramar, Havana; tel. (7) 24-4468; fax (7) 24-4140; e-mail havana@cdc.com.cu; f. 1996 in the British Virgin Islands; started lending operations in Cuba in 1997; subsidiary of the Commonwealth Devt Corpn and Grupo Nueva Banca, SA; financial services such as loans to businesses, international money transfers and leasing of equipment and machinery; Dir WILLIAM WHITE; Gen. Man. STEVEN MACQUEEN; 23 employees.

**Corporación Cuba Ron, SA:** Calle 200, No 1708, esq. 17, Reparto Atabey, Playa, Havana; tel. (7) 273-6602; fax (7) 273-6601; e-mail cubaron@cubaron.co.cu; internet www.cubaron.com; f. 1993; production, marketing and distribution of rum and other alcoholic beverages; Pres. LUIS PERDOMO HERNÁNDEZ.

**Cubalub** (Empresa Cubana de Lubricantes): Calle Oficios No 154, entre Amargura y Teniente Rey, 3°, Habana Vieja, Havana; tel. (7) 861-6512; fax (7) 867-9197; e-mail mabel@cupet.minbas.cu; manufactures and markets lubricants; Dir-Gen. MIGUEL VALDIVIA GONZÁLEZ.

**Cubapetróleo** (Cupet): Calle 11, No 511, Havana; extraction and production of petroleum; Dir FIDEL RIVERO PRIETO; Vice-Pres. JUAN FLEITES MELO.

**Electrocimex, SA:** Apdo B1, Miramar, Havana 00290; tel. (7) 33-2938; fax (7) 33-2100; f. 1968; distribution of electrical and electronic products; Man. Dir LUIS LERA; 98 employees.

**Grupo Industrial Unecamoto:** Calle E, No 12, esq. 3, Vedado, Havana; tel. (7) 33-1645; fax (7) 33-6545; e-mail unecamoto@unecamoto.com.cu; internet www.unecamoto.com.cu; produces automobile parts and accessories; Pres. CRISTÓBAL VÁZQUEZ EGAÑA.

**Grupo Refrigeración y Calderas:** Calle 31, No 19811, entre 198 y 208, La Coronela, La Lisa, Havana; tel. (7) 33-8090; fax (7) 33-8501; e-mail emilio@rc.columbus.cu; f. 1985; manufacturer of refrigerators and air-conditioning appliances; Man. Dir EMILIO MARILL FREYRE DE ANDRADE; 2,200 employees.

**Holmer Gold Mines Ltd:** Calle A, No 506, entre 21 y 23, Vedado, Apdo 5 Interior, Havana; tel. (7) 33-3966; fax (7) 66-2074; e-mail holmer@ceniai.inf.cu; internet www.holmergold.com; f. 1993; Canadian exploration and prospecting mining co; responsible for the opening of Cuba's first silver mine in 1999; Pres. and CEO Dr K. SETHU RAMAN; Gen. Sec. ED SVOBODA; 8 employees.

**TabaCuba** (Grupo Empresarial del Tabaco): Avda Independencia, entre Conill y Sta Ana, 5°, Havana; tel. (7) 82-0575; f. 2000; state-owned; regulates tobacco cultivation and cigar production in Cuba; Pres. OSCAR BASULTO TORRES.

**Unión de Empresas Constructoras Caribe (Corporación Uneca):** Calle 9A, No 614, entre 6 y 10, Miramar, Havana; tel. (7) 209-3396; fax (7) 204-1637; e-mail dmarket@uneca.co.cu; f. 1984; construction work, particularly within the tourism industry; sales of US $135m. (1995); Pres. MANUEL TOMÁS VÁZQUEZ ENRÍQUEZ.

### UTILITIES

#### Electricity

**Unión Nacional Eléctrica (UNE):** Havana; public utility; Dir-Gen. JUAN PRUNA.

#### Water

**Instituto Nacional de Recursos Hidraulicos (INRH)** (National Water Resources Institute): Virtudes 680, esq. Belascoain, Havana; regulatory body; Vice-Pres. AYMEE AGUIRRE HERNÁNDEZ.

### TRADE UNIONS

All workers have the right to become members of a national trade union according to their industry and economic branch.

The following industries and labour branches have their own unions: Agriculture, Chemistry and Energetics, Civil Workers of the Revolutionary Armed Forces, Commerce and Gastronomy, Communications, Construction, Culture, Defence, Education and Science, Food, Forestry, Health, Light Industry, Merchant Marine, Mining and Metallurgy, Ports and Fishing, Public Administration, Sugar, Tobacco and Transport.

**Central de Trabajadores de Cuba (CTC)** (Confederation of Cuban Workers): Palacio de los Trabajadores, San Carlos y Peñalver, Havana; tel. (7) 78-4901; fax (7) 55-5927; e-mail digital@trabaja.cip.cu; internet www.trabajadores.cubaweb.cu; f. 1939; affiliated to WFTU and CPUSTAL; 19 national trade unions affiliated; Gen. Sec. PEDRO ROSS LEAL; 2,767,806 mems (1996).

## Transport

The Ministry of Transportation controls all public transport.

### RAILWAYS

The total length of railways in 1998 was 14,331 km, of which 9,638 km were used by the sugar industry. The remaining 4,520 km were public service railways operated by Ferrocarriles de Cuba. All railways were nationalized in 1960. In 2001 Cuba signed an agreement with Mexico for the maintenance and repair of rolling stock.

**Ferrocarriles de Cuba:** Edif. Estación Central, Egido y Arsenal, Havana; tel. (7) 70-1076; fax (7) 33-1489; f. 1960; operates public services; Dir-Gen. FERNANDO PÉREZ LÓPEZ; divided as follows:

**División Occidente:** serves Pinar del Río, Ciudad de la Habana, Havana Province and Matanzas.

**División Centro:** serves Villa Clara, Cienfuegos and Sancti Spíritus.

**División Centro-Este:** serves Camagüey, Ciego de Avila and Tunas.

**División Oriente:** serves Santiago de Cuba, Granma, Guantánamo and Holguín.

**División Camilo Cienfuegos:** serves part of Havana Province and Matanzas.

### ROADS

In 1999 there were an estimated 60,858 km of roads, of which 4,353 km were highways or main roads. The Central Highway runs from Pinar del Río in the west to Santiago, for a length of 1,144 km. In addition to this paved highway, there are a number of secondary and 'farm-to-market' roads. A small proportion of these secondary roads is paved, but many can be used by motor vehicles only during the dry season.

### SHIPPING

Cuba's principal ports are Havana (which handles 60% of all cargo), Santiago de Cuba, Cienfuegos, Nuevitas, Matanzas, Antilla, Guayabal and Mariel. Maritime transport has developed rapidly since 1959, and at 31 December 2004 Cuba had a merchant fleet of 96 ships (with an approximate total displacement of 126,000 grt). In 2000 a US $100m. renovation and enlargement project for the port of Mariel was announced.

**Coral Container Lines, SA:** Calle Oficios 170, 1°, Habana Vieja, Havana; tel. (7) 67-0854; fax (7) 67-0850; e-mail caribe@coral.com

.cu; f. 1994; liner services to Europe, Canada, Brazil and Mexico; 11 containers; Chair. and Man. Dir EVELIO GONZÁLEZ GONZÁLEZ.

**Empresa Consignataria Mambisa:** San José No 65, entre Prado y Zulueta, Habana Vieja, Havana; tel. (7) 862-2061; fax (7) 33-8111; e-mail mercedes@mambisa.transnet.cu; shipping agent, bunker suppliers; Man. Dir MERCEDES PÉREZ NEWHALL.

**Empresa Cubana de Fletes (Cuflet):** Calle Oficios No 170, entre Teniente Rey y Amargura, Apdo 6755, Havana; tel. (7) 61-2604; e-mail antares@antares.transnet.cu; freight agents for Cuban cargo; Man. Dir CARLOS SÁNCHEZ PERDOMO.

**Empresa de Navegación Caribe (Navecaribe):** Calle San Martín, No 65, 4°, entre Agramonte y Pasco de Martí, Habana Vieja, Havana; tel. (7) 61-8611; fax (7) 33-8564; e-mail enccom@transnet.cu; f. 1966; operates Cuban coastal fleet; Dir RAMÓN DURÁN SUÁREZ.

**Empresa de Navegación Mambisa:** San Ignacio No 104, Apdo 543, Havana; tel. (7) 869-7901; fax (7) 61-0044; operates dry cargo, reefer and bulk carrier vessels; Gen. Man. GUMERSINDO GONZÁLEZ FELIÚ.

**Naviera del Caribe (Carimar):** Ofs 170, entre Amargura y Teniente Rey, 3°, Habana Vieja, Havana; tel. (7) 67-0925; fax (7) 204-8627; e-mail ftarrau@coral.com.cu.

**Naviera Frigorífica Marítima (Friomar):** 5a Avda y 240, Barlovento, Playa, Havana; tel. (7) 209-8171; fax (7) 204-5864; e-mail friocom@fishnavy.inf.cu; specializes in shipping of refrigerated cargo; Dir JORGE FERNÁNDEZ.

**Naviera Mar América:** 5a Avda y 246, Edif. No 3, 1°, Barlovento, Playa, Havana; tel. (7) 209-8076; fax (7) 204-8889; e-mail nubia@maramerica.fishnavy.inf.eu.

**Naviera Petrocost:** 5a Avda y 246, Barlovento, Playa, Havana; tel. (7) 209-8067; fax (7) 204-5113; e-mail aleida@petrocost.fishnavy.inf.cu; transports liquid cargo to domestic and international destinations.

**Naviera Poseidon:** 5a Avda y 246, Edif. No 3, 2°, Barlovento, Playa, Havana; tel. (7) 209-8073; fax (7) 204-8627; e-mail yepe@poseidon.fishnavy.inf.cu.

**Nexus Reefer:** 5a Avda y 246, Edif. No 7, 1°, Barlovento, Playa, Havana; tel. (7) 204-8205; fax (7) 204-8490; e-mail sandra@antares.fishnavy.inf.cu; merchant reefer ships; Gen. Dir QUIRINO L. GUTIÉRREZ LÓPEZ.

### CIVIL AVIATION

There are a total of 21 civilian airports, with 11 international airports, including Havana, Santiago de Cuba, Camagüey, Varadero and Holguín. Abel Santamaría International Airport opened in Villa Clara in early 2001. In January 2003 the King's Gardens International Airport in Cayo Coco was opened. The airport formed part of a new tourist 'offshore' centre. The international airports were all upgraded and expanded during the 1990s and a third terminal was constructed at the José Martí International Airport in Havana. In November 2001 three North American airlines were permitted to commence direct flights from Miami and New York to Havana.

**Aerocaribbean:** Calle 23, No 64 esq. a P, Vedado, Havana; tel. (7) 832-7584; fax (7) 336-5016; e-mail vpcr@cacsa.avianet.cu; internet www.aero-caribbean.com; f. 1982; international and domestic scheduled and charter services; Chair. JULIÁN ALVAREZ INFIESTA.

**Aerogaviota:** Avda 47, No 2814, entre 28 y 34, Reparto Kolhy, Havana; tel. (7) 203-0668; fax (7) 204-2621; e-mail vpcom@aerogaviota.avianet.cu; f. 1994; operated by Cuban air force.

**Empresa Consolidada Cubana de Aviación (Cubana):** Calle 23, No 64, esq. a Infanta, Vedado, Havana; tel. (7) 33-4949; fax (7) 33-4056; e-mail vpcom@cubana.avianet.cu; internet www.cubana.cu; f. 1929; international services to North America, Central America, the Caribbean, South America and Europe; internal services from Havana to 14 other cities; Gen. Dir HERIBERTO PRIETO.

**Instituto de Aeronáutica Civil de Cuba (IACC):** Calle 23, No 64, Plaza de la Revolución, Vedado, Havana; tel. (7) 33-4949; fax (7) 33-4553; e-mail iacc@avianet.cu; internet www.cubagob.cu/des_eco/iacc/home.htm; f. 1985; Pres. ROGELIO ACEVEDO GONZÁLEZ.

## Tourism

Tourism began to develop after 1977, with the easing of travel restrictions by the USA, and Cuba subsequently attracted European tourists. In 2000 the number of hotel rooms had increased to 34,000. In that year there were 189 hotels. A number of hotel tourism complexes were under construction in the early 2000s. In 2002 there were approximately 40,000 hotel rooms; in the same year approximately 1,686,200 tourists visited the island. The number of tourists increased to approximately 1,905,700 in 2003, when receipts from tourism totalled an estimated US $1,846m. It was feared that the tightening of restrictions on travel to Cuba by the USA in mid-2004 would have an adverse effect upon the continued development of the country's tourism sector.

**Cubanacán:** Calle 23 No 156, entre O y P, Vedado, Havana 10400; tel. (7) 833-4090; fax (7) 22-8382; e-mail com_electronic@cubanacan.cyt.cu; internet www.cubanacan.cu; f. 1987; Pres. MANUEL VILA.

**Empresa de Turismo Internacional (Cubatur):** Calle F, No 157, entre Calzada y Novena, Vedado, Havana; tel. (7) 835-4155; fax (7) 836-3170; e-mail casamatriz@cubatur.cu; internet www.cubatur.cu; f. 1968; Dir JOSÉ PADILLA.

**Empresa de Turismo Nacional (Viajes Cuba):** Calle 20, No 352, entre 21 y 23, Vedado, Havana; tel. (7) 30-0587; e-mail info@viajesmorella.com; internet www.viajes-cuba.com; f. 1981; Dir ANA ELIS DE LA CRUZ GARCÍA.

## Defence

At August 2004, according to Western estimates, Cuba's Revolutionary Armed Forces numbered 49,000 (including ready reserves serving 45 days per year to complete active and reserve units): Army 38,000, Navy 3,000 and Air Force 8,000. Cuba's paramilitary forces included 20,000 State Security troops, 6,500 border guards, a civil defence force of 50,000 and a Youth Labour Army of some 70,000. A local militia organization (Milicias de Tropas Territoriales—MTT), comprising an estimated 1m. men and women, was formed in 1980. Despite Cuban hostility, the USA maintains a base at Guantánamo Bay, which comprised 1,600 army, 510 naval, 80 marine and 65 air force personnel in 2004. In June 1993, in accordance with the unilateral decision of the then USSR in September 1991, the 3,000-strong military unit of the former USSR, which had been stationed in Cuba since 1962, was withdrawn. In December 2001 Russia closed the Lourdes military electronic surveillance base, first opened in 1964. Conscription for military service is for a two-year period from 17 years of age, and conscripts also work on the land.

**Defence Expenditure:** Expenditure on defence and internal security for 2003 was estimated at US $1,200m.

**Minister of the Revolutionary Armed Forces:** Gen. RAÚL CASTRO RUZ.

**Chief of Staff:** Gen. ALVARO LÓPEZ MIERA.

## Education

State education in Cuba is universal and free at all levels. Education is based on Marxist-Leninist principles and combines study with manual work. National schools at the pre-primary level are available for children of five years of age, and day nurseries are available for all children over 45 days old. Primary education is compulsory for children aged six to 11 years. Secondary education lasts from 12 to 17 years of age, comprising two cycles of three years each. In 1997 almost 100% of children in the appropriate age group (males 99.9%; females 99.9%) attended primary schools, while 69.9% of those in the relevant age group (males 67.2%; females 72.6%) were enrolled at secondary schools. In 2003/04 there were an estimated 273,100 students enrolled in higher education. Workers attending university courses receive a state subsidy to provide for their dependants. Courses at intermediate and higher levels lay an emphasis on technology, agriculture and teacher training. In 2002 budgetary expenditure on education was estimated at 2,790m. pesos (16.4% of total spending).

# Bibliography

For works on the Caribbean generally, see Select Bibliography (Books)

Arboleya, J. *Havana–Miami: The US–Cuban Migration Conflict*. Melbourne, Ocean Press, 1996.

Azicri, M., and Deal, E. (Eds). *Cuban Socialism in a New Century: Adversity, Survival and Renewal*. Gainesville, FL, University Press of Florida, 2005.

Azicri, M., and Kirk, J. M. *Cuba Today and Tomorrow*. Gainesville, FL, University Press of Florida, 2001.

Baez, A. C. *State Resistance to Globalisation in Cuba*. London, Pluto Press, 2004.

Balfour, S. *Castro: Profiles in Power*. London, Longman, 1995.

Basdeo, S., and Nicol, H. N. (Eds). *Canada, the United States, and Cuba: An Evolving Relationship*. Boulder, CO, Lynne Rienner Publrs, 2002.

Blight, J. A., and Welch, D. A. (Eds). *Intelligence and the Cuban Missile Crisis*. London, Frank Cass, 1998.

Brenner, P., et al. *The Cuba Reader: The Making of a Revolutionary Society*. New York, NY, Grove Press, 1998.

Calvo, H., and Declercq, K. *The Cuban Exile Movement: Dissidents or Mercenaries?* Melbourne, Ocean Press, 2000.

Centeno, M. A., and Font, M. *Towards a New Cuba?* London, Lynne Rienner Publrs, 1997.

Chaffee, W. A., and Prevost, G. (Eds). *Cuba: A Different America*. Lanham, MD, Rowman & Littlefield Publrs, 2002.

Chrisp, P. *The Cuban Missile Crisis*. London, Hodder Wayland, 2001.

Cirules, E. *The Mafia in Havana: A Caribbean Mob Story*. Melbourne, Ocean Press, 2004.

Cole, K. *Cuba: From Revolution to Development*. London, Pinter, 1998.

Coltman, L. *The Real Fidel Castro*. New Haven, CT, Yale University Press, 2003.

*Cuba's Economic Reforms: Results and Future Prospects*. London, Cuba Business, 1997.

*Cuba: Evolucíon Económica durante 1996*. UN Economic Commission for Latin America and the Caribbean, 1997.

Eckstein, S. E. *Back from the Future: Cuba Under Castro*. London, Routledge, 2003.

*Evolución de la Economía Cubana en 1996*. Havana, Centro de Estudios sobre la Economía Cubana, 1997.

Elliston, J. (Ed.). *Psywar on Cuba: The Declassified History of US Anti-Castro Propaganda*. Melbourne, Ocean Press, 1999.

Escalante, F. *The Cuba Project: CIA Covert Operations Against Cuba 1959–62*. St Paul, MN, Consortium, 2004.

Falcoff, M. *Cuba the Morning After: Normalization and its Discontents*. Washington, DC, AEI Press, 2003.

Fernandez, S. J. *Encumbered Cuba*. Gainesville, FL, University Press of Florida, 2002.

Ferrer, A. *Insurgent Cuba: Race, Nation, and Revolution, 1868–1898*. Chapel Hill, NC, University of North Carolina Press, 1999.

Franklin, J. *Cuba and the United States: A Chronological History*. Melbourne, Ocean Press, 1997.

Fuente, A. de la. *A Nation for All: Race, Inequality, and Politics in Twentieth-Century Cuba (Envisioning Cuba)*. Chapel Hill, NC, University of North Carolina Press, 2001.

García Luis, J. (Ed.). *Cuban Revolution Reader: A Documentary History of 40 Years of Revolution*. Melbourne, Ocean Press, 2000.

González, E., and McCarthy K. *Cuba After Castro: Legacies, Challenges and Impediments*. Santa Monica, CA, RAND Corpn, 2004.

González, M. *Che Guevara and the Cuban Revolution*. London, Bookmarks Publications, 2004.

Gott, R. *Cuba: A New History*. New Haven, CT, Yale University Press, 2004.

Haines, L. *Reassessing Cuba: Emerging Opportunities and Operating Challenges*. New York, NY, Economist Intelligence Unit, 1997.

Horowitz, I. L. (Ed.). *Cuban Communism, 1959–1995*, 10th edn. New Brunswick, NJ, Transaction Books, 2001.

Jenkins, G., and Haines, L. *Cuba: Prospects for Reform, Trade and Investment*. New York, NY, Economist Intelligence Unit, 1995.

Johnson, S. *The Social Transformation of Eighteenth-Century Cuba*. Gainesville, FL, University Press of Florida, 2001.

Kaplowitz, D. R. (Ed.). *Cuba's Ties to a Changing World*. Boulder, CO, Lynne Rienner Publrs, 1993.

Kirk, J. M., and Padura Fuentes, L. *Culture and the Cuban Revolution*. Gainesville, FL, University Press of Florida, 2001.

Lambie, G. *Building Cuban Democracy*. Basingstoke, Palgrave Macmillan, 2005.

Lechuga, C. *Cuba and the Missile Crisis*. Melbourne, Ocean Press, 2001.

Leonard, T. M. *Encyclopedia of Cuban-United States Relations*. Jefferson, NC, McFarland & Company, 2003.

Levine, R. *Secret Missions to Cuba*. Basingstoke, Palgrave Macmillan, 2002.

Lievesley, G. *The Cuban Revolution: Past, Present and Future*. Basingstoke, Palgrave Macmillan, 2003.

López, J. J. *Democracy Delayed: The Case of Castro's Cuba*. Baltimore, MD, Johns Hopkins University Press, 2002.

McCoy, T. *Cuba on the Verge: An Island in Transition*. New York, NY, Little, Brown USA, 2003.

Meso-Lago, C. (Ed.). *Cuba after the Cold War*. Pittsburgh, PA, University of Pittsburgh Press, 1993.

Moses, C. *Real Life in Castro's Cuba (Latin American Silhouettes)*. Wilmington, DE, Scholarly Resources, 1999.

Paris, M. L. *Embracing America*. Gainesville, FL, University Press of Florida, 2002.

Pérez, Jr, L. A. *Cuba: Between Reform and Revolution*, 2nd edn. Oxford, Oxford University Press, 1995.

*Cuba and the United States: Ties of Singular Intimacy*, 2nd edn. Athens, GA, University of Georgia Press, 1997.

*On Becoming Cuban: Identity, Nationality, and Culture*. Chapel Hill, NC, University of North Carolina Press, 1999.

*Winds of Change: Hurricanes and the Transformation of Nineteenth-Century Cuba*. Chapel Hill, NC, University of North Carolina Press, 2001.

Pérez-López, J. F. *Cuba at a Crossroads (Politics and Economics After the Fourth Party Congress)*. Gainesville, FL, University Press of Florida, 1994.

Pérez Sarduy, P., and Stubbs, J. (Eds). *AfroCuba. An Anthology of Cuban Writing on Race, Politics and Culture*. Melbourne, Ocean Press, 1993.

Prince, R., and Taylor, S. *Turning with the Enemy*. London, Channel Four Television, 1998.

Purcell, S. K., and Rothkopf, D. (Eds). *Cuba: The Contours of Change*. Boulder, CO, Lynne Rienner Publrs, 2000.

Ritter, A. R. M. (Ed.). *The Cuban Economy (Pitt Latin American Series)*. Pittsburgh, PA, University of Pittsburgh Press, 2004.

Robins, N. A. *The Culture of Conflict in Modern Cuba*. Jefferson, NC, McFarland and Co, 2002.

Robinson, E. *Last Dance in Havana: The Final Days of Fidel and the Start of the New Cuban Revolution*. New York, NY, Free Press, 2004.

Rodríguez, J. C. *The Bay of Pigs*. Melbourne, Ocean Press, 1999.

Roman, P. *People's Power: Cuba's Experience with Representative Government*. Lanham, MD, Rowman & Littlefield Publrs, 2003.

Saney, I. *Cuba: A Revolution in Motion*. London, Zed Books, 2004.

Shaffer, K. R. *Anarchism and Countercultural Politics in Early Twentieth-Century Cuba*. Gainesville, FL, University Press of Florida, 2005.

Suchlicki, J. *Cuba: From Columbus to Castro and Beyond*. London, Brassey's, 2002.

Sweig, J. E. *Inside the Cuban Revolution: Fidel Castro and the Urban Underground*. Cambridge, MA, Harvard University Press, 2004.

Thomas, H. *Cuba or the Pursuit of Freedom*, revised edn. New York, NY, First Da Capo Press, 1998.

Tulchin, J. S., and Serbin, A. *Cuba and the Caribbean (Regional Issues and Trends in the Post-Cold War Era) Latin American Silhouettes*. Wilmington, DE, Scholarly Resources, 1997.

White, M. J. *Missiles in Cuba: Kennedy, Khrushchev, Castro, and the 1962 Crisis*. Chicago, IL, Ivan R. Dee, 1997.

Whitney, R. *State and Revolution in Cuba*. Chapel Hill, NC, University of North Carolina Press, 2001.

# DOMINICA

## Geography

### PHYSICAL FEATURES

The Commonwealth of Dominica is found in the central Lesser Antilles, the northernmost of the Windward Islands in the old British West Indies. Early colonization and rule by the French has left its mark on the island, reinforced by the continued French presence in Dominica's nearest neighbours. To the north are the main islands of Guadeloupe, an overseas department of the French Republic, and to the south the island of Martinique, another such department. Dominica is the largest of the anglophone islands in the Lesser Antilles (excluding Trinidad), covering an area of 751 sq km (290 sq miles).

Formed by volcanic activity, which is still prevalent on the island, Dominica is a mountainous land, lush and fertile, the most rugged of the Lesser Antilles. Its physical geography helped the long resistance of the native Carib Amerindians to European colonization. Dominica has the highest mountain in the eastern Caribbean, Morne Diablotins, at 1,447 m (4,749 feet). The high interior is covered by dense rainforest, which covers about 60% of the island and is protected by three national parks. The southernmost, around Morne Trois Pitons (a UNESCO World Heritage Site), is considered to have the richest biodiversity in the Caribbean, but also contains five active volcanoes, fuelling about 50 fumaroles and hot springs, as well as Boiling Lake, one of the largest thermally active lakes in the world (its main rival is in New Zealand). Fertility has encouraged the wooded nature of the hills, and there is rich flora and fauna (including the sisserou or imperial parrot, which features on the country's flag, the red-necked or jacquot parrot, the forest thrush, the blue-headed hummingbird, turtles, a rare iguana and the crapaud—eaten as 'mountain chicken'—this last a large frog, which finds its main haven in Dominica since the devastating volcanic activity on Montserrat in the 1990s). The steep terrain and wet climate also encourage numerous rivers, streams and waterfalls, many of which are seasonal. Little of the landscape is farmed. There are 148 km (92 miles) of coastline, with reefs offshore, and the island itself is about 47 km long (north–south, the northern end of the island leaning more towards the north-west) and 26 km wide.

### CLIMATE

The climate is subtropical and the mountains attract rain, making the island relatively rich in water resources and often humid. The wettest month is August, and the driest months are in February–June. Average annual rainfall is high—in Roseau, the capital on the south-western coast, it is about 2,160 mm (85 ins), but in the mountains it is over some 8,640 mm (340 ins). The terrain makes the risk of flash flooding high, and the island is also in the possible path of hurricanes, during June–October particularly. Daytime temperatures average between 70°F and 85°F (21°–29°C), the hottest month being August, but it is much cooler in the highlands, especially at night.

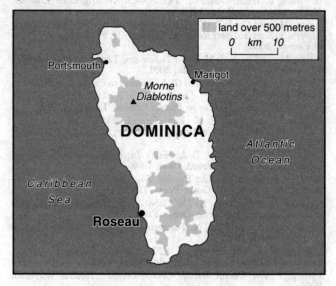

### POPULATION

Ethnically, the island population is mainly black, but there are also whites and many of mixed race, as well as a more recent Syrian community. The legacy of previous rulers survives in the only surviving Carib population of the West Indies (most living in the 15-sq-km Carib Territory, or Waitukubuli Karifuna area, on the rugged Atlantic coast) and a thriving French Creole patois as the most widely spoken local dialect. Spoken Carib has not survived. In the north-west an English patois, Kokay, is spoken by the Methodist descendants of freed slaves from Antigua. The official language, however, is English, which is widely spoken, and the main Christian denomination is Roman Catholic (80% of the population), with a further 15% adhering to a number of Protestant groups.

The total population was estimated at 70,500 in mid-2003, with about two-fifths of the population living in and around the capital, Roseau, in the south-west, in one of the 10 parishes into which the country is divided. The second town is also on the western coast, but at the northern end—Portsmouth. On the Atlantic coast, in the north-east, Marigot, near to the airport, is the main town. As mentioned above, Dominica is the only island of the Caribbean to retain some its pre-Columbian Amerindian population—about 3,000 Caribs, or those of mixed Carib descent, live on the north-eastern coast, just south of Marigot. The island is also noted for having the highest rate of centenarians in the world.

## History

### MARK WILSON

Dominica is a republic within the Commonwealth. It has a ceremonial President as head of state and a Prime Minister who leads the majority party in the unicameral House of Assembly, where elected representatives sit alongside appointed senators.

Dominica's original Carib Amerindian inhabitants knew the island as Waitukubuli—'tall is her body'—and survive in a small community at Salybia in the north-east. The island was given its present name by the explorer Christopher Columbus, who sighted it on Sunday 3 November 1493. Its mountainous topography discouraged early settlement, and it was declared neutral by England and France in 1600 and again in 1686. With the islands of Martinique and Guadeloupe in close proximity, French settlers gradually moved in during the 17th and early 18th centuries. The island was first claimed by the United

Kingdom in 1759, and changed hands several times before finally becoming a British possession in 1805. However, the French creole language, aspects of French culture and the Roman Catholic religion remain firmly established today; Dominica is a member of La Francophonie as well as the Commonwealth.

Despite the island's high rainfall and fertile soil, its broken terrain and poor communications made large-scale plantation agriculture less profitable than on some neighbouring islands. Commercial crops during the colonial period included coffee, cocoa, coconuts (for copra) and limes, as well as sugar. The island remained something of a backwater, although one administrator, Hesketh Bell, initiated significant infrastructural improvements between 1899 and 1905. Dominica formed part of the Leeward Islands Federation from 1833 until 1939.

The first elections with universal suffrage were held in 1951, with candidates supported by the Dominica Trade Union winning the majority of seats. In 1955 the Dominica Labour Party (DLP) was founded by Phyllis Shand Allfrey, a white Dominican and author of a well-known local novel, *The Orchid House*. However, an organized party system emerged more slowly than on most other islands. Dominica joined the Federation of the West Indies in 1958 along with nine other British colonies. When Jamaica and Trinidad and Tobago left in 1962, the Federation collapsed; an attempt to unite the remaining colonies as the 'little eight' was unsuccessful. Along with its neighbours, Dominica became a British Associated State in 1967. While it remained responsible for its internal affairs, the United Kingdom retained control of foreign relations and defence.

After winning an election held in 1971 the DLP was led from 1974 by Patrick John, who became the first Prime Minister upon the granting of independence on 3 November 1978. In February 1979 John announced that a 'Dominica Caribbean Free Port Authority' owned by Don Pierson, a developer from Texas, USA, would be given extensive rights over the northern third of the island. This proposal was abandoned owing to strong popular opposition. However, further disputes followed, including a strike by the civil service. On 29 May, in an attempt to disperse a large demonstration, the Defence Force shot and killed one person and left 10 injured. A general strike by the private sector then led to the resignation of several cabinet ministers, while the President, Frederick Degazon, suddenly left the island on 11 June. The former Governor, Sir Louis Cools-Lartigue, was installed as acting President, but resigned after only one day in office. Following arson attacks on the Court House and Registry shortly afterwards, the political opposition formed a Committee for National Salvation (CNS), together with private-sector, business, labour and agricultural organizations. On 21 June, in an improvised procedure, Jenner Armour was sworn in as acting President; he then appointed one of John's former ministers, Oliver Seraphin, as Prime Minister.

These turbulent events were followed in August 1979 by 'Hurricane David', which devastated homes, roads, and the banana crop, leaving 37 people dead and 5,000 injured. In September the island was struck by a second powerful storm, Hurricane Frederick.

The interim Government led by Seraphin was not an unqualified success, not least owing to disputes over the management of hurricane reconstruction efforts. At a general election held on 20 July 1980, Seraphin's followers were soundly defeated by the Dominica Freedom Party (DFP), and Eugenia Charles (later Dame Mary Eugenia Charles) became the Caribbean's first female Prime Minister. She developed a strong reputation both within the island and internationally and was an articulate supporter of the invasion of Grenada in 1983.

In March 1981 the new Government stated that members of the small Defence Force were planning a coup to restore Patrick John to office, with the support of US and Canadian 'soldiers of fortune'—members of the extremist right-wing Ku Klux Klan, some of whom were intercepted in New Orleans, Louisiana, USA. The Defence Force was disbanded in April 1982, while John was tried and acquitted of attempting to overthrow the Government in the same year. However, following a retrial in 1985 John was sentenced to a 12-year prison term; he was released before the expiry of his sentence. Meanwhile, in 1985 the DLP, the Democratic Labour Party, the United Dominica Labour Party and the Dominica Liberation Movement united to form the Labour Party of Dominica (LPD).

Although the DFP majority was reduced to a single seat in the general election of May 1990, Charles remained in office until 1995; she was replaced as leader of the DFP by Brian Alleyne, under whose leadership the party was defeated in a general election held on 12 June of that year. The Dominica United Workers' Party (UWP), led by Edison James, won 11 of the 21 seats in the House of Assembly; the DFP and the LPD secured five apiece.

Troubles in the banana industry undermined support for the UWP Government and there was considerable controversy over its spending proposals, including one for the development of an airport. In a general election on 31 January 2000, the UWP was reduced to nine seats. The LPD, led by Roosevelt (Rosie) Douglas, won 10 seats, securing 42.9% of the popular vote, and formed a coalition with the DFP, which won two seats. One UWP member subsequently defected from the party and joined the ruling coalition, strengthening the position of the new Government within the House of Assembly.

Douglas died unexpectedly of a heart attack on 1 October 2000, following an extended series of overseas visits. He was succeeded by his deputy, Pierre Charles.

In late 2001 Canada announced that Dominicans travelling to Canada would require visas. This reflected growing international concern over Dominica's economic citizenship programme, under which passports were sold to non-citizens. Following the terrorist attacks in the USA in September 2001, Dominica came under increased pressure to abandon the programme.

In June 2002 the Government announced its intention to reduce the size of the Cabinet by 1 August by abolishing three ministries and five advisory positions, in an attempt to reduce expenditure. Consequently, the Minister of Industry, Planning and Physical Development, Ambrose George, and two ministers of state resigned. The portofolios held by George, whose reputation had been severely damaged by apparent association with a money-launderer the previous year, were shared between Charles Savarin and Reginald Austrie. In July the Public Service Union organized a large-scale demonstration in the capital, Roseau, to protest against controversial tax increases introduced in a budget intended to bring about economic recovery. In November Frederick Baron, one of the two DFP members of parliament, announced that he would no longer support the Government, leaving it with backing from 11 of the 21 elected members.

With the economic situation remaining difficult, the Government was in a severely weakened condition in the first half of 2003, while the leadership of Charles was strongly questioned from within his own party. Confidence in the Prime Minister's health was shaken in 2002, when he was unable to give the budget presentation because of deep vein thrombosis, and again in February 2003 when he was taken ill at a Caribbean Community and Common Market (CARICOM) heads of government meeting in Trinidad and subsequently given an angioplasty. Charles died after a heart attack on 6 January 2004 and was succeeded by the 31-year-old Roosevelt Skerrit, who until then had been Minister of Education, Youth and Sports. Skerrit assumed direct responsibility for finance in a subsequent reshuffle, and also reappointed Ambrose George to the Cabinet as Minister of Agriculture and the Environment. Meanwhile, on 1 October 2003 Nicholas Liverpool succeeded Vernon Shaw as President. Following a visit to Beijing, Skerrit on 29 March 2004 announced that Dominica would recognize the People's Republic of China in place of Taiwan. China had agreed to provide US $6m. in budgetary support, equivalent to around 8% of annual recurrent revenue, followed by EC $300m.–$330m. in grant aid for a road, the main hospital, a school and a sports stadium; equivalent in value to some 45% of one year's gross domestic product or 1.5 years' recurrent revenue. A Chinese embassy was duly opened in Roseau in mid-June.

In spite of the country's economic difficulties and a stringent austerity programme, Skerrit re-established the Government's popularity. At a general election on 5 May 2005 the LPD took 51.7% of the popular vote, compared with 42.9% at the elections in 2000. The party won 12 of the 21 seats and a further seat was taken by a close ally of the DLP, who ran as an Independent; the

UWP secured the remaining eight seats (and 43.9% of the votes cast). The DFP was defeated in each of two constituencies in which it fielded a candidate; however, its leader, Charles Savarin, was appointed as a senator and retained his place in Skerrit's new administration as Minister of Foreign Affairs, Trade, Labour and the Public Service. Three new ministers were appointed to the re-organized Cabinet, including two non-career politicians—a businessman and a veterinary surgeon. The Prime Minister retained direct responsibility for finance, planning, Carib affairs and national security.

# Economy

## MARK WILSON

Dominica is the second smallest country in the Western hemisphere in terms of population, with 70,500 inhabitants living on its 751 sq km in mid-2003. The island has developed a modest middle-income economy, with a per-head gross domestic product (GDP) of US $3,717 in 2003. The economy has stagnated in recent years, with GDP growing at an average annual rate of 0.6% in 1990–2003; GDP contracted by 5.1% in 2002 and by a further 0.4% in 2003. In spite of a partial stabilization, the rate of unemployment rose to 25% in the latter year, according to World Bank estimates. The economic reversal in the early 2000s intensified the Government's financial difficulties; an overall deficit equivalent to around 8% of GDP in 2001 was financed partially through the accumulation of arrears. Debt rose sharply, reaching at least EC $779m. or 115% of GDP in late 2004, and included substantial amounts borrowed at commercial rates. Indeed, the Government has, since June 2002, refused to make payments on 1999 bond issues totalling US $47.5m. from two Trinidadian banks that it claims were illegally restructured to Dominica's disadvantage. The budget for 2002/03 introduced a 4% stabilization levy (repealed in the 2004/05 budget) on gross annual incomes over EC $9,000 and other measures intended to raise revenue in an attempt to reduce the deficit to 5.2% of GDP. However, major savings on public-sector staffing costs were rejected both by the Government and by trade unions.

Donor agencies, including the European Union (EU), the IMF and the World Bank, have been closely involved in the discussion of possible measures to stabilize the economy. Concessional borrowing of EC $30m. in late 2002 included assistance from Barbados, Trinidad and Tobago and Saint Vincent and the Grenadines, as well as from more obvious donor agencies. The IMF in August 2002 agreed a stand-by credit, with performance targets. However, a review of the IMF programme in March 2003 found that the island had failed to meet these targets. The IMF and World Bank were therefore unable to agree a further assistance programme by mid-year, although EU grants continued under the special programme of assistance for banana producers. The 2003/04 budget included proposals for further reductions in expenditure, including a 5% pay cut for public servants. New policy targets were set in June 2003, and were followed by a US $11.4m. line of credit under the IMF's Poverty Reduction and Growth Facility. An IMF mission in February 2004 found that the new targets had been implemented. A further positive quarterly review followed in May 2004; targets to be achieved for further reviews include a primary budget surplus equivalent to 3% of GDP by 2006/07. However, the expectation of a further 10% reduction in salary costs over two years, to be achieved mainly through reduced staff numbers, was met with some hostility from the Dominica Public Service Union; public-sector salary costs were equivalent in 2003 to 18% of GDP. The fiscal outlook benefited significantly in March 2004 from an agreement by the People's Republic of China to provide substantial grant aid in exchange for diplomatic recognition.

A third tranche of assistance was agreed with the IMF in August 2004, by which time 60% of eligible debt was covered by a restructuring programme. The recurrent deficit had been sharply reduced to 1.5% of GDP, with an increase in capital spending finance, mainly by development assistance. GDP recovered to some extent in 2004, with growth estimates of 2.0%–3.5%, and the Government was expected in mid-2005 to reverse a 5% public-sector pay cut imposed in 2003.

Dominica is a member of the Caribbean Community and Common Market, or CARICOM, which is attempting to develop a single market, in principle by 2006. It is also a member of the Organisation of Eastern Caribbean States, which links nine of the smaller Caribbean territories, while its financial affairs are supervised by the Eastern Caribbean Central Bank, headquartered in Saint Christopher and Nevis.

Agriculture accounted for 16.5% of GDP in 2003, reduced from 25.0% in 1990. The export of bananas to the United Kingdom had been the mainstay of the economy since the 1950s, and Dominica was more dependent on this activity than any other Caribbean country. In 1990 banana exports constituted 92.1% of total exports and generated 34.4% of foreign-exchange earnings, while more than one-half of the country's work-force was involved to some extent with the industry. As a result of changes in the EU's import regime, banana exports fell from 55,000 metric tons in 1993 to 12,721 tons in 2004, while there was also a steep decline in unit prices. In 2004 banana exports made up only 11.9% of merchandise exports by value and only 3.8% of foreign-exchange earnings. Significant EU grant aid was available, however, for modernizing the banana industry and for enabling economic diversification. Small farmers produce a wide variety of fruits, vegetables and livestock products for the local market, with some produce exported to Barbados, Antigua and Barbuda, and the neighbouring French Overseas Departments of Martinique and Guadeloupe.

There is a small manufacturing sector, which in 2003 accounted for 7.2% of GDP. The main exporter is Dominica Coconut Products, which makes soap from locally produced copra, as well as detergent and toothpaste.

The island has two small airports, but neither was capable of accommodating either night flights or long-haul aeroplanes. Connections to Europe and North America are made through larger regional airports such as those in Antigua and Barbuda or Barbados. The former Dominica United Workers' Party Government borrowed heavily at commercial rates to finance an ambitious airport project, which is now thought to have been neither technically nor commercially viable. The Labour Government that succeeded it in 2000 agreed EU funds to upgrade the existing airports in 2005 to allow night flights, and to improve the road connection to the capital, Roseau; the drive currently takes up to two hours.

There are few sandy beaches, and rainfall is much higher than on most Caribbean islands. These factors, as well as the lack of direct air access, have restricted the development of tourism. However, the island's wildlife and natural beauty have attracted some eco-tourists. Visitor expenditure rose from US $25m. in 1990 to $48.8m. in 1999, a year in which 70,791 tourists came to the island. A total of 79,386 visitors stayed on the island in 2004, an 8.8% increase compared with the previous year; of these, 24.3% were from North America and 12.9% from Europe; the largest groups were business and family visitors from other Caribbean islands. Stop-over tourists were easily outnumbered by the 380,608 cruise-ship passengers; however, this group spend little money ashore, and accordingly bring few economic benefits. Of some significance is the student and organizational spending of a US 'offshore' medical school, the Ross University of Medicine. A call centre opened in January 2005, with planned employment of 200.

There is a small 'offshore' financial sector, with 8,601 international business companies at the end of 2003, one international bank and four internet gambling operations. Unfortunately, the country has been accused of failing to meet internationally acceptable standards. In June 2000 Dominica was listed as a 'non-co-operative jurisdiction' on the issue of money-laundering by the Financial Action Task Force on Money Laundering

# DOMINICA

(FATF, based in Paris, France) and as a harmful tax 'haven' by the Organisation for Economic Co-operation and Development. Having made regulatory reforms, Dominica was conditionally de-listed by the FATF in October 2002, but will remain under close international scrutiny. Its problems were highlighted in November 2001, when a prominent businessman and former Vice-President of the governing Labour Party of Dominica, was arrested in Puerto Rico on money-laundering charges while on a joint business trip with the then Minister of Finance, Ambrose George. George subsequently resigned, but was reappointed to the Cabinet as Minister of Agriculture and the Environment in January 2004, and as Minister of Public Works and Public Utilities after the May 2005 election. The 'economic citizenship' programme, under which Dominican nationality can be acquired in return for a cash investment, has been in place, under various guises, since independence. It is viewed with some suspicion internationally, as it is considered open to exploitation by criminals. The programme was relaunched in June 2002, with a fee of US $75,000 for individual applicants; 650 passports were sold in 1996–2002, and retaliatory measures included the imposition of a visa requirement by Canada in December 2001.

Dominica is at risk from hurricanes. Tropical storms, while less violent, can cause serious damage to the fragile banana plants. There are also several volcanic centres, which are currently inactive; some sources suggest there is a 25% chance of a major eruption within the next 25 years. In the absence of recent eruptions, however, volcanic features such as the Boiling Lake and Valley of Desolation are attractions for the more energetic eco-tourists.

# Statistical Survey

Source (unless otherwise stated): Eastern Caribbean Central Bank; internet www.eccb-centralbank.org.

## AREA AND POPULATION

**Area:** 751 sq km (290 sq miles).

**Population:** 71,727 (males 36,434, females 35,293) at census of 12 May 2001. *2003* (mid-year estimate): 70,500 (Source: Caribbean Development Bank, *Social and Economic Indicators*).

**Density** (mid-2003): 93.9 per sq km.

**Population by Ethnic Group** (*de jure* population, excl. those resident in institutions, 1981): Negro 67,272; Mixed race 4,433; Amerindian (Carib) 1,111; White 341; Total (incl. others) 73,795 (males 36,754, females 37,041). Source: UN, *Demographic Yearbook*.

**Principal Town** (population at 1991 census): Roseau (capital) 15,853. *Mid-2003* (UN estimate, incl. suburbs): Roseau 27,401 (Source: UN, *World Urbanization Prospects: The 2003 Revision*).

**Births, Marriages and Deaths** (registrations, 2002 unless otherwise indicated): Live births 1,081 (birth rate 15.3 per 1,000); Marriages (1998) 336 (marriage rate 4.4 per 1,000); Deaths 510 (death rate 7.3 per 1,000). Source: UN, *Demographic Yearbook*.

**Expectation of Life** (WHO estimates, years at birth): 73 (males 71; females 76) in 2003. Source: WHO, *World Health Report*.

**Economically Active Population** (rounded estimates, persons aged 15 years and over, 1997): Agriculture, hunting, forestry and fishing 6,000; Fishing 100; Manufacturing 2,250; Electricity, gas and water supply 280; Construction 2,150; Wholesale and retail trade 4,050; Hotels and restaurants 980; Transport, storage and communications 1,500; Financial intermediation 540; Real estate, renting and business activities 850; Public administration, defence and social security 1,530; Education 1,260; Health and social work 1,110; Other community, social and personal service activities 930; Private households with employed persons 1,080; Not classifiable by economic activity 1,090; *Total employed* 25,690; Unemployed 7,720; *Total labour force* 33,420 (males 18,120, females 15,300) (Source: ILO). *Mid-2003* (estimates): Agriculture, etc. 8,000; Total labour force 36,000 (Source: FAO).

## HEALTH AND WELFARE

### Key Indicators

**Total Fertility Rate** (children per woman, 2003): 1.8.

**Under-5 Mortality Rate** (per 1,000 live births, 2003): 14.

**Physicians** (per 1,000 head, 1997): 0.49.

**Hospital Beds** (per 1,000 head, 1996): 2.65.

**Health Expenditure** (2002): US $ per head (PPP): 310.

**Health Expenditure** (2002): % of GDP: 6.4.

**Health Expenditure** (2002): public (% of total): 71.3.

**Access to Water** (% of persons, 2002): 97.

**Access to Sanitation** (% of persons, 2002): 83.

**Human Development Index** (2002): ranking: 95.

**Human Development Index** (2002): value: 0.743.

For sources and definitions, see explanatory note on p. vi.

## AGRICULTURE, ETC.

**Principal Crops** (FAO estimates, '000 metric tons, 2003): Sweet potatoes 1.9; Cassava 1.0; Yautia (Cocoyam) 4.6; Taro (Dasheen) 11.2; Yams 8.0; Sugar cane 4.4; Coconuts 11.5; Cabbages 0.7; Pumpkins 0.9; Cucumbers 1.7; Carrots 0.5; Other vegetables 2.7; Bananas 29.0; Plantains 5.7; Oranges 7.2; Lemons and limes 1.0; Grapefruit 17.0; Mangoes 1.9; Avocados 0.6; Other fruits 1.2.

**Livestock** (FAO estimates, '000 head, year ending September 2003): Cattle 13.4; Pigs 5.0; Sheep 7.6; Goats 9.7; Poultry 190.

**Livestock Products** (FAO estimates, '000 metric tons, 2003): Beef and veal 0.5; Pig meat 0.4; Poultry meat 0.3; Cows' milk 6.1; Hen eggs 0.2.

**Fishing** (FAO estimates, metric tons, live weight, 2003): Capture 1,100 (Skipjack tuna 51; Blackfin tuna 42; Yellowfin tuna 119; Marlins, sailfishes, etc. 75; Common dolphinfish 203); Aquaculture 3; *Total catch* 1,103.
Source: FAO.

## MINING

**Pumice** ('000 metric tons, incl. volcanic ash): Estimated production 100 per year in 1988–2004. Source: US Geological Survey.

## INDUSTRY

**Production** (preliminary, 2001, metric tons, unless otherwise indicated): Soap 9,000 (2002); Dental cream 1,562; Hard surface cleansers 2,987; Crude coconut oil 855; Coconut meal 331; Electricity 81 million kWh. Sources: IMF, *Dominica: Statistical Appendix* (October 2002), and UN, *Industrial Commodity Statistics Yearbook*.

## FINANCE

**Currency and Exchange Rates:** 100 cents = 1 Eastern Caribbean dollar (EC $). *Sterling, US Dollar and Euro Equivalents* (31 May 2005): £1 sterling = EC $4.909; US $1 = EC $2.700; €1 = EC $3.329; EC $100 = £20.37 = US $37.04 = €30.04. *Exchange Rate:* Fixed at US $1 = EC $2.70 since July 1976.

**Budget** (preliminary, EC $ million, 2002): *Revenue:* Tax revenue 162.0 (Taxes on income and profits 41.6, Taxes on domestic goods and services 32.4, Taxes on international trade and transactions 85.4, Other taxes 2.6); Other current revenue 35.3; Capital revenue 1.8; Total 199.1, excl. grants received (25.6). *Expenditure:* Current expenditure 231.4 (Wages and salaries 123.9, Goods and services 29.9, Interest payments 39.5, Transfers and subsidies 38.0); Capital expenditure and net lending 27.2; Total 258.6.

**International Reserves** (US $ million at 31 December 2004): Reserve position in IMF 0.01; Foreign exchange 42.25; Total 42.26. Source: IMF, *International Financial Statistics*.

**Money Supply** (EC $ million at 31 December 2004): Currency outside banks 37.60; Demand deposits at commercial banks 108.34; Total money (incl. others) 146.08. Source: IMF, *International Financial Statistics*.

**Cost of Living** (Retail Price Index, base: 2000 = 100): All items 101.8 in 2002; 103.3 in 2003; 105.7 in 2004. Source: IMF, *International Financial Statistics*.

**National Accounts** (EC $ million at current prices): Gross domestic product in purchasers' values 718.6 in 2001; 689.9 in 2002; 707.5 (preliminary) in 2003.

**Expenditure on the Gross Domestic Product** (preliminary, EC $ million at current prices, 2003): Government final consumption expenditure 135.14; Private final consumption expenditure 480.45; Gross fixed capital formation (incl. increase in stocks) 177.31; *Total domestic expenditure* 792.90; Exports of goods and services 340.34; *Less* Imports of goods and services 425.79; *GDP in purchasers' values* 707.45.

**Gross Domestic Product by Economic Activity** (preliminary, EC $ million at current prices, 2003): Agriculture, hunting, forestry and fishing 103.85; Mining and quarrying 4.59; Manufacturing 45.50; Electricity and water 34.40; Construction 45.86; Wholesale and retail trade 73.32; Restaurants and hotels 17.01; Transport 47.45; Communications 28.69; Finance and insurance 66.62; Real estate and housing 22.26; Government services 127.57; Other services 10.89; *Sub-total* 628.01; *Less* imputed bank service charge 47.90; *GDP at factor cost* 580.11.

**Balance of Payments** (preliminary, EC $ million, 2003): Exports of goods f.o.b. 104.73; Imports of goods f.o.b. –276.35; *Trade balance* –171.62; Exports of services 235.61; Imports of services –149.44; *Balance on goods and services* –85.45; Other income received 7.50; Other income paid –62.33; *Balance on goods, services and income* –140.28; Current transfers received 65.34; Current transfers paid –18.77; *Current balance* –93.71; Capital account (net) 52.98; Direct investment from abroad 65.89; Portfolio investment assets –10.53; Portfolio investment liabilities 10.43; Other investment (net) –20.40; Net errors and omissions 2.68; *Overall balance* 7.34.

### EXTERNAL TRADE

**Principal Commodities** (US $ million, 2002): *Imports c.i.f.:* Food products and live animals 21.9 (Meat and meat preparations 4.8; Cereals and cereal preparations 5.0); Beverages and tobacco 3.7; Mineral fuels, lubricants, etc. 11.0 (Refined petroleum products 9.5); Chemicals, etc. 14.2; Basic manufactures 19.6 (Paper products 5.7); Metal manufactures 4.7; Machinery and transport equipment 25.4 (Telecommunications and sound equipment 4.6; Road vehicles 6.8); Miscellaneous manufactured articles 15.1; Total (incl. others) 115.7. *Exports f.o.b.:* Food and live animals 15.9 (Roots and tubers 1.4; Fruit and nuts 12.2; Sauces, condiments and seasonings 1.5); Inedible crude materials 1.5 (Stone, sand and gravel 1.4); Chemicals, etc. 21.5 (Perfumes, cosmetics, toilet products, etc. 5.7; Soap 11.7; Disinfectants 1.6); Machinery and transport equipment 1.9; Total (incl. others) 41.9. Source: UN, *International Trade Statistics Yearbook*.

**Principal Trading Partners** (US $ million, 2002): *Imports c.i.f.:* Barbados 4.4; Canada 3.1; France 2.5; Guyana 1.4; Jamaica 1.6; Japan 4.7; Netherlands 2.2; Panama 1.3; Saint Lucia 4.1; Saint Vincent and the Grenadines 1.5; Trinidad and Tobago 20.4; United Kingdom 10.0; USA 42.5; Venezuela 1.5; Total (incl. others) 115.7. *Exports f.o.b.:* Antigua and Barbuda 3.5; Barbados 1.8; France 2.8; Grenada 0.6; Guyana 2.9; Jamaica 9.0; Netherlands Antilles 0.7; Saint Christopher and Nevis 1.4; Saint Lucia 1.2; Saint Vincent and the Grenadines 0.6; Trinidad and Tobago 1.9; United Kingdom 9.5; USA 3.9; Total (incl. others) 41.9. Source: UN, *International Trade Statistics Yearbook*.

### TRANSPORT

**Road Traffic** (motor vehicles licensed in 1994): Private cars 6,491; Taxis 90; Buses 559; Motorcycles 94; Trucks 2,266; Jeeps 461; Tractors 24; Total 9,985. *2000* (motor vehicles in use): Passenger cars 8,700; Commercial vehicles 3,400. Source: partly UN, *Statistical Yearbook*.

**Shipping:** *Merchant Fleet* (registered at 31 December 2004): 60 vessels (total displacement 303,866 grt) (Source: Lloyd's Register-Fairplay, *World Fleet Statistics*); *International Freight Traffic* ('000 metric tons, estimates, 1993): Goods loaded 103.2; Goods unloaded 181.2.

**Civil Aviation** (1997): Aircraft arrivals and departures 18,672; Freight loaded 363 metric tons; Freight unloaded 575 metric tons.

### TOURISM

**Tourist Arrivals** ('000): *Stop-overs:* 69.0 in 2002; 73.0 in 2003; 79.4 in 2004. *Cruise-ship passengers:* 137.0 in 2002; 177.0 in 2003; 380.6 in 2004.

**Tourism Receipts** (estimates, US $ million): 46.4 in 2001; 45.7 in 2002; 53.9 in 2003.
Source: Caribbean Development Bank, *Social and Economic Indicators*.

### COMMUNICATIONS MEDIA

**Radio Receivers** (1997): 46,000 in use.

**Television Receivers** (1999): 17,000 in use.

**Telephones** (2002): 23,700 main lines in use.

**Facsimile Machines** (1996): 396 in use.

**Mobile Cellular Telephones** (2002): 9,400 subscribers.

**Personal Computers** (2002): 7,000 in use.

**Internet Users** (2002): 12,500.

**Non-daily Newspapers** (1996): 1.
Sources: mainly UNESCO, *Statistical Yearbook*, International Telecommunication Union and UN, *Statistical Yearbook*.

### EDUCATION

**Institutions** (1994/95): Pre-primary 72 (1992/93); Primary 64; Secondary 14; Tertiary 2.

**Teachers:** Pre-primary 131 (1992/93); Primary 628 (1994/95); Secondary 269 (1994/95); Tertiary 34 (1992/93).

**Pupils** (2002): Pre-primary 3,000 (1992/93); Primary 11,025; Secondary 6,850; Tertiary 461 (1995/96).
Sources: UNESCO, *Statistical Yearbook*; Caribbean Development Bank, *Social and Economic Indicators*; UN Economic Commission for Latin America and the Caribbean, *Statistical Yearbook*.

**Adult Literacy Rate** (2001): 96.4%. Source: Secretariat of the Organisation of Eastern Caribbean States.

# Directory

## The Constitution

The Constitution came into effect at the independence of Dominica on 3 November 1978. Its main provisions are summarized below:

### FUNDAMENTAL RIGHTS AND FREEDOMS
The Constitution guarantees the rights of life, liberty, security of the person, the protection of the law and respect for private property. The individual is entitled to freedom of conscience, of expression and assembly and has the right to an existence free from slavery, forced labour and torture. Protection against discrimination on the grounds of sex, race, place of origin, political opinion, colour or creed is assured.

### THE PRESIDENT
The President is elected by the House of Assembly for a term of five years. A presidential candidate is nominated jointly by the Prime Minister and the Leader of the Opposition and on their concurrence is declared elected without any vote being taken; in the case of disagreement the choice will be made by secret ballot in the House of Assembly. Candidates must be citizens of Dominica aged at least 40 who have been resident in Dominica for five years prior to their nomination. A President may not hold office for more than two terms.

### PARLIAMENT
Parliament consists of the President and the House of Assembly, composed of 21 elected Representatives and nine Senators. According to the wishes of Parliament, the latter may be appointed by the President—five on the advice of the Prime Minister and four on the advice of the Leader of the Opposition—or elected. The life of Parliament is five years.

Parliament has the power to amend the Constitution. Each constituency returns one Representative to the House who is directly elected in accordance with the Constitution. Every citizen over the age of 18 is eligible to vote.

### THE EXECUTIVE
Executive authority is vested in the President. The President appoints as Prime Minister the elected member of the House who

# DOMINICA

commands the support of a majority of its elected members, and other ministers on the advice of the Prime Minister. Not more than three ministers may be from among the appointed Senators. The President has the power to remove the Prime Minister from office if a resolution expressing 'no confidence' in the Government is adopted by the House and the Prime Minister does not resign within three days or advise the President to dissolve Parliament.

The Cabinet consists of the Prime Minister, other ministers and the Attorney-General in an ex officio capacity.

The Leader of the Opposition is appointed by the President as that elected member of the House who, in the President's judgement, is best able to command the support of a majority of the elected members who do not support the Government.

## The Government

### HEAD OF STATE

**President:** Dr NICHOLAS LIVERPOOL (assumed office 1 October 2003).

### CABINET
(July 2005)

**Prime Minister and Minister of Finance, Planning, Carib Affairs and National Security:** ROOSEVELT SKERRIT (LPD).

**Attorney-General and Minister Legal Affairs and Immigration:** IAN DOUGLAS (LPD).

**Minister of Foreign Affairs, Trade, Labour and the Public Service:** CHARLES SAVARIN (DFP).

**Minister of Education, Human Resource Development, Sports and Youth Affairs:** VINCE HENDERSON (LPD).

**Minister of Tourism, Industry and Private-Sector Relations:** YVOR NASSIEF.

**Minister of Community Development, Gender Affairs and Information:** MATTHEW WALTER (LPD).

**Minister of Agriculture and the Environment:** COLLIN MCINTYRE.

**Minister of Public Works and Public Utilities:** AMBROSE GEORGE (LPD).

**Minister of Housing, Lands, Telecommunications, Energy and Ports:** REGINALD AUSTRIE (LPD).

**Minister of Health and Social Security:** JOHN FABIEN (LPD).

**Minister with responsibility for Carib Affairs:** KELLY GRANEAU (LPD).

### MINISTRIES

**Office of the President:** Morne Bruce, Roseau; tel. 4482054; fax 4498366; e-mail presidentoffice@cwdom.dm.

**Office of the Prime Minister:** Government Headquarters, Kennedy Ave, Roseau; tel. 4482401; fax 4485200.

All other ministries are at Government Headquarters, Kennedy Ave, Roseau; tel. 4482401.

### CARIB TERRITORY

This reserve of the remaining Amerindian population is located on the central east coast of the island. The Caribs enjoy a measure of local government and elect their chief.

**Chief:** GARNET JOSEPH.

**Waitukubuli Karifuna Development Committee:** Salybia, Carib Territory; tel. 4457336.

## Legislature

### HOUSE OF ASSEMBLY

**Speaker:** ALIX BOYD KNIGHT.

**Clerk:** ALEX F. PHILLIP.

**Senators:** 9.

**Elected Members:** 21.

### General Election, 5 May 2005

| Party | Valid votes cast | % | Seats |
|---|---|---|---|
| Labour Party of Dominica | 19,640 | 51.73 | 12 |
| Dominica United Workers' Party | 16,678 | 43.93 | 8 |
| Dominica Freedom Party | 1,211 | 3.19 | — |
| Independents | 434 | 1.14 | 1 |
| **Total** | 37,963 | 100.00 | 21 |

## Political Organizations

**A Righteous Kingdom Party (ARK):** Roseau; f. 2005; Christian; Leader HERMINA VALENTINE.

**Dominica Freedom Party (DFP):** Great George St, Roseau; tel. 4482104; Leader CHARLES SAVARIN.

**Dominica United Workers' Party (UWP):** 47 Cork St, Roseau; tel. 4485051; fax 4498448; e-mail uwp@cwdom.dm; internet www.uwp.dm; f. 1988; Political Leader EDISON JAMES; Pres. RON GREEN.

**Labour Party of Dominica (LPD):** Cork St, Roseau; tel. 4488511; f. 1985 as a merger and reunification of left-wing groups, incl. the Dominica Labour Party (DLP; f. 1961); Leader ROOSEVELT SKERRIT; Deputy Leader AMBROSE GEORGE.

## Diplomatic Representation

### EMBASSIES IN DOMINICA

**China, People's Republic:** Morne Daniel, POB 2247, Roseau; tel. 4490080; fax 4400088; Ambassador YE DABO.

**Venezuela:** 20 Bath Rd, 3rd Floor, POB 770, Roseau; tel. 4483348; fax 4486198; e-mail embven@cwdom.dm; Ambassador CARMEN MARTÍNEZ DE GRIJALVA.

## Judicial System

Justice is administered by the Eastern Caribbean Supreme Court (based in Saint Lucia), consisting of the Court of Appeal and the High Court. One of the six puisne judges of the High Court is resident in Dominica and presides over the Court of Summary Jurisdiction. The District Magistrate Courts deal with summary offences and civil offences involving limited sums of money (specified by law).

## Religion

Most of the population profess Christianity, but there are some Muslims, Bahá'ís and Jews. The largest denomination is the Roman Catholic Church (with some 85% of the inhabitants in 2003).

### CHRISTIANITY

#### The Roman Catholic Church

Dominica comprises the single diocese of Roseau, suffragan to the archdiocese of Castries (Saint Lucia). At 31 December 2003 there were an estimated 59,707 adherents in the country, representing a large majority of the inhabitants. The Bishop participates in the Antilles Episcopal Conference (currently based in Port of Spain, Trinidad).

**Bishop of Roseau:** Rt Rev. GABRIEL MALZAIRE, Bishop's House, Turkey Lane, POB 790, Roseau; tel. 4482837; fax 4483404; e-mail bishop@cwdom.dm.

#### The Anglican Communion

Anglicans in Dominica are adherents of the Church in the Province of the West Indies. The country forms part of the diocese of the North Eastern Caribbean and Aruba. The Bishop is resident in Antigua, and the Archbishop of the Province is the Bishop of the Bahamas and the Turks and Caicos Islands.

#### Other Christian Churches

**Christian Union Church of the West Indies:** Dominica Island District, 1 Rose St, Goodwill, POB 28, Roseau; tel. 4482725; e-mail cucdistrict@marpin.dm.

# DOMINICA

Other denominations include Methodist, Pentecostal, Baptist, Church of God, Presbyterian, the Assemblies of Brethren, Moravian and Seventh-day Adventist groups, and the Jehovah's Witnesses.

### BAHÁ'Í FAITH

**National Spiritual Assembly:** 79 Victoria St, POB 136, Roseau; tel. 4483881; fax 4488460; e-mail nsa_dominica@yahoo.com.

## The Press

**The Chronicle:** Wallhouse, POB 1724, Roseau; tel. 4487887; fax 4480047; e-mail thechronicle@cwdom.dm; internet www.news-dominica.com/new-index.cfm; f. 1996; Friday; progressive independent; Editor RASCHID OSMAN; Gen. Man. J. ANTHONY WHITE; circ. 3,000 (1997).

**Official Gazette:** Government Printery, Roseau; tel. 4482401, ext. 330; weekly; circ. 550.

**The Sun:** Sun Inc, 50 Independence St, POB 2255, Roseau; tel. 4484744; fax 4484764; e-mail acsun@cwdom.dm; internet www.dominicasun.com; f. 1998; Editor CHARLES JAMES.

**The Times:** Roseau; f. 2004; Friday.

**The Tropical Star:** POB 1998, Roseau; tel. 4484634; fax 4485984; e-mail tpl@cwdom.dm; weekly; circ. 3,000.

## Broadcasting and Communications

### TELECOMMUNICATIONS

#### Regulatory Authority

**Eastern Caribbean Telecommunications Authority:** POB 1886, Castries, Saint Lucia; tel. (758) 4581701; fax (758) 4581698; e-mail info@ectel.int; internet www.ectel.info; f. 2000 to regulate telecommunications in Dominica, Grenada, Saint Christopher and Nevis, Saint Lucia and Saint Vincent and the Grenadines.

#### Major Service Providers

**Cable & Wireless Dominica:** Hanover St, POB 6, Roseau; tel. 4481000; fax 4481111; e-mail pr@cwdom.dm; internet www.tod.dm; Gen. Man. CARL ROBERTS.

**Cingular Wireless Dominica:** Roseau; internet www.cingularwireless.com; joint venture between SBC Communications and BellSouth; scheduled to sell its operations and licences in the Caribbean to Digicel in 2005; Pres. and CEO STANLEY T. SIGMAN.

**Telecommunications of Dominica (TOD):** Mercury House, Hanover St, Roseau; tel. 4481024.

### BROADCASTING

#### Radio

In 2001 there were three radio stations operating in Dominica.

**Dominica Broadcasting Corporation:** Victoria St, POB 148, Roseau; tel. 4483283; fax 4482918; e-mail dbsradio@cwdom.dm; internet www.dbcradio.net; government station; daily broadcasts in English; 2 hrs daily in French patois; 10 kW transmitter on the medium wave band; FM service; programmes received throughout Caribbean excluding Jamaica and Guyana; Gen. Man. MARIETTE WARRINGTON; Programme Dir SHERMAINE GREEN-BROWN.

**Kairi FM:** Island Communications Corpn, Great George St, POB 931, Roseau; tel. 4487330; fax 4487332; e-mail kairfm@tod.dm; internet www.delphis.dm/kairi.htm; f. 1994; Owner FRANKEY BELLOT.

**Voice of Life Radio (ZGBC):** Gospel Broadcasting Corpn, Loubiere, POB 205, Roseau; tel. 4487017; fax 4400551; e-mail volradio@cwdom.dm; internet www.voiceofliferadio.com; 112 hrs weekly AM, 24 hrs daily FM; Station Man. CLEMENTINA MONRO.

#### Television

There is no national television service, although there is a cable television network serving one-third of the island.

**Marpin Telecom and Broadcasting Co Ltd:** 5–7 Great Marlborough St, POB 2381, Roseau; tel. 4484107; fax 4482965; e-mail pubrel@marpin.dm; internet www.marpin.dm; f. 1982, present name adopted in 1996; commercial; cable service; went into receivership in 2004.

## Finance

(cap. = capital; res = reserves; dep. = deposits; m. = million; amounts in East Caribbean dollars)

The Eastern Caribbean Central Bank, based in Saint Christopher, is the central issuing and monetary authority for Dominica.

**Eastern Caribbean Central Bank—Dominica Office:** Dorset House, cnr of Old St and Hodges Lane, POB 23, Roseau; tel. 4488001; fax 4488002; e-mail eccbdom@cwdom.dm; Country Dir AMBROSE SYLVESTER.

### BANKS

**Bank of Nova Scotia—Scotiabank** (Canada): 28 Hillsborough St, POB 520, Roseau; tel. 4485800; fax 4485805; e-mail scotia@cwdom.dm; Man. C. MONTE SMITH.

**FirstCaribbean International Bank (Barbados) Ltd:** Roseau; internet www.firstcaribbeanbank.com; f. 2002 following merger of Caribbean operations of Barclays Bank PLC and CIBC; Exec. Chair. MICHAEL MANSOOR; CEO CHARLES PINK.

**National Bank of Dominica:** 64 Hillsborough St, POB 271, Roseau; tel. 4484401; fax 4483982; e-mail nbd@cwdom.dm; internet www.ncbdominica.com; f. 1976, as the National Commercial Bank of Dominica; name changed as above after privatization in Dec. 2003; cap. 10.0m., res 10.0m., dep. 350.1m. (June 2003); 49% govt-owned; Chair. MILTON LAWRENCE; 4 brs.

**Royal Bank of Canada:** Dame Mary Eugenia Charles Blvd, POB 19, Roseau; tel. 4482771; fax 4485398; Man. H. PINARD.

### DEVELOPMENT BANK

**Dominica Agricultural, Industrial and Development (DAID) Bank:** cnr Charles Avenue and Rawles Lane, Goodwill, POB 215, Roseau; tel. 4482853; fax 4484903; e-mail aidbank@cwdom.dm; f. 1971; responsible to Ministry of Finance; planned privatization suspended in 1997; provides finance for the agriculture, tourism, housing, education and manufacturing sectors; cap. 9.5m. (1991); Chair. CRISPIN SORHAINDO; Man. PATRICIA CHARLES.

### STOCK EXCHANGE

**Eastern Caribbean Securities Exchange:** based in Basseterre, Saint Christopher and Nevis; e-mail info@ecseonline.com; internet www.ecseonline.com; f. 2001; regional securities market designed to facilitate the buying and selling of financial products for the eight member territories—Anguilla, Antigua and Barbuda, Dominica, Grenada, Montserrat, Saint Christopher and Nevis, Saint Lucia and Saint Vincent and the Grenadines; Gen. Man. TREVOR BLAKE.

### INSURANCE

In 2001 there were 19 insurance companies operating in Dominica. Several British, regional and US companies have agents in Roseau. Local companies include the following:

**First Domestic Insurance Co Ltd:** 19–21 King George V St, POB 1931, Roseau; tel. 4498202; fax 4485778; e-mail insurance@cwdom.dm.

**Insurance Specialists and Consultants:** 19–21 King George V St, POB 20, Roseau; tel. 4482022; fax 4485778.

**J. B. Charles and Co Ltd:** Old St, POB 121, Roseau; tel. 4482876.

**Tonge Inc Ltd:** 19–21 King George V St, POB 20, Roseau; tel. 4484027; fax 4485778.

**Windward Islands Crop Insurance Co (Wincrop):** Vanoulst House, Goodwill, POB 469, Roseau; tel. 4483955; fax 4484197; f. 1987; regional; coverage for weather destruction of, mainly, banana crops; Man. KERWIN FERREIRA; brs in Grenada, Saint Lucia and Saint Vincent.

## Trade and Industry

### DEVELOPMENT ORGANIZATIONS

**National Development Corporation (NDC):** Valley Rd, POB 293, Roseau; tel. 4482045; fax 4485840; e-mail ndc@cwdom.dm; internet www.ndcdominica.dm; f. 1988 by merger of Industrial Development Corpn (f. 1974) and Tourist Board; promotes local and foreign investment to increase employment, production and exports; promotes and co-ordinates tourism development; Chair. DESMOND CARLISLE; Gen. Man. VINCENT PHILBERT.

**Eastern Caribbean States Export Development and Agricultural Diversification Unit (EDADU):** POB 769, Roseau; tel. 4482240; fax 4485554; e-mail oecsedu@cwdom.dm; internet www

.oecs-edu.org; f. 1990 as Eastern Caribbean States Export Development Agency; reformed as above in 1997; OECS regional development org.; Exec. Dir COLIN BULLY.

### INDUSTRIAL AND TRADE ASSOCIATIONS

**Dominica Association of Industry and Commerce (DAIC):** POB 85, cnr Old St and Fields Lane, Roseau; tel. 4482874; fax 4486868; e-mail daic@marpin.dm; internet www.delphis.dm/daic.htm; f. 1972 by a merger of the Manufacturers' Association and the Chamber of Commerce; represents the business sector, liaises with the Government, and stimulates commerce and industry; 100 mems; Pres. ANTHONY BISCOMBE; CEO JEANILIA R. V. DE SMET.

**Dominica Banana Producers Ltd (DBPL):** Vanoulst House, POB 1620, Roseau; tel. 4482671; fax 4486445; e-mail dbmc@cwdom.dm; f. 1934 as Dominica Banana Growers' Association; restructured 1984 as the Dominica Banana Marketing Corpn; renamed as above in 2003; state-supported, scheduled for privatization; Chair. RICHARD CHARLES; Gen. Man. KERVIN STEPHENSON.

**Dominica Export-Import Agency (DEXIA):** Bay Front, POB 173, Roseau; tel. 4482780; fax 4486308; e-mail dexia@cwdom.dm; internet www.dexiaexport.com; f. 1986; replaced the Dominica Agricultural Marketing Board and the External Trade Bureau; exporter of Dominican agricultural products, trade facilitator and importer of bulk rice and sugar; Gen. Man. GREGOIRE THOMAS.

### EMPLOYERS' ORGANIZATION

**Dominica Employers' Federation:** 14 Church St, POB 1783, Roseau; tel. 4482314; fax 4484474; e-mail def@cwdom.dm; Pres. ACKROYD BIRMINGHAM.

### UTILITIES

#### Electricity

**Dominica Electricity Services Ltd (Domlec):** 18 Castle St, POB 1593, Roseau; tel. 4482681; fax 4485397; e-mail mansecdomlec@cwdom.dm; national electricity service; 72%-owned by the Commonwealth Development Corporation (United Kingdom); Gen. Man. MURRAY ROGERS.

#### Water

**Dominica Water and Sewerage Co Ltd (DOWASCO):** 3 High St, POB 185, Roseau; tel. 4484811; fax 4485813; e-mail dowasco@cwdom.dm; state-owned; Chair. DON CHRISTOPHER; Gen. Man. DAMIAN SHILLINGFORD.

### TRADE UNIONS

**Dominica Amalgamated Workers' Union (DAWU):** 40 Kennedy Ave, POB 137, Roseau; tel. 4483048; fax 4485787; e-mail dawu@hotmail.com; f. 1960; Gen. Sec. FEDALINE M. MOULON; 500 mems (1996).

**Dominica Association of Teachers:** 7 Boyds Ave, POB 341, Roseau; tel. 4488177; fax 4488177; e-mail dat@cwdom.dm; internet www.dateachers.4t.com; Pres. CELIA NICHOLAS; 630 mems (1996).

**Dominica Trade Union:** 70–71 Independence St, Roseau; tel. 4498139; fax 4499060; f. 1945; Pres. HAROLD SEALEY; Gen. Sec. LEO J. BERNARD NICHOLAS; 400 mems (1995).

**Media Workers' Association:** Roseau; Pres. MATTHIAS PELTIER.

**National Workers' Union:** Independence St, cnr Church St and Old Market Sq., POB 387, Roseau; tel. 4485209; fax 4481934; e-mail icss@cwdom.dm; f. 1977; Pres. RAWLINGS F. A. JEMMOTT; Gen. Sec. FRANKLIN FABIEN; 450 mems (1996).

**Public Service Union:** cnr Valley Rd and Windsor Lane, Roseau; tel. 4482102; fax 4488060; e-mail dcs@cwdom.dm; f. 1940; registered as a trade union in 1960; representing all grades of civil servants, including firemen, prison officers, nurses, teachers and postal workers; Pres. (vacant); Gen. Sec. THOMAS LETANG; 1,400 mems.

**Waterfront and Allied Workers' Union:** 43 Hillsborough St, POB 181, Roseau; tel. 4482343; fax 4480086; e-mail wawuunion@hotmail.com; f. 1965; Pres. LOUIS BENOIT; Gen. Sec. KERTIST AUGUSTIS; 1,500 mems.

## Transport

### ROADS

In 1999 there were an estimated 780 km (485 miles) of roads, of which about 50.4% was paved; there were also numerous tracks.

### SHIPPING

A deep-water harbour at Woodbridge Bay serves Roseau, which is the principal port. Several foreign shipping lines call at Roseau, and there is a high-speed ferry service between Martinique and Guadeloupe which calls at Roseau eight times a week. Ships of the Geest Line call at Prince Rupert's Bay, Portsmouth, to collect bananas, and there are also cruise-ship facilities there. There are other specialized berthing facilities on the west coast.

**Dominica Port Authority (DPA):** POB 243, Roseau; tel. 4484431; fax 4486131; e-mail domport@cwdom.dm; f. 1972; pilotage and cargo handling; Gen. Man. VINCENT ELWIN.

### CIVIL AVIATION

Melville Hall Airport, 64 km (40 miles) from Roseau, and Canefield Airport, 5 km (3 miles) from Roseau, are the two airports on the island. A EC $15m. contract to upgrade Melville Hall Airport was signed with a French construction company in August 2004. It was hoped the project would be completed by 2006. The Government redirected the funds for the development, which had been partly provided by the European Union, from the previous (Dominica United Worker's Party) administration's plan to build a brand new international airport. The regional airline, LIAT (based in Antigua and Barbuda, and in which Dominica is a shareholder), provides daily services and, with Air Caraïbe, connects Dominica with all the islands of the Eastern Caribbean, including the international airports of Puerto Rico, Antigua, Guadeloupe and Martinique.

## Tourism

The Government has designated areas of the island as nature reserves, to preserve the beautiful, lush scenery and the rich, natural heritage that constitute Dominica's main tourist attractions. Birdlife is particularly prolific, and includes several rare and endangered species, such as the Imperial parrot. There are also two marine reserves. Tourism is not as developed as it is among Dominica's neighbours, but the country is being promoted as an 'eco-tourism' and cruise destination. There were an estimated 460,000 visitors in 2004 (of whom 380,600 were cruise-ship passengers). Receipts from tourism totalled an estimated EC $145.5m. in 2003.

**National Development Corporation (NDC)—Division of Tourism:** Valley Rd, POB 293, Roseau; tel. 4482045; fax 4485840; e-mail ndctourism@cwdom.dm; internet www.ndcdominica.dm; f. 1988 following merger of Tourist Board with Industrial Devt Corpn; Dir of Tourism SHARON PASCAL.

**Dominica Hotel and Tourism Association (DHTA):** POB 384, Roseau; tel. 4403430; fax 4403433; e-mail dhta@cwdom.dm; internet www.dhta.org; Pres. ATHERTON MARTIN; Treas. MAXINE ALLEYNE.

## Defence

The Dominican Defence Force was officially disbanded in 1981. There is a police force of about 300, which includes a coastguard service. The country participates in the US-sponsored Regional Security System.

## Education

Education is free and is provided by both government and denominational schools. There are also a number of schools for the mentally and physically handicapped. Education is compulsory for 10 years between five and 15 years of age. Primary education begins at the age of five and lasts for seven years. Enrolment of children in the primary age-group was 70.7% in 1992. Secondary education, beginning at 12 years of age, lasts for five years. A teacher-training college and nursing school provide further education, and there is also a branch of the University of the West Indies on the island. In 1997 the Government announced plans to invest EC $17.9m. in a Basic Education Reform project. In February 2005 the Government announced that funding provided by Libya would pay for the construction of a primary school in Portsmouth and a high school in Goodwill, near Roseau. Estimated budgetary expenditure on schools was EC $4.1m. in 2001/02 (equivalent to 1.5% of total expenditure).

# Bibliography

For works on the Caribbean generally, see Select Bibliography (Books)

Baker, P. L. *Centring the Periphery: Chaos, Order and the Ethnohistory of Dominica*. Kingston, University of the West Indies Press, 1996.

Paravisini-Gebert, L. *Phyllis Shand Allfrey: A Caribbean Life*. Piscataway, NJ, Rutgers University Press, 1996.

# THE DOMINICAN REPUBLIC

## Geography

### PHYSICAL FEATURES

The Dominican Republic comprises almost two-thirds of the island of Hispaniola (Isla Española) in the Greater Antilles. It lies at the eastern end of the island, with Haiti to the west of the land border that runs for 360 km (224 miles), from north to south, across the widest part of Hispaniola. The next nearest neighbours are Puerto Rico, a US Commonwealth territory 120 km to the east (across the Mona Passage), and the Turks and Caicos Islands, a British territory 145 km to the north. The country has 1,288 km of coastline and a total area of 48,422 sq km (18,696 sq miles), including 350 sq km of inland waters. Although less than one-half the size of Cuba, the Dominican Republic is the second largest country of the insular Caribbean and about the same size as Costa Rica.

The Dominican Republic occupies that part of Hispaniola that tapers eastwards—in the south, this is from the Pedernales peninsula, which is just east of the border with Haiti and thrusts southwards to culminate in Cabo Beata (Blessed Cape). Like Haiti, the Dominican Republic is very rugged and mountainous (80%), but, unlike its neighbour, its hillsides have not been denuded of woodland. The highlands are cleft by fertile valleys, many of them broad, and there is one fairly extensive range of coastal plain, named for the capital, Santo Domingo, narrow to the west of the city, but broad and running to the end of the island in the east. The northern boundary of this plain is the range of hills called the Cordillera Oriental, which parallels the Atlantic coast, like the higher, north-western range known as the Cordillera Septentrional. The highest mountains of the island and, indeed, of the West Indies, are found in the Cordillera Central, where rise Pico Duarte (3,175 m or 10,420 ft) and its near twin, La Pelona. The Cordillera Central occupies the centre of the island, running eastwards out of the Massif du Nord in Haiti and eventually curving southwards to the Caribbean. A mere 85 km to the south-west of Pico Duarte is the lowest point in the Caribbean, where, 46 m below sea level, is the bitter lake, the 200-sq km Lago Enriquillo, in hot, arid surroundings between the rocky dryness of the remaining two mountain ranges, the Sierra de Neiba and, to the south of Enriquillo, the Sierra de Baharuco (a continuation of the Massif du Selle, in Haiti). The high, forested mountains of the Dominican Republic attract good rainfall, though the Cordillera Central tends to shadow the south-west of the country, but there are also some important rivers. The longest river of the Dominican Republic, watering the Ceiba Valley, is the north-westward-flowing Yaque del Norte, which exits into the sea near the border with Haiti. Other major rivers are the westward-flowing Yuna, the southward-flowing Yaque del Sur and, running into Haiti, the Artibonite. The main offshore islands, which are not significant in territorial extent, are Saona (south of the south-eastern end of the island) and the uninhabited Beata (off Cape Beata). To the west of each is a smaller island, Catalina and Alto Velo, respectively. There are also three lacustrine islands in Enriquillo and some sandy cays off the northern coast. The range of the terrain and weather conditions makes for a wide variety of environments, ranging from the arid tropical forests of the west, for instance, to the pine woodland of the highlands. Native mammals are few and endangered, notably the hutia (jutía), a small rodent, the solenodon, an insectivore, and the manatee or sea cow. Birds are more numerous and include the endemic Hispaniolan woodpecker. There are a variety of parrots, parakeets and hummingbirds.

### CLIMATE

The climate is subtropical maritime, experiencing little seasonal variation in temperature and with exposure to hurricanes for some months from the middle of the year, but particularly in September. The Dominican Republic suffers occasional flooding and more regular droughts, but generally has good rainfall,

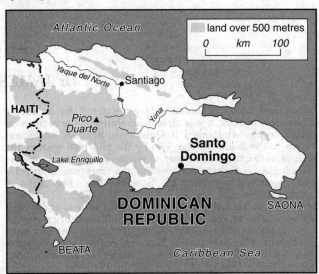

although it is very affected by the geography of the island. Rainfall is greater with altitude and with exposure to the north-eastern trade winds off the Atlantic, so the Cordillera Septentrional can receive more than 2,500 mm (98 ins) per year, on average, but the valleys between ranges can be much dryer. The average annual temperature is about 25°C (77°F), with little seasonal variation, but, again, altitude has a major influence (below-freezing temperatures have been recorded atop Pico Duarte) and the north can be noticeably cooler under the influence of weather fronts coming out of the north.

### POPULATION

Santo Domingo was the first Spanish capital in the Americas and the centre of early colonial expansion. African slaves began to arrive in the city in the 1520s, to replace the disappearing Amerindian population as labour. The proportion of black slaves in what is now called the Dominican Republic was never as high as in the French part of Hispaniola (Saint-Domingue, now Haiti); they still constituted an important part of the population and were later the object of 'liberating' incursions by their free western neighbours. Meanwhile, the mixing of Spanish and African stock produced the majority mixed-race population of today (65%), although social conditions have ensured that colour has remained an important issue into the present, with a complex vocabulary for skin colour and ethnic identity evolving. Whites and blacks each account for about 15% of the population now, because although many of the descendants of the original Spanish settlers fled at the time of independence amid fears of Haitian aggression, the ruling class considered it important to encourage new white immigration. Thus, groups of settlers from the Canary Islands (Spain) and Italy settled in the country, although there were also Syrians (known locally as Turcos), who came to establish businesses, and sugar-plantation workers from the anglophone Virgin and Leeward Islands (known as Cocolos). However, Spanish is the official language and the population remains overwhelmingly Roman Catholic, at least in nominal adherence. African traditions have also insinuated themselves into the religious scene, usually in parallel to more orthodox ecclesiastical practice. There is also an element of reconstructed Taíno (Amerindian) belief systems (the perceived legitimacy of which can also be observed in racial claims), all of which contribute to the local voodoo (vodú dominicana). Connected to this is a widespread belief in and fear of witchcraft (brujería).

The total population of the Dominican Republic, according to the results of the census conducted in October 2002, was 8.6m., making it the second most populous country of the Antilles. Social conditions are marked by income inequality and the fact that one-third of the population is under 14 years of age. There is a large immigrant Haitian population, most of whom have entered the country for work and often illegally. The capital and largest city, Santo Domingo (2.3m. in 2002), which is located midway along the southern coast, is constituted as a special National District (Distrito Nacional), in addition to the 31 provinces into which the country is divided. The second city of the republic is Santiago de los Caballeros (0.6m. in 2002), in the north-west, in the fertile and productive Ceiba Valley.

# History

## Dr JAMES FERGUSON

When Christopher Columbus sighted the island that he named Hispaniola in 1492, it was already inhabited by the indigenous Taino people, who called it Quisqueya. Spanish settlement began in the following year, and by the mid-16th century the Tainos were extinct as a separate people, having succumbed to the effects of smallpox, enslavement and repression after a 13-year insurrection against the Spanish. The importation of African slaves on a large scale began in the 1520s.

Spanish colonization of the island was piecemeal and incomplete. In the course of the 16th and 17th centuries French buccaneers established control over the empty western part, which was ceded to France as Saint-Domingue, by the Treaty of Ryswick in 1697. An English force, under Sir Francis Drake, sacked Santo Domingo in 1586 and by 1740 the area under Spanish control had a population of only 6,000, of whom only 2,000 were Spanish. Under a repopulation programme, the figure then increased to 100,000 in 40 years.

### INDEPENDENCE

At the end of the 18th century Santo Domingo became involved in the conflicts arising from the French Revolution and the Saint-Domingue slave revolt, which led to Haitian independence in 1804. France gained nominal control of the whole island through the 1795 Treaty of Basle, but it was the Haitians, under Toussaint Louverture, who succeeded in occupying Santo Domingo from 1801 to 1803. After further changes in sovereignty, Haiti invaded again in 1822, remaining until the declaration of Dominican independence in 1844.

The Haitians led further invasions, in the years following independence, and the new Republic's vulnerable position caused President Pedro Santana to request the re-establishment of Spanish dominion in 1861. Independence was regained in 1865 after a 'War of Restoration'. There followed 50 years of political and economic instability, in which governments were ephemeral, with the notable exception of the dictatorship of Ulises Heureaux (1882–99). The Republic's inability to pay foreign creditors was the pretext for the establishment, by the USA, of a customs receivership in 1905, with outright US occupation following between 1916 and 1924. A lasting result of the US occupation was the creation of a Dominican Army, the commanding general of which, Rafael Leonidas Trujillo Molina, was elected President in 1930.

### THE TRUJILLO ERA

Trujillo swiftly established a dictatorship, which lasted until his assassination in 1961. He ruled personally between 1930 and 1947, and indirectly through his brother, President Héctor Trujillo (1947–60), and through Dr Joaquín Balaguer (August 1960–January 1962). Ruling by blackmail, bribery, torture and murder, Trujillo amassed a fortune, becoming the country's largest landowner. The Trujillo family controlled two-thirds of the sugar industry, which underwent enormous expansion during his regime; their interests also covered many areas of manufacturing and commerce.

The Trujillo era witnessed a substantial programme of industrialization and public works, while the agreement to end the US customs receivership allowed Trujillo to assume the title of 'Restorer of Financial Independence'. High prices for Dominican commodities during the Second World War enabled him to liquidate the country's outstanding debt in 1947 and to introduce a national currency, the peso. Relations with Haiti, however, deteriorated, owing to the massacre of several thousand Haitian migrant labourers in 1937.

After the dictator's assassination, the titular President, Joaquín Balaguer, engineered the exile of Trujillo's son, Ramfis, and other members of the family. The Trujillo properties were seized by the state. After some months of political tension, marked by coup attempts and street disturbances, elections were held in December 1962, which gave a clear victory to the left-of-centre Partido Revolucionario Dominicano (PRD—Dominican Revolutionary Party). The PRD leader, Juan Bosch Gaviño, became President in February 1963, remaining in office for only seven months before being overthrown in a military coup, led by Col (later Gen.) Elías Wessin y Wessin. Bosch, accused of being pro-Communist, was exiled and a three-man civilian junta, led by Donald Reid Cabral, assumed power.

On 24 April 1965 followers of the PRD, supported by a group of young colonels, launched an insurrection aimed at restoring constitutional government. Fierce fighting followed between the insurgents and the armed forces under Wessin y Wessin, and on 28 April the first of 23,000 US Marines landed on the island. The Organization of American States (OAS) was subsequently requested to form a peace force and to negotiate a settlement. The intervention of the USA and the OAS resulted in an end to the fighting and the peace force withdrew in September. A provisional Government took office, under Héctor García Godoy, pending presidential and congressional elections in June 1966.

### LIMITED DEMOCRACY

The victor in the presidential contest was Balaguer, now of the Partido Reformista Social Cristiano (PRSC—Reformist Social Christian Party), who remained in office until 1978. His rule was marked by periodic outbreaks of right-wing terrorism, military coup attempts and left-wing guerrilla landings. President Balaguer encouraged closer economic links with the USA, symbolized by the powerful position obtained by Gulf and Western, the major private company in the country's sugar industry.

The PRD boycotted polls in 1970 and 1974, in protest at Balaguer's decision to seek further terms in office. Bosch resigned from the leadership of the PRD and formed his own party, the left-wing Partido de la Liberación Dominicana (PLD—Party of Dominican Liberation) in 1973. The presidential election of May 1978 was the first serious contest for 12 years. The PRD's Silvestre Antonio Guzmán Fernández was elected President, with Jacobo Majluta Azar as Vice-President. As Guzmán's victory became apparent, members of the police force attempted to stop the counting of votes. However, pressure from the US Government ensured that Guzmán was able to take office.

During his four years as President, Guzmán attempted to 'institutionalize' the armed forces, dismantling the powerful group of officers who had supported Trujillo and Balaguer. He also attempted to deal with the long-standing problem of corruption. However, in July 1982, after his successor, Salvador Jorge Blanco, also of the PRD, had been elected, Guzmán committed suicide, after learning that his daughter and son-in-law, who worked in the presidential secretariat, had been accused of corruption. Contrary to widespread fears, his suicide did not cause a military coup—a measure of his success in taming the army—and Jorge Blanco took office, as scheduled, in August.

During Jorge Blanco's presidency, as in Guzmán's, the economy suffered severe difficulties and the PRD underwent prolonged factional disputes, with the left of the party accusing the Government of abandoning its election pledges. In April 1984 at least 60 people were killed in riots, following the announcement of price increases and other austerity measures. In May 1986, as the elections approached, the PRD split into factions.

## BALAGUER AGAIN, 1986–96

The beneficiary of the PRD's disarray was Balaguer, candidate of the PRSC, who won his fifth term as President on 16 May 1986 and appointed a Cabinet comprised largely of unaffiliated technocrats. However, President Balaguer's policy of promoting rapid economic growth through a substantial public-works programme, together with incentives for foreign investors in industry and tourism, encountered increasing difficulties after 1988. The rapid price increases announced in early 1988 caused nation-wide strikes and riots, including a violent two-day general strike in June 1989. Nevertheless, Balaguer obtained a sixth term in office in the presidential election of 16 May 1990, defeating Bosch (PLD), José Francisco Peña Gómez (PRD) and Majluta, the candidate of the breakaway Partido Revolucionario Independiente (PRI—Independent Revolutionary Party), in highly controversial circumstances.

Widespread discontent at sharply rising consumer prices in 1990 forced the Government into an anti-inflationary austerity agreement with the IMF and a programme of debt rescheduling. The new policy was successful in reducing inflation and in improving the real gross domestic product (GDP) growth rate. A further eight-month IMF programme was agreed in July 1993, but inflation increased once more while growth slowed. Although 87 years old and virtually blind, Balaguer sought a seventh term in office in the presidential election of 16 May 1994. His principal opponent was the PRD's Peña Gómez, with the 84-year old Bosch standing for the PLD. Once again, the election was marked by allegations of voting irregularities, and a delayed official result gave Balaguer a narrow and bitterly contested victory over Peña Gómez. The atmosphere of crisis was only relieved by the signing of an agreement on 10 August between Balaguer and Peña Gómez, providing for Balaguer's term to last only 18 months, with a fresh election to be held in November 1995; this agreement was subsequently amended by the Congreso Nacional (National Congress) to give Balaguer a two-year term and to prohibit future presidents from seeking consecutive terms in office.

## THE PRESIDENCY OF LEONEL FERNÁNDEZ, 1996–2000

Following the volatile political atmosphere of 1995, the election of 16 May 1996 was the first for 30 years in which Balaguer did not stand, while Bosch was also absent, having resigned as PLD leader in June 1994. Another innovation was the adoption of a two-round voting system. In the first round, Peña Gómez of the PRD secured 45.9% of the ballot against 38.9% for the PLD's new candidate, Leonel Fernández Reyna, a 42-year old lawyer; the PRSC candidate, Vice-President Jacinto Peynado, won only 15.0% of the votes cast, after a campaign in which he had received only formal support from Balaguer.

In the second round, on 30 June 1996, Fernández won 51.25% of the votes, having obtained the support of PRSC voters through an alliance with their party. Peña Gómez secured 48.75% of the ballot. The majority of 71,704 votes, although narrow, was the largest in any presidential election since 1982. However, the new Government was impeded by the lack of a congressional majority, since the PLD had won only one Senate seat in the 1994 election, and 13 of the 120 seats in the Chamber of Deputies. President Fernández had promised to wage war on corruption and modernize the administration, and during 1997 he oversaw a restructuring of both the police and the judiciary. On other issues, however, notably the deteriorating state of public services, the new President made less progress. The opposition of the Congreso Nacional, which rejected the 1997 budget, played a significant role in this. This resulted in widespread disturbances and industrial action throughout 1997.

Elections to the Senate and to an enlarged Chamber of Deputies were held on 16 May 1998. The PRD (the leader of which, Peña Gómez, had died six days before) gained 83 of the 149 seats in the lower house and 24 in the Senate. President Fernández's PLD also increased its representation in the Congreso Nacional, to 49 deputies and four senators. The PRSC secured the remaining 17 seats in the lower house and two seats in the Senate. However, the PLD failed to gain a sufficient majority to secure passage of a possible constitutional amendment allowing President Fernández to seek re-election in 2000.

The remainder of President Fernández's term was characterized by bitter inter-party conflict, occasional social unrest and legislative impasses, as the Government's plans for radical reform of the taxation and tariff systems were frustrated. Some measures were implemented, including the first stages of the long-delayed privatization of the ailing state power company, Corporación Dominicana de Electricidad, in May 1999, but for the most part President Fernández concentrated on foreign-policy objectives, signing free trade agreements with the Central American Common Market (CACM) in April 1998 and with the Caribbean Community and Common Market (CARICOM) in August of that year.

## THE PRD'S RETURN

Following a period of political tension in the early months of 1999, during which candidates within the PRD, the PLD and the PRSC contested their parties' nominations for the forthcoming presidential election, due to be held in May 2000, two candidates emerged as favourites. Rafael Hipólito Mejía Domínguez won the PRD presidential nomination, while the ruling PLD selected the former Secretary of State to the Presidency, Danilo Medina Sánchez, as its nominee. In July 1999 the PRSC controversially nominated Joaquín Balaguer as its presidential candidate, with the 92-year-old former President aiming for an eighth term in office.

The campaign was the most peaceful in the Republic's history, marred only by a small number of violent incidents. It was widely assumed that the election would go to a second round, even though Mejía would probably gain a clear lead in the first. In the event, he received 49.9% of the votes cast on 16 May 2000, while Medina won 24.9% and Balaguer 24.6%. In an unprecedented move, the Junta Central Electoral allowed Mejía to declare himself the winner, although he had failed to secure the constitutionally required 50%. The PLD-PRSC alliance, which had kept the PRD out of office in 1996, failed to materialize again in 2000, probably because most PRSC activists felt that they had received little from the PLD Government over the past four years.

President Mejía's political programme was radical, including tax reform, decentralization of power and the restructuring of the public sector. Throughout his presidency he also proved to be confrontational, clashing publicly with ministers as well as the media. The PRD crusade against the previous PLD administration's alleged corruption also proved controversial, creating open hostility between the two parties. With only 73 PRD supporters out of 149 deputies in the lower house (10 of the 83 elected in May 1998 had been expelled from the party), Mejía was dependent on the PRSC and Balaguer.

Mejía stated that one of his Government's priorities was to strengthen the bilateral relationship with Haiti. In early 2001, following an increase in tensions in Haiti, a plan to protect Dominican territory was implemented, while military forces deployed on the Dominican-Haitian border were reinforced to prevent both illegal immigration and drugs-trafficking. In 2002 it was announced that US troops and equipment would be deployed on the border to prevent illegal immigration and alleged drugs-smuggling.

Some significant reforms were introduced in the first year of the Mejía administration. A package of fiscal legislation increased value-added tax (VAT) from 8% to 12%, introduced taxes on luxury goods and obliged larger companies to pay corporation taxes of 1.5% in stages throughout the year. The removal of the cumbersome 'fuel differential', the margin between the price paid by the Government and that paid by consumers, and its replacement by a fixed tax, raised petroleum prices, but there was no violent popular response. In recognition of the hardships caused to the poor by increased prices, the Government introduced a poverty mitigation programme of redistributive measures, including training, infrastructure programmes and health, education and housing reforms. The Government also announced ambitious infrastructural schemes, including a railway from the port of Haina via Santo Domingo to Santiago, and extensive new road building, to be part-funded by private interests. Social-security legislation was introduced, aimed at providing employees with improved pensions and other benefits.

However, if President Mejía enjoyed significant, albeit gradually decreasing, public popularity in 2002–04, he also attracted much criticism, especially from political opponents. It was alleged that, far from cutting the public-sector wage bill, the new Government had created over 100,000 new jobs for PRD supporters. Critics also claimed that the Government's recurrent and capital spending was outstripping revenue from taxation and that the fiscal deficit would reach crisis point. The successful floating of a US $500m. bond issue on international capital markets in September 2001 also led to allegations that the proceeds would be used for public-sector wage bills rather than capital investment.

During the course of 2002 speculation began to mount as to whether President Mejía would openly endorse a movement within the PRD calling for constitutional reform, and particularly for the right for a President to be elected to two successive terms in office. Others within the PRD, with an eye on the presidential nomination for 2004, openly opposed such constitutional changes. In January 2002 the formation of a National Constituent Assembly was approved by the Chamber of Deputies. In July the Assembly voted in favour of adopting a clause enabling a President to serve two consecutive terms. The issue of whether a presidential candidate could win an election with less than 50% of the vote remained unresolved, however. In early 2003 President Mejía finally announced that he would seek the PRD's candidacy for the 2004 election, widening the rift in the party between his followers and opponents of re-election, including Vice-President Milagros Ortiz Bosch and PRD Chairman Hatuey Decamps.

While the PRD appeared divided, the PRSC also experienced factional infighting regarding its own presidential nomination after the death of Balaguer in July 2002, with former presidential candidate Jacinto Peynado accusing his opponent, Eduardo Estrella, of engaging in irregularities while securing victory in the PRSC primaries. The PLD, for its part, supported a new bid for the presidency by former President Fernández. As the election neared, the PRD maintained its dominance but came under increasing criticism, owing to the depreciation of the peso, the resulting inflation and persistent power cuts. At the same time, a series of scandals affected the Government, including allegations of fraud made against President Mejía's former head of security. In May 2003 the collapse of the commercial bank Banco Intercontinental (Baninter), with its serious economic implications, damaged the Government as well as the PLD, with whom the President of Baninter, Ramón Báez Figueroa, was allegedly linked. The Baninter scandal was the worst financial crisis to have affected the country in recent years, owing to the involvement in illegal activity of a sum reportedly equivalent to 80% of the Government's annual budget. Meanwhile, the Government's seizure of various Baninter assets, including the *Listín Diario* newspaper, was perceived as being politically motivated, as the newspaper had been supportive of former President Fernández.

## FERNÁNDEZ'S COMEBACK

The run-up to the presidential election of 2004 was marked by social tension and controversy. Violence occurred in several cities, as protests against the Government's perceived mishandling of the economy, rising prices and power cuts ended in clashes with the police. The controversial nature of President Mejía's re-election bid also contributed to political tensions, both within the PRD and externally. After a number of high-profile quarrels between members of the PRD leadership, Mejía chose party Secretary-General Rafael Suberví Bonilla, formerly an outspoken opponent of Mejía's re-election ambitions, as his vice-presidential candidate, thus reducing some divisions within the party. Yet other senior PRD figures, such as Hatuey Decamps, remained adamantly opposed to Mejía's candidature, which was finally confirmed at a PRD convention in January 2004.

In the months preceding the election Mejía spent heavily on completing infrastructural projects in an apparent attempt to shore up support, while his opponents accused him of reckless economic mismanagement. Mejía's Government also attracted criticism from all opposition groups for its highly controversial proposed reform of the national parks system, under which large sections of environmentally sensitive and protected state-owned land would be opened to mining and tourism development. Fernández, in contrast, ran an efficient campaign, promising economic reform and an end to the country's rampant inflation and currency depreciation. Not only did he emphasize a period of steady economic growth effected during his previous incumbency, but he also impressed many with his choice of running mate, law professor and former Secretary of State for Labour Rafael Alburquerque.

Polling on 16 May 2004 took place relatively peacefully and efficiently, with a new electronic system working well. An unusually swift count of votes revealed that Fernández had secured victory in the first round, winning 57.11% of the votes. Mejía received 33.65% and Estrella 8.65%. Mejía was quick to concede defeat, calling the result a sign of the maturity of Dominican democracy. With only two PLD senators in a Senate of 32 members and 41 deputies out of a total of 150, it was apparent that Fernández would face severe challenges in enacting legislation and that he would have to form strategic alliances with the PRSC, which had 36 deputies. The Senate, however, seemed to present a stumbling block to the speedy introduction of reforms. Commentators were also quick to point out that Fernández had inherited a precarious economic situation, with high inflation, growing indebtedness, a large fiscal deficit and the prospect of continuing crisis within the bankrupt energy generation sector.

## FERNÁNDEZ'S SECOND TERM

Backed by an impressive electoral mandate, Fernández maintained his public popularity in the first year of his second presidency, promising to reform the worst aspects of Dominican political culture with measures to address corruption and clientilism. His campaign objectives included finding a solution to the long-standing energy crisis, possibly the most pressing issue for the majority of Dominicans. He also pledged to overhaul the economy, which was perceived by many to be suffering from mismanagement and excessive indebtedness from the Mejía administration. Negotiations with the IMF led to a vital stand-by agreement in early 2005, which, in turn, created the confidence among creditors to allow the country to renegotiate its debts.

Fernández's popularity was in large part owing to the crisis affecting the two main opposition parties, which were both soundly defeated in the 2004 election. The PRD was once again divided by factional interests and, in the wake of Mejía's controversial and abortive attempt to win a second consecutive term, a destructive battle for the leadership seemed inevitable. The PRSC, for its part, had yet to find a new generation of leaders to take on Balaguer's mantle. In the absence of a concerted opposition, President Fernández was able to introduce wide-ranging legislation and a potentially unpopular austerity budget without significant obstacles, despite lacking PLD majorities in both legislative chambers. With a congressional election approaching in May 2006 and the opposition parties beginning to provide a greater challenge to Fernández, the political climate of consensus seemed likely to change, however. The Government was also facing significant problems, both domestically and internationally. At home, it struggled to contain an unprecedented rise in violent crime and sought to reduce public-sector spending without exacerbating poverty among the country's most vulnerable sectors. Abroad, tensions with Haiti looked likely to increase as political violence intensified around scheduled elections there in late 2005. More pressing, however, was the issue of the free trade agreement with the USA, the Dominican Republic-Central American Free Trade Agreement (DR-CAFTA), which was to be debated in the second half of 2005 by the Dominican legislature. The Agreement was ratified by the US legislature in July of that year.

# Economy

## Dr JAMES FERGUSON

Agriculture, traditionally the mainstay of the Dominican Republic's economy, was superseded in importance by both tourism and manufacturing from the 1980s onwards, with these new sectors contributing an increasing proportion of gross domestic product (GDP) and foreign-exchange earnings. While the contribution of agriculture remained significant, especially in terms of employment, its importance to the economy as a whole declined; in 2004, according to preliminary figures, the sector's share of GDP (including livestock, forestry and fishing) was 11.7%, at constant 1970 prices, compared with 15.5% for the manufacturing sector. Government investment in infrastructural projects, largely funded by borrowing abroad, made a significant contribution to employment and to GDP. The Dominican economy is highly vulnerable to natural disasters and external shocks. In September 1998 'Hurricane Georges' killed approximately 300 people and caused damage to agricultural production estimated at US $2,000m. The terrorist attacks on the USA of 11 September 2001 had an immediate and negative impact on the Dominican economy, affecting tourism and investor confidence. In May 2004 torrential rains in the areas near the Haitian border caused mudslides and flooding, killing several hundred Dominicans and thousands of Haitians. In May 2005 heavy rains caused extensive damage to crops and infrastructure in the important agricultural area of the Cibao.

The economy suffered in the late 1980s and early 1990s from large trade and budgetary deficits, a heavy burden of debt and increases in the rate of inflation. In 1991 the adoption of an IMF-approved economic programme to restore fiscal discipline succeeded in reducing inflation and bringing the Government's accounts into surplus, while debt rescheduling and strong services earnings improved balance-of-payment and reserve positions. It was feared that inflation would increase sharply at the end of 1998, as a result of emergency imports necessitated by 'Hurricane Georges', but a rate of 4.9% was recorded over the year. The rate of inflation rose in 1999 to 6.5%, to 7.7% in 2000, and to 8.9% in 2001, largely as a consequence of higher oil prices, but fell again in 2002, to 5.3%. In 2003, however, inflation rose steeply, to 27.4%, mostly owing to the Dominican peso's depreciation against the US dollar and price increases on imported goods. The inflationary trend continued into 2004, with the rate reaching 51.4% at mid-year before showing signs of a slowdown and ending the year at 28.7%.

Government finances also fluctuated from the late 1990s: in 1996 there was a budgetary deficit of 708m. pesos, which was transformed into a surplus of 813m. pesos in 1997, and of 1,079m. pesos in 1998. In 1999, however, a deficit of 1,690m. pesos was recorded. In 2000 the Government recorded a surplus of 2,984m. pesos, but the fiscal surplus was reduced to 992m. pesos in 2001, owing to increased capital investment and rising debt commitments. In 2002 and 2003 the positive trend was reversed, with deficits of 10,859m. pesos and 19,311m. pesos, respectively, the latter representing 3.8% of GDP. The growing deficit was largely attributable to increased debt repayments, as well as increases in capital and current expenditure during a pre-election period. In the first three quarters of 2004 a surplus of 5,126m. pesos was recorded. The visible trade deficit increased steadily, from US $1,070.5m. in 1991 to $3,741.8m. in 2000. In 2001 the trade deficit narrowed slightly, to $3,451m., a 4.9% decrease in exports being offset by a 7.3% reduction in imports. The deficit fell to $2,444.0m. in 2003 and to $2,094.7m. in 2004 as recession and slow recovery dampened demand for imports.

A strong economic recovery took place throughout the 1990s, led by the construction, tourism, energy and manufacturing sectors. Despite high interest rates in 1993, which inhibited investment, growth was recorded in every year between 1992 and 2000. The strong economic performance gathered momentum in 1996, when a real GDP growth rate of 7.3% was recorded. This was based on increased activity in tourism, construction, mining, agriculture and telecommunications, together with an improvement in electricity supply. Growth 8.2% was recorded in 1997, although there was a marked deterioration in power supplies, as well as signs of diminishing business confidence, including increasing pressure on the peso-US dollar exchange rate. Tourism, construction and telecommunications were again among the leading sectors. In 1998, despite the impact of 'Hurricane Georges', real GDP increased by 7.5%. This trend continued into 1999; GDP growth of 8.8% was led by post-hurricane construction and a burgeoning telecommunications sector. In 2000 growth of 7.8% was recorded. In 2001, however, the Dominican economy experienced a significant decline, partly because of recessionary tendencies in the USA, falling tourism receipts and declining activity in the free zone sector. GDP growth for 2001 was 4.0%. Growth was stronger in 2002, standing at 4.3%. In 2003 the Dominican Republic experienced its first recession since the 1980s, as the economy contracted by 1.9%, with sectors related to domestic demand worst affected, but tourism and manufacturing expanding, owing to a depreciating peso. Modest growth of 2.0% was recorded in 2004, with communications proving the most dynamic economic sector.

According to the results of the census conducted in May 2002, the total population was 8,230,722, with a population density of 170.0 per sq km, one of the lowest in the Caribbean. The population in mid-2005 was estimated to have risen to 8,950,000. The work-force in 2002 numbered 3,943,500, of whom 634,300 (16.1%) were unemployed, although many were believed to be working in the undocumented informal sector. Unemployment was recorded at 17.5% in 2004, reflecting the impact of the recession. Approximately 1.8m. people lived in Santo Domingo (2.7m. including the suburban areas) in 2002. The annual average population growth rate between 1990 and 2003 was 1.7%.

By the 1990s deforestation was thought to have reduced the wooded area to about 6,000 sq km, the main forests being in the two mountain ridges, the northern and the southern cordilleras, both running from west to east. However, in 2000 a report by the Secretariat of State for Agriculture estimated that 13,000 sq km (or 28%) of the country were wooded. There were about 20,000 sq km of pasture and 15,000 sq km of arable land; the main agricultural areas were the Cibao and Vega Real lands in the centre of the country, where cocoa, coffee, rice and other crops were cultivated, and the south-east plains, where sugar-cane plantations and pasturage were concentrated. Mineral extraction took place near Bonao, in the centre, and in the south-west.

Clandestine emigration, mainly to the USA via Puerto Rico, was estimated at up to 40,000 attempts annually. However, increased co-operation between Puerto Rican and Dominican coast guards was reported, in 1997–2001, to have reduced the numbers. Out-migration was estimated at 3.02 persons per 1,000 in 2005. In 2003, largely owing to deteriorating economic conditions, it was reported that coastguards had intercepted 1,469 illegal migrants *en route* to Puerto Rico, a figure that increased to 2,159 in the first two months of 2004. The number of Haitians resident in the Republic remained a matter of speculation and controversy, with estimates ranging from 100,000 to 500,000. It was thought that many of these stayed and worked in the country for brief periods of time before returning to Haiti. The visit of President Leonel Fernández to Haiti in June 1998, the first by a Dominican head of state for 62 years, led to a marked improvement in relations. Relations continued to improve during the presidency of Rafael Hipólito Mejía Domínguez, and there were proposals for joint Dominican-Haitian development programmes in the border region, to be financed by the European Union (EU). The climate changed, however, with the overthrow of President Jean-Bertrand Aristide in February 2004, when supporters of the ousted Haitian leader accused the Dominican authorities of aiding the insurgents who led the rebellion. The disastrous mudslides and flooding of the following May further jeopardized proposed border development programmes. In 2005 allegations of cross-border drugs smuggling and other criminality on the part of illegal Haitian migrants created further tensions between the two countries.

## AGRICULTURE

Agriculture (including livestock, forestry and fishing) accounted for 11.7% of GDP in 2004. The sector's share of export earnings, which had been as high as 55% in 1984, decreased to less than one-half in the 1990s; in 2004 the main export commodities of sugar, coffee, cocoa and tobacco accounted for US $201.8m. from a total of $1,333.5m., excluding free zone exports. The decline in the agricultural sector was slowed by the development of new crops and by official measures to support agriculture, including the abolition of duty on imported inputs and a reduction of interest rates on loans to farmers. In the wake of 'Hurricane Georges' almost all agricultural production was badly affected; however, by 2001 there were signs of recovery, especially in crops such as citrus, coffee and cocoa, as well as such staples as rice. Flood damage in May 2004 adversely affected production of sugar and other crops, especially in the south-west.

The principal crop was sugar, although by the early 21st century it had been in decline for many years, mainly because of the inefficiency of the severely indebted state sugar company, Consejo Estatal de Azúcar (CEA). Overall, sugar exports earned US $93.5m. in 2004, compared with $170m. in 1996. The privatization of the CEA, undertaken in 1999 and 2000, has been only partly successful, with five mills leased to a Mexican consortium being closed within a year of the transfer. Five private groups, including the well-established Central Romana and Vicini companies, currently produce sugar for export and domestic consumption, although production of 524,700 metric tons in 2003 meant that the Republic needed to import 20,000 tons of refined sugar for the domestic market.

Coffee exports, which, according to UN trade figures, had reached US $170m. in value in 1995, declined throughout the late 1990s and in 2004 totalled an estimated $5.8m, falling from $16.5m. in 2003. Tobacco production increased in line with improved export marketing for cigars in the USA and the EU; export earnings reached some $100m. in 1996, but declined thereafter, standing at only $19.9m. in 2003, although a recovery, to an estimated $43.1m. was experienced in 2004. Cocoa was another important export crop, earning an estimated $59.4m. in 2004, although this was a decrease on the $77.0m. earned in the previous year. New or 'non-traditional' export crops, vegetables, cut flowers and ornamental plants, expanded considerably at the end of the 20th century, with exports reaching over $200m. in 2004. Sales of organic produce, such as bananas, continued to grow, especially in the EU. These sectors were least affected by 'Hurricane Georges', as replanting took place quickly.

Food crops included rice, maize, beans, cassava, tomatoes, bananas, plantains, mangoes and other fruit. The annual harvest of paddy rice was 609,200 metric tons in 2003, a 17% decrease on the previous year's harvest. In 2003 the country's livestock included 2.2m. head of cattle and an estimated 46.5m. chickens; the pig population was an estimated 577,500. The fish catch in that year was 21,700 tons, having increased from 13,200 tons in 2000.

## MINING

In 2004 the mining sector, dominated by ferro-nickel, contributed only an estimated 1.6% of GDP, compared with 4.5% in 1985. The sector had replaced agriculture as the main export earner between 1989 and 1990, and in 1989 ferro-nickel earnings had reached a record US $372m. Declining world prices subsequently pushed mining into second place, and throughout the 1990s revenue remained lower, except in 1994 and 1995, when a recovery in world prices pushed earnings in the latter year up to 30.0% of exports by value. Another phase of low world prices began in 1998, and exports of ferro-nickel earned only $132.1m. that year. Earnings recovered to $144m. in 1999 and a short-lived surge in world prices increased production in 2000 to $237.4m. By 2001, however, low world prices led to a fall in earnings, to $145m., a decline of almost 40%, and operations were again temporarily suspended. The slump in world prices forced the ferro-nickel mine at Bonao to close in 1998 and again in 2001 and early 2002. The mine was operated by Falconbridge Dominicana, the majority of which was owned by Falconbridge of Toronto (Canada). The mine reopened in 2003 in response to a steep increase in world demand and a doubling of US dollar nickel prices, owing in large part to the weakness of the US dollar. In 2004 the ferro-nickel exports earned $390.0m., compared with $238.7m. in the previous year.

Export values for gold and silver declined in the late 1990s, standing at a low of US $1.6m. in 1999. In 2001 the Government awarded a 25-year concession to operate the Sulfuros de Pueblo Viejo mine to Canadian-owned Placer Dome. There were also plans to develop other mines throughout the country. There was no production of gold or silver from 1999.

## ENERGY

There is no domestic petroleum production. Petroleum product imports in 2004 cost an estimated US $1,667.3m. (31% of total imports, excluding free zone activity, some of which was re-exported). Under the San José Agreement of 1980, Mexico and Venezuela were to provide 45,000 barrels per day (b/d) each, between them accounting for 90% of the Dominican Republic's petroleum imports. Under the Agreement, 20% of the cost of the petroleum was converted into a low-interest development loan. In October 2000 the signing of the Caracas energy accord gave the Dominican Republic the option to buy a further 20,000 b/d from Venezuela under San José Agreement terms. Political instability in Venezuela in 2001 and 2002 disrupted petroleum deliveries, and the Dominican oil refinery (Refidomsa) sought bids from Mexico, Colombia and Trinidad and Tobago in 2002 to make up the shortfall. A subsequent agreement signed with the Venezuelan Government in November 2004 guaranteed 10% of the country's oil needs at preferential prices. After many years of operating a so-called 'fuel differential', in which the Government bought fuel and sold it on to consumers at a profit, this was abolished in 2001 and replaced with a fixed tax on petroleum and other fuels.

The electricity-generating operations of the state concern, Corporación Dominicana de Electricidad (CDE), were severely deficient throughout the 1990s. Production fell to about one-half of peak demand at the beginning of the decade. The difficulties were caused by the frequent withdrawal from service of generating units, for repair and maintenance, together with low hydroelectric production, as a result of drought, and an inefficient distribution system, which resulted in one-half of the electricity produced being lost in transmission or through illegal connections. Private generating companies periodically reduced supplies to the CDE because of non-payment of debts.

In May 1999 the first stage of the privatization of the CDE was completed, with the sale of 50% of the shares in the CDE's two generating and three distribution companies. New Caribbean Investment, a Chilean-US consortium, paid US $177.8m. for a 50% stake in the Itabo generating company, and a consortium of the Commonwealth Development Corporation (United Kingdom) and Enron (USA) paid $144.5m. for the same proportion of the Haina generating company. Unión Fenosa (Spain) paid a combined $211.9m. for 50% shares in the northern and southern distribution companies and AES Distribución Dominicana paid $109.3m. for its stake in the eastern company. The new operators pledged substantial investment to overcome persistent supply problems and the privatization was expected to bring benefits to all of the CDE's customers, and, therefore, to the economy of the Dominican Republic as a whole. In September the generating plant at Itabo was transferred to the Gener-Coastal group, completing the transfer of the CDE's generating and distribution role to the private sector.

The privatization programme brought mixed results, with frequent conflict between energy providers and the Government. The Government has accused generators and suppliers of overcharging, while critics of privatization blamed the first Fernández administration for over-generous concessions to the private companies and creating a poor regulatory system. The companies, in turn, have criticized government ministries and departments for late payment of electricity bills. Throughout 2003 and 2004 lengthy power failures occurred across the country, frequently contributing to civil unrest and violence. The power situation worsened throughout 2004 and into 2005 as global petroleum prices rose in response to the conflict in Iraq and unrest in the Middle East. The incoming Fernández administration pledged to overhaul the energy sector, promising that contracts with generators and distributors would be revised, subsidies to consumers removed and collections improved. The crisis persisted, however, with an estimated generating deficit of 550 MW, extensive payment arrears and considerable theft of electicity in poor urban areas. The total amount owed in payment arrears to distributors and generators was estimated at US $400m. in August 2004, and higher petroleum prices threatened to exacerbate the situation.

## MANUFACTURING

Manufacturing, the Dominican Republic's largest economic sector, contributed an estimated 15.5% of GDP in 2004. Almost one-half of the contribution of the domestic manufacturing sector traditionally came from sugar refining, although this was seriously affected in 1998–2000 by uncertainties surrounding privatization and post-hurricane shortfalls in production. The sector showed signs of improvement in 2001, when growth of 7.9% was registered; however, growth slowed to 4.4% in 2002 and the sector contracted by 3.1% in 2003. An expansion of just 0.7% was recorded in 2004. Other important products included cement and other non-metallic minerals, textiles, clothing and footwear, leather goods, paper, glassware, food and drinks. The sector employed 14.1% of the active labour force in 2002. From 1990 to 2003 it demonstrated average annual growth of 3.9%. Construction contributed an estimated 10.7% of GDP in 2004, with the emphasis on private- rather than public-sector investment, although the Government's programme of infrastructural expansion, funded by foreign borrowing, contributed to the sector's growth. Following expansion of 3.2% in 2002, construction GDP decreased by 8.6% in 2003 and by 6.3% in 2004, reflecting the recessionary climate.

The Dominican Republic had a substantial free zone subsector, which numbered 513 companies in 2001, housed in 46 zones and employing 175,000 people. In 2004 the Consejo Nacional de Zonas Francas de Exportación (National Free Zones Council) reported that 116 new free zone companies had been approved, creating 18,492 new jobs. Exports from the zones in 2004 amounted to US $3,286.2m., compared with $1,765m. in 1995, representing 80% of the Republic's manufacturing export earnings. A contraction of 1.1% from the previous year's export figures reflected a declining share of the US apparel market. free zone companies enjoy 'tax holidays' of up to 20 years, duty-free raw-material imports and other benefits. In 2004 about 50% of production was accounted for by garments and textiles, while footwear, leather goods and electronic components were also produced. Since 1999 investors have warned that growing competition from Mexico is jeopardizing Dominican exports into the US market, but more recently both the Dominican Republic and Mexico have faced increasing competition in the US clothing market from the People's Republic of China. In 2004 the Dominican share of the US apparel market had fallen to 2.6%, from 3.7% in 2000.

## TRANSPORT AND INFRASTRUCTURE

Roads, totalling 17,000 km in 2005, were the main means of communication. The Mejía administration undertook an extensive road-building programme from 2000, aiming primarily to reduce traffic congestion in Santo Domingo. In 2005 the Fernández administration announced plans to construct a new subway system in the capital. In 2002 it was estimated that there were 50 vehicles per 1,000 people or 2.1m. vehicles in total. Tolls on major highways were increased by 200% in August 2002, leading to an immediate increase in public transport costs. There were several private railway companies devoted to the transport of sugar cane, the largest of which was the Central Romana network.

The country had 14 ports, of which Haina, near Santo Domingo, was the largest, handling 80% of imports. In December 2004 the country had a merchant fleet of 28 vessels with a total displacement of 13,285 grt. In 2000 the Government granted a concession to a joint-venture company for the construction of a new port and transhipment centre near the Las Americas international airport to the east of Santo Domingo, which was designed specifically for use by free zone businesses. Construction was completed in 2003. In late 2002 the Government announced plans to construct a US $1,800m. complex at Manzanillo Bay, on the north coast near the Haitian border, involving a new port, a free zone and tourist facilities. There are international airports at Santo Domingo, Puerto Plata, Barahona, La Romana, Samana and Santiago, as well as several domestic airports. In June 2002 Punta Cana airport also began receiving direct scheduled flights from the USA. The privatization of the international airports took place in 1999, with investment from Canadian and Italian consortia. Mobile cellular telephone usage has increased significantly, with 2,300,000 subscribers recorded in mid-2004, compared with 141,000 in 1997. In 2003 there were estimated to be 500,000 internet users.

## TOURISM

Tourism was the Republic's leading foreign-exchange earner, with receipts totalling about US $3,110m. in 2003. The total number of hotel rooms in 2004 was 59,416. An estimated 150,000 people were employed in the tourism industry in 2004.

In 2004 passenger arrivals by air numbered 3,772,932, an increase of 210,162, or 5.9%, on the previous year's total. The number of tourists from Europe increased throughout the 1990s, reaching about 50% of the total in 1999, but in 2001 and 2002 there was a significant decline in the number of visitors from Germany, Italy and Spain, largely owing to the weakness of the euro. In 2003 this trend was reversed as the euro gained in value against the peso, and the number of European visitors rose by 21%. In 2004 visitors from North America accounted for 36.6% of the total, while European visitors accounted for 34.1%. In US dollar terms, average daily spending by tourists declined because of the devaluation of the peso, from US $104.15 in 2002 to $101.27 in 2003.

## INVESTMENT AND FINANCE

For most years in the 1990s government accounts showed a small surplus, the largest being in 1992, when the surplus equalled 3.3% of GDP. In 1998 the Government was still able to record a budget surplus of 1,078.9m. pesos (equivalent to 0.4% of GDP) despite the damage caused by 'Hurricane Georges'. In 1999, however, the surplus turned to a deficit of 1,680.7m. pesos. In 2000 the Government again reported a fiscal surplus of 2,984m. pesos, caused, in large part, by increased revenue from income tax, sales taxes and the recently introduced fuel consumption tax, which alone generated 7,511m. pesos. A smaller surplus of 992m. pesos was recorded in 2001. In 2002, however, the Government recorded a much higher deficit of 10,859m. pesos, followed by a still bigger deficit of 19,311m. pesos in 2003, equivalent to 3.8% of GDP. The estimated deficit in 2004 was 2.7% of GDP, or 20,763m. pesos.

A scandal concerning the country's second largest commercial bank, Banco Intercontinental (Baninter), placed the financial sector under strain in early 2003, as it was revealed that over US $2,000m. had been embezzled. The Baninter affair forced the Government to intervene and reimburse investors, at a cost equivalent to approximately 20% of GDP. This, in turn, led to the intervention of the IMF, which in August 2003 agreed a $600m. stand-by agreement, designed to restore stability in the banking system and help the Government absorb the costs of compensating investors. The IMF suspended its programme in October, after the Government renationalized two electricity distribution companies without consultation, but the agreement was resumed in February 2004. After a further suspension and the election of the second Fernández Government, another US $665m. stand-by agreement was signed with the IMF in

January 2005, superseding the previous agreement. This support followed fiscal adjustment measures enacted by the new Government, including tax reforms, the removal of energy subsidies and cuts in the public-sector payroll.

In September 2001 the Government successfully placed US $500m. in sovereign bonds on the international market, allowing for increased investment in capital and other infrastructural projects. A second bond issue of $600m. took place in 2003. Private investment, external and domestic, was principally in the tourism industry, the privatized industries and telecommunications. In late 1997 a new foreign-investment law was enacted, providing for total repatriation of capital and profits, simplifying investment procedures and easing previous restrictions on the scope of foreign investment. By December 1998 foreign direct investment had reached a record $1,100m., compared with $421m. in 1997. In 1999 approximately $1,400m. of foreign investment flowed into the economy, representing some 12% of GDP. In 2001 foreign investment was estimated at $1,200m., an increase of 25% since 2000. This trend was reversed in 2002, as foreign investment fell to $916.8m., and then became increasingly negative in 2003, with investment of only $522.8m. recorded in the wake of the crisis in the banking system and declining investor confidence. Foreign direct investment fell by a further 11% in 2004 to US $463.2m. New legislation concerning a reformed monetary code was enacted in 2002, facilitating access to credit and abolishing the unpopular 4.75% commission on foreign-exchange transactions imposed on Dominican businesses.

## FOREIGN TRADE AND THE BALANCE OF PAYMENTS

The Dominican Republic's trade was in deficit from 1976 onwards and from 1989 the deficit exceeded US $1,000m. By 1998 the deficit had reached $2,617m., principally owing to a sharp rise in imports in the last quarter, as reconstruction of the damage caused by 'Hurricane Georges' began. In 1999 the deficit was $2,904m., again worsened by post-hurricane imports and reduced exports. In 2000, despite rising exports, the deficit widened still further, to $3,742m., largely owing to increases in the price of imported petroleum. However, from 2001 the deficit narrowed: to $3,451m., with a decline in exports counterbalanced by a reduced import bill; to $3,503m. and $2,156m. in 2002 and 2003, respectively, the reduction caused primarily by a fall in demand for imports; and to $2,095m. in 2004, owing to a modest increase in exports and declining domestic demand for imports.

Exports (excluding the free zones) reached a low point of US $511m. in 1993, but increased to $836m. in 1996, largely as a result of a recovery in ferro-nickel exports (including free zone transactions, total exports increased by 7.2%, to $4,053m.). Exports fluctuated thereafter. Exports in 2002 totalled $5,165m., reflecting a falling demand for free zone goods in the US market. Exports grew in 2003, to $5,471m., of which $1,064m. were non-free zone trade. In 2004, according to preliminary figures, total exports were $5,750m., of which $1,334m. were non-free zone trade.

Imports (f.o.b.—again excluding the free zones) fluctuated around a rising trend, reaching US $6,610m. ($9,479m. including free zone trade), both as a result of strong growth and high petroleum prices. Total imports fell to $6,237m. by 2002 ($8,838m., including free zones), however, reflecting economic slowdown and a fall in petroleum prices. In 2003 imports to $5,096m. (excluding free zones), again reflecting recessionary tendencies. In 2004 there was a slight increase, to $5,370m. (again, excluding free zone transactions).

The trade deficits were partially offset by earnings from tourism and free zone exports, together with remittances from Dominicans overseas (estimated at US $2,060.5m. in 2003 and rising by 6.5% in 2004). Nevertheless, the current account of the balance of payments was in deficit from 1966. In 2000 the deficit stood at $1,027m. Figures for 2001 indicated a decrease in the current-account deficit, to $741m., and the figure rose slightly to $798m. in 2002. In 2003 the current account registered a surplus of $1,036m., largely owing to the fall in imports and a 26.3% increase in the services balance, attributable to improved tourism receipts. This trend continued in 2004, when a preliminary current acount surplus of $1,399m. was registered.

The Dominican Republic's main sources of long-term development aid were the Inter-American Development Bank (IDB), the World Bank and the EU. The country was admitted to the EU's fourth Lomé Convention in 1989 (the Lomé Convention expired in February 2000 and was replaced by the Cotonou Agreement in June—see Part Three, Regional Organizations, European Union). In the first Lomé protocol (1990–95), under its national initiative programme, the Dominican Republic was allocated ECU 85m. This figure increased to ECU 106m. in the second protocol (1995–2000). Bilateral aid was also received from a variety of countries. Apart from the stand-by agreement with the IMF, the Dominican Government also renegotiated its external debt with the 'Paris Club' of foreign creditors in April 2004. In 2004 the country's total external debt was estimated at US $7,700m., a substantial increase from the figure of $5,100m. in 2001. Total international reserves stood at $806.2m. in December 2004, compared with $260.7m. in the previous year; this increase was largely owing to the peso's rally against a weaker US dollar and growing confidence in the economic competence of the new Government. In February 2005 the peso was trading at 29 to the US dollar, having recovered from a low of 48.5 at the beginning of 2004.

## CONCLUSION

Since coming to power in August 2004, the Fernández Government has done much to restore confidence in the Dominican economy after inheriting daunting problems of indebtedness and fiscal deficits in the wake of the Baninter scandal, compounded by a slowdown in economic growth. The new Government swiftly moved to reduce the high levels of inflation, to address the chronic electricity supply crisis and to lower the public-sector wage bill. Balancing fiscal authority and overtures to foreign investors with meeting campaign pledges to improve economic conditions for the majority of poor Dominicans was not expected to be an easy task. The Government introduced an austerity budget with the aim of reducing the fiscal deficit and also regained the support of the IMF, which wished to see further reforms in taxation, financial regulation and public-sector finances. The large-scale borrowing undertaken by the Mejía administration would, nevertheless, continue to create financial difficulties, as an estimated US $1,600m. of debt repayments would fall due in 2006, compared with the $859m. due in 2003. Further borrowing and debt rescheduling seemed inevitable and would create serious problems if international interest rates were to rise. The negative GDP growth recorded in 2003 proved not to be a long-term trend, as the US and European economies remained sufficiently strong to encourage investment in Dominican manufacturing and tourism, allowing a modest economic recovery (of 2.0%) in 2004. However, much would depend in the coming years on the outcome of the free trade agreement signed between the Dominican Republic and the USA in March 2004, but yet to be ratified by the Dominican legislature, bringing the country into line with the Central American countries that signed a similar agreement in the previous year. While the so-called Dominican Republic-Central American Free Trade Agreement agreement offered the Dominican economy little new access to the US market, as 91% of exports were tariff-free, it did secure existing access, which could be ended with the possible termination of the Caribbean Basin Trade Promotion Act in 2007. Conversely, the abolition of tariffs on US goods entering the Dominican Republic would adversely affect the Government's fiscal position, as 24% of tax revenue was derived from tariffs and taxes on imports.

THE DOMINICAN REPUBLIC

# Statistical Survey

Sources (unless otherwise stated): Oficina Nacional de Estadística, Edif. de Oficinas Gubernamentales, Avda México, esq. Leopoldo Navarro, Santo Domingo; tel. 682-7777; fax 685-4424; e-mail info@one.gov.do; internet www.one.gov.do; Banco Central de la República Dominicana, Calle Pedro Henríquez Ureña, esq. Leopoldo Navarro, Apdo 1347, Santo Domingo; tel. 221-9111; fax 686-7488; e-mail info@bancentral.gov.do; internet www.bancentral.gov.do.

## Area and Population

### AREA, POPULATION AND DENSITY

| | |
|---|---:|
| Area (sq km) | |
| Land | 48,137 |
| Inland water | 597 |
| Total | 48,734* |
| Population (census results) | |
| 24 September 1993 | 7,293,390 |
| 18–20 October 2002 | |
| Males | 4,297,326 |
| Females | 4,265,215 |
| Total | 8,562,541 |
| Population (official estimates)† | |
| 2003 | 8,716,500 |
| 2004 | 8,873,300 |
| Density (per sq km) at 2004 | 182.1 |

* 18,816 sq miles.
† Preliminary figures, rounded to nearest 100 persons.

### PROVINCES

| | Area (sq km) | Population (census of October 2002) | Density (per sq km) |
|---|---:|---:|---:|
| *Distrito Nacional Region* | | | |
| Distrito Nacional | 104.4 | 913,540 | 8,750.4 |
| Santo Domingo | 1,296.4 | 1,817,754 | 1,402.2 |
| *Valdesia Region* | | | |
| Peravia | 997.6 | 169,865 | 170.3 |
| San Cristóbal | 1,265.8 | 532,880 | 421.0 |
| Monte Plata | 2,632.1 | 180,376 | 68.5 |
| San José de Ocoa | 650.2 | 62,368 | 95.9 |
| *Norcentral Region* | | | |
| Espaillat | 839.0 | 225,091 | 268.3 |
| Puerto Plata | 1,856.9 | 312,706 | 168.4 |
| Santiago | 2,839.0 | 908,250 | 319.9 |
| *Nordeste Region* | | | |
| Duarte | 1,605.4 | 283,805 | 176.8 |
| María Trinidad Sánchez | 1,271.7 | 135,727 | 106.7 |
| Salcedo | 440.4 | 96,356 | 218.8 |
| Samaná | 853.7 | 91,875 | 107.6 |
| *Enriquillo Region* | | | |
| Baoruco | 1,282.2 | 91,480 | 71.3 |
| Barahona | 1,739.4 | 179,239 | 103.0 |
| Independencia | 2,006.4 | 50,833 | 25.3 |
| Pedernales | 2,074.5 | 21,207 | 10.2 |
| *Este Region* | | | |
| El Seibo | 1,786.8 | 89,261 | 50.0 |
| La Altagracia | 2,474.3 | 182,020 | 73.6 |
| La Romana | 654.0 | 219,812 | 336.1 |
| San Pedro de Macorís | 1,255.5 | 301,744 | 240.3 |
| Hato Mayor | 1,329.3 | 87,631 | 65.9 |
| *El Valle Region* | | | |
| Azua | 2,531.8 | 208,857 | 82.5 |
| Elías Piña | 1,426.2 | 63,879 | 44.8 |
| San Juan | 3,569.4 | 241,105 | 67.5 |
| *Noroeste Region* | | | |
| Dajabón | 1,020.7 | 62,046 | 60.8 |
| Monte Cristi | 1,924.4 | 111,014 | 57.7 |
| Santiago Rodríguez | 1,111.1 | 59,629 | 53.7 |
| Valverde | 823.4 | 158,293 | 192.2 |
| *Cibao Central Region* | | | |
| La Vega | 2,287.0 | 385,101 | 168.4 |
| Sánchez Ramírez | 1,196.1 | 151,179 | 126.4 |
| Monseñor Nouel | 992.4 | 167,618 | 168.9 |
| **Total** | 48,137.0* | 8,562,541 | 177.9* |

* Land area only.

### PRINCIPAL TOWNS
(population at census of October 2002)

| | | | |
|---|---:|---|---:|
| Santo Domingo DN (capital) | 2,302,759 | San Felipe de Puerto Plata | 146,882 |
| Santiago de los Caballeros | 622,101 | Higuey | 141,751 |
| San Cristóbal | 220,767 | Moca | 131,733 |
| Concepción de la Vega | 220,279 | San Juan de la Maguana | 129,224 |
| San Pedro de Macorís | 217,141 | Monseñor Nouel | 115,743 |
| La Romana | 202,488 | Baní | 107,926 |
| San Francisco de Macorís | 156,267 | Bajos de Haina | 80,835 |

### BIRTHS AND DEATHS
(UN estimates, annual averages)

| | 1985–90 | 1990–95 | 1995–2000 |
|---|---:|---:|---:|
| Birth rate (per 1,000) | 31.5 | 28.2 | 25.7 |
| Death rate (per 1,000) | 7.0 | 6.5 | 6.4 |

Source: UN, *World Population Prospects: The 2004 Revision*.

**Expectation of life** (WHO estimates, years at birth): 68 (males 65; females 72) in 2003 (Source: WHO, *World Health Report*).

### ECONOMICALLY ACTIVE POPULATION
('000 persons aged 10 years and over, national survey, April 2002)

| | Males | Females | Total |
|---|---:|---:|---:|
| Agriculture, hunting, forestry and fishing | 483.4 | 17.0 | 500.4 |
| Mining and quarrying | 8.7 | 0.0 | 8.7 |
| Manufacturing | 304.1 | 163.7 | 467.8 |
| Electricity, gas and water supply | 23.3 | 6.3 | 29.5 |
| Construction | 180.5 | 3.4 | 183.9 |
| Wholesale and retail trade | 461.3 | 255.0 | 716.3 |
| Hotels and restaurants | 74.7 | 117.7 | 192.4 |
| Transport, storage and communications | 222.0 | 20.6 | 242.6 |
| Financial intermediation | | | |
| Real estate, renting and business activities | 32.9 | 33.4 | 66.4 |
| Public administration and defence | 122.7 | 46.8 | 169.4 |
| Education | | | |
| Health and social work | 258.7 | 473.0 | 731.6 |
| Other community, social and personal service activities | | | |
| **Total employed** | 2,172.3 | 1,136.8 | 3,309.1 |
| Unemployed | 220.5 | 413.8 | 634.3 |
| **Total labour force** | 2,392.8 | 1,550.6 | 3,943.5 |

**Mid-2003** (estimates in '000): Agriculture, etc. 572; Total labour force 3,869 (Source: FAO).

# THE DOMINICAN REPUBLIC

*Statistical Survey*

## Health and Welfare

### KEY INDICATORS

| | |
|---|---:|
| Total fertility rate (children per woman, 2003) | 2.7 |
| Under-5 mortality rate (per 1,000 live births, 2003) | 35 |
| HIV/AIDS (% of persons aged 15–49, 2003) | 1.7 |
| Physicians (per 1,000 head, 2000) | 1.88 |
| Hospital beds (per 1,000 head, 1996) | 1.5 |
| Health expenditure (2002): US $ per head (PPP) | 295 |
| Health expenditure (2002): % of GDP | 6.1 |
| Health expenditure (2002): public (% of total) | 36.4 |
| Access to water (% of persons, 2002) | 93 |
| Access to sanitation (% of persons, 2002) | 57 |
| Human Development Index (2002): ranking | 98 |
| Human Development Index (2002): value | 0.738 |

For sources and definitions, see explanatory note on p. vi.

## Agriculture

### PRINCIPAL CROPS
('000 metric tons)

| | 2001 | 2002 | 2003 |
|---|---:|---:|---:|
| Rice (paddy) | 721.7 | 730.7 | 608.3 |
| Maize | 36.5 | 30.3 | 44.0 |
| Potatoes | 64.7 | 48.6 | 49.3 |
| Sweet potatoes | 32.3 | 32.3 | 33.9 |
| Cassava (Manioc) | 123.9 | 120.2 | 123.6 |
| Yautia (Cocoyam) | 39.5 | 59.4 | 79.4 |
| Sugar cane | 4,645.3 | 4,846.5 | 5,019.2 |
| Dry beans | 30.3 | 29.4 | 29.0 |
| Coconuts | 169.1 | 171.8 | 183.7 |
| Oil palm fruit* | 153.0 | 155.0 | 155.0 |
| Tomatoes | 203.0 | 154.9 | 155.0* |
| Pumpkins, squash and gourds* | 23.0 | 23.5 | 23.5 |
| Chillies and peppers, green | 24.6 | 28.7 | 30.0* |
| Dry onions | 41.2 | 42.7 | 43.0* |
| Garlic | 4.1 | 7.0 | 7.0* |
| Carrots | 26.7 | 20.8 | 21.0* |
| Cantaloupes and other melons* | 52.0 | 53.0 | 53.0 |
| Other fresh vegetables* | 35.3 | 34.3 | 35.0 |
| Bananas | 442.0 | 502.9 | 481.0* |
| Plantains* | 190.0 | 192.0 | 192.5 |
| Oranges | 66.7 | 86.1 | 88.0* |
| Mangoes* | 185.0 | 185.5 | 185.5 |
| Avocados | 111.1 | 147.5 | 150.0* |
| Pineapples | 136.9 | 108.6 | 110.0* |
| Papayas* | 24.0 | 24.5 | 25.0 |
| Other fresh fruit* | 42.2 | 43.7 | 43.7 |
| Coffee (green) | 35.5 | 49.0 | 50.0* |
| Cocoa beans | 44.9 | 49.7 | 50.0* |
| Tobacco (leaves)* | 17.5 | 18.0 | 18.5 |

* FAO estimate(s).
Source: FAO.

### LIVESTOCK
('000 head, year ending September)

| | 2001 | 2002 | 2003* |
|---|---:|---:|---:|
| Horses* | 340 | 342 | 343 |
| Mules* | 140 | 140 | 141 |
| Asses* | 150 | 150 | 151 |
| Cattle | 2,106.8 | 2,159.6 | 2,160.0 |
| Pigs | 565.5 | 577.0* | 577.5 |
| Sheep | 106.0 | 121.3 | 122.0 |
| Goats | 187.4 | 188.0 | 188.5 |
| Poultry | 47,380 | 46,000* | 46,500 |

* FAO estimate(s).
Source: FAO.

### LIVESTOCK PRODUCTS
('000 metric tons)

| | 2001 | 2002 | 2003* |
|---|---:|---:|---:|
| Beef and veal | 70.8 | 71.7 | 72.0 |
| Pig meat | 63.1 | 64.0 | 65.0 |
| Poultry meat | 203.4 | 185.3 | 185.5 |
| Cows' milk | 420.3 | 517.7 | 520.0 |
| Cheese | 2.5* | 3.7 | 3.7 |
| Butter | 1.5* | 1.8 | 1.8 |
| Hen eggs | 77.0 | 82.4 | 83.0 |
| Honey | 1.4* | 1.3 | 1.3 |
| Cattle hides (fresh)* | 8.8 | 8.9 | 8.9 |

* FAO estimate(s).
Source: FAO.

## Forestry

### ROUNDWOOD REMOVALS
('000 cubic metres, excl. bark)

| | 1983 | 1984 | 1985 |
|---|---:|---:|---:|
| Sawlogs, veneer logs and logs for sleepers | 4 | 4 | 4 |
| Other industrial wood | 3 | 3 | 3 |
| Fuel wood | 531 | 543 | 556 |
| **Total** | **538** | **550** | **563** |

**1986–2003:** Annual production as in 1985 (FAO estimates).
Source: FAO.

## Fishing

('000 metric tons, live weight)

| | 2001 | 2002 | 2003 |
|---|---:|---:|---:|
| Capture | 13.2 | 17.3 | 18.1 |
| Tilapia | 0.5 | 1.1 | 0.7 |
| Groupers, seabasses | 0.9 | 1.3 | 0.6 |
| Common carp | 0.4 | 0.6 | 0.4 |
| Snappers and jobfishes | 0.9 | 1.7 | 2.2 |
| Grunts, sweetlips | 0.4 | 0.4 | 0.2 |
| Nurse shark | 0.1 | — | 0.1 |
| Caribbean spiny lobster | 1.2 | 2.5 | 0.8 |
| Stromboid conchs | 1.4 | 2.7 | 1.7 |
| Aquaculture | 2.6 | 3.6 | 3.6 |
| Common carp | 0.4 | 0.5 | 0.5 |
| **Total catch** | **15.9** | **20.8** | **21.7** |

Source: FAO.

## Mining

('000 metric tons)

| | 2001 | 2002 | 2003 |
|---|---:|---:|---:|
| Ferro-nickel | 60.7 | 58.1 | 66.3 |
| Nickel (metal content of laterite ore) | 39.1 | 38.9 | 45.4 |
| Gypsum | 175.6 | 163.0 | 230.6 |
| Rock salt | 189.6 | 157.3 | 107.0 |

Source: US Geological Survey.

# THE DOMINICAN REPUBLIC

## Industry

### SELECTED PRODUCTS
('000 metric tons, unless otherwise indicated)

|  | 2000 | 2001 | 2002 |
|---|---|---|---|
| Wheat flour | 245 | 272 | n.a. |
| Refined sugar | 101 | 113 | 132 |
| Cement | 2,521 | 2,758 | 3,050 |
| Beer ('000 hectolitres) | 3,666 | 3,176 | 3,554 |
| Cigarettes (million) | n.a. | n.a. | 3,509 |
| Motor spirit (gasoline) | 345 | n.a. | n.a |
| Electric energy (million kWh)* | 9,316 | 10,512 | 11,510 |

* Source: UN Economic Commission for Latin America and the Caribbean, *Statistical Yearbook*.

Source: mostly UN, *Industrial Commodity Statistics Yearbook*.

**2000–03** ('000 42–gallon barrels, estimates): Motor spirit (gasoline)1,900; Jet fuel 1,800; Distillate fuel oil 2,700; residual fuel oil 4,400 (Source: US Geological Survey).

## Finance

### CURRENCY AND EXCHANGE RATES

**Monetary Units**
100 centavos = 1 Dominican Republic peso (RD $ or peso oro).

**Sterling, Dollar and Euro Equivalents** (31 May 2005)
£1 sterling = 53.110 pesos;
US $1 = 29.212 pesos;
€1 = 36.021 pesos;
1,000 Dominican Republic pesos = £18.83 = $34.23 = €27.76.

**Average Exchange Rate** (RD $ per US $)
2002  18.610
2003  30.831
2004  42.120

### BUDGET
(RD $ million)

| Revenue | 2001 | 2002 | 2003* |
|---|---|---|---|
| Tax revenue | 58,058.3 | 63,866.6 | 74,247.8 |
| Taxes on income and profits | 15,317.7 | 16,032.7 | 20,384.8 |
| Taxes on goods and services | 27,964.8 | 31,484.5 | 34,498.3 |
| Taxes on international trade and transactions | 13,415.1 | 14,773.4 | 17,573.0 |
| Other current revenue | 1,583.5 | 2,428.5 | 5,397.6 |
| Capital revenue | 213.9 | 782.9 | 39.3 |
| Total | 59,855.8 | 67,078.0 | 79,684.6 |

| Expenditure | 2001 | 2002 | 2003* |
|---|---|---|---|
| Current expenditure | 43,419.3 | 48,383.6 | 59,421.5 |
| Wages and salaries | 21,496.2 | 24,925.4 | 26,107.7 |
| Other services | 2,075.5 | 3,275.1 | 4,325.9 |
| Materials and supplies | 4,048.7 | 4,477.5 | 5,288.7 |
| Current transfers | 12,157.3 | 10,650.9 | 14,072.6 |
| Interest payments | 2,989.9 | 4,386.5 | 7,844.1 |
| Internal debt | 726.8 | 1,534.9 | 2,336.6 |
| External debt | 2,263.2 | 2,851.6 | 5,507.4 |
| Capital expenditure | 15,444.1 | 18,993.8 | 16,325.7 |
| Machines and equipment | 994.5 | 1,241.7 | 854.7 |
| Construction of works and agricultural plantations | 7,624.7 | 8,309.8 | 7,286.2 |
| Capital transfers | 6,196.0 | 8,736.3 | 7,758.4 |
| Total (incl. others) | 58,863.4 | 67,377.4 | 75,747.2 |

* Preliminary figures.

### INTERNATIONAL RESERVES
(US $ million at 31 December)

|  | 2002 | 2003 | 2004 |
|---|---|---|---|
| Gold* | 5.2 | 7.6 | 8.0 |
| IMF special drawing rights | 0.3 | 0.1 | 1.6 |
| Foreign exchange | 468.1 | 253.0 | 796.7 |
| Total | 473.6 | 260.7 | 806.2 |

* Valued at market-related prices.

Source: IMF, *International Financial Statistics*.

### MONEY SUPPLY
(RD $ million at 31 December)

|  | 2002 | 2003 | 2004 |
|---|---|---|---|
| Currency outside banks | 18,259.0 | 29,654.2 | 32,548.4 |
| Demand deposits at commercial banks | 23,764.4 | 44,902.0 | 46,350.9 |
| Total money (incl. others) | 38,469.6 | 77,558.2 | 78,555.1 |

### COST OF LIVING
(Consumer Price Index including direct taxes; base: 2000 = 100)

|  | 2002 | 2003 | 2004 |
|---|---|---|---|
| Food, beverages and tobacco | 110.7 | 140.1 | 237.0 |
| Clothing | 108.4 | 117.1 | n.a. |
| Rent | 120.6 | 165.0 | n.a. |
| All items (incl. others) | 114.6 | 146.0 | 221.2 |

Source: ILO.

### NATIONAL ACCOUNTS

**National Income and Product**
(RD $ million at current prices)

|  | 2001 | 2002 | 2003 |
|---|---|---|---|
| GDP in market prices | 366,205 | 401,883 | 509,965 |
| Net primary income from abroad | −18,531 | −21,504 | −38,501 |
| Gross national income (GNI) | 347,675 | 380,379 | 471,465 |
| Less Consumption of fixed capital | 21,972 | 24,113 | 30,598 |
| Net national income | 325,703 | 356,266 | 440,867 |

Source: IMF, *International Financial Statistics*.

**Expenditure on the Gross Domestic Product**
(RD $ million at current prices)

|  | 2001 | 2002 | 2003 |
|---|---|---|---|
| Final consumption expenditure | 311,469 | 346,146 | 405,153 |
| Households Non-profit institutions serving households | 278,622 | 307,577 | 357,828 |
| General government | 32,847 | 38,569 | 47,325 |
| Gross capital formation | 83,781 | 91,575 | 117,656 |
| Gross fixed capital formation | 82,726 | 90,417 | 116,186 |
| Changes in inventories Acquisitions, less disposals, of valuables | 1,055 | 1,158 | 1,470 |
| Total domestic expenditure | 395,250 | 437,721 | 522,809 |
| Exports of goods and services | 141,165 | 152,880 | 264,720 |
| Less Imports of goods and services | 170,209 | 187,718 | 277,563 |
| GDP in market prices | 366,205 | 401,883 | 509,965 |
| GDP at constant 1970 prices | 6,910 | 7,207 | 7,175 |

Source: IMF, *International Financial Statistics*.

# THE DOMINICAN REPUBLIC

**Gross Domestic Product by Economic Activity**
(RD $ million at constant 1970 prices, preliminary figures)

|  | 2002 | 2003 | 2004 |
|---|---|---|---|
| Agriculture, hunting, forestry and fishing | 836.1 | 814.4 | 842.5 |
| Mining and quarrying | 102.1 | 111.1 | 115.2 |
| Manufacturing | 1,144.2 | 1,109.1 | 1,116.9 |
| Electricity and water | 178.0 | 163.0 | 131.1 |
| Construction | 905.0 | 827.5 | 775.4 |
| Wholesale and retail trade, restaurants and hotels | 1,375.6 | 1,306.5 | 1,330.3 |
| Transport, storage and communications | 1,072.4 | 1,126.8 | 1,254.9 |
| Finance, insurance and real estate | 558.3 | 549.0 | 552.4 |
| General government services | 548.3 | 571.4 | 580.0 |
| Other services | 496.7 | 502.7 | 521.0 |
| **Total** | **7,216.6** | **7,081.4** | **7,219.7** |

## BALANCE OF PAYMENTS
(US $ million)

|  | 2002 | 2003 | 2004* |
|---|---|---|---|
| Exports of goods f.o.b. | 5,165.0 | 5,470.8 | 5,749.9 |
| Imports of goods f.o.b. | −8,837.7 | −7,626.8 | −7,844.6 |
| **Trade balance** | **−3,672.7** | **−2,156.0** | **−2,094.7** |
| Exports of services | 3,070.7 | 3,468.8 | 3,532.6 |
| Imports of services | −1,313.4 | −1,219.4 | −1,204.8 |
| **Balance on goods and services** | **−1,915.4** | **93.4** | **233.1** |
| Other income received | 300.4 | 340.7 | 316.2 |
| Other income paid | −1,452.2 | −1,733.8 | −1,648.1 |
| **Balance on goods, services and income** | **−3,067.2** | **−1,299.7** | **−1,098.8** |
| Current transfers (net) | 2,269.3 | 2,335.9 | 2,497.6 |
| **Current balance** | **−797.9** | **1,036.2** | **1,398.9** |
| Direct investment | 916.8 | 613.0 | 645.1 |
| Other investment (net) | −533.8 | −629.41 | −1,354.2 |
| Net errors and omissions | −139.3 | −1,566.3 | −143.6 |
| **Overall balance** | **−554.2** | **−546.5** | **546.2** |

* Preliminary figures.

# External Trade

## PRINCIPAL COMMODITIES

| Imports f.o.b. (US $ million)* | 2002 | 2003 | 2004† |
|---|---|---|---|
| Consumer goods | 3,107.6 | 2,409.0 | 2,550.2 |
| Durable goods | 784.9 | 315.3 | 375.4 |
| Foodstuffs | 288.4 | 240.5 | 214.9 |
| Petroleum products | 947.7 | 977.1 | 1,087.7 |
| Raw materials | 1,863.9 | 1,771.7 | 1,949.7 |
| Artificial plastic materials | 143.3 | 126.2 | 137.1 |
| Petroleum and petroleum products | 349.5 | 438.8 | 579.6 |
| Cast iron and steel | 188.3 | 155.4 | 199.0 |
| Capital goods | 1,265.9 | 915.2 | 870.0 |
| For transport | 237.8 | 142.0 | 78.1 |
| For industry | 232.7 | 182.1 | 229.1 |
| Machinery | 309.0 | 204.0 | 228.1 |
| **Total** (incl. others) | **6,237.3** | **5,095.9** | **5,369.9** |

* Figures exclude imports into free trade zones.
† Preliminary figures.

| Exports f.o.b. (US $ million)* | 2002 | 2003 | 2004† |
|---|---|---|---|
| Sugar and sugar cane derivatives | 99.2 | 96.9 | 93.5 |
| Raw cane sugar | 74.0 | 72.9 | 73.5 |
| Cocoa and cocoa manufactures | 67.0 | 77.0 | 59.4 |
| Cocoa beans | 60.8 | 67.6 | 52.2 |
| Tobacco and tobacco manufactures | 25.5 | 19.9 | 43.1 |
| Ferro-nickel | 156.2 | 238.7 | 390.0 |
| Other goods | 374.2 | 463.2 | 507.8 |
| Petroleum products | 98.8 | 137.7 | 218.6 |
| **Total** | **847.7** | **1,064.1** | **1,333.5** |

* Figures exclude exports from free trade zones, which totalled: US $4,317.3m. in 2002; US $4,406.8m. in 2003; US $4,416.4m. in 2004.
† Preliminary figures.

## PRINCIPAL TRADING PARTNERS
(US $ '000)

| Imports c.i.f. | 1998 | 1999 | 2000 |
|---|---|---|---|
| Argentina | 60,697 | 40,687 | 24,808 |
| Brazil | 57,863 | 66,193 | 80,719 |
| Canada | 40,593 | 40,549 | 41,350 |
| Colombia | 45,434 | 65,040 | 75,199 |
| Denmark | 36,396 | 32,633 | 50,886 |
| Germany | 51,507 | 59,488 | 87,814 |
| Italy | 43,434 | 46,340 | 57,698 |
| Japan | 148,285 | 237,190 | 266,938 |
| Korea, Democratic People's Republic | 35,139 | 21,199 | 16,025 |
| Mexico | 245,284 | 265,894 | 436,254 |
| Panama | 80,808 | 124,101 | 157,221 |
| Puerto Rico | 131,209 | 72,388 | 88,813 |
| Spain | 125,141 | 143,688 | 243,198 |
| Taiwan | 46,847 | 72,018 | 79,221 |
| USA | 1,700,855 | 2,436,210 | 2,962,044 |
| Venezuela | 403,918 | 673,960 | 979,449 |
| **Total** (incl. others) | **3,446,603** | **4,635,545** | **5,953,829** |

| Exports f.o.b. | 1998 | 1999 | 2000 |
|---|---|---|---|
| Belgium-Luxembourg | 75,774 | 71,220 | 123,293 |
| Canada | 26,490 | 15,690 | 5,160 |
| Cuba | 2,422 | 4,604 | 10,114 |
| France | 10,805 | 11,930 | 11,227 |
| Germany | 30,236 | 28,467 | 26,908 |
| Haiti | 47,601 | 67,543 | 57,936 |
| Italy | 17,935 | 9,208 | 12,892 |
| Japan | 18,461 | 8,525 | 11,949 |
| Korea, Republic | 18,618 | 18,208 | 38,517 |
| Mexico | 9,356 | 4,897 | 3,343 |
| Netherlands | 12,526 | 11,668 | 17,340 |
| Panama | 9,960 | 17,923 | 18,879 |
| Puerto Rico | 74,012 | 91,948 | 97,505 |
| Spain | 17,262 | 16,931 | 15,300 |
| United Kingdom | 2,035 | 6,518 | 10,836 |
| USA | 448,509 | 327,875 | 373,462 |
| **Total** (incl. others) | **861,990** | **743,561** | **866,054** |

Source: Inter-American Development Bank.

# Transport

## ROAD TRAFFIC
(motor vehicles in use)

|  | 1999 | 2000 | 2001 |
|---|---|---|---|
| Passenger cars ('000) | 445.9 | 455.6 | 561.3 |
| Commercial vehicles ('000) | 247.2 | 283.0 | 284.7 |

Source: UN, *Statistical Yearbook*.

## SHIPPING

**Merchant Fleet**
(registered at 31 December)

|  | 2002 | 2003 | 2004 |
|---|---|---|---|
| Number of vessels | 19 | 20 | 28 |
| Total displacement ('000 grt) | 9.2 | 12.8 | 13.3 |

Source: Lloyd's Register-Fairplay, *World Fleet Statistics*.

**International Sea-borne Freight Traffic**
('000 metric tons)

|  | 1996 | 1997 | 1998 |
|---|---|---|---|
| Goods loaded | 112 | 152 | 139 |

Source: UN, *Monthly Bulletin of Statistics*.

## CIVIL AVIATION
(traffic on scheduled services)

|  | 1997 | 1998 | 1999 |
|---|---|---|---|
| Kilometres flown (million) | 1 | 1 | 0 |
| Passengers carried ('000) | 34 | 34 | 10 |
| Passengers-km (million) | 16 | 16 | 5 |
| Total ton-km (million) | 1 | 1 | 0 |

Source: UN, *Statistical Yearbook*.

## Tourism

**ARRIVALS BY NATIONALITY**

|  | 2002 | 2003 | 2004 |
|---|---|---|---|
| Canada | 313,491 | 412,625 | 448,627 |
| France | 242,022 | 317,215 | 300,009 |
| Germany | 240,546 | 243,135 | 233,090 |
| Italy | 113,569 | 135,293 | 123,904 |
| Spain | 135,522 | 201,864 | 228,035 |
| United Kingdom | 146,257 | 171,696 | 197,964 |
| USA | 710,875 | 865,942 | 931,248 |
| **Total** (incl. others) | 3,104,709 | 3,562,770 | 3,772,932 |

**Tourism receipts** (US $ million, excl. passenger transport): 2,798 in 2001; 2,730 in 2002; 3,110 in 2003 (Source: World Tourism Organization).

## Communications Media

|  | 2001 | 2002 | 2003 |
|---|---|---|---|
| Telephones ('000 main lines in use) | 955.1 | 909.0 | 901.8 |
| Mobile cellular telephones ('000 subscribers) | 1,270.1 | 1,700.6 | 2,122.5 |
| Internet users ('000) | 186.0 | 500.0 | 800.0 |

**Daily newspapers:** 9 in 2000 (average circulation 230,000 copies).
**Non-daily newspapers:** 8 in 2000 (average circulation 215,000 copies).
**Radio receivers** ('000 in use): 1,440 in 1997.
**Television receivers** ('000 in use): 790 in 1998.
**Facsimile machines** (number in use): 2,300 in 1996.

Sources: UNESCO, *Statistical Yearbook*; UN, *Statistical Yearbook*; International Telecommunication Union.

## Education

(1996/97)

|  | Institutions | Teachers | Students Males | Students Females | Total |
|---|---|---|---|---|---|
| Pre-primary* | n.a. | 8,571 | 96,252 | 94,289 | 190,541 |
| Primary | 4,001† | 39,860 | 691,675 | 668,369 | 1,360,044 |
| Secondary: |  |  |  |  |  |
| general | 1,737 | 11,033 | 145,560* | 184,384* | 329,944* |
| teacher training† | n.a. | 86 | 549 | 743 | 1,292 |
| vocational† | n.a. | 1,211 | 9,147 | 12,356 | 21,503 |
| Higher | n.a. | 9,041 | 75,223 | 101,772 | 176,995 |

* 1997/98 figure(s).
† 1994/95 figure(s).

Source: UNESCO, *Statistical Yearbook*.

**Adult literacy rate** (UNESCO estimates): 84.4% (males 84.3%; females 84.4%) in 2002 (Source: UN Development Programme, *Human Development Report*).

# Directory

## The Constitution

The Constitution of the Dominican Republic was promulgated on 28 November 1966, and amended on 14 August 1994. Its main provisions are summarized below:

The Dominican Republic is a sovereign, free, independent state; no organizations set up by the State can bring about any act which might cause direct or indirect intervention in the internal or foreign affairs of the State or which might threaten the integrity of the State. The Dominican Republic recognizes and applies the norms of general and American international law and is in favour of and will support any initiative towards economic integration for the countries of America. The civil, republican, democratic, representative Government is divided into three independent powers: legislative, executive and judicial.

The territory of the Dominican Republic is as laid down in the Frontier Treaty of 1929 and its Protocol of Revision of 1936.

The life and property of the individual citizen are inviolable; there can be no sentence of death, torture nor any sentence which might cause physical harm to the individual. There is freedom of thought, of conscience, of religion, freedom to publish, freedom of unarmed association, provided that there is no subversion against public order, national security or decency. There is freedom of labour and trade unions; freedom to strike, except in the case of public services, according to the dispositions of the law.

The State will undertake agrarian reform, dedicating the land to useful interests and gradually eliminating the latifundios (large estates). The State will do all in its power to support all aspects of family life. Primary education is compulsory and all education is free. Social security services will be developed. Every Dominican has the duty to give what civil and military service the State may require. Every legally entitled citizen must exercise the right to vote, i.e. all persons over 18 years of age and all who are or have been married even if they are not yet 18.

### GOVERNMENT

Legislative power is exercised by Congress which is made up of the Senate and Chamber of Deputies, elected by direct vote. Senators, one for each of the 31 Provinces and one for the Distrito Nacional, are elected for four years; they must be Dominicans in full exercise of their citizen's rights, and at least 25 years of age. Their duties are to elect the President and other members of the Electoral and Accounts Councils, and to approve the nomination of diplomats. Deputies, one for every 50,000 inhabitants or fraction over 25,000 in each Province and the Distrito Nacional, are elected for four years and must fulfil the same conditions for election as Senators.

Decisions of Congress are taken by absolute majority of at least half the members of each house; urgent matters require a two-thirds' majority. Both houses normally meet on 27 February and 16 August each year for sessions of 90 days, which can be extended for a further 60 days.

Executive power is exercised by the President of the Republic, who is elected by direct vote for a four-year term. No President may serve more than one consecutive term. The successful presidential candidate must obtain at least 50% plus one vote of the votes cast; if necessary, a second round of voting is held 45 days later, with the participation of the two parties that obtained the highest number of votes. The President must be a Dominican citizen by birth or origin,

# THE DOMINICAN REPUBLIC

over 30 years of age and in full exercise of citizen's rights. The President must not have engaged in any active military or police service for at least a year prior to election. The President takes office on 16 August following the election. The President of the Republic is Head of the Public Administration and Supreme Chief of the armed forces and police forces. The President's duties include nominating Secretaries and Assistant Secretaries of State and other public officials, promulgating and publishing laws and resolutions of Congress and seeing to their faithful execution, watching over the collection and just investment of national income, nominating, with the approval of the Senate, members of the Diplomatic Corps, receiving foreign Heads of State, presiding at national functions, decreeing a State of Siege or Emergency or any other measures necessary during a public crisis. The President may not leave the country for more than 15 days without authorization from Congress. In the absence of the President, the Vice-President will assume power, or failing him, the President of the Supreme Court of Justice. The legislative and municipal elections are held two years after the presidential elections, mid-way through the presidential term.

## LOCAL GOVERNMENT

Government in the Distrito Nacional and the Municipalities is in the hands of local councils, with members elected proportionally to the number of inhabitants, but numbering at least five. Each Province has a civil Governor, designated by the Executive.

## JUDICIARY

Judicial power is exercised by the Supreme Court of Justice and the other Tribunals; no judicial official may hold another public office or employment, other than honorary or teaching. The Supreme Court is made up of at least 11 judges, who must be Dominican citizens by birth or origin, at least 35 years old, in full exercise of their citizen's rights, graduates in law and have practised professionally for at least 12 years. The National Judiciary Council appoints the members of the Supreme Court, who in turn appoint judges at all other levels of the judicial system. There are nine Courts of Appeal, a Lands Tribunal and a Court of the First Instance in each judicial district; in each Municipality and in the Distrito Nacional there are also Justices of the Peace.

Elections are directed by the Central Electoral Board. The armed forces are essentially obedient and apolitical, created for the defence of national independence and the maintenance of public order and the Constitution and Laws.

The artistic and historical riches of the country, whoever owns them, are part of the cultural heritage of the country and are under the safe-keeping of the State. Mineral deposits belong to the State. There is freedom to form political parties, provided they conform to the principles laid down in the Constitution. Justice is administered without charge throughout the Republic.

This Constitution can be reformed if the proposal for reform is supported in Congress by one-third of the members of either house or by the Executive. A special session of Congress must be called and any resolutions must have a two-thirds' majority. There can be no reform of the method of government, which must always be civil, republican, democratic and representative. Note: In July 2002 a National Constituent Assembly voted in favour of amending the Constitution to enable presidents to serve two consecutive terms; a further proposal to allow presidential candidates to win an election with less than 50% of votes cast was also under discussion.

# The Government

## HEAD OF STATE

**President:** Leonel Fernández Reyna (took office 16 August 2004).
**Vice-President:** Dr Rafael Francisco Alburquerque de Castro.

## CABINET
(July 2005)

**Secretary of State to the Presidency:** Danilo Medina.
**Secretary of State for External Relations:** Carlos Morales Troncoso.
**Secretary of State for the Interior and Police:** Franklin Almeyda.
**Secretary of State for the Armed Forces:** Rear-Adm. Sigfrido Pared Pérez.
**Secretary of State for Finance:** Vicente Bengoa.
**Secretary of State for Education:** Alejandrina Germán.
**Secretary of State for Agriculture:** Amílcar Romero.
**Secretary of State for Public Works and Communications:** Manuel de Jesús Pérez.
**Secretary of State for Public Health and Social Welfare:** Sabino Baéz.
**Secretary of State for Industry and Commerce:** Francisco Javier García.
**Secretary of State for Labour:** José Ramón Fadul.
**Secretary of State for Tourism:** Felix Jiménez.
**Secretary of State for Sport, Physical Education and Recreation:** Felipe Jay Payano.
**Secretary of State for Art and Culture:** José Rafael Lantigua.
**Secretary of State for Higher Education, Science and Technology:** Ligia Amada de Melo.
**Secretary of State for Women:** Gladys Gutiérrez.
**Secretary of State for Youth:** Manuel Crespo.
**Secretary of State for the Environment and Natural Resources:** Max Puig.
**Technical Secretary of State to the Presidency:** Juan Temistocles Montas.
**Administrative Secretary of State to the Presidency:** Luis Manuel Bonetti.
**Secretary of State without Portfolio:** Luis Inchausti.
**Secretary of State without Portfolio:** Bienvenido Perez.

## SECRETARIATS OF STATE

**Administrative Secretariat of the Presidency:** Palacio Nacional, Avda México, esq. Dr Delgado, Santo Domingo, DN; tel. 686-4771; fax 688-2100; e-mail prensa@presidencia.gov.do; internet www.presidencia.gov.do.

**Technical Secretariat of the Presidency:** Avda México, esq. Dr Delgado, Bloque B, 2°, Santo Domingo, DN; tel. 695-8028; fax 695-8432; e-mail informacion@stp.gov.do; internet www.stp.gov.do.

**Secretariat of State for Agriculture:** Autopista Duarte, Km 6.5, Los Jardines del Norte, Santo Domingo, DN; tel. 547-3888; fax 227-1268; e-mail sec.agric@verizon.net.do; internet www.agricultura.gov.do.

**Secretariat of State for the Armed Forces:** Plaza de la Independencia, Avda 27 de Febrero, esq. Luperón, Santo Domingo, DN; tel. 530-5149; fax 531-1309; internet www.secffaa.mil.do.

**Secretariat of State for Culture:** Centro de Eventos y Exposiciones, Avda George Washington, esq. Presidente Vicini Burgos, Santo Domingo, DN; tel. 221-4141; e-mail decultura@hotmail.com; internet www.cultura.gov.do.

**Secretariat of State for Education:** Avda Máximo Gómez 10, esq. Santiago, Santo Domingo, DN; tel. 688-9700; fax 689-8907; e-mail sub-administrativo@see.gov.do.

**Secretariat of State for the Environment and Natural Resources:** Avda 27 de Febrero, esq. Tiradente, Edif. Plaza Merengure, Suite 202, Ens. Naco, Santo Domingo, DN; tel. 555-5555; internet www.ceiba.gov.do.

**Secretariat of State for External Relations:** Avda Independencia 752, Santo Domingo, DN; tel. 535-6280; fax 533-5772; e-mail correspondencia@serex.gov.do; internet www.serex.gov.do.

**Secretariat of State for Finance:** Avda México 45, esq. Leopoldo Navarro, Apdo 1478, Santo Domingo, DN; tel. 687-5131; fax 682-0498; e-mail webmaster@finanzas.gov.do; internet www.finanzas.gov.do.

**Secretariat of State for Higher Education, Science and Technology:** Centro de los Héroes, Avda Enrique Jiménez Moya, esq. Juan de Dios Ventura Simó, 5°, Santo Domingo, DN; tel. 533-3881; fax 535-4694; e-mail info@seescyt.gov.do; internet www.seescyt.gov.do.

**Secretariat of State for Industry and Commerce:** Edif. de Ofs Gubernamentales, 7°, Avda Francia, esq. Leopoldo Navarro, Santo Domingo, DN; tel. 685-5171; fax 686-1973; e-mail ind.comercio@codetel.net.do; internet www.seic.gov.do.

**Secretariat of State for the Interior and Police:** Edif. de Ofs Gubernamentales, 3°, Avda Francia, esq. Leopoldo Navarro, Santo Domingo, DN; tel. 686-6251; fax 689-6599; e-mail info@seip.gov.do; internet www.seip.gov.do.

**Secretariat of State for Labour:** Centro de los Héroes, Jiménez Moya 9, Santo Domingo, DN; tel. 535-4404; fax 535-4590; e-mail secret.trabajo@codetel.net.do; internet www.set.gov.do.

**Secretariat of State for Public Health and Social Welfare:** Avda Tiradentes, esq. San Cristóbal, Ensanche La Fe, Santo Domingo, DN; tel. 541-3121; fax 540-6445; e-mail saludp@saludpublica.gov.do; internet www.saludpublica.gov.do.

# THE DOMINICAN REPUBLIC

**Secretariat of State for Public Works and Communications:** Avda Homero Fernández, esq. Horacio Blanco Fombona, Ensanche La Fe, Santo Domingo, DN; tel. 565-2811; fax 562-3382; e-mail info@seopc.gov.do; internet www.seopc.gov.do.

**Secretariat of State for Sport, Physical Education and Recreation:** Avda Ortega y Gasset, Centro Olímpico, Santo Domingo, DN; tel. 540-4010; fax 563-6586; internet www.sedefir.gov.do.

**Secretariat of State for Tourism:** Bloque D, Edif. de Ofs Gubernamentales, Avda México, esq. 30 de Marzo, Apdo 497, Santo Domingo, DN; tel. 221-4660; fax 682-3806; e-mail sectur@codetel.net.do; internet www.dominicana.com.do.

**Secretariat of State for Women:** Bloque D, Edif. de Ofs Gubernamentales, Avda México, esq. 30 de Marzo, Santo Domingo, DN; tel. 685-3755; fax 686-0911; e-mail info@sem.gov.do; internet www.sem.gov.do.

**Secretariat of State for Youth:** Avda John F. Kennedy, esq. Ortega y Gasset, Plaza Metro Politana, Santo Domingo, DN; tel. 732-7227; fax 472-8585; e-mail juventud@verizon.net.do.

## President and Legislature

### PRESIDENT

Election, 16 May 2004

| Candidate | % of votes cast |
|---|---|
| Leonel Fernández Reyna (PLD) | 57.11 |
| Rafael Hipólito Mejía Domínguez (PRD) | 33.65 |
| Rafael Eduardo Estrella Virella (PRSC) | 8.65 |
| **Total** (incl. others) | 100.00 |

### CONGRESO NACIONAL

The Congreso Nacional comprises a Senado and a Cámara de Diputados.

**President of the Senate:** ANDRES BAUTISTA (PRD).

**President of the Chamber of Deputies:** ALFREDO PACHECO (PRD).

General Election, 16 May 2002

|  | Seats | |
|---|---|---|
|  | Senate | Chamber of Deputies |
| Partido Revolucionario Dominicano (PRD) | 29 | 73 |
| Partido de la Liberación Dominicana (PLD) | 2 | 41 |
| Partido Reformista Social Cristiano (PRSC) | 1 | 36 |
| **Total** | 32 | 150 |

## Political Organizations

**Alianza por la Democracia (APD):** Santo Domingo, DN; f. 1992 by breakaway group of the PLD; split into two factions (led, respectively, by Max Puig and Nélsida Marmolejos) in 1993; Sec.-Gen. VICENTE BENGOA.

**Frente Independiente Leonel al Poder (FILA):** Santo Domingo, DN; f. 2002; established by former Pres. Leonel Fernández to support his candidacy in 2004 presidential election; Pres. VICENTE BENGOA.

**Fuerza Nacional Progresista (FNP):** Santo Domingo, DN; right-wing; Leader MARIO VINICIO CASTILLO.

**Fuerza de la Revolución:** Avda Independencia 258, Apdo 2651, Santo Domingo, DN; e-mail fr@nodo50.ix.apc.org; internet www.fuerzadelarevolucion.org; f. 1996 by merger of the Partido Comunista Dominicano, Movimiento Liberador 12 de Enero, Fuerza de Resistencia y Liberación Popular, Fuerza Revolucionaria 21 de Julio and other revolutionary groups; Marxist-Leninist; Sec.-Gen. Lic. FERNANDO PEÑA.

**Movimiento de Conciliación Nacional (MCN):** Pina 207, Santo Domingo, DN; f. 1969; centre party; 659,277 mems; Pres. Dr JAIME M. FERNÁNDEZ; Sec. VÍCTOR MENA.

**Movimiento de Integración Democrática (MIDA):** Santo Domingo, DN; tel. 687-8895; centre-right; Leader Dr FRANCISCO AUGUSTO LORA.

**Movimiento Popular Dominicano:** Santo Domingo, DN; left-wing; Leader JULIO DE PEÑA VALDÉS.

**Participación Ciudadana:** Calle Wenceslao Alvarez 8, Santo Domingo, DN; tel. 685-6200; fax 685-6631; e-mail p.ciudadana@verizon.net.do; internet www.pciudadana.com; f. 1993; Leaders JAVIER CABREJA POLANCO, RAMÓN TEJADA HOLGUÍN, JUAN BOLÍVAR DÍAZ.

**Partido Demócrata Popular:** Arz. Meriño 259, Santo Domingo, DN; tel. 685-2920; Leader LUIS HOMERO LÁJARA BURGOS.

**Partido de la Liberación Dominicana (PLD):** Avda Independencia 401, Santo Domingo, DN; tel. 685-3540; fax 687-5569; e-mail pldorg@pld.org.do; internet www.pld.org.do; f. 1973 by breakaway group of PRD; left-wing; Leader LEONEL FERNÁNDEZ REYNA; Sec.-Gen. REINALDO PARED PÉREZ.

**Partido Quisqueyano Demócrata (PQD):** Avda Bolívar 51, esq. Uruguay, Santo Domingo, DN; tel. 565-0244; e-mail pqd@verizon.net.do; f. 1968; right-wing; 600,000 mems; Pres. Dr ELÍAS WESSIN CHÁVEZ; Sec.-Gen. Lic. LORENZO VÁLDEZ CARRASCO.

**Partido Reformista Social Cristiano (PRSC):** Avda San Cristóbal, Ensanche La Fe, Apdo 1332, Santo Domingo, DN; tel. 566-7089; internet www.reformistadigital.com; f. 1964; centre-right party; Pres. ANTÚN BATLLE.

**Partido Revolucionario Dominicano (PRD):** Espaillat 118, Santo Domingo, DN; tel. 687-2193; e-mail julio.estevez@prd.partidos.com; internet www.prd.partidos.com; f. 1939; democratic socialist; mem. of Socialist International; 400,000 mems; Pres. VICENTE SÁNCHEZ BARET; Sec.-Gen. Dr RAFAEL SUBERVÍ BONILLA.

**Partido Revolucionario Independiente (PRI):** Avda Bolívar 257, Santo Domingo, DN; tel. 476-7412; f. 1985 after split by the PRD's right-wing faction; Pres. Dr TRAJANO SANTANA; Sec.-Gen. STORMI REYNOSO.

**Partido Revolucionario Social Cristiano:** Santo Domingo, DN; f. 1961; left-wing; Pres. Dr CLAUDIO ISIDORO ACOSTA; Sec.-Gen. Dr ALFONSO LOCKWARD.

**Partido de los Trabajadores Dominicanos (PTD):** Avda Duarte 69 (Altos), Santo Domingo, DN; tel. 685-7705; fax 687-4190; f. 1979; workers' party; Sec.-Gen. JOSÉ GONZÁLEZ ESPINOZA.

**Unidad Democrática (UD):** Avda Pasteur 156, Gazgue, Santo Domingo, DN; tel. 688-4222; Leader FERNANDO ALVAREZ BOGAERT.

Other parties include Unión Cívica Nacional (UCN), Partido Alianza Social Demócrata (ASD—Leader Dr JOSÉ RAFAEL ABINADER), Movimiento Nacional de Salvación (MNS—Leader LUIS JULIÁN PÉREZ), Partido Comunista del Trabajo de la República Dominicana (Sec.-Gen. RAFAEL CHALJUB MEJÍA), Partido de los Trabajadores Dominicanos Marxista–Leninista (PTD-ML—Leader IVÁN RODRÍGUEZ), Partido de Veteranos Civiles (PVC), Partido Acción Constitucional (PAC), Partido Unión Patriótica (PUP—Leader ROBERTO SANTANA), Partido de Acción Nacional (right-wing) and Movimiento de Acción Social Cristiana (ASC).

## Diplomatic Representation

### EMBASSIES IN THE DOMINICAN REPUBLIC

**Argentina:** Avda Máximo Gómez 10, Apdo 1302, Santo Domingo, DN; tel. 682-2977; fax 221-2206; e-mail embarg@aster.com.do; Ambassador JORGE J. A. ROBALLO.

**Belize:** Carretera La Isabela, Calle Proyecto 3, Arroyo Manzano 1, Santo Domingo, DN; tel. 567-7146; fax 567-7159; e-mail embelize@verizon.net.do; internet www.embelicerd.com; Ambassador R. EDUARDO LAMA S.

**Brazil:** Avda Winston Churchill 32, Edif. Franco-Acra y Asociados, 2°, Apdo 1655, Santo Domingo, DN; tel. 532-0868; fax 532-0917; e-mail contacto@embajadadebrasil.org.do; internet www.embajadadebrasil.org.do; Ambassador RONALDO EDGAR DUNLOP.

**Canada:** Apdo 2054, Santo Domingo, DN; tel. 685-1136; fax 682-2691; e-mail sdmgo@international.gc.ca; internet www.dfait-maeci.gc.ca/dominicanrepublic; Ambassador ADAM BLACKWELL.

**Chile:** Avda Anacaona 11, Mirador del Sur, Santo Domingo, DN; tel. 532-7800; fax 530-8310; e-mail embaj.chile@codetel.net.do; Ambassador RUBIO SANDOVAL CARLOS.

**China (Taiwan):** Edif. Palic, 1°, Avda Abraham Lincoln, esq. José Amado Soler, Apdo 4797, Santo Domingo, DN; tel. 562-5555; fax 563-4139; e-mail dom@mofa.gov.tw; Ambassador CHI TAI FENG.

# THE DOMINICAN REPUBLIC

*Directory*

**Colombia:** Avda Abraham Lincoln 502, 2°, Santo Domingo, DN; tel. 562-1670; e-mail c.columbia@verizon.net.do; Ambassador JORGE ENRIQUE GARAVITO DURÁN.

**Costa Rica:** Ensanche Serralles entre Abraham Lincoln y Lope de Vega, Calle Malaquías Gil 11 Altos, Santo Domingo, DN; tel. 683-7209; fax 565-6467; e-mail embarica@codetel.net.do; Ambassador EKHART PETERS SEEVERS.

**Cuba:** 808 Francisco Prats Ramírez, esq. El Millón, Santo Domingo, DN; tel. 537-2113; fax 537-9820; e-mail embadom@codetel.net.do; Ambassador OMAR R. CÓRDOVA RIVAS.

**Ecuador:** Calle Rafael Augusto Sánchez 17, Edif. Profesional Saint Michel, Of. 301, Ensanche Naco, Apdo 808, Santo Domingo, DN; tel. 563-8363; fax 563-8153; e-mail mecuador@codetel.net.do; Ambassador (vacant).

**El Salvador:** José A. Brea Peña 12, Ensanche Evaristo Morales, Santo Domingo, DN; tel. 565-4311; fax 541-7503; e-mail emb.salvador@verizon.net.do; Ambassador Dr CARLOS ERNESTO MENDOZA C.

**France:** Calle Las Damas 42, Zona Colonial, Santo Domingo, DN; tel. 695-4300; fax 695-4311; e-mail ambafrance@ambafrance-do.org; internet www.ambafrance-do.org; Ambassador JEAN-CLAUDE MOYRET.

**Germany:** Edif. Torre Piantini, 16°, Calle Gustavo Mejía Ricart, esq. Avda Abraham Lincoln, Santo Domingo, DN; tel. 542-8949; fax 542-8955; e-mail embal@verizon.net.do; internet www.santo-domingo.diplo.de; Ambassador KARL KÖHLER.

**Guatemala:** Calle Santiago 359, Gazcue, Santo Domingo, DN; tel. 689-5327; fax 689-5146; e-mail embrepublicadominicana@minex.gob.gt; Ambassador Gen. BERNA RONALDO MÉNDEZ MARA.

**Haiti:** Avda Juan Sánchez Ramírez 33, Santo Domingo, DN; tel. 686-5778; fax 686-6096; e-mail amb.haiti@codetel.net.do; Chargé d'affaires a.i. JOSEPH MATHURIN BELONY.

**Holy See:** Avda Máximo Gómez 27, Apdo 312, Santo Domingo, DN (Apostolic Nunciature); tel. 682-3773; fax 687-0287; Apostolic Nuncio Most Rev. TIMOTHY BROGLIO (Titular Archbishop of Amiternum).

**Honduras:** Calle Arístides García Mella, esq. Dolores Rodríguez Objío, Edif. El Buen Pastor VI, Apt 1-B, 1°, Mirador del Sur, Santo Domingo, DN; tel. 482-7992; fax 482-7505; e-mail e.honduras@codetel.net.do; Ambassador NERY MAGALY FÚNES.

**Israel:** Pedro Henríquez Ureña 80, Santo Domingo, DN; tel. 472-0774; fax 472-1785; e-mail info@santodomingo.mfa.gov.il; Ambassador YOAV BARON.

**Italy:** Rodríguez Objío 4, Santo Domingo, DN; tel. 682-0830; fax 682-8296; e-mail ambital@verizon.net.do; Ambassador GIORGIO SFARA.

**Jamaica:** Avda Enriquillo 61, Los Cacicazgos, Santo Domingo, DN; tel. 482-7770; fax 482-7773; e-mail emb.jamaica@verizon.net.do; internet ; Chargé d'affaires THOMAS F. ALLAN MARLEY.

**Japan:** Torre BHD, 8°, Avda Winston Churchill, esq. Luis F. Thomén, Santo Domingo, DN; tel. 567-3365; fax 566-8013; internet www.do.emb-japan.go.jp; Ambassador HARUO OKAMOTO.

**Korea, Republic:** Avda Sarasota 98, Santo Domingo, DN; tel. 532-4314; fax 532-3807; Ambassador LEE JOON-IL.

**Mexico:** Arzobispo Meriño No 265, esq. con las Mercedes, Zona Colonial, Santo Domingo, DN; tel. 687-6444; fax 687-7872; e-mail embamex@codetel.net.do; Ambassador ISABEL TÉLLEZ DE ORTEGA.

**Netherlands:** Calle Max Henriquez Ureña 50, entre Avda Winston Churchill y Abraham Lincoln, Ens. Piantini, Apdo 855, Santo Domingo, DN; tel. 262-0320; fax 565-4685; e-mail std@minbuza.nl; internet www.holanda.org.do; Ambassador B. W. SCHORTINGHUIS.

**Nicaragua:** Avda México 152, Condominio Elsa María, Apdo 1, La Esperilla, Santo Domingo, DN; tel. 563-2311; fax 565-7961; e-mail embnicaragua@codetel.net.do; Ambassador ALVARO SEVILLA SIERO.

**Panama:** Calle Enrique Henríquez 54, esq. Uruguay, Gazcue, Santo Domingo, DN; tel. 688-1043; fax 689-1273; e-mail emb.panam@codetel.net.do; Ambassador MARIO GALVEZ.

**Peru:** Edif. Curvo Of., 485 Bella Vista, Pedro A. Bobea, esq. Avda Anacaona, Santo Domingo, DN; tel. 532-6777; fax 532-6291; e-mail embaperu@codetel.net.do; Ambassador RAÚL GUTIÉRREZ.

**Spain:** Avda Independencia 1205, Apdo 1468, Santo Domingo, DN; tel. 535-6500; fax 535-1595; e-mail embespdo@mail.mae.es; Ambassador MARÍA JESÚS FIGA LÓPEZ-PALOP.

**United Kingdom:** Edif. Corominas Pepin, 7°, Avda 27 de Febrero 233, Santo Domingo, DN; tel. 472-7111; fax 472-7574; e-mail brit.emb.sadom@codetel.net.do; Ambassador ANDY ASHCROFT.

**USA:** César Nicolás Pensón, esq. Leopoldo Navarro, Santo Domingo, DN; tel. 221-2171; fax 685-6959; e-mail irc@usemb.gov.do; internet www.usemb.gov.do/index.htm; Ambassador HANS H. HERTELL.

**Uruguay:** Baltasar Brum 7, Apt 1-B, Ensanche La Esperilla, Santo Domingo, DN; tel. 682-5565; fax 687-2167; e-mail embur@codetel.net.do; Ambassador CROCI DE MULA.

**Venezuela:** Avda Anacoana 7, Mirador del Sur, Santo Domingo, DN; tel. 537-8578; fax 537-8780; e-mail embvenezuela@codetel.net.do; Ambassador FRANCISCO ALBERTO BELISARIO LANDIS.

## Judicial System

The Judicial Power resides in the Suprema Corte de Justicia (Supreme Court of Justice), the Cortes de Apelación (Courts of Appeal), the Juzgados de Primera Instancia (Tribunals of the First Instance), the municipal courts and the other judicial authorities provided by law. The Supreme Court is composed of at least 11 judges (16 in June 2005) and the Attorney-General, and exercises disciplinary authority over all the members of the judiciary. The Attorney-General of the Republic is the Chief of Judicial Police and of the Public Ministry which he represents before the Supreme Court of Justice. The Consejo Nacional de la Magistratura (National Judiciary Council) appoints the members of the Supreme Court, which in turn appoints judges at all other levels of the judicial system.

### Corte Suprema

Centro de los Héroes, Calle Juan de Dios Ventura Simó, esq. Enrique Jiménez Moya, Santo Domingo, DN; tel. 533-3118; fax 532-2906; e-mail suprema.corte@codetel.net.do; internet www.suprema.gov.do.

**President:** Dr JORGE SUBERO ISA.

**Vice-President and President of First Court:** Dr RAFAEL LUCIANO PICHARDO.

**Second Vice-President:** Dra EGLYS MARGARITA ESMURDOC.

**President of Second Court:** Dr HUGO ÁLVAREZ VALENCIA.

**Justices:** Dra MARGARITA A. TAVARES, VÍCTOR JOSÉ CASTELLANOS ESTRELLA, Dr JULIO IBARRA RÍOS, Dr EDGAR HERNÁNDEZ MEJÍA, Dra DULCE M. RODRÍGUEZ DE GORIS, Dra ANA ROSA BERGÉS DREYFOUS JUEZ, Dr JUAN LUPERÓN VÁSQUEZ, Dr JULIO ANÍBAL SUÁREZ, Dra ENILDA REYES PÉREZ, Dr JOSÉ ENRIQUE HERNÁNDEZ MACHADO, Dr PEDRO ROMERO CONFESOR, Dr DARÍO OCTAVIO FERNÁNDEZ ESPINAL.

**Attorney-General:** FRANCISCO DOMINGUEZ BRITO.

## Religion

The majority of the inhabitants belong to the Roman Catholic Church, but freedom of worship exists for all denominations. The Baptist, Evangelist and Seventh-day Adventist churches and the Jewish faith are also represented.

### CHRISTIANITY

#### The Roman Catholic Church

The Dominican Republic comprises two archdioceses and nine dioceses. At 31 December 2003 adherents represented about 86.0% of the population.

#### Bishops' Conference

Conferencia del Episcopado Dominicano, Apdo 186, Calle Isabel la Católica 55, Santo Domingo, DN; tel. 685-3141; fax 689-9454.

f. 1985; Pres. Most Rev. RAMÓN BENITO DE LA ROSA Y CARPIO (Archbishop of Santiago de los Caballeros).

**Archbishop of Santiago de los Caballeros:** Most Rev. RAMÓN BENITO DE LA ROSA Y CARPIO, Arzobispado, Duvergé 14, Apdo 679, Santiago de los Caballeros; tel. 582-2094; fax 581-3580; e-mail arzobisp.stgo@verizon.net.do.

**Archbishop of Santo Domingo:** Cardinal NICOLÁS DE JESÚS LÓPEZ RODRÍGUEZ, Arzobispado, Isabel la Católica 55, Apdo 186, Santo Domingo, DN; tel. 685-3141; fax 688-7270; e-mail arzobispado@codetel.net.do.

#### The Anglican Communion

Anglicans in the Dominican Republic are under the jurisdiction of the Episcopal Church in the USA. The country is classified as a missionary diocese, in Province IX.

**Bishop of the Dominican Republic:** Rt Rev. JULIO CÉSAR HOLGUÍN KHOURY, Santiago 114, Apdo 764, Santo Domingo, DN; tel. 688-6016; fax 686-6364; e-mail igelpidom@codetel.net.do; internet www.dominicanepiscopalchurch.org.

# THE DOMINICAN REPUBLIC

## BAHÁ'Í FAITH

**National Spiritual Assembly of the Bahá'ís of the Dominican Republic:** Cambronal 152, esq. Beller, Santo Domingo, DN; tel. 687-1726; fax 687-7606; e-mail bahai.rd.aen@codetel.net.do; f. 1961; 402 localities.

# The Press

**Dirección General de Información, Publicidad y Prensa:** Santo Domingo, DN; f. 1983; government supervisory body; Dir-Gen. LUIS GONZÁLEZ FABRA.

## DAILIES

### Santo Domingo, DN

**El Caribe:** Calle Doctor Defilló 4, Los Prados, Apdo 416, Santo Domingo, DN; tel. 683-8100; fax 544-4003; e-mail editora@elcaribe.com.do; internet www.elcaribe.com.do; f. 1948; morning; circ. 32,000; Pres. LUIS GARCÍA RECIO; Dir VÍCTOR MANUEL TEJADA.

**Diario Libre:** Avda Abraham Lincoln, esq. Max Henriquez Ureña, Santo Domingo, DN; internet www.diariolibre.com.

**Hoy:** Avda San Martín 236, Santo Domingo, DN; tel. 565-5581; fax 567-2424; internet www.hoy.com.do; f. 1981; morning; Editor LUIS M. CARDENAS; circ. 40,000.

**Listín Diario:** Paseo de los Periodistas 52, Ensanche Miraflores, Santo Domingo, DN; tel. 686-6688; fax 686-6595; e-mail webmaster@listindiario.com.do; internet www.listin.com.do; f. 1889; morning; Dir-Gen. MIGUEL FRANJUL; Editor-in-Chief FABIO CABRAL; circ. 88,050.

**La Nación:** Calle San Antonio 2, Zona Industrial de Herrera, Santo Domingo, DN; tel. 537-2444; fax 537-4865; e-mail editor.nacion@codetel.net.do; afternoon.

**El Nacional:** Avda San Martín 236, Santo Domingo, DN; tel. 565-5581; fax 565-4190; e-mail elnacional@codetel.net.do; internet www.elnacional.com.do; f. 1966; evening and Sunday; Dir RADHAMÉS GÓMEZ PEPÍN; circ. 45,000.

**El Nuevo Diario:** Ensanche Gazcue, Santo Domingo, DN; tel. 687-6205; fax 688-0763; morning; Dir PERSIO MALDONADO.

**El Siglo:** Calle San Anton 2, Zona Industrial de Herrera, Apdo 20213, Santo Domingo, DN; tel. 518-4000; fax 518-4035; e-mail elsiglo@elsiglord.com; Editorial Dir OSVALDO SANTANA.

### Santiago de los Caballeros, SD

**La Información:** Carretera Licey, Km 3, Santiago de los Caballeros, SD; tel. 581-1915; fax 581-7770; e-mail e.informacion@codetel.net.do; internet www.lainformacionrd.com; f. 1915; morning; Dir FERNANDO A. PÉREZ MEMÉN; circ. 15,000.

## PERIODICALS AND REVIEWS

**Agricultura:** Autopista Duarte, Km 6.5, Los Jardines del Norte, Santo Domingo, DN; tel. 547-3888; fax 227-1268; organ of the Secretariat of State for Agriculture; f. 1905; monthly; Dir MIGUEL RODRÍGUEZ, Jr.

**Agroconocimiento:** Apdo 345-2, Santo Domingo, DN; monthly; agricultural news and technical information; Dir DOMINGO MARTE; circ. 10,000.

**¡Ahora!:** San Martín 236, Apdo 1402, Santo Domingo, DN; tel. 565-5581; e-mail revistaahora@internet.net.do; internet www.ahora.com.do; f. 1962; weekly; Editor RAFAEL MOLINA.

**La Campiña:** San Martín 236, Apdo 1402, Santo Domingo, DN; f. 1967; Dir Ing. JUAN ULISES GARCÍA B.

**Carta Dominicana:** Avda Tiradentes 56, Santo Domingo, DN; tel. 566-0119; f. 1974; monthly; economics; Dir JUAN RAMÓN QUIÑONES M.

**Deportes:** San Martín 236, Apdo 1402, Santo Domingo, DN; f. 1967; sports; fortnightly; Dir L. R. CORDERO; circ. 5,000.

**Eva:** San Martín 236, Apdo 1402, Santo Domingo, DN; f. 1967; fortnightly; Dir MAGDA FLORENCIO.

**Horizontes de América:** Santo Domingo, DN; f. 1967; monthly; Dir ARMANDO LEMUS CASTILLO.

**Letra Grande, Arte y Literatura:** Leonardo da Vinci 13, Mirador del Sur, Avda 27 de Febrero, Santo Domingo, DN; tel. 531-2225; f. 1980; monthly; art and literature; Dir JUAN RAMÓN QUIÑONES M.

**Renovación:** José Reyes, esq. El Conde, Santo Domingo, DN; fortnightly; Dir OLGA QUISQUEYA VIUDA MARTÍNEZ.

## NEWS AGENCIES

**La Noticia:** Julio Verne 14, Santo Domingo, DN; tel. 535-0815; f. 1973; evening; Pres. JOSÉ A. BREA PEÑA; Dir BOLÍVAR BELLO.

### Foreign Press Bureaux

**Agence France-Presse (AFP)** (France): Calle Roberto Pastoriza 220, Ensanche Naco, Santo Domingo, DN; tel. 566-0379.

**Agencia EFE** (Spain): Galerías Comerciales, 5°, Of. 507, Avda 27 de Febrero, Santo Domingo, DN; tel. 567-7617; Bureau Chief ANTONIO CASTILLO URBERUAGA.

**Agenzia Nazionale Stampa Associata (ANSA)** (Italy): Leopoldo Navarro 79, 3°, Sala 17, Apdo 20324, Huanca, Santo Domingo, DN; tel. 685-8765; fax 685-8765; Bureau Chief HUMBER ANDRÉS SUAZO.

**Inter Press Service (IPS)** (Italy): Cambronal, No. 4-1, Ciudad Nueva, Santo Domingo, DN; tel. 593-5153; Correspondent VIANCO MARTÍNEZ.

**United Press International (UPI)** (USA): Carrera A. Manoguaybo 16, Manoguaybo, DN; tel. 689-7171; Chief Correspondent SANTIAGO ESTRELLA VELOZ.

The Agence Haïtienne de Presse (AHP) is also represented.

# Publishers

## SANTO DOMINGO, DN

**Arte y Cine, C por A:** San Martín 45, Santo Domingo, DN; tel. 682-0342; fax 686-8354.

**Editora Alfa y Omega:** José Contreras 69, Santo Domingo, DN; tel. 532-5577.

**Editora de las Antillas:** Santo Domingo, DN; tel. 685-2197.

**Editora Dominicana, SA:** 23 Oeste, No 3 Luperón, Santo Domingo, DN; tel. 688-0846.

**Editora El Caribe, C por A:** Calle Doctor Defilló 4, Los Prados, Apdo 416, Santo Domingo, DN; tel. 683-8100; fax 544-4003; e-mail editora@elcaribe.com.do; internet www.elcaribe.com.do; f. 1948; Dir FERNANDO FERRÁN.

**Editora Hoy, C por A:** San Martín, 236, Santo Domingo, DN; tel. 566-1147.

**Editora Listín Diario, C por A:** Paseo de los Periodistas 52, Ensanche Miraflores, Apdo 1455, Santo Domingo, DN; tel. 686-6688; fax 686-6595; f. 1889; Pres. Dr ROGELIO A. PELLERANO.

**Editorama, SA:** Calle Eugenio Contreras, No. 54, Los Trinitarios, Apdo 2074, Santo Domingo, DN; tel. 596-6669; fax 594-1421; e-mail editorama@verizon.net.do; internet www.editorama.com; f. 1970; Pres. JUAN ANTONIO QUIÑONES MARTE.

**Editorial Padilla, C por A:** Avda 27 de Febrero, Santo Domingo, DN; tel. 379-1550; fax 379-2631.

**Editorial Santo Domingo:** Santo Domingo, DN; tel. 532-9431.

**Editorial Stella:** Avda 19 de Marzo 304, Santo Domingo, DN; tel. 682-2281; fax 687-0835.

**Julio D. Postigo e Hijos:** Santo Domingo, DN; f. 1949; fiction; Man. J. D. POSTIGO.

**Publicaciones Ahora, C por A:** Avda San Martín 236, Apdo 1402, Santo Domingo, DN; tel. 565-5580; fax 565-4190; Pres. JULIO CASTAÑO.

**Publicaciones América:** Santo Domingo, DN; Dir PEDRO BISONÓ.

## SANTIAGO DE LOS CABALLEROS, SD

**Editora el País, SA:** Carrera Sánchez, Km 6½, Santiago de los Caballeros, SD; tel. 532-9511.

# Broadcasting and Communications

**Dirección General de Telecomunicaciones:** Isabel la Católica 73, Santo Domingo, DN; tel. 682-2244; fax 682-3493; government supervisory body; Dir-Gen. RUBÉN MONTAS; Dir-Gen. of Television RAMON EMILIO COLOMBO.

**Instituto Dominicano de las Telecomunicaciones (INDOTEL):** Avda Abraham Lincoln, No 962, Edif. Osiris, Santo Domingo, DN; tel. 732-5555; fax 732-3904; e-mail info@indotel.org.do; internet www.indotel.org.do; Pres. Dr JOSÉ RAFAEL VARGAS.

# THE DOMINICAN REPUBLIC

## TELECOMMUNICATIONS

**Ericsson República Dominicana:** Avda Winston Churchill esq. Víctor Garrido Puello, Edif. Empresarial Hylsa, 2°, Nivel, Santo Domingo, DN; tel. 683-7723; fax 616-0962; f. 2000; mobile cellular telephone network provider; subsidiary of Telefon AB LM Ericsson (Sweden); Country Man. ROBERT RUBIN.

**Orange Dominicana, SA:** Calle Victor Garrido Puello 23, Edif. Orange, Ens. Piantini, Santo Domingo, DN; tel. 859-1142; fax ; e-mail serviciocliente@orange.com.do; internet www.orange.com.do; f. 2000; mobile cellular telephone operator, providing GSM network coverage to 89% of the Dominican Republic population; subsidiary of Orange SA (France); Pres. FREDERICK DEBORD.

**Tricom Telecomunicaciones de Voz, Data y Video:** Avda Lope de Vega 95, Ens. Naco, Santo Domingo, DN; tel. 476-6000; fax 567-4412; e-mail sc@tricom.com.do; internet www.tricom.net; f. 1992; Pres. and CEO CARL CARLSON; Chair. RICARDO VALDEZ ALBIZU.

**Verizon Dominicana, C por A:** Avda Lincoln 1101, Apdo 1377, Santo Domingo, DN; tel. 220-1212; fax 543-1301; e-mail ayuda@verizon.net.do; internet www.verizon.net.do; f. 1930; fmrly Compañia Dominicana de Teléfonos (Codetel); name changed as above in 2004; Pres. JORGE IVAN RAMÍREZ; Gen. Man. GUILLERMO AMORE.

## BROADCASTING

### Radio

There were some 130 commercial stations in the Dominican Republic. The government-owned broadcasting network, Radio Televisión Dominicana (see Television), operates nine radio stations.

**Asociación Dominicana de Radiodifusoras (ADORA):** Paul Harris 3, Centro de los Héroes, Santo Domingo, DN; tel. 535-4057; Pres. IVELISE DE TORRES.

**Cadena de Noticias (CDN) Radio:** Calle Doctor Defilló 4, Los Prados, Apdo 416, Santo Domingo, DN; tel. 683-8100; fax 544-4003; internet www.elcaribecdn.com.do/cdnradio.

### Television

**Antena Latina, Canal 7:** Calle Gustavo Mejía Ricart, Santo Domingo, DN; internet www.antenalatina7.com; f. 1999; Pres. MIGUEL BONETTI DUBREIL; Editor-in-Chief MIGUEL ANGEL ORDONEZ.

**Cadena de Noticias (CDN) Televisión:** Calle Doctor Defilló 4, Los Prados, Apdo 416, Santo Domingo, DN; tel. 262-2100; fax 567-2671; e-mail direccion@cdn.com.do; internet www.elcaribe.com.do/cdntv.aspx; broadcasts news on Channel 37.

**Color Visión, Canal 9:** Emilio A. Morel, esq. Luis Perez, Ensanche La Fe, Santo Domingo, DN; tel. 566-5876; fax 732-9347; e-mail colorvision@codetel.net.do; internet www.colorvision.com.do; f. 1969; majority owned by Corporación Dominicana de Radio y Televisión; commercial station; Dir-Gen. MANUEL QUIROZ MIRANDA.

**Radio Televisión Dominicana, Canal 4:** Dr Tejada Florentino 8, Apdo 869, Santo Domingo, DN; tel. 689-2120; e-mail rm.colombo@codetel.net.do; internet www.rtvd.com; government station; Channel 4; Dir-Gen. RAMÓN COLOMBO; Gen. Man. AGUSTÍN MERCADO.

**Teleantillas, Canal 2:** Autopista Duarte, Km 7½, Los Prados, Apdo 30404, Santo Domingo, DN; tel. 567-7751; fax 540-4912; e-mail vbaez@corripio.com.do; f. 1979; Gen. Man. HECTOR VALENTÍN BÁEZ.

**Telecentro, Canal 13:** Avda Pasteur 204, Santo Domingo, DN; tel. 687-9161; fax 542-7582; e-mail webmaster@telecentro.com.do; Santo Domingo and east region; Pres. JOSE MIGUEL BÁEZ FIGUEROA.

**Tele-Inde Canal 13:** Avda Pasteur 101, Santo Domingo, DN; tel. 687-9161; commercial station; Dir JULIO HAZIM.

**Telemicro, Canal 5:** San Martín, Avda 27 de Febrero, Santo Domingo, DN; tel. 689-8151; fax 686-6528; e-mail programacion@telemicro.com.do; internet www.telemicro.com.do; f. 1982.

**Telesistema, Canal 11:** Avda 27 de Febrero 52, Sector Bergel, Santo Domingo, DN; tel. 563-6661; fax 472-1754; e-mail info@telesistema11.tv; internet www.telesistema11.tv; Pres. JOSÉ L. CORREPIO.

# Finance

(cap. = capital; dep. = deposits; m. = million; p.u. = paid up; res = reserves; amounts in pesos)

## BANKING

### Supervisory Body

**Superintendencia de Bancos:** 52 Avda México, esq. Leopoldo Navarro, Apdo 1326, Santo Domingo, DN; tel. 685-8141; fax 685-0859; e-mail webmaster@supbanco.gov.do; internet www.supbanco.gov.do; f. 1947; Supt RAFAEL CAMILO ABRÉU.

### Central Bank

**Banco Central de la República Dominicana:** Calle Pedro Henríquez Ureña, esq. Leopoldo Navarro, Apdo 1347, Santo Domingo, DN; tel. 221-9111; fax 687-7488; e-mail info@bancentral.gov.do; internet www.bancentral.gov.do; f. 1947; cap. 0.7m., res 92.3m., dep. 6,321.7m. (Dec. 1996); Gov. HECTOR VALDEZ ALBIZU; Vice-Gov. CLARISSA DE LA ROCHA DE TORRES.

### Commercial Banks

**Banco BHD, SA:** Avda 27 de Febrero, esq. Avda Winston Churchill, Santo Domingo, DN; tel. 243-3232; fax 541-4949; e-mail servicio@bhd.com.do; internet www.bhd.com.do; f. 1972; total assets 11,902m. (1999); Exec. Pres. LUÍS MOLINA ACHÉCAR; 35 brs.

**Banco Dominicano del Progreso, SA:** Avda John F. Kennedy 3, Apdo 1329, Santo Domingo, DN; tel. 563-3233; fax 563-2455; e-mail informacion@progreso.com.do; internet www.progreso.com.do; f. 1974; merged with Banco Metropolitano, SA, and Banco de Desarrollo Dominicano, SA, in 2000; cap. 1,004.4m., res 249.5m., dep. 11,433.5m. (Dec. 2002); Exec. Pres. PEDRO E. CASTILLO L.; 20 brs.

**Banco Gerencial y Fiduciario, SA:** Avda 27 de Febrero 50, Santo Domingo, DN; tel. 473-9400; fax 565-7569; total assets 7,876m. (1999); f. 1983; Exec. Vice-Pres. GEORGE MANUEL HAZOURY PEÑA.

**Banco Global, SA:** Avda Rómulo Betancourt 1, esq. Avda A. Lincoln, Santo Domingo, DN; tel. 532-3000; fax 535-7070; total assets 11,486m. (1999); Pres. OSCAR LAMA.

**Banco Industrial de Desarrollo e Inversión, SA:** Avda San Martín 40, esq. Dr Delgado, Santo Domingo, DN; tel. 685-3194; fax 689-3485.

**Banco Multiple Leon, SA:** Avda John F. Kennedy, esq. Tiradentes, Apdo 1408, Santo Domingo, DN; tel. 540-4441; fax 567-4854; e-mail info@leon.com.do; internet www.leon.com.do; f. 1981; fmrly Banco Nacional de Crédito, SA; became Bancrédito, SA, in 2002; name changed as above Dec. 2003; cap. 686.8m., res 181.3m., dep. 7,911.3m. (Dec. 2000); Pres. and CEO FELIPE J. MENDOZA; 38 brs.

**Banco Popular Dominicano:** Avda John F. Kennedy 20, Torre Popular, Apdo 1441, Santo Domingo, DN; tel. 544-5555; fax 544-5899; e-mail contactenos@bpd.com.do; internet www.bpd.com.do; f. 1963; cap. 4,645.8m., res 1,270.4m., dep. 61,735.9m. (Dec. 2003); Pres., Chair. and Gen. Man. MANUEL ALEJANDRO GRULLÓN; 184 brs.

**Banco de Reservas de la República Dominicana:** Isabel la Católica 201, Apdo 1353, Santo Domingo, DN; tel. 687-5366; fax 685-0602; e-mail sperdomo@brrd.com; internet www.banreservas.com.do; f. 1941; cap. 2,000.0m., res 1,326.3m., dep. 52,263.2m. (Dec. 2003); Dir DANIEL TORIBIO; Pres. and Gen. Man. MANUEL ANTONIO LARA HERNÁNDEZ; 112 brs.

**Republic Bank (DR), SA:** Avda Pastoriza 303, Ensanche Naco, Santo Domingo, DN; tel. 567-4444; fax 549-6509; e-mail info@mercantil.com.do; internet www.mercantil.com.do; f. 1985; fmrly Banco Mercantil; name changed as above in 2005 following Republic Bank's acquisition of the firm in 2003; total assets 9,579m. (1999); Pres. RONALD G. HUGGINS; 21 brs.

### Development Banks

**Banco Agrícola de la República Dominicana:** Avda G. Washington 601, Apdo 1057, Santo Domingo, DN; tel. 535-8088; fax 535-8022; e-mail bagricolabagricola.gov.do; internet www.bagricola.gov.do; f. 1945; government agricultural development bank; Pres. Lic. RAFAEL ANGELES SUÁREZ; Gen. Administrator CARLOS SEGURA FOSTER.

**Banco Continental de Desarrollo, SA:** Edif. Continental, Avda Pedro Henríquez Ureña 126, Santo Domingo, DN; tel. 472-2228; fax 472-3027; internet www.banco-continental.com; Pres. HUGO GILIANO; Man. Dir ANTON YELARI.

**Banco de Desarrollo Ademi, SA:** Avda Rafael Augusto Sánchez 33, Plaza Intercaribe, Suite 203, Apdo 2887, Santo Domingo, DN; tel. 732-4411; fax 732-4401; e-mail cgutierrez@ademi.org.do; internet www.ademi.org.do; Pres. MARGARITA DE FERRARI DE LLUBERES.

**Banco de Desarrollo Agropecuario Norcentral, SA:** Avda Independencia 801, esq. Avda Máximo Gómez, Santo Domingo, DN; tel. 686-0984; fax 687-0825.

**Banco de Desarrollo de Exportación, SA:** Fatino Falco, entre Avda Lope de Vega y Tiradentes 201, Santo Domingo, DN; tel. 566-5841; fax 565-1769.

**Banco de Desarrollo Industrial, SA:** Avda Sarasota 27, esq. La Julia, Santo Domingo, DN; tel. 535-8586; fax 535-6069.

**Banco de Desarrollo Intercontinental, SA:** Edif. Lilian, 5°, Avda Lope de Vega, esq. Gustavo Mejía Ricart, Santo Domingo, DN; tel. 544-0559; fax 563-6884.

# THE DOMINICAN REPUBLIC

**Banco Nacional de la Construcción:** Avda Alma Mater, esq. Pedro Henríquez Ureña, Santo Domingo, DN; tel. 685-9776; f. 1977; Gen. Man. LUIS MANUEL PELLERANO.

**Banco de la Pequeña Empresa, SA:** Avda Bolivar 233, entre Avda Abraham Lincoln y Avda Winston Churchill, Santo Domingo, DN; tel. 534-8383; fax 534-8385.

### Foreign Banks

**Bank of Nova Scotia** (Canada): Avda John F. Kennedy, esq. Lope de Vega, Apdo 1494, Santo Domingo, DN; tel. 567-7268; fax 567-5732; e-mail drinfo@scotiabank.com; internet www.scotiabank.com/cda/content/0,1608,CID54_LIDen,00.html; f. 1920; signed agreement with Govt in July 2003 to acquire some operations of dissolved Banco Intercontinental, SA (Baninter); Vice-Pres. and Gen. Man. LUIS FERNANDO TOBÓN; 54 brs.

**Citibank NA** (USA): Avda John F. Kennedy 1, Apdo 1492, Santo Domingo, DN; tel. 566-5611; fax 685-7535; f. 1962; Gen. Man. NICOLE REICH; 7 brs.

## STOCK EXCHANGE

**Santo Domingo Securities Exchange Inc:** Edif. Empresarial, 1°, Avda John F. Kennedy, Apdo 25144, Santo Domingo, DN; tel. 567-6694; fax 567-6697; e-mail info@bolsard.com; internet www.bolsard.com; Pres. MARINO GINEBRA.

## INSURANCE

### Supervisory Body

**Superintendencia de Seguros:** Secretaría de Estado de Finanzas, Avda México 54, esq. Leopoldo Navarro, Santo Domingo, DN; tel. 221-2606; fax 685-5096; e-mail servicio@superseguro.gov.do; f. 1969; Supt CESAR CABRERA.

### Insurance Companies

**American Life and General Insurance Co, C por A:** Edif. ALICO, 5°, Avda Abraham Lincoln, Apdo 131, Santo Domingo, DN; tel. 533-7131; fax 535-0362; e-mail caribalico@codetel.net.do; general; Gen. Man. FRANCISCO CABREJA.

**Angloamericana de Seguros, SA:** Avda Gustavo Mejía Ricard 8, esq. Hermanos Roque Martínez, Ens. El Millón, Santo Domingo, DN; tel. 227-1002; fax 227-6005; e-mail angloamericana@verizon.net.do; internet www.angloam.com.do; f. 1996; Pres. NELSON HEDI HERNÁNDEZ P.; Vice-Pres. ESTEBAN BETANCES FABRÉ.

**Atlantica Insurance, SA:** Avda 27 de Febrero 365A, 2°, Apdo 826, Santo Domingo, DN; tel. 565-5591; fax 565-4343; e-mail atlanticains@codetel.net.do; Dir-Gen. Lic. RHINA RAMÍREZ DE PERALTA; Pres. GAMALIER PERALTA.

**Bankers Security Life Insurance Society:** Calle Gustavo Mejía Ricart 61, Apdo 1123, Santo Domingo, DN; tel. 544-2626; fax 567-9389; e-mail eizquierdo@bpd.com.do; Pres. Lic. ESTELA MA. FIALLO T.

**BMI Compañía de Seguros, SA:** Avda Tiradentes 14, Edif. Alfonso Comercial, Apdo 916, Ens. Naco, Santo Domingo, DN; tel. 562-6660; fax 562-6849; e-mail gerencia@bmi.com.do; Exec. Vice-Pres. PEDRO DA CUNHA.

**Bonanza Compañía de Seguros, SA:** Avda John F. Kennedy, Edif. Bonanza, Santo Domingo, DN; tel. 565-5531; fax 566-1087; e-mail bonanza.dom@codetel.net.do; internet www.bonanza.dominicana.com.do; Pres. Lic. DARIO LAMA.

**Britanica de Seguros, SA:** Max Henríquez Ureña 35, Apdo 3637, Santo Domingo, DN; tel. 542-6863; fax 544-4542; e-mail wharper@codetel.net.do; Pres. JOHN HARPER SALETA.

**Centro de Seguros La Popular, C por A:** Gustavo Mejía Ricart 61, Apdo 1123, Santo Domingo, DN; tel. 566-1988; fax 567-9389; f. 1965; general except life; Pres. Lic. ROSA FIALLO.

**La Colonial, SA:** Avda Sarasota 75, Bella Vista, Santo Domingo, DN; tel. 508-8000; fax 508-0608; e-mail info@lacolonial.com.do; internet www.lacolonial.com.do; f. 1971; general; Pres. Dr MIGUEL FERIS IGLESIAS.

**Compañía de Seguros Palic, SA:** Avda Abraham Lincoln, esq. José Amado Soler, Apdo 1132, Santo Domingo, DN; tel. 562-1271; fax 562-1825; e-mail cia.seg.palic2@codetel.net.do; Pres. JOSÉ ANTONIO CARO GINEBRA; Exec. Vice-Pres. MILAGROS DE LOS SANTOS.

**Confederación del Canada Dominicana:** Calle Salvador Sturla 17, Ens. Naco, Santo Domingo, DN; tel. 544-4144; fax 540-4740; e-mail confedom@codetel.net.do; f. 1988; Pres. Lic. MOISES A. FRANCO.

**Federal Insurance Company:** Edif. La Cumbre, 4°, Avda Tiradentes, esq. Presidente González, Santo Domingo, DN; tel. 567-0181; fax 567-8909; e-mail eizquierdo@bpd.com.do; Pres. LUIS AUGUSTO GINEBRA H.

**General de Seguros, SA:** Avda Sarasota 55, esq. Pedro A. Bobea, Apdo 2183, Santo Domingo, DN; tel. 535-8888; fax 532-4451; e-mail general.seg@codetel.net.do; f. 1981; general; Pres. Dr FERNANDO A. BALLISTA DÍAZ.

**La Intercontinental de Seguros, SA:** Plaza Naco, 2°, Avda Tiradentes, Apdo 825, Santo Domingo, DN; tel. 227-2002; fax 472-4111; e-mail fotero@lainter.com; internet www.lainter.com; general; Pres. Lic. RAMÓN BÁEZ ROMANO.

**Magna Compañía de Seguros, SA:** Edif. Magna Motors, Avda Abraham Lincoln, esq. John F. Kennedy, Apdo 1979, Santo Domingo, DN; tel. 544-1400; fax 562-5723; e-mail jvargas@segna.com.do; f. 1974; general and life; Pres. E. ANTONIO LAMA S.; Man. MILAGROS DE LOS SANTOS.

**La Mundial de Seguros, SA:** Avda Máximo Gómez, No 31, Santo Domingo, DN; tel. 685-2121; fax 682-3269; general except life and financial; Pres. PEDRO D'ACUNHA.

**La Peninsular de Seguros, SA:** Edif. Corp. Corominas Pepín, 3°, Avda 27 de Febrero 233, Santo Domingo, DN; tel. 472-1166; fax 563-2349; e-mail peninsu@aacr.net; general; Pres. Lic. ERNESTO ROMERO LANDRÓN.

**Progreso Compañia de Seguros, SA** (PROSEGUROS): Edif. Torre Progreso, 2°, Avda Winston Churchill, esq. Ludovino Fernández, Ens. Carmelita, Santo Domingo, DN; tel. 541-7182; fax 541-7915; e-mail JuanGa@progreso.com.do; Pres. VICENZO MASTROLILLI; Exec. Vice-Pres. Lic. JUAN GARRIGÓ.

**Reaseguradora Hispaniola, SA:** Avda Gustavo Mejía Ricart, Edif. 8, 2°, esq. Hermanos Roque Martínez, Ens. El Millón, Santo Domingo, DN; tel. 548-7171; fax 548-6007; e-mail rehsa@verizon.net.do; Pres. NELSON HEDI HERNÁNDEZ P.

**SEGNA:** Avda Máximo Gómez 31, Gazcue, Santo Domingo, DN; tel. 476-7100; fax 476-3427; e-mail info@segna.com.do; internet www.segna.com; f. 1964; general; fmrly Compañía Nacional de Seguros, C por A; name changed as above in 2001; Chair. Dr MÁXIMO A. PELLERANO.

**Seguros La Antillana, SA:** Avda Abraham Lincoln, esq. Max Henríquez Ureña, Apdo 27, Santo Domingo, DN; tel. 567-4481; fax 541-2927; e-mail seguros@seguroslaantilla.com; f. 1947; general and life; Pres. Lic. OSCAR LAMA.

**Seguros BanReservas:** Avda Enrique Jiménez Moya, esq. Calle 4, Centro Technológico Banreservas, Ens. La Paz, Santo Domingo, DN; tel. 960-7200; fax 960-5148; e-mail smahfoud@segbanreservas.com; internet www.segurosbanreservas.com; Pres. Lic. MANUEL LARA HERNÁNDEZ; Exec. Vice-Pres. SIMÓN MAHFOUD MIGUEL.

**Seguros La Isleña, C por A:** Edif. Centro Coordinador Empresarial, Avda Núñez de Cáceres, esq. Guarocuya, Santo Domingo, DN; tel. 567-7211; fax 565-1448; Pres. HUÁSCAR RODRIGUEZ; Exec. Vice-Pres. MARÍA DEL PILAR RODRÍGUEZ.

**Seguros Pepín, SA:** Edif. Corp. Corominas Pepín, Avda 27 de Febrero 233, Santo Domingo, DN; tel. 472-1006; fax 565-9176; general; Pres. Dr BIENVENIDO COROMINAS PEPIN.

**Seguros Popular:** Avda Winston Churchill 1100, Torre Universal América, Apdo 1052, Santo Domingo, DN; tel. 544-7200; fax 544-7999; e-mail escribenos@seguropopular.com.do; internet www.universal.com.do; f. 1964 as La Universal de Seguros; merged with Grupo Asegurador América in 2000; general; Pres. ERNESTO IZQUIERDO.

**Seguros San Rafael, C por A:** Leopoldo Navarro 61, esq. San Francisco de Macorís, Edif. San Rafael, Apdo 1018, Santo Domingo, DN; tel. 688-2231; fax 686-2628; e-mail sanrafael@codetel.net.do; general; Admin.-Gen. RAMÓN PERALTA.

**El Sol de Seguros, SA:** Torre Hipotecaria, 2°, Avda Tiradentes 25, Santo Domingo, DN; tel. 562-6504; fax 544-3260; general; Pres. ANTONIO ALMA.

**Sudamericana de Seguros, SA:** El Conde 105, frente al Parque Colón, Santo Domingo, DN; tel. 685-0141; fax 688-8074; Pres. VINCENZO MASTROLILLI.

**Transglobal de Seguros, SA:** Avda Lope de Vega 16, esq. Andres Julio Aybar y Ensanche Piantini, Apdo 1869, Santo Domingo, DN; tel. 541-3366; fax 567-9398; e-mail transglobal@codetel.net.do; Dirs JOSÉ MANUEL VARGAS, OSCAR LAMA.

### Insurance Association

**Cámara Dominicana de Aseguradores y Reaseguradores, Inc:** Edif. Torre BHD, 5°, Luis F. Thomen, esq. Winston Churchill, Apdo 601, Santo Domingo, DN; tel. 566-0014; fax 566-2600; e-mail cadoar@verizon.net.do; internet www.cadoar.org.do; f. 1972; Pres. SIMÓN MAHFOUD MIGUEL.

# THE DOMINICAN REPUBLIC

# Trade and Industry

## GOVERNMENT AGENCIES

**Comisión Nacional de Energía (CNE):** Calle Gustavo Mejía Ricart 73, esq. Agustín Lara, 3°, Ensanche Serralles, Santo Domingo, DN; tel. 732-2000; fax 547-2073; e-mail info@cne.gov.do; internet www.cne.gov.do; f. 2001; responsible for regulation and development of energy sector; Exec. Dir Antonio Almonte Reynoso.

**Comisión para la Reforma de la Empresa Pública:** Gustavo Mejía Ricart 73, esq. Agustín Lara, Edif. Latinoamericana de Seguros, 6°, Ens. Serrallés, Santo Domingo, DN; tel. 683-3591; fax 683-3888; e-mail crepdom@codetel.net.do; internet www.crepdom.gov.do; commission charged with divestment and restructuring of state enterprises; Pres. Rafael Montilla Martínez.

**Consejo Estatal del Azúcar (CEA)** (State Sugar Council): Centro de los Héroes, Apdo 1256/1258, Santo Domingo, DN; tel. 533-1161; fax 533-7393; internet www.crepdom.gov.do/CEA/main.htm; f. 1966; management of operations contracted to private consortiums in 1999 and 2000; Dir-Gen. Víctor Manuel Báez.

**Corporación Dominicana de Empresas Estatales (CORDE)** (Dominican State Corporation): Avda General Antonio Duvergé, Apdo 1378, Santo Domingo, DN; tel. 533-5171; internet www.crepdom.gov.do/CORDE/main.htm; f. 1966 to administer, direct and develop state enterprises; Dir-Gen. Fernando Durán.

**Instituto de Estabilización de Precios (INESPRE):** Plaza Independencia, Avda Luperon, esq. 27 de Febrero, Apdo 86-2, Santo Domingo, DN; tel. 537-0020; fax 531-0198; e-mail inespre.comp@codetel.net.do; internet www.inespre.gov.do; f. 1969; price commission; Exec. Dir José Francisco Peña Guaba.

**Instituto Nacional de la Vivienda:** Avda Tiredentes, esq. Pedro Henríquez Ureña, Santo Domingo, DN; tel. 732-0600; e-mail invi@codetel.net.do; internet www.invi.gov.do; f. 1962; low-cost housing institute; Dir-Gen. Ing. Juan Antonio Vargas.

## DEVELOPMENT ORGANIZATIONS

**Centro para el Desarrollo Agropecuario y Forestal, Inc. (CEDAF):** Calle José A. Soler 50, Santo Domingo, DN; tel. 544-0616; fax 544-4727; e-mail camado@cedaf.org.do; internet www.cedaf.org.do; f. 1987 to encourage the development of agriculture, livestock and forestry; fmrly known as Fundación de Desarrollo Agropecuario, Inc. (FDA); Pres. José Miguel Bonetti; Exec. Dir Altagracia Rivera de Castillo.

**Consejo Nacional para el Desarrollo Minero:** Santo Domingo, DN; f. 2000; encourages the development of the mining sector; Exec. Dir Miguel Peña; Sec.-Gen. Pedro Vásquez.

**Departamento de Desarrollo y Financiamiento de Proyectos (DEFINPRO):** c/o Banco Central de la República Dominicana, Pedro Henríquez Ureña, esq. Leopoldo Navarro, Apdo 1347, Santo Domingo, DN; tel. 221-9111; fax 687-7488; f. 1993; associated with AID, IDB, WB, KFW; encourages economic development in productive sectors of economy, excluding sugar; authorizes complementary financing to private sector for establishing and developing industrial and agricultural enterprises and free-zone industrial parks; Dir Angel Nery Castillo Pimentel.

**Fundación Dominicana de Desarrollo** (Dominican Development Foundation): Mercedes No 4, Apdo 857, Santo Domingo, DN; tel. 688-8101; fax 686-0430; e-mail info@fdd.org.do; internet www.fdd.org.do; f. 1962 to mobilize private resources for collaboration in financing small-scale development programmes; 384 mems; Pres. Osvaldo Brugal; Exec. Dir Lic. Ada Wiscovitch.

**Instituto de Desarrollo y Crédito Cooperativo (IDECOOP):** Centro de los Héroes, Apdo 1371, Santo Domingo, DN; tel. 533-8131; fax 535-5148; f. 1963 to encourage the development of co-operatives; Dir Javier Peña Núñez.

## CHAMBERS OF COMMERCE

**Cámara Americana de Comercio de la República Dominicana:** Torre Empresarial, 6°, Avda Sarasota 20, Santo Domingo, DN; tel. 381-0777; fax 381-0286; e-mail amcham@verizon.net.do; internet www.amcham.org.do; Pres. Kevin Manning.

**Cámara de Comercio y Producción de Santo Domingo:** Arz. Nouel 206, Zona Colonial, Apdo 815, Santo Domingo, DN; tel. 682-2688; fax 685-2228; e-mail camara.sto.dgo@codetel.net.do; internet www.ccpsd.org.do; f. 1910; 1,500 active mems; Pres. José Manuel Armenteros; Exec. Dir Milagros J. Puello.

There are official Chambers of Commerce in the larger towns.

## INDUSTRIAL AND TRADE ASSOCIATIONS

**Asociación Dominicana de Hacendados y Agricultores Inc:** Santo Domingo, DN; tel. 565-0542; farming and agricultural org.; Pres. Lic. Cesario Contreras.

**Asociación de Industrias de la República Dominicana Inc:** Avda Sarasota 20, Torre Empresarial AIRD, 12°, Santo Domingo, DN; tel. 472-0000; fax 472-0303; e-mail aird@verizon.net.do; internet www.aird.org.do; f. 1962; industrial org.; Pres. Yandra Portela.

**Centro Dominicano de Promoción de Exportaciones (CEDOPEX):** Plaza de la Bandera, Avda 27 de Febrero, Apdo 199-2, Santo Domingo, DN; tel. 530-5505; fax 531-5136; e-mail emartinez@cel-rd.gov.do; internet www.cei-rd.gov.do; promotion of exports and investments; Exec. Dir Eddy Martínez Manzueta.

**Consejo Nacional de la Empresa Privada (CONEP):** Avda Sarasota 20, Torre Empresarial, 12°, Ens. La Julia, Santo Domingo, DN; tel. 472-7101; fax 472-7850; e-mail conep@conep.org.do; internet www.conep.org.do; Pres. Lic. Elena Viyella de Paliza.

**Consejo Nacional de Zonas Francas de Exportación:** Leopoldo Navarro 61, Edif. San Rafael, 5°, Apdo 21430, Santo Domingo, DN; tel. 686-8077; fax 686-8079; e-mail webmaster@cnzfe.gov.do; internet www.cnzfe.gov.do; co-ordinating body for the free trade zones; Exec. Dir Luisa Fernández Durán.

**Consejo Promotor de Inversiones** (Investment Promotion Council): Avda Abraham Lincoln, Edif. Alico, 2°, Santo Domingo, DN; tel. 532-3281; fax 533-7029; Exec. Dir and CEO Frederic Emam Zadé.

**Corporación de Fomento Industrial (CFI):** Avda 27 de Febrero, Plaza La Bandera, Apdo 1452, Santo Domingo, DN; tel. 530-0010; fax 530-1303; f. 1962 to promote agro-industrial development; Dir-Gen. Rámon Rodríguez.

**Dirección General de Minería:** Edif. de Ofs Gubernamentales, 10°, Avda México, esq. Leopoldo Navarro, Santo Domingo, DN; tel. 685-8191; fax 686-8327; e-mail direc.mineria@verizon.net.do; internet www.dgm.gov.do; f. 1947; government mining and hydrocarbon org.; Dir-Gen. Octavio López.

**Instituto Agrario Dominicano (IAD):** Avda 27 de Febrero, Santo Domingo, DN; tel. 530-8272; internet www.iad.gov.do; Dir-Gen. Ing. Agron Salvador Jiménez A.

**Instituto Dominicano de Tecnología Industrial (INDOTEC):** Avda Núñez de Cáceres, esq. Olof Palme, Santo Domingo, DN; tel. 566-8121; fax 227-8808; e-mail indotec@codetel.net.do; internet www.indotec.gov.do; Pres. Francisco Javier García; Exec. Dir Bernarda Castillo.

**Instituto Nacional del Azúcar (INAZUCAR):** Avda Jiménez Moya, Apdo 667, Santo Domingo, DN; tel. 532-5571; fax 533-2402; e-mail inst.azucar@verizon.net.do; internet www.inazucar.gov.do; sugar institute; f. 1965; Exec. Dir Severo de Jesús Ovalle.

## EMPLOYERS' ORGANIZATIONS

**Confederación Patronal de la República Dominicana (COPARDOM):** Avda Sarasota 20, Torre Empresarial AIRD, Suite 207, Santo Domingo, DN; tel. 381-4233; fax 381-4266; e-mail copardom@copardom.org.do; internet www.copardom.org.do; Pres. Virgilio Ortega Nadal.

**Consejo Nacional de Hombres de Empresa Inc:** Edif. Motorámbar, 7°, Avda Abraham Lincoln 1056, Santo Domingo, DN; tel. 562-1666; Pres. José Manuel Paliza.

**Federación Dominicana de Comerciantes:** Carretera Sánchez Km 10, Santo Domingo, DN; tel. 533-2666; Pres. Ivan García.

## MAJOR COMPANIES

**Abbott Laboratories International:** Apdo 846, Santo Domingo; tel. 542-7181; fax 922-8029; manufacturers of nutritional and pharmaceutical products.

**Barceló & Co, C por A:** Avda Ulises Heureaux 20, Villa Duarte, Santo Domingo, DN; tel. 592-2223; fax 593-4209; e-mail barcelo@barcelo.com.do; internet www.barcelo.com.do; f. 1926; manufacturers of distilled alcoholic drinks; Exec. Pres. Jose Antonio Barceló; 1,400 employees.

**Bratex Dominicana, C por A:** Zona Franca, Villa Mella, Santo Domingo, DN; tel. 568-1304; fax 568-5718; e-mail helpdesk@bratex.com.do; internet www.bratex.com.do; f. 1989; manufacturers of knitted undergarments; Pres. Peter Weinerth; 5,000 employees.

**Cartonera Hernández, SA:** Anibal Espinosa 366, Apdo 1162, Santo Domingo, DN; tel. 695-4008; fax 695-4058; e-mail geren.hdez@codetel.net.do; f. 1946; manufacturers of cardboard and packaging materials; Pres. Ricardo Hernández Elmudesi; 285 employees.

# THE DOMINICAN REPUBLIC

**Cementos Cibao, C por A:** Carretera Baitoa Km 8.5, Apdo de Correo 571, Santiago; tel. 242-7111; fax 242-7135; e-mail cementoscibao@codetel.net.do; f. 1964; manufacturers of concrete and cement blocks; Pres. HUASCAR RODRÍGUEZ; 516 employees.

**Cementos Nacionales, SA:** Avda Charles Summer 51, Apdo 285, Santo Domingo, DN; tel. 567-8811; fax 541-8880; f. 1970; manufacturers of cement and cement products; Pres. JOSÉ OSVALDO OLLER CASTRO; 457 employees.

**Cemex Dominicana, C por A:** Torre Acrópolis, 20°, Avda Winston Churchill 67, Ens. Piantini, Santo Domingo, DN; tel. 683-4901; fax 683-4949; e-mail dcedeno@cemex.com.do; internet www.cemexdominicana.com; manufacturers of cement and cement products; Pres. MIGUEL ÁNGEL TREVIÑO.

**Central Romana Corporación:** Apdo 891, La Romana, DN; tel. 523-3333; f. 1911; sugar-cane cultivation and processing; Pres. ALFONSO FANJUL; 21,200 employees.

**Cerámica Industrial del Caribe, C por A:** Autopista Duarte Km 17.5, Apdo 222, Santo Domingo, DN; tel. 560-5618; fax 560-2884; e-mail c.caribe@codetel.net.do; internet www.ceramica-carabobo.com; f. 1979; manufacturers of ceramic products; sales 450m. pesos (2000); Pres. Dr JACOBO SALAS; 450 employees.

**Cervecería Nacional Dominicana, C por A:** Avda Independencia, Santo Domingo, DN; tel. 487-3802; fax 533-5815; e-mail centro.atencionalcliente@eli.com.do; internet www.cnd.com.do; f. 1929; manufacturers of beer and malt liquor; Pres. RAFAEL MENICUCCI VILA; 2,500 employees.

**Delta Comercial, C por A:** Avda John F. Kennedy, Apdo 1376, Santo Domingo, DN; tel. 565-4421; fax 542-7582; f. 1962; manufacturers and distributors of inner tubes and tyres; Pres. JACINTO B. REYNADO; 675 employees.

**Falconbridge Dominicana, C por A:** Avda Máximo Gómez 30, 3°, Apdo 1343, Santo Domingo, DN; tel. 682-6041; fax 687-4735; f. 1971; subsidiary of Falconbridge Ltd, Canada; nickel mining and smelting; Pres. RICHARD FAUCHER; Vice-Pres. and Gen. Man. JAMES H. CORRIGAN; 1,260 employees.

**Ferretería Americana, C por A:** Avda John F. Kennedy, Km 5.5, Santo Domingo, DN; tel. 549-7777; fax 567-7063; e-mail contacto@ferreteriaamericana.com; internet www.americana.com.do; f. 1944; distributors of hardware, houseware, animal food and construction materials; Pres. LUIS GARCÍA SAN MIGUEL; 400 employees.

**Grupo M:** Caribbean Industrial Park, Santiago de los Caballeros, SD; tel. 241-7171; fax 242-7510; e-mail info@grupom.com.do; internet www.grupom.com.do; f. 1986; manufacturers of clothing; Pres. FERNANDO CAPPELLÁN; approx. 12,000 employees.

**Grupo Vicini:** Isobel la Católica 158, Zona Colonial, Santo Domingo, DN; tel. 221-8021; fax 685-7503; e-mail c.moya@codetel.net.do; internet www.inazucar.gov.do/Vic.html; f. 1883; sugar-cane cultivation and processing.

**Industria Textil del Caribe, C por A:** Isabel Aguirre, Apdo 2347, Santo Domingo, DN; f. 1957; manufacturers of cotton fabrics; Pres. PEDRO Z. BENDEK; 876 employees.

**Industrias Banilejas, C por A:** Avda Máximo Gómez 118, Apdo 942, Santo Domingo, DN; tel. 565-3121; fax 541-5465; e-mail cafe@induban.com; internet www.induban.com; f. 1962; coffee roasting and processing; Pres. MANUEL DE JESÚS PERELLO BAEZ; 877 employees.

**Industrias Textiles Puig, SA:** Anibal Espinosa 303, Apdo 954, Santo Domingo, DN; tel. 536-5800; fax 536-6579; f. 1958; producers of socks, hosiery, underwear and other clothing; Pres. JOSÉ MARÍA PUIG; Vice-Pres. DINO MARRANZINI PUIG; 343 employees.

**Induveca:** Santo Domingo, DN; f. 2000 by merger of Industrias Vegas and Mercasid; producer of meat products and edible oils.

**Interamericana Products International, SA:** Apdo 192-2, Zona Franca Industrial, Santiago de los Caballeros, SD; tel. 575-0007; fax 575-0253; e-mail info@interamericana.com.do; internet www.interamericana.com.do; manufacturers of clothing; 8,200 employees.

**León Jimenez, C por A:** Avda 27 de Febrero, Villa Progreso, Santo Domingo, DN; tel. 535-5555; fax 533-5845; f. 1985; manufacturers of cigarettes; Pres. EDUARDO A. LEÓN ASENCIO; 1,000 employees.

**Mercatti, SA:** Avda Máximo Gómez 192, Apdo 726, Santo Domingo, DN; tel. 565-2151; fax 567-0422; f. 1937; fmrly Sociedad Industrial Dominicana, C por A; producers of cooking oil, soap, detergents and margarine; Pres. MIGUEL BONETTI GUERRA; 1,100 employees.

**Rosario Dominicana, SA:** Avda Núñez de Cáceres, Edif. Indotec, 1°, Santo Domingo, DN; tel. 567-5251; fax 540-5280; f. 1972; mine exploration and development; Pres. ANTONIO IMBERT BARRERAS; 1,000 employees.

**Tabacalera, C por A:** Calle Numa Silverio 1, Villa González, Apdo 758, Santiago de los Caballeros, SD; tel. 582-3151; fax 581-3019; e-mail info@latabacalera.com; internet www.latabacalera.com; f. 1914 under the name Compañía Anónima Tabacalera; name changes as above in 2000; manufacturers of cigarettes and tobacco products; Pres. SANTIAGO REINOSO; 1,300 employees.

**Tabacalera San Luis, C por A:** Calle Gustavo Mejía Ricart 120, Ste 104, Piantini, Santo Domingo, DN; tel. 567-2201; fax 547-2862; e-mail info@deverascigars.com; internet www.deverascigars.com.

**Tabacos Dominicanos, SA:** Carretera Don Pedro, H del Caimito, Apdo 1162, Santiago de los Caballeros, SD; tel. 582-6440; fax 582-9118; manufacturers and distributors of cigarettes; Pres. HENDRIK KELNER; 800 employees.

## UTILITIES

### Electricity

**Corporación Dominicana de Empresas Eléctricas Estatales (CDEEE):** Edif. Principal CDE, Centro de los Héroes, Avda Independencia, esq. Fray C. de Utrera, Santo Domingo, DN; tel. 535-1100; fax 533-7204; e-mail info@cde.gov.do; internet electricidad-rd.gov.do; f. 1955; state electricity company; partially privatized in 1999, renationalized in 2003; Dir-Gen. RAFAEL DE JESÚS PERELLÓ; Admin. Gen. Ing. RADHAMÉS SEGURA.

**Superintendencia de Electricidad:** Avda Gustavo Mejía Ricart 73, esq. Agustín Lara, Ens. Serrallés, 5°, Edif. CREP, Santo Domingo, DN; tel. 683-2500; fax 544-1637; e-mail sielectric@sie.gov.do; internet www.sie.gov.do; f. 2001; Pres. FRANCISCO MÉNDEZ.

There are three electricity distribution companies operating in the Dominican Republic. Empresa Distribuidora de Electricidad del Norte (Ede-Norte), responsible for distribution in the north of the country, and Empresa Distribuidora de Electricidad del Sur (Ede-Sur), in the south, are owned by the government, while Empresa Distribuidora de Electricidad del Este (Ede-Este), in the east, is jointly owned by the Government and the US company AES. In early 2004 AES announced that it intended to divest its 50% stake in Ede-Este, owing to continued problems in the country's power sector.

### Water

As part of the continuing programme of repair to the country's infrastructure, in 2001 construction began on aqueducts, intended to supply water to the provinces of Bahoruco, Barahona, Duarte, Espaillat, Independencia and Salcedo.

**Instituto Nacional de Aguas Potables y Alcantarillado (INAPA):** Calle Guarocuya, Santo Domingo, DN; tel. 567-1241; e-mail info@inapa.gov.do; internet www.inapa.gov.do; Exec. Dir VICTOR DÍAZ RUA.

**Instituto Nacional de Recursos Hidráulicos:** Avda Jiménez de Moya, Centro de los Héroes, Santo Domingo, DN; tel. 532-3271; fax 508-2741; internet www.indrhi.gov.do; f. 1965; Exec. Dir SILVIO CARRASCO.

## TRADE UNIONS

**Central General de Trabajadores (CGT):** Santo Domingo, DN; tel. 688-3932; f. 1972; 13 sections; Sec.-Gen. FRANCISCO ANTONIO SANTOS; 65,000 mems.

**Central de Trabajadores Independientes (CTI):** Juan Erazo 133, Santo Domingo, DN; tel. 688-3932; f. 1978; left-wing; Sec.-Gen. RAFAEL SANTOS.

**Central de Trabajadores Mayoritarias (CTM):** Tunti Cáceres 222, Santo Domingo, DN; tel. 562-3392; Sec.-Gen. NÉLSIDA MARMOLEJOS.

**Confederación Autónoma de Sindicatos Clasistas (CASC)** (Autonomous Confederation of Trade Unions): J. Erazo 39, Santo Domingo, DN; tel. 687-8533; f. 1962; supports PRSC; Sec.-Gen. GABRIEL DEL RÍO.

**Confederación Nacional de Trabajadores Dominicanos (CNTD)** (National Confederation of Dominican Workers): Calle José de Jesús Ravelo 56, Villa Juana, 5°, Santo Domingo, DN; tel. 221-2117; fax 221-3217; e-mail cntd@codetel.net.do; internet www.acmoti.org/1.%20portal%20CNTD.htm; f. 1988 by merger; 11 provincial federations totalling 150 unions are affiliated; Sec.-Gen. MARIANO NEGRON TEJADA; 188,000 mems (est.).

**Confederación de Trabajadores Unitaria (CTU)** (United Workers' Confederation): Santo Domingo, DN; f. 1991; Pres. Dr EUGENIO PÉREZ CEPEDA.

# Transport

## RAILWAYS

In April 2001 plans were announced for the construction of a passenger and freight railway from the coastal port of Haina to Santiago, with the possibility of subsequent extension to Puerto Plata and Manzanillo.

**Dirección General de Tránsito Terrestre:** Avda Tiradente, esq. Avda San Cristóbal, Santo Domingo, DN; tel. 565-2811; e-mail dgtt_dom@hotmail.com; internet www.dgtt.gov.do; f. 1966; operated by Secretary of State for Public Works and Communications; Dir-Gen. Ing. RAFAEL T. CRESPO PEREZ.

**Ferrocarriles Unidos Dominicanos:** Santo Domingo, DN; government-owned; 142 km of track from La Vega to Sánchez and from Guayubín to Pepillo principally used for the transport of exports.

There are also a number of semi-autonomous and private railway companies for the transport of sugar cane, including:

**Ferrocarril Central Río Haina:** Apdo 1258, Haina; 113 km open.

**Ferrocarril de Central Romana:** La Romana; 375 km open; Pres. C. MORALES.

## ROADS

In 2005 there were an estimated 17,000 km of roads, of which about 6,225 km were paved. There is a direct route from Santo Domingo to Port-au-Prince in Haiti. The Mejía administration undertook an extensive road-building programme from 2000, aiming primarily to reduce traffic congestion in Santo Domingo. In July 2001 the Government announced that a Colombian-led consortium, Consorcio Dominico-Colombiano Autopista del Nordeste, was to construct a toll road between Boca and San Pedro de Macorís. In early 2002 the Government announced the construction of two further bridges over the Ozama river, following the completion of the Juan Bosch Bridge.

**Dirección General de Carreteras y Caminos Vecinales:** Santo Domingo, DN; fax 567-5470; f. 1987; government supervisory body; Dir-Gen. ELIZABETH PERALTA BRITO.

**Autoridad Metropolitana de Transporte (AMET):** Avda Expreso V Centenario, esq. Avda San Martín, Santo Domingo, DN; tel. 686-6520; fax 686-3447; e-mail info@amet.gov.do; Dir-Gen. SIGFRIDO FERNÁNDEZ FADUL.

## SHIPPING

The Dominican Republic has 14 ports, of which Santo Domingo is by far the largest, handling about 80% of imports. Construction work was in progress on the conversion of Manzanillo port into a container terminal, and on Punta Caucedo and Haina ports. A new port and transhipment centre was being built near the Las Americas international airport, which was designed specifically for use by free trade zone businesses.

A number of foreign shipping companies operate services to the island.

**Agencias Navieras B&R, SA:** Avda Abraham Lincoln 504, Apdo 1221, Santo Domingo, DN; tel. 544-2200; fax 562-3383; e-mail ops@navierasbr.com; internet www.navierasbr.com; f. 1919; shipping agents and export services; Man. JUAN PERICHE PIDAL.

**Armadora Naval Dominicana, SA:** Isabel la Católica 165, Apdo 2677, Santo Domingo, DN; tel. 689-6191; Man. Dir Capt. EINAR WETTRE.

**Autoridad Portuaria Dominicana:** Avda Máximo Gómez, Santo Domingo, DN; tel. 535-8462; Exec. Dir Prof. ARSENIO BORGES.

**Líneas Marítimas de Santo Domingo, SA:** José Gabriel García 8, Apdo 1148, Santo Domingo, DN; tel. 689-9146; fax 685-4654; Pres. C. LLUBERES; Vice-Pres. JUAN T. TAVARES.

## CIVIL AVIATION

There are international airports at Santo Domingo (Aeropuerto Internacional de las Américas José Francisco Peña Gómez and Aeropuerto Internacional de Herrera), Puerto Plata, Punta Cana, Santiago, La Romana, Samaná and Barahona (Aeropuerto Internacional María Móntez). A new international airport (Aeropuerto Internacional La Isabela) at El Higuero, near Santo Domingo, intended to replace the Aeropuerto Internacional de Herrero, was scheduled to become operational in 2004. The international airports underwent privatization in 1999–2000. Most main cities have domestic airports.

**Dirección General de Aeronáutica Civil:** Avda México, esq. Avda 30 de Marzo, Apdo 1180, Santo Domingo, DN; tel. 221-7909; fax 221-8616; e-mail aeronautica.c@verizon.net.do; internet www.dgac.gov.do; f. 1955; government supervisory body; Dir-Gen. JORGE BOTELLO FERNÁNDEZ.

**Aerochago:** Aeropuerto Internacional de las Américas, Santo Domingo, DN; tel. 549-0709; fax 549-0708; f. 1973; operates cargo and charter service in Central America and the Caribbean; Gen. Man. PEDRO RODRÍGUEZ.

**Aerolíneas Argo:** Santo Domingo, DN; f. 1971; cargo and mail services to the USA, Puerto Rico and the US Virgin Islands.

**Aerolíneas Dominicanas (Dominair):** El Sol 62, Apdo 202, Santiago; tel. 581-8882; fax 582-5074; f. 1974; owned by Aeropostal (Venezuela); scheduled and charter passenger services.

**Aerolíneas Santo Domingo:** Edif. J. P., Avda 27 de Febrero 272, esq. Seminario, Santo Domingo, DN; e-mail reservas@airsantodomingo.com.do; f. 1996; operates scheduled and charter internal, regional and international flights; Pres. HENRY W. AZAR.

**Aeromar Airlines:** Avda Winston Churchill 71, esq. Desiderio Arias, Ensanche La Julia, Santo Domingo, DN; tel. 533-4447; fax 533-4550; e-mail customer_service@aeromarairlines.com; cargo and passenger services; Pres. RAYMUNDO POLANCO.

**Caribbean Atlantic Airlines (Caribair):** Aeropuerto Internacional de Herrera, Santo Domingo,DN; tel. 542-6688; fax 567-7033; e-mail caribair@caribair.com.do; internet www.caribair.com.do; f. 1962; operates scheduled flights to Haiti, and domestic and regional charter flights;

**Compañía Dominicana de Aviación C por A:** Avda Jiménez de Moya, esq. José Contreras, Apdo 1415, Santo Domingo, DN; tel. 532-8511; fax 535-1656; f. 1944; operates on international routes connecting Santo Domingo with the Netherlands Antilles, Aruba, the USA, Haiti and Venezuela; operations suspended 1995, privatization pending; Chair. Dr RODOLFO RINCÓN; CEO MARINA GINEBRA DE BONNELLY.

# Tourism

Strenuous efforts were made to improve the tourism infrastructure, with 200m. pesos spent on increasing the number of hotel rooms by 50%, road improvements and new developments. In 2001 tourism developments were under way in the south-western province of Pedernales, in Monte Cristi and in Cap Cana. The total number of visitors to the Dominican Republic in 2004 was 3,772,932. In 2003 receipts from tourism totalled US $3,110m. There were 59,416 hotel rooms in the Dominican Republic in 2004.

**Secretaría de Estado de Turismo:** Bloque D, Edif. de Oficinas Gubernamentales, Avda México, esq. 30 de Marzo, Apdo 497, Santo Domingo, DN; tel. 221-4660; fax 682-3806; e-mail sectur@sectur.gov.do; internet www.dominicana.com.do.

**Asociación Dominicana de Agencias de Viajes:** Apdo 2097, Calle Padre Billini 263, Santo Domingo, DN; tel. 687-8984; Pres. RAMÓN PRIETO.

**Consejo de Promoción Turística:** Avda México 66, Santo Domingo, DN; tel. 685-9054; fax 685-6752; e-mail cpt@codetel.net.do; Dir EVELYN PAIEWONSKY.

# Defence

In August 2004 the Dominican Republic's armed forces numbered an estimated 24,500: Army 15,000, Navy 4,000 (including naval infantry), Air Force 5,500. There were also paramilitary forces numbering 15,000. Military service is voluntary and lasts for four years.

**Defence Expenditure:** The budget allocation for 2004 was an estimated RD $5,500m. (US $122m.).

**Secretary of State for the Armed Forces and General Chief of Staff:** Rear-Adm. JOSÉ SIGFRIDO PARED PÉREZ.

**Army Chief of Staff:** Maj.-Gen. JOSÉ RICARDO ESTRELLA FERNÁNDEZ.

**Navy Chief of Staff:** Vice-Adm. CESAR AUGUSTO DE WINDT RUIZ.

**Air Force Chief of Staff:** Maj.-Gen. NELSON MARMOLEJOS ACOSTA.

# Education

Education is, where possible, compulsory for children between the ages of six and 14 years. Primary education commences at the age of six and lasts for eight years. Secondary education, starting at 14 years of age, lasts for four years. In 1999 the total enrolment in primary, secondary and tertiary education was equivalent to 72% of the relevant age group (males 69%; females 75%). In 1997 total enrolment at primary level was equivalent to 91.3% of children in the relevant age-group (males 89.0%; females 93.6%), while secondary enrolment was equivalent to 78.5% of children in the relevant age-group (males 74.9%; females 82.1%). At the end of 1997 there were 6,424 primary schools and in 1996/97 there were an estimated 1,737 secondary schools. There were eight universities. In October 2001 the Government announced the investment of US $52m. in the construction and repair of schools. Budgetary expenditure on education in 1997 was RD $5,114.7m., representing 14.3% of total spending.

# Bibliography

For works on the Caribbean generally, see Select Bibliography (Books)

Atkins, G. P., and Wilson, L. C. *The Dominican Republic and the United States: From Imperialism to Transnationalism*. Athens, GA, University of Georgia Press, 1998.

Betances, E. *State and Society in the Dominican Republic*. Boulder, CO, Westview, 1995.

Chester, E. T. *Rag-tags, Scum, Riff–raff, and Commies: The US Intervention in the Dominican Republic, 1965–1966*. New York, NY, Monthly Review Press, 2001.

Diederich, B. *Trujillo: The Death of the Dictator*. Princeton, NJ, Markus Wiener Publrs, 2000.

Hall, M. R. *Sugar and Power in the Dominican Republic: Eisenhower, Kennedy and the Trujillos (1958–62)*. Westport, CT, Greenwood Press, 2000.

Hartlyn, J. *The Struggle for Democratic Politics in the Dominican Republic*. Chapel Hill, NC, University of North Carolina Press, 1998.

Hernandez, R. *The Mobility of Workers Under Advanced Capitalism: Dominican Migration to the United States*. New York, NY, Columbia University Press, 2002.

Hillman, R. S., and D'Agostino, T. J. *Distant Neighbors: The Dominican Republic and Jamaica in Comparative Perspectives*. New York, NY, Praeger Publrs, 1992.

Howard, D. *Coloring the Nation: Race and Ethnicity in the Dominican Republic*. Oxford, Signal Books, 2001.

Itzigsohn, J. *Developing Poverty: The State, Labor Market Deregulation, and the Informal Economy in Costa Rica and the Dominican Republic*. University Park, PA, Pennsylvania State University Press, 2000.

Martinez-Vergne, T. *Nation and Citizen in the Dominican Republic, 1880–1916*. Chapel Hill, NC, University of North Carolina Press, 2005.

Moya Pons, F. *The Dominican Republic: A National History*. Princeton, NJ, Markus Wiener Publrs, 1998.

*The Dominican Republic*. Princeton, NJ, Markus Wiener Publrs, 2002.

Peguero, V., Maslowski, P., and Grimsley, M. (Eds). *Militarization of Culture in the Dominican Republic: From the Captains General to General Trujillo (Studies in War, Society & the Military)*. Lincoln, NE, University of Nebraska Press, 2004.

Pope Atkins, G., and Wilson, L. C. *The Dominican Republic and the United States: From Imperialism to Transnationalism*. Athens, GA, University of Georgia Press, 1998.

Roorda, E. P. *The Dictator Next Door: The Good Neighbor Policy and the Trujillo Regime in the Dominican Republic, 1930–1945*. Durham, NC, Duke University Press, 1998.

Sagas, E. *Race and Politics in the Dominican Republic*. Gainesville, FL, University Press of Florida, 2000.

Turits, R. L. *Foundations of Despotism: Peasants, the Trujillo Regime and Modernity in Dominican History*. Revised edn. Palo Alto, CA, Stanford University Press, 2004.

Veeser, C. *Improving Paradise: American Capitalists and US Intervention in the Dominican Republic, 1890–1908*. New York, NY, Columbia University Press, 2003.

Wucker, M. *Why the Cocks Fight: Dominicans, Haitians and the Struggle for Hispaniola*. New York, NY, Hill and Wang Publishing, 2000.

# ECUADOR

## Geography

### PHYSICAL FEATURES

The Republic of Ecuador is in western South America, straddling the Equator, which gives the country its name. Some 965 km (about 600 miles) to the west of the mainland, but also on the Equator, is the country's Pacific territory on the Islas Galápagos (Galapagos Islands—observing their unique ecosystem was important to the development of the theory of evolution by the 19th-century British scientist Charles Darwin). Continental Ecuador was once rather more extensive, but border disputes and adverse military encounters have reduced Ecuador's territory considerably (notably in 1904–42). There have been occasional clashes subsequently, mainly with Peru—most recently in 1995, although this was resolved in 1999. Ecuador's main border concerns are now not so much territorial as preventing any extension of the conflicts in Colombia. The border with Peru, to the east and south, has been settled at some 1,420 km in extent, and that with Colombia, to the north, is 590 km, giving Ecuador a total area of 272,045 sq km (105,037 sq miles).

Ecuador stretches for 2,237 km along the west coast of South America, extending over the Andes and, in the north-east, down onto the edge of the great Amazonian plains. The country is divided, therefore, into the Costa or coastal plains, the Sierra or central highlands and the Oriente, the forested eastern slopes of the Andes descending to alluvial plains. The Costa covers about one-quarter of the country and is a rich agricultural region, with rolling, forested hills in the north and, generally, a broad lowland basin descending from the Andes to the sea. There is also tropical jungle in the south, climbing the mountain sides as wet, mossy woodland. The Sierra itself is where the chain of the Andes, the Continental Divide, forms a double range of mountains flanking a narrow, inhabited upland plateau. The great cordilleras, Occidental and Oriental, include 22 massive volcanoes among their peaks, including the highest active volcano in the world, Cotopaxi (5,897 m or 19,354 ft), signalling the seismic instability of the region. The highest point in the country is at the summit of Chimborazo (6,310 m), which is also distinguished as being the point on the surface of the earth furthest from its centre (owing to the globe being wider around the Equator). The Oriente or eastern jungles consists of the eastern, forested slopes of the Andes and the gently undulating plains, thick with tropical rainforest. In fact, trees cover over one-half of the country, with 15% classed as pastureland. The mountain heights above 3,000 m or so tend to be grassland. Wildlife includes bears, jaguars, otters, skunks and crocodiles, with a huge variety of birds, including a number of North American species that winter here. Perhaps more unique is the isolated ecosystem of the Galapagos (officially, the Archipiélago Colón), the six larger and nine smaller islands of which (mostly extinct volcanic peaks) have been declared a UNESCO World Heritage Site.

### CLIMATE

The climate is tropical at sea level, the Costa being hot and humid, with an average annual temperature of 78°F (26°C). Being on the Equator, there is little seasonal variation in temperature, although there is a wet season from December to April, with particularly heavy rains, but there is no dry season as such. On the Sierra the average temperatures range from 45°F to 70°F (7°–21°C), depending on the elevation, with Quito at 2,850 m above sea level on 13°C (55°F). However, average

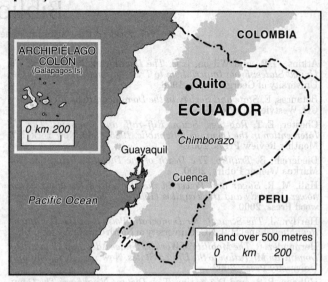

figures can obscure the contrast of warm days and chilly nights. The Oriente is even hotter and more humid than the Costa. Temperatures approach nearer 40°C (104°F) and average annual rainfall is about 2,050 mm (80 ins), falling year round.

### POPULATION

The white élite, predominantly of Spanish descent, still controls most of the wealth in Ecuador, although those of largely unmixed European descent account for little more than 10% of the population. Blacks amount to about one-half of that, with even smaller groups of Arabs and Asians. The vast bulk of the population, over four-fifths, is either Amerindian or of mixed-Amerindian (Mestizo) descent. Figures vary as to the exact proportions, but indigenous peoples could make up anywhere upward of one-fifth of the entire population. The official language, however, is Spanish, although many still speak Quechua, the language of the Incas, in daily life. There are about 700 tribes, the main ones being the Otavalos, the Salasacas, the Saraguros, the Colorados, the Cayapas, the Jivaros, the Aucas, the Yumbos, the Zaparos and the Cofan. Many retain the use of their own dialects or languages. The ancient religions, by contrast, now only persist in the more remote parts of the Oriente, although elements have been maintained in conjunction with the dominant Roman Catholic Christianity (which claims the adherence of almost 90% of the population). There is a small, but growing, Protestant Christian (2%) minority.

According to mid-year estimates in 2004, the total population was 13.0m. Almost one-half of the population lives on the Costa (predominantly Mestizos and blacks), and only slightly less on the Sierra (where most people are Amerindians). The capital is Quito, at 2,850 m the second highest capital in South America, but the largest city is Guayaquil, the main port (2.0m. at the 2001 census—compared to Quito's 1.4m.). Guayaquil is in the south, on the gulf that bears its name, while Quito is in the Sierra, towards the north. The third city, Cuenca (0.3m.), is also in the mountains, but in the south. The country is divided into 22 provinces.

# History

## SANDY MARKWICK

Archaeologists date the existence of ancient civilization in the area comprising modern-day Ecuador as far back as 3500 BC. Evidence, though limited, points to the presence of numerous distinct tribal peoples often in conflict with each other. The Esmeralda, Manta, Huancavilca and Puná peoples farmed the coastal lowlands and were also hunters, fishermen and traders. Indigenous peoples in the Andean Sierra were more tied to the land and organized in dispersed communities growing maize, beans, potatoes and squash.

In the late 15th century these various tribes of people were subsumed into the Inca empire of Tihuantinsuya as it expanded north from modern-day Peru. The Incas introduced new crops, new agricultural methods and imposed new systems of land ownership. Internal migration spread the use of Quechua at the expense of local languages and established Quito as the most important city in the Inca empire after Cusco, Peru. A standing army, a clergy and a bureaucracy reinforced Inca control.

By the time the first Spanish *conquistadores* arrived in search of land, wealth and power in 1526, the Inca empire was already disintegrating and crucially weakened by civil war following the death of the Inca Emperor Huayna Capac.

## SPANISH COLONIAL RULE AND INDEPENDENCE

The Spanish crown sponsored several exploratory voyages southwards along the Pacific coast of South America. Francisco Pizarro led the forces that conquered the Inca territory for the Spanish crown, and in 1534 Quito was captured by Sebastián de Belalcázar. In 1540 Pizarro appointed his brother Gonzalo Governor of Quito, as the colony was named.

As a result of in-fighting among rival Spanish *conquistadores*, the Spanish crown intervened to establish order by incorporating the territory into the Viceroyalty of Peru in 1544, though it was briefly transferred to the Viceroyalty of Nueva Granada in 1717. Spanish dominance was not uniform as significant areas, particularly of the coastal region and jungle lowlands, remained unconquered.

In Ecuador, the movement for independence from Spain was part of a pan-continental struggle between *criollos* (the local élite of Spanish descent) and the demands of the Spanish crown and the privileges of its local representatives. Following decades of revolts, successfully contained, colonial rule ended when Simón Bolívar's forces, under the command of Venezuelan Gen. José Antonio de Sucre, defeated the Spanish at the decisive battle of Pichincha in 1822. Ecuador was then incorporated into the short-lived Confederación de Gran Colombia. In 1830 Ecuador seceded from the Confederación and attained independence. Gen. Juan José Flores became the country's first President and dominated the politics of the new republic for the first 15 years of its existence.

## LIBERALISM VERSUS CONSERVATISM

Throughout the 19th century Ecuador endured political instability. In 1851 Gen. José María Urbina seized power. A defining characteristic of Ecuadorean politics began to emerge during his rule, which lasted until 1856: a persistent power struggle between the forces of liberalism, largely based in Guayaquil, and conservatism, based in the Sierra. This bipartisan dichotomy continued to resonate in the modern era.

Liberalism as represented by Urbina and his successor, Gen. Francisco Robles, was a modernizing, anti-clerical and pro-business influence. Urbina ended slavery and Robles partially abolished the feudal annual tributes that the indigenous population had been required to pay to landowners for centuries. Regional interests reinforced the ideological division, with modern business and trading interests concentrated in the port city of Guayaquil, while Quito was at the epicentre of the old landowning interests of the Sierra. The liberal direction of the Urbina-Robles era provoked regional opposition from local caudillos (provincial rulers), which threatened to undermine the territorial integrity of Ecuador.

In response to the liberal influence of Urbina and Robles emerged the controversial figure of Gabriel García Moreno (President in 1861–65 and 1869–75), the founder of conservatism in Ecuador. He suppressed local rebellions and expelled the Peruvian army from southern Ecuador. The conservatism of García Moreno was rooted in a fanatical adherence to contemporary Catholic theology, which emphasized personal self-discipline, hierarchy and order. In 1869 he founded the Partido Conservador (Conservative Party—PC) and attempted to install a theocratic state. García Moreno's nation-building vision also took the form of a major road- and railway-building exercise linking the capital Quito with Guayaquil.

Between 1852–90 the value of exports grew from US $1m. to $10m., partly because Ecuador was able to take advantage of trading opportunities which arose as a consequence of the attentions of Ecuador's main trading rivals—Peru, Chile and Bolivia—being diverted by the War of the Pacific (1879–83). Economic boom increased the power of the Guayaquil business élite, which was then reflected in a period of power for the Liberals between 1895 and 1925.

If García Moreno was the founder of conservatism in Ecuador, Gen. José Eloy Alfaro Delgado was the principal figure behind Ecuadorean liberalism. He seized power and founded the Partido Liberal Radical (Radical Liberal Party—PLR) in 1895, and quickly went to work stripping the Church of the powers and privileges it had enjoyed under the García Moreno regime. During this period of liberal rule, the Government tried to implement social reform to help the poor, but progress was undermined by unfavourable economic conditions. Liberal rule also failed to build and strengthen democratic institutions and the period was noted for its violent instability stemming largely from disunity within the PLR.

Political power during the early 20th century was dominated by an oligarchy of agricultural and banking interests based in the coastal region. A spendthrift attitude among money-printing bankers and agriculturalists within this liberal élite (know as *la argolla*—the ring) led to spiralling inflation. Economic conditions were exacerbated by competition from overseas producers of cocoa. The authorities ruthlessly put down a general strike in Guayaquil in 1922 and a peasant rebellion in the Sierra in 1923. In 1925, with unrest growing, a group of military social reformers known as the League of Young Officers overthrew the Government with the stated aim of ending the corruption of liberal rule.

The post-Liberal era proved to be equally unstable. From 1925 to 1948 Ecuador had 22 Governments, each one terminated by unconstitutional means. A new Constitution was introduced in 1929, which shifted the balance of power away from the executive to the legislature, thereby encouraging a plethora of minor political parties and groupings that, in combination with the Wall Street crash of 1929 and the ensuing Great Depression, increased instability. Nationalist opinion was inflamed by the signing of the Rio Protocol in 1942, in which Ecuador renounced claims to 200,000 sq km of territory to Peru. In 1948 Galo Plaza Lasso came to power at the head of a coalition of liberals and socialists. His victory began an era of 12 years of unbroken constitutional rule.

## CONSTITUTIONAL RULE, 1948–60

Galo Plaza aligned himself closely to the USA, which proved a frequent source of tension between the Government and populist opposition politicians. Galo Plaza was credited for attempts to deepen democratic rights and institutions even though the opposition largely frustrated his attempts to reform economic and development policy along pragmatic and technocratic lines. Political frustration was the price for relative stability, which was undoubtedly helped by a banana boom that saw exports grow from US $2m. in 1948 to $20m. in 1952. During this time the Government succeeded in balancing budgets and slowing inflation.

José María Velasco Ibarra, elected President for the third time (previously in 1934 and 1944), succeeded Galo Plaza in 1952. Encouraged by the relatively healthy state of the economy, Velasco increased public spending on infrastructure projects and the military, but his Government became increasingly authoritarian. He was succeeded by Camilo Ponce Enríquez in 1956, but won another term of office in 1960. However, after 14 months in office and amid a general strike and violent unrest precipitated by worsening economic conditions, the military deposed Velasco and installed Carlos Julio Arosemena Monroy as President. Nevertheless, the military was concerned about Arosemena's pro-Cuban rhetoric and, fearing the emergence of domestic left-wing insurgency, ousted the President in 1963.

## MILITARY RULE, 1963–79

The new military Government, led by a four-man junta, intended to stay in power for as long as it took to implement unpopular economic policies while defeating left-wing political activity. However, in 1966, faced with continued economic problems and growing unrest, the military stepped down. Interim Governments took office while an elected constituent assembly drafted a new constitution. In 1968 the first elections under the new Constitution returned the 75-year-old Velasco to the presidency for the fifth time, though he received just one-third of votes cast. The last Velasco Government was to be hampered by a weak mandate. Frequent resignations from his administration and economic mismanagement fuelled instability and led to an *autogolpe* in 1970, in which Velasco disbanded the Congreso Nacional (Congress) and the Supreme Court and assumed direct powers himself with the support of the military. Velasco took the opportunity to implement unpopular measures, including a devaluation of the sucre. He also introduced more controls on foreign exchange and increased import tariffs. In 1972 he was deposed by a military *coup d'état* led by Gen. Guillermo Rodríguez Lara, who became head of state.

The military regime had a nationalist outlook which underpinned its heavy investment in infrastructure and industry and the founding of the state petroleum company, Estatal Petrolera Ecuatoriana (which was replaced by Petróleos de Ecuador—PETROECUADOR—in 1989). The Government's nationalism was reflected in its petroleum policy; it took Ecuador into the Organization of the Petroleum Exporting Countries (OPEC) in 1973 and renegotiated contracts to develop petroleum concessions with foreign oil companies on more favourable terms.

The military Government tried to control inflation with low tariffs to encourage imports and absorb the spending power of the small, but growing, middle class. However, the strategy fuelled balance of payments problems. When the regime changed policy and imposed a 60% duty on luxury imports it provoked fierce opposition from Ecuador's élite, which had grown accustomed to cheap luxury goods. In 1976 a military faction led a successful and bloodless transfer of power to a three-man military junta, appointed to oversee the return to civilian rule. In 1978 voters approved a new, more progressive, Constitution in a referendum. The new Constitution acknowledged the role of the state in socio-economic development, established a legislature consisting of a single chamber, disallowed immediate re-election for serving Presidents and put an end to a literacy standard for voter eligibility. The latter was highly significant as it effectively enfranchised a large proportion of Ecuador's indigenous population.

In 1979 Jaime Roldós Aguilera, from the Concentración de Fuerzas Populares (CFP), was elected President with over 68% of the second-round votes cast. Doubts that the military would permit Roldós to take office came to nothing when he finally assumed the presidency in August 1979, though the military had put in place safeguards to prevent any investigations into human rights violation during its period in power.

## CIVILIAN GOVERNMENT RESTORED

Many of the weaknesses of civilian politics in Ecuador reemerged when democracy was restored. Regional and personal rivalries were expressed in the high number of small political parties, groupings and shifting alliances. President Roldós, who came to power with an agenda to push socio-economic reform, faced opposition not only from the traditional right, but also from within his own CFP. In May 1981 the President, his wife and the Minister of Defence were killed in a plane crash in the southern province of Loja. The Vice-President, Osvaldo Hurtado, the leader of the Partido Demócratico Cristiano (Christian Democratic Party—PDC), assumed the presidency amid a deteriorating economic situation brought about by the end of the petroleum boom. Foreign debt totalled about US $7,000m. Despite Hurtado's left-of-centre credentials, his response was to pursue IMF-approved austerity measures. He reduced government subsidies and devalued the sucre. In the 1984 presidential election the right wing united behind León Febres Cordero, a Guayaquil businessman of the Partido Social Cristiano (Social Christian Party—PSC), who narrowly defeated Rodrigo Borja Cevallos, who had broken away from the PLR to form the Izquierda Democratica (Democratic Left—ID).

Conflict between the executive and legislature threatened a number of constitutional crises. Febres Cordero's use of executive power was consistent with the long tradition in Ecuador of the caudillo rather than the constitutionally elected politician. Febres Cordero tried to implement changes that would grant him additional powers and postpone legislative elections. The President's relationship with the military was tense. His dismissal of the chief of the armed forces, Lt-Gen. Frank Vargas Pazzos, for allegations of corruption he had made against senior defence staff led to a military revolt at an air base and at Quito's international airport in 1986. Vargas was arrested and Febres Cordero subsequently ignored an amnesty granted to Vargas by the Congreso Nacional. In January 1987 paratroopers loyal to Vargas abducted the President. Vargas was given an amnesty in exchange for Febres Cordero's release.

In May 1988 Rodrigo Borja Cevallos won the presidential election, defeating Abdalá Bucaram Ortiz of the Partido Roldosista Ecuatoriano (PRE) in the second round of voting, with 46% of the votes cast. Both candidates had proposed economic nationalism and import-substitution. Borja's ID also won the largest number of seats in the Congreso and, by entering into an alliance with fringe leftist and populist parties, could look forward to the prospect of a supportive legislature. Borja pledged to defend human rights and introduce moderate socioeconomic reforms while balancing the conflicting interests of the labour movement and Guayaquil business interests. However, a concurrent programme of economic austerity provoked large public demonstrations and led to disillusion among the electorate. In mid-term legislative elections in June 1990 the ID-led coalition lost its majority in the Congreso, which resulted in legislative paralysis for the remainder of Borja's term of office.

During the Borja presidency, there was a greater mobilization of social groups representing the indigenous population. Particularly active was the Consejo de Nacionalidades Indígenas de Ecuador (Council of Indigenous Nationalities of Ecuador—CONAIE). CONAIE led a nation-wide uprising covering seven Andean provinces in 1990, during which they blocked roads and brought the countryside to a virtual standstill. The trade-union movement joined the protests of the increasingly powerful indigenous organization, which fuelled conservative fears of the Borja administration, particularly after the Government had made overtures to left-wing insurgents to end their violent struggle. In 1989 the President had successfully concluded negotiations with Alfaro Vive ¡Carajo! (Alfaro Lives, Damn It!), one of Ecuador's small guerrilla movements, after it agreed to abandon the armed struggle and enter the legal political arena.

## PRESIDENT DURÁN BALLÉN, 1992–96

These fears helped the conservative Sixto Durán Ballén of the Partido Unitario Republicano (Unitary Republican Party—PUR) to victory in the second round of the presidential election in July 1992. Durán Ballén defeated another right-wing candidate, Jaime Nebot Saadi of the PSC, with 58% of the ballot. The ID candidate was humiliated, receiving just 8.2% of votes cast in the first round in May as those associated with the labour unrest and the incumbent Government were unequivocally rejected. Durán Ballén, having promised to accelerate free-market reforms and encourage foreign investment, introduced a series of structural reforms within two months of assuming power. The Government took Ecuador into the Andean Pact free trade area and introduced a new, more liberal, foreign investment code. In November 1992 Ecuador withdrew from OPEC to allow it greater freedom in production and export of oil. Negotiations to restructure the foreign debt led to improved relations with international creditors and a stand-by facility from the IMF in 1994.

During the first half of Durán Ballén's term of office the Government depended on support from an alliance of the PUR and the PC, with additional *ad hoc* support from the PSC, to secure majority support for legislation in the Congreso. However, this period of relative legislative consensus was short-lived. Corruption charges levelled at cabinet members undermined the fragile support of the PSC and slowed down the implementation of the Government's legislative agenda. In May 1994 the opposition parties effectively took control of the Congreso following an emphatic defeat for the PUR and PC in legislative mid-term elections.

Renewed military skirmishes along the border with Peru in January 1995 temporarily stifled domestic political opposition. However, normal political discourse resumed in February after a cease-fire agreement was reached and following revelations that government representatives had paid legislators and judges to help ease the path of legislative proposals. The President survived the scandal, but it highlighted the existence of widespread corruption within the political and judicial élite. There was an increasing mobilization of the population against the policies and practices of the Durán Ballén administration in the wake of the corruption scandal. In late 1995 violence accompanied protests among inhabitants of the Galápagos Islands, who were demanding greater political and financial autonomy, and protests against labour-code reform and fuel price increases. Industrial action in the public sector led to a series of one-day national strikes and an energy crisis.

## PRESIDENTS BUCARAM AND ALARCÓN, 1996–98

The presidential elections of 1996 saw victory for the populist Abdalá Bucaram Ortiz, representing the PRE. He unexpectedly won the second round of voting, defeating Nebot Saadi of the PSC. His victory was fuelled both by a widespread disillusion with established party politics and a rejection of the structural-adjustment economic programme initiated by Durán Ballén. Bucaram's electoral platform included anti-poverty measures, such as increased subsidies for basic commodities and wage rises for public-sector workers. However, once in power, Bucaram increased the price of utilities as the Government tried to balance its budget. In early 1997 there was a general strike in response to commodity price increases, mass demonstrations and violent clashes between protesters and the security forces.

Charges of 'cronyism' were levelled against the President after family members and close friends were appointed to important cabinet positions. Eventually, Bucaram's undiplomatic style and unpredictable decision-making created such an atmosphere of instability that, on 6 February 1997, the Congreso voted by simple majority to dismiss the President on grounds of mental incapacity (a move which avoided impeachment proceedings that required a two-thirds' majority). The Congreso elected Alarcón to the presidency on 11 February, by 57 votes to two, and scheduled new elections for August 1998.

Alarcón did little to address Ecuador's fiscal crisis. Instead he reinstated the public-sector employees dismissed by his predecessor and built up support for his small, populist Frente Radical Alfarista (Alfarist Radical Front—FRA). Despite his public efforts to tackle corruption and have Bucaram extradited from Panama, whither he had fled, to face charges of misappropriating some US $90m. in government funds, Alarcón himself became embroiled in allegations of embezzlement. (After his term ended he spent several months in prison before charges were dismissed.)

Elections to a 70-member National Assembly to consider constitutional reform were brought forward to November 1997 under considerable public pressure. In February 1998 the Assembly agreed a series of reforms designed to ensure greater government stability, including the enlargement of the Congreso from 82 to 121 seats, the abolition of mid-term elections and more limited impeachment powers for the Congreso. Elections to the newly enlarged Congreso were held in May 1998. The Democracia Popular (People's Democracy—DP) became the largest single party with 32 seats, followed by the PSC with 27, the PRE with 24 and the ID with 18. In the second round of the presidential election in July, Quito mayor and DP candidate Jamil Mahuad Witt narrowly defeated the PRE's Alvaro Noboa Pontón. Mahuad took office in August and appointed a Cabinet largely composed of independents.

## PRESIDENT MAHAUD WITT, 1998–2000

A peace agreement with Peru in October 1998 increased Mahuad's popular standing. The treaty established a definitive border with Peru and granted Ecuador navigation rights on the Amazon river. Unlike previous agreements, this accord seemed to offer the opportunity of a permanent settlement, which would allow the two countries to normalize relations. However, the new administration quickly encountered Ecuador's familiar economic and political problems. The strength of the political opposition in the Congreso ensured that there was little consensus and any reforms required a tortuous process of negotiation with a multitude of disparate parties, each aiming to extract policy concessions from the Government.

Austerity measures introduced by the new Government led to a general strike in October 1998, followed by a period of violent unrest. By 1999 the economic crisis was in danger of leading to a collapse of the banking system, default on debt repayments, hyperinflation and a rapid fall in investor confidence. Protests against a 'freeze' on bank deposits, decreed in March by Mahuad in response to the falling value of the sucre, led to violent protests and the declaration of a 60-day state of emergency. The military was deployed to maintain order while the Government, aiming to secure much-needed loans from international creditors, increased its attempts to restructure the public sector in line with IMF prescriptions. The Government achieved mixed results. Ecuador received US $500m. in loans from the World Bank, the Inter-American Development Bank (IDB) and the Andean Development Corporation, but missed the deadline for signing a letter of intent with the IMF that held out the prospect of much larger loans. The threat of debt default put additional pressure on the Government to impose fiscal reform. However, under popular pressure, the Congreso rejected privatization legislation which would have allowed the Government to earn windfall revenues from the sale of state assets. In August the Mahuad administration declared a moratorium on foreign debt repayments thus making Ecuador the first country to default on its Brady bond and Eurobond commitments.

Popular protest continued despite the debt moratorium. Unions and indigenous organizations joined striking transport workers in demonstrations. The President's popularity declined precipitously. In January 2000 Mahuad announced the decision to adopt the US dollar in place of the sucre as the currency of Ecuador. The decision provoked the resignation of the Central Bank President, Pablo Better, who protested at this violation of the bank's independence. Further protests by indigenous groups received the tacit support of sections of the military who allowed thousands of protesters to occupy the Congreso building on 21 January. During the occupation CONAIE President Antonio Vargas, along with a group of army colonels (among them Lucio Gutiérrez, who would become President three years later), announced the overthrow of Mahuad in a move which illustrated the weaknesses of Ecuador's democratic institutions, despite enjoying civilian government since 1979. The military high command was divided, but the following day supported the transfer of power to Vice-President Gustavo Noboa Bejeramo, who was given the task of completing Mahuad's four-year term.

## PRESIDENT GUSTAVO NOBOA, 2000–03

Noboa was an independent, though he was associated with the PSC and another right-leaning party, the Movimiento de Integración Nacional (National Integration Movement—MIN), which was formed following a split with former President Mahuad's DP. His Cabinet included representatives of several parties and had a strong business influence.

New alliances and splits between the parties undermined the prospect of Noboa forging consistent consensus within the Congreso. Noboa frequently vetoed congressional amendments to proposed legislation, which caused delays to his reform programme. Noboa, as Mahuad before him, had to secure the passage of legislation with the help of *ad hoc* alliances to secure congressional approval. Ecuador's history of a political party system, which is only loosely based around strong ideological identification and where regional and personal rivalries are much in evidence, led to a highly fragmented political landscape that undermined the efficiency of government. However, popular weariness of further political instability gave Noboa a degree of flexibility to deepen the economic reform programme. High petroleum prices and IMF pressure in the early 21st century further encouraged neo-liberal reform.

Noboa took little time to demonstrate his intention to maintain market-orientated policies. In March 2000 the new President secured congressional approval for a Ley de Transformación Económica (Economic Transformation Law). This latest attempt to introduce structural reforms included dollarization, increased labour-market flexibility, an extended policy of privatization, as well as cuts in government spending and tax reform. Additional measures, including a timetable for the adoption of the Basel standard of capital adequacy for the banking sector, persuaded the IMF in April to extend a one-year stand-by agreement, conditionally approving US $2,000m. in multilateral aid over the following two years. In the same month the Government formally began the process of replacing the sucre with the US dollar as Ecuador's unit of currency. The immediate effect of the policy was to stabilize the exchange rate and reduce interest rates. The release of small bank depositor funds, 'frozen' since March 1999 under Noboa's predecessor, helped further to restore public confidence. Opponents of the stabilization programme were partly appeased by an increase in public-sector salaries. In July the Government proposed a second round of reforms, which included more liberal rules governing private investment in petroleum, electricity and telecommunications, as well as health and social security. However, the success of the reform package was threatened by a breakdown in relations between the DP and PSC in the legislature, following the defeat of the PSC's candidate for President of the Congreso. An impasse between the executive and the legislature was resolved in August when the Constitutional Tribunal decreed a new vote which was subsequently won by the PSC.

Opposition to economic liberalization persisted, led by the forces that had been so influential in removing Mahuad from office: CONAIE, trade unions and students. The Government faced significant unrest over its unpopular policies, which included fuel price increases and privatization plans. Anti-Government protests led by CONAIE escalated following fuel price increases in late December 2000. Protests in January and February 2001 were the most serious since similar protests led to the downfall of Mahuad one year earlier. Noboa's Government declared a national state of emergency, giving the security forces extra powers to control unrest. In contrast to Mahuad, however, Noboa maintained the support of most mainstream political parties as well as the business community and the military. His popularity declined in the wake of the protests, but not to the depths of antipathy towards Mahuad.

Opponents of the Government tried to prevent an increase in value-added tax (VAT) and the implementation of other policies imposed as part of an IMF stand-by agreement. In December 2000 the Minister of Finance and Public Credit, Luis Yturralde, resigned in protest at the conditions prescribed in the IMF agreement. His replacement, Jorge Gallardo, estimated that without the proposed increase in VAT, Ecuador's deficit would increase to 3.5%–4% of GDP, considerably more than the 1.5% target agreed with the IMF. In February 2001 Ecuador made its first six-month interest payment on 30-year Global bonds, issued in September 2000 in the aftermath of Ecuador's default on its Brady bond payments. The Congreso continued to oppose the VAT increase, particularly as petroleum revenues were increasing. In April 2001 the 'Paris Club' of Western creditor nations suspended talks with the Government until the tax issue had been resolved. The IMF followed suit shortly afterwards. When the Government succeeded in securing a 2% increase in VAT in an extraordinary congressional session, the IMF approved a US $48m. disbursement and an extension of the agreement until the end of 2001.

When the Government announced further fuel price increases as part of its budget for 2002, protests again ensued, led by CONAIE and Frente Unitario de Trabajadores (Workers' United Front—FUT), as well as an umbrella group of Amerindian organizations, student groups and trade unions, known as the Frente Popular (Popular Front—FP). The Government declared a state of emergency in the Amazon provinces of Sucumbíos and Orellana in February 2002 to counter a regional strike. Protesters blocked petroleum operations and demanded greater regional investment from central government and foreign petroleum companies. In March the conflict was resolved following a pledge from the Government to direct more state funds to local infrastructure development.

## PRESIDENT LUCIO GUTIÉRREZ BORBUA, 2003–05

Elections to the presidency and the legislature went ahead as scheduled on 20 October 2002, with the two leading presidential candidates competing in a second round of voting on 24 November. Lucio Gutiérrez Borbua, a former army colonel involved in the coup to oust President Mahuad, won 58.7% of votes cast in this run-off ballot, defeating a wealthy banana magnate, Alvaro Noboa Pontón (no relation to former President Gustavo). Gutiérrez's nationalist and leftist campaign rhetoric, targeting particularly corruption within the political establishment, earned him the support of a loose alliance of indigenous peoples' interests and left-wing groups. Principal among these was the Partido Sociedad Patriótica 21 de Enero (PSP—Patriotic Society Party), which Gutiérrez had formed with former army colleagues, the indigenous Pachakútik movement and the small left-wing Movimiento Popular Democrático (MPD—Popular Democratic Movement).

President Gutiérrez assumed office on 15 January 2003, facing an opposition-dominated Congreso Nacional. Parties of the ruling coalition accounted for just 20 seats in the 100-seat Congreso, a figure which had been reduced to 17 by mid-2003. The opposition included the larger established parties such as the PSC and ID, as well as the PRE, the DP and former presidential candidate Noboa's own Partido Renovador Institucional de Acción Nacional (National Action Institutional Renovation Party—PRIAN). This opposition bloc took control of important congressional committees, and installed ID leader Guillermo Landazuri as congressional speaker.

On taking office, President Gutiérrez moderated his rhetoric and quickly embarked on a largely orthodox programme of economic management intended to secure an IMF stand-by loan. Although the new Cabinet reflected the diverse forces that supported Gutiérrez, the key position of Minister of Finance and Public Credit went to the market-friendly Quito businessman Mauricio Pozo. The new Government immediately imposed a public-sector salary 'freeze' and reduced fuel subsidies. Gutiérrez also proposed a tax reform to increase revenues and improve the Government's fiscal position. Although not all IMF prescriptions were followed—notably, the politically sensitive subsidy on cooking gas was maintained—the Government's proposals damaged relations with its own supporters. Trade unions and popular and indigenous groups were unhappy at what they perceived to be broken promises on the part of the Government, and their reaction called into question the Government's ability to sustain its minority alliance in the Congreso or its support in the country at large. This was borne out in early August 2003 when Gutiérrez's electoral campaign ally Pachakútik was expelled from the governing coalition after failing to support his public-sector reforms. The traditional parties were not strongly opposed to Gutiérrez's IMF-supported reform efforts on ideological grounds, but they conceived the government alliance of small parties drawn from outside established politics as a threat to their influence within the state.

On 21 March 2003 the IMF approved a US $205m. stand-by loan, which paved the way for further lending. In May the World Bank announced a new four-year programme of lending, worth a projected $1,050m. Ambivalence within the Congreso led to several measures being diluted and falling short of IMF demands. The Government missed a 30 April deadline to send a labour reform bill to the Congreso, and the legislature amended measures to reform the customs service. Further proposed legislation to improve supervision of the financial sector was also rejected in May. The Government faced a delicate balancing act in attempting to satisfy IMF requirements on one hand, while trying to avoid alienating its campaign supporters on the other.

The Government's popularity was further damaged in November 2003 by a corruption scandal. Former mayor of the province of Manabí, César Fernández, a significant donor to Gutiérrez's electoral campaign and personally close to many senior figures in the Government, was arrested on suspicion of drugs-trafficking. The incident was particularly embarrassing as Gutiérrez's campaign rhetoric had focused on fighting corruption. Further isolated from his former electoral campaign allies, Gutiérrez sought a rapprochement with Ecuador's traditional parties. A cabinet reshuffle in December succeeded in shoring up political support for Gutiérrez and stemmed a deteriorating political environment that threatened his presidency. Gutiérrez appointed several members of the PSC and PRE to cabinet positions, broadening the Government's appeal in the Congreso. Furthermore, with the appointment as Minister of the Interior of Raúl Baca Carbo, a trusted politician who enjoyed good relations with indigenous organizations and trade unions, the Government went some way to appease its former allies. At the same time, Gutiérrez reappointed Mauricio Pozo to the finance portfolio, thereby reconfirming the Government's commitment to structural reform and orthodox fiscal policies agreed with the IMF. Nevertheless, the Gutiérrez administration still faced many obstacles. Despite their participation in government, the traditional parties were not formally in a coalition with President Gutiérrez. As a result, many reforms were delayed or diluted in the face of political opposition. Furthermore, the Government had yet to relinquish control of state electricity and telecommunications companies to private management and it failed to introduce tax reform at the end of 2003. In the Congreso, recognizing that it was not possible to build a sustainable majority, the Government attempted to garner *ad hoc* support.

Gutiérrez's political difficulties mounted towards the end of 2004. The ruling PSP won just 7% of the vote in local and provincial elections held in October of that year, while the PSC gained control of several coastal cities and the ID dominated in the Sierra. Emboldened by this strong electoral support the PSC, ID and Pachakutik began impeachment proceedings against Gutiérrez, accusing him of misappropriating public funds during October's elections. The President survived the attempt to depose him, which required a simple majority in the Congreso, by engaging the support of exiled former President Bucaram of the PRE and the PRIAN's Alvaro Noboa, with additional support from smaller parties. Using his new alliance, Gutiérrez launched a counter-offensive against the PSC and the ID, taking control of important congressional committees and replacing the ID President of the Congreso with leading PRE member Omar Quintana Baquerizo. In late November allies of the new governing majority were appointed as justices to the Constitutional Tribunal and the Supreme Electoral Tribunal. More controversial still, in early December the Government called an extraordinary session of the Congreso, during which it voted to expel 27 members of the Supreme Court alleged to be too closely associated with the PSC. They were replaced them with new judges with links to the PRE, PRIAN and the Government. Furthermore, any future replacements were to be appointed by other justices. This purge was widely condemned as a gross violation of the judiciary's independence and of the Constitution. In response, the opposition began to mobilize domestic and international support for its accusations of corruption, authoritarianism and 'cronyism' on the part of the governing alliance, which now effectively controlled the judiciary and electoral authorities. Among other allegations, the opposition claimed that the judicial appointments were a step towards suspending fraud charges against former President Bucaram, now a government ally.

In early 2005 the PSC called protest demonstrations against the Government, taking advantage of popular discontent over the Government's record, especially in reducing crime and providing services in Guayaquil, in order to maintain pressure on the Government over its manipulation of the judiciary. Popular protest spread to Quito and elsewhere, and demonstrations were held throughout March and April. As predicted by the opposition, at the end of March the Supreme Court announced that all charges against exiled former President Bucaram had been dismissed. Polls indicated that support for Gutiérrez had fallen to below 5%. Following a week of daily protests by thousands in the capital, on 15 April Gutiérrez declared a state of emergency and announced the dissolution of the Supreme Court; however, by this time he had lost the support both of the Congreso, which on 20 April voted to oust him for 'abandoning his post', and of the military, which had allowed protesters to reach the presidential palace on the same day. Gutiérrez (who fled to Brazil, where he claimed asylum) became the third elected President since 1997 to be removed following popular unrest.

## INTERIM GOVERNMENT, 2005–

Immediately following the removal of Gutiérrez, his Vice-President, Dr Alfredo Palacio Gonzáles, was sworn in as President. Palacio, while nominally an independent, had associations with the PSC, and his cabinet appointees similarly had links with the traditional parties. There were doubts over the constitutional legitimacy of Palacio's accession because the vote to remove Gutiérrez from office had not been approved by the required minimum of 67 legislators in the Congreso. Despite this, Palacio appeared to be consolidating his position in the weeks following his installation by forming a congressional alliance of the PSC, ID, Pachakutik, DP, MPD and assorted independents. While the direction of the new administration had yet to be clearly defined by mid-2005, there were strong suggestions of a shift away from orthodox structural reform. His appointment of Rafael Correa Delgado as Minister of the Economy and Finance suggested that oil revenues would be used for social investment rather than for debt repayment. However, a Government less likely to follow IMF prescriptions jeopardized further lending from multilateral institutions, which depended on positive assessments from the IMF. Meanwhile, the new Government had an important decision to make concerning the composition of the Supreme Court, which had been suspended since the state of emergency introduced by former President Gutiérrez in mid-April. Politically divisive decisions were also pending over supreme court rulings issued after Gutiérrez had changed its composition, the most controversial of which was the dismissal of corruption charges against Bucaram. Following this ruling Bucaram had returned to Ecuador from self-imposed exile in Panama, a move widely regarded as having catalysed the ultimately successful popular protests demanding the end of the Gutiérrez administration. As with Gutiérrez and other recent Presidents, Palacio faced the prospect of governing while having to negotiate with a fragmented Congreso in which he enjoyed limited personal support. It was possible that presidential and legislative elections, scheduled for October 2006, would be brought forward if political difficulties threatened to undermine Ecuador's governability.

## CONCLUSION

The history of politics and government in Ecuador has been turbulent. Despite a new Constitution in 1998 designed to improve governability by increasing the powers of the executive over the legislature, a fragmented party system, based around personalities and regional and ethnic interests, continued to pose a threat to coherent government, rather than providing an effective balance to executive power. Balancing policies to gain IMF approval with those to secure popular support has been an additional problem for successive Governments. These remained the dominant themes of Ecuadorean politics and raised questions about the ability of Palacio's interim Government to remain in power until the next scheduled elections. This uncertainty was in spite of the fact that the Government had assumed office in favourable economic circumstances stemming from a rise in both Ecuador's capacity to produce oil and in the export price of that commodity.

# Economy

## SANDY MARKWICK

Ecuador comprises a landmass of 272,045 sq km (105,037 sq miles), making it one of the smaller South American countries. Despite its relatively small size, Ecuador has a richly diverse geography. The Andean Sierra runs north to south through the middle of the country forming a natural barrier between the tropical lowlands of the Amazon basin to the east, the Oriente region, and the coastal lowlands, the Costa, to the west. Meanwhile, the Galapagos Islands, 1,500 km off the Pacific coast, contributes a unique range of flora and fauna to the already rich biodiversity of mainland Ecuador.

Climate varies between regions. Most of the country is tropical or subtropical while some 20% is temperate. Ecuador has some 8.29m. ha of land with potential for agricultural use. The fertile soils of the coastal plains and Oriente can support a wide range of crops. Meanwhile, the coastlines as well as freshwater rivers and lakes provide abundant opportunity for commercial fishing and seafood production. Ecuador suffers from the regular occurrence of El Niño, a periodic warming of the tropical Pacific Ocean that brings heavy rains and damages agricultural output.

Since the 2001 census counted 12.16m. Ecuadoreans, the population has risen by approximately 1.8% annually, to an estimated 13.0m. in 2004. Net emigration throughout the 1990s increased from 1998, which, along with a generally declining birth rate, slowed population growth. It is estimated that more than 1m. Ecuadoreans were working outside the country. Emigration accelerated after 2001 and continued to have an important economic impact into 2005, as growth in the formal economy struggled to absorb young entrants to the labour market and migrants to urban areas. In common with other developing countries, Ecuador had a large informal economy. Around 62% of the population lived in cities in 2004, compared with 47% in 1980. The three largest cities—Guayaquil, Quito and Cuenca—accounted for 30% of the population. Almost 50% of Ecuadoreans live in the coastal lowlands, 45% in the Sierra, while only 4.5% live in the Oriente and less than 1% in the Galapagos Islands. Ecuador has a young population, with approximately 32.4% aged under 14 years in 2004.

Most social indicators show Ecuador to be among the poorest countries in South America. Gross domestic product (GDP) per head was approximately US $2,227m. in 2005. In common with most of the region, income distribution was uneven. The poorest 20% of the population received 5.4% of national income while 33.8% went to the richest 10%. World Bank indicators suggested there had been improvements in living standards since 1980. Life expectancy increased to 71 years in 2003 from 63 years in 1980, and infant mortality declined significantly from 74 per 1,000 live births in 1980 to 27 per 1,000 live births in 2003. However, living standards declined slightly from the late 1990s, owing to economic instability. Public spending on health care has suffered as successive Governments have struggled to balance their budgets. The proportion of the urban work-force considered unemployed or underemployed was an estimated 53.6% in 2004, up from 41.5% in 2002. While high rates of emigration and the Government's narrow definition for unemployment tended to restrain official unemployment figures, it was estimated that 11.0% of the total labour force were out of work in 2004, compared with 9.8% in 2003. Unemployment increased following the completion of the Oleoducto de Crudos Pesados (OCP), a pipeline to transport heavy crude petroleum, in 2003 and in the absence of other public-works projects on a similar scale.

The GDP of Ecuador was an estimated US $30,282m. in 2004. Ecuador is the seventh largest economy in South America and also ranks seventh in terms of GDP per head in the region. In 2004 agriculture contributed an estimated 7.7% of GDP, industry contributed 33.4% and services 58.8%. Since Ecuador began producing petroleum in the early 1970s, economic growth figures have closely mirrored the performance of the petroleum sector. Following Ecuador's withdrawal from the Organization of the Petroleum Exporting Countries (OPEC) in 1992, petroleum production increased, leading to a period of strong growth until the mid-1990s. Ecuador's positive, though modest, annual growth after 1994 was interrupted in 1999 when GDP declined by 7.3% as a result of the spiralling effects of a banking crisis, devaluation and political instability, which in turn led to fiscal indiscipline, under-investment in the petroleum sector and depressed domestic demand. Other circumstances exacerbated the situation, such as the devastation wrought by the El Niño weather pattern, a credit squeeze stemming from a financial crisis in East Asia and a fall in petroleum prices. There was a recovery in 2000 when GDP increased by 2.3%, partly because of an increase in petroleum pipeline capacity and higher petroleum output. GDP increased by 5.6% in 2001, outperforming most other Latin American economies. Both the petroleum sector and non-petroleum sectors registered recovery. The construction sector grew particularly strongly, owing to the completion of postponed infrastructure projects. The economy grew by approximately 3.0% in 2002. After two quarters of economic contraction in the first half of 2003, during which output was down by 1.1%, the economy recovered strongly to register annual growth of 2.6% for the year. Recovery was led by private consumption and exports, particularly of oil, helped by the completion of the OCP. Oil-led growth increased further in 2004, contributing to overall GDP growth of 6.9% in that year.

Petroleum revenues in the 1970s stimulated domestic demand and inflationary pressures in the 1970s. Public-sector spending increases led to budget deficits and currency weakness fuelling inflation from the 1980s. By 1989 inflation reached an annual average of 76%, which, at the time, was the highest rate in Ecuador's history. Inflation in the 1990s averaged 37.4% per year, but increased to an average annual rate of 96.9% in 2000, far exceeding all other South American economies. The increase stemmed largely from an expanded monetary base as the Government bailed out banks facing liquidity crises, which in turn led to a decline in the value of the sucre, further fuelling inflation. Fears of hyperinflation led to the policy of 'dollarization', in which the sucre was replaced with the US dollar as the unit of currency. The policy was announced in January 2000 and came into effect in March. The dollarization policy had the effect of stabilizing the economy by removing exchange-rate instability and reducing the Government's scope for fiscal imprudence. Existing inflationary pressures and cuts in state subsidies meant that the rate of inflation continued to rise after the introduction of the dollarization policy, reaching 107.9% in September 2000. Subsequently, however, as the monetary base contracted and an increase in agricultural products reduced the prices of food and beverages, inflation declined to an annual average of 7.9% in 2003 and further, to 2.7% in 2004. The rate of inflation stood at 1.4% in March 2005. Despite historically low inflation rates, price rises and a fixed exchange rate undermined the competitiveness of Ecuadorean producers. Attempts to increase productivity or push for privatization or other structural reforms to boost competitiveness could expect opposition from entrenched business interests, as well as trade unions and left-wing groups.

Successive governments came under pressure to generate fiscal surpluses in order to meet its burdensome debt obligations. A fiscal deficit equivalent to 4.8% of GDP in 1998 was fuelled by the banking crisis. The deficit was reduced to 3.8% of GDP in 1999 before a surplus equivalent to 1.5% of GDP was achieved in 2000. High international oil prices helped Ecuador maintain a small fiscal surplus between 2001–02, although the position remained unhealthy and was exacerbated by the Government's undisciplined handling of the public finances. Overspending in early 2002 was followed by a drastic halt in the latter half of the year, including a failure to pay public-sector wages. A Ley Orgánica de Responsabilidad, Estabilización y Transparencia Fiscal (Fiscal Responsibility, Stability and

Transparency Law), passed in June 2002, restricted the Government to annual spending increases of no more than 3.5%, in real terms. In January 2003 the new Government of Lucio Gutiérrez Borbua addressed the fiscal shortfall by decreeing cuts in fuel subsidies and other public spending. The Gutiérrez Government also attempted to reduce the minimum tax threshold and reduce tax exemptions, but, faced with opposition in the Congreso Nacional, the proposals were not implemented. A stand-by agreement with the IMF, announced in March 2003, was an important first step in gaining access to credit from the other multilateral lending agencies and private creditors required to finance Ecuador's debt. Nevertheless, the Government faced political and popular opposition to some of the IMF-prescribed restructuring, which raised questions about its ability to fulfil its commitments and maintain access to international credit. IMF approval of the Government's economic policies was a condition for disbursements from the World Bank and the Inter-American Development Bank (IDB).

## AGRICULTURE AND FISHERIES

Close to one-third of Ecuador's landmass is used for agricultural purposes. Sectoral output grew by an estimated 0.2% in 2004, down from 1.5% in 2003 and 7.2% in 2002. The decrease is largely explained by low rainfall levels and lower prices in the period. Agriculture's contribution to GDP has been largely stable since the early 1990s. During the previous three decades, however, the importance of agriculture declined dramatically: agriculture accounted for 25% of total output in the 1960s. A combination of poor infrastructure, lack of mechanization, financing difficulties and the effects of El Niño limited productivity in the sector.

Ecuador's climatic and geographic diversity supports a wide range of crops and fisheries production. The coastal region features a modern agro-industry where land has been converted for production, largely for export, of bananas and other fruit, coffee, cocoa, rice and shrimp. Other important exports include cut flowers, the cultivation of which is concentrated in the Sierra. Staple products for domestic consumption included rice, sugar cane and plantain, grown in coastal areas, while grains, vegetables and dairy products are produced in the Sierra.

Ecuador grows more bananas than any country other than Brazil, producing 5.8m. metric tons in 2003. Bananas became the principal crop of Ecuador in the 1940s when disease and hurricanes damaged production in Central America. Banana plantations were principally located in the lowlands of Guayas province, though road construction and improved irrigation expanded the viable area of production. By 1995 the land area given to banana production slowed. Global growth in demand, particularly from the former USSR and Eastern Europe, led to heavy investment in technology. The disease resistant and higher-yielding Cavendish variety of banana replaced the traditional Gros Michel. However, increased output and decreased demand saw prices fall. Banana (including plantain) cultivation reached a high point in 1997 when Ecuador produced 7.5m. metric tons, generating export revenues of US $1,327m. A steady decline in exports was reversed in 2001, when banana production of 6.1m. tons generated exports worth $847m., equivalent to 18.2% of total export earnings. This increased to $969.3m. in 2002, some 19.2% of total export earnings and to $1,099.3m. (18.2% of export earnings) in 2003. In 2004 bananas accounted for export revenues of $1,023m., equivalent to 13.4% of the total export value, a significant decrease in the fruit's contribution to export earnings. In 1999 the World Trade Organization (WTO) ruled in favour of Ecuador in a dispute with the European Union (EU). Ecuador had long protested that the preference for trading with African, Caribbean and Pacific countries, with whom parts of the EU maintained post-colonial ties, was contrary to trade regulations. The WTO ruled that the EU pay damages of $200m. annually until 2006, by which time a non-discriminatory tariff system should be introduced. Despite the favourable ruling, the pending EU tariff on Latin American banana imports—a proposed €230 per metric ton—had the effect of deterring investment in banana growing.

Before bananas took over in the 1940s, cocoa had been Ecuador's main export crop. In the 19th century coastal plantations had produced cocoa accounting for up to three-quarters of total export earnings. Decline followed disease and the emergence of alternative sources to satisfy global demand. A revival in the sub-sector took place in the 1980s as the Government subsidized production. Production reached record levels of 131,000 metric tons in 1985. The crop is vulnerable to climate and fluctuations in demand. By 1987 production had decreased to 57,500 tons. Cocoa exports came under threat when the International Cocoa Organization reclassified the quality of Ecuadorean cocoa at a lower level. In the mid-1990s, with the help of EU funds, Ecuador embarked on a programme to improve the quality and increase production. In 1996 production increased to 93,800 tons, but damage caused by El Niño saw production suffer in the following year. Thereafter cocoa staged a recovery, with production levels increasing to 89,000 tons in 2003. Export earnings were vulnerable to world demand and prices. Ecuadorean cocoa and derivatives earned US $159m. in 2003. A delayed harvest in early 2005 resulted in a 57% fall in export volumes in January–February, compared with the same period in 2004; however, the decline was offset by higher international prices.

As with other commodities, the importance of coffee to the Ecuadorean economy has varied with international prices, supply and climatic conditions. With more than 100,000 small family growers, coffee production is fragmented and inefficient. Intermittent attempts by Latin American coffee producers to support higher international prices by limiting exports have not been sustained. El Niño severely affected coffee production in 1997, leading to a 74.7% decline in production volumes in 1996–98. Production recovered, and in 2001 Ecuador registered output of 165,000 metric tons, its largest since 1995. However, output fell significantly in 2002, to 79,000 tons, and had only recovered slightly (to 83,000 tons) in 2003. Export revenues declined after 1999, even during high production years, owing to lower demand and prices. Having earned US $105m. in export revenues in 1998, receipts from coffee fell to just $42m. in 2002. Revenues recovered somewhat in 2003, reaching $63m. Early indications in 2005 suggested that export volumes in that year would be affected by the unusual length of the dry season.

New agricultural products, including cut flowers, melons, asparagus, artichokes and strawberries gave a boost to export earnings at the beginning of the 21st century. Cut flowers earned export revenues estimated at US $346m. in 2004, up from $295m. in 2003. Ecuadorean producers took advantage of the appreciation of the Colombian peso, which ensured more competitive prices in international markets. Colombia was Ecuador's principal competitor in the flower industry. Both countries, however, were vulnerable to the strengthening real exchange rate.

Staple crops such as rice, sugar cane, potatoes, maize, soybeans, wheat, barley and cotton were important for domestic consumption, but contributed insignificant amounts to overall export earnings. Landholdings in the non-export sector tended to be smaller and have lower levels of productivity. Ecuador became self-sufficient in rice cultivation in the early 1990s. In 2003 rice production totalled 1.2m. metric tons, up from 781,000 tons in 1987. Production is concentrated in the Guayas lowlands. Ecuador relied on imports to satisfy requirements for wheat and barley. Patterns of land ownership, characterized by a predominance of subsistence smallholdings, were contributory to the shortfall in wheat production.

In 1986 Ecuador became the world's largest exporter of shrimps, mainly selling to the USA. Shrimp production, concentrated in the Guayas and Esmeraldas provinces, increased further in the 1990s. Export revenues reached a record US $886m. in 1997. At this time, Ecuador had restored its position as one of the world's most important suppliers of shrimps, which had become the country's third most important export commodity after petroleum and bananas. Despite this growth, cultivation was dramatically affected in 1998–99 by competition from Asia, environmental damage caused by El Niño, interruptions in power supply, lack of investment and disease. Export volumes and earnings were partially restored from 2000, but did not approach 1997 levels. Ecuador's shrimp fishing sector earned a provisional $324.3m. in export revenues in 2004, up from $275.7m. in 2003, with additional revenues from tuna, sardines, mackerel, anchovies and fishmeal. Export volumes in January–February 2005 declined by 25.4%, compared with the same period in 2004, largely because of a 3.4%

tariff imposed by the USA. The tariff stems from a successful anti-dumping suit brought by US and Canadian shrimp producers. Non-canned fish export earnings declined in 2004 to $68m., down from $88m. in 2003. Canned and other processed fish earned an estimated $342m. in export revenues in 2003, compared with $268m. in 2002.

## MINING AND ENERGY

Oil drilling in Ecuador began in 1917 in the Santa Elena peninsula, west of Guayaquil. However, large-scale production dates back to the late 1960s with the discovery of major reserves in Lago Agrio in the Oriente by the US consortium of Texaco-Gulf. The Trans-Andean pipeline was built linking the oilfields to a tanker terminal in the port of Esmeraldas and exports began in 1972. Ecuador joined OPEC in 1973, but the required production quotas limited exploration and proven reserves.

The Government made it easier for foreign companies to operate in the petroleum sector in 1983, following which several foreign companies signed contracts with the state oil company Corporación Estatal Petrolera Ecuatoriana (CEPE). New reserves were found in south-eastern Oriente. Ignoring OPEC production quotas, Ecuador increased production to maximize revenues when oil prices were falling. In 1989 President Rodrigo Borja formed a new state oil company, Petróleos de Ecuador (PETROECUADOR), to replace CEPE and assume greater state control over the process of production and distribution. PETROECUADOR took over the Trans-Andean pipeline, but a decline in export revenues led to a policy that encouraged greater levels of foreign investment. In 1993 President Sixto Durán Ballén introduced a new more liberal foreign investment code. The new rules led to eight new production-sharing agreements with petroleum companies in the mid-1990s.

In 1992 Durán Ballén withdrew Ecuador from membership of OPEC because of OPEC's refusal to authorize an increase in Ecuador's production quota. Durán Ballén announced a target of producing 576,000 barrels per day (b/d) by 1996, which represented an ambitious 55% increase over the 1992 production volume. In the mid-1990s petroleum discoveries almost tripled Ecuador's proven reserves and the Government signed several contracts with foreign companies for drilling and exploration. Production increased from 341,774 b/d in 1993 to 386,725 b/d in 1995, still far below government targets. Increased opposition and sensitivities to foreign participation in the petroleum sector slowed development and growth. Under-investment under President Fabián Alarcón Rivera (1997–98) contributed to a decline in production that occurred concurrently with a drop in international petroleum prices. At 370,000 b/d, production was at levels lower than any time since 1993. Production increased to 407,000 b/d in 2001, but declined to 393,273 b/d during 2002, largely as a consequence of technical problems and an industrial dispute at the Esmeraldas refinery. Poor infrastructure was another obstacle to increased production. In 2000 work was completed to expand the existing Sistema del Oleoducto Trans-Ecuatoriano (Trans-Ecuadorean Oil Pipeline System—SOTE) pipeline, which added 60,000 b/d to overall capacity. The OCP, which would double heavy crude transport capacity to 850,000 b/d, became operational in 2003. As a result, average production for that year increased to 447,000 b/d. However, in March 2004 oil shipments via the SOTE were suspended following a landslide, forcing PETROECUADOR to declare *force majeur* on its contractual obligations. Nevertheless, in 2004 Ecuador produced an average 525,500 b/d of petroleum, of which PETROECUADOR accounted for an estimated 37.4%, a declining proportion as the role played by foreign companies increased. Until 2003 PETROECUADOR had supplied over 50% of national output. However, a dispute between three private petroleum companies and the tax authorities, which may go to international arbitration, threatened the planned expansion of petroleum production. (To avoid further legal disputes, in 2004 the Congreso passed legislation specifically exempting oil exporters from value-added tax—VAT—rebates.) Continued underinvestment in PETROECUADOR and aversion to risks associated with the legal status of contracts on the part of foreign companies suggested that Ecuador's oil production potential would remain unrealized in the immediate term.

Ecuador is heavily reliant on oil. Petroleum contributed 18.2% of GDP in 2004. Petroleum and petroleum derivatives were Ecuador's most important export commodity (comprising a provisional 55.4% of total export revenue in 2004, compared with 43.2% in 2003), and the sector was the largest contributor to the central government treasury, accounting for around one-third of government revenues. The increase in petroleum's share of export earnings in 2004 was partly caused by the increase in the cost of a barrel, from an average of approximately $25 per barrel at the beginning of the year, to $31 per barrel at the end. The USA was the most significant destination for petroleum exports. Proven reserves stood at 5,100m. barrels in December 2004. Most Ecuadorean oil was medium-heavy crude, though recent discoveries have been of heavy crude.

Most mining focuses on non-metals used in construction, including limestone, sand and clay, though there are reserves of metals such as gold, silver, copper, iron, lead, zinc, uranium and magnesium. The Government opened up mining to foreign investment in 1991 in a bid to develop the sector with a more liberal mining law. Subsequently, the bureaucracy associated with investment was reduced and further reforms were introduced in 2000 granting stronger legal rights to mining companies. However, the risks surrounding exploration rights from changes in government policy and enforcement have deterred investment.

## MANUFACTURING AND CONSTRUCTION

Traditional manufacturing sectors were textiles, food and drink, tobacco, petroleum refining and cement production. Most industrial activity takes place in the Guayas and Pichincha provinces, though other areas occupied important roles in industry. Petroleum refining and wood activity take place in Esmeraldas, iron and steel in Cotopaxi, while ceramics, furniture and tyres are produced in Azuay and marine and agricultural products are manufactured in Manabí. Ecuador has established *maquiladoras* (assembly plants) at the Guayas Free Zone near Guayaquil and several other areas have been identified for further free zones.

The petroleum boom of the 1970s fuelled expansion in manufacturing, which the Government encouraged with protectionist trade initiatives designed to substitute imports with domestically produced products. Industrial development was heavily dependent on the import of capital goods. Manufacturing output increased at an annual average rate of 9.5% between 1972–82. However, the sector was characterized by recession and stagnation in the rest of the decade. The sector was restored to growth in the 1990s. Manufacturing output increased by an average 0.9% per year during the period 1990–2003. In 2004 the manufacturing and construction sector's relative contribution to GDP was 18.9% of GDP, and its share of the active work-force stood at 20.6% in 2003. In the 1990s government policy shifted significantly away from protectionism towards a more liberal trade environment in line with regional and global developments. Membership of the WTO and Andean Community of Nations (Comunidad Andina de Naciones—CAN), which was ratified under the Government of Durán Ballén (1992–96), reinforced this outlook. The elimination and reduction of tariff and non-tariff barriers has exposed domestic producers to foreign competition while, at the same time, opening up opportunities for local industry to serve foreign markets. Exports to the CAN, in particular, has led to an expansion in the chemicals, machinery, minerals, paper, printing and wood products industries. Despite new overseas markets, manufacturing continued to contribute modestly to overall export earnings. Canned fish was the single largest manufacturing export, attracting revenues of US $320m. in 2004, though revenues fell by 26% over the year. Manufacturing output (not including oil refining) grew by 3.0% in 2004, below the average for the economy as a whole. Output was marginally down compared with 2003, but an improvement on the 0.7% growth recorded in 2002. An underdeveloped stock market was an impediment to businesses seeking investment capital.

The construction sector grew by 0.5% in 2003 and by 1.2% in 2004, a considerable deceleration in growth compared with 2002 when it was the fastest growing sector, at 14.7%. The volatility is explained by the development of the OCP heavy crude pipeline, which was completed in 2003, though most of the work was carried out during 2002. The state sector accounts for most investment in construction and major public-sector infrastructure projects, such as new oil pipelines and highways, an important source of employment.

## FINANCIAL SERVICES

The regulation of financial services was overhauled in 1994. Inadequate banking-sector supervision and risk assessment as well as the importance of US dollar-denominated lending made banks vulnerable to recession or devaluation of the sucre. As well as growth, the sector experienced regular crises from the mid-1990s. Poor economic performance, declining commodity prices, exchange-rate depreciation and rising interest rates reduced bank deposits and increased bad debts. Banking collapses in 1998–99 led to a halt in banking activity, in an attempt to prevent further bankruptcies, and a deposit freeze lasting one year. Between 1998 and 2001 16 banks closed down. By mid-2000 the majority of depositors whose assets had been 'frozen' were reimbursed. A government agency, the Agencia de Garantía de Depósitos (Deposit Guarantee Agency—AGD), was pursuing the assets of former bankers and bank shareholders, many of whom had fled the country to escape criminal investigation, in order to repay remaining creditors and depositors. The influence of powerful debtors has slowed the process of debt recovery, and confidence was slow in returning to the sector. Growth in bank deposits began to return in 2001. Initially these were largely confined to current accounts and those of short maturity, but by 2005 saving deposits made up 62.8% of all bank deposits. Savings deposits were vulnerable to political turbulence and growth slowed in response to popular protests which led to the collapse of the Gutiérrez Government in April 2005. As part of the IMF stand-by agreement reached in 2003, the Government was committed to returning all remaining deposits, paying liabilities and withdrawing from the banking sector. Progress was slow as a result of the difficulties in recovering bank loans to pay depositors. Intentions to introduce Basle standards of capital adequacy on the banking sector have been frustrated by a lack of resources, though multilateral institutions were expected to provide support in improving banking supervision.

## TRANSPORT AND COMMUNICATIONS

Underinvestment left most of Ecuador's road network in poor condition. The standard of the road network was best near the coast where large-scale reconstruction followed extensive damage caused by El Niño in 1997–98. Since 1998 private companies have been able to operate concessions to build and maintain highways. After 1991 private-sector companies were authorized to manage more than 1,000 km of roads. Road traffic increased dramatically since the 1970s: in 2004 there were an estimated 1m. vehicles, compared with just 76,000 in 1971. Roads were built to open up new areas in Oriente and Costa for agriculture and settlement. There were 43,197 km of roads in 2000, of which only 18.9% were paved.

The railways offered virtually no services of significant economic benefit. The network, which was once extensive, was very limited and in disrepair. The main railway line ran between Riobamba in the Sierra and the coast, but services were irregular. The Government was seeking to privatize the railways.

Quito and Guayaquil hosted Ecuador's two main international airports. In November 2002 a private North American consortium won a 35-year concession to operate Quito's existing Mariscal Sucre airport (an upgrade was completed in March 2004) and to build a new one further from the city centre. An Argentinian-based company won a concession to undertake a more limited modernization of Guayaquil's airport. Failure to meet international safety and regulatory standards restricted opportunities for Ecuadorean carriers to serve routes to the USA. The principal airline, Transportes Aéreos Militares Ecuatorianos (TAME), served Latin American destinations as well as national routes. Ecuatoriana, which was part-privatized in 1995, subsequently went into receivership. The majority of its assets were subsequently acquired by the LAN Chile group, which launched a new airline, LAN Ecuador, in May 2003.

Ecuador had six ports capable of accommodating petroleum tankers. Of these, Guayaquil was the main trading port, handling 65% of all traffic and Esmeraldas the most important port serving the petroleum sector. The other principal ports are Manta (through which most coffee and cocoa exports are distributed), Puerto Bolívar (particularly important in supporting banana exports), Balao and La Libertad.

Ecuador had an average of just 12.6 fixed telephone lines for every 100 inhabitants in 2004, almost 50% below the Latin American regional average. Installed lines are concentrated in urban areas. Foreign investment and management is required to improve services. Plans dating back to 1998 to sell off Ecuador's two state companies in the sector, Andinatel, providing services in the Sierra, and Pacifictel, in the coastal region, were frustrated by political opposition and potential bidders' aversion to risks associated with uncertain regulations. Outright privatization plans were suspended by President Gustavo Noboa (2000–03) in favour of transferring management over to the private sector. This commitment became part of the IMF stand-by agreement of March 2003, but continued to be the subject of delays, owing to political opposition. The underdevelopment of fixed line telecommunications was partially offset by a rapid increase in mobile cellular telephone use. The Government awarded a third mobile telecommunications licence to Andinatel following an auction in February 2003. The company committed to an estimated US $520m. in investment. At the end of 2004 there were 3.5m. mobile telephone users, customers of two operators, Porta and Bell South. The generally undeveloped telecommunications infrastructure undermined the use of the internet as a tool for business, although measures to provide a legal framework to support e-commerce and fixed tariffs on internet connections helped to increase use.

## TOURISM

Tourism has been a significant sector of the national economy since the 1960s with growth in long haul, international travel. It grew to become Ecuador's fourth largest earner of foreign exchange during the 1990s, behind petroleum, bananas and shrimps. Ecuador's rich biodiversity and varied climate and landscape—Andean highlands, tropical rainforest, Pacific beaches and the Galápagos Islands—mostly within accessible journey times from main cities, make it a popular destination.

The number of tourists visiting Ecuador increased from 172,000 in 1975 to 760,000 in 2003. Tourism receipts were US $408m. in that year. Most international visitors are from neighbouring Colombia and Peru, followed by the USA and Europe. Resolution of the long-running border dispute with Peru has led to direct air links between Peru's capital, Lima, and several destinations in Ecuador. Airport capacity was an impediment to expansion of visitor numbers.

## FOREIGN INVESTMENT

Regulations governing foreign investment were liberalized from the 1980s in line with global and regional trends. In the 1980s President Febres Cordero relaxed ownership restrictions on foreign companies and raised limits on profit remittances. In 1991 President Borja opened up some sectors of the economy that had been restricted to sole or majority domestic ownership only. In 1993 President Durán Ballén introduced a further liberalization of the foreign investment code. This established equal treatment for national and international investors and opened up further sectors of the economy to foreign capital. In addition, profits could be freely repatriated and prior government approval for foreign investments was no longer required. In 1998 the remaining restrictions on foreign investment in strategic sectors such as fishing, air transport and media were abolished.

Encouraged by these reforms, together with membership of the CAN, which restricted future governments' ability to return to a less liberal investment environment, foreign direct investment (FDI) increased. From an annual average of US $108m. in 1982–92, FDI rose to $589m. in 1993–99. Since the mid-1980s, 65% of inflows have come from the USA and Canada. Europe has been the source of 18% of foreign investment. Investment in the petroleum sector accounted for more than 90% of FDI, encouraged by the introduction of production-sharing contracts introduced in the mid-1990s.

While a more liberal foreign investment regime went some way to increasing the confidence of overseas investors, this was offset by the conflict with Peru in the mid-1990s and by the 'tequila effect' (the damage caused to all Latin American economies following the devaluation of the Mexican peso in December 1994). Confidence among investors was also undermined by domestic political scandal and instability between 1997–2000 as two elected Governments failed to complete their terms of office. President Noboa succeeded in securing significant reform in 2000 with the Ley de Transformación Económica (Economic Transformation Law), the most notable feature of which was dollarization, in order to support privatization plans.

FDI in 2004 was valued at US $1,240m., a 20.3% decrease from the $1,555m. attracted in 2003. The reduced figures were a result of the completion of the OCP pipeline. Despite the fall, FDI was considerably higher than in 2000, when Ecuador attracted FDI valued at $720m. Historically, the USA has been the largest source of FDI into Ecuador of any single country, although Canada has outstripped the USA in the early 2000s with an annual average of 26.2% of total foreign investment, compared with 23.5% from the USA. In 2004 foreign investment from the USA was estimated at $304m., representing 24.5% of the total, up from $204m. in 2003, when it was at a low of 13.1% of the total. Historically, Europe accounted for 15%–20% of FDI. Italy and Spain were the leading European investors at 3.5% and 3.2%, respectively, in 2003, when 53% of foreign investment funded oil and mining projects, and 28.4% went into construction (mostly associated with the OCP project). Ecuador's total stock of FDI was estimated at $9,686m. in 2002, an increase on $8,410m. in 2001, making it one of the smallest per-head recipients of FDI in Latin America.

The contribution of the stock market to the economy was insignificant, both as a source of investment funds for local business and as a destination for portfolio investment. Banks were the principal source of financing available. Around one-third of all transactions involved central bank debt issues. Net portfolio investment represented an outflow of funds.

## DEBT

Ecuador rapidly accumulated sizeable debt in the 1970s to finance state-led industrial development. In 1970 debt was just US $242m., but had reached $4,600m. by 1980 and the debt-service ratio (debt servicing to the total value of export earnings) was 38%. Ecuador used its reserves of petroleum to finance the debt, but with a fall in international petroleum prices in 1982 and interest rate increases the country fell behind in repayments. In the 1980s infrastructure was damaged by earthquakes, which in turn detrimentally affected petroleum revenues. Lending from both commercial banks and the 'Paris Club' of creditor nations increased annually between 1987 and 1994 to unsustainable levels.

Persistent difficulties in debt repayment obligations led to the 1995 Brady Plan, in which US $7,580m. of debt with commercial banks was restructured to ensure that Ecuador had access to further commercial lending. However, a combination of a depreciating currency, low petroleum prices and a weak economy led to a further worsening of the debt position. Total public external debt increased to more than 100% of GDP in 1999, making Ecuador the first country to default on Brady and Eurobond obligations. In 2000 the Noboa Government reached an agreement with the IMF and private-sector creditors to restructure debt. The IMF provided a $2,000m. loan facility over a two-year period. However, the debt-servicing ratio remained high and Ecuador had limited access to international lending.

Ecuador's finance minister in 2001–02, Carlos Julio Emanuel, won the confidence of the international financial community by persisting with orthodox fiscal discipline in the face of political opposition. In December 2001 the IMF released funds to complete an earlier stand-by agreement, following payment of US $55m. in arrears owed to 'Paris Club' creditors. In March 2003 the new Gutiérrez Government secured a 13-month stand-by loan of $205m. from the IMF, conditional upon a programme of structural reform. This facility gave Ecuador access to further multilateral lending including a new Country Assistance Strategy, worth some $1,050m. over four years. In May 2004 this loan was extended to the end of the year. It was hoped that this new credit, in addition to high world petroleum prices, would help Ecuador finance its debt repayments. Indeed, the Government was committed to placing above-budget oil revenues and those from the new heavy crude oil pipeline into a stabilization fund to repay debt, with a target of reducing the debt/GDP ratio to 40% by the end of 2007 (the ratio was 58.5% in 2004). The Gutiérrez administration also embarked on a concerted effort to restrain public spending; its limited successes thus far included a reduction in fuel subsidies and a public-sector pay 'freeze'. The Government's IMF-approved fiscal performance target for 2003 was a non-financial primary surplus of 3.3% of GDP, requiring a primary surplus equivalent to 6.4% of GDP. Despite being on target during the first three quarters of 2003, the Government fell short, as tax revenues were lower than expected, while public-sector wages increased and the Government decided against reducing the subsidy on cooking gas. Total foreign debt was estimated at $17,200m. in 2004, more than the $15,305m. owed in 1999, although it was a marked improvement in terms of debt-service ratio, representing 20.6%, compared with 26.5% in 1999.

## FOREIGN TRADE

Liberalization of trade policy progressively made Ecuador more open in the 1990s, beginning with a reform of the Tariff Law and elimination of import quotas. The policy shift away from import substitution, which had been favoured in the 1970s, was given institutional support when Ecuador joined the WTO in 1996 and ratified the General Agreement on Trade in Services. Domestic manufacturing—with a few exceptions, such as vehicles, as permitted by the WTO—no longer enjoys the protection of high tariff barriers. Ecuador's standard tariff is below the Andean Community's agreed common external tariff, making it one of the most open in South America.

Trade liberalization, including the burgeoning new CAN markets, encouraged diversification of exports. Manufacturing exports grew significantly from the early 1990s, from 12.3% of export revenues in 1991 to an estimated 29% in 2002. Food, drink and tobacco was the most important manufacturing subsector, in particular canned fish, which earned Ecuador more revenues than any other manufacturing export. Among other growth areas in non-traditional exports were vehicles assembled for export to the rest of the Andean region. However, despite diversification Ecuador remained vulnerable, though to a lesser degree, to the volatility of international commodity markets, as exports remained dominated by primary products.

Ecuador registered a trade surplus of US $261.8m. in 2004 (imports were valued at $7,487.8m. and exports at $7,748.8m.), the first surplus since 2000. Trade deficits of $998 and $71m. were registered in 2002 and 2003, respectively. The shift into a positive trade balance in 2004 and the shrinking deficit in 2003 were led by increased volumes and prices of oil exports, which outpaced strong import growth.

Between 1999–2002 the share of consumer and capital goods imports relative to primary imports grew. In 2004 consumer goods accounted for 27.9% of all imports, while capital goods made up 26.1% and primary goods 36.0%. Consumer goods registered a 17.2% growth in 2004, following a modest 2.6% growth in 2003. In 2001 the deficit was mainly the result both of exchange-rate appreciation, which fuelled an increase in imports, and an average 21% fall in international petroleum prices in that year. Despite increases in export revenues for non-traditional goods (most notably vehicles, which earned 48.4% more in 2001 than in 2000), total export revenues fell by 9.5% in 2001. The total value of imports grew by 44.1% in the same period. Growth of 75% in capital goods imports was the result of the construction of the new petroleum pipeline and was offset by the increased capital inflows it facilitated. However, the 74.2% increase in the imports of consumer goods and the 19.7% increase in the imports of intermediate goods enhanced the deficit. Ecuador's competitiveness in international trade was undermined by inflationary pressures that were generally greater than those faced by the country's principal trading partners.

Crude petroleum, and its derivatives, was the largest export item, earning US $4,234m. in 2004 (55.4% of total export earnings), compared with $2,607m. in 2003 (43.2% of the total) and $2,055m. in 2002 (40.8% of the total). Bananas (including plantains) were the next largest export item, earning $1,023m. in 2004 (13.4% of total exports), down from $1,099m. in 2003, but an increase on the $969m. earned in 2002. The second largest non-oil export commodity was shrimps, which earned $324m. in 2004, up from $276m. in 2003, and $253m. in 2002. In 2004 42.9% of exports were destined for the USA, far outstripping other export destinations. Peru was the second most important destination, accounting for 7.9% of exports, while the largest European market was Italy, which received 4.6% of exports in 2004. In that year the USA was also the largest source of imports (20.7%), followed by Colombia (14.6%), Venezuela (6.8%) and Brazil (6.1%). Notwithstanding the trade surplus, in 2004 there was an estimated current-account deficit of $122m., equivalent to 0.4% of GDP, a reduced deficit compared with the $455m. recorded in 2003, which was equivalent to 1.5% of GDP. The repatriation of foreign investors' earnings abroad, as well as Ecuador's large interest obligations on its debt, continued to ensure consistent current-account deficits. Remittances from Ecuadoreans living abroad represented the largest source of foreign exchange earnings after petroleum, totalling a Central Bank estimate of $1,600m. in 2004. This represented a modest increase in the year. Remittances from Ecuadoreans abroad were vulnerable to economic slowdowns, particularly in the main host countries, the USA and Spain.

## CONCLUSION

The Government benefited from increased oil revenues owing to higher export volumes and international prices. However, a new episode of political instability in Ecuador in 2005 raised doubts about the direction of economic policy and the use to which these revenues would be put. Early signals from President Alfredo Palacio Gonzáles, who replaced President Gutiérrez in April 2005, suggested that the new administration would redirect some funds (specifically, from the oil stabilization fund, Fondo de Estabilización, Inversión Social y Reducción del Endeudamiento Público—FEIREP) away from debt repayment and towards increased social investment. Under the new Government there were renewed doubts about whether structural reforms advocated by the IMF would be introduced. Designed to boost the non-oil economy, these included changes to the tax regime, the introduction of private management into public-sector utilities and greater flexibility in labour markets. In the longer term, reluctance to introduce structural reforms threatened Ecuador's relationship with the IMF, whose approval was required to trigger important financing from other multilateral lending organizations.

# Statistical Survey

Sources (unless otherwise stated): Banco Central del Ecuador, Quito; Ministerio de Comercio Exterior, Industrialización, Pesca y Competitividad; Instituto Nacional de Estadística y Censos, Juan Larrea 534 y Riofrío, Quito; tel. (2) 529-858; e-mail inec1@ecnet.ec; internet www.inec.gov.ec.

## Area and Population

### AREA, POPULATION AND DENSITY

| | |
|---|---|
| Area (sq km) | 272,045* |
| Population (census results)† | |
| 25 November 1990 | 9,648,189 |
| 25 November 2001 | |
| Males | 6,018,353 |
| Females | 6,138,255 |
| Total | 12,156,608 |
| Population (UN estimate at mid-year)‡ | |
| 2002 | 12,810,000 |
| 2003 | 13,003,000 |
| 2004 | 13,040,000 |
| Density (per sq km) at mid-2004 | 47.9 |

* 105,037 sq miles.
† Excluding nomadic tribes of indigenous Indians and any adjustment for underenumeration, estimated to have been 6.3% in 1990; the 1990 total was subsequently revised to 9,697,979.
‡ Source: UN, *World Population Prospects: The 2004 Revision*.

### PROVINCES
(2001 census)

| | Area (sq km) | Population | Density (per sq km) | Capital |
|---|---|---|---|---|
| Azuay | 8,125 | 599,546 | 73.8 | Cuenca |
| Bolívar | 3,940 | 169,370 | 43.0 | Guaranda |
| Cañar | 3,122 | 206,981 | 66.3 | Azogues |
| Carchi | 3,605 | 152,939 | 42.4 | Tulcán |
| Chimborazo | 6,072 | 403,632 | 66.5 | Riobamba |
| Cotopaxi | 6,569 | 349,540 | 53.2 | Latacunga |
| El Oro | 5,850 | 525,763 | 89.9 | Machala |
| Esmeraldas | 15,239 | 385,223 | 25.3 | Esmeraldas |
| Guayas | 20,503 | 3,309,034 | 161.4 | Guayaquil |
| Imbabura | 4,559 | 344,044 | 75.5 | Ibarra |
| Loja | 11,027 | 404,835 | 36.7 | Loja |
| Los Ríos | 7,175 | 650,178 | 90.6 | Babahoyo |
| Manabí | 18,879 | 1,186,025 | 62.8 | Portoviejo |
| Morona Santiago | 25,690 | 115,412 | 4.5 | Macas |
| Napo | 11,431 | 79,139 | 6.9 | Tena |
| Orellana | 22,500 | 86,493 | 3.8 | Puerto Francisco de Orellana (Coca) |
| Pastaza | 29,774 | 61,779 | 2.1 | Puyo |
| Pichincha | 12,915 | 2,388,817 | 185.0 | Quito |
| Sucumbíos | 18,328 | 128,995 | 7.0 | Nueva Loja |
| Tungurahua | 3,335 | 441,034 | 132.2 | Ambato |
| Zamora Chinchipe | 23,111 | 76,601 | 3.3 | Zamora |
| Archipiélago de Colón (Galápagos) | 8,010 | 18,640 | 2.3 | Puerto Baquerizo (Isla San Cristóbal) |
| *Uncharted areas* | 2,289 | 72,588 | — | — |
| **Total** | 272,045 | 12,156,608 | 44.7 | |

Source: partly Stefan Helders, *World Gazetteer*.

# ECUADOR

## PRINCIPAL TOWNS
(2001 census)

| | | | | |
|---|---:|---|---:|
| Guayaquil | 1,985,379 | Ambato | 154,095 |
| Quito (capital) | 1,399,378 | Riobamba | 124,807 |
| Cuenca | 277,374 | Quevedo | 120,379 |
| Machala | 204,578 | Loja | 118,532 |
| Santo Domingo de los Colorados | 199,827 | Milagro | 113,440 |
| Manta | 183,105 | Ibarra | 108,535 |
| Portoviejo | 171,847 | Esmeraldas | 95,124 |

**Mid-2003** (UN estimate, incl. suburbs): Guayaquil 2,262,343 (Source: UN, *World Urbanization Prospects: The 2003 Revision*).

## BIRTHS, MARRIAGES AND DEATHS
(excluding nomadic Indian tribes)*

| | Registered live births† | | Registered marriages | | Registered deaths | |
|---|---:|---:|---:|---:|---:|---:|
| | Number | Rate (per 1,000) | Number | Rate (per 1,000) | Number | Rate (per 1,000) |
| 1995 | 408,983 | 23.7 | 70,480 | 6.2 | 50,867 | 4.4 |
| 1996 | 302,217 | 23.1 | 72,094 | 6.2 | 52,300 | 4.5 |
| 1997 | 288,803 | 22.8 | 66,967 | 5.6 | 52,089 | 4.4 |
| 1998 | 364,684 | 22.7 | 69,867 | n.a. | 54,357 | 4.5 |
| 1999 | 368,659 | 24.6 | 77,593 | n.a. | 55,921 | 4.5 |
| 2000 | 345,715 | 23.4 | 74,875 | n.a. | 56,240 | 4.5 |
| 2001 | 332,776 | n.a. | 67,741 | n.a. | 55,214 | 4.5 |

* Registrations incomplete.
† Figures include registrations of large numbers of births occurring in previous years. The number of births registered in the year of occurrence was: 181,268 in 1995, 182,242 in 1996, 169,869 in 1997, 199,079 in 1998, 218,108 in 1999, 202,257 in 2000 and 192,786 in 2001.

Source: partly UN, *Demographic Yearbook*.

**2002** (excluding nomadic Indian tribes): Live births registered in year of occurrence 183,792; Deaths 55,549 (Source: UN, *Population and Vital Statistics Report*).

**Marriages:** 66,208 in 2002.

**Expectation of life** (WHO estimates, years at birth): 71 (males 68; females 74) in 2003 (Source: WHO, *World Health Report*).

## ECONOMICALLY ACTIVE POPULATION
(ISIC major divisions, urban areas only, '000 persons aged 10 years and over, at November of each year, unless otherwise indicated)

| | 2001* | 2002 | 2003 |
|---|---:|---:|---:|
| Agriculture, hunting and forestry | 239.8 | 261.2 | 276.7 |
| Fishing | 42.3 | 35.6 | 46.0 |
| Mining and quarrying | 18.3 | 22.3 | 20.4 |
| Manufacturing | 610.6 | 501.5 | 487.8 |
| Electricity, gas and water | 27.8 | 14.0 | 17.0 |
| Construction | 234.9 | 240.5 | 239.9 |
| Wholesale and retail trade; repair of motor vehicles, motorcycles and personal and household goods | 1,026.7 | 971.5 | 1,000.8 |
| Hotels and restaurants | 158.2 | 147.4 | 131.3 |
| Transport, storage and communications | 244.6 | 222.3 | 233.2 |
| Financial intermediation | 33.5 | 46.4 | 51.8 |
| Real estate, renting and business activities | 158.2 | 155.0 | 154.9 |
| Public administration and defence; compulsory social security | 159.9 | 146.1 | 181.7 |
| Education | 211.4 | 238.8 | 234.4 |
| Health and social work | 99.8 | 118.6 | 116.5 |
| Other community, social and personal service activities | 154.2 | 121.0 | 174.5 |
| Private households with employed persons | 232.1 | 217.0 | 163.4 |
| Extra-territorial organizations and bodies | 1.4 | 0.4 | 1.0 |
| Activities not adequately defined | 19.6 | — | — |
| **Total employed** | 3,673.2 | 3,459.4 | 3,531.2 |
| Unemployed | 451.0 | 352.9 | 461.1 |
| **Total labour force** | 4,124.2 | 3,812.3 | 3,992.3 |
| Males | 2,380.7 | 2,267.9 | 2,353.4 |
| Females | 1,743.6 | 1,544.4 | 1,638.9 |

* At July.

Source: ILO.

# Health and Welfare

## KEY INDICATORS

| | |
|---|---:|
| Total fertility rate (children per woman, 2003) | 2.7 |
| Under-5 mortality rate (per 1,000 live births, 2003) | 27 |
| HIV/AIDS (% of persons aged 15–49, 2003) | 0.6 |
| Physicians (per 1,000 head, 2000) | 1.48 |
| Hospital beds (per 1,000 head, 1996) | 1.55 |
| Health expenditure (2002): US $ per head (PPP) | 197 |
| Health expenditure (2002): % of GDP | 4.8 |
| Health expenditure (2002): public (% of total) | 36.0 |
| Access to water (% of persons, 2002) | 86 |
| Access to sanitation (% of persons, 2002) | 72 |
| Human Development Index (2002): ranking | 100 |
| Human Development Index (2002): value | 0.735 |

For sources and definitions, see explanatory note on p. vi.

# ECUADOR

## Agriculture

**PRINCIPAL CROPS**
('000 metric tons)

|  | 2001 | 2002 | 2003 |
|---|---|---|---|
| Rice (paddy) | 1,256 | 1,285 | 1,236 |
| Barley | 26 | 29 | 27 |
| Maize | 337 | 602 | 677 |
| Potatoes | 441 | 485 | 397 |
| Cassava (Manioc) | 84 | 95 | 131 |
| Sugar cane | 5,654 | 5,670 | 5,691 |
| Dry beans | 30 | 48 | 49 |
| Soybeans (Soya beans) | 113 | 109 | 109 |
| Coconuts | 25 | 22 | 21 |
| Oil palm fruit* | 1,424 | 1,506 | 1,450 |
| Tomatoes | 29 | 48 | 73 |
| Pumpkins, squash and gourds* | 42 | 43 | 43 |
| Onions and shallots (green) | 101 | 98 | 70 |
| Carrots | 21 | 20 | 23 |
| Watermelons | 26 | 30 | 29 |
| Other vegetables and melons* | 93 | 108 | 112 |
| Bananas | 6,077 | 5,528 | 5,883 |
| Plantains | 813 | 542 | 651 |
| Oranges | 152 | 178 | 213 |
| Tangerines, mandarins, clementines and satsumas | 18 | 23 | 35 |
| Grapefruit and pomelo | 4 | 32 | 31 |
| Mangoes | 89 | 101 | 89 |
| Pineapples | 47 | 77 | 90 |
| Papayas | 17 | 20 | 23 |
| Other fresh fruit (excl. melons)* | 349 | 395 | 392 |
| Coffee (green) | 165 | 79 | 83 |
| Cocoa beans | 76 | 88 | 89 |
| Abaca (Manila hemp)* | 25 | 26 | 26 |

* FAO estimates.
Source: FAO.

**LIVESTOCK**
('000 head, year ending September)

|  | 2001 | 2002 | 2003 |
|---|---|---|---|
| Cattle | 4,657 | 4,794 | 4,977* |
| Sheep | 2,249 | 2,381 | 2,645 |
| Pigs* | 2,897 | 2,959 | 3,007 |
| Horses† | 525 | 528 | 530 |
| Goats | 273 | 278 | 279† |
| Asses† | 270 | 275 | 280 |
| Mules† | 158 | 160 | 160 |
| Chickens | 138,429 | 139,000† | 142,000† |
| Ducks† | 152 | 155 | 158 |
| Geese† | 50 | 50 | 50 |
| Turkeys† | 22 | 25 | 26 |

* Unofficial figure(s).
† FAO estimate(s).
Source: FAO.

**LIVESTOCK PRODUCTS**
('000 metric tons)

|  | 2001 | 2002 | 2003 |
|---|---|---|---|
| Beef and veal* | 189.2 | 206.2 | 232.8 |
| Mutton and lamb† | 6.3 | 6.2 | 7.7 |
| Pig meat | 139.5* | 144.9 | 154.0 |
| Goat meat | 1.3† | 1.9 | 2.5 |
| Poultry meat | 202.9* | 208.2 | 211.2 |
| Cows' milk | 2,431.1 | 2,433.2 | 2,456.5 |
| Sheep's milk† | 6.2 | 6.2 | 6.2 |
| Goats' milk† | 2.5 | 2.5 | 2.6 |
| Butter† | 5.1 | 5.3 | 4.9 |
| Cheese† | 7.7 | 8.2 | 7.3 |
| Hen eggs | 72.5 | 73.5 | 74.0 |
| Wool: greasy† | 2.2 | 2.4 | 2.6 |
| Wool: scoured† | 1.1 | 1.2 | 1.3 |
| Cattle hides (fresh)† | 31.6 | 34.4 | 34.4 |

* Unofficial figure(s).
† FAO estimate(s).
Source: FAO.

## Forestry

**ROUNDWOOD REMOVALS**
('000 cubic metres, excluding bark)

|  | 2001 | 2002 | 2003 |
|---|---|---|---|
| Pulpwood | 858 | 913 | 913 |
| Fuel wood* | 5,201 | 5,274 | 5,350 |
| **Total** | 6,059 | 6,187 | 6,263 |

* FAO estimates.
Source: FAO.

**SAWNWOOD PRODUCTION**
('000 cubic metres, including railway sleepers)

|  | 2000 | 2001 | 2002 |
|---|---|---|---|
| Coniferous (softwood) | 121 | 134 | 150 |
| Broadleaved (hardwood) | 594 | 660 | 600 |
| **Total** | 715 | 794 | 750 |

**2003:** Production as in 2002 (FAO estimates).
Source: FAO.

## Fishing

('000 metric tons, live weight)

|  | 2001 | 2002 | 2003 |
|---|---|---|---|
| Capture | 586.6 | 318.5 | 397.9 |
| Gurnards and searobins | 1.4 | 1.5 | 2.1 |
| Chilean jack mackerel | 134.0 | 0.6 | — |
| South American pilchard | 42.1 | 2.5 | 0.6 |
| Pacific thread herring | 19.9 | 11.0 | 6.9 |
| Anchoveta (Peruvian anchovy) | 2.1 | 71.0 | 33.4 |
| Pacific anchoveta | 73.5 | 18.3 | 19.5 |
| Skipjack tuna | 68.2 | 78.2 | 133.2 |
| Yellowfin tuna | 57.6 | 36.3 | 40.8 |
| Bigeye tuna | 23.4 | 21.3 | 20.2 |
| Chub mackerel | 85.4 | 17.1 | 33.3 |
| Aquaculture | 52.4 | 55.6 | 67.2 |
| Whiteleg shrimp | 45.3 | 46.7 | 57.5 |
| **Total catch** | 639.0 | 374.3 | 465.1 |

Source: FAO.

## Mining

('000 barrels, unless otherwise indicated)

|  | 2001 | 2002 | 2003 |
|---|---|---|---|
| Crude petroleum* | 148,746 | 143,758 | 152,497 |
| Natural gas (million cu m) | 1,001 | 998 | 1,039* |
| Gold (kilograms)† | 3,005 | 2,750 | 3,020 |

* Reported figure(s).
† Metal content of ore only.
Source: US Geological Survey.

# ECUADOR

## Industry

**SELECTED PRODUCTS**
('000 barrels, unless otherwise indicated)

|  | 2001 | 2002 | 2003 |
|---|---|---|---|
| Jet fuels | 1,771 | 1,797 | 1,879 |
| Motor spirit (gasoline) | 12,236 | 12,887 | 13,090 |
| Distillate fuel oils | 10,953 | 11,354 | 10,812 |
| Residual fuel oils | 11,898 | 10,742 | 8,879 |
| Liquefied petroleum gas | 2,407 | 2,060 | 2,230 |
| Crude steel ('000 metric tons) | 60 | 69 | 80 |
| Cement ('000 metric tons) | 2,920 | 3,000* | 3,100 |
| Electric energy (million kWh)† | 11,050 | 11,884 | n.a. |

* Estimate.
† Source: UN Economic Commission for Latin America and the Caribbean, *Statistical Yearbook*.

Source: mostly US Geological Survey.

## Finance

**CURRENCY AND EXCHANGE RATES**

**Monetary Units**
United States currency is used: 100 cents = 1 US dollar ($).

**Sterling and Euro Equivalents** (31 May 2005)
£1 sterling = US $1.82;
€1 = US $1.23;
US $100 = £55.00 = €81.10.

Note: Ecuador's national currency was formerly the sucre. From 13 March 2000 the sucre was replaced by the US dollar, at an exchange rate of $1 = 25,000 sucres. Both currencies were officially in use for a transitional period of 180 days, but from 9 September sucres were withdrawn from circulation and the dollar became the sole legal tender.

**BUDGET**
(US $ million)

| Revenue | 2000 | 2001 | 2002 |
|---|---|---|---|
| Petroleum revenue | 1,396 | 1,280 | 1,363 |
| Non-petroleum revenue | 1,854 | 2,566 | 3,209 |
| Taxation | 1,747 | 2,369 | 2,750 |
| Taxes on income and profits | 458 | 475 | 531 |
| Property taxes | 14 | 23 | 49 |
| Value-added tax | 836 | 1,340 | 1,529 |
| Selective excise taxes | 75 | 137 | 221 |
| Taxes on international trade | 341 | 373 | 419 |
| Import duties | 321 | 354 | 414 |
| Export duties | 17 | 18 | 5 |
| Other taxes | 23 | 21 | 1 |
| Non-tax revenue | 107 | 158 | 338 |
| Transfers | — | 58 | 122 |
| Statistical discrepancy | — | −19 | — |
| **Total** | **3,250** | **3,846** | **4,572** |

| Expenditure | 2000 | 2001 | 2002 |
|---|---|---|---|
| Wages and salaries | 703 | 1,088 | 1,673 |
| Purchases of goods and services | 169 | 122 | 318 |
| Interest payments | 1,009 | 938 | 823 |
| Current transfers | 145 | 183 | 366 |
| Other current expenditure | 542 | 300 | 351 |
| Capital expenditure | 659 | 1,438 | 1,226 |
| Fixed capital formation | 425 | 645 | 611 |
| Capital transfers | 234 | 562 | 618 |
| Other capital expenditure | — | 232 | −2 |
| **Total** | **3,227** | **4,068** | **4,757** |

Source: IMF, *Ecuador: Selected Issues and Statistical Appendix* (April 2003).

## Statistical Survey

**INTERNATIONAL RESERVES**
(US $ million at 31 December)

|  | 2002 | 2003 | 2004 |
|---|---|---|---|
| Gold* | 293.3 | 348.0 | 368.0 |
| IMF special drawing rights | 1.9 | 1.0 | 56.1 |
| Reserve position in IMF | 23.3 | 25.5 | 26.6 |
| Foreign exchange | 689.4 | 786.1 | 986.9 |
| **Total** | **1,007.9** | **1,160.6** | **1,437.6** |

* National valuation $347 per ounce at 31 December 2002; $412 per ounce at 31 December 2003; $436 per ounce at 31 December 2004.

Source: IMF, *International Financial Statistics*.

**MONEY SUPPLY**
(US $ million at 31 December)

|  | 2002 | 2003 | 2004 |
|---|---|---|---|
| Currency outside banks | 39.6 | 49.7 | 58.1 |
| Demand deposits at deposit money banks | 1,629.1 | 1,793.0 | 2,315.0 |
| **Total money** (incl. others)* | **1,794.8** | **1,985.3** | **2,510.0** |

* Includes private-sector deposits at the Central Bank.

Source: IMF, *International Financial Statistics*.

**COST OF LIVING**
(Consumer Price Index; base: 2000 = 100)

|  | 2001 | 2002 | 2003 |
|---|---|---|---|
| Food (incl. alcoholic beverages) | 131.2 | 141.7 | 145.8 |
| Fuel (excl. light) | 165.0 | 203.9 | n.a. |
| Clothing | 130.6 | 132.5 | 124.6 |
| Rent | 135.2 | 198.1 | n.a. |
| **All items** (incl. others) | **137.7** | **154.9** | **167.1** |

**2004:** Food (incl. alcoholic beverages) 146.9; All items 171.7.

Source: ILO.

**NATIONAL ACCOUNTS**
(US $ million at current prices)

**Expenditure on the Gross Domestic Product**

|  | 2002 | 2003 | 2004 |
|---|---|---|---|
| Government final consumption expenditure | 2,550 | 2,583 | 2,790 |
| Private final consumption expenditure | 16,837 | 18,473 | 19,769 |
| Changes in stocks | 1,191 | 1,329 | 1,673 |
| Gross fixed capital formation | 5,549 | 6,192 | 6,571 |
| **Total domestic expenditure** | **26,127** | **28,577** | **30,803** |
| Exports of goods and services | 5,829 | 6,461 | 8,029 |
| *Less* Imports of goods and services | 7,644 | 7,837 | 8,549 |
| **GDP at market prices** | **24,311** | **27,201** | **30,282** |

Source: mostly IMF, *International Financial Statistics*.

# ECUADOR

## Gross Domestic Product by Economic Activity

|  | 2002 | 2003 | 2004 |
|---|---:|---:|---:|
| Agriculture, hunting, forestry and fishing | 2,194.0 | 2,087.9 | 2,160.4 |
| Petroleum and other mining | 2,845.2 | 3,569.1 | 5,076.6 |
| Manufacturing (excl. petroleum refining) | 2,662.9 | 2,899.4 | 3,018.5 |
| Manufacture of petroleum derivatives | −965.9 | −1,168.8 | −1,499.8 |
| Electricity, gas and water | 433.0 | 443.6 | 482.0 |
| Construction | 1,914.2 | 2,066.0 | 2,252.2 |
| Wholesale and retail trade | 3,378.3 | 3,691.7 | 3,919.6 |
| Hotels and restaurants | 475.2 | 548.6 | 576.8 |
| Transport, storage and communications | 3,775.2 | 4,567.6 | 4,859.8 |
| Financial intermediation | 675.8 | 758.0 | 876.4 |
| Real estate, renting and business activities | 1,858.3 | 2,212.8 | 2,349.1 |
| Public administration and defence | 1,369.2 | 1,442.8 | 1,540.0 |
| Education | 1,114.3 | 1,326.4 | 1,428.3 |
| Health and social welfare | 462.8 | 548.0 | 573.6 |
| Other community, social and personal service activities | 177.6 | 218.0 | 235.7 |
| Private households with employed persons | 43.8 | 47.4 | 48.7 |
| **Sub-total** | 22,413.9 | 25,258.3 | 27,897.9 |
| *Less* Financial intermediation services indirectly measured | 800.5 | 886.8 | 909.5 |
| **Gross value added in basic prices** | 21,613.4 | 24,371.5 | 26,988.5 |
| Taxes, less subsidies, on products | 2,697.5 | 2,829.4 | 3,293.0 |
| **GDP in market prices** | 24,310.9 | 27,201.0 | 30,281.5 |

## BALANCE OF PAYMENTS
(US $ million)

|  | 2001 | 2002 | 2003 |
|---|---:|---:|---:|
| Exports of goods f.o.b. | 4,781 | 5,198 | 6,197 |
| Imports of goods f.o.b. | −5,179 | −6,196 | −6,268 |
| **Trade balance** | −397 | −998 | −71 |
| Exports of services | 862 | 923 | 898 |
| Imports of services | −1,434 | −1,632 | −1,590 |
| **Balance on goods and services** | −969 | −1,707 | −763 |
| Other income received | 48 | 30 | 27 |
| Other income paid | −1,412 | −1,335 | −1,492 |
| **Balance on goods, services and income** | −2,333 | −3,012 | −2,227 |
| Current transfers received | 1,686 | 1,712 | 1,794 |
| Current transfers paid | −47 | −58 | −22 |
| **Current balance** | −695 | −1,359 | −455 |
| Capital account (net) | −63 | 20 | 25 |
| Direct investment from abroad | 1,330 | 1,275 | 1,555 |
| Portfolio investment liabilities | −148 | — | 8 |
| Other investment assets | −1,275 | −1,394 | −904 |
| Other investment liabilities | 868 | 1,240 | −343 |
| Net errors and omissions | −276 | −4 | 184 |
| **Overall balance** | −258 | −221 | 70 |

Source: IMF, *International Financial Statistics*.

# External Trade

**PRINCIPAL COMMODITIES**
(distribution by SITC, US $ million)

| Imports c.i.f. | 2000 | 2001 | 2002 |
|---|---:|---:|---:|
| **Food and live animals** | 261.4 | 372.6 | 482.5 |
| Cereals and cereal preparations | 109.1 | 139.7 | 165.9 |
| **Crude materials (inedible) except fuels** | 134.9 | 142.9 | 127.9 |
| **Mineral fuels, lubricants, etc.** | 282.3 | 302.5 | 291.2 |
| Petroleum, petroleum products, etc. | 115.5 | 140.1 | 132.8 |
| Liquefied petroleum gases, etc. | 166.1 | 161.5 | 157.5 |
| **Chemicals and related products** | 809.3 | 968.4 | 1,038.9 |
| Medicinal and pharmaceutical products | 198.7 | 258.8 | 289.1 |
| Medicaments (incl. veterinary) | 167.4 | 220.7 | 246.3 |
| Artificial resins, plastic materials etc. | 145.0 | 161.9 | 183.3 |
| Disinfectants, insecticides, fungicides, etc. | 104.9 | 106.6 | 110.8 |
| **Basic manufactures** | 684.6 | 922.6 | 1,152.3 |
| Paper, paperboard and articles thereof | 109.0 | 122.3 | 146.7 |
| Textile yarn, fabrics, etc. | 114.0 | 138.1 | 127.8 |
| Iron and steel | 191.5 | 274.5 | 454.3 |
| Tubes, pipes and fittings | 24.7 | 76.7 | 213.5 |
| **Machinery and transport equipment** | 914.9 | 1,964.6 | 2,614.6 |
| Power-generating machinery and equipment | 62.9 | 131.0 | 119.9 |
| Machinery specialized for particular industries | 133.0 | 292.3 | 269.2 |
| General industrial machinery equipment and parts | 182.9 | 292.3 | 444.0 |
| Telecommunications and sound equipment | 109.6 | 289.3 | 374.3 |
| Other electrical machinery apparatus, etc. | 146.5 | 226.1 | 281.7 |
| Road vehicles and parts* | 184.3 | 620.3 | 923.2 |
| Passenger motor cars (excl. buses) | 72.6 | 248.0 | 399.2 |
| Motor vehicles for goods transport and special purposes | 43.7 | 188.0 | 299.0 |
| **Miscellaneous manufactured articles** | 266.5 | 241.7 | 304.9 |
| **Total** (incl. others) | 3,445.9 | 5,362.9 | 6,431.1 |

* Data on parts exclude tyres, engines and electrical parts.
Source: UN, *International Trade Statistics Yearbook*.

**2003** (US $ million): Consumer goods 1,868.4; Primary materials 2,212.4; Capital goods 1,788.6; Total imports (incl. others) 6,534.4.

**2004** (US $ million, provisional figures): Consumer goods 2,190.2; Primary materials 2,831.2; Capital goods 2,053.6; Total imports (incl. others) 7,861.1.

Source: Banco Central.

| Exports f.o.b. | 2002 | 2003* | 2004* |
|---|---:|---:|---:|
| Fresh fish, crustaceans and molluscs | 340.7 | 364.2 | 400.4 |
| Prepared fish and crustacean products | 358.0 | 411.3 | 364.7 |
| Bananas and plantains | 969.3 | 1,099.3 | 1,023.0 |
| Coffee, cocoa and their derivatives | 170.7 | 221.8 | 187.8 |
| Cut flowers and foliage | 290.3 | 295.2 | 346.4 |
| Petroleum, petroleum products etc. | 2,055.0 | 2,606.5 | 4,233.9 |
| Crude petroleum oils, etc. | 1,839.0 | 2,372.3 | 3,898.5 |
| Refined petroleum products | 216.0 | 234.2 | 335.4 |
| **Basic manufactures** | 201.0 | 305.4 | 249.2 |
| Metal manufactures | 143.0 | 239.4 | 172.4 |
| Textile manufactures | 58.0 | 66.0 | 76.8 |
| **Total** (incl. others) | 5,036.1 | 6,038.5 | 7,646.6 |

* Provisional figures.

# ECUADOR

## PRINCIPAL TRADING PARTNERS
(US $ million)

| Imports c.i.f. | 2002 | 2003 | 2004 |
|---|---|---|---|
| Argentina | 169.1 | 174.6 | 247.5 |
| Belgium | 146.8 | 99.7 | 89.0 |
| Brazil | 405.9 | 366.4 | 478.8 |
| Chile | 300.9 | 355.1 | 412.8 |
| Colombia | 902.3 | 925.8 | 1,150.4 |
| Germany | 181.5 | 177.0 | 206.4 |
| Italy | 142.5 | 111.4 | 96.8 |
| Japan | 391.6 | 273.5 | 302.3 |
| Mexico | 191.0 | 182.9 | 235.9 |
| Netherlands | 51.4 | 73.1 | 128.7 |
| Peru | 154.8 | 177.0 | 245.0 |
| Spain | 137.7 | 138.9 | 123.2 |
| Taiwan | 66.0 | 73.6 | 66.9 |
| United Kingdom | 64.2 | 47.6 | 53.6 |
| USA | 1,480.9 | 1,401.1 | 1,623.4 |
| Venezuela | 353.2 | 377.1 | 531.2 |
| **Total** (incl. others) | 6,431.1 | 6,534.4 | 7,861.1 |

| Exports f.o.b. | 2002 | 2003 | 2004 |
|---|---|---|---|
| Belgium | 71.7 | 73.5 | 75.0 |
| Chile | 74.4 | 66.6 | 123.6 |
| Colombia | 362.5 | 362.2 | 296.7 |
| Germany | 172.2 | 206.9 | 193.5 |
| Italy | 289.6 | 377.4 | 354.8 |
| Japan | 97.9 | 86.1 | 77.1 |
| Mexico | 25.7 | 48.1 | 40.9 |
| Netherlands | 86.9 | 113.1 | 123.6 |
| Peru | 374.5 | 632.9 | 602.6 |
| Spain | 65.8 | 145.8 | 111.0 |
| USA | 2,086.8 | 2,451.6 | 3,279.6 |
| Venezuela | 64.7 | 54.6 | 118.9 |
| **Total** (incl. others) | 5,036.1 | 6,038.5 | 7,646.4 |

## Transport

### RAILWAYS
(traffic)

| | 1997 | 1998 | 1999 |
|---|---|---|---|
| Passenger-kilometres (million) | 47 | 44 | 5 |
| Net ton-kilometres (million) | — | 14 | — |

**Passenger-kilometres** (million): 5 in 2000; 32 in 2001; 33 in 2002.

Source: UN, *Statistical Yearbook*.

### ROAD TRAFFIC
(motor vehicles in use at 31 December)

| | 1999 | 2000 | 2001 |
|---|---|---|---|
| Passenger cars | 532,170 | 550,448 | 529,359 |
| Buses and coaches | 9,917 | 9,183 | 8,962 |
| Lorries and vans | 51,686 | 55,580 | 54,698 |
| Road tractors | 3,630 | 5,118 | 5,588 |
| Motorcycles and mopeds | 26,641 | 25,711 | 22,574 |

Source: IRF, *World Road Statistics*.

## SHIPPING
**Merchant Fleet**
(registered at 31 December)

| | 2002 | 2003 | 2004 |
|---|---|---|---|
| Number of vessels | 182 | 211 | 206 |
| Total displacement ('000 grt) | 313.1 | 324.5 | 264.6 |

Source: Lloyd's Register-Fairplay, *World Fleet Statistics*.

**International Sea-borne Freight Traffic**
('000 metric tons)

| | 1988* | 1989* | 1990 |
|---|---|---|---|
| Goods loaded | 8,402 | 10,020 | 11,783 |
| Goods unloaded | 2,518 | 2,573 | 1,958 |

* Source: UN, *Monthly Bulletin of Statistics*.

## CIVIL AVIATION
(traffic on scheduled services)

| | 1999 | 2000 | 2001 |
|---|---|---|---|
| Kilometres flown (million) | 16 | 12 | 8 |
| Passengers carried ('000) | 1,387 | 1,319 | 1,285 |
| Passenger-km (million) | 1,388 | 1,042 | 715 |
| Total ton-km (million) | 157 | 108 | 70 |

Source: UN, *Statistical Yearbook*.

## Tourism

### FOREIGN VISITOR ARRIVALS*

| Country of residence | 2001 | 2002 | 2003 |
|---|---|---|---|
| Argentina | 12,631 | 14,265 | 15,395 |
| Chile | 19,593 | 18,571 | 16,656 |
| Colombia | 182,316 | 197,080 | 205,353 |
| France | 12,161 | 12,671 | 13,490 |
| Germany | 17,733 | 17,541 | 18,598 |
| Peru | 84,794 | 106,777 | 153,520 |
| Spain | 15,400 | 16,943 | 20,111 |
| United Kingdom | 18,968 | 17,844 | 19,554 |
| USA | 148,100 | 150,582 | 159,851 |
| Venezuela | 14,089 | 12,460 | 14,084 |
| **Total** (incl. others) | 640,561 | 682,962 | 760,776 |

* Figures refer to total arrivals (including same-day visitors), except those of Ecuadorean nationals residing abroad.

**Tourism receipts** (US $ million, incl. passenger transport): 438 in 2001; 449 in 2002; 408 in 2003.

Source: World Tourism Organization.

## Communications Media

|  | 2001 | 2002 | 2003 |
|---|---|---|---|
| Television receivers ('000 in use) | 2,900 | n.a. | n.a. |
| Telephones ('000 main lines in use) | 1,335.8 | 1,426.2 | 1,549.0 |
| Mobile cellular telephones ('000 subscribers) | 859.2 | 1,560.9 | 2,394.4 |
| Personal computers ('000 in use) | 300 | 403 | n.a. |
| Internet users ('000) | 333.0 | 537.9 | 569.7 |

**Radio receivers** ('000 in use): 5,040 in 1999.

**Facsimile machines:** 30,000 in use in 1996.

**Daily newspapers:** 36 in 2000 (average circulation 1,220,000).

Sources: UNESCO, *Statistical Yearbook*; UN, *Statistical Yearbook*; International Telecommunication Union.

## Education

(1999/2000)

|  | Institutions | Teachers | Students |
|---|---|---|---|
| Pre-primary | 5,244 | 14,686 | 209,334 |
| Primary | 18,203 | 86,598 | 1,982,636 |
| Secondary | 3,486 | 80,662 | 980,213 |
| Higher* | n.a. | 12,856 | 206,541 |

* 1990/91 figures.

Sources: UNESCO, *Statistical Yearbook*; Ministerio de Educación y Cultura.

**Adult literacy rate** (UNESCO estimates): 91.0% (males 92.3%; females 89.7%) in 2002 (Source: UN Development Programme, *Human Development Report*).

# Directory

## The Constitution

The 1945 Constitution was suspended in June 1970. In January 1978 a referendum was held to choose between two draft Constitutions, prepared by various special constitutional committees. In a 90% poll, 43% voted for a proposed new Constitution and 32.1% voted for a revised version of the 1945 Constitution. The new Constitution came into force on 10 August 1979. In November 1997 a National Constituent Assembly was elected for the purpose of reviewing the Constitution, and a new Constitution, which retained many of the provisions of the 1979 Constitution, came into force on 10 August 1998. The main provisions of the Constitution are summarized below:

### CHAMBER OF REPRESENTATIVES

The Constitution of 1998 states that legislative power is exercised by the Chamber of Representatives, which sits for a period of 60 days from 10 August. The Chamber is required to set up four full-time Legislative Commissions to consider draft laws when the House is in recess. Special sessions of the Chamber of Representatives may be called.

Representatives are elected for four years from lists of candidates drawn up by legally recognized parties. Twelve are elected nationally; two from each Province with over 100,000 inhabitants, one from each Province with fewer than 100,000; and one for every 200,000 citizens or fractions of over 150,000. Representatives are eligible for re-election.

In addition to its law-making duties, the Chamber ratifies treaties, elects members of the Supreme and Superior Courts, and (from panels presented by the President) the Comptroller-General, the Attorney-General and the Superintendent of Banks. It is also able to overrule the President's amendment of a bill that it has submitted for Presidential approval. It may reconsider a rejected bill after a year or request a referendum, and may revoke the President's declaration of a state of emergency. The budget is considered in the first instance by the appropriate Legislative Commission and disagreements are resolved in the Chamber.

### PRESIDENT

The presidential term is four years (starting from 15 January of the year following his election), and there is no re-election. The President appoints the Cabinet, the Governors of Provinces, diplomatic representatives and certain administrative employees, and is responsible for the direction of international relations. In the event of foreign invasion or internal disturbance, the President may declare a state of emergency and must notify the Chamber, or the Tribunal for Constitutional Guarantees if the Chamber is not in session.

As in other post-war Latin-American Constitutions, particular emphasis is laid on the functions and duties of the State, which is given wide responsibilities with regard to the protection of labour; assisting in the expansion of production; protecting the Indian and peasant communities; and organizing the distribution and development of uncultivated lands, by expropriation where necessary.

Voting is compulsory for every Ecuadorean citizen who is literate and over 18 years of age. An optional vote has been extended to illiterates (under 15% of the population by 1981). The Constitution guarantees liberty of conscience in all its manifestations, and states that the law shall not make any discrimination for religious reasons.

## The Government

### HEAD OF STATE

**President:** ALFREDO PALACIO GONZÁLES (assumed office 20 April 2005).
**Vice-President:** ALEJANDRO SERRANO.

### CABINET
(August 2005)

**Minister of National Defence:** Gen. (retd) ANÍBAL SOLÓN ESPINOSA.
**Minister of Government and Police:** OSCAR AYERVE.
**Minister of Foreign Affairs:** Dr ANTONIO PARRA GIL.
**Minister of the Economy and Finance:** MAGDALENA BARREIRO.
**Minister of Foreign Trade, Industrialization, Fishing and Competition:** OSWALDO MOLESTINA ZABALA.
**Minister of Labour and Employment:** Dr GALO CHIRIBOGA ZAMBRANO.
**Minister of Energy and Mines:** Ing. IVÁN RODRÍGUEZ RAMOS.
**Minister of Urban Development and Housing:** Ing. ARMANDO BRAVO NÚÑEZ.
**Minister of Education and Culture:** Dra CONSUELO YÉPEZ COSSÍO.
**Minister of Public Health:** Dr WELLINGTON SANDOVAL.
**Minister of Agriculture:** Ing. PABLO RIZZO PASTOR.
**Minister of the Environment:** Dr ANITA ALBÁN MORA.
**Minister of Tourism:** MARÍA ISABEL SALVADOR.
**Minister of Public Works and Communications:** Ing. DERLIZ PALACIOS.
**Minister of Social Welfare:** Dr ALBERTO RIGAIL.

The following are, *ex officio*, members of the Cabinet: the National Secretary of Administrative Development, the Co-ordinator of the Social Expenditure Fund (FISE), the State Comptroller-General, the State Procurator-General, the Chairman of the National Monetary Board, the General Manager of the State Bank, the General Manager of the Central Bank, the Secretary-General of the National Planning Council (CONADE), the President of the National Financial Corporation, the President of the National Modernization Council (CONAM), the Presidential Private Secretary, the Subsecretary-General of Public Administration and the Presidential Press Secretary.

### MINISTRIES

**Office of the President:** Palacio Nacional, García Moreno 1043, Quito; tel. (2) 221-6300.

# ECUADOR

**Office of the Vice-President:** Manuel Larrea y Arenas, Edif. Consejo Provincial de Pichincha, 21°, Quito; tel. (2) 250-4953; fax (2) 250-3379.

**Ministry of Agriculture:** Avda Eloy Alfaro y Amazonas, Quito; tel. (2) 255-3472; fax (2) 256-4531; internet www.mag.gov.ec.

**Ministry of Education and Cultures:** Mejía 322, Quito; tel. (2) 221-6224; fax (2) 258-0116; internet www.mec.edu.ec.

**Ministry of Energy and Mines:** Juan León Mera y Orellana, 5°, Quito; tel. (2) 255-0041; fax (2) 255-0018; e-mail menergia2@andinanet.net; internet www.menergia.gov.ec.

**Ministry of the Environment:** Avda Eloy Alfaro y Amazonas, Edif. M.A.G., Quito; tel. (2) 256-3429; fax (2) 250-0041; e-mail mma@ambiente.gov.ec.

**Ministry of the Economy and Finance:** Avda 10 de Agosto 1661 y Jorge Washington, Quito; tel. (2) 254-4500; fax (2) 253-0703; internet minfinanzas.ec-gov.net.

**Ministry of Foreign Affairs:** Avda 10 de Agosto y Carrión, Quito; tel. (2) 223-0100; fax (2) 256-4873; e-mail webmast@mmrree.gov.ec; internet www.mmrree.gov.ec.

**Ministry of Foreign Trade, Industrialization, Fishing and Competition:** Avda Eloy Alfaro y Amazonas, Quito; tel. (2) 252-7988; fax (2) 250-3549.

**Ministry of Government and Police:** Espejo y Benalcázar, Quito; tel. (2) 295-5666; fax (2) 295-8360; e-mail informacion@mingobierno.gov.ec; internet www.mingobierno.gov.ec.

**Ministry of Labour and Employment:** Clemente Ponce 255 y Piedrahita, Quito; tel. (2) 256-6148; fax (2) 250-3122; e-mail mintrab@accessinter.net; internet www.mintrab.gov.ec.

**Ministry of National Defence:** Exposición 208, Quito; tel. (2) 221-6150; fax (2) 256-9386; e-mail paginaweb@fuerzasarmadasecuador.org; internet www.fuerzasarmadasecuador.ec-gov.net.

**Ministry of Public Health:** Juan Larrea 444, Quito; tel. (2) 252-9163; fax 256-9786; e-mail msp@accessinter.net; internet www.msp.gov.ec.

**Ministry of Public Works and Communications:** Avda Juan León Mera y Orellana, Quito; tel. (2) 222-2749; fax (2) 222-3077; internet www.mop.gov.ec.

**Ministry of Social Welfare:** Quito.

**Ministry of Tourism:** Avda Eloy Alfaro 32 y Carlos Tobar, Quito; tel. (2) 250-7559; fax (2) 222-9330; e-mail promocion@turismo.gov.ec; internet www.vivecuador.com.

**Ministry of Urban Development and Housing:** Avda 10 de Agosto 2270 y Corotero, Quito; tel. (2) 223-8060; fax (2) 256-6785; e-mail mdesur2@ec-gov.net; internet www.miduvi.ec-gov.net.

**Office for Public Administration:** Palacio Nacional, García Moreno 1043, Quito; tel. (2) 251-5990.

## President and Legislature

### PRESIDENT

Elections, 20 October and 24 November 2002

| Candidate | % of votes cast in first ballot | % of votes cast in second ballot |
|---|---|---|
| Col Lucio Edwin Gutiérrez Borbua (PSP-MNPP) | 20.3 | 58.7 |
| Alvaro Fernando Noboa Pontón (PRIAN) | 17.4 | 41.3 |
| León Roldos Aguilera (PS—FA-DP-CFP) | 15.5 | — |
| Rodrigo Borja Cevallos (ID) | 14.0 | — |
| Antonio Xavier Neira Menendez (PSC) | 12.2 | — |
| **Total** (incl. others) | 100.0 | 100.0 |

### CONGRESO

#### Cámara Nacional de Representantes

**President:** WILFREDO LUCERO BOLAÑOS.

Election, 20 October 2002

| Political parties | Seats |
|---|---|
| Partido Social Cristiano (PSC) | 24 |
| Izquierda Democrática (ID) | 15 |
| Partido Roldosista Ecuatoriano (PRE) | 14 |
| Partido Renovador Institucional de Acción Nacional (PRIAN) | 10 |
| Movimiento Nuevo País-Pachakútik (MNPP) | 8 |
| Partido Sociedad Patriótica 21 de Enero (PSP) | 8 |
| Movimiento Popular Democrático (MPD) | 7 |
| Democracia Popular (DP) | 5 |
| Partido Socialista-Frente Amplio (PS—FA) | 3 |
| Independent | 1 |
| **Total** (incl. others) | 100 |

## Political Organizations

**Acción Popular Revolucionaria Ecuatoriana (APRE):** centrist; Leader Lt-Gen. FRANK VARGAS PAZZOS.

**Coalición Nacional Republicana (CNR):** Quito; f. 1986; fmrly Coalición Institucionalista Demócrata (CID).

**Concentración de Fuerzas Populares (CFP):** Quito; f. 1946; Leader GALO VAYAS; Dir Dr AVERROES BUCARAM SAXIDA.

**Democracia Popular (DP):** Calle Luis Saá 153 y Hermanos Pazmiño, Casilla 17-01-2300, Quito; tel. (2) 254-7654; fax (2) 250-2995; f. 1978 as Democracia Popular-Unión Demócrata Cristiana; Christian democrat; Leader Lic. ABSALÓN ROCHA.

**Frente Futuro de Ecuador (FFE):** f. 2002 to contest the October 2002 presidential and legislative elections; Leader ANTONIO VARGAS.

**Frente Radical Alfarista (FRA):** Quito; f. 1972; liberal; Leader IVÁN CASTRO PATIÑO.

**Izquierda Democrática (ID):** Polonia 161, entre Vancouver y Eloy Alfaro, Quito; tel. (2) 256-4436; fax (2) 256-9295; f. 1977; absorbed Fuerzas Armadas Populares Eloy Alfaro—Alfaro Vive ¡Carajo! (AVC) (Eloy Alfaro Popular Armed Forces—Alfaro Lives, Damn It!) in 1991; Leader GUILLERMO LANDAZURI; National Dir ANDRÉS VALLEJO.

**Movimiento Independiente para una República Auténtica (MIRA):** Quito; f. 1996; Leader Dra ROSALIA ARTEAGA SERRANO.

**Movimiento Nuevo País-Pachakútik (MNPP/Pachakútik):** Quito; internet www.pachakutik.org.ec; political wing of CONAIE (q.v.); represents indigenous, environmental and social groups; Leader FREDDY EHLERS ZURITA; Nat. Dir GILBERTO TALAHUA.

**Movimiento Popular Democrático (MPD):** e-mail periodicopcion@andinanet.net; internet bloquempd.tripod.com; Stalinist; Dep. Leader STALIN VARGAS MEZA.

**Partido Comunista Marxista-Leninista de Ecuador:** e-mail pcmle@bigfoot.com; internet www.geocities.com/pcmle; Sec.-Gen. CAMILO ALMEYDA.

**Partido Conservador (PC):** Wilsón 578, Quito; tel. (2) 250-5061; f. 1855; incorporated Partido Unidad Republicano in 1995; centre-right; Leader SIXTO DURÁN BALLÉN.

**Partido Demócrata (PD):** Quito; Leader Dr FRANCISCO HUERTA MONTALVO.

**Partido Liberal Radical (PLR):** Quito; f. 1895; held office from 1895 to 1944 as the Liberal Party, which subsequently divided into various factions; perpetuates the traditions of the Liberal Party; Leader CARLOS JULIO PLAZA A.

**Partido Renovador Institucional de Acción Nacional (PRIAN):** Quito; right-wing, populist; Leader ALVARO FERNANDO NOBOA PONTÓN.

**Partido Republicano (PR):** Quito; Leader GUILLERMO SOTOMAYOR.

**Partido Roldosista Ecuatoriano (PRE):** Quito; f. 1982; populist; Dir ABDALÁ BUCARAM ORTIZ.

**Partido Social Cristiano (PSC):** Carrión 548 y Reina Victoria, Casilla 9454, Quito; tel. (2) 254-4536; fax (2) 256-8562; internet www.psc.org.ec; f. 1951; centre-right party; Pres. JAIME NEBOT SAADI; Leaders Dr LEÓN FEBRES CORDERO RIVADENEIRA, Lic. CAMILO PONCE GANGOTENA, HEINZ MOELLER FREILE, Lic. PASCUAL DEL CIOPPO ARAGUNDI.

**Partido Socialista-Frente Amplio (PS-FA):** Avda Gran Colombia y Yaguachi, Quito; tel. (2) 222-1764; fax (2) 222-2184; f. 1926; Pres. Dr MANUEL SALGADO TAMAYO.

**Partido Sociedad Patriótica 21 de Enero (PSP):** Quito; internet www.sociedadpatriotica.com; contested the 2002 elections in alliance with the MNPP; Leader Col LUCIO EDWIN GUTIÉRREZ BORBUA.

# ECUADOR

**Unión Alfarista-FRA:** Quito; f. 1998; centrist; Leader CÉSAR VERDUGA VÉLEZ.

### OTHER ORGANIZATIONS

**Confederación de las Nacionalidades Indígenas de la Amazonia Ecuatoriana (CONFENIAE):** Avda 6 de Diciembre 159 y Pazmino, Of. 408, Apdo 17-01-4180, Quito; tel. (2) 543-973; fax (2) 220-325; e-mail confeniae@applicom.com; internet www.unii.net/confeniae; to represent indigenous peoples; mem. of CONAIE; Leader LUIS VARGAS.

**Confederación de Nacionalidades Indígenas de Ecuador (CONAIE)** (National Confederation of the Indigenous Population of Ecuador): Avda Los Granados 2553 y 6 de Diciembre, Quito; tel. (2) 245-2335; e-mail reincon@uio.telconet.net; internet conaie.org; f. 1986 to represent indigenous peoples; MNPP (q.v.) represents CONAIE in the legislature; Pres. LUIS MACAS; Vice-Pres. SANTIAGO DE LA CRUZ.

**Coordinadora de las Organizaciones Indígenas de la Cuenca Amazónica (COICA):** Quito; e-mail comunica@coica.org; internet www.coica.org; f. 1984; founded in Lima, Peru; moved to Quito in 1993; umbrella group of 9 orgs representing indigenous peoples of the Amazon Basin in Bolivia, Brazil, Colombia, Ecuador, French Guiana, Guyana, Suriname and Venezuela; Gen. Co-ordinator SEBASTIÃO HAJI ALVES RODRIGUES MANCHINERIS; Vice-Gen. Co-ordinator JOCELYN ROGER THERESE.

**Federación de Pueblos Indígenas, Campesinas y Negros del Ecuador (Fedepicne):** Quito; f. 1996 following split from CONAIE; formally recognized in 2003; Pres. LUIS PACHALA.

**Frente Popular (FP):** umbrella group of labour unions and other populist organizations; Pres. LUIS VILLACÍS.

### ARMED GROUPS

**Grupos de Combatientes Populares (GCP):** Cuenca; communist guerrilla grouping.

**Izquierda Revolucionaria Armada (IRA):** extreme left-wing revolutionary group opposed to international capitalism.

**Montoneros Patria Libre (MPL):** f. 1986; advocates an end to authoritarianism.

**Milicias Revolucionarias del Pueblo (MRP):** extreme left-wing grouping opposed to international capitalism.

**Partido Maoísta-Comunista 'Puka Inti':** Sec.-Gen. RAMIRO CELI.

## Diplomatic Representation

### EMBASSIES IN ECUADOR

**Argentina:** Avda Amazonas 22-147 y Roca, 8°, Quito; tel. (2) 256-2292; fax (2) 256-8177; e-mail embarge2@andinanet.net; Chargé d'affaires a.i. GONZALO TORRES CARIONI.

**Belgium:** Edif. Mansión Blanca, 10°, Avda República de El Salvador 1082 y Naciones Unidas; tel. (2) 227-6145; fax (2) 227-3910; e-mail quito@diplobel.org; Ambassador BEATRIX VAN HEMELDONCK.

**Bolivia:** Avda Eloy Alfaro 2432 y Fernando Ayarza, Casilla 17-210003, Quito; tel. (2) 244-4830; fax (2) 224-4833; e-mail embolivia-quito@andinanet.net; Ambassador Dr EDIL SANDOVAL MORÓN.

**Brazil:** Avda Amazonas 1429 y Colón, 9° y 10°, Apdo 17-01-231, Quito; tel. (2) 256-3142; fax (2) 250-4468; e-mail ebrasil@uio.satnet.net; Ambassador SERGIO AUGUSTO DE ABREU E LIMA FLORENCIO SOBRINHO.

**Canada:** Edif. Josueth Gonzales, Avda 6 de Diciembre 28-16 y Paul Rivet, 4°, Apdo 17-11-6512, Quito; tel. (2) 250-6162; fax (2) 250-3108; e-mail quito@dfait-maeci.gc.ca; internet www.dfait-maeci.gc.ca/ecuador; Ambassador BRIAN OAK.

**Chile:** Edif. Xerox, 4°, Juan Pablo Sanz 3617 y Amazonas, Quito; tel. (2) 224-9403; fax (2) 244-4470; e-mail embachileecu@trans-telco.net; Ambassador NELSON HADAD HERESY.

**China, People's Republic:** Avda Atahualpa 349 y Amazonas, Quito; tel. (2) 245-8927; fax (2) 244-4364; e-mail embchina@uio.telconet.net; Ambassador LIU YUQIN.

**Colombia:** Edif. Arista, Avda Colón 1133 y Amazonas, 7°, Apdo 17-07-9164, Quito; tel. (2) 222-8926; e-mail equito@minrelext.gov.co; Ambassador MARÍA PAULINA ESPINOZA DE LÓPEZ.

**Costa Rica:** Rumipamba 692 y República, 1°, Apdo 17-03-301, Quito; tel. (2) 225-4945; fax (2) 225-4087; e-mail embajcr@uiso.satnet.net; Ambassador LUZ ARGENTINA CALDERÓN GEI.

**Cuba:** Mercurio 365, entre La Razón y El Vengador, Quito; tel. (2) 245-6936; fax (2) 243-0594; e-mail embajada@ecuecuador.minrex.gov.cu; Ambassador ILEANA DÍAZ-ARGÜELLES ALASÁ.

**Dominican Republic:** Edif. Albatros, Avda de los Shyris 1240 y Portugal, 2°, Apdo 17-01-387-A, Quito; tel. (2) 243-4232; e-mail emrepdom@interactive.net.ec; Ambassador NESTOR JUAN CERÓN.

**Egypt:** Avda Tarqui 4-56 y Avda 6 de Diciembre, Apdo 9355, Quito; tel. (2) 222-5240; fax (2) 256-3521; Ambassador MOHAMED HANZA ELEISH.

**El Salvador:** Edif. Gabriela III, 3°, Avda República de El Salvador 733 y Portugal, Quito; tel. (2) 243-3070; fax (2) 224-2829; e-mail embelsal@uio.salnet.net; Ambassador RAFAEL ANGEL ALFARO PINEDA.

**France:** Calle Leonidas Plaza 107 y Avda Patria, CP 536, Quito; tel. (2) 256-0789; fax (2) 256-6424; e-mail francie@andinanet.net; internet www.ambafrance-ecu.org; Ambassador FRANÇOIS COUSIN.

**Germany:** Edif. Citiplaza, 13° y 14°, Avda Naciones Unidas y República de El Salvador, Casilla 17-17-536, Quito; tel. (2) 297-0820; fax (2) 297-0815; e-mail alemania@interactive.net.ec; internet www.embajada-quito.de; Ambassador SEPP JUERGEN WOELKER.

**Guatemala:** Edif. Gabriela III, 3°, Of. 301, Avda República de El Salvador 733 y Portugal, Apdo 17-03-294, Quito; tel. (2) 245-9700; e-mail embecuador@minex.gob.gt; Ambassador ANGELA GAROZ CABRERA.

**Holy See:** Avda Orellana 692, Apdo 17-07-8980, Quito; tel. (2) 250-5200; fax (2) 256-4810; e-mail nunapec@impsat.net.ec; Apostolic Nunciature; Apostolic Nuncio Most Rev. ALAIN PAUL LEBEAUPIN (Titular Archbishop of Vico Equense).

**Honduras:** Edif. World Trade Centre, Torre A, 5°, Of. 501, Avda 12 de Octubre 1942 y Luis Cordero, Apdo 17-03-4753, Quito; tel. (2) 222-3985; fax (2) 222-0441; e-mail embhquito@yahoo.com; Ambassador ANTONIO BERMUDEZ AGUILAR.

**Israel:** Edif. Plaza 2000, Avda 12 de Octubre y General Francisco Salazar, Apdo 17-21-08, Quito; tel. (2) 223-7474; fax (2) 223-8055; e-mail info@quito.mfa.gov.il; internet www.quito.mfa.il; Ambassador DANIEL SABAN.

**Italy:** Calle La Isla 111 y Humberto Alborñoz, Casilla 17-03-72, Quito; tel. (2) 256-1077; fax (2) 250-2818; e-mail ambital@ambitalquito.org; internet www.ambitalquito.org; Ambassador GIULIO CESARE PICCIRILLI.

**Japan:** Juan León Mera 130 y Avda Patria, 7°, Quito; tel. (2) 256-1899; fax (2) 250-3670; e-mail japembec@uio.satnet.net; Ambassador HIROYUKI HIRAMATSU.

**Korea, Republic:** Edif. Citiplaza, 8°, Avda Naciones Unidas y Avda de El Salvador, Quito; tel. (2) 297-0625; fax (2) 297-0630; e-mail ecemco@interactive.net.ec; Ambassador SANG-WOOK NAM.

**Mexico:** Avda 6 de Diciembre 4843 y Naciones Unidas, Casilla 17-11-6371, Quito; tel. (2) 292-3770; fax (2) 244-8245; e-mail embajadamexico@embamex.org.ec; internet www.sre.gob.mx/ecuador; Ambassador ALMA PATRICIA SORIA AYUSO.

**Netherlands:** Edif. World Trade Centre, Torre 1, 1°, Avda 12 de Octubre 1942 y Luis Cordero, Quito; tel. (2) 222-9229; fax (2) 256-7917; e-mail nlgovqui@embajadadeholanda.com; internet www.embajadadeholanda.com; Ambassador FRANCISCUS GIJSBERTUS BIJVOET.

**Panama:** Edif. ESPRO, 6°, Alpallana 505 y Whimper, Quito; tel. (2) 250-8856; fax (2) 256-5234; e-mail panaembaecuador@hotmail.com; Ambassador MATEO CASTILLERO CASTILLO.

**Paraguay:** Edif. World Trade Center, Torre A, 9°, Of. 906, Avda 12 de Octubre 1942 y Luis Cordero, Apdo 17-03-139, Quito; tel. (2) 224-5871; fax (2) 225-1446; e-mail embapar@uio.telconet.net; Ambassador (vacant).

**Peru:** Avda República de El Salvador 495 e Irlanda, Apdo 17-07-9380, Quito; tel. (2) 246-8410; fax (2) 225-2560; e-mail embpeecu@uio.satnet.net; Ambassador LUIS MARCHAND STENS.

**Russia:** Reina Victoria 462 y Ramón Roca, Quito; tel. (2) 256-6361; fax (2) 256-5531; e-mail embrusia@accessinter.net; Ambassador (vacant).

**Spain:** General Francisco Salazar E12-73 y Toledo (Sector La Floresta), Apdo 17-01-9322, Quito; tel. (2) 322-6296; fax (2) 322-7805; e-mail embespec@uio.satnet.net; internet www.embajadaespana.com.ec; Ambassador JUAN MARÍA ALZINA DE AGUILAR.

**Switzerland:** Edif. Xerox, 2°, Amazonas 3617 y Juan Pablo Sanz, Casilla 17-11-4815, Quito; tel. (2) 243-4949; fax (2) 244-9314; e-mail vertretung@qui.rep.admin.ch; internet www.eda.admin.ch/quito_emb/s/home.html; Ambassador ROBERT REICH.

**United Kingdom:** Edif. Citiplaza, 14°, Avda Naciones Unidas y República de El Salvador, Casilla 17-01-314, Quito; tel. (2) 297-0800; fax (2) 297-0809; e-mail consuio@uio.satnet.net; internet www.britembquito.org.ec; Ambassador RICHARD GEORGE LEWINGTON.

# ECUADOR

**USA:** Avda 12 de Octubre 1942 y Patria 120, Quito; tel. (2) 256-2890; fax (2) 250-2052; internet www.usembassy.org.ec; Ambassador LINDA JEWELL.

**Uruguay:** Edif. Josueth González, 9°, Avda 6 de Diciembre 2816 y Paul Rivet, Casilla 17-12-282, Quito; tel. (2) 256-3762; fax (2) 256-3763; e-mail emburugl@emburuguay.int.ec; Ambassador JOSÉ LUIS ALDABALDE ARBURUAS.

**Venezuela:** Avda Los Cabildos 115, Apdo 17-01-688, Quito; tel. (2) 226-8636; fax (2) 246-6786; e-mail embavenecua@interactive.net.ec; Ambassador OSCAR NAVAS TORTOLERO.

## Judicial System

**Attorney-General:** CECILIA ARMAS (acting).

**Supreme Court of Justice:** Palacio de Justicia, Avda 6 de Diciembre y Piedrahita 332, Quito; tel. (2) 290-0424; fax (2) 290-0425; e-mail dni-cnj@access.net.ec; internet www.justiciaecuador.gov.ec; f. 1830; Pres. (vacant).

**Higher or Divisional Courts:** Ambato, Azogues, Babahoyo, Cuenca, Esmeraldas, Guaranda, Guayaquil, Ibarra, Latacunga, Loja, Machala, Portoviejo, Quito, Riobamba and Tulcán; 90 judges.

**Provincial Courts:** There are 40 Provincial Courts in 15 districts; other courts include 94 Criminal, 219 Civil, 29 dealing with labour disputes and 17 Rent Tribunals.

**Special Courts:** National Court for Juveniles.

## Religion

There is no state religion but nearly 90% of the population are Roman Catholics. There are representatives of various Protestant Churches and of the Jewish faith in Quito and Guayaquil.

### CHRISTIANITY

#### The Roman Catholic Church

Ecuador comprises four archdioceses, 11 dioceses, seven Apostolic Vicariates and one Apostolic Prefecture. At 31 December 2003 there were an estimated 11,755,467 adherents in the country, equivalent to some 89.6% of the population.

**Bishops' Conference**
Conferencia Episcopal Ecuatoriana, Avda América 1805 y La Gasca, Apdo 17-01-1081, Quito; tel. (2) 255-8913; fax (2) 250-1429; e-mail dicuenca@etapa.com.ec.
f. 1939; statutes approved 1999; Pres. VICENTE RODRIGO CISNEROS DURÁN NAVAS (Archbishop of Cuenca).

**Archbishop of Cuenca:** VICENTE RODRIGO CISNEROS DURÁN NAVAS, Arzobispado, Manuel Vega 8-66 y Calle Bolívar, Apdo 01-01-0046, Cuenca; tel. (7) 847-234; fax (7) 844-436; e-mail dicuenca@etapaonline.ne.ec.

**Archbishop of Guayaquil:** ANTONIO ARREGUI YARZA HOLGUÍN, Arzobispado, Calle Clemente Ballén 501 y Chimborazo, Apdo 09-01-0254, Guayaquil; e-mail marregui@q.ecua.net.ec; tel. (4) 232-2778; fax (4) 232-9695.

**Archbishop of Portoviejo:** JOSÉ MARIO RUIZ NAVAS, Arzobispado, Avda Universitaria s/n, Entre Alajuela y Ramos y Duarte, Casilla 13-01-0024, Portoviejo; tel. (5) 630-404; fax (5) 634-428; e-mail arzobis@ecua.net.ec.

**Archbishop of Quito:** RAÚL VELA CHIRIBOGA, Arzobispado, Calle Chile 1140 y Venezuela, Apdo 17-01-00106, Quito; e-mail raul.vela@andinanet.net; tel. (2) 228-4429; fax (2) 258-0973.

#### The Anglican Communion

Anglicans in Ecuador are under the jurisdiction of Province IX of the Episcopal Church in the USA. The country is divided into two dioceses, one of which, Central Ecuador, is a missionary diocese.

**Bishop of Littoral Ecuador:** Rt Rev. ALFREDO MORANTE, Calle Bogotá 1010, Barrio Centenario, Apdo 5250, Guayaquil.

**Bishop of Central Ecuador:** Rt Rev. JOSÉ NEPTALÍ LARREA MORENO, Apdo 17-11-6165, Quito; e-mail ecuacen@uio.satnet.net.

#### The Baptist Church

**Baptist Convention of Ecuador:** Casilla 3236, Guayaquil; tel. (4) 238-4865; Pres. Rev. HAROLT SANTE MATA; Sec. JORGE MORENO CHAVARRÍA.

#### The Methodist Church

**United Evangelical Methodist Church:** Evangelical United Church, Rumipamba 915, Casilla 17-03-236, Quito; tel. (2) 226-5158; fax (2) 243-9576; 800 mems, 2,000 adherents.

### BAHÁ'Í FAITH

**National Spiritual Assembly of the Bahá'ís:** Apdo 869A, Quito; tel. (2) 256-3484; fax (2) 252-3192; e-mail ecua9nsa@uio.satnet.net; mems resident in 1,121 localities.

## The Press

### PRINCIPAL DAILIES

#### Quito

**El Comercio:** Avda Pedro Vicente Maldonado 11515 y el Tablón, Casilla 17-01-57, Quito; tel. (2) 267-0999; fax (2) 267-0466; e-mail contactenos@elcomercio.com; internet www.elcomercio.com; f. 1906; morning; independent; Proprs Compañía Anónima El Comercio; Pres. GUADALUPE MANTILLA DE ACQUAVIVA; Editor MARCO ARAUZ; circ. 160,000.

**La Hora:** Panamericana Norte Kilómetro 3½, Quito; tel. (2) 247-3724; fax (2) 247-5086; e-mail lahora@uio.satnet.net; internet www.lahora.com.ec; f. 1982; Nat. Pres. Dr FRANCISCO VIVANCO RIOFRÍO; Exec. Pres. FRANCISCO VIVANCO ARROYO; Gen. Editor JUANA LÓPEZ SARMIENTO.

**Hoy:** Avda Mariscal Sucre N-71345, Casilla 17-07-0969, Quito; tel. (2) 249-0888; fax (2) 249-1881; e-mail hoy@hoy.com.ec; internet www.hoy.com.ec; f. 1982; morning; independent; Man. JAIME MANTILLA ANDERSON; circ. 72,000.

**Ultimas Noticias:** Avda Pedro Vicente Maldonado 11515 y el Tablón, Casilla 17-01-57, Quito; tel. (2) 267-0999; fax (2) 267-4923; f. 1938; evening; independent; commercial; Proprs Compañía Anónima El Comercio; Dir DAVID MANTILLA CASHMORE; circ. 60,000.

#### Guayaquil

**Expreso:** Avda Carlos Julio Arosemena, Casilla 5890, Guayaquil; tel. (4) 220-1100; fax (4) 220-0291; e-mail webmaster@granasa.com.ec; internet www.diario-expreso.com; f. 1973; morning; independent; Exec. Editor MARTÍN ULLOA; circ. 60,000.

**El Extra:** Avda Carlos Julio Arosemena, Casilla 5890, Guayaquil; tel. (4) 220-1100; fax (4) 220-0291; e-mail webmaster@granasa.com.ec; internet www.diario-expreso.com; f. 1975; morning; Pres. ERROL CARTWRIGHT BETANCOURT; Editor HENRY HOLGUÍN; circ. 200,000.

**La Razón:** Avda Constitución y las Americas, Guayaquil; tel. (4) 228-0100; fax (4) 228-5110; e-mail cartas@larazonecuador.com; internet www.larazonecuador.com; f. 1965; morning; independent; Propr ROBERTO ISAÍAS DASSUM; Dir JORGE E. PÉREZ PESANTES; circ. 35,000.

**La Segunda:** Calle Colón 526 y Boyacá, Casilla 6366, Guayaquil; tel. (4) 232-0635; fax (4) 232-0539; f. 1983; morning; Propr CARLOS MANSUR; Dir VICENTE ADUM ANTÓN; circ. 60,000.

**El Telégrafo:** Avda 10 de Agosto 601 y Boyacá, Casilla 415, Guayaquil; tel. (4) 232-6500; fax (4) 232-3265; e-mail cartas@telegrafo.com.ec; internet www.telegrafo.com.ec; f. 1884; morning; independent; commercial; Proprs El Telégrafo CA; Dir CARLOS NAVARRETE CASTILLO; circ. 45,000 (weekdays), 55,000 (Sundays).

**El Universo:** Avda Domingo Comín y Alban, Casilla 09-01-531, Guayaquil; tel. (4) 249-0000; fax (4) 249-1034; e-mail editores@telconet.net; internet www.eluniverso.com; f. 1921; morning; independent; Pres. NICOLÁS PÉREZ LAPENTTI; Dir CARLOS PÉREZ BARRIGA; circ. 174,000 (weekdays), 290,000 (Sundays).

#### Cuenca

**El Tiempo:** Avda Loja y Rodrigo de Triana, Cuenca; tel. (07) 882-551; fax (07) 882-555; e-mail redaccion@eltiempo.com.ec; f. 1965; morning; independent; Dir Dr RENÉ TORAL CALLE; Editor RICARDO TELLO CARRIÓN; circ. 35,000.

**El Mercurio:** Avda las Américas (sector El Arenal); tel. (07) 880-110; fax (07) 817-266; e-mail redaccion@elmercurio.com.ec; Dir NICANOR MERCHÁN LUCO.

### PERIODICALS

#### Quito

**La Calle:** Casilla 2010, Quito; f. 1956; weekly; politics; Dir CARLOS ENRIQUE CARRIÓN; circ. 20,000.

# ECUADOR

**Cámara de Comercio de Quito:** Avda Amazona y República, Casilla 202, Quito; tel. (2) 244-3787; fax (2) 243-5862; f. 1906; monthly; commerce; Pres. ANDRÉS PÉREZ ESPINOSA; Exec. Dir ARMANDO TOMASELLI; circ. 10,000.

**Carta Económica del Ecuador:** Toledo 1448 y Coruña, Apdo 3358, Quito; f. 1969; weekly; economic, financial and business information; Pres. Dr LINCOLN LARREA B.; circ. 8,000.

**Chasqui:** Avda Diego de Almagro 32-133 y Andrade Marín, Apdo 17-01-584, Quito; tel. (2) 254-8011; fax (2) 250-2487; e-mail chasqui@ciespal.net; internet ; f. 1997; monthly; media studies; publ. of the Centro Internacional de Estudios Superiores de Comunicación para America Latina; Dir EDGAR JARAMILLO; Editor LUIS ELADIO PROAÑO.

**El Colegial:** Calle Carlos Ibarra 206, Quito; tel. (2) 221-6541; f. 1974; weekly; publ. of Student Press Association; Dir WILSON ALMEIDA MUÑOZ; circ. 20,000.

**Cosas:** Edif. Alpallana 2, 3°, Alpallana 289 y Diego Almagro, Quito; tel. (2) 250-8742; fax (2) 250-8747; e-mail redaccion@cosas.com.ec; internet www.cosas.com.ec; women's interest; Editor CRISTA RODAS; Man. CARIDAD VELA.

**Ecuador Guía Turística:** Mejía 438, Of. 43, Quito; f. 1969; fortnightly; tourist information in Spanish and English; Propr Prensa Informativa Turística; Dir JORGE VACA O.; circ. 30,000.

**Gestión:** Avda González Suárez 335 y San Ignacio, 7°, Quito; tel. (2) 250-5524; fax (2) 250-5527; e-mail info@dinediciones.com; internet www.gestion.dinediciones.com; f. 1999; economy and society; monthly; circ 15,000.

**Integración:** Solano 836, Quito; quarterly; economics of the Andean countries.

**Letras del Ecuador:** Casa de la Cultura Ecuatoriana, Avda 6 de Diciembre, Casilla 67, Quito; f. 1944; monthly; literature and art; non-political; Dir Dr TEODORO VANEGAS ANDRADE.

**El Libertador:** Olmedo 931 y García Moreno, Quito; f. 1926; monthly; Pres. Dr BENJAMÍN TERÁN VAREA.

**Mensajero:** Benalcázar 478, Apdo 17-01-4100, Quito; tel. (2) 221-9555; f. 1884; monthly; religion, culture, economics and politics; Man. OSWALDO CARRERA LANDÁZURI; circ. 5,000.

**Quince Días:** Sociedad Periodística Ecuatoriana, Los Pinos 315, Panamericana Norte Km 5½, Quito; tel. (2) 247-4122; fax (2) 256-6741; fortnightly; news and regional political analysis.

### Guayaquil

**Análisis Semanal:** Elizalde 119, 10°, Apdo 4925, Guayaquil; tel. (4) 232-6590; fax (4) 232-6842; e-mail wspurrier@ecuadoranalysis.com; internet www.ecuadoranalysis.com; weekly; economic and political affairs; Editor WALTER SPURRIER BAQUERIZO.

**Ecuador Ilustrado:** Guayaquil; f. 1924; monthly; literary; illustrated.

**El Financiero:** Casilla 6666, Guayaquil; tel. (4) 220-5051; e-mail elfinanciero@elfinanciero.com; internet www.elfinanciero.com; weekly; business and economic news; f. 1990; Dir DAVID PÉREZ MACCOLLUM.

**Generación XXI:** Aguirre 730 y Boyacá, Guayaquil; tel. (4) 232-7200; fax (4) 232-0499; e-mail webmaster@vistazo.com; internet www.generacion21.com; youth; Gen. Editor SÉBASTIEN MELIÉRES.

**Hogar:** Aguirre 724 y Boyacá, Apdo 1239, Guayaquil; tel. (4) 232-7200; f. 1964; monthly; Man. Editor ROSA AMELIA ALVARADO; circ. 35,000.

**Revista Estadio:** Aguirre 730 y Boyacá, Apdo 1239, Guayaquil; tel. (4) 327-200; fax (4) 320-499; e-mail estadio@vistazo.com; internet www.revistaestadio.com; f. 1962; fortnightly; sport; Dir-Gen. SEBASTIAN MÉLIÉRES; Editor CÉSAR TORRES TINAJERO; circ. 40,000.

**Vistazo:** Aguirre 730 y Boyacá, Casilla 09-01-1239, Guayaquil; tel. (4) 232-7200; fax (4) 232-0499; e-mail vistazo@vistazo.com; internet www.vistazo.com; f. 1957; fortnightly; general; Gen. Editor PATRICIA ESTUPIÑÁN DE BURBANO; circ. 85,000.

## NEWS AGENCIES

### Foreign Bureaux

**Agencia EFE** (Spain): Edif. La Fontana, 1°, Of. 1, Avda República de El Salvador 1058, y Naciones Unidas, Quito; tel. (2) 225-9682; fax (2) 225-5769; e-mail eferedaccion@andinanet.net; Bureau Chief EMILIO CRESPO.

**Agenzia Nazionale Stampa Associata (ANSA)** (Italy): Calle Venezuela 1013 y esq. Mejía, Of. 26, Quito; tel. (2) 258-0794; fax (2) 258-0782; Correspondent FERNANDO LARENAS.

**Associated Press (AP)** (USA): Edif. Sudamérica, 4°, Of. 44, Calle Venezuela 1018 y Mejía, Quito; tel. (2) 257-0235; Correspondent CARLOS CISTERNAS.

**Deutsche Presse-Agentur (dpa)** (Germany): Edif. Atrium, Of. 5-7, González Suárez 894 y Gonnessiat, Quito; tel. (2) 256-8986; Correspondent JORGE ORTIZ.

**Informatsionnoye Telegrafnoye Agentstvo Rossii—Telegrafnoye Agentstvo Suverennykh Stran (ITAR—TASS)** (Russia): Calle Roca 328 y 6 de Diciembre, 2°, Dep. 6, Quito; tel. (2) 251-1631; Correspondent VLADIMIR GOSTEV.

**Inter Press Service (IPS)** (Italy): Urbanización Los Arrayanes Manzanas 20, Casa 15, Calle León Pontón y Pasaje E, Casilla 17-01-1284, Quito; tel. (2) 266-2362; fax (2) 266-1977; e-mail jfrias@uio.telconet.net; Correspondent KINTTO LUCAS.

**Prensa Latina** (Cuba): Edif. Sudamérica, 2°, Of. 24, Calle Venezuela 1018 y Mejía, Quito; tel. (2) 251-9333; Bureau Chief ENRIQUE GARCÍA MEDINA.

**Reuters** (United Kingdom): Avda Amazonas 3655, 2°, Casilla 17-01-4112, Quito; tel. (2) 243-1753; fax (2) 243-2949; Correspondent JORGE AGUIRRE CHARVET.

**Xinhua (New China) News Agency** (People's Republic of China): Edif. Portugal, Avda Portugal y Avda de la República de El Salvador 730, 10°, Quito; Bureau Chief LIN MINZHONG.

## Publishers

**Artes Gráficas Ltda:** Avda 12 de Octubre 1637, Apdo 533, Casilla 456A, Quito; Man. MANUEL DEL CASTILLO.

**Casa de la Cultura Ecuatoriana:** Avdas 6 de Diciembre y Patria; tel. (2) 222-3391; e-mail info@cce.org.ec; Dir JAIME PAREDES.

**Centro de Educación Popular:** Avda América 3584, Apdo 17-08-8604, Quito; tel. (2) 252-5521; fax (2) 254-2369; e-mail centro@cedep.ec; f. 1978; communications, economics; Dir DIEGO LANDÁZURI.

**CEPLAES:** Avda 6 de Diciembre 2912 y Alpallana, Apdo 17-11-6127, Quito; tel. (2) 254-8547; fax (2) 256-6207; f. 1978; agriculture, anthropology, education, health, social sciences, women's studies; Exec. Dir ALEXANDRA AYALA.

**CIDAP:** Hermano Miguel 3-23, Casilla 01-011-943, Cuenca; tel. (7) 282-9451; fax (7) 283-1450; e-mail cidapl@cidap.org.ec; internet www.cidap.org.ec; art, crafts, games, hobbies; Dir CLAUDIO MALO GONZALES.

**CIESPAL** (Centro Internacional de Estudios Superiores de Comunicación para América Latina): Avda Diego de Almagro 32-133 y Andrade Marin, Apdo 17-01-584, Quito; tel. (2) 254-8011; fax (2) 250-2487; e-mail ejaramillo@ciespal.net; internet www.ciespal.net; f. 1959; communications, technology; Dir EDGAR JARAMILLO.

**Corporación de Estudios y Publicaciónes:** Acuna 168 y J. Agama, Casilla 17-21-0086, Quito; tel. (2) 222-1711; fax (2) 222-6256; e-mail cep@accessinter.net; f. 1963; law, public administration.

**Corporación Editora Nacional:** Roca E9–59 y Tamayo, Apdo 17-12-886, Quito; tel. (2) 255-4358; fax (2) 256-6340; e-mail cen@accessinter.net; f. 1977; archaeology, economics, education, geography, political science, history, law, literature, philosophy, social sciences; Pres. ERNESTO ALBÁN GOMEZ.

**Cromograf, SA:** Coronel 2207, Guayaquil; tel. (4) 234-6400; children's books, paperbacks, art productions.

**Ediciones Abya-Yala:** Avda 12 de Octubre 1430, CP 17-12-719, Quito; tel. (2) 250-6251; fax (2) 250-6267; e-mail admin-info@abyayala.org; internet www.abyayala.org; f. 1975; anthropology, environmental studies, languages, education, theology; Pres. Fr JUAN BOTTASSO; Dir-Gen. P. XAVIER HERRÁN.

**Ediciones Valladolid:** Colón 818 y Santa Elena, Guayaquil; tel. (4) 232-9956; e-mail edival@andinanet.net; internet edival.ecuaemail.com; f. 1996; reference; Owner and Man. CÉSAR VALLADOLID.

**Editorial Bueno:** Avda República oeste 2-23 y Teresa de Cepeda (diagonal a los Puentes República); tel. (2) 225-8959; fax (2) 437-873; e-mail buenoeditores@yahoo.com; Man. CÉSAR FERNÁNDEZ BUENO.

**Editorial Don Bosco:** Vega Muñoz 10-68 y General Torres, Cuenca; tel. (7) 283-1745; fax (7) 284-2722; e-mail edibosco@bosco.org.ec; internet www.lns.com.ec; f. 1920; Gen. Man. P. EDUARDO SANDOVAL; Deputy Man. FANNY FAJARDO Z.

**Editorial El Conejo:** 6 de Diciembre 2309 y La Niña, Quito; tel. (2) 222-7948; fax (2) 250 1066; e-mail econejo@attglobal.net; internet www.editorialelconejo.com; f. 1979; non-profit publisher of educational and literary texts; Dir ABDÓN UBIDIA.

**Editorial Edinacho SA:** Bartolomé Sánchez Lote 6 y Calle C Lotización Muñoz Carvajal, Carcelén, 5932, Quito; tel. (2) 247-0429;

fax (2) 247-0430; e-mail edinacho@ecnet.ec; f. 1985; Man. GERMÁN SEGURA; Editor Dr DALIA MARÍA NOBOA.

**Eguez-Pérez en Nombre Colectivo/Abrapalabra Editores:** América 5378, Casilla 464A, Quito; tel. and fax (2) 254-4178; f. 1990; drama, education, fiction, literature, science fiction, social sciences; Man. IVAN EGUEZ.

**Libresa SA:** Murgeon 364 y Ulloa, Quito; tel. (2) 223-0925; fax (2) 250-2992; e-mail info@libresa.com; internet www.libresa.com; f. 1979; education, literature, philosophy; Pres. FAUSTO COBA ESTRELLA; Man. JAIME PEÑA NOVOA.

**Libros Técnicos Litesa Cía Ltda:** Avda América 542, Apdo 456A, Quito; tel. (2) 252-8537; Man. MANUEL DEL CASTILLO.

**Pontificia Universidad Católica del Ecuador, Centro de Publicaciones:** Avda 12 de Octubre 1076 y Carrión, Apdo 17-01-2184, Quito; tel. (2) 252-9250; fax (2) 256-7117; e-mail puce@edu.ec; internet www.puce.edu.ec; f. 1974; literature, natural science, law, anthropology, sociology, politics, economics, theology, philosophy, history, archaeology, linguistics, languages and business; Rector Dr JOSÉ RIBADENEIRA ESPINOSA; Dir Dr MARCO VINICIO RUEDA.

**Trama Ediciones:** Edif. Marinoar PB, entre Catalina Aldaz y El Batán, Eloy Alfaro 34-85, Quito; tel. (2) 224-6315; fax (2) 224-6317; e-mail trama@trama.ec; internet www.trama.com.ec; f. 1977; architecture, design, art and tourism; Dir-Gen. ROLANDO MOYA TASQUER.

**Universidad Central del Ecuador:** Departamento de Publicaciones, Servicio de Almacén Universitario, Ciudad Universitaria, Avda América y A. Pérez Guerrero, POB 3291, Quito; tel. (2) 222-6080; fax (2) 250-1207.

**Universidad de Guayaquil:** Departamento de Publicaciones, Biblioteca General 'Luis de Tola y Avilés', Apdo 09-01-3834, Guayaquil; tel. (4) 251-6296; f. 1930; general literature, history, philosophy, fiction; Man. Dir LEONOR VILLAO DE SANTANDER.

# Broadcasting and Communications

## TELECOMMUNICATIONS

**Asociación Ecuatoriana de Radiodifusión (AER):** Edif. Atlas, 8°, Of. 802, Calle Justino Cornejo con Francisco de Orellana, Guayaquil; tel. and fax (4) 229-1783; ind. asscn; Pres. Ing. ANTONIO GUERRERO ONESSA.

**Consejo Nacional de Telecomunicaciones (CONATEL):** Avda Diego de Almagro 31-95 y Alpallana, Casilla 17-07-9777, Quito; tel. (2) 222-5614; fax (2) 222-5030; e-mail comunicacion@conatel.gov.ec; internet www.conatel.gov.ec; Pres. Dr JOSÉ PILEGGI VÉLIZ.

**Secretaría Nacional de Telecomunicaciones:** Avda Diego de Almagro 31-95 y Alpallana, Casilla 17-07-9777, Quito; tel. (2) 250-2197; fax (2) 290-1010; Secretario Nacional de Telecomunicaciones Ing. CARLOS DEL POZO CAZAR.

**Superintendencia de Telecomunicaciones (SUPTEL):** Edif. Olimpo, Avda 9 de Octubre 1645 y Berlín, Casilla 17-21-1797, Quito; tel. (2) 222-2448; fax (2) 256-6688; e-mail supertel@server.supertel .gov.ec; internet www.supertel.gov.ec; Superintendente de Telecomunicaciones IVÁN BURBANO.

### Major Service Providers

**Alegro PCS** (Telecsa, SA): Edif. Vivaldi, Amazonas 3837 y Corea, Quito; tel. and fax (2) 299-0000; internet www.alegropcs.com; co-owned by Andinatel and Pacifictel; cellular telephone provider.

**Andinatel:** Edif. Zeta, Avda Amazonas y Veintimilla, Quito; tel. (2) 256-1004; fax (2) 256-2240; e-mail ventas@andinatel.com; internet www.andinatel.com; scheduled for privatization; Exec. Pres. Ing. LUIS RECALDE.

**BellSouth Ecuador** (OTECEL, SA): Edif. BellSouth, Avda República y Esq. La Pradera, Quito; tel. (2) 222-7700; internet www .bellsouth.com.ec; f. 1997; owned by Telefónica Móviles, SA (Spain); mobile telephone services.

**Pacifictel:** Calle Panamá y Roca, Guayaquil; tel. (4) 230-8724; internet www.pacifictel.net; scheduled for privatization; Exec. Pres. (vacant).

**Porta:** Quito; e-mail callcenter@conecel.com; internet www.porta .net; f. 1993; mobile telecommunications provider; subsidiary of América Móvil group (Mexico).

## BROADCASTING

### Regulatory Authority

**Consejo Nacional de Radiodifusión y Televisión (CONARTEL):** Calle La Pinta 225 y Rábida, Quito; tel. (2) 223-3492; fax (2) 252-3188; Pres. Ing. ALOO OTTATI PINO.

### Radio

There are nearly 300 commercial stations, 10 cultural stations and 10 religious stations. The following are among the most important stations:

**Emisoras Gran Colombia:** Calle Galápagos 112 y Guayaquil, Quito; tel. (2) 244-2951; fax (2) 244-3147; f. 1943; Pres. MARIO JOSÉ CANESSA ONETO.

**Radio Católica Nacional:** Avda América 1830 y Mercadillo, Casilla 17-03-540, Quito; tel. (2) 254-1557; fax (2) 256-7309; f. 1985; Pres. ANTONIO GONZÁLEZ.

**Radio Centro:** Avda República de El Salvador 836 y Portugal, Quito; tel. (2) 244-8900; fax (2) 250-4575; f. 1977; Pres. ÉDGAR YÁNEZ VILLALOBOS.

**Radio Colón:** Avellanas E5-107 y Avda Eloy Alfaro, Casilla 17-07-9927, Quito; tel. 248-4574; fax (2) 248-5666; e-mail escucha@ radiocolon.ec; internet www.radiocolon.ec; f. 1934; Pres. Dr GERARDO CASTRO; Dir BERNARDO NUSSBAUM.

**Radio CRE Satelital** (CORTEL, SA): Edif. El Torreón, 9°, Avda Boyacá 642 y Padre Solano, Apdo 4144, Guayaquil; tel. (4) 256-4290; fax (4) 256-0386; e-mail administracion@cre.com.ec; Pres. RAFAEL GUERRERO VALENZUELA.

**Radio La Luna:** Quito; Owner PACO VELASCO.

**Radio Nacional del Ecuador:** Mariano Echeverría 537 y Brasil, Casilla 17-01-82, Quito; tel. (2) 245-9555; fax (2) 245-5266; f. 1961; state-owned; Dir ANA MALDONADO ROBLES.

**Radio Quito:** Avda 10 de Agosto 2441 y Colón, Casilla 17-21-1971, Quito; tel. (2) 250-8301; fax (2) 250-3311; f. 1940; Pres. GUADALUPE MANTILLA MOSQUERA.

**Radio Sonorama (HCAEL):** Eloy Alfaro 5400 y Los Granados, Casilla 130B, Quito; tel. (2) 244-8403; fax (2) 244-5858; f. 1975; Pres. SANTIAGO PROAÑO.

**Radio Sucre:** Joaquin Orrantia y Miguel H. Alcivar (Casa de las Americas Kennedy Norte), Guayaquil; tel. (4) 268-0588; fax (4) 268-0592; e-mail info@radiosucretv.com; internet www.radiosucre.com .ec; f. 1983; Pres. VICENTE ARROBA DITTO; Dir LUIS SÁNCHEZ MORENO.

**La Voz de los Andes (HCJB):** Villalengua 884 y Avda 10 de Agosto, Casilla 17-17-691, Quito; tel. (2) 226-6808; fax (2) 226-7263; e-mail helpdesk@hcjb.org.ec; f. 1931; operated by World Radio Missionary Fellowship; programmes in 11 languages (including Spanish and English) and 22 Quechua dialects; private, non-commercial, cultural, religious; Dir, Int. Radio CURT COLE; Gen. Man. JOHN E. BECK.

### Television

**Corporación Ecuatoriana de Televisión—Ecuavisa Canal 2:** Cerro El Carmen, Casilla 1239, Guayaquil; tel. (4) 230-0150; fax (4) 230-3677; f. 1967; Pres. XAVIER ALVARADO ROCA; Gen. Man. FRANCISCO AROSEMENA ROBLES.

**Cadena Ecuatoriana de Televisión—TC Televisión Canal 10:** Avda de las Américas, frente al Aeropuerto, Casilla 09-01-673, Guayaquil; tel. (4) 239-7664; fax (4) 228-7544; f. 1969; commercial; Pres. ROBERTO ISAÍAS; Gen. Man. JORGE KRONFLE.

**Televisora Nacional—Ecuavisa Canal 8:** Bosmediano 447 y José Carbo, Bellavista, Quito; tel. (2) 244-6472; fax (2) 244-5488; commercial; f. 1970; Pres. PATRICIO JARAMILLO.

**Televisión del Pacífico, SA—Gamavisión:** Avda Eloy Alfaro 5400 y Río Coca, Quito; tel. (2) 226-2222; fax (2) 244-0259; e-mail nicovega@gamavision.com; internet www.gamavision.com; f. 1978; Pres. NICOLÁS VEGA.

**Teleamazonas Cratel, CA:** Granda Centeno y Brasil, Casilla 17-11-04844, Quito; tel. (2) 245-1385; fax (2) 244-1620; e-mail contactenos@teleamazonas.com; f. 1974; commercial; Pres. EDUARDO GRANDA GARCÉS.

**Teleandina Canal 23:** Avda de la Prensa 3920 y Fernández Salvador, Quito; tel. (2) 259-9403; fax (2) 259-2600; f. 1991; Pres. HUMBERTO ORTIZ FLORES; Dir PATRICIO AVILES.

# Finance

(cap. = capital; res = reserves; dep. = deposits; m. = million; brs = branches; amounts in US dollars unless otherwise indicated)

**Junta Monetaria Nacional** (National Monetary Board): Quito; tel. (2) 251-4833; fax (2) 257-0258; f. 1927; Pres. FRANCISCO SWETT.

## SUPERVISORY AUTHORITY

**Superintendencia de Bancos y Seguros:** Avda 12 de Octubre 1561 y Madrid, Casilla 17-17-770, Quito; tel. (2) 255-4225; fax (2)

# ECUADOR

250-6812; e-mail alejo@e-mail.superban.gov.ec; internet www.superban.gov.ec; f. 1927; supervises national banking system, including state and private banks and other financial institutions; Superintendent ALEJANDRO MALDONADO GARCÍA.

## BANKING

### Central Bank

**Banco Central del Ecuador:** Avda Amazonas y Atahualpa, Casilla 339, Quito; tel. (2) 225-5777; fax (2) 226-9064; internet www.bce.fin.ec; f. 1927; cap. 2.5m., res 1,303.2m., dep. 1,529.0m. (Dec. 2003); Pres. JOSÉ JOUVÍN VERNAZA; Gen. Man. LEOPOLDO R. BÁEZ CARRERA; 2 brs.

### Other State Banks

**Banco Ecuatoriano de la Vivienda:** Avda 10 de Agosto 2270 y Cordero, Casilla 3244, Quito; tel. (2) 252-1311; f. 1962; Pres. Abog. JUAN PABLO MONCAGATTA; Gen. Man. Dr PATRICIO CEVALLOS MORÁN.

**Banco del Estado (BDE):** Avda Atahualpa 628 y 10 de Agosto, Casilla 17-01-00373, Quito; tel. (2) 225-0800; fax (2) 225-0320; f. 1979; Pres. Econ. CÉSAR ROBALINO; Gen. Man. LUIS MEJÍA MONTESDEOCA.

**Banco Nacional de Fomento:** Ante 107 y 10 de Agosto, Casilla 685, Quito; tel. (2) 256-0680; fax (2) 257-0286; internet www.bnf.fin.ec; f. 1928; Pres. HÉCTOR BALLESTEROS SEGARRA; 70 brs.

**Corporación Financiera Nacional (CFN):** Avda Juan León Mera 130 y Patria, Casilla 17-21-01924, Quito; tel. (2) 256-4900; fax (2) 222-3823; e-mail mbenitez@q.cfn.fin.ec; internet www.cfn.fin.ec; state-owned bank providing export credits, etc.; f. 1964; Pres. Ing. PEDRO KOHN; Financial Dir HÉCTOR SAN ANDRÉS.

### Commercial Banks

#### Quito

**Banco Amazonas, SA:** Avda Amazonas 4430 y Villalengua, Casilla 121, Quito; tel. (2) 226-0400; fax (2) 225-5123; e-mail basacomp@porta.net; internet www.bancoamazonas.com; f. 1976; affiliated to Banque Paribas; Pres. RAFAEL FERRETTI BENÍTEZ; Vice-Pres. ROBERTO SEMINARIO.

**Banco Caja de Crédito Agrícola Ganadero, SA:** Avda 6 de Diciembre 225 y Piedrahita, Quito; tel. (2) 252-8521; f. 1949; Man. HUGO GRIJALVA GARZÓN; Pres. NICOLÁS GUILLÉN.

**Banco Consolidado del Ecuador:** Avda Patria 740 y 9 de Octubre, Casilla 9150, Suc. 7, Quito; tel. (2) 256-0369; fax (2) 256-0719; f. 1981; cap. 2,874m., res 4,338m., dep. 5,545m. sucres (Oct. 1998); Chair. JAIME GILINSKI; Gen. Man. ANTONIO COY; 2 brs.

**Banco General Rumiñahui:** Avda República 720 y Eloy Alfaro, Quito; tel. (2) 250-9929; fax (2) 256-3786; e-mail bcainstitucional@bgr.com.ec; internet www.bgr.com.ec; total assets 192.0m. (Dec. 2004); Gen. Man. Gen. GUSTAVO HERRERA.

**Banco Internacional, SA:** Avda Patria E-421 y 9 de Octubre, Casilla 17-01-2114, Quito; tel. (2) 256-5547; fax (2) 256-5758; e-mail cromero@bancointernacional.com.ec; internet www.bancointernacional.com.ec; f. 1973; cap. 8,600m. (Dec. 2000); Pres. JOSÉ ENRIQUE FUSTER CAMPS; Gen. Man. RAÚL GUERRERO ANDRADE; 58 brs.

**Banco del Pichincha, CA:** Avda Amazonas 4560 y Pereira, Casilla 261, Quito; tel. (2) 298-0980; fax (2) 298-1226; internet www.pichincha.com; internet www.todo1.com; f. 1906; cap. 63.5m., res 76.4m., dep. 1,549.5m. (Dec. 2003); 61.82% owned by Exec. Pres. and Chair.; Exec. Pres. and Chair. Dr FIDEL EGAS GRIJALVA; Gen. Man. ANTONIO ACOSTA ESPINOSA; 127 brs.

**Produbanco:** Avda Amazonas 3775 y Japón, Casilla 17-03-38-A, Quito; tel. (2) 299-9000; fax (2) 244-7319; internet www.produbanco.com; f. 1978 as Banco de la Producción; name changed as above in 1996; cap. 60m., res 11.38m., dep. 542.36m. (Dec. 2004); filial of Grupo Financiero Producción; Pres. ABELARDO PACHANO BERTERO; Dir RODRIGO PAZ DELGADO; 55 brs.

**UniBanco:** Avda República 500 y Pasaje Carrión, Edif. Pucara, 1°, Quito; tel. (2) 290-7576; fax (2) 222-5000; f. 1964 as Banco de Cooperativas del Ecuador, name changed as above in 1995; Pres. SALVADOR PEDRERO; Treas. JUAN FERNANDO BERMEO.

#### Cuenca

**Banco del Austro:** Sucre y Borrero (esq.), Casilla 01-01-0167, Cuenca; tel. (7) 831-646; fax (7) 832-633; internet www.bancodelaustro.com; f. 1977; total assets 249.8m. (Sept. 2004); Pres. JUAN ELJURI ANTÓN; Gen. Man. PATRICIO ROBAYO IDROVO; 19 brs.

#### Guayaquil

**Banco Bolivariano, CA:** Junín 200 y Panamá, Casilla 09-01-10184, Guayaquil; tel. (4) 230-5000; fax (4) 256-6707; e-mail info@bolivariano.com; internet www.bolivariano.com; f. 1978; cap. 34.8m., res 16.8m., dep. 537.3m. (Dec. 2004); 61.2% owned by Tabos Investment SA; Chair. JOSÉ SALAZAR BARRAGÁN; Pres. and CEO MIGUEL BABRA LEÓN; 53 brs.

**Banco del Pacífico:** Francisco de P. Ycaza 200, Casilla 09-01-988, Guayaquil; tel. (4) 256-6010; fax (4) 232-8333; e-mail webadmin@pacifico.fin.ec; internet www.bp.fin.ec; f. 2000 by merger of Banco del Pacífico and Banco Continental; total assets 152.2m.; Exec. Pres. FÉLIX HERRERO BACHMEIER.

**Banco Industrial y Comercial—Baninco:** Pichincha 335 e Illingworth, Casilla 5817, Guayaquil; tel. (4) 232-3488; f. 1965; Pres. Ing. CARLOS MANZUR PERES; Gen. Man. GABRIEL MARTÍNEZ INTRIAGO; 2 brs.

**Banco Territorial, SA:** Panamá 814 y V. M. Rendón, Casilla 09-01-227, Guayaquil; tel. (4) 256-6695; fax (4) 256-6695; f. 1886; Pres. ROBERTO GOLDBAUM; Gen. Man. Ing. GUSTAVO HEINERT.

#### Loja

**Banco de Loja:** esq. Bolívar y Rocafuerte, Casilla 11-01-300, Loja; tel. (4) 757-1682; fax (4) 753-3019; internet www.bancodeloja.fin.ec; f. 1968; cap. 10,000m., res 4,207m., dep. 70,900m. sucres (Dec. 1996); Pres. Ing. STEVE BROWN HIDALGO; Man. Ing. LEONARDO BURNEO MULLER.

#### Machala

**Banco de Machala, SA:** Avda 9 de Mayo y Rocafuerte, Casilla 711, Machala; tel. (4) 256-6800; fax (4) 292-2744; internet www.bmachala.com; f. 1962; cap. 11.0m., res 0.4m., dep. 143.0; Pres. Dr ESTEBAN QUIROLA FIGUEROA; Exec. Pres. and Gen. Man. Dr MARIO CANESSA ONETO; 2 brs.

#### Portoviejo

**Banco Comercial de Manabí, SA:** Avda 10 de Agosto 600 y 18 de Octubre, Portoviejo; tel. (4) 265-3888; fax (4) 263-5527; f. 1980; Pres. Dr RUBÉN DARÍO MORALES; Gen. Man. ARISTO ANDRADE DÍAZ.

### Foreign Banks

**ABN AMBO Bank NV** (Netherlands): República de El Salvador 1082, Edif. Mansión Blanca, Torre Londres, 5°, Quito; tel. (2) 243-4448; fax (2) 243-2163; e-mail juan.lopez@ec.abnamro.com; internet www.wholesale.abnamro.com; f. 1960; Country Rep. SANTIAGO HIDALGO; 6 brs.

**Citibank, NA** (USA): Juan León Mera 130 y Patria, Casilla 17-01-1393, Quito; tel. (2) 256-3300; fax (2) 256-6895; f. 1959; cap. 7,000m., res 1,000m., dep. 62,000m. (Dec. 1996); Gen. Man. BENJAMÍN FRANCO; 3 brs.

**ING Bank NV** (Netherlands): Edif. Centro Financiero, Avda Amazonas 4545 y Pereira, Quito; tel. (2) 298-1650; fax (2) 298-1665.

**Lloyds TSB Bank Ltd** (United Kingdom): Avda Amazonas 3428E entre Atahualpa y Núñez de Vela, Quito; tel. (4) 693-0000; fax (4) 693-0458; e-mail ltsbec@andinanet.net; internet www.lloydstsb.com.ec; f. 1988; in succession to the Bank of London and South America, f. 1936; Country Man. PAUL McEVOY.

### 'Multibanco'

**Banco de Guayaquil, SA:** Plaza Ycaza 105 y Pichincha, Casilla 09-01-1300, Guayaquil; tel. (4) 251-7100; fax (4) 251-4406; e-mail servicios@bankguay.com; internet www.bancoguayaquil.com; f. 1923; absorbed the finance corpn FINANSUR in 1990 to become Ecuador's first 'multibanco', carrying out commercial and financial activities; cap. 145,000m., dep. 3,198,862m. sucres (Dec. 1998); Pres. DANILO CARRERA DROUET; Exec. Pres. GUILLERMO LASSO MENDOZA; 50 brs.

### Finance Corporations

**Corporación Financiera Nacional (CFN):** Juan León Mera 130 y Avda Patria, Casilla 17-21-01924, Quito; tel. (2) 256-4900; fax (2) 222-3823; internet www.cfn.fin.ec; f. 1997; development finance agency; Chair. ROSA ELVIRA MANTILLA DE VELASCO.

**Financiera Guayaquil, SA:** Carchi 702 y 9 de Octubre, 6°, Casilla 2167, Guayaquil; f. 1976; cap. 900m., res 142m. (June 1987); Gen. Man. Dr MIGUEL BABRA LYON.

### Associations

**Asociación de Bancos Privados del Ecuador:** Edif. Delta, 7°, Avda República de El Salvador 890 y Suecia, Casilla 17-11-6708, Quito; tel. (2) 246-6670; fax (2) 246-6702; e-mail echiribo@asobancos.org.ec; internet www.asobancos.org.ec; f. 1965; 36 mems; Pres. ANGELO CAPUTI; Exec. Pres. ERNESTO CHIRIBOGA BLONDET (acting).

# ECUADOR

**Asociación de Compañías Financieras del Ecuador (AFIN):** Robles 653 y Amazonas, 13°, Of. 1310-1311, Casilla 17-07-9156, Quito; tel. (2) 255-0623; fax (2) 256-7912; Pres. Ing. FRANCISCO ORTEGA.

### STOCK EXCHANGES

**Bolsa de Valores de Guayaquil:** 9 de Octubre 110 y Pichincha, Guayaquil; tel. (4) 256-1519; fax (4) 256-1871; e-mail mmurillo@bvg.fin.ec; internet www.mundobvg.com; Pres. RODOLFO KRONFLE AKEL; Dir-Gen. ENRIQUE AROSEMENA BAQUERIZO.

**Bolsa de Valores de Quito:** Edif. Londres, 8°, Avda Amazonas 540 y Carrión, Casilla 17-01-3772, Quito; tel. (2) 222-1333; fax (2) 250-0942; e-mail informacion@ccbvq.com; internet www.ccbvq.com; f. 1969; Pres. PATRICIO PEÑA; Exec. Pres. MÓNICA DE ANDERSON.

### INSURANCE

**Instituto Ecuatoriano de Seguridad Social:** Avda 10 de Agosto y Bogotá, Apdo 2640, Quito; tel. (2) 254-7400; fax (2) 250-4572; internet www.iess.gov.ec; f. 1928; various forms of state insurance provided; directs the Ecuadorean social insurance system; provides social benefits and medical service; Dir-Gen. Ing. JORGE MADERA CASTILLO.

#### Principal National Companies

**Bolívar Cía de Seguros del Ecuador, SA:** Edif. Las Cámaras, 11° y 12°, Avda Francisco de Orellana, Guayaquil; tel. (2) 268-1777; fax (2) 268-3363; e-mail ssanmiguel@seguros-bolivar.com; internet www.seguros-bolivar.com; f. 1957; Gen. Man. LEONIDAS ORTEGA AMADOR.

**Cía Reaseguradora del Ecuador, SA:** Edif. Intercambios, 1°, Junín 105 y Malecón Simón Bolívar, Casilla 09-01-6776, Guayaquil; tel. (4) 256-4461; fax (4) 256-4454; e-mail oespinoz@ecuare.fin.ec; f. 1977; Man. Dir Ing. OMAR ESPINOSA ROMERO.

**Cía de Seguros Cóndor, SA:** Plaza Ycaza 302, Apdo 09-01-5007, Guayaquil; tel. (4) 256-5300; fax (4) 256-0144; f. 1966; Gen. Man. AUGUSTO SALAME ARZUBIAGA.

**Cía de Seguros Ecuatoriano-Suiza, SA:** Avda 9 de Octubre 2101 y Tulcán, Apdo 09-01-0937, Guayaquil; tel. (4) 245-2444; fax (4) 245-3229; f. 1954; Gen. Man. LUIS F. SALAS RUBIO.

**Cía Seguros Unidos, SA:** Edif. Urania, 4°, Avda 10 de Agosto 5133 y Naciones Unidas, Casilla 17-0373, Quito; tel. (2) 252-6466; fax (2) 245-0920; Gen. Man. JUAN RIVAS DOMENECH.

**La Nacional Cía de Seguros Generales, SA:** Panamá 809 y Rendón, Guayaquil; tel. (4) 256-0700; fax (4) 256-6327; f. 1940.

**Panamericana del Ecuador, SA:** Calle Portugal E-12-72 y Avda Eloy Alfaro, Quito; tel. (2) 246-8840; fax (2) 246-9650; f. 1973; Gen. Man. GERMÁN DAVILA LEORO.

**Seguros Rocafuerte, SA:** Edif. Filanbanco, 15°, Plaza Carbo 505 y 9 de Octubre, Apdo 09-04-6491, Guayaquil; tel. (4) 232-6125; fax (4) 329-353; f. 1967; life and medical; Exec. Pres. Ing. NORMAN PICHARDO VAN DER DIJS.

**La Unión Cía Nacional de Seguros:** Urb. Los Cedros, Km 5½, Vía a la Costa, Apdo 09-01-1294, Guayaquil; tel. (4) 285-1500; fax (4) 285-1700; e-mail rgoldbaum@seguroslaunion.com; internet www.seguroslaunion.com; f. 1943; Exec. Pres. ROBERTO GOLDBAUM MORALES; Gen. Man. DAVID A. GOLDBAUM MORALES.

# Trade and Industry

### GOVERNMENT AGENCIES

**Consejo Nacional de Competitividad (CNC):** Edif. Plaza 2000, 1°, Avda 12 de Octubre 24-593 y Francisco Salazar, Quito; tel. (2) 252-1657; e-mail info@cnc.gov.ec; internet www.ecuadorcompite.gov.ec; promotes competitiveness of Ecuadorean businesses; Pres. JOYCE HIGGINGS DE GINATTA.

**Consejo Nacional de Modernización del Estado (CONAM):** Edif. CFN, 10°, Avda Juan León Mera 130 y Patria, Quito; tel. (2) 250-9432; fax (2) 222-8450; e-mail info@conam.gov.ec; internet www.conam.gov.ec; f. 1994; responsible for overseeing privatizations; Pres. Ing. RICARDO NOBOA; Exec. Dir Ing. ANTONIO PERÉ YCAZA.

**Empresa de Comercio Exterior (ECE):** Quito; f. 1980 to promote non-traditional exports; State owns 33% share in company; share capital 25m. sucres.

**Fondo de Promoción de Exportaciones (FOPEX):** Juan León Mera 130 y Patria, Casilla 163, Quito; tel. (2) 256-4900; fax (2) 256-2519; f. 1972; export promotion; Dir ELIANA SANTAMARÍA M.

**Fondo de Solidaridad:** Avda 6 de Diciembre 25-75 y Colón, Edif. Partenón, Quito; tel. (2) 220-0429; internet www.fondodesolidaridad.gov.ec; f. 1996; govt devt agency; responsibility for overseeing privatization of electricity sector; Pres. JORGE BURBANO.

**Instituto Ecuatoriano de Reforma Agraria y Colonización (IERAC):** f. 1973 to supervise the Agrarian Reform Law under the auspices and co-ordination of the Ministry of Agriculture; Dir LUIS LUNA GAYBOR.

**Superintendencia de Compañías del Ecuador:** Roca 660 y Amazonas, Casilla 687, Quito; tel. (2) 254-1606; fax (2) 256-6685; e-mail superintcias@q.supercias.gov.ec; internet www.supercias.gov.ec; f. 1964; responsible for the legal and accounting control of commercial enterprises; Supt GONZALO FABIÁN ALBUJA CHAVES; Sec.-Gen. VÍCTOR CEVALLOS VÁSQUEZ.

### DEVELOPMENT ORGANIZATIONS

**Centro Nacional de Promoción de la Pequeña Industria y Artesanía (CENAPIA):** Quito; agency to develop small-scale industry and handicrafts; Dir Econ. EDGAR GUEVARA (acting).

**Centro de Reconversión Económica del Azuay, Cañar y Morona Santiago (CREA):** Avda México entre Unidad Nacional y las Américas, Casilla 01-01-1953, Cuenca; tel. (7) 817-500; fax (7) 817-134; f. 1959; devt org.; Dir Dr JUAN TAMA.

**Consejo Nacional de Desarrollo (CONADE):** Juan Larrea y Arenas, Quito; fmrly Junta Nacional de Planificación y Coordinación Económica; aims to formulate a general plan of economic and social devt and supervise its execution; also to integrate local plans into the national; Chair. GALO ABRIL OJEDA; Sec. PABLO LUCIO PAREDES.

**Fondo de Desarrollo del Sector Rural Marginal (FODERUMA):** f. 1978 to allot funds to rural development programmes in poor areas.

**Fondo Nacional de Desarrollo (FONADE):** f. 1973; national development fund to finance projects as laid down in the five-year plan.

**Instituto de Colonización de la Región Amazónica (INCREA):** f. 1978 to encourage settlement in and economic development of the Amazon region; Dir Dr DIMAS GUZMÁN.

**Instituto Ecuatoriano de Recursos Hidráulicos (INERHI):** undertakes irrigation and hydroelectric projects; Man. Ing. EDUARDO GARCÍA GARCÍA.

**Organización Comercial Ecuatoriana de Productos Artesanales (OCEPA):** Carrión 1236 y Versalles, Casilla 17-01-2948, Quito; tel. (2) 254-1992; fax (2) 256-5961; f. 1964 to develop and promote handicrafts; Gen. Man. MARCELO RODRÍGUEZ.

**Programa Nacional del Banano y Frutas Tropicales:** Guayaquil; to promote the development of banana and tropical-fruit cultivation; Dir Ing. JORGE GIL CHANG.

**Programa Regional de Desarrollo del Sur del Ecuador (PREDESUR):** Pasaje María Eufrasia 100 y Mosquera Narváez, Quito; tel. (2) 254-4415; f. 1972 to promote the development of the southern area of the country; Dir Ing. LUIS HERNÁN EGUIGUREN CARRIÓN.

### CHAMBERS OF COMMERCE AND INDUSTRY

**Cámara de Comercio de Ambato:** Montalvo 630, Ambato; tel. and fax (3) 841-906; e-mail camcomam@uio.satnet.net; Pres. PABLO VÁSCONEZ.

**Cámara de Comercio de Cuenca:** Avda Federico Malo 1-90, 2°, Casilla 4929, Cuenca; tel. (7) 827-531; fax (7) 833-891; e-mail cccuenca@etapa.com.ec; internet www.cccuenca.com; f. 1919; 5,329 mems; Pres. JUAN PABLO VINTIMILLA.

**Cámara de Comercio Ecuatoriano-Americana** (Amcham Quito-Ecuador): tel. (2) 250-7450; fax (2) 250-4571; e-mail info@ecamcham.com; internet www.ecamcham.com; f. 1974; promotes bilateral trade and investment between Ecuador and the USA; brs in Ambato, Cuenca and Manta; Pres. MAURICIO ROBALINO; Vice-Pres. ALBERTO SANDOVAL.

**Cámara de Comercio Ecuatoriano Canadiense** (Ecuadorean-Canadian Chamber of Commerce): Quito; internet www.ecucanchamber.org; Exec. Dir PATRICIA BUSTAMANTE; Pres. BILL IBBITSON.

**Cámara de Comercio de Guayaquil:** Avda Francisco de Orellana y V. H. Sicouret, Centro Empresarial 'Las Cámaras', 2° y 3°, Guayaquil; tel. (4) 268-2771; fax (4) 268-2766; e-mail info@lacamara.org; internet www.lacamara.org; f. 1889; 31,000 affiliates; Pres. EDUARDO MARURI MIRANDA.

**Cámara de Comercio de Machala:** Edif. Cámara de Comercio, 2°, Rocafuerte y Buenavista, CP 825, Machala; tel. (7) 930-640; fax (7) 934-454; e-mail ccomach@ecua.net.ec; Pres. JOSÉ MENDIETA E.

# ECUADOR

**Cámara de Comercio de Manta:** Edif. Banco del Pichincha, 1°, Avda 2, entre Calles 10 y 11, Manta; tel. and fax (5) 621-306; e-mail cacoma@manta.telconet.net; Pres. Mariano Zambrano S.

**Cámara de Comercio de Quito:** Edif. Las Cámaras, 6°, Avda República y Amazonas, 6°, Casilla 17-01-202, Quito; tel. (2) 244-3787; fax (2) 243-5862; e-mail ccq@ccq.org.ec; internet www.ccq.org.ec; f. 1906; 12,000 mems; Pres. Blasco Peñaherrera Solah; Exec. Dir Miguel Chiriboga Torreo.

**Cámara de Industrias de Cuenca:** Edif. Cámara de Industrias de Cuenca, 12° y 13°, Avda Florencia Astudillo y Alfonso Cordero, Cuenca; tel. (7) 284-5053; fax (7) 284-0107; internet www.industriascuenca.org.ec; f. 1936; Pres. Ing. Marcejo Jaramillo Crespo.

**Cámara de Industrias de Guayaquil:** Avda Francisco de Orellana y M. Alcívar, Casilla 09-01-4007, Guayaquil; tel. (4) 268-2618; fax (4) 268-2680; e-mail caindgye@cig.org.ec; internet www.cig.org.ec; f. 1936; Pres. Alberto Dassum A.

**Federación Nacional de Cámaras de Comercio del Ecuador:** Avda Olmedo 414 y Boyacá, Guayaquil; tel. (4) 232-3130; fax (4) 232-3478; Pres. Blasco Peñaherrera Solah; Exec. Vice-Pres. Dr Roberto Illingworth.

**Federación Nacional de Cámaras de Industrias:** Avda República y Amazonas, 10°, Casilla 2438, Quito; tel. (2) 245-2994; fax (2) 244-8118; e-mail camara@camindustriales.org.ec; internet www.camindustriales.org.ec; f. 1974; Pres. Ing. Gustavo Pinto.

## INDUSTRIAL AND TRADE ASSOCIATIONS

**Centro de Desarrollo Industrial del Ecuador (CENDES):** Avda Orellana 1715 y 9 de Octubre, Casilla 2321, Quito; tel. (2) 252-7100; f. 1962; carries out industrial feasibility studies, supplies technical and administrative assistance to industry, promotes new industries, supervises investment programmes; Gen. Man. Claudio Creamer Guillén.

**Corporación de Desarrollo e Investigación Geológico-Minero-Metalúrgica (CODIGEM):** Avda 10 de Agosto 5844 y Pereira, Casilla 17-03-23, Quito; tel. (2) 225-4673; fax (2) 225-4674; e-mail prodemi2@prodeminca.org.ec; f. 1991 to direct mining exploration and exploitation; Exec. Pres. Ing. Jorge Barragán G.

**Corporación de Promoción de Exportaciones e Inversiones (CORPEI):** Edif. Centro Empresarial Las Cámaras, Avda Francisco de Orellana y Miguel H. Alcívar, 2°, Guayaquil; tel. (4) 268-1550; fax (4) 268-1551; e-mail aissa@corpei.org.ec; internet www.corpei.org.ec; f. 1997 to promote exports and investment; CEO Ricardo E. Estrada.

**Fondo Nacional de Preinversión (FONAPRE):** Jorge Washington 624 y Amazonas, Casilla 17-01-3302, Quito; tel. (2) 256-3261; f. 1973 to undertake feasibility projects before investment; Pres. Luis Parodí Valverde; Gen. Man. Ing. Eduardo Molina Grazziani.

## EMPLOYERS' ORGANIZATIONS

**Asociación de Atuneros:** Malecón s/n, Muelle Portuario de Manta 1, Manta; tel. and fax (5) 262-6467; e-mail atunec@manta.ecua.net.ec; asscn of tuna producers; Pres. Lucía Fernández de Genna.

**Asociación de Cafecultores del Cantón Piñas:** García Moreno y Abdón Calderón, Quito; coffee growers' association.

**Asociación de Comerciantes e Industriales:** Avda Boyacá 1416, Guayaquil; traders' and industrialists' association.

**Asociación de Compañías Consultoras del Ecuador:** Edif. Delta 890, 4°, República de El Salvador y Suecia, Quito; tel. (2) 246-5048; fax (2) 245-1171; e-mail acce@acce.com.ec; internet www.consultoresecuador.com; asscn of consulting cos; Pres. Rodolfo Rendón.

**Asociación de Industriales Gráficos de Pichincha:** Edif. de las Cámaras, 8°, Amazonas y República, Quito; tel. (2) 292-3141; fax (2) 245-6664; e-mail aigquito@aig.org.ec; internet www.aig.org.ec; asscn of the graphic industry; Pres. Enrique Cortéz.

**Asociación Ecuatoriana de Industriales de la Madera:** Edif. de las Cámaras, 7°, República y Amazonas, Quito; tel. (2) 226-0980; fax (2) 243-9560; e-mail secre@aima.org.ec; internet www.aima.org.ec; wood mfrs' asscn; Pres. César Alvarez.

**Asociación de Industriales Textiles del Ecuador (AITE):** Edif. Las Cámaras, 8°, Avda República y Amazonas, Casilla 2893, Quito; tel. (2) 224-9434; fax (2) 244-5159; e-mail aite@aite.org.ec; internet www.aite.com.ec; f. 1938; textile mfrs' association; 40 mems; Pres. Fernando Pérez.

**Asociación de Productores Bananeros del Ecuador (APROBANA):** Guayaquil; banana growers' association; Pres. Nicolás Castro.

**Asociación Nacional de Empresarios (ANDE):** Edif. España, 6°, Of. 67, Avda Amazonas 25–23 y Colón, Casilla 17-01-3489, Quito; tel. (2) 223-8507; fax (2) 250-9806; e-mail ande@uio.satnet.net; internet www.ande.org.ec; national employers' asscn; Pres. Pedro Villamar; Vice-Pres. Ernesto Ribadeneira Troya.

**Asociación Nacional de Exportadores de Cacao y Café (ANECAFE):** Casilla 4774, Manta; tel. (2) 229-2782; fax (2) 229-2885; e-mail anacafe@uio.satnet.net; cocoa and coffee exporters' association.

**Asociación Nacional de Exportadores de Camarones:** Pres. Luis Villacís.

**Asociación Nacional de Molineros:** 6 de Diciembre 3470 e Ignacio Bossano, Quito; tel. (2) 246-5597; fax (2) 246-4754; Exec. Dir Rafaél Callejas.

**Cámara de Agricultura:** Casilla 17-21-322, Quito; tel. (2) 223-0195; Pres. Alberto Enríquez Portilla.

**Consorcio Ecuatoriano de Exportadores de Cacao y Café:** cocoa and coffee exporters' consortium.

**Corporación Nacional de Exportadores de Cacao y Café:** Guayaquil; cocoa and coffee exporters' corporation.

**Federación Nacional de Cooperativas Cafetaleras del Ecuador (FENACAFE):** Jipijapa; tel. (4) 260-0631; e-mail orgcafex@mnb.satnet.net; coffee co-operatives' federation.

**Unión Nacional de Periodistas:** Joaquín Auxe Iñaquito, Quito; national press association.

There are several other coffee and cocoa organizations.

## STATE HYDROCARBON COMPANY

**Empresa Estatal Petróleos del Ecuador (PETROECUADOR):** Alpallana E-8-86 y Avda 6 de Diciembre, Casilla 17-11-5007, Quito; tel. (2) 256-3060; fax (2) 250-3571; e-mail rin@petroecuador.com.ec; internet www.petroecuador.com.ec; f. 1989; state petroleum co; Exec. Pres. Robert Alonso Pinzón Rojas.

## MAJOR COMPANIES

The following are some of the leading industrial and commercial companies currently operating in Ecuador.

**Acero Comercial Ecuatoriano, SA:** Avda 10 de Agosto 3653 y Maríana de Jesús, Quito; tel. (2) 252-4450; fax (2) 222-1743; e-mail acero@telconet.net; f. 1957; production of construction materials; Gen. Man. Juan Pedro Bluhm; 260 employees.

**Cementos Selva Alegre, SA:** Edif. Banco La Previsora, Of. 402, CP 6663, Amazonas y Naciones Unidas, Quito; tel. (2) 459-140; fax (2) 256-091; e-mail compras@csa.com.ec; internet www.csa.com.ec; f. 1979; manufacture of cement; Gen. Man. José Espinosa Pérez; 567 employees.

**Compañía de Cervezas Nacionales, CA:** Vía a Daule, Km 16½, Casilla 09-01-519, Guayaquil; tel. (4) 289-3088; fax (4) 289-3263; e-mail cervecerianacional@ccn.com.ec; internet www.cervecerianacional.com.ec; f. 1921; brewing; Exec. Pres. Julio Mario Santo Domingo; 440 employees.

**Fábrica de Aceites La Favorita, SA:** G. Francisco de Marcos 102 y E. Alvaro, Guayaquil; tel. (4) 241-7025; fax (4) 241-4507; e-mail valledua@jnmail.com.ec; f. 1941; production of vegetable oils; Gen. Man. Ernesto Noboa Bejarano; 480 employees.

**Hidalgo e Hidalgo, SA:** Avda 10 de Agosto y Algarrobos, Quito; tel. (2) 240-8038; fax (2) 240-0541; e-mail hidalgo@hehconstructores.com; internet hehconstructores.com; f. 1969; construction; Pres. Juan Francisco Hidalgo; Gen. Man. Julio Hidalgo González; 1,000 employees.

**Importadora El Rosado Cía Ltda:** Calle 9 de Octubre 729 y Boyacá, Guayaquil; tel. (4) 232-2000; fax (4) 232-8196; e-mail luchoweb@elrosado.com; internet www.elrosado.com; f. 1954; retailing; Dir-Gen. Johnny Czarninsky; 4,000 employees.

**Industria Ecuatoriana Productora de Alimentos (INEPACA):** Calle Malecón, Casilla 4881, Manta; tel. (5) 624-870; fax (5) 624-870; e-mail ozambra@inepaca.net; internet www.ecuadorexplorer.com/inepaca; f. 1949; fishing and processing of fish; Pres. Dr Edgar Teán; Gen. Man. Carlos E. Zárate; 987 employees.

**Industrias Ales, CA:** 10 de Agosto 8919, Quito; tel. (2) 240-2600; fax (2) 240-8344; e-mail cmera@ales.com.ec; f. 1943; manufacture of cooking oils, fats and soap; Gen. Man. Patricio Álvarez Drouet.

**La Cemento Nacional Cía, CA:** Vía a la Costa Km 7 y Km 18, Casilla 09-01-04243, Guayaquil; tel. (4) 287-1900; fax (4) 287-3482; e-mail info@lcn.com.ec; internet www.lcn.com.ec; f. 1934; manufacture of cement; Gen. Man. Patrick Bredthauer; 1,472 employees.

**La Universal, SA:** Eloy Alfaro 1101–1109 y Gómez Rendón, Guayaquil; tel. (4) 241-4009; fax (4) 241-4904; e-mail launiversal@launiversal.com.ec; internet www.launiversal.com.ec; f. 1889; manufacture of food products; Pres. Domingo Norero; Gen. Man. Fernando Guzmán; 1,050 employees.

# ECUADOR

**Lanafit, SA:** Avda 6 de Diciembre 41–245 y Tomás de Berlanga, Sector El Inca, Pichincha, Quito; tel. (2) 226-8686; fax (2) 246-7049; e-mail lanafit@textilanafit.com; internet www.textilanafit.com; f. 1953; manufacture of clothing from synthetic fibres; Gen. Man. Fuad Alberto Dassum Arméndariz; 700 employees.

**Petroproducción, SA:** Avda 6 Diciembre 4226 y Gaspar Cañero, Casilla 17-01-1006, Quito; tel. (2) 244-0333; fax (2) 244-0383; e-mail info@petroproduccion.com.ec; internet www.petroecuador.com.ec/petroprod.htm; f. 1989; state petroleum and natural gas exploration enterprise; Gen. Man. Ing. Luis Albán; 1,300 employees.

**Sociedad Agrícola e Industrial San Carlos, SA:** Primero de Mayo 813 y Los Ríos, Guayaquil; tel. (4) 229-0800; fax (4) 229-0266; e-mail xmarcos@gu.pro.ec; f. 1897; processing and refining of sugar; Pres. Mario González; Dir Xavier Marcos Stagg; 4,000 employees.

**Supermercados La Favorita, CA (SUPERMAXI):** Avigiras E-12-70 y Avda Eloy Alfaro, Casilla 17-11-04910, Quito; tel. (2) 240-1140; fax (2) 240-2499; e-mail favorita@supermaxi.com; internet www.supermaxi.com; f. 1952; Pres. Ronald Owen Wright Durán Ballén; Gen. Man. Tomás Durán Ballén; 2,475 employees.

**Tejidos Pintex, SA:** Avda de la Prensa 3741 y Manuel Herrera, Quito; tel. (2) 448-333; fax (2) 448-335; e-mail pintex@access.net.ec; internet www.tejidospintex.com; f. 1959; manufacture of textiles; Pres. Cristina Pinto Mancheno; Gen. Man. Ing. Ramiro León Paez; 420 employees.

**Textil San Pedro, SA:** Avda Napo y Pedro Pinto Guzmán 709, Apdo 17-01-3002, Quito; tel. (2) 265-5604; fax (2) 266-1596; internet sanpedro.accessinter.net; production of cotton textiles; f. 1948; Gen. Man. Pedro Pinto.

**Textiles Nacionales, SA:** Gala Plaza Lasso 73-41 y Basantes, Quito; tel. (2) 248-3980; fax (2) 248-4378; e-mail info@textilesnacionales.com; internet www.textilesnacionales.com; f. 1950; manufacture of textiles; Pres. Joseph S. Handall; 450 employees.

**Váldez SA, Cía Azucarera:** Edif. Torres del Río, Junín 114 y Malecón, Guayaquil; tel. (4) 256-3966; fax (4) 256-3248; f. 1984; processing and refining of sugar; Gen. Man. Edmundo Váldez Murillo; 3,100 employees.

## UTILITIES

### Regulatory Authorities

**Ministry of Energy and Mines:** see section on The Government (Ministries).

**Centro Nacional de Control de Energía (CENACE):** Panamericana Sur Km 17.5, Sector Santa Rosa de Cutuglagua, Casilla 17-21-1991, Quito; tel. (2) 299-2030; fax (2) 299-2031; e-mail garguello@cenace.org.ec; f. 1999; co-ordinates and oversees national energy system; Exec. Dir Gabriel Argüello Ríos.

**Comisión Ecuatoriana de Energía Atómica:** Juan Larrea 15-36 y Riofrío, Casilla 17-01-2517, Quito; tel. (2) 254-5861; fax (2) 256-3336; e-mail comecen1@comecenat.gov.ec; atomic energy commission; Exec. Dir Celiano Almeida; Head of Information Hipsy Cifuentes.

**Consejo Nacional de Electricidad (CONELEC):** Avda Amazonas 33-299 e Inglaterra, Quito; tel. (2) 226-8746; fax (2) 226-8737; e-mail conelec@conelec.gov.ec; internet www.conelec.gov.ec; f. 1999; supervises electricity industry following transfer of assets of the former Instituto Ecuatoriano de Electrificación (INECEL) to the Fondo de Solidaridad; pending privatization as six generating companies, one transmission company and 19 distribution companies; Exec. Pres. Diego Pérez Pallares.

**Directorate of Renewable Energy and Energy Efficiency (Ministry of Energy and Mines):** J. L. Mera y Orellana, Edif. MOP, 6°, Quito; tel. (2) 255-0018; fax (2) 255-0018; e-mail jvasconez@menergia.gov.ec; internet www.menergia.gov.ec; f. 1995; research and development of new and renewable energy sources; Dir Ing. José Vasconez E.

**Directorate-General of Hydrocarbons:** Avda 10 de Agosto 321, Quito; supervision of the enforcement of laws regarding the exploration and development of petroleum.

### Electricity

**Corporación Eléctrica de Guayaquil, SA:** Urb. La Garzota, Sector 3, Manzana 47; tel. (4) 224-8006; fax (4) 224-8040; major producer and distributor of electricity, mostly using oil-fired or diesel generating capacity; partially privatized in 2003.

**Empresa Eléctrica Quito, SA (EEQ):** Avda 10 de Agosto y Las Casas, Quito; tel. (2) 252-5013; fax (2) 250-3817; e-mail gerenges@eeq.com.ec; internet www.eeq.com.ec; f. 1894; produces electricity for the region around Quito, mostly from hydroelectric plants; Gen. Man. Hernán Andino Romero.

**Empresa Eléctrica Regional El Oro, SA (EMELORO):** e-mail mandrad@emeloro.gov.ec; internet www.eeq.com.ec; electricity production and generation in El Oro province; Pres. Wilson Lapo.

**Empresa Eléctrica Regional del Sur, SA (EERSSA):** internet www.eeq.com.ec; f. 1973; electricity production and generation in Loja and Zamora Chinchipe provinces; Man. Daniel Mahauad Ortega.

**Empresa Eléctrica Riobamba, SA:** Veloz 20-12 y Tarqui, Riobamba; tel. (3) 961-693; fax (3) 965-257; state-owned utility; Pres. Oswaldo García Davalos.

### Water

**Instituto Ecuatoriano de Obras Sanitarias:** Toledo 684 y Lérida, Troncal, Quito; tel. (2) 252-2738.

## TRADE UNIONS

**Frente Unitario de Trabajadores (FUT):** f. 1971; left-wing; 300,000 mems; Pres. Jaime Arciniegas Aguirre; comprises:

**Confederación Ecuatoriana de Organizaciones Clasistas (CEDOC):** Edif. Cedocut 5°, Flores 846 y Manabì, Quito; tel. (2) 295-4551; fax 295-4013; internet www.cedocut.org; f. 1938; affiliated to CMT and CLAT; humanist; Pres. Mesías Tatamuez Moreno; Vice-Pres. Wilson Álvarez Bedón; 1,065 mem orgs, 86,416 individual mems.

**Confederación Ecuatoriana de Organizaciones Sindicales Libres (CEOSL):** Avda Tarqui 15-26, 6°, Casilla 17-11-373, Quito; tel. (2) 252-2511; fax (2) 250-0836; e-mail ceosl@hoy.net; internet www.ceoslecuador.org; f. 1962; affiliated to ICFTU and ORIT; Pres. Jaime Arciniegas Aguirre; Sec.-Gen. Guillermo Touma González.

**Confederación de Trabajadores del Ecuador (CTE)** (Confederation of Ecuadorean Workers): 9 de Octubre 26-106 y Marieta de Veintimilla, Casilla 17-014166, Quito; tel. (2) 252-0456; fax (2) 252-0445; e-mail cte@punto.net.ec; internet www.cte-ecuador.org; f. 1944; admitted to WFTU and CEPUSTAL; Pres. Santiago Yagual Yagual; 1,200 affiliated unions, 76 national federations.

**Central Católica de Obreros:** Avda 24 de Mayo 344, Quito; tel. (2) 221-3704; f. 1906; craft and manual workers and intellectuals; Pres. Carlos E. Dávila Zurita.

A number of trade unions are not affiliated to the above groups. These include the Federación Nacional de Trabajadores Marítimos y Portuarios del Ecuador (FNTMPE—National Federation of Maritime and Port Workers of Ecuador) and both railway trade unions.

# Transport

## RAILWAYS

All railways are government-controlled. In 2000 the total length of track was 956 km.

**Empresa Nacional de Ferrocarriles del Estado (ENFE):** Calle Bolívar 443, Casilla 159, Quito; tel. (2) 221-6180; e-mail enfegg@andinanet.net; national network currently has only 200 miles of track; Gen. Man. Sergio Coellar.

There are divisional state railway managements for the following lines: Guayaquil–Quito, Sibambe–Cuenca and Quito–San Lorenzo.

## ROADS

There were 43,197 km of roads in 2001, of which 16.7% were paved. The Pan-American Highway runs north from Ambato to Quito and to the Colombian border at Tulcán and south to Cuenca and Loja. Major rebuilding projects were undertaken in late 1998 with finance from several development organizations to restore roads damaged by the effects of El Niño (a periodic warming of the tropical Pacific Ocean).

## SHIPPING

The following are Ecuador's principal ports: Guayaquil, Esmeraldas, Manta and Puerto Bolívar.

**Acotramar, CA:** General Gómez 522 y Coronel Guayaquil, Casilla 4044, Guayaquil; tel. (4) 240-1004; fax (4) 244-4852.

**Ecuanave, CA:** Junin 415 y Córdova, 4°, Casilla 09-01-30H, Guayaquil; tel. (4) 229-3808; fax (4) 228-9257; e-mail ecuanav@ecua.net.ec; Chair. Ing. P. Ernesto Escobar; Man. Dir A. Guillermo Serrano.

**Flota Bananera Ecuatoriana, SA:** Edif. Gran Pasaje, 9°, Plaza Ycaza 437, Guayaquil; tel. (4) 230-9333; f. 1967; owned by Govt and private stockholders; Pres. Diego Sánchez; Gen. Man. Jorge Barriga.

**Flota Mercante Grancolombiana, SA:** Guayaquil; tel. (4) 251-2791; f. 1946; with Colombia and Venezuela; on Venezuela's withdrawal, in 1953, Ecuador's 10% interest was increased to 20%; operates services from Colombia and Ecuador to European ports, US Gulf ports and New York, Mexican Atlantic ports and East Canada; offices in Quito, Cuenca, Bahía, Manta and Esmeraldas; Man. Naval Capt. J. ALBERTO SÁNCHEZ.

**Flota Petrolera Ecuatoriana (FLOPEC):** Edif. FLOPEC, Avda Amazonas 1188 y Cordero, Casilla 535-A, Quito; tel. (2) 256-4058; fax (2) 256-9794; e-mail g.general@flopec.com.ec; internet www.flopec.com.ec; f. 1972; Gen. Man. Vice-Adm. JORGE DONOSO MORAN.

**Logística Marítima, CA (LOGMAR):** Avda Córdova 812 y V. M. Rendón, 1°, Casilla 9622, Guayaquil; tel. (4) 230-7041; Pres. J. COELLOG; Man. IGNACIO RODRÍGUEZ BAQUERIZO.

**Naviera del Pacífico, CA (NAPACA):** El Oro 101 y La Ría, Casilla 09-01-529, Guayaquil; tel. (4) 234-2055; Pres. LUIS ADOLFO NOBOA NARANJO.

**Servicios Oceánicos Internacionales, SA:** Avda Domingo Comin y Calle 11, Casilla 79, Guayaquil; Pres. CARLOS VALDANO RAFFO; Man. FERNANDO VALDANO TRUJILLO.

**Transfuel, CA:** Junin 415 y Córdova, 4°, Casilla 09-01-30H, Guayaquil; tel. (4) 230-4142; Chair. Ing. ERNESTO ESCOBAR PALLARES; Man. Dir CARLOS MANRIQUE A.

**Transportes Navieros Ecuatorianos (Transnave):** Edif. Citibank, 4°–7°, Avda 9 de Octubre 416 y Chile, Casilla 4706, Guayaquil; tel. (4) 256-1455; fax (4) 256-6273; transports general cargo within the European South Pacific Magellan Conference, Japan West Coast South America Conference and Atlantic and Gulf West Coast South America Conference; Pres. Vice-Adm. YÉZID JARAMILLO SANTOS; Gen. Man. RUBÉN LANDÁZURI ZAMBRANO.

### CIVIL AVIATION

There are two international airports: Mariscal Sucre, near Quito, and Simón Bolívar, near Guayaquil.

**LAN Ecuador, SA:** Avda de las Américas s/n, Guayaquil; tel. (4) 269-2850; fax (4) 228-5433; internet www.lanecuador.com; f. 2002; commenced operations in April 2003, following acquisition of assets of Ecuatoriana by LAN Chile; scheduled daily flights between Quito, Guayaquil, Miami and New York; Dir BRUNO ARDITO.

**TAME Línea Aérea del Ecuador:** Avda Amazonas 1354 y Colón, 6°, Casilla 17-07-8736, Sucursal Almagro, Quito; tel. (2) 250-9375; fax (2) 255-4907; e-mail tamejefv@impsat.net.ec; internet www.tame.com.ec; f. 1962; fmrly Transportes Aéreos Mercantiles Ecuatorianos, SA; removed from military control in 1990, state-owned; domestic scheduled and charter services for passengers and freight; Pres. JORGE CABEZAS; Gen. Man. BOLÍVAR MORA VINTIMILLA.

The following airlines also offer national and regional services:

Aerotaxis Ecuatorianos, SA (ATESA); Cía Ecuatoriana de Transportes Aéreos (CEDTA); Ecuastol Servicios Aéreos, SA; Ecuavia Cía Ltda; Aeroturismo Cía Ltda (SAVAC).

## Tourism

Tourism has become an increasingly important industry in Ecuador, with 654,993 foreign arrivals (including same-day visitors) in 2002. Of total visitors in that year, some 29% came from Colombia, 13% were from Peru, 9% from other Latin American countries, 23% from the USA and 18% from Europe. In 2003 receipts from the tourism industry amounted to US $408m.

**Asociación Ecuatoriana de Agencias de Viajes y Turismo (ASECUT):** Edif. Banco del Pacífico, 5°, Avda Amazonas 22–94 y Veintimilla, Casilla 9421, Quito; tel. (2) 250-3669; fax (2) 250-0238; e-mail asecut@pi.pro.ec; f. 1953; Pres. ERIK RUGEL ROCA.

**Federación Hotelera del Ecuador (AHOTEC):** América 5378 y Diguja, Quito; tel. (2) 244-3425; fax (2) 245-3942; e-mail ahotec@interactive.net.ec; internet www.hotelesecuador.com.ec; Pres. JOSÉ OCHOA; Exec. Dir DIEGO UTRERAS.

**Ministry of Tourism:** Avda Eloy Alfaro 32-300 y Carlos Tobar, Quito; tel. (2) 250-7559; fax (2) 250-7565; e-mail promocion@turismo.gov.ec; internet www.vivecuador.com; Minister of Tourism MARÍA ISABEL SALVADOR.

## Defence

At 1 August 2004 Ecuador's armed forces numbered 46,500: Army 37,000, Navy 5,500 (including 1,700 marines), Air Force 4,000. Paramilitary forces included 270 coastguards. Military service lasts for one year and is selective for men at the age of 20.

**Defence Expenditure:** an estimated US $588m. in 2004.

**Chief of the Joint Command of the Armed Forces:** Rear-Adm. VÍCTOR HUGO ROSERO.

**Chief of Staff of the Army:** Gen. JORGE ZURITA RÍOS.

**Chief of Staff of the Air Force:** Lt-Gen. LUIS HERNÁN AYALA SALAZAR.

## Education

Education in Ecuador is officially compulsory for six years, to be undertaken between six and 14 years of age. All public schools are free. Private schools feature prominently in the educational system. Primary education is available for children aged between six and 12 years. In 1999 the total enrolment at primary and secondary schools was equivalent to 91% of the school-age population. In 1999/2000 a total of 1,982,636 children attended 18,203 primary schools. Secondary education, in general and specialized technical or humanities schools, is available for students aged 12 to 18. In 1999/2000 there were 3,486 secondary schools, attended by a total of 980,213 pupils. University courses last for up to six years, and include programmes for teacher training. In many rural areas, Quechua and other indigenous Amerindian languages are used in education. Total expenditure on education by the central Government was estimated at 1,957,051m. sucres (equivalent to 3.5% of GNP) in 1996.

## Bibliography

For works on South America generally, see Select Bibliography (Books)

Cruz, H., Castro, A. V., and Arnold, A. *Faith in Service: Developing Credit Unions in Ecuador.* Los Angeles, CA, Writer's Showcase Press, 2001.

Downes, R., and Marcella, G. *Security Cooperation in the Western Hemisphere: Resolving the Ecuador-Peru Conflict.* Boulder, CO, Lynne Rienner Publrs, 1999.

Gerlach, A. *Indians, Oil, and Politics: A Recent History of Ecuador (Latin American Silhouettes).* Wilmington, DE, Scholarly Resources Inc., 2003.

Herz, M., and João Pontes, N. *Ecuador Vs Peru: Peacemaking Amid Rivalry.* Boulder, CO, Lynne Rienner Publrs, 2002.

Kyle, D. *Transnational Peasants: Migrations, Networks and Ethnicity in Andean Ecuador.* Baltimore, MD, Johns Hopkins University Press, 2000.

Lane, K. *Quito 1599: City and Colony in Transition.* Albuquerque, NM, University of New Mexico Press, 2002.

Pallares, A. *From Peasant Struggles to Indian Resistance: The Ecuadorian Andes in the Late Twentieth Century.* Norman, OK, University of Oklahoma Press, 2002.

Rival, L. *Trekking Through History: The Huaorani of Amazonian Ecuador.* New York, NY, Columbia University Press, 2002.

Selverston-Scher, M. *Ethnopolitics in Ecuador: Indigenous Rights and the Strengthening of Democracy.* Boulder, CO, Lynne Rienner Publrs, 2001.

Solimano, A., and Beckerman, P. *Crisis and Dollarization in Ecuador: Stability, Growth, and Social Equity (Directions in Development).* Washington, DC, World Bank, 2002.

Striffler, S. *In the Shadows of State and Capital: The United Fruit Company, Popular Struggle, and Agrarian Restructuring in Ecuador, 1900–1995.* Durham, NC, Duke University Press, 2002.

Whitten, N. (Ed.) *Millennial Ecuador: Critical Essays on Cultural Transformations and Social Dynamics.* Iowa City, IA, University of Iowa Press, 2003.

# EL SALVADOR

## Geography

### PHYSICAL FEATURES

The Republic of El Salvador is the smallest country in Central America and the only one without a Caribbean shore. It lies on the western or Pacific side of the Central American land bridge, but itself has a southern coast. Guatemala lies to the west, further up the isthmus, beyond a 203-km (126-mile) border. Honduras is to the north and east—a definitive border demarcation, along the 342 km of frontier, was only agreed in 1992, when an International Court of Justice (ICJ—based in The Hague, Netherlands) decision was accepted, although Honduras has since complained about implementation, and many of the *bolsones* (disputed areas) remain unresolved. The ICJ referred the issue of the maritime boundary in the Gulf of Fonseca (a line was agreed in 1990 by the Honduras–Nicaragua Mixed Boundary Commission) to tripartite discussion—Nicaragua lies beyond the Gulf, in the south-east. The dispute in the Gulf is also complicated by the El Salvadorean claim to the island of Conejo, currently held by Honduras. The country has a total area of 21,041 sq km (8,124 sq miles).

El Salvador is about 260 km in length (east–west) and 140 km wide, with 307 km of coast along the Pacific Ocean. It is a land of volcanoes, and is prone to sometimes devastating earthquakes. This can make the terrain unstable and dangerous, but has also given the country rich volcanic soil suitable for growing coffee, the basis of the Salvadorean economy. The uplands consist of a double row of volcanoes and mountains, the roughly parallel and east–west coastal chain and the further inland Cordillera Apeneca, which reach their highest point in the north-west, at Cerro El Pital (2,730 m or 8,960 ft). There is also a central plateau and, beneath the highlands, falling fairly steeply into the Pacific, is a narrow coastal plain. The three main topographical areas are, therefore: a flat, tropical region in the south, some of it wetlands; the central plateau of mountains, valleys and volcanoes; and the northern lowlands formed by the valleys of the Lempa river and the Sierra Madre. In all there are 150, usually fast-flowing, rivers and three lakes. The terrain still sustains much biological diversity, despite the pressures of the densely settled human population, with, for instance, more species of trees than in all of Western Europe. There are reckoned to be large numbers of species of plants (notably orchids), butterflies, birds and fish, but, nevertheless, there are fewer than in any other Central American country. Woodland covers 17% of El Salvador (only 3% of the country remains with its natural primary forest), most of it secondary forest and scrubland, but there is an additional 9% of territory planted with coffee bushes, which are also provided with trees for shade. El Salvador has the highest rate of deforestation (just over 3% per year) and the least amount of territory protected by national parks (0.5%) in Central America.

### CLIMATE

The climate is tropical, but more temperate in the high country. The rainy season is over the summer, in May–October, the wettest month in San Salvador, the inland capital, being June. Rainfall is generally heavier on the coast, however, while the interior remains relatively dry. The average annual rainfall for the whole country is almost 1,800 mm (70 ins). Temperatures in

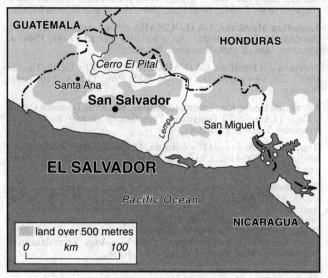

the wet season average about 28°C (82°F), whereas in the cooler dry season they range between 15°C and 23°C (59°–73°F). Greater altitude, of course, moderates these temperatures. The country is susceptible to hurricanes from the Caribbean, as well as to the vagaries of the El Niño weather phenomenon.

### POPULATION

The population is predominantly Mestizo, of mixed-race descent (90%), the rest of the population consisting of those of European stock and of indigenous Amerindians (the latter consisting mainly of the Pipil, descendants of the Aztecs, and the Lenca). A number of nationalities have settled in El Salvador over the years, but nearly all of today's population is of Spanish descent. For this reason it is no surprise to find that around 77% of the country adheres to the Roman Catholic Church, although, as elsewhere in Latin America, evangelical Protestant groups have been very active and making significant numbers of converts. The official language, and that used by nearly the entire population, is Spanish. Some speak Amerindian tongues, notably Nahua (Pipil). Informally, the people of El Salvador are sometimes known as Guanacos.

The total population was estimated to be 6.8m. in mid-2004, about two-fifths of whom live in rural areas and one-third in and around the greater metropolitan area of the capital, San Salvador, which is located in the highlands immediately above the western end of the coastal plains. There are a number of other large urban centres, but probably the two most eminent cities are Santa Ana, to the north-west of the capital, and San Miguel, in the south-east of El Salvador. Although it is the smallest country of Central America, El Salvador is second in population only to Guatemala—indeed, it is the most densely populated country on the American mainland (326.1 people per sq km in 2004). For administrative purposes, the country is divided into 14 departments.

# History

## ANDREW BOUNDS

Revised for this edition by DIEGO SÁNCHEZ-ANCOCHEA

El Salvador's history has been dominated by the land question. The country is the smallest and most densely populated on mainland America and has very few natural resources. For thousands of years before the conquest by Spain between 1524–35 indigenous (Amerindian) races populated the country. Following the conquest, Spanish settlers established large plantations of cocoa and then indigo for export, and the native population was forced on to ever smaller areas of common land where they grew traditional subsistence crops such as rice, beans and yucca. This competition for a limited land base and the practice of planting export crops over as great an area as possible shaped El Salvador's society, politics and economy.

El Salvador achieved full political independence after the end of Spanish rule and the collapse of the United Provinces of Central America, which grouped it with Honduras, Guatemala, Nicaragua and Costa Rica from 1823–29. From the mid-19th century a small group of local landowners and merchants transformed the country into a specialized producer of agricultural exports. After the collapse of world demand for indigo in the 1860s coffee emerged as the mainstay of the national commercial economy, displacing traditional systems of local food production. In 1882 all common lands were abolished and three-quarters of all land passed into the private ownership of families comprising only about 2% of El Salvador's inhabitants. The majority of the population, displaced from their traditional lands, became permanent wage labourers on the new plantations or migratory seasonal workers.

In contrast to other countries in Central America, the agro-export economy of El Salvador was created by domestic, not foreign, capital and expertise. The economy was dominated by an interlocked élite of landowners and merchants (known as the 'Fourteen Families') who controlled the state, land, capital and markets. The formal democratic procedures of a constitutional republic were maintained, but governments were effectively appointed by the oligarchy to administer power in its own interests.

## MILITARY DOMINATION OF GOVERNMENT, 1932–82

The events of 1931–32 challenged the basis of this power structure. In 1931, against a background of national and international economic depression, relatively fair elections brought to office a reformist President, Arturo Araújo. The existence of widespread popular discontent and allegations that the Partido Comunista Salvadoreño (PCS—Salvadorean Communist Party), recently established under the leadership of Agustín Farabundo Martí, was promoting armed insurrection, provoked a conservative reaction. Araújo was deposed by a military coup in December 1931 and replaced by Gen. Maximiliano Hernández Martínez. A large-scale peasant uprising in January 1932 was violently suppressed; between 10,000 and 30,000 people, among them many Amerindians, were killed in reprisals known as *la matanza* (the massacre). Farabundo Martí was arrested and executed. The landowning élite, shaken by these perceived threats to its economic interest, abdicated its control of political power to the army that had saved it.

For the next 50 years the relationship between the armed forces and the civilian oligarchy remained the central reality of the nation's power structure: the former guaranteed the privileges of the latter, while simultaneously promoting their own interests by establishing military rule as an institution. Shifts between 'conservative' and 'progressive' factions within the military led to a pattern of reform and repression. When abuses of presidential power threatened to provoke popular discontent, incumbent Presidents were removed by military coup: Gen. Hernández Martínez in 1944, Lt-Col José María Lemus in 1960 and Gen. Carlos Humberto Romero Mena in 1979. When reformist administrations were considered to be too radical, they were removed by counter-coup, as in the overthrow of the military-civilian juntas of 1944 and 1960.

The economy continued to develop after 1945. A second period of export-led agricultural growth (based mainly on sugar, cotton and cattle) was accompanied by some industrialization. The latter was as a result of the creation of the Central American Common Market (CACM), which increased market size. As in the past, this development was achieved primarily by Salvadorean capital and entrepreneurs, but the benefits accrued to a minority while the needs of the majority continued to be denied. It also benefited food processors and exporters more than producers. This created a new, more progressive class in the oligarchy that was not so dependent on land and bonded labour as the traditional coffee growers.

Between 1961 and 1979 the military leadership attempted to present its own party, the Partido de Conciliación Nacional (PCN—Party of National Conciliation), as the country's unifying force. Other parties were tolerated, but repeated electoral manipulation ensured that PCN candidates—Lt-Col Julio Adalberto Rivera, Gen. Fidel Sánchez Hernández, Col Arturo Armando Molina Barraza and Gen. Romero—gained the presidency, and so retained power in the hands of the army. From 1960 the Partido Demócrata Cristiano (PDC—Christian Democratic Party), under the leadership of José Napoleón Duarte, consistently attracted the largest electoral support for any opposition party and grew in strength steadily.

In July 1969 a 13-day war with Honduras killed 2,000 people. The catalyst for the so-called 'soccer war' was a disputed decision in the third qualifying round of the soccer world cup. The roots of the conflict were territorial disputes and migration pressures. Some 300,000 Salvadoreans had emigrated to Honduras to farm. Honduras decided to expel them as it pressed a boundary claim. The frontier dispute was settled in September 1972 when Honduras was awarded two-thirds of the disputed land by international arbitration.

The return of the emigrants, now refugees, put more pressure on El Salvador's land. The number of landless peasants grew from 30,500 in 1961 (12% of the rural population) to 167,000 in 1975 (41% of the rural population). This combined with an economic depression, caused partly by the 1973 oil shock and its global consequences and partly by the loss of the Honduran market, to put heavy pressure on the regime. In the 1972 presidential election the opposition parties united behind Duarte, but Col Molina unilaterally declared victory. Subsequent protests and an attempted coup were crushed and Duarte exiled. These events not only frustrated the hopes of fundamental reform by democratic means, but also convinced many opponents of military rule that armed insurrection was inevitable.

In 1969 ORDEN, the first of a succession of extreme right-wing terrorist groups, was established to assassinate and intimidate those in support of reform and political change. In 1970 Cayetano Carpio, Secretary-General of the PCS, broke from the party to pursue a campaign of armed insurrection. His lead was followed by a number of distinct guerrilla groups on the extreme left, and in Cuba in 1980 they co-ordinated to form the Frente Farabundo Martí de Liberación Nacional (FMLN—Farabundo Martí National Liberation Front). The FMLN also established a political wing, the Frente Democrático Revolucionario (FDR—Democratic Revolutionary Front). Reforming politicians, previously strongly opposed to armed struggle, began to support the FMLN as the only available option for the pursuit of democratic change.

The Roman Catholic Church, which had been a supporter of the conservative regime, also became more identified with the opposition after the Medellín council of bishops in 1968 called for it to transform the lives of the poor and overcome injustice. The Church began to preach a message of social justice and organized bible study groups in each parish that became one of the few forms of social gathering permitted under the military regime. These groups moulded many resistance leaders.

Gen. Romero's victory in the presidential election of 1977 followed a campaign characterized by intimidation, fraud and violent suppression of subsequent public protests. Alarmed by the implications of the overthrow of the dictatorship of Gen. Anastasio Somoza Debayle in Nicaragua in July 1979, the military ousted President Romero in October. After a series of abortive civilian-military juntas, Duarte agreed to lead a provisional government, on the condition that fundamental reforms be introduced immediately and be guaranteed by the military and by the USA.

Duarte joined the junta in March 1980. In the same month the Government, assisted by the USA, expropriated one-quarter of all agricultural land, for conversion into peasant co-operatives, and nationalized the banks and major export institutions. This was the first stage of the most important change in the nation's economy since the abolition of common lands a century before. However, civil war was already imminent. Army and 'death-squad' human rights abuses were such that the population had little faith in the Government. At the end of March Oscar Romero y Galdames, Archbishop of San Salvador, whose Sunday homilies had regularly expressed support for the cause of the poor and oppressed, was assassinated in the act of celebrating Mass. At his funeral, without provocation, soldiers fired into the crowd of mourners, which numbered more than 250,000. The Government was also under pressure from the left, which, as in 1932, favoured revolution over reform.

## THE CIVIL WAR

### The Effects of the War

In January 1980 the FMLN launched its 'Final Offensive', intended to achieve victory before the inauguration of President Ronald Reagan of the USA. It failed and the civil war continued at an increasing cost in human suffering and economic disruption. By 1992 more than 80,000 combatants and civilians had been killed, the vast majority by the armed forces; an estimated 550,000 (more than 10% of the population) had been displaced from their homes within the country, while in excess of 500,000 had fled the country as refugees. There was military stalemate. The army, relying heavily on US support, was increased in size to 55,000 military and 15,000 paramilitary personnel. However, it was ill-led and relied on a strategy of sporadic infantry attacks, supported by aerial bombings across extensive 'free-fire' zones, thus alienating the civilian population. The FMLN relied on well co-ordinated, dispersed ambush attacks, urban terrorism and economic sabotage, such as the destruction of power lines and crops and the mining of public highways, all of which increased the war's impact on civilians.

The economic cost of the conflict was so high that the collapse of the economy was only averted by direct US economic assistance. One-half of government budgets were committed to defence spending. Reduced export earnings and increased government deficits by 1987 resulted in servicing of the public-sector external debt exceeding 50% of annual export income. Unemployment and underemployment affected more than one-half of the adult population and per-head income decreased to levels of the 1960s. The crisis was exacerbated by two natural disasters: the 1986 earthquake that killed 1,500 people and the subsequent drought.

Elections for a constituent assembly in March 1982 divided power between the PDC and two major right-wing parties—the PCN and the Alianza Republicana Nacionalista (ARENA—Nationalist Republican Alliance), founded in 1981. In 1982 Alvaro Magaña Borja, a politically independent banker, was accepted by all parties as interim President. Under his guidance, in 1983 the major parties agreed to a new Constitution. This provided for a democratic political process and incorporated the essential principles of economic reform.

In May 1984 Duarte won the presidency in a direct electoral contest with the ARENA candidate, Maj. Roberto D'Aubuisson Arrieta. Elections in March 1985 gave the PDC an absolute majority in the Asamblea Nacional (National Assembly). However, Duarte failed to make progress towards a negotiated settlement of the war or social and economic reforms. In the 1988 election ARENA won control of the Asamblea Nacional and in 1989 its candidate, Alfredo Cristiani Burkard, was elected President.

### Progress Towards Peace

Formal and informal contacts between the Government and the FDR-FMLN began as early as 1980. Each attempt at a negotiated peace, however, foundered on two key issues: the future role, control and structure of the military; and the integration of the FDR-FMLN into national political life. Despite his position as Commander-in-Chief of the Armed Forces, President Duarte did not control them and so could not guarantee the FDR-FMLN's conditions. His failure to do so was demonstrated by the terror tactics of the extreme right-wing death squads, which, aided by his senior officers, were responsible for the assassination of church leaders, trade unionists and political activists.

The FDR-FMLN also considered each of the six elections held between 1982 and 1989 as invalid. On each occasion the guerrillas disrupted balloting. From 1989 the FDR leaders accepted that there had been sufficient improvement in electoral conditions to allow them openly to participate in the election, as the Convergencia Democrática (CD—Democratic Convergence), but not until 1991 did one of its five constituent parties contest national elections.

The victory of ARENA in the elections of 1989 and 1991 caused widespread expectations that the conflict would escalate. Many of its senior members were openly committed to a military solution and there was a common belief that some party leaders, such as Roberto D'Aubuisson, were involved in the death squads. After President Cristiani's assumption of power in April 1989, the FMLN launched limited, but effective, offensives in May and June, accompanied by direct attacks against the ARENA leadership, notably the assassination of the Attorney-General in April and the Minister of the Presidency in July. The reaction of the right-wing extremists, orchestrated by groups within ARENA and the military, involved the assassination of trade-union leaders and suspected FMLN sympathizers. FMLN–ARENA negotiations in September broke down almost immediately and were followed by a nation-wide November offensive by the FMLN, which posed the most formidable challenge to the Government in 10 years of civil war. The FMLN gained temporary occupation of large areas of the capital. These successes provoked major reactions by right-wing extremists: the assassination of six Jesuit priests; the less-remarked killings of alleged FDR-FMLN sympathizers; and intimidation of church and human-rights groups and of left-wing politicians. In April 1990, however, representatives of the Government and the FDR-FMLN met in Geneva, Switzerland, under the chairmanship of the UN Secretary-General. Following a series of difficult negotiations a comprehensive agreement was reached by 31 December 1991, leading to a cease-fire on 1 February 1992.

The cease-fire was brought about by both domestic and international factors. The end of the cold war in 1989 ended US–Soviet confrontation in Central America and this allowed a regional solution by reviving the Esquipulas Accord, proposed by Costa Rica and endorsed by Central American Presidents in 1987. The Accord committed all five countries to adopt specific measures to achieve regional peace. These measures included: dialogue between Governments and insurgent groups; commitments to democratic and pluralistic political systems; and cessation of support for insurgent groups from whatever source. The end of the cold war also allowed the UN to become actively involved in the task of conflict resolution in El Salvador: firstly, by sponsorship of negotiations; secondly, by the establishment of a resident UN Observer Mission in El Salvador (ONUSAL), to verify compliance with negotiated agreements. Finally, within El Salvador, the Government, and especially the military, came under intense US pressure to reach a settlement. The ARENA leadership was dominated by export and consumer-oriented business and financial interests who needed peace to continue to flourish, while the FMLN, appreciating its increasing international and regional isolation, accepted the need for a peace agreement.

This settlement was achieved in the UN-brokered agreement, announced in December 1991, and signed at Chapultepec Castle in Mexico City on 16 January 1992. The Chapultepec Accords provided a framework for the reconstruction of Salvadorean society. The peace agreement focused on the demilitarization and submission of the country to civilian control under the rule of law. On 1 February the formal cease-fire was implemented under the supervision of some 1,000 UN personnel and the National Commission for the Consolidation of Peace (COPAZ) was formally installed. COPAZ was composed of representatives from both government and guerrilla forces, as well as from all major political parties. Its aim was to supervise the enforcement of guarantees for the political integration of the FMLN.

As well as the immediate measures for the disengagement and demobilization of FMLN guerrillas, and the reform and reduction of the Salvadorean military, the Accords established a range of new civilian institutions and programmes. These included the participation of former FMLN members in a new national civilian police force (Policía Nacional Civil—PNC), which replaced the paramilitary national police. A new National Council for the Defence of Human Rights was to be supported by an independent National Judiciary Council. A Land Transfer Programme for demobilized combatants and displaced civilians envisaged the transfer of some 10% of El Salvador's agricultural land to a total beneficiary population of about 47,500 people. A tripartite forum, representing the Government, workers and the private sector, was established in order to formulate social and economic policies.

Initial progress was made possible by a widespread desire for reconciliation and a willingness to seek *concertación*, or consensus. However, mutual allegations of failure to comply with the terms of the Accords persisted throughout 1992 and resulted in the negotiation of a revised timetable for disarmament. Nevertheless, the cease-fire was observed by both sides and, on 15 December (declared National Reconciliation Day), the conflict was formally concluded. On the same day the FMLN was officially registered and recognized as a legitimate political party.

## POST-ACCORD POLITICS

Expectations over the Chapultepec Accords were only partially fulfilled. In November 1992 the Comisión de la Verdad (Truth Commission) released the names of 103 military personnel alleged to have participated in human rights abuses in the civil war. The Government, however, was at first reluctant to remove from the armed forces those personnel identified by the Commission. The FMLN was prompted to delay the demobilization of its forces. The effective operation of the new civilian police was constrained by a lack of resources. A dramatic decline in political violence and human rights violations was accompanied by increasing criminal violence. The independence and security of the judiciary, reformed on the recommendations of the Commission's report, had not yet been tested. In late 1993, following international pressure, the military personnel identified by the Commission were dismissed.

These issues posed the principal themes in the campaign leading to national elections on 20 March 1994, monitored by ONUSAL. In spite of problems in the organization of the elections, the people of El Salvador were provided with their first opportunity to express their political preferences in elections that were peaceful as well as free and fair. The three major contending parties were ARENA, the PDC and the FMLN, which in September 1993 had confirmed its political alliance with the CD, and later, in December, with the Movimiento Nacional Revolucionario (MNR—National Revolutionary Movement). Following a second-round run-off contest on 24 April, Armando Calderón Sol, the ARENA candidate, was elected President. He took office on 1 June.

Despite its success, serious divisions emerged within the FMLN in 1994. In December, two factions, the Resistencia Nacional (RN—National Resistance) and the Expresión Renovadora del Pueblo (ERP—Renewed Expression of the People), left the FMLN because of a difference in political interests. In March 1995 Joaquín Villalobos, the ERP Secretary-General, announced the formation of the Partido Demócrata (PD—Democrat Party), a centre-left grouping consisting of the ERP, the RN, the MNR and a dissident faction of the PDC. The PD co-operated with the ruling ARENA party. In June 1995 the support of five PD deputies, as well as three independents, allowed the Government to gain legislative approval for an increase in value-added tax (VAT) to 13%, which had been opposed vehemently by others. However, a year later the PD withdrew from its pact with the Government, in protest at the latter's failure to soften its neoliberal policies and boost expenditure on health and education.

Meanwhile, there was increasing dissatisfaction with the Government's failure to honour the terms of the Chapultepec Accords. Former soldiers alleged that they had not received financial compensation and other benefits promised in the 1992 agreement. In September 1994 retired soldiers occupied the parliament building and held a number of deputies hostage. The Government immediately pledged to enter into direct negotiations with the soldiers, and the siege ended peacefully. However, in January 1995 former soldiers again occupied the Asamblea Nacional and took a number of hostages. Once again the occupation ended swiftly and bloodlessly as the Government reiterated its promise to meet its obligations. However, the spectre of renewed armed conflict was still in the background.

A reduced ONUSAL contingent, known as MINUSAL, remained in El Salvador until the end of 1996. In March 1997, contrary to expectation, the ruling ARENA party lost seats to the FMLN in municipal and legislative elections. The FMLN also experienced a significant increase in support in the capital. However, the FMLN was deeply divided and unable to present a coherent alternative to ARENA nationally. In the presidential elections, held on 7 March 1999, Francisco Flores Pérez, the ARENA candidate, was elected President, with 52% of the votes cast. The election was characterized by the highest abstention rate in the country's history, with over 60% of the country's 3m. registered voters failing to turn out. Flores, a former professor of philosophy, was sworn in as President on 1 June.

Although in many respects the peace process in El Salvador had opened up the country's political arena to wider participation, and judicial and political reforms were planned, most of the population remained too overwhelmed by the daily struggle for economic survival to feel that politics had any relevance to their lives. The neoliberal reforms adopted from the beginning of the 1990s had neither delivered high rates of growth nor a substantial improvement in the living standards of a majority of the population. Moreover, the levels of criminality in the country were of greater importance to the people than the electoral contest. The number of violent deaths in El Salvador was higher after the civil war than during it. President Flores announced that his main priority was to combat crime through the joint efforts of the PNC, the judicial system, the Government and the public and the murder rate did fall during his administration. However, his first year in office was marred by a series of abductions for ransom, some apparently involving police-officers, and numerous strikes and protests over his proposals for private-sector involvement in the health service.

## THE 2000 ELECTIONS

Voter dissatisfaction with the new Government as economic growth slowed resulted in an unexpected victory for the FMLN in legislative elections held on 12 March 2000, when it became the largest single party in the Asamblea Nacional, winning 31 of the 84 legislative seats. ARENA was the second largest party, winning 29 seats. However, the PCN came third, securing 14 seats, giving the right wing a working majority in the legislature, although not guaranteeing it the 56 votes required for the approval of important legislation. Once again there was a low voter turn-out, of about 33%. ARENA's poor electoral performance initiated an upheaval within the party. The FMLN made further gains in concurrently held municipal elections, winning control of 78 of the country's 262 municipalities, compared with ARENA's 127. The FMLN candidate, Héctor Silva Aregüello, won a convincing re-election to the mayoralty of San Salvador.

## FMLN–ARENA TENSIONS

Following the 2000 elections relations between the FMLN and ARENA immediately became strained as the three right-wing parties combined to prevent the FMLN from claiming the presidency of the Asamblea Nacional, a post traditionally held by the largest party. To prevent a political impasse it was decided that the presidency was to be rotated annually among the three largest parties, the PCN, ARENA and the FMLN. In late 2000 President Flores responded to ARENA's electoral reverses by amending some of his more unpopular economic policies; he suspended the privatization of the two remaining state banks, announced new public spending in port infrastructure, and implemented a series of measures to protect the agricultural sector against cheap imports and to provide credit to the coffee sector. On 22 November President Flores unexpectedly announced that, from 1 January 2001, the US dollar would be introduced as an official currency alongside the colón, anchored to the colón at a fixed rate of exchange. The intention was to stabilize the economy, lower interest rates and encourage domestic and foreign investment. The FMLN filed a case against the legislation with the Supreme Court of Justice; however, the so-called 'Law of Monetary Integration' was approved by the Asamblea Nacional in December. While most Salvadoreans opposed the move, they soon adapted, and by 2004 99% of the cash in circulation was US dollars. Interest rates were among the lowest in Latin America, but there was no accompanying economic growth and little extra foreign investment. At the same time, exports suffered owing to El Salvador's stronger currency.

Two severe earthquakes struck El Salvador in January and February 2001, presenting the country with its most serious test since the civil war. More than 1,100 people were killed and damage was estimated at US $1,900m., representing 14% of gross domestic product (GDP) for that year. Around 1.5m. people (one-quarter of the population) were made homeless. The Government initially was overwhelmed; however, international and domestic criticism prompted President Flores to devolve much responsibility to the municipalities, declaring a state of emergency and establishing 87 medical and evacuee centres. Society grew closer through the national recovery effort and the efficient response of the police and army, which grew in popularity. Local mayors were perceived to have handled the task of reconstruction competently, often using their new responsibility to cooperate across party boundaries. Relations between ARENA and the FMLN at a national level worsened, however, and in the days immediately following the earthquake the parties continued a dispute over the budget, resulting in a delay in the distribution of aid. The housing, health and education services, and the communications sector were badly affected, while agriculture, industry and trade all suffered losses, particularly in the private sector. In March the Inter-American Development Bank pledged some $1,278.5m. in aid for economic, social and environmental development projects. In that month the US immigration authorities granted a one-year protective status to Salvadorean illegal immigrants residing in the USA, releasing those held in custody. This has since been extended a number of times, most recently until September 2005. Remittances from workers abroad had by then become of crucial importance to the economy, increasing from 10.8% of GDP in 1997 to 13.8% in 2001.

The legacy of the civil war continued to overshadow politics. The UN Secretary-General, Kofi Annan, cancelled a visit to El Salvador in February 2002 to attend celebrations marking 10 years of peace. Although Annan's office claimed that the visit had never been confirmed, both sides of the political divide had been preparing for it. However, shortly before the anniversary, the FMLN pointed out that many aspects of the accords, such as the granting of land to former guerrillas, had not been fulfilled, prompting Annan to revise his plans. President Flores had hoped that the visit would demonstrate that the wartime chapter in the country's history had finally been closed. Nevertheless, it was the FMLN which found it most difficult to leave the war behind. In April the party divided, with six moderate deputies leaving the party, including Facundo Guardado, the party's 1999 presidential candidate and leader of the reformist faction. The moderate Silva Aregüello then left the party in November, following a dispute with the hardliners in the party over his offer of mediation in a public-health workers strike. He joined the small Centro Democrático Unido (CDU—United Democratic Centre), led by another former FMLN presidential candidate, Rubén Zamora.

## THE ELECTIONS OF 2003 AND 2004

The FMLN did unexpectedly well in regional and congressional elections held on 16 March 2003. The party won 31 seats in the Asamblea Nacional, compared with ARENA's 28, in what appeared to be a protest vote against President Flores' moves towards the privatization of the health service and reservations about a free trade arrangement with the USA. The overall voter abstention rate, however, was estimated to be over 60%. The 15 seats won by the PCN continued to ensure President Flores a right-wing majority in the Asamblea and, following the elections, he announced a series of measures designed to recover voters' support, including an increase in pensions and a reduction in electricity tariffs. He also introduced the *mano dura* (iron fist) policy. This introduced strict penalties on anyone convicted of being a member of a street gang, blamed by Salvadoreans for the rampant crime rate in the country. However, the law was ruled unconstitutional by the Supreme Court, and judges and lawyers generally declined to take on cases brought under it.

The orthodox faction that controlled the FMLN selected Schafik Jorge Handal, a committed Marxist and former guerrilla leader, as its presidential candidate for the March 2004 elections. His age, 72, wartime past and hardline message went down poorly with voters. They contrasted sharply with the ARENA candidate, Elías Antonio (Tony) Saca, a 39-year-old media entrepreneur of middle-class origins. His selection confirmed ARENA's transformation from a party of the old, coffee-producing elite, to one dominated by businessmen from services such as banking and retailing. Saca won 57.7% of the vote against 35.7% for Handal. The turn-out was a record high of 67% as the public was galvanized by the fierce campaign. The PCN and a centrist coalition of the CDU and PDC, who had selected former FMLN mayor of San Salvador Silva Aregüello as their candidate, failed to poll enough votes to remain legal, which could potentially force them to dissolve and reform under new names.

## RELATIONS WITH THE USA AND THE ADVENT OF CAFTA

The US President, George W. Bush (2001–), visited El Salvador on 24 March 2002. President Bush met the leaders of five Central American countries (Belize, Guatemala, Honduras, El Salvador and Nicaragua) to offer a free trade treaty with the USA. The visit was a further sign of the historically close relations between El Salvador and the USA. Negotiations began in January 2003 and were completed by the end of the year with the creation of the Central American Free Trade Agreement (CAFTA) to which the Dominican Republic later adhered. The deal was signed in May 2004 and in December, despite the opposition of the FMLN in the legislature, El Salvador became the first country to ratify it. The introduction of CAFTA, which was intended to facilitate trade with the USA but could have high adjustment costs for the Central American countries, remained uncertain, although it was ratified by the US Congress in late July 2005. One item that remained non-negotiable during the discussions was the free movement of labour. There were an estimated 1.5m. Salvadoreans living in the USA, who found it more difficult to gain US citizenship than did Nicaraguans and Cubans, for historical reasons. The remittances they sent home were equal to 16.1% of GDP in 2004. The refugee issue remained the most important problem, as it had been throughout Salvadorean history.

## PROSPECTS FOR THE FUTURE

One of the aims of the new President in 2004 was the promotion of political dialogue between all major parties. To this end one of Saca's first acts after assuming office in June was the creation of a permanent political forum, comprising representatives from the main parties. However, his efforts soon failed. The FMLN abandoned the newly created institution in October and continued to denounce the neoliberal character of the Saca administration. At the same time, however, internal problems within the FMLN persisted, with the traditionalists remaining in con-

# Economy

## ANDREW BOUNDS

Revised for this edition by DIEGO SÁNCHEZ-ANCOCHEA

With an area of only 21,041 sq km (8,124 sq miles) and an estimated population of 6.8m. in 2004, El Salvador has one of the greatest population densities in the Western hemisphere (an estimated 321 per sq km). According to the World Bank, El Salvador is a lower-middle income country with a nominal gross national income (GNI) per head of US $2,340 in 2004. At the same time, however, El Salvador is one of the least developed countries in Latin America: in 2002 the country was ranked 103rd in the UN's Human Development Index, higher than just five other Latin American countries.

Infant mortality in 2003 was 32 per 1,000 live births, life expectancy at birth was 70.4 years and adult literacy in 2002 was 82.4% for males and 77.1% for females. Low human development is partly the result of one of the lowest levels of social spending in Latin America. According to the UN's Economic Commission for Latin America and the Caribbean (ECLAC), social spending as a proportion of gross domestic product (GDP) in 2000/01 was only 4.2%, compared with 6.2% in Guatemala, 10.0% in Honduras and 18.2% in Costa Rica.

### PHASES OF ECONOMIC DEVELOPMENT

El Salvador, like many of its neighbours in Central America and the Caribbean, has slowly moved from a primary-based economy to one that depends on manufacturing exports from free trade zones and remittances from Salvadoreans abroad. This process of change has taken place in five different phases since independence in 1821. Initially, the dominant products were indigo and cotton. From the mid-19th century coffee superseded these commodities in importance. After the abolition of common land in 1882 vast *haciendas* (plantations) worked by a seasonal peasant work-force who had lost access to common land grew up. Coffee barons branched out into finance and commerce and in the 1960s their capital helped establish a manufacturing base that exported throughout Central America.

The new process of industrialization based on the Central American Common Market (CACM) contributed to the acceleration of economic growth and opened a third stage of development. Between 1970 and 1978 GDP expanded, in real terms, at an average annual rate of 5%. However, deterioration in the price of coffee and other commodities, together with negative international conditions and the crisis of the CACM led to a severe downturn at the end of the 1970s. Between 1978 and 1982, real GDP decreased by 22.3%. With population growth averaging 2.6% annually, by 1983 GDP per head had fallen to levels comparable with those of the 1960s. From 1979 to 1982 investment, in real terms, decreased by 68% and consumption by 20%. Unemployment, combined with underemployment, was estimated to affect more than 40% of the total work-force. In an attempt to ease social tensions and avert a left-wing uprising, a new military Government nationalized the banks and the coffee industry and began breaking up large *haciendas* and handing them to worker co-operatives, initiating what can be considered the fourth phase of development. Nevertheless, these attempts failed to prevent a civil war that extended from 1979 to 1992. The war caused more than 80,000 deaths, the internal and external displacement of over 1m. people, a massive flight of capital, and economic damage estimated at more than US $2,000m. During the war external financial assistance, mainly from the USA, helped to keep the economy from sliding into recession. Between 1980 and 1990 total external financial assistance to El Salvador was in excess of $5,000m., with approximately 90% from the USA, making El Salvador the third largest recipient of assistance from the US Administration at the time. One of the main purposes of US assistance was to offset economic sabotage by the Frente Farabundo Martí de Liberación Nacional (FMLN—Farabundo Martí National Liberation Front), which particularly affected the harvesting and export of the country's main export crop, coffee, and severely disrupted transport and power transmission. Coffee exports were also affected by the government monopoly that controlled them, which paid less than market rates. Economic recovery was further impeded by the earthquake of October 1986, which caused material damage estimated at $900m. (mainly in San Salvador) and disrupted administration and public services.

El Salvador initiated the latest stage of economic development with the full adoption of the recommendations contained in the so-called 'Washington Consensus' during the administration of Alfredo Cristiani (1989–94). The new policies included deregulation, a reduction of the state's role in the economy and the attraction of foreign investment through the promotion of *maquila* textile assembly plants and a stable exchange rate with the US dollar. Important sectors of the economy were returned to private ownership, including sugar refineries, distilleries, textile mills, hotels and fish processing plants, as well as most of the banks and financial institutions. Public spending was cut and price controls and subsidies reduced or abolished. The tax system was simplified and tariffs diminished. In 1990 the IMF granted a stand-by loan agreement of US $50m., indicating international approval of the reforms.

Market reforms continued after the peace accords and throughout the 1990s under successive Alianza Republicana Nacionalista (ARENA) Governments. In 1998 the state telecommunications company, Administración Nacional de Telecomunicaciones (ANTEL), was privatized and in 1999 several electricity generating stations were sold. Private pensions were introduced in 1998. Following serious public and trade-union protests in 1999 and 2000, and an electoral reverse in March 2000, President Francisco Flores Pérez, who assumed office in June 1999, retreated from further privatization in the health and banking sectors. Another attempt to introduce private contracting to the health system prompted further protests in September 2002. In November 2000 President Flores announced the radical step of 'dollarization', or monetary integration, as he termed it, in an attempt to reactivate the economy. The legislation was approved in December and from 1 January 2001 the US dollar circulated freely with the colón at a fixed rate of 8.75. The Central Bank had maintained a fixed exchange rate of 8.76 since May 1995. The World Bank and the IMF supported the move, which aimed to integrate El Salvador into the global economy and to reduce real interest rates close to US levels, to encourage investment to expand growth.

Cristiani's reforms and demand after the end of the war led to a short-lived economic boom, partly facilitated by the expansion of remittances. Real GDP increased by as much as 6.3% in 1995, but declined by 1.7% in the following year. Between 1996 and 2004 the average annual rate of real GDP was 2.6% but, as a result of high population growth, real GDP per head stagnated, increasing by only 0.6% per year in the same period.

Frequent natural disasters were partly responsible for this uneven economic record. 'Hurricane Mitch', which struck Central America in October 1998, resulted in the loss of an estimated 8% (about US $1,760m.) of El Salvador's GDP in that year. In January and February 2001 two severe earthquakes hit the country, killing more than 1,100 people and leaving a further 1.5m. people homeless. Reconstruction was estimated at $1,900m., or 14% of GDP. Donors in March promised $1,300m., mostly in loans, and the Government diverted $150m. from the 2001 budget. The Government was also forced to borrow heavily to finance reconstruction over the following five years. External debt was expected to rise from a comparatively low 23.1% of GDP in 2000 to 38.9% of GDP in 2005. Moreover, 40,000 small businesses were damaged and 200,000 people left without temporary work in the fields, as crops and equipment were destroyed.

The earthquakes and droughts and the relative failure of economic reforms were instrumental in maintaining high levels of poverty and unemployment. Although the poverty rate fell by 15 percentage points in the 1990s it remained at 51% in 2000, according to the World Bank, while rural poverty was 55%. By the end of 1999 unemployment had reached 7.9% and remained at that level in 2003. The average income per person fell below US $100 per month in that latter year, to $96.80, emphasizing the lack of progress since 2000. At least 30% of the economically active population were underemployed and many survived by working in the parallel, illegal economy. Poor urban, as well as rural, households became increasingly dependent on remittances from family members who had migrated, especially to the USA, which helped to lessen the impact of macroeconomic adjustments. Remittances reached a record $2,547.6m. in 2004, which offset, to some extent, the expanding trade deficit.

## AGRICULTURE

Agriculture (including hunting, forestry and fishing) remained an important economic activity in the early 21st century, despite a contraction in the sector, owing to mass migration to the cities. By 2005 an estimated 59.7% of the population lived in urban areas, an increase from 50% in 1995. Agricultural growth was stagnant between 1996 and 2004. Its contribution to GDP contracted from around 14% at the end of the civil war to an estimated 9.1% in 2004. There was intense pressure for land in a country with an increasing population density. Uneven rainfall, some 84% of which occurred during May–October, was another problem. The vulnerability of subsistence farmers was highlighted in 2001 when the rains did not arrive until August. Up to 80% of crops were lost in four provinces in the east, amounting to 114,300 metric tons of beans and maize. Some 318,640 people were affected by food shortages, with international food aid given to 22,000 families.

During the war the centrist Government attempted substantial, but flawed, agricultural reform. In 1980 all plantations of more than 500 ha (20% of all agricultural land) were expropriated by the state-owned Instituto Salvadoreño de Transformación Agraria (Salvadorean Institute of Agrarian Transformation), for transfer to peasant-run co-operatives. A programme of transfer of freehold titles to tenant smallholders began in 1981, and by 1985 more than 35,000 peasants had benefited. In 1983 a statutory limit of 245 ha was placed upon the amount of land that could be owned by any Salvadorean national. The aim was to provide one-half of the landless rural population with land rights. However, this target was later reduced and at the end less than one-quarter received rights. As part of the 1992 peace settlement, an estimated 10% of agricultural land was to be distributed among 45,000 families of refugees and combatants. The programme was finally completed at the beginning of 1998, three years after the official deadline. Nevertheless, post-war Governments gave little assistance to the sector, with the exception of coffee, believing progress lay in industrialization. After the earthquakes of 2001 the Government actively discouraged replanting of crops, often locating new settlements near export processing zones, which provided jobs in the low-wage *maquila* assembly factories. Hence land reform was no panacea, with many new landowners lacking the capital and expertise necessary to take full advantage of their land. For example, the share of national coffee production by land reform co-operatives fell from 10% to 7% between 1996 and 2000 in a declining market.

Production of arabica coffee, the basis of El Salvador's economic development for a century, appeared unlikely to recover from the effects of the civil war and declining profitability. Production fell from the 1979 harvest of 4.1m. quintales (the old Spanish quintal, used in El Salvador, is equivalent to 46 kg), to 2.6m. quintales by 1988. Guerrilla sabotage caused large areas under cultivation to be destroyed or abandoned. In 1989 the international coffee agreement, which guaranteed traditional producers' quotas, collapsed. A retention scheme by the Association of Coffee Producing Countries artificially boosted prices for a while and supported production. The 1992 harvest was 3.2m. quintales (approximately 147,000 metric tons), but natural disasters and then low prices caused by rising exports from non-traditional growers such as Viet Nam adversely affected the sector. Following 'Hurricane Mitch' in 1998 the crop amounted to 117,200 tons. In 2001 coffee production fell to wartime levels of 112,200 tons and still further to 91,500 tons in 2002, remaining at this level in 2003. Export income from coffee declined even faster because of 40-year price lows in 2001 and 2002 and was determined more by the fluctuations of world prices than by output. Income fell drastically, from US $522m. (25% of total export earnings) in 1997 to $109m. in 2002 (8.9% of export earnings), before stabilizing. In 2004, following a small recuperation in international prices, coffee exports were an estimated $123m., though this was equivalent to only 3.7% of total export earnings, including *maquila* zones. A survey in 2000 found there were an estimated 22,500 producers of coffee in El Salvador, employing 150,000 people. The leading 6% of producers accounted for 72% of production while 70% of growers produced fewer than 30 bags of 60 kg per year, providing an annual income of less than $800. By 2004 the number of people working in the sector had fallen to an estimated 50,000 as many smaller growers switched to other crops. The Government gave only modest support, in the form of favourable loans to producers, demonstrating the once-powerful industry's loss of influence. El Salvador's future was thought to lie in its mountain-grown, high quality coffee, pitched at the high-margin gourmet market. However, that remained a niche field.

Output of sugar cane, another important cash crop, also declined during the civil war, and a recovery in the sector in the late 1990s proved shortlived. However, in the early 2000s, sugar production increased; in 2002 output was an estimated 476,000 metric tons, rising to an estimated 530,000 tons in 2003. Earlier in the 1990s low world prices and reductions in US import quotas adversely affected the prospects for sugar exports. The value of sugar exports fluctuated with prices, reaching US $70m. in 2001, but declining to an estimated $37m. in 2004. The Central American Free Trade Agreement (CAFTA) with the USA was expected to provide scope for growth by increasing US quotas.

The commercial fishing industry expanded considerably after the 1960s. After declines in the mid-1990s, in 1998 the total catch increased to 11,400 metric tons; however, earthquakes, pollution and the changes in climate associated with the El Niño phenomenon reduced the catch to a low of 9,900 tons in 2000 before strong growth increased this figure to 36,500 tons in 2003. Nevertheless, shrimp exports fell from US $39m. in 1996 to an estimated $5m. in 2004. In 2003 the Spanish company Grupo Calvo established an important tuna processing plant in the Gulf of Fonseca and this was expected to increase tuna exports significantly.

Non-traditional exports (including animal fodder, melons and pineapples) became more important at the end of the 20th century, particularly those to the countries of CACM. These averaged some US $300m. in the 1990s. Food production for the domestic market was dominated by the cultivation of maize, sorghum, beans and rice, primarily on smallholdings. However, El Salvador imported around 30% of its basic food needs. Production of staples declined after 1979, owing mainly to the security situation and the displacement of the population.

Nevertheless, substantial increases were achieved after 1982. The staple food of most Salvadoreans was maize (output of which was an estimated 618,000 metric tons in 2003), and beans (78,100 tons in 2003). The dry weather also affected rice production, which halved between 2000 and 2003.

## MINING AND POWER

Mining was of negligible importance in 2003, accounting for some 0.1% of the total work-force and, in 2004, an estimated 0.5% of GDP. The main minerals produced were limestone, gypsum and salt.

El Salvador still relied significantly on imported petroleum despite a drive to exploit its hydroelectric and geothermal potential in its rugged terrain and volcanoes. The River Lempa was dammed in 1950, feeding the 180 MW San Lorenzo powerstation. The development of the sector was impeded by the civil war and in 1992 installed capacity was 740 MW: 50% hydroelectric, 37% thermal and 13% geothermal. Persistently low rainfall eroded hydroelectric production, however. Deforestation, leading to erosion and silting up of rivers, also affected production. In 2003 oil accounted for 40% of the 4,077.2 kWh generated, and hydroelectric for 35%. Geothermal sources were exploited almost to the full, providing 966 kWh. Petroleum imports rose from US $175m. that year to a preliminary $226m. in 2004, equivalent to 6.9% of export receipts. Much was bought on favourable terms from Venezuela under the 2000 Acuerdo de Caracas. El Salvador became a net importer of energy in 1994.

The Government prioritized energy production as a motor of industrialization and economic recovery after the civil war. In 1998 75% of the shares were sold in four state-owned regional electricity-distribution companies. The state still owned the nation's hydroelectric dams, but competition was allowed in thermal and geothermal production, in an effort to encourage investment. That was given a boost in 2001 when construction finally began on an electricity interconnection between El Salvador and Honduras, part of a regional power network covering the whole of Central America. Funding for the project, which was due to be completed by 2006, was provided partly by the Inter-American Development Bank. It was envisaged that it would lower energy prices and encourage investment by providing economies of scale. In May 2004 the two countries agreed to build a joint hydroelectric power plant in Honduras.

## MANUFACTURING

The rapid growth of the manufacturing sector after 1960, within the CACM, increased the sector's contribution to 15% of GDP by 1979. Although the war reversed this trend temporarily, from 1983 a series of measures were adopted to revitalize the sector, including the promotion of exports within the regional market and the creation of credit lines for industrial companies in the context of the US Government's Caribbean Basin Initiative (CBI). At the same time, consideration was given to combining compensation to previous owners of expropriated agricultural land with reinvestment of this compensation in industrial enterprises. The return of business confidence and the development of regional markets in Central America contributed to the recuperation of the manufacturing sector during the post-war period. Manufacturing production grew by an annual average rate of 4% between 1996 and 2004, and exports of manufactures accounted for 53% of total exports in 1999, mainly destined for Central American markets. However, dollarization made many Salvadorean companies uncompetitive, according to the Salvadorean Association of Industrialists, and several ceased trading in 2003. In 2004 the manufacturing sector grew by only 0.7%, compared to 1.5% of the economy as a whole. The contribution of the manufacturing sector to GDP increased from 21.2% in 1996 to 23.7% in 2003, before decreasing slightly to 23.1% in 2004.

The attraction of *maquila* plants, which mainly produced clothing items for the US market, was the chief plank of the Government's job creation policy and the dominant export sector. The *maquilas* employed more than 89,000 people in 2003, and generated the equivalent of 55.2% of total exports in 2004. In May 2000 the *maquila* sector was boosted when the USA agreed to broaden the terms of the CBI in order to provide North American Free Trade Agreement (NAFTA) parity to El Salvador and 23 other Latin American countries. The enhanced CBI provided duty-free access to the US market for a number of previously excluded categories of *maquila* garments. Nevertheless, after seven years of double digit expansion, the growth of *maquila* exports slowed from 2000, suggesting that the industry had matured. In 2004 *maquila* exports earned US $1,821m., compared with $1,758m. in 2002. After importing raw materials, the value-added of *maquila* products was $441m. in 2004 (24% of the total value of *maquila* exports). It was anticpated that the approval of CAFTA might help El Salvador to remain competitive in *maquila* apparel production in the face of competition from the People's Republic of China. Whereas the CBI required the use of material from the USA, CAFTA would allow the use of material from any party to the agreement or to NAFTA. This may stimulate El Salvador's resurgent dyeing industry and domestic production and cutting industry. Allegations of poor labour conditions and worker abuse, however, left the industry vulnerable to US consumer boycott.

## TRANSPORT AND TOURISM

There were some 10,029 km of roads in 1999, of which 306 km were part of the Pan-American Highway and 1,985 km were paved. The number of vehicles on the roads tripled from 145,000 to 450,000 between 1997–2001. Improving the quality of existing roads through maintenance, rehabilitation and modernization was a major challenge for post-war El Salvador. A Vice-Ministry of Transport was established to direct and coordinate policy. The earthquakes of 2001 caused considerable damage to the road system, which, according to government estimates, would cost US $188m. to repair. In 2002 the Government established a road maintenance fund, the Fondo de Conservation Vial (Fovial), financed by a 20 US cents per gallon tax on petrol, which began to give concessions to private companies to repair the roads.

The Comisón Ejecutiva Portuaria Autónoma (CEPA) is responsible for the administration of El Salvador's only main ports, Acajutla and Cutuco, the El Salvador International Airport at Comalpa, Cuscatlán, and the national railway system, which operated 602 km of track in the late 1990s, including the 429-km Salvadorean section of the International Railways of Central America. In 2000 the rail system carried 687,300 passengers and 136,200 metric tons of freight. CEPA improved its financial situation in the 1990s, but faced the problem of recovering traffic lost during the civil war, as well as competing with other Central American ports, particularly Puerto Quetzal in Guatemala. In 2001 the Government finalized an agreement to reactivate Cutuco port on the Gulf of Fonseca. It received a US $90m. loan and was to contribute $56m. in accompanying road and other infrastructural improvements. A Japanese company was to run the port in concession. The port should reopen in 2006. Acajutla, the country's main port further north, has been in decline. Only one bidder, a multinational Latin American consortium, bid for a 25-year management concession in 2004.

El Salvador's recent bloody history and high crime rate stunted the growth of its tourism industry. Nevertheless, the country had much to offer, with Mayan temples and cities, volcanoes, mountain lakes and sandy beaches. Tourist arrivals rose from 387,052 in 1997 to 966,416 in 2004. Many of the tourist in El Salvador were Salvadoreans living in the USA returning to visit families or their home towns; in 2004 31% of all tourists went to El Salvador with the primary aim of visiting friends and family. As a result of the expansion in the number of tourists, the foreign exchange generated by the tourist sector increased steadily from US $75m. in 1997 to $425m. in 2004. In 2004 tourism became the second largest generator of foreign exchange, behind remittances, but surpassing the free trade zones. Seven countries agreed to establish a Central American Tourism Promotion Agency in Madrid, Spain, in December 2002, in an attempt to increase tourist arrivals. President Antonio Saca also prioritized tourism by creating a Ministry of Tourism when he took office in June 2004.

## INVESTMENT AND FINANCE

The cost of reconstruction following the earthquakes placed a heavy burden on the Government's finances in 2001. The public deficit of the central Government in that year was US $426m., equivalent to 3.1% of GDP, but the deficit narrowed to 2.2% of GDP in 2003 ($245.5m.) and to an estimated 0.6% in 2004. The country continued to depend on high levels of foreign aid and concessional loans to finance much-needed infrastructural development. Tax revenue rose from 10.5% of GDP in 2002 to an estimated 11.5% in 2004, still low even by Latin American standards. The Government took advantage of its low debt and healthy credit rating to borrow internationally. It also increased incentives for foreign investment as part of its programme for economic reactivation and stabilization; however, high crime levels and a violent death rate on a par with that experienced during the civil war deterred foreign investors. More recently, the introduction of the US dollar as legal tender helped to attract some investors.

In 1990–2003 the average annual rate of inflation was 6.9%. In 2000 the extension of value-added tax (VAT) contributed to a 2.3% rise in consumer prices, and the introduction of the US dollar further increased inflation to 3.7% in 2001. The annual inflation rate fell to 1.9% in 2002 and to 2.5% in 2003, but rose significantly in 2004 to 5.4%, owing to rising oil prices.

## FOREIGN TRADE AND PAYMENTS

The change in the Salvadorean economic model has been particularly reflected in the shift in the export structure. While in 1970 coffee and sugar constituted 91% of non-regional exports, they accounted for only 5% of total exports in 2004. Meanwhile, *maquilas*, tourism and, most importantly, remittances from abroad became the main generators of foreign exchange. The shift in the structure of exports coincided with an expansion of both exports and imports. According to the Central Bank, exports (including from the *maquila* zones) in 2004 totalled US $3,295m., while imports were $6,269m. Of imports, intermediary goods accounted for $2,114m., consumer goods for $1,787m. (reflecting the weight of remittance money), and capital goods for $990m. The imports for the *maquila* industries accounted for $1,378m. The principal export partner was the USA, which was the destination of 65% of Salvadorean exports and the origin of 46% of its imports (including maquila imports).

The trade surplus that El Salvador had recorded during most of the 1970s gave way to a deficit from 1981, even though the declines in export revenues were accompanied by rigorous restrictions on 'non-essential' imports. After 1992 the deficit increased. In 2004 the trade deficit stood at US $2,619m., some 16.6% of GDP. Since the 1980s the deterioration in the trade balance has been offset by remittances from abroad and by capital inflows. As a result, international reserves have steadily increased from a low of $72m. in 1981. In December 2004 international reserves amounted to $2,066.1m., enough to cover the monetary base and guarantee dollarization. Foreign direct investment increased in the late 1990s as a result of the privatization of the telecommunications sector and the sale of shares in energy companies, but fell back after the privatization programme ended. While it averaged $200m. per year in 1998–2002, it was less than $87m. in 2004.

Between 1990 and 2004 total external debt increased from US $2,148m. to $8,869.9m. as the Government borrowed to finance reconstruction after the earthquake. Almost all of this debt was incurred on a medium- to long-term basis at low interest rates, owing to El Salvador's high credit rating. In 2003 the cost of debt-servicing was estimated to be equivalent to 8.8% of the value of exports of goods and services, and external debt was 46.3% of GNI.

El Salvador has been an active participant in recent processes of regional integration. In March 2001 a free trade agreement with Mexico, Guatemala and Honduras came into effect, which, it was hoped, would gradually open up markets for industrial and agricultural products over a 12-year period. In May the Central American countries, including El Salvador, reached the basis of a deal with Mexico, the 'Plan Puebla-Panamá', to integrate the region through joint transport, industry and tourism projects. In December 2003 El Salvador and its neighbours Costa Rica, Guatemala, Honduras and Nicaragua (and later, the Dominican Republic), concluded negotiations with the USA for the creation of CAFTA. While the El Salvadorean legislature ratified the agreement in December 2004, final approval of CAFTA was still pending in mid-2005 in Costa Rica, the Dominican Republic and Nicaragua.

## OUTLOOK

By 2005 El Salvador was financially stable and had recovered from the economic crisis of the 1980s. It was also the most liberal economy in Latin America. However, many challenges remained, not least the effects of repeated natural disasters. While growth was impressive in the 1990s, GDP per head remained at pre-war levels and was not sufficient to reduce the high levels of poverty. This was accentuated by dollarization, which made the country more expensive than its neighbours for manufacturing businesses. Moreover, its balance of payments and the success of dollarization remained heavily dependent on continued high remittance flows from more than 1m. Salvadoreans working abroad. Added to the rapidly expanding *maquila* sector, this made the country heavily dependent on the US economy, something clearly highlighted during the recession of 2001. The approval of CAFTA was likely to increase dependence on the US economy still further. High levels of poverty, agricultural stagnation, environmental damage and increasing crime and violence were all issues that the country would need to confront urgently. The extent to which the advent of CAFTA and the deepening of neoliberal reforms would contribute to meeting all these development challenges remained uncertain.

# Statistical Survey

Sources (unless otherwise stated): Banco Central de Reserva de El Salvador, Alameda Juan Pablo II y 17 Avda Norte, Apdo 01-106, San Salvador; tel. 2271-0011; fax 2271-4575; internet www.bcr.gob.sv; Dirección General de Estadística y Censos, Edif. Centro de Gobierno, Alameda Juan Pablo II y Calle Guadalupe, San Salvador; tel. 2286-4260; fax 2286-2505; internet www.minec.gob.sv.

## Area and Population

### AREA, POPULATION AND DENSITY

Area (sq km)
- Land . . . . . . . . . . . . . 20,721
- Inland water . . . . . . . . . . 320
- Total . . . . . . . . . . . . . 21,041*

Population (census results)†
- 28 June 1971 . . . . . . . . . . 3,554,648
- 27 September 1992
  - Males . . . . . . . . . . . . 2,485,613
  - Females . . . . . . . . . . . 2,632,986
  - Total . . . . . . . . . . . . 5,118,599

Population (official estimates at mid-year)
- 2002 . . . . . . . . . . . . . 6,510,300
- 2003 . . . . . . . . . . . . . 6,639,000
- 2004 . . . . . . . . . . . . . 6,756,800

Density (per sq km) at mid-2004 . . . . 326.1

* 8,124 sq miles.
† Excluding adjustments for underenumeration.

### PRINCIPAL TOWNS
(official population estimates at mid-2001)*

| | | | |
|---|---|---|---|
| San Salvador (capital) | 485,847 | Mejicanos | 193,400 |
| Soyapango | 287,034 | Apopa | 179,122 |
| Santa Ana | 253,037 | Nueva San Salvador | 163,794 |
| San Miguel | 245,428 | Ciudad Delgado | 157,094 |

* Figures refer to municipios, which may each contain rural areas as well as an urban centre.

**Mid-2003** (UN estimates, incl. suburbs): San Salvador 1,423,915 (Source: UN, *World Urbanization Prospects: The 2003 Revision*).

### BIRTHS, MARRIAGES AND DEATHS

| | Registered live births | | Registered marriages | | Registered deaths | |
|---|---|---|---|---|---|---|
| | Number | Rate (per 1,000) | Number | Rate (per 1,000) | Number | Rate (per 1,000) |
| 1994 | 160,772 | 29.0 | 27,761 | 5.0 | 29,407 | 5.3 |
| 1995 | 159,336 | 28.1 | 25,308 | 4.5 | 29,130 | 5.1 |
| 1996 | 163,007 | 28.2 | 27,130 | 4.7 | 28,904 | 5.0 |
| 1997 | 164,143 | 27.8 | 23,561 | 4.0 | 29,118 | 4.9 |
| 1998 | 158,350 | 26.3 | 25,937 | 4.3 | 29,919 | 5.0 |
| 1999 | 153,636 | 25.0 | 34,306 | 5.6 | 28,056 | 4.6 |
| 2000 | 150,176 | 23.9 | 28,231 | 4.5 | 28,154 | 4.5 |
| 2001* | 138,354 | 21.6 | 29,216 | 4.6 | 29,559 | 4.6 |

* Provisional figures.

**2002** (official estimates): Live births 168,300 (birth rate 25.8 per 1,000); deaths 38,900 (death rate 6.0 per 1,000).

**Expectation of life** (WHO estimates, years at birth): 70 (males 67; females 73) in 2003 (Source: WHO, *World Health Report*).

### ECONOMICALLY ACTIVE POPULATION
(ISIC major divisions, '000 persons aged 10 years and over)

| | 2000 | 2001 | 2002 |
|---|---|---|---|
| Agriculture, hunting, forestry and fishing | 501.8 | 534.3 | 474.4 |
| Mining and quarrying | 1.5 | 3.0 | 3.5 |
| Manufacturing | 433.5 | 431.6 | 434.0 |
| Electricity, gas and water | 8.8 | 10.9 | 10.7 |
| Construction | 118.8 | 133.0 | 136.2 |
| Trade, restaurants and hotels | 610.9 | 667.4 | 688.5 |
| Transport, storage and communication | 109.4 | 113.4 | 103.4 |
| Financing, insurance, real estate and business services | 87.8 | 100.5 | 98.0 |
| Public administration, defence and social security | 123.8 | 97.5 | 100.5 |
| Education | 69.8 | 88.0 | 94.5 |
| Other community, social and personal services | 154.9 | 155.3 | 155.4 |
| Private households with employed persons | 100.4 | 115.0 | 103.2 |
| Other activities not adequately defined | 1.2 | 1.5 | 10.4 |
| **Total employed** | 2,322.7 | 2,451.3 | 2,412.8 |
| Unemployed | 173.7 | 183.5 | 160.2 |
| **Total labour force** | 2,496.4 | 2,634.8 | 2,573.0 |

Source: ILO.

## Health and Welfare

### KEY INDICATORS

| | |
|---|---|
| Total fertility rate (children per woman, 2003) | 2.9 |
| Under-5 mortality rate (per 1,000 live births, 2003) | 36 |
| HIV/AIDS (% of persons aged 15–49, 2003) | 0.7 |
| Physicians (per 1,000 head, 2002) | 1.27 |
| Hospital beds (per 1,000 head, 1996) | 1.65 |
| Health expenditure (2002): US $ per head (PPP) | 372 |
| Health expenditure (2002): % of GDP | 8.0 |
| Health expenditure (2002): public (% of total) | 44.7 |
| Access to water (% of persons, 2002) | 82 |
| Access to sanitation (% of persons, 2002) | 63 |
| Human Development Index (2002): ranking | 103 |
| Human Development Index (2002): value | 0.720 |

For sources and definitions, see explanatory note on p. vi.

## Agriculture

### PRINCIPAL CROPS
('000 metric tons)

| | 2001 | 2002 | 2003 |
|---|---|---|---|
| Rice (paddy) | 37.7 | 29.1 | 22.5 |
| Maize | 571.5 | 644.4 | 628.0 |
| Sorghum | 150.5 | 140.8 | 141.0 |
| Yautia (Cocoyam)* | 52.0 | 52.0 | 52.0 |
| Sugar cane | 4,877.2 | 4,528.2 | 4,531.5 |
| Dry beans | 74.9 | 82.6 | 83.5 |
| Coconuts | 88.3 | 24.3 | 24.3 |
| Vegetables* | 58.4 | 58.4 | 58.4 |
| Watermelons* | 75.0 | 75.0 | 75.0 |
| Bananas* | 65.0 | 65.0 | 65.0 |
| Plantains* | 66.0 | 66.0 | 66.0 |
| Other fruit* | 156.2 | 156.2 | 156.2 |
| Coffee (green) | 112.2 | 91.5 | 91.5 |

* FAO estimates.

Source: FAO.

# EL SALVADOR

## LIVESTOCK
('000 head, year ending September)

|  | 2001 | 2002 | 2003 |
|---|---|---|---|
| Horses* | 96 | 96 | 96 |
| Asses* | 3 | 3 | 3 |
| Mules* | 24 | 24 | 24 |
| Cattle | 1,216 | 1,100* | 1,000* |
| Pigs | 150 | 153 | 153 |
| Sheep* | 5 | 5 | 5 |
| Goats* | 15 | 11 | 11 |
| Chickens* | 7,200 | 8,100 | 8,100 |

* FAO estimate(s).
Source: FAO.

## LIVESTOCK PRODUCTS
('000 metric tons)

|  | 2001 | 2002 | 2003 |
|---|---|---|---|
| Beef and veal | 35.1 | 30.3 | 29.2 |
| Pig meat* | 8.7 | 8.7 | 8.7 |
| Chicken meat | 74.1 | 78.5 | 78.6* |
| Cows' milk | 383.5 | 408.0 | 393.2 |
| Cheese* | 2.4 | 2.4 | 2.4 |
| Hen eggs | 61.3 | 65.9 | 65.9* |
| Fresh cattle hides* | 6.7 | 6.7 | 6.7 |

* FAO estimate(s).
Source: FAO.

## Forestry

### ROUNDWOOD REMOVALS
('000 cubic metres, excl. bark)

|  | 2001 | 2002 | 2003 |
|---|---|---|---|
| Sawlogs, veneer logs and logs for sleepers | 682 | 682 | 682* |
| Fuel wood | 4,518 | 4,518 | 4,147 |
| **Total** | 5,200 | 5,200 | 4,829* |

* FAO estimate.
Source: FAO.

### SAWNWOOD PRODUCTION
(FAO estimates, '000 cubic metres, incl. railway sleepers)

|  | 2001 | 2002 | 2003 |
|---|---|---|---|
| **Total** (all broadleaved, hardwood) | 58 | 68 | 68 |

Source: FAO.

## Fishing

('000 metric tons, live weight)

|  | 2001 | 2002 | 2003 |
|---|---|---|---|
| Capture | 19.0 | 34.5 | 35.4 |
| Nile tilapia | 0.6 | 1.2 | 1.2 |
| Other freshwater fishes | 1.1 | 1.5 | 1.5 |
| Croakers and drums | 0.6 | 0.8 | 0.8 |
| Skipjack tuna | 4.5 | 6.8 | 4.5 |
| Yellowfin tuna | 2.2 | 3.4 | 6.6 |
| Bigeye tuna | 2.1 | 4.6 | 1.1 |
| Sharks, rays, skates, etc. | 0.8 | 1.0 | 1.0 |
| Other marine fishes | 1.7 | 3.4 | 3.3 |
| Whiteleg shrimp | 0.5 | 0.9 | 0.9 |
| Pacific seabobs | 1.5 | 0.9 | 1.1 |
| Marine molluscs | 0.8 | 0.8 | 0.6 |
| Aquaculture | 0.4 | 0.8 | 1.1 |
| **Total catch** | 19.4 | 35.2 | 36.5 |

Source: FAO.

## Mining
(metric tons, unless otherwise specified)

|  | 2001 | 2002 | 2003* |
|---|---|---|---|
| Gypsum | 5,600 | 5,600 | 5,600 |
| Limestone ('000 metric tons) | 1,425 | 1,631 | 1,190 |
| Salt (marine) | 31,610 | 31,552 | 31,366 |

* Estimates.
Source: US Geological Survey.

## Industry

### SELECTED PRODUCTS
('000 metric tons, unless otherwise indicated)

|  | 1999 | 2000 | 2001 |
|---|---|---|---|
| Raw sugar | 457 | 524 | 504 |
| Motor gasoline (petrol) | 135 | 140 | 121 |
| Kerosene | 20 | 16 | 14 |
| Distillate fuel oil | 199 | 163 | 157 |
| Residual fuel oil | 519 | 537 | 591 |
| Liquefied petroleum gas (refined) | 18 | 14 | 15 |
| Cement* | 1,031 | 1,064 | 1,174 |
| Electric energy (million kWh) | 3,732.2 | 3,504.9 | 3,755.5 |

**2002:** Raw sugar 465; Cement 1,318*; Electric energy (million kWh) 3,982.3*.

**2003:** Cement 1,390*; Electric energy (million kWh) 4,152.2*.

* Source: US Geological Survey.
Source: mostly UN, *Industrial Commodity Statistics Yearbook*.

## Finance

### CURRENCY AND EXCHANGE RATES

**Monetary Units**
100 centavos = 1 Salvadorean colón.

**Sterling, Dollar and Euro Equivalents** (31 May 2005)
£1 sterling = 15.908 colones;
US $1 = 8.750 colones;
€1 = 10.790 colones;
1,000 Salvadorean colones = £62.86 = $114.29 = €92.68.

Note: The foregoing information refers to the principal exchange rate, applicable to official receipts and payments, imports of petroleum and exports of coffee. In addition, there is a market exchange rate, applicable to other transactions. The principal rate was maintained at 8.755 colones per US dollar from May 1995 to December 2000. However, in January 2001, with the introduction of legislation making the US dollar legal tender, the rate was adjusted to $1 = 8.750 colones. Both currencies were to be accepted for a transitional period.

### CENTRAL GOVERNMENT BUDGET
(US $ million)

| Revenue* | 2001 | 2002 | 2003 |
|---|---|---|---|
| Current revenue | 1,598.6 | 1,746.0 | 1,915.2 |
| Tax revenue | 1,448.9 | 1,595.2 | 1,736.2 |
| Taxes on earnings | 431.4 | 457.2 | 502.9 |
| Import duties | 146.0 | 154.7 | 177.7 |
| Consumption of products | 49.2 | 66.5 | 61.7 |
| Value-added tax | 809.0 | 837.0 | 911.4 |
| Non-tax revenue | 149.7 | 150.8 | 179.0 |
| Capital revenue | 0.4 | 2.3 | 0.4 |
| **Total** | 1,599.0 | 1,748.3 | 1,915.6 |

# EL SALVADOR

| Expenditure† | 2001 | 2002 | 2003 |
|---|---|---|---|
| Current expenditure | 1,584.5 | 1,624.4 | 1,773.1 |
| Remunerations | 763.0 | 756.7 | 735.9 |
| Goods and services | 256.1 | 265.6 | 273.5 |
| Interest payments | 172.7 | 223.7 | 289.7 |
| Transfers | 392.6 | 378.4 | 473.9 |
| To other government bodies | 221.0 | 213.8 | 296.1 |
| To the private sector | 157.3 | 151.6 | 161.6 |
| Capital expenditure | 566.5 | 615.6 | 561.0 |
| Gross investment | 434.3 | 450.3 | 414.0 |
| **Total** | 2,151.0 | 2,240.0 | 2,334.1 |

\* Excluding grants received (US $ million): 51.4 in 2001; 45.0 in 2002; 67.1 in 2003.
† Excluding lending minus repayments (US $ million): -7.7 in 2001; 0.5 in 2002; -5.9 in 2003.

## INTERNATIONAL RESERVES
(US $ million at 31 December)

| | 2002 | 2003 | 2004 |
|---|---|---|---|
| Gold* | 117.8 | 117.8 | 138.9 |
| IMF special drawing rights | 34.0 | 37.1 | 38.8 |
| Foreign exchange | 1,588.8 | 1,905.8 | 1,888.4 |
| **Total** | 1,740.6 | 2,060.7 | 2,066.1 |

\* National valuation US $251 per troy ounce at 31 December 2002 and 2003, and $332 per troy ounce at 31 December 2004.
Source: IMF, *International Financial Statistics*.

## MONEY SUPPLY
(US $ million at 31 December)

| | 2002 | 2003 | 2004 |
|---|---|---|---|
| Currency outside banks | 60.6 | 36.4 | 35.4 |
| Demand deposits at deposit money banks | 1,027.1 | 1,085.6 | 1,218.8 |
| **Total money** (incl. others) | 1,090.4 | 1,198.6 | 1,255.8 |

Source: IMF, *International Financial Statistics*.

## COST OF LIVING
(Consumer Price Index; base: 1992 = 100)

| | 2001 | 2002 | 2003 |
|---|---|---|---|
| Food and non-alcoholic beverages | 167.9 | 169.2 | 176.4 |
| Rent, water, electricity, gas and other fuels | 175.9 | 184.5 | 184.4 |
| Clothing | 114.2 | 111.2 | 110.2 |
| Health | 180.9 | 188.8 | 195.4 |
| Transport | 116.1 | 119.8 | 121.6 |
| **All items** (incl. others) | 162.3 | 166.8 | 171.0 |

**2004:** All items 180.2.

## NATIONAL ACCOUNTS
**Expenditure on the Gross Domestic Product**
(US $ million at current prices)

| | 2002 | 2003 | 2004 |
|---|---|---|---|
| Final consumption expenditure | 14,110.9 | 14,892.6 | 16,038.4 |
| Households | 12,614.1 | 13,279.8 | 14,377.2 |
| General government | 1,496.8 | 1,612.8 | 1,661.2 |
| Gross capital formation | 2,316.8 | 2,490.7 | 2,461.5 |
| Gross fixed capital formation | 2,353.8 | 2,490.7 | 2,461.5 |
| Changes in inventories | -37.0 | 0.0 | 0.0 |
| **Total domestic expenditure** | 16,427.7 | 17,383.3 | 18,499.9 |
| Exports of goods and services | 3,773.1 | 3,986.5 | 4,310.3 |
| *Less* Imports of goods and services | 5,888.8 | 6,429.5 | 6,986.4 |
| **GDP in purchasers' values** | 14,311.9 | 14,940.3 | 15,823.9 |
| **GDP at constant 1990 prices** | 7,830.5 | 7,972.5 | 8,095.4 |

**Gross Domestic Product by Economic Activity**
(US $ million at current prices)

| | 2002 | 2003 | 2004 |
|---|---|---|---|
| Agriculture, hunting, forestry and fishing | 1,217.4 | 1,264.3 | 1,400.1 |
| Mining and quarrying | 64.6 | 69.0 | 72.1 |
| Manufacturing | 3,318.2 | 3,432.0 | 3,537.1 |
| Construction | 686.5 | 726.1 | 666.2 |
| Electricity, gas and water | 255.6 | 256.1 | 267.9 |
| Transport, storage and communications | 1,281.5 | 1,347.7 | 1,540.2 |
| Wholesale and retail trade, restaurants and hotels | 2,741.6 | 2,825.9 | 3,023.3 |
| Finance, insurance, real estate and business services | 1,227.1 | 1,273.0 | 1,374.1 |
| Owner-occupied dwellings | 1,098.4 | 1,142.9 | 1,207.8 |
| Community, social, domestic and personal services | 1,034.6 | 1,090.8 | 1,168.0 |
| Government services | 1,000.0 | 1,038.0 | 1,060.8 |
| **Sub-total** | 13,925.5 | 14,465.8 | 15,317.6 |
| Import duties and value-added tax | 979.6 | 1,079.4 | 1,119.7 |
| *Less* Imputed bank service charge | 593.0 | 604.9 | 613.5 |
| **GDP in purchasers' values** | 14,311.9 | 14,940.3 | 15,823.9 |

## BALANCE OF PAYMENTS
(US $ million)

| | 2002 | 2003 | 2004 |
|---|---|---|---|
| Exports of goods f.o.b. | 3,019.7 | 3,152.6 | 3,329.6 |
| Imports of goods f.o.b. | -4,884.7 | -5,428.0 | -5,948.8 |
| **Trade balance** | -1,865.0 | -2,275.4 | -2,619.1 |
| Exports of services | 783.3 | 853.4 | 971.6 |
| Imports of services | -1,023.0 | -1,033.4 | -1,080.4 |
| **Balance on goods and services** | -2,104.7 | -2,455.4 | -2,727.9 |
| Other income received | 159.1 | 140.4 | 144.1 |
| Other income paid | -482.5 | -562.8 | -603.5 |
| **Balance on goods, services and income** | -2,428.1 | -2,877.8 | -3,187.3 |
| Current transfers received | 2,111.1 | 2,200.2 | 2,634.4 |
| Current transfers paid | -88.2 | -85.9 | -58.6 |
| **Current balance** | -405.2 | -763.6 | -611.5 |
| Capital account (net) | 208.9 | 112.9 | 99.7 |
| Direct investment abroad | 25.7 | -18.6 | -7.4 |
| Direct investment from abroad | 470.0 | 171.8 | 465.9 |
| Portfolio investment assets | -289.2 | -263.7 | -124.8 |
| Portfolio investment liabilities | 554.8 | 452.9 | 424.8 |
| Other investment assets | -223.7 | 8.0 | -153.0 |
| Other investment liabilities | 150.4 | 888.9 | -179.1 |
| Net errors and omissions | -615.2 | -272.4 | 45.4 |
| **Overall balance** | -123.5 | 316.2 | -40.0 |

Source: IMF, *International Financial Statistics*.

# EL SALVADOR

# External Trade

## PRINCIPAL COMMODITIES
(US $ million)

| Imports c.i.f.* | 2002 | 2003† | 2004† |
|---|---|---|---|
| Live animals and animal products; vegetables, crops and related products, primary | 351.3 | 380.9 | 422.1 |
| Food, beverages and tobacco manufactures | 282.0 | 320.7 | 368.4 |
| Mineral products | 521.7 | 644.3 | 720.4 |
|    Crude petroleum oils | 175.3 | 209.4 | 225.8 |
|    Refined petroleum oils | 143.8 | 175.7 | 206.2 |
| Chemicals and related products | 506.0 | | |
|    Therapeutic and preventative medicines | 133.0 | 155.2 | 183.9 |
| Plastics, artificial resins and articles thereof | 187.3 | 217.2 | 253.7 |
| Wood pulp, paper, paperboard and articles thereof | 201.4 | 216.5 | 249.6 |
| Metals and articles thereof | 286.1 | 308.2 | 378.4 |
|    Cast iron and steel | 127.1 | 144.2 | 202.9 |
| Mechanical machinery and apparatus | 385.6 | 415.6 | 389.4 |
| Electrical machinery and appliances | 251.4 | 275.8 | 383.1 |
|    Radio and television transmitters and receivers, and parts thereof | 46.8 | 94.6 | 169.9 |
| Transport equipment | 279.6 | 337.2 | 393.0 |
| **Total** (excl. others) | 3,901.9 | 4,375.0 | 4,891.0 |

* Excluding imports into *maquila* zones (US $ million): 1,282.6 in 2002; 1,379.2 in 2003 (preliminary); 1,377.7 in 2004 (preliminary).
† Preliminary figures.

| Exports f.o.b.* | 2002 | 2003† | 2004† |
|---|---|---|---|
| Live animals and animal products, primary | 33.3 | 46.9 | 34.5 |
| Vegetables, crops and related products, primary | 142.9 | 140.3 | 167.7 |
|    Coffee, including roasted and decaffeinated | 106.9 | 105.4 | 123.4 |
| Food, beverages and tobacco manufactures | 223.0 | 239.9 | 270.9 |
|    Unrefined sugar | 44.4 | 46.6 | 37.3 |
|    Manufactured cereal products, toasted or inflated | 30.4 | 45.4 | 45.0 |
| Mineral products | 38.6 | 67.2 | 71.3 |
| Chemical products | 157.6 | 149.8 | 159.2 |
|    Therapeutic and preventative medicines | 48.3 | 58.6 | 70.4 |
| Plastics, rubber and articles thereof | 63.0 | 69.1 | 86.6 |
| Wood pulp, paper, paperboard and articles thereof | 128.2 | 122.9 | 128.9 |
|    Paper and cardboard packaging | 39.4 | 42.8 | 47.3 |
| Textiles and articles thereof | 156.8 | 156.8 | 222.9 |
|    Outer garments | 18.5 | 17.2 | 58.4 |
|    Underwear | 65.6 | 65.7 | 79.1 |
| Metals and articles thereof | 127.2 | 131.1 | 166.3 |
|    Iron and steel products, laminated | 46.8 | 49.8 | 59.4 |
|    Other iron and steel products | 27.1 | 25.1 | 47.9 |
| Electrical machinery and appliances | 56.7 | 47.3 | 72.5 |
| **Total** (incl. others) | 1,237.6 | 1,255.0 | 1,474.7 |

* Excluding exports from *maquila* zones (US $ million): 1,757.5 in 2002; 1,873.0 in 2003 (preliminary); 1,820.6 in 2004 (preliminary).
† Preliminary figures.

*Statistical Survey*

## PRINCIPAL TRADING PARTNERS
(US $ million)

| Imports c.i.f.* | 2002 | 2003† | 2004† |
|---|---|---|---|
| Brazil | 82.2 | 116.7 | 168.4 |
| Canada | 34.2 | 48.8 | 46.5 |
| Colombia | 40.1 | 65.0 | 58.2 |
| Costa Rica | 149.3 | 157.4 | 174.5 |
| Ecuador | 152.5 | 168.3 | 94.7 |
| France | 29.9 | 33.6 | 66.9 |
| Germany | 81.0 | 91.6 | 92.7 |
| Guatemala | 418.7 | 463.6 | 506.5 |
| Honduras | 155.3 | 134.9 | 153.6 |
| Italy | 42.7 | 45.7 | 36.2 |
| Japan | 136.3 | 132.3 | 134.1 |
| Mexico | 294.6 | 315.6 | 374.2 |
| Netherlands Antilles | 64.1 | 24.6 | 62.7 |
| Nicaragua | 97.5 | 111.6 | 111.5 |
| Panama | 156.2 | 126.0 | 146.3 |
| Spain | 61.9 | 66.5 | 104.7 |
| Taiwan | 40.2 | 58.5 | 47.6 |
| USA | 1,286.6 | 1,482.1 | 1,524.9 |
| Venezuela | 78.7 | 134.9 | 244.6 |
| **Total** (incl. others) | 3,901.9 | 4,375.0 | 4,891.0 |

* Excluding *maquila* zones.
† Preliminary figures.

| Exports f.o.b.* | 2002 | 2003† | 2004† |
|---|---|---|---|
| Costa Rica | 106.6 | 102.3 | 100.4 |
| Dominican Republic | 20.8 | 22.2 | 25.2 |
| Germany | 36.4 | 31.3 | 34.2 |
| Guatemala | 344.0 | 361.3 | 387.2 |
| Honduras | 176.6 | 184.9 | 205.9 |
| Mexico | 31.9 | 37.4 | 35.8 |
| Nicaragua | 113.2 | 98.1 | 128.6 |
| Panama | 45.0 | 46.3 | 46.1 |
| Spain | 5.8 | 14.6 | 37.7 |
| USA | 248.5 | 239.9 | 335.4 |
| **Total** (incl. others) | 1,237.6 | 1,255.0 | 1,474.7 |

* Excluding *maquila* zones.
† Preliminary figures.

# Transport

## RAILWAYS
(traffic)

| | 1999 | 2000 |
|---|---|---|
| Number of passengers ('000) | 543.3 | 687.3 |
| Passenger-kilometres (million) | 8.4 | 10.7 |
| Freight ('000 metric tons) | 188.6 | 136.2 |
| Freight ton-kilometres (million) | 19.4 | 13.1 |

Source: Ferrocarriles Nacionales de El Salvador.

## ROAD TRAFFIC
(motor vehicles in use at 31 December)

| | 1998 | 1999 | 2000 |
|---|---|---|---|
| Passenger cars | 187,440 | 197,374 | 207,259 |
| Buses and coaches | 34,784 | 36,204 | 37,554 |
| Lorries and vans | 166,065 | 177,741 | 189,812 |
| Motorcycles and mopeds | 32,271 | 35,021 | 37,139 |

Source: Servicio de Tránsito Centroamericano (SERTRACEN).

## SHIPPING
**Merchant Fleet**
(registered at 31 December)

| | 2002 | 2003 | 2004 |
|---|---|---|---|
| Number of vessels | 14 | 14 | 14 |
| Total displacement ('000 grt) | 5.6 | 5.6 | 5.6 |

Source: Lloyd's Register-Fairplay, *World Fleet Statistics*.

# EL SALVADOR

## CIVIL AVIATION
(traffic on scheduled services)

|  | 1999 | 2000 | 2001 |
|---|---|---|---|
| Kilometres flown (million) | 28 | 30 | 26 |
| Passengers carried ('000) | 1,624 | 1,960 | 1,692 |
| Passenger-km (million) | 5,091 | 2,829 | 2,907 |
| Total ton-km (million) | 502 | 284 | 308 |

Source: UN, *Statistical Yearbook*.

## Tourism

**TOURIST ARRIVALS BY COUNTRY OF ORIGIN**
(excluding Salvadorean nationals residing abroad)

|  | 2002 | 2003 | 2004 |
|---|---|---|---|
| Canada | 12,042 | 12,660 | 14,804 |
| Costa Rica | 27,134 | 26,495 | 29,405 |
| Guatemala | 377,329 | 329,162 | 326,437 |
| Honduras | 137,156 | 104,597 | 128,319 |
| Mexico | 21,462 | 20,198 | 22,973 |
| Nicaragua | 119,323 | 108,106 | 141,627 |
| Panama | 7,530 | 8,087 | 9,073 |
| Spain | 10,711 | 10,019 | 9,271 |
| USA | 167,765 | 179,712 | 225,910 |
| **Total** (incl. others) | 950,597 | 857,378 | 966,416 |

**Receipts from tourism** (US $ million): 342 in 2002; 373 in 2003; 425 in 2004.

Sources: Instituto Salvadoreño de Turismo.

## Communications Media

|  | 2000 | 2001 | 2002 |
|---|---|---|---|
| Telephones ('000 main lines in use) | 625 | 650 | 668 |
| Mobile cellular telephones ('000 subscribers) | 744 | 858 | 889 |
| Personal computers ('000 in use) | 100 | 140 | 163 |
| Internet users ('000) | 70 | 150 | 300 |

**Radio receivers** ('000 in use): 2,940 in 1999.
**Television receivers** ('000 in use): 1,260 in 2000.
**Daily newspapers:** 4 in 1998 (circulation 171,000).
**Non-daily newspapers:** 6 in 1996 (circulation 52,000).
**Book Production:** 663 in 1998.

Sources: UNESCO, *Statistical Yearbook*, International Telecommunication Union.

## Education
(2002, unless otherwise indicated)

|  | Institutions | Teachers* | Students Males | Students Females | Total† |
|---|---|---|---|---|---|
| Pre-primary | 4,838 | 5,116 | 112,718 | 115,129 | 228,064 |
| Primary | 5,414 | 31,921 | 659,896 | 621,141 | 1,281,693 |
| Secondary | 757 | 5,647 | 77,573 | 80,293 | 157,959 |
| Tertiary: University level* | 28 | 6,908 | 57,734 | 49,984 | 107,718 |
| Tertiary: Other higher* | 15 | 593 | 3,643 | 3,314 | 6,957 |

* 2000.
† Including unspecified gender.

Source: Ministry of Education.

**Adult literacy rate** (UNESCO estimates): 79.7% (males 82.4%; females 77.1%) in 2002 (Source: UN Development Programme, *Human Development Report*).

# Directory

## The Constitution

The Constitution of the Republic of El Salvador came into effect on 20 December 1983.

The Constitution provides for a republican, democratic and representative form of government, composed of three Powers—Legislative, Executive, and Judicial—which are to operate independently. Voting is a right and duty of all citizens over 18 years of age. Presidential and congressional elections may not be held simultaneously.

The Constitution binds the country, as part of the Central American Nation, to favour the total or partial reconstruction of the Republic of Central America. Integration in a unitary, federal or confederal form, provided that democratic and republican principles are respected and that basic rights of individuals are fully guaranteed, is subject to popular approval.

### LEGISLATIVE ASSEMBLY

Legislative power is vested in a single chamber, the Asamblea Nacional, whose members are elected every three years and are eligible for re-election. The Asamblea's term of office begins on 1 May. The Asamblea's duties include the choosing of the President and Vice-President of the Republic from the two citizens who shall have gained the largest number of votes for each of these offices, if no candidate obtains an absolute majority in the election. It also selects the members of the Supreme and subsidiary courts; of the Elections Council; and the Accounts Court of the Republic. It determines taxes; ratifies treaties concluded by the Executive with other States and international organizations; sanctions the Budget; regulates the monetary system of the country; determines the conditions under which foreign currencies may circulate; and suspends and reimposes constitutional guarantees. The right to initiate legislation may be exercised by the Asamblea (as well as by the President, through the Cabinet, and by the Supreme Court). The Asamblea may override, with a two-thirds majority, the President's objections to a Bill which it has sent for presidential approval.

### PRESIDENT

The President is elected for five years, the term beginning and expiring on 1 June. The principle of alternation in the presidential office is established in the Constitution, which states the action to be taken should this principle be violated. The Executive is responsible for the preparation of the Budget and its presentation to the Asamblea; the direction of foreign affairs; the organization of the armed and security forces; and the convening of extraordinary sessions of the Asamblea. In the event of the President's death, resignation, removal or other cause, the Vice-President takes office for the rest of the presidential term; and, in case of necessity, the Vice-President may be replaced by one of the two Designates elected by the Asamblea.

### JUDICIARY

Judicial power is exercised by the Supreme Court and by other competent tribunals. The Magistrates of the Supreme Court are elected by the Legislature, their number to be determined by law. The Supreme Court alone is competent to decide whether laws, decrees and regulations are constitutional or not.

EL SALVADOR

## The Government

### HEAD OF STATE

**President:** ELÍAS ANTONIO (TONY) SACA (assumed office 1 June 2004).
**Vice-President:** ANA VILMA DE ESCOBAR.

### CABINET
(July 2005)

**Minister of the Treasury:** GUILLERMO LÓPEZ SUÁREZ.
**Minister of Foreign Affairs:** FRANCISCO LAÍNEZ RIVAS.
**Minister of Government:** RENÉ MARIO FIGUEROA.
**Minister of the Economy:** YOLANDA MAYORA DE GAVIDIA.
**Minister of Education:** DARLYN MEZA.
**Minister of National Defence:** Gen. OTTO ROMERO ORELLANDA.
**Minister of Labour and Social Security:** JOSÉ ROBERTO ESPINAL ESCOBAR.
**Minister of Public Health and Social Welfare:** Dr JOSÉ GUILLERMO MAZA BRIZUELA.
**Minister of Agriculture and Livestock:** MARIO SALAVERRÍA NOLASCO.
**Minister of Public Works:** DAVID GUTIÉRREZ MIRANDA.
**Minister of the Environment and Natural Resources:** HUGO CÉSAR BARRERA GUERRERO.
**Minister of Tourism:** LUIS CARDENAL.

### MINISTRIES

**Ministry for the Presidency:** Avda Cuba, Calle Darío González 806, Barrio San Jacinto, San Salvador; tel. 2248-9000; fax 2248-9370; e-mail casapres@casapres.gob.sv; internet www.casapres.gob.sv.

**Ministry of Agriculture and Livestock:** Final 1a, Avda Norte y Avda Manuel Gallardo, Nueva San Salvador; tel. 2228-4443; fax 2229-9271; e-mail direccion.dgsva@mag.gob.sv; internet www.mag.gob.sv.

**Ministry of the Economy:** Edif. C1–C2, Centro de Gobierno, Alameda Juan Pablo II y Calle Guadalupe, San Salvador; tel. 2281-1122; fax 2221-5446; internet www.minec.gob.sv.

**Ministry of Education:** Edif. A, Centro de Gobierno, Alameda Juan Pablo II y Calle Guadalupe, San Salvador; tel. 2281-0044; fax 2281-0077; e-mail educacion@mined.gob.sv; internet www.mined.gob.sv.

**Ministry of the Environment and Natural Resources:** Carretera a Nueva San Salvador, Km 5.5, Calle y Col. Las Mercedes, San Salvador; tel. 2223-0444; fax 2260-3115; e-mail medioambiente@marn.gob.sv; internet www.marn.gob.sv.

**Ministry of Foreign Affairs:** Calle Circunvalación 227, Col. San Benito, San Salvador; tel. 2243-9648; fax 2243-9656; e-mail webmaster@rree.gob.sv; internet www.rree.gob.sv.

**Ministry of Government:** Centro de Gobierno, Edif. B-1, Alameda Juan Pablo II y Avda Norte 17, San Salvador; tel. 2221-3688; fax 2221-3956; e-mail DT@gobernacion.gob.sv; internet www.gobernacion.gob.sv.

**Ministry of Labour and Social Security:** Paseo Gen. Escalón 4122, San Salvador; tel. 2209-3700; fax 2209-3728; e-mail informacion@mtps.gob.sv; internet www.mtps.gob.sv.

**Ministry of National Defence:** Alameda Dr Manuel Enrique Araújo, Km 5, Carretera a Santa Tecla, San Salvador; tel. 2250-0325; e-mail fuerzaarmada@saltel.net; internet www.fuerzaarmada.gob.sv.

**Ministry of Public Health and Social Welfare:** Calle Arce 827, San Salvador; tel. 2221-0966; fax 2221-0991; e-mail webmaster@mspas.gob.sv; internet www.mspas.gob.sv.

**Ministry of Public Works:** Carretera a Santa Tecla 5.5 km, San Salvador; tel. 2223-8040; fax 2279-3723; internet www.mop.gob.sv.

**Ministry of Tourism:** San Salvador.

**Ministry of the Treasury:** 1231 Blvd Los Héroes, San Salvador; tel. 2244-3000; fax 2244-6408; e-mail webmaster@mh.gob.sv; internet www.mh.gob.sv.

## President and Legislature

### PRESIDENT

Election, 21 March 2004

| Candidates | % of votes cast |
|---|---|
| Elías Antonio (Tony) Saca (ARENA) | 57.71 |
| Schafik Jorge Handal (FMLN) | 35.68 |
| Héctor Silva Argüello (PDC/CDU) | 3.90 |
| José Rafael Machuca Zelaya (PCN) | 2.71 |
| **Total** | **100.00** |

### ASAMBLEA NACIONAL

**President:** CIRO CRUZ ZEPEDA.

General Election, 16 March 2003

| Party | % of votes cast | Seats |
|---|---|---|
| Frente Farabundo Martí para la Liberación Nacional (FMLN) | 33.8 | 31 |
| Alianza Republicana Nacionalista (ARENA) | 32.0 | 28 |
| Partido de Conciliación Nacional (PCN) | 13.1 | 15 |
| Partido Demócrata Cristiano (PDC) | 7.3 | 5 |
| Centro Democrático Unido (CDU)* | 6.4 | 5 |
| Other parties | 7.4 | — |
| **Total** | **100.0** | **84** |

*Electoral alliance comprising the Movimiento Nacional Revolucionario (MNR), the Movimiento Popular Social Cristiano (MPSC) and the Partido Social Demócrata (PSD).

## Political Organizations

**Alianza Republicana Nacionalista (ARENA):** Prolongación Calle Arce 2423, entre 45 y 47 Avda Norte, San Salvador; tel. 2260-4400; fax 2260-5918; internet www.arena.com.sv; f. 1981; right-wing; Pres. ELÍAS ANTONIO (TONY) SACA.

**Centro Democrático Unido (CDU):** Blvd Tutunichapa y Calle Roberto Masferrer 1313, Urb. Médica, San Salvador; tel. 2226-1752; fax 2225-5883; e-mail hsilva@asamblea.gob.sv; f. 1987 as Convergencia Democrática (CD); electoral alliance of the Movimiento Nacional Revolucionario (MNR), the Movimiento Popular Social Cristiano (MPSC) and the Partido Social Demócrata (PSD); changed name as above in 2000; became political party in 2001; Leader VINICIO PEÑATE; Sec.-Gen. RUBÉN IGNACIO ZAMORA.

**Concertación Socialdemócrata (CSD):** San Salvador; f. November 2004 by Facundo Guardado, Francisco Jovel and other fmr mems of the FMLN; moderate left-wing movt.

**Frente Farabundo Martí para la Liberación Nacional (FMLN):** 29 Calle Poniente 1316, Col. Layco, San Salvador; tel. 2226-5236; internet www.fmln.org.sv; f. 1980 as the FDR (Frente Democrático Revolucionario—FMLN) as a left-wing opposition front to the PDC-military coalition Government; the FDR was the political wing and the FMLN was the guerrilla front; military operations were co-ordinated by the Dirección Revolucionaria Unida (DRU); achieved legal recognition 1992; comprised various factions, including Communist (Leader SCHAFIK JORGE HANDAL), Renewalist (Leader OSCAR ORTIZ) and Terceristas (Leader MANUEL MELGAR); Co-ordinator-Gen. MEDARDO GONZÁLEZ.

**Partido de Conciliación Nacional (PCN):** 15 Avda Norte y 3a Calle Poniente 244, San Salvador; tel. 2221-3752; fax 2281-9272; e-mail czepeda@asamblea.gob.sv; f. 1961; right-wing; Leader RAFAEL MACHUCA; Sec.-Gen. CIRO CRUZ ZEPEDA.

**Partido Demócrata Cristiano (PDC):** 3a Calle Poniente 924, San Salvador; tel. 2222-6320; fax 7998-1526; e-mail pdcsal@navegante.com.sv; f. 1960; 150,000 mems; anti-imperialist, advocates self-determination and Latin American integration; Sec.-Gen. RODOLFO ANTONIO PARKER SOTO.

**Partido Nacional Liberal (PNL):** 5a Calle Poniente y 3a Avda Norte 226, San Salvador; tel. 2222-6511; Leader OSCAR SILDER CHÁVEZ HERNÁNDEZ.

**Partido Unificación Cristiana Democrática (UCD):** 7a Calle Oeste 52, Col. Los Andes, San Marcos; tel. 2213-0759; Sec.-Gen. JOSÉ ALEJANDRO DUARTE.

**Partido Unionista Centroamericana (PUCA):** San Salvador; advocates reunification of Central America; Pres. Dr GABRIEL PILOÑA ARAÚJO.

## Diplomatic Representation

### EMBASSIES IN EL SALVADOR

**Argentina:** 79 Avda Norte y 11 Calle Poniente 704, Col. Escalón, Apdo 384, San Salvador; tel. 2263-3638; fax 2263-3580; e-mail argensalv@saltel.net; Ambassador SILVIO HÉCTOR NEUMAN.

**Belize:** Calle el Bosque Norte y Calle Lomas de Candelaria, Col. Jardines de la Primera Etapa, San Salvador; tel. 2248-1423; fax 2273-6244; e-mail embelguat@guate.net; Ambassador DARWIN GABOUREL.

**Brazil:** Blvd de Hipódromo 132, Col. San Benito, San Salvador; tel. 2298-1993; fax 2279-3934; e-mail brasemb@es.com.sv; Ambassador MARÍA LUCÍA SANTOS POMPEU.

**Canada:** Centro Financiero Gigante, Torre A, Lobby 2, Alameda Roosevelt y 65 Avda Sur, Col. Escalón, San Salvador; tel. 2279-4655; fax 2279-0765; e-mail sal@dfait-maeci.gc.ca; internet www.dfait-maeci.gc.ca/elsalvador; Ambassador JAMES LAMBERT (resident in Guatemala).

**Chile:** Pasaje Bellavista 121, 9a Calle Poniente, Col. Escalón, San Salvador; tel. 2263-4285; fax 2263-4308; e-mail conchile@conchileelsalvador.com.sv; internet www.conchileelsalvador.com.sv; Ambassador RAQUEL MORALES ETCHEVERS.

**China (Taiwan):** Avda La Capilla 716, Col. San Benito, Apdo 956, San Salvador; tel. 2263-1330; fax 2263-1329; e-mail sinoemb3@sv.intercomnet.net; Ambassador HOU PING-FU.

**Colombia:** Calle El Mirador 5120, Col. Escalón, San Salvador; tel. 2263-1936; fax 2263-1942; e-mail elsalvador@minrelext.gov.co; Ambassador FABIO TORRIJOS QUINTERO.

**Costa Rica:** 85 Avda Sur y Calle Cuscatlán 4415, Col. Escalón, San Salvador; tel. 2264-3863; fax 2264-3866; e-mail embaricasal@sgsica.org; Ambassador JOSÉ DE JESÚS CONEJO AMADOR.

**Dominican Republic:** Avda República Federal de Alemania 163, Col. Escalón, San Salvador; tel. 2263-1816; fax 2263-1816; Ambassador HECTOR PEREYRA ARIZA.

**Ecuador:** 77 Avda Norte 208, Col. Escalón, San Salvador; tel. 2263-5323; fax 2263-5258; e-mail ecuador@integra.com.sv; Ambassador Dr FRANCISCO PROAÑO ARANDI.

**Egypt:** 9a Calle Poniente y 93 Avda Norte 12-97, Col. Escalón, San Salvador; tel. 2211-5788; fax 2263-2411; e-mail emebgip@telesal.net; Ambassador MEDHAT MOURSSI ELBANAN.

**France:** 1 Calle Poniente 3718, Col. Escalón, Apdo 474, San Salvador; tel. 2279-4016; fax 2298-1536; e-mail ambafrance@es.com.sv; internet www.embafrancia.com.sv; Ambassador FRANCIS ROUDIÈRE.

**Germany:** 7a Calle Poniente 3972, esq. 77a Avda Norte, Col. Escalón, Apdo 693, San Salvador; tel. 2247-0000; fax 2247-0099; e-mail zreg@sans.diplo.de; internet www.san-salvador.diplo.de; Ambassador Dr JOACHIM NEUKIRCH.

**Guatemala:** 15 Avda Norte 135, San Salvador; tel. 2271-2225; fax 2221-3019; Ambassador CARLOS ARTURO GONZÁLEZ ESTRADA.

**Holy See:** 87 Avda Norte y 7a Calle Poniente, Col. Escalón, Apdo 01-95, San Salvador (Apostolic Nunciature); tel. 2263-2931; fax 2263-3010; e-mail nunels@telesal.net; Apostolic Nuncio Most Rev. LUIGI PEZZUTO (Titular Archbishop of Torre di Proconsolare).

**Honduras:** 89 Avda Norte, entre 7 y 9 Calle Poniente, Col. Escalón, San Salvador; tel. 2263-2808; fax 2263-2296; Ambassador JAIME GÜELL BOGRÁN.

**Israel:** Centro Financiero Gigante, Torre B, 11°, Alameda Roosevelt y Avda Sur 63, San Salvador; tel. 2211-3434; fax 2211-3443; Ambassador YOSEF LIVNÉ.

**Italy:** Calle la Reforma 158, Col. San Benito, Apdo 0199, San Salvador; tel. 2223-4806; fax 2298-3050; e-mail ambitalia@ambasciatait.org.sv; internet www.ambasciatait.org.sv; Ambassador ROBERTO FALASCHI.

**Japan:** Calle Loma Linda 258, Col. San Benito, San Salvador; tel. 2224-4740; fax 2298-6685; Ambassador AKIO HOSONO.

**Korea, Republic:** 5a Calle Poniente 3970, entre 75 y 77 Avda Norte, Col. Escalón, San Salvador; tel. 2263-9145; fax 2263-0783; e-mail embcorea@mofat.go.kr; Ambassador CHOO YEON-GON.

**Mexico:** Calle Circunvalación y Pasaje 12, Col. San Benito, Apdo 432, San Salvador; tel. 2243-3190; fax 2243-0437; e-mail embamex@sv.intercom.com.sv; Ambassador PABLO RUIZ LIMÓN.

**Nicaragua:** 71 Avda Norte y 1a Calle Poniente 164, Col. Escalón, San Salvador; tel. 2263-8770; fax 2263-2292; Ambassador CECILE SABORIO COZE.

**Panama:** Avda Bungamilias 21, Col. San Francisco, San Salvador; tel. and fax 2298-0773; e-mail embpan@telesal.net; Ambassador RODERICK GUERRA BIANCO.

**Peru:** 7 Calle Poniente 4111, Col. Escalón, San Salvador; tel. 2263-3326; fax 2263-3310; e-mail embperu@telesal.net; Ambassador SERGIO KOSTRITSKY PEREIRA.

**Spain:** Calle La Reforma 167, Col. San Benito, San Salvador; tel. 2257-5700; fax 2298-0402; e-mail embajada@ambespama.com.sv; Ambassador JORGE HEVIA SIERRA.

**USA:** Blvd Santa Elena Sur, Antiguo Cuscatlán, San Salvador; tel. 2278-4444; fax 2278-1815; internet www.usinfo.org.sv; Ambassador HUGH DOUGLAS BARCLAY.

**Uruguay:** Edif. Gran Plaza 405, Blvd del Hipódromo, Col. San Benito, San Salvador; tel. 2279-1627; fax 2279-1626; Ambassador ROBERTO PABLO TOURIÑO.

**Venezuela:** 7a Calle Poniente, entre 75 y 77 Avda Norte, Col. Escalón, San Salvador; tel. 2263-3977; fax 2211-0027; e-mail venesal@amnetsal.com; Chargé d'affaires a.i. MARÍA EUGENIA SILVA.

## Judicial System

### Supreme Court of Justice

Frente a Plaza José Simeón Cañas, Centro de Gobierno, San Salvador; tel. 2271-8888; fax 2271-3767; internet www.csj.gob.sv.

f. 1824; composed of 15 Magistrates, one of whom is its President. The Court is divided into four chambers: Constitutional Law, Civil Law, Criminal Law and Litigation; Pres. AGUSTÍN GARCÍA CALDERÓN.

**Courts of Original Jurisdiction:** 201 courts throughout the country.

**Courts of Appeals:** 34 chambers composed of two Magistrates.

**Courts of Peace:** 322 courts throughout the country.

**Attorney-General of the Republic:** MARCOS GREGORIO SÁNCHEZ TREJO.

**Attorney-General for the Defence of Human Rights:** BEATRICE ALAMANNI DE CARRILLO.

## Religion

Roman Catholicism is the dominant religion, but other denominations are also permitted. In 2003 21.2% of the population belonged to Protestant churches. The Baptist Church, Seventh-day Adventists, Jehovah's Witnesses, and the Church of Jesus Christ of Latter-day Saints (Mormons) are represented.

### CHRISTIANITY

#### The Roman Catholic Church

El Salvador comprises one archdiocese and seven dioceses. At 31 December 2003, according to the Vatican, Roman Catholics represented some 77% of the total population.

##### Bishops' Conference

Conferencia Episcopal de El Salvador, 15 Avda Norte 1420, Col. Layco, Apdo 1310, San Salvador; tel. 2225-8997; fax 2226-5330; e-mail cedes.casa@telesal.net.

f. 1974; Pres. Most Rev. FERNANDO SÁENZ LACALLE (Archbishop of San Salvador).

**Archbishop of San Salvador:** Most Rev. FERNANDO SÁENZ LACALLE, Arzobispado, Avda Dr Emilio Alvarez y Avda Dr Max Bloch, Col. Médica, Apdo 2253, San Salvador; tel. 2225-0501; fax 2226-4979; e-mail arzfsl@vip.telesal.net.

#### The Anglican Communion

El Salvador comprises one of the five dioceses of the Iglesia Anglicana de la Región Central de América. La Iglesia Anglicana has some 5,000 members.

**Bishop of El Salvador:** Rt Rev. MARTÍN DE JESÚS BARAHONA PASCACIO, 47 Avda Sur, 723 Col. Flor Blanca, Apdo 01-274, San Salvador; tel. 2223-2252; fax 2223-7952; e-mail martinba@gbm.net; internet www.cristosal.org.

#### The Baptist Church

**Baptist Association of El Salvador:** Avda Sierra Nevada 922, Col. Miramonte, Apdo 347, San Salvador; tel. 2226-6287; e-mail

EL SALVADOR

asociacionbautistaabes@latinmail.com; f. 1933; Pres. Rev. MIGUEL ÁNGEL DUARTE RECINOS.

### Other Churches

**Sínodo Luterano Salvadoreño** (Salvadorean Lutheran Synod): Final 49 Avda Sur, Calle Paralela al Bulevar de los Próceres, San Salvador; tel. 2225-2843; fax 2248-3451; e-mail comunicaciones@sinodoluterano.org.sv; Pres. Bishop MEDARDO E. GÓMEZ SOTO; 12,000 mems.

## The Press

### DAILY NEWSPAPERS

#### San Miguel

**Diario de Oriente:** Avda Gerardo Barrios 406, San Miguel; internet www.elsalvador.com/DIARIOS/ORIENTE.

#### San Salvador

**Co Latino:** 23a Avda Sur 225, Apdo 96, San Salvador; tel. 2271-0671; fax 2271-0971; e-mail info@diariocolatino.com; internet www.diariocolatino.com; f. 1890; evening; Editor FRANCISCO ELÍAS VALENCIA; circ. 15,000.

**El Diario de Hoy:** 11 Calle Oriente 271, Apdo 495, San Salvador; tel. 2271-0100; fax 2271-2040; e-mail redaccion@elsalvador.com; internet www.elsalvador.com; f. 1936; morning; independent; Dir ENRIQUE ALTAMIRANO MADRIZ; circ. 115,000.

**Diario Oficial:** 4a Calle Poniente y 15e, Avda Sur 829, San Salvador; tel. 2233-7821; fax 2222-4936; e-mail diariooficial@imprentanacional.gob.sv; f. 1875; Dir RENÉ ORLANDO SANTAMARÍA COBOS; circ. 1000.

**El Mundo:** 15 Calle Poniente y 7a Avda Norte 521, San Salvador; tel. 2225-3300; fax 2222-1490; e-mail mercadeo@elmundo.com.sv; internet www.elmundo.com.sv; f. 1967; evening; Exec. Dir ERNESTO BORJA PAPINI; Gen. Man. ARTURO ARGUELLO OERTEL; circ. 40,215.

**La Prensa Gráfica:** 3a Calle Poniente 130, San Salvador; tel. 2271-1010; fax 2271-4242; e-mail lpg@gbm.net; internet www.laprensagrafica.com; f. 1915; general information; conservative, independent; Editor RODOLFO DUTRIZ; circ. 97,312 (weekdays), 115,564 (Sundays).

#### Santa Ana

**Diario de Occidente:** 1a Avda Sur 3, Santa Ana; tel. 2441-2931; internet www.elsalvador.com/DIARIOS/OCCIDENTE; f. 1910; Editor ALEX E. MONTENEGRO; circ. 6,000.

### PERIODICALS

**Cultura:** Concultura, Ministerio de Educación, 17 Avda Sur 430, San Salvador; tel. 2222-0665; fax 2271-1071; quarterly; educational; Pres. FEDERICO HERNÁNDEZ AGUILAR; Dir JOSÉ HERNÁN ARTEAGA.

**Proceso:** Universidad Centroamericana, Blvd Los Próceres, Apdo 01-168, San Salvador; tel. 2210-6600; fax 2210-6655; e-mail cidai@cidai.uca.edu.sv; f. 1980; weekly newsletter, published by the Documentation and Information Centre of the Universidad Centroamericana José Simeón Cañas; Dir LUIS ARMANDO GONZÁLEZ.

**Revista del Ateneo de El Salvador:** 13a Calle Poniente, Centro de Gobierno, San Salvador; tel. 2222-9686; f. 1912; 3 a year; official organ of Salvadorean Athenaeum; Pres. Lic JOSÉ OSCAR RAMÍREZ PÉREZ; Sec.-Gen. Lic. RUBÉN REGALADO SERMEÑO.

**Revista Judicial:** Centro de Gobierno, San Salvador; tel. 2222-4522; organ of the Supreme Court; Dir Dr MANUEL ARRIETA GALLEGOS.

### PRESS ASSOCIATION

**Asociación de Periodistas de El Salvador** (Press Association of El Salvador): Edif. Casa del Periodista, Paseo Gen. Escalón 4130, San Salvador; tel. 2263-5335; e-mail info@apes.org.sv; internet www.apes.org.sv; Pres. HERMINIA FÚNES.

### FOREIGN NEWS AGENCIES

**Agencia EFE** (Spain): San Salvador; Bureau Chief CRISTINA HASBÚN DE MERINO.

**Agenzia Nazionale Stampa Associata (ANSA)** (Italy): Calle Acaxual, 30-G, Urb. Metropolis Mejicanos, San Salvador; tel. and fax 2274-5512; e-mail reneal@hotmail.com; Bureau Chief RENÉ ALBERTO CONTRERAS.

**Associated Press (AP)** (USA): San Salvador; Correspondent ANA LEONOR CABRERA.

*Directory*

**Deutsche Presse-Agentur (dpa)** (Germany): San Salvador; Correspondent JORGE ARMANDO CONTRERAS.

**Inter Press Service (IPS)** (Italy): San Salvador; e-mail ipslatam@ipsnews.net; Correspondent PABLO IACUB.

## Publishers

**Centro de Paz (CEPAZ):** Col. Libertad, Avda Washington 405, San Salvador; tel. and fax 2226-2117; e-mail cepaz@cepaz.org.sv; internet www.cepaz.org.sv; social history.

**Clásicos Roxsil, SA de CV:** 4a Avda Sur 2–3, Nueva San Salvador; tel. 2228-1832; fax 2228-1212; e-mail roxanabe@navegante.com.sv; f. 1976; textbooks, literature; Dir ROSA VICTORIA SERRANO DE LÓPEZ.

**Distribuidora de Textos Escolares (D'TEXE):** Edif. C, Col., Paseo y Condominio Miralvalle, San Salvador; tel. 2274-2031; f. 1985; educational; Dir JORGE A. LÓPEZ HIDALGO.

**Dirección de Publicaciones e Impresos:** Ministerio de Educación, 17a Avda Sur 430, San Salvador; tel. 2271-1806; fax 2271-1071; e-mail direccion@dpi.gov.sv; f. 1953; literary and general; Dir MIGUEL ANGEL RIVERA LARIOS.

**Ediciones Thau:** Carretera a Santa Tecla, km 8.5, Antiguo Cuscatlán; tel. 2263-2149; fax 2263-2172; e-mail descobarg@ujmd.edu.sv; Dir DAVID ESCOBAR GALINDO.

**Editorial Universitaria:** Universidad 'Dr José Matías Delgado', Km 8.5, Carretera a Santa Tecla, Ciudad Merliot; tel. 2278-1011; e-mail jalas@ujmd.edu.sv; internet www.ujmd.edu.sv/seditorial.html; f. 1984.

**UCA Editores:** Apdo 01-575, San Salvador; tel. 2273-4400; fax 2273-3556; e-mail distpubli@ued.uca.edu.sv; internet www.uca.edu.sv; f. 1975; social science, religion, economy, literature and textbooks; Dir RODOLFO CARDENAL.

### PUBLISHERS' ASSOCIATION

**Cámara Salvadoreña del Libro:** Col. Flor Blanca, 47 Avda Norte y 1a Calle Poniente, Apdo 3384, San Salvador; tel. 2261-0293; fax 2261-2231; e-mail camsalibro@integra.com.sv; f. 1974; Pres. ANA MOLINA DE FAUVET.

## Broadcasting and Communications

### TELECOMMUNICATIONS

#### Regulatory Authority

**Superintendencia General de Electricidad y Telecomunicaciones (SIGET):** Sexta Décima Calle Poniente y 37 Avda Sur 2001, Col. Flor Blanca, San Salvador; tel. 2257-4438; internet www.siget.gob.sv; f. 1996; Supt Lic. JORGE NIETO.

#### Major Service Providers

**Digicel:** San Salvador; internet www.digicel.com.sv; provider of mobile telecommunications; owned by Digicel (USA).

**Telecom El Salvador:** Alameda Manuel Enrique Araujo y Calle Nueva 1, 40°, Edif. Palic, San Salvador; tel. 7800-8155; internet www.telecom.com.sv; terrestrial telecommunications network, fmrly part of Administración Nacional de Telecomunicaciones (Antel), which was divested in 1998; changed name from CTE Antel Telecom in 1999; became subsidiary of France Télécom in 2002; Pres. DOMINIQUE ST JEAN.

**Telecom Personal:** Alameda Manuel Enrique Araujo y Calle Nueva 1, 40°, Edif. Palic, San Salvador; internet www.personal.com.sv; provider of mobile telecommunications; subsidiary of Telecom El Salvador.

**Telefónica El Salvador:** San Salvador; e-mail telefonica.empresas@telefonicamail.com.sv; internet www.telefonica.com.sv; manages sale of telecommunications frequencies; fmrly Internacional de Telecomunicaciones (Intel), which was divested in 1998; controlling interest owned by Telefónica, SA (Spain); Pres. CÉSAR ALIERTA.

**Telefónica Movistar:** Torre Telefónica (Torre B de Centro Financiero Gigante), Alameda Roosevelt y 63 Avda Sur, Col. Escalón, San Salvador; tel. 2257-4000; internet www.telefonica.com.sv/movistar; provider of mobile telecommunications; 92% owned by Telefónica Móviles, SA (Spain), part of Telefónica, SA.

**Telemóvil:** Avda Roosevelt, Centro Financiero, Gigante Torre D, 9°, San Salvador; tel. 2246-9977; fax 2246-9999; e-mail servicioalcliente@tigo.com.sv; internet www.telemovil.com; provider of mobile telecommunications; subsidiary of Millicom International Cellular (Sweden).

**Tricom, SA:** San Salvador; provider of mobile telecommunications; owned by Tricom, SA (Dominican Republic).

### RADIO

**Asociación Salvadoreña de Radiodifusores (ASDER):** Avda Izalco, Bloco 6 No 33, Residencial San Luis, San Salvador; tel. 2222-0872; fax 2274-6870; e-mail asder@ejje.com; internet www.asder.com.sv; f. 1965; Pres. Dr José Luis Saca Meléndez.

**YSSS Radio Nacional de El Salvador:** Dirección General de Medios, Calle Monserrat, Plantel Ex-IVU, San Salvador; e-mail radio.elsalvador@gobernacion.gob.sv; internet www.radioelsalvador.com.sv; f. 1926; non-commercial cultural station; Dir-Gen. Jaime Vilanova.

There are 64 commercial radio stations. Radio Venceremos and Radio Farabundo Martí, operated by the former guerrilla group FMLN, were legalized in April 1992. Radio Mayavisión (operated by FMLN supporters), began broadcasting in November 1993.

### TELEVISION

**Canal 2, SA:** Carretera a Nueva San Salvador, Apdo 720, San Salvador; tel. 2223-6744; internet www.teledos.com; commercial; Pres. Boris Eserski; Gen. Man. Salvador I. Gadala María.

**Canal 4, SA:** Carretera a Nueva San Salvador, Apdo 720, San Salvador; tel. 2224-4555; owned by Telecorporación Salvadoreña (TCS); commercial; Pres. Boris Eserski; Man. Ronald Calvo.

**Canal 6, SA:** Km 6, Carretera Panamericana a Santa Tecla, San Salvador; tel. 2209-5068; fax 2209-2033; e-mail director@elnoticiero.com.sv; internet www.elnoticiero.com.sv; f. 1972; owned by Telecorporación Salvadoreña (TCS); commercial; Exec. Dir Juan Carlos Eserski; Man. Dr Pedro Leonel Moreno Monge.

**Canal 8 (Agape TV):** 1511 Calle Gerardo Barrios, Col. Cucumacuyán, San Salvador; tel. 2281-2828; fax 2271-3414; e-mail info@agapetv8.com; internet www.agapetv8.com; catholic, family channel; Pres. Flavián Mucci.

**Canal 12:** Urb. Santa Elena 12, Antiguo Cuscatlán, San Salvador; tel. 2278-0622; fax 2278-0722; internet www.canal12.com.sv; f. 1984; Dir Alejandro González.

**Canal 15:** 4a Avda Sur y 5a Calle Oriente 301, San Miguel; tel. 2661-3298; fax 2661-3298; f. 1994; Gen. Man. Joaquín Aparicio.

**Canal 19:** Final Calle Los Abetos 1, Col. San Francisco, San Salvador; e-mail serviciosmegavision@salnet.net; owned by Grupo Megavisión; children's channel; Gen. Man. Mario Cañas.

**Canal 21 (Megavisión):** 1a Calle Poniente entre 85 y 87 Avda Norte, Apdo 2789, San Salvador; tel. 2283-2121; fax 2283-2132; e-mail serviciosmegavision@salnet.net; internet www.canal21tv.com.sv; f. 1993; Pres. Oscar Antonio Safie; Dir Hugo Escobar.

**Fundación Canal 25:** Final Calle Libertad 100, Ciudad Merliot, Nueva San Salvador; tel. 2248-2525; fax 2278-8526; f. 1997; commercial; Gen. Man. Jorge Mira.

**Tecnovisión Canal 33:** Calle Arce 1120, San Salvador; tel. 2275-8888; e-mail jgomez@utec.edu.sv; internet www.utec.edu.sv; owned by the Universidad Tecnológica de El Salvador; Pres. Mauricio Loucel.

# Finance

(cap. = capital; p.u. = paid up; res = reserves; dep. = deposits; m. = million; brs = branches; amounts in colones unless otherwise stated)

### BANKING

The banking system was nationalized in March 1980. In 1994 the transfer to private ownership of six banks and seven savings and loans institutions, as part of a programme of economic reform begun in 1991, was completed.

#### Supervisory Bodies

**Superintendencia del Sistema Financiero:** 7a Avda Norte 240, Apdo 2942, San Salvador; tel. 2281-2444; fax 2281-1621; internet www.ssf.gob.sv; Pres. Luis Armando Montenegro.

**Superintendencia de Valores:** Antiguo Edif. BCR, 1a Calle Poniente y 7a Avda Norte, San Salvador; tel. 2281-8900; fax 2221-3404; e-mail info@superval.gob.sv; internet www.superval.gob.sv; Pres. Lic. Omar Ernesto Rodríguez Alemán.

#### Central Bank

**Banco Central de Reserva de El Salvador:** Alameda Juan Pablo II y 17 Avda Norte, Apdo 01-106, San Salvador; tel. 2281-8000; fax 2281-8011; e-mail comunicaciones@bcr.gob.sv; internet www.bcr.gob.sv; f. 1934; nationalized Dec. 1961; entered monetary integration process 1 Jan. 2001; cap. US $115.0m., res $29.2m., dep. $1,556.3m. (Dec. 2003); Pres. Luz María de Portillo.

#### Commercial and Mortgage Banks

**Banco Agrícola:** Blvd Constitución 100, San Salvador; tel. 2267-5787; fax 2267-5775; e-mail info@bancoagricola.com; internet www.bancoagricola.com; f. 1955; dep. US $2,456.6m., total assets $2,9991.5m. (Dec. 2003); merged with Banco Desarrollo in July 2000; acquired Banco Capital in November 2001; Pres. Rodolfo Schildkrecht; 150 brs world-wide.

**Banco Americano:** World Trade Center, Torre II, Local 109, 89 Avda Norte y Calle al Mirador, San Salvador; tel. 2245-0651; fax 2298-5251; e-mail presidencia@bamericano.com; internet www.bamericano.com; privately owned; fmrly Unibanco; cap. US $13.0m., res $0.2m., dep. $78.5m. (Dec. 2003); Pres. Ing. Ronald A. Lacayo.

**Banco de Comercio:** Centro Financiero, 25 Avda Norte y 23 Calle Poniente, San Salvador; tel. 2226-4577; fax 2225-7767; e-mail informacion@banco.com.sv; internet www.banco.com.sv; f. 1949; privately owned; scheduled to merge with Scotiabank in July 2005; total assets 9,074m. (1999); Pres. José Gustavo Belismelis Vides; 48 brs.

**Banco Cuscatlán de El Salvador, SA:** Edif. Pirámide Cuscatlán, Km 10, Carretera a Santa Tecla, Apdo 626, San Salvador; tel. 2212-3333; fax 2228-5700; e-mail cuscatlan@bancocuscatlan.com; internet www.bancocuscatlan.com; f. 1972; privately owned; cap. 90m., res 123.6m., dep. 1,531.6m. (Sept. 2003); Pres. Mauricio Samayoa Rivas; Vice-Pres. Roberto Ortiz; 31 brs.

**Banco Hipotecario:** Pasaje Senda Florida Sur, Col. Escalón, Apdo 999, San Salvador; tel. 2250-7101; fax 2250-7039; internet www.bancohipotecario.com.sv; f. 1935; cap. 104.0m., dep. 1,852m. (Dec. 1998); Pres. Ing. José Roberto Navarro Escobar; Exec. Dir Bruno Gustavo Wyld Berg; 12 brs.

**Banco Promérica:** 71 Avda Sur y Paseo Gen. Escalón 3669, Col. Escalón, San Salvador; tel. 2243-3344; fax 2245-2979; e-mail baprosal@promerica.com.sv; internet www.promerica.com.sv; privately owned; Pres. Lic. Ramiro Ortiz Gurdián.

**Banco Salvadoreño, SA (BANCOSAL):** Edif. Centro Financiero, Avda Manuel Enrique Araujo y Avda Olímpica 3550, Apdo 06-73, San Salvador; tel. 2298-4444; fax 2298-0102; e-mail info@bancosal.com; internet www.bancosal.com; f. 1885; privately owned commercial bank; cap. US $90m., dep. 1,130.2m. (June 2004); merged with Banco de Construcción y Ahorro, SA (BANCASA) in July 2000; Pres. and Chair. Lic. María Eugenia Brizuela de Ávila; Vice-Pres. Ing. Moisés Castro Maceda; 78 brs.

**Banco UNO:** Paseo Gen. Escalón y 69 Avda Sur 3563, Col. Escalón, San Salvador; tel. 2245-0055; fax 2245-0080; internet www.bancouno.com.sv; cap. US $11.4m., res $9.9m., dep. $150.6m. (Dec 2002); Pres. Ing. Albino Román.

**Financiera Calpiá:** Paseo Gen. Escalón 5438, San Salvador; tel. 2264-1194; fax 2264-1647; e-mail financiera@calpia.com.sv; Pres. Pedro Dalmau Gorrita.

#### Foreign Banks

**Citibank, NA** (USA): Edif. Palic, Calle Nueva 1, Col. Escalón, San Salvador; tel. 2211-2484; fax 2211-1842; e-mail bert.veltman@citigroup.com; internet www.citibank.com/elsalvador; Pres. Robert B. Willumstad; CEO Charles Prince.

**First Commercial Bank** (USA): Centro Comercial Gigante, Torre Telefónica, 3°, 63 Avda Sur y Alameda Roosevelt, San Salvador; tel. 2211-2121; fax 2211-2130; e-mail fcb971@saltel.net; Vice-Pres. Peter Ming-Fung Lan.

**Scotiabank** (Canada): Edif. Torre Scotiabank, Avda Olímpica 129, San Salvador; tel. 2245-1211; fax 2224-2884; internet www.scotiabank.com.sv; f. 1972; total assets 3,970.8m.; Pres. Ing. Luis Ivandic.

#### Public Institutions

**Banco de Fomento Agropecuario:** Km 10.5, Carretera al Puerto de la Libertad, Nueva San Salvador; tel. 2228-5188; fax 2228-5199; f. 1973; state-owned; cap. 605.0m., dep. 872.0m. (Oct. 1997); Pres. Lic. Raúl García Prieto; 27 brs.

**Banco Multisectorial de Inversiones:** Edif. Century Plaza, Alameda Manuel E. Araujo, San Salvador; tel. 2267-0000; fax 2267-0011; e-mail bmigada@es.com.sv; internet www.bmi.gob.sv; f. 1994; Pres. Dr Nicola Angelucci.

**Federación de Cajas de Crédito (FEDECREDITO):** 25 Avda Norte y 23 Calle Poniente, San Salvador; tel. 2209-9696; fax 2226-7161; internet www.fedecredito.com.sv; f. 1943; Pres. Lic. MACARIO ARMANDO ROSALES.

**Fondo Social Para la Vivienda:** Calle Rubén Darío, entre 15 y 17 Avda Sur, San Salvador; tel. 2271-2774; e-mail comunicaciones@fsv.gob.sv; internet www.fsv.gob.sv; Pres. Lic. ENRIQUE OÑATE.

### Banking Association

**Asociación Bancaria Salvadoreña (ABANSA):** Alameda Roosevelt 2511, entre 47 y 49 Avda Sur, San Salvador; tel. 2298-6938; fax 2223-1079; internet www.abansa.org.sv; Pres. Ing. JOSÉ GUSTAVO BELISMELIS.

### STOCK EXCHANGE

**Mercado de Valores de El Salvador, SA de CV (Bolsa de Valores):** Urb. Jardines de la Hacienda, Blvd Merliot y Avda Las Carretas, Antiguo Cuscatlán, La Libertad, San Salvador; tel. 2212-6400; fax 2278-4377; e-mail info@bves.com.sv; internet www.bves.com.sv; f. 1992; Pres. ROLANDO DUARTE SCHLAGETER.

### INSURANCE

**AIG Unión y Desarrollo, SA:** Calle Loma Linda 265, Col. San Benito, Apdo 92, San Salvador; tel. 2298-5455; fax 2298-5084; e-mail aig.elsalvador@uni-desa.com; f. 1998; following merger of Unión y Desarrollo, SA and AIG; Pres. FRANCISCO R. DE SOLA.

**Aseguradora Agrícola Comercial, SA:** Alameda Roosevelt 3104, Apdo 1855, San Salvador; tel. 2260-3344; fax 2260-5592; e-mail informacion@acasal.com.sv; internet www.acasal.com.sv; f. 1973; Pres. LUIS ALFREDO ESCALANTE SOL; Gen. Man. LUIS ALFONSO FIGUEROA.

**Aseguradora Popular, SA:** Paseo Gen. Escalón 5338, Col. Escalón, San Salvador; tel. 2263-0700; fax 2263-1246; e-mail aseposapresi@telesal.net; f. 1975; Pres. Dr CARLOS ARMANDO LAHÚD; Gen. Man. HERIBERTO PÉREZ AGUIRRE.

**Aseguradora Suiza Salvadoreña, SA (ASESUISA):** Alameda Dr M. E. Araujo, Plaza Suiza, Apdo 1490, San Salvador; tel. 2209-5000; fax 2209-5001; e-mail info@asesuisa.com; internet www.asesuisa.com; f. 1969; Pres. MAURICIO MEYER COHEN; Gen. Man. RICARDO COHEN.

**La Central de Seguros y Fianzas, SA:** Avda Olímpica 3333, Apdo 01-255, San Salvador; tel. 2279-3544; fax 2223-7647; e-mail hroque@lacentral.com.sv; internet www.lacentral.com.sv; f. 1983; Pres. EDUARDO ENRIQUE CHACÓN BORJA; Man. HECTOR ROQUE RUBALLO.

**La Centro Americana, SA:** Alameda Roosevelt 3107, Apdo 527, San Salvador; tel. 2223-6666; fax 2223-7203; internet www.lacentro.com; f. 1915; Pres. RUFINO GARAY AYALA; Gen. Man. ANTONIO PENEDO.

**Compañía Anglo Salvadoreña de Seguros, SA:** Paseo Gen. Escalón 3848, San Salvador; tel. 2263-0009; fax 2263-0106; f. 1976; Pres. JOSÉ ARTURO GÓMEZ; Vice-Pres. LETICIA FARFÁN DE GÓMEZ.

**Compañía General de Seguros, SA:** Calle Loma Linda 223, Col. San Benito, Apdo 1004, San Salvador; tel. 2279-3777; fax 2223-0719; f. 1955; Pres. JOSÉ GUSTAVO BELISMELIS VIDES; Gen. Man. ERIC PRADO.

**Internacional de Seguros, SA (Interseguros):** Centro Financiero Banco Salvadoreño, 5°, Avda Olímpica 3550, Col. Escalón, San Salvador; tel. 2238-0202; fax 2238-5727; e-mail interseguros@interseguros.com.sv; internet www.interseguros.com.sv; f. 1958; Merged with Seguros Universales in 2004; Pres. Dr ENRIQUE GARCÍA PRIETO; Gen. Man. ALEJANDRO CABRERA RIVAS.

**Seguros e Inversiones, SA (SISA):** 10.5 km Carretera Panamericana, Santa Tecla; tel. 2229-8888; fax 2229-8187; f. 1962; Pres. ALFREDO FÉLIX CRISTIANI BURKARD; Pres. EDUARDO MONTENEGRO.

## Trade and Industry

### GOVERNMENT AGENCIES AND DEVELOPMENT ORGANIZATIONS

**Consejo Nacional de Ciencia y Tecnología (CONACYT):** Col. Médica, Avda Dr Emilio Alvarez, Pasaje Dr Guillermo Rodríguez Pacas 51, San Salvador; tel. 2226-2800; fax 2225-6255; e-mail ulisest@conacyt.gob.sv; internet www.conacyt.gob.sv; f. 1992; formulation and guidance of national policy on science and technology; Exec. Dir CARLOS ROBERTO OCHOA CORDOBA.

**Corporación de Exportadores de El Salvador (COEXPORT):** Condomínios del Mediterráneo, Edif. 'A', No 23, Col. Jardines de Guadalupe, San Salvador; tel. 2243-1328; fax 2243-3159; e-mail info@coexport.com; internet www.coexport.com; f. 1973 to promote Salvadorean exports; Exec. Dir Lic. SILVIA M. CUÉLLAR.

**Corporación Salvadoreña de Inversiones (CORSAIN):** 1a Calle Poniente, entre 43 y 45 Avda Norte, San Salvador; tel. 2224-4242; fax 2224-6877; e-mail audisjus@cortedecuentas.gob.sv; Pres. ERNESTO HAREM.

**Fondo de Financiamiento y Garantía para la Pequeña Empresa (FIGAPE):** 9a Avda Norte 225, Apdo 1990, San Salvador; tel. 7771-1994; f. 1994; government body to assist small-sized industries; Pres. Lic. MARCO TULIO GUARDADO.

**Fondo Social para la Vivienda (FSV):** Calle Rubén Darío y 17 Avda Sur 455, San Salvador; tel. 2271-1662; fax 2271-2910; internet www.fsv.gob.sv; f. 1973; Pres. EDGAR RAMIRO MENDOZA JEREZ; Gen. Man. FRANCISCO ANTONIO GUEVARA.

**Instituto Salvadoreño de Transformación Agraria (ISTA):** Km 5½, Carretera a Santa Tecla, San Salvador; tel. 2224-6000; fax 2224-0259; f. 1976 to promote rural development; empowered to buy inefficiently cultivated land; Pres. JOSÉ ROBERTO MOLINA MORALES.

### CHAMBER OF COMMERCE

**Cámara de Comercio e Industria de El Salvador:** 9a Avda Norte y 5a Calle Poniente, Apdo 1640, San Salvador; tel. 2244-2000; fax 2271-4461; e-mail camara@camarasal.com; internet www.camarasal.com; f. 1915; 2,000 mems; Pres. EDUARDO OÑATE MUYSHONDT; Exec. Dir ALBERTO PADILLA AQUINO; brs in San Miguel, Santa Ana and Sonsonate.

### INDUSTRIAL AND TRADE ASSOCIATIONS

**Asociación Azucarera de El Salvador:** 103 Avda Norte y Calle Arturo Ambrogi 145, Col. Escalón, San Salvador; tel. 2263-0378; fax 2263-0361; e-mail asosugar@sal.gbm.net; internet www.asociacionazucarera.com; national sugar association, fmrly Instituto Nacional del Azúcar; Pres. MARIO SALAVERRIA; Exec. Dir Dr FRANCISCO ARMANDO.

**Asociación Cafetalera de El Salvador (ACES):** 67 Avda Norte 116, Col. Escalón, San Salvador; tel. 2223-3024; fax 2223-7471; e-mail acafesal@sal.gbm.net; f. 1930; coffee growers' asscn; Pres. Ing. EDUARDO E. BARRIENTOS.

**Asociación de Ganaderos de El Salvador:** 1a Avda Norte 1332, San Salvador; tel. 2225-7208; f. 1932; livestock breeders' asscn; Pres. Lic. CARLOS ARTURO MUYSHONDT.

**Asociación Salvadoreña de Beneficiadores y Exportadores de Café (ABECAFE):** 87a Avda Norte 720, Col. Escalón, San Salvador; tel. 2263-2834; fax 2263-2833; e-mail abecafe@intersal.com; coffee producers' and exporters' asscn; Pres. CARLOS BORGONOVO.

**Asociación Salvadoreña de Industriales:** Calles Roma y Liverpool, Col. Roma, Apdo 48, San Salvador; tel. 2279-2488; fax 2279-1880; e-mail asi@asi.com.sv; internet www.asi.com.sv; f. 1958; 400 mems; manufacturers' asscn; Pres. NAPOLEÓN GUERRERO BERRIOS; Exec. Dir Lic. JORGE ARRIAZA.

**Consejo Salvadoreño del Café (CSC):** 75 Avda Norte y 7a Calle Poniente 3876, Col. Escalón, San Salvador; tel. 2263-3787; fax 2263-3833; e-mail csc@consejocafe.org.sv; internet www.consejocafe.org.sv; f. 1989 as successor to the Instituto Nacional del Café; formulates policy and oversees the coffee industry; Exec. Dir RICARDO ESPITIA.

**Cooperativa Algodonera Salvadoreña, Ltda (COPAL):** 6a y 10a Calle Poniente y 43 Avda Sur 2305, Col. Flor Blanca, San Salvador; tel. 2298-9330; fax 2298-9331; f. 1940; 185 mems; cotton growers' asscn; Pres. LUIZ MÉNDEZ NOVOA; Vice-Pres. Lic. JOSÉ RAMIRO PARADA DÍAZ.

**UCAFES:** Avda Río Lempa, Calle Adriático 44, Jardines de Guadalupe, San Salvador; tel. 2243-2238; fax 2298-1504; union of coffee-growing co-operatives; Pres. JOAQUÍN SALAVERRÍA.

### EMPLOYERS' ORGANIZATION

There are several business associations, the most important of which is the Asociación Nacional de Empresa Privada.

**Asociación Nacional de Empresa Privada (ANEP)** (National Private Enterprise Association): 1a Calle Pte y 71a Avda Norte 204, Col. Escalón, Apdo 1204, San Salvador; tel. 2224-1236; fax 2223-8932; e-mail communicaciones.anep@telesal.net; internet www.anep.org.sv; national private enterprise association; Pres. FEDERICO COLORADO; Exec. Dir Lic. LUIS MARIO RODRÍGUEZ.

### MAJOR COMPANIES

#### Construction and Metals

**Cemento de El Salvador, SA de CV:** Avda El Espina y Blvd Sur, Urb. Madre Selva, Antigua Cuscatlán, La Libertad; tel. 2505-0000; fax 2505-0777; e-mail cessamer@cessa.com.sv; internet www.cessa

.com.sv; f. 1949; manufacturers of Portland cement; Pres. César Catani; Gen. Man. Ricardo Chávez; 400 employees.

**Conductores Eléctricos de Centro América, SA (CONELCA):** Carretera Panamérica, Km 11, Ilopango, APDO 283, San Salvador; tel. 2295-0866; fax 2295-0859; f. 1949; subsidiary of Phelps Dodge Industries Inc, USA; manufacturers of telephone cables; Pres. Mario Antonio Andino Gómez; 250 employees.

**Corporación Industrial de Centroaméricana, SA de CV (CORINCA):** Carretera a Quezaltepeque, Km 25, San Salvador; tel. 2310-2033; fax 2310-2234; internet www.corinca.com.sv; f. 1966; iron rods and wire, construction and building materials; Pres. Sergio Catani Papini; Gen. Man. Carlos Francisco Alvarado; 350 employees.

### Food and Beverages

**Grupo Calvo:** Puerto Cutuco, Punta Gorda; internet www.calvo.es; f. 2003; tuna-processing plant; Pres. José Luis Calvo; 750 employees.

**Industrias La Constancia, SA de CV (ILC):** Avda Independencia 526, Apdo 06-101, San Salvador; tel. 2231-5444; fax 2231-5152; e-mail info@laconstancia.com; internet www.laconstancia.com; f. 1906; partly owned by SABMiller PLC (United Kingdom); produces and sells beer; Pres. Roberto H. Murray Meza; Tech. Dir Rolando Caro Härter; 950 employees.

**Molinos de El Salvador, SA:** Blvd del Ejército Nacional y 50a Avda Norte, Apdo 327, San Salvador; tel. 2293-1522; fax 2293-1523; e-mail molsa@molsa.com.sv; internet www.coexport.com/molsa; f. 1959; production of wheat flour and biscuits; Pres. Alfonso Alvarez; Man. Ferit Zacarias Massis; 121 employees.

**Productos Alimenticios Diana SA de CV:** 12 Avda Sur, Soyapango, Apdo 177, San Salvador; tel. 2227-1233; fax 2227-7023; e-mail info@diana.com.sv; internet www.diana.com.sv; f. 1951; food processing; Pres. Rosy de Paredes; Gen. Man. Hugo César Barrera; 2,000 employees.

**Sello de Oro, SA, Productos Alimenticios:** 2.5 km Carretera a Jayaque, La Libertad, San Salvador; tel. 2338-4800; fax 2344-4200; e-mail info@sellodeoro.com.sv; internet www.coexport.com/pasosa; f. 1967; food and food processing; Pres. Carmen Elena del Sol; Exec. Dir José Agustín Martínez; 1,285 employees.

### Pharmaceuticals

**Droguería Santa Lucía, SA de CV:** Calle Roma 238, Col. Roma, Apdo 06-5, San Salvador; tel. 2223-8000; fax 2223-8033; e-mail ventas@drogueriasantalucia.com; internet www.drogueriasantalucia.com; AstraZeneca PLC (United Kingdom); pharmaceutical products.

**Laboratorio Lopez, SA de CV:** Blvd del Ejército Nacional, Km 5, Jurisdicción de Soyapango, San Salvador; tel. 2277-8333; fax 2277-2783; e-mail lablopez@salnet.net; f. 1948; manufacturers of pharmaceutical products; Pres. Gustavo López Rodríguez.

**Laboratorios Vijosa, SA de CV:** Calle L-3, 10 Zona Industrial Merliot, Antiguo Cuscatlán; tel. 2278-3077; fax 2278-2131; e-mail info@vijosa.com; internet www.vijosa.com; manufacturers of pharmaceutical products; Pres. Dr Víctor Jorge Saca; 130 employees.

### Textiles and Clothing

**Almacenes Simán, SA de CV:** Centro Comercial Galerías, Paseo General Escalón 3700, Col. Escalón, San Salvador; tel. 2245-3000; fax 2245-4000; e-mail contacto@siman.com.sv; internet www.siman.com.sv; f. 1921; wholesale and retail sale of clothing; Pres. Salvador Simán; 1,100 employees.

**Facalca Hiltex, SA de CV:** Km 99½, Carretera a San Salvador, Ahuachapan; tel. 2443-0033; fax 2443-0461; e-mail webmaster@facalca-hiltex.com; internet www.facalca-hiltex.com; f. 1964; thread mills; Pres. and Man. Jorge Bahait Ghia; 980 employees.

**Industrias Sintéticas de Centroamérica, SA:** Carretera Troncal del Norte, Km 12½, Apopa; tel. 2216-0055; fax 2216-0062; e-mail miguel.cabrera@navegante.com.sv; internet www.insinca.com; f. 1966; manufacturers of synthetic fibres; Pres. Félix Castillo Mayorga; 1,000 employees.

**Industrias Unidas, SA (IUSA):** Carretera Panamericana a Oeste Km 11.5, Ilopango, San Salvador; tel. 2295-0555; fax 2295-0846; e-mail ventas@iusa.com.sv; internet www.iusa.com.sv; f. 1955; manufacturers of thread; Pres. Germán Maron García; 1,600 employees.

**Textiles San Andrés, SA de CV (Hilasal):** Km 32, Carretera a Santa Ana, La Libertad, San Salvador; tel. 2338-4099; fax 2338-4064; e-mail tsanandres@es.com.sv; manufacturers of cotton goods and towelling; Pres. Pablo García Barbachano; 1,000 employees.

**Textufil, SA de CV:** 12 Avda Sur, Soyapango, Apdo 1632, San Salvador; tel. 2277-0066; fax 2227-2308; e-mail info@textufil.com; internet www.textufil.com; f. 1971; manufacturers of nylon and polyester textiles; Pres. Jorge Elías Bahaia; Man. Elías Jorge Bahaia Samour.

### Miscellaneous

**British American Tobacco El Salvador:** Alameda Roosevelt 2115, Apdo 06-113, San Salvador; tel. 2250-4444; fax 2250-4443; internet www.batcentralamerica.com; manufacture and sale of cigarettes.

**Cigarrería Morazán, SA de CV:** Blvd del Ejército Nacional, Km 7½, Jurisdicción de Soyapango, San Salvador; tel. 2277-0444; fax 2227-2534; f. 1926; cigarette manufacturer and tobacco exporter; Pres. and Gen. Man. Rafael Márquez; 150 employees.

**Compañía Química Industrial, SA de CV (COQUINSA):** 29 Calle Oriente 73, Col. La Rabida, San Salvador; tel. 2226-0137; fax 2225-8430; e-mail ventas@coquinsa.com; internet www.coquinsa.com; f. 1979; manufacturers of adhesives, detergents, disinfectants, insecticides; Pres. Manuel de J. Rodríguez.

**Distribuidora de Automóviles, SA de CV (DIDEA):** Blvd Los Héroes, Edif. DIDEA, San Salvador; tel. 2261-1133; fax 2260-3516; e-mail dideauno@es.com.sv; internet www.toyotadidea.com; f. 1919; distributor of cars and car supplies; Pres. Ricardo Poma; Dir-Gen. Casildo Quan; 350 employees.

**Empresas Adoc, SA:** Blvd Ejército Nacional, Km 4½, Final Col. Montecarlo, Apdo 687, Soyapango, San Salvador; tel. 2277-2277; fax 2277-0352; e-mail adoces@adoc.com.sv ; f. 1952; tannery, rubber, plastics, manufacturers of shoes and leather goods; retailers; Pres. Roberto Palomo; Gen. Man. Wilfredo Rosales; 4,370 employees.

**Muebles Metálicos Prado, SA de CV:** Km 7.5, Blvd del Ejército, Soyapango; tel. 2244-0000; fax 2289-1717; e-mail naguirre@prado.com.sv; internet prado.com.sv; f. 1952; manufacturers of metal furniture; Pres. Francisco José Prado Mairena; Gen. Man. Francisco José Prado Rivas; 500 employees.

**PriceWaterhouseCoopers:** Centro Profesional Presidente, Avda La Revolución y Calle Circunvalación, Apdo 695, San Salvador; tel. 2243-5844; fax 2243-3546; internet www.pwcglobal.com; accountancy and management consultancy; Pres. Mario Wilfredo López Salgado; 150 employees.

**Siemens SA:** Calle Siemens 43, Parque Industrial Santa Elena, Antiguo Cuscatlán, Apdo 1525, San Salvador; tel. 2278-3333; fax 2278-3334; e-mail siemens.elsalvador@siemens.co; internet www.siemens-centram.com; owned by Siemens AG (Germany); manufacturers of electrical equipment and machinery; PL&S Man. Roberto Huezo.

**SIGMA/Q, SA:** 8 km Blvd de Ejército, Apdo 1096, San Salvador; tel. 2263-5000; fax 2263-9404; e-mail atencionalcliente@sigmaq.com; internet www.sigmaq.com; f. 1973; manufacturers of collapsible packaging; Contact Carmen Aida de Meardi; 957 employees.

**Tabacalera de El Salvador, SA de CV:** 69 Avda Norte 213, Col. Escalón, San Salvador; tel. 2298-5888; fax 2224-3815; f. 1976; subsidiary of Philip Morris Inc, USA; manufacturers of cigarettes; Pres. Jorge Zablah Touche; 186 employees.

## UTILITIES

### Electricity

**Comisión Ejecutiva Hidroeléctrica del Río Lempa (CEL):** 9a Calle Poniente 950, San Salvador; tel. 2211-6000; fax 2222-9359; e-mail naguilar@cel.gob.sv; internet www.cel.gob.sv; f. 1948; state energy agency dealing with electricity generation and transmission, and non-conventional energy sources; scheduled for privatization; Pres. Guillermo A. Sol Bang.

**Superintendencia General de Electricidad y Telecomunicaciones (SIGET):** Col. Flor Blanca, Sexta Décima Calle Poniente 2001, San Salvador; tel. 2257-4438; e-mail jtrigueros@siget.gob.sv; internet www.siget.gob.sv; f. 1996; Supt Lic. José Luis Trigueros.

### Electricity Companies

In order to increase competition, the electricity-trading market was opened up in October 2000. Four companies (two domestic and two foreign) subsequently applied for licences to trade electricity in the wholesale market, from SIGET.

**AES Aurora:** Operates three distribution companies in El Salvador:

    **CLESA:** 23 Avda Sur y 5a Calle Oriente, Barrio San Rafael, Santa Ana; tel. 2447-6000; fax 2447-6155.

    **Compañía de Alumbrado Electric (CAESS):** Calle El Bambú, Col. San Antonio, Ayutuxtepeque, San Salvador; tel. 2207-2011; fax 2232-5012.

    **EEO:** Final 8a, Calle Poniente, Calle a Ciudad Pacífico, Plantel Jalacatal, San Miguel.

**CONEC-ES, SA de CV:** Edif. Gran Plaza 204, Blvd del Hipódromo, San Benito, San Salvador; subsidiary of Energia Global (USA); provides electricity from sustainable sources, imported from Costa Rica; Gen. Man. Vicente Machado.

**Geotérmica Salvadoreña, SA (GESAL):** Col. Utila Santa Tecla, San Salvador; tel. 2275-9385; fax 2263-9505; e-mail info@gesal.com; internet www.gesal.com.sv; electricity generation; two geothermic stations; joint US-Chilean owned; Gen. Man. José Antonio Rodríguez.

### Water

**Administración Nacional de Acueductos y Alcantarillados (ANDA):** Edif. ANDA, Final Avda Don Bosco, Col. Libertad, San Salvador; tel. 2225-3534; fax 2225-3152; internet www.anda.gob.sv; f. 1961; maintenance of water supply and sewerage systems; Pres. Manuel Enrique Arrieta; Gen. Man. Frineé Castillo de Zaldaña.

## TRADE UNIONS

**Asociación de Sindicatos Independientes (ASIES)** (Association of Independent Trade Unions): San Salvador.

**Central de Trabajadores Democráticos (CTD)** (Democratic Workers' Confederation): 1a Avda Norte y 19 Calle Poniente 12, San Salvador; tel. and fax 2235-8043; e-mail comuctd@netcomsa.com; Sec.-Gen. Amadeo García Espinoza.

**Central de Trabajadores Salvadoreños (CTS)** (Salvadoran Workers' Confederation): 224a Pasaje Espínola, Col. Manzano, San Jacinto, San Salvador; tel. 2270-5246; fax 2270-1703; f. 1966; Christian Democratic; 35,000 mems.

**Confederación General de Sindicatos (CGS)** (General Confederation of Unions): Edif. Kury, 3a Calle Oriente 226, San Salvador; tel. and fax 2222-3527; f. 1958; admitted to ICFTU/ORIT; Sec.-Gen. José Israel Huiza Cisneros; 27,000 mems.

**Confederación General del Trabajo (CGT)** (General Confederation of Workers): 4a Avda Sur y 8a Calle Oriente 240, San Salvador; tel. and fax 2222-6182; f. 1983; 20 affiliated unions; Sec.-Gen. José René Pérez; 85,000 mems.

**Confederación Unitaria de Trabajadores Salvadoreños (CUTS)** (United Salvadorean Workers' Federation): 141 Avda A, Col. San José, San Salvador; tel. and fax 2225-3756; e-mail proyectocuts@salnet.net; left-wing; Sec.-Gen. Noé Nerio.

**Coordinadora Sindical de los Trabajadores Salvadoreños (CSTS)** (Salvadorean Workers' Union): Col. San José, Blvd Universitario 2226, San Salvador; conglomerate of independent left-wing trade unions; Co-ordinator Victor Aguilar.

**Federación Campesina Cristiana de El Salvador-Unión de Trabajadores del Campo (FECCAS-UTC)** (Christian Peasant Federation of El Salvador-Union of Countryside Workers): Universidad Nacional, Apdo 4000, San Salvador; allied illegal Christian peasants' organizations.

**Federación Nacional Sindical de Trabajadores Salvadoreños (FENASTRAS)** (Salvadorean Workers' National Union Federation): 10a Avda Norte 120, San Salvador; f. 1975; left-wing; 35,000 mems in 16 affiliates.

**Unidad Nacional de Trabajadores Salvadoreños (UNTS)** (National Unity of Salvadorean Workers): Calle 27 Poniente 432, Col. Layco, Apdo 2479, Centro de Gobierno El Salvador, San Salvador; tel. 2225-7811; fax 2225-0558; f. 1986; largest trade union conglomerate; Leader Marco Tulio Lima; affiliated unions include.

  **Unidad Popular Democrática (UPD)** (Popular Democratic Unity): San Salvador; f. 1980; led by a committee of 10; 500,000 mems.

**Unión Comunal Salvadoreña (UCS)** (Salvadorean Communal Union): 2a Calle Oriente y Avda Melvyn Jones, Santa Tecla, San Salvador; tel. 2228-2023; fax 2229-1111; peasants' association; 100,000 mems; Gen. Sec. Guillermo Blanco.

**Unión Nacional Obrera-Campesina (UNOC)** (Worker-Peasant National Union): Blvd María Cristina 165, San Salvador; tel. 2225-6981; fax 2228-1434; f. 1986; centre-left labour organization; 500,000 mems.

Some unions, such as those of the taxi drivers and bus owners, are affiliated to the Federación Nacional de Empresas Pequeñas Salvadoreñas—Fenapes, the association of small businesses.

# Transport

**Comisión Ejecutiva Portuaria Autónoma (CEPA):** Edif. Torre Roble, Blvd de Los Héroes, Apdo 2667, San Salvador; tel. 2224-1133; fax 2224-0907; internet www.cepa.gob.sv; f. 1952; operates and administers the ports of Acajutla (on Pacific coast) and Cutuco (on Gulf of Fonseca) and the El Salvador International Airport, as well as Ferrocarriles Nacionales de El Salvador; Pres. Ruy César Miranda; Gen. Man. Lic. Arturo Germán Martínez.

### RAILWAYS

In 2002 there were 549 km of railway track in the country, of which 286 km were in service. The main track links San Salvador with the ports of Acajutla and Cutuco and with San Jerónimo on the border with Guatemala. The Salvadorean section of the International Railways of Central America runs from Anguiatú on the El Salvador–Guatemala border to the Pacific ports of Acajutla and Cutuco and connects San Salvador with Guatemala City and the Guatemalan Atlantic ports of Puerto Barrios and Santo Tomás de Castilla. A project to connect the Salvadorean and Guatemalan railway systems between Santa Ana and Santa Lucía (in Guatemala) is under consideration.

**Ferrocarriles Nacionales de El Salvador (FENADESAL):** Avda Peralta 903, Apdo 2292, San Salvador; tel. 2271-5632; fax 2271-5650; internet www.cepa.gob.sv/fenadesa1.htm; 562 km open; in 1975 Ferrocarril de El Salvador and the Salvadorean section of International Railways of Central America (429 km open) were merged and are administered by the Railroad Division of CEPA (q.v.); Gen. Man. Salvador Sanabria.

### ROADS

The country's highway system is well integrated with its railway services. There were some 10,029 km of roads in 1999, including: the Pan-American Highway (306 km). Following the earthquakes of early 2001, the Inter-American Development Bank (IDB) pledged some US $106.0m. to the transport sector for the restoration and reconstruction of roads and bridges.

### SHIPPING

The port of Acajutla is administered by CEPA (see above). Services are also provided by foreign lines. The port of Cutuco has been inactive since 1996; however, it was scheduled to reopen in 2006.

### CIVIL AVIATION

The El Salvador International Airport is located 40 km (25 miles) from San Salvador in Comalapa. An expansion of the airport was completed in 1998, with a second expansion phase of hotel and commercial space to be completed early in the 21st century. The former international airport at Ilopango is used for military and private civilian aircraft; there are an additional 88 private airports, four with permanent-surface runways.

**TACA International Airlines:** Local 21. Paseo Gen. Escalón y 71 Avda Norte, Centro Comercial Galerías, San Salvador; tel. 2339-9155; fax 2223-3757; internet www.taca.com; f. 1939; passenger and cargo services to Central America and the USA; Pres. Federico Bloch; Gen. Man. Ben Baldanza.

**United Airlines:** Local 14, Centro Comercial Galerías, Paseo Gen. Escalón, San Salvador; tel. 2279-3900; fax 2298-5539; e-mail patricia.mejia@ual.com; internet www.unitedelsalvador.com; f. 1910.

# Tourism

El Salvador was one of the centres of the ancient Mayan civilization, and the ruined temples and cities are of great interest. The volcanoes and lakes of the uplands provide magnificent scenery, while there are fine beaches along the Pacific coast. The civil war, from 1979 to 1992, severely affected the tourism industry. Following the earthquakes of early 2001, the IDB pledged some $3.6m. to the tourism and historical and cultural heritage sectors, for the reconstruction and renovation of recreational centres, the promotion of tourism, and the development of culture and heritage. In 2001 tourist arrivals decreased to 734,627 from 795,000 in the previous year and tourism receipts stood at US $235m. Receipts increased in subsequent years, reaching $425m. in 2004. In the same year the number of tourist arrivals was 966,416.

**Asociación Salvadoreña de Hoteles:** 123 Pasaje 1, Avda Olímpica, Col. Escalón, San Salvador; tel. 2298-3629; fax 2298-3628; e-mail asocdehoteles@navegante.com.sv; internet www.elsalhoteles.com; f. 1996; Pres. Carlos Alberto Delgado Z.; Sec. Bellyni Siguenza.

**Buró de Convenciones de El Salvador:** Edif. Olimpic Plaza, 73 Avda Sur 28, 2°, San Salvador; tel. 2224-0819; fax 2223-4912; internet www.convencioneselsalvador.com.sv; f. 1973; assists in organization of national and international events; Pres. Alfredo Morales; Exec. Dir Adriana de Gale.

**Cámara Salvadoreña de Turismo (CASATUR):** 123 Pasaje 1, Avda Olímpica, Col. Escalón, San Salvador; tel. 2279-2156; fax 2223-9775; e-mail info@casatur.com; internet www.casatur.com; f. 1978; non-profit org. concerned with promotion of tourism in El Salvador; Pres. Carlos Alberto Delgado; Exec. Dir Ena López Portillo.

**Corporación Salvadoreña de Turismo (CORSATUR):** Avda El Espino 68, Urb. Madre Selva, Santa Elena, Antigua Cuscatlán, La Libertad, San Salvador; tel. 2243-7835; fax 2243-7844; e-mail info@corsatur.gob.sv; internet www.elsalvadorturismo.gob.sv; f. 1997; Pres. LUIS CARDENAL; Marketing Man. KARLA SEQUEIRA.

**Feria Internacional de El Salvador (FIES):** Avda La Revolución 222, Col. San Benito, Apdo 493, San Salvador; tel. 2243-0244; fax 2243-3161; e-mail feria@fies.gob.sv; internet www.fies.gob.sv; f. 1965; Pres. BENJAMÍN TRABANINO LLOBELL.

**Instituto Salvadoreño de Turismo (ISTU)** (National Tourism Institute): Calle Rubén Darío 619, San Salvador; tel. 2222-5727; fax 2222-1208; e-mail informacion@istu.gob.sv; internet www.istu.gob.sv; f. 1950; Pres. MANUEL AVILÉS; Man. Dir ROLANDO GUTIÉRREZ ORIANI.

# Defence

In August 2004 El Salvador's Armed Forces numbered 15,500: Army an estimated 13,850, Navy 700, Air Force 950. Paramilitary forces numbered some 12,000, and were to be increased to 16,000. Military service is by compulsory selective conscription of males between 18 and 30 years of age and lasts for one year.

**Defence Budget:** 927m. colones (US $106m.) in 2004.

**Chief of Staff of the Armed Forces:** Maj. Gen. CARLOS EDUARDO CÁCERES FLORES.

# Education

Education in El Salvador is provided free of charge in state schools and there are also numerous private schools. Pre-primary education, beginning at four years of age and lasting for three years, and primary education, beginning at the age of seven years and lasting for nine years, are officially compulsory. In 2002 enrolment at pre-primary institutions was equivalent to 43.3% of children in the relevant age-group, while enrolment at primary schools was equivalent to 86.9% of children in the relevant age-group. Secondary education, from the age of 16, lasts for three years. In 2002 enrolment at secondary schools was equivalent to just 25.5% of students in the relevant age group. Budgetary expenditure on education by the central government in 2001 was US $459.1m., equivalent to 23.3% of total expenditure; in 2002 education expenditure rose to $461.5m.

# Bibliography

For works on Central America generally, see Select Bibliography (Books)

Byrne, H. *El Salvador's Civil War: A Study of Revolution*. Boulder, CO, Lynne Rienner Publrs, 1996.

Chislett, W. *El Salvador: A New Opportunity*. London, Euromoney, 1998.

La Comisión de la Verdad para El Salvador. *De la Locura a la Esperanza: La Guerra de 12 Años en El Salvador*. San Salvador, United Nations, 1993.

Cousens, E. M. (Ed.) et al. *Peacebuilding as Politics*. Boulder, CO, Lynne Rienner Publrs, 2001.

Eriksson, J. R. (Ed.) et al. *El Salvador: Post Conflict Reconstruction: Country Case Evaluation*. Washington, DC, World Bank, 2000.

Grenier, Y. *The Emergence of Insurgency in El Salvador*. Pittsburgh, PA, University of Pittsburgh Press, 1999.

Johnstone, I. *Rights and Reconciliation: UN Strategies in El Salvador*. Boulder, CO, Lynne Rienner Publrs, 1995.

Juhn, T. *Negotiating Peace in El Salvador: Civil–Military Relations and the Conspiracy to end the War*. London, Macmillan, 1998.

Krenn, M. L. *The Chains of Interdependence: US Policy toward Central America, 1945–1954*. Armonk, NY, M. E. Sharpe, 1996.

Ladutke, L. M. *Freedom of Expression in El Salvador: The Struggle for Human Rights and Democracy*. Jefferson NC, McFarland & Co Inc Publrs, 2004.

Lauria-Santiago, A. A. *An Agrarian Republic: Commercial Agriculture and the Politics of Peasant Communities in El Salvador, 1823–1914*. Pittsburgh, PA, University of Pittsburgh Press, 1999.

Lauria-Santiago A. A., and Binford, L. *Landscapes of Struggle: Politics, Society, and Community in El Salvador*. Pittsburgh, PA, University of Pittsburgh Press, 2004.

Lund, L., and Sepponen, C. (Eds). *Lifeline Performance: El Salvador Earthquakes of January 13 and February 13 2001*. Reston, VA, American Society of Civil Engineers, 2002.

Lungo Ucles, M. *El Salvador in the 1980s*. Philadelphia, PA, Temple University Press, 1996.

McClintock, C. *Revolutionary Movements in Latin America: El Salvador's FMLN and Peru's Shining Path*. Washington, DC, United States Institute of Peace, 1998.

Murray, K. *El Salvador: Peace on Trial*. Oxford, Oxfam, 1997.

Pearce, J. *Promised Land: Peasant Rebellion in Chalatenango, El Salvador*. London, Latin America Bureau, 1986.

Pelupessy, W. *The Limits of Economic Reform in El Salvador*. Basingstoke, Macmillan, 1997.

Popkin, M. L. *Peace Without Justice: Obstacles to Building the Rule of Law in El Salvador*. University Park, PA, Pennsylvania University Press, 2000.

Ross, D. G. *Development of Railroads in Guatemala and El Salvador, 1849–1929*. Lewiston, NY, Edwin Mellen Press, 2001.

Studemeister, M. S. *El Salvador: Implementation of the Peace Accords*. Washington, DC, US Institute of Peace, 2000.

Towell, L. *El Salvador*. New York, NY, W. W. Norton, 1997.

Williams, P. J., and Walter, K. *Militarization and Demilitarization in El Salvador's Transition to Democracy*. Pittsburgh, PA, University of Pittsburgh Press, 1998.

# THE FALKLAND ISLANDS

## Geography

### PHYSICAL FEATURES

The Falkland Islands is an Overseas Territory of the United Kingdom, claimed by Argentina as the Islas Malvinas. It is located in the South Atlantic, at the same latitude as southern Argentina and Chile, some 770 km (480 miles) north-east of Cape Horn, but 480 km from the nearest point on the South American mainland. The colony's authorities claim an economic zone around the islands and their waters, so as to regulate fishing and the exploitation of hydrocarbons, but the United Kingdom seeks agreement with Argentina on such issues. The islands cover an area more than one-half the size of Wales (United Kingdom), 12,173 sq km (4,700 sq miles).

There are several hundred islands and many more islets and rocks, with a combined area greater than that of Jamaica, but stretching over a distance of 238 km from east to west. The two main islands of East and West Falkland butterfly on either side of the intervening Falkland Sound. The coastline, 1,288 km in total length, is deeply indented and rugged. The windswept islands are hilly, clad in lichen-covered rocks, low-lying, scrubby vegetation or grassland dotted with heath and dwarf shrubs. There are some undulating and usually boggy plains. The islands reach their heights at Mt Usborne (705 m or 2,314 ft) on East Falkland and Mt Adam (700 m) on West Falkland. The natural environment is rich, despite the harsh conditions, although the last example of the only native mammal, the warrah or Falklands wolf, was killed in 1876. Introduced species of mammal, apart from sheep (of which there are some 700,000), cattle, horses and, more recently, reindeer (the only commercial herd unaffected by the Chornobyl, Ukraine, disaster), include the Patagonian fox, rats, mice, cats, rabbits and, on Staats Island, the guanaco. There are also breeding colonies of the southern sea lion, leopard, elephant and fur seals and rock-hopper, king, macaroni, gentoo and Magellanic (or jackass) penguins. In fact, over 200 species of birds have been recorded in the islands, from the tiny tussac, through two endemic species, the Falkland flightless steamer duck (logger) and the Cobb's or rock wren, to the mighty black-browed albatross.

### CLIMATE

The climate is a cold marine one, cloudy, humid, windy (there are strong westerlies) and seldom hot. Although the Falklands is at a similar latitude to the south as the British capital of London is to the north, there is no Gulf Stream to warm the islands, so temperatures range from –5°C (23°F) in July (winter) to 22°C (72°F) in January (summer). Average annual rainfall is low, but

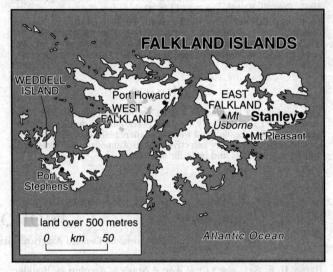

fairly evenly distributed throughout the year, at 626 mm (24 ins). It can snow in any month except the main summer months of January and February, but it seldom sticks for long. It should also be noted that the islands are sometimes directly beneath the 'hole' in the earth's ozone layer.

### POPULATION

Most of the population are of British descent (there are also some descended from Scandinavian whalers), but 96% hold British citizenship and English is the universal language. Spanish names and phrases in use stem from the mid-19th century, when farmers brought in gauchos from the South American mainland to help work the vast livestock holdings. The main religion is Christianity (Anglican, Roman Catholic and some Nonconformist Protestant denominations).

The total population at the 2001 census was 2,913, with most people (1,981) living in the capital, Stanley, in the east of northern East Falkland. There are obviously few people left in the small settlements scattered over the rest of the countryside (known locally as the 'Camp', from the Spanish for countryside). There were also 534 civilians based at the Mount Pleasant military base (the British garrison itself numbers about 1,200). There were 144 people on West Falkland and 38 on other islands (apart from East Falkland).

## History

The islands are a United Kingdom Overseas Territory. From February 1998 the British Dependent Territories were referred to as the United Kingdom Overseas Territories, following the announcement of the interim findings of a British Government review of the United Kingdom's relations with the Overseas Territories. The Governor of the islands, who is the representative of the British monarch, is advised by a six-member Executive Council; the separate post of Chief Executive (responsible to the Governor) was created in 1983. A new Constitution was introduced in 1985.

From the late 17th century the British, French and Spanish disputed sovereignty over the Falkland Islands (Islas Malvinas). Finally, in 1832 a British expedition expelled colonists from the recently independent Argentine Republic. The Falkland Islands became a British Crown Colony in 1833. Argentina continued to claim sovereignty over the islands.

In 1966 negotiations between Argentina and the United Kingdom were opened. Limited progress was made, but on 2 April 1982 Argentine forces invaded the islands, expelled the Governor and established a military governorship. A British naval task force was dispatched and the Argentine forces formally surrendered on 14 June; in the conflict about 750 Argentines, 255 British soldiers and three Falkland Islanders were killed. A Civil Commissioner (later restored to the title of Governor) resumed authority in the dependency and the British Government established a garrison of some 4,000 troops (later reduced).

The Argentine military Government refused to declare a formal cessation of hostilities until the United Kingdom agreed to negotiations on sovereignty, while the United Kingdom maintained that sovereignty was not negotiable and that the wishes of the Falkland Islanders were paramount (the Constitution of 1985 guaranteed the islanders' right to self-determination). It was only in October 1989 that the formal cessation of hostilities and the re-establishment of diplomatic relations at consular level were agreed. In February 1990 it was announced that full diplomatic relations were to be re-established and the naval protection zone around the islands was to be ended. Disputes over fishing areas and exclusion zones continued into the 1990s, but did not prevent some agreement between the United Kingdom and Argentina on the allocation of marine resources and the exploitation of hydrocarbons reserves.

In May 1999 delegations from the Falkland Islands and Argentina met in the United Kingdom for formal negotiations intended to improve relations between the two countries. Issues under discussion included co-operation in fishing and petroleum exploration, Argentine access to the islands and the resumption of air links with mainland South America. (In March Chile had ended its country's regular air services to the islands in protest at the British Government's continued detention of Gen. Pinochet.) The sovereignty of the islands was not scheduled for discussion. In July an agreement was signed by both Governments, which ended the ban on Argentine citizens visiting the islands and re-established direct flights there from Chile. Furthermore, Argentina and the islands were to co-operate in fishing conservation and in the prevention of poaching in the South Atlantic. The agreement did not affect claims to sovereignty. The arrival of the first flights carrying Argentine passengers in mid-October were greeted by protests.

In February 2000, in a meeting with the British Prime Minister, Tony Blair, the recently elected Argentine President, Fernando de la Rúa, reiterated his country's claim to sovereignty of the islands. The de la Rúa Government indicated a less friendly attitude to the Falkland islanders, in contrast to the previous administration, when it announced in July that it would not engage in any dialogue with Falkland Islands councillors that involved the British Government. In the same month the Anglo–Argentine dialogue on joint petroleum and gas exploration was suspended by mutual agreement for an indefinite period of time.

On 22 November 2001 a general election took place to elect members to a new Legislative Council. Of the eight elected, only three members had not served on the Legislative Council before.

On 3 December 2002 Howard Pearce succeeded Donald Lamont as Governor of the Falkland Islands and Commissioner of South Georgia and the Sandwich Islands. This was followed in March 2003 by the appointment of Chris Simpkins as Chief Executive; he replaced Michael Branch, who had reached the end of his three-year tenure.

In May 2003 the new Argentine President, Néstor Kirchner, promised in his inaugural speech to maintain his country's claim to the Falkland Islands; he reiterated that commitment in January 2004. In November 2003 Argentina began to demand that the increasingly frequent air-charter services flying from the Falkland Islands to Chile obtain permission to use Argentine airspace. The decision was intended to pressure the British Government into reversing its policy of not allowing Argentine airlines to fly to the Falkland Islands. The services were suspended in January 2004 after the British Government lodged objections to the Argentine demands; the situation seemed likely to damage the territory's burgeoning tourist trade. In March a British proposal for the resumption of direct charter flights from Argentina to the Falkland Islands was rejected by the Kirchner Government, which continued to demand that an Argentine carrier be authorized to benefit from the increased passenger traffic between the Falkland Islands and mainland South America. In early 2005 Argentina objected strongly to a reference to the islands included within the proposed constitutional treaty of the European Union, in relation to the United Kingdom; however, the likely unviability of the treaty ultimately rendered the issue less potent. At a session held by the Special Committee on Decolonization of the UN in mid-2005, Falkland Islands councillors appealed to be allowed to exercise the right to self-determination; the Committee requested that the Governments of Argentina and the United Kingdom resume negotiations. In July of that year it was announced that Alan Huckle would succeed Howard Pearce as Governor of the Falkland Islands from 2006.

# Economy

From 1982 the economy enjoyed a period of strong and sustained growth, notably after the introduction of the fisheries licensing scheme in 1987. The main economic activity of the islands was sheep-rearing and in the late 1990s annual exports of wool were valued at some £3.5m. From 1987, when a licensing system was introduced for foreign vessels fishing within a 150-nautical-mile conservation and management zone, the islands' annual income tripled. In 1993 it was proposed that this zone be extended to 200 miles. Revenue from licence sales totalled £25m. in 1991, but declined to £21.6m. in 1998/99, following the Argentine Government's commencement of the sale of fishing licences. The revenues, amounting to £29.4m. in 2000/2001 (including transhipments), fund social provisions and economic development programmes, including subsidies to the wool industry, which is in long-term decline, owing to the oversupply of that commodity on the international market. Significant revenue was expected from the licences for petroleum and gas exploration, issued in October 1996 under an Argentine-British agreement of the previous year, although no commercial quantities of petroleum were found in the initial phase of drilling in 1998. In July 2002 the Falkland Islands Government granted 10 petroleum exploration licences to the Falklands Hydrocarbon Consortium (comprising Global Petroleum Ltd, Hardman Resources Ltd and Falkland Islands Holdings)—now known as Falkland Oil and Gas Ltd, to an area covering 57,700 sq km to the south of the islands. The licences were granted without consultation with the Argentine Government.

The services sector has expanded rapidly during the last decade, while the importance of the agricultural sector has decreased, not least because of its reliance on direct and indirect subsidies. Tourism, and in particular eco-tourism, was developing rapidly in the early 21st century, and the number of visitors staying on the Falklands Islands had grown to some 3,000 per year, while around 40,000 tourists were expected to sail through Stanley harbour on their way to Antarctica and sub-antarctic islands such as South Georgia. The sale of postage stamps and coins also represents a significant source of income.

In 2000 the Falkland Islands recorded an estimated trade surplus of £28,041,897. Fish, most of which is purchased by the United Kingdom, Spain and Chile, is the islands' most significant export. The annual rate of inflation averaged 2.9% in 1990–2002; consumer prices increased by 0.7% in 2002. The Government's development plan for 2001–03 emphasized agricultural diversification and the promotion of tourism as its main economic aims. In early 2001 the Government brought 100 reindeer from the South Georgia Islands (with the aim of increasing the number to 10,000 over the next 20 years) in order to export venison to Scandinavia and Chile. In 2004, following the early closure for conservation reasons of the *illex* fishery and a consequential reduction in licensing fees, the Falkland Islands Government prepared an austere budget for the 2004/05 fiscal year; an approximate 5% decrease in expenditure was widely expected to account for the anticipated £10m. shortfall in income.

# THE FALKLAND ISLANDS

## Statistical Survey

Source (unless otherwise stated): The Treasury of the Falkland Islands Government, Stanley, FIQQ 1ZZ; tel. 27143; fax 27144.

### AREA AND POPULATION

**Area:** approx. 12,173 sq km (4,700 sq miles): East Falkland and adjacent islands 6,760 sq km (2,610 sq miles); West Falkland and adjacent islands 5,413 sq km (2,090 sq miles).

**Population:** 2,913 (males 1,598, females 1,315) at census of 8 April 2001. Note: Figures exclude 112 persons normally resident, but include 534 civilian personnel based at Mount Pleasant military base.

**Density** (2001): 0.24 per sq km.

**Principal Town** (2001 census): Stanley (capital), population 1,989.

**Births and Deaths** (2000): Live births 27; Deaths 6.

**Economically Active Population** (persons aged 15 years and over, 2001 census): 2,475 (males 1,370, females 1,105).

### AGRICULTURE, ETC.

**Livestock** (FAO estimates, 2003): Sheep 690,000; Cattle 4,200; Horses 1,188; Poultry 3,000.

**Livestock Products** (FAO estimates, metric tons, 2003): Beef and veal 131; Mutton and lamb 774; Cow's milk 1,500; Sheepskins (fresh) 129; Wool (greasy) 2,340; Wool (scoured) 1,520.

**Fishing** ('000 metric tons, live weight of capture, 2002): Southern blue whiting 2.5; Patagonian grenadier 9.5; Patagonian squid 43.8; Total catch (incl. others) 60.6.
Source: FAO.

### FINANCE

**Currency and Exchange Rates:** 100 pence (pennies) = 1 Falkland Islands pound (FI £). *Sterling, Dollar and Euro Equivalents* (31 May 2005): £1 sterling = FI £1.00; US $1 = 55.00 pence; €1 = 81.10 pence; FI £100 = £100.00 sterling = $181.81 = €147.44. *Average Exchange Rate* (FI £ per US dollar): 0.6672 in 2002; 0.6125 in 2003; 0.5462 in 2004. Note: The Falkland Islands pound is at par with the pound sterling.

**Budget** (FI £ million, 2000/01): *Revenue:* Operating revenue 51.1 (Sales and services 8.9, Fishing licences and transhipment 29.4, Investment income 7.2, Taxes and duties 5.6); Capital revenue 0.8; Total 51.9. *Expenditure:* Operating expenditure 32.1 (Public works 6.6, Fisheries 6.3, Health care 4.3, Education 3.2, Aviation 1.7, Police and justice 1.1, Agriculture 1.0, Central administration 2.9, other 5.0); Capital expenditure 14.2; Total 46.3.

**Cost of Living** (Consumer Price Index for Stanley; base: 2000 = 100): 101.3 in 2001; 102.0 in 2002. Source: ILO.

### EXTERNAL TRADE

**2000** (estimates): Total imports £18,958,103; Total exports £47,000,000. Fish is the principal export. Trade is mainly with the United Kingdom, Spain and Chile.

### TRANSPORT

**Shipping:** *Merchant Fleet* (at 31 December 2004): Vessels 29; Displacement 50,453 grt. Source: Lloyd's Register-Fairplay, *World Fleet Statistics*.

**Road Traffic:** 3,065 vehicles in use in 1995.

### TOURISM

**Day Visitors** (country of origin of cruise-ship excursionists, 2000/01): Germany 1,842; United Kingdom 2,042; USA 14,938; Total (incl. others) 24,000.

### EDUCATION

**2003** (Stanley): *Primary:* Teachers 18; Pupils 203, *Secondary:* Teachers 18; Pupils 160.

## Directory

### The Constitution

The present Constitution of the Falkland Islands came into force on 3 October 1985 (replacing that of 1977) and was amended in 1997. The Governor, who is the personal representative of the British monarch, is advised by the Executive Council, comprising six members: the Governor (presiding), three members elected by the Legislative Council, and two *ex-officio* members, the Chief Executive and the Financial Secretary of the Falkland Islands Government, who are non-voting. The Legislative Council is composed of eight elected members and the same two (non-voting) *ex-officio* members. One of the principal features of the Constitution is the reference in the preamble to the islanders' right to self-determination. The separate post of Chief Executive (responsible to the Governor) was created in 1983. The electoral principle was introduced, on the basis of universal adult suffrage, in 1949. The minimum voting age was lowered from 21 years to 18 years in 1977.

### The Government
(July 2005)

**Governor:** HOWARD J. S. PEARCE (took office 3 December 2002).

**Chief Executive of the Falkland Islands Government:** CHRIS SIMPKINS.

**Government Secretary:** PETER T. KING.

**Financial Secretary:** DEREK F. HOWATT.

**Attorney-General:** DAVID G. LANG.

**Military Commander:** Cdre RICHARD IBBOTSON.

### EXECUTIVE COUNCIL

The Council consists of six members (see Constitution, above).

### LEGISLATIVE COUNCIL

Comprises the Governor, two *ex-officio* (non-voting) members and eight elected members.

### GOVERNMENT OFFICES

**Office of the Governor:** Government House, Stanley, FIQQ 1ZZ; tel. 27433; fax 27434; e-mail gov.house@horizon.co.fk.

**General Office:** Secretariat, Stanley, FIQQ 1ZZ; tel. 27242; fax 27109; e-mail atomlinson@sec.gov.fk; internet www.falklands.gov.fk.

**London Office:** Falkland Islands Government Office, Falkland House, 14 Broadway, London SW1H 0BH, United Kingdom; tel. (020) 7222-2542; fax (020) 7222-2375; e-mail rep@falklands.gov.fk; internet www.falklandislands.com; f. 1983.

### Judicial System

The judicial system of the Falkland Islands is administered by the Supreme Court (presided over by the non-resident Chief Justice), the Magistrate's Court (presided over by the Senior Magistrate) and the Court of Summary Jurisdiction. The Court of Appeal for the Territory sits in England and appeals therefrom may be heard by the Judicial Committee of the Privy Council.

**Chief Justice of the Supreme Court:** JAMES WOOD.

**Judge of the Supreme Court and Senior Magistrate:** CLARE FAULDS.

THE FALKLAND ISLANDS

**Courts Administrator:** LESLEY TITTERINGTON, Ross Rd, Stanley, FIQQ 1ZZ; tel. 27271; fax 2720.

**Registrar-General:** JOHN ROWLAND, Town Hall, Stanley, FIQQ 1ZZ; tel. 27272; fax 27270.

### FALKLAND ISLANDS COURT OF APPEAL

**President:** Sir LIONEL BRETT.
**Registrar:** MICHAEL J. ELKS.

## Religion

### CHRISTIANITY

The Anglican Communion, the Roman Catholic Church and the United Free Church predominate. Also represented are the Evangelist Church, Jehovah's Witnesses, the Lutheran Church, Seventh-day Adventists and the Bahá'í faith.

#### The Anglican Communion

The Archbishop of Canterbury, the Primate of All England, exercises episcopal jurisdiction over the Falkland Islands and South Georgia.

**Rector:** Rev. ALISTAIR MCHAFFIE, The Deanery, Christ Church Cathedral, Stanley, FIQQ 1ZZ; tel. 21100; fax 21842; e-mail deanery@horizon.co.fk; internet www.horizon.co.fk/cathedral.

#### The Roman Catholic Church

**Prefect Apostolic of the Falkland Islands:** MICHAEL BERNARD MCPARTLAND, St Mary's Presbytery, 12 Ross Rd, Stanley, FIQQ 1ZZ; tel. 21204; fax 22242; e-mail stmarys@horizon.co.fk; internet www.southatlanticrcchurch.com; f. 1764; 230 adherents (2003).

## The Press

**The Falkland Islands Gazette:** Stanley, FIQQ 1ZZ; tel. 27242; fax 27109; e-mail atomlinson@sec.gov.fk; internet www.falklands.gov.fk; govt publication.

**Falkland Islands News Network:** POB 141, Stanley, FIQQ 1ZZ; tel. and fax 21182; e-mail finn@horizon.co.fk; internet www.falklandnews.com; relays news daily online and via fax as FINN(COM) Service; Man. JUAN BROCK; publishes:

**Teaberry Express:** Stanley, FIQQ 1ZZ; tel. 21182; weekly.

**Penguin News:** Ross Rd, Stanley, FIQQ 1ZZ; tel. 22684; fax 22238; e-mail pnews@horizon.co.fk; internet www.penguin-news.com; f. 1979; weekly; independent newspaper; Man. Editor JENNY COCKWELL; circ. 1,550.

## Broadcasting and Communications

### TELECOMMUNICATIONS

In 1989 Cable & Wireless PLC installed a £5.4m. digital telecommunications network covering the entire Falkland Islands. The Government contributed to the cost of the new system, which provides international services as well as a new domestic network. Further work to improve the domestic telephone system was completed in the late 1990s at a cost of £3,286,000.

**Cable & Wireless PLC:** Ross Rd, POB 584, Stanley, FIQQ 1ZZ; tel. 20801; fax 22207; e-mail info@cwfi.co.fk; internet www.horizon.co.fk; f. 1989; exclusive provider of national and international telecommunications services in the Falkland Islands under a licence issued by the Falkland Islands Government; CEO RICHARD HALL.

### BROADCASTING

#### Radio

**Falkland Islands Broadcasting Station (FIBS):** Broadcasting Studios, Stanley, FIQQ 1ZZ; tel. 27277; fax 27279; e-mail fibs.fig@horizon.co.fk; 24-hour service, financed by local Govt in association with SSVC of London, United Kingdom; broadcasts in English; Broadcasting Officer TONY BURNETT (acting); Asst Producer CORINA GOSS.

**British Forces Broadcasting Service (BFBS):** BFBS Falkland Islands, Mount Pleasant, BFPO 655; tel. 32179; fax 32193; e-mail chris.pearson@bfbs.com; internet www.bfbs.com; 24-hour satellite service from the United Kingdom; Station Man. CHRIS PEARSON; Sr Engineer ADRIAN ALMOND.

#### Television

**British Forces Broadcasting Service:** BFBS Falkland Islands, Mount Pleasant, BFPO 655; tel. 32179; fax 32193; daily four-hour transmissions of taped broadcasts from BBC and ITV of London, United Kingdom; Sr Engineer COLIN MCDONALD.

**KTV:** 16 Ross Rd West, Stanley, FIQQ 1ZZ; tel. 22349; fax 21049; e-mail kmzb@horizon.co.fk; satellite television broadcasting services; Man. MARIO ZUVIC BULIC.

## Finance

### BANK

**Standard Chartered Bank:** Ross Rd, POB 597, Stanley, FIQQ 1ZZ; tel. 21352; fax 22219; e-mail standardchartered@horizon.co.fk; branch opened in 1983; Man. N. P. HUTTON.

### INSURANCE

The British Commercial Union, Royal Insurance and Norman Tremellen companies maintain agencies in Stanley.

**Consultancy Services Falklands Ltd:** 44 John St, Stanley, FIQQ 1ZZ; tel. 22666; fax 22639; e-mail consultancy@horizon.co.uk; Man. ALISON BAKER.

## Trade and Industry

### DEVELOPMENT ORGANIZATION

**Falkland Islands Development Corporation (FIDC):** Shackleton House, Stanley, FIQQ 1ZZ; tel. 27211; fax 27210; e-mail develop@fidc.co.fk; internet www.fidc.co.fk; f. 1983; provides loans and grants; encourages private-sector investment, inward investment and technology transfer; Gen. Man. JULIAN MORRIS.

### CHAMBER OF COMMERCE

**Chamber of Commerce:** POB 378, Stanley, FIQQ 1ZZ; tel. 22264; fax 22265; e-mail commerce@horizon.co.fk; internet www.falklandislandschamberofcommerce.com; f. 1993; promotes private industry; operates DHL courier service; runs an employment agency; Pres. TIM MILLER; 70 mems.

### TRADING COMPANIES

**Falkland Islands Co Ltd (FIC):** Crozier Pl., Stanley, FIQQ 1ZZ; tel. 27600; fax 27603; e-mail fic@horizon.co.fk; internet www.the-falkland-islands-co.com; f. 1851; part of Falkland Islands Holding PLC; the largest trading co; retailing, wholesaling, shipping, insurance and Land Rover sales and servicing; operates as agent for Lloyd's of London and general shipping concerns; travel services and hoteliers; wharf owners and operators; Dir and Gen. Man. ROGER KENNETH SPINK.

**Falkland Oil and Gas Ltd** (FOGL): 56 John St, Stanley, F1QQ 1ZZ; e-mail info@fogl.co.uk; internet www.fogl.co.uk; f. 2004; Falkland Islands Holdings plc (18%), Global Petroleum (16%) and RAB Capital plc (31%); operates an offshore petroleum exploration programme with 8 licences covering 83,700 sq km; Exec. Chair. JOHN DENNIS ARMSTRONG.

### EMPLOYERS' ASSOCIATION

**Sheep Owners' Association:** Coast Ridge Farm, Fox Bay, FIQQ 1ZZ; tel. 42094; fax 42084; e-mail n.knight.coastridge@horizon.co.fk; asscn for sheep-station owners; Sec. N. KNIGHT.

### TRADE UNION

**Falkland Islands General Employees Union:** Ross Rd, Stanley, FIQQ 1ZZ; tel. 21151; f. 1943; Sec. C. A. ROWLANDS; 100 mems.

### CO-OPERATIVE SOCIETY

**Stanley Co-operative Society:** Stanley, FIQQ 1ZZ; tel. 21215; f. 1952; open to all members of the public; Man. NORMA THOM.

## Transport

### RAILWAYS

There are no railways on the islands.

# THE FALKLAND ISLANDS

### ROADS

There are 29 km (18 miles) of paved road in and around Stanley. There are 54 km (34 miles) of all-weather road linking Stanley and the Mount Pleasant airport (some of which has been surfaced with a bitumen substance), and a further 37 km of road as far as Goose Green. There are 300 km of arterial roads in the North Camp on East Falkland linking settlements, and a further 197 km of road on West Falkland. An ongoing roads network project to link remote farms is in progress. Where roads have still not been built, settlements are linked by tracks, which are passable by all-terrain motor vehicle or motor cycle except in the most severe weather conditions.

### SHIPPING

There is a ship on charter to the Falkland Islands Co Ltd which makes the round trip to the United Kingdom four or five times a year, carrying cargo. A floating deep-water jetty was completed in 1984. The British Ministry of Defence charters ships, which sail for the Falkland Islands once every three weeks. There are irregular cargo services between the islands and southern Chile and Uruguay.

The Falkland Islands merchant fleet numbered 29 vessels, with a total displacement of 50,453 grt, at December 2004; the majority of vessels registered are deep-sea fishing vessels.

**Stanley Port Authority:** c/o Department of Fisheries, POB 598, Stanley, FIQQ 1ZZ; tel. 27260; fax 27265; e-mail jclark@fisheries.gov.fk; Harbour Master J. CLARK.

#### Private Companies

**Byron Marine Ltd:** 3 'H' Jones Rd, Stanley, FIQQ 1ZZ; tel. 22245; fax 22246; e-mail info@byronmarine.co.fk; Man. LEWIS CLIFTON.

**Darwin Shipping Ltd:** Stanley, FIQQ 1ZZ; tel. 27629; fax 27626; e-mail darwin@horizon.co.fk; internet www.the-falkland-islands-co.com; subsidiary of the Falkland Islands Company Ltd; Man. EVA CLARKE.

**Falkland Islands Co. Ltd:** Crozier Pl., Stanley, FIQQ 1ZZ; tel. 27600; fax 27603; e-mail fic@horizon.co.fk; internet www.the-falkland-islands-co.com; Man. ROGER SPINK.

**Seaview Ltd:** 37 Fitzroy Rd, POB 215, Stanley, FIQQ 1ZZ; tel. 22669; fax 22670; e-mail polar.falklands@btinternet.com; internet www.fis.com/polar; f. ; Man. DICK SAWLE.

**Sulivan Shipping Services Ltd:** Davis St, Stanley, FIQQ 1ZZ; tel. 22626; fax 22625; e-mail sulivan@horizon.co.fk; internet www.sulivanshipping.com; f. 1987; provides port-agency and ground-handling services; Man. Dir JOHN POLLARD.

### CIVIL AVIATION

There are airports at Stanley and Mount Pleasant; the latter has a runway of 2,590 m, and is capable of receiving wide-bodied jet aircraft. The British Royal Air Force operates three weekly flights from the United Kingdom. The Chilean carrier LanChile operates weekly return flights from Punta Arenas.

**Falkland Islands Government Air Service (FIGAS):** Stanley Airport, Stanley, FIQQ 1ZZ; tel. 27219; fax 27309; e-mail fwallace@figas.gov.fk; f. 1948 to provide social, medical and postal services between the settlements and Stanley; aerial surveillance for Fishery Dept since 1990; operates four nine-seater aircraft to over 35 landing strips across the islands; Gen. Man. VERNON R. STEEN.

## Tourism

During the 2000/01 season some 24,000 day visitors from cruise ships (primarily US citizens, numbering 14,938) visited the islands. Wildlife photography, bird-watching and hiking are popular tourist activities. The Falkland Islands Development Corpn plans to develop the sector, which currently generates some £3m. in turnover annually.

**Falkland Islands Tourism:** Old Philomel Store, Stanley, FIQQ 1ZZ; tel. 22215; fax 22619; e-mail jettycentre@horizon.co.fk; internet www.tourism.org.fk; Man. CONNIE STEVENS.

## Defence

In August 2004 there were approximately 1,200 British troops stationed on the islands. The total cost of the conflict in 1982 and of building and maintaining a garrison for four years was estimated at £2,560m. The current annual cost of maintaining the garrison is approximately £70m. There is a Falkland Islands Defence Force, composed of 60 islanders.

## Education

Education is compulsory, and is provided free of charge, for children between the ages of five and 16 years. Facilities are available for further study beyond the statutory school-leaving age. In 2003 203 pupils were instructed by 18 teachers at the primary school in Stanley, while 160 pupils received instruction from 18 teachers at the secondary school in the capital; further facilities existed in rural areas, with six peripatetic teachers visiting younger children for two out of every six weeks (older children boarded in a hostel in Stanley). Total expenditure on education and training was estimated at £3.2m. for 2000/01.

# FRENCH GUIANA

## Geography

### PHYSICAL FEATURES

France's Overseas Department of Guiana is the easternmost and smallest of the three Guianas on the north coast of South America (until 1946 it was administered as a colony, under the name of its capital, Cayenne). Suriname (formerly Dutch Guiana) lies to the west, beyond a frontier of 510 km (317 miles), but it also claims territory in the south-east of the Department, beyond the border-marking Litani (Itany) river, as far as the Marouini (both headwaters of the Lawa and, later, the Maroni—Marowijne—which also help delineate the western border). Brazil is to the south-east and south, beyond a border of 673 km, much of it along the course of the Oiapoque (Oyapock). There are 378 km of north-east-facing Atlantic coastline. French Guiana is the only place on the American mainlands not to be independent, although it is an integral part of France. It covers an area of 83,534 sq km (32,253 sq miles).

Most of French Guiana is fairly low-lying, flat and sometimes marshy along the coast, which has a number of rocky islands off shore (notably the old penal camp of Ile du Diable—Devil's Island). Most of the Department consists of rolling, fertile plains, reaching inland to the lower, northern foothills of the Serra Tumucumaque. Hills occasionally break the monotony of the landscape, with the Chaîne Granitique in the centre of the territory, and low mountains rising in the far south. The highest point is Bellevue de l'Inini, at 851 m (2,793 ft). Most of the terrain is wooded (83%), much of it rainforest, and few people live in the swathes of wilderness inland.

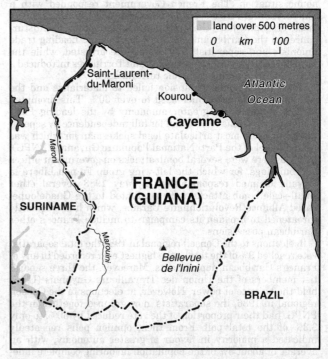

### CLIMATE

The climate is tropical, hot and humid, with little seasonal variation in temperature. Average temperatures, year round, in Cayenne, on the coast, range between 23°C and 33°C (73°–91°F). The dry season is in August–October, the wet season December–June. Average annual rainfall is over 2,500 mm (around 100 ins), with frequent heavy showers and thunderstorms able to cause flooding in the generally flat countryside.

### POPULATION

The people of the Department are French citizens, and perhaps two-thirds are black or mixed-black (mulatto)—about 17% of the population are of Haitian extraction. Some 12% are white, while the rest are mainly native Amerindian, 'East' Indian or Chinese (the latter two groups arriving when labour was needed after the abolition of slavery in the 19th century). Although there are followers of native, tribal faith systems, as well as adherents of syncretic Afro-American spiritualist groups and some Muslims, most of the population are nominally Christian. About 75% are Roman Catholic and 4% Protestant. French is the official and most widely used language, but there are speakers of Amerindian tongues and of a local Creole patois.

The total population was estimated at 183,000 in mid-2004. Most of the population live in the coastal areas, and about four-fifths in urban areas. The departmental headquarters and the largest city is Cayenne, a port on the east-central coast. To the west is Kourou, the launching centre for the European Space Agency. In the far north-west is the town of Saint-Laurent-du-Maroni. None of the inland towns is large.

## History

**PHILLIP WEARNE**

Revised for this edition by Dr JAMES FERGUSON

The land that is now French Guiana (Guyane) was first sighted by Europeans at the end of the 15th century. The French began to settle the territory in 1604, but rumours of its potential gold and diamond wealth led to frequent changes of ownership. The Dutch, British and Portuguese all occupied the area, and there were frequent border disputes before the colony was finally confirmed as French in 1817. Subsequent border disputes were settled by arbitration in 1891, 1899 and 1915. In March 1946 the colony, hitherto known as Cayenne, became an Overseas Department, like Guadeloupe and Martinique, with the same laws and administration as a department of metropolitan France. The head of state is the President of France represented locally by a Commissioner of the Republic. The Conseil général (General Council) with 19 seats and the Conseil régional (Regional Council) with 31 seats are the two local legislative houses. Both have Presidents, and the territory also sends two deputies to the French Assemblée nationale, one senator to the Sénat in Paris and one representative to the European Parliament in Strasbourg.

The discovery of gold in the basin of the Approuague river brought a brief period of prosperity in the mid-19th century, but French Guiana's chief notoriety, until 1937, was as a penal colony. After arriving, prisoners were distributed to camps scattered throughout the territory. Devil's Island became the most infamous. Prisoners mingled freely with other settlers and the indigenous population during a period of exile after serving their sentences, but few could afford to return to France.

The practice of imprisoning convicts and political prisoners in French Guiana ceased in 1937. However, the territory's reputation as a political and economic 'backwater' persisted until the 1970s, when separatist pressure and racial tension exploded in demonstrations against French rule and the deteriorating economic situation. The French Government responded with a combination of strict security measures and the allocation of more economic aid—the traditional prescription for disturbances in the Caribbean overseas departments. Leading trade unionists and separatist politicians were arrested, while the Minister for Overseas Departments and Territories introduced a wide-ranging plan for economic revitalization.

However, economic expansion failed to materialize and the rate of unemployment increased to over 30%. This prompted further demands for greater autonomy by the leading Parti Socialiste Guyanais (PSG), and for full independence by separatist groups, the most articulate legal spokesman for which was Alain Michel of the Parti National Populaire Guyanais (PNPG). In 1980 there were several bomb attacks on government offices and buildings, for which the left-wing group Fo nou Libéré la Guyane claimed responsibility. In May 1983 several other small-scale bomb attacks were attributed to the Guadeloupe-based Alliance Révolutionnaire Caraïbe, which had frequently threatened to broaden its campaign to include France's other Caribbean possessions.

In elections to the Conseil régional in 1983 the total separatist vote reached 9% of the total—the highest ever recorded in any of France's Caribbean departments. Moreover, the three separatist members of the Union des Travailleurs Guyanais (UTG) held the balance of power. However, at elections to the Conseil régional in 1986, the separatists, now grouped together in the PNPG, had their proportion of the vote reduced by 60%, to only 3.3% of the total poll. From then, opinion polls repeatedly indicated a majority in favour of greater autonomy, with an average of about 5% of the population favouring complete independence.

In the French presidential elections in April–May 1988, French Guiana followed the other Caribbean Overseas Departments in voting overwhelmingly for François Mitterrand, the Socialist candidate, who secured 60.4% of the poll. In June 1989, however, a split occurred in the PSG when Georges Othily, President of the Conseil régional, was expelled from the party for unauthorized links with the opposition, along with five other senior party figures. Analysts believed that the revolt signified growing discontent with the PSG's increasingly partisan and corrupt 10-year domination of Guianese politics. The PSG split was at least partially responsible for the party's loss of support in the 1992 elections, when it lost two seats in the Conseil général (retaining 10), although Elie Castor, leader of the PSG, remained President of the Conseil général and one of the territory's two representatives in the Assemblée nationale. In the elections to the Conseil régional, the PSG won 16 seats, but Othily supporters secured 10 seats.

Weakened by continued infighting, the decline of the PSG continued. In the 1993 elections to the Assemblée nationale a political newcomer, Christiane Taubira-Delanon of the Walwari movement, a dissident Socialist who once favoured independence, defeated the PSG's candidate, Rodolfe Alexandre, for the Cayenne-Macouria seat. Léon Bertrand of the Rassemblement pour la République (RPR), mayor of the territory's second largest town, Saint-Laurent-du-Maroni, won a convincing 52.5% of the votes cast in the second constituency, Kourou-Saint-Laurent. PSG representation in the Conseil général fell to eight seats after the March 1994 elections. Castor left the party, but a PSG member, Stéphan Phinera-Horth, was elected President of the Conseil général. Taubira-Delanon was ousted by Antoine Karam, the PSG's Secretary-General, but in June she became the first woman from the Department to enter the European Parliament. She took her place as a representative of the Energie Radicale grouping, which led the poll with 36.3% of the votes, against the right-wing government list (21.5%) and the combined list of the left-wing Rassemblement d'Outre-Mer et des Minorités (17.2%).

French Guiana's traditionally Socialist loyalties changed in the 1995 presidential election. In the first round on 23 April Jacques Chirac of the RPR took the lead, at the expense of Lionel Jospin (PS). In the second round on 7 May Chirac obtained 57% of the votes. In municipal elections in June Taubira-Delanon narrowly failed to be elected mayor of Cayenne and thus break 30 years of PSG control. In late May and early June 1997 Léon Bertrand and Taubira-Delanon were both re-elected to the Assemblée nationale. Candidates from pro-independence parties notably gained increased support, winning just over 10% of the votes cast in both constituencies.

The high rate of abstention in these elections and the re-emergence of separatist parties were symptoms of an increasing dissatisfaction with departmental politics and of rising tension between the Department and metropolitan France. These developments precipitated, and were reinforced by, escalating social unrest. In November 1996 protests in Cayenne, in support of secondary-school pupils who were boycotting classes to demand improved study conditions, degenerated into rioting and looting. Violent clashes between protesters and anti-riot police sent from metropolitan France provoked a one-day general strike in Cayenne, organized by the UTG. These tensions worsened as pupils' representatives and local politicians criticized the actions of the police, and local officials alleged that separatist groups were working to exploit the crisis for their own ends. The situation was temporarily resolved when the French Government announced administrative reform and additional funding for the education system. However, in April 1997 the arrest of five pro-independence activists suspected of setting fire to the home of the public prosecutor during the disturbances of November 1996, and the subsequent detention of five others, including leading members of the UTG and the PNPG, led to further demonstrations and riots in Cayenne. In August 1997 the release of the five original detainees, who had been held on remand since April, signalled an end to the immediate crisis.

There were elections to both the Conseil général and the Conseil régional in March 1998. The PSG retained 11 of 16 seats in the 31-seat Conseil régional, while a further 11 seats were won by other left-wing candidates, including six by the RPR and two by Walwari. The PSG's Antoine Karam defeated Georges Othily to be re-elected as the body's President. Othily was re-elected to the Sénat in September 1998. The PSG's representation in the 19-member Conseil général also declined, from eight seats to five, with an equal number won by other left-wing candidates. Independent candidates secured seven seats, a further indication of the Guianese electorate's disillusionment with traditional party politics, and André Lecante, an independent, left-wing councillor, was elected to the presidency of the Conseil général.

In January 1999 representatives from 10 separatist organizations from French Guiana, Guadeloupe and Martinique, including the PNPG and the Mouvement pour la Décolonisation et l'Emancipation Sociale (MDES), signed a joint declaration denouncing 'French colonialism'. The political and constitutional future of the Overseas Departments generated considerable debate and controversy throughout that year, especially in French Guiana. In February members of the two Councils held a congress, which recommended the replacement of the two Councils with a single body, to which added powers and responsibilities in areas such as economic development, health and education would be transferred. In October however, the French Prime Minister, Lionel Jospin, ruled out the possibility of any such merger. In December Karam co-signed a declaration along with the Presidents of the Conseil régional in both Guadeloupe and Martinique, stating their intention to propose to the French Government, legislative and constitutional amendments aimed at creating a new status of 'overseas region'. The amendments would also provide for greater financial autonomy. The declaration and subsequent announcements by Karam and his counterparts were dismissed by Jean-Jack Queyranne, Secretary of State for Overseas Departments and Territories, in February 2000 as unconstitutional and exceeding the mandate of the politicians responsible. In March, during a visit to the Department by Queyranne, rioting broke out following his refusal to meet a delegation of separatist organizations. Later that month the Conseil régional rejected, by 23 votes to seven, the reforms proposed by Queyranne in February, which included the creation of a Congress in French Guiana, as well as the extension of the Departments' powers in areas such as regional co-operation.

Nevertheless, the proposals were provisionally accepted by the Assemblée nationale in May, and were subsequently adopted, by a narrow margin, by the Sénat, following a number of modifications. In November the Assemblée nationale approved the changes and in December they were ratified by the Constitutional Council. At a referendum held in September, French Guiana voted in favour (80%) of reducing the presidential mandate from seven years to five.

In November 2000 riots took place in Cayenne. The demonstrations followed a march, organized by the UTG, demanding greater autonomy for French Guiana, as well as immediate negotiations with the new Secretary of State for Overseas Departments and Territories, Christian Paul. Protesters claimed they had been excluded from talks on French Guiana's status (Paul had invited leaders of various political parties in French Guiana to attend a meeting to be held in France in December, but the offer was rejected by MDES activists, who demanded the meeting be held in Cayenne). Nevertheless, discussions were held in mid-December in Paris at which Paul, various senior politicians from French Guiana and representatives from the PSG, the RPR, Walwari, and the Forces Démocratiques Guyanaises (FDG) were present. It was agreed that further talks were to be held between the Conseil général and the Conseil régional of French Guiana before the end of 2001, eventually to be followed by a referendum in the Department; however, no constitutional changes were to be effected before the 2002 presidential and parliamentary elections. Following a meeting of members of both Conseils in late June 2001, a series of proposals on greater autonomy, to be presented to the French Government, was agreed upon. These included: the division of the territory into four districts; the creation of a 'collectivité territoriale' (territorial community), governed by a 41-member Assembly elected for a five-year term; and the establishment of an independent executive council. Furthermore, the proposals included a request that the territory be given control over legislative and administrative affairs, as well as legislative authority on matters concerning French Guiana alone. In November the French Government announced itself to be in favour of the suggested constitutional developments, and in March 2003 the two houses of the French parliament approved constitutional changes that would allow for a referendum on proposals for greater autonomy, as occurred in Martinique and Guadeloupe. However, the electorate of French Guiana was not consulted in a referendum, a majority of 10 of the 19 members of the Conseil général voting against the proposal, provoking accusations of anti-democratic behaviour from the MDES.

At municipal elections held in March 2001, the PSG candidate for the mayorship of Cayenne, Jean-Claude Lafontaine, defeated the Walwari candidate, Christiane Taubira-Delanon. At concurrently held elections to the presidency of the Conseil général the left-wing independent candidate, Joseph Ho-Ten-You, defeated André Lecante.

In September 2001 the Secretary of State for Overseas Departments and Territories, Christian Paul, announced the establishment of a number of measures designed to improve security in the Department. Plans included a 20% increase in the police force, the creation of a small 'peace corps' and a continuous police presence in the town of Maripasoula and its surrounding region, following concerns over the security of gold prospectors in the area. Earlier that year Jacques Chirac rejected a joint request by French Guiana, Guadeloupe and Martinique that they be permitted to join, as associate members, the Association of Caribbean States.

At the first round of presidential elections, held on 21 April 2002, Christiane Taubira-Delanon obtained the greatest number of votes cast in French Guiana (52.7%), defeating both Jacques Chirac and Lionel Jospin, who came second and third, respectively. In the second round run-off election, held on 5 May, Chirac overwhelmingly defeated the Front National candidate, Jean-Marie Le Pen, securing 89.1% of the votes cast in the Department. At elections to the Assemblée nationale, held on 9 and 16 June, both deputies were re-elected, with Taubira-Delanon, representing the Walwari party, winning 65.3% of the votes cast and Léon Bertrand of the Union pour la Majorité Présidentielle (UMP) obtaining 64.5% of the ballot. In July Ange Mancini replaced Henri Masse as Prefect. In elections to the Conseil régional held in March 2004, the PSG won 17 of the council's 31 seats with 37.7% of the votes cast, while the FDG and the UMP (reconstituted as the Union pour un Mouvement Populaire) each won seven seats with 31.2% and 31.1% of the ballot, respectively. Antoine Karam was duly re-elected as President of the Conseil régional.

In a national referendum held on 29 May 2005, 60.1% of the electorate in the Department voted in favour of ratifying the proposed European Union constitution. The turn-out, however, was again very low, with only 23.1% of registered voters going to the polls. Most of the leading political parties and personalities, with the exception of Taubira-Delanon, had supported a 'oui' vote, which, however, was rejected by a significant majority of mainland French voters.

# Economy

## PHILLIP WEARNE

Revised for this edition by Dr JAMES FERGUSON

French Guiana's main exports are gold, shrimps and boats. Mineral resources and hydroelectric potential remained unexploited and although the tourist sector expanded throughout the 1980s its growth is limited by the lack of infrastructure inland. Like Martinique and Guadeloupe, the Department is highly dependent on France for its foreign trade and for aid transfers, estimated at US $500m. in 2003, to reduce the balance-of-payments deficit. The impact of the aerospace sector on the economy from the 1970s was another unsettling factor. In 2003 French Guiana's gross domestic product (GDP) was estimated at US $1,551m., equivalent per head to $8,300. GDP growth in 2001was 1.6%, while the UN estimated that between 1990 and 2001 GDP had increased an annual average of 3.1%.

In 2003 the Department recorded a trade deficit of about €529m. The value of exports was €110.8m., less than one-sixth of the total value of imports, worth €640.7m., although export earnings had more than doubled since 1991. The main market for exports was France, which received 62.8% of the total in 2003. France was also the single largest source of imports, accounting for 47.5% of their total value in that year. Other significant import sources were Trinidad and Tobago (9.6%), Italy and Japan. Other important purchasers were Switzerland (11.5%), Martinique and Guadeloupe. Road vehicles, refined petroleum products and pharmaceutical products were among the Department's principal imports.

In 1992 there was a combined deficit on the budgets of French Guiana's state, regional, departmental and communal government authorities of 696m. francs. A balanced regional budget was projected for 2000, however there was a projected deficit of some €14m. on the departmental budget in that year. The annual rate of inflation averaged 1.5% in 1990–2004; consumer prices increased by an average of 2.0% in 2003 and by 1.1% in 2004.

The agricultural sector, concentrated in forestry and fisheries, engaged 17.1% of the employed labour force in mid-2003, according to the FAO. In that year agricultural products accounted for about 20% of total export earnings, at €22.2m., although the contribution from forestry declined in the 1990s. Shrimps remained the single most important agricultural export, accounting for 11.7% of total export revenue in 2003, at

€13m. After rapid expansion in the mid-1980s shrimp production fluctuated at around 4,000 metric tons in the 1990s. In 2003 shrimp production was recorded at 3,565 tons.

The main crops grown for local consumption were cassava, vegetables, rice and sugar cane, the last for use in the making of rum. Livestock rearing was also largely for subsistence. In 2003 Guianese abattoirs produced some 2,265 metric tons of meat, mostly pork, poultry and beef. Rice, pineapples and citrus fruit continued to be cultivated for export. The value of the agricultural sector's contribution to GDP was estimated at 9% in 1999.

Timber exports declined steadily from almost 15,000 cu m in 1985 to less than 2,500 cu m in 1993. Local sales also fell sharply, from 27,549 cu m in 1991 to just 7,425 cu m by 1994. In 2001 total wood extraction was 48,122 cu m, about one-half of the 1991 figure. There were several sawmills, but exploitation of timber resources was hampered by the lack of infrastructure in the forest. Local mills produced plywood and veneers, while rosewood, satinwood and mahogany were the major hardwood products. However, by 2003 roundwood removals totalled 149,000 cu m, while sawnwood production (including railway sleepers) amounted to 15,000 cu m. The value of wood exports in that year was was €2.6m.

Industry, including construction, engaged 15.6% of the employed work-force in 1999 and contributed 18.7% of GDP, according to official sources. The mining sector was dominated by the extraction of gold, mostly in the Inini region, which involved small-scale alluvial operations and larger local and multinational mining concerns. Officially recorded gold exports were estimated at 4,857 kg in 1999. In 2002 an estimated 2,971 kg of gold were mined. In 2003 exports of gold were worth more than €55m., accounting for some 49.9% of export earnings. Crushed rock for the construction industry was the only other mineral extracted in significant quantities, but exploratory drilling of known diamond deposits began in 1995. Bauxite, kaolin and columbo-tantalite were also present in commercial quantities, in particular on the Kaw plateau and near Saint-Laurent-du-Maroni. However, low market prices and the high cost of building the infrastructure necessary for the exploitation of such reserves hampered development. Exports of all metals and metal products amounted to €59.7m. in 2003.

There was little manufacturing activity in French Guiana, except for the processing of agricultural or seafood products, mainly shrimp-freezing and rum distillation. A small quantity of sugar cane—5,400 metric tons in 2003—was processed to supply the sole rum distiller, which produced 3,072 hl of rum in 2000. Activity in the construction industry was minimal, with planning permission granted for a total of 1,100 buildings in 2003. However, the Kourou space centre added activity to the sector and employed about 1,100 people throughout the 1990s. The centre's satellite-launching activities also increased exports in transport services and imports to serve construction demands.

In 2003 the Kourou space centre was estimated to contribute 25% of French Guiana's GDP and approximately 50% of its tax revenues. Work to modernize the centre began in September 2003, with the aim of launching Russian rockets by 2006. The investment in modernization was estimated at US $295m. Only four rocket launches took place in 2003, compared with 11 the previous year, resulting in a downturn in business tourism.

Before the flooding of the Petit-Saut hydroelectric dam on the River Sinnamary in 1994 French Guiana depended heavily on imported fuels for the generation of energy. Together with existing plants, the 116-MW dam was expected to supply the Department's energy for 30 years. Imports of mineral fuels, however, still accounted for 10.3%% of total imports in 2003.

French Guiana's economic development was hindered by its location, its poor infrastructure away from the coast and the lack of a skilled indigenous work-force, which left the potential for growth in agriculture, fishing, tourism, forestry and the energy sector largely unexploited. There was also a critically high level of unemployment, especially among younger sectors of the urban work-force, while much agricultural and mining work was carried out by undocumented migrants from Haiti, Suriname and Brazil. Recorded unemployment reached 24.5% of the work-force in 2003. French Guiana's geographical characteristics, with large parts of the territory accessible only by river, made it difficult to regulate key sections of the economy, such as gold mining and forestry. There was considerable concern among environmentalists that this could have severe ecological consequences. Proposals for the creation of a national park, covering some 2.5m. ha of the south of French Guiana, with the aim of protecting an expanse of equatorial forest and due to be completed in 2000, were still being hindered in 2005 by the need to reconcile ecological concerns with economic priorities and the needs of the resident communities, notably the demands of gold prospectors. Pressure to reduce the high budget deficit increased the Department's dependence on metropolitan France, while the high demand for imported consumer goods among the relatively affluent civil-servant population tended to undermine any progress. In mid-2001 the decision by the French Government to abandon a sugar-cane plantation was met with disappointment by local politicians. The plantation of some 8,000 ha of sugar cane, with an estimated annual output of 65,000 tons of raw sugar, had been expected to generate significant employment opportunities. However, the project was considered to be too costly by the Government, which subsequently pledged to undertake a plan for the development of the agricultural sector in France. In early 2003 the French Minister of the Overseas Departments, Territories and Country announced plans to stimulate the economies of French Guiana, Guadeloupe and Martinique by introducing tax incentives for the hotel sector, to help it remain competitive in the Caribbean region, and by creating jobs, for young people in particular.

# Statistical Survey

Sources (unless otherwise stated): Institut National de la Statistique et des Etudes Economiques (INSEE), Service Régional de Guyane, ave Pasteur, BP 6017, 97306 Cayenne Cédex; tel. 5-94-29-73-00; fax 5-94-29-73-01; internet www.insee.fr/fr/insee_regions/guyane; Chambre de Commerce et d'Industrie, Hôtel Consulaire, pl. de l'Esplanade, BP 49, 97321 Cayenne Cédex; tel. 5-94-29-96-00; fax 5-94-29-96-34; internet www.guyane.cci.fr.

## AREA AND POPULATION

**Area:** 83,534 sq km (32,253 sq miles).

**Population:** 114,808 (males 59,799, females 55,009) at census of 15 March 1990; 157,213 at census of 8 March 1999. *Mid-2004* (UN estimate): 183,000 (Source: UN, *World Population Prospects: The 2004 Revision*).

**Density** (at mid-2004): 2.2 per sq km.

**Principal Towns** (population at 1999 census, provisional): Cayenne (capital) 50,594; Saint-Laurent-du-Maroni 19,211; Kourou 19,107; Matoury 18,032; Rémire-Montjoly 15,555; Mana 5,445; Macouria 5,050; Maripasoula 3,710. *Mid-2003* (UN estimate, incl. suburbs): Cayenne 56,081 (Source: UN, *World Urbanization Prospects: The 2003 Revision*).

**Births, Marriages and Deaths** (1999): Registered live births 4,907 (birth rate 31.0 per 1,000); Registered marriages 548 (marriage rate 3.5 per 1,000); Registered deaths 648 (death rate 4.1 per 1,000). *2001:* Registered marriages 553. *2002:* Registered live births 5,249; Registered deaths 656 (Source: UN, *Population and Vital Statistics Report*).

**Expectation of Life** (years at birth, 1999): Males 71.7; Females 79.2.

**Economically Active Population** (persons aged 15 years and over, 1999): Agriculture, forestry and fishing 2,888; Construction 3,256; Industry 3,524; Trade 4,573; Transport 1,616; Education, health and social services 8,990; Public administration 10,337; Other services 8,259; Total employed 43,443 (males 25,703; females 17,740). *Mid-2003* (estimates): Agriculture, etc. 13,000; Total labour force 76,000 (males 50,000, females 26,000) (Source: FAO). *December 2003:* Unemployed 12,042.

# FRENCH GUIANA

## HEALTH AND WELFARE

### Key Indicators

**Access to Water** (% of persons, 2002): 84.

**Access to Sanitation** (% of persons, 2002): 78.
For sources and definitions, see explanatory note on p. vi.

## AGRICULTURE, ETC.

**Principal Crops** (FAO estimates, '000 metric tons, 2003): Rice (paddy) 24.2; Cassava 10.4; Sugar cane 5.4; Cabbages 6.4; Tomatoes 3.8; Cucumbers and gherkins 3.7; Green beans 3.3; Other fresh vegetables 7.5; Bananas 4.5; Plantains 3.2.

**Livestock** (FAO estimates, '000 head, 2003): Cattle 9.2; Pigs 10.5; Sheep 2.6.

**Livestock Products** (FAO estimates, metric tons, 2003): Beef and veal 460; Pig meat 1,245; Poultry meat 560; Cows' milk 270; Hen eggs 460.

**Forestry** (FAO estimates, '000 cubic metres, 2003): *Roundwood Removals* (excl. bark): Sawlogs, veneer logs and logs for sleepers 51; Other industrial wood 9; Fuel wood 89; Total 149. *Sawnwood Production* (incl. railway sleepers): Total 15.

**Fishing** (metric tons, live weight, 2003): Capture 5,565 (Marine fishes 2,000—FAO estimate, Shrimps 3,565); Aquaculture 37; *Total catch* 5,602. Source: FAO.
Source: FAO.

## MINING

**Production** ('000 metric tons unless otherwise indicated, estimates, 2002): Cement 62,000; Gold (metal content of ore, kilograms) 2,971; Sand 1,500. Source: US Geological Survey.

## INDUSTRY

**Production:** Rum 3,072 hl in 2000; Electric energy 455 million kWh in 2000 (estimate). Source: partly UN, *Industrial Commodity Statistics Yearbook*.

## FINANCE

**Currency and Exchange Rates:** 100 cent = 1 euro (€). *Sterling and Dollar Equivalents* (31 May 2005): £1 sterling = €1.474; US $1 = €0.8110; €1,000 = £678.24 = $1,233.10. *Average Exchange Rate* (euros per US dollar): 1.063 in 2002; 0.886 in 2003; 0.805 in 2004. Note: The national currency was formerly the French franc. From the introduction of the euro, with French participation, on 1 January 1999, a fixed exchange rate of €1 = 6.55957 French francs was in operation. Euro notes and coins were introduced on 1 January 2002. The euro and French currency circulated alongside each other until 17 February, after which the euro became the sole legal tender. Some of the figures in this Survey are still in terms of francs.

**Budget** (million French francs, 1992): *French Government:* Revenue 706; Expenditure 1,505. *Regional Government:* Revenue 558; Expenditure 666. *Departmental Government:* Revenue 998; Expenditure 803. *Communes:* Revenue 998; Expenditure 982. *1998* (€ million): *Regional Budget:* Revenue 63; Expenditure 67. *Departmental Budget:* Revenue 137; Expenditure 15.

**Money Supply** (million French francs at 31 December 1996): Currency outside banks 3,000; Demand deposits at banks 1,621; Total money 4,621.

**Cost of Living** (Consumer Price Index; base: 2000 = 100): 103.1 in 2002; 105.2 in 2003; 106.4 in 2004. Source: ILO.

**Expenditure on the Gross Domestic Product** (million French francs at current prices, 1995): Government final consumption expenditure 4,042; Private final consumption expenditure 5,898; Increase in stocks 121; Gross fixed capital formation 2,542; *Total domestic expenditure* 12,603; Exports of goods and services 7,746; *Less* Imports of goods and services 9,577; *GDP in purchasers' values* 10,772.

**Gross Domestic Product by Economic Activity** (million French francs at current prices, 1995): Agriculture, hunting, forestry and fishing 572; Mining, quarrying and manufacturing 988; Electricity, gas and water 139; Construction 858; Trade, restaurants and hotels 1,895; Transport, storage and communications 1,241; Finance, insurance, real estate and business services 815; Public administration 1,492; Other services 2,451; *Sub-total* 10,449; *Less* Imputed bank service charge 310; *GDP at basic prices* 10,139; Taxes, less subsidies, on products 633; *GDP in purchasers' values* 10,772.

## EXTERNAL TRADE

**Principal Commodities:** *Imports* (€ million, 2003): Road vehicles 72.8; Refined petroleum products 64.3; Pharmaceutical products 27.8; Water and non-alcoholic beverages 16.5; Milk and dairy products 16.2; Road vehicle parts 14.9; Mining or construction equipment 13.2; Computers and data-processing equipment 12.3; Radios, televisions and communication equipment 11.7; Total (incl. others) 640.7. *Exports f.o.b.:* (€ million, 2003): Gold 55.3; Shrimps 13.0; Boats 9.0; Instruments of measurement and control 7.3; Metal products 4.3; Frozen fish 3.3; Rice 3.2; Wood, cut or shaped barks 2.6; Total (incl. others) 110.8. Source: Direction nationale du commerce extérieur.

**Principal Trading Partners** (€ million, 2000): *Imports c.i.f.:* Belgium-Luxembourg 10; France (metropolitan) 346; Germany 15; Italy 18; Japan 18; Martinique 11; Netherlands 12; Spain 9; Trinidad and Tobago 66; USA 12; Total (incl. others) 620. *Exports f.o.b.:* Brazil 5; France (metropolitan) 77; Guadeloupe 6; Italy 2; Martinique 4; Netherlands 3; Spain 3; Switzerland 6; USA 9; Total (incl. others) 121.

## TRANSPORT

**Road Traffic** ('000 motor vehicles in use, 2001): Passenger cars 32.9; Commercial vehicles 11.9. (Source: UN, *Statistical Yearbook*). *2002:* 50,000 motor vehicles in use.

**International Sea-borne Shipping** (traffic, 2002 unless otherwise indicated): Vessels entered 385; Goods loaded 39,047 metric tons; Goods unloaded 492,193 metric tons; Passengers carried 275,300 (1998).

**Civil Aviation** (traffic, 2002 unless otherwise indicated): Freight carried 8,167 metric tons; Passengers carried 383,889 (2003).

## TOURISM

**Tourist Arrivals by Country** (2002, rounded figures): France 40,950; Guadeloupe 5,200; Martinique 9,750; Total (incl. others) 65,000.

**Receipts from Tourism** (US $ million, incl. passenger transport): 42 in 2001; 45 in 2002.
Source: World Tourism Organization.

## COMMUNICATIONS MEDIA

**Radio Receivers** ('000 in use): 104 in 1997.

**Television Receivers** ('000 in use): 37 in 1998.

**Telephones** ('000 main lines in use): 49.2 in 1999.

**Facsimile Machines** (number in use): 185 in 1990.

**Mobile Cellular Telephones** ('000 subscribers): 81.8 in 2001.

**Personal Computers** ('000 in use): 23 in 1999.

**Internet Users** ('000): 10.1 in 2002.

**Daily Newspaper:** 1 in 1996 (average circulation 2,000 copies).
Sources: UNESCO, *Statistical Yearbook*; UN, *Statistical Yearbook*; International Telecommunication Union.

## EDUCATION

**Pre-primary** (2002/03): 43 institutions; 12,274 students.

**Primary** (2002/03): 97 institutions; 2,751 teachers (2,506 state, 245 private); 23,485 students.

**Secondary** (2002/03, unless otherwise indicated): 35 institutions (2000/01); 4,553 teachers (4,178 state, 375 private); 22,003 students.

**Higher** (2000/01): 666 students (1998/99); 42 teachers.
Source: Ministère de l'Education Nationale, de l'Enseignement Supérieur et de la Recherche, *Repères et Références Statisques sur les Enseignements, la Formation et la Recherche 2003*.

# Directory

## The Government

(July 2005)

**Prefect:** ANGE MANCINI, Préfecture, rue Fiedmont, BP 7008, 97307 Cayenne Cédex; tel. 5-94-39-45-00; fax 5-94-30-02-77; e-mail sgaer1.prefect.guyane@wanadoo.fr; internet www.guyane.pref.gouv.fr.

**President of the General Council:** PIERRE DESERT, Hôtel du Département, pl. Léopold Héder, BP 5021, 97305 Cayenne Cédex; tel. 5-94-29-55-00; fax 5-94-29-55-25.

**President of the Economic and Social Committee (CESR):** (vacant), 35 ave Léopold Héder, 97300 Cayenne; tel. 5-94-28-96-81; fax 5-94-30-73-65; e-mail cesr@cr-guyane.fr.

**President of the Culture, Education and Environment Committee (CCEE):** JEAN-PIERRE BACOT, 47 rue du XIV juillet, 97300 Cayenne; tel. 5-94-25-66-84; fax 5-94-37-94-24; e-mail ccee@cr-guyane.fr.

**Deputies to the French National Assembly:** CHRISTIANE TAUBIRA (Walwari), JULIANA RIMANE (UMP).

**Representative to the French Senate:** GEORGES OTHILY (Rassemblement Démocratique et Social Européen).

### Conseil Régional

66 ave du Général de Gaulle, BP 7025, 97300 Cayenne Cédex; tel. 5-94-29-20-20; fax 5-94-31-95-22; e-mail cabcrg@cr-guyane.fr.

**President:** ANTOINE KARAM (PSG).

**Elections, 21 and 28 March 2004**

|  | Seats |
|---|---|
| Parti Socialiste Guyanais (PSG) | 17 |
| Forces Démocratiques Guyanaises-Walwari | 7 |
| UMP | 7 |
| **Total** | **31** |

## Political Organizations

**Forces Démocratiques Guyanaises (FDG):** 41 rue du 14 Juillet, BP 403, 97300 Cayenne; tel. 5-94-30-01-55; fax 5-94-30-80-66; e-mail g.othily@senat.fr; f. 1989 by a split in the PSG; Leader GEORGES OTHILY.

**Mouvement pour la Décolonisation et l'Emancipation Sociale (MDES):** 21 rue Maissin, 97300 Cayenne; tel. 5-94-30-55-97; fax 5-94-30-97-73; e-mail webmaster@mdes.org; internet www.mdes.org; f. 1991; pro-independence party; Sec.-Gen. MAURICE PINDARD.

**Parti Socialiste (PS):** 7 rue de l'Adjudant Pindard, 97334 Cayenne Cédex; tel. 5-94-37-81-33; fax 5-94-37-81-56; Leader LÉON JEAN BAPTISTE EDOUARD; Sec. PAUL DEBRIETTE; departmental br. of the metropolitan party.

**Parti Socialiste Guyanais (PSG):** 1 Cité Césaire, 97300 Cayenne; tel. 5-94-28-11-44; f. 1956; left-wing; Sec.-Gen. ANTOINE KARAM.

**Union pour la Démocratie Française (UDF):** 29 bis ave Pasteur, 97300, Cayenne; tel. 5-94-31-74-42; f. 1979; local br. of the metropolitan party; centre-right; Leader GEORGES HABRAN-MÉRY.

**Union pour un Mouvement Populaire (UMP):** 42 rue du Docteur Barrat, 97300 Cayenne; tel. 5-94-25-24-08; fax 5-94-24-66-80; e-mail ump973@nplus.gf; internet www.u-m-p.org; f. 2002; centre-right; local br. of the metropolitan party; Leader LÉON BERTRAND; Sec. LOUIS-RÉMY BUDOC.

**Les Verts Guyane:** 64 rue Madame Payé, 97300 Cayenne; tel. 5-94-40-97-27; e-mail tamanoir.guyane@wanadoo.fr; internet guyane.lesverts.fr; green; local br. of the metropolitan party; Leader BRIGITTE WYNGAARDE.

**Walwari:** 35 rue Schoelcher, 1er étage, BP 803, 97338 Cayenne Cédex; tel. 5-94-30-31-00; fax 5-94-31-84-95; e-mail permanence-taubira@wanadoo.fr; f. 1993; left-wing; Leader CHRISTIANE TAUBIRA.

## Judicial System

**Courts of Appeal:** see Judicial System, Martinique.

**Tribunal de Grande Instance:** Palais de Justice, 9 ave du Général de Gaulle, 97300 Cayenne; Pres. M. J. FAHET; Procurators-Gen. YVES-ARMAND FRASSATIN (acting), MICHEL REDON (acting).

## Religion

### CHRISTIANITY

#### The Roman Catholic Church

French Guiana comprises the single diocese of Cayenne, suffragan to the archdiocese of Fort-de-France, Martinique. At 31 December 2003 there were an estimated 150,000 adherents in French Guiana, representing some 75% of the total population. French Guiana participates in the Antilles Episcopal Conference, currently based in Port of Spain, Trinidad and Tobago.

**Bishop of Cayenne:** Rt Rev. EMMANUEL M. P. L. LAFONT, Evêché, 24 rue Madame Payé, BP 378, 97328 Cayenne Cédex; tel. 5-94-28-98-48; fax 5-94-30-20-33; internet perso.wanadoo.fr/catholique-cayenne.

#### The Anglican Communion

Within the Church in the Province of the West Indies, French Guiana forms part of the diocese of Guyana. The Bishop is resident in Georgetown, Guyana. There were fewer than 100 adherents at mid-2000.

#### Other Churches

At mid-2000 there were an estimated 7,000 Protestants and 7,200 adherents professing other forms of Christianity.

**Assembly of God:** 1051 route de Raban, 97300 Cayenne; tel. 5-94-35-23-04; fax 5-94-35-23-05; e-mail jacques.rhino@wanadoo.fr; internet www.addguyane.com; Pres. JACQUES RHINO.

**Church of Jesus Christ of Latter-day Saints (Mormons):** Route de la Rocade, 97305 Cayenne; Br. Pres. FRANÇOIS PRATIQUE.

**Seventh-day Adventist Church:** Mission Adventiste de la Guyane, 39 rue Schoëlcher, BP 169, 97324 Cayenne Cédex; tel. 5-94-25-64-26; fax 5-94-37-93-02; e-mail mission.adventiste@wanadoo.fr; f. 1949; Pres. and Chair. NORBERT KANCEL; 2,010 mems (May 2005).

The Jehovah's Witnesses are also represented.

## The Press

**France-Guyane:** 17 rue Lallouette, 97300 Cayenne; tel. 5-94-29-70-00; fax 5-94-29-70-02; daily; Publishing Dir FRÉDÉRIC AURAND; Local Dir MARC AUBURTIN; Editor PIERRE GIRARD; circ. Mon. to Fri. 3,280, Sat. 10,000.

**La Presse de Guyane:** pl. Léopold Héder, 97300 Cayenne; tel. 5-94-29-55-55; fax 5-94-29-55-54; e-mail tchisseka@yahoo.fr; Editor-in-Chief TCHISSÉKA LOBELT; 4 a week; circ. 1,000.

**La Semaine Guyanaise:** 6 ave Pasteur, 97300 Cayenne; tel. 5-94-31-09-83; fax 5-94-31-95-20; e-mail semaine.guyanaise@nplus.gf; weekly; Dir ALAIN CHAUMET.

**Ròt Kozé:** 21 rue Maissin, 97300 Cayenne; tel. 5-94-30-55-97; fax 5-94-30-97-73; internet www.mdes.org; f. 1990; monthly; Dir MAURICE PINDARD; left-wing organ of the MDES party.

### NEWS AGENCIES

**Agence France Presse:** 22 réseau St Antoine, 97300 Cayenne; tel. and fax 5-94-29-34-72; Rep. ALEXANDRE ROZIGA.

#### Foreign Bureau

**Reuters** (United Kingdom): Impasse du 8 Mai 1945, 97300 Cayenne; tel. 5-94-30-44-26; fax 5-94-31-93-24; Rep. ALEXANDER MILES.

## Broadcasting and Communications

### TELECOMMUNICATIONS

**France Telecom:** 76 ave Voltaire, BP 8080, 97300 Cayenne; tel. 5-94-39-91-08; fax 5-94-39-91-95; local br. of national telecommunications co.

# FRENCH GUIANA

*Directory*

## BROADCASTING

**Réseau France Outre-mer (RFO):** ave le Grand Boulevard, Z.A.D. Moulin à Vent, 97354 Rémire-Montjoly; tel. 5-94-25-67-00; fax 5-94-25-67-64; internet www.rfo.fr; fmrly Société Nationale de Radio-Télévision Française d'Outre-mer; present name adopted 1999; Radio-Guyane Inter: broadcasts 18 hours daily; accounts for 46.6% of listeners (2003); Téléguyane: 2 channels, 32 hours weekly; accounts for 59.5% of viewers (2003); Pres. ANDRÉ-MICHEL BESSE; Regional Dir ANASTASIE BOURQUIN.

### Radio

Seven private FM radio stations are in operation.

**Radio Gabrielle:** rue Montravel, 97311 Roura; tel. 5-94-27-00-47; fax 5-94-28-01-28.

**Radio Ouassaille:** rue Maurice Demongeot, 97360 Mana; tel. 5-94-34-80-96.

**Radio Pagani:** ave Justin Catayee, 97355 Macouria Tonate; tel. 5-94-38-85-90.

**Radio Tout'Moun:** rue des Mandarines, 97300 Cayenne; tel. 5-94-31-80-79; fax 5-94-30-91-19; f. 1982; broadcasts 24 hours a day.

**Radio UDL** (Union Défense des Libertés): ave Félix Eboué, 97323 Saint-Laurent-du-Maroni; tel. 5-94-34-27-90; f. 1982.

**RCI Guyane:** 33 rue Madame Payé, 97300 Cayenne; tel. 5-94-30-50-00; fax 5-94-30-60-00.

**RFM 90:** Annexe Hôtel des Roches, Le Manguier, 97310 Kourou; tel. 5-94-32-07-90.

### Television

**AntenneCréole Guyane:** 31 ave Louis Pasteur, 97300 Cayenne; tel. 5-94-28-82-88; fax 5-94-29-13-08; internet www.acg.gf; private television station; accounted for 25.4% of viewers (2003); Pres. MARC HO-A-CHUCK; Gen. Man. WLADIMIR MANGACHOFF.

## Finance

(cap. = capital; res = reserves; dep. = deposits; m. = million; brs = branches; amounts in French francs)

### BANKING

#### Central Bank

**Institut d'Emission des Départements d'Outre-mer (IEDOM):** 8 rue Christophe Colomb, BP 6016, 97306 Cayenne Cédex; tel. 5-94-29-36-50; fax 5-94-30-02-76; e-mail iedomguyane@wanadoo.fr; internet www.iedom.com; f. 1959; Dir DIDIER GREBERT.

#### Commercial Banks

**Banque Française Commerciale Antilles-Guyane (BFC Antilles-Guyane):** 8 pl. des Palmistes, BP 111, 97345 Cayenne; tel. 5-94-29-11-11; fax 5-94-30-13-12; e-mail f.aujoulat@bfc-ag.com; internet www.bfc-ag.com; see chapter on Guadeloupe; f. 1985; Regional Dir PHILIPPE BISSAINTE.

**BNP Paribas Guyane SA:** 2 pl. Victor Schoëlcher, BP 35, 97300 Cayenne; tel. 5-94-39-63-00; fax 5-94-30-23-08; e-mail bnpg@bnpparibas.com; internet www.bnpparibas.com; f. 1964 following purchase of BNP Guyane (f. 1855); name changed July 2000; owned by BNP Paribas SA, 94%; BNP Paribas Martinique, 3%; BNP Paribas Guadeloupe, 3%; cap. 71.7m., res 100.0m., dep. 2,007m. (Dec. 1994); Dir and Gen. Man. FRANÇOIS DU PEUTY; Gen. Sec. GEORGES CLOUTE-CAZAALA; 5 brs.

**BRED-Banque Populaire (BRED-BP):** 5 ave du Général de Gaulle, 97300 Cayenne; tel. 5-94-25-56-80; fax 5-94-31-98-40; 3 brs.

#### Development Bank

**Société Financière pour le Développement Economique de la Guyane (SOFIDEG):** PK 3, 700 route de Baduel, BP 860, 97339 Cayenne Cédex; tel. 5-94-29-94-29; fax 5-94-30-60-44; e-mail sofideg@nplus.gf; f. 1982; bought from the Agence Française de Développement (AFD—q.v.) by BRED-BP in 2003; Dir FRANÇOIS CHEVILLOTTE.

## Trade and Industry

### GOVERNMENT AGENCIES

**Conseil Economique et Social:** 66 rue du Général de Gaulle, BPC 25, 97300 Cayenne; tel. 5-94-30-81-00; fax 5-94-30-73-65; e-mail cesrccee@netplus.gf; Pres. ROGER-MICHEL LOUPEC.

**Direction de l'Agriculture et de la Forêt (DAF):** Parc Rebard, BP 5002, 97305 Cayenne Cédex; tel. 5-94-29-63-74; fax 5-94-29-63-63; e-mail daf973@agriculture.gouv.fr; internet daf.guyane.agriculture.gouv.fr.

**Direction Régionale de l'Industrie, de la Recherche et de l'Environnement (DRIRE):** Pointe Buzaré, BP 7001, 97307 Cayenne; tel. 5-94-29-75-30; fax 5-94-29-07-34; e-mail drire-antilles-guyane@industrie.gouv.fr; internet www.ggm.drire.gouv.fr; active in industry, business services, transport, public works, tourism and distribution; Regional Dir PHILIPPE COMBE.

**Direction Régionale et Départementale des Affaires Maritimes (DRAM):** 2 bis rue Mentel, BP 6008, 97306 Cayenne Cédex; tel. 5-94-29-36-15; fax 5-94-29-36-16; e-mail stephane.gatto@equipement.gouv.fr; responsible for shipping, fishing and other maritime issues at a nat. and community level.

### DEVELOPMENT ORGANIZATIONS

**Agence Française de Développement (AFD):** Lotissement les Héliconias, route de Baduel, BP 1122, 97345 Cayenne Cédex; tel. 5-94-29-90-90; fax 5-94-30-63-32; e-mail afd.cayenne@gf.groupe-afd.org; fmrly Caisse Française de Développement; Man. GENEVIÈVE JAVALOYES.

**Agence pour la Création et le Développement des Entreprises en Guyane (ACREDEG):** pl. Schoelcher, BP 235, 97300 Cayenne; tel. 5-94-25-66-66; fax 5-94-25-43-19; e-mail acredeg@nplus.gf; internet www.acredeg.gf; f. 1998; Pres. JEAN CLAUDE SIMONEAU; Dir RAYMOND CHARPENTIER-TITY.

### CHAMBERS OF COMMERCE

**Chambre d'Agriculture:** 8 ave du Général de Gaulle, BP 544, 97333 Cayenne Cédex; tel. 5-94-29-61-95; fax 5-94-31-00-01; e-mail chambre.agriculture.973@wanadoo.fr; Pres. PATRICK LABRANCHE.

**Chambre de Commerce et d'Industrie:** Hôtel Consulaire, pl. de l'Esplanade, BP 49, 97300 Cayenne Cédex; tel. 5-94-29-96-00; fax 5-94-29-96-34; e-mail contact@guyane.cci.fr; internet www.guyane.cci.fr; Pres. JEAN-PAUL LE PELLETIER.

**Chambre de Métiers:** Jardin Botanique, blvd de la République, BP 176, 97324 Cayenne Cédex; tel. 5-94-30-21-80; fax 5-94-30-54-22; e-mail m.toulemonde@cm-guyane.fr; internet www.cm-guyane.fr; Pres. SYLVAIN LEMKI; Sec.-Gen. MYRIAM TOULEMONDE (acting).

**Jeune Chambre Economique de Cayenne:** 1 Cité A. Horth, route de Montabo, BP 1094, Cayenne; tel. 5-94-31-62-99; fax 5-94-31-76-13; f. 1960; Pres. FRANCK VERSET.

### EMPLOYERS' ORGANIZATIONS

**Coopérative des Céréales et Oléagineux de Guyane (COCEROG):** PK 24, chemin départemental 8, Sarcelles, 97360 Mana; tel. 5-94-34-20-82; fax 5-94-34-02-08.

**Groupement Régional des Agriculteurs de Guyane (GRAGE):** PK 15 route nationale 1, Domaine de Soula, 97355 Macouria; tel. 5-94-38-71-26; fax ; e-mail grage@wanadoo.fr; affiliated to the Confédération Paysanne; Pres. ALBÉRIC BENTH.

**Ordre des Pharmaciens du Département Guyane:** ave Hector Berlioz, 97310 Kourou; tel. 5-94-32-17-62 ; fax 5-94-32-17-66; Pres. LOUISE AREL-GOLITIN.

**Syndicat des Exploitants Forestiers et Scieurs de Guyane (SEFSG):** Macouria; tel. 5-94-35-26-66; fax 5-94-35-29-92; f. 1987; timber processors; Man. M. POMIES.

**Syndicat des Transformateurs du Bois de Guyane (STBG):** Menuiserie Cabassou, PK 4.5, route de Cabassou, 97354 Remire-Montjoly; tel. 5-94-31-34-49; fax 5-94-35-10-51; f. 2002; Pres. YVES ELISE; Sec. FRANÇOIS AUGER.

**MEDEF Guyane:** 27A Résidence Gustave Stanislas, Source de Baduel, BP 820, 97338 Cayenne Cédex; tel. 5-94-31-17-71; fax 5-94-30-32-13; e-mail updg@nplus.gf; f. 2005; fmrly l'Union des Entreprises de Guyane; Pres. ADRIEN AUBIN.

### MAJOR COMPANIES

**Air Liquide Spatial Guyane (ALSG):** Route de l'Espace, Ensemble de Lancement, BP 826, 97388 Kourou; tel. 5-94-33-75-69; fax 5-94-33-75-77; f. 1969; subsidiary of Air Liquide Group; manufacture of industrial gases; Gen. Man. LAURENT DU HAYES.

# FRENCH GUIANA

*Directory*

**Arianespace, SA:** BP 809, 97388 Kourou; fax 5-94-33-62-66; e-mail webmaster@arianespace.com; f. 1979; aerospace industry; Man. JEAN-CHARLES VINCENT.

**Bamyrag:** 7 Lotissement Marengo, Z. I. Collery, 97323 Cayenne; tel. 5-94-36-26-00; fax 5-94-35-14-45; distribution of fuels, metal minerals and chemical products; Pres. BERNARD HAYOT.

**CEGELEC Space:** Global Technologies, CIGMA Division, Immeuble Vercors, pl. Newton, 97310 Kourou; BP 819, 97388 Kourou; tel. 5-94-32-05-24; fax 5-94-32-31-39; supplier to the space industry; responsible for operation and maintenance of Guiana Space Centre ground infrastructure; subsidiary of CEGELEC; Man. BERNARD ASSIE; 300 employees.

**Ciments Guyanais (CIGU):** Z. I. Dégrad-des-Cannes, 97354 Rémire-Montjoly; tel. 5-94-35-54-98; fax 5-94-35-54-99; e-mail ciments-guyanais@wanadoo.fr; f. 1989; cement production; Pres. PETER HADDINOTT; Gen. Man. PATRICK VANDRESSE; 25 employees.

**Compagnie Française de Pêche Nouvelle (CFPN):** Port du Larivot, BP 834, 97338 Cayenne Cédex; tel. 5-94-35-17-77; fax 5-94-35-10-42; e-mail louis.maignan@wanadoo.fr; fishery co; Gen. Man. LOUIS MAIGNAN; 133 employees.

**Compagnie Guyanaise de Transformation des Produits de la Mer (COGUMER):** Port de Pêche du Larivot, 97351 Matoury; tel. 5-94-29-00-00; fax 5-94-30-30-46; e-mail cogumer@wanadoo.fr; internet www.cogumer.com; f. 1986 as CODEPEG; present name adopted 2003; fish processing; Pres. CHRISTIAN MADERE; Gen. Man. RENÉ GUSTAVE; 250 employees.

**Nofrayane:** 9 Parc d'Activité Cognot Matounry, BP 1166, 97300 Cayenne; tel. 5-94-35-18-65; fax 5-94-35-18-60; construction and civil engineering; Chair. M. MANTES; Dir STEVE SAINT-JEAN; 98 employees.

**Propadis:** Z. I. Collery, 97300 Cayenne; tel. 5-94-35-17-17; fax 5-94-35-19-91; distribution of food products; subsidiary of SISB since Dec. 2002; Pres. NICOLAS LECRES; Gen. Man. CHRISTIAN CHANG HING WING.

**Régulus (CSG):** Centre Spatial Guyanais, BP 73, 97372 Cayenne; tel. 5-94-35-15-00; fax 5-94-32-49-42; f. 1991; space industry; joint subsidiary of Fiat Avio (60%) and SNPE (40%); Pres. AMATO FADINI.

**SARA** (Société Anonyme de la Raffinerie des Antilles): Dégrad-des-Cannes, BP 227, 97301 Cayenne; tel. 5-94-25-50-50; fax 5-94-35-41-79; internet www.sara.mq; f. 1982; second depot opened in Kourou in 2000; Chair. CHRISITAN CHAMMAS; Regional Gen. Man. FRANÇOIS NAHAN; c. 250 employees regionally (see entry under Martinique).

**Shell SAGF** (Société Shell des Antilles et de la Guyane Française): Z. I. Pariacabo, 97310 Kourou; fax 5-94-32-33-40; distribution of petroleum products.

**SNC Dumez Guyane:** Z. I. Pariacabo, BP 817, 97310 Kourou; construction and civil engineering.

**Tanon & CIE:** PK 1.5 route de Baduel, BP 262, 97326 Cayenne; tel. 5-94-29-39-39; fax 5-94-31-37-20; e-mail guyaneautocenter@wanadoo.fr; f. 1892; distribution of motor vehicles and spare parts, mining equipment, air conditioners, agricultural and petroleum products; Pres. RAYMOND ABCHEE; Gen. Man. ANDRÉ ABCHEE; 115 employees.

## UTILITIES

### Electricity

**Électricité de France Guyane (EDF):** Blvd Jubelin, BP 6002, 97306 Cayenne; tel. 5-94-39-64-00; fax 5-94-30-10-81; state-owned; Gen. Man. MARC GIRARD.

### Water

**Société Guyanaise des Eaux (SGDE):** 2738 route de Montabo, 97305 Cayenne; tel. 5-94-30-32-32; fax 5-94-30-59-60; subsidiary of Vivendi Universal; Pres. FRANÇOIS MARTIN; Gen. Man. JACQUES FOURNET.

## TRADE UNIONS

**Centrale Démocratique des Travailleurs de la Guyane (CDTG):** 99–100 Cité Césaire, BP 383, 97328 Cayenne Cédex; tel. 5-94-31-02-32; fax 5-94-31-81-05; e-mail sg.cdtg@wanadoo.fr; Sec.-Gen. JEAN-MARC BOURETTE.

**Union Départementale Confédération Française des Travailleurs Chrétiens Guyane (UD CFTC Guyane):** BP 763, 97337 Cayenne Cédex; tel. 5-94-35-63-14; fax 5-94-35-77-30; e-mail lydie.leneveu@wanadoo.fr.

**Union Départementale Force Ouvrière de Guyane (FO):** 25 Cité Mirza, rue des Acajous, 97300 Cayenne; tel. 5-94-31-62-55; Sec.-Gen. CHRISTIAN DESFLOTS.

**Union Régionale Guyane:** 52 rue François Arago, BP 807, 97300 Cayenne; tel. 05-94-21-67-61; fax 05-94-30-89-70.

**Union des Travailleurs Guyanais (UTG):** 14 bis rue Digue Ronjon, 97300 Cayenne; tel. 5-94-30-17-67; fax 5-94-35-47-74; Sec.-Gen. CHRISTIAN RAVIN.

**UNSA Education Guyane:** 46 rue Vermont Polycarpe, 97300 Cayenne; tel. and fax 5-94-30-89-70; e-mail unsa-education-guyane@wanadoo.fr; Sec.-Gen. MARTINE NIVOIX.

# Transport

### RAILWAYS

There are no railways in French Guiana.

### ROADS

In 2004 there were 1,300 km (808 miles) of roads in French Guiana, of which 397 km were main roads. Much of the network is concentrated along the coast, although proposals for a major new road into the interior of the Department were under consideration.

### SHIPPING

Dégrad-des-Cannes, on the estuary of the river Mahury, is the principal port, handling 80% of maritime traffic in 1989. There are other ports at Le Larivot, Saint-Laurent-du-Maroni and Kourou. Saint-Laurent is used primarily for the export of timber, and Le Larivot for fishing vessels. There are river ports on the Oyapock and on the Approuague. There is a ferry service across the Maroni river between Saint-Laurent and Albina, Suriname. The rivers provide the best means of access to the interior, although numerous rapids prevent navigation by large vessels.

**Société des Transports Maritimes Guyanaise (STMG):** 515 chemin St Antoine, 97300 Cayenne; Man. ANTOINE PORRY.

**SOMARIG** (Société Maritime et Industrielle de la Guyane): Z. I. de Dégrad-des-Cannes, Rémire, BP 81, 97322 Cayenne Cédex; tel. 5-94-35-42-00; fax 5-94-35-53-44; e-mail cay.genmbox@cma-cgm.com; Man. Dir PHILIPPE BABLON.

### CIVIL AVIATION

Rochambeau International Airport, situated 17.5 km (11 miles) from Cayenne, is equipped to handle the largest jet aircraft. Access to remote inland areas is frequently by helicopter.

**Air Guyane:** Aéroport de Rochambeau, 97300 Matoury; tel. 5-94-35-03-07; fax 5-94-30-54-37; e-mail resa@airguyane.com; internet www.airguyane.com; f. 1980; operates internal and internat. services; Pres. CHRISTIAN MARCHAND.

# Tourism

The main attractions are the natural beauty of the tropical scenery and the Amerindian villages of the interior. In 2002 there were 28 hotels with some 1,300 rooms. In 2001 and 2002 some 65,000 tourist arrivals were recorded, while receipts from tourism increased from an estimated US $42m. in 2001 to $45m. in 2002.

**Comité du Tourisme de la Guyane:** 12 rue Lallouette, BP 801, 97338 Cayenne Cédex; tel. 5-94-29-65-00; fax 5-94-29-65-01; e-mail ctginfo@tourisme-guyane.com; internet www.tourisme-guyane.com; Pres. JEAN-ELIE PANELLE; Dir GEORGES EUZET.

**Délégation Régionale au Tourisme, au Commerce et à l'Artisanat pour la Guyane:** 9 rue Louis Blanc, BP 7008, 97307 Cayenne Cédex; tel. 5-94-28-92-90; fax 5-94-31-01-04; e-mail drtca973@nplus.gf; Delegate PAUL RENAUD.

**Fédération des Offices de Tourisme et Syndicats d'Initiative de la Guyane (FOTSIG):** 12 rue Lallouette, 97300 Cayenne; tel. 5-94-30-96-29; fax 5-94-31-23-41; Pres. ARMAND HILDAIRE.

**Office Culturel de la Région Guyane:** 82 ave du Général de Gaulle, BP 6007, 97306 Cayenne Cédex; tel. 5-94-28-94-00; fax 5-94-28-94-04; e-mail ocrg@nplus.gf; internet www.ocrg.gf; f. 1998; replaced the Association Régionale de Développement Culturel; Pres. MYRIAM KEREL.

# Defence

At 1 August 2004 France maintained a military force of about 3,100 in French Guiana. The headquarters is in Cayenne.

## Education

Education is modelled on the French system, and is free and compulsory for children between six and 16 years of age. Between 1980 and 1993 the number of children attending primary school increased by more than 70%, and the number of children at secondary school by 87%. This expansion placed considerable pressure on the education system. In 2002/03 there were 43 pre-primary schools and 97 primary schools; there were 35 secondary schools in 2000/01. There were 35,759 students in pre-primary and primary education in 2002/03, while in secondary education there were 22,003 (including 3,073 students in vocational education), of whom some 92% attended state schools. Higher education is provided by a branch of the Université Antilles-Guyane in Cayenne, which has faculties of law, administration and French language and literature. The university as a whole had 12,000 enrolled students in 2004. There is also an Ecole Normale for teacher training, an agricultural college and a technical college. Total expenditure on education amounted to 851m. francs in 1993. The French Government announced its decision to increase expenditure in the education sector during 2000–06, including €71m. on the construction of new school buildings. An Academy for French Guiana was established in January 1997.

## Bibliography

For works on the Caribbean generally, see Select Bibliography (Books)

Crane, J. *French Guiana*. Oxford, ABC Clio, 1999.

Mam-Lam-Fouk, S. *Histoire générale de la Guyane française*. Matoury, Ibis Rouge, 2002.

Plénet, C. *Les fonds structurels européens*. Matoury, Ibis Rouge, 2005.

Redfield, P. *Space in the Tropics: From Convicts to Rockets in French Guiana*. Berkeley, CA, University of California Press, 2000.

# GRENADA

## Geography

### PHYSICAL FEATURES

Grenada is in the Windward Islands, in the Lesser Antilles, and is considered to be the most southerly island of the eastern Caribbean. It lies about 145 km (90 miles) north of Trinidad (Trinidad and Tobago). The country includes Carriacou, the largest island of the Grenadines, and a number of smaller islands in the chain that runs north and a little east of the main island. Off shore from Carriacou, to the east, is Petit (often spelt Petite) Martinique, the most northerly of the islands of Grenada and separated from Petit St Vincent (Saint Vincent and the Grenadines) by only a narrow sea channel. Grenada is the second-smallest independent state in the Americas (after Saint Christopher and Nevis), with a total area of 344.5 sq km (133 sq miles).

The main island of Grenada is about 34 km long by 19 km wide, aligned along a north–south axis, apart from a southern tapering towards the south-west and a northern tendency to reach towards the chain of the Grenadines in the north-east. The wooded mountains march across the island following this diagonal, the land to the north and west rising (the highest point is at Mt St Catherine—840 m or 2,757 ft), the land to the south and east falling to an indented coasts of rias (drowned valleys), which provide deep harbours. Grenada is volcanic in origin, with its central highlands, but also its fertile soil, which give sustenance to tropical forests and mangrove swamps, as well as the crops that have earned it the moniker of the 'spice island of the Caribbean'. Birds and animals thrive, including armadillos and Mona monkeys originally imported from Africa, but the only unique species are the endangered hookbilled kite, a large hawk that eats tree snails and is now, world-wide, only found in the Levera National Park, and the native Grenada dove. The 121-km coastline is largely protected by reefs, particularly in the Grenadines.

The southern Grenadines form part of the country of Grenada. These islands include a group around Ronde Island, then, further north, a smaller group just to the south of the large island of Carriacou, which is 37 km north-east of Grenada itself. Carriacou is almost 34 sq km in extent, an island of low, green hills. About 4 km east of its northern end is the island of Petit Martinique, the next largest of the Grenadian Grenadines.

### CLIMATE

The climate is subtropical, tempered by north-eastern trade winds off the Atlantic. Grenada can occasionally be affected by hurricanes, but the country is generally considered to lie just to the south of the hurricane belt (in 1999 a heavy swell from a hurricane to the north caused considerable damage, for instance). There is, though, a rainy season from June to November, followed by the cooler months of December and January. Average annual rainfall is about 1,520 mm (60 ins), but in the high forest there can be over 4,000 mm per year. The average annual temperature along the coast is 28°C (82°F).

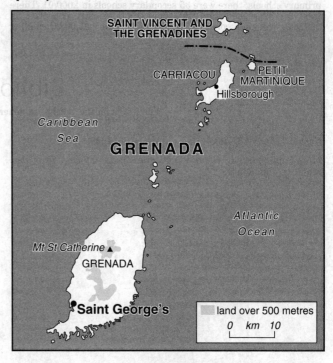

### POPULATION

Racially the population is 85% black, with 11% of mixed black–white ancestry and the rest consisting mainly of whites and 'East' Indians. Some claim to include in their ancestry traces of the original Carib and Arawak Amerindian inhabitants. Although most cultural evidence of early French rule has disappeared, just over one-half of the population is still Roman Catholic, with Nonconformist Protestant denominations (33%) and Anglicans (14%) making up the balance. The official language is English, but a French patois has survived in some areas.

According to mid-year estimates, the total population was 104,614 in 2003. Many of these people are young, with almost two-fifths under 15 years of age and a further one-quarter between 15 and 30 years old. Almost 4,000 live in the capital, the beautiful harbour city of St George's in the south-west, and over 5,000 on the islands of Carriacou and Petit Martinique. The chief town of the Grenadines is Hillsborough, on the central western shore of Carriacou. Grenada island is divided into six parishes, with Carriacou and Petit Martinique described as a dependency.

## History

### MARK WILSON

Grenada is a constitutional monarchy within the Commonwealth. Queen Elizabeth II is head of state, and is represented in Grenada by a Governor-General. There is a bicameral legislature with an elected chamber.

Grenada was known by its original Carib inhabitants as Camerhogue, but was named Concepción by Christopher Columbus on his third voyage in 1498, and later renamed La Grenade by the French, who colonized the island from 1650, meeting fierce resistance from the Caribs. The island was ceded to Great Britain by the Treaty of Paris in 1763, but was briefly recaptured by France in 1779–83. A slave revolt led by Julian Fedon in 1795–96 was the most successful in the eastern Caribbean, gaining control of almost the entire island, but was bloodily suppressed. Slavery was abolished in 1834, as in the United Kingdom's other Caribbean colonies.

Eric (later Sir Eric) Gairy dominated the island's politics in the 1950s, 1960s and 1970s. He rose to prominence in 1950, as the founder of the Grenada Manual and Mental Workers' Union, which led a series of strikes the next year. Four people were killed in violent clashes with the authorities, and his Grenada People's Party, later renamed the Grenada United Labour Party (GULP), won the first universal suffrage election, which was fought on 10 October of that year, with 71% of the popular vote. Gairy became the first Chief Minister and was fiercely supported by the rural masses, who gave him credit for increased wages and improvements in education and welfare. At the same time, he was feared and despised by the urban middle class, most of whom supported Herbert Blaize's National Party (NP), which held office in 1957–61 and 1962–67.

Grenada joined the Federation of the West Indies in 1958 along with nine other British colonies. When Jamaica and Trinidad left in 1962 the Federation collapsed, and an attempt to unite the remaining colonies as the 'little eight' was unsuccessful, while a proposal for a union with Trinidad and Tobago was not followed through. Along with its neighbours to the north, Grenada became a British Associated State in 1967, responsible for its internal affairs, with the United Kingdom retaining control of external affairs and defence.

Young left-wing opponents of Gairy, led by Maurice Bishop, in 1973 formed the New Jewel Movement (NJM—Joint Endeavour for Welfare, Education and Liberation). Grenada was the first Associated State to gain independence, on 7 February 1974. Gairy's opponents were strongly opposed to separation from the United Kingdom under what they saw as a repressive regime, and the months before were marked by protracted strikes and widespread demonstrations; Bishop's father was shot dead by the police during a demonstration on 21 January.

In elections held in 1976 an opposition alliance won 48% of the popular vote and six seats; Bishop became leader of the opposition. On 13 March 1979 the NJM removed Gairy from power, installing a People's Revolutionary Government (PRG), with Bishop as Prime Minister, but retaining the Queen as head of state. The PRG had close relations with Cuba and the USSR, while the USA saw the regime as a potential security threat. Bishop, in particular, at first enjoyed strong popular support, but this was gradually eroded both by the refusal to call elections and by economic difficulties. An important policy initiative was the construction, with Cuban assistance, of the Point Salines International Airport, replacing a short-runway facility at Pearls, which could not accommodate intercontinental flights. Although designed as a civilian facility, the USA feared that Point Salines could be used for military purposes.

In 1983 a militant wing of the NJM and the army became increasingly hostile to Bishop and his immediate supporters. In October these forces organized a coup, led by Gen. Hudson Austin. Bishop and several cabinet ministers were shot, as well as members of a crowd of demonstrators, which gathered in St George's. The USA led a military intervention, supported by seven Caribbean states, which installed an interim Government, led by Nicholas Brathwaite.

The NP, still led by Blaize, won a general election in December 1984, securing 14 of the 15 legislative seats. Blaize's Government completed the construction of the international airport and returned the country to a more normal path of development. None the less, splits and defections weakened the NP Government, as did the death of Blaize in December 1989. Blaize's successor, Ben Jones, lost a general election in March 1990; the National Democratic Congress (NDC), led by Brathwaite, won seven seats in the House of Representatives, with 34.6% of the popular vote, gaining a majority through the defection of individual members from other parties. However, the NP's main successor, the New National Party (NNP), led by Keith Mitchell, regained office in an election held on 20 June 1995, with 32.7% of the popular vote and eight parliamentary seats. After losing his parliamentary majority through further defections, Mitchell called fresh elections on 18 January 1999, in which the NNP won all 15 parliamentary seats (with 62% of the popular vote).

The NNP has at times been under attack over allegations of impropriety, including lax supervision of the 'offshore' financial sector; moreover, several prominent overseas investors have proved to be undesirable, and departed in controversial circumstances. In late 2001 Canada announced that Grenadians travelling to Canada would require visas. This reflected growing international concern over Grenada's economic citizenship programme, under which passports were sold to non-citizens. Following the terrorist attacks in the USA in September 2001, Grenada came under increased pressure to abandon the programme, eventually complying in December 2002.

The NNP narrowly won a general election held on 27 November 2003, holding eight of the 15 seats in the House of Representatives; the NDC took the remaining seven seats. Mitchell appointed a new Cabinet on 3 December 2003; the most significant change was the transferral of the agriculture portfolio to Gregory Bowen, the erstwhile Minister of Communications, Works and Public Utilities, as part of a wider plan to revitalize the ailing agricultural sector, in particular the nutmeg industry.

'Hurricane Ivan' struck Grenada on 7 September 2004, causing widespread devastation. In spite of this, the political climate remained bitterly polarized. Lawyers and trade unions campaigned vigorously in February 2005 against a proposal to appoint a controversial Jamaican lawyer, Hugh Wildman, as Attorney-General, successfully persuading the Judicial and Legal Services Commission of the Organisation of Eastern Caribbean States to oppose the appointment. The Government accepted the ruling and in March the position of Attorney-General was added to the other portfolios held by Elvin Nimrod, the Minister of Foreign Affairs and International Trade, of Carriacou and Petit Martinique Affairs, and of Legal Affairs.

Further controversy was caused by the switching of diplomatic recognition in January 2005 from Taiwan to the People's Republic of China. China had agreed in December 2004 to assist with the construction of a sports stadium, housing and other projects. A broadly similar package had previously been agreed with Taiwan, formerly the island's largest bilateral aid donor, but Mitchell argued that the destruction caused by 'Hurricane Ivan' had forced the Government to reconsider its international relationships.

# Economy

## MARK WILSON

The three-island state of Grenada, Carriacou and Petit Martinique is the second smallest nation in the Western hemisphere in area, with 345 sq km and some 104,614 inhabitants in 2003, of whom 4,900 live on Carriacou and 800 on Petit Martinique. The nation has developed a fairly prosperous middle-income economy, with a per-head gross domestic product (GDP) of US $4,172 in 2003. The unemployment rate fell from 17.0% to 12.3% in 1996–2002, while GDP increased at an average annual rate of 6.7% in 1997–2000. However, as a result of an international downturn in tourism, lower international prices for nutmeg, reduced production of some other crops, and production cutbacks in a local electronics plant, GDP contracted by 4.4% in 2001 and by a further 0.4% in 2002; a recovery in 2003 produced estimated growth of 5.7%. Reduced government revenue and increased salary costs resulted in an increase in the overall fiscal deficit from 2.9% of GDP in 2000 to 9.6% in 2001 and 19.5% in 2002, financed mainly through local commercial borrowing and an accumulation of arrears; however, the deficit was reduced to 4.9% of GDP in 2003. An IMF report, which followed Article IV consultations in January 2003, noted the need to reverse the sharp and unsustainable increase in public debt (111% of GDP at the end of that year) and to reduce the fiscal deficit.

Grenada is a member of the Caribbean Community and Common Market, or CARICOM, which is attempting to develop a single market, in principle by 2006. It is also a member of the Organisation of Eastern Caribbean States (OECS), which links nine of the smaller Caribbean territories, and the Eastern Caribbean Securities Exchange (based in Saint Christopher and Nevis), while its financial affairs are ably supervised by the Eastern Caribbean Central Bank, also headquartered in Saint Christopher and Nevis.

Agriculture and fishing made up some 9.1% of GDP in 2003. The major export crops are nutmeg and its by-product, mace, of which Grenada is the world's second most important producer, after Indonesia. The crop benefited from high international prices in 1997–2000, with the value of exports of the two products peaking at US $16.8m. in 1999, making up 25.5% of that year's exports. The same figures were $13.6m. and 35.5% in 2002. Cocoa exports are also significant to the local economy. In contrast to the other Windward Islands, banana exports are of minor importance. There is some chance of offshore oil or gas reserves, and the Government intends to negotiate Exclusive Economic Zone boundaries with Venezuela and Trinidad and Tobago.

The pleasant climate, white sand beaches and natural beauty of the islands have encouraged the growth of tourism, with the additional benefit since 1984 of direct air connections to the United Kingdom and North America. Receipts from tourism increased from US $37.5m. in 1990 to an estimated $169.5m. in 2003. The number of stop-over tourists reached some 133,700 in 2003, of whom 24.7% were from the USA, 3.9% from Canada, and 30.3% were from Europe. In the same year there were also 146,925 cruise-ship passengers, who made a smaller contribution to the economy as a result of limited onshore spending. The number of stop-over visitors was estimated to have decreased to 89,900 in 2004.

St George's University, which includes an 'offshore' medical school, has over 1,250 resident students and staff, most of them from the USA. This institution makes a substantial contribution to GDP.

Manufacturing made up some 5.8% of GDP in 2003. Industries such as the assembly of electronic components are oriented entirely to the export market, while others, such as beverages, produce for the local market with some exports to other Caribbean islands.

The Government stresses the importance of recent advances in technology for economic development. Grenada is a member of the Eastern Caribbean Telecommunications Authority, which has liberalized the regime for local and international communications. Call centres and telemarketing operations are a source of employment, but performance has been uneven.

There is a small 'offshore' financial sector, which included 859 International Business Companies and one 'offshore' bank at the end of 2004. The sector has until recently failed to meet internationally accepted regulatory standards, and Grenada was listed as a 'non-co-operative jurisdiction' by the Organisation for Economic Co-operation and Development in 2001. The main 'offshore' financial institution in the late 1990s was First International Bank, which had a close relationship with armed rebel groups in the Congo, and a balance sheet based on the reputed value of a single gemstone. The bank collapsed in July 2000, owing US $206m. to creditors in the USA and elsewhere, and with funds transferred to Saint Vincent and the Grenadines, the British Crown Dependency of Jersey and Uganda. The Prime Minister, who was then directly responsible for 'offshore'-sector supervision, had been warned by the bank's auditor of its precarious financial state. In September 2001 Grenada was added to the list of 'non-co-operative jurisdictions' compiled by the Financial Action Task Force on Money Laundering (FATF, based in Paris, France). Legislation and regulatory standards were subsequently tightened by the Grenada International Financial Services Authority, with the licences of 26 banks and trust companies revoked; as a result, Grenada was de-listed by the FATF in February 2003. None the less, international monitoring of the sector remained strict.

The islands contain several volcanic centres, although the only one with a recent history of activity is the underwater volcano of Kick 'Em Jenny to the north of the main island. Grenada was devastated by 'Hurricane Ivan' on 7 September 2004. The Organisation of Eastern Caribbean States assessed damage at US $815 million, equivalent to over twice the island's annual GDP, which contracted in 2004 by an estimated 3.2% as a result of the storm. There was extensive damage to housing stock, infrastructures and public services, particularly in the south of the island. Most notably, of the 1,700 hotel rooms, only 300 were usable in the immediate aftermath of the hurricane, and close to 90% of the nutmeg trees were uprooted or severely damaged. The hurricane disabled the main export industries, while increasing the need for reconstruction-related imports. As a consequence, the burden on the public finances was dramatically increased, while revenue was sharply reduced. Generous grant aid supported disaster relief and reconstruction efforts, with pledges totalling one-third of annual GDP, and expenditure by the Government channelled through the new Agency for Reconstruction and Development, established in order to improve public and donor confidence in accountability and transparency. However, the IMF projected a financing gap of 4.6% of GDP for 2005, rising to 10% in 2006 before decreasing to 6%–7% in 2008–09, with public debt totalling the equivalent of 130% of GDP in February 2005. Proposals for recovery included the reduction of tax exemptions, the introduction of a value added tax and extensive public-sector reform. Of the main productive sectors, hotel-based tourism was expected to be fully functional in time for the 2005–06 winter season; however, the recovery of nutmeg farming was expected to be slow at best, partly owing to the diversion of labour from agriculture to reconstruction projects that promised more immediate returns.

# Statistical Survey

Source (unless otherwise stated): Central Statistical Office, Ministry of Finance, Financial Complex, The Carenage, St George's; tel. 440-2731; fax 440-4115; e-mail director@economicaffairs.grenada.gd; internet economicaffairs.grenada.gd.

### AREA AND POPULATION

**Area:** 344.5 sq km (133.0 sq miles).

**Population:** 94,806 (males 46,637, females 48,169) at census of 12 May 1991 (excluding 537 persons in institutions and 33 persons in the foreign service); 100,895 at census of 25 May 2001 (preliminary). *2003* (mid-year estimate): 104,614 (Source: World Bank, *World Development Indicators*).

**Density** (mid-2003): 303.7 per sq km.

**Principal Town** (population at 2001 census, preliminary): St George's (capital) 3,908. *Mid-2003* (UN estimate, incl. suburbs): St George's 32,707 (Source: UN, *World Urbanization Prospects: The 2003 Revision*).

**Births and Deaths** (registrations, 2001, provisional): Live births 1,899 (birth rate 18.8 per 1,000); Deaths 727 (death rate 7.2 per 1,000).

**Expectation of Life** (WHO estimates, years at birth): 67 (males 66; females 69) in 2003. Source: WHO, *World Health Report*.

**Employment** (employees only, 1998): Agriculture, hunting, forestry and fishing 4,794; Mining and quarrying 58; Manufacturing 2,579; Electricity, gas and water 505; Construction 5,163; Wholesale and retail trade 6,324; Restaurants and hotels 1,974; Transport, storage and communications 2,043; Financing, insurance and real estate 1,312; Public administration, defence and social security 1,879; Community services 3,904; Other services 2,933; Activities not adequately defined 1,321; *Total employed* 34,789 (males 20,733, females 14,056). *Mid-2003* (estimates): Agriculture, etc. 8,000; Total labour force 37,000 (Source: FAO).

# GRENADA

## HEALTH AND WELFARE
### Key Indicators

**Total Fertility Rate** (children per woman, 2003): 3.5.

**Under-5 Mortality Rate** (per 1,000 live births, 2003): 23.

**Physicians** (per 1,000 head, 1997): 0.50.

**Hospital Beds** (per 1,000 head, 1996): 5.27.

**Health Expenditure** (2002): US $ per head (PPP): 465.

**Health Expenditure** (2002): % of GDP: 5.7.

**Health Expenditure** (2002): public (% of total): 71.0.

**Access to Water** (% of persons, 2002): 95.

**Access to Sanitation** (% of persons, 2002): 97.

**Human Development Index** (2002): ranking: 93.

**Human Development Index** (2002): value: 0.745.

For sources and definitions, see explanatory note on p. vi.

## AGRICULTURE, ETC.

**Principal Crops** (FAO estimates, '000 metric tons, 2003): Roots and tubers 3.2; Sugar cane 7.2; Pigeon peas 0.5; Coconuts 6.5; Vegetables 2.6; Bananas 4.1; Plantains 0.7; Oranges 0.9; Grapefruit and pomelos 2.0; Apples 0.5; Plums 0.7; Mangoes 1.9; Avocados 1.5; Other fruits 4.5; Cocoa beans 0.7; Nutmeg, mace and cardamons 2.7.

**Livestock** (FAO estimates, '000 head, year ending September 2003): Cattle 4.5; Pigs 5.9; Sheep 13.2; Goats 7.2; Asses 0.7; Poultry 268.

**Livestock Products** (FAO estimates, '000 metric tons, 2003): Poultry meat 0.6; Cows' milk 0.5; Hen eggs 0.9.

**Fishing** (metric tons, live weight, 2003): Red hind 203; Coney 97; Snappers and jobfishes 92; Parrotfishes 178; Blackfin tuna 335; Yellowfish tuna 749; Atlantic sailfish 171; Bigeye scad 67; Swordfish 88; Common dolphinfish 130; *Total catch* (incl. others) 2,544. Source: FAO.

## INDUSTRY

**Production** (1994): Rum 300,000 litres; Beer 2,400,000 litres; Wheat flour 4,000 metric tons (1996); Cigarettes 15m.; Electricity 118 million kWh (2001). Source: UN, *Industrial Commodity Statistics Yearbook*.

## FINANCE

**Currency and Exchange Rates:** 100 cents = 1 Eastern Caribbean dollar (EC $). *Sterling, US Dollar and Euro Equivalents* (31 May 2005): £1 sterling = EC $4.909; US $1 = EC $2.700; €1 = EC $3.329; EC $100 = £20.37 = US $37.04 = €30.04. *Exchange Rate:* Fixed at US $1 = EC $2.70 since July 1976.

**Budget** (preliminary, EC $ million 2002): *Revenue:* Tax revenue 262.4 (Taxes on income and profits 43.5, Taxes on property 18.0, Taxes on domestic goods and services 49.6, Taxes on international trade and transactions 151.3); Other current revenue 30.1; Capital revenue 2.7; Total 295.2, excluding grants received (23.5). *Expenditure:* Current expenditure 291.4 (Personal emoluments 124.6, Goods and services 57.6, Interest payments 49.7, Transfers and subsidies 59.3); Capital expenditure 237.5; Total 528.9. Source: Eastern Caribbean Central Bank.

**International Reserves** (US $ million at 31 December 2004): IMF special drawing rights 0.01; Foreign exchange 121.72; Total 121.73. Source: IMF, *International Financial Statistics*.

**Money Supply** (EC $ million at 31 December 2004): Currency outside banks 102.10; Demand deposits at deposit money banks 323.26; Total money (incl. others) 425.64. Source: IMF, *International Financial Statistics*.

**Cost of Living** (Consumer Price Index; base: 2000 = 100): 103.2 in 2001; 104.3 in 2002; 106.6 in 2003. Source: ILO.

**Expenditure on the Gross Domestic Product** (preliminary, EC $ million at current prices, 2003): Government final consumption expenditure 179.38; Private final consumption expenditure 830.15; Gross fixed capital formation 505.08; *Total domestic expenditure* 1,514.61; Exports of goods and services 504.10; *Less* Imports of goods and services 837.80; *GDP in purchasers' values* 1,180.91. Source: Eastern Caribbean Central Bank.

**Gross Domestic Product by Economic Activity** (EC $ million at current prices, 2003): Agriculture, hunting, forestry and fishing 94.25; Mining and quarrying 5.71; Manufacturing 60.58; Electricity and water 59.01; Construction 104.33; Wholesale and retail trade 108.04; Restaurants and hotels 81.49; Transport and communications 189.59; Finance and insurance 115.18; Real estate 33.83; Government services 157.89; Other services 29.56; *Sub-total* 1,039.46; *Less* Imputed bank service charge 92.39; *GDP at factor cost* 947.07. Source: Eastern Caribbean Central Bank.

**Balance of Payments** (EC $ million, 2003): Exports of goods f.o.b. 112.98; Imports of goods f.o.b. –607.69; *Trade balance* –494.71; Exports of services 391.12; Imports of services –230.11; *Balance on goods and services* –333.70; Other income received 12.40; Other income paid –150.34; *Balance on goods, services and income* –471.64; Current transfers received 88.77; Current transfers paid –24.67; *Current balance* –407.54; Capital account (net) 117.16; Direct investment from abroad 217.06; Portfolio investment assets –1.20; Portfolio investment liabilities 38.18; Other investment (net) –37.95; Net errors and omissions 42.93; *Overall balance* –31.36. Source: Eastern Caribbean Central Bank.

## EXTERNAL TRADE

**Principal Commodities** (US $ million, 2002): *Imports c.i.f.:* Food 37.6 (Meat 8.5; Dairy products and birds' eggs 6.3; Cereals 6.8); Refined petroleum products 17.5 (Motor spirit 6.9; Gas oil 9.7); Chemicals 14.4; Basic manufactures 36.6 (Metal manufactures 9.4); Machinery and transport equipment 49.7 (Power-generating machinery and equipment 6.7; Telecommunication equipment 7.5; Electric machinery 7.9; Road vehicles 9.7); Miscellaneous manufactures 30.0; Total (incl. others) 198.8 (excl. unrecorded imports). *Exports f.o.b.:* Fish 3.9; Wheat flour 4.1; Cocoa beans 1.4; Nutmeg 13.6; Flour 5.0; Chemicals 3.1; Toilet paper 1.9; Machinery and transport equipment 5.5; Miscellaneous manufactures 1.9; Total (incl. others) 38.3. Source: UN, *International Trade Statistics Yearbook*.

**Principal Trading Partners** (US $ million, 2002): *Imports c.i.f.:* Barbados 5.2; Brazil 2.4; Canada 4.8; China, People's Republic 2.3; France 3.1; Germany 5.3; Guyana 2.6; Honduras 2.2; Japan 6.4; Netherlands 2.6; Trinidad and Tobago 41.6; United Kingdom 12.6; USA 86.9; Venezuela 2.2; Total (incl. others) 198.8. *Exports f.o.b.:* Antigua and Barbuda 1.2; Argentina 0.7; Barbados 1.9; Belgium 1.7; Canada 0.4; Dominica 1.0; France 1.7; Germany 2.1; Guyana 0.4; Jamaica 0.6; Netherlands 7.4; Saint Christopher and Nevis 1.4; Saint Lucia 2.9; Saint Vincent and the Grenadines 0.5; Trinidad and Tobago 1.6; United Kingdom 0.5; USA 11.2; Total (incl. others) 38.3. Source: UN, *International Trade Statistics Yearbook*.

## TRANSPORT

**Road Traffic** ('000 motor vehicles in use, 2001): Passenger cars 15.8; Commercial vehicles 4.2. Source: UN, *Statistical Yearbook*.

**Shipping:** *Merchant Fleet* (registered at 31 December 2004) 11 vessels (total displacement 2,821 grt) (Source: Lloyd's Register-Fairplay, *World Fleet Statistics*). International Sea-borne Freight Traffic (estimates, '000 metric tons, 1995): Goods loaded 21.3; Goods unloaded 193.0. *Ship Arrivals* (1991): 1,254. *Fishing Vessels* (registered, 1987): 635.

**Civil Aviation** (aircraft arrivals, 1995): 11,310.

## TOURISM

**Visitor Arrivals** ('000): *Stop-overs:* 121.1 in 2002; 133.7 in 2003; 89.9 in 2004. *Cruise-ship Passengers:* 147.4 in 2001; 135.1 in 2002; 146.9 in 2003.

**Receipts from Tourism** (estimates, US $ million): 163.0 in 2001; 169.5 in 2002; 169.5 in 2003.
Source: Caribbean Development Bank, *Social and Economic Indicators*.

## COMMUNICATIONS MEDIA

**Radio Receivers** (1997): 57,000 in use*.

**Television Receivers** (1999): 35,000 in use*.

**Telephones** (2003): 32,600 main lines in use‡.

**Facsimile Machines** (1996): 270 in use†.

**Mobile Cellular Telephones** (2003): 42,300 subscribers‡.

**Personal Computers** (2002): 14,000 in use‡.

**Internet Users** (2003): 19,500‡.

**Non-daily Newspapers** (1996): 4; circulation 14,000*.

* Source: UNESCO, *Statistical Yearbook*.
† Source: UN, *Statistical Yearbook*.
‡ Source: International Telecommunication Union.

## EDUCATION

**Pre-primary** (1994): 74 schools; 158 teachers; 3,499 pupils.
**Primary** (1995): 57 schools; 849 teachers; 17,352 pupils (2003).
**Secondary:** 20 schools (2002); 381 teachers (1995); 10,603 pupils (2003).
**Higher** (excluding figures for the Grenada Teachers' Training College, 1993): 66 teachers; 651 students.
Source: partly Caribbean Development Bank, *Social and Economic Indicators*.

**Adult Literacy Rate:** 94.4% in 2001. Source: Secretariat of the Organisation of Eastern Caribbean States.

# Directory

## The Constitution

The 1974 independence Constitution was suspended in March 1979, following the coup, and almost entirely restored between November 1983, after the overthrow of the Revolutionary Military Council, and the elections of December 1984. The main provisions of this Constitution are summarized below:

The Head of State is the British monarch, represented in Grenada by an appointed Governor-General. Legislative power is vested in the bicameral Parliament, comprising a Senate and a House of Representatives. The Senate consists of 13 Senators, seven of whom are appointed on the advice of the Prime Minister, three on the advice of the Leader of the Opposition and three on the advice of the Prime Minister after he has consulted interests which he considers Senators should be selected to represent. The Constitution does not specify the number of members of the House of Representatives, but the country consists of 15 single-member constituencies, for which representatives are elected for up to five years, on the basis of universal adult suffrage.

The Cabinet consists of a Prime Minister, who must be a member of the House of Representatives, and such other ministers as the Governor-General may appoint on the advice of the Prime Minister.

There is a Supreme Court and, in certain cases, a further appeal lies to Her Majesty in Council.

## The Government

### HEAD OF STATE

**Monarch:** HM Queen Elizabeth II (succeeded to the throne 6 February 1952).
**Governor-General:** Sir Daniel Williams (appointed 8 August 1996).

### THE CABINET
(July 2005)

**Prime Minister and Minister of National Security, Information, Business and Private-Sector Development, Youth Development and Information Communications Technology:** Dr Keith Claudius Mitchell.

**Deputy Prime Minister and Minister of Agriculture, Lands, Forestry and Fisheries, Public Utilities, Energy and the Marketing and National Importing Board:** Sen. Gregory Bowen.

**Minister of Finance and Planning:** Anthony Boatswain.

**Minister of Foreign Affairs and International Trade, of Carriacou and Petit Martinique Affairs, and of Legal Affairs, and Attorney-General:** Sen. Elvin Nimrod.

**Minister of Communications, Works and Transport:** Clarice Modeste-Curwen.

**Minister of Education and Labour:** Claris Charles.

**Minister of Sports, Community Development and Co-operatives, with responsibility for Community Development and Co-operatives, and Minister in the Ministry of Finance with responsibility for Revenue Administration:** Roland Bhola.

**Minister of Health, Social Security and the Environment:** Ann David Antoine.

**Minister of Tourism, Civil Aviation, Culture and the Performing Arts:** Sen. Brenda Hood.

**Minister of Housing, Social Services and Gender and Family Affairs:** Yolande Bain Joseph.

**Minister of State in the Ministry of Sports, Community Development and Co-operatives, with responsibility for Sports:** Adrian Mitchell.

**Minister of State in the Prime Minister's Office with responsibility for Information, Business and Private-Sector Development and Information Communications Technology:** Einstein Louison.

**Minister of State in the Prime Minister's Office with responsibility for Youth Development:** Emmalin Pierce.

### MINISTRIES

**Office of the Governor-General:** Government House, St George's; tel. 440-6639; fax 440-6688; e-mail patogg@caribsurf.com.

**Office of the Prime Minister:** Ministerial Complex, 6th Floor, Botanical Gardens, St George's; tel. 440-2255; fax 440-4116; e-mail pmoffice@gov.gd.

**Ministry of Agriculture, Lands, Forestry and Fisheries:** Ministerial Complex, 2nd and 3rd Floors, Botanical Gardens, St George's; tel. 440-2708; fax 440-4191; e-mail agriculture@gov.gd.

**Ministry of Carriacou and Petit Martinique Affairs:** Beauséjour, Carriacou; tel. 443-6026; fax 443-6040; e-mail minccoupm@caribsurf.com.

**Ministry of Communications, Works and Transport:** Ministerial Complex, 4th Floor, Botanical Gardens, St George's; tel. 440-2181; fax 440-4122; e-mail ministerworks@caribsurf.com.

**Ministry of Education and Labour:** Ministerial Complex, Botanical Gardens, Tanteen, St George's; tel. 440-2166; fax 440-6650; e-mail psmined@yahoo.com.

**Ministry of Finance and Planning:** Financial Complex, The Carenage, St George's; tel. 440-2731; fax 440-4115; e-mail director@economicaffairs.grenada.gd; internet economicaffairs.grenada.gd.

**Ministry of Foreign Affairs and International Trade:** Ministerial Complex, 4th Floor, Botanical Gardens, St George's; tel. 440-2640; fax 440-4184; e-mail foreignaffairs@gov.gd.

**Ministry of Health, Social Security and the Environment:** Ministerial Complex, 1st and 2nd Floors, Botanical Gardens, St George's; tel. 440-2649; fax 440-4127; e-mail min-healthgrenada@caribsurf.com.

**Ministry of Housing, Social Services and Gender and Family Affairs:** Ministerial Complex, 1st and 2nd Floors, Botanical Gardens, St George's; tel. 440-7994; fax 440-7990; e-mail mhousing@hotmail.com.

**Ministry of Information:** Ministerial Complex, 6th Floor, Botanical Gardens, St George's; tel. 440-2255; fax 440-4116.

**Ministry of Legal Affairs:** Church St, St George's; tel. 440-2050; fax 440-6630; e-mail legalaffairs@caribsurf.com.

**Ministry of National Security:** Ministerial Complex, Botanical Gardens, St George's; tel. 440-2255.

**Ministry of Public Utilities and Energy:** Ministerial Complex, 4th Floor, Botanical Gardens, St George's; tel. and fax 440-2181.

**Ministry of Tourism, Civil Aviation, Culture and the Performing Arts:** Ministerial Complex, 4th Floor, Botanical Gardens, St George's; tel. 440-0366; fax 440-0443; e-mail mot@caribsurf.com; internet www.spiceisle.com.users.mot.

**Ministry of Sports, Community Development and Co-operatives:** Ministerial Complex, 2nd Floor, Botanical Gardens, St George's; tel. 440-6917; fax 440-6924; e-mail yscm@caribsurf.com.

### GOVERNMENT AGENCY

**Agency for Reconstruction and Development Inc (ARD):** Steele's Office Complex, St George's; tel. 439-5606; fax 439-5609; e-mail ard@gov.gd; f. Oct. 2004 to oversee Grenada's recovery from the effects of 'Hurricane Ivan', which struck the island in Sept., and to handle donations for reconstruction projects; comprises four depts: social recovery; economic recovery; physical infrastructure; and finance and administration; Chair. Marius St Rose; CEO Richardson Andrews.

# GRENADA

## Legislature

### PARLIAMENT

**Houses of Parliament:** Church St, St George's; tel. 440-2090; fax 440-4138.

#### Senate

**President:** Sen. Dr JOHN WATTS.
There are 13 appointed members.

#### House of Representatives

**Speaker:** LAWRENCE JOSEPH.

**General Election, 27 November 2003**

|  | Votes | % | Seats |
|---|---|---|---|
| New National Party (NNP) | 22,566 | 47.8 | 8 |
| National Democratic Congress (NDC) | 21,445 | 45.4 | 7 |
| Grenada United Labour Party (GULP) | 2,243 | 4.7 | — |
| People's Labour Movement (PLM) | 933 | 2.0 | — |
| Others | 52 | 0.0 | — |
| **Total** | **47,239** | **100.0** | **15** |

## Political Organizations

**Grenada United Labour Party (GULP):** St George's; f. 1950; merged with United Labour Congress in 2001; right-wing; Pres. WILFRED HAYES; Leader GLORIA PAYNE-BANFIELD.

**National Democratic Congress (NDC):** St George's; f. 1987 by former members of the NNP and merger of Democratic Labour Congress and Grenada Democratic Labour Party; centrist; Leader TILLMAN THOMAS; Dep. Leader GEORGE PRIME.

**New National Party (NNP):** St George's; f. 1984 following merger of Grenada Democratic Movement, Grenada National Party and National Democratic Party; centrist; Chair. LAWRENCE JOSEPH; Leader Dr KEITH MITCHELL; Dep. Leader GREGORY BOWEN.

**People's Labour Movement (PLM):** St George's; f. 1995 by former members of the NDC; fmrly known as the Democratic Labour Party; Leader Dr FRANCIS ALEXIS; Pres. Dr TERRANCE MARRYSHOW.

## Diplomatic Representation

### EMBASSIES AND HIGH COMMISSION IN GRENADA

**China, People's Republic:** St George's; Ambassador SHEN HONGSHUN.

**Cuba:** L'Anse aux Epines, St George's; tel. 444-1884; fax 444-1877; e-mail embacubagranada@caribsurf.com; Ambassador HUMBERTO RIVERO ROSARIO.

**United Kingdom:** British High Commission, Netherlands Bldg, Grand Anse, St George's; tel. 440-3536; fax 440-4939; e-mail bhcgrenada@caribsurf.com; High Commissioner JOHN WHITE (resident in Barbados).

**USA:** POB 54, St George's; tel. 444-1173; fax 444-4820; e-mail usemb_gd@caribsurf.com; internet www.spiceisle.com/homepages/usemb_gd; Ambassador MARY ELIZABETH KRAMER (resident in Barbados).

**Venezuela:** Upper Lucas St, POB 201, St George's; tel. 440-1721; fax 440-6657; e-mail embavengda@caribsurf.com; Ambassador EDNA FIGUERA CEDEÑO.

## Judicial System

Justice is administered by the West Indies Associated States Supreme Court, composed of a High Court of Justice and a Court of Appeal. The Itinerant Court of Appeal consists of three judges and sits three times a year; it hears appeals from the High Court and the Magistrates' Court. The Magistrates' Court administers summary jurisdiction.

In 1988 the OECS excluded the possibility of Grenada's readmittance to the East Caribbean court system until after the conclusion of appeals by the defendants in the Maurice Bishop murder trial (see History). Following the conclusion of the case in 1991, Parliament voted to rejoin the system, thus also restoring the right of appeal to the Privy Council in the United Kingdom.

**Attorney-General:** Sen. ELVIN NIMROD.

**Puisne Judges:** DENYS BARROW, KENNETH BENJAMIN, LYLE K. ST PAUL.

**Registrar of the Supreme Court:** ROBERT BRANCH.

**President of the Court of Appeal:** C. M. DENNIS BYRON.

**Office of the Attorney-General:** Church St, St George's; tel. 440-2050; fax 440-6630; e-mail legalaffairs@caribsurf.com.

## Religion

### CHRISTIANITY

#### The Roman Catholic Church

Grenada comprises a single diocese, suffragan to the archdiocese of Castries (Saint Lucia). The Bishop participates in the Antilles Episcopal Conference (based in Port of Spain, Trinidad and Tobago). At 31 December 2003 there were an estimated 55,888 adherents in the diocese.

**Bishop of St George's in Grenada:** Rev. VINCENT DARIUS, Bishop's House, Morne Jaloux, POB 375, St George's; tel. 443-5299; fax 443-5758; e-mail bishopgrenada@caribsurf.com.

#### The Anglican Communion

Anglicans in Grenada are adherents of the Church in the Province of the West Indies, and represented 14% of the population at the time of the 1991 census. The country forms part of the diocese of the Windward Islands (the Bishop, the Rt Rev. SEHON GOODRIDGE, resides in Kingstown, Saint Vincent).

#### Other Christian Churches

The Presbyterian, Methodist, Plymouth Brethren, Baptist, Salvation Army, Jehovah's Witness, Pentecostal (7.2% of the population in 1991) and Seventh-day Adventist (8.5%) faiths are also represented.

## The Press

### NEWSPAPERS

**Barnacle:** Frequente Industrial Park, St George's; tel. 440-5151; monthly; Editor IAN GEORGE.

**The Grenada Informer:** Market Hill, POB 622, St George's; tel. 440-1530; fax 440-4119; e-mail movanget@caribsurf.com; f. 1985; weekly; Editor CARLA BRIGGS; circ. 6,000.

**Grenada Today:** St John's St, POB 142, St George's; tel. 440-4401; internet www.belgrafix.com/gtoday98.htm; weekly; Editor GEORGE WORME.

**The Grenadian Voice:** Frequente Industrial Park, Bldg 1B, Maurice Bishop Highway, POB 633, St George's; tel. 440-1498; fax 440-4117; e-mail gvoice@caribsurf.com; internet www.grenadianvoice.com; weekly; Editor LESLIE PIERRE; circ. 3,000.

**Government Gazette:** St George's; weekly; official.

### PRESS ASSOCIATION

**Press Association of Grenada:** St George's; f. 1986; Pres. LESLIE PIERRE.

Inter Press Service (IPS) (Italy) is also represented.

## Publisher

**Anansi Publications:** Hillsborough St, St George's; tel. 440-0800; e-mail aclouden@caribsurf.com.

## Broadcasting and Communications

### TELECOMMUNICATIONS

#### Regulatory Authority

**Eastern Caribbean Telecommunications Authority:** Castries, Saint Lucia; f. 2000 to regulate telecommunications in Grenada, Dominica, Saint Christopher and Nevis, Saint Lucia and Saint Vincent and the Grenadines.

# GRENADA

*Directory*

**National Telecommunications Regulatory Commission:** Suite 8, Grand Anse Shopping Centre, POB 854, St George's; tel. 435-6872; fax 435-2132; e-mail gntrc@caribsurf.com; internet www.spiceisle.com/gntrc; Chair. LINUS SPENCER THOMAS; Dir of Telecommunications ROBERT OLIVER FINLAY.

### Major Service Providers

**Cable & Wireless Grenada Ltd:** POB 119, The Carenage, St George's; tel. 440-1000; fax 440-4134; e-mail gndinfo@caribsurf.com; internet www.candw.gd; f. 1989; until 1998 known as Grenada Telecommunications Ltd (Grentel); 30% govt-owned; Chief Exec. IAN BLANCHARD.

**Cingular Wireless Grenada:** St George's; internet www.cingularwireless.com; joint venture between SBC Communications and BellSouth; scheduled to sell its operations and licences in the Caribbean to Digicel in 2005; Pres. and CEO STANLEY T. SIGMAN.

**Digicel Grenada Ltd:** Point Salines, POB 1690, St George's; e-mail grecustcare@digicelgroup.com; internet www.digicelgrenada.com; tel. 439-4463; fax 439-4464; f. 2003; began operating cellular telephone services in Oct. 2003; owned by an Irish consortium; Chair. DENIS O'BRIEN.

**Grenada Postal Corporation:** Burns Point, St George's; tel. 440-2526; fax 440-4271; e-mail gpc@caribsurf.com; internet gndonline.com; Chair. GORDON ROBERTSON; Man. LEO ROBERTS.

## BROADCASTING

**Grenada Broadcasting Network (GBN):** Observatory Rd, POB 535, St George's; tel. 440-2446; fax 440-4180; e-mail gbn@caribsurf.com; internet www2.spiceisle.com; f. 1972; 60% privately owned, 40% govt-owned; Chair. KEN GORDON; Man. Dir RICHARD PURCELL.

### Radio

**Grenada Broadcasting Network (Radio):** see Broadcasting.

**The Harbour Light of the Windwards:** Carriacou; tel. and fax 443-7628; e-mail harbourlight@caribsurf.com; internet www.harbourlightradio.org; Station Man. RANDY CORNELIUS; Chief Engineer JOHN MCPHERSON.

**Spice Capitol Radio FM 90:** Springs, St George's; tel. 440-0162.

### Television

Television programmes from Trinidad and from Barbados can be received on the island.

**Grenada Broadcasting Network (Television):** see Broadcasting.

## Finance

(cap. = capital; res = reserves; dep. = deposits; amounts in Eastern Caribbean dollars)

The Eastern Caribbean Central Bank, based in Saint Christopher, is the central issuing and monetary authority for Grenada.

**Eastern Caribbean Central Bank—Grenada Office:** Monckton St, St George's; tel. 440-3016; fax 440-6721; e-mail eccbgnd@caribsurf.com; Country Dir TIMOTHY ANTOINE.

### BANKING

#### Regulatory Authority

**Grenada International Financial Services Authority:** Bldg 5, Financial Complex, The Carenage, St George's; tel. 440-6575; fax 440-4780; e-mail gifsa@caribsurf.com; internet gifsa-grenada.com; f. 1999; revenues 4.2m. (2002); Chair. TIMOTHY ANTOINE; Exec. Dir SHARON GRIFFITH.

#### Commercial Banks

**Grenada Co-operative Bank Ltd:** 8 Church St, POB 135, St George's; tel. 440-2111; fax 440-6600; e-mail co-opbank@caribsurf.com; f. 1932; Man. Dir and Sec. G. V. STEELE; brs in St Andrew's, St George's and St Patrick's.

**Grenada Development Bank:** Melville St, POB 734, St George's; tel. 440-2382; fax 440-6610; e-mail gdbbank@caribsurf.com; f. 1976 following merger; Chair. ARNOLD CRUICKSHANK; Man. CAMPBELL BLENMAN.

**National Commercial Bank of Grenada Ltd:** NCB House, POB 857, Grand Anse, St George's; tel. 444-2265; fax 444-5500; e-mail ncbgnd@caribsurf.com; internet www.ncbgrenada.com; f. 1979; 51% owned by Republic Bank Ltd, Port of Spain, Trinidad and Tobago; cap. 15.0m., res 29.4m., dep. 487.6m. (Sept. 2003); Chair. RONALD HARFORD; Man. Dir DANIEL ROBERTS; 9 brs.

**RBTT Bank Grenada Ltd:** Corner of Cross and Halifax Sts, POB 4, St George's; tel. 440-3521; fax 440-4153; e-mail gbcltd@caribsurf.com; internet www.rbtt.com; f. 1983; 10% govt-owned; cap. 7.4m., res 8.8m., dep. 256.6m. (Dec. 2000); Chair. PETER JULY; Man. Dir MAXIM PAZOS.

#### Foreign Banks

**Bank of Nova Scotia** (Canada): Granby and Halifax Sts, POB 194, St George's; tel. 440-3274; fax 440-4173; Man. B. ROBINSON; 3 brs.

**FirstCaribbean International Bank (Barbados) Ltd:** Church St, St George's; tel. 440-3232; internet www.firstcaribbeanbank.com; f. 2003; 87.5% owned by Barclays Bank PLC (United Kingdom) and Canadian Imperial Bank of Commerce; CEO CHARLES PINK; Exec. Dir SHARON BROWN; 4 brs.

### STOCK EXCHANGE

**Eastern Caribbean Securities Exchange:** based in Basseterre, Saint Christopher and Nevis; e-mail info@ecseonline.com; internet www.ecseonline.com; f. 2001; regional securities market designed to facilitate the buying and selling of financial products for the eight member territories—Anguilla, Antigua and Barbuda, Dominica, Grenada, Montserrat, Saint Christopher and Nevis, Saint Lucia and Saint Vincent and the Grenadines; Gen. Man. TREVOR BLAKE.

### INSURANCE

Several foreign insurance companies operate in Grenada and the other islands of the group. Principal locally owned companies include the following:

**Grenada Insurance and Finance Co Ltd:** Young St, POB 139, St George's; tel. 440-3004.

**Grenada Motor and General Insurance Co Ltd:** Scott St, St George's; tel. 440-3379.

**Grenadian General Insurance Co Ltd:** Cnr of Young and Scott Sts, POB 47, St George's; tel. 440-2434; fax 440-6618.

## Trade and Industry

### CHAMBERS OF COMMERCE

**Grenada Chamber of Industry and Commerce, Inc:** DeCaul Bldg, Mt Gay, POB 129, St George's; tel. 440-2937; fax 440-6627; e-mail gcic@caribsurf.com; internet www.spiceisle.com/homepages/gcic; f. 1921; inc 1947; 170 mems; Pres. AZAM RAHAMAN; Exec. Dir CHRISTOPHER DERIGGS.

**Grenada Manufacturing Council:** POB 129, St George's; tel. 440-2937; fax 440-6627; e-mail gcic@caribsurf.com; f. 1991 to replace Grenada Manufacturers' Asscn; Chair. CHRISTOPHER DEALLIE.

### INDUSTRIAL AND TRADE ASSOCIATIONS

**Grenada Cocoa Association:** Scott St, St George's; tel. 440-2234; fax 440-1470; e-mail gca@caribsurf.com; f. 1987 following merger; changed from co-operative to shareholding structure in late 1996; Chair. REGINALD BUCKMIRE; Man. ANDREW HASTICK.

**Grenada Co-operative Banana Society:** Scott St, St George's; tel. 440-2486; fax 440-4199; e-mail gbcs@caribsurf.com; f. 1955; a statutory body to control production and marketing of bananas; Exec. Chair. DANIEL LEWIS; Gen. Man. JOHN MARK (acting).

**Grenada Co-operative Nutmeg Association:** Lagoon Rd, POB 160, St George's; tel. 440-2117; fax 440-6602; e-mail gcna.nutmeg@caribsurf.com; internet www.grenadanutmeg.com; f. 1947; processes and markets all the nutmeg and mace grown on the island; to include the production of nutmeg oil; Chair. RAMSEY RUSH; Gen. Man. TERENCE MOORE.

**Grenada Industrial Development Corporation:** Frequente Industrial Park, Frequente, St George's; tel. 444-1035; fax 444-4828; e-mail gidc@caribsurf.com; internet www.grenadaworld.com; f. 1985; Chair. R. ANTHONY JOSEPH; Man. SONIA RODEN.

**Marketing and National Importing Board:** Young St, St George's; tel. 440-1791; fax 440-4152; e-mail mnib.com@caribsurf.com; f. 1974; govt-owned; imports basic food items, incl. sugar, rice and milk; Chair. RICHARDSON ANDREWS; Gen. Man. FITZROY JAMES.

### EMPLOYERS' ORGANIZATION

**Grenada Employers' Federation:** Mt Gay, POB 129, St George's; tel. 440-1832; e-mail gef@caribsurf.com; 60 mems.

There are several marketing and trading co-operatives, mainly in the agricultural sector.

## UTILITIES

**Public Utilities Commission:** St George's.

### Electricity

**Grenada Electricity Services Ltd (Grenlec):** Halifax St, POB 381, St George's; tel. 440-2097; fax 440-4106; e-mail grenlec@caribsurf.com; internet www.carilec.org/grenlec; generation and distribution; 90% privately owned, 10% govt-owned; Chair. G. ROBERT BLANCHARD, Jr; Gen. Man. VERNON LAWRENCE.

### Water

**National Water and Sewerage Authority:** The Carenage, POB 392, St George's; tel. 440-2155; fax 440-4107; f. 1969; Chair. MICHAEL PIERRE; Man. ALLEN MCQUIRE.

## TRADE UNIONS

**Grenada Trade Union Council (GTUC):** Green St, POB 411, St George's; tel. 440-3733; fax 440-3733; e-mail gtuc@caribsurf.com; Pres. DEREK ALLARD; Gen. Sec. RAY ROBERTS.

**Bank and General Workers' Union (BGWU):** Bain's Alley, St George's; tel. 440-3563; fax 440-0778; e-mail bgwu@caribsurf.com; Pres. DEREK ALLARD; Gen. Sec. JUSTIN CAMPBELL.

**Commercial and Industrial Workers' Union:** Bain's Alley, POB 191, St George's; tel. 440-3423; fax 440-3423; Pres. ELLIOT BISHOP; Gen. Sec. BARBARA FRASER; 492 mems.

**Grenada Manual, Maritime and Intellectual Workers' Union (GMMIWU):** c/o Birchgrove, St Andrew's; tel. 442-7724; fax 442-7724; Pres. BERT LATOUCHE; Gen. Sec. OSCAR WILLIAMS.

**Grenada Union of Teachers (GUT):** Marine Villa, POB 452, St George's; tel. 440-2992; fax 440-9019; f. 1913; Pres. MARVIN ANDALL; Gen. Sec. ELAINE MCQUEEN; 1,300 mems.

**Media Workers' Association of Grenada (MWAG):** St George's; f. 1999; Gen. Sec. RAE ROBERTS.

**Public Workers' Union (PWU):** Tanteen, POB 420, St George's; tel. 440-2203; fax 440-6615; e-mail pwu-cpsa@caribsurf.com; f. 1931; Pres. LAURETTE CLARKSON; Gen. Sec. SHIRLEY MODESTE.

**Seamen and Waterfront Workers' Union:** Ottway House, POB 154, St George's; tel. 440-2573; fax 440-7199; e-mail swwu@caribsurf.com; f. 1952; Pres. ALBERT JULIEN; Gen. Sec. LYLE SAMUEL; 350 mems.

**Technical and Allied Workers' Union (TAWU):** Green St, POB 405, St George's; tel. 440-2231; fax 440-5878; f. 1958; Pres.-Gen. Sen. CHESTER HUMPHREY; Gen. Sec. ANDRÉ LEWIS.

# Transport

## RAILWAYS

There are no railways in Grenada.

## ROADS

In 1999 there were approximately 1,040 km (646 miles) of roads, of which 61.3% were paved. Public transport is provided by small private operators, with a system covering the entire country.

## SHIPPING

The main port is St George's, with accommodation for two ocean-going vessels of up to 500 ft. A number of shipping lines call at St George's. Grenville, on Grenada, and Hillsborough, on Carriacou, are used mostly by small craft. The first phase of a project to expand the port at St George's and enable the harbour to accommodate modern super-sized cruise ships was completed in July 2003.

**Grenada Ports Authority:** POB 494, The Carenage, St George's; tel. 440-7678; fax 440-3418; e-mail grenport@caribsurf.com; internet www.grenadaports.com; Chair. WALTER ST JOHN; Gen. Man. AMBROSE PHILLIP.

## CIVIL AVIATION

The Point Salines International Airport, 10 km (6 miles) from St George's, was opened in October 1984, and has scheduled flights to most East Caribbean destinations, including Venezuela, and to the United Kingdom and North America. There is an airfield at Pearls, 30 km (18 miles) from St George's, and Lauriston Airport, on the island of Carriacou, offers regular scheduled services to Grenada, Saint Vincent and Palm Island (Grenadines of Saint Vincent).

Grenada is a shareholder in the regional airline, LIAT (Antigua and Barbuda).

**Grenada Airports Authority:** Point Salines Int. Airport, POB 385, St George's; tel. 444-4101; fax 444-4838; e-mail gaa@caribsurf.com; f. 1985; Chair. RICHARD MENEZES; CEO PHILIPPE BARIL; Gen. Man. DONALD MCPHAIL.

**Airlines of Carriacou:** Point Salines Int. Airport, POB 805, St Georges; tel. 444-1475; fax 444-2898; e-mail cayar@caribsurf.com; internet www.travelgrenada.com/aircarriacou.htm; f. 1992; national airline, operates in association with LIAT; Man. Dir ARTHUR W. BAIN.

# Tourism

Grenada has the attractions of both white sandy beaches and a scenic, mountainous interior with an extensive rainforest. There are also sites of historical interest, and the capital, St George's, is a noted beauty spot. In 2003 there were an estimated 133,700 stop-over arrivals and 146,900 cruise-ship passengers; in that year tourism earned some US $169.5m. There were approximately 1,670 hotel rooms in 1996. In 1997 a joint venture between the Government of Grenada and the Caribbean Development Bank to upgrade and market some 50 unprofitable hotels was implemented. In addition, the Ministry of Tourism, in conjunction with the Board of Tourism and the Hotel Association, announced a 10-year development plan to increase hotel capacity to 2,500 rooms.

**Grenada Board of Tourism:** Burns Point, POB 293, St George's; tel. 440-2279; fax 440-6637; e-mail gbt@caribsurf.com; internet www.grenada.org; f. 1991; Chair. RICHARD STRACHAN; Dir WILLIAM JOSEPH.

**Grenada Hotel and Tourism Association Ltd:** POB 440, St George's; tel. 444-1353; fax 444-4847; e-mail grenhota@caribsurf.com; internet www.grenadahotelsinfo.com; f. 1961; Pres. IAN DA BREO; Dirs SHEREE ANN ADAMS, LAWRENCE LAMBERT, ROYSTON HOPKIN, RUSS FIELDEN, CLIVE BARNES, COLEMAN REDHEAD, MIRIAM BEDEAU, ANN BAYNE GRIFFITH, VALENTINE FRASER, CHRISTOPHER MILLS.

# Defence

A regional security unit was formed in 1983, modelled on the British police force and trained by British officers. A paramilitary element, known as the Special Service Unit and trained by US advisers, acts as the defence contingent and participates in the Regional Security System, a defence pact with other East Caribbean states.

**Commissioner of Police:** FITZROY BEDEAU.

# Education

Education is free and compulsory for children between the ages of five and 16 years. Primary education begins at five years of age and lasts for seven years. Secondary education, beginning at the age of 12, lasts for a further five years. In 2001 a total of 19,134 children received public primary education; there were 57 primary schools in 1995. There were 20 public secondary schools in 2002, with 9,891 pupils registered, in 2001. Technical Centres have been established in St Patrick's, St David's and St John's, and the Grenada National College, the Mirabeau Agricultural School and the Teachers' Training College have been incorporated into the Technical and Vocational Institute in St George's. The Extra-Mural Department of the University of the West Indies has a branch in St George's. A School of Medicine has been established at St George's University (SGU), where a School of Arts and Sciences was also founded in 1997, while there is a School of Fishing at Victoria. In May 2003 Grenada successfully completed negotiations with the World Bank and the Organisation of Eastern Caribbean States for an EC $11.3m. loan and credit agreement for educational development. Total budgetary expenditure on education was $52.5m. in 2001 (equivalent to 12.5% of total expenditure).

# Bibliography

For works on the Caribbean generally, see Select Bibliography (Books)

Beck, R. J. *The Grenada Invasion*. Boulder, CO, Westview Press, 1993.

Brizan, G. I. *Grenada: Island of Conflict*. Grand Cayman, Caribbean Publishing, 1998.

Ferguson, J. *Grenada: Revolution in Reverse*. New York, NY, Monthly Review Press, 1990.

Heine, J. (Ed.). *A Revolution Aborted*. Pittsburgh, PA, University of Pittsburgh Press, 1991.

Pryor, F. L. *Revolutionary Grenada: A Study in Political Economy*. Westport, CT, Praeger Publrs, 1986.

Smith, C. A. *Socialist Transformation in Peripheral Economies*. Brookfield, VT, Avebury Publishing Co, 1995.

Steele, B. A. *Grenada: A History of its People*. Oxford, Macmillan Caribbean, 2003.

# GUADELOUPE
## Geography

### PHYSICAL FEATURES

The Overseas Department of Guadeloupe is an integral part of the French Republic, but lies in the Lesser Antilles. It includes not only the main islands of Guadeloupe itself (Basse-Terre and Grande-Terre) and a number of surrounding islands, which lie in the Windward Islands, but also St-Barthélemy and the northern part of the island of St-Martin (Sint Maarten), in the Leewards. All these islands lie in the north-eastern Caribbean. Guadeloupe itself is flanked by two sea lanes from the Atlantic: the Dominica Passage to the south, beyond which lies the Commonwealth of Dominica; and the wider Guadeloupe Passage to the north, beyond which, some 64 km (40 miles) to the north-west, is the southernmost of the Leeward Islands, the British dependency of Montserrat. A similar distance directly north is Antigua and Barbuda, while about one-half that distance south of Marie-Galante is the island of Dominica. The Leeward dependencies of Guadeloupe are to the north-west of the main islands, tiny St-Barthélemy about 230 km distant and St-Martin a further 35 km. In terms of international neighbours, St-Martin is only 8 km south of the British dependency of Anguilla, while St-Barthélemy is some distance to the north of the island of St Christopher (Kitts—in Saint Christopher and Nevis). However, the closest neighbour to St-Martin is the Netherlands Antilles—Saba and St Eustatius lie to the south, but, more importantly, the island is actually shared with the Dutch, who call it St Maarten. The 10.2-km border is the only land frontier in the Lesser Antilles and makes St-Martin the smallest island in the world to be divided. The French occupy the north, about 60% of the island (52 sq km or 20 sq miles). The entire territory of the Department covers 1,705 sq km, including 74 sq km of inland waters.

Guadeloupe accounts for the greatest part of the territory of the Department. The butterfly-shaped landmass is, technically, two islands, separated by a narrow channel, the Rivière Sallée. Guadeloupe originally applied only to the larger, mountainous western island, which is now known as Basse-Terre (like the first settlement and the administrative headquarters), after the only 'low shore', on the leeward side of the cliff-edged island. The broad, flat lands of the east, Grande-Terre, soon assumed commercial and agricultural significance and the name of Guadeloupe came to be applied to both 'wings'. Basse-Terre, the largest single island of Guadeloupe (848 sq km), is fertile, mountainous and volcanic, the densely forested central range reaching the highest point in the Department, at the desolate volcano of La Soufrière (1,354 m or 4,444 ft), in the south of the island. About two-thirds of the way up the eastern coast of Basse-Terre a broad peninsula reaches out to its sister island. The lower-lying Grande-Terre (about three-fifths the size of Basse-Terre), essentially a limestone plateau, extends mainly northwards, tapering, from a west–east base between the near-isthmus to Basse-Terre and the eastward-pointing Pointe des Châteaux. There are some hills in the south and mangrove swamps along the west coast, but much of the land is given to sugar cane, fruit trees and livestock. The wealth of vegetation throughout Guadeloupe includes native and widely protected forests, but indigenous wildlife has largely been wiped out—a few racoons (*ratons laveurs*) and iguanas survive on the offshore islands. The only other island of any size in the Department is 22 km south of Grande-Terre—the flat, round Marie-Galante (158 sq km). Just east of Grande-Terre is La Désirade, a rather dry island, to the south-west of which, mid-way to Marie-Galante, are the Îles de la Petite-Terre. West of Marie-Galante and just south of Basse-Terre are the Îles des Saintes (near which an important naval battle took place in 1782). These are the main offshore islands of Guadeloupe and most are hilly, though drier than the mainland. In the even drier Leewards, St-Barthélemy (St Barts) occupies only 21 sq km, but has green-clad volcanic hillsides, as well as white beaches and surrounding reefs and islets, while St-Martin

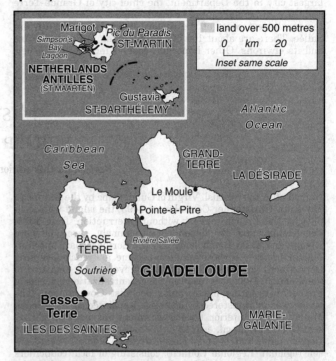

is hilly in the east, notably in the northern, French section, but low-lying in the south-west. In all, apart from Guadeloupe itself, seven islands are inhabited.

### CLIMATE

The climate is subtropical, tempered by trade winds. All the islands are in the potential path of hurricanes, and Guadeloupe itself is fairly humid. The average temperatures on the coast of the main islands range between 22°C and 30°C (72°F–86°F), but it can be about 3°C cooler inland, particularly in the mountains. Temperatures are warmer in the Leewards, but both St-Martin and St-Barthélemy normally receive about 1,100 mm (43 ins) of rain annually, which, as for all the outer islands, is less than on Guadeloupe. Most of this rain falls in September–November, although the wet season (*hivernage*) is reckoned to begin in July. The dry season (*carême*) falls in the cooler months of January–April.

### POPULATION

Around 90% of the population is black or of mixed race (mulatto), with 5% whites and smaller communities of 'East' Indians, Chinese and Lebanese, similar to many places in the West Indies. However, Les Saintes and St-Barthélemy, with permanent populations of little more than 2,000 and 5,000, respectively, are predominantly white, the former inhabited by fisherfolk of Breton descent and the latter by people of Breton, Norman and Poitevin descent. There are fewer descendants of the Swedish, who ruled St-Barthélemy for almost one century (until a referendum in 1878). Some 87% of the population is Roman Catholic, with only 1% Protestant, and the rest being Hindu or espousing pagan African religions or syncretic versions thereof. The official language is French, which most people speak, but many people generally use the Creole patois that employs West African grammatical structures with a predominantly French-derived vocabulary. On St-Barthélemy a Norman dialect of French is still sometimes in use, and English is also widely understood there. On St-Martin, with its immediate proximity to tourist-dominated St Maarten and its location, like St Barts, in the predominantly anglophone Leeward Islands, English is the language of common use.

The total population was estimated at 445,000 at mid-2004. Most of the population lives on Guadeloupe, with Grande-Terre being the more densely populated. The administrative capital, Basse-Terre, is not a large city—it is on the western coast, at the southern end, of the western 'wing' of the main island. The largest city in the Department, the chief city of Grande-Terre and the commercial capital of Guadeloupe is Pointe-à-Pitre, on the south-west coast of the eastern 'wing', near the Rivière Sallée. There are about 100,000 people living in Pointe-à-Pitre and its environs. Other important centres include: Le Gosier, just to the south-east of Pointe-à-Pitre, and Les Abymes, inland, just to the north-east; La Moule, a former capital, on Grande-Terre, but on the east coast; and Capesterre-Belle-Eau, on the south-east coast of Basse-Terre. Marie-Galante, with a total population of around 18,000, has three main settlements, the largest being Grand-Bourg, on the south-west coast. The main town on St-Barthélemy is Gustavia, its main port, in the south-west. St-Martin, with a population of about 30,000 (the largest of the offshore islands of the Department), has as its chief town Marigot, in the south-west of the French sector, on the north coast of the island, between the sea and the Simpson's Bay Lagoon.

# History

## PHILLIP WEARNE

Revised for this edition by Dr JAMES FERGUSON

Named after the Spanish Virgin of Guadeloupe by the European navigator Christopher Columbus in 1493, the island was occupied by the French, almost without interruption, from 1635. Guadeloupe and its dependencies—namely three offshore islands and Saint-Barthélemy and the northern part of Saint-Martin in the Leeward Islands—became a French Overseas Department in March 1946, but achieved some measure of autonomy in 1983 as a result of the decentralization reforms of President François Mitterrand's Socialist Government. The island was administered by a prefect, appointed by the French Ministry of the Interior. Local government was made up of a 42-seat General Council, elected for a six-year term to control the Department's budget and domestic taxation, and a 41-seat Conseil régional (Regional Council), consisting of local councillors and the two senators and four deputies Guadeloupe sends to the French Assemblée nationale.

However, progress towards greater autonomy did not prevent an upsurge of nationalist sentiment during the 1980s, witnessed by the formation of several separatist groups, making Guadeloupe the most politicized of France's Caribbean possessions. These groups comprised two broad categories: those that claimed responsibility for a series of bomb attacks on government offices and economic targets such as hotels and restaurants; and those that campaigned by lawful means. Among the former, the most active were the Alliance Révolutionnaire Caraïbe (ARC) and, during 1983 and 1984, the Groupe Libération Armée. In 1984 the ARC merged with the Mouvement Populaire pour une Guadeloupe Indépendante (MPGI), but continued its bombing campaign, using the MPGI as a legitimate 'cover' for its activities. In 1984 the leader of the ARC, Luc Reinette, was sentenced to 12 years in prison for possession of arms and conspiracy to carry out more than 20 bomb attacks in 1983–85. He escaped in June 1985, but was recaptured in July 1987 and held in Paris, where he was due to go on trial with 13 other Guadeloupe separatists. In June 1989, however, after demonstrations demanding the release of political prisoners and after 'hunger strikes' by separatist activists, among them Reinette, the Assemblée nationale granted an amnesty to those imprisoned or accused in connection with politically motivated crimes committed before July 1988 in the Overseas Departments.

The amnesty, the failure of the independence groups at the polls and the Mitterrand administration's willingness to cede more autonomy to Guadeloupe's democratic left, in firm control of the island's Conseil général (General Council) by 1988, formed the basis for a new political *modus vivendi* in Guadeloupe. The bombers, temporarily at least, changed tactics. Pressure from trade unionists, the support base of the socialist movement on the island, who attributed a further decline in the island's economy to the bombing campaign, seems to have played a crucial role.

As in other French possessions (such as New Caledonia, in the Pacific), nationalist pressure had a racial element. The original white planters represented only 5% of the population, yet they owned more than 80% of all land and property until 1945. The position of this group was consolidated, after 1946, by the arrival of French government officials and professionals. They not only altered the racial balance in favour of the whites, at a time when black national consciousness was growing, but also effectively blocked the advancement of the most able blacks and people of mixed descent by monopolizing the most desirable jobs. The problems of change in the racial balance of the island were compounded by the influx of up to 45,000 illegal immigrants from Dominica, Haiti and other neighbouring islands, attracted by the relatively high standard of living that Guadeloupeans enjoyed, and by work opportunities in the agricultural sector. Many native islanders preferred to emigrate or to seek work in the services sector rather than accept wages below the legal minimum on the land. However, salary levels remained high by Caribbean standards, being linked to those of metropolitan France. Emigration, meanwhile, continued at a rate of nearly 20,000 per year, with the result that perhaps 40% of all Guadeloupeans lived abroad by the early 1990s. Their remittances to relatives at home became a vital part of the island's economic structure.

The belief that the separatists had an influence out of all proportion to their numbers was borne out by the French presidential and parliamentary elections, which, along with local elections, took place in 1988. In all three gauges of public opinion, it was the moderate left—the Socialists—which made gains, although the independence movement claimed that the increased rate of abstention reflected growing support for their cause. For the presidency of the Republic, François Mitterrand gained 69.4% of the votes in Guadeloupe in the second round of voting in May 1988. At elections to the Assemblée nationale, held in June 1988, the Socialists made one gain, when Dominique Larifla, President of the Conseil général, defeated the sitting deputy of the conservative Union du Rassemblement du Centre (URC), a local alliance between the main national parties of the right, the Rassemblement pour la République (RPR) and the Union des Démocrates pour la République. One Socialist and one Communist were re-elected, as was Lucette Michaux-Chévry of the URC. In the elections for the Conseil général itself, the Socialist–Communist coalition increased its total by one seat, giving it 26 seats, against the 16 of the right-wing parties. Dominique Larifla was re-elected President of the Conseil général.

In March 1992 Dominique Larifla was returned as President of the Conseil général, but the Socialists suffered some reversal in the elections to the Conseil régional, held simultaneously. The right-wing grouping Objectif Guadeloupe, led by Michaux-Chévry, gained 15 of the 41 seats; two Socialist groups gained 16 between them, the Communists eight seats and the pro-independence Union Populaire pour la Libération de la Guadeloupe two seats. A split on the Socialist side meant that Michaux-Chévry, with the support of six dissident left-wingers, was able to oust the Socialist Félix Proto from the presidency of the Conseil régional. The decline of the left-wing was consolidated when a repeat of the election to the Conseil régional was required after a complaint was upheld that one party's candidates had been registered after the official deadline. Objectif Guadeloupe took a further seven seats to push its tally up to 22, while the two left-wing parties retained only 10.

At elections to the French Assemblée nationale held in March 1993 the local wing of the national Parti Socialiste (PS) lost more ground when an independent right-wing candidate, Edouard Chammougon, defeated Larifla by just 273 votes. Chammougon, mayor of Baie-Mahault, was elected despite being implicated in several corruption scandals, which involved many leading figures in the Department during the early 1990s. However, further corruption charges and allegations of abuse and misappropriation of public funds led to Chammougon's membership of the Assemblée nationale being revoked by the French Constitutional Council in November 1994, at which time Léo Andy of the 'dissident' PS was elected to his seat. In 1993 Lucette Michaux-Chévry retained her seat at the Assemblée with 79.9% of the poll, while an anti-Larifla Socialist, Frédéric Jalton, and a dissident Communist, Ernest Moutoussamy, retained the other two seats. However, the Socialist–Communist coalition retained control of the Conseil général in March 1994 and Larifla was re-elected its President. At elections to the European Parliament in June left-wing parties won the largest share of the vote, a combined 37.2%.

At the 1995 national presidential election the PS candidate, Lionel Jospin, defeated Jacques Chirac of the RPR in the second round of voting on 7 May. At municipal elections held in the following month Michaux-Chévry defeated the Communist incumbent to become mayor of the capital, Basse-Terre. She and Larifla were elected to the French Sénat in September and Philippe Chaulet of the RPR was subsequently elected to her seat in the Assemblée nationale. Jean Barfleur became Guadeloupe's first pro-independence mayor when he won Port-Louis. The defeat of the Communist candidate, Henri Bangou, was attributed to continuing divisions within the left.

At elections to the Assemblée nationale in late May and early June 1997, Moutoussamy, Andy and Chaulet all retained their seats, while Daniel Marsin, a candidate of the independent left, was elected in the Les Abymes–Pointe-à-Pitre constituency formerly held by the deceased Frédéric Jalton. In March 1998 the RPR won 25 of the 41 seats in the Conseil régional, although the PS increased its representation to 12 seats. Michaux-Chévry was re-elected to the presidency of the Conseil. In concurrent elections to the Conseil général the Socialist–Communist coalition retained a 28-seat majority, although the RPR increased its representation to eight seats. Larifla was deposed from the Conseil's presidency and replaced by Marcellin Lubeth of the Parti Progressiste Démocratique Guadeloupéen (PPDG), which had been formed in 1991, following a split in the Parti Communiste Guadeloupéen (PCG).

Guadeloupe experienced a wave of political and social unrest in 1999 surrounding the two-day visit of Prime Minister Jospin in October. In September riots broke out in Pointe-à-Pitre following the arrest and sentencing of Armand Toto, a leading member of the Union Générale des Travailleurs de la Guadeloupe (UGTG), who was accused of assaulting two policemen and threatening to kill another, while occupying the premises of a motor-vehicle company in support of a dismissed worker. Moreover, industrial action by banana producers, concerned at falling banana prices in the European market, also created widespread disruption. The workers demanded the disbursement of 100m. French francs, and additional assistance for the restructuring of their businesses, as compensation for a significant decline in banana prices. Jospin later announced that his Government would introduce an emergency plan for the banana sector.

The issue of Guadeloupe's constitutional status also arose in 1999, following a series of meetings between the Presidents of the Conseil régionaux of Guadeloupe, Martinique and French Guiana. In December Michaux-Chévry co-signed a declaration, stating the intention of the three Presidents to propose, to the French Government, legislative and constitutional amendments aimed at creating a new status of 'overseas region' and providing for greater financial autonomy. The declaration and subsequent announcements by Michaux-Chévry and her counterparts were dismissed by Jean-Jack Queyranne, the French Secretary of State for Overseas Departments and Territories, in February 2000, as unconstitutional and exceeding the mandate of the politicians responsible. However, in May 2000 a number of proposals, including the extension of the Departments' powers in areas such as regional co-operation, were provisionally accepted by the Assemblée nationale; a modified version of the proposals was subsequently adopted, by a narrow margin, by the Sénat. In November the Assemblée nationale approved the changes, and in December they were ratified by the Constitutional Council. At a referendum held in September, Guadeloupe voted overwhelmingly in favour of reducing the presidential mandate from seven years to five.

In municipal elections held in March 2001 Michaux-Chévry was re-elected mayor of Basse-Terre, despite corruption charges against her (in January Michaux-Chévry was acquitted of charges of forgery; however, she was still under investigation for charges of embezzlement). Following her election Michaux-Chévry relinquished the post to Pierre-Martin, in order to comply with regulations that no official may hold more than two elected posts simultaneously (she already held the positions of senator and President of the Conseil régional). Henri Bangou of the PPDG was also re-elected to the mayoralty of Pointe-à-Pitre. In the concurrently held election to the presidency of the Conseil général Jacques Gillot of Guadeloupe Unie, Socialisme et Réalité (GUSR) defeated Marcellin Lubeth of the PPDG.

In early June 2001 riots took place in Pointe-à-Pitre in which a number of people were injured, in protest at the arrest of the leader of the UGTG, Michel Madassamy, who had been charged in late May with vandalizing a number of shops that had remained open, in defiance of the UGTG's recommendations. The General Secretary of the UGTG subsequently called for a general strike to be held for the duration of Madassamy's incarceration. A period of severe drought, necessitating the rationing of water supplies, served to exacerbate the deteriorating social situation on the island.

Following a meeting of members of the Conseil régional and the Conseil général in late June 2001, a series of proposals on greater autonomy, to be presented to the French Government, was agreed upon. These included: the division of the territory into four districts; the creation of a 'collectivité territoriale' (territorial community), governed by a 41-member Assembly elected for a five-year term; and the establishment of an independent executive council. Furthermore, the proposals included a request that the territory be given control over legislative and administrative affairs, as well as legislative authority on matters concerning Guadeloupe alone. In November the French Government announced itself to be in favour of the suggested constitutional developments, and in March 2003 the two houses of the French parliament approved constitutional changes which would allow for a referendum on proposals for greater autonomy.

In mid-2001 the French President, Jacques Chirac, rejected a joint request by French Guiana, Guadeloupe and Martinique that they be permitted to join, as associate members, the Association of Caribbean States.

At the first round of national presidential election, held on 21 April 2002, Christiane Taubira, representing the Parti Radical de Gauche, defeated Jacques Chirac, obtaining 37.2% of local votes cast (compared with 28.9% for Chirac); Lionel Jospin won 23.1% of votes cast. At the second round of voting, held on 5 May, Chirac defeated the Front National candidate, Jean-Marie Le Pen, securing 91.3% of votes cast in the Department. At elections to the Assemblée nationale, held on 9 and 16 June, Chaulet, Marsin, Moutoussamy and Andy were all defeated; Objectif Guadeloupe emerged as the most successful political grouping, with both Gabrielle Louis-Carabin and Joël Beaugendre being elected as deputies; Eric Jalton of the PCG and Victorin Lurel of the Gauche Plurielle grouping were also successful. In August Dominique Vian replaced Jean-François Carenco as Prefect; she was suceeded in mid-2004 by Paul Girot de Langlade.

In December 2003 voting took place in the referendum to determine Guadeloupe's constitutional relationship with France. The proposed new arrangement, laid out by Jean-Pierre Raffarin's right-of-centre Government in Paris, foresaw the replacement of the Conseil général and Conseil régional with a single elected council, with the aim of streamlining administrative and political processes. Despite the enthusiastic support of Lucette Michaux-Chévry, however, an overwhelming

73.0% voted against the proposal. Supporters of the 'yes' vote claimed that voters had been misled into believing that the new constitutional status would involve cuts in French subsidies and social security payments. In Saint-Barthélemy and Saint-Martin, however, the vote was in favour of the proposed change, by 95.5% and 76.2%, respectively, meaning that the islands would no longer be communes of Guadeloupe but separate 'collectivités territoriales' with autonomous tax regimes.

In elections to the Conseil régional in March 2004 Michaux-Chévry suffered a further defeat when Victorin Lurel's left-wing coalition, Guadeloupe pour Tous, won 58.4% of votes cast and 29 of the 41 seats. The list headed by Michaux-Chévry won 12 seats and 41.6% of the votes cast; she was forced to resign as President of the Conseil régional.

In the national referendum on ratification of the proposed constitutional treaty of the European Union, conducted on 29 May 2005, 58.6% of voters in Guadeloupe were in favour of the proposal; none the less, the proposal was rejected by a majority of the national electorate. Abstention, however, stood at 77.8% of those eligible to vote on the island, reflecting the lack of interest in the issue, despite warnings from political leaders across the party spectrum that a negative result might lead to reductions in EU financing.

# Economy

### PHILLIP WEARNE

Revised for this edition by Dr JAMES FERGUSON

The structure of Guadeloupe's economy helps to explain why the separatist movement does not command greater support. The penalties that would normally be associated with huge trade deficits and low productivity do not apply in Guadeloupe as French aid and subsidies make up the difference, providing one of the highest standards of living in the Caribbean. In 2001 Guadeloupe's gross domestic product (GDP) per head was estimated at US $10,323, a figure that, it was calculated, would be dramatically reduced were France to withdraw its support of the economy. Between 1990 and 2004 the annual rate of inflation averaged 1.0%. In 2003 the annual inflation rate was 1.8% and in 2004 it was 1.4%. In December 2003 some 26.9% of the workforce was unemployed.

Guadeloupe's trade deficit rose steadily at the end of the 20th century. In 1980 the value of exports covered 14.5% of imports. By 1998 that figure had fallen to 6.6%. The deficit rose from 2,628m. francs in 1981 to 9,999.7m. francs by 1998. In 2003 the trade deficit had reached more than €1,800m., with exports representing less than 10% of imports. The main reason for this was Guadeloupe's increased consumption of high-value consumer products, such as electrical goods and cars. The principal imports in 2003 were machinery and transport equipment, food and livestock, manufactured goods and chemicals. In that year some 62.1% of imports came from metropolitan France. Imports of mineral fuels, the main source of the Department's energy, increased from 2.1% of the total in 1993 to 9.0% in 2003. The main exports in 2003 were bananas, sugar, rum, cereals and boats. In that year France took 67.0% of Guadeloupe's exports.

One possible solution to the island's deteriorating trade position was more regional integration. Closer co-operation with the Organisation of Eastern Caribbean States (OECS—a seven-state, Commonwealth-Caribbean group, using a common currency, the Eastern Caribbean dollar), was the most obvious option. At an OECS meeting in Antigua and Barbuda in 1989, it was agreed to establish a committee to examine the possibility of OECS countries using Guadeloupe and Martinique as entrepôts for exporting to Europe. Joint ventures were another possibility. However, the advent of the single market in Europe from 1993 neutralized some of the special advantages the French Overseas Departments and Territories had previously enjoyed, with all three French Caribbean possessions thenceforth being treated simply as less developed areas of Europe.

As in other states in the region, the economy is based on agriculture, tourism and some light industry, mostly the processing of food and beverages. Bananas and raw sugar were traditionally Guadeloupe's principal exports, although the volume of production fluctuated in the early and mid-1990s owing to variable climatic conditions. Production of bananas reached 148,296 metric tons in 1992, but in 1994 a combination of drought and the damage caused to plantations in September by 'Hurricane Debbie' devastated the crop. Banana production in 2003 was estimated at 115,500 tons. Banana exports, typically about 120,000 tons per year in the early 1990s, were recorded at 88,000 tons in 2003, a fall from the 97,000 tons exported in 2002. In the late 1990s Guadeloupe's banana sector was adversely affected by declining prices on the European market, while a dispute between the USA and four major Latin American producers and the European Union (EU) over the latter's banana import regime also threatened the sector (this dispute was resolved in April 2001). The proposed liberalization of EU banana imports from 2006 onwards also led many growers to abandon the crop, fearful of lower prices and increased competition. One-fifth of Guadeloupe's growers ceased harvesting bananas in the course of 2003.

The fall in international sugar prices of the mid-1980s made the production of sugar cane in Guadeloupe uneconomic. A five-year plan to provide subsidies and price guarantees to growers was largely unsuccessful in maintaining production levels, as equipment deteriorated and growers voluntarily reduced production. In 1995 Guadeloupe's worst sugar-cane harvest—except for 1990, the year after 'Hurricane Hugo'—saw production fall to less than 376,000 tons of cane and 33,000 tons of sugar, though exports of raw sugar still accounted for 11.5% of total export earnings in that year. However, in 1997 production of raw sugar recovered to 57,000 tons, when it accounted for 23.7% of total export earnings, before declining to 38,400 tons in 1998. Thereafter, however, production recovered, and stood at an estimated 64,000 tons of raw sugar in 2003, from a harvest of 820,000 tons of sugar cane.

The most promising agricultural sector remained non-traditional crops. The production of pineapples increased by 26.5% in the four years to 1990, to reach 4,660 metric tons. Production of pineapples was estimated at 7,000 tons in 2003. Melon output more than doubled in the same period, with exports increasing more than fourfold, to 2,695 tons in 1992. Following a decrease in the latter half of the 1990s, production recovered to an estimated 4,100 tons in 2003. Output of aubergines, avocados, limes and cut flowers also increased from the 1980s, although many non-traditional sectors subsequently experienced problems. Yams, sweet potatoes and plantains were the main subsistence crops. The fishing sector, traditionally underdeveloped, responded to efforts to stimulate output. Fishing, mostly at an artisanal level, fulfilled about two-thirds of domestic requirements in the 1990s. The total catch rose to some 10,114 tons by 2002. Exports rose to 176.4 tons by 1992: the recovery of shrimp exports—up to 12.8 tons in 1992—was a key factor.

The main industrial activity concerned the processing of the island's agricultural crops, particularly the refining of sugar and the distillation of rum, one of Guadeloupe's major manufactured exports. By 2002 rum production had fallen from its peak of 79,550 hl in 1989 to 67,151 hl, although exports still accounted for 7.0% of total export revenues that year. Exports in 2003 fell by 17.5%, and production contracted to 54,813 hl. From February 1996 the quotas of light rum from the African, Caribbean and Pacific (ACP) states associated with the EU under the Lomé Conventions were abolished; furthermore, the EU proposed that a phased abolition of quotas of traditional rum be in place by the end of the century. The proposals were understood to comply with requests from the French Government, which wanted to pre-empt potential competition with its rum suppliers in the overseas territories.

There was only limited activity in the textile, furniture, metal, cement, plastics and printing sub-sectors. Despite government efforts to expand the island's industrial base with the establishment of an industrial zone and free port at Jarry, and by the promotion of fiscal incentives, industry and construction continued to employ relatively few people. In 2001 some 12.8% of the economically active work-force was employed in industry. Some 5,000, or 2.4%, of the employed work-force was estimated to be engaged by the agricultural sector in mid-2003. In 2001 industrial action at Jarry and among employees of the nationalized electricity and gas providers slowed manufacturing.

Tourism remained the major source of foreign exchange, although the number of visitors declined sharply as a result of the bomb attacks by the separatist movement in the mid-1980s. Nevertheless, in 1988 tourism replaced sugar production as the Department's principal source of income and the sector continued to expand rapidly in the 1990s, with overnight stays in hotels increasing from 453,000 in 1993 to 3,233,487 in 2000. From 2000, however, the tourism sector suffered a steady slowdown, with overall passenger arrivals at the main airport falling from 2.1m. in 2000 to 1.75m. in 2003. Cruise-ship arrivals also decreased, from 300,000 in 2000 to 200,000 in 2003. Tourist arrivals numbered 438,819 in 2003. Tourism receipts in that year were estimated at €400m, and tourism was estimated to account for some 10% of local GDP. The Guadeloupean dependencies of Saint-Martin (an island shared with the Dutch territory of St Maarten, in the Netherlands Antilles) and Saint-Barthélemy were almost entirely dependent upon the tourism industry and were less affected by separatist troubles. The deregulation of air transport in 1986, which ended Air France's monopoly on flights to Guadeloupe, was a major factor in attracting more visitors to the main island. However, the decision by AOM Compagnie Aérienne Française (subsequently Air Lib) to cease flights to St Maarten in March 2001 severely affected the tourism industry on the island. Moreover, concerns were expressed in Guadeloupe at the potentially adverse effects on the tourism industry that a renewed monopoly on flights by Air France might have. These concerns became more urgent in February 2003, when Air Lib went into liquidation, reducing choice and frequency of flights. The main beneficiary of Air Lib's collapse was the French company Corsair, a subsidiary of the TUI group, which saw its share of passengers transported to Guadeloupe rise by 26% in 2003. France remained the largest source of tourists. In 2002 89.0% of arrivals came from metropolitan France or dependent territories, 6.0% from other European countries and 3.0% from North America. The number of hotels increased in line with the rise in tourism, and numbered 162 in 2002. The number of hotel rooms available in that year totalled 8,000. In late 2002 the French hotel group Accor announced that it would close its five hotels on Guadeloupe and Martinique, citing high operating costs, poor industrial relations and decreasing tourist arrivals. In early 2005 the group still operated five hotels in Guadeloupe.

As the economy was heavily dependent on France, so local commercial activity was heavily dependent on the spending power of French tourists and civil servants. More than one-half of the total salary payments made on the island went to civil servants or the French Government's contractors. Civil servants received a 40% bonus on their basic metropolitan earnings, which further increased their economic importance to the Department, in terms of purchasing power. As a result, local investment interest remained concentrated on the import-export business or in the services sector (such as the discount stores which handled imported goods), rather than in productive investment. The 1996 alignment of the social security systems of the Overseas Departments with that of metropolitan France was also of significant benefit to many Guadeloupeans. In 2003 it was estimated that social security benefits worth €515m. were paid to over 100,000 claimants, of which 40% took the form of child support. In April 2003 the Conseil régional announced the creation of a fund totalling €76m. to support local businesses and reduce reliance on French subsidies.

# Statistical Survey

Sources (unless otherwise stated): Institut National de la Statistique et des Etudes Economiques (INSEE), ave Paul Lacavé, BP 96, 97102 Basse-Terre; tel. 99-0250; internet www.insee.fr/fr/insee_regions/guadeloupe; Service de Presse et d'Information, Ministère des départements et territoires d'outre-mer, 27 rue Oudinot, 75700 Paris 07 SP, France; tel. 1-53-69-20-00; fax 1-43-06-60-30; internet www.outre-mer.gouv.fr.

## AREA AND POPULATION

**Area:** 1,705 sq km (658.3 sq miles), incl. dependencies (La Désirade, Les Saintes, Marie-Galante, Saint-Barthélemy, Saint-Martin).

**Population:** 327,002 (males 160,112, females 166,890) at census of 9 March 1982; 387,034 (males 189,187, females 197,847) at census of 15 March 1990; 422,496 at census of 8 March 1999. *Mid-2004* (UN estimate): 445,000 (Source: UN, *World Population Prospects: The 2004 Revision*)).

**Density** (at mid-2004): 261.0 per sq km.

**Principal Towns** (population at 1999 census): Les Abymes 63,054; Saint-Martin 29,078; Le Gosier 25,360; Baie-Mahault 23,389; Pointe-à-Pitre 20,948; Le Moule 20,827; Petit Bourg 20,528; Sainte Anne 20,410; Basse-Terre (capital) 12,410.

**Births, Marriages and Deaths** (1997, provisional figures): Registered live births 7,554 (birth rate 17.4 per 1,000); Registered marriages 1,936 (marriage rate 4.7 per 1,000); Registered deaths 2,441 (death rate 5.6 per 1,000). *2002:* Registered live births 7,032; Registered marriages 1,824; Registered deaths 2,614 (preliminary).

**Expectation of Life** (UN estimates, years at birth): 77.3 (males 73.6; females 80.9) in 1995–2000. Source: UN, *World Population Prospects: The 2004 Revision*.

**Economically Active Population** (persons aged 15 years and over, 1990 census): Agriculture, hunting, forestry and fishing 8,391; Industry and energy 9,630; Construction and public works 13,967; Trade 15,020; Transport and telecommunications 6,950; Financial services 2,802; Other marketable services 26,533; Non-marketable services 34,223; *Total employed* 117,516 (males 68,258, females 49,258); Unemployed 54,926 (males 25,691, females 29,235); Total labour force 172,442 (males 93,949, females 78,493). *2001* (estimates, salaried employees at 31 December): Agriculture 3,075; Industry (excl. construction) 7,798; Construction 6,521; Services 94,144; Total 111,538. *2001* (estimates, non-salaried employees at 31 December): Total 22,856. *Mid-2003* (estimates): Agriculture, etc. 5,000; Total labour force 205,000 (males 110,000, females 95,000) (Source: FAO). *December 2003:* Unemployed 44,298.

## HEALTH AND WELFARE

### Key Indicators

**Physicians** (per 1,000 head, 1996): 1.6.

**Hospital Beds** (per 1,000 head, 1996): 29.1.

For sources and definitions, see explanatory note on p. vi.

## AGRICULTURE, ETC.

**Principal Crops** (FAO estimates, '000 metric tons, 2003): Sweet potatoes 4.3; Yams 10; Other roots and tubers 4.2; Sugar cane 820.0; Cabbages 2.3; Lettuce 3.2; Tomatoes 3.1; Cucumbers and gherkins 3.9; Other fresh vegetables 15.1; Cantaloupes and other melons 4.1; Bananas 115.5; Plantains 9.2; Pineapples 7.0; Other fruits 6.7. Source: FAO.

**Livestock** (FAO estimates, '000 head, year ending September 2003): Cattle 85; Goats 28; Pigs 19; Sheep 3.2.

**Livestock Products** (FAO estimates, '000 metric tons, 2003): Beef and veal 3.3; Pig meat 1.0; Hen eggs 1.7.

**Forestry** ('000 cubic metres, 2003): *Roundwood Removals* (excl. bark): Sawlogs, veneer logs and logs for sleepers 0.3; Fuel wood 15.0; Total 15.3. *Sawnwood Production* (incl. railway sleepers): Total 1.0.

# GUADELOUPE

**Fishing** (FAO estimates, metric tons, live weight, 2003): Capture 10,100 (Common dolphinfish 700; Blackfin tuna 500; Other mackerel-like fishes 1,600; Marine fishes 7,100; Stromboid conchs 550); Aquaculture 31; *Total catch* 10,131.
Source: FAO.

## MINING

**Production** ('000 metric tons, 2002, estimates): Pumice 210.0; Salt 49. Source: US Geological Survey.

## INDUSTRY

**Production:** Raw sugar 65,200 metric tons in 2001; Rum 62,679 hl in 1998; Cement 230,000 metric tons in 2002 (estimate); Electric energy (million kWh) 1,220 in 2000 (estimate). Source: mainly UN, *Industrial Commodity Statistics Yearbook*.

## FINANCE

**Currency and Exchange Rates:** The French franc was used until the end of February 2002. Euro notes and coins were introduced on 1 January 2002, and the euro became the sole legal tender from 18 February. Some of the figures in this Survey are still in terms of francs. For details of exchange rates, see French Guiana.

**Budget:** *State Budget* (million French francs, 1990): Revenue 2,494; Expenditure 4,776. *Regional Budget* (preliminary, excl. debt rescheduling, € million, 2002): Total revenue 287 (Direct taxes 13, Indirect taxes 126, Transfers 109, Loans 38, Other 1); Total expenditure 287. *Departmental Budget* (€ million, 1998): Revenue 332; Expenditure 318.

**Money Supply** (million French francs at 31 December 1996): Currency outside banks 1,148; Demand deposits at banks 6,187; Total money 7,335.

**Cost of Living** (Consumer Price Index; base: 2000 = 100): 105.0 in 2002; 107.1 in 2003; 108.6 in 2004. Source: ILO.

**Expenditure on the Gross Domestic Product** (million French francs at current prices, 1994): Government final consumption expenditure 5,721; Private final consumption expenditure 16,779; Increase in stocks 161; Gross fixed capital formation 5,218; *Total domestic expenditure* 27,879; Exports of goods and services 912; *Less* Imports of goods and services 9,040; *GDP in purchasers' values* 19,751.

**Gross Domestic Product by Economic Activity** (million French francs at current prices, 1992): Agriculture, hunting, forestry and fishing 1,206.9; Mining, quarrying and manufacturing 1,237.5; Electricity, gas and water 307.5; Construction 1,164.4; Trade, restaurants and hotels 2,907.9; Transport, storage and communications 1,415.4; Finance, insurance, real estate and business services 2,023.0; Government services 4,769.7; Other community, social and personal services 2,587.7; Other services 348.0; *Sub-total* 17,968.0; Import duties 699.4; Value-added tax 615.7; *Less* Imputed bank service charge 1,311.2; *GDP in purchasers' values* 17,972.0 (Source: UN, *National Accounts Statistics*). *2000* (€ million): GDP in purchasers' values 5,593.

## EXTERNAL TRADE

**Principal Commodities:** *Imports c.i.f.* (US $ million, 1995): Food and live animals 302.8 (Meat and meat preparations 74.7, Dairy products and birds' eggs 51.9, Cereals and cereal preparations 55.6, Vegetables and fruit 51.4); Beverages and tobacco 88.1 (Beverages 78.8); Mineral fuels, lubricants, etc. 110.7 (Petroleum, petroleum products, etc. 52.5, Gas, natural and manufactured 58.0); Chemicals and related products 172.8 (Medicinal and pharmaceutical products 78.1); Basic manufactures 259.5 (Paper, paperboard and manufactures 39.3); Machinery and transport equipment 607.0 (Office machines and automatic data-processing equipment 40.7, Telecommunications and sound equipment 43.0, Road vehicles and parts 217.7, Other transport equipment 100.0); Miscellaneous manufactured articles 282.6 (Furniture and parts 47.2, Clothing and accessories, excl. footwear, 52.6, Printed matter 39.5); Total (incl. others) 1,901.3 (Source: UN, *International Trade Statistics Yearbook*). *Exports f.o.b.* (million French francs, 1997): Bananas 179.7; Melons and fresh papayas 33.8; Sugar 193.9; Rum 40.2; Wheaten or rye flour 35.3; Yachts and sports boats 50.7; Total (incl. others) 819.4. *2000* (€ million): *Imports:* 2,010; *Exports:* 210.

**Principal Trading Partners** (million French francs, 1997): *Imports c.i.f.:* Belgium-Luxembourg 161.8; Curaçao (Netherlands Antilles) 212.6; France (metropolitan) 6,435.7; Germany 445.5; Italy 395.0; Japan 247.6; Martinique 114.0; Spain 190.2; Trinidad and Tobago 189.3; United Kingdom 151.7; USA 315.4; Total (incl. others) 10,236.6. *Exports f.o.b.:* Belgium-Luxembourg 23.0; France (metropolitan) 497.2; French Guiana 20.6; Italy 21.7; Martinique 152.2; United Kingdom 18.7; USA 36.6; Total (incl. others) 819.4.

## TRANSPORT

**Road Traffic** ('000 motor vehicles in use, 2001): Passenger cars 117.7; Commercial vehicles 31.4. Source: UN, *Statistical Yearbook*.

**Shipping:** *Merchant Fleet* (vessels registered, '000 grt at 31 December 1992): Total displacement 6. (Source: Lloyd's Register-Fairplay, *World Fleet Statistics*). *International Sea-borne Traffic* (1999): Vessels entered 2,988; Goods loaded 428,000 metric tons; Goods unloaded 2,747,000 metric tons; Passengers carried 1,296,000.

**Civil Aviation** (2003): Passengers carried 1,761,500; Freight carried 15,000 metric tons.

## TOURISM

**Tourist Arrivals by Country** (2000): Canada 10,431; France 440,779; Italy 15,670; Switzerland 9,766; USA 92,474; Total (incl. others) 623,134. *2003:* France 386,737; Total (incl. others) 438,819.

**Receipts from Tourism** (US $ million, incl. passenger transport): 466 in 1998; 375 in 1999; 418 in 2000.
Source: World Tourism Organization.

## COMMUNICATIONS MEDIA

**Radio Receivers** ('000 in use): 113 in 1997.

**Television Receivers** ('000 in use): 118 in 1997.

**Telephones** ('000 main lines in use): 210.0 in 2001.

**Facsimile Machines** (number in use): 3,400 in 1996.

**Mobile Cellular Telephones** ('000 subscribers): 323.5 in 2002.

**Personal Computers** ('000 in use): 100 in 2001.

**Internet Users** ('000): 20.0 in 2001.

**Daily Newspaper:** 1 (estimate) in 1996 (estimated average circulation 35,000 copies).
Sources: UNESCO, *Statistical Yearbook*; UN, *Statistical Yearbook*; International Telecommunication Union.

## EDUCATION

**Pre-primary** (2002/03): 136 institutions; 23,461 students (Source: Ministère de l'Education Nationale, de l'Enseignement Supérieur et de la Recherche, *Repères et références statistiques sur les enseignements, la formation et la recherche 2003*).

**Primary** (2002/03): 220 institutions; 3,241 teachers (3,011 state, 230 private); 40,092 students (Source: Ministère de l'Education Nationale, de l'Enseignement Supérieur et de la Recherche, *Repères et références statistiques sur les enseignements, la formation et la recherche 2003*).

**Secondary** (2002/03): 4,700 teachers (4,282 state, 418 private); 52,916 students (Source: Ministère de l'Education Nationale, de l'Enseignement Supérieur et de la Recherche, *Repères et références statistiques sur les enseignements, la formation et la recherche 2003*).

**Higher** (1997): 5,800 students (Université Antilles-Guyane).

**Adult Literacy Rate:** 90.1% (males 89.7%; females 90.5%) in 1992.

# Directory

## The Government

(July 2005)

**Prefect:** PAUL GIROT DE LANGLADE, Préfecture, Palais d'Orléans, rue Lardenoy, 97109 Basse-Terre Cédex; tel. 5-90-99-39-00; fax 5-90-81-58-32; internet www.guadeloupe.pref.gouv.fr.

**President of the General Council:** Dr JACQUES GILLOT (GUSR), Hôtel du Département, blvd Félix Eboué, 97109 Basse-Terre; tel. 5-90-99-77-77; fax 5-90-99-76-00; e-mail daniel.dumirier@cg971.fr; internet www.cg971.com.

**President of the Economic and Social Committee:** GUY FRÉDÉRIC.

**Deputies to the French National Assembly:** ÉRIC JALTON (PCG), GABRIELLE LOUIS-CARABIN (UMP), JOËL BEAUGENDRE (UMP), VICTORIN LUREL (PS).

**Representatives to the French Senate:** JACQUES GILLOT (Rassemblement Démocratique et Social Européen), DANIEL MARSIN (Groupe Socialiste), LUCETTE MICHAUX-CHEVRY (UMP).

### Conseil Régional

ave Paul Lacavé, Petit-Paris, 97109 Basse-Terre; tel. 5-90-80-40-40; fax 5-90-81-34-19; internet www.cr-guadeloupe.fr.

**President:** VICTORIN LUREL (PS).

**Elections, 21 and 28 March 2004**

|  | Seats |
|---|---|
| Guadeloupe pour Tous* | 29 |
| Action Guadeloupe pour l'Initiative et le Rassemblement (AGIR)† | 12 |
| Total | 41 |

Seven other lists contested the elections: NOFWAP; Combat Ouvrier; Alternative Citoyenne; Union pour une Guadeloupe Responsable; Lé Verts Sé Nou Tout; Le Renouveau Guadeloupéen; and Priorité à l'Education et à l'Environnement.
* Comprising the PS, PPDG, GUSR and other left-wing candidates.
† Comprising the UMP, other right-wing candidates and socialist dissidents.

## Political Organizations

**Combat Ouvrier:** BP 213, 97156 Point-à-Pitre Cédex; Leaders JEAN-MARIE NOMERTIN, GÉRARD SÉNÉ; Trotskyist; sister org. of Lutte Ouvrière; mem. of the Internationalist Communist Union.

**Guadeloupe Respect:** Goyave; Leader JEAN EMMANUEL LAGUERRE; formed by dissidents from the right-wing Objectif Guadeloupe coalition.

**Guadeloupe Unie, Socialisme et Réalité (GUSR):** Pointe-à-Pitre; e-mail d.larifla@senat.fr; 'dissident' faction of the Parti Socialiste; Pres. DOMINIQUE LARIFLA.

**Konvwa pou Liberasyon Nasyon Gwadloup (KLNG):** Pointe-à-Pitre; f. 1997; pro-independence; Leader LUC REINETTE.

**Lutte Ouvrière:** Pointe-à-Pitre; internet www.lutte-ouvriere.org; extreme left-wing; sister org. of Combat Ouvrier; mem. of the Internationalist Communist Union.

**Parti Communiste Guadeloupéen (PCG):** 119 rue Vatable, 97110 Pointe-à-Pitre; tel. 5-90-88-23-07; f. 1944; Sec.-Gen. CHRISTIAN CÉLESTE.

**Parti Radical de Gauche:** Bergette, 97150 Petit Bourg; tel. 5-90-95-72-83; f. 1902; Pres. FLAVIEN FERRANT; Gen. Sec. LUCIEN ANAÏS.

**Parti Socialiste (PS):** 8 Résidence Légitimus, blvd Légitimus, 97110 Pointe-à-Pitre; tel. and fax 5-90-21-65-72; fax 5-90-83-65-20-51; e-mail fede971@parti-socialiste.fr; internet www.parti-socialiste.fr; divided into two factions to contest the March 1992 and March 1993 elections; Jt First Secs MARLÈNE MELISSE, FAVROT DAVRAIN.

**Renouveau Socialiste:** ZAC de Boisrepeaux, Abymes, Pointe-à-Pitre; e-mail d.marsin@senat.fr; f. 2003; Leader DANIEL MARSIN.

**Union pour un Mouvement Populaire (UMP):** Lotissement SIG, Sainte-Anne; tel. 5-90-99-52-99; fax 5-90-99-52-99; f. 2002; centre-right; local br. of the metropolitan party; Departmental Sec. ALDO BLAISE.

**Les Verts:** 32 rue Alsace-Lorraine, 97110 Pointe-à-Pitre; tel. 5-90-35-41-90; fax 5-90-25-02-62; internet www.les-verts.org; green; Gen. Sec. MICHÈLE MAXO; Regional Dirs HARRY DURIMEL, MARIE LINE PIRBAKAS.

Other political organizations participating in the 2004 elections included Mouvement pour la Démocratie et le Développement (MDDP), Union Populaire pour la Libération de la Guadeloupe (UPLG), Mouvman Gwadloupéyen (MG), Parti Progressiste Démocratique Guadeloupéen (PPDG), Renouveau Socialiste; and the coalitions Priorité à l'Education et à l'Environnement and Union pour une Guadeloupe Responsable.

## Judicial System

**Cour d'Appel:** Palais de Justice, 4 blvd Félix Eboué, 97100 Basse-Terre; tel. 5-90-80-63-36; fax 5-90-80-63-39; First Pres. JEAN-PIERRE ATTHENONT; Procurator-Gen. MICHEL MAROTTE; two Tribunaux de Grande Instance, four Tribunaux d'Instance.

## Religion

The majority of the population belong to the Roman Catholic Church.

### CHRISTIANITY

#### The Roman Catholic Church

Guadeloupe comprises the single diocese of Basse-Terre, suffragan to the archdiocese of Fort-de-France, Martinique. At 31 December 2003 there were an estimated 395,000 adherents, representing some 87% of the total population. The Bishop participates in the Antilles Episcopal Conference, currently based in Port of Spain, Trinidad and Tobago.

**Bishop of Basse-Terre:** Rt Rev. ERNEST MESMIN LUCIEN CABO, Evêché, pl. Saint-François, BP 369, 97100 Basse-Terre Cédex; tel. 5-90-81-36-69; fax 5-90-81-98-23; e-mail eveche@catholique-guadeloupe.info.

## The Press

**France Antilles:** 1 rue Hincelin, BP 658, 97159 Pointe-à-Pitre; tel. 5-90-90-25-25; fax 5-90-91-78-31; daily; f. 1964; subsidiary of Groupe France Antilles; Chair. PHILIPPE HERSANT; Man. Dir FRÉDÉRIC AURAND; circ. 50,000.

**Match:** 33 rue Peynier, 97110 Pointe-à-Pitre; tel. 5-90-82-18-68; fax 5-90-82-01-87; fortnightly; Dir MARIE ANTONIA JABBOUR; circ. 6,000.

**Nouvelles Etincelles Hebdo:** 119 rue Vatable, 97110 Pointe-à-Pitre; tel. 5-90-91-12-77; fax 5-90-83-69-90; f. 1944 as L'Etincelle, organ of the PCG; present name adopted 2005; weekly; Dir RAYMOND BARON; circ. 5,000.

**Le Progrès Social:** rue Toussaint L'Ouverture, 97100 Basse-Terre; tel. 5-90-81-10-41; weekly; Dir JEAN-CLAUDE RODES; circ. 5,000.

**Sept Mag Communication:** Immeuble Curaçao, voie Verte, 97122 Baie-Mahault; tel. 26-60-50; fax 26-61-87; e-mail jcanneval@outremer.com; weekly; Dir JACQUES CANNEVAL; circ. 30,000.

**TV Magazine Guadeloupe:** 1 rue Paul Lacavé, BP 658, 97169 Pointe-à-Pitre; tel. 5-90-90-25-25; weekly.

## Broadcasting and Communications

### BROADCASTING

**Réseau France Outre-mer (RFO):** BP 180, 97122 Baie Mahault Cédex; tel. 5-90-60-96-96; fax 5-90-60-96-82; internet www.rfo.fr; fmrly Société Nationale de Radio-Télévision Française d'Outre-mer; present name adopted in 1998; radio and television broadcast 24 hours daily; Pres. ANDRÉ-MICHEL BESSE; Regional Dir JEAN PHILLIPPE PASCAL; Gen. Man. FRANÇOIS GUILBEAU.

#### Radio

More than 30 private FM radio stations are in operation.

# GUADELOUPE

**Radio Actif:** rue Ferdinand Forest, 97122 Baie-Mahault; tel. 5-90-26-68-93; commercial; satellite link to Radio Monte-Carlo (Monaco and France).

**Radio Caraïbes International (RCI):** BP 1309, 97187 Point-à-Pitre Cédex; tel. 5-90-83-96-96; fax 5-90-83-96-97; two commercial stations broadcasting 24 hours daily; Dir Jean-François Ferandier-Sicard.

**Radio Saint-Martin:** Port de Marigot, 97150 Saint-Martin; commercial station broadcasting 94 hours weekly; Man. H. Cooks.

**Radio Voix Chrétiennes de Saint-Martin:** BP 103, Marigot, 97150 Saint-Martin; tel. 5-90-87-13-59; religious; Man. Fr Cornelius Charles.

### Television

**Canal Antilles:** 2 Lotissement les Jardins de Houelbourg, 97122 Baie-Mahault; tel. 5-90-26-81-79; private 'coded' station.

**TCI Guadeloupe:** Montauban, 97190 Grosier; commercial station.

## Finance

(cap. = capital; res = reserves; dep. = deposits; m. = million; brs = branches; amounts in euros unless otherwise indicated)

### BANKING

#### Central Bank

**Institut d'Emission des Départments d'Outre-mer (IEDOM):** blvd Légitimus, BP 196, 97155 Pointe-à-Pitre Cédex; tel. 5-90-93-74-00; fax 5-90-93-74-25; e-mail iedom.gpe@wanadoo.fr; internet www.iedom.com; Dir F. Roche Toussaint.

#### Commercial Banks

**Banque des Antilles Françaises:** pl. de la Victoire, BP 696, 97110 Pointe-à-Pitre Cédex; tel. 5-90-26-80-07; fax 5-90-26-74-48; internet www.bdaf.gp; f. 1853; cap. 7.5m., res 8.4m., dep. 506.1m. (Dec. 2002); Chair. Serge J. Robert; Gen. Man. Michel Sprecht; 19 brs.

**Banque Française Commerciale Antilles-Guyane (BFC Antilles-Guyane):** Immeuble BFC AG, Grand Camp-La Rocade, BP 13, 97151 Pointe-à-Pitre Cédex; tel. 5-90-21-56-70; fax 5-90-21-56-80; e-mail f.aujoulat@bfc-ag.com; internet www.bfc-ag.com; f. 1976 as br. of Banque Française Commerciale SA, separated 1984; total assets 594.6m. (2003); Chair. Pierre de Bellefon; Gen. Man. Jean Marguier.

**BNP Paribas Guadeloupe:** pl. de la Rénovation, BP 161, 97155 Pointe-à-Pitre; tel. 5-90-90-58-58; fax 5-90-90-04-07; internet www.bnpparibas.com; f. 1941; subsidiary of BNP Paribas; CEO Gérard d'Here; Gen. Sec. François Pasetti; 12 brs.

**BRED Banque Populaire (BRED-BP):** 10 rue Achille René Boisneuf, BP 35, 97110 Pointe-à-Pitre; tel. 5-90-89-67-00.

**Caisse d'Epargne de la Guadeloupe:** 20 Lotissement Plazza II, Grand Camp-La Rocade, 97142 Abymes, BP 22, 97151 Pointe-à-Pitre Cédex; tel. 5-90-93-12-12; fax 5-90-93-12-13; Pres. Daniel Nuccio.

**Crédit Agricole de la Guadeloupe:** Petit Pérou, 97176 Abymes Cédex; tel. 5-90-90-65-65; fax 5-90-90-65-89; e-mail catelnet@cr900.credit-agricole.fr; internet www.ca-guadeloupe.fr; Pres. Christian Fléreau; Gen. Man. Roger Wunschel.

**Crédit Maritime de la Guadeloupe:** 36 rue Achille René-Boisneuf, BP 292, 97175 Pointe-à-Pitre; tel. 5-90-21-08-40; fax 5-90-83-46-37; e-mail pointe-a-pitre-agence-cmm@creditmaritime.com; internet www.creditmaritime-outremer.com; Pres. Yves Joliman; Gen. Man. Philippe Beauvoir; 4 agencies.

**Société Générale de Banque aux Antilles (SGBA):** 30 rue Frébault, BP 55, 97152 Pointe-à-Pitre; tel. 5-90-25-49-77; fax 5-90-25-49-78; e-mail sgba@wanadoo.fr; internet www.sgba.fr; f. 1979; cap. 18.6m., res 0.6m., dep. 246m. (Dec. 2002); Chair. Jean-Louis Mattei; Gen. Man. Patrick Le Buffe; 5 brs.

#### Development Banks

**Banque de Développement de Petites et Moyennes Entreprises (BDPME):** c/o AFD, blvd Légitimus, BP 160, 97159 Pointe-à-Pitre Cédex; tel. 5-90-89-65-58; fax 5-90-21-04-55; Rep. Muguette Daljardin.

**Société de Crédit pour le Développement de Guadeloupe (SODEGA):** Carrefour Raizet Baimbridge, BP 54, 97152 Pointe-à-Pitre; tel. 5-90-82-65-00; fax 5-90-90-17-91; e-mail informatique@sodega.fr; internet www.sodega.fr; f. 1970; bought from the Agence Française de Développement (AFD—q.v.) by BRED-BP in 2003.

### Foreign Companies

Some 30 of the principal European insurance companies are represented in Pointe-à-Pitre, and another six companies have offices in Basse-Terre.

## Trade and Industry

### GOVERNMENT AGENCIES

**Conseil Economique et Social:** rue Peynier, 97100 Basse-Terre; tel. 5-90-41-05-15; fax 5-90-41-05-23; internet www.cr-guadeloupe.fr; Pres. Jean-Michel Penchard.

**Direction de l'Agriculture et de la Forêt (DAF):** Jardin Botanique, 97100 Basse-Terre; tel. 5-90-99-09-09; fax 5-90-99-09-10; e-mail daf@docupguadeloupe.org; Man. Gérard Chuiton.

**Direction Régionale de l'Industrie, de la Recherche et de l'Environnement (DRIRE):** 20 rue de la Chapelle, Z. I. Jarry, BP 448, 97122 Baie-Mahault; tel. 5-90-38-03-47; fax 5-90-38-03-50; e-mail henri.kaltembacher@industrie.gouv.fr; active in industry, business services, transport, public works, tourism and distribution; Departmental Co-ordinator Henri Kaltembacher.

**Direction Régionale des Affaires Maritimes (DRAM):** 1 quai Layrle, BP 473, 97164 Pointe-à-Pitre; tel. 5-90-82-03-13; fax 5-90-90-07-33; responsible for shipping, fishing and other maritime issues at a national and community level.

**Direction Régionale du Commerce Extérieur (DRCE):** see chapter on Martinique.

### DEVELOPMENT ORGANIZATIONS

**Agence de l'Environnement et de la Maîtrise de l'Energie (ADEME):** Immeuble Café Center, rue Ferdinand Forest, Z. I. Jarry, 97122 Baie-Mahault; tel. 5-90-26-78-05; fax 5-90-26-87-15; e-mail ademegua@wanadoo.fr; internet www.ademe.fr; developing energy and waste management; Man. Guy Simonnot.

**Agence Française de Développement (AFD):** blvd Légitimus, BP 160, 97159 Pointe-à-Pitre Cédex; tel. 5-90-89-65-65; fax 5-90-83-03-73; e-mail afdpointeapitre@gp.groupe-afd.org; internet www.afd.fr; fmrly Caisse Française de Développement; Man. Yves Malpel.

**Agence pour la Promotion des Investissements en Guadeloupe (APRIGA):** 12 Convenance's Center, Lieu-dit Convenance, 97122 Baie-Mahault; tel. 5-90-94-45-40; fax 5-90-95-86-47; e-mail apriga@apriga.com; internet www.apriga.com; f. 1979 as Agence pour la Promotion de l'Industrie de la Guadeloupe; Pres. Lyliane Piquion Salome.

### CHAMBERS OF COMMERCE

**Chambre d'Agriculture de la Guadeloupe:** Rond-Point Destrelland, 97122 Baie-Mahault; tel. 5-90-25-17-17; fax 5-90-26-07-22; e-mail direction@guadeloupe.chambagri.fr; Pres. Maurice Ramassamy.

**Chambre de Commerce et d'Industrie de Basse-Terre:** 6 rue Victor Hugues, 97100 Basse-Terre; tel. 5-90-99-44-44; fax 5-90-81-21-17; e-mail cci-basse-terre@wanadoo.fr; internet www.basse-terre.cci.fr; f. 1832; Pres. Gérard Théobald; 24 mems.

**Chambre de Commerce et d'Industrie de Pointe-à-Pitre:** Hôtel Consulaire, rue Félix Eboué, 97159 Pointe-à-Pitre Cédex; tel. 5-90-93-76-00; fax 5-90-90-21-87; e-mail contact@pointe-a-pitre.cci.fr; internet www.pointe-a-pitre.cci.fr; Pres. Colette Koury; Dir-Gen. Jacques Garreta; 34 full mems and 17 assoc. mems.

**Chambre de Métiers de la Guadeloupe:** route Choisy, BP 61, 97120 Saint-Claude; tel. 5-90-80-23-33; fax 5-90-80-08-93; e-mail sgr@cmguadeloupe.org; Pres. Maurice Songeons; Sec.-Gen. Serge Nirlep.

**Jeune Chambre Economique de Basse-Terre:** BP 316, 97100 Basse-Terre; tel. 5-90-81-13-73; e-mail jce.bt@laposte.net; Pres. Bernard Saulchoir.

**Jeune Chambre Economique de Pointe-à-Pitre:** BP 505, 97168 Pointe-à-Pitre Cédex; tel. 5-90-89-01-30; fax 5-90-91-72-98; e-mail jcepap2006@wanadoo.fr; Pres. Jean-Christophe Hodebar.

### EMPLOYERS' ORGANIZATIONS

**Association des Moyennes et Petites Industries (AMPI):** CWTC—Zone de Commerce International, Pointe Jarry, BP 2325, 97187 Jarry Cédex; tel. 5-90-25-06-28; fax 5-90-25-06-29; e-mail mpi.guadeloupe@wanadoo.fr; internet www.mpi-guadeloupe.com; f. 1974; Pres. Patrick Doquin; Gen. Sec Jean Joachim; 116 mem. cos.

# GUADELOUPE

**Ordre des pharmaciens du département Guadeloupe:** Rocade Forum de Grand Camp, Bâtiment A, No 1, 97142 Pointe-à-Pitre; tel. 5-90-21-66-05; fax 5-90-21-66-07; Pres. CLAUDIE ESPIAND.

**Syndicat des Producteurs-Exportateurs de Sucre et de Rhum de la Guadeloupe et Dépendances:** Z. I. de la Pointe Jarry, 97122 Baie-Mahault; BP 2015, 97191 Pointe-à-Pitre; tel. 5-90-26-62-12; fax 5-90-26-86-76; f. 1937; Pres. AMÉDÉE HUYGHUES-DESPOINTES; 4 mems.

**Union des Entreprises–Mouvement des Entreprises de France (UDE–MEDEF):** Immeuble SCI BTB, voie Principale, Z.I. de la Pointe Jarry, 97122 Baie Mahault; tel. 5-90-26-83-58; fax 5-90-26-83-67; e-mail ude.medef@medef-guadeloupe.com; Pres. CHRISTIAN VIVIES; Gen. Sec. BERTRAND JOYAU.

## MAJOR COMPANIES

**Chantiers Audebert et Cie, SARL:** Z. I. de la Pointe Jarry, 97122 Baie-Mahault; tel. 5-90-26-75-40; fax 5-90-26-75-43; internet www.chantiers-aubert.com; f. 1904; construction equipment; Chair. JEAN AUDEBERT; Gen. Man. DERRICK AUDEBERT; 100 employees.

**Colgate-Palmolive Guadeloupe:** rue Robert Fulton, 97122 Jarry Cédex; tel. 5-90-26-78-81.

**Compagnie Frigorifique de la Guadeloupe, SARL (COFRIGO):** Z. I. de la Pointe Jarry, 97122 Baie-Mahault; tel. 5-90-26-72-28; fax 5-90-26-80-91; f. 1973; manufacture of soft drinks; Pres. ALAIN HUYGUES DESPOINTES; Gen. Man. THIERRY HUYGUES DESPOINTES; 87 employees.

**Coopérative des Marins-Pêcheurs de la Guadeloupe (COMA-PEGA):** Port de Pêche de Bergevin, 97110 Ponte-à-Pitre; tel. 5-90-21-46-60; fax 5-90-91-63-78; Pres. JEAN-CLAUDE YOYOTTE.

**Guadeloupe International Paper:** Pères Blancs Baillif, 32653 Basse-Terre; manufacture of paper products; Gen. Man. L. DENZENNE; 55 employees.

**Jus de Fruits des Antilles:** Carrefour de la Lézarde—Section Colin, 97170 Petit Bourg; tel. 5-90-26-76-64; fax 5-90-26-78-21; production of fruit and vegetable juices.

**Liquoristerie Madras (LIQUOMA):** rue Eugene Freyssinet, 97122 Baie-Mahault; tel. 5-90-26-60-28; fax 5-90-26-76-69; f. 1983; production and bottling of cane syrup, rums and other beverages; Man. Dir RAYMOND BICHARA-JABOUR; 20 employees.

**Nestlé:** BP 2005, Zone Portuaire de Jarry, 97122 Baie-Mahault; tel. 5-90-32-54-44; fax 5-90-26-72-88.

**SARA** (Société Anonyme de la Raffinerie des Antilles): BP 2039, 97191 Pointe-à-Pitre; tel. 5-90-38-13-13; fax 5-90-26-70-98; internet www.sara.mq; e-mail francois.nahan@sara.mq; f. 1970; owned by TotalFinaElf 50%, Shell 24%, Esso 14.5%, Texaco 11.5%; Regional Gen. Man. FRANÇOIS NAHAN; c. 250 employees regionally (see entry in Martinique).

**Severin Industrie SARL:** Domaine de Séverin, Cadet, 97115 Sainte-Rose; f. 1929; production of alcoholic beverages; Man. JOSEPH MARSOLLE.

**Société Thermales des Eaux Capes Dolé:** Lieu-dit Dolé, 97113 Gourbeyre; tel. 5-90-92-10-92; fax 5-90-92-26-19; f. 1968; bottling of mineral water; Man. PATRICK DOQUIN; 21 employees.

**Somatco:** Z. I. de la Pointe Jarry, Impasse Augustin Fresnel, 97122 Baie-Mahault; tel. 5-90-26-71-67; fax 5-90-26-86-24; f. 1980; industrial manufacture of construction materials; Pres. CHRISTIAN BONNARDEL; 17 employees.

## UTILITIES

**EDF Guadeloupe:** BP 85, 97153 Pointe-à-Pitre; tel. 5-90-82-40-34; fax 5-90-83-30-02; e-mail marie-thérèse.fournier@edfgdf.fr; internet guadeloupe.edf.fr; electricity producer; Man. JEAN-MICHEL LEBEAU; electricity producer.

**Veolia Water-Générale des Eaux Guadeloupe:** Centre de la Guadeloupe, 7 Morne Vergain, BP 17, 91139 Abymes, Pointe-à-Pitre; tel. 5-90-89-76-76; fax 5-90-91-39-10; fmrly SOGEA.

## TRADE UNIONS

**Centrale des Travailleurs Unis (CTU):** Logement Test 14, BP 676, Bergevin, 97169 Pointe-à-Pitre; tel. 5-90-83-16-50; fax 5-90-91-78-02; affiliated to the Confédération Française Démocratique du Travail; Sec.-Gen. HENRI BERTHELOT.

**Confédération Générale du Travail de la Guadeloupe (CGTG):** 4 Cité Artisanale de Bergevin, BP 779, 97110 Pointe-à-Pitre Cédex; tel. 5-90-82-34-61; fax 5-90-91-04-00; f. 1961; Sec.-Gen. JEAN-MARIE NOMERTIN; 5,000 mems.

**Fédération Départementale des Syndicats d'Exploitants de la Guadeloupe (FDSEA):** Chambre d'Agriculture, Rond-Point de Destrellan, 97122 Baie-Mahault; tel. 5-90-26-06-47; fax 5-90-26-48-82; e-mail fdsea5@wanadoo.fr; affiliated to the Fédération Nationale des Syndicats d'Exploitants; Pres. ERIC NELSON.

**Union Départementale de la Confédération Française des Travailleurs Chrétiens:** 29 rue Victor Hugo, BP 245, 97159 Pointe-à-Pitre; tel. 5-90-82-04-01; f. 1937; Sec.-Gen. ALBERT SARKIS; 3,500 mems.

**Union Départementale des Syndicats CGT-FO:** 59 rue Lamartine, BP 687, 97110 Pointe-à-Pitre; tel. 5-90-82-86-83; fax 5-90-82-16-12; e-mail udfoguadeloupe@force-ouvriere.fr; Gen. Sec. MAX EVARISTE; 1,500 mems.

**Union des Moyennes et Petites Entreprises de Guadeloupe (UMPEG):** 17 Immeuble Patio, Grand Camp, 97142 Abymes, Pointe-à-Pitre; tel. 5-90-91-79-31; fax 5-90-93-09-18.

**Union Générale des Travailleurs de la Guadeloupe (UGTG):** rue Paul Lacavé, 97110 Pointe-à-Pitre; tel. 5-90-83-10-07; fax 5-90-89-08-70; e-mail ugtg@ugtg.org; f. 1973; confederation of pro-independence trade unions including l'Union des Employés du Commerce (UEC), l'Union des Travailleurs des Communes (UTC), l'Union des Travailleurs de l'Hôtellerie, du Tourisme et de la Restauration (UTHTR), l'Union des Travailleurs des Produits Pétroliers (UTPP), l'Union des Travailleurs de la Santé (UTS), and l'Union des Travailleurs des Télécommunications (UTT); Leader MICHEL MADASSAMY; Gen. Sec. RAYMOND GAUTHIEROT; 4,000 mems.

**Union Régionale Guadeloupe:** Immeuble Jabol, 5ème, rue de l'Assainissement, 97110 Pointe-à-Pitre; tel. 5-90-91-01-15; fax 5-90-83-08-64; e-mail m.alidor@aol.fr; internet www.unsa.org; mem. of l'Union Nationale des Syndicats Autonomes (UNSA).

**UNSA Education Guadeloupe:** Immeuble Jabol, 5ème, rue de l'Assainissement, 97110 Point-à-Pitre; tel. 5-90-91-01-15; fax 5-90-83-08-64; e-mail pelage.girard@wanadoo.fr; Sec.-Gen. GIRARD PELAGE.

# Transport

## RAILWAYS

There are no railways in Guadeloupe.

## ROADS

In 1990 there were 2,069 km (1,286 miles) of roads in Guadeloupe, of which 323 km were Routes Nationales.

## SHIPPING

The major port is at Pointe-à-Pitre, and a new port for the export of bananas has been built at Basse-Terre.

**Compagnie Générale Portuaire:** Marina Bas-du-Fort, 97110 Pointe-à-Pitre; tel. 5-90-93-66-20; fax 5-90-90-81-53; e-mail marina@marina-pap.com; internet www.marina-pap.com; port authority; Man. PHILIPPE CHEVALLIER; Harbour Master TONY BRESLAU; 1,000 berths for non-commercial traffic.

**Compagnie Générale Maritime Antilles-Guyane:** Z. I. de la Pointe Jarry, BP 92, 97100 Baie-Mahault; tel. 5-90-26-72-39; fax 5-90-26-74-62.

**Port Autonome de la Guadeloupe:** Gare Maritime, BP 485, 97165 Pointe-à-Pitre Cédex; tel. 5-90-21-39-00; fax 5-90-21-39-69; e-mail v-tarer@port-guadeloupe.com; internet www.port-guadeloupe.com; port authority; Pres. (vacant); Gen. Man. CHRISTIAN BROUTIN.

**Société Guadeloupéenne de Consignation et Manutention (SGCM):** 8 rue de la Chapelle, BP 2360, 97001 Jarry Cédex; tel. 5-90-38-05-55; fax 5-90-26-95-39; e-mail gerard.petrelluzzi@sgcm.fr; internet www.sgcm.gp; f. 1994; shipping agents, stevedoring; Chair. BERNARD AUBERY; Gen. Man. GERARD PETRELLUZZI; 17 berths.

**Société de Transport Maritimes Brudey Frères:** 78 centre St John Perse, 97110 Pointe-à-Pitre; tel. 5-90-91-60-87; fax 5-90-93-00-79; e-mail brudey.freres@wanadoo.fr; internet www.brudey-freres.fr; f. 1983; inter-island ferry service; Dir DENIS BRUDEY; 5 vessels; c. 400,000 passengers per year.

## CIVIL AVIATION

Raizet International Airport is situated 3 km (2 miles) from Pointe-à-Pitre and is equipped to handle jet-engined aircraft. There are smaller airports on the islands of Marie-Galante, La Désirade and Saint-Barthélémy.

**Air Caraïbes (CAT):** Immeuble le Caducée, Morne Vergain, 97139 Abymes, Pointe-à-Pitre; tel. 5-90-82-47-47; fax 5-90-82-47-49; e-mail info@aircaraibes.com; internet www.aircaraibes.com; f. 2000 following merger of Air St Martin, Air St Barts, Air Guadeloupe and Air Martinique; owned by Groupe Dubreuil; operates daily inter-island, regional and international services within the Caribbean, and flights to Brazil, French Guiana and Paris; CEO SERGE TSYGALNITZKY; 14 aircraft; 452,000 passengers (2003); 375 employees.

**Air Caraïbes Atlantique:** Aéroport, 97232 Le Lamentin; f. 2003; subsidiary of Air Caraïbes; services between Pointe-à-Pitre and Fort-de-France (Martinique) and Paris; Pres. FRANÇOIS HERSEN.

## Tourism

Guadeloupe is a popular tourist destination, especially for visitors from metropolitan France (who account for some 89% of tourists) and the USA. The main attractions are the beaches, the mountainous scenery and the unspoilt beauty of the island dependencies. In 2003 some 438,819 tourists visited Guadeloupe. Receipts from tourism totalled €238m. in 2002. In that year there were 162 hotels, with some 8,000 rooms.

**Comité du Tourisme:** 5 sq. de la Banque, BP 555 97166, Pointe-à-Pitre Cédex; tel. 5-90-82-09-30; fax 5-90-83-89-22; e-mail info@lesilesdeguadeloupe.com; internet www.lesilesdeguadeloupe.com; Pres. PATRICK VIAL COLLET.

**Délégation Régionale au Tourisme:** 5 rue Victor Hugues, 97100 Basse-Terre; tel. 5-90-81-10-44; fax 5-90-81-94-82; e-mail drtourisme.guadeloupe@wanadoo.fr; Dir JEAN FRANÇOIS DESBROCHES.

**Syndicat d'Initiative de Pointe-à-Pitre-:** Centre Commercial de la Marina, 97110 Pointe-à-Pitre; tel. 5-90-90-70-02; fax 5-90-90-74-70; internet www.sivap.gp; Pres. M. FORTUNE.

## Defence

At 1 August 2004 France maintained a military force of about 4,100 in the Antilles (headquartered in Martinique).

## Education

Education is free and compulsory in state schools between the ages of six and 16 years. In 2002/03 there were 43 pre-primary schools and 97 primary schools; there were 35 secondary schools in 2000/01. There were 63,553 students in pre-primary and primary education in 2002/03, while in secondary education there were 52,916 (including 8,280 students in vocational education), of whom some 90% attended state schools. Higher education is provided by a branch of the Université Antilles-Guyane at Pointe-à-Pitre, which has faculties of law, economics, sciences, medicine and Caribbean studies. There were also two teacher-training institutes, and colleges of agriculture, fisheries, hotel management, nursing, midwifery and child care. The Guadeloupe branch of the university had 5,800 enrolled students in 1997. An Academy for Guadeloupe was established in January 1997. Government expenditure on education totalled 2,842m. French francs in 1993.

## Bibliography

For works on the Caribbean generally, see Select Bibliography (Books)

Goslinga, M. *Guadeloupe*. Oxford, ABC Clio, 1999.

Jennings, E. T. *Vichy in the Tropics*. Stanford, CA, Stanford University Press, 2001.

Orizio, R. *Lost White Tribes: The End of Privilege and the Last Colonials in Sri Lanka, Jamaica, Brazil, Haiti, Namibia and Guadeloupe*. Mississauga, ON, Random House of Canada, 2001.

# GUATEMALA

## Geography

### PHYSICAL FEATURES

The Republic of Guatemala is in Central America, apart from Belize the country furthest north on the isthmus, its territory abutting into the Yucatán peninsula. Its longest border (962 km or 597 miles), therefore, is with the North American country of Mexico, which lies to the north and the north-west (Campeche is to the north, Tabasco in the north-west and Chiapas stretches west). To the south-east, further down the trunk of Central America, are El Salvador (beyond a 203-km frontier) on the Pacific Ocean and, further north, Honduras (with a 256-km border). The long Pacific coast faces south and south-west. Apart from a relatively short, north-facing shore with no natural harbours at the marshy head of the Gulf of Honduras, the country is isolated from the Caribbean coast by Belize, which lies to the east of northern Guatemala. The border with Belize is 266 km long, but Guatemala maintains claims on territory in the south of the former British colony, as well as rights of access to the Caribbean. The Organization of American States is mediating the dispute, which also gave rise to the protocols of 2000 that govern a buffer territory within the 'Lines of Adjacency', a zone 2 km wide bisected by the common border. The area of Guatemala is 108,889 sq km (42,042 sq miles).

The extent of the country is about 450 km (north–south) by 430 km, and it has about 400 km of coastline, mostly on the Pacific. Guatemala is mostly mountainous (66%), with narrow coastal plains and a rolling limestone plateau in the north. It is heavily forested (62%), woodland types varying from warm and humid, through cool and humid to warm and dry, which, together with the altitude and two coasts, contributes to the variety of flora and fauna. The most spectacular of the many bird species in the country is the national bird, the quetzal, but another noteworthy example might be the ocellated turkey, while the largest native mammal is a fresh-water sea cow (manatee) found in Lake Izabal, in the east. There are volcanoes in the mountains, indicating that the country is an area prone to seismic activity. A volcano is actually the highest point not only in Guatemala, but in all of Central America—Tajumulco, at 4,211 m or 13,821 ft—and is found at the western end of the main mountain chain, in the south-west of the country. The main mountain range of the high plateau of the south-west is the Sierra Madre, which echoes the line of the Rockies and the Andes in the continents to north and south, and it runs roughly east–west above the Pacific coast. There is a branching range, the Sierra de los Chuchumatanes, which thrusts more to the north-west, and other highlands in this area include the Sierra Chaucus, the Montanas del Mico and the Sierra de Chama. Most of the country's volcanoes are on the central plateau. The three other topographical regions are less lofty. South of the mountains is the Pacific coast, which consists of tropical savannah plains and lagoons, and is traversed by 18 short rivers. North of the central plateau and in the east of the country are the lower slopes and more easterly out-thrusts of the Continental Divide (such as the Sierra de las Minas), giving way to the often swampy Caribbean flatlands, this region dominated by three deep river valleys (of the Motagua, the Polochic and the Sarstun). Finally, north of here, occupying most of northern Guatemala and mainly beyond the Maya Mountains is the water-eroded limestone plateau of El Petén, dotted with lakes and densely cloaked in tropical forest, watered by heavy, year-round rain. The largest lakes are not in this area, however, with the 800-sq-km Izabal (also known as the Dulce gulf) in the Caribbean lowlands and the famously beautiful, 126-sq-km Atitlán in the central highlands).

### CLIMATE

The climate is tropical, hot and humid, especially on the Caribbean coast and in the El Petén lowlands, but it is cooler in the highlands. The Caribbean coast is also prone to hurricanes and

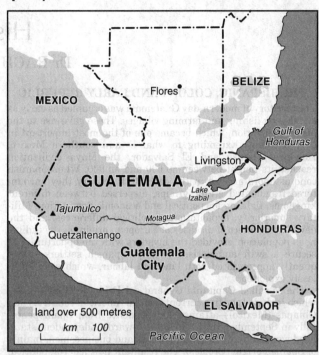

other tropical storms. Average annual rainfall varies from as low as 510 mm (20 ins) in the eastern highlands to 2,540 mm (100 ins) in the north. The average maximum and minimum temperatures in Guatemala City, the capital, which is located on the central plateau, range from 12°–23°C (54°–73°F) in January to 16°–29°C (61°–84°F) in May. May is the end of the dry season and the start of the rainy season.

### POPULATION

Guatemala has the largest surviving indigenous population in Central America. The normally dominant Mestizo population of mixed-Spanish descent (including assimilated Amerindians), here known as Ladino, accounts for barely 55% of the total. Often included with the Ladinos are the small populations of more purely Spanish descent and of other non-indigenous origins, such as Syrians and Lebanese, Asians and the Garifuna (Black Caribs, mainly of African descent, with some Amerindian ancestry, settled on the Caribbean coast). Amerindian peoples, however, account for 43% of the population. That they are concentrated in the countryside may obscure the veracity of the figure, with some claiming the Amerindians, the Mayas, account for almost two-thirds of the population. Spanish is the official language and is spoken by 60% of the population, but there is some official status for Garifuna (Carib) and around 20 of the 50 or so Mayan languages spoken in the country. The main Mayan language groups spoken in Guatemala are Quiché (Kiche), Cakchiquel, Mam and Kelchi. It is not only language that survives among the Maya, but also some traditional religious beliefs and practices, although usually syncretized with the dominant Roman Catholic Christianity. Certainly more than two-thirds of the population claim to adhere to the Church, although campaigning Protestant groups from North America gained up to one-fifth of the population over the last decades of the 20th century. There are also small groups of Orthodox Catholics ('Uniates'), Buddhists, Jews, Muslims and 'Moonies' (followers of the Unification Church).

According to the 2002 census, the people of Guatemala, sometimes informally referred to as 'Chapines', numbered 11.24m., making them the most populous nation of Central America.

According to UN estimates, at mid-2004 the population numbered 12.4m. Most of the population live in the temperate valleys of the highlands, in the south-centre of the country, around the capital. The capital is Guatemala City (Guatemala de la Asunción—the greater metropolitan area numbering some 2.5m. residents in 2002). The other chief cities of the highlands, to the north-west of the capital, are Quetzaltenango and Sololá. Livingston, on the Caribbean coast, is not a large city, but it is the chief town of the east, while Flores is the main centre of El Petén. The country is divided into 22 departments.

# History

## Dr RACHEL SIEDER

### PREHISPANIC, COLONY AND EARLY REPUBLIC

The territory of modern-day Guatemala was occupied as early as BC 2000 by fishing and farming villages. These gave rise to the Maya civilization, which became one of the most important in the continent. Extending to what is now southern Mexico, Honduras, Belize and El Salvador, the Maya civilization reached its height between AD 600 and AD 900. When Spanish conquistadors arrived in the early 16th century they encountered fragmented ethnic groups, dispersed between different kingdoms, fighting for political and economic domination. Military superiority, exploitation of political differences, and the devastating effect of diseases of European origin on the indigenous population provided the invaders with the opportunity to secure a swift and decisive victory. Spanish settlers subsequently appropriated both land and labour, which were ruthlessly exploited.

The so-called Capitanía General del Reino de Guatemala—which included the five current Central American countries and Chiapas (Mexico)—attained independence from Spain peacefully in September 1821. The move towards independent statehood was led by wealthy landowners and businessmen, mainly of Spanish origin (Creoles). The resultant new federal, political and administrative entity, the United Provinces of Central America, was dissolved in 1839, owing to internal conflicts. The current geographical territory of Guatemala was established later in the century, following the loss of Belize to the United Kingdom and Chiapas to Mexico. From the mid-19th century onwards Guatemala invested heavily in the development of coffee plantations, which had become the main source of national income by the end of the century. After the Liberal revolution of 1871, the country was integrated into the global market by the rapid expansion of agro-exports. As a result of the introduction of new legislation and the creation of a national army, Liberal governments institutionalized a more coercive role for the state in promoting agrarian capitalism. Indigenous peoples lost huge swathes of their communal lands, which were incorporated into large landholdings dedicated to the production of coffee exports. Hundreds of European settlers, particularly Germans, were attracted by the prospect of favourable grants of land and labour to aid in the development of agro-exports. The rights and obligations of rural employers and workers were codified in the late 19th century in a series of agricultural laws, which refined and intensified traditional practices of debt peonage. During the first two decades of the 20th century the Government of Manuel Estrada Cabrera opened the country to capital investment from the USA. This monopolized services such as railways, ports, electricity and maritime transport, as well as the international mail service. US monopoly capital was also used to acquire significant land holdings, as the United Fruit Company procured large tracts for the development of banana plantations.

In 1944, after 14 years in power, Gen. Jorge Ubico, the last representative of the Liberal reforms of 1871, was overthrown by a popular uprising led by nationalist members of the army and progressive intellectuals, who were inspired by US President Franklin Delano Roosevelt's 'Four Freedoms' address to the US Congress in 1941. The so-called 'October Revolution' led to presidential elections, which were won by a civilian, Juan José Arévalo. The new President introduced a series of economic and social reforms, including universal male suffrage, the establishment of a social security system for state employees and a new labour code. His successor, Col Jacobo Arbenz Guzmán, who acceded to the presidency in 1950, enhanced these measures and initiated a radical agrarian reform, which encouraged the organization of peasant leagues throughout the countryside and benefited some 100,000 families by expropriating uncultivated lands from large landowners and the United Fruit Company. At the height of the Cold War, concerned by the influence of the Partido Guatemalteco del Trabajo (PGT—the Guatemalan communist party) within the Arbenz administration, the recently created US Central Intelligence Agency (CIA) exploited discontent within the Guatemalan army and private sector and organized a military invasion from Honduras, which overthrew the Arbenz Government in June 1954. The counter-revolution reversed most of the reforms of the previous decades, began a systematic process of political persecution of government supporters, and installed a series of anti-communist regimes dominated by the armed forces. During the late 1970s the ability of these military regimes to govern deteriorated, as a result both of splits within the ruling coalition and of their lack of popular legitimacy. Elections held in 1974, 1978 and 1982 were openly fraudulent.

### GUERRILLA WARFARE AND COUNTER-INSURGENCY

Beginning in the early 1960s an armed opposition developed, originating from a schism within the army, but later garnering widespread indigenous support in the rural areas. This initiated the guerrilla and anti-guerrilla warfare that would last for more than three decades. Some 200,000 Guatemalans were either killed or disappeared during this period, largely as a result of state military operations against civilians, particularly rural Mayan communities suspected of supporting the insurgency. Mass violations of human rights under the military Government of Gen. Fernando Romeo Lucas García (1978-82) led the US Administration of Jimmy Carter (1977–81) to suspend military aid to Guatemala (this was only resumed in 1985 under Ronald Reagan). In February 1982 three different guerrilla organizations and the PGT formed a guerrilla coalition, the Unidad Revolucionaria Nacional Guatemalteca (URNG). At their height the guerrillas were operating in 18 of the country's 22 departments. Divisions within the army over the prosecution of the war also facilitated two military coups during this period. The first took place on 23 March 1982 and was led by Gen. Efrain Ríos Montt, a former Christian Democrat candidate and, latterly, evangelical preacher, who was denied the presidency in 1974, owing to electoral fraud. Gen. Ríos Montt's 15 months in office are widely regarded as having been the most violent in the country's modern history. He intensified the counter-insurgency offensive, and in the first year of his regime more than 15,000 Guatemalans were assassinated, mainly among the Mayan rural population, 70,000 fled the country, and some 500,000 were displaced internally. Hundreds of villages were eradicated, destroyed by systematic army massacres. In the rural areas all men between the ages of 16 and 60 were forced to participate in paramilitary civil defence patrols, which, at their height, comprised some 1m. peasants. 'Model villages', under strict army control, were established as detention centres for those across the country who had been displaced by war. The military operations of the guerrilla movement declined dramatically during this period, overwhelmed by the superior fire-power and ruthless tactics of the army. In August 1983 Gen. Ríos Montt was overthrown by his Minister of Defence, Gen. Oscar Humberto Mejía Víctores. The new military regime restored the promotional hierarchy that had been disrupted by the former Government, and promised a swift return to constitutional rule, while enforcing a continued programme of counter-insurgency and political repression.

## CIVILIAN RULE AND PEACE PROCESS

By the mid-1980s there was a consensus within the higher echelons of the military that Guatemala should shift to civilian rule if it was to change its international status as a pariah. On 1 July 1984 a Constituent Assembly was elected; the promulgation of a new Constitution in 1985 and the holding of general elections in the same year ushered in a period of elected civilian rule and a limited relaxation of the military's control over national affairs. However, the key institutions of the counter-insurgency, such as the civil patrols, were legalized by the new Constitution, which provided a framework for civilian government throughout the 1980s. Although the URNG boycotted the presidential elections held in November 1985, the centre-right Christian Democrat candidate, Vinicio Cerezo, won them by an ample margin. Yet, despite high hopes for the new civilian Government, Cerezo did not challenge the dominance of the military or address the country's acute socio-economic problems. Rising prices for industrial imports, coupled with falling prices for Central America's primary export products, meant that the Guatemalan economy suffered negative growth rates during the 1980s; unemployment and inflation reached unprecedented levels. Overall levels of poverty increased significantly in the second half of the decade, and by the end of the 1980s nearly 90% of the population were living below the poverty line (compared with 79% in 1980); approximately three-quarters of the population were living in extreme poverty. The social and economic crisis led to the re-emergence of social movements and civic protests, despite the harsh repressive measures that this evinced from the military. After the 1985 election the URNG proposed that the armed conflict be settled by negotiation, and in October 1987 representatives of the Government and the URNG met for the first time in Madrid, Spain, to initiate discussions. The meeting was a direct result of the regional 'Esquipulas II Accord', signed two months previously by all six Central American republics in search of a political resolution to the region's armed conflicts. On 30 March 1990 the guerrilla movement and the Government agreed to both a framework for negotiations and to a mediating role for the UN and consultative status for the National Commission for Reconciliation (Comisión Nacional de Reconciliación—CNR), a coalition of national civic groups led by the head of the Guatemalan Bishops' Conference, Rodolfo Quezada Toruño. However, the peace talks stalled within months, owing to the army's insistence upon a full demobilization of the URNG prior to a final settlement and an amnesty on human rights violations.

Presidential and congressional elections held in November 1990 registered abstention rates of 70%. No parties of the left were permitted to participate. Jorge Serrano Elías, an evangelical leader and former member of the 1982–83 Ríos Montt Government, won the second round of voting, held in January 1991. The new Government reinitiated peace negotiations with the guerrilla movement, and meetings were held in Cuernavaca, Mexico, in April. Meanwhile, the domestic political situation continued to deteriorate; human rights organizations reported some 1,750 human rights violations, including 650 extra-judicial executions, during the first nine months of the Serrano Government. The USA once again suspended military aid to Guatemala and, together with the World Bank and the European Parliament, exerted pressure upon the Government to bring an end to the political violence. In October 1992 Rigoberta Menchú, a Guatemalan Maya-K'iche' woman and indigenous human rights activist, won the Nobel Peace Prize, focusing global attention on the perpetuation of human rights violations in the country, and also signalling the emergence of an increasingly organized Mayan, indigenous, popular movement. Facing a mounting crisis of legitimacy, President Serrano, supported by a faction of the military, mounted a so-called autogolpe 'self-coup' in May 1993, attempting to suspend significant constitutional guarantees and dissolve the National Congress and the Supreme Court. However, in the face of domestic and international opposition the coup ultimately proved unsuccessful, and a few days later Serrano abandoned the country for exile in Panama. (In June 2002 an international detention order was issued for Serrano and former Secretary of the Interior Francisco Perdomo Sandoval, following the failure of attempts to extradite the men from Panama to face the criminal charges against them).

The then ombudsman for human rights, Ramiro de León Carpio, was designated interim President by the Congreso Nacional on 6 June 1993 to enable the completion of Serrano's presidential term. His cousin, Jorge Carpio Nicolle, a political leader and newspaper tycoon who had been defeated by Serrano in the second round of the 1991 election, was assassinated shortly after de León assumed power. The end of the Cold War, and the successful negotiation of settlements of conflicts in neighbouring Nicaragua and El Salvador, favoured those sectors supporting a negotiated end to the armed conflict. Yet disenchantment with domestic democratic politics persisted—a referendum to approve a series of constitutional reforms in 1994 registered an abstention rate of 85%. The hardline, populist Frente Republicano Guatemalteco (FRG), headed by the former de facto chief of state, Gen. (retd) Ríos Montt, obtained a congressional majority in elections held in the same year. In January 1994 the Government and the guerrilla movement signed a framework agreement for the resumption of peace talks in Mexico. In March a comprehensive agreement on human rights was signed following discussions held in Oslo, Norway, which was to enter into immediate effect (unlike the other accords, which would be implemented only at the end of the entire peace process) with in situ verification by the UN. Before the end of the year the guerrilla movement and the Government had signed an agreement, with UN verification, for the protection of human rights. A further agreement established conditions for the resettlement of displaced populations. Most controversial was the June 1994 accord that delineated the terms of a Commission for Historical Clarification, or truth commission, to investigate human rights violations committed during the armed conflict. Despite the efforts of domestic and international human rights organizations, the final agreement stated that the commission's report would not individualize responsibility for gross violations—this was widely interpreted as being a concession to the hardline element within the military. In March 1995 another wide-ranging agreement relating to the identity and rights of Guatemala's indigenous population (which constitutes between 50% and 60% of the total population) was signed. This was the only accord in the formulation of which civic groups played a significant role, through the consultative forum of the Civil Society Assembly (ASC), created in 1994 to replace the CNR. However, by mid-1995 peace negotiations had stalled again, a consequence of the declining strength of the de León Government and a lack of political will on the part of military and civilian elites.

Presidential elections held in November 1995 led to a second round of voting, won by Alvaro Enrique Arzú Yrigoyen of the Partido de Avanzada Nacional (PAN), a party representing the interests of the private sector. Arzú, a businessman, former mayor of Guatemala City and Minister of Foreign Affairs during the first months of Serrano's government, won a close electoral race against Alfonso Portillo of the FRG in the second round run-off held in January 1996. The rate of abstention was approximately 63%. The new administration advanced the peace process and gave new impetus to its development. Arzú appointed prominent business leaders to key government posts, securing the private sector's commitment to the peace negotiations. New peace accords, arguably some of the most controversial, were signed in 1996, before a final settlement was reached in December. The new accords included an agreement, reached in May, on socio-economic issues and the agrarian situation, committing the Government to increase social spending and to fund a land bank through which landless peasants could acquire land. Another, negotiated in September, strengthened civilian rule and defined the function of the army in a democratic society, specifying the terms by which military power would be gradually reduced and public security reformed. A number of enabling accords were signed in December, including those referring to: constitutional reform and the electoral regime; the legalization of the URNG; the conclusion of a definitive cease-fire; and the final accord for a firm and lasting peace, which was signed in Guatemala City on 29 December in the presence of UN Secretary-General, Boutros Boutros Gali.

The rebels disarmed shortly afterwards, under UN supervision, and the URNG subsequently became a legal political party.

## POST-ACCORD POLITICS

The period immediately following the achievement of the negotiated settlement was dominated by attempts to implement the reforms promised in the various accords, and by efforts to investigate past violations of human rights. In April 1998 the Roman Catholic Church's Oficina de Derechos Humanos del Arzobispado (ODHA—Archbishopric's Human Rights Office) published a report documenting gross violations of human rights that had occurred during the armed conflict, based on over 6,000 testimonies. This attributed more than 85% of war atrocities to the army and army-controlled paramilitary forces and 10% to the URNG. In February 1999 the UN Report of Historical Clarification, based on over 9,000 testimonies, went further, attributing 93% of atrocities to the army and paramilitary forces and 3% to the URNG. The UN report demonstrated that the army had committed acts of genocide (as defined by international law) legitimated by state policy during the 1981-83 period. However, the costs of such investigations were high—Juan José Gerardi Conadera, the bishop in charge of the 1998 ODHA report, was murdered just days after its publication. It took three years and the exertion of immense international pressure for domestic courts to find three army officers guilty of his extra-judicial execution. Guatemalan human rights organizations attempting to secure the prosecution of former army officers in domestic and international courts for gross violations of human rights were subject to continuing intimidation and violence.

In March 1999 the peace settlement suffered a serious setback when a referendum on a package of constitutional reforms agreed by the legislature to implement the peace agreements was rejected with a turn-out of less than 25%, slowing implementation of the settlement almost to a standstill. The FRG candidate, Alfonso Portillo Cabrera, won the 1999 presidential elections that were held on 26 December, with 68.3% of the votes cast in the second round run-off (compared with the 31.7% secured by his rival Oscar Berger Perdomo of the PAN). Although the FRG was formally committed to the implementation of the peace agreements (a key feature of the prevailing conditions by which international aid was granted), the party lacked the commitment of the Arzú administration to the process, divided as it was between moderates and hardliners, the latter allied with party founder, Ríos Montt, who maintained a powerful position as the President of the National Congress. The weakness of the public security and justice systems and the growth of organized crime resulted in governmental inability to address the increasing problem of crime and public insecurity, which had reached unprecedented levels by the end of the decade. The state-orchestrated violence of the 1980s was replaced by a wave of kidnappings, armed assaults and robberies, leading to escalating public alarm about law and order and calls for more stringent penal measures, such as the introduction of the death penalty for kidnappers, which was approved in 1996. Meanwhile, the implementation of the demilitarization agreed in the September 1996 accord met with numerous obstacles; although the size of the armed forces was reduced by one-third as stipulated, military intelligence institutions remained largely unreformed at the end of the decade, despite express provisions that they be dismantled. The UN repeatedly raised concerns that the military budget was being expanded and that the new civilian police force was becoming increasingly militarized. Consistently high levels of violence and crime with impunity led the population to resort to private justice; levels of gun ownership and homicide continued to increase and suspected criminals were subjected to mob lynchings. Registering over 80 violent deaths per 100,000 inhabitants annually, Guatemala is one of the most violent countries in the Western hemisphere.

By the end of 2000 conflict between the private sector and the ruling FRG was acute; the former was particularly opposed to the Government's attempts to raise taxes—a key demand of the IMF and bilateral donors and a necessary prerequisite for the funding of the improvements in health and education provision mandated by the peace accords. In July 2001 the legislature finally approved legislation increasing value-added tax (VAT) from 10% to 12%; the rise prompted violent, nation-wide protests. Discontent over the Portillo administration's lacklustre record in office, multiple corruption scandals involving high-ranking government officials, and poor fiscal management became increasingly manifest. Fiscal weakness, combined with divisions between the Portillo and Ríos Montt factions of the FRG, continued to undermine the administration's coherence. In February 2003 the USA added Guatemala to its list of states deemed unco-operative in the campaign against illegal drugs-trafficking. Until 2004 the country also remained on the so-called 'blacklist' of countries considered to be unco-operative in the fight against money-laundering, drawn up by the Financial Action Task Force on Money Laundering (FATF, based in Paris, France).

## THE GOVERNMENT OF OSCAR BERGER PERDOMO

Since the return to democracy in 1985 no incumbent party has ever been re-elected to office. The FRG proved no exception to this rule and the discredited party lost power in the presidential and legislative elections in November 2003. Despite the controversial approval by the Constitutional Court in July 2003 of the presidential candidacy of former dictator Ríos Montt (previously prohibited by the Constitution), the FRG fought an uninspired electoral campaign and was adversely affected by accusations of corruption and its association with human rights violations. In the first round of voting, Oscar Berger Perdomo of the centre-right alliance the Gran Alianza Nacional (GANA—Great National Alliance) won 34.3% of the votes cast. Second placed was Alvaro Colom Caballeros of the centre-left Unidad Nacional de Esperanza (UNE—National Unity of Hope). Ríos Montt attracted just 19.3% of the ballot. Following a second round of voting between the two leading candidates on 28 December, Berger emerged victorious, winning 54.1% of the vote. However, although the Berger Government represented a return to power of the business sector, the new Government proved far weaker than the Arzú administration. Berger, formerly mayor of Guatemala City for the PAN, had originally been selected as the PAN's presidential candidate. Following divisions within the party, however, in May 2003 Berger left to form his own electoral coalition, the GANA. Although Berger was successful in the presidential ballot, the GANA—a fragile alliance of small parties—did not obtain a majority in the concurrently held congressional elections, winning just 49 of the 158 seats. The FRG came second with 42 seats, followed by the UNE with 33. The PAN secured 16 seats and the Partido Unionista (PU—Unionist Party) obtained seven seats. The remaining 11 seats were shared among smaller left-wing parties.

On taking office in January 2004, the new Government faced a series of difficult challenges, including a severe funding shortfall, a looming fiscal crisis and a divided legislature. Following two weeks of negotiations, a governability pact was agreed in January between GANA, the UNE and the PAN to secure working majorities in congress. However, this promptly collapsed following accusations that both the UNE and PAN had received campaign financing from the former FRG Government. The absence of a congressional majority, combined with chronic fragmentation and instability within the parties themselves, has made it very difficult for the Government to steer its policies through the Congreso. In addition, in early 2004 one of the main parties of the GANA, the Partido Patriota, withdrew from the alliance in protest at the Government's negotiations with the FRG. Partly as a result of its weak position in the Congreso, Berger's Government pursued a minimalist agenda and by mid-2005 had made little progress in addressing social and economic inequalities, as mandated by the peace accords. In early 2004 an austerity programme was announced in response to the fiscal crisis. Nevertheless, pressure to increase taxes was resisted by the private sector and particularly by the major business organization, the Comité Coordinador de Asociaciones Agrícolas, Comerciales, Industriales y Financieras (CACIF), hitherto one of Berger's main supporters. A fiscal package was finally approved in June 2004, but fell far short of the comprehensive overhaul envisaged in the peace accords.

Confidence in state institutions had been severely eroded during the Portillo Government, and on assuming office Berger launched an anti-corruption campaign. The campaign gave rise to dozens of criminal investigations into alleged illicit activities and corrupt practices by former government officials. These measures received strong support from the USA: in 2004 a customs and immigration task force in the USA and judicial authorities in Florida began investigating money-laundering activities linked to former President Portillo and his associates. Portillo subsequently fled Guatemala for Mexico, following a ruling by the Constitutional Court in February that stripped him of the legal and political immunity afforded him as a deputy of the Central American Parliament (PARLACEN). In April a former Minister of Public Finance in the Portillo administration, Eduardo Weymann Fuentes, was arrested pending criminal investigations into his role in approving unauthorized transfers made to the superintendency of tax administration (he was released in March 2005 after being convicted on lesser charges). The superintendency had been headed by Marco Tulio Abadío, an FRG appointment, who was convicted of multiple corruption charges and eventually imprisoned after evading arrest for several months. In July former Vice-President Francisco Reyes López was also arrested and charged with fraud. The anti-corruption campaign won some much-needed support for the Government and was strongly supported by the US Administration of George W. Bush, which brought pressure to bear on its allies to address the problems of money-laundering and drugs-trafficking as part of its 'war on terror'. Relations between Guatemala and the USA, which had deteriorated significantly during Portillo's presidency, improved markedly under the Berger administration. Following a visit by the then US Secretary of Defence, Donald Rumsfeld, in 2004 the US Government ended a long-standing ban on the sale of armaments to the Guatemalan military, signalling closer military co-operation in the future—primarily in order to tackle organized crime and improve regional security.

In spite of these successes, the first year of the Berger Government saw rising social discontent, partly in response to increases in inflation. On taking office the new Government had pledged to improve social conditions, in particular the problems of violent crime and insecurity. Joint army and police patrols were introduced and a weapons amnesty was launched to reduce the number of weapons in circulation. The amnesty allowed people to swap guns for bicycles, sewing machines and other household appliances. However, high levels of crime continued to be an issue of public concern. Peasant land shortages continued to prompt demonstrations and land occupations; ensuing clashed with the police resulted in the deaths of peasants and police-officers on more than one occasion. Opposition by trade unionists and peasant organizations to the proposed Central American Free Trade Agreement (CAFTA) with the USA also led to violence in early 2005. None the less, the Agreement was ratified by the Congreso on 10 March following an agreement between the government coalition and the major opposition parties.

The Government was also dogged by protests from former members of the Patrullas de Autodefensa Civil (PAC), a pro-Government peasant militia which have been mobilized since 2002, demanding compensation for their participation in the counter-insurgency campaign. In March 2005 Berger yielded to pressure and promised to pay compensation, to the consternation of human rights groups who believed the PAC to be responsible for human rights violations. However, legal problems have thus far prevented a final resolution of the matter. Protests by indigenous peoples' organizations at the lack of proper consultation over mining concessions in indigenous communities also proved a recurrent problem for the Government. The right of consultation for indigenous peoples is guaranteed in the International Labour Organization's Convention 169 on the rights of indigenous peoples, which was ratified by Guatemala in 1995 although routinely ignored by state authorities. Overall, while governability has been secured in the medium term by the Berger Government, attempts at structural reform in line with the peace accords have delivered relatively meagre results. Crime, security and poverty continued to be major public concerns in 2005.

# Economy

## Prof. JENNY PEARCE

Revised for this edition by the editorial staff

At the beginning of the 21st century the Guatemalan economy remained dominated by agriculture, which, together with hunting, forestry and fishing, contributed an estimated 22.8% of the gross domestic product (GDP) in constant prices, in 2003 and employed 42.1% of the economically active population in 2002. At the same time, however, compared with the rest of Central America, Guatemala had a relatively highly developed industrial sector, which in 2004 contributed an estimated 18.9% of GDP and exported processed products to the USA and to the country's neighbours. Only a small proportion of the country's reserves of petroleum were exploited; copper, antimony, tungsten and nickel were also mined.

Economic growth declined in the late 1970s, after two decades of expansion, and the economy stagnated in the 1980s as war and violence escalated in Guatemala and the rest of Central America. In 1980–90 annual growth averaged 0.9%, while per-head GDP decreased by an annual average of 0.8%, in real terms. The recession had passed its worst by the mid-1980s, although per-head GDP growth remained slow until the end of the decade. Average annual growth in 1990–2003 was 3.7%, reaching its peak in 1998 with 5.0%, before steadily declining in subsequent years, falling to 2.1% in 2003. In 2004 growth was an estimated 2.7%. Average per-head growth over the same period, however, was only about one-half overall GDP growth.

Growth was stimulated by an increase in non-traditional exports, which outperformed the principal traditional export, coffee, in 1990, and was also improved by a sudden increase in infrastructure development. The non-traditional export sector showed the most dynamism in cultivated seafood (seafood-farm products), wood furniture and vegetables. The total value of non-traditional exports increased towards the end of the last century. Nevertheless, Guatemala continued to show a large trade deficit, which increased from US $521m. in 1991 to $3,127m. in 2004. During the 1990s the country had failed to diversify its export products sufficiently, had been unable to nurture skilled labour to use high technology and had not developed its manufacturing base. At the beginning of the 21st century the economy remained overly dependent on an external sector that was extremely vulnerable to world prices of the main export commodities (coffee, sugar, bananas and petroleum), which were depressed through the 1990s; coffee prices experienced record lows in 2001 and 2002.

Increasing levels of poverty remained a major constraint on economic development. It was estimated that Guatemala's poverty rate increased from 63.4% to 89.0% in 1980–90. Extreme poverty, defined as the percentage of people unable to meet basic nutritional requirements, more than doubled, from 31% in 1980, to 67% in 1990, with even higher figures among the country's majority indigenous population. However, the poverty rate subsequently decreased, falling to 56.2% in 2000, while the percentage of Guatemalans living in extreme poverty declined substantially, to 15.7%. Remittances from Guatemalans living abroad formed a large part of the economy. From 1996 remittances increased annually, becoming Guatemala's second largest hard currency inflow after commodity exports. In 2002 remittances provided more than six times the revenue generated by coffee exports and in 2004 they increased by 21% to reach a record US $2,550m. Furthermore, there was insufficient private and public investment to strengthen the productive base of the economy or to improve the social indicators. In early 2002 a severe drought in the Central American region had particularly significant repercussions in Guatemala. Many families engaged in subsistence-level farming lost their crops and poverty and malnourishment were widespread. In March 2002 the UN's World Food Programme announced an emergency plan to feed an estimated 60,000 Guatemalan children under the age of five who were severely malnourished; the Government also established more than 60 emergency feeding centres in an attempt to deal with the crisis.

A major problem for the economy was the low tax revenue, which, at around 7% of GDP at the end of the 1990s, was the lowest in Latin America. Only 20% of government revenues came from income taxes. This resulted in high (41%) government expenditure on financing the country's internal debt and forced the Government to keep interest rates high, in real terms, impeding economic growth. The peace accords signed in December 1996 committed the Government to increase tax revenue to 12% of GDP, and the issue of tax reform was a principal point of conflict during the presidencies of Alvaro Enrique Arzú Irigoyen (1996–2000) and Alfonso Portillo Cabrera (2000–04). The Portillo Government did succeed in securing the passage of a number of reforms in 2001 and 2002, but a series of corruption scandals in 2003 undermined investor confidence and slowed the progress of any further reform efforts. In 2003 tax revenue stood at 10.3% of GDP. The incoming Government of Oscar Berger Perdomo in 2004 also declared tax reform to be one of its priorities, pledging to increase tax revenue to 12% of GDP. However, the lack of an absolute majority in the Congreso meant that Berger's party, the Gran Alianza Nacional (GANA), was compelled to gain the consensus of opposition parties in order to pass legislation.

## AGRICULTURE

Agriculture, including hunting, forestry and fishing, was the main sector of the economy. Coffee, sugar, bananas, oil and cardamom were the major agricultural exports, although fluctuating world prices for many commodities had an adverse affect on exports from the 1990s onwards. Coffee, the main export crop, accounted for 11.4% of the value of total exports in 2003, compared with 50% in the mid-1980s. Coffee was followed in importance by sugar (8.1%), bananas (8.0%) and cardamom (3.0%—Guatemala was the world's largest producer, typically accounting for 90% of cardamom on international markets). Other exported agricultural products included vegetables, plants (including seeds and flowers) and fish.

Although the suspension of the International Coffee Organization quota system in 1989 was initially beneficial to Guatemala, it severely affected coffee exports in the early 1990s. There was some recovery in the mid-1990s, and in 1995, exports totalled US $539.3m. 'Hurricane Mitch', which struck Central America in October 1998, contributed to a 6% decline in coffee exports in that year. Average prices decreased steadily in 1998–2001, prompting fluctuations and an overall decline in the volume and consequently the revenue of exports. Revenue from coffee exports decreased from $586.5m. in 1998 to $261.7m. in 2002. Small producers, representing around 100,000 families and contributing 30% of overall national production, were particularly badly affected.

The question of how best to sustain international coffee prices preoccupied all exporters following the collapse of the International Coffee Agreement. Guatemala opted not to support a Central American plan to withhold coffee exports in an attempt to stimulate growth in falling international prices. The National Coffee Association (Anacafé) adopted a 'hedged' loan programme, under which producers received financial support by selling coffee beans at or above a set minimum price, which would then cover the loan in the event of depressed market conditions. Low world prices had a devastating impact on the sector in 1996; the average level at which income justified cost was around 130 cents per lb for Guatemalan growers, but prices were around 115 cents for most of that year. In 2001, with prices at an all-time low, Guatemala joined an agreement negotiated by Mexico, Colombia and the Central American nations in March, to withhold 5% of their lowest quality beans from export and put them to other uses, for example, fertilizer or fuel for industry. The plan coincided with an international scheme to withhold 20% of the export goods that had failed to prevent the plunge in coffee prices. Meanwhile, some of the nation's growers held internet auctions of quality beans.

In mid-2002 the Portillo Government established a US $100m. aid fund intended to provide coffee growers with financial assistance in an attempt to relieve the situation. However, the effects of the aid were limited by the fact that it could not be used to finance the transportation of the coffee to market; it was intended instead to enable growers to reschedule debts and to experiment with new technology. Furthermore, very little of the money was made available to those who needed it. At a meeting in early 2002 attended by coffee producers from across Central America, methods were discussed by which coffee quality could be improved and methods of production diversified in an attempt to mitigate against the effects of the crisis. Guatemalan officials petitioned the Inter-American Development Bank (IDB—based in New York, USA), USAID and the World Bank, requesting international financial assistance but their request was denied. The average price for Guatemalan coffee exports stabilized in 2001–03 before increasing in 2004, but remained well below pre-2001 prices.

From the 1980s there was a substantial decline in the area planted with cotton, a trend accelerated by the guerrilla war and high production costs, compounded by competition from other regions and from a rise in the production of synthetic fibres. Conversely, the area planted with sugar cane expanded by more than 10% in the late 1980s and early 1990s. Estimated production of sugar rose steadily in the 1990s, reaching an estimated 1.9m. metric tons in 2002. In 2003 sugar production was an estimated 1.8m. tons. Exports of sugar increased from US $220.4m. in 1996 to $316.6m. in 1998. The impact of 'Hurricane Mitch' in late 1998 adversely affected regional trade and resulted in a fall in sugar prices in the following year. Sugar exports in 1999 were worth $192.1m., a slight increase from the previous year. Exports subsequently recovered, reaching $227.1m. in 2002, before declining to $212.3m. in 2003. Guatemala remained the world's seventh largest sugar producer.

Banana export revenues stood at US $210.0m. in 2003, providing 8.0% of total exports by value. There was potential for profitable wood production, as more than one-third of the country was covered with forests, including valuable cedar and mahogany. Total roundwood removals in 2003 amounted to 16.1m. cu m, while sawnwood production was estimated at 366,000 cu m.

## MINING AND POWER

The mining sector was small, contributing an estimated 0.4% of GDP in 2004 and employing 0.2% of the work-force in 2002. The largest operation was a copper mine in Alta Verapaz Department, which began production in 1975, but had an unrealized potential of 150,000 metric tons per year. Lead and zinc output fell drastically and a nickel-mining project was suspended in 1980, after only three years of production, owing to low world prices and high production costs. Antimony and tungsten were mined and Guatemala had exploitable reserves of sulphur and marble. With the hope that guerrilla violence was over, in 1996 the Government began attempts to attract foreign investment to the mining sector. The energy and mines ministry drafted legislation to increase the profitability of mining operations for foreign investors (through reductions in government royalties for mining concessions and in the time taken to allocate permits for companies applying to move into exploitation after exploration). Australia's Broken Hill Proprietary (BHP) Co was granted a concession to explore for copper, lead, zinc and silver in Quiché and Alta Verapaz in 1996. Mining production decreased by an estimated 8.3% in 2000, before registering positive growth in the subsequent three years; the sector increased by 0.8% in 2001, by 9.8% in 2002 and by an estimated 6% in 2003. In December 2003 Glamis Gold confirmed plans to develop its Marlin gold and silver mine at a cost of US $120m. The proven and probable reserves amounted to some 25m. metric tons, which would allow production of 4,000 tons of ore per day.

The main reserves of petroleum were found in the north of the country, across the border from Mexican production areas. In January 2000 proven reserves totalled 526m. barrels, while potential reserves were believed to be 800m.–1,000m. barrels. Earnings from exports declined to US $14.6m. in 1989, providing only 1.4% of total export earnings. However, petroleum production increased during the 1990s, reaching some 25,000 barrels per day (b/d) in 1997, or $98.7m. (4.2% of total export earnings). Following a slight fall in 1998, exports continued to increase in the rest of the decade, partly owing to an increase in international prices. In 2000 petroleum exports stood at $159.3m., or 5.9% of total export value. By 2002 this figure had increased to 6.7% of the total value of exports, though exports totalled $149.4m. The Bahamas-based Basic Petroleum International commenced production from two new oil wells in 1996 and increased output from 12,000 b/d to 20,000 b/d. There were also plans to construct a pipeline connecting the company's oilfields and refinery with an existing pipeline to the Caribbean coast. Increased political stability following the final peace accord signed in December 1996 raised hopes that other foreign petroleum companies could be encouraged to conduct exploration in Guatemala. In March 1997 bidding was opened for petroleum exploration sites in two of 12 areas, with at least four foreign companies expressing interest. The optimistic plans for the petroleum sector were, however, confounded by the fall in oil prices in 1998 and the withdrawal of foreign investors. Guatemala was receiving $14.4 per barrel at the beginning of 1998; however, by the end of the year this figure had fallen to $6.8. None the less, fuel prices subsequently recovered, peaking at $21.4 per barrel in 2000. In 2003 the price of petroleum averaged $15.3 per barrel; which, combined with increased production, resulted in petroleum export revenue of $173.4m., compared with $58.3m. in 1998.

Energy was supplied principally from petroleum (44.1% in 2001) and hydroelectricity (32.9%). The Chixoy watershed is the country's main source of hydroelectric power. In late 2001 Guatemala and Mexico reached agreement, under the 'Plan Puebla–Panamá' (see below), on a project to link their electricity grids. It was hoped that the project would improve regional power infrastructure and attract investment in the sector. The privatization of the energy sector was a major objective of the Arzú Government: in July 1998 80% of the capital of the state-owned Empresa Eléctrica de Guatemala, SA was sold to foreign investors for US $520m. and in December the Instituto Nacional de Electrificación (INDE) was sold for $100m. However, the Portillo Government did much to reverse the privatization policy of its predecessor (see below).

## MANUFACTURING

Guatemala has the largest, and one of the most developed, of all national manufacturing sectors in Central America. It experienced rapid expansion in the 1960s and 1970s, under the stimulus of the Central America Common Market (CACM) and foreign investment. In the 1980s, manufacturing output was adversely affected by the contraction in demand from other Central American countries as a result of the civil wars in the region and a shortage of domestic credit. Some recovery was experienced between 1987 and 1998, by which time manufacturing contributed an estimated US $81.8m., or 13.6% of GDP. It remained at approximately this level in 2004, when it contributed an estimated 12.6% of GDP, measured at constant prices. Despite its relatively advanced industrial sector, however, insufficient investment raised a number of concerns about its future competitiveness.

The industrial sector (including construction) employed 19.2% of the work-force in 2003 and contributed an estimated 18.9% of GDP, measured in constant prices, in 2004. Small enterprises predominated producing consumer goods, such as food, beverages and tobacco; intermediate goods accounted for a further 35% and included the manufacture of building materials and textiles. The clothing assembly, or *maquila*, plants contributed to an increase in non-traditional exports from the late 1980s. Guatemala took second place, behind Costa Rica, as the region's leading apparel exporter. There were 280 apparel firms operating in Guatemala by the 1990s, 30 of which were of Korean origin; major US companies accounted for most of the rest. An estimated 80,000 workers were employed in the sector. The sector's dynamism depended considerably on special government incentives to investors and low wages for a non-unionized labour force. Attracting foreign investment remained a problem for Guatemala. There was still a belief that Guatemala lacked political and economic stability, the regulatory framework or the consumer market to attract investment. Nevertheless, the *maquila* sector continued to expand through the 1990s and into 2000, when, despite a lack of new investments, the sector grew by 22.4%. In 2001, owing to both the introduction of a business tax and to an economic slowdown in the USA, exacerbated by the effects of the terrorist attacks of September 2001, the sector experienced some contraction. A total of 38 *maquila* factories closed and a further 229 were threatened with closure. Faced with the prospect of unemployment, however, many apparel manufacturers underwent a process of reinvention and in mid-2002, as sales began to recover, prospects for further growth in the sector were thought to be positive.

Although manufacturing was traditionally orientated towards Central American markets, recession and the impact of conflict throughout the region resulted in reduced demand for Guatemalan goods. The emphasis was transferred to expanding trade with the USA. However, this tendency was reversed during the late 1980s, when Central America entered a period of greater stability. In 1996 the Arzú Government began a tariff reduction programme aimed at placing the country's manufacturing sector on an equal level with those of Costa Rica and El Salvador, where imports of raw materials were subject to recently lowered tariffs. In 1998 CACM countries agreed a joint reduction of import duties from 19% to 17% for finished goods and 12% for intermediate goods. There were further reductions in January 1999 to 15% and 10%, respectively. In 2002 39% of exports were destined for the Central America, and 30% for the USA.

The construction sector underwent a period of rapid expansion in the early 1990s, although there were suspicions that the sector was partly funded by money raised from the illegal drugs trade. Following a slowdown in the middle of decade, which fuelled fears that the expansion of the building trade had ended, with the capital area saturated with luxury construction, the sector experienced renewed dynamism at the end of the 1990s and beginning of the 2000s, expanding by 17.0% in 2000 and by 12.2% in 2001. However, from 2002 the sector contracted dramatically: by 15.3% in that year, by 2.2% in 2003 and by an estimated 15.6% in 2004. Some 207,877 people were directly employed by the construction trade (6.0% of the total labour force) in 2002 and a great many more were indirectly dependent on the industry.

## TRANSPORT AND COMMUNICATIONS

In 1999 Guatemala had 14,021 km of roads, 22% of which were paved. Paved roads increased to 35% of total roads in 2000. The Guatemalan section of the Pan-American Highway was 518.7 km long and totally asphalted. Highways are concentrated in the producing areas of the Pacific and Altiplano regions. In May 2003 the World Bank approved a loan of US $46.7m. to repair existing roads and construct new networks in rural areas. The main ports are Puerto Barrios and Santo Tomás de Castilla, on the Gulf of Mexico, and San José and Champerico, on the Pacific Ocean. There are several airports, the main international airport being La Aurora, near Guatemala City.

The Arzú Government placed much emphasis on the modernization of the country's infrastructure, which included the 1997 sale of the Ferrocarriles de Guatemala (FEGUA) rail network and the 1998 sale of a 50-year concession of the railways to the Railroad Development Corporation (USA). In June 2002 the Portillo Government announced a package of measures intended to stimulate economic growth (see below). Amongst the proposals were plans to expand La Aurora airport and to construct a new international airport on the outskirts of Guatemala City, as well as to initiate the renovation and modernization of Santa Elena airport. The shipping facilities at Puerto Quetzal and Santo Tomás de Castilla were to be expanded and several major highways and urban roads constructed; there were also plans for a thermoelectric power plant. The projects were expected to cost over US $1,100m., to be funded through loans from international aid organizations and the encouragement of private investment.

In 1998 the sale, for US $700m., of 95% of the state telecommunication company, Empresa Guatemalteca de Telecomunicaciones (Guatel), was awarded to a group of mostly domestic investors and the Mexican operator Teléfonos de México (Telmex). However, in July 2000 Arzú's successor, President Portillo, declared Guatel's sale to be damaging to national interests and referred it to the courts. The Government announced that it would not limit its revision to the sale of Guatel, but would also examine other privatizations that took place during the Arzú administration, including that of the postal service, the railway network and the electricity company. These measures, which placed in doubt many of the decisions made by the previous Government, generated much uncertainty among the Guatemalan and international business sectors. While the telecommunications service in the country improved dramatically, with the number of telephones per 1,000 people increasing from 28.7 in 1995 to 75.3 in 2002, it nevertheless remained inadequate.

## INVESTMENT AND FINANCE

Following the transfer of power to the civilian Government of President Vinicio Cerezo in 1986, a programme of austerity measures and financial reform was implemented. As a result, the economy began to show signs of revival after several years of decline. However, these positive developments were adversely affected by the decrease in coffee prices in the late 1980s and the decline in international reserves. A single exchange rate was introduced to deal with the latter in February 1989, followed in November by the flotation of the Guatemalan currency unit, the quetzal. Inflationary pressures increased in the economy as the quetzal was devalued several times, and a series of austerity measures was introduced. Although the general value of the quetzal stabilized, the rate of inflation continued to increase. The level of generalized crisis in the economy was such that by the end of the Cerezo Government, per-head income was estimated to have decreased to its 1969 level. The Government of Jorge Serrano Elías (1990–93) succeeded in stabilizing the currency and in increasing international reserves. In 1990–2003 real annual GDP grew by an average of 3.7%. The economy expanded by an estimated 2.2% in 2002 and by 2.1% in 2003. The decrease in economic growth was mainly owing to falling prices for primary export products and the resulting effects of the US economic downturn. The annual rate of inflation was over 60% in 1990, but in the course of the following decade this rate was brought under control. The average annual rate of inflation in 1990–2004 was 10.2%. Consumer prices increased by an average of 8.1% in 2002 and by 5.6% in 2003, and 7.6% in 2004, mainly owing to an increase in value-added tax (VAT) and to high global oil prices. Controlling inflation was expected to be central to monetary policy in 2005, when the rate was forecast to decline to 7.0%.

The signing of the final peace accord in December 1996 enabled Guatemala to attract international aid for reconstruction. Some US $1,900m. in grants and loans was finally agreed. The IDB promised $800m., the World Bank $400m. and the European Union (EU) $250m. The Government committed itself to raising the shortfall of the estimated $2,500m. needed to implement the peace accords by raising tax revenues to 50% above the 1995 levels by 2000. Failure to meet the pledge to raise revenue prompted increasing pressure on the Government from international organizations; the EU gave an explicit warning in June 2001. However, in 2003 tax revenue fell to less than 10% of GDP.

In December 2000 legislation was approved to allow the circulation of the US dollar and other convertible currencies, for use in a wide range of transactions, including bank transactions, the paying of salaries, and other business services, from 1 May 2001. However, many were opposed to the change, and in May a constitutional challenge to the legislation was made, arguing that it violated the constitutional requirement that the Central Bank remain in exclusive control of foreign exchange.

In June 2001 the Financial Action Task Force on Money Laundering (FATF, based in Paris, France) added Guatemala to its list of 'non-co-operative' countries in international efforts to prevent money-laundering activities. In October the Congreso Nacional approved an Anti-Money-Laundering law which classified money-laundering activities as a criminal offence, expanded banking secrecy rules to incorporate suspect financial transactions and permitted the Government to share information with other countries. The legislation also created an autonomous authority empowered to investigate suspicious financial dealings. The FATF welcomed the progress made by the Portillo and Berger Governments and in July 2004 Guatemala was finally removed from the list.

In an attempt to reduce the fragility of the banking sector, in mid-2001 the Guatemalan Monetary Board initiated a thorough review of the banking system. As a result, the courts were requested to liquidate the three banks that had been subjected to state intervention earlier in the year (following a corruption scandal), as well as two finance companies. A draft law permitting the sale of another under-performing bank, the Banco del Ejército, was approved in January 2002. Further reforms were enabled in June 2002 when the Congreso Nacional approved four bills creating a legislative framework through which the Central Bank would be able to exercise more effective control over the banking sector. The new legislation also allowed for the supervision of 'offshore' banks, for which previous banking laws had made no provision. The World Bank granted Guatemala two US $155m. loans to strengthen the newly created framework. The 11 'offshore' banks were inspected in 2003 and a licensing system was introduced.

## FOREIGN TRADE AND THE BALANCE OF PAYMENTS

The Cerezo Government attempted to alter the economy's dependence on exports of primary products (which accounted for more than 70% of export earnings), particularly coffee, by the implementation of an export-orientated strategy aimed at export diversification rather than import substitution, as had previously been the case. This policy achieved undoubted success with the development of non-traditional exports, which, by 1988, had increased by 200% in two years.

The export sector was depressed in the early 1990s, but following efforts by the new administration of President de León, exports increased. In 1997 growth in exports reached 16.3%, before slowing to 7.2% in 1998 and falling further to 6.7% in 2000. In the late 1990s import growth also accelerated to about 20% in real terms, resulting in a growing trade deficit, which stood at US $3,126.7m. in 2003, compared with $1,445.1m. in 1999. However, largely owing to remittances from citizens working abroad, the deficit on the current account of the balance of payments contracted slightly in 2004 for the third consecutive year. The USA was the principal market for Guatemalan exports, taking an estimated 29.9% in 2003. Other significant purchasers were El Salvador, Honduras, Nicaragua, Costa Rica and Mexico. The USA was also the major import source, supplying 43.8% in 2003.

In the 1990s Guatemala was threatened with exclusion from regional free trade agreements if it failed to reduce human rights abuses and corruption. The improvement of workers' rights and respect for international patents were also important areas of disagreement between the US and Guatemalan Governments. The 1996 peace accords clearly constituted a move towards improving Guatemala's standing with the international community; however, the failure to implement them adversely affected the country's status.

The prospects for free trade with Mexico were considered a potential benefit to exports; at the end of the 1990s, imports from Mexico far exceeded exports in terms of value. In April 2000 El Salvador, Guatemala and Honduras (the so-called CA-3 countries) signed a free trade agreement with Mexico, which promised greater access to the Mexican market for Guatemala, with increased bilateral trade in the future. The agreement became operational in March 2001. As predicted by many Guatemalan economic analysts, however, the agreement initially benefited the Mexican economy to the detriment of Guatemala, as in 2001 Guatemalan exports to Mexico decreased by 34.4%, while Mexican exports to Guatemala increased by 3.2%. In January 2002 Mexico withdrew a ban on imports of shoes and leather (which had not been covered by the 2000 agreement) from Guatemala following pressure from the Mexican ambassador to Guatemala. In 2003 bilateral trade increased to the benefit of both countries. In 2002 Guatemala also agreed with Costa Rica, El Salvador, Honduras and Nicaragua to eliminate all regional trade barriers; the eradication of all tariffs was perceived to be a necessary prerequisite for the implementation of free trade agreements with the USA and Canada and for the future activation of the proposed Free Trade Area of the Americas, due to come into force by 2005, negotiations towards which stalled in 2004. Negotiations on the establishment of a Central American Free Trade Agreement (CAFTA) with the USA were concluded in late 2003; and the agreement was ratified by the Congreso in March 2005. In May 2004 the incoming President Berger, together with the Presidents of El Salvador, Honduras and Nicaragua signed an agreement creating a Central American customs union. In June of that year, in an attempt to improve relations with Belize (which would not participate in the proposed customs union), the Berger Government initiated negotiations with Belize on a bilateral free trade agreement.

In May 2001 the Central American countries, including Guatemala, reached an agreement with Mexico, called the 'Plan Puebla–Panamá', to establish a series of joint transport, industry and tourism projects intended to integrate the region. In December Guatemala and Mexico agreed that they would link their electricity grids under the terms of the Plan, as part of its initiative on energy integration. Officials from the two nations agreed to co-operate on a project to construct an 80-km power line that would link a substation in Tapachula, Mexico, with a Guatemalan substation situated in Los Brillantes; the cost would be an estimated US $30m. In the same year construction also began upon the electrical interconnection of Guatemala with Panama via a transmission line that would span 1,802 km; the project was scheduled for completion in 2005 and was to cost an estimated $320m.

## PUBLIC REVENUE

In 1998, following the socio-economic component of the peace accords, President Arzú launched the proposal of a Fiscal Pact, envisaged as a multi-sectoral, long-term consensus between the principal political, economic and social bodies over the future tax regime and fiscal policies. To this end a Fiscal Pact Preparatory Committee (CPPF) was established to organize a national debate on tax policy. The CPPF's report, presented in late 1999, argued that increasing tax collection was the only means for Guatemala to meet peace accord goals and advance infrastructure development plans; the report contained 66 recommendations to be implemented by 2004. It proposed the narrowing of the fiscal deficit to 2% in 2000, attaining the 12% tax collection goal by 2000, formulating a tax policy based on taxpayers' capacity to contribute, and creating a stabilization fund to cover shortfalls in revenue stemming from business cycles. It also recommended that the Government establish mechanisms to audit the Superintendency for Tax Administration (SAT), create a body to curb tax evasion and strengthen the judiciary's capacity to process cases of tax evasion and fraud.

The incoming Government of President Portillo also cited the signing of the Fiscal Pact to be one of its priorities. One of the main agreements of the Pact was to achieve a 12% tax collection goal by 2002 (the tax collection figure was 10% in 1999, an increase of 0.5% from 1998). Such a goal seemed unattainable without the Government increasing considerably the rate of VAT, which stood at 10% in 1999. However, efforts to increase VAT to 12% were blocked in the Congreso Nacional and were protested against by business-led groups. Further attempts to raise VAT in early 2001 were again met with protests, despite growing pressure from the international finance community. However, at a congressional session in July, which was boycotted by the Partido de Avanzada Nacional (PAN), the FRG approved the controversial legislation to increase VAT. The increase came into effect on 1 August. In November the Congreso Nacional approved further legislation increasing the excise on cigarettes. In January 2002 custom duties on petroleum were also raised, followed in the next month by increases in the excises imposed on alcoholic beverages and soft drinks. Tax collection was projected to increase to 10.8% of GDP in 2002 as a result of the tax reforms that had been introduced and an improvement in taxation administration. In early 2003 President Portillo implemented by decree further unpopular tax rises on fuel, wheat and a number of other items. The target date for the 12% tax collection goal was postponed to 2004. The Berger Government also indicated its determination to implement tax reform on taking office in 2004; however, it faced opposition from other parties in the Congreso on whom it relied for a legislative majority, and from the business community.

## OUTLOOK

In June 2002 the Portillo Government announced a series of measures intended to revitalize Guatemala's stagnating economy. The 2002–04 Economic Action Plan prioritized an extensive programme of privatization and public works projects, as well as incentives to stimulate the development of the private sector and attract foreign investment. The programme was commended by the IMF and to assist its implementation it approved the release of funds totalling US $120m. in a nine-month stand-by arrangement in June 2003. However, while the proposed reforms were widely perceived to constitute an important advance in the search for a solution to the country's economic problems, recovery was believed by many to remain contingent upon the Government's ability to reduce public overspending and to attract foreign capital, as well as upon external factors. International concern over a spate of attacks on human rights workers involving members of the armed forces in 2002 and the controversial candidacy of Ríos Montt in the 2003 presidential elections also mitigated against creating a stable political climate in which economic development could be nurtured. Furthermore, in mid-2003 a report by the Consultative Group of donor countries was highly critical of the Government's failure to implement the terms of the peace accords.

The election of Oscar Berger Perdomo of the centre-right GANA to the presidency in November 2003 gave rise to optimism that the change of government would prompt inflows of much-needed foreign investment. Furthermore, the new Government's determination to address the endemic corruption at all levels of public office was also welcomed by investors, as was the fact that Berger's new Cabinet was mainly composed of businessmen. However, GANA did not obtain a majority in the Congreso, which impeded attempts to implement the promised economic reforms. Moreover, it was anticipated that reliance on the legislative support of the FRG might temper investigations into allegations of corruption within the former Government. On taking office in January 2004, President Berger had pledged to foster economic growth by means of an austere but transparent fiscal policy. The incoming administration engaged to strengthen the regulation of the financial sector and increase supervision of the banking system in an attempt to attract domestic and international investment. In April Berger declared that the fiscal deficit, which had reached US $875m., was unsustainable and could lead to an economic crisis. He presented a series of economic reforms to the Congreso that aimed to narrow the fiscal deficit to 1.8% of GDP by the end of 2005 by increasing tax revenue and decreasing budgetary expenditure. The communication and agriculture budgets were reduced by 780m. and 169m. quetzales, respectively, although the education budget was increased by 700m. quetzales. Expenditure on the armed forces was limited to 0.33% of GDP—one-half the amount allowed by the peace accords. The number of members was reduced from 27,000 to 15,500 and 13 of the 60 army bases were closed. The tax measures included alterations to exemptions from VAT and a broadening of the income-tax base. Furthermore, the unpopular Impuesto a las Empresas Mercantiles (IEMA) was replaced by a new corporation tax. However, pressure to increase taxes was resisted by the private sector, as well as by peasant organizations, trade unions and public-sector workers. Following protests by many of these groupings, in June Berger agreed to review the tax reforms with a view to decreasing the burden on low-income taxpayers, to not extending income tax to bonuses and severance pay and to not increasing the rate of VAT. A temporary fiscal package was finally approved later in June, but it was insufficient to finance the sustained increases in public spending that would be necessary to comply with the reforms laid out in the peace accords. Berger also came under pressure from former members of the

# GUATEMALA

pro-Government peasant militia (Patrullas de Autodefensa Civil—PAC), who held numerous protests throughout 2004 and early 2005 to demand compensation for their role in the civil war. Portillo had decreed the payment of US $640 to each former PAC member, but only US $215 had been disbursed before the end of his term in office. Berger agreed to pay the remainder, but was thwarted on three occasions, lastly in March 2005. Despite legislative approval for the compensation, human rights organizations successfully appealed to the Constitutional Court on the grounds that the laws were unconstitutional.

In 2005 the Government introduced a tripartite economic recovery programme called 'Vamos Guatemala' (Let's Go, Guatemala). The three principal objectives of the plan were to reduce poverty, to improve export competitiveness, and to foster economic growth, by attracting investment and stimulating private-sector participation. The Government forecast a moderate increase in economic growth from 2.6% in 2004 to 2.8% in 2005. Meanwhile, the advent of CAFTA and of the Central American customs union was expected to bolster external trade. However, in mid-2005 President Berger still faced formidable challenges and progress was expected to be restrained by the lack of public funds and the Government's minority position in the Congreso.

# Statistical Survey

Sources (unless otherwise stated): Banco de Guatemala, 7a Avda 22-01, Zona 1, Apdo 365, Guatemala City; tel. 230-6222; fax 253-4035; internet www.banguat.gob.gt ; Instituto Nacional de Estadística, Edif. América 4°, 8a Calle 9-55, Zona 1, Guatemala City; tel. 232-6136; fax 232-4790; e-mail info-ine@ine.gob.gt; internet www.segeplan.gob.gt/ine.

## Area and Population

### AREA, POPULATION AND DENSITY

| | |
|---|---:|
| Area (sq km) | |
| Land | 108,429 |
| Inland water | 460 |
| Total | 108,889* |
| Population (census results)† | |
| 17 April 1994 | 8,322,051 |
| 24 November 2002 | 11,237,196 |
| Population (UN estimates at mid-year)‡ | |
| 2003 | 11,998,000 |
| 2004 | 12,295,000 |
| Density (per sq km) at mid-2004 | 112.9 |

* 42,042 sq miles.
† Excluding adjustments for underenumeration, estimated to have been 13.7% in 1981.
‡ Source: UN, *World Population Prospects: The 2004 Revision.*

### DEPARTMENTS
(population at census of November 2002)

| | | | | |
|---|---:|---|---:|---|
| Alta Verapaz | 776,246 | Quetzaltenango | 624,716 | |
| Baja Verapaz | 215,915 | Quiché | 655,510 | |
| Chimaltenango | 446,133 | Retalhuleu | 241,411 | |
| Chiquimula | 302,485 | Sacatepéquez | 248,019 | |
| El Progreso | 139,490 | San Marcos | 794,951 | |
| Escuintla | 538,746 | Santa Rosa | 301,370 | |
| Guatemala | 2,541,581 | Sololá | 307,661 | |
| Huehuetenango | 846,544 | Suchitepéquez | 403,945 | |
| Izabal | 314,306 | Totonicapán | 339,254 | |
| Jalapa | 242,926 | Zacapa | 200,167 | |
| Jutiapa | 389,085 | | | |
| Petén | 366,735 | **Total** | **11,237,196** | |

### PRINCIPAL TOWNS
(population at census of November 2002)

| | | | |
|---|---:|---|---:|
| Guatemala City | 942,348 | Cobán | 144,461 |
| Mixco | 403,689 | Quetzaltenango | 127,569 |
| Villa Nueva | 355,901 | Escuintla | 119,897 |
| San Juan Sacatepéquez | 152,583 | Jalapa | 105,796 |
| San Pedro Carcha | 148,344 | Totonicapán | 96,392 |

### BIRTHS, MARRIAGES AND DEATHS

| | Registered live births | | Registered marriages | | Registered deaths | |
|---|---:|---:|---:|---:|---:|---:|
| | Number | Rate (per 1,000) | Number | Rate (per 1,000) | Number | Rate (per 1,000) |
| 1996 | 377,723 | 34.6 | 47,428 | 4.3 | 60,618 | 5.5 |
| 1997 | 387,862 | 36.9 | 51,908 | 4.9 | 67,691 | 6.4 |
| 1998 | 378,438 | 35.0 | 52,499 | 4.9 | 69,633 | 6.4 |
| 1999 | 409,034 | 36.9 | 62,034 | 5.6 | 65,139 | 5.9 |
| 2000 | 425,410 | 37.4 | 58,311 | 5.1 | 67,284 | 5.9 |
| 2001 | 415,410 | 35.6 | 54,722 | 4.7 | 68,041 | 5.8 |
| 2002 | 387,287 | 34.5 | 51,857 | 4.6 | 66,089 | 5.9 |
| 2003 | 375,092 | n.a. | 51,247 | n.a. | 66,695 | n.a. |

**Expectation of life** (WHO estimates, years at birth): 66 (males 64; females 69) in 2003 (Source: WHO, *World Health Report*).

### ECONOMICALLY ACTIVE POPULATION
(at census of November 2002)

| | Males | Females | Total |
|---|---:|---:|---:|
| Agriculture, forestry, hunting and fishing | 1,278,739 | 178,364 | 1,457,103 |
| Mining and quarrying | 5,313 | 756 | 6,069 |
| Manufacturing | 301,222 | 164,725 | 465,947 |
| Construction | 186,611 | 21,266 | 207,877 |
| Electricity, gas, water and sanitary services | 23,518 | 10,135 | 33,653 |
| Commerce | 343,586 | 228,114 | 571,700 |
| Transport, storage and communications | 96,410 | 16,913 | 113,323 |
| Financial and property services | 82,644 | 42,839 | 125,483 |
| Public administration and defence | 60,853 | 25,137 | 85,990 |
| Education | 42,366 | 59,796 | 102,162 |
| Community and personal services | 68,165 | 197,794 | 265,959 |
| Activities not adequately described | 18,256 | 9,875 | 28,131 |
| **Total** | **2,525,683** | **937,714** | **3,463,397** |

# GUATEMALA

## Health and Welfare

### KEY INDICATORS

| | |
|---|---:|
| Total fertility rate (children per woman, 2003) | 4.4 |
| Under-5 mortality rate (per 1,000 live births, 2003) | 47 |
| HIV/AIDS (% of persons aged 15–49, 2003) | 1.1 |
| Physicians (per 1,000 head, 1999) | 0.9 |
| Hospital beds (per 1,000 head, 1996) | 0.98 |
| Health expenditure (2002): US $ per head (PPP) | 199 |
| Health expenditure (2002): % of GDP | 4.8 |
| Health expenditure (2002): public (% of total) | 47.5 |
| Access to water (% of persons, 2002) | 95 |
| Access to sanitation (% of persons, 2002) | 61 |
| Human Development Index (2002): ranking | 121 |
| Human Development Index (2002): value | 0.649 |

For sources and definitions, see explanatory note on p. vi.

## Agriculture

### PRINCIPAL CROPS
('000 metric tons)

| | 2001 | 2002 | 2003 |
|---|---:|---:|---:|
| Maize | 1,091.5 | 1,050.1 | 1,053.6* |
| Sugar cane | 16,934.9 | 17,489.9 | 17,500.0† |
| Pulses† | 129.4 | 129.4 | 129.7† |
| Oil palm fruit† | 466.6 | 540.0 | 540.0† |
| Tomatoes | 175.0 | 187.2 | 187.2† |
| Other fresh vegetables† | 487.2 | 475.2 | 475.2 |
| Watermelons† | 126.2 | 126.2 | 126.2 |
| Cantaloupes and other melons | 188.2 | 188.2† | 188.2† |
| Bananas | 898.0 | 1,000.0† | 1,000.0† |
| Plantains | 266.5 | 268.0† | 268.0† |
| Lemons and limes | 130.8 | 142.9 | 142.9† |
| Mangoes† | 183.0 | 187.0 | 187.0 |
| Other fresh fruit† | 396.8 | 397.6 | 397.6 |
| Coffee (green)* | 275.7 | 221.8 | 210.0 |

\* Unofficial figure(s).  
† FAO estimate(s).  
Source: FAO.

### LIVESTOCK
('000 head, year ending September)

| | 2001 | 2002 | 2003 |
|---|---:|---:|---:|
| Horses* | 120 | 122 | 124 |
| Asses* | 10 | 10 | 10 |
| Mules* | 38 | 39 | 39 |
| Cattle* | 2,500 | 2,540 | 2,540 |
| Sheep | 245 | 250* | 260 |
| Pigs* | 763 | 800 | 780 |
| Goats | 112* | 112* | 112 |
| Chickens* | 24,000 | 25,500 | 27 |

\* FAO estimate(s).  
Source: FAO.

### LIVESTOCK PRODUCTS
('000 metric tons)

| | 2001 | 2002 | 2003 |
|---|---:|---:|---:|
| Beef and veal* | 62.0 | 63.0 | 63.0 |
| Pig meat* | 25.0 | 25.5 | 26.0 |
| Chicken meat | 144.0† | 155.0† | 155.0* |
| Other meat* | 3.9 | 3.9 | 4.0 |
| Cows' milk* | 270.0 | 270.0 | 270.0 |
| Cheese* | 11.3 | 11.3 | 11.3 |
| Butter* | 0.6 | 0.6 | 0.6 |
| Hen eggs† | 82.5 | 85.0 | 85.0 |
| Honey | 1.5* | 1.5* | 1.5 |
| Cattle hides* | 8.8 | 8.9 | 8.9 |

\* FAO estimate(s).  
† Unofficial figure(s).  
Source: FAO.

## Forestry

### ROUNDWOOD REMOVALS
('000 cubic metres, excl. bark)

| | 2001 | 2002 | 2003 |
|---|---:|---:|---:|
| Sawlogs, veneer logs and logs for sleepers | 425 | 492 | 492* |
| Other industrial wood | 19 | 26 | 17 |
| Fuel wood* | 14,870 | 15,207 | 15,552 |
| **Total** | 15,313 | 15,725 | 16,061 |

\* FAO estimates.  
Source: FAO.

### SAWNWOOD PRODUCTION
('000 cubic metres, incl. railway sleepers)

| | 2001 | 2002 | 2003 |
|---|---:|---:|---:|
| Coniferous (softwood) | 180* | 180* | 251 |
| Broadleaved (hardwood) | 160 | 160 | 115 |
| **Total** | 340 | 340 | 366 |

\* FAO estimate.

## Fishing

('000 metric tons, live weight)

| | 2001* | 2002 | 2003 |
|---|---:|---:|---:|
| Capture | 30.5 | 24.2 | 24.1 |
| Freshwater fishes* | 7.1 | 7.1 | 7.1 |
| Skipjack tuna | 10.7 | 10.1 | 9.9 |
| Yellowfin tuna | 5.9 | 2.4 | 3.0 |
| Bigeye tuna | 4.6 | 2.6 | 1.7 |
| Penaeus shrimps | 0.8 | 0.6 | 0.8 |
| Aquaculture | 5.1 | 8.0 | 6.3 |
| Tilapias* | 1.9 | 1.9 | 1.9 |
| Penaeus shrimps | 2.5 | 5.4 | 3.8 |
| **Total catch** | 35.6 | 32.1 | 30.5 |

\* FAO estimates.  
Source: FAO.

## Mining

('000 metric tons, unless otherwise indicated)

| | 2000 | 2001* | 2002* |
|---|---:|---:|---:|
| Crude petroleum ('000 42-gallon barrels) | 7,571 | 21,000 | 19,000 |
| Limestone | 4,532 | 2,775 | 3,040 |
| Sand and gravel | 4,158 | 4,000 | 4,000 |
| Silica sand | 173 | 170 | 170 |
| Crushed stone | 50,000 | 50,000 | 50,000 |

\* Estimates.  
Source: US Geological Survey.

## Industry

### SELECTED PRODUCTS
('000 metric tons, unless otherwise indicated)

| | 2000 | 2001 | 2002 |
|---|---:|---:|---:|
| Sugar (raw) | 883 | 1,670 | 1,912 |
| Motor spirit (petrol) | 150 | 147 | — |
| Cigarettes (million) | 4,262 | — | — |
| Tobacco (prepared leaf) | 19 | 20 | 21 |
| Cement | 2,039 | 1,976 | 2,068 |
| Electric energy (million kWh) | 6,048 | 5,856 | — |

# Finance

## CURRENCY AND EXCHANGE RATES

**Monetary Units**
100 centavos = 1 quetzal.

**Sterling, Dollar and Euro Equivalents** (31 May 2005)
£1 sterling = 13.828 quetzales;
US $1 = 7.606 quetzales;
€1 = 9.379 quetzales;
1,000 quetzales = £72.32 = $131.48 = €106.62.

**Average Exchange Rate** (quetzales per US dollar)
2002  7.8216
2003  7.9409
2004  7.9465

Note: In December 2000 legislation was approved to allow the circulation of the US dollar and other convertible currencies, for use in a wide range of transactions, from 1 May 2001.

## BUDGET
(million quetzales)

| Revenue | 2001 | 2002 | 2003 |
|---|---|---|---|
| Taxation | 15,966.2 | 19,346.7 | 20,317.7 |
| Taxes on income, profits and capital gains | 4,139.8 | 5,453.8 | 5,601.1 |
| Corporate | 2,059.7 | 2,182.3 | 2,501.9 |
| Domestic taxes on goods and services | 9,679.2 | 11,435.2 | 12,146.1 |
| Sales, turnover or value-added taxes | 6,983.1 | 8,618.4 | 9,288.6 |
| Excises | 2,177.3 | 2,249.9 | 2,244.0 |
| Taxes on international trade and transactions | 1,979.9 | 2,264.1 | 2,385.0 |
| Contributions to government employee pension and welfare funds within government | 414.2 | 468.6 | 504.7 |
| Grants | 565.7 | 414.9 | 377.1 |
| Other revenue | 1,259.2 | 525.4 | 494.9 |
| Statistical discrepancy | −14.0 | −40.0 | — |
| **Total** (incl. grants) | 18,191.3 | 20,715.6 | 21,694.4 |

| Expenditure | 2001 | 2002 | 2003 |
|---|---|---|---|
| Compensation of employees | 5,959.0 | 6,413.0 | 6,751.9 |
| Wages and salaries | 5,958.5 | 6,413.0 | 6,648.2 |
| Use of goods and services | 3,100.2 | 3,013.9 | 3,405.7 |
| Interest payments | 2,111.2 | 2,191.5 | 2,207.4 |
| Subsidies and other current transfers | 2.2 | 8.2 | 242.3 |
| Transfers to non profit-making institutions and households | 2,063.5 | 2,710.7 | 1,883.5 |
| Grants | 2,188.6 | 2,707.4 | 2,961.2 |
| To foreign governments | 482.7 | 533.9 | 742.4 |
| To other levels of national government | 1,705.8 | 2,173.5 | 2,218.8 |
| Capital | 1,705.8 | 2,173.5 | 2,218.8 |
| Other expense | 3,955.8 | 3,714.0 | 7,516.5 |
| **Total** (incl. grants) | 19,380.4 | 20,758.7 | 24,968.6 |

Source: IMF, *Government Finance Statistics Yearbook*.

## INTERNATIONAL RESERVES
(US $ million at 31 December)

| | 2002 | 2003 | 2004 |
|---|---|---|---|
| Gold* | 9.1 | 9.3 | 9.3 |
| IMF special drawing rights | 8.2 | 8.2 | 8.0 |
| Foreign exchange | 2,290.9 | 2,825.0 | 3,418.3 |
| **Total** | 2,299.1 | 2,842.5 | 3,435.6 |

* Valued at US $42 per troy ounce.

Source: IMF, *International Financial Statistics*.

## MONEY SUPPLY
(million quetzales at 31 December)

| | 2002 | 2003 | 2004 |
|---|---|---|---|
| Currency outside banks | 8,729.4 | 10,608.8 | 11,194.8 |
| Demand deposits at deposit money banks | 13,930.4 | 16,188.0 | 17,470.8 |
| **Total money** (incl. others) | 22,835.7 | 26,937.7 | 28,741.0 |

Source: IMF, *International Financial Statistics*.

## COST OF LIVING
(Consumer Price Index; base: 2000 = 100)

| | 2002 | 2003 | 2004 |
|---|---|---|---|
| Food | 121.5 | 128.5 | 141.7 |
| **All items** (incl. others) | 116.0 | 122.5 | 131.8 |

Source: ILO.

## NATIONAL ACCOUNTS

**Expenditure on the Gross Domestic Product**
(million quetzales at current prices)

| | 2002 | 2003 | 2004 |
|---|---|---|---|
| Government final consumption expenditure | 13,005 | 14,313 | 13,157 |
| Private final consumption expenditure | 156,945 | 171,951 | 191,783 |
| Increase in stocks | 6,660 | 6,746 | 5,542 |
| Gross fixed capital formation | 28,164 | 28,771 | 32,388 |
| **Total domestic expenditure** | 204,774 | 221,781 | 242,870 |
| Exports of goods and services | 31,153 | 32,920 | 36,412 |
| *Less* Imports of goods and services | 53,652 | 58,004 | 66,189 |
| **GDP in purchasers' values** | 182,275 | 196,696 | 213,093 |

Source: IMF, *International Financial Statistics*.

**Gross Domestic Product by Economic Activity**
(million quetzales at constant 1958 prices)

| | 2002 | 2003* | 2004† |
|---|---|---|---|
| Agriculture, hunting, forestry and fishing | 1,192.5 | 1,227.8 | 1,270.0 |
| Mining and quarrying | 29.2 | 29.7 | 27.0 |
| Manufacturing | 681.0 | 688.0 | 702.1 |
| Electricity, gas and water | 223.2 | 233.2 | 246.8 |
| Construction | 93.7 | 91.6 | 77.3 |
| Trade, restaurants and hotels | 1,319.2 | 1,346.3 | 1,393.3 |
| Transport, storage and communications | 552.3 | 581.9 | 631.7 |
| Finance, insurance and real estate | 265.3 | 269.2 | 275.6 |
| Ownership of dwellings | 245.4 | 252.6 | 260.2 |
| General government services | 395.8 | 378.8 | 346.4 |
| Other community, social and personal services | 311.2 | 322.3 | 334.6 |
| **Total** | 5,308.7 | 5,421.4 | 5,565.1 |

* Preliminary figures.
† Estimates.

# GUATEMALA

## BALANCE OF PAYMENTS
(US $ million)

|  | 2001 | 2002 | 2003 |
|---|---|---|---|
| Exports of goods f.o.b. | 2,864.6 | 2,818.9 | 3,048.3 |
| Imports of goods f.o.b. | −5,142.1 | −5,791.0 | −6,175.0 |
| **Trade balance** | −2,277.5 | −2,972.1 | −3,126.7 |
| Exports of services | 1,031 | 1,145.4 | 1,058.8 |
| Imports of services | −898.0 | −1,066.1 | −1,126.8 |
| **Balance on goods and services** | −2,144.5 | −2,892.8 | −3,194.7 |
| Other income received | 317.5 | 161.1 | 178.7 |
| Other income paid | −407.6 | −479.5 | −496.8 |
| **Balance on goods, services and income** | −2,234.6 | −3,211.1 | −3,512.9 |
| Current transfers received | 1,031.8 | 2,077.7 | 2,558.9 |
| Current transfers paid | −35.1 | −101.5 | −97.0 |
| **Current balance** | −1,237.9 | −1,234.9 | −1,050.9 |
| Capital account (net) | 93.3 | 124.2 | 133.8 |
| Direct investment from abroad | 455.5 | 110.6 | 115.8 |
| Portfolio investment assets | −44.9 | −38.3 | 21.7 |
| Portfolio investment liabilities | −175.4 | −107.8 | −11.0 |
| Other investment assets | 156.7 | 196.4 | 116.0 |
| Other investment liabilities | 778.1 | 1,035.6 | 913.4 |
| Net errors and omissions | 98.1 | −64.7 | 311.4 |
| **Overall balance** | 474.3 | 21.2 | 550.1 |

Source: IMF, *International Financial Statistics*.

## External Trade

### PRINCIPAL COMMODITIES
(US $ million)

| Imports c.i.f. | 2000 | 2001 | 2002 |
|---|---|---|---|
| **Food and live animals** | 494.7 | 663.5 | 681.5 |
| Cereals and cereal preparations | 150.4 | 220.9 | 260.5 |
| **Mineral fuels, lubricants, etc.** | 622.1 | 764.6 | 773.6 |
| Petroleum, petroleum products etc. | 535.8 | 673.3 | 693.0 |
| Crude petroleum oils, etc. | 169.8 | 158.2 | 114.0 |
| Refined petroleum products | 356.3 | 506.3 | 562.9 |
| Other fuel oils | 174.4 | 246.7 | n.a. |
| **Chemicals and related products** | 780.0 | 947.2 | 1,008.4 |
| Medicinal and pharmaceutical products | 185.2 | 244.1 | 258.2 |
| Medicaments (incl. veterinary medicaments) | 152.4 | 200.8 | 207.6 |
| Artificial resins, plastic materials etc. | 169.8 | 186.0 | 200.2 |
| Products of polymerizations, etc. | 146.5 | 156.6 | 168.3 |
| **Basic manufactures** | 826.7 | 912.2 | 935.5 |
| Paper, paperboards and manufactures | 230.9 | 253.3 | 268.0 |
| Paper and paperboard | 150.7 | 157.3 | 172.1 |
| Iron and steel | 174.4 | 182.4 | 177.1 |
| **Machinery and transport equipment** | 1,583.7 | 1,578.7 | 1,823.5 |
| Power-generating machinery and equipment | 157.4 | 67.9 | 150.0 |
| Machinery specialized for particular industries | 150.0 | 139.9 | 161.4 |
| Telecommunications and sound equipment | 282.6 | 263.4 | 304.2 |
| Road vehicles and parts | 499.6 | 573.5 | 636.2 |
| Passenger motor cars (excl. buses) | 213.5 | 259.5 | 295.0 |
| Motor vehicles for the transport of goods | 158.3 | 175.6 | 189.7 |
| **Miscellaneous manufactured articles** | 376.2 | 507.3 | 321.6 |
| **Total** (incl. others) | 4,882.4 | 5,597.6 | 6,074.5 |

| Exports f.o.b. | 2000 | 2001 | 2002 |
|---|---|---|---|
| **Food and live animals** | 1,406.3 | 1,127.9 | 1,079.4 |
| Cereals and cereal preparations | 74.7 | 90.9 | 82.1 |
| Vegetables and fruit | 353.5 | 300.6 | 318.9 |
| Fresh or dried fruit and nuts (excl. oil nuts) | 243.3 | 223.3 | 253.2 |
| Bananas and plantains | 178.1 | 201.2 | 240.4 |
| Sugar, sugar preparations and honey | 203.7 | 231.7 | 252.3 |
| Sugar and honey | 197.6 | 224.1 | 244.9 |
| Raw beet and cane sugars | 190.8 | 212.6 | 227.1 |
| Coffee, tea, cocoa and spices | 657.5 | 407.1 | 359.1 |
| Coffee (incl. husks and skins) and substitutes containing coffee | 575.4 | 306.9 | 261.7 |
| **Crude materials (inedible) except fuels** | 162.3 | 135.0 | 115.1 |
| **Mineral fuels, lubricants, etc.** | 162.4 | 130.1 | 154.9 |
| Crude petroleum oils, etc. | 159.3 | 100.8 | 149.4 |
| **Chemicals and related products** | 308.8 | 350.7 | 314.6 |
| Medicinal and pharmaceutical products | 85.2 | 73.2 | 73.1 |
| Medicaments (incl. veterinary) | 82.0 | 69.7 | 70.5 |
| Essential oils, perfume materials and cleansing preparations | 110.4 | 138.1 | 126.4 |
| **Basic manufactures** | 301.4 | 321.8 | 251.0 |
| **Miscellaneous manufactured articles** | 204.5 | 178.0 | 144.1 |
| **Total** (incl. others) | 2,699.4 | 2,412.6 | 2,227.5 |

### PRINCIPAL TRADING PARTNERS
(US $ million)

| Imports c.i.f. | 2000 | 2001 | 2002 |
|---|---|---|---|
| Brazil | 67.3 | 84.3 | 122.5 |
| Canada | 124.7 | 140.6 | 125.5 |
| Costa Rica | 200.9 | 232.2 | 216.3 |
| El Salvador | 313.8 | 385.0 | 333.1 |
| Germany | 126.2 | 158.9 | 171.8 |
| Honduras | 84.1 | 129.3 | 84.0 |
| Italy | 55.7 | 56.9 | 71.7 |
| Japan | 166.6 | 289.7 | 340.1 |
| Korea, Republic | 69.2 | 77.1 | 77.6 |
| Mexico | 576.0 | 595.0 | 587.4 |
| Panama | 162.0 | 69.5 | 81.9 |
| Spain | 69.3 | 83.6 | 104.6 |
| USA | 2,071.2 | 1,964.6 | 2,195.4 |
| Venezuela | 275.8 | 271.8 | 224.5 |
| **Total** (incl. others) | 5,171.4 | 5,606.6 | 6,077.7 |

| Exports f.o.b. | 2000 | 2001 | 2002 |
|---|---|---|---|
| Belgium* | 34.8 | 11.9 | 11.8 |
| Canada | 63.1 | 42.7 | 34.9 |
| Costa Rica | 126.8 | 156.3 | 119.1 |
| El Salvador | 341.1 | 477.3 | 403.6 |
| Germany | 107.8 | 58.3 | 46.1 |
| Honduras | 233.1 | 295.3 | 236.6 |
| Japan | 62.5 | 42.3 | 29.3 |
| Korea, Republic | 28.6 | 94.0 | 85.5 |
| Mexico | 120.2 | 79.0 | 76.4 |
| Netherlands | 30.7 | 15.8 | 14.2 |
| Nicaragua | 114.3 | 130.6 | 114.5 |
| Panama | 54.7 | 43.4 | 48.2 |
| Russian Federation | 20.7 | 31.3 | n.a. |
| Saudi Arabia | 47.6 | 55.9 | 46.3 |
| USA | 971.2 | 643.1 | 672.1 |
| **Total** (incl. others) | 2,699.0 | 2,412.6 | 2,227.7 |

* Includes figures for Luxembourg.

# GUATEMALA

## Transport

**RAILWAYS**
(traffic)

|  | 1994 | 1995 | 1996 |
|---|---|---|---|
| Passenger-km (million) | 991 | 0 | 0 |
| Freight ton-km (million) | 25,295 | 14,242 | 836 |

Source: UN, *Statistical Yearbook*.

**ROAD TRAFFIC**
(motor vehicles in use at 31 December)

|  | 1997 | 1998 | 1999 |
|---|---|---|---|
| Passenger cars | 470,016 | 508,868 | 578,733 |
| Buses and coaches | 9,843 | 10,250 | 11,017 |
| Lorries and vans | 34,220 | 37,057 | 42,219 |
| Motorcycles and mopeds | 111,358 | 117,536 | 129,664 |

Source: IRF, *World Road Statistics*.

**SHIPPING**
**Merchant Fleet**
(registered at 31 December)

|  | 2002 | 2003 | 2004 |
|---|---|---|---|
| Number of vessels | 11 | 9 | 10 |
| Total displacement ('000 grt) | 8.9 | 5.3 | 5.5 |

Source: Lloyd's Register-Fairplay, *World Fleet Statistics*.

**International Sea-borne Freight Traffic**
('000 metric tons)

|  | 1992 | 1993 | 1994 |
|---|---|---|---|
| Goods loaded | 2,176 | 1,818 | 2,096 |
| Goods unloaded | 3,201 | 3,025 | 3,822 |

**CIVIL AVIATION**
(traffic on scheduled services)

|  | 1997 | 1998 | 1999 |
|---|---|---|---|
| Kilometres flown (million) | 5 | 7 | 5 |
| Passengers carried ('000) | 508 | 794 | 506 |
| Passenger-km (million) | 368 | 480 | 342 |
| Total ton-km (million) | 77 | 50 | 33 |

Source: UN, *Statistical Yearbook*.

## Tourism

**TOURIST ARRIVALS BY COUNTRY OF ORIGIN**

|  | 2001 | 2002 | 2003 |
|---|---|---|---|
| Canada | 17,277 | 23,945 | 27,048 |
| Costa Rica | 28,974 | 28,918 | 29,529 |
| El Salvador | 214,114 | 228,018 | 209,745 |
| France | 15,312 | 16,119 | 18,433 |
| Germany | 20,985 | 23,559 | 27,734 |
| Honduras | 59,224 | 75,355 | 64,242 |
| Italy | 18,358 | 17,310 | 17,272 |
| Mexico | 61,326 | 65,331 | 70,732 |
| Nicaragua | 15,882 | 24,607 | 29,815 |
| Spain | 24,190 | 24,125 | 24,869 |
| USA | 193,285 | 199,614 | 209,247 |
| **Total** (incl. others) | 835,492 | 884,190 | 880,223 |

**Tourism receipts** (US $ million, incl. passenger transport): 588 in 2001; 647 in 2002; 646 in 2003.

Source: World Tourism Organization.

## Communications Media

|  | 2001 | 2002 | 2003 |
|---|---|---|---|
| Telephones ('000 main lines in use) | 756 | 846 | 944 |
| Mobile cellular telephones ('000 subscribers) | 1,146 | 1,577 | 2,035 |
| Personal computers ('000 in use) | 150 | 173 | n.a. |
| Internet users ('000) | 200 | 400 | n.a. |

Source: International Telecommunication Union.

**Radio receivers** ('000 in use): 835 in 1997.

**Television receivers** ('000 in use): 680 in 1999 (Source: UN, *Statistical Yearbook*).

**Daily newspapers** (number): 7 in 1996.

**Facsimile machines** (number in use): 10,000 in 1996.

## Education

(2003, unless otherwise indicated)

|  | Institutions | Teachers | Students |
|---|---|---|---|
| Pre-primary | 10,424* | 15,708 | 399,842 |
| Primary | 16,609 | 69,954 | 2,163,760 |
| Secondary | 3,585 | 26,372 | 444,345 |
| Tertiary | 1,946 | 16,188 | 210,225 |

* 2000 data.

**Adult literacy rate** (UNESCO estimates): 69.9% (males 77.3%; females 62.5%) in 2002 (Source: UN Development Programme, *Human Development Report*).

# Directory

## The Constitution

In December 1984 the Constituent Assembly drafted a new Constitution (based on that of 1965), which was approved in May 1985 and came into effect in January 1986. A series of amendments to the Constitution were approved by referendum in January 1994 and came into effect in April 1994. The Constitution's main provisions are summarized below:

Guatemala has a republican representative democratic system of government and power is exercised equally by the legislative, executive and judicial bodies. The official language is Spanish. Suffrage is universal and secret, obligatory for those who can read and write and optional for those who are illiterate. The free formation and growth of political parties whose aims are democratic is guaranteed. There is no discrimination on grounds of race, colour, sex, religion, birth, economic or social position or political opinions.

The State will give protection to capital and private enterprise in order to develop sources of labour and stimulate creative activity.

Monopolies are forbidden and the State will limit any enterprise which might prejudice the development of the community. The right to social security is recognized and it shall be on a national, unitary, obligatory basis.

Constitutional guarantees may be suspended in certain circumstances for up to 30 days (unlimited in the case of war).

### CONGRESS

Legislative power rests with Congress, which is made up of 158 deputies, elected according to a combination of departmental and proportional representation. Congress meets on 15 January each year and ordinary sessions last four months; extraordinary sessions can be called by the Permanent Commission or the Executive. All Congressional decisions must be taken by absolute majority of the

# GUATEMALA

members, except in special cases laid down by law. Deputies are elected for four years; they may be re-elected after a lapse of one session, but only once. Congress is responsible for all matters concerning the President and Vice-President and their execution of their offices; for all electoral matters; for all matters concerning the laws of the Republic; for approving the budget and decreeing taxes; for declaring war; for conferring honours, both civil and military; for fixing the coinage and the system of weights and measures; for approving, by two-thirds' majority, any international treaty or agreement affecting the law, sovereignty, financial status or security of the country.

## PRESIDENT

The President is elected by universal suffrage, by absolute majority for a non-extendable period of four years. Re-election or prolongation of the presidential term of office are punishable by law. The President is responsible for national defence and security, fulfilling the Constitution, leading the armed forces, taking any necessary steps in time of national emergency, passing and executing laws, international policy, nominating and removing Ministers, officials and diplomats, co-ordinating the actions of Ministers of State. The Vice-President's duties include presiding over Congress and taking part in the discussions of the Council of Ministers.

## ARMY

The Guatemalan Army is intended to maintain national independence, sovereignty and honour and territorial integrity. It is an indivisible, apolitical, non-deliberating body and is made up of land, sea and air forces.

## LOCAL ADMINISTRATIVE DIVISIONS

For the purposes of administration the territory of the Republic is divided into 22 Departments and these into 330 Municipalities, but this division can be modified by Congress to suit interests and general development of the Nation without loss of municipal autonomy. Municipal authorities are elected every four years.

## JUDICIARY

Justice is exercised exclusively by the Supreme Court of Justice and other tribunals. Administration of Justice is obligatory, free and independent of the other functions of State. The President of the Judiciary, judges and other officials are elected by Congress for five years. The Supreme Court of Justice is made up of 13 judges. The President of the Judiciary is also President of the Supreme Court. The Supreme Court nominates all other judges. Under the Supreme Court come the Court of Appeal, the Administrative Disputes Tribunal, the Tribunal of Second Instance of Accounts, Jurisdiction Conflicts, First Instance and Military, the Extraordinary Tribunal of Protection. There is a Court of Constitutionality presided over by the President of the Supreme Court.

# The Government

## HEAD OF STATE

**President:** Oscar José Rafael Berger Perdomo (took office 14 January 2004).

**Vice-President:** Eduardo Stein Barillas.

### CABINET
(July 2005)

**Minister of Foreign Affairs:** Jorge Briz Abularach.

**Minister of the Interior:** Carlos Vielman Montes.

**Minister of National Defence:** Gen. Carlos Humberto Aldana Villanueva.

**Minister of Public Finance:** Maria Antonieta del Cid de Bonilla.

**Minister of Economy:** Marcio Ronaldo Cuevas Posadas.

**Minister of Public Health and Social Welfare:** Marco Tulio Sosa Ramírez.

**Minister of Communications, Infrastructure and Housing:** Manuel Eduardo Castillo Arroyo.

**Minister of Agriculture, Livestock and Food:** Alvaro Aguilar Prado.

**Minister of Education:** María del Carmen Aceña.

**Minister of Employment and Social Security:** Jorge Francisco Gallardo Flores.

**Minister of Energy and Mines:** Luis Romeo Ortiz Peláez.

**Minister of Culture and Sport:** Manuel de Jesús Salazar Tetzahuic.

**Minister of the Environment and Natural Resources:** Juan Mario Dary Fuentes.

## MINISTRIES

**Ministry of Agriculture, Livestock and Food:** Edif. Monja Blanca, 7 Avda 12-90, Zona 13, Guatemala City; tel. 2362-4764; fax 2332-8302; e-mail magadest@intelnet.net.gt; internet www.maga.gob.gt.

**Ministry of Communications, Infrastructure and Housing:** Edif. Antiguo Cocesna, 8a Avda y 15 Calle, Zona 13, Finca Nacional La Aurora, Guatemala City; tel. 2361-0827; fax 2362-6051; e-mail relpublicas@micivi.gob.gt; internet www.civ.gob.gt.

**Ministry of Culture and Sport:** 12 Avda 11-65, Zona 1, Guatemala City; tel. 2253-0543; fax 2253-0540; internet www.minculturadeportes.gob.gt.

**Ministry of Economy:** 8a Avda 10-43, Zona 1, Guatemala City; tel. 2238-3330; fax 2238-2413; e-mail nhernandez@mail.mineco.gob.gt; internet www.mineco.gob.gt.

**Ministry of Education:** Palacio Nacional, 6a Calle 1-87, Zona 10, Guatemala City; tel. 2360-0911; fax 2361-0350; e-mail informatica@mineduc.gob.gt; internet www.mineduc.gob.gt.

**Ministry of Employment and Social Security:** Edif. NASA, 14 Calle 5-49, Zona 1, Guatemala City; tel. 2230-1361; fax 2251-3559; e-mail info@mintrabajo.gob.gt; internet www.mintrabajo.gob.gt.

**Ministry of Energy and Mines:** Diagonal 17, 29-78, Zona 11, Guatemala City; tel. 2477-0743; fax 2476-8506; e-mail informatica@mem.gob.gt; internet www.mem.gob.gt.

**Ministry of the Environment and Natural Resources:** Edif. MARN, 20 Calle 28-58, Zona 10, Guatemala City; tel. 512-2914; e-mail marnguatemala@marn.gob.gt; internet www.marn.gob.gt.

**Ministry of Foreign Affairs:** 2a Avda La Reforma 4-47, Zona 10, Guatemala City; tel. 2331-8410; fax 2331-8510; e-mail webmaster@minex.gob.gt; internet www.minex.gob.gt.

**Ministry of the Interior:** 6a Avda 4-64, Zona 4, 3°, Guatemala City; tel. 2367-5402; fax 2362-0239; internet www.mingob.gob.gt.

**Ministry of National Defence:** Antigua Escuela Politécnica, Avda La Reforma 1-45, Zona 10, Guatemala City; tel. 2360-9890; fax 2360-9909; internet www.mindef.mil.gt.

**Ministry of Public Finance:** Centro Cívico, 8a Avda y 21 Calle, Zona 1, Guatemala City; tel. 2248-5053; fax 2248-5054; e-mail info@minfin.gob.gt; internet www.minfin.gob.gt.

**Ministry of Public Health and Social Welfare:** Escuela de Enfermería, 3°, 6a Avda 3-45, Zona 1, Guatemala City; tel. 2475-2121; fax 2475-2168; e-mail info@mspas.gob.gt; internet www.mspas.gob.gt.

# President and Legislature

## PRESIDENT

**Presidential Election, 9 November and 28 December 2003**

| Candidate | First round % of votes | Second round % of votes |
|---|---|---|
| Oscar Berger Perdomo (GANA)* | 34.3 | 54.1 |
| Alvaro Colom Caballeros (UNE) | 26.4 | 45.9 |
| Gen. (retd) José Efraín Ríos Montt (FRG) | 19.3 | — |
| Leonel López Rodas (PAN) | 8.4 | — |
| Fritz García-Gallont (PU) | 3.0 | — |
| Rodrigo Asturias (URNG) | 2.6 | — |
| Jacobo Arbenz (DCG) | 1.6 | — |
| José Angel Lee (DSP) | 1.4 | — |
| **Total (incl. others)** | **100.0** | **100.0** |

*The Gran Alianza Nacional, an electoral alliance comprising the Partido Patriota, the Movimiento Reformador and the Partido Solidaridad Nacional.

## CONGRESO DE LA REPÚBLICA

**President:** Jorge Méndez Herbruger.

**Vice-Presidents:** Alejandro Baltazar Maldonado Aguirre, Edgar Leonel Rodríguez Lara, Ingrid Roxana Baldetti Elias.

# GUATEMALA

## Legislative Elections, 9 November 2003

| | Seats |
|---|---|
| Gran Alianza Nacional (GANA)* | 49 |
| Frente Republicano Guatemalteco (FRG) | 42 |
| Unidad Nacional de la Esperanza (UNE) | 33 |
| Partido de Avanzada Nacional (PAN) | 16 |
| Partido Unionista (PU) | 7 |
| Alianza Nueva Nación (ANN) | 6 |
| Unidad Revolucionaria Nacional Guatemalteca (URNG) | 2 |
| Others | 3 |
| **Total** | **158** |

*An electoral alliance comprising the Partido Patriota, the Movimiento Reformador and the Partido Solidaridad Nacional.

## Political Organizations

**Alianza Nueva Nación (ANN):** 15 Avda 5-60, Zona 1, Guatemala City; tel. 2251-2514; Sec.-Gen. ALFONSO BAUER PAZ.

**Bienestar Nacional (BIEN):** 8a Avda 6-30, Zona 2, Guatemala City; tel. 524-1448; Sec.-Gen. RUBÉN GARCÍA LÓPEZ.

**Centro de Acción Social (CASA):** Guatemala City; f. 2003; represents indigenous peoples; Leader RIGOBERTO QUEMÉ CHAY.

**Democracia Cristiana Guatemalteca (DCG):** Avda Elena 20-66, Zona 3, Guatemala City; tel. 2238-4988; fax 2337-0966; e-mail vcerezo@congreso.go.gt; f. 1955; 130,000 mems; Sec.-Gen. MARCO VINICIO CEREZO AREVALO.

**Desarrollo Integral Auténtico (DIA):** 12a Calle 'A' 2-18, Zona 1, Guatemala City; tel. and fax 2232-8044; e-mail morlain@guate.net; f. 1998; left-wing party; Sec.-Gen. JORGE LUIS ORTEGA TORRES.

**Frente Democrático Nueva Guatemala (FDNG):** left-wing faction of Partido Revolucionario; Pres. JORGE GONZÁLEZ DEL VALLE; Sec.-Gen. RAFAEL ARRIAGA.

**Frente Republicano Guatemalteco (FRG):** 3a Calle 5-50, Zona 1, Guatemala City; tel. 2238-0826; internet www.frg.org.gt; f. 1988; right-wing group; Leader Gen. (retd) JOSÉ EFRAÍN RÍOS MONTT.

**Gran Alianza Nacional (GANA):** 7a Avda, 10-40, Zona 9, Guatemala City; tel. 2331-0121; fax 2331-0144; internet www.ganaconoscarberger2003.org; f. 2003, following split with the PAN; Leader OSCAR JOSÉ RAFAEL BERGER PERDOMO; centre-right alliance comprising:

**Movimiento Reformador:** Vía 6, 4-09, Zona 4, Guatemala City; tel. 2360-0745; e-mail ruben@itelgua.com; f. 1995 as the Partido Laborista Guatemalteco; adopted present name in 2002; Sec.-Gen. ALFREDO SKINNER KLEE ARENALES.

**Partido Solidaridad Nacional (PSN):** 3a Avda 15-14, Zona 1, Guatemala City; e-mail cqa_74@hotmail.com; Sec.-Gen. JORGE FRANCISCO GALLARDO FLORES.

**Movimiento de Liberación Nacional (MLN):** Of. 10A, Condiminio Reforma, Avda Reforma 10-00, Zona 9, Guatemala City; tel. 2331-1093; fax 2331-6865; e-mail mln@wepa.com.gt; internet www.wepa.com.gt/mln; f. 1960; extreme right-wing; 95,000 mems; Sec.-Gen. ULYSSES CHARLES DENT WEISSENBERG; Public Relations Sec. JOSÉ GARAVITO GORDILLO.

**Movimiento Principios y Valores:** Guatemala City; f. 2003; Sec.-Gen. FRANCISCO BIANCHI; Sec. MANOLO BENDFELDT.

**Partido de Avanzada Nacional (PAN):** 7a Avda 10-38, Zona 9, Guatemala City; tel. 2334-1702; fax 2331-9906; internet www.pan.org.gt; Leader ALVARO ENRIQUE ARZÚ YRIGOYEN; Sec.-Gen. LEONEL ELISEO LÓPEZ RODAS.

**Partido Libertador Progresista (PLP):** 5a Calle 5-44, Zona 1, Guatemala City; tel. 2232-5548; e-mail plp@intelnet.net.gt; f. 1990; Sec.-Gen. ACISCLO VALLADARES.

**Partido Patriota (PP):** 11 Calle 11-54, Zona 1, Guatemala City; tel. 2230-6227; e-mail comunicacion@partidopatriota.org; internet www.partidopatriota.org; f. 2002; contested 2003 elections as part of GANA (q.v.); withdrew from GANA in May 2004; Sec.-Gen. OTTO PÉREZ MOLINA.

**Partido Petenero:** Guatemala City; f. 1983; defends regional interests of El Petén.

**Partido Unionista (PU):** 3a Avda 'A' 14-23, Zona 9, Guatemala City; tel. 2331-7468; fax 2331-6141; e-mail info@unionista.org; internet www.unionistas.org; f. 1917; Sec.-Gen. FRITZ GARCÍA-GALLONT.

**Transparencia:** 11 Avda 15-61, Zona 10, Guatemala City; tel. 2333-7330; contested the 2003 elections; Sec.-Gen. PEDRO CHITAY RODRÍGUES.

**Unidad Nacional Autentica (UNA):** Calzado San Juan, Col. Monserrat 1, 2°, Zona 4, Mixco; tel. 437-6816; Sec.-Gen. GERARDO VILLEDA GUERRA.

**Unidad Nacional de la Esperanza (UNE):** 2a Avda 5-11, Zona 9, Guatemala City; tel. 2232-4685; e-mail ideas@une.org.gt; internet www.une.org.gt; f. 2001 following a split within the PAN; Founder and Pres. ALVARO COLOM CABALLEROS.

**Unidad Revolucionaria Nacional Guatemalteca (URNG):** Avda Simeón Cañas 8-01, Zona 2, Guatemala City; tel. 2288-4440; fax 2254-0572; e-mail prensaurng@guate.net; internet www.urng.org.gt; f. 1982 following unification of principal guerrilla groups engaged in the civil war; formally registered as a political party in 1998, following the end of the civil war in Dec. 1996; Sec.-Gen. ALBA ESTELA MALDONADO GUEVARA.

**Unión Democrática (UD):** Of. E, 3°, Vista Hermosa II, 1a Calle 18-83, Zona 15, 01015 Guatemala City; tel. 2369-7074; fax 2369-3062; e-mail fito.paiz@ud.org.gt; internet www.ud.org.gt; f. 1983; Sec.-Gen. RODOLFO ERNESTO (FITO) PAIZ ANDRADE.

**Los Verdes (LV):** Vista Hermosa II, 1a Calle 22-08, Zona 15, Guatemala City; Sec.-Gen. RODOLFO ROSALES GARCÍA SALAS.

## Diplomatic Representation

### EMBASSIES IN GUATEMALA

**Argentina:** Edif. Europlaza 1703, 5 Avda 5-55, Zona 14, Apdo 120, Guatemala City; tel. and fax 2385-3786; e-mail embajadadeargentina@hotmail.com; internet www.embajadadeargentina.com; Ambassador CARLOS MARIO FORADORI.

**Austria:** Edif. Plaza Marítima, 4°, 6a Avda 20-25, Zona 10, Guatemala City; tel. 2368-1134; fax 2333-6180; e-mail austriabot@intelnet.net.gt; Ambassador MONIKA GRUBER-LANG.

**Belize:** Edif. El Reformador, Suite 803, 8°, Avda de la Reforma 1-50, Zona 9, Guatemala City; tel. 2334-5531; fax 2334-5536; e-mail embelguat@guate.net; internet www.embajadadebelize.org; Ambassador ALFREDO MARTINEZ.

**Brazil:** 18a Calle 2-22, Zona 14, Apdo 196-A, Guatemala City; tel. 2337-0949; fax 2337-3475; e-mail braembx@intelnet.net.gt; Ambassador RENAN PAES BARRETO.

**Canada:** Edif. Edyma Plaza, 8°, 13a Calle 8-44, Zona 10, Apdo 400, Guatemala City; tel. 2333-6102; fax 2333-6189; e-mail gtmla@dfait-maeci.gc.ca; internet www.dfait-maeci.gc.ca/guatemala; Ambassador JAMES LAMBERT.

**Chile:** 14 Calle 15-21, Zona 13, Guatemala City; tel. 2334-8273; fax 2334-8276; e-mail chilegu@intelnet.net.gt; Ambassador JORGE MOLINA VALDIVIESO.

**China (Taiwan):** 4a Avda 'A' 13-25, Zona 9, Apdo 1646, Guatemala City; tel. 2339-0711; fax 2332-2668; e-mail echina@intelnet.net.gt; Ambassador HONG-LIEN OU.

**Colombia:** Edif. Europlaza 1603, 5a Avda 5-56, Zona 14, Guatemala City; tel. 2385-3432; fax 2335-3603; e-mail embaguate@inter.net.co; Ambassador HERNANDEZ RÁMIREZ JARAMILLO.

**Costa Rica:** 1a Avda 15-52, Zona 10, Guatemala City; tel. 2363-1345; fax 2368-0506; e-mail embarica@intelnet.net.gt; Ambassador HORACIO ALVARADO BOGANTES.

**Cuba:** Avda las Americas 20-72, Zona 13, Guatemala City; tel. 2332-4066; fax 2332-5525; e-mail embagua@intelnet.net.gt; Ambassador ANGEL LORENZO ABASCAL IGLESIAS.

**Dominican Republic:** Edif. Géminis 10, Suite 804, Torre Sur, 12 Calle 1-25, Zona 10, Guatemala City; tel. 2338-2170; fax 2338-2171; e-mail embardom@intelnet.net.gt; Ambassador MIGDALIA TORRES GARCÍA.

**Ecuador:** 4 Avda 12-04, Zona 14, Guatemala City; tel. 2337-2994; fax 2368-1831; e-mail embecuad@guate.net; Ambassador MAURICIO PÉREZ MARTÍNEZ.

**Egypt:** Edif. Cobella, 5°, 5 Avda 10-84, Zona 14, Apdo 502, Guatemala City; tel. 2333-6296; fax 2368-2808; e-mail egyptemb@quetzal.net; Ambassador MOHAMED HADI MOUSTAFA EL TONSI.

**El Salvador:** 5a Avda 8-15, Zona 9, Guatemala City; tel. 2360-7660; fax 2334-2069; e-mail emsalva@pronet.net.gt; Ambassador RAFAEL ANTONIO CARBALLO ARÉVALO.

**France:** Edif. Marbella, 11°, 16a Calle 4-53, Zona 10, Apdo 971-A, Guatemala City; tel. 2337-4080; fax 2337-3180; e-mail ambfrguate@intelnet.net.gt; internet www.ambafrance.org.gt; Ambassador NORBERT CARRASCO-SAULNIER.

# GUATEMALA

**Germany:** Edif. Plaza Marítima, 2°, 20 Calle 6-20, Zona 10, Guatemala City; tel. 2364-6700; fax 2333-6906; e-mail embalemana@intelnet.net.gt; internet www.guatemala.diplo.de; Ambassador CLAUDE-ROBERT ELLNER.

**Holy See:** 10a Calle 4-47, Zona 9, Apdo 22, Guatemala City (Apostolic Nunciature); tel. 2332-4274; fax 2334-1918; e-mail nuntius@gua.net; Apostolic Nuncio Most Rev. BRUNO MUSARÒ (Titular Archbishop of Abari).

**Honduras:** Edif. Géminis 10, 12°, Torre Sur, 12 Calle 1-25, Of. 1211-1206B, Zona 10, Guatemala City; tel. 2335-3281; fax 2335-2851; e-mail embhond@intelnet.net.gt; Ambassador EDGUARDO PAZ.

**Israel:** 13a Avda 14-07, Zona 10, Guatemala City; tel. 2333-4624; fax 2333-6950; e-mail isrembgu@guaweb.net; Ambassador YAACOV PARAN.

**Italy:** 5a Avda 8-59, Zona 14, Guatemala City; tel. 2337-4557; fax 2337-0795; e-mail ambaguat@intelnet.net.gt; Ambassador PIETRO PORCARELLI.

**Japan:** Edif. Torre Internacional, 10°, Avda de la Reforma 16-85, Zona 10, Guatemala City; tel. 2367-2244; fax 2367-2245; e-mail embjpninfo@micro.com.gt; Ambassador KAGEFUMI UENO.

**Korea, Republic:** Edif. El Reformador, 7°, Avda de la Reforma 1-50, Zona 9, Apdo 1649, Guatemala City; tel. 2334-5480; fax 2334-5481; e-mail korembsy@mofat.go.kr; Ambassador KIM HONG-RAK.

**Mexico:** 2a Avda 7-57, Zona 10, Apdo 1455, Guatemala City; tel. 2420-3400; fax 2420-3410; e-mail embamexguat@itelgua.com; internet www.sre.gob.mx/guatemala; Ambassador ROSALBA OJEDA Y CARDENAS.

**Nicaragua:** 10a Avda 14-72, Zona 10, Guatemala City; tel. 2368-0785; fax 2337-4264; e-mail embaguat@terra.com.gt; Ambassador Dr JOSÉ RENÉ GUTIÉRREZ HUETE.

**Norway:** Edif. Murano Center 15°, Of. 1501, 14 Calle 3-51, Zona 10, Apdo 1764, Guatemala City; tel. 2366-5908; fax 2366-5928; e-mail ambgua@norad.no; Ambassador ROLF O. BERG.

**Panama:** 10 Avda 18-53, La Cañada, Apdo 929A, Zona 14, Guatemala City; tel. 2368-2805; fax 2333-3835; Ambassador JOSÉ ORLANDO CALVO VELÁZQUEZ.

**Peru:** 2a Avda 9-67, Zona 9, Guatemala City; tel. 2331-8558; fax 2334-3744; e-mail leprugua@concyt.gob.gt; Ambassador ALFREDO ARECCO SABLICH.

**Spain:** 6a Calle 6-48, Zona 9, Guatemala City; tel. 2379-3530; fax 2379-3533; e-mail embaespa@terra.com.gt; Ambassador JUAN LÓPEZ-DORIGA PÉREZ.

**Sweden:** 8a Avda 15-07, Zona 10, Guatemala City; tel. 2384-7300; fax 2384-7350; e-mail ambassaden.guatemala@foreign.ministry.se; internet www.swedenabroad.com/pages/general_23845.asp; Ambassador EIVOR HALKJAER.

**Switzerland:** Edif. Torre Internacional, 14°, 16 Calle 0-55, Zona 10, Apdo 1426, Guatemala City; tel. 2367-5520; fax 2367-5811; e-mail vertretung@gua.rep.admin.ch; Ambassador URS STEMMLER.

**United Kingdom:** Edif. Torre Internacional, 11°, Avda de la Reforma, 16 Calle, Zona 10, Guatemala City; tel. 2367-5425; fax 2367-5430; e-mail embassy@intelnett.com; Ambassador RICHARD LAVERS.

**USA:** Avda de la Reforma 7-01, Zona 10, Guatemala City; tel. 2331-1541; fax 2331-8885; internet guatemala.usembassy.gov; Ambassador JAMES M. DERHAM.

**Uruguay:** Edif. Plaza Marítima, 3°, Of. 341, 6a Avda 20-25, Zona 10, Guatemala City; tel. 2368-0810; fax 2333-7553; e-mail uruguate@guate.net; Ambassador ESTELLA RUBY ARMAND UGON SEPULVEDA.

**Venezuela:** Edif. Atlantis, Of. 601, 13 Calle 3-40, Zona 10, Apdo 152, Guatemala City; tel. 2366-9832; fax 2366-9838; e-mail embavene@concyt.gob.gt; Chargé d'affaires a.i. FELIX MÉNDEZ CORREA.

## Judicial System

### Corte Suprema

Centro Cívico, 21 Calle 7-70, Zona 1, Guatemala City; internet www.organismojudicial.gob.gt.

The members of the Supreme Court are appointed by the Congress.

**President of the Supreme Court:** RODOLFO DE LEÓN MOLINA.

**Members:** A. E. LÓPEZ RODRÍGUEZ, R. E. HIGUEROS GIRÓN, L. S. SECAIRA PINTO, J. F. DE MATA VELA, Dr V. M. RIVERA WÖLTKE, O. H. VÁSQUEZ OLIVA, E. R. PACAY YALIBAT, C. E. DE LEÓN CÓRDOVA, B. O. DE LEÓN REYES, L. FERNÁNDEZ MOLINA, J. G. CABRERA HURTARTE, C. G. CHACÓN TORREBIARTE.

### Civil Courts of Appeal

Ten courts, five in Guatemala City, two in Quetzaltenango, one each in Jalapa, Zacapa and Antigua. The two Labour Courts of Appeal are in Guatemala City.

**Judges of the First Instance:** Seven civil and 10 penal in Guatemala City, two civil each in Quetzaltenango, Escuintla, Jutiapa and San Marcos, one civil in each of the 18 remaining Departments of the Republic.

## Religion

Almost all of the inhabitants profess Christianity, with a majority belonging to the Roman Catholic Church. In recent years the Protestant Churches have attracted a growing number of converts.

### CHRISTIANITY

#### The Roman Catholic Church

For ecclesiastical purposes, Guatemala comprises two archdioceses, 10 dioceses and the Apostolic Vicariates of El Petén and Izabal. At 31 December 2003 adherents represented about 77% of the total population.

#### Bishops' Conference

Conferencia Episcopal de Guatemala, Secretariado General del Episcopado, Km 15, Calzada Roosevelt 4-54, Zona 7, Mixco, Apdo 1698, Guatemala City; tel. 2433-1832; fax 2433-1834; e-mail ceg@quetzal.net; internet www.iglesiacatolica.org.gt.

f. 1973; Pres. Cardinal RODOLFO QUEZADA TORUÑO (Archbishop of Guatemala City).

**Archbishop of Guatemala City:** Cardinal RODOLFO QUEZADA TORUÑO, Arzobispado, 7a Avda 6-21, Zona 1, Apdo 723, Guatemala City; tel. 2232-9707; fax 2251-5068; e-mail curiaarzobispal@intelnet.net.gt.

**Archbishop of Los Altos, Quetzaltenango-Totonicapán:** VÍCTOR HUGO MARTÍNEZ CONTRERAS, Arzobispado, 11a Avda 6-27, Zona 1, Apdo 11, 09001 Quetzaltenango; tel. 7761-2840; fax 7761-6049.

#### The Anglican Communion

Guatemala comprises one of the five dioceses of the Iglesia Anglicana de la Región Central de América.

**Bishop of Guatemala:** Rt Rev. ARMANDO GUERRA SORIA, Avda Castellana 40-06, Zona 8, Apdo 58-A, Guatemala City; tel. 2272-0852; fax 2472-0764; e-mail diocesis@infovia.com.gt; diocese founded 1967.

#### Protestant Churches

The largest Protestant denomination in Guatemala is the Full Gospel Church, followed by the Assembly of God, the Central American Church, and the Prince of Peace Church. The Baptist, Presbyterian, Lutheran and Episcopalian churches are also represented.

**The Baptist Church:** Convention of Baptist Churches of Guatemala, 12a Calle 9-54, Zona 1, Apdo 322, 01901 Guatemala City; tel. and fax 2232-4227; e-mail cibg@intelnet.net.gt; f. 1946; Pres. Lic. JOSÉ MARROQUÍN R.; 43,876 mems.

**Church of Jesus Christ of Latter-day Saints:** 12a Calle 3-37, Zona 9, Guatemala City; e-mail contactos@mormones.org.gt; internet www.mormones.org.gt; 17 bishoprics, nine chapels; Pres. GORDON B. HINCKLEY.

**Congregación Luterana La Epifanía** (Evangelical Lutheran Congregation La Epifanía): 2a Avda 15-31, Zona 10, Apdo 651, 01010 Guatemala City; tel. 2368-0301; fax 2366-4968; e-mail schweikle@web.de; Pres. MÓNICA HEGEL; 200 mems.

**Iglesia Nacional Evangélica Menonita Guatemalteca:** Guatemala City; tel. 2339-0606; e-mail AlvaradoJE@ldschurch.org; Contact JULIO ALVARADO; members 5,000.

**Divine Saviour Lutheran Church:** Zacapa; tel. 7941-0254; e-mail hogarluterano@hotmail.com; Pastor GERARDO VENANCIO VÁSQUEZ SALGUERO.

**Presbyterian Church:** Iglesia Evangélica Presbiteriana Central, 6a Avda 'A' 4-68, Zona 1, Apdo 655, Guatemala City; tel. 2232-0791; fax 2232-2832; internet www.presbiterianacentral.org; f. 1882; 36,000 mems; Pastor Rev. JOSÉ RAMIRO BOLAÑOS RIVERA.

**Union Church:** 12 Calle 7-37, Zona 9, 01009 Guatemala City; tel. 2361-2037; fax 2362-3961; e-mail unionchurch@guate.net; f. 1943; Pastor W. KARL SMITH.

# GUATEMALA

## The Press

### PRINCIPAL DAILIES

**Al Día:** Avda de la Reforma 6-64, Zona 9, Guatemala City; tel. 2339-7430; fax 2339-7435; e-mail aldia@notinet.com.gt; f. 1996; Pres. LIONEL TORIELLO NÁJERA; Dir GERARDO JIMÉNEZ ARDÓN; Editor OTONIEL MONROY HERNÁNDEZ.

**Diario de Centroamérica:** 18a Calle 6-72, Zona 1, Guatemala City; tel. 2222-4418; internet www.diariodecentroamerica.gob.gt; f. 1880; morning; official; Dir LUIS MENDIZÁBAL; circ. 15,000.

**La Hora:** 9a Calle 'A' 1-56, Zona 1, Apdo 1593, Guatemala City; tel. 2250-0447; fax 2251-7084; e-mail lahora@lahora.com.gt; internet www.lahora.com.gt; f. 1920; evening; independent; Dir OSCAR MARROQUÍN ROJAS; circ. 18,000.

**Nuestro Diario:** 15 Avda 24–51, Zona 13, Guatemala City; tel. and fax 2361-6988; e-mail opinion@nuestrodiario.com.gt; internet www.nuestrodiario.com.gt; Dir RODOLFO MÓBIL.

**El Periódico:** 15a Avda 24-51, Zona 13, Guatemala City; tel. 2362-0242; fax 2332-9761; e-mail redaccion@elperiodico.com.gt; internet www.elperiodico.com.gt; f. 1996; morning; independent; Pres. JOSÉ RUBÉN ZAMORA; Editors JUAN LUIS FONT, SYLVIA GEREDA; circ. 50,000.

**Guía Interamericana:** Guía Interamericana, 20 Calle 5-35, Zona 10, Edif. Plaza los Arcos, 3°, Guatemala City; Editor ALFREDO MAYORGA; circ. 5,000.

**Prensa Libre:** 13a Calle 9-31, Zona 1, Apdo 2063, Guatemala City; tel. 2230-5096; fax 2251-8768; e-mail nacional@prensalibre.com.gt; internet www.prensalibre.com.gt; f. 1951; morning; independent; Gen. Man. ENRIQUE SOLORZANO MOLINA; Editor GONZALO MARROQUÍN GODOY; circ. 120,000.

**Siglo Veintiuno:** 7a Avda 11-63, Zona 9, Guatemala City; tel. 2360-6724; fax 2331-9145; e-mail buzon21@sigloxxi.com; internet www.sigloxxi.com; f. 1990; morning; Dir GUILLERMO FERNÁNDEZ; Gen. Man. LUCIANA CISNEROS; circ. 65,000.

### PERIODICALS

**Amiga:** 13 Calle 9-31, Zona 1, Guatemala City; e-mail revistas@prensalibre.com.gt; health; Dir CAROLINA VÁSQUEZ.

**Control TV:** 13 Calle 9-31, Zona 1, Guatemala City; e-mail revistas@prensalibre.com.gt; Dir CAROLINA VÁSQUEZ.

**Crónica Semanal:** Guatemala City; tel. 2235-2155; fax 2235-2360; f. 1988; weekly; politics, economics, culture; Publr FRANCISCO PÉREZ.

**Especiales:** Edif. El Gráfico, 14 Avda 4-33, Zona 1, Guatemala City; e-mail moneda@guate.net; international news magazine; Dir KATIA DE CARPIO.

**Gerencia:** La Asociación de Gerentes de Guatemala, 6 Avda 1-36, Zona 14, 01014 Guatemala City; tel. 2231-1644; fax 2231-1646; e-mail agg@guate.net; internet www.nortropic.com/gerencia; f. 1967; monthly; official organ of the Association of Guatemalan Managers; Editor MARGARITA DE TARANO.

**Guatemala Business News:** 10a Calle 3-80, Zona 1, Guatemala City; monthly; Editor RODOLFO GARCÍA; circ. 5,000.

**Inforpress Centroamericana:** Guatemala City; fax 2232-9034; e-mail inforpre@inforpressca.com; internet www.inforpressca.com; f. 1972; weekly; Spanish and English; regional political and economic news and analysis; Dir ARIEL DE LEÓN.

**Magazine Business Guatemala:** 6a Avda 14-77, Zona 10, Guatemala City; consumer magazine.

**Mundo Motor:** 13 Calle 9-31, Zona 1, Guatemala City; e-mail evasquez@prensalibre.com.gt; Dir CAROLINA VÁSQUEZ.

**Revista Data Export:** Edif. Camara de Industria, 5°, 6a Ruta 9-21, Zona 4, Guatemala; monthly; Editor REGINA CEREZO; circ. 1,500.

**Revista Industria:** 6a Ruta 9-21, Zona 4, Guatemala City; tel. 2331-9191; fax 2334-1091; e-mail contactemos@industriaguate.com; internet www.industriguate.com; monthly; official organ of the Chamber of Industry; Dir OSCAR VILLAGRÁN.

**Revista Mundo Comercial:** 10a Calle 3-80, Zona 1, 01001 Guatemala City; e-mail mundo@guatemala-chamber.org; internet www.guatemala-chamber.org; monthly; business; official organ of the Chamber of Commerce; circ. 11,000.

**Tertulia:** Guatemala City; e-mail tertulia@intelnett.com; internet www.la-tertulia.net; f. 1997; women's affairs; Editor LAURA E. ASTURIAS.

**Usurarios:** 13 Calle 9-31, Zona 1, Guatemala City; e-mail revistas@prensalibre.com.gt; computing; Dir CAROLINA VÁSQUEZ.

**Viajes:** 13 Calle 9-31, Zona 1, Guatemala City; e-mail revistas@prensalibre.com.gt; tourism; Dir CAROLINA VÁSQUEZ.

**Vida Médica:** Edif. Reforma Montúfar, Torre A, Of. 1006, Avda Reforma 12-01, Zona 10, Guatemala City; tel. 2331-7679; fax 2331-7754; e-mail vidamed@infovia.com.gt; internet www.infovia.com.gt/vidamedica; health; Dir SERAPIO ALVARADO; Editorial Dir Dr CARLOS SALAZAR.

### PRESS ASSOCIATIONS

**Asociación de Periodistas de Guatemala (APG):** 14a Calle 3-29, Zona 1, Guatemala City; tel. 2232-1813; fax 2238-2781; e-mail apege@intelnet.net.gt; internet www.freewebs.com/apg; f. 1947; Pres. ILEANA ALAMILLA; Sec. WALTER HERMOSILLA.

**Cámara Guatemalteca de Periodismo (CGP):** Guatemala City; Pres. MARIO FUENTES DESTARAC.

**Círculo Nacional de Prensa (CNP):** Guatemala City; Pres. FREDY AZURDIA AZURDI.

### NEWS AGENCIES

**Inforpress Centroamericana:** Calle Mariscal 6-58, Zona 11, 0100 Guatemala City; tel. and fax 2473-1704; e-mail inforpre@guate.net; internet www.inforpressca.com/CAR; f. 1972; independent news agency; publishes two weekly news bulletins, in English and Spanish.

#### Foreign Bureaux

**ACAN-EFE** (Central America): Edif. El Centro, 8°, Of. 8-21, 9a Calle y 7a Avda, Zona 1, Guatemala City; tel. 2251-9454; fax 2251-9484; Man. ANA CAROLINA ALPÍREZ A.

**Agenzia Nazionale Stampa Associata (ANSA)** (Italy): Torre Norte, Edif. Géminis 10, Of. 805, 12a Calle 1-25, Zona 10, Guatemala City; tel. 2335-3039; e-mail ansagua@guate.net; Chief ALFONSO ANZUETO LÓPEZ.

**Deutsche Presse-Agentur (dpa)** (Germany): 5a Calle 4-30, Zona 1, Apdo 2333, Guatemala City; tel. 2251-7505; fax 2251-7505; Correspondent JULIO CÉSAR ANZUETO.

**Inter Press Service (IPS)** (Italy): Edif. El Centro, 3°, Of. 13, 7a Avda 8-56, Zona 1, Guatemala City; tel. 2253-8837; fax 2251-4736; internet www.ipslatam.net; Correspondent GEORGE RODRÍGUEZ-OTEIZA.

**United Press International (UPI)** (USA): Guatemala City; tel. and fax 2251-4258; Correspondent AMAFREDO CASTELLANOS.

## Publishers

**Cholsamaj:** 7a Avda 9-25, Zona 1, Apdo 4, Guatemala City; tel. 2232-5959; e-mail cholsamaj@micro.com; internet www.cholsamaj.org.gt; Mayan language publications.

**Ediciones Legales Comercio e Industria:** 12a Avda 14-78, Zone 1, Guatemala City; tel. 2253-5725; fax 2220-7592; Man. Dir LUIS EMILIO BARRIOS.

**Editorial Cultura:** 10A Calle 10-14, Zona 1, Guatemala City; tel. 2232-5667; fax 2230-0591; e-mail cultuarte@intelnet.net.gt; part of the Ministry of Culture and Sport.

**Editorial Nueva Narrativa:** Edif. El Patrio, Of. 108, 7a Avda 7-07, Zona 4, Guatemala City; tel. 2360-0732; fax 5704-7895; e-mail maxaraujo@intelnet.gt; Man. Dir MAX ARAÚJO A.

**Editorial Palo de Hormigo:** Calle 16-40, Zona 15, Col. El Maestro, Guatemala City; tel. 2369-2080; fax 2369-8858; e-mail juanfercif@hotmail.com; f. 1990; Man. Dir JUAN FERNANDO CIFUENTES.

**Editorial Universitaria:** Edif. de la Editorial Universitaria, Universidad de San Carlos de Guatemala, Ciudad Universitaria, Zona 12, Guatemala City; tel. and fax 2476-9628; literature, social sciences, health, pure and technical sciences, humanities, secondary and university educational textbooks.

**F & G Editores:** 31 Avda 'C' 5-54, Zona 7, 01007 Guatemala City; tel. and fax 2433-2361; e-mail fgeditor@guate.net; internet www.fygeditores.com; f. 1990 as Figueroa y Gallardo, changed name in 1993; law, literature and social sciences; Editor RAÚL FIGUEROA SARTI.

**Piedra Santa:** 5a Calle 7-55, Zona 1, Guatemala City; tel. 2324-2331; fax 2334-6801; e-mail editorialps@yahoo.com; f. 1947; children's literature, text books; Man. Dir IRENE PIEDRA SANTA.

# GUATEMALA

# Broadcasting and Communications

## TELECOMMUNICATIONS

### Regulatory Authority

**Superintendencia de Telecomunicaciones de Guatemala:** Edif. Murano Center, 16°, 14a Calle 3-51, Zona 10, 01010 Guatemala City; tel. 2366-5880; fax 2366-5890; e-mail supertel@sit.gob.gt; internet www.sit.gob.gt; f. 1996; Supt OSCAR STUARDO CHINCHILLA.

### Major Service Providers

**Telecomunicaciones de Guatemala, SA (Telgua):** Guatemala City; internet www.telgua.com.gt; fmrly state-owned Empresa Guatemalteca de Telecomunicaciones (Guatel), name changed as above to facilitate privatization; 95% share transferred to private ownership in 1998; Dir OSCAR MONTOYA.

**Telefónica MoviStar Guatemala, SA:** Blvd Los Próceres, 20-09 Torre Telefónica, 9°, Zona 10, Guatemala City; tel. 2379-7979; e-mail servicioalcliente@telefonica.com.gt; internet www.telefonica.com.gt; owned by Telefónica Group, SA; acquired BellSouth Guatemala in 2004; wireless, wireline and radio paging communications services; 298,000 customers; Pres. CÉSAR ALIERTA IZUEL.

Other service providers include: Emergia, FT & T (Telered), Cablenet, Universal de Telecomunicaciones, Comunicaciones Celulares, Telefónica Centroamérica Guatemala, Servicios de Comunicaciones Personales Inalámbricas, A-tel Communications, Cybernet de Centroamérica, Teléfonos del Norte, Americatel Guatemala, Desarrollo Integral, BNA, TTI, Optel and Concert Global Networks.

## BROADCASTING

**Dirección General de Radiodifusión y Televisión Nacional:** Edif. Tipografía Nacional, 3°, 18 de Septiembre 6-72, Zona 1, Guatemala City; tel. 2253-2539; f. 1931; government supervisory body; Dir-Gen. ENRIQUE ALBERTO HERNÁNDEZ ESCOBAR.

### Radio

There are currently five government and six educational stations, including:

**Radio Cultural TGN:** 4a Avda y 30 Calle, Zona 3, Apdo 601, Guatemala City; tel. 2471-4378; fax 2440-0260; e-mail trgn@radiocultural.com; internet www.radiocultural.com; f. 1950; religious and cultural station; programmes in Spanish and English, Cakchiquel, Kekchí, Quiché and Aguacateco; Dir ESTEBAN SYWULKA; Man. ANTHONY WAYNE BERGER.

**Radio Nacional TGW** (La Voz de Guatemala): 18a Calle 6-72, Zona 1, Guatemala City; tel. 2253-2539; internet www.radiotgw.gob.gt; government station; Dir MOÍSES JÉREZ MORALES.

There are some 80 commercial stations, of which the most important are:

**Emisoras Unidas de Guatemala:** 4a Calle 6-84, Zona 13, Guatemala City; tel. 2440-5133; fax 2440-5159; e-mail patrullajeinformativo@emisorasunidas.com; internet sites .emisorasunidas.com; f. 1964; 7 stations: Yo Sí Sideral, Supercadena, Kiss, Atmósfera, Fabustereo, Radio Estrella and La Grande; Pres. JORGE EDGARDO ARCHILA MARROQUÍN; Vice-Pres. ROLANDO ARCHILA MARROQUÍN.

**La Marca:** 30 Avda 3-40, Zona 11, Guatemala City; tel. 2434-7330; e-mail lamarca@94fm.com.gt; internet www.94fm.com.gt.

**Metro Stereo:** Guatemala City; e-mail metrored@metrostereo.net; internet www.metrostereo.net.

**Radio Panamericana:** 1a Avda 35-48, Zona 7, Guatemala City; Dir MARIA ANTONIETA DE PANIAGUA.

### Television

**Canal 5—Televisión Cultural y Educativa, SA:** 4a Calle 18-38, Zona 1, Guatemala City; tel. 2253-1913; fax 2232-7003; f. 1980; cultural and educational programmes; Dir ALFREDO HERRERA CABRERA.

**Radio-Televisión Guatemala, SA:** 30a Avda 3-40, Zona 11, Apdo 1367, Guatemala City; tel. 2434-6320; fax 2294-7492; e-mail canal3@canal3.co.gt; internet www.canal3.com.gt; f. 1956; commercial station; operates channels 3 and 10; Pres. Lic. MAX KESTLER FARNÉS; Vice-Pres. J. F. VILLANUEVA.

**Teleonce:** 20a Calle 5-02, Zona 10, Guatemala City; tel. 2368-2532; fax 2368-2221; e-mail jcof@infovia.com.gt; internet canal11y13.homestead.com/20CALLE.html; f. 1968; commercial; channel 11; Gen. Dir JUAN CARLOS ORTIZ.

**Televisiete, SA:** 30a Avda 3-40, Zona 11, Apdo 1242, Guatemala City; tel. 594-5320; fax 2369-1393; internet www.canal7.com.gt; f. 1988; commercial station channel 7; Dir ABDÓN RODRÍGUEZ ZEA.

**Trecevisión, SA:** 20a Calle 5-02, Zona 10, Guatemala City; tel. 2368-2532; e-mail jcof@canaltrece.tv; internet canal11y13.homestead.com/20CALLE.html; commercial; channel 13; Dir Ing. PEDRO MELGAR R.; Gen. Man. GILDA VALLADARES ORTIZ.

# Finance

(cap. = capital; p.u. = paid up; res = reserves; dep. = deposits; m. = million; brs = branches; amounts in quetzales)

## BANKING

**Superintendencia de Bancos:** 9a Avda 22-00, Zona 1, Apdo 2306, Guatemala City; tel. 2232-0001; fax 2232-0002; e-mail info@sib.gob.gt; internet www.sib.gob.gt; f. 1946; Supt DOUGLAS O. BORJA VIELMAN.

### Central Bank

**Banco de Guatemala:** 7a Avda 22-01, Zona 1, Apdo 365, Guatemala City; tel. 2230-6222; fax 2253-4035; e-mail webmaster@banguat.gob.gt; internet www.banguat.gob.gt; f. 1946; state-owned; cap. 41,049.4, res 94.8m., dep. 23,941.5m. (Dec. 2001); Pres. LIZARDO ARTURO SOSA LÓPEZ; Man. EDWIN HAROLDO MATUL RUANO; 751 employees.

### State Commercial Bank

**Crédito Hipotecario Nacional de Guatemala (CHN):** 7a Avda 22-77, Zona 1, Apdo 242, Guatemala City; tel. 2384-5222; fax 2238-0744; e-mail mercadeo@chn.com.gt; internet www.chn.com.gt; f. 1930; govt-owned; Pres. FREDDY ARNOLDO MUÑOZ MORÁN; Gen. Man. HUGO LEONEL CRUZ MONTERROSO; 4 brs, 72 agencies.

### Private Commercial Banks

**Banco Agromercantil de Guatemala, SA:** 7a Avda 7-30, Zona 9, 01009 Guatemala City; tel. 2338-6565; fax 2232-5406; e-mail agromercantil@bam.com.gt; internet www.agromercantil.com.gt; f. 2000 as Banco Central de Guatemala; changed name to Banco Agrícola Mercantil in 1948; name changed as above in 2000, following merger with Banco del Agro; Man. ALFONSO E. VILLA DEVOTO; 78 agencies.

**Banco de América Central, SA:** Local 6-12, 1°, 7a Avda 6-26, Zona 9, Guatemala City; tel. 2360-9440; fax 2331-8720.

**Banco Americano, SA:** 11 Calle 7-44, Zona 9, 01009 Guatemala City; tel. 2386-1700; fax 2386-1753; e-mail grufin@infovia.com.gt; internet www.bancoamericano.com.gt.

**Banco de Antigua, SA:** 5a Avda 12-35, Zona 9, Guatemala City; tel. 2420-5555; fax 2336-8205.

**Banco del Café, SA:** Avda de la Reforma 9-30, Zona 9, Apdo 831, Guatemala City; tel. 2361-3645; fax 2331-1418; e-mail bancafeonline@bancafe.com.gt; internet www.bancafe.com.gt; f. 1978; merged with Multibanco in 2000; cap. 286.6m., res 57.2m., dep. 6,143.2m. (2002); Pres. EDUARDO MANUEL GONZÁLEZ RIVERA.

**Banco de Comercio:** Edif. Centro Operativo, 6a Avda 8-00, Zona 9, Guatemala City; tel. 2339-0504; fax 2339-0555; internet www.bancomercio.com.gt; f. 1991; 33 brs.

**Banco Corporativo, SA:** 6a Avda 4-38, Zona 9, 01009 Guatemala City; tel. 2279-9999; fax 2279-9990; e-mail mercadeo@corpobanco.com.gt.

**Banco Cuscatlán de Guatemala, SA:** Edif. Céntrica Plaza, 15 Calle 1-04, Zona 10, 01010 Guatemala City; tel. 2250-2000; fax 2250-2001; internet www.bancocuscatlan.com; acquired Guatemalan assets of Lloyds TSB in Dec. 2003.

**Banco de Desarrollo Rural, SA:** Avda La Reforma 2-56, Zona 9, Guatemala City; tel. 2334-1383; fax 2360-9740; e-mail internacional4@banrural.com.gt; internet www.banrural.com.gt; f. 1971 as Banco de Desarrollo Agrícola, name changed as above in 1998; Pres. JOSE ANGEL LÓPEZ CAMPOSECO; Gen. Man. ADOLFO FERNANDO PEÑA PEREZ; 235 agencies.

**Banco de Exportación, SA (BANEX):** Avda de la Reforma 11-49, Zona 10, Guatemala City; tel. 2331-9861; fax 2332-2879; e-mail infbanex@banex.net.gt; internet www.banex.net.gt; f. 1985; cap. 322.5m., res 66.4m., dep. 1,560.6m. (Dec. 2004); Pres. ALEJANDRO BOTRÁN; Man. ROBERTO ORTEGA HERRERA; 15 brs.

**Banco Industrial, SA (BAINSA):** Edif. Centro Financiero, Torre 1, 7a Avda 5-10, Zona 4, Apdo 744, Guatemala City; tel. 2334-5111; fax 2331-9437; e-mail webmaster@bi.com.gt; internet www.bi.com.gt; f. 1964 to promote industrial development; total assets 7.91m. (1999);

# GUATEMALA

Pres. Juan Miguel Torrebiarte Lantzendorffer; Gen. Man. Lic. Diego Pulido Aragón.

**Banco Inmobilario, SA:** 7a Avda 11-59, Zona 9, Apdo 1181, Guatemala City; tel. 2339-3777; fax 2332-1418; e-mail info@bcoinmob.com.gt; internet www.bcoinmob.com.gt; f. 1958; cap. 77.6m., res 0.4m., dep. 738.6m. (Dec. 2002); Pres. Emilio Antonio Peralta Portillo; 38 brs.

**Banco Internacional, SA:** Torre Internacional, Avda Reforma 15-85, Zona 10, Apdo 2588, Guatemala City; tel. 2366-6666; fax 2366-6743; e-mail info@bco.inter.com; internet www.bancointernacional.com.gt; f. 1976; cap. 50.0m., res 15.4m., dep. 822.6m. (Dec. 1997); Pres. Juan Ruiz Skinner-Klée; Gen. Man. Juan Manuel Ventas Benitez; 35 brs.

**Banco Privado para el Desarrollo, SA:** 7a Avda 8-46, Zona 9, Guatemala City; tel. 2361-7777; fax 2361-7217; e-mail info@bancosol.com.gt.

**Banco del Quetzal, SA:** Edif. Plaza El Roble, 7a Avda 6-26, Zona 9, Apdo 1001-A, 01009 Guatemala City; tel. 2331-8333; fax 2334-0613; e-mail negocios@banquetzal.com.gt; internet banquetzal.com.gt; f. 1984; Pres. Lic. Mario Roberto Leal Pivaral; Gen. Man. Alfonso Villa Devoto.

**Banco Reformador, SA:** 7a Avda 7-24, Zona 9, 01009 Guatemala City; tel. 2362-0888; fax 2362-0847; internet www.bancoreformador.com; cap. 129.1m., res 141.5m., dep 2,128.4m. (Dec. 2001); merged with Banco de la Construcción in 2000; Pres. Miguel Aguirre; 47 brs.

**Banco SCI:** Edif. SCI Centre, Avda La Reforma 9-76, Zona 9, 01009 Guatemala City; tel. 2331-7515; fax 2339-0755; e-mail atencion@sci.net.gt; internet www.sci.com.gt; f. 1967.

**Banco de los Trabajadores:** Avda Reforma 6-20, Zona 9, 01001 Guatemala City; tel. 2339-8600; fax 2339-4750; e-mail bantrab@terra.com.gt; f. 1966; deals with loans for establishing and improving small industries as well as normal banking business; Pres. Lic. César Amilcar Bárcenas; Gen. Man. Lic. Oscar H. Andrade Elizondo.

**Banco Uno:** Edif. Unicentro, 1°, Blvd Los Próceres, 18 Calle 5-56, Zona 10, 01010 Guatemala City; tel. 2366-1777; fax 2366-1553; e-mail bancouno@gua.pibnet.com; internet www.bancouno.com.gt.

**G & T Continental:** 7a Avda 1-86, Zona 4, Guatemala City; and Plaza Continental, 6a Avda 9-08, Zona 9, Guatemala City; tel. 2331-2337; fax 2332-2682; e-mail consultas@gytcontinental.com.gt; internet www.gytcontinental.com.gt; f. 2000 following merger of Banco Continental and Banco Granai y Townson; total assets 11.4m. (2000); 130 brs.

**Vivibanco, SA:** 6a Avda 12-98, Zona 9, 01009 Guatemala City; tel. 2277-7878; fax 2277-7805; e-mail mestrada@vivibanco.com; internet www.vivibanco.com.

### Finance Corporations

**Corporación Financiera Nacional (CORFINA):** 11a Avda 3-14, Zona 1, Guatemala City; tel. 2253-4550; fax 2232-5805; e-mail corfina@guate.net; internet www.guate.net/corfina; f. 1973; provides assistance for the development of industry, mining and tourism.

**Financiera Guatemalteca, SA (FIGSA):** 1a Avda 11-50, Zona 10, Apdo 2460, Guatemala City; tel. 2338-8000; fax 2331-0873; e-mail figsa@figsa.com; internet www.figsa.com; f. 1962; investment agency; Pres. Carlos González Barrios; Gen. Man. Ing. Roberto Fernández Botrán.

**Financiera Industrial, SA (FISA):** Centro Financiero, Torre 1, 7a Avda 5-10, Zona 4, Apdo 744, Guatemala City; tel. 2334-5111; fax 2331-9437; internet www.bi.com.gt; f. 1981; subsidiary of Banco Industrial (q.v.); Pres. Carlos Arías Masselli; Gen. Man. Lic. Elder F. Calderón Reyes.

**Financiera de Inversión, SA:** 11a Calle 7-44, Zona 9, Guatemala City; tel. 2332-4020; fax 2332-4320; f. 1981; investment agency; cap. 15.0m. (June 1997); Pres. Lic. Mario Augusto Porras González; Gen. Man. Lic. José Rolando Porras González.

### Banking Association

**Asociación Bancaria de Guatemala:** Edif. Margarita 2, Of. 502, Diagonal 6, Zona 10, Guatemala City; tel. 2336-6060; fax 2336-6094; internet www.abg.org.gt; f. 1961; represents all state and private banks; Pres. Lic. Luis Lara Grojec.

### STOCK EXCHANGE

**Bolsa de Valores Nacional, SA:** 7 Avda 5-10, Zona 4, Centro Financiero, Torre II, 2°, Guatemala City; tel. 2338-4400; fax 2332-1721; e-mail bvn@bvnsa.com.gt; internet www.bvnsa.com.gt; f. 1987; the exchange is commonly owned (one share per associate) and trades stocks from private companies, government bonds, letters of credit and other securities; Pres. Juan Carlos Castillo; Gen. Man. Rolando San Román.

## INSURANCE
### National Companies

**Aseguradora La Ceiba, SA:** 20 Calle 15-20, Zona 13, Guatemala City; tel. 2379-1800; fax 2334-8167; e-mail aceiba@aceiba.com.gt; internet www.aceiba.com.gt; f. 1978; Man. Alejandro Beltranena.

**Aseguradora General, SA:** 10a Calle 3-71, Zona 10, Guatemala City; tel. 2332-5933; fax 2334-2093; e-mail generaliguate@generali.com.gt; f. 1968; Pres. Juan O. Niemann; Man. Enrique Neutze A.

**Aseguradora Guatemalteca, SA:** Edif. Torre Azul, 10° 4a Calle 7-53, Zona 9, Guatemala City; tel. 2361-0206; fax 2361-1093; e-mail aseguate@guate.net; f. 1978; Pres. Gen. Fernando Alfonso Castillo Ramírez; Man. José Guillermo H. López Cordón.

**Cía de Seguros El Roble, SA:** Torre 2, 7a Avda 5-10, Zona 4, Guatemala City; tel. 2332-1702; fax 2332-1629; e-mail rerales@elroble.com; f. 1973; Gen. Man. Hermann Giron.

**Comercial Aseguradora Suizo-Americana, SA:** 7a Avda 7-07, Zona 9, Apdo 132, Guatemala City; tel. 2334-1661; fax 2331-5495; e-mail seguros@grupocasa.com.gt; internet www.grupocasa.com.gt; f. 1946; Pres. William Bickford B.; Gen. Man. David H. Lemus Pivaral.

**Departamento de Seguros y Previsión del Crédito Hipotecario Nacional:** 7a Avda 22-77, Zona 1, Centro Cívico, Guatemala City; tel. 2250-0271; fax 2253-8584; e-mail vjsc@chn.com.gt; internet www.chn.com.gt; f. 1942; Pres. Freddy A. Muñoz Moran; Man. Hugo Cruz Monterroso.

**Empresa Guatemalteca Cigna de Seguros, SA:** 5a Avda 5-55, Zona 14, Guatemala City; tel. 2384-5454; fax 2384-5400; e-mail cigna@starnet.com.gt; f. 1951; Gen. Man. Lic. Ricardo Estrada Dardón.

**La Seguridad de Centroamérica, SA:** 7a Avda 12-23, Edif. Etísa, Plazuela España, Zona 9, Guatemala City; tel. 2285-5900; fax 2361-3026; e-mail servicios.cmg@aig.com; internet www.aig.com; f. 1967; Gen. Man. Juan Manuel Friederich; Legal Rep. Marta de Toriello.

**Seguros Alianza, SA:** 7a Avda 12-23, Edif. Etísa, 6°, Plazuela España, Zona 9, Guatemala City; tel. 2331-5475; fax 2331-0023; e-mail segualia@terra.com.gt; f. 1968; Pres. Luis Fernando Samayoa.

**Seguros Generales G & T, SA:** 7a Avda 1-84, Zona 4, Guatemala City; tel. 2334-1361; fax 2332-8970; e-mail seguros@gyt.co.gt; f. 1947; Pres. Ernesto Townson R.; Exec. Man. Mario Granai Fernández; Gen. Man. Enrique Rodríguez.

**Seguros de Occidente, SA:** 7a Calle 'A' 7-14, Zona 9, Guatemala City; tel. 2331-1222; fax 2334-2787; e-mail occidente@occidente.com.gt; internet www.occidente.com.gt; f. 1979; Pres. Lic. Pedro Aguirre; Gen. Man. Carlos Lainfiesta.

**Seguros Panamericana, SA:** Avda de la Reforma 9-00, Edif. Plaza Panamericana, Zona 9, Guatemala City; tel. 2332-5922; fax 2331-5026; e-mail sortega@exchange.palic.com; f. 1968; Pres. Frank Purvis; Gen. Man. Lic. Salvador Ortega.

**Seguros Universales, SA:** 4a Calle 7-73, Zona 9, Apdo 1479, Guatemala City; tel. 2334-0733; fax 2332-3372; e-mail seguros@universales.net; f. 1962; Man. Pedro Nolasco Sicilia.

### Insurance Association

**Asociación Guatemalteca de Instituciones de Seguros (AGIS):** Edif. Torre Profesional I, Of. 703, 4°, 6a Avda 0-60, Zona 4, Guatemala City; tel. 2335-2140; fax 2335-2357; e-mail agis@intelnet.net.gt; internet www.agis.centroamerica.com; f. 1953; 12 mems; Pres. Alejandro Beltranena; Vice-Pres. Lic. Hermann Girón.

## Trade and Industry

### DEVELOPMENT ORGANIZATIONS

**Comisión Nacional Petrolera:** Diagonal 17, 29-78, Zona 11, Guatemala City; tel. 2276-0680; fax 2276-3175; f. 1983; awards petroleum exploration licences.

**Corporación Financiera Nacional (CORFINA):** see under Finance (Finance Corporations).

**Fondo de Inversión Social (FIS):** 2a Avda 10-13, Edif. Los Arcos, Zona 10, Guatemala City; tel. 2367-2891; fax 2367-2890; e-mail comunicacion@fis.gob.gt; internet www.fis.gob.gt; f. 1993; supports community efforts to improve social and economic conditions in the country; mandate due to expire in 2006; Pres. Dr Luis A. Flores Asturias; Gen. Man. Ing. Luis A. Bolaños Zabarburú.

**Instituto de Fomento de Hipotecas Aseguradas (FHA):** Edif. Aristos Reforma, 2°, Avda Reforma 7-62, Zona 9, Guatemala City; tel. 2362-9434; fax 2362-9492; e-mail promocion@fha.com.gt;

# GUATEMALA

internet www.fha.com.gt; f. 1961; insured mortgage institution for the promotion of house construction; Pres. Lic. HOMERO AUGUSTO GONZÁLEZ BARILLAS; Man. Lic. JOSÉ SALVADOR SAMAYOA AGUILAR.

**Instituto Nacional de Administración Pública (INAP):** 5a Avda 12-65, Zona 10, Apto 2753, Guatemala City; tel. 2366-3021; fax 2366-2655; e-mail inap@inap-gt.org; internet www.inap-gt.org; f. 1964; provides technical experts to assist the Government in administrative reform programmes; provides in-service training for local and central government staff; has research programmes in administration, sociology, politics and economics; provides postgraduate education in public administration; Man. HÉCTOR TOUSSAINT CABRERA GAILLARD.

**Instituto Nacional de Transformación Agraria (INTA):** 14a Calle 7-14, Zona 1, Guatemala City; tel. 2228-0975; f. 1962 to carry out agrarian reform; current programme includes development of the 'Faja Transversal del Norte'; Pres. Ing. NERY ORLANDO SAMAYOA; Vice-Pres Ing. SERGIO FRANCISCO MORALES-JUÁREZ, ROBERTO EDMUNDO QUIÑÓNEZ LÓPEZ.

**Secretaría de Planificación y Programación (SEGEPLAN):** 9a Calle 10-44, Zona 1, Guatemala City; tel. 2251-4549; fax 2253-3127; e-mail segeplan@segeplan.gob.gt; internet www.segeplan.gob.gt; f. 1954; prepares and supervises the implementation of the national economic development plan.

## CHAMBERS OF COMMERCE AND INDUSTRY

**Comité Coordinador de Asociaciones Agrícolas, Comerciales, Industriales y Financieras (CACIF):** Edif. Cámara de Industria de Guatemala, 6a Ruta 9-21, Zona 4, Guatemala City; tel. 2231-0651; e-mail info@cacif.org.gt; internet www.cacif.org.gt; co-ordinates work on problems and organization of free enterprise; mems: 6 chambers; Pres. JORGE BRIZ; Sec.-Gen. RAFAEL POLA.

**Cámara de Comercio de Guatemala:** 10a Calle 3-80, Zona 1, Guatemala City; tel. 2253-5353; fax 2220-9393; e-mail info@camaradecomercio.org.gt; internet www.negociosenguatemala.com; f. 1894; Pres. EDGARDO WAGNER DURÁN.

**Cámara de Industria de Guatemala:** 6a Ruta 9-21, 12°, Zona 4, Apdo 214, Guatemala City; tel. 2334-0850; fax 2334-1090; e-mail cig@industriaguate.com; internet www.industriaguate.com; f. 1959; Exec. Dir OSCAR VILLAGRÁN.

**Cámara Oficial Española de Comercio de Guatemala:** Edif. Géminis, 10°, Torre Sur, Of. 1513, 12 Calle 1-25, Zona 10, Apdo 2480, Guatemala City; tel. 2335-2735; fax 2335-3380; e-mail camacoes@terra.com.gt; internet www.camacoes-guate.com; Pres. ANDRES SICILIA; Gen. Man. BEATRIZ SÁNCHEZ.

## INDUSTRIAL AND TRADE ASSOCIATIONS

**Asociación de Azucareros de Guatemala (ASAZGUA):** Edif. Europlaza, 17°, 5a Avda 5-55, Zona 14, Guatemala City; tel. 2386-2000; fax 2386-2020; e-mail asazgua@azucar.com.gt; internet www.asazgua.com.gt; f. 1957; sugar producers' asscn; 15 mems; Pres. FRATERNO VILA; Gen. Man. ARMANDO BOESCHE.

**Asociación General de Agricultores:** 9a Calle 3-43, Zona 1, Guatemala City; f. 1920; general farmers' asscn; 350 mems; Pres. DAVID ORDÓÑEZ; Man. PEDRO ARRIVILLAGA RADA.

**Asociación de Gremiales de Exportadores de Productos No Tradicionales (AGEXPRONT):** 15 Avda 14-72, Zona 13, Guatemala City; tel. 2362-2002; fax 2362-1950; internet www.export.com.gt; f. 1982; devt of export of non-traditional products.

**Asociación Nacional de Avicultores (ANAVI):** Edif. Galerías Reforma, Torre 2, 9°, Of. 904, Avda de la Reforma 8-60, Zona 9, Guatemala City; tel. 2231-1381; fax 2234-7576; f. 1964; national asscn of poultry farmers; 60 mems; Pres. Lic. FERNANDO ROJAS; Dir Dr MARIO A. MOTTA GONZÁLEZ.

**Asociación Nacional de Fabricantes de Alcoholes y Licores (ANFAL):** Guatemala City; tel. 2292-0430; f. 1947; distillers' asscn; Pres. FELIPE BOTRÁN MERINO.

**Asociación Nacional del Café—Anacafé:** Edif. Etisa, Plazuela España, Zona 9, Guatemala City; tel. 2236-7180; fax 2234-7023; e-mail sellodepureza@anacafe.org; internet www.anacafe.org; f. 1960; national coffee asscn; Pres. WILLIAM STIXRUD.

**Cámara del Agro:** 15a Calle 'A' 7-65, Zona 9, Guatemala City; tel. 2226-1473; e-mail camagro@intelnet.net.gt; f. 1973; Pres. ROBERTO CASTAÑEDA.

**Gremial de Huleros de Guatemala:** Edif. Galerias España, 7 Avda 11-63, 3°, Guatemala City; tel. 2331-8269; fax 2332-1553; e-mail gremulgt@pronet.net.gt; f. 1970; rubber producers' guild; 125 mems; Man. ALEJANDRO SOSA BROL.

## MAJOR COMPANIES

The following are some of the leading companies currently operating in Guatemala:

### Construction

**Cementos Progreso, SA:** 17 Avda 18-66, Zona 10, Guatemala City; tel. 2236-3686; fax 2338-9110; e-mail cprogreso@cempro.com; internet www.cempro.com; f. 1899; cement manufacturers; sold to the Swiss Holderbank Financiere Glaris Ltd in 2000; Pres. FREDERICK C. E. MELVILLE NOVELLA; Gen. Man. PLINIO A. HERRERA CHACON; 1,550 employees.

**Ingenieros Mayorga & Tejada:** 4a Avda 8-40, Zona 9, Guatemala City; tel. 2331-6749; fax 2332-0959; f. 1966; heavy construction and civil engineering services; Man. Dir ENRIQUE TEJADA; 800 employees.

### Food and Beverages

**Alimentos Kern de Guatemala, SA:** Km 7, Carretera al Atlántico, Zona 18, Guatemala City; tel. 2256-0537; fax 2256-2378; e-mail info@alikerns.com; internet www.alikerns.com; f. 1959; manufacturers of canned fruit juices and fruit products; Pres. JESS JOAQUÍN PARDO VIADERO; Gen. Man. MARCO ALVAREZ; 550 employees.

**Cervecería Centroamericana, SA:** 3a Avda Norte Final, Finca El Zapote, Zona 2, Guatemala City; tel. 2289-1555; fax 2289-1716; internet www.cerveceria.com.gt; f. 1886; brewery; Pres. JORGE CASTILLO LOVE; 440 employees.

**EMBOCEN, SA** (Embotelladora Central, SA): 24 Calle 6-02, Zona 11, Guatemala City; tel. 2442-3879; fax 2442-0973; e-mail embocenmail@guate.net; f. 1985; bottling factory, producers of carbonated beverages; Pres. CARLOS H. PORRAS; Gen. Man. CARLOS TRIGUEROS; 490 employees.

**Embotelladora del Pacífico, SA:** Km 166, Cutytenango-Suchitepeque, Guatemala City; tel. 2472-0884; fax 2472-0883; internet www.e-arca.com.mx; f. 1957; producers of soft drinks; acquired by El Grupo Arma in 1983; Dir FRANCISCO GARZA EGLOFF; 800 employees.

**Industria Nacional Alimenticia:** 33 Calle 6-34, Col. Las Charcas, Zona 11, Guatemala City; tel. 2476-2783; fax 2476-2862; e-mail ina@ina.com.gt; internet www.ina.com.gt; manufacturers of pasta; owned by Distribuidora Interamericana de Alimentos, S.A.; Production Man. HUGO IOLI.

**Ingenio Tulula, SA:** 19 Calle 3-97, Zona 10, Guatemala City; tel. 2367-3436; fax 2332-0004; e-mail tulula@guate.net; f. 1982; processing sugar cane; Pres. JOSÉ BOUSCAYOL; 1,600 employees.

**Kellogg de Centro América, SA:** 46 Calle 24-50, Zona 12, Guatemala City; tel. 2442-9323; fax 2434-3727; e-mail claudia.giron@kellogg.com; f. 1970; foodstuffs; Pres. WILLIAM COMSTRA.

**Pollo Campero:** Guatemala City; tel. 2333-7233; e-mail ebarillas@campero.com.gt; internet www.campero.com; f. 1971; restaurant franchise; Pres. JUAN JOSÉ GUITÉRREZ.

### Metals and Rubber

**Gran Industria de Neumáticos Centroamericanos, SA (GINSA):** 50 Calle 23-70, Zona 12, Guatemala City; tel. 2477-5412; fax 2477-5421; e-mail mirna.orellana@goodyear.com; f. 1956; subsidiary of Goodyear Tyre and Rubber Co, USA; tyre manufacturers; Pres. ADOLFO BEHRENS MOTTA; 600 employees.

**Hulera Centroamericana, SA:** 24 Calle 24-75, Zona 12, Guatemala City; tel. 2476-0364; fax 2442-3890; e-mail hucasa@infovia.com.gt; f. 1958; manufacturers of rubber goods; Gen. Man. JUAN IGNACIO TORREBIARTE; 250 employees.

**Llantas Vifrio, SA:** 42 Calle 20-64, Zona 12, Guatemala City; tel. 2476-1212; fax 2479-3017; e-mail ventas@vifrio.com; internet www.vifrio.com; f. 1967; repair and retreading of tyres; Pres. HUMBERTO SUÁREZ VALDEZ.

**SIDASA:** 10a Calle 0-52, Zona 9, Guatemala City; tel. 2334-2678; fax 2334-7149; e-mail info@sidasa.net; manufacturers of steel tubes.

### Pharmaceuticals

**Abbott Laboratories, SA:** Apdo 37-01901, Guatemala City; tel. 2291-0111; medical equipment manufacturers.

**Colgate-Palmolive Central America, SA:** Avda Ferrocarril 49-66, Zona 12, Guatemala City; tel. 2477-5511; fax 2477-5403; f. 1971; pharmaceuticals and consumer products; Pres. MICHAEL J. TANGNEY; Gen. Man. PEGGY GERICHTER; 465 employees.

### Tobacco

**British American Tobacco Central America (BATCA):** 24 Avda 3581, Zona 120, Guatemala City; tel. 2366-8787; fax 2366-8785; internet www.batca.com; f. 1928; subsidiary of BAT Industries PLC (United Kingdom); manufacturers of cigarettes; Sales Man. RICARDO SELVA; 345 employees.

# GUATEMALA

**Tabacalera Centroamericana, SA:** Km 12½, Carretera a Villa Canales, Aldea Boca del Monte, Apdo 626, Guatemala City; tel. 2449-5555; fax 2448-0154; e-mail pmi.pressoffice@pmintl.com; f. 1945; subsidiary of Philip Morris Int. Finance Corpn of the USA; manufacturers of cigarettes; Pres. MIROSLAW ZIELINSKI; Man. SUSANA DE URRUELA; 487 employees.

## Miscellaneous

**Alcatel de Guatemala, SA:** Edif. Europlaza, Torre 2, 4°, 5 Avda 5-55, Zona 14, Guatemala City; tel. 2366-1466; fax 2366-4077; subsidiary of Alcatel (France); manufacturer of telecommunications equipment.

**Fábrica de Jabones y Detergentes La Luz, SA:** Km 18, Carretera Vieja al Antigua 16–81, Zona 1, Guatemala City; tel. 594-5115; fax 594-4750; e-mail laluz@infovia.com.gt; f. 1946; producers of soap and detergents; Pres. Ing. JUAN JOSÉ URRUELA VILLACORTA; 620 employees.

**Industria Centroamericana de Vidrio, SA:** Avda Petapa 48-01, Zona 12, Apdo 1759, Guatemala City; tel. 2276-0406; producers of glass bottles and containers and tableware; Pres. EDGAR CASTILLO SINIBALDI; 670 employees.

**Industria La Popular, SA:** Vía 3 5-42, Zona 4, Guatemala City; tel. 2331-3821; fax 2331-0381; e-mail quiesa@hotmail.com; f. 1920; producers of soap and detergents; Pres. FEDERICO KONG VIELMAN; 555 employees.

**Inyectores de Plástico:** Avda Petapa, Calle 56, Zona 12, Guatemala City; tel. 2470-5700; fax 2477-4814; e-mail ventasipsa@icasa.com.gt; internet www.ipsa.com; f. 1974; subsidiary of Grupo Industrial EEC; manufacturers of plastic packaging.

**KPMG Guatemala:** 7a Avda 5-10, Zona 4, Centro Financiero, Torre 1, 16°, Guatemala City; tel. 2334-2628; fax 2331-5477; e-mail kpmg@guate.net; internet www.kpmg.com; accountants and management consultants; Sr Partner ROBERTO GARCÍA.

**Minas de Guatemala, SA:** 4a Avda 8-53, Zona 9, Guatemala City; tel. 2336-3976; f. 1969; metal ore mining; Gen. Man. RODOLFO MENDOZA TEJADA; 650 employees.

**PriceWaterhouseCoopers:** Edif. Tívoli Plaza, 6a Calle 6-38, Zona 9, Apdo 868, Guatemala City; tel. 2334-5080; fax 2331-2819; internet www.pwcglobal.com; accountants and management consultants; Partners ÓSCAR CORDÓN, CARLOS E. PARRA.

**Proquirsa:** Edif. Intecunsa, Avda 3–68, Zona 9, Apdo 441A, Guatemala City; tel. 2362-0214; e-mail gerenciaventas@proquirsa.com; internet www.proquirsa.com; distributors of industrial chemicals.

## UTILITIES

### Electricity

**Empresa Eléctrica de Guatemala, SA:** 6a Avda 8-14, Zona 1, Guatemala City; tel. 277-7000; fax 253-1746; e-mail consultas@eegsa.net; internet www.eegsa.com; f. 1972; state electricity producer; 80% share transferred to private ownership in 1998; Commercial Man. JAVIER MADRIGAL.

**Instituto Nacional de Electrificación (INDE):** Edif. La Torre, 7a Avda 2-29, Zona 9, Guatemala City; tel. (2) 34-5711; fax (2) 34-5811; e-mail gerencia.general@inde.gob.gt; internet www.inde.gob.gt; f. 1959; former state agency for the generation and distribution of hydroelectric power; principal electricity producer; privatized in 1998; CEO CARLOS COLOM; Gen. Man JORGE JUÁREZ.

## CO-OPERATIVES

**Instituto Nacional de Cooperativas (INACOP):** 13 Calle 5-16, Zona 1, Guatemala City; tel. 2234-1097; fax 2234-7536; technical and financial assistance in planning and devt of co-operatives; Pres. LUIS ANTONIO SALAZAR.

## TRADE UNIONS

**Frente Nacional Sindical (FNS)** (National Trade Union Front): Guatemala City; f. 1968 to achieve united action in labour matters; affiliated are two confederations and 11 federations, which represent 97% of the country's trade unions and whose General Secretaries form the governing council of the FNS. The affiliated organizations include:

**Federación Autónoma Sindical Guatemalteca** (Guatemalan Autonomous Trade Union Federation): Guatemala City; Gen. Sec. MIGUEL ANGEL SOLÍS.

**Federación de Obreros Textiles** (Textile Workers' Federation): Edif. Briz, Of. 503, 6a Avda 14-33, Zona 1, Guatemala City; f. 1957; Sec.-Gen. FACUNDO PINEDA.

A number of unions exist without a national centre, including the Union of Chicle and Wood Workers, the Union of Coca-Cola Workers and the Union of Workers of the Enterprise of the United Fruit Company.

**Central General de Trabajadores de Guatemala (CGTG):** 3a Avda 12-22, Zona 1, Guatemala City; tel. 2232-1010; fax 2251-3212; e-mail cgtg@guate.net.gt; f. 1987; Sec.-Gen. JOSÉ E. PINZÓN SALAZAR.

**Central Nacional de Trabajadores (CNT):** Guatemala City; f. 1968 as Confederación Nacional de Trabajadores, name changed as above in 1973; cover all sections of commerce, industry and agriculture including the public sector.

**Confederación de Trabajadores del Campo (CTC):** 12 Calle 'A', 12-44, Zona 1, Guatemala City; tel. and fax 2232-6947; e-mail centracampo@yahoo.com; Sec.-Gen. MIGUEL ANGEL LUCAS GÓMEZ.

**Federación Sindical de Trabajadores de la Alimentación Agro-Industrias y Similares de Guatemala (FESTRAS):** 6a Avda 15-41, Zona 1, 4°, Of. 6, Guatemala City; tel. and fax 2338-3075; affiliated to International Union of Food, Agricultural, Hotel, Restaurants, Catering, Tobacco and Allied Workers' Associations; Sec.-Gen. JOSÉ DAVID MORALES C.

**Unidad de Acción Sindical y Popular (UASP):** 10 Avda 'A' 5-40, Zona 1, Guatemala City; f. 1988; broad coalition of leading labour and peasant organizations; includes:

**Comité de la Unidad Campesina (CUC)** (Committee of Peasants' Unity): 31 Avda 'A' 14-46, Zona 7, Ciudad de Plata, Apdo 1002, Guatemala City; tel. and fax 2434-9754; e-mail cuc@guate.net; internet cuc.mundoweb.org.

**Confederación de Unidad Sindical de Guatemala (CUSG):** 12 Calle 'A', Zona 1, Guatemala City; tel. and fax 2232-8154; e-mail cusg@itelgua.com; f. 1983; member of ICFTU; Sec.-Gen. CARLOS ENRIQUE MANCILLA.

**Federación Nacional de Sindicatos de Trabajadores del Estado de Guatemala (FENASTEG):** 10 Avda 5-40, Zona 1, Guatemala City; tel. and fax 2232-2772; Sec. ARMANDO SÁNCHEZ.

**Sindicato de Trabajadores del Instituto Guatemalteco de Seguridad Social (STIGSS):** 11 Calle 11-15, Zona 1, Guatemala City; f. 1953.

**Unión Sindical de Trabajadores de Guatemala (UNSITRAGUA):** 9 Avda 1-43, Zona 1, Guatemala City; tel. and fax 2238-2272; e-mail unsitragua@hotmail.com; f. 1985; member unions are mostly from the private industrial sector and include STECSA, SITRALU and SCTM; Co-ordinator DANIEL VÁSQUEZ.

**Unión Guatemalteca de Trabajadores (UGT):** 13 Calle 11-40, Zona 1, Guatemala City; tel. and fax 2251-1686; e-mail ugt.guatemala@yahoo.com; Sec. Gen. CARLOS ENRIQUE MANCILLA.

# Transport

## RAILWAYS

In 1998 there were 1,390 km of railway track in Guatemala, of which some 102 km were plantation lines.

**Ferrocarriles de Guatemala (FEGUA):** 18 Calle 9-03, Zona 1, Guatemala City; tel. 2232-9270; fax 2238-3039; e-mail ferroguat@hotmail.com; internet www.quetzalnet.com/quetzalnet/fegua; f. 1968; 50-year concession to rehabilitate and operate railway awarded in 1997 to the US Railroad Devt Corpn; 782 km from Puerto Barrios and Santo Tomás de Castilla on the Atlantic coast to Tecún Umán on the Mexican border, via Zacapa, Guatemala City and Santa María. Branch lines: Santa María–San José; Las Cruces–Champerico. From Zacapa another line branches southward to Anguiatú, on the border with El Salvador; owns the ports of Barrios (Atlantic) and San José (Pacific); first 65-km section, Guatemala City–El Chile, and a further 300-km section, extending to Barrios, reopened in 1999; Interventor Dr ARTURO GRAMAJO MONDAL.

## ROADS

In 1999 there were an estimated 14,021 km of roads, of which 3,081 km were paved. The Guatemalan section of the Pan-American highway is 518.7 km long and totally asphalted. In 2002 plans were discussed for the construction of a highway between Huehuetenango and Izabal under the 'Plan Puebla–Panamá' at an estimated cost of US $292m. The highway would take five years to complete. In May 2003 the World Bank approved a loan of US $46.7m. to repair existing roads and construct new networks in rural areas.

## SHIPPING

Guatemala's major ports are Puerto Barrios and Santo Tomás de Castilla, on the Gulf of Mexico, San José and Champerico on the Pacific Ocean, and Puerto Quetzal, which was redeveloped in the late 1990s. In June 2002 the Government announced plans for the expansion of shipping facilities at Puerto Quetzal and Santo Tomás de Castilla in the near future.

**Armadora Marítima Guatemalteca, SA:** 14a Calle 8-14, Edif. Armagua, 5°, Zona 1, Apdo 1008, Guatemala City; tel. 2230-4686; fax 2253-7464; cargo services; Pres. and Gen. Man. L. R. CORONADO CONDE.

**Empresa Portuaria 'Quetzal':** Edif. Torre Azul, 1°, 4a Calle 7-53, Zona 9, Guatemala City; tel. 2334-7101; fax 2334-8172; e-mail pquetzal@terra.com.gt; internet www.puerto-quetzal.com; port and shipping co; Man. LEONEL MONTEJO.

**Empresa Portuaria Nacional Santo Tomás de Castilla:** Calle Real de la Villa, 17 Calle 16-43, Zona 10, Guatemala City; tel. 2366-9413; fax 2366-9445; internet www.empornac.gob.gt; Man. ENRIQUE SALAZAR.

**Flota Mercante Gran Centroamericana, SA:** Guatemala City; tel. 2231-6666; f. 1959; services from Europe (in association with WITASS), Gulf of Mexico, US Atlantic and East Coast Central American ports; Pres. R. S. HERRERÍAS; Gen. Man. J. E. A. MORALES.

**Líneas Marítimas de Guatemala, SA:** Edif. Plaza Marítima, 8°, 6a Avda 20-25, Zona 10, Guatemala City; tel. 2237-0166; cargo services; Pres. J. R. MATHEAU ESCOBAR; Gen. Man. F. HERRERÍAS.

Several foreign lines link Guatemala with Europe, the Far East and North America.

### CIVIL AVIATION

There are two international airports, 'La Aurora' in Guatemala City and Santa Elena Petén. In June 2002 the Government announced plans for the potential construction of a new international airport outside Guatemala City and for the renovation and modernization of the airport at Santa Elena Petén.

**Aerolíneas de Guatemala (AVIATECA):** Avda Hincapié 12-22, Aeropuerto 'La Aurora', Zona 12, Guatemala City; tel. 2331-0375; fax 2334-7846; internet www.grupotaca.com; f. 1945; internal services and external services to the USA, Mexico, and within Central America; transferred to private ownership in 1989; CEO ROBERTO KRIETE; Pres. ALFREDO SCHILDKNECHT.

**Aeroquetzal:** Avda Hincapié, Hangar EH-05, Zona 13, Guatemala City; tel. 2334-7689; fax 2232-1491; scheduled domestic passenger and cargo services, and external services to Mexico.

**Aviones Comerciales de Guatemala (Avcom):** Avda Hincapié 18, Aeropuerto La Aurora, Zona 13, Guatemala City; tel. 2331-5821; fax 2332-4946; domestic charter passenger services.

**Tikal Jets:** Avda Hincapié 18, Aeropuerto La Aurora, Zona 13, Guatemala City; tel. 2334-5631.

## Tourism

Following the end of the civil war in 1996 the number of tourist arrivals increased steadily and were recorded at some 884,190 in 2002. However, in 2003 the number of tourists decreased slightly, to 880,223. In that year receipts were an estimated US $646m.

**Instituto Guatemalteco de Turismo (INGUAT)** (Guatemala Tourist Commission): Centro Cívico, 7a Avda 1-17, Zona 4, Guatemala City; tel. 2331-1333; fax 2331-4416; e-mail informacion@inguat.gob.gt; internet www.visitguatemala.com; f. 1967; policy and planning council: 11 mems representing the public and private sectors; Pres. DANIEL MOONEY; Dir ALEJANDRO SINIBALDI APARICIO.

**Asociación Guatemalteca de Agentes de Viajes (AGAV)** (Guatemalan Association of Travel Agents): Edif. El Reformador, Avda La Reforma 1-50, Zona 9, Apdo 2735, Guatemala City; tel. 2332-0782; fax 2334-5217; e-mail agav@intelnet.net.gt; Pres. LAURA MAYORGA DE ESTRADA.

## Defence

At August 2004 Guatemala's active Armed Forces numbered an estimated 29,200: Army 27,000 (including an estimated 23,000 conscripts), Navy 1,500 (estimate, including 650 marines) and Air Force 700. There were paramilitary forces numbering 19,000.

Under the terms of a 1996 accord concluded between the Government and the opposition guerrilla forces, the Armed Forces were reduced in number by one-third in 1997. In 2003 a civic service law was introduced that allowed conscripts to choose social service instead of the formerly compulsory military service. In 2004 the Government of Oscar Berger Perdomo proposed a reduction in the size of the Armed Forces to 14,000.

**Defence Budget:** 1,300m. quetzales (US $160m.) in 2004.

**Chief of Staff of the Armed Forces:** Gen. JORGE HUGO FLORES ARANA.

## Education

Elementary education in Guatemala is free and, in urban areas, compulsory between seven and 14 years of age. Primary education begins at the age of seven and lasts for six years. Secondary education, beginning at 13 years of age, lasts for up to six years, comprising two cycles of three years each. In 2000 there were 10,424 pre-primary schools, and in 2003 there were 16,609 primary schools and 3,585 secondary schools. Net enrolment at primary schools in 2001/02 was equivalent to 85.0% of children in the relevant age group (males 86.9%; females 82.9%). In the same year secondary net enrolment was equivalent to 29.3% of those aged 13 to 19 (males 29.9%; females 28.7%). There are twelve universities, of which 11 are privately run. The inauguration of Guatemala's first indigenous university, the Universidad Maya de Guatemala, was expected to take place in 2005. Open to both indigenous and non-indigenous students, the institution would offer courses in medicine, education, Mayan Law, art, architecture, community development and agriculture. According to preliminary figures, in 2003 budgetary expenditure on education was 3,255m. quetzales.

## Bibliography

For works on Central America generally, see Select Bibliography (Books)

Chase-Dunn, C., Amaro, N., and Jonas, S. (Eds). *Globalization on the Ground: Postbellum Guatemalan Democracy and Development*. Lanham, MD, Rowman & Littlefield Publrs, 2001.

Cullather, N., and Gleijeses, P. *Secret History: The CIA's Classified Account of its Operations in Guatemala, 1952–1954*. Stanford, CA, Stanford University Press, 1999.

Dosal, P. J. *Doing Business with the Dictators: A Political History of United Fruits in Guatemala, 1899–1944*. Wilmington, DE, Scholarly Resources, 1993.

Fischer, E. F. *Cultural Logics and Global Economies: Maya Identity in Thought and Practice*. Austin, TX, University of Texas Press, 2001.

Forster, C. *The Time of Freedom: Campesino Workers in Guatemala's October Revolution*. Pittsburgh, PA, University of Pittsburgh Press, 2001.

Glebbeek, M.-L. *In the Crossfire of Democracy: Police Reform and Police Practice in Post-Civil War Guatemala (Thela Latin America Series)*. Amsterdam, Rozenberg, 2003.

Grandin, G. *The Blood of Guatemala: A History of Race and Nation*. Durham, NC, Duke University Press, 2000.

Green, D. *Guatemala: Burden of Paradise*. London, Latin American Bureau, 1992.

Hawkins, T. *Jose de Bustamante and Central American Independence: Colonial Administration in an Age of Imperial Crisis*. Tuscaloosa, AL, University of Alabama Press, 2004.

Hererra Robinson, A. *Natives, Europeans and Africans in 16th-century Santiago de Guatemala*. Austin, TX, University of Texas Press, 2003.

Little, W. E. *Mayas in the Marketplace: Tourism, Globalization, and Cultural Identity*. Austin, TX, University of Texas Press, 2004.

Lovell, W. G. *A Beauty that Hurts: Life and Death in Guatemala*. Austin, TX, University of Texas Press, 2001.

McCleary, R. M. *Dictating Democracy: Guatemala and the End of Violent Revolution*. Gainesville, FL, University Press of Florida, 1999.

McCreery, D. *Rural Guatemala 1760–1940*. Stanford, CA, Stanford University Press, 1994.

May, R. A. *Terror in the Countryside*. Athens, OH, Ohio University Press, 2001.

Menchú, R. *Crossing Borders*, (translated and edited by A. Wright). Lewiston, NY, Mellen University Press, 1998.

Montejo, V. *Voices From Exile: Violence and Survival in Modern Maya History*. Norman, OK, University of Oklahoma Press, 1999.

Montgomery, J. *Tikal: An Illustrated History*. New York, NY, Hippocrene Books, 2001.

*Poverty in Guatemala*. Washington, DC, World Bank, 2003.

Remijnse, S. *Memories of Violence: Civil Patrols and the Legacy of Conflict in Joyabaj, Guatemala*. Amsterdam, Rozenberg, 2002.

Sanford, V. *Buried Secrets: Truth and Human Rights in Guatemala*. New York, Palgrave Macmillan, 2003.

Schirmer, J. *The Guatemalan Military Project*. University Park, PA, University of Pennsylvania Press, 1998.

Schlesinger, S., and Kinzer, S. *Bitter Fruit: The Story of the American Coup in Guatemala*. Cambridge, MA, Harvard University Press, 1999.

Shea, M. E. *Culture and Customs of Guatemala*. Westport, CT, Greenwood Publishing Group, 2000.

Sieder, R. (Ed.). *Guatemala after the Peace Accords*. London, Institute of Latin American Studies, 1999.

Siekmeier, J. F. *Aid, Nationalism and Inter-American Relations—Guatemala, Bolivia and the United States 1945–1961*. Lewiston, NY, Edwin Mellen Press, 1999.

Stoll, D. *Rigoberta Menchú and the Story of All Poor Guatemalans*. Boulder, CO, Westview Press 1999.

Trudeau, R. H. *Guatemalan Politics: The Popular Struggle for Democracy*. Boulder, CO, Lynne Rienner Publrs, 1993.

Yashar, D. J. *Demanding Democracy: Reform and Reaction in Costa Rica and Guatemala, 1870s–1950s*. Stanford, CA, Stanford University Press, 1997.

# GUYANA

## Geography

### PHYSICAL FEATURES

The Co-operative Republic of Guyana is the westernmost and largest of the three Guianas that occupy that part of the north-eastern coast of South America between the Serra Tumucumaque and the Atlantic Ocean. Formerly the United Kingdom's colony of British Guiana, Guyana is bordered to the east by Suriname (formerly Dutch Guiana). The two countries share a border along the Courantyne (Corantijn) river, although Suriname disputes possession of territory between the upper reaches of the Courantyne (or Kutari) and the New River (Upper Courantyne), in the south-eastern corner of Guyana. The maritime boundary is also disputed. Venezuela, which lies to the west, claims most of northern Guyana by arguing that its border should run along the Essequibo. Brazil lies beyond the longer, southern part of the western border and in the south (Brazil has 1,119 km—695 miles—of frontier with Guyana, Venezuela has 743 km and Suriname 600 km). Guyana is the third smallest country in South America (after Suriname and Uruguay), with an area of 214,969 sq km (83,000 sq miles).

Guyana consists of a northern block, longer along the coast, with a narrower southern extension into Brazil and along the border with Suriname. Guyana's Atlantic coast, which extends for 459 km, faces north-east. Behind it, protected by a complex system of dams and dykes (except in the east, where more than just a coastal strip is still swampy), is a rich plain of alluvial mud, deposited by the Amazon and other rivers. These coastal plains vary between 8 km and 65 km in width, and are mostly below sea level. Most of the agriculture and population of the country is located here. Inland is rolling highlands, most of it clad in dense woodland. The forest region, which accounts for four-fifths of the country, covers an eroded plateau. From this, in the south-west of northern Guyana, are the Pakaraima Mountains, including the country's highest, Roraima (2,835 m or 9,304 ft), which is on the Venezuelan border, and also just north of the border with Brazil. The forest region extends into the highlands, reaching as far as where the land rises again in the far south. Here, forest tends to be displaced by savannah grasslands, such as in the Rupununi valley in the far south-west. The country is 84% wooded, and has rich farmland along the coast, all of this watered by a number of rivers (Guyana means 'land of many waters'), the main ones being the Essequibo, in the centre of the country, and its tributaries, the Demerara, the Berbice and the Courantyne.

### CLIMATE

Although the climate is characteristically hot and humid, it is relatively mild for such a low-lying area in the tropics, being moderated by the north-eastern trade winds off the Atlantic. There are two rainy seasons (May–August and November–January), during which flash floods are always a risk. The average annual rainfall in Georgetown, the capital, on the coast, is 2,280 mm (almost 90 ins). There is less rain on the higher plateau regions, with the savannah of the far south receiving about 1,525 mm per year. The average temperatures in Georgetown range from 23°C (73°F) to 31°C (88°F) all year round.

### POPULATION

Guyana has a varied and complex ethnic constitution, the result of different solutions to labour demand during the colonial

period. There are some Amerindian peoples (about 6%—mainly Caribs) in the interior still, but most of the population (51%) is now 'East' Indian, descended from indentured workers brought from the Indian subcontinent in the 19th century. Over one-third (41%) is black, descended from the African slaves whose freedom had required new labour solutions. There are also the descendants of Chinese and Portuguese workers (the latter mainly from Madeira, but who have not maintained the use of their original language). About 12% of the total population are of mixed race; in Guyana they usually form a socially distinct group maintaining closer links to the European community. English is the official language and the one most widely used by all these communities, but an English-based Creole is also spoken, as is (particularly among the older generations) Hindi, Urdu and Chinese. The Amerindians have their own languages too. Almost one-half of the population is Christian (mainly Anglicans and Roman Catholics), one-third Hindu and nearly one-10th Muslim.

The total population, according to mid-year estimates in 2003, was 751,000. Around 90% of the population lives on the coast and just over three-fifths reside in rural areas. However, about one-third of the population lives in and around the capital, Georgetown, which is located just to the east of the mouth of the Essequibo. To the east of the capital is the port of New Amsterdam, another important town, and inland is the mining town of Linden (formerly Mackenzie), with the nearby centres of Wismar and Christianborg. The country is divided into 10 regions for administrative purposes.

# History

## COLIN SMITH

Based on an earlier article by JAMES McDONOUGH

A number of Amerindian tribes had settled in Guiana for centuries, scattered throughout the territory. The name Guiana means 'land of many waters' and the rivers were an important source of food and a key communications system. In the late 16th century the Amerindians began trading with the first European power to establish a permanent presence in what is now Guyana. These Dutch traders were interested mainly in tobacco and the annatto dye used by Indians. The Dutch initially established trading posts to facilitate these activities and control supplies. They soon became permanent settlements. By 1616 they had established the first of these permanent posts at Kyk-Over-Al, on an island in the Essequibo river. In 1627, after the Dutch West India Company was established, another settlement was founded in the colony of Berbice. In 1746 a third colony was established in Demerara. These three settlements were joined to form Dutch Guiana.

The Dutch West India Company introduced the system of slavery, using Amerindians who had been captured by rival tribes. They helped in the growing of tobacco and sugar cane. The growing importance of sugar cultivation meant an increased demand for cheap labour. The Dutch then introduced Negro slaves brought from Africa. The first-large scale importation of slaves took place in the 1650s. Later, the economy was diversified through the large-scale planting of coffee and cotton. These too were grown by slave labour. During the 18th and 19th centuries the colonies were controlled variously by the British, Dutch and French. The British captured Guiana in 1796 and the colony came permanently under British rule, to be known as British Guiana. The British needed large quantities of sugar and continued the system of slavery to support this. Many more estates were established. However, bloody slave revolts in 1763 and 1823 eventually led to its abolition in 1834. The British subsequently introduced Portuguese and Chinese workers into the colony to ensure a continued supply of cheap labour. However, they soon gravitated to, and remained in, commerce, rather than in the cutting of cane. Later, indentured labourers from India also provided a large source of cheap labour. They began arriving in large numbers in 1838, working on the sugar estates for a fixed time, at the end of which they would be sent back to India. The Government found this too expensive and offered them land instead of return fares. Most agreed to this solution and helped to develop one of Guyana's most important food crops today, namely rice. Moreover, many aspects of Indian culture, such as language, food and religion, still survive in Guyana today.

In around 1680 the Dutch West India Company promulgated what is thought to have been the country's first Constitution. Under the document, the Commander of the Essequibo colony was given complete administrative authority over the land, but in the application of justice he was to be assisted by a Council consisting of the Sergeant of the Garrison and the Captains of the ships that were then in port. Later, in 1732, the Berbice region also established a Constitution, which formed the basis of government for the colony for the next 200 years. The Berbice Constitution instituted a Governor, a Council of Government and two Councils of Justice. Members of the Council of Government were selected by the Governor from a list provided by the colonial planters. Together, the Governor and Council comprised the Council of Criminal Justice. A second group, more representative of planter interests, but chaired by the Governor, made up the Council of Civil Justice. The Constitution's recall provision allowed the planters to bring charges against the Governor and his administration, and petition his removal. This provision gave the colonial planter class a curb over the colony's executive authority.

Persistent overspending by the colonial Government and a dramatic decrease in sugar prices after the First World War prompted the British, in 1928, to establish the Legislative Council, a majority of whose members were appointed by the British Colonial Office. Elected members of the Legislative Council were not to form a majority until after the extensions of the local voting franchise, which occurred in 1943 and again in 1945.

## THE BIRTH OF ETHNIC POLITICS

Thus, British Guiana became a land of six races; Amerindians, Africans, Indians (known as 'East' Indians), Chinese, Portuguese and various European strains. The races were inclined to dominate in certain professions and particular areas of national life. 'East' Indians tended to be mainly agricultural labourers working in the sugar and rice industries while Africans were concentrated in the security forces, the civil service and the teaching and nursing professions. The shopkeepers were mainly Portuguese. Although over the years some Amerindians migrated to the cities, the large majority remained in their reservations, which in the early 21st century numbered over 150, occupying a total of 15,000 sq miles. (In May 2003 over 100 Amerindian chiefs held a conference, at which it was decided to form a national council to represent Amerindians locally and overseas, and to formulate a comprehensive programme to develop indigenous communities.)

The post-Second World War period in Guyana was dominated by two personalities. In 1950 Cheddi Jagan, a US-trained dentist, and Forbes Burnham, a British-trained lawyer, formed the People's Progressive Party (PPP). Jagan was its leader and Burnham its Chairman. This was the country's first mass-based party. The two major races, Indians and Africans, continued to show great solidarity as they battled what they considered to be the autocratic rule of the British colonial power. They claimed that there was a common pattern of victimization perpetrated on both races. Burnham and Jagan both supported the popular desire for an end to colonial rule.

Cheddi Jagan proclaimed his intention to establish in what was still British Guiana a communist state in accordance with Marxist-Leninist ideology. In the PPP's 1953 election manifesto he and Burnham asserted that they were revolutionary and anti-imperialist. In the general election of that year, the first to be held under adult suffrage, the PPP won a resounding victory, winning 51% of the votes cast and 18 of the 24 seats in the colonial legislature. However, the party's strident Marxist and anti-imperialist stance caused consternation in the United Kingdom and the USA. Six months after the election the Constitution was suspended. Jagan was removed from office and an interim Government appointed. Burnham realized that this policy of antagonizing the colonial powers would only delay the granting of independence. As a result, he began to adopt a more moderate approach, which led to tensions between him and the more left-wing Jagan.

The rivalry between the two races was also gradually increasing. In 1955 Burnham, whose main support came from the Africans, broke away from the PPP and formed the People's National Congress (PNC). This left the PPP with the larger mainly 'East' Indian following. Jagan and the PPP again won elections in 1957 and 1961. The years approaching independence saw increasing racial antagonism between the two major races as further divisions began to appear over which race should lead the country into independence. The first source of conflict came in 1962 when Jagan presented a budget containing measures that were an attempt to address the serious financial problems which beset the country. Protests were held, ostensibly against the proposals contained in the budget. In reality, however, they were aimed at bringing about the downfall of the man who, it was feared, would lead the country into communism. Fidel Castro's successful revolution in Cuba in 1959 raised fears that if Jagan's Marxist-oriented party led the country to independence there could be another communist state in the Caribbean and a base for the spread of the ideology. The demonstrations soon became violent and widespread pillage, arson and racial conflicts ensued. In 1963 there was an 80-day general strike, ostensibly over the introduction of a new Labour Relations Bill. The strike was accompanied by serious clashes between strikers and strike-breakers, severe food shortages and disruption of air and sea travel.

Following the disturbances, the United Kingdom reconsidered its plans for British Guiana. Just before the 1964 general elections, the British Government changed the electoral system from 'first past the post' to proportional representation. The changes favoured Burnham's PNC. Jagan opposed the election, which he felt was an attempt to unseat him. The union controlled by the Jagan Government called a strike in the entire sugar belt. Six months of conflict ensued, during which time 176 persons died and thousands more were injured and made homeless. In the elections Jagan again won the majority of seats. However, the British Governor called on Burnham to form the new Government after he had formed a coalition with the third major party, giving them an overall majority. Two years later, on 26 May 1966, British Guiana became Guyana after it was granted its independence by the United Kingdom.

## BURNHAM'S RULE OF INDEPENDENT GUYANA

The coalition Government enjoyed relative success in the early years of independence. There was record investment, much foreign capital began to flow into Guyana, production rose rapidly and revenue was buoyant. To gain US favour, Burnham severed trade relations with Cuba, which had been strongly promoted under PPP rule. However, his original pro-capital economic plans began to falter and Burnham became increasingly left-wing and authoritarian. In February 1970 Guyana was proclaimed a Co-operative Republic, whereby the co-operative sector was to become the dominant element of the economy. In 1974, in what is known as the Declaration of Sophia, Burnham committed his party to a stronger socialist line; two years later he declared that the PNC would create a Marxist-Leninist state.

In 1980 Guyana adopted a new Constitution, which provided for a strong executive presidency. More than two years after they were due, elections were finally held in December 1980. The PNC won 41 of the 53 seats in the National Assembly and elected Burnham to the newly created position of President. A team of observers led by Lord Eric Avebury, the Secretary of the British Parliamentary Human Rights Group, were unanimous in their report of massive irregularities. Christian churches and human rights groups began to mobilize against Burnham. Tensions increased in 1980 when Walter Rodney, a distinguished historian, leader of a popular party which seemed to cross racial barriers, and vehement critic of the Burnham regime, died when a package he was carrying exploded in his lap. In August 1985 Burnham died, and was succeeded as President by Hugh Desmond Hoyte, hitherto First Vice-President and Prime Minister. Hoyte was to be the head of state until elections in 1992 and leader of his party for the next 17 years.

## THE HOYTE PRESIDENCY, 1985–92

Hoyte led the PNC to a decisive victory in the general election in December 1985, when his party won 42 of 53 elective seats in the National Assembly and received 79% of the total votes cast. The distinguishing marks of his presidency were democratic and economic transformation and the rebirth of press freedom. He was much less autocratic than his predecessor. Many of the democratic institutions that had been severely crippled by Burnham were returned and strengthened. The economy had stagnated and there were high rates of inflation and migration. Social services were on the verge of collapse and per-head income was on the decline. Hoyte promoted economic growth by reducing state bureaucracy.

Guyana became ineligible to borrow from international financial institutions because of a failure to repay its foreign debt. In 1987, after reopening negotiations with the IMF, Hoyte's Government adopted the Economic Recovery Programme (ERP), which envisaged greater participation of private enterprise in the economy, incentives to attract foreign investment, a reduction of government spending and a willingness to privatize public corporations. In 1991, after three years of decline, the country registered a 4.5% increase in gross national product (GNP).

As general elections scheduled for 1990 approached, opposition parties rallied under the Patriotic Coalition for Democracy (PCD), while civil society leaders formed the Guyanese Action for Reform and Democracy (GUARD). The trade unions mobilized under the Federation of Independent Trade Unions (FITUG). They all campaigned relentlessly for electoral and economic reform. In response, and following intervention from former US President Jimmy Carter, the Government agreed to postpone the elections and discuss several reforms. In May 1991 a survey of the entire country was undertaken, to produce a new electoral register. Meanwhile, the National Assembly moved to sanction the political agreements between President Hoyte and the opposition parties through an amendment to the country's Constitution. The PPP, which had withdrawn from the National Assembly in September 1991, returned to participate in the debate and vote on the constitutional proposals. The party subsequently remained in the National Assembly, but limited its participation to purely electoral matters. The amendment guaranteed the electoral rights of all the parties, but also extended the National Assembly's term until the end of September 1992.

## THE CHEDDI JAGAN PRESIDENCY, 1992–97

The 1992 elections, held on 5 October, were won by the PPP in coalition with CIVIC, a movement consisting principally of members of the business and professional community, thus ending the long authoritarian reign of the PNC. The Carter Center and the Commonwealth Secretariat declared that the elections were essentially free and fair, although marred by riots in Georgetown. Three demonstrators were killed by police, and another 200 arrested. After spending 28 years in political opposition, PPP leader Jagan was named President on 9 October 1992. He subsequently appointed the CIVIC leader, Samuel Hinds, as Prime Minister. The PPP/CIVIC coalition secured 28 of the national elective seats in the National Assembly (and 54% of the votes cast), compared with 23 seats for the PNC (42%). After conducting negotiations with independent legislators, the governing coalition eventually came to control 36 of the 65 seats in the National Assembly.

After his victory President Jagan attempted to rewrite the Constitution in concert with all the country's political parties. In 1995 the four political parties in the National Assembly—the PPP, the PNC, The United Force (TUF) and the Working People's Alliance (WPA)—agreed to a 1991 recommendation of the Carter Center, whereby the Chairman of the Elections Commission was to be selected from a list of nominees proposed by the opposition parties. The parties also agreed to expand the Commission to include three representatives of the ruling party and three from the opposition parties. Subsequently, PPP/CIVIC rejected the only candidate proposed by the opposition parties. This led to an electoral impasse and destroyed any possibility of electoral reform before the 1997 elections.

President Jagan expounded his economic agenda in the 1992–94 Policy Framework Paper, which expressed the new Government's strong commitment to the fundamental principles of the ERP. At the same time, the paper stressed the need for the development of the nation's human resources and efforts to reduce poverty, with 29% of the total population, and as much as 71% of the Amerindian population in the Hinterland, living in extreme poverty in 1992.

The 1992 election results had devastating consequences for the PNC. Immediately following the elections, public disagreement erupted between Hoyte and Hamilton Green, his former Prime Minister. Green was expelled from the party after a PNC disciplinary committee found him guilty of misconduct. He founded a new political group called the Forum for Democracy (subsequently renamed Good and Green Guyana). The Forum contested seats in the municipal elections, which were eventually held in August 1994, and Green was elected as mayor of Georgetown.

## THE JANET JAGAN PRESIDENCY, 1997–99

In March 1997 Cheddi Jagan died at the age of 78, after suffering a heart attack. Some 100,000 Guyanese attended his funeral, in tribute to the man who had dominated much of the country's political life for 50 years. Prime Minister Hinds was named President to complete Jagan's term. In September the PPP/CIVIC selected Janet Jagan, Cheddi Jagan's US-born wife and the Prime Minister, to be its presidential candidate in the election, scheduled to be held on 15 December. The other parties contesting the election were the PNC, led by former President Hoyte, the TUF, the Alliance for Guyana (AFG) and the Guyana Democratic Party (GDP). The PPP/CIVIC won 55% of the votes cast, compared to 41% for the PNC and 1% each for the TUF and the AFG. As a result, PPP/CIVIC secured 29 of the 53 elective seats in the National Assembly and the PNC won 22, with one each being obtained by the TUF and the AFG. The GDP obtained no national seats. With the results of the district elections, the PPP/CIVIC won a clear majority in the National Assembly and on 19 December, after counting some 90% of the votes, the Elections Commission declared Janet Jagan the President. Hoyte immediately challenged the Commission, claiming that the elections were fraudulent. Racial slurs were aimed at the new white President and the traditional racial divisions within the country again came to the fore. The PNC refused to take their seats in the National Assembly, and Hoyte called for massive demonstrations in Georgetown to protest against the election results, which quickly escalated into rioting and looting. With unrest increasing, the Government and the PNC agreed to take part in negotiations, under the auspices of the Caribbean Community and Common Market (CARICOM). In January 1998 the two sides signed the Herdmanston Accord, whereby they agreed to allow CARICOM to conduct an independent audit of the election results, and to form a commission to draft revisions to the Constitution within 18 months. New elections were to be held within three years, cutting the normal presidential term by two years. At the same time, Hoyte agreed to curtail any further demonstrations. The elections were subsequently scheduled for January 2001. In early June 1998 the CARICOM audit declared the election results fair, but public violence continued to escalate.

In July 1998, at its annual summit in Saint Lucia, CARICOM again mediated between the PPP and the PNC. The two parties signed the Saint Lucia Statement, in which they agreed to adopt measures intended to improve race relations in Guyana, to renew discussions on constitutional reform and to reinstate full legislative participation. Legislation designed to allow the PNC deputies to recover their seats in the legislature was formulated by both sides, and the PNC (except Hoyte, who continued to deny the legitimacy of Jagan's authority) rejoined the National Assembly on 14 July. In January 1999 a 20-member Constitutional Reform Commission was created, comprising members of the principal political parties and community groups. The Commission's task was to formulate recommendations for constitutional reform by mid-July 1999. In 2000 the two parties agreed to change the title of the parliamentary minority leader to Leader of the Opposition, decided on a process for appointing the head of the Elections Commission, and agreed to a constitutional change limiting the power of the presidency. The PPP, using its parliamentary majority, approved a smaller regional representation in the parliament than that advocated by the PNC.

## THE BHARRAT JAGDEO PRESIDENCY, 1999–

Despite Hoyte's promises to call for an end to the unrest, demonstrations and public violence continued in 1999. The level of discontent amongst supporters of the PNC was underlined at the end of April when public-service employees, demanding higher salaries, organized a strike lasting eight weeks. The Government accused the PNC of instigating the action, while the opposition criticized the Government for the harsh behaviour of the police in dealing with demonstrators. In August, following a mild heart attack, Janet Jagan resigned as President. Prime Minister Bharrat Jagdeo was immediately sworn into office to replace her. As President, Jagdeo committed his administration to public-service reform, the continued privatization of public enterprises, improved land surveys and major infrastructure projects such as the Berbice bridge, a deep-water port, and the Guyana–Brazil highway.

In June 2000 the Elections Commission reached an agreement on electoral reform, to be implemented for the general election, scheduled for January 2001. In October the legislature unanimously approved a constitutional amendment establishing a mixed system of proportional representation combining regional constituencies and national candidate lists, and the abolition of the Supreme Congress of the People of Guyana and the National Congress of Local Democratic Organs. However, in November the elections were postponed until 19 March 2001 in order to implement the reforms. This resulted in a dispute regarding the status of the Government in the interim.

In January 2001 a high court judge, Claudette Singh, declared the December 1997 elections to be null and void, ruling that the legislature had acted illegally in making the possession of a voter identification card a prerequisite for voting, and further alleging that there had been instances of electoral fraud. However, she subsequently ruled that the Government should remain in office until March, but that only legislation necessary for the election should be passed.

In December 2000 the National Assembly approved a constitutional amendment removing the President's immunity from prosecution and limiting his power to appoint only four ministers from outside the Assembly. The general and regional elections of 19 March 2001 were preceded by demonstrations over the late distribution of voter identification cards (despite an earlier announcement that voters would be allowed to use other forms of identification). The PPP/CIVIC gained 53.0% of the votes cast, a clear majority, as opposed to 41.6% for the PNC (which contested the elections as the PNCReform). Of the 65 parliamentary seats, the PPP/CIVIC won 34, the PNCReform garnered 27, the Guyana Action Party, in alliance with the WPA, obtained two seats, while Rise, Organize and Rebuild Guyana Movement and the TUF each secured one seat.

As had been the pattern since 1953, the voting in the election that brought President Jagdeo to power was largely along racial lines. The PNCReform contested the election in the nation's High Court, alleging numerous irregularities. The case was accompanied by protests, arson attacks and street violence across Guyana. The Carter Center, which had observed the elections, found them to be generally free and fair. On 31 March 2001 the High Court rejected the PNC challenge, ordered an immediate declaration of the official results, and the PPP/CIVIC leader, Jagdeo, was sworn in as President for a five-year term.

At his inauguration, Jagdeo pledged to convene a National Conference to discuss bipartisan co-operation. He committed his administration to working with the PNC as equals, and to incorporating the programmes of Guyana's various political parties into a National Development Strategy. He also promised to accelerate the process of constitutional reform and to expand the role of the legislative arm of government in an attempt to reassure PNC supporters. In a relatively successful attempt to reduce the intensity of the continuing violence, on 25 April 2001 Jagdeo and Hoyte announced confidence-building measures and the establishment of joint committees to examine and report on critical issues. Jagdeo's Cabinet remained largely unchanged in 2005, in spite of several alleged instances of conspicuous incompetence and misjudgement by members of government. However, the situation in Guyana was unusual in that, owing to its ethnic majority, the Government did not expect to be voted out of office. Since the present Government came into power in 1992 only four ministers have been relieved of their duties, while two have tendered their resignations.

Guyana's political history since independence in 1966 has been marked by a strong and often overruling influence of the political party in power, referred to as 'democratic centralism'. Since 1992 the country's leadership has been actively searching to find more inclusive ways of governance. The establishment of the six bipartisan committees was considered an important step in this process. The six committees dealt with: local government; national policy for land and house lots distribution and depressed communities; borders and national security; bauxite-industry resuscitation; radio monopoly; and non-partisan boards. However, in March 2002 PNCReform, the main opposition party, announced a pause in the dialogue process, pending the implementation of the outstanding decisions and recommendations made by the different committees. Frustrated by the lack of results of the dialogue process, the opposition parties jointly walked out of the National Assembly during the presentation of the 2002 budget, bringing this attempt to find more inclusive forms of governance to an abrupt halt. The PNCReform and PPP/CIVIC parties subsequently announced that they had adopted a formal policy of 'active non-co-operation'. Amid this political uncertainty, certain parts of the country, especially the Georgetown area, experienced a sharp increase in violent crime, robberies and attacks on businesses, homes and the police force, creating serious economic hardship and increasing social tension throughout the year. The number of reported murders rose by 56% in 2002, compared with the previous year, amid increased criticism of police tactics used to try to gain control of the situation.

In July 2002, on the first day of the annual CARICOM summit meeting in Georgetown, anti-Government demonstrators broke into the presidential offices. Outside the building, security forces opened fire on protesters, killing two people and wounding 15. The protests were organized by the newly-formed People's Solidarity Movement. CARICOM delegates condemned the violence and urged the PNCReform to accept the results of the 2001 elections. In the latter half of 2002 the Government introduced a number of measures in an attempt to reduce criminal activity, including an Anti-Kidnapping Bill, which received legislative approval in December. The human rights organization Amnesty International claimed that parts of the proposed legislation violated human rights and suggested that it should be reviewed.

In December 2002 Desmond Hoyte died unexpectedly following a heart attack. He was replaced as leader of the PNCReform by Robert Corbin, a lawyer from Linden with a long career in politics, in February 2003. Following several visits by a special envoy of the Commonwealth Secretary-General aimed at facilitating the dialogue between the two major political leaders, Corbin agreed to resume the dialogue process with the Government. This resulted in May in the signing of a Joint Communiqué between the President and the PNCReform and the opposition's return to the National Assembly, bringing to an end the 14-month political impasse. The Joint Communiqué allowed for the resumption of the work of the six bipartisan committees and greater inclusiveness of the opposition in parliamentary affairs, through the appointment of several new standing committees of the National Assembly. Furthermore, the two leaders agreed to strengthen the capacity of the Parliament Office and the National Assembly as a whole, to appoint an Ethnic Relations Commission, and a Disciplined Forces Commission, which would lead an inquiry into the operations of the Guyana Police Force, the Guyana Defence Force and the Fire and Prison Services.

The Inter-American Development Bank (IDB) agreed to fund a programme to improve crime management and in 2003 a UN Special Rapporteur was commissioned to investigate the incidence of racism, discrimination and xenophobia in Guyana. In his final report, the Rapporteur recommended that political leaders should demonstrate their commitment to the eradication of ethnic polarization by initiating a formal dialogue on the question of inclusive governance and by ensuring the implementation of the reforms agreed in the Communiqué. The Disciplined Forces Commission, appointed in an effort to restore professionalism and morale to the police, completed its work after 10 months. One of its main findings was that the army, police, prison service and fire service had been unable to effectively confront the recent social disorder and secure the public's safety. The report detailed perceived weaknesses and proposed 164 recommendations for improving the forces' efficiency. However, the ruling party declared that it would not be bound by the findings of the Commission. More than one year after the Commission published its report, none of the more important and significant recommendations had been implemented. The fragility of the *rapprochement* between the Government and the opposition was also demonstrated when the PNCReform warned that it would boycott elections in 2006 unless the Constitution was amended to enable power-sharing between the country's ethnic groups.

Owing to the political impasse prior to the signing of the Joint Communiqué, the constitutionally designated bodies to deal with public-sector appointments, transfers, promotions and disciplinary issues were absent for most of 2002 and 2003. However, following renewed negotiations, the members of the Public Service Commission were sworn in on 30 December 2003. Members of the other three commissions were sworn in soon after. In July 2004 the President proposed that the dialogue process between him and the opposition leader be transferred to parliament and conducted publicly.

Relations between the Government and opposition had already deteriorated in January 2004, following allegations that the Government was linked to a clandestine paramilitary group. It was claimed that the group targeted suspected criminals and persons linked to known criminals, and had been allegedly responsible for more than 40 extra-judicial killings in 2003. The Minister of Home Affairs, Ronald Gajraj, who was on leave for almost 11 months pending an inquiry into allegations that he had links with a so-called 'death squad', resumed his duties in April 2005. Although the Presidential Commission of Inquiry cleared Gajraj of involvement in the activities of the death squads, he was censured for exceeding his authority. There were immediate calls locally and internationally for his reappointment to the Cabinet to be rescinded. The opposition parties withdrew temporarily from parliament and called for the resignation of the Government. The USA, Canada, the European Union and the IDB all expressed concern about Gajraj's reinstatement and some threatened to withhold aid. At the end of April Gajraj submitted his resignation.

In 2002–05 Guyanese society was strained by ethnic tensions, underdevelopment and criminal activity. The public perception of a corrupt police force was reinforced by allegations of police involvement in extra-judicial killings and in a visa fraud scheme. The new Police Commissioner and Army Chief of Staff, who both took office in early 2004, promised wide reforms to their agencies. Heightened criminal activity also indirectly affected the economy by reducing private-sector activity, discouraging investment and increasing the external migration of skills. This echoed a World Bank report in November 2003 that maintained that the focus on ethnic security, survival and competitive prestige left little time to implement the many policies necessary to facilitate growth and development and to discourage corruption and crime. Guyana was also criticized for its police abuses, lengthy trial delays and poor prison conditions in the US Department of State's 2004 country reports on human rights practices.

In an attempt to resolve the political impasse in August 2004, former US President Jimmy Carter mediated discussions between the main political leaders. After three days of talks Carter recommended that opposition members return to the National Assembly, that the provisions of the National Development Strategy be debated in parliament, and that an independent civil society forum be established to lead a national discussion on governance.

Joey Jagan, the son of former Presidents Cheddi and Janet Jagan, and a former member of parliament for the party founded by his parents, formed a political organization, known as the Unity Party, in early 2005 and announced his intention to participate in the forthcoming general election, scheduled to be held by March 2006.

## INTERNATIONAL RELATIONS

In the early years of independence, Guyana developed close links with Cuba and became allied with the non-aligned nations. President Burnham took a strong position against the pro-apartheid Government in South Africa. However, after Ronald Reagan won the US presidency in 1980, Guyana began to modify its pro-Cuban stance in order to appease the new US Administration. International attention was focused on Guyana when, in November 1978, about 900 followers of the US-based People's Temple cult committed mass suicide, following the murder in Guyana of US Congressman Leo Ryan. Ryan had travelled to Guyana to inspect the Temple's commune at the behest of some of his constituents. After the 1992 elections, the USA approved the release of donor funds to support improvements in the country's infrastructure and other areas. In 1999 alone, a total of US $72m. was given in donor assistance, making the USA the largest bilateral donor to Guyana.

Guyana and the USA signed a number of important agreements in 2003 and 2004, including the 'Shiprider' maritime agreement to allow US law-enforcement vessels to enter Guyanese territory in pursuit of suspect vessels. Guyana's relations with its neighbouring South American states—Venezuela to the west and Suriname to the east—were, to a large extent, defined by bitter, long-running border disputes. In 1962 Venezuela declared the decision of the Paris Arbitral Tribunal null and void. The Tribunal had established the border between the two countries to the apparent satisfaction of both Venezuela and the United Kingdom, which then controlled Guyana. In repudiating the Paris decision, Venezuela laid claim to all territory west of the Essequibo river, some 130,000 sq km, or almost two-thirds of modern-day Guyana.

In 1966, just prior to granting Guyana independence, the United Kingdom and Venezuela signed the Geneva Agreement, which committed the two sides to establishing a Mixed Commission to recommend a solution to the controversy. In 1970, following a border incident, Guyana and Venezuela signed the Protocol of Port of Spain, in which the two countries agreed to reduce tensions along the border, and to work towards a peaceful resolution of the controversy. In 1982 tension between the two countries again increased, when Venezuela refused to renew the Port of Spain agreement. In November 1989 President Hoyte and his Venezuelan counterpart, Carlos Andrés Pérez, agreed to accept a UN proposal to pursue and identify a mutually acceptable means of a peaceful resolution of the controversy. Although negotiations were ongoing during the ensuing years, no agreement was ever reached on a practical means for settlement of the controversy.

In 2000 tensions increased following Guyana's agreement with the US-based Beal Aerospace Corporation to build the world's first private satellite-launching facility in the Essequibo. However, following objections by Venezuela and financial difficulties, the company abandoned the project. Nevertheless, in January 2001 Venezuela announced an oil exploration project in the same zone. In late 2001 the two countries reached agreement to resuscitate the Guyana/Venezuela High Level Bilateral Commission, which had been established in 1998.

Various Venezuelan fishing vessels were detained in 2002–04. However, in general, Guyana's relationship with Venezuela further improved and in 2003 a joint 21-point communiqué was issued by the two countries. Both countries agreed, *inter alia*, to work towards the timely completion of the remaining procedures for the sale of fuel to Guyana under the Caracas Energy Co-operation Agreement, and to co-operate more closely in the transportation, infrastructure and telecommunications sectors. In February 2004 Venezuelan President, Lt-Col (retd) Hugo Chávez Frías, undertook a state visit to Guyana, during which the two heads of state reviewed the efforts aimed at increasing bilateral co-operation.

The border problem with Suriname was also unresolved. It concerned a triangular piece of land, called the New River Triangle, located in the south-east of Guyana and bounded by the New River on the west and the Courantyne-Kutari river on the east, the Courantyne river itself and the maritime boundary offshore. The boundary with Suriname was never formally settled between the United Kingdom and the Netherlands; in 1939 the British and Dutch Governments agreed on a draft treaty, but it was never signed because of the outbreak of war in Europe. That draft treaty placed the boundary on the left bank of the Courantyne and Kutari rivers. However, in 1962 the Netherlands proposed an alternative boundary, which followed the Thalweg river, instead of the left bank of the Courantyne, and the westerly New River, instead of the Kutari. The proposal was rejected by the United Kingdom. In addition to the problems of the New River area, there were misunderstandings over the joint maritime boundary. In 1998 Guyana granted the Canadian-based CGX Energy Incorporated a concession to explore for petroleum and gas along the continental shelf. Part of this area, the Courantyne bloc, lay within territory claimed by Suriname.

In May 2000 the Government of Suriname formally protested that Guyana had violated its sovereignty and territorial integrity by granting the CGX concession. A second diplomatic note from Suriname later in the same month stated that the petroleum-exploration activity 'constituted an illegal act', and invited Guyana to begin negotiations 'in order to clarify any misunderstanding on the maritime boundary'. Guyana claimed that the exploration activities were being conducted in its own territory, but was willing to enter negotiations. In June gunboats from the Suriname navy forced CGX to remove the drill rig from the disputed area. Guyana demanded drilling by the Canadian company be allowed to resume. Both countries agreed to establish immediately a Joint Technical Committee to work towards a settlement of the dispute. A meeting was held in Port of Spain, Trinidad and Tobago, but both sides remained at an impasse, in spite of CARICOM intervention. During his first official visit to Suriname in early 2002, President Jagdeo and his Surinamese counterpart, Runaldo Venetiaan, established a Sub-Committee of the Guyana and Suriname Border Commissions to discuss further possible solutions to the dispute. Meanwhile, CGX continued its test drilling in an adjacent bloc in the Guyanese territorial waters. Progress towards a resolution was again interrupted in March 2003, when the Government of Suriname decreed that all maps circulated by diplomatic missions in its capital, Paramaribo, must include the New River Triangle. Guyana lodged a formal protest against this action. All of these efforts failed to resolve the issue. In February 2004 Guyana submitted the maritime boundary dispute with Suriname for arbitration with the International Tribunal for the Law of the Sea (ITLOS) in Hamburg, Germany. It was expected that the two countries would have to wait up to five years before the Tribunal provided a legally binding resolution of this issue. In December, at the behest of its former colony Suriname, the Netherlands denied Guyana access to its archived documents on the border issue. The British Government made its archives available to both Guyana and Suriname. Guyana subsequently filed a protest with the ITLOS and the case was submitted to court in March 2005.

Relations with Brazil improved towards the end of the 20th century. In 1989 President Hoyte signed an agreement with Brazil to start a road project, which was to provide the first land link between the two countries. Sixteen years later, that road project was still not completed. However, in 2000 Brazil did indicate a willingness to pay for the pavement of the road and for the construction of a deep-water port along Guyana's coast. The Governor of the Brazilian state of Roraima agreed to finance a bridge over the Takutu river. Between 1998 and 2002 the number of Brazilians (mostly miners) registered in Guyana increased by 52%, with a total of 522 registered in 2002. Estimates of the number of illegal Brazilian miners (some on short-term assignments) working in the remote Hinterland regions varied between 2,000 and 5,000 persons. In February 2003 Brazil and Guyana signed an international road transport agreement which was expected to facilitate increased trade and tourism between the two countries. In May of the same year the Guyana Government approved a request by the Brazilian authorities for a partial abolition of visas for both countries. The heightened co-operation between the two countries resulted in the opening of a Guyana consulate in Boa Vista in January 2003. On a visit to Guyana in February 2005, Brazil's President Luiz Inácio ('Lula') da Silva reiterated his Government's commitment to the construction of the bridge across the Takutu river between the two countries. During the visit the two countries also signed agreements on health and education, and entered into a three-year agreement aimed at facilitating improved training and qualifications for foreign service personnel in both countries. Joint-venture oil and gas exploration was also announced.

Guyana was chosen to host the 20th Rio Group summit in 2006, which 18 Latin American heads of state were expected to attend. The purpose of the Group was strengthen political and economic co-operation and to foster a more systematic approach to resolving shared problems, as well as pursuing regional integration. Guyana also participated in the third meeting of South American Presidents in Cuzco, Peru, in December 2004, which created the South American Community of Nations (Comunidad Sudamericana de Naciones), expected to become operational by 2007.

In 1967 Antigua, Barbados and Guyana were the original signatories to the document that established the Caribbean Free Trade Area, which, in 1973, became CARICOM, a grouping primarily of British Commonwealth Caribbean countries. In July 1991 CARICOM signed a trade co-operation agreement

# Economy

## COLIN SMITH

Based on an earlier article by JAMES MCDONOUGH

The Co-operative Republic of Guyana, on the southern coast of South America, remained one of the poorest countries in the Western hemisphere in 2005, as it had been for over 30 years. From 1992 to 1997, under President Cheddi Jagan, it had a per-head income of US $890. From 1997 to 2000 this declined to $840. According to World Bank estimates, per-head income remained at $840 in 2003. It was classified by the World Bank as a heavily indebted low-income country. In 2003 its gross domestic product (GDP) was $631.0m., while its external debt stood at approximately $1,083.8m. According to the UN Development Programme, in 2004 Guyana ranked 104 out of a list of 177 countries on its Human Development Index, 12 places lower than in 2003 and 23rd among 92 developing countries on its Human Poverty Index. Life expectancy in 2003 was 62 years. In 2001 the adult literacy rate was 98.6% and the combined primary, secondary and tertiary education gross enrolment ratio was 84%.

In the early 21st century Guyana faced numerous challenges, including a stagnant economy, little local and overseas investment, high unemployment and underemployment, and a relentless exodus of skilled Guyanese workers. In a report published in 2004, the IMF stated that the depletion of skills in the private and public sectors, owing to the migration of professionals, constrained the implementation of effective development programmes and ultimately inhibited the country's ability to benefit from increased donor support. The IMF also urged Guyana to address the upsurge in crime, which, it asserted, exacerbated the challenges faced in stimulating economic growth and creating jobs.

According to the results of a national population and housing census published in May 2004, the population stood at 745,150. The total number of Guyanese living abroad was estimated to be between 500,000 and 1,000,000. In 1990–2003 it was estimated that the population grew at an average annual rate of just 0.4%. The highest outward emigration was from the Berbice region. Guyana had one of the highest migration rates in the world. A World Bank report in November 2003 estimated that 77% of those with university education in the mid-1990s subsequently moved to the USA.

The country's economy is dominated by agriculture and mining and its major exports are sugar, rice, bauxite, gold and wood. Most of the country's economic activity, except for mining and forestry, and about 90% of its population are concentrated on the country's narrow coastal plain.

Estimates of the extent of the underground economy in Guyana varied in 2003 between 30% and 46% of GDP. It was estimated in 2004 that approximately US $250m. was received annually in the form of remittances sent by overseas family members. Remittances gained increasing recognition as a major source of capital; in 2004, as in previous years, they exceeded the combined total of overseas aid and foreign direct investment received by the region. Although the proportion of the population affected by poverty declined from 43% in 1993 to 35% in 2000, it remained a high percentage compared with other countries in the region. Moreover, the improvement was not uniform; the poverty rate in the rural interior actually increased from 79% in 1993 to 92% in 1999.

In addition, in 2004 it became apparent that Guyana was becoming a major transhipment point for illegal drugs from Colombia, Peru, and Bolivia, particularly for cocaine bound for Canada. In that year airport authorities intercepted 138 kgs of cocaine in 18 seizures. The police narcotics branch seized 827 kgs of cocaine between 2002 and 2004.

Since acquiring independence in 1966, Guyana's economic history can be divided into two distinct phases: the so-called 'socialist period', under the presidency of Forbes Burnham, and the market economy pursued by the Governments of Hugh Desmond Hoyte (1985–92), Dr Cheddi Jagan (1992–97), Janet Jagan (1997–99) and Bharrat Jagdeo (1999–). The policies of the socialist period led to economic disaster and a swiftly expanding external debt, while under the latter economic regime, the country experienced steady, if unspectacular, growth.

In 1970, four years after independence, the Burnham administration adopted a socialist orientation and proceeded over the next 10 years to nationalize 80% of the country's economy, including its two major export sectors: bauxite and sugar. By the end of the decade the Government had taken over 32 companies and established 12 new companies under the Guyana State Corporation, a state-owned holding company. However, after 1975, world prices for Guyana's main exports (sugar, bauxite and rice) began to fall. The Government slipped increasingly into debt, and inflation increased rapidly, reaching over 400% by 1983. Throughout the 1980s real GDP declined by approximately 6% per year. When President Burnham died in office in 1985 the Government was deeply in debt and in arrears with its international creditors, its capacity to borrow new funds exhausted.

Hoyte, the country's new President, immediately entered into negotiations with the IMF in a bid to resolve the country's debt crisis and to revive the economy. In 1987 Hoyte adopted the Economic Recovery Programme (ERP), which reversed the co-operative, socialist policies of the Burnham administration. Under the ERP the Guyanese Government committed itself to major economic reform (see Investment and Finance, below) and initiated a major programme of privatization of public assets, to reduce government employment and to raise taxes. In 1990 the Government introduced the Social Impact Amelioration Programme (SIMAP) to soften the impact of the largely IMF-imposed austerity measures on the Guyanese population.

Successive administrations continued to implement the Hoyte ERP. Guyana's economy began to show positive growth; real GDP increased by an annual average of 3.9% in 1990–2003. In 2004 the economy grew by 1.6%, reversing the negative growth of 0.6% in 2003. The annual rate of inflation, which had been reduced from over 100% in the late 1980s, stood at 5.5% in 2004, slightly higher than the rate of 4.9% recorded in 2003. President Jagdeo proposed to reform and modernize the country's civil service, provide for accurate surveying of land, expand the country's transportation and telecommunications infrastructure, and work more closely with the private sector. Progress in these areas was severely impeded, however, as a result of the uncertain economic and political climate created by a steep increase in armed robberies and attacks on businesses, and serious political disagreements between the ruling People's Progressive Party/CIVIC (PPP/CIVIC) and the main opposition party, the People's National Congress Reform (PNCReform). The signing of a joint agreement between President Jagdeo and the newly elected Leader of the Opposition, Robert Corbin, in May 2003 indicated a return to political stability and the beginning of the restoration of investor confidence. A Trade Policy Review for Guyana conducted by the World Trade Organization (WTO) in 2003 noted that Guyana continued to rely heavily on the export of a narrow range of traditional products and remained vulnerable to shifts in the level of global demand, fluctuation in prices, and the erosion of its margins of preferences in access to major foreign markets. A report by the Guyana Government Office for Investment (Go-Invest) stated that in 2004 US $46.8m. was invested in Guyana in 75 projects, of which 33 were funded by foreign investors. The projects were mainly in the fields of information and communication technology, forestry and wood products, mining, and light manufacturing.

## THE FLOODS OF 2005

In January 2005 Guyana experienced its worst flooding for more than 100 years. In that month the country witnessed more than seven times the average amount of rainfall for January: some 52 inches. Severe flooding occurred along the coast, where the majority of the population lived. Three of the country's 10 administrative regions were declared disaster areas. The UN Economic Commission for Latin America and the Caribbean (ECLAC) estimated the total damage at some US $465m. The disaster affected 37% of the population, while some 44% of the national housing stock was lost or damaged. Damage to the sugar industry was estimated at $11.5m, while the rice sector suffered losses reaching $8.0m. ECLAC also reported that commerce suffered a $72.5m. loss, manufacturing $5.5m. and tourism $4.8m. As part of the Government's short-term recovery and rehabilitation strategy, each of the 70,000 affected households received a grant of $50 as part of a $30m. relief package. However, the Government recognized that recovery from the flood was a long-term task for the country.

## AGRICULTURE

Some 500,000 ha of Guyana's total land area of about 19.7m. ha were suitable for agriculture. Over 200,000 ha were under cultivation. The country had a further 1.1m. ha of land under permanent pasture and an estimated 16.5m. ha of forests and woodlands, most of which was inaccessible. Sugar cane plantations and rice paddies occupied most of the narrow coastal plain. Much of this low coastal land had been reclaimed and converted, through the construction of dykes, into fertile estates (polders) by the Dutch West India Company during the colonial period. Successive Governments committed considerable resources to repairing and maintaining the dyke system to ensure the continued viability of the country's agricultural sector. In 2003 agriculture, including forestry and fishing, provided an estimated 37.5% of GDP. The sector employed an estimated 16.7% of the total labour force in mid-2003. Following several years of expansion, activity in the rice, fishing and forestry sectors decreased in 2002, by 10.7%, 4.0% and 8.0%, respectively. In 2003 the value of sugar exported increased by 8.1%, to US $129.2m., despite a 2.4% decline in average export prices; exports were expected to increased further in 2005 following an estimated 7.5% increase in production in 2004. Although there was a 4% increase in the volume of rice production in 2003, there was a small decline in earnings, to $45.3m., as the average export price fell. Output was projected to increase by 1.4% in 2004. The projected growth rates for 2004 for the other agricultural subsectors were: livestock 2%, fishing 0.5% and forestry 0.5%. However, the 2005 floods were expected to lower projected growth in that year.

### Sugar

Sugar was the single largest contributor to Guyana's GDP in 2003, providing 15.0% of GDP and 25.6% of export earnings. In that year approximately 7% of the labour force was employed in the sector. The state-owned Guyana Sugar Corporation (Guysuco) owned 90% of the land used for the cultivation of sugar cane (approximately 52,000 ha) and all of the country's canecrushing facilities. The rest of the sugar was grown on land owned by private farmers. From 1990 Guysuco's holdings were managed by the British company Booker Tate, the successor to Booker McConnell Ltd, whose Guyana sugar holdings were nationalized in the 1970s. Owing to initiatives taken by Booker Tate, sugar production increased steadily in 1990–99, reaching 321,438 metric tons in the later year, the highest production levels since 1978. Output subsequently fluctuated and stood at an estimated 324,940 tons in 2004, the second largest production since 1990. However, the World Bank estimated that some 25% of the cane to be harvested in 2005 was affected by the January floods and output was expected to fall as a result.

More than 90% of the sugar produced in Guyana was exported. Of its sugar exports, approximately three-quarters went to the countries of the European Union (EU), particularly to the United Kingdom, and to the USA, under quotas at rates considerably above world prices. The annual export quota to the EU was 165,000 metric tons. The US quota stood at 14,859 tons per year. Receipts from sugar exports in 2004 amounted to $136m., representing an increase of 5.9% over the previous year. The average export price of sugar of $472.3 per ton was 14% higher than in 2003. In 2003 Australia, Brazil and Thailand filed a complaint against the EU with the WTO to the effect that the subsidies that the EU paid its sugar farmers were illegal under WTO rules; in August 2004 a WTO panel ruled, on a preliminary basis, that the EU subsidies were illegal. The ruling was upheld in March 2005. The EU's sugar reform proposals were announced in June of that year. They included a 39% reduction in sugar prices over two years, beginning in mid-2006. Guyana was considering joining other Caribbean countries that would be similarly affected by the price reduction in taking legal action against the EU, on the grounds that the proposed unilateral cuts violated the provisions of the sugar protocol of the EU's Lomé Conventions and the legitimate expectations of its signatories.

However, it was hoped that the negative effects of the end of the EU's preferential regime would be mitigated by the implementation of a strategic plan geared to increase Guysuco's competitiveness, profitability and long-term viability. The centrepiece of this plan was the modernization of the Skeldon refinery, one of the largest in the country. The entire project was expected to cost US $135m., with the Government to provide $56m., the People's Republic of China $32m., the Caribbean Development Bank $25m., and Guysuco $22m., mainly from the sale of lands. The completion date was set for 2007, although the plan was subject to a feasibility study. The cost of production at the new factory would be around US 8–9 cents per pound. World prices were expected to settle in line with this figure when barriers to trade and subsidies were eventually removed. The Government also secured a loan of $25.2m. from India, on easy repayment terms, which was to be used to restructure other estates in order to improve their operational efficiency and productive capacity. Although it was considered likely that the WTO ruling and the remodelling of the Skeldon refinery would bring about the closure of three old and inefficient sugar factories, resulting in thousands of job losses, it was nevertheless anticipated that Guyana would be one of the few sugar producers in the Caribbean to survive the reforms to the EU sugar regime as a result of the country's low production costs, which were less than $0.20 per pound.

Raw sugar exports to the Caribbean ranged between 55,000 and 60,000 metric tons annually in the early 2000s; however, Guyana also hoped to gain access to the lucrative Caribbean Community and Common Market (CARICOM), which would potentially absorb up to 160,000 tons of refined sugar. In August 2003 Guysuco signed a memorandum of understanding with Trinidad and Tobago's Angostura Holdings Ltd, which should have seen Guysuco converting its molasses to rum. However, Demerara Distillers Limited (DDL) went to the courts to stop the move, claiming it had first option on molasses from Guysuco and that if the deal proceeded it would harm DDL's rum production.

### Rice

In 2003 Guyana produced approximately 501,500 metric tons of rice. Some 290,000 tons of rice were exported in 2000, mainly to the EU, which used to give preferential tariff treatment to Guyanese rice, and where Guyanese long grain and extra long grain rice were particularly popular, especially in the Netherlands, where they were sold at a premium. Nevertheless, rice producers were concerned about the future, as the country's EU tariff advantage was phased out by 2000, and they turned to Latin America for new markets.

The cultivation of rice in Guyana was dominated by 'East' Indian (Asian-descended) small-scale farmers, whose ancestors had come to Guyana as indentured workers to replace slave labour. The annual output of rice between 1960 and 1980 had averaged 256,000 metric tons. Rice production fell in the 1980s, reaching a low of 93,400 tons in 1990. Production increased thereafter, however, reaching an estimated 501,500 tons in 2003. The value of the country's rice exports was US $13m. in 1990, rising to $95m. in 1996 before declining in the rest of the decade. In 2001 and 2002 the situation in the rice sector deteriorated, with some of the major rice farmers experiencing severe financial difficulties, owing to lower commodity prices, rising production costs and the depreciation of the euro in the case of preferential markets. In 2003 rice exports earned $45.3m., 9.0% of total earnings in that year. However, a total of 127,612 ha of land were under rice cultivation in 2003, compared with 107,902 ha in 2002. On the international market, rice sold at an average of $210 per ton, compared with $410 in 1996. Exports to the EU increased to 115,000 tons in 2004, compared with 108,000 tons in 2003, and exports to non-traditional markets in 2004 more than doubled to 38,000 tons from 15,000 tons in the previous year. Although production declined, higher export volume and higher prices were responsible for exports rising in 2004 to an estimated $55.1m. Export earnings in 2005 were initially set at $58m.; however, this was revised in light of the flooding in January, when the authorities deliberately flooded thousands of acres of rice fields in an attempt to ease the pressure on a dam, the breaking of which could have caused further catastrophic flooding.

The rice industry is the largest private-sector activity in Guyana, employing over 150,000 persons directly and indirectly. One of the major factors that continued to affect rice production in 2004 was the number of mills that remained closed after 2002, mainly owing to financial constraints. A government initiative to negotiate new terms that would make it easier for farmers and millers to repay their debts met with only limited success. An EU grant of €11.7m. in 2003 and a 'soft' loan of €3.2m. in January 2004 were expected to be used by the industry for research and to improve drainage and irrigation. Guyana also announced its intention to achieve either a regional safeguard mechanism for rice or an increase in the tariff Caribbean countries have to pay to bring rice into the region from extra-regional sources.

### Fishing

An increase in the production of shrimp for export from the late 1980s led to an expansion in the fishing sector. The total export value of shrimps rose from approximately US $33.1m. in 2000 to $53.9m. in 2003. The fishing sector accounted for 6.8% of GDP in 2003. In the late 1990s the Government sought technical assistance to determine the fish biomass in its waters, in order to develop a sustainable fishing industry and to attract additional capital. Parliament approved a Fisheries Bill in 2002 aimed at developing the fisheries sector in a sustainable manner. According to a report from the UN's Food and Agriculture Organization (FAO), the fisheries sector in Guyana contributed 2.4%–2.7% to the country's GDP in 1998–2003. Fish exports were expected to rise even further following the EU certification of the country's seafood exports in January 2004. This certification allowed Guyanese seafood processors to export their products to the EU at higher prices than those previously available in the USA.

The shrimp industry provided jobs for an estimated 10,600 persons in coastal and rural communities. Shrimping is estimated to earn US $58m. annually and was the sixth largest foreign-exchange earner for Guyana. In 2003, however, the US Southern Shrimp Alliance petitioned the US Government to take action against 10 countries, including Guyana, over alleged dumping of shrimp in the US market. Guyana supplied between 50m. lbs and 60m. lbs of shrimp annually to the USA and in 2004 was the 11th largest exporter to that country.

The Government has also been actively promoting aquaculture. Areas for aquaculture production have now reached over 6,000 acres. FAO agreed to finance a project to integrate aquaculture into small rice-based farming systems to diversify production in a bid to increase income and improve nutrition. In December 2003 Guyana and Barbados signed an agreement for joint co-operation in the area where their economic exclusion zones overlap.

### Forestry

Tropical rainforest covers about 161,000 sq km, approximately three-quarters of Guyana's land area, and is composed of hundreds of hardwood species. It is probably best known with respect to forestry as the source of Green Heart (Chlorocardium rodiei), which is widely used in marine work. At the beginning of the 1990s only about 40% of the forest areas were accessible and only 10% were being exploited; a total of 3.7m. ha was allocated for commercial use by 10 large companies and 250 medium-sized and small operators, which were supplying the domestic market and some member states of CARICOM. The current commercial yield from forests in Guyana averages around 8 cu m per ha. In most of the other main producers the average yield per ha is 20 cu m–100 cu m. During the 1990s the development of the wood-processing industry was targeted as a high priority. The Government's strategy was to privatize the industry and seek outside investment. One of the first initiatives was the sale by the Hoyte administration of the state-owned logging company to Demerara Timber Ltd, a company controlled by the British Beaverbrook group, for US $16.5m.

The Jagan Government in 1997 signed a memorandum of understanding with Berjaya of Malaysia for the exploitation of 303,520 ha of forest land. Berjaya planned to invest US $150m. in the project. In the same year the Government also signed a memorandum of understanding with Solid Timber of Sarawak, Malaysia, for the exploitation of 307,567 ha of forest land. Solid Timber planned to invest $250m. in the venture. In 2000 the Jagdeo administration signed an agreement with the Chinese-owned Jilin Company Guyana, Inc, for a forest concession in the north-west district, approximately 167,125 ha in size. Jilin proposed to invest £1m. over the next three years to carry out a forest inventory and to develop a management plan. Upon acceptance of its plan, Jilin would invest up to $20m. in saw milling and other projects. The agreement represented the first large investment made by the People's Republic of China in Guyana. The main forestry products exported were plywood, sawnwood and logs.

In 1996 Guyana exported 34,000 cu m of forestry products at a value of US $10m. In 1998, however, output of forestry products declined by 24% compared with the previous year, owing mainly to the East Asian economic crisis, which reduced demand from export markets. In 1999 forestry output recovered to reach 467,000 cu m, or 6% more than the previous year. Following further increases in wood and timber export earnings in 1999 and 2000, reaching $40.9m. in the latter year, export earnings decreased in 2001 and 2002, to $37.5m. and $29.0m., respectively. In 2003 this figure recovered, albeit slightly, to $30.7m., or 6.1% of the total export value. In 2004 the industry returned a solid performance with export receipts growing by 18.9%, to $36m. Royalties levied on the production of timber in 2003 totalled $748,000, representing a 2% increase over 2002. Exports of logs increased from 35,000 cu m in 2001 to over 70,000 cu m in 2004. Few areas, such as the Kaieteur National Park, were under protection. In addition, the Iwokrama Rainforest Center and Conservation International managed a 'Conservation Concession' of some 80,000 ha. In December 2003 Iwokrama announced plans to harvest timber for green and socially responsible niche markets.

## MINING AND ENERGY

Mining, mainly of bauxite, gold and diamonds, contributed an estimated 12.9% of GDP in 2003, making it the third most important sector after agriculture and services. In 2000 the sector grew, registering a 5.9% increase over the previous year. In terms of exports, mining was even more important, with bauxite and gold alone accounting for some 34.7% of the value of Guyana's exports in 2003. The latest estimates indicated that the mining sector recorded a decline of 8.7% in 2003, following a 6.9% contraction in 2002. The country was dependent on outside investment to develop its mining industry. Foreign investment was first attracted to Guyana in the 1990s to exploit its high-grade bauxite deposits for the making of aluminium. Later in the 1990s considerable foreign investment was directed into the mining of gold, marked by the opening, in 1994, of the Omai Gold Mine operation on the Omai river by a Canadian-led partnership.

## Bauxite

Guyana's extensive bauxite deposits occur in an arcuate belt along the southern margin of the country's coastal plain. The Demerara Bauxite Company Ltd (Demba), a subsidiary of the Aluminium Company of Canada (ALCAN), started mining these bauxite deposits in 1916. The main mining centre was at Linden, 120 km (74 miles) south of Georgetown. An industrial complex was established in association with the mining operations, for the purification and drying of the mineral, with one factory producing calcium carbonate and alumina. The industry underwent a period of rapid expansion in the 1920s when foreign companies invested heavily in three operations: the Linden operation, which produced high-quality and high-value calcined bauxite, representing 31% (in 1994) of the total tonnage, but 63% of total industry sales; and the Berbice and Aroaima operations, which produced lower-value metallurgic bauxite. At the end of the 1960s Demba was producing 75% of all bauxite ore mined in Guyana, including all of the carbonated bauxite, and had a work-force of 4,500. Reynolds (Guyana) Mines, a subsidiary of the US Reynolds Metal Company, began mining bauxite around Kawakwani in the Berbice region after the Second World War, and was producing 25% of the total bauxite mined by 1975. Demba and Reynolds (Guyana) Mines were nationalized, with compensation, in 1971 and 1974, respectively, and in 1977 they were combined to form the Guyana Mining Enterprise Ltd (Guymine). The lack of a major electricity supply continued to prevent the production in Guyana of aluminium. In the late 1970s Guyana lost its market dominance when the People's Republic of China entered the high-value refractory bauxite market and gained a significant share through a substantial reduction in prices. Throughout the 1990s Guyana also suffered sharp declines in calcined, chemical-grade and metal-grade bauxite. The downturn was mainly owing to competition from Brazil and the People's Republic of China, and to shrinking markets in Europe and the USA.

In 1989 the Government relinquished some control of the bauxite industry when it entered into a joint venture with Reynolds International for the development and exploitation of new bauxite deposits in Aroaima. The jointly owned Aroaima Bauxite Company began operations in mid-1991 and by the end of that year had produced 800,000 metric tons of bauxite. In 1992 the debt-ridden and state-owned Guymine was dissolved and divided into the Linden Mining Enterprise Ltd (Linmine) and the Berbice Mining Enterprise Ltd (Bermine). In September 2002 the operations of Bermine and the Aroaima Bauxite Company were merged in an attempt to increase profits. However, the merged company, which operated without budgetary transfers, suffered a cash depletion of US $2m. at the end of 2003, placing it in a precarious financial position. On 1 August 2003 the Canadian mining company Cambior took over the management of Linmine. Cambior rehired almost 200 of the 650 Linmine workers who had been made redundant. According to a feasibility study commissioned by Cambior, sufficient investment could result in an increase in annual production to some 2.8m. tons. In December 2004 the Government and Cambior signed an agreement creating the Omai Bauxite Mining Inc, 70% owned by Cambior and the remainder by the state. The agreement committed to a two-phase programme of investment: $24m. in 2005 and a further $15m. in 2007.

Bauxite production reached an impressive 1,715,705 metric tons in 2003, but declined by 12.4% in 2004, to 1,503,416 tons. Export receipts increased from US $35.3m. in 2002 to $44.6m. in 2003. This was attributed to a 9.6% growth in the volume of exports and a 15.7% rise in the average price. In April 2004 Russian Aluminium (RUSAL) began a number of studies on the prospects for the local industry, with a view, in the first instance, to obtaining about 2m. tons of bauxite annually. RUSAL and the Government signed a memorandum of understanding for supplies of bauxite and a possible joint mining venture. In December the Guyanese Government and the Aroaima Mining Company (AMC) entered into an agreement with a subsidiary of RUSAL, the Bauxite and Alumina Mining Venture (BAMV), to form the Bauxite Company of Guyana Inc (BCGI). The new company was 90% owned by BAMV and 10% by the Government of Guyana. Under the agreement, RUSAL was to invest $20m. in the form of operating assets in the Berbice bauxite industry. RUSAL, through its subsidiary BAMV, was to lease equipment to AMC, thus allowing it to expand bauxite production from 1.3m. tons to 2.5m. tons per year. BGCI was to manage AMC until the latter was fully privatized in 2006.

In May 2004 the Prime Minister of Trinidad and Tobago announced that his Government wanted the construction of an alumina plant in Guyana to be a condition for negotiating the expansion of aluminium production in Trinidad and Tobago. Guyana began looking for investors for the plant, construction of which would greatly boost efforts to return the industry to much greater profitability. Two other initiatives could have a positive impact on the resurgence of the bauxite industry. Firstly, Brazil has shown interest in metallurgical bauxite from Guyana. Secondly, an overseas prospecting company has been studying the extensive laterite bauxite deposits in the Pakaraimas, with encouraging results.

## Gold and Diamonds

In the country's other important mining sector, gold, output rose during the last years of the 20th century. Total production of gold reached 454,485 troy oz in 1998, marking a dramatic increase for the gold industry, which had recorded a level of production of 6,800 oz in 1983. In terms of exports, gold constituted 6% of the value of all exports in 1990. This figure increased significantly to 28.6% in 1994, because Omai Gold Mines Ltd opened operations along the Omai river in 1993.

In 2003 gold exports were valued at US $130.9m., falling from $137.0m. in 1997, but still accounting for almost 26% of the total domestic export value. In 2004 gold exports earned $145.1m, benefiting from the depreciation of the US dollar that contributed to a 12.1% increase in price. Omai Gold Mines Ltd produced 319,000 troy oz of gold in 2002, but only 274,000 troy oz in 2003, owing to depleting ore resources. Production at the Omai mine for 2004 was 240,000 troy oz slightly higher than the anticipated 234,000 troy oz. By the end of 2003 the company had paid a total of $50.3m. in royalties and $39.5m. in income taxes. The company's total investment in Guyana was $252.8m. During its operation in Guyana, Omai mined 3.3m. troy oz of gold valued at $113,000m., but paid no corporation taxes because it had shown no profits. The Government gave Omai Gold Mines Ltd licences for operations in Quartz Hill and a further area of the Omai river, which, it was hoped, would extend the life of the mine. However, it was announced in September 2004 that processing operations at the Omai site were expected to end in the third quarter of 2005.

Production of diamonds, Guyana's other principal mineral resource, also increased in the 1990s, after suffering a decline in the previous decade, reaching a peak of 50,000 carats in 1993. Production of diamonds increased sharply in 2001, to 184,309 carats, with an equivalent export value of US $13.3m. Diamond production in 2003 was 412,537 carats, 63% higher than forecast. The steep increase in production was thought to be owing to the rapidly growing numbers of legal and illegal Brazilian diamond miners in remote Hinterland locations.

## Electricity

Although the potential for hydropower development in the Hinterland regions of Guyana is immense, Guyana was dependent on petroleum as its main energy source and imported most of its feed stock from Trinidad and Tobago and Venezuela. One barrier to exploiting the hydro potential is the distance from the source of power delivery. Until 1999 electricity was supplied and distributed by the Guyana Electricity Corporation (GEC) for domestic use, public lighting and small and large industries, while the sugar factories, bauxite companies and other mining and forestry companies generated their own electricity. During the 1970s, 1980s and early 1990s the country suffered a severe shortage of electrical generating capacity, leading to frequent shortages and failures of the power supply. By the mid-1990s a programme was initiated to rehabilitate distribution facilities and the GEC increased the joint generating capacity of the Garden of Eden and Kingston power-stations by 44 MW, to 90 MW. In October 1999 GEC was privatized and became Guyana Power and Light (GPL), with the Government retaining 50% of the shares.

The Government and consumer groups continued to raise concerns in the early 2000s about the steady increase in electricity tariffs, the high management costs and the apparent lack of structural investment in the sector. Poor generating capacity and the faulty distribution system of the national grid also meant that over 40% of electricity generated was lost in conversion and distribution, equivalent to an annual loss of US $16m., which was borne by consumers. Owing to unmet expectations, regulatory oversight and legal challenges over proposed tariff increases, in 2003 Americas and Caribbean Ltd sold its shares in GPL to the Government for $1. In December 2004 a new Unserved Areas Electrification Programme was launched. The $35m. project, principally funded by the Inter-American Development Bank (IDB), was intended to improve efficiency and provide power to low-income areas, with some 30,000 new connections proposed.

Over the years there has been continued interest in renewable energy development. In 2004 Enman Services of Trinidad and Tobago was examining the feasibility of developing a hydroelectric plant at Turtruba Falls with potential capacity of 1,000 MW for export of power to Brazil and for downstream industrialized use. In early 2003 the Dutch firm Delta Caribbean NV undertook a feasibility study to set up and implement the first connected wind farm in Guyana to sell power to the GPL network. With the cost of electricity rising, an increasing number of large commercial and private customers disconnected from the GPL grid and started to generate their own electricity. Meanwhile, in 2000 the Canadian-based CGX Energy Incorporated announced a find of two potentially giant oilfields off the coast of Guyana. However, the company was not able to continue test drilling in the area owing to an international boundary dispute between Guyana and Suriname (see History). Despite the border dispute, in early 2001 Venezuela granted Guyana, along with a number of other Caribbean nations, entry to the Caracas energy accord for special oil concessions.

## MANUFACTURING

In the mid-1990s the Government began implementing supportive tax reform measures and constructing industrial parks to attract private investment. Outside the state sector, there were a number of local industries and co-operatives producing consumer goods, especially food and clothing. However, in 2004 Guyana's manufacturing sector remained a small group fighting against cheap imports, excessive bureaucracy, crime, high energy costs and poor infrastructure. The sector remained stagnant in that year. There were modest increases in the production of beer and stout, aerated drinks and distilled water, but several major industries recorded lower output. Manufacturing (including power) accounted for 3.1% of GDP in 2003. Processing of bauxite, sugar, rice, shrimps and timber were the most important industries. Rum production was another traditional export industry.

## TOURISM

Guyana, with its vast, pristine rainforests and savannahs, has become an attraction for the nature traveller and eco-tourist, and several tourism resorts were developed in the interior after 1992. Although tourism does not contribute in a major way to the GDP of Guyana (less than 2% in 2001) this sector has been targeted as a viable option for economic diversification and has begun to show considerable growth. Certain areas have been selected for major development as tourist destinations. In 2002, however, the number of tourists visiting Guyana decreased as a result of the dramatic rise in violent crime and robberies. The Guyana Tourism Authority, established in 2003, began an initiative aggressively to market Guyana overseas. The results were reflected in record arrivals in the last three months of 2003 and the first six months of 2004.

Still optimistic about the growth and potential of the sector, the Government almost doubled its allocation to tourism in the 2005 budget. It was reported that there were 125,000 visitors to the country in 2004, compared with 100,911 visitors in 2003. A number of projects were underway to increase accommodation capacity. However, it was anticipated that the January floods would have a negative impact on tourist arrivals in 2005. In 2003 the sector earned US $39m.

## TRANSPORT INFRASTRUCTURE

Internal communications presented a major obstacle to the economic development of Guyana. In 1999 there were some 7,970 km (4,952 miles) of roads. Most were gravel and earth roads, which could be used in good weather. The remainder were paved, with the main roads running from Georgetown (northwestward to Charity and south-eastward to New Amsterdam). Much of the road network was confined to the coastal region.

In 1989 President Hoyte signed an agreement with Brazil for the construction of the Lethem Kurupukari road, providing the first land route of 560 km long, connecting Georgetown with the Brazilian frontier. Construction of the road, however, was slow, owing to a lack of financial resources, and by 2004 the Guyanese portion of the road had only been partly completed. However, the pace of improvements to the road increased considerably following private-sector involvement and now the journey, which once took four days, can be completed in 12 hours. A regular bus service has also been introduced and work was near completion on a bridge across the river that formed the border between Guyana and Brazil. While Guyana acknowledged that its infrastructure shortcomings were unlikely to be redressed in the short term, owing to a lack of funding, the country was nevertheless committed to the long-term rehabilitation of its transportation networks, sea defences, drainage and irrigation systems, and water and electricity supplies. A US $45m. project to upgrade the roads and bridges between the main airport to the city and from the city to the second largest town, New Amsterdam, began in 2003. The Ministry of Public Works and Communication was allocated $45m. to improve infrastructure in 2004, particularly the road network, drainage systems and sea defences.

Air transportation forms the main link within Guyana. By 1998 there were about 94 airstrips, most of which were located in the interior and catered for light aircraft. An airstrip at Ogle, close to the centre of Georgetown, was used primarily for travel between the capital and the interior. The national carrier, the Guyana Airways Corporation (GAC—known as Guyana Airways 2000 from June 1999, following the sale of a 51% stake in the company to a consortium of local businesses), suspended operations in June 2001. In 2003 privately funded expansion of the Ogle Aerodrome began at a budgeted cost of US $660,000. The airport was expected to provide direct links to the Caribbean.

Guyana is served mainly by the port of Georgetown, which is located at the mouth of the Demerara river. Another important port is that at Linden on the Demerara, which serves as a transit point for bauxite products from the Demerara region. The port facilities at Georgetown were constructed initially to handle the colony's large sugar exports. New Amsterdam's port was improved for the transhipment of bauxite ore, taken to the coast by river from Linden and Everton.

The transport and communications industry contributed 9.3% of GDP in 2003, which was more than the combined contributions of manufacturing and construction. In that year the sector grew by 3.5%, driven largely by the expansion in landline, cellular and information technology infrastructure. The shipping industry saw a sharp decline in trading activities in the late 1970s and 1980s. The launching of the IMF-imposed ERP in 1987 changed the sector's fortunes and the movement of cargo to and from Port Georgetown between 1993 and 1997 was greater than at any other time in the country's history. The combined value of imports and exports exceeded US $1,147m. in 1996, with the shipping industry being responsible for 90% of the transportation. In 2002 President Jagdeo also committed his administration to building a deep-water port, although a site had yet to be selected by mid-2005. The high costs of shipping from Guyana continued to be a serious impediment to the economy as the average cost for imports into the USA was 11.6% of the value of the goods, compared with an average of 3.9% from other countries. The inadequacies of the port facilities and harbours increased the cost of production of exports by an average of 30%. The shipping industry faced further financial difficulties when it moved to comply with the International Ship and Port Facility Security (ISPS) code. Following the terrorist attacks in the USA in September 2001, new procedures for improving the safety and security of maritime and port facilities were mandated by the International Maritime Organisation.

Compliance with the new regulations was estimated to cost the local industry around US $25m.

## INVESTMENT AND FINANCE

Following excessive borrowing from the mid-1960s, in 1985 the IMF formally declared Guyana ineligible for further assistance. After long negotiations with the IMF, the Hoyte administration in 1987 announced the IMF-imposed ERP, approved by the National Assembly in the following year. The ERP was a complete repudiation of the Government's 'socialist' policies: it demanded reductions in government spending, curbs on government intervention in the economy, the establishment of a 'floating' (free) exchange rate for the Guyanese dollar, promotion of foreign investment and the privatization of state enterprises.

In 1990 a settlement was finally reached, enabling Guyana to restructure accumulated debt. Guyana received further debt relief in May 1993, when the 'Paris Club' of Western creditor nations and Trinidad and Tobago rescheduled Guyana's debt under the 'Enhanced Toronto Terms'. Furthermore, in 1996 these parties agreed to cancel 67% of the country's short-term bilateral debt. In 1992, when Cheddi Jagan took power, Guyana's external debt stood at US $2,000m., but by the end of 2003 it had fallen to $1,447m. In May 1999 the IMF and the World Bank granted Guyana $440m. in debt reduction under the heavily indebted poor countries (HIPC) initiative. The total assistance provided to Guyana reduced the country's external debt burden by some 24%. It was estimated that this reduction would, ultimately, translate into debt-service relief of some 3% of GDP per year in 1999–2003 and some 2% per year in 2004–09.

Upon successful completion of a series of pre-arranged conditions in November 2000, Guyana qualified for additional debt relief under the enhanced HIPC initiative, totalling US $590m. One of the conditions required Guyana to reduce the size of the civil service and introduce slower wage increases for public-sector employees. Another stipulated the establishment of an independent tax collection agency. The Government eventually obtained a waiver from the international institutions on the shrinking of the public service because of the high political and ethnic tension and the potential for serious unrest. The enhanced HIPC initiative called for the development of the Poverty Reduction Strategy Paper (PRSP), a blueprint for promoting economic growth and attacking poverty. Another condition was the signing of the SIMAP Third Stage (SIMAP III) loan agreement with the IDB in 2001.

In February 2002 Guyana and the IMF agreed on the implementation of a new three-year programme under the Poverty Reduction Growth Facility (PRGF). In September 2002 the IMF approved a US $73m. poverty reduction credit for Guyana and in December the IDB approved a $64m. debt-relief package. In the same month the EU formalized a major €38.9m. grant for the country and the World Bank approved a $16.8m. zero-interest credit to support Guyana's implementation of the poverty reduction strategy, targeted at stimulating economic growth, strengthening governance, and providing improved access to health care and education. In December 2003 the boards of the IMF and World Bank confirmed that Guyana had met the 'completion point' of the enhanced HIPC initiative, thus qualifying for additional debt relief of $334m. in net present value terms over 10 years. In June 2005, in advance of a summit meeting of the Group of Eight (G-8) industrialized nations in Gleneagles, Scotland, G-8 finance ministers announced that Bolivia was one of 18 countries to have satisfied all the criteria under the HIPC initiative to qualify for immediate debt relief.

Guyana also received debt relief from bilateral creditors. The People's Republic of China agreed in 2003 to cancel three loans totalling US $21.3m., while India cancelled the outstanding balance of about $0.5m. of credit. In June 2004 the Dutch Government confirmed cancellation of debt totalling $4.7m., representing 100% of the stock of commercial debt, including the accrued interest up to 1 December 2003. In July 2004 Canada cancelled $1.4m. of commercial debt, representing 100% of Guyana's debt to Canada. This bought the total debt reduction to Guyana by the Canadian Government to approximately $5.5m. In September France also cancelled all of the $4.4m. debt owed by Guyana and in December Russia wrote off $16.2m. of debts. By mid-2005 only 20% of the country's revenue was used to service its international debt.

Earlier, in September 2003, the IMF completed the first review of Guyana's performance under the PRGF arrangement. As a result of the favourable review, the IMF released US $8.2m. in funds. In December the World Bank disbursed $13.2m. under the Poverty Reduction Support Credit (PRSC). In July 2004 the IMF completed its second review of Guyana's performance under its $73m. PRGF. The favourable review entitled Guyana to the release of a further $8.8m. However, the IMF expressed concern about debt sustainability in light of a number of planned new projects, including the construction of a football stadium for the 2007 World Cup, the rehabilitation of a number of sugar factories, and a bridge across the Berbice river. The review highlighted the risk that the estimated cost of these projects would breach fiscal conditions specified under the HIPC initiative. There was also concern about the lack of growth. The IMF had forecast growth in 2004 to increase to 2.5%, primarily owing to the privatization of the bauxite sector and increased sugar production. However, before the January floods, it was predicted that the economy would contract by 0.3% in 2005, largely as a result of the closure of Omai Gold Mine, followed by modest expansion in subsequent years. Following the third review of the PRGF in January 2005, the IMF released a further $14.1m., and agreed to extend the facility until 2006. In December 2004 Guyana and China signed an agreement for a $32m. concessional loan to be used for the construction of a co-generating plant and the upgrading of the Skeldon sugar refinery (see above).

Guyana's first stock exchange was opened on 30 June 2003. Trade was facilitated through four brokers. The shares of 12 public companies registered with the Guyana Securities Council as reporting issuers were traded on the exchange. In June 2004 the IDB approved a 'soft' loan of US $23m. for the health sector. It was to be used to support improvement in the effectiveness, quality and equity of the distribution of health care nationally. In November the National Assembly approved legislation granting the Bank of Guyana power to take control of troubled financial institutions.

## FOREIGN TRADE AND THE BALANCE OF PAYMENTS

Guyana's exports grew significantly during the 1990s and early 2000s, increasing from US $493m. in 1991 to $506m. in 2003. Imports also rose in the same period from $244m. in 1991, to a high of $628m. in 1997, and standing at $572m. in 2003. Traditionally, Guyana's trade strategy hinged on the preservation of preferential access to the EU market for rice and sugar. Additionally, Guyana had trade agreements with Venezuela, Colombia, Cuba, Canada and the Dominican Republic through CARICOM and bilateral trading agreements with the People's Republic of China, Brazil, Thailand, Argentina and Jordan. Through the enhanced Caribbean Basin Initiative some of Guyana's commodities enjoyed duty-free access to the USA.

In 2001 the principal destinations for the country's exports were Canada (42.9% of total exports), the USA (25.2%) and the United Kingdom (9.4%). The principal origins of its imports were the USA (38.1% of total imports), Trinidad and Tobago (7.6%), the United Kingdom (6.2%) and Japan (5.1%). In 2003 the main imports were intermediate goods, including fuel and lubricants, consumer goods and capital goods. A survey of major export categories from 1990 to 2002 indicated that improvements were registered in the exports of non-traditional products such as shrimp, timber, diamonds, garments and furniture. An important factor inhibiting exports was the high cost of shipping. A new fuel-marking scheme to prevent the smuggling of fuel resulted in an additional $1m. per month in additional revenues. However, although many culprits were identified, not many were prosecuted.

In 2003 the trade deficit stood at US $58.8m.; according to official estimates, this was expected to increase to $92.7m. in 2004. The economic crisis in East Asia, bad weather brought on by the El Niño phenomenon, and continued public violence contributed to an increase in the current-account deficit in the late 1990s, which totalled $102m. in 1998. However, by 2003 the current-account deficit had decreased to $88m. A deficit of

$119m. was forecast for 2004. The country's international reserves stood at $231.84m. at the end of 2004.

## CONCLUSION

Guyana experienced positive growth for most of the 1990s, while at the same time reducing inflation and controlling spending. The country implemented most of the structural reforms demanded by the IMF and World Bank, thus regaining the trust of its international creditors. Nevertheless, despite generous debt-reduction packages from the international financial community, Guyana's economic progress was still impeded by a large external debt, which would necessitate continued austerity measures. At the same time, the country faced falling prices for its export commodities, especially gold, bauxite and sugar, and rising prices for imported commodities, such as petroleum and manufactured goods. Despite the negative growth forecast for 2005, the Government maintained political and financial stability and kept inflation in single figures. Futhermore, partly as a result of the disciplined wages policy, the non-financial public-sector deficit remained well within IMF limits in 2005. The Government reiterated its commitment to introducing value-added tax by 2006. However, an IMF review in 2005 expressed concern that the weak growth in recent years reflected deep-seated structural weaknesses and inefficiencies. It warned that although economic growth was expected to strengthen in the medium term, there were considerable risks, particularly in relation to high oil prices, the planned liberalization of the EU sugar regime, and the general election, scheduled to be held early in 2006. The IMF recommended that the Government focus on strengthening tax collection and limiting public spending to ensure its financial viability and consistency with the debt strategy.

# Statistical Survey

Sources (unless otherwise stated): Bank of Guyana, 1 Church St and Ave of the Republic, POB 1003, Georgetown; tel. 226-3261; fax 227-2965; e-mail communications@solutions2000.net; internet www.bankofguyana.org.gy; Bureau of Statistics, Ministry of Finance, Main and Urquhart Sts, Georgetown; tel. 227-1114; fax 226-1284.

## AREA AND POPULATION

**Area:** 214,969 sq km (83,000 sq miles).

**Population:** 758,619 (males 375,481, females 382,778) at census of 12 May 1980; 701,704 (males 344,928, females 356,776) at census of 12 May 1991; 751,000 in 2003 (official estimate at mid-year).

**Density** (2003): 3.5 per sq km.

**Ethnic Groups** (official estimates, 1999): 'East' Indians 51%, Africans 41%, Amerindians 6%, Portuguese and Chinese 2%, Others less than 1%.

**Regions** (estimated population, 1986): Barima–Waini 18,500; Pomeroon–Supenaam 42,000; Essequibo Islands–West Demerara 102,800; Demerara–Mahaica 310,800; Mahaica–Berbice 55,600; East Berbice–Corentyne 149,000; Cuyuni–Mazaruni 17,900; Potaro–Siparuni 5,700; Upper Takutu–Upper Essequibo 15,300; Upper Demerara–Berbice 38,600.

**Principal Towns** (official estimates, 1991): Georgetown (capital) 151,679; Linden 28,560; New Amsterdam 18,460; Corriverton 13,429 (Source: Government Information Agency). *Mid-2003* (UN estimate, incl. suburbs): Georgetown 231,007 (Source: UN, *World Urbanization Prospects: The 2003 Revision*).

**Births, Marriages and Deaths** (official estimates): Birth rate 21.8 per 1,000 in 1995–2000, 23.6 per 1,000 in 2001; Marriage rate 7.3 per 1,000 in 2001; Crude death rate 7.4 per 1,000 in 1995–2000, 6.6 per 1,000 in 2001.

**Expectation of Life** (WHO estimates, years at birth): 62 (males 61; females 64) in 2003. Source: WHO, *World Health Report*.

**Economically Active Population** (persons between 15 and 65 years of age, 1997): Agriculture, forestry and fishing 66,789; Mining and quarrying 7,299; Manufacturing 27,869; Electricity, gas and water 2,547; Construction 16,545; Trade, repair of motor vehicles and personal and household goods 43,056; Restaurants and hotels 1,597; Transport, storage and communications 20,154; Financial intermediation 4,334; Real estate, renting and business services 7,885; Public administration, defence and social security 15,219; Education 6,659; Health and social work 2,988; Other community, social and personal service activities 4,765; Private households with employed persons 8,146; Extra-territorial organizations and bodies 294; Activities not adequately defined 3,701; Total employed 239,847. *Mid-2003* (estimates): Agriculture, etc. 55,000; Total labour force 329,000 (Source: FAO).

## HEALTH AND WELFARE
### Key Indicators

**Total Fertility Rate** (children per woman, 2003): 2.3.

**Under-5 Mortality Rate** (per 1,000 live births, 2003): 69.

**HIV/AIDS** (% of persons aged 15–49, 2003): 2.5.

**Physicians** (per 1,000 head, 2000): 0.48.

**Hospital Beds** (per 1,000 head, 1996): 3.87.

**Health Expenditure** (2002): US $ per head (PPP): 227.

**Health Expenditure** (2002): % of GDP: 5.6.

**Health Expenditure** (2002): public (% of total): 76.3.

**Access to Water** (% of persons, 2002): 83.

**Access to Sanitation** (% of persons, 2002): 70.

**Human Development Index** (2002): ranking: 104.

**Human Development Index** (2002): value: 0.719.

For sources and definitions, see explanatory note on p. vi.

## AGRICULTURE, ETC.

**Principal Crops** (FAO estimates unless otherwise indicated, '000 metric tons, 2003): Rice (paddy) 501.5 (unofficial estimate); Cassava (Manioc) 29; Roots and tubers 11; Sugar cane 3,000; Coconuts 45; Vegetables 45; Bananas 17; Plantains 17; Other fruit 34.

**Livestock** (FAO estimates, '000 head, year ending September 2003): Horses 2.4; Asses 1; Cattle 110; Sheep 130; Pigs 20; Goats 79; Chickens 21,300.

**Livestock Products** (FAO estimates, '000 metric tons, 2003): Beef and veal 1.8; Sheep and goat meat 0.8; Pig meat 0.5; Chicken meat 23.7; Cows' milk 30; Hen eggs 1.5.

**Forestry** (FAO estimates, '000 cubic metres, 2003): *Roundwood Removals:* Sawlogs, veneer logs and logs for sleepers 165, Pulpwood 112, Other industrial wood 15, Fuel wood 870; Total 1,162; *Sawnwood Production:* Total (all broadleaved) 38.

**Fishing** ('000 metric tons, live weight, 2003): Capture 59.7 (Marine fishes 35.7; Atlantic seabob 19.2; Whitebelly prawn 2.2); Aquaculture 0.6 (FAO estimate); *Total catch* 60.3 (FAO estimate). Note: Figures exclude crocodiles: the number of spectacled caimans caught in 2003 was 3,124.
Source: FAO.

## MINING

**Production** (2003): Bauxite 1,715,705 metric tons; Gold 12,172 kilograms.

## INDUSTRY

**Selected Products** (2003, unless otherwise indicated): Raw sugar 302,378 metric tons; Rice 355,019 metric tons; Bauxite 1,715,705 metric tons; Raw gold 102,662 oz; Rum 145,586 hectolitres (2002); Beer 108,550 hectolitres (2002); Electric energy 549.3m. kWh.

## FINANCE

**Currency and Exchange Rates:** 100 cents = 1 Guyana dollar ($ G). Sterling, *US Dollar and Euro Equivalents* (29 April 2005): £1 sterling = $ G382.400; US $1 = $ G200.000; €1 = $ G259.140; $ G1,000 = £2.62 = US $5.00 = €3.86. *Average Exchange Rate:* ($ G per US $): 190.7 in 2002; 193.9 in 2003; 198.3 in 2004.

# GUYANA

**Budget** (estimates, $ G million, 2004): *Revenue:* Tax revenue 44,460.3 (Income tax 20,657.3, Consumption tax 16,968.2, Trade taxes 4,631.3, Other tax 2,203.5); Other current revenue 3,439.8; Total 47,900.1. *Expenditure:* Current expenditure 50,032.3 (Personnel emoluments 17,716.4, Other goods and services 12,971.6, Interest 7,237.6, Transfers 12,106.7); Capital expenditure 23,838.6; Total (excl. lending minus repayments) 73,870.9.

**International Reserves** (US $ million at 31 December 2004): IMF special drawing rights 7.15; Foreign exchange 224.70; *Total* 231.84. Source: IMF, *International Financial Statistics*.

**Money Supply** ($ G million at 31 December 2004): Currency outside banks 19,546; Demand deposits at commercial banks 17,052; Total money (including also private-sector deposits at the Bank of Guyana) 36,604. Source: IMF, *International Financial Statistics*.

**Cost of Living** (Consumer Price Index; base: 1994 = 100): 183.3 in 2001; 196.4 in 2002; 206.0 in 2003.

**Expenditure on the Gross Domestic Product** ($ G million at current prices, 2003): Government final consumption expenditure 37,928; Private final consumption expenditure 71,615; Gross fixed capital formation (incl. changes in stocks) 50,473; *Total domestic expenditure* 160,016; Net exports of goods and services –15,952; *GDP in purchasers' values* 144,064.

**Gross Domestic Product by Economic Activity** ($ G million at current prices, 2003): Agriculture 35,463 (Sugar 18,448, Rice 8,621, Livestock 2,979); Forestry 2,411; Fishing 8,389; Mining and quarrying 15,930; Manufacturing (incl. utilities) 3,874; Engineering and construction 6,199; Distribution 4,996; Transport and communication 11,502; Rented dwellings 5,087; Financial services 4,400; Other services 2,201; Government 22,809; *GDP at factor cost* 123,261; Indirect taxes, *Less* Subsidies 20,803; *GDP in purchasers' values* 144,064.

**Balance of Payments** (budget estimates, US $ million, 2003): Exports of goods f.o.b. 512.8; Imports of goods f.o.b. –571.7; *Trade balance* –58.8; Exports of services (net) –69.8; *Balance on goods and services* –128.6; Current transfers received (net) 40.3; *Current balance* –88.3; Capital transfer 43.8; Medium and long-term capital (net) 38.5; Short-term capital (net) –27.3; Net errors and omissions 24.5; *Overall balance* –8.9.

## EXTERNAL TRADE

**Principal Commodities** (US $ million, 2003): *Imports c.i.f.:* Capital goods 116.1; Consumer goods 149.4; Fuel and lubricants 147.2; Other intermediate goods 158.3; Total (incl. others) 571.7. *Exports f.o.b.:* Bauxite 44.6; Sugar 129.2; Rice 45.3; Gold 130.9; Shrimps 53.9; Timber 30.7; Total (incl. others, excl. re-exports) 505.6.

**Principal Trading Partners** ($ G million, 2001, rounded figures): *Imports:* USA 45,200; Trinidad and Tobago 9,000; United Kingdom 7,400; Japan 6,100; Venezuela 4,500; Netherlands 3,900; China 3,200; Canada 1,600; Total (incl. others) 118,700. *Exports:* Canada 56,300; USA 33,200; United Kingdom 12,300; Jamaica 4,300; Netherlands 2,600; Trinidad and Tobago 2,400; Barbados 2,300; Total (incl. others) 131,300.

## TRANSPORT

**Road Traffic** ('000 vehicles in use, 2001): Passenger cars 61.3; Commercial vehicles 15.5. Source: UN, *Statistical Yearbook*.

**Shipping:** *International Sea-borne Freight Traffic* ('000 metric tons, estimates, 1990): Goods loaded 1,730; Goods unloaded 673 (Source: UN, *Monthly Bulletin of Statistics*). *Merchant Fleet* (at 31 December 2004): Vessels 106; Displacement 33,233 grt (Source: Lloyd's Register-Fairplay, *World Fleet Statistics*).

**Civil Aviation** (traffic on scheduled services, 2001): Kilometres flown (million) 1; Passengers carried ('000) 48; Passenger-km (million) 475; Total ton-km (million) 17. Source: UN, *Statistical Yearbook*.

## TOURISM

**Tourist Arrivals** ('000): 99.3 in 2001; 104.3 in 2002; 100.9 in 2003.

**Tourism Receipts** (US $ million, excluding passenger transport): 61 in 2001; 49 in 2002; 39 in 2003.
Source: World Tourism Organization.

## COMMUNICATIONS MEDIA

**Radio Receivers** (1999): 400,000 in use.

**Television Receivers** (2000): 70,000 in use.

**Telephones** (2003): 80,400 main lines in use.

**Facsimile Machines** (1990): 195 in use.

**Mobile Cellular Telephones** (2002): 87,300 subscribers.

**Personal Computers** (2002): 24,000 in use.

**Internet Users** (2002): 125,000.

**Daily Newspapers** (1998): 2; estimated circulation 56,750.

**Non-daily Newspapers** (2000): 4; estimated circulation 47,700.

**Book Production** (1997): 25.
Sources: mainly UNESCO, *Statistical Yearbook*; UN, *Statistical Yearbook*; International Telecommunication Union.

## EDUCATION

**Pre-primary** (1999/2000): Institutions 320; Teachers 2,218; Students 36,995.

**Primary** (1999/2000): Institutions 423; Teachers 3,951; Students 105,800.

**General Secondary** (1999/2000): Institutions 70; Teachers 1,972; Students 36,055.

**Special Education** (1999/2000): Institutions 6; Teachers 64; Students 617.

**Technical and Vocational** (1999/2000): Institutions 6; Teachers 215; Students 4,662.

**Teacher Training** (1999/2000): Institutions 1; Teachers 297; Students 1,604.

**University** (1999/2000): Institutions 1; Teachers 371; Students 7,496.

**Private Education** (1999/2000): Institutions 7; Teachers 120; Students 1,692.
Source: Ministry of Education.

**Adult Literacy Rate** (UNESCO estimates): 98.6% (males 99.0%; females 98.2%) in 2001. Source: UN Development Programme, *Human Development Report*.

# Directory

## The Constitution

Guyana became a republic, within the Commonwealth, on 23 February 1970. A new Constitution was promulgated on 6 October 1980, and amended in 1998, 2000 and 2001. Its main provisions are summarized below:

The Constitution declares the Co-operative Republic of Guyana to be an indivisible, secular, democratic sovereign state in the course of transition from capitalism to socialism. The bases of the political, economic and social system are political and economic independence, involvement of citizens and socio-economic groups, such as co-operatives and trade unions, in the decision-making processes of the State and in management, social ownership of the means of production, national economic planning and co-operativism as the principle of socialist transformation. Personal property, inheritance, the right to work, with equal pay for men and women engaged in equal work, free medical attention, free education and social benefits for old age and disability are guaranteed. Additional rights include equality before the law, the right to strike and to demonstrate peacefully, the right of indigenous peoples to the protection and preservation of their culture, and a variety of gender and work-related rights. Individual political rights are subject to the principles of national sovereignty and democracy, and freedom of expression to the State's duty to ensure fairness and balance in the dissemination of information to the public. Relations with other countries are guided by respect for human rights, territorial integrity and non-intervention.

### THE PRESIDENT

The President is the supreme executive authority, Head of State and Commander-in-Chief of the armed forces, elected for a five-year term of office, with no limit on re-election. The successful presidential candidate is the nominee of the party with the largest number of votes in the legislative elections. The President may prorogue or dissolve the National Assembly (in the case of dissolution, fresh elections must be held immediately) and has discretionary powers to postpone elections for up to one year at a time for up to five years. The President may be removed from office on medical grounds, or for

# GUYANA

violation of the Constitution (with a two-thirds' majority vote of the Assembly), or for gross misconduct (with a three-quarters' majority vote of the Assembly if allegations are upheld by a tribunal).

The President appoints a First Vice-President and Prime Minister who must be an elected member of the National Assembly, and a Cabinet of Ministers, which may include four non-elected members and is collectively responsible to the legislature. The President also appoints a Leader of the Opposition, who is the elected member of the Assembly deemed by the President most able to command the support of the opposition.

### THE LEGISLATURE

The legislative body is a unicameral National Assembly of 65 members (66 in special circumstances), elected by universal adult suffrage in a system of proportional representation; 40 members are elected at national level, and a further 25 are elected from regional constituency lists. The Assembly passes bills, which are then presented to the President, and may pass constitutional amendments.

### LOCAL GOVERNMENT

Guyana is divided into 10 Regions, each having a Regional Democratic Council elected for a term of up to five years and four months, although it may be prematurely dissolved by the President.

### OTHER PROVISIONS

Impartial commissions exist for the judiciary, the public service and the police service. An Ombudsman is appointed, after consultation between the President and the Leader of the Opposition, to hold office for four years.

## The Government

### HEAD OF STATE

**President:** BHARRAT JAGDEO (sworn in 11 August 1999; re-elected 19 March 2001).

### CABINET
(July 2005)

**Prime Minister and Minister of Public Works and Communication:** SAMUEL A. HINDS.
**Minister of Parliamentary Affairs in the Office of the President:** PANDIT REEPU DAMAN PERSAUD.
**Minister of Foreign Affairs:** Dr RUDY INSANALLY.
**Minister of Foreign Trade in the Ministry of Foreign Affairs:** CLEMENT ROHEE.
**Minister of Finance:** SAISNARINE KOWLESSAR.
**Minister of Agriculture and of Fisheries, Crops and Livestock:** SATYADEOW SAWH.
**Minister of Legal Affairs and Attorney-General:** DOODNAUTH SINGH.
**Minister of Information:** PREM MISIR.
**Minister of Education:** Dr HENRY BENFIELD JEFFREY.
**Minister of Health:** Dr LESLIE RAMSAMMY.
**Minister of Housing and Water:** SHAIK BAKSH.
**Minister of Labour, Human Services and Social Security:** Dr DALE BISNAUTH.
**Minister of Tourism, Industry and Commerce:** MANZOOR NADIR.
**Minister of Amerindian Affairs:** CAROLYN RODRIGUES.
**Minister of Local Government and Regional Development:** HARRIPERSAUD NOKTA.
**Minister of Public Service Management:** JENNIFER WESTFORD.
**Minister of Transport and Hydraulics:** CARL ANTHONY XAVIER.
**Minister of Culture, Youth and Sports and Acting Minister of Home Affairs:** GAIL TEIXEIRA.
**Secretary to the Cabinet:** Dr ROGER LUNCHEON.

### MINISTRIES

**Office of the President:** New Garden St, Bourda, Georgetown; tel. 225-1330; fax 227-3050; e-mail opmed@sdnp.org.gy; internet www.op.gov.gy.
**Office of the Prime Minister:** Wight's Lane, Georgetown; tel. 227-3101; fax 226-7573; e-mail pmoffice@sdnp.org.org.gy; internet www.pm.gov.org.gy.
**Ministry of Agriculture:** Regent and Vlissengen Rds, POB 1001, Georgetown; tel. 227-5527; fax 227-3638; e-mail moa@sdnp.org.gov.gy; internet www.sdnp.org.gy/minagri.
**Ministry of Amerindian Affairs:** Thomas and Quamina Sts, Georgetown; e-mail ministryofamerindian@networksgy.com.
**Ministry of Culture, Youth and Sports:** 71 Main St, North Cummingsburg, Georgetown; tel. 226-8542; fax 226-8549; e-mail psmincys@guyana.net.gy.
**Ministry of Education:** 26 Brickdam, Stabroek, POB 1014, Georgetown; tel. 226-3094; fax 225-5570; e-mail moegyweb@yahoo.com; internet www.sdnp.org.gy/minedu.
**Ministry of Finance:** Main St, Kingston, Georgetown; tel. 227-1114; fax 226-1284; e-mail guyanadmd@solutions2000.net.
**Ministry of Fisheries, Crops and Livestock:** Regent Rd, Bourda, Georgetown; tel. 226-1565; fax 227-2978; e-mail minfcl@sdnp.org.gy; internet www.sdnp.org.gy/minagri.
**Ministry of Foreign Affairs:** Takuba Lodge, 254 South Rd, Bourda, Georgetown; tel. 226-1607; fax 225-9192; e-mail minfor@sdnp.org.gy; internet www.sdnp.org.gy/minfor.
**Ministry of Health:** Brickdam, Stabroek, Georgetown; tel. 226-1560; fax 225-4505; e-mail moh@sdnp.org.gy; internet www.sdnp.org.gy/moh.
**Ministry of Home Affairs:** 6 Brickdam, Stabroek, Georgetown; tel. 225-7270; fax 226-2740.
**Ministry of Housing and Water:** 41 Brickdam, Stabroek, Georgetown; tel. 225-7192; fax 227-3455; e-mail mhwps@sdnp.org.gy.
**Ministry of Information:** Area B, Homestretch Ave, Durban Park, Georgetown; tel. 227-1101; fax 226-4003.
**Ministry of Labour, Human Services and Social Security:** 1 Water St, Stabroek, Georgetown; tel. 225-0566; fax 226-6076; e-mail khadoo@networksgy.com; internet www.sdnp.org.gy/mohss.
**Ministry of Legal Affairs and Office of the Attorney-General:** 95 Carmichael St, Georgetown; tel. 225-3663; fax 226-9721.
**Ministry of Local Government and Regional Development:** De Winkle Bldg, Fort St, Kingston, Georgetown; tel. 225-8621; fax 226-5070; e-mail mlgrdps@telsnetgy.net.
**Ministry of Public Service Management:** 164 Waterloo St, North Cummingsburg, Georgetown; tel. 227-1193; fax 227-2700; e-mail psm@sdnp.org.gy; internet www.sdnp.org.gy/psm.
**Ministry of Public Works and Communication:** Georgetown.
**Ministry of Tourism, Industry and Commerce:** 229 South Rd, Lacytown, Georgetown; tel. 226-2505; fax 225-4370; e-mail ministry@mintic.gov.gy; internet www.sdnp.org.gy/mtti.
**Ministry of Transport and Hydraulics:** Wights Lane, Kingston, Georgetown; tel. 226-1875; fax 225-8395; e-mail minoth@networksgy.com.

## President and Legislature

### NATIONAL ASSEMBLY

**Speaker:** RALPH RAMKARRAN.
**Deputy Speaker:** CLARISSA RIEHL.
**Election, 19 March 2001**

| Party | No. of seats | | |
| --- | --- | --- | --- |
| | Regional | National | Total |
| People's Progressive Party/CIVIC (PPP/CIVIC) | 11 | 23 | 34 |
| People's National Congress Reform (PNCReform) | 13 | 14 | 27 |
| Guyana Action Party/Working People's Alliance (GAP/WPA) | 1 | 1 | 2 |
| Rise, Organize and Rebuild Guyana Movement (ROAR) | — | 1 | 1 |
| The United Force (TUF) | — | 1 | 1 |
| **Total** | 25 | 40 | 65 |

Under Guyana's system of proportional representation, the nominated candidate of the party receiving the most number of votes was elected to the presidency. Thus, on 23 March 2001 the candidate of the PPP/CIVIC alliance, BHARRAT JAGDEO, was declared President-elect, defeating HUGH DESMOND HOYTE of the PNCReform. JAGDEO was inaugurated as President on 31 March.

# GUYANA

## Political Organizations

**CIVIC:** New Garden St, Georgetown; internet www.ppp-civic.org; social/political movement of businessmen and professionals; allied to PPP; Leader SAMUEL ARCHIBALD ANTHONY HINDS.

**Guyana Action Party (GAP):** Georgetown; allied to WPA; Leader PAUL HARDY.

**Guyana Democratic Party (GDP):** Georgetown; f. 1996; Leaders ASGAR ALLY, NANDA K. GOPAUL.

**Guyana National Congress (GNC):** Georgetown.

**Guyana People's Party (GPP):** Georgetown; f. 1996; Leader MAX MOHAMED.

**Guyana Republican Party (GRP):** Paprika East Bank, Essequibo; f. 1985; right-wing; Leader LESLIE PRINCE (resident in the USA).

**Horizon and Star (HAS):** Georgetown.

**Justice For All Party (JFAP):** 73 Robb and Wellington Sts, Lacytown, Georgetown; tel. 226-5462; fax 227-3050; e-mail sharma@guyana.net.gy; internet www.jfa-gy.com; Leader CHANDRANARINE SHARMA.

**National Democratic Front (NDF):** Georgetown; Leader JOSEPH BACCHUS.

**National Front Alliance:** Georgetown.

**Democratic Labour Movement (DLM):** 34 Robb and King Sts, 4th Floor, Lacytown, POB 10930, Georgetown; f. 1983; democratic-nationalist; Pres. PAUL NEHRU TENNASSEE.

**People's Democratic Movement (PDM):** Stabroek House, 10 Croal St, Georgetown; tel. 226-4707; fax 226-3002; f. 1973; centrist; Leader LLEWELLYN JOHN.

**People's Progressive Party (PPP):** Freedom House, 41 Robb St, Lacytown, Georgetown; tel. 227-2095; fax 227-2096; e-mail ppp@guyana.net.gy; internet www.ppp-civic.org; f. 1950; Marxist-Leninist; allied to CIVIC; Gen. Sec. DONALD RAMOTAR.

**National Independence Party:** Georgetown; Leader SAPHIER HUSSIEN.

**People's National Congress Reform (PNCReform):** Congress Place, Sophia, POB 10330, Georgetown; tel. 225-7852; fax 225-6055; e-mail pnc@guyana-pnc.org; internet www.guyanapnc.org; f. 1955 after a split with the PPP; Reform wing established in 2000; PNC Leader ROBERT H. O. CORBIN; Chair. VINCENT ALEXANDER.

**Rise, Organize and Rebuild Guyana Movement (ROAR):** 186 Parafield, Leonora, West Coast Demerara, POB 101409, Georgetown; tel. 068-2452; e-mail guyroar@hotmail.com; f. 1999.

**The United Force (TUF):** 95 Robb and New Garden Sts, Bourda, Georgetown; tel. 226-2596; fax 225-2973; e-mail manzoornadir@yahoo.com; f. 1960; right-wing; advocates rapid industrialization through govt partnership and private capital; Leader MANZOOR NADIR.

**United People's Party:** 77 Winter Pl., Brickdam, Georgetown; tel. 227-5217; fax 227-5166; e-mail unitedguyana@yahoo.com.

**Unity Party:** 77 Hadfield St, Georgetown; tel. 227-6744; fax 227-6745; e-mail unityp@networksgy.com; f. 2005; promotes private enterprise and coalition politics; Pres. CHEDDI (JOEY) JAGAN, Jr.

**Working People's Alliance (WPA):** Walter Rodney House, 80 Croal St, Stabroek, Georgetown; tel. and fax 225-3679; internet www.guyanacaribbeanpolitics.com/wpa/wpa.html; originally popular pressure group, became political party 1979; independent Marxist; allied to GAP; Collective Leadership Dr CLIVE THOMAS, Dr RUPERT ROOPNARINE.

## Diplomatic Representation

### EMBASSIES AND HIGH COMMISSIONS IN GUYANA

**Brazil:** 308 Church St, Queenstown, POB 10489, Georgetown; tel. 225-7970; fax 226-9063; e-mail guibrem@solutions2000.net; Ambassador NEY DO PRADO DIEGUEZ.

**Canada:** High and Young Sts, POB 10880, Georgetown; tel. 227-2081; fax 225-8380; e-mail grgtn@international.gc.ca; internet www.dfait-maeci.gc.ca/guyana; High Commissioner BRUNO PICARD.

**China, People's Republic:** Botanic Gardens, Mandella Ave, Georgetown; tel. 227-1651; fax 231-6602; e-mail chinaemb_gy@mfa.gov.cn; Ambassador SHEN QING.

**Cuba:** 46 High St, Georgetown; tel. 225-1881; fax 226-1824; e-mail emguyana@networksgy.com; Ambassador JOSÉ MANUEL INCLAN EMBADE.

**India:** Bank of Baroda Bldg, 10 Ave of the Republic, POB 101148, Georgetown; tel. 226-3996; fax 225-7012; High Commissioner AVINASH C. GUPTA.

**Russia:** 3 Public Rd, Kitty, Georgetown; tel. 227-1738; fax 227-2975; e-mail reing@networks.gy.com; Ambassador VLADIMIR STEPANOVICH STARIKOV.

**Suriname:** 304 Church St, POB 10508, Georgetown; tel. 226-7844; fax 225-0759; e-mail surnemb@gol.net.gy; Ambassador MANORMA SOEKNANDAN.

**United Kingdom:** 44 Main St, POB 10849, Georgetown; tel. 226-5881; fax 225-3555; e-mail enquiries@britain-in-guyana.org; internet britain-in-guyana.org; High Commissioner STEPHEN HISCOCK.

**USA:** 100 Young and Duke Sts, POB 10507, Kingston, Georgetown; tel. 225-4900; fax 225-8497; Ambassador ROLAND W. BULLEN.

**Venezuela:** 296 Thomas St, South Cummingsburg, Georgetown; tel. 226-1543; fax 225-3241; e-mail embveguy@gol.net.gy; Ambassador FRANÇOIS PULVENIS.

## Judicial System

The Judicature of Guyana comprises the Supreme Court of Judicature, which consists of the Court of Appeal and the High Court (both of which are superior courts of record), and a number of Courts of Summary Jurisdiction.

The Court of Appeal, which came into operation in 1966, consists of the Chancellor as President, the Chief Justice, and such number of Justices of Appeal as may be prescribed by the National Assembly.

The High Court of the Supreme Court consists of the Chief Justice as President of the Court and Puisne Judges. Its jurisdiction is both original and appellate. It has criminal jurisdiction in matters brought before it on indictment. A person convicted by the Court has a right of appeal to the Guyana Court of Appeal. The High Court of the Supreme Court has unlimited jurisdiction in civil matters and exclusive jurisdiction in probate, divorce and admiralty and certain other matters. Under certain circumstances, appeal in civil matters lies either to the Full Court of the High Court of the Supreme Court, which is composed of no fewer than two judges, or to the Guyana Court of Appeal. On 4 November 2004 the National Assembly approved legislation recognizing the Caribbean Court of Justice (CCJ) as Guyana's highest court of appeal. The CCJ was inaugurated in Port of Spain, Trinidad and Tobago, on 16 April 2005.

A magistrate has jurisdiction to determine claims where the amount involved does not exceed a certain sum of money, specified by law. Appeal lies to the Full Court.

**Chancellor of Justice:** CARL SINGH (acting).

**Chief Justice:** CARL SINGH (acting).

**High Court Justices:** NANDRAM KISSOON, CLAUDETTE SINGH.

**Attorney-General:** DOODNAUTH SINGH.

## Religion

### CHRISTIANITY

**Guyana Council of Churches:** 26 Durban St, Lodge, Georgetown; tel. 225-3020; e-mail bishopedghill@hotmail.com; f. 1967 by merger of the Christian Social Council (f. 1937) and the Evangelical Council (f. 1960); 15 mem. churches, 1 assoc. mem.; Chair. Bishop JUAN A. EDGHILL; Sec. Rev. NIGEL HAZEL.

#### The Anglican Communion

Anglicans in Guyana are adherents of the Church in the Province of the West Indies, comprising eight dioceses. The Archbishop of the Province is the Bishop of the North Eastern Caribbean and Aruba, resident in St John's, Antigua. The diocese of Guyana also includes French Guiana and Suriname. In 1996 the estimated membership in the country was 118,000.

**Bishop of Guyana:** Rt Rev. RANDOLPH OSWALD GEORGE, The Church House, 49 Barrack St, POB 10949, Georgetown 1; tel. and fax 226-4183; e-mail dioofguy@networksgy.com; internet www.anglican.bm/G/01.html.

#### The Baptist Church

**The Baptist Convention of Guyana:** POB 10149, Georgetown; tel. 226-0428; 1,823 mems.

#### The Lutheran Church

**The Evangelical Lutheran Church in Guyana:** Lutheran Courts, Berbice, POB 88, New Amsterdam; tel. and fax 333-6479;

# GUYANA

e-mail lcg@guyana.net.gy; f. 1947; 11,000 mems; Pres. Rev. ROY K. THAKURDYAL.

### The Roman Catholic Church

Guyana comprises the single diocese of Georgetown, suffragan to the archdiocese of Port of Spain, Trinidad and Tobago. At 31 December 2003 adherents of the Roman Catholic Church comprised about 11% of the total population. The Bishop participates in the Antilles Episcopal Conference Secretariat, currently based in Port of Spain, Trinidad.

**Bishop of Georgetown:** FRANCIS DEAN ALLEYNE, Bishop's House, 27 Brickdam, POB 10720, Stabroek, Georgetown; tel. 226-4469; fax 225-8519; e-mail rcbishop@networksgy.com.

### Other Christian Churches

Other denominations active in Guyana include the African Methodist Episcopal Church, the African Methodist Episcopal Zion Church, the Church of God, the Church of the Nazarene, the Ethiopian Orthodox Church, the Guyana Baptist Mission, the Guyana Congregational Union, the Guyana Presbyterian Church, the Hallelujah Church, the Methodist Church in the Caribbean and the Americas, the Moravian Church and the Presbytery of Guyana.

### BAHÁ'Í FAITH

**National Spiritual Assembly:** 220 Charlotte St, Bourda, Georgetown; tel. and fax 226-5952; e-mail nsaguy@networksgy.gy; internet www.sdnp.org.gy/bahai; incorporated in 1976.

### HINDUISM

Hindus constituted an estimated 35% of the population.

**Hindu Religious Centre:** 162 Lamaha St, POB 10576, Georgetown; tel. 225-7443; f. 1934; Hindus account for about one-third of the population; Gen. Sec. CHRISHNA PERSAUD.

### ISLAM

Muslims in Guyana comprised approximately 10% of the population.

**The Central Islamic Organization of Guyana (CIOG):** M.Y.O. Bldg, Woolford Ave, Thomas Lands, POB 10245, Georgetown; tel. 225-8654; fax 227-2475; e-mail ciog@sdnp.org.gy; internet www.sdnp.org.gy/ciog; Pres. Alhaji FAZEEL M. FEROUZ; Gen. Sec. MUJTABA NASIR.

**Guyana United Sad'r Islamic Anjuman:** 157 Alexander St, Kitty, POB 10715, Georgetown; tel. 226-9620; f. 1936; 120,000 mems; Pres. Haji A. HAFIZ RAHAMAN.

## The Press

### DAILIES

**Guyana Chronicle:** 2A Lama Ave, Bel Air Park, POB 11, Georgetown; tel. 226-3243; fax 227-5208; e-mail gm@guyanachronicle.com; internet www.guyanachronicle.com; f. 1881; govt-owned; also produces weekly *Sunday Chronicle* (tel. 226-3243); Editor-in-Chief SHARIEF KHAN; circ. 23,000 (weekdays), 43,000 (Sundays).

**Kaieteur News:** 24 Saffon St, Charlestown; tel. 225-8458; fax 225-8473; e-mail kaieteur_kaieteur@yahoo.com; internet www.kaieteurnews.com; f. 1994; independent; Editor W. HENRY SKERRETT; Publr GLEN LALL; circ. 30,000.

**Stabroek News:** 46–47 Robb St, Lacytown, Georgetown; tel. 227-4080; fax 225-4637; e-mail stabroeknews@stabroeknews.com; internet www.stabroeknews.com; f. 1986; also produces weekly *Sunday Stabroek*; liberal independent; Editor-in-Chief DAVID DE CAIRES; Editor ANAND PERSAUD; circ. 13,500 (weekdays), 26,000 (Sundays).

### WEEKLIES AND PERIODICALS

**The Catholic Standard:** 293 Oronoque St, Queenstown, POB 10720, Georgetown; tel. 226-1540; f. 1905; weekly; Editor COLIN SMITH; circ. 10,000.

**Diocesan Magazine:** 144 Almond and Oronoque Sts, Queenstown, Georgetown; quarterly.

**Guyana Business:** 156 Waterloo St, POB 10110, Georgetown; tel. 225-6451; f. 1889; organ of the Georgetown Chamber of Commerce and Industry; quarterly; Editor C. D. KIRTON.

**Guyana Review:** 143 Oronoque St, POB 10386, Georgetown; tel. 226-3139; fax 227-3465; e-mail guyrev@networksgy.com; internet www.guyanareview.com; f. 1993; monthly.

**Guynews:** Georgetown; monthly.

**Mirror:** Lot 8, Industrial Estate, Ruimveldt, Greater Georgetown; tel. 226-2471; fax 226-2472; e-mail ngmirror@guyana.net.gy; internet www.mirrornewsonline.com; owned by the New Guyana Co Ltd; Sundays; Editor ROBERT PERSAUD; circ. 25,000.

**New Nation:** Congress Pl., Sophia, Georgetown; tel. 226-7891; f. 1955; organ of the People's National Congress; weekly; Editor FRANCIS WILLIAMS; circ. 26,000.

**The Official Gazette of Guyana:** Guyana National Printers Ltd, Lot 1, Public Rd, La Penitence; weekly; circ. 450.

**Thunder:** Georgetown; f. 1950; organ of the People's Progressive Party; quarterly; Editor RALPH RAMKARRAN; circ. 5,000.

### PRESS ASSOCIATION

**Guyana Press Association (GPA):** Georgetown; Pres. DENIS CHABROL; Vice-Pres. JULIA JOHNSON.

### NEWS AGENCIES

**Guyana Information Agency:** Area B, Homestretch Ave, D'Urban Backlands, Georgetown; tel. 225-3117; fax 226-4003; e-mail gina@gina.gov.gy; internet www.gina.gov.gy; f. 1993; Dir PREM MISIR.

### Foreign Bureaux

**Xinhua (New China) News Agency** (People's Republic of China): 52 Brickdam, Stabroek, Georgetown; tel. 226-9965.

Associated Press (USA) and Informatsionnoye Telegrafnoye Agentstvo Rossii—Telegrafnoye Agentstvo Suverennykh Stran (ITAR—TASS) (Russia) are also represented.

## Publishers

**Guyana Free Press:** POB 10386, Georgetown; tel. 226-3139; fax 227-3465; e-mail guyrev@networksgy.com; books and learned journals.

**Guyana National Printers Ltd:** 1 Public Rd, La Penitence, POB 10256, Greater Georgetown; tel. 225-3623; e-mail gnpl@guyana.net.gy; f. 1939; govt-owned printers and publishers; privatization pending.

**Guyana Publications Inc:** 46/47 Robb St, Lacytown, Georgetown; tel. 225-7473; fax 225-4637; e-mail stabroeknews.com; internet www.stabroeknews.com; Man. Dir DOREEN DECAIRES.

## Broadcasting and Communications

### TELECOMMUNICATIONS

The telecommunications sector was restructured and opened up to competition in 2002.

**Guyana Telephones and Telegraph Company (GT & T):** 79 Brickdam, POB 10628, Georgetown; tel. 226-7840; fax 226-2457; internet www.gtt.co.gy; f. 1991; fmrly state-owned Guyana Telecommunications Corpn; 80% ownership by Atlantic Tele-Network (USA); CEO SONITA JAGAN; Chair. CORNELIUS PRIOR.

**Trans-World Telecom Guyana (TWT Guyana):** 56 High St, POB 101845, Georgetown; tel. 223-6531; fax 223-6532; e-mail info@celstarguyana.net; internet www.twtcaribbean.com/guyana; f. 1999; acquired Cel Star Guyana in 2003; GSM cellular telecommunications network; mem. of Trans-World Telecom Caribbean group; COO PIERRE STRASSER; Vice-Pres. GREGORY LIBERTINY.

### BROADCASTING

In May 2001 the Government implemented the regulation of all broadcast frequencies. Two private stations relay US satellite television programmes.

**National Communications Network (NCN):** Homestretch Ave, D'Urban Park, Georgetown; tel. 227-1566; fax 226-2253; e-mail ncnfeedback@ncnguyana.com; internet www.ncnguyana.com; f. 2004 following merger of Guyana Broadcasting Corpn (f. 1979) and Guyana Television and Broadcasting Co (f. 1993); govt-owned; operates three radio channels and six TV channels; CEO DESMOND MOHAMED SATTAUR; Editor-in-Chief WILFRED CAMERON.

### Radio

**National Communications Network (NCN):** see Broadcasting; operates three channels:

**Hot FM**.

**Radio Roraima**.

**Voice of Guyana**.

# GUYANA

## Television

**National Communications Network:** see Broadcasting; TV network covers channels 8, 11, 13, 15, 21 and 26.

# Finance

(cap. = capital; res = reserves; dep. = deposits; m. = million; brs = branches; amounts in Guyana dollars)

## BANKING

### Central Bank

**Bank of Guyana:** 1 Church St and Ave of the Republic, POB 1003, Georgetown; tel. 226-3250; fax 227-2965; e-mail communications@bankofguyana.org.gy; internet www.bankofguyana.org.gy; f. 1965; cap. 1,000m., res 4,062.7m., dep. 86,482.3m. (Dec. 2003); central bank of issue; acts as regulatory authority for the banking sector; Gov. LAWRENCE T. WILLIAMS.

### Commercial Banks

**Citizens' Bank Guyana Inc:** 201 Camp and Charlotte Sts, Lacytown, Georgetown; tel. 226-1705; fax 227-1719; e-mail citizens@guyana.net.gy; f. 1994; 51% owned by Banks DIH; Chair. CLIFFORD REIS; Man. Dir ALAN PARRIS.

**Demerara Bank Ltd:** 230 Camp and South Sts, Georgetown; tel. and fax 225-0610; e-mail banking@demerarabank.com; internet www.demerarabank.com; f. 1994; cap. 450m., res 167m., dep. 12,491m. (Sept. 2003); Chair. YESU PERSAUD; Man. Dir AHMAD M. KHAN.

**Guyana Americas Merchant Bank (GBTI):** GTBI Bldg, 138 Regent St, Lacytown, Georgetown; tel. 223-5193; fax 223-5195; e-mail gambi@networksgy.com; f. 2001; merchant bank.

**Guyana Bank for Trade and Industry Ltd:** 47–48 Water St, POB 10280, Georgetown; tel. 226-8430; fax 227-1612; e-mail banking@gbtibank.com; internet www.gbtibank.com; f. 1987 to absorb the operations of Barclays Bank; cap. 800m., res 601.5m., dep. 21,724.6m. (Dec. 2003); CEO RADHA KRISHNA SHARMA; Dir JOHN TRACEY; 6 brs.

**National Bank of Industry and Commerce Ltd (NBIC):** Promenade Court, 155–156 New Market St, Georgetown; tel. 223-7938; fax 227-4506; e-mail email@nbicltd.com; internet www.nbicgy.com; f. 1984; 51% owned by Republic Bank Ltd, Port of Spain, Trinidad and Tobago; acquired Guyana National Co-operative Bank in 2003; cap. 300m., res 2,899.3m., dep. 48,589.7m. (Sept. 2003); Man. Dir MICHAEL ARCHIBALD; 5 brs.

### Foreign Banks

**Bank of Baroda** (India): 10 Ave of the Republic, POB 10768, Georgetown; tel. 226-4005; fax 225-1691; e-mail bobinc@networksgy.com; f. 1908; Chief Man. P. SAVID.

**Bank of Nova Scotia** (Canada): 104 Carmichael St, POB 10631, Georgetown; tel. 225-9222; fax 225-9309; e-mail ian.cooper@scotiabank.com; internet www.scotiabank.com; Man. IAN COOPER; 5 brs.

## STOCK EXCHANGE

**Guyana Association of Securities Companies and Intermediaries (GASCI):** Hand-in-Hand Bldg, 1–4 Ave of the Republic, Georgetown; tel. 223-6175; e-mail info@gasci.com; internet www.gasci.com; f. 2003; Chair. CHANDRA GAJRAJ.

## INSURANCE

**Demerara Mutual Life Assurance Society Ltd:** Demerara Life Bldg, 63 Robb St and Ave of the Republic, POB 10409, Georgetown; tel. 231-3636; fax 225-8995; e-mail demlife@demeraramutual.com; internet www.demeraramutual.com; f. 1891; Chair. RICHARD B. FIELDS; CEO KEITH N. CHOLMONDELEY.

**Diamond Fire and General Insurance Inc:** High St, Kingston, Georgetown; tel. 223-9771; fax 223-9770; e-mail diamondins@solutions2000.net; f. 2000; privately-owned; Man. TARA CHANDRA; cap. $G 100m.

**Guyana Co-operative Insurance Service:** 47 Main St, Georgetown; tel. 225-9153; f. 1976; 67% owned by the Hand-in-Hand Group; Area Rep. SAMMY RAMPERSAUD.

**Guyana and Trinidad Mutual Life Insurance Co Ltd:** 27–29, Robb and Hinck St, Georgetown; tel. 225-7910; fax 225-9397; e-mail gtmgroup@gtm-gy.com; f. 1925; Chair. HAROLD B. DAVIS; Man. Dir R. E. CHEONG; affiliated company: Guyana and Trinidad Mutual Fire Insurance Co Ltd.

**Hand-in-Hand Mutual Fire and Life Group:** Hand-in-Hand Bldg, 1–4 Ave of the Republic, POB 10188, Georgetown; tel. 225-0462; fax 225-7519; f. 1865; fire and life insurance; Chair. J. A. CHIN; Gen. Man. K. A. EVELYN.

### Insurance Association

**Insurance Association of Guyana:** 54 Robb St, Bourda, POB 10741, Georgetown; tel. 226-3514; f. 1968.

# Trade and Industry

## GOVERNMENT AGENCIES

**Environmental Protection Agency, Guyana:** University of Guyana Campus, Turkeyen, Georgetown; tel. 022-5783; fax 022-2442; e-mail epa@epaguyana.org; internet www.epaguyana.org; f. 1988 as Guyana Agency for the Environment, renamed 1996; formulates, implements and monitors policies on the environment; Exec. Dir DOORGA PERSAUD.

**Guyana Energy Agency (GEA):** 295 Quamina St, POB 903, Georgetown; tel. 226-0394; fax 226-5227; e-mail ecgea@sdnp.org.gy; internet www.sdnp.org.gy/gea; f. 1998 as successor to Guyana National Energy Authority; CEO JOSEPH O'LALL.

**Guyana Office for Investment (Go-Invest):** 190 Camp and Church Sts, Georgetown; tel. 225-0653; fax 225-0655; e-mail goinvest@goinvest.gov.gy; internet www.goinvest.gov.gy; f. 1994; Chair. BHARRAT JAGDEO; CEO GEOFFREY DA SILVA.

**Guyana Public Communications Agency:** Georgetown; tel. 227-2025; f. 1989; Exec. Chair. KESTER ALVES.

**'New' Guyana Marketing Corporation:** 87 Robb and Alexander Sts, POB 10810, Georgetown; tel. 226-8255; fax 227-4114; e-mail newgmc@networksgy.com; internet www.agrinetguyana.org.gy/ngmc; Chair. GEOFFY DA SILVA; Gen. Man. NIZAM HASSAN.

## DEVELOPMENT ORGANIZATION

**Institute of Private Enterprise Development (IPED):** 254 South Rd, Bourda, Georgetown; tel. 225-8949; fax 226-4675; internet www.ipedgy.com; f. 1986 to help establish small businesses; total loans provided $ G870m. (2003); Chair. YESU PERSAUD; Exec. Dir Dr LESLIE CHIN.

## CHAMBER OF COMMERCE

**Georgetown Chamber of Commerce and Industry:** 156 Waterloo St, Cummingsburg, POB 10110, Georgetown; tel. 225-5846; fax 226-3519; internet www.georgetownchamberofcommerce.org; f. 1889; 122 mems; Pres. EDWARD BOYER.

## INDUSTRIAL AND TRADE ASSOCIATIONS

**Guyana Rice Development Board:** 117 Cowan St, Georgetown; tel. 225-8717; fax 225-6486; e-mail grdb@gol.net.gy; f. 1994 to assume operations of Guyana Rice Export Board and Guyana Rice Grading Centre; Gen. Man. JAGNARINE SINGH.

**National Dairy and Development Programme (NDDP):** REPAHA Compound, Agriculture Rd, Mon Repos, East Coast Demerara; tel. 220-6556; e-mail nichanw@yahoo.com; f. 1984; aims to increase domestic milk and beef production; Dir Dr NICHOLAS WALDRON.

## EMPLOYERS' ASSOCIATIONS

**Consultative Association of Guyanese Industry Ltd:** 157 Waterloo St, POB 10730, Georgetown; tel. 226-4603; fax 227-0725; e-mail cagi@guyana.net.gy; f. 1962; 193 mems, 3 mem. asscns, 159 assoc. mems; Chair. YESU PERSAUD; Exec. Dir DAVID YANKANA.

**Forest Products Association of Guyana:** 157 Waterloo St, Georgetown; tel. 226-9848; e-mail fpasect@sdnp.org.gy; internet www.fpaguyana.org; f. 1944; 47 mems; Pres. L. J. P. WILLEMS; Exec. Officer WARREN PHOENIX.

**Guyana Manufacturers' Association Ltd:** 157 Waterloo St, North Cummingsburg, Georgetown; tel. 227-4295; fax 227-5615; e-mail gma_guyana@yahoo.com; internet www.gy-gma.org; f. 1967; 190 mems; Pres. NORMAN MCLEAN.

**Guyana Rice Producers' Association (GRPA):** 126 Parade & Barrack St, Georgetown; tel. 226-4411; fax 223-7249; e-mail grpa.riceproducers@networksgy.com; f. 1946; non-govt org.; 18,500 mems; Gen. Sec. DHARAMKUMAR SEERAJ.

## MAJOR COMPANIES

The following are some of the major companies operating in Guyana:

## Food and Beverages

**Banks DIH Ltd:** The Rotunda, Thirst Park, Ruimveldt, Georgetown; tel. 226-2491; fax 226-6523; internet www.banksdih.com; f. 1848; brewers and soft drinks and snacks manufacturers; sales of $ G6,706m. (1995/96); Chair. and Man. Dir CLIFFORD B. REIS; Vice-Chair. KATHLEEN D'AGUIAR; 1,500 employees.

**Chin's Manufacturing Industries Ltd:** Area K, Le Ressouvenir, ECD; tel. 220-2818; fax 220-3592; e-mail chinsagency@yahoo.com; internet www.geocities.com/chinsmfg; Man. Dir COMPTON CHIN.

**Demerara Distillers Ltd:** Diamond, East Bank, Demerara; tel. 265-6000; fax 265-3367; internet www.demrum.com; f. 1952; producer of alcoholic and non-alcoholic beverages; sales of G $10.6m. (2003); Exec. Chair. YESU PERSAUD; 1,165 employees.

**Edward Beharry & Co Ltd:** 191 Charlotte St, Lacytown, Georgetown; tel. 227-0632; fax 225-6062; internet www.beharrygroup.com/manufacturing.html; producer of confectionery, condiments and pasta.

**Guyana Sugar Corpn Inc (Guysuco):** Ogle Estate, POB 10547, ECD; tel. 222-6030; fax 222-6048; e-mail info@GuySuCo.com; internet www.guysuco.com; f. 1976; from 1990 managed by Booker Tate (United Kingdom); sugar production; scheduled for privatization; Chair. RONALD ALLI; Dir KEITH WARD.

**National Milling Co of Guyana Inc:** Agricola, East Bank, Demerara; tel. 225-2990; e-mail info@namilcoflour.com; internet www.namilcoflour.com; subsidiary of Seaboard Corporation (USA); flour millers; Man. Dir DONALD FRANKE.

## Forestry and Timber

**A. Mazaharally and Sons Ltd:** 22 Wight's Lane, Kingston, Georgetown; tel. 226-0442; fax 226-4151; logs supplier.

**Barama Co. Ltd:** Land of Canaan, East Bank, Demerara; tel. 225-4555; e-mail barama@samling.com.my; internet www.samling.com.my; subsidiary of Samling (Malaysia); Man. Dir JIM KEYLON; 1,400 employees.

**Demerara Timbers Ltd (DTL):** 1 Water St and Battery Rd, Kingston, Georgetown; tel. 225-3835; fax 227-1663; owned by Prime Group Holdings Ltd (British Virgin Islands); CEO LU KUI SAN.

**OREU Timber and Trading Co Ltd:** 695 Pennylane St, South Riumveldt Gardens, Georgetown, Guyana; tel. 231-7780; e-mail oreutimbers@solutions2000.net; tel. 227-3103; f. 1982; logging, timber trading, forest industry consultants; Chair. OREU C. BENJAMIN.

**Toolsie Persaud Ltd:** 1–4 Lombard St, Georgetown; tel. 226-4071; fax 226-2554; e-mail tpl@tpl-gy.com; internet www.tpl-gy.com; f. 1949; logging and quarrying co and manufacturer of construction materials; Pres. TOOLSIE PERSAUD; Man. Dir DAVID PERSAUD; 650 employees.

**Willems Timber and Trading Co Ltd:** 7 Water St, Werk-en-Rust, POB 10443, Georgetown; tel. 226-9252; fax 226-0983; production of timber and lumber; Dir JOHN WILLEMS.

## Mining

**Bauxite Company of Guyana Inc (BCGI):** Aroaima; f. 2004 as jt venture between Govt and Russian Aluminium Company (RUSAL); 90% owned by RUSAL subsidiary, the Bauxite and Alumina Mining Venture (BAMV).

**Aroaima Bauxite Company (ABC):** Aroaima; f. 1989 as jt venture between the Govt and Reynolds International (USA); began mining of bauxite in 1999; merged with Bermine in 2002; managed by BCGI from 2005 until its scheduled privatization in 2006.

**Guyana Oil Company (GUYOIL):** 166 Waterloo St, Georgetown; tel. 225-7161; fax 225-2320; e-mail guyoilmd@networksgy.com; internet guyoil.com; state-owned; petroleum exploration and production; Marketing Man. ALWYN APPIAH.

**Omai Bauxite Mining Inc:** POB 32217, Linden; tel. (444) 6415; fax (444) 6103; internet www.cambior.com; f. 1992; fmrly state-owned Linden Mining Enterprises Ltd, name changed as above upon full privatization in 2004; mining of bauxite; Chair. C. P. PLUMMER; CEO H. JAMES; 300 employees.

## Miscellaneous

**A. H. & L. Kissoon Group of Companies:** 80 Camp and Robb Sts, Georgetown; tel. 226-0967; fax 227-5265; e-mail AHLGROUP@kissoon-furniture.com; internet www.kissoon-furniture.com; construction of wooden furniture and housing, rice cultivation, cattle rearing.

**Alesie Guyana (Group Management):** 79 Cowan St, Kingston, Georgetown; tel. 226-4601; fax 226-2038; e-mail InfoGuyana@alesierice.com; exporters of rice.

**Brass Aluminium and Cast Iron Foundry Ltd:** 11–14 West Ruimveldt, Greater Georgetown; tel. 225-7531; fax 225-4341; producer of ferrous and non-ferrous metals.

**Colgate-Palmolive (Guyana) Ltd:** Ruimveldt, East Bank, Georgetown; tel. 226-2663; fax 225-6792; subsidiary of Colgate-Palmolive Co (USA); manufacturer of toothpaste and domestic cleaning products.

**Continental Group of Companies:** 9–12 Industrial Site, Georgetown; tel. 226-4041.

**Courts Guyana Ltd:** 25–26 Main St, POB 10481, Georgetown; tel. 225-5886; fax 227-8751; e-mail d.burgess@courtsguyana.com; internet www.courtsguyana.com; Man. MOHAMED ABU ZAMAN.

**Deloitte & Touche:** 77 Brickdam, Stabroek, POB 10506, Georgetown; tel. 226-3226; fax 225-7578; accountancy and management consultancy; Sr Partner RAMESTIWAR LAL.

**Demerara Tobacco Co Ltd:** Eping Ave, Bel Air Park, POB 10262, Greater Georgetown; tel. 226-5190; fax 226-9322; f. 1975; importer of cigarettes; Chair. CHARLES R. QUINTIN; 200 employees.

**Denmor Garments (Manufacturers) Inc:** 101 Regent St, Georgetown; tel. 225-7630; fax 226-1939; clothing manufacturer; Dir DENNIS MORGAN MUDLIER; 1,000 employees.

**Gafsons Group of Companies:** Lot 1–2, Area X, Plantation Houston, Georgetown; tel. 226-3666; fax 227-8763; e-mail nil@guyana.net.gy; Chair. ABDOOL SATTAUR GAFOOR.

**Guyana National Engineering Corp Ltd:** 1–9 Lombard St, Charlestown, POB 10520, Georgetown, Demerara-Mahaica; tel. 226-3291; fax 225-8525; state-owned; ship repairs, ship building, aluminium castings; CEO CLAUDE SAUL; Gen. Man. M. F. BASCOM; 1,150 employees.

**KPMG:** 8 Church St, Georgetown; tel. 227-8825; fax 227-8824; internet www.kpmg.com; f. 1993; accountancy and management consultancy; Sr Partner NIZAM ALI; 12 employees.

**Laparkan Holdings Ltd:** 34 Water St; tel. 225-6870; e-mail lpkadmin@inetguyana.net; Chair. GLEN KHAN.

**New GPC Inc:** Al Farm, East Bank, Georgetown, Demerara; tel. 265-4261; fax 265-2229; e-mail limacol@newgpc.com; internet www.newgpc.com; fmrly Guyana Pharmaceutical Corpn Ltd, privatized Dec. 1999; manufacturer of pharmaceuticals and cosmetics; Exec. Chair. Dr R. RAMROOP; Company Sec. D. A. PIERRE.

**Guyana Refrigerators Ltd:** 15A Water and Holmes Sts, POB 10392, Georgetown; tel. 225-4934; fax 227-0302; manufacturer of refrigerators, freezers and ice buckets.

**Guyana Stores Ltd (GSL):** 19 Water St, Georgetown, Demerara-Mahaica; tel. 226-6171; e-mail guyanastores@telsnetgy.com; f. 1976; offered for privatization in 1999; retailers (supermarket, department store, pharmacies), wholesalers of hardware, motor vehicle sales concessions; Chair. PAUL CHAN-A-SUE; 1,100 employees.

**Ram & MacRae:** 157C Waterloo St, POB 10148, Georgetown; tel. 226-1072; fax 226-4221; f. 1985; accountants; Man. Partner CHRISTOPHER L. RAM.

**G & C Sanata Company Inc:** Industrial Site, Ruimveldt, Georgetown; tel. 231-7273-6; fax 227-8197; e-mail Sanata@networksgy.com; fabrics manufacturer; Man. Dir CHEN RONG.

**Torginol (Guyana) Ltd:** Industrial Site, Ruimveldt, East Bank, Georgetown, Demerara; tel. 226-4041; fax 225-3568; internet www.continentalgy.com/continentalpaints.htm; manufacturer of paints; owned by The Continental Group of Companies.

# UTILITIES

## Electricity

**Guyana Power and Light Inc (GPL):** 40 Main St, POB 10390, Georgetown; tel. 225-4618; fax 227-1978; e-mail enquiries@gplinc.com; internet www.gplinc.com; f. 1999; fmrly Guyana Electricity Corpn; state-owned; Chair. RONALD ALLI; CEO ROBIN SINGH.

## Water

**Guyana Water Inc (GWI):** 10 Fort St, Georgetown; tel. 225-0471; fax 225-0478; e-mail gwi@networks.gy.com; f. 2002, following merger of Guyana Water Authority (GUYWA) and Georgetown Sewerage and Water Comm.; operated by Severn Trent Water International (United Kingdom); Man. Dir MICHAEL KENT.

# CO-OPERATIVE SOCIETIES

**Chief Co-operatives Development Officer:** Ministry of Labour, Human Services and Social Security, 1 Water and Cornhill Sts, Stabroek, Georgetown; tel. 225-8644; fax 227-1308; e-mail coopdept@telsnet.gy.net; f. 1948; Dir CLIVE NURSE.

# GUYANA

In October 1996 there were 1,324 registered co-operative societies, mainly savings clubs and agricultural credit societies, with a total membership of 95,950.

## TRADE UNIONS

**Trades Union Congress (TUC):** Critchlow Labour College, Woolford Ave, Non-pareil Park, Georgetown; tel. 226-1493; fax 227-0254; e-mail gtuc@guyana.net.gy; f. 1940; national trade union body; 22 affiliated unions; 70,000 mems; merged with the Federation of Independent Trade Unions in Guyana in 1993; Pres. NORRIS WITTER; Gen. Sec. LINCOLN LEWIS.

**Amalgamated Transport and General Workers' Union:** 46 Urquhart St, Georgetown; tel. 226-6243; fax 225-6602; Pres. RICHARD SAMUELS; Gen. Sec. VICTOR JOHNSON.

**Association of Masters and Mistresses:** c/o Critchlow Labour College, Georgetown; tel. 226-8968; Pres. GANESH SINGH; Gen. Sec. T. ANSON SANCHO.

**Clerical and Commercial Workers' Union (CCWU):** Clerico House, 140 Quamina St, South Cummingsburg, POB 101045, Georgetown; tel. 225-2827; fax 227-2618; e-mail ccwu@guyana.net.gy; Pres. ROY HUGHES; Gen. Sec. GRANTLEY L. CULBARD.

**Guyana Bauxite and General Workers' Union:** 180 Charlotte St, Georgetown; tel. 225-4654; Pres. CHARLES SAMPSON; Gen. Sec. LEROY ALLEN.

**Guyana Labour Union:** 198 Camp St, Cummingsburg, Georgetown; tel. 227-1196; fax 225-0820; e-mail glu@solutions2000.net; Pres. SAMUEL WALKER; Gen. Sec. CARVILLE DUNCAN; 6,000 mems.

**Guyana Local Government Officers' Union:** Woolford Ave, Georgetown; tel. 227-2131; fax 227-6905; e-mail daleantford@yahoo.com; Pres. ANDREW GARNETT; Gen. Sec. DALE BERESFORD.

**Guyana Mining, Metal and General Workers' Union:** 56 Wismar St, Linden, Demerara River; tel. 204-6822; Pres. VERNON SEMPLE; Gen. Sec. LESLIE GONSALVES; 5,800 mems.

**Guyana Postal and Telecommunication Workers' Union:** 310 East St, Georgetown; tel. 226-5255; fax 225-1633; e-mail dawnc_edwards@hotmail.com; Pres. BENJAMIN NEDD; Gen. Sec. DAWN EDWARDS.

**Guyana Teachers' Union:** Woolford Ave, POB 738, Georgetown; tel. 226-3183; fax 227-0403; Pres. LANCELOT BAPTISTE; Gen. Sec. SHIRLEY HOOPER.

**National Mining and General Workers Union:** 10 Church St, New Amsterdam, Berbice; tel. 203-3496; Pres. CYRIL CONWAY; Gen. Sec. MARILYN GRIFFITH.

**National Union of Public Service Employees:** 4 Fort St, Kingston, Georgetown; tel. 227-1491; Pres. ROBERT JOHNSON; Gen. Sec. PATRICK QUINTYNE.

**Printing Industry and Allied Workers' Union:** c/o Guyana TUC, Georgetown; tel. 226-8968; Gen. Sec. LESLIE REECE.

**Public Employees' Union:** Regent St, Georgetown; Pres. REUBEN KHAN.

**Union of Agricultural and Allied Workers (UAAW):** 10 Hadfield St, Werk-en-Rust, Georgetown; tel. 226-7434; Pres. JEAN SMITH; Gen. Sec. SEELO BAICHAN.

**University of Guyana Workers' Union:** POB 841, Turkeyen, Georgetown; tel. 222-3586; e-mail adeolaplus@yahoo.com; supports Working People's Alliance; Pres. Dr ADELOA JAMES; Sec. AUDREY IRVING.

**Guyana Agricultural and General Workers' Union (GAWU):** 59 High St and Wight's Lane, Kingston, Georgetown; tel. 227-2091; fax 227-2093; e-mail gawu@networksgy.com; Pres. KOMAL CHAND; Gen. Sec. SEEPAUL NARINE; 20,000 mems.

**Guyana Public Service Union (GPSU):** 160 Regent and New Garden Sts, Georgetown; tel. 225-0518; fax 226-5322; e-mail gpsu@guyana.net.gy; internet www.guyanapsu.org; Pres. PATRICK YARDE; Gen. Sec. SURENDRA PERSAUD (acting); 11,600 mems.

**National Association of Agricultural, Commercial and Industrial Employees (NAACIE):** 64 High St, Kingston, Georgetown; tel. 227-2301; f. 1946; Pres. KENNETH JOSEPH; Gen. Sec. KAISREE TAKECHANDRA; c. 2,000 mems.

# Transport

## RAILWAY

There are no public railways in Guyana.

**Linmine Railway:** Mackenzie, Linden; tel. 444-2279; fax 444-6699; e-mail orin_barnwell@cambior.com; bauxite transport; 15 km of track, Coomaka to Linden; Dept Head ORIN BARNWELL.

## ROADS

The coastal strip has a well-developed road system. In 1999 there were an estimated 7,970 km (4,952 miles) of paved and good-weather roads and trails. In September 2001 a European Union-funded road improvement programme between Crabwood Creek and the Guyana–Suriname Ferry Terminal was completed; the project was intended to help integrate the region. In the same year construction began on a bridge across the Takutu river, linking Guyana to Brazil; construction was suspended in 2002, owing to difficulties with planning regulations and financing, but was expected to resume in 2004. The US $40m. rehabilitation of the Mahaica–Rosignol road, partly funded by the Inter-American Development Bank, began in 2003. In the same year work began on a complementary project to remove and relocate both the Mahaica and Mahaicony bridges. Construction of a bridge over the Berbice river was to begin in early 2004.

## SHIPPING

Guyana's principal ports are at Georgetown and New Amsterdam. The port at Linden serves for the transportation of bauxite products. A ferry service is operated between Guyana and Suriname. Communications with the interior are chiefly by river, although access is hindered by rapids and falls. There are 1,077 km (607 miles) of navigable rivers. The main rivers are the Mazaruni, the Potaro, the Essequibo, the Demerara and the Berbice. In 2000 the Brazilian Government announced that it was to finance the construction of both a deep-water port and a river bridge.

**Transport and Harbours Department:** Battery Rd, Kingston, Georgetown; tel. 225-9350; fax 227-8545; e-mail t&hd@solutions2000.net; Gen. Man. WILLIAM JOSEPH.

**Shipping Association of Guyana Inc:** 5–9 Lombard St, Georgetown; tel. 226-1448; fax 225-0849; e-mail gnsc@futurenetgynet.com; f. 1952; non-governmental forum; Pres. CLINTON WILLIAMS; Vice-Pres. TARACHANDRA KHELAWAN; members:

**Guyana National Industrial Company Inc (GNIC):** 2–9 Lombard St, Charlestown, POB 10520, Georgetown; tel. 225-8428; fax 225-8526; e-mail gnicadmin@futurenetgy.com; internet gnicgy.tripod.com/home.htm; metal foundry, ship building and repair, agents for a number of international transport cos; privatized 1995; CEO CLINTON WILLIAMS; Gen. Man. MICHAEL FORDE.

**Guyana National Shipping Corporation Ltd:** 5–9 Lombard St, La Penitence, POB 10988, Georgetown; tel. 225-0849; fax 225-3815; e-mail gnsc@guyana.net.gy; internet www.gnsc.com; govt-owned; Man. Dir M. F. BASCOM.

**John Fernandes Ltd:** 24 Water St, POB 10211, Georgetown; tel. 227-3344; fax 226-1881; e-mail chris@jf-ltd.com; internet www.jf-ltd.com; ship agents, pier operators and stevedore contractors; Chair. and CEO CHRIS FERNANDES.

## CIVIL AVIATION

The main airport, Cheddi Jaggan International Airport, is at Timehri, 42 km (26 miles) from Georgetown. In 1998 there were some 94 airstrips.

**Roraima Airways:** 101 Cummings St, Bourda, Georgetown; tel. 225-9648; fax 225-9646; e-mail ral@roraimaairways.com; internet www.roraimaairways.com; f. 1992; flights to Venezuela and 4 domestic destinations; Owner GERALD GOUVEIA.

**Trans Guyana Airways:** Ogle Aerodrome, Ogle, East Coast Demerara; tel. 222-2525; fax ; e-mail commercial@transguyana.com; internet www.transguyana.com; f. 1956; internal flights to 22 destinations.

**Universal Airlines:** 65 Main St, Georgetown; tel. 226-9262; fax 226-9264; e-mail RES@Universal-Airlines.com; internet www.universal-airlines.com; f. 2001; Owners CHANDRAMATTIE HARPAUL, RAMASHREE SINGH.

# Tourism

Despite the beautiful scenery in the interior of the country, Guyana has limited tourist facilities, and began encouraging tourism only in the late 1980s. During the 1990s Guyana began to develop its considerable potential as an eco-tourism destination. However, tourist arrivals declined towards the end of the decade. The total number of visitors to Guyana in 2003 was 100,911 and expenditure by tourists amounted to US $39m.

**Tourism and Hospitality Association of Guyana:** 157 Waterloo Street, Georgetown; tel. 225-0807; fax 225-0817; e-mail thag@networksgy.com; internet www.exploreguyana.com; f. 1992; Pres. Capt. GERALD GOUVEIA; Exec. Dir MAUREEN PAUL.

## Defence

The armed forces are united in a single service, the Combined Guyana Defence Force, which consisted of some 1,600 men (of whom 1,400 were in the army, 100 in the air force and about 100 in the navy) at 1 August 2004. The Guyana People's Militia, a paramilitary reserve force, totalled about 1,500. The President is Commander-in-Chief.

**Defence Budget:** An estimated $ G1,000m. (US $5.8m.) in 2004.

**Chief-of-Staff:** Brig.-Gen. EDWARD COLLINS.

## Education

Education is free and compulsory for children aged between six years and 15 years of age. Children receive primary education for a period of six years, followed by secondary education, beginning at 12 years of age. Secondary education in a general secondary school lasts for up to seven years, comprising an initial cycle of five years, followed by a cycle of two years. Alternatively, children may remain at primary school or a Community High School for an additional four-year period. In 1999/2000 there were 320 nursery schools/classes and 423 primary schools/classes. Net enrolment at primary schools in 1999 was equivalent to 98.4% of children in the relevant age group (males 99.7%; females 97.1%). Gross enrolment at secondary schools in that year was equivalent to 75.2% of children in the relevant age group (males 71.6%; females 78.9%).

Higher education is provided by five technical and vocational schools, one teacher-training college, and one school for home economics and domestic crafts. Training in agriculture is provided by the Guyana School of Agriculture, at Mon Repos. The Burrowes School of Art offers education in fine art. The University of Guyana at Turkeyen has faculties of natural sciences, social sciences, arts, medicine, law, agriculture, technology and education. The University's Institute of Distance and Continuing Education offers training in a broad range of subjects ranging from home management to psychology and industrial relations. In 1999/2000 13,762 students were enrolled in higher education.

Expenditure on education by the central Government in 2004 was estimated at $ G14,500m., and represented 14.5% of total government expenditure in that year, compared to an allocation of 4.4% of the national budget in 1990. In 2002 the Inter-American Development Bank approved a US $30m. loan to assist with the modernization of basic education in Guyana. The Government proposed to build an additional 25 schools in 2004.

## Bibliography

Abrams, O. *Metegee: The History and Culture of Guyana*. Eldorado Publications, 1998.

Bartilow, H. A. *The Debt Dilemma: IMF Negotiations in Jamaica, Grenada and Guyana*. Warwick, Warwick University Caribbean Studies, 1997.

Colchester, M. *Guyana: Fragile Frontier*. Kingston, Ian Randle Publishers, 1997.

Cruickshank, J. G. *Scenes from the History of the Africans in Guyana*. Georgetown, Guyana Free Press, 1999.

Egoume-Bossogo, P., Faal, E., Nallari, R., Weisman, E. *Guyana: Experience With Macroeconomic Stabilization, Structural Adjustment, and Poverty Reduction*. Ottowa, ON, Renouf Publishing Co Ltd, 2003.

Gafar, J. *Guyana: From State Control to Free Markets*. Hauppauge, NY, Nova Science Publrs Inc, 2003.

Graham Burnett, D. *Masters of All They Surveyed: Exploration, Geography and a British El Dorado*. Chicago, IL, University of Chicago Press, 2001.

Granger, D. G. (Ed.). *Emancipation*. Georgetown, Guyana Free Press, 1999.

*Guyana's Military Veterans: Promises, Problems and Prospects*. Georgetown, Guyana Free Press, 1999.

*Guyana General and Regional Elections, 19 March 2001: the Report of the Commonwealth Observer Group*. London, Commonwealth Secretariat Group, 2001.

Hoyte, H. D. *Guyana's Economic Recovery: Leadership, Will-power and Vision. Selected Speeches of Hugh Desmond Hoyte*. Georgetown, Guyana Free Press, 1997.

Irving, B. *Guyana: a Composite Monograph*. Hato Rey, Puerto Rico, Inter American University Press, 1972.

Joseph, C. L. *Anglo-American Diplomacy and the Re-Opening of the Guyana—Venezuela Boundary Controversy, 1961–66*. Georgetown, Guyana Free Press, 1998.

Mars, P., Young, A. L. *Caribbean Labor and Politics: Legacies of Cheddi Jagan and Michael Manley*. Wayne State University Press, Detroit, MI, 2004.

McGowan, W. F., et al (Eds). *Themes in African—Guyanese History*. Georgetown, Guyana Free Press, 1998.

*The Demerara Revolt, 1823*. Georgetown, Guyana, Free Press, 1998.

Mars, J. R. *Deadly Force, Colonialism and the Rule of Law*. Westport, CT, Greenwood Press, 2002.

Mohamed, I. A. *Guyana's Approach: From Singapore to Seattle: World Trade Negotiations: Pushing for a Development Round of Negotiations through Process of Review, Repair and Reform of the World Trade Organization*. Georgetown, Ministry of Foreign Affairs, 2000.

Mitchell, W. B., Bibbiana, W. A., DuPre, C. E., et al. *Area Handbook for Guyana*. Washington, DC, US Government Printing Office, 1969.

Morrison, A. *Justice: The Struggle for Democracy in Guyana 1952–1992*. Georgetown, Red Thread Women's Press, 1998.

Munslow, B. *Guyana: Microcosm of Sustainable Development Challenges*. Aldershot, Hampshire, Ashgate Publishing Ltd, 1998.

Peake, L., and Peake, A. *Gender, Ethnicity and Poverty in Guyana*. London, Routledge, 1999.

Premdas, R. R. *Ethnic Conflict and Development: The Case of Guyana (Research in Ethnic Relations)*. Brookfield, VT, Avebury, 1995.

Rabe, S. G. *U.S. Intervention in British Guiana: A Cold War Story (The New Cold War History)*. University of North Carolina Press, Chapel Hill, NC, 2005.

Ramcharan, B. G. *The Guyana Court of Appeal: the Challenges of the Rule in a Developing Country*. London, Cavendish Publishing Ltd, 2002.

Seecomar, J. *Contributions Towards the Resolution of Conflict in Guyana*. Leeds, Peepal Tree Press, 2002.

Singh, J. N. *Guyana: Democracy Betrayed*. Kingston, Jamaica, Kingston Publrs, 1996.

# HAITI

## Geography

### PHYSICAL FEATURES

The Republic of Haiti lies at the western end of the island of Hispaniola, which it shares with the larger Dominican Republic (beyond a 360-km or 224-mile eastern border, partly along the Pedernales river in the south and the Massacre in the north). The land border was first set by treaty between France and Spain at Ryswick (Rijswijk, Netherlands) in 1697, and most recently revised in 1936. Cuba lies 80 km beyond the Windward Passage in the north-west and Jamaica twice that distance to the west of the south-western peninsula. The country includes a number of offshore islands, but also claims Navassa Island. This is a scrubby, uninhabited, 5-sq km (2-sq miles), coral and limestone rock, some 65 km west of Haiti and 160 km south of the USA's Guantánamo Bay naval base in southern Cuba. Navassa is currently administered as an unincorporated territory and wildlife reserve by the USA. To the north, Haiti is less than 100 km from the southernmost Bahamas and about 140 km from the Turks and Caicos Islands, a British dependency. Haiti includes only 190 sq km of inland waters, but has 1,771 km of coastline. At 27,750 sq km (10,714 sq miles), the country is the same size as the US state of Massachusetts and somewhat smaller than Belgium.

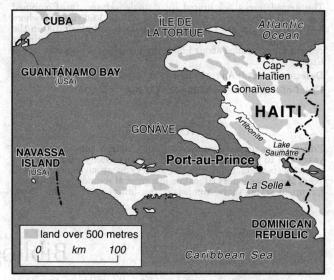

Mainland Haiti has a long, westward-extending peninsula in the south of the country, while the centre and the north broaden north-westwards to form the other arm of the Gulf of Gonâve. At the head of the Gulf, in the south-east, lies the capital, Port-au-Prince. The Gulf embraces the barren island of Gonâve, at some 60 km in length and 15 km in width, the largest of Haiti's offshore territories. The next largest is the Ile de la Tortue, or Tortuga, 40 km in length and 12 km off Port-de-Paix on the north coast (it was the main Caribbean pirate centre in the 17th century, then the base for the French acquisition of western Hispaniola). Stretching east of here run the northern coastal plains (about 150 km in length, 30 km wide and covering some 2,000 sq km), the main area of flat land in the country, apart from the drier Cul-de-Sac in the south, which extends eastwards from Port-au-Prince to brackish Lake Saumâtre (20 km long and 6 km–14 km wide) in the south-east. The fertile, well-watered northern plains had attracted the original colonial plantations and, earlier, piratical outcasts and renegades, who hunted the wild cattle roaming the plains, cooking over wood fires known as *boucans*—hence the word buccaneer. The other main areas of flat land in Haiti are in the valley of the Artibonite and on the south coast, the densely populated Léogâne plains. However, it is the highlands that are more typical of the country. Most of Haiti is mountainous and, in fact, it is considered the most mountainous country in the Caribbean. The Massif du Nord lies behind the coastal plains of the north, running north-westwards into the dry and drought-afflicted peninsula and, in the other direction, over the border of the Dominican Republic as the Cordillera Central. The highest point in the country, however, is Morne de la Selle (2,680 m or 8,796 ft), which rises from the south-eastern massif—the mountains then continue westwards, along the peninsula, where, towards the end, is the Massif de la Hotte. The centre of Haiti is dominated by the elevated eastern central plateau and a number of other ranges, extensions of the Massif du Nord or the Dominican Sierra de Neibe. Cleaving through the central highlands is the valley of the Artibonite (some 800 sq km of plains land), the main river of Haiti and the trunk of the largest drainage system. The river, 400 km in length, enters the country from the Dominican Republic (it also forms part of the border), but it arose in Haiti as the Libón. Although deep and strong, and prone to flooding in the wet seasons, at drier times of the year the river can shrink dramatically and even cease to flow in some places. The most important tributary of the Artibonite is the 95-km Guayamouc. The main river of the north, at 150 km in length, is the oddly named Trois Rivières. Flooding of the often swift-flowing rivers in steep terrain is made worse by the extensive soil erosion caused by deforestation. The widespread poverty in the country has contributed to the wiping out of Haiti's once-extensive tree cover. There are only really two places where woodland survives, both high and inaccessible (one on the Massif de la Selle, the other around the Pic de Macaya, at the end of the south-western peninsula). These places shelter some remnant riches of Haiti's devastated environment, which has been denuded of forest and rendered prone to soil erosion and drought.

### CLIMATE

The climate is subtropical, but parts of Haiti are semi-arid, owing to the eastern mountains shadowing the rain brought by the Atlantic trade winds. The northern plains, the mountains of the north-east and the highlands of the south-western peninsula receive the most rain. The driest and most barren areas are the coastal strip east of Port-au-Prince and the north-western peninsula. The recurring problem of drought does not preclude severe storms during the June–October hurricane season, which can bring flooding, and the country also suffers from occasional earthquakes. There are two rainy seasons, April–May and September–October, with average annual rainfall, unevenly distributed, ranging between 1,400 mm (55 ins) and 2,000 mm (79 ins). It is humid on the coast, but the heat is made bearable by morning and evening breezes. Temperatures on the coast vary between about 20°C (68°F) and 35°C (95°F), being slightly hotter in April–September and cooler in the hills at any time of year.

### POPULATION

The people of Haiti are black (95%), the inheritors of the world's only successful slave revolt, which drove out or killed most of its white, planter aristocracy in the early 19th century (the French colony was known as Saint-Domingue). However, there is still a white and mulatto minority, most of whom remain disproportionately wealthy and powerful. Although race and colour have remained important prejudices bedevilling Haitian society, it is language that has had the widest effect. The élite continued to use French after the revolution, and to disparage the lower classes' constant use of Creole (Créole), perpetuating the old colonial tongue as the speech of formal use and as the official language into the present. However, only about 15% of the population speak French and, although they usually also speak Creole, only about 5% can speak both fluently, crippling a society that has long permitted education and advancement only on the basis of French. Only since the 1990s has Creole begun to make official advances, the culmination of a movement to dignify the native culture that only began in the 1920s. Creole, now also an official language, is a tongue that shares many words with French, but has a very different (West African) grammatical structure, its origins either in a pidgin used between slaves of various African nationalities and their French masters or in a French maritime-trade dialect. Another influence imported from West Africa into Haitian culture is in the field of religion, with voodoo (vaudou), as it is labelled, practised by over one-half of the population. This does not preclude continued adherence to the official religion, Roman Catholic Christianity, as voodoo adapted the African veneration of family and ancestral spirits with local conditions under the French. Haitians distinguish between the benign practice of specialists and witchcraft, but, as the belief system is essentially family based, there is a limit to sorcery, as there is to organization or political exploitation, as under the Duvaliers. The Roman Catholic Church is the main official religious affiliation, and still claims the loyalty of 65% of the population. However, its identity with the ruling establishment helped the growth of Protestant churches, which are now adhered to by 16% (mainly Baptists—10%, and Pentecostalists—4%). The Protestants, in contrast to the uneasy, *de facto* accommodation by the Roman Catholic Church, tend to be fierce in their rejection of voodoo.

The population of Haiti totalled approximately 7.9m., according to results of the census conducted in July 2003. At mid-2004 the population was estimated at approximately 8.4m. There is a high mortality rate from AIDS. Moreover, about four-fifths of the population live in considerable poverty and Haiti is described as the poorest country in the Western hemisphere (and one of the poorest in the world) under many definitions. Port-au-Prince, the capital, had an estimated population of some 2.0m. in mid-2003 (including its suburbs of Carrefour and Delmas). The third city, Gonaïves (known as the 'city of independence'), is also on the Gulf, but in the north-eastern corner. The second city of Haiti, and the largest port, is Cap-Haïtien (as Cap-Français, it was the colonial capital—now usually referred to as Cap or, in Creole, Okap), on the central northern coast. The fourth city is Les Cayes (Okay), on the south coast, towards the end of the south-western peninsula. Haiti is administered as nine departments.

# History

## GREG CHAMBERLAIN

The western half of the island of Hispaniola was first colonized by the French in the late 17th century and its borders with the Spanish part of the island were agreed in the Treaty of Ryswick in 1697. Haiti became the world's first independent black republic on New Year's Day 1804 after a bloody 12-year rebellion by African-descended slaves led by Toussaint Louverture, which was one of the most dramatic by-products of the French Revolution. Haiti's first ruler, Jean-Jacques Dessalines, and his successor, Alexandre Pétion (1807–18), divided the estates of the departed French among the freed peasantry, laying the economic, social and psychological foundations of modern Haiti.

Despite this early establishment of national identity, the attainment of meaningful independence proved to be a struggle. Economic development was hampered until the early 20th century by repayment of a massive indemnity of 90m. francs demanded by France in exchange for recognition, which did not come from the USA until 1862. Greed and the legacy of violence and colour division between blacks and the ruling mulattos produced increasingly unstable and chaotic government. In the late 19th century Haiti's economic dependence led to growing involvement of US and European interests in the country's affairs. In 1915 the USA, fearing European (especially German) economic and political rivalry in its Caribbean sphere of influence, invaded Haiti, officially to restore order. There followed a 19-year occupation of what a senior US official had once called 'a public nuisance at our doors'.

US military rule brought racial segregation and forced labour. This provoked sharp resistance and between 1918 and 1920 some 3,000 rebels and their leader, Charlemagne Péralte, were killed. The occupation entrenched the mulatto minority in power and aroused strong nationalist and socialist sentiments among black intellectuals and liberal mulattos, which erupted into open discontent in 1946 when President Elie Lescot was overthrown. The country's first trade unions and left-wing parties were founded and the election of the liberal Dumarsais Estimé gave the emerging black bourgeoisie some political power. Estimé was overthrown in 1950 by the army commander, Gen. Paul Magloire, whose rule saw relative political peace, infrastructure development and social and economic advances.

### 'PAPA DOC' AND 'BABY DOC' DUVALIER

In 1957 Dr François ('Papa Doc') Duvalier, a black nationalist intellectual, was elected President. He resolved to end the mulattos' political power even if he could not break their dominance of the economy. He also sought to limit the influence of the army, the Roman Catholic Church and the USA, which had traditionally been strong in Haitian society. More blacks entered the Government and a lumpen black militia, the National Security Volunteers (popularly known as the Tontons Macoutes, a Creole expression for 'bogeymen'), was created to counterbalance the Army, which was purged and tamed. Duvalier appointed the first Haitian head of the local Roman Catholic Church, until then controlled by French Breton clergy, many of whom Duvalier deported.

Partly in response to opposition, Duvalier's rule soon degenerated into violence and killings, particularly of mulattos. By 1965 tens of thousands of Haitians had been murdered, tortured or had fled to North America, Africa and Europe. Duvalier expelled the US ambassador in 1963 and the USA suspended aid to Haiti. However, US governments continued to regard the regime favourably for not aligning the country with neighbouring Cuba. Duvalier declared himself 'President-for-Life' in 1964. A dozen attempts at invasion or internal rebellion were crushed and when Duvalier died in April 1971, he bequeathed power to his 19-year-old son, Jean-Claude ('Baby Doc') Duvalier, also as President-for-Life. The young man's inexperience and initial lack of interest in his legacy was a problem, but rival factions kept him in office as a pliable arbiter of their quarrels. The USA resumed economic and military aid.

Repression eased, however, and dissidence appeared from 1977 onwards, partly encouraged by US President Jimmy Carter's human rights policies. The press criticized government failures, symbolized by famine and the flight of thousands of 'boat people' to the USA, although criticism of the Duvaliers and their estimated US $500m. private fortune remained taboo. The regime tolerated the foundation of two small political parties and a trade union federation, held parliamentary elections (without opposition candidates) and freed some political prisoners.

In November 1980 the Government arrested and deported all its main press critics and political opponents. Leadership of the opposition was assumed by progressive Roman Catholic priests and the Church's grass-roots organizations (Ti Legliz). The Church-run Radio Soleil became the country's most popular station, using the national language, Creole, rather than the official French favoured by the élite. The President's marriage in 1980 to a mulatto, Michèle Bennett, angered many Duvalierists, who saw it as a sign that the old élite was re-establishing itself.

### THE END OF THE DUVALIER ERA

The first anti-Government demonstrations in 20 years broke out in 1984, when food warehouses were looted. After the USA threatened to reduce aid, the Government promised to lift restrictions on political activity. However, the killing of four schoolchildren by police in Gonaïves in November 1985 prompted a revolt and the regime lost control of the provinces. The US Government, fearing radicalization of the revolt, had the Duvaliers flown into exile in France on 7 February 1986.

A civilian-military junta, led by the Army commander, Gen. Henri Namphy, assumed power with the approval of the USA and leading Duvalierists. Freedom of expression was restored and formal democracy promised. Hundreds of Macoutes were killed during several months of violent *déchoukaj* (uprooting) of traces of the old regime. However, its top officials escaped punishment.

By mid-1987 regime-sanctioned violence had resumed and about 140 peasants were massacred near the town of Jean-Rabel. A new Constitution banned leading Duvalierists from contesting elections for 10 years. Haiti's first attempt at free elections, on 29 November, was thwarted when Duvalierist gangs attacked polling stations, killing dozens of people. The election was cancelled in mid-vote and the country's main sources of aid (Canada, France and the USA) suspended their funding in protest.

The Army organized widely boycotted new elections in January 1988 and a university professor, Leslie Manigat, was declared to have won the presidency. He tried to eradicate the Army's extensive smuggling activities and dismissed Gen. Namphy as the Army commander. However, Manigat was deposed in June and Namphy resumed control. In September pro-Namphy Duvalierists massacred a dozen people during a church service conducted by a popular radical priest, Fr Jean-Bertrand Aristide. Disgusted young soldiers overthrew Namphy and appointed Gen. Prosper Avril as President.

Avril promised to restore democracy. Rank-and-file troops overthrew their commanders, but their radical leaders were soon arrested. Meanwhile, conservative Church leaders engineered the expulsion of Fr Aristide from his order, the Salesians. Avril declared a state of siege in 1990 and the USA and a united opposition forced him to flee into exile in March. A Supreme Court judge, Ertha Pascal-Trouillot, was appointed provisional President.

## THE ARISTIDE EXPERIMENT

Haiti's first free presidential elections took place on 16 December 1990. Aristide won 67% of the votes cast, compared with only 14% for a US-backed conservative, Marc Bazin. In January 1991 the Duvaliers' former Minister of the Interior, Roger Lafontant, who had been barred from the election, took outgoing President Pascal-Trouillot hostage and tried to seize power. Street protests, accompanied by the death of nearly 100 people, many by 'necklacing' (killing with burning rubber tyres), prompted the Army to suppress the coup.

Fr Aristide assumed the presidency on 7 February 1991. His huge popular support was not matched by administrative experience and he distanced himself from the political groups that had supported him. His impatience with parliament and his failure firmly to condemn 'necklacing' also lost him support from the traditional ruling class. However, his Government reduced corruption and political violence, and won overwhelming favour with foreign aid donors. The adulation of the black majority and Aristide's advocacy of social justice greatly alarmed the old élite. His attempts to reduce the defence budget and make the armed forces and police accountable to civilian authorities also met opposition. On 30 September the police chief, Maj. Michel François, supported by the Army, led by Gen. Raoul Cédras, overthrew the President. Hundreds of Aristide's supporters were killed and he was sent into exile. Foreign aid was again suspended and the Organization of American States (OAS) imposed economic sanctions in an attempt to force his reinstatement.

## INTERNATIONAL ISOLATION

In February 1992 parliament agreed, at OAS prompting, to a phased restoration of Aristide's presidency, but the deal was rejected by the Army. In June Bazin, whom Aristide had defeated in the 1990 election, was appointed Prime Minister, in the hope that he could end Haiti's world-wide diplomatic isolation. Meanwhile, violations of the sanctions, notably by imports of petroleum, strengthened the regime and the killing of Aristide supporters continued.

New sanctions resulted in an agreement at Governor's Island (New York, USA) in July 1993, under which the Army again agreed to Aristide's phased return to power. Aristide named a Prime Minister, Robert Malval, but the Army prevented him from governing and its agents murdered a leading Aristide supporter, Antoine Izméry, during a church service in September. In October the US President, Bill Clinton, ordered a ship bringing the first 200 US troops of a UN peace-keeping force not to dock for fear of attacks by pro-Army demonstrators. Two days later the Minister of Justice, Guy Malary, was assassinated. The UN authorized a blockade of Haiti by US warships.

Political killings, notably by the army-sponsored paramilitary group, Front Revolutionnaire pour l'Avancement et le Progès d'Haïti (FRAPH), increased. About 4,000 supporters of Aristide were estimated to have been murdered during the Army's rule. The main consideration in US policy, however, was the tens of thousands of Haitian boat people fleeing to the USA from poverty and political repression. In May 1992 the US President, George Bush, ordered all those intercepted at sea to be summarily returned to Haiti. In May 1994 his successor, Bill Clinton, having declared full support for Aristide under pressure from black US politicians, agreed to resume offshore hearings for asylum applications. Most of the resulting daily flood of thousands of boat people were held at the US naval base at Guantánamo, in Cuba. Further UN sanctions were imposed in May, after the Army appointed the 81-year-old head of the Supreme Court, Émile Jonassaint, as President. In July the regime expelled UN observers, after which all air links with Haiti were suspended and the UN authorized an invasion to depose the military regime. On 18 September a mission led by former US President Jimmy Carter and Gen. Colin Powell negotiated the eventual departure abroad of the regime's leaders. A total of 21,000 US troops began landing the next day, followed by several hundred from other countries. The Army and the police were neutralized and Aristide returned to Haiti from the USA on 15 October to resume office.

## THE RETURN TO CIVILIAN RULE

President Aristide soon criticized the USA for not systematically disarming the former soldiers, police and other agents of the dictatorship. Describing the Army as a 'cancer', he declared it dissolved in April 1995. Violent crime, much of it by former soldiers, increased and retaliatory lynchings occurred. The USA, Canada and France trained a new police force to replace the Army. In March a 7,000-strong UN peace-keeping force took over from the US soldiers. It was steadily reduced and the last 1,200 foreign troops left in November 1997.

Aristide had named as his Prime Minister a businessman, Smarck Michel, who pledged to modernize the economy. Foreign assistance was made conditional on the privatization of state bodies and a reduction in the number of government employees. Opposition to this by Aristide prompted Michel to resign in October 1995. He was replaced by the Minister of Foreign Affairs, Claudette Werleigh.

Despite growing discontent over government inaction, parliamentary elections in mid-1995 produced a landslide victory for the government coalition led by Aristide's Organisation Politique Lavalas (OPL). The opposition boycotted the poll as 'fraudulent'. The OPL candidate, René Préval, won the presidential election in December 1995, gaining 88% of the vote in a low turnout (28%). He succeeded Aristide on 7 February 1996, named an agronomist, Rosny Smarth, as Prime Minister and pledged to enact a diluted version of the austerity programme.

The ruling party split in November 1996, when Aristide founded Fanmi Lavalas (FL), accusing the Government of having lost contact with its 'grass-roots' supporters. The split between the FL and the remaining OPL faction deepened in April 1997, when accusations of fraud in legislative by-elections led to parliamentary and political deadlock.

Demonstrations and strikes, as well as increasing violence, led to Smarth's resignation as premier in June 1997, and parliament rejected Préval's nominees to replace him. Disbursement of hundreds of millions of dollars in foreign aid was held up. Préval announced in January 1999 that he would no longer recognize the legislature and appointed as Prime Minister Jacques Édouard Alexis, the Minister of Education. Using street gangs to exert pressure, the FL began a campaign to gain control of the police force, which it accused of brutality. Aristide, who retained strong influence over his former protégé Préval, forced the resignation in October of Robert Manuel, the Secretary of State for Public Security, who fled abroad.

## THE 2000 ELECTIONS

Legislative and municipal elections were held on 21 May 2000 after several postponements, during which the FL obtained changes in the electoral law in its favour. The vote produced a clear majority for the FL and observers initially said the election had been conducted satisfactorily. However, the OAS subsequently revealed that the first-round votes for the Sénat had been counted incorrectly, exaggerating the number of seats won by the FL and disregarding about 1.5m. votes for candidates of the fragmented opposition. The President of the Conseil Electoral Provisoire (CEP), Léon Manus, fled abroad, saying he had been threatened by Préval and Aristide when he refused to endorse the inaccurate count. The FL officially won 73 seats in the 83-member Chambre des Députés, and 18 of the 19 seats contested in the 27-member Sénat. The USA, the European Union (EU) and international organizations denounced the results as fraudulent and maintained their aid suspension. The election campaign had been conducted amid growing violence, in which opposition offices were burned down, the country's leading radio journalist, Jean Dominique, was murdered and the head of the national chamber of commerce fled abroad. The political situation and the police force were also seriously undermined by the large sums of money involved in cocaine-trafficking through Haiti to the USA. However, in November 37 people, including senior army officers and the ex-dictator, Gen. Cédras, were sentenced, *in absentia*, to life imprisonment with hard labour for the 1994 murder of 15 people in the shanty town of Raboteau, in Gonaïves.

Aristide won the presidential election of 26 November 2000 with 92% of the votes cast, against six minor candidates. The opposition, now a coalition of 15 small groups known as the Convergence Démocratique (CD), boycotted the vote. The Government estimated the turn-out to be 61%, although journalists claimed it was nearer 10%.

## THE NEW GOVERNMENT

In December 2000, under pressure from the USA, Aristide pledged to implement a package of political reforms. However, the CD continued to call for a complete rerun of all the elections held in 2000 and announced a symbolic alternative government, headed by an elderly former Minister of Justice, Gérard Gourgue, who demanded the re-establishment of the dissolved Army. Aristide appointed a close aide, Jean-Marie Chérestal, as Prime Minister and a Cabinet that included his former election rival, Bazin, and several former Duvalierists. It was announced that early legislative elections would be held to satisfy international and opposition criticism and restore suspended foreign aid. In June 2001 seven senators elected in the disputed May 2000 elections resigned in a conciliatory gesture. However, arrests of opposition supporters and pro-Government violence continued. Hostility towards the independent media grew, culminating in December in the murder of a radio journalist by government supporters in Petit-Goâve.

About 20 armed men briefly seized the presidential palace without resistance on 17 December 2001, after which thousands of people, mostly recruited by so-called 'grass-roots organizations', took to the streets to resist what the Government claimed was a coup attempt. Opposition homes and offices were looted and burned, as well as one of the country's most important historical archives. An OAS commission of inquiry concluded that there had been no attempted coup.

In January 2002 the OAS urged Aristide to create 'a climate of security' so that negotiations between the Government and the opposition could resume. Corruption scandals and quarrels within the FL resulted in the resignation of Chérestal in January. He was succeeded by the President of the Sénat, Yvon Neptune, who was also the chief FL spokesman.

Neptune could not resolve the political impasse. Gang warfare, partly provoked by government supporters and fuelled by illegal drugs-trafficking, increased, especially in the capital's main slum, Cité Soleil, and anti-poverty protests against the Government and Aristide became more frequent. In mid-2001 a judge said he would charge the powerful FL senator, Dany Toussaint, a former soldier and police chief, with the April 2000 murder of journalist Jean Dominique. However, after death threats from Toussaint's supporters, he fled abroad.

In September 2002 the OAS recommended resumption of foreign aid if the Government curbed and disarmed its violent supporters, compensated their victims, reformed the police force and moved towards holding elections. Anti-Aristide demonstrations broke out in most large towns at the end of the year. The CD was eclipsed in late 2002 by a new body, the 'Group of 184' business and civil society organizations, led by businessman André Apaid, although it attracted negligible public support, despite widespread discontent with the Government. In May a group of armed men, apparently former soldiers, sabotaged the country's main hydroelectric plant at Lake Péligre, cutting off supply for two days.

Meanwhile, President Aristide called on France to pay US $21,700m. to Haiti to mark the bicentenary of independence in 2004, representing the current value of the 90m.-francs independence indemnity. France rejected the demand, saying 'bad governance' was the cause of Haiti's decline.

Opposition protests continued throughout 2003 and were mostly broken up by police and government supporters, including armed gangs of '*chimères*' ('hotheads'). Amiot Métayer, the strongman of the city of Gonaïves and leader of the rebel 'Cannibal Army', was murdered in September and his followers accused the Government of being responsible. The killing triggered weeks of protests in the city, the burning of public buildings and failed government attempts to regain control.

In December 2003 pro-Government thugs invaded the state university and severely assaulted the rector. The Minister of Education resigned in protest, followed by two more ministers and other regime figures. Opposition demonstrations became almost daily and pro-opposition radio stations were attacked in retaliation. The bicentenary celebration in January 2004 was boycotted by most Haitian intellectuals and foreign leaders in protest against government repression, though South African President Thabo Mbeki attended.

## THE DEPARTURE OF ARISTIDE AND CONTINUING INSTABILITY

Mediation efforts to break the long deadlock failed as the civil and political opposition (now united as the Plateforme Démocratique) hardened its position, demanding Aristide's resignation as a condition for any settlement or elections. This ruled out a church proposal that an interim government be formed, an offer by Aristide to hold parliamentary elections within six months, new mediation efforts by the OAS and a plan formulated by the Caribbean Community and Common Market (CARICOM), supported by the USA and other countries, for a power-sharing government, police reform and disbanding of the *chimères*.

The Cannibal Army rebels (now the Gonaïves Resistance Front) formally seized control of Gonaïves on 5 February 2004 and other insurgents took over a dozen further towns. A small but well-armed group of rebels led by a former police chief, Guy Philippe, and including a former leader of the FRAPH, Louis-Jodel Chamblain, returned from exile in the Dominican Republic to support the insurgents and seized Cap-Haïtien on 22 February.

Aristide appealed in vain for international help to fight the rebels, while France and the USA called on him to step down. He claimed that he would 'rather die' but secretly negotiated with US officials to leave the country. After formally resigning, Aristide left Haiti before dawn on 29 February 2004 on a US aircraft, and flew to the Central African Republic, where the USA had arranged for him to go. Once there, he claimed he had been 'kidnapped' by the USA and forced to leave Haiti.

The USA and France landed troops in Haiti hours after Aristide's departure and as the UN Security Council voted its approval. The US-led Multinational Interim Force (MIF) of

3,400 troops was replaced in June 2004 by a UN peace-keeping force (the UN Stabilization Mission in Haiti—MINUSTAH) led by Brazil

A council of 'elders' chose retired UN economist Gérard Latortue to serve as Prime Minister under a figurehead President (Chief Justice Boniface Alexandre). Latortue, who took office on 12 March 2004 and had lived outside Haiti for 40 years, named a Government of mostly non-political figures, but including former Army commander Gen. Hérard Abraham and excluding any FL supporters.

CARICOM protested that Aristide's departure had taken place under US pressure, accused the UN of not defending him, and said it would not recognize the Latortue Government. Jamaica admitted Aristide as a refugee in March 2004, but in May he left for exile in South Africa.

Latortue, meanwhile, praised the ex-rebels as 'freedom fighters' and many Aristide supporters were arrested, including former Prime Minister Neptune. When the Government announced a disarmament drive in July 2004, the former soldiers responded with armed demonstrations and occupied several provincial towns in August and September. The Government offered them 10 years of salary arrears, but payment of some of this (in 2005) did not diminish their calls for revival of the Army and the Government's overthrow. They and armed gangs remained in effective control of some parts of the country in 2005.

In mid-September 2004 flash floods devastated the northwestern city of Gonaïves, killing about 3,000 people and leaving 200,000 homeless. The Government's failure to provide swift help and the obstruction of international relief efforts by looters and gangs controlling the city brought Latortue under heavy criticism.

Violence and killings, mostly in the capital, increased sharply in early October 2004 after Aristide supporters, who had been staging large demonstrations, and gangs of *chimères* launched a concerted campaign against the Government, to which the police responded harshly.

Aristide, along with 'drugs-lords' and other interests, was suspected (and accused by US officials) of orchestrating and funding the violence. In April 2005 he claimed that, since his departure from the country, 10,000 of his followers had been killed in a 'black holocaust' carried out by the USA and France. Human rights organizations, however, estimated the dead on all sides at about 1,500 by mid-2005.

UN troops were criticized for failing to disarm the population and eradicate the violence, especially in the Port-au-Prince slums, but MINUSTAH was slow to reach its full strength because countries were reluctant to contribute troops and police. By December 2004 more troops had arrived and the force took a more aggressive stance, regaining control of some parts of the country from the ex-soldiers in March 2005 and from slum gangs in the capital. The ex-soldiers' leader, Rémissainthe Ravix, was killed by police in April, followed by several gang leaders in Port-au-Prince. The UN Security Council renewed MINUSTAH's mandate in June, for eight months, and added a further 1,000 soldiers and police, bringing the total number of troops to 7,700. Meanwhile, the US Government, at Latortue's request, ended its 13-year ban on arms sales to Haiti.

The justice system continued to lack credibility and drew strong local and international criticism. In August 2004 Chamblain, along with a notorious former Army captain, Jackson Joanis, was cleared of involvement in the 1993 murder of Antoine Izméry. Sixteen people convicted in 2000 for the 1994 murders in Raboteau were cleared by the Supreme Court in May 2005 on technicalities. Brutality and executions by the police, which hundreds of ex-soldiers joined in 2005, were deplored by all sides, including MINUSTAH. Kidnappings for large ransoms of businessmen and members of their families, as well as other prominent people and foreigners (some were tortured or killed), grew steadily in the capital from March 2005. Some of those responsible declared themselves Aristide supporters. In May the number of these kidnappings increased dramatically. Many people who received threats fled the country. In July a prominent journalist, Jacques Roche, was kidnapped and murdered and the Government accused Aristide supporters of being responsible. The Government, meanwhile, refused to release former Prime Minister Neptune, who was accused (but not charged for 11 months) of ordering a massacre of Aristide opponents, despite reports by a UN inspector that no massacre had taken place. International pressure, including from the UN and the US Government, contributed to the forced resignation in June 2005 of the hardline Minister of Justice and Public Security, Bernard Gousse.

Plans for presidential and legislative elections to be held in November 2005 were being seriously hampered by disorganization, corruption and divisions in the electoral council, as well as by very low voter registration. Some urged that the vote be delayed owing to these factors and to the growing violence in the country. The FL, whose supporters continued to stage demonstrations, also said it would boycott the elections while Neptune (who went on hunger strike in mid-2005) and scores of its leaders and activists remained in prison without trial and while Aristide remained in exile. However, some in the party favoured participation and a priest, Fr Gérard Jean-Juste, was emerging as a possible new leader. Non-participation by the FL would seriously undermine the legitimacy of the elections. Former rebel leader Guy Philippe was prominent among about 100 presidential candidates. The mainstream social-democratic coalition, the Grand Parti Socialiste Haïtien (GPSH), launched in May 2004, named a former senator, Serge Gilles, as its candidate. About 150 parties were competing for parliamentary seats.

The pre-election political situation was complicated by growing criticism of Latortue's Government from all quarters, including the traditional élite, which initially supported him but then accused him of being ineffective and of failing to obtain disbursement of the majority of a promised US $1,000m. in foreign aid. However, UN, US and EU officials continued to support Latortue and the planned elections.

# Economy

## GREG CHAMBERLAIN

Most of Haiti's 8.4m. inhabitants (mid-2004 estimate) are among the poorest in the Americas and the contrast in wealth between rulers and ruled is stark. In the late 18th century Haiti was France's richest colony. By the end of the 20th the country had few material resources and was racked by drought, hunger, severe deforestation (96% in 2000, up from 91% in 1987) and soil erosion (some 10,000 ha of land and 30m. trees are lost each year), aggravated by the international sanctions imposed during the 1991–94 military dictatorship and political paralysis since 1997. Many consider the country an ecological disaster with little chance of recovery.

These factors, together with high population density (303 per sq km in mid-2004) and primitive production methods, ensured the steady decline of agriculture, the main economic sector. At the end of the 20th century nearly 40% of Haitians lived in towns, compared with 12% half a century earlier. Since 1970 the population of Port-au-Prince, the capital, has tripled in size, with two-thirds of its inhabitants living in slums. Haiti is heavily dependent on foreign aid, including food relief, mainly from the USA. By 2000 more than one-10th of the population was being fed by international relief organizations. In 2003 the UN Development Programme (UNDP) stated that 56% of Haitians were malnourished. Along with Afghanistan and Somalia, the country had the world's worst per-head food deficit, which also contributed to political tensions.

At the start of the 21st century unemployment and underemployment affected 85% of the labour force. Some 97% of Haitians were outside the formal economy. Per-head gross domestic product (GDP) was estimated in 2003 at US $364, with 76% living in poverty (less than $2 per day) and 55% in extreme poverty (less than $1). The wealthiest 4% of Haitians accounted for nearly two-thirds of gross national income (GNI). Illiteracy was estimated at 47% in 2004 (unofficial estimates were far higher). Life expectancy in that year was 51.8 years. Some 42% of children under five were malnourished and one-quarter of them died of preventable diseases. Only one-third of the population had access to health care and fast-diminishing supplies of drinking water, which was eight times more expensive than in developed countries. Malaria (the incidence of which more than tripled between 1994 and 2002), tuberculosis and HIV/AIDS were rampant. Haiti's place in UNDP's Human Development Index declined further in 2004, to 153rd out of a total of 177 countries.

Factories, virtually all North American 'offshore' assembly operations (mainly clothing), provided low-paying jobs in Port-au-Prince. There was little other foreign investment. Tourism virtually died out in the 1980s owing to political instability and publicity about poverty and the high incidence of HIV/AIDS. However, some cruise-ship tourism resumed in the mid-1990s, with visits to the north coast resort of Labadie. In 1997 the Government began a US $18m. project to develop tourism near Jacmel, in the south-east.

The minimum wage was doubled in 2003, to 70 gourdes (worth about US $1.70); this was, however, a fall of 30% in real terms since the last increase in 1995. Enforcement was difficult because of high unemployment and weak trade unions. Pressure to keep the minimum wage low came from owners in the assembly sector. In 2004 Haiti was listed by the non-governmental organization Transparency International as the most corrupt country in the world (tied with Bangladesh).

Since the early 1960s Haitians have responded to their bleak prospects by emigrating, mostly to the USA and Canada. More than 1m. Haitians also live in the Dominican Republic, one-half of them illegally and many cutting sugar cane in virtual slavery, while tens of thousands provide cheap labour in the Bahamas, the Turks and Caicos Islands and French Guiana. In 2003 remittances from this diaspora amounted to some US $900m. (about one-quarter of GDP). The outflow of most skilled people condemned those who remained in Haiti to even worse economic conditions.

From 1980 the economy steadily declined, mainly owing to political turmoil and extensive smuggling. In 1991 the advent of the country's first freely elected President, Fr Jean-Bertrand Aristide, revived business confidence and reforms began. However, Aristide's overthrow, after seven months, ended these economic hopes. An international trade embargo was imposed in an attempt to force Aristide's restoration. This was undermined by smuggling, especially of petroleum from Europe and the Dominican Republic. Nearly 100,000 jobs were lost, however, prices rose dramatically, poverty deepened and health conditions worsened in the countryside, to which some 300,000 people fled from the capital, while many of the rich became even wealthier from their control of smuggling. Thousands of others left the country as 'boat people'. The economy contracted by nearly one-third between 1991 and 1994.

The restoration to power of Aristide in 1994 raised hopes of a substantial economic revival. These quickly faded as government incapacity and political deadlock hampered foreign and domestic efforts and about US $500m. in foreign aid was suspended. Meanwhile, the smuggling of cocaine through Haiti from Colombia to the USA was playing an increasingly large role in the economy. About 8% of the cocaine arriving in the USA in 2003 had passed through Haiti.

The collapse of the Aristide Government in 2004 was accompanied by heavy looting in the capital, which severely damaged the economy. The new Prime Minister, Gérard Latortue, a retired senior UN economist, embarked on serious economic and administrative reforms, largely conditional on foreign assistance. Pledges of more than US $1,000m. worth of aid were received in July. The plans were set back, however, by severe flooding in the south-east in May, which killed more than 2,000 people, and by even worse flooding three months later in Gonaïves, where 3,000 people died. Both illustrated the grave consequences of chronic erosion and the Government's inability to provide relief supplies. GDP contracted by 3.8% in 2003/04 and was two-thirds of its 1981 value. Growing political violence undermined efforts to restore economic confidence and slowed the reforms, as well as the arrival of foreign aid, as did the fact that the country's historically predatory élite had regained control of the state. Civil society and the poor complained they were not being consulted over the planned reforms in 2004.

While technical results (especially fiscal discipline and business incentives) impressed international financial institutions in 2004 and 2005, little progress was made in reducing poverty and upgrading infrastructure. The business climate was further damaged in 2005 by the hundreds of ransom kidnappings of wealthy or middle-class Haitians, and many fled the country, transferring millions of dollars abroad.

## AGRICULTURE

Nearly two-thirds of Haiti's working population were engaged in farming and fishing, but the sector produced less than one-third of the country's GDP (26.2% in 2003/04), about 5% of total exports (5.6% in 2004) and less than one-half of domestic food requirements. Only about 20% of Haiti's land was considered arable, but, as a result of population pressure, one-half of total land area was under cultivation, mostly in small and inefficient plots. About one-quarter of the peasantry had no land and the poorest 60% owned only 1% of it. Tenure was difficult to prove and land seizures and disputes were, with rural taxes and lack of access to water, a cause of political unrest. The disruptions caused by the 1991–94 embargo and military dictatorship reduced farm production by 40% and sent the peasantry into serious debt. The Government began a modest agrarian reform programme in 1996.

Owing to erosion, drought, primitive farming methods, poor maintenance and population density, soil fertility was low. Only 7% of the arable land was irrigated, most of it in the main valley, the Artibonite, and the Port-au-Prince and Les Cayes plains. Poverty and the country's mountainous terrain made mechanization impractical.

Production of coffee, long the main cash crop, fell dramatically, from 450,000 bags in 1985 to officially only 10,000 in 2003, mainly owing to smuggling into the Dominican Republic in recent years. A US Agency for International Development (USAID) project, begun in 1990, organized growers into cooperatives and increased production, as well as helping reforestation. Sugar cane was once the second cash crop, but falling quality and cheaper imports forced closure of all the country's major sugar factories, although one, at Darbonne, near Léogane, was reopened with Cuban help in 2001. Cocoa and essential oils were exported.

The main food crops, apart from sugar, were maize, rice, sorghum, millet, bananas, avocados, vegetables and fruit. Good rains in 2004 increased maize production by 17.5%, but output of rice fell by 10%. The lowering of tariffs under IMF-imposed reforms in 1994 increased the inflow of cheap foreign rice and about 75% of rice was being imported by 2005, destroying the livelihood of tens of thousands of small producers. The influx of relief aid in 2004 further depressed local food production. In 2001 Cuban experts advised the undeveloped fishing industry; the first national fishing congress was held in December 2000.

## MINING AND POWER

Charcoal from local timber accounted for more than two-thirds of the energy used in Haiti. This aggravated deforestation, which was accelerated by the oil fuel embargo imposed on the 1991–94 dictatorship.

Some of the country's electricity came from the ageing Péligre hydroelectric plant above the Artibonite valley and more from several thermal plants in the Port-au-Prince area and the Saut-Mathurine dam near Les Cayes. Less than 10% of the population (45% in the capital and 3% in rural areas) had electricity. Many assembly industries, along with wealthier citizens, relied on private generators, which provided about one-third of the national total. More than one-half of the current produced by the state electricity company, Electricité d'Haïti, was stolen or illegally resold. By 2004 overall production had dropped to one-10th of the capital's needs. The Latortue Government improved supply as one its priorities, through a US $22m. programme with USAID, but extensive blackouts returned when this ended in March 2005 and were worsened by high prices of oil for electricity generation.

The mining sector virtually disappeared in 1983, when the US-owned Reynolds Company closed its bauxite mine at Miragoâne because of low world prices. Marble and limestone are extracted. Copper and gold deposits in the north have occasionally been mined.

## MANUFACTURING

Owing to Haiti's extreme poverty, the domestic market for finished goods is small, and local factories produce only such staples as cooking oil, detergents, household utensils, shoes, beverages, cigarettes, cement and flour. There is also strong competition from smuggled goods.

More than 100 mostly US-owned factories assembling light industrial products (mostly clothing) for re-export operated on the outskirts of the capital. They were attracted to Haiti by the lowest labour costs in the region and the virtual absence of trade unions. They provided 23,000 jobs (mainly for women) in 2005, a fall of 3,000 from 2004, continuing a decline dating from the political instability after the fall of the Duvaliers in 1986, which saw many firms move to more stable neighbouring countries with more skilled and literate workers. The sector's contribution to government revenue was insignificant, as the companies were exempt from taxation for up to 15 years and free to repatriate profits.

In mid-2003 a Haitian-Dominican free trade zone along the northern border near the town of Ouanaminthe began operating, but disputes about poor working conditions were not settled until early 2005. A proposed US measure, the Haiti Economic Recovery and Opportunity Act (HERO), to give duty-free access to the USA for Haitian clothing exports, thus providing a major boost to the Haitian economy, was sidelined by the US Congress in September 2004. Another branch of manufacturing was handicrafts, including paintings, furniture and carvings.

## TRANSPORT AND INFRASTRUCTURE

Only about one-quarter of the 4,160 km (2,585 miles) of roads were paved in 2003 and these had been deteriorating for several years. A single railway near the capital, used only to transport sugar, closed in the early 1990s.

There were direct air links to North America, Europe and the rest of the Caribbean, but few regular services to provincial airfields, most of which were grass strips. The Government announced an airport modernization plan in 2002 and a new airport was built in 2003 in the main southern town of Les Cayes.

A container terminal in Port-au-Prince handled most of the country's foreign trade. Provincial ports and the porous frontier with the Dominican Republic were the main centres of imported contraband.

Fewer than 5% of Haitians had fixed-line telephones and service was poor, with frequent sabotage and theft of lines. By the early 2000s the state-owned telephone company, Téléco, was being rivalled by private mobile telephone companies and internet service providers.

## INVESTMENT AND FINANCE

Most public expenditure was on government operations and personnel. There was little capital investment and public services were minimal. The USA was the principal foreign investor in Haiti, almost exclusively in the assembly sector. Foreign direct investment fell from US $30m. in 1999 to only $5m. in 2003.

Foreign aid accounted for much of the national budget, despite Haiti's notable inability to absorb it. Most of the aid came from the European Union (EU), the USA and Canada. In 1991, under President Aristide, nearly US $500m. was promised, only to be suspended after the military coup later that year. When Aristide returned to power in 1994, the international community pledged more than $1,000m. to restore the shattered economy. The IMF made it conditional on a structural-adjustment programme involving privatization of nine major state bodies, most of them overstaffed and in deficit, and a 20% cut in the bloated government work-force, as well as decentralization. Aristide later opposed the reforms and negotiations with the IMF were suspended in 1995.

In 1996 President René Préval amended the structural-adjustment programme to one of partial privatization. But by 2003 only two state bodies—a flour mill and a cement company—had been sold, and only about one-half of the total foreign aid promised had been disbursed owing to the legislature's inaction. Charges of electoral fraud in May 2000 once again resulted in the suspension of most foreign assistance. In July 2004, soon after Aristide's departure, a donors' conference pledged more than US $1,000m. in aid and approved the Government's recovery plan, though only 2% of the funding was for the key sector of environmental restoration. By March 2005 disorganization and fears of political violence had prevented disbursement of 80% of the aid, but a further donors' conference that month, in Cayenne, French Guiana, renewed the pledges on the basis of more specific projects and disbursement speeded up, notably after settlement of $52.6m. in arrears to the World Bank. The renewed aid was reflected in a 34% increase in the national budget for 2004/05.

The post-Duvalier regimes made some attempt at tax reform and monopolies were ended. However, state revenue halved under the 1991–94 military dictatorship. Two-thirds of Haiti's foreign debt was cancelled by the main creditor countries in 1994, but parliamentary disputes prevented approval of a national budget between 1996 and 2000. About 40% of potential tax revenue was being lost through corruption or evasion. In 2003 total tax revenue was still only 8% of GDP, two-thirds of it from customs duties, but collection improved in 2004. The value of the gourde steadily declined after 1986, falling by about 50% in 2002–03, when it stabilized. Civil servants were given a 30% pay rise in October 2004 in an effort to fight corruption, but the new salary levels were a 40% decrease in real terms since 2000.

## FOREIGN TRADE

Haiti's principal export in 2004 was light manufactures (84.7% of total exports) produced by 'offshore' assembly industries, followed by agricultural exports (5.6%), mainly coffee (1.2%). The main imports in that year were food and live animals (32.6% of all imports), manufactured goods (27.1%) and fuel (17.1%). In 2003 the USA took 84% of Haiti's exports and provided 52% of imports. Trade figures were deceptive owing to large-scale smuggling, estimated to account for two-thirds of imports.

## CONCLUSION

Chronic corruption and lack of material and human resources, as well as state-sponsored violence in a deeply divided society, hindered efforts to establish political and economic stability from the mid-1990s. The crisis worsened when a stubborn traditional *élite* regained control in 2004 with the overthrow of President Aristide and found itself confronting the poor majority of Haitians and their violent agents. This *élite* and their quarrelling politicians had no coherent political or economic programme and elections, scheduled to take place in late 2005, were not expected to change this situation. Undemocratic government, political violence and extreme poverty have also long isolated Haiti from its more peaceful and prosperous neighbours, to which millions of Haitians have emigrated and from which thousands are routinely deported. However, diplomatic relations with Cuba were resumed in 1996 after a 32-year gap and hundreds of Cuban doctors came to Haiti to work as volunteers, but Haiti's admission to the Caribbean Community and Common Market in 2002 was suspended owing to the political crisis. Prospects for a real transformation of Haitian society and for economic improvement and political stability were poor. With few viable institutions and scant natural resources, Haiti's hopes for progress in 2005 still rested on carefully targeted and co-ordinated international aid and, if they could be marshalled, the skills and relative wealth of the country's large North American diaspora.

# Statistical Survey

Sources (unless otherwise stated): Banque de la République d'Haïti, angle rues du Magasin d'État et des Miracles, BP 1570, Port-au-Prince; tel. 299-1200; fax 299-1045; e-mail webmaster@brh.net; internet www.brh.net; Ministère de l'Economie et des Finances, Palais des Ministères, rue Monseigneur Guilloux, Port-au-Prince; Institut Haitien de Statistique et d'Informatique.

## Area and Population

### AREA, POPULATION AND DENSITY

| | |
|---|---:|
| Area (sq km) | 27,750* |
| Population (census results)† | |
| 30 August 1982 | 5,053,792 |
| 7 July 2003‡ | |
| Males | 3,832,980 |
| Females | 4,096,068 |
| Total | 7,929,048 |
| Population (UN estimate at mid-year)§ | |
| 2004 | 8,407,000 |
| Density (per sq km) at mid-2004 | 303.0 |

* 10,714 sq miles.
† Excluding adjustment for underenumeration.
‡ Preliminary results.
§ Source: UN, *World Population Prospects: The 2004 Revision*.

### DEPARTMENTS
(preliminary figures, 2003 census)

| | Population | Capital |
|---|---:|---|
| Ouest | 3,093,699 | Port-au-Prince |
| Sud-Est | 449,585 | Jacmel |
| Nord | 773,546 | Cap-Haïtien |
| Grand'Anse | 603,894 | Jérémie |
| Nord-Est | 300,493 | Fort Liberté |
| L'Artibonit (Artibonite) | 1,070,397 | Gonaïves |
| Centre | 565,043 | Hinche |
| Sud | 627,311 | Les Cayes |
| Nord-Ouest | 445,080 | Port-de-Paix |
| **Total** | **7,929,048** | — |

### PRINCIPAL TOWNS
(estimated population at mid-1999)

| | | | |
|---|---:|---|---:|
| Port-au-Prince (capital) | 990,558 | Delmas | 284,079 |
| Carrefour | 336,222 | | |

Source: UN Statistics Division.

**Mid-2003** (UN estimate, incl. suburbs): Port-au-Prince 1,961,455 (Source: UN, *World Urbanization Prospects: The 2003 Revision*).

### BIRTHS AND DEATHS
(UN estimates)

| | 1985–90 | 1990–95 | 1995–2000 |
|---|---:|---:|---:|
| Crude birth rate (per 1,000) | 41.6 | 33.4 | 31.6 |
| Crude death rate (per 1,000) | 16.1 | 15.8 | 14.6 |

Source: UN, *World Population Prospects: The 2004 Revision*.

**Expectation of life** (WHO estimates, years at birth): 53 (males 52; females 54) in 2003 (Source: WHO, *World Health Report*).

## ECONOMICALLY ACTIVE POPULATION
(official estimates, persons aged 10 years and over, mid-1990)

| | Males | Females | Total |
|---|---:|---:|---:|
| Agriculture, hunting, forestry and fishing | 1,077,191 | 458,253 | 1,535,444 |
| Mining and quarrying | 11,959 | 12,053 | 24,012 |
| Manufacturing | 83,180 | 68,207 | 151,387 |
| Electricity, gas and water | 1,643 | 934 | 2,577 |
| Construction | 23,584 | 4,417 | 28,001 |
| Trade, restaurants and hotels | 81,632 | 271,338 | 352,970 |
| Transport, storage and communications | 17,856 | 2,835 | 20,691 |
| Financing, insurance, real estate and business services | 3,468 | 1,589 | 5,057 |
| Community, social and personal services | 81,897 | 73,450 | 155,347 |
| Activities not adequately defined | 33,695 | 30,280 | 63,975 |
| **Total employed** | 1,416,105 | 923,356 | 2,339,461 |
| Unemployed | 191,333 | 148,346 | 339,679 |
| **Total labour force** | 1,607,438 | 1,071,702 | 2,679,140 |

Source: ILO, *Yearbook of Labour Statistics*.

**Mid-2003** (estimates in '000): Agriculture, etc. 2,213; Total labour force 3,645 (Source: FAO).

## Health and Welfare

### KEY INDICATORS

| | |
|---|---:|
| Total fertility rate (children per woman, 2003) | 3.9 |
| Under-5 mortality rate (per 1,000 live births, 2003) | 118 |
| HIV/AIDS (% of persons aged 15–49, 2003) | 5.6 |
| Physicians (per 1,000 head, 1998) | 0.25 |
| Hospital beds (per 1,000 head, 1996) | 0.71 |
| Health expenditure (2002): US $ per head (PPP) | 83 |
| Health expenditure (2002): % of GDP | 7.6 |
| Health expenditure (2002): public (% of total) | 39.4 |
| Access to water (% of persons, 2002) | 71 |
| Access to sanitation (% of persons, 2002) | 34 |
| Human Development Index (2002): ranking | 153 |
| Human Development Index (2002): value | 0.463 |

For sources and definitions, see explanatory note on p. vi.

## Agriculture

### PRINCIPAL CROPS
('000 metric tons)

| | 2001 | 2002* | 2003* |
|---|---:|---:|---:|
| Rice (paddy) | 103 | 104 | 105 |
| Maize | 180 | 185 | 198 |
| Sorghum | 80 | 85 | 95 |
| Sweet potatoes | 174 | 175 | 175 |
| Cassava (Manioc) | 332 | 335 | 340 |
| Yams | 197 | 198 | 199 |
| Other roots and tubers* | 41 | 42 | 41 |
| Sugar cane | 1,008 | 1,010 | 1,050 |
| Fresh vegetables (incl. melons)* | 199 | 200 | 204 |
| Bananas | 290 | 295 | 300 |
| Plaintains* | 280 | 285 | 283 |
| Mangoes | 250 | 260 | 261 |
| Other fruit* | 149 | 162 | 170 |

* FAO estimates.

Source: FAO.

# HAITI

## LIVESTOCK
(FAO estimates, '000 head, year ending September)

|  | 2001 | 2002 | 2003 |
|---|---|---|---|
| Horses | 501 | 501 | 501 |
| Mules | 82 | 82 | 83 |
| Asses | 215 | 215 | 220 |
| Cattle | 1,440 | 1,450 | 1,455 |
| Pigs | 1,001 | 1,001 | 1,002 |
| Sheep | 152 | 153 | 154 |
| Goats | 1,942 | 1,943 | 1,944 |
| Poultry | 5,905 | 6,005 | 6,075 |

Source: FAO.

## LIVESTOCK PRODUCTS
(FAO estimates, '000 metric tons)

|  | 2001 | 2002 | 2003 |
|---|---|---|---|
| Beef and veal | 40.5 | 42.0 | 42.5 |
| Goat meat | 6.5 | 6.5 | 6.5 |
| Pig meat | 31.2 | 29.1 | 29.1 |
| Horse meat | 5.6 | 5.6 | 5.7 |
| Poultry meat | 8.3 | 8.6 | 9.1 |
| Cows' milk | 42.0 | 42.5 | 44.0 |
| Goats' milk | 24.4 | 24.8 | 25.2 |
| Hen eggs | 4.2 | 4.3 | 4.4 |
| Cattle hides | 5.8 | 6.2 | 6.3 |

Source: FAO.

## Forestry

### ROUNDWOOD REMOVALS
(FAO estimates, '000 cubic metres, excl. bark)

|  | 2001 | 2002 | 2003 |
|---|---|---|---|
| Sawlogs, veneer logs and logs for sleepers* | 224 | 224 | 224 |
| Other industrial wood* | 15 | 15 | 15 |
| Fuel wood | 1,971 | 1,971 | 1,978 |
| Total | 2,210 | 2,210 | 2,217 |

* Output assumed to be unchanged since 1971.

Source: FAO.

### SAWNWOOD PRODUCTION
('000 cubic metres, incl. railway sleepers)

|  | 1969 | 1970 | 1971 |
|---|---|---|---|
| Coniferous (softwood) | 5 | 8 | 8 |
| Broadleaved (hardwood) | 10 | 5 | 6 |
| Total | 14 | 13 | 14 |

**1972–2003:** Annual production as in 1971 (FAO estimates).

Source: FAO.

## Fishing
(metric tons, live weight)

|  | 1997 | 1998 | 1999* |
|---|---|---|---|
| Freshwater fishes* | 500 | 500 | 500 |
| Marine fishes* | 4,000 | 4,000 | 3,800 |
| Marine crabs | 71 | 59 | 50 |
| Caribbean spiny lobster* | 200 | 200 | 200 |
| Natantian decapods* | 150 | 150 | 150 |
| Stromboid conchs* | 380 | 350 | 300 |
| Total catch | 5,301 | 5,259 | 5,000 |

* FAO estimates.

Note: Figures exclude corals and madrepores (FAO estimates, metric tons): 13 in 1997, 10 in 1998, 10 in 1999.

**2000–03:** Annual catch as in 1999 (FAO estimates).

Source: FAO.

## Industry

### SELECTED PRODUCTS
(metric tons, unless otherwise indicated, year ending 30 September)

|  | 1999/2000 |
|---|---|
| Edible oils | 38,839.6 |
| Butter | 2,972.2 |
| Margarine | 2,387.4 |
| Cornflour | 104,542.6 |
| Soap | 30,069.9 |
| Detergent | 4,506.1 |
| Beer ('000 cases of 24 bottles) | 784.5 |
| Beverages ('000 cases of 24 bottles) | 1,807.7 |
| Rum ('000 750ml bottles) | 2,009.5 |
| Electric energy (million kWh) | 697.6 |

## Finance

### CURRENCY AND EXCHANGE RATES

**Monetary Units**
100 centimes = 1 gourde.

**Sterling, Dollar and Euro Equivalents** (31 May 2005)
£1 sterling = 70.365 gourdes;
US $1 = 38.703 gourdes;
€1 = 47.724 gourdes;
1,000 gourdes = £14.21 = $25.84 = €20.95.

**Average Exchange Rate** (gourdes per US $)
2002  29.251
2003  42.367
2004  38.352

Note: The official rate of exchange was maintained at US $1 = 5 gourdes until September 1991, when the central bank ceased all operations at the official rate, thereby unifying the exchange system at the 'floating' free market rate.

### BUDGET
(million gourdes, year ending 30 September)

| Revenue* | 2002 | 2003 | 2004 |
|---|---|---|---|
| Current receipts | 7,826 | 10,746 | 12,457 |
| Internal receipts | 5,587 | 7,462 | 8,769 |
| Customs | 2,069 | 2,762 | 3,481 |
| Transfers from public enterprises | 2 | — | — |
| Total | 7,828 | 10,746 | 12,457 |

| Expenditure | 2002 | 2003 | 2004 |
|---|---|---|---|
| Current expenditure | 8,864 | 11,156 | 11,037 |
| Wages and salaries | 3,480 | 3,862 | 4,126 |
| Operations | 3,888 | 4,845 | 4,517 |
| Interest on public debt | 729 | 1,050 | 1,150 |
| External debt | 353 | 561 | 662 |
| Internal debt | 376 | 489 | 488 |
| Transfers and subsidies | 767 | 1,399 | 1,261 |
| Capital expenditure | 1,908 | 3,928 | 4,960 |
| Total | 10,772 | 15,084 | 15,996 |

* Excluding grants received (million gourdes): 113 in 2002; 171 in 2003; 1,848 in 2004.

Source: IMF, *Haiti: Selected Issues* (June 2005).

### INTERNATIONAL RESERVES
(US $ million at 31 December)*

|  | 2002 | 2003 | 2004 |
|---|---|---|---|
| IMF special drawing rights | 0.5 | 0.3 | 0.2 |
| Reserve position in IMF | 0.1 | 0.1 | 0.1 |
| Foreign exchange | 81.1 | 61.6 | 114.1 |
| Total | 81.7 | 62.0 | 114.4 |

* Excluding gold (valued at market-related prices, US $ million): 6.6 in 1989.

Source: IMF, *International Financial Statistics*.

# HAITI

*Statistical Survey*

## MONEY SUPPLY
(million gourdes at 31 December)

|  | 2002 | 2003 | 2004 |
|---|---|---|---|
| Currency outside banks | 8,687.5 | 9,843.2 | 10,218.3 |
| Demand deposits at commercial banks | 4,394.7 | 6,142.5 | 7,1455.1 |
| **Total money** (incl. others) | 13,501.5 | 16,222.1 | 17,975.8 |

Source: IMF, *International Financial Statistics*.

## COST OF LIVING
(Consumer Price Index, year ending 30 September; base: 2000 = 100, metropolitan areas)

|  | 2001 | 2002 | 2003 |
|---|---|---|---|
| Food | 115.5 | 127.4 | 174.2 |
| Clothing and footwear | 112.6 | 123.4 | 155.1 |
| Rent | 111.3 | 121.9 | 170.1 |
| **All items** (incl. others) | 114.0 | 125.3 | 174.5 |

**2004:** All items 214.3.

Source: ILO.

## NATIONAL ACCOUNTS
(million gourdes, year ending 30 September)

**Expenditure on the Gross Domestic Product**
(at current prices)

|  | 2001/02 | 2002/03 | 2003/04 |
|---|---|---|---|
| Final consumption expenditure | 92,388 | 120,528 | 141,243 |
| Gross capital formation | 23,412 | 36,732 | 38,312 |
| **Total domestic expenditure** | 115,800 | 157,260 | 179,555 |
| Exports of goods and services | 11,403 | 18,945 | 20,194 |
| *Less* Imports of goods and services | 33,363 | 56,589 | 59,755 |
| **GDP in purchasers' values** | 93,840 | 119,616 | 139,994 |
| **GDP at constant 1987 prices** | 12,930 | 12,992 | 12,502 |

Source: IMF, *International Financial Statistics*.

**Gross Domestic Product by Economic Activity**
(at constant 1986/87 prices)

|  | 2001/02* | 2002/03† | 2003/04‡ |
|---|---|---|---|
| Agriculture, hunting, forestry and fishing | 3,326.4 | 3,336.8 | 3,157.7 |
| Mining and quarrying | 13.9 | 14.1 | 13.4 |
| Manufacturing | 998.6 | 1,003.5 | 999.8 |
| Electricity, gas and water | 61.0 | 62.9 | 64.2 |
| Construction | 955.5 | 969.4 | 941.8 |
| Trade, restaurants and hotels | 3,471.9 | 3,509.1 | 3,257.9 |
| Transport, storage and communication | 762.6 | 774.7 | 781.4 |
| Business services | 1,531.6 | 1,535.1 | 1,522.5 |
| Other services | 1,400.2 | 1,380.5 | 1,335.8 |
| **Sub-total** | 12,521.7 | 12,586.1 | 12,074.5 |
| *Less* Imputed bank service charge | 512.2 | 515.5 | 496.7 |
| Import duties | 920.0 | 921.0 | 924.0 |
| **GDP in purchasers' values** | 12,929.5 | 12,991.6 | 12,501.8 |

* Revised figures.
† Preliminary figures.
‡ Official estimates.

## BALANCE OF PAYMENTS
(US $ million, year ending 30 September)

|  | 2002 | 2003 | 2004 |
|---|---|---|---|
| Exports of goods f.o.b. | 273.2 | 330.4 | 372.7 |
| Imports of goods f.o.b. | −982.6 | −1,115.8 | −1,182.6 |
| **Trade balance** | −709.4 | −785.4 | −809.9 |
| Exports of services | 163.7 | 130.9 | 131.7 |
| Imports of services | −256.3 | −283.0 | −336.6 |
| **Balance on goods and services** | −802.0 | −937.5 | −1,014.8 |
| Other income (net) | −15.2 | −14.3 | −13.6 |
| **Balance on goods, services and income** | −817.2 | −951.8 | −1,028.4 |
| Current transfers received | 784.2 | 948.0 | 1,044.1 |
| **Current balance** | −33.0 | −3.7 | 15.6 |
| Capital flows (net) | −8.0 | 25.3 | −4.6 |
| Direct investment (net) | 4.7 | 7.8 | 5.9 |
| Other investment assets* | 3.1 | −46.8 | 29.0 |
| Investment liabilities† | −35.3 | 6.5 | −12.9 |
| **Overall balance** | −68.5 | −10.9 | 33.0 |

* Excludes commercial banks' foreign currency deposits.
† Includes short-term capital and net errors and omissions.

Source: IMF, *Haiti: Selected Issues* (June 2005).

# External Trade

## PRINCIPAL COMMODITIES
(US $ million, year ending 30 September)

| Imports c.i.f. | 2002 | 2003 | 2004 |
|---|---|---|---|
| Food and others* | 345.2 | 362.7 | 415.2 |
| Mineral fuels, lubricants, etc. | 157.3 | 196.5 | 218.0 |
| Machinery and transport equipment | 159.8 | 165.3 | 173.2 |
| Manufactured goods | 293.9 | 346.9 | 344.2 |
| **Total** (incl. others) | 1,054.2 | 1,199.8 | 1,271.7 |

| Exports f.o.b. | 2002 | 2003 | 2004 |
|---|---|---|---|
| Agricultural exports | 19.9 | 20.9 | 21.0 |
| Coffee | 2.9 | 3.6 | 4.3 |
| Light manufactures† | 273.0 | 310.6 | 315.5 |
| Domestic inputs | 17.4 | 17.6 | 18.2 |
| Imported inputs | 255.6 | 293.0 | 297.3 |
| **Total** (incl. others) | 310.5 | 369.0 | 372.7 |

* Including beverages, oils and fats, and pharmaceutical products.
† Includes valuation and classification adjustments made by Banque de la République d'Haïti.

Source: IMF, *Haiti: Selected Issues* (June 2005).

## PRINCIPAL TRADING PARTNERS
(US $ million, year ending 30 September)*

| Imports c.i.f. | 1989/90 | 1990/91 | 1991/92 |
|---|---|---|---|
| Belgium | 3.4 | 3.7 | 2.9 |
| Canada | 22.0 | 31.9 | 15.2 |
| France | 24.5 | 32.4 | 17.2 |
| Germany, Federal Republic | 14.6 | 19.2 | 10.0 |
| Japan | 23.6 | 31.2 | 17.7 |
| Netherlands | 11.2 | 13.9 | 8.7 |
| United Kingdom | 5.6 | 6.7 | 4.2 |
| USA | 153.1 | 203.2 | 126.7 |
| **Total** (incl. others) | 332.2 | 400.5 | 277.2 |

# HAITI

| Exports f.o.b.† | 1989/90 | 1990/91 | 1991/92 |
|---|---|---|---|
| Belgium | 15.9 | 19.5 | 6.0 |
| Canada | 4.5 | 4.7 | 2.3 |
| France | 17.4 | 21.6 | 6.1 |
| Germany, Federal Republic | 5.4 | 6.6 | 2.4 |
| Italy | 16.5 | 20.7 | 8.7 |
| Japan | 2.4 | 2.9 | 0.9 |
| Netherlands | 3.4 | 4.3 | 1.4 |
| United Kingdom | 2.3 | 2.3 | 0.7 |
| USA | 78.3 | 96.3 | 39.7 |
| **Total** (incl. others) | 163.7 | 198.7 | 74.7 |

* Provisional.
† Excluding re-exports.

Source: Administration Générale des Douanes.

## Transport

**ROAD TRAFFIC**
('000 motor vehicles in use)

| | 1994 | 1995 | 1996 |
|---|---|---|---|
| Passenger cars | 30.0 | 49.0 | 59.0 |
| Commercial vehicles | 30.0 | 29.0 | 35.0 |

**1999** ('000 motor vehicles in use): Passenger cars 93.0; Commercial vehicles 61.6.

Source: UN, *Statistical Yearbook*.

**SHIPPING**

**Merchant Fleet**
(registered at 31 December)

| | 2002 | 2003 | 2004 |
|---|---|---|---|
| Number of vessels | 5 | 5 | 5 |
| Total displacement ('000 grt) | 1.3 | 1.3 | 1.3 |

Source: Lloyd's Register-Fairplay, *World Fleet Statistics*.

**International Sea-borne Freight Traffic**
('000 metric tons)

| | 1988 | 1989 | 1990 |
|---|---|---|---|
| Goods loaded | 164 | 165 | 170 |
| Goods unloaded | 684 | 659 | 704 |

Source: UN, *Monthly Bulletin of Statistics*.

**CIVIL AVIATION** (international flights, 1995): Passengers arriving 367,900; Passengers departing 368,330.

## Tourism

**TOURIST ARRIVALS BY COUNTRY OF ORIGIN**

| | 1999 | 2000 | 2001 |
|---|---|---|---|
| Canada | 15,955 | 14,752 | 14,953 |
| Dominican Republic | 6,769 | 7,034 | 7,401 |
| France | 6,318 | 6,420 | 6,473 |
| Jamaica | 3,069 | 3,531 | 3,351 |
| USA | 92,543 | 92,921 | 93,065 |
| **Total** (incl. others) | 143,362 | 140,492 | 141,632 |

**Receipts from tourism** (US $ million, excl. passenger transport): 105 in 2001; 112 in 2002; 93 in 2003.

Source: World Tourism Organization.

## Communications Media

| | 2001 | 2002 | 2003 |
|---|---|---|---|
| Telephones ('000 main lines in use) | 80.0 | 130.0 | 140.0 |
| Mobile cellular telephones ('000 subscribers) | 91.5 | 140.0 | 320.0 |
| Internet users ('000) | 30.0 | 80.0 | 150.0 |

**Personal computers** ('000 in use): 2.0 in 1999.

Source: International Telecommunication Union.

**Radio receivers** ('000 in use): 415 in 1997.

**Television receivers** ('000 in use): 42 in 1999.

**Daily newspapers:** 4 in 1996 (total circulation 20,000 copies).

**Book production:** 340 titles published in 1995.

Sources (unless otherwise indicated): UNESCO, *Statistical Yearbook*; UN, *Statistical Yearbook*.

## Education

(1994/95)

| | Institutions | Teachers | Students |
|---|---|---|---|
| Pre-primary | n.a. | n.a. | 230,391* |
| Primary | 10,071 | 30,205 | 1,110,398 |
| Secondary | 1,038 | 15,275 | 195,418 |
| Tertiary | n.a. | 654* | 6,288* |

* 1990/91 figure.

**Adult literacy rate** (UNESCO estimates): 51.9% (males 53.8%; females 50.0%) in 2002 (Source: UN Development Programme, *Human Development Report*).

# Directory

## The Constitution

The Constitution of the Republic of Haiti, which was approved by the electorate in a referendum held in March 1987, provided for a system of power-sharing between a President (who may not serve two consecutive five-year terms), a Prime Minister, a bicameral legislature (comprising a chamber of deputies elected for four years and a senate whose members serve six-year terms, one-third of whom are elected every two years) and regional assemblies. The army and the police were no longer to be a combined force. The death penalty was abolished. Official status was given to the Creole language spoken by Haitians and to the folk religion, voodoo (vaudou). The Constitution was suspended after a military *coup d'état* in June 1988. It was restored when the military ruler, Brig.-Gen. Prosper Avril, fled in March 1990 and an interim President was appointed, pending a presidential election in December 1990. Fr Jean-Bertrand Aristide was elected President, but was deposed in September 1991 by a military coup. In October a new President and Government were installed by the army. In June 1992 the presidency was declared to be vacant, but in May 1994 a pro-military faction of the Senate declared the head of the Supreme Court, Émile Jonassaint, provisional President. Following US mediation, US forces (officially an international peace-keeping force) arrived on the island in September. In October Jonassaint and Lt-Gen. Raoul Cédras, the Commander-in-Chief of the armed forces, resigned. In the same month President Aristide returned to Haiti, to begin the restoration of constitutional government. He declared the army dissolved in April 1995. The constitutional amendment formally abolishing it was due to be passed by the legislature elected in May and July 2000. However, the amendment was never presented to the legislature and, after the fall of Aristide in February 2004, there were calls for the revival of the army. The head of the Supreme Court, Boniface Alexandre, succeeded Aristide, and a Council of Elders appointed Gérard Latortue as Prime Minister. The Council was to expand into an informal consultative body until new legislative elections were held. The Government and Haiti's political parties agreed in April to

# HAITI

suspend the constitutional requirement that new elections be held within 90 days of a presidential resignation. The Government later announced that elections would be held in November 2005.

## The Government

### HEAD OF STATE

**Acting President:** BONIFACE ALEXANDRE (assumed office on 29 February 2004 following the resignation of Jean-Bertrand Aristide).

### CABINET
(July 2005)

**Prime Minister:** GÉRARD LATORTUE.
**Minister of Agriculture, Natural Resources and Rural Development:** PHILIPPE MATHIEU.
**Minister of Culture and Communications:** MAGALIE COMEAU DENIS.
**Minister of Economy and Finance:** HENRI BAZIN.
**Minister of the Environment:** YVES-ANDRÉ WAINRIGHT.
**Minister of Foreign Affairs and Religion:** Gen. (retd) HÉRARD ABRAHAM.
**Minister of Haitians Residing Abroad:** ALIX BAPTISTE.
**Minister of Health:** JOSETTE BIJOUX.
**Minister of the Interior and National Security:** PAUL GUSTAVE MAGLOIRE.
**Minister of Justice and Public Security:** HENRI MARGE DORLÉANS.
**Minister of National Education, Youth and Sport:** PIERRE BUTEAU.
**Minister of Planning and External Co-operation:** ROLAND PIERRE.
**Minister of Public Works, Transport and Communications:** FRITZ ADRIEN.
**Minister of Social Affairs:** FRANCK CHARLES.
**Minister of Trade and Industry:** FRITZ KÉNOL.
**Minister of Women's Affairs and Rights:** ADELINE MAGLOIRE CHANCY.

### MINISTRIES

**Office of the President:** Palais National, rue de la République, Port-au-Prince; tel. 222-3024; e-mail webmestre@palaisnational.info; internet www.palaisnational.info/pnh/accueil.php.
**Office of the Prime Minister:** Villa d'Accueil, Delmas 60, Musseau, Port-au-Prince; tel. 245-0007; fax 245-1624.
**Ministry of Agriculture, Natural Resources and Rural Development:** BP 1441, Damien, Port-au-Prince; tel. 250-7558.
**Ministry of Culture and Communications:** 4 rue Nagny, Port-au-Prince; tel. 221-1716; e-mail dg1@haiticulture.org; internet www.haiticulture.org.
**Ministry of Economy and Finance:** Palais des Ministères, rue Monseigneur Guilloux, Port-au-Prince; tel. 222-7113; fax 223-1247.
**Ministry of the Environment:** Haut Turgeau 181, Port-au-Prince; tel. 245-7572; fax 245-7360; e-mail dgmde@rehred-haiti.net; internet www.rehred-haiti.net.
**Ministry of Foreign Affairs and Religion:** blvd Harry S Truman, Cité de l'Exposition, Port-au-Prince; tel. 222-8482; fax 223-1668; e-mail webmaster@maehaitiinfo.org; internet www.maehaitiinfo.org.
**Ministry of Haitians Residing Abroad:** 87 ave Jean-Paul II, Turgeau, Port-au-Prince; tel. 244-4321; fax 245-3400; internet haiti2004lakay.com; f. 1995.
**Ministry of Health:** Palais de Ministères, Port-au-Prince; tel. 222-1583; fax 222-4066.
**Ministry of the Interior and Local Government:** Palais des Ministères, Port-au-Prince; tel. 223-4491; fax 222-4429.
**Ministry of Justice:** ave Charles Sumner 19, Port-au-Prince; tel. 245-1626.
**Ministry of National Education, Youth and Sport:** rue Dr Audain, Port-au-Prince; tel. 222-1036; fax 223-7887.
**Ministry of Planning and External Co-operation:** Palais des Ministères, Port-au-Prince; tel. 222-4148; fax 223-4193.

**Ministry of Public Works:** Palais des Ministères, rue Mgr Guilloux, Port-au-Prince; tel. 222-2528; fax 222-3240; e-mail bmministre@haititptc.org; internet www.haititptc.org.
**Ministry of Social Affairs:** 16 rue de la Révolution, Port-au-Prince; tel. 222-1244; fax 223-8084.
**Ministry of Trade and Industry:** rue Légitime, Champ-de-Mars, BP 200, Port-au-Prince; tel. 222-1628; fax 223-8402.
**Ministry of Women's Affairs and Rights:** Delmas 31, rue Louverture et Biassou 2, Port-au-Prince; tel. 249-5913.

## President and Legislature

### PRESIDENT

**Presidential Election, 26 November 2000**

| Candidates | % of votes |
|---|---|
| Jean-Bertrand Aristide (FL) | 91.7 |
| Jean-Arnold Dumas | 2.0 |
| Evan Nicolas | 1.6 |
| Serge Sylvain | 1.3 |
| Calixte Dorisca | 1.3 |
| Jacques Philippe Dorce | 1.1 |
| Paul Arthur Fleurival | 1.0 |
| **Total** (incl. others) | 100.0 |

### LEGISLATURE

In January 2004, owing to the country's failure to hold legislative elections, the mandates of all 83 members of the Chamber of Deputies and 12 members of the Senate expired, leaving Haiti effectively without a legislature and entitling the President to rule by decree.

#### Sénat
(Senate)

**President:** (vacant).

**Elections, 21 May, 9 July and 26 November 2000**

| | Seats |
|---|---|
| La Fanmi Lavalas (FL) | 26* |
| Pati Louvri Baryè (PLB) | 1 |
| **Total** | 27 |

*In June 2001 seven FL senators resigned their seats.

#### Chambre des Députés
(Chamber of Deputies)

**President:** (vacant).

**Elections, 21 May, 9 July and 26 November 2000**

| | Seats |
|---|---|
| La Fanmi Lavalas (FL) | 73 |
| Mouvement Chrétien pour Batir une Nouvelle Haïti (MOCHRENA) | 3 |
| Espace de Concertation | 2 |
| Pati Louvri Baryè (PLB) | 2 |
| Koordinasyon Resistans Grandans (KOREGA-ESCANP) | 1 |
| Organisation du Peuple en Lutte (OPL) | 1 |
| Independent | 1 |
| **Total** | 83 |

## Political Organizations

**Action Démocratique pour Bâtir Haïti (ADEBHA):** Delmas 60, 16, Port-au-Prince; tel. 558-1623; e-mail camilleleblanc@hotmail.com; f. 2004; Pres. CAMILLE LEBLANC.

**Alliance pour la Libération et l'Avancement d'Haïti (ALAH):** Haut Turgeau 95, BP 13350, Port-au-Prince; tel. 245-0446; fax 257-4804; e-mail reynoldgeorges@yahoo.com; f. 1975; Leader REYNOLD GEORGES.

**L'Alternative pour le Changement (AC):** 88 Chemin des Dalles, Port-au-Prince; tel. 557-7388; f. 2000; Leader GÉRARD BLOT.

**Congrès National des Mouvements Démocratiques (KONAKOM):** 101 Bois Verna, Port-au-Prince; tel. 245-6228; f. 1987; social-democratic; Leader VICTOR BENOÎT.

# HAITI — Directory

**Espace de Concertation:** f. 1999; centre-left coalition of five parties; Leaders EVANS PAUL, MICHA GAILLARD.

**La Fanmi Lavalas:** blvd 15 Octobre, Tabarre, Port-au-Prince; tel. 256-7208; f. 1996 by Jean-Bertrand Aristide; formed a coalition with the MOP, the OPL and the PLB.

**Front National pour le Changement et la Démocratie (FNCD):** f. 1990; social-democratic; Leader EVANS PAUL.

**Front de Reconstruction Nationale (FRN):** Gonaïves; f. 2004; Sec.-Gen. GUY PHILIPPE.

**Grand Front de Centre Droit (GFCD):** 21 blvd Harry S Truman, Cité de l'Exposition, Port-au-Prince; tel. 245-6251; e-mail hdr@mdnhaiti.org; internet www.gfcd.org; f. 2003; centre-right alliance; Pres. HUBERT DE RONCERAY.

**Grand Parti Socialiste Haïtien (GPSH):** Port-au-Prince; f. 2004; mem. of the GFCD; Leader SERGE GILLES.

**Jeunesse Pouvoir Populaire (JPP):** 410 rue Tiremasse, Port-au-Prince; tel. 558-1647; f. 1997; Leader RENÉ CIVIL.

**Konvansyon Inite Demokratik (KID):** 14 rue Camille Leon, Port-au-Prince; tel. 245-0185; f. 1986; fmrly Konfederasyon Inite Demokratik; Leader EVANS PAUL.

**Koordinasyon Resistans Grandans (KOREGA-ESCANP):** e-mail crb@maf.org; regionally based; radical left; Leader Fr JOACHIM SAMEDI.

**Mobilisation pour le Développement National (MDN):** c/o CHISS, 33 rue Bonne Foi, BP 2497, Port-au-Prince; tel. 222-3829; e-mail info@mdnhaiti.org; internet www.mdnhaiti.org; f. 1986; Pres. HUBERT DE RONCERAY; Sec.-Gen. MAX CARRE.

**Mouvement Chrétien pour une Nouvelle Haïti (MOCHRENHA):** rue M 7 Turgeau, Port-au-Prince, Carrefour; tel. 401-3120; e-mail mochrenha@hotmail.com; internet www.mochrenha.com; f. 1991; Leaders LUC MÉSADIEU, GILBERT N. LÉGER.

**Mouvement pour l'Instauration de la Démocratie en Haïti (MIDH):** 114 ave Jean Paul II, Port-au-Prince; tel. 245-8377; f. 1986; centre-right; Pres. MARC BAZIN.

**Mouvement National et Patriotique du 28 Novembre (MNP-28):** f. 1991; Leader DÉJEAN BÉLIZAIRE.

**Mouvement pour l'Organisation du Pays (MOP):** 9 rue Stella, Delmas 31, Port-au-Prince; tel. 249-3408; f. 1946; centre party; Leader JEAN MOLIÈRE.

**Mouvement Patriotique pour le Sauvetage National (MPSN):** f. 1998; right-wing coalition comprising 7 parties; Leader HUBERT DE RONCERAY.

**Mouvement pour la Reconstruction Nationale (MRN):** f. 1991; Leader JEAN-ENOL BUTEAU.

**Mouvman Konbit Nasyonal (MKN):** Leader VOLVICK RÉMY JOSEPH.

**Nouveau Parti Communiste Haïtien (NPCH):** Port-au-Prince; e-mail contact@npch.net; internet www.npch.net; Marxist-Leninist.

**Organisation du Peuple en Lutte (OPL):** 105 ave Lamartinière, Bois Verna, Port-au-Prince; tel. 245-4214; e-mail info@oplpeople.com; internet www.oplpeople.com/home.html; f. 1991 as Organisation Politique Lavalas; name changed as above 1998; Leader SAUVEUR PIERRE-ÉTIENNE.

**Parti Agricole et Industrie National (PAIN):** f. 1956; Spokesman TOUSSAINT DESROSIERS.

**Parti des Démocrates Haïtiens (PADEMH):** Leader JEAN-JACQUES CLARK PARENT.

**Parti Démocratique et Chrétien d'Haïti (PDCH):** 127 rue du Magasin de l'Etat, Port-au-Prince; tel. 550-7282; f. 1979; Christian Democrat party; Leaders OSNER FÉVRY, JOACHIN PIERRE.

**Parti pour un Développement Alternatif (PADH):** Leader GÉRARD DALVIUS.

**Parti Nationale Démocratique Progressiste d'Haïti (PNDPH):** Port-au-Prince; Pres. TURNEB DELPÉ.

**Parti National Progressiste Révolutionnaire (PANPRA):** 5 rue Marcelin, Port-au-Prince; tel. 257-5359; f. 1989; social-democratic.

**Parti National des Travailleurs (PNT):** Port-au-Prince.

**Parti Populaire Nationale (PPN):** 11 rue Capois, Port-au-Prince; tel. 222-6513; f. 1987 as Assemblée Populaire Nationale (APN); name changed as above in 1999; radical left; Sec.-Gen. BEN DUPUY.

**Parti Revolutionnaire Démocratique d'Haïti (PRDH):** fmrly Mouvement Démocratique pour la Libération d'Haïti (MODELH); Leader FRANÇOIS LATORTUE.

**Parti Social Chrétien d'Haïti (PSCH):** BP 84, Port-au-Prince; f. 1979; Leader GRÉGOIRE EUGÈNE.

**Pati Louvri Baryè (PLB):** f. 1992 by Renaud Bernardin; Sec.-Gen. FRANÇOIS PIERRE-LOUIS.

**Rassemblement des Démocrates Chrétiens (RDC):** 177 rue du Centre, Port-au-Prince; tel. 234-4214; Leader EDDY VOLEL.

**Rassemblement des Démocrates Nationalistes et Progressistes (RDNP):** 234 route de Delmas, Delmas, Port-au-Prince; tel. 246-3313; f. 1979; centre party; Sec.-Gen. LESLIE FRANÇOIS MANIGAT.

**Union Démocrates Patriotiques (UDP):** 30 rue Geffrard, Port-au-Prince; tel. 256-1953; Leader ROCKFELLER GUERRE.

**Union Patriotique:** f. 2002; coalition of four opposition parties; Sec.-Gen. LESLIE FRANÇOIS MANIGAT.

## Diplomatic Representation

### EMBASSIES IN HAITI

**Argentina:** 8 rue Mangones, Berthé, Pétionville, Port-au-Prince; tel. 257-5725; fax 256-8227; e-mail embarghaiti@hainet.net; Chargé d'affaires a.i. MARIO JOSÉ PINO.

**Bahamas:** 12 rue Goulard, Place Boyer, Pétionville; tel. 257-8782; fax 256-5729; e-mail bahamasemb@hainet.net; Ambassador Dr EUGENE NEWRY.

**Brazil:** 34 rue Lamarre, Pétionville, BP 6140, Port-au-Prince; tel. 256-9662; fax 510-6111; e-mail brasemb1@transnethaiti.com; Ambassador WALTER MOREIRA.

**Canada:** Route de Delmas, entre Delmas 71 et 75, BP 826, Port-au-Prince; tel. 249-9000; fax 249-9920; e-mail prnce@international.gc.ca; internet www.dfait-maeci.gc.ca/haiti; Ambassador CLAUDE BOUCHER.

**Chile:** 2 rue Coutilien, Musseau, Port-au-Prince; tel. 256-7960; fax 257-0623; e-mail myoung@minrel.gov.cl; Ambassador ALAIN MARCEL YOUNG DEBEUF.

**China (Taiwan):** 16 rue Léon Nau, Pétionville, BP 655, Port-au-Prince; tel. 257-2899; fax 256-8067; e-mail hti@ms90.url.com.tw; Ambassador YANG CHENG-TA.

**Cuba:** 18 rue Marion, Peguy Ville, POB 15702, Port-au-Prince; tel. 256-3812; fax 257-8566; e-mail ecuhaiti@hainet.net; Ambassador ROLANDO GÓMEZ GONZÁLEZ.

**Dominican Republic:** rue Panaméricaine 121, BP 56, Pétionville, Port-au-Prince; tel. 257-0568; fax 221-8718; e-mail embrepdomhai@yahoo.com; Ambassador JOSÉ SERULLE RAMIA.

**France:** 51 pl. des Héros de l'Indépendance, BP 1312, Port-au-Prince; tel. 222-0952; fax 223-5675; e-mail ambafrance@hainet.net; Ambassador THIERRY BURKARD.

**Germany:** 2 impasse Claudinette, Bois Moquette, Pétionville, BP 1147, Port-au-Prince; tel. 257-7280; fax 257-4131; e-mail germanem@haitelonline.com; Ambassador Dr GORDON KRICKE.

**Holy See:** rue Louis Pouget, Morne Calvaire, BP 326, Port-au-Prince; tel. 257-6308; fax 257-3411; e-mail nonciature@haitiworld.com; Apostolic Nuncio Most Rev. MARIO GIORDANA (Titular Archbishop of Minora).

**Jamaica:** 141 rue Pavée, Edif. Firestone, BP 1065, Port-au-Prince; tel. 248-0589; fax 226-0613; Ambassador PETER BLACK.

**Japan:** Villa Bella Vista, 2 Impasse Tulipe Desprez, BP 2512, Port-au-Prince; tel. 245-3333; fax 245-8834; Ambassador HARUO OKOMOTO.

**Mexico:** Delmas 60, 2, BP 327, Port-au-Prince; tel. 257-8100; fax 256-6528; e-mail embmxhai@yahoo.com; Ambassador ANACELIA PÉREZ CHARLES.

**Spain:** 54 rue Pacot, State Liles, BP 386, Port-au-Prince; tel. 245-4410; fax 245-3901; e-mail ampespht@mail.mae.es; Ambassador PAULINO GONZÁLEZ FERNÁNDEZ-CORUGEDO.

**USA:** 5 blvd Harry S Truman, BP 1761, Port-au-Prince; tel. 222-0200; fax 223-9038; internet usembassy.state.gov/haiti; Ambassador JAMES B. FOLEY.

**Venezuela:** blvd Harry S Truman, Cité de l'Exposition, BP 2158, Port-au-Prince; tel. 222-0973; e-mail venhtamb@compa.net; Ambassador MARCO REQUENA.

## Judicial System

Law is based on the French Napoleonic Code, substantially modified during the presidency of François Duvalier.

Courts of Appeal and Civil Courts sit at Port-au-Prince and the three provincial capitals: Gonaïves, Cap Haïtien and Port de Paix. In principle each commune has a Magistrates' Court. Judges of the

HAITI — Directory

Supreme Court and Courts of Appeal are appointed by the President.

### Supreme Court
Port-au-Prince; tel. 222-3212.
Pres. (vacant); Vice-Pres. PRADEL PÉAN.

**Citizens' Rights Defender:** NECKER DESSABLES.

## Religion

Roman Catholicism and the folk religion voodoo (vaudou) are the official religions. There are various Protestant and other denominations.

### CHRISTIANITY

#### The Roman Catholic Church

For ecclesiastical purposes, Haiti comprises two archdioceses and seven dioceses. At 31 December 2003 adherents represented some 65.4% of the population.

#### Bishops' Conference

Conférence Episcopale de Haïti, angle rues Piquant et Lammarre, BP 1572, Port-au-Prince; tel. 222-5194; fax 223-5318; e-mail ceh56@hotmail.com.
f. 1977; Pres. Rt Rev. HUBERT CONSTANT (Archbishop of Cap-Haïtien).

**Archbishop of Cap-Haïtien:** Most Rev. HUBERT CONSTANT, Archevêché, rue 19–20 H, BP 22, Cap-Haïtien; tel. 262-0071; fax 262-1278.

**Archbishop of Port-au-Prince:** Most Rev. FRANÇOIS-WOLFF LIGONDÉ, Archevêché, rue Dr Aubry, BP 538, Port-au-Prince; tel. 222-2045; e-mail archeveche.pap@globalsud.com.

#### The Anglican Communion

Anglicans in Haiti fall under the jurisdiction of a missionary diocese of Province II of the Episcopal Church in the USA.

**Bishop of Haiti:** Rt Rev. JEAN ZACHE DURACIN, Eglise Episcopale d'Haïti, BP 1309, Port-au-Prince; fax 257-3412; e-mail epihaiti@globalsud.net.

#### Protestant Churches

**Baptist Convention:** BP 20, Cap-Haïtien; tel. 262-0567; e-mail conventionbaptiste@yahoo.com; f. 1964; Pres. Rev. GÉDÉON EUGÈNE.

**Evangelical Lutheran Church of Haiti:** 144 rue Capitale, BP 15, Les Cayes; tel. 286-3398; f. 1975; Pres. (vacant).

Other denominations active in Haiti include Methodists and the Church of God 'Eben-Ezer'.

## The Press

### DAILIES

**Le Matin:** 6 rue du Quai, BP 367, Port-au-Prince; tel. 222-2040; fax 223-2551; f. 1908; French; independent; Editor DUMAYRIC CHARLIER; circ. 5,000.

**Le Nouvelliste:** 198 rue du Centre, BP 1013, Port-au-Prince; tel. 223-2114; fax 223-2313; f. 1898; evening; French; independent; Editor CARLO DÉSINOR; circ. 6,000.

### PERIODICALS

**Ayiti Fanm:** 3 bis, rue Sapotille, Port-au-Prince; tel. 245-1930; Creole; Dir CLORINDE ZÉPHYR.

**Bon Nouvèl:** 103 rue Pavée, Étage Imprimerie La Phalange, BP 1594, Port-au-Prince; tel. 223-9186; fax 222-8105; e-mail bonnouvel@rehred-haiti.net; internet www.rehred-haiti.net/membres/bonouvel/; f. 1967; monthly; Creole; Dir JEAN HOET.

**Bulletin de Liaison:** Centre Pedro-Arrupe, BP 1710, Port-au-Prince; tel. 245-2360; e-mail dcm3@georgetown.edu; internet liaison.lemoyne.edu; f. 1996; 4 a year; Editors ANDRÉ CHARBONNEAU, ALFRED DORVIL, DONALD MALDARI.

**Haïti en Marche:** 74 bis, rue Capois, Port-au-Prince; tel. and fax 221–8596; e-mail pub@haitienmarche.com; internet www.haitienmarche.com; f. 1987; weekly; Editor MARCUS GARCIA.

**Haïti Observateur:** 98 ave John Brown, Port-au-Prince; tel. 228-0782; f. 1971; weekly; Editor LÉO JOSEPH.

**Haïti Progrès:** 11 rue Capois, Port-au-Prince; tel. 222-6513; fax 222-7022; e-mail editor@haiti-progres.com; internet www.haiti-progres.com; f. 1983; weekly; French, English and Creole; Dir KIM IVES.

**Jounal Libète:** BP 13441, Delmas, Port-au-Prince; tel. 245-7766; fax 245-7760; e-mail jyurfie@acn.com; f. 1990; weekly; Creole; Dir-Gen. JEAN-YVES URFIÉ; Editor GARY BÉLIZAIRE.

**Liaison:** 9 rue Chochotte, Babiole, BP 2481, Port-au-Prince; tel. 245-1186; fax 245-3629; e-mail havainfo@acn.com; French; available in Creole as *Aksyon*; Dir IVES-MARIE CHANEL.

**Le Messager du Nord-Ouest:** Port de Paix; weekly.

**Le Moniteur:** rue du Centre, BP 214 bis, Port-au-Prince; tel. 222-1744; f. 1845; 2 a week; French; the official gazette; circ. 2,000.

**Optique:** Institut Français d'Haiti, BP 1316, Port-au-Prince; tel. 245-7766; monthly; arts.

**Le Septentrion:** Cap-Haïtien; weekly; independent; Editor NELSON BELL; circ. 2,000.

**Superstar Détente:** 3 ruelle Chériez, Port-au-Prince; tel. 245-3450; fax 222-6329; cultural magazine; Dir CLAUDEL VICTOR.

### NEWS AGENCIES

**Agence Haïtienne de Presse (AHP):** 6 rue Fernand, Port-au-Prince; tel. 245-7222; fax 245-5836; e-mail ahp@yahoo.com; internet www.ahphaiti.org; f. 1989; publishes daily news bulletins in french and english; Dir-Gen. GEORGES VENEL REMARAIS.

**AlterPresse:** 38 Delmas 8, BP 19211, Port-au-Prince; tel. 554-1882; e-mail alterpresse@medialternatif.org; internet www.alterpresse.org; f. 2001; independent; Dir GOTSON PIERRE.

#### Foreign Bureaux

**Agence France-Presse (AFP):** 177 rue du Centre, Cour Buromatic, Port-au-Prince; tel. 222-3469; fax 222-3759.

**Agencia EFE** (Spain): Port-au-Prince; tel. 255-9517; Correspondent HEROLD JEAN-FRANÇOIS.

**Associated Press (AP)** (USA): Port-au-Prince; e-mail hib@igc.apc.com.

**Inter Press Service** (Italy): 16 rue Malval, Turgeau, Port-au-Prince; tel. 260-5512; fax 260-5513; e-mail ipsnoramcarib@ipsnews.net; internet ipsnews.org/new_focus/haiti/index.asp; Correspondent JANE REGAN.

**Prensa Latina** (Cuba): Port-au-Prince; tel. 246-5149; internet www.prensa-latina.com; Correspondent JACQUELÍN TELEMAQUE.

**Reuters** (United Kingdom): Port-au-Prince; Correspondent GUYLER C. DELVA.

## Publishers

**Editions des Antilles:** route de l'Aéroport, Port-au-Prince.

**Editions Caraïbes S.A.:** 57 rue Pavée, BP 2013, Port-au-Prince; tel. 222-0032; Man. PIERRE J. ELIE.

**Editions du Soleil:** BP 2471, rue du Centre, Port-au-Prince; tel. 222-3147; education.

**L'Imprimeur Deux:** Le Nouvelliste, 198 rue du Centre, Port-au-Prince.

**Maison Henri Deschamps—Les Entreprises Deschamps Frisch, SA:** 25 rue Dr Martelly Seïde, BP 164, Port-au-Prince; tel. 223-2215; fax 223-4976; e-mail entdeschamps@gdfhaiti.com; f. 1898; education and literature; Man. Dir JACQUES DESCHAMPS, Jr; CEO HENRI R. DESCHAMPS.

**Natal Imprimerie:** rue Barbancourt, Port-au-Prince; Dir ROBERT MALVAL.

**Théodore Imprimerie:** rue Dantes Destouches, Port-au-Prince.

## Broadcasting and Communications

### TELECOMMUNICATIONS

**Conseil National des Télécommunications (CONATEL):** 16 ave Marie Jeanne, Cité de l'Exposition, BP 2002, Port-au-Prince; tel. 222-0300; fax 223-9229; e-mail conatel@haitiwww.haiticonatel.org; f. 1969; govt communications licensing authority; Dir-Gen. JEAN MICHEL BOISROND.

**Digicel Haiti:** Port-au-Prince; internet www.digicelhaiti.com; f. 2005; mobile telephone network provider; sole company licensed to operate a GSM network within Haiti; Group Chair. DENIS O'BRIEN.

# HAITI

**Télécommunications d'Haïti (Téléco):** blvd Jean-Jacques Dessalines, BP 814, Port-au-Prince; tel. 245-2200; fax 223-0002; e-mail info@haititeleco.com; internet www.haititeleco.com; Dir-Gen. RENÉ MÉRONEY.

## BROADCASTING

### Radio

**Radio Antilles International:** 175 rue du Centre, BP 2335, Port-au-Prince; tel. 223-0696; fax 222-0260; f. 1984; independent; Dir-Gen. JACQUES SAMPEUR.

**Radio Cacique:** 5 Bellevue, BP 1480, Port-au-Prince; tel. 245-2326; f. 1961; independent; Dir JEAN-CLAUDE CARRIÉ.

**Radio Canal du Christ:** 175 rue du Centre, Port-au-Prince; tel. 223-9917; fax 222-0260; e-mail canalchrist93.5@mcm.net; f. 1998; Dir JACQUES SAMPEUR.

**Radio Caraïbes:** 19 rue Chavannes, Port-au-Prince; tel. 223-0644; fax 223-4955; e-mail caraibesfm@netcourrier.com; f. 1949; independent; Dir PATRICK MOUSSIGNAC.

**Radio Céleste:** 106 rue de la Réunion, Port-au-Prince; tel. 222-4714; fax 222-6636; f. 1991; Dir JEAN-EDDY CHARLEUS.

**Radio Galaxie:** 17 rue Pavée, Port-au-Prince; tel. 223-9942; fax 223-9941; e-mail rgalaxie@hotmail.com; f. 1990; independent; Dir YVES JEAN-BART.

**Radio Ginen:** 9 bis, Delmas 31, Port-au-Prince; tel. 249-1738; e-mail feedback@radyoginen.com; internet www.radyoginen.com; f. 1994; Dir LUCIEN BORGES.

**Radio Haïti Inter:** Delmas 66A, 522, en face de Delmas 91, BP 737, Port-au-Prince; tel. 257-3111; f. 1935; independent; ceased broadcasting indefinitely in Feb. 2003; Dir MICHÈLE MONTAS.

**Radio Kadans FM:** 3 rue Neptune, Delmas 65, Port-au-Prince; tel. 249-4040; fax 245-2672; f. 1991; Dir LIONEL BENJAMIN.

**Radio Kiskeya:** 7 rue Pavée, Port-au-Prince; tel. 222-6002; fax 223-6204; f. 1994; Dir SONY BASTIEN.

**Radio Lakansyèl:** 285 route de Delmas, Port-au-Prince; tel. 246-2020; independent; Dir ALEX SAINT-SURIN.

**Radio Lumière:** Côte-Plage 16, BP 1050, Port-au-Prince; f. 1959; tel. 234-0330; internet www.radiolumiere.org; f. 1959; Protestant; independent; Dir VARNEL JEUNE.

**Radio Magic Stéreo:** 346 route de Delmas, Port-au-Prince; tel. 245-5404; f. 1991; independent; Dir FRITZ JOASSIN.

**Radio Mélodie:** 74 bis, rue Capois, Port-au-Prince; tel. 221-8567; fax 221-1323; e-mail melodiefm@hotmail.com; f. 1998; Dir MARCUS GARCIA.

**Radio Metropole:** 18 Delmas 52, BP 62, Port-au-Prince; tel. 246-2626; fax 249-2020; internet www.metropolehaiti.com; f. 1970; independent; Dir-Gen. RICHARD WIDMAIER.

**Radio Nationale d'Haïti:** 174 rue du Magasin de l'Etat, BP 1143, Port-au-Prince; tel. 223-5712; fax 223-5911; f. 1977; govt-operated; Dir-Gen. MICHEL FAVARD.

**Radio Plus:** 85 rue Pavée, BP 1174, Port-au-Prince; tel. 222-1588; fax 223-2288; f. 1988; independent; Dir GUY CÉSAR.

**Radio Port-au-Prince:** Stade Sylvio Cator, BP 863, Port-au-Prince; f. 1979; independent; Dir GEORGE L. HÉRARD.

**Radio Sans Souci:** 16 rue Malval, Turgeau, Port-au-Prince; tel. 260-5512; fax 262-5444; e-mail sanssoucifm@mediacom-ht.com; internet www.radiosanssouci.com; f. 1998; Dir IVES-MARIE CHANEL.

**Radio Signal FM:** 127 rue Louverture, Pétionville, BP 391, Port-au-Prince; tel. 298-4370; fax 298-4372; e-mail signalfm@netcourrier.com; internet www.signalfmhaiti.com; f. 1991; independent; Dir-Gen. MARIO VIAUD.

**Radio Soleil:** BP 1362, Archevêché de Port-au-Prince; tel. 222-3062; fax 222-3516; e-mail contact@radiosoleil.com; internet www.radiosoleil.com; f. 1978; Catholic; independent; educational; broadcasts in Creole and French; Dir Fr ARNOUX CHÉRY.

**Radio Solidarité:** 6 rue Fernand, Port-au-Prince; tel. 244-0469; fax 244-6698; e-mail solidarite@haitiwebs.com; internet www.haitiwebs.com/radiosolidarite/; Dir VENEL REMARAIS.

**Radio Superstar:** 38 rue Safran, Delmas 68, Port-au-Prince; tel. 257-7219; fax 257-3015; e-mail superstar1029fmfr@yahoo.fr; f. 1987; independent; Dir-Gen. ALBERT CHANCY.

**Radio Timoun:** 27 bis, rue Camille Léon, Port-au-Prince; tel. 245-1099; f. 1996; Dir MARIE KEMLY PERCY.

**Radio Tropic FM:** 6 ave John Brown, Port-au-Prince; tel. 223-6565; f. 1991; independent; Dir GUY JEAN.

**Radio Tropicale Internationale:** Delmas 27–29, Dubois Shopping Center, Port-au-Prince; tel. 249-1646; f. 1994; Dir JOEL BORGELLA.

**Radio Vision 2000:** 184 ave John Brown, Port-au-Prince; tel. 245-4914; f. 1991; Dir LÉOPOLD BERLANGER.

### Television

**Galaxy 2:** 6 rue Henri Christophe, Jacmel; tel. 288-2324; f. 1989; independent; Dir MILOT BERQUIN.

**PVS Antenne 16:** 137 rue Monseigneur Guilloux, Port-au-Prince; tel. and fax 222-1277; f. 1988; independent; Dir-Gen. RAYNALD DELERME.

**Télé Eclair:** 526 route de Delmas, Port-au-Prince; tel. 256-4505; fax 256-3828; f. 1996; independent; Dir PATRICK ANDRÉ JOSEPH.

**Télé Express Continentale:** rue de l'Eglise, Jacmel; tel. 288-2246; fax 288-2191; f. 1985; independent; Dirs JEAN-FRANÇOIS VERDIER, JACQUES JEAN-PIERRE.

**Télé Haïti:** blvd Harry S Truman, BP 1126, Port-au-Prince; tel. 222-3887; fax 222-9140; f. 1959; independent; pay-cable station with 33 channels; in French, Spanish and English; Dir MARIE CHRISTINE MOURRAL BLANC.

**Télé Smart:** Hinche; tel. 277-0347; f. 1998; independent; Dir MOZART SIMON.

**Télé Timoun:** blvd 15 Octobre, Tabarre; tel. 250-1924; fax 250-9972; f. 1996; independent.

**Télémax:** 3 Delmas 19, Port-au-Prince; tel. 246-2002; fax 246-1155; f. 1994; independent; Dir ROBERT DENIS.

**Télévision Nationale d'Haïti:** Delmas 33, BP 13400, Port-au-Prince; tel. 246-0200; fax 246-0693; e-mail info@tnh.ht; internet www.tnh.ht; f. 1979; govt-owned; cultural; 4 channels in Creole, French and Spanish; administered by four-mem. board; Dir JACQUES PRICE JEAN.

**Trans-America:** ruelle Roger, Gonaïves; tel. 274-0113; f. 1990; independent; Dir-Gen. HÉBERT PELISSIER.

**TV Magik:** 16 rue Conty, Jacmel; tel. 288-2456; f. 1992; independent; Dirs LOUIS ANTONIN BLAISE, RICHARD CYPRIEN.

**TVA:** rue Liberté, Gonaïves; independent; cable station with three channels; Dir-Gen. GÉRARD LUC JEAN-BAPTISTE.

# Finance

(cap. = capital; m. = million; res = reserves; dep. = deposits; brs = branches; amounts in gourdes)

## BANKING

### Central Bank

**Banque de la République d'Haïti:** angle rues du Magasin d'État et des Miracles, BP 1570, Port-au-Prince; tel. 299-1200; fax 299-1045; e-mail brh_adm@brh.net; internet www.brh.net; f. 1911 as Banque Nationale de la République d'Haïti; name changed as above in 1979; bank of issue; cap. 50m., res 4,084.0m., dep. 9,986.4m. (Sept. 2001); Gov. RAYMOND MAGLOIRE; Dir-Gen. CHARLES CASTEL.

### Commercial Banks

**Banque Industrielle et Commerciale d'Haïti:** 158 rue Dr Aubry, Port-au-Prince; tel. 299-6800; fax 299-6804; f. 1974.

**Banque Nationale de Crédit:** angle rues du Quai et des Miracles, BP 1320, Port-au-Prince; tel. 299-4081; fax 299-4045; f. 1979; cap. 25m., dep. 729.9m. (Sept. 1989); Pres. GUITEAU TOUSSAINT; Dir-Gen. LEVÈQUE VALBRUN.

**Banque Populaire Haïtienne:** angle rues des Miracles et du Centre, Port-au-Prince; tel. 299-6000; fax 222-4389; e-mail bphinfo@brh.net; f. 1973; state-owned; cap. and res 22m., dep. 614m. (31 Dec. 2001); 3 brs; Dir-Gen. ERNST GILLES; Pres. RODNÉE DESCHINEAUX.

**Banque de l'Union Haïtienne:** angle rues du Quai et Bonne Foi, BP 275, Port-au-Prince; tel. 299-8500; fax 223-2852; e-mail buh@buhsa.com; internet www.buhsa.com; f. 1973; cap. 30.1m., res 6.2m., dep. 1,296.7m. (Sept. 1997); Pres. OSWALD J. BRANDT II; 9 brs.

**Capital Bank:** 149–151 rue des Miracles, BP 2464, Port-au-Prince; tel. 299-6500; fax 299-6519; e-mail capitalbank@brh.net; f. 1985; fmrly Banque de Crédit Immobilier, SA; Pres. BERNARD ROY; Gen. Man. LILIANE C. DOMINIQUE.

**PROMOBANK** (Banque de Promotion Commerciale et Industrielle, SA): angle des rues Faubert et Rigaud, Pétionville, Port-au-Prince; tel. 299-8000; fax 299-8125; e-mail info@promointer.net; internet www.promointer.net; f. 1974 as B.N.P. Haïti, name changed as above in 1994; cap. 60.4m., res 16.4m., dep. 1,183.4m. (Dec. 1998); Pres. and Chair. MICHAEL GAY; Gen. Man. JOSSELINE COLIMON-FÉTHIÈRE.

**Société Caribéenne de Banque, SA (SOCABANK):** 37 rue Pavée, BP 80, Port-au-Prince; tel. 299-7000; fax 299-7036; e-mail socabankcard@usa.net; internet www.socabank.net; f. 1994; Pres. JEAN-CLAUDE NADAL; Dir-Gen. CHARLES CLERMONT.

**Sogebank, SA** (Société Générale Haïtienne de Banque, SA): route de Delmas BP 1315, Delmas; tel. 229-5000; fax 229-5022; internet www.sogebank.com; f. 1986; cap. 79.5m.; Pres. GERARD MOSCOSO; 7 brs.

**Sogebel** (Société Générale Haïtienne de Banque d'Espargne et de Logement): route de l'Aéroport, BP 2409, Delmas; tel. 229-5353; fax 229-5352; f. 1988; cap. 15.1m., dep. 249.9m.; Gen. Man. CLAUDE PIERRE-LOUIS; 2 brs.

**Unibank:** 94 place Geffard, BP 46, Port-au-Prince; tel. 299-2300; fax 229-2332; e-mail info@unibankhaiti.com; internet www.unibankhaiti.com; f. 1993; cap. 100m., res 17.5m., dep. 3,366m. (Sept. 1999); Pres. F. CARL BRAUN; Dir-Gen. FRANCK HELMCKE; 20 brs.

### Foreign Banks

**Bank of Nova Scotia** (Canada): 360 blvd J. J. Dessalines, BP 686, Port-au-Prince; tel. 299-3000; fax 229-3204; Man. CHESTER A. S. HINKSON; 3 brs.

**Citibank, NA** (USA): 242 route de Delmas, BP 1688, Port-au-Prince; tel. 299-3200; fax 299-3227; internet www.citibank.com/haiti; Dir GLADYS M. COUPET.

### Development Bank

**Banque Haïtienne de Développement:** 20 ave Lamartinière, Port-au-Prince; tel. 245-4422; fax 244-3737; f. 1998; Dir-Gen. YVES LEREBOURS.

### INSURANCE

#### National Companies

**L'Atout Assurance, SA:** 77 rue Lamarre, Port-au-Prince; tel. 223-9378; Dir JEAN EVEILLARD.

**Compagnie d'Assurances d'Haïti, SA (CAH):** étage Dynamic Entreprise, route de l'Aéroport, BP 1489, Port-au-Prince; tel. 250-0700; fax 250-0236; e-mail info@groupedynamic.com; internet www.groupedynamic.com/cah.php; f. 1978; merged with Multi Assurances SA in 2000; subsidiary of Groupe Dynamic SA; Group Chair. and CEO PHILIPPE R. ARMAND.

**Excelsior Assurance, SA:** rue 6, no 24, Port-au-Prince; tel. 245-8881; fax 245-8598; Dir-Gen. EMMANUEL SANON.

**Générale d'Assurance, SA:** Champ de Mars, Port-au-Prince; tel. 222-5465; fax 222-6502; f. 1985; Dir-Gen. ROLAND ACRA.

**Haïti Sécurité Assurance, SA:** 16 rue des Miracles, BP 1754, Port-au-Prince; tel. 223-2118; Dir-Gen. WILLIAM PHIPPS.

**International Assurance, SA (INASSA):** angle rues des Miracles et Pétion, Port-au-Prince; tel. 222-1058; Dir-Gen. RAOUL MÉROVÉ-PIERRE.

**MAVSA Multi Assurances, SA:** étage Dynamic Enterprise, route de l'Aéroport, BP 1489, Port-au-Prince; tel. 250-0700; fax 250-0236; e-mail info@groupedynamic.com; internet www.groupedynamic.com/mavsa.php; f. 1992; subsidiary of Groupe Dynamic SA; credit life insurance and pension plans; Group Chair. and CEO PHILIPPE R. ARMAND.

**National Assurance, SA (NASSA):** 25 rue Ferdinand Canapé-Vert, Port-au-Prince, HT6115; tel. 245-9800; fax 245-9701; e-mail nassa@nassagroup.com; Dir-Gen. FRITZ DUPUY.

**Office National d'Assurance Vieillesse (ONA):** Champ de Mars, rue Piquant, Port-au-Prince; tel. 223-9034; Dir-Gen. JEAN RONALD JOSEPH.

**Société de Commercialisation d'Assurance, SA (SOCOMAS):** Etage Complexe STELO, 56 route de Delmas, BP 636, Port-au-Prince; tel. 246-4768; fax 246-4874; Dir-Gen. JEAN DIDIER GARDÈRE.

#### Foreign Companies

**Les Assurances Léger, SA** (France): 40 rue Lamarre, BP 2120, Port-au-Prince; tel. 222-3451; fax 223-8634; Pres. GÉRARD N. LÉGER.

**Cabinet d'Assurances Fritz de Catalogne** (USA): angle rues du Peuple et des Miracles, BP 1644, Port-au-Prince; tel. 222-6695; fax 223-0827; Dir FRITZ DE CATALOGNE.

**Capital Life Insurance Company Ltd** (Bahamas): angle rues du Peuple et des Miracles, BP 1644, Port-au-Prince; tel. 222-6695; fax 223-0827; Agent FRITZ DE CATALOGNE.

**Groupement Français d'Assurances** (France): Port-au-Prince; Agent ALBERT A. DUFORT.

**National Western Life Insurance** (USA): 13 rue Pie XII, Cité de l'Exposition, Port-au-Prince; tel. 223-0734; Agent VORBE BARRAU DUPUY.

#### Insurance Association

**Association des Assureurs d'Haïti:** 153 rue des Miracles, Port-au-Prince; tel. 223-0796; fax 223-8634; Dir FRITZ DE CATALOGNE.

## Trade and Industry

### GOVERNMENT AGENCY

**Conseil de Modernisation des Entreprises Publiques (CMEP):** Palais National, Port-au-Prince; tel. 222-4111; fax 222-7761; f. 1996; oversees modernization and privatization of state enterprises.

### DEVELOPMENT ORGANIZATIONS

**Fonds de Développement Industriel (FDI):** Immeuble PROMO-BANK, 4 étage, ave John Brown et rue Lamarre, BP 2597, Port-au-Prince; tel. 222-7852; fax 222-8301; e-mail FDI-Finance@globelsud.net; f. 1981; Dir ROOSEVELT SAINT-DIC.

**Société Financière Haïtienne de Développement, SA (SOFIHDES):** 11 blvd Harry S Truman, BP 1399, Port-au-Prince; tel. 222-8904; fax 222-8997; f. 1983; industrial and agro-industrial project financing, accounting, data processing, management consultancy; cap. 7.5m. (1989); Dir-Gen. FAUBERT GUSTAVE; 1 br.

### CHAMBERS OF COMMERCE

**Chambre de Commerce et d'Industrie d'Haïti (CCIH):** blvd Harry S Truman, Cité de l'Exposition, BP 982, Port-au-Prince; tel. 223-0786; fax 222-0281; e-mail ccih@compa.net; internet www.intervision2000.com/iv2-trop/index.html; f. 1895; Pres. REGINALD BOULOS.

**Chambre de Commerce et d'Industrie Haïtiano-Américaine (HAMCHAM):** 6 rue Oge, Pétionville, Port-au-Prince; tel. 511-3024; e-mail csjean@hamcham.org; f. 1979; Exec. Dir CHANTAL SALOMON-JEAN.

**Chambre de Commerce et d'Industrie des Professions du Nord:** BP 244, Cap-Haïtien; tel. 262-2360; fax 262-2895.

**Chambre Franco-Haïtienne de Commerce et d'Industrie (CFHCI):** Le Plaza Holiday Inn, 10 rue Capois, Champ de Mars, Port-au-Prince; tel. 223-8424; fax 223-8131; e-mail cfhci@haitiworld.com; internet www.inhaiti.com/forms/cfhci-home.html; f. 1987; Pres. PHILIPPE LAHENS; Dir JOSETTE NAZON.

### INDUSTRIAL AND TRADE ORGANIZATIONS

**Association des Exportateurs de Café (ASDEC):** rue Barbancourt, BP 134, Port-au-Prince; tel. 249-2919; fax 249-2142; Pres. HUBERT DUPORT.

**Association des Industries d'Haïti (ADIH):** 199 route de Delmas, entre Delmas 31 et 33, étage Galerie 128, BP 2568, Port-au-Prince; tel. 246-4509; fax 246-2211; e-mail adih@acn2.net; f. 1980; Pres. (vacant).

**Association Nationale des Distributeurs de Produits Pétroliers (ANADIPP):** Centre Commercial Dubois, route de Delmas, Bureau 401, Port-au-Prince; tel. 246-1414; fax 245-0698; f. 1979; Pres. MAX ROMAIN.

**Association Nationale des Importateurs et Distributeurs de Produits Pharmaceutiques (ANIDPP):** blvd Harry S Truman, Port-au-Prince; tel. 222-0268; fax 222-7887; e-mail anidpp@direcway.com; Pres. RALPH EDMOND.

**Association des Producteurs Agricoles (APA):** BP 1318, Port-au-Prince; tel. 246-1848; fax 246-0356; f. 1985; Pres. REYNOLD BONNEFIL.

**Association des Producteurs Nationaux (APRONA):** c/o Mosaïques Gardère, ave Hailé Sélassié, Port-au-Prince; tel. and fax 511-8611; e-mail frantzgardere@hotmail.com; Pres. FRANTZ GARDÈRE.

### MAJOR COMPANIES

**AGC Apparel and Garment Contractors:** route de l'Aéroport, Port-au-Prince; tel. 246-2772; fax 246-1417; f. 1952; fmrly Alpha Electronics; manufacture of garments; Pres. ANDRÉ APAID; 970 employees.

**Brasserie Nationale d'Haïti, SA (BRANA):** ave Hailé Sélassié, BP 1334, Port-au-Prince; tel. 250-1501; fax 250-1300; e-mail brana@

branahaiti.com; internet www.prestigebeer.com; f. 1973; brewery and soft-drinks bottler; Pres. and CEO MICHAEL MADSEN; 145 employees.

**Cimenterie Nationale, SEM (CINA):** Km 25, route Nationale 1, Fond Mombin, Port-au-Prince; tel. 298-3234; e-mail csantamaria@cinahaiti.com; controlled since 1997 by a Colombian-Swiss consortium; cement manufacturers; Pres. CARLOS ESTEBAN SANTA MARIA; 345 employees.

**Compagnie Nationale de Ciment, SA (CNC):** 24 route de l'Aéroport, BP 745, Port-au-Prince; tel. 250-1819; e-mail PGAR7695@aol.com; cement distributor; Pres PATRICK GARDÈRE, PETER OESTROM.

**Etablissement Raymond Flambert, SA:** 430 route de Delmas, BP 896, Port-au-Prince; tel. 246-2605; fax 246-0110; e-mail erf@haitelonline.com; internet www.erfhaiti.com; manufacturers of building materials; Pres. ALEX FLAMBERT; 245 employees.

**Groupe Dynamic, SA:** étage Dynamic Enterprise, route de l'Aéroport, BP 1489, Port-au-Prince; tel. 250-0700; fax 250-0236; e-mail info@groupedynamic.com; internet www.groupedynamic.com; holding company with 12 subsidiaries across the financial, insurance, medical, tourism, transportation and building sectors; Chair. and CEO PHILIPPE R. ARMAND.

**Haïti Metal, SA:** BP 327, Port-au-Prince; tel. 240-412; fax 241-175; f. 1955; aluminium producers; Pres. RAYMOND L. ROY; 300 employees.

**Industries Nationales Réunies, SA:** Delmas 6 & 8, Port-au-Prince; tel. 222-0153; manufacturers of plastic products; Pres. ANDRÉ APAID; 455 employees.

**Laboratoires 4C:** Delmas 71, BP 44, Port-au-Prince; tel. 246-2207; fax 246-5332; e-mail cchti@dnetwork.net; f. 1952; manufacturers of pharmaceuticals and paper products; Pres. and Gen. Man. MAURICE ACRE; 450 employees.

**Les Moulins d'Haïti, SEM:** 1 route Nationale, Laffiteau, BP 15509, Pétionville; tel. 298-3616; fax 244-8050; e-mail ddaines@lmh-ht.com; fmrly La Minoterie d'Haïti; privatized 1999; now a semi-public co; 70% owned by two US agribusiness cos, 30% state-owned; flour milling; Pres. CHRISTIAN FUCINA.

**Modern Business Systems & Equipment, SA:** rue des Miracles, Port-au-Prince; tel. 222-5374; f. 1975; industrial and commercial equipment; Man. Dir GERARD LELIO JOSEPH.

**SA Filature et Corderia d'Haïti (SAFICO):** Diquini 63, Port-au-Prince; tel. 234-0523; e-mail micama45@yahoo.com; f. 1952; sisal processor, mattress and foam manufacturer; sales US $2m. (2001); Dir-Gen. THOMAS ADAMSON; 150 employees.

## UTILITIES

### Electricity

**Electricité d'Haïti:** rue Dante Destouches, Port-au-Prince; tel. 222-4600; state energy utility company; Dir-Gen. CHARLES ALBERT JACQUES.

    **Péligre Hydroelectric Plant:** Artibonite Valley.

    **Saut-Mathurine Hydroelectric Plant:** Les Cayes.

### Water

**Service Nationale d'Eau Potable (SNEP):** 48 route de Delmas, Port-au-Prince; Dir-Gen. PÉTION ROY.

## TRADE UNIONS

**Association des Journalistes Haïtiens (AJH):** f. 1954; Sec.-Gen. GUYLER C. DELVA.

**Batay Ouvriye** (Workers' Struggle): BP 13326, Delmas, Port-au-Prince; tel. 222-6719; e-mail batay@batayouvriye.org; internet www.batayouvriye.org; f. 2002; independent umbrella organization providing a framework for various autonomous trade unions and workers' associations.

**Centrale Autonome des Travailleurs Haïtiens (CATH):** 93 rue des Casernes, Port-au-Prince; tel. 222-4506; f. 1980; Sec.-Gen. FIGNOLE SAINT-CYR.

**Confédération Ouvriers Travailleurs Haïtiens (KOTA):** 155 rue des Césars, Port-au-Prince.

**Confédération Nationale des Educateurs Haïtiens (CNEH):** rue Berne 21, Port-au-Prince; tel. 245-1552; fax 245-9536; e-mail lana14@caramail.com; f. 1986.

**Confédération des Travailleurs Haïtiens (CTH):** f. 1989; Sec.-Gen. JACQUES BELZIN.

**Fédération Haïtienne de Syndicats Chrétiens (FHSC):** BP 416, Port-au-Prince; Pres. LÉONVIL LEBLANC.

**Fédération des Ouvriers Syndiques (FOS):** angle rues Dr Aubry et des Miracles 115, BP 371, Port-au-Prince; tel. 222-0035; f. 1984; Pres. JOSEPH J. SÉNAT.

**Organisation Générale Indépendante des Travailleurs et Travailleuses d'Haïti (OGITH):** 121, 2–3 étage, angle route Delmas et Delmas 11, Port-au-Prince; tel. 249-0575; e-mail pnumas@yahoo.fr; f. 1988; Gen. Sec. PATRICK NUMAS.

**Syndicat des Employés de l'EDH (SEEH):** c/o EDH, rue Joseph Janvier, Port-au-Prince; tel. 222-3367.

**Union Nationale des Ouvriers d'Haïti (UNOH):** Delmas 11, 121 bis, Cité de l'Exposition, BP 3337, Port-au-Prince; f. 1951; Pres. MARCEL VINCENT; Sec.-Gen. FRITZNER ST VIL; 3,000 mems from 8 affiliated unions.

A number of unions are non-affiliated and without a national centre, including those organized on a company basis.

# Transport

### RAILWAYS

The railway service, for the transportation of sugar cane, closed during the early 1990s.

### ROADS

In 1999, according to International Road Federation estimates, there were 4,160 km (2,585 miles) of roads, of which 24.3% was paved. There are all-weather roads from Port-au-Prince to Cap-Haïtien, on the northern coast, and to Les Cayes, in the south

### SHIPPING

Many European and American shipping lines call at Haiti. The two principal ports are Port-au-Prince and Cap-Haïtien. There are also 12 minor ports.

**Autorité Portuaire Nationale:** blvd La Saline, BP 616, Port-au-Prince; tel. 222-1942; fax 223-2440; e-mail apnpap@hotmail.com; f. 1978; Dir-Gen. JEAN EVENS CHARLES.

### CIVIL AVIATION

The international airport, situated 8 km (5 miles) outside Port-au-Prince, is the country's principal airport, and is served by many international airlines linking Haiti with the USA and other Caribbean islands. There is an airport at Cap-Haïtien, and smaller airfields at Jacmel, Jérémie, Les Cayes and Port-de-Paix.

**Office National de l'Aviation Civile (OFNAC):** Aéroport International Mais Gate, BP 1346, Port-au-Prince; tel. 246-0052; fax 246-0998; e-mail lpierre@ofnac.org; Dir-Gen. JEAN-LEMERQUE PIERRE.

**Air Haïti:** Aéroport International, Port-au-Prince; tel. 246-3311; f. 1969; began cargo charter operations 1970; scheduled cargo and mail services from Port-au-Prince to Cap-Haïtien, San Juan (Puerto Rico), Santo Domingo (Dominican Republic), Miami and New York (USA).

**Caribintair:** Aéroport International, Port-au-Prince; tel. 246-0778; scheduled domestic service and charter flights to Santo Domingo (Dominican Republic) and other Caribbean destinations.

**Haiti Air Freight, SA:** Aéroport International, BP 170, Port-au-Prince; tel. 246-2572; fax 246-0848; cargo carrier operating scheduled and charter services from Port-au-Prince and Cap-Haïtien to Miami (USA) and Puerto Rico.

**Haiti International Airlines:** Delmas 65, Rue Zamor 2, Port-au-Prince; tel. 434-7201; f. 1996; scheduled passenger and cargo services from Port-au Prince to Miami and New York (USA); Pres. and Chair. KHAN RAHMAN.

**Tropical Airways d'Haïti, SA:** 76 rue Panaméricaine, Pétionville, Port-au-Prince; tel. 250-3420; fax 250-3422; internet www.tropical-haiti.com; f. 1998; domestic and regional services; Pres. PHILIP GORNAIL.

# Tourism

Tourism was formerly Haiti's second largest source of foreign exchange. However, as a result of political instability, the number of cruise ships visiting Haiti declined considerably, causing a sharp decline in the number of tourist arrivals. With the restoration of democracy in late 1994, the development of the tourism industry was identified as a priority by the Government. In 2001 stop-over tourists totalled 141,632, while cruise-ship excursionists numbered 246,221 in 1998. Receipts from tourism in 2003 totalled US $93m.

**Secrétariat d'Etat au Tourisme:** 8 rue Légitime, Champ de Mars, Port-au-Prince; tel. 221-5960; fax 222-8659; e-mail info@

haititourisme.org; internet www.haititourisme.org; Sec. of State for Tourism Harold Florentino Latortue.

**Association Haïtienne des Agences de Voyages:** 17 rue des Miracles, Port-au-Prince; tel. 222-8855; fax 222-2054.

**Association Touristique d'Haïti:** rue Lamarre, BP 2562, Port-au-Prince; tel. 257-4647; fax 257-4134; Pres. Dominique Carvonis; Dir Elizabeth Silvera Ducasse.

## Defence

In November 1994, following the return to civilian rule, measures providing for the separation of the armed forces from the police force were approved by the legislature. In December President Aristide ordered the reduction of the armed forces to 1,500. In that month two commissions were established for the restructuring of the armed forces and the formation of a new 4,000-strong civilian police force (later enlarged to 6,000). In 1995 the armed forces were effectively dissolved, although officially they remained in existence pending an amendment to the Constitution providing for their abolition; such an amendment was, however, never presented to the legislature. In August 2004 the civilian police force numbered an estimated 2,000. There was also a coast guard of 30. In June 2004, following the resignation of President Aristide and his departure from the country in February of that year, a UN security force—the UN Stabilization Mission in Haiti (MINUSTAH)—assumed peace-keeping responsibilities in the country from a US-led Multinational Interim Force (authorized by the UN Security Council to maintain order in the country following Aristide's departure). MINUSTAH, which was to remain in the country for an initial period of six months, comprised 6,700 troops and a civilian police force of 1,622. In June 2005 the mandate of MINUSTAH was extended by a further eight months and the number of troops increased to 7,700. The security budget for 2003 was an estimated US $23m.

**Director-General of the Police Nationale:** Léon Charles.

## Education

Some 80% of education is provided by private or missionary schools and the rest by the State. Learning is based on the French model, and French is used as the language of instruction, although most Haitians speak only Creole. Primary education, which normally begins at six years of age and lasts for six years, is officially compulsory. Secondary education usually begins at 12 years of age and lasts for a further six years, comprising two cycles of three years each. In 1997 primary enrolment included only 19.4% of children in the relevant age-group (males 18.9%; females 19.9%). Enrolment at secondary schools in 1997 was equivalent to only 34.2% of children in the relevant age-group (males 35.2%; females 33.2%). In 1999 combined enrolment in primary, secondary and tertiary education was 52%. Some basic adult education programmes, with instruction in Creole, were created in the late 1980s in an attempt to address the problem of adult illiteracy. Higher education is provided by 18 technical and vocational centres, 42 domestic-science schools, and by the Université d'Etat d'Haïti, which has faculties of law, medicine, dentistry, science, agronomy, pharmacy, economics, veterinary medicine and ethnology. Government expenditure on education in 1990 was 216m. gourdes, or 20.0% of total government expenditure.

## Bibliography

For works on the Caribbean generally, see Select Bibliography (Books)

Avril, P. *Haiti (1995–2000): The Black Book on Insecurity.* Boca Raton, FL, Universal Publrs, 2004.

Brown, G. S. *Toussaint's Clause: The Founding Fathers and the Haitian Revolution.* Jackson, MS, University Press of Mississippi, 2005.

Catanese, A. *Haitians: Migration and Diaspora.* Boulder, CO, Westview Press, 1999.

Centre Haïtien de Presse. www.cybermedia-ht.com.

Chin, P., Flounders, S., and Dunkel, G. (Eds). *Haiti: A Slave Revolution — 200 Years After 1804.* New York, NY, International Action Center, 2004.

Chomsky, N., Farmer, P., and Goodman, A. *Getting Haiti Right This Time: The US and the Coup.* Monroe, ME, Common Courage Press, 2004.

Corbett, B. www.webster.edu/-corbetre/haiti/haiti.html.

Cousens, E. M. (Ed.) et al. *Peacebuilding as Politics.* Boulder, CO, Lynne Rienner Publrs, 2001.

Dash, J. M. *Culture and Customs of Haiti.* London, Greenwood Press, 2001.

Delince, K. *Les forces politiques en Haïti: manuel d'histoire contemporaine.* Paris, Editions Karthala, 1993.

Diederich, B., and Burt, A. *Papa Doc: Haiti and Its Dictator.* Princeton, NJ, Markus Wiener Publications, 1998.

Dubois, L. *Avengers of the New World: The Story of the Haitian Revolution.* Cambridge, MA, Belknap Press, 2004.

Fatton, Jr, R. *Haiti's Predatory Republic: The Unending Transition to Democracy.* Boulder, CO, Lynne Rienner Publrs, 2002.

Ferguson, J. *Papa Doc, Baby Doc.* Oxford, Blackwell Publrs, 1998.

Fischer, S. *Modernity Disavowed: Haiti and the Cultures of Slavery in the Age of Revolution.* Durham, NC, Duke University Press, 2004.

Geggus, D. P. (Ed.). *The Impact of the Haitian Revolution in the Atlantic World.* Columbia, SC, University of South Carolina Press, 2001.

*Haitian Revolutionary Studies.* Bloomington, IN, Indiana University Press, 2002.

Gibbons, E. D. *Sanctions in Haiti: Human Rights and Democracy under Assault.* New York, NY, Praeger Publrs, 1999.

Gray, O. *Economic Implications of CARICOM for Haiti.* Lewiston, NY, Edwin Mellen Press, 2003.

Haïti-Info. haiti-info.com.

Haiti Press Network. www.haitipressnetwork.com.

Haiti-Référence. www.haiti-reference.com.

Heinl, R. D., and Gordon, N. *Written in Blood.* New York, NY, University Press of America, 1995.

Kumar, C. *Building Peace in Haiti.* Boulder, CO, Lynne Reinner Publrs, 1998.

*Diasporic Citizenship: Haitian Americans in Transnational America.* Basingstoke, Palgrave Macmillan, 1998.

Matthewson, T. *A Proslavery Foreign Policy: Haitian-American Relations During the Early Republic.* Westport, CT, Praeger Publrs, 2003.

Pamphile, L. D. *Haitians and African Americans: A Heritage of Tragedy and Hope.* Gainesville, FL, University Press of Florida, 2002.

Renda, M. *Taking Haiti: Military Occupation and the Culture of US Imperialism, 1915–1940.* Chapel Hill, NC, University of North Carolina Press, 2001.

Rhodes, L. *Democracy and the Role of the Haitian Media.* Lewiston, NY, Edwin Mellen Press, 2001.

Rotberg, R. I. *Haiti Renewed: Political and Economic Prospects.* Washington, DC, Brookings Institute Press, 1997.

Stotzky, I. P. *Silencing the Guns in Haiti: the Promise of Deliberative Democracy.* Chicago, IL, University of Chicago Press, 1997.

Von Hippel, K. *Democracy by Force.* Cambridge, Cambridge University Press, 1999.

# HONDURAS

## Geography

### PHYSICAL FEATURES

The Republic of Honduras is a Central American country, which sits on a north-facing Caribbean coast and, on its east side, tapers southwards to a short Pacific coast, on the Gulf of Fonseca. Its territory includes a number of offshore islands and cays in the Caribbean, but also in the Gulf of Fonseca. El Salvador, which lies in the south-west of the country, occupying the Pacific coast west of the Gulf of Fonseca, has claims on the island of Conejo in the Gulf. Competing maritime claims in the Gulf have been referred to a tripartite commission with Nicaragua. The exact demarcation of the Honduras–El Salvador border (342 km or 212 miles) remains unresolved in many areas, despite a decision by the International Court of Justice (ICJ—based in The Hague, Netherlands) in 1992. Guatemala lies beyond a 256-km border in the west and north-west. Belize also lies to the north-west, but beyond the Gulf of Honduras (its Sapodilla Cays are claimed by Honduras). The longest border, across the widest part of the country, running south-westwards from the Caribbean (Atlantic) to the Pacific coasts, is the 922-km frontier with Nicaragua, which lies to the south-east. The maritime boundary between the two countries has been the subject of dispute since Nicaragua filed claims with the ICJ in 1999. The demarcation of the maritime boundary with the Cayman Islands, a British dependency in the Caribbean, has not yet been settled. Honduras, which is slightly larger than Guatemala, is the second largest country in Central America (after Nicaragua), covering an area of 112,492 sq km (43,433 sq miles), including about 200 sq km of inland waters.

Honduras is dominated by its central highlands, widest and highest in the west, towering above the narrow Pacific coastal plains and above the broader northern lowlands and plains on the Caribbean. About three-quarters of the country is mountainous, largely consisting of extinct volcanoes and their outflows (the whole country is prone to usually mild earthquakes), while the shores along the southern Gulf of Fonseca amount to only 124 km and the Caribbean coast extends for 644 km. The northern lowlands include out-thrust ranges from the Continental Divide and river valleys, notably those of the Ulua and the Agua, running down to the coastal plains. The North Coast region is agricultural and well populated. Other important rivers include the Guayape and the Patuca, the latter dominating the second northern topographical region, the flat, hot and humid Mosquito Coast (the name a corruption of the local Miskito people), densely clad in rainforest and scantily inhabited. This region, in the far north-east of the country, includes extensive wetlands, especially around the Caratasca lagoon. Honduras encompasses only the northern part of the Mosquito Coast (Mosquitia—the rest is Nicaragua's eastern shore), which begins at Cabo Gracias a Dios, named in gratitude by the Genoese (Italian) navigator exploring for the Spanish monarchy, Christopher Columbus, when he rounded it and escaped the 'deep waters' (in Spanish) that gave Honduras its name. The central highlands, crowded in the west and dominating the south, consist of two main ranges, the Central American Cordillera and the Volcanic Highlands. The highest point in Honduras is in the west, at Celaque, the loftiest peak of which is Cerro de las Minas (2,870 m or 9,419 ft). The far south, beneath the heights where the national capital, Tegucigalpa, sprawls, is the fertile strip of coastal plain (only about 24 km in width) along the Gulf of Fonseca. Offshore are a number of islands belonging to Honduras, including the disputed Conejo and volcanic cones of Tigre and Zacate Grande, for instance. Strewn over a wider area are the Caribbean islands and cays of Honduras, notably the Bay Islands (Islas de la Bahía—mostly jungle-clad volcanic cones), just off the North Coast, and, further out, the Swan Islands. This vast territory includes a varied natural environment, little of it free from risk, be it mining damage to part of the country's extensive river system or to Lake

Yojoa, an important source of fresh water, or the encroachments of urban expansion and agriculture. Deforestation throughout Honduras is a problem, with the accompanying hazards of soil erosion, etc., but just over one-half of the country is still wooded, more than anywhere else in Central America. The surviving forests host a huge variety of plant, bird and animal life, such as butterflies, jaguars, and white-faced, spider and howler monkeys. The insular environments are often more unusual, with the Bay Islands, for instance, noted for features such as its reef system (the most diverse in the Caribbean after Jamaica), the yellow-naped parrot or the spiny-tailed iguana found only on Utila.

### CLIMATE

The climate is tropical in the lowlands and temperate in the mountainous interior. Both climatic zones have a wet season, which is in April–October. In the north of Honduras, on the North Coast and in the hill country between there and the highest uplands, average annual rainfall varies from 1,780 mm (70 ins) to about 2,550 mm, while along the Pacific coastal plains the range is around 1,550–2,050 mm. The Caribbean coasts are very susceptible to hurricanes and flooding. The average minimum and maximum temperatures in Tegucigalpa range from 4°–27°C (39°–81°F) in February to 12°–33°C (54°–91°F) in May.

### POPULATION

The people of Honduras, who refer to themselves as 'Catrachos', are mainly (90%) Mestizo (Ladino), with 7% Amerindian, 2% black and 1% white. Most Ladinos are of predominantly Spanish descent, and Spanish is the official and overwhelmingly the most widely spoken language. Some English is spoken on the North Coast and in the Bay Islands, by blacks and Anglo-Antilleans descended from those arriving in the country from elsewhere in the Caribbean about one century ago. Two centuries ago the main black population had arrived in this part of Central America, when the British deported the Black Caribs (mixed descendants of Amerindians and escaped slaves of African origin) from St Vincent (now part of Saint Vincent and the Grenadines); they speak Garifuna (Carib), originally an Amerindian tongue. Generally, however, such racial identities are preserved in cultural vestiges rather than language—the six dominant Amerindian language groups native to the region are Lenca, Jicaque or Tol, Paya, Chorti, Miskito and Sumo. Another legacy of the dominant Spanish influence in Honduras is that in

2003 some 80% of the population still adhered to the Roman Catholic Church, although about 10% were Protestant.

Most of the 6.9m. (official estimate at mid-2003) population lives in central and western upland valleys, and on the North Coast. There are considerably fewer people in the north-east and in the south. The people of Honduras are poor—there are very pronounced levels of income inequality—and only two-fifths are urbanized. Moreover, since the civil conflicts in neighbouring countries during the 1970s and 1980s, some 50,000 legal refugees, as well as many illegal migrants, have moved into Honduras. These are split between Salvadorean Mestizos in the west and Sumo and Miskito Amerindians in the north-east. The capital of Honduras is Tegucigalpa, in the centre-south of the country, in the highlands. The second city, in the north-west, is San Pedro Sula, which controls the important Ulea valley. La Ceiba, actually on the North Coast, was famous as the headquarters of the Standard Fruit Co, the US banana company that once dominated the country, whereas the Copán valley, in the far west, was the centre for the even earlier rule of the Mayas. Choluteca is the focus for southern Honduras, the Pacific coast. There are 18 departments.

# History

## HELEN SCHOOLEY
Revised for this edition by SANDY MARKWICK

The history of Honduras has been strongly influenced by a series of geographical factors. The country has enjoyed relatively few natural resources in proportion to the size of its population, is prone to natural disasters, and at the turn of the 21st century was reckoned as one of the poorest states in Latin America. It has suffered from endemic political instability, and despite many attempts at regional integration, Honduras has often come into conflict with its Central American neighbours. Its principal economic activity has helped to earn it the epithet 'banana republic', and the cultivation of bananas also provided the basis for another major theme of the country's history in the 20th century, the dominant influence of the USA in the nation's internal politics and regional stance. Major developments in the country's history have often been precipitated as much from events outside its borders as from internal politics.

Before the Spanish conquest Honduras lay on the southern edge of the Mayan civilization, which was already struggling with internal disturbances. Socio-economic pressures caused by famine, disease and soil erosion contributed to political divisions. After the Spanish adventurer Hernán Cortés had established his power in Mexico City, he turned his attention south and arrived in Honduras in 1525, enticed by greatly exaggerated rumours of mineral wealth there. The significant deposits of silver were largely exhausted by the 18th century, and the economic mainstay subsequently became subsistence farming. In 1549 Honduras became a province of the Kingdom of Guatemala, which was in turn part of the Viceroyalty of New Spain (Mexico). The existing rivalries between the provinces were fostered by Spain to help it retain its imperial power.

## INDEPENDENCE

In 1822 the newly independent Mexico annexed Central America into its short-lived empire. The following year the provinces seceded and formed the Central American Federation. In 1827 internal conflict between the two main political factions, the Liberals, who supported the Federation, and the Conservatives, who opposed it, escalated into civil war. The Liberal Honduran leader, Francisco Morazán, emerged as the President, but in 1838 a peasant revolt in Guatemala, led by the conservative Rafael Guerrera, brought the Federation to an end, and Honduras became an independent state.

The new country had internal divisions of its own, retaining the Liberal–Conservative divide, which continued to dominate political life throughout the 19th and 20th centuries. The Conservatives held power for most of the first four decades of independence, and the first national capital was the city of Comayagua. The Liberals returned to government in 1876 under Marco Aurelio Soto, and remained there for 56 years, consolidating their dominance with the removal of the capital to Tegucigalpa. Despite the existence of the political tendencies (not yet fully formalized into parties), government was relatively weak, and was subject to considerable influence from other countries. Soto was both installed and removed in 1881 with the active participation of Guatemala, and in 1907 Nicaraguan forces helped to replace President Policarpo Bonilla with Miguel Dávila. The most powerful foreign influence, however, was that of the USA. In the late 19th and early 20th centuries vast tracts of land were granted to US companies, in particular the United Fruit Company (UFCO), on very advantageous terms, often in exchange for political support. By 1913 UFCO controlled two-thirds of Honduras' banana exports. A political crisis arose in 1911–12, in which President Dávila was overthrown and replaced by Manuel Bonilla in what was essentially a struggle between UFCO and another US banana enterprise.

## DICTATORSHIP AND MILITARY RULE, 1932–80

The US Great Depression had a severe impact on the Honduran economy with its dependence on exports to the USA, and, as in Guatemala and El Salvador, contributed to the rise of dictatorship in the 1930s. Gen. Tiburcio Carías Andino, leader of the conservative Partido Nacional (PN), was elected President in 1932, and proceeded to hold office for 16 years without further election. He was finally persuaded to resign in 1948, and was succeeded by Dr Juan Manuel Galvez of the PN, the sole candidate. After inconclusive elections held in 1954, Vice-President Julio Lozano Díaz dissolved the Congress, abrogated the Constitution and declared himself head of state. Lozano was in turn overthrown in a bloodless coup two years later. After constituent elections in which the Partido Liberal (PL) achieved a substantial majority, the PL leader, Dr José Ramón Villeda Morales, took office in 1957. His moderate social reforms aroused opposition from the traditional ruling class and he was overthrown in 1963 in a military coup led by Gen. Oswaldo López Arellano. A new Constitution was approved in June 1965 and López Arellano was installed as President.

Elections were held again in 1971, and were won by the PN candidate, Dr Ramón Ernesto Cruz Uclés. This brief return to civilian rule followed a humiliating defeat for the Honduran military in the war that had broken out against El Salvador in 1969. Triggered by events after two football (soccer) World Cup qualifying matches, it became known as the 'Football War', although it had its roots in long-standing economic grievances between the two countries. Full-scale hostilities lasted less than two weeks, but a final peace agreement was not signed until 1980.

Gen. López Arellano returned to power in a coup in 1972, but in 1975 was overthrown by Col (later Gen.) Juan Melgar Castro. Only a few weeks before the coup López Arellano had been named in a bribery scandal involving United Brands (as UFCO had become in 1970), to which the company later admitted. Melgar Castro's attempts to introduce a comprehensive land-reform programme provoked opposition from landed interests and he was removed from power in 1978, in a coup led by Gen. Policarpo Paz García.

## RETURN TO CIVILIAN GOVERNMENT

During the 1980s Honduras became an important player in US policy in Central America. Ironically, the return to civilian government marked a sharp increase in actual military power, because the armed forces had greater, if not always overt, US support and were consequently better equipped than they had been previously. Under US encouragement, constituent elections were held in 1980, resulting in a surprise victory for the Liberals. PL leader Roberto Suazo Córdova was elected President in 1981 and a new Constitution was promulgated in January 1982, but an amendment transferred the post of Commander-in-Chief from the President to the head of the armed forces (then Gen. Gustavo Adolfo Alvarez Martínez). Gen. Alvarez was widely regarded as the most powerful man in the country and, under his command, the army carried out a 'dirty war' against 'subversives'; the attendant allegations of human rights abuses would continue for the next two decades. (In 1993 a human rights commission attributed responsibility for 184 'disappearances' to special units directly answerable to Gen. Alvarez and his successors.) Gen. Alvarez was removed from power in 1984 by a group of junior officers disenchanted with his authoritarian policies, and he was assassinated in 1989.

The presidential election of November 1985 was won by José Simeón Azcona del Hoyo of the PL, under a new system that awarded victory to the leading candidate of the party with the greatest number of votes. (As each party fielded more than one candidate, the winner might not be the most popular individual candidate.) Azcona took office in January 1986, the first time for 55 years in which one freely elected president had succeeded another. The voting procedure was simplified for the 1989 elections, and the presidential ballot was won by Rafael Callejas Romero of the PN (who had polled the highest individual number of votes in 1985).

## HUMAN RIGHTS AND SOCIAL ISSUES

Lacking a majority in the Asamblea Nacional, the Callejas administration also encountered opposition on many fronts. The armed forces, resenting attempts to prosecute some of its members for human rights abuses, continued to demonstrate that their influence outweighed the country's political and judicial institutions, and made unsuccessful coup attempts in 1991 and 1993. Meanwhile, the Government's economic structural-adjustment programme raised vociferous protests not only from trade unions and peasant groups, but also from private business and the Roman Catholic Church. The Church was concerned principally with the matter of land distribution, and denounced the amassing of large estates by foreign companies, a development justified by the Government on the grounds of agricultural modernization. In addition, several indigenous Amerindian groups protested over the Government's failure to fulfil land-rights agreements and to improve living and working conditions. A series of accords providing for the return of some 7,000 ha of land to the indigenous communities, along with many infrastructure projects on their land, was reached in 1994, but conflicted with vested interests.

Human rights was a prominent issue in the elections of November 1993, with the two main parties accusing each other of complicity in past abuses. The campaign also exposed the regular practice of purchasing favourable press coverage. The PL retained its majority and its candidate, Carlos Roberto Reina Idiaquez, was elected President. Reina, a former President of the Inter-American Court of Human Rights (part of the Organization of American States—OAS), set out to reform the judicial system, to curb the power of the army and to combat political corruption. Under Reina the post of Commander-in-Chief was returned to the civilian authority (the defence minister), the Military High Council was abolished and a new national police force was created outside direct military control. Despite these measures, the armed forces generally managed to evade the judicial process, and in 1998 the independent Comité para la defensa de Derechos Humanos en Honduras (CODEH—Committee for the Defence of Human Rights in Honduras) claimed that paramilitary 'death squad' activity had doubled since 1995.

In the 1997 elections the PL retained power and Carlos Roberto Flores Facussé became President. In 1999 he appointed Edgardo Dumas as the country's first civilian defence minister. Dumas opened an audit of the military pension fund, which had wide-ranging interests through the economy and allegedly operated an extensive network of corruption. The armed forces' economic penetration was so thorough that it constituted one of the principal business interests in the country. In November 2001 an investigation reported that five former Commanders-in-Chief had diverted some US $8m. in government revenues between 1986 and 1997. The appointment of Dumas angered the military command, and in July 1999 President Flores removed four senior officers and redeployed 33 more in an effort to reassert political supremacy over the armed forces. Tension between the civilian and military authorities increased with the announcement in August that a series of graves had been discovered at a former US-built military base at El Aguacate. The camp had been used for training right-wing Nicaraguan guerrillas (known as 'Contras') and allegedly for the detention and torture of suspected left-wing activists. The graves were estimated to contain about one-half of the 184 'disappeared'.

Protests over the land rights issue intensified, and were increasingly accompanied by demands from environmental groups. From 1996 regular demonstrations were staged in Tegucigalpa calling for a full investigation into the murders of over 40 indigenous leaders, who had come into conflict with cattle ranchers, logging interests, energy companies and tourism developers. In March 2000 four leaders of the Chortí people were shot dead by the private security guards of a landowner in the west of the country, and in September the Chortí held a protest in Copán over the minimal implementation of the 1994 accords.

The country also faced a range of socio-economic problems, and ranked as the poorest in Central America. In 2001 a survey indicated that 79% of the population lived in poverty, and 56% were classified as destitute (according to the UN's Economic Commission for Latin America and the Caribbean), with contingent poor standards of housing and health; there were several outbreaks of cholera in the 1990s. The infant mortality rate was high, and, of surviving children, 43% had no access to school education, with many living precariously on the margins of society. Street children were at risk of summary brutality and execution, with over 1,500 killed between 1998 and 2002, mostly by police and security forces. Many of the street children became involved in child prostitution, allied with a rise in 'sex tourism', and also in criminal gangs and drugs-related crime. The country had a rising rate of violent crime; by 2001 the city of San Pedro Sula had the highest murder rate, with 95 per 100,000 head of population, and the incidence of kidnapping increased sharply in 2001 and 2002, particularly targeted at the business community. The seriously overcrowded prisons often experienced violent riots by inmates, and some 90% of those in detention had not been convicted. The number of Hondurans suffering from AIDS rose from just over 2,500 in 1993 to over 14,000 in 1999, with over 70% of cases occurring in the 19–35 age group.

The country's socio-economic problems have been compounded by frequent natural disasters, of which one of the most severe was 'Hurricane Mitch' in October 1998. There was major damage to the national infrastructure and environment, as well as to health and job prospects; this prompted an increase in illegal immigration north to Mexico and the USA. Furthermore, the arrival of international aid focused attention on the country's inadequate provision of housing and services and on the high level of corruption.

## THE GOVERNMENT OF RICARDO MADURO JOEST

The general election held on 25 November 2001 resulted in a surprise victory for the PN. That party's candidate, Ricardo Maduro Joest, was elected President with 52% of the votes cast, and the party won the largest number of seats (61) in the Asamblea Nacional for the first time since the restoration of civilian rule in 1980. A number of smaller parties, which had previously been seen as politically irrelevant, also made significant gains. In June 2002 the Partido Demócrata Cristiano de Honduras (PDCH), which had won three seats in the elections, joined the government coalition to give the Maduro administration a majority in the legislature. Maduro, a prominent businessman and former Governor of the Central Bank, undertook to pursue the reforms advocated by the IMF, including acceleration of the privatization programme. A number of fiscal reforms were subsequently introduced in 2002 and 2003, including the extension of sales tax to a wider range of basic goods and an increase in income tax. President Maduro was also committed to reducing spending on public-sector wages, a policy that led to strikes among teachers and health workers in early 2005.

While previous administrations had been criticized for failing to address the issue of crime, the new President, who took office in January 2002, undertook to give it a high priority. In March 2003 the Government announced the formation of a special commission to establish responsibility for the murder of street children, and in August increased sentences for crimes by gang members were introduced. There had been a marked rise in violent gang warfare since 1997, when the USA had begun to deport Hondurans convicted of offences; by 2002 the number of such deportees had reached 4,680. The most prominent gang was Mara 18, which had grown out of the civil wars of the 1980s and which operated throughout Central America and within the USA. The Honduran police force lacked training and resources and, at only 7,000 strong, proved unequal to combat the rising crime level. The new police force was also adjusting to a new penal code, passed in 2002, which introduced the concept of 'innocent until found guilty'; however, many officers were still unfamiliar with new procedures. Since November 2000 army units had been deployed to help them patrol the streets, a measure that alarmed many human rights groups. Many people died in gun battles between gang members and the police, and there were also many reports of innocent bystanders being arrested and killed. The authorities alleged that many of the street children killed by security forces (with 549 such fatalities in 2002 alone) were connected with gangs, but the human rights group Casa Alianza claimed many had been shot indiscriminately. Criminal activity included extortion, with armed robberies in the streets and the levying of an illegal 'tax' to enter areas of the capital, as well as illegal drugs-trafficking. In order to counter the growing links between narcotics and political corruption, President Maduro announced restrictions on parliamentary immunity. There were also renewed prison riots, and in April 2003, in a disturbance at La Ceiba prison, 69 inmates were shot dead (61 of whom were members of Mara 18) by security forces, some apparently after they had surrendered. Despite the Government's focus on crime, there was a continuing public perception that violence was on the increase.

In response to domestic and international accusations that the Government was neglecting to support indigenous rights and enforce land-reform legislation, in September 2003 President Maduro created a 'National Dialogue', a discussion forum involving several sectors of society and opposition groupings designed to reach a consensus on policy. There were many complaints that no judicial proceedings were taken against landowners accused of intimidating farmers and occupying their ancestral lands. Earlier in 2003 leaders of an environmental campaign reported having received death threats from timber companies, and there continued to be claims of intimidation against human rights activists in general.

Elections to the presidency and the Asamblea Nacional were scheduled for 27 November 2005. In line with a new electoral law, the PL and PN primaries took place simultaneously, in February of that year. The opposition PL selected Manuel Zelaya Rosales, the former head of a social investment fund, as its presidential candidate. The incumbent President of the Asamblea Nacional, Porfirio Lobo Sosa, was selected to represent the ruling PN. Maduro was ineligible to stand again as the Constitution prohibited Presidents from serving a second term in office. There was broad consensus between the two leading candidates to continue with the economic programme agreed with the IMF within its Poverty Reduction and Growth Facility (PRGF).

## RELATIONS WITH CENTRAL AMERICA AND THE USA

Honduras was drawn into the Central American conflicts of the 1980s when right-wing Nicaraguan Contras, with US support, began using its territory as a base for military operations and sabotage missions against Sandinista-ruled Nicaragua. The legacy of the conflict was a powerful and arguably over-manned and over-equipped armed forces, which needed to redefine a role in the 1990s, while also facing a drastic cut in US military aid. In addition, the virtual demise of the few and relatively small guerrilla groups formed in the 1980s eliminated an obvious internal threat. Another long-term consequence of the conflict was the presence of some 30,000 landmines along the border with Nicaragua, mostly planted by the Contras. A clearance programme launched in 1995 was disrupted by 'Hurricane Mitch' in 1998, which also caused a large-scale displacement of many of these devices.

The advent of peace in Central America, and the electoral defeat of the Sandinistas in April 1990, led to substantially improved relations between the Central American states, although the traditional sources of conflict resurfaced thereafter. Once again domestic troubles and the economic burdens imposed by natural disasters plagued renewed attempts at regional integration. One notable attempt was the formation of the Central American Integration System in 1991, which helped to foster a rapid expansion of the Central American Common Market (CACM).

The 1980 peace agreement with El Salvador (see above) had left territorial issues unresolved, both along the border and in the Gulf of Fonseca, which have caused intermittent tension. In 1986 the territorial dispute was referred to the International Court of Justice (ICJ) at The Hague, Netherlands, which, in 1992, awarded about two-thirds of the disputed territories to Honduras. In 1998 the two countries undertook to complete the border demarcation, but relations remained uneasy, and in August 2001 Honduras expelled two Salvadorean diplomats, alleging that they had been involved with naval espionage. El Salvador gave a fresh impetus to negotiations in October 2002 by challenging the ICJ ruling of 1992.

In November 1999 the Government provoked a new dispute with Nicaragua when it asked the Asamblea Nacional to ratify an agreement, signed with Colombia in 1986, delineating maritime boundaries. The Nicaraguan Government protested that this move threatened its national maritime claims, and both countries posted troops on their joint border, as well as introducing high tariffs on each other's goods. The border area was demilitarized in February 2000 and, following OAS mediation, the two nations agreed to establish a maritime exclusion zone in the disputed area. After a year of relative calm, relations deteriorated in February 2001 when the Nicaraguan Government accused Honduras of violating the agreement reached the previous year by holding military exercises in the border region. The tension was diffused by OAS-mediated talks held in June, but in August Nicaragua again protested, on this occasion accusing Honduras of planning an attack on its territory. In September 2002 the Honduran Government protested over the sale by the Nicaraguan authorities of oil-drilling rights in the disputed region. In March 2003, however, the Nicaraguan legislature voted to suspend import taxes on Honduran goods, imposed in 1999. A final ruling on the region by the ICJ was still pending in mid-2005.

During the 1980s relations with the USA had revolved chiefly around security matters. This issue dwindled in the 1990s, although the human rights investigations alleged the involvement of US military advisers in the conduct of the 'dirty war'. Bilateral relations predominantly reverted to economic considerations, among which was the question of the thousands of illegal Honduran immigrants currently living in the USA. Following the devastation caused by 'Hurricane Mitch' in 1998, Honduran immigrants to the USA were granted Temporary Protection Status. This was extended in successive years, most recently until July 2006. There were an estimated 81,000 Hondurans living in the USA in 2005. Continuing a tradition of close alliance with the USA, in June 2003 Honduras agreed to send up to 400 troops to join the peace-keeping effort in Iraq, serving under Spanish command. However, when the incoming Spanish Government announced in March 2004 that it would be withdrawing its forces from Iraq in June, Honduras also decided to withdraw its troops.

# Economy

**PHILLIP WEARNE**

Revised for this edition by SANDY MARKWICK

With a national income per head estimated at just US $1,056 in 2004, Honduras remained one of the poorest and least developed nations in Latin America. Over one-half of the population were considered to be living below the poverty line and the country also posted one of the highest inequality ratings in the region. Despite some success in efforts to diversify the country's economic base in the 1990s, Honduras remained overly dependent on primary agricultural and fisheries exports such as coffee, bananas, meat, shrimp and lobster, as well as on large quantities of external finance. Dependence on agriculture also made the country vulnerable to serial natural disasters.

Real annual average gross domestic product (GDP) growth in 1980–90 was 2.3%. By 1990 the economy was stagnant, expanding by a mere 0.1%. Nevertheless, growth increased in 1991 and the country sustained steady rates of growth in the mid-1990s. However, the devastation caused by 'Hurricane Mitch' in October 1998 interrupted the economic growth trend. GDP increased, in real terms, at an average annual rate of 3.1% in 1990–2003; the negative rate posted in 1999 was offset by growth of nearer 5% in 2000. The rate then stabilized: in 2001 the economy grew by 2.6%, in 2002 by 2.7% and in 2003 by 3.5%. GDP grew by an estimated 5.0% in 2004, underpinned by confidence generated by debt reduction and free trade agreements. Growth in 2004 was led by the agricultural sector, particularly coffee and bananas, followed by manufacturing and tourist-driven commerce, particularly restaurants and hotels.

Economic growth could not keep pace with population growth in the last decade of the 20th century and caused a steady rise in unemployment. The population increased by an annual average of 2.8% during 1990–2003, with the number of inhabitants totalling an estimated 6.9m. in mid-2003. Although the official unemployment rate was just 3.8% in 2002, it was estimated that more than 35% of the labour force were actually jobless, or underemployed. Some 60% of the agricultural workforce were believed to be underemployed as a result of the pressure on land, the increased mechanization of export agriculture, and the seasonal nature of coffee, banana, sugar and fruit production. The maintenance of the fixed exchange rate, at two lempiras (the Honduran currency unit) per US dollar, and a reduction in the money supply kept the annual inflation rate below 5% between 1984 and 1988. However, massive devaluations of the lempira in 1990–91, combined with more gradual ones thereafter, took the annual inflation rate to a record 34% in 1991. By 2000 the rate of inflation had fallen to 6.0%, but progressive devaluation of the lempira increased the pressure on domestic prices, and in the period 2001–04 annual inflation averaged 8.0%. Annual inflation remained stable, averaging 9.2% in 2004, although it had increased to 9.4% by April 2005.

After stable growth in the 1970s the economy declined in the 1980s, as a result of world recession, rising petroleum prices, falling international agricultural commodity prices and the Central American civil wars. In the mid-1980s a fall in international petroleum prices and an increase in coffee revenues, combined with reductions in public-sector expenditure and increased inflows of US aid, renewed creditors' confidence in the economy. However, by 1989 the position had been reversed. As imports soared and coffee prices fell drastically, the World Bank declared the country ineligible for further credits. This followed President José Simeón Azcona del Hoyo's (1986–90) unilateral suspension of all interest payments on the country's US $3,200m. foreign debt. The Government of President Rafael Leonardo Callejas Romero (1990–94) steadily recovered the situation, adhering to the economic prescriptions of the US Agency for International Development (USAID) more closely. An effective 100% devaluation of the lempira against the US dollar in March 1990 began the process.

By mid-1990 the credibility of the new Government had been rewarded by a stand-by agreement with the IMF, a second structural-adjustment accord with the World Bank and a US $247m. bridging loan. In 1992 Honduras signed an Extended Structural Adjustment Facility (ESAF) loan agreement with the IMF, but deterioration in the fiscal accounts meant that the Government was soon missing targets and the incoming Government of President Carlos Roberto Reina Idiaquez (1994–98) was forced to repeat the initial measures of its predecessor. In February 1994 the tax base was widened, the lempira devalued and public expenditure further decreased, in an effort to reduce a fiscal deficit of $326m., equivalent to 10.6% of GDP.

By late 1995 the Honduran economy was recovering. The rise in the rate of inflation was slowing and the fiscal deficit had been reduced to less than 4% of GDP. As a result, by early 1996 the Inter-American Development Bank (IDB) had made some US $160m. available for the modernization of the economy and the World Bank approved a further $60m. In May 1997 the IMF announced a partial agreement for the third-year disbursement of funds under its ESAF agreement with Honduras.

Following 'Hurricane Mitch', the IMF approved a Poverty Reduction and Growth Facility (PRGF) worth US $215m. in March 1999 to aid economic reconstruction. Honduras was further aided by a three-year deferral of bilateral debt-service payments to the 'Paris Club' of Western creditor countries. Subsequently, the country was formally approved for debt relief under the heavily indebted poor countries (HIPC) initiative. Under this programme, in July 2000 the IMF and the World Bank announced a reduction in debt-service payments, contingent on the Government's economic management. The IMF did not immediately renew the PRGF, which expired in December 2002, dissatisfied with the Government's progress towards IMF targets. A central objective of the IMF programme was a reduction of the fiscal deficit to below 3% of GDP by 2005. The deficit was reduced from 10.7% in 2001 to 5.4% in 2003, following which, in February 2004, the IMF approved a three-year $108m. PRGF, while at the same time maintaining pressure on the Government to extend reform. Also in support of these structural policies, the IDB approved a funding for Honduras totalling almost $600m. in April of that year. The fiscal deficit fell further, to 3.1% of GDP in 2004, and with forecasts for a further reduction in 2005 to within the IMF target, in early April 2005 the IMF announced that Honduras had reached 'completion point' within the HIPC initiative. This entitled Honduras to immediate debt relief amounting to $1,200m. over 10 years. In June, in advance of a summit meeting of the Group of Eight (G-8) industrialized nations in Gleneagles, Scotland, this decision was endorsed by G-8 finance ministers. Total external debt rose from $5,600m. in 2000 to an estimated $5,800m. in 2004, but at 77% of GDP, the debt was more manageable compared with a debt equivalent to nearly 93% of GDP in 2000.

## AGRICULTURE

Agriculture remained the most important sector of the economy. In 2003 the sector (including hunting, forestry and fishing) employed 37.4% of the economically active population. Agriculture accounted for 13.3% of GDP in 2004, compared with 23% in 1997, a decline which reflected long-term trends and the extent of the devastation caused by 'Hurricane Mitch'. The sector remained underdeveloped—of a total land area of 11.2m. ha, only an estimated 1.7m. ha were utilized for arable farming, with a further 1.5m. ha under pasture. Despite the country's relatively low density of population, land shortages were a persistent problem. In 1999 some 650,000 peasant farmers in Honduras were estimated to be landless. As a result, disturbances caused by the unofficial occupation of unused or underutilized agricultural land had become a marked feature of life in rural Honduras.

The rapid development of cash-crop farming, most notably coffee and bananas, as well as relatively stable world commodity prices, allowed for an average annual growth in agricultural production of 5.8% during the 1960s. Although the average annual rate of growth was reduced to 2.4% during the 1970s, significant gains were made in livestock farming and sugar production. Adverse weather conditions, falling commodity prices and the onset of world recession reduced the growth rate markedly during the 1980s. In the early 1990s the sector began to recover. However, 'Hurricane Mitch', which affected the agricultural sector more adversely than any other, ended the revival. Coffee and banana production were particularly badly affected. Losses in the sector were estimated at US $200m. in 1998, and at a further $500m. in 1999, when the agriculture sector contracted by 8.5%. Some 70% of the banana crop was destroyed and coffee production fell markedly at a time when prices were already experiencing record lows. In 2000 the sector increased by an estimated 9.5%, but decreased, albeit slightly (0.1%), in 2001. Modest growth, averaging around 3.5% in 2002–03, was followed by more significant growth, of 7.1%, in 2004.

In November 1983 the effective control of the banana industry by two US conglomerates, United Brands Company (formerly the United Fruit Company) and Standard Fruit Company, was broken when agreements were reached for the sale of a large proportion of the national crop to local trading companies. This, combined with a government export-incentive scheme in the mid-1980s, an increase in demand following the collapse of the Communist bloc in Eastern Europe in the early 1990s, and the implementation of the free market within the European Community (EC, known as the European Union—EU—from 1993) in 1992, acted as a stimulant to banana production. In 1990 the Anglo-Irish banana company Fyffes began financing new, independent banana co-operatives, in a further challenge to the monopoly of the two US companies. Nevertheless, Chiquita Brands International Inc (formerly United Brands Company) and Dole Food Company (formerly Standard Fruit Company) remained the dominant forces in the industry.

In the first half of the 1990s export earnings from bananas experienced a decline. Following a brief recovery in the middle of the decade, in 1997 earnings fell back as export volumes decreased by more than 15%. Figures for 1999 reflected the disastrous consequences of 'Hurricane Mitch', with the destruction of more than 70% of the total crop. Banana export earnings fell dramatically to US $38m., compared with $220m. in the previous year. In 2000–01 a recovery in the sector occurred and exports increased to $204.6m. in 2001; exports declined to $132.4m. in 2003, but recovered to $208.3m. in 2004. Export volumes ranged from 38.5m. 40-lb boxes in 1992 to just 6.8m. boxes in 1999, but stood at 29m. boxes in 2004.

In the mid-1980s the volume of coffee exports increased significantly under the combined stimuli of greatly improved international prices and the success of government efforts to eradicate diseases such as coffee rust. However, the improvement was negated by the consequences of the collapse of the International Coffee Agreement in July 1989. The disastrous fall in prices that followed did not begin to be reversed until 1994, when world shortages produced the highest coffee prices in a decade. In 1994 exports of 2.2m. 46-kg bags earned $200.1m.; the vagaries of the market were fully illustrated the following year, when a minimal rise in export volume produced an increase in earnings to $349.3m. Although 'Hurricane Mitch' had less impact on the coffee sector than the banana industry, the devastation, combined with low prices, cut earnings to just $256.1m. in 1999 following earnings of $429.8m. the previous year. Earnings partially recovered in 2000, to $339.4m., before more than halving again, to just $160.7m., in 2001, as prices declined to their lowest real levels for more than a century. A slight improvement was recorded in 2002 and 2003, when coffee exports earned a preliminary $182.5m. and $183.3m. Earnings increased significantly in 2004, amounting to $251.8m. In 2000 Central American coffee producers, along with Mexico, Colombia and Brazil, agreed to retain up to 10% of their stocks in order to allow the price to rise, although the arrangement proved difficult to enforce.

Poor prices and the consequent lower production caused the sugar, cotton and tobacco sectors to decrease dramatically in importance as export crops in the 1980s. By 2002 sugar exports were worth a mere US $16.8m. However, growth in some non-traditional agricultural sectors and substantial growth in the seafood industry during the 1990s helped to offset the deficit. Melon exports significantly increased in both volume and value in the late 1990s, reaching $44m. in 1998, before falling to an estimated $27.8m. in 2002. Earnings recovered slightly thereafter, reaching $34.0m. in 2004. Pineapple exports declined from $19m. in 1999 to just $8.7m. in 2002, before increasing to $24.4m. in 2004. Exports of shellfish, principally lobster and shrimp, more than tripled in little more than a decade. In 2004 shellfish exports earned the country $194.2m., with shrimp representing the vast majority of that figure. The shrimp farming industry, in particular, although badly hit by 'Hurricane Mitch', recovered rapidly.

The timber industry, which was nationalized in 1974 and placed under the control of the Corporación Hondureña de Desarrollo Forestal (COHDEFOR—Honduran Corporation of Forestry Development), encountered a series of problems. Like other state corporations, COHDEFOR incurred huge financial losses and its principal sawmill, the Bonito timber project in Olancho province, was eventually sold to a US citizen for only 5m. lempiras, representing a loss of 18m. lempiras. This followed devastating fires at the beginning of the 1980s, as a consequence of which the value of timber exports declined from US $44.7m. in 1982 to just $19.0m. in 1995, a level at which exports stabilized in the mid-1990s. In 2004 wood exports earned Honduras $32.9m. Almost all timber exports were pine and other softwoods. An estimated 2.5m. ha of Honduras were believed to be in need of reforestation, following the loss of more than 30% of the country's forests after 1970, at a rate of some 88,000 ha per year. The IDB loan agreement of April 2004 (see below) included $55m. for sustainable forestry development.

## MINING AND POWER

Although no comprehensive geological survey had been undertaken, Honduras was reported to have substantial reserves of tin, iron ore, coal, pitchblende and antimony and exploitable reserves of gold. In recent years there has been an increase in interest in the sector from foreign companies, following the liberalization of the mining code, and even though the mining and quarrying sector contributed only 1.8% of GDP in 2004, the increased activity was reflected in some revival of exports. In 2003 gold, silver, zinc and lead exports earned the country some US $96.6m.

Activity focused on the extraction of lead, zinc, silver and gold. Gold transformed from a statistically insignificant export commodity in 1999 to the most important mineral export, earning US $80.5m. in 2002, before declining slightly to $66.6m. in 2003. Gold exports earned $73.6m. in 2004. Such dramatic shifts reflect the limited extent of mineral exploitation in Honduras. Zinc exports increased dramatically in the mid-1990s, more than doubling export revenues in 1997 (to $53.8m.), although this fell at the end of the decade and into the 2000s. Production of lead similarly increased, with exports nearly doubling to 13.6m. lbs in 1997, earning $4.6m., before falling. In 2000 exports of 10.8m. lbs earned the country $3.3m. Production increased by 50% in 2001, but there was no increase in earnings. In 2004 combined lead and zinc exports generated $45.2m. in earnings. Silver production earnings experienced similar fluctuations. Earnings and production peaked in 1998, with Honduras exporting 1.5m. troy oz to earn $7.8m. In 1999 exports fell to 1.1m. troy oz, earning $5.4m., and there was a further fall in 2000 to 976,000 troy oz, which produced earnings of $4.8m. In 2003 silver exports of 1.4m. troy oz earned $5.6m. In 2001 Entremares Honduras, SA (a subsidiary of Glamis Golds Ltd of the USA), the Canadian company Geomaque Explorations and the Argentine concern Coviasa began mining operations in Honduras, joining the Canadian-owned American Pacific, which worked the El Mochito lead, zinc and copper mines.

Honduras had no exploited oil reserves. Petroleum accounted for 15.9% of imports in 2003. The power industry centred on hydroelectric generation, and in the 1970s two major hydroelectric systems, the El Cajón dam (capacity 292 MW) and the

Río Lindo–Yojoa system (capacity 285 MW), were developed. By the 1990s the combination of under-investment, mismanagement and natural disasters had brought the sector into crisis. Declining levels of water in both hydroelectric systems were blamed on deforestation, caused by logging and the clearing of woodland for agriculture in order to cope with the country's growing land crisis. In 1997–98 a drought caused by the El Niño weather phenomenon reduced El Cajón's generating capacity by 50%. The floods caused by 'Hurricane Mitch' in 1998 then wrought havoc with the power-distribution network, despite raising water levels and thus increasing power-generation capacity.

From 1994 successive Governments attempted, with some success, to resolve the problems of the power sector. Electricity prices were increased several times and the state-owned Empresa Nacional de Energía Eléctrica (ENEE—National Electrical Energy Company) intensified efforts to collect more than 120m. lempiras in debts. In February 1996 a new 40-MW thermal electricity plant was inaugurated. However, plans for the development of other plants were postponed, and continued under-investment in the transmission network meant that more than 20% of the power generated was lost. The damage caused by 'Hurricane Mitch' proved a major obstacle to efforts to overhaul the distribution network to major cities. In April 1999 the Governments of Honduras and El Salvador agreed to construct a regional electricity grid using a grant of US $30m. from the IDB. The transmission cable came into operation in May 2002. In 2000 the country generated some 3,930 GWh, with state-owned hydroelectric plants generating more than two-thirds of the total, and state and privately owned thermal plants generating the remainder. National consumption was put at 3,548 GWh in 2002, with demand increasing by an average of 7.5% per year in 1999–2002. By 2003 this rising demand prompted the ENEE to arrange for an additional 60 MW per day to be imported from Costa Rica in order to ease the strain on domestic supplies. In mid-2003 President Ricardo Maduro Joest commissioned a new authority to develop the country's first new hydroelectric projects since 1985.

## MANUFACTURING

Owing to the establishment of *maquila* (offshore assembly) plants producing goods for re-export, the Honduran manufacturing sector underwent a substantial transformation in the 1980s and 1990s. According to the Central Bank, earnings from *maquila* plants totalled US $830.7m. in 2004, and the sector employed around 115,000 people. This represented a recovery in the sector, which had been contracting from 2001. In that year some 220 *maquila* plants accounted for $560.8m. in value-added export earnings, down from $662m. in the previous year, and employed an estimated 94,000 people. The rate of growth had been slowing in the 1990s, with little diversification taking place from the main industry, textile assembly, and competition from the Caribbean impacting the relative attraction of the country. The manufacturing sector's contribution to GDP remained relatively stable as the decline in traditional industries offset the growth in the offshore assembly plants. Manufacturing GDP increased at an average annual rate of 4.0% during 1991–2003, but growth slowed somewhat, with the extension of North American Free Trade Agreement (NAFTA) privileges to Honduran textiles failing to offset the marked slowdown in demand from the USA, the country's major market. The sector's contribution to overall GDP was 20.5% in 2004.

Traditional manufacturing activity was based largely on agro-forestry products for both export and domestic consumption and included food processing, beverages, textiles, furniture making, cigarettes, sugar refining, seafood and meat processing, and paper and pulp processing. Cement, textile fabric, beer, soft drinks, wheat flour and rum were some of the major products. Despite some initial damage to plants and markets as a result of 'Hurricane Mitch', much of the domestic manufacturing sector benefited substantially as a result of the reconstruction programme. Cement production recovered to 32.7m. bags in 2004, compared with 28.8m. bags in 2002. Fabric production increased by more than 20% in 2001, to 119m. yards, and then increased further, to 206m. yards, in 2004. Soft-drink output rose by 7.1%, to 1,454m. bottles, in 2002, before registering marginal growth in 2003 and a 9.6% decline, to 1,329m. bottles, in 2004.

In the 1990s it was the free-market and privatization policies of the Governments of Presidents Callejas and Reina that dominated prospects in the manufacturing sector. During 1991 several factories, including a cement and textile plant, were sold, following the liquidation of the Corporación Nacional de Inversiones (CONADI—National Investment Corporation), the government-established industrial development agency. Reductions in tariff protection in 1991 adversely affected industry, but, for exporters at least, the successive currency devaluations helped to increase competitiveness.

Another major stimulus was the creation of five free trade zones from 1976 onwards. The success of these led to the creation of six private free trade zones from 1987. All of the free trade zones were concentrated near the coastal cities of San Pedro Sula and Puerto Cortés. Textile and assembly companies predominated in the sector. Most companies operated under the customs provisions of the USA's Caribbean Basin Initiative (CBI), which allowed for the duty-free importation of clothes assembled from US cloth. Electronics, furniture and metal-manufacturing assembly plants all increased in number in the 1990s, albeit from a very low base, and the country of origin of the investment diversified.

For the *maquila* sector, 1999 and 2000 were the most successful years in more than a decade. With 'Hurricane Mitch' devastating traditional agricultural exports such as bananas and coffee, export earnings of some US $545m. in 1999, then $662m. in 2000, made the *maquila* sector Honduras' most important foreign-exchange earner for the first time. The sector was further boosted in July 2000 by the extension of NAFTA import-duty parity to Honduran-assembled goods. The extension of these provisions reduced the import tariff on Honduran-assembled goods into the US market by 15% and allowed the sector to diversify into textile operations such as dyeing and cutting. The move encouraged the Government to target $700m. in new investment into the sector over the subsequent five years, which, it hoped, would create up to 80,000 jobs. In November 2000 the Formosa Industrial Park, financed by the Republic of China (Taiwan), opened. However, the US economic slowdown that began in 2001 badly affected the sector, with 32 *maquila* plants closing and employment levels falling, with an estimated 10,000 jobs lost, for the first time in more than a decade.

## FOREIGN TRADE AND PAYMENTS

Honduras recorded a persistently large current-account deficit. By 1994 this deficit had reached US $343m., although there was some improvement in the late 1990s. In 1999, however, with exports affected by 'Hurricane Mitch', the deficit increased to $625m. In 2000 this figure fell to $276m., before increasing to $339m. in 2001. A slight improvement, to $242m., was recorded in 2002, but the deficit grew again, to $279m., in 2003 and further, to $391m., in 2004.

The trade deficit has always fluctuated in line with the prices of major exports and the demand for imports; however, it increased dramatically after 1990. In 2001 the trade deficit stood at US $832.6m. and continued to rise in subsequent years, reaching $987.2m. in 2003. Export earnings (f.o.b.) rose by more than 10% per year in the mid-1990s, but fell back substantially in the wake of 'Hurricane Mitch'. In 2000 export earnings were $1,436.2m., before decreasing to $1,324,4m. in 2002. This figure increased to an estimated $1,343.3m. in 2003, and to an estimated $1,553,9m. in 2004. In 2003 imports (c.i.f.) totalled an estimated $3,275.6m.

In the mid-1980s the widening trade gap was partially offset by a sharp rise in aid from the USA, but subsequent falls in both aid and export revenue forced the Government to yield to pressure from the IMF for a devaluation, which took place in 1990. Throughout the 1990s the fiscal situation continued to deteriorate, and there were further devaluations in 1994 in order to secure a US $500m. loan agreement from the World Bank and the IMF. Multilateral and bilateral creditor support subsequently remained strong, despite the significant weakening of government finances in the wake of 'Hurricane Mitch'. The most obvious demonstration of this was the country's achievement of HIPC status in mid-2000. The agreement promised debt-service relief of some $900m. in return for continued economic reform and adherence to an IMF- and World Bank-agreed PRGF programme. The programme allowed for an increase in social spending, subject to an increase in national revenue. The debt relief secured by the Government after reducing the fiscal deficit and reaching 'completion point' within the HIPC initiative in 2005 enabled further investment in poverty reduction.

Improved terms from creditors in the 1990s was contingent upon investment liberalization, which helped dramatically increase foreign investment inflows during that decade. In 2000 direct foreign investment increased to US $282m. (a 19% rise on the previous year), before falling to $176m. in 2002. In 2003 direct foreign investment increased to $247m.

External debt more than doubled in the 1980s, reaching US $3,700m. in 1990. However, rescheduling, renegotiation and relief steadily reduced the Honduran debt-service ratio (debt service as a percentage of earnings from the export of goods and services) from 35.3% in 1990 to 13.0% 10 years later. By 2003 total debt stocks had risen to $5,641m., while the debt-service ratio had declined to 1.6%.

By 2005 it was increasingly clear that successive Honduran Governments had enjoyed some success in their efforts to diversify the economy and the country's sources of foreign exchange. Non-traditional agricultural and fishery exports soared, earnings from the *maquila* industry exceeded those from both coffee and bananas, and invisibles, such as tourism and remittances from Hondurans living abroad (estimated at US $1,250m. in 2004), were making increasingly important contributions to the current account. Revenues from tourism grew dramatically, rising from $31.8m. in 1992 to $341m. in 2003, when some 610,535 tourists visited the country; in 2004 some 384,500 day-trippers visited the country, with a further 672,000 making overnight visits. Tourist expenditure increased to $410m. in 2004, supporting employment for 93,500 Hondurans. More than 88% of the mainland arrivals were from North and Central America.

There was less success in the diversification of the direction of trade, with dependence on the USA, the country's principal market, remaining strong. In 2004 41.5% of the country's exports went to the USA, compared with just over 54% a decade earlier. The major trade development was the growth of intra-regional trade; El Salvador and Guatemala took 10.9% and 7.3%, respectively, of the country's total exports in 2004. Germany and Belgium were the main destinations for exports outside the Americas. Reliance on the USA for imports gradually declined slightly, yet still remained strong. In 2004 34.6% of the country's total imports came from the USA, compared with 47% a decade earlier. Other important suppliers were regional neighbours, with Guatemala, El Salvador, Mexico and Costa Rica collectively accounting for 22.3% of the total in 2004.

Following four years of negotiations, in April 2000 the Central American countries of El Salvador, Guatemala and Honduras signed a free trade agreement with Mexico, which promised greater access to the Mexican market and increased bilateral trade. The 'Plan Puebla–Panamá' of May 2001 provided for regional integration through joint transport, industry and tourism projects. Honduras was also party to the establishment of the Central American Free Trade Agreement (CAFTA) with the USA, which was agreed in December 2003 and ratified by the Honduran legislature in March 2005. The Agreement would consolidate trade privileges, which currently come under the auspices of the CBI (which expires in 2008) and was expected to produce many *maquila* jobs. The Agreement was ratified by the US House of Representatives in late July 2005. In a bilateral initiative, Honduras and Guatemala signed a customs-union agreement in April 2004.

## CONCLUSION

President Maduro's Government pursued policies in line with IMF and World Bank programmes, in particular, curbing public expenditure and diversifying the country's narrow economic base. The Government made a commitment to alleviate the country's high level of poverty. Debt relief under the HIPC initiative promised to relieve pressure on the country's finances. Meanwhile, the potential implementation of CAFTA was expected to improve export performance in the near future. However, the country continued to be vulnerable to natural disasters, depressed international commodity prices and any downturn in the US economy. The lasting effects of 'Hurricane Mitch' continued to impede the crucial agricultural sector, while high international oil prices fuelled inflation. Furthermore, many economic and social problems, including widespread crime and corruption, remained.

# Statistical Survey

Sources (unless otherwise stated): Department of Economic Studies, Banco Central de Honduras, Avda Juan Ramón Molina, 1a Calle, 7a Avda, Apdo 3165, Tegucigalpa; tel. 237-2270; fax 238-0376; e-mail jreyes@bch.hn; internet www.bch.hn; Instituto Nacional de Estadística, Edif. Gómez, Blvd Suyapa, Col. Florencia Sur, Apdo 9412, Tegucipgalpa; e-mail info@ine.online.hn; internet www.ine-hn.org.

Note: Although the metric system is in force, some old Spanish measures are still used, including: 25 libras = 1 arroba; 4 arrobas = 1 quintal (46 kg).

## Area and Population

### AREA, POPULATION AND DENSITY

| | |
|---|---:|
| Area (sq km) | 112,492* |
| Population (census results)† | |
| 29 May 1988 | 4,614,377 |
| 1 August 2001 | |
|   Males | 3,230,958 |
|   Females | 3,304,386 |
|   Total | 6,535,344 |
| Population (official estimates at mid-year) | |
|   2001 | 6,530,300 |
|   2002 | 6,694,800 |
|   2003 | 6,860,800 |
| Density (per sq km) at mid-2003 | 61.0 |

\* 43,433 sq miles.
† Excluding adjustments for underenumeration, estimated to have been 10% at the 1974 census.

### PRINCIPAL TOWNS
(estimated population, '000 at mid-2001)

| | | | |
|---|---:|---|---:|
| Tegucigalpa (captial) | 1,089.2 | Siguatepeque | 53.7 |
| San Pedro Sula | 490.6 | Puerto Cortés | 36.0 |
| El Progreso | 115.0 | Juticalpa | 34.8 |
| La Ceiba | 111.2 | Santa Rosa de Copán | 28.6 |
| Choluteca | 101.6 | Tela | 26.6 |
| Comayagua | 77.4 | Olanchito | 23.9 |
| Danlí | 68.8 | | |

**Mid-2003** (UN estimate, incl. suburbs): Tegucigalpa 1,006,592 (Source: UN, *World Urbanization Prospects: The 2003 revision*).

### BIRTHS AND DEATHS
(UN estimates)

| | 1985–90 | 1990–95 | 1995–2000 |
|---|---:|---:|---:|
| Birth rate (per 1,000) | 39.4 | 37.1 | 33.4 |
| Death rate (per 1,000) | 7.4 | 6.7 | 6.2 |

Source: UN, *World Population Prospects: The 2004 Revision*.

**Expectation of life** (WHO estimates, years at birth): 67 (males 65; females 69) in 2003 (Source: WHO, *World Health Report*).

# HONDURAS

## ECONOMICALLY ACTIVE POPULATION
('000 persons)

|  | 2001 | 2002 | 2003 |
|---|---|---|---|
| Agriculture, hunting, forestry and fishing | 844.8 | 931.2 | 906.3 |
| Mining and quarrying | 5.1 | 4.6 | 5.8 |
| Manufacturing | 367.0 | 337.3 | 381.2 |
| Electricity, gas and water | 11.5 | 8.8 | 9.6 |
| Construction | 123.6 | 120.8 | 122.8 |
| Trade, restaurants and hotels | 462.7 | 507.0 | 495.9 |
| Transport, storage and communications | 71.7 | 71.6 | 79.1 |
| Financing, insurance, real estate and business services | 66.5 | 57.4 | 73.3 |
| Community, social and personal services | 335.8 | 312.4 | 352.1 |
| **Total** | **2,288.7** | **2,351.1** | **2,426.1** |

# Health and Welfare

## KEY INDICATORS

| | |
|---|---|
| Total fertility rate (children per woman, 2003) | 3.7 |
| Under-5 mortality rate (per 1,000 live births, 2003) | 41 |
| HIV/AIDS (% of persons aged 15–49, 2003) | 1.8 |
| Physicians (per 1,000 head, 1997) | 0.83 |
| Hospital beds (per 1,000 head, 1996) | 1.06 |
| Health expenditure (2002): US $ per head (PPP) | 156 |
| Health expenditure (2002): % of GDP | 6.2 |
| Health expenditure (2002): public (% of total) | 51.2 |
| Access to water (% of persons, 2002) | 90 |
| Access to sanitation (% of persons, 2002) | 68 |
| Human Development Index (2002): ranking | 115 |
| Human Development Index (2002): value | 0.672 |

For sources and definitions, see explanatory note on p. vi.

# Agriculture

## PRINCIPAL CROPS
('000 metric tons)

|  | 2001 | 2002 | 2003 |
|---|---|---|---|
| Maize | 516 | 392 | 502 |
| Sorghum | 75 | 43 | 52 |
| Sugar cane | 3,451 | 3,451† | 5,363 |
| Dry beans | 59 | 50 | 69 |
| Oil palm fruit | 669 | 736 | 740† |
| Tomatoes | 50 | 53* | 53† |
| Other vegetables* | 140 | 142 | 142 |
| Melons* | 114 | 182 | 145 |
| Bananas | 516 | 965 | 965* |
| Plantains* | 260 | 260 | 260 |
| Oranges | 142* | 167 | 167* |
| Pineapples | 67* | 62 | 62* |
| Other fruit* | 71 | 77 | 77 |
| Coffee (green) | 206 | 182† | 150† |

\* FAO estimate(s).
† Unofficial estimate.

Source: FAO.

## LIVESTOCK
('000 head, year ending September)

|  | 2001 | 2002 | 2003 |
|---|---|---|---|
| Cattle | 1,875* | 1,860 | 2,403* |
| Sheep† | 14 | 14 | 13 |
| Goats† | 32 | 32 | 32 |
| Pigs | 538* | 538 | 478* |
| Horses† | 180 | 180 | 181 |
| Mules† | 70 | 70 | 70 |
| Asses† | 23 | 23 | 23 |
| Poultry | 18,000† | 18,648 | 18,700† |

\* Unofficial figure.
† FAO estimate(s).

Source: FAO.

## LIVESTOCK PRODUCTS
('000 metric tons)

|  | 2001 | 2002 | 2003 |
|---|---|---|---|
| Beef and veal | 55.3 | 62.0* | 57.0 |
| Pig meat | 9.7 | 10.5† | 9.6† |
| Poultry meat | 66.3 | 66.3† | 74.0† |
| Cows' milk | 593.8 | 595.5† | 597.0† |
| Cheese† | 9.0 | 9.2 | 9.0 |
| Butter† | 4.4 | 4.5 | 4.5 |
| Hen eggs† | 44.4 | 44.4 | 40.0 |
| Cattle hides† | 9.5 | 9.5 | 9.5 |

\* Unofficial estimate.
† FAO estimate(s).

Source: FAO.

# Forestry

## ROUNDWOOD REMOVALS
('000 cubic metres, excl. bark)

|  | 2001 | 2002 | 2003 |
|---|---|---|---|
| Sawlogs, veneer logs and logs for sleepers | 832 | 971 | 801 |
| Fuel wood* | 8,720 | 8,710 | 8,703 |
| **Total \*** | **9,552** | **9,681** | **9,504** |

\* FAO estimates.

Source: FAO.

## SAWNWOOD PRODUCTION
('000 cubic metres, incl. railway sleepers)

|  | 2001 | 2002 | 2003 |
|---|---|---|---|
| Coniferous (softwood) | 412 | 470 | 421 |
| Broadleaved (hardwood) | 7 | — | — |
| **Total** | **419** | **470** | **421** |

Source: FAO.

# Fishing

('000 metric tons, live weight)

|  | 2001 | 2002 | 2003 |
|---|---|---|---|
| Capture* | 18.9 | 11.4 | 10.8 |
| Marine fishes | 4.2 | 4.9 | 5.1 |
| Caribbean spiny lobster | 0.9 | 0.8 | 1.0 |
| Penaeus shrimps | 5.2 | 4.7 | 3.5 |
| Stromboid conches | 0.7 | 0.8 | 1.0 |
| Aquaculture | 12.1 | 14.6 | 20.0 |
| Nile tilapia | 1.2 | 2.0 | 3.5 |
| Penaeus shrimps | 10.9 | 12.6 | 16.5 |
| **Total catch\*** | **31.0** | **26.0** | **30.8** |

\* FAO estimates.

Source: FAO.

# Mining

(metal content)

|  | 2000 | 2001 | 2002 |
|---|---|---|---|
| Lead (metric tons) | 4,805 | 6,750 | 8,128 |
| Zinc (metric tons) | 31,226 | 48,485 | 46,339 |
| Silver (kilograms) | 31,958 | 46,831 | 52,877 |
| Gold (kilograms) | 878 | 4,574 | 4,984 |

Source: US Geological Survey.

# Industry

## SELECTED PRODUCTS

|  | 2002 | 2003 | 2004* |
|---|---:|---:|---:|
| Raw sugar ('000 quintales) | 7,220 | 6,588 | 8,145 |
| Cement ('000 bags of 42.5 kg) | 28,803 | 29,827 | 32,700 |
| Cigarettes ('000 packets of 20) | 300,491 | 308,042 | 320,556 |
| Beer ('000 12 oz bottles) | 260,007 | 255,173 | 257,198 |
| Soft drinks ('000 12 oz bottles) | 1,453,703 | 1,457,876 | 1,329,206 |
| Wheat flour ('000 quintales) | 2,566 | 2,498 | 2,637 |
| Fabric ('000 yards) | 135,290 | 187,750 | 206,427 |
| Liquor and spirits ('000 litres) | 11,169 | 12,444 | 12,799 |
| Vegetable oil and butter ('000 libras) | 169,664 | 177,148 | 177,649 |
| Electric energy (million kWh) | 4,494.9 | 4,845.2 | n.a. |

* Preliminary figures.

# Finance

## CURRENCY AND EXCHANGE RATES

**Monetary Units**
100 centavos = 1 lempira.

**Sterling, Dollar and Euro Equivalents** (31 May 2005)
£1 sterling = 34.284 lempiras;
US $1 = 18.857 lempiras;
€1 = 23.252 lempiras;
1,000 lempiras = £29.17 = $53.03 43.01.

**Average Exchange Rate** (lempiras per US $)
2002   16.4334
2003   17.3453
2004   18.2062

## BUDGET
(million lempiras)

| Revenue | 2001 | 2002 | 2003* |
|---|---:|---:|---:|
| Current revenue | 17,848.9 | 19,776.0 | 22,175.1 |
| Taxes | 16,083.1 | 17,229.0 | 19,632.4 |
| Non-tax revenue | 1,765.8 | 2,547.0 | 2,542.7 |
| Capital revenue | 31.4 | — | — |
| Transfers | 1,657.8 | 1,302.3 | 1,328.1 |
| Other revenue | 188.4 | 62.8 | 58.8 |
| **Total** | 19,726.5 | 21,141.1 | 23,562.1 |

| Expenditure | 2001 | 2002 | 2003* |
|---|---:|---:|---:|
| Current expenditure | 17,400.7 | 19,593.3 | 22,823.0 |
| Consumption expenditure | 12,680.9 | 14,223.1 | 15,441.1 |
| Interest and other charges | 1,186.4 | 1,296.8 | 1,469.9 |
| Internal debt | 300.7 | 314.4 | 406.2 |
| External debt† | 885.6 | 982.4 | 1,063.7 |
| Transfers | 3,533.5 | 4,073.3 | 5,912.0 |
| Capital expenditure | 6,744.7 | 5,433.7 | 6,594.2 |
| Real investment | 2,824.6 | 2,448.9 | 2,795.0 |
| Transfers | 3,920.1 | 2,984.8 | 3,799.2 |
| **Total**‡ | 24,145.4 | 25,027.0 | 29,417.2 |

* Preliminary figures.
† Excludes debt-servicing relief.
‡ Excluding net lending (million lempiras): 821.0 in 2001; 1,262.0 in 2002; 671.6 in 2003 (preliminary).

Source: Ministry of Finance, *Memoria Secretaria de Finanzas*.

## CENTRAL BANK RESERVES
(US $ million at 31 December)

|  | 2002 | 2003 | 2004 |
|---|---:|---:|---:|
| Gold* | 7.11 | 9.26 | 9.93 |
| IMF special drawing rights | 0.47 | 0.11 | 0.90 |
| Foreign exchange | 1,511.80 | 1,417.10 | 1,956.90 |
| Reserve position in IMF | 11.73 | 12.82 | 13.40 |
| **Total** | 1,531.11 | 1,439.29 | 1,981.13 |

* National valuation: $339 per troy ounce at 31 December 2002; $441 per troy ounce at 31 December 2003; $473 per troy ounce at 31 December 2004.

Source: IMF, *International Financial Statistics*.

## MONEY SUPPLY
(million lempiras at 31 December)

|  | 2002 | 2003 | 2004 |
|---|---:|---:|---:|
| Currency outside banks | 5,549 | 6,448 | 7,639 |
| Demand deposits at commercial banks | 7,658 | 9,315 | 9,840 |
| **Total money** (incl. others) | 14,224 | 17,251 | 20,019 |

Source: IMF, *International Financial Statistics*.

## COST OF LIVING
(Consumer Price Index, December; base: 1999 = 100)

|  | 2002 | 2003 | 2004 |
|---|---:|---:|---:|
| Food | 120.1 | 126.1 | 137.8 |
| Alcohol and tobacco | 135.7 | 151.5 | 161.9 |
| Rent, water, fuel and power | 141.3 | 153.2 | 169.0 |
| Clothing and footwear | 131.3 | 140.7 | 147.7 |
| Health | 154.9 | 165.0 | 176.4 |
| Transport | 135.5 | 146.3 | 167.9 |
| Communications | 93.5 | 94.5 | 97.2 |
| Culture and recreation | 121.8 | 127.8 | 135.2 |
| Education | 158.2 | 180.9 | 203.4 |
| Restaurants and hotels | 128.7 | 136.8 | 147.5 |
| **All items** (incl. others) | 129.5 | 138.3 | 151.0 |

## NATIONAL ACCOUNTS
(million lempiras at current prices)

### Expenditure on the Gross Domestic Product

|  | 2001* | 2002* | 2003† |
|---|---:|---:|---:|
| Government final consumption expenditure | 13,792 | 14,925 | 16,209 |
| Private final consumption expenditure | 72,198 | 82,063 | 91,627 |
| Increase in stocks | 5,756 | 3,710 | 4,261 |
| Gross fixed capital formation | 23,525 | 23,992 | 28,124 |
| **Total domestic expenditure** | 115,271 | 124,691 | 140,221 |
| Exports of goods and services | 37,481 | 41,203 | 46,684 |
| *Less* Imports of goods and services | 53,720 | 57,770 | 66,440 |
| **GDP in purchasers' values** | 99,032 | 108,124 | 120,465 |
| **GDP at constant 1978 prices** | 7,324 | 7,523 | 7,785 |

* Preliminary figures.
† Estimates.

# HONDURAS

## Gross Domestic Product by Economic Activity

|  | 2001* | 2002* | 2003† |
|---|---|---|---|
| Agriculture, hunting, forestry and fishing | 12,122 | 12,895 | 13,701 |
| Mining and quarrying | 1,591 | 1,794 | 1,998 |
| Manufacturing | 17,540 | 19,640 | 21,980 |
| Electricity, gas and water | 3,728 | 4,263 | 5,052 |
| Construction | 4,269 | 3,966 | 4,922 |
| Wholesale and retail trade, restaurants and hotels | 10,870 | 12,050 | 13,412 |
| Transport, storage and communications | 5,096 | 5,643 | 6,354 |
| Finance, insurance and real estate | 9,441 | 10,406 | 11,458 |
| Owner-occupied dwellings | 5,201 | 5,840 | 6,552 |
| Public administration and defence | 6,139 | 7,095 | 7,478 |
| Other services | 10,561 | 12,177 | 13,577 |
| **GDP at factor cost** | 86,558 | 95,769 | 106,484 |
| Indirect taxes, *less* subsidies | 12,474 | 12,355 | 13,981 |
| **GDP in purchasers' values** | 99,032 | 108,124 | 120,465 |

\* Preliminary figures.
† Estimates.

## BALANCE OF PAYMENTS
(US $ million)

|  | 2001 | 2002 | 2003 |
|---|---|---|---|
| Exports of goods f.o.b. | 1,935.5 | 1,973.6 | 2,078.2 |
| Imports of goods f.o.b. | −2,768.1 | −2,809.2 | −3,065.4 |
| **Trade balance** | −832.6 | −835.6 | −987.2 |
| Exports of services | 487.2 | 530.3 | 576.2 |
| Imports of services | −626.6 | −601.9 | −653.2 |
| **Balance on goods and services** | −972.0 | −907.2 | −1,064.2 |
| Other income received | 88.0 | 66.8 | 56.7 |
| Other income paid | −259.2 | −249.7 | −240.0 |
| **Balance on goods, services and income** | −1,143.2 | −1,090.0 | −1,247.5 |
| Current transfers received | 894.0 | 946.6 | 1,072.2 |
| Current transfers paid | −89.9 | −98.9 | −104.0 |
| **Current balance** | −339.1 | −242.3 | −279.2 |
| Capital account (net) | 36.7 | 23.6 | 21.0 |
| Direct investment from abroad | 189.5 | 175.5 | 198.0 |
| Portfolio investment assets | −3.6 | −3.8 | −4.1 |
| Other investment assets | −102.3 | 24.2 | −77.6 |
| Other investment liabilities | 76.5 | −56.6 | −136.3 |
| Net errors and omissions | 68.3 | 60.5 | 76.1 |
| **Overall balance** | −74.0 | −18.8 | −202.1 |

Source: IMF, *International Financial Statistics*.

# External Trade

## PRINCIPAL COMMODITIES
(US $ million, preliminary figures)

| Imports c.i.f. | 2001 | 2002 | 2003 |
|---|---|---|---|
| Vegetables and fruit | 141.7 | 151.5 | 155.4 |
| Mineral fuels and lubricants | 395.1 | 413.5 | 519.2 |
| Chemicals and related products | 397.3 | 445.1 | 457.0 |
| Plastic and manufactures | 169.6 | 184.3 | 196.9 |
| Paper, paperboard and manufactures | 158.7 | 166.9 | 176.6 |
| Textile yarn, fabrics and manufactures | 87.7 | 96.7 | 99.3 |
| Metal and manufactures | 291.3 | 201.3 | 246.3 |
| Food products | 282.8 | 285.9 | 302.2 |
| Machinery and electrical appliances | 463.8 | 481.2 | 559.4 |
| Transport equipment | 276.8 | 275.7 | 241.8 |
| **Total** (incl. others) | 2,941.8 | 2,981.1 | 3,275.6 |

| Exports f.o.b. | 2002 | 2003 | 2004 |
|---|---|---|---|
| Bananas | 172.4 | 132.4 | 208.3 |
| Coffee | 182.5 | 183.3 | 251.8 |
| Gold | 80.5 | 66.6 | 73.6 |
| Lead and zinc | 36.8 | 39.7 | 45.2 |
| Palm oil | 30.2 | 53.9 | 53.1 |
| Shellfish | 172.4 | 191.8 | 194.2 |
| Soaps and detergents | 26.1 | 33.7 | n.a. |
| Wood | 33.7 | 31.0 | 32.9 |
| **Total** (incl. others) | 1,324.4 | 1,343.3 | 1,553.9 |

## PRINCIPAL TRADING PARTNERS
(US $ million, preliminary figures)

| Imports c.i.f. | 2001 | 2002 | 2003 |
|---|---|---|---|
| Brazil | 29.9 | 43.6 | 66.0 |
| Colombia | 30.2 | 34.3 | 36.9 |
| Costa Rica | 115.2 | 148.4 | 164.0 |
| El Salvador | 183.5 | 165.3 | 182.0 |
| Germany | 43.7 | 34.3 | 40.7 |
| Guatemala | 244.9 | 249.4 | 268.1 |
| Japan | 140.9 | 97.6 | 114.0 |
| Mexico | 149.9 | 169.3 | 162.1 |
| Nicaragua | 31.4 | 45.9 | 48.7 |
| Spain | 51.3 | 33.1 | 48.0 |
| Trinidad and Tobago | 33.5 | 79.6 | 26.5 |
| USA | 1,097.9 | 1,104.6 | 1,227.3 |
| Venezuela | 23.9 | 57.4 | 48.6 |
| **Total** (incl. others) | 2,941.8 | 2,981.1 | 3,275.6 |

| Exports f.o.b. | 2001 | 2002 | 2003 |
|---|---|---|---|
| Belgium | 70.5 | 41.0 | 41.1 |
| Canada | 43.9 | 31.2 | 29.5 |
| Costa Rica | 35.8 | 30.6 | 32.8 |
| El Salvador | 135.0 | 157.0 | 150.5 |
| France | 14.0 | 9.8 | 13.0 |
| Germany | 57.4 | 83.9 | 64.1 |
| Guatemala | 129.3 | 83.9 | 78.0 |
| Italy | 14.6 | 19.6 | 8.5 |
| Japan | 30.0 | 20.2 | 15.1 |
| Mexico | 7.4 | 11.2 | 36.4 |
| Netherlands | 19.2 | 23.9 | 50.4 |
| Nicaragua | 26.5 | 19.1 | 43.0 |
| Spain | 32.8 | 39.3 | 42.9 |
| United Kingdom | 21.6 | 31.8 | 39.3 |
| USA | 571.4 | 616.5 | 595.3 |
| **Total** (incl. others) | 1,324.4 | 1,321.2 | 1,332.3 |

# Transport

## ROAD TRAFFIC
(licensed vehicles in use)

|  | 2001 | 2002 | 2003 |
|---|---|---|---|
| Passenger cars | 345,931 | 369,303 | 386,468 |
| Buses and coaches | 20,380 | 21,814 | 22,514 |
| Lorries and vans | 81,192 | 86,893 | 91,230 |
| Motorcycles and bicycles | 36,828 | 39,245 | 41,852 |

## SHIPPING

**Merchant Fleet**
(registered at 31 December)

|  | 2002 | 2003 | 2004 |
|---|---|---|---|
| Number of vessels | 1,155 | 1,143 | 1,094 |
| Total displacement ('000 grt) | 933.2 | 812.5 | 784.1 |

Source: Lloyd's Register-Fairplay, *World Fleet Statistics*.

# HONDURAS

**International Sea-borne Freight Traffic**
('000 metric tons)

|  | 1988 | 1989 | 1990 |
|---|---|---|---|
| Goods loaded | 1,328 | 1,333 | 1,316 |
| Goods unloaded | 1,151 | 1,222 | 1,002 |

Source: UN, *Monthly Bulletin of Statistics*.

## CIVIL AVIATION
(traffic on scheduled services)

|  | 1993 | 1994 | 1995 |
|---|---|---|---|
| Kilometres flown (million) | 4 | 5 | 5 |
| Passengers carried ('000) | 409 | 449 | 474 |
| Passenger-km (million) | 362 | 323 | 341 |
| Total ton-km (million) | 50 | 42 | 33 |

Source: UN, *Statistical Yearbook*.

## Tourism

**TOURIST ARRIVALS BY COUNTRY OF ORIGIN**

|  | 2001 | 2002 | 2003 |
|---|---|---|---|
| Canada | 9,000 | 8,844 | 10,324 |
| Costa Rica | 16,854 | 17,845 | 19,621 |
| El Salvador | 95,338 | 110,931 | 130,547 |
| Guatemala | 78,717 | 86,068 | 99,894 |
| Italy | 10,210 | 11,684 | 10,020 |
| Mexico | 10,579 | 10,948 | 13,149 |
| Nicaragua | 77,774 | 96,532 | 94,174 |
| Spain | 7,939 | 7,028 | 7,335 |
| USA | 148,554 | 137,953 | 161,954 |
| **Total** (incl. others) | 517,914 | 549,500 | 610,535 |

**Receipts from tourism** (US $ million, incl. passenger transport): 260 in 2001; 305 in 2002; 341 in 2003.

Source: World Tourism Organization.

## Communications Media

|  | 2001 | 2002 | 2003 |
|---|---|---|---|
| Television receivers ('000 in use)* | 640 | n.a. | n.a. |
| Telephones ('000 main lines in use) | 310.6 | 322.5 | 334.4 |
| Mobile cellular telephones ('000 subscribers) | 237.6 | 326.5 | 351.3 |
| Personal computers ('000 in use)* | 80 | 91 | n.a. |
| Internet users ('000) | 90.0 | 168.6 | 185.4 |
| Daily newspapers | 4 | 4 | 4 |
| Weekly newspapers | 3 | 3 | 3 |

* Source: International Telecommunication Union.

**Radio receivers** ('000 in use): 2,450 in 1997 (Source: UN, *Statistical Yearbook*).

## Education

(2003)

|  | Institutions | Teachers | Students |
|---|---|---|---|
| Pre-primary | 5,357 | 15,232 | 197,408 |
| Primary (grades 1 to 6) | 11,115 | 42,788 | 1,255,859 |
| Secondary (grades 7 to 9) |  |  | 290,100 |
| High school | 871 | 16,435 | 454,489 |
| University level | 8 | 3,707 | 75,643 |

**Adult literacy rate** (UNESCO estimates): 80.0% (males 80.2%; females 79.8%) in 2002 (Source: UN Development Programme, *Human Development Report*).

# Directory

## The Constitution

Following the elections of April 1980, the 1965 Constitution was revised. The new Constitution was approved by the National Assembly in November 1982, and amended in 1995. The following are some of its main provisions:

Honduras is constituted as a democratic Republic. All Hondurans over 18 years of age are citizens.

### THE SUFFRAGE AND POLITICAL PARTIES

The vote is direct and secret. Any political party that proclaims or practises doctrines contrary to the democratic spirit is forbidden. A National Electoral Council will be set up at the end of each presidential term. Its general function will be to supervise all elections and to register political parties. A proportional system of voting will be adopted for the election of Municipal Corporations.

### INDIVIDUAL RIGHTS AND GUARANTEES

The right to life is declared inviolable; the death penalty is abolished. The Constitution recognizes the right of habeas corpus and arrests may be made only by judicial order. Remand for interrogation may not last more than six days, and no-one may be held incommunicado for more than 24 hours. The Constitution recognizes the rights of free expression of thought and opinion, the free circulation of information, of peaceful, unarmed association, of free movement within and out of the country, of political asylum and of religious and educational freedom. Civil marriage and divorce are recognized.

### WORKERS' WELFARE

All have a right to work. Day work shall not exceed eight hours per day or 44 hours per week; night work shall not exceed six hours per night or 36 hours per week. Equal pay shall be given for equal work. The legality of trade unions and the right to strike are recognized.

### EDUCATION

The State is responsible for education, which shall be free, lay, and, in the primary stage, compulsory. Private education is liable to inspection and regulation by the State.

### LEGISLATIVE POWER

Deputies are obliged to vote, for or against, on any measure at the discussion of which they are present. The National Assembly has power to grant amnesties to political prisoners; approve or disapprove of the actions of the Executive; declare part or the whole of the Republic subject to a state of siege; declare war; approve or withhold approval of treaties; withhold approval of the accounts of public expenditure when these exceed the sums fixed in the budget; decree, interpret, repeal and amend laws, and pass legislation fixing the rate of exchange or stabilizing the national currency. The National Assembly may suspend certain guarantees in all or part of the Republic for 60 days in the case of grave danger from civil or foreign war, epidemics or any other calamity. Deputies are elected in the proportion of one deputy and one substitute for every 35,000 inhabitants, or fraction over 15,000. Congress may amend the basis in the light of increasing population.

### EXECUTIVE POWER

Executive power is exercised by the President of the Republic, who is elected for four years by a simple majority of the people. No President may serve more than one term.

### JUDICIAL POWER

The Judiciary consists of the Supreme Court, the Courts of Appeal and various lesser tribunals. The nine judges and seven substitute judges of the Supreme Court are elected by the National Assembly for a period of four years. The Supreme Court is empowered to declare laws unconstitutional.

# HONDURAS

## THE ARMED FORCES
The Armed Forces are declared by the Constitution to be essentially professional and non-political. The President exercises direct authority over the military.

## LOCAL ADMINISTRATION
The country is divided into 18 Departments for purposes of local administration, and these are subdivided into 290 autonomous Municipalities; the functions of local offices shall be only economic and administrative.

## The Government

### HEAD OF STATE
**President:** Ricardo Maduro Joest (assumed office 27 January 2002).

**Vice-President:** Vicente Williams Agasse.

### CABINET
(July 2005)

**Minister of the Interior and Justice:** Jorge Ramón Hernández Alcerro.
**Minister in the Office of the President:** Ramón Medina.
**Minister of Foreign Affairs:** Mario Fortín.
**Minister of Industry and Commerce:** Norman García.
**Minister of Finance:** William Chong Wong.
**Minister of National Defence:** Federico Brevé Travieso.
**Minister of Public Security:** Oscar Alvarez Guerrero.
**Minister of Labour and Social Welfare:** Germán Leitzelar Vidaurreta.
**Minister of Health:** Elías Lizardo Zelaya.
**Minister of Public Education:** Roberto Martínez Lozano.
**Minister of Public Works, Transport and Housing:** Jorge Carranza Díaz.
**Minister of Culture, Art and Sports:** Arnoldo Avilés.
**Minister of Agriculture and Livestock:** Mariano Jiménez Talavera.
**Minister of Natural Resources and Environment:** Patricia Panting Galo.
**Minister of Tourism:** Thierry de Pierrefeu Midence.
**Minister of Strategy and Communication:** Ramón Medina Luna.
**Minister of International Co-operation:** Brenie Liliana Matute.

### MINISTRIES

**Office of the President:** Palacio José Cecilio del Valle, Blvd Juan Pablo II, Tegucigalpa; tel. 232-6282; fax 231-0097; internet www.casapresidencial.hn.

**Ministry of Agriculture and Livestock:** Tegucigalpa; tel. 235-7388; e-mail infoagro@sag.gob.hn; internet www.sag.gob.hn.

**Ministry of Culture, Art and Sports:** Avda La Paz, Apdo 3287, Tegucigalpa; tel. 236-9643; fax 236-9532; e-mail binah@sdnhon.org.hn; internet www.secad.gob.hn.

**Ministry of Finance:** 5a Avda, 3a Calle, Tegucigalpa; tel. 222-1278; fax 238-2309; e-mail despacho@sefin.gob.hn; internet www.sefin.gob.hn.

**Ministry of Foreign Affairs:** Centro Cívico Gubernamental, Antigua Casa Presidencial, Blvd Kuwait, Contiguo a la Corte Suprema de Justicia, Tegucigalpa; tel. 234-1962; fax 234-1484; internet www.sre.hn.

**Ministry of Health:** 4a Avda, 3a Calle, Tegucigalpa; tel. 222-8518; fax 238-4141; internet www.secsalud.hn.

**Ministry of Industry and Commerce:** Edif. Salame, 5a Avda, 4a Calle, Tegucigalpa; tel. 238-2025; fax 237-2836; internet www.sic.gob.hn.

**Ministry of the Interior and Justice:** Residencia La Hacienda, Calle La Estancia, Tegucigalpa; tel. 232-1892; fax 232-0226; e-mail atencionalpublico@gobernacion.gob.hn; internet www.gobernacion.gob.hn.

**Ministry of International Co-operation:** Edif. El Sol, Col. Puerta del Sol, 1 c. atrás del Blvd La Hacienda, Tegucigalpa; tel. 239-5545; fax 239-5277; e-mail webmaster@setco.gob.hn; internet www.setco.gob.hn.

**Ministry of Labour and Social Welfare:** Edif. Olympus (STSS), Col. Puerta del Sol, Contiguo a SETCO, Intersección Bulevares Villa Olímpica, La Hacienda, Tegucigalpa; tel. 235-3455; fax 235-3456; e-mail mtrabajohonduras@yahoo.com.

**Ministry of National Defence:** 5a Avda, 4a Calle, Tegucigalpa; tel. 238-3427; fax 238-0238.

**Ministry of Natural Resources and Environment:** 100 m al sur del Estadio Nacional, Apdo 1389, Tegucigalpa; tel. 235-7833; fax 232-6250; e-mail sdespacho@serna.gob.hn; internet www.serna.gob.hn.

**Ministry of Public Education:** 1a Avda, 2a y 3a Calle 201, Comayagüela, Tegucigalpa; tel. 222-8571; fax 237-4312; e-mail info@se.gob.hn; internet www.se.gob.hn.

**Ministry of Public Security:** 5a Avda, 4a Calle, Tegucigalpa; tel. 220-4323; fax 220-4352.

**Ministry of Public Works, Transport and Housing:** Barrio La Bolsa, Comayagüela, Tegucigalpa; tel. 225-0489; fax 225-2227; internet www.soptravi.gob.hn.

**Ministry of Strategy and Communication:** Tegucigalpa.

**Ministry of Tourism:** Edif. Europa, Col. San Carlos, Apdo 3261, Tegucigalpa; tel. and fax 222-2124; e-mail tourisminfo@iht.hn; internet www.letsgohonduras.com.

## President and Legislature

### PRESIDENT
Election, 25 November 2001

| Candidate | Votes cast | % of votes |
| --- | ---: | ---: |
| Ricardo Maduro Joest (PN) | 1,137,734 | 52.21 |
| Rafael Piñeda Ponce (PL) | 964,590 | 44.26 |
| Olban F. Valladares (PINU) | 31,666 | 1.45 |
| Matías Funes (PUD) | 24,102 | 1.11 |
| Marco Orlando Iriarte (PDCH) | 21,089 | 0.97 |
| **Total*** | 2,179,181 | 100.00 |

*Excluding 23,927 blank ballots, 81,959 spoilt votes and 1,163,213 abstentions.

### ASAMBLEA NACIONAL
**President:** Porfirio Lobo Sosa.

General Election, 25 November 2001

| | Seats |
| --- | ---: |
| Partido Nacional (PN) | 61 |
| Partido Liberal (PL) | 55 |
| Partido de Unificación Democrática (PUD) | 5 |
| Partido Innovación y Unidad—Social Democracia (PINU) | 4 |
| Partido Demócrata Cristiano de Honduras (PDCH) | 3 |
| **Total** | **128** |

## Political Organizations

**Partido Demócrata Cristiano de Honduras (PDCH):** Col. San Carlos, Tegucigalpa; tel. 236-5969; fax 236-9941; e-mail pdch@hondutel.hn; internet www.pdch.hn; legally recognized in 1980; Pres. Dr Hernán Corrales Padilla; Sec.-Gen. Saúl Escobar Andrade.

**Partido Innovación y Unidad—Social Democracia (PINU—SD):** 2a Avda, entre 9 y 10 calles, Comaygüela; tel. 220-4224; fax 220-4232; e-mail pinu-sd@sdnhon.org.hn; internet www.pinu.org; f. 1970; legally recognized in 1978; Pres. Bernard Martínez.

**Partido Liberal (PL):** Col. Miramonte, Tegucigalpa; tel. 232-0520; fax 232-0797; f. 1891; factions within the party include the Movimiento Pinedista (Leader Dr Rafael Pinda Ponce), the Movimiento LIBRE (Leader Jaime Rosenthal Oliva, and the Movimiento Esperanza Liberal (Leader Manuel Zelaya); has a youth organization called the Frente Central de Juventud Liberal de Honduras (Pres. Eduardo Raina García); Pres. Dr Rafael Piñeda Ponce; Sec.-Gen. Jaime Rosenthal Oliva.

**Partido Nacional (PN):** Paseo el Obelisco, Comayagüela, Tegucigalpa; tel. 237-7310; fax 237-7365; e-mail partidonacional@partidonacional.hn; f. 1902; traditional right-wing party; internal opposition tendencies include Movimiento Democratizador Nacionalista (MODENA), Movimiento de Unidad y Cambio (MUC), Movi-

# HONDURAS

miento Nacional de Reivindicación Callejista (MONARCA) and Tendencia Nacionalista de Trabajo; Pres. RAFAEL LEONARDO CALLEJAS ROMERO; Sec.-Gen. JORGE CARRANZA.

**Partido de Unificación Democrática (PUD):** Barrio La Plazuela, Avda Cervantes, Tegucigalpa; tel. and fax 238-2498; e-mail colectivoparlud@hotmail.com; f. 1993; left-wing coalition comprising Partido Revolucionario Hondureño, Partido Renovación Patriótica, Partido para la Transformación de Honduras and Partido Morazanista; Sec.-Gen. HERMILIO SOTO.

## Diplomatic Representation

### EMBASSIES IN HONDURAS

**Argentina:** Calle Palermo 302, Col. Rubén Darío, Apdo 3208, Tegucigalpa; tel. 232-3376; fax 231-0376; e-mail ehond@mrecic.gov.ar; Ambassador ALFREDO WALDO FORTI.

**Brazil:** Col. Palmira, Calle República del Brasil, Apdo 341, Tegucigalpa; tel. 221-4432; fax 236-5873; e-mail brastegu@sigmanet.hn; Ambassador SERGIO LUIZ BEZERRA CAVALCANTI.

**Chile:** Calle Oslo C-4242, Col. Lomas del Guijarro, Tegucigalpa; tel. 232-2114; fax 239-7925; e-mail echilehn@123.hn; Ambassador MARÍA CORREA MARIN.

**China (Taiwan):** Col. Lomas del Guijarro, Calle Eucaliptos 3750, Apdo 3433, Tegucigalpa; tel. 239-5837; fax 232-5103; e-mail hnd@mofa.gov.tw; Ambassador TIEN-DER YOU.

**Colombia:** Edif. Palmira, 4°, Col. Palmira, Apdo 468, Tegucigalpa; tel. 232-9709; fax 232-8133; e-mail ehonduras@minrelext.gov.co; Ambassador JUAN ANTONIO RANGEL.

**Costa Rica:** Residencial El Triángulo, 1 Calle, Lomas del Guijarro, Apdo 512, Tegucigalpa; tel. 232-1768; fax 232-1876; e-mail embacori@multivisionhn.net; Ambassador EDGAR GARCÍA MIRANDA.

**Cuba:** Col. Florencia Sur, Avda Principal 4313, Tegucigalpa; tel. 239-0610; fax 235-7624; e-mail oficuba@cybertelh.hn; Ambassador ELIS ALBERTO GONZÁLEZ POLANCO.

**Dominican Republic:** Calle Principal frente al Banco Continental, Col. Miramontes, Tegucigalpa; tel. 239-0130; fax 239-1594; e-mail embadom@compunet.hn; Ambassador ELADIO KNIPPING VICTORIA.

**Ecuador:** Casa 2968, Sendero Senecio, Col. Castaños Sur, Apdo 358, Tegucigalpa; tel. 221-1049; fax 235-4074; e-mail mecuahon@multivisionhn.ne; Ambassador FRANCISCO MARTÍNEZ SALAZAR.

**El Salvador:** Col. Rubén Darío, 2a Avda y 5a Calle No 620, Apdo 1936, Tegucigalpa; tel. 239-0901; fax 239-7009; e-mail embasalva@cablecolor.hn; Ambassador SIGIFREDO OCHOA PÉREZ.

**France:** Col. Palmira, Avda Juan Lindo, Callejón Batres 337, Apdo 3441, Tegucigalpa; tel. 236-6800; fax 236-8051; e-mail ambafrance@cablecolor.hn; Ambassador FRÉDÉRIC BASAGUREN.

**Germany:** Edif. Paysen, 3°, Blvd Morazán, Apdo 3145, Tegucigalpa; tel. 232-3161; fax 239-9018; e-mail embalema@multivisionhn.net; internet www.tegucigalpa.diplo.de; Ambassador Dr THOMAS BRUNS.

**Guatemala:** Col. Las Minitas, Calle Arturo López Rodezno 2421, Tegucigalpa; tel. 232-5018; fax 232-1580; e-mail embhonduras@minex.gob.gt; Ambassador ERWIN FERNANDO GUZMÁN OVALLE.

**Holy See:** Palacio de la Nunciatura Apostólica, Col. Palmira, Avda Santa Sede 412, Apdo 324, Tegucigalpa; tel. 232-6613; fax 239-8869; e-mail nunciatureateg@multivisionhn.net; Apostolic Nuncio Most Rev. ANTONIO ARCARI (Titular Archbishop of Ceciti).

**Italy:** Col. Montecarlo, Avda Enrique Tierno Galván, Apdo U-9093, Tegucigalpa; tel. 236-6391; fax 236-5659; e-mail ambtegu@multivisionhn.net; internet www.ambitaliahn.org; Ambassador ESTEFANO MARÍA CACCIAGUERRA RANGHIERI.

**Japan:** Col. San Carlos, Calzada Rep. Paraguay, Apdo 3232, Tegucigalpa; tel. 236-2628; fax 236-6100; internet www.hn.emb-japan.go.jp; Ambassador MASAMI TAKEMOTO.

**Mexico:** Col. Lomas de Guijarro, Col. Eucalipto 1001, Tegucigalpa; tel. 232-4039; fax 232-4719; e-mail embamexhon@newcom.hn; Ambassador WALTERIO ASTIÉ BURGOS.

**Nicaragua:** Col. Tepeyac, Bloque M-1, Avda Choluteca 1130, Apdo 392, Tegucigalpa; tel. 231-1977; fax 231-1412; e-mail embanic@multivisionhn.net; Ambassador JULIO DELGADO MENDOZA.

**Panama:** Edif. Palmira, 2°, Col. Palmira, Apdo 397, Tegucigalpa; tel. 239-5508; fax 232-8147; e-mail ephon@multivisionhn.net; Ambassador ROBERT JOVANÉ.

**Peru:** Col. La Reforma, Calle Principal 2618, Tegucigalpa; tel. 221-0596; fax 236-6070; e-mail embajdadelperu@cablecolor.hn; Ambassador GUSTAVO OTERO ZAPATA.

**Spain:** Col. Matamoros, Calle Santander 801, Apdo 3221, Tegucigalpa; tel. 236-6875; fax 236-8682; e-mail embesphn@correo.mae.es; Ambassador AUGUSTIN NÚÑEZ MARTÍNEZ.

**USA:** Avda La Paz, Apdo 3453, Tegucigalpa; tel. 236-9320; fax 236-9037; internet honduras.usembassy.gov; Ambassador LARRY PALMER.

**Venezuela:** Col. Rubén Darío, 2116 Circuito Choluteca, Apdo 775, Tegucigalpa; tel. 232-1879; fax 232-1016; e-mail emvenezue@hondutel.hn; Ambassador MARÍA SALAZAR SANABRIA.

## Judicial System

Justice is administered by the Supreme Court (which has 15 judges), five Courts of Appeal and departmental courts (which have their own local jurisdiction).

Tegucigalpa has two Courts of Appeal which have jurisdiction (1) in the department of Francisco Morazán, and (2) in the departments of Choluteca Valle, El Paraíso and Olancho.

The Appeal Court of San Pedro Sula has jurisdiction in the department of Cortés; that of Comayagua has jurisdiction in the departments of Comayagua, La Paz and Intibucá; and that of Santa Bárbara in the departments of Santa Bárbara, Lempira and Copán.

**Supreme Court:** Edif. Palacio de Justicia, contiguo Col. Miraflores, Centro Cívico Gubernamental, Tegucigalpa; tel. 233-9208; fax 233-6784.

**President of the Supreme Court of Justice:** VILMA CECILIA MORALES MONTALVÁN.

**Attorney-General:** LEÓNIDAS ROSA BAUTISTA.

## Religion

The majority of the population are Roman Catholics; the Constitution guarantees toleration to all forms of religious belief.

### CHRISTIANITY

#### The Roman Catholic Church

Honduras comprises one archdiocese and six dioceses. At 31 December 2003 some 80% of the population were adherents.

#### Bishops' Conference

Conferencia Episcopal de Honduras, Blvd Estadio Suyapa, Apdo 3121, Tegucigalpa; tel. 229-1111; fax 229-1144; e-mail ceh@unicah.edu.

f. 1929; Pres. Cardinal OSCAR ANDRÉS RODRÍGUEZ MARADIAGA (Archbishop of Tegucigalpa).

**Archbishop of Tegucigalpa:** Cardinal OSCAR ANDRÉS RODRÍGUEZ MARADIAGA, Arzobispado, 3a y 2a Avda 1113, Apdo 106, Tegucigalpa; tel. 237-0353; fax 222-2337; e-mail orodriguez@unicah.edu.

#### The Anglican Communion

Honduras comprises a single missionary diocese, in Province IX of the Episcopal Church in the USA.

**Bishop of Honduras:** Rt Rev. LLOYD ALLEN, Apdo 586, San Pedro Sula; tel. 556-6155; fax 556-6467; e-mail emmanuel@anglicano.hn.

#### The Baptist Church

**Baptist Convention of Honduras:** Apdo 2176, Tegucigalpa; tel. and fax 236-6717; e-mail conibah@sigmanet.hn; Pres. Pastor ALEXIS SALVADOR VIDES; 21,355 mems.

#### Other Churches

**Iglesia Cristiana Luterana de Honduras** (Christian Lutheran Church of Honduras): Apdo 2861, Tegucigalpa; tel. and fax 225-4464; fax 225-4893; e-mail iclh@123.hn; Pres. Rev. J. GUILLERMO FLORES V.; 1,000 mems.

### BAHÁ'Í FAITH

**National Spiritual Assembly:** Sendero de los Naranjos 2801, Col. Castaños, Apdo 273, Tegucigalpa; tel. 232-6124; fax 231-1343; e-mail bahaihon@bahaihon.org; internet www.bahaihon.org; 40,000 mems resident in more than 500 localities.

HONDURAS — *Directory*

## The Press

### DAILIES

**La Gaceta:** Tegucigalpa; f. 1830; morning; official govt paper; circ. 3,000.

**El Heraldo:** Avda los Próceres, Frente Instituto del Tórax, Barrio San Felipe, Apdo 1938, Tegucigalpa; tel. 236-6000; internet www.elheraldo.hn; f. 1979; morning; independent; Pres. JORGE CANAHUATI LARACH; Editor CARLOS MAURICIO FLORES; circ. 45,000.

**El Nuevo Día:** 3a Avda, 11–12 Calles, San Pedro Sula; tel. 52-4298; fax 57-9457; f. 1994; morning; independent; Pres. ABRAHAM ANDONIE; Editor ARMANDO CERRATO; circ. 20,000.

**El Periódico:** Carretera al Batallón, Tegucigalpa; tel. 234-3086; fax 234-3090; f. 1993; morning; Pres. EMIN ABUFELE; Editor OSCAR ARMANDO MARTÍNEZ.

**La Prensa:** 3a Avda, 6a–7a Calles No 34, Apdo 143, San Pedro Sula; tel. 53-3101; fax 53-0778; e-mail correos@laprensa.com; internet www.laprensahn.com; f. 1964; morning; independent; Pres. JORGE CANAHUATI LARACH; Exec. Dir NELSON EDGARDO FERNÁNDEZ; circ. 50,000.

**El Tiempo:** 1a Calle, 5a Avda 102, Santa Anita, Apdo 450, San Pedro Sula; tel. 53-3388; fax 53-4590; e-mail tiempo@continental.hn; internet www.tiempo.hn; f. 1960; morning; left-of-centre; Pres. JAIME ROSENTHAL OLIVA; Editor MANUEL GAMERO; circ. 35,000.

**La Tribuna:** Col. Santa Bárbara, Comayagüela, Apdo 1501, Tegucigalpa; tel. 233-1138; fax 233-1188; e-mail tribuna@david.intertel.hn; internet tribuna.icomstec.com; f. 1977; morning; independent; Dir ADÁN ELVIR FLORES; Gen. Man. MANUEL ACOSTA MEDINA; circ. 45,000.

### PERIODICALS

**Cambio Empresarial:** Apdo 1111, Tegucigalpa; tel. 237-2853; fax 237-0480; monthly; economic, political, social; Editor JOAQUÍN MEDINA OVIEDO.

**El Comercio:** Cámara de Comercio e Industrias de Tegucigalpa, Blvd Centroamérica, Apdo 3444, Tegucigalpa; tel. 232-4200; fax 232-0759; f. 1970; monthly; commercial and industrial news; Dir-Gen. Lic. HÉCTOR MANUEL ORDÓÑEZ.

**Hablemos Claro:** Edif. Torre Libertad, Blvd Suyapa, Tegucigalpa; tel. 232-8058; fax 239-7008; e-mail abrecha@hondutel.hn; internet www.hablemosclaro.com; f. 1990; weekly; Editor RODRIGO WONG ARÉVALO; circ. 9,000.

**Hibueras:** Apdo 955, Tegucigalpa; Dir RAÚL LANZA VALERIANO.

**Honduras This Week:** Centro Comercial Villa Mare, Blvd Morazán, Apdo 1323, Tegucigalpa; tel. 239-3654; fax 232-2300; e-mail hondweek@hondutel.hn; internet www.hondurasthisweek.com; f. 1988; weekly; English language; tourism, culture and the environment; Man. Editor GLADYS ACOSTA; Editor MARIO GUTIÉRREZ MINERA.

**El Libertador:** Tegucigalpa; Dir JHONY LAGOS.

**Pregonero Evangélico:** Tegucigalpa; e-mail mcm@sdnhon.org.hn; internet www.mcmhn.org/pregonero; f. 2000; quarterly; Christian magazine.

### PRESS ASSOCIATION

**Asociación de Prensa Hondureña:** Casa del Periodista, Avda Gutemberg 1525, Calle 6, Barrio El Guanacaste, Apdo 893, Tegucigalpa; tel. and fax 378-345; f. 1930; Pres. GUILLERMO PAGÁN SOLORZANO; Sec. Gen. FELA ISABEL DUARTE.

### FOREIGN NEWS AGENCIES

**Agencia EFE** (Spain): Edif. Jiménez Castro, 5°, Of. 505, Tegucigalpa; tel. 22-0493; Bureau Chief ARMANDO ENRIQUE CERRATO CORTÉS.

**Agenzia Nazionale Stampa Associata (ANSA)** (Italy): Edif. JS, Cubículo 202, Blvd Tegucigalpa, Tegucigalpa; tel. 239-3943; Correspondent RAÚL MONCADA.

**Deutsche Presse-Agentur (dpa)** (Germany): Edif. Jiménez Castro, Of. 203, 4a Calle y 5a Avda, No 405, Apdo 3522, Tegucigalpa; tel. 237-8570; Correspondent WILFREDO GARCÍA CASTRO.

**Inter Press Service (IPS)** (Italy): Apdo 228, Tegucigalpa; tel. 232-5342; Correspondent JUAN RAMÓN DURÁN.

## Publishers

**Centro Editorial:** San Pedro Sula; tel. and fax 558-1282; e-mail escoto@globalnet.hn; Dir JULIO ESCOTO.

**Ediciones Ramses:** Edif. Torres Fiallos, Avda Jerez, Apdo 5600, Tegucigalpa; tel. 220-4248; fax 220-0833; e-mail servicioalcliente@edicionesramses.hn; internet edicionesramses.galeon.com; educational material.

**Editora Fuego Nuevo:** Col. Florencia Sur, Blvd Suyapa, Tegucigalpa; tel. 232-4638; fax 232-4964; e-mail mirna_detorres@yahoo.com; Dir MYRNA LANZA GONZÁLEZ.

**Editorial Pez Dulce:** 143 Paseo La Leona, Barrio La Leona, Tegucigalpa; tel. and fax 222-1220; e-mail pezdulce@yahoo.com.

**Guaymuras:** Apdo 1843, Tegucigalpa; tel. 237-5433; fax 238-4578; e-mail ediguay@123.hn; f. 1980; Dir ISOLDA ARITA MELZER.

**Universidad Nacional Autónoma de Honduras:** Blvd Suyapa, Tegucigalpa; tel. 231-4601; fax 231-4601; f. 1847.

## Broadcasting and Communications

### TELECOMMUNICATIONS

#### Regulatory Authority

**Comisión Nacional de Telecomunicaciones (Conatel):** Col. Modelo, 6a Avda Suroeste, Apdo 15012, Tegucigalpa; tel. 234-8600; fax 234-8611; e-mail conatel@conatel.hn; internet www.conatel.hn; Pres. MARLON RAMSSÉS TÁBORA.

#### Major Service Providers

**Empresa Hondureña de Telecomunicaciones (Hondutel):** Apdo 1794, Tegucigalpa; tel. 221-6555; fax 236-7795; e-mail miguel.velez@hondutelnet.hn; internet www.hondutel.hn; scheduled for partial privatization by 2005; Gen. Man. ALONSO VALENZUELA.

**Multifon:** Tegucigalpa; f. 2003; subsidiary of MultiData; awarded govt contract with UT Starcom (q.v.) for fixed telephone lines in 2003; Pres. JOSÉ RAFAEL FERRARI.

**UT Starcom** (USA): Tegucigalpa; awarded govt contract with Multifon (q.v.) for fixed telephone lines in 2003; Pres. and CEO HONG LIANG LU; Vice-Pres. (Latin America) RENÉ MÉNDEZ.

### BROADCASTING

#### Radio

**Estereo McIntosh:** La Ceiba, Atlantida; tel. 440-0326; fax 440-0325; e-mail McIntosh@psinet.hn; internet www.psinet.hn/mcintosh; commercial channel.

**HRN, La Voz de Honduras:** Blvd Suyapa, Apdo 642, Tegucigalpa; internet www.radiohrn.hn; commercial station; f. 1933; broadcasts 12 channels; 23 relay stations; Gen. Man. NOEMI VALLADARES.

**Power FM:** Tegucigalpa; tel. 552-4898; e-mail xavier@powerfm.hn; internet www.powerfm.hn; Gen. Man. XAVIER SIERRA.

**Radio América:** Col. Alameda, frente a la Droguería Mandofer, Apdo 259, Tegucigalpa; tel. 232-8338; fax 232-1009; internet www.hondutel.hn; commercial station; broadcasts Radio San Pedro, Radio Continental, Radio Monderna, Radio Universal, Cadena Radial Sonora, Super Cien Stereo, Momentos FM Stereo and 3 regional channels; f. 1948; 13 relay stations; Gen. Man. ARTURO VARELA.

**Radio Club Honduras:** Salida Chamelecon, Apdo 273, San Pedro Sula; tel. 556-6173; e-mail hr2rch@mayanet.hn; internet www.qsl.net/hr2rch; amateur radio club; Pres. NORMA LEIVA.

**Radio Esperanza:** La Esperanza, Intibucá; tel. 783-0025; fax 783-0644; e-mail radioesperanza1@hotmail.com; internet www.honducontact.com/Radio%20Esperanza.htm; Dir J. M. DEL CID.

**Radio la Voz del Atlántico:** 12 Calle, 2–3 Avda, Barrio Copen, Puerto Cortés; tel. 665-5166; fax 665-2401; e-mail atlantico@sescomnet.com; internet radioatlantico.8m.com.

**Radio Nacional de Honduras:** Avda La Paz, contiguo a la Secretaría de Cultura, Artes y Deportes, Tegucigalpa; tel. 236-7551; fax 236-7400; f. 1976; official station, operated by the Govt; Gen. Man. ROLANDO SARMIENTO.

**La Voz de Centroamérica:** 9a Calle, 10a Avda 64, Apdo 120, San Pedro Sula; tel. 52-7660; fax 57-3257; f. 1955; commercial station; Gen. Man. NOEMI SIKAFFY.

#### Television

**Televicentro:** Edif. Televicentro, Blvd Suyapa, Apdo 734, Tegucigalpa; e-mail mercadeoventas@televicentro.hn; internet www.televicentro.hn; f. 1987; 11 relay stations.

# HONDURAS

**Canal 5:** tel. 232-7835; fax 232-0097; f. 1959; Gen. Man. José Rafael Ferrari.

**Telecadena 7 y 4:** tel. 239-2081; fax 232-0097; f. 1959; Pres. José Rafael Ferrari; Gen. Man. Rafael Enrique Villeda.

**Telesistema Hondureño, Canal 3 y 7:** tel. 232-7064; fax 232-5019; f. 1967; Gen. Man. Rafael Enrique Villeda.

**VICA Television:** 9a Calle, 10a Avda 64, Barrio Guamilito, Apdo 120, San Pedro Sula; tel. 552-4478; fax 557-3257; e-mail info@mayanet.hn; internet www.vicatv.hn; f. 1986; operates regional channels 2, 9 and 13; Pres. Blanca Sikaffy.

## Finance

(cap. = capital; res = reserves; dep. = deposits; m. = million; brs = branches; amounts in lempiras unless otherwise stated)

### BANKING

#### Central Bank

**Banco Central de Honduras (BANTRAL):** Avda Juan Ramón Molina, 7a Avda y 1a Calle, Apdo 3165, Tegucigalpa; tel. 237-2270; fax 237-1876; e-mail eanariba@mail.bch.hn; internet www.bch.hn; f. 1950; bank of issue; cap. 218.9m., res 477.8m., dep. 22,679.5m. (Dec. 2002); Pres. María Elena Mondragón; 4 brs.

#### Commercial Banks

**Banco Atlántida, SA (BANCATLAN):** Plaza Bancatlán, Blvd Centroamérica, Apdo 3164, Tegucigalpa; tel. 232-1050; fax 232-6120; e-mail webmaster@bancatlan.hn; internet www.bancatlan.hn; f. 1913; cap. 1,000m., res 344.6m., dep. 10,825.3m. (Dec. 2004); Exec. Pres. Guillermo Bueso; Exec. Vice-Pres Gustavo Oviedo, Ildoira G. de Bonilla; 17 brs.

**Banco Continental, SA (BANCON):** Edif. Continental, 3a Avda 7, entre 2a y 3a Calle, Apdo 390, San Pedro Sula; tel. 550-0880; fax 550-8580; e-mail pattyr@continental.hn; internet www.bancon.hn; f. 1974; cap. 250m., res 134.7m., dep. 1,081.0m. (Dec. 2002); Pres. Jaime Rosenthal Oliva; 41 brs.

**Banco de las Fuerzas Armadas, SA (BANFFAA):** Centro Comercial Los Castaños, Blvd Morazán, Apdo 877, Tegucigalpa; tel. 232-4208; fax 239-4252; e-mail webmaster@banffaa.hn; internet www.banffaa.hn; f. 1979; cap. 10m., res 33.2m., dep. 428.1m. (Dec. 1992); Pres. Luis Alfonso Discua Elvir; Gen. Man. Carlos Rivera Xatruch; 38 brs.

**Banco Grupo El Ahorro Hondureño (BGA):** Intersección Blvd Suyapa y Blvd Juan Pablo Segundo, Apdo 344, Tegucigalpa; tel. 232-0909; fax 232-0743; e-mail bga@bancobga.com; internet www.bancobga.com; f. 2000 following a merger of Banco del Ahorro Hondureño and Banco La Capitalizadora Hondureña; cap. 704.1m., res 82.7m., dep. 8,077.8m. (Dec. 2003); 125 brs; Pres. Alberto Vallarino Clément.

**Banco de Honduras, SA:** Blvd Suyapa, Col. Loma Linda Sur, Tegucigalpa; tel. 232-6122; fax 232-6167; e-mail patricia.ferro@citicorp.com; internet www.bancodehonduras.citibank.com; f. 1889; total assets 15,106m. (1999); Gen. Man. Maximo R. Vidal; 2 brs.

**Banco Mercantil, SA:** Blvd Suyapa, frente a Emisoras Unidas, Apdo 116, Tegucigalpa; tel. 232-0006; fax 239-4509; internet www.bamernet.hn; Pres. José Lamas; Vice-Pres. Manuel Villeda Toledo.

**Banco de Occidente, SA (BANCOCCI):** 6a Avda, Calle 2–3, Apdo 3284, Tegucigalpa; tel. 237-0310; fax 237-0486; e-mail bancocci@cybertelh.hn; f. 1951; cap. and res 69m., dep. 606m. (June 1994); Pres. and Gen. Man. Jorge Bueso Arias; Vice-Pres. Emilio Medina R.; 146 brs.

**Banco de los Trabajadores, SA (BANCOTRAB):** 3a Avda, 13a Calle, Comayagüela, Apdo 3246, Tegucigalpa; tel. 238-0017; fax 238-0077; e-mail marcio@btrab.com; f. 1967; cap. and res US $6.6m., dep. $43.1m. (Dec. 1992); Pres. Rolando del Cid Velásquez; 13 brs.

#### Development Banks

**Banco Centroamericano de Integración Económica:** Edif. Sede BCIE, Blvd Suyapa, Apdo 772, Tegucigalpa; tel. 228-2182; fax 228-2183; e-mail cmartine@bcie.hn; internet www.bcie.org; f. 1960 to finance the economic development of the Central American Common Market and its mem. countries; mems Costa Rica, El Salvador, Guatemala, Honduras, Nicaragua; cap. and res US $1,020.0m. (June 2003); Exec. Pres. Harry Brautigan.

**Banco Financiera Centroamericana, SA (FICENSA):** Edif. La Interamericana, Blvd Morazán, Apdo 1432, Tegucigalpa; tel. 238-1661; fax 238-1630; e-mail rrivera@ficensa.com; internet www.ficensa.com; f. 1974; private org. providing finance for industry, commerce and transport; Pres. Oswaldo López Arellano; Gen. Man. Roque Rivera Ribas.

**Banco Hondureño del Café, SA (BANHCAFE):** Calle República de Costa Rica, Blvd Juan Pablo II, Col. Lomas del Mayab, Apdo 583, Tegucigalpa; tel. 232-8370; fax 232-8782; e-mail bcaferhu@hondutel.hn; f. 1981 to help finance coffee production; owned principally by private coffee producers; cap. 119.2m., res 77.6m., dep. 1,379.9m. (Dec. 2002); Pres. Miguel Alfonso Fernández Rápalo; Gen. Man. René Ardón Matute; 50 brs.

**Banco Municipal Autónomo (BANMA):** 6a Avda, 6a Calle, Tegucigalpa; tel. 237-8503; fax 237-6946; e-mail bancomunicipal@tutopia.com; f. 1963; Pres. Justo Pastor Calderón; 2 brs.

**Banco Nacional de Desarrollo Agrícola (BANADESA):** 4a Avda y 5a Avda, 13a y 14a Calles, Barrio Concepción, Apdo 212, Tegucigalpa; tel. 237-2201; fax 237-5187; e-mail info@banadesa.hn; f. 1980; govt development bank (transfer to private ownership pending); loans to agricultural sector; cap. 34.5m., res 42.7m., dep. 126.9m. (March 1993); Gen. Man. Enrique Alberto Castellon; 35 brs.

#### Banking Associations

**Asociación Hondureña de Instituciones Bancarias (AHIBA):** Edif. AHIBA, Blvd Suyapa, Apdo 1344, Tegucigalpa; tel. 235-6770; fax 239-0191; e-mail ahiba@ahiba.hn; f. 1957; 21 mem. banks; Exec. Dir María Lydia Solano.

**Comisión Nacional de Bancos y Seguros (CNBS):** Calle 6 y 7, Avda Juan Ramón Molina, Tegucigalpa; tel. 238-0580; fax 237-6232; internet www.cnbs.gov.hn; Pres. Ana Cristina Mejía de Pereira.

### STOCK EXCHANGE

**Bolsa Hondureña de Valores:** Edif. Martínez Valenzuela, 1°, 2a Calle, 3a Avda, Apdo 161, San Pedro Sula; tel. 553-4410; fax 553-4480; e-mail bhvsps@bhv.hn; internet www.bhv.hn; Pres. José Rubén Mendoza.

### INSURANCE

**American Home Assurance Co:** Edif. Los Castaños, 4°, Blvd Morazán, Apdo 3220, Tegucigalpa; tel. 232-3938; fax 232-8169; f. 1958; Mans Leonardo Moreira, Edgar Wagner.

**Aseguradora Hondureña, SA:** Edif. El Planetario, 4°, Col. Lomas de Guijarro Sur, Calle Madrid, Avda Paris, Apdo 312, Tegucigalpa; tel. 232-2729; fax 231-0982; e-mail gerencia@aseguradora.hn; internet www.laaseguradora.com.hn; f. 1954; Pres. Gerardo Corrales.

**Compañía de Seguros El Ahorro Hondureño, SA:** Edif. Trinidad, Avda Colón, Apdo 3643, Tegucigalpa; tel. 237-8219; fax 237-4780; e-mail pbetanco@eahsa.hn; internet www.seguroselahorro.hn; f. 1917; Pres. Doris Patricia Hernández Valeriano.

**Interamericana de Seguros, SA:** Col. Los Castaños, Apdo 593, Tegucigalpa; tel. 232-7614; fax 232-7762; internet www.ficohsa.hn; f. 1957; part of Grupo Financiero Ficohsa; Pres. Leonel Giannini.

**Pan American Life Insurance Co (PALIC):** Edif. PALIC, Avda República de Chile 804, Tegucigalpa; tel. 220-5757; fax 232-3907; e-mail palic@david.intertel.hn; f. 1944; Vice-Pres. Felix Martínez; Gen. Man. Alberto Agurcia.

**Previsión y Seguros, SA:** Edif. Grupo Financiero IPM, 4°, Blvd Centroamérica, Apdo 770, Tegucigalpa; tel. 232-4119; fax 232-4113; internet www.previsahn.com; f. 1982; Gen. Man. Gerardo A. Rivera.

**Seguros Atlántida:** Edif. Sonisa, Costado Este Plaza Bancatlán, Tegucigalpa; tel. 232-4014; fax 232-3688; e-mail info@seatlan.com; internet www.seatlan.com; f. 1986; Pres. Guillermo Bueso; Gen. Man. Juan Miguel Orellana.

**Seguros Continental, SA:** Edif. Continental, 4°, 3a Avda SO, 2a y 3a Calle, Apdo 320, San Pedro Sula; tel. 550-0880; fax 550-2750; e-mail seguros@continental.hn; f. 1968; Pres. Jaime Rosenthal Oliva; Gen. Man. Mario R. Solís.

**Seguros Crefisa:** Edif. Ficensa, 1°, Blvd Morazán, Apdo 3774, Tegucigalpa; tel. 238-1750; fax 238-1714; e-mail gerencia@crefisa.hn; internet www.crefisa.hn; f. 1993; Pres. Oswaldo López Arellano; Gen. Man. Mario Batres Piñeda.

#### Insurance Association

**Cámara Hondureña de Aseguradores (CAHDA):** Edif. Casa Real, 3°, Col. San Carlos, Tegucigalpa; tel. 221-5354; fax 221-5356; e-mail cahda@gbm.hn; internet www.cahda.org; f. 1974; Man. José Luis Moncada Rodríguez.

# Trade and Industry

## GOVERNMENT AGENCIES

**Fondo Hondureño de Inversión Social (FHIS):** Antiguo Edif. I.P.M., Col. Godoy, Comayagüela, Apdo 3581, Tegucigalpa; e-mail dir.ejectutiva@fhis.hn; internet www.fhis.hn; tel. 234-5231; fax 534-5255; social investment fund; Dir LEONI YUWAI.

**Fondo Social de la Vivienda (FOSOVI):** Col. Florencia, Tegucigalpa; tel. 239-1605; social fund for housing, urbanization and devt; Gen. Man. MARIO MARTÍ.

**Secretaria Técnica del Consejo Superior de Planificación Económica (CONSUPLANE):** Edif. Bancatlán, 3°, Apdo 1327, Comayagüela, Tegucigalpa; tel. 22-8738; f. 1965; national planning office; Exec. Sec. FRANCISCO FIGUEROA ZÚÑIGA.

## DEVELOPMENT ORGANIZATIONS

**Consejo Hondureño de la Empresa Privada (COHEP):** Edif. 8, Calle Yoro, Col. Tepeyac, Apdo 3240, Tegucigalpa; tel. 235-3336; fax 235-3345; e-mail consejo@cohep.com; internet www.cohep.com; f. 1968; represents 52 private-sector trade associations; Pres. JOSÉ MARÍA AGURCIA.

**Corporación Hondureña de Desarrollo Forestal (COHDEFOR):** Salida Carretera del Norte, Zona El Carrizal, Comayagüela, Apdo 1378, Tegucigalpa; tel. 223-7383; fax 223-3348; internet www.cohdefor.hn; f. 1974; control and management of the forestry industry; Gen. Man. GUSTAVO MORALES.

**Dirección Ejecutiva de Fomento a la Minería (DEFOMIN):** Edif. DEFOMIN, 3°, Blvd Miraflores, Apdo 981, Tegucigalpa; tel. 232-6721; fax 232-8635; promotes the mining sector; Dir-Gen. SANDRA MARLENE PINTO.

**Instituto Hondureño del Café (IHCAFE):** Edif. El Faro, Col. las Minitas, Apdo 40-C, Tegucigalpa; tel. 237-3131; e-mail ihcafe@cafesdehonduras.com; f. 1970; coffee devt programme; Gen. Man. FERNANDO D. MONTES M.

**Instituto Hondureño de Mercadeo Agrícola (IHMA):** Apdo 727, Tegucigalpa; tel. 235-3193; fax 235-5719; f. 1978; agricultural devt agency; Gen. Man. TULIO ROLANDO GIRÓN ROMERO.

**Instituto Nacional Agrario (INA):** Col. La Almeda, 4a Avda, entre 10a y 11a Calles, No 1009, Apdo 3391, Tegucigalpa; tel. 232-4893; fax 232-7398; agricultural devt programmes; Exec. Dir ERASMO PORTILLO FERNÁNDEZ.

## CHAMBERS OF COMMERCE

**Cámara de Comercio e Industrias de Copán:** Edif. Comercial Romero, 2°, Barrio Mercedes, Santa Rosa de Copán; tel. 662-0843; fax 662-1783; e-mail info@camaracopan.com; internet www.camaracopan.com; f. 1940; Pres. EUDOCIO LEIVA AMAYA.

**Cámara de Comercio e Industrias de Cortés (CCIC):** Col. Trejo, 17a Avda, 10a Calle, Apdo 14, San Pedro Sula; tel. 53-0761; fax 553-3777; e-mail ccic@ccichonduras.org; internet www.ccichonduras.org; f. 1931; 812 mems; Pres. OSCAR GALEANO; Dir RAÚL REINA.

**Cámara de Comercio e Industrias de Tegucigalpa:** Blvd Centroamérica, Apdo 3444, Tegucigalpa; tel. 232-4200; fax 232-0159; e-mail ccit@ccit.hn; internet www.ccit.hn; Pres. AMÍLCAR BULNES.

**Federación de Cámaras de Comercio e Industrias de Honduras (FEDECAMARA):** Edif. Castañito, 2°, 6a Avda, Col. los Castaños, Apdo 3393, Tegucigalpa; tel. 232-1870; fax 232-6083; e-mail fedecamara@sigmanet.hn; f. 1948; 1,200 mems; Pres. JOSÉ NOLASCO; Exec. Dir JUAN MANUEL MOYA.

**Fundación para la Inversión y Desarrollo de Exportaciones (FIDE)** (Foundation for Investment and Export Development): Col. La Estancia, Plaza Marte, final del Blvd Morazán, POB 2029, Tegucigalpa; tel. 221-6303; fax 221-6318; internet www.hondurasinfo.hn/eng/fide/fide.asp; f. 1984; private, non-profit agency; Pres. VILMA SIERRA DE FONSECA.

**Honduran American Chamber of Commerce (Amcham Honduras):** Commercial Area Hotel Honduras Maya, POB 1838, Tegucigalpa; tel. 232-6035; fax 232-2031; e-mail amcham1@quikhonduras.com; internet www.amchamhonduras.org; f. 1981; Pres. ROBERTO ALVAREZ GUERRERO.

## INDUSTRIAL AND TRADE ASSOCIATIONS

**Consejo Hondureño de la Empresa Privada (COHEP):** Edif. 8, Col. Tepeyac, Calle Yoro, Apdo 3240, Tegucigalpa; tel. 235-3336; fax 235-3345; e-mail consejo@cohep.com; internet www.cohep.com; f. 1968; umbrella org. representing 54 industrial and trade asscns; Pres. JOSÉ MARÍA AGURCIA.

**Asociación Hondureña de Productores de Café (AHPROCAFE)** (Coffee Producers' Association): Edif. AHPROCAFE, Avda La Paz, Apdo 959, Tegucigalpa; tel. 236-8286; fax 236-8310; e-mail ahprocafe@123.hn; Pres. FREDY ESPINOZA MONDRADON; Gen. Man. PEDRO MENDOZA FLORES.

**Asociación Nacional de Acuicultores de Honduras** (National Fish Farmers' Association of Honduras): Pasaje Sarita Rubinstein 11, 2°, Apdo 229, Choluteca; tel. 882-0986; fax 882-3848; e-mail andahn@hondutel.hn; f. 1986; 136 mems; Pres. ISMAEL WONG; Exec. Dir FRANCISCO AVALOS.

**Asociación Nacional de Exportadores de Honduras (ANEXHON)** (National Association of Exporters): Local de la C.C.I.C, Tegucigalpa; tel. 553-3029; fax 557-0203; e-mail Roberto@itsa.com; comprises 104 private enterprises; Pres. ROBERTO PANAYOTTI.

**Asociación Nacional de Industriales (ANDI)** (National Association of Manufacturers): Edif. Fundación Covelo, 3°, Col. Castaño Sur, Blvd Morazán, Apdo 3447, Tegucigalpa; tel. 232-2221; fax 221-5199; e-mail andi@andi.hn; internet www.andi.hn; Pres. ADOLFO FACUSSÉ; Exec. Sec. GRACO PAREDES HERRERA.

**Federación Nacional de Agricultores y Ganaderos de Honduras (FENAGH)** (Farmers and Livestock Breeders' Association): Apdo 3209, Tegucigalpa; tel. 239-1303; fax 231-1392; Pres. ROBERTO GALLARDO LARDIZÁBAL.

## MAJOR COMPANIES

The following are some of the major companies currently operating in Honduras.

**Aquacultura Fonseca:** Carretera a Orocuina, desvío a Linaca, Apdo 181/255, Choluteca; tel. 882-0099; fax 882-2579; e-mail hond@seajoy.com; internet www.seajoy.com; seafood exporters; owned by Seajoy, Miami, FL, USA; Gen. Man. ISMAEL WONG C.

**Azucarera La Grecia, SA de CV:** Apdo 32, Choluteca; tel. 882-0629; fax 882-0633; f. 1974 as Azucarera Central; processing and refining of sugar cane; Man. Lic. GUILLERMO LIPPMANN.

**Breakwater Resources:** Apdo 342, San Pedro Sula; tel. 659-3051; fax 659-3059; e-mail mochito@breakwater.hn; internet www.breakwater.ca; Canadian mining co, owner/operator of El Mochito mine; Chair. NED GOODMAN; CEO GARTH MACRAE.

**Cementos del Norte, SA de CV:** Río Bijao, Choloma Cortés, Apdo 132; tel. 669-3640; fax 669-3639; cement producers; 470 employees.

**Cervecería Hondureña, SA:** Carretera a Puerto Cortés, Apdo 86, San Pedro Sula; tel. 553-3310; fax 552-2845; f. 1915; brewery and soft drink manufacturers; Man. FERNANDO A. PEÑABAD; 1,150 employees.

**Compañía Azucarera Choluteca, SA de CV:** Zona de los Mangos, Choluteca; tel. 882-0530; fax 882-0554; e-mail achsa@hondudata.hn; f. 1967; sugar-cane refining; Pres. SERGIO R. SALINAS S.; Gen. Man. BRAULIO CRUZ; 180 employees.

**Compañía Azucarera Hondureña, SA:** 3 Avda 36, Apdo 552, San Pedro Sula; tel. 574-8090; fax 574-8093; e-mail sugar@netsys.hn; f. 1938; cultivation and refining of sugar cane; Gen. Man. CHARLES HEYER; 560 employees.

**Compañia Hulera Sula:** 3.5km, Carretera a Puerto Cortés, Frente al Seguro Social, Apdo 202, San Pedro Sula; tel. 551-2832; fax 551-3718; e-mail hulesula@globalnet.hn; manufacturer of plastics and rubber products; Gen. Man. HÉCTOR GUILLÉN.

**Corporación Lady Lee:** Centro Comercial Megaplaza 2km, Autopista al Aeropuerto, Apdo 948, San Pedro Sula; tel. 553-1642; fax 552-6426; e-mail ladylee@ladylee.hn; department stores; Gen. Man. RAYMOND MAALOUF.

**Derivados de Metal:** 3km después de Choloma, Quebrada Seca Choloma, Apdo 797, Cortés; tel. 669-3529; fax 669-3066; e-mail demesa@netsys.hn; steel manufacturer; Gen. Man. MAHFUZ EMIL HAWIT MEDRANO.

**Droguería Pharma Internacional:** Casa 1812, Calle 11, 2a Avda, Col. Alameda, Tegucigalpa; tel. 239-6499; fax 239-3222; e-mail pharmaint@multivisionhn.net; manufacturers of pharmaceutical products; Gen. Man. AURELIO NEMBRINI.

**Entremares Honduras, SA:** Valle de Siria, Francisco Morazán; subsidiary of Glamis Golds Ltd of the USA; mining co.

**Gabriel Kafti:** Barrio La Bolsa, Comayaguela; tel. 225-1675; fax 225-3792; e-mail gksa@hondutel.hn; coffee producer; Gen. Man. MIGUEL OSCAR KAFATI.

**Hilos y Mechas, SA de CV:** Carretera a Puerto Cortés, Apdo 118, San Pedro Sula; fax 552-2441; e-mail himesa@himesa.hn; internet www.himesa.hn; f. 1953; manufacturers of cotton fabrics, twine and thread; Gen. Man. EDUARDO HANDAL; 856 employees.

# HONDURAS

**Lácteos de Honduras SA de CV (LACTHOSA):** Carretera a Puerto Cortés, Apdo 140, San Pedro Sula; tel. 566-0055; fax 566-3917; e-mail sula@hn2.com; dairy products and fruit juices; Gen. Man. JULIO MONTESI PALMA; 1,000 employees.

**Leche y Derivados, SA (LEYDE):** 8km Carretera a Tela, Apdo 95, La Ceiba; tel. 441-1859; fax 441-0108; f. 1973; milk and dairy products; Pres. JOSÉ BONANO; 1,200 employees.

**Palao Williams & Company:** Col. Lomas del Mayab, Calle Hibueras, Avda Cotán 3359, Tegucigalpa; tel. 232-0799; fax 231-3709; e-mail dpca@david.intertel.hn; accountants; Gen. Man. OSCAR HERNAN CASTILLO.

**Procesadora de Tabaco SA:** 2506 Blvd del Sur, Apdo 130, San Pedro Sula; tel. 556-9113; fax 556-9410; e-mail protabsa@sulanet.net; f. 1996; tobacco maufacturer; Gen. Man. CÉSAR LÓPEZ PÉREZ; 1,000 employees.

**Químicas Handal de Centroamérica SA de CV:** Km 2.6 carretera a Puerto Cortés, Choloma, Apdo 559, Cortés; tel. 551-8184; fax 669-3484; e-mail info@quimicashandal.com; internet www.quimicashandal.com; f. 1967; manufactures shoe care products; Gen. Man. FUAD HANDAL.

**Químicas Magna, SA de CV:** Edif. Conjunto Químicas Dinant, Barrio Morazán, Tegucigalpa; tel. 231-0777; fax 232-3729; f. 1962; producers of agricultural chemicals, soaps, cosmetics and coffee; Man. FABIO TÁBORA.

**Tabacalera Hondureña, SA:** Carretera Chamelecon, Zona El Cacao, Apdo 64, San Pedro Sula; tel. 556-6161; fax 556-6189; f. 1928; subsidiary of British-American Tobacco Ltd, United Kingdom; cigarette manufacturers; Man. GRACO PAREDES; 280 employees.

**Tela Railroad Company:** Edif. Banco del País, 5°, Blvd Suyapa, Apdo 155, Tegucigalpa; tel. 235-8084; fax 235-8083; Honduras' largest banana producers; a subsidiary of Chiquita Brands International; Vice-Pres. FERNANDO SÁNCHEZ.

**Textiles Río Lindo, SA de CV:** Col. San José del Pedregal, Calle Principal, Apdo 211, Tegucigalpa; tel. 245-5411; fax 245-5433; e-mail kfacusse@riolindo.hn; internet www.riolindo.hn; f. 1950; manufacturers of textiles; CEO ADOLFO J. FACUSSÉ; Gen. Man. KAREN FACUSSÉ; 850 employees.

## UTILITIES

### Electricity

**AES Honduras:** Tegucigalpa; tel. 556-5563; fax 556-5567; e-mail carlospineda@aes.com; subsidiary of AES Corpn (USA); CEO CARLOS LARACH; Gen. Man. CARLOS V. PINEDA.

**Empresa Nacional de Energía Eléctrica (ENEE)** (National Electrical Energy Co): Edif. Autobanco Atlántida, 5°, Calle Real Comayagüela, Apdo 99, Tegucigalpa; tel. 238-5977; fax 237-9881; e-mail dirplan@enee.hn; internet www.enee.hn; f. 1957; state-owned electricity co; scheduled for privatization; Pres. GILBERTO RAMOS DUBÓN; Man. ANGELO BOTAZZI.

**Luz y Fuerza de San Lorenzo, SA (LUFUSA):** Tegucigalpa; generates thermoelectric power.

## TRADE UNIONS

**Central General de Trabajadores de Honduras (CGTH)** (General Confederation of Labour of Honduras): Barrio La Granja, antiguo Local CONADI, Comayagüela, Apdo 1236, Tegucigalpa; tel. 225-2509; fax 225-2525; e-mail cgt@david.intertel.hn; f. 1970; legally recognized from 1982; attached to Partido Demócrata Cristiano; Sec.-Gen. DANIEL A. DURÓN.

**Federación Auténtica Sindical de Honduras (FASH):** Barrio La Granja, antiguo Local CONADI, Apdo 1236, Comayagüela, Tegucigalpa; tel. 225-2509; affiliated to CGTH, CCT, CLAT, CMT; Sec.-Gen. JOSÉ HUMBERTO LARA ENAMORANDO.

**Federación Sindical del Sur (FESISUR):** Barrio La Ceiba, 1 c. al norte del Instituto Santa María Goretti, Apdo 256, Choluteca; tel. 882-0328; affiliated to CGT, CLAT, CMT; Pres. REINA DE ORDÓÑEZ.

**Unión Nacional de Campesinos (UNC)** (National Union of Farmworkers): antiguo Local CONADI, Barrio La Granja, Comayagüela, Tegucigalpa; tel. 225-1005; linked to CLAT; Sec.-Gen. VÍCTOR MANUEL CAMPO.

**Confederación Hondureña de Cooperativas (CHC):** 3001 Blvd Morazán, Edif. I.F.C., Apdo 3265, Tegucigalpa; tel. 232-2890; fax 231-1024; Pres. JOSÉ R. MORENO PAZ.

**Confederación de Trabajadores de Honduras (CTH)** (Workers' Confederation of Honduras): Edif. Beige, 2°, Avda Juan Ramón Molina, Barrio El Olvido, Apdo 720, Tegucigalpa; tel. 238-3178; fax 237-8575; e-mail cthhn@yahoo.com; f. 1964; affiliated to CTCA, ORIT, CIOSL, FIAET and ICFTU; Pres. WILFREDO GALEAS ANGEL MEZA; Sec.-Gen. REINA DINORA ACEITUNO; 200,000 mems; comprises the following federations:

**Asociación Nacional de Campesinos Hondureños (ANACH)** (National Association of Honduran Farmworkers): Edif. Chávez Mejía, 2°, Calle Juan Ramón Molina, Barrio El Olvido Tegucigalpa; tel. 238-0558; f. 1962; affiliated to CTH, ORIT, CIOSL; Pres. BENEDICTO CÁRCAMO MEJÍA; 80,000 mems.

**Federación Central de Sindicatos de Trabajadores Libres de Honduras (FECESITLIH)** (Honduran Federation of Free Trade Unions): antiguo Edif. EUKZKADI, 3a Avda 3 y 4, Calles 336, Comayagüela, Tegucigalpa; tel. 237-3955; affiliated to CTH, ORIT, CIOSL; Pres. ROSA ALTAGRACIA FUENTES.

**Federación Sindical de Trabajadores Nacionales de Honduras (FESITRANH)** (Honduran Federation of Farmworkers): 10a Avda, 11a Calle, Barrio Los Andes, San Pedro Sula, Apdo 245, Cortés; tel. 57-2539; f. 1957; affiliated to CTH; Pres. MAURO FRANCISCO GONZÁLES.

**Sindicato Nacional de Motoristas de Equipo Pesado de Honduras (SINAMEQUIPH)** (National Union of HGV Drivers): Avda Juan Ramón Molina, Barrio El Olvido, Tegucigalpa; tel. 237-4415; affiliated to CTH, IFF; Pres. ERASMO FLORES.

**Confederación Unitaria de Trabajadores de Honduras (CUTH):** Barrio Bella Vista, 10 Calle, 8 y 9 Avda, Casa 829, Tegucigalpa; tel. and fax 220-4732; e-mail cuth@123.hn; f. 1992; Sec.-Gen. ISRAEL SALINAS.

**Asociación Nacional de Empleados Públicos de Honduras (ANDEPH)** (National Association of Public Employees of Honduras): Barrio Los Dolores, Avda Paulino Valladares, frente Panadería Italiana, atrás Iglesia Los Dolores, Tegucigalpa; tel. 237-4393; Pres. FAUSTO MOLINA CASTRO.

**Federación Unitaria de Trabajadores de Honduras (FUTH):** Barrio La Granja, contiguo Banco Atlántida, Casa 3047, frente a mercadito la granja, Apdo 1663, Comayagüela, Tegucigalpa; tel. 225-1010; f. 1981; linked to left-wing electoral alliance Frente Patriótico Hondureño; Pres. JUAN ALBERTO BARAHONA MEJÍA; 45,000 mems.

**Federación de Cooperativas de la Reforma Agraria de Honduras (FECORAH):** Casa 2223, antiguo Local de COAPALMA, Col. Rubén Darío, Tegucigalpa; tel. 232-0547; fax 225-2525; f. 1970; legally recognized from 1974; Pres. Ing. WILTON SALINAS.

# Transport

## RAILWAYS

The railway network is confined to the north of the country and most lines are used for fruit cargo. There are 995 km of railway track in Honduras, of which 349 km are narrow gauge. Only 256 km of track were in use in 2003.

**Ferrocarril Nacional de Honduras** (National Railway of Honduras): 1a Avda entre 1a y 2a Calle, Apdo 496, San Pedro Sula; tel. and fax 552-8001; f. 1870; govt-owned; Gen. Man. M. A. QUINTANILLA.

**Tela Railroad Co:** La Lima; tel. 56-2037; Pres. RONALD F. WALKER; Gen. Man. FREDDY KOCH.

**Vaccaro Railway:** La Ceiba; tel. 43-0511; fax 43-0091; fmrly operated by Standard Fruit Co.

## ROADS

In 2004 there were an estimated 13,720 km of roads in Honduras, of which 2,970 km were paved. A further 3,156 km of roads have been constructed by the Fondo Cafetero Nacional and some routes have been built by COHDEFOR in order to facilitate access to coffee plantations and forestry development areas. In November 2000 the World Bank approved a US $66.5m. loan to repair roads and bridges damaged or destroyed by 'Hurricane Mitch' in 1998. In March 2003 the Central American Bank for Economic Integration approved funding, worth $22.5m., for the construction of a highway from Puerto Cortés to the Guatemalan border.

**Dirección General de Caminos:** Barrio La Bolsa, Comayagüela, Tegucigalpa; tel. 225-1703; fax 225-0194; e-mail miayapew@mailcity.com; f. 1915; highways board; Dir MARCIO ALVARADO ENAMORADO.

## SHIPPING

The principal port is Puerto Cortés on the Caribbean coast, which is the largest and best-equipped port in Central America. Other ports include Tela, La Ceiba, Trujillo/Castilla, Roatán, Amapala and San Lorenzo; all are operated by the Empresa Nacional Portuaria. There are several minor shipping companies. A number of foreign shipping lines call at Honduran ports.

**Empresa Nacional Portuaria** (National Port Authority): Apdo 18, Puerto Cortés; tel. 55-0192; fax 55-0968; e-mail gerencia@enp.hn; internet www.enp.hn; f. 1965; has jurisdiction over all ports in Honduras; a network of paved roads connects Puerto Cortés and San Lorenzo with the main cities of Honduras, and with the principal cities of Central America; Gen. Man. ROBERTO VALENZUELA SIMÓN.

### CIVIL AVIATION

Local airlines in Honduras compensate for the deficiencies of road and rail transport, linking together small towns and inaccessible districts. There are four international airports: Golosón airport in La Ceiba, Ramón Villeda Morales airport in San Pedro Sula, Toncontín airport in Tegucigalpa and Juan Manuel Gálvaz airport in Roatán. In 2001 it was announced that San Francisco Airport, USA, was to invest some US $150m. in the four airports over two years. In 2000 plans for a new airport inside the Copán Ruinas archaeological park, 400 km west of Tegucigalpa, were announced.

**AeroHonduras:** Edif. Corporativo, Hotel Real Clarion, Col. Alameda, 1521 Avda Juan Manuel Gálvez, Apdo 1861, Tegucigalpa; tel. 235-3737; fax 232-5005; internet www.aerohonduras.com; f. 2002 as Sol Air, name changed as above in 2004; flights to Managua, Nicaragua, and Miami, FL, USA; Pres. RICARDO MARTÍNEZ.

**Atlantic Airlines de Honduras:** Tegucigalpa; tel. 440-2343; fax 440-2347; e-mail atlantic@caribe.hn; f. 2001; affiliated to Atlantic Airlines (Nicaragua); scheduled domestic and international services.

**Isleña Airlines:** Avda San Isidro, frente al Parque Central, Barrio El Iman, Apdo 402, La Ceiba; tel. 443-0179; e-mail info@flyislena.com; internet www.flyislena.com; subsidiary of TACA, El Salvador; domestic service and service to the Cayman Islands; Pres. and CEO ARTURO ALVARADO WOOD.

## Tourism

Tourists are attracted by the Mayan ruins, the fishing and boating facilities in Trujillo Bay and Lake Yojoa, near San Pedro Sula, and the beaches on the northern coast. There is an increasing ecotourism industry. In May 2001 the tomb of a Mayan king was discovered near to the Copán Ruinas archaeological park; it was expected to become an important tourist attraction. Honduras received around 610,535 tourists in 2003, when tourism receipts totalled US $341m.

**Instituto Hondureño de Turismo:** Edif. Europa, 5°, Col. San Carlos, Apdo 3261, Tegucigalpa; tel. and fax 222-2124; e-mail tourisminfo@iht.hn; internet www.letsgohonduras.com; f. 1972; dept of the Secretaría de Cultura y Turismo; Dir-Gen. THIERRY DE PIERREFEU.

**Asociación Hotelera y Afines de Honduras (AHAH):** Hotel Escuela Madrid, Suite 402, Col. 21 de Octubre-Los Girasoles, Tegucigalpa; tel. 221-4579; fax 221-1778; e-mail staynhonduras@123.hn; Pres. ANASTASIO ANASTASSIU; Exec. Dir KAREN BONILLA.

**Asociación Nacional de Agencias de Viajes y Turismo de Honduras:** Blvd Morazán, frente a McDonald's, Tegucigalpa; tel. 236-9455; e-mail travelex@multivisiohn.net; Pres. SCARLETT DE MONCADA.

**Asociación de Tour Operadores de Honduras:** Col. Los Alamos, III Etapa, Bloque BA-1, Cortes, San Pedro Sula; tel. 551-4391; fax 551-4393; e-mail mayanct@hondutel.hn; f. 1997; Pres. JOSÉ GILBERTO ARITA.

## Defence

In August 2004 the Honduran Armed Forces numbered 12,000: Army 8,300, Navy 1,400 and Air Force some 2,300. Paramilitary Public Security and Defence Forces numbered 8,000. Military service was ended in 1995. From January 1999 the post of Commander-in-Chief was abolished and the military was brought under the authority of the President. The military budget was substantially reduced in the late 1990s. In August 2004 some 587 US troops (Army 382, Air Force 205) were based in Honduras.

**Defence Budget:** 950m. lempiras (US $52m.) in 2004.

**Chairman of the Joint Chiefs of Staff:** Brig.-Gen. ROMEO ORLANDO VÁSQUEZ VELÁSQUEZ.

**Chief of Staff (Army):** Col RENE OLIVA SAUCEDA.

**Chief of Staff (Air Force):** Flight Col MANUEL ENRIQUE CÁCERES DÍAZ.

**Chief of Staff (Navy):** Capt. JOSÉ EDUARDO ESPINAL PAZ.

## Education

Primary education, beginning at six years of age and comprising three cycles of three years, is officially compulsory and is provided free of charge. Secondary education is not compulsory and begins at the age of 15. It lasts for a further three years. In 2002 the enrolment at primary schools was 99.0% of the relevant age-group, while enrolment at secondary schools in that year was equivalent to only 42.0% of children in the appropriate age-group. There are eight universities, including the Autonomous National University in Tegucigalpa. Estimated spending on education in 2003 was 8,783m. lempiras, representing 26.7% of the total budget. In December 2000 the Inter-American Development Bank approved a US $29.6m. loan to initiate an expansion and reform of the education system.

## Bibliography

For works on Central America generally, see Select Bibliography (Books)

Binns, J. R. *The United States in Honduras*. Jefferson, NC, McFarland & Co, 2000.

Bradshaw, S., and Linneker, B. *Challenging Women's Poverty: Perspectives on Gender and Poverty Reduction Strategies from Nicaragua and Honduras (CIIR Briefing)*. London, Catholic Institute for International Relations, 2004.

De Coster, J., and Anson R. (Ed.). *Profile of the Maquila Apparel Industry in Honduras*. Wilmslow, Textiles Intelligence Ltd, 2003.

Douglass, J. G. *Hinterland Households: Rural Agrarian Diversity in Northwest Honduras*. Boulder, CO, University of Colorado Press, 2002.

Euraque, D. A. *Reinterpreting the Banana Republic: Region and State in Honduras, 1870–1972*. Chapel Hill, NC, University of North Carolina Press, 1996.

Fasquelle, R. P. *Perfil de un nuevo discurso político*. San Pedro Sula, Centro Editorial, 1992.

Loker W. M. *Changing Places: Environment, Development and Social Change in Rural Honduras*. Durham, NC, Carolina Academic Press, 2004.

Thorpe, A. *Agrarian Modernisation in Honduras*. Lewiston, NY, Edwin Mellen Press, 2001.

Turck, M., and Black, N. J. *Honduras: Hunger and Hope*. Parsipanny, NJ, Dillon Press, 1999.

USA Ibp. *Honduras Foreign Policy and Government Guide*. Milton Keynes, Lightning Source UK Ltd, 2003.

*Honduras Customs, Trade Regulations And Procedures Handbook*. Milton Keynes, Lightning Source UK Ltd, 2005.

# JAMAICA

## Geography

### PHYSICAL FEATURES

Jamaica is in the Caribbean Sea, about 145 km (90 miles) south of eastern Cuba and 160 km west of south-western Haiti. The small, uninhabited US island of Navassa falls mid-way between Jamaica and Haiti, and the Cayman Islands, a British dependency, lies 290 km (180 miles) to the north-west. Jamaica lies between two of the main sea lanes to Panama, the Cayman Trench to the north and the Jamaica Channel to the east. Jamaica has an area of 10,991 sq km (4,244 sq miles), contained within 1,022 km of coastline and including 160 sq km of inland waters.

Jamaica is the third largest island of the Greater Antilles (after Cuba and Hispaniola), being 235 km from east to west and 82 km from north to south at its widest. The length of the island is dominated by ranges of mountains, reaching their height at Blue Mountain Peak (2,256 m or 7,404 ft) in the east, but falling away in the west. The mountains stretch north and south in a series of spurs and gullies, and the heights are luxuriantly forested. The terrain is, therefore, mainly mountainous, relieved only by narrow and discontinuous coastal plains. The soil is fertile and, although cultivation has made an impact, there are still vast tracts of native vegetation—for instance, some 3,000 species of flowering plant (827 of which are endemic) and 550 varieties of fern. There are several species of reptile (including the crocodile, the Jamaican iguana, which was thought to be extinct until 1990, and five types of snake, all harmless and rare, such as the yellow snake or Jamaican boa) and few large mammals (the mongoose, an introduced pest, wild boar in the mountains and the endangered hutia or coney, as well as the Pedro seal and a small number of sea cows, the manatee). Life in the air is obviously more adaptable on Jamaica, with at least 25 species of bat (including a fish-eating one) and many butterflies and birds. There are a number of indigenous creatures, such as the extremely rare Jamaican butterfly (a black-and-yellow swallowtail, the largest butterfly of the Americas and the second largest in the world), the red-billed streamertail hummingbird or doctor bird (the national bird) and some parrots.

### CLIMATE

The climate is subtropical, hot and humid on the coast, particularly, but more temperate inland, mainly owing to elevation. The average, annual temperature on the coast is 27°C (81°F). The thermometer can sometimes reach readings of 32°C (90°F) in the height of summer (July–August), but never falls below 20°C (68°F). However, in the mountains winter temperatures can fall as low as 7°C (45°F). The hurricane season is July–November and the rainy season October–November, although rain need not be uncommon from May onwards (average annual rainfall is 78 inches or 1,980 mm) and is more copious in the mountains.

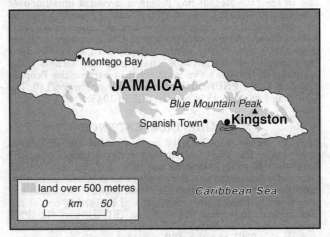

### POPULATION

The population is mostly black, 90.9% being of African descent, with those of mixed race at 7.3%, 'East' Indians at 1.3% and Chinese and whites at 0.2% each. The original settlers from Europe, who displaced the Taino Amerindians, were soon outnumbered by the slaves they brought in as agricultural labour, whose descendants are now Jamaica's dominant ethnic group. The vast majority of these slaves were brought in by the British, the main colonial power, but some earlier groups joined the original Spanish settlers in resisting the British occupation of the island, and their descendants became known as the Maroons, some of whom still maintain a distinct identity in the west. British institutions and culture have, therefore, been influenced by other traditions, notably West African, giving Jamaican culture its modern vibrancy. Thus, English is the official language, but an English patois is generally spoken (using many Ashanti words) and, while most people are Christian, there are influences of other traditions in religion too. Protestant denominations have the largest number of followers (Church of God claims 24% of the total population, Adventists 11% and Pentecostalists 10%), but there are also Anglicans, Roman Catholics and Ethiopian Orthodox adherents, and 35% of people confess other religions (apart from some Jews, Muslims and followers of various Asian belief systems, these are mainly spiritualist cults or the indigenous Rastafarian religion, a pan-Africanist movement). Some early adaptations of Christianity, such as Pocomania and Revival, also still survive.

Estimates for December 2003 put the total population of Jamaica at about 2.6m., but there are far more people of Jamaican descent living abroad. About 0.6m. people live in the capital, Kingston, which is found at the eastern end of the southern coast. The second city is Montego Bay, right across the island, in the north-west. For administrative purposes, Jamaica is divided into 14 parishes.

## History

### Dr JAMES FERGUSON

Jamaica was sighted and visited by the navigator Christopher Columbus in 1494, but Spanish settlement did not take place until 1509. The existing inhabitants of the island, the Taino people, had arrived in AD 650–1000 and had established small villages, based upon fishing and subsistence agriculture. The indigenous population was rapidly exterminated by the European colonists, who were in search of gold. Finding little precious metal, the Spanish imported African slaves and developed cattle-raising ranches and small-scale sugar plantations.

The island was an insignificant part of the Spanish Empire and was easily captured by an English expedition in 1655. Introducing a much greater numbers of slaves, the English transformed Jamaica into a major sugar-exporting colony over the next two centuries. The wealth of the planters contrasted greatly with the harsh conditions endured by the slaves, and there were several major slave rebellions as well as an extended war between the planters and the Maroons (escaped slaves) during the 18th century.

Slavery was abolished in 1834, but social conflict persisted, and in 1865 the Morant Bay Rebellion, a violent uprising over land rights, led to the removal of the island's Assembly and the imposition of direct rule from the British Government. This Crown Colony status persisted into the 20th century, but was gradually replaced by greater self-government, especially after the social unrest of the 1930s. That period also witnessed the founding of Jamaica's two main political parties: the People's National Party (PNP) and the Jamaica Labour Party (JLP). The PNP was formed in 1938 and the JLP in 1943.

## TOWARDS INDEPENDENCE

The two parties were both led by charismatic and influential individuals, the PNP by the University of Oxford-educated barrister, Norman Manley, and the JLP by his cousin, Alexander Bustamante, a fiery orator and professional money-lender of little formal education. Both political organizations also had affiliated trade unions, the JLP evolving out of the Bustamante Industrial Trade Union (BITU), and the PNP linking itself to the National Workers' Union (NWU).

Both the PNP and the JLP appealed to the impoverished majority of Jamaicans, and both parties advocated independence from the United Kingdom. From the outset, however, the JLP was more conservative, pro-USA and vigorously anti-communist than the PNP, which adopted social-democratic policies, inspired by the moderate socialism of the British Labour Party. The parties were different in leadership style too, as Bustamante prided himself on his theatrical speeches and rapport with the working-class electorate, while Manley was more restrained and intellectual in approach.

Jamaican politics and society were henceforth to be divided, sometimes bitterly, between the two parties. However, the path towards independence was smooth and peaceful, as political conflict was confined to electoral and constitutional methods. The year after the formation of the JLP, in 1944, a new Constitution was introduced, ushering in universal adult suffrage. In elections that year the JLP won a majority of seats, campaigning on the slogan 'Bread and Butter', while the PNP put emphasis on the need for independence. The more populist JLP won again in the 1949 general election.

As the move towards British withdrawal from the Caribbean gathered pace, Jamaica's independence became increasingly inevitable. However, there were differences between the PNP and the JLP, the former favouring the federal model of government proposed by the United Kingdom, which was to bring together the English-speaking colonies of the Caribbean into a single state. Bustamante, for his part, was more ambivalent, suspecting that Jamaica's relative economic strength would be diluted to the detriment of its citizens in a federal arrangement with smaller, poorer territories.

Under the aegis of Manley, whose PNP won the legislative elections of 1955, Jamaica became a key member of the West Indies Federation, which was established in 1958. From the outset, however, there were tensions between the larger, more developed islands, particularly between Jamaica and Trinidad and Tobago. As Jamaica argued about its level of representation within the Federation and its right to determine its own economic policy, the federal ideal became tarnished and Manley's popularity at home waned. Seizing his opportunity, Bustamante called for Jamaica's withdrawal from the Federation and a referendum on the issue, which the JLP duly won in September 1961. With Jamaica's retreat from the federal experiment and the ensuing collapse of the West Indies Federation, the prospect of Jamaica's independence as a unitary state was imminent. On 6 August 1962, following elections in the previous April, Bustamante oversaw the final transition to independence.

## COLD WAR POLITICS

The JLP triumphed again in elections in 1967, taking the credit for a period of sustained economic growth that was owing to high commodity prices. Bustamante had retired in 1964, to be replaced by Donald Sangster and then, when Sangster suddenly died, by Hugh Shearer. Maintaining a conservative outlook during a period of political ferment and student radicalism, the JLP faced growing opposition in its second term in office and was damaged by allegations of corruption. The so-called 'Rodney riots' in 1968, caused by the expulsion of the Marxist 'Black Power' advocate Walter Rodney, further weakened the JLP's position. As Shearer's popularity declined, the PNP, under the leadership of Michael Manley (Norman Manley's son), gathered impetus.

The elections of 1974 returned the PNP to power with a large majority and a mandate for radical change. The party now openly embraced what it called 'socialism with a human face', or 'democratic socialism', and pledged to nationalize key sectors of the economy dominated by foreign or private capital, to raise minimum wages and the living standards of the poor, and to follow a foreign policy based on 'Third Worldism' and closer links with Cuba. These policies alienated the Jamaican business élite and, more importantly, the USA, which viewed Manley's version of socialism with distrust. It was alleged, both inside Jamaica and beyond, that the USA, alarmed at perceived political radicalism in the Caribbean, was intent on destabilizing the Manley Government.

In the event, it was the 1973 oil crisis and the ensuing global recession that put most pressure on Jamaica's socialist experiment. While Manley honoured his election pledges, raising wages and improving social provision, he also sought to extract a better deal from foreign companies mining bauxite in Jamaica. The decision to impose a higher tax on bauxite exports produced little benefit as aluminium companies, severely impacted by world-wide recession, reduced their activities. Economic downturn also affected foreign investment in other areas of the economy, and unemployment increased dramatically.

Even amid the turmoil, the PNP won a second term of office in 1976, but economic conditions continued to deteriorate. Political violence between PNP and JLP militants had reached new heights during the election, when a state of emergency had been declared, and tourism and foreign investment began to decline. Faced with open US hostility and a serious balance-of-payments crisis, the PNP Government was forced in 1978 to seek assistance from the IMF, which, in return for short-term financial support, insisted on public-sector cuts and the abolition of many social programmes. After little improvement in the economy was achieved, Manley broke with the IMF. By 1980, as new elections approached, the economy was in danger of collapse and social turmoil was increasing—over 1,000 died in violence surrounding the elections.

A resurgent JLP, led by Edward Seaga, won the 1980 elections, securing 51 of the 60 seats in the House of Representatives. Seaga, a former Minister of Finance, espoused a free-market approach and rejected Manley's social reformism. Close relations with the USA were restored, Cuba was isolated and Jamaica participated in the October 1983 'intervention' to overthrow a Marxist regime in Grenada. Seaga's political rhetoric was stridently anti-communist, designed to appeal to the conservative Administration of President Ronald Reagan (1981–89) in the USA. Capitalizing on his post-Grenada popularity, Seaga won a 'snap' election in December 1983, which the PNP boycotted. However, even without an opposition, Seaga struggled to revitalize the economy, facing mounting popular opposition as he abolished price controls on basic foods and oversaw a sharp rise in unemployment. Furthermore, as Jamaica suffered from the effects of a recession in the USA and austerity schemes seemed to deliver few benefits, Seaga went against IMF recommendations and adopted an expansionary budget. Arrears in debt payments to the IMF, however, forced Seaga to accept further unpopular conditions, such as limits on public-sector pay. Despite spending beyond IMF-approved limits in 1988 in an attempt to regain some popularity, the JLP rapidly lost ground to the PNP, which won local elections in 1986.

## THE RETURN OF THE PNP

Elections in February 1989, which were delayed because of the chaos caused by 'Hurricane Gilbert' in September 1988, returned the PNP to power with 45 seats in the House of

Representatives. The party leader was still Michael Manley, but, after nine years out of power, he was a very different political figure, having abandoned most of the radical rhetoric that had earned him such hostility in the 1970s. Instead, he advocated social consensus and a market-driven economic policy, as well as close relations with the USA. Manley also honoured ongoing agreements with the IMF, obtaining fresh loans in 1989 and 1990 in return for devaluing the Jamaican dollar. Further measures aimed at deregulating the financial sector and international trade took place, including the flotation of the Jamaican dollar and the removal of foreign-exchange controls. Yet despite such liberalizing reforms, the Manley administration wrestled with foreign-exchange shortages and a legacy of excessive debt, inherited from the Seaga Government.

In March 1992 Michael Manley announced his retirement on grounds of ill health, and in the ensuing party leadership election Percival J. Patterson, the former Deputy Prime Minister, emerged the victor. Under Patterson's leadership the policy of stringent economic management continued, with the result that inflation lessened and the exchange rate stabilized.

Conscious of his own popularity and factional disarray within the JLP, Patterson opted for an early election in March 1993, in which the PNP polled an overwhelming 60% of the votes, albeit on a low turn-out. The election was to some degree marred by violence and allegations of irregularities, leading to the installation in 1996 of a computerized electoral register. The JLP's defeat also led to increased infighting among senior members and criticism of Seaga's leadership, with the result that in 1995 former JLP Chairman and finance spokesman Bruce Golding left the party to lead a new organization, the National Democratic Movement (NDM). Golding's defection, along with several other leading JLP members, led to sporadic violence between NDM and JLP partisans, but the NDM failed to make an immediate impact.

The PNP Government's second term was characterized by continuing attempts to control exchange rates and inflation, with high interest rates adopted as a means to increase foreign-exchange reserves and reduce inflation. However, these deflationary policies, which had achieved their objective by the end of 1997, created other problems, notably zero or negative growth in gross domestic product (GDP), which was lower at the end of the 1990s than at the beginning of the decade. Agriculture and manufacturing were badly affected, the former by drought, the latter by cheaper regional competition, but both also by high interest rates, which deterred investment.

Despite the PNP Government's economic difficulties, the party was returned to power on a third and unprecedented consecutive occasion in the election of December 1997, winning 56% of the votes, compared with the JLP's 39% and the NDM's 5%. The polling, based on the new electoral register, was judged fair and free, while the death toll in the accompanying political violence was reduced to two people.

## SOCIAL PROBLEMS AND VIOLENCE

If the 1997 election marked a relative success story in terms of violence, social problems and a disproportionately high murder rate continued to blight Jamaicans' daily life and the image of the island overseas. In 1999 the British Government announced a £2.9m. programme of assistance aimed at reforming and modernizing the Jamaica Constabulary Force, and in 2000 Patterson announced that a specialized police force had been formed to combat organized crime. In the same year an investigation took place into widespread allegations of corruption in the police force, involving complicity in drugs-trafficking and the illicit recording of ministerial telephone conversations. The British Government insisted in January 2001 that the release of a shipment of arms and ammunition, intended for the police, was dependent on training in the non-lethal apprehension of suspects. In 2004 the British Government dispatched a team of police advisers in order to strengthen key areas of local policing. 'Operation Kingfish', implemented in October 2004, was proclaimed a success in April 2005 by the Minister of National Security, Peter Phillips, who claimed that organized criminal gangs had been disrupted by the seizure of drugs and arms, as well as by 154 arrests.

Despite such measures, the murder rate continued to rise, reaching 1,138 in 2001, one of the worst per-head rates in the world. Many of these violent deaths were connected to the illegal drugs business and to related feuds within the deprived inner-city areas of Kingston. However, many were also blamed on the police, who, according to the human rights group Amnesty International, were responsible for over 600 killings in 1999–2002. The number of murders overall dropped slightly in 2002, to 1,045, with 133 individuals reportedly killed by police personnel. Of those killed, some 400 were thought to have been murdered in inter-party violence in the weeks preceding elections in October 2002. These killings occurred in spite of a pact signed in June by Seaga and Patterson, condemning political violence and calling for non-belligerent, nation-building debates. Political violence rose again in the aftermath of the October elections, with 100 people reportedly murdered. Amnesty International estimated that 975 people were murdered in 2003, of whom at least 113 were victims of police shootings. In 2004 a record 1,445 people died as a result of violent crime, and in the first five months of 2005 alone there were 630 murders. In May 2005 the Private Sector Organization of Jamaica (PSOJ) issued a 13-point declaration, calling on all politicians to dissociate themselves from criminal elements.

Jamaica's involvement in the international narcotics trade also damaged the island's image and international relations. In 1997, after the threat of economic sanctions from the US Administration of Bill Clinton (1993–2001), Jamaica signed the Shiprider Agreement, whereby US Coast Guard and other anti-drugs personnel, with the permission of the Jamaican authorities, were allowed to enter Jamaica's territorial waters in pursuit of suspected drugs-smugglers. Several high-profile interceptions of supplies of cocaine and marijuana, mainly originating from Colombia, were reported. None the less, the activities of so-called 'Yardie' gangs operating in Jamaica, Canada, the United Kingdom and the USA continued to cause alarm, as did the frequent interception of drugs-traffickers entering the United Kingdom from Jamaica. In 2001 the British Deputy High Commissioner in Kingston remarked that more than one in 10 Jamaicans arriving at British airports were carrying illegal drugs. That same year an estimated 21 people were killed in London, United Kingdom, in what were believed to be 'Yardie'-related feuds. In January 2002 the British Government announced that Jamaican nationals would henceforth require a visa to enter the United Kingdom. The British authorities cited not only drugs-smuggling but also illegal immigration as reasons for this new measure. In September 2004 it was estimated that there were 2,000 Jamaican nationals in British jails, mostly connected with the drugs trade and other 'Yardie'-related criminality.

## RECENT DEVELOPMENTS

In the legislative election of 16 October 2002, the PNP defied the opinion polls and won a historic fourth term of office, taking 34 of the 60 seats in the House of Representatives. The small 'third' parties once again failed to make any headway against the two traditional parties. The PNP's success was judged to be owing in part to the Government's recently increased spending on housing and infrastructure, but also to public distrust of the JLP. Nevertheless, the Government's majority was radically reduced from 38 to eight, while electoral turn-out of only 57% was the lowest since the introduction of universal suffrage in 1944. While the election campaign itself involved many violent deaths, the voting process was relatively peaceful.

The PNP Government began its fourth term in office under considerable pressure from the JLP, and facing the problems of a large fiscal deficit, growing debt and pressure on the Jamaican dollar. Reducing the rate of violent crime also remained a considerable challenge. Local elections in June 2003 confirmed the Government's unpopularity, with the JLP winning 130 of the 227 contested local government seats, compared with the PNP's 97. Political apathy was also reflected in the voter turn-out of 37%. The PNP's standing among voters was further dented by its austerity budget for 2003/04, which included sharp tax rises in order to meet debt commitments. In the face of widespread opposition, the Government was forced to withdraw certain proposals, including a widening of the range of items liable to the general consumption tax. Nevertheless, the JLP was unable to capitalize on dissatisfaction with the Government, which in a March 2004 opinion poll was estimated at 61%. Instead, the opposition was damaged by public divisions within its leadership and growing speculation over the future of its veteran leader, Edward Seaga. In July 2004 Seaga finally announced his intention to step down as leader in November. Following acrimonious divisions within the JLP over the leadership election, Golding, who had rejoined the JLP in 2001, was finally elected unopposed in February 2005. He subsequently won the safe JLP Western Kingston constituency, vacated by Seaga, in a by-election in April, allowing him to act as official Leader of the Opposition.

Given the lack of effective opposition, the PNP Government encountered few significant political obstacles in its first three years in office. It began to adopt a more conciliatory stance in 2004, by signing in February a Memorandum of Understanding with the Jamaica Confederation of Trades Unions. The agreement stipulated that the public-sector unions would accept wage increases limited to 3% per year over two years in exchange for the Government's pledge not to introduce wide-scale redundancies. The Government was also in consensus talks with the PSOJ, developing a 'Partnership for Progress' that involved policies on tax and pensions reforms, as well as privatization programmes. These developments in the first half of 2004 suggested that the PNP Government was under no immediate threat and might be able to introduce vital reforms with regard to the fiscal deficit, debt and interest rates. The 2005/06 budget was approved by the Parliament without significant opposition in April 2005, aiming at a balanced budget through an increase in the general consumption tax and a lowering of interest rates to reduce domestic debt payments.

While reiterating its support for enhanced integration within the Caribbean Community and Common Market (CARICOM), the PNP Government also endorsed the long-mooted proposal for a Caribbean Court of Justice to replace the Privy Council (based in the United Kingdom) as a final court of appeal in Jamaica. Despite opposition demands for a referendum on such an important constitutional reform, after gaining approval in the Senate and the House of Representatives, in June 2003 the Government ratified the establishment of the Court, to be based in Trinidad and Tobago. In early 2005, however, following a ruling by the Privy Council itself, it emerged that a referendum on the proposed change was constitutionally necessary, as amendments to existing procedures regarding the appointment and tenure of judges had to be approved by referendum. In addition, the PNP advocated the replacement of the British monarch with an elected president as head of state and pledged a referendum on this change during the current term of Parliament, although little was heard on this issue thereafter. The PNP's high-profile support for the ousted Haitian President, Jean-Bertrand Aristide, in 2004 created some friction with the US Administration, as Patterson offered Aristide temporary asylum in Jamaica after implicitly accusing the USA of endorsing the Haitian leader's overthrow. Patterson's long-awaited retirement was expected to be announced at the PNP party conference in September 2005, opening up the prospect of a fiercely fought and potentially divisive leadership contest. Several leading PNP politicians were expected to compete for the party leadership, including PNP Vice-Presidents Portia Simpson-Miller and Peter Phillips.

# Economy
## Dr JAMES FERGUSON

With a substantial mining sector, an extensive tourism industry and an established manufacturing base, Jamaica has a diversified economy by Commonwealth Caribbean standards. Once the main sugar-producing colony in the English-speaking Caribbean, the island gradually abandoned its dependence on monoculture in the first part of the 20th century, developing a diversified rural economy based on other export crops, such as bananas, citrus fruits and coffee, as well as food crops for domestic consumption. From 1952 onwards bauxite and alumina production became an important part of the Jamaican economy, and from the 1960s, with the advent of mass tourism, hotels, restaurants and associated services grew in importance. Prior to and following independence (achieved in 1962), successive Governments also attempted to reduce reliance on imported manufactures by encouraging local industries to produce goods for the domestic and export markets.

Despite its relative diversification, however, Jamaica has remained extremely vulnerable to external economic factors, such as unstable commodity prices, competition from other producer nations and the vagaries of the tourism industry. Natural disasters and political unrest have also had a markedly negative impact on economic performance. During periods of growth, particularly in the USA, the Jamaican economy has tended to expand, buoyed by investment, expanding tourism and demand for commodities. In times of recession, conversely, it has suffered. During the 1960s Jamaica's gross domestic product (GDP) grew by an average of almost 6% per year, in real terms, owing to bauxite exports and foreign investment. From 1974 to 1980, however, the economy contracted sharply, with GDP declining by 16% cumulatively as a consequence of the global oil crisis, rising petroleum prices and recession in Europe and the USA. The 1980s witnessed some economic growth, notably in 1987, when GDP increased by 6.2%, but the destructive impact of 'Hurricane Gilbert' in 1988 reversed the trend. Overall average annual growth in 1980–90 was 1.6%, but at the end of the 1980s the economy had still not recovered to the level of 1974. During the 1990s Jamaica fared even worse, with an average annual growth of 0.9% in 1990–2003. This was, in part, the result of a strict economic reform programme, intended to counter Jamaica's indebtedness and overspending by the Government, as well as a consequence of low bauxite and alumina prices. Recessionary trends continued throughout the late 1990s, with negative growth in 1997–99. In 2000 modest growth of 0.8% was recorded, with growth of 1.5% in 2001 and 1.1% in 2002. In 2003 slightly higher growth of 2.3% was recorded, while in 2004 growth slowed once more, to 1.2%, owing, in large part, to damage caused by 'Hurricane Ivan' in September 2004. GDP per head was estimated to have decreased, in real terms, by an annual average of 0.2% in 1990–2003, but sustained, albeit limited, growth from 2000 marked Jamaica's best period of expansion since the 1980s.

During the 1990s the population increased at an average annual rate of 1.0%. In July 2005 the population was estimated at 2.7m. and population density was estimated at 248.5 per sq km. According to the 2001 census, 52.1% of the population lived in urban areas, of which some 61.0% resided in Kingston and the surrounding urban parish of St Andrew. Spanish Town in St Catherine and Montego Bay on the north-west coast were the two biggest towns outside Kingston. In 2005 net emigration from the country was estimated at 4.07 persons per 1,000.

During the economic troubles of the 1980s unemployment was often extremely high, reaching levels of between 20% and 30%. By 2004 official unemployment was recorded at some 15.0% of the work-force. Many more Jamaicans are underemployed or work in the unregulated informal sector, either as casual farm workers or in service jobs. Unemployment has usually affected women and those under 25 most severely. Despite consistently high levels of unemployment and declining per-head GDP throughout the 1990s, the IMF calculated that the percentage of Jamaicans defined as living in poverty had fallen from 44.6% in 1991 to 16.1% in 2004.

## AGRICULTURE

Agriculture, including forestry and fishing, comprised 5.3% of GDP in 2003. The sector was adversely affected by droughts during 1997–98 and 2000, by 'Hurricane Michelle' in October 2001 and by 'Hurricane Ivan' in 2004. Persistent flooding and droughts from 2002 compounded problems, notably among coffee producers. The sector provided work for some 19.4% of the engaged work-force in 2002, although much work is seasonal and casual in nature. According to the Statistical Institute of Jamaica, agricultural production decreased by 8.3% in 2002, largely owing to drought and flooding, but output increased in 2003, especially towards the end of the year, with domestic crop production rising by 21% and export crop production by 7%. Hurricane damage in September 2004 accounted for an estimated 35% fall in production in the fourth quarter of that year. In 2002 the Inter-American Development Bank (IDB) agreed a loan worth some US $22m. to modernize the Ministry of Agriculture.

The sugar industry, historically the most important sector of Jamaica's economy, experienced a crisis in 1985, following decades of decline. In that year the Government contracted the management of the state sugar corporation to the British Booker Tate company. In 1993 the Government sold the five state-owned sugar mills to private consortia, but in 1998 was forced to repurchase the mills for a nominal J $1 in order to rescue the Sugar Company of Jamaica from bankruptcy. In 2001 it was estimated that the Government had spent some J $3,000m. annually since 1998 in subsidizing the sugar industry. The Government has come under pressure to reduce the 40,000-strong work-force, but was unwilling to run too strong a political risk. In December 2002, following its re-election in October, the People's National Party (PNP) Government announced the closure of the loss-making Hampden sugar factory, which had been under state control after being put into receivership in 1999. In 2004 and 2005 a number of labour disputes in state-owned and private plantations disrupted production. Sugar production has gradually declined since the 1990s, recovering to 209,825 metric tons in 2000, before declining to 153,542 tons in 2003. Production recovered in 2004, to 181,000 tons, before the damage inflicted by 'Hurricane Ivan', which was expected to reduce yields in 2005. Sugar export earnings, underpinned by preferential quota arrangements with the European Union (EU) and the USA, fell from US $98m. in 1999 to US $66m. in 2002, partly owing to the depreciation of the euro, in which EU exports are paid, against the US dollar. The subsequent recovery of the euro maintained sugar earnings at US $66m. in 2003, despite reduced export volumes. However, these preferential arrangements were due to be phased out by the EU over the next decade. One private-sector business, Appleton Estates, meanwhile completed a US $14m. modernization programme in late 2002, trebling potential annual output to 60,000 tons and enabling it to compete in the liberalized EU market. The proposed reduction of guaranteed EU prices from 2006 would mean further hardships for the sector, although the EU announced an eight-year support programme for Caribbean sugar producers in 2004.

Banana production and exports were badly disrupted by 'Hurricane Gilbert' in 1988, but recovered in the 1990s, reaching an export volume of 89,000 metric tons in 1996. Exports fell to 40,000 tons in 2002, with export earnings declining to US $18m. (from US $45m. in 1996). Exports further declined in volume in 2003, albeit slightly, to 39,000 tons and earning US $19m., while drought in early 2004, a reported outbreak of Moko disease and hurricane damage were responsible for stopping exports altogether from September 2004 until June 2005. Falling world prices, together with uncertainty over the future of the industry, reduced investment and acreage in the banana sector, with the prospect of reduced preferences into the EU market deterring farmers from further investment. In 2001, after pressure from the USA, the EU agreed to end its preferential treatment of bananas from Africa-Caribbean-Pacific (ACP) exporters and to introduce a transitional system of licences based on historic trade patterns. A tariff-only system was to enter into force on 1 January 2006. In mid-2002 the EU agreed to provide the country with US $4.3m. to improve the banana industry, offering a further US $5.5m. in 2003.

Coffee production and exports have also fluctuated owing to adverse weather conditions, with exports reaching a record high of almost 3,000 metric tons in 1996, before falling to 1,000 tons in 1998. Exports recovered to 1,700 tons in 2000, but were severely affected by flooding in late 2001. The value of coffee exports in 2001 was US $30.7m., a fall of 7% on the previous year. 'Hurricane Ivan' in September 2004 reportedly destroyed 60% of coffee plantations. Non-traditional agricultural products for export showed strong growth in the mid-1990s, but declined thereafter, owing to drought and disease, falling in value from US $409m. in 1997 to US $302m. in 2001. Exports in 2004 earned US $183.3m. Cocoa, citrus fruits, pimento and coconuts are also exported.

A programme of land reform and investment undertaken in the 1990s with the aim of increasing food production enjoyed some success, but farmers have been affected by droughts and hurricanes, with production of sweet potatoes, yams and vegetables declining between 1999 and 2000 and increasing only slightly in 2001. A marked increase in production was recorded in the second half of 2003 after improved weather conditions, but drought in early 2005 caused a fall in production. Further pressure on domestic farming came from a reduction in tariffs resulting from an agreement with the World Trade Organization. Imported chicken and dairy goods, in particular, reduced prices for consumers, but also for farmers.

In 2001 6% of Jamaica's land area, or 77,000 ha, was estimated to consist of undisturbed natural forest. Other areas of forest have come under increasing pressure from developers and farmers.

## MINING AND POWER

The mining sector is dominated by the bauxite/alumina industry. Mining and quarrying contributed 4.4% of GDP in 2003 and employed 0.4% of the work-force in 2002, while alumina and bauxite exports together earned US $758m. in 2003, or 60% of total export earnings. In 2004 Jamaica was the world's third largest producer of bauxite.

Low world prices in the early 1980s led to a crisis in the bauxite industry, but production recovered in the early 1990s, before industrial disputes and operating problems depressed output. From a total of 12.6m. metric tons in 1998, production fell to 11.1m. tons in 2000, largely because of a fire at the Kaiser Aluminium refinery in Texas, USA. The reopening of the Kaiser plant in 2001 led to increased production of 12.1m. tons in that year. Thereafter, increased world demand led to increased production of 13.4m. tons in 2003. Alumina production was also static in the 1980s, before rising in the 1990s, reaching 3.6m. tons in 2000. Production declined to 3.5m. tons in 2001, as a result of industrial action in the Jamalco processing plant, before rising to 4.0m. tons in 2004. Bauxite mining and alumina refining were carried out by three major companies: Alcan of Canada (taken over by Glencore of Switzerland and renamed in 2001); Jamalco, a joint venture between Alcoa of the USA and the Jamaican Government; and Alumina Partners of Jamaica (Alpart), which was jointly owned by Kaiser (of the USA) and Norsk Hydro (of Norway). In early 2002 the Minister of Mining and Energy announced that the Government wished to renegotiate the bauxite levy in force since the 1970s, stating that it was prepared to replace a tax on production with a tax on profits in return for increased investment. This was in response to complaints by mining companies that the levy rendered Jamaica uncompetitive as it taxed production irrespective of profitability. An agreement was announced later in 2002 whereby the Government replaced the bauxite levy imposed on Alcoa with a tax on profits and Alcoa, in turn, pledged to expand its Jamalco refinery. In December 2004 Alcoa reached an agreement with the Government for a US $800m. expansion of the Jamalco plant, doubling the refinery's annual capacity to almost 3m. tons. Alcoa was to increase its share of ownership to 70%, while the Government was to retain control of the remaining 30%.

Jamaica is a petroleum-importing country, with one-third of fuel required for the bauxite industry. In 2003 the cost of fuel imports was some US $830m., representing nearly one-quarter of all imports. Oil-fired electricity plants produced 94% of the power generated in 2002, which amounted to 6,289m. kWh. The remainder was hydroelectric power. In May 2002 a number of foreign companies expressed their interest in investing in a liquefied natural gas plant which, it was hoped, would provide

Jamaica with an alternative source of energy. Petroleum exploration in Jamaica did not produce any exploitable finds and official policy was directed towards energy conservation and the development of alternative energy sources. In 2004 the state-owned Petroleum Corporation of Jamaica (PCJ) offered exploration licences to foreign companies interested in exploring for oil deposits. PCJ also signed an agreement with the Venezuelan state oil corporation, Petróleos de Venezuela, SA (PDVSA), to expand Jamaica's oil-refining capacity to 60,000 b/d, enabling Jamaica to export some refined products.

## MANUFACTURING

Jamaica has had a developed manufacturing sector since the 1950s. The sector accounted for 12.8% of GDP in 2003 and employed 6.9% of the work-force in 2002. Manufacturing has declined in importance since the 1980s, with its share of GDP declining from 19.6% in 1988.

In the 1960s and 1970s Jamaica followed the import substitution model of industrialization, with government incentives and tariffs intended to encourage manufactures for the local market. The Jamaica Labour Party (JLP) Government (1980–89) shifted the emphasis towards export-orientated manufacturing within its structural adjustment policy. Reducing import restrictions and attracting foreign investment with tax incentives, the JLP administration oversaw the establishment of free-trade zones in Kingston, Montego Bay and Spanish Town, as well as the creation of a promotional organization, now known as Jamaica Promotions Ltd (JAMPRO), to increase export manufacturing. The PNP has subsequently pursued a similar policy of export-orientated manufacturing.

Traditional manufacturing revolved around food and drinks and tobacco products, building materials, chemicals, fertilizers and metals. Cement, mostly for the domestic market, is an important manufactured item, with production totalling 607,682 metric tons in 2003. Following the imposition of a 40% customs duty on imports from outside the Caribbean Community and Common Market (CARICOM), production by the Caribbean Cement Company, Jamaica's only producer, grew even more sharply, reaching 573,900 tons in the first nine months of 2004.

From the mid-1980s the export-orientated manufacturing sector grew rapidly, especially within the free trade zones. The most significant investors in this sector were companies from the USA, Taiwan and Hong Kong. Under the terms of the Caribbean Basin Initiative and the Section 807 textile and garment facility, Jamaica was a beneficiary of preferential entry arrangements for garment exports into the US market. In 1994 32,000 people were employed in the garment and textile sector, but by 2004 this number had dropped to fewer than 10,000. The decline in textile exports was largely attributable to the high cost of borrowing and to competition from other low-wage competitors, notably Mexico. Other competitors included the Dominican Republic, Haiti and Honduras, all of which offered foreign investors lower wage rates. Earnings from garments exports declined from US $227.6m. in 1997 to US $88.5m. in 2001 and continued to decline thereafter. Apart from textiles, Jamaica-based assembly plants produce some electronic components, chemicals and processed foods.

## TRANSPORT AND COMMUNICATIONS

Jamaica's transport infrastructure is well-developed by regional standards. There are two international airports, at Kingston and Montego Bay, serviced by a range of international and charter airlines, including the national carrier, Air Jamaica. The principal port is at Kingston, a major transhipment centre for the Caribbean region.

The railway network totals 339 km (211 miles), but is used exclusively for the transportation of bauxite. Passenger services were suspended in 1992, and, after a series of talks with investors concerning the rehabilitation of the network, in early 2005 the Government signed a memorandum of understanding with China National Machinery and Equipment Corporation for the reconstruction of the railway for commercial and passenger use.

In 2004 there were an estimated 18,700 km (11,620 miles) of roads, of which 70.1% were paved. The Government undertook a five-year programme of road improvements in 1999, with the emphasis on creating tolled highways and eliminating congestion. In February 2002 the Government signed an agreement with the French construction company Bouygues, by which the firm was to build the first section of the tolled Highway 2000 between Kingston and Montego Bay. Other important roadworks around Montego Bay, Ocho Rios and Negril have been completed or are near completion. In 2004 there were some 225,000 registered motor vehicles, many of them imported used cars from Japan.

From the mid-1980s Jamaica's telecommunications system was substantially modernized and improved. In 1999 the Government approved the liberalization of the telecommunications sector, with the issuing of mobile cellular telephone licences commencing soon after. In 2004 there were some 1.8m. subscribers to fixed line and mobile telephone services, with an estimated 2m. subscribers by mid-2005. The cellular sector is dominated by Oceanic Digital Jamaica and Digicel Jamaica, a subsidiary of the Irish company Mossel, while the British Cable & Wireless company, traditionally the main supplier, remained in competition. In 2004 the US company AT&T acquired a licence from the Jamaican Government to introduce a wireless telecommunications service. In mid-2005 there were an estimated 2.8m. internet subscribers in Jamaica, with access to 21 servers.

## TOURISM

From the 1970s tourism began to supplant agriculture and mining as Jamaica's main source of foreign currency. In 2004 the sector was estimated to employ 85,000 people directly, with twice as many employed in ancillary and 'informal' sector work related to tourism.

Stop-over tourist arrivals increased steadily in the 1980s, averaging an annual 11.3% increase over the decade and reaching a total of 840,777 in 1990. Growth was slower in the 1990s, with arrivals rising to 1.3m. in 2000. Cruise-ship arrivals, however, increased at a much greater rate, with numbers expanding from 133,400 in 1980 to 764,341 in 1999. In 2003 the total number of tourists visiting Jamaica was 1,350,284, with a further 1,132,300 arriving on cruise ships. In that year the average length of stay was 10.1 nights. In 2004, despite the adverse effects of 'Hurricane Ivan', stop-over tourist arrivals increased by 4.8%, to 1,414,786, while cruise ship arrivals fell by 2.9%. Tourism receipts in 2003 amounted to US $1,621m.

The Jamaican tourism industry has frequently faced problems related to global recession, adverse publicity and competition from other Caribbean and long-haul destinations. Harassment of tourists has been a long-standing problem and caused the withdrawal of some cruise-ship companies in 2001. Moreover, 2001 was a particularly problematic year, with widely reported violence in Kingston, the impact of the 11 September terrorist attacks in the USA and a decline in the US economy, upon which Jamaica is highly dependent (some 70% of tourists came from the USA in 2004).

## INVESTMENT AND FINANCE

From the 1970s successive Governments have been confronted with fiscal deficits, caused in large part by overspending on state-sector salaries and inadequate tax collection. The PNP Government led by Michael Manley substantially increased government spending on welfare, job creation schemes and infrastructure, and was obliged to seek the financial support of the IMF. The austerity programme proposed by the IMF, including wide-scale redundancies, reductions in welfare spending and deregulation of the state sector, proved politically unacceptable to the PNP Government. The succeeding JLP Government reopened negotiations with the IMF, securing a three-year agreement with balance-of-payments assistance worth some US $650m. However, political pressures forced the JLP Government briefly to abandon its IMF-approved austerity policies in the wake of social unrest in 1985 and to pursue more expansionary policies. A further series of arrangements followed, and the PNP Government maintained relations with the IMF following its return to power in 1989. IMF balance-of-payments assistance continued until 1995, with a three-year facility of US $153m. In 2000 the Government announced that it would enter into a Staff Monitored Programme (SMP) with the IMF. This arrangement, while involving neither disbursements from the IMF nor any politically sensitive conditions, included policy advice, the setting of targets and access to loans from other multilateral agencies such as the IDB and World Bank.

Despite the intervention of the IMF since the 1980s, successive Governments have struggled with fiscal deficits and mounting debt. In 2001/02 the fiscal deficit was calculated at J $19,241m. The deficit widened considerably in the following year to J $35,840m., before declining to J $28,259 in 2003/04. Government targets to reduce the fiscal deficit in 2004/05 were damaged by the negative impact of 'Hurricane Ivan', and the estimated deficit for that year stood at 4.5% of GDP. The 2005/06 budget aimed to eliminate the deficit through increased indirect taxation and public-sector pay restraint.

As part of the IMF plan to liberalize the Jamaican economy, the Jamaican dollar was devalued in 1989, having stood at J $5.50 to the US dollar since 1985. After a further devaluation in 1990, the Jamaican dollar was floated that same year, becoming freely convertible as part of a series of measures aimed at deregulating the economy and making Jamaican exports more competitive. Commercial banks were also allowed to trade in foreign currency. In September 1991 the complete abolition of exchange controls was announced, and by December the exchange rate had declined to J $21.49 to the US dollar. The rate continued to fall throughout the 1990s, reaching J $39.62 to the US dollar at the end of 1995. A temporary strengthening of the currency occurred in 1996, owing to an increased supply of 'hard' currency, but the rate declined from 1998 to 2002, reaching J $47.29 to the US dollar at the close of the later year. The intervention of the Bank of Jamaica in mid-2003 could not halt the currency's further decline, to J $60.52 to the US dollar by the end of that year. Nevertheless, at the close of 2004 the exchange rate still stood at J $61.45 to the US dollar and the value of the Jamaican dollar remained relatively steady in the period up to April 2005, as international reserves increased owing to stronger exports.

In 1997 the Government was obliged to intervene as Jamaica's seven largest commercial banks faced insolvency, creating a Financial Sector Adjustment Co (FINSAC), which compensated depositors in the failed companies. By late 2000 FINSAC was estimated to have left the Government with debts of J $30,000m. In early 2002, however, the Government announced that it had reached an agreement with the Beal Bank of Texas, whereby Beal would collect monies owed on FINSAC's bad debt portfolio, paying the Government a sliding share of funds recovered. The sale of one of the country's two largest banks, the National Commercial Bank, to a Canadian fund management company, AIC Group, was announced in March. Later that year the Government announced that it intended to close down FINSAC; however, a number of legal suits concerning unpaid debts and liabilities remained unresolved.

Low tax-collection rates have contributed to Jamaica's fiscal deficits, with only an estimated 45% of property taxes collected in 2001, and high levels of income-tax evasion. Only 10% of direct taxes owed by self-employed workers were paid in 2004, according to the Andrew Young School of Policy Studies, while the informal sector was extremely difficult to tax. The new tax regime introduced in the 2003/04 budget increased revenue by 30% over the preceding financial year, to J $119,100m., but, nevertheless, was 2.6% below the target set for the period in question. The 2004/05 budget saw a significant shift towards indirect taxation in an attempt to narrow the fiscal deficit.

The recessionary climate from 1997 enabled the Government to achieve one key target: a reduction in inflation. In 1996 inflation was recorded at 26.4%, declining to 6.0% in 1999. After rising to 8.2% in 2000, largely owing to increased oil prices, inflation fell again to 7.0% in 2001 and 7.1% in 2002. In 2003, however, inflation rose to 10.3%, owing largely to the depreciation in the Jamaican dollar and increased taxation on consumer goods. Inflation rose in 2004, to 13.6%. Interest rates have also followed a generally downward trend, with the commercial bank lending rate falling from 44.2% in 1996 to 18.3% in 2004.

Jamaica has been forced to depend on external financial support since the 1970s. In 2001 grants of J $3,236m. were recorded, falling to J $600m. in 2002/03 and J $400m. in 2003/04. Substantial assistance is provided by the World Bank, the IDB, the Barbados-based Caribbean Development Bank, the EU and bilateral donors such as Kuwait. In May 2005 the World Bank announced plans for a US $150m. programme of poverty alleviation and development initiatives scheduled to take place between 2006 and 2009. Jamaica's external debt was estimated to have reached US $6,200m. in early 2005, with a debt-service ratio of 16.5% of GDP. Interest repayments on foreign loans rose from J $14,700m. in 2002/03 to J $16,100m. in 2003/04. The Government successfully issued a US $300m. sovereign bond in May 2005, reflecting growing confidence in Jamaica's economic prospects. Of greater concern, however, was the Government's internal debt, estimated at J $447,000m. at the end of 2004 and subject to high domestic interest rates.

## FOREIGN TRADE AND BALANCE OF PAYMENTS

Jamaica's foreign-trade balance has been in deficit for three decades, but between 1992 and 1996 the deficit was offset by revenues from tourism, remittances from Jamaicans overseas and capital inflows attracted by high interest rates. In 2001 the merchandise trade deficit stood at US $1,618.2m., as declining exports of alumina, sugar and bananas were compounded by a 2.3% rise in imports, largely owing to capital goods such as telecommunications equipment. The trade deficit in 2002 and 2003 rose still higher, to US $1,870.5m. and US $1,943.8m., respectively. In 2004 exports were estimated at US $1,635.8m., reflecting improved performance in the mining sector, while imports were valued at US $3,519.7m., resulting in a deficit of US $1,883.9m.

The deficit on the current account of the balance of payments more than doubled in 2001, rising from US $367.4m. in 2000 to US $758.8m. in 2001, then to US $1,074.2m. in 2002, before falling in 2003 to US $761.4m. Positive capital movements offset this deficit, with official capital inflows rising to US $511m. in 2004, owing to successful bond issues and increased borrowing, while private capital inflows increased by 34.9%, to US $786m., owing to investment in telecommunications, mining and tourism. Remittance payments from Jamaicans overseas were solid in 2004, accounting for much of the US $1,178.1m. recorded under current transfers. At the beginning of 2005 net international reserves (excluding gold) stood at US $1,858m., compared with US $1,195m. at the end of 2003.

## CONCLUSION

While recording modest GDP growth rates in 2001–04, the Jamaican economy was still below 1990 levels and remained highly vulnerable to external shocks, such as decreasing tourism receipts and low commodity prices. While inflation was still under control and interest rates much lower than in the 1990s, this was achieved at the cost of stagnation in most productive sectors and a continuing fiscal deficit. A growing deficit, financed by increased borrowing on international capital markets and from domestic sources, was likely to increase an already high level of indebtedness, which would become critical in the event of international interest rate rises. Manufacturing remained uncompetitive in the face of cheaper production in Asia, and agriculture, though undergoing a recovery in 2003–04, was threatened by falling world prices and adverse weather conditions. Tourism, by contrast, experienced a revival, while increased investment in the bauxite/alumina sector signalled a projected increase in production and export earnings. The country faced the challenges of increased trade liberalization, as well as the gradual removal of trade preferences for sugar and bananas currently offered by the EU. In the short term, Jamaica's prospects for reversing a long period of recession would depend on the recovery of the US economy, improved tourism conditions and avoiding a renewed debt crisis.

# Statistical Survey

Sources (unless otherwise stated): Statistical Institute of Jamaica, 7 Cecelio Ave, Kingston 10; tel. 926-5311; fax 926-1138; e-mail info@statinja.com; internet www.statinja.com; Jamaica Information Service, 58a Half Way Tree Rd, POB 2222, Kingston 10; tel. 926-3740; fax 926-6715; e-mail jis@jis.gov.jm; internet www.jis.gov.jm; Bank of Jamaica, Nethersole Pl., POB 621, Kingston; tel. 922-0752; fax 922-0854; e-mail info@boj.org.jm; internet www.boj.org.jm.

## Area and Population

### AREA, POPULATION AND DENSITY

| | |
|---|---:|
| Area (sq km) | 10,991* |
| Population (census results) | |
| 7 April 1991 | 2,314,479 |
| 10 September 2001 | |
| Males | 1,283,547 |
| Females | 1,324,085 |
| Total | 2,607,632 |
| Population (official estimates at 31 December) | |
| 2001 | 2,612,500 |
| 2002 | 2,624,700 |
| 2003 | 2,641,600 |
| Density (per sq km) at 31 December 2003 | 240.3 |

* 4,243.6 sq miles.

### PARISHES

| | Area (sq km) | Population (census of 2001) | Parish capitals (with population at 1991 census) |
|---|---:|---:|---|
| Kingston | 22 | 651,880 | Kingston M.A. (587,798) |
| St Andrew | 431 | | |
| St Thomas | 743 | 91,604 | Morant Bay (9,185) |
| Portland | 814 | 80,205 | Port Antonio (13,246) |
| St Mary | 611 | 111,466 | Port Maria (7,651) |
| St Ann | 1,213 | 166,762 | St Ann's Bay (10,518) |
| Trelawny | 875 | 73,066 | Falmouth (7,245) |
| St James | 595 | 175,127 | Montego Bay (83,446) |
| Hanover | 450 | 67,037 | Lucea (6,002) |
| Westmoreland | 807 | 138,947 | Savanna La Mar (16,553) |
| St Elizabeth | 1,212 | 146,404 | Black River (3,675) |
| Manchester | 830 | 185,801 | Mandeville (39,430) |
| Clarendon | 1,196 | 237,024 | May Pen (46,785) |
| St Catherine | 1,192 | 482,308 | Spanish Town (92,383) |
| **Total** | **10,991** | **2,607,632** | — |

### PRINCIPAL TOWNS
(population at census of 7 April 1991)

| | | | | |
|---|---:|---|---|---:|
| Kingston (capital) | 587,798 | | Montego Bay | 83,446 |
| Spanish Town | 92,383 | | May Pen | 46,785 |
| Portmore | 90,138 | | Mandeville | 39,430 |

Source: Thomas Brinkhoff, *City Population* (internet www.citypopulation.de).

Mid-2003 (UN estimate, incl. suburbs): Kingston 575,105 (Source: UN, *World Urbanization Prospects: The 2003 Revision*).

## BIRTHS, MARRIAGES AND DEATHS*

| | Registered live births | | Registered marriages | | Registered deaths | |
|---|---:|---:|---:|---:|---:|---:|
| | Number | Rate (per 1,000) | Number | Rate (per 1,000) | Number | Rate (per 1,000) |
| 1992 | 56,276 | 23.5 | 13,042 | 5.6 | 13,225 | 5.5 |
| 1993 | 58,627 | 24.0 | 14,352 | 5.9 | 13,927 | 5.7 |
| 1994 | 57,404 | 23.2 | 15,171 | 6.1 | 13,503 | 5.5 |
| 1995 | 57,607 | 23.0 | 16,515 | 6.6 | 12,776 | 5.1 |
| 1996 | 59,194 | 23.4 | 19,170 | 7.6 | 16,926 | 6.7 |
| 1997 | 59,385 | 23.3 | 21,502 | 8.4 | 15,087 | 5.9 |
| 1998 | 56,937 | 22.1 | 24,131 | 9.4 | 15,967 | 6.2 |
| 1999 | 56,911 | 22.0 | 26,871 | 10.4 | 17,353 | 6.7 |

**2000:** Birth rate 20.9 per 1,000; Death rate 6.3 per 1,000.

**2001:** Birth rate 20.6 per 1,000; Death rate 6.2 per 1,000.

**2002:** Birth rate 20.0 per 1,000; Death rate 6.5 per 1,000.

**2003:** Birth rate 19.3 per 1,000; Death rate 6.4 per 1,000.

* Data are tabulated by year of registration rather than by year of occurrence, and 1992–99 figures have not been revised to take account of 2001 census results.

Source: partly UN, *Demographic Yearbook*.

**Expectation of life** (WHO estimates, years at birth): 73 (males 71; females 74) in 2003 (Source: WHO, *World Health Report*).

### ECONOMICALLY ACTIVE POPULATION
('000 persons aged 14 years and over)

| | 2000 | 2001 | 2002 |
|---|---:|---:|---:|
| Agriculture, forestry and fishing | 194.5 | 195.2 | 182.4 |
| Mining and quarrying | 4.7 | 5.2 | 3.9 |
| Manufacturing | 65.8 | 69.1 | 64.6 |
| Electricity, gas and water | 5.6 | 6.4 | 6.3 |
| Construction | 82.4 | 80.5 | 90.4 |
| Trade, restaurants and hotels | 204.8 | 214.7 | 209.0 |
| Transport, storage and communications | 57.6 | 60.6 | 62.9 |
| Financing, insurance, real estate and business services | 55.9 | 46.4 | 60.4 |
| Community, social and personal services | 261.4 | 261.9 | 259.1 |
| Activities not adequately defined | 2.9 | 2.3 | 3.3 |
| **Total employed** | **935.6** | **942.3** | **942.3** |
| Males | 552.5 | 556.6 | 546.3 |
| Females | 383.1 | 385.7 | 396.0 |

Source: ILO.

## Health and Welfare

### KEY INDICATORS

| | |
|---|---:|
| Total fertility rate (children per woman, 2003) | 2.3 |
| Under-5 mortality rate (per 1,000 live births, 2003) | 20 |
| HIV/AIDS (% of persons aged 15–49, 2003) | 1.2 |
| Physicians (per 1,000 head, 1996) | 1.40 |
| Hospital beds (per 1,000 head, 1996) | 2.12 |
| Health expenditure (2002): US $ per head (PPP) | 234 |
| Health expenditure (2002): % of GDP | 6.0 |
| Health expenditure (2002): public (% of total) | 57.4 |
| Access to water (% of persons, 2002) | 93 |
| Access to sanitation (% of persons, 2002) | 80 |
| Human Development Index (2002): ranking | 79 |
| Human Development Index (2002): value | 0.764 |

For sources and definitions, see explanatory note on p. vi.

# JAMAICA

## Agriculture

**PRINCIPAL CROPS**
('000 metric tons)

|  | 2001 | 2002 | 2003 |
|---|---|---|---|
| Sweet potatoes | 25 | 20 | 25 |
| Yams | 158 | 148 | 151 |
| Other roots and tubers | 42 | 40* | 42* |
| Sugar cane | 2,231 | 1,966 | 1,776 |
| Coconuts | 170* | 170* | 170 |
| Cabbages | 22 | 19 | 29 |
| Tomatoes | 24 | 19 | 19* |
| Pumpkins, squash and gourds | 36 | 36* | 36* |
| Carrots | 20 | 19 | 19* |
| Other vegetables (incl. melons)* | 94 | 92 | 93 |
| Bananas* | 130 | 130 | 130 |
| Plantains | 25 | 21 | 20 |
| Oranges* | 140 | 140 | 140 |
| Lemons and limes* | 24 | 24 | 24 |
| Grapefruit and pomelos* | 42 | 42 | 42 |
| Pineapples | 20 | 21 | 21* |
| Other fruit* | 90 | 90 | 90 |
| Pimento, allspice* | 10 | 10 | 10 |

* FAO estimate(s).
Source: FAO.

**LIVESTOCK**
(FAO estimates, '000 head, year ending September)

|  | 2001 | 2002 | 2003 |
|---|---|---|---|
| Horses | 4 | 4 | 4 |
| Mules | 10 | 10 | 10 |
| Asses | 23 | 23 | 23 |
| Cattle | 400 | 400 | 430 |
| Pigs | 180 | 150 | 180 |
| Sheep | 1 | 1 | 1 |
| Goats | 440 | 440 | 440 |
| Poultry | 13,000 | 13,000 | 11,000 |

Source: FAO.

**LIVESTOCK PRODUCTS**
('000 metric tons)

|  | 2001 | 2002 | 2003 |
|---|---|---|---|
| Beef and veal | 13.4 | 14.3 | 14.5* |
| Goat meat* | 1.7 | 1.6 | 1.6 |
| Pig meat | 6.4 | 5.5 | 5.5* |
| Poultry meat | 83.0† | 83.8 | 81.3 |
| Cows' milk* | 28.5 | 28.5 | 28.5 |
| Hen eggs | 7.9* | 7.9* | 5.8 |
| Honey* | 1 | 1 | 1 |
| Cattle hides* | 1.3 | 1.4 | 1.4 |

* FAO estimate(s).
† Unofficial figure.
Source: FAO.

## Forestry

**ROUNDWOOD REMOVALS**
(FAO estimates, '000 cubic metres, excl. bark)

|  | 2001 | 2002 | 2003 |
|---|---|---|---|
| Sawlogs, veneer logs and logs for sleepers | 132 | 132 | 132 |
| Other industrial wood | 150 | 150 | 150 |
| Fuel wood | 591 | 584 | 577 |
| **Total** | 874 | 867 | 860 |

Source: FAO.

**SAWNWOOD PRODUCTION**
('000 cubic metres, incl. railway sleepers)

|  | 1996 | 1997 | 1998 |
|---|---|---|---|
| Coniferous (softwood) | 3 | 3 | 3 |
| Broadleaved (hardwood) | 61 | 62 | 63 |
| **Total** | 64 | 65 | 66 |

**1999–2003:** Annual production as in 1998 (FAO estimates).
Source: FAO.

## Fishing

('000 metric tons, live weight)

|  | 2001* | 2002 | 2003 |
|---|---|---|---|
| Capture | 6.6 | 7.7* | 8.7* |
| Marine fishes | 5.6 | 6.6* | 7.6* |
| Caribbean spiny lobster | 0.5 | 0.5* | 0.5* |
| Aquaculture | 4.5 | 6.2 | 3.0 |
| Nile tilapia | 4.5 | 6.0 | 2.5 |
| **Total catch*** | 11.2 | 13.9 | 11.7 |

* FAO estimate(s).
Source: FAO.

## Mining

('000 metric tons)

|  | 2001 | 2002 | 2003 |
|---|---|---|---|
| Bauxite* | 12,139 | 13,139 | 13,443 |
| Alumina | 3,542 | 3,630 | 3,844 |
| Crude gypsum | 320 | 165 | 162 |
| Salt | 19.1 | 19.0† | n.a. |

* Dried equivalent of crude ore.
† Estimate.
Source: partly US Geological Survey.

## Industry

**SELECTED PRODUCTS**

|  | 2001 | 2002 | 2003 |
|---|---|---|---|
| Sugar (metric tons) | 205,128 | 174,949 | 153,542 |
| Molasses (metric tons) | 86,983 | 79,653 | 72,631 |
| Rum ('000 litres) | 23,705 | 24,727 | 25,519 |
| Beer and stout ('000 litres) | 78,566 | 77,551 | 75,189 |
| Fuel oil ('000 litres) | 576,736 | 671,826 | 511,900 |
| Gasoline (petrol) ('000 litres) | 181,045 | 271,215 | 130,229 |
| Kerosene, turbo and jet fuel ('000 litres) | 94,105 | 122,488 | 81,902 |
| Auto diesel oil ('000 litres) | 231,451 | 302,497 | 172,340 |
| Cement ('000 metric tons) | 596,248 | 621,831 | 607,682 |
| Concrete ('000 cu metres) | 86,820 | 140,314 | 176,815 |

**Electrical energy** (million kWh): 6,656 in 2001 (Source: UN, *Industrial Commodity Statistics Yearbook*).

## Finance

**CURRENCY AND EXCHANGE RATES**

**Monetary Units**
100 cents = 1 Jamaican dollar (J $).

**Sterling, US Dollar and Euro Equivalents** (31 May 2005)
£1 sterling = J $111.955;
US $1 = J $61.578;
€1 = J $75.932;
J $1,000 = £8.93 = US $16.24 = €13.17.

**Average Exchange Rate** (J $ per US $)
2002   48.416
2003   57.741
2004   61.197

# JAMAICA

*Statistical Survey*

## BUDGET*
(J $ million, year ending 31 March)

| Revenue† | 1999 | 2000 | 2001 |
|---|---|---|---|
| Tax revenue | 74,634 | 84,659 | 90,985 |
|   Taxes on income, profits and capital gains | 32,185 | 35,456 | 35,495 |
|     Individual | 16,681 | 16,386 | 19,750 |
|     Corporate | 9,051 | 8,825 | 6,680 |
|   Domestic taxes on goods and services | 29,451 | 34,285 | 38,561 |
|     General sales, turnover or value-added taxes | 26,392 | 30,717 | 34,625 |
|   Taxes on international trade | 6,919 | 7,503 | 8,501 |
| Other current revenue | 22,777 | 29,260 | 27,423 |
|   Entrepreneurial and property income | 8,626 | 9,875 | 9,954 |
|   Administrative fees, non-industrial and incidental sales | 6,606 | 8,260 | 9,365 |
|   Fines and forfeits | 2,590 | 2,891 | 2,808 |
| Capital revenue | 601 | 5,063 | 4,595 |
| **Total** | 98,012 | 118,982 | 123,004 |

| Expenditure | 1999 | 2000 | 2001 |
|---|---|---|---|
| General public services | 11,164 | 26,992 | 13,698 |
| Defence | 1,802 | 1,896 | 2,212 |
| Public order and safety | 8,143 | 8,677 | 10,693 |
| Education | 17,470 | 4,877 | 21,628 |
| Health | 7,048 | 3,743 | 8,222 |
| Social security and welfare | 1,182 | 1,697 | 1,295 |
| Housing and community amenities | 2,565 | 2,572 | 2,741 |
| Recreational, cultural and religious affairs and services | 167 | 260 | 276 |
| Economic affairs and services | 10,689 | 11,969 | 11,258 |
|   Agriculture, forestry, fishing and hunting | 2,327 | 2,048 | 2,291 |
|   Transport and communications | 3,383 | 3,576 | 3,577 |
| Other purposes | 52,278 | 60,245 | 63,500 |
| **Total expenditure** | 112,506 | 122,929 | 135,523 |
| Current‡ | 97,945 | 105,461 | 120,192 |
| Capital | 14,561 | 17,468 | 15,331 |

* Figures refer to consolidated accounts of the central Government.
† Excluding grants received (J $ million): 797 in 1999; 3,236 in 2001.
‡ Including interest payments (J $ million): 41,911 in 1999; 43,335 in 2000; 51,550 in 2001.

Source: IMF, *Government Finance Statistics Yearbook*.

**Budget** (J $ million, 2004/05): *Revenue:* Tax revenue 150,481.6; Non-tax revenue 9,824.5; Capital development fund transfers (incl. bauxite levy) 2,479.1; Capital revenue 4,533.8; Total 167,319.0 (excl. grants). *Expenditure:* Recurrent expenditure 188,382.0 (Personnel emoluments 63,516.8, Programmes 32,081.0; Interest 92,784.2); Capital expenditure and net lending 11,105.9; Total 199,487.9.

## INTERNATIONAL RESERVES
(US $ million at 31 December, excluding gold)

| | 2002 | 2003 | 2004 |
|---|---|---|---|
| IMF special drawing rights | 0.9 | 0.1 | 0.1 |
| Foreign exchange | 1,644.2 | 1,194.8 | 1,846.4 |
| **Total** | 1,645.1 | 1,194.9 | 1,846.5 |

Source: IMF, *International Financial Statistics*.

## MONEY SUPPLY
(J $ million at 31 December)

| | 2002 | 2003 | 2004 |
|---|---|---|---|
| Currency outside banks | 20,399 | 23,186 | 26,683 |
| Demand deposits at commercial banks | 39,020 | 39,871 | 48,010 |
| **Total money** | 59,419 | 63,057 | 74,693 |

Source: IMF, *International Financial Statistics*.

## COST OF LIVING
(Consumer Price Index; base: 2000 = 100)

| | 2001 | 2002 | 2003 |
|---|---|---|---|
| Food (incl. beverages) | 103.4 | 109.7 | 120.2 |
| Fuel and power | 110.7 | 114.9 | 129.9 |
| Clothing and footwear | 104.2 | 107.2 | 112.9 |
| Rent | 129.9 | 145.5 | 157.1 |
| **All items** (incl. others) | 107.0 | 114.6 | 126.4 |

**2004:** All items 143.6.
Source: ILO.

## NATIONAL ACCOUNTS
(J $ million at current prices)

**Expenditure on the Gross Domestic Product**

| | 2001 | 2002 | 2003 |
|---|---|---|---|
| Government final consumption expenditure | 59,146 | 66,734 | 71,015 |
| Private final consumption expenditure | 266,035 | 293,869 | 345,412 |
| Increase in stocks | 408 | 664 | 769 |
| Gross fixed capital formation | 108,323 | 129,497 | 140,427 |
| **Total domestic expenditure** | 433,913 | 490,764 | 557,623 |
| Exports of goods and services | 144,614 | 147,948 | 191,473 |
| *Less* Imports of goods and services | 205,484 | 229,947 | 278,656 |
| **GDP in purchasers' values** | 373,042 | 408,765 | 470,440 |
| **GDP at constant 1996 prices** | 227,070 | 229,536 | 234,730 |

Source: IMF, *International Financial Statistics*.

**Gross Domestic Product by Economic Activity**

| | 2001 | 2002 | 2003 |
|---|---|---|---|
| Agriculture, forestry and fishing | 22,892.1 | 22,678.9 | 24,250.5 |
| Mining and quarrying | 14,820.1 | 15,689.5 | 20,153.6 |
| Manufacturing | 48,250.0 | 51,279.2 | 59,105.4 |
| Electricity and water | 12,489.9 | 13,577.6 | 17,191.9 |
| Construction | 35,013.7 | 38,590.2 | 43,766.0 |
| Wholesale and retail trade | 72,426.8 | 78,605.2 | 88,496.3 |
| Transport, storage and communication | 43,408.2 | 51,093.7 | 56,539.3 |
| Finance and insurance services | 23,275.6 | 22,177.2 | 35,501.7 |
| Real estate and business services | 21,529.6 | 24,292.0 | 27,200.3 |
| Producers of government services | 43,694.8 | 49,593.3 | 54,643.4 |
| Household and private non-profit services | 2,100.2 | 2,261.5 | 2,326.3 |
| Other services | 24,936.5 | 26,510.5 | 31,289.2 |
| **Sub-total** | 364,837.4 | 396,348.9 | 460,463.9 |
| Value-added tax | 23,337.3 | 26,590.3 | 33,866.7 |
| *Less* Imputed bank service charge | 15,132.2 | 14,173.9 | 23,890.4 |
| **GDP in purchasers' values** | 373,042.6 | 408,765.3 | 470,440.2 |

# JAMAICA

## BALANCE OF PAYMENTS
(US $ million)

|  | 2001 | 2002 | 2003 |
|---|---|---|---|
| Exports of goods f.o.b. | 1,454.4 | 1,309.1 | 1,385.6 |
| Imports of goods f.o.b. | −3,072.6 | −3,179.6 | −3,329.4 |
| **Trade balance** | −1,618.2 | −1,870.5 | −1,943.8 |
| Exports of services | 1,897.0 | 1,912.2 | 2,131.6 |
| Imports of services | −1,513.9 | −1,597.5 | −1,566.9 |
| **Balance on goods and services** | −1,235.1 | −1,555.8 | −1,379.1 |
| Other income received | 218.2 | 221.0 | 217.6 |
| Other income paid | −656.0 | −826.3 | −789.0 |
| **Balance on goods, services and income** | −1,672.9 | −2,161.1 | −1,950.5 |
| Current transfers received | 1,090.7 | 1,337.9 | 1,523.5 |
| Current transfers paid | −176.6 | −251.0 | −334.4 |
| **Current balance** | −758.8 | −1,074.2 | −761.4 |
| Capital account (net) | −22.3 | −16.9 | 0.1 |
| Direct investment abroad | −89.0 | −73.9 | −116.3 |
| Direct investment from abroad | 613.9 | 481.1 | 720.7 |
| Portfolio investment assets | −39.3 | −351.3 | −1,105.2 |
| Portfolio investment liabilities | 69.7 | 155.8 | 819.6 |
| Other investment assets | −215.5 | −164.9 | −308.8 |
| Other investment liabilities | 1,320.7 | 859.6 | 302.1 |
| Net errors and omissions | −14.4 | −55.1 | 14.1 |
| **Overall balance** | 865.0 | −239.8 | −435.1 |

Source: IMF, *International Financial Statistics*.

## External Trade

### PRINCIPAL COMMODITIES
(US $ million)

| Imports c.i.f. | 2001 | 2002 | 2003 |
|---|---|---|---|
| Foods | 476.8 | 480.1 | 487.4 |
| Mineral fuels and lubricants | 615.5 | 636.8 | 830.1 |
| Chemicals | 383.5 | 375.4 | 444.1 |
| Manufactured goods | 467.0 | 459.5 | 486.4 |
| Machinery and transport equipment | 883.3 | 1,036.9 | 878.9 |
| Miscellaneous manufactured articles | 394.5 | 399.4 | 384.0 |
| **Total** (incl. others) | 3,402.6 | 3,570.5 | 3,678.9 |

| Exports f.o.b. | 2001 | 2002 | 2003 |
|---|---|---|---|
| Foods | 224.9 | 206.3 | 219.6 |
| Beverages and tobacco | 48.4 | 56.2 | 53.9 |
| Crude materials (excl. fuels) | 742.7 | 716.1 | 785.6 |
| Chemicals | 68.5 | 59.4 | 57.7 |
| Miscellaneous manufactured articles | 98.1 | 30.4 | 22.3 |
| **Total** (incl. others) | 1,223.1 | 1,117.3 | 1,196.0 |

### PRINCIPAL TRADING PARTNERS
(US $ million)

| Imports c.i.f. | 2001 | 2002 | 2003 |
|---|---|---|---|
| Canada | 97.6 | 112.6 | 97.9 |
| CARICOM* | 433.0 | 398.6 | 469.3 |
| Latin America | 361.0 | 447.8 | 388.8 |
| United Kingdom | 102.5 | 93.1 | 149.6 |
| Other European Union | 214.3 | 281.5 | 235.2 |
| USA | 1,525.8 | 1,546.6 | 1,632.0 |
| **Total** (incl. others) | 3,402.6 | 3,570.5 | 3,678.9 |

| Exports f.o.b. | 2001 | 2002 | 2003 |
|---|---|---|---|
| Canada | 191.1 | 157.1 | 192.1 |
| CARICOM* | 50.6 | 48.7 | 50.9 |
| Latin America | 12.0 | 14.7 | 6.8 |
| United Kingdom | 157.1 | 134.3 | 153.2 |
| Other European Union | 202.7 | 213.9 | 205.2 |
| Norway | 91.6 | 93.5 | 44.4 |
| USA | 380.2 | 313.4 | 344.4 |
| **Total** (incl. others) | 1,223.1 | 1,117.3 | 1,196.0 |

* Caribbean Community and Common Market.

## Transport

### RAILWAYS
(traffic)

|  | 1988 | 1989 | 1990 |
|---|---|---|---|
| Passenger-km ('000) | 36,146 | 37,995 | n.a. |
| Freight ton-km ('000) | 115,076 | 28,609 | 1,931 |

Source: Jamaica Railway Corporation.

### ROAD TRAFFIC
('000 motor vehicles in use)

|  | 1999 | 2000 | 2001 |
|---|---|---|---|
| Passenger cars | 140.4 | 129.4 | 129.4 |
| Commercial vehicles | 56.6 | 57.6 | 65.2 |

Source: UN, *Statistical Yearbook*.

### SHIPPING
**Merchant Fleet**
(registered at 31 December)

|  | 2002 | 2003 | 2004 |
|---|---|---|---|
| Number of vessels | 11 | 16 | 23 |
| Total displacement ('000 grt) | 75.4 | 56.8 | 131.2 |

Source: Lloyd's Register-Fairplay, *World Fleet Statistics*.

**International Sea-borne Freight Traffic**
(estimates, '000 metric tons)

|  | 1989 | 1990 | 1991 |
|---|---|---|---|
| Goods loaded | 7,711 | 8,354 | 8,802 |
| Goods unloaded | 5,167 | 5,380 | 5,285 |

Source: Port Authority of Jamaica.

### CIVIL AVIATION
(traffic on scheduled services)

|  | 1999 | 2000 | 2001 |
|---|---|---|---|
| Kilometres flown (million) | 35 | 32 | 46 |
| Passengers carried ('000) | 1,670 | 1,922 | 1,946 |
| Passenger-km (million) | 3,495 | 4,087 | 4,412 |
| Total ton-km (million) | 377 | 400 | 471 |

Source: UN, *Statistical Yearbook*.

## Tourism

### VISITOR ARRIVALS BY COUNTRY OF ORIGIN

|  | 2002 | 2003 | 2004 |
|---|---|---|---|
| Canada | 89,570 | 87,908 | 105,623 |
| Europe | 173,412 | 211,011 | 241,925 |
| USA | 859,347 | 904,666 | 996,131 |
| **Total** (incl. others) | 1,266,366 | 1,350,284 | 1,414,786 |

**Visitor expenditure** (US $ million, incl. passenger transport): 1,494 in 2001; 1,482 in 2002; 1,621 in 2003.

Sources: partly World Tourism Organization.

## Communications Media

|  | 2000 | 2001 | 2002 |
|---|---|---|---|
| Television receivers ('000 in use) | 500 | 510 | n.a. |
| Telephones ('000 main lines in use) | 429 | 376 | 367 |
| Mobile cellular telephones ('000 subscribers) | 367 | 700 | 1,400 |
| Personal computers ('000 in use) | 110* | 130* | 141 |
| Internet users ('000) | 60* | 100 | 600 |

**Radio receivers** ('000 in use): 1,215 in 1997.

**Facsimile machines** (number in use): 1,567 in 1992†.

**Daily newspapers:** 3 in 1996 (circulation 158,000).

* Estimate.
† Year ending 1 April.

Sources: International Telecommunication Union; UN, *Statistical Yearbook*; UNESCO, *Statistical Yearbook*.

## Education

(2003/04)

|  | Institutions* | Teachers | Students |
|---|---|---|---|
| Pre-primary | 2,137† | 6,021 | 123,520 |
| Primary | 355 | 10,500 | 304,771 |
| Secondary | 161 | 11,900 | 236,949 |
| Tertiary | 15 | 1,051 | 11,600 |

* Excludes 349 all-age schools and 88 primary and junior high schools.
† Includes 2,008 community-operated basic schools.

Source: Ministry of Education, Youth and Culture.

**Adult literacy rate** (UNESCO estimates): 87.6% (males 83.9%; females 91.4%) in 2002 (Source: UN Development Programme, *Human Development Report*).

# Directory

## The Constitution

The Constitution came into force at the independence of Jamaica on 6 August 1962. Amendments to the Constitution are enacted by Parliament, but certain entrenched provisions require ratification by a two-thirds' majority in both chambers of the legislature, and some (such as a change of the head of state) require the additional approval of a national referendum.

### HEAD OF STATE

The Head of State is the British monarch, who is locally represented by a Governor-General, appointed by the British monarch, on the recommendation of the Jamaican Prime Minister in consultation with the Leader of the Opposition party.

### THE LEGISLATURE

The Senate or Upper House consists of 21 Senators, of whom 13 will be appointed by the Governor-General on the advice of the Prime Minister and eight by the Governor-General on the advice of the Leader of the Opposition. (Legislation enacted in 1984 provided for eight independent Senators to be appointed, after consultations with the Prime Minister, in the eventuality of there being no Leader of the Opposition.)

The House of Representatives or Lower House consists of 60 elected members called Members of Parliament.

A person is qualified for appointment to the Senate or for election to the House of Representatives if he or she is a citizen of Jamaica or other Commonwealth country, of the age of 21 or more and has been ordinarily resident in Jamaica for the immediately preceding 12 months.

### THE PRIVY COUNCIL

The Privy Council consists of six members appointed by the Governor-General after consultation with the Prime Minister, of whom at least two are persons who hold or who have held public office. The functions of the Council are to advise the Governor-General on the exercise of the Prerogative of Mercy and on appeals on disciplinary matters from the three Service Commissions.

### THE EXECUTIVE

The Prime Minister is appointed from the House of Representatives by the Governor-General, and is the leader of the party that holds the majority of seats in the House of Representatives. The Leader of the party is voted in by the members of that party. The Leader of the Opposition is voted in by the members of the Opposition party.

The Cabinet consists of the Prime Minister and not fewer than 11 other ministers, not more than four of whom may sit in the Senate. The members of the Cabinet are appointed by the Governor-General on the advice of the Prime Minister.

### THE JUDICATURE

The Judicature consists of a Supreme Court, a Court of Appeal and minor courts. Judicial matters, notably advice to the Governor-General on appointments, are considered by a Judicial Service Commission, the Chairman of which is the Chief Justice, members being the President of the Court of Appeal, the Chairman of the Public Service Commission and three others.

### CITIZENSHIP

All persons born in Jamaica after independence automatically acquire Jamaican citizenship and there is also provision for the acquisition of citizenship by persons born outside Jamaica of Jamaican parents. Persons born in Jamaica (or persons born outside Jamaica of Jamaican parents) before independence who immediately prior to independence were citizens of the United Kingdom and colonies also automatically become citizens of Jamaica.

Appropriate provision is made which permits persons who do not automatically become citizens of Jamaica to be registered as such.

### FUNDAMENTAL RIGHTS AND FREEDOMS

The Constitution includes provisions safeguarding the fundamental freedoms of the individual, irrespective of race, place of origin, political opinions, colour, creed or sex, subject only to respect for the rights and freedoms of others and for the public interest. The fundamental freedoms include the rights of life, liberty, security of the person and protection from arbitrary arrest or restriction of movement, the enjoyment of property and the protection of the law, freedom of conscience, of expression and of peaceful assembly and association, and respect for private and family life.

## The Government

### HEAD OF STATE

**Monarch:** HM Queen ELIZABETH II (succeeded to the throne 6 February 1952).

**Governor-General:** Sir HOWARD FELIX HANLAN COOKE (appointed 1 August 1991).

### PRIVY COUNCIL OF JAMAICA

KENNETH SMITH, DONALD MILLS, DENNIS LALOR, JAMES KERR, ELSA LEO RHYNIE, HEADLEY CUNNINGHAM.

### CABINET
(July 2005)

**Prime Minister and Minister of Defence:** PERCIVAL JAMES PATTERSON.

**Minister of Finance and Planning:** Dr OMAR DAVIES.

**Minister of Labour and Social Security:** HORACE DALLEY.

**Minister of Industry and Tourism:** ALOUN N'DOMBET-ASSAMBA.

**Minister of Local Government, Community Development and Sport:** PORTIA SIMPSON-MILLER.

**Minister of National Security and Leader of Government Business in the House of Representatives:** Dr PETER PHILLIPS.

**Minister of Justice and Attorney-General:** ARNOLD J. NICHOLSON.

**Minister of Agriculture:** ROGER CLARKE.

JAMAICA                                                                                                                                    Directory

Minister of Foreign Affairs and Foreign Trade: KEITH (K. D.) KNIGHT.
Minister of Health: JOHN JUNOR.
Minister of Education, Youth and Culture: MAXINE HENRY-WILSON.
Minister of Transport and Works: ROBERT PICKERSGILL.
Minister of Water and Housing: DONALD BUCHANAN.
Minister of Commerce, Science and Technology: PHILLIP PAULWELL.
Minister of Land and Environment: DEAN PEART.
Minister of Information and Leader of Government Business in the Senate: BURCHELL WHITEMAN.
Minister of Development: PAUL ROBERTSON.

### MINISTRIES

**Office of the Governor-General:** King's House, Hope Rd, Kingston 10; tel. 927-6424; fax 978-6025; e-mail kingshouse@cwjamaica.com.

**Office of the Prime Minister:** Jamaica House, 1 Devon Rd, POB 272, Kingston 6; tel. 927-9941; fax 929-0005; e-mail pmo@opm.gov.jm.

**Ministry of Agriculture:** Hope Gardens, POB 480, Kingston 6; tel. 927-1731; fax 927-1904; e-mail psoffice@moa.gov.jm; internet www.moa.gov.jm.

**Ministry of Commerce, Science and Technology:** PCJ Bldg, 36 Trafalgar Rd, Kingston 10; tel. 929-8990; fax 960-1623; e-mail admin@mct.gov.jm; internet www.mct.gov.jm.

**Ministry of Development:** Cabinet Office, 1 Devon Rd, POB 272, Kingston 6; tel. 927-9941; fax 929-8405; e-mail info@cabinet.gov.jm; internet www.cabinet.gov.jm.

**Ministry of Education, Youth and Culture:** 2A National Heroes Circle, Kingston 4; tel. 922-1400; fax 922-1837; internet www.moec.gov.jm.

**Ministry of Finance and Planning:** 30 National Heroes Circle, Kingston 4; tel. 922-8600; fax 922-7097; e-mail info@mof.gov.jm; internet www.mof.gov.jm.

**Ministry of Foreign Affairs and Foreign Trade:** 21 Dominica Dr., POB 624, Kingston 5; tel. 926-4220; fax 929-6733; e-mail mfaftjam@cwjamaica.com; internet www.mfaft.gov.jm.

**Ministry of Health:** Oceana Hotel Complex, 2–4 King St, Kingston; tel. 967-1092; fax 967-7293; e-mail junorj@moh.gov.jm; internet www.moh.gov.jm.

**Ministry of Industry and Tourism:** 64 Knutsford Blvd, Kingston 5; tel. 920-4924; fax 920-4944; e-mail mts@cwjamaica.com.

**Ministry of Information:** Kingston.

**Ministry of Justice and Attorney-General's Department:** 2 Oxford Rd, Kingston 5; tel. 906-2414; fax 922-5109; internet www.moj.gov.jm.

**Ministry of Labour and Social Security:** 1F North St, POB 10, Kingston; tel. 922-8000; fax 922-6902; e-mail mlss@netcomm-jm.com; internet www.minlab.gov.jm.

**Ministry of Land and Environment:** 16A Halfway Tree Rd, Kingston 5; tel. 920-3273; fax 929-7349; e-mail mehsys@hotmail.com.

**Ministry of Local Government, Community Development and Sport:** 85 Hagley Park Rd, Kingston 10; tel. 754-0994; fax 754-0210; e-mail communications@mlgcd.gov.jm; internet www.mlgycd.gov.jm.

**Ministry of National Security:** Mutual Life Bldg, North Tower, 2 Oxford Rd, Kingston 5; tel. 906-4908; fax 906-1724; e-mail enquiries@mns.gov.jm; internet www.mns.gov.jm.

**Ministry of Transport and Works:** 138H Maxfield Ave, Kingston 10; tel. 754-1900; fax 920-8763; e-mail ps@mtw.gov.jm; internet www.mtw.gov.jm.

**Ministry of Water and Housing:** 25 Dominica Dr., Kingston 5; tel. 754-0973; fax 754-2853; e-mail genefa@cwjamaica.com; internet www.mwh.gov.jm.

## Legislature

### PARLIAMENT

**Houses of Parliament:** Gordon House, 81 Duke St, POB 636, Kingston; tel. 922-0200; e-mail slewis@mail.infochan.com.

### Senate

**President:** SYRINGA MARSHALL-BURNETT.
**Vice-President:** NAVEL FOSTER CLARKE.
The Senate has 19 other members.

### House of Representatives

**Speaker:** MICHAEL PEART.
**Deputy Speaker:** ONEL T. WILLIAMS.

**General Election, 16 October 2002**

|  | % of votes cast | Seats |
|---|---|---|
| People's National Party (PNP) | 52.2 | 34 |
| Jamaica Labour Party (JLP) | 47.2 | 26 |
| **Total** (incl. others) | 100.0 | 60 |

## Political Organizations

**Jamaica Labour Party (JLP):** 20 Belmont Rd, Kingston 5; tel. 929-1183; fax 968-0873; e-mail info@thejlp.com; internet www.thejlp.org; f. 1943; supports free enterprise in a mixed economy and close co-operation with the USA; Leader ORRETT BRUCE GOLDING; Chair. ROBERT PICKERSGILL; Gen. Sec. KARL SAMUDA.

**Jamaica Alliance Movement (JAM):** Kingston; f. 2001; Rastafarian; Pres. ASTOR BLACK.

**Jamaican Alliance for National Unity (JANU):** Kingston; f. 2002; mem. of the New Jamaica Alliance; Chair. Rev. AL MILLER.

**National Democratic Movement (NDM):** 72 Half Way Tree Rd, Kingston 10; e-mail ndmjamaica@yahoo.com; internet www.ndm4jamaica.org; f. 1995; advocates a clear separation of powers between the central executive and elected representatives; supports private investment and a market economy; mem. of the New Jamaica Alliance; Gen. Sec. MICHAEL WILLIAMS; Chair. PETER TOWNSEND (acting).

**Natural Law Party:** c/o 21st Century Integrated Medical Centre, Shop OF3, Overton Plaza, 49 Union St, Montego Bay; tel. 971-9107; fax 971-9109; e-mail nlp@cwjamaica.com; f. 1996; Leader Dr LEO CAMPBELL.

**People's National Party (PNP):** 89 Old Hope Rd, Kingston 5; tel. 978-1337; fax 927-4389; e-mail information@pnpjamaica.com; internet www.pnpjamaica.com; f. 1938; socialist principles; affiliated with the National Workers' Union; Leader PERCIVAL J. PATTERSON; Gen. Sec. BURCHELL WHITEMAN; Vice-Pres PETER PHILLIPS, PORTIA SIMPSON-MILLER.

**Republican Party of Jamaica (RPJ):** Kingston; mem. of the New Jamaica Alliance; Leader DENZIL TAYLOR.

**United People's Party (UPP):** 6 Trinidad Terrace, Kingston 5; tel. 929-6429; fax 968-4615; e-mail uppjam@cwjamaica.com; f. 2001; Pres. ANTOINETTE HAUGHTON CARDENAS; Gen. Sec. HORACE MATTHEWS.

## Diplomatic Representation

### EMBASSIES AND HIGH COMMISSIONS IN JAMAICA

**Argentina:** Dyoll Life Bldg, 6th Floor, 40 Knutsford Blvd, Kingston 5; tel. 926-5588; fax 926-0580; e-mail embargen@kasnet.com; Ambassador GONZALO FERNÁNDEZ MEDRANO.

**Brazil:** First Life Bldg, 3rd Floor, 64 Knutsford Blvd, Kingston 5; tel. 929-8607; fax 929-1259; e-mail brasking@infochan.com; Ambassador CÉZAR AUGUSTO DE SOUZA LIMA AMARAL.

**Canada:** 3 West Kings House Rd, POB 1500, Kingston 10; tel. 926-1500; fax 511-3491; e-mail kngtn@dfait-maeci.gc.ca; internet www.kingston.gc.ca; High Commissioner CLAUDIO VALLE.

**Chile:** Island Life Centre, 5th Floor, South Sixth St, Lucia Ave, Kingston 5; tel. 968-0260; fax 968-0265; e-mail chilejam@cwjamaica.com; Ambassador OSCAR ALFONSO SILVA NAVARRO.

**China, People's Republic:** 8 Seaview Ave, POB 232, Kingston 6; tel. 927-3871; fax 927-6920; e-mail chinaemba@cwjamaica.com; Ambassador ZHAO ZHENYU.

**Colombia:** Victoria Mutual Bldg, 3rd Floor, 53 Knutsford Blvd, Kingston 5; tel. 929-1701; fax 968-0577; Ambassador Dr FRANCIS JAMES KENT.

# JAMAICA

**Costa Rica:** Belvedere House, Beverly Dr., Hopedale, Old Hope Rd, Kingston 6; tel. 927-5988; fax 978-3946; e-mail cr_emb_jam14@hotmail.com; Ambassador JOYCELYN SAWYERS ROYAL.

**Cuba:** 9 Trafalgar Rd, Kingston 10; tel. 978-0931; fax 978-5372; e-mail embacubajam@cwjamaica.com; Ambassador GISELA GARCÍA RIVERA.

**Dominican Republic:** 32 Earls Court, Kingston 8; tel. 755-4155; fax 755-4156; e-mail domemb@cwjamaica.com; Ambassador ARSENIO JIMENEZ-POLANCO.

**France:** 13 Hillcrest Ave, POB 93, Kingston 6; tel. 978-0210; fax 927-4998; e-mail frenchembassy@cwjamaica.com; internet ambafrance-jm.org; Ambassador FRANCIS HURTUT.

**Germany:** 10 Waterloo Rd, POB 444, Kingston 10; tel. 926-6728; fax 929-8282; e-mail germanemb@cwjamaica.com; Ambassador Dr CHRISTIAN HAUSMANN.

**Haiti:** 2 Munroe Rd, Kingston 6; tel. 927-7595; fax 978-7638; Ambassador JEAN-GABRIEL AUGUSTIN.

**Honduras:** 7 Lady Kay Dr., Norbrook, Kingston 8; tel. 931-5248; fax 941-6470; e-mail emhonjam@kasnet.com; Ambassador CARLOS AUGUSTO MATUTE RIVERA.

**India:** 27 Seymour Ave, POB 446, Kingston 6; tel. 927-3114; fax 978-2801; e-mail hicomindkin@cwjamaica.com; High Commissioner INDER VIR CHOPRA.

**Japan:** Mutual Life Centre, North Tower, 6th Floor, 2 Oxford Rd, POB 8104, Kingston 5; tel. 929-3338; fax 968-1373; Ambassador HIROSHI SAKURAI.

**Mexico:** PCJ Bldg, 36 Trafalgar Rd, Kingston 10; tel. 926-6891; fax 929-7995; e-mail mexicoj@kasnet.com; Ambassador BENITO ANDIÓN.

**Nigeria:** 5 Waterloo Rd, POB 94, Kingston 10; tel. 926-6400; fax 968-7371; High Commissioner F. A. UKONGA.

**Panama:** 1 St Lucia Ave, Spanish Court, Suite 26, Kingston 5; tel. 968-2928; fax 960-1618; Ambassador FRANKLIN ANTONIO BARRETT TAIT.

**Peru:** 23 Barbados Ave, Kingston 5; tel. 920-5027; fax 920-4360; e-mail embaperu-kingston@rree.gob.pe; Ambassador LUIS SÁNDIGA CABRERA .

**Russia:** 22 Norbrook Dr., Kingston 8; tel. 924-1048; fax 925-8290; e-mail Rusembja@colis.com; Ambassador EDUARD MALAYAN.

**Saint Christopher and Nevis:** 11A Opal Ave, POB 157, Golden Acres, Kingston 7; tel. 944-3861; fax 945-0105; High Commissioner CEDRIC HARPER.

**South Africa:** First Life Bldg, 7th Floor, 60 Knutsford Blvd, Kingston 5; tel. 960-3750; fax 929-0240; High Commissioner THANDYISE CHILIZA.

**Spain:** The Towers, 9th Floor, 25 Dominica Dr., Kingston 5; tel. 929-6710; fax 929-8965; e-mail jamesp@jamweb.net; Ambassador RAFAEL ADOLFO JOVER Y DE MORA FIGUEROA.

**Trinidad and Tobago:** First Life Bldg, 3rd Floor, 60 Knutsford Blvd, Kingston 5; tel. 926-5730; fax 926-5801; e-mail t&thckgn@infochan.com; High Commissioner DENNIS FRANCIS.

**United Kingdom:** 28 Trafalgar Rd, POB 575, Kingston 10; tel. 510-0700; fax 511-5303; e-mail bhckingston@cwjamaica.com; internet www.britishhighcommission.gov.uk/jamaica; High Commissioner PETER J. MATHERS.

**USA:** Mutual Life Centre, 2 Oxford Rd, Kingston 5; tel. 929-4850; fax 935-6019; e-mail opakgn@state.gov; internet kingston.usembassy.gov; Chargé d'affaires THOMAS C. (CLIFF) TIGHE.

**Venezuela:** PCJ Bldg, 3rd Floor, 36 Trafalgar Rd, Kingston 10; tel. 926-5510; fax 926-7442; Ambassador ROCIO MANEIRO GONZÁLEZ.

## Judicial System

The judicial system is based on English common law and practice. Final appeal is to the Judicial Committee of the Privy Council in the United Kingdom, although in 2001 the Jamaican Government signed an agreement to establish a Caribbean Court of Justice to fulfil this function.

Justice is administered by the Privy Council, Court of Appeal, Supreme Court (which includes the Revenue Court and the Gun Court), Resident Magistrates' Court (which includes the Traffic Court), two Family Courts and the Courts of Petty Sessions.

**Judicial Service Commission:** Office of the Services Commissions, 63–67 Knutsford Blvd, Kingston 5; advises the Governor-General on judicial appointments, etc.; chaired by the Chief Justice.

**Attorney-General:** ARNOLD J. NICHOLSON.

### SUPREME COURT
(Public Bldg E, 134 Tower St, POB 491, Kingston; tel. 922-8300; fax 967-0669; e-mail webmaster@sc.gov.jm; internet www.sc.gov.jm)

**Chief Justice:** LENSLEY H. WOLFE.

**Senior Puisne Judge:** H. F. COOKE.

**Master:** CHRISTINE MCDONALD.

**Registrar:** AUDRE LINDO.

### COURT OF APPEAL
(POB 629, Kingston; tel. 922-8300)

**President:** I. X. FORTE.

**Registrar:** G. P. LEVERS.

## Religion

### CHRISTIANITY

There are more than 100 Christian denominations active in Jamaica. According to the 2001 census, the largest religious bodies were the Church of God (whose members represented 24% of the population), Seventh-day Adventists (11% of the population), Pentecostalist (10%), Baptists (7%) and Anglicans (4%). Other denominations include Jehovah's Witnesses, the Methodist and Congregational Churches, United Church, the Church of the Brethren, the Ethiopian Orthodox Church, the Disciples of Christ, the Moravian Church, the Salvation Army and the Society of Friends (Quakers).

**Jamaica Council of Churches:** 14 South Ave, Kingston 10; tel. and fax 926-0974; e-mail jchurch@cwjamaica.com; f. 1941; 10 member churches and three agencies; Gen. Sec. HARRIS CUNNINGHAM.

#### The Anglican Communion

Anglicans in Jamaica are adherents of the Church in the Province of the West Indies, comprising eight dioceses. The Archbishop of the Province is the Bishop of the North East Caribbean and Aruba. The Bishop of Jamaica, whose jurisdiction also includes Grand Cayman (in the Cayman Islands), is assisted by three suffragan Bishops (of Kingston, Mandeville and Montego Bay). The 2001 census recorded that some 4% of the population were Anglicans.

**Bishop of Jamaica:** Rt Rev. ALFRED C. REID, Church House, 2 Caledonia Ave, Kingston 5; tel. 952-4963; fax 952-2933.

#### The Roman Catholic Church

Jamaica comprises the archdiocese of Kingston in Jamaica (also including the Cayman Islands), and the dioceses of Montego Bay and Mandeville. At 31 December 2003 the estimated total of adherents in Jamaica was 79,222, representing about 3% of the total population. The Archbishop and Bishops participate in the Antilles Episcopal Conference (currently based in Port of Spain, Trinidad and Tobago).

**Archbishop of Kingston in Jamaica:** Most Rev. LAWRENCE ALOYSIUS BURKE, Archbishop's Residence, 21 Hopefield Ave, POB 43, Kingston 6; tel. 927-9915; fax 927-4487; e-mail rcabkgn@cwjamaica.com.

#### Other Christian Churches

**Assembly of God:** Evangel Temple, 3 Friendship Park Rd, Kingston 3; tel. 928-2995; Sec. Pastor WILSON.

**Baptist Union:** 6 Hope Rd, Kingston 10; tel. 926-7820; fax 968-7832; e-mail info@jbu.org.jm; internet www.jbu.org.jm; 40,000 mems; Pres. Rev. JONATHAN HEMMINGS; Gen. Sec. Rev. KARL JOHNSON.

**Church of God in Jamaica:** 35 Hope Rd, Kingston 10; tel. 927-5990.

**First Church of Christ, Scientist:** 17 National Heroes Circle, Kingston 4; tel. 967-3814.

**Methodist Church (Jamaica District):** 143 Constant Spring Rd, POB 892, Kingston 8; tel. and fax 925-6768; e-mail jamaicamethodist@cwjamaica.com; f. 1789; 15,820 mems; Pres. Rev. Dr BYRON CHAMBERS; Synod Sec. Rev. CATHERINE GALE.

**Moravian Church in Jamaica:** 3 Hector St, POB 8369, Kingston 5; tel. 928-1861; fax 928-8336; e-mail moravianchja@colis.com; f. 1754; 30,000 mems; Pres. Rev. Dr LIVINGSTONE THOMPSON.

**Seventh-day Adventist Church:** 56 James St, Kingston; tel. 922-7440; f. 1901; 205,000 mems; Pres. Dr PATRICK ALLEN.

**United Church in Jamaica and the Cayman Islands:** 12 Carlton Cres., POB 359, Kingston 10; tel. 926-6059; fax 929-0826; e-mail churchunited@hotmail.com; internet www.ucjci.netfirms.com; f. 1965 by merger of the Congregational Union of Jamaica (f. 1877) and the Presbyterian Church of Jamaica and Grand Cayman to become United Church of Jamaica and Grand Cayman; merged

# JAMAICA

with Disciples of Christ in Jamaica in 1992 when name changed as above; 20,000 mems; Moderator RODERICK HEWITT; Gen. Sec. Rev. MAITLAND EVANS.

### RASTAFARIANISM

Rastafarianism is an important influence in Jamaican culture. The cult is derived from Christianity and a belief in the divinity of Ras (Prince) Tafari Makonnen (later Emperor Haile Selassie) of Ethiopia. It advocates racial equality and non-violence, but causes controversy in its use of 'ganja' (marijuana) as a sacrament. The 2001 census recorded 24,020 Rastafarians (0.9% of the total population). Although the religion is largely unorganized, there are some denominations.

**Haile Selassie Rastafari Royal Ethiopian Judah Coptic Church:** Balcombe Dr., Waterhouse, Kingston; not officially incorporated, on account of its alleged use of marijuana; Leader Abuna BONGO BLACKHEART.

### BAHÁ'Í FAITH

**National Spiritual Assembly:** 208 Mountain View Ave, Kingston 6; tel. 927-7051; fax 978-2344; incorporated in 1970.

### ISLAM

At the 2001 census there were an estimated 5,000 Muslims.

### JUDAISM

The 2001 census recorded some 350 Jews.

**United Congregation of Israelites:** K. K. Shaare Shalom Synagogue, POB 540, Kingston 6; tel. and fax 927-7948; e-mail ainsley@cwjamaica.com; internet www.ujcl.org; f. 1655; c. 250 mems; Dir AINSLEY COHEN HENRIQUES.

## The Press

### DAILIES

**Daily Gleaner:** 7 North St, POB 40, Kingston; tel. 922-3400; fax 922-6223; e-mail feedback@jamaica-gleaner.com; internet www.jamaica-gleaner.com; f. 1834; morning; independent; Chair. and Man. Dir OLIVER CLARKE; Editor-in-Chief GARFIELD GRANDISON; circ. 50,000.

**Daily Star:** 7 North St, POB 40, Kingston; tel. 922-3400; fax 922-6223; e-mail feedback@jamaica-gleaner.com; internet www.jamaica-gleaner.com; f. 1951; evening; Editor-in-Chief GARFIELD GRANDISON; Editor CLAIRE CLARKE-GRANT; circ. 45,000.

**Jamaica Herald:** 86 Hagley Park Rd, Kingston 10; tel. 937-7304; Man. Editor FRANKLIN MCKNIGHT.

**Jamaica Observer:** 40 Beechwood Ave, Kingston 5; tel. 920-8136; fax 926-7655; e-mail feedback@jamaicaobserver.com; internet www.jamaicaobserver.com; f. 1993; Chair. GORDON 'BUTCH' STEWART.

### PERIODICALS

**Catholic Opinion:** Roman Catholic Chancery Office, 21 Hopefield Ave, POB 43, Kingston 6; tel. 927-9915; fax 927-4487; e-mail rcabkgn@cwjamaica.com; 6 a year; religious; circulated in the Sunday Gleaner; Editor Very Rev. Fr MICHAEL LEWIS; circ. over 500,000.

**Children's Own:** 7 North St, POB 40, Kingston; tel. 922-3400; fax 922-6223; e-mail feedback@jamaica-gleaner.com; internet www.jamaica-gleaner.com; weekly during term time; Editor-in-Chief GARFIELD GRANDISON; circ. 120,000.

**Jamaica Churchman:** 2 Caledonia Ave, Kingston 5; tel. 926-6608; quarterly; Editor BARBARA GLOUDON; circ. 7,000.

**Jamaica Journal:** 12 East St, Kingston; tel. 929-0620; fax 926-1147; e-mail ioj.jam@mail.infochan.com; internet www.instituteofjamaica.org.jm; f. 1967; 3 a year; literary, historical and cultural review; publ. by Institute of Jamaica Publs Ltd; Chair. of Editorial Cttee BARBARA GLOUDON.

**Mandeville Weekly:** 31 Ward Ave, Mandeville, Manchester; tel. 961-0118; fax 961-0119; internet www.eyegrid.com/mandevilleweekly; Chief Editor ANTHONY FRECKLETON.

**Sunday Gleaner:** 7 North St, POB 40, Kingston; tel. 922-3400; fax 922-6223; e-mail feedback@jamaica-gleaner.com; internet www.jamaica-gleaner.com; weekly; Editor-in-Chief GARFIELD GRANDISON; circ. 100,000.

**Sunday Herald:** 17 Norwood Ave, Kingston 5; tel. 906-7572; fax 908-4044; e-mail mainsection@sundayheraldjamaica.com; internet www.sundayheraldjamaica.com; f. 1997; weekly; Consulting Editor FRANKLIN MCKNIGHT; Exec. Editor R. CHRISTINE KING.

*Directory*

**Sunday Observer:** 2 Fagan Ave, Kingston 8; tel. 920-8136; fax 926-7655; internet www.jamaicaobserver.com; weekly; Chair. GORDON 'BUTCH' STEWART.

**Teen Herald:** 86 Hagley Park Rd, Kingston 10; tel. 758-5275; fax 901-9335; e-mail teenherald@yahoo.com.

**The Visitor Vacation Guide:** 4 Cottage Rd, POB 1258, Montego Bay; tel. 952-5256; fax 952-6513; Editor LLOYD B. SMITH.

**Weekend Star:** 7 North St, POB 40, Kingston; tel. 922-3400; fax 922-6223; e-mail feedback@jamaica-gleaner.com; internet www.jamaica-gleaner.com; f. 1951; weekly; Editor-in-Chief GARFIELD GRANDISON; Editor CLAIRE CLARKE-GRANT; circ. 80,000.

**The Western Mirror:** 4 Cottage Rd, POB 1258, Montego Bay; tel. 952-5253; fax 952-6513; e-mail westernmirror@mail.infochan.com; internet www.westernmirror.com; f. 1980; 2 a week; Man. Dir and Editor LLOYD B. SMITH; circ. 20,000.

**West Indian Medical Journal:** Faculty of Medical Sciences, University of the West Indies, Kingston 7; tel. 927-1214; fax 927-1846; e-mail wimj@uwimona.edu.jm; f. 1951; quarterly; Editor EVERARD N. BARTON; circ. 2,000.

**X-News Jamaica:** 86 Hagley Park Rd, Kingston 10; tel. 937-7304; fax 901-7667; e-mail comments@xnewsjamaica.com; internet www.xnewsjamaica.com; f. 1993; weekly.

### PRESS ASSOCIATION

**Press Association of Jamaica (PAJ):** Kingston 8; tel. 925-7836; f. 1943; Pres. DESMOND RICHARDS.

#### Foreign Bureaux

Associated Press (USA), Caribbean Media Corpn and Inter Press Service (Italy) are represented in Jamaica.

## Publishers

**Ian Randle Publishers (IRP):** 11 Cunningham Ave, POB 686, Kingston 6; tel. 978-0745; fax 978-1156; e-mail ian@ianrandlepublishers.com; internet www.ianrandlepublishers.com; f. 1991; history, gender studies, politics, sociology; Publr IAN RANDLE; Finance Man. CARLENE RANDLE.

**Jamaica Publishing House Ltd:** 97 Church St, Kingston; tel. 967-3866; fax 922-5412; e-mail jph@mail.infochan.com; f. 1969; wholly-owned subsidiary of Jamaica Teachers' Asscn; educational, English language and literature, mathematics, history, geography, social sciences, music; Chair. WOODBURN MILLER; Man. ELAINE R. STENNETT.

**LMH Publishing Ltd:** 7 Norman Rd, Suite 10, LOJ Industrial Complex, Kingston CSO; tel. 938-0005; fax 759-8752; e-mail lmhbookpublishing@cwjamaica.com; internet lmhpublishingjamaica.com; f. 1970; educational textbooks, general, travel, fiction, non-fiction, children's books; Chair. L. MICHAEL HENRY; Man. Dir DAWN CHAMBERS-HENRY.

**University of the West Indies Press (UWI Press):** 1A Aqueduct Flats, Mona, Kingston 7; tel. 977-2659; fax 977-2660; e-mail cuserv@cwjamaica.com; f. 1992; Caribbean history, culture and literature, gender studies, education and political science; Man. Editor SHIVAUN HEARNE; Gen. Man. LINDA SPETH.

**Western Publishers Ltd:** 4 Cottage Rd, POB 1258, Montego Bay; tel. 952-5253; fax 952-6513; e-mail westernmirror@mail.infochan.com; f. 1980; Man. Dir and Editor-in-Chief LLOYD B. SMITH.

### GOVERNMENT PUBLISHING HOUSE

**Jamaica Printing Services:** 77 Duke St, Kingston; tel. 967-2250; internet www.jps1992.com; Chair. EVADNE STERLING; Man. RALPH BELL.

## Broadcasting and Communications

### TELECOMMUNICATIONS

The telecommunications sector became fully liberalized on 1 March 2003. The sector was regulated by the Office of Utilities Regulation (see Utilities).

**Cable & Wireless Jamaica Ltd:** 7 Cecilio Ave, Kingston 10; tel. 926-9450; fax 929-9530; f. 1989; in 1995 merged with Jamaica Telephone Co Ltd and Jamaica International Telecommunications Ltd, name changed as above 1995; 79% owned by Cable & Wireless (United Kingdom); Pres. GARY BARROW.

**Digicel:** Kingston; internet www.digiceljamaica.com; mobile cellular telephone operator; owned by Irish consortium, Mossel

JAMAICA

(Jamaica) Ltd; f. 2001; Chair. DENIS O'BRIEN; CEO (Jamaica) DAVID HALL.

**Oceanic Digital Jamaica:** Kingston; e-mail dkcastaldi@oceanicdigital.com; internet www.oceanicdigital.com; mobile cellular telephone operator; fmrly Centennial Digital; Oceanic Digital, 49% stockholder, bought remaining 51% in August 2002.

### BROADCASTING

#### Radio

**Independent Radio:** 6 Bradley Ave, Kingston 10; tel. 968-4880; fax 968-9165; commercial radio station; broadcasts 24 hrs a day on FM; Gen. Man. NEWTON JAMES.

**IRIE FM:** 1B Derrymore Rd, Kingston 10; tel. 968-5013; fax 968-8332; e-mail iriefmmarket@cwjamaica.com; internet www.iriefm.net; f. 1991; commercial radio station owned by Grove Broadcasting Co; plays only reggae music.

**Island Broadcasting Services Ltd:** 41B Half Way Tree Rd, Kingston 5; tel. 968-8115; fax 929-1345; commercial; broadcasts 24 hrs a day on FM; Exec. Chair. NEVILLE JAMES.

**KLAS-FM 89:** 81 Knutsford Blvd, Kingston 5; tel. 929-1344; f. 1991; commercial radio station.

**Love FM:** 12 Carlton Cres., Kingston 10; tel. 968-9596; f. 1997; commercial radio station, broadcasts religious programming on FM; owned by Religious Media Ltd.

**Radio Jamaica Ltd (RJR):** Broadcasting House, 32 Lyndhurst Rd, POB 23, Kingston 5; tel. 926-1100; fax 929-7467; e-mail rjr@radiojamaica.com; internet www.radiojamaica.com; f. 1947; commercial, public service; 3 channels.

**FAME FM:** internet www.famefm.fm; broadcasts on FM, island-wide 24 hrs a day; Exec. Producer FRANCOIS ST. JUSTE.

**Radio 2 FM:** internet www.radio92fm.com; broadcasts on FM, island-wide 24 hrs a day; Media Services Man. DONALD TOPPING.

**RJR 94 FM:** internet www.rjr94fm.com; broadcasts on AM and FM, island-wide 24 hrs a day; Exec. Producer NORMA BROWN-BELL.

**ZIP 103 FM:** 1B Derrymore Rd, Kingston 10, Jamaica; tel. 819-7699; fax 960-0523; e-mail zip103fm@cwjamaica.com; internet www.ZIPFM.net; f. 2002; commercial radio station; Dir JUDITH BODLEY.

Other stations broadcasting include Hot 102 FM, Power 106 FM, Roots FM and TBC FM.

#### Television

**Creative TV (CTV):** Kinston; tel. 967-4482; fax 924-9432; internet www.creativetvjamaica.com; operated by Creative Production & Training Centre Limited (CPTC); local cable channel; regional cultural, educational and historical programming; CEO Dr HOPETON DUNN.

**CVM Television:** 69 Constant Sprint Rd, Kingston 10; tel. 931-9400; fax 931-9417; e-mail wsmith@cvmtv.com; internet www.cvmtv.com.

**Love Television:** Kingston; internet www.love101.org; f. 1997; religious programming; owned by National Religious Media Ltd.

**Television Jamaica Limited (TVJ):** 5–9 South Odeon Ave, POB 100, Kingston 10; tel. 926-5620; fax 929-1029; e-mail tvjadmin@cwjamaica.com; internet www.televisionjamaica.com; f. 1959 as Jamaica Broadcasting Corporation; privatized 1997, name changed as above; subsidiary of RJR Communications Group; island-wide VHF transmission 24 hrs a day.

## Finance

(cap. = capital; p.u. = paid up; res = reserves; dep. = deposits; m. = million; brs = branches; amounts in Jamaican dollars)

### BANKING

#### Central Bank

**Bank of Jamaica:** Nethersole Pl., POB 621, Kingston; tel. 922-0750; fax 922-0828; e-mail info@boj.org.jm; internet www.boj.org.jm; f. 1960; cap. 4.0m., res 2,396.4m., dep. 112,307.9m. (Dec. 2003); Gov. DERICK MILTON LATIBEAUDIÈRE.

#### Commercial Banks

**Bank of Nova Scotia Jamaica Ltd** (Canada): Scotiabank Centre Bldg, cnr Duke and Port Royal Sts, POB 709, Kingston; tel. 922-1000; fax 924-9294; f. 1967; cap. 1,463.6m., res 9,612.6m., dep. 91,314.6m. (Oct. 2003); Chair. R. H. PITFIELD; Man. Dir WILLIAM E. CLARKE; 35 brs.

*Directory*

**Citibank, NA** (USA): 63–67 Knutsford Blvd, POB 286, Kingston 5; tel. 926-3270; fax 929-3745; internet www.citibank.com/jamaica; Vice-Pres. PETER MOSES.

**FirstCaribbean International Bank Ltd** (Canada/United Kingdom): 23–27 Knutsford Blvd, POB 762, Kingston 5; tel. 929-9310; fax 960-2837; internet www.firstcaribbeanbank.com; 51% owned by Canadian Imperial Bank of Commerce and Barclays Bank PLC; cap. 96.7m., res 1,025.2m., dep. 16,058.5m. (Oct. 2003); Chair. MICHAEL MANSOOR; Man. RAYMOND CAMPBELL; 12 brs.

**National Commercial Bank Jamaica Ltd:** 'The Atrium', 32 Trafalgar Rd, POB 88, Kingston 10; tel. 929-9050; fax 929-8399; internet www.jncb.com; f. 1977; merged with Mutual Security Bank in 1996; cap. 2,466.8m., res 6,552.6m., dep. 121,525.6m. (Sept. 2003); Chair. MICHAEL LEE-CHIN; Man. Dir AUBYN HILL; 37 brs.

**RBTT Bank Jamaica Ltd:** 17 Dominica Dr., Kingston 5; tel. 960-2340; fax 960-2332; internet www.rbtt.com; f. 1993 as Jamaica Citizens Bank Ltd; name changed to Union Bank of Jamaica Ltd; acquired by Royal Bank of Trinidad and Tobago in 2001 and name changed as above; Chair. Dr OWEN JEFFERSON; Man. Dir MICHAEL E. A. WRIGHT; 614 brs.

#### Development Banks

**Development Bank of Jamaica Ltd:** 11A–15 Oxford Rd, POB 466, Kingston 5; tel. 929-6124-7; fax 929-6055; e-mail dbank@cwjamaica.com; replaced Jamaica Development Bank; f. 1969; provides funds for medium- and long-term devt-orientated projects in the agricultural, tourism, industrial, manufacturing and services sectors through financial intermediaries; Man. Dir KINGSLEY THOMAS.

**Jamaica Mortgage Bank:** 33 Tobago Ave, POB 950, Kingston 5; tel. 929-6350; fax 968-5428; e-mail jmb@cwjamaica.com; internet www.jamaicamortgagebank.com; f. 1971 by the Jamaican Govt and the US Agency for Int. Devt; govt-owned statutory org. since 1973; intended to function primarily as a secondary market facility for home mortgages and to mobilize long-term funds for housing devts in Jamaica; also insures home mortgage loans made by approved financial institutions, thus transferring risk of default on a loan to the Govt; Man. Dir PETER THOMAS; Gen. Man. EVERTON HANSON.

**Pan Caribbean Financial Services:** 60 Knutsford Blvd, Kingston 5; tel. 929-5583-4; fax 926-4385; e-mail options@gopancaribbean.com; internet www.gopancaribbean.com; fmrly Trafalgar Development Bank, name changed as above in December 2002; Chair. RICHARD O. BYLES; Pres. DONOVAN H. PERKINS.

#### Other Banks

**National Export-Import Bank of Jamaica Ltd:** 48 Duke St, POB 3, Kingston; tel. 922-9690; fax 922-9184; e-mail eximjam@cwjamaica.com; internet www.eximbankja.com; replaced Jamaica Export Credit Insurance Corpn; finances import and export of goods and services; Chair. Dr OWEN JEFFERSON; Deputy Chair. PAUL THOMAS.

**National Investment Bank of Jamaica Ltd:** 11 Oxford Rd, POB 889, Kingston 5; tel. 960-9691; fax 920-0379; e-mail nibj@infochan.com; internet www.nibj.com; Chair. DAVID COORE; Pres. REX JAMES.

#### Banking Association

**Jamaica Bankers' Association:** 39 Hope Rd, POB 1079, Kingston 10; tel. 927-6238; fax 927-5137; e-mail jbainfo@jba.org.jm; internet www.jba.org.jm; f. 1973; Pres. WILLIAM CLARKE.

#### Financial Sector Adjustment Company

**FINSAC Ltd:** 76 Knutsford Blvd, POB 54, Kingston 5; tel. 906-1809; fax 906-1822; e-mail info@finsac.com; internet www.finsac.com; f. 1997; state-owned; intervenes in the banking and insurance sectors to restore stability in the financial sector; Chair. KENNETH RATTRAY.

### STOCK EXCHANGE

**Jamaica Stock Exchange Ltd:** 40 Harbour St, Kingston; tel. 967-3271; fax 922-6966; internet www.jamstockex.com; f. 1968; 32 listed cos (2004); Chair. ROY JOHNSON.

### INSURANCE

**Office of the Superintendent of Insurance:** 51 St Lucia Ave, POB 800, Kingston 5; tel. 926-1790; fax 968-4346; f. 1972; regulatory body; Supt ERROL MCLEAN (acting).

**Jamaica Association of General Insurance Companies:** 3-3A Richmond Ave, Kingston 10; tel. 929-8404; e-mail jagic@cwjamaica.com; internet www.jagonline.com; Man. GLORIA M. GRANT; Chair. LESLIE CHUNG.

#### Principal Companies

**British Caribbean Insurance Co Ltd:** 36 Duke St, POB 170, Kingston; tel. 922-1260; fax 922-4475; e-mail bricar@cwjamaica

JAMAICA                                                                                                      *Directory*

.com; internet www.bciconline.com; f. 1962; general insurance; Man. Dir LESLIE W. CHUNG.

**General Accident Insurance Co Jamaica Ltd:** 58 Half Way Tree Rd, Kingston 10; tel. 929-8451; fax 929-1074; e-mail genac@cwjamaica.com; internet www.genac.com; f. 1981; Gen. Man. SHARON E. DONALDSON.

**Globe Insurance Co of the West Indies Ltd:** 19 Dominica Dr., POB 401, Kingston 5; tel. 926-3720; fax 929-2727; e-mail admin@globeins.com; internet www.globeins.com; f. 1962; Man. Dir EVAN THWAITES.

**Guardian Life:** 12 Trafalgar Rd, Kingston 5; tel. 978-8815; fax 978-4225; internet www.guardianholdings.com; pension and life policies; Chair. ARTHUR LOK JACK; Pres. and CEO DOUGLAS CAMACHO.

**Insurance Co of the West Indies Ltd (ICWI):** 2 St Lucia Ave, POB 306, Kingston 5; tel. 926-9182; fax 929-6641; Chair. DENNIS LALOR; CEO KENNETH BLAKELEY.

**Island Life Insurance Co:** 6 St Lucia Ave, Kingston 5; tel. 968-6874; e-mail ceo@islandlife-ja-com.jm; 64% owned by Barbados Mutual Life Assurance Co; 26.5% owned by FINSAC; merged with Life of Jamaica Ltd in 2001.

**Jamaica General Insurance Co Ltd:** 9 Duke St, POB 408, Kingston; tel. 922-6420; fax 922-2073; Man. Dir A. C. LEVY.

**Life of Jamaica Ltd:** 28–48 Barbados Ave, Kingston 5; tel. 960-8705; fax 929-4730; f. 1970; life and health insurance, pensions; 76% owned by Barbados Mutual Life Assurance Co; merged with Island Life Insurance Co in 2001; bought First Life Insurance in May 2005; Pres. R. D. WILLIAMS.

**NEM Insurance Co (Jamaica) Ltd:** NEM House, 9 King St, Kingston; tel. 922-1460; fax 922-4045; e-mail info@nemjam.com; internet www.nemjam.com; fmrly the National Employers' Mutual General Insurance Asscn; Man. Dir ERROL ZIADIE.

## Trade and Industry

### GOVERNMENT AGENCY

**Jamaica Information Service (JIS):** 58A Half Way Tree Rd, POB 2222, Kingston 10; tel. 926-3741; fax 920-7427; e-mail jis@jis.gov.jm; internet www.jis.gov.jm; f. 1963; information agency for govt policies and programmes, ministries and public-sector agencies; CEO CARMEN E. TIPLING.

### DEVELOPMENT ORGANIZATIONS

**Agricultural Development Corpn (ADC) Group of Companies:** Mais House, Hope Rd, POB 552, Kingston; tel. 977-4412; fax 977-4411; f. 1989; manages and develops breeds of cattle, provides warehousing, cold storage, offices and information for exporters and distributors of non-traditional crops and ensures the proper utilization of agricultural lands under its control; Chair. Dr ASTON WOOD; Gen. Man. DUDLEY IRVING.

**Jamaica Promotions Corpn (JAMPRO):** 18 Trafalgar Rd, Kingston 10; tel. 978-7755; fax 946-0090; e-mail jampro@investjamaica.com; internet www.investjamaica.com; f. 1988 by merger of Jamaica Industrial Development Corpn, Jamaica National Export Corpn and Jamaica Investment Promotion Ltd; trade and investment promotion agency; Pres. PATRICIA FRANCIS; Chair. JOSEPH A. MATALON.

**Planning Institute of Jamaica:** 10–16 Grenada Way, Kingston 5; tel. 906-4463; fax 906-5011; e-mail doccen@mail.colis.com; internet www.pioj.gov.jm; f. 1955 as the Central Planning Unit; adopted current name in 1984; formulates policy on and monitors performance in the fields of the economy and and social, environmental and trade issues; publishing and analysis of social and economic performance data; Dir-Gen. WESLEY HUGHES.

**Urban Development Corpn:** The Office Centre, 8th Floor, 12 Ocean Blvd, Kingston; tel. 922-8310; fax 922-9326; e-mail info@udcja.com; internet www.udcja.com; f. 1968; responsibility for urban renewal and devt within designated areas; Chair. Dr VINCENT LAWRENCE; Gen. Man. MARJORIE CAMPBELL.

### CHAMBERS OF COMMERCE

**Associated Chambers of Commerce of Jamaica:** 7–8 East Parade, POB 172, Kingston; tel. 922-0150; f. 1974; 12 associated Chambers of Commerce; Pres. RAY CAMPBELL.

**Jamaica Chamber of Commerce:** 7–8 East Parade, POB 172, Kingston; tel. 922-0150; fax 924-9056; e-mail jamcham@cwjamaica.com; internet www.jcc.org.jm; f. 1779; Chair. AVIS HENRIQUES; 450 mems.

### INDUSTRIAL AND TRADE ASSOCIATIONS

**Cocoa Industry Board:** Marcus Garvey Dr., POB 1039, Kingston 15; tel. 923-6411; fax 923-5837; e-mail cocoajam@cwjamaica.com; f. 1957; has statutory powers to regulate and develop the industry; owns and operates four central fermentaries; Chair. JOSEPH SUAH; Man. and Sec. NABURN NELSON.

**Coconut Industry Board:** 18 Waterloo Rd, Half Way Tree, Kingston 10; tel. 926-1770; fax 968-1360; f. 1945; 9 mems; Chair. Dr RICHARD JONES; Gen. Man. JAMES S. JOYLES.

**Coffee Industry Board:** Marcus Garvey Dr., POB 508, Kingston 15; tel. 923-5850; fax 923-7587; e-mail coffeeboard@jamaicancoffee.gov.jm; internet www.jamaicancoffee.gov.jm; f. 1950; 9 mems; has wide statutory powers to regulate and develop the industry; Chair. RICHARD DOWNER.

**Jamaica Bauxite Institute:** Hope Gardens, POB 355, Kingston 6; tel. 927-2073; fax 927-1159; f. 1975; adviser to the Govt in the negotiation of agreements, consultancy services to clients in the bauxite/alumina and related industries, laboratory services for mineral and soil-related services, Pilot Plant services for materials and equipment testing, research and development; Chair. CARLTON DAVIS; Gen. Man. PARRIS LYEW-AYEE.

**Jamaica Export Trading Co Ltd (JETCO):** 188 Spanish Town Rd, AMC Complex, Kingston 11; tel. 937-1798; fax 937-6547; e-mail jetcoja@mail.infochan.com; internet www.exportjamaica.org; f. 1977; export trading in non-traditional products, incl. spices, fresh produce, furniture, garments, processed foods, minerals, etc.; Man. Dir HERNAL HAMILTON.

**Sugar Industry Authority:** 5 Trevennion Park Rd, POB 127, Kingston 5; tel. 926-5930; fax 926-6149; e-mail sia@cwjamaica.com; internet www.jamaicasugar.org; f. 1970; statutory body under portfolio of Ministry of Agriculture; responsible for regulation and control of sugar industry and sugar marketing; conducts research through Sugar Industry Research Institute; Exec. Chair. DERICK HEAVEN.

**Trade Board:** 107 Constant Spring Rd, Kingston 8; tel. 969-0478; fax 969-6513; e-mail tboard@colis.com; Admin. JEAN MORGAN.

### EMPLOYERS' ORGANIZATIONS

**All-Island Banana Growers' Association Ltd:** Banana Industry Bldg, 10 South Ave, Kingston 4; tel. 922-5492; fax 922-5497; f. 1946; 1,500 mems (1997); Chair. A. A. POTTINGER; Sec. I. CHANG.

**All-Island Jamaica Cane Farmers' Association (AIJCFA):** 4 North Ave, Kingston Gardens, Kingston 4; tel. 922-3010; fax 922-2077; e-mail allcane@cwjamaica.com; f. 1941; registered cane farmers; 27,000 mems; Pres. ALAN RICKARDS; Man. ADITER MILLER.

**Banana Export Co (BECO):** 1A Braemar Ave, Kingston 10; tel. 978-8762; fax 978-6096; f. 1985 to replace Banana Co of Jamaica; oversees the devt of the banana industry; Chair. Dr MARSHALL MCGOWAN HALL.

**Citrus Growers' Association Ltd:** Bog Walk, Linstead; tel. 922-8230; f. 1944; 13,000 mems; Chair. C. L. BENT.

**Jamaica Association of Sugar Technologists:** c/o Sugar Industry Research Institute, Kendal Rd, Mandeville; tel. 962-2241; fax 962-1288; f. 1936; 275 mems; Chair. KARL JAMES; Pres. GILBERT THORNE.

**Jamaica Exporters' Association (JEA):** 39 Hope Rd, Kingston 10; tel. 927-6238; fax 927-5137; e-mail infojea@exportja.org; internet www.exportjamaica.org; Pres. Dr ANDRE GORDON; Exec. Dir PAULINE GRAY.

**Jamaica Livestock Association:** Newport East, POB 36, Kingston; tel. 922-7130; fax 923-5046; e-mail jlapurch@cwjamaica.com; internet www.jlaltd.com; f. 1941; 7,584 mems; Chair. Dr JOHN MASTERTON; Man. Dir and CEO HENRY J. RAINFORD.

**Jamaica Manufacturers' Association Ltd:** 85a Duke St, Kingston; tel. 922-8880; fax 922-9205; e-mail jma@toj.com; internet www.jma.com.jm; f. 1947; 400 mems; Pres. CLARENCE CLARK.

**Jamaica Producers' Group Ltd:** 6A Oxford Rd, POB 237, Kingston 5; tel. 926-3503; fax 929-3636; e-mail cosecretary@jpjamaica.com; f. 1929; fmrly Jamaica Banana Producers' Asscn; Chair. C. H. JOHNSTON; Man. Dir Dr MARSHALL HALL.

**Private Sector Organization of Jamaica (PSOJ):** 39 Hope Rd, POB 236, Kingston 10; tel. 927-6238; fax 927-5137; e-mail psojinfo@psoj.org; internet www.psoj.org; federative body of private business individuals, cos and asscns; Pres. BEVERLY LOPEZ; CEO LOLA FONG WRIGHT.

**Small Businesses' Association of Jamaica (SBAJ):** 2 Trafalgar Rd, Kingston 5; tel. 927-7071; fax 978-2738; e-mail sbaj@anbell.net; internet www.sbaj.org.jm; Chair. ADOLPH BROWN; Man. ALBERT HUIE.

# JAMAICA

**Sugar Manufacturing Corpn of Jamaica Ltd:** 5 Trevennion Park Rd, Kingston 5; tel. 926-5930; fax 926-6149; established to represent the sugar manufacturers in Jamaica; deals with all aspects of the sugar industry and its by-products; provides liaison between the Govt, the Sugar Industry Authority and the All-Island Jamaica Cane Farmers' Asscn; 9 mems; Chair. CHRISTOPHER BOVELL; Gen. Man. DERYCK T. BROWN.

## MAJOR COMPANIES

### Chemicals and Pharmaceuticals

**Alkali Group of Cos:** 259 Spanish Town Rd, POB 200, Kingston 11; tel. 923-6131; fax 923-4947; e-mail abe@alkaligroup.com; internet www.alkaligroup.com; f. 1960; holding co comprising Powertrac, Industrial Chemical Company, Leder Mode Ltd and Tanners Ltd; Chair. BARCLAY EWART; Group Administrator ANDREW BROWN; 800 employees.

**Colgate Palmolive Co (Jamaica) Ltd:** 216 Marcus Garvey Dr., POB 4, Kingston 11; tel. 923-5691; fax 923-4355; f. 1938; subsidiary of Colgate-Palmolive Co (USA); manufacturer of toiletries; Man. Dir TREVOR OTTEY; 165 employees.

**Federated Pharmaceutical Co Ltd:** 1 Bell Rd, Kingston 11; tel. 932-7236; fax 922-0183; f. 1958; manufacturer of pharmaceuticals and cosmetics; Chair. JAMES LINDSAY; Gen. Man. CYRIL BRIDGE; 210 employees.

### Construction

**Ashtrom Building Services Ltd:** Mandela Highway, Central Village, Kingston; tel. 984-2395; fax 984-3210; f. 1970; production and installation of pre-constructed commercial buildings; Chair. HOWARD HAMILTON; Man. YORAME KEREN; 455 employees.

**B. & H. Structures Ltd:** 3 Lady Huggins Ave, Kingston 8; tel. 931-9237; fax 969-8443; f. 1975; civil and structural engineering; Man. STAFFORD HYDE; 455 employees.

**Caribbean Cement Co Ltd:** 28 Barbados Ave, POB 448, Kingston 5; tel. 928-6232; fax 928-7381; e-mail info@caribcement.com; internet www.caribcement.com; f. 1947; manufacturer of cement; sales of J $2,918m. (1999); Chair. BRIAN YOUNG; CEO Dr ROLLIN BERTRAND; 318 employees.

**Hardware and Lumber Ltd:** 697 Spanish Town Rd, Kingston 11; tel. 923-8912; fax 923-8629; e-mail handl@cwjamaica.ca.com; f. 1969; wholesale distribution of construction materials; sales of J $1,217m. (1999); Chair. RICHARD BYLES; Man. Dir A. ANTHONY HOLNESS; 197 employees.

### Food and Beverages

**Dairy Industries (Jamaica) Ltd:** 111 Washington Blvd, POB 336, Kingston 11; tel. 934-8272; fax 934-1852; e-mail dairy@cwjamaica.cim; f. 1964; 50% owned by Grace Kennedy and Co PLC and 50% owned by Fonterra Co-op Group Ltd (New Zealand); manufacturer of dairy products, contract packer of powdered drinks; Chair. D. ORANE; Gen. Man. ANDREW HO; 100 employees.

**Desnoes and Geddes Ltd:** 214 Spanish Town Rd, POB 190, Kingston 11; tel. 923-9291; fax 923-8599; f. 1918; brewery and soft-drinks bottlers; producers of Red Stripe lager and Dragon Stout; sales of J $4,615m. (1995); Chair. PATRICK H. O. ROUSSEAU; Pres. JOHN IRVINE; 667 employees.

**Eastern Banana Estates Ltd:** 6A Oxford Rd, Kingston 5; tel. 926-3503; fax 926-3636; f. 1983; subsidiary of the Jamaica Producers' Group Ltd; producers and exporters of bananas; Chair. C. H. JOHNSTON; Man. Dir MARSHALL HALL; 800 employees.

**Grace Kennedy and Co PLC:** 73 Harbour St, POB 86, Kingston; tel. 922-3440; fax 922-7567; e-mail gracefoods@gkco.com; internet www.gracekennedy.com; f. 1922; holding co concerned with food-processing and wholesale distribution, manufacturing, financial services, maritime activities, information technology; over 65 subsidiaries and related cos; sales of J $44,466m. (2003); Chair. and CEO DOUGLAS ORANE; 2,200 employees.

**Jamaica Broilers Group Ltd:** 15 McCooks Pen, St Catherine, C.S.O; tel. 943-4376; fax 943-4955; e-mail mchristian@jabgl.com; f. 1958; manufacturer of animal feed, producer of poultry, beef, tilapia and hatching eggs; sales of J $6,870.74m. (2003); cap. $339m.; Pres. and CEO ROBERT E. LEVY; Chair. R. DANVERS WILLIAMS; 970 employees.

**Jamaica Flour Mills Ltd:** 24 Trafalgar Rd, POB 28, Kingston; tel. 928-7221; fax 928-7348; e-mail garnett_williams@admworld.com; f. 1966; milling of grain, including flour; Man. Dir JAMES GILL; Gen. Man. GARNETT WILLIAMS; 104 employees.

**Jamaica Standard Products Co Ltd:** POB 2, Williamsfield, Manchester; tel. 963-4211; fax 963-4309; e-mail sanco@colis.com.jm; internet www.caribplace.com/foods/jspcl.htm; f. 1942; sales J $60m. (1995); production and export of coffee; Man. Dir JOHN O. MINOTT; 232 employees.

**National Rums Jamaica Ltd:** 25 Dominica Dr., Kingston; tel. 926-7548; fax 926-7499; f. 1979; manufacturer of distilled alcoholic drinks; 51% govt-owned; Man. Dir R. EVON BROWN; 139 employees.

**Nestlé-JMP Jamaica Ltd:** 60 Knutsford Blvd, POB 281, Kingston; tel. 926-1300; fax 926-7388; f. 1986; subsidiary of Nestlé (Switzerland); manufacturer of milk products; Chair. FELIPE SILVA; Gen. Man. JAMES RAWLE; 456 employees.

**Salada Foods:** 20 Bell Rd, POB 71, Kingston 11; tel. 923-7114; fax 923-5336; e-mail info@saladafoodsjamaica.com; internet www.saladafoodsjamaica.com; coffee processing; Chair. VINCENT CHEN; Man. Dir ROBERT PARKINS.

**Trelawny Sugar Company:** Trelawny Estate, Clark's Town, Trelwany; fax 954-2436; f. 1978; production and processing of sugar cane; Pres. LIVINGSTONE MORRISON.

**Walker's Wood Caribbean Foods ltd:** Walkerswood PO, St Ann; tel. 917-2318; fax 917-2648; e-mail partners@walkerswood.com; internet www.walkerswood.com; manufacturer of food products; sales of J $50m. (1996); Chair. RHODERICK EDWARDS; Man. Dir WOODROW MITCHELL; 65 employees.

**J. Wray and Nephew Ltd:** 234 Spanish Town Rd, POB 39, Kingston 11; tel. 923-6141; fax 923-8619; f. 1960; sugar plantation and rum distillery; owns Appleton Estates; Man. Dir WILLIAM McCONNELL; Gen. Man. TANYA MILLER; 2,400 employees.

### Mining and Power

**Alumina Partners of Jamaica** (Alpart): Spur Tree Post Office, Manchester; tel. 962-3251; fax 962-3532; f. 1966; 65% owned by Kaiser Aluminium (USA), 35% owned by Norsk Hydro Aluminium (Norway); bauxite mining and processing, aluminium products; Gen. Man. DARREL HARRIMAN; 1,400 employees.

**Jamalco** (Alcoa Minerals of Jamaica, Inc): Clarendon Parish, Clarendon; tel. 986-2561; fax 986-9637; internet www.alcoa.com/jamaica; f. 1959 as Alcoa Minerals of Jamaica; 50% owned by Alcoa (USA), 50% by the Govt of Jamaica; alumina and bauxite mining; Man. Dir JEROME T. MAXWELL.

**Petroleum Corpn of Jamaica (PCJ):** 36 Trafalgar Rd, POB 579, Kingston 10; tel. 929-5380; fax 929-2409; e-mail ica@pcj.com; internet www.pcj.com; f. 1979; state-owned; owns and operates petroleum refinery; holds exploration rights to local petroleum and gas reserves; Chair. JOHN COOKE.

**Petrojam Ltd:** 96 Marcus Garvey Dr., POB 241, Kingston; tel. 923-8611; fax 923-0384; e-mail wlw@petrojam.com; internet www.pcj.com/petrojam; f. 1964 by Esso, bought by Govt in 1982; wholly-owned subsidiary of PCJ; operates sole oil refinery in Jamaica; Chair. PAUL THOMAS; Man. Dir Dr RAYMOND WHITE; 145 employees.

**Petroleum Co of Jamaica Ltd (PETCOM):** 695 Spanish Town Rd, Kingston 11; tel. 934-6682; fax 934-6690; e-mail petcom@cwjamaica.com; internet www.pcj.com/petcom; f. 1985; wholly owned subsidiary of PCJ; markets gasoline, lubricants and petrochemicals and operates service stations; Chair. BARBARA CLARKE.

**West Indies Alumina Co** (Windalco): Kirkvine PO, Manchester; tel. 962-3141; fax 962-0606; internet www.windalco.com; f. 1943 as Alcan Jamaica; 7% Govt-owned, 93% acquired by Glencore Alumina Jamaica Ltd (Switzerland) in 2001; bauxite mining and processing, production of calcinated alumina; operates Kirkvine and Ewarton refineries; Man. Dir MICHAEL COLLINS; 1,470 employees.

### Miscellaneous

**Berger Paints Jamaica Ltd:** 256 Spanish Town Rd, POB 8, Kingston 11; tel. 923-9116; fax 923-5129; e-mail bergerja@infochem.com; f. 1952; subsidiary of UB International Ltd (United Kingdom); manufacturer of paints; sales J $764m. (2000). Chair. Dato ABDUL GHANI BIN YUSOF; Man. Dir WARREN McDONALD; 125 employees.

**Broadway Import Export Co Ltd:** 325 Spanish Town Rd, Kingston 11; tel. 937-3188; f. 1968; manufacturer and distributor of household linen; Man. JOSEPHINE JONES; 341 employees.

**Caribbean Brake Products Ltd:** 11 Bell Rd, POB 66, Kingston 11; tel. 923-7236; fax 923-6352; e-mail cbpsales@toj.com; f. 1959; production and distribution of automobile components; Chair. GORDON A. STEWART; Man. Dir PHILIP N. CRIMARCO; 170 employees.

**Caribbean Casting and Engineering Ltd:** 138 Spanish Town Rd, POB 163, Kingston 11; tel. 923-6558; fax 923-9538; f. 1970; producer of cast-iron products and machinery used for the manufacture of sugar; Man. Dir DENNIS FLETCHER; 123 employees.

**Carreras Group:** Twickenham Park, POB 100, Spanish Town; tel. 984-3051; fax 984-6571; e-mail jacig@infochan.com; internet www.batcentralamerica.com; acquired Cigarette Co of Jamacia in 2004; 50.4% owned by British American Tobacco plc (United Kingdom);

manufactures tobacco products incl. 'Matterhorn' and 'Craven A' brands; Chair. GEORGE ASHENHEIM; Man. Dir MICHAEL BERNARD; 129 employees.

**Ciboney Hotels Ltd:** 39–43 Barbados Ave, New Kingston, Kingston 5; tel. 929-6198; fax 929-2230; e-mail ciboney@infochan.com; f. 1991; public co; hotel management; sales of J $128m. (2001); Chair. PATRICK HYLTON; CEO GEOFFREY MESSADO; 567 employees.

**Courts Jamaica:** 79–81A Slipe Rd, Cross Roads, Kingston 5; tel. 926-2110; fax 929-0887; e-mail charter@courts.com.jm; internet www.courts.com.jm; subsidiary of Courts (Furnishers) Ltd, United Kingdom; furniture retailers; Chair. BRUCE COHEN; Man. Dir R. HAYDEN SINGH.

**Goodyear (Jamaica) Ltd:** 8 Oliver Rd, Kingston 8; tel. 924-6130; fax 924-6372; f. 1945; subsidiary of Goodyear Tire and Rubber Co (USA); manufacturer of automobile tyres; sales of J $717m. (2000); Man. Dir PETER GRAHAM; 223 employees.

**Seprod Group of Companies:** 3 Felix Blvd, Kingston; tel. 922-1220; fax 922-6948; e-mail corporate@seprod.com; internet www.seprod.com; f. 1940; manufacturer and distributor of soap, detergents, edible oils and fats, animal feeds; processors of grain, cereals, glycerine, etc.; sales of J $2,165m. (2000); Chair. A. DESMOND BLADES; CEO BRYON E. THOMPSON; 900 employees.

**West Indies Pulp and Paper Ltd:** 19 West Kings House Rd, Kingston 10; tel. 926-7423; fax 929-6726; f. 1968; manufacturer of paper products; sales of J $672m. (1999); Chair. PAUL GEDDES; Man. Dir MICHAEL PICKERSGILL; 288 employees.

## UTILITIES

### Regulatory Authority

**Office of Utilities Regulation (OUR):** PCJ Resource Centre, 3rd Floor, 36 Trafalgar Rd, Kingston 10; tel. 929-6672; fax 929-3635; e-mail office@our.org.jm; internet www.our.org.jm; f. 1995; regulates provision of services in the following sectors: water, electricity, telecommunications, public passenger transportation, sewerage; Dir-Gen. J. PAUL MORGAN.

### Electricity

**Jamaica Public Service Co (JPSCo):** Dominion Life Bldg, 6 Knutsford Blvd, POB 54, Kingston 5; tel. 926-3190; fax 968-5341; e-mail media@jpsco.com; internet www.jpsco.com; responsible for the generation and supply of electricity to the island; 80% sold to Mirant Corpn (USA) in March 2001; Pres. and CEO CHARLES MATTHEWS.

### Water

**National Water Commission:** LOJ Centre, 28–48 Barbados Ave, Kingston 5; tel. 929-5430; fax 929-1329; e-mail pr@nwc.com.jm; internet www.nwcjamaica.com; f. 1980; statutory body; provides potable water and waste water services; Chair. RICHARD BYLES.

**Water Resources Authority:** Hope Gardens, POB 91, Kingston 7; tel. 927-0077; fax 977-0179; e-mail commander@cwjamaica.com; internet www.wra-ja.org; f. 1996; manages, protects and controls allocation and use of water supplies; Man. Dir BASIL FERNANDEZ.

## TRADE UNIONS

**Bustamante Industrial Trade Union (BITU):** 98 Duke St, Kingston; tel. 922-2443; fax 967-0120; e-mail bitu@cwjamaica.com; f. 1938; Pres. HUGH SHEARER; Gen. Sec. GEORGE FYFFE; 60,000 mems.

**Jamaica Confederation of Trade Unions (JCTU):** 1A Hope Blvd, Kingston 6; tel. 977-5170; fax 977-4575; e-mail jctu@cwjamaica.com; Sec.-Gen. LLOYD GOODLEIGH.

**National Workers' Union of Jamaica (NWU):** 130–132 East St, POB 344, Kingston 16; tel. 922-1150; e-mail nwyou@cwjamaica.com; f. 1952; affiliated to the International Confederation of Free Trade Unions, etc.; Pres. CLIVE DOBSON; Gen. Sec. LLOYD GOODLEIGH; 10,000 mems.

**Trades Union Congress of Jamaica:** 25 Sutton St, POB 19, Kingston; tel. 922-5313; fax 922-5468; affiliated to the Caribbean Congress of Labour and the International Confederation of Free Trade Unions; Pres. E. SMITH; Gen. Sec. HOPETON CRAVEN; 20,000 mems.

### Principal Independent Unions

**Caribbean Union of Teachers:** 97 Church St, Kingston; tel. 922-1385; fax 922-3257; e-mail jta@cwjamaica.com; Pres. BYRON FARQUHARSON; Gen. Sec. ADOLPH CAMERON.

**Jamaica Association of Local Government Officers:** 15A Old Hope Rd, Kingston 5; tel. 929-5123; fax 960-4403; e-mail jalgo@cwjamaica.com; Pres. STANLEY THOMAS.

**Jamaica Civil Service Association:** 10 Caledonia Ave, Kingston 5; tel. 968-7087; fax 926-2042; e-mail jacisera@cwjamaica.com; Pres. WAYNE JONES; Sec. DENHAM WHILBY.

**Jamaica Federation of Musicians:** 5 Balmoral Ave, Kingston 10; tel. 926-8029; fax 929-0485; e-mail jafedmusic@cwjamaica.com; internet jafedmusic.tripod.com; f. 1958; Pres. HEDLEY H. G. JONES; Sec. CARL AYTON; 2,000 mems.

**Jamaica Teachers' Association:** 97 Church St, Kingston; tel. 922-1385-7; fax 922-1385; e-mail jta@cwjamaica.com; Gen. Sec. Dr ADOLPH CAMERON.

**Jamaica Union of Public Officers and Public Employees:** 4 Northend Pl., Kingston 10; tel. 929-1354; Pres. FITZROY BRYAN; Gen. Sec. NICKELLOH MARTIN.

**Jamaica Workers' Union:** 3 West Ave, Kingston 4; tel. 922-3222; fax 967-3128; Pres. CLIFTON BROWN; Gen. Sec. MICHAEL NEWTON.

**Union of Schools, Agricultural and Allied Workers (USAAW):** 2 Wildman St, Kingston; tel. 967-2970; f. 1978; Pres. DWAYNE BARNETT.

**Union of Technical, Administrative and Supervisory Personnel:** 108 Church St, Kingston; tel. 922-2086; Pres. ANTHONY DAWKINS; Gen. Sec. REG ENNIS.

**United Portworkers' and Seamen's Union (UPWU):** Kingston.

**United Union of Jamaica:** 35A Lynhurst Rd, Kingston; tel. 960-4206; Pres. JAMES FRANCIS; Gen. Sec. WILLIAM HASFAL.

**University and Allied Workers' Union (UAWU):** 50 Lady Musgrave Rd, Kingston; tel. 927-7968; fax 927-9931; e-mail jacisera@cwjamaica.com; affiliated to the WPJ; Pres. Prof. TREVOR MUNROE.

There are also some 30 associations registered as trade unions.

# Transport

## RAILWAYS

There are about 339 km (211 miles) of railway, all standard gauge, in Jamaica. The government-subsidized Jamaica Railway Corpn operated 207 km of the track until freight and passenger services were suspended in 1992 owing to falling revenues not meeting maintenance costs. Negotiations towards the privatization of the Jamaica Railway Corpn between the Government and the Indian/Canadian consortium Railtech broke down in late 2003. In early 2005 the Government signed a memorandum of understanding with China National Machinery and Equipment Corpn for the reconstruction of the railway system, including the provision of passengers coaches and engines.

**Jamaica Railway Corpn (JRC):** 142 Barry St, POB 489, Kingston; tel. 922-6620; fax 922-4539; f. 1845 as Jamaica Railway Co, the earliest British colonial railway; transferred to JRC in 1960; govt-owned, but autonomous, statutory corpn until 1990, when it was partly leased to Alcan Jamaica Co Ltd (subsequently West Indies Alumina Co) as the first stage of a privatization scheme; 207 km of railway; Chair. W. TAYLOR; Gen. Man. OWEN CROOKS.

**Alcoa Railroads:** Alcoa Minerals of Jamaica Inc, May Pen PO; tel. 986-2561; fax 986-2026; 43 km of standard-gauge railway; transport of bauxite; Supt RICHARD HECTOR; Man. FITZ CARTY (Railroad Operations and Maintenance).

**Kaiser Jamaica Bauxite Co Railway:** Discovery Bay PO, St Ann; tel. 973-2221; 25 km of standard-gauge railway; transport of bauxite; Gen. Man. TIM DAMON.

## ROADS

Jamaica has a good network of tar-surfaced and metalled motoring roads. According to estimates by the International Road Federation, there were 18,700 km of roads in 2004, of which 70.1% were paved. In 2001 a consortium of two British companies, Kier International and Mabey & Johnson, was awarded a contract to supply the materials for and construct six road bridges in Kingston and Montego Bay, and a further 20 bridges in rural areas. In the same year the Inter-American Development Bank approved a US $24.5m. loan to improve road maintenance. In March 2002 construction of the first part of a 230-km highway system linking major cities was scheduled to begin.

## SHIPPING

The principal ports are Kingston, Montego Bay and Port Antonio. The port at Kingston has four container berths, and is a major transhipment terminal for the Caribbean area. Jamaica has interests in the multinational shipping line WISCO (West Indies Shipping Corpn—based in Trinidad and Tobago). Services are also provided by most major foreign lines serving the region. In January

2004 a US $35m. plan to expand Kingston's container-handling capacity by 25% was announced.

**Port Authority of Jamaica:** 15–17 Duke St, Kingston; tel. 922-0290; fax 924-9437; e-mail pajmktg@infochan.com; internet www.seaportsofjamaica.com; f. 1966; Govt's principal maritime agency; responsible for monitoring and regulating the navigation of all vessels berthing at Jamaican ports, for regulating the tariffs on public wharves, and for the devt of industrial Free Zones in Jamaica; Pres. and Chair. NOEL HYLTON.

**Kingston Free Zone Co Ltd:** 27 Shannon Dr., POB 1025, Kingston 15; tel. 923-5274; fax 923-6023; e-mail kfzclsvc@infochan.com; internet www.portjam.com/f_zones.htm; f. 1976; subsidiary of Port Authority of Jamaica; management and promotion of an export-orientated industrial free trade zone for cos from various countries; Gen. Man. CLAUDE FLETCHER.

**Montego Bay Free Zone:** POB 1377, Montego Bay; tel. 979-8696-8; fax 979-8088; e-mail clients-mbfz@jadigiport.com; internet www.portjam.com; Gen. Man. CLAUDE FLETCHER.

**Shipping Association of Jamaica:** 4 Fourth Ave, Newport West, POB 1050, Kingston 15; tel. 923-3491; fax 923-3421; e-mail jfs@jashipco.com; internet www.seaportsofjamaica.com/tsaj; f. 1939; 63 mems; an employers' trade union which regulates the supply and management of stevedoring labour in Kingston; represents members in negotiations with govt and trade bodies; Pres. HENRY MARAGH; Gen. Man. TREVOR RILEY.

### Principal Shipping Companies

**Jamaica Freight and Shipping Co Ltd (JFS):** 80–82 Second St, Port Bustamante, POB 167, Kingston 13; tel. 923-9271; fax 923-4091; e-mail cshaw@toj.com; cargo services to and from the USA, Caribbean, Central and South America, the United Kingdom, Japan and Canada; Exec. Chair. CHARLES JOHNSTON; Man. Dir GRANTLEY STEPHENSON.

**Portcold Ltd:** 122 Third St, Newport West, Kingston 13; tel. 923-7425; fax 923-5713; Chair. and Man. Dir ISHMAEL E. ROBERTSON.

### CIVIL AVIATION

There are two international airports linking Jamaica with North America, Europe, and other Caribbean islands. The Norman Manley International Airport is situated 22.5 km (14 miles) outside Kingston. The Donald Sangster International Airport is 5 km (3 miles) from Montego Bay. A J $800m. programme to expand and improve the latter was under consideration.

**Airports Authority of Jamaica:** Victoria Mutual Bldg, 53 Knutsford Blvd, POB 567, Kingston 5; tel. 926-1622; fax 926-0356; e-mail aaj@cwjamaica.com; internet www.aaj.com.jm; Chair. DENNIS MORRISON; Pres. EARL RICHARDS.

**Civil Aviation Authority:** 4 Winchester Rd, POB 8998, Kingston 10; tel. 960-3948; fax 920-0194; e-mail jcivav@jcaa.gov.jm; internet www.jcaa.gov.jm; f. 1996; Dir-Gen. Col TORRANCE LEWIS.

**Air Jamaica Ltd:** 72–76 Harbour St, Kingston; tel. 922-3460; fax 922-0107; internet www.airjamaica.com; f. 1968; privatized in 1994, Govet reacquired in 2004; services within the Caribbean and to Canada (in asscn with Air Canada), the USA and the United Kingdom; Exec. Chair Dr VINCENT M. LAWRENCE; CEO CHRISTOPHER ZACCA.

**Air Jamaica Express:** Tinson Pen Aerodrome, Kingston 11; tel. 923-6664; fax 937-3807; internet www.airjamaica.com/express; previously known as Trans-Jamaican Airlines; internal services between Kingston, Montego Bay, Negril, Ocho Rios and Port Antonio and services to the Cayman Islands, the Bahamas, the Dominican Republic and Cuba; Man. Dir PAULO MOREIRA.

**Air Negril:** Montego Bay; tel. 940-7741; fax 940-6491; e-mail negriljeff@yahoo.com; internet caribbean-travel.com/airnegril; domestic charter services.

## Tourism

Tourists, mainly from the USA, visit Jamaica for its beaches, mountains, historic buildings and cultural heritage. In 2004 there were an estimated 1,414,786 visitors (excluding cruise-ship passengers). Tourism receipts were estimated to be US $1,621m. in 2003. In 2001 there were some 24,007 rooms in all forms of tourist accommodation.

**Jamaica Tourist Board (JTB):** 64 Knutsford Blvd, Kingston 5; tel. 929-9200; fax 929-9375; e-mail info@visitjamaica.com; internet www.visitjamaica.com; f. 1955; a statutory body set up by the Govt to promote all aspects of the tourist industry; Chair. DENNIS MORRISON; Dir of Tourism PAUL PENNICOOK.

**Jamaica Hotel and Tourist Association (JHTA):** 2 Ardenne Rd, Kingston 10; tel. 926-3635; fax 929-1054; e-mail info@jhta.org; internet www.jhta.org; f. 1961; trade asscn for hoteliers and other cos involved in Jamaican tourism; Pres. HORACE PETERKIN; Exec. Dir CAMILLE NEEDHAM.

## Defence

In August 2004 the Jamaican Defence Force consisted of 2,830 men on active service. This included an army of 2,500, a coastguard of 190 and an air wing of 140 men. There were reserves of some 953.

**Defence Budget:** an estimated J $3,000m. (US $50m.) in 2004.

**Chief of Staff:** Rear Adm. H. M. LEWIN.

## Education

Primary education was compulsory in certain districts, and free education was ensured. The education system consisted of a primary cycle of six years, followed by two secondary cycles of three and four years, respectively. In 2002 enrolment at primary schools included 97% of children in the relevant age-group. In the same academic year enrolment at secondary schools was equivalent to 79% of children in the relevant age-group. Higher education was provided by the College of Arts, Science and Technology, the College of Agriculture and the University of the West Indies, which had five faculties (arts and general studies, natural sciences, social sciences, medicine and a school of education) situated at its Mona campus, in Kingston. Government spending on education in 2005/06 was budgeted at some J $34,000m., representing some 9.8% of total planned expenditure.

# Bibliography

For works on the Caribbean generally, see Select Bibliography (Books)

Besson, J., and Mintz, S. W. *Martha Brae's Two Histories*. Raleigh, NC, University of North Carolina Press, 2002.

Harrison, M. *King Sugar: Jamaica, the Caribbean and the World Sugar Industry*. New York, NY, New York University Press, 2001.

Harriott, A. *Understanding Crime in Jamaica*. Kingston, University of the West Indies Press, 2004.

Hillman, R. S., and D'Agostino, T. J. *Distant Neighbors: The Dominican Republic and Jamaica in Comparative Perspectives*. New York, NY, Praeger Publrs, 1992.

Ingram, K. E. *Jamaica* (World Bibliographical Series). Oxford, ABC Clio, 1997.

Johnson, A. S. *Jamaican Leaders*. Kingston, Teejay, 2001.

King, C. L. *Michael Manley and Democratic Socialism: Political Leadership and Ideology in Jamaica*. San José, CA, Resource Publications, 2003.

Kirton, C. *Jamaica: Debt and Poverty*. Oxford, Oxfam, 1992.

Lundy, P. *Debt and Adjustment: Social and Environmental Consequences in Jamaica*. Aldershot, Ashgate, 1999.

Manderson, P., et al. *The Story of the Jamaican People*. Kingston, Ian Randle Publrs, 1997.

Manley, M. *The Politics of Change: A Jamaican Testament*, revised edn. Washington, DC, Howard University Press, 1990.

*Jamaica: Struggle in the Periphery*. London, Writers' and Readers' Publishing Co-operative Society, 1982.

Mars, P., and Young, A. L. *Caribbean Labor and Politics: Legacies of Cheddi Jagan and Michael Manley*. Detroit, MI, Wayne State University Press, 2004.

Miller, E. *Jamaica in the Twenty-First Century: Contending Issues*. Kingston, Grace Kennedy Foundation, 2001.

Mason, P. *Jamaica in Focus*. London, Latin American Bureau, 2000.

Monteith, K., and Richards, G. (Eds). *Jamaica in Slavery and Freedom: History, Heritage and Culture*. Kingston, University of the West Indies Press, 2001.

Payne, A. J. *Politics in Jamaica*, revised edn. Kingston, Ian Randle Publrs, 1994.

Patterson, P. J. *A Jamaica Voice in Caribbean and World Politics*. Kingston, Ian Randle Publrs, 2002.

Persaud, R. B., and Cox, R. W. *Counter-Hegemony and Foreign Policy: The Dialectics of Marginalized and Global Forces in Jamaica*. Albany, NY, SUNY, 2001.

Weston, A., and Viswanathan, U. (Eds). *Jamaica after NAFTA: Trade Options and Sectoral Strategies*. Ottawa, ON, North-South Institute, 1998.

# MARTINIQUE

## Geography

### PHYSICAL FEATURES

The Overseas Department of Martinique is an integral part of France, but is located in the Windward Islands, in the Lesser Antilles, 6,856 km (4,261 miles) from the French capital, Paris. Its immediate neighbours are the two anglophone Windward nations of Dominica (25 km to the north—the islands of Guadeloupe, which are also French, lie beyond that) and Saint Lucia (37 km to the south). The Department, comprising the island of Martinique and its few offshore islets, covers an area of 1,100 sq km (425 sq miles), including 40 sq km of inland waters. This makes Martinique the largest single island of the Windwards or the Leewards (though smaller than the combined island of Guadeloupe, Basse-Terre and Grande-Terre together).

Martinique is about 80 km in length and 32 km at its widest, aligned more towards the north-west than along a straight north–south axis, its long, thin shape, broader in the north, distorted by two peninsulas. In the south-west there is a broad abutment of land, the north littoral of which forms the southern shore of the main bay on the leeward coast. The bay is named after the capital, Fort-de-France, which is on the north shore. Further north, but on the more rugged, eastern coast, thrusting out into the Atlantic, is the thinner Caravelle peninsula. South of these two features, the coastline (350 km around the whole island) is deeply indented and eroded, whereas the north is less so, having been more recently added to by volcanic action. The great volcano of Mt Pelée (1,397 m or 4,585 ft) dominates the north, while to its south the twin peaks (pitons) of Carbet achieve heights just below its own. The mountains in the south are much lower, with the land tending to fall away southwards, from the central, raised Lamentin plain of low, rounded hills and gentle valleys. The soil is extremely fertile, the north dominated by extensive rainforest (and banana and pineapple plantations) and the Lamentin plain by sugar cane. Native fauna is now scarce, but the flora remains rich and varied (the Carib name for Martinique meant 'island of flowers'—see History), the rainforest consisting of mahogany trees, mountain palms, bamboo and many other types of tree, as well as fostering flowering plants and orchids, and over 1,000 species of fern. This luxuriance flourishes on the productive emissions of former eruptions by Pelée—the last were in 1902, the one on 8 May being the most famously devastating, wiping out the then capital of St-Pierre and all but one of its 26,000 inhabitants.

### CLIMATE

The climate is a humid, subtropical one, moderated by the trade winds (alizés) from the Atlantic (Antilles means 'breezy islands'). Precipitation is higher in Martinique than in many of the Caribbean islands, owing to its mountains. There is a rainy season from June to November, brought by the same weather conditions that can bring hurricanes to the region—the latter only occasionally hit Martinique. A more usual natural hazard is the risk of flooding, while an unusual one is the volcano, which is now dormant. The average annual temperature is about 26°C (79°F), but it can be much cooler at altitude.

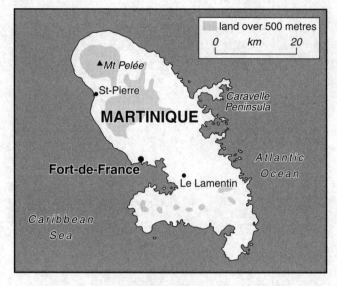

### POPULATION

Most of the population are of African descent, some 90% of the population being either black or of mixed descent (mainly black, white and Asian) and only 5% white (the descendants of the original white colonial settlers are known as *békés*). The rest of the population is mainly Asian—Indian or Chinese. The Roman Catholic Church claims the adherence of 78% of the population, with the rest being mainly Hindu or practitioners of pagan rites adapted by those brought from Africa in the 17th and 18th centuries. Everyone speaks French, and many a Creole patois thereof. The hybrid culture, although increasingly French since Martinique became a department, has always flourished and the island has long been considered the most sophisticated part of the French Antilles. Famous products and inhabitants of the island range from the rum invented by a Dominican monk, through the 'three queens' (Madame de Maintenon, mistress of Louis XIV, Marie Rose Tascher de la Pagerie, better known as Napoléon Bonaparte's Empress Josephine, and her cousin, Aimée Dubuc de Rivery, who married an Ottoman sultan and gave birth to his successor) to Aimé Césaire (politician, poet and a pioneer of the literary 'négritude' movement), one of the many writers from Martinique.

The total population was estimated at 394,000 in mid-2004. The capital and largest city is Fort-de-France (originally Fort-Royale, then République-Ville), on the west coast. The old administrative capital, St-Pierre, is further north up the coast. Inland from Fort-de-France, at the western end of the central plain of the same name, is Le Lamentin. There are also a number of other large towns not too far from Fort-de-France, and others on the east coast. Most people live in the capital or on the central plains.

# History

**PHILLIP WEARNE**

Revised for this edition by Dr JAMES FERGUSON

Martinique's name is either a corruption of the Amerindian (Carib) name of Madinina ('island of flowers') or a derivation of Saint Martin. The navigator Christopher Columbus sighted the island in 1493 or 1502—the date is disputed. It was first settled by the French in 1635, despite the hostility of the local Caribs, and was occupied with little interruption thereafter. Like Guadeloupe, Martinique was made an Overseas Department of France in 1946, its people becoming French citizens. The island's Governor was replaced with a Prefect and an elected Conseil général (General Council) was constituted. Thereafter, the French Government's policy of assimilation created a strongly French society, bound by linguistic, cultural and economic ties to metropolitan France. The island enjoyed a better infrastructure and a higher standard of living than its immediate Caribbean neighbours, but in consequence it also became heavily dependent on France. For many years economic power remained concentrated in the hands of the *békés* (descendants of white colonial settlers), who still owned most of the agricultural land and controlled the lucrative import-export market in the last decade of the 20th century. This led to little incentive for innovation or self-sufficiency and fostered resentment of lingering colonial attitudes.

The evolution of Martinique's political system was based on the French Government's response to the growth in nationalist sentiment during the latter half of the 20th century. In 1960 the mandate of the island's Conseil général was broadened, to permit discussion of political as well as administrative issues. In 1974 Martinique was granted regional status, as were Guadeloupe and French Guiana, and an indirectly elected Conseil régional (Regional Council), with some control over the local economy, was established. In the early 1980s the Socialist Government of President François Mitterrand tried to curb the continued growth of nationalist pressure and the threat of civil disturbances by instituting a policy of greater decentralization. The two local Conseils were given increased control over taxation, the local police and the economy.

In the first direct election to the Conseil régional in February 1983 the Department's left-wing parties, which articulated nationalist sentiments while supporting the French Government's policy of decentralization, gained 21 of the 41 seats. That success weakened the threat posed by militant separatist challengers inside and outside the left-wing parties. The most vocal of the separatist parties, the Mouvement Indépendantiste Martiniquais (MIM), won less than 3% of the votes cast.

One of the principal campaigners for greater autonomy, from the 1940s, was the veteran socialist writer and poet Aimé Césaire, leader of the Parti Progressiste Martiniquais (PPM). Mayor of Fort-de-France, Césaire held a seat in the French Assemblée nationale from 1945 until March 1993, when he was succeeded by Camille Darsières, the General Secretary of the PPM. With Louis-Joseph Dogué of the Fédération Socialiste de la Martinique, the local wing of the national Parti Socialiste (PS), standing down in the same election, the Gaullist Rassemblement pour la République (RPR) was able to increase its representation to three of the four seats reserved for Martinique in the Assemblée nationale, through the election of André Lesueur, mayor of Rivière-Salée, Pierre Petit and Anicet Turinay.

The 1993 election results in Martinique reflected the increase in support for the right wing seen throughout France, reversing the trend of the late 1980s towards consolidation of the left-wing vote. In the 1988 presidential elections François Mitterrand received 71% of votes cast on the island. Later in that year, the left wing secured control of the Conseil général with a one-seat majority. Emile Maurice, from the conservative RPR, was elected President of the Conseil général, for his seventh term. However, Maurice was defeated in the 1992 elections, and replaced by the PPM candidate, Claude Lise. The left-wing predominance in local politics was further enhanced by the results of elections to the Conseil régional at the same time. Although the right-wing grouping was the largest single list, with 16 of the 41 seats, the parties of the left had a working majority.

Elections in March 1994 brought little change in the composition of the Conseil général. The PPM retained 10 seats, the RPR seven and the centrist Union pour la Démocratie Française (UDF) three seats, with Claude Lise re-elected President of the Conseil général. However, at elections to the European Parliament in June, the conservative government list secured the greatest proportion of the votes cast (37%). A combined list of the parties of the left, the Rassemblement d'Outre-Mer et des Minorités, came second with 20% of the votes.

At the first round of voting in the 1995 national presidential election, which took place on 23 April, Martinique was the only French Overseas Possession in which the candidate of the PS, Lionel Jospin, received the greatest proportion of the valid votes cast. Jospin, supported by all the parties of the left, won the presidency at the second round on 7 May, which was contested against Jacques Chirac of the RPR. At municipal elections in June the PPM retained control of Martinique's principal towns. In September Lise was elected to the Sénat, while the incumbent PPM representative, Rodolphe Désiré, was returned to office.

At elections to the Assemblée nationale in May and June 1997, Turinay and Pierre Petit of the RPR were re-elected, together with Camille Darsières of the PPM. Alfred Marie-Jeanne, the First Secretary and a founding member of the MIM, was elected in the Le François-Le Robert constituency (hitherto held by the RPR). At elections to the Conseil régional in March 1998 the left retained a majority. The MIM increased its representation to 13 seats, while the PPM held seven seats, the RPR six and the UDF five. Marie-Jeanne was elected to the presidency of the Conseil régional. In concurrent elections to the Conseil général the parties of the left again performed well, increasing their representation from 26 to 29 seats, with right-wing candidates securing 14 seats and independents two. Claude Lise was re-elected President of the Conseil général.

Martinique was adversely affected by industrial action in 1998 and 1999, with strikes occurring among banana- and automobile-sector workers. The crisis in the banana industry was caused by falling prices in the European market; however, the two-month strike was ended in January 1999, when a pay agreement was reached. There was further conflict in the sector in October, however, when, prior to a two-day visit by French Prime Minister Lionel Jospin, banana producers occupied the headquarters of the French naval forces for several days, demanding the disbursement of exceptional aid to compensate for the adverse effect on their industry of a dramatic decline in prices on the European market. Marie-Jeanne, who was opposed to the limited nature of the Government's plans for institutional reform, refused to participate in the events organized for Jospin's visit. The Prime Minister announced an emergency plan for the banana sector and agreed, in principle, to a proposal for greater autonomy for the local authorities in conducting relations with neighbouring countries and territories. The dispute at the Toyota motor company, where workers were demanding substantial pay increases and a reduction in working hours, lasted five months and involved secondary action and blockades by trade unionists, but was eventually settled in November 1999.

The issue of Martinique's constitutional status also arose in 1999, following a series of meetings between the Presidents of the Conseils régionaux of Martinique, Guadeloupe and French Guiana. In December Marie-Jeanne co-signed a declaration, stating the intention of the three Presidents to propose, to the

French Government, legislative and constitutional amendments aimed at creating a new status of 'overseas region'. The declaration and subsequent announcements by Marie-Jeanne and his counterparts were dismissed by Jean-Jack Queyranne, Secretary of State for Overseas Affairs, in February 2000 as unconstitutional and exceeding the mandate of the politicians responsible. However, in May a number of proposals, including the extension of the Departments' powers in areas such as regional co-operation, were provisionally accepted by the Assemblée nationale; a modified version of the proposals was subsequently adopted, by a narrow margin, by the Sénat. In November the Assemblée approved the proposals and in December they were ratified by the Conseil constitutionnel. Following a meeting of members of the Conseil régional and the Conseil général in late June 2001, a series of proposals on greater autonomy, to be presented to the French Government, was agreed upon. These included: the division of the territory into four districts; the creation of a 'collectivité territoriale' (territorial community), governed by a 41-member assembly elected for a five-year term; and the establishment of an independent executive council. Furthermore, the proposals included a request that the territory be given control over legislative and administrative affairs, as well as legislative authority on matters concerning Martinique alone. In November the French Government announced itself to be in favour of the suggested constitutional developments, and in March 2003 the two houses of the French parliament approved constitutional changes that would allow for a referendum on proposals for greater autonomy.

At a referendum held in September 2000, Martinique voted overwhelmingly in favour (91%) of reducing the presidential mandate from seven to five years. At municipal elections held in March 2001 the PPM retained control of the majority of municipalities (including Fort-de-France, where Aimé Césaire, retiring as mayor after 56 years, was succeeded by a fellow PPM member, Serge Letchimy, who defeated Alfred Marie-Jeanne). In the concurrent election to the Conseil général, Claude Lise was re-elected as President. In mid-2001 the French President, Jacques Chirac, rejected a joint request by French Guiana, Guadeloupe and Martinique that they be permitted to join, as associate members, the Association of Caribbean States.

At the first round of national presidential elections, held on 21 April 2002, Jacques Chirac obtained 33.0% of votes cast on the island; Lionel Jospin came second, with 29.2% of the votes. At the second round, held on 5 May, Chirac emerged victorious, overwhelmingly defeating the Front National candidate, Jean-Marie Le Pen, who secured only 3.9% of votes cast in Martinique. At elections to the Assemblée nationale, held on 9 and 18 June, only the MIM candidate, Alfred Marie-Jeanne, was re-elected; the PS representative, Louis-Joseph Manscour, Alfred Almont of the Union pour un Mouvement Populaire (UMP—the new grouping of the centre-right) and Pierre-Jean Samot of the left-wing Bâtir le Pays Martinique were also successful. Samot was ordered to resign by the Conseil constitutionnel in Paris in March 2003 for receiving campaign funds for the 2002 elections from his party, which was not officially registered at the time. He was replaced by Philippe Edmond-Mariette, also of Bâtir le Pays Martinique, who won a by-election in May 2003.

The referendum on Martinique's constitutional status and the proposed new 'collectivité territoriale' framework was held in December 2003. Despite the explicit support of Marie-Jeanne for the new arrangement, which would have replaced the Conseil général and Conseil régional with a single elected council, 50.48% of those voting opposed the change. The rejection of the reform plan, proposed by the centre-right Government of Jean-Pierre Raffarin, was widely seen as symptomatic of fears that the French Government would seek to reduce subsidies and social security payments under any new constitutional arrangement.

In elections to the Conseil régional held in March 2004, a joint list comprising Marie-Jeanne's MIM won 28 of the 41 seats, with 53.8% of the votes cast, while the PPM and other left-wing candidates won only nine seats and 30.6% of the ballot. Marie-Jeanne was subsequently elected President of the Conseil régional.

In the national referendum on ratification of the proposed constitutional treaty of the European Union, conducted on 29 May 2005, Martinique voted overwhelmingly in favour of the proposal (69.0%). At odds with the mainland French rejection of the proposed constitution, Martinique's endorsement was also overshadowed by a massive rate of abstention, with 71.6% of voters ignoring the poll. Neither the pro-constitution centre-right and centre-left, nor Marie-Jeanne's anti-constitution, pro-independence supporters, could claim credit for the outcome.

# Economy

## PHILLIP WEARNE

Revised for this edition by Dr JAMES FERGUSON

Like Guadeloupe, Martinique's economy is closely tied to that of France. Aid and subsidies, in various forms, are necessary to balance a huge deficit between visible exports and imports. In 2003 the trade deficit was €1,570m., with export earnings worth only approximately 18% of the total value of imports. Some 73% of exports in 2003 went to France; Guadeloupe and French Guiana together accounted for a further 20%. The European Union (EU) countries of Belgium, Luxembourg and the United Kingdom are also significant purchasers. The main exports were bananas, refined petroleum products and rum. France was also the single largest source of imports, accounting for 63.0% of their total value in 2003. The bulk of the remainder came from other members of the EU, the USA and Venezuela. The principal imports in 2003 included machinery and transport equipment (especially road vehicles), food and live animals, manufactured goods, chemicals and mineral fuels. Agriculture was the primary economic activity, with sugar, bananas, fruit, vegetables and some flowers being the principal crops. There was some light industry, the largest export being rum. Tourism was a major source of convertible-currency revenue. Consumer prices increased by an annual average of 2.0% in 1990–2004; consumer prices increased by 2.1% in 2003 and in 2004. Per-head gross domestic product (GDP) was estimated at approximately US $10,723 in 2001.

## AGRICULTURE

The sugar industry was Martinique's original source of prosperity. However, it was dealt a devastating blow by the volcanic eruption of Mt Pelée in 1902. Bananas then became the major export. Banana exports grew steadily in the late 1980s to reach 215,980 metric tons by 1990. From the 1990s the volume of exports fluctuated, owing to variable climatic conditions. A significant decline in prices on the European market, and an ongoing dispute between the USA and four Latin American countries and the EU over the latter's banana import regime also threatened Martinique's banana-growing sector. By 2000 production had recovered to 316,194 tons, but by 2003 output had fallen to 290,000 tons. Exports in that year reached 264,959 tons, accounting for around 40% of total export revenues.

Sugar cane remained the island's major agricultural crop, despite low world prices, under-investment and the diversification of some cane-growing land to the cultivation of other crops. By 1982 local production proved insufficient to supply domestic demand, causing a reversal of the policy of neglect and a dramatic increase in the cane harvest, to 217,000 metric tons by 1989. In the early 1990s production declined, but recovered again, to reach 212,000 tons in 1996; in 2003 the cane harvest totalled an estimated 207,000 tons. Output of sugar fluctuated accordingly,

# MARTINIQUE

and stood at 5,200 tons in 2003. Local consumption accounted for virtually all the harvest, with about one-third going on the production of rum, the island's major manufactured product. Output of rum historically varied according to the supply of sugar, but in the early 1990s fell precipitously. In 1990 it declined by almost 20%, to 84,828 hl. Production declined further, to 69,458 hl by 1998; however, by 2002 output had recovered to 91,629 hl, before falling to 80,731 hl in 2003.

In the 1980s agricultural diversification became official government policy and it contributed to efforts to increase export earnings and reduce the cost of food imports. Pineapples, avocados and aubergines became significant export crops, and flowers and citrus fruits, particularly limes, were also shipped abroad. The most dramatic growth was in the cultivation of melons and pineapples, according to the UN's Food and Agriculture Organization (FAO) estimates, output of the former reaching an estimated 2,700 metric tons in 2003, and production of the latter rising to 20,800 tons in the same year. However, only melon exports were having a major economic impact by 2003, when 1,654 tons were shipped overseas. Some 265 tons of taro, 11 tons of pineapples, 21 tons of avocados and 882,000 cut flowers were exported in 1998.

With less land available for pasture than on Guadeloupe, a significant proportion of the island's meat and dairy products had to be imported, although the local administration claimed some success in boosting livestock production. In 2003, according to FAO estimates, some 2,300 metric tons of beef and veal, 1,700 tons of pig meat and 2,200 tons of cows' milk were produced. This was produced from a livestock population put at 25,000 cattle, 34,000 sheep, 35,000 pigs and 17,000 goats. In 2003 it was estimated that 46% of fresh meat consumed was locally produced. The total fishing catch declined sharply in the late 1980s, to only 3,314 tons in 1989, but by the early 2000s had recovered, totalling 6,300 tons in 2003. The absence of a full marketing structure was cited as the main obstacle to further development of the sector.

## INDUSTRY

Industry (including construction and public works) employed some 10.6% of the total labour force in 1998 and accounted for an estimated 14% of GDP in 2000, but the sector remained underdeveloped. Industry's total contribution to the economy was somewhat inflated by the petroleum refinery, processing crude petroleum imported from Venezuela, Trinidad and Tobago and even Saudi Arabia. By 2003 exports of refined petroleum products accounted for 26.0% of the total value of exports. Energy was derived mainly from mineral fuels, imports of which accounted for 12.0% of the total value of imports in 2003. Martinique generated an estimated 1,151m. kWh of energy in 2001.

Other industrial activity was generally concentrated on food and drink processing, in particular fish and fruit canning, rum distillation, soft-drink manufacture and sugar refining. Some 44,678 hl of rum were exported in 2003, a fall of 9% compared with 2002's figure and a significant decrease on the 53,324 hl exported in 1991. Most of the rum exports were destined for France. The island did, however, boast a polyethelyne plant and a cement factory. The latter produced an estimated 220,000 metric tons of cement in 2002, virtually all of which was used locally. There were also some small wood-furniture manufacturers, construction material producers and a paper-carton outlet.

## TOURISM

Martinique's tourist attractions are its beaches and coastal scenery, its mountainous interior and the historic towns of Fort-de-France and St-Pierre. From the early 1990s tourism has remained one of the most important sources of foreign exchange, with a steady growth in numbers of arrivals. Most visitors were from Europe, although proportionally the number of visitors from the USA grew most. The hotel industry experienced rapid expansion, with the total number of rooms offered rising from 3,735 in 1991 to 5,115 by 2002. The number of tourist arrivals excluding cruise-ship passengers and same-day visitors totalled 453,159 in 2003. In that year some 78.9% of arrivals were from metropolitan France and 4% from European countries other than France. Tourism receipts totalled US $247m. in that year. In late 2002 the French hotel group Accor announced that it would close its five hotels on Guadeloupe and Martinique, citing high operating costs, poor industrial relations and decreasing tourist arrivals, but in early 2005 the group still owned two hotels in Martinique. The liquidation of the French airline Air Lib in February 2003 had a further negative impact on the sector.

## EMPLOYMENT

As in Guadeloupe, relatively high wages—tied to those of metropolitan France—coupled with the high levels of aid and imports from France served to restrict economic development. The descendants of the original colonial settlers, or *békés*, were reinforced by professionals, whose spending power, together with that of the foreign tourists, simply encouraged the development of the services sector and import businesses. On the other hand, the native black population was reinforced by illegal immigrants from Saint Lucia, Dominica and Haiti.

Thus, as the more labour-intensive agricultural sector contracted, emphasis on the services sector increased and unemployment rose, particularly as more young people entered the job market. The lack of job prospects encouraged extensive emigration, to France and other Caribbean islands. Some 22.5% of the labour force was unemployed in 2003. Dependency on French subsidies increased in the first years of the 21st century, with social security payments worth €466m. paid to 91,589 Martinicans in 2003.

# Statistical Survey

Sources (unless otherwise stated): Institut national de la statistique et des études économiques (INSEE), Service Régional de Martinique, Centre Delgrès, blvd de la Pointe des Sables, Les Hauts de Dillon, BP 641, 97262 Fort-de-France Cédex; tel. 60-73-60; fax 60-73-50; internet www.insee.fr/fr/insee_regions/martinique; Ministère des départements et territoires d'outre-mer, 27 rue Oudinot, 75700 Paris 07 SP; tel. 1-53-69-20-00; fax 1-43-06-60-30; internet www.outre-mer.gouv.fr.

## AREA AND POPULATION

**Area:** 1,100 sq km (424.7 sq miles).

**Population:** 359,579 (males 173,878, females 185,701) at census of 15 March 1990; 381,427 at census of 8 March 1999. *Mid-2004* (UN estimate): 394,000 (Source: UN, *World Population Prospects: The 2004 Revision*).

**Density** (at mid-2004): 358.2 per sq km.

**Principal Towns** (at 1999 census): Fort-de-France (capital) 94,049; Le Lamentin 35,460; Le Robert 21,240; Schoelcher 20,845; Sainte-Marie 20,098; Le François 18,559; Saint-Joseph 15,785; Ducos 15,240. *Mid-2003* (UN estimate, incl. suburbs): Fort-de-France 93,138 (Source: UN, *World Urbanization Prospects: The 2003 Revision*).

**Births, Marriages and Deaths** (2001): Registered births 5,774 (birth rate 14.9 per 1,000); Registered marriages 1,571; Registered deaths 2,734 (death rate 7.1 per 1,000). *2002:* Registered births 5,391; Registered marriages 1,511; Registered deaths 2,659 (preliminary).

**Expectation of Life** (years at birth, 1997): Males 75.2; Females 81.7.

**Economically Active Population** (persons aged 15 years and over, 1998): Agriculture and fishing 7,650; Industry 7,103; Construction and public works 10,405; Trade 16,196; Transport 4,383; Financial services and real estate 3,354; Business services 8,376;

# MARTINIQUE

Public services 14,179; Education 14,991; Health and social security 10,676; Administrative services 18,742; *Total employed* 116,055 (males 62,198, females 53,857); Unemployed 48,537 (males 22,628, females 25,909); *Total labour force* 164,592 (males 84,826, females 79,766). *Mid-2003:* Agriculture, etc. 6,000; Total labour force 187,000 (Source: FAO). *December 2003:* Unemployed 37,751.

## HEALTH AND WELFARE
### Key Indicators

**Number of Physicians** (1995): 680.

**Number of Hospital Beds** (1995): 2,100.

For definitions, see explanatory note on p. vi.

## AGRICULTURE, ETC.

**Principal Crops** (FAO estimates, '000 metric tons, 2003): Yams 7.5; Other roots and tubers 13.3; Sugar cane 207; Lettuce 7.8; Tomatoes 6.1; Cucumbers and gherkins 4; Other fresh vegetables 8.7; Bananas 290; Plantains 16; Pineapples 20.8.

**Livestock** (FAO estimates, '000 head, year ending September 2003): Cattle 25; Sheep 34; Pigs 35; Goats 17.

**Livestock Products** (FAO estimates, '000 metric tons, 2003): Beef and veal 2.3; Pig meat 1.7; Poultry meat 1.0; Cows' milk 2.2; Hen eggs 1.5.

**Forestry:** Roundwood removals ('000 cu m, excluding bark, 1999): Sawlogs, veneer logs and logs for sleepers 2; Fuel wood 10; Total 12. *2001–03:* Production as in 1999 (FAO estimates).

**Fishing** ('000 metric tons, live weight, 2003): Capture 6.2—FAO estimate (Clupeoids 4,000; Common dolphinfish 220—FAO estimate; Other marine fishes 1,080—FAO estimate; Caribbean spiny lobster 190; Clams, etc. 700); Aquaculture 100; *Total catch* 6,300—FAO estimate. Source: FAO.

## MINING

**Production** (estimates, '000 metric tons, unless otherwise indicated, 2002): Pumice 130; Salt 200. Source: US Geological Survey.

## INDUSTRY

**Production** ('000 metric tons, 2000, unless otherwise indicated): Pineapple juice 3.2 (1994); Canned or bottled pineapples 18.4 (1994); Raw sugar 5.3 (2002); Rum (hl) 91,629 (2002); Motor spirit (petrol) 155 (estimate); Kerosene 134 (estimate); Gas-diesel (distillate fuel) oils 162 (estimate); Residual fuel oils 278 (estimate); Liquefied petroleum gas 22 (estimate); Cement 220 (estimate, 2002); Electric energy (million kWh) 1,085 (estimate). Source: partly UN, *Industrial Commodity Statistics Yearbook*.

## FINANCE

**Currency and Exchange Rates:** The French franc was used until the end of 2001. Euro notes and coins were introduced on 1 January 2002, and the euro became the sole legal tender from 18 February. Some of the figures in this Survey are still in terms of francs. For details of exchange rates, see French Guiana.

**Budget** (forecasts, million French francs, 2001): *Revenue:* Tax revenue 836.9 (Departmental taxes 332.0, Fuel tax 295.0, Transfer taxes, etc. 58.0, Motor vehicle tax 68.0, Fiscal subsidy 53.0); Other current revenue 886.6 (Refunds of social assistance 65.0, Operational allowance 315.0, Decentralization allowance 477.0); Capital revenue 499.5 (EU development funds 71.0, Capital allowances 59.0, Other receipts 101.4, Borrowing 270.0); Total 2,223.0. *Expenditure:* Current expenditure 1,482.2 (Finance service 57.1, Permanent staff 394.7, General administration 65.1, Other indirect services 69.0, Administrative services 108.4, Public health 49.9, Social assistance 503.6, Support costs of minimum wage 99.8, Economic services 114.7); Capital expenditure 740.8 (Road system 139.5, Networks 47.9, Education and culture 111.5, Other departmental programmes 101.6, Other public bodies 83.7, Other programmes 96.3, Non-programme expenditure 162.3); Total 2,223.0. Note: Figures refer only to the departmental budget.

**State Budget** (million francs, 1998): Revenue 4,757, Expenditure 8,309.

**Regional Budget** (€ million, 2002, excl. debt rescheduling): Total revenue 247 (Direct taxes 22, Indirect taxes 107, Transfers 79, Loans 38, Other 1); Total expenditure 247.

**Money Supply** (million French francs at 31 December 1998): Currency outside banks 924; Demand deposits at banks 6,330; Total money 7,254.

**Cost of Living** (Consumer Price Index; base: 2000 = 100): 104.2 in 2002; 106.4 in 2003; 108.6 in 2004. Source: ILO.

**Expenditure on the Gross Domestic Product** (million French francs at current prices, 1994): Government final consumption expenditure 6,962; Private final consumption expenditure 20,133; Increase in stocks 20; Gross fixed capital formation 5,102; *Total domestic expenditure* 32,217; Exports of goods and services 1,439; *Less* Imports of goods and services 9,150; *GDP in purchasers' values* 24,506. *2000* (€ million): GDP in purchasers' values 5,496.

**Gross Domestic Product by Economic Activity** (million French francs at current prices, 1992): Agriculture, hunting, forestry and fishing 1,106.3; Mining, quarrying and manufacturing 1,770.6; Electricity, gas and water 483.6; Construction 1,145.3; Trade, restaurants and hotels 4,022.1; Transport, storage and communications 1,427.6; Finance, insurance, real estate and business services 2,590.1; Government services 5,416.0; Other community, social and personal services 3,576.4; Other services 330.5; *Sub-total* 21,868.5; Import duties 791.0; Value-added tax 640.9; *Less* Imputed bank service charge 1,207.1; *GDP in purchasers' values* 22,093.4. Source: UN, *National Accounts Statistics*.

## EXTERNAL TRADE

**Principal Commodities:** *Imports c.i.f.* US $ million, 1995): Food and live animals 319.6 (Meat and meat preparations 82.8, Dairy products and birds' eggs 53.6, Fish and fish preparations 38.6, Cereals and cereal preparations 45.6, Vegetables and fruit 49.1); Beverages and tobacco 52.0 (Beverages 45.0); Mineral fuels, lubricants, etc. 148.0 (Petroleum and petroleum products 146.2); Chemicals and related products 189.5 (Medicinal and pharmaceutical products 83.9); Basic manufactures 260.1 (Paper, paperboard and manufactures 45.1); Machinery and transport equipment 637.6 (Power-generating machinery and equipment 62.8, General industrial machinery, equipment and parts 83.2, Telecommunications and sound equipment 41.1, Road vehicles and parts 240.7, Ships and boats 50.3); Miscellaneous manufactured articles 288.5 (Furniture and parts 46.2, Clothing and accessories, excl. footwear 62.6); Total (incl. others) 1,969.8 (Source: UN, *International Trade Statistics Yearbook*). *Exports f.o.b.* (million French francs, 1997): Bananas 462.2; Rum 126.7; Flavoured or sweetened water 78.8; Refined petroleum products 201.9; Yachts and sports boats 74.4; Total (incl. others) 1,263.3. *1997* (million French francs): *Imports:* 9,947.0; *Exports:* 1,239.0. *1998* (million French francs): *Imports:* 10,046.8; *Exports:* 1,701.4. *1999* (million French francs): *Imports:* 10,605.6; *Exports:* 1,715.1. *2000* (million French francs): *Imports:* 1,759; *Exports:* 294. *2001* (€ million): *Imports:* 1,885.5; *Exports:* 308.9. *2002* (€ million): *Imports:* 1,824.2; *Exports:* 324.8. Sources: Direction Générale des Douanes, Chambre de Commerce et d'Industrie de la Martinique.

**Principal Trading Partners** (€ million, 2000): *Imports c.i.f.*: France (metropolitan) 1,117; Germany 69; Italy 55; Japan 36; Netherland Antilles 41; USA 36; Venezuela 102; Total (incl. others) 1,759; *Exports f.o.b.*: Belgium-Luxembourg 8; France (metropolitan) 170; French Guiana 11; Guadeloupe 63; United Kingdom 10; Total (incl. others) 294.

## TRANSPORT

**Road Traffic** ('000 motor vehicles in use, 1995): Passenger cars 95.0; Commercial vehicles 21.5. Source: UN, *Statistical Yearbook*.

**Shipping:** *Merchant Fleet* (vessels registered '000 grt at 31 December, 1992): 1 (Source: Lloyd's Register of Shipping). *International Sea-borne Traffic* (2002): Goods loaded 848,000 metric tons; Goods unloaded 1,943,000 metric tons (Source: Direction Départementale de l'Equipement).

**Civil Aviation** (203): Passengers carried 1,519,000; Freight carried 13,354 metric tons. Source: Chambre de Commerce et d'Industrie de la Martinique.

## TOURISM

**Tourist Arrivals by Country** (excl. same-day visitors and cruise-ship arrivals, 2003): France 357,726; Guadeloupe 40,668; French Guiana 10,619; Total (incl. others) 453,159.

**Receipts from Tourism** (US $ million, incl. passenger transport): 245 in 2001; 237 in 2002; 247 in 2003.
Source: World Tourism Organization.

## COMMUNICATIONS MEDIA

**Radio Receivers** ('000 in use): 82 in 1997.

**Television Receivers** ('000 in use): 62 in 1999.

**Telephones** ('000 main lines in use): 172.0 in 2001.

**Facsimile Machines** (number in use): 5,200 in 1997.

MARTINIQUE

**Mobile Cellular Telephones** ('000 subscribers): 319.9 in 2002.

**Personal Computers** ('000 in use): 52 in 2001.

**Internet Users** ('000): 40 in 2001.

**Daily Newspaper:** 1 (estimate) in 1996 (estimated average circulation 30,000 copies).
Sources: UNESCO, *Statistical Yearbook*; UN, *Statistical Yearbook*; International Telecommunication Union.

### EDUCATION

**Pre-primary** (2002/03): 85 institutions; 19,041 students (Source: Ministère de l'Education Nationale, de l'Enseignement Supérieur et de la Recherche, *Repères et références statistiques sur les enseignements, la formation et la recherche 2003*).

**Primary** (2002/03): 188 institutions; 1,971 teachers (1,863 state, 108 private); 31,096 students (Source: Ministère de l'Education Nationale, de l'Enseignement Supérieur et de la Recherche, *Repères et références statistiques sur les enseignements, la formation et la recherche 2003*).

**Secondary** (2002/03, unless otherwise indicated): 78 institutions (2001/02); 2,024 teachers (1,887 state, 137 private); 49,426 students (Source: Ministère de l'Education Nationale, de l'Enseignement Supérieur et de la Recherche, *Repères et références statistiques sur les enseignements, la formation et la recherche 2003*).

**Vocational** (1998/99): 15 institutions (9 state, 6 private); 7,661 students (7,101 state, 560 private) (Source: Ministère de l'Education Nationale, de l'Enseignement Supérieur et de la Recherche, *Repères et références statistiques sur les enseignements, la formation et la recherche 2003*).

**Higher** (2004): 12,000 students (Université Antilles-Guyane).

# Directory

## The Government

(July 2005)

**Prefect:** YVES DASSONVILLE, Préfecture, 82 rue Victor Sévère, BP 647–648, 97262 Fort-de-France Cédex; tel. 5-96-63-18-61; fax 5-96-71-40-29.

**President of the General Council:** CLAUDE LISE (PPM), Conseil Général de la Martinique, blvd Chevalier Sainte-Marthe, 97200 Fort-de-France Cédex; tel. 5-96-55-26-00; fax 5-96-73-59-32; internet www.cg972.fr.

**Deputies to the French National Assembly:** LOUIS-JOSEPH MANSCOUR (PS), ALFRED ALMONT (UMP), PHILIPPE EDMOND-MARIETTE (Bâtir le Pays Martinique), ALFRED MARIE-JEANNE (MIM).

**Representatives to the French Senate:** SERGE LARCHER (Groupe Socialiste), CLAUDE LISE (Groupe Socialiste).

### Conseil Régional

Hôtel de Région, rue Gaston Deferre, BP 601, 97200 Fort-de-France Cédex; tel. 5-96-59-63-00; fax 5-96-72-68-10; e-mail service.communication@cr-martinique.fr; internet www.cr-martinique.fr.

**President:** ALFRED MARIE-JEANNE (MIM).

**Elections, 21 and 28 March 2004**

| | Seats |
|---|---:|
| Groupe des Patriotes—MIM CNCP* | 28 |
| Convergences Martiniquaises Union de la Gauche† | 9 |
| Forces Martiniquaises de Progrès | 4 |
| Total | 41 |

*A joint list comprising the Mouvement Indépendantiste Martiniquais, the Conseil National des Comités Populaires and the Alliance pour le Pays Martinique.
†A joint list comprising the Parti Progressiste Martiniquais and other left-wing candidates.

## Political Organizations

**Bâtir le Pays Martinique:** Fort-de-France; f. 1998; left-wing; split from the Parti Communiste Martiniquais; Leader PIERRE-JEAN SAMOT.

**Combat Ouvrier:** BP 821, 97258 Fort-de-France Cédex; Trostskyist; mem. of the Communist Internationalist Union; Leader GHISLAINE JOACHIM-ARNAUD.

**Conseil National des Comités Populaires (CNCP):** 97200 Fort-de-France; e-mail robert.sae@wanadoo.fr; pro-independence party; contested the 2004 regional elections in alliance with the MIM (q.v.); Leader ROBERT SAÉ.

**Fédération Socialiste de la Martinique (FSM):** 52 rue du Capitaine Pierre-Rose, 97200 Fort-de-France; tel. 5-96-60-14-88; e-mail fede972@parti-socialiste.fr; internet www.parti-socialiste.fr; local br. of the **Parti Socialiste (PS)**; Sec.-Gen. JEAN CRUSOL.

**Forces Martiniquaises de Progrès:** 97200 Fort-de-France; f. 1998 to replace the local br. of the Union pour la Démocratie Française; Pres. MIGUEL LAVENTURE.

**Mouvement des Démocrates et Écologistes pour une Martinique Souveraine (MODEMAS):** Fort-de-France; f. 1992; Pres. GARCIN MALSA.

**Mouvement Indépendantiste Martiniquais (MIM):** Fort-de-France; f. 1978; pro-independence party; First Sec. ALFRED MARIE-JEANNE.

**Mouvement Populaire Franciscain:** Fort-de-France; left-wing; Leader MAURICE ANTISTE.

**Osons Oser:** Fort-de-France; f. 1998; right-wing; affiliated with the metropolitan Union pour un Mouvement Populaire (UMP); Pres. PIERRE PETIT.

**Parti Progressiste Martiniquais (PPM):** Quartier Trénelle, 97200 Fort-de-France; tel. 5-96-71-86-83; f. 1957; left-wing; Leader MADELEINE DE GRANDMAISON; Sec.-Gen. PIERRE SUÉDILE.

**Parti Radical de Gauche:** 27 ancienne route de Schoelcher, 97233 Schoellcher; tel. 5-96-47-54-97; e-mail didier.saint-louis@wanadoo.fr; f. 1902; Pres. DIDIER SAINT-LOUIS; Gen. Sec. ROLAND ROSILLETTE.

**Les Verts Martinique:** Lot D, 54 rue Madinina, Cluny, 97200 Fort-de-France; tel. 5-96-73-07-45; fax 5-96-71-58-21; Leader LOUIS-LÉONCE LECURIEUX-LAFFERONNAY.

## Judicial System

**Cour d'Appel de Fort-de-France:** Palais de Justice, pl. Légitimée Défense, BP 634, 97262 Fort-de-France Cédex; tel. 5-96-70-62-62; fax 5-96-63-52-13; e-mail ca-fort-de-france@justice.fr; highest court of appeal for Martinique and French Guiana; First Pres. RENÉ SALOMON; Procurator-Gen. GÉRARD LOUBENS; Chief Clerk MARIETTE BELHUMEUR.

There are two Tribunaux de Grande Instance, at Fort-de-France and Cayenne (French Guiana), and three Tribunaux d'Instance (two in Fort-de-France and one in Cayenne).

## Religion

The majority of the population belong to the Roman Catholic Church.

### CHRISTIANITY

#### The Roman Catholic Church

Martinique comprises the single archdiocese of Fort-de-France, with an estimated 297,515 adherents (some 78% of the total population) at 31 December 2003. The Archbishop participates in the Antilles Episcopal Conference, currently based in Port of Spain, Trinidad and Tobago.

**Archbishop of Fort-de-France and Saint-Pierre:** Rev. MAURICE MARIE-SAINTE (designate), Archevêché, 5–7 rue du Révérend Père Pinchon, BP 586, 97207 Fort-de-France Cédex; tel. 5-96-63-70-70; fax 5-96-63-75-21; e-mail archeveche-martinique@wanadoo.fr.

#### Other Churches

Among the denominations active in Martinique are the Assembly of God, the Evangelical Church of the Nazarene and the Seventh-day Adventist Church.

MARTINIQUE                                                                                                                    *Directory*

## The Press

**Antilla:** 60 Jambette Beauséjour, 97200 Fort-de-France; tel. 5-96-75-48-68; fax 5-96-75-58-46; e-mail antilla@wanadoo.fr; weekly; Dir ALFRED FORTUNE.

**Combat Ouvrier:** BP 213, 97156 Pointe-à-Pitre Cédex; e-mail combatouvrier@fr.fm; internet www.chez.com/combatouvrier; weekly; communist; Dir M. G. BEAUJOUR.

**France Antilles:** pl. Stalingrad, 97200 Fort-de-France; tel. 5-96-59-08-83; fax 5-96-60-29-96; f. 1964; subsidiary of Groupe France Antilles; daily; Dir HENRI MERLE; circ. 30,000 (Martinique edition).

**Justice:** rue André Aliker, 97200 Fort-de-France; tel. 5-96-71-86-83; fax 5-96-63-13-20; e-mail ed.justice@wanadoo.fr; weekly; organ of the PCM; Dir FERNAND PAPAYA; circ. 8,000.

**Le Naif:** route Phare, 97200 Fort-de-France; tel. 5-96-61-62-55; weekly; Owner CAMILLE CHAUVET.

**Le Progressiste:** c/o PPM, Quartier Trénelle, 97200 Fort-de-France; tel. 5-96-72-68-56; fax 5-96-71-88-01; weekly; organ of the PPM; Dir PAUL GABOURG; circ. 13,000.

**TV Magazine:** Immeuble Gouya, Z. I. Californie, BP 1064, 97200 Fort-de-France; tel. 5-96-42-51-28; fax 5-96-42-98-94; e-mail tv.mag@media-antilles.fr; f. 1989; weekly.

## Broadcasting and Communications

### BROADCASTING

**Réseau France Outre-mer (RFO):** La Clairière, BP 662, 97263 Fort-de-France; tel. 5-96-59-52-00; fax 5-96-63-29-88; internet www.rfo.fr; fmrly Société Nationale de Radio-Télévision Française d'Outre-mer; present name adopted 1998; broadcasts 24 hours of radio programmes daily and 37 hours of television programmes weekly; Pres. ANDRÉ-MICHEL BESSE; Regional Dir MARIJOSÉ ALIE-MONTHIEUX; Editor-in-Chief GÉRARD LE MOAL.

#### Radio

There are some 40 licensed private FM radio stations.

**Radio Caraïbe International (RCI):** 2 blvd de la Marne, 97200 Fort-de-France Cédex; tel. 5-96-63-98-70; fax 5-96-63-26-59; internet www.fwinet.com/rci.htm; commercial station broadcasting 24 hours daily; Dir YANN DUVAL.

#### Television

**ATV Antilles Télévision:** 28 rue Arawaks, 97200 Fort-de-France; tel. 5-96-75-44-44; fax 5-96-75-55-65; commercial station; Dir DANIEL ROBIN.

**Canal Antilles:** Centre Commerciale la Galléria, 97232 Le Lamentin; tel. 5-96-50-57-87; private commercial station.

## Finance

(cap. = capital; res = reserves; dep. = deposits; m. = million; brs = branches; amounts in French francs)

### BANKING

#### Central Bank

**Institut d'Emission des Départements d'Outre-mer (IEDOM):** 1 blvd du Général de Gaulle, BP 512, 97206 Fort-de-France Cédex; tel. 5-96-59-44-00; fax 5-96-59-44-04; e-mail agence@iedom-martinique.fr; internet www.iedom.com; Dir C. APANON.

#### Major Commercial Banks

**Banque des Antilles Françaises:** 28–34 rue Lamartine, BP 582, 97207 Fort-de-France Cédex; tel. 5-96-60-72-72; fax 5-96-60-72-54; internet www.bdaf.gp; f. 1853; Dir ALBERT CLERMONT.

**BNP Paribas Martinique:** 72 ave des Caraïbes, BP 588, 97200 Fort-de-France; tel. 5-96-59-46-00; fax 5-96-63-71-42; internet martinique.bnpparibas.net; f. 1941; subsidiary of BNP Paribas; 12 brs.

**BRED Banque Populaire:** 5 pl. Monseigneur Romero, 97200 Fort-de-France; tel. 5-96-63-77-63; internet www.bred.fr; Chair. STEVE GENTILI.

**Crédit Agricole:** rue Case Nègre, pl. d'Armes, 97232 Le Lamentin Cédex 2; tel. 5-96-66-59-39; fax 5-96-66-59-67; internet www.ca-martinique.fr; f. 1950; Pres. GUY RANLIN; Gen. Man. PASCAL DURIEUX.

**Crédit Martiniquais:** 17 rue de la Liberté, 97200 Fort-de-France; tel. 5-96-59-93-00; fax 5-96-60-29-30; f. 1922; associated since 1987 with Chase Manhattan Bank (USA), and since 1990 with Mutuelles du Mans Vie (France); cap. 185.4m. (1998); Administrator ALAIN DENNHARDT; 10 brs.

**Société Générale de Banque aux Antilles:** 19–21 rue de la Liberté, BP 408, 97200 Fort-de-France; tel. 5-96-72-82-82; internet www.sgba.fr; f. 1979; cap. 15m.; Dir ODETTE MAZARIN; 4 brs.

#### Development Bank

**Société de Crédit pour le Développement de Martinique (SODEMA):** 12 blvd du Général de Gaulle, BP 575, 97242 Fort-de-France Cédex; tel. 5-96-72-87-72; fax 5-96-72-87-70; e-mail sodema@compuserve.com; f. 1970; part of the Agence Française de Développement (AFD—q.v.); cap. 25m.; medium- and long-term finance; Gen. Man. JACKIE BATHANY.

### INSURANCE

**Caraïbe Assurances:** 11 rue Victor Hugo, BP 210, 97202 Fort-de-France; tel. 5-96-63-92-29; fax 5-96-63-19-79.

**Groupement Français d'Assurances Caraïbes (GFA Caraïbes):** 46–48 rue Ernest Desproges, 97205 Fort-de-France; tel. 5-96-59-04-04; fax 5-96-73-19-72.

**La Nationale (GAN):** 30 blvd Général de Gaulle, BP 185, Fort-de-France; tel. 5-96-71-30-07; Reps MARCEL BOULLANGER, ROGER BOULLANGER.

**Le Secours:** 74 ave Duparquet, 97200 Fort-de-France; tel. 5-96-70-03-79; Dir Y. ANGANI.

## Trade and Industry

### GOVERNMENT AGENCIES

**Conseil Economique et Social:** Hôtel de la Région, ave Gaston Deferre, Plateau Roy Cluny, BP 601, 97200 Fort-de-France; tel. 5-96-59-63-00; fax 5-96-59-64-43; Pres. RENÉ FABIEN.

**Direction de la Santé et du Développement Social (DSDS):** Centre d'Affaires AGORA, Etang Z'abricots, Pointe des Grives, BP 658, 97263, Fort-de-France Cédex; tel. 5-96-39-42-43; fax 5-96-60-60-12; e-mail josiane.pinville@sante.gouv.fr; internet martinique.sante.gouv.fr.

**Direction Régionale du Commerce Extérieur Antilles-Guyane (DRCE):** c/o Préfecture SGAR, BP 647, 97262 Fort-de-France Cédex; tel. 5-96-39-49-90; fax 5-96-60-08-14; e-mail drceantilles@missioneco.org; internet www.missioneco.org/antilles-guyane; Regional Man. MARTIN LAMMERT.

**Direction Régionale de l'Industrie, de la Recherche et de l'Environnement (DRIRE):** see chapter on French Guiana.

### DEVELOPMENT ORGANIZATIONS

**Agence Française de Développement (AFD):** Immeuble AFD/IEDOM, 12 blvd du Général de Gaulle, BP 804, 97244 Fort-de-France Cédex; tel. 5-96-59-44-73; fax 5-96-59-44-88; e-mail afdfdf@wanadoo.fr; internet www.afd.fr; fmrly Caisse Française de Développement; Man. JEAN-YVES CLAVEL.

**Secrétariat Général pour les Affaires Régionales (SGAR)—Bureau de la Coopération Régionale:** Préfecture, 97262 Fort-de-France; e-mail jean-charles.barrus@martinique.pref.gouv.fr; tel. 5-96-39-49-78; fax 5-96-39-49-59; successor to the Direction de l'Action Economique Régionale (DAER); research, documentation, and technical and administrative advice on investment in industry and commerce; Chief JEAN-CHARLES BARRUS.

### CHAMBERS OF COMMERCE

**Chambre d'Agriculture:** pl. d'Armes, BP 312, 97286 Le Lamentin; tel. 5-96-51-75-75; fax 5-96-51-93-42; internet paris.apca.chambagri.fr; Pres. GUY OVIDE-ETIENNE; Dir NICAIRE MONROSE.

**Chambre de Commerce et d'Industrie de la Martinique:** 50 rue Ernest Desproge, BP 478, 97241 Fort-de-France Cédex; tel. 5-96-55-28-00; fax 5-96-60-66-68; e-mail ccim.doi@martinique.cci.fr; internet www.martinique.cci.fr; f. 1907; Pres. CLAUDE POMPIÈRE; Sec. JOSEPH DE JAHAM.

**Chambre des Métiers de la Martinique:** 2 rue du Temple, Morne Tartenson, BP 1194, 97200 Fort-de-France; tel. 5-96-71-32-22; fax 5-96-70-47-30; f. 1970; Pres. CHRISTIAN CAYOL; 8,000 mems.

# MARTINIQUE

## EMPLOYERS' ORGANIZATIONS

**Banalliance:** Centre d'Affaires le Baobab, rue Léon Gontran Damas, 97232 Le Lamentin; tel. 5-96-57-42-42; fax 5-96-57-35-18; f. 1996; Pres. DANIEL DISIER; 220 mems.

**Banamart:** Quartier Bois Rouge, 97224 Ducos; tel. 5-96-42-43-44; fax 5-96-51-47-70; f. 2005; formed by merger of SICABAM and GIPAM; represents banana producers.

**Compagnie Bananière de la Martinique (COBAMAR):** Immeuble les Palétuviers Z. I. Lézarde, 97232 Le Lamentin; tel. 5-96-30-00-50; fax 5-96-57-14-03; internet www.cobamar.fr; f. 1993; Pres. MARCEL FABRE; 200 mems.

**Ordre des Médecins de la Martinique:** 80 rue de la République, 97200 Fort-de-France; tel. 5-96-63-27-01; fax 5-96-60-58-00; Pres. Dr RENÉ LEGENDRI.

**Ordre des Pharmaciens de la Martinique:** BP 587, 97207 Fort-de-France Cédex; tel. 5-96-52-23-67; fax 5-96-52-20-92; Pres. JEAN BIGNON.

## MAJOR COMPANIES

**Bellonie Bourdillon Successeurs (BBS):** BP 10, Z. I. Génipa, 97215 Rivière Salée; tel. 5-96-56-82-82; fax 5-96-56-82-83; e-mail info@rhumdemartinique.com; internet www.rhumdemartinique.com; f. 1919; rum producer, markets other spirits and wines; Pres. XAVIER THIEBELIN; Gen. Man. BENOÎT PILLET; 161 employees.

**Biometal, SA:** Usine de Robert, Parc d'activité du Robert, 97231 Le Robert; tel. 5-96-65-14-44; fax 5-96-65-10-01; e-mail mbellemare@biometal.com; f. 1979; manufacture of steel products; Pres. and Dir.-Gen. LIONEL DE LAGUARIGUE; Man. GILLES DE REYNAL DE SAINT MICHEL; 70 employees.

**COMIA:** pl. d'Armes, BP 266, 97232 Le Lamentin; tel. 5-96-66-61-62; fax 5-96-51-40-21; e-mail info@comia.fr; f. 1978; cooked meats; Dir ALEX BILAS; c. 40 employees.

**Delattre-Levivier Martinique:** Bassin de Radoub, BP 708, 97207 Fort-de-France Cédex; tel. 5-96-72-64-64; fax 5-96-60-61-02; e-mail bruno.rossovich@wanadoo.fr; industrial maintenance and metalwork, ship repair; Dir BRUNO ROSSOVICH; c. 40 employees.

**Denel (Ex Royal SA):** Usine Dénel, 97213 Gros-Morne; tel. 5-96-67-51-23; fax 5-96-67-67-56; e-mail info@denelmartinique.com; internet www.denelmartinique.com; f. 1932; food processing, fruit juices and preserves; CEO ALAIN HUYGHUES DESPOINTES; Gen. Mans LAURENT HUYGHUES DESPOINTES, PHILIPPE VOURCH; 49 employees.

**Distillerie Dillon, SA:** 9 route Chateauboeuf, BP 212, 97257 Fort-de-France Cédex; tel. 5-96-75-20-20; fax 5-96-75-30-33; e-mail info@rhum-dillon.com; internet www.rhum-dillon.com; f. 1967; rum producer; Man. PASCAL RENARD; 45 employees; 80 planters.

**Esso Antilles Guyane, SA:** pl. d'Armes, BP 272, 97285 Le Lamentin Cédex 2; tel. 5-96-66-90-60; fax 5-96-51-17-87; f. 1965; distribution of petroleum and petroleum products; Pres. JEAN FRANÇOIS DUSSOULIER; 35 employees.

**Nestlé:** Z. I. pl. d'Armes, 97232 Lamentin; tel. 5-96-51-04-84; fax 5-96-51-66-20.

**Prochimie, SA:** BP 233, 97284 Le Lamentin Cédex 2; tel. 5-96-50-32-82; fax 5-96-50-22-48; e-mail prochimie@wanadoo.fr; f. 1972; domestic and sanitary products and paper; Pres. MARCEL PLISSONNEAU DUQUENE; 50 employees.

**SAEM PSRM Le Galion** (S.A. d'Economie Mixte de Production Sucrière et Rhumière de La Martinique): Usine Le Galion, 97220 La Trinité; tel. 5-96-58-20-65; fax 5-96-58-34-40; f. 1984; sugar refinery; rum business managed by COFEPP; Man. J. M. TOTO; 91 employees.

**SARA** (Société Anonyme de la Raffinerie des Antilles): Californie, BP 436, 97292 Le Lamentin, Cédex 2; tel. 5-96-50-18-94; fax 5-96-50-00-15; e-mail christine.ransay@sara.mq; internet www.sara.mq; f. 1969; TotalFinaElf 50%, Shell 24%, Esso 14.5%, Texaco 11.5%; depots in French Guiana and Guadeloupe; processes 800,000 metric tons of crude oil annually; Regional Gen. Man. FRANÇOIS NAHAN; c. 250 employees regionally.

**Siapoc, SA** (Société Industrielle Antillaise de Peintures et de Produits Chimiques): Zone de Californie, Acajou, 97232 Le Lamentin; tel. 5-96-50-54-14; fax 5-96-50-09-11; e-mail stesiapoc@siapoc.org; f. 1965; paints; Man. MARCEL PRESENT; 65 employees.

**SISAL, SA** (Industrielle du Siège de L'ameublement et de la Literie): Cocotte Canal, 97224 Ducos; tel. 5-96-56-32-32; fax 5-96-56-33-19; f. 1983; furniture manufacturer; Pres. FELIX SYLVIUS; 35 employees.

**SMPA:** Z. I. pl. d'Armes, 97232 Le Lamentin; tel. 5-96-30-00-14; fax 5-96-51-70-43; e-mail eursulet@sasi.fr; f. 1987; industrial bakery products and frozen foods; Man. EMMANUEL URSULET; 35 employees.

**Socara, SARL** (Société Caraibe de Representation Importation Exportation): 2 ave des Arawaks, BP 560, 97242 Fort-de-France Cédex; tel. 5-96-75-04-04; fax 5-96-75-04-76; e-mail ddesgrottes@socara-antilles.com; f. 1948; fruit juices, wines, beer, spirits; Dir NICOLAS CHABROL; 32 employees.

**Société d'Embouteillage de l'Eau Minérale Didier (SEEMD):** PK 9 route de Didier, 97200 Fort-de-France; tel. 5-96-64-07-88; fax 5-96-64-01-69; mineral water; CEO JEAN-LUC GARCIN; Dir MIGUEL GOSSELIN; 41 employees.

**SOFECA, SA** (Solaire Filtration Epuration Conditionnement d'Air): Quartier Californie, 97232 Le Lamentin; tel. 5-96-50-30-00; fax 5-96-50-03-11; f. 1977; refrigeration, air-conditioning, solar heating, catering equipment; Man. CHARLES BLANCANEAUX; 27 employees.

**SOMES** (Société Martiniquaise des Eaux de Source): Quartier Champflore, 97260 Morne Rouge; tel. 5-96-52-52-52; fax 5-96-52-30-55; e-mail somes@wanadoo.fr; f. 1976; carbonated drinks bottler and distributor; Pres. BERTRAND CLERC; 49 employees.

## UTILITIES

**EDF Martinique:** Pointe des Carrières, BP 573, 97242 Fort-de-France Cédex 01; tel. 5-96-59-20-00; fax 5-96-60-29-76; e-mail edf-services-martinique@edfgdf.fr; internet www.edf.fr/martinique; f. 1975; electricity supplier; successor to Société de Production et de Distribution d'Electricité de la Martinique (SPDEM); Dir GÉRARD BELLANGER; 160,000 customers.

## TRADE UNIONS

**Centrale Démocratique Martiniquaise du Travail (CDMT):** Maison des Syndicats, Jardin Desclieux, 97200 Fort-de-France; tel. 5-96-70-19-86; fax 5-96-71-32-25; Sec.-Gen. NICOLE ELANA.

**Confédération Générale du Travail de la Martinique (CGTM):** Maison des Syndicats, Jardin Desclieux, 97200 Fort-de-France; tel. 5-96-60-45-21; f. 1961; affiliated to World Federation of Trade Unions; Sec.-Gen. GHISLAINE JOACHIM-ARNAUD.

**Fédération Départementale des Syndicats d'Exploitants Agricoles de la Martinique** (FDSEA): Immeuble Chambre d'Agriculture, pl. d'Armes, 97232 Le Lamentin; tel. 5-96-51-61-46; fax 5-96-57-05-43; e-mail fdsea.martinique@wanadoo.fr; affiliated to the Fédération Nationale des Syndicats d'Exploitants Agricoles; Pres. LOUIS-DANIEL BERTOME.

**Union Départementale CFTC Martinique** (UD-CFTC): Maison des Syndicats, Jardin Desclieux, 97200 Fort-de-France; tel. 5-96-60-95-10; fax 5-96-60-39-10.

**Union Départementale Force Ouvrière Martinique (UD-FO):** rue Bouillé, BP 1114, 97248 Fort-de-France Cédex; tel. 5-96-70-07-04; fax 5-96-70-18-20; e-mail udfomartinique@force-ouvriere.fr; affiliated to International Confederation of Free Trade Unions; Sec.-Gen. MARIE-ALICE MEDEUF-ANDRIEUX.

**Union Régionale Martinique:** Maison de Syndicats, Jardin Desclieux, Salles 5–7, 97200 Fort-de-France; tel. 5-96-70-16-80.

**UNSA Education Martinique (SE-UNSA):** Maison des Syndicats, Salles 6–7, Jardin Desclieux, 97200 Fort-de-France; tel. 5-96-72-64-74; fax 5-96-70-16-80; e-mail unsa972@wanadoo.fr; Sec.-Gen. MICHEL CRISPIN.

# Transport

## RAILWAYS

There are no railways in Martinique.

## ROADS

There were 2,077 km (1,291 miles) of roads in 1998, of which 261 km were motorways and first-class roads.

## SHIPPING

**Direction des Concessions Services Portuaires:** quai de l'Hydro Base, BP 782, 97244 Fort-de-France Cédex; tel. 5-96-59-00-00; fax 5-96-71-35-73; e-mail port@martinique.cci.fr; port services management; Dir FRANTZ THODIARD; Operations Man. VICTOR EUSTACHE.

**Direction Départementale des Affaires Maritimes:** blvd Chevalier de Sainte-Marthe, BP 620, 97261 Fort-de-France Cédex; tel. 5-96-71-90-05; fax 5-96-63-67-30; Dir FRANÇOIS NIHOUL.

**Alcoa Steamship Co, Alpine Line, Agdwa Line, Delta Line, Raymond Witcomb Co, Moore MacCormack, Eastern Steamship Co:** c/o Etablissements Ren, Cottrell, Fort-de-France.

**American President Lines:** c/o Compagnie d'Agence Multiples Antillaise (CAMA), 44 rue Garnier Pages, 97205 Fort-de-France Cédex; tel. 5-96-71-31-00; fax 5-96-63-54-40.

**CMA-CGM CGM Antilles-Guyane:** ave Francois Mitterrand, BP 574, 97242 Fort-de-France Cédex; tel. 5-96-55-32-00; fax 5-96-63-08-87; e-mail fdf.rjoseph-alexandre@cma-cgm.com; internet www.cma-cgm.com; also represents other passenger and freight lines; Man. Dir JEAN-CHARLES CREN.

**Compagnie de Navigation Mixte:** Immeuble Rocade, La Dillon, BP 1023, 97209 Fort-de-France; Rep. R. M. MICHAUX.

### CIVIL AVIATION

Martinique's international airport is at Le Lamentin, 6 km from Fort-de-France.

**Direction des Services Aéroportuaires:** BP 279, 97285 Le Lamentin; tel. 5-96-42-16-00; fax 5-96-42-18-77.

**Air Caraïbes:** see chapter on Guadeloupe (Civil Aviation).

## Tourism

Martinique's tourist attractions are its beaches and coastal scenery, its mountainous interior, and the historic towns of Fort-de-France and Saint-Pierre. In 2002 there were 122 hotels, with some 5,115 rooms. In 2002 some 446,689 tourists visited the island, increasing to 453,159 in 2003. During the same period receipts from tourism increased from an estimated US $237m. to $247m.

**Comité Martiniquais du Tourisme:** Immeuble Beaupré, Pointe de Jaham, 97233 Schoelcher; tel. 5-96-61-61-77; fax 5-96-61-22-72; internet www.touristmartinique.com; Sec.-Gen. ROBERT CONRAD.

**Délégation Régionale au Tourisme:** 41 rue Gabriel Périé 97200 Fort-de-France; tel. 5-96-63-18-61; Dir GILBERT LECURIEUK.

**Fédération Martiniquaise des Offices de Tourisme et Syndicats d'Initiative (FMOTSI):** Maison du Tourisme Vert, 9 blvd du Général de Gaulle, BP 491, 97207 Fort-de-France Cédex; tel. 5-96-63-18-54; fax 5-96-70-17-61; f. 1984; Pres. VICTOR GRANDIN.

**Office Départemental du Tourisme de la Martinique:** 2 rue Ernest Desproges, BP 520, 97206 Fort-de-France; tel. 5-96-63-79-60; fax 5-96-73-66-93.

## Defence

At 1 August 2004 France maintained a military force of about 4,100 and a gendarmerie in the Antilles. The headquarters is in Fort-de-France.

## Education

There is free and compulsory education in government schools for children aged between six and 16 years. In 2002/03 there were 50,137 pupils in pre-primary and primary education, while in secondary education there were 49,426 students (including 7,891 students in vocational education), of whom some 92% attended state schools. Higher education in law, French language and literature, human sciences, economics, medicine and Creole studies is provided by a branch of the Université Antilles-Guyane. The University as a whole had 12,000 enrolled students in 2004. There are two teacher-training institutes and colleges of agriculture, fisheries, hotel management, nursing, midwifery and child care. Separate Academies for Martinique, French Guiana and Guadeloupe were established in January 1997, replacing the single Academy for the Antilles-Guyane, which was based in Fort-de-France. Departmental expenditure on education and culture was estimated at 111.5m. French francs in 2001.

## Bibliography

For works on the Caribbean generally, see Select Bibliography (Books)

Crane, J. *Martinique* (World Bibliographical Series). Oxford, ABC Clio, 1995.

Laguerre, M. S. *Urban Poverty in the Caribbean (French Martinique as a Social Laboratory)*. New York, NY, Palgrave, 1990.

Scarth, A. *La Catastrophe: The Eruption of Mount Pelée*. Oxford University Press, 2002.

# MEXICO

## Geography

### PHYSICAL FEATURES

The United Mexican States is the southernmost of the three great federations of North America, a republic that narrows south and east towards the great land bridge of Central America. Mexico is the smallest of the continental North American countries, being about one-fifth the size of the USA or Canada, but it is the third largest country of Latin America (it is less than one-quarter the size of Brazil and 70% of Argentina, but the next country in area, Peru, is only two-thirds its size) and the most northerly. The longest border (3,152 km or 1,959 miles) is with the USA, which lies to the north and north-east, while its south-eastern frontier is with the Central American countries of Guatemala (956 km), on the Pacific side, and Belize (193 km), on the Caribbean side. The country's nearest insular Caribbean neighbour is Cuba, some 210 km to the east of the Yucatán peninsula, in the south. The eastern coast is along the Gulf of Mexico and on the Caribbean, while the western coast, almost double in length, is on the Pacific Ocean. In total, there are about 11,122 km of coastline. Mexico covers an area of 1,964,375 sq km (758,449 sq miles), including 5,127 sq km of islands (the latter an area almost equivalent in size to Trinidad and Tobago).

Mexico has a diverse and crumpled topography, covering a vast area (it is the 13th largest country in the world). It is at its widest in the north, along the US border, where it is about 2,000 km from east to west. It is at its narrowest, 210 km from north to south, on the Tehuantepec isthmus, where there is a break in the Continental Divide, just before the country broadens westwards into Central America. Some two-thirds of Mexico is mountainous and one-half of it above 1,500 m (about 5,000 ft). Most of it is dry and sere, although the extent and variety of the terrain makes for regional contrasts. The Pacific coast runs roughly from north-west to south-east, before curving into a south-facing shore and continuing into Guatemala. The more northerly section of this western coast is dominated by the lowlands of the Sonora Desert, shielded from the immediate presence of the Pacific by the 1,300-km peninsula of Baja California, joined to the rest of Mexico in the far north-west, then thrusting southwards, parallel to the main coast on the other shore of what is known in Mexico as the Sea of Cortez (Cortés—Gulf of California). On the other side of the country, the western shore curves out of the north-east, circling around the Gulf of Mexico and into the north-thrusting abutment of the Yucatán peninsula, which separates the Gulf from the Caribbean. The flat, forested Yucatán has a relatively short eastern coast on the Caribbean, below which is Belize. Much of the interior of the country is an elevated plateau, the Mesa Central, which occupies about 60% of the territory of Mexico and contains most of its mountains, population and historical remains. A number of other distinct regions can be identified—the narrow coastal plains of the Pacific and of the Gulf of Mexico, the peninsulas of Baja California and Yucatán, and two further areas of upland, the southern highlands and the Chiapas highlands.

The central plateau is flanked, to west and east, by two mountain ranges, each rising steeply from their respective coastal plains, the Sierra Madre Occidental and the Sierra Madre Oriental. The western (Occidental) mountains are generally higher, above narrower coastal plains. Both ranges continue the thrust of the Rockies from the north, and merge some 240 km south of Mexico City (Ciudad de México), the capital, in an area of towering, inactive volcanoes and some of the country's highest peaks—such as Potacatépetl and Ixaticcíhuatl (just south-east of Mexico City) or the highest, Orizaba (5,610 m or 18,412 ft), between there and the Gulf city of Veracruz. The average elevation of the central plateau itself is about 900 m in the north and 2,400 m in the south. The plateau is interrupted by heights and broad basins, consisting of desert in the north (*bolsones*) and areas of settlement amid the rolling hills of the south. One such basin in the south of the plateau is the site of Mexico City and is known as the Valley of Mexico.

Just as the Sierra Madre is a part of the Continental Divide and a continuation of the Rockies, the mountainous interior of Baja (Lower) California is a continuation of the Coast Ranges in the USA. The northern connection to 'mainland' Mexico is interrupted by a deep cleft (continuing in the Sea of Cortez) in which the Laguna Salada is the lowest point in the country, being 10 m below sea level. Sparsely inhabited and arid, the peninsula most resembles the desert of the northern shores on the other side of the Sea of Cortez. Mid-way down the long inlet of the Pacific begin the coastal plains, about 50 km in width and extending southwards to just below Tepic (which lies north-west of Guadalajara). These plains are extensively irrigated and used for agriculture, being one of the limited fertile areas of the country (only 13% of the national territory is cultivated). Thereafter, the few patches of coastal plain are much narrower, as the southern highlands meet the sea, but widen again beyond Tehuantepec. The southern highlands are the Sierra Madre del Sur, which begin on the coast to the west of the Valley of Mexico and swing south of the central highlands, pushing south-eastwards to tail away at the Tehuantepec isthmus. The terrain consists of steep mountains, deep valleys (hot and dry inland) and high plateaux. The mountains often plunge straight into the Pacific and the coast is rugged, and sometimes called the 'Mexican Riviera'. Beyond Tehuantepec the Sierra Madre and the Continental Divide resume in the Chiapas highlands, continuing into Guatemala. The coastal plains also resume here, beneath the mountains, but the landscape is very different here in the south. Chiapas receives more rainfall than anywhere else in Mexico and is thinly inhabited, leaving forest-cover still extensive. Beyond the highlands, east of the central plateau, are the limestone lowlands and coastal plains along the Gulf of Mexico. In the north, near the US border along the Río Bravo del Norte (known as the Rio Grande in the USA), there are about 280 km between the sea and the steep mountain wall, but this narrows to only a few kilometres near Veracruz (which is east of Mexico City). Beyond here the plains widen again into the lowlands of the Isthmus of Tehuantepec, before thrusting north and east beyond Tabasco into the Yucatán peninsula. Much of the plains immediately on the Gulf consist of lagoons and swampy lowlands. Yucatán itself is broad and flat, without surface rivers, being dry and scrubby in the north-west, but much wetter and covered in dense rainforest in the south, the jungle often cloaking the ruins of the old Maya civilization.

# MEXICO

The country is largely dry and also has few large rivers. There are the Grijalva and the Usumacinta, for instance, which rise in Guatemala and empty into the Gulf of Mexico. The most important river system is that of the Grande de Santiago and the Lerma, which empty into the Pacific. The two rivers are interrupted by the country's largest lake, Chapala, which is 80 km long and 13 km wide—Mexico has relatively few lakes, although many lagoons and enclosed bodies of salt water contribute to inland waters of 49,510 sq km. This contributes to the variety of ecosystems to be found in the country, which still enjoys a diverse range of living species, both plant and animal. However, human settlement, in the north particularly, has had an adverse effect on the natural environment, with a continuing high rate of deforestation (this and the lack of clean water have both been declared national-security issues by the Government—as an example of the seriousness, the depletion of groundwater reserves in the Valley of Mexico is causing subsidence). Nevertheless, about one-quarter of the country is forested (mainly in the south), and this helps to shelter the range of species found in Mexico—30,000 plants, 1,000 birds and 1,500 mammals, reptiles and amphibians. About 15% of these species are reckoned to be unique to Mexico. The human population being less numerous and less widespread in the south, the ecological damage here is considerably less than in the north, with the environment similar to Central America, still inhabited by animals such as tapirs, jaguars and monkeys. Even in the north, in the high Sierra Madre, animals such as bears, deer, coyotes and mountain lions can still be found. The continually rising population and the fact that most of Mexico's profitable natural resources are subsoil, however, means a continuing threat to the environment, even in the hitherto neglected desert areas. It is the north of the country that is dominated by desert foliage, with extensive grasslands and hardwood forests in the highlands, while the wetter south is more typically rainforest.

## CLIMATE

The climate is as varied as the topography, ranging from tropical to temperate, mainly depending on altitude. Most of the country is south of the Tropic of Cancer, which crosses the tip of the Baja California peninsula. It is also generally dry, particularly in the north-west, but the south receives copious rainfall. Mexico mostly experiences two seasons, wet and dry, the latter (November–May) being the cooler. The central plateau is milder than the rest of the country, the evenings not only cool but even cold at times during the winter. Most rain is between June and September, ranging from about 300 mm (12 ins) in the north to 500–650 mm in the south. However, the deleterious effects human settlement can have even on weather is amply demonstrated in the disruption to the traditional rainfall patterns in the Valley of Mexico, owing to industrial pollution. The north-west of the country is the driest part, particularly on the Baja California peninsula. The northern part of the peninsula has a climate similar to that of southern California in the USA, with dry, warm summers and mild winters, while the south of the peninsula can get extremely hot. Here and in the Sonora is the area of lowest rainfall (about 130 mm annually). The coastal plains of the Pacific tend to have a little more rain and rather stronger storm patterns. Further down this coast, on the Riviera region, there is a definite tropical monsoon climate, although the southern highlands hinterland tends to be drier. In Chiapas the tropical conditions are even more pronounced, with average annual rainfall reaching 2,030 mm, but many places getting as much as double that amount. The coast of the Gulf of Mexico is muggy, with average rainfall about twice that of Baja California and increasing significantly south of Tampico (mid-way between Veracruz and the US border). Summer temperatures are high, but rain in autumn and winter (September–February), a slightly different wet season, is helped by brisk, cold northerlies. Yucatán is hot and humid in its jungle interior, alleviated by proximity to the coasts. The north of the peninsula is dry, but precipitation levels in the south approach those of Chiapas. There are two main wet seasons, in April–May and September–January. September–October can bring hurricanes, to which Caribbean, Gulf and Pacific coasts are all prone (natural hazards also come from the earthquakes common in a volcanic region such as this, with tsunamis liable to hit the Pacific coast). The average maximum and minimum temperatures in Mexico City range from 6°–19°C (43°–66°F) in January to 12°–26°C (54°–79°F) in May.

## POPULATION

The population of Mexico is predominantly Mestizo (60%), of mixed Spanish and Amerindian descent, and full-blood Amerindian (anywhere between 15% and 30%, depending on which figures are given credence), the balance made up of whites (9%—of mixed descent from Spanish and other immigrant, mainly European, groups), with small black, mixed-black and Asian groups. The early adoption of Spanish and Roman Catholicism by the native Amerindians has made for a more uniform culture than might otherwise have been the case. Spanish remains the official language, although 7% of the population speak only one of at least 62 indigenous tongues (mainly Nahuatl or, in the south, Mayan languages). English is widely spoken along the northern border. About 90% of the population are at least nominally Roman Catholic, despite the official anticlericalism of the regime for most of the 20th century, and 6% are Protestant.

At mid-2004 the total population of Mexico was an estimated 105.3m. After decades of rural exodus, about three-quarters of this population is urbanized, with nearly one-fifth of the total living in or around Mexico City. The capital is just north of where the Sierra Madre Occidental and the Sierra Madre Oriental converge, in central Mexico. The population of Mexico City at the time of the 2000 census was 8.61m., although, in addition, some cities on its outskirts are virtually suburbs (such as Ecatepec—1.62m. and Nezahualcóyotl—1.23m.). According to UN estimates, the population of Mexico City, including suburbs, was 18.7m. in 2003. The second city of the country is to the west-north-west of the capital, about three-quarters of the way to the coast, Guadalajara (1.65m.). Puebla, not far to the south-east of Mexico City, had a 2000 population of 1.27m., while the border cities of Ciudad Juárez (next to the US city of El Paso, in south-western Texas) and Tijuana (in the far north-west, south of California) had 1.19m. and 1.15m., respectively. The other two cities with over 1m. people were Monterrey (1.11m.), south-west of the Texas border, on the edge of the Gulf lowlands and the Sierra Madre Oriental, and León (1.02m.), north-east of Guadalajara. Chihuahua, to the south of Ciudad Juárez, is the main city of the northern central plateau and Merida the chief city of the Yucatán. Mexico is a federal republic, consisting of 31 states and the Federal District (Distrito Federal) encompassing Mexico City.

# History

## SANDY MARKWICK

### SPANISH CONQUEST

The area comprising modern Mexico was home to some of the most advanced pre-Columbian civilizations in the Americas. Evidence suggests human settlement dates back thousands of years. The most sophisticated societies were those of the Mayans, centred on the Yucatán peninsula, at its height between AD 700 and 900, and the Aztecs, centred in the central Valle de México (Valley of Mexico), location of present-day Mexico City, at its height between AD 1300 and 1500.

The meeting of European and indigenous cultures resonates powerfully in modern Mexico, as it does in much of Latin America. Some commentators see in the authoritarianism of modern Mexico, only recently challenged by genuine multi-party democracy, not just contemporary influences, but a persistent theme that can be traced back to the hierarchical organization of Aztec society. The modern population of Mexico is predominantly of mixed European and indigenous ethnicity (Mestizo), though large populations of Mexican Amerindians continue to live throughout Mexico, typically occupying the poorest strata of rural and urban society.

Invading Europeans led by Hernán Cortés destroyed the Aztec empire, led by its last emperor, Cuauhtémoc, in 1521; following the conquest, diseases brought from Europe severely reduced the indigenous population. Mexico City was subsequently founded on the ruins of the Aztec capital of Tenochtitlan. Mexico was ruled by a governing committee (audencia) in Mexico City with executive and legislative powers; it was also part of the larger territory of New Spain, incorporating much of Central America, the Caribbean and the Philippines, under the nominal control of the King of Spain's viceroy. Colonial society divided along racial and class lines and was built on the wealth derived largely from Mexico's silver resources.

## INDEPENDENCE

The liberal philosophies behind the French and American revolutions influenced opinion in favour of independence in Mexico from the late 18th century. Loyalty to the Spanish crown was called into question following Napoleon's invasion of the Iberian peninsula in 1808. The French emperor had ousted Carlos IV and installed his brother, Joseph Bonaparte, as king. The ensuing breakdown in Spanish royal authority in Mexico exacerbated tensions between Mexican-born élites (Criollos) and their counterparts from Spain (Peninsulares), leading to conflict and an 11-year independence struggle. On 16 September 1810 a radical Criollo priest, Miguel Hidalgo, formally proclaimed independence from his base in Guanajuato. The date is celebrated as Independence Day.

The insurrection met with partial success, but by 1820 had failed to defeat the colonial forces in Mexico City and the two principal leaders of the pro-independence forces, Hidalgo and his successor, José María Morelos, had been executed. However, the seizure of power in Spain by liberals had the effect of provoking conservative loyalists, led by Augustín de Iturbide, to turn against Spain and unite with the rebels behind independence. Spain signed the Treaty of Córdoba in September 1821, recognizing Mexican independence under the rule of a Mexican monarch, with equal privileges for Criollos and Peninsulares, and with the Roman Catholic Church afforded privileges as the official religion.

Iturbide was appointed 'emperor' as he forcibly annexed most of Spanish-speaking Central America, though his empire was short-lived. A garrison commander, Antonio López de Santa Ana, led a republican revolt, which received significant support and led to Iturbide's abdication. Central American territories subsequently declared their independence from Mexico (the southern state of Chiapas, which had been part of Guatemala, opted to remain part of Mexico).

### Federal Republic, 1824–36
A constitutional congress drew up the Constitution of 1824 and established a federal republic of 19 states and four territories. Superficially, the Constitution had the liberal appearance of the US Constitution, with separation of powers between the states, the executive, legislature and judiciary. Where it departed from a liberal influence was in privileges granted to the military and the Roman Catholic Church, concessions secured by conservatives within the constitutional congress.

For a decade political conflict in the Federalist Republic set federalists against conservatives. The former received support from liberal Criollos and Mestizos and feared control from a conservative Mexico City, while the conservatives were centralists and traditionalists supported by the military and the Roman Catholic Church. Liberal rule under the federal republic was brought to an end with the assumption of dictatorial powers by the provincial ruler (Caudillo) López de Santa Ana. He had been elected President in 1833 after his defeat of earlier conservative revolts, but subsequently abandoned liberalism. López de Santa Ana dominated politics either as President or as the power behind the scenes until 1855.

### Santa Ana and Territorial Loss
López de Santa Ana drew up the new Constitution of 1836, with power concentrated in the hands of the President. Regional Caudillos were appointed to run the former states as military districts. The new regime's authoritarianism and nationalism brought it into conflict with the English-speaking settlers of Texas. Texans declared independence from Mexico in 1836 and, with the support of volunteers from the USA, defeated López de Santa Ana's forces. Texan independence lasted until 1845, when it became the 28th state of the USA.

War between Mexico and the USA broke out in 1846 over the disputed status of Texas (the Mexican legislature had never ratified López de Santa Ana's treaty recognizing Texan independence), culminating in the occupation of Mexico City by US forces. The 1848 Treaty of Guadalupe Hidalgo established the border between Mexico and the USA at the Río Bravo del Norte (Rio Grande), accepted Texas' incorporation into the USA and saw more than one-half of its territory—a large part of present-day south-western USA, including California, Nevada and Utah—ceded to the USA. López de Santa Ana sold further territory to the USA—southern New Mexico and Arizona—in the Gadsden Purchase of 1854. This last episode enraged a group of reformists led by Benito Juárez, a lawyer and Zapotec Amerindian. Inspired by the liberal European political philosophers, such as Jean-Jacques Rousseau and John Stuart Mill, this group planned the overthrow of López de Santa Ana. Support for the conspirators grew and led to López de Santa Ana's resignation in 1855.

### The Reform and Emperor Maximilian
A period of liberal rule ensued under the provisional Government of Juan Ruíz de Alvarez, charged with governing until a new constitution could be drafted. A series of reform laws saw the powers of the Church curtailed, slavery abolished and civil liberties guaranteed. These reforms were incorporated into the new Constitution of 1857, but they polarized Mexico, leading to a civil war, known as the War of the Reform, from 1858–61. Conservatives, supported by the military and the Church, dissolved the Congreso (Congress) while Juárez formed a liberal 'Government-in-exile'. The liberals defeated the conservative forces at the end of 1860 and entered the capital on 1 January 1861. Juárez won a presidential election and, faced with a stagnant economy and the demands of foreign creditors to repay loans, declared a moratorium on foreign debt repayments. In response, Spain, the United Kingdom and France sent a joint military expedition to enforce repayment. The United Kingdom and Spain subsequently withdrew their forces, but the French continued the occupation with the support of the defeated conservatives. The French lost a disastrous battle at Puebla on 5 May 1862, which is commemorated by the Cinco de Mayo national holiday. The French finally captured the capital in 1863 and, with Napoleon III hoping to gain territory in the Americas, imposed Archduke Maximilian of Austria, from the house of Habsburg, as the Emperor of Mexico.

When the French, facing a threat from the Prussians, recalled their forces, Maximilian was left isolated, not just against Juárez and his supporters (which by now included the USA, after its own Civil War had ended), but also against the conservatives, who were disillusioned with Maximilian's refusal to repeal liberal reforms. Liberals and conservatives united to defeat Maximilian in June 1867, after which he was executed, and Juárez returned to power.

The republican forces under the newly elected Juárez restored the Constitution of 1857 and embarked on a programme to develop communications, natural resources and education. Juárez died in 1872 and was replaced by another liberal, Sebastián Lerdo de Tejada. When Lerdo tried to seek re-election he was deposed during an uprising by José de la Cruz Porfirio Díaz, who had earlier staged an unsuccessful revolt against Juárez. Porfirio Díaz was to rule largely uninterrupted from 1876 to 1910, during a period that came to be known as the Porfiriato.

### The Porfiriato
Díaz pursued a strategy of modernization and economic development based on the integration of Mexico into the global economy through export-orientated growth. Díaz created a modern infrastructure, including an expanded railway network, to support the development of agriculture and minerals. The Porfiriato was

characterized by economic growth, modernization and political stability, although achieved at the cost of social division and authoritarianism. Wealth was concentrated into the hands of a narrow élite, political opposition was not tolerated and Díaz flouted the clause in the Constitution that prohibited re-election. The Government concentrated land ownership in the hands of a few rich landowners (*hacendados*), creating widespread rural landlessness.

## THE MEXICAN REVOLUTION, 1910–20

Díaz's attempt to ensure his re-election (for the seventh time) in 1910 sparked a rebellion led by Francisco Madero, a wealthy northerner who represented the political resentment towards Díaz's monopoly of power. Other revolts broke out against Díaz, including those led by 'Pancho' Villa in the northern state of Chihuahua and Emiliano Zapata, a peasant leader from Morelos state. Zapata expressed the discontent among poor, rural, indigenous Mexicans, particularly in the south. Díaz resigned in the face of the rebellions in May 1911.

Madero became President in November 1911, but found it difficult to satisfy the expectations of those revolutionary forces who were calling for social justice as well as political reform and nationalist rhetoric based on indigenous heritage. Zapata turned his rebellion against Madero, demanding land reform, while other revolutionary and counter-revolutionary forces also plotted against the President. A commander formerly loyal to Madero, Victoriano Huerta, joined counter-revolutionary forces under a nephew of Díaz, deposed Madero and installed himself as President. Madero was assassinated in February 1913. Opposition to Huerta stemmed from numerous northern Caudillos, including Villa in Chihuahua, Alvaro Obregón in Sonora and Venustiano Carranza in Coahuila, and from Zapata in the south. Huerta tried to maintain control through repressive use of federal forces, but the economy was in ruins and conflict continued. Huerta resigned in July 1914.

A conference of military leaders organized by Carranza revealed a broad division between those principally concerned with the restoration of the Constitution, led by Carranza and Obregón, and those fighting for land reform, led by Zapata and Villa. At one point all four leaders established governments and claimed legitimacy: Carranza based in Veracruz; Obregón in Mexico City; Villa in Guanajuato; and Roque González Garza in Cuernavaca, supported by Zapata. This period of civil war ended in victory for Carranza. His Government was recognized by the USA, though Villa continued to resist. Carranza held the Congress of Querétaro on 1 December 1916 to draw up a new constitutional settlement. (Mexican Presidents are inaugurated on 1 December to commemorate the Querétaro meeting.) The Constitution of 1917 that emerged from Querétaro was a liberal democratic charter. It departed from the 1857 Constitution by giving greater powers to the executive, although the clause prohibiting re-election was maintained, and by incorporating revolutionary goals: civil liberties; labour rights and land reform; subordination of the military; secularization of education; and strict observance of the electoral calendar. It established a federal system of 31 states and a Federal District (Distrito Federal) comprising Mexico City.

With the new Constitution in place, Carranza took office in May 1917, following electoral victory. Civil war was to continue for another three years. Carranza restored expropriated land to former landowners rather than legally recognize land seizures or extend land reform. This ensured that insurrection persisted, particularly from the Zapatistas in Morelos. Carranza ordered Zapata's assassination in 1919. However, Carranza provoked further military opposition when he tried to install an ally as a successor to allow him to maintain power behind the scenes. Adolfo de la Huerta and Plutarco Elías Calles led an army from the north, invoking the Constitution. Carranza left for exile, but was assassinated en route. Obregón was elected to a new four-year term in 1920. Villa ended resistance in the north by signing a peace treaty with the federal Government.

## POST-REVOLUTIONARY SETTLEMENT AND THE FOUNDING OF A STATE PARTY

Under Obregón's presidency (1920–24), land redistribution proceeded cautiously, disappointing radicals in the process. His Government furthered revolutionary goals through a large-scale education policy in rural areas, in which it sought to integrate indigenous groups into Mexican society. During this time the Government commissioned Diego Rivera and other renowned muralists to teach the history and cultivate the myths of the revolution through visual imagery.

Obregón nominated his interior minister Calles as his successor in 1924. Calles' presidency was more radical than his predecessor's: he redistributed 3.2m. ha of land, three times as much as had been redistributed under Obregón, and also implemented the anti-clerical provisions of the Constitution in closing Church-run schools, banning religious processions and forcing priests to register with the Government in order to perform duties. The policies provoked fierce revolt among militant supporters of the Roman Catholic Church, violence which was met in kind by government forces. By 1929 the Church had been forced to accept government terms. During this period a religious militant assassinated former President Obregón, who had won elections in 1928 (thereby flouting the constitutional clause prohibiting re-election) and was due to assume office in that year. Calles effectively remained in power behind three weak Presidents until 1934, in a period known as the Maximato.

In 1929 Calles established the Partido Nacional Revolucionario (PNR—National Revolutionary Party) providing institutional support for his political ambitions. Previous elections had been fought using *ad hoc* parties established for specific electoral campaigns. The new PNR was designed as a permanent organization and as an official institutional expression of the revolution. Under the Maximato, Calles and his Governments shifted to the right, as land redistribution was halted and independent labour organizations suppressed. The President came under pressure from the left within the PNR and, to avoid a split in the party, nominated Lázaro Cárdenas, a populist Governor of Michoacán, as candidate in the 1934 presidential elections.

### The Presidency of Lázaro Cárdenas, 1934–40

The Cárdenas presidency gave significant new impetus to the revolutionary vision. Cárdenas redistributed close to 18m. ha of land, eclipsing the combined efforts of all previous Governments. The new President displayed his independence from Calles by removing many of his supporters from their positions in the federal bureaucracy. The military demonstrated its loyalty in suppressing a rebellion of Calles' conservative supporters in San Luis Potosí. Cárdenas is also noted for his expropriation of foreign oil companies in 1938 and the formation of a national oil company, Petróleos Mexicanos (PEMEX). The initiative damaged relations with the USA and led to under-investment in the industry, but ensured Cárdenas' place in Mexican history as a hero of the nationalist left.

Cárdenas reorganized the PNR into the Partido de la Revolución Mexicana (PRM—Party of the Mexican Revolution) and in doing so integrated key sectors of Mexican society into the party in official 'corporatist' sectors. Labour was represented by the newly formed Confederación of Trabajadores Mexicanos (CTM—Confederation of Mexican Workers). Other quasi-official organizations represented rural labourers and small farmers and urban employees of the state bureaucracy. These organizations did not emerge from the grassroots of sectoral organization, but instead were imposed from above by the political hierarchy to distribute benefits while simultaneously controlling and suppressing demands. The corporatist organizations enjoyed the benefits conferred on them as a result of their quasi-official status, which independent organizations did not.

## THE MEXICAN 'MIRACLE', 1940–81

The period 1940–81 was one of steady, sustained economic growth. Following Cárdenas, Mexican Presidents over the next half-century focused on ensuring that the state took the lead in economic and industrial development. Domestic industry was sheltered behind protectionist tariff barriers, and governments used their authority through the CTM to suppress workers' wage demands. Public investment was directed towards modernizing infrastructure and developing strategic industries such as oil and petrochemicals. (Mexico had become the 15th largest economy in the world by 1980.) All Mexican Governments up to this time had shared this broad strategy, varying principally in the degree of nationalist tone and their enthusiasm for rural development and land redistribution. Avila Camacho (1940–46), Miguel Alemán (1946–52) and Adolfo Ruíz Cortines (1952–58) were more conservative Presidents than Cárdenas, slowing land reform and promoting private investment by keeping wages low. By contrast, Adolfo López Mateos (1958–64) represented a shift back to the left, with a reinvigoration of land redistribution, while also maintaining tight control of the unions. The combination of economic growth and relative political stability was unique in Latin America during this period and often referred to as the Mexican 'miracle'.

The political system that underpinned the Mexican miracle for half a century and evolved during the post-revolutionary period under Calles, and then Cárdenas, was unique. One of its principle features was the enormous power provided de facto to the President. While the Constitution of 1917 provided for a separation of powers between the executive, legislature and judiciary, in reality the President dominated everything. The Congreso was a pliant body, little more than a rubber stamp for presidential ambitions. The judiciary was nominally independent, but in fact expressed the will of the executive. The 31 state governors were not accountable to an electorate, but to the gran elector, the President himself, to whom they had to show loyalty, dependent as they were on the federal treasury. The President controlled state resources and could use them as punishment or reward. The enormous powers of the President were limited only by the fixed electoral calendar and the constitutional clause prohibiting re-election. Even then the incumbent could choose his successor by nominating the ruling party's presidential candidate. The nomination was not the result of any internal party democracy.

The party was a vital feature of the system and a central pillar of Mexico's relative stability in the post-war decades. The Partido Revolucionario Institucional (PRI—Institutional Revolutionary Party), as the party founded by Calles in 1929 came to be known, was not established as a traditional political party, as the term is understood in modern democracies. The aim of the party was not to compete with other parties, but to include all the factions behind the revolution and to provide a mechanism for the peaceful resolution of disputes. The party was a co-ordinator of the electoral process, an election-winning machine dominated by successive Presidents, which ensured victory through privileged access to state resources, patronage and, where necessary, fraud and repression.

Leaders sought democratic legitimacy through the electoral process, but for most of the post-revolutionary 20th century genuine electoral competition was non-existent. The authoritarian reality behind a democratic and constitutional façade inspired the Peruvian novelist Mario Vargas Llosa to refer to the Mexican political system as the 'perfect dictatorship'. The PRI 'won' all elections at state and federal level and the great majority at municipal level too. The PRI encouraged some semi-official 'opposition' parties and occasionally conceded defeat in municipal elections to the opposition, largely to boost democratic legitimacy. Genuine opposition, for example from the conservative Partido de Acción Nacional (PAN—National Action Party), tended to be limited regionally, in this case to the pro-business north and the Roman Catholic south-east. PRI candidates wielded significant advantages over opposition candidates in their proximity to the apparatus and largesse of the state, including funds and police, which they used in pursuit of electoral support. If that was not enough, the executive's dominance over the electoral commission ensured PRI victories unless the regime felt it prudent to concede defeat.

The most important political battles took place in relative obscurity within the PRI itself rather than at the polling stations. The unity of the PRI was maintained by imposing obstacles and disincentives to party resignations, such as electoral failure or repression, and incentives to remain loyal, such as jobs in the state bureaucracy or the promise of political power in the future. The fixed calendar, strictly adhered to, allowed for a circulation of competing governing élites. Government opponents could bide their time in the knowledge that a new administration with new supporters would be in power at the end of the sexenio (six-year presidential term). A large state bureaucracy and widespread patronage, with a large turnover of benefits handed out at the outset of each sexenio, ensured that many millions of Mexicans had a stake in the system. This encouraged loyalty to the system. It was also the source of one of its weaknesses. There was a pattern of crises accompanying transitions between Governments, as outgoing administrations, made up of multiple patron-client relationships about to lose power, did not necessarily behave in the national interest.

While presidential power backed up by a dominant state party made the Mexican system undoubtedly authoritarian, it was also socially inclusive. Mexican Presidents were heirs of a revolutionary and nationalist struggle. The legacy was not just rhetoric: the economic growth generated was used to pursue social justice (to varying degrees, according to the administration). This included land redistribution to communal landholdings (ejidos), subsidies for the poor, job creation, and investment in health and education. The institutional expression of the socially inclusive character of the Mexican system was the PRI-affiliated corporatist sectoral organizations representing the labour unions, small farmers, landless peasants and employees of urban public-sector bureaucracies. The system distributed benefits to those groups within these sectors that were affiliated with the PRI in return for control, and discriminated against any independent organizations. In this way, the system secured the loyalty of potential opponents. Rather than allow independent unions to thrive, the Government created 'official' unions and incorporated them into the PRI and ensured that they received preferential treatment. Similarly, the peasantry was organized from above, in agrarian co-operatives. The Government's control of land titles, loans, seeds and fertilizer ensured these unions' dependence, and discouraged the emergence of any alternative autonomous organization.

### Cracks in the System

Under President Gustavo Díaz Ordaz (1964–70), the authoritarianism of the system was exposed to the world when the Government responded to mounting student unrest with brutal repression. Hundreds of protesters were killed in Tlatelolco Square in October 1968, in response to protests from students and intellectuals, who had been demanding greater openness. Mexican society had changed significantly since the days of Calles and Cárdenas. The steady growth, stability and economic development promoted by PRI Governments transformed Mexico from a predominantly uneducated rural society to a predominantly educated and urban one, and PRI authoritarianism began to appear anachronistic.

In response to the threat of unrest, the Government responded with a tactic that previous Governments had used successfully when faced with a threat to stability: it co-opted into the regime the very group that had threatened it. In 1970 Luis Echeverría, who as interior minister was responsible for the actions of the security forces, became President (1970–76) and defused urban protest by integrating disaffected students into the system and conferring some of its benefits to them. Echeverría released imprisoned students, brought other student leaders into prominent positions in his Government and gave jobs in an expanded bureaucracy to the educated middle-class youth.

### Debt Crisis and the End of the 'Miracle'

Under President Luis López Portillo (1976–82) the seeds were sown for a crisis that would change the development model and threaten the system of government. López Portillo benefited from the discovery of massive new oil reserves, particularly in the Gulf of Mexico. By 1981 Mexico had tripled its oil production to become the world's fourth largest producer, earning huge revenues as the price of crude increased. Mexico used these reserves as collateral to borrow extensively from willing foreign banks, the coffers of which were filled with the deposits of oil producers in the Middle East. When the petroleum price slumped in 1981 and interest rates increased, it meant that by 1982 Mexico was faced with a US $10,000m. current-account deficit and servicing obligations on a $90,000m. foreign debt, equivalent to 45% of export earnings. The Government devalued the peso and, to control the flight of capital overseas, nationalized the banking sector.

President Miguel de la Madrid Hurtado (1982–88) inherited an economy in crisis, in which the old statist and protectionist development model was no longer appropriate. Instead the Government chose spending cuts and austerity measures in order to keep credit lines open from the international financial community, as well as trade liberalization. In 1986 Mexico joined the General Agreement on Tariffs and Trade, the forerunner to the World Trade Organization (WTO), and so began a process of progressively reducing tariffs and eliminating non-tariff barriers.

The new economic environment had important political reverberations. The efficiency of the system in co-opting sufficient numbers and sufficient interests behind the President came under severe strain. The system was no longer delivering steady economic growth. The austerity measures threw the inequality within Mexican society into sharp relief; rising inflation, public spending cuts and devaluation affected the poor the worst. During de la Madrid's administration, real wages declined by 40%. The PRI and the political system were under pressure. The authoritarian model could suppress opposition while it was delivering economic growth, but began to come under intense pressure after economic crisis in the 1980s. It also opened up regional divisions between the pro-business north and Mexico City. Support for the PAN in the north grew. The Government, less able now to use economic policy to shore up loyalty, began tentative moves towards greater political pluralism. Such a move was not born out of democratic conviction, but in the hope of channelling opposition without threatening the PRI. The PAN won municipal mayoralties and seats in state legislatures in 1983 and 1985, respectively.

Economic crisis, inequality and the shift towards the free market in government policy resulted in a left-wing breakaway from the PRI, led by Cuauhtémoc Cárdenas (son of nationalist hero and former President Cárdenas). He stood as an independent candidate in the 1988 elections, campaigning for a return to the principles of the Mexican revolution against the PRI candidate, the technocrat, Carlos Salinas de Gortari.

## THE ELECTION OF CARLOS SALINAS

The year 1988 proved a watershed in Mexican politics. For the first time, a candidate from the PRI faced a serious challenge for the presidency. Salinas was declared the winner with just over 50% of votes cast (PRI presidential candidates had consistently polled over 80%), amid widespread allegations of fraud. For the first time, the PRI held less than a two-thirds' majority in the Congreso, requiring it to negotiate with opposition parties to enact constitutional reform.

President Salinas extended the fundamental structural transformation of the Mexican economy through privatization, trade liberalization and modernization of the financial system. There was also a significant political element to his economic policies. The manner of his 'victory' strongly influenced the path of his presidency. The election demonstrated a loss of support for the PRI and the party's system, and the dubious nature of the count called into question the legitimacy of the Government. From the outset, Salinas aimed to regain support. He was determined to win back the support of the middle classes through exchange-rate and price stability; liberalization of trade, together with a fixed, overvalued, exchange rate, was designed to reduce inflation, as much through cheap imports as innate Mexican competitiveness.

The policy opened a trade gap, which, along with debt-servicing requirements, fuelled a current-account deficit. Salinas attempted to finance the deficit by attracting capital through privatization. Between 1986 and 1994 the Government sold the state's interest in more than 700 companies, including all its commercial banks, airlines, television networks, telephone companies, ports, industrial processing plants, mines and steel-production facilities. The process generated revenues of US $22,000m. The Government lifted the tight restrictions that faced foreign investors. The new rules fuelled a boom in the Mexican stock exchange and in the growth of the *maquiladora* sector (tax-free assembly plants predominantly located along the border with the USA, to which products were exported).

President Salinas' policies brought some macroeconomic success, but the benefits were not evenly distributed. Annual inflation fell from 159% in 1987 to 6.7% in 1994, garnering significant support from the urban middle classes, who enjoyed a consumer boom brought about by cheap imports. However, growth rates remained modest, averaging 2.6% in the period 1988–94. Unemployment and underemployment remained high, fuelling high rates of emigration to the USA. Joining the North American Free Trade Agreement (NAFTA) in 1994 consolidated the radical shift in Mexican trading policy. Salinas turned around decades of nationalist-driven protectionism to direct the economy towards integration into the international economy. Salinas promised political reform, but delivered a highly selective version, which did little to threaten the PRI's hold on power, at least in the short term.

In July 1989 the PRI conceded the governorship of the state of Baja California Norte to the PAN. This was unprecedented, the first time the opposition had won any gubernatorial election. However, it represented only a limited political opening: state governors were reliant on the federal treasury, and Salinas could afford to 'allow' a PAN victory, knowing that it would give the impression of increasing pluralism without being an immediate threat to PRI control. While the PRI allowed a PAN victory in the north, it appeared to be engaging in ballot rigging and control of the electoral authorities to prevent a victory for Cárdenas' party, by now known as the Partido de la Revolución Democrática (PRD—Party of the Democratic Revolution), in his home state of Michoacán.

President Salinas had to work with the PAN in the Congreso. He was forced to accept a succession of electoral-reform measures—limits were placed on campaign financing and proportional representation was extended in the Cámara Federal de Diputados (Federal Chamber of Deputies). On their own neither measure marked a major watershed in electoral democracy in Mexico, but each brought it a little closer, leaving the Government less bargaining power when negotiating concessions from the opposition. Greater openness encouraged the opposition, and the process of democratization was gathering a momentum of its own, no longer under the control of the Government. Meanwhile, the PAN's strategy was to stop questioning the legitimacy of the Government. Instead, it co-operated with Salinas, who had adopted many of the policies that the party had advocated.

### Political Instability

President Salinas' neo-liberal policies and the PRI's continued grip on power provoked the emergence of an armed guerrilla group inspired by the revolutionary peasant leader Emiliano Zapata, known as the Ejército Zapatista de Liberación Nacional (EZLN—Zapatista Army of National Liberation). A small but well-organized group, consisting of several hundred indigenous guerrillas, led by the enigmatic 'Subcomandante Marcos', the EZLN briefly took over several municipalities in the southern state of Chiapas on 1 January 1994, the day that Mexico's membership of NAFTA came into force. The modern-day Zapatistas blamed the Government's liberal free-market policies, and NAFTA in particular, for enriching big business and the agricultural industry at the expense of the rural poor. The movement inspired the left elsewhere in Mexico and emphasized Mexico's history of authoritarian domination of indigenous communities. Another shattering blow to Mexico's image of stability came in March with the assassination of the PRI's presidential candidate, Luis Donaldo Colosio, during a campaign rally in Tijuana. It was the first assassination of a national PRI figure since the killing of President Álvaro Obregón in 1928. Few accepted the official version of events surrounding the killing, which held that a lone gunman was responsible. Following so shortly after the Zapatista uprising, the event raised fears of a breakdown in governance in Mexico. Former education minister, Ernesto Zedillo Ponce de León, succeeded Colosio as PRI presidential candidate. Fraud allegations in the 1994 elections were not on the scale of the 1988 ballot and the vote was considered relatively free and fair, aided by electoral reforms that had recently been put in place.

However, political violence did not subside. Just a month after Zedillo's victory, in September 1994, José Ruiz Massieu, another senior PRI official and former brother-in-law of Salinas, was assassinated. Salinas' brother, Raúl, was subsequently convicted for the murder of Massieu (and for money-laundering on a massive scale), while the victim's own brother and Deputy Attorney-General, Mario Ruiz Massieu, was arrested for obstructing the investigation into the crime. These two crimes were high-profile examples of what many feared was a growing willingness for disputes within élite circles to be settled violently instead of through negotiation. Whoever was behind Colosio's murder (drugs-traffickers and anti-reformist PRI 'dinosaurs' are most often blamed), the fact that the dispute was resolved violently appeared to reflect a breakdown of the system. This was partly ascribed to the new, increasingly liberal economy, which had weakened the basis of authoritarian rule. Many of the state's former assets were now controlled by the private sector (often by close associates of Salinas), reducing the opportunities for patronage and undermining the mechanisms for the peaceful settlement of disputes among élites. There were now fewer state-owned companies, jobs, seats in the legislature or governorships to offer. An opaque system of authoritarian control was being eroded, but as yet it had not been replaced with a new and transparent democratic order.

The increase in drugs-trafficking activity and the consequent influx of so-called 'narco-dollars' worsened law and order in a country where law enforcement had traditionally been weak or corrupt. Owing to its geography and history of state corruption, Mexico was a conduit for the shipment of cocaine from Colombia to markets in the USA. The opportunities for increased revenues from shipping Colombian cocaine compromised the security forces and state sector further. Traffickers laundered their revenues through legitimate business, greater opportunities for which presented themselves through privatization.

## ERNESTO ZEDILLO: STABILIZATION AND REFORM

The election of Zedillo to the presidency in 1994 was arguably the first post-authoritarian election in Mexico. Although the PRI continued to enjoy advantages over opposition parties in terms of media coverage, financial resources, state access and organizational strength, there was genuine doubt about the outcome of the ballot, owing to high voter turn-out and greater independent scrutiny of electoral practices.

Zedillo took office in December 1994 and was immediately embroiled in crisis management, lasting into the first half of 1995. Throughout the Salinas presidency an overvalued peso had been maintained to support a political imperative, that of low inflation and a boom in consumer spending. There were few opportunities for a corrective devaluation which would not have had damaging political implications. Salinas had had to consider mid-term elections, state elections and the close US congressional vote on Mexico's entry into NAFTA. Perhaps so as not to undermine his ambition of heading the WTO after leaving office, or perhaps simply to help preserve his place in history, Salinas did not take the opportunity to help his successor by devaluing the peso before he left office. In the mean time, Mexico had become highly dependent on short-term, speculative investment from overseas to finance the deficit accumulated under Salinas. Such investment was highly vulnerable to foreign interest rates and political instability, of which there had been no shortage in 1994. With fears of an impending devaluation, many investors switched to US-dollar-denominated short-term treasury bonds (*tesobonos*). When devaluation came on 20 December, the Government found itself unable to meet its US $29,000m. *tesobono* obligations, payable in 1995.

Zedillo's Government was criticized for mishandling the devaluation. Initially, the Government attempted to raise the ceiling at which the peso was traded by 15%, but this merely aggravated investors already nervous with low foreign-exchange reserves. Pressure on the peso forced the Government to float it freely against the US dollar. Inflation rocketed from 6% to 50%, businesses were forced into bankruptcy and consumers suffered as Mexico descended into deep recession. Many people lost jobs and many more were pushed into poverty. Only international intervention on a massive scale (a US-led rescue package amounting to US $50,000m.) prevented further decline. Zedillo's emergency economic plan involved stricter fiscal and monetary policies and price and wage restraints agreed by business groups and trade unions. With the help of export performance, growth was restored in 1997.

President Zedillo played a significant role in advancing the cause of political reform. In a government system characterized by a divide between the constitutional theory and the informal reality, Zedillo's contribution was not so much in pushing through electoral reform legislation, although among the new rules agreed by the main parties in 1996 was the highly significant establishment of full independence of the federal electoral body. Instead, it was his willingness to step back from using the informal powers at the disposal of the presidency that advanced political reform in Mexico. In doing so he allowed the other institutions and office holders, including municipal mayors, state governors and judges, to realize their potential as alternative sources of power and to begin to accept the accountability that accompanies it. Early in his administration Zedillo appointed an opposition PAN member as Attorney-General, and gave autonomy to the central bank, as part of his vision for greater transparency in public finances. Some saw Zedillo's non-intervention in disputes that previous Presidents would have settled as a sign of weakness. A more generous assessment suggests that Zedillo recognized the need to allow nominally independent institutions to mature, free from interference from the executive.

## END OF THE PRI ERA

In the mid-term congressional elections of July 1997, the PRI lost its absolute majority in the lower house for the first time, so beginning the development of the Congreso into a genuine check on executive power. Simultaneously, the PRD's Cuauhtémoc Cárdenas won the mayoral election in Mexico City. Zedillo also agreed to internal reforms to the PRI, allowing party nominations for President to be chosen in primary elections. Zedillo was the first incumbent President to agree to forego his right to nominate the PRI candidate.

In a presidential election held on 2 July 2000, Vicente Fox Quesada, the candidate of the PAN-led Alianza por el Cambio (Alliance for Change), of which the Partido Verde Ecologista de México (PVEM—Green Ecologist Party of Mexico) was also a member, defeated Francisco Labastida Ochoa of the PRI. Fox received 43.5% of the valid votes cast, while Labastida secured 36.9%. The result brought 71 years of uninterrupted dominance by the PRI and its predecessor ruling parties to an end. By 2000 10 states were controlled by governors belonging to opposition parties.

Vicente Fox was a charismatic agro-industrialist and former President of Coca-Cola Mexico, who had been elected Governor of Guanajuato state in 1995. Only a constitutional amendment of 1993, negotiated by the PAN with Fox in mind, allowed him to stand. For the first time a presidential candidate whose parents were not born in Mexico was entitled to hold the presidency. Fox aimed to apply reforms that he had introduced in Guanajuato state on a national level. Examples included the devolution of powers such as tax collection and responsibility for services to districts and municipalities to encourage local autonomy in a country long accustomed to lobbying for funds from central government. Fox's Government, which took office in December 2000, unlike those of his predecessors, contained ministers without party affiliations, and from the private sector. Fox himself showed independence from his own party, with which he had an intermittently tense relationship. This approach has meant that he has been unable automatically to rely on the support of PAN legislators.

## THE GOVERNMENT OF VICENTE FOX, 2000–

The Alianza por el Cambio formed the largest bloc in the Cámara Federal de Diputados, although its 224 seats fell short of an absolute majority (the PRI secured 209 seats). In the Senado (Senate) the alliance held 51 of the 128 seats, compared with the PRI's 60. Fox was prevented from pursuing a more ambitious agenda because of his minority support in the Congreso and the need to seek consensus with the PRI, still the largest single party in the legislature. In particular, Fox's plans for reform of the labour market and liberalization of the energy sector faced opposition from the Congreso.

While President Fox was experiencing difficulties in implementing his agenda, the Congreso was maturing into a genuine institutional check on the executive, and developed dramatically from its typically pliant existence under PRI Governments. In 2001 the legislature blocked two of the Government's key initiatives (on tax reform and indigenous rights). Relations between the executive and the legislature reached a nadir in April 2002, when the Congreso refused permission for President Fox to travel to the USA and Canada—Presidents require congressional permission to travel, but it had never previously been denied. Deputies from the PRI, the PRD and the PVEM were angered at budget cuts and Fox's adversarial approach to Cuba.

Scandals involving funding of presidential campaigns embroiled both the PAN and the PRI, seriously undermining attempts at reconciliation between the parties. In January 2003 the federal electoral court (Tribunal Electoral del Poder Judicial de la Federación—TEPJF) ruled that banking secrecy laws did not apply in investigations into the fund-raising activities of political parties. The ruling led to the conviction and fine of the PRI for receiving US $45m. in illegal campaign financing from the oil-workers' union, the Sindicato de Trabajadores Petroleros de la República Mexicana (STPRM). In October 2003 both the PAN and the PVEM were fined for receiving illegal funding from Amigos de Fox, responsible for Fox's presidential campaign.

In June 2002 the Government published its Programa Nacional de Financiamiento del Desarrollo (Pronafide—National Financing for Development Programme), which outlined its plans for the remainder of its term in office. The main fiscal aim of Pronafide was to reduce the budget deficit and cut the public-sector borrowing requirement. It aimed to do this by increasing federal tax revenues, including eradicating certain income tax and value-added tax (VAT) exemptions, as well as increasing state and municipal tax revenues to reduce the federal tax burden; however, attempts to reform the tax system were blocked by Congreso.

The Fox Government gave new impetus to efforts to improve Mexico's human rights record. In July 2002 the Office of the UN High Commissioner for Human Rights (OHCHR) opened an office in Mexico City, while the Special Prosecutor's office launched an investigation into the repression of protest movements in 1968. In November 2003 the Supreme Court of Justice ruled that prosecution for murder could proceed, even in cases where no body had been found, thereby paving the way for prosecution relating to the 'dirty war' of the 1970s. However, in August 2003 a report by the human rights organization Amnesty International, *Muertes Intolerables*, criticized the Government for inefficiency and negligence in investigating the murders of an estimated 370 women in Juárez over the past 10 years. Fox had some success in defusing the conflict with the EZLN in Chiapas. The Government withdrew the Army and freed Zapatista prisoners. However, in August 2001 constitutional reforms guaranteeing indigenous rights were approved by the Congreso only after significant changes weakening the proposals had been made, causing the EZLN to suspend all contact with the Government. An unresolved stalemate prevailed in Zapatista-dominated areas of Chiapas. In June 2005 the Zapatista leadership appeared to be preparing for a change in tactics, possibly to include participation in electoral politics.

The Government's focus on human rights extended into foreign policy. Fox abandoned a policy of non-intervention in foreign affairs that had been pursued for decades by successive PRI Governments. The shift saw Mexico actively promoting human rights and democracy, which, along with closer relations with the USA, led to a deterioration of the historically good relations between Mexico and Cuba. After Mexico voted in favour of a UN motion to censure Cuba for its human rights record in April 2004, the Fox Government was accused by the Cuban leader, Fidel Castro, of interference in his country's affairs. In response, Mexico temporarily recalled its ambassador from Cuba. Fox's main foreign-affairs priority was to strengthen relations with Mexico's NAFTA partners, the USA and Canada, while defending the interests of the millions of Mexicans working, legally and illegally, in the USA. On the surface, President Fox and his US counterpart, George W. Bush, shared a pragmatic, pro-business outlook. Bush's first foreign trip as President was to Mexico, and Fox was the first head of state to visit Bush in Washington, DC, USA, which reflected the importance of the relationship. In March 2002, at a conference in Monterrey, Fox and Bush announced the so-called 'smart border' declaration, which set out to increase security on the joint border, as well as facilitating the safe and legal passage of migrants and goods.

The PAN's electoral success instilled new life into party politics in Mexico. New parties were successfully registered by the Código Federal Electoral (Cofipe—Federal Electoral Code) with the aim of contesting congressional mid-term elections in July 2003, during which the 500-seat Cámara Federal de Diputados, as well as six state governorships, went to the vote. While the PRI's monopoly of federal government was over, the party remained the largest, with the most powerful electoral machine in Mexico. Election campaigning limited legislative progress in the first half of 2003. Labour reforms, as well as reforms in the electricity and telecommunications sectors, were postponed until the new Congreso had been elected.

In the mid-term congressional elections held on 6 July 2003, the ruling PAN polled approximately 30.5% of votes, trailing the PRI's 35.0%; the PRD made the largest gains in terms of seats. The results reduced the PAN's representation in the lower house to 153 seats, with the PRI winning 224 seats and the PRD 95. However, the balance of forces was largely unchanged, as the PAN's status as a minority party in the Congreso continued, and the informal PRI-PVEM alliance fell short of a working majority. The fall in support for the PAN reflected protest and dissatisfaction at the record of the Government, particularly in the economy, which had been largely stagnant during the first three years of Fox's presidency. However, it also showed that the PRI, as the traditional party of government, was better equipped to organize nation-wide election campaigns. The mid-term election results and the tensions between Fox and his own party left the President increasingly isolated from the Congreso.

## THE BEGINNING OF THE END OF THE FOX PRESIDENCY

Although presidential elections were not due until 2006, the first signs of unofficial campaigning were placing obstacles in the way of the Government's legislative agenda in 2004. There were indications that President Fox's structural reform programme would lose support from presidential hopefuls concerned that, for example, liberalization of the electricity and telecommunications sectors, or fiscal and labour reform, would lose them important areas of support. One area of reform where it was hoped that consensus would lead to congressional approval, however, was fiscal decentralization. The Government established the Convención Nacional Hacendaria (CNH—National Finance Convention), made up of multi-party representatives from municipal, state and federal levels. From February to July 2004 the CNH discussed the transfer of powers from federal government to the states and municipalities in areas of revenue, expenditure, debt and public ownership. The *status quo* ensured that fiscal power and responsibility were concentrated in the Secretariat of State for Finance and Public Credit, with lower tiers of government relying on federal transfers. It was hoped that decentralization would lead to greater responsibility and accountability at lower levels of government, thereby rejuvenating the democratic process and reducing patronage and control from the centre. The proposals to give states greater responsibility extended to a national scale reforms introduced by President Fox when he was Governor of Guanajuato, when powers were devolved to district level. In mid-2005 some progress had been made in submitting more than 300 proposals to a series of committees for analysis and technical evaluation. However, despite the progress and multi-party involvement in the CNH, there was no guarantee that Congreso would approve its recommendations. As President Fox entered his last year in office before the election, his ability to forge the necessary consensus for reform was considerably weakened. The political parties were entering a long election campaign phase not conducive to cross-party initiatives, which threatened proposed electoral reform as well as reform of public finances. Conflict arose in late 2004 over whether the Government had the right to veto a budget already approved by the Congreso.

With polls in early 2005 indicating an uncertain race for the presidency involving three leading candidates, the main parties—the PRI, PAN and PRD—were manoeuvring in preparation for the presidential and congressional elections in July 2006. In February 2005 the PRD had received an electoral fillip by winning the governorship in the historic PRI stronghold of Guerrero. Political tensions increased in April after the PRI and PAN voted to remove the immunity enjoyed by the PRD mayor of Mexico City, Andrés Manuel López Obrador, who was accused of contempt of court in a 2001 planning dispute. López Obrador was a popular figure and a leading contender for the PRD nomination (particularly after Cuauhtémoc Cárdenas announced in early July that he would not be seeking the PRD's candidacy). In May all charges against López Obrador were dropped in the face of widespread support for the mayor, who emerged stronger from the episode, while the PAN and the PRI were accused of cynically attempting to remove him from the ballot in 2006 rather than promoting, as both parties claimed, accountability among public officials. López Obrador resigned as mayor in late July to focus on securing the PRD nomination. In the mean time, he began stressing fiscal prudence and economic pragmatism in an attempt to widen his appeal and to win the confidence of the business community.

Disappointment with the Government of President Fox disadvantaged the presidential prospects of the former Secretary of the Interior, Santiago Creel Miranda Semblanza, who was the PAN's most likely candidate for the 2006 ballot. Creel resigned on 1 June 2005 in order to campaign for the PAN's nomination, and was replaced by Carlos Abascal Carranza, previously Secretary of Labour and Social Welfare, who was seen as one of the more conservative members of Fox's Cabinet. On 23 June Alberto Cárdenas resigned as Secretary of the Environment and Natural Resources, also with the intention of competing for the PAN's nomination.

The PRI was expected to mount a strong challenge to the presidency, although there were divisions within the party over the choice of candidate for the presidential ballot. The two leading candidates for the PRI nomination in mid-2005 were the outgoing Governor of Estado de México, Arturo Montiel Rojas, and party President Roberto Madrazo Pintado, although Madrazo faced considerable opposition from a faction of the PRI. Facing genuinely contested elections and greater internal democracy, there were signs that the nature of the PRI was beginning to change: no longer exclusively a mechanism designed to mobilize for electoral victory controlled from above, it was hoped that the PRI in the 21st century would show more of the signs of a party in a democracy, challenging for political power through appeal to its centre-left ideology, rather than relying on fraudulently gained or privileged access to state power. An early indication of the relative fortunes of the main parties came in July 2005 in the gubernatorial election in Estado de México, the most populous state and traditionally a bellwether reflecting national voting trends. The PRI's candidate, Enrique Peña Nieto, won easily, with just under 50% of the votes cast. The PRI also won the governorship of Nayarit, held on the same day.

# Economy

## SANDY MARKWICK

Occupying an area of 1.96m. sq km (758,449 sq miles), Mexico is the 13th largest country in the world. It shares a 3,152 km-long border with the USA in the north, and in the south is bounded by Guatemala and Belize. The Gulf of Mexico and the Caribbean Sea lie to the east, and the Pacific Ocean and the Sea of Cortés (Gulf of California) to the west. The climate and topography are extremely varied: the tropical southern region and coastal lowlands are hot and wet, while the highlands of the centre are temperate and much of the north and west is arid desert. Although conditions are not ideally suited to agriculture, Mexico is among the world's leading producers of a number of crops. By contrast, the country's forestry and fishing resources are underdeveloped. Extensive mineral potential also remains largely unrealized, although this does not apply to petroleum reserves, extraction of which began at the start of the 20th century and accelerated in the 1970s, following important discoveries.

The population of Mexico was an estimated 106m. in 2005, making it the world's 11th most populous country. A population rate of approximately 1.5% per year is significantly lower than in recent decades, reflecting improving education and health care. The economy became increasingly industrialized in the second half of the 20th century. Successive Governments encouraged this trend with fiscal incentives and protection against imports. Supported by high government spending, considerable overseas investment and massive foreign borrowing, as well as by petroleum discoveries and increases in petroleum prices, the Mexican economy grew by over 6% per year, on average, between 1958 and 1982. However, rapid growth was accompanied by high fiscal deficits, a growing rate of inflation and increasing trade deficits. By 1982 Mexico's external debt totalled US $90,000m. Domestic investors, concerned about the Government's ability to manage the economy, withdrew massive amounts of capital from the country. With its foreign-exchange reserves all but exhausted, Mexico was forced to suspend debt-service payments. In late 1982 the incoming Government of President Miguel de la Madrid Hurtado was forced to implement austerity measures in order to secure the much-needed support of multilateral lending agencies and thus reschedule the debt. In 1983–88 consumer prices increased by almost 4,000%, while annual gross domestic product (GDP) growth averaged just 0.1%.

The Government of President Carlos Salinas de Gortari (1988–94) continued and advanced the structural-adjustment process. Deregulation loosened the constraints on market forces, encouraged foreign investment and reduced the role of the state in numerous sectors of the economy, including finance, agriculture, transport and communications. The state's shares in many state-owned companies were sold to the private sector. In 1992 import licences were required for fewer than 2% of all imports, while the average tariff was some 11%, compared with 13% in 1986. Part of President Salinas' motives for trade liberalization were political: cheaper consumer imports restored the support of a significant proportion of the middle classes who had abandoned the ruling Partido Revolucionario Institucional (PRI—Institutional Revolutionary Party) in 1988. Trade liberalization was consolidated by the coming into effect, on 1 January 1994, of the North American Free Trade Agreement (NAFTA) between Mexico, the USA and Canada. Trade liberalization and an exchange-rate band system succeeded in reducing inflation from 113.9% in 1988 to around 7% in December 1994, when Salinas was succeeded as President by Ernesto Zedillo Ponce de León. The economy looked stronger, partly because of broadening the tax base and more efficient tax collection, but also because of wage and price restraints, which were the result of a pact with labour and business leaders. Another reason for the economic improvement was a reduction in domestic interest rates. The country's admittance into the Organisation for Economic Co-operation and Development in June 1994 was intended to have a similar effect to that of admittance into NAFTA, by committing the Mexican economy to a free-market programme and ensuring that capital receipts remained substantial. Encouraged by the new regulatory environment, foreign direct investment increased to US $8,000m. by 1994. Private investment was mainly responsible for the modest annual average GDP growth in the early 1990s.

However, President Salinas' economic successes were based on an exchange-rate policy which, ultimately, proved unsustainable. The rate at which the peso was traded with the US dollar was allowed to 'float' (i.e., be made freely convertible) within a band, the floor of which was fixed at 3.0562 pesos to US $1, while the ceiling depreciated daily. President Salinas maintained this policy in spite of peso appreciation in real terms, rapidly increasing balance-of-payments problems and a current-account deficit which rose to $29,662m. in 1994 (almost 8% of GDP). To cover the current-account shortfall, the Government maintained real interest rates in order to continue to attract capital inflows, which were under threat because of fears of political instability in Mexico and because of higher interest rates in the USA. In 1994 dollar-linked bonds, *tesobonos*, were introduced to strengthen inflows. The capital-account deficit was largely financed by volatile, speculative portfolio inflows. Domestic political instability resulted in a loss of confidence in the currency during 1994. There was a massive withdrawal of investment funds, from $28,400m. to less than $8,200m., and a rapid depletion of foreign-exchange reserves, from $24,886m. at the end of 1993 to $6,101m. one year later.

Overvaluation of the peso led to a devaluation crisis within just three weeks of President Zedillo assuming power in December 1994. With foreign reserves low, the Government opted for a limited devaluation by lifting the exchange-band ceiling to 4.10 pesos to US $1, but this failed to ease pressure on the peso. Two days later the Government allowed the peso to float freely. At the end of 1994 there were 5.33 pesos to $1 and the currency continued to weaken in 1995. Investor nervousness was exacerbated by the Government's delay in producing a credible plan to control devaluation. The crisis led to fears that the Government would be unable to repay $29,000m. of short-term *tesobono* debt due in early 1995 and that the banking system would collapse because of increases in external debt-servicing costs.

In late January 1995 the US Government's announcement of a rescue programme succeeded in preventing default on debt payments and in supporting the Mexican economy until it could develop firmer foundations. The US Government pledged US $20,000m., while the Bank for International Settlements and the IMF agreed to extend a further $17,800m. Other financial institutions also agreed credit facilities. This rescue programme, the largest since the US Marshall Plan (European Reconstruction Program) to aid Europe following the Second World War, was agreed after the Zedillo administration committed itself to emergency stabilization measures, consisting of strict control of the money supply, reductions in public spending in order to create a public-sector surplus, increased taxes and wage constraints. The Government maintained its priority of reducing inflation by means of strict monetary policies.

The austerity measures succeeded in improving the balance of payments, but the Mexican economy entered a deep recession, contracting by 6.2% in 1995. The annual rate of inflation, having averaged 18% per year in 1991–95, increased to 35.0% in 1995. The Government continued its strict monetary policies into 1996, to re-establish credibility and to regain the confidence of the financial markets and the IMF. Mexico began refinancing its public-sector external debt, which had increased by US $15,636m. in 1995, to $95,167m. The successful issue of a floating-rate note supported by petroleum exports helped to raise $7,000m. towards repaying the US Treasury. Mexico was able to resume borrowing in international financial markets in June 1996.

The economy began to recover in 1996, led by the manufacturing, construction, mining, communications and transport sectors. Export revenues increased in 1996 and 1997 and a trade deficit of US $18,464m. in 1994 became a surplus of $7,089m. in 1995 after the devaluation of the peso and domestic recession. Mexico maintained a trade surplus until mid-1997. Thereafter, it has registered consistent deficits, in part owing to low petroleum prices. Inflation fell from 35.0% in 1995 to 15.9% in 1998. After a consumer price increase of 16.6% in 1999, annual inflation declined steadily to 5.0% in 2002, above the government target because of peso depreciation and increases in state-managed food prices, although the underlying trend was downward. The inflation rate fell below 4%, a record low, at the end of 2003. The rate increased to an annual average of 4.7% in 2004. Annual inflation fell slightly to 4.6% in May 2005. The higher inflation rate was a result of high international commodity prices and strong domestic demand.

President Zedillo adhered to prudent monetary and fiscal policies throughout his term in office. Public spending was restricted to meet budget targets and foreign-exchange reserves restored, increasing from US $6,101m. at the end of 1994 to $44,384m. in 2001. The current-account deficit doubled from 1997 to 1998, declined slightly in 1999, but increased in 2000, fuelled by interest payments on public debt. However, as a proportion of GDP the deficit was not a cause for alarm. The current-account deficit has fallen since 2000 (see below).

GDP growth slowed to 4.9% in 1998 and to 3.7% in 1999, principally owing to turmoil in Asian markets and currency devaluations in Brazil and Russia, although Mexico benefited from sustained demand from the USA. Strong growth, of 6.6%, was restored in 2000, led by commerce, restaurants and hotels, and again by the transport and communications sector. Fears of traditional fiscal indiscipline on the part of the incumbent Government in an election year proved unfounded, helped partly by the autonomy of the central bank. The new President, Vicente Fox Quesada, who assumed office in December 2000, was the first head of state to come from outside the traditional ruling PRI. While this marked a watershed in modern Mexican history, the implications for immediate economic policy were few. President Fox, from the conservative Partido Acción Nacional (PAN—National Action Party), renewed the Government's broad commitment to macroeconomic orthodoxy and free-market reform. The Fox administration's first budget proposal included a strict fiscal policy, while monetary policy remained rigorous under the continuing stewardship of the central bank governor, Guillermo Ortiz Martínez. Following strong growth in 2000, GDP contracted in 2001 by an estimated 0.2%, largely owing to reduced demand from the USA, leading to a significant fall in export revenues, as well as lower investment and public spending. The poorest performing sectors were construction and manufacturing. Growth was restored in 2002, but remained sluggish, at an overall 0.7%. Sectoral performance was very uneven, with agriculture, mining, manufacturing and commerce still in recession. Exports were the main source of growth in the economy. The economy grew by 1.2% in 2003, despite continued recession in the manufacturing sector, owing to stronger output in services, construction and agriculture. Growth of 4.4% in 2004 led to GDP of US $676,500m. Output was fuelled by private consumption, and underpinned by remittances from overseas workers, a credit expansion and by fixed investment and a positive trade balance.

In spite of its recovery in the latter half of the 1990s, Mexico remained vulnerable to external conditions, particularly to changes in interest and growth rates in the USA, the source of most of the country's trade and investment. In addition, a decline in world petroleum prices in 1998 (dramatically reversed subsequently) threatened the country's fiscal and trade accounts. The structural reforms of the 1980s caused widespread social hardship, which persisted at the beginning of the 21st century. There was a marked deterioration in income distribution between 1984 and 1992, and in the late 1990s extreme poverty affected around 20% of the population, the problem being particularly acute in the rural south. At the depth of the recession in July 1995 more than 500,000 people became unemployed. Job creation accompanied economic recovery, but underemployment remained a significant issue. There were hopes that the stable transition to the first non-PRI Government represented a maturing democracy that would underpin economic stability.

## EMPLOYMENT

Linked with the inequality of income distribution in Mexico was the problem of unemployment. As Mexico's population was predominantly a young one—some 31% of the population was under 14 years of age in mid-2005—the labour market came under intense pressure to absorb approximately 0.7m. new additions to the work-force each year. The fact that the economy consistently did not achieve adequate growth rates for more than a decade was reflected in high unemployment rates and the existence of a thriving informal economy.

The official rate of urban unemployment was 4.2% in April 2005 and averaged 3.7% in 2003, the highest rate since 1997. Economic growth led to an additional 538,000 jobs in the formal sector in 2004, but this was not sufficient to reduce unemployment. Official unemployment figures concealed the problem of underemployment. In 2003 an average of 20.5% of the workforce worked fewer than 35 hours per week. This was the highest rate since 1998. The informal sector was very large, although difficult to quantify accurately.

## AGRICULTURE, LIVESTOCK, FORESTRY AND FISHING

Mexico's complicated topography and wide variation of climates restricted the area that could be cultivated to about 21% of its total territory. By contrast, 57% was pasture or brushland while around 17% was forested. About 16% of the arable land was irrigated. In 2004 the agricultural sector (including livestock, forestry and fishing) engaged 16.4% of the employed work-force. Agriculture accounted for a stable 4.0% of GDP in 2004, compared with 5.8% in 1993 and 20% in 1950. The decline in its share of GDP was largely a result of post-Second World War industrialization, but it also reflected low growth rates. In 1900–2003 the sector achieved an average growth rate of 1.9% per year, in real terms. There were several reasons for such poor performances, including decapitalization, inadequate transport facilities, low world prices, adverse weather conditions and unpaid debts. One fundamental problem was the *ejido* land-holding system, which arose from the agrarian reforms of the 1930s. This system allowed for individual plots to be cultivated on communally owned land. As the land was communally owned there was no provision for the individual plots to be sold or leased, although they could be passed on to descendants. The result was an increasingly uneconomical fragmentation of the 105m. ha of land in communal ownership (in 2001). In 1992 the Government amended the Constitution in order to allow *ejido* land to be rented or sold, or to be used as collateral security for raising finance.

Other impediments to agricultural growth were trade liberalization and the decrease in subsidies, which caused output of many crops to fall. In 1993 the Government sought to address this problem by ending price supports for basic grains and by introducing a new scheme that subsidized farmers according to the area of land they owned. Indebtedness was made worse by devaluation. Faced with demonstrations by farmers, the authorities also arranged for the restructuring of overdue bank loans, but debts and high interest rates continued to restrain growth. In 1995 President Zedillo introduced a programme of direct cash subsidies for the agricultural sector and followed this with measures designed to stimulate output, including debt rescheduling for producers and increases in financing and credit from state agencies.

Agricultural output grew by 3.9% in 2003 and by 4.0% in 2004. Increased competition from producers in the USA and Canada was expected to increase pressure on the sector. On 1 January 2003 tariffs were lifted on trade in agricultural products representing some 30% of agricultural imports, in accordance with progressive liberalization under NAFTA. While devaluation resulted in a rare surplus being achieved in 1995, Mexico regularly recorded trade deficits in agricultural products. In 2003 the total agricultural trade deficit was US $5,603m., representing a 29% decrease from the deficit in 2002. Crop production accounted for just under 50% of agricultural output in 1999. Livestock products contributed some 30% and forestry products 21%. The contribution of fisheries to sectoral output was marginal.

Sugar cane and maize were the principal crops produced by Mexico, in terms of volume, in 2005. Mexico has produced a stable annual average of more than 40.0m. metric tons of sugar cane since the mid-1990s. Annual maize production averaged around 20m. tons, occupying 37% of harvested land. Production was an estimated 19.7m. tons in 2003. Other staples included sorghum (6.5m. tons in 2003), wheat, stable at 2.8m. tons in 2003, and dry beans (1.4m. tons in 2003). Barley, rice and soya beans were also grown. In the late 1990s coffee was the most valuable export crop; however, in late 1998 production in the main coffee-growing state of Chiapas was adversely affected by extensive flooding associated with the El Niño weather phenomenon (a warm ocean current which appears periodically in the Pacific). Coffee production then fluctuated, earning US $162m. in 2002, down from $227m. in 2001. Coffee was surpassed by red tomatoes as the most important single export crop (2.1m. tons produced in 2003). The contribution of agricultural produce to total export revenues declined steadily after 1995, when revenues of $4,016m. represented 5.0% of total export revenues. Agricultural exports were valued at $5,683.9m. in 2004, equivalent to 3.0% of total export revenues.

Fruit and vegetables increased their sales substantially in the first half of the 1990s, earning some US $2,251.3m. in 1995, compared with $935m. in 1988. The increase came about mainly owing to an expansion in exports to the USA. Mexico's fruit and vegetable production included citrus fruits, strawberries, mangoes, apples, pears, melons, pineapples and tomatoes. Owing to weak world prices, rising production costs and adverse climatic conditions, in 1992 cotton (lint) production collapsed to an estimated 33,000 metric tons, compared with 202,000 tons in 1991. Although it recovered in the late 1990s, by 2003 output stood at just 65,000 tons. Tobacco cultivation was yet another sector adversely affected by bad weather, with production fluctuating, reaching 51,000 tons by 1999, before falling back to 22,000 tons in 2003.

In the livestock sub-sector, cattle-rearing was the most important activity. Total meat production, including poultry, amounted to some 4.9m. metric tons in 2003. Production of cows' milk steadily increased from the late 1980s and reached some 9.8m. tons in 2003. Cattle stocks were recorded at an estimated 30.8m. head in that year. From the 1980s there was a sharp increase in poultry stocks, which was reflected by rising production of white meat (roughly 2.2m. tons in 2003, compared with 399,200 tons in 1980) and eggs (1.9m. in 2003, compared with 644,400 in 1980). Another product of the livestock sector was honey, an estimated 56,000 tons of which were produced in 2003.

Neither the transport infrastructure nor the landownership system encouraged long-term investment in forestry. Some 5.6% of forested areas were managed. Exploitation depleted Mexico's forested land at an annual rate of 1.4% in 1981–90. The sector faced further damage to output from poor planning and the lack of an infrastructure for effectively tackling the regular problems of El Niño and forest fires. In 2001 the Government established the Comisión Nacional Forestal (CONAFOR—National Forestry Commission) to encourage sustainable development of the forestry sector.

Although it has long Caribbean and Pacific coastlines and extensive inland waters, in the 21st century Mexico had yet to develop a modern fishing industry of any real importance. The annual catch increased gradually until 2003, when it stood at an estimated 1.5m. metric tons. Leading varieties of fish caught were tuna (tunny), Californian pilchard (sardines), squid and anchovy. Exports of fish (including shellfish) earned just US $584m. in 2002, down from the $731m. earned in 2001, because of lower prices and volumes for shrimp, the main fishing export. There are around 103,000 vessels in Mexico's fishing fleet operating out of 62 fishing ports.

## MINING AND POWER

Petroleum was by far the most important product of the extractive sector. Mexico was one of the world's largest crude oil-producing countries and the state oil monopoly, Petróleos Mexicanos (PEMEX), was one of the largest oil companies in the world. However, towards the end of the 20th century its contribution to the Mexican economy declined as world petroleum prices fell. Between 1986 and 1998 petroleum export revenues declined by an estimated 61%, from the equivalent of 5.7% of GDP to just 1.7%. In 1990 petroleum accounted for 35.6% of export earnings. Oil earnings fluctuated significantly with volatile international prices. Earnings from oil increased after 1999, when total earnings were US $9,605m., as a result of dramatic increases in the oil price in 2000. Having declined to their lowest level in two decades in December 1998, at $7.67 per barrel (p/b), prices for Mexican crude recovered, reaching $29.27 p/b in March 2000; however, prices declined in 2001 to as low as $12 p/b in December. By March 2002, however, prices had recovered to over $20 p/b, helping Mexico record a 13.1% increase in output in 2002 and exports of $14,477m. (9.0% of the value of total exports). In 2003 prices for Mexican crude increased further, to an average of $25 p/b, leading to export earnings of $18,654m. (11.3% of total export earnings). In 2004 Mexico earned $23,667m. from oil exports, equivalent to 12.6% of total exports. The increased earnings were predominantly the result of higher prices rather than volumes. An average price of Mexican crude was $33.3 p/b in the fourth quarter of 2004, compared with $25.03 p/b in the equivalent period of 2003. The trend continued into 2005. In June 2005 Mexican crude was $46.36 p/b.

Despite the importance of oil to the economy and the monopoly exercised by PEMEX, the company faced financial loss preventing it from pursuing further investment in exploration and development which was necessary to maintain current production levels. Oil revenues from PEMEX accounted for 33.4% of total government revenues in 2003. In the same year PEMEX recorded a loss (of US $3,700m.) after tax for the sixth consecutive year. PEMEX's tax bill for 2003 was $33,800m. Lobbying by PEMEX to reduce the company's tax contributions and thereby allow it to invest in development and exploration was frustrated by the Government's failure to implement sufficient fiscal reform. In the mean time, PEMEX expansion was financed by debt, causing a degree of nervousness among international lenders.

Petroleum production (and liquid gas equivalent) increased slightly during the previous decade, from 2.67m. barrels per day (b/d) in 1993 to a record 3.37m. b/d in 2003, a 6.1% increase over production in 2002. Production increased marginally in 2004, to an average of 3.38m. b/d. PEMEX enjoyed a monopoly on petroleum production, conferred by the Constitution. However, foreign petroleum companies did operate under service contracts with PEMEX, while US and Canadian concerns had the opportunity to enter into 'performance' contracts with the company. Plans to sell petrochemicals assets proceeded slowly, owing to investor concern about low petroleum prices and domestic opposition to privatization in the industry. Around 52% of Mexico's petroleum came from offshore sites in the Campeche Sound (Gulf of Mexico) and 39% from the Chicontepec region. Mexico's total hydrocarbons reserves stood at 50,032m. barrels (and equivalent) in 2003, a slight decline over previous years. Proven reserves amounted to 18,900m. barrels in 2004, down from an average 20,077m. barrels in 2003, itself a sharp decline from 30,837m. barrels in 2002. Total gas reserves were equivalent to 13,766m. barrels in 2003, down from 16,236m. barrels in 2001. Output of natural gas averaged 4,573m. cu ft per day (129m. cu m) in 2004, compared with an average of 4,623m. cu ft in 1998–2003. In March 2002 the Government announced the discovery of three gas fields in Veracruz and Campeche states and the Gulf of Mexico, which, it was hoped, would boost reserves by as much as 25% and reduce imports. Mexico hoped to continue to increase gas production in order to become a net exporter. A decline in exploratory activity in the 1980s had led to a diminution of reserves in the 1990s.

Apart from petroleum, Mexico produced an impressive amount of other minerals. It was the largest producer of silver in the world (3.1m. kg in 2004) and was also a leading source of fluorite, celestite, and sodium sulphate, as well as bismuth, graphite, antimony, arsenic, barite, sulphur and copper. It also produced iron ore, lead, zinc and coal. However, despite the wealth of resources, the mining sector was not a major force in the economy, accounting for just 0.3% of exports in 2003. In 2004 mining contributed 1.3% of total GDP and employed 1.0% (including the electricity sector) of the economically active population. Mining output increased by 3.7% in 2003 in response to increased global demand, a rate that was sustained into the first quarter of 2004. The sector was restrained by outdated technology, poor world prices and lack of finance. However, government regulations acted as a deterrent to private investment. The Salinas Government, therefore, focused on increasing private investment: it made available a large proportion of the country's mining reserves, relaxed curbs on foreign investment, lowered taxes and extended concession terms. The country registered a deficit in its trade in metallic and non-metallic minerals. Mining products earned an estimated US $500m. in export revenues in 2003, a increase over the $389m. earned in 2002.

Unlike mining, the utilities sector expanded relatively rapidly in the early 1990s, with growth averaging 3.5% per year in 1990–93. Utilities (electricity, gas and water) grew by 1.1% in 2003, compared with a 3.5% expansion experienced in 2002. The sector contributed 1.3% of GDP in 2004. Electricity production rose steadily, doubling in output to 228,873m. kWh in 1987–2000. Mexico generated 181,806m. kWh in 2003. Of Mexico's installed capacity of electric power, 62.0% came from oil-fired plants, 25.0% from hydroelectric plants, 6.8% from coal-fired plants, 2.2% from geothermal plants and 3.5% from the Laguna Verde nuclear power-station. National distribution of electricity was uneven. In March 2003 the state-owned power utility, Comisión Federal de Electricidad (CFE), awarded a contract to build the El Cajón hydroelectric plant to a consortium led by ICA, Mexico's largest construction company. The plant in Nayarit state, due to be completed in 2007, was expected to provide generating capacity of 750 MW.

In this sector too, with the challenge of expanding output to meet rising demand, from the late 1980s the Government gradually increased the opportunities for private enterprise, including foreign competition, both to collaborate with the CFE in installing new plants, and to produce power on its own account. However, political opposition delayed the introduction of comprehensive measures to increase private-sector involvement in the electricity industry. Opposition parties obstructed President Zedillo's attempts to change the Constitution to allow the full participation of private investors in the sector. President Fox renewed efforts to liberalize the electricity sector from late 2000. His proposals were similar to those of his predecessor, including ending the CFE's monopoly and creating a free market for electricity surpluses. However, in April 2002 the Senado and Supreme Court rejected, respectively, a reform proposal and a change decreed by Fox a year earlier, both of which were designed to increase the participation of private companies in the electricity sector. The supreme court ruling cast constitutional doubt on foreign investment in electricity generation. Fox responded by submitting a constitutional reform to the Congreso in August 2002, allowing private companies access to the transmission and distribution infrastructure in order to sell electricity freely.

## MANUFACTURING

The manufacturing sector began to be developed after the Second World War, and for four decades it enjoyed a considerable degree of protection from outside competition. However, from the 1980s it was increasingly subjected to such competition, as trade barriers were dismantled. The results were mixed. Manufacturing exports grew quite strongly, but many sectors were unable to meet the challenge of competition on either the export or the domestic market. This was particularly the case with small and medium-sized enterprises, which accounted for the majority of manufacturing businesses and for more than 90% of official employment in the sector. The manufacturing sector contributed 17.9% of GDP in 2004, compared with 19.5% in 2003. Excluding minerals, manufacturing generated 83.9% of total export earnings in 2004. The manufacturing sector recovered from three years of recession in 2004, recording annual growth of 3.8%. Recovery stemmed from increased US and domestic demand. In 1990–2003 the sector had averaged annual growth of 3.9%.

The most important manufacturing branch, in terms of its contribution to sectoral GDP and employment, was metal products, machinery and equipment, which accounted for 29.4% of manufacturing output and 25.8% of sectoral employment in 2003. This branch of manufacturing grew at an annual average rate of 10.3% in 1995–2000 and outperformed other manufacturing areas in 2000, recording 12.2% growth. However, this sub-sector performed relatively poorly during 2001–03, compared with the rest of manufacturing, recording contractions of 6.2%, 2.2% and 6.9% in 2001, 2002 and 2003, respectively. Significant recovery in 2004 was fuelled by strong US demand and led to growth of 28.9%. The metal products sub-sector was led by the automotive industry, which expanded rapidly from the mid-1990s as transnational corporations established operations in Mexico, in order to take advantage of NAFTA and of growth in the Mexican market.

The food, drink and tobacco sub-sector produced 26.7% of manufacturing output in 2003, employing 27.7% of the manufacturing work-force. The sub-sector recorded growth of 29.5% in 2004 compared with just 1.2% in 2003 and 1.8% in 2002. Growth stemmed from increased domestic demand. Key industries in this sub-sector, which drew on domestically produced raw materials, included grain-milling and bakeries, sugar processing, fruit and vegetable processing, cigarettes and beers and spirits. The chemicals industry (15.1% of sectoral GDP in 2003) was another major sub-sector, of which the petrochemicals division was a particularly dynamic component, but which also encompassed plastics, rubber and pharmaceuticals. One of the main traditional industrial activities was textiles and clothing, developed following the success of locally grown cotton and, later, of petroleum resources. In fact, in the 1990s synthetic fibres output far exceeded that of natural fibres. The industry traditionally had a significant export market, but the domestically orientated segment had suffered considerably from import competition. The chemicals sector grew by 1.8% and 3.8% in 2003 and 2004, respectively.

Mexico had a significant iron and steel industry, which used mainly national iron-ore resources to produce over 8m. metric tons of steel per year in the 1980s. The industry was privatized in 1989, after which production increased above overall rates of GDP, while the work-force was reduced by 40%. In 2001 Mexico produced an estimated 13.3m. metric tons of steel, down from the 15.6m. tons produced in 2000. There was concern that Mexico's reserves of iron ore would last only a further 25 years at existing rates of steel production.

Mexico also developed silver-, copper-, lead-, zinc- and tin-processing facilities, while non-metallic mineral resources provided the basis for glass and cement industries. Mexico was one of the world's largest manufacturers of cement. The other main industrial activities were paper, printing and publishing, and wood and cork manufactures.

In 1965 the Government began to plan an 'in-bond' industry, allowing temporary imports of inputs (parts), which were then assembled for duty-free exports. Originally, the provisions only covered the northern border areas, but they were later extended to the whole country. This in-bond, or *maquila* (assembly plant), sector, which took advantage of Mexico's low wage rates, prospered from the early 1980s. By 2000 there were 3,667 in-bond plants in operation, employing some 1.3m. workers. The main activities with which the *maquila* industries were concerned were electrical and electronic machinery and equipment, transport equipment and textiles. The sector generated an appreciable amount of foreign exchange and was one of the principal destinations for overseas foreign investment in the 1990s. Recession in manufacturing in 2001 was felt particularly in the export-orientated *maquila* sector, which contracted by 8.9% in 2001. By 2003 employee numbers in *maquilas* had fallen to 1.1m., working in 2,829 plants. In 2004 *maquilas* earned US $87,548m. in export revenues, representing 55.5% of overall manufacturing revenues. However, since the *maquila* sector was orientated on the processing of imported inputs from the USA, its outlays were also high, which left net earnings in 2004 at $18,924m.

## CONSTRUCTION

The construction industry grew by a steady 5.0% in 1999–2000, but declined by 4.5% in 2001, owing to a combination of government spending cuts and falling private-sector investment. Growth was restored to the sector in 2002 (1.7%) and continued in 2003 and 2004 with expansion of 3.4% and 5.3%, respectively, owing to a combination of public-sector investment in infrastructure as well as low interest rates and a wider availability of mortgages encouraging private-sector house-building. Public-sector investment accounted for 61% of construction activity in 2001, down from 65% in 2000. The sector contributed 5.6% of GDP in 2004 and employed 6.5% of the active labour force.

## TRANSPORT AND COMMUNICATIONS

Road transport was the chief means of conveying passengers and freight in Mexico. In 2003 there were an estimated 348,529 km (216,478 miles) of roads in the total network, of which 32.6% were paved. In order to support the development of the export trade and to spread the benefits of economic growth more evenly, the Government pursued extensive road-building projects in the 1990s. It also attempted to encourage the private sector to participate in road building, by granting concessionaires the right to build, maintain and collect the fees from toll roads. An innovative scheme for raising finance by issuing bonds backed by toll receipts was introduced. President Fox announced a plan to offer new concessions in 2003, but progress was slow and had yet to achieve results by mid-2005.

The railway system, covering 26,655 km in 2003, was owned by the state-owned company, Ferrocarriles Nacionales de México, until 2001, when it was sold off. Inadequate investment had resulted in the deterioration of the service, with the volume of freight traffic decreasing by 13% between 1986 and 1992, to 49.8m. metric tons, before recovering to 80.4m. tons by 1999. In 2003 there was 85.2m. tons of freight traffic. Passenger numbers decreased by 40% between 1986 and 1992, and continued to decline dramatically, from 14.7m. in 1992 to 0.27m. by 2003. However, numbers were considered likely to increase with new investment. The Government had slowly opened the railway system to private-sector involvement as it prepared to link the Mexican system with that of the USA and Canada, under NAFTA.

Mexico had 97 maritime ports, but the majority of cargo shipments were handled by just nine of them. The need for port facilities to meet international standards induced the Government to offer their management to the private sector. In the early 1990s US $700m. was spent on port development, much of it from the private sector. In the mid-1990s the management of the ports of Altamira, Acapulco, Guaymas, Tampico, Lázaro Cárdenas, Manzanillo and Veracruz was transferred to the private sector.

Mexico boasts one of the largest networks of airports in the world, with nearly every town or city of over 50,000 inhabitants having its own airport. In 2004 there were 57 international airports, 35 of which were used by 97% of total passengers. There were numerous national air companies, of which Aerovías de México (Aeroméxico) and Compañía Mexicana de Aviación, SA de CV (Mexicana) were the two largest and together controlled around three-quarters of the domestic market. Formerly state-owned, the companies were bought out by a coalition of banks in the late 1980s under the auspices of a new government-owned holding company, Cintra. However, Cintra's monopolistic arrangement was reviewed by the Federal Competition Commission in 1999. Plans to sell off Cintra by 2004 were delayed; the Government postponed advancing the process owing to poor demand. The Government also planned to allow the private sector to build and operate airports on 50-year, renewable contracts. In December 1998 a consortium including Danish, French and Spanish investors successfully bid in the first auction for management of airports, gaining control of nine southern airports, including Cancún. In August 1999 Mexican and Spanish investors won control of Pacific coast airports, including Guadalajara and Tijuana. The location for a proposed new international airport serving Mexico City was unresolved in early 2005. In 2002 the Government abandoned plans to build on land east of the city, following violent protests from local farmers. In the mean time the Government proposed to expand existing airports in central Mexico to ease the pressure on Mexico City's Benito Juárez airport.

The state-owned telephone company, Teléfonos de México, SA de CV (Telmex), was privatized in 1990. The number of telephone lines increased from 5.4m. in 1990 to an estimated 16.3m. in 2003. Further liberalization opened long-distance and cellular services to competition in 1996. In 2005 there were an estimated 35m. mobile cellular telephone subscribers. In 1997 Telmex received a provisional licence to provide services in the USA, in conjunction with US telecommunications company Sprint. The local market was opened to competition in 1998. However, Mexico and the USA have been in dispute over the rules governing competition and access to a market which is still dominated by Telmex. The latter restricted access to local markets for competitors following disputes over interconnection fees and debt payments. These financial disputes were resolved in January 2001, but the extra market access subsequently made available to competitors was insufficient to stop the USA challenging Mexico in the World Trade Organization (WTO).

## TOURISM

Tourism was an important foreign-exchange earner. Mexico recorded surplus revenue flows from tourism amounting to US $3,794m. in 2004, a marked increase from a surplus of just $7m. in 1990. There were an estimated 98.7m. visits from abroad made to Mexico in 2004, including 81m. day trips, of which 90% originated in the USA and Canada. Receipts in the first four months of 2005 were 26.2% higher than in the equivalent period of 2004. Significant travel by Mexicans abroad reduced the beneficial effects of overseas visitors' contribution to Mexico's 'invisible earnings' account. Mexico's attractions range from beach resorts such as Acapulco and Bahías de Huatulco on the Pacific coast and Cancún on the Caribbean coast, to a number of pre-Columbian sites including Teotihuacan and Chichén Itzá, as well as various colonial cities. A significant category of both inbound and outbound tourism was the cross-border day trip. From the late 1980s the Government made a concerted effort to encourage further 'open-skies' policy (a deregulated civil aviation market), while investment rules were liberalized for domestic and foreign investors alike. In 2003 tourism in Mexico supported employment for some 1.7m. people.

## FINANCE

The financial sector underwent dramatic changes from the early 1980s. In 1982 the Government nationalized the banking sector in an attempt to help overcome the debt crisis. A process of rationalization ensued, with the number of commercial banks being reduced from 58 to 18. They were reprivatized in 1991–92, and were generally sold for a high price, reflecting their profitability. However, operating systems had become cumbersome and outmoded and the new owners had to invest in modernization.

An added incentive to increase the banking sector's competitiveness came in 1993, when the Government began to authorize the establishment of new domestic banks. Following the ratification of NAFTA in 1994, US and Canadian banks were allowed access to the Mexican financial sector, albeit with certain restrictions. In addition, investors from countries other than the USA and Canada could also gain entry to the Mexican market if they operated through a North American subsidiary. Financial services sectoral GDP grew by 4.6% in 2004, following growth of 4.3% in 2003.

The object of the financial-sector reforms was both to increase domestic savings and to reduce borrowing costs. Before the bank privatizations began, the rules on lending had been relaxed. This, combined with success in reducing the fiscal deficit, lowered interest rates from 95% in 1987 to 14% by the end of 1993. In April 1994 interest rates in particular, and the setting of monetary policy in general, became the preserve of an autonomous central bank, Banco de México (BANXICO). The Government believed that by giving BANXICO independence it was guaranteeing that price stability would be maintained. The Government intervened to support ailing commercial banks and to prevent a collapse in the banking sector owing to 'bad' debts, which multiplied following devaluation in late 1994. In early 1998 the Government submitted to Congreso proposals to grant additional powers of autonomy to BANXICO. However, a further crisis in the financial sector was prompted following an investigation by the US authorities, which in May resulted in the arrest of several prominent bankers, accused of money-laundering and cultivating links with drugs cartels.

The Government established the Fondo Bancario de Protección al Ahorro (Fobaproa—Banking Fund for the Protection of Savings) to assume bad banking debts in exchange for new capital injections from shareholders. By late 1999 Fobaproa had absorbed US $89,000m. of liabilities, of which only an estimated 20% of the value would be recovered. After much political wrangling in the Congreso, in 1999 a new agency was approved, the Instituto de Protección al Ahorro Bancario (IPAB—Bank Savings Protection Institute). IPAB continued to prop up bankrupt banks, notably the third largest bank, Banca Serfin, which it sold to Spanish bank Banco Santander Central Hispano in May 2000. IPAB also introduced new insurance quotas for banks to cover deposits and auctioned the rights to manage and recover loans accumulated by Fobaproa.

## FOREIGN TRADE AND THE BALANCE OF PAYMENTS

The economic adjustments made necessary by the debt crisis of the 1980s led to a distinct improvement in the foreign-trade balance, with surpluses being recorded each year between 1982 and 1989. However, from 1990 the balance of trade was in deficit once again. Trade liberalization measures and a foreign-exchange policy that tended to allow the peso to become overvalued were partly to blame. The recovery in domestic growth rates and weak commodity prices also had a negative impact. The trade deficit amounted to US $18,464m. in 1994, compared with a surplus of $405m. in 1989. Following devaluation of the peso in 1994, exports performed better than imports. In 1995 Mexico registered a trade surplus of $7,089m. This surplus declined slightly to $6,531m. in 1996, before falling more dramatically to $623m. in 1997, as the growth in non-petroleum exports stimulated demand for imports of intermediate goods. While the value of exports increased by 22% in 1996–98, the value of imports increased by 40% over the same period, resulting in the trade balance falling into deficit once again in 1998. Thereafter the deficit continued to increase, reaching $9,955m. in 2001, the widest deficit since 1994. Despite a decline of $6,182m. in the value of imports in 2001 from the previous year, the value of exports declined further over the same period, by $7,908m. The deficit narrowed to $7,633m. in 2002 and to $5,780m. in 2003, largely owing to higher oil prices and lower demand for imports. A rapid acceleration of import spending widened the trade deficit significantly in 2004, to $8,530m.

Petroleum was traditionally the most important export, but it was surpassed in the mid-1990s by manufactures. Crude petroleum and petroleum products accounted for just 12.6% of the value of total exports in 2004, compared with 83.9% for manufactures (including *maquila* manufactures, which made up 46.2% of total exports). Agriculture (including livestock, forestry and fisheries) accounted for 3.0% of total exports in 2004, registering a sharp decline from 5.0% in 1995.

Imports consisted mainly of intermediate products required by the manufacturing sector. In 2004 these intermediate products totalled US $149,289m., or 75.9% of total imports. Imports of capital goods, which rose during the 1990s to support the increase in investment activity, amounted to $22,599m. in 2004. In that year this sector accounted for 11.5% of total imports. In 2003 imports of consumer goods amounted to $25,359m., representing 12.9% of total imports.

The vast majority of Mexico's foreign trade was, unsurprisingly, with the USA, a pattern which was reinforced by NAFTA. In 2004 the USA was the destination for an estimated 87.7% of Mexico's exports and the origin of 56.5% of imports. In 1997 Mexico displaced Japan as the USA's second most important trading partner. The next largest source of imports after the USA was the People's Republic of China, accounting for 7.3% in 2004, followed by Japan (5.4%) and Germany (3.6%). The second largest single export market for Mexican goods was Canada, which received 1.7% of exports in 2004, followed by Spain (1.1%). Although NAFTA would also bring about a strengthening of trade flows with Canada, the Mexican Government endeavoured to prevent excessive dependence on its northern neighbours. To this end it entered into free trade agreements with Bolivia, Chile, Colombia, Costa Rica and Venezuela. The Government also sought to foster closer trading relations with countries outside Latin America.

As well as a trade deficit, Mexico had a perennial shortfall on its 'invisibles' or services account. The largest contributors to this deficit were interest payments on the foreign debt, which led to massive current-account deficits in the balance of payments. At the same time there was substantial new foreign borrowing by the public- and private-sector alike. The result was a rapid increase in the size of the foreign debt. During the 1980s the authorities embarked on various debt-rescheduling exercises with commercial-bank creditors. The most fruitful was an agreement reached in 1989, adopting the idea of debt relief proposed by Nicholas Brady, the US Treasury Secretary, under the so-called Brady Plan. Although the Brady Plan required new borrowing from the World Bank, the IMF and Japan to provide a certain amount of collateral, it did lead to a reduction in the stock of external debt. However, in the early 1990s the debt increased once more and reached US $166,000m. in 1999. Meanwhile, the current-account deficit declined. A continuing deficit on the invisibles account offset a trade surplus to produce a current-account deficit of $2,330m. in 1996, which was covered by new debt and foreign investment. In 1996–98 the current-account deficit increased, reaching $15,724m. in the later year, as imports increased faster than exports. The current-account deficit declined to $14,017m. in 1999, then increased to $18,185m. in 2000. It decreased slightly, to $18,003m., in 2001, before declining to $13,072m. in 2002, principally because improved oil earnings reduced the trade deficit. The current-account deficit fell further still, to $6,613m., in 2003, but rose slightly, to $7,813m., in 2004. Deficits in trade, services and income accounts were offset by a 24% increase in worker remittances in 2004, to $16,613m. These transfers have grown in importance, to account for 2.5% of GDP in 2004, compared with 1.2% of GDP in 2000. Most of these transfers related to earnings sent by Mexicans working in the USA.

Contributing to the current-account deficit along with the trade deficit, were the services and income balances, amounting to deficits of US $5,689m. and $10,622m., respectively, in 2004. The bulk of the income deficit was made up of net interest on public debt. At the end of 2004 total public- and private-sector foreign debt, as estimated by the central bank, amounted to an estimated $130,500m., a further overall decline since 1999 as liabilities of both commercial banks and the public sector were reduced. Mexico registered a capital-account surplus of $13,692m. in 2004, down from $18,044m. in 2003, $21,100m. in 2002 and a record $24,407m. in 2001. The smaller surplus stemmed from a decline in the value of assets and was in spite of a 46% increase in foreign direct investment (FDI) inflows, which amounted to $16,602m. in 2004. Financial services were the target of 29.9% of FDI inflows during 2004, with the acquisition of Bancomer by Spain's Banco Bilbao Vizcaya Argentaria (BBVA) alone accounting for $4,200m. Manufacturing was the principal target of FDI, accounting for 52% in 2004. This reflected a shift in the direction of FDI over the previous decade. In 1990 32% of FDI was directed towards industrial projects and 59% towards services. The principal source of FDI was the USA, which provided 48% of the total in 2004. Spain provided 34.7% in that year. The total stock of FDI was an estimated $165,000m. at the end of 2003. Portfolio investment totalled $7,565m. in 2004, up from $3,864m. in 2003, and a further considerable increase compared with levels in 2000, when $401m. was invested in the money and equity markets. In the first quarter of 2005 portfolio investment increased by 95% over the same period in 2004. The increase in portfolio investment was largely explained by higher interest rates attracting international investors in the latter half of 2004.

## CONCLUSION

Unquestionably, there was a dramatic transformation in the Mexican economy in the last two decades of the 20th century. The role of the state was steadily reduced and opportunities for private and foreign investment increased. Low growth in the 1990s saw the poorer sections of society, already adversely affected by the previous decade's austerity measures, steadily become poorer. Political instability in 1994 caused a deterioration in economic conditions. The new peso came under pressure, leading to the devaluation crisis of December and recession and economic crisis in 1995. However, the devaluation of the peso restored export growth and the imposition of strict austerity measures ensured that the economy was more solidly based. Assisted by substantial financial aid, the economy began to recover from 1996. In 1997–98 growth returned to the Mexican economy and inflation was brought under control, with few adverse effects from the Asian financial crisis and the decline in world petroleum prices as Mexico strengthened its economic links with its North American neighbours. While NAFTA and other free trade agreements with Latin American and Pacific countries were expected to continue to provide an impetus to exports and to investment, it was the Government's hope that this investment would be increasingly directed to productive, rather than speculative, assets. The maintenance of political calm was crucial to Mexico's economic future. The defeat of the ruling PRI in the presidential elections of 2000 represented a profound watershed in Mexican history. Investor confidence in Mexican political stability was encouraged by an orderly transfer of power in December 2000 and macroeconomic stability. However, political weakness prevented the Government of President Vicente Fox from achieving economic and state sector reforms which might have underpinned more dynamic or consistent growth, and Mexico also remained vulnerable to fluctuations in the US economy.

# Statistical Survey

Sources (unless otherwise stated): Dirección General de Estadística, Instituto Nacional de Estadística, Geografía e Informática (INEGI), Edif. Sede, Avda Prolongación Héroe de Nacozari 2301 Sur, 20270 Aguascalientes, Ags; tel. (14) 918-1948; fax (14) 918-0739; internet www.inegi.gob.mx; Banco de México, Avda 5 de Mayo 1, Col. Centro, Del. Cuauhtémoc, 06059 México, DF; tel. (55) 5237-2000; fax (55) 5237-2370; internet www.banxico.org.mx.

## Area and Population

### AREA, POPULATION AND DENSITY

| | |
|---|---:|
| Area (sq km) | |
| Continental | 1,959,248 |
| Islands | 5,127 |
| Total | 1,964,375* |
| Population (census results)† | |
| 5 November 1995 | 91,158,290 |
| 14 February 2000 | |
| Males | 47,592,253 |
| Females | 49,891,159 |
| Total | 97,483,412 |
| Population (official estimates at mid-year)‡ | |
| 2002 | 103,039,964 |
| 2003 | 104,213,503 |
| 2004 | 105,349,837 |
| Density (per sq km) at mid-2004‡ | 53.6 |

* 758,449 sq miles.
† Including adjustment for underenumeration (90,855 in 1995 and 1,730,016 in 2000).
‡ Source: Presidencia de la República, *Cuarto Informe de Gobierno*.

### ADMINISTRATIVE DIVISIONS
(at census of 14 February 2000)

| States | Area (sq km)* | Population | Density (per sq km) | Capital |
|---|---:|---:|---:|---|
| Aguascalientes (Ags) | 5,623 | 944,285 | 167.9 | Aguascalientes |
| Baja California (BC) | 71,540 | 2,487,367 | 34.8 | Mexicali |
| Baja California Sur (BCS) | 73,937 | 424,041 | 5.7 | La Paz |
| Campeche (Camp.) | 57,718 | 690,689 | 12.0 | Campeche |
| Chiapas (Chis) | 73,680 | 3,920,892 | 53.2 | Tuxtla Gutiérrez |
| Chihuahua (Chih.) | 247,490 | 3,052,907 | 12.3 | Chihuahua |
| Coahuila (de Zaragoza) (Coah.) | 151,447 | 2,298,070 | 15.2 | Saltillo |
| Colima (Col.) | 5,629 | 542,627 | 96.4 | Colima |
| Distrito Federal (DF) | 1,485 | 8,605,239 | 5,794.8 | Mexico City |
| Durango (Dgo) | 123,364 | 1,448,661 | 11.7 | Victoria de Durango |
| Guanajuato (Gto) | 30,617 | 4,663,032 | 152.3 | Guanajuato |

# MEXICO

## Statistical Survey

| States—continued | Area (sq km)* | Population | Density (per sq km) | Capital |
|---|---|---|---|---|
| Guerrero (Gro) | 63,618 | 3,079,649 | 48.4 | Chilpancingo de los Bravos |
| Hidalgo (Hgo) | 20,855 | 2,235,591 | 107.2 | Pachuca de Soto |
| Jalisco (Jal.) | 78,624 | 6,322,002 | 80.4 | Guadalajara |
| México (Méx.) | 22,332 | 13,096,686 | 586.5 | Toluca de Lerdo |
| Michoacán (de Ocampo) (Mich.) | 58,672 | 3,985,667 | 67.9 | Morelia |
| Morelos (Mor.) | 4,894 | 1,555,296 | 317.8 | Cuernavaca |
| Nayarit (Nay.) | 27,861 | 920,185 | 33.0 | Tepic |
| Nuevo León (NL) | 64,206 | 3,834,141 | 59.7 | Monterrey |
| Oaxaca (Oax.) | 93,348 | 3,438,765 | 36.8 | Oaxaca de Juárez |
| Puebla (Pue.) | 34,246 | 5,076,686 | 148.2 | Heroica Puebla de Zaragoza |
| Querétaro (de Arteaga) (Qro) | 11,659 | 1,404,306 | 120.4 | Querétaro |
| Quintana Roo (Q.Roo) | 42,544 | 874,963 | 20.6 | Ciudad Chetumal |
| San Luis Potosí (SLP) | 61,165 | 2,299,360 | 37.6 | San Luis Potosí |
| Sinaloa (Sin.) | 57,334 | 2,536,844 | 44.2 | Culiacán Rosales |
| Sonora (Son.) | 179,527 | 2,216,969 | 12.3 | Hermosillo |
| Tabasco (Tab.) | 24,747 | 1,891,829 | 76.4 | Villahermosa |
| Tamaulipas (Tam.) | 80,155 | 2,753,222 | 34.3 | Ciudad Victoria |
| Tlaxcala (Tlax.) | 3,988 | 962,646 | 241.4 | Tlaxcala de Xicohténcatl |
| Veracruz-Llave (Ver.) | 71,856 | 6,908,975 | 96.2 | Jalapa Enríquez |
| Yucatán (Yuc.) | 39,675 | 1,658,210 | 41.8 | Mérida |
| Zacatecas (Zac.) | 75,412 | 1,353,610 | 17.9 | Zacatecas |
| **Total** | 1,959,248 | 97,483,412 | 49.8 | — |

* Excluding islands.

### PRINCIPAL TOWNS
(population at census of 14 February 2000)

| | | | |
|---|---|---|---|
| Ciudad de México (Mexico City, capital) | 8,605,239 | Querétaro | 536,463 |
| Guadalajara | 1,646,183 | Torreón | 502,964 |
| Ecatepec de Morelos (Ecatepec) | 1,621,827 | San Nicolás de los Garzas | 496,879 |
| Heroica Puebla de Zaragoza (Puebla) | 1,271,673 | Santa María Chimalhuacán (Chimalhuacán) | 482,530 |
| Nezahualcóyotl | 1,225,083 | Atizapán de Zaragoza | 467,544 |
| Ciudad Juárez | 1,187,275 | Tlaquepaque | 458,674 |
| Tijuana | 1,148,681 | Toluca de Lerdo (Toluca) | 435,125 |
| Monterrey | 1,110,909 | Cuautitlán Izcalli | 433,830 |
| León | 1,020,818 | Victoria de Durango (Durango) | 427,135 |
| Zapopan | 910,690 | Tuxtla Gutiérrez | 424,579 |
| Naucalpan de Juárez (Naucalpan) | 835,053 | Veracruz Llave (Veracruz) | 411,582 |
| Tlalnepantla de Baz (Tlalnepantla) | 714,735 | Reynosa | 403,718 |
| Guadalupe | 669,842 | Benito Juárez (Cancún) | 397,191 |
| Mérida | 662,530 | Matamoros | 376,279 |
| Chihuahua | 657,876 | Jalapa Enríquez (Xalapa) | 373,076 |
| San Luis Potosí | 629,208 | Villahermosa | 330,846 |
| Acapulco de Juárez (Acapulco) | 620,656 | Mazatlán | 327,989 |
| Aguascalientes | 594,092 | Cuernavaca | 327,162 |
| Saltillo | 562,587 | Valle de Chalco (Xico) | 322,784 |
| Morelia | 549,996 | Irapuato | 319,148 |
| Mexicali | 549,873 | Tonalá | 315,278 |
| Hermosillo | 545,928 | Nuevo Laredo | 308,828 |
| Culiacán Rosales (Culiacán) | 540,823 | | |

**2003** (UN estimate, including suburbs): Ciudad de México 18,660,221 (Source: UN, *World Urbanization Prospects: The 2003 Revision*).

### BIRTHS, MARRIAGES AND DEATHS

| | Registered live births | | Registered marriages | | Registered deaths | |
|---|---|---|---|---|---|---|
| | Number | Rate (per 1,000) | Number | Rate (per 1,000) | Number | Rate (per 1,000) |
| 1996 | 2,707,718 | 24.4 | 670,523 | 6.9 | 436,321 | 4.5 |
| 1997 | 2,698,425 | 23.7 | 707,840 | 7.5 | 440,437 | 4.7 |
| 1998 | 2,668,428 | 23.0 | n.a. | n.a. | 444,665 | n.a. |
| 1999 | 2,769,089 | 22.3 | 743,856 | n.a. | 443,950 | n.a. |
| 2000 | 2,798,339 | 21.1 | 707,422 | 7.0 | 437,667 | 4.2 |
| 2001 | 2,767,610 | 20.5 | 665,434 | 6.5 | 443,127 | 4.2 |
| 2002 | 2,699,084 | 19.9 | n.a. | 6.0 | 459,687 | 4.5 |
| 2003 | 2,655,894 | 19.3 | n.a. | 5.6 | 472,140 | 4.5 |

**2004:** Birth rate 18.8 per 1,000; Death rate 4.5 per 1,000.

**Expectation of life** (UN estimates, years at birth): 74 (males 72; females 77) in 2003 (Source: WHO *World Health Report*).

### ECONOMICALLY ACTIVE POPULATION
(sample surveys, '000 persons aged 12 years and over, April–June)

| | 2002 | 2003 | 2004 |
|---|---|---|---|
| Agriculture, hunting, forestry and fishing | 7,206.7 | 6,813.6 | 6,937.9 |
| Mining, quarrying and electricity | 340.2 | 352.2 | 409.2 |
| Manufacturing | 7,132.8 | 6,991.5 | 7,350.7 |
| Construction | 2,533.1 | 2,748.4 | 2,741.8 |
| Trade | 7,451.4 | 7,688.3 | 8,147.7 |
| Services | 11,851.3 | 12,197.0 | 12,851.5 |
| Transport and communications | 1,831.4 | 1,865.5 | 1,888.7 |
| Public sector | 1,802.7 | 1,829.8 | 1,816.2 |
| Activities not adequately defined | 152.4 | 147.0 | 162.5 |
| **Total employed** | 40,302.0 | 40,633.2 | 42,306.1 |
| Unemployed | 783.7 | 882.5 | 1,092.7 |
| **Total labour force** | 41,085.7 | 41,515.7 | 43,398.8 |
| Males | 26,888.1 | 27,277.0 | 28,013.5 |
| Females | 14,197.6 | 14,238.6 | 15,385.2 |

## Health and Welfare

### KEY INDICATORS

| | |
|---|---|
| Total fertility rate (children per woman, 2003) | 2.5 |
| Under-5 mortality rate (per 1,000 live births, 2003) | 28 |
| HIV/AIDS (% of persons aged 15–49, 2003) | 0.3 |
| Physicians (per 1,000 head, 2002) | 1.5 |
| Hospital beds (per 1,000 head, 2001) | 1.1 |
| Health expenditure (2002): US $ per head (PPP) | 550 |
| Health expenditure (2002): % of GDP | 6.1 |
| Health expenditure (2002): public (% of total) | 44.9 |
| Access to water (% of persons, 2002) | 91 |
| Access to sanitation (% of persons, 2002) | 77 |
| Human Development Index (2002): ranking | 53 |
| Human Development Index (2002): value | 0.802 |

For sources and definitions, see explanatory note on p. vi.

# MEXICO

## Agriculture

**PRINCIPAL CROPS**
('000 metric tons)

|  | 2001 | 2002 | 2003 |
|---|---|---|---|
| Wheat | 3,275 | 3,236 | 2,750 |
| Rice (paddy) | 227 | 227 | 191 |
| Barley | 762 | 737 | 1,109 |
| Maize | 20,134 | 19,299 | 19,652 |
| Oats | 89 | 64 | 147 |
| Sorghum | 6,567 | 5,206 | 6,462 |
| Potatoes | 1,628 | 1,483 | 1,735 |
| Sugar cane | 47,250* | 45,635 | 45,126 |
| Dry beans | 1,063 | 1,549 | 1,400 |
| Chick-peas | 326 | 235 | 240* |
| Soybeans (Soya beans) | 122 | 87 | 76 |
| Groundnuts (in shell) | 120 | 75 | 75* |
| Coconuts* | 1,100 | 959 | 959 |
| Safflower seed | 111 | 53 | 213 |
| Cottonseed | 152 | 68 | 102 |
| Cabbages | 216 | 197 | 197* |
| Lettuce | 213 | 228 | 243 |
| Tomatoes | 2,183 | 1,990 | 2,148 |
| Cauliflower* | 200 | 200 | 200 |
| Pumpkins, squash and gourds* | 560 | 470 | 560 |
| Cucumbers and gherkins* | 420 | 433 | 435 |
| Chillies and green peppers | 1,871 | 1,784 | 1,854 |
| Green onions and shallots | 1,029 | 1,131 | 1,131* |
| Dry onions* | 102 | 100 | 100 |
| Carrots | 356 | 378 | 378* |
| Green corn* | 191 | 186 | 186 |
| Bananas | 2,028 | 1,886 | 2,027 |
| Oranges | 4,035 | 3,844 | 3,970† |
| Tangerines, mandarins, clementines and satsumas | 365 | 360* | 360* |
| Lemons and limes | 1,594 | 1,725 | 1,825 |
| Grapefruit and pomelos | 320 | 269 | 257† |
| Apples | 443 | 428 | 488† |
| Peaches and nectarines | 176 | 198 | 224 |
| Strawberries | 131 | 142 | 150 |
| Grapes | 436 | 363 | 457 |
| Watermelons | 970 | 858 | 970 |
| Cantaloupes and other melons* | 510 | 510 | 510 |
| Mangoes | 1,577 | 1,523 | 1,503 |
| Avocados | 940 | 901 | 1,040 |
| Pineapples | 626 | 660 | 721 |
| Papayas | 873 | 876 | 956 |
| Coffee (green) | 303 | 313 | 311 |
| Cocoa beans | 47 | 46 | 48 |
| Pimento and allspice* | 55 | 55 | 55 |
| Cotton (lint) | 97 | 43 | 65 |
| Tobacco (leaves) | 41 | 22 | 22 |

* FAO estimate(s).
† Unofficial figure.
Source: FAO.

**LIVESTOCK**
('000 head, year ending September)

|  | 2001 | 2002 | 2003 |
|---|---|---|---|
| Horses* | 6,255 | 6,255 | 6,260 |
| Mules* | 3,280 | 3,280 | 3,280 |
| Asses* | 3,260 | 3,260 | 3,260 |
| Cattle | 30,621 | 30,700* | 30,800* |
| Pigs | 17,584 | 18,000* | 18,100* |
| Sheep | 6,165 | 6,260* | 6,560* |
| Goats | 8,702 | 9,600* | 9,500* |
| Chickens* | 497,600 | 540,000 | 540,000 |
| Ducks* | 8,100 | 8,100 | 8,100 |
| Turkeys* | 5,850 | 5,850 | 5,850 |

* FAO estimate(s).
Source: FAO.

**LIVESTOCK PRODUCTS**
('000 metric tons)

|  | 2001 | 2002 | 2003 |
|---|---|---|---|
| Beef and veal | 1,445 | 1,468 | 1,496 |
| Mutton and lamb | 36 | 38 | 40 |
| Goat meat | 39 | 42 | 42 |
| Pig meat | 1,058 | 1,070 | 1,043 |
| Horse meat* | 79 | 79 | 79 |
| Poultry meat | 1,976 | 2,123 | 2,204 |
| Cows' milk | 9,472 | 9,658 | 9,842 |
| Goats' milk | 140 | 146 | 148 |
| Butter | 15 | 14 | 14 |
| Cheese | 153 | 152 | 140 |
| Evaporated and condensed milk* | 156 | 158 | 158 |
| Hen eggs | 1,892 | 1,901 | 1,882 |
| Cattle hides* | 176 | 176 | 176 |
| Honey | 59 | 59 | 56 |

* FAO estimates.
Source: FAO.

## Forestry

**ROUNDWOOD REMOVALS**
('000 cubic metres, excl. bark)

|  | 2001 | 2002 | 2003 |
|---|---|---|---|
| Sawlogs, veneer logs and logs for sleepers | 6,176 | 6,176* | 6,176* |
| Pulpwood | 1,028 | 1,028* | 1,028* |
| Other industrial wood | 216 | 216* | 216* |
| Fuel wood* | 37,736 | 37,913 | 38,090 |
| **Total** | 45,156 | 45,333 | 45,510 |

* FAO estimate(s).
Source: FAO.

**SAWNWOOD PRODUCTION**
('000 cubic metres, incl. railway sleepers)

|  | 1999* | 2000* | 2001 |
|---|---|---|---|
| Coniferous (softwood) | 2,904 | 2,904 | 2,904* |
| Broadleaved (hardwood) | 206 | 206 | 483 |
| **Total** | 3,110 | 3,110 | 3,387 |

* FAO estimate(s).
**2002–03:** Production assumed to be unchanged from 2001 (FAO estimates).
Source: FAO.

## Fishing

('000 metric tons, live weight)

|  | 2001 | 2002 | 2003* |
|---|---|---|---|
| Capture | 1,398.6 | 1,450.6 | 1,450.0 |
| Tilapias | 60.3 | 54.9 | 54.5 |
| California pilchard (sardine) | 609.8 | 624.8 | 620.0 |
| Yellowfin tuna | 135.5 | 148.6 | 165.8 |
| Marine shrimps and prawns | 57.5 | 54.6 | 54.6 |
| American cupped oyster | 48.6 | 47.6 | 47.6 |
| Jumbo flying squid | 73.7 | 115.9 | 100.0 |
| Aquaculture | 76.1 | 73.7 | 73.7 |
| Whiteleg shrimp | 48.0 | 45.9 | 45.9 |
| **Total catch** (incl. others) | 1,474.7 | 1,524.3 | 1,523.7 |

* FAO estimates.
Note: Figures exclude aquatic plants ('000 metric tons, capture only): 46.9 in 2001; 30.1 in 2002; 30.0 in 2003 (FAO estimate). Also excluded are aquatic mammals and crocodiles (recorded by number rather than by weight), shells and corals. The number of Morelet's crocodiles caught was: 3,643 in 2001; 1,588 in 2002; 1,037 in 2003. The catch of marine shells (metric tons) was: 363 in 2001 (FAO estimate); 265 in 2002; 249.7 in 2003 (FAO estimate).
Source: FAO.

MEXICO                                                                                                           Statistical Survey

## Mining

(metric tons, unless otherwise indicated)

|  | 2002 | 2003 | 2004 |
|---|---:|---:|---:|
| Antimony* | 153 | 434 | 503 |
| Arsenic* | 1,946 | 1,729 | 1,828 |
| Barytes | 163,621 | 287,451 | 306,668 |
| Bismuth* | 1,126 | 1,064 | 1,014 |
| Cadmium* | 1,389 | 1,639 | 1,618 |
| Coal | 6,370,874 | 6,648,257 | 6,450,594 |
| Coke | 1,451,094 | 1,462,106 | 1,445,052 |
| Copper* | 314,820 | 303,765 | 352,286 |
| Crude petroleum ('000 barrels per day) | 3,177 | 3,371 | 3,383 |
| Celestite | 94,016 | 130,329 | 87,610 |
| Diatomite | 48,029 | 53,395 | 59,818 |
| Dolomite | 457,665 | 565,896 | 1,158,929 |
| Feldspar | 332,101 | 346,315 | 364,166 |
| Flourite | 622,478 | 756,258 | 842,698 |
| Gas (million cu ft per day) | 4,423 | 4,498 | 4,573 |
| Gold (kg)* | 23,596 | 22,177 | 24,496 |
| Graphite | 13,885 | 8,730 | 14,769 |
| Gypsum | 3,549,550 | 3,779,659 | 4,840,099 |
| Iron* | 5,965,427 | 6,759,198 | 6,889,538 |
| Lead* | 138,749 | 144,297 | 141,578 |
| Manganese* | 88,358 | 114,550 | 135,893 |
| Molybdenum* | 3,427 | 3,524 | 3,731 |
| Silica | 1,778,714 | 1,689,042 | 2,055,940 |
| Silver* | 3,146,257 | 2,945,710 | 3,093,366 |
| Sulphur | 887,035 | 1,051,968 | 1,121,546 |
| Wollastonite | 29,197 | 31,234 | 28,224 |
| Zinc* | 431,663 | 412,255 | 384,338 |

* Figures for metallic minerals refer to metal content of ores.

## Industry

**SELECTED PRODUCTS**
('000 metric tons, unless otherwise indicated)

|  | 2000 | 2001 | 2002 |
|---|---:|---:|---:|
| Wheat flour | 2,538 | 2,611 | 2,619 |
| Other cereal flour | 1,179 | 1,678 | 1,614 |
| Raw sugar | 2,531 | 3,018 | 2,736 |
| Beer ('000 hectolitres) | 59,851 | 61,632 | 63,530 |
| Soft drinks ('000 hectolitres) | 126,460 | 130,050 | 127,530 |
| Cigarettes (million units) | 56,383 | 56,057 | 54,704 |
| Cotton yarn (pure and mixed) | 21 | 19 | 17 |
| Tyres ('000 units)* | 16,780 | 13,533 | 11,628 |
| Cement | 33,429 | 32,239 | 33,478 |
| Gas stoves—household ('000 units) | 3,973 | 4,021 | 4,510 |
| Refrigerators—household ('000 units) | 2,049 | 2,071 | 2,222 |
| Washing machines—household ('000 units) | 1,720 | 1,636 | 1,657 |
| Lorries, buses, tractors, etc. ('000 units) | 554 | 529 | 504 |
| Passenger cars ('000 units) | 1,294 | 1,273 | 1,247 |
| Electric energy (million kWh) | 228,873 | 226,686 | n.a. |

* Tyres for road motor vehicles.

Source: UN, *Industrial Commodity Statistics Yearbook*.

## Finance

**CURRENCY AND EXCHANGE RATES**

**Monetary Units**
100 centavos = 1 Mexican nuevo peso.

**Sterling, Dollar and Euro Equivalents** (31 May 2005)
£1 sterling = 19.823 nuevos pesos;
US $1 = 10.903 nuevos pesos;
€1 = 13.444 nuevos pesos;
1,000 Mexican nuevos pesos = £50.45 = $91.72 = €74.38

**Average Exchange Rate** (nuevos pesos per US $)
2002    9.6560
2003   10.7890
2004   11.2860

Note: Figures are given in terms of the nuevo (new) peso, introduced on 1 January 1993 and equivalent to 1,000 former pesos.

**BUDGET***
(million new pesos)

| Revenue | 2001 | 2002 | 2003 |
|---|---:|---:|---:|
| Taxation | 654,870 | 728,284 | 766,123 |
| Income taxes | 285,523 | 318,380 | 336,546 |
| Value-added tax | 208,408 | 218,442 | 254,437 |
| Excise tax | 110,689 | 136,257 | 117,762 |
| Import duties | 28,902 | 27233 | 26,975 |
| Other revenue | 284,244 | 261,070 | 367,061 |
| Royalties | 203,752 | 158,507 | n.a. |
| Petroleum royalties | 187,607 | 139,842 | n.a. |
| **Total revenue** | **939,115** | **989,353** | **1,133,184** |

| Expenditure | 2001 | 2002 | 2003 |
|---|---:|---:|---:|
| Programmable expenditure | 631,774 | 745,104 | 832,215 |
| Current expenditure | 539,425 | 607,691 | 697,025 |
| Personal services | 93,132 | 101,735 | 367,736 |
| Transfers, subsidies and aid | 416,333 | 471,416 | 285,537 |
| Other current expenditure | 29,960 | 34,541 | 43,752 |
| Capital expenditure | 92,348 | 137,412 | 135,190 |
| Non-programmable expenditure | 365,177 | 379,348 | 400,926 |
| Interest and fees | 166,826 | 158,543 | 159,657 |
| Revenue sharing | 196,931 | 214,910 | 225,380 |
| Other | 1,419 | 5,895 | 15,890 |
| **Total expenditure** | **996,951** | **1,124,451** | **1,233,141** |

* Figures refer to the consolidated accounts of the central Government, including government agencies and the national social security system. The budgets of state and local governments are excluded.

Source: Dirección General de Planeación Hacendaria.

**INTERNATIONAL RESERVES***
(US $ million at 31 December)

|  | 2002 | 2003 | 2004 |
|---|---:|---:|---:|
| IMF special drawing rights | 392 | 433 | 465 |
| Reserve position in the Fund | 308 | 782 | 898 |
| Foreign exchange | 49,895 | 57,740 | 62,778 |
| **Total** | **50,594** | **58,956** | **64,141** |

* Excluding gold reserves ($357 million at 30 September 1989).

Source: IMF, *International Financial Statistics*.

**MONEY SUPPLY**
(million new pesos at 31 December)

|  | 2002 | 2003 | 2004 |
|---|---:|---:|---:|
| Currency outside banks | 232,082 | 263,387 | 300,982 |
| Demand deposits at deposit money banks | 364,664 | 415,909 | 435,392 |
| **Total money** | **596,746** | **679,296** | **736,374** |

Source: IMF, *International Financial Statistics*.

**COST OF LIVING**
(Consumer Price Index; base: 2000 = 100)

|  | 2002 | 2003 | 2004 |
|---|---:|---:|---:|
| Food, beverages and tobacco | 109.6 | 115.1 | 122.9 |
| Clothing and footwear | 108.9 | 109.8 | n.a. |
| Rent (incl. fuel and light) | 113.3 | 120.2 | n.a. |
| **All items** (incl. others) | **111.7** | **116.8** | **122.3** |

Source: ILO.

# MEXICO

## Statistical Survey

### NATIONAL ACCOUNTS

**National Income and Product**
(million new pesos at current prices)

|  | 1999 | 2000 | 2001 |
|---|---|---|---|
| Compensation of employees | 1,434,759 | 1,718,147 | 1,892,584 |
| Operating surplus | 2,309,085 | 2,737,206 | 2,830,568 |
| **Domestic factor incomes** | 3,743,844 | 4,455,353 | 4,723,152 |
| Consumption of fixed capital | 461,860 | 523,432 | 562,454 |
| **Gross domestic product at factor cost** | 4,205,704 | 4,980,785 | 5,285,606 |
| Indirect taxes | 404,370 | 528,205 | 564,561 |
| Less Subsidies | 16,388 | 17,617 | 21,577 |
| **GDP in purchasers' values** | 4,593,685 | 5,491,373 | 5,828,591 |

Source: ECLAC, *Statistical Yearbook for Latin America and the Caribbean*.

**Expenditure on the Gross Domestic Product**
('000 million new pesos at current prices)*

|  | 2001 | 2002 | 2003 |
|---|---|---|---|
| Government final consumption expenditure | 683.38 | 758.49 | 855.87 |
| Private final consumption expenditure | 4,044.88 | 4,319.87 | 4,672.27 |
| Increase in stocks | 53.01 | 85.72 | 35.10 |
| Gross fixed capital formation | 1,161.95 | 1,209.68 | 1,304.29 |
| **Total domestic expenditure** | 5,943.22 | 6,373.76 | 6,867.53 |
| Exports of goods and services | 1,598.52 | 1,677.56 | 1,920.55 |
| Less Imports of goods and services | 1,730.39 | 1,794.95 | 2,033.31 |
| **GDP in purchasers' values** | 5,811.35 | 6,256.38 | 6,754.77 |
| **GDP at constant 1993 prices** | 1,602.71 | 1,613.21 | 1,633.08 |

* Quarterly data seasonally adjusted at annual rates; figures are rounded to the nearest 10 million new pesos.

Source: IMF, *International Financial Statistics*.

**Gross Domestic Product by Economic Activity**
(provisional figures, million new pesos at current prices)

|  | 2002 | 2003 | 2004 |
|---|---|---|---|
| Agriculture, forestry and fishing | 226,397 | 243,080 | 281,392 |
| Mining and quarrying | 77,207 | 82,512 | 91,523 |
| Manufacturing | 1,068,603 | 1,123,213 | 1,253,500 |
| Electricity, gas and water | 81,881 | 79,687 | 93,144 |
| Construction | 292,180 | 326,319 | 388,947 |
| Trade, restaurants and hotels | 1,148,997 | 1,270,197 | 1,418,090 |
| Transport, storage and communications | 611,602 | 645,750 | 724,264 |
| Finance, insurance, real estate and business services | 769,222 | 824,536 | 895,732 |
| Community, social and personal services | 1,547,672 | 1,728,288 | 1,848,760 |
| **Sub-total** | 5,823,761 | 6,323,582 | 6,995,352 |
| Less Imputed bank service charge | 84,778 | 75,036 | 84,778 |
| **GDP at factor cost** | 5,738,983 | 6,248,545 | 6,910,904 |
| Indirect taxes, *less* subsidies | 528,491 | 646,448 | 724,002 |
| **GDP in purchasers' values** | 6,267,474 | 6,894,993 | 7,634,926 |

### BALANCE OF PAYMENTS
(US $ million)

|  | 2002 | 2003 | 2004 |
|---|---|---|---|
| Exports of goods f.o.b. | 161,046 | 164,766 | 188,627 |
| Imports of goods f.o.b. | −168,679 | −170,546 | −197,156 |
| **Trade balance** | −7,633 | −5,780 | −8,530 |
| Exports of services | 12,740 | 12,712 | 14,090 |
| Imports of services | −17,660 | −18,142 | −19,779 |
| **Balance on goods and services** | −12,553 | −11,209 | −14,219 |
| Other income received | 4,405 | 5,662 | 4,455 |
| Other income paid | −15,176 | −14,908 | −15,077 |
| **Balance on goods, services and income** | −23,324 | −20,455 | −24,842 |
| Current transfers received | 10,287 | 13,880 | 17,108 |
| Current transfers paid | −35 | −37 | −80 |
| **Current balance** | −13,072 | −6,613 | −7,813 |
| Direct investment abroad | −930 | −1,784 | −2,240 |
| Direct investment from abroad | 15,129 | 11,373 | 16,602 |
| Portfolio investment assets | 1,134 | 91 | 1,718 |
| Portfolio investment liabilities | −632 | 3,864 | 7,565 |
| Other investment assets | 11,601 | 8,627 | −4,066 |
| Other investment liabilities | −3,377 | −4,126 | −5,888 |
| Net errors and omissions | −2,493 | −1,613 | −1,774 |
| **Overall balance** | 7,359 | 9,817 | 4,104 |

Source: IMF, *International Financial Statistics*.

## External Trade

**PRINCIPAL COMMODITIES**
(distribution by SITC, US $ million)

| Imports f.o.b. | 1999 | 2000 | 2001 |
|---|---|---|---|
| **Food and live animals** | 5,909.8 | 7,050.7 | 8,314.3 |
| **Crude materials (inedible) except fuels** | 4,156.7 | 4,902.1 | 4,688.6 |
| **Mineral fuels, lubricants and related materials** | 3,181.4 | 5,618.3 | 5,635.8 |
| **Chemicals and related products** | 12,309.6 | 14,892.5 | 15,269.7 |
| **Basic manufactures** | 25,444.1 | 33,422.8 | 31,941.1 |
| Textile yarn, fabrics, etc. | 4,928.6 | 6,252.6 | 6,045.0 |
| **Machinery and transport equipment** | 71,585.5 | 95,897.0 | 98,634.5 |
| Power-generating machinery and equipment | 5,410.6 | 6,890.4 | 6,598.1 |
| Machinery specialized for particular industries | 4,844.5 | 5,386.7 | 4,710.1 |
| General industrial machinery and equipment and parts | 8,309.7 | 10,281.1 | 10,366.1 |
| Office machines and automatic data-processing equipment | 4,357.5 | 5,774.9 | 8,374.2 |
| Telecommunications and sound equipment | 6,898.4 | 9,771.8 | 10,017.6 |
| Other electrical machinery, apparatus, etc. | 27,002.1 | 36,388.3 | 38,020.7 |
| Switchgear, resistors, printed circuits, switchboards, etc. | 6,472.5 | 8,486.5 | 9,644.6 |
| Thermionic valves, tubes, etc. | 10,106.0 | 14,279.5 | 14,467.6 |
| Electronic microcircuits | 5,604.8 | 8,937.7 | 8,669.5 |
| Road vehicles and parts* | 12,157.1 | 18,964.4 | 18,309.5 |
| Parts and accessories for cars, buses, lorries, etc.* | 8,230.1 | 12,054.8 | 11,192.7 |
| **Miscellaneous manufactured articles** | 18,022.6 | 23,294.9 | 23,763.6 |
| Articles of plastic materials, etc. | 5,613.0 | 7,265.2 | 7,265.3 |
| **Total** (incl. others) | 146,064.6 | 190,790.5 | 190,365.2 |

* Data on parts exclude tyres, engines and electrical parts.

# MEXICO

*Statistical Survey*

| Exports f.o.b. | 1999 | 2000 | 2001 |
|---|---|---|---|
| **Food and live animals** | 5,999.6 | 6,472.1 | 6,419.4 |
| **Mineral fuels, lubricants, etc.** | 9,731.0 | 16,052.8 | 12,639.3 |
| Petroleum, petroleum products, etc. | 9,604.8 | 15,964.1 | 12,493.3 |
| Crude petroleum oils, etc. | 8,858.8 | 14,878.5 | 11,597.9 |
| **Chemicals and related products** | 4,402.5 | 5,253.1 | 5,326.2 |
| **Basic manufactures** | 12,120.2 | 13,848.2 | 12,810.9 |
| **Machinery and transport equipment** | 81,258.9 | 98,281.3 | 95,497.4 |
| Power-generating machinery and equipment | 5,731.6 | 6,222.4 | 6,047.3 |
| General industrial machinery equipment and parts | 4,475.4 | 5,171.9 | 5,074.3 |
| Office machines and automatic data-processing equipment | 9,760.3 | 11,756.7 | 13,188.4 |
| Automatic data-processing machines and units | 6,398.9 | 8,137.8 | 9,692.5 |
| Telecommunications and sound equipment | 14,379.5 | 19,221.1 | 19,142.2 |
| Colour television receivers | 5,156.1 | 5,727.4 | 6,239.0 |
| Other electrical machinery apparatus, etc. | 21,756.8 | 26,063.9 | 22,285.9 |
| Equipment for distributing electricity | 6,008.2 | 6,719.8 | 5,949.7 |
| Insulated electric wire, cable etc. | 5,977.2 | 6,673.9 | 5,898.4 |
| Road vehicles and parts* | 23,383.4 | 27,898.1 | 27,825.2 |
| Passenger motor cars (excl. buses) | 12,407.5 | 16,296.7 | 15,297.4 |
| Goods vehicles (lorries and trucks) | 4,101.5 | 4,815.7 | 6,447.6 |
| Parts and accessories for cars, buses, lorries, etc.* | 5,107.9 | 5,812.5 | 5,579.3 |
| **Miscellaneous manufactured articles** | 19,379.9 | 22,607.6 | 22,454.4 |
| Clothing and accessories (excl. footwear) | 7,772.9 | 8,631.3 | 8,012.0 |
| **Total** (incl. others) | 136,262.8 | 166,191.7 | 158,684.6 |

* Data on parts exclude tyres, engines and electrical parts.

Source: UN, *International Trade Statistics Yearbook*.

**2002** (US $ million, preliminary figures): *Imports f.o.b.*: Agricultural products 4,871.9; Mineral products 1,967.0; Manufactured goods 160,622.8; Total (incl. others) 168,678.9; *Exports f.o.b.*: Agricultural products 4,214.5; Mineral products 15,196.9; Manufactured goods 141.634.5; Total (incl. others) 161,046.0.

**2003** (US $ million, preliminary figures): *Imports f.o.b.*: Agricultural products 5,464.9; Mineral products 3,124.2; Manufactured goods 160,975.4; Total (incl. others) 170,545.8; *Exports f.o.b.*: Agricultural products 5,035.6; Mineral products 19,098.7; Manufactured goods 140,632.1; Total (incl. others) 164,766.2.

**2004** (US $ million, preliminary figures): *Imports f.o.b.*: Total 196,809.7; *Exports f.o.b.*: Agricultural products 5,683.9; Mineral products 24,567.4; Manufactured goods 157,747.3; Total (incl. others) 187,998.6.

## PRINCIPAL TRADING PARTNERS*
(US $ million)

| Imports c.i.f. | 2002 | 2003 | 2004 |
|---|---|---|---|
| Argentina | 687.3 | 867.1 | 1,109.8 |
| Brazil | 2,565.0 | 3,267.4 | 4,343.8 |
| Canada | 4,480.3 | 4,120.5 | 5,334.1 |
| Chile | 1,010.2 | 1,081.9 | 1,463.8 |
| China, People's Republic | 6,274.4 | 9,400.6 | 14,459.4 |
| China (Taiwan) | 4,250.1 | 2,509.1 | 3,502.1 |
| France | 1,806.8 | 2,015.4 | 2,397.6 |
| Germany | 6,065.8 | 6,218.2 | 7,154.1 |
| Italy | 2,171.1 | 2,473.9 | 2,820.9 |
| Japan | 9,348.6 | 7,595.1 | 10,624.1 |
| Korea, Republic | 3,910.0 | 4,112.9 | 5,262.7 |
| Malaysia | 1,993.2 | 2,760.6 | 3,399.9 |
| Singapore | 1,555.0 | 1,337.8 | 2,225.5 |
| Spain | 2,223.9 | 2,288.0 | 2,853.3 |
| Thailand | 838.8 | 987.4 | 1,270.0 |
| United Kingdom | 1,349.8 | 1,242.2 | 1,460.5 |
| USA | 106,921.9 | 105,724.0 | 111,319.0 |
| **Total** (incl. others) | 168,678.9 | 170,545.8 | 197,156.5 |

| Exports f.o.b. | 2002 | 2003 | 2004 |
|---|---|---|---|
| Aruba† | 166.1 | 780.5 | 1,443.8 |
| Canada | 2,991.3 | 3,041.8 | 3,298.7 |
| Germany | 1,159.1 | 1,715.2 | 1,609.3 |
| Spain | 1,393.7 | 1,512.4 | 2,000.3 |
| USA | 142,167.3 | 144,557.2 | 165,423.2 |
| **Total** (incl. others) | 161,046.0 | 164,766.4 | 188,626.5 |

* Imports by country of origin; exports by country of destination.
† Estimates.

# Transport

## RAILWAYS
(traffic)

| | 2001 | 2002 | 2003 |
|---|---|---|---|
| Passengers carried ('000) | 242 | 237 | 270 |
| Passenger-kilometres (million) | 67 | 69 | 78 |
| Freight carried ('000 tons) | 76,182 | 80,451 | 85,232 |
| Freight ton-kilometres (million) | 46,614 | 52,432 | 54,813 |

Source: Dirección General de Planeación, Secretaría de Comunicaciones y Transportes.

## ROAD TRAFFIC
(estimates, vehicles in use at 31 December)

| | 1998 | 1999 | 2000 |
|---|---|---|---|
| Passenger cars | 9,378,587 | 9,842,006 | 10,443,489 |
| Buses and coaches | 108,690 | 109,929 | 111,756 |
| Lorries and vans | 4,403,953 | 4,639,860 | 7,931,590 |

Source: IRF, *World Road Statistics*.

## SHIPPING

**Merchant Fleet**
(registered at 31 December)

| | 2002 | 2003 | 2004 |
|---|---|---|---|
| Number of vessels | 658 | 654 | 687 |
| Total displacement ('000 grt) | 937.2 | 972.7 | 1,008.0 |

Source: Lloyd's Register-Fairplay, *World Fleet Statistics*.

**Sea-borne Shipping**
(domestic and international freight traffic, '000 metric tons)

| | 2002 | 2003 | 2004 |
|---|---|---|---|
| Goods loaded | 160,894 | 175,923 | 177,915 |
| Goods unloaded | 87,203 | 88,816 | 88,093 |

Source: Coordinación General de Puertos y Marina Mercante.

## CIVIL AVIATION
(traffic on scheduled services)

|  | 2001 | 2002 | 2003 |
|---|---|---|---|
| Kilometres flown (million) | 363 | n.a. | n.a. |
| Passengers carried ('000) | 38,282 | 37,256 | 39,276 |
| Passenger-km (million) | 29,621 | n.a. | n.a. |
| Freight carried ('000 tons) | 459.2 | 488.2 | 497.0 |

Sources: UN, *Statistical Yearbook* and Secretaría de Comunicaciones y Transportes.

## Tourism

|  | 2001 | 2002 | 2003 |
|---|---|---|---|
| Tourist arrivals ('000) | 19,810 | 19,667 | 18,665 |
| Border tourists ('000) | 9,659 | 9,784 | 8,312 |
| Total expenditure (US $ million) | 8,401 | 8,858 | 9,457 |

Source: Secretaría de Turismo de México.

## Communications Media

|  | 2001 | 2002 | 2003 |
|---|---|---|---|
| Telephone ('000 main lines in use) | 13,533.0 | 14,941.6 | 16,311.1 |
| Mobile cellular telephones ('000 subscribers) | 20,136.0 | 25,928.3 | 30,097.7 |
| Personal computers ('000 in use) | 6,900 | 8,363 | n.a. |
| Internet users ('000) | 3,500.0 | 12,250.3 | n.a. |

**Radio receivers** ('000 in use): 31,000 in 1997.

**Television receivers** ('000 in use): 28,000 in 2000.

**Facsimile machines** ('000 in use): 285 in 1997.

**Daily newspapers** (2000): Number 311; Average circulation 9,251,000.

**Non-daily newspapers** (2000): Number 26; Average circulation 614,000.

**Books published** (titles): 6,952 in 1998.

Sources: International Telecommunication Union; UNESCO Institute for Statistics; UNESCO, *Statistical Yearbook*; UN, *Statistical Yearbook*.

## Education

(2003)

|  | Institutions | Teachers | Students* |
|---|---|---|---|
| Pre-primary | 76,108 | 169,081 | 3,742.6 |
| Primary | 99,034 | 559,499 | 14,781.3 |
| Secondary (incl. technical) | 30,337 | 331,563 | 5,780.4 |
| Intermediate: professional/technical | 1,626 | 31,557 | 359.9 |
| Intermediate: Baccalaureate | 10,312 | 210,565 | 3,083.8 |
| Higher (incl. post-graduate) | 4,585 | 24,1236 | 2,250.5 |

* Figures are in thousands.

Source: partly Secretaría de Educación Pública.

**Adult literacy rate** (UNESCO estimates): 90.5% (males 92.6%; females 88.7%) in 2002 (Source: UN Development Programme, *Human Development Report*).

# Directory

## The Constitution

The present Mexican Constitution was proclaimed on 5 February 1917, at the end of the revolution, which began in 1910, against the regime of Porfirio Díaz. Its provisions regarding religion, education and the ownership and exploitation of mineral wealth reflect the long revolutionary struggle against the concentration of power in the hands of the Roman Catholic Church and the large landowners, and the struggle which culminated, in the 1930s, in the expropriation of the properties of the foreign petroleum companies. It has been amended from time to time.

### GOVERNMENT

#### The President and Congress

The President of the Republic, in agreement with the Cabinet and with the approval of the Congreso de la Unión (Congress) or of the Permanent Committee when the Congreso is not in session, may suspend constitutional guarantees in case of foreign invasion, serious disturbance, or any other emergency endangering the people.

The exercise of supreme executive authority is vested in the President, who is elected for six years and enters office on 1 December of the year of election. The presidential powers include the right to appoint and remove members of the Cabinet and the Attorney-General; to appoint, with the approval of the Senado (Senate), diplomatic officials, the higher officers of the army, and ministers of the supreme and higher courts of justice. The President is also empowered to dispose of the Armed Forces for the internal and external security of the federation.

The Congreso is composed of the Cámara Federal de Diputados (Federal Chamber of Deputies) elected every three years, and the Senado whose members hold office for six years. There is one deputy for every 250,000 people and for every fraction of over 125,000 people. The Senado is composed of two members for each state and two for the Distrito Federal. Regular sessions of the Congreso begin on 1 September and may not continue beyond 31 December of the same year. Extraordinary sessions may be convened by the Permanent Committee.

The powers of the Congreso include the right to: pass laws and regulations; impose taxes; specify the criteria on which the Executive may negotiate loans; declare war; raise, maintain and regulate the organization of the Armed Forces; establish and maintain schools of various types throughout the country; approve or reject the budget; sanction appointments submitted by the President of the Supreme Court and magistrates of the superior court of the Distrito Federal; approve or reject treaties and conventions made with foreign powers; and ratify diplomatic appointments.

The Permanent Committee, consisting of 29 members of the Congreso (15 of whom are deputies and 14 senators), officiates when the Congreso is in recess, and is responsible for the convening of extraordinary sessions of the Congreso.

#### The States

Governors are elected by popular vote in a general election every six years. The local legislature is formed by deputies, who are changed every three years. The judicature is specially appointed under the Constitution by the competent authority (it is never subject to the popular vote).

Each state is a separate unit, with the right to levy taxes and to legislate in certain matters. The states are not allowed to levy inter-state customs duties.

#### The Federal District

The Distrito Federal consists of Mexico City and several neighbouring small towns and villages. The first direct elections for the Head of Government of the Distrito Federal were held in July 1997; hitherto a Regent had been appointed by the President.

## EDUCATION

According to the Constitution, the provision of educational facilities is the joint responsibility of the federation, the states and the municipalities. Education shall be democratic, and shall be directed to developing all the faculties of the individual students, while imbuing them with love of their country and a consciousness of international solidarity and justice. Religious bodies may not provide education, except training for the priesthood. Private educational institutions must conform to the requirements of the Constitution with regard to the nature of the teaching given. The education provided by the states shall be free of charge.

## RELIGION

Religious bodies of whatever denomination shall not have the capacity to possess or administer real estate or capital invested therein. Churches are the property of the nation; the headquarters of bishops, seminaries, convents and other property used for the propagation of a religious creed shall pass into the hands of the state, to be dedicated to the public service of the federation or of the respective state. Institutions of charity, provided they are not connected with a religious body, may hold real property. The establishment of monastic orders is prohibited. Ministers of religion must be Mexican; they may not criticize the fundamental laws of the country in a public or private meeting; they may not vote or form associations for political purposes. Political meetings may not be held in places of worship.

A reform proposal, whereby constitutional restrictions on the Catholic Church were formally ended, received congressional approval in December 1991 and was promulgated as law in January 1992.

## LAND AND MINERAL OWNERSHIP

Article 27 of the Constitution vests direct ownership of minerals and other products of the subsoil, including petroleum and water, in the nation, and reserves to the Federal Government alone the right to grant concessions in accordance with the laws to individuals and companies, on the condition that they establish regular work for the exploitation of the materials. At the same time, the right to acquire ownership of lands and waters belonging to the nation, or concessions for their exploitation, is limited to Mexican individuals and companies, although the State may concede similar rights to foreigners who agree not to invoke the protection of their governments to enforce such rights.

The same article declares null all alienations of lands, waters and forests belonging to towns or communities made by political chiefs or other local authorities in violation of the provisions of the law of 25 June 1856*, and all concessions or sales of communally-held lands, waters and forests made by the federal authorities after 1 December 1876. The population settlements which lack ejidos (state-owned smallholdings), or cannot obtain restitution of lands previously held, shall be granted lands in proportion to the needs of the population. The area of land granted to the individual may not be less than 10 hectares of irrigated or watered land, or the equivalent in other kinds of land.

The owners affected by decisions to divide and redistribute land (with the exception of the owners of farming or cattle-rearing properties) shall not have any right of redress, nor may they invoke the right of amparo† in protection of their interests. They may, however, apply to the Government for indemnification. Small properties, the areas of which are defined in the Constitution, will not be subject to expropriation. The Constitution leaves to the Congreso the duty of determining the maximum size of rural properties.

In March 1992 an agrarian reform amendment, whereby the programme of land-distribution established by the 1917 Constitution was abolished and the terms of the ejido system of tenant farmers were relaxed, was formally adopted.

Monopolies and measures to restrict competition in industry, commerce or public services are prohibited.

A section of the Constitution deals with work and social security.

On 30 December 1977 a Federal Law on Political Organizations and Electoral Procedure was promulgated. It includes the following provisions:

Legislative power is vested in the Congreso de la Unión which comprises the Cámara Federal de Diputados and the Senado. The Cámara shall comprise 300 deputies elected by majority vote within single-member electoral districts and up to 100 deputies (increased to 200 from July 1988) elected by a system of proportional representation from regional lists within multi-member constituencies. The Senado comprises two members for each state and two for the Distrito Federal, elected by majority vote.

Executive power is exercised by the President of the Republic of the United Mexican States, elected by majority vote.

Ordinary elections will be held every three years for the federal deputies and every six years for the senators and the President of the Republic on the first Sunday of July of the year in question. When a vacancy occurs among members of the Congreso elected by majority vote, the house in question shall call extraordinary elections, and when a vacancy occurs among members of the Cámara elected by proportional representation it shall be filled by the candidate of the same party who received the next highest number of votes at the last ordinary election.

Voting is the right and duty of every citizen, male or female, over the age of 18 years.

A political party shall be registered if it has at least 3,000 members in each one of at least half the states in Mexico or at least 300 members in each one of at least half of the single-member constituencies. In either case the total number of members must be no less than 65,000. A party can also obtain conditional registration if it has been active for at least four years. Registration is confirmed if the party obtains at least 1.5% of the popular vote. All political parties shall have free access to the media.

In September 1993 an amendment to the Law on Electoral Procedure provided for the expansion of the Senado to 128 seats, representing four members for each state and the Distrito Federal, three to be elected by majority vote and one by proportional representation.

* The Lerdo Law against ecclesiastical privilege, which became the basis of the Liberal Constitution of 1857.
† The Constitution provides for the procedure known as juicio de amparo, a wider form of habeas corpus, which the individual may invoke in protection of his constitutional rights.

# The Government

## HEAD OF STATE

**President:** VICENTE FOX QUESADA (took office 1 December 2000).

## CABINET
(July 2005)

**Secretary of the Interior:** CARLOS MARÍA ABASCAL CARRANZA.

**Secretary of Foreign Affairs:** Dr LUIS ERNESTO DERBEZ BAUTISTA.

**Secretary of Finance and Public Credit:** FRANCISCO GIL DÍAZ.

**Secretary of National Defence:** Gen. GERARDO CLEMENTE RICARDO VEGA GARCÍA.

**Secretary of the Navy:** Adm. MARCO ANTONIO PEYROT GONZÁLEZ.

**Secretary of the Economy:** FERNANDO CANALES CLARIOND.

**Secretary of Social Development:** JOSEFINA EUGENIA VÁZQUEZ MOTA.

**Secretary of Public Security and Judicial Services:** RAMÓN MARTÍN HUERTA.

**Secretary of Public Function:** EDUARDO ROMERO RAMOS.

**Secretary of Communications and Transport:** PEDRO CERISOLA Y WEBER.

**Secretary of Labour and Social Welfare:** FRANCISCO JAVIER SALAZAR SÁENZ.

**Secretary of the Environment and Natural Resources:** JOSÉ LUIS LUEGE TAMARGO.

**Secretary of Energy:** FERNANDO ELIZONDO BARRAGÁN.

**Secretary of Agriculture, Livestock, Rural Development, Fisheries and Food:** JAVIER USABIAGA ARROYO.

**Secretary of Public Education:** Dr REYES S. TAMEZ GUERRA.

**Secretary of Health:** Dr JULIO JOSÉ FRENK MORA.

**Secretary of Tourism:** RODOLFO ELIZONDO TORRES.

**Secretary of Agrarian Reform:** FLORENCIO SALAZAR ADAME.

**Attorney-General:** DANIEL FRANCISCO CABEZA DE VACA HERNÁNDEZ.

## SECRETARIATS OF STATE

**Office of the President:** Los Pinos, Puerta 1, Col. San Miguel Chapultepec, 11850 México, DF; tel. (55) 5091-1100; fax (55) 5277-2376; e-mail vicente.fox.quesada@presidencia.gob.mx; internet www.presidencia.gob.mx.

**Secretariat of State for Agrarian Reform:** Edif. de Avda Heroica, 1°, Escuela Naval Militar 701, Col. Presidentes Ejidales, 04470 México, DF; tel. (55) 5695-6776; fax (55) 5695-6368; e-mail sra@sra.gob.mx; internet www.sra.gob.mx.

**Secretariat of State for Agriculture, Livestock, Rural Development, Fisheries and Food:** Avda Municipio Libre 377, A°, Col. Santa Cruz Atoyac, Del. Benito Juárez, 03310 México, DF; tel. (55) 9183-1000; fax (55) 9183-1018; e-mail contacto@sagarpa.gob.mx; internet www.sagarpa.gob.mx.

# MEXICO

**Secretariat of State for Communications and Transport:** Avda Xola y Universidad, Col. Narvarte, Del. Benito Juárez, 03020 México, DF; tel. (55) 5519-7456; fax (55) 5519-0692; e-mail buzon_sct@sct.gob.mx; internet www.sct.gob.mx.

**Secretariat of State for the Economy:** Alfonso Reyes 30, Col. Hipódromo Condesa, 06170 México, DF; tel. (55) 5729-9291; fax (55) 5729-9358; e-mail fcanales@economia.gob.mx; internet www.economia.gob.mx.

**Secretariat of State for Energy:** Insurgentes Sur 890, 17°, Col. del Valle, 03100 México, DF; tel. (55) 5000-6030; fax (55) 5000-6222; e-mail felizondo@energia.gob.mx; internet www.energia.gob.mx.

**Secretariat of State for the Environment and Natural Resources:** Blvd Adolfo Ruíz Cortines 4209, Col. Jardines en la Montaña, Tlalpan, 14210 México, DF; tel. (55) 5628-0602; fax (55) 5628-0643; e-mail secretario@semarnat.gob.mx; internet www.semarnat.gob.mx.

**Secretariat of State for Finance and Public Credit:** Palacio Nacional, Primer Patio Mariano, 3°, Of. 3045, Col. Centro, Del. Cuauhtémoc, 06000 México, DF; tel. (55) 5542-2213; fax (55) 9158-1142; e-mail secretario@hacienda.gob.mx; internet www.shcp.gob.mx.

**Secretariat of State for Foreign Affairs:** Avda Ricardo Flores Magón 2, Col. Guerrero, Del. Cuauhtémoc, 06995 México, DF; tel. (55) 5063-3000; fax (55) 5782-4109; e-mail comment@sre.gob.mx; internet www.sre.gob.mx.

**Secretariat of State for Health:** Lieja 7, 1°, Col. Juárez, Del. Cuauhtémoc, 06600 México, DF; tel. (55) 5286-2383; fax (55) 5553-7917; e-mail jfrenk@salud.gob.mx; internet www.salud.gob.mx.

**Secretariat of State for the Interior:** Bucareli 99, 1°, Col. Juárez, 06069 México, DF; tel. (55) 5592-1141; fax (55) 5546-5350; internet www.gobernacion.gob.mx.

**Secretariat of State for Labour and Social Welfare:** Edif. A, 4°, Anillo Periférico Sur 4271, 4°, Col. Fuentes del Pedregal, 14149 México, DF; tel. (55) 5645-3965; fax (55) 5645-5594; e-mail correo@stps.gob.mx; internet www.stps.gob.mx.

**Secretariat of State for National Defence:** Manuel Avila Camacho, esq. Avda Industria Militar, 3°, Col. Lomas de Sotelo, Del. Miguel Hidalgo, 11640 México, DF; tel. (55) 5557-5571; fax (55) 5557-1370; e-mail comsoc@mail.sedena.gob.mx; internet www.sedena.gob.mx.

**Secretariat of State for the Navy:** Eje 2 Ote, Tramo Heroica, Escuela Naval Militar 861, Col. Los Cipreses, Del. Coyoacán, 04830 México, DF; tel. (55) 5624-6500; fax (55) 5679-6411; e-mail srio@semar.gob.mx; internet www.semar.gob.mx.

**Secretariat of State for Public Education:** Dinamarca 84, 5°, Col. Juárez, 06600 México, DF; tel. (55) 5510-2557; fax (55) 5329-6873; e-mail educa@sep.gob.mx; internet www.sep.gob.mx.

**Secretariat of State for Public Function:** Insurgentes Sur 1735, 10°, Col. Guadalupe Inn, Del. Alvaro Obregón, 01020 México, DF; tel. (55) 3003-3000; fax (55) 5662-4763; e-mail sactel@funcionpublica.gob.mx; internet www.secodam.gob.mx.

**Secretariat of State for Public Security and Judicial Services:** Londres 102, 7°, Col. Juárez, 06600 México, DF; e-mail buzon@ssp.gob.mx; internet www.ssp.gob.mx.

**Secretariat of State for Social Development:** Avda Reforma 116, Col. Juárez, Del. Cuauhtémoc, 06600 México, DF; tel. (55) 5328-5000; fax (55) 5271-8862; e-mail demandasocial@sedesol.gob.mx; internet www.sedesol.gob.mx.

**Secretariat of State for Tourism:** Presidente Masarik 172, Col. Chapultepec Morales, 11587 México, DF; tel. (55) 3002-6300; fax (55) 1036-0789; e-mail relizondo@sectur.gob.mx; internet www.sectur.gob.mx.

**Office of the Attorney-General:** Avda Paseo de la Reforma 211–213, Col. Cuauhtémoc, Del. Cuauhtémoc, 06500 México, DF; tel. (55) 5626-9600; fax (55) 5626-4447; e-mail ofproc@pgr.gob.mx; internet www.pgr.gob.mx.

## State Governors

(July 2005)

**Aguascalientes:** Luís Armando Reynoso (PAN).
**Baja California:** Eugenio Elorduy Walther (PAN).
**Baja California Sur:** Narciso Agúndez (PRD).
**Campeche:** Jorge Carlos Hurtado Valdez (PRI).
**Chiapas:** Pablo Salazar Mendiguchía (Ind.).
**Chihuahua:** José Reyes Baeza (PRI).
**Coahuila (de Zaragoza):** Enrique Martínez y Martínez (PRI).
**Colima:** Jesús Silverio Cavazos Ceballos (PRI-PT-PVEM).
**Durango:** Ismael Hernández (PRI).
**Guanajuato:** Juan Carlos Romero Hicks (PAN).
**Guerrero:** Zeferino Torreblanco (PRD).
**Hidalgo:** Miguel Angel Osorio Chong (PRI).
**Jalisco:** Lic. Francisco Ramirez Acuña (PAN).
**México:** Enrique Peña Nieto (PRI) (Governor-elect).
**Michoacán (de Ocampo):** Lazaro Cárdenas Batel (PRD).
**Morelos:** Lic. Sergio Estrada Cajigal Ramírez (PAN).
**Nayarit:** Ney González Sánchez (PRI) (Governor-elect).
**Nuevo León:** Lic. José Natividad González Parás (PRI-PVEM-PLM-Fuerza Ciudadana).
**Oaxaca:** Ulises Ruíz (PRI).
**Puebla:** Mario Marín Torres (PRI).
**Querétaro (de Arteaga):** Francisco Garrido Patrón (PAN).
**Quintana Roo:** Félix González Canto (PRI).
**San Luis Potosí:** Marcelo de los Santos Fraga (PAN).
**Sinaloa:** Jesús Aguilar Padilla (PRI).
**Sonora:** José Eduardo Robinson Bours Castelo (PRI-PVEM).
**Tabasco:** Manuel Andrade Díaz (PRI).
**Tamaulipas:** Eugenio Hernández Flores (PRI).
**Tlaxcala:** Mariano González Zarur (PRI).
**Veracruz-Llave:** Fidel Herrera (PRI).
**Yucatán:** Patricio Patrón Laviada (PAN-PRD-PT-PVEM).
**Zacatecas:** Amalia García (PRD).
**Head of Government of the Distrito Federal:** Alejandro Encinas (PRD).

## President and Legislature

### PRESIDENT

Election, 2 July 2000

| Candidate | Number of votes | % of votes |
|---|---:|---:|
| Vicente Fox Quesada (Alianza por el Cambio*) | 15,988,740 | 43.47 |
| Francisco Labastida Ochoa (PRI) | 13,576,385 | 36.91 |
| Cuauhtémoc Cárdenas Solórzano (Alianza por México) | 6,259,048 | 17.02 |
| Others | 957,455 | 2.60 |
| **Total** | **36,781,628** | **100.00** |

*An alliance of the PAN and the PVEM.

### CONGRESO DE LA UNIÓN

#### Senado

**Senate:** Xicoténcatl 9, Centro Histórico, 06010 México, DF; tel. (55) 5130-2200; internet www.senado.gob.mx.

**President:** Diego Fernández de Cevallos.

Elections, 2 July 2000

| Party | Seats |
|---|---:|
| Partido Revolucionario Institucional (PRI) | 60 |
| Partido Acción Nacional (PAN)* | 46 |
| Partido de la Revolución Democrática (PRD)† | 15 |
| Partido Verde Ecologista de México (PVEM)* | 5 |
| Convergencia por la Democracia (CD)† | 1 |
| Partido del Trabajo (PT)† | 1 |
| **Total** | **128** |

*Contested the elections jointly as the Alianza por el Cambio.
†Contested the elections as part of the Alianza por México.

#### Cámara Federal de Diputados

**Federal Chamber of Deputies:** Avda Congreso de la Unión 66, Col. El Parque, Del. Venustiano Carranza, 15969 México, DF; tel. (55) 5628-1300; internet www.cddhcu.gob.mx.

**President:** Manlio Fabio Beltrones Rivera.

MEXICO
*Directory*

**Elections, 6 July 2003**

| Party | Seats |
|---|---|
| Partido Revolucionario Institucional (PRI) | 224 |
| Partido Acción Nacional (PAN) | 153 |
| Partido de la Revolución Democrática (PRD) | 95 |
| Partido Verde Ecologista de México (PVEM) | 17 |
| Partido del Trabajo (PT) | 6 |
| Convergencia por la Democracia (CD) | 5 |
| **Total** | **500** |

## Political Organizations

To retain legal political registration, parties must secure at least 1.5% of total votes at two consecutive federal elections. Several of the parties listed below are no longer officially registered but continue to be politically active.

**Convergencia:** Louisiana 113, Col. Nápoles, 03810 México, DF; tel. (55) 5543-8517; e-mail diconvergencia@prodigy.net.mx; internet www.convergencia.org.mx; f. 1995 as Convergencia por la Democracia; Pres. DANTE DELGADO RANNAURO; Sec.-Gen. ALEJANDRO CHANONA BURGUETE.

**Democracia Social:** San Borja 416, Col. del Valle, 03100 México, DF; tel. (55) 5559-2875; e-mail correos@democraciasocial.org.mx; f. 1999; Pres. GILBERTO RINCÓN GALLARDO.

**Fuerza Ciudadana:** Rochester 94, Col. Nápoles, 03810 México, DF; tel. (55) 5534-4628; e-mail info@fuerzaciudadana.org.mx; internet www.fuerzaciudadana.org.mx; f. 2002; citizens' asscn; National Co-ordinator JORGE ALCOCER VILLANUEVA; Sec.-Gen. EMILIO CABALLERO URDIALES.

**Nueva Alianza:** México, DF; f. 2005 by members of the Sindicato Nacional de Trabajadores de la Educación (SNTE, see Trade Unions); Pres. MIGUEL ANGEL JIMÉNEZ.

**Partido Acción Nacional (PAN):** Avda Coyoacán 1546, Col. del Valle, México, DF; tel. (55) 5200-4000; e-mail correo@cen.pan.org.mx; internet www.pan.org.mx; f. 1939; democratic party; 150,000 mems; Pres. MANUEL ESPINO BARRIENTOS; Sec.-Gen. ARTURO GARCÍA PORTILLO.

**Partido Alianza Social (PAS):** Édison 89, Col. Tabacalera, 06030 México, DF; tel. (55) 5592-5688; fax (55) 5566-1665; Pres. GUILLERMO CALDERÓN DOMÍNGUEZ; Sec.-Gen. ADALBERTO ROSAS LÓPEZ.

**Partido Auténtico de la Revolución Mexicana (PARM):** Pueblo 286, 1°, Col. Roma, 06700 México, DF; tel. (55) 5514-9676; f. 1954 to sustain the ideology of the Mexican Political Constitution of 1917; 191,500 mems; Pres. CARLOS GUZMÁN PÉREZ.

**Partido de Centro Democrático (PCD):** Amores 923, Col. del Valle, Del. Benito Juárez, 03100 México, DF; tel. (55) 5575-3101; fax (55) 5575-8888; e-mail pcdcen@pcd2000.org.mx; centrist party; f. 1997; Leader MANUEL CAMACHO SOLÍS.

**Partido del Frente Cardenista de Reconstrucción Nacional (PFCRN):** Avda México 199, Col. Hipódromo Condesa, 06170 México, DF; f. 1972; Marxist-Leninist; fmrly Partido Socialista de los Trabajadores; 132,000 mems; Pres. RAFAEL AGUILAR TALAMANTES; Sec.-Gen. GRACO RAMÍREZ ABREU.

**Partido Liberal México (PLM):** México, DF; f. 2002; Pres. C. SALVADOR ORDAZ MONTES DE OCA.

**Partido México Posible—La Nueva Política:** Dr Vértiz 1200, Col. Letrán Valle, 03650 México, DF; tel. (55) 5243-6061; e-mail correo@mexicoposible.org.mx; internet www.mexicoposible.org.mx; f. 2002; focus on the rights of women and indigenous peoples, and ecological issues; Pres. PATRICIA MERCADO CASTRO; Sec.-Gen. WILFRIDO ISAMÍ SALZAR RULE.

**Partido Popular Socialista de México (PPS):** Avda Alvaro Obregón 185, Col. Roma, Del. Cuauhtémoc, 06797 México, DF; tel. (55) 5511-0184; fax (55) 5514-9498; e-mail ejesusantoniocarlos@msm.com; internet www.ppsdemexico.org.mx; f. 1948; left-wing party; Sec.-Gen. JESÚS ANTONIO CARLOS HERNÁNDEZ.

**Partido de la Revolución Democrática (PRD):** Monterrey 50, Col. Roma, 54879 México, DF; tel. (55) 5689-7895; e-mail consenal@prd.org.mx; internet www.prd.org.mx; f. 1989 by the Corriente Democrática (CD) and elements of the Partido Mexicano Socialista (PMS); centre-left; Pres. LEONEL COTA MONTAÑO; Sec.-Gen. CARLOS NAVARRETE RUIZ.

**Partido Revolucionario Institucional (PRI):** Insurgentes Norte 59, Edif. 2, subsótano, Col. Buenavista, 06359 México, DF; tel. (55) 5729-9600; fax (55) 5546-3552; internet www.pri.org.mx; f. 1929 as the Partido Nacional Revolucionario, but is regarded as the natural successor to the victorious parties of the revolutionary period; broadly based and centre govt party; Pres. ROBERTO MADRAZO PINTADO; Sec.-Gen. ELBA ESTHER GORDILLO MORALES; groups within the PRI include: the Corriente Crítica Progresista, the Corriente Crítica del Partido, the Corriente Constitucionalista Democratizadora, Corriente Nuevo PRI XIV Asamblea, Democracia 2000, México Nuevo, Galileo and Unidad Democrática.

**Partido Social Demócrata Mexicano (PSDM):** Edisón 89, Col. Revolución, 06030 México, DF; tel. (55) 5592-5688; fax (55) 5535-0031; f. 1975 as Partido Demócrata Mexicano; adopted current name in 1998; Christian Democrat party; 450,000 mems; Pres. BALTASAR IGNACIO VALADEZ MONTOYA.

**Partido de la Sociedad Nacionalista (PSN):** Magdalena 117, Col. del Valle, 03100 México, DF; tel. (55) 5682-5960; e-mail psn@psn.org.mx; internet www.psn.org.mx; Pres. GUSTAVO RIOJAS SANTANA.

**Partido del Trabajo (PT):** Avda Cuauhtémoc 47, Col. Roma, 06700 México, DF; tel. and fax (55) 5525-8419; internet www.pt.org.mx; f. 1991; labour party; Leader ALBERTO ANAYA GUTIÉRREZ.

**Partido Verde Ecologista de México (PVEM):** Medicina 74, esq. Avda Copilco–Universidad, Del. Coyoacán, 04360 México, DF; tel. and fax (55) 5658-7172; e-mail pve@infosel.net.mx; internet www.pvem.org.mx; f. 1987; ecologist party; Leader BERNARDO DE LA GARZA HERRERA.

The following parties are not legally recognized:

**Partido Popular Revolucionario Democrático:** e-mail pdprepr@hotmail.com; internet www.pengo.it/PDPR-EPR; f. 1996; political grouping representing the causes of 14 armed peasant orgs, including the EPR and the PROCUP.

**Partido Revolucionario Obrerista y Clandestino de Unión Popular (PROCUP):** internet www.pengo.it/PDPR-EPR/; peasant org.

Illegal organizations active in Mexico include the following:

**Ejército Popular Revolucionario (EPR):** e-mail pdprepr@hotmail.com; internet www.pengo.it/PDPR-EPR/; f. 1994; left-wing guerrilla group active in southern states, linked to the Partido Popular Revolucionario Democrático (q.v.).

**Ejército Revolucionario Popular Insurgente (ERPI):** f. 1996; left-wing guerrilla group active in Guerrero, Morelos and Oaxaca; Leader JACOBO SILVA NOGALES.

**Ejército Zapatista de Liberación Nacional (EZLN):** internet www.ezln.org; f. 1993; left-wing guerrilla group active in the Chiapas region; Leader 'Subcomandante INSURGENTE MARCOS'.

**Frente Democrático Oriental de México Emiliano Zapata (FDOMEZ):** peasant org.

## Diplomatic Representation

### EMBASSIES IN MEXICO

**Algeria:** Sierra Madre 540, Col. Lomas de Chapultepec, Del. Miguel Hidalgo, 11000 México, DF; tel. (55) 5520-6950; fax (55) 5540-7579; e-mail embjargl@iwm.com.mx; Ambassador ABDELKADER TAFFAR.

**Angola:** Gaspar de Zúñiga 226, Col. Lomas de Chapultepec, Sección Virreyes, Del. Miguel Hidalgo, 11000 México, DF; tel. (55) 5202-4421; fax (55) 5540-0503; e-mail info@embangolamex.org; Ambassador JOSÉ JAIME FURTADO GONCALVEZ.

**Argentina:** Blvd Manuel Avila Camacho 1, 7°, Edif. Scotiabank Inverlat, Col. Lomas de Chapultepec, Del. Miguel Hidalgo, 11009 México, DF; tel. (55) 5520-9430; fax (55) 5540-5011; e-mail embajadaargentina@prodigy.net.mx; Ambassador OSCAR GUILLERMO GALIE.

**Australia:** Rubén Darío 55, Col. Polanco, Del. Miguel Hidalgo, 11580 México, DF; tel. (55) 1101-2200; fax (55) 1101-2201; e-mail dima-mexico.city@dfat.gov.au; internet www.mexico.embassy.gov.au; Ambassador NEIL ALLAN MULLES.

**Austria:** Sierra Tarahumara 420, Col. Lomas de Chapultepec, Del. Miguel Hidalgo, 11000 México, DF; tel. (55) 5251-1606; fax (55) 5245-0198; e-mail mexiko-ob@bmaa.gv.at; internet www.embajadadeaustria.com.mx; Ambassador Dr WERNER DRUML.

**Belgium:** Alfredo Musset 41, Col. Polanco, Del. Miguel Hidalgo, 11550 México, DF; tel. (55) 5280-0758; fax (55) 5280-0208; e-mail mexico@diplobel.org; internet www.diplobel.org/mexico; Ambassador MICHEL DELFOSSE.

**Belize:** Bernardo de Gálvez 215, Col. Lomas de Chapultepec, Del. Miguel Hidalgo, 11000 México, DF; tel. (55) 5520-1274; fax (55) 5520-6089; e-mail embelize@prodigy.net.mx; Ambassador SALVADOR AMÍN FIGUEROA.

# MEXICO

**Bolivia:** Paseo de la Reforma 45, 4°, Col. Tabacalera, Del. Cuauhtémoc, 06030 México, DF; tel. (55) 5703-0983; fax (55) 5703-0994; e-mail embajada@embol.org.mx; Ambassador Guido Rafael Capra Jemio.

**Brazil:** Lope de Armendáriz 130, Col. Lomas Virreyes, Del. Miguel Hidalgo, 11000 México, DF; tel. (55) 5201-4531; fax (55) 5520-4929; e-mail embrasil@brasil.org.mx; internet www.brasil.org.mx; Ambassador Luiz Augusto de Araujo Castro.

**Bulgaria:** Paseo de la Reforma 1990, Col. Lomas de Chapultepec, Del. Miguel Hidalgo, 11000 México, DF; tel. (55) 5596-3283; fax (55) 5596-1012; e-mail ebulgaria@yahoo.com; Ambassador Ivan Christov.

**Canada:** Schiller 529, Col. Polanco, Del. Miguel Hidalgo, 11560 México, DF; tel. (55) 5724-7900; fax (55) 5724-7985; e-mail embajada@canada.org.mx; internet www.canada.org.mx; Ambassador Gaëtan Lavertu.

**Chile:** Andrés Bello 10, 18°, Col. Polanco, Del. Miguel Hidalgo, 11560 México, DF; tel. (55) 5280-9681; fax (55) 5280-9703; e-mail echilmex@prodigy.net.mx; internet www.embajadadechile.com/mx; Ambassador Eurardo Aninat Ureta.

**China, People's Republic:** Avda San Jerónimo 217B, Del. Álvaro Obregón, 01090 México, DF; tel. (55) 5616-0609; fax (55) 5616-0460; e-mail embchina@data.net.mx; internet www.embajadachina.org.mx; Ambassador Ren Jingyu.

**Colombia:** Paseo de la Reforma 379, 1°, 5° y 6°, Col. Cuauhtémoc, Del. Cuauhtémoc, 06500 México, DF; tel. (55) 5525-0277; fax (55) 5208-2876; e-mail emcolmex@prodigy.net.mx; internet www.colombiaenmexico.org; Ambassador Luis Guillermo Giraldo Hurtado.

**Costa Rica:** Río Po 113, Col. Cuauhtémoc, Del. Cuauhtémoc, 06500 México, DF; tel. (55) 5525-7764; fax (55) 5511-9240; e-mail embcrica@ri.redint.com; Ambassador Ronald Gurdián Marchena.

**Côte d'Ivoire:** Tennyson 57, Col. Polanco, Del. Miguel Hidalgo, 11560 México, DF; tel. (55) 5280-8573; fax 5282-2954; Ambassador Yao Charles Koffi.

**Cuba:** Presidente Masaryk 554, Col. Polanco, Del. Miguel Hidalgo, 11560 México, DF; tel. and fax (55) 5280-8039; e-mail cancilleria@embacuba.com.mx; internet www.embacuba.com.mx; Ambassador Jorge Alberto Bolaños Suárez.

**Cyprus:** Sierra Gorda 370, Col. Lomas de Chapultepec, Del. Miguel Hidalgo, 11000 México, DF; tel. (55) 5202-7600; fax (55) 5520-2693; e-mail chipre@att.net.mx; Ambassador Antonis Mandritis.

**Czech Republic:** Cuvier 22, Col. Nueva Anzures, Del. Miguel Hidalgo, 11590 México, DF; tel. (55) 5531-2777; fax (55) 5531-1837; e-mail mexico@embassy.mzv.cz; internet www.czechembassy.org; Ambassador Vladimire Eisenbruck.

**Denmark:** Tres Picos 43, Col. Chapultepec Morales, Del. Miguel Hidalgo, 11580 México, DF; tel. (55) 5255-3405; fax (55) 5545-5797; e-mail mexamb@um.dk; internet www.danmex.org; Ambassador Søren Haslund.

**Dominican Republic:** República de Guatemala 84, Centro Histórico, Del. Cuauhtémoc, 06020 México, DF; tel. and fax (55) 5542-3553; e-mail embadomi@data.net.mx; internet www.embajadadominicana.com.mx; Ambassador Pablo Mariñez Alvarez.

**Ecuador:** Tennyson 217, Col. Polanco, Del. Miguel Hidalgo, 11560 México, DF; tel. (55) 5545-3141; fax (55) 5254-2442; e-mail mecuamex@prodigy.net.mx; Ambassador Reynaldo Eduardo Huerta Ortega.

**Egypt:** Alejandro Dumas 131, Col. Polanco, Del. Miguel Hidalgo, 11560 México, DF; tel. (55) 5281-0823; fax (55) 5282-1294; e-mail embofegypt@prodigy.net.mx; Ambassador Mamdouh Shawky Moustafa.

**El Salvador:** Temístocles 88, Col. Polanco, Del. Miguel Hidalgo, 11560 México, DF; tel. and fax (55) 5281-5725; e-mail embesmex@webtelmex.net.mx; Ambassador Francisco Flor Imendia Maza.

**Finland:** Monte Pelvoux 111, 4°, Col. Lomas de Chapultepec, Del. Miguel Hidalgo, 11000 México, DF; tel. (55) 5540-6036; fax (55) 5540-0114; e-mail finmex@prodigy.net.mx; internet www.finlandia.org.mx; Ambassador Ilkka Heiskanen.

**France:** Campos Elíseos 339, Col. Polanco, Del. Miguel Hidalgo, 11560 México, DF; tel. (55) 9171-9700; fax (55) 9171-9703; e-mail webmaster@francia.org.mx; internet www.francia.org.mx; Ambassador Alain Le Gourriérec (from Sept. 2005).

**Germany:** Lord Byron 737, Col. Polanco, Del. Miguel Hidalgo, 11560 México, DF; tel. (55) 5283-2200; fax (55) 5281-2588; e-mail info@embajada-alemana.org.mx; internet www.embajada-alemana.org.mx; Ambassador Eberhard Kolsch.

**Greece:** Sierra Gorda 505, Col. Lomas de Chapultepec, Del. Miguel Hidalgo, 11010 México, DF; tel. (55) 5520-2070; fax (55) 5202-4080; e-mail grecemb@prodigy.net.mx; Ambassador Alexander Migliaressis.

**Guatemala:** Explanada 1025, Col. Lomas de Chapultepec, Del. Miguel Hidalgo, 11000 México, DF; tel. (55) 5540-7520; fax (55) 5202-1142; e-mail embaguate@mexis.com; Ambassador Manuel Arturo Soto Aguirre.

**Haiti:** Presa Don Martín 53, Col. Irrigación, Del. Miguel Hidalgo, 11500 México, DF; tel. (55) 5557-2065; fax (55) 5395-1654; e-mail ambadh@mail.internet.com.mx; Ambassador Idalbert Pierre-Jean.

**Holy See:** Juan Pablo II 118, Col. Guadalupe Inn, Del. Alvaro Obregón, 01020 México, DF; tel. (55) 5663-3999; fax (55) 5663-5308; e-mail nuntiusmex@infosel.net.mx; Apostolic Nuncio Most Rev. Giuseppe Bertello (Titular Archbishop of Urbisaglia).

**Honduras:** Alfonso Reyes 220, Col. Condesa, Del. Cuauhtémoc, 06170 México, DF; tel. (55) 5211-5747; fax (55) 5211-5425; e-mail emhonmex@mail.internet.com.mx; Ambassador Elisa Mercedes Pineda Pineda.

**Hungary:** Paseo de las Palmas 2005, Col. Lomas de Chapultepec, Del. Miguel Hidalgo, 11000 México, DF; tel. (55) 5596-0523; fax (55) 5596-2378; e-mail secretaria@embajadahungria.com.mx; internet embajadahungria@vantel.net.mx; Ambassador György Tibor Herczsg.

**India:** Musset 325, Col. Polanco, Del. Miguel Hidalgo, 11550 México, DF; tel. (55) 5531-1050; fax (55) 5254-2349; e-mail indembmx@prodigy.net.mx; internet www.indembassy.org; Ambassador Surinder Singh Gill.

**Indonesia:** Julio Verne 27, Col. Polanco, Del. Miguel Hidalgo, 11560 México, DF; tel. (55) 5280-6363; fax (55) 5280-7062; e-mail kbrimex@prodigy.net.mx; Ambassador Ahwil Luthan.

**Iran:** Paseo de la Reforma 2350, Col. Lomas Altas, Del. Miguel Hidalgo, 11950 México, DF; tel. (55) 9172-2691; fax (55) 9172-2694; e-mail iranembmex@hotmail.com; Ambassador Mohammad Roohi Sefat.

**Iraq:** Paseo de la Reforma 1875, Col. Lomas de Chapultepec, Del. Miguel Hidalgo, 11000 México, DF; tel. (55) 5596-0933; fax (55) 5596-0254.

**Ireland:** Cerrada Blvd M. Avila Camacho 76, 3°, Col. Lomas de Chapultepec, Del. Miguel Hidalgo, 11000 México, DF; tel. (55) 5520-5803; fax (55) 5520-5892; e-mail embajada@irlanda.org.mx; Ambassador Dermot Brangan.

**Israel:** Sierra Madre 215, Col. Lomas de Chapultepec, Del. Miguel Hidalgo, 11000 México, DF; tel. (55) 5201-1500; fax (55) 5201-1555; e-mail embisrael@prodigy.net.mx; Ambassador David Dadonn.

**Italy:** Paseo de las Palmas 1994, Col. Lomas de Chapultepec, Del. Miguel Hidalgo, 11000 México, DF; tel. (55) 5596-3655; fax (55) 5596-2472; e-mail info@embitalia.org.mx; internet www.embitalia.org.mx; Ambassador Felice Scauso.

**Jamaica:** Schiller 326, 8°, Col. Chapultepec Morales, Del. Miguel Hidalgo, 11570 México, DF; tel. (55) 5250-6804; fax (55) 5250-6160; e-mail embjamaicamex@infosel.net.mx; Ambassador Vilma Kathleen McNish.

**Japan:** Paseo de la Reforma 395, Apdo 5-101, Col. Cuauhtémoc, Del. Cuauhtémoc, 06500 México, DF; tel. (55) 5211-0028; fax (55) 5207-7743; e-mail embjapmx@mail.internet.com.mx; internet www.mx.emb-japan.go.jp; Ambassador Noriteru Fukushima.

**Korea, Democratic People's Republic:** Eugenio Sue 332, Col. Polanco, Del. Miguel Hidalgo, 11550 México, DF; tel. (55) 5545-1871; fax (55) 5203-0019; e-mail dpkoreaemb@prodigy.net.mx; Ambassador So Jae Myong.

**Korea, Republic:** Lope de Armendáriz 110, Col. Lomas Virreyes, Del. Miguel Hidalgo, 11000 México, DF; tel. (55) 5202-9866; fax (55) 5540-7446; e-mail coremex@prodigy.net.mx; Ambassador Kyu-hyung Cho.

**Lebanon:** Julio Verne 8, Col. Polanco, Del. Miguel Hidalgo, 11560 México, DF; tel. (55) 5280-5614; fax (55) 5280-8870; e-mail embalib@prodigy.net.mx; Ambassador Nouhad Mahmoud.

**Malaysia:** Calderón de la Barca 215, Col. Polanco, 11550 México, DF; tel. (55) 5254-1118; fax (55) 5254-1295; e-mail mwmexico@infosel.net.mx; Ambassador Mohammed Abdul Halim bin Abdul Rahman.

**Morocco:** Paseo de las Palmas 2020, Col. Lomas de Chapultepec, Del. Miguel Hidalgo, 11000 México, DF; tel. (55) 5245-1786; fax (55) 5245-1791; e-mail sifamex@infosel.net.mx; internet www.marruecos.org.mx; Chargé d'affaires a.i. Noureddine Khalifa.

**Netherlands:** Edif. Calakmul 7°, Avda Vasco de Quiroga 3000, Col. Santa Fe, Del. Alvaro Obregón, 01210 México, DF; tel. (55) 5258-9921; fax (55) 5258-8138; e-mail nlgovmex@nlgovmex.com; internet www.paisesbajos.com.mx; Ambassador Jan-Jaap van de Velde.

# MEXICO — Directory

**New Zealand:** Edif. Corporativo Polanco 4°, Jaime Balmes 8, Col. Polanco, Del. Miguel Hidalgo, 11510 México, DF; tel. (55) 5283-9460; fax (55) 5283-9480; e-mail kiwimexico@compuserve.com.mx; Ambassador George Robert Furness Troup.

**Nicaragua:** Prado Norte 470, Col. Lomas de Chapultepec, Del. Miguel Hidalgo, 11000 México, DF; tel. (55) 5540-5625; fax (55) 5520-6961; e-mail embanic@prodigy.net.mx; Ambassador Leopoldo Ramírez Eva.

**Nigeria:** Paseo de las Palmas 1875, Col. Lomas de Chapultepec, Del. Miguel Hidalgo, 11000 México, DF; tel. (55) 5596-1274; fax (55) 5245-0105; e-mail nigembmx@att.net.mx; Ambassador Iyorwuese Hagher.

**Norway:** Avda de los Virreyes 1460, Col. Lomas Virreyes, Del. Miguel Hidalgo, 11000 México, DF; tel. (55) 5540-3486; fax (55) 5202-3019; e-mail emb.mexico@mfa.no; internet www.noruega.org.mx; Ambassador Helge Skaara.

**Pakistan:** Hegel 512, Col. Chapultepec Morales, Del. Miguel Hidalgo, 11570 México, DF; tel. (55) 5203-3636; fax (55) 5203-9907; Ambassador Khalid Aziz Babar.

**Panama:** Sócrates 339, Col. Polanco, Del. Miguel Hidalgo, 11560 México, DF; tel. (55) 5280-7857; fax (55) 5280-7586; e-mail embpanmx@prodigy.net.mx; internet www.embpanamamexico.com; Ambassador Ricardo José Aleman Alfaro.

**Paraguay:** Homero 415, 1°, esq. Hegel, Col. Polanco, Del. Miguel Hidalgo, 11570 México, DF; tel. (55) 5545-0405; fax (55) 5531-9905; e-mail embapar@prodigy.net.mx; Ambassador José Félix Fernández Estigarribia.

**Peru:** Paseo de la Reforma 2601, Col. Lomas Reforma, Del. Miguel Hidalgo, 11000 México, DF; tel. (55) 5570-2443; fax (55) 5259-0530; e-mail embaperu@prodigy.net.mx; Ambassador Alfredo Arosemena Ferreyros.

**Philippines:** Sierra Gorda 175, Col. Lomas de Chapultepec, Del. Miguel Hidalgo, 11000 México, DF; tel. (55) 5202-8456; fax (55) 5202-8403; e-mail ambamexi@mail.internet.com.mx; Ambassador Justo O. Orros, Jr.

**Poland:** Cracovia 40, Col. San Angel, Del. Alvaro Obregón, 01000 México, DF; tel. (55) 5550-4700; fax (55) 5616-0822; e-mail embajadadepolonia@prodigy.net.mx; Ambassador Wojciech Tomaszewski.

**Portugal:** Avda Alpes 1370, Lomas de Chapultepec, Del. Miguel Hidalgo, 11000 México, DF; tel. (55) 5520-7897; fax (55) 5520-4688; e-mail embpomex@prodigy.net.mx; internet www.portugalenmexico.com.mx; Ambassador Francisco Henriques da Silva.

**Romania:** Sófocles 311, Col. Polanco, Del. Miguel Hidalgo, 11560 México, DF; tel. (55) 5280-0197; fax (55) 5280-0343; e-mail ambromaniei@prodigy.net.mx; internet www.gilbert.ro; Ambassador Vasile Dan.

**Russia:** José Vasconcelos 204, Col. Condesa, Del. Cuauhtémoc, 06140 México, DF; tel. (55) 5273-1305; fax (55) 5273-1545; e-mail embrumex@mail.internet.com.mx; Chargé d'affaires a.i. Alexander Dogadin.

**Saudi Arabia:** Paseo de las Palmas 2075, Col. Lomas de Chapultepec, Del. Miguel Hidalgo, 11000 México, DF; tel. (55) 5596-0173; fax (55) 5520-3160; e-mail saudiemb@prodigy.net.mx; Chargé d'affaires a.i. Ali Ahmad Alghamdi.

**Serbia and Montenegro:** Montañas Rocallosas Ote 515, Col. Lomas de Chapultepec, Del. Miguel Hidalgo, 11000 México, DF; tel. (55) 5520-0524; fax (55) 5520-9927; e-mail ambayumex@att.met.mx; Ambassador Milisav Paic.

**Slovakia:** Julio Verne 35, Col. Polanco, Del. Miguel Hidalgo, 11560 México, DF; tel. (55) 5280-6669; fax (55) 5280-6294; e-mail eslovaquia@prodigy.net.mx; Ambassador Branislav Hitka.

**South Africa:** Andrés Bello 10, Edif. Forum, Col. Polanco 9°, Del. Miguel Hidalgo, 11560 México, DF; tel. and fax (55) 5282-9260; e-mail safrica@prodigy.net.mx; Ambassador Malcolm Grant Ferguson.

**Spain:** Galileo 114, esq. Horacio, Col. Polanco, Del. Miguel Hidalgo, 11550 México, DF; tel. (55) 5282-2271; fax (55) 5282-1520; e-mail embaes@prodigy.net.mx; Ambassador María Cristina Barrios y Almazor.

**Sweden:** Paseo de las Palmas 1375, Col. Lomas de Chapultepec, Del. Miguel Hidalgo, 11000 México, DF; tel. (55) 9178-5010; fax (55) 5540-3253; e-mail info@suecia.com.mx; internet www.suecia.com.mx; Ambassador Ewa Polano.

**Switzerland:** Paseo de las Palmas 405, 11°, Torre Óptima, Col. Lomas de Chapultepec, Del. Miguel Hidalgo, 11000 México, DF; tel. (55) 5520-3003; fax (55) 5520-8685; e-mail vertretung@mex.rep.admin.ch; internet www.eda.admin.ch/mexico_emb/s/home.html; Ambassador Gian Federico Pedotti.

**Thailand:** Paseo de la Reforma 930, Col. Lomas de Chapultepec, Del. Miguel Hidalgo, 11000 México, DF; tel. (55) 5540-4551; fax (55) 5540-4817; e-mail thaimex@prodigy.net.mx; Ambassador Ravee Hongsaprabhas.

**Turkey:** Monte Libano 885, Col. Lomas de Chapultepec, Del. Miguel Hidalgo, 11000 México, DF; tel. (55) 5282-5446; fax (55) 5282-4894; e-mail turkem@mail.internet.com.mx; Ambassador Ahmet Sedat Banguoglu.

**Ukraine:** Paseo de la Reforma 730, Col. Lomas de Chapultepec, Del. Miguel Hidalgo, 11000 México, DF; tel. and fax (55) 5282-4789; e-mail ukrainembasy@mexis.com; Ambassador Olexander Taranenko.

**United Kingdom:** Río Lerma 71, Col. Cuauhtémoc, Del. Cuauhtémoc, 06500 México, DF; tel. (55) 5242-8500; fax (55) 5242-8517; e-mail ukinmex@att.net.mx; internet www.embajadabritanica.com.mx; Ambassador Giles Paxman (from Oct. 2005); Chargé d'affaires a.i. Vijay Rangarajan.

**USA:** Paseo de la Reforma 305, Del. Cuauhtémoc, 06500 México, DF; tel. (55) 5080-2000; fax (55) 5080-2150; internet www.usembassy-mexico.gov; Ambassador Antonio O. Garza, Jr.

**Uruguay:** Hegel 149, 1°, Col. Chapultepec Morales, Del. Miguel Hidalgo, 11560 México, DF; tel. (55) 5531-0880; fax (55) 5545-3342; e-mail uruazte@ort.org.mx; Ambassador Juan Delgado Genta.

**Venezuela:** Schiller 326, Col. Chapultepec Morales, Del. Miguel Hidalgo, 11570 México, DF; tel. (55) 5203-4233; fax (55) 5203-5072; e-mail venez-mex@embajadadevenezuela.com.mxn; Ambassador Hely Vladimir Villegas Poljak.

**Viet Nam:** Sierra Ventana 255, Col. Lomas de Chapultepec, Del. Miguel Hidalgo, 11000 México, DF; tel. (55) 5540-1632; fax (55) 5540-1612; e-mail dsqvn@terra.com.mx; Ambassador Le Van Thinh.

# Judicial System

The principle of the separation of the judiciary from the legislative and executive powers is embodied in the 1917 Constitution. The judicial system is divided into two areas: the federal, dealing with federal law, and the local, dealing only with state law within each state.

The federal judicial system has both ordinary and constitutional jurisdiction and judicial power is exercised by the Supreme Court of Justice, the Electoral Court, Collegiate and Unitary Circuit Courts and District Courts. The Supreme Court comprises two separate chambers: Civil and Criminal Affairs, and Administrative and Labour Affairs. The Federal Judicature Council is responsible for the administration, surveillance and discipline of the federal judiciary, except for the Supreme Court of Justice.

In 2005 there were 172 Collegiate Circuit Courts (Tribunales Colegiados), 62 Unitary Circuit Courts (Tribunales Unitarios) and 285 District Courts (Juzgados de Distrito). Mexico is divided into 29 judicial circuits. The Circuit Courts may be collegiate, when dealing with the derecho de amparo (protection of constitutional rights of an individual), or unitary, when dealing with appeal cases. The Collegiate Circuit Courts comprise three magistrates with residence in the cities of México, Toluca, Naucalpan, Guadalajara, Monterrey, Hermosillo, Puebla, Boca del Río, Xalapa, Torreón, Saltillo, San Luis Potosí, Villahermosa, Morelia, Mazatlán, Oaxaca, Mérida, Mexicali, Guanajuato, León, Chihuahua, Ciudad Juárez, Cuernavaca, Ciudad Victoria, Ciudad Reynosa, Tuxtla Gutiérrez, Tapachula, Acapulco, Chilpancingo, Querétaro, Zacatecas, Aguascalientes, Tepic, Durango, La Paz, Cancún, Tlaxcala and Pachuca. The Unitary Circuit Courts comprise one magistrate with residence mostly in the same cities as given above.

### SUPREME COURT OF JUSTICE

**Suprema Corte de Justicia de la Nación:** Pino Suárez 2, Col. Centro, 06065 México, DF; tel. (55) 5522-0096; fax (55) 5522-0152; internet www.scjn.gob.mx.

**Chief Justice:** Mariano Azuela Güitrón.

#### First Chamber—Civil and Criminal Affairs

**President:** José de Jesús Gudiño Pelayo.

#### Second Chamber—Administrative and Labour Affairs

**President:** Guillermo I. Ortiz Mayagoitia.

# Religion

## CHRISTIANITY

### The Roman Catholic Church

The prevailing religion is Roman Catholicism, but the Church, disestablished in 1857, was for many years, under the Constitution of 1917, subject to state control. A constitutional amendment, promulgated in January 1992, officially removed all restrictions on the Church. For ecclesiastical purposes, Mexico comprises 14 archdioceses, 67 dioceses, five territorial prelatures and two eparchates (both directly subject to the Holy See). An estimated 90% of the population are adherents.

### Bishops' Conference

Conferencia del Episcopado Mexicano (CEM), Edif. S. S. Juan Pablo II, Prolongación Ministerios 24, Col. Tepeyac Insurgentes, Apdo 118-055, 07020 México, DF; tel. (55) 5781-8462; e-mail segcem@cem.org.mx; internet www.cem.org.mx.

Pres. JOSÉ GUADALUPE MARTÍN RÁBAGO (Bishop of León).

**Archbishop of Acapulco:** FELIPE AGUIRRE FRANCO, Arzobispado, Quebrada 16, Apdo 201, 39300 Acapulco, Gro; tel. and fax (744) 482-0763; e-mail buenpastor@acabtu.com.mx.

**Archbishop of Antequera/Oaxaca:** JOSÉ LUIS CHÁVEZ BOTELLO, Independencia 700, Apdo 31, 68000 Oaxaca, Oax.; tel. (951) 64822; fax (951) 65580; e-mail antequera@oax1.telmex.net.mx.

**Archbishop of Chihuahua:** JOSÉ FERNÁNDEZ ARTEAGA, Arzobispado, Avda Cuauhtémoc 1828, Apdo 7, 31020 Chihuahua, Chih.; tel. (614) 10-3202; fax (614) 10-5621; e-mail curiao1@chih1.telmex.net.mx.

**Archbishop of Durango:** HÉCTOR GONZÁLEZ MARTÍNEZ, Arzobispado, 20 de Noviembre 306 Poniente, Apdo 116, 34000 Durango, Dgo; tel. (618) 114242; fax (618) 128881; e-mail arqdgo@logicnet.com.mx.

**Archbishop of Guadalajara:** Cardinal JUAN SANDOVAL IÑIGUEZ, Arzobispado, Liceo 17, Apdo 1-331, 44100 Guadalajara, Jal.; tel. (33) 614-5504; fax (33) 658-2300; e-mail arzgdl@arquinet.com.mx; internet www.arquidiocesisgdl.org.mx.

**Archbishop of Hermosillo/Sonora:** JOSÉ ULISES MACÍAS SALCEDO, Arzobispado, Dr Paliza 81, Apdo 1, 83260 Hermosillo, Son.; tel. (658) 13-2138; fax (658) 13-1327; e-mail obispo@rtn.uson.mx.

**Archbishop of Jalapa:** SERGIO OBESO RIVERA, Arzobispado, Avda Manuel Avila Camacho 73, Apdo 359, 91000 Jalapa, Ver.; tel. (228) 12-0579; fax (228) 817-5578; e-mail arzobispadoal_xal@infosel.net.mx.

**Archbishop of Mexico City:** Cardinal NORBERTO RIVERA CARRERA, Curia del Arzobispado de México, Durango 90, 5°, Col. Roma, Apdo 24433, 06700 México, DF; tel. (55) 5208-3200; fax (55) 5208-2894; e-mail arzobisp@arzobispadomexico.org.mx; internet www.arzobispadomexico.org.mx.

**Archbishop of Monterrey:** FRANCISCO ROBLES ORTEGA, Zuazua 1100, Apdo 7, 64000 Monterrey, NL; tel. (81) 8345-2466; fax (81) 8345-3557; e-mail curia@sdm.net.mx; internet www.arquidiocesismty.org.mx.

**Archbishop of Morelia:** ALBERTO SUÁREZ INDA, Arzobispado, Costado Catedral, Frente Avda Madero, Apdo 17, 58000 Morelia, Mich; tel. (443) 120523; fax (443) 123744; e-mail arzobispo@arquimorelia.org.mx; tel. www.arquimorelia.org.mx.

**Archbishop of Puebla de los Angeles:** ROSENDO HUESCA PACHECO, Avda 2 Sur 305, Apdo 235, 72000 Puebla, Pue.; tel. (222) 32-4591; fax (222) 46-2277; e-mail rhuesca@mail.cem.org.mx.

**Archbishop of San Luis Potosí:** LUIS MORALES REYES, Arzobispado, Francisco Madero 300, Apdo 1, 78000 San Luis Potosí, SLP; tel. (444) 812-4555; fax (444) 812-7979; e-mail adiosclp@mail.cem.org.mx.

**Archbishop of Tlalnepantla:** RICARDO GUÍZAR DÍAZ, Arzobispado, Avda Juárez 42, Apdo 268, 54000 Tlalnepantla, Méx.; tel. (55) 5565-3944; fax (55) 5565-3944; e-mail rguizar@mail.cem.org.mx; internet www.arqtlalnepantla.org.

**Archbishop of Yucatán:** EMILIO CARLOS BERLIE BELAUNZARÁN, Arzobispado, Calle 58 501, 97000 Mérida, Yuc.; tel. (999) 928-6214; fax (999) 923-7983; e-mail aryu@sureste.com; internet www.arquidiocesisdeyucatan.org.

### The Anglican Communion

Mexico is divided into five dioceses, which form the Province of the Anglican Church in Mexico, established in 1995.

**Bishop of Cuernavaca:** MARTINIANO GARCÍA MONTIEL, Minerva 1, Col. Las Delicias, 62330 Cuernavaca, Mor.; tel. and fax (777) 315-2870; e-mail diovca@edsa.net.mx.

**Bishop of Mexico City and Primate of the Anglican Church in Mexico:** CARLOS TOUCHE-PORTER, La Otra Banda 46, Avda San Jerónimo 117, Col. San Ángel, 01000 México, DF; tel. (55) 5616-2205; fax (55) 5616-2205; e-mail diomex@adetel.net.mx.

**Bishop of Northern Mexico:** MARCELINO RIVERA, Simón Bolívar 2005 Nte, Col. Mitras Centro, 64460 Monterrey, NL; tel. (81) 333-0922; fax (81) 348-73625; e-mail diocesisdelnorte@att.net.mx.

**Bishop of South-Eastern Mexico:** BENITO JUÁREZ MARTÍNEZ, Avda de las Américas 73, Col. Aguacatl, 91130 Jalapa, Ver.; tel. and fax (932) 144387; e-mail dioste99@aol.com; internet diosemexico.org.

**Bishop of Western Mexico:** LINO RODRÍGUEZ-AMARO, Javier J. Gamboa 255, Col. Sector Juárez 44100, Guadalajara, Jal.; tel. (33) 615-5070; fax (33) 615-4413; e-mail diocte@vianet.com.mx.

### Protestant Churches

**Federación Evangélica de México.**

**Iglesia Evangélica Luterana de México:** POB 1-1034, 44100 Guadalajara, Jal.; tel. (52) 3639-7253; e-mail dtrejocoria@hotmail.com; Pres. ENCARNACIÓN ESTRADA; Pres. DANIEL TREJO CORIA; 1,500 mems.

**Iglesia Metodista de México, Asociación Religiosa:** Miravelle 209, Col. Albert 03570, México, DF; tel. (55) 5539-3674; e-mail prenapro@iglesia-metodista.org.mx; internet www.iglesia-metodista.org.mx; f. 1930; 55,000 mems; Pres. Bishop RAÚL ROSAS GONZÁLEZ; 370 congregations; comprises six episcopal areas; Bishop (México) MOISÉS VALDERRAMA GÓMEZ; Bishop (North-West) JAIME VÁSQUEZ OLMEDA; Bishop (North-Central) JUAN MILTON VELASCO LEGORRETA, Bishop (East) RAÚL ROSAS GONZÁLEZ; Bishop (South) BASILIO HERRERA LÓPEZ, Bishop (South-East) PEDRO MORENO CANO.

**National Baptist Convention of Mexico:** Tlalpan 1035-A, Col. Américas Unidas, 03610 México, DF; tel. and fax (55) 5539-7720; e-mail webmaster@cnbm.org.mx; internet www.cnbm.org.mx; f. 1903; Pres. GILBERTO GUTIÉRREZ LUCERO.

## BAHÁ'Í FAITH

**National Spiritual Assembly of the Bahá'ís of Mexico:** Emerson 421, Col. Chapultepec Morales, 11570 México, DF; tel. (55) 5545-2155; fax (55) 5255-5972; e-mail bahaimex@mx.inter.net; internet www.bahaimex.org; mems resident in 978 localities.

## JUDAISM

The Jewish community numbered some 40,000 in 2004.

**Comité Central de la Comunidad Judía de México:** Cofre de Perote 115, Lomas Barrilaco, 11010 México, DF; tel. (55) 5520-9393; fax (55) 5540-3050; Pres. SIMÓN NISSAN ROVERO.

# The Press

## DAILY NEWSPAPERS

### México, DF

**La Afición:** Ignacio Mariscal 23, Apdo 64 bis, Col. Tabacalera, 06800 México, DF; tel. (55) 5546-4780; fax (55) 5546-5852; e-mail opino@aguila.el-universal.com.mx; internet www.milenio.com/deportes; f. 1930; sport; Pres. Lic. JUAN FRANCISCO EALY ORTIZ; Gen. Man. ANTONIO GARCÍA SERRANO; circ. 85,000.

**La Crónica de Hoy:** Grupo Editorial Convergencia, SA de CV, Balderas 33, 6°, Col. Centro, 06040 México, DF; tel. and fax (52) 5512-3429; internet www.cronica.com.mx; Pres. JORGE KAHWAGI GASTINE; Editorial Dir PABLO HIRIART LE BERT.

**Cuestión:** Laguna de Mayrán 410, Col. Anáhuac, 11320 México, DF; tel. (55) 5260-0499; fax (55) 5260-3645; e-mail contacto@cuestion.com.mx; internet www.cuestion.com.mx; f. 1980; midday; Dir-Gen. Lic. ALBERTO GONZÁLEZ PARRA; circ. 48,000.

**El Día:** Insurgentes Norte 1210, Col. Capultitlán, 07370 México, DF; tel. (55) 5729-2155; fax (55) 5537-6629; e-mail cduran@servidor.unam.mx; f. 1962; morning; Dir-Gen. JOSÉ LUIS CAMACHO LÓPEZ; circ. 50,000.

**Diario de México:** Chimalpopoca 38, Col. Obrera, 06800 México, DF; tel. (55) 5442-6501; fax (55) 5588-4289; e-mail dirgral@diariodemexico.com.mx; internet www.diariodemexico.com.mx; f. 1948; morning; Dir-Gen. FEDERICO BRACAMONTES GÁLVEZ; Dir RAFAEL LIZARDI DURÁN; circ. 76,000.

**El Economista:** Avda Coyoacán 515, Col. del Valle, 03100 México, DF; tel. (55) 5326-5444; fax (55) 5687-3821; internet www

# MEXICO

.economista.com.mx; f. 1988; financial; Pres. José Gómez Cañibe; Dir-Gen. Luis Enrique Mercado Sánchez.

**Esto:** Guillermo Prieto 7, 1°, 06470 México, DF; tel. and fax (55) 5591-0866; e-mail esto@oem.com.mx; internet www.esto.com.mx; f. 1941; morning; sport; Pres. and Dir-Gen. Lic. Mario Vázquez Raña; Dir Carlos Trapaga Barrientos; circ. 400,000, Mondays 450,000.

**Excélsior:** Paseo de la Reforma 18 y Bucareli 1, Apdo 120 bis, Col. Centro, 06600 México, DF; tel. (55) 5566-2200; fax (55) 5566-0223; e-mail foro@excelsior.com.mx; internet www.excelsior.com.mx; f. 1917; morning; independent; Pres. Rafael de la Huerta Reyes; Dir José Manuel Nava Sánchez; Gen. Man. Juventino Olivera López; circ. 200,000.

**El Financiero:** Lago Bolsena 176, Col. Anáhuac entre Lago Peypus y Lago Onega, 11320 México, DF; tel. (55) 5227-7600; fax (55) 5254-6427; e-mail pilar@elfinanciero.com.mx; internet www.elfinanciero.com.mx; f. 1981; financial; Dir-Gen. Pilar Estandía de Cárdenas; circ. 147,000.

**El Heraldo de México:** Dr Lucio, esq. Dr Velasco, Col. Doctores, 06720 México, DF; tel. (55) 5578-7022; fax (55) 5578-9824; e-mail heraldo@iwm.com.mx; internet www.heraldo.com.mx; f. 1965; morning; Dir-Gen. Gabriel Alarcón Velázquez; circ. 209,600.

**La Jornada:** Avda Cuauhtémoc 1236, Col. Santa Cruz Atoyac, Del. Benito Juárez, 03310 México, DF; tel. (55) 9183-0300; internet www.jornada.unam.mx; f. 1984; morning; Dir-Gen. Lic. Carmen Lira Saade; Gen. Man. Lic. Jorge Martínez Jiménez; circ. 106,471, Sundays 100,924.

**El Milenio:** México, DF; internet www.milenio.com; publishes 6 other newspapers in Mexico, and a weekly news magazine, *Milenio Semanal*.

**Novedades:** Balderas 87, esq. Morelos, Col. Centro, 06040 México, DF; tel. (55) 5518-5481; fax (55) 5521-4505; internet www.novedades.com.mx; f. 1936; morning; independent; Pres. and Editor-in-Chief Romulo O'Farrill, Jr; Vice-Pres. José Antonio O'Farrill Avila; circ. 42,990, Sundays 43,536.

**Ovaciones:** Lago Zirahuén 279, 20°, Col. Anáhuac, 11320 México, DF; tel. (55) 5328-0700; fax (55) 5260-2219; e-mail ovaciones@televisa.com.mx; f. 1947; morning and evening editions; Pres. and Dir-Gen. Lic. Alberto Ventosa Aguilera; circ. 130,000; evening circ. 100,000.

**La Prensa:** Basilio Badillo 40, Col. Tabacalera, 06030 México, DF; tel. (55) 5228-9947; fax (55) 5521-8209; e-mail prensa@oem.com.mx; internet www.la-prensa.com.mx; f. 1928; morning; Pres. and Dir-Gen. Lic. Mario Vázquez Raña; Dir Mauricio Ortega Camberos; circ. 270,000.

**Reforma:** Avda México Coyoacán 40, Col. Sta Cruz Atoyac, 03310 México, DF; tel. (55) 5628-7100; fax (55) 5628-7188; internet www.reforma.infosel.com; f. 1993; morning; Pres. and Dir-Gen. Alejandro Junco de la Vega Elizondo; circ. 94,000.

**El Sol de México:** Guillermo Prieto 7, 20°, Col. San Rafael, 06470 México, DF; tel. (55) 5566-1511; fax (55) 5535-5560; e-mail info@oem.com.mx; internet www.elsoldemexico.com.mx; f. 1965; morning and midday; Pres. and Dir-Gen. Lic. Mario Vázquez Raña; Dir Pilar Ferreira García; circ. 76,000.

**El Universal:** Bucareli 8, Apdo 909, Col. Centro, Del. Cuauhtémoc, 06040 México, DF; tel. (55) 5709-1313; fax (55) 5510-1269; e-mail redaccio@servidor.unam.mx; internet www.el-universal.com.mx; f. 1916; morning; independent; Pres. and Dir-Gen. Lic. Juan Francisco Ealy Ortiz; circ. 165,629, Sundays 181,615.

**Unomásuno:** Gabino Barreda 86, Col. San Rafael, México, DF; tel. (55) 1055-5500; fax (55) 5598-8821; e-mail cduran@servidor.unam.mx; internet www.unomasuno.com.mx; f. 1977; morning; left-wing; Pres. Naim Libien Kaui; Dir José Luis Rojas Ramírez; circ. 40,000.

## PROVINCIAL DAILY NEWSPAPERS

### Baja California

**El Mexicano:** Carretera al Aeropuerto s/n, Fracc. Alamar, Apdo 2333, 22540 Tijuana, BC; tel. (664) 21-3400; fax (664) 21-2944; f. 1959; morning; Dir and Gen. Man. Eligio Valencia Roque; circ. 80,000.

**El Sol de Tijuana:** Rufino Tamayo 4, Zona Río, 22320 Tijuana, BC; tel. (664) 34-3232; fax (664) 34-2234; e-mail soltij@oem.com.mx; internet www.oem.com.mx; f. 1989; morning; Pres. and Dir-Gen. Lic. Mario Vázquez Raña; circ. 50,000.

**La Voz de la Frontera:** Avda Madero 1545, Col. Nueva, Apdo 946, 21100 Mexicali, BC; tel. (686) 53-4545; fax (686) 53-6912; e-mail lavoz@oem.com.mx; internet www.oem.com.mx; f. 1964; morning; independent; Pres. and Dir-Gen. Mario Vázquez Raña; Gen. Man. Lic. Mario Valdés Hernández; circ. 65,000.

### Chihuahua

**El Diario:** Publicaciones del Chuviscar, SA de CV, Avda Universidad 1900, Col. San Felipe, Chihuahua, Chih.; tel. (614) 429-0700; internet www.diario.com.mx; www.eldiariodechihuahua.com.mx; f. 1976; Pres. Osvaldo Rodríguez Borunda.

**El Heraldo de Chihuahua:** Avda Universidad 2507, Apdo 1515, 31240 Chihuahua, Chih.; tel. (614) 13-9339; fax (614) 13-5625; e-mail elheraldo@buzon.online.com.mx; f. 1927; morning; Pres. and Dir-Gen. Lic. Mario Vázquez Raña; Dir Lic. Javier H. Contreras; circ. 27,520, Sundays 31,223.

### Coahuila

**La Opinión:** Blvd Independencia 1492 Ote, Apdo 86, 27010 Torreón, Coah.; tel. (871) 13-8777; fax (871) 13-8164; internet www.editoriallaopinion.com.mx; f. 1917; morning; Pres. Francisco A. González; circ. 40,000.

**El Siglo de Torreón:** Avda Matamoros 1056 Pte, Apdo 19, 27000 Torreón, Coah.; tel. (871) 12-8600; fax (871) 16-5909; internet www.elsiglodetorreon.com.mx; f. 1922; morning; Pres. Olga De Juambelz y Horcasitas; circ. 38,611, Sundays 38,526.

**Vanguardia:** Blvd Venustiano Carranza 1918, República Oriente, 25280 Saltillo, Coah.; tel. (844) 411-0835; e-mail hola@vanguardia.com.mx; internet www.vanguardia.com.mx.

### Colima

**Diario de Colima:** Avda 20 de Noviembre 380, 28060 Colima, Col.; tel. (312) 12-5688; internet www.diariodecolima.com; f. 1953; Dir-Gen. Héctor Sánchez de la Madrid; Dir Esteban Cortes Rojas.

### Guanajuato

**El Nacional:** Carretera Guanajuato–Juventino Rosas, km 9.5, Apdo 32, 36000 Guanajuato, Gto; tel. (477) 33-1286; fax (477) 33-1288; f. 1987; morning; Dir-Gen. Arnoldo Cuéllar Ornelas; circ. 60,000.

### Jalisco

**El Diario de Guadalajara:** Calle 14 2550, Zona Industrial, 44940 Guadalajara, Jal.; tel. (33) 612-0043; fax (33) 612-0818; f. 1969; morning; Pres. and Dir-Gen. Luis A. González Becerra; circ. 78,000.

**El Informador:** Independencia 300, Apdo 3 bis, 44100 Guadalajara, Jal.; tel. (33) 614-6340; fax (33) 614-4653; internet www.informador.com.mx; f. 1917; morning; Editor Jorge Álvarez del Castillo; circ. 50,000.

**El Occidental:** Calzada Independencia Sur 324, Apdo 1-699, 44100 Guadalajara, Jal.; tel. (33) 613-0690; fax (33) 613-6796; f. 1942; morning; Dir Lic. Ricardo del Valle del Peral; circ. 49,400.

### México

**ABC:** Avda Hidalgo Ote 1339, Centro Comercial, 50000 Toluca, Méx.; tel. (722) 179880; fax (722) 179646; e-mail miled1@mail.miled.com; internet www.miled.com; f. 1984; morning; Pres. and Editor Miled Libien Kaui; circ. 65,000.

**Diario de Toluca:** Allende Sur 209, 50000 Toluca, Méx.; tel. (722) 142403; fax (722) 141523; f. 1980; morning; Editor Anuar Maccise Dib; circ. 22,200.

**El Heraldo de Toluca:** Salvador Díaz Mirón 700, Col. Sánchez Colín, 50150 Toluca, Méx.; tel. (722) 173453; fax (722) 122535; f. 1955; morning; Editor Alberto Barraza Sánchez A; circ. 90,000.

**El Mañana:** Avda Hidalgo Ote 1339, Toluca, Méx.; tel. (722) 179880; fax (722) 178402; e-mail miled1@mail.miled.com; internet www.miled.com; f. 1986; morning; Pres. and Editor Miled Libien Kaui; circ. 65,000.

**Rumbo:** Allende Sur 205, Toluca, Méx.; tel. (722) 142403; fax (722) 141523; f. 1968; morning; Editor Anuar Maccise Dib; circ. 10,800.

**El Sol de Toluca:** Santos Degollado 105, Apdo 54, 50050 Toluca, Méx.; tel. (722) 150340; fax (722) 147441; f. 1947; morning; Pres. and Dir-Gen. Lic. Mario Vázquez Raña; circ. 42,000.

### Michoacán

**La Voz de Michoacán:** Blvd del Periodismo 1270, Col. Arriaga Rivera, Apdo 121, 58190 Morelia, Mich.; tel. (443) 327-3712; fax (443) 327-3728; e-mail lavoz@mail.giga.com; internet www.voznet.com.mx; f. 1948; morning; Dir-Gen. Lic. Miguel Medina Robles; circ. 50,000.

### Morelos

**El Diario de Morelos:** Morelos Sur 817, Col. Las Palmas, 62000 Cuernavaca, Mor.; tel. (777) 14-2660; fax (777) 14-1253; internet www.diariodemorelos.com.mx; morning; Dir-Gen. Federico Bracamontes; circ. 47,000.

MEXICO  *Directory*

### Nayarit

**Meridiano de Nayarit:** E. Zapata 73 Pte, Apdo 65, 63000 Tepic, Nay.; tel. (321) 20145; fax (321) 26630; internet www.meridiano.com.mx; f. 1942; morning; Dir Dr DAVID ALFARO; circ. 60,000.

**El Observador:** Allende 110 Oeste, Despachos 203–204, 63000 Tepic, Nay.; tel. (321) 24309; fax (321) 24309; morning; Pres. and Dir-Gen. Lic. LUIS A. GONZÁLEZ BECERRA; circ. 55,000.

### Nuevo León

**ABC:** Platón Sánchez Sur 411, 64000 Monterrey, NL; tel. (81) 344-4480; fax (81) 344-5990; e-mail abc2000@mexis.com; f. 1985; morning; Pres. GONZALO ESTRADA CRUZ; Dir-Gen. GONZALO ESTRADO TORRES; circ. 40,000, Sundays 45,000.

**El Diario de Monterrey:** Eugenio Garza Sada 2245 Sur, Col. Roma, Apdo 3128, 647000 Monterrey, NL; tel. (81) 359-2525; fax (81) 359-1414; internet www.diariodemonterrey.com; f. 1974; morning; Dir-Gen. Lic. FEDERICO ARREOLA; circ. 80,000.

**El Norte:** Washington 629 Ote, Apdo 186, 64000 Monterrey, NL; tel. (81) 345-3388; fax (81) 343-2476; internet www.elnorte.com.mx; f. 1938; morning; Man. Dir Lic. ALEJANDRO JUNCO DE LA VEGA; circ. 133,872, Sundays 154,951.

**El Porvenir:** Galeana Sur 344, Apdo 218, 64000 Monterrey, NL; tel. (81) 345-4080; fax (81) 345-7795; internet www.elporvenir.com.mx; f. 1919; morning; Dir-Gen. JOSÉ GERARDO CANTÚ ESCALANTE; circ. 75,000.

**El Sol:** Washington 629 Ote, Apdo 186, 64000 Monterrey, NL; tel. (81) 345-3388; fax (81) 343-2476; internet www.infosel.com.mx; f. 1922; evening (except Sundays); Man. Dir Lic. ALEJANDRO JUNCO DE LA VEGA; circ. 45,300.

### Oaxaca

**El Imparcial:** Armenta y López 312, Apdo 322, 68000 Oaxaca, Oax.; tel. (951) 516-2812; fax (951) 516-0050; internet www.imparoax.com.mx; f. 1951; morning; Dir-Gen. Lic. BENJAMÍN FERNÁNDEZ PICHARDO; circ. 17,000, Sundays 20,000.

### Puebla

**El Sol de Puebla:** Avda 3 Ote 201, Apdo 190, 72000 Puebla, Pue.; tel. (222) 42-4560; fax (222) 46-0869; internet www.oem.com.mx; f. 1944; morning; Pres. and Dir-Gen. Lic. MARIO VÁZQUEZ RAÑA; circ. 67,000.

### San Luis Potosí

**El Heraldo:** Villerías 305, 78000 San Luis Potosí, SLP; tel. (444) 812-3312; fax (444) 812-2081; e-mail heraldsl@prodigy.net.mx; internet www.elheraldodesanluis.com.mx; f. 1954; morning; Dir-Gen. ALEJANDRO VILLASANA MENA; circ. 60,620.

**Pulso:** Galeana 485, 78000 San Luis Potosí, SLP; tel. (444) 812-7575; fax (444) 812-3525; internet www.pulsoslp.com.mx; morning; Dir-Gen. MIGUEL VALLADARES GARCÍA; circ. 60,000.

**El Sol de San Luis:** Avda Universidad 565, Apdo 342, 78000 San Luis Potosí, SLP; tel. and fax (444) 812-4412; f. 1952; morning; Pres. and Dir-Gen. Lic. MARIO VÁZQUEZ RAÑA; Dir JOSÉ ANGEL MARTÍNEZ LIMÓN; circ. 60,000.

### Sinaloa

**El Debate de Culiacán:** Madero 556 Pte, 80000 Culiacán, Sin.; tel. (667) 16-6353; fax (667) 15-7131; e-mail redaccion@debate.com.mx; internet www.debate.com.mx; f. 1972; morning; Dir ROSARIO I. OROPEZA; circ. 23,603, Sundays 23,838.

**Noroeste Culiacán:** Grupo Periodicos Noroeste, Angel Flores 282 Ote, Apdo 90, 80000 Culiacán, Sin.; tel. (667) 713-2100; fax (667) 712-8006; e-mail cschmidt@noroeste.com.mx; internet www.noroeste.com.mx; f. 1973; morning; Pres. MANUEL J. CLOUTHIER; Editor RODOLFO DIAZ; circ. 35,000.

**El Sol de Sinaloa:** Blvd G. Leyva Lozano y Corona 320, Apdo 412, 80000 Culiacán, Sin.; tel. (667) 13-1621; f. 1956; morning; Pres. Lic. MARIO VÁZQUEZ RAÑA; Dir JORGE LUIS TÉLLEZ SALAZAR; circ. 30,000.

### Sonora

**El Imparcial:** Sufragio Efectivo y Mina 71, Col. Centro, Apdo 66, 83000 Hermosillo, Son.; tel. (658) 59-4700; fax (658) 17-4483; e-mail impar@imparcial.com.mx; internet www.imparcial.com.mx; f. 1937; morning; Pres. and Dir-Gen. JOSÉ SANTIAGO HEALY LOERA; circ. 32,083, Sundays 32,444.

### Tabasco

**Tabasco Hoy:** Avda de los Ríos 206, 86035 Villahermosa, Tab.; tel. (993) 16-3333; fax (993) 16-2135; internet www.tabascohoy.com.mx; f. 1987; morning; Dir-Gen. MIGUEL CANTÓN ZETINA; circ. 52,302.

### Tamaulipas

**El Bravo:** Morelos y Primera 129, Apdo 483, 87300 Matamoros, Tamps; tel. (871) 160100; fax (871) 162007; e-mail elbravo@riogrande.net.mx; f. 1951; morning; Pres. and Dir-Gen. JOSÉ CARRETERO BALBOA; circ. 60,000.

**El Diario de Nuevo Laredo:** González 2409, Apdo 101, 88000 Nuevo Laredo, Tamps; tel. (867) 128444; fax (867) 128221; internet www.diario.net; f. 1948; morning; Editor RUPERTO VILLARREAL MONTEMAYOR; circ. 68,130, Sundays 73,495.

**Expresión:** Calle 3A y Novedades 1, Col. Periodistas, 87300 Matamoros, Tamps; tel. (871) 174330; fax (871) 173307; morning; circ. 50,000.

**El Mañana de Reynosa:** Prof. Lauro Aguirre con Matías Canales, Apdo 14, 88620 Ciudad Reynosa, Tamps; tel. (899) 921-9950; fax (899) 924-9348; internet www.elmananarey.com.mx; f. 1949; morning; Dir-Gen. HERIBERTO DEANDAR MARTÍNEZ; circ. 65,000.

**El Mundo:** Ejército Nacional 201, Col. Guadalupe, 89120 Tampico, Tamps; tel. (833) 134084; fax (833) 134136; f. 1918; morning; Dir-Gen. ANTONIO MANZUR MARÓN; circ. 54,000.

**La Opinión de Matamoros:** Blvd Lauro Villar 200, Apdo 486, 87400 Matamoros, Tamps; tel. (871) 123141; fax (871) 122132; e-mail opinion1@tamps1.telmex.net.mx; f. 1971; Pres. and Dir-Gen. JUAN B. GARCÍA GÓMEZ; circ. 50,000.

**Prensa de Reynosa:** Matamoros y González Ortega, 88500 Reynosa, Tamps; tel. (899) 23515; fax (899) 223823; f. 1963; morning; Dir-Gen. FÉLIX GARZA ELIZONDO; circ. 60,000.

**El Sol de Tampico:** Altamira 311 Pte, Apdo 434, 89000 Tampico, Tamps; tel. (833) 12-3566; fax (833) 12-6986; internet www.oem.com.mx; f. 1950; morning; Dir-Gen. Lic. RUBÉN DÍAZ DE LA GARZA; circ. 77,000.

### Veracruz

**Diario del Istmo:** Avda Hidalgo 1115, 96400 Coatzacoalcos, Ver.; tel. (921) 48802; fax (921) 48514; e-mail info@istmo.com.mx; internet www.diariodelistmo.com; f. 1979; morning; Dir-Gen. JAÍR BENJAMÍN ROBLES BARAJAS; circ. 64,600.

**El Dictamen:** 16 de Septiembre y Arista, 91700 Veracruz, Ver.; tel. (229) 311745; fax (229) 315804; f. 1898; morning; Pres. CARLOS ANTONIO MALPICA MARTÍNEZ; circ. 38,000, Sundays 39,000.

**La Opinión:** Ver.; internet www.laopinion.com.mx; Man. Dir SILVIA GIBB GUERRERO.

### Yucatán

**Diario de Yucatán:** Calle 60, No 521, 97000 Mérida, Yuc.; tel. (999) 23-8444; fax (999) 42-2204; internet www.yucatan.com.mx; f. 1925; morning; Dir-Gen. CARLOS R. MENÉNDEZ NAVARRETE; circ. 54,639, Sundays 65,399.

**El Mundo al Día:** Calle 62, No 514A, 97000 Mérida, Yuc.; tel. (999) 23-9933; fax (999) 24-9629; e-mail nmerida@cancun.novenet.com.mx; f. 1964; morning; Pres. ROMULO O'FARRILL, Jr; Gen. Man. Lic. GERARDO GARCÍA GAMBOA; circ. 25,000.

**Por Esto!:** Calle 60, No 576 entre 73 y 71, 97000 Mérida, Yuc.; tel. (999) 24-7613; fax (999) 28-6514; internet www.poresto.net; f. 1991; morning; Dir-Gen. MARIO R. MENÉNDEZ RODRÍGUEZ; circ. 26,985, Sundays 28,727.

### Zacatecas

**Imagen:** Avda Revolución 24, Col. Tierra y Libertad, Guadalupe, Zac.; tel. and fax (492) 923-4412; internet www.imagenzac.com.mx; Dir-Gen. EUGENIO MERCADO.

## SELECTED WEEKLY NEWSPAPERS

**Bolsa de Trabajo:** San Francisco 657, 9A, Col. del Valle, 03100 México, DF; tel. (55) 5536-8387; f. 1988; employment; Pres. and Dir-Gen. MÓNICA ELÍAS CALLES; circ. 30,000.

**El Heraldo de León:** Hermanos Aldama 222, Apdo 299, 37000 León, Gto; tel. (477) 713-1194; fax (477) 714-3464; e-mail heraldo@el-heraldo-bajio.com; internet www.el-heraldo-bajio.com; f. 1957; Pres. and Dir-Gen. MAURICIO BERCÚN LÓPEZ; circ. 85,000.

**Segundamano:** Insurgentes Sur 619, Col. Nápoles, Del. Benito Juárez, 03810 México, DF; tel. (55) 1107-1715; e-mail soporte@segundamano.com.mx; internet www.segundamano.com.mx; f. 1986; Dir-Gen. LUIS MAGAÑA MAGAÑA; circ. 105,000.

**Zeta:** Avda las Américas 4633, Fraccionamiento El Paraíso, Tijuana, BC; tel. (681) 69-13; fax (621) 00-25; e-mail zeta@zetatijuana.com; internet www.zetatijuana.com; f. 1980; news magazine; Dir JESÚS BLANCORNELOS.

MEXICO
*Directory*

## GENERAL INTEREST PERIODICALS

**Car and Driver:** Alabama 113, Col. Nápoles, 03810 México, DF; tel. (55) 5523-5201; fax (55) 5536-6399; f. 1999; monthly; Pres. PEDRO VARGAS G.; circ. 80,000.

**Casas & Gente:** Ediarte, SA de CV, Amsterdam 112, 06100 México, DF; tel. (55) 5286-7794; fax (55) 5211-7112; e-mail informac@casasgente.com; internet www.casasgente.com; 10 a year; interior design; Dir-Gen. NICOLÁS H. SÁNCHEZ-OSORIO.

**Conozca Más:** Vasco de Quiroga 2000, Col. Santa Fe, 01210 México, DF; tel. (55) 5261-2600; fax (55) 5261-2704; f. 1990; monthly; scientific; Dir EUGENIO MENDOZA; circ. 90,000.

**Contenido:** Darwin 101, Col. Anzures, 11590 México, DF; tel. (55) 5531-3162; fax (55) 5545-7478; f. 1963; monthly; popular appeal; Dir ARMANDO AYALA A.; circ. 124,190.

**Cosmopolitan (México):** Vasco de Quiroga 2000, Col. Santa Fe, 01210 México, DF; tel. (55) 5261-2600; fax (55) 5261-2704; f. 1973; monthly; women's magazine; Dir SARA MARÍA CASTANY; circ. 260,000.

**Expansión:** Avda Constituyentes 956, Lomas Altas, 11950 México, DF; tel. and fax (55) 9177-4190; e-mail quien@expansion.com.mx; internet www.expansion.com.mx; fortnightly; business and financial; Editor ALBERTO BELLO.

**Fama:** Avda Eugenio Garza Sada 2245 Sur, Col. Roma, Apdo 3128, 64700 Monterrey, NL; tel. (81) 359-2525; fortnightly; show business; Pres. JESÚS D. GONZÁLEZ; Dir RAÚL MARTÍNEZ; circ. 350,000.

**Impacto:** Avda Ceylán 517, Col. Industrial Vallejo, Apdo 2986, 02300 México, DF; tel. (55) 5587-3855; fax (55) 5567-7781; f. 1949; weekly; politics; Man. and Dir-Gen. JUAN BUSTILLOS OROZCO; circ. 115,000.

**Kena Especiales:** Romero de Terreros 832, Col. del Valle, 03100 México, DF; tel. (55) 5543-1032; fax (55) 5669-3465; e-mail armonia@netsevice.com.mx; f. 1977; fortnightly; women's interest; Dir-Gen. Lic. LILIANA MORENO G.; circ. 100,000.

**Letras Libres:** Presidente Carranza 210, Col. Coyoacán, 04000 México, DF; tel. (55) 5554-8810; fax (55) 5658-0074; e-mail correo@letraslibres.com; internet www.letraslibres.com; monthly; culture; Chief Editor JULIO TRUJILLO.

**Marie Claire:** Editorial Televisa, SA de CV, Vasco de Quiroga 2000, Col. Santa Fe, 01210 México, DF; tel. (55) 5261-2600; fax (55) 5261-2704; f. 1990; monthly; women's interest; Editor LOUISE MERELES; circ. 80,000.

**Mecánica Popular:** Vasco de Quiroga 2000, Edif. E, Col. Santa Fe, Deleg. Alvaro Obregón, 01210 Mexico, DF; tel. (55) 5261-2600; fax (55) 5261-2705; e-mail mecanica.popular@siedi.spin.com.mx; f. 1947; monthly; crafts and home improvements; Dir ANDRÉS JORGE; circ. 247,850.

**Men's Health:** Vasco de Quiroga 2000, Col. Santa Fe, 01210 México, DF; tel. (55) 5261-2645; fax (55) 5261-2733; e-mail mens.health@editorial.televisa.com.mx; f. 1994; monthly; health; Editor JUAN ANTONIO SEMPERE; circ. 130,000.

**Muy Interesante:** Vasco de Quiroga 2000, Col. Santa Fe, 01210 México, DF; tel. (55) 5261-2600; fax (55) 5261-2704; f. 1984; monthly; scientific devt; Dir PILAR S. HOYOS; circ. 250,000.

**Proceso:** Fresas 7, Col. del Valle, 03100 México, DF; tel. (55) 5629-2090; fax (55) 5629-2092; f. 1976; weekly; news analysis; Pres. JULIO SCHERER GARCÍA; circ. 98,784.

**La Revista Peninsular:** Calle 35, 489 x 52 y 54, Centro, Mérida, Yuc.; e-mail revista@sureste.com; internet www.larevista.com.mx; weekly; news and politics; Dir-Gen. RODRIGO MENÉNDEZ CÁMARA.

**Selecciones del Reader's Digest:** Avda Prolongación Paseo de la Reforma 1236, 10°, Col. Santa Fe, 05348 México, DF; tel. (55) 5351-2200; f. 1940; monthly; Editor AUDÓN CORIA; circ. 611,660.

**Siempre!:** Vallarta 20, Col. Tabacalera, 06030 México, DF; tel. and fax (55) 5566-1804; e-mail mauricio@siempre.com.mx; internet www.siempre.com.mx; f. 1953; weekly; left of centre; Dir Lic. BEATRIZ PAGÉS REBOLLAR DE NIETO; circ. 100,000.

**Tele-Guía:** Vasco de Quiroga 2000, Col. Santa Fe, 01210 México, DF; tel. (55) 5261-2600; fax (55) 5261-2704; f. 1952; weekly; television guide; Editor MARÍA EUGENIA HERNÁNDEZ; circ. 375,000.

**Tiempo Libre:** Holbein 75 bis, Col. Nochebuena, 03720 México, DF; tel. (55) 5611-7332; fax (55) 5611-3874; e-mail info@tiempolibre.com.mx; internet www.tiempolibre.com.mx; f. 1980; weekly; entertainment guide; Dir-Gen. ANGELES AGUILAR ZINSER; circ. 95,000.

**Tú:** Vasco de Quiroga 2000, Col. Santa Fe, 01210 México, DF; tel. (55) 5261-2600; fax (55) 5261-2730; f. 1980; monthly; Editor MARÍA ANTONIETA SALAMANCA; circ. 275,000.

**TV y Novelas:** Vasco de Quiroga 2000, Col. Santa Fe, 01210 México, DF; tel. (55) 5261-2600; fax (55) 5261-2704; f. 1982; weekly; television guide and short stories; Dir JESÚS GALLEGOS; circ. 460,000.

**Ultima Moda:** Morelos 16, 6°, Col. Centro, 06040 México, DF; tel. (55) 5518-5481; fax (55) 5512-8902; f. 1966; monthly; fashion; Pres. ROMULO O'FARRILL, Jr; Gen. Man. Lic. SAMUEL PODOLSKY RAPOPORT; circ. 110,548.

**Vanidades:** Vasco de Quiroga 2000, Col. Santa Fe, 01210 México, DF; tel. (55) 5261-2600; fax (55) 5261-2704; f. 1961; fortnightly; women's magazine; Dir SARA MARÍA BARCELÓ DE CASTANY; circ. 290,000.

**Visión:** Homero 411, 5°, Col. Polanco, 11570 México, DF; tel. (55) 5531-4914; fax (55) 5531-4915; e-mail 74174.3111@compuserve.com; offices in Santafé de Bogotá, Buenos Aires and Santiago de Chile; f. 1950; fortnightly; politics and economics; Gen. Man. ROBERTO BELLO; circ. 27,215.

**Vogue (México):** Grupo Idéas de México, SA de CV, México, DF; tel. (55) 5095-8066; fax (55) 5530-2828; internet www.vogue.com.mx; f. 1999; monthly; women's fashion; circ. 208,180.

## SPECIALIST PERIODICALS

**Boletín Industrial:** Goldsmith 37-403, Col. Polanco, 11550 México, DF; tel. (55) 5280-6463; fax (55) 5280-3194; e-mail bolind@viernes.iwm.com.mx; internet www.bolind.com.mx; f. 1983; monthly; Dir-Gen. HUMBERTO VALADÉS DÍAZ; circ. 36,000.

**Comercio:** Río Tíber 87, 06500 México, DF; tel. (55) 5514-0873; fax (55) 5514-1008; f. 1960; monthly; business review; Dir RAÚL HORTA; circ. 40,000.

**Gaceta Médica de México:** Academia Nacional de Medicina, Unidad de Congresos del Centro Médico Nacional Siglo XXI, Bloque B, Avda Cuauhtémoc 330, Col. Doctores, 06725 México, DF; tel. (55) 5578-2044; fax (55) 5578-4271; e-mail gacetamx@starnet.net.mx; f. 1864; every 2 months; medicine; Editor Dr LUIS BENÍTEZ; circ. 20,000.

**Manufactura:** Avda Constituyentes 956, esq. Rosaleda, Col. Lomas Altas, 11950 México, DF; tel. (55) 9177-4100; e-mail dluna@expansion.com.mx; internet www.manufacturaweb.com; f. 1994; monthly; industrial; Dir-Gen. DAVID LUNA ARELLANO; circ. 29,751.

**Negobancos** (Negocios y Bancos): Bolívar 8-103, Apdo 1907, Col. Centro, 06000 México, DF; tel. (55) 5510-1884; fax (55) 5512-9411; e-mail nego_bancos@mexico.com; f. 1951; fortnightly; business, economics; Dir ALFREDO FARRUGIA REED; circ. 10,000.

## ASSOCIATIONS

**Asociación Nacional de Periodistas y Comunicadores, A.C.:** Luis G. Obregón 17, Of. 209, Col. Centro, 06020 México, DF; tel. (55) 5341-1523; Pres. MOISÉS HUERTA.

**Federación de Asociaciones de Periodistas Mexicanos (Fapermex):** Avda de las Vegas 111, Col. Colinas de Tarango, 01610 México, DF; tel. (55) 5643-4238; e-mail fapermex@fapermex.com; internet www.fapermex.com; Pres. TEODORO RENTERÍA ARRÓYAVE; Sec.-Gen. HILDA LUISA VALDEMAR Y LIMA; 88 mem. asscns; c. 9,000 mems.

**Federación Latinoamericana de Periodistas (FELAP):** Nuevo Leon 144, 1°, Col. Hipódromo Condesa, 06170 México, DF; tel. (55) 5286-6055; fax (5) 286-6085.

## NEWS AGENCIES

**Agencia de Información Integral Periodística, SA (AIIP):** Tabasco 263, Col. Roma, Del. Cuauhtémoc, 06700 México, DF; tel. (55) 8596-9643; fax (55) 5235-3468; e-mail aiipsa@axtel.net; internet www.aiip.com.mx; f. 1987; Dir-Gen. MIGUEL HERRERA LÓPEZ.

**Agencia Mexicana de Información (AMI):** Avda Cuauhtémoc 16, Col. Doctores, 06720 México, DF; tel. (55) 5761-9933; e-mail info@red-ami.com; internet www.ami.com.mx; f. 1971; Dir-Gen. JOSÉ LUIS BECERRA LÓPEZ; Gen. Man. EVA VÁZQUEZ LÓPEZ.

**Notimex, SA de CV:** Morena 110, 3°, Col. del Valle, 03100 México, DF; tel. (55) 5420-1100; fax (55) 5682-0005; e-mail comercial@notimex.com.mx; internet www.notimex.com.mx; f. 1968; services to press, radio and television in Mexico and throughout the world; Dir-Gen. Dr JORGE MEDINA VIEDAS.

### Foreign Bureaux

**Agence France-Presse (AFP):** Torre Latinoamericana, 9°, Eje Central y Madero 1, 06007 México, DF; tel. (55) 5518-5494; fax (55) 5510-4564; e-mail redaccion.mexico@afp.com; Bureau Chief PAUL RUTLER.

**Agenzia Nazionale Stampa Associata (ANSA)** (Italy): Emerson 150, 2°, Col. Chapultepec Morales, 11570 México, DF; tel. (55) 5255-3696; fax (55) 5255-3018; e-mail ansamexico@prodigy.net.mx; Bureau Chief MARCO BRANCACCIA.

**Associated Press (AP)** (USA): Paseo de la Reforma 350, Col. Juárez 06600, México, DF; tel. (55) 5080-3400; fax (55) 5208-2684; e-mail apmexico@ap.org; Bureau Chief ELOY O. AGUILAR.

# MEXICO — Directory

**Deutsche Presse-Agentur (dpa)** (Germany): Avda Cuauhtémoc 16-301, Col. Doctores, 06720 México, DF; tel. (55) 5578-4829; fax (55) 7561-0762; e-mail info@dpa.com.mx; Bureau Chief KLAUS BLUME.

**EFE** (Spain): Lafayette 69, Col. Anzures, 11590 México, DF; tel. (55) 5545-8256; fax (55) 5254-1412; e-mail direccion@efe.com.mx; Bureau Chief MANUEL FUENTES GARCÍA.

**Informatsionnoye Telegrafnoye Agentstvo Rossii—Telegrafnoye Agentsvo Suverennykh Stran (ITAR—TASS)** (Russia): Monte Líbano 965, Col. Lomas de Chapultepec 11000, México, DF; tel. (55) 5202-4831; fax (55) 5202-4879; e-mail itartass@prodigy.net; Bureau Chief IGOR VARLAMOV.

**Inter Press Service (IPS)** (Italy): Avda Cuauhtémoc 16-403, Col. Doctores, Del. Cuauhtémoc, 06720 México, DF; tel. (55) 5578-0417; fax (55) 5578-2094; e-mail mex@ipservespanol.org; Chief Correspondent DIEGO CEVALLOS ROJAS.

**Jiji Tsushin-Sha** (Japan): Sevilla 9, 2°, Col. Juárez, Del. Cuauhtémoc, 06600 México, DF; tel. (55) 5528-9651; fax (55) 5511-0062; Bureau Chief FUJIO IKEDA.

**Kyodo Tsushin** (Japan): Cerro Dios de Hacha 66, Col. Romero de Terreros 04310, México, DF; tel. (55) 5554-7199; fax (55) 5658-2957; e-mail kyodonews@mexis.com; Bureau Chief MASAHARU NANAMI.

**Maghreb Arabe Presse** (Morocco): Miguel de Cervantes Saavedra 448, 4°, Col. Irrigación, 11500 México, DF; tel. (55) 1997-2558; fax (55) 1997-6198; e-mail mohammedtanji@hotmail.com; Correspondent MOHAMMED TANJI.

**Prensa Latina** (Cuba): Edif. B, Dpto 504, Insurgentes Centro 125, Col. San Rafael, Del. Cuauhtémoc, 06470 México, DF; tel. (55) 5546-6015; fax (55) 5592-0570; e-mail plenmex@mail.internet.com.mx; Chief Correspondent Lic. AISSA GARCÍA GOREIA.

**Reuters Ltd** (United Kingdom): Manuel Ávila Camacho 36, Edif. Torre Esmeralda 11, 19°, Col. Lomas de Chapultepec 11000, México, DF; tel. (55) 5282-7000; fax (55) 5282-7171; e-mail mexicocity.newsroom@reuters.com; internet about.reuters.com/latam; Bureau Chief KIERAN MICHAEL MURRAY.

**Viet Nam News Agency (VNA)** (Viet Nam): Río Pánuco 180, Col. Cuauhtémoc 06500, México, DF; tel. (55) 5514-9013; fax (55) 5514-1015; e-mail vnamex@prodigy.net.mx; Correspondent PHAM PHOI.

**Xinhua (New China) News Agency** (People's Republic of China): Francisco I. Madero 17, Col. Tlacopac, 01040 México, DF; tel. (55) 5662-8548; fax (55) 5662-9028; e-mail xinhuamx@xinhuanet.com; Bureau Chief SONG XINDE.

### FOREIGN CORRESPONDENTS' ASSOCIATION

**Asociación de Corresponsales Extranjeros en México (ACEM):** Avda Cuauhtémoc 16, 1°, Col. Doctores, Del. Cuauhtémoc, 06720 México, DF; tel. (55) 5588-3241; fax (55) 5588-6382.

# Publishers

### MÉXICO, DF

**Aguilar, Altea, Taurus, Alfaguara, SA de CV:** Avda Universidad 767, Col. del Valle, 03100 México, DF; tel. (55) 5688-8966; fax (55) 5604-2304; e-mail sealtiel@santillana.com.mx; f. 1965; general literature; Dir SEALTIEL ALATRISTE.

**Arbol Editorial, SA de CV:** Avda Cuauhtémoc 1430, Col. Sta Cruz Atoyac, 03310 México, DF; tel. (55) 5688-4828; fax (55) 5605-7600; e-mail editorialpax@maxis.com; f. 1979; health, philosophy, theatre; Man. Dir GERARDO GALLY TEOMONFORD.

**Artes de México y del Mundo, SA de CV:** Plaza Río de Janeiro, Col. Roma, 06700 México, DF; tel. (55) 5525-5905; fax (55) 5525-5925; e-mail artesdemexico@artesdemexico.com; internet www.artesdemexico.com; f. 1988; art, design, poetry.

**Editorial Avante, SA de CV:** Luis G. Obregón 9, 1°, Apdo 45-796, Col. Centro, 06020 México, DF; tel. (55) 5510-8804; fax (55) 5521-5245; e-mail editorialavante@editorialavante.com.mx; internet www.editorialavante.com.mx; f. 1948; educational, drama, linguistics; Man. Dir Lic. MARIO A. HINOJOSA SAENZ.

**Editorial Azteca, SA:** Calle de la Luna 225–227, Col. Guerrero, 06300 México, DF; tel. (55) 5526-1157; fax (55) 5526-2557; f. 1956; religion, literature and technical; Man. Dir ALFONSO ALEMÓN JALOMO.

**Cía Editorial Continental, SA de CV (CECSA):** Renacimiento 180, Col. San Juan Tlihuaca, Azcapotzalco, 02400 México, DF; tel. (55) 5561-8333; fax (55) 5561-5231; e-mail info@patriacultural.com.mx; f. 1954; business, technology, general textbooks; Pres. CARLOS FRIGOLET LERMA.

**Ediciones de Cultura Popular, SA:** Odontología 76, Copilco Universidad, México, DF; f. 1969; history, politics, social sciences; Man. Dir URIEL JARQUÍN GALVEZ.

**Editorial Diana, SA de CV:** Arenal No 24, Edif. Norte, Ex-Hacienda Guadalupe, Chimalistac, Del. Álvaro Obregón, 01050 México, DF; tel. (55) 5089-1220; fax (55) 5089-1230; e-mail jlr@diana.com.mx; internet www.editorialdiana.com.mx; f. 1946; general trade and technical books; Pres. and CEO JOSÉ LUIS RAMÍREZ.

**Edamex, SA de CV:** Heriberto Frias 1104, Col. del Valle, 03100 México, DF; tel. (55) 5559-8588; fax (55) 5575-0555; e-mail info@edamex.com; internet www.edamex.com; arts and literature, sport, journalism, education, philosophy, food, history, children's, health, sociology; Dir-Gen. MONICA COLMENARES.

**Ediciones Era, SA de CV:** Calle del Trabajo 31, Col. La Fama, Tlalpan, 14269 México, DF; tel. (55) 5528-1221; fax (55) 5606-2904; e-mail edicionesera@laneta.apc.org; internet www.edicionesera.com.mx; f. 1960; general and social science, art and literature; Gen. Man. NIEVES ESPRESATE XIRAU.

**Editorial Everest Mexicana, SA:** Calzada Ermita Iztapalapa 1631, Col. Barrio San Miguel del Iztapalapa, 09360 México, DF; tel. (55) 5685-1966; fax (55) 5685-3433; f. 1980; general textbooks; Gen. Man. JOSÉ LUIS HUIDOBRO LEÓN.

**Espasa Calpe Mexicana, SA:** Pitágoras 1139, Col. del Valle, 03100 México, DF; tel. (55) 5575-5022; f. 1948; literature, music, economics, philosophy, encyclopaedia; Man. FRANCISCO CRUZ RUBIO.

**Fernández Editores, SA de CV:** Eje 1 Pte México-Coyoacán 321, Col. Xoco, 03330 México, DF; tel. (55) 5605-6557; fax (55) 5688-9173; f. 1943; children's literature, textbooks, educational toys, didactic material; Man. Dir LUIS GERARDO FERNÁNDEZ PÉREZ.

**Editorial Fondo de Cultura Económica, SA de CV:** Carretera Picacho-Ajusco 227, Col. Bosques del Pedregal, 14200 México, DF; tel. (55) 5227-4672; fax (55) 5227-4640; e-mail editorial@fce.com.mx; f. 1934; economics, history, philosophy, children's books, science, politics, psychology, sociology, literature; Dir Lic. MIGUEL DE LA MADRID.

**Editorial Grijalbo, SA de CV:** Calzada San Bartolo-Naucalpan 282, Col. Argentina, Apdo 17-568, 11230 México, DF; tel. (55) 5358-4355; fax (55) 5576-3586; e-mail diredit@grijalbo.com.mx; internet www.randomhousemondadori.com.mx; f. 1954; owned by Mondadori (Italy); general fiction, history, sciences, philosophy, children's books; Man. Dir AGUSTÍN CENTENO RÍOS.

**Nueva Editorial Interamericana, SA de CV:** Cedro 512, Col. Atlampa, Apdo 4-140, 06450 México, DF; tel. (55) 5541-6789; fax (55) 5541-1603; f. 1944; medical publishing; Man. Dir RAFAEL SÁINZ.

**Distribuidora Intermex, SA de CV:** Lucio Blanco 435, Azcapotzalco, 02400 México, DF; tel. (55) 5230-9500; fax (55) 5230-9516; e-mail pmuhechi@televisa.com.mx; f. 1969; romantic fiction; Gen. Dir Lic. ALEJANDRO PAILLÉS.

**McGraw-Hill Interamericana de México, SA de CV:** Cedro 512, Col. Atlampa Cuauhtémoc, 06450 México, DF; tel. (55) 5576-7304; fax (55) 5628-5367; e-mail mcgraw-hill@infosel.net.mx; internet www.mcgraw-hill.com.mx; education, business, science; Man. Dir CARLOS RIOS.

**Editorial Joaquín Mortiz, SA de CV:** Insurgentes Sur 1162, 3°, Col. del Valle, 03100 México, DF; tel. (55) 5575-8585; fax (55) 5559-3483; f. 1962; general literature; Man. Dir Ing. HOMERO GAYOSO ANIMAS.

**Editorial Jus, SA de CV:** Plaza de Abasolo 14, Col. Guerrero, 06300 México, DF; tel. (55) 526-0616; fax (55) 5529-0951; f. 1938; history of Mexico, law, philosophy, economy, religion; Man. TOMÁS G. REYNOSO.

**Ediciones Larousse, SA de CV:** Dinamarca 81, Col. Juárez, 06600 México, DF; tel. (55) 5208-2005; fax (55) 5208-6225; f. 1965; Man. Dir DOMINIQUE BERTÍN GARCÍA.

**Editora Latino Americana, SA:** Guatemala 10-220, México, DF; popular literature; Dir JORGE H. YÉPEZ.

**Editorial Limusa, SA de CV:** Balderas 95, 1°, Col. Centro, 06040 México, DF; tel. (55) 5521-2105; fax (55) 5512-2903; e-mail limusa@noriega.com.mx; internet www.noriega.com.mx; f. 1962; science, general, textbooks; Pres. CARLOS NORIEGA MILERA.

**Editorial Nuestro Tiempo, SA:** Avda Universidad 771, Despachos 103–104, Col. del Valle, 03100 México, DF; tel. (55) 5688-8768; fax (55) 5688-6868; f. 1966; social sciences; Man. Dir ESPERANZA NACIF BARQUET.

**Editorial Oasis, SA:** Avda Oaxaca 28, 06700 México, DF; tel. (55) 5528-8293; f. 1954; literature, pedagogy, poetry; Man. MARÍA TERESA ESTRADA DE FERNÁNDEZ DEL BUSTO.

# MEXICO

**Editorial Orión:** Sierra Mojada 325, 11000 México, DF; tel. (55) 5520-0224; f. 1942; archaeology, philosophy, psychology, literature, fiction; Man. Dir SILVA HERNÁNDEZ BALTAZAR.

**Editorial Patria, SA de CV:** Renacimiento 180, Col. San Juan Tlihuaca, Azcapotzalco, 02400 México, DF; tel. (55) 5561-6042; fax (55) 5561-5231; e-mail info@patriacultural.com.mx; f. 1933; fiction, general trade, children's books; Pres. CARLOS FRIGOLET LERMA.

**Editorial Planeta Mexicana, SA de CV:** Clavijero 70, Col. Esperanza, México, DF; tel. (55) 5533-1250; internet www.editorialplaneta.com.mx; general literature, non-fiction; part of Grupo Planeta (Spain); Man. Dir JOAQUIN DIEZ-CANEDO.

**Editorial Porrúa Hnos, SA:** Argentina 15, 5°, 06020 México, DF; tel. (55) 5702-4574; fax (55) 5702-6529; e-mail servicios@porrua.com; internet www.porrua.com; f. 1944; general literature; Dir JOSÉ ANTONIO PÉREZ PORRÚA.

**Editorial Posada, SA de CV:** Eugenia 13, Despacho 501, Col. Nápoles, 03510 México, DF; tel. (55) 5682-0660; f. 1968; general; Dir-Gen. CARLOS VIGIL ZUBIETA.

**Editorial Quetzacoatl, SA:** Medicina 37, Local 1 y 2, México, DF; tel. (55) 5548-6180; Man. Dir ALBERTO RODRÍGUEZ VALDÉS.

**Medios Publicitarios Mexicanos, SA de CV:** Eugenia 811, Eje 5 Sur, Col. del Valle, 03100 México, DF; tel. (55) 5523-3342; fax (55) 5523-3379; e-mail editorial@mpm.com.mx; internet www.mpm.com.mx; f. 1958; advertising media rates and data; Man. FERNANDO VILLAMIL AVILA.

**Reverté Ediciones, SA de CV:** Río Pánuco 141A, 06500 México, DF; tel. (55) 5533-5658; fax (55) 5514-6799; e-mail 101545.2361@compuserve.com; f. 1955; science, technical, architecture; Man. RAMÓN REVERTÉ MASCÓ.

**Salvat Mexicana de Ediciones, SA de CV:** Presidente Masarik 101, 5°, 11570 México, DF; tel. (55) 5250-6041; fax (55) 5250-6861; medicine, encyclopaedic works; Dir GUILLERMO HERNÁNDEZ PÉREZ.

**Siglo XXI Editores, SA de CV:** Avda Cerro del Agua 248, Col. Romero de Terreros, Del. Coyoacán, 04310 México, DF; tel. (55) 5658-7999; fax (55) 5658-7599; e-mail informes@sigloxxieditores.com.mx; internet www.sigloxxieditores.com.mx; f. 1966; art, economics, education, history, social sciences, literature, philology and linguistics, philosophy and political science; Dir-Gen. Lic. JAIME LABASTIDA OCHOA; Gen. Man. Ing. JOSÉ MARÍA CASTRO MUSSOT.

**Editorial Trillas, SA:** Avda Río Churubusco 385 Pte, Col. Xoco, Apdo 10534, 03330 México, DF; tel. (55) 5688-4233; fax (55) 5601-1858; e-mail trillas@ovinet.com.mx; internet www.trillas.mx; f. 1954; science, technical, textbooks, children's books; Man. Dir FRANCISCO TRILLAS MERCADER.

**Universidad Nacional Autónoma de México:** Dirección General de Fomento Editorial, Avda del Iman 5, Ciudad Universitaria, 04510 México, DF; tel. (55) 5622-6581; fax (55) 5665-2778; f. 1935; publications in all fields; Dir-Gen. ARTURO VELÁZQUEZ JIMÉNEZ.

## ESTADO DE MÉXICO

**Pearson Educación de México, SA de CV:** Calle 4, 25, Fraccionamiento Industrial Alce Blanco, 53370 Naucalpan de Juárez, Méx.; tel. (55) 5387-0700; fax (55) 5358-6445; internet www.pearson.com.mx; f. 1984; educational books under the imprints Addison-Wesley, Prentice Hall, Allyn and Bacon, Longman and Scott Foresman; Pres. STEVE MARBAN.

## ASSOCIATIONS

**Cámara Nacional de la Industria Editorial Mexicana:** Holanda 13, Col. San Diego Churubusco, 04120 México, DF; tel. (55) 5688-2011; fax (55) 5604-3147; e-mail cepromex@caniem.com; internet www.caniem.com; f. 1964; Pres. JOSÉ ÁNGEL QUINTANILLA; Gen. Man. GUILLERMO COCHRAN.

**Instituto Mexicano del Libro, AC:** México, DF; tel. (55) 5535-2061; Pres. KLAUS THIELE; Sec.-Gen. ISABEL RUIZ GONZÁLEZ.

**Organización Editorial Mexicana, SA:** Guillermo Prieto 7, 06470 México, DF; tel. (55) 5566-1511; fax (55) 5566-0694; e-mail info@oem.com.mx; internet www.oem.com.mx; Pres. Lic. MARIO VÁZQUEZ RAÑA.

**Prensa Nacional Asociada, SA (PRENASA):** Insurgentes Centro 114-411, 06030 México, DF; tel. (55) 5546-7389.

# Broadcasting and Communications

## TELECOMMUNICATIONS

### Regulatory Authorities

**Comisión Federal de Telecomunicaciones (Cofetel):** Bosque de Radiatas 44, 4°, Col. Bosques de las Lomas, 05120 México, DF; tel. (55) 5261-4000; fax (55) 5261-4000; e-mail información@cft.gob.mx; internet www.cofetel.gob.mx; Pres. JORGE ARREDONDO MARTÍNEZ.

**Dirección General de Politica de Telecomunicaciones:** Eje Central Lazaro Cardeñas 567, Torre Telecomunicaciones, 15° Ala Sur, Col. Narvarte, Del. Benito Juárez, 03028 México, DF; tel. (55) 5519-1993; fax (55) 5530-1816; e-mail buzon_dgpt@sct.gob.mx; Dir Ing. LEONEL LÓPEZ CELAYA.

**Dirección General de Telecomunicaciones:** Lázaro Cárdenas 567, 11°, Ala Norte, Col. Narvarte, 03020 México, DF; tel. (55) 5519-9161; Dir-Gen. Ing. ENRIQUE LUENGAS H.

### Principal Operators

**Alestra:** Paseo de las Palmas 405, Col. Lomas de Chapultepec, 11000 México, DF; internet www.alestra.com.mx; 49% owned by AT&T; Dir-Gen. ROLANDO ZUBIRÁN SHETLER.

**America Movil, SA de CV (Telcel):** Lago Alberto, 366 Col. Anáhuac, 11320 México, DF; internet www.telcel.com; CEO DANIEL HAJJ.

**Avantel:** Liverpool 88, Col. Juárez, 06600 México, DF; e-mail webmaster@avantel.com.mx; internet www.avantel.net.mx; f. 1994; Dir-Gen. OSCAR RODRÍGUEZ MARTÍNEZ.

**Carso Global Telecom, SA de CV:** Insurgentientes Sur 3500, Col. Peña Pobre, 14060 México DF; tel. (55) 5726-3686; fax (55) 5238-0601; internet www.cgtelecom.com.mx; Chair. CARLOS SLIM DOMIT.

**Grupo Iusacell, SA de CV:** Avda Prolongación Paseo de la Reforma 1236, Col. Santa Fe, 05438 México, DF; e-mail webmaster@iusacell.com.mx; internet www.iusacell.com.mx; f. 1992; operates mobile cellular telephone network; 74% owned by Móvil Access; Dir-Gen. GUSTAVO GUZMÁN SEPÚLVEDA.

**Telecomunicaciones de México (TELECOMM):** Torre Central de Telecomunicaciones, Eje Central Lázaro Cárdenas 567, 11°, Ala Norte, Col. Narvarte, 03020 México, DF; tel. (55) 5629-1166; fax (55) 5559-9812; internet www.sct.gob.mx; govt-owned; Dir-Gen. ANDRÉS FIGUEROA COBIÁN.

**Telefónica México:** Prolongación Paseo de la Reforma 1200, Lote B-2, Col. Santa Fe, Cruz Manca, Casilla 05348, México, DF; tel. (55) 1616-5000; internet www.telefonicamoviles.com.mx; operates mobile telephone service Telefónica Móviles (MoviStar), telecommunications co Telefónica Data; controls Pegaso, Cedetel, Norcel, Movitel and Baja Celular; owned by Telefónica, SA (Spain).

**Teléfonos de México, SA de CV (Telmex):** Parque Via 198, Of. 701, Col. Cuauhtémoc, 06599 México, DF; tel. (55) 5222-5462; fax (55) 5545-5500; internet www.telmex.com.mx; Dir-Gen. CARLOS SLIM DOMIT.

## BROADCASTING

### Regulatory Authorities

**Cámara Nacional de la Industria de Radio y Televisión (CIRT):** Horacio 1013, Col. Polanco Reforma, Del. Miguel Hidalgo, 11550 México, DF; tel. (55) 5726-9909; fax (55) 5545-6767; e-mail cirt@cirt.com.mx; internet www.cirt.com.mx; f. 1942; Pres. ALEJANDRO GARCÍA GAMBOA; Dir CESAR HERNÁNDEZ ESPEJO.

**Dirección General de Radio, Televisión y Cine (RTC):** Secretaría de Gobernación, Bucareli 99, Col. Juaréz, Del. Cuahtémoc, 06600 México, DF; internet www.rtc.gob.mx; tel. (55) 5566-0262.

**Dirección de Normas de Radiodifusión:** Eugenia 197, 1°, Col. Narvarte, 03020 México, DF; tel. (55) 5590-4372; e-mail amilpg@sct.gob.mx; internet www.sct.gob.mx; licence-issuing authority; Dir Dr ALFONSO AMILPAS.

**Instituto Mexicano de Televisión:** Anillo Periférico Sur 4121, Col. Fuentes del Pedregal, 14141 México, DF; tel. (55) 5568-5684; Dir-Gen. Lic. JOSÉ ANTONIO ALVAREZ LIMA.

### Radio

There were around 1,337 radio stations in Mexico. Among the most important commercial networks are:

**ARTSA:** Avda de Los Virreyes 1030, Col. Lomas de Chapultepec, 11000 México, DF; tel. (55) 5202-3344; fax (55) 5202-6940; Dir-Gen. Lic. GUSTAVO ECHEVARRÍA ARCE.

**Corporación Mexicana de Radiodifusión:** Tetitla 23, esq. Calle Coapa, Col. Toriello Guerra, 14050 México, DF; tel. (55) 5424-6380;

# MEXICO

fax (55) 5666-5422; e-mail comentarios@cmr.com.mx; internet www.cmr.com.mx; Pres. ENRIQUE BERNAL SERVÍN; Dir-Gen. OSCAR BELTRÁN.

**Firme, SA:** Gauss 10, Col. Nueva Anzures, 11590 México, DF; tel. (55) 55250-7788; fax (55) 5250-7788; Dir-Gen. LUIS IGNACIO SANTIBÁÑEZ.

**Grupo Acir, SA:** Monte Pirineos 770, Col. Lomas de Chapultepec, 11000 México, DF; tel. (55) 5540-4291; fax (55) 5540-4106; f. 1965; comprises 140 stations; Pres. FRANCISCO IBARRA LÓPEZ.

**Grupo Radio Centro, SA de CV:** Constituyentes 1154, Col. Lomas Atlas, Del. Miguel Hidalgo, 11950 México, DF; tel. (55) 5728-4947; fax (55) 5259-2915; f. 1965; comprises 100 radio stations; Pres. ADRIÁN AGUIRRE GÓMEZ; Dir-Gen. Ing. GILBERTO SOLIS SILVA.

**Grupo Siete Comunicácion:** Montecito 38, 31°, Of. 33, México, DF; tel. (55) 5488-0887; e-mail jch@gruposiete.com.mx; internet www.gruposiete.com.mx; f. 1997; Pres. Lic. FRANCISCO JAVIER SÁNCHEZ CAMPUZANO.

**Instituto Mexicano de la Radio (IMER):** Mayorazgo 83, 2°, Col. Xoco, 03330 México, DF; tel. (55) 5628-1730; f. 1983; Dir-Gen. CARLOS LARA SUMANO.

**MVS Radio Stereorey y FM Globo:** Mariano Escobedo 532, Col. Anzures, 11590 México, DF; tel. (55) 5203-4574; fax (55) 5255-1425; e-mail vargas@data.net.mx; f. 1968; Pres. Lic. JOAQUÍN VARGAS G; Vice-Pres. Lic. ADRIÁN VARGAS G.

**Núcleo Radio Mil:** Prolongación Paseo de la Reforma 115, Col. Paseo de las Lomas, Santa Fe, 01330 México, DF; tel. (55) 5258-1200; e-mail radiomil@rnm.com.mx; internet www.nrm.com.mx; f. 1942; comprises seven radio stations; Pres. and Dir-Gen. Lic. E. GUILLERMO SALAS PEYRÓ.

**Organización Radio Centro:** Artículo 123, No 90, Col. Centro, 06050 México, DF; tel. (55) 5709-2220; fax (55) 512-8588; nine stations in Mexico City; Pres. MARÍA ESTHER GÓMEZ DE AGUIRRE.

**Organización Radiofónica de México, SA:** Tuxpan 39, 8°, Col. Roma Sur, 06760 México, DF; tel. (55) 5264-2025; fax (55) 5264-5720; Pres. JAIME FERNÁNDEZ ARMENDÁRIZ.

**Radio Cadena Nacional, SA (RCN):** Lago Victoria 78, Col. Granada, 11520 México, DF; tel. (55) 2624-0401; e-mail loregonzalez@rcn.com.mx; internet www.rcn.com.mx; f. 1948; Pres. RAFAEL C. NAVARRO ARRONTE; Dir-Gen. SERGIO FAJARDO ORTIZ.

**Radio Comerciales, SA de CV:** Avda México y López Mateos, 44680 Guadalajara, Jal.; tel. (33) 615-0852; fax (33) 630-3487; 7 major commercial stations.

**Radio Educación:** Angel Urraza 622, Col. del Valle, 03100 México, DF; tel. (55) 1500-1015; fax (55) 1500-1053; e-mail direccion@radioeducacion.edu.mx; internet www.radioeducacion.edu.mx; f. 1968; Dir-Gen. LIDIA CAMACHO.

**Radio Fórmula, SA:** Privada de Horacio 10, Col. Polanco, 11560 México, DF; tel. (55) 282-1016; Dir Lic. ROGERIO AZCARRAGA.

**Radiodifusoras Asociadas, SA de CV (RASA):** Durango 331, 2°, Col. Roma, 06700 México, DF; tel. (55) 5553-6620; fax (55) 5286-2774; f. 1956; Pres. JOSÉ LARIS ITURBIDE; Dir-Gen. JOSÉ LARIS RODRÍGUEZ.

**Radiodifusores Asociados de Innovación y Organización, SA:** Emerson 408, Col. Chapultepec Morales, 11570 México, DF; tel. (55) 5203-5577; fax (55) 5545-2078; Dir-Gen. Lic. CARLOS QUIÑONES ARMENDÁRIZ.

**Radiópolis, SA de CV:** owned by Grupo Televisa and Grupo Prisa; owns 5 radio stations; affiliated to Radiorama, SA de CV (q.v.) in 2004; Dir-Gen. RAÚL RODRÍGUEZ GONZÁLEZ.

**Radiorama, SA de CV:** Reforma 56, 5°, 06600 México, DF; tel. (55) 5566-1515; fax (55) 5566-1454; Dir JOSÉ LUIS C. RESÉNDIZ.

**Representaciones Comerciales Integrales:** Avda Chapultepec 431, Col. Juárez, 06600 México, DF; tel. (55) 5533-6185; Dir-Gen. ALFONSO PALMA V.

**Sistema Radio Juventud:** Pablo Casals 567, Prados Providencia, 44670 Guadalajara, Jal.; tel. (33) 641-6677; fax (33) 641-3413; f. 1975; network of several stations including Estereo Soul 89.9 FM; Dirs ALBERTO LEAL A., J. JESÚS OROZCO G., GABRIEL ARREGUI V.

**Sistema Radiofónico Nacional, SA:** Baja California 163, Of. 602, 06760 México, DF; tel. (55) 5574-0298; f. 1971; represents comercial radio networks; Dir-Gen. RENÉ C. DE LA ROSA.

**Sociedad Mexicana de Radio, SA de CV (SOMER):** Gutenberg 89, Col. Anzures, 11590 México, DF; tel. (55) 5255-5297; fax (55) 5545-0310; Dir-Gen. EDILBERTO HUESCA PERROTIN.

Radio Insurgente, the underground radio station of the Ejército Zapatista de Liberación Nacional (EZLN—Zapatistas), is broadcast from south-eastern Mexico. Programmes can be found on www.radioinsurgente.org.

## Television

There are around 468 television stations. Among the most important are:

**Asesoramiento y Servicios Técnicos Industriales, SA (ASTISA):** México, DF; tel. (55) 5585-3333; commercial; Dir ROBERTO CHÁVEZ TINAJERO.

**MVS** (Multivisión): Blvd Puerto Aéreo 486, Col. Moctezuma, 15500 México, DF; tel. (55) 5764-8100; internet www.mvs.com; subscriber-funded.

**Once TV:** Carpio 475, Col. Casco de Santo Tomás, 11340 México, DF; tel. (55) 5356-1111; fax (55) 5396-8001; e-mail canal11@vmredipn.ipn.mx; f. 1959; Dir-Gen. ALEJANDRA LAJOUS VARGAS.

**Tele Cadena Mexicana, SA:** Avda Chapultepec 18, 06724 México, DF; tel. (55) 5535-1679; commercial, comprises about 80 stations; Dir Lic. JORGE ARMANDO PIÑA MEDINA.

**Televisa, SA de CV:** Edif. Televicentro, Avda Chapultepec 28, Col. Doctores, 06724 México, DF; tel. (55) 5709-3333; fax (55) 5709-3021; e-mail webmaster@televisa.com.mx; internet www.televisa.com; f. 1973; commercial; began broadcasts to Europe via satellite in Dec. 1988 through its subsidiary, Galavisión; 406 affiliated stations; Chair. and CEO EMILIO AZCÁRRAGA JEAN; Vice-Pres. ALEJANDRO BURILLO AZCÁRRAGA.

**Televisión Azteca, SA de CV:** Anillo Periférico Sur 4121, Col. Fuentes del Pedregal, 14141 México, DF; tel. (55) 5420-1313; fax (55) 5645-4258; e-mail webtva@tvazteca.com; internet tvazteca.todito.com; f. 1992; assumed responsibility for former state-owned channels 7 and 13; Pres. RICARDO B. SALINAS PLIEGO; CEO PEDRO PADILLA LONGORIA.

**Televisión de la República Mexicana:** Mina 24, Col. Guerrero, México, DF; tel. (55) 5510-8590; cultural; Dir EDUARDO LIZALDE.

As a member of the Intelsat international consortium, Mexico has received communications via satellite since the 1960s. The launch of the Morelos I and Morelos II satellites, in 1985, provided Mexico with its own satellite communications system. The Morelos satellites were superseded by a new satellite network, Solidaridad, which was inaugurated in early 1994. In late 1997 Mexico's three satellites (grouped in a newly formed company, SatMex) were transferred to private ownership.

# Finance

(cap. = capital; res = reserves; dep. = deposits; m. = million; amounts in new pesos unless otherwise stated)

### BANKING

The Mexican banking system is comprised of the Banco de México (the central bank of issue), multiple or commercial banking institutions and development banking institutions. Banking activity is regulated by the Federal Government.

Commercial banking institutions are constituted as *Sociedades Anónimas*, with wholly private social capital. Development banking institutions exist as *Sociedades Nacionales de Crédito*, participation in their capital is exclusive to the Federal Government, notwithstanding the possibility of accepting limited amounts of private capital. In 2005 there were 34 commercial and development banks operating in Mexico and 71 foreign banks maintained offices.

All private banks were nationalized in September 1982. By July 1992, however, the banking system had been completely returned to the private sector. Legislation removing all restrictions on foreign ownership of banks received congressional approval in 1999.

#### Supervisory Authority

**Comisión Nacional Bancaria y de Valores (CNBV)** (National Banking and Securities Commission): Avda Insurgentes Sur 1971, Torre Norte, Sur y III, Col. Guadalupe Inn, Del. Alvaro Obregón, 01020 México, DF; tel. and fax (55) 5724-6000; e-mail info@cnbv.gob.mx; internet www.cnbv.gob.mx; f. 1924; govt commission controlling all credit institutions in Mexico; Pres. JONATHAN DAVIS ARZAC.

#### Central Bank

**Banco de México (BANXICO):** Avda 5 de Mayo 1, Col. Centro, Del. Cuauhtémoc, 06059 México, DF; tel. (55) 5237-2000; fax (55) 5237-2070; e-mail comsoc@banxico.org.mx; internet www.banxico.org.mx; f. 1925; currency issuing authority; became autonomous on 1 April 1994; cap. 4,869.0m., res 15,000.0m., dep. 526,808.0m. (Dec. 2003); Gov. Dr GUILLERMO ORTIZ MARTÍNEZ; 6 brs.

#### Commercial Banks

**Banca Serfín, SA:** Mod 409, 4°, Prolongación Paseo de la Reforma 500, Col. Lomas de Santa Fe, 01219 México, DF; tel. (55) 5259-8860;

fax (55) 5257-8387; internet www.serfin.com.mx; f. 1864; merged with Banco Continental Ganadero in 1985; transferred to private ownership in Jan. 1992; acquired by Banco Santander Central Hispano (Spain) in Dec. 2000; cap. 6,874.9m., res 2,982.7m., dep. 96,919.7m. (Dec. 2003); Chair. and Pres. CARLOS GÓMEZ; CEO and Gen. Man. ADOLFO LAGOS ESPINOSA; 554 brs.

**Banco del Bajío, SA:** Avda Manuel J. Clouthier 508, Col. Jardines del Campestre, 37128 León, Gto; tel. (477) 710-4600; fax (477) 710-4693; e-mail internacional@bancobajio.com.mx; internet www.bancobajio.com.mx; f. 1994; cap. 1,285.1m., res −17.0m., dep. 10,995.4m. (Dec. 2003); Pres. SALVADOR OÑATE; Gen. Man. CARLOS DE LA CERDA SERRANO.

**Banco Mercantil del Norte, SA (BANORTE):** Avda Morones Prieto 2312 Pte, 2°, Col Lomas de San Francisco, 64710 Monterrey, NL; tel. (81) 3319-7200; fax (81) 3319-5216; internet www.banorte.com; f. 1899; merged with Banco Regional del Norte in 1985; cap. 5,351.8m., res 1,332.8m., dep. 179,725.6m. (Dec. 2002); Chair. ROBERTO GONZÁLEZ BARRERA; CEO LUIS PEÑA KEGEL; 457 brs.

**Banco Nacional de México, SA (Banamex):** Roberto Medellín 800, Col. Santa Fe, 01210 México, DF; tel. (55) 5720-7091; fax (55) 5920-7323; internet www.banamex.com; f. 1884; transferred to private ownership in 1991; merged with Citibank México, SA in 2001; cap. 25,551.0m., res 22,922.0m., dep. 305,063.0m. (Dec. 2003); CEO MANUEL MEDINA MORA; 1,260 brs.

**BBVA Bancomer, SA:** Centro Bancomer, Avda Universidad 1200, Col. Xoco, 03339 México, DF; tel. (55) 5621-3434; fax (55) 5621-3230; internet www.bancomer.com.mx; f. 2000 by merger of Bancomer (f. 1864) and Mexican operations of Banco Bilbao Vizcaya Argentaria (Spain); privatized in 2002; cap. 67,283.2m., res −22,306.5m., dep. 358,537.3m. (Dec. 2002); Chair. RICARDO GUAJARDO TOUCHÉ.

**HSBC México:** Paseo de la Reforma 156, Col. Juárez, Del. Cuauhtémoc, 06600 México, DF; tel. (55) 5721-2222; fax (55) 5721-2393; internet www.hsbc.com.mx; f. 1941; bought by HSBC (UK) in 2002; name changed from Banco Internacional, SA (BITAL) in 2004; cap. 14,962.0m., res 662.1m., dep. 141,533.3m. (Dec. 2002); Gen. Man. and CEO ALEXANDER FLOCKHART; 1,400 brs.

**Scotiabank Inverlat, SA:** Blvd Miguel Avila Camacho 1, 18°, Col. Polanco, Del. Miguel Hidalgo, 11009 México, DF; tel. (55) 5728-1000; fax (55) 5229-2157; internet www.inverlat.com.mx; f. 1977 as Multibanco Comermex, SA; changed name to Banco Inverlat, SA 1995; 55% holding acquired by Scotiabank Group (Canada) and name changed as above 2001; cap. 2,957.8m., res 1,240.1m., dep. 73,860.3m. (Dec. 2002); CEO ANATOL VON HANN; 371 brs.

### Development Banks

**Banco Nacional de Comercio Exterior, SNC (BANCOMEXT):** Camino a Santa Teresa 1679, Jardines de Pedegral, Del. Álvaro Obregón, 14210 México, DF; tel. (55) 5481-6000; fax (55) 5449-9030; internet www.bancomext.com; f. 1937; cap. 19,324.0m., res −12,152.0m., dep. 44,544.0m. (Dec. 2003); Man. Dir ENRIQUE VILATELA RIBA.

**Banco Nacional de Crédito Rural, SNC (BANRURAL):** Avda Baja California 261, Col. Hipódromo Condesa, Del. Cuauhtémoc, 06170 México, DF ; tel. (55) 5273-1300; fax (55) 5584-2664; e-mail contacto@banrural.gob.mx; internet www.banrural.gob.mx; f. 1975; provides financing for agriculture and normal banking services; in liquidation March 2004; Dir-Gen. ALFREDO GÓMEZ AGUIRRE; 187 brs.

**Banco Nacional del Ejército, Fuerza Aérea y Armada, SNC (BANJERCITO):** Avda Industria Militar 1055, Col. Lomas de Sotelo, 11200 México, DF; tel. and fax (55) 5626-6290; internet www.banjercito.com.mx; f. 1947; Dir-Gen. Gral-Bgda FERNANDO MILLÁN VILLEGAS; 51 brs.

**Banco Nacional de Obras y Servicios Públicos, SNC (BANOBRAS):** Avda Javier Barros Sierra 515, Col. Lomas de Santa Fe, México, DF; tel. (55) 5270-1200; fax (55) 5723-6108; e-mail bneumann@banobras.gob.mx; internet www.banobras.gob.mx; f. 1933; govt-owned; cap. 8,176.4m., res 1,743.1m., dep. 118,041.0m. (Dec. 2001); Chair. Dr GUILLERMO ORTIZ MARTÍNEZ.

**Nacional Financiera, SNC:** Insurgentes Sur 1971, Torre IV, 13°, Col. Guadalupe Inn, 01020 México, DF; tel. (55) 5325-6700; fax (55) 5661-8418; e-mail info@nafin.gob.mx; internet www.nafin.com; f. 1934; cap. 6,777.2m, res 253.9m., dep. 253,542.4m. (Dec. 2003); Chair. JOSÉ ANGEL GURRIA TREVINO; Pres. CARLOS SALES GUTIÉRREZ; 32 brs.

### Foreign Bank

**Dresdner Bank Mexico, SA:** Bosque de Alisos 47A, 4°, Col. Bosques de las Lomas, 05120 México, DF; tel. (55) 5258-3170; fax (55) 5258-3199; e-mail mexico@dbla.com; f. 1995; Man. Dir LUIS NIÑO DE RIVERA.

### BANKERS' ASSOCIATION

**Asociación de Banqueros de México:** 16 de Setiembre 27, Col. Centro Histórico, 06000 México, DF; tel. (55) 5722-4305; internet www.abm.org.mx; f. 1928; fmrly Asociación Mexicano de Bancos; Pres. MANUEL MEDINA MORA; Dir-Gen. Lic. ADOLFO RIVAS MARTIN DEL CAMPO; 52 mems.

### STOCK EXCHANGE

**Bolsa Mexicana de Valores, SA de CV:** Paseo de la Reforma 255, Col. Cuauhtémoc, 06500 México, DF; tel. (55) 5726-6600; fax (55) 5591-0642; e-mail cinforma@bmv.com.mx; internet www.bmv.com.mx; f. 1894; Pres. Lic. MANUEL ROBELDA GONZALES DE CASTILLA; Dir-Gen. Ing. GERARDO FLORES DEUCHLER.

### INSURANCE

#### México, DF

**ACE Seguros:** Bosques de Alisos, 47-A, 1°, Col. Bosques de las Lomas, 5120 México, DF; tel. (5) 258-5800; fax (5) 258-5899; e-mail info@acelatinamerica.com; f. 1990; fmrly Seguros Cigna.

**Aseguradora Cuauhtémoc, SA:** Manuel Avila Camacho 164, 11570 México, DF; tel. (55) 5250-9800; fax (55) 5540-3204; f. 1944; general; Exec. Pres. JUAN B. RIVEROLL; Dir-Gen. JAVIER COMPEÁN AMEZCUA.

**Aseguradora Hidalgo, SA:** Presidente Masarik 111, Col. Polanco, Del. Miguel Hidalgo, 11570 México, DF; f. 1931; life; Dir-Gen. JOSÉ GÓMEZ GORDOA; Man. Dir HUMBERTO ROQUE VILLANUEVA.

**ING Comercial América—Seguros:** Insurgentes Sur 3900, Col. Tlalpan, 14000 México, DF; tel. (55) 5169-2500; internet www.comercialamerica.com.mx; f. 1936 as La Comercial; acquired by ING Group in 2000; life, etc.; Pres. GLENN HILLIARS; Dir-Gen. Ing. ADRIÁN PÁEZ.

**La Nacional, Cía de Seguros, SA:** México, DF; f. 1901; life, etc.; Pres. CLEMENTE CABELLO; Chair. Lic. ALBERTO BAILLERES.

**Pan American de México, Cía de Seguros, SA:** México, DF; f. 1940; Pres. Lic. JESS N. DALTON; Dir-Gen. GILBERTO ESCOBERA PAZ.

**Royal & SunAlliance Mexico:** Blvd Adolfo López Mateos 2448, Col. Altavista, 01060 México, DF; tel. (55) 5723-7999; fax (55) 5723-7941; e-mail omar.antonio@mx.royalsun.com; internet www.royalsun.com.mx; f. 1941; acquired Seguros BBV-Probursa in 2001; general, except life.

**Seguros América Banamex, SA:** Avda Revolución 1508, Col. Guadalupe Inn, 01020 México, DF; f. 1933; Pres. AGUSTÍN F. LEGORRETA; Dir-Gen. JUAN OROZCO GÓMEZ PORTUGAL.

**Seguros Azteca, SA:** Insurgentes 102, México, DF; f. 1933; general including life; Pres. JUAN CAMPO RODRÍGUEZ.

**Seguros Constitución, SA:** Avda Revolución 2042, Col. La Otra Banda, 01090 México, DF; tel. (55) 5550-7910; f. 1937; life, accident; Pres. ISIDORO RODRÍGUEZ RUIZ; Dir-Gen. ALFONSO DE ORDUÑA Y PÉREZ.

**Seguros el Fénix, SA:** México, DF; f. 1937; Pres. VICTORIANO OLAZÁBAL E.; Dir-Gen. JAIME MATUTE LABRADOR.

**Seguros Internacional, SA:** Abraham González 67, México, DF; f. 1945; general; Pres. Lic. GUSTAVO ROMERO KOLBECK.

**Seguros de México, SA:** Insurgentes Sur 3496, Col. Peña Pobre, 14060 México, DF; tel. (55) 5679-3855; f. 1957; life, etc.; Dir-Gen. Lic. ANTONIO MIJARES RICCI.

**Seguros La Provincial, SA:** México, DF; f. 1936; general; Pres. CLEMENTE CABELLO; Chair. ALBERTO BAILLERES.

**Seguros La República, SA:** Paseo de la Reforma 383, México, DF; f. 1966; general; 43% owned by Commercial Union (United Kingdom); Pres. LUCIANO ARECHEDERRA QUINTANA; Gen. Man. JUAN ANTONIO DE ARRIETA MENDIZÁBAL.

#### Guadalajara, Jal.

**Nueva Galicia, Compañía de Seguros Generales, SA:** Guadalajara, Jal.; f. 1946; fire; Pres. SALVADOR VEYTIA Y VEYTIA.

#### Monterrey, NL

**Seguros Monterrey Aetna, SA:** Avda Diagonal Sta Engracia 221 Ote, Col. Lomas de San Francisco, 64710 Monterrey, NL; tel. (81) 319-1111; fax (81) 363-0428; f. 1940; casualty, life, etc.; Dir-Gen. FEDERICO REYES GARCÍA.

**Seguros Monterrey del Círculo Mercantil, SA, Sociedad General de Seguros:** Padre Mier Pte 276, Monterrey, NL; f. 1941; life; Gen. Man. CARMEN G. MASSO DE NAVARRO.

## MEXICO

### Insurance Association

**Asociación Mexicana de Instituciones de Seguros, AC:** Fco I Madero 21, Col. Tlacopac San Angel, 01140 México, DF; tel. (55) 5662-6161; e-mail aglez@amis.com.mx; internet www.amis.com.mx; f. 1946; all insurance cos operating in Mexico are mems; Pres. José Luis Llamosas Portilla; Dir-Gen. Recaredo Arias Jiménez.

# Trade and Industry

### GOVERNMENT AGENCIES

**Comisión Federal de Protección Contra Riesgos Sanitarios (COFEPRIS):** Monterrey 33, Esq. Oaxaca, Col. Roma, Del. Cuauhtémoc, 06700 Mexico, DF; tel. (55) 5280-5200; e-mail contacto_cofepris@salud.gob.mx; internet www.cofepris.gob.mx; f. 2003; pharmaceutical regulatory authority; Sec.-Gen. Alejandra Olguín Ramírez.

**Comisión Nacional Forestal (CONAFOR):** Carretera a Nogales s/n, Esq. Periférico Pte. 5°, San Juan de Ocotán, 45019 Zapopan, Jal.; tel. (33) 3777-7077; fax (33) 3770-7078; e-mail transparencia@conafor.gob.mx; internet www.conafor.gob.mx; f. 2001; Dir-Gen. Ing. Manuel Agustín Reed Segovia.

**Comisión Nacional de Precios:** Avda Juárez 101, 17°, México 1, DF; tel. (55) 5510-0436; f. 1977; national prices commission; Dir-Gen. Jesús Sánchez Jiménez.

**Comisión Nacional de los Salarios Mínimos (CNSM):** Avda Cuauhtémoc 14, 2°, Col. Doctores, 06720 México 7, DF; tel. (55) 5761-5778; fax (55) 5578-5775; f. 1962, in accordance with Section VI of Article 123 of the Constitution; national commission on minimum salaries; Pres. Lic. Basilio González-Nuñez; Tech. Dir Alida Bernal Cosío.

**Consejo Mexicano de Comercio Exterior (CONCE):** Lancaster 15, Col. Juárez, 06600 México, DF; tel. (52) 5231-7100; fax (55) 5321-7109; e-mail comce@comce.org.mx; internet www.comce.org.mx; f. 1999 to promote international trade; Chair. Federico Sada González.

**Consejo Nacional de Comercio Exterior, AC (CONACEX):** Avda Parque Fundidora 501, Of. 95E, Edif. CINTERMEX, Col. Obrera, 64010 Monterrey, NL; tel. (81) 369-0284; fax (81) 369-0293; e-mail conacex@technet.net.mx; f. 1962 to promote national exports; Chair. Ing. Javier Prieto de la Fuente; Pres. Lic. Juan Manuel Quiroga Lam.

**Instituto Nacional de Investigaciones Nucleares (ININ):** Centro Nuclear de México, Carretera México–Toluca km 36.5, 52045 Ocoyoacac, Méx.; tel. (55) 5329-7200; fax (55) 5329-7298; e-mail webmaster@nuclear.inin.mx; internet www.inin.mx; f. 1979 to plan research and devt of nuclear science and technology, as well as the peaceful uses of nuclear energy, for the social, scientific and technological devt of the country; administers the Secondary Standard Dosimetry Laboratory, the Nuclear Information and Documentation Centre, which serves Mexico's entire scientific community; operates a tissue culture laboratory for medical treatment; the 1 MW research reactor which came into operation, in 1967, supplies part of Mexico's requirements for radioactive isotopes; also operates a 12 MV Tandem van de Graaff. Mexico has two nuclear reactors, each with a generating capacity of 654 MW; the first, at Laguna Verde, became operational in 1989 and is administered by the Comisión Federal de Electricidad (CFE); Dir-Gen. José Raúl Ortíz Magaña.

**Instituto Nacional de Pesca** (National Fishery Institute): Pitágoras 1320, Col. Santa Cruz Atoyac, Del. Juárez, 03310 México, DF; tel. (55) 5688-1469; fax (55) 5604-9169; e-mail compean@inp.sagarpa .gob.mx; internet inp.sagarpa.gob.mx; f. 1962; Dir Guillermo Compean Jiménez.

**Procuraduría Federal del Consumidor (Profeco):** Dr Carmona y Valle 11, Col. Doctores, 06720 México, DF; tel. (55) 5761-3021; internet www.profeco.gob.mx; f. 1975; consumer protection; Procurator Carlos Arce Macías.

**Servicio Geológico Mexicano:** Blvd Felipe Angeles, Carretera México–Pachuca, km 93.5, Col. Venta Prieta, 42080 Pachuca, HI; tel. (771) 771-4016; fax (771) 771-3938; e-mail dirgeneral@coremisgm .gob.mx; internet www.coremisgm.gob.mx; f. 1957; govt agency for the devt of mineral resources; Dir-Gen. Ing. Francisco José Escandón Valle.

### DEVELOPMENT ORGANIZATIONS

**Centro de Investigación para el Desarollo, AC (CIDAC)** (Centre of Research for Development): Jaime Balmes 11, Edif. D-2, Col. Los Morales Polanco, 11510 México, DF; tel. (55) 5985-1010; fax (55) 5985-1030; e-mail info@cidac.org.mx; internet www.cidac.org; f. 1984; researches economic and political development.

**Comisión Nacional de las Zonas Aridas:** Blvd Venustiano Carranza 1623, Col. República, 25280 Saltillo, Coah.; e-mail uenlace@conaza.gob.mx; internet www.conaza.gob.mx; f. 1970; commission to co-ordinate the devt and use of arid areas; Dir-Gen. Lic. Eduardo Terrazas Ramos.

**Fideicomiso de Fomento Mineiro (Fifomi):** Puente de Tecamachalco 26, 2°, Col. Lomas de Chapultepec, Del. Miguel Hidalgo, 11000 México, DF; tel. (55) 5202-0968; e-mail pguerra@fifomi.gob .mx; internet www.fifomi.gob.mx; trust for the development of the mineral industries; Dir-Gen. Pedro Guerra Menéndez.

**Fideicomisos Instituídos en Relación con la Agricultura (FIRA):** Km 8, Antigua Carretera Pátzucuaro, 58341 Morelia, Mich.; tel. (443) 322-2390; fax (443) 327-7860; e-mail webmaster@correo.fira.gob.mx; internet www.fira.gob.mx; Dir Francisco Meré Palafox; a group of devt funds to aid agricultural financing, under the Banco de México, comprising:

**Fondo de Garantía y Fomento para la Agricultura, Ganadería y Avicultura (FOGAGA):** f. 1954.

**Fondo Especial para Financiamientos Agropecuarios (FEFA):** f. 1965.

**Fondo Especial de Asistencia Técnica y Garantía para Créditos Agropecuarios (FEGA):** f. 1972.

**Fondo de Garantía y Fomento para las Actividades Pesqueras (FOPESCA).**

**Fondo de Operación y Financiamiento Bancario a la Vivienda (FOVI):** Ejército Nacional 180, Col. Anzures, 11590 México, DF; tel. (55) 5263-4500; fax (55) 5263-4541; e-mail jmartinez@fovi .gob.mx; internet www.fovi.gob.mx; f. 1963 to promote the construction of low-cost housing through savings and credit schemes; devt fund under the Banco de México; Dir-Gen. Lic. Manuel Zepeda Payeras.

**Instituto Mexicano del Petróleo (IMP):** Eje Central Lázaro Cárdenas 152, Col. Bernardo Atepehuacan, Del. Gustavo A. Madero, 07730 México, DF; tel. (55) 9175-6000; fax (55) 9175-8000; e-mail sabugalp@imp.mx; internet www.imp.mx; f. 1965 to foster devt of the petroleum, chemical and petrochemical industries; Dir Gustavo A. Chapela Castañares.

### CHAMBERS OF COMMERCE

**Confederación de Cámaras Nacionales de Comercio, Servicios y Turismo (CONCANACO-SERVYTUR)** (Confederation of National Chambers of Commerce, Services and Tourism): Balderas 144, 3°, Col. Centro, 06079 México, DF; tel. (55) 5772-9300; fax (55) 5709-1152; e-mail gerardo@concanacored.com; internet www .concanacored.com; f. 1917; Pres. Raúl Alejandro Padilla Orozco; Dir-Gen. Lic. Ana María Amezcua Ríos; comprises 283 regional Chambers.

**Cámara de Comercio, Servicios y Turismo Ciudad de México (CANACO)** (Chamber of Commerce, Services and Tourism of Mexico City): Paseo de la Reforma 42, 3°, Col. Centro, Apdo 32005, 06048 México, DF; tel. (55) 5592-2677; fax (55) 5592-2279; internet www.ccmexico.com.mx; f. 1874; 50,000 mems; Pres. Manuel Tron Campos; Dir-Gen. Lic. Eduardo García Villaseñor.

**Cámara Nacional de la Industria de Transformación (CANACINTRA):** Avda San Antonio 256, Col. Ampliación Nápoles, Del. Juárez, México, DF; tel. (55) 5482-3000; internet contacto@canacintra-digital.com.mx; internet www.canacintra.org.mx; representa majority of smaller manufacturing businesses; Pres. Cuauhtémoc Martínez García; Dir-Gen. Angélica Rubí Rubí.

Chambers of Commerce exist in the chief town of each state as well as in the larger centres of commercial activity. There are also international Chambers of Commerce.

### CHAMBERS OF INDUSTRY

The 46 National Chambers, 19 Regional Chambers, 3 General Chambers and 40 Associations, many of which are located in the Federal District, are representative of the major industries of the country.

#### Central Confederation

**Confederación de Cámaras Industriales de los Estados Unidos Mexicanos (CONCAMIN)** (Confed. of Industrial Chambers): Manuel María Contreras 133, 7°, Col. Cuauhtémoc, 06500 México, DF; tel. (55) 5140-7800; fax (55) 5140-7831; e-mail webmaster@concamin.org.mx; internet www.concamin.org.mx; f. 1918; represents and promotes the activities of the entire industrial sector; Pres. Léon Halkin Bider; Dir-Gen. Gerardo Barrios Espinosa; 108 mem. orgs.

MEXICO                                                                                                                           *Directory*

## INDUSTRIAL AND TRADE ASSOCIATIONS

**Asociación Nacional de Importadores y Exportadores de la República Mexicana (ANIERM)** (National Association of Importers and Exporters): Monterrey 130, Col. Roma, Del. Cuauhtémoc, 06700 México, DF; tel. (55) 5584-9522; fax (55) 5584-5317; e-mail anierm@anierm.org.mx; internet www.anierm.org.mx; f. 1944; Pres. RODRIGO GUERRA B.; Vice-Pres. Lic. HUMBERTO SIMONEEN ARDILA.

**Comisión de Fomento Minero (COFOMI):** Puente de Tecamachalco 26, Lomas de Chapultepec, 11000 México, DF; tel. (55) 5540-3400; fax (55) 5202-0342; f. 1934 to promote the devt of the mining sector; Dir Lic. LUIS DE PABLO SERNA.

**Comisión Nacional de Inversiones Extranjeras (CNIE):** Blvd Avila Camacho 1, 11°, 11000 México, DF; tel. (55) 5540-1426; fax (55) 5286-1551; f. 1973; commission to co-ordinate foreign investment; Exec. Sec. Dr CARLOS CAMACHO GAOS.

**Comisión Nacional de Seguridad Nuclear y Salvaguardias (CNSNS):** Dr Barragán 779, Col. Narvarte, Del. Juárez, 03020 México, DF; tel. (55) 5095-3200; fax (55) 5095-3295; e-mail swaller@cnsns.gob.mx; f. 1979; nuclear regulatory agency; Dir-Gen. JUAN EIBENSCHUTZ HARTMAN.

**Comisión Petroquímica Mexicana:** México, DF; to promote the devt of the petrochemical industry; Tech. Sec. Ing. JUAN ANTONIO BARGÉS MESTRES.

**Consejo Empresarial Mexicano para Asuntos Internacionales (CEMAI):** Homero 527, 7°, Col. Polanco, 11570 México, DF; tel. (55) 5250-7033; fax (55) 5531-1590.

**Consejo Mexicano del Café (CMCAFE):** José María Ibarrarán 84, 1°, Col. San José Insurgentes, Del. Juárez, 03900 México, DF; tel. and fax (55) 5611-9075; e-mail cmc@sagar.gob.mx; internet www.cmcafe.org.mx; f. 1993; devt of coffee sector.

**Consejo Nacional de la Industria Maquiladora de Exportación (CNIME):** Ejército Nacional 418, 12°, Of. 1204, Col. Chapultepec Morales, Del. Miguel Hidalgo, 11570 México, DF; tel. and fax (55) 5250-6093; internet www.cnime.org.mx; f. 1975; Pres. ENRIQUE CASTRO SEPTIEN; Sec. ENRIQUE HURTADO.

**Instituto Nacional de Investigaciones Forestales y Agropecuarios (INIFAP)** (National Forestry and Agricultural Research Institute): Serapio Rendón 83, Col. San Rafael, Del. Cuauhtémoc, 06470 México, DF; tel. (55) 5140-1674; fax (55) 5566-3799; internet www.inifap.gob.mx; f. 1985; conducts research into plant genetics, management of species and conservation; Dir-Gen. PEDRO BRAJCICH GALLEGOS.

## EMPLOYERS' ORGANIZATIONS

**Consejo Coordinador Empresarial (CCE):** Lancaster 15, Col. Juárez, 06600 México, DF; tel. (55) 5229-1100; fax (55) 5592-3857; e-mail sistemas@cce.org.mx; internet www.cce.org.mx; f. 1974; co-ordinating body of private sector; Pres. JOSÉ LUIS BARRAZA; Dir FRANCISCO CALDERÓN.

**Consejo Mexicano de Hombres de Negocios (CMHN):** México, DF; f. 1963; represents leading businesspeople; affiliated to CCE; Pres. ANTONIO DEL VALLE RUIZ.

## STATE HYDROCARBONS COMPANY

**Petróleos Mexicanos (PEMEX):** Avda Marina Nacional 329, Col. Huasteca, 11311 México, DF; tel. (55) 1944-2500; fax (55) 5531-6354; internet www.pemex.com; f. 1938; govt agency for the exploitation of Mexico's petroleum and natural gas resources; Dir-Gen. LUIS RAMÍREZ CORZO; 106,900 employees.

## MAJOR COMPANIES

### Mining and Metals

**Altos Hornos de México, SA:** Prolongación Juárez s/n, Edif. GAN Modulo II, Col. La Loma, Monclova, Coah.; tel. (866) 649-3400; fax (866) 633-2390; e-mail ventas@ahmsa.com; internet www.ahmsa.com.mx; f. 1942; fmr state-owned iron and steel foundry and rolling mill; privatized in the early 1990s; subsidiary of Grupo Imsa; Dir-Gen. LUIS ZAMUDIO MIECHIELSEN; 17,000 employees (incl. subsidiaries).

**Empresas Frisco, SA de CV:** Jaime Balmes 11, 5°, Col. Los Morales Polanco, 11510 México, DF; tel. (55) 5626-7799; fax (55) 5557-1591; internet www.gcarso.com.mx; f. 1973; mining, railway and chemical co; part of Grupo Carso, SA de CV; Man. Dir JESÚS GUTIÉRREZ BASTIDA; 2,807 employees (2002).

**Grupo Imsa:** Avda Batallón de San Patricio 111, 66269 San Pedro García, NL; tel. (81) 8153-8300; fax (81) 8153-8400; internet www.grupoimsa.com; f. 1936; steel, aluminium; Chair. EUGENIO CLARIOND GARZA; CEO and Dir EUGENIO CLARIOND REYES RETANA; 16,373 employees (2001).

**Grupo Industrial Saltillo, SA de CV:** Chiapas 375, Col. República, 25280 Saltillo, Coah.; tel. (844) 111-000; fax (844) 158-096; internet www.gis.com.mx; f. 1966; ceramics, iron, aluminium producers; Chair. ISIDRO LÓPEZ DEL BOSQUE; 9,881 employees in 2003.

**Grupo México, SA de CV:** Avda Baja California 200, Col. Roma Sur, Del Cuauhtémoc, 06760 México, DF; tel. (55) 5080-0050; fax (55) 5574-7677; internet www.gmexico.com; f. 1901; began operations in Mexico as Asarco (USA); holding co with interests in extraction and processing of metallic ores, and transportation; Chair. and CEO GERMAN LARREA MOTA VELASCO; 23,769 employees in 2002.

**Grupo Simec, SA de CV:** Avda Lázaro Cárdenas 601, Col. La Nogalera, 44440 Guadalajara, Jal.; tel. (33) 1057-5757; fax (33) 1057-5726; e-mail mktcsg@sidek.com.mx; steel producers; Man. Dir LUIS GARCÍA LIMÓN; 1,386 employees in 2001.

**Hylsamex:** Avda Munich 101, 66452 San Nicolás de los Garza, NL; tel. (81) 8865-1224; e-mail webmaster@hylsamex.com.mx; internet www.hylsamex.com.mx; 42% owned by Techint (Argentina); steel; Dir-Gen. ALEJANDRO M. ELIZONDO BARRAGÁN.

**Industrias Peñoles, SA de CV:** Moliere 222, Col. Polanco, 11540 México, DF; tel. (55) 5279-3000; fax (55) 5279-3514; internet www.penoles.com.mx; f. 1969; mining co; Chair. ALBERTO BAILLÈRES; 9,081 employees.

**Nacional de Cobre, SA de CV (Nacobre):** Poniente 134, No 719, Col. Industria Vallejo, 02300 México, DF; tel. (55) 5728-5300; fax (55) 5728-5369; internet www.nacobre.com.mx; f. 1951; copper, brass, aluminium, plastics producers; part of Grupo Carso, SA de CV; Gen. Man. ALEJANDRO OCHOA ABARCA; 715 employees.

**Tubacero, SA de CV:** Avda Guerrero 3729 Norte, Col. del Norte, 64500 Monterrey, NL; tel. (81) 8305-5555; fax (81) 8305-5550; e-mail sistemas@tubacero.com; internet www.tubacero.com; f. 1943; manufacturers of piping; Dir-Gen. LEÓN GUTIÉRREZ VELA; 299 employees.

**Tubos de Acero de México, SA (TAMSA):** Km 433.7 Carretera México–Veracruz, Vía Xalapa, 91697 Veracruz, Ver.; tel. (229) 989-1100; fax (229) 989-1120; e-mail elenah@tamsa.com.mx; internet www.tamsa.com.mx; f. 1952; manufacturers of seamless steel tubes and fittings; Chair. PAOLO ROCCA; 2,500 employees.

### Motor Vehicles

**BMW de México, SA:** Plaza Arquímedes 130, 10°, Col. Polanco, 11560 México, DF; tel. (55) 5282-8700; fax (55) 5282-8731; e-mail lineadeatencion@bmw.com.mx; internet www.bmw.com.mx; f. 1994; subsidiary of Bayerische Motoren Werke AG of Germany; motor-vehicles and motor-parts manufacturers; Man. Dir FRANZ BAUMGARTNER.

**Daimler Chrysler de México, SA:** Paseo de la Reforma 1240, Col. Santa Fé, 05709 México, DF; tel. (55) 5729-1442; fax (55) 5729-1461; internet www.daimlerchrysler.com.mx; f. 1972; subsidiary of the Chrysler Corpn, USA; automobile assembly; Man. Dir MILES BRIJANT, III; 11,070 employees.

**Ford Motor Company, SA de CV:** Paseo de la Reforma 333, Col. Cuauhtémoc, 06500 México, DF; tel. (55) 5326-0000; fax (55) 5525-3840; internet www.ford.com.mx; f. 1925 in Mexico; subsidiary of Ford Motor Co, USA; manufacturers of motor-vehicle, truck and tractor parts; Chair. KATHLEEN LEGOCKI; 7,765 employees.

**General Motors de México, SA de CV:** Avda Ejército Nacional 843, Col. Granada, 11520 México, DF; tel. (55) 5329-0800; fax (55) 5625-3335; internet www.gm.com.mx; f. 1931 in Mexico; subsidiary of General Motors Corpn of the USA; automobile assembly; Pres. and Dir-Gen. TROY CLARKE; 10,445 employees.

**Mercedes-Benz México, SA de CV:** División Automovil, Edif. Corporativo, Km 23.7, Carretera Santiago Tianguistengo, Santiago Tianguistenco, 52600 México, DF; tel. (713) 279-2400; fax (713) 279-2493; internet www.mercedes-benz.com.mx; f. 1968; subsidiary of DaimlerChrysler de México; Man. Dir JOSÉ DIERE; 900 employees.

**Nissan Mexicana, SA de CV:** Avda Insurgentes Sur 1958, Col. Florida, 01030 México, DF; tel. (55) 5661-6120; fax (55) 5628-2690; internet www.nissan.com.mx; f. 1961 in Mexico; subsidiary of Nissan Motors Co Ltd, Japan; automobile assembly plant; Dir-Gen. HIROSHI YOSHIOKA; 2,987 employees.

**Transmisiones y Equipos Mecánicos, SA de CV:** Avda 5 de Febrero 2115, Zona Industrial Benito Juárez, 76120 Querétaro, Qro; tel. (442) 211-7300; e-mail azarete.tremec@spicer.com; internet www.tremec.com.mx; f. 1964; makers of motor-vehicle components; Pres. BERNARDO QUINTANA ISAAC; 1,800 employees.

**Volkswagen de Mexico, SA de CV:** Autopista México–Puebla km 116, San Lorenzo Almecatla, 72008 Cuautlancingo, Pue.; tel. (222) 308-111; fax (222) 308-468; e-mail contacto@vw.com.mx; internet www.vw.com.mx; f. 1964; subsidiary of Volkswagen AG of Germany; manufacture of motor vehicles; Chair. BERND LEISSNER; 16,000 employees.

# MEXICO

## Food and Drink, etc.

**Coca-Cola Femsa, SA de CV:** Guillermo González Camarena 600, Centro de Cuidad Santa Fe, 01210 México, DF; tel. (55) 5081-5100; fax (55) 5292-3474; internet www.cocacola-femsa.com.mx; f. 1991; subsidiary of Coca-Cola Export Co, USA, and FEMSA, SA de CV; soft-drink manufacturer; CEO Carlos Salazar Lomelin; 14,457 employees in 2002.

**Fomento Económico Mexicano, SA de CV (FEMSA):** General Anaya 601 Pte., Col. Bella Vista, 64410 Monterrey, NL; tel. (81) 8326-6000; fax (81) 8328-6080; e-mail comunicacion@femsa.com; internet www.femsa.com; f. 1991; beer and soft-drink producers; Pres. and CEO José Antonio Fernández Carbajal; 41,656 employees in 2002.

**Gruma, SA:** Calzada del Valle 407 Oeste, Col. del Valle, 66220 San Pedro Garza García, NL; tel. (81) 8399-3300; fax (81) 8399-3359; e-mail acruz@gruma.com; internet www.gruma.com; f. 1949; tortilla and corn-flour products manufacturers and distributors; Pres. Roberto González Barrera; 15,585 employees in 2001.

**Grupo Bimbo, SA:** Paseo de la Reforma 1000, Col. Peña Blanca, Santa Fe, Del. Alvaro Obregon, 01210 México, DF; tel. (55) 5268-6585; fax (55) 5258-6847; e-mail prensa@grupobimbo.com; internet www.gibsa.com; f. 1945; bread, confectionery and canned food manufacturers; Dir-Gen. Daniel Servitje Montull; 71,000 employees in 2003.

**Grupo Modelo, SA de CV:** Campo Eliseos 400, 18°, Col. Lomas de Chapultepec, 01100 México, DF; tel. (55) 9138-9990; fax (55) 5280-6718; e-mail invelrations@gmodelo.com.mx; internet www.gmodelo.com.mx; f. 1925; beer producers; Chair. Antonio Fernández Rodríguez; 48,445 employees in 2001.

**Grupo Sanborns:** Calvario 106, Col. Tlalpan, 14000 México, DF; tel. (55) 5325-9900; fax (55) 5325-9941; internet www.sanborns.com.mx; operates a chain of restaurants and manufactures confectionery; part of Grupo Carso, SA de CV; Dir.-Gen. Carlos Slim Domit; 35,659 employees in 2001.

**Molinos Azteca y Juper, SA de CV:** Calle 7 1057, Zona Industrial Guadalajara, 44940, Jal.; tel. (33) 3645-6308; fax (33) 3645-2393; internet www.molinosazteca.com; f. 1950; agro-industrial subsidiary of Gruma, SA; Gen. Man. Gerardo Gómez; 250 employees.

**Savia, SA de CV:** Río Sena 500, Col. del Valle Oriente, 66220 San Pedro Garza García, NL; tel. (81) 8173-5500; fax (81) 8173-5508; internet www.savia.com.mx; food and real estate; Chair. and CEO Alfonso Romo Garza; 8,000 employees.

## Electrical Goods

**Ericsson Telecom, SA de CV:** Punta Santa Fé, Torre A, 5–11°, Prolongación Paseo de la Reforma 1015, Col. Santa Fé, Del. Alvaro Obregón, 01376 México, DF; tel. (55) 1103-0000; fax (55) 5726-2333; internet www.ericsson.com.mx; f. 1904; subsidiary of Telefonaktiebolaget L. M. Ericsson of Sweden; makers of telecommunications equipment; Man. Dir Gerhard Skladal; 3,565 employees.

**Grupo Condumex, SA de CV:** Miguel de Cervantes Saavedra 255, Col. Amplicaión Granada México, 11520 México, DF; tel. (55) 5328-5868; fax (55) 5255-1026; internet www.condumex.com.mx; f. 1952; electrical equipment; part of Grupo Carso, SA de CV; Pres. Julio Gutiérrez Trujillo; 18,000 employees.

**IBM de México, SA de CV:** Alfonso Napoles Gandara 3111, Corporativo Santa Fé, 01210 México, DF; tel. (55) 5267-2754; fax (55) 5267-2754; internet www.ibm.com.mx; f. 1927; manufacturers of computers and office equipment; Pres. and Dir-Gen. José Decurnex; 2,100 employees.

## Cement and Construction

**Cemento Cruz Azul:** Torres Adalid 517, Col. del Valle, 03100 México, DF; tel. (55) 5687-2030; fax (55) 5682-6773; e-mail webmaster@cruzazul.com.mx; internet www.cruzazul.com.mx; f. 1881; co-operative manufacturers of cement; Dir-Gen. Guillermo Alvarez Cuevas.

**Cemex, SA de CV:** Avda Ricardo Margáin Zozaya 325, 66265 San Pedro Garza García, NL; tel. (81) 8888-8888; fax (81) 8888-4417; internet www.cemex.com; f. 1906; manufacturers and distributors of cement, concrete and building materials; Pres. and CEO Lorenzo H. Zambrano; 26,679 employees (2004).

**Corporación GEO, SA de CV:** Margaritas 433, Col. Guadalupe Chimalistac, 01050 México, DF; tel. (55) 5480-5000; fax (55) 5554-6064; internet www.casasgeo.com; construction and real estate; Pres. Luis Orvañanos Lascurain; 10,379 employees in 2001.

**Empresas ICA Sociedad Controladora, SA:** Edif. D, 4°, Minería 145, 11800 México, DF; tel. (55) 5272-9991; fax (55) 5271-1607; e-mail comunicacion@ica.com.mx; internet www.ica.com.mx; f. 1947; holding co with interests in the construction industry; Pres. Bernardo Quintana Isaac; 11,482 employees in 2001.

**Grupo Empresarial Maya, SA de CV:** Avda Constitución 444 Pte, 64900 Monterrey, NL; tel. (81) 8345-2000; fax (81) 8345-2025; f. 1987; cement manufacturers; Pres. Marcelo Zambrano; 1,990 employees.

**Grupo Sidek-Situr, SA de CV:** Circ. Agustín Yañez 2343, 3°, Col. Moderna, 44100 Guadalajara, Jal.; tel. (33) 678-5985; fax (33) 678-5920; e-mail sidek@sidek.com.mx; internet www.sidek.com.mx; f. 1989; hotel and resort construction; Chair. Luis Rebollar Corona; 4,717 employees.

**Holcim Apasco:** Campos Eliseos 345, Col. Polanco Chapultepec, Del. Miguel Hidalgo, 11560 México, DF; tel. (55) 5724-0000; fax (55) 5724-0288; internet www.holcim.com; f. 1963; manufacture and distribution of construction materials; Dir Alejandro Carrillo; 2,500 employees.

**Tolmex, SA de CV:** Avda Constitución 444 Pte, 64900 Monterrey, NL; tel. (81) 345-2000; fax (81) 345-2425; f. 1989; cement and ready-made concrete manufacturers; Dir-Gen. Lorenzo Zambrano; 4,400 employees.

## Pharmaceutical

**Armstrong Laboratorios, SA de CV:** Avda Periférico Sur 6677, Col. Ejidos de Tepepán, Xochimilco, México, DF; tel. (55) 5629-9950; fax (55) 5641-5400; internet www.armstronglaboratorios.com.mx; f. 1972 in Mexico; part of Bagó Laboratorios.

**Grupo Casa Saba, SA de CV:** Paseo de la Reforma 215, Lomas de Chapultepec, Del. Miguel Hidalgo, 11000 México, DF; tel. (55) 5284-6600; internet www.casasaba.com; f. 1944; name changed from Grupo Casa Autrey in 2000; pharmaceutical co; Pres. Isaac Saba Raffoul; Dir-Gen. Manuel Saba Ades; 5,700 employees.

**Grupo Roche Syntex de México, SA de CV:** Cerrada de Bezares 9, Col. Lomas de Bezares, 11910 México, DF; tel. (55) 5258-5000; fax (55) 5258-5472; internet www.roche.com.

**Laboratorios Liomont:** Adolfo López Mateos 68, Cuajimalpa, 05000 México, DF; tel. (55) 5814-1200; fax (55) 5812-1074; e-mail direccioncomercial@liomont.com.mx; internet www.liomont.com; f. 1938; 1,050 employees.

**Merck Sharp & Dohme—MSD, SA de CV:** Avda San Jerónimo 369, 8°, Col. Tizapán San Angel, 01090 México, DF; tel. (55) 5481-9708; internet www.msd.com.mx; f. 1932; prescription medications and vaccinations.

**Sanofi-Synthélabo:** Col. Parque Industrial Cuamatla, Cuautitlán Izcalli, 54730 México, DF; tel. (55) 5062-3300; fax (55) 5872-0433; internet www.sanofi-synthelabo.com.mx.

**Schering-Plough:** Avda 16 de Septiembre 301, Xaltocan, Xochimilco, 16090 México, DF; tel. (55) 5728-4444.

**Sicor de México, SA de CV:** Vicente Guerrero 3, 2°, Col. El Mirador, Tlalnepantla, México, DF; tel. (55) 5239-9039; internet www.sicor.com.mx; part of Sicor group.

## Retail

**Controladora Comercial Mexicana, SA de CV (CCM):** Adolfo López Mateos 201, Col. Santa Cruz Acatalán, 53140 Naucalpan de Juárez, Méx; tel. (55) 5270-9000; fax (55) 5371-7302; internet www.comerci.com.mx; f. 1944; retail traders; CEO Carlos González Nova; 31,995 employees in 2001.

**El Puerto de Liverpool, SA de CV:** Torcuato Tasso 241, 2°, Col. Chapultepec Morales, 11570 México, DF; tel. (55) 5262-9999; fax (55) 5254-5688; internet www.liverpool.com.mx; f. 1944; retail traders; Chair. Max Michel; Pres. Enrique Bremond; 23,000 employees in 2001.

**Far-Ben, SA de CV** (Farmacias Benavides): Avda Fundadores 935, Col. Valle del Mirador, 64750 Monterrey, NL; tel. (81) 359-6215; fax (81) 359-5150; e-mail adriang@benavides.com.mx; internet www.benavides.com.mx; f. 1917; retail chemists (pharmacies); Chair. Jaime M. Benavides Pompa; Dir-Gen. Walter Westphal Urrieta; 8,300 employees.

**Grupo Corvi:** Pico de Tolima 29, Jardines en la Montaña, 14210 México, DF; tel. (55) 5628-5100; fax (55) 5645-1581; e-mail bvillasenor@infosel.net.mx; internet www.grupocorvi.com.mx; retail distribution and confectionary; Pres. Benjamín Villaseñor Costa.

**Grupo Elektra, SA de CV:** Avda Insurgentes Sur 3579, Col. Tlalpán La Joya, 14000 México, DF; tel. (55) 1720-7000; fax (55) 5629-9234; internet www.grupoelektra.com.mx; f. 1950; retail and consumer finance; Chair. Ricardo B. Salinas Pliego; Dir-Gen. Javier Sarro Cortina; 24,328 employees.

**Grupo Gigante, SA de CV:** Ejército Nacional 769, Del. Miguel Hidalgo, 11520 México, DF; tel. (55) 5269-8000; fax (55) 5269-8308; internet www.gigante.com.mx; f. 1962; retail traders; Chair. Angel Losada Moreno; 36,849 employees in 2001.

**Organización Soriana, SA de CV:** Alejandro de Rodas 3102-A, Col. Las Cumbres, 8 Sector, 64610 Monterrey, NL; tel. (81) 8329-9000; fax (81) 8329-9128; e-mail pmejia@soriana.com.mx; internet www.soriana.com.mx; f. 1968; holding co with interests in the grocery trade; Pres. Francisco J. Martín Bringas; 35,659 employees in 2001.

**Sears Roebuck de México, SA de CV:** San Luis Potosí 214, Col. Romao, 06700 México, DF; tel. (55) 5247-7500; fax (55) 5584-6848; internet www.sears.com.mx; f. 1947; subsidiary of Sears Roebuck and Co of the USA; department stores; Pres. Thurmon A. Williams Stewart; 10,000 employees.

**Wal-Mart de México, SA de CV:** Blvd Manuel Avila Camacho 641, Col. Periodistas, 11220 México, DF; tel. (55) 5557-1133; internet www.walmartmexico.com.mx; subsidiary of Walmart Inc of the USA; retail traders; Chair. Cesareo Fernández; 112,294 employees.

### Miscellaneous

**Alfa, SA de CV:** Avda Gómez Morín 1111 sur, Col. Carrizalejo, 66254 San Pedro Garza García, NL; tel. (81) 8748-1111; fax (81) 8748-2552; internet www.alfa.com.mx; holding co with interests in steel, petrochemicals, food products and telecommunications; Chair. and CEO Dionisio Garza Medina; 42,069 employees in 2004.

**Berol, SA de CV:** Vía Dr Gustavo Baz 309, Col. La Loma, 54060 Tlalnepantia, Edo de México; tel. (55) 5729-3400; fax (55) 5729-3433; internet www.berol.com.mx; f. 1970; stationery manufacturers; Pres. Carlos Moreno Rivas; 600 employees.

**Celanese Mexicana, SA de CV (CelMex):** Tecoyotitla 412, Col. Guadalupe Chimalistac, 01050 México, DF; tel. (55) 5480-9100; fax (55) 5480-9324; e-mail aorduna3@celanese.com.mx; internet www.celanese.com.mx; f. 1944; holding co with interests in chemicals and packaging manufacturing; CEO Francisco Puente; 2,464 employees.

**Compañía Industrial de San Cristobal, SA de CV (CISCSA):** Manuel María Contreras 133, 3°, Col. San Rafael, 06470 México, DF; tel. (55) 5516-3000; fax (55) 5326-2262; f. 1951; holding co for paper products; Chair. A. Patrón Luján; Gen. Man. C. J. Booth; 2,700 employees.

**CYDSA, SA y Subsidiarias:** Avda Ricardo Margáin Zozaya 565-B, Parque Corporativo Santa Engracia, 66267 San Pedro Garza García, NL; tel. (81) 8152-4500; fax (81) 8152-4813; internet www.cydsa.com; f. 1945; manufacturers of synthetic fibres, textiles, chemicals and packaging materialso; Pres. Ing. Tomás González Sada; Dir-Gen. José de Jesús Montemayor Castillo; 10,838 employees.

**Desc, SA de CV:** Paseo de los Tamarindos 400B, 28°, Bosque de las Lomas, 05120 México, DF; tel. (55) 5261-8000; fax (55) 5261-8096; e-mail desc@mail.desc.com.mx; internet www.desc.com.mx; f. 1973; holding co with interests in auto parts, food, chemicals and real estate; Pres. and CEO Fernando Senderos Mestre; 19,079 employees in 2001.

**Empaques Ponderosa, SA de CV:** Río Sena Oeste 500, Col. del Valle, 66220 San Pedro Garza García, NL; tel. and fax (81) 8173-5510; fax (81) 8173-5508; e-mail amurrieta@empaq.com.mx; f. 1989; cardboard manufacturers; Chair. José Manuel Martínez González; 858 employees.

**Grupo Carso, SA de CV:** Insurgentes 3500, POB 03, Col. Pena Pobre, 14060 Mexico, DF; tel. (55) 5202-8838; fax (55) 5238-0601; internet www.gcarso.com.mx; f. 1980; holding co with interests in retail, food, mining, electricals, tobacco; Pres. Carlos Slim Domit; 30,840 employees.

**Industrias John Deere de México, SA de CV:** Blvd Díaz Ordáz 500, San Pedro Garza García, NL; tel. (81) 8288-1212; internet www.deere.com/es_MX/ag/; f. 1955; subsidiary of Deere and Co of the USA; farming machinery and equipment makers; Pres. Agustín Santamarina Vázquez; 1,215 employees.

**Internacional de Cerámica, SA de CV:** Avda Carlos Pacheco 7200, 31080 Chihuahua, Chih.; fax (614) 429-1166; internet www.interceramic.com; f. 1978; makers of floor tiles; Chair. Oscar E. Almeida Chabre; Dir-Gen. Victor D. Almeida García; 2,900 employees.

**Kimberly-Clark de México, SA de CV:** José Luis Lagrange 103, 3°, Col. Los Morales, 11510 Mexico, DF; tel. (55) 5282-7300; fax (55) 5282-7282; e-mail kcm.informacion@kcc.com; internet www.kimberly-clark.com.mx; f. 1955; subsidiary of Kimberly Clark Corpn of the USA; paper manufacturers; Pres. and Dir-Gen. Claudio Xavier González Laporte; 7,700 employees.

**Nadro, SA de CV:** Vasco de Quiroga 3100, Col. Centro de Ciudad Santa Fe, 01210 México, DF; tel. (55) 5292-4343; internet www.nadro.com.mx; f. 1943 as Nacional de Drogas; distribution of pharmaceuticals and beauty products; Chair. and CEO Pablo Escandón Cusi.

**Vitro, SA de CV:** Avda Ricardo Margáin Zozaya 440, Col. Valle del Campestre, 66265 San Pedro Garza García, NL; tel. (81) 8863-1300; fax (81) 8863-7839; internet www.vitro.com; f. 1909; manufacturers of glass, glass bottles and containers; Pres. and CEO Federico Sada G.; 33,378 employees in 2001.

## UTILITIES

### Regulatory Authorities

**Comisión Nacional del Agua (CNA):** Avda Insurgentes Sur 2146, Col. Copilco el Bajo, Del. Coyoacán, 04340 México, DF; tel. (55) 5550-7607; fax (55) 5550-6721; e-mail direccion@cna.gob.mx; internet www.cna.gob.mx; commission to administer national water resources; Dir-Gen. Cristobal Jaime Jaquez.

**Comisión Reguladora de Energía (CRE):** Horacio 1750, Col. Polanco, Del. Miguel Hidalgo, 11510 México, DF; tel. (55) 5283-1500; internet www.cre.gob.mx; f. 1994; commission to control energy policy and planning; Pres. Dionisio Pérez-Jácome; Exec. Sec. Francisco J. Valdes López.

**Secretariat of State for Energy:** see section on The Government (Secretariats of State).

### Electricity

**Comisión Federal de Electricidad (CFE):** 2 Sec. del Bosque de Chapultepec, Museo Tecnológico, Del. Miguel Hidalgo, México, DF; tel. (55) 5229-4400; fax (55) 5553-5321; e-mail alfredo.elias@cfe.gob.mx; internet www.cfe.gob.mx; state-owned power utility; Dir-Gen. Alfredo Elías Ayub.

**Luz y Fuerza del Centro:** Melchor Ocampo 171, Col. Tlaxpana, Del. Miguel Hidalgo, 11379 México, DF; tel. (55) 5140-0040; fax (55) 5140-0300; e-mail ldpablo@inter01.lfc.gob.mx; internet www.lfc.gob.mx; operates electricity network in the centre of the country; Dir-Gen. Luis de Pablo Serna.

### Gas

**Gas Natural México (GNM):** Monterrey, NL; e-mail sugerencias@gnm.com.mx; internet www.gasnaturalmexico.com.mx; f. 1994 in Mexico; distributes natural gas in the states of Tamaulipas, Aguascalientes, Coahuila, San Luis Potosí, Guanajuato, Nuevo León and México and the in Distrito Federal; subsidiary of Gas Natural (Spain).

**Petróleos Mexicanos (PEMEX):** see State Hydrocarbons Company, above; distributes natural gas.

## TRADE UNIONS

**Congreso del Trabajo (CT):** Avda Ricardo Flores Magón 44, Col. Guerrero, 06300 México 37, DF; tel. (55) 5583-3817; f. 1966; trade union congress comprising trade union federations, confederations, etc.; Pres. Lic. Héctor Valdés Romo.

**Confederación Regional Obrera Mexicana (CROM)** (Regional Confederation of Mexican Workers): República de Cuba 60, México, DF; f. 1918; Sec.-Gen. Ignacio Cuauhtémoc Paleta; 120,000 mems, 900 affiliated syndicates.

**Confederación Revolucionaria de Obreros y Campesinos de México (CROC)** (Revolutionary Confederation of Workers and Farmers): Hamburgo 250, Col. Juárez, Del. Cuauhtémoc, 06600 México, DF; tel. (55) 5208-5449; e-mail crocmodel@hotmail.com; internet www.croc.org.mx; f. 1952; Sec.-Gen. Isias González Cuevas; 4.5m. mems in 32 state federations and 17 national unions.

**Confederación Revolucionaria de Trabajadores (CRT)** (Revolutionary Confederation of Workers): Dr Jiménez 218, Col. Doctores, México, DF; f. 1954; Sec.-Gen. Mario Suárez García; 10,000 mems; 10 federations and 192 syndicates.

**Confederación de Trabajadores de México (CTM)** (Confederation of Mexican Workers): Insurgente Norte 59, Edif. 2, 2°, Col. Buena Vista, México, DF; tel. (55) 5703-3137; fax (55) 5705-0966; f. 1936; admitted to ICFTU; Sec.-Gen. Joaquín Gamboa; 5.5m. mems.

**Federación Obrera de Organizaciones Femeniles (FOOF)** (Workers' Federation of Women's Organizations): Vallarta 8, México, DF; f. 1950; women workers' union within CTM; Sec.-Gen. Hilda Anderson Nevárez; 400,000 mems.

**Federación Nacional de Sindicatos Independientes** (National Federation of Independent Trade Unions): Isaac Garza 311 Ote, 64000 Monterrey, NL; tel. (8) 375-6677; internet www.fnsi.org.mx; f. 1936; Sec.-Gen. Jacinto Padilla Valdez; 230,000 mems.

**Federación de Sindicatos de Trabajadores al Servicio del Estado (FSTSE)** (Federation of Unions of Government Workers): Gómez Farías 40, Col. San Rafael, 06470 México, DF; f. 1938; Sec.-Gen. Lic. Joel Ayala; 2.5m. mems; 80 unions.

**Frente Unida Sindical por la Defensa de los Trabajadores y la Constitución** (United Union Front in Defence of the Workers and

# MEXICO

the Constitution): f. 1990 by more than 120 trade orgs to support the implementation of workers' constitutional rights.

**Unión General de Obreros y Campesinos de México, Jacinto López (UGOCM-JL)** (General Union of Workers and Farmers of Mexico, Jacinto López): José María Marroquí 8, 2°, 06050 México, DF; tel. (55) 5518-3015; f. 1949; admitted to WFTU/CSTAL; Sec.-Gen. JOSÉ LUIS GONZÁLEZ AGUILERA; 7,500 mems, over 2,500 syndicates.

**Unión Nacional de Trabajadores (UNT)** (National Union of Workers): Villalongen 50, Col. Cuauhtémoc, México, DF; tel. (55) 5140-1425; fax (55) 5703-2583; e-mail secretariageneral@strm.org.mx; internet www.unt.org.mx; f. 1998; Sec.-Gen. FRANCISCO HERNÁNDEZ JUÁREZ.

A number of major unions are non-affiliated; they include:

**Federación Democrática de Sindicatos de Servidores Públicos** (Democratic Federation of Public Servants): México, DF; f. 2005.

**Frente Auténtico de los Trabajadores (FAT).**

**Pacto de Unidad Sindical Solidaridad (PAUSS):** comprises 10 independent trade unions.

**Sindicato Nacional de Trabajadores Mineros, Metalúrgicos y Similares de la República Mexicana** (Industrial Union of Mine, Metallurgical and Related Workers of the Republic of Mexico): Avda Dr Vertiz 668, Col. Narvarte, 03020 México, DF; tel. (55) 5519-5690; f. 1933; Sec.-Gen. NAPOLEON GÓMEZ URRUTIA; 86,000 mems.

**Sindicato Nacional de Trabajadores de la Educación (SNTE):** Venezuela 44, Col. Centro, México, DF; tel. (55) 5702-0005; fax (55) 5702-6303; f. 1943; teachers' union; Pres. ELBA ESTHER GORDILLO MORALES; Sec.-Gen. TOMÁS VÁZQUEZ VIGIL; 1.3m. mems.

   **Coordinadora Nacional de Trabajadores de la Educación (CNTE):** dissident faction; Leader TEODORO PALOMINO.

**Sindicato de Trabajadores Petroleros de la República Mexicana (STPRM)** (Union of Petroleum Workers of the Republic of Mexico): Zaragoza 15, Col. Guerrero, 06300 México, DF; tel. (55) 5546-0912; close links with PEMEX; Sec.-Gen. CARLOS ROMERO DESCHAMPS; 110,000 mems; includes:

   **Movimiento Nacional Petrolero:** reformist faction; Leader HEBRAÍCAZ VÁSQUEZ.

**Sindicato de Trabajadores Ferrocarrileros de la República Mexicana (STFRM)** (Union of Railroad Workers of the Republic of Mexico): Avda Ricardo Flores Magón 206, Col. Guerrero, México 3, DF; tel. (55) 5597-1011; f. 1933; Sec.-Gen. VÍCTOR F. FLORES MORALES; 100,000 mems.

**Sindicato Unico de Trabajadores Electricistas de la República Mexicana (SUTERM)** (Sole Union of Electricity Workers of the Republic of Mexico): Río Guadalquivir 106, Col. Cuauhtémoc, 06500 México, DF; tel. (55) 5207-0578; Sec.-Gen. LEONARDO RODRÍGUEZ ALCAINE.

**Sindicato Unico de Trabajadores de la Industria Nuclear (SUTIN):** Viaducto Río Becerra 139, Col. Nápoles, 03810 México, DF; tel. (55) 5523-8048; fax (55) 5687-6353; e-mail sutin@nuclear.inin.mx; internet www.prodigyweb.net.mx/sutin; Sec.-Gen. RICARDO FLORES BELLO.

**Unión Obrera Independiente (UOI):** non-aligned.

The major agricultural unions are:

**Central Campesina Independiente (CCI):** México, DF; e-mail cencci@prodigy.net.mx; Dir RAFAEL GALINDO JAIME.

**Confederación Nacional Campesina (CNC):** Mariano Azuela 121, Col. Santa María de la Ribera, México, DF; Sec.-Gen. Lic. BEATRIZ PAREDES RANGEL.

**Confederación Nacional Ganadera:** Calzada Mariano Escobedo 714, Col. Anzures, México, DF; tel. (55) 5203-3506; Pres. Ing. CÉSAR GONZÁLEZ QUIROGA; 300,000 mems.

**Consejo Agrarista Mexicano:** 09760 Iztapalapa, México, DF; Sec.-Gen. HUMBERTO SERRANO.

**Unión Nacional de Trabajadores Agriculturas (UNTA).**

## Transport

Road transport accounts for about 98% of all public passenger traffic and for about 80% of freight traffic. Mexico's terrain is difficult for overland travel. As a result, there has been an expansion of air transport and there were 67 international and national airports in 2004. In 2002 plans to build a new airport in the capital were postponed after conflict over the proposed site. International flights are provided by a large number of national and foreign airlines.

Mexico has 140 seaports, 29 river docks and a further 29 lake shelters. More than 85% of Mexico's foreign trade is conducted through maritime transport. In the 1980s the Government developed the main industrial ports of Tampico, Coatzacoalcos, Lázaro Cárdenas, Altamira, Laguna de Ostión and Salina Cruz in an attempt to redirect growth and to facilitate exports. The port at Dos Bocas, on the Gulf of Mexico, was one of the largest in Latin America when it opened in 1999. A 300-km railway link across the isthmus of Tehuantepec connects the Caribbean port of Coatzacoalcos with the Pacific port of Salina Cruz.

In 1992, as part of an ambitious divestment programme, the Government announced that concessions would be offered for sale to the private sector, in 1993, to operate nine ports and 61 of the country's airports. The national ports authority was to be disbanded, responsibility for each port being transferred to Administraciones Portuarias Integrales (APIs). In 1998 plans were announced for public share offerings in 35 airports. From 1997 the national railway system underwent privatization granted under 50-year concessions.

**Secretariat of State for Communications and Transport:** see section on The Government (Secretariats of State).

**Aeropuertos y Servicios Auxiliares (ASA):** Avda 602 No 161, Col. San Juan de Aragón, Del. V. Carranza, 15620 México, DF; tel. (55) 5786-9526; fax (55) 5786-9709; internet www.asa.gob.mx; Dir-Gen. ERNESTO VALESCO LEÓN.

**Caminos y Puentes Federales (CAPUFE):** e-mail contacto@capufe.gob.mx; internet www.capufe.gob.mx; Dir-Gen. MANUEL ZUBIRIA MAQUEO.

### STATE RAILWAYS

In 2003 there were 26,655 km of main line track. In 2003 the railway system carried an estimated 270,000 passengers and 54,813m. freight ton-km. Ferrocarriles Nacionales de México (FNM), government-owned since 1937, was liquidated in 2001 following a process of restructuring and privatization. In July 2003 plans were announced for the construction of suburban train system for the Valle de México.

**Ferrocarril del Noreste:** Avda Manuel L. Barragán 4850, Col. Hidalgo, 64281 Monterrey, NL; tel. (81) 8305-7931; fax (81) 8305-7766; e-mail tfm@tfm.com.mx; internet www.tfm.com.mx; concession awarded to Transportación Ferroviaria Mexicana (TFM) in 1997; 4,251 km of line, linking Mexico City with the ports of Lázaro Cárdenas, Veracruz, Tampico, Altamira and Matamoros and the US border at Nuevo Laredo; Dir-Gen. M. MOHAR.

**Ferrocarril Pacifico-Norte:** Avda Baja California 200, Col. Roma Sur, 06760 México, DF; internet www.ferromex.com.mx; 50-year concession awarded to Grupo Ferroviario Mexicano, SA, (GFM) commencing in 1998; owned by Grupo México, SA de CV; operates through wholly-owned subsidiary Ferrocarril Mexicano, SA de CV (FERROMEX); 7,500 km of track and Mexico's largest rail fleet; Dir of Operations Ing. LORENZO REYES RETANA.

**Ferrocarril del Sureste (Ferrosur):** Jaime Balmes 11, 4°, Col. Los Morales Polanco, 11510 México, DF; tel. (55) 5387-6500; fax (55) 5387-6533; 50-year concession awarded to Grupo Tribasa in 1998; 66.7% sold to Empresas Frisco, SA de CV, in 1999, owned by Grupos Carso, SA de CV; Dir GUILLERMO MUÑOZ LARA.

**Servicio de Transportes Eléctricos del Distrito Federal (STE):** Avda Municipio Libre 402, Col. San Andrés Tetepilco, México, DF; tel. (55) 5539-6500; fax (55) 5672-4758; e-mail infoste@df.gob.mx; internet www.ste.df.gob.mx; suburban tram route with 17 stops upgraded to light rail standard to act as a feeder to the metro; also operates bus and trolleybus networks; Dir-Gen. Dra FLORENCIA SERRANIA SOTO.

**Sistema de Transporte Colectivo (Metro):** Delicias 67, 06070 México, DF; tel. (55) 5709-1133; fax (55) 5512-3601; internet www.metro.df.gob.mx; f. 1967; the first stage of a combined underground and surface railway system in Mexico City was opened in 1969; 10 lines, covering 158 km, were operating, in 1998, and five new lines, bringing the total distance to 315 km, are to be completed by 2010; the system is wholly state-owned and the fares are partially subsidized; Dir-Gen. Dr JAVIER GONZÁLEZ GARZA.

### ROADS

In 2003 there were an estimated 348,529 km of roads, of which 32.6% were paved. The construction of some 4,000 km of new four-lane toll highways, through the granting of govt concessions to the private sector, was undertaken during 1989–93. In mid-1997 the Government announced that it would repurchase almost one-half of the road concessions granted in an attempt to stimulate road construction.

Long-distance buses form one of the principal methods of transport in Mexico, and there are some 600 lines operating services throughout the country.

MEXICO

**Dirección General de Autotransporte Federal:** Calzada de las Bombas 411, 11°, Col. San Bartolo Coapa, 04800 México, DF; tel. (55) 5684-0757; co-ordinates long distance bus services.

### SHIPPING

At the end of 2004 Mexico's registered merchant fleet numbered 687 vessels, with a total displacement of 1,007,998 grt. The Government operates the facilities of seaports. In 1989–94 US $700m. was spent on port development, much of it from the private sector. In 1994–95 management of several ports were transferred to the private sector.

**Coordinación General de Puertos y Marina Mercante (CGPMM):** Avda Nuevo León 210, Col. Hipódromo, 06100 México, DF03310 México, DF; tel. (55) 5723-9300; e-mail cgpmmweb@sct.gob.mx; Dir César Patricio Reyes Roel.

**Port of Acapulco:** Puertos Mexicanos, Malecón Fiscal s/n, Acapulco, Gro.; tel. (744) 22067; fax (744) 31648; Harbour Master Capt. René F. Novales Betanzos.

**Port of Coatzacoalcos:** Administración Portuaria Integral de Coatzacoalcos, SA de CV, Interior recinto portuario s/n Coatzacoalcos, 96400 Ver.; tel. (921) 214-6744; fax (921) 214-6758; e-mail apicoa@apicoatza.com; internet www.apicoatza.com; Dir-Gen. Ing. Gilberto António Ríos Ruíz.

**Port of Dos Bocas:** Administración Portuaria Integral de Dos Bocas, SA de CV, Carretera Federal Puerto Ceiba–Paraíso 414, Col. Quintín Arzuz Paraíso, 86600 Tabasco, Tab.; tel. (933) 353-2744; e-mail dosbocas@apidosbocas.com; internet www.apidosbocas.com.

**Port of Manzanillo:** Administración Portuaria Integral de Manzanillo, SA de CV, Avda Tte Azueta 9, Col. Burócrata, 28250 Manzanillo, Col.; tel. (314) 331-1400; fax (314) 332-1005; e-mail comercializacion@apimanzanillo.com.mx; internet www.apimanzanillo.com.mx; Dir-Gen. Capt. Héctor Mora Gómez.

**Port of Tampico:** Administración Portuaria Integral de Tampico, SA de CV, Edif. API de Tampico, 1°, Recinto Portuario, 89000 Tampico, Tamps.; tel. (833) 212-4660; fax (833) 212-5744; e-mail apitam@puertodetampico.com.mx; internet www.puertodetampico.com.mx; Gen. Dir Ing. Rafael Meseguer Lima.

**Port of Veracruz:** Administración Porturia Integral de Veracruz, SA de CV, Marina Mercante 210, 7°, 91700 Veracruz, Ver.; tel. (229) 32-1319; fax (229) 32-3040; e-mail portverc@infosel.net.mx; internet apiver.com; privatized in 1994; Port Dir Juan José Sánchez Esquda.

**Petróleos Mexicanos (PEMEX):** Edif. 1917, 2°, Avda Marina Nacional 329, 44°, Col. Anáhuac, 11300 México, DF; tel. (55) 5531-6053; Dir-Gen. J. R. Moctezuma.

**Transportación Marítima Mexicana, SA de CV:** Avda de la Cúspide 4755, Col. Parques del Pedregal, Del. Tlalpan, 14010 México, DF; tel. (55) 5652-4111; fax (55) 5665-3566; internet www.tmm.com.mx; f. 1955; cargo services to Europe, the Mediterranean, Scandinavia, the USA, South and Central America, the Caribbean and the Far East; Pres. Juan Carlos Merodio; Dir-Gen. Javier Segovia.

### CIVIL AVIATION

There were 67 airports in Mexico in 2004, of which 57 were international. Of these, México, Guadalajara, Monterrey and Cancún registered the highest number of operations.

**Aerocalifornia:** Aquiles Serdán 1955, 23000 La Paz, BCS; tel. (612) 26655; fax (612) 53993; e-mail aeroll@aerocalifornia.uabcs.mx; f. 1960; regional carrier with scheduled passenger and cargo services in Mexico and the USA; Chair. Paul A. Arechiga.

**Aerocancún:** Edif. Oasis 29, Avda Kukulcan, esq. Cenzontle, Zona Hotelera, 77500 Cancún, Q. Roo; tel. (988) 32475; fax (988) 32558; charter services to the USA, South America, the Caribbean and Europe; Dir-Gen. Javier Maranon.

**Aeroliteral:** e-mail comentarios@aerolitoral.com; internet www.aerolitoral.com.mx; operates internal flights, and flights to the USA.

**Aeromar, Transportes Aeromar:** Hotel Maria Isabel Sheraton, Paseo de la Reforma 325, Local 10, México, DF; tel. (55) 5514-2248; e-mail web.aeromar@aeromar.com.mx; internet www.aeromar.com.mx; f. 1987; scheduled domestic passenger and cargo services; Dir-Gen. Juan I. Steta.

**Aerovías de México (Aeroméxico):** Paseo de la Reforma 445, 3°, Torre B, Col. Cuauhtémoc, 06500 México, DF; tel. (55) 5133-4000; fax (55) 5133-4619; internet www.aeromexico.com; f. 1934 as Aeronaves de México, nationalized 1959; fmrly Aeroméxico until 1988, when, following bankruptcy, the Govt sold a 75% stake to private investors and a 25% stake to the Asociación Sindical de Pilotos de México; services between most principal cities of Mexico and the USA, Brazil, Peru, France and Spain; Dir-Gen. Gilberto Perez-alonso; Pres. Andres Consesa.

*Directory*

**Aviacsa:** Aeropuerto Internacional, Zona C, Hangar 1, Col. Aviación General, 15520 México, DF; tel. (55) 5716-9005; fax (55) 5758-3823; internet www.aviacsa.com; f. 1990; operates internal flights, and flights to the USA.

**Azteca Lineas Aéreas:** e-mail info@aazteca.com; internet www.aazteca.com.mx; f. 2000; domestic flights.

**Mexicana (Compañía Mexicana de Aviación, SA de CV):** Avda Xola 535, Col. del Valle, 03100 México, DF; tel. (55) 5448-3000; fax (55) 5448-3129; e-mail dirgenmx@mexicana.com.mx; internet www.mexicana.com; f. 1921; owned by state holding co Cintra; scheduled to be sold by 2006; international services between Mexico City and the USA, Central America and the Caribbean; domestic services; CEO Emilio Romano Mussali.

**Click Mexicana:** Avda Xola 535, Col. Del Valle, 03100 México, DF; tel. (55) 5284-3132; internet www.clickmx.com; f. 2005; owned by Mexicana; budget airline operating internal flights; CEO Isaac Volin Bolok.

## Tourism

Tourism remains one of Mexico's principal sources of foreign exchange. Mexico received an estimated 18.7m. foreign visitors in 2003, and receipts from tourism in that year were estimated at US $9,457m. More than 90% of visitors come from the USA and Canada. The country is famous for volcanoes, coastal scenery and the great Sierra Nevada (Sierra Madre) mountain range. The relics of the Mayan and Aztec civilizations and of Spanish Colonial Mexico are of historic and artistic interest. Zihuatanejo, on the Pacific coast, and Cancún, on the Caribbean, were developed as tourist resorts by the Government. In 1998 there were 392,402 hotel rooms in Mexico. The government tourism agency, FONATUR, encourages the renovation and expansion of old hotels and provides attractive incentives for the industry. FONATUR is also the main developer of major resorts in Mexico.

**Secretariat of State for Tourism:** see section on The Government (Secretariats of State).

**Fondo Nacional de Fomento al Turismo (FONATUR):** Insurgentes Sur 800, 17°, Col. del Valle, 03100 México, DF; tel. (55) 5448-4200; internet www.fonatur.gob.mx; f. 1956 to finance and promote the devt of tourism; Dir Lic. John McCarthy.

## Defence

In August 2004 Mexico's regular Armed Forces numbered 192,770: Army 144,000 (including some 60,000 conscripts), Navy 37,000 (including naval air force—1,100—and marines—8,700) and Air Force 11,770. There were also 300,000 reserves. There is a rural defence militia numbering 11,000. Military service, on a part-time basis, is by a lottery and lasts for one year.

**Defence Budget:** 31,800m. new pesos in 2004.

**Chief of Staff of National Defense:** Gen. Humberto Alfonso Guillermo Aguilar.

**Superintendant and Comptroller of the Army and Air Force:** Gen. Salvador Leonardo Bejarano Gómez.

**Commander of the Air Force:** Gen. Manuel Victor Estrada Ricardez.

**Navy Chief of Staff:** Vice-Adm. Alberto Castro Rosas.

## Education

State education in Mexico is free and compulsory at primary and secondary level. Primary education lasts for six years between the ages of six and 11. Secondary education lasts for up to six years. Children aged four years and over may attend nursery school. In 2000/01 99% of children in the relevant age-group were enrolled in primary education (males 99%; females 101%), while UNESCO estimated that 60% of children in the relevant age-group were enrolled in secondary schools (males 57%; females 62%). In 2003 there were an estimated 76,108 nursery schools, 99,034 primary schools, 42,275 secondary and intermediate schools and 4,585 institutes of higher education. In 2001/02 there were 17,552 nursery and primary schools for the indigenous population. However, in spite of the existence of more than 80 indigenous languages in Mexico, there were few bilingual secondary schools. Total enrolment at primary schools was 14,781,327 students in 2003; in that year 5,780,437 students were enrolled at all levels of secondary education. In 2003 there were an estimated 4,585 institutes of higher education, attended by 2,250,461 students. Federal expenditure on education in 2003 was an estimated 265,238.1m. new pesos (equivalent to 21.4% of total central government expenditure). In November 2002 the Congreso de la Unión approved legislation establishing a lower limit on education expenditure equivalent to 8% of GDP.

# Bibliography

Babb, S. L. *Managing Mexico: Economists from Nationalism to Neoliberalism*. Princeton, NJ, Princeton University Press, 2002.

Bartra, R. *Agrarian Structure and Political Power in Mexico*. Baltimore, MD, Johns Hopkins University Press, 1993.

*Blood, Ink and Culture: Miseries and Splendours of the Post-Mexican Condition*. Raleigh, NC, Duke University Press, 2002.

Beatty, E. *Institutions and Investment: The Political Basis of Industrialization in Mexico before 1911*. Stanford, CA, Stanford University Press, 2001.

Bortz, J., and Haber, S. (Eds). *The Mexican Economy, 1870–1930*. Stanford, CA, Stanford University Press, 2002.

Brown, J. C. *Oil and Revolution in Mexico*. Berkeley, CA, University of California Press, 1993.

Cameron, M. A., and Tomlin, B. W. *The Making of NAFTA*. Ithaca, NY, Cornell University Press, 2002.

Camp, R. A. *Politics in Mexico*, 2nd Edn. Oxford, Oxford University Press, 1996.

Castaneda, J. G. *The Mexican Shock*. New York, NY, New Press, 1995.

*Perpetuating Power: How Mexican Presidents are Chosen*. New York, NY, New Press, 2000.

Chand, V. K. *Mexico's Political Awakening*. Notre Dame, IN, University of Notre Dame Press, 2000.

Chappell Lawson, J. *Building the Fourth Estate: Democratization and the Rise of a Free Press in Mexico*. Berkeley, CA, University of California Press, 2002.

Cockcroft, J. D. *Mexico: Class Formation, Capital Accumulation and the State*. New York, NY, Monthly Review Press, 1983.

Collier, G. A., and Quaratiello, E. L. *Basta! Land and the Zapatista Rebellion in Chiapas*. Oakland, CA, Institute for Food and Development Policy, 1999.

Corchabo, A., and Schwartz, G. *Mexico: Experiences with Pro-Poor Expenditure*. Washington, DC, International Monetary Fund, 2002.

Day, S. A. *Staging Politics in Mexico: The Road to Neoliberalism*. Lewisburg, PA, Bucknell University Press, 2004.

Ellingwood, K. *Hard Line: Life and Death on the US–Mexico Border*. New York, NY, Pantheon, Random House, 2004.

Flores de la Pena, H., et al. *Bases para la planeación económica y social de México*. Madrid, Editores Siglo XXI, 2002.

Fuentes, C. *A New Time for Mexico*. London, Bloomsbury, 1997.

Garber, P. M. (Ed.). *The Mexico–US Free Trade Agreement*. London, MIT Press, 1993.

Guardino, P. F. *Peasants, Politics and the Formation of Mexico's National State*. Stanford, CA, Stanford University Press, 2001.

Hodges, D. C., and Gandy, R. *Mexico: The End of the Revolution*. New York, NY, Praeger Publrs, 2001.

Krauze, E. *Mexico: Biography of Power. A History of Modern Mexico, 1810–1996*. London, HarperCollins, 1997.

Murphy, R. D., and Feltenstein, A. *Private Costs and Public Infrastructure: The Mexican Case*. Washington, DC, International Monetary Fund, 2001.

Orme, Jr, W. A. *Continental Shift, Free Trade and the New North America*. Washington, DC, The Washington Post, 1993.

Otero, G. (Ed.). *Mexico in Transition: Neoliberal Globalism, the State and Civil Society*. London, Zed Books, 2004.

Prescott, W. H. *History of the Conquest of Mexico*. London, Phoenix Press, 2002.

Preston, J., and Dillon, S. *Opening Mexico: The Making of a Democracy*. New York, NY, Farrar, Straus and Giroux, 2004.

Reding, A. 'The next Mexican revolution', in *World Policy Journal*, Vol. XIII, No 3, 1996.

Richmond, D. W. *The Mexican Nation: Historical Continuity and Modern Change*. Paramus, NJ, Prentice Hall, 2001.

Riding, A. *Mexico: Inside the Volcano*. London, I. B. Tauris, 1987.

Rubio, L. *Políticas económicas del México contemporáneo*. México, DF, Fondo de Cultura Económica, 2001.

Simon, J. *Endangered Mexico: An Environment on the Edge*. London, Latin America Bureau, 1998.

Suchlicki, J. *Mexico: From Montezuma to the Fall of the PRI*. London, Brassey's UK, 2001.

Thomas, H. *The Conquest of Mexico*. London, Hutchinson, 1993.

Vincent, T. G. *The Legacy of Vincente Guerrero, Mexico's First Black Indian President*. Gainesville, FL, University Press of Florida, 2002.

Weintraub, S. *A Marriage of Convenience*. Oxford, Oxford University Press, 1990.

Wise, T. A., Salazar, H., and Carlsen, L. (Eds). *Confronting Globalization: Economic Integration and Popular Resistence in Mexico*. Bloomfield, CT, Kumarian Press, 2003.

# MONTSERRAT

## Geography

### PHYSICAL FEATURES

The devastated island of Montserrat is a British Overseas Territory located in the Leeward Islands, in the eastern Caribbean. The nearest other polity to the colony is Antigua and Barbuda, with Antigua island some 43 km (27 miles) to the north-east, although its uninhabited outpost of Redonda is only 24 km to the north-west. Beyond Redonda lies Nevis, the smaller unit of the federation of Saint Christopher (St Kitts) and Nevis. The main islands of the French department of Guadeloupe lie 64 km to the south-east, where the arc of the Lesser Antilles continues. Montserrat is 102 sq km (39 sq miles) in extent, only slightly larger than Anguilla (but much more mountainous). Volcanic activity has increased its area somewhat since the mid-1990s.

Montserrat is a roughly pear-shaped island about 19 km long (north–south) and 11 km wide, with a mountainous terrain provided by three groups of highlands, steadily rising towards the south: Silver Hill in the far north; the Centre Hills; and the great Soufrière Hills, which dominate the south. This last range reaches its height at Chance's Peak (914 m—3,000 ft) and now also dominates the island because of the renewed volcanic activity centred here. Previously distinguished by verdant, forest-clad heights, occasionally scarred by areas of sulphurous springs (soufrières), the Soufrière Hills resumed a more active volcanic state for the first time in over 350 years on 18 July 1995. The series of eruptions resulted in the removal of government from the capital, Plymouth (on the south-western coast), and the evacuation of other towns (including the airport, across the island from Plymouth) from an Exclusion Zone covering over one-half of the island by April 1996. People remain generally forbidden to visit this Zone, although a Daytime Entry Zone (volcanic conditions permitting) now moderates this, operating in the west, as far as just to the north of the site of Plymouth.

Eruptions in the Soufrière Hills steadily increased in seriousness up to September 1996, but the deadly pyroclastic flows took their worst toll in June 1997 (19 people died, in the Exclusion Zone, despite official warnings, and the air and sea ports were finally closed) and, thereafter, the flows destroyed the centre of the abandoned capital. The largest flow occurred on 26 December 1997. Pyroclastic flows are a particularly fast-moving and deadly form of volcanic emission, making the Exclusion Zone a fundamental safety feature during eruptions. Dome collapses fuel the most devastating flows, and these have continued into the 2000s. The largest dome collapse of the eruption was on 27 July 2000, but most of the flow was down the Tar River (a seasonal stream before 1995), in the east of the Zone. Another major conduit is the valley of the White River, the delta of which is also expanding the south of the island, on the south-west coast. The south of Montserrat is now dominated by seven active volcanoes, which have devastated the landscape, emptying it of life and rendering it uninhabitable for at least one decade more.

Prior to the eruption of the Soufrière Hills, the richest land was in the south, and most people lived in the south-west. The greenery on the hills and coasts had confirmed the sobriquet of the 'Emerald Isle of the Caribbean', earned by Montserrat's Irish heritage. Its forestland sheltered a rich bird and animal life, including the unique black and gold Montserrat oriole and the terrestrial frog ('mountain chicken'), the latter being only otherwise found in Dominica, where it is known as the crapaud. It is as yet uncertain quite how devastating or prolonged the effect of the eruptions has been on Montserrat's natural environment, but, as about one-third of the island has been rendered into a bleak and desolate rock-scape, it is severe. The central and northern highlands retain their vegetation, but the north was always drier than the south and woodland less extensive. The mountainous nature of the island has contributed to a rugged coastline, particularly on the Atlantic or weather coast, and

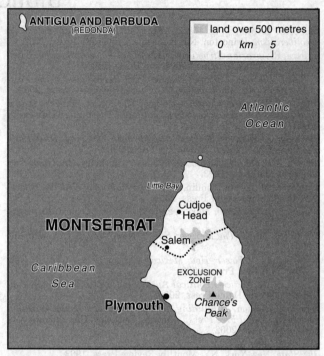

there are no natural harbours, although the more sheltered western coast offers some anchorages (port facilities at Plymouth were finally destroyed by the volcanic activity of the mid-1990s and Little Bay now serves as the main port).

### CLIMATE

The climate is subtropical, tempered by the trade winds and by altitude. There is little seasonal variation in temperature, with average monthly maximums ranging between 24°C and 31°C (76°F–88°F). Rainfall averages some 1,530 mm (60 ins) per year, or more with altitude, and most occurs in November–January. Humidity is low, but the island is in the hurricane belt, the last time it was seriously affected being in 1989.

### POPULATION

The population is now predominantly black or of mixed race, but Montserrat was originally noted as the only centre of Roman-Catholic Irish settlement in the West Indies (in the 17th century). Other white families joined the Irish ones, but the main addition to the population was African slaves, whose descendants now preserve the legacy of Irish names, phrases and, allegedly, a touch of brogue in the local accent (English is the official language). There is still a large Roman Catholic community in Montserrat, although the Anglican Church is also well represented and there are many other Christian denominations in a traditionally religious society.

Owing to volcanic activity from 1995 and the evacuation of the south, the total population declined from its normal level of about 11,000 (10,639 at the 1991 census) to less than 3,000 in 1998 (a mid-year estimate in 2004 put the population at 4,690). The largest emigrant communities are in Antigua and Barbuda and in the United Kingdom. Many people left during 1997 particularly, although some began to return from 1998, especially as new facilities began to be built in the north. The capital is still officially Plymouth, but this has been largely destroyed by the volcano and lies within the Exclusion Zone. Government buildings have been established at Brades Estate in the Carr's

Bay area, in the north-west. This is on the coastal lowlands, just south of the new port facilities at Little Bay and to the north of Cudjoe Head Village, on the heights above. Cudjoe Head and Salem were the original two settlements of any size in the north of the island, in the area now outside the Exclusion Zone. Salem is just north of the boundary, also on hills near the west coast.

Montserrat was originally divided into three parishes, but virtually the entire Parish of St Anthony (in the south and south-west) is now in the Exclusion Zone, as is most of the eastern Parish of St George. This leaves only the northern, hillier (and originally more sparsely populated) parts of St George's and the Parish of St Peter in the north and north-west.

# History

## MARK WILSON

Montserrat is a United Kingdom Overseas Territory. A Governor, who is the representative of the British monarch, has important reserve powers, including responsibility for national security and defence, the civil service, the judiciary and certain financial matters. A Chief Minister is responsible to the Legislative Council, similar in function to a parliament, which contains a majority of elected members. The Governor presides over an Executive Council, similar in function to a cabinet, which includes the Chief Minister and the other ministers. A constitutional review was concluded in 2005.

Few traces remain of the original Amerindian inhabitants, who knew the island as Alliougana, or land of the prickly bush. Christopher Columbus sighted the island in 1493, and named it after the Spanish monastery of St Ignatius of Loyola. The first British settlers arrived in 1632, most of them Irish Catholics, resettling from Saint Christopher (St Kitts) or from the North American colony of Virginia. To this day, Montserrat is sometimes known as the Emerald Isle of the Caribbean, and has received some limited reconstruction assistance from the Irish Government. For most of its colonial history small slave plantations grew sugar and other crops, Montserrat being most famous for its limes and its cotton; however, the island was never an economic powerhouse.

From 1871 to 1956 Montserrat formed part of the Leeward Islands Federation. Unlike the larger islands of the eastern Caribbean, it did not join the short-lived Federation of the West Indies in 1958–62, and did not progress to associated statehood, retaining instead its status as a United Kingdom dependent territory, at first under an Administrator and then under a Governor.

In the first universal suffrage elections, held in 1952, all the seats were won by the Montserrat Labour Party, linked to the Montserrat Trades and Labour Union. This pattern was repeated in 1955, but broken in 1958. Since then, party allegiances in this small community have been fluid, overridden on most occasions by personal loyalties. The People's Liberation Movement, led by John Osborne, won all seven elected seats in a 1978 election, and remained the majority party with five seats in 1983 and four in 1987, when Osborne sought a fresh mandate during a dispute with the United Kingdom authorities. Osborne's relations with the United Kingdom and with the Governor had deteriorated sharply as the result of a dispute over the management of the island's 'offshore' financial sector. Several hundred 'offshore' banks had been allowed to operate with no physical presence on the island and no effective supervision; most of these were later closed down. There was also some concern over Osborne's proposal to introduce casino gambling. In response, the British Government proposed a revised constitution, with increased powers of control over the financial sector for the Governor.

While this dispute was in progress, the island was devastated in September 1989 by 'Hurricane Hugo'. Most of the housing stock was severely damaged, public buildings were badly damaged, and telephone and overhead electricity-distribution systems had to be completely reconstructed. The disaster underlined the advantages of a continuing link with the United Kingdom, and somewhat weakened Osborne's internal political position. Along with his three ministers, British police detectives questioned him in June in connection with a fraud investigation. He lost his majority in September 1991, and the National Progressive Party (NPP), whose leader, Reuben Meade, became Chief Minister, won elections on 8 October.

The Soufrière Hills volcano, which had been inactive since before the time of European settlement, came to life in July 1995. The southern part of the island was temporarily evacuated in August and December of that year, and what turned out to be a permanent evacuation was ordered on 1–3 April 1996. Twenty-three emergency shelters were established in the north of the island, initially housing 1,381; the last residents moved out of emergency accommodation only in 2001. By the time of the April 1996 evacuation, around 3,000 residents had left the island; soon afterwards, the British Government agreed Montserratians would be granted two years' stay in the United Kingdom, with the right to work or to obtain income support and housing benefits; however, no assistance with travel was offered for low-income islanders.

A general election was held in November 1996, despite the volcanic emergency. With a low turn-out, the NPP held one seat, with two each for the People's Progressive Alliance and the Movement for National Reconstruction, and two for independent candidates. Bertrand Osborne formed a coalition Government, and, in contrast to some of Montserrat's previous Chief Ministers, enjoyed consistently good relations with the United Kingdom.

In March and April 1997 the pace of volcanic activity quickened. There were flows of hot ash down the sides of the volcano, and ash clouds extended up to 12.5 km into the atmosphere. The remaining inhabitants of the southern part of the island, including the former capital, Plymouth, were forced to leave their homes. Most of the south has since been covered by a thick layer of volcanic ash, while volcanic deposits have at one point extended the coastline out to sea by 1.5 km. However, not everyone complied with instructions to stay out of the danger area at all times, and 19 people were killed on 25 June, some of whom were visiting closed areas to tend crops or livestock. The port was destroyed, while the airport lay within the danger zone. Some emergency accommodation was provided in the north, but two-thirds of the population left the island altogether, moving either to the United Kingdom, to Antigua and Barbuda or to other Caribbean islands. At one point, it appeared that the island might need to be evacuated altogether.

There was considerable public disquiet both locally and in the United Kingdom at what was seen as a lack of assistance from the British Government. After public protests over conditions in emergency accommodation, Osborne resigned in August 1997, and was replaced as Chief Minister by the more vociferous David Brandt, who had been a key member of John Osborne's Government during the 'offshore' banking crisis. Brandt immediately moved into a bitter dispute with the British authorities, which escalated after remarks perceived as insensitive were made by the British Secretary of State for International Development, Clare Short, about Montserratian demands for 'golden elephants'.

Since the entire population now lived in the north of the island, the old constituency system was no longer relevant. Brandt's Government lost its majority in February 2001, and elections were held on 2 April under a new 'at-large' voting system, without constituencies. John Osborne returned to office as Chief Minister, as leader of the New People's Liberation Movement, which won seven of the nine elective seats and 52% of the popular vote, with the NPP taking the remaining two seats. In contrast to his earlier period in office, Osborne has in general enjoyed a harmonious relationship with the British Government and its local representatives.

In February 2002 the Organisation of Eastern Caribbean States (OECS) countries, including Montserrat, agreed to allow nationals of member states to travel freely within the OECS area and to remain in a foreign territory within the area for up to six months. It was also planned to introduce a common OECS passport by mid-2004. In May 2002 the British Overseas Territories Act, having received royal assent in the United Kingdom in February, came into force and granted British citizenship to the people of its Overseas Territories, including Montserrat. Under the new law, Montserratians would be able to hold British passports and work in the United Kingdom and elsewhere in the European Union.

The Government's majority was reduced to a single seat in February 2003, when the Minister of Communications and Works, Lowell Lewis, resigned his post, and, along with a party colleague, crossed the floor of the Legislative Council to sit as an independent. A former government backbencher, John Wilson, was appointed to the Executive Council in his place. The Chief Minister was forced to take a 10-week leave from December 2004, owing to ill health, forcing a postponement of the annual budget presentation; however, he subsequently returned to work, and the political situation appeared stable.

The level of volcanic activity increased again from September 2002, and a small area containing 300 houses was taken out of the 'safe' zone. Most of those affected were wealthy expatriates with winter homes; however, 80 Montserratians were placed in emergency accommodation. A further massive eruption, reported to be the largest since 1995, occurred on 12 July 2003, causing damage to the supply of water and electricity, and the destruction of numerous buildings in the Salem area, which is on the periphery of the 'safe' zone. There was another major eruption of ash on 2 March 2004; this was followed by a scientific assessment that suggested that a nine-year phase of volcanic activity was moving towards a close. A section of the Exclusion Zone was subsequently opened for access, and the island administration applied to the United Kingdom for £10m. in assistance for ash clearance and infrastructural rehabilitation. In July the USA informed the 200 Montserratians granted emergency residence during the volcanic emergency that they should plan to return home. However, the volcano was far from quiet, with renewed activity occurring in May 2005. In May 2004 Deborah Barnes-Jones was inaugurated as Governor, replacing Anthony Longrigg. The next election was constitutionally due by April 2006, but it was possible that it would be held earlier.

# Economy

## MARK WILSON

Montserrat is an Overseas Territory of the United Kingdom in the eastern Caribbean, with an area of 102 sq km and a population of 4,690 in mid-2004; this figure is much lower than the total of 11,581 in 1994, just before the volcano became active, but considerably more than the low point of 2,850 reached in 1998. A proportion of the increased population were migrant workers, however (thus some of the money earned by them was sent abroad in the form of remittances). In spite of the volcanic emergency, the island is reasonably prosperous, with a per-head gross domestic product (GDP) of some US $8,797 in 2003, and an economy supported by substantial grant aid from the United Kingdom, although this was expected gradually to diminish from 2005. GDP was estimated to have grown by 4.3% in 2004 and was projected at 6.5% for 2006.

Montserrat is a member of the Caribbean Community and Common Market, or CARICOM, which is attempting to develop a single market and economy, in principle, by 2006. It is also a member of the Organisation of Eastern Caribbean States, which links nine of the smaller Caribbean territories, while the Eastern Caribbean Central Bank, headquartered in Saint Christopher (St Kitts) and Nevis, supervises its financial affairs. The territory also uses the services of the regional stock exchange, the Eastern Caribbean Securities Exchange (also based in Saint Christopher and Nevis), established in 2001.

The southern two-thirds of the island forms an Exclusion Zone around the Soufrière Hills volcano, which has been active since July 1995, and is expected to remain active for several decades. Before the recent volcanic devastation, there was a small manufacturing sector, including a rice mill and an electronics assembly plant. These, as well as most tourist accommodation and much of the transport, utilities and infrastructure, were in what is now the Exclusion Zone. Since 1997 there has been no significant manufacturing or commercial agriculture. There is a small 'offshore' financial sector, which, at the end of 2003, included 22 International Business Companies and 11 'offshore' banks, now better regulated than in the mid-1980s. In mid-2005 the first major export was made since the volcanic eruption: a quantity of aggregate from a quarry in Little Bay was shipped to Antigua.

Economic activity has centred since 1997 on the construction of new facilities in the north of the island, supported largely by British development grants. The Government's own recurrent revenue was expected in 2005 to contribute only EC $32.7m. to a total current expenditure of $80.1m. The remainder of the current budget and the entire capital programme were funded by grants from the United Kingdom and the European Union (EU). The United Kingdom spent £59m. in Montserrat in 1995–98. A three-year reconstruction programme, funded by British budgetary assistance of £75m., was agreed in January 1999. A £55m. programme for 2001–04 was followed by a £40m. programme for 2004/05–06/07, announced in June 2004 as part of a new Country Policy Plan. The temporary capital is at Brades, formerly a small village with few services. Facilities completed in the north include a small port, a housing development, fuel-storage facilities, a power plant, a water-supply system, schools, a small hospital, a police headquarters, a fire station, sheltered housing for the elderly and a new building for the volcanic observatory. Privately funded projects have included a small shopping centre. The major projects in progress in 2005 were the construction of a new port and an airport, funded by the EU and the United Kingdom, with a 600-m runway, completed in July, capable of harbouring small aircraft for flights to neighbouring islands. A new port building was completed in 2003, with work on a jetty extension to commence on completion of the airport project. A new power-station was also to be built, at Brades, and a plan proposed for the use of geothermal power to generate electricity.

A number of interesting proposals were made with the intention of revitalising the island's economy. An encouraging prospect is a specialized tourism industry based on the volcano itself, with facilities for visiting scientists, students and the merely curious. A number of visitors already make day side-trips while staying in Antigua and Barbuda or other islands. For stop-over visitors, there were 60 rooms in hotels and guest houses in mid-2002, and there were 9,569 stop-over tourist arrivals in 2004; further hotel, golf and scuba-diving developments were also planned. An 'offshore' medical school opened in 2004, with the intention of making use of Montserrat's status as an overseas territory of an EU member state; two others have been licensed. If a substantial number of resident students were to be attracted, their spending would contribute significantly to the island's foreign-exchange earnings. In addition, the Fédération internationale de football association (FIFA) made funding

MONTSERRAT

available for an international football-training facility, complete with a stadium, pool, gymnasium and accommodation, for use out of season by overseas teams. Irish-based investors have also expressed an interest in plans to quarry volcanic ash from the south of the island, for use as raw material by the cement and ceramics industries, should the south of the island be declared safe. Another proposal was a graduate school of disaster studies, to be operated in collaboration with British and US universities. Besides volcanic activity and hurricanes, the island is at some risk from earthquakes, landslides and tsunamis.

# Statistical Survey

Sources (unless otherwise stated): Government Information Service, Media Centre, Chief Minister's Office, Old Towne; tel. 491-2702; fax 491-2711; Eastern Caribbean Central Bank, POB 89, Basseterre, Saint Christopher; internet www.eccb-centralbank.org; OECS Economic Affairs Secretariat, *Statistical Digest*.

## AREA AND POPULATION

**Area:** 102 sq km (39.5 sq miles).

**Population:** 10,639 (males 5,290, females 5,349) at census of 12 May 1991; 4,482 at census of 12 May 2001 (Source: UN, *Population and Vital Statistics Report*). 2004 (estimate): 4,690.

**Density** (2004): 46.0 per sq km.

**Principal Towns:** Plymouth, the former capital, was abandoned in 1997. Brades is the interim capital.

**Births and Deaths** (1986): 200 live births (birth rate 16.8 per 1,000); 123 deaths (death rate 10.3 per 1,000). 2003: Crude birth rate 9.6 per 1,000; Crude death rate 12.3 per 1,000 (Source: Caribbean Development Bank, *Social and Economic Indicators*).

**Expectation of Life** (years at birth, estimates): 78.5 (males 76.4; females 80.8) in 2004. Source: Pan-American Health Organization.

**Employment** (1992): Agriculture, forestry and fishing 298; Mining and manufacturing 254; Electricity, gas and water 68; Wholesale and retail trade 1,666; Restaurants and hotels 234; Transport and communication 417; Finance, insurance and business services 242; Public defence 390; Other community, social and personal services 952; *Total* 4,521 (Source: *The Commonwealth Yearbook*). 1998 (estimate): Total labour force 1,500.

## HEALTH AND WELFARE

**Physicians** (per 1,000 head, 1999): 0.18.

**Hospital Beds** (per 1,000 head, 2003): 3.3.

**Health Expenditure** (public,% of GDP, 2000): 7.7.

**Health Expenditure** (public, % of total, 1995): 67.0.

**Access to Water** (% of persons, 2002): 100.

**Access to Sanitation** (% of persons, 2002): 96.
Source: Pan-American Health Organization.
For definitions, see explanatory note on p. vi.

## AGRICULTURE, ETC.

**Principal Crops** (FAO estimate, metric tons, 2003): Vegetables 475; Fruit (excl. melons) 710.

**Livestock** (FAO estimates, '000 head, 2003): Cattle 9.7; Sheep 4.7; Goats 7.0; Pigs 1.2.

**Livestock Products** (FAO estimates, '000 metric tons, 2003): Beef and veal 0.7; Cows' milk 2.3.

**Fishing** (FAO estimate, metric tons, live weight, 2003): Total catch 50 (all marine fishes).
Source: FAO.

## INDUSTRY

**Electric Energy** (million kWh): 15 in 1999 (estimate); 12 in 2000 (estimate); 12 in 2001. Source: UN, *Industrial Commodity Statistics Yearbook*.

## FINANCE

**Currency and Exchange Rates:** 100 cents = 1 East Caribbean dollar (EC $). Sterling, US Dollar and Euro Equivalents (31 May 2005): £1 sterling = EC $4.909; US $1 = EC $2.700; €1 = EC $3.329; EC $100 = £20.37 = US $37.04 = €30.04. *Exchange Rate*: Fixed at US $1 = EC $2.70 since July 1976.

**Budget** (EC $ million, preliminary figures, 2004): *Revenue:* Revenue from taxation 30.3 (Taxes on income and profits 12.8; Taxes on domestic goods and services 1.0; Taxes on international trade and transactions 13.3); Non-tax revenue 2.0; Total 32.3 (excl. grants 92.7). *Expenditure:* Current expenditure 83.0 (Personal emoluments 25.9, Goods and services 29.8, Interest payments 0.2; Transfers and subsidies 27.1); Capital expenditure 32.8; Total 115.8.

**International Reserves** (US $ million at 31 December 2004): Foreign exchange 14.10. Source: IMF, *International Financial Statistics*.

**Money Supply** (EC $ million at 31 December 2004): Currency outside banks 12.96; Demand deposits at deposit money banks 31.51; *Total money* 44.47. Source: IMF, *International Financial Statistics*.

**Cost of Living** (consumer price index; base: previous year = 100): 103.5 in 2002; 101.2 in 2003; 103.6 in 2004. Source: partly Caribbean Development Bank, *Social and Economic Indicators*.

**Expenditure on the Gross Domestic Product** (EC $ million at current prices, 2003, provisional figures): Government final consumption expenditure 58.32; Private final consumption expenditure 71.88; Gross fixed capital formation 63.66; *Total domestic expenditure* 193.86; Export of goods and services 39.44; *Less* Imports of goods and services 128.79; *GDP at market prices* 104.51.

**Gross Domestic Product by Economic Activity** (EC $ million at current prices, 2003, provisional figures): Agriculture, forestry and fishing 1.35; Mining and quarrying 0.05; Manufacturing 0.62; Electricity, gas and water 6.24; Construction 15.64; Wholesale and retail trade 4.22; Restaurants and hotels 0.70; Transport 6.89; Communications 3.61; Banks and insurance 9.32; Real estate and housing 12.13; Government services 30.84; Other services 7.14; *Sub-total* 98.75; *Less* Financial intermediation services indirectly measured 7.13; *Gross value added at basic prices* 91.62; Taxes, less subsidies, on products 12.89; *GDP at market prices* 104.51.

**Balance of Payments** (EC $ million, 2003): Goods (net) –55.6; Services (net) –20.5; *Balance on goods and services* –76.1; Income (net) –4.9; *Balance on goods, services and income* –81.0; Current transfers (net) 62.8; *Current balance* –18.2; Capital account (net) 31.9; Direct investment (net) 6.6; Portfolio investment (net) 0.2; Public sector long term investment –0.9; Commercial banks –12.6; Other investment assets 0.9; Other investment liabilities (incl. net errors and omissions) –3.7; *Overall balance* 4.2.

## EXTERNAL TRADE

**Principal Commodities** (US $ million, 2003): *Imports c.i.f.:* Food and live animals 3.3; Beverages and tobacco 1.4; Crude materials (inedible) except fuels 0.8; Mineral fuels, lubricants, etc. 4.4; Chemicals 1.4; Manufactured goods 4.7; Machinery and transport equipment 9.2; Miscellaneous manufactured articles 3.1; Total (incl. others) 28.4. *Exports f.o.b.:* Mineral fuels, lubricants, etc. 0.7; Manufactured goods 0.1; Machinery and transport equipment 0.5 (Complete digital data-processing machines 0.3); Miscellaneous manufactured articles 0.3; Total (incl. others) 1.8.

**Principal Trading Partners** (US $ million, 2003): *Imports c.i.f.:* Barbados 0.7; Canada 0.6; Dominica 0.3; Japan 1.3; Netherlands 0.3; Trinidad and Tobago 1.0; United Kingdom 5.7; USA 16.6; Total (incl. others) 28.4. *Exports f.o.b.:* Antigua and Barbuda 0.8; St Christopher and Nevis 0.1; United Kingdom 0.1; USA 0.6; Total (incl. others) 1.8. Source: UN Statistics Division.

## TOURISM

**Tourist Arrivals** (preliminary figures, 2004): Stay-over arrivals 9,569 (USA 1,871, Canada 334, United Kingdom 2,663, Caribbean 4,389, Others 312); Excursionists 5,106; Cruise-ship passengers 363; *Total visitor arrivals* 15,038.

**Tourism Receipts** (EC $ million): 23.4 in 2002; 19.8 in 2003; 23.2 in 2004 (preliminary figure).

## TRANSPORT

**Road Traffic** (vehicles in use, 1990): Passenger cars 1,823; Goods vehicles 54; Public service vehicles 4; Motorcycles 21; Miscellaneous 806.

## MONTSERRAT

**Shipping:** ('000 metric tons, 1990): *International Freight Traffic:* Goods loaded 6; Goods unloaded 49. Source: UN, *Monthly Bulletin of Statistics.*

**Civil Aviation** (1985): Aircraft arrivals 4,422; passengers 25,380; air cargo 132.4 metric tons.

### COMMUNICATIONS MEDIA

**Radio Receivers** (1997): 7,000 in use.

**Television Receivers** (1999): 3,000 in use.

**Telephones** (2000): 2,811 main lines in use.

**Mobile Cellular Telephones** (2000): 489 subscribers.

**Non-daily Newspapers** (1996): 2 (estimated circulation 3,000).
Sources: UNESCO, *Statistical Yearbook*; International Telecommunication Union.

### EDUCATION

**Pre-primary** (1993/94): 12 schools; 31 teachers; 407 pupils.

**Primary** (1993/94, unless otherwise indicated): 11 schools; 85 teachers; 460 pupils (2003).

**Secondary:** 2 schools (1989); 80 teachers (1993/94); 308 pupils (2003).

Sources: UNESCO, *Statistical Yearbook*; Caribbean Development Bank, *Social and Economic Indicators.*

# Directory

The eruption in June 1997 of the Soufrière Hills volcano, and subsequent volcanic activity, rendered some two-thirds of Montserrat uninhabitable and destroyed the capital, Plymouth. Islanders were evacuated to a 'safe zone' in the north of the island.

## The Constitution

The present Constitution came into force on 19 December 1989 and made few amendments to the constitutional order established in 1960. The Constitution now guarantees the fundamental rights and freedoms of the individual and grants the Territory the right of self-determination. Montserrat is governed by a Governor and has its own Executive and Legislative Councils. The Governor retains responsibility for defence, external affairs (including international financial affairs) and internal security. The Executive Council consists of the Governor as President, the Chief Minister and three other Ministers, the Attorney-General and the Financial Secretary. The Legislative Council consists of the Speaker (chosen from outside the Council), nine elected, two official and two nominated members. Owing to the disruption caused by evacuation from the south of the island, the 2001 general election was conducted according to a new 'at-large' voting system, without constituencies, but still choosing nine members of the legislature.

## The Government

**Governor:** DEBORAH BARNES-JONES (took office on 10 May 2004).

### EXECUTIVE COUNCIL
(July 2005)

**President:** DEBORAH BARNES-JONES (The Governor).

**Official Members:**

**Attorney-General:** ESCO HENRY-GREER.

**Financial Secretary:** JOHN SKERRIT.

**Chief Minister and Minister of Finance, Economic Development, Trade, Tourism and Media:** JOHN A. OSBORNE.

**Minister of Agriculture, Lands, Housing and the Environment:** ANN DYER-HOWE.

**Minister of Communications and Works:** JOHN WILSON.

**Minister of Education, Health and Community Services:** IDABELLE MEADE.

**Clerk to the Executive Council:** CLAUDETTE WEEKES.

### MINISTRIES

**Office of the Governor:** Lancaster House, Olveston; tel. 491-2688; fax 491-8867; e-mail govoff@candw.ag; internet www.montserrat-newsletter.com; internet www.gov.ms/index/governor.htm.

**Office of the Chief Minister:** Government Headquarters, POB 292, Brades; tel. 491-3378; fax 491-6780; e-mail ocm@gov.ms.

**Ministry of Agriculture, Lands, Housing and the Environment:** Government Headquarters, POB 292, Brades; tel. 491-2546; fax 491-9275; e-mail malhe@gov.ms; internet www.malhe.gov.ms.

**Ministry of Communications and Works:** Woodlands; tel. 491-2521; fax 491-3475; e-mail mcw@gov.ms; internet www.gov.ms/commsworks.

**Ministry of Education, Health and Community Services:** Government Headquarters, Brades; tel. 491-2880; fax 491-3131; e-mail mehcs@gov.ms; internet www.mehcs.gov.ms.

**Ministry of Finance, Economic Development, Trade, Tourism and Media:** Government Headquarters, POB 292, Brades; tel. 491-2777; fax 491-2367; e-mail minfin@gov.ms; internet minfin@gov.ms.

### LEGISLATIVE COUNCIL

**Speaker:** Dr JOSEPH MEADE.

**Election, 2 April 2001**

| Party | Seats |
|---|---|
| New People's Liberation Movement (NPLM) | 7 |
| National Progressive Party (NPP) | 2 |

There are also two *ex-officio* members (the Attorney-General and the Financial Secretary) and two nominated members.

## Political Organizations

**Movement for National Reconstruction (MNR):** f. 1996; Leader AUSTIN BRAMBLE.

**National Progressive Party (NPP):** tel. 491-2444; f. 1991; Leader REUBEN T. MEADE.

**New People's Liberation Movement (NPLM):** f. 2001 as successor party to People's Progressive Alliance; Leader JOHN A. OSBORNE.

## Judicial System

Justice is administered by the Eastern Caribbean Supreme Court (based in Saint Lucia), the Court of Summary Jurisdiction and the Magistrate's Court. A revised edition of the Laws of Montserrat came into force on 15 April 2005, following five years of preparation by a Law Revision Committee.

**Puisne Judge (Montserrat Circuit):** NEVILLE L. SMITH.

**Magistrate:** CLIFTON WARNER, Govt HQ, Brades; tel. 491-4056; fax 491-8866; e-mail magoff@candw.ag.

**Registrar:** SONYA YOUNG, Govt HQ, Brades; tel. 491-2129; fax 491-8866; e-mail courtreg@candw.ag.

## Religion

### CHRISTIANITY

**The Montserrat Christian Council:** St Peter's, POB 227; tel. 491-4864; fax 491-2813; e-mail 113057.1074@compuserve.com; Chair. Rev. SINCLAIR WILLIAMS.

#### The Anglican Communion

Anglicans are adherents of the Church in the Province of the West Indies, comprising eight dioceses. Montserrat forms part of the diocese of the North Eastern Caribbean and Aruba. The Bishop is resident in The Valley, Anguilla.

# MONTSERRAT

### The Roman Catholic Church
Montserrat forms part of the diocese of St John's-Basseterre, suffragan to the archdiocese of Castries (Saint Lucia). The Bishop is resident in St John's, Antigua and Barbuda.

### Other Christian Churches
There are Baptist, Methodist, Pentecostal and Seventh-day Adventist churches and other places of worship on the island.

## The Press

**Montserrat Newsletter:** Unit 8, Farara Plaza, Brades; tel. 491-2688; fax 491-8867; e-mail monmedia@candw.ag; internet www.montserrat-newsletter.com; government information publication; Publicity Officer RICHARD ASPIN.

**The Montserrat Reporter:** POB 306, Davy Hill; tel. 491-4715; fax 491-2430; e-mail editor@montserratreporter.org; internet www.themontserratreporter.com; weekly on Fridays; circ. 2,000; Editor BENNETTE ROACH.

**The Montserrat Times:** POB 28; tel. 491-2501; fax 491-6069; weekly on Fridays; circ. 1,000.

## Broadcasting and Communications

### TELECOMMUNICATIONS

**Cable & Wireless:** POB 219, Sweeney's; tel. 491-1000; fax 491-3599; e-mail venus.george@cwni.cwplc.com; internet www.cwmontserrat.com.

### BROADCASTING

Prior to the volcanic eruption of June 1997 there were three radio stations operating in Montserrat. Television services can also be obtained from Saint Christopher and Nevis, Puerto Rico and from Antigua and Barbuda (ABS).

#### Radio

**Gem Radio Network:** Barzey's, POB 488; tel. 491-5728; fax 491-5729; f. 1984; commercial; Station Man. KEVIN LEWIS; Man. Dir KENNETH LEE.

**Radio Antilles:** POB 35/930; tel. 491-2755; fax 491-2724; f. 1963; in 1989 the Govt of Montserrat, on behalf of the OECS, acquired the station; has one of the most powerful transmitters in the region; commercial; regional; broadcasts in English and French; Chair. Dr H. FELLHAUER; Man. Dir KRISTIAN KNAACK; Gen. Man. KEITH GREAVES.

**Radio Montserrat (ZJB):** POB 51, Sweeney's; tel. 491-2885; fax 491-9250; e-mail zjb@gov.ms; internet www.zjb.gov.ms; f. 1952; first broadcast 1957; govt station; Station Man. ROSE WILLOCK; CEO LOWELL MASON.

#### Television

**Cable Television of Montserrat Ltd:** POB 447, Olveston; tel. 491-2507; fax 491-3081; Man. SYLVIA WHITE.

## Finance

The Eastern Caribbean Central Bank, based in Saint Christopher and Nevis, is the central issuing and monetary authority for Montserrat.

**Eastern Caribbean Central Bank—Montserrat Office:** Farara Plaza, POB 484, Brades; tel. and fax 491-6877; e-mail eccbmni@candw.ag; internet www.eccb-centralbank.org; Resident Rep. CHARLES T. JOHN.

**Financial Services Commission:** POB 188, Phoenix House, Brades; tel. 491-6887; fax 491-9888; e-mail enquiries@fscmontserrat.org; internet www.fscmontserrat.org; f. 2002; the Commission consists of the Commissioner and three other members appointed by the Governor; Chair. C. T. JOHN.

### BANKING

**Bank of Montserrat Ltd:** POB 10, St Peters; tel. 491-3843; fax 491-3163; e-mail bom@candw.ag; Man. ANTON DOLDRON.

**Montserrat Building Society:** POB 101, Brades; tel. 491-2391; fax 491-6127; e-mail mbsl@candw.ms.

**Royal Bank of Canada:** POB 222, Brades; tel. 491-2426; fax 491-3991; e-mail rbcmont@candw.ms; Man. J. R. GILBERT.

**St Patrick's Co-operative Credit Union Ltd:** POB 33, Brades; tel. 491-666; fax 491-6566.

### STOCK EXCHANGE

**Eastern Caribbean Securities Exchange:** based in Basseterre, Saint Christopher and Nevis; e-mail info@ecseonline.com; internet www.ecseonline.com; f. 2001; regional securities market designed to facilitate the buying and selling of financial products for the eight member territories—Anguilla, Antigua and Barbuda, Dominica, Grenada, Montserrat, Saint Christopher and Nevis, Saint Lucia and Saint Vincent and the Grenadines; Gen. Man. TREVOR E. BLAKE.

### INSURANCE

**Caribbean Alliance Insurance Co Ltd:** POB 185, Brades; tel. 491-2103; fax 491-6013; e-mail ismcall@candw.ag.

**Insurance Services (Montserrat) Ltd:** POB 185, Sweeney's; tel. 491-2103; fax 491-6013.

**NAGICO:** Ryan Investments, Brades; tel. 491-9301; fax 491-3403; e-mail ryaninvestments@candw.ag.

**Nemwil:** POB 287, Brades; tel. 491-3813; fax 491-3814.

**United Insurance Co Ltd:** Jacquie Ryan Enterprises Ltd, POB 425, Brades; tel. 491-2055; fax 491-3257; e-mail united@candw.ms.

## Trade and Industry

### GOVERNMENT AGENCY

**Montserrat Economic Development Unit:** POB 292; tel. 491-2066; fax 491-4632; e-mail devunit@gov.ms; internet www.devunit.gov.ms.

### CHAMBER OF COMMERCE

**Montserrat Chamber of Commerce and Industry:** Vue Pointe Hotel, Old Towne, POB 384, Brades; tel. 491-3640; fax 491-3639; e-mail chamber@candw.ag; internet www.montserratcci.com; refounded 1971; 31 company mems, 26 individual mems; Pres. KENNETH A. CASSELL.

### UTILITIES

#### Electricity

**Montserrat Electricity Services Ltd (MONLEC):** POB 16, St John's; tel. 491-3148; fax 491-3143; e-mail monlec@candw.ag; domestic electricity generation and supply.

#### Gas

**Grant Enterprises and Trading:** POB 350, Brades; tel. 491-9654; fax 491-4854; e-mail granten@candw.ag; domestic gas supplies.

#### Water

**Montserrat Water Authority:** POB 324; tel. 491-2527; fax 491-4904; e-mail mwa@candw.ag; f. 1972; domestic water supplies; Chair. ALIC TAYLOR; Gen. Man. EMILE DU BERRY.

### TRADE UNIONS

**Montserrat Allied Workers' Union (MAWU):** POB 245, Dagenham, Plymouth; tel. 491-6049; fax 491-6145; e-mail bramblehl@candw.ag; f. 1973; private-sector employees; Gen. Sec. HYLROY BRAMBLE; 1,000 mems.

**Montserrat Civil Service Association:** POB 468, Plymouth; tel. 491-3797; fax 491-2367.

**Montserrat Seamen's and Waterfront Workers' Union:** tel. 491-6335; fax 491-6335; f. 1980; Sec.-Gen. CHEDMOND BROWNE; 100 mems.

**Montserrat Union of Teachers:** POB 460; tel. 491-7034; fax 491-5779; e-mail hcb@candw.ag; f. 1978; Pres. GREGORY JULIUS (acting); Gen. Sec. HYACINTH BRAMBLE-BROWNE; 46 mems.

## Transport

The eruption in June 1997 of the Soufrière Hills volcano, and subsequent volcanic activity, destroyed much of the infrastructure in the southern two-thirds of the island, including the country's principal port and airport facilities, as well as the road network.

# MONTSERRAT

## ROADS

Prior to the volcanic eruption of June 1997 Montserrat had an extensive and well-constructed road network. There were 203 km (126 miles) of good surfaced main roads, 24 km (15 miles) of secondary unsurfaced roads and 42 km (26 miles) of rough tracks. In 2004 some EC $3.5m. was allocated to a programme of road reinstatment, which was to be completed in 2005.

## SHIPPING

The principal port at Plymouth was destroyed by the volcanic activity of June 1997. An emergency jetty was constructed at Little Bay in the north of the island. Regular transhipment steamship services are provided by Harrison Line and Nedlloyd Line. The Bermuth Line and the West Indies Shipping Service link Montserrat with Miami, USA, and with neighbouring territories. A twice-daily ferry service is in operation between Montserrat and Antigua.

**Port Authority of Montserrat:** Little Bay; tel. 491-2791; fax 491-8063; Man. ROOSEVELT JEMMOTTE.

**Montserrat Shipping Services:** POB 46, Carr's Bay; tel. 491-3614; fax 491-3617.

## CIVIL AVIATION

The main airport, Blackburne at Trants, 13 km (8 miles) from Plymouth, was destroyed by the volcanic activity of June 1997. A helicopter port at Gerald's in the north of the island was completed in 2000. A new, temporary international airport at Gerald's, financed at a cost of EC $42.6m. by the European Union and the British Department for International Development, was completed in 2005; a new terminal building was opened in February and the airport officially opened in July, to be serviced by Windward Islands Airways International (WINAIR). In the longer term, the Government intended to construct a permanent international airport at Thatch Valley. Montserrat is linked to Antigua by a helicopter service, which operates three times a day. The island is also a shareholder in the regional airline, LIAT (based in Antigua and Barbuda).

**Montserrat Airways Ltd:** tel. 491-6494; fax 491-6205; charter services.

**Montserrat Aviation Services Ltd:** POB 257, Nixon's, Cudjoe Head; tel. 491-2533; fax 491-7186; f. 1981; handling agent.

# Tourism

Since the 1997 volcanic activity, Montserrat has been marketed as an eco-tourism destination. Known as the 'Emerald Isle of the Caribbean', Montserrat is noted for its Irish connections, and for its range of flora and fauna. In 2004 there were 9,569 stay-over tourist arrivals. In that year some 46% of tourist arrivals were from Caribbean countries, 28% from the United Kingdom, 20% from the USA and 3% from Canada. A large proportion of visitors are estimated to be Montserrat nationals residing overseas. In 2004 earnings from the sector amounted to a preliminary EC $23.2m.

**Montserrat Tourist Board:** POB 7, Brades; tel. 491-2230; fax 491-7430; e-mail info@montserrattourism.ms; internet www.visitmontserrat.com; f. 1961; Chair. JOHN RYAN; Dir of Tourism ERNESTINE CASSELL.

# Education

Education, beginning at five years of age, is compulsory up to the age of 14. In 1993 there were 11 primary schools, including 10 government schools. Secondary education begins at 12 years of age, and comprises a first cycle of five years and a second, two-year cycle. In 1989 there was one government secondary school and one private secondary school. In addition, in 1993 there were 12 nursery schools, sponsored by a government-financed organization, and a Technical College, which provided vocational and technical training for school-leavers. There was also an extra-mural department of the University of the West Indies in Plymouth. In 2002 the European Union and the United Kingdom contributed a total of EC $6m. to the construction of a further education college, the Montserrat Community College, which was completed in late 2003. A nursery school was constructed in 2002. Three 'offshore' medical schools were licensed in 2003. Education was allocated EC $7.0m. in the 2005 budget.

# THE NETHERLANDS ANTILLES

## Geography

### PHYSICAL FEATURES

The Netherlands Antilles is part of the tripartite Kingdom of the Netherlands. The dependency has been known informally as the 'Antilles of the Five' since Aruba separated in 1986, and is now composed of the island territories of Curaçao and Bonaire in the southern Caribbean and, in the north-eastern Caribbean, of the Dutch (southern) half of the island of Sint Maarten (Saint-Martin), Sint Eustatius (Statia—from the original Spanish name, St Anastasia) and Saba. The Netherlands Antilles' Caribbean colleague in the Kingdom, Aruba, is 68 km (42 miles) to the west of Curaçao. These two islands, together with Bonaire known as the 'ABC islands', lie off the coast of Venezuela, and are the westernmost extension of the southern Lesser Antilles. The Venezuelan mainland is about 55 km south of Curaçao and 80 km south of Bonaire (Bonaire is actually closer to the offshore Venezuelan Antilles, specifically, the Islas Las Aves, to the east). Curaçao and Bonaire, themselves only about 35 km apart, together with Aruba, are known as the *Benedenwindse Eilands* ('Leeward Islands'). The other islands of the Dutch Caribbean, about 900 km to the north-east, are grouped in the *Bovenwindse Eilands* ('Windward Islands'—the Dutch adopted this terminology from the Spanish, at variance with the anglophone tradition, which terms the surrounding islands in the north-eastern Caribbean as the Leeward Islands). The *Bovenwindse Eilands* are sometimes known as the 'three Ss' (Saba, St Eustatius or Statia, and St Maarten). St Eustatius is about 20 km north-west of St Kitts (Saint Christopher and Nevis). Saba lies 27 km to the north-west of St Eustatius. St Maarten lies 56 km to the north of St Eustatius and 45 km from Saba. On St Maarten is the only land border (10.2 km) in the Lesser Antilles. The north of the island (St-Martin—as well as the island of St-Barthélemy, some 20 km to the south-east) lies in France and is part of the Overseas Department of Guadeloupe. The British dependency of Anguilla lies just to the north of the Franco-Dutch island, with the Virgin Islands some distance to the west. The five island territories of the Netherlands Antilles, formerly known as Curaçao and Dependencies (then including Aruba), together cover an area of 800 sq km (309 sq miles).

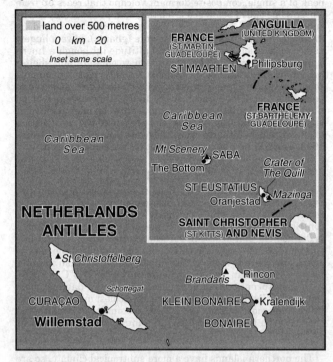

Curaçao is the largest island of the Netherlands Antilles, with an area of 444 sq km. It is about 61 km long, undulating in width (at most, 14 km) south-eastwards from the north-western tip (North Point) to East Point. Volcanic in origin, reef development has added to the island, which remains surrounded by coral barriers. The northern coasts face the weather, while a complex littoral of drowned valleys and inlets, still home to some mangrove wetlands, is more characteristic of the leeward shore. The largest enclosed bay is the Schottegat, on the south coast, which is where the port of Willemstad, the capital, has its Annabaai harbour, inland from the original site of the city. The interior (cunucu) is hilly (the highest point being St Christoffelberg, at 345 m or 1,132 ft, in the north-west), but dry, with vegetation being scrubby and drought resistant. The natural environment is similar to that in Aruba and Bonaire, but has eroded more from human development, although it is distinguished by fauna such as the blue iguana and the small Curaçao deer (a white-tail deer believed to have been imported by the original Amerindian Arawaks in the 14th century).

Bonaire pushes south-eastwards, before thrusting south, the gentler, south-western, leeward shore forming a rough crescent, which shelters the small limestone island of Klein Bonaire, some 750 m off the west coast. Bonaire covers an area of 288 sq km, and is about 40 km long and 11 km at its widest. Like Curaçao, Bonaire is a reef-fringed island originally of volcanic origin, but it is much lower, most of the south being only about 2 m above sea level. Hills are confined to the north (reaching their height at Mt Brandaris—238 m). Water retention of the soil is poor and this, combined with low rainfall, has added the feature of salinas (extremely saline lakes separated from the sea by coral-rubble barriers), which attracted the original Dutch settlers, who wished to harvest the salt. The saltpans also shelter one of the largest colonies of Caribbean flamingos in the Americas. The island's natural environment, including its economically important marine environment (Bonaire is considered to be one of the world's three leading diving destinations), has been long and extensively protected. Among the multitude of other bird species found on the island are the indigenous Bonaire green parrot and the rare yellow-shouldered amazon, while the only native mammal is also airborne—the bat, of which eight species have been identified. Reptiles include the iguana and the anolis. Bonaire is the Dutch island least densely populated by people.

The island of St Martin (Sint Maarten or Saint-Martin) is of volcanic origin, but is fairly flat and comprised primarily of a coralline limestone base. The dry, often scrubby landscape is dotted with salt ponds and other inland waters, notably the great Simpson's Bay Lagoon, which dominates the low-lying, west-pointing peninsula that extends the island in the south-west. The highest point is in the centre of the hillier, eastern bulk of the island, at Pic du Paradis (424 m), and lies in the French sector. At 86 sq km, St Maarten is the smallest island in the world to be divided between two nations—the southern part of the island is Dutch, covers 34 sq km (as finally agreed in the 1839 revision of the original 1648 Franco-Dutch Treaty of Mount Concordia) and is known as Sint Maarten. St Maarten has a slightly larger population than St-Martin and is heavily developed.

St Eustatius (Statia) covers 21 sq km and, like Saba, is of a more explicit volcanic origin than the other three islands. In the south it rises to the rainforest-filled crater of an extinct volcano, The Quill. The highest point of the crater rim, Mazinga, is 600 m above sea level. To the north-west are less lofty heights, culminating in the Boven at the northern end of Statia, with central plains connecting the two highlands. Although much of the native forest disappeared during the period of economic expansion that ended towards the end of the 18th century, when the island was known as the 'Golden Rock' and slave-worked plantations added to its immense entrepôt wealth, Statia still has a diverse plant life and hosts a number of bird species. Butterflies abound and animal life includes the threatened Antilles iguana, land crabs and tree frogs. The island is rich in historical remains.

At 13 sq km, Saba is the smallest of the Dutch West Indian islands, but reaches their highest point (indeed, the highest point in the whole Kingdom of the Netherlands), at Mt Scenery (866 m—2,842 ft). Rising steeply from the sea, the island is the peak of a single, composite dormant volcano that rears up from a base some 600 m below sea level, the main peak surrounded by subsidiary peaks and domes. The green hillsides, scored by drainage valleys known locally as guts (ghauts), have a hugely varied vegetation, dependent upon altitude (although a hurricane in 1999 severely damaged much woodland), and shelter numerous birds and some animals, mainly reptilian, among them the green iguana, the endemic anole lizard and a non-poisonous red-bellied racer snake (only otherwise known on Statia). The mountain terrain permits only one inlet at which it is possible to land from the sea, and prevented the construction of a road (which, hitherto, was believed to be unachievable) until 1943. There is no human settlement on the coast as such.

## CLIMATE

The climate of Curaçao and Bonaire is tropical and the two islands are south of the main hurricane belt. The islands also receive little rain, while the fairly constant Atlantic breezes contribute to a high evaporation rate, ensuring that the climate is defined as semi-arid. On Curaçao average annual rainfall is 570 mm (22 ins—although this varies considerably from year to year), and there is more in November–December than in other months. The mean annual air temperature is 27.5°C (81.5°F), with a nocturnal minimum of about 26°C and a daytime maximum of 30°C (79°–86°F). January is the coolest month. Temperatures, and most weather conditions, are similar in Bonaire, but there is even less precipitation (an annual average of just over 490 mm) and it falls far more unevenly, with much more in the north.

The northern islands have a more subtropical climate, so are slightly less hot. The islands receive more rainfall, but lie firmly in the hurricane belt. Average annual rainfall in all three island territories is about 1,100 mm, but the loftier altitudes of Saba and St Eustatius can receive considerably more. Temperature too can vary with altitude, with averages most affected on Saba, which is higher than the other islands and where settlement begins at a greater altitude. Saba is used to 25°–28°C (77°–82°F) during the day, but thermometric readings can fall as low as 16°C (61°F) at night. The other two islands are warmer, but remain milder, though more humid, than the southern islands.

## POPULATION

Most of the population are of a mixed black descent (85%), including some claiming descent from the original Amerindians, while the rest are other mestizos, whites and Asians. Roman Catholicism is the largest of the Christian denominations, but Protestants are also well represented, with proportions varying from island to island and the Roman Catholics more dominant on the southern islands. There are also some Jews in the Netherlands Antilles, with Curaçao boasting the oldest synagogue in continuous use in the Americas. Curaçao and Bonaire are also where the cosmopolitan islanders developed the native Papiamento dialect, an eclectic mix of Portuguese, Spanish, Dutch and English evolved from a slave pidgin. In the northern islands English tends to be more generally spoken, and is the first language for many, while Spanish is spoken widely in the south and mainly by immigrant workers from the Dominican Republic in the north. Dutch, however, is the official language of the dependency.

The total population was estimated at 180,870 in January 2004. About 74.4% of this total lives in Curaçao, 17.4% in St Maarten, 6.1% in Bonaire, 1.3% in St Eustatius and 0.8% in Saba. St Maarten is the most densely populated part of the Netherlands Antilles and Bonaire the least. The capital of the Netherlands Antilles is Willemstad, on Curaçao, where most of that island's population lives (134,003 at mid-2003). The second city of the dependency is Philipsburg, the chief town of St Maarten, where it is located in the south-west of the island, between an inland lake and the Great Bay on the southern coast. The government headquarters and main settlement on Bonaire is Kralendijk (known locally as Playa), on the coast opposite Klein Bonaire, while the second town of the island is the older and more northerly inland settlement of Rincon. Oranjestad, on the west coast of St Eustatius, just to the north-west of The Quill, was once the main port of the entire region, but remains the chief town of Statia. The main settlement on Saba is The Bottom, with fewer than 400 residents.

# History

### RENÉ VAN DONGEN

Revised for this edition by the editorial staff

## EARLY OCCUPATION

The first inhabitants of the island of Curaçao were small groups of Amerindians (most likely from the Caquetíos tribe), who had crossed over from Venezuela. Some of their artefacts found on Curaçao can be dated back to 2500 BC. Another group of Amerindians (most likely Arawaks) had settled on Aruba, Bonaire and Curaçao around AD 900. These Amerindians clashed several times with groups of Caribs who were roaming around the Caribbean Sea. A few Amerindian artefacts were found on the 'S' Islands (near Norman Estate on St Maarten), which greatly resembled Amerindian artefacts found in Venezuela. By the time the first European settlers set foot on the 'S' Islands, the Amerindians had disappeared.

The first written source of information about the Netherlands Antilles was from the Italian sailor Amerigo Vespucci. He travelled through the Caribbean in 1499 together with Alonso de Ojeda and the cartographer Juan de la Cosa, and reported an encounter with Amerindians on the islands of Bonaire and Curaçao. Between 1499 and 1526 these Amerindians were captured as slaves by the Spaniards and put to work on other islands. The Spaniards considered Bonaire and Curaçao to be *islas inútiles*, or 'useless islands', and used them mainly to graze sheep, goats and horses. They also started to grow fruit trees and tobacco on the islands. They took no interest in St Maarten, Saba and St Eustatius, perhaps also because the influence of the French and English was much stronger in the northern part of the Caribbean.

## UNDER THE RULE OF THE WEST INDIAN COMPANY

Dutch sailors belonging to the West-Indische Compagnie (WIC—West Indian Company) began visiting the islands of Bonaire and St Maarten from around 1624, with the sole purpose of collecting salt in the natural saltpans there. The Dutch captured St Maarten, St Eustatius and Saba in 1630, 1635 and 1640, respectively. Curaçao was captured by the Dutch in 1634, without experiencing any resistance from the Spanish. Bonaire came under Dutch rule for the first time in 1636.

Like so many territories in the Caribbean, the ownership and administration of the islands was far from stable. The British and the Dutch governed Curaçao and Bonaire for alternating periods between the early 17th and the early 19th century. The government of St Maarten, St Eustatius and Saba changed hands several times between the Dutch, French and British (and, in the case of St Maarten, the Spanish, between 1633 and

1648). St Eustatius changed flag 22 times before the Dutch claimed uncontested sovereignty in 1816. The most famous Governor of one of the five Antillean islands was Peter Stuyvesant, who was Director-General of New Holland and Curaçao from 1647 until 1664, before he moved on to the newly established Dutch colony of New Amsterdam, which later changed name and ownership and became New York, USA. The most influential Governor was Jan Doncker (1673–79), who brought prosperity to Curaçao. Under his leadership trade and agriculture (important to feed the growing number of slaves) flourished.

From the beginning the interest of the Dutch in the Caribbean was confined to (legal and illegal) trade. Anything that would fetch a price or could find a buyer either in Europe, in other parts of the Caribbean, on the South American mainland and, during the American War of Independence (1775–83), on the North American continent was traded. The range of goods transported varied from salt (an essential preservative) and lumber (Bonaire and Curaçao), to sugar, tobacco and cotton (St Maarten, St Eustatius and Saba) and slaves.

Slave trade to the Caribbean commenced in 1501, by Spanish and Portuguese slavers. The Dutch WIC began the transportation of slaves from West Africa to Brazil in 1637. Curaçao became the centre of the Dutch slave trade in 1648 after the Spanish traders lost ground to the Portuguese in Brazil. In 1675 Curaçao achieved the status of 'free trade zone', whereby traders from all nations could come to buy or sell. Between 1685 and 1713 the majority of the slaves transported from West Africa into the Caribbean passed through Curaçao. In the 18th century St Eustatius took over this role from Curaçao, supplying slaves to plantations in Suriname and the counties of Essequibo, Demerara and Berbice (located in what is now known as Guyana). In the 17th century the Dutch traders transported 100,000 slaves from Africa to the Americas. Together with the 400,000 slaves they expatriated in the 18th century, this represents 5% of the total Atlantic slave trade that took place from the 16th century until its abolition in the 19th century.

Although the islands of Aruba, Bonaire and Curaçao were not really suited for the cultivation of any crops, planters came to these islands in the 17th century. The majority of these white planters originated from the Protestant south-west of the Netherlands (Zeeland) and were referred to as 'high Protestants'. Also, Jews from Portugal and Portuguese Jews from former Dutch colonies in Brazil arrived in Curaçao to buy land and establish plantations. Together with these successful white settlers came Europeans who had left or had been forced to leave Europe for a variety of other reasons. These settlers made up the group of so-called 'low Protestants', or 'poor whites'. Small groups of refugees from Colombia and Venezuela also took up residence on the islands.

The Protestant traders from Zeeland were also the first Europeans to colonize the 'S' islands. As British and French influence was more pronounced in this part of the Caribbean, a mixture of white Europeans (mainly English, Irish and Scottish and, on St Maarten, some French and English-Americans) occupied these islands in the 17th and 18th centuries. These settlers also brought knowledge about the cultivation of sugar cane, a crop that was introduced in 1650.

During the American War of Independence the island of St Eustatius became well known for supplying the rebels from the American colonies (later to become the USA) with goods, which enabled them to continue the war against the British. This continued supply to the rebels through St Eustatius, as well as the island's official recognition of the flag of the rebel American colonies in 1776, forced the British to declare war on the Dutch Republic. The conflict that followed was the beginning of the end of the success of Dutch trading in the Caribbean in general, and the shipment and transhipment of goods and slaves through St Eustatius in particular.

## 19TH CENTURY: SEARCHING FOR A NEW DIRECTION

The rule of the Dutch WIC came to an end in 1828 and the Dutch Crown assumed responsibility for the islands. By that time the commercial value of the Dutch West Indies had lessened considerably, particularly compared with the attractions of the Dutch East Indies (now Indonesia). The Dutch territories were the last to abolish slavery in 1863, after the slave trade had ceased to exist in the Dutch colonies in 1818. The main reasons at that time to abolish slavery were the static or declining profits in the sugar industry. The growers of cane sugar could not compete with large producers such as Cuba and the discovery of beet sugar in 1747 also meant that Europe was no longer dependent on imports of cane sugar from the Caribbean.

As a result, the old plantocracy system slowly began to decline in the second half of the 19th century. After the abolition of slavery, many former slaves stayed on as wage-labourers on the plantations or in the saltpans. Others obtained or rented small pieces of land and became small farmers. One of the very few new economic opportunities that were created during this period was the mining of newly discovered deposits of phosphate on Little Curaçao. Trade with Venezuela became very difficult at the end of the 19th century, when the Venezuelan President, Antonio Guzmán Blanco, implemented regulations in order to protect Venezuela's economy.

In the early part of the 19th century the Dutch Crown possessed three colonies in the West Indies: the colony of Suriname, the colony of Curaçao (including Little Curaçao, Aruba and Bonaire) and the islands of St Maarten, St Eustatius and Saba, which were grouped together as one colony. In order to stimulate unity between these three colonies, a form of government was attempted, with a Governor-General, seated in Paramaribo, Suriname, overlooking all three colonies. In practice, however, it proved impossible to govern the colonies in unison. Therefore, in 1845 the islands now forming the Netherlands Antilles, together with Aruba, were placed under one Governor (based in Curaçao) for the first time. The Governor was appointed by the King and, later, by the Dutch Government, which remained the major source of funding and thus retained a considerable amount of influence.

## CHANGE IN FORTUNE

A sharp change in fortune for Curaçao came with the discovery of significant petroleum deposits in and around Lake Maracaibo, in Venezuela, at the turn of the 20th century. Because oil tankers could not pass over a large sandbar at the entrance to the shallow lake and since the Royal Dutch-Shell Company regarded Venezuela as politically unstable, a petroleum refinery was built on Curaçao in 1915. The refining industry was very successful and attracted labour from the other islands of the Netherlands Antilles and many other countries.

The people of the Netherlands Antilles had to wait until 1936 before they were given some form of self-governance and it was not until 1948 that the first real parliament (called the Staten) for the Netherlands Antilles was elected and appointed. The movement for (more) independence from the Netherlands started long before the Second World War, but gained real momentum with the foundation of the Democratische Partij (DP—Democratic Party) in 1944, which campaigned actively for independence from the Netherlands. Similar parties were established in Aruba. Following the first Round Table Conference in The Hague, Netherlands, in 1948 and the second Conference in 1954, the Statute of the Kingdom of the Netherlands was produced. This important document provided a legal framework for greater autonomy of the Netherlands Antilles and Suriname, except for matters pertaining to foreign affairs, defence, security and migration. It also created the opportunity for the Netherlands Antilles to introduce its own currency, the Netherlands Antillean guilder or florin, and its own system of taxation and social security. Efraín Jonckheer became the first Prime Minister of the Netherlands Antilles in 1954, a position he held until 1968. The statute regulations with Suriname became redundant in 1975 when Suriname gained independence.

Although the petroleum industry brought wealth and prosperity to the Netherlands Antilles in general and to Curaçao and Bonaire in particular, not everybody profited in equal terms. This became very apparent in May 1969 when the high level of unemployment caused by the contraction of the refining industry provoked serious riots in Willemstad, necessitating the dispatch of Dutch marines to restore order. The riots revealed the existence of racial tensions but also led to a new sense of awareness, especially for the Afro-Antillean part of the population. It also resulted in a decrease in influence of the 'old white élite' (white Protestants, Portuguese Jews and Catholic leaders) and an increase in the influence of the labour movement. In the Netherlands the riots and its effects confirmed the views of many Dutch politicians that the Netherlands Antilles should be given independence as soon as possible.

## TRYING OUT NEW FORMS OF CO-OPERATION

The independence movement in the Netherlands Dependencies gained momentum in Aruba when the Movimentu Electoral di Pueblo (MEP—People's Electoral Movement) gained large support under the charismatic leadership of Gilberto (Betico) Croes in the 1970s. The MEP was determined to leave the Antillean federation and, by doing so, free itself from control by Willemstad. Croes argued that Aruba was unfairly treated by Curaçao and Aruba's budgetary contributions were excessive. The MEP's cause was endorsed by a referendum in Aruba in 1977, in which 82% of voters supported separation from the Netherlands Antilles. In 1983 it was agreed that Aruba would leave the federation in 1986 and obtain a *status aparte*, on the condition that Aruba would gain full independence by 1996. This clause was abandoned in 1993 and Aruba's *status aparte* remained more or less permanent.

Although some politicians favoured the idea of a *status aparte* for Curaçao in the early 1990s, a referendum in Curaçao in November 1993 revealed that 75% of the island's inhabitants wished to remain part of the Netherlands Antilles. On Bonaire, Saba and St Eustatius a large majority of the people voted, in October 1994, in favour of remaining part of the Federation (88%, 91% and 86%, respectively). In the simultaneously held referendum on St Maarten, however, only 60% of those voting were in favour of maintaining the *status quo*. Moreover, in a subsequent referendum, held on 23 June 2000, only 4% of the people of St Maarten favoured maintaining the *status quo*. Some 69% favoured obtaining *status aparte* within the Kingdom of the Netherlands, 14% favoured complete independence and 12% preferred a restructuring of the Antilles of the Five.

On 8 October 2004 the Jesurun Commission, established by the Dutch and Antillean Governments and headed by Edsel Jesurun, a former Governor of the territory, recommended the dissolution of the Netherlands Antilles; support for the federation had, it was argued, virtually disintegrated on most of the islands. The Commission proposed that Curaçao and St Maarten should become autonomous states within the Netherlands (i.e., have *status aparte*), while Saba, Bonaire and St Eustatius should be directly administered by the Dutch Government. In September and November, respectively, in official referendums, a majority of voters on both Bonaire (59%) and Saba (86%) strongly favoured becoming part of the Netherlands. Further referendums on the constitutional futures of Curaçao and St Eustatius took place on 8 April 2005: 68% of participants in Curaçao favoured *status aparte*, in line with the recommendations of the Commission, 23% voted for closer ties with the Netherlands and 5% voted for complete independence—there was a 54% voter turn-out. In St Eustatius, from a voter turn-out of 55%, 76% of the electorate favoured remaining part of the Netherlands Antilles, while 20% voted for closer ties with the Netherlands and 1% preferred to seek complete independence for the island. The Dutch Government was understood to be willing to review its relationship with the islands on the basis of the results. Subsequently, an inter-island constitutional conference in late April decided to abolish the federation by 2007. St Eustatius, the only island to favour by referendum the *status quo*, refused to sign the agreement; none the less, it was decided that Curaçao and St Maarten would be granted *status aparte* while Bonaire and Saba would become *koninkrijkseilanden* (royal islands). The agreement was expected to be formalized by October after talks with the Dutch Government.

There was a strong tradition of migration between the islands of the Antilles. For instance, when Royal Dutch-Shell opened its petroleum refineries on Bonaire and Curaçao, many skilled and unskilled workers from St Eustatius, Saba and St Maarten went there to take up jobs. The first wave of migration from the Netherlands Antilles to the Netherlands occurred in the 1960s and 1970s, following large-scale redundancies in the petroleum industry as a result of automation and scale reductions. After the closure of the main refineries on Curaçao in 1985 the number of emigrants to the Netherlands more than doubled. Between 1999 and 2001 approximately 20,000 Antilleans migrated to the Netherlands. However, more Dutch people were expected to migrate to the Netherlands Antilles, following an agreement signed in July 2000 between the Governments of the Netherlands and the Netherlands Antilles, making it easier for Dutch people to work and live in the Netherlands Antilles.

Since the introduction of direct flights to and from Amsterdam, the Netherlands, a dramatic increase in the level of drugs-trafficking was seen in 2002 in Bonaire—there were 131 drugs-related cases in that year compared with 30 cases during 2001. In order to curb the rapidly growing number of drugs-couriers on flights between Amsterdam and the Netherlands Antilles, the Dutch Government demanded the use of body-scanners at Curaçao airport, to assist with the speedy detection of narcotics-smuggling. This measure met with considerable resistance on the island; a similar reaction greeted a Dutch proposal to ban known drugs-traffickers from Antillean airlines.

The island of St Maarten experienced great economic development in the 1970s under the leadership of Claude Wathey, the leader of the Democratische Partij—Sint Maarten (DP—StM), who led the campaign for the island's secession. However, by the 1990s there was growing evidence to support allegations of gross economic mismanagement, drugs-trafficking and corruption by Wathey and other senior DP—StM figures. In 1991 popular protests against Wathey forced him to resign and in 1994 he was convicted of charges of perjury and forgery and sentenced to 18 months' imprisonment. In 1992 the Dutch Government imposed 'higher supervision', which limited the island Government's power to make decisions on expenditure. That regime remained in place until October 1995, when it was rescinded in order to facilitate reconstruction efforts following the destruction caused in the previous month by Hurricanes 'Luis' and 'Marilyn'.

At elections in January 1998 the governing Partido Antía Restrukturá (PAR—Restructured Antilles Party), in office since 1994, lost four of its eight seats in the 22-seat legislature and won only 19% of the total votes cast (compared with 39% in 1994). A new coalition Government, comprising the Partido Nashonal di Pueblo (PNP—National People's Party) and the newly formed Partido Laboral Krusado Popular (PLKP—Popular Workers' Crusade Party) took office on 1 June 1998, under the leadership of Susanne F. C. Camelia-Römer of the PNP. A committee advising the Government produced a National Recovery Plan (Plan Nasional), in which very stringent measures were proposed in order to revitalize the Antillean economy.

Nevertheless, since the Plan Nasional failed to show immediate visible results, in October 1999 the Government of Camelia-Römer was forced to resign. The new coalition Government, which took office in November 1999, again under the leadership of PAR leader Miguel A. Pourier, consisted of nine coalition parties. An important agreement between the Government of the Netherlands Antilles and the IMF was signed in September 2000 and assessed in March 2001. However, negotiations for the design and implementation of a multi-year adjustment programme were not concluded when elections took place in January 2002.

The elections of 18 January 2002 were won by the Frente Obrero i Liberashon 30 di mei (FOL—Workers' Liberation Front of 30 May), led by Anthony Godett, which had campaigned against the stringent measures imposed by the IMF. The FOL won five seats in the 22-seat Staten (and 23.0% of the popular vote). Pourier's party, the PAR, won four legislative seats and 20.4% of the popular vote. The PNP won three seats (13.4% of the ballot), the PLKP, the DP—StM and the Unión Patriótico Bonairiano (UPB—Patriotic Union of Bonaire) each won two seats (12.1%, 5.5% and 3.6%, respectively), and the Windward Islands People's Movement (WIPM), the Democratische Partij—Statia (DP—StE), the Democratische Partij—Bonaire (DP—B) and the National Alliance (comprising the National Progressive Party and the St Maarten Patriotic Alliance) each gained one legislative seat. None the less, despite the party's victory, attempts by the FOL to form a coalition Government failed because of allegations of corruption and mismanagement of funds by party leaders. Eventually, in June 2002 a coalition Government, which included representatives of the PAR, the

PNP, the PLKP, DP—StM, the UPD and the DP—StE, took office, under the leadership of the new PAR leader, Etienne Ys, replacing Pourier's caretaker Government.

Local elections in Curaçao in May 2003 were also won by the FOL under the leadership of Anthony Godett, who had been detained by police in April for alleged corrupt activities. The victory of Godett's party, known for its independent and assertive attitude towards the Netherlands and the IMF, was expected to impact negatively on the implementation of IMF measures and the political relationship with the Netherlands, which had improved under the PAR leadership. Later in the same month Prime Minister Ys' cabinet resigned to allow the FOL to attempt to form a fresh governing coalition in the light of new local coalition alliances. In late July Ben Komproe of the FOL was sworn in as Prime Minister on a temporary basis, since his party leader, Godett, could not be approved for the post until he was cleared of corruption charges. On 11 August Mirna Luisa Godett, the sister of the FOL leader, was elected by the party to the post of Prime Minister, despite not being a member of the Staten. Komproe was appointed Minister of Justice.

Following the conviction in December 2003 of Anthony Godett (and 16 other party members, business leaders and officials) on fraud, embezzlement and corruption charges, in January 2004 the governing coalition almost lost control of the legislature when the WIPM and the UPB withdrew their support. The Government managed to retain its majority, however, when the DP—B and the National Alliance agreed in principle to join the coalition later in the same month. Although the National Alliance later reversed its decision, the governing coalition retained a small majority of seats (12 out of 22). Meanwhile, as anticipated, relations between the Netherlands Antilles and the Dutch Government worsened following the advent of the FOL-dominated Government; the situation reached a nadir when in January the Prime Minister accused Dutch officials of spying on her Government after local justice officials met two visiting Dutch ministers without her knowledge.

The Government lost its parliamentary majority on 5 April 2004 when four parties (the PNP, the PLKP, DP—B and the DP—StE) withdrew from the FOL-led coalition after failing to effect the resignation of the Minister of Justice, Ben Komproe. Among other complaints, the coalition partners accused Komproe of allowing the FOL's main political donor, Nelson Monte, who was serving a custodial sentence for corruption, to stay in a luxury hospital rather than be jailed. After the Staten voted in favour of a 'no confidence' motion against the Government, the Prime Minister (in addition to three other cabinet members) resigned. The Governor, Fritz Goedgedrag, assumed the task of forming a new government; a seven-party coalition (consisting of the PAR, the PNP, the PLKP, the UPB, the DP—StM, the DP—StE and the WIPM) led by Etienne Ys was duly appointed in June and charged with the responsibility of repairing the territory's relationship with the Netherlands and reducing the burdensome national debt. Meanwhile, in July Anthony Godett was sentenced to 15 months in jail on corruption-related charges. In October Komproe, who had been arrested in September and was under investigation for corruption and forgery crimes, died from apparent complications relating to recent gastric surgery.

In May 2005 the Netherlands announced plans to introduce legislation to require citizens of the Netherlands Antilles aged between 16 and 24 years to find work or begin studies within three months of arriving in the Netherlands, or face deportation; it was estimated by the Dutch Government that immigrants from the Netherlands Antilles aged under 25 years were four times more likely to turn to crime than the general Dutch population. Ys described the policy as discriminatory and threatened to take the Dutch Government to court.

# Economy

## RENÉ VAN DONGEN
Revised for this edition by the editorial staff

The mainstays of the economy of the Netherlands Antilles are the tourism industry, 'offshore' finance activities and the petroleum-refining and transport industry. Tourism and the 'offshore' financial sectors were the main cause of economic expansion from the 1960s. No reliable long-term economic data were available for the Netherlands Antilles, but between 1996 and 2001 gross domestic product (GDP) declined. Limited positive economic growth of 0.4%, 1.4% and 2.1% was reported for 2002, 2003 and 2004, respectively. The disappointing economic performance was caused by a combination of factors: the damage inflicted by hurricanes 'Luis' and 'Marilyn' in September 1995; the subsequent implementation of stricter monetary policy by the Government of Miguel A. Pourier; large debt accumulation; uncertainty over a tax treaty with the USA, which depressed activity in the 'offshore' sector; the global economic slowdown, exacerbated by the September 2001 terrorist attacks on the USA; and a number of internal political problems. In 2002–04 the long formation periods of three successive Governments and the consequent lack of a central programme, coupled with the sluggish world economy and weak international tourism market, damaged confidence, leading to only slight increases in real GDP. Continued low-level growth was anticipated in 2005 and 2006. Total external debt in 2003 was NA Fl. 768.9m. (14.3% of GDP), owed mostly to the Netherlands, and total debt was NA Fl. 3,834.9m. in the same year (71.4% of GDP). The national debt was reported to have increased to around NA Fl. 4,986m. in mid-2005. Financial aid from the Netherlands was likely to continue to remain important to the islands' economies.

The fight against corruption and fraud by civil servants had been a priority since 1980, but was intensified in the 1990s with the implementation of a legal framework to prevent corrupt activities and to screen civil servants before they took up senior positions. In December 2002 civil servants belonging to the Partido Antío Restrukturá (Restructured Antilles Party) and Partido Nashonal di Pueblo (National People's Party) were arrested and charged for alleged corrupt activities, and in April 2003 the leader of the Frente Obrero i Liberashon 30 di mei, Anthony Godett, was detained by the police on charges of alleged fraudulent and corrupt practices. Godett was sentenced to jail for 15 months on charges of corruption in July 2004.

In 2004 the annual rate of inflation in the Netherlands Antilles was 1.3%. The figure is highly dependent on the inflation rates of the islands' main trading partners (Venezuela, the USA and the Netherlands) and on world petroleum prices. The rate of unemployment decreased from 16.2% in 1998 to 12.9% in 2000, mainly owing to the vast migration to the Netherlands; however, the rate was estimated to have increased to 15.3% by September 2004; 16% of the Curaçao work-force was unemployed in that year. Figures from the 2001 census showed that the rate of unemployment in St Maarten stood at 12.2%, with the rate of youth unemployment at 24.1%. Most employment opportunities existed in the trade sector, where at least middle, and sometimes higher, levels of education were required. The internal market was very limited and relied heavily on imports. In 2002 the principal source of imports was Venezuela (60.4% of the total), followed by the USA (13.4%), the Netherlands and Iraq. Some 22.1% of the Netherlands Antilles' exports went to the USA in 2002 and 10.0% went to Venezuela, with smaller quantities being exported to the Bahamas, El Salvador, Honduras and the Netherlands.

The secession of Aruba from the six-island Antillean federation in 1986 was followed by economic disruption in the region, in the form of 'capital flight' (the withdrawal of funds and investments), mostly by local citizens. The 'Antilles of the Five' had to provide Aruba with nearly one-third of the assets and

27% of the gold and foreign reserves of the former federation of six islands.

Following its election in 1994, the Pourier administration negotiated a structural-adjustment programme with the IMF, which was intended to reduce the fiscal deficit over a period of four years, from 1996. Measures included the privatization of loss-making government services, a reduction in the number of civil servants (over 60% of the work-force was employed by the Government), a freeze on salaries and a reduction in holiday allowance in 1996–98, and a reform of civil-service pensions. Furthermore, petrol and motor-fuel duties and utility prices were to be increased, and a 6% sales tax was to be introduced. Public- and private-sector workers reacted to the announced austerity measures by organizing a series of strikes in November 1995. The fiscal deficit, which had increased from US $16.3m. in 1994 to $32m. in 1995, fell sharply in 1997 as a result of the adjustment programme. By 1999, however, it had risen to $53.1m., mainly because of reduced tax, income inefficiencies in the system of tax collection and the overall economic recession.

Although the Government of Susanne F. C. Camelia-Römer (which took office in June 1998) was instrumental in the design of a National Recovery Plan (Plan Nasional), the government coalition parties were unable to reach consensus on the necessary measures for a structural reduction of expenditure. This caused the collapse of the governing coalition in October 1999. A new Government, appointed in the following month and once again led by Pourier, enthusiastically began the implementation of the Plan Nasional, under the watchful eyes of the IMF, the Inter-American Development Bank (IDB) and the Dutch Government. In its first month in office the new coalition administration dismissed 850 civil servants and replaced the sales tax with a turnover tax, which was raised from 2% to 5% in October 1999. In order to address the economic crisis, an important piece of legislation was approved by the Staten in July 2000 for a period of one year (until August 2001). This special legislation enabled the Government of the Netherlands Antilles to accelerate considerably the processing of tax and finance laws.

The agreement signed in September 2000 between the IMF and the Government of the Netherlands Antilles sought to simplify the tax system, restructure the pension schemes and create a more flexible labour market. Furthermore, agreement was reached on a maximum limit for the fiscal deficit for 2001, the dismissal of a further 394 civil servants and the implementation of more stringent controls on the contracting of consultants. Measures to encourage foreign investment were also announced (including the elimination of market protection and the simplification of import procedures and bureaucracy for new investors). The IMF's first assessment of the efficacy of these measures in March 2001 showed that the Government of the Netherlands Antilles needed to implement even more stringent measures to keep the 2001/02 budget deficit within the agreed limits. The Netherlands approved the release of €125m. in additional funds in 2003, and the Dutch and Netherlands Antillean Governments agreed on a more prominent role for the central bank in the monitoring and implementation of the IMF targets. During rounds of the Article IV consultation visits in March 2003, the IMF team concluded that although the Netherlands Antilles was experiencing a period of 'fragile growth', there was a continued need for financial reforms and savings, especially in the health-care sector, government institutions and limited companies. Total fiscal debt was equivalent to 6.3% of GDP for 2004 (US $189.5m.) and had to be reduced to a maximum of 3% of GDP, according to the IMF team.

## TOURISM

Although the most important growth in the tourism industry of the Dutch Caribbean took place on Aruba, the sector was also a major contributor to the economy of the Netherlands Antilles; the tourism sector was the largest employer (28.1% in 1992) after the public sector. From 1995 foreign-exchange earnings from tourism increased rapidly, with a total of NA Fl.1,906.5m. earned in 2004. However, the total number of stop-over tourists in the Netherlands Antilles decreased from a peak of 825,812 in 1991 to 779,325 in 2004. Easily the largest total decrease during 1991–2004 was suffered by Saba (of 61%), while St Maarten also experienced a decline (of 10%) over the same period; the number of stop-over visitors to Curaçao, Bonaire and St Eustatius increased slightly over the same period. St Maarten still attracted the largest number of stop-over tourists: 475,032 in 2004. The total number of cruise passengers visiting the Netherlands Antilles (excluding Saba and St Eustatius) in 2004 was the highest ever for the Antilles of the Five: 1,621,178. In particular, Bonaire was becoming increasingly popular as a cruise-ship destination: 53,343 cruise passengers visited in 2004, compared with 10,688 in 1995. Some 52.0% of the tourists staying over on St Maarten in 2003 were from the USA and Canada, 9.3% from other parts of the Caribbean and 3.0% from the Netherlands. Most of the tourists (41.3%) staying over on Curaçao in 2003 came from Europe (primarily from the Netherlands), with 20.3% from the USA and Canada and 17.9% from South America (especially Venezuela). The total number of visitors from Europe (in particular from the Netherlands) decreased in 2001 and 2002 owing to the depreciation of the euro against the US dollar; however, the number of stop-over tourists from Europe to Curaçao, Bonaire and St Maarten rallied strongly in 2003, increasing by 14.6%.

The main attractions of Curaçao were the (largely man-made) beaches and the modern and numerous shopping facilities. Curaçao continued to invest in its tourism industry, with several large hotel complexes and a golf course under construction in the early 21st century. At the entrance to the Curaçao harbour, a 'mega-pier' was built, aimed at accommodating large cruise ships. The project was funded with assistance from the European Union, the Dutch Government and local resources. With additional major new tourism investments planned (Brionplein Hotel, Kura Hulanda, Cornelisbaai, Jan Thiel Baai) in the early 2000s, the prospects for the sector on Curaçao were bright.

St Maarten, with its fine beaches and exceptional duty-free cruise-ship port (and large airport) had a highly developed tourism industry, accounting for 76.0% of total visitors to the Netherlands Antilles in 2004. The destruction caused to St Maarten by hurricanes 'Luis' and 'Marilyn' in 1995 and by 'Hurricane Georges' in September 1998 resulted in a decline in the sector in the late 1990s. In 1999 the tourism sector of the Windward Islands again suffered from the impact of two hurricanes ('José' and 'Lenny'), incurring a decline in numbers of stop-over and cruise tourists in 1999 and closing down some hotels for most of 2000. On the neighbouring islands of Saba and St Eustatius the tourism industry was small but very important for the local economy. Prospects for the tourism industry on St Maarten for 2005 and beyond were expected to continue to improve, with most of the room capacity restored. The number of cruise tourists visiting St Maarten reached record levels in 2003 and 2004, with the completion of the Wathey Pier in Grand Bay.

## FINANCIAL SERVICES AND FOREIGN TRADE

The second largest source of foreign-exchange income in the Netherlands Antilles was the international financial and business services sector. From the early 1970s banking and financial services in the Netherlands Antilles changed from an almost completely locally owned industry to a sophisticated global network. By the late 1990s many foreign banks were well established on the islands and were specializing in 'offshore' banking facilities. In 1988 a major international trade centre opened, capitalizing upon years of accumulated financial expertise. The financial-intermediation sector employed 6.4% of the working population of Curaçao in 2000–02 and contributed 17.3% of GDP in 2003. The 'tax haven' facilities established on Curaçao in the 1960s offered an attractive incentive for foreign companies to channel their money through the island. Interest on overseas investments was tax-free for all US companies registered in Curaçao. Furthermore, a tax treaty signed in 1963 between the USA and the Netherlands Antilles made it possible for US companies to spend money banked in Curaçao in other countries without having to pay tax in the recipient country. These favourable tax treaties were abrogated in the late 1980s, leading to an immediate decline of Curaçao as a tax haven.

From the late 1980s the finance industry concentrated on the promotion of trade brokerage, counter-trade finance, trusts and private companies, over 40,000 of which were registered in 1995. Most transactions in the 'offshore' finance sector were legal, but money-laundering, involving drugs cartels, became a serious problem by the mid-1990s. Legislation was introduced in 1995 aimed at combating financial crime. In 1998 the Bank of the Netherlands Antilles conducted a number of checks at several international credit institutions to ensure that sufficient efforts were being undertaken to deter and detect use of their facilities for money-laundering purposes. The Bank received assistance in this matter from the Financial Action Task Force on Money Laundering (FATF, based in Paris, France) and the Caribbean Financial Action Task Force (CFATF).

In 1998–2000 the international financial and business services sector continued to perform poorly. This weak performance was reflected by a decline in both income received for services rendered and profit taxes transferred to the Government. However, the introduction, in November 1998, of legislation to promote proper functioning of the stock-exchange markets, the introduction of a new tax regime (New Fiscal Regime—NFR) and the conclusion of a new tax arrangement for the Netherlands laid the groundwork for improvements in this sector.

Although new jurisdiction for the adequate supervision of financial institutions was introduced and much effort and capital were invested to comply with international standards of effective banking supervision, in May 2000 the Financial Stability Forum (FSF) categorized the Netherlands Antilles in group III (the lowest group). The FSF report focused mainly on licensing and admission processes for financial institutions, local supervisory infrastructure and instruments used to combat and prevent money-laundering. In June 2000 the Organisation for Economic Co-operation and Development (OECD) included the Netherlands Antilles on a list of countries deemed to be operating 'unco-operative tax havens' and urged the Government to improve further the accountability and transparency of the financial services sector. In April 2002 OECD removed the Netherlands Antilles from the list after favourably assessing the Government's legislative amendments. In the same month, an agreement was signed with the USA, pledging to share information on tax matters, with the aim of combating money-laundering and associated criminal activities The sector achieved impressive growth during 2002–04.

The free trade zones at Schottegat Harbour and at Curaçao's Hato International Airport recorded a marked increase in re-exports and the number of visits in 1998. Goods destined for a market outside the Netherlands Antilles could be traded in these free trade zones, without having to pay import duty. The increase in merchandise exports for 1998 was related almost entirely to the increased re-exporting activities of some of the main free zone companies on Curaçao and was the main factor responsible for an improved deficit on the current account of the balance of payments in the same year. After five consecutive years of expansion, free zone activities deteriorated significantly in 1999, mainly owing to set-backs in re-exports, increased competition from other free zones in the region and the economic downturn in some of the main export markets. Although the number of free zone visits decreased in 2000 and 2001, larger quantities of merchandise were bought. Furthermore, legislation on e-commerce and on the establishment of economic zones (so-called 'E-zones') was adopted by the Staten in 2000 in order to promote business development.

## PETROLEUM INDUSTRY

The refining industry began on Curaçao in 1915, with the construction of a plant by the Royal Dutch-Shell Company. Most of the crude petroleum for the refineries came from the nearby Venezuelan field and the refined products were mainly shipped to the USA. At its peak, in the 1950s, the industry produced 400,000 barrels per day (b/d), accounting for about one-third of employment and 40% of gross national income. During the 1970s the industry diversified into petroleum transhipment and petroleum storage. Huge terminals were built on Curaçao to handle supertankers carrying African and Middle Eastern crude petroleum and petroleum products. The terminal at Bullen Bay on Curaçao remained one of the world's largest into the 1990s. Since US ports on the eastern seaboard could not handle heavy tankers, the deep-water facilities of Curaçao and Aruba were ideal. Smaller terminals were also built on Bonaire and St Eustatius, and by 1980 the revenue from transhipping almost equalled that of refining. However, the refining industry suffered a reverse when, in the early 1980s, the US federal authorities permitted the construction of large offshore terminals on the Gulf of Mexico and the USA's eastern coasts.

The situation deteriorated further in the 1980s, as the new US terminals often fed into US refineries. As the price of petroleum fell and with reductions in supplies by Venezuela, as a result of production quotas by the Organization of the Petroleum Exporting Countries (OPEC), the Netherlands Antilles' refining industry became unprofitable. In 1985 the Government purchased the plant for a nominal one guilder (plus US $47m. for stocks and machinery), after which it was leased to the Venezuelan state petroleum company, Petróleos de Venezuela, SA (PDVSA), for 10 years. In 1987 PDVSA agreed to spend $25m. to adapt the refinery to process heavy crude petroleum. In 1991 capacity increased to 470,000 b/d, from the original 320,000 b/d, mostly destined for markets in Latin American and the Caribbean. In 1995 PDVSA extended the lease for a further 20 years, ensuring that Curaçao would not be entirely dependent upon Dutch financial support. PDVSA planned to invest at the plant through a build-operate-own scheme. Upgrading programmes begun in the late 1990s were intended to secure the refinery's long-term viability. In 2002 petroleum accounted for 68.4% of imports and 94.7% of exports. Results for 2000 and 2001 for the petroleum sector were favourable, with an increase in the amount of petroleum refined and a reduction in the refinery's operating costs; however, social unrest and strikes in Venezuela forced the refinery to close down operations in December 2002–March 2003, adversely affecting the local economy.

## TRANSPORT

The performance of the national carrier of the Netherlands Antilles, Antilliaanse Luchtvaart Maatschappij (ALM—Antillean Airlines), deteriorated from 1999 and negotiations on the privatization of the airline (which changed its name to Dutch Caribbean Airlines—DCA in 2002) finally failed in late 2004 when the company was declared bankrupt. Regional and international transit arrivals at Curaçao's airport increased in the late 1990s, particularly following the establishment of a free trade zone there in 1998.

The port of Curaçao was one of the largest natural ports in the world and played an important role in Caribbean petroleum transhipment. After a slight increase in harbour activities in Curaçao and St Eustatius in 1998, performance for all harbours in the Netherlands Antilles deteriorated in 1999. For the third quarter of 2000, Curaçao's ship-repair industry grew, following a decline in the corresponding quarter of 1999. Furthermore, the number of ships piloted into the harbour increased in the early 2000s, in particular, owing to an increase in the number of tankers, cruise ships and other types of vessels using the port.

## MANUFACTURING AND AGRICULTURE

The Netherlands Antilles has a very small, highly protected manufacturing sector. Its main sub-sectors were food processing, the production of Curaçao liqueur and the manufacture of paper, plastics and textiles. There was also a dry dock at Curaçao. Manufacturing contributed 5.5% of GDP in 2003 and employed 8.7% of the economically active population on Curaçao in 2000–02. The mining sector only employed 0.2% of the working population on Curaçao during the same period. The poor quality of the soils, combined with water shortages (especially on Curaçao and Bonaire) restricted agricultural development. The major crops and products used to be aloes, charcoal, goat meat, goatskins, sorghum and divi-divi nuts (nuts from the *Casealpinia coriaria* tree used in the tanning industry). Aloe (*Aloe barbadensis*) was used in the pharmaceutical industry and Bonaire was a major exporter until 1973, when production ended.

By the early 2000s cultivation of groundnuts, beans, fresh vegetables and tropical fruit was very limited. The Government of the Netherlands Antilles imposed restrictions on the imports of certain products (notably cucumbers, hot peppers and aubergines) in an attempt to stimulate domestic production. Irrigation with water from wells and pipe water took place on a very limited scale. Fresh water on Bonaire and Curaçao was an # THE NETHERLANDS ANTILLES

expensive commodity (it was manufactured in large desalination plants), making cultivation using large-scale irrigation unlikely. Agriculture, forestry and fishing employed 1.0% of the Curaçao working population in 2000–02.

## OUTLOOK

The Netherlands Antilles experienced great difficulties in adapting to altered economic fortunes at the turn of the millennium, despite great efforts to generate new business and to diversify its activities. Curaçao suffered from high salary levels, a vestige of the petroleum-refining industry, and only moderate productivity levels. For example, in 1993 labour charges in the large but financially troubled Curaçao dry-dock company were US $20 per hour, compared to an average $12 in Western Europe. However, prospects for Curaçao for 2005 and beyond appeared more positive. Estimates from the island's Chamber of Commerce and Industry suggested that large-scale investment in the tourism industry, petroleum refineries and the energy sector had returned. In addition, the Caribbean Rim Investment Initiative (an OECD programme, launched in 2001) aimed to maximize foreign investment in the Caribbean and Latin America.

The economic situation on the island of Bonaire was at least as serious as on Curaçao at the turn of the century. St Maarten demonstrated great economic resilience after being hit by several hurricanes in the 1990s. That island's economic success, as well as the results of the referendum that took place on 23 June 2000 (see History), again raised the issue of St Maarten's proposed *status aparte* within the Antilles of the Five. Developments on St Eustatius and Saba remained stable, with tourism the most important economic activity. The Governments of Bonaire, St Maarten and Saba experienced severe financial problems in 2002–03 and were forced to implement drastic economic austerity measures in exchange for continued financial support from the central Government in Curaçao.

# Statistical Survey

Sources (unless otherwise stated): Centraal Bureau voor de Statistiek, Fort Amsterdam, Willemstad, Curaçao; tel. (9) 461-1031; fax 461-1696; internet www.central-bureau-of-statistics.an; Bank van de Nederlandse Antillen, Simon Bolivar Plein 1, Willemstad, Curaçao; tel. (9) 434-5500; fax (9) 461-5004; e-mail info@centralbank.an; internet www.centralbank.an.

## AREA AND POPULATION

**Area** (sq km): Curaçao 444; Bonaire 288; St Maarten (Dutch sector) 34; St Eustatius 21; Saba 13; Total 800 (309 sq miles).

**Population:** 189,474 at census of 27 January 1992 (excluding adjustment for underenumeration, estimated at 3.2%); 175,653 (males 82,521, females 93,132) at census of 29 January 2001; 180,870 at 1 January 2004 (estimate). *By Island* (2001 census): Curaçao 130,627; Bonaire 10,791; St Maarten (Dutch sector) 30,594; St Eustatius 2,292; Saba 1,349.

**Density** (per sq km, 2001 census): Curaçao 294.2; Bonaire 37.5; St Maarten (Dutch sector) 899.8; St Eustatius 109.1; Saba 103.8; Total 219.6. *1 January 2004:* Total 226.1.

**Principal Town:** Willemstad (capital), population (UN estimate, incl. suburbs): 134,003 at mid-2003. Source: UN, *World Urbanization Prospects: The 2003 Revision.*

**Births, Marriages and Deaths** (2003): Registered live births 2,512; Registered marriages 748 (marriage rate 4.2 per 1,000); Registered deaths 1,374 (death rate 7.7 per 1,000).

**Expectation of Life** (years at birth): 76.7 (males 73.6; females 79.5) in 2004. Source: Pan-American Health Organization.

**Economically Active Population** (sample survey, Curaçao only, persons aged 15 years and over, average 2000–02): Agriculture, forestry and fishing 487; Mining and quarrying 102; Manufacturing 4,271; Electricity, gas and water 825; Construction 3,567; Wholesale and retail trade, repairs 9,072; Hotels and restaurants 3,491; Transport, storage and communications 3,200; Financial intermediation 3,137; Real estate, renting and business activities 4,080; Public administration, defence and social security 4,481; Education 2,276; Health and social work 4,217; Other community, social and personal services 3,748; Private households with employed persons 2,095; Extra-territorial organizations and bodies 8; *Total employed* 49,056 (males 25,206, females 23,850); Unemployed 9,056; *Total labour force* 58,112.

## HEALTH AND WELFARE

**Total Fertility Rate** (children per woman, 2004): 2.0.

**Under-5 Mortality Rate** (per 1,000 live births, 2002): 14.2.

**Physicians** (per 1,000 head, 1999): 1.4.

**Hospital Beds** (per 1,000 head, 1996): 6.15.

**Health Expenditure:** % of GDP (1995): 4.5.
Source: Pan American Health Organization.
For definitions, see explanatory note on p. vi.

## AGRICULTURE, ETC.

**Livestock** (FAO estimates, '000 head, year ending September 2003): Asses 2.6; Cattle 0.6; Pigs 2.4; Goats 13.3; Sheep 8.9; Poultry 135.

**Livestock Products** (FAO estimates, metric tons, 2003): Pig meat 180; Poultry meat 300; Cows' milk 410; Hen eggs 510.

**Fishing** (metric tons, live weight, 2003): Skipjack tuna 12,084; Yellowfin tuna 6,667; Bigeye tuna 3,203; *Total catch* (incl. others) 23,070 (FAO estimate).
Source: FAO.

## MINING

**Production** ('000 metric tons, 2003, estimate): Salt 500. Source: US Geological Survey.

## INDUSTRY

**Production** ('000 metric tons, 2001, unless otherwise indicated): Jet fuel 915; Kerosene 46 (estimate); Residual fuel oils 5,115; Lubricating oils 395 (estimate); Petroleum bitumen (asphalt) 1,025 (estimate); Liquefied petroleum gas 100; Motor spirit (petrol) 1,750; Aviation gasoline 14; Distillate fuel oils (gas-diesel oil) 2,528 (estimate); Sulphur (recovered) 30 (2002); Electric energy (million kWh) 1,125.
Sources: UN, *Industrial Commodity Statistics Yearbook*, and US Geological Survey.

## FINANCE

**Currency and Exchange Rates:** 100 cents = 1 Netherlands Antilles gulden (guilder) or florin (NA Fl.). *Sterling, Dollar and Euro Equivalents* (31 May 2005): £1 sterling = NA Fl. 3.254; US $1 = NA Fl. 1.790; €1 = NA Fl. 2.207; NA Fl. 100 = £30.73 = $55.87 = €45.31. *Exchange Rate:* In December 1971 the central bank's mid-point rate was fixed at US $1 = NA Fl. 1.80. In 1989 this was adjusted to $1 = NA Fl. 1.79. The US dollar also circulates on St Maarten.

**Central Government Budget** (NA Fl. million, 2004): *Revenue:* Tax revenue 577.2 (Taxes on goods and services 420.7, Taxes on international trade and transactions 127.0, Other taxes 29.5); Non-tax revenue 79.8; Grants (from other levels of government, excluding overseas development aid) 35.3; Total 692.3. *Expenditure:* Wages and salaries 292.9; Other goods and services 107.2; Interest payments 142.0; Subsidies 1.7; Current transfers 324.3; Capital expenditure (incl. transfers and net lending) 30.3; Total 898.4. *Total General Government Budget* (incl. island governments, NA Fl. million, 2004): *Revenue:* Tax revenue 1,205.0; Non-tax revenue 144.3; Total 1,349.2. *Expenditure:* Current expenditure 1,625.6; Capital expenditure 64.7; Total 1,690.3.

**International Reserves** (US $ million at 31 December 2004): Gold (national valuation) 152; Foreign exchange 415; Total 567. Source: IMF, *International Financial Statistics*.

**Money Supply** (NA Fl. million at 31 December 2004): Currency outside banks 231.3; Demand deposits at commercial banks 1,096.8; Total (incl. others) 1,406.5. Source: IMF, *International Financial Statistics*.

**Cost of Living** (Consumer Price Index; base: 2000 = 100): All items 102.2 in 2002; 104.3 in 2003; 105.7 in 2004. Source: IMF, *International Financial Statistics*.

**Expenditure on the Gross Domestic Product** (million NA Fl. at current prices, 2003): Final consumption expenditure 4,142.7 (Government 1,146.3, Households and non-profit institutions serving households 2,996.4); Gross fixed capital formation 1,266.2; Changes in inventories 3.4; *Total domestic expenditure* 5,412.3; Exports of goods and services 4,451.9; *Less* Imports of goods and services 4,496.6; Statistical discrepancy 5.9; *GDP in market prices* 5,373.6.

**Gross Domestic Product** (million NA Fl. at current prices, 2003): Agriculture, fishing, mining, etc. 40.1; Manufacturing 268.0; Electricity, gas and water 198.8; Construction 204.9; Wholesale and retail trade 735.3; Hotels and restaurants 212.5; Transport, storage and communications 437.1; Financial intermediation 836.4; Real estate, renting and business activities 462.0; Public administration, defence, etc. 348.9; Education 508.4; Health care and social services 444.0; Other community, social and personal services 137.2; Private households with employed persons 10.6; *Gross value added at basic prices* 4,844.4; Taxes, less subsidies, on products 529.2; *Gross domestic product in market prices* 5,373.6.

**Balance of Payments** (US $ million, 2003): Exports of goods f.o.b. 681.5; Imports of goods f.o.b. –1,684.4; *Trade balance* –1,002.9; Exports of services 1,705.7; Imports of services –811.6; *Balance on goods and services* –108.8; Other income received 89.8; Other income paid –97.1; *Balance on goods, services and income* –116.0; Current transfers received 399.1; Current transfers paid –276.3; *Current balance* 6.8; Capital transfers (net) 26.2; Direct investment abroad 0.9; Direct investment from abroad –80.6; Portfolio investment assets –0.8; Portfolio investment liabilities 5.0; Other investment assets –119.5; Other investment liabilities 143.0; Net errors and omissions 46.3; *Overall balance* 27.1. Source: IMF, *International Financial Statistics*.

### EXTERNAL TRADE

**Principal Commodities** (US $ million, 2002): *Imports c.i.f.*: Food and live animals 145.5; Petroleum, petroleum products, etc. 1,552.1 (Crude petroleum 1,354.4); Basic manufactures 110.4; Machinery and transport equipment 126.3 (Road vehicles 72.9); Total (incl. others) 2,268.5. *Exports f.o.b.*: Refined petroleum products 1,609.0; Total (incl. others) 1,699.2. Source: UN, *International Trade Statistics Yearbook*.

**Principal Trading Partners** (US $ million, 2002): *Imports c.i.f.*: Colombia 30.3; Germany 39.0; Iraq 151.1; Japan 36.7; Netherlands 187.4; USA 305.1; Venezuela 1,370.9; Total (incl. others) 2,268.5. *Exports f.o.b.*: Antigua and Barbuda 22.0; Aruba 22.7; Bahamas 129.9; Belize 26.1; Canada 68.6; Colombia 18.2; Cuba 61.5; El Salvador 80.0; Guatemala 66.5; Guyana 69.7; Haiti 34.2; Honduras 74.2; Netherlands 73.7; Nicaragua 36.3; Panama 68.0; Suriname 24.1; United Arab Emirates 40.6; USA 375.0; Venezuela 169.9; Total (incl. others) 1,699.2. Source: UN, *International Trade Statistics Yearbook*.

### TRANSPORT

**Road Traffic** (Curaçao and Bonaire, motor vehicles registered, excluding government-owned vehicles, 2003): Passenger cars 57,252; Lorries 13,660; Buses 467; Taxis 227; Other cars 166; Motorcycles 1,348.

**Shipping:** *International Freight Traffic* (Curaçao, '000 metric tons, excl. petroleum, 1997): Goods loaded 215.2; Goods unloaded 516.7. *Merchant Fleet* (registered at 31 December 2004): Number of vessels 214; Total displacement 1,661,631 grt (Source: Lloyd's Register-Fairplay, *World Fleet Statistics*).

### TOURISM

**Tourist Arrivals:** *Stop-overs*: 668,425 in 2002; 728,402 in 2003; 779,325 in 2004. *Cruise-ship Passengers* (Bonaire, Curaçao and St Maarten only): 1,416,288 in 2002; 1,495,713 in 2003; 1,621,178 in 2004.

**Tourism Receipts** (NA Fl. million, incl. passenger transport): 1,683.5 in 2002; 1,761.0 in 2003; 1,906.5 in 2004.

### COMMUNICATIONS MEDIA

**Radio Receivers** (1997): 217,000 in use.

**Television Receivers** (1999): 71,000 in use.

**Telephones** (2001, UN estimate): 81,000 main lines in use.

**Mobile Cellular Telephones** (1998): 16,000 subscribers.

**Internet Users** (1999, UN estimate): 2,000.

**Daily Newspapers** (1996): 6 titles (estimated circulation 70,000 copies per issue).

Sources: UNESCO, *Statistical Yearbook*; UN, *Statistical Yearbook*; International Telecommunication Union.

### EDUCATION

**Pre-primary** (2000/01): 6,811 pupils; 316 teachers.

**Primary** (2000/01): 22,140 pupils; 1,022 teachers.

**Secondary** (2000/01): 13,392 pupils; 1,167 teachers.

**English Language Secondary** (2000/01): 377 pupils.

**Vocational** (2000/01): 1,747 pupils.

**Special Education** (2000/01): 2,337 pupils; 178 teachers.

**Teacher Training** (2000/01): 133 students; 22 teachers.

**University** (2000/01): 795 students; 131 teachers.

**Adult Literacy Rate** (UNESCO estimates): 96.5% (males 96.5%; females 96.5%) in 2000.

# Directory

## The Constitution

The form of government for the Netherlands Antilles is embodied in the Charter of the Kingdom of the Netherlands, which came into force on 20 December 1954. The Netherlands, the Netherlands Antilles and, since 1986, Aruba each enjoy full autonomy in domestic and internal affairs and are united on a basis of equality for the protection of their common interests and the granting of mutual assistance.

The monarch of the Netherlands is represented in the Netherlands Antilles by the Governor, who is appointed by the Dutch Crown for a term of six years. The central Government of the Netherlands Antilles appoints a Minister Plenipotentiary to represent the Antilles in the Government of the Kingdom. Whenever the Netherlands Council of Ministers is dealing with matters coming under the heading of joint affairs of the realm (in practice mainly foreign affairs and defence), the Council assumes the status of Council of Ministers of the Kingdom. In that event, the Minister Plenipotentiary appointed by the Government of the Netherlands Antilles takes part, with full voting powers, in the deliberations.

A legislative proposal regarding affairs of the realm and applying to the Netherlands Antilles as well as to the 'metropolitan' Netherlands is sent, simultaneously with its submission, to the Staten Generaal (the Netherlands parliament) and to the Staten (parliament) of the Netherlands Antilles. The latter body can report in writing to the Staten Generaal on the draft Kingdom Statute and designate one or more special delegates to attend the debates and furnish information in the meetings of the Chambers of the Staten Generaal. Before the final vote on a draft the Minister Plenipotentiary has the right to express an opinion on it. If he disapproves of the draft, and if in the Second Chamber a three-fifths' majority of the votes cast is not obtained, the discussions on the draft are suspended and further deliberations take place in the Council of Ministers of the Kingdom. When special delegates attend the meetings of the Chambers this right devolves upon the delegates of the parliamentary body designated for this purpose.

The Governor has executive power in external affairs, which he exercises in co-operation with the Council of Ministers. He is assisted by an advisory council, which consists of at least five members appointed by him.

Executive power in internal affairs is vested in the nominated Council of Ministers, responsible to the Staten. The Netherlands Antilles Staten consists of 22 members, who are elected by universal adult suffrage for four years (subject to dissolution). Each island forms an electoral district. Curaçao elects 14 members, Bonaire three members, St Maarten three members and Saba and St Eustatius one member each. In the islands where more than one member is elected, the election is by proportional representation. Inhabitants have the right to vote if they have Dutch nationality and have reached 18 years of age. Voting is not compulsory. Each island territory also elects its Island Council (Curaçao 21 members, Bonaire 9, St Maarten 7, St Eustatius and Saba 5), and its internal affairs are managed by an executive council, consisting of the Gezaghebber (Lieutenant-Governor), and a number of commissioners. The central Government of the Netherlands Antilles has the right to annul any local island decision which is in conflict with the public interest or the Constitution. Control of the police, communications, monetary affairs, health and education remain under the jurisdiction of the central Government.

On 1 January 1986 Aruba acquired separate status (*status aparte*) within the Kingdom of the Netherlands. However, in economic and monetary affairs there is a co-operative union between Aruba and the Antilles of the Five, known as the 'Union of the Netherlands Antilles and Aruba'.

## The Government

### HEAD OF STATE

**Queen of the Netherlands:** HM Queen BEATRIX.

**Governor:** Dr FRITZ M. DE LOS SANTOS GOEDGEDRAG.

### COUNCIL OF MINISTERS
(July 2005)

The Government comprised a seven-party coalition of the Partido Antía Restrukturá (PAR), Partido Nashonal di Pueblo (PNP), Partido Laboral Krusado Popular (PLKP), the Unión Patriótico Bonairiano (UPB), the Democratische Partij—Sint Maarten (DP—StM), the Democratic Partij—Statia (DP—StE) and the Windward Islands People's Movement (WIPM).

**Prime Minister, Minister of General Affairs and Foreign Relations:** ETIENNE YS.

**Deputy Prime Minister and Minister for Economic Affairs and Labour:** ERROL A. COVA.

**Minister of Constitutional and Internal Affairs:** RICHARD F. GIBSON.

**Minister of Finance:** ERSILIA T. M. DE LANNOY.

**Minister of Justice:** NORBERTO V. RIBEIRO.

**Minister of Education, Culture, Youth and Sports:** MARITZA D. SILBERIE.

**Minister of Transport and Telecommunications:** OMAYRA LEE-FLANGE.

**Minister of Public Health and Social Development:** JOAN THEODORA-BREWSTER.

**Minister Plenipotentiary and Member of the Council of Ministers of the Realm of the Netherlands Antilles:** PAUL R. J. COMENENCIA.

**Attorney-General of the Netherlands Antilles:** DICK A. PIAR.

### GEZAGHEBBERS
(Lieutenant-Governors)

**Bonaire:** RICHARD N. HART, Wilhelminaplein 1, Kralendijk, Bonaire; tel. 717-5330; fax 717-5100; e-mail gezag@bonairelive.com.

**Curaçao:** LISA DINDIAL, Centraal Bestuurskantoor, Concordiastraat 24, Willemstad, Curaçao; tel. (9) 461-2900.

**Saba:** ANTOINE J. M. SOLAGNIER, The Bottom, Saba; tel. 416-3215; fax 416-3274; e-mail antoine@solagnier.com.

**St Eustatius:** IRWIN E. TEMMER, Oranjestad, St Eustatius; tel. 318-2213.

**St Maarten:** FRANKLYN E. RICHARDS, Central Administration, Secretariat, Clem Labega Sq., POB 1121, Philipsburg, St Maarten; tel. 542-6085; fax 542-4172; e-mail cabgov@sintmaarten.net; internet www.sintmaarten.net/gis.

### MINISTRIES

**Office of the Governor:** Fort Amsterdam 2, Willemstad, Curaçao; tel. (9) 461-2000; fax (9) 461-1412; e-mail kabinet@kgna.an.

**Ministry of Constitutional Affairs and Internal Affairs:** Willemstad, Curaçao.

**Ministry for Economic Affairs and Labour:** Scharlooweg 106, Willemstad, Curaçao; tel. (9) 465-6236; fax (9) 465-6316; e-mail info .DEZ@ibm.net.

**Ministry of Education, Culture, Youth and Sports:** Boerhavestraat 16, Otrobanda, Willemstad, Curaçao; tel. (9) 462-4777; fax (9) 462-4471.

**Ministry of Finance:** Pietermaai 17, Willemstad, Curaçao; tel. (9) 432-8000; fax (9) 461-3339; e-mail g.d.dirfin@curinfo.an.

**Ministry of General Affairs and Foreign Relations:** Plasa Horacio Hoyer 9, Willemstad, Curaçao; tel. (9) 461-1866; fax (9) 461-1268.

**Ministry of Justice:** Willhelminaplein, Willemstad, Curaçao; tel. (9) 463-0299; fax (9) 465-8083.

**Ministry of Public Health and Social Development:** Santa Rosaweg 122, Willemstad, Curaçao; tel. (9) 736-3530; fax (9) 736-3531; e-mail vornil@cura.net.

**Ministry of Transport and Telecommunications:** Fort Amsterdam 17, Willemstad, Curaçao; tel. (9) 461-3988.

**Office of the Minister Plenipotentiary of the Netherlands Antilles:** Antillenhuis, Badhuisweg 173–175, POB 90706, 2509LS The Hague, the Netherlands; tel. (70) 3066111; fax (70) 3066110; e-mail info@antillenhuis.nl; internet www.antillenhuis.nl.

## Legislature

### STATEN

**Speaker:** D. A. S. LUCIA (PNP).

**General Election, 18 January 2002**

| Party | % of votes | Seats |
|---|---|---|
| Frente Obrero i Liberashon 30 di mei | 23.0 | 5 |
| Partido Antía Restrukturá | 20.6 | 4 |
| Partido Nashonal di Pueblo | 13.4 | 3 |
| Partido Laboral Krusado Popular | 12.1 | 2 |
| Unión Patriótico Bonairiano | 3.6 | 2 |
| Democratic Party—St Maarten | 5.5 | 2 |
| Democratische Partij—Bonaire | 2.6 | 1 |
| National Alliance* | 4.8 | 1 |
| Democratic Party—Statia | 0.5 | 1 |
| Windward Islands People's Movement | 0.5 | 1 |
| **Total** (incl. others) | 100.0 | 22 |

*Comprising the St Maarten Patriotic Alliance and the National Progressive Party.

## Political Organizations

**Democratische Partij—Bonaire (DP—B)** (Democratic Party—Bonaire): Kaya America 13A, POB 294, Kralendijk, Bonaire; tel. 717-5923; fax 717-7341; f. 1954; also known as Partido Democratico Boneriano; liberal; Leader JOPIE ABRAHAM.

**Democratische Partij—Curaçao (DP—C)** (Democratic Party—Curaçao): Neptunusweg 28, Willemstad, Curaçao; f. 1944; Leader RAYMOND BENTOERA.

**Democratische Partij—Sint Maarten (DP—StM):** Tamarind Tree Dr. 4, Union Rd, Cole Bay, St Maarten; tel. 543-1166; fax 542-4296; Leader SARAH WESCOTT-WILLIAMS.

**Democratische Partij—Statia (DP—StE):** Oranjestad, St Eustatius; Leader KENNETH VAN PUTTEN.

**Frente Obrero i Liberashon 30 di mei (FOL)** (Workers' Liberation Front of 30 May): Mayaguanaweg 16, Willemstad, Curaçao; tel. (9) 461-8105; internet www.fol.an; f. 1969; socialist; Leaders ANTHONY GODETT, RIGNALD LAK, EDITHA WRIGHT.

**Movimentu Antiyas Nobo (MAN)** (Movement for a New Antilles): Landhuis Morgenster, Willemstad, Curaçao; tel. (9) 468-4781; internet www.man.an; f. 1971; socialist; Leader DOMINICO (DON) F. MARTINA.

**National Progressive Party:** Willemstad, Curaçao; contested the 2002 elections as the National Alliance with the St Maarten Patriotic Alliance (q.v.).

**Nos Patria** (Our Fatherland): Willemstad, Curaçao; Leader CHIN BEHILIA.

**Partido Antía Restrukturá (PAR)** (Restructured Antilles Party): Fokkerweg 28, Willemstad, Curaçao; tel. (9) 465-2566; fax (9) 465-2622; e-mail par@partidopar.com; internet www.partidopar.com; f. 1993; social-Christian ideology; Leader ETIENNE YS.

**Partido Kousa Akshan Sosial (KAS):** Santa Rasaweg Naast 156, Willemstad, Curaçao; tel. (9) 747-2660; Leader BERNARD S. A. DEMEI.

**Partido Laboral Krusado Popular (PLKP):** Schouwburgweg 44, Willemstad, Curaçao; tel. (9) 737-0644; fax (9) 737-0831; internet www.cura.net/krusada; f. 1997; progressive; Leader ERROL A. COVA.

**Partido Nashonal di Pueblo (PNP)** (National People's Party): Winston Churchillweg 133, Willemstad, Curaçao; tel. (9) 869-6777; fax (9) 869-6688; internet www.pnp.an; f. 1948; also known as Nationale Volkspartij; Social Christian Party; Pres. MARIA LIBERIA-PETERS; Leader SUSANNE F. C. CAMELIA-RÖMER.

# THE NETHERLANDS ANTILLES

**Partido Obrero di Bonaire** (Bonaire Workers' Party): Kralendijk, Bonaire.

**Partido Union den Reino Ulandés (PURU):** Binnenweg 11, Willemstad, Curaçao; Leader FREDDY I. ANTERSUN.

**People's Democratic Party (PDP):** Philipsburg, St Maarten; tel. 542-2696; Leader MILLICENT DE WEEVER.

**People's Progressive Party:** Philipsburg, St Maarten.

**Saba United Democratic Party (SUDP):** Saba; tel. 416-3311; fax 416-3434; Leader STEVE HASSELL.

**Saint Eustatius Alliance (SEA):** Oranjestad, St Eustatius; Leader INGRID WHITFIELD.

**Serious Alternative People's Party (SAPP):** St Maarten; Leader JULIAN ROLLOCKS.

**St Maarten Patriotic Alliance (SPA):** Frontstraat 69, Philipsburg, St Maarten; tel. 543-1064; fax 543-1065; contested the 2002 elections as the National Alliance with the National Progressive Party (q.v.); Leader VANCE JAMES, Jr.

**Social Independiente (SI):** Willemstad, Curaçao; f. 1986 by fmr PNP mems in Curaçao; formed electoral alliance with FOL for 1990 election; Leader GEORGE HUECK.

**Unión Patriótico Bonairiano (UPB)** (Patriotic Union of Bonaire): Kaya Sabana 22, Kralendijk, Bonaire; tel. 717-8906; fax 717-5552; 2,134 mems; Christian-democratic; Leader RAMONSITO T. BOOI; Sec.-Gen. C. V. WINKLAAR.

**Windward Islands People's Movement (WIPM):** Windwardside, POB 525, Saba; tel. 416-2244; Chair. and Leader WILL JOHNSTON; Sec.-Gen. DAVE LEVENSTONE.

## Judicial System

Legal authority is exercised by the Court of First Instance (which sits in all the islands) and in appeal by the Joint High Court of Justice of the Netherlands Antilles and Aruba. The members of the Joint High Court of Justice sit singly as judges in the Courts of First Instance. The Chief Justice of the Joint High Court of Justice, its members (a maximum of 30) and the Attorneys-General of the Netherlands Antilles and of Aruba are appointed for life by the Dutch monarch, after consultation with the Governments of the Netherlands Antilles and Aruba.

### Joint High Court of Justice

Wilhelminaplein 4, Willemstad, Curaçao; tel. (9) 463-4111; fax (9) 461-8341; e-mail hofcur@cura.net.

**Chief Justice of the Joint High Court:** Dr LUIS ALBERTO JOSÉ DE LANNOY.

**Attorney-General of the Netherlands Antilles:** DICK A. PIAR.

**Secretary-Executive of the Joint High Court:** M. E. N. ROJER-DE FREITAS (acting).

## Religion

### CHRISTIANITY

Most of the population were Christian, the predominant denomination being Roman Catholicism. According to the 1992 census, Roman Catholics formed the largest single group on four of the five islands: 82% of the population of Bonaire, 81% on Curaçao, 65% on Saba and 41% on St Maarten. On St Eustatius the Methodists formed the largest single denomination (31%). Of the other denominations, the main ones were the Anglicans and the Dutch Reformed Church. There were also small communities of Jews, Muslims and Bahá'ís.

**Curaçaose Raad van Kerken** (Curaçao Council of Churches): Barenblaan 11, Willemstad, Curaçao; tel. (9) 737-3070; fax (9) 736-2183; f. 1958; six member churches; Chair. IDA VISSER; Exec. Sec. PAUL VAN DER WAAL.

### The Roman Catholic Church

The Netherlands Antilles and Aruba together form the diocese of Willemstad, suffragan to the archdiocese of Port of Spain (Trinidad and Tobago). At 31 December 2003 the diocese numbered an estimated 224,809 adherents (about 78% of the total population). The Bishop participates in the Antilles Episcopal Conference, currently based in Trinidad and Tobago.

**Bishop of Willemstad:** Rt Rev. LUIGI ANTONIO SECCO, Bisdom, Breedestraat 31, Otrobanda, Willemstad, Curaçao; tel. (9) 462-5857; fax (9) 462-7437; e-mail bisdomwstad@curinfo.an.

### The Anglican Communion

Saba, St Eustatius and St Maarten form part of the diocese of the North Eastern Caribbean and Aruba, within the Church in the Province of the West Indies. The Bishop is resident in The Valley, Anguilla.

### Other Churches

**Iglesia Protestant Uni** (United Protestant Church): Fortkerk, Fort Amsterdam, Willemstad, Curaçao; tel. (9) 461-1139; fax (9) 465-7481; f. 1825 by union of Dutch Reformed and Evangelical Lutheran Churches; Pres. D. J. LOPES; 3 congregations; 11,280 adherents.

**Methodist Church:** Oranjestad, St Eustatius.

Other denominations active in the islands include the Moravian, Apostolic Faith, Wesleyan Holiness and Norwegian Seamen's Churches, the Baptists, Calvinists, Jehovah's Witnesses, Evangelists, Seventh-day Adventists, the Church of Christ and the New Testament Church of God.

### JUDAISM

**Reconstructionist Shephardi Congregation Mikvé Israel-Emanuel:** Hanchi di Snoa 29, POB 322, Willemstad, Curaçao; tel. (9) 461-1067; fax (9) 465-4141; e-mail board@snoa.com; internet www.snoa.com; f. 1732 on present site; about 350 mems.

**Orthodox Ashkenazi Congregation Shaarei Tsedek:** Leliweg 1A, Willemstad, Curaçao; tel. (9) 737-5738; 100 mems.

## The Press

**Algemeen Dagblad:** Daphneweg 44, POB 725, Willemstad, Curaçao; tel. (9) 747-2200; fax (9) 747-2257; e-mail adcarib@cura.net; internet www.ad-caribbean.com; daily; Dutch; Editor NOUD KÖPER.

**Amigoe:** Kaya Fratumam di Skirpiri z/n, POB 577, Willemstad, Curaçao; tel. (9) 767-2000; fax (9) 767-4084; e-mail management@amigoe.com; internet www.amigoe.com; f. 1884; Christian; daily; evening; Dutch; Dir INGRID DE MAAIJER-HOLLANDER; Editor-in-Chief MICHAEL WILLEMSE; circ. 12,000.

**Bala:** Noord Zapateer nst 13, Willemstad, Curaçao; tel. (9) 467-1646; fax (9) 467-1041; daily; Papiamento.

**Beurs- en Nieuwsberichten:** A. M. Chumaceiro Blvd 5, POB 741, Willemstad, Curaçao; tel. (9) 465-4544; fax (9) 465-3411; f. 1935; daily; evening; Dutch; Editor L. SCHENK; circ. 8,000.

**Bonaire Holiday:** POB 569, Curaçao; tel. (9) 767-1403; fax (9) 767-2003; f. 1971; tourist guide; English; 3 a year; circ. 95,000.

**Bonaire Reporter:** Kaya Gob. Debrot 200-6, Bonaire; tel. and fax 717-8988; e-mail reporter@bonairereporter.com; internet bonairereporter.com; English; weekly.

**The Business Journal:** Indjuweg 30A, Willemstad, Curaçao; tel. (9) 461-1367; fax (9) 461-1955; monthly; English.

**Colors:** Liberty Publications, Curaçao; tel. and fax (9) 869-6066; e-mail colors@curacao-online.net; internet www.curacao-online.net/colors; f. 1998; general interest magazine; 4 a year; Publr TIRZAH Z. B. LIBERT.

**De Curaçaosche Courant:** Frederikstraat 123, POB 15, Willemstad, Curaçao; tel. (9) 461-2766; fax (9) 462-6535; f. 1812; weekly; Dutch; Editor J. KORIDON.

**Curaçao Holiday:** POB 569, Curaçao; tel. (9) 767-1403; fax (9) 767-2003; f. 1960; tourist guide; English; 3 a year; circ. 300,000.

**Daily Herald:** Bush Rd 22, POB 828, Philipsburg, St Maarten; tel. 542-5253; fax 542-5913; e-mail editorial@thedailyherald.com; internet www.thedailyherald.com; daily; English.

**Extra:** W. I. Compagniestraat 41, Willemstad, Curaçao; tel. (9) 462-4595; fax (9) 462-7575; daily; morning; Papiamento; Man. R. YRAUSQUIN; Editor MIKE OEHLERS; circ. 20,000.

**Newsletter of Curaçao Trade and Industry Association:** Kaya Junior Salas 1, POB 49, Willemstad, Curaçao; tel. (9) 461-1210; fax (9) 461-5422; f. 1972; monthly; English and Dutch; economic and industrial paper.

**Nobo:** Scherpenheuvel w/n, POB 323, Willemstad, Curaçao; tel. (9) 467-3500; fax (9) 467-2783; daily; evening; Papiamento; Editor CARLOS DAANTJE; circ. 15,000.

**Nos Isla:** Refineria Isla (Curazao) SA, Emmastad, Curaçao; 2 a month; Papiamento; circ. 1,200.

**La Prensa:** W. I. Compagniestraat 41, Willemstad, Curaçao; tel. (9) 462-3850; fax (9) 462-5983; e-mail laprensa@laprensacur.com; internet www.laprensacur.com; f. 1929; daily; evening; Papiamento; Man. R. YRAUSQUIN; Editor SIGFRIED RIGAUD; circ. 10,750.

# THE NETHERLANDS ANTILLES

**Saba Herald:** The Level, Saba; tel. 416-2244; f. 1968; monthly; news, local history; Editor WILL JOHNSON; circ. 500.

**St Maarten Guardian:** Vlaun Bldg, Pondfill, POB 1046, Philipsburg, St Maarten; tel. 542-6022; fax 542-6043; e-mail guardian@sintmaarten.net; f. 1989; daily; English; Man. Dir RICHARD F. GIBSON; Man. Editor JOSEPH DOMINIQUE; circ. 4,000.

**St Maarten Holiday:** POB 569, Curaçao; tel. (9) 767-1403; fax (9) 767-2003; f. 1968; tourist guide; English; 3 a year; circ. 175,000.

**Teen Times:** c/o The Daily Herald, Bush Rd 22, POB 828, Philipsburg, St Maarten; tel. 542-5597; e-mail info@teentimes.com; internet www.teentimes.com; for teenagers by teenagers; sponsored by The Daily Herald; English; Editor-in-Chief MICHAEL GRANGER.

**Ultimo Noticia:** Frederikstraat 123, Willemstad, Curaçao; tel. (9) 462-3444; fax (9) 462-6535; daily; morning; Papiamento; Editor A. A. JONCKHEER.

**La Unión:** Rotaprint NV, Willemstad, Curaçao; weekly; Papiamento.

### NEWS AGENCIES

**Algemeen Nederlands Persbureau (ANP)** (Netherlands): Panoramaweg 5, POB 439, Willemstad, Curaçao; tel. (9) 461-2233; fax (9) 461-7431; Representative RONNIE RENS.

**Associated Press (AP)** (USA): Roodeweg 64, Willemstad, Curaçao; tel. (9) 462-6586; Representative ORLANDO CUALES.

## Publishers

**Curaçao Drukkerij en Uitgevers Maatschappij:** Willemstad, Curaçao.

**Ediciones Populares:** W. I. Compagniestraat 41, Willemstad, Curaçao; f. 1929; Dir RONALD YRAUSQUIN.

**Drukkerij Scherpenheuvel NV:** Scherpenheuvel, POB 60, Willemstad, Curaçao; tel. (9) 467-1134.

**Drukkerij de Stad NV:** W. I. Compagniestraat 41, Willemstad, Curaçao; tel. (9) 462-3566; fax (9) 462-2175; e-mail kenrick@destad.an; f. 1929; Dir KENRICK A. YRAUSQUIN.

**Holiday Publications:** POB 569, Curaçao; tel. (9) 767-1403; fax (9) 767-2003.

**Offsetdrukkerij Intergrafia NV:** Essoweg 54, Willemstad, Curaçao; tel. (9) 464-3180.

## Broadcasting and Communications

### TELECOMMUNICATIONS

**Curaçao Telecom:** Schottegatweg Oost 19, Willemstad, Curaçao; tel. 736-1056; fax 736-1057; internet www.curacaotelecom.com; f. 1999; bought by Digicel (Ireland) in 2005; telephone and internet services; Chair. DENIS O'BRIEN.

**East Caribbean Cellular NV (ECC):** 13 Richardson St, Philipsburg, St Maarten; tel. 542-2100; fax 542-5675; e-mail info@eastcaribbeancellular.com; internet www.eastcaribbeancellular.com; f. 1989.

**Servicio de Telekomunikashon (SETEL):** F. D. Rooseveltweg 337, POB 3177, Willemstad, Curaçao; tel. (9) 833-1222; fax (9) 868-2596; e-mail setel@curinfo.an; internet www.curinfo.an; f. 1979; telecommunications equipment and network provider; state-owned, but expected to be privatized; Pres. ANGEL R. KOOK; Man. Dir JULIO CONSTANSIA; 400 employees.

**Smitcoms NV:** Dr A. C. Wathey Cruise & Cargo Facility, St Maarten; tel. 542-9140; fax 542-9141; e-mail matthews@sintmaarten.net; internet smitcomsltd.com; f. 2000; international telephone network provider; Man. Dir CURTIS K. HAYNES.

**St Maarten Telephone Co (TelEm):** C. A. Cannegieter St 17, POB 160, Philipsburg, St Maarten; tel. 542-2278; fax 543-0101; e-mail lpeters@telem.an; internet www.sinmaarten.net; f. 1975; local landline and value-added services, also operates TelCell digital cellular service; 15,000 subscribers; Man. Dir CURTIS K. HAYNES.

**United Telecom Services (UTS):** Schouwburgweg 22, POB 103, Willemstad, Curaçao; tel. (9) 777-0101; fax (9) 777-1238; e-mail info@antele.com; f. 1908; fmrly called Antelecom NV; Chair. DAVID DICK; Man. Dir HENDRIK J. EIKELENBOOM.

### BROADCASTING

#### Radio

**Easy 97.9 FM:** Arikokweg 19A, Willemstad, Curaçao; tel. (9) 462-3162; fax (9) 462-8712; e-mail radio@easyfm.com; internet www.easyfm.com; Dir KEVIN CARTHY.

**Radio Caribe:** Ledaweg 35, Brievengat, Willemstad, Curaçao; tel. (9) 736-9555; fax (9) 736-9569; f. 1955; commercial station; programmes in Dutch, English, Spanish and Papiamento; Dir-Gen. C. R. HEILLEGGER.

**Radio Curom** (Curaçaose Radio-Omroep Vereniging): Roodeweg 64, POB 2169, Willemstad, Curaçao; tel. (9) 462-6586; fax (9) 462-5796; f. 1933; broadcasts in Papiamento; Dir ORLANDO CUALES.

**Radiodifusión Boneriana NV:** Kaya Gobernador Debrot 2, Kralendijk, Bonaire; tel. 717-8273; fax 717-8220; e-mail vdb@vozdibonaire.com; internet www.vozdibonaire.com; f. 1980; Owner FELICIANO DA SILVA PILOTO.

**Voz di Bonaire (PJB2)** (Voice of Bonaire): broadcasts in Papiamento, Spanish and Dutch.

**Radio Exito:** Wolkstraat 15, Willemstad, Curaçao; tel. (9) 462-5577; fax (9) 462-5580.

**Radio Hoyer NV:** Plasa Horacio Hoyer 21, Willemstad, Curaçao; tel. (9) 461-1678; fax (9) 461-6528; e-mail hoyer@cura.net; internet www.radiohoyer.com; f. 1954; commercial; two stations: Radio Hoyer I (mainly Papiamento, also Spanish) and II (mainly Dutch, also English) in Curaçao; Man. Dir HELEN HOYER.

**Radio Korsou FM:** Bataljonweg 7, POB 3250, Willemstad, Curaçao; tel. (9) 737-3012; fax (9) 737-2888; e-mail master@korsou.com; internet www.korsou.com; 24 hrs a day; programmes in Papiamento and Dutch; Gen. Man. ALAN H. EVERTSZ.

**Laser 101 (101.1 FM):** tel. (9) 737-7139; fax (9) 737-5215; e-mail master@laser101.com; internet www.laser101.fm; 24 hours a day; music; English and Papiamento; Gen. Man. ALAN H. EVERTSZ.

**Radio Paradise:** ITC Bldg, Piscadera Bay, POB 6103, Curaçao; tel. (9) 463-6103; fax (9) 463-6404; Man. Dir J. A. VISSER.

**Radio Tropical:** Willemstad, Curaçao; Dir DWIGHT RUDOLPHINA.

**Ritme FM (PJB4):** broadcasts in Dutch.

**Trans World Radio (TWR):** Kaya Gouverneur N. Debrotweg 64, Kralendijk, Bonaire; tel. 717-8800; fax 717-8808; e-mail 800am@twr.org; internet www.twrbonaire.com; f. 1964; religious, educational and cultural station; programmes to South, Central and North America, Caribbean in six languages; Pres. Dr DAVID TUCKER, Jr; Station Dir RICHARD FULLER.

**Voice of St Maarten (PJD2 Radio):** Plaza 21, Backstreet, POB 366, Philipsburg, St Maarten; tel. 542-2580; fax 542-4905; also operates PJD3 on FM (24 hrs); commercial; programmes in English; Gen. Man. DON R. HUGHES.

**Voice of Saba (PJF1):** The Bottom, POB 1, Saba; studio in St Maarten; tel. 546-3213; also operates The Voice of Saba FM; Man. MAX W. NICHOLSON.

There is a relay station for Radio Nederland on Bonaire.

#### Television

**Antilliaanse Televisie Mij NV** (Antilles Television Co): Berg Arraret, POB 415, Willemstad, Curaçao; tel. (9) 461-1288; fax (9) 461-4138; f. 1960; operates Tele-Curaçao (fmrly operated Tele-Aruba); commercial; govt-owned; also operates cable service, offering programmes from US satellite television and two Venezuelan channels; Dir JOSÉ M. CIJNTJE; Gen. Man. NORMAN K. RICHARDS.

**Leeward Broadcasting Corporation—Television:** Philipsburg, St Maarten; tel. (5) 23491; transmissions for approx. 10 hours daily.

Five television channels can be received on Curaçao, in total. Relay stations provide Bonaire with programmes from Curaçao, St Maarten with programmes from Puerto Rico, and Saba and St Eustatius with programmes from St Maarten and neighbouring islands. Curaçao has a publicly owned cable television service, TDS.

# Finance

(cap. = capital; res = reserves; dep. = deposits; m. = million; brs = branches; amounts in Netherlands Antilles guilders unless otherwise stated)

## BANKING

### Central Bank

**Bank van de Nederlandse Antillen** (Bank of the Netherlands Antilles): Simon Bolivar Plein 1, Willemstad, Curaçao; tel. (9) 434-5500; fax (9) 461-5004; e-mail info@centralbank.an; internet centralbank.an; f. 1828 as Curaçaosche Bank, name changed as above 1962; cap. 30.0m., res 111.4m., dep. 688.6m. (Dec. 2002); Chair. RALPH PALM; Pres. Dr EMSLEY D. TROMP; 2 brs on St Maarten and Bonaire.

### Commercial Banks

**ABN AMRO Bank NV:** Kaya Flamboyan 1, POB 3144, Willemstad, Curaçao; tel. (9) 763-8000; fax (9) 737-0620; f. 1964; Gen. Man. H. V. IGNACIO; 6 brs.

**Banco di Caribe NV:** Schottegatweg Oost 205, POB 3785, Willemstad, Curaçao; tel. (9) 432-3000; fax (9) 461-5220; e-mail info@bancodicaribe.com; internet www.bancodicaribe.com; f. 1973; dep. 941.4m., total assets 1,025.3m. (Dec. 2004); Chair. W. J. CURIEL; CEO and Gen. Man. Dir E. DE KORT; Dir K. ABRAHAM; 5 brs.

**Banco Mercantil CA (Banco Universal):** Abraham de Veerstraat 1, POB 565, Willemstad, Curaçao; tel. (9) 461-8241; fax (9) 461-1824; f. 1988; Gen. Man. FRANK GIRIGORI.

**Banco de Venezuela NV:** POB 131, c/o Amicorp NV, Bronsweg 8A, Willemstad, Curaçao; tel. (9) 434-3500; fax (9) 434-3533; f. 1993; Man. Dirs H. P. F. VON AESCH, R. YANES, V.E. BORBERG.

**Bank of Nova Scotia NV** (Canada): Backstreet 64, POB 303, Philipsburg, St Maarten; tel. 542-3317; fax 542-2562; f. 1969; Man. ROBERT G. JUDD.

**Barclays Bank plc** (United Kingdom): 29 Front St, POB 941, Philipsburg, St Maarten; tel. 542-3511; fax 542-4531; f. 1959; Man. EDWARD ARMOGAN (offices in Saba and St Eustatius).

**Chase Manhattan Bank NA** (USA): Chase Financial Center, Vlaun Bldg, Cannegieter Rd (Pondfill) and Mullet Bay Hotel, POB 921, Philipsburg, St Maarten; tel. 542-3726; fax 542-3692; f. 1971; Gen. Man. K. BUTLER.

**CITCO Banking Corporation NV:** Kaya Flamboyan 9, POB 707, Willemstad, Curaçao; tel. (9) 732-2322; fax (9) 732-2330; e-mail cbc@citco.com; f. 1980 as Curaçao Banking Corporation NV; Man. Dir and Gen. Man. R. F. IRAUSQUIN; Man. Dir A. A. HART.

**Fortis Bank (Curaçao) NV:** Berg Arrarat 1, POB 3889, Willemstad, Curaçao; tel. (9) 463-9300; fax (9) 461-3769; internet www.fortisbank.com; f. 1952 as Pierson, Heldring and Pierson (Curaçao) NV, became Meespierson (Curaçao) NV in 1993, name changed as above in 2000; international banking/trust company; Man. Dir GREGORY ELIAS.

**Giro Curaçao NV:** Scharlooweg 35, Willemstad, Curaçao; tel. (9) 433-9999; fax (9) 461-7861; Gen. Dir L. C. BERGMAN; Financial Dir H. L. MARTHA.

**ING Bank NV** (Internationale Nederlanden Bank NV): Kaya W. F. G. (Jombi) Mensing 14, POB 3895, Willemstad, Curaçao; tel. (9) 732-7000; fax (9) 732-7502; f. 1989 as Nederlandse Middenstandsbank NV, name changed as above 1992; Gen. Man. MARK SCHNEIDERS.

**Maduro & Curiel's Bank NV:** Plaza Jojo Correa 2–4, POB 305, Willemstad, Curaçao; tel. (9) 466-1100; fax (9) 466-1130; e-mail info@mcb-bank.com; internet www.mcb-bank.com; f. 1916 as NV Maduro's Bank; merged with Curiel's Bank in 1931; affiliated with Bank of Nova Scotia NV, Toronto; cap. 50.2m., res 119.3m., dep. 3,079.7m. (Dec. 2003); Chair. N. D. HENRÍQUEZ; Man. Dirs WILLIAM H. L. FABRO, RON GOMES CASSERES; 25 brs.

**Orco Bank NV:** Dr Henry Fergusonweg 10, POB 4928, Willemstad, Curaçao; tel. (9) 737-2000; fax (9) 737-6741; e-mail info@orcobank.com; internet www.orcobank.com; f. 1986; cap. 30.7m., res 27.5m., dep. 523.9m. (Dec. 1999); Chair. E. L. GARCIA; Man. Dir I. D. SIMON; 1 br.

**Rabobank Curaçao NV:** Zeelandia Office Park, Kaya W. F. G. (Jombi), Mensing 14, POB 3876, Willemstad, Curaçao; tel. (9) 465-2011; fax (9) 465-2066; e-mail l.an.curacao.ops@rabobank.com; internet www.rabobank.com; f. 1978; cap. US $53.0m., res US $17.8m., dep. US $4,535.2m. (Dec. 2003); Chair. S. SCHAT; Gen. Man. J. S. KLEP.

**RBTT Bank NV:** Kaya Flamboyan 1, Willemstad; tel. (9) 763-8000; fax (9) 763-8449; e-mail info@tt.rbtt.com; internet www.rbtt.com; f. 1997 as Antilles Banking Corpn; name changed to RBTT Bank Antilles in 2001; name changed as above in 2002; Pres. RODNEY S. PRASAD; Chair. PETER J. JULY; 4 brs.

**SFT Bank NV:** Schottegatweg Oost 44, POB 707, Willemstad, Curaçao; tel. (9) 732-2900; fax (9) 732-2902.

**Windward Islands Bank Ltd:** Clem Labega Square 7, POB 220, Philipsburg, St Maarten; tel. 542-2313; fax 542-4761; affiliated to Maduro and Curiel's Bank NV; f. 1960; cap. and res 3.6m., dep. 53.6m. (Dec. 1984); Man. Dirs VICTOR P. HENRÍQUEZ, W. G. H. STRIJBOSCH.

### 'Offshore' Banks
(without permission to operate locally)

**ABN AMRO Bank Asset Management (Curaçao) NV:** Kaya Flamboyan 1, POB 3144, Willemstad, Curaçao; tel. (9) 736-6755; fax (9) 736-9246; f. 1976; Man. D. M. VROEGINDEWEY.

**Abu Dhabi International Bank NV:** Kaya W. F. G. (Jombi), Mensing 36, POB 3141, Willemstad, Curaçao; tel. (9) 461-1299; fax (9) 461-5392; internet www.adibwash.com; f. 1981; cap. US $20.0m., res $30.4m., dep. $329.3m. (Dec. 2001); Pres. QAMBAR AL MULLA; Man. Dir NAGY S. KOLTA.

**Banco Caracas NV:** Kaya W. F. G. (Jombi) Mensing 36, POB 3141, Willemstad, Curaçao; tel. (9) 461-1299; fax (9) 461-5392; f. 1984; Pres. GEORGE L. REEVES.

**Banco Consolidado NV:** Handelskrade 12, POB 3141, Willemstad, Curaçao; tel. (9) 461-3423; f. 1978.

**Banco Latino NV:** De Ruyterkade 61, POB 785, Willemstad, Curaçao; tel. (9) 461-2987; fax (9) 461-6163; f. 1978; cap. US $25.0m., res $12.3m., dep. $450.8m. (Nov. 1992); Chair. Dr GUSTAVO GÓMEZ LÓPEZ; Pres. FOLCO FALCHI.

**Banco Provincial Overseas NV:** Santa Rosaweg 51–55, POB 5312, Willemstad, Curaçao; tel. (9) 737-6011; fax (9) 737-6346; Man. E. SUARES.

**Banque Artesia Curaçao NV:** Castorweg 22–24, POB 155, Willemstad, Curaçao; tel. (9) 461-8061; fax (9) 461-5151; f. 1976 as Banque Paribas Curaçao NV; name changed as above 1998.

**Caribbean American Bank NV:** POB 6087, TM1 10, WTC Bldg, Piscadera Bay, Willemstad, Curaçao; tel. (9) 463-6380; fax (9) 463-6556; Man. Dir Dr MARCO TULIO HENRÍQUEZ.

**F. Van Lanschot Bankiers (Curaçao) NV:** Schottegatweg Oost 32, POB 4799, Willemstad, Curaçao; tel. (9) 737-1011; fax (9) 737-1086; f. 1962; Man. A. VAN GEEST.

**First Curaçao International Bank NV:** Office Park Zeelandia, Kaya W. F. G. (Jombi) Mensing 18, POB 299, Willemstad, Curaçao; tel. (9) 737-2100; fax (9) 737-2018; f. 1973; cap. and res US $55m., dep. $244m. (1988); Pres. and CEO J. CH. DEUSS; Man. M. NEUMAN-ROUIRA.

**Toronto Dominion (Curaçao) NV:** c/o SCRIBA NV, Polarisweg 31–33, POB 703, Willemstad, Curaçao; tel. (9) 461-3199; fax (9) 461-1099; f. 1981; Man. E. L. GOULDING.

**Union Bancaire Privée (TDB):** J. B. Gorsiraweg 14, POB 3889, Willemstad, Curaçao; tel. (9) 463-9300; fax (9) 461-4129.

Other 'offshore' banks in the Netherlands Antilles include American Express Overseas Credit Corporation NV, Banco Aliado NV, Banco del Orinoco NV, Banco Mercantil Venezolano NV, Banco Principal NV, Banco Provincial International NV, Banunion NV, CFM Bank NV, Citco Banking Corporation NV, Compagnie Bancaire des Antilles NV, Deutche Bank Finance NV, Ebna Bank NV, Exprinter International Bank NV, Integra Bank NV, Lavoro Bank Overseas NV, Lombard-Atlantic Bank NV, Middenbank (Curaçao) NV, Netherlands Caribbean Bank NV, Noro Bank NV, Premier Bank International NV.

### Development Banks

**Ontwikkelingsbank van de Nederlandse Antillen NV:** Schottegatweg Oost 3C, POB 267, Willemstad, Curaçao; tel. (9) 747-3000; fax (9) 747-3320; e-mail obna@curinfo.an; f. 1981.

**Stichting Korporashon pa Desaroyo di Korsou (KORPDEKO):** Breedestraat 29C, POB 656, Willemstad, Curaçao; tel. (9) 461-6699; fax (9) 461-3013.

### Other Banks

**Postspaarbank van de Nederlandse Antillen:** Waaigatplein 7, Willemstad, Curaçao; tel. (9) 461-1126; fax (9) 461-7561; f. 1905; post office savings bank; Chair. H. J. J. VICTORIA; cap. 21m.; 20 brs.

**Spaar- en Beleenbank van Curaçao NV:** MCB Salinja Bldg, Schottegatweg Oost 130, Willemstad, Curaçao; tel. (9) 466-1585; fax (9) 466-1590.

There are also several mortgage banks and credit unions.

THE NETHERLANDS ANTILLES — *Directory*

### Banking Associations

**Association of International Bankers in the Netherlands Antilles (IBNA):** Chumaceiro Blvd 3, POB 220, Curaçao; tel. (9) 461-5367; fax (9) 461-5369; e-mail info@ibna.an; internet www.ibna.an; Pres. HANS F. C. BLANKVOORT.

**Bonaire Bankers' Association:** POB 288, Kralendijk, Bonaire.

**Curaçao Bankers' Association (CBA):** A. M. Chumaceiro Blvd 3, Willemstad, Curaçao; tel. (9) 465-2486; fax (9) 465-2476; e-mail florisela.bentoera@an.rbtt.com; f. 1972; Pres. RODNEY PRASAD; Sec. FLORISELA BENTOERA.

**Federashon di Kooperativanan di Spar i Kredito Antiyano (Fekoskan):** Curaçaostraat 50, Willemstad, Curaçao; tel. (9) 462-3676; fax (9) 462-4995; e-mail fekoskan@attglobal.net.

**International Bankers' Association in the Netherlands Antilles:** Scharlooweg 55, Willemstad, Curaçao.

**The Windward Islands Bankers' Association:** Clem Labega Square, Philipsburg, St Maarten; tel. 542-2313; fax 542-4761.

### INSURANCE

**Amersfoortse Antillen NV:** Kaya W. F. G. Mensing 19, Willemstad, Curaçao; tel. (9) 461-6399; fax (9) 461-6709.

**Aseguro di Kooperativa Antiyano (ASKA) NV:** Scharlooweg 15, Willemstad, Curaçao; tel. (9) 461-7765; fax (9) 461-5991; accident and health, motor vehicle, property.

**Ennia Caribe Schaden NV:** J. B. Gorsiraweg 6, POB 581, Willemstad, Curaçao; tel. (9) 434-3800; fax (9) 434-3873; e-mail mail@ennia.com; f. 1948; general; life insurance as Ennia Caribe Leven NV; Pres. DONALD BAKHUIS; Man. Dir ALBARTUS WILLEMSEN.

**ING Fatum:** Cas Coraweg 2, Willemstad, Curaçao; tel. (9) 777-7777; fax (9) 461-2023; f. 1904; property insurance.

**MCB Group Insurance NV:** MCB Bldg Scharloo, Scharloo, Willemstad, Curaçao; tel. (9) 466-1370; fax (9) 466-1327.

**Netherlands Antilles and Aruba Assurance Company (NA&A) NV:** Pietermaai 135, Willemstad, Curaçao; tel. (9) 465-7146; fax (9) 461-6269; accident and health, motor vehicle, property.

**Seguros Antilliano NV:** S. b. N. Doormanweg/Reigerweg 5, Willemstad, Curaçao; tel. (9) 736-6877; fax (9) 736-5794; general.

A number of foreign companies also have offices in Curaçao, mainly British, Canadian, Dutch and US firms.

### Insurance Association

**Insurance Association of the Netherlands Antilles (NAVV):** c/o Ing Fatum, POB 3002, Cas Coraweg 2, Willemstad, Curaçao; tel. (9) 777-7777; fax (9) 736-9658; Pres. R. C. MARTINA-JOE.

## Trade and Industry

### DEVELOPMENT ORGANIZATIONS

**Curaçao Industrial and International Trade Development Company NV (CURINDE):** Emancipatie Blvd 7, Landhuis Koninsplein, Curaçao; tel. (9) 737-6000; fax (9) 737-1336; e-mail info@curinde.com; internet www.curinde.com; f. 1980; state-owned; manages the harbour free zone, the airport free zone and the industrial zone; Man. Dir E. R. SMEULDERS.

**Foreign Investment Agency Curaçao (FIAC):** Scharlooweg 174, Curaçao; tel. (9) 465-7044; fax (9) 461-5788; e-mail fiac@curinfo.an.

**World Trade Center Curaçao:** POB 6005, Piscadera Bay, Curaçao; tel. (9) 463-6100; fax (9) 462-4408; e-mail info@wtccuracao.com; Man. Dir JOSÉ VICENTE SANCHES PIÑA.

### CHAMBERS OF COMMERCE

**Bonaire Chamber of Commerce and Industry:** Princess Mariestraat, POB 52, Kralendijk, Bonaire; tel. 717-5595; fax 717-8995.

**Curaçao Chamber of Commerce and Industry:** Kaya Junior Salas 1, POB 10, Willemstad, Curaçao; tel. (9) 461-3918; fax (9) 461-5652; e-mail businessinfo@curacao-chamber.an; internet www.curacao-chamber.an; f. 1884; Chair. HERMAN BEHR; Exec. Dir PAUL R. J. COMENENCIA.

**St Maarten Chamber of Commerce and Industry:** C. A. Cannegieterstraat 11, POB 454, Philipsburg, St Maarten; tel. 542-3590; fax 542-3512; e-mail coci@sintmaarten.net; f. 1979; Exec. Dir J. M. ARRINDELL VAN WINDT.

### INDUSTRIAL AND TRADE ASSOCIATIONS

**Association of Industrialists of the Netherlands Antilles (ASINA):** Kaya Junior Salas 1, Willemstad, Curaçao; tel. (9) 461-2353; fax (9) 465-8040; tel. asina@cura.net; f. 1981; Pres. E. ZIMMERMAN.

**Bonaire Trade and Industry Asscn** (Vereniging Bedrijfsleven Bonaire): POB 371, Kralendijk, Bonaire.

**Curaçao Exporters' Association (CEA):** c/o Seawings NV, Maduro Plaza z/n CEA, POB 6049, Curaçao; tel. (9) 733-1591; fax (9) 733-1599; e-mail albert.elens@seawings-curacao.com; f. 1903; Dir ALBERT ELENS.

**Curaçao International Financial Services Asscn (CIFA):** Chumaceiro Blvd 3, POB 220, Curaçao; tel. (9) 461-5371; fax (9) 461-5378; e-mail info@cifa.an; internet www.cifa.an; Chair. EDSEL L. DORAN.

**Curaçao Trade and Industry Asscn** (Vereniging Bedrijfsleven Curaçao—VBC): Kaya Junior Salas 1, POB 49, Willemstad, Curaçao; tel. (9) 461-1210; fax (9) 461-5652; e-mail vbc1@cura.net; f. 1944; Pres. B. KOOYMAN; Exec. Dir R. P. J. LIEUW.

### MAJOR COMPANIES

The following are some of the leading industrial and commercial companies currently operating in the Netherlands Antilles:

**ADM Milling Co (Netherlands Antilles) NV:** Brionwerf/Emancipatie Blvd w/n, POB 290, Willemstad, Curaçao; tel. (9) 461-6627; fax (9) 461-6116; e-mail admcur@interneeds.net; f. 1971; previously called Continental Milling Co (Netherlands Antilles) NV; millers for bakeries, producers of animal feeds; Gen. Man. BRYAN G. POOL; 37 employees.

**Antillean Paper and Plastic Co NV:** Industrial Park Brievengat, POB 3505, Curaçao; tel. (9) 737-6422; fax (9) 737-2424; f. 1976; plastics, paper; subsidiaries produce plastic containers and soaps; Mans HUBERT VAN GRIEKEN, E. J. HALABI, J. B. M. KOOL; 56 employees.

**Antillean Soap Co NV:** Industrial Park Brievengat, Willemstad, Curaçao; tel. (9) 737-7177; fax (9) 737-7191; f. 1976; manufactures powder and liquid soap, disinfectants, abrasives, industrial cleaners; Man. Dir EDWARD J. HALABI; 87 employees.

**Antilliaanse Brouwerij NV:** Rijkseenheid Blvd w/n, POB 465, Willemstad, Curaçao; tel. (9) 461-2944; fax (9) 461-2035; e-mail info@amstelcuracao.com; internet www.amstelcuracao.com; f. 1958; manufacture and distribution of beer and soft drinks; Man. Dir JOHANNES BRUNING; 121 employees.

**Antilliaanse Emballagefabriek NV:** Westwerf, POB 3247, Curaçao; f. 1961; manufactures packaging materials such as cans, foils and plastic bags, solvents, margarine and milk powder; Man. E. A. BRUSSEN; 25 employees.

**Antilliaanse Verffabrick NV:** Asteroidenweg z/n, POB 3944, Curaçao; tel. (9) 37-9866; fax (9) 37-2048; f. 1956; manufactures paint products; Man. E. VAN ARKEL; 47 employees.

**Betonbouw NV:** Baai Macolaweg 8, POB 3884, Curaçao; tel. (9) 461-1293; fax (9) 461-2354; e-mail beton@cura.net; f. 1975; construction; Man. Dir CO KLEYN; 300 employees.

**BOPEC** (Bonaire Petroleum Corporation NV): Kralendijk, POB 117, Bonaire; tel. 717-8177; fax 717-8266; e-mail bopec@bonairelive.com; oil terminal with storage capacity of 10.1m. barrels; bought by PDVSA (Venezuela) in 1989; Gen. Man. ERROL RIENHART.

**Building Depot Curaçao NV:** Industrial Park Zeelandia, POB 3712, Curaçao; tel. (9) 461-3233; fax (9) 461-8732; internet www.building-depot.net; e-mail bdinfo@building-depot.net; f. 1977; sale of construction materials, hardware and household items; Man. RODNEY M. LUCIA; 107 employees.

**Cargill Salt Bonaire NV:** Bonaire; internet www.cargillsalt.com; salt production and export; has operated the Solar Salt Works since 1997 when Cargill Salt acquired the North American assets of Akzo Nobel Salt Inc.

**Caribbean Bottling Co Ltd:** Kaminda André J. E. Kusters 6, POB 302, Willemstad, Curaçao; tel. (9) 461-2488; fax (9) 465-1377; e-mail info@pop.an; internet www.pop.an; f. 1948; produces carbonated drinks; Man. A. MADURO; 110 employees.

**Caribbean Metal Products Inc:** Industrial Park Brievengat, POB 4054, Curaçao; tel. (9) 737-0633; fax (9) 737-0106; f. 1989; manufactures masonry nails, galvanized common nails and umbrella nails; Man. R. LUCIA; 17 employees.

**Carnefco Group:** Kaya Buena Vista w/n, POB 3121, Willemstad, Curaçao; tel. (9) 869-2121; fax (9) 869-2486; f. 1973; ship maintenance, grit-blasting, wet-blasting, high-pressure water blasting, internal coatings, etc.; Man. Dir W. VOS; 60 employees.

# THE NETHERLANDS ANTILLES

**Curaçao Beverage Bottling Co NV:** Rijkseenheid Blvd 1, POB 95, Willemstad, Curaçao; tel. (9) 463-3311; fax (9) 461-1310; e-mail info@cocacola.an; internet www.fria.com; f. 1938; bottlers and manufacturers of soft drinks, many under licence; Man. Dir TIBOR LUCKMANN; 105 employees.

**Curaçao Mining Co Inc:** Nieuwpoort z/n, POB 3078, Curaçao; tel. (9) 767-3400; fax (9) 767-6721; f. 1875; produces phosphate, limestone pebbles in five sizes, sand; Man. H. DE VOOGD; 70 employees.

**Curaçao Oil NV (CUROIL):** A. M. Chumaceiro Blvd 15, POB 3927, Curaçao; tel. (9) 432-0000; fax (9) 461-3335; e-mail curoil@curoil.com; internet www.curoil.com; f. 1985; fuel and petroleum lubricants marketing company; supplies automotive, marine, aviation and industrial fuels and lubricants; Man. Dir M. J. NICOLINA; 80 employees.

**Kooyman NV:** Kaya W. F. G. (Jombi) Mensing 44, POB 3062, Willemstad, Curaçao; tel. (9) 461-3433; fax (9) 461-5806; e-mail bkooyman@kooyman.an; internet www.kooymannv.com; f. 1939; building supplies, hardware, steel/aluminium goods, glass, timber goods, etc.; Man. Dir BASTIAAN KOOIJMAN; 300 employees.

**Lovers Industrial Corporation NV:** Industrial Park Brievengat, Curaçao; tel. (9) 737-0499; fax (9) 737-1747; e-mail lovers@curinfo.an; f. 1984; produces fruit juices, ice cream, frozen yoghurt; Man. OSWALD C. VAN DER DIJS; 180 employees.

**Otto Senior NV:** Scharlooweg 25–27, Willemstad, Curaçao; tel. (9) 461-1701; fax (9) 461-5978; e-mail dserphos@cura.net; internet www.businesscuracao.com/ritz; f. 1938; 'Ritz' trade name; dairy products, fruit juices, processed foods, catering supplies; Mans I. F. SERPHOS, D. I. SERPHOS, R. H. SERPHOS; 112 employees.

**Plastico NV:** Industrial Park Brievengat, POB 3561, Curaçao; tel. (9) 737-9568; fax (9) 737-5843; e-mail info@plastico-nv.com; internet www.plastico-nv.com; f. 1985; manufactures polythene pipes for water-distribution and irrigation purposes; Man. L. LIEUW-SJONG; 20 employees.

**Refinería Isla (Curazao) SA:** Margrietlaan, Emmastad, POB 3843, Curaçao; tel. (9) 466-2273; fax (9) 466-2488; e-mail jhdsisl@ccopl.pdv.com; f. 1985; refinery established in 1915 by Royal Dutch-Shell Co, but from 1985 operated by PDVSA (Venezuela); petroleum products; Pres. JOAQUIN TREDINICK; Man. JAVIER HERNÁNDEZ; 1,360 employees.

**Refinería di Korsou NV:** Ara Hilltop Office Complex, POB 3627, Curaçao; tel. (9) 461-1050; fax (9) 461-1250; f. 1985; owner co of the fmr Shell Curaçao Refinery and the former Curaçao Oil Terminal; Man. H. PARISIUS; 16 employees.

**Softex Products NV:** Industrial Park Brievengat, POB 3795, Curaçao; tel. (9) 737-7811; fax (9) 737-7903; e-mail softex@cura.net; f. 1976; produces paper towels, napkins, etc.; Man. P. LIEUW; 35 employees.

**Vasos Antillanos NV:** Industrial Park Brievengat, Curaçao; tel. (9) 737-7488; fax (9) 737-2424; f. 1986; produces plastic cold-drink cups; Man. H. VAN GRIEKEN; 20 employees.

**WIMCO** (West India Mercantile Co Ltd): POB 74, Willemstad, Curaçao; tel. (9) 461-1833; fax (9) 461-1627; e-mail info@wimco-nv.com; internet www.wimco-nv.com; f. 1928; wholesale and retail of household electrodomestic goods; Man. Dir VICTOR HENRÍQUEZ; 90 employees.

## UTILITIES

### Electricity and Water

**Aqualectra Production NV (KAE):** Rector Zwijsenstraat 1, POB 2097, Curaçao; tel. (9) 433-2200; fax (9) 462-6685; e-mail mgmt@aqualectra.com; internet www.aqualectra.com; Dir S. MARTINA.

**GEBE NV:** Pond Fill, W. J. A. Nisbeth Rd, POB 123, St Maarten; tel. 542-2213; fax 542-4810; f. 1961; Man. Dir J. A. LAMBERT.

**Water & Energiebedrijf Bonaire (WEB) NV:** Carlos Nicolaas 3, Kralendijk; tel. 717-8244.

## TRADE UNIONS

**Algemene Bond van Overheidspersoneel (ABVO)** (General Union of Civil Servants): POB 3604, Willemstad, Curaçao; tel. (9) 737-6097; fax (9) 737-3145; e-mail abvo_na@cura.net; internet www.abvoinforma.org; f. 1936; Pres. R. H. IGNACIO; Sec. W. E. CALMES; 4,000 mems.

**Algemene Federatie van Bonaireaanse Werknemers (AFBW):** Kralendijk, Bonaire.

**Central General di Trahado di Corsow (CGTC)** (General Headquarters for Workers of Curaçao): POB 2078, Willemstad, Curaçao; tel. (9) 737-6097; fax (9) 737-3145; e-mail abvo-na@cura.net; f. 1949; Sec.-Gen. ROLAND H. IGNACIO.

**Curaçaosche Federatie van Werknemers** (Curaçao Federation of Workers): Schouwburgweg 44, Willemstad, Curaçao; f. 1964; Pres. WILFRED SPENCER; Sec.-Gen. RONCHI ISENIA; 204 affiliated unions; about 2,000 mems.

**Federashon Bonaireana di Trabou (FEDEBON):** Kaya Krabè 6, Nikiboko, POB 324, Bonaire; tel. and fax 717-8845; Pres. GEROLD BERNABELA.

**Petroleum Workers' Federation of Curaçao:** Willemstad, Curaçao; tel. (9) 737-0255; fax (9) 737-5250; affiliated to Int. Petroleum and Chemical Workers' Fed.; f. 1955; Pres. R. G. GIJSBERTHA; approx. 1,500 mems.

**Sentral di Sindikatonan di Korsou (SSK)** (Confederation of Curaçao Trade Unions): Schouwburgweg 44, POB 3036, Willemstad; tel. (9) 737-0794; 6,000 mems.

**Sindikato di Trahado den Edukashon na Korsou (SITEK)** (Curaçao Schoolteachers' Trade Union): Landhuis Stenen Koraal, Willemstad, Curaçao; tel. (9) 468-2902; fax (9) 469-0552; 1,234 mems.

**Windward Islands' Federation of Labour (WIFOL):** Pond Fill, Long Wall Rd, POB 1097, St Maarten; tel. 542-2797; fax 542-6631; e-mail wifol@sintmaarten.net; Pres. THEOPHILUS THOMPSON.

# Transport

## RAILWAYS

There are no railways.

## ROADS

All the islands have a good system of all-weather roads. There were 590 km of roads in 1992, of which 300 km were paved.

## SHIPPING

Curaçao is an important centre for the refining and transhipment of Venezuelan and Middle Eastern petroleum. Willemstad is served by the Schottegat harbour, set in a wide bay with a long channel and deep water. Facilities for handling containerized traffic at Willemstad were inaugurated in 1984. A Mega Cruise Facility, with capacity for the largest cruise ships, has been constructed on the Otrobanda side of St Anna Bay. Ports at Bullen Bay and Caracas Bay also serve Curaçao. St Maarten is one of the Caribbean's leading ports for visits by cruise ships and in January 2001 new pier facilities were opened which could accommodate up to four cruise ships and add more cargo space. Each of the other islands has a good harbour, except for Saba, which has one inlet, equipped with a large pier. In May 2002 the Netherlands provided NA Fl. 9.6m. for the repair of Saba's port, which sustained severe hurricane damage in 1999. Many foreign shipping lines call at ports in the Netherlands Antilles.

**Curaçao Ports Authority:** Werf de Wilde, POB 3266, Willemstad, Curaçao; tel. (9) 461-4422; fax (9) 461-3907; e-mail cpamanag@cura.net; internet curports.com; Man. Dir RICHARD LÓPEZ-RAMÍREZ.

**Curaçao Shipping Association (SVC):** c/o Dammers & van der Heide (Antilles) Inc, Kaya Flamboyan 11, Willemstad, Curaçao; tel. (9) 737-0600; fax (9) 737-3875; Pres. K. PONSEN.

**St Maarten Ports Authority:** J. Yrausquin Blvd, POB 146, Philipsburg, St Maarten; tel. 542-2307; fax 542-5048; e-mail smpa1shh@sintmaarten.net; internet www.portofstmaarten.com; Man. Dir ROMMEL CHARLES.

### Principal Shipping Companies

**Caribbean Cargo Services NV:** Jan Thiel w/n, POB 442, Willemstad, Curaçao; tel. (9) 467-2588.

**Curaçao Dry-dock Co Inc:** POB 3012, Curaçao; tel. (9) 733-0000; fax (9) 736-5580; e-mail marketing@cdmnv.com; internet www.cdmnv.com; f. 1958; Man. Dir MARIO RAYMOND EVERTSZ.

**Curaçao Ports Authority (CPA) NV:** Werf de Wilde z/n, POB 689, Curaçao; tel. (9) 434-5999; fax (9) 461-3907; e-mail cpamanag@cura.net; internet www.curports.com; Man. Dir RICHARD LÓPEZ-RAMÍREZ.

**Curaçao Ports Services Inc NV (CPS):** Curaçao Container Terminal, POB 170, Curaçao; tel. (9) 461-5079; fax (9) 461-6536; e-mail cps@ibm.net; Man. Dir KAREL JAN O. ASTER.

**Dammers & van der Heide, Shipping and Trading (Antilles) Inc:** Kaya Flamboyan 11, POB 3018, Willemstad, Curaçao; tel. (9) 737-0600; fax (9) 737-3875; e-mail general@dammers-curacao.com; internet www.dammers-curacao.com; f. 1964; Man. Dir J. J. PONSEN.

**Gomez Transport NV:** Zeelandia, Willemstad, Curaçao; tel. (9) 461-5260; fax (9) 461-3358; e-mail gomez-shipping@ibm.net; Man. FERNANDO DA COSTA GÓMEZ.

**Hal Antillen NV:** De Ruyterkade 63, POB 812, Curaçao.

**Intermodal Container Services NV:** Fokkerweg 30, Willemstad, POB 3747, Curaçao; tel. (9) 461-3330; fax (9) 461-3432; Mans A. R. BEAUJON, N. N. HARMS.

**Kroonvlag Curaçao NV:** Maduro Plaza, POB 231, Curaçao; tel. (9) 737-6900; fax (9) 737-1266; e-mail hekro@cura.net.

**Lagendijk Maritime Services:** POB 3481, Curaçao; tel. (9) 465-5766; fax (9) 465-5998; e-mail ims@ibm.net.

**S. E. L. Maduro & Sons (Curaçao) Inc:** Maduro Plaza, POB 3304, Willemstad, Curaçao; tel. (9) 733-1501; fax (9) 733-1506; e-mail hmeijer@madurosons.com; Man. Dir H. MEIJER; Vice-Pres. R. CORSEN.

**St Maarten Port Services:** POB 270, Philipsburg, St Maarten; tel. 542-2304.

**Anthony Veder & Co NV:** Zeelandia, POB 3677, Curaçao; tel. (9) 461-4700; fax (9) 461-2576; e-mail anveder@ibm.net; Man. Dir JOOP VAN VLIET.

### CIVIL AVIATION

There are international airports at Curaçao (Dr Albert Plesman, or Hato, 12 km from Willemstad), Bonaire (Flamingo Field) and St Maarten (Princess Juliana, 16 km from Philipsburg); and airfields for inter-island flights at St Eustatius and Saba. In 1998 a free trade zone was inaugurated at the international airport on Curaçao. The second phase of a US $118m. project to expand Princess Juliana Airport commenced in June 2004. Financing was secured for the construction of new passenger terminal building at Dr Albert Plesman Airport in September 2003. The national carrier of the Netherlands Antilles, known as Dutch Caribbean Airlines (DCA) from 2002, was declared bankrupt in late 2004.

**Windward Islands Airways International (WIA—Winair) NV:** Princess Juliana Airport, POB 2088, Philipsburg, St Maarten; tel. 545-2568; fax 545-4229; e-mail info@fly-winair.com; internet www.fly-winair.com; f. 1961; govt-owned since 1974; scheduled and charter flights throughout north-eastern Caribbean; Man. Dir EDWIN HODGE.

## Tourism

Tourism is a major industry on all the islands. The principal attractions for tourists are the white, sandy beaches, marine wildlife and diving facilities. There are marine parks in the waters around Curaçao, Bonaire and Saba. The numerous historic sites are of interest to visitors. The largest number of tourists visit St Maarten, Curaçao and Bonaire. In 2004 stop-over visitors totalled some 779,325 (of whom 61.0% were on St Maarten). In the same year 1,621,178 cruise-ship passengers visited St Maarten, Curaçao and Bonaire (of whom 83.2% were on St Maarten).

**Bonaire Tourism Corporation:** Kaya Grandi 2, Kralendijk, Bonaire; tel. 717-8322; fax 717-8408; e-mail info@tourismbonaire.com; internet www.infobonaire.com; Dir RONELLA CROES.

**Curaçao Tourism Development Bureau (CTDB):** Pietermaai 19, POB 3266, Willemstad, Curaçao; tel. (9) 434-8200; fax (9) 461-2305; e-mail info@ctbd.net; internet www.curacao-tourism.com; f. 1989; Dir JAMES HEPPLE.

**Saba Tourist Office:** Windwardside, POB 527, Saba; tel. 416-2231; fax 416-2350; e-mail iluvsaba@unspoiledqueen.com; internet www.sabatourism.com; Dir GLENN C. HOLM.

**St Eustatius Tourist Office:** Fort Oranje Straat z/n, Oranjestad, St Eustatius; tel. and fax 318-2433; e-mail euxtour@goldenrock.net; internet www.statiatourism.com; Dir ALIDA FRANCIS.

**St Maarten Tourist Bureau:** Vineyard Office Park, W. G. Buncamper Rd 33, Philipsburg, St Maarten; tel. 5422-337; fax 542-2734; e-mail info@st-maarten.com; internet www.st-maarten.com; Dir CORNELIUS DE WEEVER.

### HOTEL ASSOCIATIONS

**Bonaire Hotel and Tourism Association:** Kralendijk, Bonaire; e-mail info@bonhata.org; internet www.bonhata.org; Man. Dir JACK CHALK.

**Curaçao Hospitality and Tourism Association (CHATA):** POB 6115, Kurason Komèrsio, Curaçao; tel. (9) 465-1005; fax (9) 465-1052; e-mail information@chata.org; internet www.chata.org; f. 1967 as Curaçao Hotel Asscn; Pres. ROLF SPRECHER.

**St Maarten Hospitality and Trade Association:** W. J. A. Nisbeth Rd 33A, POB 486, Philipsburg, St Maarten; tel. 542-0108; fax 542-0107; e-mail info@shta.com; internet www.shta.com; Pres. EMIL LEE.

## Defence

Although defence is the responsibility of the Netherlands, compulsory military service is laid down in an Antilles Ordinance. The Governor is the Commander-in-Chief of the armed forces in the islands, and a Dutch contingent is stationed in Willemstad, Curaçao. The Netherlands also operates a Coast Guard Force (to combat organized crime and drugs-smuggling), based at St Maarten and Aruba. In May 1999 the US air force and navy began patrols from a base on Curaçao to combat the transport of illegal drugs.

**Commander of the Navy:** Brig.-Gen. R. L. ZUIDERWIJK.

## Education

Education was made compulsory in 1992. The islands' educational facilities are generally of a high standard. The education system is the same as that of the Netherlands. Dutch is used as the principal language of instruction in schools on the 'Leeward Islands', while English is used in schools on the 'Windward Islands'. Instruction in Papiamento (using a different spelling system from that adopted by Aruba) has been introduced in primary schools. Primary education begins at six years of age and lasts for six years. Secondary education lasts for a further five years. The University of the Netherlands Antilles, sited on Curaçao, had 795 students in 2000/01. In April 2002 the Netherlands Government made more than €12.7m. available for improvements to education provision in the Netherlands Antilles. In 1995 local government expenditure on education in the 'Antilles of the Five' was NA Fl. 178.9m. (19.3% of total spending by the island governments).

## Bibliography

For works on the Caribbean generally, see Select Bibliography (Books)

Brown, E. *Suriname and the Netherlands Antilles*. Lanham, MD, Scarecrow Press, 1992.

Ferguson, J. *Eastern Caribbean*. New York, NY, Latin American Bureau, 1997.

Keinders, A. *Politieke Geschiedenis van de Nederlandse Antillen en Aruba, 1950–93*. Zutphen, Walburg Press, 1993.

Klomp, A. *Politics on Bonaire*. Assen, Van Gorcum, 1986.

Koulen, I. *Netherlands Antilles and Aruba: A Research Guide*. London, ICP Publishing Ltd, 1987.

Schaap, C. D. *Fighting Money Laundering*. Dordrecht, Kluwer Law International, 1998.

Schoenhais, K. *Netherlands Antilles and Aruba*, World Bibliographical Series. Oxford, ABC Clio, 1993.

# NICARAGUA

## Geography

### PHYSICAL FEATURES

The Republic of Nicaragua spans the Central American isthmus and is the largest country lying between Mexico and Colombia. Nicaragua tapers westwards from its long eastern coast on the Caribbean, to a south-west-facing Pacific coast, but also tapers southwards from the long, north-western border (922 km) with Honduras, to the shorter, more east–west border (309 km) with Costa Rica (on the latter frontier, the question of Costa Rican navigation rights on the San Juan river has currently become vexatious). In the far north-west Nicaragua looks across the mouth of the Gulf of Fonseca at El Salvador—the two countries and Honduras are holding tripartite discussions on the delimitation of maritime boundaries in the Gulf. Such boundaries also concern Nicaragua in the Caribbean, where it has challenged Honduras and Colombia over some 50,000 sq km (19,300 sq miles) of maritime territory, owing to the Colombian possession of a number of islands and cays lying to the east of the Nicaraguan coast (San Andrés lies 180 km offshore, included administratively with Providencia and a number of other satellites). Nicaraguan territory also includes some islands and islets, notably the Corn Islands, some 70 km off the Caribbean coast near Bluefields, and the Miskito cays further north. The total area of the country is 130,373 sq km, including 10,034 sq km of inland waters.

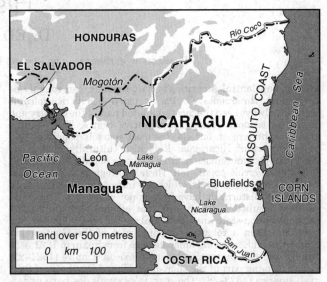

Nicaragua is about 440 km from north to south and 450 km, at its widest, from east to west. Like most of the Central American countries, it consists of a central highlands divide between a Pacific coastal region and a wider Caribbean (Atlantic) coastal region. In Nicaragua, which is dominated by the great lake that bears its name, in the south-west of the country, the Pacific lowlands are rather more involved topographically. Rising immediately above the narrow plains along the shore are the Coastal or Diriamba Highlands, which are volcanic and, mainly in the north, still active. These heights subside into the Rivas isthmus as they head south, separating Lake Nicaragua from the ocean (by no more than 17 km at one point). Between this line and the central mountain uplands lies the Great Rift, fertile central lowlands dominated by two great lakes—Managua, on the south-western shores of which is the eponymous national capital, and, to the south of it, the 8,157-sq-km Nicaragua, one of the largest freshwater lakes in the world. The Pacific lowlands broaden southwards around the lakes, following the Rift, sufficient to allow Lake Nicaragua actually to empty into the Caribbean (along the San Juan). This is possible because the central mountain region tapers southwards, as the main thrust of the Continental Divide becomes more diffuse. The highlands, therefore, mainly rise in the north-centre of the country, the highly dissected region contrasting forested ranges with fertile basins and valleys. The highest point is Mogoton (2,438 m—8,002 ft), in the north-west. Finally, the eastern or Caribbean region is dominated by the Mosquito Coast (Costa de Mosquitos—an erroneous adaptation of the name of the local Amerindians, the Miskitos) and its hinterland. This involves the lower slopes of the central uplands merging into an extensive, alluvial plain with gentle valleys, giving way to an often marshy coast of shallow bays, lagoons and salt marshes. The eastern lowlands tend to be savannah, but the coast is dominated by dense rainforest. In fact, about 27% of Nicaragua is forested, and 46% is classified as pastureland. This terrain is well watered, not just by copious rainfall in most places, but by many rivers—the longest is the 780-km Coco, which forms much of the eastern part of the Honduran border. The varied ecosystems common to the countries of Central America (owing to the contrasting topography and its position as the land bridge between two continents) shelter in Nicaragua the usual array of wildlife—numerous bird species, big cats, a variety of reptiles, anteaters, monkeys, etc.—but the country's wide maritime shelf also allows it to boast the richest marine fauna in the Caribbean.

### CLIMATE

The climate is tropical in the lowlands and cooler in the highlands. Nicaragua is very susceptible to hurricanes, with volcanoes and earthquakes adding to nature's destructive potential, so the country is particularly prone to landslides. This is not helped by deforestation and the consequent soil erosion. Generally, it is slightly cooler and much wetter in the east than in the Pacific west. The rainy season is May–January, and average annual rainfall in Managua, located in the western lowlands, is 1,140 mm (44 ins). This figure can be more than doubled in the east. The average temperatures in Managua range from 20°C (68°F) to 34°C (93°F) over the year.

### POPULATION

Most Nicaraguans ('Nicas') are Mestizo (77%), of mixed race, but there are fairly substantial white (10%) and black (9%) communities, with the balance consisting of indigenous Amerindians (who may be underenumerated). The black population, which is concentrated in the east, mainly along the Caribbean coast, is generally of Jamaican descent or Black Carib (escaped African slaves who married into a Carib Amerindian population, preserving a Carib or Garifuna language, but were deported from St Vincent island—now in Saint Vincent and the Grenadines—in 1797 to an island off the Honduran coast). Most of the population (96%) use Spanish, the official language, but, as well as Garifuna, a Creole and English are also used in the east (apart from those descended from an anglophone background, there are traces of the time when the United Kingdom claimed a protectorate over the Mosquito Coast). The indigenous languages of the country are mainly of the Miskito, Sumo and Rama groups, in the east, with others spoken by the small numbers of Amerindians in the west, the Monimbo and Subtiaba. About 80% of the population are Roman Catholic and the remainder mostly Evangelical Protestant.

According to official estimates for July 2004, the total population numbered 5.6m. Almost one-half of this population is under 15 years of age, 56% are urbanized (the highest rate of urbanization in Central America—the same as Panama) and it is one of the poorest populations in the Western hemisphere. Most people live in the Pacific lowlands and in the adjacent high country. In mid-2003 the capital, Managua, had an estimated population of 1.1m., with the second city, León, to the north-west, totalling over 120,000 inhabitants in 1995. The country is divided into 15 departments, with the Mosquito Coast now consisting of two autonomous regions (Atlántico Norte and Atlántico Sur).

# History

## Dr ED BROWN

Based on an earlier essay by LUCIANO BARACCO and revised by the editorial staff

Nicaragua, Central America's largest republic, can roughly be divided into three zones. The Pacific region with its rich volcanic soils has always been the most populated part of the country and forms its economic and administrative centre. The Atlantic region is the largest, but also the most sparsely populated and isolated part of Nicaragua. It is also culturally distinctive, having important minorities from a variety of indigenous groups and a black/creole population. These two regions are separated by the Central Highlands, which are important for coffee cultivation and historically have been major zones of political and military conflict. Nicaragua did not emerge as a fully independent nation state until 1838. During the colonial period, the area had been a relatively isolated part of the Guatemalan Audiencia and, following independence from Spain in 1821, was briefly annexed by the Iturbidean Mexican Empire (1821–23) before forming part of the short-lived United Provinces of Central America (1823–38). The country's boundaries have undergone a number of alterations since independence: Guanacaste was ceded to Costa Rica in 1858; in the 20th century the islands of San Andreas and Providencia were ceded to Colombia, and a large piece of territory in north-eastern Nicaragua was granted to Honduras by the International Court of Justice in the 1960s.

### NICARAGUA, 1838–1936

The first decades of Nicaraguan independence are known as the 'Age of Anarchy', owing to successive civil wars between the Liberal élites of León and the Conservatives of Granada. With these two cities acting like autonomous states, the emergence of a national state apparatus was severely impeded. In 1854 Leonese Liberals recruited a group of North American mercenaries, led by William Walker, to fight their Granadan rivals. After defeating the Conservatives, Walker turned on his Liberal patrons, declared himself President, reintroduced slavery and established English as the official language of the republic. However, his ambitions of regional domination united the rest of Central America against him and he was eventually removed from office. The Walker episode was a turning point for the Nicaraguan élite. The Liberals, weakened by their association with the American mercenary, compromised with Conservative Presidents over the next 30 years. The stability that this brought saw the first coherent nation-building project of Nicaragua's history as the country's infrastructure began to be developed and the judicial and institutional basis of the state was enhanced; both of these facilitated the introduction of several new export crops. It is interesting to note that this nation-building impulse was born under a period of Conservative rule, suggesting the inadequacies of traditional interpretations which depict the Conservatives as reactionary opponents of the modernizing impulses of the Liberals.

While the foundations of the modern Nicaraguan state were laid in the 30 years of Conservative hegemony, it was the Liberal regime of José Santos Zelaya (1893–1909) that marked the definitive insertion of the Nicaraguan economy into the international division of labour as a supplier of primary commodities (primarily coffee, but also bananas, gold and timber). This expansion was supported by further developments of the national road, telegraph and rail systems and an expanded state role in the promotion of the emerging export economy. Zelaya also introduced a Constitution separating the Church from the State, modernized the army, gained control of the Atlantic Coast from the United Kingdom and encouraged overseas capital investment. However, the courting of European and Japanese capital for the construction of an inter-oceanic canal in southern Nicaragua provoked a US-supported rebellion by Nicaragua's Conservatives, which forced Zelaya into exile in 1909 and ushered in 25 years of intermittent US occupation. Marines were stationed in Nicaragua from 1912 to 1925 and again from 1926 to 1933 in order to protect a series of Conservative presidents from Liberal rebellions. During this period Nicaragua virtually became a colony of the USA. The incipient Nicaraguan state became enmeshed within a financial system where profligate loans were guaranteed by turning over Nicaragua's customs revenues, the national railroads and the national bank to US interests. Meanwhile, the Bryan-Chamorro Treaty of 1916 granted the USA exclusive rights to the construction of any canal within Nicaraguan territory (thereby safeguarding the Panamanian canal from any rival Nicaraguan project).

The second period of occupation saw the USA modify their support for the Conservatives in an effort to find a solution that would allow the USA to continue to dominate Nicaragua, but without the constant need to have troops on the ground. As a result, an attempt was made to foment a longer-term agreement between the Liberals and the Conservatives. Such an agreement was reached in 1927 and included provisions for US-supervised elections to be held in 1927, together with the continued presence of US troops in Nicaragua until a new 'non-partisan' National Guard could be trained. Only one Liberal lieutenant, Augusto César Sandino, rejected these measures and, with a 300-strong 'Defending Army of National Sovereignty', continued to fight a sustained guerrilla war against the US forces, their Conservative puppets and the Liberal leadership, whom Sandino now saw as little more than traitors. Learning from early defeats in open warfare, Sandino and his followers quickly developed very successful guerrilla tactics and the US forces became embroiled in a protracted conflict.

The US marines eventually left following the election of the Liberal Juan Bautista Sacasa in 1932, leaving the National Guard to carry on the struggle against Sandino. In February 1933, however, Sandino signed a new peace accord with the Sacasa Government, which guaranteed that US troops would not return, granted an amnesty to Sandino's followers and created a new department in the north of the country, which they would settle in and police themselves. The Sandinistas, however, continued to be harassed by a uncontrollable National Guard, while Sacasa was becoming increasingly concerned about the aspirations of the Guard's director, Anastasio Somoza García. These fears proved well grounded when Sandino was murdered on Somoza's orders on the 21 February 1934. Somoza eventually overthrew Sacasa in May 1936. Highly dubious presidential elections were held in December of the same year and on 1 January 1937 Somoza was installed as President of Nicaragua. He and his sons after him were to stay in control of Nicaragua until 1979.

### THE SOMOZA DYNASTY

Anastasio Somoza García proved to be a skilful manipulator of successive US Administrations. He utilized his control of the National Guard, clever political manoeuvring and strategies of accommodation with the traditional oligarchy to sustain his position within Nicaragua. Following his assassination in 1956 by the poet Rigoberto López Pérez, Anastasio Somoza García was succeeded as President by his son, Luís Somoza Debayle, while his youngest son, Anastasio Somoza Debayle (or 'Tachito'), became Commander-in-Chief of the National Guard. Luís pursued a much lower profile and more humanitarian style of leadership, opening up the political system, at least ostensibly, and instigating a range of social programmes under the US-sponsored Alliance for Progress. On the death of Luís in 1967, his brother assumed the presidency and, in spite of lacking the political skill of his father, operated a more repressive form of dictatorship.

Under the Somozas, the Nicaraguan economy expanded considerably as they facilitated the successive evolution of a series of further agro-export crops (including beef, sugar and especially cotton, although coffee continued to be important) and a certain level of industrial development. The size of the state apparatus also grew in order to promote and service the rapidly growing economy. However, their control of the National Guard, their close ties with the US Administration, as well as rampant nepotism and cronyism enabled the Somozas to ensure that the state essentially acted to service their own business interests (and those of their closest allies), rather than national economic and social development. As such, the Somozas were able rapidly to construct a business empire that virtually monopolized the most modern sectors of the Nicaraguan economy. It is estimated that by 1979 the Somozas owned around 40% of Nicaragua's economy.

The regime finally began to fragment in the 1970s as the brutality, corruption and political weakness of 'Tachito' grew increasingly apparent. As organized opposition to the dictatorship grew, Somoza's response became ever more repressive. His abuse of humanitarian aid, following an earthquake that destroyed Managua in 1972, and the assassination of the Conservative opposition leader, Pedro Joaquín Chamorro, in 1978, helped to unite internal opposition to the regime (see below). Moreover, Somoza's excesses were also leading to international condemnation and, most importantly, the partial withdrawal of support from the Administration of US President Jimmy Carter (1977–81). Facing international isolation, a growing popular insurrection and a disintegrating National Guard, Anastasio Somoza Debayle resigned the presidency in July 1979. He fled to Paraguay, where he was assassinated in September 1980.

## THE SANDINISTA NATIONAL LIBERATION FRONT

The overthrow of the Somoza dictatorship represented one of the most broadly based and popular insurrections in Latin America's history. Its origins lay in 1961 when, inspired by the Cuban revolution, a group of students from the University of León (together with surviving members of Sandino's Defending Army) formed the Frente Sandinista de Liberación Nacional (FSLN—Sandinista National Liberation Front). Despite the fact that many of its founders were Marxists, the FSLN rapidly broadened its ideological base, and its politics represented a mix of nationalism, radical Christianity and a brand of socialism that owed more to cultural practices peculiar to Latin America than to the USSR. By the end of the 1970s, the FSLN had become the hegemonic group within an extremely broad revolutionary alliance (which also included the non-Somocista business élite, sections of the Roman Catholic Church, left-wing activists and intellectuals) and had succeeded in mobilizing large numbers of the urban poor and the rural peasantry.

The Somoza regime, however, remained firmly in control of the country until the mid-1970s, inflicting serious losses on the FSLN in the late 1960s and early 1970s. By 1974 most of the FSLN's founders were dead or imprisoned, many of its underground networks had been destroyed, and the organization was internally divided. However, the dictator's increasingly brutal and arbitrary repression over the ensuing years contributed to growing unrest across Nicaraguan society, which enabled the FSLN to attract increasing numbers of new recruits. The murder of Chamorro in January 1978 provoked mass demonstrations against Somoza and was followed by a national strike and a series of uprisings in the major cities in September. Following the Government's brutal suppression of these demonstrations, the FSLN prepared for the launching of a final offensive in April and June of 1979. Sensing the imminent fall of Somoza, US President Carter attempted to negotiate his replacement with members of the non-Sandinista opposition. However, owing to the growing popularity of the FSLN, and the refusal of Somoza to accept this proposal, the plan failed. By late July Somoza had been removed and Nicaragua was ruled by a Government of National Reconstruction dominated by the FSLN.

## THE SANDINISTA GOVERNMENT, 1979–90

The guiding principles of the new regime were the pursuit of political pluralism, a mixed economy, popular participation, and a non-aligned foreign policy. The state was to function according to the 'logic of the majority', with priority given to the interests of the poor. The five-member governing Junta consisted of two members of the anti-Somoza bourgeoisie and three Sandinistas, among them the General Secretary of the FSLN, Daniel Ortega Saavedra, who became *de facto* President of Nicaragua. The 1974 Constitution was abrogated, the bicameral Congreso Nacional (National Congress) was dissolved and the National Guard was replaced with the Sandinista People's Army. In May 1980 a 47-seat Council of State was convened (increased to 51 seats in May 1981) to act as a legislative body. The new assembly consisted of representatives of various mass organizations active within Nicaraguan civil society. Although the Sandinistas enjoyed an institutionalized parliamentary majority, the non-Sandinista business sectors and constitutional opposition parties were also represented. The new Constitution guaranteed a range of civil liberties common to liberal democracies, abolished the death penalty and established an independent judiciary.

Elections to the presidency and to a new 96-seat Asamblea Nacional (National Assembly) were held on 4 November 1984. The FSLN presidential candidate, Ortega, received 67% of the votes cast, and the FSLN won 61 seats in the Asamblea Nacional. Most national and international observers endorsed the legitimacy of the elections (with the significant exception of the USA) and they were undoubtedly indicative of the high level of popular support enjoyed by the Sandinistas in the mid-1980s, particularly among the beneficiaries of wide-ranging agricultural, health and educational reforms.

Under the Sandinista Government, Nicaragua was characterized by a series of popular consultations and Asamblea Nacional debates, which led to the promulgation of a new Constitution on 9 January 1987. Prominent amongst its provisions was the granting of autonomy to the indigenous population of the Atlantic coast. This reflected a recognition of the errors of earlier heavy-handed centralizing policies that had prompted armed resistance by various indigenous groups, including the Miskito Yatama (Mother Earth) group. Initially, the Sandinistas had promoted innovative forms of governance across the country organized at a local community level; although the management demands of the intensifying military conflict against the Contras (counter-revolutionary forces) gradually led to more authoritarian tendencies from the mid-1980s.

Economically, the Sandinista regime pursued a number of redistributive policies, which included subsidies on basic goods and nationalization (although the public sector never accounted for more than 40% of the economy during the Sandinista period and the private sector, despite its frequently antagonistic attitude towards the revolution, was directly encouraged through incentives). An agrarian reform programme was also undertaken, which gradually redistributed lands taken first from the Somozas and later from other large landowners deemed not to be using their land productively. Initially, a significant proportion of this occurred through the creation of large state farms but, as the policy progressed, the majority was granted to landless peasants, both in the form of co-operatives and individual land holdings. By the end of the 1980s more than 50% of Nicaragua's agricultural land had been affected by the land reform programme. Furthermore, in the urban areas, many unauthorized inhabitants of unoccupied land or premises were made legal and received materials with which to construct houses.

For the first few years of the Sandinista regime, the economy performed relatively well, especially given the fact that the whole continent was contending with a debt crisis. By 1988, however, the effects of the US trade embargo, imposed in 1985, and the ongoing conflict with the Contras had severely undermined the original Sandinista economic model. Production levels and earnings from the important agricultural export sector collapsed (owing to the combination of labour shortages, the disruption to production caused by the war and dramatic declines in the prices of the major commodities), leading to unsustainable increases in the national debt. The combination of this worsening external situation with rapidly expanding budget deficits caused by the escalating costs of defence (over one-half of the national budget was spent on defence by 1985) gradually developed into a severe economic crisis. The problems were compounded by the Government's attempts to narrow the deficit by printing money, thereby accelerating a hyperinflationary spiral. That same year, an economic adjustment plan was instituted, which reduced public-sector employment and state subsidies and devalued the currency. The initiation of peace negotiations also resulted in significant reductions in military expenditure. These measures had some success in combating inflation but, by the time of the elections in early 1990, the parlous state of the economy remained a key problem facing the Nicaraguan people.

## OPPOSITION AND THE WAR

The period of relative political stability following the Sandinistas' insurrection was short-lived. Ronald Reagan's inauguration as US President in 1981 marked the beginning of a US campaign to destabilize the revolution. The US Government claimed that the FSLN were totalitarian communists who directly threatened US security interests and other countries in the region, as well as the liberties of the Nicaraguan people. On this basis, Reagan adopted an active policy of resourcing, arming and training counter-revolutionary forces (Contras). The most important of these, initially consisting of around 2,000 former National Guards, were based in Honduras and were organized into the Movimiento Democrático Nicaragüense (MDN—Nicaraguan Democratic Movement) and the Fuerzas Democráticas Nicaragüenses (FDN—Nicaraguan Democratic Forces). A smaller force, the Alianza Revolucionaria Democrática (ARDE—Democratic Revolutionary Alliance), led by a disillusioned Sandinista commander, Edén Pastora Gómez, was based in Costa Rica.

By 1985 the Contras had recruited up to 10,000 members, including a significant number of peasants disaffected by the Sandinistas' agrarian reform programme. With the support of the Honduran army and financial aid from the USA (as well as the direct involvement of US personnel in some operations), the Contras were able to organize numerous incursions into Nicaragua from the early 1980s. Their targets included health facilities and schools, as well as the country's infrastructure and productive activities. It was estimated that nearly 45,000 Nicaraguans were killed, wounded or abducted in the war and about one-quarter of a million people forced to flee the major zones of Contra activity. In 1986 increasing US congressional opposition to the funding of the Contras and growing international censure provoked elements of the Reagan Administration to become involved in the illegal transfer to the Contras of the proceeds of clandestine sales of military equipment to Iran. The eventual exposure of these actions, known as the Iran–Contra scandal, contributed to the strengthening of a congressional ban on aid to the Contras.

Despite their destructive impact on the Sandinista economic project, the failure of the Contras to achieve a decisive military victory, combined with the end of US aid, prompted them to enter into peace negotiations. The first in a series of agreements, known as the Esquipulas Peace Accords, was signed in 1987. These accords laid the foundations for a general disarmament process, with international observation by the Organization of American States (OAS). Subsequent accords (Esquipulas II and III) led to further disarmament and provided the framework for the 1990 elections.

## THE 1990 ELECTIONS

The presidential, legislative and municipal elections of February 1990 were a landmark in Nicaraguan history. Recognized to have been free and fair by international observers, the Sandinistas abided by the results. Such peaceful transitions of power were rare in Nicaraguan history. The ballot produced an unexpected victory for the Unión Nacional Opositora (UNO—National Opposition Union), a hastily prepared coalition of 14 anti-Sandinista parties. The UNO's presidential candidate, Violeta Barrios de Chamorro, widow of the assassinated Pedro Joaquín Chamorro, won 55% of the votes cast, compared with Ortega's 41%. The opposition alliance also gained an absolute majority in the Asamblea Nacional and control of 101 of 132 municipal councils. The belief that a Sandinista victory would mean the continuation of war and economic hardship, the unpopularity of military conscription and the increasingly hierarchical attitude of the Sandinista leadership were the principal causes of the Government's electoral defeat.

The new head of state, Violeta Chamorro, a former member of the Government of National Reconstruction in the 1980s, was seen as a figure of national reconciliation. On assuming the presidency she avoided confrontation with the FSLN by reappointing Sandinista Gen. Humberto Ortega as Commander-in-Chief of the Armed Forces, and by agreeing to uphold the 1987 Constitution. Chamorro's actions led to tensions with more extremist factions within the UNO coalition, led by the Vice-President, Virgilio Godoy Reyes, and the newly elected mayor of Managua, Arnoldo Alemán Lacayo. This infighting within the governing coalition led to the collapse of the UNO as a coherent political force in early 1993. Following the failure of the UNO, a new consensus was reached between a number of former UNO deputies, the FSLN, the Partido Conservador de Nicaragua (PCN—Conservative Party of Nicaragua), and the President. Right-wing UNO deputies reformed as the Alianza Política Opositora (Political Opposition Alliance) in February 1993 and boycotted the Asamblea Nacional, demanding the dismissal of Gen. Ortega and of the Minister to the Presidency, Antonio Lacayo Oyanguren.

## THE UNO GOVERNMENT

Although the victory of the UNO coalition generated fears of revenge-taking and political chaos, owing to the extremism of some of its component parties, government policy quickly dispelled such concerns. The architect of government policy was Antonio Lacayo, the son-in-law of Chamorro and *de facto* prime minister. Lacayo represented Nicaragua's modernizing bourgeoisie and sought pragmatically to implement a range of market-orientated economic reforms, while recognizing the continued influence of the FSLN. However, a number of factors prevented the successful implementation of Lacayo's 'centrist solution'. Chamorro's conciliatory stance towards the Sandinistas led to the suspension of US aid on the advice of the Chairman of the US Senate's Foreign Relations Committee, Jesse Helms. Moreover, vital loans from the IMF were contingent on the implementation of economic-liberalization measures and public-expenditure reductions, which provoked industrial action by opposition-led trade unions and popular organizations. By mid-1991 Lacayo's strategic policy of market-led modernization had been replaced by short-term crisis management. Further economic reforms were enacted only after lengthy negotiations with the Sandinistas.

## THE FSLN IN OPPOSITION

Despite their electoral defeat, the Sandinistas remained the largest single party in the Asamblea Nacional and the most experienced, organized political party in Nicaragua. They continued to exert informal influence over the police and the army, both of which were led by prominent Sandinistas. The FSLN also exercised considerable influence among the trade unions and civil associations. However, the party's popularity among the wider population suffered considerably from what became known as 'la piñata', a process whereby large amounts of state and government property was transferred to the leadership of the FSLN during the two-month transition period between the FSLN and UNO Governments.

In 1994 internal divisions within the FSLN over its co-operation with the Chamorro Government led to the formation of two factions. The Izquierda Democrática Sandinista (Sandinista Democratic Left), led by Daniel Ortega, supported a more orthodox revolutionary outlook and carried the support of mass organizations, the trade unions and the urban and rural poor. This faction was opposed by the more socio-democratic, 'renewalist' faction, led by a former Vice-President and leader of the FSLN in the Asamblea Nacional, Sergio Ramírez Mercado. The Ortega faction emerged successful from the ensuing political struggle, and in January 1995 Ramírez left the FSLN to form the Movimiento de Renovación Sandinista (MRS—Sandinista Renewal Movement). The majority of the Sandinista deputies moved to the MRS, leaving the FSLN with just eight deputies. However, the overwhelming majority of the grassroots membership remained loyal to the FSLN.

## OPPOSITION TO THE CHAMORRO GOVERNMENT

As well as constitutional opposition, the Chamorro presidency was challenged by an increasing number of armed opposition groups. Following the 1990 elections, the OAS supervised a progressive demobilization of Contra forces, and the Nicaraguan army was reduced from 90,000 to 15,000 troops. The resettlement into civilian life of both former Contras and former Sandinista soldiers proved problematic, however, and promises guaranteed in the Esquipulas Accords to provide land, social services and credit to facilitate this process were not fulfilled.

As a result, the Government faced its first major military crisis in July 1993, following the occupation of the town of Estelí by a pro-Sandinista armed group. The security forces recaptured the town by force, resulting in 22 deaths. In August a rearmed group of former Contras, Frente Norte 3-80 (FN—Northern Front 3-80), kidnapped a number of deputies who had gone to the northern town of Quilalí to negotiate its disarmament. The FN demanded the resignations of Gen. Humberto Ortega and Antonio Lacayo. In response, former Sandinista soldiers kidnapped a group of right-wing deputies in Managua, including Vice-President Godoy. A peaceful resolution was negotiated by the OAS and a general amnesty was granted. In late 1993 Gen. Ortega was replaced as Commander-in-Chief of the Armed Forces by Gen. Joaquín Cuadra Lacayo, also a Sandinista. Nevertheless, re-armed groups continued to be a serious problem in parts of northern Nicaragua until as late as 2001. In 1994–95 violence also erupted sporadically among the student population, in protest at the Government's failure to spend 6% of the national budget on higher education, as stated in the Constitution.

A further source of tension for the UNO Government arose from the political question of how to deal with the unconsolidated legal status of the businesses, houses and agricultural land that had been nationalized or redistributed during the revolutionary years. Many owners who had fled to the USA following the revolution had returned to Nicaragua following the elections and were making vocal demands for the return of their property. These demands were frequently backed by the US embassy, since many of these individuals had adopted US citizenship. This evolved into a complicated political predicament for the Chamorro Government as it sought to respond to these sometimes violent demands without invalidating the redistribution processes of the 1980s, which would have precipitated a descent into further social and political instability. Eventually, following mediation from former US President Carter, a compromise law dealing with these issues was passed, in the form of Law 209, the law of property stability. Despite the legal protection that this gave to those who had benefited from redistribution, the lack of access of the poorest sectors to legal assistance contributed to a gradual reconcentration of property ownership in the country.

## THE 1996 ELECTIONS

Presidential and legislative elections were held on 20 October 1996. Owing to a constitutional amendment passed in the previous year that prohibited close relatives of a serving President from standing in a presidential election, Lacayo was unable to present his candidacy. In the absence of a prominent UNO nominee, the anti-Sandinista platform was filled by the Alianza Liberal (AL—Liberal Alliance). This coalition, under the candidacy of Arnoldo Alemán Lacayo, mayor of Managua from 1990, brought together a range of liberal parties, including Somoza's former Partido Liberal Nacionalista (PLN—Nationalist Liberal Party) and Alemán's Partido Liberal Constitucionalista (PLC—Constitutional Liberal Party), although the PLC would eventually come to dominate the coalition. Owing to the discovery of various procedural irregularities, including significant quantities of missing votes, the provisional results of the presidential contest, which indicated a victory for Alemán, were alleged to be fraudulent by the FSLN and several other parties. However, although the election process was undoubtedly characterized by administrative incompetence and logistical difficulties, international observers suggested that there was no indication of a systematic attempt to manipulate the poll and declared the ballot to have been generally free and fair. Although many of the provisional results were subsequently revised, the Consejo Supremo Electoral (CSE—Supreme Electoral Council) declared that these anomalies had not affected the overall result. Alemán was declared the winner, with 51% of the valid votes cast, compared with 38% for the FSLN candidate, former President Daniel Ortega. The remaining 21 presidential candidates shared the remainder of the votes. In the legislative elections, the AL also gained a majority of seats in the Asamblea Nacional and control of 92 municipalities.

## THE ALEMÁN ADMINISTRATION

In contrast to the conciliatory agenda of Violeta Chamorro, Arnoldo Alemán had an openly anti-Sandinista agenda. Indeed, Alemán's association with the Somoza family prompted some to label him a Somocista. He immediately adopted a more abrasive style of government, with somewhat predictable consequences. The property issue, for example, returned to the centre of the political stage. The eviction of families occupying land granted to them by the Sandinistas under the new Government's desire to return land to its former owners prompted the FSLN to organize national protests in April 1997. The issue was eventually settled via direct negotiations between the FSLN and the Government in 1997, resulting in the passing of the definitive law of urban and rural reformed property. Meanwhile, unemployment continued to increase, despite Alemán's campaign promise to create 100,000 jobs per year, and the establishment of a multi-party commission to seek solutions to social hardship produced few tangible benefits. In June 1997 the Government convened a 'national dialogue', involving representatives of more than 50 political parties and civic organizations, to examine poverty, unemployment and property-rights issues. However, it was boycotted by the FSLN and anti-Government demonstrations continued, including, from late June, violent clashes between security forces and students.

During 1998 continuing economic and social problems were accompanied by an increase in allegations of corruption against the Government. The work of the Comptroller-General, Agustín Jarquín Anaya, was systematically obstructed by Alemán, who was, together with members of his family, the subject of a number of investigations by Jarquín's office. In February 1999 allegations against the President gained credence with the publication of a report by the Comptroller-General's office, which revealed that Alemán's personal wealth had increased by some 900% during his terms of office as mayor of Managua and as President. However, the campaign of obstruction against Jarquín was aided by senior figures from the FSLN, who were also under investigation concerning their acquisition of former state assets during the transition to the UNO Government in 1990. This culminated in Jarquín being briefly imprisoned in November 1999 on charges of illegal use of state funds to employ an investigative journalist to examine Alemán's finances.

## 'HURRICANE MITCH'

The weakness of Nicaragua's public institutions was amply illustrated following the impact of 'Hurricane Mitch' in October 1998. The hurricane devastated the north-western parts of the country, killing almost 3,000 people and causing massive infrastructural damage. Initially, Alemán refused to declare a state of emergency and questioned the severity of the disaster. When the full consequences of the disaster became known, international donors adopted a strategy of delivering humanitarian aid directly to non-governmental organizations (NGOs), owing to concerns over the possible misappropriation of funds by the Government. In response, the President threatened to impose a tax of up to 40% on all aid entering the country, a threat universally condemned and subsequently withdrawn. The aftermath of the hurricane also exposed an increasing distance between the FSLN and the social organizations from which it traditionally drew its core support. Ortega, along with President Alemán, was criticized by NGOs for using the disaster for political purposes; the apparent lack of input by the FSLN leadership in the mass mobilization of volunteers illustrated the extent of the FSLN's transformation from a mass social movement to a parliamentary party.

The confused response to this disaster and the subsequent inefficiency of the Government's reconstruction plans brought criticism from various quarters. The Stockholm Consultative Group, established by international donors to co-ordinate reconstruction efforts in consultation with the Nicaraguan Government and NGOs, approved the Government's plans for posthurricane reconstruction in April 1999, but became increasingly critical of the Alemán administration's repeated inability to present more detailed proposals on how aid would be spent. Growing accusations of corruption and incompetence in the allocation of post-hurricane reconstruction funds, as well as the attacks on the Comptroller-General's office, led some donors to suspend a number of aid programmes. Nicaragua's accelerated inclusion in the World Bank and IMF heavily indebted poor countries (HIPC) initiative was, nevertheless, approved on 18 December 2000, but Nicaragua's participation in the lengthy process (which was dependent upon the Government meeting specific indicators set by the international institutions) was frequently put at risk by the economic mismanagement and lack of transparency of the Alemán Government.

## THE PLC-FSLN PACT

During 1998 the FSLN maintained strong vocal opposition to Alemán's Government, but the party's apparent lack of long-term strategy, combined with failure in the March 1998 Atlantic Coast local elections, led to renewed divisions within the FSLN leadership. Although Ortega was re-elected Secretary-General of the party in May, opposition to his leadership intensified when he subsequently entered into negotiations with President Alemán on political co-operation with the PLC. These negotiations, concluded in October 1999, centred on changes to the Constitution and to the electoral system. It was agreed that the PLC and FSLN would appoint their own members to important institutions such as the CSE and the Supreme Court, and that elaborate procedures for the granting of legal status to political parties would be established in order to make it more difficult for smaller parties to function, thereby institutionalizing a limited two-party system, consisting of the PLC and the FSLN. The principal changes to the electoral laws included the following provisions: each party must obtain 75,000 signatures in order to gain legal recognition (in the case of an alliance, 75,000 signatures must be collected for each party included in the alliance); each party must gain 4% of the popular vote in order to retain legal status after an election; and all parties must have a presence in all 150 municipalities. Moreover, the threshold for a first-round victory in the presidential elections was reduced from 45% to 35%, albeit on the condition that the victor have a 5% lead over the second placed candidate; and all former Presidents were granted a life seat in the Asamblea Nacional (and, with it, comprehensive parliamentary immunity). Other reforms transformed the Comptroller-General's office into a collegiate body of five Comptrollers, chosen by the PLC and FSLN, thus effectively marginalizing the existing Comptroller-General, Jarquín, who, as a result, resigned in June 2000.

## OPPOSITION TO THE PLC-FSLN PACT

The PLC-FSLN pact was initiated in the context of a growing crisis within both the Sandinista party and the PLC. In March 1998 Daniel Ortega's stepdaughter, Zoilamérica Narváez Murillo, accused him of having abused her as a child. Ortega claimed that the accusations were politically motivated and that Narváez was manipulated by his political rivals within the FSLN. In March 2000 Narváez finally took the matter to the Nicaraguan courts. Ortega's decision to avoid trial by invoking his parliamentary immunity deepened the divisions in the FSLN, with a number of leading figures being removed from prominent positions owing to their criticism of Ortega's refusal to answer the charges. Also in early 2000 Alemán was implicated in the 'narcojet scandal', a complex affair involving a stolen plane, drugs-trafficking and the falsification of government documents. He similarly refused to respond to these accusations nor to the growing public unrest over the alleged misuse of post-hurricane reconstruction funds. As a result, the opportunist PLC-FSLN pact deepened divisions within both parties, with the PLC losing the support of several of its deputies in the Asamblea Nacional, and the FSLN expelling some of its senior figures, while others resigned.

The European Union, Scandinavian Governments and the USA all expressed concern regarding the impact of the PLC-FSLN pact on the transparency of government administration. In June 2000 the countries constituting the Stockholm Consultative Group informed the Alemán Government that although they would fulfil their existing aid commitments, no additional aid would be forthcoming. In particular, the Group pointed out that the failure of the new Comptroller-General's office to investigate clearly documented cases of corruption involving government officials played a significant part in its decision to suspend further aid.

The impact of the constitutional changes was first explicitly felt in the municipal elections of November 2000, when the CSE denied legal status to all political parties except for the FSLN, the PLC, the PCN, the Camino Cristiano Nicaragüense (CCN—Nicaraguan Christian Way), and two small Atlantic Coast indigenous parties. New political initiatives, such as the founding of the Movimiento Unidad Nacional (MUN—National Unity Movement) by the retired army general Joaquín Cuadra and the formation of the Partido Liberal Democrático (PLD—Democratic Liberal Party) by Alemán's former Minister of National Defence, José Antonio Alvarado, were left without any way of engaging in the formal political process. The decision to deny legal status to the new parties caused widespread outrage and was seen as blatant political manipulation on the part of the FSLN and the PLC. This had the effect of further polarizing Nicaraguan politics and deepening the cynicism of the electorate towards the political process. Partly as a result, the municipal elections of November 2000 were marked by a high rate of abstention, and the result was widely portrayed as a 'technical draw'. The FSLN did qualitatively better than the PLC, winning in 11 of the 17 departmental capitals, including Managua, while the PLC won control of more municipalities (97, compared with the FSLN's 49).

## THE 2001 ELECTIONS

The presidential and congressional elections of 4 November 2001 produced another victory for the PLC. In the presidential ballot, former Vice-President Enrique Bolaños Geyer gained 56.3% of votes cast, compared with the 42.3% polled by the FSLN candidate, Daniel Ortega. Jorge Saborio, representing the PCN, attracted just 1.4% of the votes cast. In concurrently held elections to the Asamblea Nacional, the PLC won 47 seats, the FSLN 42 seats, and the PCN one seat. Although opinion polls had predicted both an FSLN victory and a high rate of abstention, during the final months of campaigning the PLC gained important support and some 92% of the electorate voted. This higher than expected turn-out and the decisive margin of the PLC's victory helped the legitimacy of the electoral process, despite the adverse impacts of the constitutional changes. In accordance with electoral reforms resulting from the PLC-FSLN pact described above, Alemán and Ortega, as outgoing President and second placed presidential candidate, respectively, were both given seats in the Asamblea Nacional, increasing the PLC's and FSLN's legislative representation by one seat each. As a further consequence of the electoral reforms initiated by the pact, which stipulated that any party receiving less than 4% of the total votes cast would cease to have legal status, the PCN was prevented from contesting the elections to the Atlantic Coast Regional Councils, which were held in March 2002. The PLC was also successful in these ballots, winning a majority of seats in the South Atlantic Regional Council, and an equal number of seats as the FSLN in the northern assembly. The indigenous party YATAMA also made significant gains, to become the third largest party on the Coast.

Having been excluded from participating in the 2001 elections as a result of the PLC-FSLN pact, many of the smaller parties had given their support to the FSLN (including demobilized Contras and the MRS, who returned to the Sandinista mainstream for the first time after the split in 1995). Among the party's other supporters was Antonio Lacayo. For the FSLN, the

2001 election results represented a third successive defeat for its candidate, Daniel Ortega. The failure of Ortega to win the presidency, despite the support of such a broad alliance (known as the Convergencia), was attributed by some to a heightened awareness of terrorism following the terrorist attacks on the USA on 11 September 2001. His opponents made much of his connections with the Libyan leader, Col Muammar al-Qaddafi, who was suspected by the US Government of maintaining links with terrorist networks. The final weeks of the election campaign also saw direct expressions of support for Bolaños from the US Administration of George W. Bush (2001-) through the statements of his Secretary of State for the Western Hemisphere, Lino Gutiérrez, and paid adverts taken out in the Nicaraguan media by Bush's brother Jeb, the Governor of the US state of Florida. Ortega's continued reluctance to answer the sexual abuse charges against him (he finally surrendered his immunity in December 2001), as well as his close association with the Alemán administration through the FSLN-PLC pact, also damaged his political reputation. Reformers within the party were hopeful that a third electoral defeat and a new impetus towards reform sparked by the formation of the Convergencia, which had survived beyond the elections, could lead towards a change of leadership and direction within the FSLN. Nevertheless, Ortega and those close to him remained in very firm control of the party.

## THE BOLAÑOS GOVERNMENT

President Enrique Bolaños Geyer's Government has faced a considerable number of challenges since taking office in January 2002. The political situation remained volatile and his fortunes ebbed and flowed in relation to those of the other two main political players, Alemán and Ortega. His first challenge upon assuming the presidency was to determine how to deal with Alemán, who seemingly intended to dominate the Government through his former Vice-President. Bolaños' position was particularly weak because most of the PLC deputies owed their positions to Alemán's control of PLC party structures and later proved fiercely loyal to him. The extent of Alemán's influence over the PLC was demonstrated by his election to the presidency of the Asamblea Nacional as soon as Bolaños took office. Nevertheless, from the outset, Bolaños surprised many observers by exerting his independence from Alemán and resolutely pursuing his own political agenda. The most important of these was a sustained effort to deal with the problem of corruption (in 2002 Nicaragua was rated as the third most corrupt country in Latin America by the German-based NGO Transparency International). The first step in this direction was the appointment of a Cabinet composed of party members with few direct links to the Alemán faction of the PLC, many of whom had been critical of the former President's poor record on corruption. The first direct move was to seek the collaboration of the US Department of Justice in the prosecution of Byron Jérez (a close associate of Alemán and former Director of the General Tax Office), who was accused of embezzling state funds and 'laundering' them in US banks. In June 2003 Jérez was found guilty of defrauding the Government of more than US $400,000. However, he was unexpectedly declared innocent of many of the charges in December and released from prison on bail.

Bolaños' campaign to distance himself from his predecessor intensified in March 2002 when the Attorney-General issued charges against Alemán himself in connection with fraud involving the state television company (Sistema Nacional de Televisión Canal 6) and, in May, in connection with the state telecommunications company, Enitel (Empresa Nicaragüense de Telecomunicaciones). Several attempts by Bolaños to remove Alemán's congressional immunity were thwarted by Alemán loyalists within the Asamblea Nacional. However, this was finally achieved on 12 December, and the former President was placed under house arrest. Nevertheless, the case proved to be drawn-out and highly politicized, and was compounded by the fact that the presiding Supreme Court judge in the case, Juana Méndez, was an Ortega loyalist. In August 2003, under Méndez's direction, Alemán was moved to a prison in Managua, before being released back to house arrest on his ranch in November. On 7 December, however, he was finally sentenced to 20 years' imprisonment on charges of money-laundering, fraud and theft of state property. Although he was initially allowed to serve this sentence at his ranch, he was once more confined to prison in mid-March 2004 after the Supreme Court rejected his appeal to serve his sentence under house arrest on grounds of ill health. His transfer back to jail was broadcast live on Nicaraguan television.

The convoluted process of Alemán's trial and imprisonment reflected the complex political interplay between Alemán loyalists in the PLC and Bolaños, as well as between the Government and the FSLN (and Ortega in particular), and the Bush Administration in the USA. In the initial months of Bolaños' administration, Alemán's continued control of the PLC votes in the legislature meant that the new President was forced to negotiate with the FSLN in order to pursue the case against his predecessor. This alienated the pro-Alemán legislative bloc of the PLC still further, leading eventually to its definitive break with the Government in March 2003. Meanwhile, the US Administration was alarmed by the negotiations with the FSLN and began urging Bolaños to adopt a more hardline, anti-Sandinista stance, even if it meant some sort of rehabilitation for Alemán (although it was also made abundantly clear that there could be no return to Nicaraguan politics for the latter). Bolaños, whose main remaining claim to legitimacy had been through the clear international support for his Government's fight against corruption, appeared to support the US stance. One of his main strategies to this end was an attempt in mid-2003 to depoliticize the system by which judges were appointed, a move designed to break Ortega's power in the judiciary. To do this, he needed the support of Alemán loyalists in the legislature. However, it was considered that they would demand too high a price in terms of the exoneration of their leader. Ultimately, US intervention merely seemed to unite, albeit temporarily, Alemán loyalists and Sandinistas against the Bolaños Government. The pro-Alemán PLC and the FSLN continued to block the proposed judicial reforms throughout most of 2004. Nevertheless, in mid-October the Asamblea Nacional finally approved the creation of an independent judicial council to appoint judges.

In essence, the central difficulty of Bolaños' presidency was his need, despite the rhetoric from the USA about anti-Sandinista unity, to reach some sort of *rapprochement* with the FSLN if he was to have any hope of delivering effective government in Nicaragua (and thus the fulfilment of his wider international commitments) or of developing a liberal alternative free from the influence of Alemán. As a result, in April 2004 Ortega and Bolaños announced that they had reached a new understanding. This consensus saw Bolaños agree to the restructuring of the directorate of the Asamblea, from which the FSLN had initially been excluded. President Bolaños also made attempts to mould a new political alternative to the PLC and the FSLN. In May he launched an electoral alliance with the PCN, known as the Alianza por la República (APRE—Alliance for the Republic). The anti-corruption crusade did win some victories: in 2004 Alemán was imprisoned, greater controls had been placed upon the activities of state functionaries and resources appeared to be used more effectively. Some of the worst elements of the constitutional changes resulting from the PLC-FSLN pact were also reversed. For example, in January 2003 the CSE recognized that the removal of the legal status of most political parties under the 2000 electoral law had been unconstitutional.

Nevertheless, President Bolaños' political woes worsened in 2004. In late March the President, Vice-President Rizo and 31 other senior members of the PLC were accused of illegal campaign financing during the 2001 presidential election. Seven PLC members were arrested, including party President, Jorge Castillo Quant. In October the Comptroller-General, Juan Gutiérrez, requested that Bolaños be removed from power and fined two months' wages for withholding information regarding the financing of his electoral campaign. Supporters of the President claimed the request was politically motivated, as the office of the Comptroller-General was controlled by the FSLN and the PLC. At the invitation of Bolaños, a delegation arrived from the OAS to investigate the Comptroller-General's findings.

The FSLN won a decisive victory in the local elections of 7 November 2004, securing control of 84 of the 151 municipalities, including Managua, while the PLC won power in 57 municipalities and the APRE obtained control of six local councils. The following day a two-thirds' majority in the Asamblea Nacional voted in support of constitutional amendments limiting presidential powers and increasing legislative ones. The reforms would require the President to seek legislative ratification of key appointments, such as ministers, ambassadors, the chief prosecutor and banking superintendent; the amendments would also enable the Asamblea to remove officials deemed to be incompetent and, most controversially, overturn the presidential veto by a simple majority vote rather than the existing two-thirds' majority. Furthermore, on 24 November the PLC and the FSLN approved a law to transfer control of the state energy, water and telecommunications services from the President to the one regulatory body, the Superintendencia de Servicios Públicos (Sisep). President Bolaños appealed to the Supreme Court in December, contending that the proposed reforms were unconstitutional and that such an attempt to redefine the powers of the executive and legislative branches of government exceeded the remit of the Asamblea Nacional. Bolaños also asserted that the legislation would engender irreconcilable tension between the Government and the legislature. Nevertheless, in January 2005 the Asamblea ratified the amendments. Bolaños agreed to promulgate the reforms in return for a pledge from the opposition that it would work towards a consensus with the executive on such matters as the budget and social security reform. At the end of March the Supreme Court ruled that the reforms were valid, despite the Central American Court of Justice (CCJ) declaring them to be illegal.

Internal divisions within the FSLN arose in early 2005 ahead of the presidential and congressional elections that were scheduled for November 2006. On 18 January 2005 the FSLN ruled that any prospective presidential nominees should have been party members for at least 10 years. The rule was intended to prevent Herty Lewites, the popular former mayor of Managua, from challenging Ortega in the party's primary election. Lewites had represented another party in the 1996 mayoral elections. In spite of the ruling, Lewites and a number of his supporters were expelled from the party at the end of February, accused of 'abandoning anti-imperialist principles'. On 5 March the FSLN's national congress duly selected Ortega as its presidential candidate. It was widely expected that Lewites, who enjoyed widespread popular support, would, nevertheless, stand as a dissident FSLN candidate in the presidential ballot.

On 26 April 2005 industrial action by transport workers caused serious disruption in Managua, and thousands of protesters gathered outside the presidential palace calling on Bolaños to resign. The demonstration by transport workers and students was provoked by a 20% increase in bus fares. Bolaños, who subsequently agreed to provide a US $1.2m. subsidy to the transport sector and repeal the increment, accused the FSLN of inciting the protests. There were also two protests in mid-2005 in support of Bolaños, organized by the Red por Nicaragua, a coalition of social groups, businesspeople and minor political parties. The protesters demanded an end to the political pact between the pro-Alemán PLC faction and the FSLN.

Tensions were heightened in June 2005 when Bolaños issued a presidential decree compelling the police to comply with the CCJ ruling that invalidated the transfer of control of the public utilities from the President to the newly created Sisep. The President subsequently ordered the police to prevent the new executives nominated by the legislature from entering the offices of the public utilities. In response to the decree, the Comptroller-General once more called for President Bolaños to be impeached. Despite the arrival on 29 June of a second OAS mission to facilitate dialogue between the two sides, the conflict between the President and the Asamblea Nacional remained at an impasse in mid-2005.

# Economy

## PHILLIP WEARNE

Revised for this edition by the editorial staff

Nicaragua has experienced a period of gradual economic recovery since the mid-1990s, despite the huge set-back of the devastation caused by 'Hurricane Mitch' in October 1998, although few Nicaraguans appear to have enjoyed the fruits of that recovery. Real gross domestic product (GDP) increased by 3.7% in 1998, by 7.0% in 1999 (reflecting the considerable inflow of international resources in aid of reconstruction) and by 4.1% in 2000. The rate of growth subsequently slowed: in 2001 GDP increased by 3.0% and in 2002 the rate was 0.8%. However, the economic growth rate improved thereafter: in 2003 GDP increased by an estimated 2.3% and in 2004 the rate was an estimated 5.1%. The average annual rate of inflation was brought under greater control, falling from 13.0% in 1998 to 3.9% in 2002, although it rose again to 5.2% in 2003 and to 8.4% in 2004. Exports nearly doubled in the six years to 2000 (rising from US $351.7m. in 1994 to $629.4m. in 2000), but fell back thereafter and stood at $604.5m. in 2003, before recovering to $755.6m. in the following year.

Nicaragua's recovery in the latter half of the 1990s followed one of the most prolonged and profound economic crises ever witnessed in Latin America. Between the late 1970s and the early 1990s the Nicaraguan economy was beset by a prolonged civil war, the international depression of the early 1980s, the imposition of economic sanctions by the USA and a certain amount of government mismanagement. After the revolution of 1979, which overthrew the dictator Anastasio Somoza Debayle, the size of the state sector increased dramatically. However, the Sandinista Government also actively encouraged the expansion of the private sector, which continued to account for more than 50% of economic activity throughout its time in office. The early revolutionary years saw an attempt to restructure Nicaraguan society and a series of important social reforms, including a wide-ranging agrarian reform programme, were embarked upon. Despite the onset of world recession and the downward pressure on agricultural export prices, the initial restructuring and rehabilitation of the economy following the revolution saw a small increase in growth in the early 1980s. However, the subsequent US-funded 'Contra' war against the Sandinista Government and the economic boycott by multilateral lending agencies caused great damage to the Nicaraguan economy, resulting in a decrease in agricultural output, growing trade and budget deficits and, eventually, spiralling inflation levels (which peaked at a staggering 38,000% in 1988). GDP decreased, in real terms, by an annual average of 6.1% between 1985 and 1994 (with a particularly severe contraction of 18.6% in the two years from 1988). It was not until 1992 that positive growth, of 0.3%, was recorded. The cumulative impact of the economic contraction of the late 1980s and early 1990s was devastating. By 1994 economic output was more than 60% below what it had been in 1977.

All these factors served to accentuate a traditional dependence on imports and external financing (after 1990 the flows of international aid into Nicaragua expanded markedly) and, productively, economic performance remained excessively reliant on the country's narrow range of traditional agricultural commodity exports, the most important of which were bananas, coffee, cotton, meat and sugar. By 1994 average GDP per head in Nicaragua was just US $414, making the country the second poorest in the Western hemisphere, after Haiti. By 2003 average Nicaraguan GDP per head, at just $496, remained pitifully low.

## AGRICULTURE

Nicaragua's agricultural sector was, along with the rest of the economy, deeply depressed in the 1980s and early 1990s. In addition to all of the normal variables affecting agricultural production in Central America (weather, labour shortages, fluctuating world commodity prices and crop disease), Nicaragua has had to face the additional burden of the impacts of war in the country's most important agricultural zones, the effects of which have lasted long beyond the cessation of hostilities. Immediately after the revolution the state took over more than 1m. ha of land, of which about 70% was converted into state farms, with the remainder transferred to peasant co-operatives. Large areas of underutilized and idle land were then expropriated during the second stage of the Government's land-reform programme in mid-1981. From 1986 the emphasis was increasingly placed upon production from co-operatives, or individual smallholders, with the Government continuing to exercise considerable control of the sector through its monopoly purchases of export crops.

The Government of President Violeta Barrios de Chamorro (1990–97) adopted a pragmatic approach, privatizing a number of large state farms, but declining to reverse the Sandinista land reforms. However, the demands of resettling former Contras, who were given land from former state farms, took priority. The administration of President Arnoldo Alemán Lacayo, which took power in January 1997, endorsed this flexibility, reaching an agreement with the Sandinista opposition, which sanctioned the land and property distribution to legitimize beneficiaries of the Sandinista reforms. At the same time, the new Government reinforced an arbitration mechanism to adjudicate the claims of many larger property owners who insisted that they had been victims of asset seizures by senior Sandinistas. Instead of returning properties to these claimants, both the Chamorro and Alemán Governments made generous compensation payments to those who had been expropriated. These individuals received approximately US $900m. in government funding between 1990 and 2003. These payments had a massive impact upon the country's internal debt, representing more than one-half of total internal debt in 2003. Despite all of these payments, the property question continued to be a major political and economic issue in Nicaragua. Some saw this as a major disincentive to significant private investment in agriculture, although by 2004 there were some signs that the problem had begun to recede.

With the exception of the obvious impacts of 'Hurricane Mitch' on the sector in 1998 and 1999, agricultural production showed a modest recovery from the mid-1990s, although it was impeded by a scarcity of inputs, credit and continuing political instability in many rural areas. Nevertheless, the sector (including forestry and fishing) remained the mainstay of the economy, employing an estimated 43.4% of the economically active population in 2001 and (including hunting, fishing and forestry) accounting for an estimated 18.3% of GDP in 2004. Despite the negative impacts of 'Hurricane Mitch', the recovery became more consistent after 1999. Production levels in 2003 were, for example, 80% higher than in 1999 in the case of maize, 50.8% higher for beans and 50.6% for rice. Production levels of beans and maize increased in every year after 1999, as did rice output (with the exception of a fairly sharp fall in 2001, which was followed by equally dramatic improvements in 2002 and 2003). These increases did not, however, simply reflect a continuous expansion in the area harvested of these crops, but also improvements in productivity. For example, the areas harvested of rice, maize and sorghum all decreased in 2001, but overall production levels of sorghum and maize still grew in that year.

The output and exchange earnings of export crops have been similarly variable. During the revolutionary years, production of the country's major export crop, coffee, had declined steadily from 1984, reaching a nadir in 1990. From that point, however, there was throughout the 1990s a reasonably consistent expansion in the total area of coffee harvested, which continued in the 21st century (with the exception of the Mitch-affected year of 1999). Production levels, however, were more variable. Production increased in every year (with the exception of 1996) in the 1990s, reaching a peak of 91,800 metric tons in 1999. However, levels declined in every year thereafter (although the 59,700 tons produced in 2003 was still greater than production had been during any year between 1990 and 1996).

Coffee producers and workers have faced considerable difficulties in recent years. Coffee export earnings reflected not only the production levels, but also fluctuations in the international price of coffee. An improvement in earnings to US $173m. in 1998 was the result of recovering production levels, yielded by the expansion programme begun several years earlier, rather than any rise in international prices. By 2002 international coffee prices had reached their lowest ever level (although there were finally some signs of a recovery in 2004). In response, production began to falter and export earnings fell to $135.3m. in 1999. After recovering to $170.9m. in 2000, in 2001 and 2002 earnings fell again, to $98.6m. and $73.6m., respectively, their lowest level for 10 years. Earnings recovered to $85.5m. in 2003, and further, to $126.8m., in 2004. The whole sector has been decimated by these recent problems, farms have been abandoned and labourers in the traditional coffee growing areas have faced unemployment and starvation.

Cotton fared even worse than coffee and by 1998 the industry had virtually ceased to exist. In 1991 export earnings from cotton had been as high as US $44.4m., representing the production of 30,000 metric tons of cotton lint. In just two years, however, that figure had fallen to as little as $400,000. Some recovery occurred around 1995, with an increase in world prices stimulating planting, but another cycle of low prices had a predictable effect on the area planted and export earnings. By 1998 cotton exports had fallen to $300,000, less than one-10th the level of the previous year. By 2000 earnings from cotton stood at less than $100,000 and production levels did not increase thereafter. In 2003 just 1,000 tons were produced, according to estimates by the UN's Food and Agriculture Organization. Cotton dominated the best agricultural lands of the North Pacific region from the end of the Second World War and the decline of the industry left many of the communities in the regions around León and Chinandega, areas harshly affected by 'Hurricane Mitch', with considerable economic problems.

The other major traditional agricultural exports were sugar and bananas (and beef, which is considered separately below). Sugar production increased more or less steadily from the early 1990s, reaching an estimated 346,000 metric tons in 2003. Despite this relatively stable level of output, the volatility of international prices produced a complex pattern of sugar export earnings over this period. Low prices in the early 1990s resulted in a fall in earnings. They then steadily increased to reach US $53.2m. by 1997, but fell back to $30.4m. by 1999. There then followed a sharp improvement, with sugar exports earning $33.9m. in 2001. Earnings fell in 2002 and 2003 to $28.6m. and $25.7m., respectively, before recovering to $36.8m. in 2004. Export earnings from bananas have followed a similarly complex pattern. Falling international prices led to a fall in production levels in 1993, with earnings declining to a paltry $5.5m. in that year. The next three years saw a gradual improvement, with earnings recovering to $21.7m. by 1996. Thereafter, however, there was a precipitous decline, with earnings reaching a mere $10.1m. in 2000 (this reflected a fall in production from 96,797 tons in 1996 to just 56,700 tons in 2000). There was some recovery after 2000, with production averaging around 60,000 tons and earnings stabilizing at around $11.5m. in the early years of the 21st century. Banana production stood at 61,091 tons in 2004, while banana exports earned $10.7m.

The increasing export of less traditional products has been one of the success stories of the Nicaraguan economy in recent years. In 2000–04, for example, groundnut exports earned more foreign exchange than any other agricultural commodity apart from coffee (exports reached US $39.7m. in 2004). Similarly, the export of beans, for many years seen as merely a staple of the peasant sector, expanded rapidly. Earnings from bean exports to neighbouring countries increased from around $13.0m. in 2001 to $18.8m. in 2004.

In the mid-2000s the livestock industry, like other sectors, had yet to recover fully from the shortage of foreign exchange for foreign machinery and other essential products it had suffered during the Contra war. Compounding this, the sector was excluded from government credit schemes in the early 1990s and it seemed unlikely to recover its former position in terms of export importance. Some sub-sectors did, none the less, gradually show some improvement. From 1998 beef and veal export earnings increased annually, with earnings reaching US $110.4m. in 2004. Furthermore, the export of meat products has been complemented in recent years by the expanded export of livestock to neighbouring countries, which, in the early 2000s, generated earnings of around $25m. per year.

Nicaragua possessed substantial supplies of timber, including considerable reserves of hardwoods such as mahogany, cedar, rosewood, caoba and oak. In the early 2000s roundwood production averaged about 6m. cu m annually, an increase on the 4m. cu m averaged in the 1990s. Concern about overexploitation in the mid-1990s led the Asamblea Nacional (National Assembly) to ban further logging concessions in 1996, although many environmental organizations continued to be seriously concerned about illegal logging.

The development of the fishing industry was a major priority around the turn of the 21st century. Three fish-processing plants were rehabilitated and commercial fishing of crab, crayfish, shrimp, lobster and tuna was encouraged. By 2000 the total shrimp catch had more than doubled in just seven years, to 16.1m. lbs. Similarly, lobster production reached 4.4m. lbs in 2000. Production then decreased slightly and, as a result, earnings from lobster and shrimp, which reached US $118.8m. in 2000, fell back to just $78.6m. in 2002, and to $69.1m. in 2003. An improvement, to $80.6m., was recorded in 2004, some 10.7% of total export earnings.

## MINING AND POWER

The mining sector was completely nationalized in 1979. However, after the end of the Sandinista regime, successive Governments ceded the rights of exploitation to private companies under long-term lease agreements. This policy met with some success: an estimated US $70m. of private investment was directed into the sector in the 1990s. The contribution of the mining sector to overall GDP reflected the investment, rising from 0.6% in 1994 to 1.1% in 2004. The sector encompasses salt, marble and quarried stone, but the real export value lies in Nicaragua's nine gold and silver mines, three of which were in production. Increases in gold production were in line with sectoral performance, a 51% expansion resulting in output of 121,800 troy oz in 1998. The sector recorded a further rise in 1999, with output of 143,000 troy oz, but there was then a fall to 118,100 troy oz in 2000. Output recovered temporarily in 2001 and 2002, to 123,500 troy oz and 126,500 troy oz, respectively, before another fall, to 108,260 troy oz, was provisionally registered in 2003. Silver production increased by almost 50% in 1997, rising to 34,600 troy oz, before surging further, to 62,500 troy oz, in 1998. Output fell to 51,100 troy oz in 2000, but recovered to 81,400 troy oz in 2001 and 70,700 troy oz in 2002 with the reopening of the El Mojote mine in Chontales. In 2003 output fell to a provisional 65,710 troy oz. Some forecasts saw exports trebling in the coming years, as some of five other closed gold and silver mines returned to production.

Dependence on imported energy sources, particularly petroleum, proved a major problem for successive Nicaraguan Governments. During much of its time in office, the Sandinista Government relied on cheap or bartered petroleum from Mexico or the former USSR. However, Mexico suspended supplies when even 'soft-term' payments became overdue and in 1987 the USSR announced large reductions in supplies, causing severe energy problems which compounded the general economic malaise of the time. In the early 2000s, as the country's credit rating was somewhat restored, much of Nicaragua's petroleum needs were met by Venezuela and Mexico, under the concessionary terms of the 1980 San José Agreement. Despite ambitious diversification efforts, in particular the inauguration of the Momotombo geothermal plant in 1984 and the Asturias hydroelectric scheme five years later, petroleum and related imports remained crucial at the beginning of the 21st century, costing the country as much as US $425.9m. in 2004, compared with the $143.3m. required in 1998. In 2003 hydroelectricity provided less than 12% of Nicaragua's total production of 2,063.9 GWh, down from as much as nearly 25% in 1996. Geothermal generation has followed a similar pattern, falling from as much as 15% of total output in 1996 to zero in 2000 and beyond.

Net electrical generating capacity was estimated to be about 546 MW in 2001, an increase of almost 40% on what had been available in 1997–98, largely a result of a number of new private power-plants coming on stream. The new capacity was expected to alleviate the frequent power shortages so common in the 1990s when supply covered just one-half of consumer demand. Foreign companies had been required to restrict output to 30 MW per year until 1999, but in that year the US company Amfels was given permission to increase capacity at its Puerto Sandino plant to 57 MW. At the same time, Ormat Industries of Israel signed a 15-year operating contract for the Momotombo geothermal plant with the state power company, Empresa Nicaragüense de Electricidad (ENEL). The Israeli firm pledged to invest an initial US $15m., with up to $30m. to follow, and announced plans to increase the plant's capacity from 13 MW to 70 MW. In 1999 ENEL was divided into three generating companies (one hydroelectric and two thermal plants) and two distribution companies, in preparation for privatization, for which tenders were solicited in May 2000. Meanwhile, ENEL continued to increase spending on the rehabilitation and repair of existing plant and transmission lines and a regional electricity grid, the core of which already existed in the import and export of power to Honduras and Panama. The dilapidated state of the power-distribution system in Nicaragua was responsible for the loss of up to 25% of all power generated during the 1990s. The distribution companies were sold to the Spanish company Unión Fenosa in November 2000, which came under intense scrutiny from sectors of civil society unhappy with its pricing policies and failure to fulfil promised infrastructural investments. However, government attempts to sell the generating companies failed when, in July 2003, the Supreme Court annulled the sale of power company HIDROGESA in the previous year, owing to alleged irregularities. The expansion of private generating capacity meant that about 89% of the country's total output of 2,063.9 GWh in 2003 was supplied by the private sector, with ENEL's output of 455.9 GWh only 31% of its 1999 figure.

## MANUFACTURING AND INDUSTRY

Traditionally, manufacturing activity in Nicaragua centred on the processing and packing of local agricultural produce, although some heavy industries, such as chemical and cement production, also existed. In 2004 manufacturing accounted for an estimated 19.2% of GDP. The origins of the sector lie in the rapid expansion of import-substitution industries that occurred during the 1960s. This was, however, short-lived and the average annual rate of manufacturing growth decreased from 11.4% between 1960 and 1970 to 2.5% between 1970 and 1982. The situation worsened during the 1980s as political strife, hyperinflation, and shortages of raw materials, imported machinery and skilled personnel took full effect. The Contra war seriously disrupted manufacturing activity in that, while the shifting of the economy onto a permanent war footing stimulated the production of consumer goods for the armed forces and of construction materials for the repair of roads and buildings, other import-dependent industries suffered from the reduction in consumer spending and from Nicaragua's chronic shortage of foreign exchange caused by the US trade embargo. By 1985 the private sector had contracted by about 20% and was operating at little more than 50% of capacity, while ambitious state enterprises had failed to yield the hoped-for production levels. After a brief respite in 1986–87, manufacturing was again badly affected by the 1988–89 recession. Industrial production was estimated to have contracted by some 20% in 1989 alone.

The contraction persisted into the early 1990s and, although it has subsequently recovered (the sector grew by an annual average of 2.0% between 1990 and 2003), by 2004 the sector had still to reach the levels attained 20 years earlier. The Chamorro Government privatized some of the state agro-business projects initiated by the Sandinistas, including the US $500m. Malacatoya sugar mill, the Chiltepe dairy project and a fruit and vegetable processing and canning factory in the Sebaco valley. At the same time, the end of the war, a sharp fall in import tariffs and renewed access to the US market provided new opportunities in manufacturing, with many Nicaraguan exiles returning to take advantage of them. However, new export opportunities were not enough to counteract continued contraction in the domestic market, particularly in production sectors such as beverages, processed foods and cigarettes. As Nicaragua entered the new millennium, sector-wide figures improved, and manufacturing output grew by 2.8% in 2000 and 2.6% in 2001 (with the building materials sub-sector expanding by nearly 40%, owing to post-Mitch reconstruction activities).

The pace of growth slowed in 2002, with sectoral growth of 1.6%. However, economic expansion improved thereafter, with growth of 2.5% in 2003, and a robust increase of 6.6% in 2004, which reflected a 35.7% increase in sugar refining. A glance at the experiences of particular sub-sectors, however, suggested a more complex picture. In 2000–03 certain foodstuffs (meat, biscuits, milk), beer, leather goods, construction materials and certain plastics underwent a considerable expansion. Some of these improvements reflected the acceleration of existing trends, but in other cases they followed prior decreases. Other industries, however, contracted over the same period (examples included rum, some canned goods and most paper goods). The varied performance reflected a major restructuring of the Nicaraguan manufacturing sector, following economic liberalization efforts. An importation tariff reform brought cheap imports and, unable to compete, some traditional manufacturing businesses closed, diversified or formed joint ventures with foreign firms or returning exiles attracted into the country by a more favourable foreign investment environment.

The most sustained expansion in the sector has been seen in the *maquila*, or 'offshore' assembly, sector. The Las Mercedes free trade zone, the country's first, doubled its total exports to US $80m. in 1995, although subsequent growth was more modest. However, in 2000 and 2001 the country's free trade zones replicated some of their earlier spectacular growth, with value added rising from $203m. to $300m. and $380m., respectively. Although the principal product was clothing assembled from US-imported textiles, plants also produced footwear, aluminium frames and jewellery, with companies from Hong Kong and Taiwan augmenting those from the USA. The number of jobs in the free trade zones expanded from around 20,000 in 1999 to over 50,000 in 2003. The attraction of more of these enterprises formed the key component of the Government's national economic development plan, with an assumption that more US companies would be attracted to Nicaragua following the implementation of the Central American Free Trade Agreement (CAFTA), agreed in late 2003 but still awaiting ratification by the Nicaraguan legislature in mid-2005. Analysts remained divided as to the probability of sustained growth in this area and the ability of the sector to generate sufficient employment to offset the likely agricultural losses that CAFTA would produce.

Much of the investment in the construction sector during the 1980s was strategic. By the end of the decade, however, the industry contracted as external assistance ended. The Chamorro Government's reconstruction programme reversed the decline in the early 1990s, but sectoral growth fluctuated in the rest of the decade. However, the assistance given to the sector by reconstruction efforts in the wake of 'Hurricane Mitch', combined with several large-scale projects in the tourism and transport sectors, dramatically expanded construction in 1999–2001, when it accounted for more than 6.2% of annual GDP. In 2001 the sector employed some 102,300 people, equivalent to 6.0% of the economically active population. In 1999 the sector expanded by more than 60%, with a further expansion of more than 10% the following year. However, by 2003 the rate of growth had slowed to 3.9%, before rising to 10.7% in the following year. In 2004 the sector contributed 6.4% of GDP.

## PUBLIC FINANCE AND PAYMENTS

The Governments of Presidents Violeta Chamorro and Arnoldo Alemán Lacayo had some success in controlling what had become one of the most chronic current-account budget deficits of any state. Access to external funds from the USA and multilateral lending agencies, a dramatic decrease in inflation, severe reductions in public spending, as well as privatization of state enterprises, improved tax collection and the retirement or renegotiation of some of the country's massive foreign debt combined to help reduce the budget deficit from the average 25% of GDP it had reached in the 1980s. By 1998 the budget deficit (before grants) had fallen to the equivalent of 4.8% of GDP. The next year, however, saw an increase to 13.2% of GDP and by 2001 it had reached around 19.0%, as election year spending and the consequences of 'Hurricane Mitch' took a particularly heavy toll. Much of this deficit was financed by foreign grants and concessionary loans, which, in turn, contributed substantially to an expenditure programme that, by 2001, had reached nearly 40% of GDP. In 2002 and 2003, however, the budget deficit was dramatically reduced to 4.1% and 2.3% of GDP, respectively, as a result of the severe fiscal measures that the administration of Enrique Bolaños Geyer (2002–) was forced to impose in order to be eligible for access to the World Bank and IMF's heavily indebted poor countries (HIPC) initiative. In 2004, according to preliminary figures, the budget deficit decreased further, to 1,035m. córdobas, equivalent to 1.4% of GDP.

Many of the policies of the Chamorro Government were in fact implemented by the previous administration. Although the Sandinista Government had attempted to offset the increase in state spending on defence in the 1980s by increasing taxation on profits, workers' remittances and non-essential imports, the Government was forced to implement a number of austerity measures in 1988–89, in an attempt to solve the growing budget crisis. The measures included large devaluations, the creation of new currencies, wide-ranging price increases and budget reductions that involved a large number of job losses in the public sector. There were also job losses under the Chamorro Government, as attempts were made to reduce the deficit as part of its stabilization programme of 1991. This trend continued during the Alemán Government (1997–2002), as well as the current Bolaños administration. As a result, the number of public-sector workers in Nicaragua fell from 218,703 in 1990 to only 70,699 by 2002, a dramatic reduction of 67% in public employment.

One of the achievements of the economic policy followed by the last three Governments was the control of inflation. By 2002, for example, the rate was just 3.9%, down from 13.0% in 1998 (by 2004 the annual rate had risen to 8.4%). Tax revenues rose by 28% in 1998, 14.9% in 1999, and some 12.0% in 2000. A decrease of 2.8% in 2001 was followed by rises of 4% in 2002 and 8.7% in 2003. This trend was explained by different factors, among them a substantial rise in import tax and general sales tax revenue as reconstruction efforts stimulated demand, and an important increase in the taxpayers' base. Capital spending in 1999 was double that of the previous year, but was nominally at least covered by the international donations and grants made available in the wake of 'Hurricane Mitch' (this level of spending was at least maintained every year thereafter, with the exception of 2002).

During the 1980s the growing budget deficits produced by the war effort and growing balance of payments deficits had forced the Sandinista Government to increase its foreign borrowing. As a result, the external debt of US $1,600m., which was inherited from the Somoza dictatorship in 1979, had increased to more than $9,300m. by 1989. The main problem was Nicaragua's restricted access to multilateral loans owing to US opposition. Consequently, Nicaragua became dependent on bilateral credits from Western states and countries that were members of the Council for Mutual Economic Assistance (CMEA). Following initial rescheduling, a moratorium on public-sector debt-service payments had, by September 1982, become inevitable. In 1983 Nicaragua failed to repay $45m. and requested the renegotiation of $180m. in interest payments due in 1984. Unable to pay either interest or principal on its debt, Nicaragua entered a state of 'passive default'. However, in 1988–89 the Sandinista Government implemented many of the measures traditionally demanded by the IMF as part of its own austerity programme.

These measures prepared the way for the implementation of the new policies of the Chamorro Government, which was able to take advantage of much more favourable circumstances. Thus, in 1992, with the USA agreeing to waive payment of US $259.5m. in bilateral debt and the 'Paris Club' of Western creditor nations reducing an $830m. debt to $207m., the new Government was able to pay arrears to the World Bank and the Inter-American Development Bank (IDB). This, in turn, opened a series of loan opportunities. The IMF ended its 12-year boycott of Nicaragua by approving a $55.7m. stand-by loan and the IDB approved a loan to support adjustment programmes in the trade and financial sectors. Restructuring of debt payments with various major bilateral creditors such as Mexico, Russia (as the principal successor state of the USSR) and several European countries followed in 1992. However, in 1993 Nicaragua again defaulted on repayment.

In 1994 the Government reached agreement with the IMF on an enhanced structural-adjustment loan and redoubled its efforts to secure debt renegotiation and cancellation. The following year the Government secured the purchase of more than 80% of its commercial debt at eight cents in the dollar, effectively paying US $112m. to cancel $1,400m. in debt. This followed the restructuring of debts owed to the Paris Club. In 1998 the Club agreed to a further two-year postponement of $201m. in debt-service payments. In April 1996 Nicaragua finally reached agreement with Russia over its $3,500m. debt to the former USSR, with 95% of the outstanding amount being forgiven. Later that year Mexico followed suit, agreeing to waive 91% of the $1,100m. that it was owed. Agreement was also reached with the IMF in 1998 on a second enhanced structural-adjustment loan. In November 1998 a number of countries pardoned or softened terms on bilateral debt they were owed, in the wake of 'Hurricane Mitch'. Cuba, Austria and France, in particular, waived a total of $157m. in bilateral debt. France then waived a further $90m. in outstanding debt in February 2000. In March 2001 Spain assumed most of Nicaragua's $500m. debt to Guatemala, writing off some $399m. of the total in the process. In April 2003 France, Germany and Spain cancelled a further $263m. in bilateral debt.

In September 1999 Nicaragua was declared eligible for inclusion in the HIPC debt-relief initiative, which would eventually make the cancellation of 80%–90% of the country's foreign debt possible. In December 2000 the IMF and World Bank declared that the country had fulfilled the necessary conditions to enter the initiative. In 2002 debt-service waivers reached US $206m., almost double the amount of the previous year, following the implementation of a severe fiscal programme by the Bolaños Government, while in 2003 the total relief was $214m. In January 2004, after the fulfilment of all obligations, Nicaragua finally attained HIPC status, which included the forgiveness of 80% of the country's public external debt (totalling $6,400m. in 2003). In addition, the country only had to pay 10% of the total servicing of all World Bank loans contracted between 2001 and 2023, and would also receive a debt release of some $106m. on servicing IMF loans entered into between 2002 and 2009. The institutions were also encouraging Nicaragua's bilateral creditors to follow suit. Finally, in 2004 Nicaragua obtained a $21m. loan from the IMF and a $70m. loan from the World Bank, both under preferential terms. Then, in June 2005, Nicaragua was among 18 countries to be granted 100% debt relief on multilateral debt agreed by the Group of Eight (G-8) leading industrialized nations, subject to the approval of the lenders.

This significant change in Nicaragua's financial circumstances was expected, at least in principle, to increase the resources that could be allocated to social programmes and public services designed to alleviate the massive poverty and unemployment in the country. It would not, however, despite the insistence on the targeting of poverty central to the whole HIPC process, automatically lead to a significant improvement in the living conditions of the majority of Nicaraguans.

## FOREIGN TRADE

Until the early 1990s Nicaragua's export trade remained dominated by agricultural commodities, mainly coffee, cotton, sugar, beef, bananas and, increasingly, seafood. Import trade was (and continued to be) dominated by petroleum, other raw materials, non-durable consumer goods and machinery. The enormous cost of the war, increased purchases of petroleum and reduced income from exports (chiefly owing to low prices for coffee, sugar and cotton and the loss of the US market because of the embargo) resulted in an increasingly bleak trade deficit by the end of the 1980s. From the early 1990s, however, there was substantial improvement, with revenue from exports (f.o.b.) rising steadily to reach US $666.6m. by 1997. The damage caused to export crops by 'Hurricane Mitch' resulted in a fall in revenues in 1998, to $552.8m., with a further decrease to an estimated $508.5m. in 1999; however, export revenues recovered to $629.4m. in 2000, before falling again to $605.0m. in 2001, and to $561.0m. in 2002. Export earnings subsequently increased to $604.5m. in 2003 and to $755.6m. in the following year. Although export growth was not particularly dynamic, as mentioned earlier, there was a significant transformation in the relative composition of export trade at the beginning of the 21st century. In 2000 traditional exports (coffee, cotton, sugar, bananas, meat, etc.) represented 63.5% of total earnings; by 2003 these products represented only 51.7%. Non-traditional exports, led by industrial production in the *maquila* zones and non-traditional agricultural exports such as groundnuts and beans, increased dramatically in importance over the same period.

Imports increased in the 1990s, reaching a level of US $1,491.7m. by 1998. In 1999, fuelled by the reconstruction and repair needs resulting from 'Hurricane Mitch', imports increased to $1,723.1m., falling slightly, to $1,720.6m., in 2000. The imports bill remained static in 2001 and 2002, but increased again in 2003 and 2004 to an estimated $1,879.4m. and $2,212.3m., respectively. These levels of importation, coupled with the lack of dynamism in the export sector, produced a burgeoning trade deficit. In 1995 this stood at $527m., but by 1999 this figure had practically doubled to $1,071m. Some improvement in the trade balance was recorded in 2000 and 2001 (when the deficit shrank to $897m.); by 2003 the deficit had increased to $972m.

The only major new sources of income that helped to offset the growing trade deficit were income from tourism and remittances from Nicaraguan citizens living abroad, mainly in the USA and Costa Rica. By 2003 income from tourism had reached US $155m., while arrivals increased by 11.5%, to reach 525,775. There was a massive increase in remittances in the early 21st century. In 1999 migrants sent $345m. back to Nicaragua. In 2003 remittances reached $788m.

Fighting in the border regions disrupted Nicaragua's trade with neighbouring Central American countries during the 1980s. Non-traditional exports to the Central American Common Market (CACM) suffered most, decreasing by more than 75% between 1982 and 1985. The irregular availability of foreign exchange was the principal problem, but the continued economic recession throughout the region was also a contributory factor. Trade with Argentina and Brazil increased, owing to the extension of trade credits for Nicaragua's prime agricultural exports. With the exception of bananas, Nicaragua's trade with the USA fell dramatically. However, with the US trade embargo lifted in 1990, trade with Nicaragua's most natural trading partner rapidly began to improve. In 1993 exports to the USA represented 48.1% of Nicaragua's total exports, while the value of goods from the USA rose to 26.6% of total imports. By 1999, however, with total imports increasing more quickly than exports, these figures had moved into balance, with trade with the USA accounting for 38.3% of exports and 24.9% of imports. Thereafter there was a gradual decrease in both figures, which stood at 34.9% and 22.2%, respectively, in 2004. This reflected the growing proportion of trade carried out with Nicaragua's Central American neighbours. In 2003 CACM states accounted for 36.9% of Nicaragua's exports and 22.8% of the country's imports, while in 2004 their share of exports declined to 32.9%, while imports increased slightly to 23.0%. El Salvador became Nicaragua's second largest export market, taking 17.3% of export goods in 2003 and 14.5% in 2004.

In May 2001 the Central American countries, including Nicaragua, reached an accord with Mexico, known as the 'Plan Puebla–Panamá', under which the region would be integrated through joint transport, industry and tourism projects. Throughout 2003 Nicaragua joined Costa Rica, El Salvador, Guatemala and Honduras in negotiations with the USA towards CAFTA, a treaty that was signed by the Presidents of Central America and the USA in May 2004 in Washington, DC. The final approval of the treaty by the Nicaraguan legislature was yet to happen in mid-2005. Important debates remained ongoing about the potential benefits and damages of this agreement for the Nicaraguan economy and for its society in general. The main concern was that any economic growth that free trade would promote would disproportionately benefit a very small group of people.

## CONCLUSION

Repeated promises from those promoting neo-liberal reforms that the Nicaraguan economy would improve went largely unfulfilled after the 1980s. However, by 2005 there was at least a general consensus that, despite the huge set-backs caused by 'Hurricane Mitch', there had been some recovery from the economic abyss of the late 1980s. The economy maintained some momentum in 2003 but, as elsewhere in the region, given the world economic slowdown in 2001–02, it could not repeat the 7.0% growth of 1999, and growth of only 2.3% was registered in that year. However, growth increased in 2004 to 5.1%, underpinned by a strong recovery in agriculture, particularly in coffee and sugar production, in addition to good performances in the livestock, manufacturing and construction sectors.

Although the Bolaños administration had succeeded in meeting the IMF's criteria for inclusion in the HIPC initiative, the economic indicators in 2005 were not particularly encouraging: the unemployment rate had fallen from 10.5% of the labour force to 6.5% in 2004, but fiscal and trade deficits remained unsustainable and social deprivation was still at an unacceptably high level. The political impasse in the Asamblea Nacional in 2003–05 threatened to undermine President Bolaños' ability to implement his economic policy agenda throughout the remainder of his term in office. Indeed, in early 2005 increased budget expenditure approved by the legislature and its failure to implement structural reforms led to the loss of financing support under the IMF's Poverty Reduction and Growth Facility (PRGF). The best prospect for the amelioration of the country's chronic economic problems remained debt relief, although the degree to which any resources were likely to be transferred to productive purposes or the social sector remained to be seen.

Longer-term economic strategies revolved around an attempt to attract foreign investment through Nicaragua's integration into regional trade initiatives. It was anticipated that the potential implementation of CAFTA and the proposed Central American customs union would advance such an aim. Export earnings were also likely to benefit from improved commodity prices. It was anticipated that the trade deficit would be further offset by rising levels of remittances. According to official forecasts, GDP was expected to expand by 3.8% in 2005, but this was dependent on the resumption of the PRGF and the ratification of CAFTA.

# Statistical Survey

Sources (unless otherwise stated): Banco Central de Nicaragua, Carretera Sur, Km 7, Apdos 2252/3, Zona 5, Managua; tel. (2) 65-0500; fax (2) 65-2272; e-mail bcn@cabcn.gob.ni; internet www.bcn.gob.ni; Instituto Nacional de Estadísticas y Censos (INEC), Las Brisas, Frente Hospital Fonseca, Managua; tel. (2) 66-2031; internet www.inec.gob.ni.

## Area and Population

### AREA, POPULATION AND DENSITY

| | |
|---|---:|
| Area (sq km) | |
|   Land | 120,340 |
|   Inland water | 10,034 |
|   Total | 130,373* |
| Population (census results) | |
|   20 April 1971 | 1,877,952 |
|   25 April 1995 | |
|     Males | 2,147,105 |
|     Females | 2,209,994 |
|     Total | 4,357,099 |
| Population (official estimates at mid-year) | |
|   2002 | 5,341,883 |
|   2003 | 5,482,340 |
|   2004 | 5,626,492 |
| Density (per sq km) at mid-2004 | 46.8† |

* 50,337 sq miles.
† Land area only.

### ADMINISTRATIVE DIVISIONS
(official estimates at mid-2004)

| | Area (sq km) | Population | Density (per sq km) | Capital |
|---|---:|---:|---:|---|
| *Departments:* | | | | |
| Chinandega | 4,822.4 | 452,190 | 93.8 | Chinandega |
| León | 5,138.0 | 402,710 | 78.4 | León |
| Managua | 3,465.1 | 1,413,257 | 407.9 | Managua |
| Masaya | 610.8 | 324,855 | 531.9 | Masaya |
| Carazo | 1,081.4 | 182,640 | 168.9 | Jinotepe |
| Granada | 1,039.7 | 196,275 | 188.8 | Granada |
| Rivas | 2,161.8 | 172,119 | 79.6 | Rivas |
| Estelí | 2,229.7 | 220,521 | 98.9 | Estelí |
| Madriz | 1,708.2 | 137,111 | 80.3 | Somoto |
| Nueva Segovia | 3,491.3 | 217,444 | 62.3 | Ocotal |
| Jinotega | 9,222.4 | 305,818 | 33.2 | Jinotega |
| Matagalpa | 6,803.9 | 497,931 | 73.2 | Matagalpa |
| Boaco | 4,176.7 | 173,444 | 41.5 | Boaco |
| Chontales | 6,481.3 | 186,672 | 28.8 | Juigalpa |
| Río San Juan | 7,540.9 | 97,825 | 13.0 | San Carlos |
| *Autonomous Regions:* | | | | |
| Atlántico Norte (RAAN) | 32,819.7 | 256,440 | 7.8 | Bilwi |
| Atlántico Sur (RAAS) | 27,546.3 | 389,240 | 14.1 | Bluefields |
| **Total** | **120,339.5** | **5,626,492** | **46.8** | — |

### PRINCIPAL TOWNS
(population at 1995 census)

| | | | |
|---|---:|---|---:|
| Managua (capital) | 864,201 | Granada | 71,783 |
| León | 123,865 | Estelí | 71,550 |
| Chinandega | 97,387 | Tipitaga | 67,925 |
| Masaya | 88,971 | Matagalpa | 59,397 |

**Mid-2003** (UN estimate, including suburbs): Managua 1,097,611 (Source: UN, *World Urbanization Prospects: The 2003 Revision*).

### BIRTHS AND DEATHS
(UN estimates, annual averages)

| | 1985–90 | 1990–95 | 1995–2000 |
|---|---:|---:|---:|
| Birth rate (per 1,000) | 39.8 | 386.0 | 32.4 |
| Death rate (per 1,000) | 8.4 | 6.3 | 5.5 |

Source: UN, *World Population Prospects: The 2004 Revision*.

**2001:** Registered births 103,593 (19.9 per 1,000); Registered deaths 10,071 (1.9 per 1,000).

**2002:** Registered births 103,643 (19.4 per 1,000); Registered deaths 10,830 (2.0 per 1,000).

**Expectation of life** (WHO estimates, years at birth): 70 (males 68; females 73) in 2003 (Source: WHO, *World Health Report*).

# NICARAGUA

## ECONOMICALLY ACTIVE POPULATION
('000 persons)

|  | 1999 | 2000 | 2001 |
|---|---|---|---|
| Agriculture, forestry and fishing | 655.3 | 711.8 | 739.0 |
| Mining and quarrying | 11.7 | 9.4 | 9.6 |
| Manufacturing | 125.3 | 127.8 | 131.6 |
| Electricity, gas and water | 5.8 | 5.9 | 6.1 |
| Construction | 88.1 | 97.3 | 102.3 |
| Trade, restaurants and hotels | 259.2 | 268.3 | 279.8 |
| Transport and communications | 49.7 | 51.2 | 52.9 |
| Financial services | 20.1 | 21.8 | 22.6 |
| Government services | 67.5 | 65.0 | 63.5 |
| Other services | 261.5 | 278.8 | 294.3 |
| **Total employed** | 1,544.2 | 1,637.3 | 1,701.7 |
| Unemployed | 184.7 | 178.0 | 198.7 |
| **Total labour force** | 1,728.9 | 1,815.3 | 1,900.4 |

**2002:** Total employed 1,976.2; Unemployed 135.6; Total labour force 2,111.8.

**2003:** Total employed 1,917.0; Unemployed 160.5; Total labour force 2,077.4.

**2004:** Total employed 1,973.1; Unemployed 138.0; Total labour force 2,111.1.

## Health and Welfare

### KEY INDICATORS

| | |
|---|---|
| Total fertility rate (children per woman, 2003) | 3.7 |
| Under-5 mortality rate (per 1,000 live births, 2003) | 38 |
| HIV/AIDS (% of persons aged 15–49, 2003) | 0.2 |
| Physicians (per 1,000 head, 1997) | 0.86 |
| Hospital beds (per 1,000 head, 1996) | 1.48 |
| Health expenditure (2002): US $ per head (PPP) | 206 |
| Health expenditure (2002): % of GDP | 7.9 |
| Health expenditure (2002): public (% of total) | 49.1 |
| Access to water (% of persons, 2002) | 81 |
| Access to sanitation (% of persons, 2002) | 66 |
| Human Development Index (2002): ranking | 118 |
| Human Development Index (2002): value | 0.667 |

For sources and definitions, see explanatory note on p. vi.

## Agriculture

### PRINCIPAL CROPS
('000 metric tons)

|  | 2001 | 2002 | 2003 |
|---|---|---|---|
| Rice (paddy) | 237.4 | 283.9 | 290.6 |
| Maize | 419.9 | 499.5 | 523.7 |
| Sorghum | 88.9 | 117.8 | 99.5 |
| Cassava (Manioc) | 51.0 | 52.0 | 52.0 |
| Sugar cane | 3,144.6 | 3,119.4 | 3,602.8 |
| Dry beans | 176.8 | 196.9 | 202.9 |
| Groundnuts (in shell) | 80.6 | 60.4 | 81.9 |
| Oil palm fruit* | 53.0 | 53.0 | 53.0 |
| Bananas | 63.9 | 56.4 | 59.0 |
| Plantains* | 38.0 | 40.0 | 40.5 |
| Oranges* | 70.0 | 66.0 | 70.0 |
| Pineapples* | 45.0 | 46.0 | 48.0 |
| Coffee (green) | 66.8 | 60.2 | 59.7 |

* FAO estimates.
Source: FAO.

### LIVESTOCK
('000 head, year ending September)

|  | 2001 | 2002 | 2003 |
|---|---|---|---|
| Cattle | 3,300 | 3,350 | 3,500 |
| Pigs* | 410 | 430 | 440 |
| Goats* | 6.6 | 6.7 | 6.8 |
| Horses* | 248 | 250 | 260 |
| Asses* | 8.7 | 8.8 | 8.9 |
| Mules* | 46.0 | 46.5 | 47.0 |
| Poultry* | 15,200 | 15,500 | 16,200 |

* FAO estimates.
Source: FAO.

### LIVESTOCK PRODUCTS
('000 metric tons)

|  | 2001 | 2002 | 2003 |
|---|---|---|---|
| Beef and veal | 54.1 | 60.1 | 65.6 |
| Pig meat | 6.0 | 6.3 | 6.5 |
| Horse meat* | 1.9 | 1.9 | 2.0 |
| Poultry meat | 55.4 | 56.1 | 61.5 |
| Cows' milk | 247.1 | 263.5 | 281.1 |
| Butter | 0.5 | 0.6 | 0.6 |
| Cheese* | 21.9 | 23.2 | 23.5 |
| Hen eggs | 20.2 | 22.4 | 22.3 |
| Cattle hides* | 8.1 | 8.7 | 8.7 |

* FAO estimates.
Source: FAO.

## Forestry

### ROUNDWOOD REMOVALS
('000 cubic metres, excl. bark)

|  | 2001 | 2002 | 2003 |
|---|---|---|---|
| Sawlogs, veneer logs and logs for sleepers | 93 | 124 | 124* |
| Fuel wood* | 5,791 | 5,827 | 5,866 |
| **Total** | 5,884 | 5,951 | 5,990* |

* FAO estimate(s).
Source: FAO.

### SAWNWOOD PRODUCTION
('000 cubic metres, incl. railway sleepers)

|  | 2001 | 2002 | 2003* |
|---|---|---|---|
| Coniferous | 55 | 16 | 16 |
| Broadleaved | 10 | 29 | 29 |
| **Total** | 65 | 45 | 45 |

* FAO estimates.
Source: FAO.

## Fishing

('000 metric tons, live weight)

|  | 2001 | 2002 | 2003 |
|---|---|---|---|
| Capture | 19.5 | 16.4 | 15.3 |
| Snooks | 1.4 | 1.4 | 1.5 |
| Snappers | 1.9 | 2.2 | 2.2 |
| Yellowfin tuna | 3.5 | n.a. | n.a. |
| Common dolphinfish | 1.4 | 1.4 | 0.7 |
| Caribbean spiny lobsters | 3.9 | 4.3 | 3.9 |
| Penaeus shrimp | 4.2 | 4.0 | 3.7 |
| Aquaculture | 5.8 | 6.1 | 7.0 |
| Whiteleg shrimp | 5.6 | 6.0 | 6.9 |
| **Total catch** | 25.3 | 22.5 | 22.3 |

Source: FAO.

## Mining

|  | 2001 | 2002 | 2003* |
|---|---|---|---|
| Gold ('000 troy ounces) | 123.5 | 126.5 | 108.3 |
| Silver ('000 troy ounces) | 81.4 | 70.7 | 65.7 |
| Sand ('000 cubic metres) | 401.4 | 256.0 | 259.1 |
| Limestone ('000 cubic metres) | 231.2 | 290.0 | 186.6 |
| Gypsum ('000 metric tons) | 34.4 | 28.2 | 14.8 |

* Provisional figures.

## Industry

**SELECTED PRODUCTS**
('000 barrels, unless otherwise indicated)

|  | 2001 | 2002* | 2003† |
|---|---|---|---|
| Raw sugar ('000 metric tons) | 354 | 363 | 346 |
| Liquid gas | 219 | 194 | 230 |
| Motor spirit | 905 | 872 | 855 |
| Kerosene | 451 | 383 | 359 |
| Diesel | 1,578 | 1,485 | 1,436 |
| Fuel oil | 3,174 | 2,726 | 2,668 |
| Bitumen (asphalt) | 57 | 35 | 71 |
| Cement ('000 metric tons) | 595 | 549 | 577 |
| Cardboad boxes ('000 sq m) | 9,879 | 8,212 | 4,673 |
| Electric energy (million kWh) | 2,614 | n.a. | n.a. |
| Soap (kgs) | 10,000 | 10,200 | n.a. |
| Rum ('000 litres) | 8,595 | 7,216 | 8,345 |

* Preliminary figures.
† Estimates.

## Finance

**CURRENCY AND EXCHANGE RATES**

**Monetary Units**
100 centavos = 1 córdoba oro (gold córdoba).

**Sterling, Dollar and Euro Equivalents** (29 April 2005)
£1 sterling = 31.726 gold córdobas;
US $1 = 16.593 gold córdobas;
€1 = 21.500 gold córdobas;
1,000 gold córdobas = £31.52 = $60.27 = €46.51.

**Average Exchange Rate** (gold córdobas per US dollar)
2002   14.25
2003   15.11
2004   15.94

Note: In February 1988 a new córdoba, equivalent to 1,000 of the former units, was introduced, and a uniform exchange rate of US $1 = 10 new córdobas was established. Subsequently, the exchange rate was frequently adjusted. A new currency, the córdoba oro (gold córdoba), was introduced as a unit of account in May 1990 and began to be circulated in August. The value of the gold córdoba was initially fixed at par with the US dollar, but in March 1991 the exchange rate was revised to $1 = 25,000,000 new córdobas (or 5 gold córdobas). On 30 April 1991 the gold córdoba became the sole legal tender.

**BUDGET**
(million gold córdobas)

| Revenue* | 2002 | 2003 | 2004† |
|---|---|---|---|
| Taxation | 7,738.9 | 9,422.4 | 11,252.5 |
| Income tax | 1,609.8 | 2,447.9 | 3,176.0 |
| Value-added tax | 3,355.9 | 3,812.9 | 4,575.1 |
| Taxes on petroleum products | 1,399.4 | 1,566.4 | 1,618.4 |
| Taxes on imports | 641.9 | 628.2 | 684.4 |
| Other revenue | 824.4 | 728.6 | 983.1 |
| **Total** | 8,563.3 | 10,151.0 | 12,235.6 |

| Expenditure‡ | 2002 | 2003 | 2004† |
|---|---|---|---|
| Compensation of employees | 3,443.5 | 3,834.4 | 4,178.0 |
| Goods and services | 1,332.6 | 1,246.3 | 1,468.3 |
| Interest payments | 1,286.8 | 1,918.9 | 1,478.1 |
| Current transfers | 2,504.2 | 3,050.1 | 4,240.0 |
| Social security contributions | 157.1 | 223.2 | 227.4 |
| Other expenditure | 360.7 | 300.5 | 339.9 |
| **Total** | 9,085.0 | 10,573.5 | 11,931.6 |

* Excluding grants received (million gold córdobas): 1,522.3 in 2002; 2,079.0 in 2003; 2,373.6 in 2004 (preliminary).
† Preliminary figures.
‡ Excluding net acquisition of non-financial assets (million gold córdobas): 2,416.3 in 2002; 3,420.2 in 2003; 4,252.6 in 2004 (preliminary).

**INTERNATIONAL RESERVES***
(US $ million at 31 December)

|  | 2002 | 2003 | 2004 |
|---|---|---|---|
| IMF special drawing rights | 0.03 | 0.06 | 0.50 |
| Foreign exchange | 448.10 | 502.00 | 667.70 |
| **Total** | 448.13 | 502.06 | 668.20 |

* Excluding gold reserves (US $ million at 31 December): 4.10 in 1993.

Source: IMF, *International Financial Statistics*.

**MONEY SUPPLY**
(million gold córdobas at 31 December)

|  | 2002 | 2003 | 2004 |
|---|---|---|---|
| Currency outside banks | 2,085.8 | 2,506.6 | 3,103.3 |
| Demand deposits at commercial banks | 1,280.5 | 1,696.0 | 1,702.3 |
| **Total money** (incl. others) | 3,368.1 | 4,208.3 | 4,806.7 |

Source: IMF, *International Financial Statistics*.

**COST OF LIVING**
(Consumer Price Index; base: 2000 = 100)

|  | 2001 | 2002 | 2003 |
|---|---|---|---|
| Food (incl. beverages) | 108.6 | 111.8 | 115.9 |
| Clothing (incl. footwear) | 102.4 | 104.6 | 106.4 |
| Rent, fuel and light | 98.3 | 102.7 | 108.7 |
| **All items** (incl. others) | 107.4 | 111.6 | 117.4 |

Source: ILO.

**NATIONAL ACCOUNTS**
(million gold córdobas at current prices)

**Expenditure on the Gross Domestic Product**

|  | 2002 | 2003* | 2004† |
|---|---|---|---|
| Government final consumption expenditure | 6,124.0 | 6,614.0 | 7,575.2 |
| Private final consumption expenditure | 51,384.8 | 56,402.0 | 64,343.8 |
| Increase in stocks | 695.8 | 1,008.7 | 1,752.6 |
| Gross fixed capital formation | 14,261.8 | 15,334.2 | 18,870.4 |
| **Total domestic expenditure** | 72,466.4 | 79,359.0 | 92,542.0 |
| Exports of goods and services | 12,846.4 | 15,202.3 | 19,120.9 |
| *Less* Imports of goods and services | 27,936.4 | 31,887.5 | 39,059.6 |
| **GDP in purchasers' values** | 57,376.3 | 62,673.8 | 72,603.3 |
| **GDP at constant 1994 prices** | 28,087.5 | 28,721.2 | 30,199.9 |

* Preliminary figures.
† Estimates.

# NICARAGUA

## Gross Domestic Product by Economic Activity

|  | 2002 | 2003* | 2004† |
|---|---|---|---|
| Agriculture, hunting, forestry and fishing | 9,809.9 | 10,435.4 | 12,388.1 |
| Mining and quarrying | 603.3 | 585.7 | 736.2 |
| Manufacturing | 10,279.2 | 11,184.3 | 12,995.5 |
| Electricity, gas and water | 1,307.8 | 1,507.7 | 1,694.0 |
| Construction | 2,977.1 | 3,358.1 | 4,342.7 |
| Wholesale and retail trade | 8,029.9 | 8,695.5 | 9,967.1 |
| Transport and communications | 3,085.4 | 3,459.5 | 3,967.2 |
| Finance, insurance, real estate and business services | 6,973.7 | 7,689.2 | 9,044.8 |
| Other private services | 4,110.1 | 4,426.7 | 4,935.5 |
| Government services | 6,163.6 | 6,812.5 | 7,737.0 |
| **Sub-total** | 53,340.1 | 58,154.7 | 67,808.2 |
| Net taxes on products | 6,141.4 | 7,001.2 | 8,106.3 |
| *Less* Imputed bank service charge | 2,105.2 | 2,482.1 | 3,311.2 |
| **GDP in purchasers' values** | 57,376.3 | 62,673.8 | 72,603.3 |

\* Preliminary figures.
† Estimates.

## BALANCE OF PAYMENTS
(US $ million)

|  | 2001 | 2002 | 2003 |
|---|---|---|---|
| Exports of goods f.o.b. | 910.9 | 916.7 | 1,049.1 |
| Imports of goods f.o.b. | −1,808.2 | −1,853.1 | −2,021.2 |
| **Trade balance** | −897.3 | −936.4 | −972.1 |
| Exports of services | 223.4 | 226.2 | 249.0 |
| Imports of services | −353.1 | −335.5 | −372.1 |
| **Balance on goods and services** | −1,027.0 | −1,045.7 | −1,095.2 |
| Other income received | 14.7 | 9.2 | 6.7 |
| Other income paid | −255.0 | −209.6 | −209.9 |
| **Balance on goods, services and income** | −1,267.3 | −1,246.1 | −1,298.4 |
| Current transfers received | 482.5 | 462.4 | 518.9 |
| **Current balance** | −784.8 | −783.7 | −779.5 |
| Direct investment from abroad | 150.2 | 203.9 | 201.3 |
| Other investment assets | −60.4 | 2.9 | −16.0 |
| Other investment liabilities | −31.5 | −220.6 | −194.5 |
| Net errors and omissions | −61.2 | 135.1 | 26.4 |
| **Overall balance** | −493.0 | −414.2 | −500.7 |

Source: IMF, *International Financial Statistics*.

# External Trade

## PRINCIPAL COMMODITIES
(US $ million)

| Imports c.i.f. | 2002 | 2003 | 2004* |
|---|---|---|---|
| Consumer goods | 571.4 | 631.6 | 733.7 |
| Non-durable consumer goods | 452.0 | 489.4 | 569.8 |
| Durable consumer goods | 119.4 | 142.2 | 163.9 |
| Petroleum, mineral fuels and lubricants | 253.7 | 328.3 | 425.9 |
| Crude petroleum | 149.5 | 194.4 | 235.7 |
| Mineral fuels and lubricants | 103.8 | 133.5 | 190.2 |
| Intermediate goods | 508.9 | 557.6 | 646.1 |
| Primary materials and intermediate goods for agriculture and fishing | 55.8 | 60.2 | 67.6 |
| Primary materials and intermediate good for industry | 365.1 | 403.2 | 452.2 |
| Construction materials | 88.0 | 94.2 | 126.3 |
| Capital goods | 414.8 | 359.3 | 404.6 |
| For agriculture and fishing | 27.8 | 19.5 | 17.5 |
| For industry | 237.9 | 221.1 | 250.4 |
| For transport | 149.1 | 118.7 | 136.7 |
| Miscellaneous | 4.9 | 2.6 | 1.9 |
| **Total** | 1,753.7 | 1,879.4 | 2,212.3 |

\* Preliminary figures.

| Exports f.o.b. | 2002 | 2003* | 2004* |
|---|---|---|---|
| Coffee | 73.6 | 85.5 | 126.8 |
| Groundnuts | 24.2 | 28.4 | 39.7 |
| Cattle on hoof | 23.3 | 25.9 | 35.9 |
| Beans | 18.2 | 20.1 | 18.8 |
| Bananas | 11.0 | 12.0 | 10.7 |
| Raw tobacco | 4.9 | 7.2 | 7.3 |
| Lobster | 45.5 | 36.1 | 43.4 |
| Shrimp | 33.1 | 33.0 | 37.2 |
| Gold | 35.0 | 35.0 | 45.2 |
| Meat and meat products | 78.0 | 83.8 | 110.4 |
| Refined sugars, etc. | 28.6 | 25.7 | 36.8 |
| Cheese | 13.4 | 20.5 | 22.2 |
| Instant coffee | 8.9 | 7.6 | 9.1 |
| Wood products | 17.7 | 13.4 | 12.7 |
| Chemical products | 15.7 | 21.9 | 26.7 |
| Refined petroleum | 9.2 | 7.8 | 8.0 |
| Porcelain products | 9.9 | 10.2 | 11.3 |
| **Total** (incl. others) | 561.0 | 604.5 | 755.6 |

\* Preliminary figures.

## PRINCIPAL TRADING PARTNERS
(US $ million)

| Imports c.i.f. | 2002 | 2003 | 2004* |
|---|---|---|---|
| Canada | 19.2 | 15.7 | 21.4 |
| Costa Rica | 151.3 | 168.6 | 189.2 |
| Ecuador | 18.3 | 26.0 | 38.1 |
| El Salvador | 102.7 | 91.9 | 103.6 |
| Germany | 27.6 | 42.1 | 41.7 |
| Guatemala | 136.8 | 137.3 | 157.4 |
| Honduras | 9.7 | 32.4 | 51.5 |
| Japan | 93.8 | 81.9 | 95.5 |
| Mexico | 112.2 | 158.8 | 163.0 |
| Panama | 22.4 | 18.9 | 16.6 |
| Spain | 32.4 | 26.8 | 29.2 |
| Sweden | 21.2 | 32.8 | 6.5 |
| Taiwan | 19.1 | 17.1 | 16.0 |
| USA | 475.1 | 462.5 | 491.9 |
| Venezuela | 196.1 | 183.2 | 320.1 |
| **Total** (incl. others) | 1,753.7 | 1,879.4 | 2,212.3 |

\* Preliminary figures.

| Exports f.o.b. | 2002 | 2003* | 2004* |
|---|---|---|---|
| Belgium | 3.5 | 3.8 | 11.7 |
| Canada | 19.4 | 21.3 | 34.9 |
| Costa Rica | 48.4 | 49.2 | 50.6 |
| El Salvador | 86.7 | 104.4 | 109.3 |
| France | 9.2 | 6.0 | 13.0 |
| Germany | 13.0 | 9.5 | 14.1 |
| Guatemala | 23.2 | 25.9 | 32.3 |
| Honduras | 38.4 | 43.5 | 56.6 |
| Italy | 15.3 | 19.0 | 17.6 |
| Mexico | 21.4 | 27.9 | 39.9 |
| Puerto Rico | 16.1 | 15.2 | 19.4 |
| Russia | 13.4 | 14.5 | 1.2 |
| Spain | 11.2 | 15.9 | 23.6 |
| USA | 205.2 | 201.9 | 263.4 |
| **Total** (incl. others) | 561.0 | 604.5 | 755.6 |

\* Preliminary figures.

# Transport

## RAILWAYS
(traffic)

|  | 1990 | 1991 | 1992 |
|---|---|---|---|
| Passenger-km (million) | 3 | 3 | 6 |

**Freight ton-km** (million): 4 in 1985.

Source: UN, *Statistical Yearbook*.

# NICARAGUA

## ROAD TRAFFIC
(motor vehicles in use)

|  | 2000 | 2001 | 2002 |
|---|---|---|---|
| Cars | 61,357 | 70,372 | 83,168 |
| Buses and coaches | 5,460 | 6,078 | 6,947 |
| Goods vehicles | 87,358 | 95,986 | 108,308 |
| Motorcycles and mopeds | 23,857 | 26,654 | 28,973 |

## SHIPPING
**Merchant fleet**
(registered at 31 December)

|  | 2002 | 2003 | 2004 |
|---|---|---|---|
| Number of vessels | 26 | 26 | 27 |
| Total displacement ('000 grt) | 3.6 | 3.6 | 5.0 |

Source: Lloyd's Register-Fairplay, *World Fleet Statistics*.

**International Sea-Borne Freight Traffic**
('000 metric tons)

|  | 1997 | 1998 | 1999 |
|---|---|---|---|
| Imports | 1,272.7 | 1,964.7 | 1,180.7 |
| Exports | 329.3 | 204.5 | 183.6 |

**Total freight traffic** ('000 metric tons): 2,215.9 in 2000; 2,363.0 in 2001; 2,093.8 in 2002.

## CIVIL AVIATION
(traffic on scheduled services)

|  | 1998 | 1999 | 2000 |
|---|---|---|---|
| Kilometres flown (million) | 1.2 | 0.8 | 0.8 |
| Passengers carried ('000) | 52 | 59 | 61 |
| Passenger-km (million) | 93 | 67 | 72 |
| Freight ton-km (million) | n.a. | 0.5 | 0.5 |

Source: UN Economic Commission for Latin America and the Caribbean.

## Tourism

**TOURIST ARRIVALS BY COUNTRY OF ORIGIN**

|  | 2001 | 2002 | 2003 |
|---|---|---|---|
| Canada | 11,138 | 9,800 | 13,124 |
| Costa Rica | 62,055 | 57,824 | 76,659 |
| El Salvador | 71,886 | 69,691 | 73,806 |
| Guatemala | 38,311 | 36,964 | 40,132 |
| Honduras | 118,282 | 111,947 | 107,365 |
| Panama | 10,720 | 10,545 | 11,988 |
| USA | 88,375 | 97,863 | 117,156 |
| **Total** (incl. others) | 482,869 | 471,622 | 525,775 |

**Tourism receipts** (US $ million, incl. passenger transport): 138 in 2001; 138 in 2002; 155 in 2003.

Sources: World Tourism Organization.

## Communications Media

|  | 2001 | 2002 | 2003 |
|---|---|---|---|
| Telephones ('000 main lines in use) | 158.6 | 171.6 | 205.0 |
| Mobile cellular telephones ('000 subscribers) | 156.0 | 239.9 | 466.7 |
| Personal computers ('000 in use) | 50 | 150 | n.a. |
| Internet users ('000) | 50 | 90 | n.a. |

**Radio receivers** ('000 in use): 1,240 in 1997.
**Television receivers** ('000 in use): 350 in 2000.
**Daily newspapers:** 4 in 1996 (average circulation 135,000 copies).

Sources: UNESCO, *Statistical Yearbook*; International Telecommunication Union.

## Education

(2002, unless otherwise indicated)

|  |  |  | Students | | |
|---|---|---|---|---|---|
|  | Institutions | Teachers | Males | Females | Total |
| Pre-primary | 5,980 | 3,672* | 88,916 | 88,618 | 177,534 |
| Primary | 8,251 | 21,020* | 471,656 | 451,735 | 923,391 |
| Secondary: general | 1,249 | 5,970* | 160,399 | 203,613 | 364,012 |
| Tertiary: university level | 35 | 3,630† | n.a. | n.a. | 70,925† |
| Tertiary: other higher | 73 | 210† | 8,615 | 9,898 | 18,513 |

\* 1996 figure.
† 2001 figure.

Sources: UNESCO, *Statistical Yearbook*; Ministry of Education, Culture and Sports.

**Adult literacy rate** (UNESCO estimates): 76.7% (males 76.8%; females 76.6%) in 2002 (Source: UN Development Programme, *Human Development Report*).

# Directory

## The Constitution*

Shortly after taking office on 20 July 1979, the Government of National Reconstruction abrogated the 1974 Constitution. On 22 August 1979 the revolutionary junta issued a 'Statute on Rights and Guarantees for the Citizens of Nicaragua', providing for the basic freedoms of the individual, religious freedom and freedom of the press and abolishing the death penalty. The intention of the Statute was formally to re-establish rights which had been violated under the deposed Somoza regime. A fundamental Statute took effect from 20 July 1980 and remained in force until the Council of State drafted a political constitution and proposed an electoral law. A new Constitution was approved by the National Constituent Assembly on 19 November 1986 and promulgated on 9 January 1987. Amendments to the Constitution were approved by the Asamblea Nacional (National Assembly) in July 1995 and January 2000. The following are some of the main points of the Constitution.

Nicaragua is an independent, free, sovereign and indivisible state. All Nicaraguans who have reached 16 years of age are full citizens.

### POLITICAL RIGHTS

There shall be absolute equality between men and women. It is the obligation of the State to remove obstacles that impede effective participation of Nicaraguans in the political, economic and social life of the country. Citizens have the right to vote and to be elected at elections and to offer themselves for public office. Citizens may organize or affiliate with political parties, with the objective of

# NICARAGUA

participating in, exercising or vying for power. The supremacy of civilian authority is enshrined in the Constitution.

## SOCIAL RIGHTS

The Nicaraguan people have the right to work, to education and to culture. They have the right to decent, comfortable and safe housing, and to seek accurate information. This right comprises the freedom to seek, receive and disseminate information and ideas, both spoken and written, in graphic or any other form. The mass media are at the service of national interests. No Nicaraguan citizen may disobey the law or prevent others from exercising their rights and fulfilling their duties by invoking religious beliefs or inclinations.

## LABOUR RIGHTS

All have a right to work, and to participate in the management of their enterprises. Equal pay shall be given for equal work. The State shall strive for full and productive employment under conditions that guarantee the fundamental rights of the individual. There shall be an eight-hour working day, weekly rest, vacations, remuneration for national holidays and a bonus payment equivalent to one month's salary, in conformity with the law.

## EDUCATION

Education is an obligatory function of the State. Planning, direction and organization of the secular education system is the responsibility of the State. All Nicaraguans have free and equal access to education. Private education centres may function at all levels.

## LEGISLATIVE POWER

The Asamblea Nacional exercises Legislative Power through representative popular mandate. The Asamblea Nacional is composed of 90 representatives elected by direct secret vote by means of a system of proportional representation, of which 70 are elected at regional level and 20 at national level. The number of representatives may be increased in accordance with the general census of the population, in conformity with the law. Representatives shall be elected for a period of five years. The functions of the Asamblea Nacional are to draft and approve laws and decrees; to decree amnesties and pardons; to consider, discuss and approve the General Budget of the Republic; to elect judges to the Supreme Court of Justice and the Supreme Electoral Council; to fill permanent vacancies for the Presidency or Vice-Presidency; and to determine the political and administrative division of the country.

## EXECUTIVE POWER

The Executive Power is exercised by the President of the Republic (assisted by the Vice-President), who is the Head of State, Head of Government and Commander-in-Chief of the Defence and Security Forces of the Nation. The election of the President (and Vice-President) is by equal, direct and free universal suffrage in secret ballot. Should a single candidate in a presidential election fail to secure the necessary 35% of the vote to win outright in the first round, a second ballot shall be held. Close relatives of a serving President are prohibited from contesting a presidential election. The President shall serve for a period of five years and may not serve for two consecutive terms. All outgoing Presidents are granted a seat in the Asamblea Nacional.

## JUDICIAL POWER

The Judiciary consists of the Supreme Court of Justice, Courts of Appeal and other courts of the Republic. The Supreme Court is composed of at least seven judges, elected by the Asamblea Nacional, who shall serve for a term of six years. The functions of the Supreme Court are to organize and direct the administration of justice. There are 12 Supreme Court justices, appointed for a period of seven years.

## LOCAL ADMINISTRATION

The country is divided into regions, departments and municipalities for administrative purposes. The municipal governments shall be elected by universal suffrage in secret ballot and will serve a six-year term. The communities of the Atlantic Coast have the right to live and develop in accordance with a social organization which corresponds to their historical and cultural traditions. The State shall implement, by legal means, autonomous governments in the regions inhabited by the communities of the Atlantic Coast, in order that the communities may exercise their rights.

*In January 2000 a constitutional amendment established a constitutional assembly to effect further reform of the Constitution, following amendments in 1995 and 2000.

# The Government

## HEAD OF STATE

**President:** ENRIQUE BOLAÑOS GEYER (took office 10 January 2002).

**Vice-President:** JOSÉ RIZO CASTELLÓN.

## CABINET
(July 2005)

**Minister of Foreign Affairs:** NORMAN CALDERA CARDENAL.

**Minister of Government:** JULIO VEGA PASQUIER.

**Minister of National Defence:** AVIL RAMÍREZ.

**Minister of Finance and Public Credit:** MARIO ARANA SEVILLA.

**Minister of Development, Industry and Trade:** AZUCENA CASTILLO DE SOLANO.

**Minister of Labour:** Dr VIRGILIO JOSÉ GURDIÁN CASTELLÓN.

**Minister of the Environment and Natural Resources:** ARTURO HARDING LACAYO.

**Minister of Transport and Infrastructure:** PEDRO SOLÓRZANO CASTILLO.

**Minister of Agriculture and Forestry:** JOSÉ AUGUSTO NAVARRO FLORES.

**Minister of Health:** MARGARITA GURDIÁN LÓPEZ.

**Minister of Education, Culture and Sports:** MIGUEL ANGEL GARCÍA GUTIÉRREZ.

**Minister of the Family:** IVANIA DEL SOCORRO TORUÑO PADILLA.

**Secretary to the Presidency:** ERNESTO LEAL.

There are, in addition, five Secretaries of State.

## MINISTRIES

**Ministry of Agriculture and Forestry:** Km 8½, Carretera a Masaya, Managua; tel. (2) 76-0235; e-mail prensa@magfor.gob.ni; internet www.magfor.gob.ni.

**Ministry of Development, Industry and Trade:** Edif. Central, Km 6, Carretera a Masaya, Apdo 8, Managua; tel. (2) 78-8702; fax (2) 70-095; e-mail webmaster@mific.gob.ni; internet www.mific.gob.ni.

**Ministry of Education, Culture and Sports:** Complejo Cívico Camilo Ortega Saavedra, Managua; tel. (2) 65-1451; e-mail rivash@mecd.gob.ni; internet www.mecd.gob.ni.

**Ministry of the Environment and Natural Resources:** Km 12½, Carretera Norte, Apdo 5123, Managua; tel. (2) 33-1111; fax (2) 63-1274; e-mail cap@marena.gob.ni; internet www.marena.gob.ni.

**Ministry of the Family:** Managua; tel. (2) 78-1620; e-mail webmaster@mifamilia.gob.ni; internet www.mifamilia.gob.ni.

**Ministry of Finance and Public Credit:** Frente a la Asamblea Nacional, Apdo 2170, Managua; tel. (2) 22-6530; fax (2) 22-6430; e-mail webmaster@mhcp.gob.ni; internet www.hacienda.gob.ni.

**Ministry of Foreign Affairs:** Del Cine González al Sur sobre Avda Bolivar, Managua; tel. (2) 44-8000; fax 28-5102; e-mail despacho.ministro@cancilleria.gob.ni; internet www.cancilleria.gob.ni.

**Ministry of Government:** Apdo 68, Managua; tel. (2) 28-2284; fax (2) 22-2789; e-mail webmaster@migob.gob.ni; internet www.migob.gob.ni.

**Ministry of Health:** Complejo Cívico Camilo Ortega Saavedra, Managua; tel. (2) 89-7164; e-mail secretaria@minsa.gob.ni; internet www.minsa.gob.ni.

**Ministry of Labour:** Estadio Nacional, 400 m al Norte, Apdo 487, Managua; tel. (2) 28-2028; fax (2) 28-2103.

**Ministry of National Defence:** Casa de la Presidencia, Managua; tel. (2) 66-3580; fax (2) 28-7911; internet www.midef.gob.ni.

**Ministry of Tourism:** Hotel Intercontinental 1C, Managua; tel. (2) 22-3333; internet www.intur.gob.ni.

**Ministry of Transport and Infrastructure:** Frente al Estadio Nacional, Apdo 26, Managua; tel. (2) 28-2061; fax (2) 22-5111; e-mail webmaster@mti.gob.ni; internet www.mti.gob.ni.

# President and Legislature

## PRESIDENT
Election, 4 November 2001

| Candidate | Votes | % of total |
|---|---|---|
| Enrique Bolaños Geyer (Partido Liberal Constitucionalista) | 1,216,863 | 56.3 |
| Daniel Ortega Saavedra (Frente Sandinista de Liberación Nacional) | 915,417 | 42.3 |
| Alberto Saborío (Partido Conservador de Nicaragua) | 29,933 | 1.4 |
| **Total** | **2,162,213** | **100.0** |

## ASAMBLEA NACIONAL
(National Assembly)

**Asamblea Nacional**
Avda Bolívar, Contiguo a la Presidencia de la República, Managua; e-mail webmaster@correo.asamblea.gob.ni; internet www.asamblea.gob.ni.

**President:** SANTOS RENÉ NÚÑEZ TÉLLEZ.
**First Vice-President:** CARLOS WILFREDO NAVARRO MOREIRA.
**Second Vice-President:** MIRNA DEL ROSARIO ROSALES AGUILAR.
**Third Vice-President:** GABRIEL RIVERA ZELEDÓN.

Election, 4 November 2001

| Party | Seats |
|---|---|
| Partido Liberal Constitucionalista (PLC) | 47 |
| Frente Sandinista de Liberación Nacional (FSLN) | 42 |
| Partido Conservador de Nicaragua (PCN) | 1 |
| **Total** | **90\*** |

\* In addition to the 90 elected members, supplementary seats in the Asamblea Nacional are awarded to the unsuccessful candidates at the presidential election who were not nominated for the legislature but who received, in the presidential poll, a number of votes at least equal to the average required for one of the 70 legislative seats decided at a regional level. On this basis, the FSLN obtained one additional seat in the Asamblea Nacional. A legislative seat is also awarded to the out-going President. Thus, the PLC also gained an additional seat, bringing the total number of seats in the Asamblea Nacional to 92.

# Political Organizations

**Acción Nacional Conservadora (ANC):** Costado Oeste SNTV, Managua; tel. (2) 66-8755; f. 1956; Pres. Dr FRANK DUARTE TAPIA.

**Alianza por la República (APRE):** Casa 211, Col. Los Robles, Funeraria Monte de los Olivos 1.5 c. al norte, Managua; f. 2004; Pres. MIGUEL LÓPEZ BALDIZÓN.

**Movimiento Democrático Nicaragüense (MDN):** Casa L-39, Ciudad Jardín Bnd, 50 m al sur, Managua; tel. (2) 43898; f. 1978; Leader ROBERTO SEQUEIRA GÓMEZ.

**Partido Conservador de Nicaragua (PC):** Colegio Centroamérica, 500 m al sur, Managua; tel. (2) 67-0484; internet www.partidoconservador.org.ni; f. 1992 following merger between Partido Conservador Demócrata (PCD) and Partido Socialconservadurismo; Pres. MARIO RAPPACCIOLI MCGREGOR.

**Partido Social Cristiano (PSC):** Ciudad Jardín, Pizza María, 1 c. al Lago, Managua; tel. (2) 22-026; f. 1957; 42,000 mems; Pres. ABEL REYES.

**Camino Cristiano Nicaragüense (CCN):** Managua; Pres. GUILLERMO ANTONIO OSORNO MOLINA.

**Frente Sandinista de Liberación Nacional (FSLN)** (Sandinista National Liberation Front): Costado Oeste Parque El Carmen, Managua; tel. and fax (2) 66-8173; internet www.fsln.org.ni; f. 1960; led by a 15-member directorate; embraces Izquierda Democrática Sandinista 'orthodox revolutionary' faction, led by Daniel Ortega Saavedra; 120,000 mems; Gen. Sec. DANIEL ORTEGA SAAVEDRA.

**Movimiento Renovador Sandinista (MRS):** Tienda Katty 1 c. abajo, Apdo 24, Managua; tel. (2) 78-0279; fax (2) 78-0268; f. 1995; former faction of Frente Sandinista de Liberación Nacional; Pres. DORA MARÍA TÉLLEZ.

**Partido Indígena Multiétnico (PIM):** Residencial Los Robles, de Farmacentro 1 c. al este, 80 varas al sur, Managua; Pres. CARLA WHITE HODGSON.

**Partido Liberal Constitucionalista (PLC):** Semáforos Country Club 100 m al este, Apdo 4569, Managua; tel. (2) 78-8705; fax (2) 78-1800; e-mail plc@ibw.com.ni; internet www.plc.org.ni; f. 1967; Pres. JORGE CASTILLO QUANT; Nat. Sec. Dr NOEL RAMÍREZ SÁNCHEZ.

**Partido Liberal Independiente (PLI):** Ciudad Jardín, H-4, Calle Principal, Managua; tel. (2) 44-3556; fax (2) 48-0012; f. 1944; Leader VIRGILIO GODOY REYES.

**Partido Liberal Independiente de Unidad Nacional (PLIUN):** Munich, 2½ c. arriba, Managua; tel. (2) 61-672; f. 1988; splinter group of PLI; Pres. EDUARDO CORONADO PÉREZ; Sec.-Gen. CARLOS ALONSO.

**Partido Movimiento de Unidad Costeña (PAMUC):** Bilwi Puerto Cabeza; Pres. KENNETH SERAPIO HUNTER.

**Partido Neo-Liberal (Pali):** Cine Dorado, 2 c. al sur, 50 m arriba, Managua; tel. (2) 66-5166; f. 1986; Pres. Dr RICARDO VEGA GARCÍA.

**Partido de los Pueblos Costeños (PPC):** Bluefields; f. 1997; multi-ethnic; Pres. HUGO SUJO.

**Partido Regional Nueva Alternativa (PARNA):** Managua.

**Partido Resistencia Nicaragüense (PRN):** Edif. VINSA, frente a Autonica, Carretera Sur, Managua; tel. and fax (2) 70-6508; e-mail salvata@ibw.com.ni; f. 1993; nationalist party; Pres. SALVADOR TALAVERA ALANIZ.

**Partido Social Demócrata (PSD):** Frente al Teatro Aguerri, Managua; tel. (2) 28-1277; f. 1979; Pres. ADOLFO JARQUÍN ORTEL; Sec.-Gen. Dr JOSÉ PALLAIS ARANA.

**Partido Socialista (PS):** Hospital Militar, 100 m al norte, 100 m al oeste, 100 m al sur, Managua; tel. (2) 66-2321; fax (2) 66-2936; f. 1944; social democratic party; Sec.-Gen. Dr GUSTAVO TABLADA ZELAYA.

**Partido Unionista Centroamericano (PUCA):** Cine Cabrera, 1a c. al este, 20 m al norte, Managua; tel. (2) 27-472; f. 1904; Pres. BLANCA ROJAS ECHAVERRY.

**Unión Demócrata Cristiana (UDC):** De Iglesia Santa Ana, 2 c. abajo, Barrio Santa Ana, Apdo 3089, Managua; tel. (2) 66-2576; f. 1976 as Partido Popular Social Cristiano; name officially changed as above in Dec. 1993; Pres. AGUSTÍN JARQUÍN.

**Yatama** (Yapti Tasba Masraka Nanih Aslatakanka): Of. de Odacan, Busto José Martí, 1 c. al este y ½ c. al norte, Managua; tel. (2) 28-1494; Atlantic coast Miskito organization; Leader BROOKLYN RIVERA BRYAN.

# Diplomatic Representation

## EMBASSIES IN NICARAGUA

**Argentina:** Semáforos de Villa Fontana, 2 c. abajo, 1 al sur, 1 abajo, 75 varas oeste, Casa 133, Apdo 703, Managua; tel. (2) 83-7066; fax (2) 70-2343; e-mail embargentina@teranet.com.ni; Ambassador HORACIO ALBERTO AMOROSO.

**Brazil:** Km 7¾, Carretera Sur, Quinta los Pinos, Apdo 264, Managua; tel. (2) 65-0035; fax (2) 65-2206; e-mail ebrasil@ibw.com.ni; Ambassador RICARDO DRUMMOND DE MELLO.

**Chile:** Entrada principal los Robles, 1 c. abajo, 1 c. al sur, Apdo 1289, Managua; tel. (2) 78-0619; fax (2) 70-4073; e-mail echileni@cablenet.com.ni; Ambassador CARLOS GONZÁLEZ MÁRQUEZ.

**China (Taiwan):** Planes de Altamira, 19–20, frente a la cancha de tenis, Apdo 4653, Managua 5; tel. (2) 77-1333; fax (2) 67-4025; e-mail embchina@ibw.com.ni; Ambassador MING-TA HUNG.

**Colombia:** 2da Entrada a las Colinas, 1 c. arriba, ½ c. al lago, Casa 97, Managua; tel. (2) 76-2149; e-mail emanagua@minrelext.gov.co; Ambassador MELBA MARTÍNEZ LÓPEZ.

**Costa Rica:** Edif. Car, 3°, 4.5km Carretera a Masaya, Managua; tel. (2) 70-3779; fax (2) 70-3780; e-mail info@embajadadecostarica.com; Ambassador RODRIGO CARRERAS JIMÉNEZ.

**Cuba:** Carretera a Masaya, 3a Entrada a las Colinas, Managua; tel. (2) 76-0742; fax (2) 76-0166; e-mail embacuba@cablenet.com.ni; Chargé d'affaires a.i. MANUEL GUILLOT PÉREZ.

**Denmark:** De la Plaza España 1 c. abajo, 2 c. al lago, ½ c. abajo, Apdo 4942, Managua; tel. (2) 68-0250; fax (2) 68-8095; e-mail mgaamb@um.dk; Ambassador THOMAS SCHJERBECK.

**Dominican Republic:** Reparto Las Colinas, Prado Ecuestre 100, con Curva de los Gallos, Apdo 614, Managua; tel. (2) 76-2029; fax (2) 76-0654; e-mail embdom@alfanumeric.com.ni; Ambassador HÉCTOR DARÍO FREITES CAMINERO.

# NICARAGUA

**Ecuador:** De los Pipitos 1½ c. abajo, Apdo C-33, Managua; tel. (2) 68-1098; fax (2) 66-8081; e-mail ecuador@ibw.com.ni; Ambassador María del Carmen González Cabal.

**El Salvador:** Reparto Las Colinas, Avda del Campo y Pasaje Los Cerros 142, Apdo 149, Managua; tel. (2) 76-0712; fax (2) 76-0711; e-mail embelsa@cablenet.com.ni; Ambassador José Roberto Francisco Imendia.

**Finland:** Suc. Jorge Navarro, Apdo 2219, Managua; tel. (2) 66-3415; fax (2) 66-3416; e-mail sanomat.mgu@formin.fi; Ambassador Inger Hirvelä López (non-resident).

**France:** Iglesia el Carmen 1½ c. abajo, Apdo 1227, Managua; tel. (2) 22-6210; fax (2) 28-1057; e-mail ambafrance-mnga@tmx.com.ni; internet www.ambafrance-ni.org; Ambassador Jean-Pierre Lafosse.

**Germany:** Bolonia, de la Rotonda El Güegüense, 1½ c. al lago, Contiguo a Optica Nicaragüense, Apdo 29, Managua; tel. (2) 66-3917; fax (2) 66-7667; e-mail alemania@ibw.com.ni; internet www.managua.diplo.de; Ambassador Gregor Koebel.

**Guatemala:** Km 11½, Carretera a Masaya, Apdo E-1, Managua; tel. (2) 79-9609; fax (2) 79-9610; e-mail embnicaragua@minex.gob.gt; Ambassador Jorge Rolando Echeverría Roldán.

**Holy See:** Apostolic Nunciature, Km 10.8, Carretera Sur, Apdo 506, Managua; tel. (2) 65-8657; fax (2) 65-7416; e-mail nuntius@cablenet.com.ni; Apostolic Nuncio Most Rev. Jean-Paul Gobel (Titular Archbishop of Galazia in Campania).

**Honduras:** Reparto San Juan 312, del Gimnasio Hércules 1 c. al sur y 1½ arriba, Apdo 321, Managua; tel. (2) 78-4133; fax (2) 78-3043; e-mail embhonduras@ideay.net.ni; Ambassador Jorge Milla Reyes.

**Italy:** Rotonda El Güegüense, 1 c. al norte, Apdo 2092, ½ c. abajo, Managua 4; tel. (2) 66-2961; fax (2) 66-3987; e-mail embitaliasegr@cablenet.com.ni; internet www.ambitaliamanagua.org; Ambassador Dr Alberto Boniver.

**Japan:** Plaza España, 1 c. abajo y 1 c. al lago, Bolonia, Apdo 1789, Managua; tel. (2) 66-8668; fax (2) 66-8566; internet www.ni.emb-japan.go.jp; Ambassador Kunio Shimizu.

**Korea, Democratic People's Republic:** Managua; Ambassador Ri Kang Se.

**Libya:** Mansión Teodolinda, 1 c. al sur, ½ c. abajo, Managua; Sec. of the People's Bureau Mohamed Majdoub.

**Mexico:** Contiguo a Optica Matamoros, Km 4½, Carretera a Masaya, Apdo 834, Managua; tel. (2) 78-1859; fax (2) 78-2886; e-mail embamex@cablenet.com.ni; Ambassador Columba Calvo Varga.

**Netherlands:** Col. los Robles III etapa, de Plaza el Sol 1 c. al sur 1½ c. al oeste, Apdo 3688, Managua; tel. (2) 70-4505; fax (2) 70-0399; e-mail mng@minbuza.nl; internet www.embholanda.org.ni; Ambassador Kees Rade.

**Norway:** Plaza España, Apdo 2090, Correo Central, Managua; tel. (2) 66-4199; fax (2) 66-3303; e-mail emb.managua@norad.no; internet www.noruega.org.ni; Ambasssador Idar Johansen.

**Panama:** Casa 93, Reparto Mántica, del Cuartel General de Bommberos 1 c. abajo, Apdo 1, Managua; tel. (2) 66-2224; fax (2) 66-8633; e-mail mbdpma@hotmail.com; Ambassador Dr Daniel Silvera Serracín.

**Peru:** Casa 325, Barrio Bolonia, del Hospital Militar 'Alejandró Dávila', 1 c. al lago y 2 c. abajo, Apdo 211, Managua; tel. (2) 66-8677; fax (2) 66-1408; e-mail peru1@ibw.com.ni; Ambassador Eduardo Cartillo Hernández.

**Russia:** Reparto Las Colinas, Calle Vista Alegre 214, Entre Avda Central y Paseo del Club, Apdo 249, Managua; tel. (2) 76-0374; fax (2) 76-0179; e-mail rossia@ibw.com.ni; Ambassador Igor Dyakonov.

**Spain:** Avda Central 13, Las Colinas, Apdo 284, Managua; tel. (2) 76-0966; fax (2) 76-0937; e-mail embespni@correo.mae.es; Ambassador Jaime Lacadena Higuera.

**Sweden:** Plaza España, 1 c. abajo, 2 c. al lago y ½ c. al oeste, Apdo 2307, Managua; tel. (2) 66-2762; fax (2) 66-6778; e-mail ambassaden.managua@sida.se; internet www.suecia.org.ni; Ambassador Eva Zetterber.

**USA:** Km 4½, Carretera Sur, Apdo 327, Managua; tel. (2) 66-6010; fax (2) 66-3861; e-mail ConsularManagua@state.gov; Chargé d'affaires Oliver Garza; Ambassador Paul A. Trivelli (from Sept. 2005).

**Venezuela:** Edif. Málaga, 2°, Plaza España, Módulo A-13, Apdo 406, Managua; tel. (2) 72-0267; fax (2) 72-2265; e-mail embaveznica@cablenet.com.ni; Ambassador Miguel Gómez.

## Judicial System

### The Supreme Court

Km 7½, Carretera Norte, Managua; tel. (2) 33-0083; fax (2) 33-0581; internet www.csj.gob.ni.

Deals with both civil and criminal cases, acts as a Court of Cassation, appoints Judges of First Instance, and generally supervises the legal administration of the country.

**President:** Dr José Manuel Martínez Sevilla.
**Vice-President:** Dr Carlos Guerra Gallardo.
**Attorney-General:** Alberto Novoa.

## Religion

All religions are tolerated. Almost all of Nicaragua's inhabitants profess Christianity, and the great majority belong to the Roman Catholic Church. The Moravian Church predominates on the Caribbean coast.

### CHRISTIANITY

#### The Roman Catholic Church

Nicaragua comprises one archdiocese, six dioceses and the Apostolic Vicariate of Bluefields. At 31 December 2003 there were an estimated 5,213,778 adherents, representing about 80% of the total population.

**Bishops' Conference**
Conferencia Episcopal de Nicaragua, Ferretería Lang 1 c. al norte, l c. al este, Zona 3, Las Piedrecitas, Apdo 2407, Managua; tel. (2) 66-6292; fax (2) 66-8069; e-mail cen@tmx.com.ni.

f. 1975; statute approved 1987; Pres. Cardinal Miguel Obando y Bravo (Archbishop of Managua).

**Archbishop of Managua:** Leopoldo José Brenes Solórzano, Arzobispado, Apdo 3058, Managua, tel. (2) 77-1754; fax (2) 76-0130; e-mail mob@adm.unica.edu.ni.

#### The Anglican Communion

Nicaragua comprises one of the five dioceses of the Iglesia Anglicana de la Región Central de América.

**Bishop of Nicaragua:** Rt Rev. Sturdie W. Downs, Apdo 1207, Managua; tel. (2) 22-5174; fax (2) 22-6701; e-mail episcnic@tmx.com.ni.

#### Protestant Churches

**Baptist Convention of Nicaragua:** Apdo 2593, Managua; tel. (2) 25785; fax (2) 24131; e-mail cbn@ibw.com.ni; f. 1917; 155 churches, 20,000 mems (2003); Pres. Rev. Pablo García Morales (2002–05); Gen. Sec. Rev. Elías González Argüello.

**The Nicaraguan Lutheran Church of Faith and Hope:** Apdo 151, Managua; tel. (2) 66-4467; fax (2) 66-4609; e-mail luterana@ibw.org.ni; f. 1994; 4,000 mems (2000); Pres. Rev. Victoria Cortez Rodríguez.

## The Press

### NEWSPAPERS AND PERIODICALS

**Avance:** Ciudad Jardín 0-30, Apdo 4231, Managua; tel. (2) 49-5047; f. 1972; weekly publication of the Partido Comunista de Nicaragua; circ. 10,000.

**Bolsa de Noticias:** Col. Centroamérica 852, Apdo VF-90, Managua; tel. (2) 70-0546; fax (2) 77-4931; e-mail grupoese@tmx.com.ni; internet www.grupoese.com.ni/BolsadeNoticias; f. 1974; daily; Dir María Elsa Suárez García.

**Confidencial:** De la Iglesia El Carmen 1 c. al lago, ½ c. abajo, Managua; tel. (2) 68-0129; fax (2) 68-4650; e-mail revista@confidencial.com.ni; internet www.confidencial.com.ni; weekly; political analysis; Editor Carlos F. Chamorro.

**La Gaceta, Diario Oficial:** Semáforos de Plaza Inter, 1 c. arriba, 1½ c. al lago, Managua; tel. (2) 28-3791; e-mail lagaceta@ibw.com.ni; f. 1912; morning; official.

**Novedades:** Pista P. Joaquín Chamorro, Km 4, Carretera Norte, Apdo 576, Managua; daily, evening.

**Nuevo Diario:** Pista P. Joaquín Chamorro, Km 4, Carretera Norte, Apdo 4591, Managua; tel. (2) 49-1190; fax (2) 49-0700; e-mail info@elnuevodiario.com.ni; internet www.elnuevodiario.com.ni; f. 1980;

# NICARAGUA

morning, daily; independent; Editor Xavier Chamorro Cardenal; circ. 45,000.

**El Observador Económico:** Antiguo Hospital el Retiro, 2 c. al lago, Apdo 2074, Managua; tel. (2) 66-8708; fax (2) 66-8711; e-mail amc@elobservadoreconomico.com; internet www.elobservadoreconomico.com; Dir-Gen. Alejandro Martínez Cuenca.

**El Popular:** Managua; tel. (2) 66-2936; monthly; official publication of the Partido Socialista Nicaragüense.

**La Prensa:** Km 4½, Carretera Norte, Apdo 192, Managua; tel. (2) 49-8405; fax (2) 49-6926; e-mail info@laprensa.com.ni; internet www.laprensa.com.ni; f. 1926; morning, daily; independent; Pres. Jaime Chamorro Cardenal; Editor Eduardo Enríquez; circ. 30,000.

**Prensa Proletaria:** Managua; tel. (2) 22-594; fortnightly; official publication of the Movimiento de Acción Popular Marxista-Leninista.

**Revista Ambiente:** 27 Avda 6901, Managua; tel. and fax (2) 66-8206; fax (2) 66-8206.

**Revista del Pensamiento Centroamericano:** Apdo 2108, Managua; quarterly; centre-left; Editor Xavier Zavala Cuadra.

**Revista 7 Días:** Altamira de lo Vicky, 5½ al lago, Managua; e-mail 7dias@ibw.com.ni; internet www.7dias.com.ni.

**La Semana Cómica:** Centro Comercial Bello Horizonte, Módulos 7 y 9, Apdo SV-3, Managua; tel. (2) 44-909; e-mail bmejia@lasemanacomica.com; internet www.lasemanacomica.com; f. 1980; weekly; Dir Róger Sánchez; circ. 45,000.

**El Socialista:** Managua; fortnightly; official publication of the Partido Revolucionario de los Trabajadores.

**Tiempos del Mundo:** Apdo 3525, Managua; tel. (2) 70-3418; fax (2) 70-3419; e-mail tiempos@tdm.com.ni; internet www.tdm.com.ni; f. 1996; weekly; Gen. Man. Takuya Ishii; circ. 5,000.

**La Tribuna:** Detrás del Banco Mercantil, Plaza España, Apdo 1469, Managua; tel. (2) 66-9282; fax (2) 66-5167; e-mail tribuna@latribuna.com.ni; internet www.latribuna.com.ni; f. 1993; morning, daily; Dir Haroldo J. Montealegre; Gen. Man. Mario González.

**Vision Sandinista:** Managua; f. 1980; weekly; official publication of the Frente Sandinista de Liberación Nacional.

### Association

**Unión de Periodistas de Nicaragua (UPN):** Apdo 4006, Managua; Pres. Carlos Salgado.

### NEWS AGENCIES

#### Foreign Bureaux

**Agencia EFE** (Spain): Ciudad Jardín S-22, Apdo 1951, Managua; tel. (2) 24-928; Bureau Chief Filadelfo Martínez Flores.

**Deutsche Presse-Agentur (dpa)** (Germany): Apdo 2095, Managua; tel. (2) 78-1862; fax (2) 78-1863; Correspondent José Esteban Quezada.

**Informatsionnoye Telegrafnoye Agentstvo Rossii—Telegrafnoye Agentstvo Suverennykh Stran (ITAR—TASS)** (Russia): Col. Los Robles, Casa 17, Managua; internet www.itar-tass-com; Correspondent Aleksandr Trushin.

**Notimex** (Mexico): Reparto San Juan, opp. corner from Hercules Gymnasium, Managua; tel. (2) 78-4540; fax (2) 67-0413.

**Prensa Latina** (Cuba): Casa 280, de los Semáforos del Portón de Telcor de Villa Fontana, 25 m al este, 2 c. al lago, Managua; tel. (2) 72-697; e-mail platina@ibw.com.ni; internet www.prensa-latina.org; Correspondent Raúl García Alvarez.

**United Press International** (USA): 165 Col. del Periodista, Managua; tel. and fax (2) 78-1712.

**Xinhua (New China) News Agency** (People's Republic of China): De Policlínica Nicaragüense, 80 m al sur, Apdo 5899, Managua; tel. (2) 62-155; Bureau Chief Liu Riuchang.

## Publishers

**Academia Nicaragüense de la Lengua:** Calle Central, Reparto Las Colinas, Apdo 2711, Managua; f. 1928; languages; Dir Pablo Antonio Cuadra; Sec. Julio Ycaza Tigerino.

**Editora de Arte SA:** 53 Reparto Los Robles III, Managua; tel. (2) 78-5854.

**Editorial Impresora Comercial:** Julio C. Orozco L., 9 C.S.E. Entre 27a y 28a Avda, Apdo 10-11, Managua; tel. (2) 49-2258; art.

**Editorial Nueva Nicaragua:** Paseo Salvador Allende, Km 3½, Carretera Sur, Apdo 073, Managua; fax (2) 66-6520; f. 1981; Pres. Dr Sergio Ramírez Mercado; Dir-Gen. Roberto Díaz Castillo.

**Editorial Rodríguez:** Iglesia Santa Faz, 1½ c. abajo, Blvd Costa Rica, Apdo 4702, Managua.

**Editorial Unión:** Avda Central Norte, Managua; travel.

**Editorial Universitaria Centroamérica:** Col. Centroamérica, K-752, Managua.

**Librería y Editorial, Universidad Nacional de Nicaragua:** León; tel. (2) 33-1950; education, history, sciences, law, literature, politics.

**Librería Hispaniamericana (HISPAMER):** Costado Este de la UCA, Apdo A-221, Managua; e-mail hispamer@hispamer.com.ni; internet www.hispamer.com.ni; f. 1991.

# Broadcasting and Communications

## TELECOMMUNICATIONS

### Regulatory Bodies

**Instituto Nicaragüense de Telecomunicaciones y Correos (Telcor):** Edif. Telcor, Avda Bolivar diagonal a Cancillería, Apdo 2264, Managua; tel. (2) 22-7350; fax (2) 22-7554; e-mail Webmaster@telcor.gob.ni; internet www.telcor.gob.ni; scheduled to be succeeded by Sisep (q.v.) in late 2005; Dir-Gen. Marta Julia Lugo.

**Superintendencia de Servicios Públicos (Sisep):** Managua; f. 2004 by the Asamblea Nacional to oversee the telecommunications, energy and water sectors; scheduled to succeed Telcor, the INE and the INAA (qq.v.) in late 2005; Supt Víctor Manuel Guerrero.

### Major Service Providers

**ALO (PCS de Nicaragua):** Managua; internet www.pcsdigital.com.ni; mobile cellular telephone operator; awarded licence in 2002.

**BellSouth Nicaragua** (Telefonía Celular de Nicaragua, SA): Km 6½, Carretera a Masaya, Managua; tel. (2) 77-0731; internet www.bellsouth.com.ni; fmrly Nicacel; owned by Telefónica Móviles, SA (Spain); mobile cellular telephone provider.

**Empresa Nicaragüense de Telecomunicaciones (Enitel):** Villafontana 2°, Apdo 232, Managua; tel. (2) 77-3057; fax (2) 70-2128; internet www.enitel.com.ni; f. 1925; 40% stake sold to Swedish-Honduran consortium in 2001; further 49% sold to Mexican operator América Móvil in 2004; CEO Dr Carlos Ramos Fones.

## BROADCASTING

### Radio

**Radio Cadena de Oro:** Altamira 73, Managua; tel. (2) 67-0035; fax (2) 78-1220; f. 1990; Dir Allan David Tefel Alba.

**Radio Católica:** Altamira D'Este 621, 3°, Apdo 2183, Managua; tel. (2) 78-0836; fax (2) 78-2544; e-mail catolica@ibw.com.ni; f. 1961; controlled by Conferencia Episcopal de Nicaragua; Dir Fr José Bismarck Carballo; Gen. Man. Alberto Carballo Madrigal.

**Radio Corporación:** Ciudad Jardín Q-20, Apdo 2442, Managua; tel. (2) 49-1619; fax (2) 44-3824; internet www.rc540.com.ni; Dir José Castillos Osejo; Man. Fabio Gadea Mantilla.

**Radio Estrella:** Sierritas de Santo Domingo, Frente al Cementerio, Apdo UNICA 104, Managua; tel. (2) 76-0241; fax (2) 76-0062; e-mail radiosm@radioestrelladelmar.com; internet www.radioestrelladelmar.com.

**Radio Mundial:** 36 Avda Oeste, Reparto Loma Verde, Apdo 3170, Managua; tel. (2) 66-6767; fax (2) 66-4630; commercial; Pres. Manuel Arana Valle; Dir-Gen. Alma Rosa Arana Hartig.

**Radio Nicaragua:** Villa Fontana, Contiguo a Enitel, Apdo 4665, Managua; tel. (2) 27-2330-1; fax (2) 67-1448; e-mail radio.nicaragua@netport.com.ni; internet www.radionicaragua.com.ni; f. 1960; government station; Dir-Gen. Ramón Rodríguez Salinas.

**Radio Noticias:** Col. Robles 92, 4°, Apdo A-150, Managua; tel. (2) 49-5914; fax (2) 49-6393; e-mail fuentes@ibw.com.ni; Dir Agustín Fuentes Sequeira.

**Radio Ondas de Luz:** Costado Sur del Hospital Bautista, Apdo 607, Managua; tel. and fax (2) 49-7058; f. 1959; religious and cultural station; Pres. Guillermo Osorno Molina; Dir Eduardo Gutiérrez Narváez.

**Radio Sandino:** Paseo Tiscapa Este, Contiguo al Restaurante Mirador, Apdo 4776, Managua; tel. (2) 28-1330; fax (2) 62-4052; f. 1977; station controlled by the Frente Sandinista de Liberación Nacional; Pres. Bayardo Arce Castaño; Dir Conrado Pineda Aguilar.

# NICARAGUA

**Radio Tiempo:** Reparto Pancasan 217, 7°, Apdo 2735, Managua; tel. (2) 78-2540; f. 1976; Pres. Ernesto Cruz; Gen. Man. Mariano Valle Peters.

**Radio Universidad:** Avda Card, 3 c. abajo, Apdo 2883, Managua; tel. (2) 78-4743; fax (2) 77-5057; f. 1984; Dir Luis López Ruiz.

**Radio Ya:** Pista de la Resistencia, Frente a la Universidad Centroamericana, Managua; tel. (2) 78-5600; fax (2) 78-6000; internet www.nuevaya.com.ni; f. 1990; Dir Dennis Schwartz.

There are some 50 other radio stations.

### Television

**Nicavisión, Canal 12:** Bolonia Dual Card, 1 c. abajo, ½ c. al sur, Apdo 2766, Managua; tel. (2) 66-0691; fax (2) 66-1424; e-mail info@tv12-nic.com; internet www.tv12-nic.com; f. 1993; Dir Mariano Valle Peters.

**Nueva Imagen, Canal 4:** Montoya, 1 c. al sur, 2 c. arriba, Managua; tel. (2) 28-1310; fax (2) 22-4067; Pres. Dionisio Marenco; Gen. Man. Orlando Castilllo.

**Sistema Nacional de Televisión Canal 6 (SNTV Canal 6):** Km 3½, Carretera Sur, Contiguo a Shell, Las Palmas, Apdo 1505, Managua; tel. (2) 66-4958; fax (2) 66-1520; state-owned; Dir Walter René Pérez.

**Televicentro de Nicaragua, SA, Canal 2:** Casa del Obrero, 6½ c. al Sur, Apdo 688, Managua; tel. (2) 68-2222; fax (2) 66-3688; e-mail canal2@canal2.com.ni; internet www.canal2.com.ni; f. 1965; Pres. Octavio Sacasa.

**Televisora Nicaragüense, SA, Telenica 8:** De la Mansión Teodolinda, 1 c. al sur, ½ c. abajo, Bolonia, Apdo 3611, Managua; tel. (2) 66-5021; fax (2) 66-5024; e-mail cbriceno@nicanet.com.ni; f. 1989; Pres. Carlos A. Briceño Lovo.

**Televisión Internacional, Canal 23:** Casa L-852, Col. Centroamérica, Managua; tel. (2) 68-7466; fax (2) 68-0625; e-mail canal23@ibw.com.ni; f. 1993; Pres. César Riguero.

**Ultravisión de Nicaragua, SA:** Casa 567, Rotonda los Cocos, Altamira, Managua; tel. (2) 77-3524; Pres. Criseyda Olivas Vega.

# Finance

(cap. = capital; res = reserves; dep. = deposits; m. = million; amounts in gold córdobas unless otherwise stated)

### BANKING

All Nicaraguan banks were nationalized in July 1979. Foreign banks operating in the country are no longer permitted to secure local deposits. All foreign exchange transactions must be made through the Banco Central or its agencies. Under a decree issued in May 1985, the establishment of private exchange houses was permitted. In 1990 legislation allowing for the establishment of private banks was enacted.

### Supervisory Authority

**Superintendencia de Bancos y de Otras Instituciones Financieras:** Edif. SBOIF, Km 7, Carretera Sur, Apdo 788, Managua; tel. (2) 65-1555; fax (2) 65-0965; e-mail correo@sibiof.gob.ni; internet www.superintendencia.gob.ni; f. 1991; Supt Alfonso J. Llanes Cardenal.

### Central Bank

**Banco Central de Nicaragua:** Carretera Sur, Km 7, Apdos 2252/3, Zona 5, Managua; tel. (2) 65-0500; fax (2) 65-0561; e-mail bcn@bcn.gob.ni; internet www.bcn.gob.ni; f. 1961; bank of issue and govt fiscal agent; cap. 10.6m., res 21.1m., dep. 15,176.9m. (Dec. 2002); Pres. Dr Mario Alonso Icabalceta; Dir Mario J. Flores Loáisiga.

### Private Banks

**Banco de América Central (BAC):** Pista Sub-Urbana, Frente a Lotería Popular, Managua; tel. (2) 67-0220; fax (2) 67-0224; e-mail info@bancodeamericacentral.com; internet www.bancodeamericacentral.com; f. 1991; total assets 10,516m. (1999); Pres. Carlos Pellas Chamorro; Gen. Man. Carlos Matus Tapia.

**Banco de Crédito Centroamericano (BANCENTRO):** Edif. BANCENTRO, Km 4 Carretera a Masaya, Managua; tel. (2) 78-2777; fax (2) 78-6001; e-mail bancentro@net.com.ni; internet www.bancentro.com.ni; f. 1991; total assets 5,249m. (1999); Pres. Roberto J. Zamora Llanes; Gen. Man. Carlos A. Briceño Ríos.

**Banco Uno SA:** Plaza España, Rotonda el Güegüense 20 m al oeste, Managua; tel. (2) 78-7171; fax (2) 77-3154; e-mail info@banexpo.com.ni; internet www.bancouno.com.ni; dep. 2,372m. (Dec. 2002); fmrly Banco de la Exportación (BANEXPO), present name adopted in Nov. 2002; Dir Adolfo Argüello Lacayo.

### STOCK EXCHANGE

**Bolsa de Valores de Nicaragua:** Edif. Oscar Pérez Cassar, Centro BANIC, Km 5½, Carretera Masaya, Apdo 121, Managua; tel. (2) 78-3830; fax (2) 78-3836; e-mail info@bolsanic.com; internet bolsanic.com; f. 1993; Pres. Dr Raúl Lacayo Solórzano; Gen. Man. Carolina Solórzano de Barrios.

### INSURANCE

#### State Company

**Instituto Nicaragüense de Seguros y Reaseguros (INISER):** Centro Comercial Camino de Oriente, Km 6, Carretera a Masaya, Apdo 1147, Managua; tel. (2) 66-6772; fax (2) 66-5636; e-mail iniser@iniser.com.ni; internet www.iniser.com.ni; f. 1979 to assume the activities of all the pre-revolution national private insurance companies; Exec. Pres. Manuel R. Gurdián Ubago.

#### Private Companies

**Compañía de Seguros del Pacifico, SA:** Bolonia Rotonda Güegüense, 2½ c. al norte, Managua; tel. (2) 66-9208; f. 1997; Vice-Pres. Sergio Raskosky Holmann.

**Metropolitana Compañía de Seguros, SA:** Reparto Serrano Plaza El Sol, 400 m al norte, Managua; tel. (2) 78-8538; Pres. Dr Leonel Argüello Ramírez; Sec. Leopoldo J. Arosemena.

**Seguros América, SA:** Managua; f. 1996; Pres. Carlos F. Pellas Chamorro; Sec. Danilo Manzanares Enríquez.

**Seguros Centroamericanos, SA (Segurossa):** Edif. Oscar Pérez Cassar, Centro Banic, Managua; tel. (2) 70-3505; fax (2) 70-3558; f. 1996; Pres. Roberto Zamora Llanes; Gen. Man. Narciso Arellano Suárez.

# Trade and Industry

### GOVERNMENT AGENCIES

**Empresa Nicaragüense de Alimentos Básicos (ENABAS):** Salida a Carretera Norte, Apdo 1041, Managua; tel. (2) 23082; fax (2) 26185; f. 1979; controls trading in basic foodstuffs; Dir Mariano Vega Noguera.

**Instituto Nicaragüense de Apoyo a la Pequeña y Mediana Empresa (INPYME):** De la Shell Plaza el Sol, 1 c. al sur, 300 m abajo, Apdo 449, Managua; tel. (2) 77-0599; fax (2) 77-0598; internet www.inpyme.gob.ni; supports small and medium-sized enterprises; Exec. Dir Harold Antonio Rocha Solís.

**Instituto Nicaragüense de Reforma Agraria (INRA):** Km 8, Entrada a Sierrita Santo Domingo, Managua; tel. (2) 73210; f. 1991; agrarian reform; Dir (vacant).

**Instituto Nicaragüense de Seguridad Social (INSS):** Frente Cementerio San Pedro, Edif. INSS, Apdo 1649, Managua; tel. (2) 22-7454; fax (2) 22-7445; f. 1955; social security and welfare; Dir Edna Balladares; Deputy Dir Emilio Selva Tapia.

**Instituto Nicaragüense de Tecnología Agropecuaria (INTA):** Managua; tel. 278-0469; fax 278-1259; e-mail mapache@inta.gob.ni; internet www.inta.gob.ni; f. 1993; Pres. Luis A. Osorio; Dir-Gen. Dr Noel Pallais Checa.

**MEDEPESCA:** Km 6½, Carretera Sur, Managua; tel. (2) 67-3490; state fishing agency; Dir Ing. Emilio Olivares.

### DEVELOPMENT ORGANIZATIONS

**Asociación de Productores y Exportadores de Nicaragua (APEN):** Del Hotel Intercontinental, 2 c. al sur y 2 c. abajo, Bolonia, Managua; tel. (2) 66-5038; fax (2) 66-5039; Gen. Man. Jorge Brenes.

**Cámara de Industrias de Nicaragua:** Rotonda el Güegüense, 300 m al sur, Apdo 1436, Managua; tel. (2) 66-8847; fax (2) 66-1891; e-mail cadin@cadin.org.ni; internet www.cadin.org.ni; Pres. Gabriel Pasos Lacayo.

**Cámara Nacional de la Mediana y Pequeña Industria (CONAPI):** Plaza 19 de Julio, Frente a la UCA, Apdo 153, Managua; tel. (2) 78-4892; fax (2) 67-0192; e-mail conapi@nicarao.org.ni; Pres. Flora Vargas; Gen. Man. Uriel Argañel C.

**Cámara Nicaragüense de la Construcción (CNC):** Bolonia de Aval Card, 2 c. abajo, 50 varas al Sur, Managua; tel. (2) 26-3363; fax (2) 66-3327; e-mail cncsecre@nicarao.org.ni; f. 1961; construction industry; Pres. Alejandro Terán.

# NICARAGUA

**Instituto Nicaragüense de Fomento Municipal (INIFOM):** Edif. Central, Carretera a la Refinería, entrada principal residencial Los Arcos, Apdo 3097, Managua; tel. (2) 66-6050; fax (2) 66-6050; internet www.inifom.gob.ni; Pres. CARLOS MIGUEL DUARTE AREAS.

## CHAMBERS OF COMMERCE

**Cámara de Comercio de Nicaragua:** Rotonda El Güegüense 300 m al sur, 20 m al oeste; tel. (2) 68-3505; fax (2) 68-3600; e-mail comercio@ibw.com.ni; f. 1892; 530 mems; Pres. MARCO A. MAYORGA L.; Gen. Man. MARIO ALEGRÍA.

**Cámara de Comercio Americana de Nicaragua:** Semáforos ENEL Central, 500 m al sur, Apdo 2720, Managua; tel. (2) 67-3099; fax (2) 67-3098; e-mail amcham@ns.tmx.com.ni; f. 1974; Pres. HUMBERTO CORRALES M.

**Cámara Oficial Española de Comercio de Nicaragua:** Restaurante la Marseilleisa, ½ c. arriba, Los Robles, Apdo 4103, Managua; tel. (2) 78-9047; fax (2) 78-9088; e-mail camacoesnic@cablenet.com.ni; Pres. JOSÉ ESCALANTE ALVARADO; Sec.-Gen. AUXILIADORA MIRANDA DE GUERRERO.

## EMPLOYERS' ORGANIZATIONS

**Asociación de Productores de Café Nicaragüenses:** coffee producers; Vice-Pres. DANILO GONZÁLEZ.

**Consejo Superior de la Empresa Privada (COSEP):** De Telcor Zacarías Guerra, 1 c. abajo, Apdo 5430, Managua; tel. (2) 28-2030; fax (2) 28-2041; e-mail cosep@nic.gbm.net; f. 1972; private businesses; consists of Cámara de Industrias de Nicaragua (CADIN), Unión de Productores Agropecuarios de Nicaragua (UPANIC), Cámara de Comercio, Cámara de la Construcción, Confederación Nacional de Profesionales (CONAPRO), Instituto Nicaragüense de Desarrollo (INDE); mem. of Coordinadora Democrática Nicaragüense; Vice-Pres. ROBERTO TERÁN; Sec. ORESTES ROMERO ROJAS.

**Instituto Nicaragüense de Desarrollo (INDE):** Camas Lunes 1 c. al oeste, Calle 27 de Mayo, Apdo 2598, Managua; tel. and fax (2) 68-1900; f. 1963; private business org.; 650 mems; Pres. Ing. GABRIEL SOLORZANO.

**Unión de Productores Agropecuarios de Nicaragua (UPANIC):** Reparto San Juan No 300, Detrás del Ginmasio Hércules, Managua; tel. (2) 78-3382; fax (2) 78-3291; e-mail upanic@ibw.com.ni; private agriculturalists' asscn; Pres. OSCAR ALEMÁN; Exec. Sec. ALEJANDRO RASKOSKY.

## MAJOR COMPANIES

**AMANCO:** Km 3½, Carretera Sur, Apdo 2964, Managua; tel. (2) 66-1551; fax (2) 66-2534; e-mail amanconi@gruponueva.com; internet www.amanco.com; f. 1967; fmrly NICALIT; makers of asbestos products; Pres. FERNANDO MONTES OROZCO; Gen. Man. CAMILO LÓPEZ; 300 employees.

**Café Soluble, SA:** Km 8½, Carretera Norte, Apdo 429, Managua; tel. (2) 33-1122; fax (2) 33-1110; e-mail info@cafesoluble.com; internet www.cafesoluble.com; f. 1958; instant-coffee manufacturers; 200 employees.

**Comercializadora del Banano Nicaragüense, SA (COMBANISA):** De la Vichy, 75 m al Sur, Altamira del Este, Managua; tel. (2) 67-8311; fax (2) 67-3633; f. 1979; wholesale fruit trading; Gen. Man. BLANCO OLIVIA LÓPEZ; 4,000 employees.

**Compañía Cervecera de Nicaragua, SA:** Km 6½, Carretera Norte, de la Cruz Lorena 600 varas. al Lago, Managua; tel. (2) 48-5080; fax (2) 48-5120; e-mail jrosales@victoria.com.ni; internet www.victoria.com.ni; brewery; f. 1926; Gen. Man. JAIME ROSALES PASQUIER; 700 employees.

**Compañía Licorera de Nicaragua, SA:** Camino Oriente, Casilla 1494, Managua; tel. (2) 78-3270; fax (2) 77-3649; e-mail info@flordecana.com; internet www.cln.com.ni; f. 1890; producers of rum; Chair. CARLOS PELLAS.

**Corporación Nicaragüense del Banano (BANANIC):** Altamira del Este, Managua; tel. (2) 67-8312; fax (2) 77-3633; f. 1988; fruit growers and sellers; Gen. Man. BLANCO OLIVIA LÓPEZ; 6,000 employees.

**Embotelladora Nacional, SA (ENSA):** Km 7½, Carretera Norte, Apdo 471, Managua; tel. (2) 33-1300; fax (2) 63-1320; e-mail mcaldera@pepsicentroamerica.com; f. 1944; bottlers of carbonated beverages; Pres. CARLOS ENRIQUE CASTILLO MONGE; Gen. Man. MILTON CALDERA CARDENAL; 700 employees.

**Grupo Premio Coffee SA:** 48 Reparto San Martin, Managua; tel. (2) 663984; fax (2) 666048; e-mail ecastro77@cafepremio.com; internet www.cafepremio.com; Dir EDGAR CASTRO.

**Industria Nacional de Clavos y Alambres de Púas, SA (INCA):** Apdo 14, Masaya, Masaya; tel. (2) 62-2605; f. 1961; producers of nails, wire and meshwork; Man. Dir RAMÓN ORLANDO GARCÍA PÉREZ; 560 employees.

**Kraft Foods Nicaragua/Nabisco de Nicaragua, SA:** Barrio Altagracia Contiguo al Edif. El Eskimo, Managua; tel. (2) 49-2740; fax (2) 48-0704; e-mail Manuel.Oliver@kraftla.com; f. 1949; subsidiary of Nabisco of the USA; makers of biscuits and puddings; Commercial Dir MANUEL OLIVER.

**Metales y Estructuras, SA (METASA):** Km 28, Carretera Norte, Tipitapa, Managua; tel. (2) 21-1124; f. 1958; production of metals; Pres. PEDRO BLANDON MORENO; 498 employees.

**Molinos de Nicaragua, SA (MONISA):** Final Calle Inmaculada, Apdo 45, Granada; tel. and fax (5) 52-2291; e-mail monisa@ibw.com.ni; internet www.monisa.com; f. 1964; flour producers; Gen. Man. CARLOS GERMAN SEQUEIRA.

**Monte Rosa:** Chinandega; sugar production; owned by Pantaleon Sugar Holdings, Guatemala; Gen. Man. FRANCISCO BALTODANO.

**Narciso Salas y Asociados Ernst & Young:** Apdo 2446, Managua; tel. (2) 26-4591; fax (2) 66-0436; f. 1972; accountants; Man. Partner NARCISO SALAS CHÁVEZ.

**Nicaragua Sugar Estate Ltd:** Ingenio San Antonio, Chichigalpa, Chinandega; tel. (2) 78-3270; fax (2) 67-2874; e-mail gerencia-general@nicaraguasugar.com.ni; internet www.nicaraguasugar.com; f. 1890; sugar refinery; Pres. CARLOS PELLAS CHAMORRO; Gen. Man. XAVIER ARGÜELLO BARILLAS; 3,500 employees.

**Plásticos de Nicaragua, SA (PLASTINIC):** Km 44½, Carretera Sur, Apdo 27, Jinotepe; tel. 532-2575; fax 532-3478; e-mail plastinic@plastinic.com; f. 1968; plastic-bag manufacturers; Gen. Man. MIRIAM LACAYO; 245 employees.

**PriceWaterhouseCoopers:** Apdo 2697, Managua; tel. (2) 70-9639; fax (2) 70-9665; accountants and management consultants; Partner OSCAR CUADRA.

**Sacos Macen (MACEN):** Km 14½, Carretera a los Brasiles, Managua; tel. (2) 69-9213; fax (2) 69-9217; e-mail macen@interlink.com.ni; f. 1957; producers of sacking, string, linings and hammocks; Mans ALBERTO MACGREGOR, ADOLFO MACGREGOR, DONALD SPENCER; 200 employees.

**Siemens, SA:** Carreterra Norte, Apdo 1049, Managua; tel. (2) 49-1111; fax (2) 49-1849; e-mail siemens.nic@siemens.com.mx; internet www.siemens-centram.com; manufacturer of electrical equipment and machinery; subsidiary of Siemens AG, Germany; Man. Dir LUIS ADOLFO GABUARDI.

**Tabacalera Nicaragüense, SA (TACUNISA):** Km 7½, Carretera Norte, Apdo 1049, Managua; tel. (2) 63-1900; fax (2) 63-1642; e-mail gtanic@ibw.com.ni; f. 1934; cigarette manufacturers; Gen. Man. MIGUEL TRIVELLI; 1,100 employees.

## UTILITIES

### Regulatory Bodies

**Comisión Nacional de Energía:** Managua; Exec. Sec. RAÚL SOLÓRZANO.

**Instituto Nicaragüense de Acueductos y Alcantarillados (INAA):** De la Mansión Teodolinda, 3 c. al sur, Bolonia, Apdo 1084, Managua; tel. (2) 66-7882; fax (2) 66-7917; e-mail inaa@inaa.gob.ni; internet www.inaa.gob.ni; f. 1979; water regulator; scheduled to be succeeded by Sisep (q.v.) in late 2005; Exec. Pres. JORGE HAYN VOGL.

**Instituto Nicaragüense de Energía (INE):** Edif. Petronic, 4°, Managua; tel. (2) 28-1142; fax (2) 28-2049; internet www.ine.gob.ni; scheduled to be succeeded by Sisep (q.v.) in late 2005; Pres. OCTAVIO SALINAS MORAZÁN.

**Superintendencia de Servicios Públicos (Sisep):** see Telecommunications—Regulatory Bodies.

### Electricity

**Unidad de Reestructuración de la Empresa Nicaragüense de Electricidad (URE):** Altamira d'Este 141, de la Vicky 1 c. abajo, 1 c. al sur, Managua; tel. (2) 70-9989; fax (2) 78-2284; e-mail ure@ibw.com.ni; internet www.ure.gob.ni; f. 1999 to oversee the privatization of the state-owned distribution and generation companies of ENEL; Exec. Dir SALVADOR QUINTANILLA.

**Empresa Nicaragüense de Electricidad (ENEL):** Ofs Centrales, Pista Juan Pablo II y Avda Bolívar, Managua; tel. (2) 67-4159; fax (2) 67-2686; e-mail relapub@ibw.com.ni; Pres. MARIO MONTENEGRO; responsible for planning, organization, management, administration, research and development of energy resources; split into two distribution companies and three generation companies in 1999:

**DISNORTE, SA:** electricity distribution company; distributes some 802 GWh; sold to Unión Fenosa (Spain) in 2000.

**DISSUR, SA:** electricity distribution company; distributes some 658 GWh; sold to Unión Fenosa (Spain) in 2000.

**GECSA:** electricity generation company; 79 MW capacity thermal plant; almost obsolete and therefore difficult to privatize, GECSA was likely to be retained for emergency purposes.

**GEOSA:** electricity generation company; 112 MW capacity thermal plant; sold to Coastal Power International (USA) in Jan. 2002.

**HIDROGESA:** electricity generation company; 94 MW capacity hydroelectric plant; privatized in 2002; however, sale annulled in July 2003 owing to alleged irregularities.

### CO-OPERATIVES

**Cooperativa de Algodoneros de Managua, RL:** De la Iglesia Recolección 1 c. al norte, León; tel. (3) 11-2450; cotton-growers.

**Empresa Cooperativa de Productores Agropecuarios:** Managua; represents 13,000 mems from 48 affiliated co-operatives; Chair. DANIEL NÚÑEZ.

### TRADE UNIONS

**Asociación Nacional de Educadores de Nicaragua (ANDEN):** Managua; e-mail anden@guegue.com.ni; Sec.-Gen. JOSÉ ANTONIO ZEPEDA; 19 affiliates, 15,000 mems.

**Asociación de Trabajadores del Campo (ATC)** (Association of Rural Workers): Apdo A-244, Managua; tel. (2) 23-2221; e-mail atcnic@ibw.com.ni; f. 1977; Gen. Sec. EDGARDO GARCÍA; 52,000 mems.

**Central Sandinista de Trabajadores (CST):** Iglesia del Carmen, 1 c. al oeste, ½ c. al sur, Managua; tel. (2) 65-1096; fax (2) 40-1285; e-mail cts/cor@alfamumeric.com.ni; Sec.-Gen. ROBERTO GONZÁLEZ GAITÁN.

**Central de Trabajadores de Nicaragua (CTN)** (Nicaraguan Workers' Congress): De la Iglesia del Carmen, 1 c. al sur, ½ c. arriba y 75 varas al sur, Managua; tel. (2) 68-3061; fax (2) 652056; e-mail ctn@alfanumeric.com.ni; f. 1962; mem. of Coordinadora Democrática Nicaragüense; Sec.-Gen. CARLOS HUEMBES.

**Confederación de Acción y Unidad Sindical (CAUS)** (Confederation for Trade Union Action and Unity): Semáforos de Rubenia, 2 c. abajo y 2 c. al lago, arrio Venezuela, Managua; tel. and fax (2) 44-2587; f. 1973; trade-union wing of Partido Comunista de Nicaragua; Sec.-Gen. EMILIO MÁRQUEZ.

**Confederación General de Trabajadores Independientes (CGT(I))** (Independent General Confederation of Labour): Centro Comercial Nejapa, 1 c. arriba y 3 c. al lago, Managua; tel. (2) 22-5195; fax (2) 28-7505; f. 1953; Sec.-Gen. NILO M. SALAZAR; 4,843 mems (est.) from six federations with 40 local unions, and six non-federated local unions.

**Confederación de Unificación Sindical (CUS)** (Confederation of United Trade Unions): Casa Q3, del Colegio la Tenderi 2½ c. arriba, Ciudad Jardín, Managua; tel. (2) 48-3681; fax (2) 40-1330; e-mail sindicatocus@yahoo.com; f. 1972; affiliated to the Inter-American Regional Organization of Workers, etc.; mem. of Coordinadora Democrática Nicaragüense; Sec.-Gen. JOSÉ ESPINOZA.

**Federation Enrique Schmidt (FESC):** Managua; e-mail fschmidt@tmx.com.ni; communications and postal workers' union.

**Federación de Trabajadores de la Salud (FETSALUD)** (Federation of Health Workers): Optica Nicaragüense, 2 c. arriba ½ c. al sur, Apdo 1402, Managua; tel. and fax (2) 66-3065; e-mail fntsid@ibw.com.ni; Dir DAVE GODSON; 25,000 mems.

**Federación de Trabajadores Nicaragüenses (FTN):** workers' federation; Leader DOMINGO PÉREZ.

**Federación de Transportadores Unidos Nicaragüense (FTUN)** (United Transport Workers' Federation of Nicaragua): De donde fue el Vocacional, esq. este, 30 m al sur, Apdo 945, Managua; f. 1952; Pres. MANUEL SABALLOS; 2,880 mems (est.) from 21 affiliated associations.

**Frente Nacional de los Trabajadores (FNT)** (National Workers' Front): Residencial Bolonia, de la Optica Nicaragüense, 2 c. arriba, 30 varas al sur, Managua; tel. and fax (2) 66-3065; e-mail fnt@ibw.com.ni; f. 1979; affiliated to Frente Sandinista de Liberación Nacional; Leader Dr GUSTAVO PORRAS CORTÉS; Sec.-Gen. JOSÉ A. BERMÚDEZ.

**Unión Nacional de Agricultores y Ganaderos (UNAG)** (National Union of Agricultural and Livestock Workers): Contiguo edif. Julia Pasos, Reparto Las Palmas, 3½ km Carretera Sur, Managua; tel. (2) 66-1675; fax (2) 66-2135; e-mail unag@unag.org.ni; internet www.unag.org.ni; f. 1981; Pres. ALVARO FIALLOS OYANGUREN.

**Unión Nacional de Caficultores de Nicaragua (UNCAFENIC)** (National Union of Coffee Growers of Nicaragua): Reparto San Juan, Casa 300, Apdo 3447, Managua; tel. (2) 78-2586; fax (2) 78-2587; Pres. FREDDY TORRES.

**Unión Nacional de Empleados (UNE):** Managua; e-mail cocentrafemenino@xerox.com.ni; f. 1978; public sector workers' union; Sec.-Gen. DOMINGO PÉREZ; 18,000 mems.

**Unión de Productores Agropecuarios de Nicaragua (UPANIC)** (Union of Agricultural Producers of Nicaragua): Reparto San Juan, Casa 300, Apdo 2351, Managua; tel. (2) 78-3382; fax (2) 78-2587; Pres. MANUEL ALVAREZ.

## Transport

### RAILWAYS

**Ferrocarril de Nicaragua:** Plantel Central Casimiro Sotelo, Del Parque San Sebastián, 5 c. al lago, Apdo 5, Managua; tel. (2) 22-2160; fax (2) 22-2542; f. 1881; govt-owned; main line from León via Managua to Granada on Lake Nicaragua (132 km), southern branch line between Masaya and Diriamba (44 km), northern branch line between León and Río Grande (86 km) and Puerto Sandino branch line between Ceiba Mocha and Puerto Sandino (25 km); total length 287 km; reported to have ceased operations in 1994; Gen. Man. N. ESTRADA.

### ROADS

In 2000 there were an estimated 19,032 km of roads, of which 2,093 km were paved. Of the total, only some 9,000–10,000 km were accessible throughout the entire year. Some 8,000 km of roads were damaged by 'Hurricane Mitch', which struck in late 1998. The Pan-American Highway runs for 384 km in Nicaragua and links Managua with the Honduran and Costa Rican frontiers and the Atlantic and Pacific Highways connecting Managua with the coastal regions.

### SHIPPING

Corinto, Puerto Sandino and San Juan del Sur, on the Pacific, and Puerto Cabezas, El Bluff and El Rama, on the Caribbean, are the principal ports. Corinto deals with about 60% of trade. In 2001 the US-based company Delasa was given a 25-year concession to develop and modernize Puerto Cabezas port. It was to invest some US $200m.

**Empresa Portuaria Nacional (EPN):** Apdo 2727–3570, Managua; tel. (2) 22-3827; fax (2) 66-3488; e-mail epn_puertos@epn.com.ni; internet www.epn.com.ni; Pres. ROBERTO ZELAYA BLANCO.

   **Administración Portuaria de Bluefields-Bluff:** tel. (8) 22-2632; e-mail puertobluff@epn.com.ni.

   **Administración Portuaria de Corinto:** De Telcor, 1 c. al oeste, Corinto; tel. (3) 42-2768; e-mail puertocorinto@epn.com.ni; f. 1956.

   **Administración Portuaria de San Juan del Sur:** tel. (4) 58-2336; e-mail puertosanjuan@epn.com.ni.

   **Administración Portuaria de Puerto Cabezas:** tel. (2) 82-2331.

   **Administración Portuaria de Sandino:** tel. (3) 12-2212; e-mail puertosandino@epn.com.ni.

### CIVIL AVIATION

The principal airport is the Augusto Sandino International Airport, in Managua. There are some 185 additional airports in Nicaragua.

**Atlantic Airlines:** Estatua José Martí, 150 m este, Managua; tel. (2) 22-3037; fax (2) 28-5614; e-mail reservaciones@atlanticairlines.com.ni; internet www.atlanticairlines.com.ni; scheduled domestic servies, charters, cargo transportation and courier services; Gen. Man. LUIS ARÉVALO.

**La Costeña:** Managua International Airport, Managua; tel. (2) 63-2142; fax (2) 63-1281; e-mail jcaballero@lacostena.com.ni; Gen. Man. ALFREDO CABALLERO.

## Tourism

In 2003 tourist arrivals totalled 525,775 and receipts from tourism totalled US $155m.

**Instituto Nicaragüense de Turismo (INTUR):** Del Hotel Intercontinental Managua, 1 c. al sur, 1 c. al oeste, Managua; tel. (2) 22-6460; fax (2) 22-6610; e-mail promocion@intur.gob.ni; internet www.visit-Nicaragua.com; f. 1998; Pres. LUCÍA SALAZAR DE ROBELO; Vice-Pres. LEDA SÁNCHEZ DE PARRALES.

**Asociación Nicaragüense de Agencias de Viajes y Turismo (ANAVYT):** Edif. Policlínica Nicaragüense, Reparto Bolonia, Apdo 1045, Managua; tel. (2) 66-9742; fax (2) 66-4474; e-mail aeromund@cablenet.com.ni; f. 1966; Pres. Ana María Rocha C.

**Cámara Nacional de Turismo (CANATUR):** Contiguo al Ministerio de Turismo, Apdo 2105, Managua; tel. (2) 66-5071; fax (2) 66-5071; e-mail canatur@munditel.com.ni; f. 1976; Pres. Miguel Romero.

## Defence

In August 2004 Nicaragua's professional Armed Forces numbered an estimated 14,000: Army 12,000, Navy 800 (estimated) and Air Force 1,200. Conscription was introduced in September 1983, but was abolished in April 1990. In 1995 the armed forces were renamed the Ejército de Nicaragua, following a constitutional amendment removing their Sandinista affiliation. There is a voluntary military service which lasts 18–36 months.

**Defence Budget:** 504m. gold córdobas (US $32m.) in 2004.
**Commander-in-Chief:** Gen. Moisés Omar Hallesleven.

## Education

From 1979 primary and secondary education in Nicaragua were provided free of charge. Primary education, which is officially compulsory, begins at seven years of age and lasts for six years. Secondary education, beginning at the age of 13, lasts for up to five years, comprising a first cycle of three years and a second of two years. In 2001/02 the total enrolment at primary schools was equivalent to 81.9% of the relevant age-group (males 81.6%; females 82.2%). Secondary enrolment in that year was equivalent to 37.0% of children in the relevant age-group (males 34.0%; females 40.1%). There are many commercial schools and four universities. In 2001 70,925 students attended universities. In 2001 expenditure on education accounted for 13.0% of total government expenditure.

## Bibliography

For works on Central America generally, see Select Bibliography (Books)

*Agriculture in Nicaragua: Promoting Competitiveness and Stimulating Broad-based Growth.* Washington, DC, World Bank, 2003.

Babb, F. E. *After Revolution: Mapping Gender and Cultural Politics in Neoliberal Nicaragua.* Austin, TX, University of Texas Press, 2002.

Bickham Mendez, J. *From the Revolution to the Maquiladoras: Gender, Labor, and Globalization in Nicaragua.* Durham, NC, Duke University Press, 2005.

*Binational Study: The State of Migration Flows Between Costa Rica and Nicaragua—An Analysis of Economic and Social Implications for Both Countries.* Geneva, Intergovernmental Committee for Migration, 2003.

Charlip, J. A. *Cultivating Coffee.* Columbus, OH, Ohio University Press, 2002.

Chavez Metoyer, C. *Women and the State in Post-Sandinista Nicaragua.* Boulder, CO, Lynne Rienner Publrs, 1999.

Close, D. *Nicaragua: The Chamorro Years.* Boulder, CO, Lynne Rienner Publrs, 1998.

*Undoing Democracy: The Politics of Electoral Caudillismo.* Lanham, MD, Lexington Books, 2004.

Cruz, C. *Political Culture and Institutional Development in Costa Rica and Nicaragua: World-Making in the Tropics.* Cambridge, Cambridge University Press, 2005.

Dye, D. R. *Observing the 2001 Nicaraguan Elections: Final Report.* Atlanta, GA, Carter Center, 2002.

Edmisten, P. *Nicaragua Divided: La Prensa and the Chamorro Legacy.* Gainesville, FL, University Press of Florida, 1990.

Field, L. W. *The Grimace of MacHo Raton: Artisans, Identity and Nation in Late Twentieth-Century Western Nicaragua.* Durham, NC, Duke University Press, 1999.

Gambone, M. D. *Eisenhower, Somoza and the Cold War in Nicaragua.* New York, NY, Praeger Publrs, 1997.

Heyck, D. L. D. *Life Stories of the Nicaraguan Revolution.* London, Routledge, 1990.

Isbester, K. *Still Fighting: The Nicaraguan Women's Movement, 1977–2000.* Pittsburgh, PA, University of Pittsburgh Press, 2001.

Kodrich, K. *Tradition and Change in the Nicaraguan Press: Newspapers and Journalists in a New Democratic Era.* Lanham, MD, University Press of America, 2002.

Leiken, R. S. *Why Nicaragua Vanished: A Story of Reporters and Revolutionaries.* Lanham, MD, Rowman and Littlefield Publrs, 2003.

MacAulay, N. *The Sandino Affair.* Micanopy, FL, Wacahoota Press, 1998.

Marti i Puig, S. *The Origins of the Peasant-Contra Rebellion in Nicaragua, 1979–87.* London, Institute of Latin American Studies, 2001.

Moltaván Belliz, C. A. *Hurricane Mitch and the Impact of the NGOs on Indigenous Miskito Communities in Río Coco, North Atlantic Autonomous Region, Nicaragua.* Managua, Universidad de las Regiones Autónomas de la Costa Caribe Nicaragüense, 2002.

Morley, M. H. *Washington, Somoza and the Sandinistas: State and Regime in US Policy towards Nicaragua, 1969–1981.* Cambridge, Cambridge University Press, 1994.

*Washington, Somoza and the Sandinistas.* Cambridge, Cambridge University Press, 2002.

Orzoco, M. *International Norms and Mobilization of Democracy: Nicaragua in the World.* Burlington, VA, Ashgate, 2002.

Pastor, R. *Not Condemned to Repetition: The United States and Nicaragua.* Boulder, CO, Westview Press, 2002.

Prevost, G., and Vanden, H. E. (Eds) *The Undermining of the Sandinista Revolution.* New York, NY, Palgrave Macmillan, 1999.

Smith, H. *Nicaragua: Self-Determination and Survival.* London, Pluto Press, 1992.

Walker, T. W. (Ed.). *Nicaragua Without Illusions: Regime Transition and Structural Adjustment in the 1990s.* Wilmington, DE, Scholarly Resources, 1997.

# PANAMA

## Geography

### PHYSICAL FEATURES

The Republic of Panama is often not included with Central America, its early history being related to Colombia rather than to the countries of the isthmus, but, geographically, it does occupy the narrowest part of the great land bridge connecting South and North America. In shape, Panama is a sinuous east–west land corridor, with the Caribbean to the north and the Pacific to the south. To the east is the South American country of Colombia, from which Panama was hewn in 1903, and to the west Costa Rica. The Costa Rican border is about 330 km (205 miles) in length and the Colombian border 225 km, but the country has an even more extensive coastline, of 2,490 km. From 1903 the USA held 'sovereign rights' over 1,432 sq km (553 sq miles) of Panamanian territory, the Canal Zone that flanked the route of the transisthmian waterway for 8 km (5 miles) on either side. However, the lease was negotiated to an early end on 31 December 1999, when Panama resumed full sovereignty over all of its national territory. The country covers 75,517 sq km, making it larger than El Salvador, Belize or Costa Rica, and a little smaller than Scotland (United Kingdom).

The shape of Panama contributes to its irregular coast, dotted with many islands and islets offshore. In the east a northward loop encloses the Gulf of Panama, while the southward loop defines the Mosquito Gulf in the Caribbean. On the south coast, a great promontory of land jutting further into the Pacific from the southernmost part of the arc is split into the semi-arid Azuero peninsula and the much smaller Las Palmas peninsula to the west, forming the eastern arm of the Gulf of Chiriquí. The largest Pacific islands are Coiba (used as a prison island), just south-west of the Las Palmas peninsula, and Isla del Rey in the Archipiélago de las Perlas (Pearl Islands), in the Gulf of Panama. The San Blas chain of coral atolls in the north-east, parallel to the coast, are the abode of the indigenous Kuna (Kuna Yala). The bulk of the country consists of a discontinuous mountain spine, steep, rugged mountains, interspersed with upland plains and rolling hills, flanked by coastal plains of varying width. From west to east the country measures about 650 km, but it narrows to as little as 48 km from north to south, at roughly the point where the Canal crosses. There are 2,210 sq km of inland waters, much of them artificially restrained, including the Bayono lake on Chepo river and Gatún lake, one of the largest artificial reservoirs in the world, built as part of the Canal. The Canal, built between 1904 and 1914 by the USA, is about 82 km in length, raising and lowering vessels 26 m by means of six pairs of locks. It crosses a low seat of land, situated in a gap between the western and eastern mountain ranges that form the backbone of the country. This region is known as the central isthmus or the transit zone, and consists of coastal plains and a highland interior. About one-half of the population, 90% of Panama's industry and all the transisthmian links are situated here.

To the east of the transit zone is the sparsely populated and barely developed territory of Darién, covering one-third of Panama with the largest area of rainforest in the Americas outside the Amazon basin. The central mountains here continue nearer the north coast of the isthmus, as the Serranía de San Blas and the Serranía del Darién, with more mountains to the south. Densely wooded and containing the Tuira river, the longest in Panama, Darién is the home of the Choco and other Amerindians. The region contributes the bulk of the country's woodland (most of the northern coast of Panama is also densely forested), which in all accounts for about 44% of the nation's total area. Despite problems of deforestation (immediate concerns with this are focused on soil erosion and the consequent threat of silting in the Canal) and of mining, Panama is still home to a considerable variety of flora and fauna, with more than 2,000 species of tropical plants, numerous native and migratory birds, and many animals common to both South and

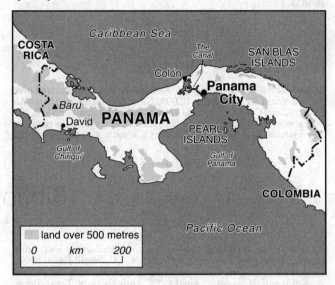

Central America, as well as some unique species (such as the golden tree frog or the giant tree sloth). Biodiversity is less rich west of the Canal and south of the mountains, much of this area having long been cleared for arable farming or ranching. The Continental Divide here forms higher mountains than in the east, where the Cordillera Central reaches its heights in the Serranía de Tabasará, and the highest peak in Panama, here and not in Darién, is the Baru volcano of Chiriquí (3,475 m or 11,405 ft), in the west. Such volcanoes, none now active in the country, have given the fertile soil that is so widely cultivated in the west of the country. Agriculture dominates Chiriquí—generally the area south of the mountains and extending onto the rolling hills of Las Palmas. Between here and the Canal, along the north-western shores of the Gulf of Panama and in the lowlands west of the transit zone, is the old, settled heartland of central Panama. The final topographical area is the forested north-west, the Atlantic region, between the mountains and the north coast. This region, centred on Bocas del Toro, is the indigenous home of the Guaymí (Ngöbe-Buglé) and of the mainly black descendants of West Indian immigrants, who came to work the banana plantations that once flourished along the Caribbean.

### CLIMATE

Panama has a tropical maritime climate, making it hot, humid and cloudy, with a prolonged rainy season (May–January) over the summer and autumn. Prevailing winds are off the Atlantic for most of the year, but change to south-westerlies during the autumn. The Caribbean side of the country is wetter, with Azuero the driest part. The northern mountain slopes receive average rainfall of 2,970 mm (117 ins) per year, with the drier Pacific coast generally on about 1,650 mm annually. The average temperatures in Colón, at the northern, Caribbean end of the Canal, range from the lowest minimum of 20°C to the highest maximum of 32°C (68°–90°F). At greater altitudes the average temperature can be about 19°C (66°F). Panama falls outside either the Atlantic hurricane belt or the Eastern Pacific one.

### POPULATION

The people of Panama (Panameños) are mainly of mixed Spanish and Amerindian race, the Mestizos; they constitute some 63% of the population. There are also large communities of predominantly black descent (14%) and Spanish descent (10%), as well as some of mixed-black descent (mulatto—5%) and a number of Amerindian peoples (5%). There are also some of Middle Eastern and Asian descent. More recently, there has been a largely illegal influx of Colombians, either seeking employment or political refuge, particularly in the deserted borderlands. The official language, and the one in general use, is Spanish, but, after long years of US rule along the Canal and immigration from North America and the West Indies, English is widely spoken (being the main language for up to 14% of the population) and many Panamanians are reckoned to be bilingual. There is also an English-based Creole still used, while a French-based Creole (San Miguel Creole) is virtually extinct. Some indigenous Amerindian languages are also spoken, by the Choco and Chibchan groups. The vast majority of people are still Roman Catholic, but in recent years membership of Protestant groups has increased, while the largest non-Christian community is Muslim. There are also some Jews. Traditional, indigenous belief systems have usually syncretized with Roman Catholic practice.

The total population at mid-2004 was estimated at 3.2m. Most of these people live in the region of the Canal, in the transit zone of the central isthmus, and 56% of the national total is urbanized (as high a rate as in Nicaragua, the most urbanized country of Central America proper). The largest city and the national capital is Panama City (Panamá), the greater metropolitan area of which had a population of around 1m. Panamá lies at the southern end of the Canal, on the Gulf that bears its name, and was an early viceregal capital for the Spanish (until replaced by Bogotá, Colombia) in the Americas. The second city of the country, Colón, with its own collection of suburbs, lies at the northern end of the Canal and constitutes the world's second largest free zone for trade and industry. David, the capital of the agricultural, western province of Chiriquí, located on the Pacific coastal plains, is the most populous city outside the central isthmus. The country is divided into nine provinces, with the more recent formation of three territories constituted for various Amerindian groups.

# History

## HELEN SCHOOLEY
Revised for this edition by KATHARINE MURISON

Geographically, Panama is part of Central America, but, as a result of Spanish colonial divisions and the construction of the Panama Canal through its territory, its history was very different from that of its northern neighbours. The question of the Canal dominated the country's history and, as a result, Panama gained political independence, first from Spain and then from Colombia, only to experience 150 years of US involvement in its national life. The USA was concerned less with the country's internal politics than with the protection of US interests in the operation of the Canal, and with US security policy throughout Latin America. The cession of the Canal to Panama on 31 December 1999 marked a decline in US involvement in both the political life and, coincidentally, in the national economy of Panama. The Spanish arrived in Panama in 1502, and made it the centre of viceregal government over an area stretching as far south as Peru from 1533 to 1751, when the seat of government was moved to Santafé de Bogotá (Colombia). In 1821, along with the rest of Central America, Panama declared independence from Spain, but instead of joining the Central American Federation (1823–38) it opted for incorporation into the Federation of Gran (Greater) Colombia, which also included Venezuela and Ecuador, both of which seceded in 1830, while Panama, despite a number of revolts, continued under Colombian rule for a further 70 years.

## CONSTRUCTION OF THE CANAL

The idea of constructing a transisthmian route was originated by the Spanish, who hoped to build a trading passage for Peruvian silver. The idea gained renewed prominence in the 1840s as a result of the California gold-rush, and led to the construction of the Panama railway in 1850–55 by a US company. The USA directly intervened in the country's internal politics on five occasions between 1860 and 1902, to protect the railway under the terms of a US-Colombian treaty of 1846. The original contract to build a canal was held by the French Panama Canal Co, but the work begun in 1879 proved so much more complex and costly, in terms of human lives, than anticipated that it was suspended in 1888. The Canal was finished by US interests and finally opened in 1914.

## INDEPENDENCE

In 1903 Panama achieved independence from Colombia, largely at the instigation of the USA. Earlier in that year the US Government had negotiated a treaty with Colombia, gaining the canal concession for the sum of US $10m., together with a 100-year lease on a 1,432-sq-km strip of Panamanian territory (known as the Canal Zone), extending for 8 km on either side of the Panama Canal route. The Colombian Congress raised objections to the treaty, causing a revolt in Panama, which then received US assistance in declaring independence in November. The USA concluded a similar treaty with the new Panamanian Government, according Panama sovereignty and the USA 'sovereign rights', creating an ambiguity that was to cause considerable disagreement between the two countries in the following years. The USA retained the right to military intervention to protect the Canal and, within the Zone, it had its own military bases, police force, laws, currency and postal service; it also maintained a direct role in Panama's internal political life until 1918. After more than two decades of relative stability in Panama, constitutional government was disrupted in the 1940s with a progression of bloodless coups and disputed elections. Adolfo de la Guardia (President in 1941 and in 1949–50) pursued a series of right-wing policies, and was impeached and banned from public life in 1950. Antonio Remón was elected President in 1952 and enacted a programme of moderate reforms until his assassination in 1955. During this period the USA established the School of the Americas (SOA), a military base in Panama, which acquired a formidable reputation for its training of anti-insurgency forces from several Latin American countries; the SOA was moved to the USA in 1984.

## THE TORRIJOS ERA, 1968–81

After 12 years of uninterrupted elected government, political turbulence erupted again in 1968. Arias Madrid won the presidential election, but after 11 days in office he was deposed in a military coup led by the National Guard under the command of Col (later Gen.) Omar Torrijos Herrera. (The National Guard, which wielded considerable political influence in the country, functioned as the national defence force, since the formation of an army had been proscribed in the 1904 Constitution.) Freedoms of the press, and of speech and assembly were suspended for one year, and party political activity was banned from February 1969 until October 1978. The National Guard retained power, first with military, and subsequently with civilian, appointees. Torrijos took the executive title of Chief of Government in 1972, and legislative power was vested in a 505-member Asamblea Nacional de Representantes de Corregimientos (National Assembly of Community Representatives), elected in August on a non-party basis. Under Torrijos, Panama enjoyed a greater degree of internal stability; his Government adopted more broadly left-wing sympathies in foreign policy and undertook considerable agrarian reform, while its most important achievement was the negotiation of a new Canal Treaty with the USA.

### The Revision of the Canal Treaty

Despite revisions of the Canal Treaty in 1936 and 1955, which included an increase in the annual rent paid by the USA, the 1903 Treaty continued to be a focus for anti-US sentiment within the country. After repeated pressure from Panama and a number of other Latin American countries, negotiations on a new treaty began in 1973. Two draft treaties were signed in September 1977, in which the USA agreed to cede the Canal to the Government of Panama at noon on 31 December 1999. Prior to that, there would be a phased withdrawal of US troops, with Panama eventually taking control of all US military bases in the Canal Zone. The Zone was to be abolished and renamed the Canal Area. The Panama Canal Co would be replaced by a nine-member Panama Canal Commission, a non-profit US government agency with a board of Panamanian and US directors approved by the US Senate, on which the USA was to retain majority representation until 1989. In addition, Panama and the USA were to be jointly responsible for guaranteeing the Canal's permanent neutrality. The treaties were approved by a national referendum in Panama in October 1977, but opposition within the US Congress delayed their ratification in the USA until October 1979.

In 1978 Gen. Torrijos announced plans to return Panama to elected government. He resigned as Chief of Government in October (retaining the post of National Guard Commander), when a newly elected Asamblea Nacional endorsed his nominee, Dr Arístides Royo Sánchez, as President for a six-year term. However, Gen. Torrijos maintained his hold on power when, at elections in August to the 19-seat Consejo Nacional de Legislación (National Legislative Council—an upper house that also contained 38 members nominated by the Asamblea Nacional), the Partido Revolucionario Democrático (PRD—Revolutionary Democratic Party, led by Torrijos) obtained 10 seats. The elections were boycotted by Arias Madrid's Partido Panameñista Auténtico (PPA—Authentic Panamanian Party). In July 1981 Gen. Torrijos was killed in an aeroplane crash, which was assumed to be accidental, although there were allegations of US Central Intelligence Agency (CIA) involvement.

## THE RISE AND FALL OF NORIEGA

After the death of Gen. Torrijos, relations between the presidency and the National Guard deteriorated, especially after the appointment as Commander, in March 1982, of Gen. Rubén Darío Paredes, a much keener advocate of pro-US foreign policy than President Royo. In July Gen. Paredes forced Royo from office, and the First Vice-President, Ricardo de la Espriella, became President. In the following year Paredes withdrew from the National Guard in order to contest the 1984 presidential election and was replaced by Gen. Manuel Antonio Noriega Morena. Dr Jorge Illueca, who took over as President in February 1984, was also highly critical of the USA and its alleged violations of the new Canal Treaties.

Constitutional amendments introduced in 1983 provided for the direct election of the President and Consejo Nacional de Legislación, replaced the 505-member Asamblea Nacional de Representantes de Corregimientos with a 67-seat Asamblea Legislativa (Legislative Assembly), reduced the presidential term of office to five years and prevented members of the National Defence Forces (as the National Guard was renamed) from standing as political candidates. In the May 1984 elections the PRD, in coalition with five other parties, won a majority in the Asamblea Legislativa and its candidate, Nicolás Ardito Barletta, was elected President. President Ardito resigned in September 1985, however, amid rumours that the National Defence Forces had assisted his election; it was then alleged that Gen. Noriega had removed him in order to disrupt an investigation into the murder, in the same month, of a leading politician, Hugo Spadáfora, in which the armed forces' command was implicated. Ardito Barletta was succeeded by the First Vice-President, Eric Delvalle, who promised 'a return to Torrijista principles', but faced strong opposition to his economic policies. His main problems, however, were the growing power of Gen. Noriega and Panamanian claims that the USA was trying to renege on its commitment to withdraw from the Canal area and also failing to hand over Panama's rightful share of profits.

Despite mounting criticism of Gen. Noriega within Panama, US pressure for an investigation into allegations against him merely increased anti-US sentiment. In February 1988 President Delvalle attempted to dismiss Noriega, following his indictment in the USA on drugs-trafficking charges. Instead, Noriega deposed Delvalle and replaced him with Manuel Solís Palma. Delvalle went into hiding, but continued to be recognized as head of state by the opposition and the US Administration. The May 1989 elections were contested entirely on the issue of Noriega's continuance in power, between two hastily formed coalitions: the government Coalición de Liberación Nacional (COLINA—National Liberation Coalition) and the Alianza Democrática de Oposición Civilista (ADOC—Civic Opposition Democratic Alliance). There were reports of substantial fraud on the part of COLINA and, when the ADOC presidential candidate, Guillermo Endara Galimany, claimed victory, the counting was suspended and the whole election annulled. As ADOC refused to accept the annulment, COLINA formed a provisional Government.

### US Military Intervention

In December 1989 Gen. Noriega was declared Head of Government. He announced that Panama was at war with the USA, and on 20 December US forces intervened to overthrow him. Endara was installed as President, US economic sanctions (in force since 1987) were ended and full diplomatic relations restored. Although Noriega himself had few allies abroad, the US action was condemned by most Latin American countries, and by the UN General Assembly, as a violation of Panamanian sovereignty. The official death toll resulting from the intervention was 527, excluding 27 US troops; the economic cost to Panama was estimated to be at least US $2,000m.

Noriega eventually gave himself up in return for an assurance that he would receive a fair trial and would not face the death penalty. He was flown to the USA, found guilty on charges relating to drugs-trafficking and money-laundering, and in July 1992 was sentenced to 40 years' imprisonment (later reduced to 30 years, after an appeal by his lawyers that he had given years of service to the USA as an 'asset' of the CIA). By August 2001 he had received prison sentences in Panama totalling 98 years, on charges that included murder, corruption and drugs-trafficking.

## RESTORATION OF THE DEMOCRATIC PROCESS

In order to regain the confidence of the international community, a new Asamblea Legislativa was formed in February 1990, based on the results of the May 1989 elections; ADOC was awarded 51 seats and COLINA six, while fresh elections were held for the remaining seats. The National Defence Forces were dissolved and a new Public Force created, consisting of the National Police, the National Air Force and the National Maritime Service. President Endara's Government, however, lacked domestic confidence, and was criticized for its failure to address the problems of political corruption, widespread poverty and rising violent crime, largely linked to the illegal drugs trade. Furthermore, in 1993 various ethnic groups (which formed about 8% of the population) increased their demands for land rights and accused the Government of failing to honour a commitment, made in February 1992, to prevent settlers encroaching on indigenous Amerindian lands.

The PRD returned to power after the May 1994 elections, although it failed to secure an outright majority in either the presidential or legislative polls. The party's presidential candidate, Ernesto Pérez Balladares, won 33.2% of the votes cast, ahead of Mireya Moscoso de Gruber (widow of former President Arias Madrid) of the Partido Arnulfista (PA—Arnulfist Party, the leading member of the ADOC coalition) with 29.1%. In the enlarged 72-seat legislature the PRD reached an agreement with minority parties to command just 36 seats. President Pérez's social and economic policies were highly controversial. The unions opposed the Government's attempts to liberalize the labour market and to implement a privatization programme, and the President's apparent endorsement of a system of political favours damaged public confidence in the administration. A draft constitutional amendment allowing the President to stand for re-election was passed by the legislature in late 1997, but was rejected by a ratio of nearly two to one in a referendum held in August 1998.

In the mid-1990s protests by indigenous groups increased, forcing greater concessions from the Government. The Kuna Indians and the Guaymí (also known as the Ngöbe-Buglé) were granted a degree of autonomy in June 1996 and March 1997, respectively. The agreement allowed them self-rule but not, as the Guaymí had demanded, any powers over either mineral rights or exploitation.

## TRANSFER OF THE CANAL TO PANAMANIAN SOVREIGNTY

The final US base in the Canal Area was closed at the end of November 1999, and the Canal was formally transferred to Panamanian sovereignty on 31 December. The Canal also became a fully commercial operation; under US control it had been run on a non-profit basis. The official ceremony was held on 14 December, and the US delegation was headed by former President Jimmy Carter. The absence of US President Bill Clinton was interpreted in Panama as a diplomatic snub, and a concession to US right-wing interests still opposed to the handover. The administration of the Canal was assumed by an 11-member Autoridad del Canal de Panamá (ACP—Panama Canal Authority). In early 2000 a five-year canal-modernization project was announced, including the development of technology to raise capacity, a general improvement of facilities, the construction of a second bridge, the widening of the narrowest section, at Culebra Cut, and work to deepen certain sections. The loss of US personnel was a serious blow to the economy of the region, and by 2002 a group of local Panamanians had launched a campaign for the return of US interests.

## THE MOSCOSO PRESIDENCY, 1999–2004

The presidential election of 2 May 1999 was a contest between representatives of traditional forces in national politics, with Mireya Moscoso de Gruber standing for the PA-led Unión por Panamá (UPP—Union for Panama) alliance against PRD nominee Martín Torrijos Espino, son of Gen. Torrijos, the candidate of its Nueva Nación (NN—New Nation) grouping. Both candidates promised to continue with free-market reforms, while seeking to mitigate their economic consequences. Moscoso's unexpected victory was perceived as an endorsement of her more populist style and greater emphasis on social justice. Moscoso was inaugurated as the country's first female President in September, and formed a new 'Government of national unity', comprising members of the Partido Solidaridad (PS—Solidarity Party), the Partido Liberal Nacional (PLN—National Liberal Party), the Partido Demócrata Cristiano (PDC—Christian Democratic Party) and the Partido Renovación Civilista (PRC—Civil Renewal Party). In the Asamblea Legislativa the largest party was the NN, with 41 seats, obliging Moscoso to negotiate with six minority parties to achieve a working parliamentary majority, albeit of just one seat. The majority was eliminated in September 2000, but was re-established in September 2002. By this time the Government was facing severe economic pressures and Moscoso needed the majority in order to introduce fiscal and social security reforms.

Moscoso's election campaign included pledges to increase social spending, suspend the privatization programme and increase tariffs on agricultural imports. However, an economic downturn in 2000, and an even more marked decline in 2001, triggered increasing discontent with her Government. There was a series of civilian protests, over a range of social, economic and political issues, including rising unemployment, health and education provision and allegations of bribery among members of the Asamblea Legislativa. In August 2001 riots in protest at rising unemployment broke out in Colón, a town which had suffered economically from the departure of the US staff from the Canal Area and which was afflicted by a particularly high crime rate. As in many other Latin American countries, there was a marked increase in violent crime, apparently linked to drugs-trafficking, and a rising prison population. Prisons were over-populated by up to five times their capacity, provoking sporadic prison riots.

A national dialogue forum, comprising members of the main political parties, business and trade-union leaders, was established by Moscoso to discuss the reform of the tax and social welfare system. Although a preliminary agreement was reached in July 2002, business leaders still objected to the Government's proposed increases to income and corporate tax. The withdrawal of the opposition PRD from the forum and the defection of a number of opposition deputies to the governing coalition in late 2002 enabled the President to proceed with some legislation; nevertheless, plans to reform regulations governing income and corporate tax were omitted from the final agreement concluded by the forum in January 2003.

Following the discovery of human remains in a former military barracks in Tocumen, in December 2000 President Moscoso announced the establishment of a Truth Commission to investigate human rights abuses under former military leaders. Human rights groups estimated the number of 'disappearances' under the military dictatorships during 1968–89 to be some 160. Little effort had been made to investigate these cases, and there were allegations that the US authorities wished to avoid further scrutiny of their association with Noriega. The offices of a human rights organization in Panama City were raided in 2002 and much documentary evidence was taken, including files regarding regional arms-trafficking and the mistreatment of children. In March 2004 the Asamblea Legislativa approved the creation of a Special Prosecutor's Office to investigate crimes committed during the military dictatorships. The Truth Commission had reportedly documented 110 cases of 'disappearances' and murders during the period of military rule, and had located the remains of 40 victims.

## THE 2004 ELECTIONS

The presidential election held on 2 May 2004 signalled a move away from recent trends, with a decisive victory for Martín Torrijos Espino of the PRD, with 47.4% of the votes cast. The PS candidate, former President Endara, won 30.9% of the ballot, while José Miguel Alemán, representing the PA-led Visión de País alliance, only received 16.4% of the votes cast. It appeared that Torrijos, who had the support of former presidential candidate and popular entertainer Rubén Blades, had succeeded in attracting much of the younger electorate with his promises to address the problems of corruption and human rights abuses, to introduce judicial reform and to improve free trade with the USA (in line with the current regional initiative). His appeal rested on both the enduring popularity of his father, Gen. Omar Torrijos, and his image as representing a change from the traditional elements of the country's political scene. The PRD also performed well in concurrent elections to the enlarged Asamblea Legislativa, securing 42 of the 78 seats, while the PA won 16 seats and the PS nine seats.

## THE TORRIJOS PRESIDENCY

Torrijos succeeded in securing legislative approval for various constitutional reforms even before assuming the presidency on 1 September 2004. The amendments, adopted in late July, included a reduction in the number of seats in the Asamblea Legislativa to 71 from 2009, the shortening of the transitional period following elections from four months to two, the abolition of parliamentary immunity from prosecution, and the creation of a constitutional assembly to consider future changes to the Constitution.

During his inaugural address President Torrijos was severely critical of Moscoso's administration, pledging to ensure that those guilty of corruption would be brought to justice. The newly appointed Minister of Finance and the Treasury, Ricaurte Vásquez Morales, later claimed that the fiscal deficit was far higher than that calculated by the previous Government, partly owing to its failure to separate canal finances from government revenue. At the beginning of November 2004 the Government announced that it was pursuing Vásquez's predecessor, Norberto Delgado, on charges of illicit enrichment and corruption. Meanwhile, a Consejo Nacional de Transparencia contra la Corrupción (National Transparency Council against Corruption) was formed within the Ministry of the Presidency, and audits of major government offices and other public institutions were initiated.

In mid-October 2004 Torrijos encountered the first significant demonstration of opposition to his intention to reform the social security fund, the Caja de Seguro Social (CSS), when some 5,000 people participated in a march organized by the newly created Frente Nacional por la Defensa de la Seguridad Social en este Gobierno (FRENADESSO) to protest against the mooted privatization of the institution. The Government denied that it was considering fully privatizing the CSS, which was burdened by an estimated deficit of at least US $2,500m. The march was also attended by farmers opposed to the proposed enlargement of the Panama Canal, which was no longer able to accommodate the new largest container vessels. A referendum was to be held on the issue once more detailed plans had been drafted. Although the country relied heavily on the revenue from the Canal, there was considerable opposition to the accompanying relocation of communities required for its expansion, many of whom were among the poorest in the country.

Fiscal reform legislation aimed at reducing the fiscal deficit from the current 5% of gross domestic product to 1% by the end of Torrijo's mandate in 2009 was approved by the Asamblea Legislativa in January 2005, despite strong opposition from the business sector and trade unions. Corporate taxation was to be increased and the public-sector work-force reduced. In May the Government finally presented its long-awaited plans to address the CSS's financial difficulties. The proposals, which included an increase in the retirement age and an increase in employers' and employees' contributions, prompted a series of demonstrations, led by FRENADESSO, which claimed that the Government had failed to consult the trade unions about the reforms. The organization called a strike, but this was not widely supported. After the Government had made a number of concessions to opponents of the proposals, the reforms to the CSS were approved by the Asamblea Legislativa at the beginning of June, despite ongoing protests, particularly by teachers and medical workers. However, on 26 June Torrijos agreed to suspend the pension reform law for 90 days pending negotiations with representatives of FRENADESSO and the Roman Catholic Church.

At mid-2005 the opposition remained largely divided, despite the formation, in December 2004, of the Coordinadora Nacional de Oposición, a loose coalition of four disparate parties, including the PLN and the PS. Endara, perhaps the most prominent opposition figure, had split from the PS shortly after the 2004 elections and later established a new party, but it lacked legislative representation. Following the PA's poor performance in the elections, it was renamed the Partido Panameñista (PP—Panamanian Party) in January 2005. In a further attempt to revitalize the party, Moscoso resigned as President of the PP in March, to be replaced by Marco Ameglio.

## INTERNATIONAL RELATIONS

Political wrangling over the possibility and benefits of a continued US military presence from 2000 shaped Panama's relations with the USA in the 1990s, and relations between the two countries were further complicated by developments in Panama's relationship with Colombia. One legacy of the Central American civil wars of the 1980s was a vast surplus of arms and ammunition in the region, and Panama became a point of passage for arms-trafficking to Colombia and drugs-trafficking to the USA. The presence of left-wing Colombian guerrillas just inside the Panamanian border, in the province of Darién, had been tacitly accepted since 1993, although under President Pérez Panama appeared to offer more assistance to Colombia's right-wing paramilitary groups. In September 1999 there was a sharp increase in activity by these groups, which some Panamanians alleged were part of an orchestrated scheme to strengthen the case for a continued US military presence. Cross-border incursions continued, and there were death threats against the local population. Colombian efforts to contain the guerrillas and the drugs trade tended to increase the number of refugees fleeing to Panama (estimated at some 8,000 between 1996 and 2000), although by the end of 2001 a series of repatriation initiatives had reduced the number to around 1,700. Investigating reports that some refugees had been forcibly repatriated by Panamanian security forces, the human rights organization Amnesty International expressed concern in 2003 for the safety of the refugees once in Colombia.

Although plans were announced to turn a number of former US bases in Panama into tourist developments, some of these were affected by the presence of US chemical and biological weapons and munitions left behind. US forces had tested chemical and biological weapons in Panama during 1930–68, including napalm and 'Agent Orange' (the defoliant used by the USA during the Viet Nam War). In February 2000 the USA maintained that there were insufficient technological resources available to them at that time to ensure the safe removal of the weapons. The Panamanian Government declared its intention to pursue the matter through the machinery of the Chemical Weapons Convention, to which both countries were signatories. Nevertheless, the two Governments subsequently concluded an agreement regarding stricter military patrols of air and maritime borders to combat trafficking of illegal drugs.

Both the staging of the 10th Ibero-American Summit in Panama in November 2000, and the establishment in Panama in 2001 of the temporary headquarters for negotiations regarding the creation of the Free Trade Area of the Americas, were seen as an indication of the country's growing independence from the USA. Since the ending of the Central American civil wars relations between countries in the region had improved, and in 1997 the countries agreed to co-operate in a major environmental initiative, including the promotion of organic agriculture and ecotourism. Although the USA remained the Canal's principal customer, Panama sought to strengthen trading links with the rest of Latin America and with Europe, while moderating its relations with Asia. Like other Central American countries, Panama accorded diplomatic recognition to the Republic of China (Taiwan), receiving considerable investment in return; however, the country also extended links to the People's Republic of China (PRC). The PRC was the third greatest user of the Canal, accounted for one-third of exports to Panama's duty-free re-export zone, and actively encouraged Panama to end its recognition of Taiwan.

After taking office in September 2004, President Torrijos made clear his intention to enhance Panama's trading relations within Latin America, and in November Panama was invited to join the Group of Three, comprising Colombia, Mexico and Venezuela, which sought to abolish trade barriers between its members. Furthermore, in June 2005 Panama was officially invited to become an associate member of the Mercado Común del Sur (Mercosur—Southern Common Market). Meanwhile, in April 2004 Panama and the USA commenced negotiating a free trade agreement. It had been hoped that the two countries would conclude the talks by September, but an eighth round ended in February 2005 without agreement on a number of sensitive issues, including trade in agricultural products and US access to contracts for the development of the Panama Canal. Negotiations remained in abeyance in mid-2005, as the office of the US Trade Representative focused its efforts on gaining congressional approval of an agreement with other Central American countries and the Dominican Republic.

Panama's foreign relations were adversely affected by a reputation for political corruption, money-laundering and providing shelter to allegedly corrupt or ruthless foreign politicians. There was also particular concern in the USA over apparent use of Panama as a conduit for illegal drugs and immigrants bound for North America. In January 2004 Panamanian security forces co-operated with US officials in an operation to arrest a Colombian national in Panama who was wanted in both Colombia and the USA on charges of drugs-trafficking and money-laundering. It was also widely suspected that the practice of allowing foreign vessels to sail under the Panamanian 'flag of convenience' was used as a cover for illegal activity; Panama had the largest shipping registry in the world. In May 2004 the Panamanian authorities agreed to permit US officials to board any ships sailing under the Panamanian flag that they suspected of carrying weapons of mass destruction.

Relations between Panama and Cuba were severely strained in late August 2004 when, during her last week in office, outgoing President Moscoso issued a pardon to four prisoners who had been convicted by a Panamanian court of charges related to an alleged plot to assassinate the Cuban leader, Fidel Castro Ruz, during his visit to Panama in November 2000. In

# Economy

## PHILLIP WEARNE

Revised for this edition by the editorial staff

response, the Cuban Government, which had sought the extradition of the four, severed diplomatic ties with Panama. On the following day the Venezuelan Government, which had requested the extradition of one of the men in connection with the bombing of a Cuban aeroplane in 1976, also suspended diplomatic relations with Panama. Following the inauguration of President Torrijos in September, the new Minister of Foreign Affairs, Samuel Lewis Navarro, denounced the pardons granted by Moscoso and successfully normalized relations with Venezuela. Restoring full ties with Cuba proved more difficult, however, although in December consulates in the respective countries were reopened.

Panama's geographic location enabled it to develop as one of the most important shipping crossroads and entrepôts in the world. The country's most famous asset was the 82 km-long Panama Canal, which traversed the Darién isthmus, thus linking the Pacific Ocean with the Caribbean Sea and enabling shipping to avoid the lengthy Cape Horn route around the South American landmass. The Canal itself, however, diminished in importance with the advent of supertankers and freighters, as the largest of the modern oil and bulk-cargo tankers could not use it. However, with the completion of the widening of the narrowest part of the Canal in 2001 and the rapid growth in cruise-ship traffic in both the Caribbean and the Pacific, the Panama Canal has gained prominence in terms of income and strategic importance in recent years. Panama's continued role as a 'land bridge' was also reinforced with the opening, in October 1982, of a transisthmian pipeline to carry petroleum deemed economically impractical for transit in the usual way.

For a relatively small country, Panama possessed abundant natural resources, including high-quality fishing grounds, mineral deposits, forests and, above all, a topography and climate that were ideal for the development of hydroelectric and thermoelectric power. Substantial reserves of gold, copper and coal were underexploited and, apart from some manufacturing in the Colón Free Zone (CFZ—the second largest free trade zone in the world, after Hong Kong), the primary and secondary sectors of the productive economy were also grossly underdeveloped. Panama was thus traditionally a services-based economy, reliant on revenues from the Canal, ship registration, free trade zone transactions and contributions from 'offshore' banking activities. In 2004 services accounted for some 75.1% of gross domestic product (GDP); some of the largest individual subsectors were transport, storage and communications (which contributed an estimated 18.3% of GDP), and insurance, real estate and business services (which accounted for some 16.4% of GDP). In the same year agriculture (including hunting, forestry and fishing) and industry (including manufacturing) together accounted for only 24.9% of GDP.

Panama's currency was, effectively, the US dollar, although a nominal local currency, the balboa, existed at par with the dollar. The country's banknote supply was thus determined exclusively by trading relations and capital flows. Balance-of-payment surpluses automatically increased the money supply, while deficits caused it to dwindle. The country's central bank, the Banco Nacional de Panamá, could only influence the credit-creation constituent of the money supply (although it could issue local coinage) and the Government was unable to use currency devaluation or revaluation as an instrument of economic management.

The Government's economic policies were largely dictated by the IMF, the World Bank and the Inter-American Development Bank (IDB), all of which were involved in structural-readjustment programmes from 1985. One of the principal aims of these programmes was to correct Panama's lack of international competitiveness by removing distortions and inefficiencies in the tax regime and labour market. Import substitution in industry and agriculture was considered an important means of closing the gap between an over-regulated domestic economy and an under-regulated international services sector. To this end Panama joined the World Trade Organization in September 1997 and began to dismantle its protective import-tariff regime. From the mid-1980s there was some effort to diversify the Panamanian services-orientated economic model. Revenues from the Panama Canal were by no means guaranteed and the credibility of the 'offshore' banking sector was damaged by its perceived association with drugs-traffickers. Furthermore, the CFZ's exclusive advantages were being increasingly challenged by the creation of free trade ports and zones all over Latin America. It seemed clear, therefore, that the Panamanian economic base had to adapt to avoid a steady decline. From the mid-1980s the contribution of the services sector to the Panamanian economy remained fairly steady; however, there was considerable diversification within the sector itself, with tourism a particularly robust growth area.

Economic policies originating in the 1980s were designed to curb the budget deficit by reducing subsidies, rationalizing employment in the public sector and stimulating export growth, particularly in agriculture, by eliminating bureaucracy and offering better incentives. The initial implementation of such policies, however, produced a high level of political instability in Panama. This led to the resignation of President Ardito Barletta in September 1985 and the ousting of his successor, Eric Arturo Delvalle, by the then Commander of the Panamanian Defence Forces, Gen. Manuel Antonio Noriega Morena, in February 1988. There could be little doubt that the economic situation, worsened immeasurably by the US economic boycott of 1988–89, also played a crucial role in the downfall of Noriega himself, who was ousted by, and surrendered to, US troops during a brief but devastating invasion of the country in December 1989.

Panama withstood the general global recession of the early 1980s fairly well until 1983–84. In 1983 GDP grew by only 0.4%, then contracted by the same amount in 1984, before a recovery in the services sector and manufacturing output from the CFZ stimulated growth in 1985–87. US economic sanctions provoked a precipitous 15.6% decline in 1988 and there was a contraction of 0.4% in 1989, before the lifting of US sanctions and a more realistic economic policy in 1990 and 1991 stimulated economic growth. The recovery was maintained in the early 1990s, before falling to just 1.8% in 1995. There was, however, a subsequent revival, and steady progress was maintained until 2000. However, the growth rate subsequently declined, particularly, as elsewhere in the region, in 2001 when the economy grew by just 0.6%, before recovering slightly to expand by an estimated 2.2% in 2002. Improved economic expansion of 4.3% and an estimated 6.2% was recorded in 2003 and 2004, respectively. The rate of increase in consumer prices remained low into the early 2000s. Consumer prices remained stagnant in 2001, and in 2002 the annual increase in inflation was 1.0%, rising to 1.4% in 2003. The rate was an estimated 1.5% in 2004.

Panama had one of the highest foreign debts per head of population in the world (partly because it needed to borrow the money that it could not print). At the end of 2003 total external debt was an estimated US $8,770m., almost three-quarters of which was public debt. High real interest rates significantly increased the debt-service ratio (annual amortization and interest, expressed as a proportion of foreign-exchange earnings on goods and non-factor services) on Panama's external debt during the 1980s, and in 1988 the country suspended all debt-service payments. By the end of 1990 arrears to the IMF, the World Bank and the IDB had reached $610m. Despite a series of financial reforms in 1991–92, the situation continued to deteriorate. Total arrears were estimated at $3,000m. in early 1994, before the Government intensified its efforts to reach an agreement with commercial and multilateral creditors.

In May 1995 a Brady Plan debt-restructuring agreement was announced (the initiative on debt relief originally proposed by the then US Treasury Secretary, Nicholas Brady, in 1989), which covered a total of US $3,230m. in principal and interest arrears. The agreement, whereby Panama exchanged debt for new bonds with virtually all its commercial creditors, reduced the country's debt by more than $400m. and opened the way for fresh credit. In 1997 the country further reduced its debt by renegotiating more than $300m. in petroleum-supply debts incurred with Mexico and Venezuela under the San José Agreement of 1980. This restructuring gave Panama access to concessionary finance from the IMF, the World Bank and the IDB and allowed the country access to international capital markets. Panama took full advantage of this and since March 1999, when it sold $500m. worth of 30-year global bonds, using 40% of the receipts to buy back more foreign debt, has raised money to cancel debt. In January 2001 the country sold $750m. worth of 10-year global bonds, and in July and November 2002 it issued a total of $580m. worth of paper in its Global 2012 bond series.

## AGRICULTURE, FORESTRY AND FISHING

Agriculture and fishing were vitally important to the success of the economic diversification effort in Panama. This fact was recognized in the country's 1986 development plan, which set out to reverse a steady economic decline originating in the early 1970s. In 2004 19.1% of the working population were employed in agriculture, forestry and fishing, but the sector contributed just 7.5% of GDP. About 23% of Panama's 7.6m. ha of land were cultivated, while another 21% were permanent pasture land and 44% were forest and woodland (more than one-half of which has been declared protected park, woodland and forest in the past 25 years).

Bananas, melons, sugar cane and coffee were the principal export crops. In 2003 Panama's exports of bananas, formerly the country's leading export commodity, earned an estimated US $105.2m., compared with $109.4m. in the previous year and $122.2m. in 2001. These figures represented yet another step in the decade-old decline in the relative importance of banana exports; in 1992 bananas accounted for 44.2% of total export earnings, in 2003 they represented some 13.0%. The decline was variously attributed to falling international prices, industrial action in the sector, the adverse effects of the El Niño weather phenomenon (the warm current that periodically appears in the Pacific Ocean, altering normal weather patterns) and, perhaps most crucially, the quotas imposed by the European Union (EU) on banana exports from Latin America. However, from 2006 the EU quota system was scheduled to be replaced by an increase in import tariffs from €75 per metric ton to €230 per ton. Representatives of the industry claimed that the new tariffs would destabilize the economy and substantially increase unemployment. In the late 1990s two subsidiaries of a US company, Chiquita Brands, were responsible for about 80% of Panama's banana production and exports, with private Panamanian producers accounting for the remainder. In 2000 the sector employed some 15,000 people directly, with a further 60,000 people dependent on the industry.

Traditionally, sugar was the second highest agricultural export earner. However, revenues declined dramatically from the late 1980s as a result of depressed world prices and dramatic annual reductions in the USA's sugar import quota. Revenue decreased from US $41.3m. in 1983 to $10.2m. in 1989, although it recovered thereafter, reaching $28.7m. in 1997, following significant increases in Panama's share of the US sugar import quota in the mid-1990s. In 1999 export revenue declined dramatically, to $14.5m., when it accounted for just 2.0% of export revenue. In 2000 export revenue recovered, with the value reaching $19.9m., before decreasing again in 2001 to just $13.9m. (approximately 2% of the value of total exports). Following a slight recovery in 2002, when exports earned $15.1m., earnings from sugar declined to an estimated $12.8m. in 2003, equivalent to 1.6% of total earnings. Melon cultivation increased substantially in the early 21st century. Export earnings from melons rose from $25.6m. in 2001 to an estimated $47.3m. in 2003, equivalent to 5.9% of total export revenue. Coffee production increased to 13,900 metric tons in 2002, after remaining stable at about 11,000 tons throughout the 1990s; however, earnings fluctuated owing to international price movements. In 1995 earnings more than doubled to $33.4m., before declining to just $12.1m. in 2003, or 1.5% of total exports.

Successive Governments recognized the urgent need to increase agricultural production in their agreements with the IMF and the World Bank. A five-year agricultural development plan announced in 1985 intended to counteract the effects of underinvestment resulting from price controls on rice, meat, potatoes and dairy products. Diversification of both crops and markets was given a high priority, and new crops such as African palm (oil palm), cocoa, coconuts, various winter vegetables and tropical fruits were promoted. However, the need to boost production was somewhat undermined by the move to abolish the high import tariffs that had protected Panamanian farmers for many years, which caused many of those who catered for domestic consumption to be forced out of business from the mid-1980s. Production did, however, increase thereafter, but annual average expansion of 2.3% in 1990–2003 masked some fundamental structural changes in favour of export-driven production. In 2001 the agricultural sector expanded by 6.4%; however, growth in the sector slowed to 2.0% and 1.5% in 2002 and 2003, respectively. The livestock sector, the epitome of the new export-driven policy, underwent particularly rapid expansion, averaging 4.0% annual growth in the late 1980s and early 1990s. However, the sector subsequently suffered from credit problems and export restrictions, and export earnings from livestock products fell dramatically as a result. The sector, however, recovered quite dramatically after the mid-1990s, with meat exports totalling $21.4m. in 2002, of which beef exports contributed almost three-quarters.

Panama was one of the world's largest exporters of shrimps, with exports of shellfish more than doubling in the 1990s. In 2001 exports of shrimps earned US $70.1m. (equivalent to 8.6% of total export earnings); however, this figure fell in 2002, to $58.0m., and further in 2003, to an estimated $57.6m. (7.1% of exports). In the same period fresh and frozen fish and fillets, including yellowfin tuna, almost doubled in importance, from $148.1m. in 2001 (18.3% of total exports), to $261.9m. in 2003 (32.4% of exports).

## MINING AND ENERGY

Despite valuable mineral deposits that included gold, silver, copper and coal, mineral extraction was traditionally limited to clay, limestone and salt for local consumption. In 2004 mining contributed only an estimated 1.2% of GDP, engaging less than 0.1% of the employed labour force. With a return to political and economic stability in the mid-1990s, and studies indicating potential export earnings from gold, silver and copper of some US $500m. per year, Panama's mining potential began to attract the interest of foreign investors. However, in 2005 the sector remained relatively undeveloped, with potential investors discouraged by low world metal prices, the high cost of initial exploratory and extractive activity, and in some areas environmental opposition to development. Panama had estimated reserves of 3,600m. metric tons of proven and probable reserves of 0.6% grade copper ore at one deposit alone, the world's 12th largest copper deposit, at Petaquilla, in the western Chiriquí province.

Gold-mining in Panama tended to be small-scale, but gold-mining concessions were more attractive to foreign investors owing to the lower capital costs. Nevertheless, at the beginning of the 21st century Panama had only one gold mine in production, the Santa Rosa mine in Veraguas province. Following an initial investment of US $25m. in 1995, the Santa Rosa mine yielded some 57,000 troy oz in 1997, producing $20m. in export revenue. However, output fell steadily thereafter, with export earnings put at just $12m. in 2000. In 2003 RNC Gold Inc increased its share of the Cerro Quema gold- and silver-mining concession from 10% to 60%. In the following year RNC invested $14m. in the construction of an open mine at Cerro Quema. Production of gold was scheduled to begin in late 2005, with an anticipated annual yield of 48,000 troy oz.

Panama had enormous hydroelectric and thermoelectric potential and aimed eventually to eliminate petroleum-powered electricity generation. The opening in 1984 of the La Fortuna (now Edwin Fabrega) hydroelectricity plant seemed at the time to bring this aim closer. The plant cost US $500m. and, with a capacity of 300 MW, was by far the largest of the country's three hydroelectricity plants. However, the value of the plant, in terms of savings on petroleum imports, was reduced by the decrease in international petroleum prices in the late 1980s and by fluctuating rainfall levels, which led to a crisis in the hydroelectricity sector. Nevertheless, by 2001 2,575 MW, or 53% of the country's total electricity output, was hydroelectric, with demand growing at an average of 6% per year. In 2001 electricity generation rose by more than 4%, an increase almost entirely owing to an improved performance in the thermoelectric sector and a result that outstripped the 3.4% rise in consumption in that year. In 2003 the Estí hydroelectric project increased installed capacity by 120 MW, while the thermoelectric Pedregal power installation added a further 53.4 MW of generating capacity.

During periods of low hydroelectric output Panama was dependent on power generated from imported petroleum products. The cost of petroleum imports, which had been as high as US $350m. in the early 1980s, fell to less than $125m. by 1994, owing to increased hydroelectric capacity and lower world petroleum prices. However, the subsequent poor performance of the hydroelectricity sector and a rise in the price of crude petroleum resulted in the cost of fuel imports doubling between 2001 and 2003, to $360.0m. Despite the increase in production at powerstations in Estí and Pedregal from 2003, oil import costs were expected to continue to rise.

## MANUFACTURING AND CONSTRUCTION

Manufacturing accounted for about 8.1% of GDP and engaged an estimated 9.4% of the country's employed labour force in 2004. The sector was based on agricultural processing and light manufacturing, particularly food and beverages, clothing, household goods and construction materials. Growth in the sector fluctuated considerably, successively depressed by a stagnant internal market, political instability and foreign competition, and then buoyed by the removal or neutralization of such impediments. In 1990–2003 manufacturing GDP increased by an annual average of 2.7%; however, it contracted by an estimated 1.6% in 2003, before increasing by an estimated 2.6% in 2004. Industrial activity was concentrated in the corridor of land running between Panama City and Colón (roughly following the line of the Canal), known as the CFZ. This was by far the most important manufacturing area in Panama, covering 400 ha and accommodating some 2,000 companies, with net exports from the CFZ accounting for around 7% of GDP while employing less than 1% of Panama's work-force. Most of the imports are from the Far East, destined for markets in South and Central America.

Ambitious expansion plans for the CFZ, costing more than US $65m., were initiated by the Government of President Arturo Endara Galimany (1989–94). There was a corresponding increase in confidence and private-sector investment, with total trade—the sum of imported and re-exported goods—increasing from $5,100m. in 1990 to $11,197m. in 1998. Nevertheless, in the late 1990s competition from other free trade zones and the increasing removal of trade barriers within the region, combined with the severe financial crises that crippled Latin America and Asia, threatened the long-term competitiveness of the CFZ, which responded by emphasizing its geographical location and seeking to reposition itself in the free trade zone market as a transhipment hub. In 2004 the CFZ generated approximately $11m. in imports and exports. Trade in the CFZ increased by 19.0% in that year.

The construction sector suffered a sharp decline in 2001 and 2002, contracting by 21.8% and 7.1%, respectively. However, in 2003 the sector expanded by 32.5%, owing, in part, to the conversion or restoration of a large number of buildings in the Canal Area (known as the Canal Zone until 1979) and to the start of work on a second bridge over the Canal. A further increase, of an estimated 16.0%, was recorded in 2004. In that year the sector contributed an estimated 4.8% of GDP and engaged an estimated 7.5% of the employed population.

## TRANSPORT AND TOURISM

Economic sanctions against the Noriega regime severely damaged Panama's ship registration industry; however, despite competition, it subsequently recovered well. Panama's open registry fleet grew by 4.5%–5.0% per year in the 1980s and by 1990 the number of vessels registered, including tankers, was 12,149. However, although the country's shipping registry remained the largest in the world, by 2004 this figure had fallen to 6,477, owing to fierce competition from other countries. Earnings from ship-registration fees also declined. Nevertheless, this represented a recovery from the disastrous US sanctions of the late 1980s, when Panamanian ships were banned from US ports. Thousands of ships transferred to other registers before the Endara Government waived re-registration fees for a year in 1990, in an effort to recover market share. Further measures enacted in 1995 included offering volume discounts of up to 50% in fees to shipowners whose registry was over 100,000 grt in total. Most shipping remained foreign-owned, reflecting the preferential tax treatment available to shipping companies in Panama. Cristóbal and Balboa, ports in the Canal Area, could accommodate ocean-going freighters and passenger ships, following their redevelopment in 1996. Such investment made Panama one of the busiest container transhipment locations in the world. The Manzanillo International Terminal, Evergreen and Panamá ports were the three largest container ports in Latin America. Late 2000 saw the successful inauguration of Panama's first cruise-ship terminals, at both ends of the Canal.

In 1999 the road network totalled 11,400 km, of which about 35% was paved, and there were an estimated 420,000 vehicles in use in 2002. The Pan-American Highway ran for 545 km in Panama, from the Costa Rican border through Panama City, to Chepo. Three railways served the banana plantations and other agricultural areas in the western parts of Bocas del Toro and Chiriquí, which bordered Costa Rica. On ratification of the 1977 Canal Treaties, Panama also acquired control of the Panama Railroad, which connected Panama City and Colón. The 83-km railroad reopened on 1 July 2001, operating daily passenger and cargo container services. It was operated by Ferrocarril de Panamá, a subsidiary of Kansas City Southern Railway (USA). The Tocumen (formerly Omar Torrijos) International Airport was officially opened in 1978 and, as a consequence of increased tourist arrivals, a project to expand the facilities at a cost of US $20m. was announced in 2004.

Two long, varied coastlines with good beaches and 700 tropical islands offered vast tourism potential. The southern (Pacific) coast of Panama offered some of the best deep-sea fishing in the world. Other tourist attractions included the mountains and volcanic scenery, the ruins of the original Panama City and the Panama Canal. Although tourism declined steadily throughout the 1980s, a strong recovery was established in the early 1990s. The number of visits by foreigners rose by 5.7%, to 1.21m., in 2001, with those defined as being tourists, as opposed to stopover arrivals, rising from 221,677 in 1990 to 534,208 in 2003. Receipts from tourism rose steadily from US $538m. in 1999 to $809m. in 2003. The construction of six hotels and 120 tourist villas in the former US military base at Fort Amador, at the entrance to the Canal, was expected eventually to increase the country's hotel capacity by one-third; the $450m., four-stage scheme was begun in 1997. The most spectacular growth in recent years was in cruise-ship arrivals following the completion of two new cruise-ship terminals in late 2000. In the six months to March 2001 60 cruise ships docked, while the Canal earned $21.3m. in tolls from some 234 cruise-ship transits in the same period.

## THE CANAL AREA

The Canal Area (known as the Canal Zone until 1979) is a strip of land, 16 km wide, between the Pacific and Caribbean coasts, running north-west to south-east. The Canal itself is 82 km long, and raises or lowers ships through 26 m by means of six pairs of locks. An average passage takes about eight hours. The Canal can accommodate ships with a maximum draught of 12.0 m and beams of 32.3 m. Improvements to the Canal in the early 21st century increased the transit capacity to 43 vessels per day. In 2004 traffic in the Canal increased by 7.1%. Canal operations—tolls, transit-related services and sales of surplus water and electricity—accounted for 7.2% of Panama's GDP in 2003.

Almost 70% of all traffic through the Canal either originated from, or was destined for, the USA (moving between Asia and the east coast of the USA). Japan was the second most regular user of the Canal, followed by the People's Republic of China, Japan and Chile. In 1996, following a report by the US Army Corps of Engineers that listed more than 1,000 maintenance repairs and equipment upgrades requiring immediate attention, the Panama Canal Commission (PCC, replaced by the Autoridad del Canal de Panamá, ACP—Panama Canal Authority—in December 1999) initiated a $1,000m. modernization programme. The programme, which was expected to take up to six years to complete and to increase capacity by one-fifth, was financed entirely from the Canal's revenue, and significantly increased employment in the area. The results of this programme have already been seen. The Culebra Cut widening project—the expansion of the narrowest part of the Canal from 152 m to 192 m—was completed in December 2001. The project allowed two vessels to pass at the same time and increased the capacity of the Canal by some 20%. New lock configurations were also commissioned and two new tugboats were added to the Canal's fleet in December 2001. In 2005 the ACP was expected to announce plans to widen the Canal to allow the passage of larger commercial container vessels. While widely considered to be vital to the future of the Canal and therefore the country's commercial sector, concerns were raised relating to the environmental consequences of the expansion and the possible relocation of farmers in communities bordering the Canal. Under the Canal Treaties of 1977, which came into force in October 1979, the neutrality of the Canal Area was guaranteed, so as to ensure the continuous and clear transit of traffic. Panama administered the Canal from January 2000, although the USA reserved the right to protect the Canal by military force if necessary, and assumed a majority on the nine-member PCC in 1990.

The vessels that used the Canal were predominantly bulk cargo carriers carrying grain, petroleum and related products. The number of transits and the cargo figures remained virtually static from 1995. However, revenue from the Canal rose, owing to increases in transit charges that were introduced in 1995 and 2003. In the 2004 fiscal year (October to September for the Canal) canal transit fees earned Panama some US $757.7m. (compared to US $460m. in 1995), derived from some 14,035 commercial transits of 200.2m. long tons. Other services, notably the sale of surplus electricity and water from the Canal Area, earned a further $238.7m.

## EXTERNAL TRADE AND FINANCE

Panama regularly incurred a large deficit on its merchandise trade account as a result of its heavy dependence on imported fuel and 'invisibles' (banking, ship registration, Canal fees and re-exports). This was, however, partially offset by a surplus on transactions in services. The trade deficit grew inexorably from the late 1980s, and reached US $2,807.8m. in 1999. There was a slight improvement in the deficit in 2000, to $1,189.8m., and a more significant decrease, to $825.8m., in 2001; however, the deficit then increased to $1,176.4m. in 2002 before decreasing to $1,112.7m. in 2003. In 2004 the imports rose steeply to $7,470.9m. (from $6,161.6m. in the previous year), resulting in an increased deficit of $1,585.3m. The principal imports were primary materials and intermediate goods, which were valued at $671.3m. in 2003, non-durable consumer goods ($384.0m.) and transport and telecommunications equipment ($330.1m.).

Panama tended to be heavily dependent on capital inflows, such as IMF assistance. This was partly because its unusually liberal economic system made it particularly vulnerable to lower world-trading activity during periods of recession. Furthermore, because of its use of the US dollar, Panama was unable to resort to currency devaluation in order to correct trading imbalances. Between 1981 and 1986 the debt-service ratio (debt-servicing compared with the total value of exports of goods and services) fluctuated between 43% and 55%, but declined steadily thereafter following renegotiation and rescheduling. Nearly a decade of debt restructuring, including debt forgiveness, repurchasing and exchanging commercial bank debt for bonds, beginning with a Brady bond deal in 1995, enabled the country to reduce its liabilities steadily in recent years. With access to concessional loans and international capital markets, the Panamanian Government was able to issue bonds worth more than $1,500m. to buy back debt in 1998–2003. At the end of the later year Panama's external debt stood at $8,770m., of which $8,296m. was long-term debt. In that year the cost of debt-servicing was equivalent to 11.3% of the value of exports of goods and services.

Panama traditionally encouraged foreign investment, a policy that was intensified after 1990, as part of efforts to improve the economy and counterbalance the legacy of sanctions and the US invasion. Many restrictions were ended as part of the Endara Government's privatization programme. Panama exercised no exchange controls, and transfers of funds were never prevented. There were no restrictions on the transfer of profits, dividends, interest, royalties or fees, nor on the repatriation of capital nor the repayment of principal. A 10% withholding tax was levied on dividends from operations in Panama (excluding the CFZ), but Panama did not levy tax on income earned in 'offshore' financial dealings. In 1997–98 foreign capital inflows increased to an annual average of more than US $1,200m. Although this figure was somewhat distorted by the purchase of a 49% stake in the state telecommunications monopoly, Instituto Nacional de Telecomunicaciones (INTEL), by Cable & Wireless (United Kingdom) for $652m., the incidence of such major investments was increasing. For example, the sale of toll-road concessions to two Mexican companies in 1998 secured future inward investment in excess of $600m., while the sale of electricity assets in the same year yielded $603m. Other important sales and joint-venture investments, including many tourism and infrastructure developments linked to redevelopment of military and Canal sites following the withdrawal of US personnel, were expected to help maintain a healthy level of capital investment. However, the completion of a number of major investment projects in 2000 caused a decline in foreign direct investment from $517m. to $393m. According to the Government, the level of investment recovered to $512.6m. in the following year and continued to rise, reaching $1,012.3m. in 2004.

From the late 1960s onwards Panama developed its potential as an international finance centre, based on the full transferability of its currency, the country's favourable tax laws and the absence of state controls. The 'offshore' business, foreign exchange, money and reinsurance markets expanded in the early 1980s and in 2001 the sector accounted for an estimated 11% of GDP. However, the 'offshore' banking sector experienced difficulties caused by political and economic instability and radical changes in business in the late 20th century.

By December 2004 the net consolidated assets of the banking sector were estimated to be US $40,149m. This represented substantial increase from the $34,020m. recorded in 2002, but was still well below the levels of the mid-1980s, when total assets amounted to $49,000m. In December 2004 the number of banks registered was 73, significantly fewer than the 118 registered in 1982. As general financial liberalization eroded Panama's competitiveness in the 1990s, competitors in Latin American and the Caribbean were establishing themselves. The relaxation of financial restrictions globally meant that many banks in the USA and Europe began to deal directly with clients in Latin America, denying Panama's banks one of their principal roles. In June 2000 the Organisation of Economic Co-operation and Development (OECD, based in Paris, France) identified Panama as meeting the technical criteria for being a tax 'haven'. Furthermore, in the same month the Financial Action Task Force on Money Laundering (FATF, also based in Paris) included the country on a list of those jurisdictions considered to be 'non-co-operative' in international efforts to prevent money-laundering and the financing of terrorist organizations. Panama protested against its inclusion on both lists and subsequently modified its legislation to introduce greater legal and administrative transparency in the financial sector. In June 2001 the country was removed from the FATF list and in April 2002 Panama met OECD criteria in committing to improve the transparency of its tax and regulatory systems and establish effective exchange of information for tax matters by 31 December 2005.

From 1983 Panama's fiscal policies required IMF approval. However, such approval was difficult to secure, principally owing to the Government's difficulty in achieving IMF targets for the budget deficit. Successful rescheduling and negotiations of new loans from the IMF were achieved in the mid-1980s. However, these were gained not only as a result of budgetary austerity, but also because of the Government's commitment to certain reforms opposed by the trade unions and by the private sector. Satisfaction with the Government's efforts at economic restructuring was signalled by the resumption of lending to the country by the World Bank in December 1986. However, in 1987 the Government began to withhold payments to bilateral creditors and by March 1988 the country's IMF agreement had lapsed, with no new accord negotiated to replace it. By the end of that year accumulated interest and principal arrears on public-sector debt were estimated at US $1,400m.

In 1991 the Endara Government had rescheduled US $520m. in bilateral debts with the 'Paris Club' of Western creditor nations. After paying part of the $610m. in arrears to the IMF, the World Bank and the IDB, the country became eligible for further credits during 1992–93. In February 1992 the IMF disbursed $50.4m. in support of the Government's economic programme. The trend was consolidated in April 1995 when Panama signed a debt-rescheduling agreement based on the Brady Plan. The accord covered $2,000m. in principal arrears and $1,500m. in interest arrears, offering creditors a variety of options with the IMF, the World Bank and the IDB, which were all actively supporting the agreement.

Thereafter, Panama repurchased or exchanged US $1,220m. of this Brady-bond debt for its own 30-year government bonds. In December 1997 the IMF approved a credit of $162m., under the Extended Fund Facility, in support of the Government's economic programme for 1998–2000. In March 2000 Panama began negotiations with other Central American countries on a free trade accord. In May 2001, following further extensive discussions, the Central American countries, including Panama, reached an agreement with Mexico to establish the 'Plan Puebla–Panamá' and in March 2002 Panama and El Salvador signed a free trade agreement. In August 2003 Panama signed a free trade accord with Taiwan and in April 2005 a free trade agreement was established between Panama and Singapore. In mid-2005 negotiations were ongoing towards a bilateral free trade agreement with the USA.

## CONCLUSION

The administration of President Martín Torrijos Espino faced three major challenges upon assuming office in September 2004: the ongoing problem of the fiscal deficit; reform of the social security fund, Caja de Seguro Social (CSS); and the proposed widening of the Canal. Although the previous Government of Mireya Moscoso de Gruber (1999–2004) had made progress in reducing the deficit, further reform was needed. In January 2005, despite opposition from the business sector, the Asamblea Legislativa approved legislation that aimed to reduce the deficit from 5% to 1% of GDP by the end of Torrijo's mandate in 2009. The Government's attempts to reform the CSS in 2005 met with much public opposition, however, and in late June President Torrijos was compelled temporarily to suspend implementation of the reforms to allow further dialogue with opposition groups. A referendum on the planned expansion of the Canal was expected to be held in late 2005. It was considered imperative for the issue of social security reform to be resolved and greater financial stability to be established before the Government could embark on ambitious investment projects such as the widening of the Canal.

The Panamanian economy recovered well from the world-wide economic slowdown of 2001, expanding by 2.2% in 2002, by 4.3% in 2003 and by an impressive 6.2% in 2004, according to official estimates. Although still high, the unemployment rate fell from 13.1% of the economically active population in 2003 to an estimated 11.8% in 2004. Growth in 2004 was underpinned by strong expansion in external trade, ongoing public works projects, and pre-electoral spending in the first half of the year. However, the introduction by the incoming Torrijos Government of a fiscal reform programme, and lower capital investment resulting from these budgetary constraints, as well as a more hostile global trade environment, were expected adversely to affect economic growth in 2005, when the economy was forecast to expand by a more modest 3.5%.

# Statistical Survey

Sources (unless otherwise stated): Dirección de Estadística y Censo, Contraloría General de la República, Avda Balboa y Federico Boyd, Apdo 5213, Panamá 5; tel. 210-4800; fax 210-4801; e-mail cgrdec@contraloria.gob.pa; internet www.contraloria.gob.pa; Viceministerio del Comercio Exterior, Plaza Edison, 3°, Avda El Paical, Apdo 55-2359, Panamá; e-mail secomex@mici.gob.pa; internet www.vicomex.gob.pa.

Note: The former Canal Zone was incorporated into Panama on 1 October 1979.

## Area and Population

### AREA, POPULATION AND DENSITY

| | |
|---|---:|
| Area (sq km) | 75,517* |
| Population (census results) | |
| 13 May 1990 | 2,329,329 |
| 14 May 2000 | |
| Males | 1,432,566 |
| Females | 1,406,611 |
| Total | 2,839,177 |
| Population (official estimates at mid-year) | |
| 2002 | 3,060,090 |
| 2003 | 3,116,277 |
| 2004 | 3,172,360 |
| Density (per sq km) at mid-2004 | 42.0 |

* 29,157 sq miles.

### ADMINISTRATIVE DIVISIONS
(population at census of May 2000)

| Province | Population | Capital (and population)* |
|---|---:|---|
| Bocas del Toro | 89,269 | Bocas del Toro (9,916) |
| Chiriquí | 368,790 | David (124,280) |
| Coclé | 202,461 | Penonomé (72,448) |
| Colón | 204,208 | Colón (42,133) |
| Comarca Emberá | 8,246 | — |
| Comarca Kuna Yala | 32,446 | — |
| Comarca Ngöbe-Buglé | 110,080 | — |
| Darién | 40,284 | Chepigana (27,461) |
| Herrera | 102,465 | Chitré (42,467) |
| Los Santos | 83,495 | Las Tablas (24,298) |
| Panamá | 1,388,357 | Panamá (708,438) |
| Veraguas | 209,076 | Santiago (74,679) |
| **Total** | **2,839,177** | — |

* Population of district in which capital is located.
Note: Population figures include the former Canal Zone.

# PANAMA

*Statistical Survey*

## PRINCIPAL TOWNS
(population at 2000 census)

| | | | |
|---|---|---|---|
| Panamá (Panama City, capital) | 463,093 | Pacora | 57,232 |
| San Miguelito | 291,769 | Santiago | 55,146 |
| Tocumen | 81,250 | La Chorrera | 54,823 |
| David | 76,481 | Colón | 52,286 |
| Nuevo Arraiján | 63,753 | Changuinola | 45,063 |
| Puerto Armuelles | 60,102 | Pedregal | 45,033 |

## BIRTHS, MARRIAGES AND DEATHS

| | Registered live births | | Registered marriages* | | Registered deaths | |
|---|---|---|---|---|---|---|
| | Number | Rate (per 1,000)† | Number | Rate (per 1,000)† | Number | Rate (per 1,000)† |
| 1996 | 63,401 | 23.7 | 10,206 | 3.8 | 11,161 | 4.2 |
| 1997 | 68,009 | 25.0 | 10,357 | 4.1 | 12,179 | 4.5 |
| 1998 | 62,351 | 22.6 | 10,415 | 4.1 | 11,824 | 4.3 |
| 1999 | 64,248 | 22.9 | 10,388 | 3.9 | 11,938 | 4.2 |
| 2000 | 64,839 | 22.7 | 10,430 | 3.9 | 11,841 | 4.1 |
| 2001 | 63,900 | 21.3 | 9,687 | 3.6 | 12,442 | 4.1 |
| 2002 | 61,671 | 20.2 | 9,558 | 3.1 | 12,428 | 4.1 |
| 2003 | 61,753 | 19.8 | 10,310 | 3.3 | 13,248 | 4.3 |

\* Excludes tribal Indian population.
† Based on official mid-year population estimates.

**Expectation of life** (WHO estimates, years at birth): 75 (males 73; females 78) in 2003 (Source: WHO, *World Health Report*).

## EMPLOYMENT
('000 persons aged 15 years and over, August of each year)

| | 2002 | 2003 | 2004 |
|---|---|---|---|
| Agriculture, hunting and forestry | 222.2 | 228.3 | 215.1 |
| Fishing | 12.4 | 11.7 | 13.7 |
| Mining and quarrying | 1.6 | 1.0 | 0.7 |
| Manufacturing | 100.2 | 105.8 | 112.7 |
| Electricity, gas and water supply | 8.6 | 8.8 | 8.3 |
| Construction | 71.9 | 79.9 | 90.4 |
| Wholesale and retail trade; repair of motor vehicles, motorcycles and personal and household goods | 195.6 | 196.4 | 209.6 |
| Hotels and restaurants | 47.9 | 53.4 | 60.8 |
| Transport, storage and communications | 81.9 | 85.9 | 89.0 |
| Financial intermediation | 22.5 | 21.5 | 24.8 |
| Real estate, renting and business activities | 42.1 | 44.6 | 53.8 |
| Public administration and defence; compulsory social service | 70.5 | 74.5 | 73.9 |
| Education | 62.5 | 65.1 | 67.8 |
| Health and social work | 39.1 | 38.2 | 43.8 |
| Other community, social and personal service activities | 70.4 | 65.7 | 63.2 |
| Private households with employed persons | 62.0 | 64.0 | 69.2 |
| Extra-territorial organizations and bodies | 0.2 | 1.0 | 0.7 |
| **Total employed** | 1,111.7 | 1,146.0 | 1,197.6 |
| Unemployed | 173.3 | 172.0 | 159.7 |
| **Total labour force** | 1,285.0 | 1,318.0 | 1,357.3 |

# Health and Welfare

## KEY INDICATORS

| | |
|---|---|
| Total fertility rate (children per woman, 2003) | 2.7 |
| Under-5 mortality rate (per 1,000 live births, 2003) | 24 |
| HIV/AIDS (% of persons aged 15–49, 2003) | 0.9 |
| Physicians (per 1,000 head, 2000) | 1.68 |
| Hospital beds (per 1,000 head, 1996) | 2.21 |
| Health expenditure (2002): US $ per head (PPP) | 576 |
| Health expenditure (2002): % of GDP | 8.9 |
| Health expenditure (2002): public (% of total) | 71.7 |
| Access to water (% of persons, 2002) | 91 |
| Access to sanitation (% of persons, 2002) | 72 |
| Human Development Index (2002): ranking | 61 |
| Human Development Index (2002): value | 0.791 |

For sources and definitions, see explanatory note on p. vi.

# Agriculture

## PRINCIPAL CROPS
('000 metric tons)

| | 2001 | 2002 | 2003* |
|---|---|---|---|
| Rice (paddy) | 279.3 | 245.2 | 250.0 |
| Maize | 37.2 | 75.6 | 78.0† |
| Sugar cane | 1,789.0 | 1,440.6 | 1,500.0 |
| Roots and tubers | 76.1 | 79.9* | 85.5 |
| Fresh vegetables (incl. melons)* | 103.5 | 140.8 | 142.8 |
| Bananas | 489.4 | 500.0* | 550.0 |
| Plantains | 100.0* | 105.0* | 106.0 |
| Oranges | 35.0* | 46.6 | 47.0 |
| Other fruit (excl. melons)* | 55.8 | 57.3 | 59.8 |
| Coffee (green) | 12.4 | 13.9 | 14.0 |
| Tobacco (leaves)* | 2.1 | 2.2 | 2.2 |

\* FAO estimate(s).
† Unofficial figure.
Source: FAO.

## LIVESTOCK
('000 head, year ending September)

| | 2001 | 2002 | 2003* |
|---|---|---|---|
| Horses* | 168 | 170 | 175 |
| Mules* | 4 | 4 | 4 |
| Cattle | 1,533 | 1,533 | 1,550 |
| Pigs | 312 | 303 | 305 |
| Goats | 5 | 6 | 6 |
| Chickens | 14,133 | 13,894 | 14,000 |
| Ducks* | 210 | 215 | 220 |
| Turkeys* | 27 | 30 | 33 |

\* FAO estimates.
Source: FAO.

## LIVESTOCK PRODUCTS
('000 metric tons)

| | 2001 | 2002 | 2003* |
|---|---|---|---|
| Beef and veal | 56.7 | 53.7 | 54.0 |
| Pig meat | 19.0 | 18.2 | 18.2 |
| Poultry meat | 76.8 | 88.7 | 89.0 |
| Cows' milk | 170.6 | 178.1 | 180.0 |
| Cheese | 7.9 | 10.0 | 10.5 |
| Hen eggs | 12.7 | 26.3 | 27.0 |

\* FAO estimates.
Source: FAO.

## Forestry

**ROUNDWOOD REMOVALS**
('000 cubic metres, excluding bark)

|  | 2001 | 2002 | 2003* |
|---|---|---|---|
| Sawlogs, veneer logs and logs for sleepers | 42 | 63 | 63 |
| Pulpwood | 31 | — | — |
| Other industrial wood | — | 90 | 90 |
| Fuel wood* | 1,264 | 1,248 | 1,234 |
| **Total*** | 1,337 | 1,401 | 1,387 |

* FAO estimates.
Source: FAO.

**SAWNWOOD PRODUCTION**
('000 cubic metres, incl. railway sleepers)

|  | 2000 | 2001 | 2002 |
|---|---|---|---|
| Coniferous (softwood) | 2 | 2* | 2* |
| Broadleaved (hardwood) | 46 | 40 | 24 |
| **Total** | 48 | 42* | 26* |

* FAO estimate.
**2003:** Production as in 2002 (FAO estimates).
Source: FAO.

## Fishing

('000 metric tons, live weight)

|  | 2001 | 2002 | 2003 |
|---|---|---|---|
| Capture | 256.4 | 305.1 | 223.4 |
| Snappers and jobfishes | 22.3 | 26.6 | 13.1 |
| Pacific thread herring | 29.0 | 48.2 | 55.7 |
| Pacific anchoveta | 129.1 | 160.4 | 78.6 |
| Skipjack tuna | 5.8 | 7.6 | 11.5 |
| Yellowfin tuna | 12.3 | 20.4 | 28.7 |
| Marine fishes | 52.4 | 41.5 | 12.8 |
| Aquaculture | 3.1 | 4.3 | 6.2 |
| **Total catch** | 259.5 | 309.4 | 229.7 |

Note: Figures exclude crocodiles. The number of spectacled caimans caught was: 11,700 in 2001; 13,298 in 2002; 14,694 in 2003.
Source: FAO.

## Industry

**SELECTED PRODUCTS**
('000 metric tons, unless otherwise indicated)

|  | 1999 | 2000 | 2001 |
|---|---|---|---|
| Salt* | 22.5 | 22.5 | 22.5 |
| Sugar | 176.7 | 156.0 | 146.3 |
| Beer (million litres) | 146.1 | 139.9 | 126.7 |
| Evaporated, condensed and powdered milk | 23.8 | 27.1 | 26.3 |
| Frozen fish | 18.1 | 11.4 | 22.6 |
| Footwear ('000 pairs) | 507.8 | 242.1 | 112.2 |
| Motor spirit—petrol | 298 | 274 | 281 |
| Electricity (million kWh, net) | 4,929 | 4,836 | 5,124 |

**2002:** Salt 22.5*; Sugar 149.9; Evaporated, condensed and powdered milk 29.7.
* Estimated data from US Geological Survey.
Source (unless otherwise specified): mainly UN, *Industrial Commodity Statistics Yearbook*.

## Finance

**CURRENCY AND EXCHANGE RATES**
**Monetary Units**
100 centésimos = 1 balboa (B).
**Sterling, Dollar and Euro Equivalents** (31 May 2005)
£1 sterling = 1.818 balboas;
US $1 = 1.000 balboas;
€1 = 1.233 balboas;
100 balboas = £55.00 = $100.00 = €81.10.
**Exchange Rate:** The balboa's value is fixed at par with that of the US dollar.

**BUDGET**
(million US $)

| Revenue* | 2000 | 2001 | 2002† |
|---|---|---|---|
| Tax revenue | 1,120 | 1,039 | 1,051 |
| Income tax | 494 | 454 | 452 |
| Taxes on foreign trade | 202 | 172 | 184 |
| Taxes on domestic transactions | 371 | 351 | 359 |
| Other direct taxes | 53 | 61 | 56 |
| Other current revenue | 989 | 990 | 926 |
| Panama Canal | 142 | 133 | 152 |
| Transfers from balance of public sector | 341 | 252 | 212 |
| Capital revenue | 1 | 59 | 90 |
| **Total revenue** | 2,110 | 2,088 | 2,067 |

| Expenditure | 2000 | 2001 | 2002† |
|---|---|---|---|
| Current expenditure | 1,964 | 1,967 | 1,973 |
| Wages and salaries‡ | 641 | 672 | 696 |
| Goods and services | 178 | 169 | 147 |
| Pensions and transfers | 616 | 560 | 568 |
| Social Security Agency | 346 | 298 | 280 |
| Interest payments | 485 | 500 | 508 |
| Other current expenditure | 45 | 66 | 54 |
| Capital expenditure | 277 | 320 | 332 |
| Fixed capital formation | 223 | 245 | 284 |
| Capital transfers | 54 | 75 | 49 |
| **Total** | 2,241 | 2,287 | 2,305 |

* Excluding grants received (US $ million): 4 in 2000; 1 in 2001.
† Preliminary figures.
‡ Including severance payments and payments of wages outstanding.

**INTERNATIONAL RESERVES**
(US $ million at 31 December*)

|  | 2002 | 2003 | 2004 |
|---|---|---|---|
| IMF special drawing rights | 1.1 | 0.8 | 0.9 |
| Reserve position in IMF | 16.1 | 17.6 | 18.4 |
| Foreign exchange | 1,165.7 | 992.5 | 611.4 |
| **Total** | 1,182.8 | 1,011.0 | 630.6 |

* Excludes gold, valued at US $476,000 in 1991–93.
Source: IMF, *International Financial Statistics*.
Note: US treasury notes and coins form the bulk of the currency in circulation in Panama.

**COST OF LIVING**
(Consumer Price Index, Panamá (Panama City); base: 1987 = 100)

|  | 2001 | 2002 | 2003 |
|---|---|---|---|
| Food (incl. beverages) | 112.8 | 112.1 | 113.5 |
| Rent, fuel and light | 122.1 | 125.8 | 129.5 |
| Clothing (incl. footwear) | 113.5 | 113.2 | 114.5 |
| Medical and health care | 126.4 | 124.5 | 122.6 |
| Transport and communications | 106.6 | 113.4 | 114.5 |
| **All items** (incl. others) | 114.3 | 115.5 | 117.1 |

# PANAMA

*Statistical Survey*

## NATIONAL ACCOUNTS
(million balboas at current prices)

### National Income and Product

| | 2001 | 2002 | 2003 |
|---|---|---|---|
| Compensation of employees | 4,125.0 | 4,180.0 | 4,258.7 |
| Operating surplus | 5,911.6 | 6,162.8 | 6,490.9 |
| **Domestic factor incomes** | 10,036.6 | 10,342.8 | 10,749.6 |
| Consumption of fixed capital | 958.2 | 981.7 | 1,004.2 |
| **Gross domestic product (GDP) at factor cost** | 10,994.8 | 11,324.5 | 11,753.8 |
| Indirect taxes | 875.7 | 1,002.8 | 1,154.8 |
| *Less* Subsidies | 63.0 | 55.0 | 46.2 |
| **GDP in purchasers' values** | 11,807.5 | 12,272.3 | 12,862.4 |
| *Less* Net factor income paid to the rest of the world | 910.1 | 565.7 | 1,047.5 |
| **Gross national product** | 10,897.4 | 11,706.6 | 11,814.9 |
| *Less* Consumption of fixed capital | 958.2 | 981.7 | 1,004.2 |
| **National income in market prices** | 9,939.2 | 10,724.9 | 10,810.7 |
| Other current transfers from abroad (net) | 161.1 | 170.9 | 175.2 |
| **National disposable income** | 10,100.3 | 10,895.8 | 10,985.9 |

### Expenditure on the Gross Domestic Product

| | 2001 | 2002 | 2003 |
|---|---|---|---|
| Government final consumption expenditure | 1,646.1 | 1,819.3 | 1,845.3 |
| Private final consumption expenditure | 7,276.3 | 7,885.2 | 7,784.6 |
| Increase in stocks | 288.3 | 268.1 | 249.4 |
| Gross fixed capital formation | 1,794.2 | 1,664.8 | 2,207.3 |
| **Total domestic expenditure** | 11,004.9 | 11,637.4 | 12,086.6 |
| Exports of goods and services | 8,586.5 | 8,278.9 | 8,231.4 |
| *Less* Imports of goods and services | 7,783.9 | 7,643.9 | 7,455.6 |
| **GDP in purchasers' values** | 11,807.5 | 12,272.4 | 12,862.4 |
| **GDP at constant 1996 prices** | 11,436.2 | 11,691.1 | 12,196.2 |

Source: IMF *International Financial Statistics*.

### Gross Domestic Product by Economic Activity
(million balboas at constant 1996 prices)

| | 2002 | 2003* | 2004† |
|---|---|---|---|
| Agriculture, hunting, forestry and fishing | 877.5 | 918.0 | 940.4 |
| Mining and quarrying | 89.3 | 112.3 | 150.2 |
| Manufacturing | 999.4 | 983.1 | 1,008.6 |
| Electricity, gas and water | 381.5 | 388.1 | 422.1 |
| Construction | 390.0 | 516.8 | 599.6 |
| Wholesale and retail trade | 1,668.1 | 1,656.1 | 1,839.4 |
| Hotels and restaurants | 277.4 | 307.7 | 333.0 |
| Transport, storage and communications | 1,803.2 | 2,047.6 | 2,285.6 |
| Financial services | 1,011.9 | 949.6 | 925.0 |
| Insurance, real estate and business services | 1,867.5 | 1,936.2 | 2,058.2 |
| Social welfare and private health | 118.7 | 126.2 | 133.7 |
| Education (private) | 84.8 | 85.7 | 87.6 |
| Other community, social and personal service activities | 388.4 | 383.9 | 409.6 |
| Government services | 1,171.3 | 1,201.4 | 1,219.4 |
| Employed persons in private households | 93.1 | 95.0 | 99.6 |
| **Sub-total** | 11,222.1 | 11,707.6 | 12,512.0 |
| *Less* Financial intermediation services indirectly measured | 288.2 | 307.0 | 320.1 |
| **Gross value added in basic prices** | 10,933.9 | 11,400.6 | 12,191.9 |
| Import duties and other taxes | 809.0 | 838.5 | 812.0 |
| *Less* Grants and subsidies | 51.8 | 42.9 | 46.5 |
| **GDP in market prices** | 11,691.1 | 12,196.2 | 12,957.4 |

* Provisional figures.
† Preliminary estimates.

## BALANCE OF PAYMENTS
(US $ million)*

| | 2002 | 2003 | 2004 |
|---|---|---|---|
| Exports of goods f.o.b. | 5,283.8 | 5,048.9 | 5,885.6 |
| Imports of goods f.o.b. | –6,460.2 | –6,161.6 | –7,470.9 |
| **Trade balance** | –1,176.4 | –1,112.7 | –1,585.3 |
| Exports of services | 2,290.6 | 2,556.6 | 2,725.8 |
| Imports of services | –1,263.6 | –1,302.5 | –1,430.4 |
| **Balance on goods and services** | –149.4 | 141.4 | –289.9 |
| Other income received | 919.5 | 769.5 | 786.0 |
| Other income paid | –1,136.3 | –1,589.4 | –1,828.4 |
| **Balance on goods, services and income** | –366.2 | –678.5 | –1,332.3 |
| Current transfers received | 242.0 | 301.8 | 323.3 |
| Current transfers paid | –29.5 | –60.5 | –95.3 |
| **Current balance** | –153.7 | –437.2 | –1,104.3 |
| Direct investment from abroad | 56.9 | 791.5 | 1,012.3 |
| Portfolio investment assets | 159.7 | –59.3 | –605.2 |
| Portfolio investment liabilities | 367.5 | 139.6 | 775.9 |
| Other investment assets | 3,342.3 | 464.1 | –889.3 |
| Other investment liabilities | –3,660.0 | –1,310.5 | 713.8 |
| Net errors and omissions | 729.8 | 257.3 | –299.4 |
| **Overall balance** | 842.5 | –154.5 | –396.1 |

* Including the transactions of enterprises operating in the Colón Free Zone.

Source: IMF, *International Financial Statistics*.

# External Trade

## PRINCIPAL COMMODITIES
(US $ million)

| Imports c.i.f. | 1999 | 2000 | 2001 |
|---|---|---|---|
| Food products, beverages and tobacco | 374.8 | 373.3 | 342.3 |
| Mineral products | 408.8 | 628.8 | 612.8 |
| Petroleum, petroleum products and related materials | 390.4 | 617.6 | 605.4 |
| Chemicals and related products | 382.1 | 380.3 | 361.0 |
| Medicinal and pharmaceutical products | 125.1 | 128.4 | 129.3 |
| Manufactured goods | 533.7 | 485.3 | 398.8 |
| Machinery and transport equipment | 1,266.5 | 994.7 | 813.4 |
| Office machines and automatic data-processing equipment | 120.6 | 106.6 | 93.4 |
| Telecommunications, sound recording and reproducing equipment | 156.1 | 131.8 | 161.2 |
| Other electrical machinery, apparatus, appliances and parts | 161.2 | 146.5 | 126.6 |
| Road vehicles | 477.8 | 337.3 | 228.6 |
| **Total (incl. others)** | 3,515.2 | 3,378.3 | 2,963.6 |

**Total imports c.i.f.** (US $ million): 3,069.9 in 2002; 3,069.1 in 2003 (preliminary).

| Exports f.o.b. | 2001 | 2002 | 2003* |
|---|---|---|---|
| Sugar | 13.9 | 15.1 | 12.8 |
| Bananas | 122.2 | 109.4 | 105.2 |
| Melons | 25.6 | 40.7 | 47.3 |
| Coffee | 11.1 | 9.2 | 12.1 |
| Shrimps | 70.1 | 58.0 | 57.6 |
| Fresh and frozen fish and fillets (incl. yellowfin tuna) | 148.1 | 169.5 | 261.9 |
| Clothing | 14.3 | 13.3 | 11.2 |
| Petroleum | 191.5 | 148.7 | n.a. |
| **Total (incl. others)** | 809.5 | 755.7 | 808.0 |

* Preliminary figures.

# PANAMA

*Statistical Survey*

## PRINCIPAL TRADING PARTNERS
(US $ '000)

| Imports c.i.f. | 2001 | 2002 | 2003 |
|---|---|---|---|
| Canada | 26,129 | 20,035 | 20,192 |
| Colombia* | 167,889 | 179,627 | 123,191 |
| Costa Rica | 103,656 | 127,822 | 151,133 |
| Ecuador | 236,720 | 99,116 | 7,003 |
| Germany | 51,052 | 45,747 | 49,211 |
| Guatemala | 57,748 | 62,953 | 71,097 |
| Japan | 128,035 | 164,610 | 193,231 |
| Korea, Republic | 53,148 | 63,536 | 63,803 |
| Mexico | 118,346 | 112,440 | 119,155 |
| Netherlands Antilles† | 22,192 | 52,013 | 89,613 |
| Spain | 60,772 | 41,617 | 48,693 |
| Trinidad and Tobago | 1,876 | 38,551 | 19,550 |
| United Kingdom | 20,429 | 18,590 | 31,815 |
| USA | 965,096 | 1,016,191 | 1,066,132 |
| Venezuela | 154,104 | 127,675 | 84,877 |
| **Total** (incl. others) | **2,963,585** | **3,035,737** | **3,124,885** |

| Exports f.o.b. | 2001 | 2002 | 2003 |
|---|---|---|---|
| Belgium-Luxembourg | 36,515 | 32,778 | 24,340 |
| Colombia* | 6,199 | 5,749 | 8,574 |
| Costa Rica | 38,959 | 36,167 | 33,466 |
| Dominican Republic | 7,799 | 15,774 | 8,264 |
| Ecuador | 4,814 | 3,903 | 3,883 |
| El Salvador | 6,033 | 7,675 | 9,403 |
| Germany | 25,832 | 6,443 | 3,636 |
| Guatemala | 17,171 | 15,916 | 13,254 |
| Honduras | 20,075 | 33,560 | 13,362 |
| Hong Kong | 6,360 | 5,220 | 5,122 |
| Italy | 22,137 | 7,563 | 9,517 |
| Japan | 12,067 | 10,111 | 6,039 |
| Mexico | 15,397 | 14,821 | 12,060 |
| Nicaragua | 41,098 | 20,900 | 24,815 |
| Portugal | 13,827 | 28,627 | 27,469 |
| Puerto Rico | 13,434 | 12,506 | 12,954 |
| Spain | 18,123 | 21,867 | 45,593 |
| Sweden | 30,306 | 44,239 | 48,262 |
| Taiwan | 11,699 | 4,495 | 6,618 |
| USA | 388,607 | 348,266 | 402,604 |
| **Total** (incl. others) | **809,537** | **755,747** | **798,747** |

* Excluding San Andrés island.
† Curaçao only.

## Transport

### RAILWAYS
(traffic)

| | 2001 | 2002 |
|---|---|---|
| Passenger-km (million)* | 24,576 | 35,693 |
| Freight ton-km (million)† | 4,896 | 20,665 |

* Panama Railway and National Railway of Chiriquí.
† Panama Railway only.

Source: UN, *Statistical Yearbook*.

### ROAD TRAFFIC
(motor vehicles in use)

| | 2000 | 2001 | 2002 |
|---|---|---|---|
| Cars | 223,433 | 219,372 | 224,504 |
| Buses and coaches | 16,865 | 15,558 | 16,371 |
| Lorries and vans | 75,454 | 73,139 | 74,247 |

Source: IRF, *World Road Statistics*.

## SHIPPING

### Merchant Fleet
(registered at 31 December)

| | 2002 | 2003 | 2004 |
|---|---|---|---|
| Number of vessels | 6,247 | 6,302 | 6,477 |
| Total displacement ('000 grt) | 124,729.1 | 125,721.7 | 131,451.7 |

Source: Lloyd's Register-Fairplay, *World Fleet Statistics*.

### International Sea-borne Freight Traffic
('000 metric tons)

| | 2001 | 2002 | 2003 |
|---|---|---|---|
| Goods loaded | 108,456 | 110,556 | 99,516 |
| Goods unloaded | 84,864 | 99,288 | 76,152 |

### Panama Canal Traffic

| | 2002 | 2003 | 2004 |
|---|---|---|---|
| Transits | 13,185 | 13,154 | 14,035 |
| Cargo (million long tons) | 187.8 | 188.3 | 200.2 |

Source: Panama Canal Authority.

### CIVIL AVIATION
(traffic on scheduled services)

| | 2001 | 2002 | 2003 |
|---|---|---|---|
| Kilometres flown (million) | 40 | 38 | 43 |
| Passengers carried ('000) | 1,115 | 1,048 | 1,264 |
| Passengers-km (million) | 3,004 | 2,974 | 3,371 |
| Total ton-km (million) | 25 | 22 | 20 |

Source: UN Economic Commission for Latin America and the Caribbean, *Statistical Yearbook*.

## Tourism

### VISITOR ARRIVALS BY COUNTRY OF ORIGIN

| | 2001 | 2002 | 2003 |
|---|---|---|---|
| Argentina | 9,403 | 7,196 | 9,363 |
| Canada | 14,652 | 15,897 | 15,167 |
| Chile | 6,006 | 6,444 | 7,116 |
| Colombia | 80,972 | 88,049 | 93,821 |
| Costa Rica | 56,846 | 57,550 | 54,290 |
| Dominican Republic | 8,843 | 7,702 | 6,291 |
| Ecuador | 19,171 | 22,049 | 20,357 |
| El Salvador | 7,585 | 7,992 | 9,837 |
| Germany | 4,494 | 4,441 | 4,941 |
| Guatemala | 12,937 | 14,205 | 14,284 |
| Honduras | 9,012 | 8,508 | 9,846 |
| Italy | 4,510 | 7,213 | 5,378 |
| Jamaica | 8,491 | 4,720 | 4,620 |
| Mexico | 20,447 | 20,784 | 24,398 |
| Nicaragua | 12,905 | 12,045 | 11,755 |
| Peru | 8,378 | 10,643 | 10,789 |
| Puerto Rico | 6,666 | 7,254 | 7,526 |
| Spain | 7,947 | 9,298 | 9,549 |
| USA | 112,585 | 116,103 | 132,898 |
| Venezuela | 17,606 | 14,357 | 14,827 |
| **Total** (incl. others) | **482,040** | **499,643** | **534,208** |

**Tourism receipts** (US $ million, incl. passenger transport): 674 in 2001; 721 in 2002; 809 in 2003.

Sources: World Tourism Organization.

## Communications Media

|  | 2001 | 2002 | 2003 |
|---|---|---|---|
| Telephones ('000 main lines in use) | 430 | 377 | 387 |
| Mobile cellular telephones ('000 subscribers) | 600 | 475 | 834 |
| Personal computers ('000 in use) | 110 | 110 | n.a. |
| Internet users ('000) | 90 | 120 | n.a. |

**Radio receivers** ('000 in use): 815 in 1997.

**Television receivers** ('000 in use): 550 in 2000.

**Daily newspapers:** 7 in 1997.

Sources: UNESCO, *Statistical Yearbook*; UN, *Statistical Yearbook*; International Telecommunication Union.

## Education

(2001/02, unless otherwise indicated)

|  | Institutions | Teachers | Pupils |
|---|---|---|---|
| Pre-primary | 1,662 | 3,466 | 64,929 |
| Primary | 3,116 | 15,058 | 408,249 |
| Secondary | 442 | 12,327 | 244,097 |
| University | 24 | 4,972* | 117,624 |

* 2000 figure.

Sources: Ministry of Education; UNESCO, *Statistical Yearbook*.

**Adult literacy rate** (UNESCO estimates): 92.3% (males 92.9%; females 91.7%) in 2002 (Source: UN Development Programme, *Human Development Report*).

# Directory

## The Constitution

Under the terms of the amendments to the Constitution, implemented by the adoption of Reform Acts No. 1 and No. 2 in October 1978, and by the approval by referendum of the Constitutional Act in April 1983, the 67 (later 78) members of the unicameral Asamblea Legislativa (Legislative Assembly) are elected by popular vote every five years. Executive power is exercised by the President of the Republic, who is also elected by popular vote for a term of five years. Two Vice-Presidents are elected by popular vote to assist the President. The President appoints the Cabinet. The armed forces are barred from participating in elections.

## The Government

### HEAD OF STATE

**President:** MARTÍN TORRIJOS ESPINO (took office 1 September 2004).
**First Vice-President:** SAMUEL LEWIS NAVARRO.

### THE CABINET
(July 2005)

**Minister of the Interior and Justice:** HÉCTOR ALEMÁN.
**Minister of Foreign Affairs:** SAMUEL LEWIS NAVARRO.
**Minister of Public Works:** CARLOS VALLARINO.
**Minister of Finance and the Treasury and Chairman of the Panama Canal Authority:** RICAURTE VÁSQUEZ MORALES.
**Minister of Agricultural Development:** LAURENTINO CORTIZO.
**Minister of Commerce and Industry:** ALEJANDRO FERRER.
**Minister of Public Health:** CAMILO ALLEYNE.
**Minister of Labour and Social Welfare:** REYNALDO RIVERA.
**Minister of Education:** JUAN BOSCO BERNAL.
**Minister of Housing:** BALBINA HERRERA.
**Minister of the Presidency:** UBALDINO REAL.
**Minister of Youth, Women, Family and Childhood:** LEONOR CALDERÓN.

### MINISTRIES

**Office of the President:** Palacio Presidencial, Valija 50, Panamá 1; tel. 227-4062; fax 227-0076; internet www.presidencia.gob.pa.

**Ministry of Agricultural Development:** Edif. 576, Calle Manuel E. Melo, Altos de Curundú, Apdo 5390, Panamá 5; tel. 232-6254; fax 232-5044; e-mail infomida@mida.gob.pa; internet www.mida.gob.pa.

**Ministry of Commerce and Industry:** El Paical, 3°, Avda Ricardo J. Alfaro, Plaza Edison, Apdo 9658, Panamá 4; tel. 360-0600; fax 360-0663; e-mail uti@mici.gob.pa; internet www.mici.gob.pa.

**Ministry of Education:** Edif. Poli y Los Rios, Avda Justo Arosemena, Calles 26 y 27, Apdo 2440, Panamá 3; tel. 211-4400; fax 262-9087; e-mail meduc@meduc.gob.pa; internet www.meduc.gob.pa.

**Ministry of Finance and the Treasury:** Edif. Ogawa, Vía España, Apdo 5245, Panamá 5; tel. 269-4369; fax 264-7755; e-mail mhyt@mhyt.gob.pa; internet www.mef.gob.pa.

**Ministry of Foreign Affairs:** Altos de Ancón, Complejo Narciso Garay, Panamá 4; tel. 227-0013; fax 227-4725; e-mail prensa@mire.gob.pa; internet www.mire.gob.pa.

**Ministry of Housing:** Avda Ricardo J. Alfaro, Edif. Plaza Edison, 4°, Apdo 5228, Panamá 5; tel. 279-9200; fax 321-0028; e-mail webmaster@mivi.gob.pa; internet www.mivi.gob.pa.

**Ministry of the Interior and Justice:** Avda Central, entre calle 2da y 3era, San Felipe, Apdo 1628, Panamá 1; tel. 212-2000; fax 212-2126; e-mail informa@gobiernoyjusticia.gob.pa; internet www.gobiernoyjusticia.gob.pa.

**Ministry of Labour and Social Welfare:** Avda Ricardo J. Alfaro, Plaza Edison, 5°, Apdo 2441, Panamá 3; tel. 260-9087; fax 260-4466; e-mail mitrabs2@sinfo.net; internet www.mitradel.gob.pa.

**Ministry of the Presidency:** Palacio de Las Garzas, Corregimiento de San Felipe, Apdo 2189, Panamá 1; tel. 227-9663; fax 227-4119; e-mail ofasin@presidencia.gob.pa; internet www.presidencia.gob.pa.

**Ministry of Public Health:** Apdo 2048, Panamá 1; tel. 225-6080; fax 212-9202; e-mail ministro@minsa.gob.pa; internet www.minsa.gob.pa.

**Ministry of Public Works:** Edif. 1019, Curundú, Zona 1, Apdo 1632, Panamá 1; tel. 232-5333; fax 232-8776; e-mail info@mop.gob.pa; internet www.mop.gob.pa.

**Ministry of Youth, Women, Family and Childhood:** Avda Ricardo J. Alfaro, 4°, Edison Plaza, Apdo 680-50, El Dorado, Panamá; tel. 279-0702; fax 279-0665; e-mail minjumnfa@sinfo.net; internet www.minjumnfa.gob.pa.

## President and Legislature

### PRESIDENT

Election, 2 May 2004

| Candidate | Votes | % of votes |
|---|---|---|
| Martín Torrijos Espino (Patria Nueva*) | 711,447 | 47.44 |
| Guillermo Endara Galimany (Partido Solidaridad) | 462,766 | 30.86 |
| José Miguel Alemán (Visión de País†) | 245,845 | 16.39 |
| Ricardo A. Martinelli Berrocal (Cambio Democrático) | 79,595 | 5.31 |
| **Total** | 1,499,653 | 100.00 |

* Electoral alliance comprising the Partido Revolucionario Democrático and the Partido Popular.
† Electoral alliance comprising the Partido Arnulfista, the Movimiento Liberal Republicano Nacionalista and the Partido Liberal Nacional.

### ASAMBLEA LEGISLATIVA
(Legislative Assembly)

**President:** JERRY V. WILSON N.

# PANAMA

### General Election, 2 May 2004

| Affiliation/Party | % of votes | Seats |
|---|---|---|
| Patria Nueva | | |
|   Partido Revolucionario Democrático (PRD) | 37.9 | 42 |
|   Partido Popular (PP) | 6.0 | 1 |
| Visión de País | | |
|   Partido Arnulfista (PA) | 19.3 | 16 |
|   Movimiento Liberal Republicano Nacionalista (MOLIRENA) | 8.6 | 4 |
|   Partido Liberal Nacional (PLN) | 5.2 | 3 |
| Partido Solidaridad | 15.7 | 9 |
| Cambio Democrático (CD) | 7.4 | 3 |
| **Total** | **100.0** | **78** |

## Political Organizations

**Cambio Democrático (CD):** Parque Lefevre, Plaza Carolina, arriba de la Juguetería del Super 99, Panamá; tel. 217-2643; fax 217-2645; e-mail cambio.democrático@hotmail.com; formally registered 1998; Pres. RICARDO A. MARTINELLI BERROCAL; Sec.-Gen. GIACOMO TAMBURELLI.

**Movimiento Liberal Republicano Nacionalista (MOLIRENA):** Calle Venezuela, Casa No 5, entre Vía España y Calle 50, Panamá; tel. 213-5928; fax 265-6004; formally registered 1982; conservative; contested the 2004 elections as part of the Visión de País electoral alliance; Pres. JESÚS L. ROSAS ABREGO; Sec.-Gen. MIGUEL CÁRDENAS SANDOVAL.

**Movimiento Nacional por la Defensa de la Soberanía (Monadeso):** Panamá; left-wing umbrella group; Leader CONRADO SANJUR.

**Partido Panameñista (PP):** Avda Perú y Calle 38E, No 37–41, al lado de Casa la Esperanza, Apdo 9610, Panamá 4; tel. 227-1267; f. 1990 by Arnulfista faction of the Partido Panameñista Auténtico as Partido Arnulfista (PA); contested the 2004 elections as part of the Visión de País electoral alliance; name changed as above in Jan. 2005; Pres. MARCO AMEGLIO; Sec.-Gen. CARLOS RAÚL PIAD H.

**Partido Liberal Nacional (PLN):** Vía Fernandez de Córdoba, Vista Hermosa, Plaza Córdoba, antigua Ersa, Local 6–7, Panamá; tel. 229-7523; fax 229-7524; e-mail pln@sinfo.net; internet www.sinfo.net/liberal-nacional; f. 1979; mem. of Liberal International, and founding mem. of Federación Liberal de Centroamérica y el Caribe (FELICA); contested the 2004 elections as part of the Visión de País electoral alliance; 40,645 mems; Pres. ANÍBAL GALINDO; Sec. ABRAHAM WILLIAMS.

**Partido Popular (PP):** Avda Perú, frente al Parque Porras, Apdo 6322, Panamá 5; tel. 227-3204; fax 227-3944; e-mail pdc@cwpanama.net; f. 1960 as Partido Demócrata Cristiano, name changed as above in Sept. 2001; contested the 2004 elections as part of the Patria Nueva electoral alliance; Pres. RUBÉN AROSEMENA VALDÉS.

**Partido Revolucionario Democrático (PRD):** Calle 42 Bella Vista, entre Avda Perú y Avda Cuba, bajando por el teatro Bella Vista, Panamá 9; tel. 225-1050; e-mail prdpanama@yahoo.com; f. 1979; supports policies of late Gen. Omar Torrijos Herrera; combination of Marxists, Christian Democrats and some business interests; contested the 2004 elections as part of the Patria Nueva electoral alliance; Pres. HUGO H. GUIRAUD; Sec.-Gen. MARTÍN TORRIJOS ESPINO.

**Partido Solidaridad:** Edif. Maheli, Avda Ramón Arias, esq. con la Vía Transístmica, Panamá; tel. 261-2966; fax 261-5083; formally registered 1993; Pres. SAMUEL LEWIS GALINDO; Vice-Pres. JOSÉ RAÚL MULINO; Sec.-Gen. JORGE RICARDO FABREGA.

**Vanguardia Moral de la Patria:** Vía España, esq. con Vía Porras, Panamá; tel. 212-7300; f. 2004; Pres. GUILLERMO ENDARA; Gen. Sec. Dr JOHN HOGER CASTRELLON.

## Diplomatic Representation

### EMBASSIES IN PANAMA

**Argentina:** Edif. del Banco de Iberoamérica, 7°, Avda 50 y Calle 53, Apdo 1271, Panamá 1; tel. 264-6561; fax 269-5331; e-mail embargen@c-com.net.pa; Ambassador ERNESTO MARIO PFIRTER.

**Bolivia:** Calle Eric Arturo del Valle, Bella Vista 1, Panamá; tel. 269-0274; fax 264-3868; e-mail emb_bol_pan@cwpanama.net; internet www.emboliviapanama.com.pa; Ambassador CARLOS AGUIRRE BASTOS.

**Brazil:** Edif. El Dorado, 1°, Calle Elvira Méndez y Avda Ricardo Arango, Urb. Campo Alegre, Apdo 4287, Panamá 5; tel. 263-5322; fax 269-6316; e-mail embrasil@embrasil.org.pa; Ambassador LUÍZ TUPY CALDAS DE MOURA.

**Canada:** Edif. World Trade Center, Galería Comercial, 1°, Urb. Marbella, Apdo 0832-2446, Panamá; tel. 264-7115; fax 263-8083; e-mail panam@international.gc.ca; internet www.dfait-maeci.gc.ca/panama; Ambassador DAVID ADAM.

**Chile:** Edif. Banco de Boston, 11°, Calle Elvira Méndez y Vía España, Apdo 7341, Panamá 5; tel. 223-9748; fax 263-5530; e-mail echilepa@cw.panama.net; internet www.embachilepanama.com; Ambassador JAIME ROCHA MANRIQUE.

**Colombia:** Edif. World Trade Center, Of. 1802, Calle 53, Urb. Marbella, Panamá; tel. 264-9644; fax 223-1134; e-mail epanama@minrelext.gov.co; Ambassador GINA BENEDETTI DE VELEZ.

**Costa Rica:** Edif. Plaza Omega, 3°, Calle Samuel Lewis, Apdo 8963, Panamá; tel. 264-2980; fax 264-4057; e-mail embarica@cwp.net.pa; Ambassador VERA VIOLETA CASTRO CASTRO.

**Cuba:** Avda Cuba y Ecuador 33, Apdo 6-2291, Bellavista, Panamá; tel. 227-5277; fax 225-6681; e-mail embacuba@cableonda.net; Ambassador CARLOS RAFAEL ZAMORA RODRÍGUEZ (recalled in August 2004).

**Dominican Republic:** Casa 40A, Calle 75, Apdo 6250, Panamá 5; tel. 270-3884; fax 270-3886; e-mail embajdom@sinfo.net; Ambassador RODOLFO LEYBA POLANCO.

**Ecuador:** Edif. Torre 2000, 6°, Calle 50, Marbella, Bellavista, Panamá; e-mail eecuador@cwpanama.net; tel. 264-2654; fax 223-0159; Ambassador ZOILA NAVAS PERALTA.

**Egypt:** Calle 55, No 15, El Cangrejo, Apdo 7080, Panamá 5; tel. 263-5020; fax 264-8406; Ambassador SAFIA IBRAHIM AMIEN.

**El Salvador:** Edif. Metropolis, 4°, Avda Manuel Espinosa Batista, Panamá; tel. 223-3020; fax 264-1433; e-mail embasalva@cwpanama.net; Ambassador AÍDA ELENA MINERO.

**France:** Plaza de Francia 1, Las Bovedas, San Felipe, Apdo 869, Panamá 1; tel. 211-6200; fax 211-6201; internet www.ambafrance-pa.org; Ambassador CHRISTOPHE PHILBERT.

**Germany:** Edif. World Trade Center, 20°, Calle 53E, Marbella, Apdo 0832-0536, Panamá 5; tel. 263-7733; fax 223-6664; e-mail germpanama@cwp.net.pa; internet www.panama.diplo.de; Ambassador BORUSSO VON BLÜCHER.

**Guatemala:** Edif. Altamira, Of. 925, Vía Argentina, El Cangrejo, Corregimiento de Bella Vista, Panamá 9; tel. 269-3475; fax 223-1922; Ambassador GUISELA ATALIDA GODÍNEZ DE GUTIÉRREZ.

**Haiti:** Edif. Dora Luz, 2°, Calle 1, El Cangrejo, Apdo 442, Panamá 9; tel. 269-3443; fax 223-1767; e-mail ambhaiti@panama.c-com.net; Chargé d'affaires a.i. JOSEPH ETIENNE.

**Holy See:** Punta Paitilla, Avda Balboa y Vía Italia, Apdo 4251, Panamá 5 (Apostolic Nunciature); tel. 269-2102; fax 264-2116; e-mail nuncio@sinfo.net; Apostolic Nuncio Most Rev. GIAMBATTISTA DIQUATTRO (Titular Archbishop of Giru Mons).

**Honduras:** Edif. Bay Mall, 1°, 112 Avda Balboa, Apdo 8704, Panamá 5; tel. 264-5513; fax 224-5513; e-mail ehpan@cableonda.net; Ambassador RAÚL CARDONA LÓPEZ.

**India:** Avda Federico Boyd y Calle 51, Belle Vista, Apdo 8400, Panamá 7; tel. 264-3043; fax 264-2855; e-mail indempan@c-com.net.pa; internet www.indempan.org; Ambassador ASHOK TOMAR.

**Italy:** Torre Banco Exterior, 25°, Avda Balboa, Apdo 2369, Panamá 9; tel. 225-8950; fax 227-4906; e-mail panitamb@cwp.net.pa; Ambassador MARCO ROCCA.

**Japan:** Calle 50 y 60E, Obarrio, Apdo 1411, Panamá 1; tel. 263-6155; fax 263-6019; e-mail taiship2@sinfo.net; internet www.panama.emb-japan.go.jp; Ambassador SHUJI SHIMOKOJI.

**Korea, Republic:** Edif. Plaza, planta baja, Calle Ricardo Arias y Calle 51E, Campo Alegre, Apdo 8096, Panamá 7; tel. 264-8203; fax 264-8825; e-mail panama@mofat.go.kr; Ambassador TAE YOUNG MOON.

**Libya:** Avda Balboa y Calle 32 (frente al Edif. Atalaya), Apdo 6-894 El Dorado, Panamá; tel. 227-3342; fax 227-3886; Chargé d'affaires ABDULMAJID MILUD SHAHIN.

**Mexico:** Edif. Torre ADR, 10°, Avda Samuel Lewis y Calle 58, Urb. Obarrio, Corregimiento de Bella Vista, Panamá; tel. 263-4900; fax 263-5446; e-mail embamexpan@cwpanama.net; Ambassador JOSÉ IGNACIO PIÑA ROJAS.

**Nicaragua:** Quarry Heights, 16°, Ancon, Apdo 772, Zona 1, Panamá; tel. 211-2113; fax 211-2116; e-mail embapana@sinfo.net; Ambassador XAVIER ENRIQUE SARRIA ABAUNZA.

# PANAMA

**Peru:** Edif. World Trade Center, 12°, Calle 53 Marbella, Apdo 4516, Panamá 5; tel. 223-1112; fax 269-6809; e-mail embaperu@pananet.com; Ambassador JOSÉ ANTONIO BELLINA ACEVEDO.

**Russia:** Torre IBC, 10°, Avda Manuel Espinosa Batista, Apdo 6-4697, El Dorado, Panamá; tel. 264-1408; fax 264-1588; e-mail emruspan@sinfo.net; Ambassador EVGENY ROSTISLAVOVICH VORONIN.

**Spain:** Calle 53 y Avda Perú (frente a la Plaza Porras), Apdo 1857, Panamá 1; tel. 227-5122; fax 227-6284; e-mail embespa@cwpanama.net; Ambassador GERARDO ZALDÍVAR MIQUELARENA.

**United Kingdom:** Torre Swiss Bank, 4°, Urb. Marbella, Calle 53, Apdo 0816-07946, Panamá 1; tel. 269-0866; fax 223-0730; e-mail britemb@cwp.net.pa; Ambassador JAMES MALCOLM.

**USA:** Avda Balboa y Calle 38, Apdo 6959, Panamá 5; tel. 207-7000; fax 227-1964; e-mail usembisc@cwp.net.pa; Ambassador LINDA E. WATT; Ambassador WILLIAM ALAN EATON (from late 2005).

**Uruguay:** Edif. Los Delfines, Of. 8, Avda Balboa, Calle 50E Este, Apdo 8898, Panamá 5; tel. 264-2838; fax 264-8908; e-mail urupanam@cwpanama.net; Ambassador DOMINGO SCHIPANI.

**Venezuela:** Torre Hong Kong Bank, 5°, Avda Samuel Lewis, Apdo 661, Panamá 1; tel. 269-1014; fax 269-1916; e-mail embvenp@c-com.net.pa; Ambassador FLAVIO GRANADOS POMENTA.

## Judicial System

The judiciary in Panama comprises the following courts and judges: Corte Suprema de Justicia (Supreme Court of Justice), with nine judges appointed for a 10-year term; 10 Tribunales Superiores de Distrito Judicial (High Courts) with 36 magistrates; 54 Jueces de Circuito (Circuit Judges) and 89 Jueces Municipales (Municipal Judges).

Panama is divided into four judicial districts and has seven High Courts of Appeal. The first judicial district covers the provinces of Panamá, Colón, Darién and the region of Kuna Yala and contains two High Courts of Appeal, one dealing with criminal cases, the other dealing with civil cases. The second judicial district covers the provinces of Coclé and Veraguas and contains the third High Court of Appeal, located in Penonomé. The third judicial district covers the provinces of Chiriquí and Bocas del Toro and contains the fourth High Court of Appeal, located in David. The fourth judicial district covers the provinces of Herrera and Los Santos and contains the fifth High Court of Appeal, located in Las Tablas. Each of these courts deals with civil and criminal cases in their respective provinces. There are two additional special High Courts of Appeal. The first hears maritime, labour, family and infancy cases; the second deals with antitrust cases and consumer affairs.

### Corte Suprema de Justicia
Edif. 236, Ancón, Calle Culebra, Apdo 1770, Panamá 1; tel. 262-9833; e-mail prensa@organojudicial.gob.pa; internet www.organojudicial.gob.pa.

**President of the Supreme Court of Justice:** JOSÉ ANDRÉS TROYANO PEÑA.

**Attorney-General:** ANA MATILDE GÓMEZ DE RUILOBA.

## Religion

The Constitution recognizes freedom of worship and the Roman Catholic Church as the religion of the majority of the population.

### CHRISTIANITY

#### The Roman Catholic Church

For ecclesiastical purposes, Panama comprises one archdiocese, five dioceses, the territorial prelature of Bocas del Toro and the Apostolic Vicariate of Darién. There were an estimated 1,802,175 adherents at 31 December 2003, equivalent to 85% of the population.

**Bishops' Conference**
Conferencia Episcopal de Panamá, Secretariado General, Apdo 870933, Panamá 7; tel. 223-0075; fax 223-0042.

f. 1958; statutes approved 1986; Pres. Rt Rev. OSCAR MARIO BROWN JIMÉNEZ (Bishop of Santiago de Veraguas).

**Archbishop of Panamá:** Most Rev. JOSÉ DIMAS CEDEÑO DELGADO, Arzobispado Metropolitano, Calle 1a Sur Carrasquilla, Apdo 6386, Panamá 5; tel. 261-0002; fax 261-0820; e-mail asccn4@keops.utp.ac.pa.

### The Baptist Church

**The Baptist Convention of Panama** (Convención Bautista de Panamá): Apdo 6212, Panamá 5; tel. 264-5585; fax 259-5485; internet www.panamabaptist.com; f. 1959; Pres. Justo Pastor CASTILLO; Exec. Sec. ESMERALDA DE TUY.

### The Anglican Communion

Panama comprises one of the five dioceses of the Iglesia Anglicana de la Región Central de América.

**Bishop of Panama:** Rt Rev. JULIO MURRAY, Edif. 331-A, Calle Culebra, Apdo R, Balboa; tel. 212-0062; fax 262-2097; e-mail anglipan@sinfo.net; internet www.panama.anglican.org.

### BAHÁ'Í FAITH

**National Spiritual Assembly of the Bahá'ís:** Apdo 815-0143, Panamá 15; tel. 231-1191; fax 231-6909; e-mail panbahai@cwpanama.net; internet www.pa.bahai.org; mems resident in 550 localities; National Sec. EMELINA RODRÍGUEZ.

## The Press

### DAILIES

**Crítica Libre:** Vía Fernández de Córdoba, Apdo B-4, Panamá 9A; tel. 261-0575; fax 230-0132; e-mail esotop@epasa.com; internet www.critica.com.pa; f. 1925; morning; independent; Pres. ROSARIO ARIAS DE GALINDO; Dir JUAN PRITSIOLAS; circ. 40,000.

**La Estrella de Panamá:** Calle Alejandro Duque, Vía Transistmica y Frangipani, Panamá; tel. 222-0900; fax 227-1026; f. 1853; morning; independent; Pres. ALEJANDRO A. DUQUE; Editor JAMES APARICIO; circ. 20,000.

**El Panamá América:** Vía Ricardo J. Alfaro, al lado de la USMA, Apdo B-4, Panamá 9A; tel. 230-1666; fax 230-1035; e-mail director@epasa.com; internet www.epasa.com; f. 1958; morning; independent; Pres. ROSARIO ARIAS DE GALINDO; Editor OCTAVIO AMAT; circ. 25,000.

**La Prensa:** Avda 12 de Octubre, Hato Pintado, Apdo 6-4586, El Dorado, Panamá; tel. 222-1222; fax 221-7328; e-mail editor@prensa.com; internet www.prensa.com; f. 1980; morning; independent; closed by Govt 1988–90; Pres. RICARDO ALBERTO ARIAS; Editor GILBERTO SUCRE; circ. 38,000.

**La República:** Vía Fernández de Córdoba, Apdo B-4, Panamá 9A; tel. 261-0813; evening; circ. 5,000.

**El Siglo:** Calle 58 Obarrio, Panama; tel. 264-3921; fax 269-6954; e-mail redaccion@elsiglo.com; internet www.elsiglo.com; f. 1985; morning; acquired by Geo-Media SA in 2001; Pres. Dr NIVIA ROSSANA CASTRELLÓN; Editor OCTAVIO COGLEY; circ. 30,000.

**El Universal:** Avda Justo Arosemena, entre Calles 29 y 30, Panamá; tel. 225-7010; fax 225-6994; e-mail eluniver@sinfo.net; f. 1995; Pres. TOMÁS GERARDO DUQUE ZERR; Editor MILTON HENRÍQUEZ; circ. 16,000.

### PERIODICALS

**Análisis:** Edif. Señorial, Calle 50, Apdo 8038, Panamá 7; tel. 226-0073; fax 226-3758; e-mail mrognoni@revistaanalisis.com; internet www.revistaanalisis.com; monthly; economics and politics; Dir MARIO A. ROGNONI.

**El Camaleón:** Calle 58E, No 12, Apdo W, Panamá 4; tel. 269-3311; fax 269-6954; internet www.noticiasdenavarra.com/camaleon; weekly; satire; Editor JAIME PADILLA BÉLIZ; circ. 80,000.

**Diálogo Social:** Calle 71 Este Bis, Barrio Carrasquilla, Apdo 9A-192, Panamá; tel. 229-1542; fax 261-0215; e-mail ccspanama@cwpanama.net; internet www.hri.ca/partners/ccs; f. 1967; published by the Centro de Capacitación Social; monthly; religion, economics and current affairs; Pres. CELIA SANJUR; circ. 3,000.

**Dirección de Estadística y Censo:** Avda Balboa y Federico Boyd, Apdo 0816-01521, Panamá 5; tel. 210-4800; fax 210-4801; e-mail cgrdec@contraloria.gob.pa; internet www.contraloria.gob.pa/dec; f. 1941; published by the Contraloría General de la República; statistical survey in series according to subjects; Controller-General DANI KUZNIECKY; Dir of Statistics and Census LUIS ENRIQUE QUESADA.

**FOB Colón Free Zone:** Apdo 6-3287, El Dorado, Panamá; tel. 225-6638; fax 225-0466; e-mail focusint@sinfo.net; internet www.colonfreezone.com; annual; bilingual trade directory; circ. 35,000.

**Focus on Panama:** Apdo 6-3287, Panamá; tel. 225-6638; fax 225-0466; e-mail focusint@sinfo.net; internet www.focuspublicationsint.com; f. 1970; 2 a year; visitors' guide; separate English and Spanish editions; Dir KENNETH JONES; circ. 70,000.

# PANAMA

*Directory*

**Informativo Industrial:** Apdo 6-4798, El Dorado, Panamá 1; tel. 230-0482; fax 230-0805; monthly; organ of the Sindicato de Industriales de Panamá; Pres. GASPAR GARCÍA DE PAREDES.

**Maga:** Avda Edif. 2, Avda Manuel E. Batista, Panamá; tel. 223-2388; e-mail jlevi@keops.utp.ac.pa; monthly; literature, art and sociology; Dir ENRIQUE JARAMILLO LEVI.

**Sucesos:** Calle 58E, No 12, Apdo W, Panamá 4; tel. 269-3311; fax 269-6854; 3 a week; Editor JAIME PADILLA BÉLIZ; circ. 50,000.

## PRESS ASSOCIATION

**Sindicato de Periodistas de Panamá:** Avda Gorgas 287, Panamá; tel. 214-0163; fax 214-0164; e-mail sindiperpana@yahoo.com; f. 1949; Sec.-Gen. JAIME BEITIA.

## FOREIGN NEWS BUREAUX

**Agencia EFE** (Spain): Edif. Comosa, 22°, Avda Samuel Lewis y Calle Manuel María Icaza, Apdo 479, Panamá 9; tel. 223-9014; fax 264-8442; e-mail cenaca@sinfo.net; Bureau Chief HUGO FABIÁN ORTIZ DURÁN.

**Agenzia Nazionale Stampa Associata (ANSA)** (Italy): Edif. Banco de Boston, 17°, Vía España 601, Panamá; tel. 269-6623; e-mail panansa@sinfo.net; Dir LUIS LAMBOGLIA.

**Central News Agency** (Taiwan): Apdo 6-693, El Dorado, Panamá; tel. 223-8837; Correspondent HUANG KWANG CHUN.

**Deutsche Press-Agentur (dpa)** (Germany): Panamá; tel. 233-0396; fax 233-5393.

**Inter Press Service (IPS)** (Italy): Panamá; tel. 225-1673; fax 264-7033; Correspondent SILVIO HERNÁNDEZ.

**Reuters** (United Kingdom): Edif. Banco de Boston, Of. 504, Calle Elvira, Apdo 2523, Panamá 9; tel. 263-8285.

**Xinhua (New China) News Agency** (People's Republic of China): Vía Cincuentenario 48, Viña del Mar, Panamá 1; tel. 226-4501; Dir HU TAIRAN.

# Publishers

**Editora 'La Estrella de Panamá':** Avda 9A Sur 7-38, Apdo 159, Panamá 1; tel. 222-0900; f. 1853.

**Editora Panamá América (EPASA):** Avda Ricardo J. Alfaro, al lado de la USMA, Apdo B-4, Zona 9A, Panamá; tel. 230-7777; fax 230-0136; internet www.epasa.com; Pres. ROSARIO ARIAS DE GALINDO; Gen. Man. RAMÓN R. VALLARINO A.

**Editora Renovación, SA:** Vía Fernández de Córdoba, Apdo B-4, Panamá 9A; tel. 261-2300; newspapers; govt-owned; Exec. Man. ESCOLÁSTICO CALVO.

**Editora Sibauste, SA:** Panamá; tel. 269-0983; Dir ENRIQUE SIBAUSTE BARRÍA.

**Editorial Litográfica, SA (Edilito):** Panamá 10; tel. 224-3087; Pres. EDUARDO AVILES C.

**Editorial Universitaria:** Vía José de Fábrega, Panamá; tel. 264-2087; f. 1969; history, geography, law, sciences, literature.

**Focus Publications:** Apdo 6-3287, El Dorado, Panamá; tel. 225-6638; fax 225-0466; e-mail focusint@sinfo.net; internet www.focuspublicationsint.com; f. 1970; guides, trade directories and yearbooks.

**Industrial Gráfica, SA:** Vía España entre Calles 95 y 96 (al lado de Orange Crush), Apdo 810014, Panamá 10; tel. 224-3994; Pres. EDUARDO AVILES C.

**Publicar Centroamericana, SA:** Edif. Banco de Boston, 7°, Vía España 200 y Calle Elvira Méndez, Apdo 4919, Panamá 5; tel. 223-9655; fax 269-1964.

## GOVERNMENT PUBLISHING HOUSE

**Editorial Mariano Arosemena:** Instituto Nacional de Cultura, Apdo 662, Panamá 1; tel. 211-4000; fax 211-4016; e-mail comunicacion@inac.gob.pa; internet www.inac.gob.pa; f. 1974; division of National Institute of Culture (INAC); literature, history, social sciences, archaeology; Dir LESLIE MOCK.

# Broadcasting and Communications

## REGULATORY AUTHORITY

**Ente Regulador de los Servicios Públicos:** Vía España, Edif. Office Park, Apdo 4931, Panamá 5; tel. 278-4500; fax 278-4600; e-mail webmaster@ersp.gob.pa; internet www.enteregulador.gob.pa; f. 1996; state regulator with responsibility for television, radio, telecommunications, water and electricity; Pres. JOSÉ GALÁN PONCE; Dirs CARLOS RODRÍGUEZ BETHANCOURT, NILSON ESPINO.

## TELECOMMUNICATIONS

**Dirección Nacional de Medios de Comunicación Social:** Avda 7A Central y Calle 3A, Apdo 1628, Panamá 1; tel. 262-3197; fax 262-9495; Dir EDWIN CABRERA.

### Major Service Providers

**Cable & Wireless Panama:** Apdo 659, Panamá 9A; e-mail cwp@cwpanama.com; internet www.cwpanama.com.pa; 49% govt-owned, 49% owned by Cable & Wireless; major telecommunications provider; lost monopoly in Jan. 2003; Vice-Pres. ENRIQUE GARCÍA.

**Movistar:** Edif. Magna, Area Bancaria, Calle 51 Este y Manuel M. Icaza, Panamá; tel. 265-0955; f. 1996 as BellSouth Panamá SA; acquired by Telefónica Móviles, SA (Spain) in Oct. 2004; name changed as above in April 2005; mobile telephone services; Gen. Man. CLAUDIO HIDALGO.

**Optynex Telecom SA:** Calle 40 Bella Vista, Edif. 2-79, Entre Avda México y Avda Chile, Apdo 0832-2650, Panamá; tel. 380-0000; fax 380-0099; e-mail info@optynex.com; internet www.optynex.com; f. 2002; Gen. Man. ERIC MEYER.

## BROADCASTING

### Radio

**Asociación Panameña de Radiodifusión:** Apdo 7387, Estafeta de Paitilla, Panamá; tel. 263-5252; fax 226-4396; Pres. ALESSIO GRONCHI; Vice-Pres. RICARDO A. BUSTAMANTE.

In 2004 there were 109 AM (Medium Wave) and 181 FM stations registered in Panama. Most stations are commercial.

**La Mega 98.3 FM:** Casa 35, Calle 50 y 77 San Francisco, Panamá; tel. 270-3242; fax 226-1021; e-mail ventas@lamegapanama.com; internet www.lamegapanama.com; f. 2000.

**Omega Stereo:** Panamá; e-mail omegaste@omegastereo.com; internet www.omegastereo.com; f. 1981; Pres. GUILLERMO ANTONIO ADAMES.

**RPC Radio:** Panamá; e-mail rpcradio@medcom.com.pa; internet www.rpcradio.com; f. 1949; broadcasts news, sports and commentary; Man. LUIS EDUARDO QUIROS.

**SuperQ:** Panamá; tel. 227-0366; e-mail exitosa@sinfo.net; internet www.superqpanama.com; f. 1984; Pres. G. ARIS DE ICAZA.

**WAO 97.5:** Edif. Plaza 50, 2°, Calle 50 y Vía Brasil, Panamá; tel. 223-8348; fax 223-8351; internet www.wao975.com; Gen. Man. ROGELIO CAMPOS.

### Television

**Fundación para la Educación en la Televisión—FETV (Canal 5):** Vía Ricardo J. Alfaro, Apdo 6-7295, El Dorado, Panamá; tel. 230-8000; fax 230-1955; e-mail comentarios@fetv.org; internet www.fetv.org; f. 1992; Pres. JOSÉ DIMAS CEDEÑO DELGADO; Dir MANUEL SANTIAGO BLANQUER I PLANELLS; Gen. Man. MARÍA EUGENIA FONSECA M.

**Medcom:** Calle 50, No 6, Apdo 116, Panamá 8; tel. 210-6700; fax 210-6797; e-mail murrutia@medcom.com.pa; internet www.rpctv.com; f. 1998 by merger of RPC Televisión (Canal 4) and Telemetro (Canal 13); commercial; also owns Cable Onda 90, and RPC radio; Pres. FERNANDO ELETA; CEO NICOLÁS GONZÁLEZ-REVILLA.

**RTVE_Panama (Canal 11):** Curundu, Area Revertida, Avda Omar Torrijos, al lado del MOP, Panamá; tel. 232-8558; fax 223-2921; f. 1978; educational and cultural; Dir-Gen. CARLOS AGUILAR NAVARRO.

**Televisora Nacional—TVN (Canal 2):** Vía Bolívar, Apdo 6-3092, El Dorado, Panamá; tel. 236-2222; fax 236-2987; e-mail tvn@tvn-2.com; internet www.tvn-2.com; f. 1962; Dir JAIME ALBERTO ARIAS.

In 2005 there were 133 authorized television channels broadcasting in Panama.

# Finance

(cap. = capital; res = reserves; dep. = deposits; m. = million; amounts in balboas, unless otherwise stated)

## BANKING

In February 1998 new banking legislation was approved, providing for greater supervision of banking activity in the country, including the creation of a Banking Superintendency. In 2004 a total of 73 banks operated in Panama.

# PANAMA

**Superintendencia de Bancos** (Banking Superintendency): Torre HSBC, 18°, Apdo 2397, Panamá 1; tel. 206-7800; fax 264-9422; internet www.superbancos.gob.pa; f. 1970 as Comisión Bancaria Nacional (National Banking Commission) to license and control banking activities within and from Panamanian territory; Comisión Bancaria Nacional superseded by Superintendencia de Bancos in June 1998 with enhanced powers to supervise banking activity; Supt DELIA CÁRDENAS.

## National Bank

**Banco Nacional de Panamá:** Torre BNP, Vía España, Apdo 5220, Panamá 5; tel. 263-5151; fax 269-0091; e-mail bnpvalores@cwp.net.pa; internet www.banconal.com.pa; f. 1904; govt-owned; cap. 500.0m., dep. 2,846.8m., total assets 3,453.2m. (Dec. 2002); Dirs JOSÉ F. JELENSKY, VIRGILIO E. SOSA G., CARLOS EDUARDO CARRIZO ALBA, ANTONIO DOMÍNGUEZ; Man. BOLIVAR PARIENTE C.; 53 brs.

## Savings Banks

**Banco General, SA:** Calle Aquilino de la Guardia, Apdo 4592, Panamá 5; tel. 227-3200; fax 265-0210; e-mail info@bgeneral.com; internet www.bgeneral.com; f. 1955; cap. 300.0m., res 11.3m., dep. 2,057.1m., total assets 2,626.5m. (2004); purchased Banco Comercial de Panamá (BANCOMER) in 2000; Chair. and CEO FEDERICO HUMBERT; Exec. Vice-Pres. and Gen. Man. RAÚL ALEMÁN Z.; 35 brs.

**Banco Panameño de la Vivienda (BANVIVIENDA):** Casa Matriz-Bella Vista, Avda Chile y Calle 41, Apdo 8639, Panamá 5; tel. 227-4020; fax 227-5433; e-mail bpvger@pty.com; internet www.banvivienda.com; f. 1981; cap. 12.4m., dep. 128.0m., total assets 150.3m. (2004); Pres. ORLANDO SANCHEZ AVILES; Dir MARIO L. FÁBREGA AROSEMENA; 3 brs.

**Caja de Ahorros:** Vía España y Calle Thays de Pons, Apdo 1740, Panamá 1; tel. 205-1000; fax 269-3674; e-mail atencionalcliente@cajadeahorros.com.pa; internet www.cajadeahorros.com.pa; f. 1934; govt-owned; cap. 129.0m., dep. 671.8m., total assets 1,096.3m. (2004); Pres. ROGELIO ALEMÁN; Gen. Man. EUDORO JAÉN E.; 37 brs.

## Domestic Private Banks

**Banco Atlántico (Panamá), SA:** Edif. Banco Iberóamerica, Calle 50 y Calle 53, Apdo 6553, Panamá 5; tel. 263-5656; fax 269-1616; e-mail ibergeren@pan.gbm.net; internet www.bancoatlantico.com.pa; f. 1975 as Banco de Iberoamerica, SA, current name adopted in 2000; Pres. JOSÉ M. CHIMENO CHILLÓN; Gen. Man. RUBÉN FABREGAT BRACCO; 4 brs.

**Banco Continental de Panamá, SA:** Calle 50 y Avda Aquilino de la Guardia, Apdo 135, Panamá 9A; tel. 215-7000; fax 215-7134; e-mail bcp@bcocontinental.com; internet www.bbvabancocontinental.com; f. 1972; merged with Banco Internacional de Panamá in Jan. 2002; cap. 813.2m., res 189.1m., dep. 11,288.0m. (Dec. 2003); Pres. PEDRO BRESCIA CAFFERATA; Gen. Man. JOSÉ ANTONIO COLOMER GUIU; 7 brs.

**Banco Mercantil del Istmo, SA:** Calle Manuel M. Icaza, Panamá; tel. 205-5306; fax 263-6262; e-mail mguerra@banistmo.com; internet www.banistmo.com; f. 1967 as Banco de Santander y Panamá; current name adopted in 1992; cap. 26.2m., res 0.2m., dep. 308.2m., total assets 378.8m. (Dec. 2002); Pres. SAMUEL LEWIS GALINDO; Gen. Man. MANUEL JOSÉ BARREDO MARTÍNEZ.

**Banco Panamericano, SA (PANABANK):** Edif. Panabank, Casa Matriz, Calle 50, Apdo 1828, Panamá 1; tel. 263-9266; fax 269-1537; e-mail gerencia@panabank.com; internet www.panabank.com; f. 1983; Exec. Vice-Pres. GUIDO J. MARTINELLI, Jr.

**Global Bank Corporation:** Torre Global Bank, Calle 50, Apdo 55-1843, Paitilla, Panamá; tel. 206-2000; fax 263-3518; e-mail global@pan.gbm.net; internet www.globalbank.com.pa; f. 1994; total assets 458.6m. (June 2001); Pres. JORGE VALLARINO S.

**Multi Credit Bank Inc:** Edif. Prosperidad, planta baja, Vía España 127, Apdo 8210, Panamá 7; tel. 269-0188; fax 264-4014; e-mail banco@grupomulticredit.com; internet www.grupomulticredit.com; f. 1990; total assets 343.9m. (Dec. 2000); Gen. Man. MOISÉS D. COHEN M.

**Primer Banco del Istmo, SA:** Edif. Bancoistmo, Calle 50 y 77, San Francisco, Panamá; tel. 270-0015; fax 270-1952; internet www.banistmo.com; f. 1984 as Banco del Istmo; current name adopted in 2000 following merger of Primer Grupo Nacional and Banco del Istmo; merged with Banco de Latinoamerica in Sept. 2002; cap. US $480m., res $58m., dep. $2,869m. (Dec. 2003); Chair. JOSÉ RAÚL ARIAS GARCÍA DE PAREDES; Gen. Man. L. J. MONTAGUE BELANGER.

**Towerbank International Inc:** Edif. Tower Plaza, Calle 50 y Beatriz M. de Cabal, Apdo 6-6039, Panamá; tel. 269-6900; fax 269-6800; e-mail towerbank@towerbank.com; internet www.towerbank.com; f. 1971; cap. 32.0m., res 0.04m., dep. 316.5m. (Dec. 2003); Pres. SAM KARDONSKI; Gen. Man. GIJSBERTUS ANTONIUS DE WOLF.

## Foreign Banks

Principal Foreign Banks with General Licence

**BAC International Bank (Panamá), Inc** (USA): Avda de la Guardia, planta baja, Apdo 6-3654, Panamá; tel. 213-0822; fax 269-3879; e-mail rcucalon@bacbank.com; internet www.bacbank.com; Dir and Sec.-Gen. BARNEY VAUGHAN.

**BANCAFE (Panamá), SA** (Colombia): Avda Manuel María Icaza y Calle 52E, No 18, Apdo 384, Panamá 9A; tel. 264-6066; fax 263-6115; e-mail bancafe@bancafe-panama.com; internet www.bancafe-pa.com; f. 1966 as Banco Cafetero; current name adopted in 1995; cap. 27.6m., dep. 295.5m. (Dec. 1994); Pres. JORGE CASTELLANOS RUEDA; Gen. Man. JAIME DE GAMBOA GAMBOA; 2 brs.

**Banco Bilbao Vizcaya Argentaria (Panama), SA** (Spain): Torre BBVA, Avda Balboa, Apdo 8673, Panamá 5; tel. 227-0973; fax 227-3663; e-mail fperezp@bbvapanama.com; internet www.bbvapanama.com; f. 1982; cap. 8.7m., res 56.6m., dep. 648.6m. (Dec. 2002); Chair. MANUEL ZUBIRÍA PASTOR; Gen. Man. FRANCISCO JAVIER LEJARRAJA.

**Banco de Bogotá, SA** (Colombia): Avda Aquilino de la Guardia 48, Apdo 4599, Panamá 5; tel. 264-6000; fax 263-8037; e-mail banbogo@sinfo.net; internet www.bancobogota-panama.com; f. 1967; cap. 2,254m., res 858,248m., dep. 4,697,328m. (Dec. 2002); merged with Banco del Comercio, SA (Colombia) in 1994; Gen. Man. FABIO RIAÑO.

**Banco do Brasil, SA:** Edif. Interseco, planta baja, Calle Elvira Méndez 10, Apdo 87-1123, Panamá 7; tel. 263-6566; fax 269-9867; e-mail bdbrasil@bbpanama.com; internet www.bancodobrasil.com.br; f. 1973; cap. and res 52.3m., dep. 1,248.1m., total assets 1,320.2m. (Dec. 1993); Gen. Man. LUIZ EDUARDO JACOBINA; 1 br.

**Banco Internacional de Costa Rica, SA:** Calle Manual M. Icaza 25, Apdo 600, Panamá 1; tel. 263-6822; fax 263-6393; e-mail informacion@bicaspan.net; internet www.bicsa.com; f. 1976; total assets 15,382m. (1999); Gen. Man. JOSÉ FRANCISCO ULATE.

**Banco Latinoamericano de Exportaciones (BLADEX)** (Multinational): Casa Matriz, Calles 50 y Aquilino de la Guardia, Apdo 6-1497, El Dorado, Panamá; tel. 210-8500; fax 269-6333; e-mail infobla@blx.com; internet www.blx.com; f. 1979; groups together 254 Latin American commercial and central banks, 22 international banks and some 3,000 New York Stock Exchange shareholders; cap. US $280.0m., res $153.3m., dep. $1,875.7m. (Dec. 2003); CEO JOSÉ CASTAÑEDA; COO JAIME RIVERA.

**Banco Unión, SACA** (Venezuela): Torre Hongkong Bank, 1°, Avda Samuel Lewis, Apdo 'A', Panamá 5; tel. 264-9133; fax 263-9985; e-mail bcounion@orbi.net; internet www.bancunion.com; f. 1974; cap. and res 15.6m., dep. 177.1m. (Dec. 1993); Gen. Man. MARÍA M. DE MÉNDEZ; 1 br.

**Bancolombia (Panama), SA:** Calle Manuel María Icaza 11, Apdo 8593, Panamá 5; tel. 263-6955; fax 269-1138; e-mail mdebetan.bicpma@mail.bic.com.co; internet www.bancolombiapanama.com; f. 1973; current name adopted in 1999; cap. US $14.0m., res $20.7m., dep. $1,057.5m. (Dec. 2003); Gen. Man. MARÍA ISABEL URIBE.

**Bank of China:** Apdo 87-1056, Panamá 7; tel. 263-5522; fax 223-9960; e-mail heqr@bank-of-china.com; f. 1994; Gen. Man. QUINGBO WANG.

**Bank of Nova Scotia** (Canada): Edif. P. H. Scotia Plaza, Avda Federico Boyd y Calle 51, Apdo 7327, Panamá 5; tel. 263-6255; fax 263-8636; e-mail scotiabk@sinfo.net; Gen. Man. TERENCE S. MCCOY.

**Bank of Tokyo-Mitsubishi Ltd** (Japan): Vía España y Calle Aquilino de la Guardia, Apdo 1313, Panamá 1; tel. 263-6777; fax 263-5269; f. 1973; cap. and res 8.3m., dep. 1,574.0m., total assets 1,615.8m. (March 1998); Gen. Man. SHIGEYUKI ONISHI.

**Bank Leumi Le-Israel, BM** (Israel): Edif. Grobman, planta baja, Calle Manuel M. Icaza 10, Apdo 6-4518, El Dorado, Panamá; tel. 263-9377; fax 269-2674; Gen. Man. URI ROM.

**Banque Sudameris, SA** (Multinational): Avda Balboa y Calle 41, Apdo 1847, Panamá 9A; tel. 227-2777; fax 227-5828; e-mail banque.sudameris@sudameris.com.pa; internet www.sudameris.com.pa; Gen. Man. JULIO A. CORTÉS; 2 brs.

**BNP Paribas SA** (France): Edif. Omanco, Vía España 200, Apdo 1774, Panamá 1; tel. 263-6600; fax 263-6970; e-mail bnpparibas.panama@americas.bnpparibas.com; internet www.bnpparibas.com; f. 1948; name changed as above in 2002; cap. 10.0m., res 6.7m., dep. 194.9m., total assets 224.0m. (Dec. 1999); Gen. Man. CHRISTIAN GIRAUDON.

**Citibank NA** (USA): Plaza Panama Bldg, Calle 50, Apdo 555, Panamá 9A; tel. 210-5900; fax 210-5901; internet www.citibank.com.pa; f. 1904; Gen. Man. FRANCISCO CONTO; 10 brs.

**Credicorp Bank, SA:** Apdo 833-0125, Panamá; tel. 210-1111; fax 210-0069; e-mail sistemas@plazapan.com; internet www.credicorpbank.com; f. 1993; Gen. Man. CARLOS GUEVARA.

# PANAMA

**Dai Ichi Kangyo Bank, SA** (Japan): Edif. Plaza Internacional, Vía España, Apdo 2637, Panamá 9A; tel. 269-6111; fax 269-6815; f. 1979; cap. and res 2.0m., dep. 216.1m., total assets 218.4m. (Sept. 1991); Gen. Man. GAKUO FUKUTOMI.

**Dresdner Bank Lateinamerika AG** (Germany): Torre Dresdner Bank, Calles 50 y 55E, Apdo 5400, Panamá 5; tel. 206-8100; fax 206-8109; e-mail panama@dbla.com; internet www.dbla.com; f. 1971; affiliated to Dresdner Bank AG; Gen. Mans HENNING HOFFMEYER, BERND KLEINWORTH.

**HSBC Bank USA:** Apdo 9A-76, Panamá 9A; tel. 263-5855; fax 263-6009; e-mail hsbcpnm@sinfo.net; internet www.hsbcpnm.com; dep. 477m. (1999); in 2000 it acquired the 11 branch operations of the Chase Manhattan Bank, with assets of US $752m.; Gen. Man. JOSEPH L. SALTERIO; 3 brs.

**International Commerical Bank of China** (Taiwan): Calles 50 y 56E, Apdo 4453, Panamá 5; tel. 263-8565; fax 263-8392; Gen. Man. SHOW-LOONG HWANG.

**Korea Exchange Bank** (Republic of Korea): Torre Global Bank, Calle 50, Apdo 8358, Panamá 7; tel. 269-9966; fax 264-4224; Gen. Man. KWANG-SUCK KOH.

### Principal Foreign Banks with International Licence

**Atlantic Security Bank:** Apdo 6-8934, El Dorado, Panamá; tel. 269-5944; fax 215-7302; internet www.credicorpnet.com; f. 1984; Gen. Man. JORGE PONCE MENDOZA.

**Banco Alemán Platina, SA:** Panamá; tel. 269-2666; fax 269-0910; e-mail baplatina@balpa.com; internet www.bancoalemanplatina.com; f. 1965 as Banco Alemán-Panameño; current name adopted in 1993; cap. US $42.0m., res $10.3m., dep. $244.6m. (Aug. 2001); Pres. DANIEL PFISTER; Gen. Man. RALF FISCHER.

**Banco de la Nación Argentina:** Edif. World Trade Center 501, Calle 53, Urb. Marbella, Panamá; tel. 269-4666; fax 269-6719; e-mail bna@panama.phoenix.net; internet www.bna.com.ar; f. 1977; Man. OLGA SOLÍS.

**Banco de Occidente (Panama), SA:** Calle 50 y Aquilino de la Guardia, Apdo 6-7430, El Dorado, Panamá; tel. 263-8144; fax 269-3261; e-mail boccipan@pty.com; internet www.bancoccidente.com.pa; f. 1982; cap. US $6.3m., res $6.0m., dep. $199.0m. (Dec. 2002); Pres. EFRAÍN OTERO ALVAREZ.

**Banco del Pacífico (Panama), SA:** Calle Aquilino de la Guardia y Calle 52, Apdo 6-3100, El Dorado, Panamá; tel. 263-5833; fax 263-7481; e-mail bpacificopanama@pacifico.fin.ec; internet www.bancodelpacifico.com.pa; f. 1980; Pres. ANA ESCOBAR.

**Banco de la Provincia de Buenos Aires** (Argentina): Torre Banco Continental, 26°, Calle 50, Apdo 6-4592, El Dorado, Panamá; tel. 215-7703; fax 215-7718; e-mail bpbapma@psianet.pa; internet www.bapro.com; f. 1982; total assets 15,000m. (Dec. 2000); Gen. Man. JUAN CARLOS STURLESI.

**Bancrédito (Panama), SA:** Plaza Regency, 22°, Apdo 0832-1700, Panamá; tel. 223-2977; fax 264-6781; internet www.bancredito.com; f. 1998; Gen. Man. CARLOS HUMBERTO ROJAS MARTÍNEZ.

**BNP Paribas:** Edif. Omanco, Vía España 200, Apdo 0816-07547, Panamá; tel. 264-8555; fax 263-5004; e-mail bnpparibas@americas.bnpparibas.com.pa; internet www.bnpparibas.com.pa; f. 1972; Gen. Man. THIERRY DINGREVILLE.

**Popular Bank Ltd Inc:** Apdo 0816-00265, Panamá; tel. 269-4166; fax 269-1309; e-mail gversari@bpt.com.pa; f. 1983 as Banco Popular Dominicano (Panama), SA; current name adopted in 2003; cap. 24.4m., res 9.3m., dep. 338.8m. (Dec. 2004); Pres. RAFAEL A. RODRÍGUEZ; Gen. Man. GIANNI VERSARI.

**UBS (Panama), SA:** Calle 53 Este, Edif. Marbella Swiss Tower, Apdo 0834-61, Panamá 9A; tel. 206-7100; fax 206-7100; internet www.ubs.com; f. 1968; Chair. MARTIN WIRZ.

### Banking Association

**Asociación Bancaria de Panamá (ABP):** Torre Hong Kong Bank, 15°, Avda Samuel Lewis, Apdo 4554, Panamá 5; tel. 263-7044; fax 223-5800; e-mail abp@orbi.net; internet www.asociacionbancaria.com; f. 1962; 79 mems; Pres. JORGE E. VALLARINO S.; Exec. Vice-Pres. MARIO DE DIEGO, Jr.

### STOCK EXCHANGE

**Bolsa de Valores de Panamá:** Edif. Vallarino, planta baja, Calles Elvira Méndez y 52, Apdo 87-0878, Panamá; tel. 269-1966; fax 269-2457; e-mail bvp@pty.com; internet www.panabolsa.com; f. 1960; Gen. Man. ROBERTO BRENES PÉREZ.

### INSURANCE

**Aseguradora Mundial, SA:** Edif. Aseguradora Mundial, Avda Balboa y Calle 41, Apdo 8911, Panamá 5; tel. 207-7600; fax 207-8787; e-mail info@mundial.com; internet www.amundial.com; general; f. 1937; Pres. MANUEL JOSÉ PAREDES L.

**ASSA Cía de Seguros, SA:** Edif. ASSA, Avda Nicanor de Obarrio (Calle 50), Apdo 5371, Panamá 5; tel. 269-0443; fax 263-9623; e-mail assamercadeo@assanet.com; internet www.assanet.com; f. 1973; Pres. LORENZO ROMAGOSA; Gen. Man. PABLO DE LA HOYA.

**Cía Nacional de Seguros, SA:** Edif. No 62, Calle 50, Apdo 5303, Panamá 5; tel. 205-0300; fax 223-1146; e-mail conase@conase.net; f. 1957; Gen. Man. RAÚL MORRICE.

**CONASE (Cía Internacional de Seguros, SA):** Avda Cuba y Calles 35 y 36, Apdo 1036, Panamá 1; tel. 227-4000; internet www.conase.com; f. 1910; Pres. RICHARD A. FORD; Gen. Man. MANUEL A. ESKILDSEN.

**La Seguridad de Panamá, Cía de Seguros, SA:** Edif. American International, Calle 50 esq. Aquilino de la Guardia, Apdo 5306, Panamá 5; tel. 263-6700; f. 1986; Gen. Man. MARIELA OSORIO.

## Trade and Industry

**Colón Free Zone (CFZ):** Avda Roosevelt, Apdo 1118, Colón; tel. 445-1033; fax 445-2165; e-mail zonalibre@zolicol.org; internet www.colonfreezone.com; f. 1948 to manufacture, import, handle and re-export all types of merchandise; some 1,800 companies were established in 2005; well-known international banks operate in the CFZ, where there are also customs, postal and telegraph services; the main exporters to the CFZ are Japan, the USA, Hong Kong, Taiwan, the Republic of Korea, Colombia, France, Italy and the United Kingdom; the main importers from the CFZ are Brazil, Venezuela, Mexico, Ecuador, the Netherlands Dependencies, Bolivia, the USA, Chile, Argentina and Colombia; the total area of the CFZ was 485.3 ha; Gen. Man. NILDA IRIS QUIJANO.

### GOVERNMENT AGENCY

**Instituto Panameño de Comercio Exterior (IPCE):** 3 Edison Plaza, Avda El Paical, Apdo 55-2339, Paitilla, Panamá; tel. 225-7244; fax 225-2193; f. 1984; foreign trade and investment promotion organization; Dir KENIA JAÉN RIVERA.

### CHAMBERS OF COMMERCE

**Cámara Oficial Española de Comercio:** Calle 33E, Apdo 1857, Panamá 1; tel. 225-1487; fax 225-0626; internet www.caespan.com.pa; Pres. EDELMIRO GARCÍA VILLA VERDE; Sec.-Gen. ATILIANO ALFONSO MARTÍNEZ.

**Cámara de Comercio, Industrias y Agricultura de Panamá:** Avda Cuba y Ecuador 33A, Apdo 74, Panamá 1; tel. 227-1233; fax 227-4186; e-mail infocciap@panacamara.com; internet www.panacamara.com; f. 1915; Pres. AUGUST E. SIMONS; First Vice-Pres. JOSÉ JAVIER RIVERA; 1,300 mems.

### INDUSTRIAL AND TRADE ASSOCIATIONS

**Cámara Panameña de la Construcción:** Calle Aquilino de la Guardia No 19, Apdo 0816-02350, Panamá 5; tel. 265-2500; fax 265-2571; e-mail finanzas@capac.org; internet www.capac.org; represents interests of construction sector; Pres. Ing. MANUEL R. VALLARINO.

**Codemín:** Panamá; tel. 263-7475; state mining org.; Dir JAIME ROQUEBERT.

**Corporación Azucarera La Victoria:** Transístmica, San Miguelito; tel. 229-4794; state sugar corpn; scheduled for transfer to private ownership; Dir Prof. ALEJANDRO VERNAZA.

**Corporación para el Desarrollo Integral del Bayano:** Avda Balboa, al lado de la estación del tren, Estafeta El Dorado, Panamá 2; tel. 232-6160; f. 1978; state agriculture, forestry and cattle-breeding corpn.

**Dirección General de Industrias:** Edif. Plaza Edison, 3°, Apdo 9658, Panamá 4; tel. 360-0720; govt body which undertakes feasibility studies, analyses and promotion; Dir-Gen. LUCÍA FUENTES DE FERGUSON; Nat. Dir of Business Development FRANCISCO DE LA BARRERA.

**Sindicato de Industriales de Panamá:** Vía Ricardo J. Alfaro, Entrada Urb. Sara Sotillo, Apdo 6-4798, Estafeta El Dorado, Panamá; tel. 230-0169; fax 230-0805; e-mail sip@cableonda.net; internet www.industriales.org; f. 1945; represents and promotes activities of industrial sector; Pres. GASPAR GARCÍA DE PAREDES.

### EMPLOYERS' ORGANIZATIONS

**Asociación Panameña de Ejecutivos de Empresas (APEDE):** Edif. APEDE, Calle 42, Bella Vista y Avda Balboa, Apdo 1331,

Panamá 1; tel. 227-3511; fax 227-1872; e-mail apede@sinfo.net; internet www.apede.org; Pres. ENRIQUE DE OBARRIO.

**Consejo Nacional para el Desarrollo de la Pequeña Empresa:** Ministry of Commerce and Industry, Apdo 9658, Panamá 4; tel. 227-3559; fax 225-1201; f. 1983; advisory and consultative board to the Ministry of Commerce and Industry.

**Consejo Nacional de la Empresa Privada (CONEP):** Avda Morgan, Balboa, Ancón, Casa 302, Apdo 1276, Panamá 1; tel. 211-2672; fax 211-2964; e-mail www.conep.org.pa; Pres. JUAN F. KIENER.

## MAJOR COMPANIES

### Beverages and Tobacco

**Cervecería Latina:** f. 2002; production of Pilsner beer.

**Compañía Embotelladora Coca Cola de Panamá, SA:** Parque Industrial San Cristóbal, Calle Santa Rosa, Panamá 9A; tel. 236-0700; fax 260-3504; e-mail info@cocacolapanama.com; f. 1913; bottlers of soft drinks; Pres. JOAQUÍN J. VALLARINO; 500 employees.

**Grupo Cervecería Nacional:** Carrera Transísmica y Vía Ricardo J. Alfaro, El Dorado, Apdo 6-1393, Panamá 1; tel. 279-5800; fax 279-5861; internet www.grupocn.com; f. 1909; brewing; Pres. JULIO MARIO SANTO DOMINGO; Exec. Pres. RICARDO JANSON CALHOUN; 400 employees.

**TANASEC Panama:** La Rinconada, Vía Tocumén, Apdo 0839-0714, Panamá 13; tel. 220-7077; fax 220-3877; e-mail eloy.collado@pmintl.com; internet www.philipmorris.com; f. 1955; manufacture of cigarettes; merged with Tabacalera Nacional in 2001; Dir ELOY COLLADO; Man. DAVID FONG; 200 employees.

**Varela Hermanos, SA:** Calle 85, Urb. Industrial Los Angeles, entre Coagro y Fósforos El Gallo, Panamá; tel. 217-3111; fax 217-3627; e-mail varela@varehelahermanos.com; internet www.varelahermanos.com; f. 1950; distillation and bottling of liquors; Dir JUAN CARLOS PINO; 300 employees.

### Food Products

**Compañía Azucarera La Estrella, SA:** La Locería y Nata, Calle Arturo del Valle, Apdo 8404, Panamá 7; tel. 236-2577; fax 236-1308; e-mail calesa@cerco.net; f. 1949; sugar mill and refinery; Chair. ERIC A. DELVALLE; 2,390 employees.

**Cooperativa de Servicios Múltiples de Puerto Armuelles (COOSEMUPAR):** Puerto Armuelles, Chinquiniqua; fmrly a division of the US Chiquita Brands co; bought by an employees' co-operative in March 2003.

**Industrias Lácteas, SA:** Vía Simón Bolívar, Apdo 4362, Panamá 5; tel. 229-1122; fax 261-6883; f. 1956; dairy products; Man. JOSÉ LUIS GARCÍA DE PAREDES; 780 employees.

**Nestlé Panamá, SA, and Kraft Foods, SA:** Calle 69, 74D, Urb. La Loma, Panamá 9A; tel. 229-1333; fax 229-1982; f. 1937; subsidiary of Nestlé Co, SA (Switzerland); manufacture and wholesale of food products; Gen. Man. LUIS CARBARCOS GIL; 760 employees (Nestlé Panamá, SA).

**Productos Alimenticios Pascual, SA:** Avda José Agustín Arango, Apdo 8422, Panamá 7; tel. 217-2133; fax 233-2825; e-mail pashnos@sinfo.net; f. 1946; food and food-processing; Pres. VICENTE ESTEVEZ PASCUAL; Gen. Man. JUAN ALBERTO PASCUAL; 448 employees.

### Textile Manufactures

**Promedias, SA:** Local 3, al lado de Herrería Santiago, Panamá; tel. 261-3649; fax 261-5548; internet www.promedias.com.pa; manufacturers of socks and tights.

**Tejidos y Confecciones, SA:** Edif. Durex Carrasquilla, Calle 2A, Panamá; Apdo X, Panamá 9A; tel. 263-8888; fax 264-5022; e-mail sales@teyco.com; internet www.teyco.com; f. 1964; manufacturers of men's and children's clothing; Pres. MAYER ATTIE CHAYO; Dir RAMY ATTIE; 500 employees.

### Miscellaneous

**Cemento Panamá, SA:** Edif. Cemento Panamá, Avda Manuel Batista, Apdo 1755, Panamá 1; tel. 229-3011; fax 229-3151; e-mail esanchez@incem.com; internet www.incem.com; f. 1943; manufacturers of portland cement; Pres. JUAN MANUEL RUISECO; Dir-Gen. EVARISTO SÁNCHEZ; 90 employees.

**Compañía Atlas, SA:** Edif. 40, Calle 16½, Apdo 6-1092, El Dorado, Panamá; tel. 236-0066; fax 236-0044; e-mail info@ciatlas.com; internet www.atlastore.com; f. 1949; holding co with interests in the wholesale manufacture, import and distribution of stationery; Gen. Man. ROBERTO C. HENRÍQUEZ; 110 employees.

**Deloitte & Touche:** Edif. Banco de Boston 20, Calle Elvira Méndez y Via España, Panamá 5; tel. 263-9900; fax 269-2386; e-mail info@deloitte.com.pa; internet www.deloitte.com.pa; accountancy and management consultancy; Chair. ANTONIO R. BURÓN; CEO MIGUEL HERAS.

**Grupo Corcione:** Duplex 10, Calle 59, San Francisco, Apdo 0816-05715, Zona 5, Panamá; tel. 263-3617; fax 269-6057; e-mail ventascor@cableonda.net; internet www.gcorcione.com; construction contractors; Pres. NICOLÁS CORCIONE.

**International Paint:** Calle 69, Casa 115, San Francisco; tel. 270-0148; fax 270-0164; wholesalers of protective coatings; owned by Akzo Nobel NV, Netherlands.

**KPMG:** 54 Calle Nicanor A. de Obarrio, Apdo 5307, Panamá 5; tel. 263-5677; fax 263-9852; e-mail pa-fminformation@kpmg.com.pa; internet www.kpmg.com.pa; accountancy and management consultancy; Dir EDUARDO CHOY.

**Lindo y Maduro, SA:** Apdo 5300, Panamá 5; tel. 227-0100; fax 227-2935; e-mail linduro@cwpanama.net; manufacturers and distributors of non-durable goods and perfumery; Pres. RALPH J. LINDO; Man. MARÍA VICTORIA DELGADO; 60 employees.

**El Machetazo SA:** Avda Peru y Calle 29, Apdo 2587-3, Panamá; clothes retailer; f. 1967; Man. JUAN RAMON CABRERA; 389 employees.

**Melo y Cía, SA:** Apdo 333, Panamá 1; tel. 221-0033; fax 224-2311; e-mail grupomelo@grupomelo.com; internet www.grupomelo.com; vendors of veterinary, agricultural and agrochemical products, building materials and household goods; Pres. ARTURO DONALDO MELO S.; Gen. Man. FEDERICO MELO K.; 400 employees.

**Petroterminal de Panamá, SA:** Edif. Scotia Plaza, 1°, Calle 51 y Federico Boyd, Panamá; tel. 263-7777; fax 263-9949; e-mail info@petroterminal.com; internet www.petroterminal.com; f. 1977; petroleum and gas field services and petroleum storage facilities; Dir-Gen. LUIS ROQUEFORT; 80 employees.

**RNC Gold Inc:** Azuero Peninsula, Los Santos; e-mail info@rncgold.com; internet www.rncgold.com; Canadian-owned; operates Cerro Quema gold mine; Chair. and CEO RANDY J. MARTIN; Pres. THOMAS LOUGH.

**Syntex Corporation:** Edif. Bank of America, 9°, Calle 50, Panamá; tel. 263-5255; fax 441-4568; f. 1944; wholesale import and distribution of pharmaceuticals; Man. PAUL FREIMAN; 32 employees.

**H. Tzanetatos Inc.:** Via Domingo Díaz 100, Apdo 6025, Tocumen; tel. 220-1977; fax 220-5122; e-mail emelinar@sinfo.net; general wholesaler.

## UTILITIES

### Regulatory Authority

**Ente Regulador de los Servicios Públicos:** Vía España, Edif. Office Park, Apdo 4931, Panamá 5; tel. 278-4500; fax 278-4600; e-mail webmaster@ersp.gob.pa; internet www.enteregulador.gob.pa; f. 1996; state regulator with responsibility for water, electricity, broadcasting and telecommunications; Pres. JOSÉ GALÁN PONCE; Dirs CARLOS RODRÍGUEZ BETHANCOURT, NILSON ESPINO.

### Electricity

**Instituto de Recursos Hidráulicos y Electrificación (IRHE):** Edif. Poli, Avda Justo Arosemena y 26E, Apdo 5285, Panamá 5; tel. 262-6272; state org. responsible for the national public electricity supply; partial divestment of generation and distribution operations completed in Dec. 1998; transmission operations remain under complete state control; Dir-Gen. Dr FERNANDO ARAMBURÚ PORRAS.

### Water

**Instituto de Acueductos y Alcantarillados Nacionales (IDAAN)** (National Waterworks and Sewage Systems Institute): Panamá; Dir-Gen. ELIDA Dí.

## TRADE UNIONS

**Central General Autónoma de Trabajadores de Panamá (CGTP):** Casa 15, Calle Tercera Perejil, Vía España, Panamá; tel. 269-9741; fax 223-5287; e-mail cgtpan@cwpanama.net; fmrly Central Istmeña de Trabajadores; Sec.-Gen. MARIANO E. MENA.

**Confederación Nacional de Unidad Sindical Independiente (CONUSI):** 0421B Calle Venado, Ancón, Apdo 830344, Zona 3, Panamá; tel. 212-3865; fax 212-2565; e-mail conusipanama@hotmail.com; Sec.-Gen. GABRIEL E. CASTILLO C.

**Confederación de Trabajadores de la República de Panamá (CTRP)** (Confederation of Workers of the Republic of Panama): Calle 31, entre Avdas México y Justo Arosemena 3-50, Apdo 8929, Panamá 5; tel. 225-0293; fax 225-0259; e-mail ctrp@sinfo.net; f. 1956; admitted to ICFTU/ORIT; Sec.-Gen. GUILLERMO PUGA; 62,000 mems from 13 affiliated groups.

**Consejo Nacional de Trabajadores Organizados (CONATO)** (National Council of Organized Labour): Edif. 777, 2°, Balboa-

Ancón, Panamá; tel. and fax 228-0224; e-mail conato@cwpanama.net; Co-ordinator PEDRO HURTADO; 150,000 mems.

**Convergencia Sindical:** Casa 2490, Balboa Corregimiento de Ancón, Calle Bonparte, Apdo 10536, Panamá; tel. and fax 314-1615; e-mail conversind@cwpanama.net; Sec.-Gen. LUIS GONZÁLEZ.

**Federación Nacional de Asociaciones de Empleados y Servidores Públicos (FENASEP)** (National Federation of Associations of Public Employees): Galerías Alvear, 2°, Oficina 301, Vía Argentina, Apdo 66-48, Zona 5, Panamá; tel. and fax 269-1316; e-mail fenasep@sinfo.net; f. 1984; Sec.-Gen. LEANDRO ÁVILA.

A number of unions exist without affiliation to a national centre.

## Transport

### RAILWAYS

In 1998 there were an estimated 485 km of track in Panama. In 2000 a US $75m. project to modernize the line between the ports at either end of the Panama Canal began. In July 2001 the 83-km transisthmian railway, originally founded in 1855, reopened.

**Chiriquí Land Co:** Panamá; tel. 770-7243; fax 770-8064; operates two lines: the Northern Line (Almirante–Guabito in Bocas del Toro Province) and the Southern Line (Puerto Armuelles–David in Chiriquí Province); purchased by the Govt in 1978; Gen. Man. FRED JOHNSON.

**Ferrocarril Nacional de Chiriquí:** Apdo 12B, David City, Chiriquí; tel. 775-4241; fax 775-4105; 126 km linking Puerto Armuelles and David; Gen. Man. M. ALVARENGA.

**Ferrocarril de Panamá:** Apdo 2023, Estafeta de Balboa, Panamá; tel. 232-6000; fax 232-5343; govt-owned; 83 km linking Panama City and Colón, running parallel to Panama Canal; operation on concession by Kansas City Southern (USA); modernization programme completed in 2001; operates daily passenger and cargo service; Dir-Gen. VÍCTOR MANUEL DANELO.

### ROADS

In 1999 there were an estimated 11,400 km of roads, of which some 3,944 km were paved. The two most important highways are the Pan-American Highway and the Boyd-Roosevelt or Trans-Isthmian, linking Panama City and Colón. The Pan-American Highway to Mexico City runs for 545 km in Panama and was being extended towards Colombia. There is also a highway to San José, Costa Rica.

### SHIPPING

The Panama Canal opened in 1914. In 1984 more than 4% of all the world's seaborne trade passed through the waterway. It is 82 km long, and ships take an average of eight hours to complete a transit. In 2004 some 14,035 transits were recorded. The Canal can accommodate ships with a maximum draught of 12 m, beams of up to approximately 32.3 m (106 ft) and lengths of up to about 290 m (950 ft), roughly equivalent to ships with a maximum capacity of 65,000–70,000 dwt. In 2000 a five-year modernization project was begun. The project included: a general improvement of facilities; the implementation of a satellite traffic-management system; the construction of a bridge; and the widening of the narrowest section of the Canal, the Culebra Cut (which was completed in 2001). Plans were also announced to construct a 203-ha international cargo-handling platform at the Atlantic end of the Canal, including terminals, a railway and an international airport. Terminal ports are Balboa, on the Pacific Ocean, and Cristóbal, on the Caribbean Sea.

**Autoridad del Canal de Panamá (ACP):** Balboa Heights, Panamá; tel. 272-3202; fax 272-2122; e-mail info@pancanal.com; internet www.pancanal.com; in Oct. 1979 the Panama Canal Commission, a US govt agency, was established to perform the mission, previously accomplished by the Panama Canal Co, of managing, operating and maintaining the Panama Canal; the Autoridad del Canal de Panamá (ACP) was founded in 1997 and succeeded the Panama Canal Commission on 31 December 1999, when the waterway was ceded to the Govt of Panama; the ACP is the autonomous agency of the Govt of Panama; there is a Board of 11 mems; Chair. RICAURTE VÁSQUEZ; Administrator ALBERTO ALEMÁN ZUBIETA; Dep. Administrator MANUEL BENITEZ.

**Autoridad Marítima de Panamá:** Edif. 5534, Diablo Heights, Ancón, Apdo 8062, Panama 7; tel. 232-5528; fax 232-5527; e-mail ampadmin@amp.gob.pa; internet www.amp.gob.pa; f. 1998 to unite and optimize the function of all state institutions with involvement in maritime sector; Administrator RUBÉN AROSEMENA VALDÉS.

**Autoridad de la Región Interoceánica (ARI):** Panamá; tel. 228-5668; fax 228-7488; e-mail ari@sinfo.net; internet www.ari.gob.pa; f. 1993; administers the land and property of the former Canal Zone following their transfer from US to Panamanian control from 2000; legislation approved in early 1995 transferred control of the ARI to the Pres. of the Republic; Pres. GUSTAVO GARCÍA DE PAREDES.

**Panama City Port Authority and Foreign Trade Zone 65:** Apdo 15095, Panamá; FL 32406, USA; tel. 767-3220; e-mail wstubbs@portpanamacityusa.com; internet www.portpanamacityusa.com; Exec. Dir WAYNES STUBBS.

There are deep-water ports at Balboa and Cristóbal (including general cargo ships, containers, shipyards, industrial facilities); Coco Solo (general cargo and containers); Bahía Las Minas (general bulk and containers); Vacamonte (main port for fishing industry); Puerto Armuelles and Almirante (bananas); Aguadulce and Pedregal (export of crude sugar and molasses, transport of fertilizers and chemical products); and Charco Azul and Chiriquí Grande (crude oil).

The Panamanian merchant fleet was the largest in the world in December 2004, numbering 6,477 vessels with total displacement of 131.5m. gross registered tons. In November 2000 construction was completed on the largest container terminal in Latin America, in Balboa.

### CIVIL AVIATION

Tocumen (formerly Omar Torrijos) International Airport, situated 19 km (12 miles) outside Panamá (Panama City), is the country's principal airport and is served by many international airlines. A project to expand the airport's facilities, at a cost of US $20m., was announced in 2004. The France Airport in Colón and the Rio Hato Airport in Coclé province have both been declared international airports. There are also 11 smaller airports in the country.

**Aerolíneas Pacífico Atlántico, SA (Aeroperlas):** Apdo 6-3596, El Dorado, Panamá; tel. 315-7500; fax 315-0331; e-mail info@aeroperlas.com; internet www.aeroperlas.com; f. 1970; fmrly state-owned, transferred to private ownership in 1987; operates scheduled regional and domestic flights to 16 destinations; in 2000 initiated international flights; Pres. GEORGE F. NOVEY; Gen. Man. EDUARDO STAGG.

**Compañía Panameña de Aviación, SA (COPA):** Avda Justo Arosemena 230 y Calle 39, Apdo 1572, Panamá 1; tel. 227-2522; fax 227-1952; e-mail proquebert@mail.copa.com.pa; internet www.copaair.com; f. 1947; scheduled passenger and cargo services from Panamá (Panama City) to Central America, South America, the Caribbean and the USA; Chair. ALBERTO MOTTA; Exec. Pres. PEDRO O. HEILBRON.

## Tourism

Panama's attractions include Panamá (Panama City), the ruins of Portobelo and 800 sandy tropical islands, including the resort of Contadora, one of the Pearl Islands in the Gulf of Panama, and the San Blas Islands, lying off the Atlantic coast. In 2003 the number of visitors stood at 534,208. Income from tourism was some US $809m. in that year. In 2000 there were some 5,700 hotel rooms in Panama.

**Instituto Panameño de Turismo (IPAT):** Centro de Convenciones ATLAPA, Vía Israel, Apdo 4421, Panamá 5; tel. 226-7000; fax 226-3483; e-mail ggral@ns.ipat.gob.pa; internet www.ipat.gob.pa; f. 1960; Dir-Gen. RUBÉN BLADES.

**Asociación Panameña de Agencias de Viajes y Turismo (APAVIT):** Bella Vista Local 24, Calle 51, Apdo 55-1000 Paitilla, Panamá 3; tel. 264-3526; fax 264-5355; e-mail apavit@cableonda.net; internet www.apavitpanama.org; f. 1957; Pres. AIDA QUIJANO J.; Vice-Pres. ERICK GOLDONI.

## Defence

In 1990, following the overthrow of Gen. Manuel Antonio Noriega Morena, the National Defence Forces were disbanded and a new Public Force (numbering an estimated 11,800 men at 1 August 2004), comprising the National Police (11,000 men), the National Air Service (an estimated 400 men) and the National Maritime Service (an estimated 400 men), was created. The new force was representative of the size of the population and affiliated to no political party.

**Security Expenditure:** an estimated 100m. balboas (US $100m.) in 2003.

**Executive Secretary of National Defence Council:** RAMIRO JARVIS.

**Director of National Police:** GUSTAVO A. PÉREZ A.

**Director of National Air Service:** Cmmdr JAIME I. FÁBREGA.

**Director-General of National Maritime Service:** Rear Adm. RICARDO TRAAD PORRAS.

## Education

The education system in Panama is divided into elementary, secondary and university schooling, each of six years' duration. Education is free up to university level and is officially compulsory between six and 15 years of age. Primary education begins at the age of six and secondary education, which comprises two three-year cycles, at the age of 12. In 2001/02 the enrolment at primary schools was 99.0% of children in the relevant age-group (males 99.2%; females 98.8%), while secondary enrolment was 62.4% of children in the relevant age-group (males 59.6%; females 65.4%). There are four official universities and 11 private ones, including one that specializes in distance learning. Of total budgetary expenditure by the central Government in 2004, an estimated 505.8m. balboas was allocated to education.

## Bibliography

For works on Central America generally, see Select Bibliography (Books)

Coniff, M. L. *Panama and the United States*. Atlanta, GA, University of Georgia Press, 2001.

Dinges, J. *Our Man in Panama: How General Noriega Fooled the United States and Made Millions in Drugs and Arms*. New York, NY, Random House, 1990.

Dudley Gold, S. *The Panama Canal Transfer: Controversy at the Crossroads*. Austin, TX, Raintree/Steck Vaughn, 1999.

Espino, O. D. *How Wall Street Created a Nation: J. P. Morgan, Teddy Roosevelt and the Panama Canal*. New York, NY, Four Walls Eight Windows, 2001.

Guevara Mann, C. *Panamanian Militarism: A Historical Interpretation*. Athens, OH, Ohio University Press, 1996.

Harding II, R. C. *Military Foundations of Panamanian Politics*. Piscataway, NJ, Transaction Publrs, 2001.

Johns, C. J., and Ward Johnson, P. *State Crime, the Media and the Invasion of Panama*. London, Praeger Publrs, 1994.

Kempe, F. *Divorcing the Dictator: America's Bungled Affair with Noriega*. New York, NY, G. P. Putnam's Sons, 1990.

Langstaff, E. *Panama*. Oxford, ABC Clio, 2000.

Major, J. *Prize Possession: The United States and the Panama Canal, 1903–1979*. Cambridge, Cambridge University Press, 1993.

McCullough, D. *Path Between the Seas: Creation of the Panama Canal, 1870–1914*. New York, NY, Simon and Schuster, 1999.

Pearcy, T. L. *We Answer Only to God: Politics and the Military in Panama, 1903–1947*. Albuquerque, NM, University of New Mexico Press, 1998.

Pérez, O. J. (Ed.). *Post-Invasion Panama*. Lexington, MA, Lexington Books, 2000.

Rodríguez, J. C. *The Panama Canal: Its History, Its Political Aspects, and Financial Difficulties*. Honolulu, HI, University Press of the Pacific, 2002.

Rudolf, G. *Panama's Poor: Victims, Agents and Historymakers*. Gainesville, FL, University Press of Florida, 1999.

Sandoval Forero, E. A., and Salazar Pérez, R. *Lectura Crítica del Plan Puebla Panama*. Buenos Aires, Libros En Red, 2002.

Sosa, J. B. *In Defiance: The Battle Against Gen. Noriega Fought from Panama's Embassy in Washington*. Washington, DC, The Francis Press, 1999.

Snapp, J. S. *Destiny by Design: The Construction of the Panama Canal*. Pacific Heritage Press, 2000.

Taw, J. M. *Operation Just Cause: Lessons for Operations Other than War*. Skokie, IL, Rand McNally, 1996.

USA Ibp. *Doing Business And Investing in Panama*. Milton Keynes, Lightning Source UK Ltd, 2005.

*Panama Canal Handbook: Organization And Business Activity*. Milton Keynes, Lightning Source UK Ltd, 2005.

Ward, C. *Imperial Panama: Commerce and Conflict in Isthmian America, 1550–1800*. Alberquerque, NM, University of New Mexico Press, 1993.

# PARAGUAY

## Geography

### PHYSICAL FEATURES

The Republic of Paraguay is one of only two land-locked countries in South America—although, unlike Bolivia, it has never had a coastline. The country did, however, suffer immense loss from war in the 1860s, although in the 1930s it secured control over much of the Chaco after war with Bolivia. Paraguay is located in the south-central part of the continent, with Bolivia to the north-west, beyond a 750-km (466-mile) border. Argentina lies to the south-west, behind a 1,880-km frontier which abuts its neighbour in south-eastern Paraguay, while Brazil (1,290 km) is to the north-east. Paraguay is about the same size as the US state of California, covering an area of 406,752 sq km (157,048 sq miles).

Paraguay consists of two rough rectangles of territory on either side of the north–south river of the same name. The slightly larger block (with the south-west corner sliced off), covering about 60% of the territory, is to the west of the river. Offset slightly to the south is the smaller, original block of territory (a block with a much smaller block removed from the north-east corner), east of the Paraguay river. The two regions, known as Paraguay Occidental and Paraguay Oriental or Paraguay proper, have very different landscapes. The west is dominated by the Gran Chaco, an infertile alluvial plain extending through Paraguay and into Bolivia and Argentina. The region receives little rain, but is poorly drained and prone to flooding, so the rough prairie of dry grass and shadeless trees is patched with reedy marshes or thorny scrub. To the south and east, in the Oriental region, are some grassy plains, but rather more of broad, fertile valleys and rolling, wooded hills. The Paraná plateau thrusts down from the north, to create a highland of between 300 m (985 ft) and 600 m, its sharp crest running down the centre of Paraguay proper, highest in the west, but dropping abruptly to the fertile grassy foothills that roll down to the Paraguay, while descending more gently in the east towards the River Paraná. The Paraná forms the entire eastern (much of it through the reservoir behind the Itaipú dam) and southern border of Paraguay Oriental. Its tributary, the Paraguay, joins the Paraná only after cutting across the country and then forming a west-facing border with Argentina. The confluence of the Paraguay and the Paraná, at 46 m above sea level, is the lowest point in Paraguay. The highest point is in the centre of south-eastern Paraguay. It is Cerro Pero (or Cerro Tres Kandu), at 842 m. In total, just over one-half of the country is pastureland, while almost one-third is forested. The terrain is not favourable to many of the larger mammals, although there are jaguars, tapirs, armadillos and anteaters, for instance, but exotically plumaged birds are common (parrots, toucans, black ducks, etc.) and the grasslands are the natural habitat of the rhea or American ostrich.

### CLIMATE

Paraguay is bisected by the Tropic of Capricorn, and about two-thirds of the country experiences a mild, subtropical climate. On the Chaco plains of the Occidental region it is hotter, more humid and drier; temperatures can often reach 38°C (100°F). The far north and west is semi-arid. Rainfall on the plains averages about 815 mm (32 ins) per year, falling heavily in the summer (November–May) and virtually not at all in winter. By contrast, the eastern forests receive about 1,525 mm annually, with Asunción, the national capital, on the Paraguay river, getting 1,120 mm. The average temperatures range from 17°C (63°F) in July to 27°C (81°F) in January. The prevailing summer

wind is the hot, north-eastern sirocco, while in winter it is the cold pampero from the south.

### POPULATION

Paraguay has one of the more homogenous populations of South America, 95% of them being Mestizos. Otherwise, there are some unassimilated Amerindians (notably Guaraní of the eastern forests), some full-blood Spanish (mainly in the cities) and small groups originating from Japan, Italy, Portugal, Canada and Germany (the last mainly the religious minority the Mennonites). Spanish is joined by Guaraní as an official language, as the latter is commonly spoken by about 90% of the population. Spanish is spoken by about three-quarters of the population, with almost one-half of Paraguayans being completely bilingual. Guaraní was the official language for many of the early years of independence. However, the overwhelming majority of people, and most Mestizos, remain at least nominally Roman Catholic. There are other Christians, with the Evangelical Protestants active here as elsewhere in South America, but the largest Protestant minority is that of the Mennonites (based near Filadélfia, in the heart of the Chaco plains).

The total population, according to UN estimates at mid-2004, was 6.0m. Two-thirds of the population are urban and most live in western Paraguay proper. The east is very sparsely populated. The capital and largest city is Asunción, a port on the Paraguay, located where the river flows out of the north, having bisected the country that bears its name, and continues south, forming the border with Argentina. Asunción, which had a population of 510,910 in 2002, is in a capital district, which, together with 17 departments, forms the local administration of the country. The next largest city, not even one-quarter of the size, is Ciudad del Este (previously known as Puerto Presidente Stroessner), on the Paraná, near the Itaipú dam.

# History

## Prof. PETER CALVERT

Paraguay was already well populated when the Spanish first arrived in 1524. In 1537 the Spanish founded Asunción. The city enjoyed a brief period of importance until the foundation of Buenos Aires, in what is now Argentina, in 1580, when the seat of regional government was moved to the new port city. In the absence of important resources that were of interest to the Spanish, Paraguay remained economically undeveloped throughout the colonial period, and was politically and economically dependent on Buenos Aires. The indigenous Indians (Amerindians) established good relations with the Spanish and many intermarried; however, they retained their own language, Guaraní.

Jesuit missionaries soon arrived in the country. Indians converted to Christianity were resettled in missions, each farming the surrounding land. They built churches and created a unique, theocratic society. In 1767 the Jesuits were expelled and, for the first time, the Indians were directly exposed to Spanish rule. In 1810 Buenos Aires declared self-government and on 14 May 1811 Paraguay became the first Spanish territory in the Americas to achieve independence.

The newly independent country was governed by Dr José Gaspar Rodríguez de Francia. Dr Francia, known as 'El Supremo', closed Paraguay's borders to the outside world, thus ensuring Paraguayan sovereignty, despite Argentine plans for annexation. He proclaimed Guaraní the sole language and ruled alone until his death in 1840. His successor, Carlos Antonio López (President, 1842–62), reopened Paraguay's borders. No less a dictator than his predecessor, López encouraged trade, built Paraguay's first railway and abolished slavery. He also gave Paraguay the institution which was to dominate its politics from then on, the army. López's son, Francisco Solano López, succeeded him as President-for-life, and during his rule Paraguay was defeated in the War of the Triple Alliance (1865–70) by the combined forces of Argentina, Brazil and Uruguay. The War, which became known in Paraguay as the 'National Epic' (*Epopeya Nacional*), resulted in the loss of 90% of the country's male population and left the economy devastated. Marshal López himself perished and was buried on the battlefield of Cerro Cora.

International rivalry for control of a defeated Paraguay began with the emergence of national political parties; Anglo-Argentine capital supported the Partido Liberal (Liberal Party), while Brazilian interests supported the Partido Colorado (Red Party), forerunner of the present Asociación Nacional Republicana—Partido Colorado (National Republican Association—Colorado Party). The conservative Colorados remained in power between 1870 and 1904. The 1883 Law of Sale of Public Land enclosed land that had previously been accessible to all and turned it into vast private estates. Peasants were either forced to leave or to work for a pittance. In 1904 a revolution brought the Liberals to power, but achieved little else. The period from 1870 to 1940 was marked by a tumultuous series of coups and counter-coups, as rival groups within each party vied for political control of the riches to be gained from widespread foreign ownership of the national territory.

Of the 45 Presidents who governed between 1870 and 1979, all achieved power by force or by fraud and most were ousted by violence or the threat of it. Factionalism, inability to compromise and the relative weakness of the parties in government left the army as the central institution in Paraguayan politics. Paraguay's historic need to defend its borders justified the role of the armed forces and led to one of the highest ratios in the world of military and police to population.

### THE CHACO WAR AND ITS AFTERMATH

Paraguay gradually extended its control west of the River Paraguay, into the arid Chaco region. In response, Bolivia, which also claimed possession of the region, sent troops into the disputed territory. In the ensuing Chaco War (1932–35), Paraguay defeated Bolivia and, under the terms of a 1938 peace treaty, brokered by Argentina, was awarded three-quarters of the disputed territory.

Dissatisfaction with the Liberal war effort, however, led to the overthrow, by a military coup, of the Government of Eusebio Ayala in February 1936. The coup, led by Col Rafael Franco and supported by war veterans, brought to power the reformist Government of the Partido Revolucionario Febrerista (PRF—February Revolution Party). Although the Government did manage to seize and redistribute some land, its tenure was short-lived and the Febreristas, as Franco's followers soon became known, were overthrown in 1937 by army officers loyal to the Liberals.

Following a two-year interim presidency, Marshal José Félix Estigarribia (the Liberal leader in the Chaco War) became President. Estigarribia was a reformist nationalist, popular with both the military and peasants. Nevertheless, his restoration of political freedoms was met with generalized unrest, including strikes, attacks by the press and conspiracies by some military cliques, among which the Febrerista movement survived. Estigarribia therefore declared himself a temporary dictator, repressed opposition and announced a developmentalist land programme which included land expropriation. A new corporatist Constitution, which came into force in August 1940, strengthened executive powers and permitted the President to serve a second term. Estigarribia did not benefit from these constitutional changes, however, as both he and his wife were killed in an aeroplane crash only three weeks after the Constitution was approved.

Estigarribia's former Minister of War, Gen. Higinio Morínigo, assumed power following Estigarribia's death. Initially, Morínigo was regarded as a reasonably benevolent autocrat who faced the unenviable task of balancing opposing political forces to retain control of the nation. However, he soon assumed absolute powers, banning all political parties, repressing the activities of the trade-union movement and dissolving the legislature. Eminent Liberals were forced into exile and Febrerista uprisings were suppressed.

The Allied victory in the Second World War, however, gave rise to a military movement in 1946, which was directed against the Axis sympathizers behind Morínigo. Exiles were allowed to return and some conservative Colorados, as well as young, developmentalist Febrerista officers, were invited to join a Colorado-PRF coalition Government, under Morínigo's nominal control. From 1946 there was an increase in public unrest. Threatened by a growth in popular political activity and the emergence of a strong left-wing movement, Morínigo excluded the PRF from the Government and openly sided with the Colorados. An attempted coup late in 1946 was followed by the disintegration of the coalition Government. Declaring a state of siege, Morínigo formed a new military cabinet in 1947; on the rebel side, Liberals and Communists joined Febrerista forces, led by Col Rafael Franco, in a civil war which broke out in March 1947 and which divided the armed forces, with some four-fifths of officers defecting to the rebels. The Colorados triumphed, partly owing to support from the Argentine Government of Gen. Juan Domingo Perón Sosa.

The defeat of the rebels gave the Colorados control of the army and thus of the country. A number of coups followed, the first of which, in 1948, removed Morínigo from power. Presidential elections were held in the same year, but the only candidate was the Colorado, Juan Natalicio González. President González was supported by an all-Colorado legislature; Paraguay had become a one-party state. Although Morínigo went into exile, factional infighting continued. There were uprisings by Colorado officers against the González Government. Eventually, González too fled abroad. Another Colorado faction, led by Dr Federico Chávez, assumed power.

In October 1951 President Chávez appointed Gen. Alfredo Stroessner Mattiauda, a veteran of the Chaco War, as

Commander-in-Chief of the Armed Forces. Paraguay's economy began to deteriorate, however, and as inflation rose, so did political opposition. On 4 May 1954 Stroessner deposed Chávez in a military coup and in the July presidential elections Stroessner, a Colorado candidate, was elected unopposed to complete Chávez's term of office.

## THE RULE OF GENERAL STROESSNER

Gen. Stroessner immediately established a personal dictatorship. Restrictions were placed on all political activities and the Febrerista and Liberal opposition groups were ruthlessly suppressed. In 1956 Stroessner forced his principal rival within the Partido Colorado, Epifanio Méndez Fleitas, into exile. A state of siege was imposed. For over 30 years, until 1987, this state of siege was renewed every 60 days to comply with constitutional requirements. The unaccustomed sense of order that existed during Stroessner's dictatorship was advantageous to both domestic and foreign companies, and his commitment to the IMF austerity measures contributed to the stabilization of the national currency, the guaraní, by 1957. In 1958 Stroessner, as the sole candidate of the only permitted party, was re-elected President and continued to be re-elected in this way every five years until 1988. Opposition continued, but it originated mostly from outside Paraguay; attacks from Argentina by exiles were repelled in 1959 and 1960.

Stroessner's command of the armed forces and of the economy contributed to the strength of his position. Increasing confidence in the economy in the 1960s led to the encouragement of some limited political activity. Stroessner encouraged the pretence of democracy by allowing a dissident wing of the Partido Liberal, the Renovación (Renovation) wing, to participate in controlled legislative elections, in which it received one-third of the seats in the legislature. This did not in any way diminish the President's personal dominance of Paraguayan politics, since he controlled the ruling Partido Colorado, which held the remaining two-thirds of the parliament. In 1959 some 400 Colorado politicians who opposed Stroessner had been imprisoned or had fled into exile, where they formed the Movimiento Popular Colorado (MOPOCO—Colorado Popular Movement), under the leadership of Méndez Fleitas. The President then reorganized the purged Partido Colorado in order to facilitate the entrenchment of an authoritarian style of government. By 1967 the Partido Colorado constituted only members loyal to Stroessner. In that year the President changed the Constitution of 1940 to permit his legal re-election to a fourth term of office.

In the late 1960s the overtly autocratic nature of Stroessner's regime encouraged criticism from the Roman Catholic Church in Paraguay, which, in turn, resulted in popular unrest. However, the upturn in the economy experienced in the 1970s contained much of the opposition. The majority of opposition parties boycotted the presidential and legislative elections of February 1983, enabling Stroessner to obtain more than 90% of the votes cast in the presidential poll, and in August he formally took office for a seventh five-year term. In the mid-1980s the question of who would succeed Stroessner became increasingly important; his son was considered a likely candidate. In April 1987 the President announced that the state of siege was to be ended, since extraordinary security powers were no longer necessary to maintain peace. His decision to seek re-election in the February 1988 presidential elections, for an eighth consecutive term, precipitated the final crisis of Stroessner's regime. Although it was announced that he had received 88.6% of the votes cast, opposition leaders complained of electoral malpractice, and denounced his re-election as fraudulent. On 3 February 1989 Stroessner was overthrown in a coup, led by his son-in-law, Gen. Andrés Rodríguez, the second-in-command of the armed forces. (In September 2004 arrest warrants were issued for Stroessner and some 30 retired military personnel, including former Chief of Staff Alejandro Fretes Davalos. The warrants related to three Paraguayan nationals who 'disappeared' in Argentina, allegedly as a result of 'Plan Condor'—an intelligence operation to eliminate opponents of the Latin American military dictatorships in the 1970s.)

## DEMOCRATIZATION

On 1 May 1989 Gen. Rodríguez was elected President, as the official candidate of the Partido Colorado, with 74% of the votes cast. The Partido Colorado, having won 73% of the votes in the congressional elections, automatically took two-thirds of the seats in both the lower house of the legislature, the Cámara de Diputados (Chamber of Deputies), and the Senado (Senate—48 and 24 seats, respectively). However, the process of democratization continued. In Paraguay's first ever municipal elections, which took place on 26 May 1991, the Colorados won only 43% of the votes cast in 154 of the 206 municipalities, compared with 29% taken by the opposition Partido Liberal Radical Auténtico (PLRA—Authentic Radical Liberal Party). The Government's liberalizing austerity programme accelerated the formation of both new trade-union and peasant movements, which carried out a series of illegal land occupations. On 20 June 1992 a new Constitution was promulgated.

Luis María Argaña Ferraro, leader of the conservative Movimiento de Reconciliación Colorado (MRC—Movement of Colorado Reconciliation) faction, was nominated in December 1992 as the Partido Colorado's candidate for the presidential elections due in August 1993. His nomination, however, was reversed following pressure from President Rodríguez and from the Commander of the First Corps, Gen. Lino César Oviedo Silva, who had political ambitions of his own. Argaña took refuge in Brazil, urging his supporters not to vote for his rival. On 9 May 1993, however, the new official candidate, Juan Carlos Wasmosy, a former business associate of President Rodríguez but a political novice, won the presidential election with 40% of the votes cast. International observers agreed that the elections were generally fair, despite the partisanship of Gen. Oviedo, and Wasmosy thus became the first civilian President of Paraguay for 39 years.

## THE GOVERNMENT OF PRESIDENT WASMOSY

Wasmosy was inaugurated as President on 15 August 1993. There was widespread concern over the composition of Wasmosy's first Council of Ministers, many of whom had served in the administrations of Rodríguez and Stroessner. Despite the new President's apparent desire to restrict the influence of the military, the appointment on 18 August of Gen. Oviedo as Commander of the Army provoked further criticism. As a result, the PLRA refused to co-operate with the Colorados and there were violent demonstrations outside the Congreso Nacional (Congress) building. In September 1994 President Wasmosy carried out a reshuffle of the command of the armed forces, strengthening the position of Gen. Oviedo, but tensions remained high between the two men and, on 22 April 1996, the President finally requested that Oviedo resign. Oviedo had begun to campaign for the leadership of the Partido Colorado and hoped to succeed Wasmosy as President in 1998. Oviedo refused to step down and, with the support of some 5,000 troops, in turn demanded the resignation of the President, who sought asylum in the US embassy. On 24 April the President agreed to a compromise by which Oviedo would retire from active service in exchange for the offer of the post of defence minister.

The Congreso, however, refused to ratify Oviedo's appointment, which was withdrawn. The new Commander of the Army, Gen. Oscar Díaz Delmas, did not intervene, and three days later Argaña was again elected President of the Partido Colorado. A purge of senior military commanders followed. On 13 June 1996 Oviedo was arrested and detained on a charge of sedition, of which he was cleared by the Courts of Appeal on 7 August and freed. In September he narrowly defeated Argaña in the election to become the Partido Colorado's presidential candidate. Following Oviedo's victory in the primary election, the President attempted to exclude him from the presidential contest. In October an arrest order was issued against Oviedo on charges of making inflammatory public statements. Nevertheless, the general eluded capture until mid-December, when he surrendered.

In March 1998 the Special Military Tribunal convened by President Wasmosy found Oviedo guilty of rebellion and sentenced him to 10 years' imprisonment and dishonourable discharge. In April the Supreme Electoral Tribunal annulled Oviedo's presidential candidacy. The nomination of the Partido Colorado was assumed by Raúl Cubas Grau, a wealthy engineer.

Argaña was to be the new candidate for the vice-presidency. Despite public protests, the elections were held as planned on 10 May 1998. Cubas Grau obtained 55.4% of the votes cast, ahead of the candidate of the Alianza Democrática (Democratic Alliance), Domingo Laíno, who received 43.9% of the votes.

## DEMOCRATIZATION FALTERS

Cubas Grau assumed office on 15 August 1998 and a new Council of Ministers, including two pro-Oviedo generals, was sworn in on the same day. On 18 August the President issued a decree commuting Oviedo's prison sentence to time already served. The new Congreso immediately voted to condemn the decree and to initiate proceedings to impeach the President for unconstitutional behaviour. On 2 December the Supreme Court ruled the decree unconstitutional. Three days later the Argaña faction-controlled central apparatus of the Colorados expelled Oviedo from the party. While the Congreso was unable to muster the two-thirds' majority support necessary to impeach President Cubas Grau, the country remained in effective political deadlock, a situation exacerbated by the fact that the economy had been in recession since 1997.

On 23 March 1999, however, the political impasse ended dramatically when Vice-President Argaña was assassinated in the capital, Asunción, by three men in military uniform. The 66-year-old Vice-President and his supporters had just succeeded in regaining control of the Colorado headquarters, from which they had been expelled on 14 March by supporters of Oviedo and President Cubas Grau, who were immediately accused by supporters of Argaña of being at least the 'moral instigators' of the assassination. Large crowds took to the streets to demand the President's resignation, and the situation was further inflamed when six protesters, demonstrating outside the Congreso building on 25 March, were killed, apparently by an official at the finance ministry who was filmed firing at the crowds from a building near the Congreso.

On 28 March 1999, however, hours before the Congreso was due to vote on his impeachment, President Cubas Grau resigned and fled to Brazil, where he was granted political asylum. (In February 2002 Cubas Grau surrendered to the Paraguayan authorities to face trial for the killings of the protesters.) At the same time, Oviedo was granted asylum in Argentina. The President of the Congreso, pro-Argaña Colorado senator Luis González Macchi, became Head of State for the remainder of Cubas Grau's presidential term (which was scheduled to end in 2003). Hoping to overcome the disagreements existing within the Partido Colorado and between the Government and opposition in the legislature, on 30 March 1999 he announced the composition of a multi-party Government of National Unity. On 12 July Oviedo was officially expelled from the Partido Colorado. His arrest was ordered three days later, but the Argentine authorities twice refused to extradite him, and on 9 December he left Argentina to avoid being extradited by the new Government of Fernando de la Rúa.

On 18 May 2000 rebellious soldiers thought to be sympathetic to Oviedo seized the First Cavalry Division barracks and other strategic points. The coup was swiftly suppressed by the Government, which declared a 30-day nation-wide state of emergency, assuming extraordinary powers which resulted in the arrest of more than 70 people, mostly members of the security forces. (The subsequent trial of 18 military officials accused of involvement in the attempted coup collapsed, in judicially controversial circumstances, in May 2003.) However, on 11 June Oviedo himself was arrested in Foz do Iguaçu, Brazil, by the Brazilian authorities. In March 2001 the Brazilian chief prosecutor ruled that Oviedo could be extradited; however, in December Brazil's Supreme Court rejected the ruling, stating that Oviedo was a victim of political persecution, and released him from imprisonment.

The PLRA withdrew from the government coalition in February 2000 when it became clear that, contrary to the national unity agreement, the Partido Colorado would contest the elections for a new Vice-President. In April the Colorados nominated Félix Argaña, son of the late Vice-President, to be their party's candidate. However, at the elections, which were held on 13 August, the 53-year Colorado monopoly on power was ended when Argaña was narrowly defeated (by less than 1% of the votes cast) by the PLRA candidate, Julio César Franco.

In February 2001 the Minister of Education and Culture, Oscar Nicanor Duarte Frutos, resigned in order to campaign for the presidency of the Partido Colorado, which he secured, with almost 50% of the vote, in elections in early May. At the same time, the Government was further undermined by tensions between the President and the Vice-President; Franco, as the most senior elected government official, demanded the resignation of President González Macchi in the interest of democratic legitimacy. In an attempt to restore confidence, in March González Macchi carried out a cabinet reshuffle in which José Antonio Moreno Rufinelli was appointed to the post of Minister of Foreign Affairs, replacing Juan Esteban Aguirre. On the same day, some 10,000 farmers marched through Asunción to protest against the Government's economic policy and to demand greater action on rural issues. Furthermore, at the end of March, several thousand rural workers who had been camping outside the parliament building in protest at depressed cotton prices returned home after securing government concessions. Another corruption scandal emerged in early May 2001 that forced the President of the Central Bank of Paraguay, Washington Ashwell, to resign over his alleged involvement in the fraudulent transfer of US $16m. to a US bank account. Later that month, in response to allegations that the President had been a beneficiary of the misappropriated funds, opposition parties launched a bid to impeach him. However, the PLRA was forced to drop the charges when, on 6 September, it failed to secure the necessary two-thirds' majority in the Congreso. In April 2002 González Macchi was formally charged with involvement in the corruption scandal.

In late January 2002 the credibility of the Government was further undermined after two leaders of the left-wing party Movimiento Patria Libre (MPL—Free Homeland Movement) alleged that they had been illegally detained for 13 days and tortured by the police, with the knowledge of government ministers, as part of an investigation into a kidnapping case. In response to the allegations, the head of the national police force and his deputy, as well as the head of the judicial investigations department, were dismissed. Following sustained public and political pressure, the Minister of the Interior, Julio César Fanego, and the Minister of Justice and Labour, Silvio Ferreira Fernández, resigned soon afterwards, although both protested their innocence. In addition, the national intelligence agency, the Secretaría Nacional de Informaciones, was disbanded. The Cámara de Diputados issued a statement assigning some responsibility for the detention of the MPL leaders to the President and the Attorney-General, and describing the event as 'state terrorism'. Later that month 2,000 police officers were deployed to prevent a demonstration by peasants protesting against such 'state terrorism' from reaching government buildings.

During the first half of 2002 an alliance of farmers, trade unions and left-wing organizations staged mass protests throughout the country, calling for an end to the Government's free-market economic policies. The protests succeeded in reversing some of the Government's policies; most notably, the planned privatization of the telecommunications company Corporación Paraguaya de Comunicaciones (COPACO) was suspended in June. Nevertheless, the protests continued, and in July González Macchi declared a state of emergency, which was lifted two days later, after clashes between anti-Government protesters and the security forces resulted in two fatalities in Ciudad del Este. A further mass demonstration in September, organized by opposition parties, including Oviedo's Unión Nacional de Ciudadanos Éticos (UNACE) grouping, resulted in more clashes with the security forces.

No sooner was the state of emergency lifted than dissident legislators of the ruling Partido Colorado were reported to have initiated impeachment proceedings against President González Macchi on corruption charges. Meanwhile, despite limited attempts to bring the economy under control, both the fiscal deficit and, in consequence, government borrowing had continued to rise. On 29 June 2002 the Congreso had to raise the state guarantee of private bank deposits when Banco Alemán, the second largest private bank in the country, appeared to be in difficulties. (The bank was subsequently liquidated in September 2002.) Furthermore, the IMF refused to give assistance until the Congreso had agreed to increase the standard rate of value-added tax (VAT) and introduced a special tax on luxury items. However, the legislature failed to approve the necessary reforms, postponing the tax reform package until December; as a result, on 22 October James Spalding resigned as Minister of Finance.

On 5 December 2002 the Cámara de Diputados approved a proposal to impeach President González Macchi on five charges of corruption and failure to fulfil the duties of the presidency. Supported by the pro-Oviedo faction of the Partido Colorado and by the opposition PLRA, this was the third such proposal since July. González Macchi found his position as President ever more tenuous ahead of the presidential election that was due to be held in April 2003. In a bid to avoid proceedings against him, he offered to step down from office early, immediately following the April elections (instead of in August, when the inauguration of the new head of state would take place). However, on 11 February 2003 the Senado voted against approving the charges against him, by 25 votes to 18, with one abstention and one absence, thus falling short of the two-thirds' majority required by the Constitution.

### THE GOVERNMENT OF PRESIDENT DUARTE FRUTOS

The legislative and presidential elections of 27 April 2003 passed off without incident. The ruling Asociación Nacional Republicana—Partido Colorado extended its unbroken 56-year hold on power as its candidate, the 46-year-old lawyer and former minister Oscar Nicanor Duarte Frutos, won 37.1% of the votes cast in the presidential ballot, compared with the 24.0% secured by the PLRA candidate and former Vice-President (until October 2002), Julio César Franco. A businessman, Pedro Nicolás Fadul, of the Patria Querida (Beloved Fatherland) movement, came third, with 21.3% of the votes, and Guillermo Sánchez Guffanti, a radical populist representing Oviedo's UNACE party, came fourth, with 13.5% of the ballot. Concurrent elections to the 80-seat Cámara de Diputados gave the Colorados 37 seats, the PLRA 21 seats, the Patria Querida 10 seats, UNACE 10 seats and the Partido País Solidario (Party for a Country of Solidarity) two seats. The Colorados also won the most seats in the 45-seat Senado (16), followed by the PLRA with 12 and the Patria Querida with eight seats.

The new President, who took office on 15 August 2003, advocated free-market policies and pledged to tackle corruption, to put the public finances in order by renegotiating Paraguay's external debt and to restore the country's credibility internationally. He also promised to take personal charge of the customs service. President Duarte also promised that the sale of state-owned assets would remain a lesser priority than other measures to improve the Government's finances, although it remained unclear what these would be. The IMF initially remained reluctant to revive its lending programme to Paraguay in view of the Congreso's record of hostility towards reform measures, raising fears that the incoming administration faced the risk of an early financial crisis. However, the President soon showed he was serious in his determination to address the problem of corruption, announcing his intention to reform the judiciary in August. Furthermore, in October the Minister of the Interior, Roberto Eudez González Segovia, a close ally of the President, was forced to resign after being accused of corrupt practices. As a result of the new Government's policies, in December the World Bank granted the country a structural adjustment loan worth US $30m. to fund the proposed reforms, and in the same month the IMF approved a stand-by loan of $73m. on condition that structural reform continue.

Together with moves to overhaul the unpopular Supreme Court, the anti-corruption campaign won President Duarte Frutos unprecedented popularity, with polls in late 2003 showing support for him of over 60%. However, in January 2004 the President arrived back from his annual holiday in Brazil surrounded by heavy security, reportedly because an assassination attempt had been planned. Later in the same month farmers' organizations demanded the resignation of the Minister of the Interior after police had opened fire on a group of people demonstrating against crop-spraying, killing two farmers; a large number of officials were later disciplined. The President also continued to encounter congressional opposition to his reform measures. In spite of this opposition, in March reform of the Supreme Court was achieved with the replacement of two-thirds of its nine justices; however, many observers were disappointed that the new appointments were simply distributed between the three main political parties. Later in the year a proposed amendment to the Constitution to allow presidential re-election failed, and by the end of 2004 the President's popularity had fallen, though at 44% it was still high compared with many of his predecessors. Meanwhile, in June Oviedo returned from exile in Brazil, apparently hoping both to avoid conviction for the various charges he faced, including sedition against the Wasmosy administration, and to relaunch his political career. Instead he was arrested at the airport and taken to military prison to begin a 10-year sentence. However, in October he was provisionally acquitted of charges of sedition and in January 2005 he was also found not guilty of charges relating to the discovery of an arms cache. Despite these rulings in Oviedo's favour, several other charges were pending against him in 2005, and in April the Supreme Court ruled that he would have to serve his 10-year term of imprisonment. In late June several thousand of his supporters protested in Asunción against his detention.

In one of the worst tragedies of its kind in Latin America, on 1 August 2004 a fire at a supermarket near Asunción killed nearly 500 people and injured hundreds more. Following the disaster, the Director of the National Emergency Commission, Manuel Sarquis, was dismissed by President Duarte Frutos. Meanwhile, a spate of kidnappings led to the replacement of the interior minister, Orlando Fiorotto Sánchez, with Nelson Mora, hitherto the Attorney-General, following the kidnap and murder in October of the 11-year-old son of a businessman. The head of the national police was also dismissed. Previously, on 21 September Cecilia Cubas, the 32-year old daughter of former President Cubas Grau, was abducted in a paramilitary-style operation, apparently by a large criminal syndicate. Despite a ransom allegedly having been paid by her parents, on 16 February 2005 her body was found in an underground chamber at a house near Asunción. The six people subsequently arrested in connection with the kidnap and murder were alleged to have connections to the MPL and to the Fuerzas Armadas Revolucionarias de Colombia—Ejército del Pueblo, which was believed to be using Paraguay as a transhipment centre for drugs-trafficking. Allegations of serious investigative shortcomings by the authorities prompted the dismissal of Nelson Mora, who was replaced by the popular Rogelio Benítez, a former mayor of Encarnación. Thirty-two senior police officials were also dismissed amid criticism of the deteriorating security situation in the country. On 7 March Colombia and Paraguay signed an agreement to co-operate on security and drugs-trafficking issues, in support of the new three-year security plan for Paraguay announced by President Duarte Frutos in the previous month. The plan met with vociferous and occasionally violent protests by peasants' groups, owing primarily to fears that a greater military presence in rural areas would lead to increased repression of their campaign for agrarian reform, which had involved illegal land occupations since 2003.

# Economy
## Prof. PETER CALVERT

Paraguay is an agricultural country but its economy is determined to a large extent by its geographical position, which enables a relatively large informal sector to benefit from smuggling. This informal sector is reputed to be as large as the formal sector. In recent years Paraguay has also suffered greatly from overspill from the economic crisis in Argentina. The country has a land area of 406,752 sq km (157,048 sq miles) and is one of the smaller republics of South America. Together with Bolivia, it is one of the continent's two land-locked countries, with Brazil to the north and east, Argentina to the south and Bolivia to the west. According to July 2005 estimates, the country had a population of 6,348,884 and, therefore, a population density of approximately 15.6 persons per sq km. The country is divided into two distinct geographical regions by the River Paraguay, which joins the River Paraná at an altitude of 46 m above sea level. The majority of the population lives within 160 km of the capital, Asunción, to the east of the river. This region consists of grassy plains and low hills and has a temperate climate. It is divided into two zones by a ridge of hills, to the east of which lies the Paraná plateau, which ranges from 600 m to 2,300 m in height. West of the ridge lie gently rolling hills. To the west of the Paraguay river is the Chaco region, an arid, marshy plain extending to the foothills of the Andes and to the country's border with Bolivia. The Chaco accounts for 61% of Paraguay's land area but is inhabited by less than 2% of the population.

In the 1980s Paraguay had one of the highest population growth rates in Latin America. It was still projected to be high, at 2.5%, in 2005. In that year the estimated birth rate was 29.4 births per 1,000 inhabitants and the death rate was forecast at 4.5 deaths per 1,000. The infant mortality rate per 1,000 live births was projected in 2005 at 25.6, compared with 27.7 in 2004, and estimates of life expectancy at birth had improved further to 74.9 years (females 77.6, males 72.4). The climate was generally healthy, though water pollution presented a health risk for many urban residents and in 2002 only around 83% of the population had access to an improved water source. In 2000 there was one doctor for every 1,170 inhabitants.

Owing to the fertility of eastern Paraguay, economic activity was widespread but unevenly distributed. Historically, the main concentration of activity was around Asunción, with a secondary, more recent, economic zone based in the industrial park, Parque Industrial Oriente, 23 km from Ciudad del Este (formerly Puerto Presidente Stroessner). The Chaco region accounted for less than 3% of economic activity. In 2001 the percentage of the population below the national poverty line was 36%, representing more than one in every three Paraguayans. However, according to the evidence of a survey published by the Government's own statistical service, this figure would include some 70% of Paraguayans in rural areas (only 56% of the population is urban) living on or near the margin of subsistence. Spending on social welfare in the Stroessner era (1954–89) was very low, resulting in a still poorly funded education system, inadequate health care and limited sanitation. The literacy rate, however, was relatively impressive by mid-2005, estimated at 94%, most Paraguayans being bilingual in Spanish and the indigenous official language, Guaraní. In 2003 the World Bank estimated gross national income (GNI) at $5,836.2m. and GNI per head at US $1,110 ($4,740 by purchasing-power parity), ranking Paraguay 127th in the world, among the World Bank's lower middle-income countries. However, on the UN Development Programme's Human Development Index Paraguay ranked significantly higher, at 89th.

## AGRICULTURE

Agriculture (including hunting, forestry and fishing) was fundamental to the Paraguayan economy, accounting for a a provisional 27.2% of gross domestic product (GDP) in 2003 and virtually all export earnings. During 1990–2003 the agricultural sector grew by an annual average of 1.9%; real agricultural GDP increased by 2.2% in 2002 and by 3.3% in 2003. In 2003 the sector (including forestry) engaged an estimated 53% of the employed labour force, according to the UN Food and Agriculture Organization (FAO). Some 9m. ha of land was classified as arable, of which only some 30% was cultivated. There had been little progress on the promised agrarian reform; the ownership of land, therefore, remained one of the most unequal in Latin America. The 1991 agricultural census showed that 351 landowners controlled some 40% of the country's arable land. There was a large subsistence sector, including more than 200,000 families. A further 200,000 rural families were 'landless'.

Paraguay is largely self-sufficient in basic foodstuffs. Maize, cassava (manioc) and wheat are the main food crops. According to the FAO, in 2004 the largest area, 2,064,300 ha, was planted with oil crops, of which 775,050 metric tons were harvested. In the same year 681,500 ha of cereal crops were planted, yielding a total of 1,379,000 tons, of which wheat accounted for 360,000 tons and maize 870,000 tons. Some 260,000 ha was cultivated with cassava, of which 3.9m. tons were produced.

The main products grown for export were soya beans, oilseeds, cotton and sugar cane. Production of soya and cotton increased dramatically with the colonization of the eastern border region in the 1970s, and the area planted continued to expand at the beginning of the 21st century. Production of soybeans was estimated to have reached a new record of 4.2m. metric tons in 2003, falling back to an estimated 3.8m. tons in 2004, although Paraguay remained the world's fourth largest exporter. The value of exports, moreover, had risen from US $59m. in 2002 to an estimated $516m. in 2003. In order to increase productivity and the pest- and drought-resistance of their crops, Paraguayan farmers agreed in March 2005 to pay royalties in the 2004/05 crop year to a US-based corporation, Monsanto, for its genetically modified (GM) soybeans, at a rate of $2.82 per sack of seed, sufficient to plant roughly 1 ha.

Production of cotton lint declined in 2002, to an estimated 41,000 metric tons from more than double that (97,200 tons) in the previous year. In 2003 270,000 ha were dedicated to fibre crops and an estimated 57,000 tons of cotton lint produced. However, exports of cotton fibres in that year were worth US $64.9m., compared with $37.9m. in 2002, justifying the hope that in 2003 a recovery in world prices for both soya and cotton would improve both production and export revenues. In 2003 some 1.77m. tons of oilseeds were exported at a total estimated value of $530m., making them marginally the country's most valuable export crop. However, only 871 tons of processed oils were exported, bringing in only $213,000 in earnings. According to FAO estimates, in 2003 sugar cane covered 62,205 ha and 3.33m. tons were produced. Sugar exports in 2003 also rose, from $1.2m. to $11.9m. Among tree crops, three were of particular note in 2004: 48,000 tons of tung nuts were grown on 9,000 ha; 3,100 tons of green coffee were produced on 3,000 ha of land; and 82,000 tons of maté, sometimes known as Paraguayan tea, was grown on 31,000 ha.

The gently undulating plains of eastern Paraguay are good ranching country and beef exports make up the largest remaining part of the formal sector. Beef, pork, eggs and milk are produced for domestic consumption. Stocks estimated by FAO in 2003 included 8.8m. head of cattle, 362,600 sheep, 108,000 goats and 3.25m. pigs, as well as 15.6m. chickens. Some 360,000 horses were also recorded. Like land ownership, possession of livestock in Paraguay was unevenly distributed, with nearly two-thirds (58%) of cattle owned by 1% of the producers. Cattle-ranching used to be the most important sector of the Paraguayan economy, but all meat-packing plants were closed in 1981 as a result of the European Union's (EU) Common Agricultural Policy (CAP) and the Lomé Convention (which was succeeded by the Cotonou Agreement in 2000), which excluded Paraguay. Only 2,812 metric tons of fresh beef and other beef products were exported in 2003, with a total value of US $2.6m. The problem was that cattle-ranching was not only subject to the changing fortunes of the Argentine and Brazilian markets, but affected by local outbreaks of foot-and-mouth disease, such as the reported cases in the Chaco in early 2003. The fishing catch in 2003 totalled an estimated 25,110 tons, almost all of which was for domestic consumption.

Paraguay's once abundant forest resources continue to be severely depleted by competitive logging for export. The country was officially estimated to have lost 0.5% of its forest cover per year between 1990 and 2000. The damage was most extreme in eastern Paraguay, where less than one-quarter was still forested by the late 1990s. Transport costs acted as a deterrent in the Chaco region, where some 10.5m. ha of primary forest cover remained. Given the continued illegal logging trade and the absence of a systematic reforestation programme, it was not surprising that production showed a steep decline compared with 1999. The World Bank had predicted that the primary forest cover of eastern Paraguay would completely disappear by 2005. Official figures almost certainly understated the amount of damage done and the quantities involved; they showed that in 2003 the country produced 4.0m. cu m of industrial roundwood and 5.8m. cu m of fuelwood, including charcoal, for domestic consumption. In 2003 sawnwood production was an estimated 550,000 cu m, of which 163,000 cu m was exported, valued at US $13.4m. Exports of wood and wood products totalled an estimated $52.9m. in that year.

## MINING AND ENERGY

Paraguay has few proven mineral resources. From 1964 limestone was mined for the manufacture of portland cement by the state-owned Industria Nacional de Cemento at Vallemí, in the department of Concepción. Deposits of bauxite, copper, iron ore and manganese were known to exist. Deposits of uranium have been found in both the east and west of the country.

Total electricity consumption in 2002 was estimated by the US Energy Information Administration (EIA) at 2,500m. kWh. Paraguay, bordered by two of the great rivers of South America (the Paraguay and Paraná), had abundant potential hydroelectric generating capacity, which was developed to create substantial revenue. Installed capacity in 1999 was 12,600 MW. Of the estimated 48,400m. kWh generated in 2002, 45,900m. kWh was exported, primarily to Brazil. The most important source of power was the Paraguayan-Brazilian Itaipú project on the River Paraná, with an installed generating capacity of 13,300 MW, scheduled to increase to 14,000 MW on completion of two new turbines. The first stage was inaugurated in late 1982 and the final turbine of the original plans came into operation in May 1991. Under the terms of the agreement Paraguay and Brazil were each to receive one-half of the energy generated; any that they could not consume was then to be offered to the other country at a preferential rate. Paraguay, however, did not benefit from the arrangement as much as was expected, as its electrical supply grid is on the Argentine and not the Brazilian standard, and persistent financial crises meant that the Brazilian Government was often late in settling its account.

A similar agreement for a hydroelectric project with an installed capacity of 2,760 MW (later upgraded to 3,100 MW) was signed by Paraguay and Argentina in 1983. The Yacyretá project was delayed and exceeded its budget, but the final turbine came 'on stream' in February 1998. However, later that year cracks in the dam walls were discovered, which led to three turbines being taken out of production, and plans to raise the water level from 240 ft above sea level to 272 ft have been delayed because of insufficient capital—the World Bank having refused in January 2000 to fund the project for environmental reasons. None the less, in April 2004 the Argentine and Paraguayan Governments pledged to raise the water level sufficiently by 2007.

Following the Chaco War (1932–35) there were repeated hopes that petroleum deposits might be found in that part of the Chaco plain adjoining the southern Bolivian oilfields. These hopes were reactivated by the discovery of petroleum deposits in the northern Argentine province of Formosa in 1984. However, exploratory wells in the Chaco proved to be dry, and in July 1996 the US company Phillips Petroleum abandoned its search. During 2003 Paraguay consumed some 23,000 barrels per day (b/d) of petroleum products, all of which were imported. The Petróleos Paraguayos (PETROPAR) refinery at Villa Elisa had a capacity of only 7,500 b/d. Total energy use in 2001 amounted to 715 kg of oil equivalent per inhabitant. In June 2005 the Presidents of the four Mercosur (Mercado Común del Sur— Southern Common Market) member countries (Paraguay, Argentina, Brazil and Uruguay) signed an agreement creating an 'energy ring' (*anillo energético*), which would facilitate the supply and storage of natural gas from the Camisea field in Peru.

## MANUFACTURING

During 1990–2003 the manufacturing sector grew by an average of only 1.2% per year. According to provisional figures, industry (including mining, manufacturing, construction and power) accounted for 24.2% of GDP and manufacturing for 13.6% of GDP in 2003. The sector was dominated by the processing of agricultural inputs, including sugar and wood products. There were also many small companies engaged in import substitution for the domestic market, particularly in the cement, textiles and beverages sub-sectors. From 1995 the country's membership of Mercosur brought significant export advantages, but also increased competition from Argentine and Brazilian imports. In August 2001 the official rate of manufacturing unemployment was 15.3%.

There were two steel mills. The formerly state-owned Aceros del Paraguay (ACEPAR) was privatized in 1997; a second, built by a Brazilian company, Ioscape, to process ore from Corumbá (Brazil), began operations in 1994. There was some production of metal goods and machinery. In the 1990s there was substantial investment in the production of cotton yarn and paper.

## COMMERCE

The main feature of the economic upturn of the 1970s was the joint development with Brazil of the Itaipú hydroelectric project on the River Paraná, which involved a good deal of construction work and attracted many Brazilian settlers. The opening of road links with Brazil stimulated trade, both legal and illegal. Although inflation was high, the second half of the 1970s generated annual GDP growth rates of more than 10%. This growth ended abruptly, however, when the Brazilian economy entered a period of crisis in 1982. GDP growth fell to 1.6% per year, and although work continued at Itaipú, the huge revenues anticipated from the sale of Paraguay's share of its power to Brazil's developing south did not materialize.

In the late 1980s Paraguay, taking advantage of its geographical position, was able to re-establish itself as an entrepôt for intra-regional trade. A busy commercial sector was well established, engaged in the import of consumer goods from the USA, Japan and other Asian countries, for re-export to neighbouring countries. The services sector accounted for a provisional 48.5% of GDP in 2003 and grew by an annual average of 0.9% during 1990–2003.

Paraguay's informal economy was believed to be as extensive as its formal one. Vastly improved road links with Brazil had contributed to Paraguay becoming a major centre for contraband activities. Much of the informal economy consisted of profit from the smuggling of genuine and counterfeit clothing, valuable electronic goods and luxury items such as watches, perfume, spirits and tobacco into Argentina and Brazil. Paraguay was also a major illicit producer of marijuana (cannabis) for the international market, and in recent years played an increasingly significant role in the transhipment of cocaine from Colombia to the USA and Europe.

One of the consequences of this 'black' economy was inflation. In 2002 consumer prices rose by 10.4%. The rate increased to 14.3% in 2003, but fell back to 4.4% in 2004. Since the Brazilian devaluation in 1998, Paraguay has operated a 'managed float' of its currency. At the end of December 2003 the exchange rate to the US dollar was 6,079 guaraníes, little changed since the end of December 2002. By the end of December 2004 the rate had improved slightly, to 5,967 guaraníes to the dollar, but fell back to 6,235 at the end of May 2005.

## TRANSPORT AND COMMUNICATIONS

Historically, waterways provided the main mode of transport for foreign trade in Paraguay, although from the 1980s road transport became more important and by 1999 the leading point of exit for exports of cereals and vegetables was Ciudad del Este (869,373 metric tons). The country had 3,100 km of waterways, with ports at the capital, Asunción, San Antonio, Encarnación and Villeta. In 2004 the merchant fleet consisted of 44 vessels, with a total displacement of 44,300 grt, including two oil tankers. The country had free-port facilities at Nueva Palmira in Uruguay, although in the 1990s services on the Paraguay–Paraná network became both irregular and expensive. From 1994 Paraguay was linked with the Brazilian port of Santos on the Atlantic Ocean by the Hidrovía project, a network of canals and waterways sponsored by Mercosur.

Paraguay's principal road network was a triangle linking Asunción, Encarnación and Ciudad del Este. The Trans-Chaco Highway linked Asunción to the Bolivian border and there were links with Argentina and Brazil via the international bridge over the River Paraná at Ciudad del Este. In 1999, of the estimated 29,500 km of roads, 14,986 km were paved. Excessive reliance on road-hauled container traffic continued to cause congestion on the bridge at Ciudad del Este, despite the opening, in 1995, of a new container port at Hernandarias on the banks of Lake Itaipú.

The state-owned railway, Ferrocarril Presidente Carlos Antonio López, owned a 370-km (274-mile) line linking Asunción with Encarnación, on the Argentine border. In 1998 a regular steam service operated only as far as Ypacaraí, 35 km from Asunción. A number of other lines, mostly privately owned, made up the remainder of the nominally operated 441 km of track.

There were 12 airports in 2004 with paved runways, three of which had runways of more than 3,047 m. The main international airport, Aeropuerto Internacional Silvio Pettirossi, is situated 15 km from Asunción. The state-owned international airline, Líneas Aéreas Paraguayas (LAPSA), was privatized in October 1994; in 1997 its name changed to Transportes Aéreos del Mercosur (TAM Paraguay) and 80% of ownership was transferred to Transportes Aéreos of Brazil. Aerolíneas Paraguayas (ARPA) operates daily flights between Paraguay's major cities.

The state-owned telephone service, Administración Nacional de Telecomunicaciones (ANTELCO), renamed the Corporación Paraguaya de Comunicaciones (COPACO) from December 2001, had a long history of overstaffing and low productivity. In November 1995 (when the number of people waiting for telephone lines was twice as many as the number of existing subscribers) ANTELCO signed an agreement with the German company Siemens to modernize the telephone system. By 2003, however, the network, covering 273,200 subscribers, was still inadequate, although two mobile cellular telephone companies were also in operation, reaching an estimated 1.8m. mobile subscribers. In 2001 there were 990,000m. televisions, although given Paraguay's central role in the regional smuggling of electrical goods, this figure seemed rather low. In December 2003 there were four internet service providers and 120,000 users, a penetration rate of only 2.2%.

## BANKING AND FINANCIAL SERVICES

There was a rapid expansion in the provision of banks and finance companies in the country after 1990. The expansion was generally believed to be directly related to the reorientation of the drugs trade from Colombia's Andean highlands to Paraguay and the subsequent increase in illegal funds flowing through the country.

In 1995 a major banking crisis occurred, which had a significant recessionary impact on the economy. In April a US $3.8m. fraud was uncovered in the Central Bank of Paraguay. The scandal involved a well-established practice under which obligatory commercial-bank deposits lodged with the Central Bank as part of legal reserve requirements were lent for short periods on the flourishing parallel, or 'black', market, through non-registered finance houses. Here they earned rates of return of over 30% for the high-ranking bank officials involved. Following the disclosure of the arrangement, Banco Comercial Paraguayo (BANCOPAR) and Banco General were required to return missing funds to the Central Bank at short notice. The banking superintendency intervened to rescue both banks, but the crisis continued and by the end of 1995 four banks, 10 finance houses, two mortgage companies and two private pension schemes were experiencing difficulties. Both the Central Bank and the IMF argued strongly against further subsidies for these institutions and eventually the Government agreed to their closure, while strengthening the regulatory regime in the banking sector. Fears that the 1995 banking crisis had not been overcome were confirmed in June 1997 when both Ahorros Paraguayos, a mortgage company, and its parent company, Banco Unión, collapsed as a result of the earlier crisis. Further, apparently isolated, banking crises in more recent years were attributed to speculation and uncertainty surrounding the guaraní's deteriorating value against the major global currencies. Moreover, in June 2003 an official investigation was launched into allegations that some $30m. in financing for the Lebanese militant group Hezbollah had been 'laundered' via the Ciudad del Este region, further damaging the reputation of Paraguay's financial services industry.

Gross domestic investment fell by 4.3% in 2000 and by 16.0% in 2001. Moreover, owing to the loss of confidence in the country's economy, foreign direct investment in Paraguay had already fallen by some 56% in 1997. Concern about political instability continued to make overseas investors wary thereafter. Total foreign direct investment in 2001 was US $79.2m. In 2002 Paraguay was rated one of the most corrupt countries in the world by the non-governmental organization Transparency International.

## GOVERNMENT FINANCE AND INVESTMENT

In 2000 the economy contracted by 0.4%, and GDP per head fell by 2.9%. GDP increased by 2.7% in 2001, although GDP per head grew by only 0.2%. Real GDP fell in 2002, by 2.3% (a decline per head of 4.3%), but rose thereafter, by 2.6% in 2003 and by 2.9% in 2004. It was projected to further expand in 2005, by 2.7%. Total public-sector expenditure in 2004 was US $1,129m. This represented about 11% of GDP. The tax burden stood at only 9% of GDP (compared with 35% in Brazil) and tax evasion was estimated at 60%–70%. Despite this, a fiscal surplus (of an estimated 72,000m. guaraníes) was achieved in 2004.

Unlike many lower middle-income countries, Paraguay was not heavily indebted. At the end of 2004 total external debt reportedly stood at US $3,239m., compared with $3,210m. in 2003. The cost of debt-servicing in 2003 was $310m., down from $337m. in 2002. Total debt as a percentage of GNI was 53.2% in 2003, and the cost of debt-servicing was equivalent to 9.9% of the total value of exports of goods and services. Reserves had fallen significantly from the third quarter of 1997 onwards as the Central Bank was obliged to sell foreign exchange to arrest speculative attacks on the local currency, caused by persistent political uncertainty. Total foreign-exchange reserves, including gold, were $1,167m. at the end of 2004.

## FOREIGN TRADE AND BALANCE OF PAYMENTS

In 2001–03 Paraguay's most important trading partner was Brazil, which accounted for around one-third of all exports imports. However, there were substantial differences both between exports and imports and from one year to the next. The main exports in 2003 were soya beans, animal feed, cotton, meat, edible oils and electricity. After Brazil (34.2%), the main destinations for exports in 2003 were Uruguay (19.6%), Switzerland and Liechtenstein (7.8%) and Argentina (5.3%). Paraguay imported road vehicles, consumer goods, tobacco (much of which was illegally re-exported), fuels and electrical machinery. The main sources of imports, after Brazil (32.5%), in 2003 were Argentina (21.6%), the USA (3.8%) and Japan (3.5%).

In 2003 the trade deficit stood at US $260.2m., a slight increase on the $279.9m. recorded in the previous year. In 1992–2000 the trade balance had followed a trend of gently rising export levels being overtaken by a steep and continuing increase in the value of imports, a pattern that led to a widening deficit on the balance of trade. The trade deficit reached $678.2m. in that later year, falling to $613.9m. in 2001 and to $279.9m. in 2002. It was uncertain, however, what the impact of Paraguay's considerable informal trading activity was on this pattern. In the first half of 2003 higher international prices for Paraguay's main agricultural commodities led the value of exported commodities to increase by 38%, resulting in a current-account surplus of $146.0m. in that year as a whole, compared with a surplus of $73.2m. in 2002 and a current-account deficit of $266.4m. in 2001.

## CONCLUSION

Paraguay's geographical position placed its economy at a competitive disadvantage while also offering opportunities, which the informal sector was quick to seize. The country's main economic problems remained the extent of its informal sector, a lack of confidence among investors, as well as the vagaries of climate, a dependency on agriculture and the low intrinsic value of the country's principal agricultural exports. Successive Governments attempted to resolve at least some of these problems and Paraguay benefited substantially from its incorporation into Mercosur. Political instability, however, left Governments with little option but to delay the reform measures recommended by the IMF (most notably, the privatization of the telecommunications company COPACO was suspended in June 2002 following widespread protests) and, therefore, to retain the problems created by an under-funded and inefficient public sector. Reform of the public banking sector also remained a priority in order to increase investor confidence, which had waned from 2004, as demonstrated by the depreciation of the guaraní. This remained the main challenge facing the Government of President Duarte Frutos. His administration succeeded in eliminating the fiscal deficit by the end of 2004, at least according to the IMF's projections. It did this by increasing some tax rates, revising others and introducing a new tax on road vehicles; this was also successful in reducing inflation. In March 2005 the IMF, in its fourth review of the stand-by agreement with Paraguay, gave a broadly positive assessment of the Government's economic management, while encouraging further reform. Nevertheless, the Congreso remained unwilling to enact the necessary legislation and political uncertainty continued to impede economic recovery.

# Statistical Survey

Sources (unless otherwise stated): Dirección General de Estadística, Encuestas y Censos, Naciones Unidas esq. Saavedra, Fernando de la Mora, Zona Norte; tel. (21) 511016; fax (21) 508493; internet www.dgeec.gov.py; Banco Central del Paraguay, Avda Federación Rusay Sargento Marecos, Casilla 861, Asunción; tel. (21) 61-0088; fax (21) 60-8149; e-mail ccs@bcp.gov.py; internet www.bcp.gov.py; Secretaría Técnica de Planificación, Presidencia de la República, Iturbe y Eligio Ayala, Asunción.

## Area and Population

### AREA, POPULATION AND DENSITY

| | |
|---|---:|
| Area (sq km) | 406,752* |
| Population (census results) | |
| 26 August 1992 | 4,152,588 |
| 28 August 2002 | |
| Males | 2,627,831 |
| Females | 2,555,249 |
| Total | 5,183,080 |
| Population (UN estimates at mid-year)† | |
| 2003 | 5,878,000 |
| 2004 | 6,017,000 |
| Density (per sq km) at mid-2004 | 14.8 |

* 157,048 sq miles.
† Source: UN, *World Population Prospects: The 2004 Revision*.

### DEPARTMENTS
(census of August 2002)

| | Area (sq km) | Population | Density (per sq km) | Capital |
|---|---:|---:|---:|---|
| Alto Paraguay (incl. Chaco) | 82,349 | 13,250 | 0.2 | Fuerte Olimpo |
| Alto Paraná | 14,895 | 559,769 | 37.6 | Ciudad del Este |
| Amambay | 12,933 | 115,320 | 8.9 | Pedro Juan Caballero |
| Asunción | 117 | 510,910 | 4,366.8 | — |
| Boquerón (incl. Nueva Asunción) | 91,669 | 43,480 | 0.5 | Doctor Pedro P. Peña |
| Caaguazú | 11,474 | 443,311 | 38.6 | Coronel Oviedo |
| Caazapá | 9,496 | 139,080 | 14.6 | Caazapá |
| Canindeyú | 14,667 | 140,250 | 9.6 | Salto del Guairá |
| Central | 2,465 | 1,362,650 | 552.8 | Asunción |
| Concepción | 18,051 | 178,900 | 9.9 | Concepción |
| Cordillera | 4,948 | 233,170 | 47.1 | Caacupé |
| Guairá | 3,846 | 178,130 | 46.3 | Villarrica |
| Itapúa | 16,525 | 459,480 | 27.8 | Encarnación |
| Misiones | 9,556 | 102,230 | 10.7 | San Juan Bautista |
| Ñeembucú | 12,147 | 76,730 | 6.3 | Pilar |
| Paraguarí | 8,705 | 224,850 | 25.8 | Paraguarí |
| Presidente Hayes | 72,907 | 82,030 | 1.1 | Pozo Colorado |
| San Pedro | 20,002 | 319,540 | 16.0 | San Pedro |
| **Total** | **406,752** | **5,183,080** | **12.7** | — |

### PRINCIPAL TOWNS
(population at 2002 census, incl. rural environs)

| | | | |
|---|---:|---|---:|
| Asunción (capital) | 510,910 | Lambaré | 119,830 |
| Ciudad del Este* | 222,109 | Fernando de la Mora | 113,990 |
| San Lorenzo | 203,150 | Caaguazú | 100,132 |
| Luque | 185,670 | Encarnación | 97,000 |
| Capiatá | 154,520 | Pedro Juan Caballero | 88,530 |

* Formerly Puerto Presidente Stroessner.

### BIRTHS, MARRIAGES AND DEATHS
(UN estimates, annual averages)

| | 1985–90 | 1990–95 | 1995–2000 |
|---|---:|---:|---:|
| Birth rate (per 1,000) | 36.6 | 34.1 | 31.3 |
| Death rate (per 1,000) | 6.7 | 6.0 | 5.4 |

Source: UN, *World Population Prospects: The 2004 Revision*.

**2002:** Registered live births 46,012; Marriages 16,100; Deaths 19,416.
**2003:** Registered live births 45,669; Marriages 17,717; Deaths 19,593.
**Expectation of life** (WHO estimates, years at birth): 72 (males 69; females 75) in 2003 (Source: WHO, *World Health Report*).

### ECONOMICALLY ACTIVE POPULATION
(household survey, August–December 2000)

| | Males | Females | Total |
|---|---:|---:|---:|
| Agriculture, hunting, forestry and fishing | 689,825 | 212,497 | 902,322 |
| Mining and quarrying | 2,500 | — | 2,500 |
| Manufacturing | 186,322 | 99,864 | 286,186 |
| Electricity, gas and water | 10,432 | 3,300 | 13,732 |
| Construction | 107,340 | — | 107,340 |
| Trade, restaurants and hotels | 281,928 | 306,039 | 587,967 |
| Transport, storage and communications | 62,788 | 11,461 | 74,249 |
| Financing, insurance, real estate and business services | 55,836 | 28,950 | 84,786 |
| Community, social and personal services | 174,747 | 326,571 | 501,318 |
| Activities not adequately described | 208 | — | 208 |
| **Total employed** | **1,571,926** | **988,682** | **2,560,608** |

# Health and Welfare

## KEY INDICATORS

| | |
|---|---|
| Total fertility rate (children per woman, 2003) | 3.8 |
| Under-5 mortality rate (per 1,000 live births, 2003) | 29 |
| HIV/AIDS (% of persons aged 15–49, 2003) | 0.5 |
| Physicians (per 1,000 head, 2000) | 1.17 |
| Hospital beds (per 1,000 head, 1996) | 1.34 |
| Health expenditure (2002): US $ per head (PPP) | 343 |
| Health expenditure (2002): % of GDP | 8.4 |
| Health expenditure (2002): public (% of total) | 38.1 |
| Access to water (% of persons, 2002) | 83 |
| Access to sanitation (% of persons, 2002) | 78 |
| Human Development Index (2002): ranking | 89 |
| Human Development Index (2002): value | 0.751 |

For sources and definitions, see explanatory note on p. vi.

# Agriculture

## PRINCIPAL CROPS
('000 metric tons)

| | 2001 | 2002 | 2003 |
|---|---|---|---|
| Wheat | 359 | 359 | 360* |
| Rice (paddy) | 106 | 105 | 105* |
| Maize | 947 | 867 | 870* |
| Sorghum | 43 | 40 | 44* |
| Sweet potatoes | 131 | 124 | 125* |
| Cassava (Manioc) | 3,568 | 4,430 | 3,900* |
| Sugar cane | 2,396 | 3,210 | 3,300* |
| Dry beans | 53 | 54 | 54* |
| Soybeans (Soya beans) | 3,511 | 3,300 | 4,205† |
| Oil palm fruit* | 125 | 126 | 127 |
| Sunflower seed | 40 | 36 | 31† |
| Tung nuts | 45 | 46* | 48* |
| Cottonseed | 177† | 74† | 108* |
| Tomatoes | 55 | 58 | |
| Dry onions* | 34 | 34 | 35 |
| Carrots | 22 | 29 | 29* |
| Other vegetables | 51 | 48 | 50* |
| Watermelons* | 115 | 115 | 115 |
| Cantaloupes and other melons* | 28 | 28 | 30 |
| Bananas | 64 | 61 | 64* |
| Oranges | 209 | 207 | 210* |
| Tangerines, mandarins, clementines and satsumas | 24 | 23 | 25* |
| Grapefruit and pomelo | 48 | 47 | 47* |
| Mangoes* | 29 | 29 | 29 |
| Pineapples | 41 | 42* | 43* |
| Maté | 69 | 81 | 82* |

* FAO estimate(s).
† Unofficial figure.
Source: FAO.

## LIVESTOCK
('000 head, year ending September)

| | 2001 | 2002 | 2003 |
|---|---|---|---|
| Cattle | 9,889 | 9,260 | 8,810 |
| Horses | 358 | 358* | 360* |
| Pigs* | 3,150 | 3,200 | 3,250 |
| Sheep | 406 | 410 | 362† |
| Goats | 124 | 125 | 108† |
| Chickens | 15,350 | 15,504 | 15,550* |
| Ducks* | 715 | 720 | 725 |
| Geese* | 80 | 85 | 90 |
| Turkeys* | 95 | 95 | 100 |

* FAO estimate(s).
† Unofficial figure.
Source: FAO.

## LIVESTOCK PRODUCTS
('000 metric tons)

| | 2001 | 2002 | 2003 |
|---|---|---|---|
| Beef and veal | 200 | 205 | 215 |
| Pig meat* | 151 | 154 | 156 |
| Poultry meat* | 59 | 59 | 59 |
| Cows' milk | 331 | 375 | 380* |
| Hen eggs* | 70 | 72 | 73 |
| Cattle hides* | 38 | 36 | 34 |

* FAO estimate(s).
Source: FAO.

# Forestry

## ROUNDWOOD REMOVALS
(FAO estimates, '000 cubic metres, excluding bark)

| | 2001 | 2002 | 2003 |
|---|---|---|---|
| Sawlogs, veneer logs and logs for sleepers | 3,515 | 3,515 | 3,515 |
| Other industrial wood | 529 | 529 | 529 |
| Fuel wood | 5,646 | 5,743 | 5,843 |
| **Total** | 9,690 | 9,787 | 9,887 |

Source: FAO.

## SAWNWOOD PRODUCTION
('000 cubic metres, including railway sleepers)

| | 1995 | 1996 | 1997 |
|---|---|---|---|
| **Total** (all broadleaved) | 400 | 500 | 550 |

**1998–2003:** Annual production as in 1997 (FAO estimates).
Source: FAO.

# Fishing

(FAO estimates, unless otherwise indicated, '000 metric tons, live weight)

| | 1996 | 1997 | 1998 |
|---|---|---|---|
| Capture | 22.0* | 28.0* | 25.0 |
| Characins | 8.0 | 10.0 | 9.0 |
| Freshwater siluroids | 10.0 | 13.0 | 12.0 |
| Other freshwater fishes | 4.0 | 5.0 | 4.0 |
| Aquaculture | 0.4 | 0.4 | 0.1* |
| **Total catch** | 22.4 | 28.4 | 25.1 |

* Official figure.
Note: Figures exclude crocodiles, recorded by number rather than by weight. The number of spectacled caimans caught was: 503 in 1997; 4,445 in 1998; 0 in 1999; 9,750 in 2000; 3,793 in 2001; 8,373 in 2002; 3,781 in 2003.

**1999–2003:** Capture data as in 1998 (FAO estimates).
Source: FAO.

# Industry

## SELECTED PRODUCTS
('000 metric tons, unless otherwise indicated)

| | 2000 | 2001 | 2002 |
|---|---|---|---|
| Soya bean oil* | 138 | 177 | 219 |
| Sugar (raw)† | 135 | 149 | 170 |
| Hydraulic cement‡ | 650 | 650 | 650§ |

* Unofficial figures.
† FAO estimates.
‡ Data from US Geological Survey.
§ Estimate.

Source (unless otherwise indicated): FAO.

# Finance

## CURRENCY AND EXCHANGE RATES

**Monetary Units**
100 céntimos = 1 guaraní (G).

**Sterling, Dollar and Euro Equivalents** (31 May 2005)
£1 sterling = 11,335.8 guaraníes;
US $1 = 6,235.0 guaraníes;
€1 = 7,688.4 guaraníes;
100,000 guaraníes = £8.82 = $16.04 = €13.01

**Average Exchange Rate** (guaraníes per US dollar)
2002   5,716.3
2003   6,424.3
2004   5,974.6

## BUDGET
('000 million guaraníes)

| Revenue | 2002 | 2003* | 2004† |
|---|---|---|---|
| Taxation | 2,923 | 3,676 | 4,813 |
| Value-added tax | 1,235 | 1,570 | 1,995 |
| Import and export duties | 507 | 728 | 841 |
| Other | 122 | 101 | 841 |
| Non-tax revenue | 2,118 | 2,318 | 2,280 |
| Capital revenues | 7 | 7 | 3 |
| **Total** | 5,048 | 6,001 | 7,095 |

| Expenditure | 2002 | 2003* | 2004† |
|---|---|---|---|
| Current expenditure | 4,766 | 4,981 | 5,329 |
| Goods and services | 373 | 408 | 527 |
| Wages and salaries | 2,582 | 2,724 | 2,938 |
| Interest payments | 456 | 489 | 508 |
| Transfers | 1,326 | 1,334 | 1,299 |
| Pensions and benefits | 913 | 943 | 872 |
| Other | 29 | 26 | 57 |
| Capital expenditure and net lending | 1,281 | 1,165 | 1,694 |
| Net lending | −6 | −88 | −47 |
| **Total** | 6,047 | 6,146 | 7,023 |

* Estimates.
† Projections.

Source: IMF, *Paraguay: 2004 Article IV Consultation and Second Review Under the Stand-By Arrangement and Requests for Waiver and Modifications of Performance Criteria-Staff Report; Staff Statement; Public Information Notice and Press Release on the Executive Board Discussion; and Statement by the Executive Director for Paraguay* (February 2005).

## INTERNATIONAL RESERVES
(US $ million at 31 December)

| | 2002 | 2003 | 2004 |
|---|---|---|---|
| Gold* | 12.14 | 14.51 | n.a. |
| IMF special drawing rights | 113.18 | 125.75 | 133.62 |
| Reserve position in IMF | 29.20 | 31.91 | 33.35 |
| Foreign exchange | 486.82 | 811.19 | 1,000.17 |
| **Total** | 641.35 | 983.36 | 1,167.14 |

* National valuation $347 per troy ounce at 31 December 2002; $415 per troy ounce at 31 December 2003.

Source: IMF, *International Financial Statistics*.

## MONEY SUPPLY
('000 million guaraníes at 31 December)

| | 2002 | 2003 | 2004 |
|---|---|---|---|
| Currency outside banks | 1,451.20 | 1,814.57 | 2,116.69 |
| Demand deposits at commercial banks | 1,273.39 | 1,918.52 | 2,610.19 |
| **Total money** (incl. others) | 2,759.53 | 3,788.14 | 4,784.12 |

Source: IMF, *International Financial Statistics*.

## COST OF LIVING
(Consumer Price Index for Asunción; base: 2000 = 100)

| | 2001 | 2002 | 2003 |
|---|---|---|---|
| Food (incl. beverages) | 103.8 | 114.4 | 139.3 |
| Housing (incl. fuel and light) | 111.6 | 125.4 | 137.8 |
| Clothing (incl. footwear) | 103.1 | 108.4 | 117.0 |
| **All items** (incl. others) | 107.3 | 118.5 | 135.4 |

Source: ILO.

**2004:** All items 141.3.

## NATIONAL ACCOUNTS
('000 million guaraníes at current prices)

**National Income and Product**

| | 1999 | 2000 | 2001 |
|---|---|---|---|
| Compensation of employees | 7,764.3 | 8,340.1 | 9,030.2 |
| Operating surplus* | 12,736.9 | 14,467.8 | 14,699.2 |
| **Domestic factor incomes** | 20,501.2 | 22,807.9 | 23,729.4 |
| Consumption of fixed capital | 1,889.6 | 2,051.9 | 2,142.8 |
| **Gross domestic product (GDP) at factor cost** | 22,390.7 | 24,859.8 | 25,872.1 |
| Indirect taxes, *less* subsidies | 1,753.6 | 2,061.2 | 2,246.7 |
| **GDP in purchasers' values** | 24,144.3 | 26,921.0 | 28,118.8 |
| Factor income received from abroad, *less* factor income paid abroad | 132.2 | 380.7 | 164.2 |
| **Gross national product** | 24,276.5 | 27,301.7 | 28,283.0 |
| *Less* Consumption of fixed capital | 1,889.6 | 2,051.9 | 2,142.8 |
| **National income in market prices** | 22,386.9 | 25,249.8 | 26,140.2 |

* Obtained as a residual.

Source: UN Economic Commission for Latin America and the Caribbean.

**Expenditure on the Gross Domestic Product**

| | 2001 | 2002 | 2003* |
|---|---|---|---|
| Government final consumption expenditure | 2,479.9 | 2,470.2 | 2,679.3 |
| Private final consumption expenditure | 24,604.3 | 27,527.0 | 34,044.8 |
| Increase in stocks | 265.5 | 292.3 | 315.7 |
| Gross fixed capital formation | 5,295.0 | 5,812.5 | 7,373.6 |
| **Total domestic expenditure** | 32,644.7 | 36,102.1 | 44,413.4 |
| Exports of goods and services | 6,170.1 | 9,822.8 | 12,524.1 |
| *Less* Imports of goods and services | 10,696.0 | 13,948.0 | 18,131.9 |
| **GDP in purchasers' values** | 28,118.8 | 31,977.0 | 38,805.5 |
| **GDP at constant 1982 prices** | 1,157.0 | 1,130.1 | 1,159.0 |

* Provisional figures.

**Gross Domestic Product by Economic Activity**

| | 2001 | 2002 | 2003* |
|---|---|---|---|
| Agriculture, hunting, forestry and fishing | 6,663.1 | 7,546.6 | 10,570.8 |
| Mining and quarrying | 85.8 | 87.9 | 105.2 |
| Manufacturing | 4,011.6 | 4,475.0 | 5,274.9 |
| Construction | 1,310.9 | 1,432.8 | 1,793.1 |
| Electricity, gas and water | 1,783.2 | 2,041.8 | 2,229.9 |
| Trade, finance and insurance | 6,772.5 | 8,189.4 | 9,998.8 |
| Transport, storage and communications | 1,486.1 | 1,679.6 | 1,862.6 |
| Government services | 1,588.5 | 1,570.9 | 1,714.2 |
| Real estate and housing | 687.6 | 695.3 | 771.7 |
| Other services | 3,729.5 | 4,257.6 | 4,484.2 |
| **GDP in purchasers' values** | 28,118.8 | 31,976.9 | 38,805.5 |

* Provisional figures.

# PARAGUAY

*Statistical Survey*

## BALANCE OF PAYMENTS
(US $ million)

|  | 2001 | 2002 | 2003 |
|---|---|---|---|
| Exports of goods f.o.b. | 1,889.7 | 1,858.0 | 2,260.5 |
| Imports of goods f.o.b. | −2,503.6 | −2,137.9 | −2,520.7 |
| **Trade balance** | −613.9 | −279.9 | −260.2 |
| Exports of services | 555.2 | 568.4 | 589.8 |
| Imports of services | −390.0 | −349.8 | −348.1 |
| **Balance on goods and services** | −448.7 | −61.3 | −18.5 |
| Other income received | 255.9 | 193.5 | 165.1 |
| Other income paid | −240.1 | −174.9 | −165.2 |
| **Balance on goods, services and income** | −432.9 | −42.7 | −18.6 |
| Current transfers received | 168.0 | 117.5 | 166.1 |
| Current transfers paid | −1.5 | −1.6 | −1.5 |
| **Current balance** | −266.4 | 73.2 | 146.0 |
| Capital account (net) | 15.0 | 4.0 | 15.0 |
| Direct investment abroad | −5.8 | −5.5 | −5.5 |
| Direct investment from abroad | 84.2 | 9.3 | 90.8 |
| Portfolio investment assets | 0.7 | — | — |
| Portfolio investment liabilities | −0.1 | −0.1 | −0.4 |
| Other investment assets | 64.4 | −9.5 | 202.3 |
| Other investment liabilities | 4.8 | 60.8 | −69.6 |
| Net errors and omissions | 52.9 | −257.9 | −145.8 |
| **Overall balance** | −50.2 | −125.7 | 232.8 |

Source: IMF, *International Financial Statistics*.

## External Trade
(excl. border trade)

### PRINCIPAL COMMODITIES
(US $ million)

| Imports f.o.b. | 2001 | 2002 | 2003* |
|---|---|---|---|
| Food and live animals | 57.3 | 30.9 | 30.7 |
| Beverages and tobacco | 111.9 | 67.8 | 56.8 |
| Mineral fuels | 302.8 | 239.2 | 327.0 |
| Chemical products | 105.5 | 141.9 | 219.3 |
| Pharmaceuticals | 53.9 | 38.1 | 38.7 |
| Road vehicles | 110.7 | 59.4 | 58.6 |
| Transport equipment and accessories | 66.3 | 55.4 | 68.1 |
| Paper, paperboard and pulp | 83.4 | 68.4 | 67.9 |
| Textiles and textile products | 51.8 | 33.1 | 67.9 |
| Base metals and metal goods | 81.0 | 60.1 | 69.4 |
| Motors, general industrial machinery equipment and parts | 336.0 | 281.7 | 281.6 |
| Agricultural equipment and vehicles | 35.0 | 41.2 | 91.4 |
| **Total** (incl. others) | 1,590.0 | 1,251.8 | 1,598.9 |

| Exports f.o.b. | 2001 | 2002 | 2003* |
|---|---|---|---|
| Meat and derivatives | 65.1 | 72.7 | 61.2 |
| Cereals | 47.3 | 35.9 | 85.9 |
| Oleaginous seeds | 364.5 | 350.4 | 529.9 |
| Vegetable oils | 44.9 | 79.0 | 95.3 |
| Feeding stuff for animals | 96.2 | 120.1 | 129.4 |
| Leather | 48.2 | 47.5 | 45.1 |
| Wood and wooden products | 63.1 | 49.8 | 52.9 |
| Cotton fibres | 90.5 | 37.9 | 64.9 |
| **Total** (incl. others) | 1,167.4 | 1,132.8 | 1,496.1 |

* Preliminary figures.

## PRINCIPAL TRADING PARTNERS
(US $ million)

| Imports c.i.f. | 2001 | 2002 | 2003* |
|---|---|---|---|
| Argentina | 478.6 | 309.4 | 402.7 |
| Bahamas | 42.8 | 28.5 | 54.4 |
| Brazil | 563.6 | 477.6 | 605.1 |
| Chile | 42.0 | 19.3 | 32.7 |
| France (incl. Monaco) | 21.3 | 17.6 | 15.5 |
| Germany | 50.3 | 45.6 | 34.2 |
| Italy | 25.3 | 13.6 | 13.7 |
| Japan | 86.2 | 49.6 | 65.5 |
| Korea, Republic | 24.2 | 14.6 | 17.7 |
| Spain | 21.0 | 16.2 | 33.5 |
| Switzerland-Liechtenstein | 36.6 | 59.1 | 56.3 |
| Taiwan | 18.5 | 16.8 | 16.1 |
| United Kingdom | 35.1 | 22.0 | 33.9 |
| USA | 118.4 | 77.6 | 71.2 |
| Uruguay | 69.3 | 58.4 | 59.1 |
| **Total** (incl. others) | 1,988.8 | 1,510.2 | 1,862.0 |

| Exports f.o.b. | 2001 | 2002 | 2003* |
|---|---|---|---|
| Argentina | 60.8 | 34.7 | 66.4 |
| Brazil | 277.9 | 353.0 | 424.9 |
| Chile | 61.5 | 49.1 | 12.6 |
| Colombia | 10.5 | 3.0 | 1.4 |
| France (incl. Monaco) | 4.8 | 8.0 | 10.4 |
| Germany | 13.1 | 8.8 | 7.2 |
| Hong Kong | 12.0 | 11.8 | 9.2 |
| India | 28.5 | 2.8 | 0.1 |
| Italy | 42.1 | 34.9 | 40.4 |
| Japan | 11.8 | 8.1 | 5.4 |
| Netherlands | 29.4 | 19.2 | 12.5 |
| Spain | 10.8 | 7.7 | 8.7 |
| Switzerland-Liechtenstein | 34.4 | 33.4 | 97.1 |
| Taiwan | 11.6 | 11.0 | 11.5 |
| United Kingdom | 2.6 | 1.1 | 1.5 |
| USA | 29.3 | 37.4 | 44.1 |
| Uruguay | 180.0 | 165.1 | 243.1 |
| Venezuela | 7.4 | 9.7 | 6.5 |
| **Total** (incl. others) | 990.2 | 950.6 | 1,241.5 |

* Preliminary figures.

Source: UN, *International Trade Statistics Yearbook*.

## Transport

### RAILWAYS
(traffic)

|  | 1988 | 1989 | 1990 |
|---|---|---|---|
| Passengers carried | 178,159 | 196,019 | 125,685 |
| Freight (metric tons) | 200,213 | 164,980 | 289,099 |

Source: UN, *Statistical Yearbook*.

**Passenger-kilometres:** 3.0 million per year in 1994–96.

**Freight ton-kilometres:** 5.5 million in 1994.

Source: UN Economic Commission for Latin America and the Caribbean.

### ROAD TRAFFIC
(vehicles in use)

|  | 1999 | 2000 |
|---|---|---|
| Cars | 267,587 | 274,186 |
| Buses | 8,991 | 9,467 |
| Lorries | 41,329 | 42,992 |
| Vans and jeeps | 134,144 | 138,656 |
| Motorcycles | 6,872 | 8,825 |

Source: Organización Paraguaya de Cooperación Intermunicipal.

# PARAGUAY

## SHIPPING

**Merchant Fleet**
(registered at 31 December)

|  | 2002 | 2003 | 2004 |
|---|---|---|---|
| Number of vessels | 46 | 44 | 44 |
| Total displacement ('000 grt) | 47.5 | 45.1 | 44.3 |

Source: Lloyd's Register-Fairplay, *World Fleet Statistics*.

## CIVIL AVIATION
(traffic on scheduled services)

|  | 2001 | 2002 | 2003 |
|---|---|---|---|
| Kilometres flown (million) | 5.5 | 5.8 | 6.5 |
| Passengers carried ('000) | 281.0 | 268.6 | 313.0 |
| Passenger-km (million) | 294.3 | 278.6 | 324.5 |

Source: UN Economic Commission for Latin America and the Caribbean, *Statistical Yearbook*.

## Tourism

**ARRIVALS BY NATIONALITY**

|  | 2001 | 2002 | 2003 |
|---|---|---|---|
| Argentina | 170,575 | 160,758 | 177,741 |
| Brazil | 60,193 | 43,134 | 40,651 |
| Chile | 5,964 | 4,694 | 6,262 |
| Uruguay | 5,351 | 4,939 | 5,775 |
| USA | 8,695 | 9,012 | 9,210 |
| **Total** (incl. others) | 278,672 | 250,423 | 268,175 |

**Tourism receipts** (US $ million, incl. passenger transport): 87 in 2001; 73 in 2002; 81 in 2003.

Source: World Tourism Organization.

## Communications Media

|  | 2001 | 2002 | 2003 |
|---|---|---|---|
| Telephones ('000 main lines in use) | 288.8 | 273.2 | 273.2 |
| Mobile cellular telephones ('000 subscribers) | 1,150.0 | 1,667.0 | 1,770.3 |
| Personal computers ('000 in use) | 150 | 200 | n.a. |
| Internet users ('000) | 60 | 100 | 120 |

**Television receivers** ('000 in use): 1,200 in 1997.
**Radio receivers** ('000 in use): 925 in 1997.
**Facsimile machines** (number in use): 1,691 in 1992.
**Daily newspapers:** 5* in 1996 (average circulation 213,000* copies).
**Non-daily newspapers:** 2 in 1988 (average circulation 16,000* copies).
**Book production:** 152 titles (incl. 23 pamphlets) in 1993.
* Estimate.

Sources: UNESCO, *Statistical Yearbook*; UN, *Statistical Yearbook*; International Telecommunication Union.

## Education

(1999, unless otherwise indicated)

|  | Institutions | Teachers*† | Students |
|---|---|---|---|
| Pre-primary schools | 4,071 | 4,188 | 123,597 |
| Primary | 7,456 | 24,526 | 1,036,770 |
| Secondary: general | 1,844 | 21,052 | 260,500 |
| Secondary: vocational | 305 | | 31,136 |
| Tertiary: university level‡ | 2 | n.a. | 37,009 |
| Tertiary: other higher‡ | 109 | 1,135 | 20,283 |

* 1998 figures.
† Full-time teachers only.
‡ Excluding private universities.

Source: partly UNESCO Institute for Statistics.

**Adult literacy rate:** 91.6% (males 93.1%; females 90.2%) in 2002.

# Directory

## The Constitution

A new Constitution for the Republic of Paraguay came into force on 22 June 1992, replacing the Constitution of 25 August 1967.

### FUNDAMENTAL RIGHTS, DUTIES AND FREEDOMS

Paraguay is an independent republic whose form of government is representative democracy. The powers accorded to the legislature, executive and judiciary are exercised in a system of independence, equilibrium, co-ordination and reciprocal control. Sovereignty resides in the people, who exercise it through universal, free, direct, equal and secret vote. All citizens over 18 years of age and resident in the national territory are entitled to vote.

All citizens are equal before the law and have freedom of conscience, travel, residence, expression, and the right to privacy. The freedom of the press is guaranteed. The freedom of religion and ideology is guaranteed. Relations between the State and the Catholic Church are based on independence, co-operation and autonomy. All citizens have the right to assemble and demonstrate peacefully. All public- and private-sector workers, with the exception of the Armed Forces and the police, have the right to form a trade union and to strike. All citizens have the right to associate freely in political parties or movements.

The rights of the indigenous peoples to preserve and develop their ethnic identity in their respective habitat are guaranteed.

### LEGISLATURE

The legislature (Congreso Nacional—National Congress) comprises the Senado (Senate) and the Cámara de Diputados (Chamber of Deputies). The Senado is composed of 45 members, the Cámara of 80 members, elected directly by the people. Legislation concerning national defence and international agreements may be initiated in the Senado. Departmental and municipal legislation may be initiated in the Cámara. Both chambers of the Congreso are elected for a period of five years.

### GOVERNMENT

Executive power is exercised by the President of the Republic. The President and the Vice-President are elected jointly and directly by the people, by a simple majority of votes, for a period of five years. They may not be elected for a second term. The President and the Vice-President govern with the assistance of an appointed Council of Ministers. The President participates in the formulation of legislation and enacts it. The President is empowered to veto legislation sanctioned by the Congreso, to nominate or remove ministers, to direct the foreign relations of the Republic, and to convene extraordinary sessions of the Congreso. The President is Commander-in-Chief of the Armed Forces.

### JUDICIARY

Judicial power is exercised by the Supreme Court of Justice and by the tribunals. The Supreme Court is composed of nine members who are appointed on the proposal of the Consejo de la Magistratura, and has the power to declare legislation unconstitutional.

## The Government

### HEAD OF STATE

**President:** Oscar Nicanor Duarte Frutos (took office 15 August 2003).

**Vice-President:** Luis Alberto Castiglioni Soria.

# PARAGUAY

## COUNCIL OF MINISTERS
(July 2005)

**Minister of the Interior:** ROGELIO BENÍTEZ.

**Minister of Foreign Affairs:** LEILA RACHID DE COWLES.

**Minister of Finance:** ERNST FERDINAND BERGEN SCHMIDT.

**Minister of Industry and Commerce:** RAÚL JOSÉ VERA BOGADO.

**Minister of Public Works and Communications:** JOSÉ ALBERTO ALDERETE RODRÍGUEZ.

**Minister of National Defence:** ROBERTO EUDEZ GONZÁLEZ SEGOVIA.

**Minister of Public Health and Social Welfare:** Dra MARÍA TERESA LEÓN MENDARO.

**Minister of Justice and Labour:** RUBÉN CANDIA AMARILLA.

**Minister of Agriculture and Livestock:** GUSTAVO RUIZ DÍAZ.

**Minister of Education and Culture:** BLANCA OVELAR DE DUARTE.

### MINISTRIES

**Ministry of Agriculture and Livestock:** Presidente Franco 472, Asunción; tel. (21) 44-9614; fax (21) 49-7965.

**Ministry of Education and Culture:** Chile, Humaitá y Piribebuy, Asunción; tel. (21) 44-3078; fax (21) 44-3919; internet www.paraguaygobierno.gov.py/mec.

**Ministry of Finance:** Chile 128 esq. Palmas, Asunción; tel. (21) 49-2599; fax (21) 49-2599; e-mail info@hacienda.gov.py; internet www.hacienda.gov.py.

**Ministry of Foreign Affairs:** Juan E. O'Leary y Presidente Franco, Asunción; tel. (21) 49-4593; fax (21) 49-3910; internet www.mre.gov.py.

**Ministry of Industry and Commerce:** Avda España 323, Asunción; tel. (21) 20-4638; fax (21) 21-3529; internet www.mic.gov.py.

**Ministry of the Interior:** Estrella y Montevideo, Asunción; tel. (21) 49-3661; fax (21) 44-6448; internet www.ministeriodelinterior.gov.py.

**Ministry of Justice and Labour:** G. R. de Francia y Estados Unidos, Asunción; tel. (21) 49-3515; fax (21) 20-8469; e-mail mjt@conexion.com.py.

**Ministry of National Defence:** Avda Mariscal López y Vice-Presidente Sánchez, Asunción; tel. (21) 20-4771; fax (21) 21-1583.

**Ministry of Public Health and Social Welfare:** Avda Pettirossi y Brasil, Asunción; tel. (21) 20-7328; fax (21) 20-6700; internet www.mspbs.gov.py.

**Ministry of Public Works and Communications:** Oliva y Alberdi, Asunción; tel. (21) 44-4411; fax (21) 44-4421; internet www.mopc.gov.py.

## President and Legislature

### PRESIDENT

**Election, 27 April 2003**

| Candidate | % of votes |
| --- | --- |
| Oscar Nicanor Duarte Frutos (Partido Colorado) | 37.1 |
| Julio César Ramón Franco Gómez (PLRA) | 24.0 |
| Pedro Nicolás Fadul Niella (Patria Querida) | 21.3 |
| Guillermo Sánchez Guffanti (UNACE) | 13.5 |
| Total (incl. others) | 100.0 |

### CONGRESO NACIONAL
(National Congress)

**President of the Senado:** Dr CARLOS FILIZZOLA.

**President of the Cámara de Diputados:** OSCAR SALOMÓN.

**General Election, 27 April 2003**

| Party | Cámara de Diputados | Senado |
| --- | --- | --- |
| Partido Colorado | 37 | 16 |
| Partido Liberal Radical Auténtico | 21 | 12 |
| Patria Querida | 10 | 8 |
| Unión Nacional de Ciudadanos Eticos | 10 | 7 |
| Partido País Solidario | 2 | 2 |
| Total | 80 | 45 |

## Political Organizations

**Asociación Nacional Republicana—Partido Colorado** (National Republican Association—Colorado Party): Casa de los Colorados, 25 de Mayo 842, Asunción; tel. (21) 45-2543; fax (21) 49-7857; f. 19th century, ruling party since 1940; principal factions include: Movimiento de Reconciliación Colorada; Coloradismo Unido, led by Dr ANGEL ROBERTO SEIFART; Coloradismo Democrático, led by BLÁS RIQUELME; Acción Democrática Republicana, led by CARLOS FACETTI MASULLI; Frente Republicano de Unidad Nacional, led by WÁLTER BOWER MONTALTO; 947,430 mems (1991); Pres. HERMINIO CÁCERES; Vice-Pres. CÁNDIDO AGUILERA.

**Encuentro Nacional (EN):** Vice Presidente Sánchez y Herrera 2283, Asunción; tel. (21) 22-3510; fax (21) 61-0699; e-mail parenac@pla.net.py; coalition comprising factions of PRF, PDC, Asunción Para Todos and a dissident faction of the Partido Colorado; formed to contest presidential and legislative elections of May 1993; f. 1991; Pres. LUIS TORALES KENNEDY; Vice-Pres. Dr SECUNDINO NÚÑEZ.

**Independiente en Acción:** Irrazabal 857, Asunción; tel. (21) 20-1375; Pres. MIGUEL RAFAEL OTAZÚ MONTANARO; Sec.-Gen. AURORA ALMADA FLORES.

**Movimiento Patria Libre (MPL):** 15 de Agosto 1939, Asunción; tel. (21) 37-2384; left-wing; Asst. Sec.-Gen. ANUNCIO MARTI MÉNDEZ.

**Partido Blanco:** Asunción; tel. (21) 55-4068; Pres. GREGORIO SEGOVIA SILVERA; Vice-Pres. EDGAR A. ORTIGOZA CARDOZO.

**Partido Comunista Paraguayo (PCP):** Asunción; f. 1928; banned 1928–46, 1947–89; Sec.-Gen. ANANÍAS MAIDANA.

**Partido Demócrata Cristiano (PDC):** Colón 871, Casilla 1318, Asunción; internet www.pdc.org.py; f. 1960; 20,500 mems; Pres. Dr LUIS M. ANDRADA NOGUÉS; Vice-Pres. Dr JOSÉ V. ALTAMIRANO.

**Partido Frente Amplio Paraguayo:** Antequera 764, esq. Fulgencio R. Moreno, Asunción; tel. (21) 44-1389; Sec.-Gen. VÍCTOR BAREIRO ROA.

**Partido Humanista Paraguayo:** Fulgencio R. Moreno 584, esq. Paraguari, Asunción; tel. (21) 44-2625; internet www.humanista.org.py; f. 1985; recognized by the Supreme Tribunal of Electoral Justice in March 1989; campaigns for the protection of human rights and environmental issues; Gen. Sec. NICOLÁS SERVÍN.

**Partido Liberal Radical Auténtico (PLRA):** Azara y General Santos 2486, Asunción; tel. (21) 20-1337; fax (21) 20-4869; f. 1978; centre party; Leader JULIO CÉSAR ('YOYITO') FRANCO GÓMEZ.

**Partido País Solidario:** Avda 5, esq. Méjico, Asunción; tel. (21) 39-1271; Pres. Dr CARLOS FILIZZOLA; Vice-Pres. JORGE GIUCICH.

**Partido Revolucionario Febrerista (PRF):** Casa del Pueblo, Manduvira 552, Asunción; tel. (21) 494041; e-mail partyce@mixmail.com; f. 1951; social democratic party; affiliated to the Socialist International; Pres. OSCAR MONTIEL GALVÁN; Vice-Pres. MIRTA TORRES ANTÚNEZ.

**Partido de los Trabajadores (PT):** Asunción; f. 1989; Socialist.

**Partido Unión Nacional de Ciudadanos Eticos (UNACE):** Eusebio Ayala 4135, esq. Corrales, Asunción; e-mail loviedo@unace.org.py; internet www.unace.org.py ; f. 2002; left-wing; breakaway faction of the Partido Colorado; Pres. ENRIQUE GONZÁLEZ QUINTANA; Vice-Pres. Dr CARLOS ROGER CABALLERO FIORI.

**Patria Querida:** Asunción; f. 2002; recognized by the Supreme Tribunal of Electoral Justice in March 2004; Leader PEDRO NICOLÁS FADUL NIELLA.

**Unidad Popular:** Azara 2843, esq. Rodó, Asunción; tel. (21) 21-5059; recognized by the Supreme Tribunal of Electoral Justice in March 2004; Pres. JUAN DE DIOS ACOSTA MENA.

### OTHER ORGANIZATIONS

**Federación Nacional Campesina de Paraguay (FNC):** Nangariry 1196, esq. Cacique Cará Cará, Asunción; tel. (21) 51-2384;

# PARAGUAY

grouping of militant peasants' orgs; Sec.-Gen. ODILÓN ESPÍNOLA; Asst Sec.-Gen. MARCIAL GÓMEZ.

**Frente en Defensa de los Bienes Públicos y el Patrimonio Nacional:** Asunción; left-wing grouping of orgs opposed to privatization; Co-ordinator GABRIEL ESPÍNOLA.

**Frente Nacional de Lucha por la Soberanía y la Vida:** Asunción; left-wing grouping of orgs campaigning for agrarian reform and opposed to privatization; Co-ordinator LUIS AGUAYO.

## Diplomatic Representation

### EMBASSIES IN PARAGUAY

**Argentina:** Avda España esq. Avda Perú, Casilla 757, Asunción; tel. (21) 21-2320; fax (21) 21-1029; e-mail embarpy@supernet.com.py; internet www.embajada-argentina.org.py; Ambassador RAFAEL EDGARDO ROMÁ.

**Bolivia:** América 200 y Mariscal Lopez, Asunción; tel. (21) 22-7213; fax (21) 21-0440; e-mail embolivia-asuncion@webmail.com.py; Ambassador ALFREDO SEOANE FLORES.

**Brazil:** Col Irrazábal esq. Eligio Ayala, Casilla 22, Asunción; tel. (21) 21-4466; fax (21) 21-2693; e-mail acesar@embajadabrasil.org.py; internet www.embajadabrasil.org.py; Ambassador WALTER PECLY MOREIRA.

**Chile:** Capital Emilio Nudelman 351, Asunción; tel. (21) 61-3855; fax (21) 66-2755; e-mail echilepy@conexion.com.py; Ambassador JUAN EDUARDO BURGOS SANTANDER.

**China (Taiwan):** Avda Mariscal López 1143 y Mayor Bullo, Casilla 503, Asunción; tel. (21) 21-3362; fax (21) 21-2373; e-mail giopy@telesurf.com.py; e-mail emorc@highway.com.py; internet www.roc-taiwan.org.py; Ambassador BING-FAN YEN.

**Colombia:** Calle Coronel Brizuela esq. Ciudad del Vaticano, Asunción; tel. (21) 22-9888; fax (21) 22-9703; e-mail easuncio@minrelext.gov.co; Ambassador CARLOS ALBERTO BERNAL ROMÁN.

**Costa Rica:** Carlos Díaz León 3245, casi Escurra Barrio Herrera, Asunción; tel. and fax (21) 67-5297; fax (21) 67-3750 ; e-mail embarica@uninet.com.py; Ambassador FERNANDO JOSÉ GUARDIA ALVARADO.

**Cuba:** Luis Morales 757, esq. Luis León y Luis Granado, Barrio Jara, Asunción; tel. (21) 22-2763; fax (21) 21-3879; e-mail embacuba@cmm.com.py; Ambassador IRMA A. GONZÁLEZ CRUZ.

**Ecuador:** Justo Román y Julio C. Escobar, esq. Barrio Manorá, Casilla 13162, Asunción; tel. (21) 61-4814; fax (21) 61-4813; e-mail mecuapy@conexion.com.py; Ambassador FRANCISCO SUÉSCUM OTATTI.

**France:** Avda España 893, Calle Pucheu, Casilla 97, Asunción; tel. (21) 21-2449; fax (21) 21-1690; e-mail chancellerie@ambafran.gov.py; internet www.ambafran.gov.py; Ambassador DENIS VÈNE.

**Germany:** Avda Venezuela 241, Casilla 471, Asunción; tel. (21) 21-4009; fax (21) 21-2863; e-mail aaasun@pla.net.py; internet www.pla.net.py/embalem; Ambassador Dr HORST-WOLFRAM KERLL.

**Holy See:** Calle Ciudad del Vaticano 350, casi con 25 de Mayo, Casilla 83, Asunción (Apostolic Nunciature); tel. (21) 21-5139; fax (21) 21-2590; e-mail nunapos@conexion.com.py; Apostolic Nuncio Most Rev. ANTONIO LUCIBELLO (Titular Archbishop of Thurio).

**Italy:** Quesada 5871 con Bélgica, Asunción; tel. (21) 61-5620; fax (21) 61-5622; e-mail ambitalia@cmm.com.py; internet www.embajadadeitalia.org.py; Ambassador BENEDETTO AMARI.

**Japan:** Avda Mariscal López 2364, Casilla 1957, Asunción; tel. (21) 60-4616; fax (21) 60-6901; e-mail japoncul@rieder.net.py; internet www.py.emb-japan.go.jp; Ambassador TOSHIHIRO TAKAHASHI.

**Korea, Republic:** Avda Rep. Argentina Norte 678 esq. Pacheco, Casilla 1303, Asunción; tel. (21) 60-5606; fax (21) 60-1376; e-mail paraguay@mofat.go.kr; Ambassador CHUNG YOUNG-KOO.

**Mexico:** Avda España, esq. San Rafael, Casilla 1184, Asunción; tel. (21) 616-8200; fax (21) 6-2500; e-mail embamex@embamex.com.py; internet www.embamex.com.py; Ambassador ANTONIO VILLEGAS.

**Panama:** Piribeduy 765, casi Ayolas, Casilla 873, Asunción; tel. (21) 44-3522; fax (21) 44-6192; e-mail embapana@conexion.com.py; Ambassador ROBERTO MORENO OLIVARES.

**Peru:** Feliciano Marecos 441, casi Agustín Barrios y España, Manorá, Casilla 433, Asunción; tel. (21) 60-0226; fax (21) 60-7327; e-mail embperu@embperu.com.py; Ambassador ENRIQUE PALACIOS REYES.

**Spain:** Edif. S. Rafael, 5° y 6°, Yegros 437, Asunción; tel. (21) 49-0686; fax (21) 44-5394; e-mail embesppy@correo.mae.es; Ambassador EDUARDO DE QUESADA FERNÁNDEZ DE LA PUENTE.

**Switzerland:** Edif. Parapití, 4°, Ofs 419–423, Juan E. O'Leary 409 y Estrella, Casilla 552, Asunción; tel. (21) 44-8022; fax (21) 44-5853; e-mail vertretung@asu.rep.admin.ch; Ambassador URS STEMMLER.

**USA:** Avda Mariscal López 1776, Casilla 402, Asunción; tel. (21) 21-3715; fax (21) 21-3728; e-mail paraguayusembassy@state.gov; internet asuncion.usembassy.gov; Ambassador JOHN FRANCIS KEANE.

**Uruguay:** Guido Boggiani 5832, 3°, Asunción; tel. (21) 66-4244; fax (21) 60-1335; e-mail embauru@telesurf.com.py; Ambassador CARLOS ERNESTO ORLANDO BONET.

**Venezuela:** Mariscal Estigarribia 1023 con Estados Unidos, Asunción; tel. (21) 66-4682; tel. (21) 66-4683; fax (21) 66-4683; e-mail bolivar@pla.net.py; internet www.embvenezuela.org.py; Ambassador JOSÉ HUERTA CASTILLO.

## Judicial System

The Corte Suprema de Justicia (Supreme Court of Justice) is composed of nine judges appointed on the recommendation of the Consejo de la Magistratura (Council of the Magistracy).

**Corte Suprema de Justicia:** Palacio de Justicia, Asunción; Members VÍCTOR MANUEL NÚÑEZ (President), MIGUEL O. BAJAC, ANTONIO FRETES, SINDULFO BLANCO, ALICIA PUCHETA DE CORREA, WILDO RIENZI GALEANO, JOSÉ V. ALTAMIRANO AQUINO, CÉSAR A. GARAY ZUCCOLILLO, JOSÉ R. TORRES KIRMSER.

### Consejo de la Magistratura

Palacio de Justicia, Asunción.

Members RUBÉN DARÍO ROMERO (President), MARIO SOTO ESTIGARRIBIA (Vice-President), GUILLERMO DELMÁS FRESCURA, RODOLFO IRÚN ALAMANNI, RAÚL BATTILANA NIGRA, MARCELINO GAUTO BEJERANO, EUSEBIO RAMÓN AYALA, ANTONIO FRETES.

**Attorney-General:** OSCAR LATORRE.

Under the Supreme Court are the Courts of Appeal, the Tribunal of Jurors and Judges of First Instance, the Judges of Arbitration, the Magistrates (Jueces de Instrucción), and the Justices of the Peace.

## Religion

The Roman Catholic Church is the established religion, although all sects are tolerated.

### CHRISTIANITY

#### The Roman Catholic Church

For ecclesiastical purposes, Paraguay comprises one archdiocese, 11 dioceses and two Apostolic Vicariates. At 31 December 2003 there were an estimated 5,241,609 adherents in the country, representing about 91.5% of the total population.

#### Bishops' Conference

Conferencia Episcopal Paraguaya, Calle Alberdi 782, Casilla 1436, 1209 Asunción; tel. (21) 49-0920; fax (21) 49-5115; e-mail cep@infonet.com.py.

f. 1977; statutes approved 2000; Pres. Rt Rev. CATALINO CLAUDIO GIMÉNEZ MEDINA (Bishop of Caacupé).

**Archbishop of Asunción:** Most Rev. EUSTAQUIO PASTOR CUQUEJO VERGA, Arzobispado, Avda Mariscal López 130 esq. Independencia Nacional, Casilla 654, Asunción; tel. (21) 44-5551; fax (21) 44-4150.

#### The Anglican Communion

Paraguay constitutes a single diocese of the Iglesia Anglicana del Cono Sur de América (Anglican Church of the Southern Cone of America). The Presiding Bishop of the Church is the Bishop of Northern Argentina.

**Bishop of Paraguay:** Rt Rev. JOHN ELLISON, Iglesia Anglicana, Avda España casi Santos, Casilla 1124, Asunción; tel. (21) 20-0933; fax (21) 21-4328; e-mail iapar@sce.cnc.una.py; internet www.anglicanos.net.

#### The Baptist Church

**Baptist Evangelical Convention of Paraguay:** Casilla 1194, Asunción; tel. (21) 22-7110; Exec. Sec. Lic. RAFAEL ALTAMIRANO.

### BAHÁ'Í FAITH

**National Spiritual Assembly of the Bahá'ís of Paraguay:** Eligio Ayala 1456, Apdo 742, Asunción; tel. (21) 22-0250; fax (21) 22-5747; e-mail bahai@uninet.com.py; Sec. MIRNA LLAMOSAS DE RIQUELME.

# The Press

## DAILIES

**ABC Color:** Yegros 745, Apdo 1421, Asunción; tel. (21) 49-1160; fax (21) 415-1310; e-mail azeta@abc.com.py; internet www.abc.com.py; f. 1967; independent; Propr ALDO ZUCCOLILLO; circ. 45,000.

**El Día:** Avda Mariscal López 2948, Asunción; tel. (21) 60-3401; fax (21) 66-0385; e-mail eldia@infonet.com.py; internet www.infonet.com.py/eldia; Dir HUGO OSCAR ARANDA; circ. 12,000.

**La Nación:** Avda Zavala Cué entre 2da y 3ra, Fernando de la Mora, Asunción; tel. (21) 51-2520; fax (21) 51-2535; e-mail redaccion@lanacion.com.py; internet www.lanacion.com.py; f. 1995; Dir-Gen. OSVALDO DOMÍNGUEZ DIBB; circ. 10,000.

**Noticias:** Avda Artigas y Avda Brasilia, Casilla 3017, Asunción; tel. (21) 29-2721; fax (21) 29-2716; e-mail alebluth@diarionoticias.com; internet www.diarionoticias.com.py; f. 1985; independent; Dir ALEJANDRO BLUTH; circ. 20,000.

**Patria:** Tacuari 443, Asunción; tel. (21) 92011; f. 1946; Colorado Party; Dir JUAN RAMÓN CHÁVEZ; circ. 8,000.

**Popular:** Avda Mariscal López 2948, Asunción; tel. (21) 60-3401; fax (21) 60-3400; e-mail popular@mm.com.py; internet www.diariopopular.com.py; Dir JAVIER PIROVANO PEÑA; circ. 28,000.

**Última Hora:** Benjamín Constant 658, Asunción; tel. (21) 49-6261; fax (21) 44-7071; e-mail ultimahora@uhora.com.py; internet www.ultimahora.com; f. 1973; independent; Dir DEMETRIO ROJAS; circ. 30,000.

## PERIODICALS

**Acción:** Casilla 1072, Asunción; tel. (21) 37-0753; e-mail cepag@uninet.com.py; internet www.uninet.com.py/accion; monthly; Dir JOSÉ MARÍA BLANCH.

**La Opinión:** Boggiani esq. Luis Alberto de Herrera, Asuncíon; tel. (21) 50-7501; fax (21) 50-2297; weekly; Dir FRANCISCO LAWS; Editor BERNARDO NERI.

**TeVeo:** Santa Margarita de Youville 250, Santa María, Asunción; tel. (21) 67-2079; fax (21) 21-1236; e-mail sugerencias@teveo.com.py; internet www.teveo.com.py; weekly; society.

**Tiempo 14:** Mariscal Estigarribia 4187, Asunción; tel. (21) 60-4308; fax (21) 60-9394; weekly; Dir HUMBERTO RUBÍN; Editor ALBERTO PERALTA.

## NEWS AGENCIES

**Agencia Paraguaya de Noticias (APN):** Asunción; e-mail apn@supernet.com.py; internet www.supernet.com.py/usuarios/apn.

### Foreign Bureaux

**Agence France-Presse (AFP):** Herrera 195, 8°, Of. 802, Asunción; tel. (21) 49-4520; fax (21) 44-3725; Correspondent HUGO RUIZ OLAZAR.

**Agencia EFE** (Spain): Yegros 437, Asunción; tel. (21) 49-2730; fax (21) 49-1268; Bureau Chief LUCIO GÓMEZ-OLMEDO.

**Agenzia Nazionale Stampa Associata (ANSA)** (Italy): Edif. Interexpress, 4°, Of. 403, Luis Alberto de Herrera 195, Asunción; tel. (21) 44-9286; fax (21) 44-2986; e-mail ansaasu@rieder.net.py.

**Associated Press (AP)** (USA): Calle Caballero 742, Casilla 264, Asunción; tel. (21) 60-6334.

**Deutsche Presse-Agentur (dpa)** (Germany): Edif. Segesa, Of. 705, Oliva 309, Asunción; tel. (21) 45-0329; fax (21) 44-8116; Correspondent EDUARDO ARCE.

**Inter Press Service (IPS)** (Italy): Edif. Segesa, 3°, Of. 5, Oliva 393 y Alberdi, Asunción; tel. and fax (21) 44-6350; Legal Rep. CLARA ROSA GAGLIARDONE.

TELAM (Argentina) is also represented in Paraguay.

# Publishers

**La Colmena, SA:** Asunción; tel. (21) 20-0428; Dir DAUMAS LADOUCE.

**Dervish SA, Editorial:** Avda Mariscal López 1735, CP 1584, Asunción; tel. (21) 21-1729; fax (21) 22-2580; e-mail dervish@dervish.com.py; f. 1989; Co-ordinator JORGELINA MIGLIORISI; Vice-Pres. and Dir JANINE GIANI PATTERSON.

**Ediciones Diálogo:** Calle Brasil 1391, Asunción; tel. (21) 20-0428; f. 1957; fine arts, literature, poetry, criticism; Man. MIGUEL ANGEL FERNÁNDEZ.

**Ediciones Nizza:** Eligio Ayala 1073, Casilla 2596, Asunción; tel. (21) 44-7160; medicine; Pres. Dr JOSÉ FERREIRA MARTÍNEZ.

**Editorial Comuneros:** Cerro Corá 289, Casilla 930, Asunción; tel. (21) 44-6176; fax (21) 44-4667; e-mail rolon@conexion.com.py; f. 1963; social history, poetry, literature, law; Man. OSCAR R. ROLÓN.

**Editorial Quijote:** Mall Excelsior, Chile y Manduvirá, Asunción; tel. (21) 49-4445; fax (21) 44-7677; e-mail guasti@quijote.com.py; internet www.quijote.com.py; f. 1983; Gen. Man. GUSTAVO GUASTI.

**Librería Intercontinental:** Caballero 270, Asunción; tel. (21) 49-6991; fax (21) 48-721; e-mail agatti@pla.net.py; internet www.libreriaintercontinental.com.py; political science, law, literature, poetry; Dir ALEJANDRO GATTI VAN HUMBEECK.

**R. P. Ediciones:** Eduardo Víctor Haedo 427, Ascunción; tel. (21) 49-8040; Man. RAFAEL PERONI.

## ASSOCIATION

**Cámara Paraguaya del Libro:** Nuestra Señora de la Asunción 697 esq. Eduardo Víctor Haedo, Asunción; tel. (21) 44-4104; fax (21) 44-7053; Pres. PABLO LEÓN BURIAN; Sec. EMA DE VIEDMA.

# Broadcasting and Communications

## TELECOMMUNICATIONS

**Comisión Nacional de Telecomunicaciones (CONATEL):** Edif. San Rafael, 2° Yegros 437 y 25 de Mayo, Asunción; tel. (21) 44-0020; fax (21) 49-8982; Pres. VÍCTOR ALCIDES BOGADO.

**Corporación Paraguaya de Comunicaciones (COPACO):** Edif. Morotí, 1°–2°, esq. Gen. Bruguéz y Teodoro S. Mongelos, Casilla 2042, Asunción; tel. (21) 20-3800; fax (21) 20-3888; internet www.copaco.com.py; fmrly Administración Nacional de Telecomunicaciones (ANTELCO); changed name as above in December 2001 as part of the privatization process; privatization suspended in June 2002; Gen. Man. EDGAR PINEDA.

## BROADCASTING

### Radio

**Radio Arapysandú:** Avda Mariscal López y Capitán del Puerto San Ignacio, Misiones; tel. (82) 2374; fax (82) 2206; f. 1982; Dir EMILIO BOTTINO.

**Radio Asunción:** Avda Artígas y Capitán Lombardo 174, Asunción; tel. (21) 29-5375; fax (21) 29-2718; Dir MIGUEL G. FERNÁNDEZ.

**Radio Cáritas:** Kubitschek y 25 de Mayo, Asunción; tel. (21) 21-3570; fax (21) 20-4161; f. 1936; station of the Franciscan order; medium-wave; Pres. Most Rev. EUSTAQUIO PASTOR CUQUEJO VERGA (Archbishop of Asunción); Dir MARIO VELÁZQUEZ.

**Radio Cardinal:** Río Paraguay 1334 y Guariníes Lambaré, Casilla 2532, Asunción; tel. (21) 31-0555; fax (21) 31-0557; f. 1991; Pres. NÉSTOR LÓPEZ MOREIRA.

**Radio City:** Edif. Líder III, Antequera 652, 9°, Asunción; tel. (21) 44-3324; fax (21) 44-4367; f. 1980; Dir GREGORIO RAMÓN MORALES.

**Radio Concepción:** Coronel Panchito López 241, entre Schreiber y Profesor Guillermo A. Cabral, Casilla 78, Concepción; tel. (31) 42318; fax (31) 42254; f. 1963; medium-wave; Dir SERGIO E. DACAK; Gen. Man. JULIÁN MARTÍ IBÁÑEZ.

**Radio Emisoras Paraguay:** Avda General Santos 2525 y 18 de Julio, Asunción; tel. (21) 31-0644; FM; Gen. Man. GUILLERMO HEISECKE.

**Radio Encarnación:** General Artigas 798 y Caballero, Encarnación; tel. (71) 3345; fax (71) 4099; medium- and short-wave; Dir BIENVENIDA KRISTOVICH.

**Radio Guairá:** Presidente Franco 788 y Alejo García, Villarica; tel. (541) 42130; fax (541) 42385; f. 1950; medium-, long- and short-wave; Dir LÍDICE RODRÍGUEZ DE TRAVERSI.

**Radio Itapirú S.R.L.:** Avda San Blás esq. Coronel Julián Sánchez, Ciudad del Este, Alto Paraná; tel. (61) 57-2206; fax (61) 57-2210; f. 1969; Gen. Man. ANTONIO ARANDA ENCINA.

**Radio La Voz de Amambay:** 14 de Mayo y Cerro León, Pedro Juan Caballero, Amambay; tel. (36) 72537; f. 1959; Gen. Man. LUIS ROLÓN PEÑA.

**Radio Nacional del Paraguay:** Blas Garay 241 y Iturbe, Asunción; tel. (21) 39-0374; fax (21) 39-0376; medium- and short-wave and FM; Dir GILBERTO ORTÍZ BAREIRO.

**Radio Ñandutí:** Choferes del Chaco y Carmen Soler, Asunción; tel. (21) 60-4308; fax (21) 60-6074; internet www.infonet.com.py/holding/nanduam/; f. 1962; Dirs HUMBERTO RUBÍN, GLORIA RUBÍN.

PARAGUAY                                                                                                                    *Directory*

**Radio Nuevo Mundo:** Coronel Romero 1181 y Flórida, San Lorenzo, Asunción; tel. (21) 58-6258; fax (21) 58-2424; f. 1972; Dir Julio César Pereira Bobadilla.

**Radio Oriental:** Avda Mariscal López 450 y San Lorenzo, Caaguazú; tel. (522) 42790; Dir Hector Omar Yinde.

**Radio Primero de Marzo:** Avda General Perón y Concepción, Casilla 1456, Asunción; tel. (21) 31-1564; fax (21) 33-3427; Dir-Gen. Angel Guerreros A.

**Radio Santa Mónica FM:** Avda Boggiani y Herrera, 3°, Asunción; tel. (21) 50-7501; fax (21) 50-9494; f. 1973; Exec. Dir Gustavo Canatta.

**Radio Uno:** Avda Mariscal López 2948, Asunción; tel. (21) 61-2151; f. 1968 as Radio Chaco Boreal; Dir Benjamín Livieres.

**Radio Venus:** República Argentina y Souza, Asunción; tel. (21) 61-0151; e-mail venus@infonet.com.py; internet www.venus.com.py; f. 1987; Dir Angel Aguilera.

**Radio Ysapy:** Independencia Nacional 1260, 1°, Asunción; tel. (21) 44-4037; Dir José Cabriza Salvioni.

### Television

**Teledifusora Paraguaya—Canal 13:** Chile 993, Asunción; tel. (21) 33-2823; fax (21) 33-1695; f. 1980; Gen. Man. Nestor López Moreiga.

**Televisión Cerro Corá—Canal 9:** Avda Carlos A. López 572, Asunción; tel. (21) 42-4222; fax (21) 48-0230; f. 1965; commercial; Dir Gen. Ismael Hadid.

**Televisora del Este:** San Pedro, Calle Pilar, Area 5, Ciudad del Este; tel. (61) 8859; commercial; Dir Lic. Jalil Safuan; Gen. Man. A. Villalba V.

**Televisión Itapúa—Canal 7:** Encarnación; tel. (71) 20-4450; commercial; Dir Lic. Jalil Safuan; Station Man. Jorge Mateo Granada.

# Finance

(cap. = capital; res = reserves; dep. = deposits; m. = million; amounts in guaraníes, unless otherwise indicated)

### BANKING

**Superintendencia de Bancos:** Edif. Banco Central del Paraguay, Avda Federación Rusa y Avda Sargento Marecos, Barrio Santo Domingo, Asunción; tel. (21) 60-8011; fax (21) 60-8149; e-mail supban@bcp.gov.py; internet www.bcp.gov.py/supban/principal .htm; Supt Rodrigo Ortíz.

### Central Bank

**Banco Central del Paraguay:** Avda Federación Rusa y Cabo 1° Marecos, Casilla 861, Barrio Santo Domingo, Asunción; tel. (21) 61-0088; fax (21) 60-8149; e-mail informaciones@bcp.gov.py; internet www.bcp.gov.py; f. 1952; Pres. Dr Mónica Luján Pérez Dos Santos; Gen. Man. Darío Rolando Arréllaga Yaluk (acting).

### Development Banks

**Banco Nacional de Fomento:** Independencia Nacional y 25 de Mayo, Asunción; tel. (21) 44-4440; fax (21) 44-6056; e-mail correo@ bnf.gov.py; internet www.bnf.gov.py; f. 1961 to take over the deposit and private banking activities of the Banco del Paraguay; Pres. Lic. Germán Hugo Rojas Irigoyen; Sec.-Gen. Alfredo Maldonado Gómez; 52 brs.

**Crédito Agrícola de Habilitación:** Caríos 362 y Willam Richardson, Asunción; tel. (21) 56-9010; fax (21) 55-4956; e-mail cah@quanta.com.py; f. 1943; Pres. Ing. Agr. Walberto Ferreira.

**Fondo Ganadero:** Avda Mariscal López 1669 esq. República Dominicana, Asunción; tel. (21) 29-4361; fax (21) 44-6922; internet www .fondogan.gov.py; f. 1969; govt-owned; Pres. Guillermo Serratti G.

### Commercial Banks

**Banco Amambay, SA:** Avda Aviadores del Chaco, entre San Martín y Pablo Alborno, Asunción; tel. (21) 60-8831; fax (21) 60-8813; e-mail bcoama@bcoamabancoamambay.com.py; internet www .bancoamambay.com.py; f. 1992; Pres. Guiomar de Gásperi; Gen. Man. Hugo Portillo Sosa.

**Banco Continental, SAECA:** Estrella 621, Casilla 2260, Asunción; tel. (21) 44-2002; fax (21) 44-2001; e-mail contil@connexion.com.py; f. 1980; cap. US $4.2m., res $2.5m., dep. $44.6m. (Dec. 2001); Chair. Guillermo Gross Brown; First Vice-Pres. Javier González Pérez.

**Banco Finamérica SA:** Chile y Oliva, Casilla 1321, Asunción; tel. (21) 49-1021; fax (21) 44-5199; f. 1988; Pres. Dr Guillermo Heisecke Velázquez; Gen. Man. Enrique Fernández Romay.

**Interbanco, SA:** Oliva 349 esq. Chile y Alberdi, Asunción; tel. (21) 49-4992; fax (21) 41-71372; e-mail interban@conexion.com.py; internet www.interbanco.com.py; f. 1978; owned by Unibanco (Brazil); cap. 19,000.0m., res 55,864.0m., dep. 812,840.4m. (Dec. 2002); Pres. Sergio Zappa; Gen. Man. Carlos Eduardo Castro; 5 brs.

**Multibanco, SAECA:** Ayolas 482 esq. Oliva, Asunción; tel. (21) 44-7066; fax (21) 48-8496; internet www.multibanco.com.py; f. 1976 as Financiera Asunción, SA; adopted present name in 1995; under Central Bank administration from June 2003; Pres. Pedro Daniel Miraglio.

### Foreign Banks

**ABN AMRO Bank NV** (Netherlands): Alberdi y Estrella, Casilla 1180, Asunción; tel. (21) 49-0001; fax (21) 49-1734; e-mail clientes@ py.abnamro.com; internet www.abnamronet.com; f. 1965; Gen. Man. Peter Baltussen.

**Banco Asunción, SA:** Palma esq. 14 de Mayo, Asunción; tel. (21) 41-77000; fax (21) 41-77222; e-mail bancoasuncion@bancoasuncion .com.py; internet www.bancoasuncion.com.py; f. 1964; major shareholder Banco Central Hispano (Spain); Exec. Pres. Lisardo Peláez Acero; 2 brs.

**Banco Bilbao Vizcaya Argentaria Paraguaya, SA** (Spain): Yegros 435 y 25 de Mayo, Casilla 824, Asunción; tel. (21) 49-2072; fax (21) 44-7874; e-mail bbva.paraguay@bbva.com.py; f. 1961 as Banco Exterior de España, SA; renamed Argentaria Banco Exterior in 1999, name changed to above in 2000; Pres. and Gen. Man. Angel Soria Tabuenca; 5 brs.

**Banco do Estado de São Paulo SA (BANESPA)** (Brazil): Independencia Nacional esq. Fulgencio R. Moreno, Casilla 2211, Asunción; tel. (21) 49-4981; fax (21) 49-4985; e-mail banesspa@infonet .com.py; Pres. and Dir Gabriel Jaramillo Sanint.

**Banco de la Nación Argentina:** Chile y Palma, Asunción; tel. (21) 44-8566; fax (21) 44-4365; e-mail bnaasn@bna.com.py; f. 1942; Man. Aldo Dario Paviotti; 3 brs.

**Banco del Paraná, SA:** Chile esq Eduardo V. Haedo, Apdo 2298, Asunción; tel. (21) 44-6691; fax (21) 49-8909; e-mail banco.parana@ bancodelparana.com.py; internet www.bancodelparana.com.py; f. 1980; cap. US $12.6m., dep. $32.3m. (Dec. 2001); 90.79% owned by Banestado (Brazil); Pres. Antonio Carlos Genovese; Dir Miramar Bottini Filho; 5 brs.

**Banco Sudameris Paraguay, SA:** Independencia Nacional y Cerro Corá, Casilla 1433, Asunción; tel. (21) 44-8670; fax (21) 44-4024; e-mail gerencia@sudameris.com.py; internet www.sudameris .com.py; f. 1961; savings and commercial bank; subsidiary of Banque Sudameris; cap. 41,659.1m., res 24,067.3m., dep. 1,033,524.7m. (Dec. 2002); Pres. Carlos González Taboada; Man. Ignacio Jaquotot; 5 brs.

**Citibank NA** (USA): Estrella, esq. Chile, Asunción; tel. (21) 494-951; fax (21) 444-820; e-mail citservi.paraguay@citicorp.com; internet www.citibank.com/paraguay; f. 1958; dep. 55,942.2m.; total assets 1,316,258.1m. (Dec. 1997); Gen. Man. Henry Comber.

**Lloyds Bank PLC** (United Kingdom): Palma esq. Juan E. O'Leary, Apdo 696, Asunción; tel. (21) 491-7000; fax (21) 491-7414; e-mail lbpyitec@conexion.com.py; f. 1920; Man. Stuart R. C. Duncan.

### Banking Associations

**Asociación de Bancos del Paraguay:** Jorge Berges 229 esq. EEUU, Asunción; tel. (21) 21-4951; fax (21) 20-5050; e-mail abp .par@pla.net.py; mems: Paraguayan banks and foreign banks with brs in Asunción; Pres. Celio Tunholi.

**Cámara de Bancos Paraguayos:** 25 de Mayo esq. 22 de Setiembre, Asunción; tel. (21) 22-2373; Pres. Miguel Angel Larreinegabe.

### STOCK EXCHANGE

**Bolsa de Valores y Productos de Asunción SA:** Estrella 540, Asunción; tel. (21) 44-2445; fax (21) 44-2446; internet www.bvpasa .com.py; f. 1977; Pres. Jorge Daniel Pecci Miltos; Gen. Man. Hugo Eugenio Salinas Valdés.

### INSURANCE

**La Agrícola SA de Seguros Generales:** Mariscal López 5377 y Consejal Vargas, Asunción; tel. (21) 609-509; fax (21) 609-606; e-mail agricola@rieder.net.py; f. 1982; general; Pres. Dr Vicente Osvaldo Bergues; Gen. Man. Carlos Alberto Levi Sosa.

**ALFA SA de Seguros y Reaseguros:** Yegros 944, Asunción; tel. (21) 44-9992; fax (21) 44-9991; e-mail alfa.seg@conexion.com.py; Pres. Arnaldo Ramírez González.

## PARAGUAY

**América SA de Seguros y Reaseguros:** Alberdi esq. Manduvirá, 1°, Asunción; tel. (21) 49-1713; fax (21) 44-8036; Pres. EDUARDO NICOLÁS BO.

**Aseguradora Paraguaya, SA:** Israel 309 esq. Rio de Janeiro, Casilla 277, Asunción; tel. (21) 21-5086; fax (21) 22-2217; e-mail asepasa@asepasa.com.py; f. 1976; life and risk; Pres. GERARDO TORCIDA CONEJERO.

**Atalaya SA de Seguros Generales:** Independencia Nacional 565, 1°, esq. Azara y Cerro Corá, Asunción; tel. (21) 49-2811; fax (21) 49-6966; e-mail ataseg@telesurf.com.py; f. 1964; general; Pres. KARIN M. DOLL.

**Cenit de Seguros, SA:** Ayolas 1082 esq. Ibáñez del Campo, Asunción; tel. (21) 49-4972; fax (21) 44-9502; e-mail cenitsa@rieder.net.py; Pres. Dr FELIPE OSCAR ARMELE BONZI.

**Central SA de Seguros:** Edif. Betón I, 1° y 2°, Eduardo Víctor Haedo 179, Independencia Nacional, Casilla 1802, Asunción; tel. (21) 49-4654; fax (21) 49-4655; e-mail censeg@conexion.com.py; f. 1977; general; Pres. MIGUEL JACOBO VILLASANTI; Gen. Man. Dr FÉLIX AVEIRO.

**El Comercio Paraguayo SA Cía de Seguros Generales:** Alberdi 453 y Oliva, Asunción; tel. (21) 49-2324; fax (21) 49-3562; f. 1947; life and risk; Dir Dr BRAULIO OSCAR ELIZECHE.

**La Consolidada SA de Seguros y Reaseguros:** Chile 719 y Eduardo Víctor Haedo, Casilla 1182, Asunción; tel. (21) 44-5788; fax (21) 44-5795; f. 1961; life and risk; Pres. Dr JUAN DE JESÚS BIBOLINI; Gen. Man. Lic. JORGE PATRICIO FERREIRA FERREIRA.

**Fénix SA de Seguros y Reaseguros:** Iturbe 823 y Fulgencio R. Moreno, Asunción; tel. (21) 49-5549; fax (21) 44-5643; e-mail fenix@quanta.com.py; Pres. VÍCTOR MARTÍNEZ YARYES.

**Grupo General de Seguros y Reaseguros, SA:** Edif. Grupo General, Jejuí 324 y Chile, 3°, Asunción; tel. (21) 49-7897; fax (21) 44-9259; e-mail general_de_seguros@ggeneral.com.py; Pres. JORGE OBELAR LAMAS.

**La Independencia de Seguros y Reaseguros, SA:** Edif. Parapatí, 1°, Juan E. O'Leary 409 esq. Estrella, Casilla 980, Asunción; tel. (21) 44-7021; fax (21) 44-8996; e-mail la_independencia@par.net.py; f. 1965; general; Pres. REGINO MOSCARDA; Gen. Man. JUAN FRANCISCO FRANCO LÓPEZ.

**Intercontinental SA de Seguros y Reaseguros:** Iturbe 1047 con Teniente Fariña, Altos, Asunción; tel. (21) 49-2348; fax (21) 49-1227; e-mail mvmodica@yahoo.com; f. 1978; Pres. Dr JUAN MODICA; Gen. Man. LUIS SANTACRUZ.

**Mapfre Paraguay, SA:** Avda Mariscal López 910 y General Aquino, Asunción; tel. (21) 44-1983; fax (21) 49-7441; e-mail mafpy@conexion.com.py; Pres. LUIS MARÍA ZUBIZARRETA.

**La Meridional Paraguaya SA de Seguros:** Iturbe 1046, Teniente Fariña, Asunción; tel. (21) 49-8827; fax (21) 49-8826; Pres. JUAN PASCUAL BURRÓ MUJICA.

**Mundo SA de Seguros:** Estrella 917 y Montevideo, Asunción; tel. (21) 49-2787; fax (21) 44-5486; e-mail mundosa@par.net.py; f. 1970; risk; Pres. JUAN MARTÍN VILLALBA DE LOS RÍOS; Gen. Man. BLÁS MARCIAL CABRAL BARRIOS.

**La Paraguaya SA de Seguros:** Estrella 675, 7°, Asunción; tel. (21) 49-1367; fax (21) 44-8235; f. 1905; life and risk; Pres. JUAN BOSCH.

**Patria SA de Seguros y Reaseguros:** General Santos 715, Asunción; tel. (21) 22-5250; fax (21) 21-4001; e-mail patria@conexion.com.py; f. 1968; general; Pres. Dr MARCOS PERERA R.

**Porvenir SA de Seguros y Reaseguros:** Rca Argentina 222, Asunción; tel. (21) 61-3132; Pres. Dr JOSÉ DE JESÚS TORRES AGUILERA.

**La Previsora SA de Seguros Generales:** Estrella 1003, 5°, Asunción; tel. (21) 49-2442; fax (21) 49-4791; f. 1964; general; Pres. RUBÉN ODILIO DOMECQ M.; Man. JORGE ENRIQUE DOMECQ F.

**El Productor SA de Seguros y Reaseguros:** Ind. Nacional 811, 8°, Asunción; tel. (21) 49-1576; fax (21) 49-1599; Pres. REINALDO PAVÍA.

**Real Paraguaya de Seguros, SA:** Edif. Banco Real, 1°, Estrella esq. Alberdi, Casilla 1442, Asunción; tel. (21) 49-3171; fax (21) 49-8129; f. 1974; general; Pres. PETER M. BALTUSSEN.

**Rumbos SA de Seguros:** Estrella 851, Ayolas, Casilla 1017, Asunción; tel. (21) 44-9488; fax (21) 44-9492; f. 1960; general; Pres. Dr ANTONIO SOLJANCIC; Man. Dir ROBERTO GÓMEZ VERLANGIERI.

**La Rural del Paraguay SA de Seguros:** Avda Mariscal López 1082 esq. Mayor Bullo, Casilla 21, Asunción; tel. (21) 49-1917; fax (21) 44-1592; e-mail larural@larural.com.py; f. 1920; general; Pres. YUSAKU MATSUMIYA; Gen. Man. EDUARDO BARRIOS PERINI.

**Seguros Chaco SA de Seguros y Reaseguros:** Mariscal Estigarribia 982, Casilla 3248, Asunción; tel. (21) 44-7118; fax (21) 44-9551; e-mail segucha@conexion.com.py; f. 1977; general; Pres. EMILIO VELILLA LACONICH; Exec. Dir ALBERTO R. ZARZA TABOADA.

**Seguros Generales, SA (SEGESA):** Edif. SEGESA, 1°, Oliva 393 esq. Alberdi, Casilla 802, Asunción; tel. (21) 49-1362; fax (21) 49-1360; e-mail segesa@conexion.com.py; f. 1956; life and risk; Pres. CÉSAR AVALOS.

**El Sol del Paraguay, Cía de Seguros y Reaseguros, SA:** Cerro Corá 1031, Asunción; tel. (21) 49-1110; fax (21) 21-0604; f. 1978; Pres. Dr ANGEL ANDRÉS MATTO; Vice-Pres. JALIL SAFUÁN.

**Universo de Seguros y Reaseguros, SA:** Edif. de la Encarnación, 9°, 14 de Mayo esq. General Díaz, Casilla 788, Asunción; tel. (21) 44-8530; fax (21) 44-7278; f. 1979; Pres. BRAULIO GONZÁLEZ RAMOS.

**Yacyretá SA de Seguros y Reaseguros:** Oliva 685 esq. Juan E. O'Leary y 15 de Agosto, Asunción; tel. (21) 45-2374; f. 1980; Pres. OSCAR A. HARRISON JACQUECT.

### Insurance Association

**Asociación Paraguaya de Cías de Seguros:** 15 de Agosto esq. Lugano, Casilla 1435, Asunción; tel. (21) 44-6474; fax (21) 44-4343; e-mail apcs@uninet.com.py; f. 1963; Pres. Dr EMILIO VELILIA LACONICH; Sec. D. GERARDO TORCIDA CONEJERO.

## Trade and Industry

### GOVERNMENT AGENCIES

**Consejo Nacional para las Exportaciones:** Asunción; f. 1986; founded to eradicate irregular trading practices; Dir ERNST FERDINAND BERGEN SCHMIDT (Minister of Industry and Commerce).

**Consejo de Privatización:** Edif. Ybaga, 10°, Presidente Franco 173, Asunción; fax (21) 44-9157; responsible for the privatization of state-owned enterprises; Exec. Dir RUBÉN MORALES PAOLI.

**Instituto Nacional de Tecnología y Normalización (INTN)** (National Institute of Technology and Standardization): Avda General Artigas 3973 y General Roa, Casilla 967, Asunción; tel. (21) 29-0160; fax (21) 29-0266; e-mail intn@intn.gov.py; internet www.intn.gov.py; national standards institute; Dir-Gen. LILIAN MARTÍNEZ DE ALONSO.

**Instituto de Previsión Social:** Constitución y Luis Alberto de Herrera, Casilla 437, Asunción; tel. (21) 22-5719; fax (21) 22-3654; f. 1943; responsible for employees' welfare and health insurance scheme; Pres. PEDRO FERREIRA.

### DEVELOPMENT ORGANIZATIONS

**Secretaría Técnica de Planificación del Desarrollo Económico y Social:** Edif. AYFRA, 3°, Presidente Franco y Ayolas, Asunción; tel. (21) 45-0422; fax (21) 49-6510; e-mail webmarketing@stp.gov.py; govt body responsible for overall economic and social planning; Exec. Sec. CARLOS LUIS FILIPPI SANABRÍA; Sec.-Gen. Lic. OSVALDO MARTINEZ ORTEGA.

**Acuerdo Ciudadano** (Articulación de la Sociedad Civil): República de Siria 35, Asunción; tel. (21) 20-7757; fax (21) 20-2918; e-mail info@acuerdociudadano.org.py; internet www.acuerdociudadano.org.py; f. 2001; grouping of social devt orgs; Gen. Co-ordinator PASCUAL RUBIANI YANHO.

**AFS:** Azara 2242, con 22 de Setiembre, Asunción; tel. (21) 44-2369; fax (21) 49-3277; e-mail info-paraguay@afs.org; internet www.afs.org.py; educational and social devt; Exec. Dir VICTORIA VILLALBA.

**Alter Vida** (Centro de Estudios y Formación para el Ecodesarrollo): Itapúa 1372, esq. Primer Presidente y Río Monday, Barrio Trinidad, Asunción; tel. (21) 29-8842; fax (21) 29-8845; e-mail info@altervida.org.py; internet www.altervida.org.py; f. 1985; ecological devt; Dir BEATRIZ CHASE.

**Asociación Rural del Paraguay (ARP):** Ruta Transchaco Km. 14; tel. (21) 75-4412; e-mail ania@arp.org.py; internet www.arp.org.py; grouping of agricultural cos and farmers; Dir M. R. ALONSO.

**Centro de Información y Recursos para el Desarrollo (CIRD):** tel. (21) 22-6071; fax (21) 21-2540; e-mail cird@cird.org.py; internet www.cird.org.py; f. 1988; information and resources for devt orgs; Exec. Pres. AGUSTÍN CARRIZOSA.

**Consejo Nacional de Coordinación Económica:** Presidencia de la República, Paraguayo Independiente y Juan E. O'Leary, Asunción; responsible for overall economic policy; Sec. FULVIO MONGES OCAMPOS.

**Consejo Nacional de Desarrollo Industrial** (National Council for Industrial Development): Asunción; national planning institution.

# PARAGUAY

**Cooperación Empresarial y Desarrollo Industrial (CEDIAL):** Edif. UIP, 2°, Cerro Corá 1038, esq. Estados Unidos y Brasil, Asunción; tel. (21) 23-0047; e-mail cedial@cedial.org.py; internet www.cedial.org.py; promotes commerce and industrial devt; f. ; Gen. Man. HERNÁN RAMÍREZ.

**Instituto de Bienestar Rural (IBR):** Tacuary 276, Asunción; tel. (21) 44-0578; fax (21) 44-6534; responsible for rural welfare and colonization; Pres. ANTONIO IBÁÑEZ.

**Instituto Paraguayo del Indígena (INDI):** Don Bosco 745, Casilla 1575, Asunción; tel. (21) 49-3737; fax (21) 44-7154; f. 1981; responsible for welfare of Indian population; Pres. OLGA ROJAS DE BAEZ.

**ProParaguay:** Edif. Ayfra, 12°, Pdte Franco y Ayolas, Asunción; tel. (21) 49-3625; fax (21) 49-3862; e-mail ppy@proparaguay.gov.py; internet www.proparaguay.gov.py; f. 1991; responsible for promoting investment in Paraguay and the export of national products; Dir-Gen. LUIS MORINIGO GANCHI.

**Red Rural de Organizaciones Privadas de Desarrollo:** Manuel Domínguez 1045, Asunción; tel. (21) 22-9740; e-mail redrural@telesurf.com.py; internet www.redrural.org.py; f. 1989; co-ordinating body for rural development orgs; Gen. Co-ordinator IDALINA GÓMEZ; Sec. HEBE GONZÁLEZ.

## CHAMBERS OF COMMERCE

**Cámara Nacional de Comercio y Servicios de Paraguay:** Estrella 540-550, Asunción; tel. (21) 49-3321; fax (21) 44-0817; e-mail info@ccparaguay.com.py; internet www.ccparaguay.com.py; f. 1898; fmrly Cámara y Bolsa de Comercio; name changed 2002; Pres. RAUL RICARDO DOS SANTOS; Gen. Man. Lic. MIGUEL RIQUELME OLAZAR.

**Cámara de Comercio Paraguayo-Americano** (Paraguayan-American Chamber of Commerce): Edif. El Faro Internacional, 4°, Of. 1, General Díaz 521, Asunción; tel. and fax (21) 44-2135; e-mail pamchamb@conexion.com.py; internet www.pamcham.com.py; f. 1981; c. 120 mem cos.

## EMPLOYERS' ORGANIZATIONS

**Asociación de Empresas Financieras del Paraguay (ADEFI):** Edif. Ahorros Paraguayos, Torre II, 6°, Of. 05, General Díaz 471, Asunción; tel. Tel.: (21) 44-82 98; fax (21) 49-8071; e-mail adefi@conexion.com.py; internet www.adefi.org.py; f. 1975; grouping of financial cos; Pres. BELTRÁN MACCHI SALÍN.

**Asociación Paraguaya de la Calidad:** Lugano 627 y 15 de Agosto, 2°, Asunción; tel. (21) 44-7348; fax (21) 45-0705; e-mail apc@conexion.com.py; internet www.apc.org.py; f. 1988; grouping of cos to promote quality of goods and services; Pres. JORGE MIGUEL BRUNOTTE.

**Federación de la Producción, Industria y Comercio (FEPRINCO):** Edif. Union Club, Palma 751, 3°, esq. O'Leary y Ayolas, Asunción; tel. (21) 44-6634; fax (21) 44-6638; e-mail feprinco@quanta.com.py; organization of private-sector business executives; Pres. MIGUEL ANGEL CARRIZOSA GALLIANO.

**Unión Industrial Paraguaya (UIP):** Cerro Corá 1038, entre Estados Unidos y Brasil, Casilla 782, Asunción; tel. (21) 21-2556; fax (21) 21-3360; e-mail uip@uip.org.py; internet www.uip.org.py; f. 1936; organization of business entrepreneurs; Pres. Ing. GUILLERMO STANLEY.

## MAJOR COMPANIES

**Azucarera Paraguaya, SA (AZPA):** Avda General Artigas 552, Asunción; tel. (21) 21-3778; fax (21) 21-3150; e-mail informes@azpa.com.py; internet www.azpa.com.py; f. 1905; refining and wholesale distribution of cane sugar, alcohol and carbon dioxide; Pres. RAÚL HOECKLE; Dir-Gen. JAN MARC BOSCH; 700 employees.

**Consorcio de Ingeniería Electromecanica, SA (CIE):** Artigas 3443, Casilla 2078, Asunción; tel. (21) 64-2850; fax (21) 64-4130; e-mail ventas@cie.com.py; internet www.cie.com.py; f. 1978; manufacture of sheet metal work; Pres. HUGO ARANDA NÚÑEZ; 800 employees.

**Empresa Distribuidora Especializada, SA (EDESA):** Prof Conradi 1690, esq. Avda Eusebio Ayala 1690 Asunción; tel. (21) 50-1652; fax (21) 50-8549; e-mail edesa@pla.net.py; internet www.edesa.com.py; f. 1981; wholesale distribution of durable goods; Pres. RAÚL ALBERTO DÍAZ DE ESPADA; 450 employees.

**Fenix, SA:** Avda Boggiani 5086, Asunción; tel. (21) 66-0517; fax (21) 66-3375; e-mail fenix@quanta.com.py; f. 1968; import and manufacture of clothing and general merchandise; Pres. Dr ROLANDO NIELLA; 1,800 employees.

**Grandes Tiendas La Riojana, SA:** Avda Mariscal Estigarribia 165/171, Asunción; tel. (21) 49-2211; fax (21) 44-6698; e-mail riojana@rieder.net.py; department stores; CEO LÁZARO MORGA LACALLE; 480 employees.

**Industrializadora Guaraní, SA:** Madame Lynch y Sucre, Asunción; tel. (21) 67-3385; fax (21) 50-1761; f. 1967; manufacture of canned and bottled soft drinks; Man. NERY LOVERA PÉREZ RAMÍREZ; 1,170 employees.

**La Vencedora, SA:** Manduvirá y Alberdi, 6°, Asunción; tel. (21) 49-2225; fax (21) 44-8036; e-mail info@lavencedora.com; internet www.lavencedora.com; f. 1895; cigarette manufacturers; Pres. EDUARDO NICOLÁS BO; Man. Dir MARCUS NICOLÁS BO; 200 employees.

**Las Palmas:** Avda República Argentina 2154, Asunción; tel. (21) 55-3438; fax (21) 55-3403; f. 1943; production of cotton and manufacture of cotton linens; Pres. and Gen. Man. Dr ALEJANDRO GONZÁLEZ; 965 employees.

**Paraguay Refrescos, SA:** 3°, Ruta Ñemby Barcequillo Km 3, San Lorenzo; tel. (21) 50-4121; fax (21) 50-3608; f. 1964; manufacture of soft drinks; owned by Quilmes SA of Argentina; Pres. CARLOS MIGUENS BEMBERG; 700 employees.

**Petróleos Paraguayos, SA (PETROPAR):** Chile 753 casi Eduardo V. Haedo, Edif. Oga Rape, 9°, Asunción; tel. (21) 44-8503; fax (21) 448-503; e-mail contactenos@petropar.gov.py; internet www.petropar.gov.py; f. 1986; petroleum refining; Pres. ANGEL MARÍA RECALDE; 400 employees.

**Scavone Hermanos, SA:** Santa Ana 431 y Avda España, Asunción; tel. (21) 60-8171; fax (21) 66-1480; e-mail preshsa@scavonehnos.com.py; f. 1905; manufacture and retail distribution of pharmaceuticals; Pres. MARÍA LUISA SCAVONE DE MERA; 530 employees.

**Shell Paraguay Ltda:** Edif. Citibank, 3°, Avda España esq. Brasilia, Asunción; tel. (21) 49-1111; fax (21) 44-9253; f. 1953; wholesale distribution of petroleum products; Dir-Gen. ADRIAN HEARLE; 119 employees.

**Tecno Electric, SA:** Tte Primero Demetrio Araujo Miño y Smo Sacramento, Asunción; tel. (21) 29-0080; fax (21) 29-2863; e-mail tesa@tecnoelectric.com.py; f. 1968; manufacture of electrical equipment; Pres. GUIDO BOETTNER BALANSA; 200 employees.

**Vargas Peña Apezteguia y Compañía, SAIC:** Quesada 5240 y Cruz del Chaco, Asunción; tel. (21) 60-2841; fax (21) 60-0262; e-mail abgvm@rieder.net.py; internet www.vargaspena.com.py; f. 1977; production of cotton and edible oils; Pres. JOSÉ MARÍA HERNAN VARGAS PEÑA APEZTEGUIA; 600 employees.

## UTILITIES

### Electricity

**Administración Nacional de Electricidad (ANDE):** Avda España 1268, Asunción; tel. (21) 21-1001; fax (21) 21-2371; e-mail ande@ande.gov.py; internet www.ande.gov.py; f. 1949; national electricity board, privatization plans cancelled in June 2002; Pres. MARTÍN GONZÁLEZ GUGGIANI; Sec.-Gen. LUÍS OSVALDO VIVEROS GÓMEZ.

**Itaipú Binacional:** Centro Administrativo, Ruta Internacional Km 3.5, Avda Señor Rodríguez 150, Ciudad del Este; tel. (61) 57-2600; e-mail itaipu@itaipu.gov.br; internet www.itaipu.gov.br; f. 1974; jtly owned by Paraguay and Brazil; hydroelectric power-station on Brazilian-Paraguayan border; 1.3m. GWh of electricity produced in 2004; Dir-Gen. (Paraguay) Dr VÍCTOR BERNAL GARAY.

### Water

**Empresa de Servicios Sanitarios del Paraguay (ESSAP):** José Berges 516, entre Brasil y San José, Asunción; tel. (21) 21-0319; fax (21) 21-5061; fmrly Corporación de Obras Sanitarias (CORPOSANA); responsible for public water supply, sewage disposal and drainage; privatization plans suspended in 2002; Pres. CARLOS ANTONIO LÓPEZ RODRÍGUEZ.

## TRADE UNIONS

**Central Nacional de Trabajadores (CNT):** Piribebuy 1078, Asunción; tel. (21) 44-4084; fax (21) 49-2154; e-mail cnt@telesurf.com.py; Sec.-Gen. EDUARDO OJEDA; 80,000 mems.

**Central de Sindicatos de Trabajadores del Estado Paraguayo (Cesitep):** Asunción; Pres. REINALDO BARRETO MEDINA.

**Central Unitaria de Trabajadores (CUT):** San Carlos 836, Asunción; tel. (21) 44-3936; fax (21) 44-8482; f. 1989; Pres. ALAN FLORES; Sec.-Gen. JORGE ALVARENGA.

**Confederación Paraguaya de Trabajadores (CPT)** (Confederation of Paraguayan Workers): Yegros 1309–33 y Simón Bolívar, Asunción; tel. (21) 44-4921; fax (21) 20-5070; e-mail sixto10@telesurf.com.py; f. 1951; Pres. SIXTO ALONSO MENDOZA; Sec.-Gen. PATROCINIO CARMONA; 43,500 mems from 189 affiliated groups.

**Coordinadora Agrícola de Paraguay (CAP):** Asunción; farmers' organization; Pres. HÉCTOR CRISTALDO.

**Organización de Trabajadores de Educación del Paraguay (OTEP):** Avda del Pueblo 845, con Ybyra Pyta, barrio Santa Lucía, Lambaré; tel. and fax (21) 55-5525; e-mail otepsn@highway.com.py.

## Transport

### RAILWAYS

**Ferrocarriles del Paraguay:** México 145, Casilla 453, Asunción; tel. (21) 44-3273; fax (21) 44-2733; e-mail ferroca@rieder.net.py; internet www.ferrocarriles.com.py; f. 1854; state-owned since 1961; 376 km of track; scheduled for privatization; Pres. Lauro Manuel Ramírez López.

### ROADS

In 1999 there were an estimated 29,500 km of roads, of which 14,986 km were paved. The Pan-American Highway runs for over 700 km in Paraguay and the Trans-Chaco Highway extends from Asunción to Bolivia.

### SHIPPING

**Administración Nacional de Navegación y Puertos (ANNP)** (National Shipping and Ports Administration): Cólon y El Paraguayo Independiente, Asunción; tel. (21) 49-5086; fax (21) 49-7485; e-mail annp@mail.pla.net.py; internet www.annp.gov.py; f. 1965; responsible for ports services and maintaining navigable channels in rivers and for improving navigation on the Rivers Paraguay and Paraná; Pres. Milciades Rabery Ocampos.

### Inland Waterways

**Flota Mercante Paraguaya SA (FLOMEPASA):** Estrella 672-686, Casilla 454, Asunción; tel. (21) 44-7409; fax (21) 44-6010; boats and barges up to 1,000 tons displacement on Paraguay and Paraná rivers; cold storage ships for use Asunción–Buenos Aires–Montevideo; Pres. Capt. Aníbal Gino Pertile R.; Commercial Dir Dr Emigdio Duarte Sostoa.

### Ocean Shipping

**Compañía Paraguaya de Navegación de Ultramar, SA:** Presidente Franco 625, 2°, Casilla 77, Asunción; tel. (21) 49-2137; fax (21) 44-5013; f. 1963 to operate between Asunción, US and European ports; 10 vessels; Exec. Pres. Juan Bosch B.

**Navemar S.R.L.:** B. Constant 536, 1°, Casilla 273, Asunción; tel. (21) 49-3122; 5 vessels.

**Transporte Fluvial Paraguayo S.A.C.I.:** Edif. de la Encarnación, 13°, 14 de Mayo 563, Asunción; tel. (21) 49-3411; fax (21) 49-8218; e-mail tfpsaci@tm.com.py; Admin. Man. Daniella Charbonnier; 1 vessel.

### CIVIL AVIATION

The major international airport, Aeropuerto Internacional Silvio Pettirossi, is situated 15 km from Asunción. A second international airport, Aeropuerto Internacional Guaraní, 30 km from Ciudad del Este, was inaugurated in 1996.

### National Airlines

**Transportes Aéreos del Mercosur (TAM Mercosur):** Aeropuerto Internacional Silvio Pettirossi, Hangar TAM/ARPA, Luque, Asunción; tel. (21) 49-1039; fax (21) 64-5146; e-mail tammercosur@uninet.com.py; internet www.tam.com.py; f. 1963 as Líneas Aéreas Paraguayas (LAP), name changed as above in 1997; services to destinations within South America; 80% owned by TAM Linhas Aéreas (Brazil); Pres. Miguel Candia.

**Aerolíneas Paraguayas (ARPA):** Terminal ARPA, Aeropuerto Internacional Silvio Pettirossi, Asunción; tel. (21) 21-5072; fax (21) 21-5111; f. 1994; domestic service; wholly-owned by Transportes Aéreos del Mercosur (TAM Mercosur).

## Tourism

Tourism is undeveloped, but, with recent improvements in infrastructure, efforts were being made to promote the sector. Tourist arrivals in Paraguay in 2003 totalled 268,175. In that year tourism receipts were US $81m.

**Secretaría Nacional de Turismo:** Palma 468, Asunción; tel. (21) 49-4110; fax (21) 49-1230; e-mail infosenatur@senatur.gov.py; internet www.senatur.gov.py; f. 1998; Exec. Sec. María Evangelista Troche de Gallegos.

## Defence

At 1 August 2004 Paraguay's Armed Forces numbered 10,010, of which 1,900 were conscripts. There was an army of 7,600, and an air force of 1,100. The navy, which operates on the rivers, had 1,400 members, including 900 marines. There were 14,800 men in the paramilitary police force, including 4,000 conscripts. Military service, which is compulsory, lasts for 12 months in the army and two years in the navy.

**Defence Budget:** 300,000m. guaraníes in 2004.

**Commander-in-Chief of the Armed Forces:** President of the Republic.

**Commander of the Armed Forces:** Gen. José Kanazawa.

**Chairman of the Joint Chiefs of Staff:** Gen. Expedito Garrigoza.

**Commander of the Army:** Gen. Rubén Alberto Alviso González.

**Commander of the Navy:** Vice-Adm. Miguel Angel Caballero Dellaloggia.

**Commander of the Air Force:** Gen. Roberto Vera.

## Education

Education in Paraguay is, where possible, compulsory for six years, to be undertaken between six and 12 years of age, but there are insufficient schools, particularly in the remote parts of the country. Primary education commences at the age of six and lasts for six years. Secondary education, beginning at 12 years of age, lasts for a further six years, comprising two cycles of three years each. In 1996 91% of children in the relevant age-group attended primary schools (males 91%; females 91%), while secondary enrolment in the same year was 38% (males 37%; females 39%). There were 7,456 primary schools and 2,149 secondary schools in 1999. There is one state and one Roman Catholic university in Asunción. Education spending amounted to 777,652m. guaraníes (some 18.6% of total government expenditure) in 1996.

## Bibliography

For works on South America generally, see Select Bibliography (Books)

Ferradas, C. A. *Power in the Southern Cone Borderlands: An Anthology of Development Practice*. Westport, CT, Bergin & Garvey, 1998.

Hernandez, R. E., and Henderson, J. D. *Paraguay*. Broomall, PA, Mason Crest Publrs, 2003.

Jermyn, L. *Paraguay*. New York, NY, Benchmark Books, 2000.

Kolinsky, C., and Nickson, R. A. *Historical Dictionary of Paraguay*. Lanham, MD, Rowman & Littlefield Publrs, 2000.

Lambert, P., and Nickson, A. *The Transition to Democracy in Paraguay*. Basingstoke, Palgrave, 1997.

Leuchars, C. *To the Bitter End: Paraguay and the War of the Triple Alliance*. Westport, CT, Greenwood Publishing Group, 2002.

Nickson, R. A. *Paraguay*. Oxford, ABC Clio, 1999.

*Paraguay Country Study Guide (World Country Study Guide Library)*. Washington, DC, International Business Publications, USA, 2000.

*Paraguay Foreign Policy and Government Guide*. Washington, DC, International Business Publications, USA, 2000.

Turner, B. *Politics and Peasant-State Relations in Paraguay*. Lanham, MD, University Press of America, 1993.

# PERU
## Geography

### PHYSICAL FEATURES

The Republic of Peru is the third largest country in South America and is located on the west coast of the continent, astride the Andes and descending into the Amazonian plains. The country's longest border is with Brazil (1,560 km or 969 miles), which lies to the east of central Peru. Colombia (1,496 km of border) lies to the north-east, and Ecuador (1,420 km) to the north. Peru has 2,414 km of Pacific coastline, which includes the westernmost bulge of the South American continent, but then runs generally south-eastwards. In the south-east, there is a short southern border with Chile (160 km) and a much longer eastern border with Bolivia (900 km). Only in the late 1990s did Peru settle its outstanding border disputes with Ecuador and Chile, although the maritime boundary with the latter remains unresolved. Slightly smaller than the US state of Alaska, in total Peru covers 1,285,216 sq km (496,225 sq miles), including 5,130 sq km of inland waters.

Apart from some offshore islands, Peru consists of the Costa (the Pacific littoral and the foothills of the Continental Divide), the Sierra of the high Andes and the north-eastern Montaña or Selvas (the wooded lower slopes of the mountains, the high Selvas, and the rainforests of the Amazonian plains, the low Selvas). The coastal plains and lowlands stretch the length of the country, some 65–160 km in width, accounting for some 10% of the country. There are few natural harbours, while inland are dry, flat plains and sand dunes near the Sechura Desert, rising to the foothills. This country is an extension of the extremely arid Atacama Desert of Chile, and so dry that only 10 of the 52 rivers that leave the Andes for the Pacific have sufficient flow to make it to the sea. Dotted along its length are about 40 'oases' suitable for farming. Parallel to the coast, and covering about 30% of the country, is the broad Sierra region. The Andes here consist of three ranges, the main one being the Cordillera Occidental, the one closest to the sea. The mountain ranges, which narrow in width from about 400 km in the south to 240 km in the north, are interspersed with lofty plateaux and deep valleys and gorges. The average height of the region is 3,660 m (about 1,180 ft), but in the mountains north of Lima, the coastal capital of Peru, Nevada de Huascarán reaches 6,768 m (22,213 ft). In the south-east, the interior plateau is broad enough to contain Lake Titicaca, the highest navigable lake in the world, which is shared with Bolivia. Directly west of here, but rather closer to the Pacific seashore to the south-west, is the source of the Amazon (which drains into the Atlantic, far to the east, 6,516 km later), which is usually cited as Lake McIntyre. Those rivers that do not drain into the Amazon or directly into the Pacific drain into Lake Titicaca, which drains through Desaguadero into Lake Poopó in Bolivia. On the rainier eastern slopes of the Andes the rivers have carved deep valleys and sharp crests, and it is this that forms the main barrier between the highlands and the Amazon basin. These forested slopes of the high Selvas in the north-east broaden into the flat, tropical jungle of the Amazon basin, the region (60% of the country's land area) collectively being known as the Montaña. The Montaña, the eastern strip of foothills and the north-east, is largely unexplored and reaches a maximum width of 965 km in the north, where the rainforest continues into Brazil. About one-half of the country is wooded and one-fifth is pasture, the dense rainforest dominating the east, but with more varied tropical vegetation in the centre and east (and still greater variety at altitude). There is some volcanic activity in the mountains, which are prone to earthquakes and landslides, with tsunamis and flooding occurrences at lower levels.

### CLIMATE

The climate is tropical in the lowlands, but it can be arctic on the highest mountains—there is permanent snow and ice on heights over 5,000 m. Agriculture is possible up to 4,400 m, with the

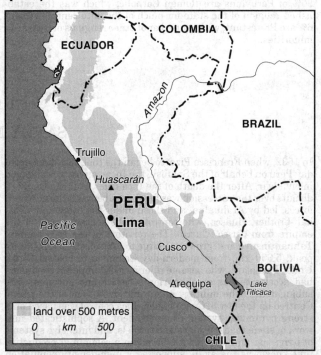

country able to grow a great range of crops and vegetation types at different altitudes. The Costa has an equable climate (temperatures averaging about 20°C—68°F—all year round), cooled by a major offshore sea current, but little rainfall gets past the mountains. For much of the coastal region average rainfall can be only about 50 mm (2 ins) per year, with slightly more in the north and less in the south. The grasslands of the foothills or on the western slopes of the Andes survive, when there is no rain, in the mists of the clouds that often cloak the heights. The interior and eastern parts of the Sierra, however, can receive heavy rainfall in October–April. The south-east Sierra, in Cusco (Cuzco), for instance, gets average annual rainfall of 815 mm, but it is the exposed eastern slopes that get most (more than 2,500 mm in places). Minimum and maximum temperatures, on average, range seasonally between –7°C and 21°C (20°–70°F). Although it is much hotter than the other two major regions of Peru, as is to be expected for a region concentrated in the north and east, it is the Montaña that gets most rainfall (mainly November–April). It is very hot, particularly lower down, and very humid. Total rainfall where the land is beginning to rise can be as much as 3,800 mm or so per year, although much of the water drains back into the lower Montaña from which it was originally evaporated by the prevailing north-easterlies. Normal patterns can, of course, be disrupted, notably during that periodic feature of the climate known as El Niño. This weather phenomenon can have severe repercussions for Peru, as it brings the heavy rainfall normally delivered to the western Pacific to the east, to a region unsuited to such conditions, which can result in widespread destruction.

### POPULATION

There are pronounced class and ethnic divisions, originating in the colonial period, that persist to this day, with the white, urban élite dominating the largely Amerindian and Mestizo countryside. Most of the indigenous natives are descendants of the imperial Incas, who were based in the Peruvian highlands. About 45% of the population are Amerindian, mainly Quechua-speaking, with some Aymará in the south and about 100 other, isolated groups in the east. A further 37% are of mixed Amerindian and white (mainly Spanish) stock, the Mestizos. About 15% are of unmixed white descent, mainly living in Lima and elsewhere along the Costa, with the rest being black, mixed-black or of Arab, Japanese or Chinese ancestry. Spanish, the principal language of 70% of the population, was the sole official language until 1975, when it was joined by Quechua (of which 28 dialects are spoken in Peru), followed by Aymará. More than 80% of Peruvians are Roman Catholic, which was the established religion of the state for much of the 20th century. About 6% are Protestant, with a number of other religious or non-faith minorities.

The total population was officially estimated at 27.5m. at mid-2004. Almost three-quarters of these people are urbanized, the largest city by far being the national capital, Lima (with over 7.6m.), which is on the central coast. Other important centres include Callao, the port of Lima, and the larger cities of Arequipa (the country's second city, but with only about one-10th the population of Lima), to the south and inland from the coast, and Trujillo, to the north. High in the Sierra, to the north of Arequipa, is the ancient Inca city of Cusco (Cuzco). The country is divided into 24 regions and one province (Callao).

# History

## SANDY MARKWICK

In 1532, when Francisco Pizarro began the conquest of modern-day Peru on behalf of the Spanish, the Inca empire was engaged in civil war. After the death of the Inca ruler, Huayna Capac, a dispute over the succession was fought out between the northern forces, led by Atahualpa, and those under the command of his half-brother, Huásca, who controlled the southern part of the empire from Cusco (Cuzco). The territory of the Inca empire, Tahuantinsuyo, had grown too large to control, stretching, at its peak, 5,000 km, from modern-day Colombia to central Chile. Under Pachacuti, who became ruler in 1432, imperial expansion had accelerated as neighbouring societies and cultures were subjugated by the military might and despotism of the Incas. Centred on the fertile Cusco region, Inca control was based on a strong pyramidal system of government. A sophisticated network of stone roads, bridges and tunnels, and intensive systems of irrigation and agriculture, based on a communal productive unit known as the ayllu, supported a high-density population (estimates vary between 12m. and 32m.). Maize and potatoes were the principal crops. The metals worked were gold, silver and bronze.

Pizarro's forces captured Atahualpa and killed some 7,000 of his troops after luring them into a trap in Cajamarca. The conquistadors, fearing the arrival of an Inca army to rescue Atahualpa, demanded riches from him before carrying out his public execution. The Cusco region fell to Pizarro's forces in 1533, and in 1544 the Viceroyalty of Peru was formally established, with its capital in Lima. Sporadic resistance continued until the Incas were finally defeated in 1572, when Tupac Amarú I, the last Inca emperor, was captured at the fall of Vilcabamba, taken to Cusco and beheaded. The Spanish colonial administration, having completed its military conquest (despite intermittent uprisings), embarked on a repressive programme of Christianization. Indigenous Indian (Amerindian) religious traditions survived, however, either assimilated into the new Roman Catholic order or concealed from the Church hierarchy. During the colonial period, the Andean tradition of communalism was weakened, as mining for export became the dominant economic activity, and the new criollo (Creole) élite of Spanish origin established large agricultural estates (haciendas) on the most fertile lands to supply produce for mining and coastal towns. The defeat of the Incas by the Spanish greatly influenced the character of modern Peru, where sharp divisions follow ethnic lines and long-established resentments occasionally escalate into violence.

## INDEPENDENCE

At the end of the 18th century an independence movement grew among the criollos, who resented their inferior status as well as the restrictive trade regulations and high taxes imposed by the Spanish Government. Peru declared independence from Spain on 28 July 1821, helped by the republican forces of the Argentine General, José de San Martín. The republicans, assisted by the United Kingdom, defeated Spain at the decisive Battle of Ayacucho in 1824. Political instability characterized the first half-century of independence. In the aftermath of the war, political office was held by regional military leaders (caudillos) who, having fought against a common enemy—the Spanish—now fought each other in the contest for central power. After independence Peru was highly regionalized, with economic power vested in the hacienda and political control exercised by hacienda owners in alliance with local caudillos. The weakness of the central state prevented successive Governments from consolidating national political control and forced them to work in co-operation with regional leaders to rule at local level. Governments in Lima were unstable and represented a series of shifting alliances between caudillos and regional groupings of the criollo élite of merchants and landowners. Between 1826 and 1865 some 34 Presidents, including 27 military officers, held office. One such contender for power was a Bolivian President, Gen. Andrés Santa Cruz, who, allied with southern Peruvian landowners, proclaimed the Peruvian-Bolivian Confederation in 1836, with its capital in Lima. Chile, viewing the Confederation as a threat to its interests in the Pacific, declared war. The Confederation was finally defeated and, subsequently, dissolved by Chilean forces in 1839.

## THE GUANO YEARS

The emergence of the guano trade ended a period of economic stagnation. International demand for the fertilizing properties of guano (the accumulated nitrate-rich deposits of bird colonies on the offshore islands) escalated with the development of capital-intensive commercial agriculture in Europe. The Peruvian Government established a state monopoly over the exploitation of guano. However, despite average annual economic growth rates of over 5% between 1840 and 1878, resulting largely from the trade, the success of the guano industry was not fully exploited. Revenue from guano was spent on the expansion of the bureaucracy, the military and the servicing of the Government's foreign debt, which had accumulated to finance imports. The sharp distinction remained intact between the colonial character of production on large estates and the self-sufficient nature of peasant economies, perpetuating deep ethnic and class divisions and preventing the emergence of a strong internal market, which would have provided Peru's productive sectors with the domestic stimulus for growth. Internal demand among élite consumers, which developed after the dramatic expansion of guano exports, could not be satisfied by the domestic productive sector and resulted in increased imports. The most lucrative years of the guano trade came to an end in the 1870s, as the result of competition from cheaper nitrates and synthetic fertilizers. Peru was under threat of bankruptcy, with a huge external debt of US $35m. and no access to foreign loans. The crisis culminated in another military defeat of Peru and Bolivia by Chile, in the War of the Pacific (1879–83), which resulted from a dispute over nitrate resources in the Atacama desert. Chile occupied much of Peru, including Lima, in 1881. Under the terms of the peace settlement, Chile annexed the province of Tarapac as well as the Bolivian department of Antofagasta.

## THE RISE OF A COASTAL OLIGARCHY

Post-war reconstruction led to the expansion of coastal haciendas and the rise of an oligarchy of merchants, financiers and landowners, better adapted to a new capitalist age, whose interests centred on the production and export of primary products. The coastal oligarchy (comprising some 50 extended families), allied to foreign capital and semi-feudal landowners of the sierra, exercised widespread influence and control over government (until as late as 1968, when a modernizing military regime seized power—see below). Raw material prices grew at an average of 7% per year between 1890 and 1929. Peru's reconstruction after the War of the Pacific also received substantial support from the so-called Grace Contract, which was finally approved in 1889. The Contract, negotiated by a British businessman, Michael Grace, provided for the cancellation of Peru's national debt in return for the Government ceding control over its railway network to the largely British-owned Peruvian Corporation. Sugar and cotton became the principal exports, though wool, rubber, silver, copper and petroleum each accounted for more than 15% of total export earnings at various times between 1890 and 1929.

Peru's first modern political party, the Partido Civil (Civil Party), was created in 1871 to represent the interests of the new oligarchy. The *civilistas*, as the party's sympathizers were known, opposed military government and rejected the unproductive use of resources from guano. In 1872 Manuel Pardo was elected as the first *civilista* President. The Partido Civil, guided by liberal, *laissez-faire* doctrines, and the rival Partido Democrático (Democratic Party), which was more pro-Church and conservative, led Peru through a period of export-orientated development and rapid economic and social change during the next half-century. Formal parliamentary democracy was, in practice, extremely limited, with no more than some 3% of the total population enfranchised. During this period, however, the military was professionalized and, at least between 1895 and 1919, was brought under civilian control. The Government was reorganized to conform better to the demands of a modern export economy. Democratic government was interrupted by the dictatorship of Augusto Bernardino Legua, which lasted from 1919 until 1930. From this period, marked by economic prosperity and relative political stability, the structure of contemporary Peru emerged. The advance of capitalism saw peasants migrate to the cities, while traditional haciendas and small-scale mining operations were transformed into more modern agro-industrial or mining complexes. During this export-led expansion, which was allied to foreign capital, the USA replaced the United Kingdom as Peru's main trading partner and source of direct investment.

## THE RISE OF APRA

The export-led model of the Peruvian economy was seriously affected by the collapse of world commodity prices in 1929. From this time there was a growing realization that export dependency was fundamental to the country's underdevelopment. The debate over methods of resolving the problem of development was led by the reformist Víctor Raúl Haya de la Torre and the revolutionary José Carlos Mariátegui. De la Torre founded the left-wing and populist Alianza Popular Revolucionaria Americana (APRA—American Popular Revolutionary Alliance, also known as the Partido Aprista Peruano—PAP). Mariátegui founded the Partido Comunista Peruano (PCP—Peruvian Communist Party) in 1928. The emergence of APRA reflected the increasing radicalism among the Peruvian masses as world economic crisis in 1929 led to depression. APRA was a left-wing, populist and anti-imperialist organization claiming to represent peasants, workers and the progressive middle classes whose interests were not being served by the governing élite. When Legua's regime was deposed by a military coup led by Col Luis Sánchez Cerro, a popular military caudillo, the oligarchical élite, conscious of the threat from APRA, gave its support to the military. The take-over inaugurated a period of political instability in national politics between the 1930s and the 1960s, as an emerging national industrial bourgeoisie and Peru's popular sectors challenged the old alliance of the Peruvian agrarian oligarchy and foreign capital that had dominated the Peruvian state. This was reflected in the increase in popularity of APRA, which became an important political force and often had the largest representation in the Congreso (Congress). APRA was prevented from forming a government by an alliance of old oligarchical and military forces whose tactics included electoral fraud, repression (APRA was declared illegal in 1931–45 and 1948–56) and finally, in July 1962, a pre-emptive *coup d'état* (which annulled the elections of the previous month).

## MILITARY GOVERNMENT, 1968–80

In October 1968 a left-wing military regime led by the army Chief of Staff, Gen. Juan Velasco, supplanted President Fernando Belaúnde Terry of the reformist Acción Popular (AP—Popular Action). Belaúnde had been elected President in June 1963, as the candidate of both the AP and of the Partido Demócrata Cristiano (PDC—Christian Democratic Party). Until 1968 government in Peru had continued broadly to reflect the priorities of the old export-orientated élites. However, a developmentalist ideology, which emerged from within the Peruvian armed forces, coincided more closely with the priorities of industrial interests. Velasco and other ranking officers in the army held the old élites responsible for Peru's underdevelopment and assumed the role of modernizers. The military Government introduced land reform and nationalized banks, telecommunications, the railways, electricity production, fisheries and heavy industry. Large haciendas were partitioned and converted into agricultural co-operatives. This policy of nationalization was presented as a nationalist, socialist and anti-imperialist struggle, partly in order to inspire popular support for the reforms.

As opposition to the Velasco regime increased, attention focused on the Government's restrictions on education, its controls on currency and imports, and the failure of its agrarian reform. Agricultural co-operatives proved to be poorly conceived, corrupt and badly managed, resulting in rebellions in rural areas. Velasco was replaced, in August 1975, by Gen. Francisco Morales Bermúdez, in an internal military coup. The new Government abandoned the former regime's anti-imperialist image and pledged a return to civilian democracy. A new Constitution, which, for the first time, granted universal adult suffrage, was adopted in July 1979. In May 1980 military rule came to an end and Belaúnde was elected President for a second term, winning 45.4% of the votes cast. The Constitution came into effect in July.

## SENDERO LUMINOSO AND THE GUERRILLA WAR

On the eve of presidential elections in May 1980 the Maoist insurgent group, Sendero Luminoso (Shining Path), initiated a strategy of 'people's war', with the burning of ballot boxes in the rural community of Chuschi, in Ayacucho department. Subsequently, Sendero Luminoso extended its influence to most other departments. By the mid-1990s the movement's destructive campaign and the army's counter-insurgency were estimated to have caused some 30,000 deaths. The origins of Sendero Luminoso lay in the worsening of relations between the USSR and the People's Republic of China in the 1960s. Sendero Luminoso left the pro-Soviet PCP in 1970 under the leadership of its founder and ideological force, Abimael Guzmán Reynoso (also known as 'Chairman Gonzalo'). Sendero Luminoso won endorsement from some sectors of the peasantry by exploiting Amerindian traditions and centuries-old resentment against the central Government and foreign conquerors. It went on to penetrate shanty-town civic groups, trade unions, universities and teacher-training colleges and expanded its influence through Huamanga University in Ayacucho, the underdeveloped region chosen for the group's inception as a peasant movement.

The rise of Sendero Luminoso was assisted by the incomplete and largely unsuccessful land reform under Velasco, and by a fragmentation among the parties of the left over the issue of reform. Sendero Luminoso emphasized revolutionary violence as the fundamental mechanism for obtaining political change. Its use of extreme brutality brought a response in kind from the army. The movement regarded as a legitimate target any body or person that collaborated with the Peruvian state, participated in state-organized activities or was involved in interests allied to it. Sendero Luminoso's strategy was to mobilize the peasantry initially, and only focus on urban areas once its power and influence in the countryside was consolidated. Sendero Luminoso grew to represent a genuine threat to the country's internal security, particularly since it increased the risk of a military take-over.

## GARCÍA LEADS APRA TO GOVERNMENT

The Belaúnde Government became increasingly unpopular, owing, in particular, to the measures it imposed to deal with Peru's considerable economic problems and its ineffectiveness in combating Sendero Luminoso. The presidential election of April 1985 was won by the APRA candidate, Alan García, with 46% of the votes cast. The army accepted García's victory and demonstrated that it was prepared to adhere to the Constitution. President García introduced radical economic policies, including a limit on debt repayments, in order to control inflation. However, the economy continued to deteriorate and guerrilla violence from Sendero Luminoso and a second insurgent group, the Movimiento Revolucionario Tupac Amarú (MRTA—Tupac Amarú Revolutionary Movement), continued to increase. Support for the Izquierda Unida (IU—United Left, a coalition of seven left-wing parties formed in 1980) grew and trade unions organized a series of general strikes in 1987 and 1988. Rightwing opposition to the Government also increased as a result of the decision, announced in July 1987, to nationalize the financial sector. A 'freedom movement', Libertad (Liberty), expressing opposition to the nationalization plans, was formed under the leadership of the novelist Mario Vargas Llosa.

## INFLUENCE OF THE MILITARY RESTORED UNDER FUJIMORI

Alberto Fujimori, an independent candidate, was the unexpected victor in the presidential election of June 1990, and he was inaugurated in July of that year. An agricultural economist of Japanese descent, he had never previously held political office. Fujimori's main rival in the election was Vargas Llosa, representing the centre-right Frente Democrático alliance (FREDEMO—Democratic Front, established in early 1988 by the AP, Libertad and the Partido Popular Cristiano—Popular Christian Party). Fujimori's victory, following a second round of voting in which he gained 57% of the votes cast, reflected a widespread disillusionment with traditional party politics. Fujimori lacked the benefit of a party machine when he entered politics, and therefore established the Cambio 90 (C90—Change 90) party to contest the election. The party consisted largely of members of Protestant movements and centrists dissatisfied with the inefficiency and corruption of APRA and other traditional parties. Although C90 lacked a coherent ideology or programme, Fujimori won the election because he represented a more populist alternative to Vargas Llosa, who was identified with a Lima-based élite. During the election campaign Fujimori had advocated 'hard work, honesty and technology' and promised an economic-reform programme that was markedly less austere than the one proposed by Vargas Llosa. However, shortly after taking office, Fujimori broke his campaign pledges and introduced an uncompromisingly orthodox reform programme, which became known as the 'Fujishock'. He reduced state subsidies, freed the exchange rate, and, in December, began negotiations with international lending agencies. In addition, President Fujimori liberalized the economy to encourage foreign investment and embarked on an ambitious privatization programme involving all the state's productive ventures.

The lack of a coherent organizational support base meant that the army's loyalty was crucial. As soon as Fujimori came to power he attempted to win the support of the army in order to reduce the risk of a military coup and to secure a powerful political ally in the event of widespread unrest. He gave the army increased powers and autonomy in the counter-insurgency campaign and appointed army generals to his cabinet (Consejo de Ministros—Council of Ministers). On 5 April 1992 President Fujimori launched an *autogolpe*, an incumbent's coup, when he suspended the 1979 Constitution, dissolved parliament, suspended the authority of the judiciary and assumed wide powers of decree, with the endorsement of the armed forces. He justified his actions by claiming that the traditional political élite was blocking the reforms necessary to combat the economic crisis and ultra-leftist insurgencies. When President Fujimori suspended the legislature he appealed for, and received, mass popular approval, although the political class in Peru denounced his shift towards authoritarianism, unconstitutionality and demagoguery. The mainstream opposition maintained that he had forfeited his position and, in accordance with the 1979 Constitution, declared First Vice-President Máximo San Román Cáceres to be the new President. Similarly, the USA and other foreign powers urged the early restoration of democratic government.

Following the suspension of international financial aid to Peru, which threatened Fujimori's radical economic-reform programme, the President was forced to hold elections to the Congreso Constituyente Democrático (CCD—Democratic Constituent Congress), a constituent assembly, in November 1992. The pro-Fujimori alliance of C90 and the Nueva Mayoría (NM—New Majority) received some 40% of the votes cast, and won 44 seats in the 80-seat chamber. APRA and the AP boycotted the CCD elections, as did Libertad and small leftist parties. Of the significant political parties, only the Partido Popular Cristiano participated, winning eight seats. The assembly was inaugurated in December. In early January 1993 the CCD exonerated Fujimori for his coup and confirmed him as head of state.

The final draft of the proposed new Constitution was approved by the CCD in early September 1993. It was endorsed by the electorate, with 52.2% of the votes cast, in a national referendum held in October. The new Constitution consolidated President Fujimori's position by at least partially restoring his democratic legitimacy and by permitting him to be re-elected for a successive five-year term. All presidential decrees issued during the interim period were declared ratified until revised or revoked by the CCD. Foreign aid to Peru, which had been withdrawn in protest against the coup, was largely restored. The CCD acted as the legislature until the end of President Fujimori's existing term of office.

President Fujimori consistently protected the armed forces from accusations of human rights abuses. The extent to which the military influenced government policy was revealed by an amnesty law in June 1995. Unexpectedly, the legislation was hastily passed by the Congreso, without a committee stage to allow the opportunity for debate. The law extended amnesty to the military for human rights offences committed during the previous 15 years of counter-insurgency. The protection given to the military by Fujimori was motivated, in part, by a desire to head off any threat of a coup, although the President was not a weak party in the relationship. The President's power *vis-à-vis* the military was consolidated by a change in the law, granting him the power to appoint all generals and admirals, whereas, under previous Governments, the armed forces had always appointed their own high command.

### Border War with Ecuador

The armed forces' support for President Fujimori was tested by a short, undeclared border war with Ecuador in January–February 1995, in which more than 100 soldiers were killed. A cease-fire was brokered in Montevideo, Uruguay, in March by the guarantors of the 1942 Rio Protocol (Argentina, Brazil, Chile and the USA), which ended an earlier war with Ecuador in 1941. This short stretch of unmarked border in the Condor mountain range remained a source of tension between Peru and Ecuador, with occasional accusations from both sides of incursions and skirmishes, until in October 1998 in Brasília, Brazil, Peru and Ecuador signed a peace deal to end the long-standing dispute. The accord recognized Peruvian claims regarding the delineation of the border, but granted Ecuador navigation rights in Peruvian Amazonia. In addition, Peru ceded 1 sq km of territory around a former Ecuadorean outpost, the burial site of many Ecuadorean soldiers killed during the fighting of 1995. While opponents of the accord in Peru protested vociferously that the ceding of territory to Ecuador was incompatible with national dignity and security, its broad acceptance in both countries raised hopes that the peace would endure. The physical demarcation of the frontier, which was begun in mid-January 1999, was completed by May of that year.

## CAPTURE OF GUZMÁN

The capture, on 12 September 1992, of Abimael Guzmán and other members of Sendero Luminoso's Central Committee (the organization's highest decision-making body) and, subsequently, many others from among its senior leaders inflicted serious damage on the movement from which it seemed unlikely to recover. Guzmán's arrest, in a raid on a house in Lima, came at a crucial time, when Sendero Luminoso had been escalating its terrorist campaign in Lima and planning further offensives as part of its 'sixth military plan'. A large car bomb in the Lima suburb of Miraflores in July had killed 21 and injured 250 in the worst single terrorist incident in 12 years of guerrilla war. The bombing was designed to cause indiscriminate civilian casualties and to intimidate the wealthy and the middle class of Lima, which had largely escaped the worst effects of terrorism. The tactic embodied the change in strategy adopted after Sendero Luminoso's First Congress in 1988, in which it shifted its focus from the countryside to the cities, despite not having sufficient control in rural areas. This move away from orthodox Maoism proved Sendero Luminoso's undoing, as it caused the authorities to redouble their efforts to combat the organization while giving them a more manageable geographic focus in which to search for key guerrilla leaders. After Guzmán's arrest the overall scale of guerrilla violence decreased by approximately 50% in the latter half of 1992, and violence continued to decline before stabilizing at a significantly reduced rate. Sendero Luminoso experienced a devastating decline in the geographical extent of its influence, which became confined to shanty towns in Lima and coca-growing areas in the Huallaga valley and central Sierra, and other isolated areas.

In September 1993 Guzmán offered to commence peace negotiations with the Government, in what was either an extraordinary capitulation or a tactical retreat with the long-term aim of rebuilding the organization. Following Guzmán's offer of peace negotiations, the number of guerrillas surrendering in order to take advantage of the lenient sentences offered as part of the Government's Repentance Law increased significantly. Guzmán's advances to the Government gave rise to divisions within Sendero Luminoso. Although he succeeded in persuading imprisoned followers to support his demand for peace, an extremist faction, which controlled much of the organization outside prison, opposed him, and remained committed to the popular war. This faction, under the leadership of Oscar Ramírez Durand, continued to wage guerrilla insurgency, although attacks were smaller and less frequent than in previous years. Despite intermittent violence, Sendero Luminoso was no longer considered to represent a serious threat to the stability of the Government or to the state long before the capture of Durand by the security forces in July 1999.

## FUJIMORI'S SECOND TERM

President Fujimori was re-elected in April 1995 for a second five-year term of office, with 64.4% of the votes cast. His main challenger, former UN Secretary-General Javier Pérez de Cuéllar, supported by an independent movement, Unión Por el Perú (UPP—Union for Peru), received 21% of the ballot and conceded the popular mandate that the vote gave to Fujimori. The restoration of economic growth and success against the insurgencies in Fujimori's first term of office ensured the President's popularity. Meanwhile, the traditional parties had lost their support base and the opposition was divided. Fujimori intended to pursue the liberal, free-market reforms introduced since his accession to the presidency in 1990, although he pledged to use funds from privatization to alleviate poverty. However, divisions within the administration over economic policy widened as a result of the contraction of Peru's economy and a growth in inflation, after three years of rapid economic growth.

The threat of left-wing insurgency re-emerged dramatically in December 1996, with the seizure of the Japanese ambassador's residence in Lima and some 600 hostages by 14 MRTA guerrillas. The guerrillas, led by Néstor Cerpa Cartolini, released most of the hostages within days, but continued to hold the remaining 72 hostages, among them high-level government officials, foreign diplomats and businessmen, until the residence was stormed by Peruvian commandos in April 1997, resulting in the deaths of all 14 guerrillas and one hostage. Cerpa had demanded the release of MRTA prisoners, including the group's leader, Víctor Polay Campos, a ransom payment and safe passage; the raid appeared to mark an end to MRTA insurgent activity.

President Fujimori's style of government continued to be personal and autocratic, with little in the way of checks and balances to his rule. His NM/C90 supporters enjoyed an absolute majority in the new 120-seat Congreso, which made the passage of legislation and even constitutional reform straightforward. In addition, opposition political parties were in disarray, the judiciary was discredited, the trade unions largely defeated, regional and municipal authorities powerless and the Council of Ministers generally weak and subservient. The judiciary and prosecution services were widely accused of becoming tools of the executive. Civilian democracy was weakened, moreover, by the lack of engaging political parties and civic organizations through which popular demands could be articulated. The President's approach was populist, to appeal directly to the public, rather than seek agreement and compromise with rival political groups. He travelled widely throughout Peru, addressing problems at local level.

Disaffection with President Fujimori's authoritarian tendencies became evident from mid-1997, following a series of controversies, including allegations of the clandestine recording by the security services of the telephone conversations of opposition politicians. A vote by the pliant Congreso in August 1996 allowed Fujimori to stand for re-election to a third presidential term in 2000, despite strong opposition, which focused on the Constitution's clearly worded statute against any third consecutive term of office.

During 1998 President Fujimori repeatedly altered the composition of his administration in an attempt to consolidate his position and restore public confidence in his leadership. In August the President moved to reinforce his control of the levers of state power and strengthen his democratic credentials with the unexpected dismissal of Gen. Nicolás Hermoza Ríos, who had come to be seen as one of a ruling triumvirate, along with President Fujimori and his security chief, Vladimir Montesinos. President Fujimori's popularity continued to wane and a major reshuffle of the Council of Ministers in January 1999 could not reverse the trend. With the administration increasingly perceived as authoritarian in attitude and inflexibly liberalizing in its handling of a deteriorating economy, the early signs were not encouraging for President Fujimori's anticipated attempt to secure a third successive term of office in the presidential election scheduled for April 2000.

## FUJIMORI WINS DISCREDITED ELECTIONS FOR THIRD TERM

The first round of elections was held on 9 April 2000. Official results gave Fujimori just less than the 50% of votes required for an outright victory. Fujimori's main rival was an economist, Alejandro Toledo, of Perú Posible (PP—Possible Peru), who received just over 40% of the votes cast. Toledo's campaign focused on what he referred to as Fujimori's record of authoritarianism and abuse of power, rather than on any significant economic or political differences. All other candidates registered less than 4% of the votes cast. Toledo came from relative obscurity, having himself received less than 4% of votes cast as a candidate in the 1995 presidential election, to eclipse the challenges of the mayor of Lima, Alberto Andrade, and Luis Casteñeda Lossio, the former head of the Social Security Institute, of Solidaridad Nacional (SN—National Solidarity), who failed to muster nation-wide support. Toledo's mestizo (mixed race) ethnicity and impoverished background attracted support from many poor Peruvians. The vote was marred by allegations, made by the opposition candidates, of large-scale irregularities in the vote count. These allegations received widespread support from international observers and representatives of foreign governments, including, in particular, that of the USA. Complaints made by opposition candidates included government manipulation of the media to deny the opposition equal access, the use of state resources (including the military and the intelligence services) to promote Fujimori's campaign, and the alleged falsification of more than 1m. signatures by members of Perú

2000 (a new movement formed to support the incumbent's re-election campaign) to facilitate registration of Fujimori's candidacy. Suspicions of electoral fraud provoked street protests by opposition supporters. Legislative elections, also held on 9 April, resulted in 52 seats for Fujimori's Perú 2000 alliance, short of the 61 needed for an absolute majority. Toledo's PP was the second largest group, with 29 seats.

A second presidential ballot, between Fujimori and Toledo, was scheduled for 28 May 2000, despite efforts by Toledo and the Organization of American States (OAS) to postpone the vote for 10 days to ensure that the conditions for fair elections were in place. Toledo withdrew from the election and called on his supporters to spoil their ballot papers in protest at electoral fraud. The Jurado Nacional Electoral (JNE—National Electoral Board) refused the postponement request and, as a result, the OAS issued a statement that the elections could not be considered free and fair. The OAS and other international observers from the US-based Carter Center and the European Union withdrew their missions. Fujimori won an uncontested election (although the JNE ruled that Toledo remained a candidate and refused to remove his name from the ballot papers), with 51% of votes cast, while Toledo received 18%, and 31% of voters either spoiled their ballot papers or left them blank. The result was denounced as invalid by the political opposition. For its part, the OAS decided against the imposition of economic sanctions, instead dispatching a mission to Peru to explore options for strengthening democracy. Fujimori's inauguration on 28 July was accompanied by violent protests in Lima.

However, Fujimori's term of office was prematurely curtailed by the disclosure, in September 2000, of hundreds of videos, recorded covertly, that showed Vladimiro Montesinos, Fujimori's controversial security adviser, bribing, among others, an opposition member of the Congreso, a Supreme Court judge, television executives and an election official. In response, Fujimori declared that new elections would be organized, in which he would not participate, and that the national intelligence service, which Montesinos headed, would be disbanded. Montesinos' close links with the military high command led to fears of a military coup, but these concerns subsided after the military publicly supported Fujimori's decision. Demonstrations, led by Toledo, demanded the immediate resignation of Fujimori and the arrest of Montesinos, who fled to Panama. At the same time, 10 congressional representatives of Perú 2000 defected, thus depriving Fujimori of his majority in the legislature. In October the Congreso approved OAS-mediated proposals preparing the way for power to be transferred from Fujimori to his successor (as well as the dissolution of the Congreso to make way for a newly elected legislature) in mid-2001. The political crisis deepened in late October 2000, when Montesinos returned from Panama, after failing to secure political asylum there. At the same time, the first Vice-President, Francisco Tudela, resigned in protest at government attempts to tie new elections to a military amnesty, and middle-ranking officers in southern Peru staged an isolated rebellion demanding the resignation of Fujimori. Under pressure, the Government launched an investigation into Montesinos' activities, considering allegations including torture and murder, as well as corruption, gun-running and money-laundering. By this time Montesinos had again gone into hiding. Fujimori's attempts to consolidate political support focused on distancing himself from Montesinos, and included the staging of a highly public operation, led in person by Fujimori himself, to arrest the former intelligence chief and to dismiss his allies in the military high command. Despite his efforts, Fujimori failed to assuage his critics and, with Montesinos' whereabouts still unknown, a number of Fujimori's supporters defected, leading to a shift in power in the Congreso, which appointed Valentin Paniagua, a moderate opposition legislator, as President of the legislature. Paniagua immediately reinstated judges dismissed for challenging the constitutionality of Fujimori's third consecutive term of office.

## THE PRESIDENCY OF ALEJANDRO TOLEDO

Fujimori resigned from the presidency during a visit to Japan in November 2000, prompting speculation that he would live there in self-imposed exile. The Japanese authorities confirmed that Fujimori was entitled to Japanese citizenship and could stay indefinitely. The Congreso, meanwhile, rejected his resignation, and instead voted to dismiss him, declaring him 'morally unfit' to hold office. It appointed Paniagua as interim President of Peru, and he immediately named former UN Secretary-General and former presidential candidate Javier Pérez de Cuéllar as Prime Minister. The new President, lacking a democratic mandate to pursue a legislative agenda, pledged to act only as an 'administrator' overseeing new elections in April 2001. Paniagua continued the process of purging the military of Montesinos' influence in order to ensure the impartiality of the military in the transition towards a newly elected Government. In June 2001 Paniagua set up a Comisión de Verdad y Reconciliación (CVR—Truth and Reconciliation Commission) to investigate political violence in the preceding two decades. When the CVR delivered its final report in August 2003, it estimated that a total of 69,000 people had been killed during the insurgency and counter-insurgency, 30% of whom had died at the hands of the security forces. The end of the Fujimori era was widely embraced as an opportunity to restore the rule of law and the integrity of democratic institutions.

The principal candidates for the elections scheduled for 8 April 2001 were Alejandro Toledo, former President Alan García, who had returned from exile to contest the poll, and the conservative Lourdes Flores Nano of the Unidad Nacional (National Unity) alliance. Toledo won 36.5% of the first-round vote, García 25.8% and Flores 24.3%, thereby eliminating Flores from a second round of elections, which was held on 3 June. Toledo emerged victorious, polling 53.1%, ahead of García's 46.9%. The PP gained 45 seats in the concurrently held legislative elections but, together with allied parties, failed to secure a majority in the 120-seat Congreso, leaving the prospect of coalition government. Two days before his inauguration, President-elect Toledo announced a new, broad-based, 15-member cabinet, headed by an economist and lawyer, Roberto Dañino. Dañino's appointment was unexpected and reflected a shift in the balance of the Government towards technocrats and away from Toledo's political allies in the PP. Technocrats were already well represented in the cabinet in the form of the Minister of Economy and Finance, Pedro Pablo Kuczynski, a former IMF economist. The fiscal orthodoxy of both Kuczynski and Dañino was seen as a restriction on Toledo's populist tendencies, which gained for the Government a greater confidence among international investors.

Meanwhile, in June 2001 Venezuelan military intelligence officers, assisted by the US Federal Bureau of Investigation and the Peruvian police force, arrested Montesinos in Caracas, Venezuela. (In July 2002 Montesinos was convicted of abuse of authority and sentenced to nine years' imprisonment. A further 57 indictments were subsequently brought against him, and in March 2003 he was sentenced to a further five years in prison following the first of these trials.) In late June the interim Government of Pérez de Cuéllar had increased diplomatic pressure on Japan to allow the extradition of Fujimori by recalling its ambassador to Japan. However, there was no bilateral extradition treaty between the two countries and the prospect of Fujimori returning to face corruption charges seemed remote.

At least 31 deaths were attributed to the Sendero Luminoso in 2001, showing that while its ability to wage guerrilla war had been severely damaged, the movement had not completely disappeared. In late August President Toledo ordered more than 100 army and police posts to be reopened in remote parts of the jungle where the guerrilla organization were thought to be active. The resurgence of the drugs trade, partly displaced from Colombia, was believed to be financing the Sendero Luminoso's activities. In March 2002, three days before US President George W. Bush was due to arrive in Peru (the first such visit by a serving US head of state), a car bomb killed nine people and injured at least 40 more near the US embassy in Lima. The attack was attributed to a radical wing of the Sendero Luminoso. In April the Toledo Government increased the defence budget by US $203.5m. to combat terrorist activity. In 2002 US aid to Peru was tripled to $156m., which included substantial funds to support alternative development strategies, in order to help combat the illegal drugs trade.

During his first year in office, President Toledo's initially high popularity began to decline. Economic growth was insufficient to deliver on the Government's populist election promises to create jobs. Owing to the unpopularity of plans for a series of major privatizations, anti-Government demonstrations were frequent in 2002, and in June, following a week of violent demonstrations in Arequipa which left more than 100 people wounded, a state of emergency was declared. The unrest was the first serious episode of protest in what was to become a common feature under Toledo's presidency. In May 2003 the President declared a 30-day state of emergency following violent unrest by public-sector workers demanding improved pay and conditions. Toledo's attempts to build a broad cross-party consensus in the Congreso in support of his policy of structural reform and combating corruption were seriously undermined by a series of scandals throughout that year, which cast doubt on his credibility. The President was accused of using his position to influence the courts in a paternity suit. Other scandals saw Toledo accused of manipulating the media and of nepotism.

The Government's loss of popularity was reflected in poor results for the PP in regional and municipal elections in November 2002. Toledo's promise of decentralization led to the creation of 25 new regional authorities, each with its own president and council (Lima was not designated a region, but its mayor was afforded the same status as a regional president). Of the new regional governments, 13 were won by APRA, while independent candidates won a further eight. The PP and its coalition partner, the Frente Independiente Moralizador (FIM—Independent Morality Front), won just one region apiece. The results indicated that Toledo was unable to draw on a political base with a nation-wide presence. APRA's success in the regional elections raised the prospect of its leading figure, Alan García, launching another presidential campaign for the election scheduled for 2006. However, the performance of all regional governments seemed likely to be limited by lack of resources and clarity about the division of powers between the central Government and the regions.

The tension between satisfying President Toledo's political constituency in the PP and appointing the most capable, often independent, ministers became a theme of his presidency. Cabinet reshuffles during Toledo's administration were generally attempts to shore up political support in response to popular unrest and, therefore, tended to represent a shift to the left, at the expense of capable technocrats. Following the June 2002 unrest the Minister of Economy and Finance, Kuczynski, resigned and the privatization process was suspended. The respected Minister of the Interior, Gino Costa, resigned in January 2003 and was replaced by the PP's Alberto Sanabría. Wide-ranging cabinet changes in July, after the unrest of the previous two months, failed to improve Toledo's political fortunes.

The Government's stated objectives continued to focus on promoting economic growth through pro-market management of the economy, including structural reform and fiscal prudence. A new IMF stand-by agreement setting fiscal targets for the period 2004–06 looked set to underpin this policy for the remainder of President Toledo's term. The inauguration of a fifth new cabinet in February 2004, in contrast to earlier reshuffles, was designed to bring greater technical expertise and independence to the Government, as well as to distance President Toledo from a scandal involving his former legal adviser and intelligence chief, César Almeyda. Kuczynski was reappointed to the economy and finance ministry; his return to government was expected to give renewed impetus to the administration's privatization efforts, though Kuczynski held discussions with the opposition immediately after assuming his position, suggesting he had become more conscious of political realities than during his previous spell at the Ministry of Economy and Finance.

Despite favourable economic indicators, the Government's political weaknesses were exacerbated in July 2004 when the Congreso elected Antero Flores-Aráoz, of the opposition Unidad Nacional, as its new speaker, defeating the governing PP's candidate, Luis Solari. The defeat prevented President Toledo from controlling the legislative agenda. Popular support for President Toledo remained very low in 2004–05. Despite impressive economic growth, significant job creation remained elusive and his political and personal integrity had been undermined by a series of damaging allegations. Charges that Toledo was closely involved in the fraudulent collection of signatures as part of the process to register the PP for the 2000 legislative elections continued to undermine the President's authority in 2005. The President appeared before a congressional commission to answer questions on the matter in March and was judged culpable in early April. The Congreso, however, ruled in late May (by 57 votes to 47) not to impeach the President on the basis of this judgment. Other sources of unrest included coca growers in Cusco and Puno, whose violent protests in late 2004 were aimed at the Government's coca-eradication programme, and cotton farmers demanding compensation for losses incurred as a result of lower trade tariffs attracting cheaper US imports. In January 2005 a revolt by around 150 army recruits led by a retired army major occurred in the southern Andes. The uprising led to the death of seven people, including four police-officers. While the insurgents' demand for Toledo's resignation might have received some sympathy from the wider population, their espousal of extreme right-wing nationalist beliefs and their recourse to violence were widely condemned.

In a cabinet reorganization in February 2005 President Toledo attempted to strengthen his position by appointing three new PP ministers. However, continued low levels of popularity (approval rates were running at around 10%), along with the lack of political support in the Congreso, threatened to bring about the President's resignation before the scheduled end of his term in 2006. Meanwhile, the opposition itself remained deeply divided and with no clear leadership. In August Toledo's appointment of the FIM leader, Fernando Olivera, as Minister of Foreign Affairs prompted the resignation of prime minister Carlos Ferrero Costa and of the housing minister, Bruce Montes de Oca. Considering his position untenable, Olivera, who, controversially, had supported the legalization of coca cultivation in several provinces, resigned two days later. In the ensuing cabinet reshuffle, Kuczynski was appointed President of the Council of Ministers.

From late 2004 former President Fujimori made public his intention to contest the presidency in 2006, despite the sustained efforts of the Peruvian authorities to extradite him from Japan to face charges of corruption and other abuses of power. In February 2005 the Government announced its intention to seek Fujimori's extradition through the International Court of Justice. In mid-June the Japanese authorities announced that the extradition process against Fujimori would not proceed in the absence of further evidence against the former head of state.

Also in mid-June 2005, the UN Office on Drugs and Crime (UNODC) reported that coca cultivation in Peru had increased by some 14% during 2004. None the less, on 21 June the department of Cusco legalized the cultivation of coca for traditional use, a measure condemned by the central Government and by the USA as likely to increase the amount of coca grown for cocaine production. Despite its criticism, the Toledo Government did not intervene to reverse the move, fearing further social unrest and demands for greater regional autonomy. However, wary of allowing Cusco to set a precedent, the Government refused to accept a measure legalizing coca cultivation adopted in mid-July in the department of Puno.

# Economy

## SANDY MARKWICK

Peru is the third largest country in South America in terms of geographical area (1.29m. sq km, or just under 500,000 sq miles) and, with a population estimated at 27.5m. in mid-2004, the fifth most populous. Peru's economy was ranked seventh in Latin America both in terms of gross domestic product (GDP), at an estimated US $69,000m. in 2004, and also in terms of GDP per head (estimated at $2,491 in the same year).

Population growth averaged 1.8% per year between 1995–2000, declining slightly to a projected annual average of 1.5% between 2000–05. Peru's population is young, with about one-third under 15 years old. The birth rate fell from 46 per 1,000 in 1960 to 23.0 per 1,000 in 2003, by which time life expectancy had risen to 70 years (from 47 years in 1960). However, health and nutritional standards were poor, as reflected by a high (although diminishing) under-five mortality rate of 34 per 1,000 live births in 2003. The geographical distribution of the population shifted dramatically in favour of the coastal region after 1940, as migrants sought improved standards of living, better employment prospects and access to infrastructure in the cities concentrated along the coastal strip. The proportion of Peruvians living in cities was projected to rise from 46% in 1960 to an estimated 72.6% by mid-2005. The process of urbanization accelerated as guerrilla insurgency intensified in the 1980s, and slowed as the conflict in the countryside all but disappeared in the mid-1990s. Economic activity was highly concentrated in the capital city, Lima.

From the 1950s successive Governments pursued a strategy of state-sponsored industrialization replacing the prevailing *laissez-faire* economy characterized by little direct government participation and few regulations. Thereafter, Governments attempted to nurture domestic manufacturing by preventing an influx of cheap foreign goods through the imposition of high import tariffs. Industry grew at a faster pace than other sectors of the economy. Broadly, Peru experienced positive growth in GDP until the late 1970s. Real GDP increased by an annual average of 5.3% in the 1960s and 4.5% in the 1970s, but a decrease in the value of exports, an increase in imports and an expensive nationalization programme prompted a recession in 1977–78. Between 1980 and 1985 GDP fell, in real terms, at an annual average rate of 0.5%, signalling the worst economic performance since the 1930s. Average annual inflation doubled from 51% between 1975 and 1980 to 102% between 1980 and 1985. President Alan García (1985–90) stimulated demand-led growth by increasing real wages and extending import controls to encourage import substitution. However, by 1988 a balance-of-payments deficit, an overvalued exchange rate, fiscal difficulties and rising inflation led the Government to adopt a more orthodox free-market approach to economic management until President García, concerned about the outcome of the April 1990 elections, allowed wages to rise, causing an increase in inflation.

On taking office in 1990, García's successor, President Alberto Fujimori (1990–2000), implemented a rigid stabilization policy aimed at combating inflation by reducing the fiscal deficit. Subsidies were eliminated, causing the rate of inflation to rise in the short term, by dramatically raising the prices of public services. However, the measures resulted in a reduced annual inflation rate of 11.1% by 1995; the rate decreased steadily to 3.5% in 1999. Historically low inflation rates continued after the end of the Fujimori Government. Inflation rose slightly in 2000, to 3.8%, before falling to an annual average of 2.0% in 2001 and to just 0.2% during 2002, the lowest rate for decades. From 2002 the inflation rate increased a little, largely owing to fuel price increases and climatic conditions, which affected agriculture and the price of basic foodstuffs. Despite the increase, modest domestic demand ensured inflation remained within official targets. The average rate of inflation in 2003 was 2.3% and 3.6% in 2004. Reduced inflationary pressures in the first half of 2005 saw annual inflation at 2.0% in April of that year. Output from the huge Camisea gas project began to reduce energy costs, while the price of imported goods was kept down by a weak US dollar.

High growth rates were recorded in the mid-1990s, but stringent economic stabilization policies, implemented in response to a widening current-account deficit on the balance of payments, led to reduced demand and a slower growth rates for the remainder of the decade. The decline was partly owing to high interest rates and weakened demand in international markets brought about by the appearance of El Niño (a warm surface current from the western Pacific that periodically replaces cold water and brings heavy rains, causing considerable damage to crops) in 1997–98 and financial crises in Asia, Brazil and Russia. In 2001 the economy went into recession as a result of depressed global demand. Recovery later in the year, aided by the opening of the Antamina copper and zinc mine for commercial production in October, saw overall GDP grow in 2001 by 0.3%. The economy grew by 4.9% in 2002, fuelled by a particularly strong performance in mining, as the Antamina mine went into its first full year of production. Mining continued to be the main driver behind growth in 2003, when the economy increased by 3.8%, but contributions from the utilities and construction sectors made the performance of the economy more evenly spread. In 2004 the economy grew by an estimated 4.8%, owing in part to an expansion of exports and investment in mining and energy. Growth was consistent across most sectors of the economy. Some 9.7% of the work-force was unemployed at the end of 2003, while a much higher percentage, estimated at around 40%, was thought to be underemployed. A large proportion of the urban work-force, around 50% by some estimates, was employed in the informal sector, living a hand-to-mouth existence beyond the reach of the tax authorities and unable to take advantage of benefits available to workers in the formal economy.

## AGRICULTURE

Irrigation and a temperate climate along Peru's arid coastline supported traditional crops of rice, sugar and cotton from the 1860s, but they increasingly supported a modern and varied agro-industrial sector of non-traditional agricultural produce such as asparagus, cocoa and fruit. There are extensive pasture lands in the Andean Sierra, although some crops, notably maize, barley and potatoes, are cultivated in deeper valley basins, while tea, coffee and coca are grown on lower eastern-facing slopes. The tropical forests east of the Andes have enormous agricultural potential, but a lack of transport and infrastructure meant that they remained largely unexploited. Of the 1.3m. ha under cultivation in Peru, 53% was in the Andean Sierra, 30% in the coastal region and 17% in the eastern rainforests. There is little room for expansion in the sierra region.

Land reform introduced under the regime of Gen. Juan Velasco between 1968 and 1975 turned huge privately owned coastal haciendas into co-operatives alongside single family smallholdings (*minifundios*). The reforms, however, were largely unsuccessful. Increasing inefficiency and growing demand converted Peru into a net importer of cotton, sugar and rice, which it had traditionally exported in large quantities. Attempts by subsequent Presidents to reactivate agriculture were limited, owing to a lack of resources to provide credit to farmers and the overvaluation of the currency. President Fujimori introduced free-market reforms into the sector by declaring land to be a freely tradable commodity, exchanged under market conditions. Strict landholding limits were relaxed and subsidized credit, tax rebates and export credits were removed. Foreign investors in agriculture faced the same rules as Peruvian nationals. President Fujimori's policy was designed to promote higher agricultural output more efficiently, by favouring medium-sized commercial farms producing for export, at the expense of traditional smallholdings.

The contribution of agriculture to GDP declined after the 1950s, when it accounted for more than 20% of GDP. In 2004 the sector (including livestock, but excluding fishing) accounted for an estimated 8.8% of GDP. However, the sector grew by an annual average of 4.7% in 1990–2003. Agriculture in Peru, particularly production on non-irrigated land, was vulnerable to unpredictable weather conditions. The El Niño offshore current, in 1982–83, 1992 and 1997–98 caused widespread rainfall and flooding in some parts of Peru, and drought in others, resulting in reduced output. Agricultural output recovered from the effects of El Niño in 1993 because of strong domestic demand. The severe impact of El Niño was felt again in 1997 and particularly in 1998, when the agricultural sector grew by just 0.5%. Strong growth was restored in 1999, when output increased by 10.1%, with substantial year-on-year increases in the production of cotton, rice, coffee and potatoes. More modest growth of 6.6% was recorded in 2000, led particularly by the production of maize and cotton. A combination of bad weather, price increases for inputs and difficulties in raising capital led to a reduction in output, of 0.5%, in 2001. Strong growth in 2002, of 5.7%, was based on significant increases in sugar cane, potato and poultry production. Sectoral growth slowed in 2003, to 4.6%, largely owing to drought conditions adversely affecting potato production. In 2004 output was reported to have declined by 1.1%, owing to unfavourable climatic conditions which particularly affected the production of sugar cane, rice and maize. Output recovered in the fourth quarter of 2004 with annual growth of 2.5%.

Modern productive commercial agro-industry in Peru was located in the coastal region, where irrigation projects enhanced the agricultural potential of the coast. The cultivation of high-price export crops could be maintained all year: mangoes, bananas, passion fruit and lemons and limes on the northern coast; asparagus, broccoli, green beans, snow-peas (mangetouts) and grapes in the central coastal area; and beans, garlic, onions and oregano in the south. Sugar estates, concentrated in the northern coastal area, were able to produce all year round with careful control of irrigation. Cotton required less water and could be grown where supplies were seasonal. The importance of cotton and sugar as export crops diminished, although they continued to rank second and third in terms of legal agricultural export earnings. Cotton exports benefited from the removal of tariff barriers when entering the US market as part of the Andean Trade Promotion and Drug Eradication Act (ATPDEA, the successor agreement to the Andean Trade Preference Act), which came into effect in October 2002. Peru became a net importer of sugar because of rising domestic consumption, inefficient methods of production, weak prices and a reduction in the US sugar quota. A dramatic fall in the export earnings of traditional crops saw non-traditional crops overtake sugar, cotton and rice for the first time in 1992. However, exports of traditional crops again earned more in 1994–95 and in 1997, when non-traditional produce earned US $340m., compared with the $472m. earned by traditional crops. The tendency towards non-traditional crops as a foreign-exchange earner was restored in 1999, with earnings of $406m. outpacing the $282m. earned by traditional crops. In 2004 non-traditional crops earned 71.1% of all agricultural export revenues. The best performers in the new range of crops were asparagus, mangoes, mandarin oranges, grapes, apples and bananas. These non-traditional crops were also boosted by the ATPDEA. During 1990–2005 it was estimated that cultivation began of some 400 new export crops, with paprika alone projected to generate some $80m. in export revenues in 2005, compared with just $6m. in 2000.

Agricultural production in the Sierra was based on predominantly subsistence-orientated smallholdings and large estates for livestock. The major crops were maize, potatoes, barley and wheat. Low productivity, scarce capital and limited land resources perpetuated rural poverty. In the eastern rainforests, new road construction was followed by internal migration and agricultural colonization. In this area there was a variety of crops produced for the domestic market, including cassava, rice, bananas, oranges, tea, cacao (cocoa plant), beef, rubber and oil palm. Coffee, Peru's most valuable legal agricultural export commodity, was produced in this region. The land area devoted to coffee increased from 76,000 ha in 1960, producing an output of 32,000 metric tons, to an estimated 230,000 ha in 2003, producing an unofficial 172,000 tons. Most coffee production was exported. Lower international prices in the early 2000s meant that, despite the rise in total coffee production, export earnings declined in 2001 to just $180.2m. Earnings increased to $187.9m. in 2002 as higher volumes offset lower prices, reduced to an average of $1,122 per metric ton. Prices rose to $1,204 in 2003, but reduced volumes saw Peru earn just $181.1m. from coffee exports.

Coca is grown on lower, eastern-facing slopes where the Sierra descends to meet the eastern rainforests. Coca is a traditional crop used for centuries in its leaf form as a stimulant by Amerindians of the Sierra. The role of coca as the raw material for cocaine resulted in a dramatic increase in cultivation after the 1960s, as world demand for cocaine expanded. In 1970 an estimated 5,000 ha of land was under coca cultivation. This figure increased, by some estimates, to over 100,000 ha by the mid-1990s (with potential production of an estimated 460m. metric tons). Peru was the world's second largest supplier, after Colombia, of coca for the production of cocaine. Crop-substitution programmes and dwindling prices led to a reduction in coca-leaf production in the late 1990s, though production increased in 2000–04, in spite of the destruction of an estimated 15% of the coca crop under cultivation in 2003. The total area under coca cultivation was estimated by the UN Office on Drugs and Crime (UNODC) to have increased by some 14% in 2004, largely as a result of displacement of coca cultivation operations from Colombia. In June 2005 the department of Cusco adopted legislation legalizing the cultivation of coca (see History).

## FISHING

An abundance of fish in the cold Humboldt current, and investment since the 1960s in industrial processing plants, made Peru one of the world's largest exporters of fish products. The anchoveta (Peruvian anchovy) was used to produce fishmeal for animal feed and fertilizers and was the most important species fuelling the fishmeal industry. Other species, principally sardines and pilchards, were caught for human consumption. Peru's main fishing ports are Chimbote, Tambo de Mora, Pisco, Callao, Supe and Ilo.

In 1995 the Government imposed restrictions on fishing in order to replenish stocks, owing to fears of over-exploitation. Despite smaller catches in 1995–96, earnings were stable because of higher international prices. The El Niño current seriously damaged fisheries in 1997–98, but, despite the small catch, export earnings increased by 21.5% in 1997 because of high world market prices and increased canning capacity. Fisheries' production and exports, including fishmeal and fish oil, recovered strongly in 1999 and the total catch increased by more than 25% in 2000, to 10.7m. metric tons, fuelled by a significantly improved anchovy catch. Following a decrease in the catch in 2001 (8.0m. tons), fisheries output increased to 8.8m. tons; however, this figure masked highly inconsistent growth during the year. High monthly year-on-year growth rates during 2002 were explained by low production levels in 2001 stemming from occasional fishing bans to counter over-fishing. Fishing output declined to 6.1m. tons in 2003, owing to a ban on anchovy fishing imposed early in the year. The Government was developing a plan to finance a reduction in the fishing fleet in order to replenish anchovy and sardine stocks. The sector recovered in 2004, growing by an estimated 28.3% following the suspension of the ban and the largest anchovy catch since 2000. In 2003 fishing accounted for 0.6% of total GDP, compared with 0.9% in the previous year.

Unexploited stocks of mackerel, shrimp, trout and turbot, as well as frozen fish, provided opportunities for diversification in the industry. Foreign investment was increasing in specific areas of fisheries and Peru granted concessions to foreign fishing operations to exploit Peruvian waters for species not consumed locally. Japanese and Korean boats fished for giant squid. Fishing contracts were awarded on the basis of open and competitive international tender. Furthermore, in 1995 the Government sold fishmeal plants belonging to the state fishing concern, PescaPerú.

## MINING AND ENERGY

Mining is an important contributor to export earnings, owing to Peru's varied and considerable mineral wealth. In 2004 the mining sector (excluding petroleum) accounted for 54.8% of Peru's total export revenue, earning some US $6,880.5m. Copper, gold and zinc ranked respectively as first, second and fourth most significant export earners across all sectors in 2004. Copper regained its position as the most important export, having been second to gold since 1998. In 2004 Peru earned an estimated $2,446m. from copper exports and $2,361m. from gold. In 2003 the mining sector contributed 6.6% of GDP, up from an average of 5.4% per year during 1998–2002. Despite the importance of mining, there was potential for even greater production. Peru was among the world's leading mining countries, yet less than one-fifth of its reserves were under exploitation.

The US-owned mining companies Southern Peru Copper Corporation (SPCC), Cerro de Pasco Corporation and Marcona Mining Company played the major role in the expansion of the mining industry, helped by the 1950 Mining Code, which fostered a favourable investment climate. However, by the 1970s the mining industry had begun to stagnate. The state mining company, MineroPerú (Empresa Minera del Perú), was created in 1970 to participate directly in areas of production where foreign investment was not forthcoming. Those foreign mining companies that failed to invest and were considered to form part of the old power bloc were nationalized. In 1974 Peru took over the Cerro de Pasco complex under its new name, Centromín. The Marcona Mining Co was nationalized in 1975 and its name changed to HierroPerú (known as Shougang Hierro Perú following its reprivatization in 1992). At the same time, the Government negotiated terms with SPCC for one of the world's largest investments in the copper industry.

In 1991 President Fujimori declared the mining sector, heavily undercapitalized, to be in a state of crisis. Successive Governments had overburdened the industry as a source of tax revenue. Currency overvaluation damaged the export sector as a whole. Hyperinflation, restrictions on profit remittances and an oversized state bureaucracy under President García saw the mining sector deteriorate further. The sector as a whole, supported by the Government, sought salvation through foreign investment, and many large companies subsequently entered into discussions to explore joint-venture possibilities. Until the early 1990s foreign investment in the mining sector was relatively limited. Best estimates indicate that total foreign investment in Peru over the 15 years to 1992 amounted to only US $700m. The most important foreign company in the mining sector was SPCC, which consistently produced around two-thirds of Peru's copper. It was one of the world's 10 largest copper-production companies and owned and operated two huge open-pit mines in southern Peru, at Cuajone and Toquepala. In 1997 SPCC began a two-stage expansion project to increase production at the Cuajone mine. Total investment was projected to reach some $1,000m.

High international demand fuelled copper production of a 868,600 metric tons in 2004, a 39% increase over output in 2003, following a 2.7% decline in production in 2003. Output doubled between 1998–2004. Copper production was boosted in 2002 by the first full year of production of Antamina.

Peru was one of the world's largest suppliers of zinc; in 2004 1.2m. metric tons were produced, mostly for export. Centromín dominated zinc production in the early 1990s, accounting for 35% of the total, which was extracted largely from its Cerro de Pasco mine. Revenues from zinc sales were US $576.8m. in 2004, an increase on the $528.7m. earned in 2003. Centromín was also the main producer of lead. In 2004 Peru's total lead output was 306,200 tons, an 8.1% increase on the previous year. Despite this, export earnings were 4.5% lower, at an estimated $201m., owing to lower prices. In May 1999 a Peruvian-owned zinc producer, Volcán Compañía Minera, bid successfully for the huge Cerro de Pasco zinc and lead mine, which had been renamed Paragsha. The Volcán-Paragsha operation became the largest zinc producer in Latin America. In 2002 Antamina offset lower production from the SPCC and the Tintaya mine to give a boost to overall annual national copper and zinc production by 16% and 15%, respectively. The sole producer of iron ore was Shougang Hierro Perú of the People's Republic of China. Iron ore production in 2004 amounted to 4.2m. tons, representing a decline from the 1994 peak of 4.7m. tons, although it was higher than in 2000 when 2.8m. tons were produced. Exports of iron ore amounted to $94.1m. in 2003, up 13.6% from in the previous year.

In 2004 Peru produced 173,200 kg of gold, an increase of just 0.2% in output over the previous year, which followed an 13.5% increase over the year before. Silver production increased by 17.2% in 2004, to 3.1m. kg. Silver production contrasted with other major minerals, as the sector was dominated by small and medium-sized mines. Peru earned US $191m. in silver exports in 2003, an increase of 10.0% on the previous year.

After high growth in the late 1970s, petroleum production declined from its highest point of 204,000 barrels per day (b/d) in 1980 to an estimated 94,120 b/d in 2004 (consumption, however, was estimated at 191,000 b/d in that year). The decline in production in recent years stemmed from the depletion of light crude reserves in the northern jungle, which had the further effect of making production of some heavy crudes no longer economical. Peru was a net petroleum importer from 1992, when it was forced to import increasing amounts of light crude petroleum because of rising domestic demand, declining production and a lower quality of extracted crude.

Encouraged by President Fujimori's liberal hydrocarbons law, the petroleum sector, like mining, expanded in the 1990s. Some 90% of output stemmed from the operations of four companies: Occidental Petroleum and Petrotech of the USA and Pluspetrol and Pecom, these last predominantly Argentine concerns. New investment in the petroleum sector was slow initially, but the pace increased in the mid-1990s; 20 exploration and drilling contracts were signed in 1996. In 1993 Occidental signed a further US $34m. contract with the Peruvian Government for exploration in the Amazon basin.

Petroleum was drilled in three principal areas. Most crude was extracted from fields in the eastern jungle (66.7% in 2002). Northern coastal fields (17.8%) and the continental shelf comprised the other two key zones. In the late 1990s petroleum production continued to be dominated by the state company, PetroPerú (Empresa de Petróleos de Perú), and Occidental, although new investments were diversifying the operators. The Fujimori Government initiated the gradual privatization of PetroPerú in 1996. The first parts to be sold were the La Pampilla petroleum refinery, which was purchased by a consortium comprising Repsol of Spain, Yacimientos Petrolíferos Fiscales (YPF) of Argentina and Mobil. At the same time, Pluspetrol won an option on a block for petroleum extraction.

In February 2000 the contract for first-phase exploration of natural gas and hydrocarbon deposits at Camisea was won by a consortium of Pluspetrol, Hunt Oil of the USA and SK Corporation of the Republic of Korea. The consortium has the right to develop the field for 40 years. The second and third phases of the project for transport and distribution were awarded, respectively, to Techint of Argentina and Tractebel of Belgium. The state-owned electricity company, ElectroPerú, guaranteed a domestic market for the fuel. Gas flows to Lima began on schedule in August 2004. In June 2005 the Presidents of the four Mercosur (Mercado Común del Sur—Southern Common Market) member countries (Argentina, Brazil, Paraguay and Uruguay) signed an agreement creating an 'energy ring' (*anillo energético*), which would facilitate the supply and storage of natural gas from the Camisea field. Gas exports from the project were not, however, expected before 2007.

Peru has a huge potential to secure hydroelectric power because of its abundance of steep running rivers. Electricity generation rose steadily in the latter half of the 1990s, from production of 17,440m. kWh in 1995 to 22,983m. kWh in 2003. However, installed hydroelectric capacity represented only a fraction of the potential. Hydroelectric plants at Santa Eulalia, Marcapomachocha and the Mantaro valley produced Lima's electricity supplies. Hydroelectric power accounted for 80.9% of the total electricity generated, rendering the electricity supply vulnerable to climatic conditions. A lack of rainfall in 1991 resulted in severe rationing and an estimated loss of US $150m. per month in production. The electricity network in Peru suffered from a lack of investment. In 1994 only 25% of homes were supplied with electricity. At that time, annual consumption per head was 610 kWh, one of the lowest levels in Latin America. In 2000 an estimated 22.4% of the population were still without power, despite new investment. The Government introduced progressive electricity pricing to benefit the poor. The costs of delivering electricity to low-use households was, in part, subsidized by higher-use consumers. The Fujimori Government encouraged the private sector to generate and supply electricity to fulfil demand and embarked on the privatization of electricity assets in 1995, beginning with Cahua, a 40-MW hydroelectric plant, and Edegel, the 700-MW generating unit of former state company Electrolima. Violent protests followed the sale by the Toledo administration of two Arequipa-based generating companies, Egasa and Egesur, to Tractebel in June 2002. The protests led Tractebel to withdraw from the purchase and forced the Government to cancel plans to auction further electricity concessions.

## MANUFACTURING AND CONSTRUCTION

In the early 1970s manufacturing output grew at an annual rate of 10%, although at a long-term cost to efficiency, as industry was protected by high tariff barriers. In the late 1970s growth slowed. When protectionism was reduced under President Fernando Belaúnde Terry from 1980, manufacturing output declined dramatically. The Government returned to its policy of protectionist growth in 1984. Under President Fujimori's leadership manufacturing was exposed to greater foreign competition and underwent a process of restructuring. However, there was a wide divergence in performance between industries and volatility year on year. Average annual growth in 1990–2003 was 3.2%. The manufacturing sector grew by 4.9% in 2003 and by some 6.1% in 2004, primarily owing to the boost given to textiles and leathers by ATPDEA. The sector contributed an estimated 14.6% to GDP in 2004, a stable figure since 1998, but lower than its contribution to GDP during the mid-1990s, which averaged over 20%.

Prior to the military regime, manufacturing largely comprised the processing of agricultural and mineral products. A policy of import-substitution industrialization, introduced under the Velasco regime, encouraged the development of domestic industry by raising the cost of imports. The military's developmentalist outlook saw the beginnings of heavy industry, including petroleum refining, chemicals, non-ferrous metals and electrical industries.

The principal manufacturing sectors in Peru were food, metal-working, steel, textiles, chemicals, cement, automobile assembly, fish processing and petroleum refining. The food industry was heavily dependent on agricultural performance. The main products were processed fish, coffee, cocoa and sugar. The textile industry, accounting for about 10% of industrial output, was the most significant non-traditional exporter.

The construction sector went into recession in the mid-1990s, but recovered strongly in 1997 with growth of 18.9%. Subsequently sectoral growth slowed dramatically and in 2000 output contracted by 3.8% as a result of low levels of both public and private investment. Recession in the sector continued into 2001 (sectoral GDP declined by 6.0%); however, by the end of 2001 strong growth had returned. The construction sector grew by an estimated 4.7% in 2004, slightly stronger growth than the 4.0% experienced in 2003, but weaker than the 7.9% in the previous year, when the Camisea gas project was nearing completion. Growth in the sector was underpinned by the Government's investment in Mivivienda, a low-cost housing programme.

## TRANSPORT AND COMMUNICATIONS

In 2005 Peru had a little over 2,000 km of railway track, run by the state national railway enterprise, Empresa Nacional de Ferrocarriles del Perú (Enafer-Perú), following nationalization in 1972. The railways were largely used for transporting minerals. The Central Railway (Ferrocarril del Centro del Perú) connected Lima to Huancayo (Junín), with a branch to the mining operations at Cerro de Pasco. The line rose to a high point of 4,775 m above sea level. The Southern Railway (Ferrocarril del Sur del Perú ENAFER, SA) ran from Matarani, through Arequipa, to Juliaca, with branches running to Puno and Cusco. The Cusco line ran through the Urubamba valley, bypassing Machu Picchu, to Quillabamba. Passenger use declined in the 1990s, both in terms of total numbers and in distances travelled. However, in the late 1990s the Government began offering concessions to run Enafer's more commercially-viable routes to private operators. In 2000 a private consortium, Ferrocarriles del Perú, won a 30-year concession to manage the Southern, Central and South-eastern railways. Private investment and improved services doubled passenger numbers from 793,000 in 1999 to 1.43m. by 2001. The use of railways for cargo also increased in 2000, with 6,964m. metric tons transported that year, compared with 5,055m. tons in 1999. In June 2005 it was announced that the world's first train powered by compressed natural gas had been inaugurated by Ferrocarril Central Andino, SA.

The road system was greatly improved after 1960. However, until the early 1990s the network lacked adequate investment to ensure basic maintenance or expansion. However, from 1993, as a result of new investment from the Inter-American Development Bank (IDB) and the Government, road repairs were begun. The major road links were: the coastal Pan-American Highway, which linked Peru to Ecuador and Chile; the Central Highway from Lima to Pucallpa, which ran alongside the Ucayali river via Oroya, Huanuco and Tingo María; the northern Trans-Andean Highway from Olmos to Yurimaguas on the Huallaga river; and the Carretera Marginal de la Selva, which was built to provide access to new settlements in the east. The Fujimori Government declared road-building central to its aims of integration, development and pacification. It was hoped that repairs to the Central Highway between Huanuco, Tingo María and Pucallpa would reduce transport costs and encourage coca-growers to cultivate alternative crops. About 30% of Peru's 78,000-km road network was paved or semi-paved, though much of it was in poor condition. The Government was encouraging private-sector investment to extend and improve the road network. In 2004 there were five main road-building projects under offer to the private sector, in the form of concessions totalling approximately US $620m. in capital investment.

There were 37 ports in Peru, of which Lima's neighbouring port of Callao was by far the most important. The deregulation of ports under President Fujimori reduced costs and increased competitiveness with ports in neighbouring Ecuador and Chile. River transport was important in the Amazon region. The port of Iquitos was accessible to ocean-going shipping from the Atlantic Ocean. River traffic extended from there as far as Pucallpa and Yurimaguas, where new ports were constructed in the 1980s. In September 2004, the Autoridad Porturia Nacional (APN—National Port Authority) approved a development plan including private-sector investment. Callao was set to be the first beneficiary, in 2005. Air services were well developed and particularly important in eastern Peru, where the road and railway networks were less extensive than those in other areas. Peru has five international airports and a further 50 domestic airports, in addition to hundreds of airfields. The Government awarded a contract to manage the largest airport, the Jorge Chávez airport in Lima, to a consortium including Frankfurt airport (Germany) and US construction firm Bechtel. Concessions were planned for the management of the other international airports in Cuzco, Tacna, Iquitos and Arequipa.

Telefónica of Spain bought a controlling interest in Peru's two former state telecommunications companies, Empresa Nacional de Telecomunicaciones del Perú (ENTEL PERU) and Compañía Peruana de Teléfonos (CPT), in 1994. The Government's objective was to modernize the service and to expand the provision of telephone lines, which, at 2.5 lines per 100 inhabitants in 1994, was the worst in Latin America. Improved communications were considered to be a fundamental requirement for developing poor provincial communities. Telefónica was obliged to increase Peru's existing 637,000 lines nearly three-fold within 10 years and to guarantee that each town of 500 inhabitants or more had telephone provision. In 2003 there were 1.8m. fixed lines and 10.8 lines per 100 inhabitants. In 1996 the telecommunications market regulator also obliged Telefónica to provide a cellular-communications infrastructure in the provinces. In the late 1990s the sector was opened to greater competition. The new entrants into the mobile telecommunications market improved services and reduced prices, leading to a rapid increase in users, from 820,700 in 1999 to 3.8m. in 2004. Internet access in the home is limited to a small minority of households, a large and growing number of Peruvians, estimated at more than 2m. in 2005, used the internet regularly via public access points.

## GOVERNMENT FINANCE AND INVESTMENT

Investment declined in the 1980s, particularly under President García, when there was virtually no foreign investment, owing to restrictions on profit remittances, and little external funding. The ratio of domestic investment to GDP decreased during the 1980s as corporate profits declined, private assets were sent overseas and government savings became negative. During the 1980s domestic savings were the main source of investment, with loans to the Government making up most of the remainder. The level of central government investment declined from 18% of total expenditure to 3% and annual direct private-sector investments declined from US $125m. in 1981 to $34m. in 1990. Under President Fujimori foreign investment increased dramatically. A new liberal foreign-investment regime lifted restrictions on remittances and provided foreign investors with the same opportunities as national investors. In addition, the Government began an ambitious privatization programme, which aimed to sell virtually all of the state's productive ventures. The sale of the state iron company, HierroPerú, to the Chinese Shougang Corporation in 1992 was the first major privatization. The Government intended to use a proportion of the revenue received for anti-poverty programmes and job-creation schemes, although a large part was to be used for debt-servicing. The sale of state-owned assets generated $7,180m. in 1991–97 and there were further inflows of foreign direct investment (FDI), averaging $1,791m. in 1998–99. FDI inflows then fell significantly, to $662m., in 2000; most of the most attractive assets had already been sold off and investors remained cautious in the aftermath of the international crisis and in the context of political instability in Peru. New FDI inflows increased to $1,317m. in 2003, when the total stock of FDI tied up in projects in Peru amounted to approximately $12,000m., lower than most other economies in Latin America. Spain and the United Kingdom were the main sources of new foreign investment, ahead of the USA. Spanish investment has predominantly been channelled into the Telefónica business while British investment was divided between the mining, communications and finance sectors. The Camisea gas project and the Antamina copper and zinc mine were expected to support stable overall FDI levels. FDI was not expected to increase significantly, because of a policy shift by the Toledo Government away from further privatizations, a response to popular opposition. The 2003 budget anticipated revenues from privatization of $350m., but the negative experience of Tractebel in attempting to acquire the state electricity generating companies Egasa and Egesur was expected to discourage other foreign investors. Foreign-exchange reserves, which were negative in 1988, stood at $12,176m. at the end of 2004.

The García administration's policy of extending subsidies while reducing taxes required the Government to finance a wide public-sector deficit and gave rise to hyperinflation at the end of the 1980s. Central government revenues fell from 14.1% of GDP in 1985 to 6.5% in 1989. In 1989 government expenditure was equivalent to 12.8% of GDP. President Fujimori's policy of maintaining fiscal equilibrium as part of a wider monetarist policy of reducing public-sector wages, delaying payments and decreasing investments lowered the public-sector deficit. In addition, Fujimori introduced reforms to increase revenues and simplify tax collection. When Fujimori assumed office, tax collection amounted to 4.4% of GDP. Following improvements in the efficiency of the tax authority as well as increases in consumption, this figure increased through the 1990s. In 2003 taxes represented 12.9% of GDP, although confidence in the integrity of the tax authority remained low. A new tax on banking transactions was introduced in March 2004, which helped to reduce the non-financial public-sector deficit to an estimated 1.1% of GDP, from an estimated 1.7% in 2003.

Consistently high inflation resulted in the introduction of two new currencies between 1985 and 1991. Firstly, the inti replaced the sol in 1985 at a rate of one inti per 1,000 sols. In 1991 the new sol replaced the inti at a rate of one new sol per 1m. intis. Between 1978 and 1985 the sol had been gradually devalued by means of a 'crawling-peg' mechanism of mini-devaluations (i.e. the exchange rate changed in response to market pressure, but only by limited amounts over set periods). Multiple exchange rates, designed to favour priority imports and manufacturing exports over traditional exports, were introduced in the mid-1980s. These were simplified to a two-rate system under President García, with a series of crawling-peg mini-devaluations for the official rate and a floating rate for all other transactions. President Fujimori further simplified the exchange-rate mechanism by adopting a free-floating, single-rate system. The new sol appreciated in real terms against the US dollar from 1993 as capital inflows maintained a balance-of-payments surplus. The exchange rate was affected by the inflow of dollars from the illegal drugs trade. High real interest rates tended to maintain the overvaluation of the new sol. The strength of the sol helped control inflation, but weakened exports. A lack of confidence in the new sol, along with fears that the problem of inflation had not yet been fully solved, encouraged the 'dollarization' of the economy, made possible by President Fujimori's deregulation of the financial system. In 1993 more than 80% of bank deposits were made in dollars. Financial reform implemented by the Fujimori Government gave foreign commercial banks equal status with local private banks. In order to support export competitiveness in 2004 and 2005, the Central Bank intervened on several occasions by buying US dollars. At the end of May 2005 new sols were exchanged at a rate of 3.25 to the dollar.

## FOREIGN DEBT

Heavy public spending in the mid-1970s resulted in a rapid accumulation of foreign debt. Foreign debt increased further in the 1980s, because of the need to meet large debt-service payments and finance large public-sector deficits. In 1982 Peru secured a three-year extended fund facility from the IMF. However, the country's failure to meet its commitments resulted in the suspension of the agreement one year later. In 1984 a 15-month stand-by loan agreement with the IMF collapsed after three months. Furthermore, Peru failed to meet bilateral commitments to member countries of the 'Paris Club' of creditor nations, following a rescheduling of debts worth US $1,046m. in 1984. In 1985 President García announced that a maximum of 10% of export earnings would be reserved for debt-servicing and that no agreement with the IMF would be sought. Payment arrears affected multilateral as well as commercial-bank debt. The IMF declared Peru ineligible for further lending in 1986, closely followed by the IDB and the World Bank. Foreign creditor banks ruled Peru ineligible for rescheduling negotiations. On taking office, President Fujimori immediately sought to re-establish good relations with the international financial community. In September 1990 he agreed to resume debt repayments and introduced a programme of economic restructuring. A support group, including the USA and Japan, provided a $2,000m. 'bridging' loan to enable Peru to clear part of its arrears. In 1991 Peru rescheduled payment of $6,660m. in principal debts, interest and arrears owed to the Paris Club. The IMF agreed to support the Government's economic reforms, thereby enabling credits from the World Bank and the IDB. Agreements with the Paris Club in July 1996 rescheduled $7,000m. in bilateral debts contracted before 1983. The agreement was to reduce the annual servicing bill by one-half, to $450m. over 20 years. In October 1995 Peru announced a debt-restructuring scheme for debts amounting to $10,560m. ($4,400m. principal and $6,160m. interest and arrears) with commercial-bank creditors. The scheme, implemented in March 1997, normalized relations between Peru and the international financial community and reopened private- and public-sector access to international capital markets. By 2001 the proportion of more costly short-term debt to long-term debt had declined, making Peru less vulnerable to external shocks. At the end of 2004 Peru's total external debt amounted to an estimated $29,600m., or 42.9% of GDP, a healthier figure than that in 2003, when debt represented some 50% of GDP. In February 2004 the IMF approved a new two-year stand-by credit for Peru, committing the Government to reducing the non-financial public-sector deficit to 1% of GDP in 2005/06. The IMF gave Peru a broadly positive review of its performance to date in November 2004, while issuing warnings about risk to inflation and fiscal targets from public-sector wage pressures. Political opposition to further privatizations underlined the need for further borrowing to service the debt. At the same time, the Government embarked on a series of bond issues, which raised $2,500m. in 2002–03, in order to reduce the fiscal deficit. The bond issues were further evidence of Peru's rehabilitation in the international financial markets.

## FOREIGN TRADE AND BALANCE OF PAYMENTS

The slow increase in the value of exports from the 1970s reflected Peru's vulnerability to international prices of raw materials. Exports declined and stagnated in the 1980s, from their highest figure of US $3,916m. in 1980 to $2,691m. in 1988. The value of exports increased slightly after 1988, owing to changes in government trading policy and more favourable metal prices. An overvalued new sol, recession and weak commodity prices resulted in retarded export growth during the early 1990s, but buoyancy returned in 1994 because of larger volumes and higher commodity prices. However, the value of exports declined in 1998, owing to the effects of El Niño, low commodity prices and diminished overseas demand. Annual export revenue growth of 8.2% in 2002, fuelled by 17% in mineral exports, had helped Peru record its first trade surplus (of $306m.) since 1990 in that year. Exports grew by 17.1% in 2003, fuelled by the commencement of the Antamina mine's operations, high gold production and gold price increases. After minerals the next best earners in that year were traditional fisheries, non-traditional agriculture, livestock and textiles. As a result, the trade surplus increased to $731m. An export boom in 2004 increased annual earnings to an estimated $12,546m. Growth was spread across traditional and non-traditional export products. Among traditional exports, minerals were the best performing sub-sector, representing 55% of total export earnings. Strong demand, notably for copper from China, underpinned high commodity prices, while production volumes also increased. Fisheries products, coffee and sugar were also notable strong performers from among traditional exports. Non-traditional exports were led by chemicals, base metals, textiles and agriculture, which benefited from greater competitiveness in the US market as a result of ATPDEA. In 2004 the trade surplus increased significantly, to $2,728m. The improving trade balance from 1999 helped to reduce the current-account deficit to US $71m. in 2004 (equivalent to 0.1% of GDP), from a record $4,314m. in 1995.

In 1970 primary products (excluding fuel) accounted for 98% of all exports. Manufacturing amounted to only 1.3% and fuels 0.7%. The composition of exports thereafter shifted away from traditional commodities to non-traditional exports, principally finished and agro-industrial products. However, traditional exports recovered in 1994, when they represented 70% of the total. In 2004 traditional exports contributed 72.6% of total export revenues. In the same year raw materials and intermediate goods accounted for 54.6% of total import costs, representing the largest share. This was followed by capital goods, which accounted for 24.1%. The relative importance of consumer-product imports increased, owing to the Government's trade liberalization policies. Consumer imports were valued at US $1,980.2m. in 2004, equivalent to 20.2% of the total value of imports, compared with $289m. in 1988, which was equivalent to 11% of the total cost of imports.

In the early 2000s the USA continued to be Peru's main trading partner, although its importance had declined over several decades. In 1970 the USA was the source of just under one-third of all imports and the destination for just over one-third of exports. The USA accounted for 24.1% of imports and 25.9% of exports in 2003. In October 2002 tariffs on the export of 6,000 products to the USA from Peru and other Andean countries had been removed as part of the ATPDEA. After the USA, the next most important export markets in 2002 were Switzerland (7.9%), the People's Republic of China (7.0% of exports) and Japan (4.7%). Trade with Asia increased following a 1997 agreement with the Asia-Pacific Economic Co-operation (APEC). From 1970 Peru diversified trading partners, creating a greater role for other Latin American countries as sources of imports. In 2004 Chile, accounting for 6.1% of imports, Brazil (5.5%) and Colombia (5.5%) were the next most important sources of imports after the USA.

## TOURISM

The number of tourists visiting Peru grew steadily from an estimated 134,000 in 1970 to 373,000 in 1980. After a decline in 1980–83, numbers reached almost 360,000 by 1988, when the industry accounted for earnings of US $448m. A decrease after 1988 resulted from a combination of guerrilla violence and an outbreak of cholera. The number of visitors to Peru in 1992 was 217,000, the lowest since 1972. However, a dramatic improvement in Peru's image after 1992, as the threat from cholera and Sendero Luminoso guerrillas diminished, improved the fortunes of the industry. In 2003 there were 933,643 international visitors, generating revenues of $959m. Tourists were attracted by Peru's Inca heritage, Amerindian cultural traditions and varied topography. The Amazonian rainforest and the Andean Sierra are rich in flora and fauna. The most popular places to visit were the former Inca capital, Cusco (which retained original Inca structures alongside Spanish colonial architecture), the spectacular Inca-period city of Machu Picchu and the mysterious earth designs at Nazca. The gold-filled tombs of the ancient Moche culture of Sipán, excavated and exhibited in museums world-wide, greatly enhanced the tourism industry. Despite these attractions, underdeveloped hotel availability was an obstacle to further growth.

The Peruvian Government continued to promote the expansion of the tourism sector. Greater competition between airlines had been introduced and the Government was investing increased sums to improve infrastructure for tourism, including airports and railways. Furthermore, the Government encouraged foreign investment in hotels and hotel construction. In the early 2000s, however, underdeveloped infrastructure and services remained an obstacle to the expansion of the sector. Investment in hotels increased throughout the 1990s, but infrastructure remained insufficient to meet the Government's proposed expansion.

## CONCLUSION

There was a significant economic revival during the early 2000s. Inflation was reduced and controlled and real GDP growth restored. However, the challenge for the Government was to distribute the benefits of restored economic growth, in order to reduce poverty and avoid social unrest. Job creation remained an acute problem, with almost 50% of the work-force either unemployed or underemployed. The administration of President Toledo faced growing opposition after 2002, in part reflecting frustration at the slow pace of jobs creation and poverty reduction, as well as a lack of confidence in his administration caused by a number of allegations of corruption and political interference in the judiciary and media. Despite presiding over significant economic growth since 2002, Toledo's popularity remained at very low levels in mid-2005, making his Government, already weakened by opposition control of the Congreso, weaker still. With elections scheduled to be held in 2006, political considerations also threatened to undermine the Government's fiscal reforms.

# Statistical Survey

Sources (unless otherwise stated): Banco Central de Reserva del Perú, Jirón Antonio Miró Quesada 441–445, Lima 1; tel. (1) 4276250; fax (1) 4275880; e-mail webma; internet www.bcrp.gob.pe; Instituto Nacional de Estadística e Informática, Avda General Garzón 658, Jesús María, Lima; tel. (1) 4334223; fax (1) 4333140; e-mail infoinei@inei.gob.pe; internet www.inei.gob.pe.

## Area and Population

### AREA, POPULATION AND DENSITY
(excluding Indian jungle population)

| | |
|---|---:|
| Area (sq km) | |
| Land | 1,280,086 |
| Inland water | 5,130 |
| Total | 1,285,216* |
| Population (census results)† | |
| 12 July 1981 | 17,005,210 |
| 11 July 1993 | |
| Males | 10,956,375 |
| Females | 11,091,981 |
| Total | 22,048,356 |
| Population (official estimates at mid-year) | |
| 2002 | 26,748,972 |
| 2003 | 27,148,101 |
| 2004 | 27,546,574 |
| Density (per sq km) at mid-2004 | 21.5‡ |

* 496,225 sq miles.
† Excluding adjustment for underenumeration, estimated at 4.1% in 1981 and 2.35% in 1993.
‡ Land area only.

### REGIONS
(30 June 2003)

| | Area (sq km) | Population (estimates) | Density (per sq km) | Capital |
|---|---:|---:|---:|---|
| Amazonas | 39,249 | 435,556 | 11.1 | Chachapoyas |
| Ancash | 35,877 | 1,123,410 | 31.3 | Huaráz |
| Apurimac | 20,896 | 470,719 | 22.5 | Abancay |
| Arequipa | 63,345 | 1,113,916 | 17.6 | Arequipa |
| Ayacucho | 43,814 | 561,029 | 12.8 | Ayacuchu |
| Cajamarca | 33,318 | 1,515,827 | 45.5 | Cajamarca |
| Callao* | 147 | 799,530 | 5439.0 | Callao |
| Cusco | 72,104 | 1,223,248 | 17.0 | Cusco (Cuzco) |
| Huancavelica | 22,131 | 451,508 | 20.4 | Huancavelica |
| Huánuco | 36,887 | 822,804 | 22.3 | Huánuco |
| Ica | 21,328 | 698,437 | 32.7 | Ica |
| Junín | 44,197 | 1,260,773 | 28.5 | Huancayo |
| La Libertad | 25,500 | 1,528,448 | 59.9 | Trujillo |
| Lambayeque | 14,231 | 1,131,467 | 79.5 | Chiclayo |
| Lima | 34,802 | 7,880,039 | 226.4 | Lima |
| Loreto | 368,852 | 919,505 | 2.5 | Iquitos |
| Madre de Dios | 85,183 | 102,174 | 1.2 | Puerto Maldonado |
| Moquegua | 15,734 | 160,232 | 10.2 | Moquegua |
| Pasco | 25,320 | 270,987 | 10.7 | Cerro de Pasco |
| Piura | 35,892 | 1,660,952 | 46.3 | Piura |
| Puno | 71,999 | 1,280,555 | 17.8 | Puno |
| San Martín | 51,253 | 767,890 | 15.0 | Moyabamba |
| Tacna | 16,076 | 301,960 | 18.8 | Tacna |
| Tumbes | 4,669 | 206,578 | 44.2 | Tumbes |
| Ucayali | 102,411 | 460,557 | 4.5 | Pucallpa |
| **Total** | **1,285,216** | **27,148,101** | **21.1** | — |

* Province.

### PRINCIPAL TOWNS
(estimated population of towns and urban environs at 1 July 1998)

| | | | |
|---|---:|---|---:|
| Lima (capital) | 7,060,600* | Piura | 308,155 |
| Arequipa | 710,103 | Huancayo | 305,039 |
| Trujillo | 603,657 | Chimbote | 298,800 |
| Callao | 515,200† | Cusco (Cuzco) | 278,590 |
| Chiclayo | 469,200 | Pucallpa | 220,866 |
| Iquitos | 334,013 | Tacna | 215,683 |

* Metropolitan area (Gran Lima) only.
† Estimated population of town, excluding urban environs, at mid-1985.

**Mid-2004** (official estimate, metropolitan area): Lima 8,049,619.

### BIRTHS AND DEATHS*

| | Live births | | Deaths | |
|---|---:|---:|---:|---:|
| | Number | Rate (per 1,000) | Number | Rate (per 1,000) |
| 1996 | 656,435 | 27.1 | 160,045 | 6.6 |
| 1997 | 652,467 | 26.4 | 160,830 | 6.5 |
| 1998 | 648,075 | 25.8 | 161,615 | 6.4 |
| 1999 | 642,874 | 25.2 | 162,457 | 6.4 |
| 2000 | 636,064 | 24.5 | 163,263 | 6.3 |
| 2001 | 630,947 | 24.0 | 164,296 | 6.2 |
| 2002 | 626,714 | 23.4 | 165,467 | 6.2 |
| 2003 | 623,521 | 23.0 | 166,777 | 6.1 |

* Data are estimates and projections based on incomplete registration, but including an upward adjustment for under-registration.

**Marriages:** 78,946 in 1997 (marriage rate 3.2 per 1,000); 60,730 in 1998 (marriage rate 2.4 per 1,000) (Source: UN, *Demographic Yearbook*).

**Expectation of life** (WHO estimates, years at birth): 70 (males 68; females 73) in 2003 (Source: WHO, *World Health Report*).

### ECONOMICALLY ACTIVE POPULATION
('000 persons aged 14 and over, urban areas, July–September)

| | 1999 | 2000 | 2001 |
|---|---:|---:|---:|
| Agriculture, hunting and forestry | 355.0 | 456.4 | 620.7 |
| Fishing | 65.3 | 25.8 | 47.1 |
| Mining and quarrying | 31.3 | 52.5 | 45.9 |
| Manufacturing | 897.6 | 963.5 | 956.4 |
| Electricity, gas and water | 41.3 | 28.1 | 20.4 |
| Construction | 378.2 | 299.5 | 341.3 |
| Wholesale and retail trade; repair of motor vehicles, motorcycles and personal and household goods | 2,077.0 | 2,060.5 | 2,124.5 |
| Hotels and restaurants | 470.6 | 431.9 | 593.8 |
| Transport, storage and communications | 618.2 | 639.0 | 641.0 |
| Financial intermediation | 76.3 | 66.6 | 47.9 |
| Real estate, renting and business activities | 407.5 | 413.4 | 342.6 |
| Public administration and defence; compulsory social security | 349.0 | 346.7 | 298.1 |
| Education | 551.5 | 479.7 | 552.1 |
| Health and social work | 165.1 | 169.7 | 213.5 |
| Other services | 372.9 | 368.2 | 406.2 |
| Private households | 353.8 | 326.8 | 366.3 |
| Not classifiable | — | — | 1.9 |
| **Total employed** | **7,211.2** | **7,128.4** | **7,619.9** |
| Unemployed | 624.9 | 566.5 | 651.5 |
| **Total labour force** | **7,836.1** | **7,694.9** | **8,271.4** |

Source: ILO.

# PERU

## Health and Welfare

**KEY INDICATORS**

| | |
|---|---|
| Total fertility rate (children per woman, 2003) | 2.8 |
| Under-5 mortality rate (per 1,000 live births, 2003) | 34 |
| HIV/AIDS (% of persons aged 15–49, 2003) | 0.5 |
| Physicians (per 1,000 head, 1999) | 1.17 |
| Hospital beds (per 1,000 head, 1996) | 1.47 |
| Health expenditure (2002): US $ per head (PPP) | 226 |
| Health expenditure (2002): % of GDP | 4.4 |
| Health expenditure (2002): public (% of total) | 49.9 |
| Access to water (% of persons, 2002) | 81 |
| Access to sanitation (% of persons, 2002) | 62 |
| Human Development Index (2002): ranking | 85 |
| Human Development Index (2002): value | 0.752 |

For sources and definitions, see explanatory note on p. vi.

## Agriculture

**PRINCIPAL CROPS**
('000 metric tons)

| | 2001 | 2002 | 2003 |
|---|---|---|---|
| Wheat | 181.9 | 186.7 | 190.6 |
| Rice (paddy) | 2,028.7 | 2,118.6 | 2,135.7 |
| Barley | 177.4 | 199.7 | 193.7 |
| Maize | 1,314.3 | 1,292.0 | 1,356.9 |
| Potatoes | 2,681.8 | 3,298.0 | 3,151.4 |
| Sweet potatoes | 254.0 | 224.5 | 193.7 |
| Cassava (Manioc) | 859.0 | 891.1 | 913.8 |
| Sugar cane* | 8,000.0 | 9,100.0 | 9,550.0 |
| Dry beans | 61.0 | 62.8 | 59.4 |
| Oil palm fruit | 175.8 | 149.0 | 180.4 |
| Cottonseed | 81.5 | 76.0 | 77.0 |
| Cabbages | 48.3 | 34.6 | 31.9 |
| Asparagus | 184.1 | 181.2 | 187.2 |
| Tomatoes | 188.5 | 129.9 | 148.9 |
| Pumpkins, squash and gourds | 97.8 | 90.9 | 90.0 |
| Green chillies and peppers | 60.9 | 52.0 | 58.0 |
| Dry onions | 415.7 | 458.2 | 472.8 |
| Garlic | 63.9 | 62.9 | 57.9 |
| Green peas | 83.6 | 81.1 | 83.7 |
| Green broad beans | 65.8 | 66.0 | 61.9 |
| Carrots | 145.0 | 155.8 | 157.2 |
| Green corn | 363.0 | 393.9 | 395.0 |
| Other fresh vegetables | 206.9 | 204.5 | 209.3 |
| Plantains | 1,557.7 | 1,570.0 | 1,600.0 |
| Oranges | 278.5 | 292.4 | 305.8 |
| Tangerines, mandarins clementines and satsumas | 129.5 | 133.2 | 161.2 |
| Lemons and limes | 204.0 | 254.3 | 250.0 |
| Apples | 138.0 | 123.5 | 134.4 |
| Grapes | 127.7 | 136.1 | 138.0 |
| Watermelons | 61.5 | 52.7 | 60.0 |
| Mangoes | 144.9 | 181.4 | 160.0 |
| Avocados | 93.4 | 94.3 | 95.0† |
| Pineapples | 149.3 | 155.9 | 150.0 |
| Papayas | 158.8 | 173.2 | 170.0 |
| Cotton (lint)† | 47.0 | 42.0 | 52.0 |
| Coffee (green) | 159.9 | 169.6 | 172.0† |

* FAO estimates.
† Unofficial figure.
Source: FAO.

**LIVESTOCK**
('000 head, year ending September)

| | 2001 | 2002 | 2003 |
|---|---|---|---|
| Horses* | 700 | 710 | 720 |
| Mules* | 260 | 270 | 280 |
| Asses* | 570 | 580 | 600 |
| Cattle | 4,962 | 4,990 | 5,046 |
| Pigs | 2,781 | 2,849 | 2,851 |
| Sheep | 14,253 | 14,025 | 13,995 |
| Goats | 1,998 | 1,942 | 1,942 |
| Poultry | 84,634 | 90,685 | 91,118 |

* FAO estimates.
Source: FAO.

**LIVESTOCK PRODUCTS**
('000 metric tons)

| | 2001 | 2002 | 2003 |
|---|---|---|---|
| Beef and veal | 137.8 | 141.5 | 144.9 |
| Mutton and lamb | 31.8 | 31.8 | 32.3 |
| Pig meat | 85.4 | 84.9 | 85.7 |
| Poultry meat | 571.3 | 609.4 | 636.0 |
| Cows' milk | 1,115.0 | 1,194.3 | 1,226.1 |
| Hen eggs | 152.2 | 181.6 | 181.8 |
| Wool (greasy) | 12.9 | 11.6 | 11.6 |

Source: FAO.

## Forestry

**ROUNDWOOD REMOVALS**
('000 cubic metres, excluding bark)

| | 2001 | 2002 | 2003 |
|---|---|---|---|
| Sawlogs, veneer logs and logs for sleepers | 1,081 | 1,194 | 1,062 |
| Other industrial wood | 1 | 0 | 0 |
| Fuel wood | 7,299 | 7,335 | 9,074* |
| **Total** | 8,381 | 8529 | 10,136 |

* FAO estimate.
Source: FAO.

**SAWNWOOD PRODUCTION**
('000 cubic metres, including railway sleepers)

| | 2001 | 2002 | 2003 |
|---|---|---|---|
| Coniferous (softwood) | 3* | 5 | 6 |
| Broadleaved (hardwood) | 503 | 621 | 551 |
| **Total** | 506 | 626 | 556 |

* FAO estimate.
Source: FAO.

## Fishing

('000 metric tons, live weight)

| | 2001 | 2002 | 2003 |
|---|---|---|---|
| Capture | 7,982.9 | 8,763.0 | 6,089.7 |
| Chilean jack mackerel | 723.7 | 154.2 | 217.7 |
| Anchoveta (Peruvian anchovy) | 6,358.2 | 8,104.7 | 5,347.2 |
| Aquaculture | 7.6 | 11.6 | 13.8 |
| **Total catch** | 7,990.5 | 8,774.6 | 6,103.5 |

Note: Figures exclude aquatic plants ('000 metric tons): 5.5 (capture 5.5, aquaculture 0.0) in 2001; 6.2 in 2002 (all capture); 7.9 in 2003 (all capture). Also excluded are aquatic mammals, recorded by number rather than by weight. The number of toothed whales caught was: 26 in 2001; 195 in 2002; 0 in 2003.

Source: FAO.

## Mining

('000 metric tons, unless otherwise indicated)*

| | 2002 | 2003 | 2004 |
|---|---|---|---|
| Crude petroleum ('000 barrels) | 35,355.8 | 33,342.6 | 34,448.0 |
| Copper | 642.8 | 625.3 | 868.6 |
| Lead | 273.9 | 283.2 | 306.2 |
| Zinc | 1,045.4 | 1,171.0 | 1,209.0 |
| Iron ore | 3,105.0 | 3,540.7 | 4,247.2 |
| Gold (metric tons) | 152.4 | 172.9 | 173.2 |
| Silver (metric tons) | 2,528.0 | 2,611.0 | 3,059.8 |

* Figures for metallic minerals refer to metal content only.
Source: Ministry of Energy and Mines.

## Industry

**SELECTED PRODUCTS**
('000 metric tons, unless otherwise indicated)

|  | 1999 | 2000 | 2001 |
|---|---|---|---|
| Canned fish | 63.6 | 77.2 | 79.0* |
| Wheat flour | 903 | 907 | 1,001* |
| Raw sugar† | 617 | 747 | 786 |
| Beer ('000 hectolitres) | 6,168 | 5,706 | 5,296* |
| Cigarettes (million) | 3,580 | 3,605 | 3,310* |
| Rubber tyres ('000)‡ | 1,067 | 1,208 | 1,351 |
| Motor spirit (petrol, '000 barrels) | 9,449 | 9,299 | 8,646 |
| Kerosene ('000 barrels) | 4,910 | 5,239 | 5,441 |
| Distillate fuel oils ('000 barrels) | 13,622 | 12,371 | 13,476 |
| Residual fuel oils ('000 barrels) | 17,437 | 18,382 | 19,256 |
| Cement | 3,799 | 3,684 | 3,589 |
| Crude steel§ | 510 | 510 | 510 |
| Copper (refined) | 185.7 | 193.3 | 196.4 |
| Lead (refined) | 121.1‖ | 116.4‖ | 116.0 |
| Zinc (refined) | 197.2 | 199.9 | 201.8 |
| Electric energy (million kWh) | 17,366.2 | 18,327.7 | 19,213.3 |

* Estimate.
† Data from FAO.
‡ Data from UN Economic Commission for Latin America and the Caribbean. Excludes tyres for bicycles and other non-motorized cycles.
§ US Geological Survey estimates.
‖ Includes secondary metal production.

**2002** ('000 barrels): Motor spirit 8,414; Kerosene 5,464; Distillate fuel oils 13,706; Residual fuel oils 17,576.

**2003** ('000 barrels): Motor spirit 8,408; Kerosene 3,945; Distillate fuel oils 13,866; Residual fuel oils 17,836.

Source (unless otherwise indicated): partly UN, *Industrial Commodity Statistics Yearbook*.

## Finance

**CURRENCY AND EXCHANGE RATES**

**Monetary Units**
100 céntimos = 1 nuevo sol (new sol).

**Sterling, Dollar and Euro Equivalents** (31 May 2005)
£1 sterling = 5.92 new soles;
US $1 = 3.25 new soles;
€1 = 4.01 new soles;
100 new soles = £16.91 = $30.74 = €24.93.

**Average Exchange Rate** (new soles per US $)
2002    3.517
2003    3.478
2004    3.413

Note: On 1 February 1985 Peru replaced its former currency, the sol, by the inti, valued at 1,000 soles. A new currency, the nuevo sol (equivalent to 1m. intis), was introduced in July 1991.

**GENERAL BUDGET**
(million new soles)

| Revenue | 1999 | 2000 | 2001 |
|---|---|---|---|
| Taxation | 21,483 | 22,376 | 22,626 |
| Taxes on income, profits, etc. | 5,072 | 5,130 | 5,630 |
| Taxes on payroll and work-force | 1,000 | 1,038 | 847 |
| Domestic taxes on goods and services | 14,475 | 15,418 | 15,341 |
| Value-added tax | 11,029 | 11,996 | 11,808 |
| Excises | 3,446 | 3,421 | 3,533 |
| Import duties | 2,848 | 2,913 | 2,740 |
| Other taxes | 229 | 648 | 902 |
| Adjustment to tax revenue | −2,142 | −2,769 | −2,834 |
| Other current revenue | 3,810 | 5,038 | 4,121 |
| Resources of ministries and other non-tax revenues | 3,548 | 4,893 | 4,088 |
| Interest on privatization funds | 262 | 145 | 33 |
| Capital revenue | 624 | 530 | 291 |
| **Total** (incl. grants) | 25,916 | 27,944 | 27,039 |

| Expenditure | 1999 | 2000 | 2001 |
|---|---|---|---|
| Current non-interest expenditure | 21,797 | 23,757 | 23,853 |
| Labour services | 11,778 | 12,394 | 12,566 |
| Wages and salaries | 7,774 | 8,180 | 8,385 |
| Pensions | 3,281 | 3,418 | 3,432 |
| Social security contributions | 724 | 796 | 750 |
| Goods and non-labour services | 6,210 | 7,068 | 7,067 |
| Transfers | 3,808 | 4,294 | 4,219 |
| Private sector | 1,095 | 1,140 | 821 |
| Non-financial public sector | 2,714 | 3,154 | 3,397 |
| Interest payments | 3,680 | 4,076 | 4,059 |
| Capital expenditure | 5,900 | 5,232 | 4,467 |
| Gross capital formation | 5,652 | 4,749 | 3,898 |
| **Total** | 31,378 | 33,065 | 32,378 |

Source: IMF, *Peru—Statistical Appendix* (March 2003).

**INTERNATIONAL RESERVES**
(US $ million at 31 December)

|  | 2002 | 2003 | 2004 |
|---|---|---|---|
| Gold* | 386.7 | 462.7 | 488.7 |
| IMF special drawing rights | 0.7 | 0.4 | 0.4 |
| Foreign exchange | 9,338.3 | 9,776.4 | 12,176.0 |
| **Total** | 9,725.7 | 10,239.5 | 12,665.1 |

* National valuation $347 per troy ounce at 31 December 2002; $415 per troy ounce at 31 December 2003; $438 per troy ounce at 31 December 2004.

Source: IMF, *International Financial Statistics*.

**MONEY SUPPLY**
(million new soles at 31 December)

|  | 2002 | 2003 | 2004 |
|---|---|---|---|
| Currency outside banks | 5,615 | 6,370 | 8,036 |
| Demand deposits at commercial and development banks | 8,359 | 8,193 | 9,942 |
| **Total money** (incl. others) | 22,049 | 21,351 | 23,728 |

Source: IMF, *International Financial Statistics*.

**COST OF LIVING**
(Consumer Price Index, Lima metropolitan area; base: 2000 = 100)

|  | 2002 | 2003 | 2004 |
|---|---|---|---|
| Food (incl. beverages) | 100.2 | 101.0 | 106.6 |
| Rent | 103.7 | 110.5 | 115.5 |
| Clothing (incl. footwear) | 103.9 | 104.6 | 105.9 |
| **All items** (incl. others) | 102.2 | 104.5 | 108.3 |

Source: ILO.

**NATIONAL ACCOUNTS**
(million new soles at current prices)

**National Income and Product**

|  | 2001 | 2002 | 2003 |
|---|---|---|---|
| Compensation of employees | 46,306 | 47,792 | 49,949 |
| Operating surplus | 110,828 | 118,880 | 126,250 |
| **Domestic factor incomes** | 157,134 | 166,672 | 176,199 |
| Consumption of fixed capital | 14,086 | 14,409 | 15,111 |
| **Gross domestic product (GDP) at factor cost** | 171,220 | 181,081 | 191,310 |
| Indirect taxes | 17,093 | 17,576 | 19,437 |
| **GDP at market prices** | 188,313 | 198,657 | 210,747 |

Source: UN Economic Commission for Latin America and the Caribbean.

# PERU

## Expenditure on the Gross Domestic Product

|  | 2001 | 2002 | 2003 |
|---|---|---|---|
| Government final consumption expenditure | 20,214 | 20,386 | 21,359 |
| Private final consumption expenditure | 135,876 | 142,534 | 149,611 |
| Increase in stocks | 275 | 2,177 | 1,906 |
| Gross fixed capital formation | 35,132 | 35,128 | 37,748 |
| **Total domestic expenditure** | 191,497 | 200,225 | 210,624 |
| Exports of goods and services | 30,128 | 32,811 | 37,270 |
| *Less* Imports of goods and services | 33,312 | 34,379 | 37,146 |
| **GDP in purchasers' values** | 188,313 | 198,657 | 210,747 |
| **GDP at constant 1994 prices** | 121,104 | 126,980 | 131,757 |

Source: IMF, *International Financial Statistics*.

## Gross Domestic Product by Economic Activity

|  | 2001 | 2002 | 2003 |
|---|---|---|---|
| Agriculture, hunting and forestry | 13,218.1 | 12,731.2 | 13,315.8 |
| Fishing | 1,318.2 | 1,592.2 | 1,235.0 |
| Mining and quarrying | 10,021.5 | 11,272.5 | 12,703.9 |
| Manufacturing | 27,230.2 | 28,658.0 | 30,065.0 |
| Electricity and water | 4,447.7 | 4,327.6 | 4,679.7 |
| Construction | 9,389.6 | 10,515.8 | 11,475.4 |
| Wholesale and retail trade | 25,442.7 | 27,561.0 | 28,123.7 |
| Restaurants and hotels | 7,779.7 | 8,226.5 | 8,679.5 |
| Transport and communications | 15,386.0 | 15,768.9 | 16,819.0 |
| Government services | 13,908.1 | 14,585.2 | 15,978.3 |
| Finance, insurance, real estate, business and other services | 43,208.1 | 46,368.2 | 49,025.5 |
| **Sub-total** | 171,349.7 | 181,606.9 | 192,100.9 |
| Import duties | 2,755.0 | 14,364.2 | 2,549.7 |
| Other taxes on products | 13,146.3 | 2,465.8 | 16,096.8 |
| **GDP in purchasers' values** | 187,251.0 | 198,436.9 | 210,747.4 |

## BALANCE OF PAYMENTS
(US $ million)

|  | 2002 | 2003 | 2004 |
|---|---|---|---|
| Exports of goods f.o.b. | 7,723 | 8,986 | 12,546 |
| Imports of goods f.o.b. | −7,417 | −8,255 | −9,818 |
| **Trade balance** | 306 | 731 | 2,728 |
| Exports of services | 1,544 | 1,679 | 1,844 |
| Imports of services | −2,530 | −2,609 | −2,801 |
| **Balance on goods and services** | −680 | −200 | 1,771 |
| Other income received | 337 | 282 | 322 |
| Other income paid | −1,827 | −2,364 | −3,628 |
| **Balance on goods, services and income** | −2,170 | −2,281 | −1,535 |
| Current transfers received | 1,052 | 1,227 | 1,470 |
| Current transfers paid | −8 | −6 | −6 |
| **Current balance** | −1,127 | −1,061 | −71 |
| Capital account (net) | −95 | −93 | −94 |
| Direct investment abroad | — | −60 | — |
| Direct investment from abroad | 2,156 | 1,377 | 1,802 |
| Portfolio investment assets | −280 | −1,435 | −448 |
| Portfolio investment liabilities | 1,724 | 1,211 | 1,239 |
| Other investment assets | 5 | 328 | 120 |
| Other investment liabilities | −1,581 | −361 | −244 |
| Net errors and omissions | 208 | 655 | 153 |
| **Overall balance** | 1,010 | 561 | 2,457 |

Source: IMF, *International Financial Statistics*.

# External Trade

## PRINCIPAL COMMODITIES
(distribution by SITC, US $ million)

| Imports c.i.f. | 2000 | 2001 | 2002 |
|---|---|---|---|
| **Food and live animals** | 742.8 | 808.0 | 795.1 |
| Cereals and cereal preparations | 335.7 | 365.2 | 364.4 |
| **Mineral fuels, lubricants, etc.** | 1,159.6 | 973.5 | 1,037.8 |
| Petroleum, petroleum products, etc. | 1,058.7 | 873.2 | 919.7 |
| Crude petroleum oils, etc. | 614.4 | 595.2 | 646.4 |
| Refined petroleum products | 436.5 | 268.1 | 264.2 |
| **Chemicals and related products** | 1,109.6 | 1,178.2 | 1,245.5 |
| Artificial resins and plastics | 299.6 | 295.2 | 318.7 |
| Polymerization and copolymerization products | 214.1 | 206.6 | 229.1 |
| **Basic manufactures** | 1,133.9 | 1,144.8 | 1,303.9 |
| Paper, paperboard and pulp | 229.5 | 235.3 | 250.2 |
| Iron and steel | 239.8 | 250.2 | 388.5 |
| **Machinery and transport equipment** | 2,410.7 | 2,291.7 | 2,083.0 |
| Machinery specialized for particular industries | 386.3 | 316.1 | 303.3 |
| General industrial machinery equipment and parts | 377.3 | 351.3 | 373.1 |
| Office machines and automatic data-processing equipment | 278.0 | 261.5 | 229.5 |
| Telecommunications and sound equipment | 367.1 | 408.3 | 371.2 |
| Parts and accessories for telecommunications and sound equipment | 249.7 | 286.7 | 219.1 |
| Other electrical machinery apparatus, etc. | 313.1 | 336.0 | 254.7 |
| Road vehicles and parts (excl. tyres, engines and electrical parts) | 513.2 | 461.1 | 408.7 |
| **Miscellaneous manufactured articles** | 591.1 | 600.7 | 685.7 |
| **Total** (incl. others) | 7,415.0 | 7,315.9 | 7,493.0 |

| Exports f.o.b. | 2000 | 2001 | 2002 |
|---|---|---|---|
| **Food and live animals** | 1,634.2 | 1,631.2 | 1,693.6 |
| Fish and fish preparations | 185.8 | 206.4 | 172.4 |
| Vegetables and fruit | 249.5 | 292.2 | 365.6 |
| Coffee, tea, cocoa and spices | 248.0 | 210.0 | 226.3 |
|   Coffee (not roasted); husks and skins | 223.7 | 180.2 | 187.9 |
| Feeding stuff for animals (excl. unmilled cereals) | 902.0 | 858.9 | 848.0 |
|   Flours and meals of meat, fish, etc. (unfit for human consumption) | 873.8 | 838.3 | 819.9 |
| **Crude materials (inedible) except fuels** | 963.9 | 1,020.0 | 1,277.5 |
| Metalliferous ores and metal scrap | 775.6 | 860.0 | 1,092.8 |
|   Ores and concentrates of base metals | 669.0 | 725.1 | 959.1 |
|     Copper ores and concentrates | 140.5 | 218.9 | 425.4 |
|     Zinc ores and concentrates | 347.8 | 355.7 | 338.4 |
| **Mineral fuels, lubricants, etc.** | 403.7 | 414.0 | 472.0 |
| Petroleum, petroleum products, etc. | 403.6 | 414.0 | 271.8 |
|   Refined petroleum products | 276.1 | 294.1 | 308.5 |
| **Basic manufactures** | 1,746.8 | 1,520.1 | 1,505.8 |
| Non-ferrous metals | 1,457.0 | 1,213.9 | 1,209.7 |
|   Copper and copper alloys | 866.2 | 800.7 | 829.2 |
|     Unwrought copper and alloys | 788.6 | 741.3 | 759.9 |
|       Refined copper (excl. master alloys) | 724.4 | 697.1 | 710.8 |
|   Zinc and zinc alloys | 223.3 | 180.6 | 131.6 |
| **Miscellaneous manufactured articles** | 634.8 | 657.9 | 718.6 |
| Clothing and accessories | 504.0 | 506.3 | 530.1 |
|   Cotton undergarments, not elastic or rubberized | 334.3 | 302.9 | 317.4 |
| **Non-monetary gold (excl. gold ores and concentrates)** | 1,144.2 | 1,166.1 | 1,467.4 |
| **Total** (incl. others) | 6,866.0 | 6,825.6 | 7,490.4 |

Source: UN, *International Trade Statistics Yearbook*.

## PRINCIPAL TRADING PARTNERS
(US $ million)

| Imports c.i.f. | 2000 | 2001 | 2002 |
|---|---|---|---|
| Argentina | 333.5 | 455.0 | 593.7 |
| Brazil | 377.6 | 327.7 | 489.5 |
| Canada | 218.0 | 148.9 | 123.5 |
| Chile | 395.2 | 428.5 | 419.0 |
| China, People's Republic | 288.9 | 353.6 | 463.4 |
| Colombia | 399.7 | 378.6 | 456.5 |
| Ecuador | 329.9 | 348.2 | 436.4 |
| France (incl. Monaco) | 129.6 | 137.4 | 115.8 |
| Germany | 217.0 | 222.9 | 231.0 |
| Italy | 123.2 | 134.1 | 140.8 |
| Japan | 485.9 | 429.1 | 411.1 |
| Korea, Republic | 220.5 | 256.5 | 228.9 |
| Mexico | 237.6 | 248.1 | 275.3 |
| Nigeria | 68.6 | 123.3 | 128.8 |
| Spain | 176.9 | 175.2 | 165.3 |
| United Kingdom | 93.5 | 90.6 | 73.4 |
| USA | 1,736.9 | 1,691.5 | 1,440.5 |
| Venezuela | 621.7 | 372.7 | 245.6 |
| **Total** (incl. others) | 7,415.0 | 7,315.9 | 7,493.0 |

| Exports f.o.b. | 2000 | 2001 | 2002 |
|---|---|---|---|
| Belgium | 108.3 | 107.4 | 103.1 |
| Bolivia | 95.4 | 97.7 | 90.3 |
| Brazil | 221.3 | 227.1 | 193.5 |
| Canada | 123.2 | 143.1 | 140.2 |
| Chile | 262.7 | 281.9 | 251.4 |
| China, People's Republic | 442.7 | 426.3 | 596.9 |
| Colombia | 144.4 | 150.4 | 156.7 |
| Ecuador | 97.3 | 120.4 | 135.4 |
| Germany | 215.5 | 207.8 | 251.3 |
| Italy | 121.6 | 139.5 | 174.1 |
| Japan | 325.4 | 383.0 | 372.6 |
| Korea, Republic | 137.6 | 110.6 | 168.1 |
| Mexico | 150.5 | 127.9 | 128.5 |
| Netherlands | 132.7 | 78.4 | 126.7 |
| Panama | 47.3 | 75.1 | 48.5 |
| Spain | 187.4 | 202.6 | 231.4 |
| Switzerland-Liechtenstein | 549.6 | 305.6 | 563.3 |
| Thailand | 74.9 | 70.3 | 25.9 |
| United Kingdom | 578.9 | 923.2 | 864.3 |
| USA | 1,919.9 | 1,693.6 | 1,917.0 |
| Venezuela | 110.9 | 145.2 | 113.9 |
| **Total** (incl. others) | 6,866.0 | 6,825.6 | 7,490.4 |

Source: UN, *International Trade Statistics Yearbook*.

# Transport

## RAILWAYS
(traffic)*

| | 2000 | 2001 | 2002 |
|---|---|---|---|
| Passenger-km (million) | 107 | 122 | 98 |
| Freight ton-km (million) | 877 | 1,148 | 1,008 |

* Including service traffic.

Source: UN, *Statistical Yearbook*.

## ROAD TRAFFIC
(motor vehicles in use)

| | 2000 | 2001 | 2002 |
|---|---|---|---|
| Passenger cars | 716,931 | 750,610 | 791,862 |
| Buses and coaches | 44,820 | 47,452 | 45,089 |
| Lorries and vans | 372,398 | 382,664 | 400,015 |

Source: IRF, *World Road Statistics*.

## SHIPPING

**Merchant Fleet**
(registered at 31 December)

| | 2002 | 2003 | 2004 |
|---|---|---|---|
| Number of vessels | 719 | 718 | 726 |
| Total displacement ('000 grt) | 240.3 | 223.6 | 226.8 |

Source: Lloyd's Register-Fairplay, *World Fleet Statistics*.

**International Sea-borne Freight Traffic**
('000 metric tons)

| | 2000* | 2001 | 2002 |
|---|---|---|---|
| Goods loaded | 6,504 | 6,616 | 6,111 |
| Goods unloaded | 6,900 | 7,152 | 8,260 |

* Approximate figures extrapolated from monthly averages.

Source: UN, *Monthly Bulletin of Statistics*.

# PERU

## CIVIL AVIATION
(traffic on scheduled services)

|  | 2001 | 2002 | 2003 |
|---|---|---|---|
| Kilometres flown (million) | 38 | 38 | 40 |
| Passengers carried ('000) | 1,844 | 2,092 | 2,233 |
| Passenger-km (million) | 2,627 | 2,340 | 2,443 |
| Total ton-km (million) | 116 | 100 | 114 |

Source: UN Economic Commission for Latin America and the Caribbean, *Statistical Yearbook*.

## Tourism

### ARRIVALS BY NATIONALITY

|  | 2001 | 2002 | 2003 |
|---|---|---|---|
| Argentina | 36,416 | 34,912 | 39,242 |
| Bolivia | 47,407 | 58,356 | 60,849 |
| Brazil | 23,744 | 24,945 | 29,016 |
| Canada | 19,868 | 21,572 | 21,995 |
| Chile | 107,994 | 97,724 | 138,856 |
| Colombia | 26,220 | 32,022 | 30,895 |
| Ecuador | 33,632 | 58,255 | 54,206 |
| France | 33,681 | 37,571 | 39,820 |
| Germany | 30,174 | 31,558 | 33,390 |
| Italy | 18,623 | 22,530 | 21,922 |
| Japan | 14,711 | 17,737 | 20,823 |
| Spain | 26,938 | 31,224 | 30,847 |
| United Kingdom | 40,743 | 42,363 | 46,747 |
| USA | 186,459 | 197,944 | 203,072 |
| Venezuela | 18,408 | 15,253 | 12,930 |
| **Total** (incl. others) | 801,334 | 865,602 | 933,643 |

**Tourism receipts** (US $ million, incl. passenger transport): 818 in 2001; 838 in 2002; 959 in 2003.

Source: World Tourism Organization.

## Communications Media

|  | 2001 | 2002 | 2003 |
|---|---|---|---|
| Telephones ('000 main lines in use) | 1,571.0 | 1,656.6 | 1,839.2 |
| Mobile cellular telephones ('000 subscribers) | 1,793.3 | 2,306.9 | 2,908.8 |
| Personal computers ('000 in use) | 1,250 | 1,149 | n.a. |
| Internet users ('000) | 2,000 | 2,400 | 2,850 |

**Television receivers** ('000 in use): 3,800 in 2000.

**Radio receivers** ('000 in use): 6,650 in 1997.

**Facsimile machines** ('000 in use, estimate): 15 in 1995.

**Book production** (titles): 612 in 1996.

**Daily newspapers**: 74 in 1996.

Sources: UNESCO, *Statistical Yearbook*; International Telecommunication Union.

## Education

(2003, incl. adult education, at documented institutions only)

|  | Institutions | Teachers | Pupils |
|---|---|---|---|
| Nursery | 16,211 | 41,718 | 763,252 |
| Primary | 34,600 | 182,369 | 4,225,086 |
| Secondary | 10,278 | 152,047 | 2,505,956 |
| Higher: universities* | 78 | 33,177 | 435,637 |
| Higher: other tertiary | 1,066 | 27,478 | 389,223 |
| Special | 407 | 3,331 | 24,672 |
| Vocational | 1,893 | 11,438 | 248,003 |

* Figures for 2000.

Source: Ministerio de Educación del Perú.

**Adult literacy rate** (UNESCO estimates): 85.0% (males 91.3%; females 80.3%) in 2002 (Source: UN Development Programme, *Human Development Report*).

# Directory

## The Constitution

In 1993 the Congreso Constituyente Democrático (CCD) began drafting a new constitution to replace the 1979 Constitution. The CCD approved the final document in September 1993, and the Constitution was endorsed by a popular national referendum that was conducted on 31 October. The Constitution was promulgated on 29 December 1993.

### EXECUTIVE POWER

Executive power is vested in the President, who is elected for a five-year term of office by universal adult suffrage; this mandate is renewable once. The successful presidential candidate must obtain at least 50% of the votes cast, and a second round of voting is held if necessary. Two Vice-Presidents are elected in simultaneous rounds of voting. The President is prohibited from serving two consecutive terms. The President is competent to initiate and submit draft bills, to review laws drafted by the legislature (Congreso) and, if delegated by the Congreso, to enact laws. The President is empowered to appoint ambassadors and senior military officials without congressional ratification, and retains the right to dissolve parliament if two or more ministers have been censured or have received a vote of 'no confidence' from the Congreso. In certain circumstances the President may, in accordance with the Council of Ministers, declare a state of emergency for a period of 60 days, during which individual constitutional rights are suspended and the armed forces may assume control of civil order. The President appoints the Council of Ministers.

### LEGISLATIVE POWER

Legislative power is vested in a single-chamber Congreso (removing the distinction in the 1979 Constitution of an upper and lower house) consisting of 120 members.* The members of the Congreso are elected for a five-year term by universal adult suffrage. The Congreso is responsible for approving the budget, for endorsing loans and international treaties and for drafting and approving bills. It may conduct investigations into matters of public concern, and question and censure the Council of Ministers and its individual members. Members of the Congreso elect a Standing Committee, to consist of not more than 25% of the total number of members (representation being proportional to the different political groupings in the legislature), which is empowered to make certain official appointments, approve credit loans and transfers relating to the budget during a parliamentary recess, and conduct other business as delegated by parliament.

### ELECTORAL SYSTEM

All citizens aged 18 years and above, including illiterate persons, are eligible to vote. Voting in elections is compulsory for all citizens aged 18–70, and is optional thereafter.

### JUDICIAL POWER

Judicial power is vested in the Supreme Court of Justice and other tribunals. The Constitution provides for the establishment of a National Council of the Judiciary, consisting of nine independently elected members, which is empowered to appoint judges to the Supreme Court. An independent Constitutional Court, comprising seven members elected by the Congreso for a five-year term, may interpret the Constitution and declare legislation and acts of government to be unconstitutional.

The death penalty may be applied by the Judiciary in cases of terrorism or of treason (the latter in times of war).

Under the Constitution, a People's Counsel is elected by the Congreso with a five-year mandate which authorizes the Counsel to defend the constitutional and fundamental rights of the individual. The Counsel may draft laws and present evidence to the legislature.

According to the Constitution, the State promotes economic and social development, particularly in the areas of employment, health, education, security, public services and infrastructure. The State recognizes a plurality of economic ownership and activity, supports free competition, and promotes the growth of small businesses. Private initiative is permitted within the framework of a social market economy. The State also guarantees the free exchange of foreign currency.

* In April 2003 the Congreso voted to restore a bicameral legislature. A new lower chamber would comprise 150 members, elected for a five-year term, and would include at least one representative from each electoral district. A 50-member Senate would also be created.

## The Government

### HEAD OF STATE

**President:** Dr ALEJANDRO TOLEDO MANRIQUE (took office 28 July 2001).

**Vice-Presidents:** Dr DAVID WAISMAN RJAVINSTHI, (vacant).

### COUNCIL OF MINISTERS
(August 2005)

**President of the Council of Ministers:** PEDRO PABLO KUCZYNSKI.
**Minister of Foreign Affairs:** OSCAR MAÚRTUA DE ROMAÑA.
**Minister of Defence:** MARCIANO RENGIFO RUIZ.
**Minister of the Interior:** RÓMULO PIZARRO TOMASIO.
**Minister of Justice:** ALEJANDRO TUDELA CHOPITEA.
**Minister of Economy and Finance:** FERNANDO ZAVALA LOMBARDI.
**Minister of Labour and Employment:** JUAN SHEPUT.
**Minister of International Trade and Tourism:** ALFREDO FERRERO DIEZ CANSECO.
**Minister of Transport and Communications:** JOSÉ ORTIZ RIVERA.
**Minister of Housing, Construction and Sanitation:** RUDECINDO VEGA CARREAZO.
**Minister of Health:** PILAR MAZZETTI SOLER.
**Minister of Agriculture:** MANUEL MANRIQUE UGARTE.
**Minister of Energy and Mines:** GLODOMIRO SÁNCHEZ MEJÍA.
**Minister of Production:** DAVID LEMOR BEZDIN.
**Minister of Education:** JAVIER SOTA NADAL.
**Minister for the Advancement of Women and Social Development:** ANA MARÍA ROMERO LOZADA LAUEZZARI.

### MINISTRIES

**Office of the President of the Council of Ministers:** 28 de Julio 878, Miraflores, Lima; tel. (1) 4469800; fax (1) 4449168; e-mail webmaster@pcm.gob.pe; internet www.pcm.gob.pe.

**Ministry for the Advancement of Women and Social Development:** Jirón Camaná 616, Lima 1; tel. (1) 4289800; fax (1) 4261665; e-mail postmaster@mimdes.gob.pe; internet www.mimdes.gob.pe.

**Ministry of Agriculture:** Avda Salaverry s/n, Jesús María, Lima 11; tel. (1) 4310424; fax (1) 4310109; e-mail postmast@minag.gob.pe; internet www.minag.gob.pe.

**Ministry of Defence:** Avda Arequipa 291, Lima 1; tel. (1) 4335150; fax (1) 4333636; e-mail webmaster@mindef.gob.pe; internet www.mindef.gob.pe.

**Ministry of Economy and Finance:** Jirón Junín 339, 4°, Circado de Lima, Lima 1; tel. (1) 4273930; fax (1) 4282509; e-mail postmaster@mef.gob.pe; internet www.mef.gob.pe.

**Ministry of Education:** Avda Van Develde 160, (cuadra 33 Javier Prado Este), San Borja, Lima 41; tel. (1) 4353900; fax (1) 4370471; e-mail postmaster@minedu.gob.pe; internet www.minedu.gob.pe.

**Ministry of Energy and Mines:** Avda Las Artes Sur 260, San Borja, Apdo 2600, Lima 41; tel. (1) 4752969; fax (1) 4750689; internet www.minem.gob.pe.

**Ministry of Foreign Affairs:** Jirón Lampa 535, Lima 1; tel. (1) 3112402; fax (1) 3112406; internet www.rree.gob.pe.

**Ministry of Health:** Avda Salaverry cuadra 8, Jesús María, Lima 11; tel. (1) 4310408; fax (1) 4310093; e-mail webmaster@minsa.gob.pe; internet www.minsa.gob.pe.

**Ministry of Housing, Construction and Sanitation:** Avda Paseo de la República 3361, San Isidro, Lima; tel. (1) 2117930; e-mail webmaster@vivienda.gob.pe; internet www.vivienda.gob.pe.

**Ministry of the Interior:** Plaza 30 de Agosto 150, San Isidro, Lima 27; tel. (1) 2242406; fax (1) 2242405; e-mail ofitel@mininter.gob.pe; internet www.mininter.gob.pe.

**Ministry of International Trade and Tourism:** Calle 1 Oeste 50, Urb. Corpac, San Isidro, Lima 27; tel. (1) 2243345; fax (1) 2243362; e-mail webmaster@mincetur.gob.pe; internet www.mincetur.gob.pe.

**Ministry of Justice:** Scipión Llona 350, Miraflores, Lima 18; tel. (1) 4222654; fax (1) 4223577; e-mail webmaster@minjus.gob.pe; internet www.minjus.gob.pe.

**Ministry of Labour and Employment:** Avda Salaverry 655 cuadra 8, Jesús María, Lima 11; tel. (1) 4332512; fax (1) 4230741; e-mail webmaster@mintra.gob.pe; internet www.mintra.gob.pe.

**Ministry of the Presidency:** Avda Paseo de la República 4297, Lima 1; tel. (1) 4465886; fax (1) 4470379; internet www.peru.gob.pe.

**Ministry of Production:** Calle 1 Oeste 60, Urb. Corpac, San Isidro, Lima 27; tel. (1) 2243333; fax (1) 2243237; internet www.produce.gob.pe.

**Ministry of Transport and Communications:** Avda 28 de Julio 800, Lima 1; tel. (1) 4330010; fax (1) 4339378; internet www.mtc.gob.pe.

## Regional Presidents
(July 2005)

**Amazonas:** MIGUEL REYES CATALINO CONTRERAS (APRA).
**Ancash:** RICARDO NARVÁEZ SOTO (APRA).
**Apurímac:** LUIS BELTRÁN BARRA PACHECO (APRA).
**Arequipa:** DANIEL ERNESTO VERA BALLÓN (APRA).
**Ayacucho:** WERNER OMAR QUEZADA MARTÍNEZ (APRA).
**Cajamarca:** LUIS FELIPE PITA GASTELUMENDI (APRA).
**Callao:** ROGELIO CANCHES (PP).
**Cusco:** CARLOS CUARESMA SÁNCHEZ (FIM).
**Huancavelica:** SALVADOR ESPINOZA HUAROC (Ind.).
**Huánuco:** LUZMILA TEMPLO CONDEZO (Ind.).
**Ica:** VICENTE TELLO CÉSPEDES (APRA).
**Junín:** MANUEL DUARTE VELARDE (Ind.).
**La Libertad:** HOMERO BURGOS LIVEROS (APRA).
**Lambayeque:** YEHUDE SIMON MUNARO (UPP).
**Lima:** MIGUEL ANGEL MUFARECH (APRA).
**Loreto:** ROBINSON RIVADENEYRA (Ind.).
**Madre de Díos:** RAFAEL EDWIN RÍOS LÓPEZ (Ind.).
**Moquegua:** MARÍA CRISTALA CONSTANTINIDES (SP).
**Pasco:** VÍCTOR RAÚL ESPINOZA SOTO (Ind.).
**Piura:** CÉSAR TRELLES LARA (APRA).
**Puno:** DANIEL ANÍBAL JIMÉNEZ SARDÓN (Ind.).
**San Martín:** MAX HENRY RAMÍREZ GARCÍA (APRA).
**Tacna:** JULIO ANTONIO ALVA CENTURIÓN (APRA).
**Tumbes:** IRIS MEDINA FEIJOÓ (APRA).
**Ucayali:** EDWIN VÁSQUEZ LÓPEZ (Ind.).

## President and Legislature

### PRESIDENT

**Presidential Election, 8 April and 3 June 2001**

| Candidate | First round % of votes | Second round % of votes |
|---|---|---|
| Alejandro Toledo (PP) | 36.51 | 53.08 |
| Alan García (APRA) | 25.78 | 46.92 |
| Lourdes Flores (Unidad Nacional) | 24.30 | — |
| Fernando Olivera Vega (FIM) | 9.85 | — |
| Carlos Boloña (Solución Popular) | 1.69 | — |
| Ciro A. Gálvez (Renacimiento Andino) | 0.81 | — |
| Marco A. Arrunategui (Proyecto País) | 0.75 | — |
| Ricardo Noriega (Todos por la Victoria) | 0.31 | — |
| **Total** | **100.00** | **100.00** |

### CONGRESO

**President:** ANTERO FLORES-ARÁOZ.

# PERU

## General Election, 8 April 2001

| Parties | Seats |
|---|---|
| Perú Posible (PP) | 45 |
| Alianza Popular Revolucionaria Americana (APRA) | 27 |
| Unidad Nacional | 17 |
| Frente Independiente Moralizador (FIM) | 12 |
| Unión por el Perú (UPP) | 6 |
| Somos Perú (SP) | 4 |
| Acción Popular (AP) | 3 |
| Cambio 90-Nueva Mayoría (C90-NM) | 3 |
| Renacimiento Andino | 1 |
| Solución Popular | 1 |
| Todos por la Victoria | 1 |
| **Total** | **120** |

## Political Organizations

**Acción Popular (AP):** Paseo Colón 218, Lima 1; tel. (1) 3321965; fax (1) 3321965; e-mail webmaster@accionpopular.org.pe; internet www.accionpopular.org.pe; f. 1956; 1.2m. mems; liberal; Leader FERNANDO BELAÚNDE TERRY; Sec.-Gen. LUIS ALBERTO VELARDE YAÑEZ.

**Alianza Popular Revolucionaria Americana (APRA):** Avda Alfonso Ugarte 1012, Lima 5; tel. (1) 4281736; internet www.apra.org.pe; f. in Mexico 1924, in Peru 1930 as Partido Aprista Peruano (PAP); legalized 1945; democratic left-wing party; Sec.-Gen. JORGE DEL CASTILLO; 700,000 mems.

**Cambio 90 (C90):** Jr Santa Isabel 590, Urb. Colmenares, Pueblo Libre, Lima; tel. (1) 9441739; group of independents formed to contest the 1990 elections, entered into coalition with Nueva Mayoría (q.v.) in 1992 to contest elections to the CCD, local elections in 1993, the presidential and congressional elections in 1995 and 2000 and the 2001 congressional elections; Leader PABLO CORREA.

**Coordinación Democrática (CODE):** Lima; f. 1992 by dissident APRA members; Leader JOSÉ BARBA CABALLERO.

**Frente Independiente Moralizador (FIM):** Pancho Fierro 133, San Isidro, Lima; tel. (1) 4220583; internet www.congreso.gob.pe/grupo_parlamentario/fim/inicio.htm; f. 1990; right-wing; Pres. LUIS FERNANDO OLIVERA VEGA; Sec.-Gen. GONZALO CARRIQUIRY BLONDET.

**Frente Nacional de Trabajadores y Campesinos (FNTC/FRENATRACA):** Lima; tel. (1) 4272868; f. 1968; left-wing party; Pres. Dr RÓGER CÁCERES VELÁSQUEZ; Sec.-Gen. Dr EDMUNDO HUANQUI MEDINA.

**Frente Popular Agrícola del Perú (Frepap):** Avda Morro Solar 1234; Santiago de Surco, Lima; tel. (1) 2753847; f. 1989; Leader EZEQUIEL ATAUCUSI GAMONAL.

**Nueva Mayoría:** Lima; f. 1992; group of independents, including former cabinet ministers, formed coalition with Cambio 90 in 1992 in order to contest elections to the CCD, local elections in 1993, the presidential and congressional elections in 1995 and 2000 and the 2001 congressional elections; Leader JAIME YOSHIYAMA TANAKA.

**Partido Comunista Peruano (PCP):** Plaza Ramón Castilla 67, Lima; tel. and fax (1) 3306106; e-mail unidad@ec-red.com; internet www.pcp.miarroba.com; f. 1928; Pres. JORGE DEL PRADO.

**Partido Demócrata Cristiano (PDC):** Avda España 321, Lima 1; tel. (1) 4238042; f. 1956; 95,000 mems; Chair. CARLOS BLANCAS BUSTAMANTE.

**Partido Democrático Descentralista (PDD):** Huancayo; e-mail pddperu@yahoogroups.com; internet www.pddperu.net; f. 2002; left-wing; regionalist; Leader JAVIER DIEZ CANSECO.

**Partido Popular Cristiano (PPC):** Avda Alfonso Ugarte 1484, Lima; tel. (1) 4238723; fax (1) 4238721; e-mail estflores@terra.com.pe; f. 1967; splinter group of Partido Demócrata Cristiano; 250,000 mems; Pres. Dr ANTERO LOURDES FLORES-NANO; Vice-Pres RAÚL CASTRO STAGNARO, JAVIER BEDOYA DE VIVIANCO, XAVIER BARRÓN CEBRERROS, ALEJANDRO CASTAGNOLA PINILLOS.

**Partido Verde del Perú** (Movimiento Independiente verde del Perú en Formación): Avda Jorge Chávez 654, 4°, Lima 33; e-mail allparuna@yahoo.es; internet www.unii.net/allparuna/peruverde.html; ecologist.

**Perú Ahora:** Plaza Bolognesi 600, Breña, Lima 5; tel. (1) 3309230; e-mail partidopolitico@peruahora.org; internet www.peruahora.org; f. 1998; relaunched in 2003 by fmr mems of Perú Posible; Sec.-Gen. LUIS GUERRERO FIGUEROA.

**Perú Posible (PP):** Avda Faustino Sánchez Carrión, Lima; tel. (1) 4620303; e-mail sgpp@mixmail.com; internet www.peruposible.org.pe/indice.htm; f. 1994 to support Toledo's candidacy for the 1995 presidential election in alliance with CODE, contested the 2000 and 2001 presidential and congressional elections unaligned; Leader ALEJANDRO TOLEDO; Sec.-Gen. LUIS SOLARI.

**Proyecto País:** Nicolás de Piérola 917, Of. 211, Plaza San Martín, Lima; tel. (1) 3312696; fax (1) 2222785; e-mail info@proyectopais.org.pe; internet www.proyectopais.org.pe; f. 1998; Leader MARCO ANTONIO ARRUNATEGUI CEVALLOS.

**Renacimiento Andino:** Calle Real 583, 2°, Huancayo; tel. (4) 214620; fax (4) 217480; e-mail webmaster@renacimientoandino.org.pe; internet www.renacimientoandino.org.pe; f. 2001; Leader CIRO ALFREDO GÁLVEZ HERRERA.

**Renovación:** c/o Edif. Complejo 510, Avda Abancay 251, Lima; tel. (1) 4264260; fax (1) 4263023; f. 1992; Leader RAFAEL REY REY.

**Solidaridad Nacional (SN):** Armando Blondet 106, San Isidro, Lima; tel. (1) 2218948; e-mail fsandoval@psn.org.pe; internet www.psn.org.pe; f. 1999; centre-left; Pres. LUIS CASTAÑEDA LOSSIO; Sec.-Gen. MARCO ANTONIO PARRA SÁNCHEZ.

**Solución Popular:** Avda Guzmán Blanco 240, Of. 1001, Lima; tel. (1) 4263017; Leader CARLOS BOLOÑA.

**Somos Perú (SP):** Avda Arequipa 3990, Miraflores, Lima; tel. (1) 4219363; e-mail postmaster@somosperu.org.pe; internet www.somosperu.org.pe; f. 1998; Leader ALBERTO ANDRADE; Sec.-Gen. EDUARDO CARHUAICRA MEZA.

**Todos por la Victoria:** Avda Grau 122, La Victoria, Lima; tel. (1) 4242446; f. 2001; Leader RICARDO MANUEL NORIEGA SALAVERRY.

**Unión de Izquierda Revolucionaria (UNIR)** (Union of the Revolutionary Left): Jirón Puno 258, Apdo 1165, Lima 1; tel. (1) 4274072; f. 1980; Chair. Sen. ROLANDO BREÑA PANTOJA; Gen. Sec. JORGE HURTADO POZO.

**Unidad Nacional:** Calle Ricardo Palma 1111, Miraflores, Lima; tel. (1) 2242773; f. 2000; centrist alliance; Leader LOURDES FLORES.

**Unión por el Perú (UPP):** Pablo de Olavide 270, San Isidro, Lima; tel. (1) 4403227; e-mail pradobus@hotmail.com; independent movement; f. 1994 to contest presidential and legislative elections; Leader Dr ALDO ESTRADA CHOQUE.

**Vamos Vecino:** Lima; f. 1998; Leader ABSALON VÁSQUEZ.

Other parties include Frente Obrero Campesino Estudiantil y Popular (FOCEP), Partido Unificado Mariateguista (PUM), Partido Comunista del Perú—Patria Roja and Movimiento Amplio País Unido.

### ARMED GROUPS

**Movimiento Nacionalista Peruano (MNP)** (Movimiento Etnocacerista): Pasaje Velarde 188, Of. 204, Lima; tel. (1) 3311074; e-mail movnacionalistaperuano@yahoo.es; internet mnp.tripod.com.pe; ultra-nationalist paramilitary group; Pres. Dr ISAAC HUMALA NÚÑEZ; Leaders of paramilitary wing Maj. (retd) ANTAURO IGOR HUMALA TASSO (arrested January 2005 following an armed uprising in Andahuaylas), Lt-Col (retd) OLLANTA MOISES HUMALA TASSO.

**Sendero Luminoso** (Shining Path): f. 1970; began armed struggle 1980; splinter group of PCP; based in Ayacucho; advocates the policies of the late Mao Zedong and his radical followers, including the 'Gang of Four' in the People's Republic of China; founder Dr ABIMAEL GUZMÁN REYNOSO (alias Comandante Gonzalo—arrested September 1992); current leaders MARGIE CLAVO PERALTA (arrested March 1995), PEDRO DOMINGO QUINTEROS AYLLÓN (alias Comrade Luis—arrested April 1998).

**Sendero Rojo** (Red Path): dissident faction of Sendero Luminoso opposed to leadership of Abimael Guzmán; Leader FILOMENO CERRÓN CARDOSO (alias Comrade Artemio).

**Movimiento Revolucionario Tupac Amarú (MRTA):** f. 1984; began negotiations with the Government to end its armed struggle in September 1990; Leader VÍCTOR POLAY CAMPOS (alias Comandante Rolando—arrested in 1992).

## Diplomatic Representation

### EMBASSIES IN PERU

**Argentina:** Arequipa 121, Lima 1; tel. (1) 4339966; fax (1) 4330769; e-mail embajada@terra.com.pe; Ambassador JORGE ALBERTO VÁZQUEZ.

**Austria:** Avda Central 643, 5°, San Isidro, Lima 27; tel. (1) 4420503; fax (1) 4428851; e-mail lima-ob@bmaa.gv.at; Ambassador GERHARD DOUJAK.

**Bolivia:** Los Castaños 235, San Isidro, Lima 27; tel. (1) 4402095; fax (1) 4402298; e-mail jemis@emboli.attla.com.pe; Ambassador ELOY AVILA ALBERDI.

**Brazil:** Avda José Pardo 850, Miraflores, Lima; tel. (1) 4215660; fax (1) 4452421; e-mail embajada@embajadabrasil.org.pe; internet www.embajadabrasil.org.pe; Ambassador ANDRÉ MATTOSO MAIA AMADO.

**Canada:** Calle Libertad 130, Miraflores, Lima 18; tel. (1) 4444015; fax (1) 4444347; e-mail lima@dfait-maeci.gc.ca; internet www.dfait-maeci.gc.ca/peru; Ambassador GENEVIÈVE DES RIVIÈRES.

**Chile:** Avda Javier Prado Oeste 790, San Isidro, Lima; tel. (1) 6112211; fax (1) 6112223; e-mail embajada@embachileperu.com.pe; internet www.embachileperu.com.pe; Ambassador JUAN PABLO LIRA BIANCHI.

**China, People's Republic:** Jirón José Granda 150, San Isidro, Apdo 375, Lima 27; tel. (1) 2220841; fax (1) 4429467; e-mail chinaemb_pe@mfa.gov.cn; internet www.embajadachina.org.pe; Ambassador MAI GUOYAN.

**Colombia:** Avda J. Basadre 1580, San Isidro, Lima 27; tel. (1) 4410954; fax (1) 4419806; e-mail elima@minrelext.gov.co; Ambassador HÉCTOR JOSÉ QUINTERO ARREDONDO.

**Costa Rica:** Calle Baltazar La Torre 828, San Isidro, Lima; tel. (1) 2642999; fax (1) 2642799; e-mail costarica@terra.com.pe; Ambassador JULIO SUÑOL LEAL.

**Cuba:** Coronel Portillo 110, San Isidro, Lima; tel. (1) 2642053; fax (1) 2644525; e-mail embacuba@ecuperu.minrex.gov.cu; Ambassador ROGELIO SIERRA DÍAZ.

**Czech Republic:** Baltazar La Torre 398, San Isidro, Lima 27; tel. (1) 2643374; fax (1) 2641708; e-mail lima@embassy.mzv.cz; internet www.mfa.cz/lima; Chargé d'affaires a.i. KAMILA HRABÁKOVÁ.

**Dominican Republic:** Calle Tudela y Varela 360, San Isidro, Lima; tel. (1) 4219765; fax (1) 2220639; e-mail embdomperu@terra.com.pe; Ambassador MIGUEL FERSOBE PICHARDO.

**Ecuador:** Las Palmeras 356 y Javier Prado Oeste, San Isidro, Lima 27; tel. (1) 2124171; fax (1) 4220711; e-mail embajada@mecuadorperu.org.pe; internet www.mecuadorperu.org.pe; Ambassador Dr LUIS VALENCIA RODRÍGUEZ (designate).

**Egypt:** Avda Jorge Basadre 1470, San Isidro, Lima 27; tel. (1) 4402642; fax (1) 4402547; e-mail emb-egypt@amauta.rcp.net.pe; Ambassador SHADIA AHMED SHOUKRI.

**El Salvador:** Avda Javier Prado 2108, San Isidro, Lima 27; tel. (1) 4403500; fax (1) 2212561; e-mail embajadasv@terra.com.pe; internet www.embajadaelsalvador.org.pe; Ambassador RAÚL SOTO-RAMÍREZ.

**Finland:** Avda Víctor Andrés Belaúnde 147, Edif. Real Tres, Of. 502, San Isidro, Lima; tel. (1) 2224466; fax (1) 2224463; e-mail sanomat.lim@formin.fi; internet www.finlandiaperu.org.pe; Ambassador KIMMO PULKKINEN.

**France:** Avda Arequipa 3415, Lima 27; tel. (1) 2158400; fax (1) 2158410; e-mail france.presse@ambafrance-pe.org; internet www.ambafrance-pe.org; Ambassador PIERRE CHARRASSE.

**Germany:** Avda Arequipa 4210, Miraflores, Lima 18; tel. (1) 2125016; fax (1) 4226475; e-mail kanzlei@embajada-alemana.org.pe; internet www.embajada-alemana.org.pe; Ambassador ROLAND ERNEST-AUGUST KLIESOW.

**Greece:** Avda Principal 190, 6°, Urb. Santa Catalina, Lima 13; tel. (1) 4761548; fax (1) 4761329; e-mail emgrecia@terra.com.pe; Ambassador VASSILOS SIMANTIRAKIS.

**Guatemala:** Inca Ripac 309, Lima 27; tel. (1) 4602078; fax (1) 4635885; e-mail popolvuh@amauta.rcp.net.pe; Ambassador OLGA MARÍA AGUJA SUÑIGA.

**Holy See:** Avda Salaverry 6a cuadra, Apdo 397, Lima 100 (Apostolic Nunciature); tel. (1) 4319436; fax (1) 4315704; e-mail nunciatura@speedy.com.pe; Apostolic Nuncio Most Rev. RINO PASSIGATO (Titular Archbishop of Nova Caesaris).

**Honduras:** Avda Larco 930, Miraflores, Lima 18; tel. (1) 4442345; fax (1) 2425017; e-mail embhonpe@speedy.com.pe; Chargé d'affaires a.i. GRACIA LAZO DE VELÁSQUEZ.

**Hungary:** Calle Alfredo Roldán 124, San Isidro, , Lima 18; tel. (1) 4223069; fax (1) 4223093; e-mail oficinalima@embajadadehungaria.com; Ambassador JÓZSEF KOSÁRKA.

**India:** Avda Salaverry 3006, San Isidro, Lima 27; tel. (1) 4602289; fax (1) 4610374; e-mail hoc@indembassy.org.pe; internet www.indembassy.org.pe; Ambassador RIEWAD V. WARJRI.

**Indonesia:** Avda Las Flores 334, San Isidro, Lima; tel. (1) 222-0308; fax (1) 222-2684; Ambassador GUSTI NGURAH SWETJA.

**Israel:** Edif. El Pacifico, 6°, Plaza Washington, Natalio Sánchez 125, Santa Beatriz, Lima; tel. (1) 4334431; fax (1) 4338925; e-mail info@lima.mfa.gov.il; internet lima.mfa.gov.il; Ambassador ORY NOY.

**Italy:** Avda Gregorio Escobedo 298, Apdo 0490, Lima 11; tel. (1) 4632727; fax (1) 4635317; e-mail segretaria@italembperu.org.pe; internet www.italembperu.org.pe; Ambassador SERGIO BUSETTO.

**Japan:** Avda San Felipe 356, Apdo 3708, Jesús María, Lima 11; tel. (1) 2181130; fax (1) 4630302; internet www.pe.emb-japan.go.jp; Ambassador YUBUN NARITA.

**Korea, Democratic People's Republic:** Los Nogales 227, San Isidro, Lima; tel. (1) 4411120; fax (1) 4409877; e-mail embcorea@hotmail.com; Ambassador YU CHANG UN.

**Korea, Republic:** Avda Principal 190, 7°, Lima 13; tel. (1) 4760815; fax (1) 4760950; e-mail korembj-pu@mofat.go.kr; Ambassador CHUNG JIN-HO.

**Malaysia:** Calle 41 894, 3°, Urb. Córpac, San Isidro, Lima 27; tel. (1) 4220297; fax (1) 4410795; e-mail embmalperu@lullitec.com.pe; Ambassador ABDUL JALIL BIN HARON.

**Mexico:** Avda Jorge Basadre 710, esq. Los Ficus, San Isidro, Lima; tel. (1) 2211100; fax (1) 4404740; e-mail info@mexico.org.pe; internet www.mexico.org.pe; Ambassador ANTONIO GUILLERMO VILLEGAS VILLALOBOS.

**Morocco:** Calle Manuel Ugarte y Morosco 790, San Isidro, Lima; tel. (1) 2643323; fax (1) 2640006; e-mail sifamlim@chavin.rcp.net.pe; Ambassador MAHMOUD RMIKI.

**Netherlands:** Avda Principal 190, 4°, Urb. Santa Catalina, La Victoria, Lima; tel. (1) 4150660; fax (1) 4150689; e-mail info@nlgovlim.com; internet www.nlgovlim.com; Ambassador PAUL W. A. SCHELLEKENS.

**Nicaragua:** Calle Parque Ramón Castilla 305, Urb. Aurora, Miraflores, Lima; tel. (1) 4459274; fax (1) 4468554; Chargé d'affaires a.i. MARITZA ROSALES GRANERA.

**Panama:** Alvarez Calderón 738, San Isidro, Lima 27; tel. (1) 4413652; fax (1) 4419323; e-mail panaemba@amauta.rcp.net.pe; Ambassador ROBERTO DÍAZ HERRERA.

**Paraguay:** Alcanfores 1286, Miraflores, Lima; tel. (1) 4474762; fax (1) 4442391; e-mail embaparpe@terra.com.pe; Ambassador Dra JULIA VELILLA LACONICH.

**Poland:** Avda Salaverry 1978, Jesús María, Lima 11; tel. (1) 4713925; fax (1) 4714813; e-mail ambrplima@telefonica.net.pe; Ambassador ZDZISLAW SOSNICKI.

**Portugal:** Avda Central 643, 4°, Lima 27; tel. (1) 4409905; fax (1) 4429655; e-mail limaportugal@hotmail.com; Ambassador MÁRIO ALBERTO LINO DA SILVA.

**Romania:** Avda Jorge Basadre 690, San Isidro, Lima; tel. (1) 4224587; fax (1) 4210609; e-mail ambrom@terra.com.pe; Ambassador STEFAN COSTIN.

**Russia:** Avda Salaverry 3424, San Isidro, Lima 27; tel. (1) 2640036; fax (1) 2640130; e-mail embrusa@amauta.rcp.net.pe; Ambassador ANATOLY P. KUZNETSOV.

**Serbia and Montenegro:** Carlos Porras Osores 360, Apdo 18-0392, San Isidro, Lima 27; Apdo 0392, Lima 18; tel. (1) 4212423; fax (1) 4212427; e-mail yugoembperu@amauta.rcp.net.pe; Chargé d'affaires a.i. MILIVOJ SUCEVIĆ.

**South Africa:** PO Box 27-013 L27, Lima; tel. (1) 4409996; fax 4223881; e-mail saemb@amauta.rcp.net.pe; Ambassador Dr C. J. STREETER.

**Spain:** Jorge Basadre 498, San Isidro, Lima 27; tel. (1) 2125155; fax (1) 4410084; e-mail embesppe@correo.mae.es; Ambassador JULIO ALBI DE LA CUESTA.

**Switzerland:** Avda Salaverry 3240, San Isidro, Lima 27; tel. (1) 2640305; fax (1) 2641319; e-mail vertretung@lim.rep.admin.ch; internet www.eda.admin.ch/lima_emb/s/home.html; Ambassador BEAT LOELIGER.

**Ukraine:** Calle José Dellepiani 470, San Isidro, Lima; tel. (1) 264-2884; fax (1) 264-2892; e-mail emb-pe@mfa.gov.ua; Ambassador IGOR GRUSHKO.

**United Kingdom:** Torre Parque Mar, 22°, Avda José Larco 1301, Miraflores, Lima; tel. (1) 6173000; fax (1) 6173100; e-mail belima@fco.gov.uk; internet www.britemb.org.pe; Ambassador RICHARD RALPH.

**USA:** Avda La Encalada 17, Surco, Lima 33; tel. (1) 4343000; fax (1) 6182397; internet usembassy.state.gov/lima; Ambassador JAMES CURTIS STRUBLE.

**Uruguay:** Calle José D. Anchorena 84, San Isidro, Lima; tel. (1) 2640099; fax (1) 2640112; e-mail uruinca@embajada-uruguay.com; Ambassador JUAN BAUTISTA ODDONE SILVEIRA.

**Venezuela:** Avda Arequipa 298, Lima; tel. (1) 4334511; fax (1) 4331191; e-mail embavene@millicom.com.pe; Chargé d'affaires a.i. MAGALY GARCÍA RUIZ.

PERU — *Directory*

## Judicial System

The Supreme Court consists of a President and 17 members. There are also Higher Courts and Courts of First Instance in provincial capitals. A comprehensive restructuring of the judiciary was implemented during the late 1990s.

### SUPREME COURT

**Corte Suprema**
Palacio de Justicia, 2°, Avda Paseo de la República, Lima 1; tel. (1) 4284457; fax (1) 4269437.

**President:** Dr Walter Humberto Vásquez Vejarano.

**Attorney-General:** Nelly Calderón Navarro.

## Religion

### CHRISTIANITY

#### The Roman Catholic Church

For ecclesiastical purposes, Peru comprises seven archdioceses, 18 dioceses, 11 territorial prelatures and eight Apostolic Vicariates. At 31 December 2003 92.0% of the country's population (an estimated 29.5m.) were adherents of the Roman Catholic Church.

#### Bishops' Conference

Conferencia Episcopal Peruana, Jirón Estados Unidos 838, Apdo 310, Lima 100; tel. (1) 4631010; fax (1) 4636125; e-mail sgc@iglesiacatolica.org.pe.

f. 1981; statutes approved 1987, revised 1992 and 2000; Pres. Mgr José Hugo Garaycoa Hawkins (Bishop of Tacna y Moquegua).

**Archbishop of Arequipa:** José Paulino Ríos Reynoso, Arzobispado, Moral San Francisco 118, Apdo 149, Arequipa; tel. (54) 234094; fax (54) 242721; e-mail arzobispadoaqp@planet.com.pe.

**Archbishop of Ayacucho or Huamanga:** Luis Abilio Sebastiani Aguirre, Arzobispado, Jirón 28 de Julio 148, Apdo 30, Ayacucho; tel. and fax (64) 812367; e-mail arzaya@mail.udep.edu.pe.

**Archbishop of Cusco:** Juan Antonio Ugarte Pérez, Arzobispado, Herrajes, Hatun Rumiyoc s/n, Apdo 148, Cusco; tel. (84) 225211; fax (84) 222781; e-mail arzobisp@terra.com.pe.

**Archbishop of Huancayo:** Pedro Ricardo Barreto Jimeno, Arzobispado, Jirón Puno 430, Apdo 245, Huancayo; tel. (64) 234952; fax (64) 239189; e-mail arzohyo@hotmail.com.

**Archbishop of Lima:** Cardinal Juan Luis Cipriani Thorne, Arzobispado, Jirón Carabaya, Plaza Mayor, Apdo 1512, Lima 100; tel. (1) 4275980; fax (1) 4271967; e-mail arzolim@terra.com.pe; internet www.arzobispadodelima.org.

**Archbishop of Piura:** Oscar Rolando Cantuarias Pastor, Arzobispado, Libertad 1105, Apdo 197, Piura; tel. and fax (74) 327561; e-mail ocordova@upiura.edu.pe.

**Archbishop of Trujillo:** Héctor Miguel Cabrejos Vidarte, Arzobispado, Jirón Mariscal de Orbegozo 451, Apdo 42, Trujillo; tel. (44) 256812; fax (44) 231473; e-mail arztrujillo@terra.com.pe.

#### The Anglican Communion

The Iglesia Anglicana del Cono Sur de América (Anglican Church of the Southern Cone of America), formally inaugurated in April 1983, comprises seven dioceses, including Peru. The Presiding Bishop of the Church is the Bishop of Northern Argentina.

**Bishop of Peru:** Rt Rev. Harold William Godfrey, Apdo 18-1032, Miraflores, Lima 18; tel. and fax (1) 4229160; e-mail wgodfrey@amauta.rcp.net.pe.

#### The Methodist Church

There are an estimated 4,200 adherents of the Iglesia Metodista del Perú.

**President:** Rev. Jorge Figueroa, Baylones 186, Lima 5; Apdo 1386, Lima 100; tel. (1) 4245970; fax (1) 4318995; e-mail iglesiamp@computextos.com.pe.

#### Other Protestant Churches

Among the most popular are the Asamblea de Dios, the Iglesia Evangélica del Perú, the Iglesia del Nazareno, the Alianza Cristiana y Misionera and the Iglesia de Dios del Perú.

### BAHÁ'Í FAITH

**National Spiritual Assembly of the Bahá'ís of Peru:** Horacio Urteaga 827, Jesús María, Apdo 11-0209, Lima 11; tel. (1) 4316077; fax (1) 4333005; e-mail bahai@terra.com.pe; mems resident in 220 localities; Nat. Sec. María Loreto Jara de Roeder.

## The Press

### DAILIES

#### Lima

**El Bocón:** Jirón Jorge Salazar Araoz 171, Urb. Santa Catalina, Apdo 152, Lima 1; tel. (1) 4756355; fax (1) 4758780; internet www.elbocon.com.pe; f. 1994; football; Editorial Dir Jorge Estéves Alfaro; circ. 90,000.

**El Comercio:** Empresa Editora 'El Comercio', SA, Jirón Antonio Miró Quesada 300, Lima; tel. (1) 4264676; fax (1) 4260810; e-mail editorweb@comercio.com.pe; internet www.elcomercioperu.com.pe; f. 1839; morning; Editor Juan Carlos Luján; Dir-Gen. Alejandro Miró Quesada G., Francisco Miró Quesada C.; circ. 150,000 weekdays, 220,000 Sundays.

**Expreso:** Jirón Antonio Elizalde 753, Lima; tel. (1) 6124000; fax (1) 4447125; e-mail webmaster@expreso.com.pe; internet www.expreso.com.pe; f. 1961; morning; conservative; Pres. Manuel Ulloa; Dir Carlos Espá; circ. 100,000.

**Extra:** Jirón Libertad 117, Miraflores, Lima; tel. (1) 4447088; fax (1) 4447117; e-mail extra@expreso.com.pe; f. 1964; evening edition of *Expreso*; Dir Carlos Sánchez; circ. 80,000.

**Gestión:** Avda Salaverry 156, Miraflores, Lima 18; tel. (1) 4776919; fax (1) 4476569; e-mail gestion@gestion.com.pe; internet www.gestion.com.pe; f. 1990; Gen. Editor Julio Lira; Gen. Man. Oscar Romero Caro; circ. 131,200.

**Ojo:** Jirón Jorge Salazar Araoz 171, Urb. Santa Catalina, Apdo 152, Lima; tel. (1) 4709696; fax (1) 4761605; internet www.ojo.com.pe; f. 1968; morning; Editorial Dir Agustín Figueroa Benza; circ. 100,000.

**El Peruano** (Diario Oficial): Avda Alfonso Ugarte 873, Lima 1; tel. (1) 3150400; fax (1) 4245023; e-mail gbarraza@editoraperu.com.pe; internet www.elperuano.com.pe; f. 1825; morning; official State Gazette; Editorial Dir Gerardo Barraza Soto; circ. 27,000.

**La República:** Jirón Camaná 320, Lima 1; tel. (1) 4276455; fax (1) 2511029; e-mail otxoa@larepublica.com.pe; internet www.larepublica.com.pe; f. 1982; left-wing; Dirs Gustavo Mohme Seminario, Gustavo Gorriti; circ. 50,000.

**Perú 21:** Jirón Miró Quesada 247, 6°, Lima; tel. (1) 311-6500; fax (1) 311-6391; e-mail director@peru21.com; internet www.peru21.com; independent; Editor Augusto Álvarez Rodrich.

#### Arequipa

**Arequipa al Día:** Avda Jorge Chávez 201, IV, Centenario, Arequipa; tel. (54) 223566; fax (54) 217810; f. 1991; Editorial Dir Carlos Meneses Cornejo.

**Correo de Arequipa:** Calle Bolívar 204, Arequipa; tel. (54) 235150; e-mail diariocorreo@epensa.com.pe; internet www.correoperu.com.pe; Dir Aldo Mariátegui; circ. 70,000.

**El Pueblo:** Sucre 213, Apdo 35, Arequipa; tel. (54) 211500; fax (54) 213361; f. 1905; morning; independent; Editorial Dir Eduardo Laime Valdivia; circ. 70,000.

#### Chiclayo

**La Industria:** Tacna 610, Chiclayo; tel. (74) 237952; fax (74) 227678; internet www.laindustria.com.pe; f. 1952; Dir Julio Alberto Ortiz Cerro; circ. 20,000.

#### Cusco

**El Diario del Cusco:** Centro Comercial Ollanta, Avda El Sol 346, Cusco; tel. (84) 229898; fax (84) 229822; e-mail buzon@diariodelcusco.com; internet www.diariodelcusco.com; morning; independent; Exec. Pres. Washinton Alosilla Portillo; Gen. Man. José Fernandez Núñez.

#### Huacho

**El Imparcial:** Avda Grau 203, Huacho; tel. (34) 2392187; fax (34) 2321352; e-mail elimparcial1891@hotmail.com; f. 1891; evening; Dir Adán Manrique Romero; circ. 5,000.

#### Huancayo

**Correo de Huancayo:** Jirón Cusco 337, Huancayo; tel. (64) 235792; fax (64) 233811; evening; Editorial Dir Rodolfo Orosco.

**La Opinión Popular:** Huancayo; tel. (64) 231149; f. 1922; Dir Miguel Bernabé Suárez Osorio.

### Ica

**La Opinión:** Avda Los Maestros 801, Apdo 186, Ica; tel. (56) 235571; f. 1922; evening; independent; Dir Gonzalo Tueros Ramírez.

**La Voz de Ica:** Castrovirreyna 193, Ica; tel. and fax (56) 232112; e-mail lavozdeica1918@infonegocio.net.pe; f. 1918; Dir Atilio Nieri Boggiano; Man. Mariella Nieri de Macedo; circ. 4,500.

### Pacasmayo

**Diario Últimas Noticias:** Ancash 691, San Pedro de Lloc, Pacasmayo; fax (44) 9651477; e-mail ultimasnoticias@pacasmayo.net; internet www.pacasmayo.net/ultimasnoticias; f. 1973; morning; independent; Editor María del Carmen Ballena Razuri; circ. 3,000.

### Piura

**Correo:** Zona Industrial Manzana 246, Lote 6, Piura; tel. (74) 321681; fax (74) 324881; Editorial Dir Rolando Rodrich Arango; circ. 12,000.

**El Tiempo:** Ayacucho 751, Piura; tel. (74) 325141; fax (74) 327478; e-mail direccion@eltiempo.com.pe; internet www.eltiempo.com.pe; f. 1916; morning; independent; Dir Luz María Helguero; circ. 18,000.

### Tacna

**Correo:** Jirón Hipólito Unanue 636, Tacna; tel. (54) 711671; fax (54) 713955; Editorial Dir Rubén Collazos Romero; circ. 8,000.

### Trujillo

**La Industria:** Gamarra 443, Trujillo; tel. (44) 234720; fax (44) 427761; e-mail industri@united.net.pe; internet www.unitru.edu.pe/eelitsa; f. 1895; morning; independent; Gen. Man. Isabel Cerro de Burga; circ. 8,000.

## PERIODICALS AND REVIEWS

### Lima

**Alerta Agrario:** Avda Salaverry 818, Lima 11; tel. (1) 4336610; fax (1) 4331744; f. 1987 by Centro Peruano de Estudios Sociales; monthly review of rural problems; Dir Bertha Consiglieri; circ. 100,000.

**The Andean Report:** Pasaje Los Pinos 156, Of. B6, Miraflores, Apdo 531, Lima; tel. (1) 4472552; fax (1) 4467888; e-mail egriffis@peruviantimes.com; internet www.perutimes.com; f. 1975; weekly newsletter; economics, trade and commerce; English; Publisher Eleanor Griffis de Zúñiga; circ. 1,000.

**Caretas:** Jirón Huallaya 122, Portal de Botoneros, Plaza de Armas, Lima 1; Apdo 737, Lima 100; tel. (1) 4289490; fax (1) 4262524; e-mail info@caretas.com.pe; internet www.caretas.com.pe; weekly; current affairs; Editor Enrique Zileri Gibson; circ. 90,000.

**Cosas:** Calle Recaveren 111, Miraflores, Lima 18; tel. (1) 2411178; fax (1) 4473776; internet www.cosasperu.com; weekly; society; Editor Elizabeth Dulanto.

**Debate:** Apdo 671, Lima 100; tel. (1) 2425656; fax (1) 4455946; f. 1980; every 2 months; Editor Gonzalo Zegarra-Huilandvich.

**Debate Agrario:** Avda Salaverry 818, Lima 11; tel. (1) 4336610; fax (1) 4331744; e-mail cepes@cepes.org.pe; f. 1987 by Centro Peruano de Estudios Sociales; every 4 months; rural issues; Dir Fernando Eguren L.

**Gente:** Eduardo de Habich 170, Miraflores, Lima 18; tel. (1) 4465046; fax (1) 4461173; e-mail correo@genteperu.com; internet www.genteperu.com; f. 1958; weekly; circ. 25,000.

**Hora del Hombre:** Lima; tel. (1) 4220208; f. 1943; monthly; cultural and political journal; illustrated; Dir Jorge Falcón.

**Industria Peruana:** Los Laureles 365, San Isidro, Apdo 632, Lima 27; f. 1896; monthly publication of the Sociedad de Industrias; Editor Rolando Celi Burneo.

**Lima Times:** Pasaje Los Pinos 156, Of. B6, Miraflores, Apdo 531, Lima 100; tel. (1) 4469120; fax (1) 4467888; e-mail perutimes@amauta.rcp.net.pe; internet www.perutimes.com; f. 1975; monthly; travel, cultural events, general news on Peru; English; Editor Eleanor Griffis de Zúñiga; circ. 10,000.

**Mercado Internacional:** Lima; tel. (1) 4445395; business.

**Monos y Monadas:** Lima; tel. (1) 4773483; f. 1981; fortnightly; satirical; Editor Nicolás Yerovi; circ. 17,000.

**Oiga:** Pedro Venturo 353, Urb. Aurora, Miraflores, Lima; tel. (1) 4475851; weekly; right-wing; Dir Francisco Igartua; circ. 60,000.

**Ollanta:** e-mail ollantaprensa@yahoo.com; internet ollantaprensa.tripod.com.pe; fortnightly; published by the Movimiento Nacionalista Peruano (MNP—'Movimiento Etnocacerista'); Dir Maj. (retd) Antauro Igor Humala Tasso (arrested January 2005 following an armed uprising in Andahuaylas).

**Onda:** Jorge Vanderghen 299, Miraflores, Lima; tel. (1) 4227008; f. 1959; monthly cultural review; Dir José Alejandro Valencia-Arenas; circ. 5,000.

**Orbita:** Parque Rochdale 129, Lima; tel. (1) 4610676; weekly; f. 1970; Dir Luz Chávez Mendoza; circ. 10,000.

**Perú Económico:** Apdo 671, Lima 100; tel. (1) 2425656; fax (1) 4455946; f. 1978; monthly; Editor Gonzalo Zegarra-Huilandvich.

**QueHacer:** León de la Fuente 110, Lima 17; tel. (1) 6138300; fax (1) 6138308; e-mail qh@desco.org.pe; internet www.desco.org.pe/qh/qh-in.htm; f. 1979; 6 a year; supported by Desco research and development agency; Editor Martín Paredes; Dir Mónica Pradel; circ. 5,000.

**Runa:** Lima; f. 1977; monthly; review of the Instituto Nacional de Cultura; Dir Mario Razzeto; circ. 10,000.

**Semana Económica:** Apdo 671, Lima 100; tel. (1) 2425656; fax (1) 4455946; f. 1985; weekly; Editor Gonzalo Zegarra-Huilandvich.

**Unidad:** Jirón Lampa 271, Of. 703, Lima; tel. (1) 4270355; weekly; Communist; Dir Gustavo Esteves Ostolaza; circ. 20,000.

**Vecino:** Avda Petit Thouars 1944, Of. 15, Lima 14; tel. (1) 4706787; f. 1981; fortnightly; supported by Yunta research and urban publishing institute; Dirs Patricia Córdova, Mario Zolezzi; circ. 5,000.

## NEWS AGENCIES

### Government News Agency

**Agencia de Noticias Peruana** (Andina): Jirón Quilca 556, Lima; tel. (1) 3306341; fax (1) 4312849; e-mail webmaster_anidina@editoraperu.com.pe; internet www.andina.com.pe; f. 1981; state-owned news agency; Dir Gerardo Barraza Soto.

### Foreign Bureaux

**Agence France-Presse (AFP):** F. Masías 544, San Isidro, Lima; tel. (1) 4214012; fax (1) 4424390; e-mail yllorca@amautarcp.net.pe; Bureau Chief Yves-Claude Llorca.

**Agencia EFE** (Spain): Manuel González Olaechea 207, San Isidro, Lima; tel. (1) 4412094; fax (1) 4412422; Bureau Chief Francisco Rubio Figueroa.

**Agenzia Nazionale Stampa Associata (ANSA)** (Italy): Avda Gen. Córdoba 2594, Lince, Lima 14; tel. (1) 4225130; fax (1) 4229087; Correspondent Alberto Ku-King Maturana.

**Deutsche Presse-agentur (dpa)** (Germany): Schell 343, Of. 707, Miraflores, Apdo 1362, Lima 18; tel. (1) 4441437; fax (1) 4443775; Bureau Chief Gonzalo Ruiz Tovar.

**Inter Press Service (IPS)** (Italy): Daniel Olaechea y Olaechea 285, Lima 11; tel. and fax (1) 4631021; Correspondent Abraham Lama.

**Prensa Latina** (Cuba): Edif. Astoria, Of. 303, Avda Tacna 482, Apdo 5567, Lima; tel. (1) 4233908; Correspondent Luis Manuel Arce Isaac.

**Reuters** (United Kingdom): Avda Paseo de la República 3505, 4°, San Isidro, Lima 27; tel. (1) 2212111; fax (1) 4418992; Man. Eduardo Hilgert.

**Xinhua (New China) News Agency** (People's Republic of China): Parque Javier Prado 181, San Isidro, Lima; tel. (1) 4403463; Bureau Chief Wang Shubo.

## PRESS ASSOCIATIONS

**Asociación Nacional de Periodistas del Perú:** Jirón Huancavélica 320, Apdo 2079, Lima 1; tel. (1) 4270687; fax (1) 4278493; e-mail anp@amauta.rcp.net.pe; internet ekeko2.rcp.net.pe/anp/; f. 1928; 8,800 mems; Pres. Roberto Marcos Mejía Alarcón.

**Federación de Periodistas del Perú (FPP):** Avda Abancay 173, Lima; tel. (1) 4284373; f. 1950; Pres. Pablo Truel Uribe.

# Publishers

**Asociación Editorial Bruño:** Avda Arica 751, Breña, Lima 5; tel. (1) 4244134; fax (1) 4251248; f. 1950; educational; Man. Federico Díaz Pinedo.

**Biblioteca Nacional del Perú:** Avda Abancay 4a cuadra, Apdo 2335, Lima 1; tel. (1) 4287690; fax (1) 4277331; e-mail dn@binape.gob.pe; internet www.binape.gob.pe; f. 1821; general non-fiction, directories; National Dir Sinesio López Jiménez.

# PERU

**Colección Artes y Tesoros del Perú:** Calle Centenario 156, Urb. Las Laderas de Melgarejo, La Molina, Lima 12; tel. (1) 3493128; fax (1) 3490579; e-mail acarulla@bcp.com.pe; f. 1971; Dir ALVARO CARULLA.

**Ediciones Médicas Peruanas, SA:** Lima; f. 1965; medical; Man. ALBERTO LOZANO REYES.

**Editora Normas Legales SA:** La Santa María 173, San Isidro, Lima; tel. (1) 2212598; fax (1) 2212598; e-mail enormaslegales@terra.com.pe; law textbooks; Man. JAVIER SANTA MARÍA SILVE.

**Editorial Book City:** Calle José R. Pizarro 1260 (espalda cuadra 11 de La Mar), Pueblo Libre, Lima; tel. (1) 2613266; general interest, juvenile, reference and literature; Man. MIRTHA YI YANG.

**Editorial Colegio Militar Leoncio Prado:** Avda Costanera 1541, La Perla, Callao; f. 1946; textbooks and official publications; Man. OSCAR MORALES QUINA.

**Editorial Cuzco SA:** Calle 5 Marzo, Jirón Lote 3, Urb. Las Magnolias, Surco, Lima; tel. (1) 4453261; e-mail ccuzco@camaralima.org.pe; law; Man. SERGIO BAZÁN CHACÓN.

**Editorial D.E.S.A.:** General Varela 1577, Breña, Lima; f. 1955; textbooks and official publications; Man. ENRIQUE MIRANDA.

**Editorial Desarrollo, SA:** Ica 242, 1°, Apdo 3824, Lima; tel. and fax (1) 4286628; f. 1965; business administration, accounting, auditing, industrial engineering, English textbooks, dictionaries, and technical reference; Dir LUIS SOSA NÚÑEZ.

**Editorial Horizonte:** Avda Nicolás de Piérola 995, Lima 1; tel. (1) 4279364; fax (1) 4274341; e-mail damonte@terra.com.pe; f. 1968; social sciences, literature, politics; Man. HUMBERTO DAMONTE.

**Editorial Labrusa, SA:** Los Frutales Avda 670-Ate, Lima; tel. (1) 4358443; fax (1) 4372925; f. 1988; literature, educational, cultural; Gen. Man. ADRIÁN REUILLA CALVO; Man. FEDERICO DÍAZ TINEO.

**Editorial Milla Batres, SA:** Lima; f. 1963; history, literature, art, archaeology, linguistics and encyclopaedias on Peru; Dir-Gen. CARLOS MILLA BATRES.

**Editorial Navarrete SRL-Industria del Offset:** Manuel Tellería 1842, Apdo 4173, Lima; tel. (1) 4319040; fax (1) 4230991; Man. LUIS NAVARRETE LECHUGA.

**Editorial Océano Peruana SA:** Avda Salaverry 2890, San Isidro, Lima; tel. (1) 2613999; fax (1) 4618628; e-mail ocelibros@oceano.com.pe; general interest and reference.

**Editorial Peisa:** Avda 2 de Mayo 1285, San Isidro, Lima; tel. (1) 4410473 ; fax (1) 2215988; e-mail burtech@yahoo.com; fiction and scholarly; Man. BENJAMIN URTECHO.

**Editorial Salesiana:** Avda Brasil 218, Apdo 0071, Lima 5; tel. (1) 4235225; f. 1918; religious and general textbooks; Man. Dir Dr FRANCESCO VACARELLO.

**Editorial Santillana:** Avda San Felipe 731, Jesús María, Lima; tel. (1) 4610277; fax (1) 2181014; e-mail santillana@santillana.com.pe; internet www.santillana.com; literature, scholarly and reference; Man. ANA CECILIA HALLO.

**Editorial Universo, SA:** Lima; f. 1967; literature, technical, educational; Pres. CLEMENTE AQUINO; Gen. Man. Ing. JOSÉ A. AQUINO BENAVIDES.

**Fundación del Banco Continental para el Fomento de la Educación y la Cultura (EDUBANCO):** Avda República de Panamá 3055, San Isidro, Apdo 4687, Lima 27; tel. (1) 2111000; fax (1) 2112479; f. 1973; Pres. PEDRO BRESCIA CAFFERATA; Man. FERNANDO PORTOCARRERO.

**Industrial Gráfica, SA:** Jirón Chavín 45, Breña, Lima 5; fax (1) 4324413; f. 1981; Pres. JAIME CAMPODONICO V.

**INIDE:** Van de Velde 160, Urb. San Borja, Lima; f. 1981; owned by National Research and Development Institute; educational books; Editor-in-Chief ANA AYALA.

**Librerías ABC, SA:** Avda Paseo de la República 3440, Local B-32, Lima 27; tel. (1) 4422900; fax (1) 4422901; f. 1956; history, Peruvian art and archaeology; Man. Dir HERBERT H. MOLL.

**Librería San Pablo:** Jirón Callao 198, Lima 1; tel. (1) 3795336; fax (1) 4593842; e-mail admlima@paulinas.org.pe; internet www.paulinas.org.pe; f. 1981; religious and scholastic texts; Man. Sister MARÍA GRACIA CAPALBO.

**Librería Studium, SA:** Lima; tel. (1) 4326278; fax (1) 4325354; f. 1936; textbooks and general culture; Man. Dir EDUARDO RIZO PATRÓN RECAVARREN.

**Pablo Villanueva Ediciones:** Lima; f. 1938; literature, history, law, etc.; Man. AUGUSTO VILLANUEVA PACHECO.

**Pontificia Universidad Católica del Perú:** Fondo Editorial, Plaza Francia 1164, Lima; tel. (1) 330710; fax (1) 3307405; e-mail feditor@pucp.edu.pe; internet www.pucp.edu.pe; Dir of Admin. AUGUSTO EGUIGUREN PRAELI.

**Sociedad Bíblica Peruana, AC:** Avda Petit Thouars 991, Apdo 14-0295, Lima 100; tel. (1) 4335815; fax (1) 4336389; internet www.members.tripod.com/sbpac; f. 1821; Christian literature and bibles; Gen. Sec. PEDRO ARANA-QUIROZ.

**Universidad Nacional Mayor de San Marcos:** Of. General de Editorial, Avda República de Chile 295, 5°, Of. 508, Lima; tel. (1) 4319689; f. 1850; textbooks, education; Man. Dir JORGE CAMPOS REY DE CASTRO.

## PUBLISHING ASSOCIATION

**Cámara Peruana del Libro:** Avda Cuba 427, esq. Jesús María, Apdo 10253, Lima 11; tel. (1) 4729516; fax (1) 2650735; e-mail cp-libro@amauta.rep.net.pe; internet www.pl.org.pe; f. 1946; 102 mems; Pres. CARLOS A. BENVIDES AGUIJE; Exec. Dir LOYDA MORÁN BUSTAMANTE.

# Broadcasting and Communications

## TELECOMMUNICATIONS

### Regulatory Authorities

**Dirección de Administración de Frecuencias:** Ministerio de Transportes y Comunicaciones, Avda 28 de Julio 800, Lima 1; tel. (1) 4331990; e-mail dgcdir@mtc.gob.pe; manages and allocates radio frequencies; Dir JOSÉ VILLA GAMBOA.

**Dirección General de Telecomunicaciones:** Ministerio de Transportes y Comunicaciones, Avda 28 de Julio 800, Lima 1; tel. (1) 4330752; fax (1) 4331450; e-mail dgtdir@mtc.gob.pe; Dir-Gen. MIGUEL OSAKI SUEMITSU.

**Instituto Nacional de Investigación y Capacitación de Telecomunicaciones (INICTEL):** Avda San Luis 17, esq. Bailetti, San Borja, Lima 41; tel. (1) 3360993; fax (1) 3369281; e-mail postmaster@inictel.gob.pe; Pres. MANUEL ADRIANZEN.

**Organismo Supervisor de Inversión Privada en Telecomunicaciones (OSIPTEL):** Calle de la Prosa 136, San Borja, Lima 41; tel. (1) 2251313; fax (1) 4751816; e-mail sid@osiptel.gob.pe; internet www.ospitel.gob.pe; body established by the Peruvian Telecommunications Act to oversee competition and tariffs, to monitor the quality of services and to settle disputes in the sector; Pres. ANA MARÍA YSHIKAWA NAKASHIMA.

### Major Service Providers

**BellSouth Perú:** Edif. Banco Continental, República de Panamá 3055, San Isidro, Lima 27; tel. (1) 19552000; internet www.bellsouth.com.pe; f. 1997; 97% owned by Telefónicas Móviles, SA (Spain); mobile telephone services; Pres. JUAN SACA; Exec. Vice-Pres. FABIO COELHO; 900,000 customers.

**Telefónica del Perú, SA:** Avda Arequipa 1155, Santa Beatriz, Lima 1; tel. (1) 2101013; fax (1) 4705950; e-mail mgarcia@tp.com.pe; Exec. Pres. MANUEL GARCÍA G.

**Telefónica MoviStar:** Juan de Arona 786, San Isidro, Lima; tel. (1) 19817000; internet www.telefonicamoviles.com.pe; f. 1994, 98% bought by Telefónicas Móviles, SA (Spain) in 2000; mobile telephone services; 1.8m. customers.

## BROADCASTING

In 1999 there were 1,425 radio stations and 105 television stations in Peru.

### Regulatory Authorities

**Asociación de Radio y Televisión del Perú (AR&TV):** Avda Roma 140, San Isidro, Lima 27; tel. (1) 4703734; Pres. HUMBERTO MALDONADO BALBÍN; Dir DANIEL LINARES BAZÁN.

**Coordinadora Nacional de Radio:** Santa Sabina 441, Urb. Santa Emma, Apdo 2179, Lima 100; tel. (1) 5640760; fax (1) 5640059; e-mail postmaster@cnr.org.pe; internet www.cnr.org.pe; f. 1978; Pres. RODOLFO AQUINO RUIZ.

**Instituto Nacional de Comunicación Social:** Jirón de la Unión 264, Lima; Dir HERNÁN VALDIZÁN.

**Unión de Radioemisoras de Provincias del Perú (UNRAP):** Mariano Carranza 754, Santa Beatriz, Lima 1.

### Radio

**Radio Agricultura del Perú, SA—La Peruanísima:** Casilla 625, Lima 11; tel. (1) 4246677; e-mail radioagriculturadelperu@yahoo.com; f. 1963; Gen. Man. LUZ ISABEL DEXTRE NÚÑEZ.

**Radio América:** Montero Rosas 1099, Santa Beatriz, Lima 1; tel. (1) 2653841; fax (1) 2653844; e-mail kcrous@americatv.com.pe; f. 1943; Dir-Gen. KAREN CROUSILLAT.

**Radio Cadena Nacional:** Los Angeles 129, Miraflores, Lima; tel. (1) 4220905; fax (1) 4221067; Pres. MIGUEL DÍEZ CANSECO; Gen. Man. CÉSAR LECCA ARRIETA.

**Cadena Peruana de Noticias:** Gral Salaverry 156, Miraflores, Lima; tel. (1) 4461554; fax (1) 4457770; e-mail webmastercpn@gestion.com.pe; internet www.cpnradio.com.pe; f. 1996; Pres. MANUEL ROMERO CARO; Gen. Man. OSCAR ROMERO CARO.

**Radio Cutivalú, La Voz del Desierto:** Jirón Ignacio de Loyola 300, Urb. Miraflores, Castilla, Piura; tel. (74) 342802; fax (74) 343370; e-mail cutivalu@cipcaorg.pe; f. 1986; Pres. FRANCISCO MUGUIRO IBARRA; Dir RODOLFO AQUINO RUIZ.

**Emisoras 'Cruz del Perú':** Victorino Laynes 1402, Urb. Elio, Lima 1; tel. (1) 4521028; Pres. FERNANDO CRUZ MENDOZA; Gen. Man. MARCO CRUZ MENDOZA M.

**Emisoras Nacionales:** León Velarde 1140, Lince, Lima 1; tel. (1) 4714948; fax (1) 4728182; Gen. Man. CÉSAR COLOMA R.

**Radio Inca del Perú:** Pastor Dávila 197, Lima; tel. (1) 2512596; fax (1) 2513324; e-mail corporacion@corporacionradial.com.pe; f. 1951; Gen. Man. ABRAHAM ZAVALA CHOCANO.

**Radio Nacional del Perú:** Avda Petit Thouars 447, Santa Beatriz, Lima; tel. (1) 4331712; fax (1) 4338952; Pres. LUIS ALBERTO MARAVÍ SÁENZ; Gen. Man. CARLOS PIZANO PANIAGUA.

**Radio Panamericana:** Paseo Parodi 340, San Isidro, Lima 27; tel. (1) 4226787; fax (1) 4221182; e-mail mad@radiopanamericana.com; internet www.radiopanamericana.com; f. 1953; Dir RAQUEL DELGADO DE ALCÁNTARA.

**Radio Programas del Perú** (GRUPORPP): Avda Paseo de la República 38667, San Isidro, Lima; tel. (1) 4338720; Pres. MANUEL DELGADO PARKER; Gen. Man. HUGO DELGADO NACHTIGALL.

**Radio Santa Rosa:** Jirón Camaná 170, Apdo 206, Lima; tel. (1) 4277488; fax (1) 4269219; e-mail santarosa@viaexpresa.com.pe; f. 1958; Dir P. JUAN SOKOLICH ALVARADO.

**Sonograbaciones Maldonado:** Mariano Carranza 754, Santa Beatriz, Lima; tel. (1) 4715163; fax (1) 4727491; Pres. HUMBERTO MALDONADO B.; Gen. Man. LUIS HUMBERTO MALDONADO.

### Television

**América Televisión, Canal 4:** Jirón Montero Rosas 1099, Santa Beatriz, Lima; tel. (1) 2657361; fax (1) 2656979; e-mail infoamerica@americatv.com.pe; internet www.americatv.com.pe; Gen. Man. MARISOL CROUSILLAT.

**ATV, Canal 9:** Avda Arequipa 3570, San Isidro, Lima 27; tel. (1) 2118800; fax (1) 4427636; e-mail andinatelevision@atv.com.pe; internet www.atv.com.pe; f. 1983; Gen. Man. MARCELLO CÚNEO LOBIANO.

**Frecuencia Latina, Canal 2:** Avda San Felipe 968, Jesús María, Lima; tel. (1) 4707272; fax (1) 4714187; internet www.frecuencialatina.com.pe; Pres. BARUCH IVCHER.

**Global Televisión, Canal 13:** Gen. Orbegoso 140, Breña, Lima; tel. (1) 3303040; fax (1) 4238202; f. 1989; Pres. GENARO DELGADO PARKER; Gen. Man. RAFAEL LEGUÍA.

**Nor Peruana de Radiodifusión, SA:** Avda Arequipa 3520, San Isidro, Lima 27; tel. (1) 403365; fax (1) 419844; f. 1991; Dir FRANCO PALERMO IBARGUENGOITIA; Gen. Man. FELIPE BERNINZÓN VALLARINO.

**Panamericana Televisión SA, Canal 5:** Avda Alejandro Tirado 217, Santa Beatriz, Lima; tel. (1) 4113201; fax (1) 4703001; e-mail fanchorena@pantel.com.pe; internet www.24horas.com.pe; Pres. RAFAEL RAVETTINO FLORES; Gen. Man. FREDERICO ANCHORENA VÁSQUEZ.

**Cía Peruana de Radiodifusión, Canal 4 TV:** Mariano Carranza y Montero Rosas 1099, Santa Beatriz, Lima; tel. (1) 4728985; fax (1) 4710099; f. 1958; Dir JOSÉ FRANCISCO CROUSILLAT CARREÑO.

**RBC Televisión, Canal 11:** Avda Manco Cápac 333, La Victoria, Lima; tel. (1) 4310169; fax (1) 4331237; Pres. FERNANDO GONZÁLEZ DEL CAMPO; Gen. Man. JUAN SÁENZ MARÓN.

**Radio Televisión Peruana, Canal 7:** Avda José Galvez 1040, Santa Beatriz, Lima 1; tel. (1) 4718000; fax (1) 4726799; f. 1957; Pres. LUIS ALBERTO MARAVÍ SÁENZ; Gen. Man. CARLOS ALBERTO PIZZANO P.

**Cía de Radiodifusión Arequipa SA, Canal 9:** Centro Comercial Cayma, R2, Arequipa; tel. (54) 252525; fax (54) 254959; e-mail crasa@ibm.net; f. 1986; Dir ENRIQUE MENDOZA NÚÑEZ; Gen. Man. ENRIQUE MENDOZA DEL SOLAR.

**Uranio, Canal 15:** Avda Arequipa 3570, 6°, San Isidro, Lima; e-mail agamarra@atv.com.pe; Gen. Man. ADELA GAMARRA VÁSQUEZ.

# Finance

In April 1991 a new banking law was introduced, which relaxed state control of the financial sector and reopened the sector to foreign banks (which had been excluded from the sector by a nationalization law promulgated in 1987).

### BANKING

(cap. = capital; res = reserves; dep. = deposits; m. = million; amounts in new soles)

**Superintendencia de Banca y Seguros:** Los Laureles 214, San Isidro, Lima 27; tel. (1) 2218990; fax (1) 4417760; e-mail mostos@sbs.gob.pe; internet www.sbs.gob.pe; f. 1931; Supt JUAN JOSÉ MARTHANS LEÓN; Sec.-Gen. NORMA SOLARI PRECIADO.

#### Central Bank

**Banco Central de Reserva del Perú:** Jirón Antonio Miró Quesada 441-445, Lima 1; tel. (1) 6132000; fax (1) 4275880; e-mail webmaster@bcrp.gob.pe; internet www.bcrp.gob.pe; f. 1922; refounded 1931; cap. 172.0m., res 171.8m., dep. 25,814.9m. (Dec. 2003); Pres. JAVIER SILVA RUETE; Gen. Man. RENZO ROSSINI; 7 brs.

#### Other Government Banks

**Banco de la Nación:** Avda Canaval y Moreyra 150, San Isidro, Lima 1; tel. (1) 4405858; fax (1) 4223451; e-mail imagen@bn.com.pe; internet www.bn.com.pe; f. 1966; cap. 674.1m., res 239.3m., dep. 6,673.9m. (Dec. 2003); conducts all commercial banking operations of official government agencies; Pres. KURT BURNEO FARFÁN; Gen. Man. PEDRO ERNESTO MENÉNDEZ RICHTER; 368 brs.

**Corporación Financiera de Desarrollo (COFIDE):** Augusto Tamayo 160, San Isidro, Lima 27; tel. (1) 4422550; fax (1) 4423374; e-mail postmaster@cofide.com.pe; internet www.cofide.com.pe; f. 1971; also owners of Banco Latino; Pres. AURELIO LORET DE MOLA BÖHME; Gen. Man. MARCO CASTILLO TORRES; 11 brs.

#### Commercial Banks

**Banco de Comercio:** Avda Paseo de la República 3705, San Isidro, Lima; tel. (1) 4229800; fax (1) 4405458; e-mail postmaster@bancomercio.com.pe; internet www.bancomercio.com.pe; f. 1967; fmrly Banco Peruano de Comercio y Construcción; cap. 68.8m., res 0.2m., dep. 445.8m. (Dec. 2003); Chair. OSCAR BALLÓN JARGAS; Gen. Man. CARLOS MUJICA CASTRO; 14 brs.

**Banco de Crédito del Perú:** Calle Centenario 156, Urb. Las Laderas de Melgarejo, Apdo 12-067, Lima 12; tel. (1) 3132000; internet www.viabcp.com; f. 1889; cap. 1,226.4m., res 698.4m., dep. 18,850.7m., total assets 22,941.2m. (Dec. 2003); Pres. and Chair. DIONISIO ROMERO SEMINARIO; 217 brs.

**Banco Interamericano de Finanzas, SA:** Avda Rivera Navarrete 600, San Isidro, Lima 27; tel. (1) 2113000; fax (1) 2212489; internet www.bif.com.pe; f. 1991; cap. 105.9m., res 6.3m., dep. 1,555.3m. (Dec. 2003); Pres. FRANCISCO ROCHE; Gen. Man. and CEO RAÚL BALTAR.

**Banco Sudamericano:** Avda Camino Real 815, San Isidro, Lima 27; tel. (1) 6161111; fax (1) 6161112; e-mail e-servicios@bansud.com.pe; internet www.sudamericano.com; f. 1981; cap. 147.9m., res 9.4m., dep. 2,038.7m. (Dec. 2003); Pres. ROBERTO CALDA CAVANNA; Gen. Man. RAFAEL VENEGAS VIDAURRE; 9 brs.

**Banco del Trabajo:** Avda Paseo de la República 3587, 4°, San Isidro, Lima; tel. (1) 4219000; fax (1) 4212521; e-mail informes@bantra.com.pe; internet www.bantra.com.pe; Chair. CARLOS ENRIQUE CARRILLO QUIÑONES; Gen. Man. MAX JULIO CHION LI; 46 brs.

**Banco Wiese Sudameris:** Avda Dionisio Derteano 102, San Isidro, Apdo 1235, Lima; tel. (1) 2116000; fax (1) 2116886; e-mail wiesenet@wiese.com.pe; internet www.bws.com.pe; f. 1943; cap. 186.0m., dep. 7,434.4m., total assets 10,531.6m. (Dec. 2003); taken over by Govt in Oct. 1987; returned to private ownership in Oct. 1988 as Banco Wiese Ltdo; 71.11% owned by Banca Intesa SpA (Italy); Chair. RAÚL BARRIOS; Gen. Man. CARLOS GONZÁLEZ TABOADA.

**BBVA Banco Continental:** Avda República de Panamá 3055, San Isidro, Lima 27; tel. (1) 2111000; fax (1) 2111788; internet www.bbvabancocontinental.com; f. 1951; cap. 813.2m., res 339.1m., dep. 11,288.0m. (Dec. 2003); merged with BBVA of Spain in 1995; 92.01% owned by Holding Continental, SA; Pres. and Chair. PEDRO BESCIA CAFFERATA; Gen. Man. JOSÉ ANTONIO COLOMER GUIU; 190 brs.

**INTERBANK** (Banco Internacional del Perú): Carlos Villarán 140, Urb. Santa Catalina, Lima 13; tel. (1) 2192000; fax (1) 2192336; e-mail krubin@intercorp.com.pe; internet www.interbank.com.pe; f. 1897; commercial bank; cap. 286.1m., res 84.1m., dep. 4,570.5m. (Dec. 2003); Chair. and Pres. CARLOS RODRÍGUEZ-PASTOR; Gen. Man. ISMAEL BENAVIDES FERREYROS; 90 brs.

PERU                                                                                                                                   *Directory*

#### Foreign Banks

**Banco do Brasil, SA:** Avda Camino Real 348, 9°, Torre El Pilar, San Isidro 27, Lima; tel. (1) 2124230; fax (1) 4424208; e-mail bblima@infonegocio.net.pe; closed to public.

**Citibank NA** (USA): Torre Real, 5°, Avda Camino Real 456, Lima 27; tel. (1) 4214000; fax (1) 4409044; internet www.citibank.com/peru/homepage/index-e-htm; f. 1920; cap. US $47,520m., res $1,936m., dep. $139,627m. (Nov. 1997); Vice-Pres. GUSTAVO MARÍN; 1 br.

#### Banking Association

**Asociación de Bancos del Perú:** Calle 41, No 975, Urb. Corpac, San Isidro, Lima 27; tel. (1) 2241718; fax (1) 2241707; e-mail earroyo@asbanc.com.pe; f. 1929; refounded 1967; Pres. JOSÉ NICOLINI LORENZONI; Gen. Man. JUAN KLINGENBERGER LOMELLINI.

### STOCK EXCHANGE

**Bolsa de Valores de Lima:** Pasaje Acuña 106, Lima 100; tel. (1) 4260714; fax (1) 4267650; internet www.bvl.com.pe; f. 1860; Exec. Pres. RAFAEL D'ANGELO SERRA.

### REGULATORY AUTHORITY

**Comisión Nacional Supervisora de Empresas y Valores (CONASEV):** Santa Cruz 315, Miraflores, Lima; tel. (1) 4416620; fax (1) 4428401; internet www.conasev.gob.pe; f. 1968; regulates the securities market; responsible to Ministry of Economy and Finance; Pres. FABIOLA BARRIGA SAN MIGUEL.

### INSURANCE

#### Lima

**Altas Cumbres Cía de Seguros de Vida, SA:** Avda Paseo de la República 3587, San Isidro, Lima; tel. (1) 4428228; fax (1) 2213313; e-mail asalazar@altascumbres.com.pe; internet www.altascumbres.com.pe; f. 1999; life; Pres. CARLOS CARRILLO QUIÑONES; Gen. Man. ALFREDO SALAZAR DELGADO.

**Generali Perú, Cía de Seguros y Reaseguros:** Jirón Antonio Miró Quesada 191, Apdo 1751, Lima 100; tel. (1) 3111000; fax (1) 3111004; e-mail borlandini@generali-peru-com.pe; internet www.generali-peru.com.pe; f. 1896; Pres. RAFFAELE TIANO SAMBO; Gen. Man. BRUNO ORLANDINI ALVAREZ-CALDERÓN.

**Interseguro Compañía de Seguros de Vida, SA:** Avda Pardo y Aliaga 640, 4°, San Isidro, Lima; tel. (1) 2223233; fax (1) 2223222; e-mail juan.vallejo@intercorp.com.pe; f. 1998; life; Pres. FELIPE MORRIS GUERINONI; Gen. Man. JUAN CARLOS VALLEJO BLANCO.

**Invita Seguros de Vida, SA:** Torre Wiese, Canaval y Moreyra 532, San Isidro, Lima; tel. (1) 2222222; fax (1) 2211683; e-mail dcosta@invita.com.pe; internet www.invita.com.pe; f. 2000; life; fmrly Wiese Aetna, SA; Pres. CARIDAD DE LA PUENTE WIESE; Gen. Man. DULIO COSTA OLIVERA.

**Mapfre Perú Cía de Seguros:** Avda 28 de Julio 873, Miraflores, Apdo 323, Lima 100; tel. (1) 4444515; fax (1) 4469599; e-mail fmarco@mapfreperu.com; internet www.mapfreperu.com; f. 1994; general; fmrly Seguros El Sol, SA; Pres. Dr FRANCISCO JOSÉ MARCO ORENES.

**Pacífico, Cía de Seguros y Reaseguros:** Avda Arequipa 660, Lima 100; tel. (1) 4333626; fax (1) 4333388; e-mail arodrigo@pps.com.pe; f. 1943; general; Pres. CALIXTO ROMERO SEMINARIO; Gen. Man. ARTURO RODRIGO SANTISTEVAN.

**La Positiva Cía de Seguros y Reaseguros, SA:** Esq. Javier Prado Este y Francisco Masías 370, San Isidro, Lima; tel. (1) 2110000; fax (1) 2110020; e-mail jaimep@lapositiva.com.pe; internet www.lapositiva.com.pe; f. 1947; Pres. Ing. JUAN MANUEL PEÑA ROCA; Gen. Man. JAIME PÉREZ RODRÍGUEZ.

**Rimac Internacional, Cía de Seguros:** Las Begonias 475, 3°, San Isidro, Lima; tel. (1) 4218383; fax (1) 4210570; e-mail jortecho@rimac.com.pe; internet www.rimac.com.pe; f. 1896; acquired Seguros Fénix in 2004; Pres. Ing. PEDRO BRESCIA CAFFERATA; Gen. Man. PEDRO FLECHA ZALBA.

**SECREX, Cía de Seguro de Crédito y Garantías:** Avda Angamos Oeste 1234, Miraflores, Apdo 0511, Lima 18; tel. (1) 4424033; fax (1) 4423890; e-mail ciaseg@secrex.com.pe; internet www.secrex.com.pe; f. 1980; Pres. Dr RAÚL FERRERO COSTA; Gen. Man. JUAN A. GIANNONI MURGA.

**Sul América Compañía de Seguros, SA:** Jirón Sinchi Roca 2728, Lince, Lima; tel. (1) 2150515; fax (1) 4418730; e-mail lavila@sulamerica.com.pe; internet www.sulamerica.com.pe; f. 1954; part of Sul América, SA (Brazil); Pres. RAÚL BARRIOS ORBEGOSO; Gen. Man. LUIS MIGUEL AVILA MERINO.

#### Insurance Association

**Asociación Peruana de Empresas de Seguros (APESEG):** Arias Araguez 146, Miraflores, Lima 100; tel. (1) 4442294; fax (1) 4468538; e-mail rda@apeseg.org.pe; internet www.apeseg.org.pe; f. 1904; Pres. RAÚL BARRIOS ORBEGOSO; Gen. Man. RAÚL DE ANDREA DE LAS CARRERAS.

## Trade and Industry

### GOVERNMENT AGENCIES

**Agencia de Promoción de la Inversión Privada (ProInversión):** Avda Paseo de la República 3361, 9°, San Isidro, Lima 27; tel. (1) 6121200; fax (1) 2212942; e-mail gvillegas@proinversion.gob.pe; internet www.proinversion.gob.pe; f. 2002 to promote economic investment; Dir RENÉ CORNEJO DÍAZ; Gen. Sec. GUSTAVO VILLEGAS DEL SOLAR.

**Empresa Nacional de la Coca, SA (ENACO):** Avda Arequipa 4528, Miraflores, Lima; tel. (1) 442-3746; fax (1) 447-1667; e-mail hramos@enaco.com.pe; internet www.enaco.com.pe; f. 1949; agency with exclusive responsibility for the purchase and resale of legally produced coca and the promotion of its derivatives; Pres. HÉCTOR RAMOS LÓPEZ; Gen. Man. ERVER RAFAEL CÓRDOVA PALIZA.

**Fondo Nacional de Compensación y Desarrollo Social (FONCODES):** Avda Paseo de la República 3101, San Isidro, Lima; tel. (1) 4212102; fax (1) 4218026; e-mail consultas@foncodes.gob.pe; internet www.foncodes.gob.pe; f. 1991; responsible for social development and eradicating poverty; Exec. Dir Dr ALEJANDRO NARVÁEZ LICERAS.

**Instituto Nacional de Recursos Forestales (INRENA):** Calle Diecisiete 355, Urb. El Palomar, Lima; tel. (1) 225-2113; fax (1) 224-3218; e-mail jefatura@inrena.gob.pe; internet www.inrena.gob.pe; f. 1992; promotes sustainable development of Amazon rain forest; Pres. LEONCIO ALVAREZ VÁSQUEZ; Gen. Man. FÉLIX AMADEO RIVERA LECAROS.

**Perupetro:** Luis Aldana 320, San Borja, Lima; tel. (1) 4759590; fax (1) 4757722; e-mail admweb@perupetro.com.pe; internet www.perupetro.com.pe; f. 1993; responsible for promoting investment in hydrocarbon exploration and exploitation; Chair. ANTONIO CUETO DUTHURBURU; CEO JOSÉ CHAVEZ CÁCERES .

### DEVELOPMENT ORGANIZATIONS

**Acción Comunitaria del Perú:** Avda Domingo Orue 165, Surquillo, Lima 34; tel. (1) 2220202; fax (1) 2224166; e-mail accion@accion.org.pe; internet www.accion.org.pe; f. 1969; promotes economic, social and cultural devt through improvements in service provision; Dir CARLOS CULQUICHICÓN.

**Asociación de Exportadores (ADEX):** Javier Prado Este 2875, San Borja, Lima 41; Apdo 1806, Lima 1; tel. (1) 3462530; fax (1) 3461879; e-mail postmaster@adexperu.org.pe; internet www.adexperu.org.pe; f. 1973; exporters' asscn; Pres. LUIS VEGA MONTEFERRI; Gen. Man. ALVARO BARRENECHEA; 600 mems.

**Asociación Kallpa para la Promoción Integral de la Salud y el Desarrollo:** Jirón Rospigliosi 105, Barranco, Lima 4; tel. (1) 4455521; fax (1) 2429693; e-mail postmast@kallpa.org.pe; internet www.kallpa.org.pe; health devt for youths; Pres. ARIELA LUNA FLORES.

**Asociación Nacional de Centros de Investigación, Promoción Social y Desarrollo:** Pablo Bermúdez 234, Jesús María, Lima; tel. (1) 4411063; fax (1) 4411227; e-mail postmaster@anc.org.pe; internet www.anc.org.pe; umbrella grouping of devt orgs; Pres. LUIS SIRUMBAL; Exec. Dir FEDERICO ARNILLAS L.

**Asociación para la Naturaleza y Desarrollo Sostenible (ANDES):** Calle Ruinas 451, Cusco; tel. (8) 4245021; e-mail andes@andes.org.pe; internet www.andes.org.pe; devt org. promoting the culture, education and environment of indigenous groups.

**Sociedad Nacional de Industrias (SNI)** (National Industrial Association): Los Laureles 365, San Isidro, Apdo 632, Lima 27; tel. (1) 4218830; fax (1) 4422573; e-mail sni@sni.org.pe; internet www.sni.org.pe; f. 1896; comprises permanent commissions covering various aspects of industry including labour, integration, fairs and exhibitions, industrial promotion; its Small Industry Committee groups over 2,000 small enterprises; Pres. ROBERTO NESTA BRERO; Gen. Man. SERGIO MAZURÉ; 90 dirs (reps of firms); 2,500 mems; 60 sectorial committees.

**Centro de Desarrollo Industrial (CDI):** c/o SNI; tel. (1) 4218881; fax (1) 4213132; e-mail cdi@sni.org.pe; internet www.cdi.org.pe; f. 1986; supports industrial development and programmes to develop industrial companies; Exec. Dir LUIS TENORIO PUENTES.

PERU *Directory*

## CHAMBERS OF COMMERCE

**Cámara de Comercio de Lima** (Lima Chamber of Commerce): Avda Gregorio Escobedo 398, Jesús María, Lima 11; tel. (1) 4633434; fax (1) 4632837; e-mail perured@camaralima.org.pe; internet www.camaralima.org.pe; f. 1888; Pres. GRACIELA FERNÁNDEZ-BACA DE VALDEZ; 3,500 mems.

**Cámara Nacional de Comercio, Producción y Servicios (PERUCAMARAS):** Avda Gregorio Escobedo 396, Jesús María, Lima 11; e-mail administracion@perucam.com; internet www.perucamaras.com; national asscn of chambers of commerce; Pres. SAMUEL GLEISER KATZ; Gen. Man. JOSÉ MARTÍN TELLO.

There are also Chambers of Commerce in Arequipa, Cusco, Callao and many other cities.

## EMPLOYERS' ORGANIZATIONS

**Asociación Automotriz del Perú:** Dos de Mayo 299, Apdo 1248, San Isidro, Lima 27; tel. (1) 4404119; fax (1) 4428865; e-mail aap@terra.com.pe; f. 1926; association of importers of motor cars and accessories; 360 mems; Pres. CARLOS BAMBARÉN GARCÍA-MALDONADO; Gen. Man. CÉSAR MARTÍN BARREDA.

**Asociación de Ganaderos del Perú** (Association of Stock Farmers of Peru): Pumacahua 877, 3°, Jesús María, Lima; f. 1915; Gen. Man. Ing. MIGUEL J. FORT.

**Consejo Nacional del Café:** Lima; representatives of Government and industrial coffee growers; Pres. ENRIQUE ALDAVE.

**Sociedad Nacional de Minería y Petróleo:** Francisco Graña 671, Magdalena del Mar, Lima 17; tel. (1) 4601600; fax (1) 4601616; e-mail postmaster@snmpe.org.pe; internet www.snmpe.org.pe; f. 1940; Pres. JOSÉ MIGUEL MORALES DASSO; Sec.-Gen. KLAUS HUYS JACOBI; association of companies involved in mining, petroleum and energy.

**Sociedad Nacional de Pesquería (SNP):** Javier Prado Oeste 2442, San Isidro, Lima 27; tel. (1) 2612970; fax (1) 2617912; e-mail snpnet@terra.com.pe; internet www.snp.org.pe; f. 1952; private-sector fishing interests; Pres. RAÚL ALBERTO SÁNCHEZ SOTOMAYOR.

## MAJOR COMPANIES

The following is a selection of the principal industrial companies operating in Peru.

### Metals, Mining and Petroleum

**Centromín, SA** (Empresa Minera del Centro del Perú): Edif. Solgas, Avda Javier Prado Este 2175, San Borja, Apdo 2142, Lima 41; tel. (1) 4759057; fax (1) 4769756; e-mail postmaster@centromin.com.pe; internet www.centromin.com.pe; f. 1902; fmr state-owned mining corpn, transferred to private ownership in 1997; Pres. RICARDO GIESECKE SARA LAFOSSE; Gen. Man. JUAN CARLOS BARCELLOS MILLA; 11,527 employees.

**Compañía de Minas Buenaventura, SA:** Avda Carlos Villarán 790, Santa Catalina, La Victoria, Lima 13; tel. (1) 4192500; fax (1) 4717349; e-mail ddominguez@buenaventura.com.pe; internet www.buenaventura.com; f. 1953; mining of silver ores; Pres. ALBERTO BENAVIDES DE LA QUINTANA; Gen. Man. ROQUE BENAVIDES GANOZA; 1,400 employees.

**Compañía Minera Antamina, SA:** La Floresta 497, 4°, Urb. Chacarilla del Estanque, San Borja, Lima 41; tel. (1) 2173000; fax (1) 2173095; e-mail comcorp@antamina.com; internet www.antamina.com; mine produces copper, lead, zinc and molybdenum; Pres. and CEO RICK PAULING; 1,433 employees world-wide.

**Compañía Minera Atacocha, SA:** Avda Javier Prado Oeste 980, San Isidro, Lima; tel. (1) 6123600; fax (1) 6123610; internet www.atacocha.com.pe; f. 1936; mining of lead and zinc; Pres. CARLOS GUILLÉN SANZ; Gen. Man. JUAN JOSÉ HERRERA TÁVARA; 1,270 employees.

**Compañía Minera del Madrigal, SA:** Morelli 181, 3°, Lima 41; tel. (1) 4414700; fax (1) 4751349; f. 1967; copper mining; Man. MIGUEL ACLEN; 760 employees.

**Compañía Minera Milpo, SA:** San Borja Norte 523, Lima 41; tel. (1) 7105500; e-mail comunicaciones@milpo.com; internet www.milpo.com; f. 1946; lead, silver and zinc mining; sales of US $62.5m. (2003); Pres. IVO UCOVICH DORSNER; Gen. Man. ABRAHAM CHAHUAN A.; 314 employees.

**Compañía Rex, SA:** Avda Alfreo Mendiola 1879, San Martín de Porres, Lima 31; tel. (1) 5342143; fax (1) 5342295; f. 1958; clay mining and production of clay and ceramic goods; Gen. Man. FRANCISCO ARANETA LAVINZ; 445 employees.

**Corporación Aceros Arequipa, SA:** Avda Enrique Meiggs 297, Parque Internacional de la Industria y Comercio, Callao, Lima; tel. (1) 5171800; fax (1) 5622436; internet www.acerosarequipa.com; f. 1966; iron and steel manufacturer; fmrly ACERSA; sales of 379m. new soles (2001); Exec. Pres. RICARDO CILLONIZ OBERTI; 792 employees.

**Corporación Minera Nor Perú, SA:** Avda República de Panamá 3055, 11°, San Isidro, Lima 27; tel. (1) 2222988; fax (1) 4410570; e-mail postmast@cmnpsa-01.com.pe; subsidiary of Pan American Silver Corpn; copper, gold, silver, lead and zinc mining; Chair. JAVIER NUÑEZ CARVALLO; Gen. Man. MARIO DEL RÍO A.; 1,070 employees.

**Empresa Minera Especial Tintaya, SA:** Avda San Martin 301, Urb. Vallecito, Cusco; tel. (54) 246000; fax (54) 246592; owned since 1994 by Magma Copper Co, USA; copper mining; Pres. MIKE ANGLIN; 1,456 employees.

**Empresa Siderúrgica del Perú:** Avda Tacna 543, Lima; fax (1) 4330807; f. 1971; processing of steel; Gen. Man. CÉSAR GARAY GHILARDI; 4,195 employees.

**Minera Arcata, SA:** Pasaje el Carmen 180, Urb. El Vivero de Monterrico, Santiago de Surco, Lima; tel. (1) 3172000; fax (1) 2212747; f. 1961; silver mining; owned by Garrison Corpn; Pres. and CEO JUAN INCHAUSTEGUI VARGAS; 463 employees.

**Occidental Petroleum Corpn of Peru:** Los Forestales 910, Urb. Los Ingenieros, La Molina, Lima 1; tel. (1) 3480600; fax (1) 3480500; internet www.oxy.com; f. 1970; principal shareholder Occidental Petroleum Corpn of the USA; petroleum and natural-gas exploration and extraction; Man. LUIS MOREYRA FERREYROS; 970 employees.

**PetroPerú (Empresa de Petróleos de Perú, SA):** Avda Paseo de la República 3361, San Isidro, Lima 27; tel. (1) 2117800; fax (1) 6145000; internet www.petroperu.com; f. 1948; state-owned petroleum refining co; transfer to private ownership commenced in 1996; Pres. JORGE KAWAMURA; Gen. Man. ANTONIO CUETO; 1,500 employees.

**Pluspetrol Exploración y Producción, SA:** Avda República de Panamá 3055,8°, San Isidro, Lima; tel. (1) 411-7100; fax (1) 411-7120; e-mail rrhh-cv-peru@pluspetrol.com.ar; internet www.pluspetrol.net; oil and gas exploration and production; Chair. and Pres. LUIS ALBERTO REY.

**Shougang Hierro Perú, SA:** Avda República de Chile 262, Jesús María, Lima 1; tel. (1) 3304600; fax (1) 3305136; e-mail lima@shp.com.pe; internet www.shougang.com.pe; f. 1993; owned by Shougang Corpn (People's Republic of China); mining, processing and shipment of iron ore; sales of 281m. new soles (1999); Chair. JIN YONG HUN; Gen. Man. WANG BAO JUN; 1,988 employees.

**Southern Peru Copper Corporation (SPCC):** Avda Caminos del Inka 171, Surco, Lima; tel. (511) 3721414; fax (511) 3720262; internet www.southernperu.com; f. 1952; copper mining; owned by Grupo México (54.1%), Cerro Trading Co, and Phelps Dodge; sales of 1,833m. new soles (1999); Chair. and CEO GERMÁN LARREA MOTA-VELASCO; 3,554 employees.

**Volcán Compañía Minera, SA:** Avda Gregorio Escobedo 710, Jesús María, Lima; tel. (1) 2194000; fax (1) 2619716; e-mail contact@volcan.com.pe; internet www.volcan.com.pe; f. 1943; lead, zinc and silver mining; owns 495 mining concessions; Chair. FRANCISCO MOREYRA GARCÍA SAYAN; 3,000 employees.

### Food and Drink

**Alicorp, SA:** Calle Chinchón 980, San Isidro, Lima; tel. (1) 4422552; fax (1) 4216642; e-mail psacchi@alicorp.com.pe; internet www.alicorp.com.pe; f. 1946; manufacturers of edible oils, lard and soaps; fmrly Compañía Oleaginosa del Perú, SA; sales of 1,537.3m. nuevos soles in 2001; part of Grupo Romero; Pres. DIONISIO ROMERO; Gen. Man. LESLIE PIERCE DIEZ CANSERO; 2,500 employees.

**Arturo Field y La Estrella Ltda, SA:** Avda Venezuela 2470, Lima 1; tel. (1) 4317510; fax (1) 4247184; f. 1864; production of confectionery; Pres. NICANOR ARTEAGA DOMÍNGUEZ; 560 employees.

**Backus y Johnston, SAA** (Unión de Cervecerías Peruanas): Avda Oscar R. Benavides 3866, Bellavista, Callao 2, Apdo 256; tel. (1) 4518040; fax (1) 4519118; e-mail cobackus@backus.com.pe; internet www.backus.com.pe; f. 1879; beverages and bottling corpn; also owns Cervecería San Juan, SAA (q.v.); Chair. ELIAS BENTÍN PERAL; 1,457 employees.

**Cervecería San Juan, SAA:** Avda Felipe Pardo y Aliaga 0666, Lima; tel. (64) 571131; fax (64) 573790; internet www.backus.com.pe; f. 1971; owned by Backus y Johnston, SAA (q.v.); brewery; Pres. ELIAS BENTÍN PERAL; Gen. Man. WALTER PASACHE CARBAJAL.

**Compañía Cervecera del Sur del Perú, SA (Cervesur):** Variante de Uchumayo 1801, Castilla 43, Arequipa; tel. (54) 470000; fax (54) 449602; e-mail postmaster@cervesur.com; internet www.cervesur.com.pe; f. 1898; brewery; sales of 462m. new soles (1999); Pres. VICTOR MONTORI ALFARO; 866 employees.

**Del Mar, SA:** Aristides Aljovin 690, Lima 18; tel. (1) 4457829; fax (1) 4471350; f. 1978; canning and processing of fish; Gen. Man. SYLVIA HERRERA AREVALO; 900 employees.

**D'Onofrio, SA:** Avda Venezuela 2580, Lima 1; tel. (1) 3365065; fax (1) 3365065; e-mail joscoa@donofrio.com.pe; f. 1933; manufacturers of desserts, confectionery and biscuits; owned by Nestlé Corpn of Switzerland; Pres. VITO RODRÍGUEZ RODRÍGUEZ; Gen. Man. ALBERTO HAITO MOARRI; 1,254 employees.

**Empresa de la Sal, SA (EMSAL):** Avda Nestor Gambetta, Km 8.5 Carretera Ventanilla, Callao; tel. (1) 5770669; fax (1) 5770685; internet www.quimpac.com.pe; f. 1969; acquired in 1994 by Quimpac; salt production; Chair. MARCOS FISHMAN; Gen. Man. JOSÉ CARLOS DE LOS RÍOS; 500 employees.

**Flota Pesquera Peruana, SA:** Avda Argentina 4090, Callao; tel. (1) 4299808; fax (1) 4640170; f. 1986; fishing co; Gen. Man. JORGE LAINES DE LA CRUZ; 564 employees.

### Rubber and Cement

**Cementos Lima, SA:** Avda Atocongo 2440, Villa María del Triunfo, Lima; tel. (1) 9541900; fax (1) 9541297; e-mail postmaster@cementolima.com.pe; internet www.cementoslima.com.pe; f. 1967; cement producers; Pres. JAIME RIZO-PATRÓN; Gen. Man. CARLOS UGAS D.; 500 employees.

**Cementos Norte Pacasmayo, SA:** Pasaje El Carmen 180, Urb. El Vivero de Monterrico, Santiago de Surco, Lima 33; tel. (1) 3172000; fax (1) 4375009; f. 1974; cement producers; Pres. ALBERTO BEECK ULLOA; Man. Dir LINO ABRAM CABALLERINO; 495 employees.

**Lima Caucho, SA:** Carretera Central 349, Km 1, Santa Anita, Lima 3; tel. (1) 3623845; fax (1) 3624069; internet www.limacaucho.com.pe; f. 1955; manufacturers of tyres and industrial rubber products; Dir JAVIER EDUARDO ALVA GUERRERO; 300 employees.

### Textiles and Clothing

**Compañía Industrial Nuevo Mundo, SA:** Jirón José Celendón 750, Lima; tel. (1) 3368110; fax (1) 3368193; e-mail nmcom@qnet.com.pe; f. 1949; manufacturers of textiles; sales of 123.9m. new soles (2001); 700 employees.

**Consorcio Textil del Pacífico, SA:** Avda Argentina 2400, Lima 1; tel. (1) 3368429; fax (1) 3368431; f. 1993; textiles and clothing manufacturer; Pres. MICHAEL MICHEL STAFFORD; Gen. Man. OLAF HEIN CHRISTIANI; 1,350 employees.

**Fábrica Nacional Textil el Amazonas:** Avda Argentina 1440–1448, Lima 1; tel. (1) 3367946; fax (1) 3367944; f. 1943; yarn mills; Man. GIANNO FAVIO GERBOLINI ISOLA; 1,000 employees.

**Michell y Compañía, SA:** Avda Juan de la Torre 101, San Lázaro, Arequipa; tel. (54) 202525; fax (54) 202626; e-mail michell@michell.com.pe; internet www.michell.com.pe; f. 1957; yarn mills; Exec. Pres. MICHAEL MICHELL STAFFORD; Dir JUAN PEPPER PASTOR; 419 employees.

**Universal Textil, SA:** Avda Venezuela 2505, Apdo 554, Lima 1; tel. (1) 3375260; fax (1) 3375270; e-mail jseminario@unitex.com.pe; internet www.universaltextil.com.pe; f. 1952; manufacturers of synthetic fabrics for outerwear; part of Romero group; sales of 66.0m. nuevos soles (2003); Chair. DIONISIO ROMERO, JAVIER SEMINARIO; Gen. Man. GEORGE R. SCHOFIELD BONELLO; 850 employees.

### Miscellaneous

**Bayer, SA:** Avda Paseo de la República 3074, 10° y 11°, San Isidro, Lima; tel. (1) 211-3800; fax (1) 4213381; internet www.bayerandina.com; f. 1969; chemicals, plastics and pharmaceuticals manufacturer; sales of 121m. nuevos soles (2001); Pres. HENNING VON KOSS; Gen. Man. CARLOS CORNEJO DE LA PIEDRA; 150 employees.

**Indeco, SA:** Avda Universitaria 683, Lima 1; tel. (1) 4642570; fax (1) 4521266; e-mail postmaster@indeco.com.pe; internet www.indeco.com.pe; f. 1952; manufacturers of electrical cables; sales of 160.9m. nuevos soles (2001); Pres. ERNESTO BAERTL MONTORI; Gen. Man. JUAN ENRIQUE RIVERA; 247 employees.

**Industrias Eletroquímicas, SA (Ieqsa):** Avda Elmer Faucett 1920, Callao, Lima; tel. (1) 572-4444; fax (1) 572-0118; e-mail export@ieqsa.com.pe; internet www.ieqsa.com.pe; f. 1963; manufacturers of batteries; sales of US $36m. (2003); Dir-Gen. RAÚL MUSSO; Pres. CARLOS GLIKSMAN; 400 employees.

**Industrias Pacocha, SA:** Francisco Graña 155, Urb. Santa Catalina, La Victoria, Lima; tel. (1) 4111600; fax (1) 4762424; f. 1971; manufacturers of detergents, soaps, fats and vegetable oils; Man. Dir EDUARDO MONTERO ARAMBURÚ; Gen. Man. MOISÉS DANNON LEVY; 675 employees.

**Ingenieros Constratistas Cosapi, SA:** Nicolás Arriola 500, Lima 13; tel. (1) 2113500; fax (1) 2248645; e-mail postmaster@cosapi.com.pe; internet www.cosapi.com.pe; f. 1967; engineering and construction; Pres. WALTER PIAZZA TANGUIS; 4,350 employees.

**Nissan Motors del Perú, SA:** Avda Tomás Valle 601, San Martín de Porres, Lima 31; tel. (1) 5342248; fax (1) 5342326; internet www.nissan.com.pe; f. 1957 as Maquinarias, SA; subsidiary of Nissan Motors of Japan; automobile assembly plant; Man. YOKI TASHANUKI; 435 employees.

**Paramonga Chemical–Paper Complex:** Avda Nestor Gambetta 8585, Callao; tel. (1) 5770700; fax (1) 5770273; e-mail quimpac@panasa.com.pe; internet www.quimpac.com.pe; f. 1898 as W. A. Grace—Sociedad Paramonga LTDA; pulp mills; Man. CARLOS ORAMS BASADRE; 3,000 employees.

**Tabacalera Nacional, SA:** Avda La Molina 140, Ate Vitarte, Lima 3; tel. (1) 4360388; fax (1) 4370066; f. 1964; acquired by British American Tobacco Peru Holdings Ltd in 2004; cigarette manufacturers; sales of 94.2m. new soles (2001); Pres. MANUEL ISABAL ROCA; Man. Dir JULIO CAIPO GUERRERO; 205 employees.

## UTILITIES

### Regulatory Authority

**Comisión de Tarifas Eléctricas (CTE):** Avda Canadá 1470, San Borja, Lima 41; tel. (1) 2240487; fax (1) 2240491; e-mail info@cte.org.pe; internet www.cte.org.pe; 5-mem. autonomous agency controlling tariffs.

### Electricity

**Electrolima, SA:** Jirón Zorritos 1301, Lima 5; tel. (1) 4324153; fax (1) 4323042; internet www.electrolima.com; f. 1906; produces and supplies electricity for Lima and the surrounding districts; state-owned; Gen. Man. Dr DARÍO CUERVO VILLAFAÑE.

**ElectroPerú:** Prolongación Pedro Miotta 421, San Juan de Miraflores, Lima 29; tel. (1) 4660506; fax (1) 4663448; internet www.electroperu.com; state-owned; Pres. Ing. GUILLERMO CASTILLO JUSTO; Gen. Man. Ing. IVÁN LA ROSA ALZAMORA.

**Empresa Regional de Servicio Público de Electricidad Norte (Electronorte, SA):** Vicente de la Vega 318, Chiclayo; tel. (74) 231580; fax (74) 227751; Pres. JORGE RODRÍGUEZ RODRÍGUEZ; Gen. Man. RICARDO ARRESE PÉREZ.

**Sociedad Eléctrica del Sur-Oeste, SA (SEAL):** Consuelo 310, Arequipa; tel. (54) 212946; fax (54) 213296; e-mail seal@sealperu.com; internet www.sealperu.com; f. 1905; Pres. ALFREDO LLOSA BARBER; Gen. Man. AMÉRICO PORTUGAL AMPUERO.

## TRADE UNIONS

The right to strike was restored in the Constitution of July 1979. In 1982 the Government recognized the right of public employees to form trade unions.

**Central Unica de Trabajadores Peruanos (CUTP):** Lima; f. 1992; Pres. JULIO CÉSAR BAZÁN; comprises:

**Confederación General de Trabajadores del Perú (CGTP):** Plaza 2 de Mayo 4, Lima 1; tel. (1) 4314738; e-mail cgtp@cgtp.org.pe; internet www.cgtp.org.pe; f. 1968; Pres. MARIO HUAMÁN RIVERA; Sec.-Gen. JUAN JOSÉ GORRITI VALLE.

**Confederación Nacional de Trabajadores (CNT):** Avda Iquitos 1198, Lima; tel. (1) 4711385; affiliated to the PPC; 12,000 mems (est.); Sec.-Gen. ANTONIO GALLARDO EGOAVIL.

**Confederación de Trabajadores del Perú (CTP):** Jirón Ayacucho 173, CP 3616, Lima 1; tel. (1) 4261310; e-mail ctp7319@hotmail.com; affiliated to APRA; Sec.-Gen. ELÍAS GRIJALVA ALVARADO.

**Confederación Intersectorial de Trabajadores Estatales (CITE)** (Union of Public Sector Workers): Lima; tel. (1) 4245525; f. 1978; Sec.-Gen. ALAVARO COLE; Asst Sec. OMAR CAMPOS; 600,000 mems.

**Federación de Empleados Bancarios (FEB)** (Union of Bank Employees): Jirón Miró Quesada 260, 7°, Lima; tel. (1) 7249570; e-mail febperu@terra.com.pe; Sec.-Gen. HÉCTOR PÉREZ PÉREZ.

**Federación Nacional de Trabajadores Mineros, Metalúrgicos y Siderúrgicos (FNTMMS)** (Federation of Peruvian Mineworkers): Jirón Callao 457, Of. 311, Lima; tel. (1) 4277554; Sec.-Gen. PEDRO ESCATE SULCA; 70,000 mems.

**Movimiento de Trabajadores y Obreros de Clase (MTOC):** Lima.

**Sindicato Unico de Trabajadores de Educación del Perú (SUTEP)** (Union of Peruvian Teachers): Camaná 550, Lima; tel. (1) 4276677; fax (1) 4268692; Sec.-Gen. NÍLVER LÓPEZ AMES.

Independent unions, representing an estimated 37% of trade unionists, include the Comité para la Coordinación Clasista y la Unificación Sindical, Confederación de Campesinos Peruanos (CCP) and the Confederación Nacional Agraria (Pres. MIGUEL CLEMENTE ALEGRE).

**Confederación Nacional de Comunidades Industriales (CONACI):** Lima; co-ordinates worker participation in industrial management and profit-sharing.

PERU — *Directory*

The following agricultural organizations exist:

**Confederación Nacional de Productores Agropecuarios de las Cuencas Cocaleras del Perú (CONPACCP):** Lima; coca-growers' confederation; Leader ELSA MALPARTIDA.

**Consejo Unitario Nacional Agrario (CUNA):** f. 1983; represents 36 farmers' and peasants' organizations, including:

  **Confederación Campesina del Perú (CCP):** radical left-wing; Pres. ANDRÉS LUNA VARGAS; Sec. HUGO BLANCO.

  **Organización Nacional Agraria (ONA):** organization of dairy farmers and cattle-breeders.

## Transport

### RAILWAYS

In 2000 there were some 2,123 km of track. A programme to develop a national railway network (Sistema Nacional Ferroviario) was begun in the early 1980s, aimed at increasing the length of track to about 5,000 km initially. The Government also plans to electrify the railway system and extend the Central and Southern Railways.

**Ministerio de Transportes y Comunicaciones:** see section on The Government (Ministries).

**Consorcio de Ferrocarriles del Perú:** following the privatization of the state railway company, Enafer, the above consortium won a 30-year concession in July 1999 to operate the following lines:

**Empresa Minera del Centro del Perú SA—División Ferrocarriles (Centromín-Perú SA)** (fmrly Cerro de Pasco Railway): Edif. Solgas, Avda Javier Prado Este 2175, San Borja, Apdo 2412, Lima 41; tel. (1) 4761010; fax (1) 4769757; 212.2 km; acquired by Enafer-Perú in 1997; Pres. HERNÁN BARRETO; Gen. Man. GUILLERMO GUANILO.

**Ferrocarril Central Andino, SA:** Jirón Brasil, esq. San Fernando-Chosica, Lima; tel. (1) 3612828; fax (1) 3610380; e-mail ferrocarrilcentral@fcca.com.pe; internet www.ferroviasperu.com.pe; f. 1999.

**Ferrocarril del Centro del Perú** (Central Railway of Peru): Ancash 201, Apdo 301, Lima; tel. (1) 4276620; fax (1) 4281075; 591 km open; Man. ADRIEL ESTRADA FARFAN.

**Ferrocarril del Sur del Perú ENAFER, SA** (Southern Railway): Avda Tacna y Arica 200, Apdo 194, Arequipa; tel. (54) 215350; fax (54) 231603; 915 km open; also operates steamship service on Lake Titicaca; Man. C. NORIEGA.

**Tacna–Arica Ferrocarril** (Tacna–Arica Railway): Avda Aldarracín 484, Tacna; 62 km open.

**Ferrocarril Pimentel** (Pimentel Railway): Pimentel, Chiclayo, Apdo 310; 56 km open; owned by Empresa Nacional de Puertos; cargo services only; Pres. R. MONTENEGRO; Man. LUIS DE LA PIEDRA ALVIZURI.

#### Private Railways

**Ferrocarril Ilo–Toquepala–Cuajone:** Apdo 2640, Lima; 219 km open, incl. five tunnels totalling 27 km; owned by the Southern Peru Copper Corpn for transporting copper supplies and concentrates only; Pres. ÓSCAR GONZÁLEZ ROCHA; Gen. Man. WILLIAM TORRES.

**Ferrocarril Supe–Barranca–Alpas:** Barranca; 40 km open; Dirs CARLOS GARCÍA GASTAÑETA, LUIS G. MIRANDA.

### ROADS

There were an estimated 78,000 km of roads in Peru, of which approximately 30% was paved or semi-paved. The most important highways are: the Pan-American Highway (3,008 km), which runs southward from the Ecuadorean border along the coast to Lima; Camino del Inca Highway (3,193 km) from Piura to Puno; Marginal de la Selva (1,688 km) from Cajamarca to Madre de Dios; and the Trans-Andean Highway (834 km), which runs from Lima to Pucallpa on the River Ucayali via Oroya, Cerro de Pasco and Tingo María.

### SHIPPING

Most trade is through the port of Callao but there are 13 deep-water ports, mainly in northern Peru (including Salaverry, Pacasmayo and Paita) and in the south (including the iron-ore port of San Juan). There are river ports at Iquitos, Pucallpa and Yurimaguas, aimed at improving communications between Lima and Iquitos, and a further port is under construction at Puerto Maldonado.

**Empresa Nacional de Puertos, SA (Enapu):** Avda Contralmirante Raygada 110, Callao; tel. (1) 4299210; fax (1) 4691011; e-mail enapu@inconet.net.pe; internet www.enapu.gob.pe; f. 1970; government agency administering all coastal and river ports; Pres. LUIS E. VARGAS CABALLERO; Gen. Man. OTTO BOTTGER R.

**Asociación Marítima del Perú:** Avda Javier Prado Este 897, Of. 33, San Isidro, Apdo 3520, Lima 27; tel. and fax (1) 4221904; f. 1957; association of 20 international and Peruvian shipping companies; Pres. LUIS FELIPE VILLENA GUTIÉRREZ.

**Consorcio Naviero Peruano, SA:** Avda Central 643, San Isidro, Apdo 18-0736, Lima 1; tel. (1) 4116500; fax (1) 4116599; e-mail cnp@cnpsa.com; f. 1959.

**Naviera Humboldt, SA:** Edif. Pacífico–Washington, 9°, Natalio Sánchez 125, Apdo 3639, Lima 1; tel. (1) 4334005; fax (1) 4337151; e-mail postmast@sorcomar.com.pe; internet www.humbolt.com.pe; f. 1970; cargo services; Pres. AUGUSTO BEDOYA CAMERE; Man. Dir LUIS FREIRE R.

**Agencia Naviera Maynas, SA:** Avda San Borja Norte 761, San Borja, Lima 41; tel. (1) 4752033; fax (1) 4759680; e-mail lima@navieramaynas.com.pe; f. 1996; Pres. R. USSEGLIO D.; Gen. Man. ROBERTO MELGAR B.

**Naviera Universal, SA:** Calle 41 No 894, Urb. Corpac, San Isidro, Apdo 10307, Lima 100; tel. (1) 4757020; fax (1) 4755233; Chair. HERBERT C. BUERGER.

**Petrolera Transoceánica, SA (PETRANSO):** Víctor Maúrtua 135, Lima 27; tel. (1) 4422007; fax (1) 4403922; internet www.petranso.com.pe; Gen. Man. JUAN VILLARÁN.

A number of foreign lines call at Peruvian ports.

### CIVIL AVIATION

Of Peru's 294 airports and airfields, the major international airport is Jorge Chávez Airport near Lima. Other important international airports are Coronel Francisco Secada Vignetta Airport, near Iquitos, Velasco Astete Airport, near Cusco, and Rodríguez Ballón Airport, near Arequipa.

**Corporación Peruana de Aeropuertos y Aviación Comercial:** Aeropuerto Internacional Jorge Chávez, Callao; tel. (1) 5750912; fax (1) 5745578; internet www.corpac.gob.pe; f. 1943; Pres. LEOPOLDO PFLUCKER LLONA; Gen. Man. ROBERT MCDONALD ZAPFF.

#### Domestic Airlines

**Aero Condor:** Juan de Arona 781, San Isidro, Lima; tel. (1) 4425215; fax (1) 2215783; internet www.aerocondor.com.pe; domestic services; Pres. CARLOS PALACÍN FERNÁNDEZ.

**LAN Perú, SA:** Lima; tel. (1) 2138200; internet www.lan.com; f. 1999; operations temporarily suspended in Oct. 2004; Exec. Vice-Pres. ENRIQUE CUETO P.

**Nuevo Continente:** Avda José Pardo 605, Lima 18; tel. (1) 2414816; fax (1) 2413074; internet www.aerocontinente.com.pe; f. 1992; domestic services; fmrly Aero Continente; operations suspended in July 2004; acquired by Vuela Perú in November 2004; Pres. LUPE L. Z. GONZALES.

## Tourism

Tourism is centred on Lima, with its Spanish colonial architecture, and Cusco, with its pre-Inca and Inca civilization, notably the 'lost city' of Machu Picchu. Lake Titicaca, lying at an altitude of 3,850 m above sea level, and the Amazon jungle region to the north-east are also popular destinations. From the mid-1990s there was evidence of a marked recovery in the tourism sector, which had been adversely affected by health and security concerns. In 2003 Peru received 933,643 visitors. Receipts from tourism generated US $959m. in that year.

**Comisión de Promoción del Perú (PromPerú):** Edif. Mitinci, Calle Uno Oeste, 13°, Urb. Corpac, San Isidro, Lima 27; tel. (1) 2243279; fax (1) 2243323; e-mail postmaster@promperu.gob.pe; internet www.peru.org.pe; f. 1993; Head of Tourism MARÍA DEL PILAR LAZARTE CONROY; Exec. Sec. MARIELA AUSEJO VIDAL.

## Defence

At 1 August 2004 Peru's Armed Forces numbered 80,000: Army 40,000, Navy 25,000, Air Force 15,000. Paramilitary forces numbered 77,000. There were 188,000 army reserves. Military service was selective and lasted for two years.

**Defence Budget:** 3,100m. new soles for defence and domestic security in 2004.

**President of the Joint Command of the Armed Forces:** Gen. AURELIO CROVETTO YAÑEZ.

**Commander of the Army:** Gen. José Graham Ayllón.
**Commander of the Air Force:** Gen. Orlando Denegri Ayllón.
**Commander of the Navy:** Adm. José Luis Noriega Lores.

## Education

Education in Peru is based on a series of reforms introduced after the 1968 revolution. The educational system is divided into three levels: the first level is for children up to six years of age in either nurseries or kindergartens. Basic education is provided at the second level. It is free and, where possible, compulsory between six and 15 years of age. Primary education lasts for six years. Secondary education, beginning at the age of 12, is divided into two stages, of two and three years respectively. Total enrolment at primary schools, in 1997, was equivalent to 91% of children in the relevant age-group, while an estimated 55% of children in the relevant age-group attended secondary schools. Higher education includes the pre-university and university levels. There were 78 universities in 2000. There is also provision for adult literacy programmes and bilingual education. Total central government expenditure on education was estimated at 2.9% of GDP in 2000. Budget proposals for 2005 allocated some US $2,600m. to education.

## Bibliography

For works on South America generally, see Select Bibliography (Books)

Arce, M. *Market Reform in Society: Post-crisis Politics and Economic Change in Authoritarian Peru.* University Park, PA, Pennsylvania State University Press, 2005.

Clayton, L. A. *Peru and the United States.* Athens, GA, University of Georgia Press, 1999.

Bennett, J. M. *Sendero Luminoso in Context.* Lanham, MD, Scarecrow Press, 1998.

Conaghan, C. *Fujimori's Peru: Deception in the Public Sphere.* Pittsburgh, PA, University of Pittsburgh Press, 2005.

Cook, N. C. *Demographic Collapse: Indian Peru 1520–1620.* Cambridge, Cambridge University Press, 2002.

Crabtree, J., and Thomas, J. (Eds) *Fujimori's Peru.* London, Institute of Latin American Studies, 1998.

Daeschner, J. *The War of the End of Democracy: Mario Vargas Llosa vs Alberto Fujimori.* Lima, Peru Reporting, 1993.

Durand, F. *Business and Politics in Peru: The State and the National Bourgeoisie.* Oxford, Westview Press, 1994.

García-Bryce, I. *Crafting the Republic: Lima's Artisans and Nation-building in Peru, 1821–1879.* Albuqurque, NM, University of New Mexico Press, 2004.

Herz, M., and João Pontes, N. *Ecuador vs Peru: Peacemaking Amid Rivalry.* Boulder, CO, Lynne Rienner Publrs, 2002.

Kimura, R. *Alberto Fujimori of Peru: The President Who Dared to Dream.* Woodstock, NY, Beekman Publrs, 1998.

Klaren, P. *Peru: Society and Nationhood in the Andes.* Oxford, Oxford University Press, 1999.

McClintock, C., and Vallas, F. *The United States and Peru.* London, Routledge, 2002.

Paredes, C., and Sachs, J. *Peru's Path to Recovery: A Plan for Economic Stabilization and Growth.* Washington, DC, Brookings Institution Press, 2004.

Parodi, J., and Conaghan, C. *To Be A Worker: Identity and Politics in Peru.* Chapel Hill, NC, University of North Carolina Press, 2000.

*Peru Foreign Policy and Government Guide.* New York, NY, International Business Publications, 2000.

Prescott, W. H. *History of the Conquest of Peru.* London, Phoenix Press, Revised edn, 2005.

Roberts, K. *Deepening Democracy?: The Modern Left and Social Movements in Chile and Peru.* Stanford, CA, University of California Press, 1999.

Sheahan, J. *Searching for a Better Society: The Peruvian Economy from 1950.* University Park, PA, Penn State University Press, 1999.

Silverblatt, I. *Modern Inquisitions: Peru and the Colonial Origins of the Civilized World.* Durham, NC, Duke University Press, 2004.

Starn, O., Degregori, C., and Kirk, R. *The Peru Reader: History, Culture, Politics.* Durham, NC, Duke University Press, Revised edn, 2005.

Stern, S. *Shining and Other Paths: War and Society in Peru, 1980–95.* Durham, NC, Duke University Press, 1998.

Strong, S. *Shining Path, the World's Deadliest Revolutionary Force.* London, Fontana, 1993.

Taylor, L. *Maoism in the Andes: Sendero Luminoso and the Contemporary Guerrilla Movement in Peru.* Liverpool, Centre for Latin American Studies, 1998.

# PUERTO RICO

## Geography

### PHYSICAL FEATURES

The Commonwealth (Estado Libre Asociado) of Puerto Rico is a US territory based on the smallest and easternmost island of the Greater Antilles. Puerto Rico and its offshore islands comprise a Commonwealth Territory in voluntary association with the USA since 1952, but a colonial possession of the North American country since its military victory against Spain in 1898. Puerto Rico was also known as Borinquén by the Spanish, after the Amerindian name for the island, Boriquén or Boriken. To the east is more US territory, the island of St Thomas in the Virgin Islands being 64 km from the main island of Puerto Rico, although the Isla de Culebra and its own offshore islands lie mid-way between the two. About 15 km to the south-west of Culebra is Vieques, which itself only lies 11 km off the south-eastern coast of Puerto Rico island. In the west, Puerto Rico lies on the strategic Mona Passage, which separates it from the Dominican Republic on the island of Hispaniola. The two islands are only 120 km apart at the narrowest part of this sea lane from the Atlantic into the Caribbean. Their territories come closer only owing to Puerto Rico possessing the small, now-uninhabited island of Mona (80 km west of the port of Mayagüez). Puerto Rico, which has 501 km of coastline and a number of fine, natural harbours, has an area about the same as that of Cyprus, at 8,959 sq km (3,459 sq miles), including about 145 sq km of inland waters.

The island of Puerto Rico is roughly rectangular in shape, with a missing south-eastern corner. It is almost 180 km in length (east–west) and nearly 60 km wide, an island of high, central peaks, surrounded by coastal lowlands, except in the west, where the mountains are sheer to the sea. The rugged mountain range, running from east to west, is known as the Cordillera Central and reaches 1,338 m (4,391 ft) at Cerro de Punta, north of the city of Ponce. Parts of the mountains are densely vegetated and there are fairly extensive protected woodland areas. For instance, there is the unique dry-forest vegetation at Guánica (700 plant species, of which 48 are endangered and 16 exist only there) or the main reserve on the island, the El Yunque tropical rainforest, which is a bird sanctuary and the home of the few remaining Puerto Rican parrots. In the 1930s about 90% of Puerto Rico was under agriculture but, as a result of post-war industrialization, the rural population migrated and forest coverage grew from 10% in the 1940s to more than 40% at the turn of the 21st century; urban areas occupied 14% of the island. To the north of the Cordillera Central is a coastal belt, where the limestone has been formed into karst country of conical hills and holes by water erosion, very different to the ancient volcanic peaks. There are also many rock caverns, with the Camuy underground river system, the third largest such in the world. Rain-catching highlands, from which many small rivers spring (falling steeply to the sea, as waterfalls over cliffs in the more rugged terrain), ensure that the island is well watered.

The largest offshore island of Puerto Rico, and the first leading east into the chain of the Lesser Antilles, is Vieques. It is about 34 km long and 6 km wide, and in the past has also been called Graciosa and then Crab Island. Two-thirds of the hilly island was owned until May 2003 by the US navy; upon the withdrawal of naval personnel, the land was ceded to the US Department of the Interior. To the north of the eastern end of Vieques, and directly east of north-eastern Puerto Rico, is

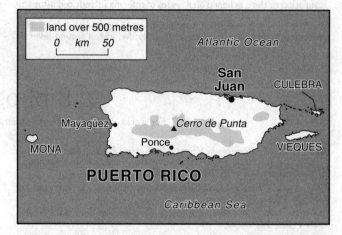

Culebra and its cluster of smaller satellites, much of which is a nature reserve. Culebra, which bears a closer resemblance to the Virgin Islands, is some 11 km long and 5 km wide.

### CLIMATE

The climate is a mild, subtropical marine one. The south of the island is often in the rain shadow of the central highlands, making it drier. Average annual rainfall is good, being about 2,300 mm (90 ins) on the north coast, and about one-half that on the south coast. Rain falls regularly from May, but mostly in July–October, which also coincides with the hurricane season—the word hurricane is derived from the name of the local Amerindian weather god, Juracan. There is little seasonal variation in temperature. Atlantic trade winds moderate the heat of the higher summer (July–August) averages of up to 30°C (86°F), while in winter the thermometer readings drop to a range of 21°–26°C (70°–79°C).

### POPULATION

Racially, Puerto Rico has more in common with Latin America than the anglophone black West Indies. White people account for 80.5% of the total population, most of them being of Spanish descent, with blacks at 8.0%, Amerindians 0.4% and Asians 0.2%; the remaining population is of mixed race. Another sign of the island's Spanish past is that Roman Catholics make up almost three-quarters of the total population, and there are few non-Christian religions on the island. Over one century of rule by the USA, however, has had a significant impact on the culture, not least that English has joined Spanish as an official language.

The total population of 3.9m. at mid-2004 (official provisional estimate) was about 80% urbanized. The capital, San Juan (with a population of 0.4m. at the April 2000 census), has one of the best natural harbours in the Caribbean. It is located in the north-east of Puerto Rico, about one-third of the way along the northern coast. Traditionally, the second city is Ponce (named for Juan Ponce de León, the Spanish nobleman who led the colonization of the island and who sought the mythical fountain of eternal youth here), on the south coast, and the third city Mayagüez, on the west coast. The territory is divided into 78 municipalities for local administration.

# History

**Prof. PETER CALVERT**

Based on an earlier article by JAMES McDONOUGH

## INTRODUCTION

Puerto Rico was discovered by a Spanish expedition, led by the navigator Christopher Columbus, in 1493, and named by him San Juan Bautista. Known to its indigenous Taino inhabitants as Boriquén, it takes its modern name from the name given in 1509 to its capital, Puerto Rico, now known as San Juan. The excellence of its harbour and its strategic position commanding the Mona Passage made it an important Spanish military outpost. However, the island tended to be neglected by Spain in favour of richer possessions, and it was only after trade had been opened up by royal decree in 1815 that a significant coffee and sugar plantation economy developed. The slave trade was ended in 1835, but slavery on the island was not abolished until 1873. A movement for autonomy emerged slowly after the change of government in Spain in 1868, but it was not until 1897 that Spain granted its colony a Carta Autonómica (Autonomy Charter). However, Puerto Rico's new-found autonomy abruptly ended just a year later when, following its capture by US forces during the Spanish–American War, the Caribbean possession was ceded to the USA by the Treaty of Paris.

In 1900 the island was reorganized as a US Territory and a civil government replaced the military government that had ruled since 1898. As a Territory, Puerto Rico became subject to most laws of the US Congress, and the US President appointed the Governor and members of the Island's Executive Council, which functioned as an upper house of the legislative branch, the lower house being elected by popular vote. In 1917 Congress extended US citizenship to the island's inhabitants over the objections of the island's lower house, and provided for the popular election of the members of an upper house or Senate. The island's Governor, however, continued to be appointed by the US President.

Further internal self-government was achieved under President Harry S Truman in 1947, when Congress approved a law giving the people of Puerto Rico the right to elect their own Governor, and a year later they chose as their first elected Governor the charismatic Luis Muñoz Marín. Now recognized as the 'father of modern Puerto Rico', it was he who persuaded the US Government to approve and in large measure to fund 'Operation Bootstrap' (see Economy), a programme to industrialize Puerto Rico and raise its standard of living towards that of the rest of the USA. In 1950, in yet another move towards greater internal autonomy, Congress approved Law 600, allowing Puerto Rico to draft its own constitution, although this was subject to congressional review. This process culminated in 1952, when, in a special referendum, the people of Puerto Rico approved the island's first Constitution under US rule. Puerto Rico was given the status of a 'Commonwealth' (in Spanish 'Estado Libre Asociado', or associated free state) in its relation to the USA, and the following year, 1953, when the island became self-governing, the UN removed it from its list of territories due for decolonization. Although Puerto Rico gained wide powers of organization over its internal affairs, the island has remained US sovereign territory and most federal laws have continued to apply to it, with the important exception of taxation. Critics maintain that Commonwealth status, although more liberal than earlier forms of US rule, affords fewer powers of self-government than Puerto Rico enjoyed under Spain's Carta Autonómica.

## DOMESTIC POLITICS

Since 1898 Puerto Ricans have been divided on the question of the island's political relationship with the USA. The island's two dominant parties both favour continued strong links with the USA. The Partido Popular Democrático (PPD—Popular Democratic Party) supports the existing Commonwealth status with 'enhancements', such as greater autonomy, while the Partido Nuevo Progresista (PNP—New Progressive Party) favours Puerto Rico's inclusion as a state of the USA. The island's third and smallest party, the Partido Independentista Puertorriqueño (PIP—Puerto Rican Independence Party) campaigns for independence.

Historically, various clandestine, pro-independence forces have operated in Puerto Rico outside the electoral process. During the 1940s and 1950s the most influential of these groups was the Partido Nacionalista (Nationalist Party), led by the charismatic Pedro Albizu Campos. The Partido Nacionalista was responsible for an uprising in Puerto Rico in 1950, involving an armed attack on La Fortaleza, and a simultaneous, but unsuccessful attempt on the life of President Truman in Washington, DC, USA. It also carried out an armed assault on members of the US Congress in 1954. In the 1970s another pro-independence group, the Partido Socialista Puertorriqueño (PSP—Puerto Rican Socialist Party) succeeded in raising the question of Puerto Rico's status before the UN's Decolonization Committee and the Conference of Non-Aligned Nations. With the active participation of Cuba, the Decolonization Committee approved a resolution recognizing Puerto Rico's inalienable right to self-determination and independence. However, the USA was able to prevent the UN General Assembly from returning Puerto Rico to the list of political dependencies from which it had been removed in 1953, following the approval of the Commonwealth Constitution. In the early 1980s the most important clandestine pro-independence group was the Ejército Popular Boricua (Puerto Rican Popular Army), known as 'Los Macheteros'. This clandestine group claimed responsibility for armed attacks in the USA and Puerto Rico against military targets. However, in 1985 the group fragmented, following a series of raids by the US Federal Bureau of Investigation (FBI), in both Puerto Rico and the USA. Seventeen members of Los Macheteros, including most of its leadership, were arrested and imprisoned. The group's most important leader, Filiberto Ojeda Ríos, escaped and was still being sought by US authorities; however, although the group is believed still to exist, it has not recently been active. Other insurgent and dissident groups active from time to time have included the Armed Forces for National Liberation (FALN), the Armed Forces of Popular Resistance and the Volunteers of the Puerto Rican Revolution.

The PPD, under the leadership of Muñoz Marín, dominated Puerto Rican electoral politics from 1940 to 1968. Muñoz Marín voluntarily retired from the leadership of the party in 1964, after serving four terms as Governor. He nominated his successor, then Secretary of State, Roberto Sánchez Vilella, who was elected Governor in 1964. A division in the PPD allowed the gubernatorial candidate of the newly formed PNP, Luis A. Ferré, to win the governorship in 1968. In the gubernatorial election held in 1972 the PPD returned to power under the leadership of Senate President Rafael Hernández Colón. He was defeated in 1976 by the PNP Mayor of San Juan, Carlos Romero Barceló, who had taken over the party's leadership following Ferré's defeat in 1972. Romero Barceló won again in 1980 by a mere 3,500 votes, in the narrowest election win in Puerto Rican history. The PPD, however, won a majority in both the House of Representatives and the Senate, thus preventing Romero Barceló from holding a referendum to determine whether people favoured US statehood for Puerto Rico.

During 1981, dissatisfaction with Romero Barceló's leadership developed. The murder of two independence activists, allegedly by Puerto Rican police, brought into question the role of Governor Romero Barceló in the affair. Romero Barceló faced additional problems from his own party, when the Mayor of San Juan, Hernán Padilla Ramírez, left the PNP over the issue of internal party democracy and formed a new political grouping, the Partido de Renovación Puertorriqueño (PRP—Puerto Rican Renewal Party). Padilla entered the 1984 gubernatorial election as the PRP candidate and received some 70,000 votes (about 4% of the total votes cast). The division in the PNP was enough to ensure the election of former Governor Hernández Colón, the PPD candidate. Hernández Colón dedicated his second term in office to the resolution of economic problems. In 1986 he successfully persuaded the US Congress to retain a special tax benefit for US corporations operating in Puerto Rico, although this was subsequently rescinded in 1996 (see Economy).

The Mayor of San Juan, Baltasar Corrada del Río, assumed control of the PNP following Romero Barceló's electoral defeat in 1984. He was the party's candidate in the 1988 election, but was defeated by Hernández Colón by some 50,000 votes. For the first time in 20 years the PNP lost its traditional bastion of power, the island's capital city of San Juan, by a mere 300 votes, to the PPD candidate, Héctor Luis Acevedo. At his third gubernatorial inauguration, in January 1989, Hernández Colón announced that he would seek congressional approval for a status referendum to be held in mid-1991; subsequently, the leaders of the island's three political parties formally petitioned the US Congress to approve legislation to authorize and implement such a referendum. In late 1989, after the US House of Representatives had approved a non-binding status plebiscite bill for Puerto Rico, the US Senate's Energy and Natural Resources Committee, which had jurisdiction in the Senate over territorial matters, defeated the proposed legislation in a dramatic tied vote, thus ending the decolonization initiative. In Puerto Rico the defeat of the plebiscite measure was attributed to the reluctance of the US Congress to approve legislation that could lead to statehood for Puerto Rico.

Following the defeat of the status legislation in the US Congress, Hernández Colón's administration, which held a majority in both legislative chambers, introduced legislation establishing a charter of 'democratic rights'. The charter included guarantees of US citizenship regardless of future changes in the island's constitutional status, and made Spanish the only official language of Puerto Rico, abrogating a 1902 law that had established both Spanish and English as the island's official languages—by this time one-fifth of the inhabitants had English as their first language. The PPD also approved legislation, opposed by the PNP, to hold a plebiscite to amend the island's Constitution by adding six 'principles of self-determination'. The electorate rejected the proposed amendment by a vote of 53.6% against to 45.4% in favour. The referendum represented a major defeat for Governor Hernández Colón, and a victory for the new PNP leader, Dr Pedro J. Rosselló, a 48 year-old physician and formerly prominent tennis player. On 8 January 1992, one year before his gubernatorial term expired, Hernández Colón resigned as leader of the PPD, a position that he had occupied almost continuously since 1969. He was succeeded by Victoria Muñoz Mendoza, the daughter of the former Governor, Luis Muñoz Marín.

In the 1992 election campaign the pro-statehood PNP candidate, Rosselló, pledged to reduce crime, privatize the island's public-health system and lower taxes for the middle classes and small businesses. He also promised to hold a referendum on the three traditional status options: statehood, enhanced Commonwealth or independence. His rival from the PPD, Victoria Muñoz, pledged to reform government bureaucracy and to encourage economic development. In the gubernatorial election, held on 3 November, Rosselló secured 50% of the votes cast, defeating Muñoz, who won 46% of the ballot. Former Governor Romero Barceló was also elected as the island's Resident Commissioner in the US Congress. At the same time, the PNP won control of both the Senate and the House of Representatives, as well as 58 of the island's 78 municipalities. The PPD had suffered a major political reverse.

### THE STATUS ISSUE

Upon taking office in January 1993, the first bill passed by the PNP-dominated legislature restored English as an official language of the island. Rosselló moved to fulfil his election promise to reduce crime by mobilizing the National Guard in several low-income residential areas of San Juan with high crime rates. He also moved to privatize the island's public-health system by subsidizing private health insurance for the poor and selling or renting out the Government's health-care facilities. Legislation presented by Rosselló to enable a status plebiscite to be held was overwhelmingly endorsed in the island legislature: on 14 November, 48.6% of the electorate voted for the retention of Commonwealth status, 46.3% supported statehood and 4% advocated independence; some 73% of registered voters participated. The results were disappointing for the governing PNP, which had hoped to win a mandate to urge the US Congress to grant the island statehood. Those in favour of continued Commonwealth status, which included the PPD, were equally disappointed that their formula had not received a clear majority and shocked to see their share of the vote, approximately 60% in the 1967 plebiscite, decline to less than 50%. For the 1996 election the PPD nominated the Mayor of San Juan, Héctor Acevedo, to oppose Rosselló, who was standing for a second term. On 5 November Rosselló was re-elected by an even greater margin than in 1992, with 51% of the votes cast, compared with 44% for Acevedo. The PNP retained control of both houses of the island legislature and won 54 of the 78 island municipalities. Romero Barceló was also re-elected to the post of Resident Commissioner, soundly defeating his PPD opponent, Celeste Benítez.

In November 1994 the Republicans won control of both the US House of Representatives and the Senate. In the following year Don Young, newly appointed Chairman of the House Resources Committee, with jurisdiction over Puerto Rico, introduced a bill that required Puerto Rico to hold periodic plebiscites until the issue of the island's status was resolved. The legislation was withdrawn in late 1996 at the insistence of Romero Barceló, after it had been amended to make English the sole official language of a future US state of Puerto Rico. It was revived in 1997, but was opposed by the PPD, owing to the bill's categorization of the Commonwealth as a territory of the USA. The PPD argued that when the Commonwealth was established in 1952, Puerto Rico ceased to be a colonial dependency of the USA, an interpretation that has been continually challenged by both the PNP and the PIP. In March 1998, however, the bill passed the US House of Representatives by one vote. The law authorized a plebiscite to be held in 1998 to allow Puerto Rico to choose between Commonwealth status, independence and statehood. In the event of the electorate voting for the Commonwealth formula, a plebiscite was to be held every 10 years until the island chose either independence or statehood. However, the US Senate leadership opposed the plebiscite and the bill was not voted on in the upper house. Subsequently, Governor Rosselló held another Puerto Rican-sponsored plebiscite in December 1998, which allowed a choice between five options: maintaining the existing Commonwealth status, defined as territorial; independence; statehood; free association with the USA (whereby the USA would yield sovereignty over Puerto Rico); or 'none of the above'. The PPD campaigned for the last of the five options, 'none of the above', which won the plebiscite with 50.2% of the vote, compared with 46.5% in favour of statehood, 2.5% in favour of independence, and less than 1% supporting free association or the existing Commonwealth status.

In June 2000, at the insistence of Rosselló, US President Bill Clinton (1993–2001) met with representatives of the island's three major political parties to discuss a new formula for resolving the status issue. However, the Republican congressional leadership boycotted the meeting and the Senate Majority Leader, Trent Lott, stated that the US legislature was not interested in dealing with Puerto Rico's status during the current session of Congress. On 7 November 2000 Sila María Calderón Serra of the PPD, who had been elected mayor of San Juan in 1996, became the first woman to be elected Governor of Puerto Rico. She gained 48.6% of the votes cast, while the PNP candidate, Carlos I. Pesquera Morales, won 45.7%. Aníbal Acevedo Vilá of the PPD was elected to the post of Resident Commissioner with 49.2% of the votes cast, defeating the incumbent, Romero Barceló. The PPD also gained control of the Senate, the House of Representatives and 46 of the island's 78 municipalities. The simultaneous election of the Republican President George W. Bush and a Republican majority in both houses of the US Congress blocked further moves on the status issue. In July 2002 Governor Calderón announced the creation of a 'status committee' to resolve the issue; the committee was to consist of representatives of the PPD, PNP and PIP, including former Governors Hernández Colón and Romero Barceló. However, the PNP leadership rejected the proposal and the status issue continued to divide local politicians.

Also in July 2002 Governor Calderón launched a non-partisan campaign on the mainland under the slogan '¡Que Nada Nos Detenga!' ('Let Nothing Hold Us Back!') to encourage Puerto Ricans in the USA to become more actively engaged in community politics. Voter registration and participation rates on the island were 95% and 86%, respectively; however, both these rates fell to about 40% when Puerto Ricans moved to the USA. In May 2003 Calderón announced that she would not seek re-election in the gubernatorial election of November 2004; Aníbal Acevedo Vilá was expected to contest the post for the PPD, and Pedro Rosselló for the PNP.

Several prominent members of the PNP who had served in the Rosselló administration went on trial in 2002, and in December Victor Fajardo was sentenced to 12 years' imprisonment after being convicted of diverting state funds to his party. In January 2003 the former Speaker of the House of Representatives under Rosselló, Edison Misla Aldorano, was found guilty of extortion, money-laundering and perverting the course of justice; a former member of Rosselló's staff was sentenced to 18 months in prison in the same month.

In July 2004, in the face of rising crime rates and drugs-related violence, Governor Calderón mobilized 500 National Guard to patrol public events, releasing police officers for operations in high-crime areas, such as housing projects. It was envisaged that if levels of violence did not decrease during the first phase of deployment, the police and National Guard would begin joint patrols; as a last resort, troops would be stationed in housing projects. Crime was a major issue during the 2004 election campaign and ways to address the issue were prominent in both gubernatorial candidates' manifestos.

## VIEQUES

In April 1999 a US navy bomb accidentally fell on an observation tower on the small island of Vieques, which lies off the eastern coast of Puerto Rico, killing a Puerto Rican civilian who worked for the US Navy. Since the end of the Second World War the US Navy had been using most of Vieques as a live firing range, despite the fact that it was also home to some 10,000 civilian inhabitants. (It was also used as the scene of military exercises, notably those that preceded in US intervention in Grenada in 1983.) The killing of the civilian prompted angry public protests in Puerto Rico, which ultimately led to the illegal occupation of the firing range and a general outcry for the Navy to leave Vieques. In January 2000 Rosselló reached an agreement with President Clinton to allow the Navy to continue bombing practice with inert or dummy ordnance until a referendum could be held on Vieques to allow the population to choose between the Navy staying and renewing live bombing practice, or leaving the island altogether by May 2003. The US Administration was also to provide immediate development aid to Vieques of US $40m., which would increase to $90m. if residents agreed to the resumption of live ammunition testing. In return, Rosselló promised to help the federal authorities remove the protesters, who had been occupying the firing range since the bombing incident. In May 2000, in anticipation of the recommencement of (dummy) ammunition testing, protests were held on Vieques. US Federal Government agents were deployed on the island forcibly to remove the protesters. Further protests were held throughout 2000.

Governor Calderón owed her success in 2000 in part to denouncing corruption in the Rosselló administration and in part to her promise to resolve the Vieques question. While mayor of San Juan, Calderón had questioned Rosselló's Vieques agreement with President Clinton and, following her election as Governor, she renounced the agreement reached between her predecessor and President Clinton, and called for the immediate withdrawal of the US Navy from the island, claiming that bombing over the years had caused serious health problems for the residents.

Military exercises were resumed on the island in May 2001. Following legal challenges and protests, operations were suspended, but they resumed again in June, prompting further protests. In early July the new US President, George W. Bush, announced that the US Navy would leave the island by May 2003, as planned. However, a non-binding referendum on the issue, organized by the Calderón Government, took place on 29 July, in which 68% of islanders voted for an immediate cessation of bombing. The US Navy's activities received endorsement from only 2% of voters. The binding referendum, agreed by Rosselló and Clinton in 2000, was still scheduled to take place in November, but was eventually cancelled by the US Congress, which ordered the US military to remain on the island, claiming that further tests were necessary as part of the Bush Administration's 'war on terror'.

Tests resumed in early April 2002, amid protests from members of the PIP, five of whom were arrested on charges of trespassing on US Navy property. Later that month, 10 US marines based on Vieques were attacked outside a café in San Juan; although those responsible were not caught, there was widespread belief that the attack was a result of anti-US sentiment. The US Navy carried out its final bombing exercises in February 2003 and withdrew from Vieques on 1 May, as promised. On that date, the Navy announced the end of its tenure on Vieques, ceding the land it owned to the US Department of the Interior to become a wildlife reserve. The Puerto Rican Government subsequently negotiated with the US Navy over cleaning up the island: after decades of bombing, contamination by heavy metals and other pollutants was identified in some two dozen sites. The base had provided jobs for around 2,000 people locally and contributed some US $300m. to the Puerto Rican economy. To compensate for the loss in tax revenue, the US Government was to pay $1.2m. per year to each of the municipalities during a period of redevelopment. The establishment of a local redevelopment authority meant that Vieques also qualified for federal redevelopment aid. Beyond more general proposals to encourage recreation and ecotourism, there were also plans to convert the base hospital and airport facilities for civilian use. The last remaining base, at Ceiba, was closed on 31 March 2004.

## THE 2004 ELECTIONS

In the gubernatorial election held on 2 November 2004, the PPD's candidate was Aníbal Acevedo Vilá of the PPD, the island's delegate to the US Congress, while the PNP's nominee was former Governor Pedro Rosselló. The closeness of the result led to a recount and a court challenge, but on 2 January 2005 Acevedo Vilá was officially declared the winner. He obtained 963,303 votes (48.4% of the votes cast), compared with 959,737 votes (48.2% of the ballot) for Rosselló. Rubén Berríos Martínez, once again representing the PIP, attracted 54,551 votes (2.7%). Acevedo Vilá took office as Governor on the same day. However, in the ballot to elect Puerto Rico's non-voting delegate in the US Congress, the pro-statehood PNP's Luis Fortuño won with 48.6% of the votes cast, ahead of the 48.0% attracted by Roberto Prats Palerm of Acevedo Vilá's PDP. Rosselló's PNP also took control of the legislature from the PDP, winning 17 senate seats (and 43.4% of the votes), compared with the PDP's nine seats (40.3%) and the PIP's one seat (9.4%). In the House of Representatives the PNP secured 32 seats and 46.3% of the ballot, while the PPD won 18 (43.1%) and the PIP again secured just one seat (9.7%).

In mid-March 2005 Governor Acevedo Vilá presented budget proposals for 2006 that included a spending reduction of US $370m. and the elimination of some 23,000 government jobs. While the proposals were rejected by the House of Representatives, by mid-year the members had still not presented an alternative budget. Meanwhile, in the same month there was widespread opposition to the decision by the US Territorial District Court of Puerto Rico to impose the death sentence on two convicted murderers. Capital punishment had been banned in the Territory in 1930, a decision that had been upheld in 2000 by a ruling of the Supreme Court of Puerto Rico that it violated the island's Constitution. However, the ruling was subsequently overturned by the US Court of Appeals, which found that Puerto Rico was subject to US federal law and that the death penalty was applicable in certain cases; this decision was upheld by the US Supreme Court. In early April Acevedo Vilá wrote to the US Attorney-General requesting that the death penalty should not apply to residents of Puerto Rico. In May the jury serving on the trial moved to sentence the two men to life imprisonment.

Acevedo Vilá had promised to summon a constitutional assembly to discuss Puerto Rico's status in 2005. He favoured maintaining the Commonwealth system with changes to allow for greater autonomy, particularly in economic development. Following the 2004 election members of the island's main political parties agreed a tripartite status bill. The proposed legislation scheduled a referendum for 10 July 2005, when Puerto Rican voters would vote for or against a petition urging the US Congress and President to provide Puerto Rico with 'non-colonial and non-territorial' status options and to pledge that the results would be honoured. Although the legislation was approved by both chambers and Governor Acevedo Vilá gave the impression that he would ratify it, he refused to do so. A hastily prepared substitute bill included an amendment stating that the US Congress was fully committed to legislation that would allow the Puerto Rican electorate to choose a mechanism to determine status, either by means of a constituent assembly or through a request for a direct congressionally binding referendum, in the event that the US Government did not commit itself to a process of free self-determination before the end of 2006. Although it was unanimously approved by the Puerto Rican congress on 10 April that legislation was also vetoed by Governor Acevedo Vilá as 'deceptive', because it did not guarantee that the method favoured by the Governor would be adopted.

# Economy

## Prof. PETER CALVERT

Based on an earlier article by JAMES MCDONOUGH

## INTRODUCTION

Puerto Rico has been a US territory since 1898, and, though it has full internal self-government, its economy is closely bound to that of the USA. It is a mountainous island measuring 177 km (110 miles) long by 56 km (35 miles) wide, with a relatively flat and narrow coastal belt. It is the smallest of the three Caribbean islands making up the Greater Antilles. At mid-2004 Puerto Rico's population was estimated at 3.9m. inhabitants, with a population density of around 434.8 per sq km, making the island one of the most densely populated areas on earth.

Until the 1950s Puerto Rico's economy was based on agriculture; in the 19th and early 20th centuries its principal cash crops were coffee, sugar and tobacco. However, in the 1940s the Government decided to seek economic growth through industrialization. The governing Partido Popular Democrático (PPD—Popular Democratic Party) decided to seek external capital, mainly from the USA, to spur economic growth. US capital was encouraged to invest in manufacturing facilities on the island through a unique combination of low wages supported by massive local and federal tax exemptions. The results of 'Operation Bootstrap' were immediate and astonishing, and soon became known throughout the world. Real gross national product (GNP) increased by 68% in the 1950s and by 90% in the 1960s. The average annual growth in GNP was approximately 6% during these two decades. Living standards rose accordingly, with personal income per head rising from US $342 per year in 1950, to $1,511 per year in 1971 and to $16,320 per year in 2001. By 1970 manufacturing constituted approximately 40% of the island's gross domestic product (GDP). By 2003/04 manufacturing accounted for 45.1% of GDP, but only 18.6% of the active work-force. At that time services constituted 54.4% of GDP and employed 79.5% of the engaged work-force. In 1970 unemployment declined to 10% of the labour force, astonishing progress given the island's traditional agricultural economy with its massive seasonal unemployment and widespread underemployment, and was still around 12% in 2005.

The extraordinary growth produced by 'Operation Bootstrap' ended following the petroleum crisis of 1973. The dramatic increases in world oil prices imposed by the Organization of the Petroleum Exporting Countries (OPEC) severely impeded Puerto Rico's capacity to sustain high economic growth, dependent as it was on imported oil and gas to meet all but a tiny fraction of its energy needs. At the same time, wage increases and reduced US tariffs on foreign products hampered Puerto Rico's ability to attract US capital, compared with other low-wage economies. Governor Rafael Hernández Colón, during his first term in office (1973–77), attempted to combat the economic slowdown through aggressive government intervention in the economy. Under his leadership, the Government purchased the Puerto Rico Telephone Company, the two major shipping lines serving Puerto Rico and US ports, and most of the island's sugar mills. Nevertheless, these efforts did not slow the relative economic decline and Puerto Rico's growth rate dropped to an annual average of 1.8% between 1974 and 1984. For the first time since 'Operation Bootstrap' began, the island experienced a decline in real GNP in 1975, 1982 and 1983. Economic disaster was avoided only by a massive increase in federal (US Government) funds, which offset the decline in the productive sectors.

Governor Hernández Colón won his second term in office in the 1984 gubernatorial election on the strength of his pledge to make job creation his main priority. His new term of office coincided with the recovery of the USA from the 1981 recession, which was accompanied by a fall in world petroleum prices. Nevertheless, he was faced with a serious challenge from the US Administration of President Ronald Reagan (1981–89), which sought to eliminate the special federal tax incentives for US investment in Puerto Rico. In response, Hernández Colón proposed an ambitious 'twin-plant' programme to promote industrial development in the Caribbean. His idea was to use the capital generated by the operations of US companies in Puerto Rico to invest in secondary Caribbean plants that would feed the companies' Puerto Rico operations. His programme won the approval of the Reagan Administration, which was deeply committed to assisting the Caribbean region under the Caribbean Basin Initiative, and efforts to eliminate the federal tax incentives for Puerto Rico were subsequently postponed. The success of Hernández Colón's policies was reflected in an annual average growth rate in real GDP of 3.8% in 1984–88. However, economic growth slowed considerably during Hernández Colón's third term in office (1989–93), mainly because of the advent of another world recession. Puerto Rico's real GDP grew by only 2.0% per year during this term, which in turn led to an increase in unemployment which remained high throughout the early 1990s.

In 1993 the incoming administration of Governor Pedro J. Rosselló (1993–2001) structured a new economic development programme in which the private sector was to be the primary vehicle for economic development. In 1994 the Governor introduced his 'new economic model', advocating reduced government regulation of businesses, the privatization of public enterprises and greater promotion of tourism, and the encouragement of local investment and local industry. In 1990–2001 Puerto Rico's GDP grew, in real terms, by an average of 4.2% per year, as the island mirrored the general success of the US economy over the same period. During the period total employment increased by 169,000 jobs, and unemployment decreased from 16.8% to 11.3%.

In keeping with its privatization policy, the Rosselló administration sold the government-owned shipping company, a major state-owned hotel property, the public sugar corporation and majority control of the state-owned Puerto Rico Telephone Company. However, in 1996, in a major reversal for the island, the US Congress approved legislation ending the island's tax credit (see below), the powerful federal tax incentive used to attract manufacturing investment from US corporations to Puerto Rico, which the Reagan Administration had tried to eliminate in the 1980s. Since the elimination of the tax incentive, Puerto Rico has lost some 38,000 manufacturing jobs. At mid-2005 total manufacturing employment was reported to be 118,000 (down from an historic high of 172,000 jobs in 1995). The PPD administration of Governor Sila María Calderón (2001–05) faced a slowing economy. Growth slowed in 2001–03, largely owing to the slowdown in the US economy, but recovered in 2004. GNP grew by 1.7% in 2001, by 2.0% in 2002, by 1.7% in 2003 and by 2.7% in 2004. The unemployment rate also increased slightly, to 12.0% in 2003, and net migration in 2004 was negative, at –1.38 per 1,000 people. In 2002 consumer prices increased by 6.2% and in 2003 by 7.4%. In 2004 GDP was estimated at US $68,950m. and per-head GDP at $17,700, ranking Puerto Rico with the world's higher-income countries. As the currency is the US dollar, purchasing power parity figures are identical.

## AGRICULTURE

Only 3.9% of the land area of the country was regarded by FAO as arable in 2001, with a further 5.5% under permanent crops and 23.5% permanent pasture. Some 400 sq km of arable land were irrigated. In 2003/04 only 2.1% of the active work-force worked on the land and agriculture contributed only an estimated 0.5% of GDP. Sugar had further declined in importance; in 2003 an estimated 320,000 metric tons were produced on 8,300 ha. Some 12,800 tons of green coffee were grown on 30,450 ha. Bananas and plantains accounted for a further 4,200 ha and 9,500 ha, respectively, producing 50,000 tons of bananas and 82,000 tons of plantains. Ranching and dairy farming (with a total of 390,000 head of cattle in 2003) had become the mainstay of the agricultural sector. Other farm animals included 118,000 pigs and 12.5m. chickens. In 2003 the total fish catch amounted to 3,188 tons, much of which was processed for export.

## MANUFACTURING

The growth of manufacturing in Puerto Rico was almost entirely the result of special tax benefits from both the US and Puerto Rican Governments. Under US law (Section 936 of the US Tax Code), the income of a subsidiary of a US corporation operating a manufacturing facility in Puerto Rico was eligible to receive a federal tax credit, which practically exempts the corporation from the payment of US income taxes on its Puerto Rican earnings. The Puerto Rican Government in 1948 matched the federal possessions tax credit with its own equally generous tax incentive programme. The combined incentives created a powerful lure to attract US corporate investment to Puerto Rico. In the early years of 'Operation Bootstrap', the island attracted the labour-intensive textiles and apparel industries. However, over the years, and as the federal incentive grew more generous, Puerto Rico manufacturing became more capital intensive and diversified, marked by substantial investment in sectors such as pharmaceuticals, scientific instruments, computers, microprocessors, medical products and electrical goods.

After 1996, when the US Congress repealed Section 936, Puerto Rico's industrialization programme began to decline. Nevertheless, the full impact of the credit's elimination will not be felt until the scheduled end of the 10-year 'grace' period, granted to companies operating in Puerto Rico at the time of the repeal of the tax credit, in December 2005. Growth in the manufacturing sector in the late 1990s resulted mainly from the expansion of those US companies (largely from the pharmaceutical sector) that were converting their legal structures to become Controlled Foreign Corporations (CFCs). Under US tax law, a CFC does not have to pay federal taxes on income earned in a foreign jurisdiction as long as the money remains outside the USA. For the purpose of US tax law, Puerto Rico is considered a foreign tax jurisdiction, so CFCs pay only royalties to their parent companies plus 10% in local taxes. This provision has been particularly attractive to pharmaceutical companies and other large corporations with global manufacturing and distribution networks. Since 1996 some 57 large manufacturing firms operating in Puerto Rico have reincorporated as CFCs.

Governor Calderón committed her administration to securing new tax incentives from the US Congress, at a time at which Congress remained committed to phasing out existing support by 2006. In 2001 Calderón sought US congressional approval to allow Puerto Rican CFCs to repatriate their earnings to their US corporate shareholders at a preferential 10% tax rate. The Republican-controlled Congress was not receptive to any change until February 2004, when Puerto Rico's then Resident Commissioner in the USA, Aníbal Acevedo Vila (who was elected Governor in November 2004), supported a 'compromise' bill to provide companies in the USA with a one-year period of a very low income tax rate on CFC funds. The 35% corporate income tax rate would be reduced to 5.25% for the year in the case of 'repatriated' funds. There was, however, a serious risk involved in providing a strong incentive for US-based companies to move capital out of the Commonwealth quickly.

## COMMERCE

Wholesale and retail trade played a significant role in Puerto Rico's economy. In 2003/04 the sector contributed 12.1% of total GDP. With 259,000 workers in 2003/04, commerce represented the third largest source of employment (21.0% of the active population) on the island, after services and government. In that year trade generated US $9,581.7m. in total revenues. Shopping centres dominated the retail trade. The island's first 'megastore' was built in the mid-1990s, consisting of several businesses that had previously operated individually, but were now gathered under one roof. This phenomenon spread and could be seen in the incorporation of pharmacies and banking service centres into department stores, as well as in fast-food establishments.

## FINANCE

Finance, insurance and real estate contributed an estimated 16.4% to the island's GDP in 2003/04. This thriving industry is represented by a broad spectrum of financial services, from large commercial banks and investment brokerage houses, to small loan companies and money transfer outlets. In 2001 bank assets totalled US $62,120m. and deposits were more than $35,000m. Brokered deposits by CFCs were reported to have risen from $2,600m. in 1995 to some $6,300m. in 2004. In 2003/04 the finance, insurance and real-estate sector employed a total of 43,000 workers, some 3.5% of the active work-force.

## TOURISM AND TRANSPORTATION

Tourism forms an increasingly important, but still small, part of Puerto Rico's economy. In 2002 the tourism industry constituted approximately 5.5% of GDP. Approximately 14,000 people, or 1.2% of the work-force, is employed by the tourism industry. During the early years of 'Operation Bootstrap' the Government utilized a combination of direct government and private investment to create the necessary infrastructure for tourism. It invested directly in the construction of hotels, but contracted with US corporations for the hotel management. In 1959 the industry was favourably affected by the Cuban Revolution, which led to the US economic embargo and the diversion of much of Cuba's US tourism to other Caribbean destinations. The island's traditional emphasis has been on relatively expensive casinos and hotels located along the beaches. However, in the late 1970s the Government developed a unique system of country inns to move tourism out of San Juan, and improved transportation facilities, including the expansion of the airport and new tourism piers in San Juan's harbour. In 2003 tourism revenue totalled US $2,677m., compared with $2,486m. in the previous year. San Juan is the world's second largest home port for cruise-ship passengers; it has seven piers and can dock 12 ships at one time. In 2002 the island received approximately 1.28m. cruise-ship visitors. However, this was a marked fall from the 1.36m. cruise-ship visitors in 2001. In 2003 some 3.2m. tourists visited the island.

The development of tourism was a priority for the Rosselló administration, which extended additional tax credits to the sector to encourage hotel construction. At the same time, Rosselló created a new subsidiary of the Government Development Bank to encourage and promote investment in new hotels. As a result, the number of hotel rooms increased every year—except in 1998, when 'Hurricane Georges' resulted in the closure of several major hotels—under his administration, reaching

11,915 by mid-2000. Tourism received a further set-back owing to the general impact of the terrorist attacks on the USA of 11 September 2001, from which it subsequently recovered.

The Rosselló administration was particularly successful in decentralizing tourism outside San Juan. New projects were built or planned east of San Juan, extending along the coast to Fajardo, and new tourism enclaves were appearing along the west coast between San Juan and Aguadilla. In 1999 the Government announced plans to create an ambitious tourism and trade centre, to be known as the Golden Triangle, on the Isla Grande peninsula in San Juan. The Calderón administration continued with the project, signing a US $200m. agreement with LCOR, Inc, for the construction of two hotels, the trade centre and a retail and entertainment area, but work proceeded slowly. In March 2004 Hotel Sol Meliá, one of Europe's largest hotel chains, inaugurated the first all-inclusive resort on the island, the Paradisus, in Rio Grande, and further resorts and hotels under construction throughout the island brought the total number of rooms to 13,000 in that year. Construction of new hotels elsewhere was ongoing in 2005.

The island's transport facilities are excellent. Of Puerto Rico's three international airports, Luis Muñoz Marín International Airport, in San Juan, was served by 45 US and international airlines. In 2001 the airport served approximately 4.7m. passengers and moved 215,603 metric tons of freight. The airport offered daily direct air services between San Juan and more than seven US cities, and regular scheduled services to other Caribbean islands and major Latin American and European cities, though it suffered a reverse in late 1997, when American Airlines, the island's biggest passenger carrier, transferred its Caribbean hub. There are 24,431 km of roads, 94% of which are paved, 96 km of narrow-gauge railway and 10 major ports apart from San Juan itself, regarded by many as the finest in the Caribbean.

In 1996 construction began of a 'super-aqueduct' to bring water from the western part of the island to the San Juan metropolitan area, where traditional water supplies were insufficient to handle the city's growing population. In the same year the Government also began the construction of a 16.6-km (10.3-mile), mass transit system for the San Juan area. The first phase of the Urban Train was inaugurated in January 2005. The US $2,150m. project was made possible by an initial $300m. commitment by the US Government, and the Urban Train's corridor, Ciudad Red, was expected to bring in $400m. in public- and private-sector investment, 16,000 direct jobs, 1,080 housing units and 373,000 sq ft of commercial and office space. An extension to Carolina and Minillas was proposed in March 2005, at a projected cost of $900m. In 2000 plans were also announced for the construction of a deep-water 'megaport' in Guayanilla on the south coast. The Calderón administration broadened the reach of the port to include the neighbouring ports of Ponce and Peñuelas. The planned 'Port of the Américas' at Ponce would provide world-class, deep-water docking and onshore warehouse space and factory sites for assembly and re-export of semi-finished goods. Construction of the port was ongoing in 2005, at a projected cost of $700m., and was scheduled to be operational by 2006–07. The Puerto Rican Government intended to capture a significant part of the Caribbean transhipment market, which had grown by 273% since 1995. Despite optimism surrounding the project, however, the Government has not fully implemented its ambitious infrastructure proposals and construction has been subject to delays.

## THE PUBLIC SECTOR

The government sector was the second largest area of employment. In 2003/04 the sector employed approximately 275,000 people, or 22.3% of the total labour force. The sector accounted for US $7,388.5m., or an estimated 9.3%, of Puerto Rico's GDP. One major reason for Puerto Rico's ability to maintain such high public-sector employment was the economic aid it received from the USA. Net federal disbursements to Puerto Rico were estimated to have increased from $2,900m. in 1979 to $8,900m. in 2002, the latter an increase of 3.7% over the previous year. Three-quarters of this (75.8%) consisted of earned benefits, e.g. social security, Medicare and veterans payments; the balance consisted of the National Assistance Program and scholarships.

Consequently, in 2002 Puerto Ricans paid $1,724.19 per person in taxes, compared with an average of $6,702.42 for other US citizens.

The Rosselló administration opened the government-owned and -operated electricity system, the Electrical Energy Authority of Puerto Rico (Autoridad de Energía Eléctrica de Puerto Rico), to private co-generators. In 2001 a 507-MW, liquefied natural gas (LNG) co-generation plant, a desalination plant and an LNG storage facility, located on the southern coast of the island, became operational. The water from the desalination plant was used to cool the turbines and the excess sold as fresh water to the Government's Aqueduct and Sewer Authority. In early 2002 the French utilities group Suez won a contract worth US $4,100m. to supply the island's water and treat its waste water over a period of 10 years.

## ENERGY

In mid-2001 the USA's AES Corporation opened a US $850m., coal-burning, co-generation plant. Like the LNG co-generation plant, it has a contract with the Electrical Energy Authority of Puerto Rico to sell its output to the government network. Construction of the plants was encouraged by the Government, in an attempt to diversify the island's electrical generation, which hitherto had been completely dependent on imported petroleum as its fuel source. In 2002 oil consumption was estimated at 223,000 barrels per day; refining capacity, however, was only 114,400 in 2004. Puerto Rico has no indigenous sources of petroleum or, for that matter, of natural gas, of which 200,000m. cu ft were imported and consumed in 2002. Electricity consumption in 2002 was 22,100m. kWh, 99.2% of which came from the burning of fossil fuel and 0.8% from hydropower. In April 2005 the Electrical Energy Authority of Puerto Rico announced that it was to build two new gas-fired power-stations with a combined capacity of 971 MW.

## PUBLIC FINANCE

Puerto Rico is constitutionally required to operate within a balanced budget. The island's Constitution provided for a limitation on the amount of general debt that could be issued, equal to 15% of the average annual revenues raised over the previous two years. Historically, Puerto Rico kept within its constitutional bounds; in mid-2002 the Commonwealth had a 6.7% borrowing margin, well within its 15% limit. At 30 June 2004 the provisional figure for total public debt—including the debt of the central government, public corporations and municipalities—amounted to US $33,942.7m. The island suffered a reverse in early 2002 when its bonds were reclassified at a lower level. This meant that the Commonwealth would have to pay a higher interest rate on outstanding debts, resulting from operational budget deficits for two years, which the former administration had concealed.

## TRADE

Although Puerto Rico for import/export purposes is a distinct entity, the Commonwealth lies within the customs barrier of the USA. Consequently, exports and imports are dominated by the USA. Puerto Rico is the fifth largest exporter in the western hemisphere in terms of the total value of its exports, ahead of such larger economies as Chile, Argentina, Colombia and Venezuela. Between 1997 and 2003 exports doubled in value. In 2002/03 the island exported goods valued at US $55,175m., equivalent to 73.7% of GDP. Imports grew more slowly, reaching $33,750m. in 2002/03, giving the island an increasingly positive balance of trade, much of which was accounted for by exports of pharmaceuticals. Imports were driven mainly by raw materials used by the island's factories. From 1982 until the present the island has operated with an annual trade surplus, which stood at $14,732.9m. in 2003/043. The percentage of exports to the USA remained relatively constant from the 1950s to the end of the 1990s, at around 90% of the total. However, goods shipped from the USA to Puerto Rico decreased from 92% of total imports in 1950 to 53.5% in 2001.

# PUERTO RICO

After the USA (86.4%), in 2003 Puerto Rico's biggest external markets were the Netherlands (2.1%) and Belgium (2.0%). With respect to imports, next to the USA (48.9%) in 2003 came Ireland (20.7%) and Japan (3.9%). The Calderón administration sought to increase trade to five target areas: Florida and the east coast of the USA; the US Hispanic/Latino market (some 39m. people); Mexico (a North American Free Trade Agreement—NAFTA—trading partner); the Dominican Republic, the island's fourth largest trading partner; and the Greater Caribbean area. The island Government believed it was competitive internationally in food and beverage production, contract manufacturing, technology, construction and services—particularly engineering and architecture—environmental engineering and finance. Puerto Rico intended to strengthen bilateral relations with other important economies, including Costa Rica, Panama, Chile and the European Union, and to become more active in regional economic organizations.

# Statistical Survey

Source (unless otherwise stated): Puerto Rico Planning Board, POB 41119, San Juan, 00940-1119; tel. (787) 723-6200; internet www.jp.gobierno.pr.

## Area and Population

### AREA, POPULATION AND DENSITY

| | |
|---|---:|
| Area (sq km) | 8,959* |
| Population (census results) | |
| 1 April 1990 | 3,522,037 |
| 1 April 2000 | |
| Males | 1,833,577 |
| Females | 1,975,033 |
| Total | 3,808,610 |
| Population (official estimates at mid-year) | |
| 2002 | 3,859,000 |
| 2003 | 3,879,000 |
| 2004 | 3,895,000† |
| Density (per sq km) at mid-2004 | 434.8 |

* 3,459 sq miles.
† Provisional figure.

### PRINCIPAL TOWNS
(population at census of 1 April 2000)

| | | | | |
|---|---:|---|---:|
| San Juan (capital) | 421,958 | Caguas | 88,680 |
| Bayamón | 203,499 | Guaynabo | 78,806 |
| Carolina | 168,164 | Mayagüez | 78,647 |
| Ponce | 155,038 | Trujillo Alto | 50,841 |

Source: Bureau of the Census, US Department of Commerce.

### BIRTHS, MARRIAGES AND DEATHS

| | Registered live births | | Registered marriages | | Registered deaths | |
|---|---:|---:|---:|---:|---:|---:|
| | Number | Rate (per 1,000) | Number | Rate (per 1,000)* | Number | Rate (per 1,000) |
| 1996 | 63,259 | 17.0 | 32,572 | 11.7 | 29,871 | 8.0 |
| 1997 | 64,214 | 16.9 | 31,493 | 11.0 | 29,119 | 7.7 |
| 1998 | 60,518 | 15.8 | 26,390 | 9.3 | 29,990 | 7.8 |
| 1999 | 59,684 | 15.4 | 27,255 | 9.3 | 29,145 | 7.5 |
| 2000 | 59,460 | 15.5 | 25,980 | 8.9 | 28,550 | 7.6 |
| 2001 | 55,983 | 14.6 | 28,598 | 7.4 | 28,794 | 7.5 |
| 2002 | 52,871 | 13.7 | 25,645 | 6.6 | 28,098 | 7.3 |

* Rates calculated using estimates of population aged 15 years and over.

**2003** (revised figures, rounded): Births 51,000 (Birth rate 13.1 per 1,000); Deaths 28,000 (Death rate 7.3 per 1,000).

**2004** (provisional figures, rounded): Births 53,000 (Birth rate 13.8 per 1,000); Deaths 28,000 (Death rate 7.4 per 1,000).

Source: mainly Department of Health, Commonwealth of Puerto Rico.

**Expectation of life** (UN estimates, years at birth): 74.9 (males 70.4; females 79.6) in 1995–2000 (Source: UN, *World Population Prospects: The 2002 Revision*).

### ECONOMICALLY ACTIVE POPULATION
('000 persons aged 16 years and over)

| | 2001/02 | 2002/03 | 2003/04 |
|---|---:|---:|---:|
| Agriculture, forestry and fishing | 23 | 25 | 26 |
| Manufacturing | 139 | 136 | 139 |
| Construction | 86 | 83 | 90 |
| Trade | 240 | 257 | 259 |
| Transportation | 30 | 28 | 27 |
| Communication | 18 | 15 | 16 |
| Finance, insurance and real estate | 42 | 43 | 43 |
| Other public utilities | 14 | 14 | 13 |
| Services | 316 | 335 | 348 |
| Government | 261 | 274 | 275 |
| **Total employed*** | **1,170** | **1,211** | **1,234** |

* Includes sectors employing fewer than 2,000 people, not listed separately.

Source: Department of Labor and Human Resources Statistics, *Household Survey*.

**Unemployed:** 147 in 2001; 166 in 2002; 167 in 2003 (Source: ILO).

**Total labour force:** 1,297 in 2001; 1,356 in 2002; 1,393 in 2003 (Source: ILO).

## Health and Welfare

### KEY INDICATORS

| | |
|---|---:|
| Total fertility rate (children per woman, 2002) | 1.9 |
| Under-5 mortality rate (per 1,000 live births, 2002) | 12.2 |
| Physicians (per 1,000 head, 1999) | 1.75 |
| Health expenditure (1994): % of GDP | 6.0 |

Source: Pan-American Health Organization.
For definitions see explanatory note on p. vi.

## Agriculture

### PRINCIPAL CROPS
('000 metric tons)

| | 1997 | 1998 | 1999* |
|---|---:|---:|---:|
| Sugar cane | 307.4 | 244.4 | 320.0 |
| Tomatoes | 9.9 | 4.2 | 5.0 |
| Pumpkins, squash and gourds | 14.0* | 11.1 | 11.0 |
| Bananas | 38.2 | 49.3 | 50.0 |
| Plantains | 76.0* | 92.5 | 82.0 |
| Oranges | 16.1 | 25.7 | 25.7 |
| Mangoes | 17.4 | 17.4* | 17.4 |
| Pineapples | 19.2 | 14.2 | 15.0 |
| Coffee (green) | 11.6 | 13.4 | 12.8 |

* FAO estimate(s).

**2000–03** (FAO estimates): data assumed to be unchanged from 1999.

**Oranges** (FAO estimates, '000 metric tons): 25.7 in 2001; 25.5 in 2002; 25.5 in 2003.

Source: FAO.

# PUERTO RICO

## LIVESTOCK
('000 head, year ending September)

|  | 1998 | 1999* | 2000* |
|---|---|---|---|
| Cattle | 387.0 | 390.0 | 390.0 |
| Sheep | 16.4 | 16.0 | 16.0 |
| Goats | 9.2 | 9.0 | 9.0 |
| Pigs | 114.8 | 118.0 | 118.0 |
| Horses | 25.0* | 25.0 | 26.0 |
| Poultry | 12,487 | 12,500 | 12,500 |

* FAO estimate(s).

**2001–03** (FAO estimates): Data assumed to be unchanged from 2000.

Source: FAO.

## Fishing

(metric tons, live weight)

|  | 2001 | 2002 | 2003 |
|---|---|---|---|
| Capture | 3,794 | 2,529 | 2,919 |
| Snappers and jobfishes | 744 | 512 | 603 |
| Seerfishes | 124 | 90 | 94 |
| Marine fishes | 173 | 126 | 104 |
| Caribbean spiny lobster | 190 | 158 | 196 |
| Stromboid conchs | 1,643 | 931 | 1,141 |
| Aquaculture | 414 | 441 | 269 |
| Penaeus shrimps | 205 | 225 | 69 |
| **Total catch** | 4,208 | 2,970 | 3,188 |

Source: FAO.

## Industry

### SELECTED PRODUCTS
(year ending 30 June)

|  | 1995/96 | 1996/97 | 1997/98* |
|---|---|---|---|
| Distilled spirits ('000 proof gallons) | 25,343 | 36,292 | 33,471 |

* Preliminary.

**Electric energy** (million kWh): 20,140.8 in 1998/99; 21,459.8 in 1999/2000; 22,132.2 in 2000/01.

**Cement** ('000 metric tons): 1,660 in 2000; 1,550 in 2001; 1,540 in 2002 (estimate) (Source: US Geological Survey).

**Beer** ('000 hectolitres): 317 in 1997; 263 in 1998; 259 in 1999 (Source: UN, *Industrial Commodity Statistics Yearbook*).

## Finance

### CURRENCY AND EXCHANGE RATES

**Monetary Units**
United States currency: 100 cents = 1 US dollar (US $).

**Sterling and Euro Equivalents** (31 May 2005)
£1 sterling = US $1.8181;
€1 = US $1.2331;
$100 = £55.00 = €81.09.

### BUDGET
(US $ '000, year ending 30 June)

| Revenue | 1998/99 | 1999/2000 | 2000/01 |
|---|---|---|---|
| Taxation | 6,136,230 | 6,791,200 | 6,417,856 |
| Income taxes | 4,413,860 | 4,967,138 | 4,536,840 |
| Excise taxes | 1,714,444 | 1,736,539 | 1,788,992 |
| Service charges | 457,454 | 617,020 | 645,806 |
| Intergovernmental transfers | 3,435,765 | 2,971,528 | 3,807,049 |
| Other receipts | 162,228 | 383,548 | 270,711 |
| **Total*** | 10,360,557 | 10,854,821 | 11,208,442 |

* Including interest payments (US $ '000): 97,880 in 1998/99; 91,525 in 1999/2000; 67,020 in 2000/01.

| Expenditure | 1998/99 | 1999/2000 | 2000/01 |
|---|---|---|---|
| General government | 526,629 | 853,040 | 739,009 |
| Public safety | 1,103,606 | 1,310,322 | 1,623,362 |
| Health | 625,475 | 972,757 | 954,563 |
| Aid to municipalities | 318,664 | 373,016 | 222,721 |
| Public housing and welfare | 2,485,092 | 2,102,410 | 2,315,899 |
| Education | 2,272,903 | 2,436,267 | 2,308,479 |
| Economic development | 314,897 | 337,255 | 170,937 |
| Capital outlays | 642,016 | 853,597 | 1,020,344 |
| Debt service | 794,336 | 860,964 | 1,011,468 |
| **Total** | 9,083,618 | 10,079,628 | 10,366,782 |

**Total expenditure** (US $ million): 10,508.4 in 2001/02; 11,465.4 in 2002/03; 11,299.8 in 2003/04.

Source: Department of the Treasury, Commonwealth of Puerto Rico.

### COST OF LIVING
(Consumer Price Index; base: 2000 = 100)

|  | 2001 | 2002 | 2003 |
|---|---|---|---|
| Food (incl. beverages) | 114.1 | 127.8 | 145.8 |
| Fuel and light | 101.2 | 100.6 | 105.6 |
| Rent | 102.6 | 104.4 | 106.0 |
| Clothing (incl. footwear) | 98.1 | 97.0 | 95.5 |
| **All items** (incl. others) | 107.0 | 113.6 | 122.5 |

Source: ILO.

### NATIONAL ACCOUNTS
(US $ million at current prices, year ending 30 June)

**Expenditure on the Gross Domestic Product**

|  | 1997/98 | 1998/99 | 1999/2000* |
|---|---|---|---|
| Government final consumption expenditure | 7,098.9 | 7,486.3 | 7,208.2 |
| Private final consumption expenditure | 32,194.2 | 34,620.1 | 36,592.7 |
| Increase in stocks | 31.2 | 484.5 | 347.2 |
| Gross fixed capital formation | 9,293.7 | 12,057.0 | 12,560.6 |
| **Total domestic expenditure** | 48,618.0 | 54,647.9 | 56,708.7 |
| Exports of goods and services *Less* Imports of goods and services | 5,515.1 | 5,391.4 | 6,441.0 |
| **GDP in purchasers' values** | 54,133.1 | 60,039.3 | 63,149.7 |
| **GDP at constant 1954 prices** | 9,252.1 | 9,915.8 | 10,276.8 |

* Preliminary.

**Gross Domestic Product by Economic Activity**

|  | 2001/02 | 2002/03 | 2003/04* |
|---|---|---|---|
| Agriculture | 276.5 | 313.9 | 434.7 |
| Manufacturing | 31,242.9 | 32,500.6 | 34,077.5 |
| Construction and mining† | 1,647.6 | 1,614.3 | 1,740.7 |
| Transportation and public utilities‡ | 4,948.3 | 5,205.4 | 5,349.5 |
| Trade | 8,622.8 | 9,005.3 | 9,581.7 |
| Finance, insurance and real estate | 11,211.9 | 12,425.4 | 13,024.2 |
| Services | 7,078.5 | 7,257.0 | 7,899.2 |
| Government | 6,302.8 | 7,005.8 | 7,388.5 |
| **Sub-total** | 71,331.3 | 75,327.7 | 79,496.0 |
| Statistical discrepancy | 292.3 | –493.3 | –653.7 |
| **Total** | 71,623.5 | 74,834.4 | 78,842.2 |

* Preliminary.
† Mining includes only quarries.
‡ Includes radio and television broadcasting.

PUERTO RICO

## BALANCE OF PAYMENTS
(US $ million, year ending 30 June)

|  | 2001/02 | 2002/03 | 2003/04 |
|---|---|---|---|
| Merchandise exports f.o.b. | 49,610.1 | 56,334.7 | 59,449.4 |
| Merchandise imports f.o.b. | −36,740.7 | −42,512.8 | −44,716.5 |
| **Trade balance** | 12,869.4 | 13,821.9 | 14,732.9 |
| Investment income received | 1,025.9 | 1,011.2 | 1,185.4 |
| Investment income paid | −28,553.7 | −29,392.1 | −30,683.8 |
| Services and other income (net) | 1,175.7 | 1,107.1 | 1,243.1 |
| Unrequited transfers (net) | 9,723.5 | 10,144.1 | 10,245.9 |
| Net interest of Commonwealth and municipal governments | −416.2 | −421.0 | −378.9 |
| **Current balance** | −4,175.4 | −3,728.8 | −3,655.4 |

## External Trade

### PRINCIPAL COMMODITIES
(US $ million, year ending 30 June)*

| Imports | 2000/01 | 2001/02 | 2002/03 |
|---|---|---|---|
| Mining products | 225.1 | 315.5 | 800.1 |
| Manufacturing products | 27,679.5 | 27,360.2 | 31,608.1 |
| Food, beverages and tobacco | 2,567.2 | 2,338.8 | 2,274.1 |
| Clothing and textiles | 1,016.7 | 960.4 | 931.4 |
| Paper, printing and publishing | 733.5 | 700.9 | 691.3 |
| Chemical products | 10,503.0 | 12,324.5 | 15,111.2 |
| Petroleum refining and related products | 2,309.3 | 1,592.7 | 1,923.1 |
| Rubber and plastic products | 546.8 | 521.9 | 637.9 |
| Primary metal products | 398.5 | 341.4 | 401.8 |
| Machinery, except electrical | 1,280.9 | 1,073.6 | 1,153.3 |
| Computer, electronic and electrical products | 3,507.7 | 2,889.3 | 3,432.0 |
| Transport equipment | 2,160.1 | 2,189.7 | 2,345.8 |
| **Total** (incl. others) | 29,149.3 | 29,984.6 | 33,749.7 |

| Exports | 2000/01 | 2001/02 | 2002/03 |
|---|---|---|---|
| Manufacturing products | 46,442.3 | 46,722.9 | 54,690.1 |
| Food, beverages and tobacco | 3,842.0 | 3,832.1 | 3,265.4 |
| Clothing and textiles | 680.8 | 621.2 | 537.3 |
| Chemical products | 30,767.1 | 33,307.1 | 39,603.9 |
| Machinery, except electrical | 680.4 | 508.7 | 616.8 |
| Computer, electronic and electrical products | 7,308.3 | 5,389.0 | 6,886.7 |
| **Total** (incl. others) | 46,900.8 | 47,172.3 | 55,175.3 |

* Figures refer to recorded transactions only.

### PRINCIPAL TRADING PARTNERS
(US $ million)*

|  | 1994/95 | | 1995/96 | |
|---|---|---|---|---|
|  | Imports | Exports | Imports | Exports |
| Belgium | 53.2 | 165.2 | 122.9 | 205.9 |
| Canada | 187.5 | 74.2 | 197.8 | 75.5 |
| Dominican Repub. | 664.2 | 693.7 | 768.1 | 677.4 |
| France | 130.3 | 86.1 | 194.1 | 103.6 |
| Germany | 207.0 | 280.8 | 175.7 | 233.8 |
| Ireland | 180.8 | 27.4 | 278.8 | 19.6 |
| Italy | 203.8 | 58.3 | 279.0 | 56.7 |
| Japan | 1,278.5 | 91.1 | 1,222.8 | 184.5 |
| Korea, Repub. | 216.8 | 43.8 | 236.7 | 61.1 |
| Mexico | 202.2 | 90.7 | 181.3 | 106.0 |
| Netherlands | 93.1 | 157.2 | 100.2 | 119.1 |
| United Kingdom | 708.3 | 178.0 | 560.8 | 159.0 |
| USA | 12,158.1 | 21,106.9 | 11,909.3 | 20,148.6 |
| US Virgin Islands | 307.0 | 164.8 | 366.8 | 137.3 |
| Venezuela | 424.0 | 43.1 | 509.5 | 35.8 |
| **Total** (incl. others) | 18,816.6 | 23,811.3 | 19,060.9 | 22,944.4 |

* Recorded trade only.

**1996/97** (US $ million): Imports: USA 13,317.8; Total (incl. others) 21,387.4. Exports: USA 21,187.3; Total (incl. others) 23,946.8.

**1997/98** (US $ million): Imports: USA 13,225.9; Total (incl. others) 21,797.5. Exports: USA 27,397.4; Total (incl. others) 30,272.9.

## Transport

### ROAD TRAFFIC
(vehicles in use)

|  | 1987/88 | 1988/89 | 1989/90 |
|---|---|---|---|
| Cars: private | 1,322,069 | 1,289,873 | 1,305,074 |
| Cars: for hire | 14,814 | 11,033 | 10,513 |
| Trucks: private | 15,790 | 13,273 | 12,577 |
| Trucks: for hire | 4,131 | 3,933 | 3,283 |
| Light trucks | 176,583 | 174,277 | 189,705 |
| Other vehicles | 75,155 | 74,930 | 60,929 |
| **Total** | 1,608,542 | 1,567,319 | 1,582,081 |

**2001:** Passenger cars 2,075,521; Trucks 33,803; Other vehicles (incl. motorcycles) 25,163 (Source: Federal Highway Administration, US Department of Transportation, *Highway Statistics 2001*).

### SHIPPING
(year ending 30 June)

|  | 1985/86 | 1986/87 | 1987/88 |
|---|---|---|---|
| Passengers arriving | 29,559 | 59,089 | 33,737 |
| Passengers departing | 33,683 | 63,987 | 35,627 |
| Cruise visitors | 448,973 | 584,429 | 723,724 |

**Cruise visitors** (calendar year): 1,300,075 in 2000; 1,349,630 in 2001.

**Freight handled** (short tons): 10,231,435 in 2000; 9,729,644 in 2001.

Source: Puerto Rico Ports Authority.

### CIVIL AVIATION
(year ending 30 June)

|  | 1988/89 | 1989/90 | 1990/91 |
|---|---|---|---|
| Passengers arriving | 4,064,762 | 4,282,324 | 4,245,137 |
| Passengers departing | 4,072,828 | 4,297,521 | 4,262,154 |
| Freight (tons)* | 173,126 | 208,586 | 222,172 |

* Handled by the Luis Muñoz Marin International Airport.

**Passengers arriving** (calendar year): 4,688,477 in 2001.

**Passengers departing** (calendar year): 4,707,829 in 2001.

**Freight handled** (calendar year): 235,880 tons in 2000; 215,603 tons in 2001.

Source: Puerto Rico Ports Authority.

## Tourism

|  | 2001 | 2002 | 2003 |
|---|---|---|---|
| Total visitors ('000) | 3,551.2 | 3,087.1 | 3,238.3 |
| From USA | 2,616.2 | 2,212.9 | 2,454.3 |
| From US Virgin Islands | 18.8 | 17.5 | 16.2 |
| From other countries | 916.2 | 856.7 | 767.8 |
| Expenditure ($ million) | 2,728 | 2,486 | 2,677 |

## Communications Media

|  | 1999 | 2000 | 2001 |
|---|---|---|---|
| Television receivers ('000 in use) | 1,270 | 1,290 | n.a. |
| Telephones ('000 main lines in use) | 1,294.7 | 1,299.3 | 1,329.5 |
| Mobile cellular telephones ('000 subscribers) | 813.8 | 926.4 | 1,211.1 |
| Internet users ('000) | 200 | 400 | 600 |

**Radio receivers** ('000 in use): 2,840 in 1997.

**Facsimile machines** ('000 in use): 543 in 1993.

**Daily newspapers** (1996): 3; average circulation ('000 copies) 475.

**Non-daily newspapers** (1988 estimates): 4; average circulation ('000 copies) 106.

Sources: UNESCO, *Statistical Yearbook*; UN, *Statistical Yearbook*; International Telecommunication Union.

## Education

|  | 1993/94 | 1994/95 | 1995/96 |
|---|---|---|---|
| Total number of students | 933,183 | 934,406 | 947,249 |
| Public day schools | 631,460 | 621,370 | 627,620 |
| Private schools (accredited)* | 140,034 | 145,864 | 148,004 |
| University of Puerto Rico† | 53,935 | 54,353 | 62,341 |
| Private colleges and universities | 107,754 | 112,819 | 109,284 |
| Number of teachers‡ | 39,816 | 40,003 | 39,328 |

* Includes public and private accredited schools not administered by the Department of Education.
† Includes all university-level students.
‡ School teachers only.

**2000:** Students at public day schools 606,487; Total number of students 909,998.

**Adult literacy rate** (UNESCO estimates): 94.1% (males 93.9%; females 94.4%) in 2002 (Source: UNESCO).

# Directory

## The Constitution

### RELATIONSHIP WITH THE USA

On 3 July 1950 the Congress of the United States of America adopted Public Law No. 600, which was to allow 'the people of Puerto Rico to organize a government pursuant to a constitution of their own adoption'. This Law was submitted to the voters of Puerto Rico in a referendum and was accepted in the summer of 1951. A new Constitution was drafted in which Puerto Rico was styled as a commonwealth, or estado libre asociado, 'a state which is free of superior authority in the management of its own local affairs', though it remained in association with the USA. This Constitution, with its amendments and resolutions, was ratified by the people of Puerto Rico on 3 March 1952, and by the Congress of the USA on 3 July 1952; and the Commonwealth of Puerto Rico was established on 25 July 1952.

Under the terms of the political and economic union between the USA and Puerto Rico, US citizens in Puerto Rico enjoy the same privileges and immunities as if Puerto Rico were a member state of the Union. Puerto Rican citizens are citizens of the USA and may freely enter and leave that country.

The Congress of the USA has no control of, and may not intervene in, the internal affairs of Puerto Rico.

Puerto Rico is exempted from the tax laws of the USA, although most other federal legislation does apply to the island. Puerto Rico is represented in the US House of Representatives by a non-voting delegate, the Resident Commissioner, who is directly elected for a four-year term. The island has no representation in the US Senate.

There are no customs duties between the USA and Puerto Rico. Foreign products entering Puerto Rico—with the single exception of coffee, which is subject to customs duty in Puerto Rico, but not in the USA—incur the same customs duties as would be paid on their entry into the USA.

The US social security system is extended to Puerto Rico, except for unemployment insurance provisions. Laws providing for economic co-operation between the Federal Government and the States of the Union for the construction of roads, schools, public health services and similar purposes are extended to Puerto Rico. Such joint programmes are administered by the Commonwealth Government.

Amendments to the Constitution are not subject to approval by the US Congress, provided that they are consistent with the US federal Constitution, the Federal Relations Act defining federal relations with Puerto Rico and Public Law No. 600. Subject to these limitations, the Constitution may be amended by a two-thirds' vote of the Puerto Rican Legislature and by the subsequent majority approval of the electorate.

### BILL OF RIGHTS

No discrimination shall be made on account of race, colour, sex, birth, social origin or condition, or political or religious ideas. Suffrage shall be direct, equal and universal for all over the age of 18. Public property and funds shall not be used to support schools other than State schools. The death penalty shall not exist. The rights of the individual, of the family and of property are guaranteed. The Constitution establishes trial by jury in all cases of felony, as well as the right of habeas corpus. Every person is to receive free elementary and secondary education. Social protection is to be afforded to the old, the disabled, the sick and the unemployed.

### THE LEGISLATURE

The Legislative Assembly consists of two chambers, the members of which are elected by direct vote for a four-year term. The Senate is composed of 27 members, who must be over 30 years of age. The House of Representatives is comprised of 51 members, of whom 40 are elected on a constituency basis, and a further 11 are at large members, elected by proportional representation. Representatives must be over 25 years of age. The Constitution guarantees the minority parties additional representation in the Senate and the House of Representatives, which may fluctuate from one quarter to one third of the seats in each House.

The Senate elects a President and the House of Representatives a Speaker from their respective members. The sessions of each house are public. A majority of the total number of members of each house constitutes a quorum. Either house can initiate legislation, although bills for raising revenue must originate in the House of Representatives. Once passed by both Houses, a bill is submitted to the Governor, who can either sign it into law or return it, with his reasons for refusal, within 10 days. If it is returned, the Houses may pass it again by a two-thirds' majority, in which case the Governor must accept it.

The House of Representatives, or the Senate, can impeach one of its members for treason, bribery, other felonies and 'misdemeanours involving moral turpitude'. A two-thirds' majority is necessary before an indictment may be brought. The cases are tried by the Senate. If a Representative or Senator is declared guilty, he is deprived of his office and becomes punishable by law.

### THE EXECUTIVE

The Governor, who must be at least 35 years of age, is elected by direct suffrage and serves for four years. Responsible for the execution of laws, the Governor is Commander-in-Chief of the militia and has the power to proclaim martial law. At the beginning of every regular session of the Assembly, in January, the Governor presents a report on the state of the treasury, and on proposed expenditure. The Governor chooses the Secretaries of Departments, subject to the approval of the Legislative Assembly. These are led by the Secretary of State, who replaces the Governor at need.

### LOCAL GOVERNMENT

The island is divided into 78 municipal districts for the purposes of local administration. The municipalities comprise both urban areas and the surrounding neighbourhood. They are governed by a mayor and a municipal assembly, both elected for a four-year term.

## The Government

### HEAD OF STATE

**Governor:** ANÍBAL ACEVEDO VILÁ (took office 2 January 2005).

# PUERTO RICO

## EXECUTIVE
(July 2005)

**Secretary of State:** Marisara Pont Marchese.
**Secretary of the Interior:** Aníbal José Torres.
**Secretary of Justice:** Roberto Sánchez Ramos.
**Secretary of the Treasury:** Juan Carlos Méndez.
**Secretary of Education:** Gloria E. Baquero Lleras.
**Secretary of Transportation and Public Works:** Gabriel Alcaraz Emmanuelli.
**Secretary of Health:** Rosa Pérez Perdomo.
**Secretary of Agriculture:** José Orlando Fabré Laboy.
**Secretary of Housing:** Jorge Rivera Jiménez.
**Secretary of Natural and Environmental Resources:** Javier Vélez Arocho.
**Secretary of Consumer Affairs:** Alejandro García Padilla.
**Secretary of Sports and Recreation:** David E. Bernier Rivera.
**Secretary of Economic Development and Commerce:** Jorge P. Silva Puras.
**Attorney-General:** Salvador Antonetti Stutts.
**Resident Commissioner in Washington:** Luis Fortuño.

## GOVERNMENT OFFICES

**Office of the Governor:** La Fortaleza, POB 9020082, PR 00909-0082; tel. (787) 721-7000; fax (787) 724-0942; internet www.fortaleza.gobierno.pr.

**Department of Agriculture:** POB 10163, Santurce, PR 00908-1163; tel. (787) 721-2120; fax (787) 723-9747; internet www.agricultura.gobierno.pr.

**Department of Consumer Affairs:** POB 41059, Minillas Station, San Juan, PR 00940-1059; tel. (787) 721-0940; fax (787) 726-007; internet www.daco.gobierno.pr.

**Department of Economic Development and Commerce:** 355 Avda Roosevelt, Suite 401, Hato Rey, PR 00918; internet www.ddecpr.com; tel. (787) 765-2900; fax (787) 753-6874.

**Department of Education:** Avda Teniente César González, esq. Calle Juan Calaf, Urb. Industrial Tres Monjitas, Hato Rey, PR 00917; POB 190759, San Juan, PR 00919-0759; tel. (787) 759-2000; fax (787) 250-0275; internet eduportal.de.gobierno.pr.

**Department of the Family:** POB 11398, Santurce, San Juan, PR 00910; tel. (787) 722-7400; fax (787) 722-7910; internet www.familia.gobierno.pr.

**Department of Health:** POB 70184, San Juan, PR 00936-8184; tel. (787) 766-1616; fax (787) 250-6547; e-mail webmaster@salud.gov.pr; internet www.salud.gov.pr.

**Department of Housing:** 606 Avda Barbosa, Juan C. Cordero, San Juan, 00928-1365; Apdo 21365, Rio Piedras, PR 00928; tel. (787) 274-2525; fax (787) 758-9263; e-mail mcardona@vivienda.gobierno.pr; internet www.vivienda.gobierno.pr.

**Department of Justice:** POB 9020192, San Juan, PR 00902-0192; tel. (787) 721-2900; fax (787) 724-4770; internet www.justicia.gobierno.pr; includes the Office of the Attorney-General.

**Department of Labor and Human Resources:** Edif. Prudencio Rivera Martínez, 505 Avda Muñoz Rivera, Hato Rey, PR 00918; tel. (787) 754-2159; fax (787) 753-4201; e-mail lortiz@dtrh.gobierno.pr; internet www.dtrh.gobierno.pr.

**Department of Natural and Environmental Resources:** Pda 3 1/2 Avda Muñoz Rivera, Puerta de Tierra, San Juan; POB 906660 Puerta de Tierra Station, San Juan, PR 00906-6600; tel. (787) 724-8774; fax (787) 723-4255; e-mail webmaster@drna.gobierno.pr; internet www.drna.gobierno.pr.

**Department of Recreation and Sports:** Avda Fernández Juncos, esq. Calle Bolívar, 1611 Antiguo Edif. del Fondo del Seguro del Estado, San Juan, 00902-3207; POB 9023207, Rio Piedras, PR 00909; tel. (787) 721-2800; fax (787) 728-0313; e-mail fgandara@drd.gobierno.pr; internet www.drd.gobierno.pr.

**Department of State:** Apdo 9023271, San Juan, PR 00902-3271; tel. (787) 721-1768; fax (787) 723-3304; e-mail estado@gobierno.pr; internet www.estado.gobierno.pr.

**Department of Transportation and Public Works:** POB 41269, San Juan, PR 00940; tel. (787) 722-2929; fax (787) 728-8963; e-mail servciud@act.dtop.gov.pr; internet www.dtop.gobierno.pr.

**Department of the Treasury:** POB 9024140, San Juan, PR 00902-4140; tel. (787) 721-2020; fax (787) 723-6213; e-mail infoserv@hacienda.gobierno.pr; internet www.hacienda.gobierno.pr.

## Gubernatorial Election, 2 November 2004

| Candidate | Votes | % |
|---|---:|---:|
| Aníbal Acevedo Vilá (PPD) | 963,303 | 48.40 |
| Pedro Rosselló González (PNP) | 959,737 | 48.22 |
| Rubén Berríos Martínez (PIP) | 54,551 | 2.74 |
| **Total** (incl. others)* | 1,990,372 | 100.00 |

*Including 4,960 blank votes and 4,042 spoilt votes.

# Legislature

## LEGISLATIVE ASSEMBLY

### Senate

**President of the Senate:** Kenneth McClintock.

Election, 2 November 2004

| Party | Seats |
|---|---:|
| PNP | 17 |
| PPD | 9 |
| PIP | 1 |
| **Total** | **27** |

### House of Representatives

**Speaker of the House:** José Aponte Hernández.

Election, 2 November 2004

| Party | Seats |
|---|---:|
| PNP | 32 |
| PPD | 18 |
| PIP | 1 |
| **Total** | **51** |

# Political Organizations

**Frente Socialista:** 103 Américo Miranda, POB 70359, San Juan, PR 00936; e-mail internacional@frentesocialista.org; internet members.tripod.com/frentesocialistapr; f. 2001; mem. orgs include the Partido Revolucionario de los Trabajadores Puertorriqueños (PRTP-Macheteros) and.

**Movimiento Socialista de Trabajadores (MST):** Apdo 22699, Estación UPR, San Juan, PR 00931-2699; e-mail info@bandera.org; internet www.bandera.org; f. 1982 by merger of the Movimiento Socialista Popular and Partido Socialista Revolucionario; mainly composed of workers and university students; pro-independence.

**Movimiento Independentista Nacional Hostosiano (MINH):** f. 2004 by merger of the Congreso Nacional Hostosiano and Nuevo Movimiento Independentista (fmr mems of the Partido Socialista Puertorriqueño); pro-independence; Co-president Julio Muriente Pérez.

**Partido Independentista Puertorriqueño (PIP)** (Puerto Rican Independence Party): 963 F. D. Roosevelt Ave, Hato Rey, San Juan PR 00920-2901; e-mail pipnacional@independencia.net; internet www.independencia.net; f. 1946; advocates full independence for Puerto Rico as a socialist-democratic republic; Leader Rubén Berríos Martínez; c. 6,000 mems.

**Partido Nuevo Progresista (PNP)** (New Progressive Party): POB 1992, Fernández Zuncos Station, San Juan 00910-1992; tel. (787) 289-2000; e-mail dannyls@caribe.net; internet www.pnp.org; f. 1967; advocates eventual admission of Puerto Rico as a federated state of the USA; Pres. Pedro Rosselló; Sec.-Gen. Thomas Rivera Schatz; c. 225,000 mems.

**Partido Popular Democrático (PPD)** (Popular Democratic Party): Comité Central PPD, Puerta de Tierra, San Juan; POB 9065788 San Juan, PR 00906-5788; e-mail lherrero@ppdpr.net; internet www.ppdpr.net; f. 1938; supports continuation and improvement of the present Commonwealth status of Puerto Rico; Pres. and Leader Aníbal Acevedo Vilá; c. 950,000 mems.

**Pro Patria National Union:** seeks independence for Puerto Rico; encourages assertion of distinct Puerto Rican citizenship as recognized by US Supreme Court in Nov. 1997; Leader Fufi Santori.

# PUERTO RICO

**Puerto Rican Republican Party:** 1629 Avda Piñero, Suite 203, San Juan, PR 00920; tel. (787) 793-8084; e-mail cchardon@goppr.org; internet www.goppr.org; Chair. Dr TIODY DE JESÚS DE FERRÉ; Exec. Dir ANNIE J. MAYOL.

**Puerto Rico Democratic Party:** POB 5788, San Juan, PR 00906; tel. (787) 722-4952; Pres. WILLIAM MIRANDA MARÍN.

**Refundación Comunista:** Organización RC, POB 13362, San Juan, PR 00908-3362; e-mail refundacionpcp@hotmail.com; internet refundacioncomunista.tripod.com; f. 2001; Marxist-Leninist; pro-independence; maintains close relations with the Frente Socialista.

## Judicial System

The Judiciary is vested in the Supreme Court and other courts as may be established by law. The Supreme Court comprises a Chief Justice and six Associate Justices, appointed by the Governor with the consent of the Senate. The lower Judiciary consists of Superior and District Courts and Municipal Justices equally appointed.

There is also a US Federal District Court, whose judges are appointed by the President of the USA. Judges of the US Territorial District Court are appointed by the Governor.

### Supreme Court of Puerto Rico

POB 2392, Puerta de Tierra, San Juan, PR 00902-2392; tel. (787) 724-3551; fax (787) 725-4910; e-mail buzon@tribunales.gobierno.pr; internet www.tribunalpr.org.

**Chief Justice:** FEDERICO HERNÁNDEZ DENTON.

**Justices:** FRANCISCO REBOLLO LÓPEZ, JAIME B. FUSTER BERLINGERI, EFRAÍN E. RIVERA PÉREZ, LIANA FIOL MATA, ANABELLE RODRÍGUEZ RODRÍGUEZ.

### US Territorial District Court for Puerto Rico

Federico Degetau Federal Bldg, Carlos Chardón Ave, Hato Rey, PR 00918; tel. (787) 772-3011; internet www.prd.uscourts.gov.

**Judges:** JOSÉ A. FUSTÉ (Chief Judge), JUAN M. PÉREZ-GIMÉNEZ, CARMEN C. CEREZO, HECTOR M. LAFITTE, SALVADOR E. CASELLAS, DANIEL R. DOMÍNGUEZ, JAY A. GARCÍA-GREGORY.

## Religion

About 73% of the population belonged to the Roman Catholic Church at the end of 2003. The Protestant churches active in Puerto Rico include the Episcopalian, Baptist, Presbyterian, Methodist, Seventh-day Adventist, Lutheran, Mennonite, Salvation Army and Christian Science. There is a small Jewish community.

### CHRISTIANITY

#### The Roman Catholic Church

Puerto Rico comprises one archdiocese and four dioceses. At 31 December 2003 there were 2,800,581 adherents.

##### Bishops' Conference of Puerto Rico

POB 40682, San Juan, PR 00940-0682; tel. (787) 728-1650; fax (787) 728-1654; e-mail ceppr@coqui.net.

f. 1960; Pres. Rt Rev. ROBERTO OCTAVIO GONZÁLEZ NIEVES (Archbishop of San Juan de Puerto Rico).

**Archbishop of San Juan de Puerto Rico:** ROBERTO OCTAVIO GONZÁLEZ NIEVES, Arzobispado, Calle San Jorge 201, Santurce, POB 00902-1967; tel. (787) 725-4975; fax (787) 723-4040; e-mail cancilleria@arqsj.org.

#### Other Christian Churches

**Episcopal Church of Puerto Rico:** POB 902, St Just, PR 00978; tel. (787) 761-9800; fax (787) 761-0320; e-mail iep@spiderlink.net; internet www.iepanglicom.org; diocese of the Episcopal Church in the USA, part of the Anglican Communion; Leader Bishop Rt Rev. DAVID ANDRÉS ALVAREZ.

**Evangelical Council of Puerto Rico:** Calle El Roble 54, Apdo 21343, Río Piedras, San Juan, PR 00928; tel. (787) 765-6030; fax (787) 765-5977; f. 1954; 6 mem. churches; Pres. Rev. HÉCTOR SOTO; Exec. Sec. Rev. HERIBERTO MARTÍNEZ RIVERA.

### BAHÁ'Í FAITH

**National Spiritual Assembly:** POB 11603, San Juan, PR 00910-2703; tel. (787) 763-0982; fax (787) 753-4449.

### JUDAISM

**Jewish Community Center:** 903 Ponce de León Ave, San Juan, PR 00907; tel. (787) 724-4157; fax (787) 722-4157; f. 1953; conservative congregation with 250 families; Rabbi GABRIEL TRYDMAN. There is also a reform congregation with 60 families.

## The Press

Puerto Rico has high readership figures for its few newspapers and magazines, as well as for mainland US periodicals. Several newspapers have a large additional readership among the immigrant communities in New York.

### DAILIES
(m = morning; s = Sunday)

**El Nuevo Día:** Parque Industrial Amelia, Carretera 164, Guaynabo, Puerto Rico; POB 9067512, San Juan, PR 00906-7512; tel. (787) 641-8000; fax (787) 641-3924; e-mail laferre@elnuevodia.com; internet www.endi.com; f. 1970; Chair. and Editor ANTONIO LUIS FERRÉ; Pres. MARÍA EUGENIA FERRÉ RANGEL; Dir LUIS ALBERTO FERRÉ RANGEL; circ. 214,441 (m), 246,765 (s).

**Primera Hora:** Parque Industrial Amelia, Calle Diana Lote 18, Guaynabo, PR 00966; POB 2009, Cataño, PR 00963-2009; tel. (787) 641-5454; fax (787) 641-4472; internet www.primerahora.com; Pres. and Editor ANTONIO LUIS FERRÉ; Dir JORGE CABEZAS; Gen. Man. JUAN MARIO ÁLVAREZ CARTAÑA; circ. 133,483 (m), 92,584 (Sat.).

**The San Juan Star:** POB 364187, San Juan, PR 00936-4187; tel. (787) 782-4200; fax (787) 783-5788; internet www.thesanjuanstar.com; f. 1959; English; Pres. and Publr GERARD ANGULO; Gen. Man. SALVADOR HASBÚN; circ. 50,000.

**El Vocero de Puerto Rico:** Apdo 7515, San Juan, PR 00906-7515; tel. (787) 721-2300; fax (787) 722-0131; e-mail opinion@vocero.com; internet www.vocero.com; f. 1974; Publr and Editor GASPAR ROCA; circ 143,150 (m), 123,869 (Sat.).

### PERIODICALS

**Buena Salud:** 1700 Fernández Juncos Ave, San Juan, PR 00909; tel. (787) 728-7325; f. 1990; monthly; Editor IVONNE LONGUEIRA; circ. 59,000.

**Caribbean Business:** 1700 Fernández Juncos Ave, San Juan, PR 00909-2938; POB 12130, San Juan, PR 00914-0130; tel. (787) 728-9300; fax (787) 726-1626; e-mail cbeditor@casiano.com; internet www.casiano.com/html/cb.html; f. 1973; weekly; business and finance; Editor ELISABETH ROMÁN; circ. 45,000.

**Educación:** c/o Dept of Education, POB 190759, Hato Rey Station, San Juan, PR 00919; f. 1960; 2 a year; Spanish; Editor JOSÉ GALARZA RODRÍGUEZ; circ. 28,000.

**La Estrella de Puerto Rico:** 165 Calle París, Urb. Floral Park, Hato Rey, PR 00917; tel. (787) 754-4440; fax (787) 754-4457; e-mail editor@estrelladepr.com; internet www.estrelladepr.com; f. 1983; weekly; Spanish and English; Editor-in-Chief FRANK GAUD; circ. 123,500.

**Imagen:** 1700 Fernández Juncos Ave, Stop 25, San Juan, PR 00909-2999; tel. (787) 728-4545; fax (787) 728-7325; e-mail imagen@casiano.com; internet www.casiano.com/html/imagen.html; f. 1986; monthly; women's interest; Editor ANNETTE OLIVERAS; circ. 71,000.

**Qué Pasa:** Loiza St Station, POB 6338, San Juan, PR 00914; tel. (787) 728-3000; fax (787) 728-1075; internet www.casiano.com/html/quepasa.html; f. 1948; quarterly; English; publ. by Puerto Rico Tourism Co; official tourist guide; Editor YAHIRA CARO; circ. 120,000.

**Resonancias:** POB 9024184, San Juan, PR 00902-4184; tel. (787) 721-0901; e-mail revista@icp.gobierno.pr; internet www.icp.gobierno.pr; f. 2000; biannual; Spanish; music, Puerto Rican culture and music, folk; Editor GLORIA TAPIA; circ. 3,000.

**Revista Colegio de Abogados de Puerto Rico:** POB 9021900, San Juan, PR 00902-1900; tel. (787) 721-3358; fax (787) 725-0330; e-mail abogados@prtc.net; f. 1914; quarterly; Spanish; law; Editor Lic. ALBERTO MEDINA; circ. 10,000.

**Revista del Instituto de Cultura Puertorriqueña:** POB 9024184, San Juan, PR 00902-4184; tel. (787) 721-0901; e-mail revista@icp.gobierno.pr; internet www.icp.gobierno.pr; f. 1958; biannual; Spanish; arts, literature, history, theatre, Puerto Rican culture; Editor GLORIA TAPIA; circ. 3,000.

**La Semana:** Calle Cristóbal Colón, esq. Ponce de León, Casilla 6527, Caguas 00725; tel. (787) 743-6537; e-mail lasemana@lasemana.com; internet www.lasemana.com/principal.html; weekly; f. 1963; Spanish; Gen. Man. MARJORIE M. RIVERA RIVERA.

# PUERTO RICO                                                                                   *Directory*

**La Torre:** POB 23322, UPR Station, San Juan, PR 00931-3322; tel. (787) 758-0148; fax (787) 753-9116; e-mail ydef@hotmail.com; f. 1953; publ. by University of Puerto Rico; quarterly; literary criticism, linguistics, humanities; Editor J. Martínez; circ. 1,000.

**Vea:** POB 190240, San Juan, PR 00919-0240; tel. (787) 721-0095; fax (787) 725-1940; e-mail editor@veavea.com; f. 1969; weekly; Spanish; TV, films and celebrities; Editor Enrique Pizzi; circ. 92,000.

**El Visitante:** POB 41305, San Juan, PR 00940-1305; tel. (787) 728-3710; fax (787) 268-1748; e-mail director@elvisitante.biz; internet www.elvisitante.biz; f. 1975; weekly; publ. by the Puerto Rican Conf. of Roman Catholic Bishops; Dir José R. Ortiz Valladares; Editor Rev. Efraín Zabala; circ. 58,500.

### FOREIGN NEWS BUREAUX

**Agencia EFE** (Spain): Cobian's Plaza, Suite 214, Santurce, PR 00910; tel. (787) 723-6023; fax (787) 725-8651; Dir Elías García.

**Associated Press** (USA): Metro Office Park 8, 1 St 108, Guaynabo, PR 00968.

## Publishers

**Ediciones Huracán Inc:** 874 Baldorioty de Castro, San Juan, PR 00925; tel. (787) 763-7407; fax (787) 753-1486; e-mail edhucan@caribe.net; f. 1975; textbooks, literature, social studies, history; Pres. Carmen Rivera-Izcoa.

**Editorial Académica, Inc:** 67 Santa Anastacia St, El Vigía, Río Piedras, PR 00926; tel. (787) 760-3879; f. 1988; regional history, politics, government, educational materials, fiction; Dir Fidelio Calderón.

**Editorial Coquí:** POB 21992, UPR Station, San Juan, PR 00931.

**Editorial Cordillera, Inc:** POB 192363, San Juan, PR 00919-2363; tel. (787) 767-6188; fax (787) 767-8646; e-mail info@editorialcordillera.com; internet www.editorialcordillera.com; f. 1962; Pres. Patricia Gutiérrez; Sec. and Treas. Adolfo R. López.

**Editorial Cultural Inc:** POB 21056, Río Piedras, San Juan, PR 00928; tel. (787) 765-9767; e-mail cultural@editorialcultural.com; f. 1949; general literature and political science; Dir Francisco M. Vázquez.

**Editorial Edil, Inc:** POB 23088, UPR Station, Río Piedras, PR 00931; tel. (787) 753-9381; fax (787) 250-1407; e-mail editedil@coqui.net; internet www.editorialedil.com; f. 1967; univ. texts, literature, technical and official publs; Man. Dir and Publr Consuelo Andino Ortiz.

**Librería y Tienda de Artesanias Instituto de Cultura Puertorriqueña:** POB 9024184, San Juan, PR 00902-4184; tel. (787) 724-4295; fax (787) 723-0168; e-mail www@icp.gobierno.pr; internet www.icp.gobierno.pr; f. 1955; literature, history, poetry, music, textbooks, arts and crafts; Man. Dir Maira Piazza.

**University of Puerto Rico Press (EDUPR):** POB 23322, UPR Station, Río Piedras, San Juan, PR 00931-3322; tel. (787) 250-0550; fax (787) 753-9116; f. 1932; general literature, children's literature, Caribbean studies, law, philosophy, science, educational; Dir Carlos D'Alzina.

## Broadcasting and Communications

### TELECOMMUNICATIONS

**Junta Reglamentadora de Telecomunicaciones de Puerto Rico:** Edif. Capital Center II, 235 Suite 1001, Avda Arterial Hostos, San Juan, PR 00918-1453; tel. (787) 756-0804; fax (787) 756-0814; e-mail correspondencia@jrtpr.gobierno.pr; internet www.jrtpr.gobierno.pr; Pres. Miguel Reyes Dávila.

**Puerto Rico Telephone Co (PRTC):** GPO Box 360998, San Juan, PR 00936-0998; tel. (787) 782-8282; fax (787) 774-0037; internet www.telefonicapr.com; provides all telecommunications services in Puerto Rico; majority control transferred from govt to private-sector GTE Corpn in March 1999; Pres. and CEO Jon E. Slater.

### BROADCASTING

There were 120 radio stations and 15 television stations operating in early 2002. The only non-commercial stations are the radio station and the two television stations operated by the Puerto Rico Department of Education. The US Armed Forces also operate a radio station and three television channels.

**Asociación de Radiodifusores de Puerto Rico** (Puerto Rican Radio Broadcasters' Asscn): Caparra Terrace, Delta 1305, San Juan, PR 00920; tel. (787) 783-8810; fax (787) 781-7647; e-mail prbroadcasters@centennialpr.net; internet www.radiodifusores.com; f. 1947; 90 mems; Pres. José A. Martínez Giraud; Exec. Dir José A. Ribas Dominicci.

## Finance

(cap. = capital; res = reserves; dep. = deposits; brs = branches; amounts in US dollars)

### BANKING

#### Government Bank

**Government Development Bank for Puerto Rico** (Banco Gubernamental de Fomento para Puerto Rico—BGF): POB 42001, San Juan, PR 00940-2001; tel. (787) 728-9200; fax (787) 268-5496; e-mail gdbcomm@prstar.net; internet www.gdb-pur.com; f. 1942; an independent govt agency; acts as fiscal (borrowing) agent to the Commonwealth Govt and its public corpns and provides long- and medium-term loans to private businesses; cap. 20.5m., res 46.8m., dep. 5,816.5m. (June 2001); Pres. Héctor Méndez Vázques; Chair. Juan Agosto Alicea; 350 employees.

**Autoridad para el Financiamiento de la Vivienda de Puerto Rico:** POB 213365, Hato Rey, San Juan, PR 00928-1365; tel. (787) 274-0000; fax (787) 764-8680; f. 1961; fmrly Banco y Agencia de Financiamiento de la Vivienda de Puerto Rico until Aug. 2001; subsidiary of the Government Development Bank of Puerto Rico; finance agency; helps low-income families to purchase houses; Exec. Dir José Cestero Casanova.

#### Commercial Banks

**Banco Bilbao Vizcaya Argentaria Puerto Rico:** 15th Floor, Torre BBVA, 258 Muñoz Rivera Ave, San Juan 00918; POB 364745, San Juan, PR 00936-4745; tel. (787) 777-2000; fax (787) 777-2999; internet www.bbvapr.com; f. 1967 as Banco de Mayagüez; taken over by Banco Occidental in 1979; merged with Banco Bilbao Vizcaya, S.A. in 1988; named changed from BBV Puerto Rico in 2000; cap. 138.7m., res. 215.6m., dep. 4,640.3m. (Dec. 2002); Pres. Antonio Uguina; 65 brs; 1,300 employees.

**Banco Popular de Puerto Rico:** Popular Center, 209 Muñoz Rivera Ave, San Juan, PR 00918; tel. (787) 765-9800; fax (787) 758-2714; internet www.bppr.com; f. 1893; cap. 8.0m., res 1,504m., dep. 19,595m. (Dec. 2002); Chair., Pres. and CEO Richard L. Carrión; 193 brs; 10,959 employees.

**Banco Santander Puerto Rico:** 207 Ponce de León Ave, Hato Rey, PR 00919; POB 362589, San Juan 00936-0062; tel. (787) 759-7070; fax (787) 767-7913; internet www.santanderpr.com; f. 1976; cap. 106.2m., res 6,383.1m., dep. 6,656.0m. (Dec. 2002); Pres. José Ramon Gonzalez; Chair. and CEO Monica Aparicio; 64 brs; 1,700 employees.

**Scotiabank de Puerto Rico:** Plaza Scotiabank, 273 Ponce de León Ave, esq. Calle Méjico, Hato Rey, PR 00918; POB 362230, San Juan 00936-2230; tel. (787) 758-8989; fax (787) 766-7879; f. 1910; cap. 23.2m., res 143.5m., dep. 1,164.3m. (Dec. 2002); Chair. Bruce R. Birmingham; Pres. and CEO Ivan A. Méndez; 13 brs; 451 employees.

#### Savings Banks

**FirstBank Puerto Rico:** First Federal Bldg, 1519 Ponce de León Ave, POB 9146, Santurce, PR 00908-0146; tel. (787) 729-8200; fax (787) 729-8139; internet www.firstbancorppr.com; f. 1948; cap. and res 368.3m., dep. 3,363.0m. (Dec. 2000); Pres. Angel Alvarez-Pérez; 45 brs.

**Oriental Bank and Trust:** Ave Fagot, esq. Obispado M-26, Ponce; tel. (787) 259-0000; fax (787) 259-0700; e-mail jllantin@orientalfg.com; internet www.orientalonline.com; total assets 2,039m. (June 2001); Chair., Pres. and CEO José Enrique Fernández.

**Ponce Federal Bank, FSB:** Villa esq. Concordia, POB 1024, Ponce, PR 00733; tel. (787) 844-8100; fax (787) 848-5380; f. 1958; Pres. and CEO Hans H. Hertell; 19 brs.

**R & G Corporation:** POB 2510, Guaynabo, PR 00970; tel. (787) 766-6677; fax (787) 766-8175; internet www.rgonline.com; total assets 4,676m. (Dec. 2001); Chair., Pres. and CEO Víctor J. Galán.

#### US Banks in Puerto Rico

**Chase Manhattan Bank NA:** 254 Muñoz Rivera Ave, Hato Rey, PR 00918; tel. (787) 753-3400; fax (787) 766-6886; Man. Dir and Gen. Man. Robert C. Dávila; 1 br.

**Citibank NA:** 252 Ponce de León Ave, San Juan, PR 00918; tel. (787) 753-5619; fax (787) 766-3880; Gen. Man. Horacio Igust; 14 brs.

# PUERTO RICO

## SAVINGS AND LOAN ASSOCIATIONS

**Caguas Central Federal Savings of Puerto Rico:** POB 7199, Caguas, PR 00626; tel. (787) 783-3370; f. 1959; assets 800m.; Pres. Lorenzo Muñoz Franco.

**Westernbank Puerto Rico:** 19 West McKinley St, Mayagüez, PR 00680; tel. (787) 834-8000; fax (787) 831-5958; internet www.wbpr .com; cap. and res 250.6m.; dep. 2,610.4m. (Dec. 2000); Chair. and CEO Lic. Frank C. Stipes; 31 brs.

## Banking Organization

**Puerto Rico Bankers' Association:** 208 Ponce de León Ave, Suite 1014, San Juan, PR 00918; tel. (787) 753-8630; fax (787) 754-6077; e-mail info@abpr.com; internet www.abpr.com; Pres. Juan A. Net; Exec. Vice-Pres. Arturo L. Carrión.

## INSURANCE

**Atlantic Southern Insurance Co:** POB 362889, San Juan, PR 00936-2889; tel. (787) 767-9750; fax (787) 764-4707; internet www .atlanticsouthern.com; f. 1945; Chair. Diane Bean Schwartz; Pres. Ramón L. Galanes.

**Caribbean American Life Assurance Co:** 273 Ponce de Léon Ave, Suite 1300, Scotiabank Plaza, San Juan, PR 00917; tel. (787) 250-1199; fax (787) 250-7680; internet www.calac.com; Pres. Iván C. López.

**Cooperativa de Seguros Multiples de Puerto Rico:** POB 3846, San Juan, PR 00936; internet www.segurosmultiples.com; general insurance; Pres. Edwin Quiñones Suárez.

**La Cruz Azul de Puerto Rico:** Rd 1, Río Piedras, San Juan, PR 00927; tel. (787) 272-9898; fax (787) 272-7867; internet www .cruzazul.comhealth; Exec. Dir Marks Vidal.

**Great American Life Assurance Co of Puerto Rico:** POB 363786, San Juan, PR 00936-3786; tel. (787) 758-4888; fax (787) 766-1985; e-mail galifepr@galifepr.com; known as General Accident Life Assurance Co until 1998; Pres. Artura Carión; Sr Vice-Pres. Edgardo Diaz.

**National Insurance Co.:** POB 366107, San Juan, PR 00936-6107; tel. (787) 758-0909; fax (787) 756-7360; internet www.nicpr.com; f. 1961; subsidiary of National Financial Group; Chair., Pres. and CEO Carlos M. Benítez Jr.

**Pan American Life Insurance Co:** POB 364865, San Juan, PR 00936-4865; tel. (787) 724-5354; fax (787) 723-3860; e-mail mmunozguren@panamericanlife.com; internet www .panamericanlife.com; Gen. Man. Maite Muñozguren.

**Puerto Rican–American Insurance Co:** POB 70333, San Juan, PR 00936-8333; tel. (787) 250-5214; fax (787) 250-5371; f. 1920; total assets 119.9m. (1993); Chair. and CEO Rafael A. Roca; Pres. Rodolfo E. Criscuolo.

**Security National Life Insurance Co:** POB 193309, Hato Rey, PR 00919; tel. (787) 753-6161; fax (787) 758-7409; Pres. Carlos Fernández.

**Universal Insurance Group:** Calle 1, Lote 10, 3°, Metro Office Park, Guaynabo; POB 2145, San Juan, PR 00922-2145; tel. (787) 793-7202; fax (787) 782-0692; internet www.universalpr.com; f. 1972; comprises Universal Insurance Co, Eastern America Insurance Agency and Caribbean Alliance Insurance Co; Chair. and CEO Luis Miranda Casañas.

There are numerous agents, representing Puerto Rican, US and foreign companies.

# Trade and Industry

## DEVELOPMENT ORGANIZATION

**Puerto Rico Industrial Development Co (PRIDCO):** POB 362350, San Juan, PR 00936-2350; 355 Roosevelt Ave, Hato Rey, San Juan, PR 00918; tel. (787) 758-4747; fax (787) 764-1415; internet www.pridco.com; public agency responsible for the govt-sponsored industrial development programme; Exec. Dir Jorge Silva Puras.

## CHAMBERS OF COMMERCE

**Chamber of Commerce of Puerto Rico:** 100 Calle Tetuán, POB 9024033, San Juan, PR 00902-4033; tel. (787) 721-6060; fax (787) 723-1891; f. 1913; 1,800 mems; Pres. Ricardo D'Acosta; Exec. Vice-Pres. Edgardo Bigas.

**Chamber of Commerce of Bayamón:** POB 2007, Bayamón, PR 00619; tel. (787) 786-4320; 350 mems; Pres. Iván A. Marrero; Exec. Sec. Angelica B. de Remírez.

**Chamber of Commerce of Ponce and the South of Puerto Rico:** 65 Calle Isabel, POB 7455, Ponce, PR 00732-7455; tel. (787) 844-4400; fax (787) 844-4705; e-mail info@camarasur.org; internet www.camarasur.org; f. 1885; 550 mems; Pres. Dr Ernesto Córdova.

**Chamber of Commerce of Río Piedras:** San Juan; f. 1960; 300 mems; Pres. Neftalí González Pérez.

**Chamber of Commerce of the West of Puerto Rico Inc:** POB 9, Mayagüez, PR 00681; tel. (787) 832-3749; fax (787) 832-4287; e-mail ccopr@coqui.net; internet www.ccopr.com; f. 1962; 300 mems; Pres. Julio César Sanabria (designate); Exec. Dir Brenda Gil Enseñat (designate).

**Official Chamber of Commerce of Spain:** POB 894, San Juan, PR 00902; tel. (787) 725-5178; fax (787) 724-0527; f. 1966; promotes Spanish goods; provides information for Spanish exporters and Puerto Rican importers; 300 mems; Pres. Manuel García; Gen. Sec. Antonio Trujilo.

**Puerto Rico/United Kingdom Chamber of Commerce:** 1509 Calle López Landrón, Suite 100, San Juan, Puerto Rico 00911; tel. (787) 721-0160; fax (787) 721-7333; e-mail iancourt1@cs.com; 120 mems; Chair. Dr Ian Court.

## INDUSTRIAL AND TRADE ASSOCIATIONS

**Home Builders' Association of Puerto Rico:** 1605 Ponce de León Ave, Condominium San Martín, Santurce, San Juan, PR 00909; tel. (787) 723-0279; 150 mems; Pres. Franklin D. López; Exec. Dir Wanda I. Navajas.

**Pharmaceutical Industry Association of Puerto Rico (PIA-PR):** City View Plaza, Suite 407, Guaynabo, PR 00968; tel. (787) 622-0500; fax (787) 622-0503; e-mail info@piapr.com; internet www .piapr.com; Chair. César Simich; Sec. Edgardo Fábregas; 17 mem. cos.

**Puerto Rico Farm Bureau:** Cond. San Martín, 16054 Ponce de León Ave, Suite 403, San Juan, PR 00909-1895; tel. (787) 721-5970; fax (787) 724-6932; f. 1925; over 1,500 mems; Pres. Antonio Alvarez.

**Puerto Rico Manufacturers' Association (PRMA):** POB 195477, San Juan, PR 00919-5477; tel. (787) 759-9445; fax (787) 756-7670.

**Puerto Rico United Retailers Center:** POB 190127, San Juan, PR 00919-0127; tel. (787) 641-8405; fax (787) 641-8406; e-mail mhernandez@centrounido.org; internet www.centrounido.org; f. 1891; represents small and medium-sized businesses; 20,000 mems; Pres. Enid Toro; Gen. Man. Ignacio T. Veloz.

## MAJOR COMPANIES

The following is a selection of some of the principal industrial and commercial companies operating in Puerto Rico.

### Construction

**Aireko Construction Corpn:** POB 2128, San Juan 00922–2128; tel. (787) 653-600; fax (787) 793-5063; e-mail aireko@aireko.com; internet www.aireko.com; f. 1963; management and construction of commercial, industrial and institutional buildings; sales of US $105.5m. (2003); Pres. Lorenzo Dragoni; 1,200 employees.

**Bermúdez & Longo, S.E.:** Rd 845, Km 0.5, Cupey Bajo, Río Piedras; POB 191213, San Juan, Puerto Rico 00919-1213; tel. (787) 761-3030; fax (787) 755-0020; e-mail lfeliciano@bermudez-longo .com; internet www.bermudez-longo.com; f. 1962; electrical and mechanical contractors; sales of US $108.3m. (2004); Co-Chairs Juan J. Bermúdez, Adriel Longo; Pres. and CEO Jaime Vásquez; 825 employees.

**Cemex Puerto Rico:** POB 364487, San Juan, PR 00936-4487; tel. (787) 783-3000; fax (787) 781-8850; internet www.cemexpuertorico .com; f. 2002; cement hydraulics; subsidiary of Cemex, SA, Mexico; fmrly Puerto Rican Cement Co Inc.; Chair. Lorenzo Zambrano; 1,060 employees.

**F. & R. Construction Co & Affiliates:** University Gardens Urb., 1010 Harvard and Interamerican Sts, San Juan, PR 00927; tel. (787) 753-7010; fax (787) 767-8454; e-mail estimate@fyrconstruction.com; f. 1972; sales of US $103.2m. (2004); Pres. Jaime Fullana Olivencia; 718 employees.

### Electronics and Computers

**Hewlett-Packard Puerto Rico Co:** Carretera 110, Km 5.1, Barrio Aguacate, POB 4048, Aguadilla 00605; tel. (787) 890-6000; fax (787) 890-6262; f. 1980; subsidiary of Hewlett-Packard Co, USA; mfrs of computer hardware; Man. Lucy Crespo; 1,700 employees.

**Intel Puerto Rico Inc.:** South Industrial Park 00771, Las Piedras 00771; tel. (787) 733-8080; fax (787) 733-8020; f. 1982; subsidiary of Intel Corpn, USA; mfrs of semiconductors and computer parts; Chair. Gordon Moore.

**Microsoft Puerto Rico Inc.:** Humacao Industrial Park, Rd 3, Km 77.8, Humacao 00792; tel. (787) 850-1600; fax (787) 852-7076; f. 1990; subsidiary of Microsoft Corpn, USA; mfrs of computer software; Pres. Rodolfo Acevedo; 100 employees.

**Sensormatic Electronics:** Cortec Bldg, Suite 104, Corporate Office Park, Guaynabo, 009680; tel. (787) 782-7373; fax (787) 782-0920; f. 1989; subsidiary of Tyco International, USA; mfrs of electronic goods, incl. surveillance equipment; Chair. Ronald Assaf; Gen. Man. Ángel Rosa; 1,500 employees.

### Food and Beverages

**Bacardi Caribbean Corpn:** Carretera 165, Km 2.6, Cataño 00962; tel. (787) 788-1500; fax (787) 788-0340; internet www.bacardi.com; f. 1992; distillers and distributors of rum; Pres. Angel O. Torres; 1,200 employees.

**Ballester Hermanos Inc.:** POB 364548, San Juan, PR 00936-4548; tel. (787) 788-4110; fax (787) 788-6460; e-mail contact@ballesterhermanos.com; internet www.ballesterhermanos.com; f. 1914; food and beverage distributors; sales of US $156.5m. (2004); Chair., Pres. and CEO Alfonso F. Ballester; 200 employees.

**Bumble Bee Seafoods, LLC:** Marina Station, POB 3268, Mayaguez, PR 00681-3268; tel. (787) 834-3450; fax (787) 265-6130; e-mail ramosd@bumblebee.com; subsidiary of Connors Bros, USA; seafood cannery; Gen. Man. Zulma Rivera; 300 employees.

**Destilería Serrallés Inc.:** Main Rd 1, Mercedita, San Juan 00715; POB 198 Mercedita, PR 00715-0198; tel. (787) 840-1000; fax (787) 840-1155; e-mail donq@donq.com; internet www.donq.com; f. 1949; mfrs of distilled alcoholic beverages; sales of US $118.9m. (2004); Pres. Félix J. Serrallés; 376 employees.

**B. Fernández & Hermanos Inc.:** Urb. Industrial Luchetti 305, Carretera 5, Bayamón; tel. (787) 792-7272; fax (787) 288-7291; f. 1888; food and beverage distributors; sales of US $180m. (2004); Pres. and CEO José Teixidor; 240 employees.

**Goya de Puerto Rico Inc.:** Carretera 28, Esq. Carretera 5, Urb. Industrial Luchetti, Bayamón, PR 00961; POB 601467, Bayamón 00960; tel. (787) 740-9000; fax (787) 740-5040; e-mail olgaluz@luztirada.com; internet www.goya.com; f. 1949; mfrs of canned fruit and vegetables; subsidiary of Goya Foods, USA; sales of US $105m. (2004); Pres. Carlos Unanue; 500 employees.

**Holsum de Puerto Rico, Inc.:** Carretera 2, Km 20.1, Barrio Candelaria, Toa Baja, PR 00951-8282; tel. (787) 798-8282; fax (787) 251-2060; e-mail ramon.calderon@holsumpr.com; internet www.holsum.com; f. 1958; sales of US $107.5m. (2004); mfrs and distributors of bakery products; Pres Ramón Calderón Rivera; 875 employees.

**Méndez & Co Inc.:** Carretera 20, Km 2.4, Guaynabo, PR 00969; tel. (787) 793-8888; fax (787) 783-4085; e-mail mdz@mendezcopr.com; internet www.mendezcopr.com; f. 1912; food and beverage distributors; sales of US $246.0m. (2004); Chair. Salustiano 'Tito' Alvarez Méndez; Pres. and CEO José Arturo Alvarez; 503 employees.

**Northwestern Selecta, Inc.:** Matadero Ave between A and C, Puerto Nuevo, PR 00920; POB 10718, Caparra Heights Station, San Juan, PR 00922-0718; tel. (787) 781-1950; fax (787) 781-1125; e-mail enunezjr@northwesternselecta.com; internet www.northwesternselecta.com; f. 1980; meat and fish processors and distributors; affiliate of Northwestern Meat Inc, USA; sales of US $210.0m. (2004); Pres. Elpidio Nuñez, Jr; 300 employees.

**Plaza Provision Co:** POB 363328, San Juan, PR 00936-3328; tel. (787) 781-2070; fax (787) 781-2210; e-mail hvelez@plazaprovision.com; internet www.plazaprovision.com; f. 1907; food and beverage distributors; sales of US $266.0m. (2004); Chair. James N. Cimino; Pres. and CEO James M. Finn; 410 employees.

**Puerto Rico Supplies Co Inc.:** C and B Sts, Luchetti Industrial Park, Bayamón, PR 00959-4390; POB 11908, San Juan, PR 00922-1908; tel. (787) 780-4043; fax (787) 780-4390; e-mail eperez@prsupplies.com; internet www.prsupplies.com; f. 1945; tobacco and food distributors; sales of US $183.2m. (2004); Chair. and CEO Stanley Pasarell; Pres. Edwin Pérez; 265 employees.

**José Santiago, Inc:** Marginal Carretera 5, Km 4.4, Urb. Industrial Luchetti, Bayamon; POB 191795, San Juan, PR 00919-1795; tel. (787) 288-8835; fax (787) 288-8809; internet www.josesantiago.com; f. 1902; food and beverage distributors; sales of US $101m. (2004); Pres. José E. Santiago; 300 employees.

**V. Suárez & Co, Inc:** El Horreo de V. Suárez, Highway 165 and Buchanan, Guaynabo, PR 00968; POB 364588, San Juan, PR 00936; tel. (787) 792-1212; fax (787) 474-0735; e-mail jdavila@vsuarez.com; internet www.vsuarez.com; f. 1943; food and beverage distributors; sales of US $545.0m. (2004); Chair. Diego Suárez Sánchez; Pres. and CEO Diego Suárez, Jr; 494 employees.

**Packers Provision Co of Puerto Rico:** Mercado Central, Edif. C, Zona Portuaria, Puerto Nuevo, PR 00920; tel. (787) 783-0011; fax (787) 782-7134; e-mail packers@packersprovision.com; internet www.packersprovision.com; f. 1974; acquired by V. Suárez & Co, Inc in 2004; meat packers, distributors of frozen and refrigerated products; sales of US $195.0m. (2004); Pres. Guillermo García Lamadrid; c. 600 employees.

### Pharmaceuticals, Biotechnology and Medical Supplies

**Abbott Laboratories (Puerto Rico) Inc:** Carretera 2, Km 58.0, Cruce Davila, Barceloneta, PR 00617; POB 278, Barceloneta 00617; tel. (787) 846-6900; fax (787) 846-5132; f. 1968; subsidiary of Abbott Laboratories, USA; mfrs of pharmaceuticals; Man. Harry Rodríguez; 2,400 employees.

**Amgen Puerto Rico, Inc:** Carretera 31, Km 24.5, POB 4060, Juncos, PR 00777; tel. (787) 734-2000; fax (787) 734-6161; internet www.amgen.com; mfrs of biotechnology and pharmaceuticals; subsidiary of Amgen Inc, USA; Vice-Pres. and Gen. Man. Madhu Balachandran; c. 1,000 employees at 2 locations.

**Bristol-Myers Squibb:** POB 364707, San Juan, PR 00936-4707; tel. (787) 774-2800; fax (787) 850-6764; internet www.bms.com; f. 1970; Pres. and Gen. Man. Edda Guerrero; over 2,000 employees at 4 locations.

**Glaxo Wellcome Puerto Rico Inc:** Carretera 172, Km 9.1, Barrio Sertenejas, POB 11975, Cidra, PR 00739-1975; tel. (787) 250-3700; fax (787) 250-3857; e-mail marian.a.lon@gsk.com; internet www.gsk.com; subsidiary of GlaxoSmithKline, United Kingdom; Vice-Pres. and Gen. Man. Dr Marian Lon.

**IPR Pharmaceuticals, Inc:** South Main St, Sabana Gardens Industrial Park, POB 1967, Carolina, PR 00984-1967; tel. (787) 750-5353; fax (787) 750-5332; internet www.astrazeneca.com; subsidiary of AstraZeneca, United Kingdom; Pres. and Gen. Man. Rubén Freyre.

**Johnson & Johnson:** HC-02, POB 19250, Gurabo, PR 00778-9629; tel. (787) 272-7651; fax (787) 272-7691; comprises 14 subsidiaries of Johnson & Johnson, USA; mfrs of pharmaceuticals, biotechnology and medical supplies; Vice-Pres. (Manufacturing Operations) Edgardo J. Fábregas.

**Lilly del Caribe Inc:** 65th Infantry Rd, Carretera 3, Km 12.6, POB 1198, Carolina, PR 00986-1198; tel. (787) 257-5561; fax (787) 251-5429; f. 1985; subsidiary of Eli Lilly & Co, USA; mfrs of pharmaceuticals; Pres. and Gen. Man. María Crowe; 1,000 employees in 3 locations.

**Medtronic Puerto Rico Operations Co (MPROC):** Carretera 149, Km 56.3, POB 6001, Villalba, PR 00766; tel. (787) 847-3500; fax (787) 848-4415; f. 1974; subsidiary of Medtronic, Inc, USA; medical supplies mfrs, particularly electrodes for use in pacemakers; Gen. Man. Manuel Santiago; 2,000 employees.

**Merck Sharp & Dohme (I.A) Corpn** (Merck Sharp & Dohme Química de Puerto Rico): 65th Infantry Rd, Carretera 3, Km 12.6, Carolina, PR 00986; POB 3689, Carolina, PR 00984-3689; tel. (787) 474-8234; internet www.msd.com.pr; f. 1953; subsidiary of Merck & Co, Inc, USA; mfrs of pharmaceuticals; Vice-Pres. and Gen. Man. Daneris Fernández; more than 2,000 employees in 4 locations.

**MOVA Pharmaceutical Corpn:** Villa Blanca Industrial Park, State Rd 1, Km 34.8, Caguas, PR 00725; POB 8639, Caguas, PR 00726; tel. (787) 746-8500; fax (787) 258-6405; e-mail jose.casellas@movapharm.com; internet www.movapharm.com; f. 1986; acquired by Patheon, Canada, in 2004; mfrs and distributors of pharmaceuticals; sales of US $146m. (2004); Pres.and CEO Joaquín B. Viso; 1,617 employees.

**Pfizer Global Manufacturing:** Carretera 689, Km 1.9, Barrio Carmelita, Vega Baja, PR 00693; POB 786, Vega Baja, PR 00694-0786; tel. (787) 654-2100; fax (787) 858-7966; e-mail info@pfizereducador.com; internet www.pfizereducador.com; subsidiary of Pfizer Inc, USA; Vice-Pres. and Gen. Man. Carlos H. del Río; over 5,500 employees in 5 locations.

**Schering-Plough Products, LLC:** Carretera Estatal 686, Km 0.5, POB 486, Manatí, PR 00674-0486; tel. (787) 854-2700; fax (787) 854-5127; internet www.schering-plough.com; subsidiary of Schering-Plough Corpn, USA; Gen. Man. Frank Rodríguez; c. 1,300 employees in 3 locations.

**Wyeth Pharmaceuticals:** Barrio Amelia, Centro Guaynabo, PR 00968; tel. (787) 782-3838; fax (787) 781-2820; internet www.wyeth.com; f. 1973; mfrs of pharmaceuticals and biotechnology; subsidiary of Wyeth, USA; Vice-Pres. and Gen. Man. Hector M. Cabrera; 3,300 employees in 4 locations.

### Miscellaneous

**Almacenes Pitusa Inc.:** Urb. Valencia, 357 Calle Guipuzcoa, San Juan 00923; tel. (787) 641-8200; fax (787) 641-8278; f. 1976; department stores; sales of US $265.0m. (2004); Pres. Israel Kopel; 1,300 employees.

# PUERTO RICO

**Bella Group Corpn:** POB 190816, San Juan, PR 00918-0816; tel. (787) 620-7010; fax (787) 620-7030; internet www.bellainternational.com; f. 1963; sales of US $212.9m. (2004); motor vehicle distributors; Chair. and CEO JERONIMO ESTEVE-ABRIL; Pres. CARLOS A. LÓPEZ-LAY; 421 employees.

**Borschow Hospital & Medical Supplies, Inc:** Calle Calaf 68, Tres Monjitas Industrial Park, Hato Rey, PR 00918-1302; tel. (787) 754-2300; fax (787) 250-4658; e-mail bhms@borschow.com; internet www.borschow.com; f. 1951; pharmaceutical product and medical supplies distributors; sales US $300.8m. (2004); Pres. and CEO JON BORSCHOW; 250 employees.

**Chevron Phillips Chemical Puerto Rico Core Inc:** Ruta 710, Km 1.3, Barrio Las Mareas, Guayama, PR 00785; tel. (787) 864-1515; fax (787) 864-1545; internet www.cpchem.com; f. 1967; subsidiary of Chevron Phillips Co, USA; mfrs of chemicals; Pres. RICHARD C. KLETT; 66 employees.

**Colgate-Palmolive (Puerto Rico) Inc:** Puente de Jobos, POB 540, Guayama 00784; tel. (787) 723-5625; fax (787) 864-5053; f. 1988; subsidiary of Colgate-Palmolive Co, USA; mfrs of toiletries and cleaning products; Man. VÍCTOR SUÁREZ; 90 employees.

**Empresas Cordero Badillo Inc:** Avda Ponce de Léon 56, Barrio Amelia, Guaynabo, PR 00962; POB 458, Cataño, PR 00963-0458; tel. (787) 749-1400; fax (787) 749-1500; f. 1967; supermarkets; sales of US $365.0m. (2004); Chair., Pres. and CEO ATILANO CORDERO BADILLO; 2,100 employees.

**Empresas Fonalledas Inc:** POB 71450, San Juan, PR 00936-8550; tel. (787) 767-1525; f. 1890; retail and property devt; sales of US $275.0m. (2004); Pres. and CEO JAIME FONALLEDAS, Jr.

**Hilton International of Puerto Rico (Caribe Hilton International Hotel):** Los Rosales, San Juan 00901; tel. and fax (787) 721-0303; e-mail info@caribehilton.com; internet www.caribehilton.com; f. 1981; hotel management; Man. RAÚL BUSTAMANTE; 875 employees.

**Supermercados Mr Special Inc:** Carretera 114, Km 0.3, Avda Santa Teresa Jornet, POB 3389, Mayaguez, PR 00681; tel. (787) 834-2695; fax (787) 833-9843; e-mail presidente@mrspecialpr.com; internet www.mrspecialpr.com; f. 1966; supermarket chain; sales of US $213.0m. (2004); Pres. SANTOS ALONSO MALDONADO; 1,108 employees.

## UTILITIES

### Electricity

**Autoridad de Energía Eléctrica de Puerto Rico** (Puerto Rico Electric Power Authority): POB 364267, San Juan, PR 00936-4267; tel. (787) 289-3434; fax (787) 289-4690; e-mail director@prepa.com; internet www.aeepr.com; govt-owned electricity corpn, opened to private co-generators in the mid-1990s; installed capacity of 4,389 MW; Dir HÉCTOR ROSARIO.

## TRADE UNIONS

**American Federation of Labor–Congress of Industrial Organizations (AFL–CIO):** San Juan; internet www.afl-cio.org; Regional Dir AGUSTÍN BENÍTEZ; c. 60,000 mems.

**Central Puertorriqueña de Trabajadores (CPT):** POB 364084, San Juan, PR 00936-4084; tel. (787) 781-6649; fax (787) 277-9290; f. 1982; Pres. FEDERICO TORRES MONTALVO.

**Confederación General de Trabajadores de Puerto Rico:** 620 San Antonio St, San Juan, PR 00907; f. 1939; Pres. FRANCISCO COLÓN GORDIANY; 35,000 mems.

**Federación del Trabajo de Puerto Rico (AFL-CIO):** POB S-1648, San Juan, PR 00903; tel. (787) 722-4012; f. 1952; Pres. HIPÓLITO MARCANO; Sec.-Treas. CLIFFORD W. DEPIN; 200,000 mems.

**Puerto Rico Industrial Workers' Union, Inc:** POB 22014, UPR Station, San Juan, PR 00931; Pres. DAVID MUÑOZ HERNÁNDEZ.

**Sindicato Empleados de Equipo Pesado, Construcción y Ramas Anexas de Puerto Rico, Inc** (Construction and Allied Trades Union): Calle Hicaco 95, Urb. Milaville, Río Piedras, San Juan, PR 00926; f. 1954; Pres. JESÚS M. AGOSTO; 950 mems.

**Sindicato de Obreros Unidos del Sur de Puerto Rico** (United Workers' Union of South Puerto Rico): POB 106, Salinas, PR 00751; f. 1961; Pres. JOSÉ CARABALLO; 52,000 mems.

**Unión General de Trabajadores de Puerto Rico:** Apdo 29247, Estación de Infantería, Río Piedras, San Juan, PR 00929; tel. (787) 751-5350; fax (787) 751-7604; f. 1965; Pres. JUAN G. ELIZA-COLÓN; Sec.-Treas. OSVALDO ROMERO-PIZARRO.

**Unión de Trabajadores de la Industría Eléctrica y Riego de Puerto Rico (UTIER):** POB 13068, Santurce, San Juan, PR 00908; tel. (787) 721-1700; e-mail utier@coqui.net; internet www.utier.org; Pres. RICARDOS SANTOS RAMOS; Sec.-Treas. LUIS MERCED; 6,000 mems.

## Transport

In January 2004 a 17-km urban railway (Tren Urbano), capable of carrying some 300,000 passengers per day, was inaugurated in greater San Juan. The railway took eight years to build and cost some US $2,150m. An extension to Carolina and Minillas was proposed in March 2005, at a projected cost of $900m.

**Ponce and Guayama Railway:** Aguirre, PR 00608; tel. (787) 853-3810; owned by the Corporación Azucarera de Puerto Rico; transports sugar cane over 96 km of track route; Exec. Dir A. MARTÍNEZ; Gen. Supt J. RODRÍGUEZ.

### ROADS

The road network totalled 24,431 km (15,181 miles) in 2003, of which some 94% was paved. A modern highway system links all cities and towns along the coast and cross-country. A highways authority oversees the design and construction of roads, highways and bridges. In April 2002 it was announced that some US $585.6m. was to be invested in projects to improve or expand the road network.

**Autoridad de Carreteras:** Centro Gubierno Minillas, Edif. Norte, Avda de Diego 23, POB 42007, Santurce, San Juan, PR 00940-2007; tel. (787) 721-8787; fax (787) 727-5456; internet www.dtop.gov.pr/act/actmain.html; Dir Dr FERNANDO FACUNDO.

### SHIPPING

There are 11 major ports on the island, the principal ones being San Juan, Ponce and Mayagüez. Other ports include Guayama, Guayanilla, Guánica, Yabucoa, Aguirre, Aguadilla, Fajardo, Arecibo, Humacao and Arroyo. San Juan, one of the finest and longest all-weather natural harbours in the Caribbean, is the main port of entry for foodstuffs and raw materials and for shipping finished industrial products. In 2001 it handled 9.7m. tons of cargo. Passenger traffic is limited to tourist cruise vessels. Work on the US $700m. Las Américas 'megaport' was ongoing in 2005. In April a high-speed ferry service was launched, connecting San Juan to the islands of Vieques and Culebra.

**Autoridad de los Puertos** (Puerto Rico Ports Authority): POB 362829, San Juan, PR 00936-2829; tel. (787) 729-8805; fax (787) 722-7867; internet www.prpa.gobierno.pr; manages and administers all ports and airports; Exec. Dir FERNANDO J. BONILLA.

### CIVIL AVIATION

There are 11 airports on the island, the principal ones are at San Juan (Carolina), Aguadilla, Ponce and Mayagüez. There are also six heliports.

## Tourism

An estimated 3.2m. tourists visited Puerto Rico in 2003, when revenue from this source was estimated at US $2,677m. More than three-quarters of all tourist visitors were from the USA mainland. In 2001 there were approximately 12,353 hotel rooms in Puerto Rico, with an occupancy rate of 67%. In April 2002 several major tourist projects were announced, with the aim of increasing the number of hotel rooms to 13,795 by 2004.

**Compañía de Turismo** (Puerto Rico Tourism Co): Edif. La Princesa, 2 Paseo La Princesa, POB 9023960, San Juan, PR 00902-3960; tel. (787) 721-2400; fax (787) 722-6238; internet www.prtourism.com; f. 1970; Dir TERESTELLA GONZÁLEZ DENTON.

## Defence

The USA is responsible for the defence of Puerto Rico. In 2003 the US Navy withdrew from Puerto Rico closing its bases at Roosevelt Roads and on the island of Vieques. Puerto Rico has a paramilitary National Guard of some 11,000 men, which is funded mainly by the US Department of Defense. At January 2005, as part of the ongoing 'war on terror', some 2,300 Puerto Rican National Guard and Puerto Rican US Army reservists were serving under US command, mainly in Iraq, out of a total of 7,600 to have done so since September 2001. The National Guard have also been deployed to support domestic police operations.

## Education

The public education system is centrally administered by the Department of Education. Education is compulsory for children between six and 16 years of age. In the academic year 2000/01 there were 606,487 pupils attending public day schools and 128,961 pupils attending private schools. The 12-year curriculum, beginning at five years of age, is subdivided into six grades of elementary school, three years at junior high school and three years at senior high school. Vocational schools at the high-school level and kindergartens also form part of the public education system. Instruction is conducted in Spanish, but English is a required subject at all levels. In 2004 there were five universities. The State University system consists of three principal campuses and six regional colleges. In 2003 there were 199,842 students in higher education of whom some 63% were enrolled at private universities and colleges. In 2000/01 US $2,308,479m. of general government expenditure was allocated to education (22.3% of total expenditure).

# Bibliography

For works on the Caribbean generally, see Select Bibliography (Books)

Barreto, A. A. *Vieques, the Navy and Puerto Rican Politics*. Gainesville, FL, University Press of Florida, 2002.

Briggs, L. *Reproducing Empire: Race, Sex, Science and U.S. Imperialism in Puerto Rico*. Berkeley, CA, and London, University of California Press, 2002.

Camara Fuentes, L. R. *The Phenomenon of Puerto Rican Voting (New Directions in Puerto Rican Studies*. Gainesville, FL, University Press of Florida, 2004.

*Constructing a Colonial People: Puerto Rico and the United States, 1898–1932*. Boulder, CO, Westview Press, 1999.

Dietz, J. L. *Economic History of Puerto Rico: Institutional Change and Capitalist Development*. Princeton, NJ, Princeton University Press, 1987.

*Puerto Rico, Negotiating Development and Change*. Boulder, CO, L. Rienner, 2003.

Duany, J. *The Puerto Rican Nation on the Move*. Chapel Hill, NC, University of North Carolina Press, 2002.

Duffy Burnett, C., and Marshall, B. (Eds). *Foreign in a Domestic Sense: Puerto Rico, American Expansion, and the Constitution*. American Encounters/Global Interactions, Durham, MC, Duke University Press, 2001.

Fernández, R. *The Disenchanted Island: Puerto Rico and the United States in the Twentieth Century*. London, Praeger Publrs, 1993.

Fernández, R., Méndez Méndez, S., and Cueto, G. *Puerto Rico Past and Present: An Encyclopedia*. Westport, CT, Greenwood Press, 1998.

González, J. L. *El país de cuatro pisos*. Río Piedras, Ediciones Piedras, 1980.

Lewis, G. K. *Puerto Rico: Freedom and Power in the Caribbean*. Oxford, James Currey Publrs, and Kingston, Ian Randle Publrs, 2004.

Malavet, P. A. *America's Colony: The Political and Cultural Conflict Between the United States and Puerto Rico*. New York, NY, New York University Press, 2004.

McCaffrey, K. T. *Military Power and Popular Protest: the U.S. Navy in Vieques, Puerto Rico*. New Brunswick, NJ, and London, Rutgers University Press, 2002.

Monge, J. T. *Puerto Rico: The Trials of the Oldest Colony in the World*. New Haven, Yale, CT, Universal Press, 1999.

Negrón-Muntaner, F. *Boricua Pop: Puerto Ricans and the Latinization of American Culture*. New York, NY, and London, New York University Press, 2004.

Rivera Ramos, E. *The Legal Construction of Identity: The Judicial and Social Legacy of American Colonialism in Puerto Rico*. Washington, DC, American Psychological Asscn, 2001.

Romberg, R. *Witchcraft and Welfare: Spiritual Capital and the Business of Magic in Modern Puerto Rico*. Austin, University of Texas Press, 2003.

Schmidt-Nowara, C. *Empire and Antislavery; Spain, Cuba and Puerto Rico, 1833–1874*. Pittsburgh, PA, University of Pittsburgh Press, 1999.

# SAINT CHRISTOPHER* AND NEVIS

## Geography

### PHYSICAL FEATURES

Saint Christopher and Nevis, a federation of two of the Leeward Islands, is in the Lesser Antilles. The larger island, St Christopher (usually known as St Kitts), is separated from Nevis to the south-east by a 3-km (2-mile) channel called The Narrows. Rather more distance (20 km) separates Saint Christopher from the next island in the chain, St Eustatius (Netherlands Antilles), to the north-west. South-east of Nevis, about 40 km away, is Redonda, a small and uninhabited island dependency of Antigua and Barbuda (the main islands of which are to the east), and 24 km beyond that is Montserrat. All of these islands lie in the Caribbean Sea. Saint Christopher covers 176.4 sq km (68 sq miles) and Nevis 93.3 sq km, giving a total area of 269.4 sq km, the smallest for any sovereign state in the Americas.

Saint Christopher is 37 km (23 miles) long and tapers south-eastwards, forming a low-lying peninsula (which widens at the end and is dotted with salt ponds) pointing towards the more globular island of Nevis. Both islands are of volcanic origins, with mountainous interiors. The main landmass of Saint Christopher is dominated by mountains grouped in three ranges, highest in the north-west, where Mt Liamuiga (formerly Mt Misery, but now renamed with the old Carib name for the island) reaches 1,156 m (3,794 ft). A narrow, sea-flanked ridge connects the main part of the island to the flat lands around the Great Salt Pond. Nevis is also lofty, cone-shaped, its central Nevis Peak rising to 985 m. Both islands are fertile and green, although much of the original forest has long since disappeared, except on the higher slopes. Native wildlife suffered not only from the intensive cultivation of sugarcane, but also from the introduction of the mongoose, which has had a serious effect in several Caribbean islands. Other immigrants include green vervet monkeys, originally imported by the French from West Africa (also found in Barbados), and some deer on Saint Christopher. Monkeys are said to outnumber Kittitians, and donkeys Nevisians. Fauna as well as flora will benefit from the recent expansion in woodland, as agriculture contracts and the authorities make efforts to enhance the environment.

### CLIMATE

The climate is subtropical marine and in the hurricane belt. The average annual temperature is 26°C (79°F), with sea breezes keeping the islands relatively cool, particularly during the driest months of December–March. Temperatures rarely exceed 33°C (91°F) or fall below 17°C (63°F), even in the cooler heights—or on Nevis (despite its name coming from the Spanish for snow). The hurricane season is in July–October. There is relatively little humidity. Average annual rainfall on Saint Christopher is about 1,400 mm (55 ins) and on Nevis some 1,220 mm (48 ins).

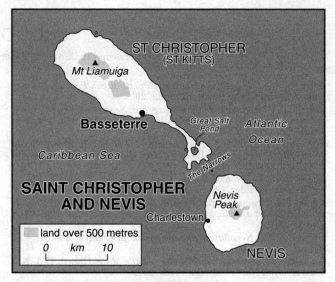

### POPULATION

Saint Christopher and Nevis is important in the history of the old British West Indies, with Saint Christopher the 'mother' colony and Nevis long one of the wealthiest. There are still some traces of the French influence on Saint Christopher (the island was at one time partitioned, the French holding the north and the south, and the peninsula being neutral), notably military ruins built in an age of great-power competition—and to protect against the doomed struggle of the native Caribs resisting European encroachment. However, the cultural legacy on both islands is predominantly British, although most of the population (more than 90%) is now black, descended from the African slaves brought to work the plantations. The official language is English and the leading Christian denomination is Anglican. Roman Catholics and Nonconformist Protestants (Methodists established themselves in the islands early) are also represented, and there are even some Orthodox Christians, as well as some Rastafarians.

The total population of Saint Christopher and Nevis was 47,300 in mid-2003. About five-sixths of the people live on Saint Christopher, and almost one-third of these in or around the capital city, Basseterre, in the south-west of the island. The chief town of Nevis, located on the central western coast, is the old settlement of Charlestown (even so, it was preceded as the capital by Jamestown, to the north, which was completely destroyed in 1690, by an earthquake and tidal wave). The country is split into 14 parishes for administrative purposes.

## History

### MARK WILSON

Saint Christopher (also known as St Kitts) and Nevis is a constitutional monarchy within the Commonwealth. Queen Elizabeth II is head of state, and is represented in Saint Christopher and Nevis by a Governor-General. There is a bicameral legislature with an elected chamber representing both islands, to which the federal Government is responsible. Nevis has a five-member Assembly, and an administration, headed by a Premier, which manages the island's affairs; Saint Christopher, on the other hand, is managed directly by the federal Government.

Few traces remain of the islands' original Amerindian inhabitants. The islands were visited and named by Christopher Columbus in 1493 as San Cristóbal and Nieves. British settlement on Saint Christopher dates from 1623, making the island the first British possession in the Caribbean; Nevis was settled five years later. However, the larger island was later parti-

---

*While this island is officially called Saint Christopher as part of the state, the name is usually abbreviated to St Kitts.

tioned, with France taking the centre, and Great Britain the extreme north and south, an arrangement that came to an end with the Treaty of Utrecht in 1713. The islands were administered together with Anguilla and the British Virgin Islands from 1816 until 1871, and then as part of the Leeward Islands Federation.

Poverty among the rural working class was extreme in the first part of the 20th century, and in 1935 resulted in a bitter sugar workers' strike; several strikers were killed by police. Political life on the island of Saint Christopher was dominated in the 1950s, 1960s and 1970s by Robert Llewellyn Bradshaw, leader of the St Kitts-Nevis Labour Party (SKNLP), which, in spite of its name, has never had a substantial following outside the larger island. Bradshaw's most solid political support was among the sugar workers of Saint Christopher, who benefited under his leadership from increased wages and improved education and welfare services. However, his relations with the urban middle class and with the populations of Nevis and Anguilla were in general very poor.

With the Leeward Islands Federation disbanded in 1957, Saint Christopher-Nevis-Anguilla joined the Federation of the West Indies in 1958 along with nine other British colonies. When Jamaica and Trinidad and Tobago left in 1962, the Federation collapsed, and an attempt to unite the remaining colonies as the 'little eight' was unsuccessful. Along with its neighbours, Saint Christopher-Nevis-Anguilla became a British Associated State in 1967, responsible for its internal affairs, with the United Kingdom retaining control of external affairs and defence. However, this arrangement was fiercely resisted by Anguilla, which feared domination by its larger neighbour, and broke away from the three-island grouping; a British commissioner administered that island separately from 1969.

The Bradshaw Government's poor relations with the island of Nevis led to the smaller island delaying the move to full independence, originally scheduled for 1980. In an election held that year, the SKNLP lost its overall majority, and a new Government was formed by a coalition of the People's Action Movement on Saint Christopher (PAM) and the Nevis Reformation Party (NRP), which took the three Nevis seats. Together, they agreed an arrangement which granted a high degree of autonomy to Nevis, while Saint Christopher, where the SKNLP retained a local majority, remained under the direct control of a federal Government. On this basis, the islands moved to independence on 19 September 1983.

The PAM remained in office for 15 years. However, the latter part of this period was overshadowed by widespread reports that the Government had failed to control major cocaine-traffickers, who had established powerful positions in local politics and business.

On Nevis, the leading position of the NRP came under challenge from the Concerned Citizens Movement (CCM), led by Vance Amory, which won island Assembly elections on 1 June 1992, taking three of the five seats. The CCM has in general favoured independence for Nevis, a position about which the NRP has, in recent years, been ambivalent.

In a general election held on 29 November 1993 the SKNLP, led by Dr Denzil Douglas, gained 54% of the popular vote and four of the eight seats on the main island, with the remainder taken by the PAM. On Nevis, one seat was taken by the NRP while the CCM, which was not willing to form an alliance with either of the Saint Christopher parties, won two seats. The leader of the PAM, Dr Kennedy A. Simmonds, remained in office with a minority Government, an arrangement that was fiercely opposed by SKNLP supporters on Saint Christopher. There was rioting in the capital, Basseterre, in December, which was countered by a state of emergency and a curfew, but was controlled after intervention by the Barbados-based Regional Security System. In November 1994 the two parties agreed that fresh elections should be held within a year.

Vincent Morris, the son of the Deputy Prime Minister Sidney Morris, disappeared on 2 October 1994, and was found dead in a burnt-out car on 12 November; this incident was believed to be linked to narcotics-trafficking. A police superintendent who was leading the investigation was shot dead while driving to work on 13 October. Two other sons of Sidney Morris were arrested on firearms charges in November. Although Sidney Morris resigned from his post, the complex scandal that ensued further damaged the standing of the PAM Government. Four police officers from the United Kingdom took control of the local force in May 1995.

Elections were held on 3 July 1995, and resulted in a clear victory for the SKNLP, which took seven of the eight seats on Saint Christopher with 58% of the popular vote on that island, with PAM retaining one Saint Christopher seat, and the Nevis seats split as they had been the previous year. The SKNLP, still led by Douglas, was returned for a second term on 6 March 2000, winning 53.9% of the popular vote on Saint Christopher, and all eight of that island's seats.

On Nevis, the CCM won a second term in office in 1997. A proposal to secede from the federation received unanimous support in the Nevis assembly, although the NRP later changed its position. The island administration exercised its constitutional option to call a referendum on secession, which was held on 10 August 1998. On a 57.8% voter turn-out, the proposal received 61.7% of the votes cast, fewer than the required two-thirds' majority. The federal Government has also proposed constitutional reforms, including replacement of the Queen as head of state with a president, and a separate assembly and island administration for Saint Christopher. However, any change would require a two-thirds' majority on each island, a difficult target in view of the strongly partisan nature of political debate in Saint Christopher and Nevis. In 2003 a parliamentary select committee on constitutional reform was asked to undertake further consultations. In April Amory announced that the CCM would seek to hold a second referendum on the issue of separation.

In elections to the Nevis Island Assembly in September 2001 the CCM, still led by Amory, strengthened its control of the legislature, gaining a total of four elective seats. The NRP, under the leadership of Joseph Parry, took the remaining one seat. The island assembly voted in June 2003 to hold another secession referendum; this initiative was opposed however both by the Caribbean Community and Common Market (CARICOM) and the USA, and by July 2005 no date had been set for the vote.

At a general election on 25 October 2004 the SKNLP won seven of the eight seats on Saint Christopher, with 60.4% of the popular vote on that island; the PAM took the remaining seat. On Nevis the CCM again secured two of the three seats in the National Assembly, with 54.0% of the votes cast on the smaller island; the remaining seat was won by the NRP. Commonwealth and CARICOM observer teams made several criticisms of the electoral process, and the Government subsequently promised reforms. Challenges by the PAM to the declared result in three marginal seats remained before the courts in mid-2005.

Narcotics-trafficking and violent crime remained serious concerns for the SKNLP Government; the per-head murder rate in 2004 (of 22 per 100,000) was extremely high by international standards and ahead of much larger Caribbean countries, such as Guyana and Trinidad and Tobago. Two alleged major drugs-traffickers have successfully resisted extradition to the USA since 1996, and remained at liberty in 2005.

# Economy

## MARK WILSON

Saint Christopher and Nevis is the smallest country in the Western hemisphere in terms of both area and population, with an area of 269.4 sq km and 47,300 inhabitants in mid-2003, of whom approximately 10,000 lived on Nevis. The islands have developed a prosperous middle-income economy, with a per-head gross domestic product (GDP) of US $7,807 in 2003. GDP grew at an average rate of 3.8% per year in 1990–2003; the pace of growth slackened to 1.7% in 2001, with a contraction of 0.4% in 2002 and growth of only 0.5% in 2003, largely owing to the slowdown in international demand for tourism; the rate of growth recovered to 4.0% in 2004. There was some concern over the overall fiscal deficit, which averaged 13.2% of GDP in 1999–2002, increasing to 16.4% in the latter year, and over the level of external debt-service payments, which were equivalent to 8.7% of GDP and 25.7% of the Government's current revenue in 2004. Total government and government-guaranteed debt, including borrowing by Nevis, was estimated in the February 2005 budget speech at 171% of GDP. However, since the Government adopted a stabilization programme in 2003, the overall deficit was reduced to 7.4% of GDP in 2004, and the Government expected the current high level of foreign direct investment in hotel construction further to benefit the balance of payments and government revenue in the next few years.

Saint Christopher and Nevis is a member of the Caribbean Community and Common Market, or CARICOM, which is attempting to develop a single market, in principle by 2006. It is also a member of the Organisation of Eastern Caribbean States, which links nine of the smaller Caribbean territories. The Eastern Caribbean Central Bank, headquartered just outside the capital, Basseterre, supervises its financial affairs.

The natural beauty, beaches and climate of both islands provide the basis for a prosperous tourism industry, which has grown rapidly in recent years. Visitor expenditure increased from US $57.7m. in 1990 to $106.9m. in 2004. On both islands, there are large modern resorts, but several former plantation houses have also been converted to small luxury hotels, while historic sites such as the Brimstone Hill fortress are important tourist attractions. A 900-room hotel with a casino and golf course opened in February 2003 and was expected to increase arrivals significantly. The number of stop-over arrivals reached 93,190 in 1998, but decreased in subsequent years as a result of hurricane damage to hotels as well as uncertain international demand in 2001–03. However, there were some 120,100 stop-over tourist arrivals in 2004; in 2002 47.2% of stop-over tourists were from North America and 8.1% from the United Kingdom.

There were also 260,100 cruise-ship passengers in 2004, who made only a modest economic contribution because of low per-head spending.

Agriculture and fishing made up only 3.2% of GDP in 2004, a proportion that was expected to decline further. The sugar industry, which dominated the economy of Saint Christopher until the 1960s, was to close after the 2005 harvest. Output in 2004 was 14,157 metric tons, well below the 2003 peak figure of 31,375 tons. In spite of a guaranteed European Union sugar price, which in recent years had been at least three times world market levels, the sugar industry survived only as a result of government subsidies, which were equivalent to 4.0% of GDP by 2004. Large areas formerly used for sugar have been redesignated for tourism, and the Government was expected to make provision for retraining of sugar industry employees.

Manufacturing comprised 8.7% of GDP in 2003. Industries such as brewing produce mainly for the local market, while others, including the assembly of electronic components, are orientated entirely to exports. The telecommunications sector has been liberalized; the Government viewed call centres and telemarketing as important growth areas for the economy.

There is a small 'offshore' financial sector on Nevis, which, by 2004, had registered 15,000 International Business Companies and 950 trusts. The division of regulatory powers between the island administration on Nevis and the federal Government has at times been unclear. Partly for this reason, Saint Christopher and Nevis was, in June 2000, listed as a 'non-co-operative jurisdiction' by the Financial Action Task Force on Money Laundering (FATF, based in Paris, France); however, the country was removed from this list in 2002 after instituting stricter regulatory controls. In April 2003 a report indicated that the islands' 'offshore' banking operations had been halved as a result of the FATF action and the subsequent financial reforms. Income is also derived from the sale of citizenship, a programme that has received considerable criticism, both locally and internationally. The Ross University School of Veterinary Medicine, which caters mainly to students from the USA, makes a substantial contribution to the economy, and the International University of Nursing opened in May 2005.

The islands are in the heart of the hurricane belt, and have been damaged by several storms in recent years, most recently by 'Georges' in September 1998, and 'José' and 'Lenny' in 1999. Mount Liamuiga (on Saint Christopher) and Nevis Peak are both volcanic centres, although they are not presently active.

# Statistical Survey

Source (unless otherwise stated): St Kitts and Nevis Information Service, Government Headquarters, Church St, POB 186, Basseterre; tel. 465-2521; fax 466-4504; e-mail skninfo@caribsurf.com; internet www.stkittsnevis.net.

## AREA AND POPULATION

**Area** (sq km): 269.4 (Saint Christopher 176.1, Nevis 93.3).

**Population:** 40,618 (males 19,933, females 20,685) at census of 12 May 1991; 45,841 (males 22,784, females 23,057) at census of 14 May 2001 (provisional results). *2003:* 47,300 (mid-year estimate). Sources: partly UN, *Population and Vital Statistics Report* and Caribbean Development Bank, *Social and Economic Indicators*.

**Density** (mid-2003): 175.6 per sq km.

**Principal Town** (estimated population incl. suburbs, mid-2003): Basseterre (capital) 13,262. Source: UN, *World Urbanization Prospects: The 2003 Revision*.

**Births and Deaths** (2000): Registered live births 838 (estimated birth rate 20.7 per 1,000); Registered deaths 356 (estimated death rate 8.8 per 1,000). *2003:* Crude birth rate 15.6 per 1,000; Crude death rate 7.6 per 1,000 (Source: Caribbean Development Bank, *Social and Economic Indicators*).

**Expectation of life** (years at birth): 70 (males 69; females 72) in 2003. Source: WHO, *World Health Report*.

**Employment** (labour force survey, 1994): Sugar cane production/manufacturing 1,525; Non-sugar agriculture 914; Mining and quarrying 29; Manufacturing (excl. sugar) 1,290; Electricity, gas and water 416; Construction 1,745; Trade (except tourism) 1,249; Tourism 2,118; Transport and communications 534; Business and general services 3,708; Government services 2,738; Other statutory bodies 342; *Total* 16,608 (Saint Christopher 12,516, Nevis 4,092). Source: IMF, *St Kitts and Nevis: Recent Economic Developments* (August 1997).

# SAINT CHRISTOPHER AND NEVIS

## HEALTH AND WELFARE
### Key Indicators

**Total Fertility Rate** (children per woman, 2003): 2.3.

**Under-5 Mortality Rate** (per 1,000 live births, 2003): 22.

**Physicians** (per 1,000 head, 1997): 1.17.

**Hospital Beds** (per 1,000 head, 1996): 6.36.

**Health Expenditure** (2002): US $ per head (PPP): 667.

**Health Expenditure** (2002): % of GDP: 5.5.

**Health Expenditure** (2002): public (% of total): 62.1.

**Access to Water** (% of persons, 2002): 99.

**Access to Sanitation** (% of persons, 2002): 96.

**Human Development Index** (2002): ranking: 39.

**Human Development Index** (2002): value: 0.844.
For sources and definitions, see explanatory note on p. vi.

## AGRICULTURE, ETC.

**Principal Crops** (FAO estimates, '000 metric tons, 2003): Roots and tubers 1.0; Sugar cane 193.0; Coconuts 1.0; Vegetables and melons 0.6; Fruits and berries 1.4.

**Livestock** (FAO estimates, '000 head, year ending September 2003): Cattle 4.3; Sheep 14.0; Goats 14.4; Pigs 4.0; Poultry 60.

**Livestock Products** (FAO estimates, '000 metric tons, 2003): Beef and veal 0.1; Pig meat 0.3; Poultry meat 0.1; Hen eggs 0.2.

**Fishing** (FAO estimates, metric tons, live weight, 2003): Snappers 16; Tuna-like fishes 13; Needlefishes, etc. 34; Flyingfishes 42; Bigeye scad 41; Common dolphinfish 40; Stromboid conchs 36; *Total catch* 370.
Source: FAO.

## INDUSTRY

**Production:** Raw sugar 21,000 metric tons in 2002; Electric energy 100 million kWh in 2001. Source: UN, *International Commodity Statistics Yearbook*.

## FINANCE

**Currency and Exchange Rates:** 100 cents = 1 Eastern Caribbean dollar (EC $). *Sterling, US Dollar and Euro Equivalents* (31 May 2005): £1 sterling = EC $4.909; US $1 = EC $2.700; €1 = EC $3.329; EC $100 = £20.37 = US $37.04 = €30.04. *Exchange Rate*: Fixed at US $1 = EC $2.70 since July 1976.

**Budget** (preliminary, EC $ million, 2002): *Revenue:* Revenue from taxation 212.1 (of which, Taxes on income and profits 62.0; Taxes on domestic goods and services 41.3; Taxes on international trade and transactions 104.3); Other current revenue 75.5; Capital revenue 6.5; Foreign grants 30.5; Total 324.6. *Expenditure:* Current expenditure 315.1 (Personal emoluments 142.9, Goods and services 71.9; Public debt charges 67.1, Transfers 33.3); Capital expenditure and net lending 132.5; Total 447.6. Source: Eastern Caribbean Central Bank, *Report and Statement of Accounts*.

**International Reserves** (US $ million at 31 December 2004): Reserve position in IMF 0.13; Foreign exchange 78.34; Total 78.47. Source: IMF, *International Financial Statistics*.

**Money Supply** (EC $ million at 31 December 2004): Currency outside banks 44.61; Demand deposits at deposit money banks 162.41; Total money (incl. others) 207.21. Source: IMF, *International Financial Statistics*.

**Cost of Living** (Consumer Price Index; base: 2000 = 100): 102.1 in 2001; 104.2 in 2002; 106.6 in 2003. Source: IMF, *International Financial Statistics*.

**Gross Domestic Product** (EC $ million at current factor cost): 802.1 in 2002; 821.0 in 2003; 889.6 in 2004 (preliminary). Source: Eastern Caribbean Central Bank.

**Expenditure on the Gross Domestic Product** (EC $ million at current prices, 2003): Government final consumption expenditure 176.23; Private final consumption expenditure 497.60; Gross fixed capital formation 469.73; *Total domestic expenditure* 1,143.56; Exports of goods and services 428.30; *Less* Imports of goods and services 587.47; *GDP in purchasers' values* 984.39. Source: Eastern Caribbean Central Bank.

**Gross Domestic Product by Economic Activity** (EC $ million at current factor cost, 2003): Agriculture, hunting, forestry and fishing 24.57; Mining and quarrying 2.32; Manufacturing 76.67; Electricity and water 23.72; Construction 129.80; Wholesale and retail trade 104.01; Restaurants and hotels 56.62; Transport 64.26; Communications 47.25; Finance and insurance 131.85; Real estate and housing 23.20; Government services 156.49; Other community, social and personal services 36.21; *Sub-total* 876.97; *Less* Imputed bank service charge 55.95; *Total in basic prices* 821.02. Source: Eastern Caribbean Central Bank.

**Balance of Payments** (EC $ million, 2003): Exports of goods f.o.b. 169.08; Imports of goods f.o.b. −383.62; *Trade balance* −214.54; Exports of services 259.22; Imports of services −203.85; *Balance on goods and services* −159.17; Other income received (net) −117.91; *Balance on goods, services and income* −277.08; Current transfers received (net) 44.65; *Current balance* −232.43; Capital account (net) 14.14; Direct investment from abroad (net) 138.43; Portfolio investment (net) 120.95; Other investment (net) −66.53; Net errors and omissions 22.85; *Overall balance* −2.59. Source: Eastern Caribbean Central Bank.

## EXTERNAL TRADE

**Principal Commodities** (US $ million, 2001): *Imports c.i.f.:* Food and live animals 27.3; Mineral fuels, lubricants, etc. 14.2 (Refined petroleum products 12.1); Chemicals 13.0; Basic manufactures 42.5 (Iron and steel manufactures 6.8); Machinery and transport equipment 51.9 (Road vehicles 9.5); Total (incl. others) 189.2. *Exports f.o.b.:* Food and live animals 7.2 (Raw sugar 6.5); Basic manufactures 0.9 (Metal manufactures 0.9); Machinery and transport equipment 20.2 (Telecommunications equipment 1.3; Other electrical machinery 18.2); Miscellaneous manufactures 2.0 (Printed matter 1.0); Total (incl. others) 31.0. Source: UN, *International Trade Statistics Yearbook*.

**Principal Trading Partners** (US $ million, 2001): *Imports:* Barbados 4.1; Canada 21.6; France 2.1; Japan 4.7; Netherlands 2.5; Trinidad and Tobago 23.4; United Kingdom 15.6; USA 95.6; Total (incl. others) 189.2. *Exports* (excl. re-exports): Dominica 0.4; United Kingdom 7.3; USA 22.1; Total (incl. others) 30.9. Source: UN, *International Trade Statistics Yearbook*.

## TRANSPORT

**Road Traffic** (registered motor vehicles): 11,352 in 1998; 12,432 in 1999; 12,917 in 2000.

**Shipping:** *Arrivals* (2000): 1,981 vessels. *International Sea-borne Freight Traffic* ('000 metric tons, 2000): Goods loaded 24.7; Goods unloaded 234.2.

**Civil Aviation** (aircraft arrivals): 24,800 in 1998; 23,500 in 1999; 19,400 in 2000.

## TOURISM

**Tourist Arrivals** ('000): 236.2 (69.0 visitor arrivals, 167.2 cruise-ship passengers) in 2002; 236.9 (90.6 visitor arrivals, 146.3 cruise-ship passengers) in 2003; 380.2 (120.1 visitor arrivals, 260.1 cruise-ship passengers) in 2004.

**Visitor Expenditure** (estimates, US $ million): 57.1 in 2002; 75.4 in 2003; 106.9 in 2004.
Source: Caribbean Development Bank, *Social and Economic Indicators*.

## COMMUNICATIONS MEDIA

**Radio Receivers** ('000 in use, 1997): 28.

**Television Receivers** ('000 in use, 1999): 10.

**Telephones** ('000 main lines in use, 2002): 23.5.

**Facsimile Machines** (1996): 450.

**Mobile Cellular Telephones** (subscribers, 2002): 5,000.

**Personal Computers** (2002): 9,000.

**Internet Users** (2002): 10,000.

**Non-daily Newspapers** (2000): Titles 4; Circulation 34,000 (1996).
Sources: mainly UNESCO, *Statistical Yearbook*; UN, *Statistical Yearbook*; International Telecommunication Union.

# SAINT CHRISTOPHER AND NEVIS

## EDUCATION

**Pre-primary** (2000/01): 77 schools; 170 teachers; 2,819 pupils.
**Primary** (2000/01): 23 schools; 302 teachers; 5,608 pupils (2001).
**Secondary** (2000/01): 7 schools; 365 teachers; 4,445 pupils (2001).
**Tertiary** (2000/01): 1 institution; 64 teachers (1999/2000); 1,235 students.

**Adult Literacy Rate:** 97.8% in 2002. Source: UN Development Programme, *Human Development Report*.

# Directory

## The Constitution

The Constitution of the Federation of Saint Christopher and Nevis took effect from 19 September 1983, when the territory achieved independence. Its main provisions are summarized below

### FUNDAMENTAL RIGHTS AND FREEDOMS

Regardless of race, place of origin, political opinion, colour, creed or sex, but subject to respect for the rights and freedoms of others and for the public interest, every person in Saint Christopher and Nevis is entitled to the rights of life, liberty, security of person, equality before the law and the protection of the law. Freedom of conscience, of expression, of assembly and association is guaranteed, and the inviolability of personal privacy, family life and property is maintained. Protection is afforded from slavery, forced labour, torture and inhuman treatment.

### THE GOVERNOR-GENERAL

The Governor-General is appointed by the British monarch, whom the Governor-General represents locally. The Governor-General must be a citizen of Saint Christopher and Nevis, and must appoint a Deputy Governor-General, in accordance with the wishes of the Premier of Nevis, to represent the Governor-General on that island.

### PARLIAMENT

Parliament consists of the British monarch, represented by the Governor-General, and the National Assembly, which includes a Speaker, three (or, if a nominated member is Attorney-General, four) nominated members (Senators) and 11 elected members (Representatives). Senators are appointed by the Governor-General; one on the advice of the Leader of the Opposition, and the other two in accordance with the wishes of the Prime Minister. The Representatives are elected by universal suffrage, one from each of the 11 single-member constituencies.

Every citizen over the age of 18 years is eligible to vote. Parliament may alter any of the provisions of the Constitution.

### THE EXECUTIVE

Executive authority is vested in the British monarch, as Head of State, and is exercised on the monarch's behalf by the Governor-General, either directly or through subordinate officers. The Governor-General appoints as Prime Minister that Representative who, in the Governor-General's opinion, appears to be best able to command the support of the majority of the Representatives. Other ministerial appointments are made by the Governor-General, in consultation with the Prime Minister, from among the members of the National Assembly. The Governor-General may remove the Prime Minister from office if a resolution of 'no confidence' in the Government is passed by the National Assembly and if the Prime Minister does not resign within three days or advise the Governor-General to dissolve Parliament.

The Cabinet consists of the Prime Minister and other Ministers. When the office of Attorney-General is a public office, the Attorney-General shall, by virtue of holding that office, be a member of the Cabinet in addition to the other Ministers. The Governor-General appoints as Leader of the Opposition in the National Assembly that Representative who, in the Governor-General's opinion, appears to be best able to command the support of the majority of the Representatives who do not support the Government.

### CITIZENSHIP

All persons born in Saint Christopher and Nevis before independence who, immediately before independence, were citizens of the United Kingdom and Colonies automatically become citizens of Saint Christopher and Nevis. All persons born in Saint Christopher and Nevis after independence automatically acquire citizenship, as do those born outside Saint Christopher and Nevis after independence to a parent possessing citizenship. There are provisions for the acquisition of citizenship by those to whom it is not automatically granted.

### THE ISLAND OF NEVIS

There is a Legislature for the island of Nevis which consists of the British monarch, represented by the Governor-General, and the Nevis Island Assembly. The Assembly consists of three nominated members (one appointed by the Governor-General in accordance with the advice of the Leader of the Opposition in the Assembly, and two appointed by the Governor-General in accordance with the advice of the Premier) and such number of elected members as corresponds directly with the number of electoral districts on the island.

There is a Nevis Island Administration, consisting of a premier and two other members who are appointed by the Governor-General. The Governor-General appoints the Premier as the person who, in the Governor-General's opinion, is best able to command the support of the majority of the elected members of the Assembly. The other members of the Administration are appointed by the Governor-General, acting in accordance with the wishes of the Premier. The Administration has exclusive responsibility for administration within the island of Nevis, in accordance with the provisions of any relevant laws.

The Nevis Island Legislature may provide that the island of Nevis is to cease to belong to the Federation of Saint Christopher and Nevis, in which case this Constitution would cease to have effect in the island of Nevis. Provisions for the possible secession of the island contain the following requirements: that the island must give full and detailed proposals for the future Constitution of the island of Nevis, which must be laid before the Assembly for a period of at least six months prior to the proposed date of secession; that a two-thirds majority has been gained in a referendum which is to be held after the Assembly has passed the motion.

## The Government

### HEAD OF STATE

**Monarch:** HM Queen ELIZABETH II.

**Governor-General:** Sir CUTHBERT MONTROVILLE SEBASTIAN (took office 1 January 1996).

### CABINET
(July 2005)

**Prime Minister and Minister of Finance, Technology and Sustainable Development, Tourism, Sports and Culture:** Dr DENZIL LLEWELLYN DOUGLAS.

**Deputy Prime Minister and Minister of Education, Youth, Social and Community Development and Gender Affairs:** SAM TERRENCE CONDOR.

**Minister of Public Works, Utilities, Transport and Posts:** Dr EARL ASIM MARTIN.

**Minister of National Security, Justice, Immigration and Labour:** GERALD ANTHONY DWYER ASTAPHAN.

**Minister of Health:** RUPERT EMMANUEL HERBERT.

**Minister of Foreign Affairs, International Trade, Industry and Commerce:** TIMOTHY SYLVESTER HARRIS.

**Minister of Housing, Agriculture, Fisheries and Consumer Affairs:** CEDRIC ROY LIBURD.

**Attorney-General and Minister of Legal Affairs:** DELANO FRANK BART.

**Minister of State in the Office of the Prime Minister with responsibility for Finance, Technology and Sustainable Development:** Sen. NIGEL ALEXIS CARTY.

**Minister of State in the Office of the Prime Minister with responsibility for Tourism, Sports and Culture:** Sen. RICHARD SKERRITT.

# SAINT CHRISTOPHER AND NEVIS

## MINISTRIES

**Office of the Governor-General:** Government House, Basseterre; tel. 465-2315.

**Government Headquarters:** Church St, POB 186, Basseterre; tel. 465-2521; fax 465-1001; internet www.stkittsnevis.net.

**Prime Minister's Office and Ministry of Finance, Technology and Sustainable Development, Tourism, Sports and Culture:** Church St, POB 186, Basseterre; tel. 465-2521; fax 465-1001; e-mail sknpmoffice@caribsurf.com; internet www.fsd.gov.kn.

**Attorney-General's Office and Ministry of Legal Affairs:** Church St, POB 186, Basseterre; tel. 465-2521; fax 465-5040; e-mail attnygenskn@caribsurf.com.

**Ministry of Education, Youth, Social and Community Development and Gender:** Church St, POB 186, Basseterre; tel. 465-2521; fax 465-2556; e-mail minelsc@caribsurf.com.

**Ministry of Foreign Affairs, International Trade, Industry and Commerce:** Church St, POB 186, Basseterre; tel. 465-2521; fax 465-1778; e-mail mintica@thecable.net.

**Ministry of Health:** Church St, POB 186, Basseterre; tel. 465-2521; fax 465-1316; e-mail minhwa@caribsurf.com.

**Ministry of Housing, Agriculture, Fisheries and Consumer Affairs:** Church St, POB 186, Basseterre; tel. 465-2521; fax 465-2635; e-mail minafclh@caribsurf.com.

**Ministry of National Security, Justice, Immigration and Labour:** Church St, POB 186, Basseterre; tel. 465-2521; fax 465-8244; e-mail mwaskn@caribsurf.com.

**Ministry of Public Works, Utilities, Transport and Posts:** Church St, POB 186, Basseterre; tel. 466-7032; fax 465-9475.

### NEVIS ISLAND ADMINISTRATION

**Premier:** VANCE W. AMORY.

There are also two appointed members.

**Administrative Centre:** Bath Hotel, Nevis; tel. 469-5521; fax 469-1207; e-mail nevisinfo@nevisweb.kn; internet www.nevisweb.kn.

## Legislature

### NATIONAL ASSEMBLY

**Speaker:** MARCELLA LIBURD.

Elected members: 11. Nominated members: 3. *Ex–officio* members: 1.

**Election, 25 October 2004**

| Party | % of votes | Seats |
|---|---|---|
| St Kitts-Nevis Labour Party | 50.6 | 7 |
| Concerned Citizens' Movement | 8.8 | 2 |
| People's Action Movement | 31.7 | 1 |
| Nevis Reformation Party | 7.5 | 1 |
| **Total** (incl. others) | 100.0 | 11 |

### NEVIS ISLAND ASSEMBLY

Elected members: 5. Nominated members: 3.

Elections to the Nevis Island Assembly took place in September 2001. The Concerned Citizens' Movement took four seats, and the Nevis Reformation Party retained one seat.

## Political Organizations

**Concerned Citizens' Movement (CCM):** Charlestown, Nevis; alliance of four parties; Leader VANCE W. AMORY.

**Nevis Reformation Party (NRP):** Government Rd, POB 480, Charlestown, Nevis; tel. 469-0630; f. 1970; Leader JOSEPH PARRY; Sec. LEVI MORTON.

**People's Action Movement (PAM):** Basseterre; e-mail exec@pamskb.com; internet www.pamdemocrat.org; f. 1965; Political Leader LINDSAY GRANT; Deputy Leader SHAWN RICHARDS.

**St Kitts-Nevis Labour Party (SKNLP):** Masses House, Church St, POB 239, Basseterre; tel. 465-5347; fax 465-8328; internet www.sknlabourparty.org; f. 1932; socialist party; Chair. TIMOTHY HARRIS; Leader Dr DENZIL LLEWELLYN DOUGLAS.

**United National Empowerment Party (UNEP):** Basseterre; e-mail bhsor34@yahoo.com; internet www.sknunep.org; Pres. Dr HENRY L. O. STOGUMBER BROWNE.

## Diplomatic Representation

### EMBASSIES IN SAINT CHRISTOPHER AND NEVIS

**China (Taiwan):** Taylor's Range, POB 119, Basseterre; tel. 465-2421; fax 465-7921; e-mail rocemb@caribsurf.com; Ambassador MARIETTA KAO LIAU.

**Cuba:** 34 Bladen Housing Devt, POB 600, Basseterre; tel. 466-3374; fax 465-8072; e-mail cubaask@thecable.net; Ambassador ORLANDO ALVAREZ.

**Venezuela:** Delisle St, POB 435, Basseterre; tel. 465-2073; fax 465-5452; e-mail frontado@caribsurf.com; Ambassador MIRIAM TROCONIS LUZARDO.

Diplomatic relations with other countries are maintained at consular level, or with ambassadors and high commissioners resident in other countries of the region, or directly with the other country.

## Judicial System

Justice is administered by the Eastern Caribbean Supreme Court, based in Saint Lucia and consisting of a Court of Appeal and a High Court. One of the nine puisne judges of the High Court is responsible for Saint Christopher and Nevis and presides over the Court of Summary Jurisdiction. The Magistrates' Courts deal with summary offences and civil offences involving sums of not more than EC $5,000. In 1998 the death penalty was employed in Saint Christopher and Nevis for the first time since 1985.

**Puisne Judge:** NEVILLE SMITH.

**Magistrates' Office:** Losack Rd, Basseterre; tel. 465-2926.

## Religion

### CHRISTIANITY

**St Kitts Christian Council:** Victoria Rd, POB 48, Basseterre; tel. 465-2167; e-mail stgeorgessk@hotmail.com; Chair. Archdeacon VALENTINE HODGE.

**St Kitts Evangelical Association:** Princess St, Basseterre.

#### The Anglican Communion

Anglicans in Saint Christopher and Nevis are adherents of the Church in the Province of the West Indies. The islands form part of the diocese of the North Eastern Caribbean and Aruba. The Bishop is resident in The Valley, Anguilla.

#### The Roman Catholic Church

The diocese of Saint John's-Basseterre, suffragan to the archdiocese of Castries (Saint Lucia), includes Anguilla, Antigua and Barbuda, the British Virgin Islands, Montserrat and Saint Christopher and Nevis. At 31 December 2003 the diocese contained an estimated 15,423 adherents. The Bishop participates in the Antilles Episcopal Conference (currently based in Port of Spain, Trinidad and Tobago).

**Bishop of Saint John's-Basseterre:** Rt Rev. DONALD JAMES REECE (resident in St John's, Antigua).

#### Other Churches

There are also communities of Methodists, Moravians, Seventh-day Adventists, Baptists, Pilgrim Holiness, the Church of God, Apostolic Faith and Plymouth Brethren.

## The Press

**The Democrat:** Cayon St, POB 30, Basseterre; tel. 465-2091; fax 465-0857; internet www.pamskb.com/democrat; f. 1948; weekly on Saturdays; organ of PAM; Dir Capt. J. L. WIGLEY; Editor FITZROY P. JONES; circ. 3,000.

**The Labour Spokesman:** Masses House, Church St, POB 239, Basseterre; tel. 465-2229; fax 466-9866; e-mail skn.union@caribsurf.com; internet www.sknlabourparty.org/spokesman; f. 1957; Wednesdays and Saturdays; organ of St Kitts-Nevis Trades and Labour Union; Editor DAWUD BYRON; Man. WALFORD GUMBS; circ. 6,000.

**The Leeward Times:** Old Hospital Rd, Charlestown, POB 535, Nevis; tel. 469-1409; fax 469-0662; e-mail hbramble@caribsurf.com; Editor HOWELL BRAMBLE.

**The St Kitts and Nevis Observer:** Lozack Rd, Basseterre; tel. 466-4994; fax 466-4995; e-mail observsk@caribsurf.com; weekly; Bureau Chief JOHN PERKER.

## SAINT CHRISTOPHER AND NEVIS

**FOREIGN NEWS AGENCIES**

Associated Press (USA) and Inter Press Service (IPS) (Italy) are represented in Basseterre.

## Publishers

**Caribbean Publishing Co (St Kitts-Nevis) Ltd:** Dr William Herbert Complex, Frigate Bay Rd, POB 745, Basseterre; tel. 465-5178; fax 466-0307; e-mail lsk-sales@caribsurf.com.

**MacPennies Publishing Co:** 10A Cayon St East, POB 318, Basseterre; tel. 465-2274; fax 465-8668.

## Broadcasting and Communications

### TELECOMMUNICATIONS

#### Regulatory Authority

**Eastern Caribbean Telecommunications Authority:** based on Castries, Saint Lucia; f. 2000 to regulate telecommunications in Saint Christopher and Nevis, Dominica, Grenada, Saint Lucia and Saint Vincent and the Grenadines.

#### Service Providers

**Cable:** Basseterre; f. 1983; granted a licence in May 2002 to provide fixed public telecommunications network and services.

**Cable & Wireless St Kitts and Nevis:** Cayon St, POB 86, Basseterre; tel. 465-1000; fax 465-1106; internet www.candw.kn; f. 1985; fmrly St Kitts and Nevis Telecommunications Co Ltd (SKANTEL); 65% owned by Cable & Wireless plc; 17% state-owned; Chair. LEE L. MOORE; Gen. Man. K. C. RICKY WENT.

**Cariglobe:** Basseterre; granted a licence in May 2002 to provide mobile telecommunications network.

**Cingular Wireless St Kitts and Nevis:** Basseterre; internet www.cingularwireless.com; joint venture between SBC Communications and BellSouth; scheduled to sell its operations and licences in the Caribbean to Digicel in 2005; Pres. and CEO STANLEY T. SIGMAN.

### BROADCASTING

#### Radio

**Choice FM:** Wellington Rd, Needsmust, Basseterre; tel. 466-1891; fax 466-1892; e-mail choicefm@caribsurf.com; Man. VINCENT HERBERT.

**Goodwill Radio FM 104.5:** POB 98, Lodge Project; tel. 465-7795; fax 465-9556; Gen. Man. DENNIS HUGGINS-NELSON.

**Radio One (SKNBC):** Bakers Corner, POB 1773, Basseterre; tel. 466-0941; fax 465-0406; e-mail gwa_house_skb@hotmail.com; Man. GUS WILLIAMS.

**Radio Paradise:** Bath Plain, POB 423, Nevis; tel. 469-1994; fax 469-1642; owned by US co (POB A, Santa Ana, CA 92711); religious; Man. ARTHUR GILBERT.

**Sugar City Rock FM:** Greenlands, Basseterre; tel. 466-1113; e-mail sugarcityrock@hotmail.com; Gen. Man. VAL THOMAS.

**Trinity Broadcasting Ltd:** Bath Plain Rd, Charlestown, Nevis; tel. 469-0285; fax 469-1723; Dir ARTHUR GILBERT.

**Voice of Nevis (VON) Radio 895 AM:** Bath Plain, Bath Village, POB 195, Charlestown, Nevis; tel. 469-1616; fax 469-5329; e-mail vonradio@caribsurf.com; internet www.skbee.com/vonlive.html.

**WINN FM:** Unit c24, The Sands, Newtown Bay Rd, Basseterre; tel. 466-9586; fax 466-7904; e-mail info@winnfm.com; internet www.winnfm.com; owned by Federation Media Group; Chair. MICHAEL KING.

**ZIZ Radio and Television:** Springfield, POB 331, Basseterre; tel. 465-2622; fax 465-5624; e-mail zbc@caribsurf.com; internet www.zizonline.com; f. 1961; television from 1972; commercial; govt-owned; Gen. Man. WINSTON MCMAHON.

#### Television

**ZIZ Radio and Television:** see Radio.

## Finance

(cap. = capital; res = reserves; dep. = deposits; brs = branches)

### BANKING

#### Central Bank

**Eastern Caribbean Central Bank (ECCB):** Headquarters Bldg, Bird Rock, POB 89, Basseterre; tel. 465-2537; fax 465-9562; e-mail eccbinfo@caribsurf.com; internet www.eccb-centralbank.org; f. 1965 as East Caribbean Currency Authority; expanded responsibilities and changed name 1983; responsible for issue of currency in Anguilla, Antigua and Barbuda, Dominica, Grenada, Montserrat, Saint Christopher and Nevis, Saint Lucia and Saint Vincent and the Grenadines; res EC $121.7m., dep. EC $1,041.9m., total assets EC $1,740.5m. (March 2004); Gov. and Chair. Sir K. DWIGHT VENNER; Country Dir WENDELL LAWRENCE.

#### Local Banks

**Bank of Nevis Ltd:** Main St, POB 450, Charlestown, Nevis; tel. 469-5564; fax 469-5798; e-mail bon@caribsurf.com; internet www.bankofnevis.com.

**RBTT Bank (SKN) Ltd:** Main and Chappel Sts, POB 673, Charlestown, Nevis; tel. 469-5277; fax 469-1493; internet www.rbtt.com; f. 1955 as Nevis Co-operative Banking Co Ltd; acquired by Royal Bank of Trinidad and Tobago (later known as RBTT) in 1996; Group Chair. PETER J. JULY.

**St Kitts-Nevis Anguilla National Bank Ltd:** Central St, POB 343, Basseterre; tel. 465-2204; fax 466-1050; e-mail national_bank@sknanb.com; internet www.sknanb.com; f. 1971; Govt of St Kitts and Nevis 51%; cap. EC $81.0m., res EC $51.2m., dep. EC $848.0m. (June 2003); Chair. RUBLIN AUDAIN; Man. Dir EDWIN W. LAWRENCE; 5 brs.

#### Foreign Banks

**Bank of Nova Scotia:** Fort St, POB 433, Basseterre; tel. 465-4141; fax 465-8600; Man. W. A. CHRISTIE.

**FirstCaribbean International Bank (Barbados) Ltd:** Basseterre; internet www.firstcaribbeanbank.com; f. 2002 following merger of Caribbean operations of Barclays Bank PLC and CIBC; Exec. Chair MICHAEL MANSOOR; CEO CHARLES PINK.

**Royal Bank of Canada:** cnr Bay and Fort St, POB 91, Basseterre; tel. 465-2259; fax 465-1040.

#### Development Bank

**Development Bank of St Kitts and Nevis:** Church St, POB 249, Basseterre; tel. 465-2288; fax 465-4016; e-mail dbskn@caribsurf.com; internet www.skndb.com; f. 1981; cap. EC $8.0m., res EC $1.8m., dep. EC $2.5m.; Chair. JOSEPH LLEWELYN EDMEADE; Gen. Man. AUCKLAND HECTOR.

### STOCK EXCHANGE

**Eastern Caribbean Securities Exchange Ltd:** Bird Rock, POB 94, Basseterre; tel. 466-7192; fax 465-3798; e-mail info@ecseonline.com; internet www.ecseonline.com; f. 2001; regional securities market designed to facilitate the buying and selling of financial products for the eight member territories—Anguilla, Antigua and Barbuda, Dominica, Grenada, Montserrat, Saint Christopher and Nevis, Saint Lucia and Saint Vincent and the Grenadines; Gen. Man. TREVOR BLAKE.

### INSURANCE

**National Caribbean Insurance Co Ltd:** Central St, POB 374, Basseterre; tel. 465-2694; fax 465-3659; subsidiary of St Kitts-Nevis Anguilla National Bank Ltd.

**St Kitts-Nevis Insurance Co Ltd:** Central St, POB 142, Basseterre; tel. 465-2845; fax 465-5410.

Several foreign companies also have offices in Saint Christopher and Nevis.

## Trade and Industry

### GOVERNMENT AGENCIES

**Central Marketing Corpn (CEMACO):** Pond's Pastire, Basseterre; tel. 465-2326; fax 465-2326; Man. MAXWELL GRIFFIN.

**The Department of Planning and Development:** The Cotton House, Market St, Charlestown, Nevis; tel. 469-5521; fax 469-1273; e-mail planevis@caribsurf.com.

**Frigate Bay Development Corporation:** Frigate Bay, POB 315, Basseterre; tel. 465-8339; fax 465-4463; promotes tourist and residential developments.

**Investment Promotion Agency:** Investment Promotion Division, Ministry of Tourism, Commerce and Consumer Affairs, Pelican Mall, Bay Rd, POB 132, Basseterre; tel. 465-4040; fax 465-6968; f. 1987.

**St Kitts Sugar Manufacturing Corpn (SSMC):** St Kitts Sugar Factory, POB 96, Basseterre; tel. 466-8503; merged with National Agricultural Corpn in 1986; expected to be dissolved following the announcement of the closure of the sugar industry in March 2005; Gen. Man. J. E. S. ALFRED.

**Social Security Board:** Bay Rd, POB 79, Basseterre; tel. 465-2535; fax 465-5051; e-mail ssbdirof@caribsurf.com; f. 1977; Dir SEPHLIN LAWRENCE.

### CHAMBER OF COMMERCE

**St Kitts-Nevis Chamber of Industry and Commerce:** Horsford Rd, Fortlands, POB 332, Basseterre; tel. 465-2980; fax 465-4490; e-mail sknchamber@caribsurf.com; internet www.stkittsnevischamber.org; incorporated 1949; 140 mems; Pres. ANTHONY ABOURIZK; Exec. Dir WENDY PHIPPS.

### EMPLOYERS' ORGANIZATIONS

**Building Contractors' Association:** Anthony Evelyn Business Complex, Paul Southwell Industrial Park, POB 1046, Basseterre; tel. 465-6897; fax 465-5623; e-mail sknbca@caribsurf.com; Pres. ANTHONY E. EVELYN.

**Nevis Cotton Growers' Association Ltd:** Charlestown, Nevis; Pres. IVOR STEVENS.

**Small Business Association:** Anthony Evelyn Business Complex, Paul Southwell Industrial Park, POB 367, Basseterre; tel. 465-8630; fax 465-6661; e-mail sb-association@caribsurf.com; Pres. EUSTACE WARNER.

### UTILITIES

**Nevis Electricity Company Ltd (Nevlec):** Charlestown, Nevis; Gen. Man. EDGAR WIGGINS.

### TRADE UNIONS

**St Kitts-Nevis Trades and Labour Union:** Masses House, Church St, POB 239, Basseterre; tel. 465-2229; fax 466-9866; e-mail sknunion@caribsurf.com; f. 1940; affiliated to Caribbean Maritime and Aviation Council, Caribbean Congress of Labour, International Federation of Plantation, Agricultural and Allied Workers and International Confederation of Free Trade Unions; associated with St Kitts-Nevis Labour Party; Pres. LEE L. MOORE; Gen. Sec. STANLEY R. FRANKS; about 3,000 mems.

**United Workers' Union (UWU):** Market St, Basseterre; tel. 465-4130; associated with People's Action Movement.

## Transport

### RAILWAYS

There are 58 km (36 miles) of narrow-gauge light railway on Saint Christopher, serving the sugar plantations. The railway, complete with new trains and carriages, was restored and developed for tourist excursions and opened in late 2002.

**St Kitts Scenic Railway:** Basseterre; tel. 465-7263; e-mail scenicreservations@caribsurf.com; internet www.stkittsscenicrailway.com; f. 2002.

**St Kitts Sugar Railway:** St Kitts Sugar Manufacturing Corpn, POB 96, Basseterre; tel. 465-8099; fax 465-1059; e-mail agronomy@caribsurf.com; Gen. Man. J. E. S. ALFRED.

### ROADS

In 1999 there were 320 km (199 miles) of road in Saint Christopher and Nevis, of which approximately 136 km (84 miles) are paved. In July 2001 the Caribbean Development Bank loaned US $3.75m. to the Nevis Government for a road improvement scheme.

### SHIPPING

The Government maintains a commercial motor-boat service between the islands, and numerous regional and international shipping lines call at the islands. A deep-water port, Port Zante, was opened at Basseterre in 1981. In June 2003 Government of Kuwait agreed to provide a loan of EC $15m. to help fund the development of the cruise-ship facilities at Port Zante.

**St Kitts Air and Sea Ports Authority:** Administration Bldg, Deep Water Port, Bird Rock, Basseterre; tel. 465-8121; fax 465-8124; f. 1993 to combine St Kitts Port Authority and Airports Authority; Gen. Man. SIDNEY OSBORNE; Airport Man. EDWARD HUGHES; Sea Port Man. CARL BRAZIER-CLARKE.

#### Shipping Companies

**Delisle Walwyn and Co Ltd:** Liverpool Row, POB 44, Basseterre; tel. 465-2631; fax 465-1125; e-mail delwal@caribsurf.com; internet www.delisleco.com.

**Sea Atlantic Cargo Shipping Corpn:** Main St, POB 556, Charlestown, Nevis.

**Tony's Ltd:** Main St, POB 564, Charlestown, Nevis; tel. 469-5953; fax 469-5413.

### CIVIL AVIATION

Robert Llewellyn Bradshaw (formerly Golden Rock) International Airport, 4 km (2½ miles) from Basseterre, is equipped to handle jet aircraft and is served by scheduled links with most Caribbean destinations, the United Kingdom, the USA and Canada. Saint Christopher and Nevis is a shareholder in the regional airline, LIAT (see chapter on Antigua and Barbuda). Vance W. Amory International Airport (formerly Newcastle Airfield), 11 km (7 miles) from Charlestown, Nevis, has regular scheduled services to St Kitts and other islands in the region. A new airport, Castle Airport, was opened on Nevis in 1998. In September 2002 a US $5.9m. project to construct a new passenger terminal at Vance W. Amory Airport was completed.

**St Kitts Air and Sea Ports Authority:** see Shipping.

#### Private Airlines

**Air St Kitts-Nevis:** Vance W. Amory International Airport, Newcastle, Nevis; tel. 469-9241; fax 469-9018.

**Caribbean Star Airlines:** Robert Llewellyn Bradshaw International Airport; f. 2000; relocated from Antigua in 2003; operates regional services; Pres. and CEO PAUL MOREIRA.

**Nevis Express Ltd:** Vance W. Amory International Airport, Newcastle, Nevis; tel. 469-9756; fax 469-9751; e-mail reservations@nevisexpress.com; internet www.nevisexpress.com; passenger and cargo charter services to all Caribbean destinations; St Kitts–Nevis shuttle service.

## Tourism

The introduction of regular air services to the US cities of Miami and New York has opened up the islands as a tourist destination. Visitors are attracted by the excellent beaches on Saint Christopher and the spectacular mountain scenery of Nevis, the historical Brimstone Hill Fortress National Park on Saint Christopher and the islands' associations with Lord Nelson and Alexander Hamilton. In 2004 there were an estimated 260,100 cruise-ship passengers and 120,100 stop-over visitors. Receipts from tourism were estimated at EC $288.6m. There were 1,508 rooms in hotels and guest houses in Saint Christopher and Nevis in 1999. The National Assembly was expected to approve the Cricket Cup Tourism Accommodation Act in 2005; the Act was designed to encourage the construction and refurbishment of hotels and other tourist accommodation for the 2007 Cricket World Cup, hosted by several Caribbean nations, including St Kitts.

**Nevis Tourism Authority:** Main St, POB 917, Charlestown; tel. 469-7550; fax 469-7551; e-mail nta2001@caribsurf.com; internet www.nevisisland.com; Dir HELEN KIDD.

**St Kitts-Nevis Department of Tourism:** Pelican Mall, Bay Rd, POB 132, Basseterre; tel. 465-4040; fax 465-8794; e-mail mintitcc@stkittstourism.kn; internet www.stkitts-nevis.com; Permanent Sec. HILARY WATTLEY.

**St Kitts-Nevis Hotel and Tourism Association:** Liverpool Row, POB 438, Basseterre; tel. 465-5304; fax 465-7746; e-mail stkitnevhta@caribsurf.com; internet www.stkittsnevishta.org; f. 1972; Pres. SAM NG'ALLA; Exec. Dir VAL HENRY.

## Defence

The small army was disbanded by the Government in 1981, and its duties were absorbed by the Volunteer Defence Force and a special tactical unit of the police. In July 1997 the National Assembly approved legislation to re-establish a full-time defence force. Coastguard operations were to be brought under military command; the defence force was also to include cadet and reserve forces. St Christopher and Nevis participates in the US-sponsored Regional

# SAINT CHRISTOPHER AND NEVIS

Security System, comprising police, coastguards and army units, which was established by independent Eastern Caribbean states in 1982. Budgetary expenditure on national security in 1998 was approximately EC $23.8m.

## Education

Education is compulsory for 12 years between five and 17 years of age. Primary education begins at the age of five, and lasts for seven years. Secondary education, from the age of 12, generally comprises a first cycle of four years, followed by a second cycle of two years. In 1993 enrolment at all levels of education was estimated to be equivalent to 78% of the school-age population. There are 30 state, eight private and five denominational schools. There is also a technical college. Budgetary expenditure on education by the central Government in 1998 was projected to be EC $25m. (6.7% of total government expenditure). In September 2000 a privately financed 'offshore' medical college, the Medical University of the Americas, opened in Nevis with 40 students registered. The Ross University of School of Veterinary Medicine and the International University of Nursing also operated on Saint Christopher. A Basic Education Project funded by the Caribbean Development Bank was in 2003 complemented by a $18.8m. Secondary Education Project, which was to include the construction of a new school in Saddlers.

# SAINT LUCIA

## Geography

### PHYSICAL FEATURES

Saint Lucia is the second largest of the Windward Islands, and is located in the eastern Caribbean, between the French department of Martinique and the fellow Commonwealth state of Saint Vincent and the Grenadines. Martinique lies 34 km away, north of the Saint Lucia Channel (a sea-lane from the Atlantic into the Caribbean), while the main island of St Vincent lies 42 km to the south, across the St Vincent Passage. The total area of the country is 616 sq km (238 sq miles), most of which consists of the main island itself, although there are a few small islands off shore, such as the uninhabited Maria Islands in the south-east.

The island of Saint Lucia is rugged and volcanic, bulked around a great barrier range of mountains along the backbone of the island, the Barre de l'Isle. The highlands are loftiest towards the south, reaching their highest point at Morne Gimie (950 m or 3,118 ft) in the south-west, although the peaks considered most emblematic of the island lie still further to the south-west. Here twin mountain horns rear above the spa town of Soufrière, jungle-clad volcanic plugs known as the Pitons (Petit Piton and, to its south, Gros Piton), steep cones plunging straight into the sea. Elsewhere on the island there are places where the highland terrain gives way to broad, fertile valleys, while the rich soil generally makes for a verdant landscape. The native rainforest has suffered since European colonization, particularly in the later 20th century, but is now protected and is still home to a rich variety of flora and fauna (deforestation was also affecting water supply). Like other islands in the Windwards, Saint Lucia is home to animals such as the iguana, the fer de lance (the only poisonous snake on the island), the manicou, the rarely seen agouti and the historically introduced mongoose. There are some endemic species, such as nine types of flamboyant tree, the pygmy gecko, the Saint Lucia tree lizard and, on the protected Maria Islands, a ground lizard and a grass snake. Bird species include the Saint Lucia parrot, the endangered Saint Lucia oriole and Saint Lucia black finch, and the Semper's warbler, which may now be extinct. On the rougher Atlantic coast turtles lay their eggs and other birds nest.

### CLIMATE

The climate is subtropical marine, moderated by the eastern and north-eastern trade winds. The island lies within the hurricane belt, and storms can cause flooding and mudslides in the steep terrain. Most rain falls between May and November, with annual averages varying considerably in different parts of the island, mainly owing to altitude. The range is between 1,540 mm and 3,540 mm (60–138 ins). Temperatures are fairly constantly around 27°C (80°F), though sometimes slightly lower in the drier months or in the heights.

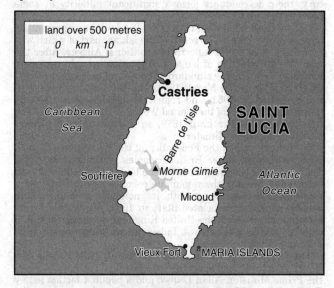

### POPULATION

The long years of alternating French and British rule have contributed to a rich cultural legacy, and the indigenous product flourishes with hybrid vigour. The descendants of the black slaves employed under both colonial powers (about 90% of the population is now black—the rest are of mixed race, 'East' Indian or white) use English as their official language, but a French patois or Creole (known as Kweyol locally) is widespread. Some even claim descent from the original Carib residents. Similarly, the system of government and law follows the British model, but about 68% of the population are Roman Catholic (about 7% are Protestant—especially Methodists and Baptists—and 3% are Anglican, with a few Hindu, Muslim and Jewish residents too). Certainly the island has produced some of the Caribbean's foremost writers, artists and thinkers—the country has the highest per-head rate of Nobel laureates ever (Sir Arthur Lewis, Economics, 1979, and Derek Walcott, Literature, 1992).

Estimates put the total population of Saint Lucia at 162,300 in mid-2004. About two-fifths of this total live in and around the island's capital, Castries, which is sited on the north-western coast. Other important towns include Micoud, on the south-eastern coast, across the island from Soufrière, and Vieux Fort, the main industrial centre, by the airport, near the southern tip of the island. Gros Islet is north of Castries, towards the end of the island, and Dennery is north of Micoud, where the road across the Barre de l'Isle reaches the east coast. For administrative purposes the island is divided into quarters, although, confusingly, there are 11 of these.

## History

### MARK WILSON

Saint Lucia is a constitutional monarchy within the Commonwealth. Queen Elizabeth II is head of state, and is represented in Saint Lucia by a Governor-General. There is a bicameral legislature with an elected chamber.

A small number of Saint Lucians trace their ancestry to the island's original Amerindian inhabitants, who knew the island as Hiwanarau or Hewanorra. There is no mention of Saint Lucia in Christopher Columbus's logbook, although a local tradition holds that he discovered and named the island on 13 December 1502, St Lucy's day. The first successful European settlement was French. The island changed hands on 14 occasions before passing to British rule in 1814, but extended periods of French rule in the 17th and 18th centuries established a French Creole as the main spoken language, and the Roman Catholic religion as the faith of the majority of the population. The island's hybrid legal code also contains elements of the pre-Revolutionary French legal tradition.

As in the United Kingdom's other Caribbean colonies, slavery was abolished in 1834. After this date, some former slaves established small peasant farms. Saint Lucia received significant numbers of Indian indentured labourers in the 19th century; their descendants form a significant minority in some parts of the island.

During the Second World War, the USA built an important air base at Vieux Fort in the south of the island, part of which was later developed as Hewanorra International Airport. After the war, the introduction of banana cultivation brought a modest increase in rural living standards. The Saint Lucia Labour Party (SLP) was formed in March 1949, won the first universal suffrage election in 1951, and remained in office until its 1964 defeat at the hands of the United Workers' Party (UWP), which then retained power until 1979, with John (later Sir John) Compton as Prime Minister.

Saint Lucia joined the Federation of the West Indies in 1958 along with nine other British colonies. When Jamaica and Trinidad and Tobago left in 1962, the Federation collapsed, and an attempt to unite the remaining colonies as the 'little eight' was unsuccessful. Along with its neighbours, Saint Lucia became a British Associated State in 1967, responsible for its internal affairs, with the United Kingdom retaining control of external affairs and defence. Independence was granted on 22 February 1979.

The SLP won a general election held in July 1979, but the new Government was torn apart by fierce factional rivalries between the Prime Minister, Allan Louisy, and a populist faction led by George Odlum. The Government was further damaged by allegations of mismanagement. After a Government defeat in Parliament, Winston Cenac replaced Louisy as Prime Minister. However, factional struggles continued; a series of protests, including an organized short-term closure of private-sector businesses in 1982, led to the formation of an interim Government led by Michael Pilgrim, which presided over fresh elections in that year. The UWP regained power with a convincing majority; Odlum fought the election as leader of the newly formed Progressive Labour Party, but was beaten into third place by the mainstream SLP.

Compton's majority was reduced to a single seat in the 1987 elections. Another election held immediately afterwards produced exactly the same result; however, the Government's position was strengthened when an SLP member crossed the floor.

After a total of 29 years in office, Compton resigned as Prime Minister in March 1996, in preparation for retirement. He was succeeded as Prime Minister by Dr Vaughan Lewis, a former Director-General of the Organisation of Eastern Caribbean States. The SLP at the same time chose a new leader, Dr Kenny Anthony, a lawyer with an academic post in the University of the West Indies. An election on 23 May 1997 resulted in an overwhelming victory for the SLP, which took 16 of the 17 seats and 60.0% of the popular vote. A disunited UWP did not perform well in opposition, and was not strengthened by the return of Compton from retirement, nor the formation of an alliance with Odlum, who had, for the second time in his career, broken with an SLP Government. In spite of the continuing troubles of the banana industry and in a poor year for tourism, the SLP won a second comfortable election victory on 3 December 2001, winning 14 of the House of Assembly's 17 seats, although its share of the popular vote fell to 54.2%; the UWP increased its strength to three seats.

The removal of the Minister of Home Affairs and Gender Relations, Sarah Flood-Beaubrun, in January 2004 over her vehement opposition to a liberalization of abortion law and her subsequent move to the opposition benches as an independent, barely weakened the Government. The opposition, meanwhile, remained chronically divided. Marius Wilson resigned from his post as leader of the opposition in February 2003 and sat in Parliament as an independent, leaving the UWP with only two seats. He was replaced by Arsene James, who was not considered a particularly effective parliamentary performer and was in turn succeeded in April 2004 by Marcus Nicholas. Nicholas received the support of Flood-Beaubrun and Wilson, but not of the UWP itself, leaving the former party of government with James as its sole loyal supporter in the legislature. Sir John Compton, at 79 years of age, on 13 March 2005 again emerged from retirement, replacing his chosen successor Dr Vaughn Lewis as leader of the UWP, although, like Lewis, he did not hold a parliamentary seat. Legislative elections were due by April 2007, but were expected to be held before that date. Drugs transhipment and violent crime remained serious concerns for the Government; the estimated per-head murder rate of 21 per 100,000 in 2004 was extremely high by international standards.

# Economy

## MARK WILSON

With 162,300 inhabitants on 616.3 sq km in mid-2004, Saint Lucia has a somewhat larger population than its neighbours in the Organisation of Eastern Caribbean States (OECS). With a per-head gross domestic product (GDP) of US $4,411 in 2002, it was a reasonably prosperous middle-income economy. However, GDP grew at an annual rate of only 1.6% in 1993–2003, largely because of the severe problems of the banana industry; GDP contracted by 0.4% in 2000 and, with tourism also affected by an international decline in demand, by 4.3% in 2001. Limited growth of 0.5% was achieved in 2002 and, with tourism recovering, GDP grew by 3.6% in 2003 and an estimated 5.4% in 2004. The fiscal position deteriorated from 1999, with an overall surplus equivalent to 3.4% of GDP declining to a deficit that reached 6.4% in 2003; meanwhile, unemployment rose from 16.5% in 2000 to 21% in 2004. Total public-sector debt was equivalent to 70% of GDP at the end of 2004, an increase from 44% in 1999; none the less, the ratio was still below levels recorded in most neighbouring islands.

Saint Lucia is a member of the Caribbean Community and Common Market, or CARICOM, which is attempting to develop a single market, in principle by 2006. It also houses the secretariat of the OECS, which links nine of the smaller Caribbean territories, while the Eastern Caribbean Central Bank, based in Saint Christopher and Nevis, supervises its financial affairs.

Agriculture (including hunting and forestry) and fishing comprised only 5.5% of GDP in 2004, down from 14.7% in 1990; however, the sector still employed 19.5% of the working population in 2000. Banana exports to the United Kingdom under protected market access arrangements were an economic mainstay from the 1950s, but the industry suffered a steep decline in the 1990s, as the trade privileges of Caribbean producers in the European market were eroded. In 1992 income from banana exports was US $68.1m., equivalent to 53.6% of merchandise exports and 21.2% of foreign-exchange earnings. In 2004 exports totalled only $19.9m., and in 2003 were equivalent to an estimated 4.0% of foreign-exchange earnings. However, significant grant aid from the European Union was available for irrigation and technical support, as well as for economic diversification and social support. Small farmers produce a wide variety of fruits, vegetables and livestock products for the local market, with some produce exported to neighbouring islands.

Saint Lucia's natural attractions and direct air links to North America and the United Kingdom have encouraged the development of tourism, which is the principal source of foreign exchange. Tourism receipts increased from EC $415.3m. in 1990 to an estimated $848.6m. in 2004. In that year the island attracted some 298,431 stop-over tourists, more than any of its OECS neighbours and above the previous peak figure of 285,400 recorded in 2000. There were also some 481,279 cruise-ship passengers, although this group had much lower per-head spending. In 2003 41.0% of tourist arrivals were from North America and 32.8% were from Europe. Tourism in Saint Lucia is reliant to a great extent on all-inclusive properties, where meals and other facilities are pre-paid; one Jamaican all-inclusive group had three properties in operation, with a fourth under construction in 2005. These and other hotel investments provided strong support to the construction industry. In mid-2005 tourism investments in progress or under development had a total value of close to US $1,000m., or 140% of GDP, with a total of more than 2,100 rooms, villas and condominium units. Yachting is an important sub-sector of tourism. The Jazz Festival in May is also a significant attraction, as is the annual Carnival, now held in July in place of its traditional pre-Lenten date. However, the high rate of violent crime, much of it drugs-related, was a potential threat to tourism.

Manufacturing comprised 4.5% of GDP in 2004, down from 8.2% in 1990. A brewery exports some of its product to the regional market. Another major enterprise—a cardboard box manufacturer—has been severely damaged by the decline in the banana industry, its major market. Saint Lucia houses the headquarters of the Eastern Caribbean Telecommunications Authority; the telecommunications sector has been liberalized, and the Government saw telemarketing and informatics as areas with great potential for economic growth.

The 'offshore' financial services sector was established only in 2000 and was better regulated than those of most other OECS members. Saint Lucia, in contrast to several of its neighbours, was not listed in 2000 by the Financial Action Task Force on Money Laundering (based in Paris, France) as 'non-co-operative' in the control of money-laundering. The scale of activity remained small, however, with 1,438 International Business Companies, 17 insurance companies and five 'offshore' banks in 2005.

The island is at some risk from hurricanes, as well as from less powerful tropical storms which can cause serious damage to the banana industry. There is a volcanic centre at Soufrière in the south of the island, which has no recent history of threatening activity and has some importance as a tourist attraction.

# Statistical Survey

Source (unless otherwise indicated): St Lucian Government Statistics Department, Block A, Government Bldgs, Waterfront, Castries; tel. 452-6653; fax 451-8254; e-mail statsdept@candw.lc; internet www.stats.gov.lc.

## AREA AND POPULATION

**Area:** 616.3 sq km (238 sq miles).

**Population:** 135,685 (males 65,988, females 69,697) at census of 12 May 1991; 158,147 (males 77,264, females 80,883) at census of 22 May 2001. *2004:* 162,300 (mid-year estimate). *By District* (2001 census, preliminary figures): Castries 60,390; Anse La Raye 5,954; Canaries 1,741; Soufrière 7,337; Choiseul 5,993; Laborie 7,329; Vieux-Fort 14,561; Micoud 15,892; Dennery 12,537; Gros Islet 19,409. Source: partly Caribbean Development Bank, *Social and Economic Indicators.*

**Density** (mid-2004): 263.3 per sq km.

**Principal Town** (estimated population incl. suburbs, mid-2003): Castries (capital) 13,687. Source: UN, *World Urbanization Prospects: The 2003 Revision.*

**Births, Marriages and Deaths** (2001): Registered live births 2,729 (birth rate 17.3 per 1,000); Registered marriages 449 (marriage rate 2.8 per 1,000); Registered deaths 980 (death rate 6.2 per 1,000). *2003:* Crude birth rate 14.7 per 1,000; Crude death rate 6.5 per 1,000 (Source: Caribbean Development Bank, *Social and Economic Indicators*).

**Expectation of Life** (years at birth): 72 (males 69; females 75) in 2003. Source: WHO, *World Health Report.*

**Economically Active Population** (persons aged 15 years and over, labour survey for July–December 2000): Agriculture, hunting and forestry 11,660; Fishing 900; Manufacturing 6,610; Electricity, gas and water 530; Construction 6,460; Wholesale and retail trade; repair of motor vehicles, motorcycles and personal and household goods 11,090; Hotels and restaurants 6,140; Transport, storage and communications 4,540; Financial intermediation 870; Real estate, renting and business activities 1,450; Public administration and defence; compulsory social security 8,180; Education 1,390; Health and social work 450; Other community, social and personal service activities 1,130; Private households with employed persons 1,390; Activities not adequately defined 1,580; *Total employed* 64,370 (males 35,620, females 28,750); Unemployed 13,630; *Total labour force* 78,000.

## HEALTH AND WELFARE
### Key Indicators

**Total Fertility Rate** (children per woman, 2003): 2.3.

**Under-5 Mortality Rate** (per 1,000 live births, 2003): 18.

**Physicians** (per 1,000 head, 1998): 5.18.

**Hospital Beds** (per 1,000 head, 1996): 3.38.

**Health Expenditure** (2002): US $ per head (PPP): 306.

**Health Expenditure** (2002): % of GDP: 5.0.

**Health Expenditure** (2002): public (% of total): 68.4.

**Access to Water** (% of persons, 2002): 98.

**Access to Sanitation** (% of persons, 2002): 89.

**Human Development Index** (2002): ranking: 71.

**Human Development Index** (2002): value: 0.777.

For sources and definitions, see explanatory note on p. vi.

## AGRICULTURE, ETC.

**Principal Crops** (FAO estimates, '000 metric tons, 2003): Cassava 1.0; Yams 4.5; Other roots and tubers 5.7; Coconuts 14.0; Vegetables 1.0; Bananas 120.0; Plantains 1.3; Grapefruit 3.0; Mangoes 28.0; Other fruits 5.7.

**Livestock** (FAO estimates, '000 head, year ending September 2003): Cattle 12.4; Sheep 12.5; Goats 9.8; Pigs 15.0; Horses 1.0; Asses 0.5; Mules 1.0; Poultry 240.

**Livestock Products** (FAO estimates, '000 metric tons, 2003): Beef and veal 0.5; Pig meat 0.7; Poultry meat 1.2; Cows' milk 1.0; Hen eggs 0.5.

**Fishing** (metric tons, live weight, 2003): Capture 1,462 (Snappers 57; Wahoo 169; Skipjack tuna 132; Blackfin tuna 169; Yellowfin tuna 139; Flyingfishes 75; Common dolphinfish 286; Stromboid conches 48); Aquaculture 2 (estimate); Total catch 1,464 (estimate). Figures exclude aquatic plants and mammals.
Source: FAO.

## INDUSTRY

**Production** (preliminary, 2003 unless otherwise stated): Electric energy 203.6 million kWh; Copra 1,162 metric tons; Coconut oil (unrefined) 734,000 litres (2002); Coconut oil (refined) 2,288,000 litres; Coconut meal 829,000 kg; Rum 9,000 hectolitres (revised figure, 1999). Sources: IMF, *St Lucia: Statistical Appendix* (December 2004), and UN, *International Commodity Statistics Yearbook.*

## FINANCE

**Currency and Exchange Rates:** 100 cents = 1 Eastern Caribbean dollar (EC $). *Sterling, US Dollar and Euro Equivalents* (31 May 2005): £1 sterling = EC $4.909; US $1 = EC $2.700; €1 = EC $3.329; EC $100 = £20.37 = US $37.04 = €30.04. *Exchange Rate:* Fixed at US $1 = EC $2.70 since July 1976.

**Budget** (EC $ million, 2003): *Revenue:* Tax revenue 441.8 (Taxes on income and profits 105.5; Taxes on domestic goods and services 174.9; Taxes on international trade and transactions 120.8); Other current revenue 36.2; Total 478.1 (excl. grants 36.3). *Expenditure:* Current expenditure 436.0 (Personal emoluments 211.6; Goods and services 75.6; Interest payments 45.4; Transfers 68.2); Capital expenditure and net lending 179.6; Total 615.6. Source: IMF, *St Lucia: Statistical Appendix* (December 2004).

**International Reserves** (US $ million at 31 December 2004): IMF special drawing rights 2.34; Foreign exchange 130.19; Total 132.53. Source: IMF, *International Financial Statistics.*

# SAINT LUCIA

**Money Supply** (EC $ million at 31 December 2004): Currency outside banks 99.16; Demand deposits at deposit money banks 407.67; Total money (incl. others) 506.84. Source: IMF, *International Financial Statistics*.

**Cost of Living** (Consumer Price Index; base: 2000 = 100): 101.7 in 2002; 102.7 in 2003; 107.4 in 2004. Source: IMF, *International Financial Statistics*.

**Gross Domestic Product** (EC $ million at current factor cost): 1,529.3 in 2002; 1,598.4 in 2003; 1,689.0 in 2004 (preliminary). Source: Eastern Caribbean Central Bank.

**Expenditure on the Gross Domestic Product** (EC $ million at current prices, 2003): Government final consumption expenditure 457.97; Private final consumption expenditure 1,246.50; Gross capital formation 408.40; *Total domestic expenditure* 2,112.87; Exports of goods and services 1,096.12; *Less* Imports of goods and services 1,296.43; *GDP at market prices* 1,912.56. Source: Eastern Caribbean Central Bank.

**Gross Domestic Product by Economic Activity** (EC $ million at current factor cost, 2003): Agriculture, hunting, forestry and fishing 84.14; Mining and quarrying 6.03; Manufacturing 78.08; Electricity and water 85.02; Construction 116.76; Wholesale and retail trade 195.21; Restaurants and hotels 233.49; Transport 200.56; Communications 141.67; Banking and insurance 169.63; Real estate and housing 100.07; Government services 241.48; Other services 82.92; *Sub-total* 1,735.06; *Less* Imputed bank service charge 136.64; *Total in basic prices* 1,598.42. Source: Eastern Caribbean Central Bank.

**Balance of Payments** (EC $ million, 2003): Exports of goods f.o.b. 189.05; Imports of goods f.o.b. -932.90; *Trade balance* -743.85; Exports of services 907.07; Imports of services -363.53; *Balance on goods and services* -200.31; Other income received (net) -127.46; *Balance on goods, services and income* -327.77; Current transfers received 78.64; Current transfers paid -43.01; *Current balance* -292.14; Capital account (net) 51.12; Direct investment from abroad (net) 268.89; Portfolio investment (net) 168.99; Other investment (net) -193.38; Net errors and omissions 32.52; *Overall balance* 36.00. Source: Eastern Caribbean Central Bank.

## EXTERNAL TRADE

**Principal Commodities** (US $ million): *Imports c.i.f.* (2003): Food 70.9; Beverages and tobacco 13.1; Fuels 41.6; Chemicals 28.3; Basic manufactures 58.6; Machinery and transport equipment 121.8; Miscellaneous manufactured articles 57.0; Total (incl. others) 400.6. *Exports f.o.b.* (2002): Food and live animals 23.7 (Bananas and plantains 22.2); Beverages and tobacco 8.7 (Beer 7.0); Mineral fuels, lubricants, etc. 4.6 (Refined petroleum products 4.6); Basic manufactures 5.8 (Paper products 3.8); Machinery and transport equipment 12.2 (Telecommunications equipment 2.3; Electric machinery, etc. 3.4); Miscellaneous manufactured articles 4.9 (Clothing 2.5); Total (incl. others) 62.0. Sources: UN, *International Trade Statistics Yearbook* and IMF, *St Lucia: Statistical Appendix* (December 2004).

**Principal Trading Partners** (US $ million, 2002): *Imports c.i.f.*: Barbados 8.0; Canada 9.0; China, People's Republic 7.6; France 7.8; Germany 3.8; Jamaica 3.3; Japan 10.6; Netherlands 3.6; Sweden 5.3; Trinidad and Tobago 47.7; United Kingdom 27.5; USA 134.7; Total (incl. others) 314.8. *Exports f.o.b.*: Antigua and Barbuda 1.1; Barbados 6.0; Dominica 3.3; France 1.3; Grenada 1.9; Saint Christopher and Nevis 0.8; Saint Vincent and the Grenadines 1.7; Trinidad and Tobago 7.3; United Kingdom 23.3; USA 12.6; Total (incl. others) 62.0. Source: UN, *International Trade Statistics Yearbook*.

## TRANSPORT

**Road Traffic** (registered motor vehicles, 2001): Goods vehicles 8,972; Taxis and hired vehicles 1,894; Motorcycles 757; Private vehicles 22,453; Passenger vans 3,387; Total (incl. others) 39,416.

**Shipping:** *Arrivals* (1999): 2,328 vessels. *International Sea-borne Freight Traffic* ('000 metric tons, 1999): Goods loaded 117; Goods unloaded 666.

**Civil Aviation** (aircraft movements): 41,702 in 1996; 41,160 in 1997; 42,040 in 1998.

## TOURISM

**Tourist Arrivals** ('000): 622.7 (235.5 visitor arrivals, 387.2 cruise-ship passengers) in 2002; 670.1 (276.9 visitor arrivals, 393.2 cruise-ship passengers) in 2003; 779.7 (298.4 visitor arrivals, 481.3 cruise-ship passengers) in 2004.

**Tourism Receipts** (estimates, US $ million): 210.0 in 2002; 282.1 in 2003; 314.3 in 2004.
Source: Caribbean Development Bank, *Social and Economic Indicators*.

## COMMUNICATIONS MEDIA

**Radio Receivers** ('000 in use, 1997): 111.

**Television Receivers** ('000 in use, 1999): 56.

**Telephones** ('000 main lines in use, 2002): 51.1.

**Mobile Cellular Telephones** (subscribers, 2002): 14,300.

**Personal Computers** ('000 in use, 2002): 24.

**Internet Users** ('000, 2001): 13.

**Non-daily Newspapers** (1996): Titles 5; Circulation 34,000.
Sources: UN, *Statistical Yearbook*; UNESCO, *Statistical Yearbook*; International Telecommunication Union.

## EDUCATION

**Pre-primary** (state institutions only, 2000/01): 106 schools; 359 teachers; 4,275 pupils.

**Primary** (state institutions only, 2000/01): 82 schools; 1,052 teachers; 27,175 pupils (2002).

**General Secondary** (state institutions only, 2000/01): 18 schools; 678 teachers; 12,655 pupils (2002).

**Special Education** (state institutions only, 2000/01): 5 schools; 39 teachers; 227 students.

**Adult Education** (state institutions only, 2000/01): 19 centres; 80 facilitators; 729 learners.

**Tertiary** (state institutions, including part-time, 2000/01): 127 teachers; 1,403 students.
Source: partly Caribbean Development Bank, *Social and Economic Indicators*.

**Adult Literacy Rate:** 94.8% in 2002. Source: UN Development Programme, *Human Development Report*.

# Directory

## The Constitution

The Constitution came into force at the independence of Saint Lucia on 22 February 1979. Its main provisions are summarized below:

### FUNDAMENTAL RIGHTS AND FREEDOMS

Regardless of race, place of origin, political opinion, colour, creed or sex but subject to respect for the rights and freedoms of others and for the public interest, every person in Saint Lucia is entitled to the rights of life, liberty, security of the person, equality before the law and the protection of the law. Freedom of conscience, of expression, of assembly and association is guaranteed and the inviolability of personal privacy, family life and property is maintained. Protection is afforded from slavery, forced labour, torture and inhuman treatment.

### THE GOVERNOR-GENERAL

The British monarch, as Head of State, is represented in Saint Lucia by the Governor-General.

### PARLIAMENT

Parliament consists of the British monarch, represented by the Governor-General, the 11-member Senate and the House of Assembly, composed of 17 elected Representatives. Senators are appointed by the Governor-General: six on the advice of the Prime Minister, three on the advice of the Leader of the Opposition and two acting on his own deliberate judgement. The life of Parliament is five years.

Each constituency returns one Representative to the House who is directly elected in accordance with the Constitution.

At a time when the office of Attorney-General is a public office, the Attorney-General is an *ex-officio* member of the House.

Every citizen over the age of 21 is eligible to vote.

# SAINT LUCIA

Parliament may alter any of the provisions of the Constitution.

## THE EXECUTIVE

Executive authority is vested in the British monarch and exercisable by the Governor-General. The Governor-General appoints as Prime Minister that member of the House who, in the Governor-General's view, is best able to command the support of the majority of the members of the House, and other Ministers on the advice of the Prime Minister. The Governor-General may remove the Prime Minister from office if the House approves a resolution expressing 'no confidence' in the Government, and if the Prime Minister does not resign within three days or advise the Governor-General to dissolve Parliament.

The Cabinet consists of the Prime Minister and other Ministers, and the Attorney-General as an *ex-officio* member at a time when the office of Attorney-General is a public office.

The Leader of the Opposition is appointed by the Governor-General as that member of the House who, in the Governor-General's view, is best able to command the support of a majority of members of the house who do not support the Government.

## CITIZENSHIP

All persons born in Saint Lucia before independence who immediately prior to independence were citizens of the United Kingdom and Colonies automatically become citizens of Saint Lucia. All persons born in Saint Lucia after independence automatically acquire Saint Lucian citizenship, as do those born outside Saint Lucia after independence to a parent possessing Saint Lucian citizenship. Provision is made for the acquisition of citizenship by those to whom it is not automatically granted.

# The Government

## HEAD OF STATE

**Monarch:** HM Queen ELIZABETH II.

**Governor-General:** Dame PEARLETTE LOUISY (took office 17 September 1997).

### CABINET
(July 2005)

**Prime Minister, Minister of Finance, International Financial Services and Economic Affairs and of Information:** Dr KENNY DAVIS ANTHONY.

**Deputy Prime Minister and Minister of Education, Human Resource Development, Youth and Sports:** MARIO F. MICHEL.

**Minister of Commerce, Tourism, Investment and Consumer Affairs:** PHILLIP J. PIERRE.

**Minister of Agriculture, Forestry and Fisheries:** IGNATIUS JEAN.

**Minister of Health, Human Services, Family Affairs and Gender Relations:** DAMIAN E. GREAVES.

**Minister of Physical Development, Environment and Housing:** THEOPHILUS FERGUSON JOHN.

**Minister of Communications, Works, Transport and Public Utilities:** FELIX FINNISTERRE.

**Minister of Labour Relations, Public Service and Co-operatives:** VELON L. JOHN.

**Minister of Home Affairs and Internal Security:** Sen. CALIXTE GEORGE.

**Minister of Foreign Affairs, International Trade and Civil Aviation:** Sen. PETRUS COMPTON.

**Minister of Social Transformation, Culture and Local Government:** MENISSA RAMBALLY.

**Attorney-General and Minister of Justice:** Sen. VICTOR PHILIP LA CORBINIERE.

### MINISTRIES

**Office of the Prime Minister:** Greaham Louisy Administrative Bldg, 5th Floor, Waterfront, Castries; tel. 468-2111; fax 453-7352; e-mail pmoffice@candw.lc; internet www.stlucia.gov.lc.

**Ministry of Agriculture, Forestry and Fisheries:** NIS Bldg, 5th Floor, Waterfront, Castries; tel. 468-4210; fax 453-6314; e-mail admin@candw.lc; internet www.slumaffe.org.

**Ministry of Commerce, Tourism, Investment and Consumer Affairs:** Ives Heraldine Rock Bldg, 4th Floor, Block B, Waterfront, Castries; tel. 468-4202; fax 451-6986; e-mail mitandt@candw.lc; internet www.commerce.gov.lc.

**Ministry of Communications, Works, Transport and Public Utilities:** Williams Bldg, Bridge St, Castries; tel. 468-4300; fax 453-2769; e-mail min_com@candw.lc.

**Ministry of Education, Human Resource Development, Youth and Sports:** Francis Compton Bldg, Waterfront, Castries; tel. 468-5203; fax 453-2299; e-mail mineduc@candw.lc; internet www.education.gov.lc.

**Ministry of Finance, International Financial Services and Economic Affairs:** Financial Centre, Bridge St, Castries; tel. 468-5503; fax 452-6700; e-mail minfin@gosl.gov.lc.

**Ministry of Foreign Affairs, International Trade and Civil Aviation:** Conway Business Centre, Waterfront, Castries; tel. 468-4501; fax 452-7427; e-mail foreign@candw.lc.

**Ministry of Health, Human Services, Family Affairs and Gender Relations:** Chaussee Rd, Castries; tel. 452-2859; fax 452-5655; e-mail health@candw.lc.

**Ministry of Home Affairs and Internal Security:** Erdiston's Pl., Manoel St, Castries; tel. 452-3772; fax 453-6315.

**Ministry of Labour Relations, Public Service and Co-operatives:** Greaham Louisy Administrative Bldg, 2nd Floor, Waterfront, Castries; tel. 468-2202; fax 453-1305.

**Ministry of Physical Development, Environment and Housing:** Greaham Louisy Administrative Bldg, 3rd Floor, Waterfront, Castries; tel. 468-4402; fax 452-2506; e-mail sde@planning.gov.lc.

**Ministry of Social Transformation, Culture and Local Government:** Greaham Louisy Administrative Bldg, 4th Floor, Waterfront, Castries; tel. 468-5101; fax 453-7921.

**Attorney-General's Office and Ministry of Justice:** Francis Compton Bldg, Waterfront, Castries; tel. 468-3200; fax 458-1131; e-mail atgen@candw.lc.

# Legislature

## PARLIAMENT

### Senate

The Senate has nine nominated members and two independent members.

**President:** HILFORD DETERVILLE.

### House of Assembly

**Speaker:** JOSEPH BADEN ALLAIN.

**Clerk:** DORIS BAILEY.

**Election, 3 December 2001**

| Party | % of votes | Seats |
| --- | --- | --- |
| Saint Lucia Labour Party | 54.2 | 14 |
| United Workers' Party | 36.6 | 3 |
| National Alliance | 3.5 | — |
| Others | 5.7 | — |
| **Total** | **100.0** | **17** |

# Political Organizations

**Committee for Meaningful Change and Reconstruction:** Castries; f. 2004.

**National Alliance (NA):** Castries; f. 2001; opposition electoral alliance.

**National Development Movement (NDM):** Castries; f. 2004; Leader AUSBERT D'AUVERGNE.

**Organization for National Empowerment (ONE):** Castries; f. 2004; Leader SARAH FLOOD-BEAUBRUN; Chair. THOMAS ROSERIE (acting).

**Saint Lucia Freedom Party:** Castries; f. 1999; campaigns against the political establishment; Spokesman MARTINUS FRANÇOIS.

**Saint Lucia Labour Party (SLP):** Tom Walcott Bldg, 2nd Floor, Jeremie St, POB 427, Castries; tel. 451-8446; fax 451-9389; e-mail slp@candw.lc; internet www.geocities.com/~slp; f. 1946; socialist party; Leader Dr KENNY DAVIS ANTHONY; Chair. TOM WALCOTT.

**United Workers' Party (UWP):** 9 Coral St, POB 1550, Castries; tel. 451-9103; fax 451-9207; e-mail unitworkers@netscape.net; internet www.uwpstlucia.org; f. 1964; right-wing party; Chair. STEPHENSON KING; Leader Sir JOHN COMPTON.

# SAINT LUCIA

## Diplomatic Representation

### EMBASSIES AND HIGH COMMISSION IN SAINT LUCIA

**China, People's Republic:** Cap Estate, Gros Islet, POB GM 999, Castries; tel. 452-0903; fax 452-9495; e-mail chinaemb_lc@mfa.gov.cn; Ambassador GU HUAMING.

**Cuba:** Rodney Heights, Gros Islet, POB 2150, Castries; tel. 458-4665; fax 458-4666; e-mail embacubasantalucia@candw.lc; Ambassador VÍCTOR DANIEL RAMÍREZ PEÑA.

**France:** French Embassy to the OECS, Vigie, Castries; tel. 455-6060; fax 455-6056; e-mail frenchembassy@candw.lc; Ambassador BERNARD VENZO.

**United Kingdom:** Francis Compton Bldg, Waterfront, POB 227, Castries; tel. 452-2484; fax 453-1543; e-mail britishhc@candw.lc; High Commissioner JOHN WHITE (resident in Barbados).

**Venezuela:** Vigie, POB 494, Castries; tel. 452-4033; fax 453-6747; e-mail vembassy@candw.lc; Ambassador MAIGUALIDA APONTE.

## Judicial System

### SUPREME COURT

**Eastern Caribbean Supreme Court:** Waterfront, Government Bldgs, POB 1093, Castries; tel. 452-2574; fax 452-5475; e-mail appeal@candw.lc; the West Indies Associated States Supreme Court was established in 1967 and was known as the Supreme Court of Grenada and the West Indies Associated States from 1974 until 1979, when it became the Eastern Caribbean Supreme Court. Its jurisdiction extends to Anguilla, Antigua and Barbuda, the British Virgin Islands, Dominica, Grenada (which rejoined in 1991), Montserrat, Saint Christopher and Nevis, Saint Lucia and Saint Vincent and the Grenadines. It is composed of the High Court of Justice and the Court of Appeal. The High Court is composed of the Chief Justice and 13 High Court Judges. The Court of Appeal is presided over by the Chief Justice and includes three other Justices of Appeal. Jurisdiction of the High Court includes fundamental rights and freedoms, membership of the parliaments, and matters concerning the interpretation of constitutions. Following the inauguration of the Caribbean Court of Justice in April 2005, appeals from the Court of Appeal would no longer be made to the Judicial Committee of the Privy Council, based in the United Kingdom, but to the new regional court.

**Chief Justice:** C. M. DENNIS BYRON.

## Religion

### CHRISTIANITY

#### The Roman Catholic Church

Saint Lucia forms a single archdiocese. The Archbishop participates in the Antilles Episcopal Conference (currently based in Port of Spain, Trinidad and Tobago). At 31 December 2003 there were an estimated 100,243 adherents, equivalent to some 68% of the population.

**Archbishop of Castries:** KELVIN EDWARD FELIX, Archbishop's Office, POB 267, Castries; tel. 452-2416; fax 452-3697; e-mail archbishop@candw.lc.

#### The Anglican Communion

Anglicans in Saint Lucia are adherents of the Church in the Province of the West Indies, comprising eight dioceses. The Archbishop of the West Indies is the Bishop of Nassau and the Bahamas. Saint Lucia forms part of the diocese of the Windward Islands (the Bishop is resident in Kingstown, Saint Vincent).

#### Other Christian Churches

**Seventh-Day Adventist Church:** St Louis St, POB 117, Castries; tel. 452-4408; e-mail adventist@candw.lc; Pastor THEODORE JARIA.

**Trinity Evangelical Lutheran Church:** Gablewoods Mall, POB 858, Castries; tel. 458-4638; fax 450-3382; e-mail spiegelbergs@candw.lc; Reverend TOM SPIEGELBERG.

Baptist, Christian Science, Methodist, Pentecostal and other churches are also represented in Saint Lucia.

## The Press

**The Catholic Chronicle:** POB 778, Castries; f. 1957; monthly; Editor Rev. PATRICK A. B. ANTHONY; circ. 3,000.

**The Crusader:** 19 St Louis St, Castries; tel. 452-2203; fax 452-1986; f. 1934; weekly; circ. 4,000.

**The Mirror:** Bisee Industrial Estate, Castries; tel. 451-6181; fax 451-6197; e-mail mirror@candw.lc; internet www.stluciamirroronline.com; f. 1994; weekly on Fridays; Man. Editor GUY ELLIS.

**One Caribbean:** POB 852, Castries; e-mail dabread@candw.lc; internet www.onecaribbean.com; Editor D. SINCLAIR DABREO.

**The Star:** Rodney Bay Industrial Estate, Gros Islet, POB 1146, Castries; tel. 450-7827; fax 450-8694; e-mail starpub@candw.lc; internet www.stluciastar.com; weekly; Propr RICK WAYNE.

**The Vanguard:** Hospital Rd, Castries; fortnightly; Editor ANDREW SEALY; circ. 2,000.

**Visions of St Lucia Tourist Guide:** 7 Maurice Mason Ave, Sans Souci, POB 947, Castries; tel. 453-0427; fax 452-1522; e-mail visions@candw.lc; internet www.stlucia.com/visions; f. 1989; official tourist guide; published by Island Visions Ltd; annual; Chair. and Man. Dir ANTHONY NEIL AUSTIN; circ. 100,000.

**The Voice of St Lucia:** Odessa Bldg, Darling Rd, POB 104, Castries; tel. 452-2590; fax 453-1453; f. 1885; 2 a week; circ. 8,000.

**The Weekend Voice:** Odessa Bldg, Darling Rd, POB 104, Castries; tel. 452-2590; fax 453-1453; weekly on Saturdays; circ. 8,000.

### PRESS ORGANIZATION

**Eastern Caribbean Press Council (ECPC):** Castries; f. 2003; independent, self-regulating body designed to foster and maintain standards in regional journalism, formed by 14 newspapers in the Eastern Caribbean and Barbados; Chair. Lady MARIE SIMMONS.

### NEWS AGENCY

**Caribbean Media Corporation:** Bisee Rd, Castries; tel. 453-7162; e-mail admin@cananews.com; internet www.cananews.com; f. by merger of Caribbean News Agency and Caribbean Broadcasting Union.

### FOREIGN NEWS AGENCY

**Inter Press Service (IPS)** (Italy): Hospital Rd, Castries; tel. 452-2770.

## Publishers

**Caribbean Publishing Co Ltd:** American Drywall Bldg, Vide Boutielle Highway, POB 104, Castries; tel. 452-3188; fax 452-3181; e-mail publish@candw.lc; f. 1978; publishes telephone directories and magazines.

**Crusader Publishing Co Ltd:** 19 St Louis St, Castries; tel. 452-2203; fax 452-1986.

**Island Visions:** Sans Soucis, POB 947, Castries; tel. 453-0427; fax 452-1522; e-mail visions@candw.lc; internet www.sluonestop.com/visions.

**Mirror Publishing Co Ltd:** Bisee Industrial Estate, POB 1782, Castries; tel. 451-6181; fax 451-6503; e-mail mirror@candw.lc; internet www.stluciamirroronline.com; Man. Editor GUY ELLIS.

**Star Publishing Co:** Massade Industrial Park, Gros Islet, POB 1146, Castries; tel. 450-7827; fax 450-8694; e-mail starpub@candw.lc.

**Voice Publishing Co Ltd:** Odessa Bldg, Darling Rd, POB 104, Castries; tel. 452-2590; fax 453-1453.

## Broadcasting and Communications

### TELECOMMUNICATIONS

#### Regulatory Authorities

**Eastern Caribbean Telecommunications Authority (ECTEL):** Castries; internet ectel.int; f. 2000 to regulate telecommunications in Saint Lucia, Dominica, Grenada, Saint Christopher and Nevis and Saint Vincent and the Grenadines; Dir (Saint Lucia) EMBERT CHARLES.

**National Telecommunications Regulatory Commission (NTRC):** Global Tile Bldg, Bois d'Orange, Gros Islet, POB 690,

Castries; tel. 458-2035; fax 453-2558; e-mail ntrc_slu@candw.lc; internet www.ectel.int/lca; f. 2000; regulates the sector in conjunction with ECTEL; Chair. ELDON MATHURIN.

### Major Service Providers

**Cable & Wireless (St Lucia) Ltd:** Bridge St, POB 111, Castries; tel. 453-9000; fax 453-9700; e-mail talk2us@candw.lc; internet www.candw.lc; also operates a mobile cellular telephone service; CEO FRED WALCOTT; in February 2001 Cable & Wireless announced it was to cease operations in Saint Lucia when its current licence expired and relocate its regional call centres to Barbados and Anguilla.

**Cingular Wireless St Lucia:** Castries; internet www.cingularwireless.com; joint venture between SBC Communications and BellSouth; scheduled to sell its operations and licences in the Caribbean to Digicel in 2005; Pres. and CEO STANLEY T. SIGMAN.

**Digicel:** Rodney Bay, Gros Islet, POB 791, Castries; tel. 456-3400; fax 450-3872; e-mail slucustomercare@digicelgroup.com; internet www.digicelstlucia.com; f. 2003; mobile cellular telephone operator, owned by an Irish consortium; Chair. DENIS O'BRIEN.

**Saint Lucia Boatphone Ltd:** POB 2136, Gros Islet; tel. 452-0361; fax 452-0394; e-mail boatphone@candw.lc.

### BROADCASTING

#### Radio

**Gem Radio Network:** POB 1146, Castries; tel. 459-0609.

**Radio Caribbean International:** 11 Mongiraud St, POB 121, Castries; tel. 452-2636; fax 452-2637; e-mail rci@candw.lc; internet www.candw.lc/homepage/rci.htm; operates Radio Caraïbes; English and Creole services; broadcasts 24 hrs; Pres. H. COQUERELLE; Station Man. WINSTON FOSTER.

**Radyo Koulibwi:** POB 20, Morne Du Don, Castries; tel. 451-7814; fax 453-1983; f. 1990.

**Saint Lucia Broadcasting Corporation:** Morne Fortune, POB 660, Castries; tel. 452-2337; fax 453-1568; govt-owned; Man. KEITH WEEKES.

**Radio 100-Helen FM:** Morne Fortune, POB 621, Castries; tel. 451-7260; fax 453-1737.

**Radio Saint Lucia (RSL):** Morne Fortune, POB 660, Castries; tel. 452-7415; fax 453-1568; English and Creole services; Chair. VAUGHN LOUIS FERNAND; Man. KEITH WEEKES.

#### Television

**Cablevision:** George Gordon Bldg, Bridge St, POB 111, Castries; tel. 453-9311; fax 453-9740.

**Catholic TV Broadcasting Service:** Micoud St, Castries; tel. 452-7050.

**Daher Broadcasting Service Ltd:** Vigie, POB 1623, Castries; tel. 453-2705; fax 452-3544; Man. Dir LINDA DAHER.

**Helen Television System (HTS):** National Television Service of St Lucia, POB 621, The Morne, Castries; tel. 452-2693; fax 454-1737; e-mail hts@candw.lc; internet www.htsstlucia.com; f. 1967; commercial station; Man. Dir LINFORD FEVRIERE; Prog. Dir VALERIE ALBERT.

**National Television Network (NTN):** Castries; f. 2001; operated by the Government Information Service; provides information on the operations of the public sector.

## Finance

(cap. = capital; dep. = deposits; m. = million; brs = branches)

### BANKING

The Eastern Caribbean Central Bank, based in Saint Christopher, is the central issuing and monetary authority for Saint Lucia.

**Eastern Caribbean Central Bank—Saint Lucia Office:** Financial Centre, 3rd Floor, Bridge St, POB 295, Castries; tel. 452-7449; fax 453-6022; e-mail eccbslu@candw.lc; Country Dir TREVOR BRAITHWAITE.

### Local Banks

**Bank of Saint Lucia Ltd:** Financial Centre, 5th Floor, 1 Bridge St, POB 1862, Castries; tel. 456-6000; fax 456-6720; e-mail bankofstlucia@candw.lc; f. 1981; fmrly National Commercial Bank of St Lucia Ltd; dep. EC $480.6m., total assets EC $556.4m. (Dec. 2000); 31.3% state-owned; parent co is East Caribbean Holding Co Ltd; Chair. VICTOR A. EUDOXIE; Man. Dir MARIUS ST ROSE; 7 brs.

**First National Bank Saint Lucia Ltd:** 21 Bridge St, POB 168, Castries; tel. 452-2880; fax 453-1630; e-mail coopbank@candw.lc; internet www.coopbank.com; inc 1937 as Saint Lucia Co-operative Bank Ltd; name changed as above 1 January 2005; commercial bank; share cap. EC $4.2m., total assets EC $154.4m. (Dec. 1999); Pres. FERREL CHARLES; Man. Dir C. CARLTON GLASGOW; 3 brs.

**RBTT Bank Caribbean (SLU) Ltd:** 22 Micoud St, POB 1531, Castries; tel. 452-2265; fax 452-1668; e-mail rbttslu.isd@candw.lc; internet www.rbtt.com/caribbank.htm; f. 1985 as Caribbean Banking Corpn Ltd, in March 2002 name changed as above; owned by R and M Holdings Ltd; Chair. PETER JULY; Country Man. EARL P. CRICHTON; 4 brs.

### Development Bank

**Saint Lucia Development Bank:** Financial Centre, 1 Bridge St, POB 368, Castries; tel. 457-7532; fax 457-7299; f. 1981; merged with former Agricultural and Industrial Development Bank; provides credit for agriculture, industry, tourism, housing and workforce training; Man. Dir DANIEL GIRARD.

### Foreign Banks

**Bank of Nova Scotia Ltd** (Canada): 6 William Peter Blvd, POB 301, Castries; tel. 456-2100; fax 456-2130; e-mail bns@candw.lc; Man. S. COZIER; 3 brs.

**FirstCaribbean International Bank (Barbados) Ltd:** Castries; internet www.firstcaribbeanbank.com; f. 2002 following merger of Caribbean operations of Barclays Bank PLC and CIBC; Exec. Chair. MICHAEL MANSOOR; CEO CHARLES PINK.

**Royal Bank of Canada:** Laborie St and William Peter Blvd, POB 280, Castries; tel. 452-2245; fax 452-7855; Man. JOHN MILLER.

### 'Offshore' Bank

**Bank Crozier International Ltd:** Crozier Bldg, 21 Brazil St, Castries; e-mail stlucia@bankcrozier.com; internet www.bankcrozier.com; tel. 455-7600; fax 455-6009; f. 2001; CEO and Chair. PETER JOHANSSON; Gen. Man. PATRICK HOLMES.

### STOCK EXCHANGE

**Eastern Caribbean Securities Exchange:** based in Basseterre, Saint Christopher and Nevis; e-mail info@ecseonline.com; internet www.ecseonline.com; f. 2001; regional securities market designed to facilitate the buying and selling of financial products for the eight member territories—Anguilla, Antigua and Barbuda, Dominica, Grenada, Montserrat, Saint Christopher and Nevis, Saint Lucia and Saint Vincent and the Grenadines; Gen. Man. TREVOR BLAKE.

### INSURANCE

Local companies include the following:

**Caribbean General Insurance Ltd:** Laborie St, POB 290, Castries; tel. 452-2410; fax 452-3649.

**Eastern Caribbean Insurance Ltd:** Laborie St, POB 290, Castries; tel. 452-2410; fax 452-3649; e-mail gci.ltd@candw.lc.

**Saint Lucia Insurances Ltd:** 48 Micoud St, POB 1084, Castries; tel. 452-3240; fax 452-2240.

**Saint Lucia Motor and General Insurance Co Ltd:** 38 Micoud St, POB 767, Castries; tel. 452-3323; fax 452-6072.

## Trade and Industry

### DEVELOPMENT ORGANIZATION

**National Development Corporation (NDC):** Block B, 1st Floor, Govt Bldgs, The Waterfront, POB 495, Castries; tel. 452-3614; fax 452-1841; e-mail devcorp@candw.lc; internet www.stluciandc.com; f. 1971 to promote the economic development of Saint Lucia; owns and manages four industrial estates; br. in New York, USA; Chair. MATTHEW BEAUBRUN; Gen. Man. JACQUELINE EMMANUEL-ALBERTINIE.

### CHAMBER OF COMMERCE

**Saint Lucia Chamber of Commerce, Industry and Agriculture:** Vide Bouteille, POB 482, Castries; tel. 452-3165; fax 453-6907; e-mail chamber@candw.lc; internet www.sluchamber.com.lc; f. 1884; 150 mems; Pres. GUY MAYERS.

### INDUSTRIAL AND TRADE ASSOCIATIONS

**Saint Lucia Banana Corporation (SLBC):** Castries; f. 1998 following privatization of Saint Lucia Banana Growers' Asscn (f. 1967); Chair. (vacant); Sec. FREEMONT LAWRENCE.

# SAINT LUCIA

**Saint Lucia Industrial and Small Business Association:** 2nd Floor, Ivy Crick Memorial Bldg, POB 312, Castries; tel. 453-1392; Pres. Leo Clarke; Exec. Dir Dr Urban Seraphine.

## EMPLOYERS' ASSOCIATIONS

**Saint Lucia Agriculturists' Association Ltd:** Mongiraud St, POB 153, Castries; tel. 452-2494; fax 453-2693; e-mail kseverin@slaa.net; internet www.slaa.net; distributor and supplier of agricultural, industrial and organic products; exporter of cocoa; Chair. Cuthbert Phillips; CEO Kerde M. Severin.

**Saint Lucia Coconut Growers' Association Ltd:** Manoel St, POB 259, Castries; tel. 452-2360; fax 453-1499; Chair. Johannes Leonce; Man. N. E. Edmunds.

**Saint Lucia Employers' Federation:** Morgan's Bldg, Maurice Mason Ave, POB 160, Castries; tel. 452-2190; fax 452-7335.

**Saint Lucia Fish Marketing Corpn:** POB 91, Castries; tel. 452-1341; fax 451-7073; e-mail slfmc@candw.lc.

**Saint Lucia Marketing Board (SLMB):** Conway, POB 441, Castries; tel. 452-3214; fax 453-1424; e-mail slmb@candw.lc; Chair. David Demaque; Man. Michael Williams.

**Windward Islands Banana Development and Exporting Co (Wibdeco):** POB 115, Castries; tel. 452-2651; f. 1994 in succession to the Windward Islands Banana Growers' Association (WINBAN); regional organization dealing with banana development and marketing; jointly owned by the Windward governments and island banana associations; Chair. Eustace Monrose.

## UTILITIES

### Electricity

**Caribbean Electric Utility Services Corpn (CARILEC):** 16–19 Orange Park Centre, POB 2056, Bois D'Orange, Gros Islet; tel. 452-0140; fax 452-0142; e-mail admin@carilec.org; internet www.carilec.org.

**St Lucia Electricity Services Ltd (LUCELEC):** Lucelec Bldg, John Compton Highway, POB 230, Castries; tel. 457-4400; fax 457-4409; e-mail lucelec@candw.lc; internet www.lucelec.com; f. 1964; Chair. Marius St Rose; Man. Dir Trevor Louisy.

### Water

**Water and Sewerage Authority (WASA):** L'Anse Rd, POB 1481, Castries; tel. 452-5344; fax 452-6844; e-mail wasco@candw.lc; f. 1999; a bill to privatize WASA was passed by Parliament in February 2005; Chair. Gordon Charles; Man. Dir John C. Joseph.

## TRADE UNIONS

**National Farmers' Union (NFU):** St Louis St, Castries; Pres. Peter Josie; 3,500 mems.

**National Workers' Union:** POB 713, Castries; tel. 452-3664; represents daily-paid workers; affiliated to World Federation of Trade Unions; Pres. Tyrone Maynard; Gen. Sec. George Goddard; 3,000 mems (1996).

**Saint Lucia Civil Service Association:** POB 244, Castries; tel. 452-3903; fax 453-6061; e-mail csa@candw.lc; f. 1951; Pres. Francis Raphael; Gen. Sec. James Perineau; 2,381 mems.

**Saint Lucia Media Workers' Association:** Castries; provides training for media workers in partnership with the Government.

**Saint Lucia Nurses' Association:** POB 819, Castries; tel. 452-1403; fax 453-0960; f. 1947; Pres. Lilia Harracksingh; Gen. Sec. Esther Felix.

**Saint Lucia Seamen, Waterfront and General Workers' Trade Union:** 24 Chaussee Rd, POB 166, Castries; tel. 452-1669; fax 452-5452; e-mail seamen@candw.lc; f. 1945; affiliated to International Confederation of Free Trade Unions (ICFTU), International Transport Federation (ITF) and Caribbean Congress of Labour; Pres. Michael Hippolyte; Sec. Crescentia Phillips; 1,000 mems.

**Saint Lucia Teachers' Union:** POB 821, Castries; tel. 452-4469; fax 453-6668; e-mail sltu@candw.lc; internet www.findsltu.org; f. 1934; Pres. Urban Dolor; Gen.-Sec. Kentry D. J. Pierre.

**Saint Lucia Workers' Union:** Reclamation Grounds, Conway, Castries; tel. 452-2620; f. 1939; affiliated to ICFTU; Pres. George Louis; Sec. Titus Francis; 1,000 mems.

**Vieux Fort General and Dock Workers' Union:** New Dock Rd, POB 224, Vieux Fort; tel. 454-6193; fax 454-5128; f. 1954; Pres. Joseph Griffith; 846 mems (1996).

# Transport

## RAILWAYS

There are no railways in Saint Lucia.

## ROADS

In 2000 there was an estimated total road network of 910 km, of which 150 km were main road and 127 km were secondary roads. In that year only 5.2% of roads were paved. The main highway passes through every town and village on the island. A construction project of a coastal highway, to link Castries with Cul de Sac Bay, was completed in February 2000. Internal transport is handled by private concerns and controlled by Government. A major five-year road maintenance programme was commenced in 2000 to rehabilitate the estimated 60% of roads which were on the point of collapse. The programme was to be partially funded by the Caribbean Development Bank. In May 2002 it was announced that Agence Française de Développement was to provide US $18.5m. to repair 20 km of tertiary and agricultural roads.

## SHIPPING

The ports at Castries and Vieux Fort have been fully mechanized. Castries has six berths with a total length of 2,470 ft (753 m). The two dolphin berths at the Pointe Seraphine cruise-ship terminal have been upgraded to a solid berth of 1,000 ft (305 m) and one of 850 ft (259 m). A project to upgrade the port at Vieux Fort to a full deep-water container port was completed in 1993. The port of Soufrière has a deep-water anchorage, but no alongside berth for ocean-going vessels. There is a petroleum transhipment terminal at Cul de Sac Bay. In 2004 481,279 cruise-ship passengers called at Saint Lucia. Regular services are provided by a number of shipping lines, including ferry services to neighbouring islands.

**Saint Lucia Air and Sea Ports Authority:** Manoel St, POB 651, Castries; tel. 452-2893; fax 452-2062; e-mail frenchg@slaspa.com; internet www.slaspa.com; Chair. Trevor Braithwaite; Gen. Man. Vincent Hippolyte.

**Saint Lucia Marine Terminals Ltd:** POB 355, Vieux Fort; tel. 454-8742; fax 454-8745; e-mail slumarterm@candw.lc; f. 1995; private port management co.

## CIVIL AVIATION

There are two airports in use: Hewanorra International (formerly Beane Field near Vieux Fort), 64 km (40 miles) from Castries, which is equipped to handle large jet aircraft; and George F. L. Charles Airport, which is at Vigie, in Castries, and which is capable of handling medium-range jets. Saint Lucia is served by scheduled flights to the USA, Canada, Europe and most destinations in the Caribbean. The country is a shareholder in the regional airline LIAT (see chapter on Antigua and Barbuda).

**Saint Lucia Air and Sea Ports Authority:** see Shipping.

**Air Antilles:** Laborie St, POB 1065, Castries; f. 1985; designated as national carrier of Grenada in 1987; flights to destinations in the Caribbean, the United Kingdom and North America; charter co.

**Caribbean Air Transport:** POB 253, Castries; f. 1975 as Saint Lucia Airways; local shuttle service, charter flights.

**Eagle Air Services Ltd:** Vigie Airport, POB 838, Castries; tel. 452-1900; fax 452-9683; e-mail eagleairslu@candw.lc; internet www.eagleairslu.com; charter flights; Man. Dir Capt. Ewart F. Hinkson.

**Helenair Corpn Ltd:** POB 253, Castries; tel. 452-1958; fax 451-7360; e-mail helenair@candw.lc; internet www.stluciatravel.com.lc/helenair.htm; f. 1987; charter and scheduled flights to major Caribbean destinations; Man. Arthur Neptune.

# Tourism

Saint Lucia possesses spectacular mountain scenery, a tropical climate and sandy beaches. Historical sites, rich birdlife and the sulphur baths at Soufrière are other attractions. Visitor arrivals totalled 779,710 in 2004. Tourist receipts in the same year were an estimated US $314.3m. North America is the principal market (38.9% of total stop-over visitors in 2003), followed by the United Kingdom (with 29.1%). There were an estimated 4,000 hotel rooms in late 2004; increasing this figure by at least 75% before the 2007 Cricket World Cup, hosted by several Caribbean states, including Saint Lucia, was a government priority.

**Saint Lucia Tourist Board:** Sureline Bldg, Top Floor, Vide Bouteille, POB 221, Castries; tel. 452-4094; fax 453-1121; e-mail slutour@candw.lc; internet www.stlucia.org; 2 brs overseas; Chair. Costello Michel; Dir Peter Hilary Modeste.

# SAINT LUCIA

**Saint Lucia Hotel and Tourism Association (SLHTA):** POB 545, Castries; tel. 452-5978; fax 452-7967; e-mail slhta@candw.lc; internet www.stluciatravel.com.lc; f. 1963; Pres. ANTHONY BOWEN; Exec. Vice-Pres. TERENCE GUSTAVE.

## Defence

The Royal Saint Lucia Police Force, which numbers about 300 men, includes a Special Service Unit for purposes of defence. Saint Lucia participates in the US-sponsored Regional Security System, comprising police, coastguards and army units, which was established by independent East Caribbean states in 1982. There are also two patrol vessels for coastguard duties.

## Education

Education is compulsory for 10 years between five and 15 years of age. Primary education begins at the age of five and lasts for seven years. Secondary education, beginning at 12 years of age, lasts for five years, comprising a first cycle of three years and a second cycle of two years. Free education is provided in more than 90 government-assisted schools. Facilities for industrial, technical and teacher-training are available at the Sir Arthur Lewis Community College at Morne Fortune, which also houses an extra-mural branch of the University of the West Indies. In May 2002 it was announced that an additional US $19m. will be invested in the education system during September 2002–September 2006. The project, to build two new secondary schools and renovate existing ones, will be partially funded by the World Bank and UNESCO. Some EC $132.7m. was allocated to the Ministry of Education, Human Resource Development, Youth and Sports in the 2004/05 budget.

# SAINT VINCENT AND THE GRENADINES

## Geography

### PHYSICAL FEATURES

Saint Vincent and the Grenadines is in the Windward Islands and consists of the main island of Saint Vincent itself and 32 other islands and cays to its south. The country is surrounded by other Commonwealth states in its position on the Antillean chain that separates the Caribbean Sea, to the west, from the Atlantic Ocean, to the east. The Lesser Antilles are here beginning to arc more towards the south-west as they head towards the South American mainland. Saint Lucia lies 47 km (29 miles) to the north, and a little east, while yet another Commonwealth country, Grenada, is southwards along the chain. The large islands of Saint Vincent and Grenada, which are about 100 km apart, are connected by the many smaller islands of the Grenadines, which are split between the two countries. Only a narrow channel separates Petit St Vincent from Petit Martinique, the northernmost island of Grenada. About 160 km to the east, beyond the main arc of the Lesser Antilles, lies Barbados. Saint Vincent and the Grenadines is the third smallest independent state in the Western hemisphere, with an area of 389 sq km (150 sq miles), 88% of which is accounted for by the main island of Saint Vincent itself.

The main island of Saint Vincent is 29 km from north to south and almost 18 km at its widest. Like the rest of the Windwards, Saint Vincent has rugged eastern Atlantic coasts and gentler western and south-western shores. The interior, however, is generally mountainous and steep, formed on the largely sunken volcanic ridge that runs north from Grenada and through the connecting archipelago. The Grenadines are largely coralline, although the larger ones are hilly, while the main island is steep and lofty. Saint Vincent is dominated by the central mass of the steep Morne Garu range, which runs from south to north and thrusts side spurs to the east and west coasts, but the highest point is to the north, in the Waterloo mountains, where the Soufrière volcano reaches 1,234 m (4,050 ft). This is an active volcano, which last erupted in April 1979, but without the loss of life that accompanied the eruptions of 1902 or 1812. Apart from in the immediate vicinity of the volcano, the island is fertile and productive, the central mountains being covered in rainforest and the lower hillsides and few flatter areas planted with crops, notably bananas and arrowroot (the island is the world's leading producer of the latter, while economic dependence on the fluctuating prices of the former has led to the alternative, but illicit, cultivation of marijuana). The woodland areas are home to a number of plant, animal and bird species, most notable among the last being the protected Saint Vincent parrot and the whistling warbler, both of which are unique to the island. Marine life is especially rich among the reefs and islands of the Grenadines.

The largest of the Grenadines, and the one closest to Saint Vincent itself (14 km), is Bequia, which covers about 18 sq km, its hills wooded with fruit and nut trees. There are a number of smaller islands to the south, such as Isle à Quatre, while to the east of them are Baliceaux and the smaller Battowia, which stand at the end of the main chain of the Grenadines, which run south-west from here towards Union Island. The main islands in this chain are Mustique (5 km in length and over 2 km wide), then, beyond some smaller islands and a wider than average sea channel, Canouan (almost 8 sq km in area, and 40 km south of Saint Vincent), followed by Mayreau and the Tobago Cays, and Union Island (the main island of the south, 64 km from Saint Vincent—5 km in length and almost 2 km in width, or just over 8 sq km). Union Island has hills that reach 305 m. To the east, and a little south, the Vincentian Grenadines end at Petit St Vincent, a resort island.

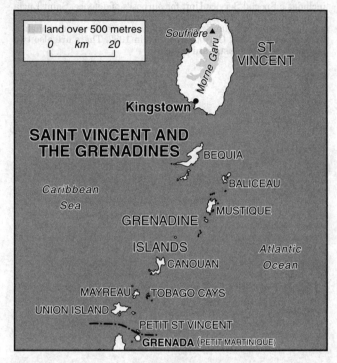

### CLIMATE

The climate is subtropical, and the country is on the edge of the hurricane belt (the southern boundary has been determined to fall just south of Canouan), although swell from storms passing to the north can still cause damage on the coasts. The hurricane season coincides with the rainy season (roughly July–October), which contributes substantially to annual rainfall, which is high, averaging some 2,050 mm (80 ins) on the coast, but up to 3,850 mm in the hilly interior of the main island. The Grenadines, being lower and less forested than Saint Vincent, receive slightly less rainfall than the main island (around 1,500 mm). Average temperatures range between about 25°C and 27°C (77°–81°F).

### POPULATION

The population is a heterogeneous mix, but of predominantly black African descent. Some 66% are counted as black, 19% as mixed race (mainly black-white), 6% as 'East' Indian, 4% as white and 2% as Carib Amerindian. The Caribs are so-called Black Caribs, descended from the union of escaped black slaves with native Caribs, the latter also fiercely resisting the British and French, but more susceptible to disease and cultural isolation. Some 5,000 of the continuously rebellious Black Caribs were deported to an island off Honduras in 1797, and the descendants of the remainder are now generally confined to the north of the island. The main religion is Christianity, with the Church in the Province of the West Indies being the largest single denomination (with somewhere between one-third and one-half of the population being Anglican), followed by the Methodist (around one-quarter) and Roman Catholic (about one-10th) Churches. The city of Kingstown is the seat of both the Anglican and Roman Catholic bishops. There are also a number of other Protestant groups, including the Seventh-day Adventists, and some adherents of non-Christian traditions. A local blend of African and Christian traditions produced a faith the observers of which were known as Spiritual Baptists. These 'Shakers' (not to be confused with the millenarian sect founded in the United Kingdom in the 18th century), like the 'Shouters' of Trinidad (Trinidad and Tobago) emerged in the 1830s, in the aftermath of emancipation, but were banned between 1912 and 1965. English is both the official and the spoken tongue, although there is a French patois or Creole still used in some areas.

In mid-2004 the total population of Saint Vincent and the Grenadines was some 118,000. About one-quarter of the population lives in and around Kingstown, the capital on the south-western coast, with the 2001 census putting the numbers actually residing within the city boundaries as 13,526. Some 8% of the population lives on the Grenadines. The main towns of the smaller islands are Port Elizabeth on Bequia, Charlestown on Canouan and Ashton (and, to a lesser extent, Clifton) on Union Island. The Grenadines are described as dependencies, while Saint Vincent itself is divided into six administrative parishes.

# History

## MARK WILSON

Saint Vincent and the Grenadines is a constitutional monarchy within the Commonwealth. Queen Elizabeth II is head of state, and is represented by a Governor-General. There is a bicameral legislature with an elected chamber.

Saint Vincent's Amerindian inhabitants knew the island as Hairoun. It was given its present name by Christopher Columbus in 1498, although fierce Amerindian resistance delayed European settlement. In 1675 a Dutch ship carrying African slaves was shipwrecked to the south of the island. The human cargo reached land, where they intermarried with the local population, producing a mixed-race Black Carib community. There was fierce resistance to European colonization and when settlements were established, the island changed hands on several occasions between the United Kingdom and France, finally passing to the United Kingdom in 1783. A rebellion by the Black Caribs in 1795–96 succeeded in gaining control of most of the island, but was eventually suppressed. Most of the Black Caribs were deported in 1797 to the island of Roatan, off the coast of Honduras. Their descendants now form the Garifuna community of Belize. Smaller groups of Black Caribs remain in Saint Vincent, and are concentrated in the north-east of the island. As in the rest of the British Caribbean, slavery was abolished in 1834. Poverty remained extreme, and there was some immigration of indentured Portuguese and Indian labourers after emancipation.

The first universal suffrage elections were in 1951, and were won by the loosely organized Workers', Peasants' and Rate-payers' Union. After the elections, Ebenezer Joshua founded the People's Political Party (PPP), while Milton Cato formed the Saint Vincent Labour Party (SVLP) in 1954. The SVLP won a general election in 1967, holding office with Cato as Prime Minister until 1972, and again during 1974–84. In the 1972 election, each of the main parties won six seats on the main island. James Mitchell took the Grenadines seat as an independent, and persuaded the PPP to join a Government in which he was Prime Minister, an arrangement which lasted for two years. However, in 1974 the PPP and SVLP joined forces to bring down the Government. In the election that followed, Mitchell again won a single seat, but Cato became Prime Minister, with Joshua as Minister of Finance. Joshua's wife, Ivy, sat on the opposition benches, and she, rather than Mitchell, was appointed Leader of the Opposition.

Saint Vincent and the Grenadines joined the Federation of the West Indies in 1958 along with nine other British colonies. When Jamaica and Trinidad and Tobago left in 1962, the Federation collapsed, and an attempt to unite the remaining colonies as the 'little eight' was unsuccessful. Two years later than its neighbours, Saint Vincent and the Grenadines became a British Associated State in 1969, responsible for its internal affairs, with the United Kingdom retaining control of external affairs and defence, before moving to full independence on 27 October 1979.

James (from 1995 Sir James) Mitchell, won a general election in July 1984 as leader of the National Democratic Party (NDP), which took 12 of the 15 parliamentary seats, with the Cato Government by then widely perceived to be tainted by corruption. The NDP Government won all 15 seats in a May 1989 election, and 12 in February 1994, with an alliance of the opposition Labour Party and Movement for National Unity taking three. By 1998 the opposition parties had merged to form the Unity Labour Party (ULP) and were able to mount a strong challenge to the NDP Government, which was itself by now widely accused of corruption, and was further damaged by the serious problems facing the banana industry. The ULP took 54.6% of the vote in an election on 15 June, but the NDP held onto eight of the 15 legislative seats, some with very narrow minorities and amid accusations of electoral malpractice.

The political climate became confrontational thereafter and a serious controversy developed over a move in April 2000 to increase the pensions paid to retired politicians. In May the Caribbean Community and Common Market (CARICOM) brokered an agreement under which fresh elections were to be held by March 2001, giving Mitchell time to retire as Prime Minister in October 2000 at the age of 69; in August his party chose the Minister of Finance, Arnhim Eustace, as his successor. At legislative elections duly held on 28 March 2001 the ULP took 57% of the popular vote and 12 seats; its leader, Ralph Gonsalves, thus became Prime Minister.

In June 2001 Gonsalves announced his intention to establish a commission to review the Constitution. A 25-member body was appointed in December 2002, and was due to complete its work by the end of 2004; a referendum was expected to be held on any proposed reforms. The commission submitted its report, to be used as a basis for further discussion, in February 2005. Its proposals included: replacing the Queen as head of state with an indirectly elected president; the establishment of a non-legislative National Advisory Council of Elders; and the introduction of a revamped single-chamber National Assembly, to include three appointed senators and seven civil-society representatives. Meanwhile, in June 2002 the Governor-General, Sir Charles Antrobus, died. Sir Frederick Ballantyne succeeded Antrobus in September.

In March 2004 the parliamentary opposition boycotted the House of Assembly and established an 'alternative parliament' at the NDP's headquarters; Eustace claimed that the parliamentary Speaker, Hendrick Alexander, had consistently demonstrated a pro-Government bias. In late February Eustace had led NDP members out of Parliament after the Prime Minister was allowed to present an amendment to a motion on a day when opposition business should have taken precedence. Despite this, the main challenge facing Gonsalves' Government in 2005 was the increasingly prominent issue of violent crime. The per-head murder rate of 26 per 100,000 in 2004 was extremely high by international standards and was the third highest in the English-speaking Caribbean after Jamaica and Belize; it was widely held that the Government had devoted insufficient funds to attempting to resolve the problem. Moreover, Saint Vincent remained one of the largest cultivators of marijuana in the region and in 2003 the Government reported that 40% of prison inmates were jailed for drugs-related offences. Another potential problem for the Government arose after Japan pledged to provide US $6m. for the construction of a fish market. It was claimed that, in return for Japanese financial aid, Saint Vincent and the Grenadines would support the pro-whaling bloc at meetings of the International Whaling Commission; Gonsalves refuted the allegation, pointing out that the

Government had abstained on crucial votes relating to the provision of whale sanctuaries in the South Atlantic and the Pacific.

Gonsalves effected a minor reorganization of his Cabinet in May 2005: the most notable change was the transferral of the erstwhile Minister of Transport, Works and Housing, Julian Francis, to a new role in the Office of the Prime Minister related to the creation of employment opportunities for young people, the lack of which was regarded as being closely related to the high rate of crime on the islands. Legislative elections were due to be held by March 2006, but were widely expected at an earlier date.

# Economy

## MARK WILSON

Saint Vincent and the Grenadines is the third smallest country in the western hemisphere in area, with 389.3 sq km and 118,000 inhabitants in mid-2004. Most of the population live on the main island. Of the Grenadine islands to the south, the most significant are Bequia, Mustique and Union Island. The islands have developed a modest middle-income economy, but the per-head gross domestic product (GDP) of US $3,613 in 2003 was the lowest in the Eastern Caribbean. GDP grew at an average rate of 4.0% per year in 1995–2000, but contracted by 0.1% in 2001 as international demand for tourism decreased and the problems of the banana industry intensified; this was followed by growth of 2.0% in 2002 and 3.6% in 2003, stimulated by a recovery in tourism. The current-account deficit on the balance of payments widened from 9.2% of GDP in 2000 to 18.1% in 2003, while public-sector debt increased from the equivalent of 49% of GDP in 1998 to 80% in 2004. With the Government adopting a policy of counter-cyclical spending, the overall fiscal balance moved from a surplus of 0.1% of GDP in 2000 to a deficit of 3.3% in 2003. International donor agencies urged the Government strictly to maintain its restraint of current expenditure and to widen the tax base with the aim of protecting its capital-investment programme. A fiscal-consolidation plan was developed in 2004 by the finance ministry, the Eastern Caribbean Central Bank (ECCB) and the Barbados-based Caribbean Technical Assistance Centre (CARTAC). In addition, the Unity Labour Party Government was attempting to reschedule or reduce debt incurred by the previous administration for the failed Ottley Hall marina, which comprised some 30% of the country's total debt.

Saint Vincent and the Grenadines is a member of the Caribbean Community and Common Market, or CARICOM, which is attempting to develop a single market, in principle by 2006. It is also a member of the Organisation of Eastern Caribbean States (OECS), which links nine of the smaller Caribbean territories, while the ECCB, based in Saint Christopher and Nevis, supervises its financial affairs.

Agriculture (including hunting, forestry and fishing) comprised some 8.5% of GDP in 2003, down from 21.2% in 1990. Banana exports to the United Kingdom under protected market access arrangements were an economic mainstay from the 1950s, but the industry suffered a steep decline in the 1990s, as the trade privileges of Caribbean producers in the European market were eroded. In 1990 income from banana exports was US $45.5m., equivalent to 52.2% of merchandise exports and 34.2% of foreign exchange earnings. By 2003 exports had diminished to an estimated $9.8m., equivalent to 24.6% of merchandise exports and an 6% of foreign-exchange earnings. However, significant grant aid from the European Union was available for irrigation and technical support, as well as for economic diversification and social assistance, and there were some hopes that the industry would survive. Small farmers produce a wide variety of fruits, vegetables and livestock products for the local market, and significant quantities are exported to Trinidad and Tobago and Barbados. Illegal cultivation of marijuana for export to other islands plays a significant role in the parallel economy.

The E. T. Joshua Airport on Saint Vincent cannot accommodate intercontinental flights, and inbound passengers must transfer to a short-haul flight in Barbados or Saint Lucia. The lack of direct air connections has hampered the growth of tourism. The main island has few white sand beaches, and most tourism development has taken place on the Grenadines, where a dry climate, clear water and fine beaches have proved powerful attractions for luxury development, and difficulty of public access has been turned into an asset for those who seek seclusion. Yachting is an important sub-sector, but an ambitious marina project at Ottley Hall on the main island led only to an increase in government commercial borrowing. Tourism receipts increased from EC $111m. in 1992 to an estimated $222m. in 2003. In 2004 the islands attracted 86,727 stop-over tourists, a 10.4% increase over the previous year. There were also 74,657 cruise-ship passengers, although this group had much lower per-head spending. In 2003 31.1% of stop-over tourists were from North America and 27.4% were from Europe.

Manufacturing comprised some 6.2% of GDP in 2003, down from 9.5% in 1992. The most important facility is a mill, which produces flour, milled rice and animal feed for the local and Eastern Caribbean markets. Flour exports totalled some US $6.8m. in 2002 and comprised 16.1% of total merchandise exports. The flour trade was protected from non-OECS imports, but would be threatened by producers on larger islands should the CARICOM single market become established. Other enterprises include a brewery, and 'enclave' industries producing electronic components for export.

Saint Vincent and the Grenadines is a member of the Eastern Caribbean Telecommunications Authority (based in Saint Lucia) and has liberalized its telecommunications regime. The Government sees telemarketing and informatics as a potentially important source of employment. However, call centres opened from 2001 did not appear to have been as successful as had originally been anticipated.

There is a small 'offshore' financial services sector; 11 'offshore' banks and 6,276 International Business Companies were registered at the end of 2004. A package of financial legislation passed in 1996 was viewed with some concern by international regulators, and Saint Vincent and the Grenadines was in 2000 listed as a 'non-co-operative jurisdiction' by the Financial Action Task Force on Money Laundering (FATF, based in Paris, France). Following some regulatory and legislative reforms, including the establishment of a Financial Intelligence Unit in May 2002, Saint Vincent and the Grenadines was delisted by the FATF in June 2003. On 18 November 2001 the island's leading 'offshore banker', Thierry Nano, was able to leave the island in time to evade a US request for his extradition on money-laundering charges. Opposition politicians accused the Government of failing to prevent Nano's escape and, in February 2002, Arnhim Eustace, the opposition leader, criticized the Government's decision to take control of the two banks owned by Nano, instead of liquidating them.

Saint Vincent and the Grenadines is close to the southern fringe of the hurricane belt, and was affected by 'Hurricane Allen' in 1980, 'Hurricane Emily' in 1987 and by the fringe of 'Hurricane Ivan' in 2004. The Soufrière volcano in the north of the island erupted in 1902 and 1979, causing several thousand deaths on the first occasion. There are other volcanic centres in the Grenadines, and there is some long term risk from the underwater volcano of Kick 'Em Jenny, just north of Grenada.

# Statistical Survey

Sources (unless otherwise stated): Statistical Office, Ministry of Finance, Planning, Economic Development, Labour and Information, Government Bldgs, Kingstown; tel. 456-1111; e-mail statssvg@vincysurf.com.

## AREA AND POPULATION

**Area:** 389.3 sq km (150.3 sq miles). The island of Saint Vincent covers 344 sq km (133 sq miles).

**Population:** 97,845 at census of 12 May 1980; 106,499 (males 53,165, females 53,334) at census of 12 May 1991; 109,022 at preliminary census count of May 2001. *2004:* 118,000 (mid-year estimate). Source; partly UN, *World Population Prospects: The 2004 Revision.*

**Density** (mid-2004): 303.1 per sq km.

**Principal Town:** Kingstown (capital), population 13,526 at preliminary census count of May 2001. *Mid-2003* (UN estimate, incl. suburbs): Kingstown 29,382 (Source: UN, *World Urbanization Prospects: The 2003 Revision*).

**Births and Deaths** (registrations, 2000): Live births 2,149 (birth rate 19.2 per 1,000); Deaths 698 (death rate 6.2 per 1,000). *2003:* Birth rate 18.5 per 1,000; Death rate 7.6 per 1,000 (Source: Caribbean Development Bank, *Social and Economic Indicators*).

**Expectation of Life** (years at birth): 70 (males 68; females 72) in 2003. Source: WHO, *World Health Report.*

**Economically Active Population** (persons aged 15 years and over, 1991 census): Agriculture, hunting, forestry and fishing 8,377; Mining and quarrying 98; Manufacturing 2,822; Electricity, gas and water 586; Construction 3,535; Trade, restaurants and hotels 6,544; Transport, storage and communications 2,279; Financing, insurance, real estate and business services 1,418; Community, social and personal services 7,696; *Total employed* 33,355 (males 21,656, females 11,699); Unemployed 8,327 (males 5,078, females 3,249); *Total labour force* 41,682 (males 26,734, females 14,948). Source: ILO, *Yearbook of Labour Statistics.*

## HEALTH AND WELFARE
### Key Indicators

**Total Fertility Rate** (children per woman, 2003): 2.2.

**Under-5 Mortality Rate** (per 1,000 live births, 2003): 27.

**Physicians** (per 1,000 head, 1997): 0.88.

**Hospital Beds** (per 1,000 head, 1996): 1.85.

**Health Expenditure** (2002): US $ per head (PPP): 340.

**Health Expenditure** (2002): % of GDP: 5.9.

**Health Expenditure** (2002): public (% of total): 65.5.

**Access to Water** (% of persons, 2000): 93.

**Access to Sanitation** (% of persons, 2000): 96.

**Human Development Index** (2002): ranking: 87.

**Human Development Index** (2002): value: 0.751.

For sources and definitions, see explanatory note on p. vi.

## AGRICULTURE, ETC.

**Principal Crops** ('000 metric tons, 2003): Maize 0.6; Cassava 0.7; Sweet potatoes 1.1; Yams 2.0; Other roots and tubers 9.8 (FAO estimate); Sugar cane 20.0; Coconuts 2.6; Vegetables 4.3 (FAO estimate); Bananas 50.0; Plaintains 1.5 (FAO estimate); Oranges 1.5; Lemons and limes 1.2; Apples 1.2; Mangoes 1.5; Other fruits 0.8.

**Livestock** ('000 head, year ending September 2003): Cattle 5.0; Sheep 12.0; Goats 7.0; Pigs 9.2; Asses 1.3 (FAO estimate); Poultry 125.

**Livestock Products** ('000 metric tons, 2003): Pig meat 0.6 (FAO estimate); Cows' milk 1.2; Hen eggs 0.6.

**Fishing** (metric tons, live weight, 2003): Albacore 1,555; Yellowfin tuna 568; Other tuna-like fishes 1,682; *Total catch* (incl. others) 4,782.

Source: FAO.

## INDUSTRY

**Selected Products** ('000 metric tons, 2003, unless otherwise stated): Copra 2 (FAO estimate); Raw sugar 2 (FAO estimate); Rum 1,000 hectolitres (2001); Electric energy 107 million kWh (IMF estimate, 2001). Sources: FAO, and UN, *International Commodity Statistics Yearbook.*

## FINANCE

**Currency and Exchange Rates:** 100 cents = 1 Eastern Caribbean dollar (EC $). *Sterling, US Dollar and Euro Equivalents* (31 May 2005): £1 sterling = EC $4.909; US $1 = EC $2.700; €1 = EC $3.329; EC $100 = £20.37 = US $31.33 = €30.04. *Exchange rate:* Fixed at US $1 = EC $2.70 since July 1976.

**Budget** (estimates, EC $ million, 2002): *Revenue:* Revenue from taxation 252 (Taxes on income 80, Taxes on goods and services 49, Taxes on international trade and transactions 121); Other current revenue 43; Capital revenue 1; Foreign grants 16; Total 312. *Expenditure:* Personal emoluments 146; Other goods and services 59; Interest payments 28; Transfers and subsidies 49; Capital expenditure and net lending 65; Total 348. Source: IMF, *St Vincent and the Grenadines: Statistical Appendix* (February 2003).

**International Reserves** (US $ million at 31 December 2004): Reserve position in IMF 0.78; Foreign exchange 74.20; Total 74.98. Source: IMF, *International Financial Statistics.*

**Money Supply** (EC $ million at 31 December 2004): Currency outside banks 64.39; Demand deposits 257.46; Total money 321.85. Source: IMF, *International Financial Statistics.*

**Cost of Living** (Consumer Price Index; base: 2000 = 100): 101.6 in 2002; 101.9 in 2003; 104.9 in 2004. Source: IMF, *International Financial Statistics.*

**Gross Domestic Product** (EC $ million at current factor cost): 805.5 in 2002; 836.4 in 2003; 949.0 in 2004 (preliminary). Source: Eastern Caribbean Central Bank.

**Expenditure on the Gross Domestic Product** (EC $ million at current prices, 2003): Government final consumption expenditure 199.26; Private final consumption expenditure 647.39; Gross capital formation 343.90; *Total domestic expenditure* 1,190.55; Exports of goods and services 475.56; *Less* Imports of goods and services 650.61; *GDP at market prices* 1,015.50. Source: Eastern Caribbean Central Bank.

**Gross Domestic Product by Economic Activity** (EC $ million at current factor cost, 2003): Agriculture, hunting, forestry and fishing 75.11; Mining and quarrying 1.97; Manufacturing 54.90; Electricity and water 52.84; Construction 101.88; Wholesale and retail trade 158.88; Restaurants and hotels 16.05; Transport 121.72; Communications 45.39; Banking and insurance 61.35; Real estate and housing 19.89; Government services 161.69; Other services 16.98; *Sub-total* 888.65; *Less* Imputed bank service charge 52.30; *Total in basic prices* 836.35. Source: Eastern Caribbean Central Bank.

**Balance of Payments** (EC $ million, 2003): Exports of goods f.o.b. 105.59; Imports of goods f.o.b. −477.68; *Trade balance* −372.09; Exports of services 369.97; Imports of services −172.93; *Balance on goods and services* −175.05; Other income received (net) −42.23; *Balance on goods, services and income* −217.28; Current transfers received 65.41; Current transfers paid −32.33; *Current balance* −184.20; Capital account (net) 14.10; Direct investment from abroad (net) 118.21; Portfolio investment (net) 53.24; Other investment (net) −35.16; Net errors and omissions 20.47; *Overall balance* −13.34. Source: Eastern Caribbean Central Bank.

## EXTERNAL TRADE

**Principal Commodities** (estimates, US $ million, 2002): *Imports:* Food and live animals 43.6; Mineral fuels, lubricants, etc. 17.3; Chemicals and related products 18.2; Basic manufactures 34.7; Machinery and transport equipment 34.8; Total (incl. others) 181.9. *Exports:* Bananas 12.9; Eddoes and dasheens 1.8; Flour 6.8; Rice 4.5; Total domestic exports (incl. others) 40.4; Re-exports 1.8; Total 42.2. Source: IMF, *St Vincent and the Grenadines: Statistical Appendix* (February 2003).

**Principal Trading Partners** (estimates, US $ million, 2002): *Imports:* Barbados 8.1; Canada 5.3; Germany 1.9; Guyana 4.1; Japan 6.7; Netherlands 2.2; Trinidad and Tobago 38.1; United Kingdom 15.6; USA 69.1; Total (incl. others) 181.9. *Exports:* Antigua and Barbuda 2.8; Barbados 3.6; Dominica 1.5; Jamaica 1.3; Saint Lucia 3.7; Trinidad and Tobago 4.3; United Kingdom 16.9; USA 1.4; Total (incl. others) 42.2. Source: IMF, *St Vincent and the Grenadines: Statistical Appendix* (February 2003).

# SAINT VINCENT AND THE GRENADINES

## TRANSPORT

**Road Traffic** (motor vehicles in use, 2002): Private cars 10,504; Buses and coaches 1,150; Lorries and vans 3,019; Road tractors 89.
Source: International Road Federation, *World Road Statistics*.

**Shipping:** *Arrivals* (2000): Vessels 1,007. *International Sea-borne Freight Traffic* ('000 metric tons, 2000): Goods loaded 54; Goods unloaded 156. *Merchant Fleet* (vessels registered at 31 December 2004): Number 1,182; Total displacement 6,324,289 grt (Source: Lloyd's Register-Fairplay, *World Fleet Statistics*).

**Civil Aviation** (visitor arrivals): 94,030 in 2000; 85,735 in 2001 (preliminary); 90,879 in 2002 (estimate). Source: IMF, *St Vincent and the Grenadines: Statistical Appendix* (February 2003).

## TOURISM

**Tourist Arrivals** ('000): 142.7 (77.6 visitor arrivals, 65.1 cruise-ship passengers) in 2002; 143.5 (78.5 visitor arrivals, 65.0 cruise-ship passengers) in 2003; 161.4 (86.7 visitor arrivals, 74.7 cruise-ship passengers) in 2004.

**Tourism Receipts** (estimates, US $ million): 80.2 in 2001; 82.0 in 2002; 82.4 in 2003.
Source: Caribbean Development Bank, *Social and Economic Indicators*.

## COMMUNICATIONS MEDIA

**Radio Receivers** ('000 in use, 2000): 100.
**Television Receivers** ('000 in use, 2000): 50.
**Telephones** ('000 main lines in use, 2003): 32.4.
**Facsimile Machines** (1996): 1,500.
**Mobile Cellular Telephones** (subscribers, 2003): 62,900.
**Personal Computers** ('000 in use, 2002): 14.
**Internet Users** ('000, 2002): 7.0.
**Non-daily Newspapers** (2000): Titles 8; Circulation 50,000.
Sources: mainly UNESCO, *Statistical Yearbook*; UN, *Statistical Yearbook*; and International Telecommunication Union.

## EDUCATION

**Pre-primary** (1993/94): 97 schools; 175 teachers; 2,500 pupils.
**Primary** (2000): 60 schools; 987 teachers; 19,279 pupils (2002).
**Secondary** (2000): 21 schools; 406 teachers; 7,909 pupils (2002).
**Teacher Training** (2000): 1 institution; 10 teachers; 107 students.
**Technical College** (2000): 1 institution; 19 teachers; 187 students.
**Community College** (2000): 1 institution; 13 teachers; 550 students.
**Nursing College** (2000): 1 institution; 6 teachers; 60 students.
Source: partly Caribbean Development Bank, *Social and Economic Indicators*.

**Adult Literacy Rate:** 83.1% in 2002. Source: UN Development Programme, *Human Development Report*.

# Directory

## The Constitution

The Constitution came into force at the independence of Saint Vincent and the Grenadines on 27 October 1979. The following is a summary of its main provisions

### FUNDAMENTAL RIGHTS AND FREEDOMS

Regardless of race, place of origin, political opinion, colour, creed or sex, but subject to respect for the rights and freedoms of others and for the public interest, every person in Saint Vincent and the Grenadines is entitled to the rights of life, liberty, security of the person and the protection of the law. Freedom of conscience, of expression, of assembly and association is guaranteed and the inviolability of a person's home and other property is maintained. Protection is afforded from slavery, forced labour, torture and inhuman treatment.

### THE GOVERNOR-GENERAL

The British Monarch is represented in Saint Vincent and the Grenadines by the Governor-General.

### PARLIAMENT

Parliament consists of the British monarch, represented by the Governor-General, and the House of Assembly, comprising 15 elected Representatives (increased from 13 under the provisions of an amendment approved in 1986) and six Senators. Senators are appointed by the Governor-General—four on the advice of the Prime Minister and two on the advice of the Leader of the Opposition. The life of Parliament is five years. Each constituency returns one Representative to the House who is directly elected in accordance with the Constitution. The Attorney-General is an *ex-officio* member of the House. Every citizen over the age of 18 is eligible to vote. Parliament may alter any of the provisions of the Constitution.

### THE EXECUTIVE

Executive authority is vested in the British monarch and is exercisable by the Governor-General. The Governor-General appoints as Prime Minister that member of the House who, in the Governor-General's view, is the best able to command the support of the majority of the members of the House, and selects other Ministers on the advice of the Prime Minister. The Governor-General may remove the Prime Minister from office if a resolution of 'no confidence' in the Government is passed by the House and the Prime Minister does not either resign within three days or advise the Governor-General to dissolve Parliament.

The Cabinet consists of the Prime Minister and other Ministers and the Attorney-General as an *ex-officio* member. The Leader of the Opposition is appointed by the Governor-General as that member of the House who, in the Governor-General's view, is best able to command the support of a majority of members of the House who do not support the Government.

### CITIZENSHIP

All persons born in Saint Vincent and the Grenadines before independence who, immediately prior to independence, were citizens of the United Kingdom and Colonies automatically became citizens of Saint Vincent and the Grenadines. All persons born outside the country after independence to a parent possessing citizenship of Saint Vincent and the Grenadines automatically acquire citizenship, as do those born in the country after independence. Citizenship can be acquired by those to whom it would not automatically be granted.

## The Government

### HEAD OF STATE

**Monarch:** HM Queen ELIZABETH II.

**Governor-General:** Sir FREDERICK BALLANTYNE (took office 2 September 2002).

### CABINET
(July 2005)

**Prime Minister and Minister of Finance, Planning, Economic Development, Labour and Information:** Dr RALPH E. GONSALVES.

**Deputy Prime Minister and Minister of Transport, Works and Housing:** LOUIS STRAKER.

**Minister of National Security, the Public Service, Seaports and Airports and Air Development:** VINCENT BEACHE.

**Minister of Foreign Affairs and of Commerce and Trade:** MIKE BROWNE.

**Minister of Social Development, the Family, Gender Affairs and Ecclesiastical Affairs:** GIRLYN MIGUEL.

**Minister of Agriculture, Forestry and Fisheries:** SELMON WALTERS.

**Minister of Tourism and Culture:** RENÉ BAPTISTE.

**Minister of Health and the Environment:** Dr DOUGLAS SLATER.

**Minister of Telecommunications, Science, Technology and Industry:** Dr JERROL THOMPSON.

**Minister of Education, Youth and Sports:** CLAYTON BURGIN.

**Attorney-General:** JUDITH S. JONES-MORGAN.

# SAINT VINCENT AND THE GRENADINES

## MINISTRIES

**Office of the Governor-General:** Government Bldgs, Kingstown; tel. 456-1401.

**Office of the Prime Minister:** Administrative Centre, Kingstown; tel. 456-1703; fax 457-2152; e-mail pmosvg@caribsurf.com.

**Office of the Attorney-General:** Government Bldgs, Kingstown; tel. 457-2807.

**Ministry of Agriculture, Forestry and Fisheries:** Richmond Hill, Kingstown; tel. 457-1410; fax 457-1688; e-mail agrimin@caribsurf.com.

**Ministry of Commerce and Trade:** Government Bldgs, Kingstown; tel. 456-1223; fax 457-2880; e-mail mtrade@caribsurf.com.

**Ministry of Education, Youth and Sports:** Government Bldgs, Kingstown; tel. 457-1104; e-mail minedsvg@vincysurf.com.

**Ministry of Finance, Planning, Economic Development, Labour and Information:** Government Bldgs, Kingstown; tel. 456-1111.

**Ministry of Foreign Affairs:** Administrative Centre, Kingstown; tel. 456-2060; fax 456-2610; e-mail svgforeign@caribsurf.com.

**Ministry of Health and the Environment:** Government Bldgs, Kingstown; tel. 457-2586; fax 457-2684.

**Ministry of National Security, the Public Service, Seaports and Airports and Air Development:** Government Bldgs, Kingstown.

**Ministry of Social Development, the Family, Gender Affairs and Ecclesiastical Affairs:** Government Bldgs, Kingstown; tel. 426-2949.

**Ministry of Telecommunications, Science, Technology and Industry:** Government Bldgs, Kingstown; tel. 457-2039.

**Ministry of Tourism and Culture:** Government Bldgs, Kingstown; tel. 457-1502; fax 451-2425; e-mail tourism@caribsurf.com.

**Ministry of Transport, Works and Housing:** Government Bldgs, Kingstown.

## Legislature

### HOUSE OF ASSEMBLY

**Senators:** 6.
**Elected Members:** 15.
**Election, 28 March 2001**

| Party | % of votes | Seats |
| --- | --- | --- |
| Unity Labour Party | 57.0 | 12 |
| New Democratic Party | 40.9 | 3 |
| People's Progressive Movement | 2.1 | — |
| Total | 100.0 | 15 |

## Political Organizations

**Canouan Progressive Movement:** Kingstown; Leader TERRY BYNOE.

**New Democratic Party (NDP):** Murray Rd, POB 1300, Kingstown; tel. 457-2647; f. 1975; democratic party supporting political unity in the Caribbean, social development and free enterprise; Pres. ARNHIM EUSTACE; Sec.-Gen. STUART NANTON; 7,000 mems.

**People's Progressive Movement (PPM):** Kingstown; f. 2000 following split from ULP; Leader KEN BOYEA.

**People's Working Party:** Kingstown; Leader BURTON WILLIAMS.

**Unity Labour Party (ULP):** Beachmont, Kingstown; tel. 457-2761; fax 456-2811; e-mail ulpweb@aol.com; internet www.ulp.org; f. 1994 by merger of Movement for National Unity and the Saint Vincent Labour Party; moderate, social-democratic party; Leader Dr RALPH E. GONSALVES.

In 2000 the **Organization in Defence of Democracy** (ODD), an umbrella group of trade unions and non-governmental organizations, was formed to protest against government policy.

## Diplomatic Representation

### EMBASSIES AND HIGH COMMISSION IN SAINT VINCENT AND THE GRENADINES

**China (Taiwan):** Murray Rd, POB 878, Kingstown; tel. 456-2431; fax 456-2913; e-mail rocemsvg@caribsurf.com; Ambassador ELIZABETH Y. F. CHU.

**United Kingdom:** Granby St, POB 132, Kingstown; tel. 457-1701; fax 456-2750; e-mail bhcsvg@caribsurf.com; High Commissioner resident in Barbados; High Commissioner JOHN WHITE (resident in Barbados).

**Venezuela:** Baynes Bros Bldg, Granby St, POB 852, Kingstown; tel. 456-1374; fax 457-1934; e-mail lvccsvg@caribsurf.com; Ambassador TIBISAY URDANETA.

## Judicial System

Justice is administered by the Eastern Caribbean Supreme Court, based in Saint Lucia and consisting of a Court of Appeal and a High Court. Two Puisne Judges are resident in Saint Vincent and the Grenadines. There are five Magistrates, including the Registrar of the Supreme Court, who acts as an additional Magistrate.

**Puisne Judges:** FREDERICK BRUCE-LYLE, LOUISE ESTHER BLENMAN.

### Office of the Registrar of the Supreme Court

Registry Dept, Court House, Kingstown; tel. 457-1220; fax 457-1888. Registrar COLEEN MCDONALD.

**Chief Magistrate:** SIMONE CHURAMAN.

**Magistrates:** SHARON MORRIS-CUMMINGS, CARL JOSEPH, HILARY SAMUEL.

**Director of Public Prosecutions:** COLIN WILLIAMS.

## Religion

### CHRISTIANITY

**Saint Vincent Christian Council:** Melville St, POB 445, Kingstown; tel. 456-1408; f. 1969; four mem. churches; Chair. Mgr RENISON HOWELL.

### The Anglican Communion

Anglicans in Saint Vincent and the Grenadines are adherents of the Church in the Province of the West Indies, comprising eight dioceses. The Archbishop of the West Indies is the Bishop of Nassau and the Bahamas, and is resident in Nassau. The diocese of the Windward Islands includes Grenada, Saint Lucia and Saint Vincent and the Grenadines.

**Bishop of the Windward Islands:** Rt Rev. SEHON GOODRIDGE, Bishop's Court, POB 502, Kingstown; tel. 456-1895; fax 456-2591; e-mail diocesewi@vincysurf.com.

### The Roman Catholic Church

Saint Vincent and the Grenadines comprises a single diocese (formed when the diocese of Bridgetown-Kingstown was divided in October 1989), which is suffragan to the archdiocese of Castries (Saint Lucia). The Bishop participates in the Antilles Episcopal Conference, currently based in Port of Spain, Trinidad and Tobago. At 31 December 2003 there were an estimated 10,073 adherents in the diocese, comprising about 8% of the population.

**Bishop of Kingstown:** Rt Rev. ROBERT RIVAS, Bishop's Office, POB 862, Edinboro, Kingstown; tel. 457-2363; fax 457-1903; e-mail rcdok@caribsurf.com.

### Other Christian Churches

The Methodists, Seventh-day Adventists, Baptists and other denominations also have places of worship.

### BAHÁ'Í FAITH

**National Spiritual Assembly:** POB 1043, Kingstown; tel. 456-4717.

SAINT VINCENT AND THE GRENADINES — Directory

## The Press

### DAILY

**The Herald:** Blue Caribbean Bldg, Kingstown; tel. 456-1242; fax 456-1046; e-mail herald@caribsurf.com; internet www.heraldsvg.com; daily; internationally distributed.

### SELECTED WEEKLIES

**The Independent Weekly:** 85 Sharpe St, Kingstown; tel. 457-2866; Man. Editor CONLEY ROSE.

**Justice:** Kingstown; weekly; organ of the United People's Movement; Editor RENWICK ROSE.

**The New Times:** POB 1300, Kingstown; f. 1984; Thursdays; organ of the New Democratic Party.

**The News:** McCoy St, POB 1078, Kingstown; tel. 456-2942; fax 456-2941; e-mail thenews@caribsurf.com; weekly.

**Searchlight:** Cnr Bay and Egmont St, POB 152, Kingstown; tel. 456-1558; fax 457-2250; e-mail search@caribsurf.com; weekly on Fridays.

**The Star:** POB 854, Kingstown.

**The Vincentian:** Paul's Ave, POB 592, Kingstown; tel. 457-7430; f. 1919; weekly; owned by the Vincentian Publishing Co; Man. Dir EGERTON M. RICHARDS; Editor-in-Chief TERRANCE PARRIS; circ. 6,000.

**The Westindian Crusader:** Kingstown; tel. 456-9315; fax 456-9315; e-mail crusader@caribsurf.com; weekly.

### SELECTED PERIODICALS

**Caribbean Compass:** POB 175, Bequia; tel. 457-3409; fax 457-3410; e-mail compass@caribsurf.com; internet www.caribbeancompass.com; marine news; monthly; free distribution in Caribbean from Puerto Rico to Panama; circ. 12,000.

**Government Bulletin:** Government Information Service, Kingstown; tel. 456-3410; circ. 300.

**Government Gazette:** POB 12, Kingstown; tel. 457-1840; f. 1868; Govt Printer HAROLD LLEWELLYN; circ. 492.

**Unity:** Middle and Melville St, POB 854, Kingstown; tel. 456-1049; fortnightly; organ of the United Labour Party.

## Publishers

**CJW Communications:** Frenches Gate, Kingstown; tel. 456-2942; fax 456-2941.

**Great Works Depot:** Commission A Bldg, Granby St, POB 1849, Kingstown; tel. 456-2057; fax 457-2055; e-mail gwd@caribsurf.com.

**The Vincentian Publishing Co Ltd:** St George's Place; POB 592, Kingstown; tel. 456-1123; Man. Dir EGERTON M. RICHARDS.

## Broadcasting and Communications

### TELECOMMUNICATIONS

#### Regulatory Authority

**Eastern Caribbean Telecommunications Authority:** based in Castries, Saint Lucia; f. 2000 to regulate telecommunications in Saint Vincent and the Grenadines, Dominica, Grenada, Saint Christopher and Nevis and Saint Lucia.

#### Major Service Providers

**Cable & Wireless (WI) Ltd:** Halifax St, POB 103, Kingstown; tel. 457-1901; fax 457-2777; e-mail svdinfo@caribsurf.com; CEO DARYL JACKSON.

**Cable & Wireless Caribbean Cellular:** Halifax St, Kingstown; tel. 457-4600; fax 457-4940; cellular telephone service.

**Cingular Wireless Saint Vincent and the Grenadines:** Kingstown; internet www.cingularwireless.com; joint venture between SBC Communications and BellSouth; scheduled to sell its operations and licences in the Caribbean to Digicel in 2005; Pres. and CEO STANLEY T. SIGMAN.

**Digicel:** Suite KO59, cnr Granby and Sharpe Sts, Kingstown; tel. 453-3000; fax 453-3010; e-mail customercaresvg@digicelgroup.com; internet www.digicelsvg.com; f. 2003; mobile cellular phone operator, owned by an Irish consortium; Chair. DENIS O'BRIEN.

### BROADCASTING

**National Broadcasting Corporation of Saint Vincent and the Grenadines:** Dorsetshire Hill, POB 617, Kingstown; tel. 456-1078; fax 456-1015; e-mail svgbc@caribsurf.com; internet www.nbcsvg.com; govt-owned; Chair. ST CLAIR LEACOCK.

#### Radio

**Radio 705:** National Broadcasting Corpn, Richmond Hill, POB 705, Kingstown; tel. 457-1111; fax 456-2749; e-mail nbcsvg@caribsurf.com; internet www.nbcsvg.com; commercial; broadcasts BBC World Service (United Kingdom) and local programmes.

#### Television

**National Broadcasting Corporation of Saint Vincent and the Grenadines:** see Broadcasting.

**SVG Television:** Dorsetshire Hill, POB 617, Kingstown; tel. 456-1078; fax 456-1015; e-mail svgbc@caribsurf.com; broadcasts US and local programmes; Chief Engineer R. P. MACLEISH.

Television services from Barbados can be received in parts of the islands.

## Finance

(cap. = capital; res = reserves; dep. = deposits; m. = million; brs = branches)

### BANKING

The Eastern Caribbean Central Bank, based in Saint Christopher, is the central issuing and monetary authority for Saint Vincent and the Grenadines.

**Eastern Caribbean Central Bank—Saint Vincent and the Grenadines Office:** Granby St, POB 839, Kingstown; tel. 456-1413; fax 456-1412; e-mail eccbsvg@caribsurf.com; Country Dir MAURICE EDWARDS.

#### Regulatory Authority

**Financial Intelligence Unit (FIU):** POB 1826, Kingstown; tel. 451-2070; fax 457-2014; e-mail svgfiu@vincysurf.com; internet www.stvincentoffshore.com/fin_intl_unit.htm; f. 2002; Dir SHARDA SINANAN-BOLLERS.

#### Principal Banks

**First Saint Vincent Bank Ltd:** Lot 112, Granby St, POB 154, Kingstown; tel. 456-1873; fax 456-2675; f. 1988; fmrly Saint Vincent Agricultural Credit and Loan Bank; Man. Dir O. R. SYLVESTER.

**National Commercial Bank (SVG) Ltd:** Bedford St, POB 880, Kingstown; tel. 457-1844; fax 456-2612; e-mail info@ncbsvg.com; internet www.ncbsvg.com; f. 1977; govt-owned; share cap. EC $14.0m., dep. EC $338.4m., total assets EC $366.4m. (June 2000); Chair. RICHARD JOACHIM; Man. DIGBY AMBRIS; 8 brs.

**RBTT Bank Caribbean Ltd:** 81 South River Rd, POB 81, Kingstown; tel. 456-1501; fax 456-2141; internet www.rbtt.com; f. 1985 as Caribbean Banking Corporation Ltd; Chair. PETER J. JULY.

**Saint Vincent Co-operative Bank:** Cnr Long Lane (Upper) and South River Rd, POB 886, Kingstown; tel. 456-1894; fax 457-2183.

#### Foreign Banks

**Bank of Nova Scotia Ltd** (Canada): 76 Halifax St, POB 237, Kingstown; tel. 457-1601; fax 457-2623; Man. S. K. SUBRAMANIAM.

**FirstCaribbean International Bank (Barbados) Ltd:** Kingstown; internet www.firstcaribbeanbank.com; f. 2002 following merger of Caribbean operations of Barclays Bank PLC and CIBC; Exec. Chair. MICHAEL MANSOOR; CEO CHARLES PINK.

### STOCK EXCHANGE

**Eastern Caribbean Securities Exchange:** e-mail info@ecseonline.com; internet www.ecseonline.com; based in Basseterre, Saint Christopher and Nevis; f. 2001; regional securities market designed to facilitate the buying and selling of financial products for the eight member territories—Anguilla, Antigua and Barbuda, Dominica, Grenada, Montserrat, Saint Christopher and Nevis, Saint Lucia and Saint Vincent and the Grenadines; Gen. Man. TREVOR BLAKE.

### INSURANCE

A number of foreign insurance companies have offices in Kingstown. Local companies include the following:

# SAINT VINCENT AND THE GRENADINES

**Abbott's Insurance Co:** Cnr Sharpe and Bay St, POB 124, Kingstown; tel. 456-1511; fax 456-2462.

**BMC Agencies Ltd:** Sharpe St, POB 1436, Kingstown; tel. 457-2041; fax 457-2103.

**Durrant Insurance Services:** South River Rd, Kingstown; tel. 457-2426.

**Haydock Insurances Ltd:** Granby St, POB 1179, Kingstown; tel. 457-2903; fax 456-2952.

**Metrocint General Insurance Co Ltd:** St George's Place, POB 692, Kingstown; tel. 456-1821.

**Saint Hill Insurance Co Ltd:** Bay St, POB 1741, Kingstown; tel. 457-1227; fax 456-2374.

**Saint Vincent Insurances Ltd:** Lot 69, Grenville St, POB 210, Kingstown; tel. 456-1733; fax 456-2225; e-mail vinsure@caribsurf.com.

### 'OFFSHORE' FINANCIAL SECTOR

Legislation permitting the development of an 'offshore' financial sector was introduced in 1976 and revised in 1996 and 1998. International banks are required to have a place of business on the islands and to designate a licensed registered agent. International Business Companies registered in Saint Vincent and the Grenadines are exempt from taxation for 25 years. Legislation also guarantees total confidentiality. By December 2004 the 'offshore' financial centre comprised 6,276 International Business Companies and 11 banks; in addition, by May 2003, there were five mutual funds, 413 trusts, six mutual-fund managers and three international insurance companies.

**International Financial Services Authority (IFSA):** Browne's Business Centre, 2nd Floor, Grenville St, POB 356, Kingstown; tel. 456-2577; fax 457-2568; e-mail info@stvincentoffshore.com; internet www.stvincentoffshore.com; f. 1996; Chair. CHRISTIAN IVOR MARTIN; Offshore Finance Inspector and CEO S. LOUISE MITCHELL.

**Saint Vincent Trust Service:** Trust House, 112 Bonadie St, POB 613, Kingstown; tel. 457-1027; fax 457-1961; e-mail trusthouse@saint-vincent-trust.com; internet www.saint-vincent-trust.com; br. in Liechtenstein; Pres. BRYAN JEEVES.

## Trade and Industry

### DEVELOPMENT ORGANIZATION

**Saint Vincent Development Corporation (Devco):** Grenville St, POB 841, Kingstown; tel. 457-1358; fax 457-2838; e-mail devco@caribsurf.com; f. 1970; finances industry, agriculture, fisheries, tourism; Chair. SAMUEL GOODLUCK; Man. CLAUDE M. LEACH.

### CHAMBER OF COMMERCE

**Saint Vincent and the Grenadines Chamber of Industry and Commerce (Inc):** Hillsboro St, POB 134, Kingstown; tel. 457-1464; fax 456-2944; e-mail svgcic@caribsurf.com; internet www.svgcic.com; f. 1925; Pres. MARTIN LABORDE; Exec. Dir LEROY ROSE.

### INDUSTRIAL AND TRADE ASSOCIATION

**Saint Vincent Marketing Corporation:** Upper Bay St, POB 872, Kingstown; tel. 457-1603; fax 456-2673; f. 1959; Man. M. DE FREITAS.

### EMPLOYERS' ORGANIZATIONS

**Saint Vincent Arrowroot Industry Association:** Upper Bay St, Kingstown; tel. 457-1511; f. 1930; producers, manufacturers and sellers; 186 mems; Chair. GEORGE O. WALKER.

**Saint Vincent Banana Growers' Association:** Sharpe St, POB 10, Kingstown; tel. 457-1605; fax 456-2585; f. 1955; over 7,000 mems; Chair. LESLINE BEST; Gen. Man. HENRY KEIZER.

**Saint Vincent Employers' Federation:** Middle St, POB 348, Kingstown; tel. 456-1269; e-mail svef@caribsurf.com; Dir ST CLAIR LEACOCK.

### UTILITIES

#### Electricity

**Saint Vincent Electricity Services Ltd (VINLEC):** Paul's Ave, POB 856, Kingstown; tel. 456-1701; fax 456-2436; internet www.vinlec.com; Chair. CLAUDE SAMUEL; CEO JOEL HUGGINS.

#### Water

**Central Water and Sewerage Authority (CWSA):** New Montrose, POB 363, Kingstown; tel. 456-2946; fax 456-2552; e-mail cwsa@caribsurf.com; f. 1961.

### CO-OPERATIVES

There are 26 Agricultural Credit Societies, which receive loans from the Government, and five Registered Co-operative Societies.

### TRADE UNIONS

**Commercial, Technical and Allied Workers' Union (CTAWU):** Lower Middle St, POB 245, Kingstown; tel. 456-1525; fax 457-1676; f. 1962; affiliated to CCL, ICFTU and other international workers' organizations; Pres. ALICE MANDEVILLE; Gen. Sec. LLOYD SMALL; 2,500 mems.

**National Labour Congress:** POB 1290, Kingstown; tel. 457-1950; five affiliated unions; Chair. FITZ JONES.

**National Workers' Movement:** Grenville St, POB 1290, Kingstown; tel. 457-1950; fax 456-2858; e-mail natwok@caribsurf.com; Gen. Sec. NOEL C. JACKSON.

**Public Services Union of Saint Vincent and the Grenadines:** McKies Hill, POB 875, Kingstown; tel. 457-1950; fax 456-2858; e-mail psuofsvg@caribsurf.com; f. 1943; Pres. CONRAD SAYERS; Exec. Sec. ROBERT I. SAMUEL; 738 mems.

**Saint Vincent Union of Teachers:** POB 304, Kingstown; tel. 457-1062; e-mail svgtu@caribsurf.com; f. 1952; members of Caribbean Union of Teachers affiliated to FISE; Pres. TYRONE BURKE; 1,250 mems.

## Transport

### RAILWAYS

There are no railways in the islands.

### ROADS

In 2002 there was an estimated total road network of 829 km (515 miles), of which 580 km (360 miles) was paved.

### SHIPPING

The deep-water harbour at Kingstown can accommodate two ocean-going vessels and about five motor vessels. There are regular motor-vessel services between the Grenadines and Saint Vincent. Geest Industries, formerly the major banana purchaser, operated a weekly service to the United Kingdom. Numerous shipping lines also call at Kingstown harbour. Some exports are flown to Barbados to link up with international shipping lines. A new marina and shipyard complex at Ottley Hall, Kingstown, was completed during 1995. A new container port at Campden Park, near Kingstown, was opened in the same year. A new dedicated Cruise Terminal opened in 1999, permitting two cruise ships to berth at the same time.

**Saint Vincent and the Grenadines Port Authority:** POB 1237, Kingstown; tel. 456-1830; fax 456-2732; e-mail svgport@caribsurf.com; internet www.svgpa.com.

### CIVIL AVIATION

There is a civilian airport, E. T. Joshua Airport, at Arnos Vale, situated about 3 km (2 miles) south-east of Kingstown. The islands of Mustique and Canouan have landing strips for light aircraft only, although a project to upgrade Canouan Airport began in May 1997. In late 1999 the Government began the process of raising finance for the extension of E. T. Joshua Airport in order to permit the use of long-haul jet aircraft. The project was still in progress in 2005. In April 2002 Kuwait provided a US $8m. loan for the expansion of the islands' international airport facilities.

**American Eagle:** POB 1232, E. T. Joshua Airport, Arnos Vale; tel. 456-5555; fax 456-5616.

**Mustique Airways:** POB 1232, E. T. Joshua Airport, Arnos Vale; tel. 456-4380; fax 456-4586; e-mail info@mustique.com; internet mustique.com; f. 1979; charter flights; Chair. JONATHAN PALMER.

**Saint Vincent and the Grenadines Air (1990) Ltd** (SVG Air): POB 39, E. T. Joshua Airport, Arnos Vale; tel. 457-5124; fax 457-5077; e-mail info@svgair.com; internet www.svgair.com; f. 1990; charter and scheduled flights; CEO J. E. BARNARD.

## Tourism

The island chain of the Grenadines is the country's main tourist asset. There are superior yachting facilities, but the lack of major air links with countries outside the region has resulted in a relatively slow development for tourism. In 2004 Saint Vincent and the Grenadines received 74,700 cruise-ship passengers and 86,700 stop-over tourists. Tourist receipts were estimated to total US $82.4m. in 2003. There were 1,762 hotel rooms on the islands in 2001.

**Department of Tourism:** Cruise Ship Terminal, POB 834, Kingstown; tel. 457-1502; fax 451-2425; e-mail tourism@caribsurf.com; internet www.svgtourism.com; Dir VERA-ANN BRERETON.

**Saint Vincent and the Grenadines Hotel and Tourism Association:** E. T. Joshua Airport; tel. 458-4379; fax 456-4456; e-mail office@svghotels.com; internet www.svghotel.com; Exec. Dir DAWN SMITH.

## Defence

Saint Vincent and the Grenadines participates in the US-sponsored Regional Security System, comprising police, coastguards and army units, which was established by independent Eastern Caribbean states in 1982. Since 1984, however, the paramilitary Special Service Unit has had strictly limited deployment. Government current expenditure on public order and safety was estimated at EC $20.8m. (10.2% of total current expenditure) in 1998.

## Education

Free primary education, beginning at five years of age and lasting for seven years, is available to all children in government schools, although it is not compulsory and attendance is low. There are 60 government, and five private, primary schools. Secondary education, beginning at 12 years of age, comprises a first cycle of five years and a second, two-year cycle. However, government facilities at this level are limited, and much secondary education is provided in schools administered by religious organizations, with government assistance. There are also a number of junior secondary schools. There is a teacher-training college and a technical college. In 1994/95 76% and 24% of children in the relevant age-groups were attending primary and secondary schools, respectively. Current expenditure on education by the central Government was a projected EC $46.4m. in 1998 (22.7% of total current expenditure). The Goverment was committed to achieving universal secondary education by 2010.

# SOUTH GEORGIA AND THE SOUTH SANDWICH ISLANDS

South Georgia, an island of 3,592 sq km (1,387 sq miles), lies in the South Atlantic Ocean, about 1,300 km (800 miles) east-south-east of the Falkland Islands. The South Sandwich Islands, which have an area of 311 sq km (120 sq miles), lie about 750 km (470 miles) south-east of South Georgia.

The United Kingdom annexed South Georgia and the South Sandwich Islands in 1775. With a segment of the Antarctic mainland and other nearby islands (now the British Antarctic Territory), they were constituted as the Falkland Islands Dependencies in 1908. Argentina made formal claim to South Georgia in 1927, and to the South Sandwich Islands in 1948. In 1955 the United Kingdom unilaterally submitted the dispute over sovereignty to the International Court of Justice (based in the Netherlands), which decided not to hear the application in view of Argentina's refusal to submit to the Court's jurisdiction. South Georgia was the site of a British Antarctic Survey base (staffed by 22 scientists and support personnel) until it was invaded in April 1982 by Argentine forces, who occupied the island until its recapture by British forces three weeks later. The South Sandwich Islands were uninhabited until the occupation of Southern Thule in December 1976 by about 50 Argentines, reported to be scientists. Argentine personnel remained until removed by British forces in June 1982.

In mid-1989 it was reported that the British Government was considering the imposition of a conservation zone extending to 200 nautical miles (370 km) around South Georgia, similar to that which surrounds the Falkland Islands, in an attempt to prevent the threatened extinction of certain types of marine life.

Under the provisions of the South Georgia and South Sandwich Islands Order of 1985, the islands ceased to be governed as dependencies of the Falkland Islands on 3 October 1985. The Governor of the Falkland Islands is, *ex officio*, Commissioner for the territory.

In May 1993, in response to the Argentine Government's decision to commence the sale of fishing licences for the region's waters, the British Government announced an extension, from 12 to 200 nautical miles, of its territorial jurisdiction in the waters surrounding the islands, in order to conserve crucial fishing stocks.

In September 1998 the British Government announced that it would withdraw its military detachment from South Georgia in 2000, while it would increase its scientific presence on the island with the installation of a permanent team from the British Antarctic Survey to investigate the fisheries around the island for possible exploitation. The small military detachment finally withdrew in March 2001. The British garrison stationed in the Falkland Islands would remain responsible for the security of South Georgia and the South Sandwich Islands. In July 2005 it was announced that Alan Huckle would succeed Howard Pearce as Governor of the Falkland Islands and Commissioner of South Georgia and the South Sandwich Islands from 2006.

**Commissioner:** HOWARD PEARCE (assumed office 3 December 2002).
**Assistant Commissioner and Director of Fisheries:** HARRIET HALL (Stanley, Falkland Islands).

# SURINAME

## Geography

### PHYSICAL FEATURES

The Republic of Suriname is on the north coast of South America and is the smallest country on the continent. Until independence in 1975 Suriname (usually rendered Surinam in English) was known as Dutch or Netherlands Guiana, and it is flanked by the smaller territory of French Guiana (part of France) on the east and Guyana (formerly British Guiana) on the west. Brazil lies beyond a 597-km (371-mile) border in the south. The Atlantic coast is 386 km in extent, while the border with French Guiana is 510 km and that with Guyana is 600 km. However, Suriname disputes the current course of the western and eastern borders and maintains territorial claims on its neighbours on the Guianese coast. Suriname claims part of south-western French Guiana, between the Itany (Litani—the current border) and the Marowijne (Marouini) rivers. In the west, the country claims territory in south-eastern Guyana as far as the New River and not along the current upper reach of the Corantijn (Courantyne) used as a border (Koetari or Kutari). Beyond the mouth of the Corantijn there is also a dispute with Guyana about maritime lines of demarcation. The country covers an area of 163,265 sq km (63,037 sq miles).

The north of Suriname consists of a typically Guianan coastal strip, with rich, alluvial soil made usable and habitable by extensive dykes and irrigation systems. These plains can be as much as 2 m below sea level, so, where not protected or reclaimed, the coastlands tend to be swampy mangrove country. The Atlantic littoral, up to 80 km in width, accounts for about 16% of the country, and is where most of the population lives. Between the plains and the densely forested interior is an intermediate plateau with a landscape alternating tracts of savannah with dunes or woodland. This gives way to a more solid forest cover, a region that accounts for about three-quarters of the country, as rolling hills climb into more mountainous terrain. The highlands of the forest region rear relatively abruptly once away from the coastal belt and its immediate hinterland, so that the country's highest point, at Juliana Top (1,230 m or 4,037 ft) is located near the centre of Suriname. The mountain is just beyond the north end of the Eilerts de Haan Gebergste (a ridge thrusting up from the south and marking off the south-west) and to the west of the Wilhelmina Gebergste; the main range of the south-east is the Oranje Gebergste. Much of this territory is dense tropical rainforest, which hosts a considerable diversity of flora and fauna (although deforestation and the pollution from small-scale mining threaten the environment—the country's timber and mineral reserves, notably bauxite, are the most lucrative natural resources). Finally, in the far south-west, there is a region of high savannah. This entire landscape is laced with numerous rivers, some feeding large reservoirs. For instance, the Professor W. J. van Blommestein Meer, in the north-east, on the edge of the forested region, is one of the largest reservoirs in the world. Dams contain a number of other lakes, and inland waters cover 1,800 sq km of the country's territory. The main rivers, apart from the Corantijn (700 km in length—most of it marking the border with Guyana) and the Marowijne (720 km—also defining much of a border, but with French Guiana, and including its upper reaches, such as the Lawa and its tributaries), are the Suriname, obviously, and the Coppename and the Saramacca.

### CLIMATE

Suriname experiences high rainfall, humidity and temperatures, although its tropical climate is tempered by the Atlantic trade winds. There are two wet seasons, April–August and November–February, although neither wet nor dry seasons are absolute. The average annual precipitation in Paramaribo, the capital, which is on the coast, is over 2,200 mm (86 ins). Inland, rainfall is more like 1,500 mm per year. Average temperatures

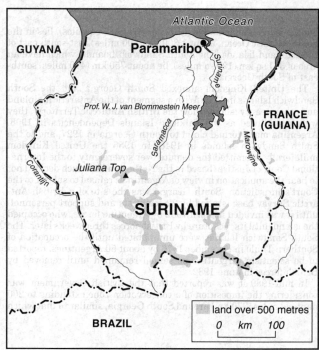

range between 23°C and 32°C (73°–90°F), but there is little seasonal variation.

### POPULATION

Suriname is one of the most racially diverse countries in the Americas or, indeed, in the world. Recent estimates put the 'Hindustani' (descendants of indentured workers brought from the Indian subcontinent in the 19th century, generally referred to as 'East' Indians in the Caribbean) population at 37% of the total, Creoles (of black and mixed-black descent) 31%, the Indonesian-descended 'Javanese' 15%, the predominantly black Boschnegers (Bush Negroes or Maroons—descendants of slaves who escaped into freedom in the mountainous interior in the 17th and 18th centuries, preserving more of their African heritage than the Creoles) 10%, indigenous Amerindians 3%, Chinese 2% and whites 1%. Dutch is the official language for all these groups, but many others are also still spoken—not only a Hindi dialect (known as Hindustani—Sarnami Hindi) and Javanese, but also some English-based Creoles, the main dialect being known as Sranan Tongo or Taki-Taki (Djukka and Saramaccan are others); English itself is widely known in the cities or among the better educated. The cosmopolitan nature of Surinamese society is also seen in the country's variety of religious affiliations—while just under one-half the people are Christian (the largest denominations are the Roman Catholics, with 23% of the population, and the Moravian Brethren, with 15%, followed by the Dutch Reformed and other Protestants), 27% are Hindu and 22% Muslim. Others practise native or adapted African tribal beliefs, although these are often not considered incompatible with more orthodox faiths.

Most of the population live on the coast and about 65% are classed as urban, with somewhere between one-third and one-half in or around the capital city. Creoles tend to dominate the urban areas and Hindustanis the countryside. However, over recent decades many people have emigrated to the Netherlands. The total population was estimated at 487,024 in mid-2004. Paramaribo, the country's only major urban area, its main seaport and the national capital, is east of centre on the coast, in the north of Suriname. There are 10 districts.

# History

## JAMES MCDONOUGH
Revised for this edition by the editorial staff

In the 15th century the only inhabitants of Suriname were Carib, Arawak and Awarao Indians (Amerindians). Another tribe, the Surinas, who inhabited the country at an earlier time, is considered the source of the name Suriname. In 1499 Alonso de Ojeada, a Spanish lieutenant serving the Italian navigator Amerigo Vespucci, landed on the north-eastern coast of South America, which was called Guiana by the Amerindians. The Spanish claimed possession of the coast, but no actual settlement was attempted. During the next century the Dutch began establishing trading posts along the Commewijn and Corantijn rivers (now in Suriname), and later along the Essequibo and Berbice (now in Guyana). The French were attempting to establish settlements along the Cayenne River (now in French Guiana). During this period, lumbering and tobacco farming were the chief commercial activities. It was the English, however, who founded the first successful colony in Suriname, as the result of an expedition financed by Lord Francis Willoughby, the colonial Governor of the flourishing but overcrowded English sugar island of Barbados. The group of English planters and their slaves established a large settlement on the Suriname River, near what is now Paramaribo. The British Crown ceded its Suriname colony to the Netherlands in the Treaty of Breda (1667), in exchange for the colony of Nieuw Amsterdam (now New York, USA). The colony remained under Dutch rule for the next 300 years, except for two brief periods of British control in 1799–1802 and 1804–14, during the Napoleonic wars.

The territory became known as Dutch Guiana, and was flanked to the west by British Guiana (now Guyana) and to the east by French Guiana. The colony was administered by a Governor, with the assistance of the Political Council, the members of which were appointed by the Governor, following nomination by the colonial planter class. In 1828 the administration of all Dutch West Indies colonies was centralized under a Governor-General, stationed in Suriname, who reported directly to the Colonial Office in the Netherlands. During this early period the colony flourished on the basis of large, Dutch-owned sugar plantations, worked by African slave labour. Between 1650 and 1820 some 300,000 West African slaves were brought into Suriname. Nevertheless, the plantations suffered a continual labour drain owing to escaping slaves, who would seek refuge from the authorities in the vast and underdeveloped interior of the country. By 1728 these runaway slaves, known as 'Boschnegers' (Bush Negroes) or maroons, had established a number of settlements based on African tribal customs and were warring with the white plantation owners and the colonial authorities. Expeditions were sent into the jungle to subdue them, but without success. Finally, in 1761 the Dutch signed a treaty with the Boschnegers, guaranteeing their liberty and supplying them with yearly shipments of arms. In return, the Boschnegers promised to return all future runaway slaves and never to appear in Paramaribo in armed groups of more than six persons. From that time, the Boschnegers led an isolated, independent life in the Suriname interior.

The abolition of slavery in neighbouring British Guiana in 1834 and in French Guiana in 1848 produced a period of unrest among Suriname's slaves. These events led King William III of the Netherlands finally to abolish slavery in 1863. To solve the problem of labour shortage created by the abolition of slavery, the Dutch turned to overseas contract or indentured labour. Between 1873 and 1917 some 37,000 indentured labourers were brought to Suriname from India. A similar influx of contract labourers, numbering about 33,000, were brought from the Dutch East Indies (now Indonesia) between 1893 and 1939. Furthermore, the Dutch encouraged the immigration of Chinese, Portuguese and, later, Lebanese workers. Suriname's ethnic and racial make-up reflected the plantation colony's historic need for cheap labour. The census in 1980 recorded that 34.70% of Suriname's population were Creole (urban dwellers of African descent), 33.49% were of Indian descent (known locally as Hindustani), 16.33% were Indonesian-descended 'Javanese', 9.55% Boschneger, 3.10% Amerindian and 1.55% Chinese. The country also had European and other minorities, amounting to about 1.28% of the population. Suriname comprised seven different ethnic groups, speaking more than 15 languages. It was the most fragmented country in the Latin American and Caribbean region and among the 20 most fragmented countries in the world. This ethnic fragmentation also existed in the political arena, where the majority of political parties were organized along ethnic lines.

## INDEPENDENCE AND POLITICAL DEVELOPMENT

In 1866 the Koloniale Staten (Colonial Assembly, also known as the Staten van Suriname), was established. A representative body with limited local power, its members were elected from a small group of colonial planters, who were extended the franchise on the basis of a poll tax. While ultimate power continued to reside in The Hague, the Koloniale Staten remained the principal administrative body in Suriname until the colony gained independence. In the 20th century the exploitation of Suriname's large bauxite reserves and the cultivation of rice replaced sugar as the principal foreign-exchange earner, although the Dutch Government found it necessary to subsidize an ever-increasing share of the colony's budget. In 1950 the Dutch Government granted Suriname internal self-government. Then in 1954, Suriname became an overseas territory of the Dutch 'Tripartite Kingdom', composed of the Netherlands, the Netherlands Antilles (now the Netherlands Antilles and Aruba) and Suriname. Full and complete independence was granted to Suriname by the Dutch on 25 November 1975.

Local political parties began forming during the Second World War, at the time of the promise of local autonomy. Further political participation was stimulated by the introduction of universal suffrage for the general election of 1949. The Nationale Partij Suriname (NPS—Suriname National Party), representing the country's Creole population, won the majority of the seats in the Koloniale Staten, under the leadership of Johan Pengel. During the 1950s the Verenigde Hindostaanse Partij (VHP—United Hindustani Party), representing the Hindustani population, gained prominence under the leadership of Jaggernath Lachmon. The NPS and VHP formed an alliance during the 1960s, which gave Suriname a long period of stability under the principle of ethnic *verbroedering* (fraternization). In the 1973 general election the NPS formed the Nationale Partij Kombinatie (NPK—Combined National Party) alliance with three other parties and won 22 of 39 seats in the Koloniale Staten. Henck Arron, who had replaced Pengel as party leader upon the latter's death in 1970, was named Prime Minister and was in power when the country gained independence in 1975.

Dutch aid, worth 3,500m. guilders, gave considerable support to the economy of the new republic. However, the international economic recession of the mid-1970s, brought on by the petroleum crisis, and the fall of the world price of bauxite, caused growing concern that Suriname would be unable to promote economic development, despite the country's large natural resources and relatively small population. Moreover, more than 40,000 persons, mostly the well-educated and well-trained, emigrated to the Netherlands on the eve of independence, in order to qualify for Dutch citizenship. A series of strikes underlined the growing dissatisfaction of the people, while corruption scandals involving cabinet ministers undermined the Government. Nevertheless, in the general election of 1977, the NPK again won 22 seats, with the remaining 17 seats going to a left-wing opposition front, led by Lachmon and the VHP.

## MILITARY TAKE-OVER, 1980–87

On 25 February 1980 the Armed Forces took control of government in a *coup d'état*. The coup followed the civilian Government's refusal to recognize demands of members of the military to form a trade union. The take-over was led by a junior army officer, Sgt-Maj. (later Lt-Col) Désiré (Desi) Bouterse, who seized power in alliance with the left-wing Partij Nationalistiche Republiek (PNR—Nationalist Republican Party). Dr Henk Chin-A-Sen, a PNR leader, was chosen as Prime Minister, presiding over a PNR-assembled Government and the eight-member Nationale Militaire Raad (NMR—National Military Council) named by Bouterse. In August 1980, following a disagreement over policy, Bouterse strengthened his control over the Government by dissolving the legislature and declaring a state of emergency. In March 1981 Sgt-Maj. Wilfred Hawker led an unsuccessful Hindustani-inspired, right-wing coup against Bouterse. In December Bouterse launched the Revolutionaire Volksfront (Revolutionary People's Front) and in February 1982 Chin-A-Sen, who earlier had been named President, was dismissed along with his civilian Government. In March, a second coup attempt by Sgt-Maj. Hawker failed, resulting in his execution.

As a result of the coup attempt, Bouterse declared a state of siege and imposed martial law. However, to prevent the Netherlands from suspending aid under the terms of the independence treaty, the military regime appointed a 12-member Council of Ministers with a civilian majority, and Henry Neyhorst, a moderate economist, became Prime Minister. The failure to solve Suriname's economic difficulties lost Bouterse the support of the left-wing groups and the trade unions, and soon the country was plagued by strikes, demonstrations, and calls for an end to military rule. Bouterse promised to hold elections for a constituent assembly to draft a new constitution. On 8 December 1982, members of the armed forces burned down Paramaribo offices of the Bouterse opposition. In the ensuing disturbances some 15 leading politicians, trade unionists, lawyers, journalists and academics were killed, in what became known as the 'December Murders'.

In response to the December Murders, the Dutch Government suspended its large aid programme to the country. The USA and the European Community (EC—known as the European Union from 1993) immediately followed suit and the Council of Ministers resigned. Bouterse, however, retained the loyalty of the 3,000-man military by dismissing two-thirds of the officer corps. In February 1983 he formed a new civilian-military Council of Ministers, with Dr Errol Alibux (a former Minister of Social Affairs) as Prime Minister. The new Government was composed of two left-wing parties, the Progressieve Arbeiders en Landbouwers Unie (PALU—Progressive Workers' and Farm Labourers' Union) and the Revolutionaire Volkspartij (Revolutionary Peoples' Party).

In foreign affairs, the Bouterse Government followed a non-alignment policy, establishing close relations with Cuba and Libya, to balance the historically close links with the USA and the Netherlands. These moves alarmed both the French and the US authorities. The French saw potential danger to its Kourou space centre, the launching site for the European Ariane rocket, which was located close to the Surinamese border in French Guiana. The USA, determined to stop the spread of 'Communist' governments in the western hemisphere, was very wary of Suriname's growing ties with Cuba and the large presence of Cuban advisers in the country. George Schultz, former US Secretary of State under President Ronald Reagan (1981–89), revealed in a 1993 memoir that the US Administration had been ready to intervene militarily as a result of the 1982 December Murders; however, US plans for the military overthrow of the Bouterse regime were abandoned after the Netherlands refused to participate.

## CIVILIAN RESTORATION

In August 1984 the state of emergency imposed in 1982 was lifted, as the military Government began to move the country towards civilian rule. In December plans were announced for the formation of a supreme deliberating council, the Topberaad, consisting of representatives of the trade unions, the business sector and Standvaste, a new movement Bouterse had established in November 1983 as a political power base. The Topberaad met in January 1985, with the main task of drafting a new constitution. In March 1987 a draft document consisting of some 186 articles was completed. The Constitution was approved by referendum in September of that year, and a general election was held in November, for the first time in eight years.

In July 1987 Standvaste was reconstituted as the Nationale Democratische Partij (NDP—National Democratic Party), under the leadership of Jules Wijdenbosch. Wijdenbosch was Prime Minister in the last Bouterse-appointed Cabinet of Ministers prior to the November election. In August the three major opposition parties, the Creole NPS, the Hindustani VHP and the Kaum-Tani Persuatan Indonesia (KTPI—Javanese Indonesian Farmers' Union), formed an electoral alliance, the Front voor Demokratie en Ontwikkeling (FDO—Front for Democracy and Development). With the restoration of electoral politics, Suriname's ethnic parties, which had dominated the political scene prior to the 1980 military coup, returned to prominence. At the November election the FDO won a decisive victory, taking 40 seats in the new 51-seat National Assembly, with the PALU. the Progressieve Bosneger Partij (PBP—Progressive Boschneger Party) and the NDP sharing out the remaining seats. The National Assembly took over political control in January 1988 and unanimously elected Ramsewak Shankar of the VHP to the presidency. Henck Arron of the NPS was elected Vice-President and Prime Minister, heading a 14-member Cabinet of Ministers.

## THE BOSCHNEGER REVOLT

The return to constitutional rule did not end Suriname's internal conflicts. From 1986 the military was fighting against a Boschneger insurgency in the interior of the country, which threatened the successful move to constitutional government. The insurgency was led by Ronnie Brunswijk, a Boschneger who was once a member of the presidential bodyguard. He claimed that plans to develop the interior of the country violated the autonomy of the Boschneger society, which had been guaranteed by the 1761 treaty and subsequent agreements signed between the Boschneger and the colonial Dutch. In 1987 Brunswijk's Surinamese Liberation Army (SLA—popularly known as the Jungle Commando) attacked economic targets, causing severe disruption, including the closure of the main bauxite smelting and refining plants.

Bouterse retaliated against the insurgency with raids into the interior; the rebels claimed that the army massacred Boschnegers in several interior villages. The army also moved to arm about 1,000 Amerindians, leading to armed clashes between the Boschneger and the Tucayana Amerindians. As a result of the fighting, some 10,000 Boschnegers took refuge in French Guiana and the French reinforced the border with Suriname with paratroopers and legionnaires, refusing to let the Suriname military pursue the insurgents across the border.

In June 1988 negotiations began between the Government and representatives of the SLA. Bouterse's willingness to negotiate was attributed by many to the announcement that there would be an amnesty for personnel from both sides involved in the conflict, which the SLA claimed would prevent investigation of alleged abuses of human rights by the army. Nevertheless, in July 1989 the SLA and the Suriname Government signed a peace accord at Kourou (French Guiana). The provisions of the so-called Kourou Accord included a general amnesty for those involved in the conflict, the ending of the state of emergency established in December 1986, the incorporation of members of the SLA into the national police and measures to provide for the safe return of the Surinamese refugees in French Guiana. However, the Accord failed when Bouterse vetoed the clause demanding the integration of Brunswijk's fighting force into the national police, and the Amerindians refused to abide by its terms. In addition, the Boschneger refugees refused to move back to Suriname, fearing reprisals by either the army or the Amerindians once they left the protection of French territory.

The Kourou Accord included proposals for a Consultative Council for the Development of the Interior. Much of the interior's infrastructure had been destroyed and development suspended during the insurgency, but it was not until late 1995 that the Council was appointed. The Government failed, however, to consult the Boschneger and Amerindian representatives about the granting of gold and timber concessions on their land. The Boschneger and Amerindians complained that small-scale mining operations, mainly by illegal Brazilian gold miners, were disrupting tribal and community life. They were concerned in particular about the damaging effects on the food chain by the gold miners' widespread use of mercury.

## THE MILITARY INTERREGNUM

In 1990 the US Department of State noted, in its annual report to the US Congress on human rights, that the Surinamese military had 're-established itself as the dominant political force in the country'. Only a few of the 120 new laws required to implement the Constitution had been passed by the Assembly, and the Constitutional Court, which was to interpret the Constitution and rule on human rights issues, had not been established. Moreover, the Government had not taken steps to deprive the military of such powers as the investigation and detention of civilians, the issue of visas and the supervision of customs and immigration at airports and harbours. Of concern to the USA was the growing military involvement in the international trafficking of illicit drugs. By the end of 1991 Western intelligence sources reported that Suriname had become a major centre of the illegal drugs trade, serving as a transhipment point for increasing quantities of cocaine intended for Europe and the USA. The country also served as a transhipment point for the sale of illegal arms to the Colombian drugs cartels. Sources in the USA and in Suriname alleged that Bouterse and the army were behind the illegal trade in drugs and arms.

In early 1990 President Shankar's Government renewed contacts with the SLA, following the failure of the Kourou Accord. With a presidential guarantee of safety, Ronnie Brunswijk travelled to Paramaribo to negotiate. However, Bouterse violated the guarantee and arrested Brunswijk. Although Brunswijk was later released on the insistence of President Shankar, the action of the military showed the weakness of the civilian Government and eventually led to its downfall. On 24 December the military overthrew the Shankar Government and installed leaders of the NDP in the executive. Jules Wijdenbosch held the posts of Vice-President, Prime Minister and Minister of Finance until the elections of May 1991. In August 1992 the new administration signed an accord with the SLA, finally ending the insurgency. The former rebels recognized the Government's authority over the entire country, while the Government promised to honour the rights of the Boschnegers, including their right to engage in gold prospecting and forestry, and to join the army. Furthermore, the Government gave assurances that programmes for economic development and social welfare would be undertaken in the interior. Nevertheless, in March 2005 former guerrillas claimed that the Government had not fulfilled its pledge to provide them with jobs and medical care.

## THE FIRST VENETIAAN GOVERNMENT

A general election was held on 25 May 1991, monitored by a delegation from the Organization of American States (OAS). The Nieuw Front coalition (NF—New Front, formerly the FDO), consisting of the dominant NPS, the VHP, the KTPI and the Surinaamse Partij van de Arbeid (SPA—Suriname Labour Party), won 30 of the 51 seats in parliament. Bouterse's NDP won 12 seats and the Democratisch Alternatief 1991 (DA '91—Democratic Alternative 1991) the remaining nine seats. The NPS leader, Runaldo Ronald Venetiaan, was elected President and Jules Adjodhia, Prime Minister.

In March 1992 the Government requested that the National Assembly remove references in the Constitution that allowed the army to act in a way that contravened the proper democratic functioning of the State. The action to curb the military was taken as a measure designed to improve relations with the Netherlands, Suriname's main international benefactor. With the restoration of democracy in 1987, the Dutch Government had renewed aid to Suriname, but under more restrictive conditions than those imposed at the time of independence. The Dutch agreed to disburse funds for only specific projects, a policy which limited the amount of overall aid. In early 1990 some US $700m., which had accumulated over the period of outright military control, had yet to be disbursed by the Netherlands. In 1992 the Netherlands agreed to renew economic assistance, but required the Government to implement the IMF's structural-adjustment programme, a stringent monetary policy that included reduced public spending, increased taxes and the removal of food and fuel subsidies. The structural-adjustment programme, implemented in 1994, proved highly successful and by 1995 the depreciation of the Suriname guilder had been halted and Central Bank reserves had reached a healthy $100m. However, the economic reforms caused widespread hardship and the Government became increasingly unpopular.

## THE WIJDENBOSCH GOVERNMENT

The results of the elections to the National Assembly, held on 23 May 1996, represented a reverse for the ruling NF, winning, as it did, fewer seats than in 1991, while Bouterse's NDP increased its legislative representation. The election was also important because Amerindians were elected to the Assembly for the first time. In an attempt to secure broader support, the NDP chose Jules Wijdenbosch, instead of Bouterse, as its candidate to contest the presidency. In the National Assembly's first vote for President, Venetiaan gained more support than Wijdenbosch, but not the two-thirds' majority necessary to win the election outright. Responsibility for electing the President then passed to the Vereinigde Volksvergadering (United People's Assembly), a body comprising national, regional and local representatives. With only a simple majority required, Wijdenbosch was elected President. He was inaugurated on 14 September. The NF alliance disintegrated, with the VHP and the KTPI joining an NDP-led coalition, on condition that Bouterse should not hold office in the new administration. The new coalition Government was appointed on 20 September, comprising representatives of four different political groupings.

The Wijdenbosch Government soon became characterized by internal political crisis and increasing pressure from diverse opposition groups. In August 1997, the dismissal of the finance minister, Motilal Mungra, following his outspoken criticism of the President's extravagant use of public funds, including the purchase of a presidential yacht, prompted Mungra's Beweging voor Vernieuwing en Democratie (BVD—Movement for Renewal and Democracy) and two other small parties to announce their withdrawal from the governing coalition. Following negotiations with the three parties concerned and other smaller opposition parties, President Wijdenbosch was able to secure sufficient support to maintain his Government's parliamentary majority. However, the Government's instability was apparent and the withdrawal of support for the administration by four members of the National Assembly precipitated a further reorganization of the coalition in February 1998. President Wijdenbosch also drew accusations of political corruption when he revealed, under considerable pressure from local and international human rights groups, that a five-member committee, appointed to investigate past human rights abuses, in particular the 1982 December Murders, was being led by a former Bouterse lawyer.

In July 1998 the President attracted further criticism, following his controversial appointment of a new President of the Court of Justice and Prosecutor-General, without consultation with, and disregarding the objections of, the sitting justices. The members of the Court refused to recognize the appointments; however, despite the objections, President Wijdenbosch named additional justices in December, and in May 1999 the appointed President of the Court of Justice swore in himself and then the new justices. Meanwhile, the Government's mismanagement of the economy was causing increasing public unrest. A strike by petroleum workers in May 1998 was followed by widespread industrial action in June, in support of the Trade and Manufacturers' Association's demands for the resignation of the Government in favour of a non-political administration.

## THE RE-ELECTION OF VENETIAAN

In early 1999 the economic situation became extremely grave, with spiralling inflation caused by an ever-widening budget deficit and a decline in the international price of bauxite, by far Suriname's most important source of foreign-exchange earnings. The Dutch Government continued to withhold the US $300m. in aid that had accumulated after it suspended payments in 1998, stating that the beneficial use of the funds by the Surinamese Government could not be assured. Under pressure from the Netherlands, in April 1999 President Wijdenbosch dismissed Bouterse, a precursor to his dismissal of his entire Cabinet of Ministers in the following month in an attempt to avoid demands for his own resignation. However, on 31 May some 30,000 protesters gathered in Paramaribo to demand President Wijdenbosch's removal, while a general strike paralysed the country. On 1 June the National Assembly passed a vote of 'no confidence' in the Government by 27 votes to 14 (with 10 abstentions). The vote, however, fell short of the two-thirds' majority needed to remove the President from office. President Wijdenbosch refused to resign, but did agree to hold new elections on 25 May 2000, one year earlier than was constitutionally required.

In the election of 25 May 2000 the NF, led by Venetiaan, secured 33 of the 51 seats in the National Assembly. The Millenium Combinatie (MC—Millennium Alliance, an alliance that included the NDP) won 10 seats, and the Democratisch Nationaal Platform 2000 (DNP—National Democratic Platform 2000), which had been formed by President Wijdenbosch in an apparent attempt to distance himself from Bouterse, took three seats. Nevertheless, Bouterse, as an NDP candidate, won a seat in the National Assembly, even as a Dutch appeals court upheld an earlier drugs-trafficking conviction (see below). As the NF narrowly failed to win the two-thirds' majority to appoint a new President directly, it immediately began coalition negotiations with smaller parties. Following the conclusion of these negotiations, on 4 August 2000 Venetiaan was elected to the presidency for the second time, winning 37 of the 51 votes cast in the National Assembly. The conflict between the Government and the judiciary precipitated by President Wijdenbosch's controversial and unilateral judicial appointments was resolved, when the disputed judges resigned. President Venetiaan also took steps to establish the independent Constitutional Court required by the 1987 Constitution.

In its electoral campaign, the NF had pledged to revitalize the faltering economy; soon after taking office, the new administration instituted a series of economic reforms aimed at reversing the failed economic policies of the Wijdenbosch administration. However, while the Venetiaan administration made progress in stabilizing the economy, tensions within the coalition and the impatience of the populace impeded its efforts.

## DECEMBER MURDERS INVESTIGATION

Following the election of Venetiaan, there were calls for the new Government to investigate the December Murders, before the 18-year statute of limitations expired in December 2000. In October the country's highest court, the Court of Justice, began hearings on the December Murders, in response to a request by relatives of the victims. The Court heard testimony from the victims' relatives, human rights activists, as well as the prosecutor's office, which had not yet made any investigation into the case. Following an order from the Court of Justice, an examining judge called for a full investigation into the Murders, including the involvement of 36 individuals. In addition, in late 2002 the Court ordered the exhumation of the remains of 15 of the murder victims. A series of burglaries at the homes of the Minister of Justice and a leading judge in April 2003 were thought to be connected to the investigation.

In early 2002 Errol Alibux, a former Prime Minister in the Wijdenbosch administration and suspect in the December Murders case, was charged with defrauding the Government while in office. The National Assembly subsequently approved a motion instigating court proceedings against him and on 5 November 2003 Alibux was sentenced to one year's imprisonment. In December 2004, following a four-year investigation, a military court indicted Bouterse and 25 other suspects for the December Murders. The trials were expected to begin in 2005.

## THE 2005 ELECTIONS

In March 2005, in advance of the indirect presidential ballot in June that was to follow the legislative election in May, the NDP formally nominated Bouterse as its candidate. The nomination prompted the USA to threaten to sever diplomatic relations with Suriname in the event of Bouterse being re-elected President. The NDP accused the USA of political interference and lodged a formal complaint with the Caribbean Community and Common Market (CARICOM) and the OAS. The other main contenders for the presidency were former President Jules Wijdenbosch of the DNP 2000 and the incumbent Venetiaan, who was seeking a third term in office. Early forecasts suggested that the ruling pro-reform coalition was generally expected to perform well in the legislavive election (and thus secure the presidency) and, moreover, enjoyed the support of the Governments of the Netherlands and the USA. Indeed, at the legislative election of 25 May, in which some 74% of the electorate participated, the NF retained its position as the largest party in the National Assembly, securing 23 of the 51 seats and 39.4% of the total votes cast. However, this was 10 seats fewer than in the previous legislature. The NDP performed well, obtaining 15 seats and 22.2% of the votes, while the Volksalliantie Voor Vooruitgang (VVV)—an alliance forged by the DNP 2000—secured just five seats and 13.8% of the votes cast. The A-Combinatie, a coalition including the Algemene Bevrijdings- en Ontwikkelingspartij, led by former guerrilla leader Ronnie Brunswijk, also secured five seats (7.2%), while the A1, a pro-business grouping containing the Democratisch Alternatief 1991, the Democraten van de 21 and the Politieke Vleugel van de FAL obtained the remaining three seats (5.9%).

The major parties subsequently entered into negotiations to form alliances with smaller groupings in an attempt to garner the two-thirds' parliamentary majority needed directly to appoint a presidential nominee. Bouterse and Brunswijk were both elected to the National Assembly. However, in June 2005 Bouterse withdrew his presidential candidacy in favour of the NDP's vice-presidential candidate, Rabin Parmessar. At the first vote in the National Assembly on 19 July no candidate garnered the requisite two-thirds' majority (34 votes); Venetiaan secured 27 votes and Parmessar obtained 20 votes. A second round of voting was scheduled to take place on 21 July; however, this was postponed after the NF alleged that Pamessar still held Dutch nationality, and was therefore ineligible for the presidency. The second round, eventually held on 26 July, was similarly unsuccessful, each candidate securing the same number of votes as previously. Responsibility for electing the new head of state subsequently passed to the 891-member United People's Assembly. With only a simple majority required, on 3 August Venetiaan won 560 votes, compared with 315 votes for Parmessar. Venetiaan was inaugurated for a third term in office on 12 August. Despite securing a convincing majority in the final round of voting, it was likely that the fragmented nature of the legislature would require Venetiaan to rely on coalition-building in order to implement new policies.

## INTERNATIONAL RELATIONS

Relations with the Government of the Netherlands deteriorated rapidly under President Wijdenbosch, principally owing to his administration's continued links with Bouterse. In 1997 the President appointed Bouterse an adviser to the Government of Suriname, a cabinet-level position, despite an ongoing investigation by the Dutch Government into drugs-trafficking allegations against the former dictator. In March 1999 the Dutch authorities began legal proceedings against Bouterse, and on 16 July a Dutch court convicted Bouterse *in absentia* of leading a Surinamese cartel that had attempted to smuggle about two metric tons of cocaine seized at Dutch and Belgian ports and airports in 1989–97. Bouterse received a sentence of 16 years' imprisonment (later reduced to 11 years) and a US $2.2m. fine. The Dutch Government secured a warrant from the International Criminal Police Commission (Interpol) for Bouterse's arrest on drugs-trafficking charges with hopes of detaining him in a third country, since the Surinamese Constitution barred extradition of its nationals. The Attorney-General of the Netherlands filed further charges (this time for torture resulting in death) against Bouterse in January 2000. The new suit concerned the December Murders in 1982 and arose because of a complaint filed by relatives of the victims.

As a result of the efforts shown by Venetiaan's Government to investigate the murders committed during the Bouterse military regime, in October 2000 the Dutch Government agreed to resume aid to Suriname, suspended since 1998. Despite an incident in February 2002, when a hand grenade was thrown at the house of the Dutch ambassador, Rudolf Treffers, relations between the two countries began to improve. In June Dutch forensic specialists assisted the Surinamese authorities in the search for crucial information regarding the case. In June 2003 authorities in the Netherlands Antilles arrested the son of Bouterse, Dino Bouterse, alleging that he had attempted to enter the Dutch dependency on a false passport. He was suspected of participating in an earlier raid on the premises of the Surinamese Central Intelligence and Protection Services, in which a large number of weapons were stolen.

In February 2001 President Venetiaan announced his intention to seek the amendment of an article in the Constitution that banned the extradition of Surinamese citizens to other countries for trial. Following a series of meetings in early 2002 between the Minister of Foreign Affairs, Marie Levens, and her Dutch counterpart, a Returned Emigration Committee was established to oversee the voluntary repatriation of Surinamese with Dutch nationality without the loss of social benefits. Proposals to make it easier for Dutch senior citizens to visit Suriname by ending temporary visa requirements for visitors over 60 years of age were also discussed.

Successive Governments made efforts to combat illegal drugs-trafficking, in order to appease both its European and its US allies. In June 1997 the President installed a commission to monitor the drugs trade. In June 1998 the Government signed the Anti-Drugs Strategy for the Western Hemisphere that had been prepared by the OAS. In January 1999 new legislation was passed, providing for heavier sentences for drugs-trafficking. Finally, the Government prepared a 'Drugs Masterplan 1997–2002', which, among other things, proposed that money-laundering be made a criminal offence. Notwithstanding these and other measures, the Government was unable effectively to stem the tide of drugs-trafficking; according to official estimates, roughly 26,000 kgs of cocaine, with a street value of slightly over US $1,000m., were shipped to Europe each year. In March 2001 Surinamese police made the largest ever seizure of drugs, confiscating a consignment of 1,198 kgs of cocaine; three Surinamese, three Brazilians and a Colombian were sentenced to 14 years' imprisonment in the following March after being convicted on charges relating to the seizure. In October 2002 the Dutch Government provided the security forces with specialist training and equipment, in addition to funds for the construction of a new police headquarters. Surinamese and foreign nationals continued to be arrested for drugs offences and in June 2004 three soldiers and a Dutchman were sentenced to between seven and 10 years' imprisonment for their involvement in the production of the illegal stimulant, MDMA, also known as ecstasy, in a laboratory that had been discovered in Paramaribo in the previous year. In early 2004 the Dutch and Surinamese Governments agreed to co-operate on intelligence-gathering and to increase security on both passenger and cargo flights from Suriname. On 25 October 2004 customs officials in the two countries reached an agreement to share information in an attempt to reduce tax evasion on imports from the Netherlands.

## Maritime Border Dispute

The Wijdenbosch administration came into conflict with Guyana over the two countries' common border. The present boundary between Guyana and Suriname was based on a draft treaty agreed, but never ratified, between the United Kingdom and the Netherlands in 1939. Under the draft treaty the boundary between the two countries was established on the left bank of the Corantijn and Kutari rivers. In 1962 the Netherlands questioned Guyana's sovereignty over an area of land that protrudes from Guyana into Suriname. The Dutch proposed a modification of the treaty, favouring a boundary that followed the Thalweg, instead of the left bank of the Corantijn, and the westerly New River, instead of the Kutari. The British Government, however, refused to reopen the issue.

In June 1998 a new border dispute erupted when Guyana granted the Canadian-based CGX Energy Inc a concession to explore for petroleum and gas along the continental shelf off its coastline. Part of the area, designated the Corantijn block, lay within waters claimed by Suriname. The Surinamese Government made a formal protest against the CGX concession in May 2000, claiming it violated Suriname's sovereignty and territorial integrity. The Guyanese Government maintained that the exploration activities were being conducted in Guyanese territory, but indicated its willingness to attend talks. In early June gunboats of the Suriname navy forced CGX to remove the drill rig from the disputed waters. At a meeting held on 6 June in Port of Spain, Trinidad and Tobago, representatives from the two countries agreed that a Joint Technical Committee be established to resolve the dispute, but both sides remained at an impasse. Further talks, held later in the month in Jamaica, also ended in deadlock, and, with an agreement between the two countries unlikely, CGX Energy Inc abandoned the area. In September 2000 Guyana reported an alleged intrusion into its territory by Surinamese soldiers.

Negotiations made new progress following the election of Venetiaan in 2000, and in 2001 the Presidents of both countries made a declaration of their commitment to peace and co-operation. In January 2002 President Bharrat Jagdeo of Guyana addressed the Surinamese National Assembly on the possibility of a joint oil exploration project in the disputed territory; however, opposition members boycotted the speech in protest at the proposals. In July representatives of the Guyanese and Surinamese border commissions met and discussed ways in which the development of oil resources in the disputed maritime areas could be shared. However, a decree issued by Venetiaan's Government in March 2003 that maps of the country circulated by diplomatic missions in Paramaribo should include the disputed territory, once again jeopardized the negotiations. Guyana responded with a formal protest and dispatched a naval patrol to the Corantijn river. On 24 February 2004 Guyana referred the maritime boundary dispute to arbitration under the provisions of Article 287 of the UN Convention on the Law of the Sea. Representatives of the two countries participated in talks with the President of the International Tribunal on the Law of the Sea, Judge Dolliver Nelson, in Hamburg, Germany, on 12 and 29 May. Guyana also requested a number of interim measures that would allow gas and oil exploration to continue. It was anticipated that a provisional ruling would be followed by a substantive hearing within three years.

# Economy

## JAMES MCDONOUGH

Revised for this edition by the editorial staff

Suriname occupies 163,265 sq km (63,037 sq miles) on the north-east coast of South America, lying between Guyana to the west and French Guiana to the east. Suriname's economy is based on bauxite and agriculture. Bauxite, or rather its derivative, alumina, from which aluminium is made, generally accounted for some 70% of the total value of Suriname's exports and about 15% of gross domestic product (GDP). As a result, Suriname's economy was susceptible to 'boom and bust' cycles caused by variations in the international price of alumina. The tax receipts that Suriname received from the export of alumina and aluminium (hardly any unrefined bauxite was exported from Suriname) provided the revenue to support the large civil service, which employed close to 50% of the working population of about 100,000 in 2000. Both the civil service and the system of public enterprises required reform. Both were large and unwieldy, and the largely unprofitable public enterprises impeded private-sector development. The interruption of Dutch aid in 1982–86 and in the late 1990s, and the six-year civil war in the 1980s, inhibited economic and social development. Exacerbating the country's economic problems were the poor fiscal policies adopted by the Wijdenbosch administration in 1996–2000. These policies included borrowing from the Centrale Bank van Suriname (Central Bank of Suriname) to cover the budget deficit. Inflation soared during those years, reaching 98.9% in 1999. While the IMF welcomed the Venetiaan Government's attempts to restore fiscal policy, it condemned a 60% increase in public-sector wages in 2002. Further adding to Suriname's economic difficulties was a decline in the world price of alumina in the late 1990s. As a result, Suriname's GDP declined by 2.4% in 1999 and by 0.1% in 2000. However, in 2001 and 2002 economic growth recovered to 4.5% and 3.0%, respectively. In 2002 Suriname's gross national income per head, according to World Bank estimates, was approximately US $1,990. It was officially estimated that in 1999 up to one-half of the population was living in poverty.

## AGRICULTURE

In the 18th century Suriname's economy was based on the sugar industry, but by the 19th century coffee, cocoa and cotton were the country's main commodities. With the decline of large-scale plantation agriculture in the 20th century the former contract labourers from India and Indonesia were induced to remain in Suriname by the offer of free land. The Government distributed small plots of land for the growing of rice along the country's rich coastal plain. In the 1990s rice was the country's principal agricultural export, followed by bananas. The coastal polders (land that had been drained) remained the country's focus of agricultural activity and settlement. At least 70% of Suriname's population, estimated to total some 487,024 in mid-2004, lived on the estuarine lands of the Suriname River, within 25 km of the capital, Paramaribo, while a further 15% lived along the coastal plain.

The majority of Suriname's 58,000 ha of cultivated land, which represented only 0.4% of the country's total land area, was on the coastal plain. One-half of the cultivated area was in the polders close to Paramaribo, between the Commewijne and Saramacca estuaries. The country's agricultural potential was far from being realized, partly because of the inaccessibility of the interior savannahs, but also because of the unequal pattern of land tenure. In the 1980s, of the total number of land holdings, 46% were less than 2 ha in area and a further 27% were between 2 ha and 4 ha. At the other end of the scale, one-half of the agricultural land was occupied by 139 large holdings averaging 370 ha. Nevertheless, during 1980–85 agricultural production increased by an average of 6.1% per year. From 1986–90, as a result of the disruption caused by guerrilla activity, production declined by 6.4% per year. There was a brief recovery at the start of the 1990s, but the agricultural sector registered declines of 1.0% in 1996 and 5.6% in 1998, owing to a fall in paddy-rice prices, which led to a sharp reduction in the cultivated acreage. The sector expanded by 2.6% per year in 1999–2001; however, after a steep increase of 10.8% in 2001, the sector contracted by 0.7% in 2002, before recovering slightly to growth of 1.3% in 2003. In 2003 agriculture (including hunting, forestry and fishing) contributed an estimated 10.4% of GDP and in 1999 employed an estimated 6.4% of the working population, which represented a notable diminution in the importance of the sector during the 1990s. In January 2005 some US $23m. was granted by the Dutch Government to assist the diversification of the agricultural sector and to improve rural roads and irrigation systems.

In the early 21st century rice remained the crop of greatest commercial value in Suriname. About 50% of all cultivated land was devoted to rice, chiefly in the western polders of the Nickerie district. Much of the rice was produced by Hindustanis and Javanese on plots of less than 1 ha, located on the older polders. On the new polders, land holdings were typically of 80 ha or more and cultivation was mechanized and well managed. The fully mechanized rice farm at Wageningen was one of the largest in the world. The cultivated land area dedicated to rice fluctuated between 40,000 ha and around 50,000 ha in 1988–2002; in 2003 it stood at 52,425 ha. The annual output of paddy rice declined after 1985, however, owing to a lack of government resources to rebuild and expand the country's decaying canal system. Rice production reached a low of 163,400 metric tons in 2002. Milled rice production was forecast to increase substantially from 2003 to over 350,000 tons, benefiting from a US $3.4m. grant from the Inter-American Development Bank (IDB) and a further US $15m. from the European Union, intended to aid the restructuring of the drainage and irrigation system. Rice output in 2003 increased only marginally, to 193,685 tons, while export earnings declined to $9.1m. from $14.2m. the previous year.

Bananas and plantains (in Suriname usage, an eating banana is referred to as a *bacove*, while *banaan* is a cooking variety), which together comprised the next most important export crop, were grown on plantations in the Paramaribo region. In April 2002 the Government closed its banana plantation company, Surland, claiming that it could no longer afford to pay the workers' wages. Consequently, in 2002, according to government figures, banana production declined to an estimated 8,071m. metric tons, from 43,139 tons in the previous year. In 2003 this figure decreased still further to 1,278 tons, although the UN's Food and Agriculture Organization still estimated output in that year at 43,000 tons. Plaintain output remained fairly stable in 2003 at 11,843 tons. However, Surland was subsequently restructured with the aid of a US $6m. loan from the IDB and production and exports resumed in March 2004.

Livestock received little attention and the output of livestock products was insufficient for local needs. The few cattle that were reared by small-scale farmers were not of high quality and were used more as draught animals than for the production of either beef or milk. In 2003 there were 137,000 head of cattle. Some efforts were made by the Government to breed stock of better quality and there were experiments to cross the Holstein breed of cattle with the Santa Gertrudi. On the initiative of the Suriname Aluminium Company (Suralco), a scientifically operated dairy farm was established at Moengo. The ideal location for cattle would be the interior savannah, but the lack of access roads has, so far, prevented such development. In 2003 the country produced 1,860 metric tons of beef.

Fishing played a small, though significant, role in the economy. There was a modern fishing industry, located in Paramaribo. This industry was dominated by Japanese and Korean companies, which exported most of their catch to the USA and Canada. Shrimps were the most important single fisheries export, although the industry suffered a decline in the early 1990s. In 1991 the USA banned shrimp imports from Suriname because the country refused to take the necessary measures to protect the sea turtle, an endangered species, from the fishing practices of the Suriname fishermen. Nevertheless, the sector remained an important source of export revenue, with shrimps contributing an estimated US $36.9m. in 2003, or 5.8% of total export earnings.

More than 80% of Suriname is covered by forest, making it one of the most densely forested countries in the world. The first administration of President Runaldo Ronald Venetiaan (1991–96) moved to exploit this resource: in January 1994 the Government granted Mitra Usaha Sejati Abadi, a private Indonesian logging company, a timber concession for 125,000 ha in western Suriname, the maximum amount permitted without the approval of the National Assembly. The Government then formulated plans to grant concessions to three Asian companies that would permit over 3m. ha (over 7.4m. acres) to be logged. These plans provoked protests from local interest groups, international environmental organizations and many in the country's legislature. The indigenous Amerindian peoples and the 'Boschnegers' (Bush Negroes) protested against the proposed concessions, maintaining that they infringed upon the economic zones promised under the 1992 peace accord that the Government had signed with various insurgent groups. Environmentalists urged the Government to consider alternative strategies, such as the exploitation of 'eco-tourism'. As a result of the opposition, President Jules Wijdenbosch (1996–2000) announced in January 1997 that Suriname would not proceed with the huge logging concessions. In 2003 lumber contributed US $1.1m. to export revenue, equivalent to less than 0.2% of total exports by value. This represented a significant decrease from the previous year, when lumber exports totalled $5.4m.

## MINING

Mining, dominated by the extraction of bauxite, was the single most important economic activity in Suriname. In 2003 mining employed 3.5% of the population and contributed an estimated 8.6% of GDP. Bauxite was mined from deposits found along the northern edge of the central plateau, close to the Cottica river (the Moengo deposit) and the Suriname river (the Paranam and Onverwacht deposits). In Suriname the mining of bauxite and its refining into alumina and aluminium was controlled by two multinational companies, one US and the other Dutch. Mining began in 1915, by Suralco, a wholly owned subsidiary of the Aluminum Company of America (Alcoa), the world's largest producer of aluminium. In 1939 Billiton Maatschappij Suriname (BMS), a subsidiary of the Royal Dutch Shell-owned Billiton Company, initiated bauxite mining operations in Suriname's Para District, some 35 km south of Paramaribo. In 1983 BMS bought 46% of Suralco's Paranam refinery on the bank of the Suriname river, 100 km from its mouth, and Suralco purchased 24% of BMS's bauxite mining operations. In 1997 BMS opened the Lelydorp III deposit in the Paranam area. The Paranam refinery included installations for the extraction of alumina (1.7m. metric tons per year) from bauxite ore, as well as a smelter for the production of aluminium using alumina. However, in 1999 Suralco closed the smelter, which had a capacity of 30,000 tons per year. The closure was influenced by the high relative cost of the smelter and low rainfall affecting power generation. The company used hydroelectric power generated by the Brokopondo-Afobaka dam on the Suriname river to run the manufacturing operations. The dam was built by Suralco in the 1960s at a cost of US $150m., creating a 1,560-sq km lake, one of the largest artificial lakes in the world. BMS estimated that it would have enough bauxite for the Paranam refinery until 2006. In the early 2000s BHP Billiton, as BMS became known following a merger with Australian natural resources company, sought more sites in and outside Suriname to continue to supply the Paranam facility beyond the depletion of its current bauxite reserves. In early 2005 Alcoa World Alumina and Chemicals (AWAC) completed a project to increase production at the refinery in Paranam by 250,000 metric tons, to 2.2m. tons per year.

The bauxite industry was heavily taxed and traditionally accounted for over 40% of the Government's revenue. Suriname's traditional markets for bauxite derivatives were the USA, Canada and Norway. Suriname's annual bauxite output increased steadily in the 1990s, reaching 4.4m. tons in 2001, before falling back to 4.2m. tons in 2003. Suriname's alumina production remained fairly constant during the 1990s, with an average annual production of about 1.5m. tons. This figure rose to an average of 1.9m. tons between 1998–2002, and stood at 2.0m. tons in 2003. In 2003 alumina constituted 52.6% of the value of Suriname's exports, generating income of US $335.8m. In 2000–03 bauxite revenue was used to repay in full the loans secured to build bridges over the Coppename and Suriname rivers. In early 2003 BHP Billiton and Alcoa secured an agreement with the Government for the joint exploration of potential bauxite extraction sites and permission to build an aluminium-smelting plant and hydroelectric dam in the Backhuis mountains. In mid-June 2003, however, the Government filed a lawsuit against BHP Billiton, claiming that the company had neglected to pay taxes due on its 1994 profits.

Other mineral resources included iron ore, nickel, platinum, tin, copper, manganese, diamonds and gold. Gold and diamonds were extracted in small quantities from the river beds by private prospectors. In 1994 the Government signed an agreement with Golden Star Resources Inc (USA) and Cambior (Canada), envisaging the large-scale exploitation of gold in the central-eastern part of the country. Cambior began construction at the Gross Rosebel gold mine in late 2002 and commercial production began there in February 2004. Rosebel was scheduled to process 4.6m. metric tons of ore in 2004, for production of 245,000 troy oz of gold, at an average mine operating cost of US $184 per oz. The total reserves of the mine area were estimated at 42.9m. metric tons of ore. Exports of gold increased from 129,100 troy oz in 2001 to 385,500 troy oz in 2003, while over the same period export earnings from gold increased four-fold, from $35.0m. to $140.3m.

Petroleum was discovered in the Saramacca district in 1981, by the Gulf Oil Corporation of the USA. As a result, a Suriname State Oil Company (Staatsolie) was formed to exploit the reserves. Suriname exported small quantities of crude petroleum and imported refined petroleum products, as the country lacked refining capacity. Output in the early 2000s was 12,000 b/d, and regional geology suggested additional potential. Staatsolie was actively seeking international joint-venture partners, but with little success. In June 2001 a consortium of foreign oil companies halted exploration in waters off Suriname after deciding it was too high risk an operation. Undeterred, in 2003 Staatsolie announced plans to tender 19 offshore exploration blocks and in April 2004 Staatsolie signed a 30-year contract for joint exploration and production for an offshore block in the Guyana basin with Repsol YPF.

## MANUFACTURING

The industrial sector was dominated by the production of alumina. However, the country manufactured some foodstuffs (flour, margarine, cattle fodder), tobacco products, beverages, construction materials, clothing and furniture, using chiefly local raw materials, but imported machinery. Manufacturing, aside from alumina, grew only marginally in the 1990s owing to shortages caused by lack of foreign exchange, and increased competition from the Caribbean Community and Common Market (CARICOM) countries. In February 1995 Suriname was granted full membership of CARICOM, the first member from outside the English-speaking countries of the region. Suriname accepted CARICOM economic obligations on 1 January 1996 and on 1 January 1997 CARICOM's maximum common external tariff was reduced to 25%. Manufacturing employed an estimated 9.7% of the working population in 2003 and contributed an estimated 4.4% of GDP.

## TRANSPORT AND COMMUNICATIONS INFRASTRUCTURE

Suriname's first 10-Year Development Plan, covering the period 1955–65, placed emphasis on improving the infrastructure. Projects included the modernization of Paramaribo's port and Zanderij airport. In addition, a coastal road was constructed from Albina to Nieuw Nickerie, and another from Paramaribo, 110 km along the Suriname river, to the Brokopondo dam. This latter road formed the first part of a projected highway to Brazil. There were two railway lines, one a narrow-gauge line between Onverwacht and Brownsweg, the other a 70-km track between the bauxite deposits in the west, and Apoera on the Corantijn river, opened in 1980. However, owing to the slow progress of plans to mine bauxite in the Backhuis mountains, the latter railway remained inoperative. Despite the major improvements made in 1955–65, the infrastructure remained only minimally developed in Suriname. In 2000 there were an estimated 4,492 km of roads, mainly in the north of the country. In the late 1980s a ferry link between Suriname and Guyana, across the Corantijn, was opened. Relative to Suriname's population and level of economic development, there was a large number of telephones, with 79,800 in use in 2003.

## FOREIGN TRADE AND BALANCE OF PAYMENTS

Exports were dominated by alumina and gold, which accounted for almost three-quarters of Suriname's total export earnings. The remaining exports included rice, crude petroleum, shrimps and timber products. Imports consisted largely of machinery and transport equipment, as well as manufactured goods, and mineral fuels and lubricants. Imports declined steeply in the 1980s and early 1990s, owing to official regulation and a shortage of foreign exchange, and the balance of trade was generally in Suriname's favour as a result. Imports began to recover in 1996, and remained strong until 1999, although exports decreased in the same period. From 1996 to 2000, the value of imports increased from US $426.3m. to $526.5m., while the total value of exports increased from $434.3m. in 1996 to $514.0m. in the same period. Both imports and exports declined in 2001, recovered somewhat in 2002, and rose substantially in 2003; in that year imports totalled $703.9m. while export earnings amounted to $638.5m. The trade balance was in deficit until 1998; however, exports subsequently increased, resulting in a trade surplus of $153.0m. in 2000. The trade surplus then declined to $47.4m. in 2002 and to $29.8m. in 2003. In 2004, according to preliminary figures, the surplus increased to $67.8m. Total public external debt grew from $195.7m. in 2000 to $235.6m. by the end of 2004. In the same period the country's international reserves rose from $127.4m. to $136.9m.

In 2003 the principal destinations for Suriname's exports were the USA (21.0% of the total), Norway (16.5%) and France (9.1%). In the same year, imports came chiefly from the USA (30.1%), the Netherlands (17.8%) and Trinidad and Tobago (11.9%).

## INVESTMENT AND FINANCE

Prior to independence the country's two 10-Year Development Plans, beginning in 1955, served to stimulate economic growth. During this period the Brokopondo dam was built to supply hydroelectric power for the bauxite industry, the irrigation of polders was perfected and the country's infrastructure improved. Scientific techniques were introduced in agriculture, and social programmes, particularly in education, were funded. In the 1980s Suriname's economy began to decline. Official capital imports came to a virtual halt in 1982, when the Netherlands suspended its development co-operation because of the political murders by the military regime, which had taken power in 1980. A reduction in foreign capital investment and in exports, owing to the weakness of the world market for alumina and aluminium products, was reflected in a 20% fall in government revenues. Nevertheless, the military Government increased expenditure excessively, doubling the level of spending during the 1980s. The number of public employees increased by one-fifth, with the result that by 1994 nearly one-half of the active labour force consisted of civil servants. By contrast, spending on development projects collapsed. Over 50% of total government spending in the early 1990s was estimated to go on wages, with another 30% used to buy materials. The Government was able to devote a mere 2% to development projects.

Suriname's budget deficit increased from the equivalent of barely 5% of gross national product in 1980 to over 25% in 1992. The Government financed its expenditures first with international reserves and then by printing money, which by 1994 had precipitated a 'hyperinflationary' crisis, with the annual rate of inflation at 368.5%. Subsequently, the implementation of a stringent austerity programme and an increase in revenues caused by a rise in the world price of bauxite derivatives brought about dramatic improvement in the country's economic situation. By 1996 the depreciation of the Surinamese guilder had been halted, international reserves had reached almost US $100m. and the country was experiencing deflation. In that year there was a budget surplus equivalent to 2% of GDP. However, in the late 1990s a decline in revenues from exports of alumina and aluminium, along with increased government spending and a relaxation of fiscal controls by the Wijdenbosch administration, combined to recreate the conditions of an economic crisis. In 1998 the budget deficit increased to an estimated Sf 50,800m., equivalent to 11.1% of GDP, financed mainly through domestic borrowing. In 1999 a reduction in government expenditure reduced the deficit to 8.8% of GDP. Nevertheless, inflation, which had been brought under control in 1996, increased to 98.9% in 1999. This figure fell to 59.3% in 2000 and inflation continued to fall in 2001 and 2002, to 38.7% and 15.5%, respectively, owing to the new Government's ending of central bank credit. In 2003 consumer prices increased by an average of 23.0%. However, the annual inflation rate was an estimated 9.1% in 2004, and was forecast to remain at this level in 2005.

The guilder depreciated rapidly, despite a 43% devaluation on 1 January 1999. The differential between the official and parallel exchange rate, which had all but been eliminated with the devaluation, reached an average of 82% in 1999–2000. In an attempt to halt the economic decline, the second Venetiaan administration, which assumed power in August 2000, devalued the official exchange rate by 89%. In late July 2002 the Central Bank further devalued the guilder against the US dollar, from 2,200 guilders to 2,500 guilders. As well as ending government borrowing from the Central Bank, the new Government also eliminated subsidies on petroleum products, substantially increased electricity and water rates, rationalized the list of price controls on 12 basic food items, and increased the tax on cigarettes, alcohol and soft drinks. President Venetiaan also dismissed the President of the Central Bank for financing the previous administration's budget deficits. A new banking supervisory act was passed in January 2003 that aimed to strengthen the powers of the Central Bank. One key measure was the introduction of a formal licensing system that gave the Central Bank the exclusive authority to issue bank licences. In July of that year the Government announced that the Surinamese guilder (Sf) was to be replaced with the Surinamese dollar (SRD). The Surinamese dollar was introduced in January 2004 at an initial exchange rate of 2.8 to the US dollar.

## FINANCIAL AID

Suriname's fractured political spectrum and its weak financial situation combined to deny the country access to significant financial aid through the major international lending institutions. Suriname had no loans from the IMF or the World Bank. It had a series of small loans with the IDB, mainly to upgrade its financial and tax operations. Most of its other loans came from bilateral arrangements with individual countries, including Brazil, Japan and the Netherlands. The Framework Treaty of 1975, by which Suriname gained its independence from the Netherlands, provided for substantial amounts of economic aid to assist the development of the new republic. According to the Treaty, the Dutch were to provide US $1,700m. in the form of outright grants to implement a general development programme, and 500m. guilders as import guarantees for a period of 10–15 years. An additional 300m. guilders were to be available for matching grants to be financed from savings in Suriname. However, the Dutch suspended the treaty aid in 1982, as a result of the 'December Murders' (see History).

Upon the approval of a new Constitution in 1987, the Dutch agreed to resume their aid, but on a limited basis. Prior to 1989 some US $100m. had been disbursed, but about $700m., which had accumulated during the six years of military rule, remained undisbursed. The Netherlands insisted that aid would be provided only on a project-by-project basis until the Suriname Government implemented the IMF's structural-adjustment programme, considered necessary to correct the hyperinflationary crisis of the early 1990s and provide long-term stability for the economy. In 1998 the aid payments were suspended once again, owing to disagreements relating to the attempts of the Dutch Government to arrest Lt-Col Désiré (Desi) Bouterse, Suriname's former military ruler, and to Bouterse's continued presence in government. In 2000, with the advent of a new administration, the Dutch Government revised the structure of its aid to Suriname in support of sectoral priorities rather than individual projects. The Venetiaan administration was not in favour of that approach, and the two Governments began negotiations on the issue.

## OUTLOOK

Suriname's economic prospects, improved in the early 21st century. The economy grew by 5.3% in 2003 and by an estimated 4.6% in 2004, mainly owing to the expansion in the mining sector and some recovery in manufacturing and agriculture. However, following a small fiscal surplus equivalent to 0.1% of GDP in 2003 the central government recorded a deficit of 1.8% of GDP in 2004, largely owing to a rise in capital expenditure and a decline in domestic fuel taxes. The deficit was forecast to expand to 2.8% of GDP in 2005, despite a forecast expansion in revenue from the alumina sector. Indeed, the strength of the economy was highly dependent on the world price of alumina and the overall health of the world economy. According to the IMF, Suriname needed to rationalize the civil service, encourage privatization and institute a public-sector investment programme. The Government also needed to promote agricultural development and exports. In a report published in March 2005, the IMF recommended that over the medium term the Government should work to limit the country's economic vulnerability by widening the fiscal revenue base, reducing dependency on the

# SURINAME

mining sector and reforming domestic fuel taxes. The opening of the Rosebel gold mine in 2004 increased employment and contributed substantially to fiscal revenue, however. Further foreign direct investment was expected in 2005, particularly in agriculture, mining and oil, and GDP growth was predicted to remain strong, with an increase of 4.8% forecast for 2005.

# Statistical Survey

Sources (unless otherwise stated): Algemeen Bureau voor de Statistiek, Kromme Elleboogstraat 10, POB 244, Paramaribo; tel. 473927; fax 425004; e-mail info@statistics-suriname.org; internet www.statistics-suriname.org; Ministry of Trade and Industry: Nieuwe Haven, POB 557, Paramaribo; tel. 475080; fax 477602.

## AREA AND POPULATION

**Area:** 163,265 sq km (63,037 sq miles).

**Population:** 355,240 (males 175,814, females 179,426) at census of 1 July 1980; 487,024 (males 244,931, females 241,084, not known 1,009) at census of 2 August 2004.

**Density** (census of 2004): 3.0 per sq km.

**Ethnic Groups** (1980 census, percentage): Creole 34.70; Hindustani 33.49; Javanese 16.33; Bush Negro 9.55; Amerindian 3.10; Chinese 1.55; European 0.44; Others 0.84.

**Administrative Districts** (population at 2 August 2004): Paramaribo 243,640; Wanica 86,072; Nickerie 36,611; Coronie 2,809; Saramacca 16,135; Commewijne 24,657; Marowijne 16,641; Para 18,958; Brokopondo 13,299; Sipaliwini 28,202.

**Principal Towns** (estimated population at 1 July 1996): Paramaribo (capital) 205,000; Lelydorp 15,600; Nieuw Nickerie 11,100. Source: Thomas Brinkoff, *City Population* (internet www.citypopulation.de).

**Births and Deaths** (2000): Registered live births 9,804 (birth rate 22.5 per 1,000); Registered deaths 3,090 (death rate 7.1 per 1,000).

**Expectation of Life** (WHO estimates, years at birth): 66 (males 63; females 69) in 2003. Source: WHO, *World Health Report*.

**Economically Active Population** ('000 persons aged 15–66 years, January–June 1999): Total employed 72,834 (males 47,756, females 25,078); Unemployed 11,812 (males 5,361, females 6,451); Total labour force 84,646 (males 53,117, females 31,529). 2003 (preliminary estimates): Mining and quarrying 2,276; Manufacturing 6,269; Utilities 1,769; Construction 1,266; Trade 6,480; Transport, storage and communication 2,102; Finance and insurance 1,514; Insurance 315; Other services 2,557; Government 40,129; Total employed 64,678. Source: IMF, *Suriname: Selected Issues and Statistical Appendix* (March 2005).

## HEALTH AND WELFARE

### Key Indicators

**Total Fertility Rate** (children per woman, 2003): 2.4.

**Under-5 Mortality Rate** (per 1,000 live births, 2003): 39.

**HIV/AIDS** (% of persons aged 15–49, 2003): 1.7.

**Physicians** (per 1,000 head, 2000): 0.45.

**Hospital Beds** (per 1,000 head, 1996): 3.74.

**Health Expenditure** (2002): US $ per head (PPP): 385.

**Health Expenditure** (2002): % of GDP: 8.6.

**Health Expenditure** (2002): public (% of total): 41.8.

**Access to Water** (% of persons, 2002): 92.

**Access to Sanitation** (% of persons, 2002): 93.

**Human Development Index** (2002): ranking: 67.

**Human Development Index** (2002): value: 0.780.

For sources and definitions, see explanatory note on p. vi.

## AGRICULTURE, ETC.

**Principal Crops** (FAO estimates, '000 metric tons, 2003): Rice (paddy) 195.0; Roots and tubers 5; Sugar cane 120; Coconuts 9; Vegetables 22; Bananas 43; Plantains 11; Oranges 10; Other citrus fruit 5.

**Livestock** (FAO estimates, '000 head, 2003): Cattle 137; Sheep and Goats 15; Pigs 25; Chickens 3,800.

**Livestock Products** (FAO estimates, '000 metric tons, 2003): Beef and veal 2; Pig meat 1; Poultry meat 6; Cows' milk 9; Hen eggs 3.

**Forestry** (FAO estimates, '000 cu metres, 2003): *Roundwood Removals:* Sawlogs, veneer logs and logs for sleepers 148; Other industrial wood 7; Fuel wood 44; Total 199. *Sawnwood Production:* Total (incl. railway sleepers) 56.

**Fishing** (FAO estimates, '000 metric tons, 2002): Capture 28.2 (Marine fishes 11.6; Penaeus shrimps 2.4; Atlantic seabob 13.9); Aquaculture 0.2; *Total catch* 28.4.
Source: FAO.

## MINING

**Production** (2002, unless otherwise indicated): Crude petroleum ('000 barrels) 4,500; Bauxite ('000 metric tons, 2003) 4,215. Sources: US Geological Survey and IMF, *Suriname: Selected Issues and Statistical Appendix* (November 2003 and March 2005).

## INDUSTRY

**Production** ('000 metric tons, unless otherwise indicated, 2001): Gold-bearing ores (kg) 300; Gravel and crushed stone 85; Distillate fuel oil 26; Residual fuel oils 258; Cement 60; Alumina 1,900; Beer of barley 16; Coconut oil 0.82; Palm oil 0.22; Cigarettes 483 million (1996); Plywood ('000 cubic metres) 3; Electricity (million kWh) 1,870. Sources: mainly UN, *Industrial Commodity Statistics Yearbook* and FAO. 2002: Alumina 1,903. 2003: Alumina 2,005. Source: IMF, *Suriname: Selected Issues and Statistical Appendix* (November 2003 and March 2005).

## FINANCE

**Currency and Exchange Rates:** 100 cents = 1 Surinamese dollar. *Sterling, Dollar and Euro Equivalents* (31 May 2005): £1 sterling = 4.982 Surinamese dollars; US $1 = 2.740 Surinamese dollars; €1 = 3.379 Surinamese dollars; 1,000 Surinamese dollars = £200.74 = $364.96 = €295.97. *Average Exchange Rate* (Surinamese dollars per US $): 2.515 in 2002; 2.625 in 2003; 2.734 in 2004. *Note:* Between 1971 and 1993 the official market rate was US $1 = 1.785 guilders. A new free market rate was introduced in June 1993, and a unified, market-determined rate took effect in July 1994. A mid-point rate of US $1 = 401.0 guilders was in effect between September 1996 and January 1999. A new currency, the Surinamese dollar, was introduced on 1 January 2004, and was equivalent to 1,000 old guilders. Some data in this survey are still presented in terms of the former currency.

**Budget** (million Surinamese dollars, 2003): *Revenue:* Direct taxation 308.7; Indirect taxation 429.5 (Domestic taxes on goods and services 175.2, Taxes on international trade 251.2, Other taxes (incl. bauxite levy) 3.0); Non-tax revenue 119.1; Total 857.3 (excl. grants 62.3). *Expenditure:* Wages and salaries 406.0; Current transfers 139.1; Goods and services 203.6; Interest payments 65.2; Capital 98.7; Total 912.6 (excl. net lending 10.2). Source: IMF, *Suriname: Selected Issues and Statistical Appendix* (March 2005).

**International Reserves** (US $ million at 31 December 2004): Gold (national valuation) 7.52; IMF special drawing rights 1.89; Reserve position in IMF 9.51; Foreign exchange 118.00; Total 136.92. Source: IMF, *International Financial Statistics*.

**Money Supply** (million Surinamese dollars at 31 December 2004): Currency outside banks 246,826; Demand deposits at deposit money banks 414,364; Total money (incl. others) 701,316. Source: IMF, *International Financial Statistics*.

**Cost of Living** (Consumer Price Index for Paramaribo area; base: October–December 2000 = 100): 138.6 in 2001; 160.1 in 2002; 196.9 in 2003. Source: IMF, *International Financial Statistics*.

**National Income and Product** (Sf million at current prices, 1998): Compensation of employees 213,371; Operating surplus 109,143; *Domestic factor incomes* 322,514; Consumption of fixed capital 38,697; *GDP at factor cost* 361,210; Indirect taxes 53,644; *Less* subsidies 7,726; *GDP in purchasers' values* 407,128; Net factor income from abroad –244; *Gross national product* 406,884; *Less* consumption of fixed capital 38,697; *National income in market prices* 368,187. Source: UN Economic Commission for Latin America and the Caribbean, *Statistical Yearbook*.

# SURINAME

*Directory*

**Expenditure on the Gross Domestic Product** ('000 Surinamese dollars at current prices, preliminary figures, 2003): Public consumption 814,013; Private consumption 1,971,593; Public investment 47,667; Private investment 674,337; Exports of goods and non-factor services 654,589; *Less* Imports of goods and non-factor services 1,508,804; *GDP in purchasers' values* 2,653,396. Source: IMF, *Suriname: Selected Issues* (March 2005).

**Gross Domestic Product by Economic Activity** ('000 Surinamese dollars at current prices, preliminary figures, 2003): Agriculture, hunting and forestry 244,096; Mining and quarrying 203,579; Manufacturing 126,754; Electricity, gas and water 70,693; Construction 80,586; Trade, restaurants and hotels 290,181; Transport, storage and communications 182,011; Finance 293,107; Government 426,983; Other community, social and personal services 52,168; Informal sector 387,398; *Sub-total* 2,357,558; *Less* Imputed bank service charge 78,769; *GDP at factor cost* 2,278,789; Taxes on products, less subsidies 374,607; *GDP in purchasers' values* 2,653,396. Source: IMF, *Suriname: Selected Issues* (March 2005).

**Balance of Payments** (US $ million, preliminary figures, 2004): Exports of goods f.o.b. 880.6; Imports of goods f.o.b. −812.8; *Trade balance* 67.8; Exports of services 64.5; Imports of services −208.4; *Balance on goods and services* −76.1; Other income (net) −138.9; *Balance on goods, services and income* −215.0; Current transfers received (net) 69.3; *Current balance* −145.7; Capital account (net) 23.0; Financial account (net) 146.0; Net errors and omissions 7.6; Overall balance 30.9. Source: IMF, *Staff Report for the 2004 Article IV Consultation* (March 2005).

## EXTERNAL TRADE

**Principal Commodities** (US $ million, 2003): *Imports c.i.f.:* Food and live animals 84.2; Mineral fuels, lubricants, etc. 96.9; Chemicals 61.8; Manufactured goods 116.3; Machinery and transport equipment 239.2; Total (incl. others, excl. re-exports) 703.9. *Exports:* Alumina 335.8; Gold 140.3; Shrimp and fish 36.9; Crude oil 34.7; Rice 9.1; Total (incl. others) 638.5. Source: IMF, *Suriname: Selected Issues* (March 2005).

**Principal Trading Partners** (US $ million, 2003): *Imports:* Belgium 14.6; Brazil 18.6; China, People's Republic 48.4; Germany 20.7; Italy 9.3; Japan 43.1; Netherlands 125.5; Trinidad and Tobago 83.8; United Kingdom 18.2; USA 219.9; Total (incl. others) 703.9. *Exports:* Barbados 9.8; France 58.2; Iceland 26.5; Japan 12.1; Netherlands 23.9; Norway 105.6; Trinidad and Tobago 11.1; USA 134.3; Total (incl. others) 638.5. Source: IMF, *Suriname: Selected Issues* (March 2005).

## TRANSPORT

**Road Traffic** (registered motor vehicles, 2000 estimates): Passenger cars 61,365; Buses and coaches 2,393; Lorries and vans 20,827; Motorcycles and mopeds 30,598.

**Shipping:** *International Sea-borne Freight Traffic* (estimates, '000 metric tons, 2001): Goods loaded 2,306; Goods unloaded 1,212. *Merchant Fleet* (registered at 31 December 2004): Number of vessels 13; Total displacement 5,229 grt Source: Lloyd's Register-Fairplay, *World Fleet Statistics.*

**Civil Aviation** (traffic on scheduled services, 2001): Kilometres flown (million) 5; Passengers carried ('000) 203; Passenger-km (million) 898; Total ton-km (million) 103. Source: UN, *Statistical Yearbook.*

## TOURISM

**Tourist Arrivals** (number of non-resident arrivals at airports, '000): 72 in 1999; 72 in 2000; 68 in 2001 (preliminary figure).

**Tourism Receipts** (US $ million, excl. passenger transport): 26 in 2001; 17 in 2002; 18 in 2003.
Source: World Tourism Organization.

## COMMUNICATIONS MEDIA

**Radio Receivers** (1997): 300,000 in use.

**Television Receivers** (2000): 110,000 in use.

**Telephones** (2003): 79,800 main lines in use.

**Facsimile Machines** (1996): 800 in use.

**Mobile Cellular Telephones** (2003): 168,100 subscribers.

**Personal Computers** (2002): 20,000 in use.

**Internet Users** (2002): 20,000.

**Daily Newspapers** (2001): 3.

**Non-daily Newspapers** (2000): 10.
Sources: mainly UNESCO, *Statistical Yearbook*; UN, *Statistical Yearbook*; International Telecommunication Union.

## EDUCATION

**Pre-primary** (2001/02): 637 teachers; 15,746 pupils.

**Primary** (2001/02, incl. special education): 308 schools; 3,159 teachers (excl. special education); 65,611 pupils.

**General Secondary** (2001/02, incl. teacher-training): 141 schools; 2,056 teachers (1994/95); 39,858 pupils.

**University** (2001/02): 1 institution; 2,949 students.

**Other Higher** (2001/02): 3 institutions; 1,456 students (1999/2000).

**Adult Literacy Rate** (UNESCO projections): 94.0% in 2002.

# Directory

## The Constitution

The 1987 Constitution was approved by the National Assembly on 31 March and by 93% of voters in a national referendum in September.

### THE LEGISLATURE

Legislative power is exercised jointly by the National Assembly and the Government. The National Assembly comprises 51 members, elected for a five-year term by universal adult suffrage. The Assembly elects a President and a Vice-President and has the right of amendment in any proposal of law by the Government. The approval of a majority of at least two-thirds of the number of members of the National Assembly is required for the amendment of the Constitution, the election of the President or the Vice-President, the decision to organize a plebiscite and a People's Congress and for the amendment of electoral law. If it is unable to obtain a two-thirds' majority following two rounds of voting, the Assembly may convene a People's Congress and supplement its numbers with members of local councils. The approval by a simple majority is sufficient in the People's Congress.

### THE EXECUTIVE

Executive authority is vested in the President, who is elected for a term of five years as Head of State, Head of Government, Head of the Armed Forces, Chairman of the Council of State, the Cabinet of Ministers and the Security Council.

The Government comprises the President, the Vice-President and the Cabinet of Ministers. The Cabinet of Ministers is appointed by the President from among the members of the National Assembly. The Vice-President is the Prime Minister and leader of the Cabinet, and is responsible to the President.

In the event of war, a state of siege, or exceptional circumstances to be determined by law, a Security Council assumes all government functions.

### THE COUNCIL OF STATE

The Council of State comprises the President (its Chairman) and 14 additional members, composed of two representatives of the combined trade unions, one representative of the associations of employers, one representative of the National Army and 10 representatives of the political parties in the National Assembly. Its duties are to advise the President and the legislature and to supervise the correct execution by the Government of the decisions of the National Assembly. The Council may present proposals of law or of general administrative measures to the Government. The Council has the authority to suspend any legislation approved by the National Assembly which, in the opinion of the Council, is in violation of the Constitution. In this event, the President must decide within one month whether or not to ratify the Council's decision.

# SURINAME

# The Government

## HEAD OF STATE

**President:** RUNALDO RONALD VENETIAAN (assumed office 12 August 2000; re-elected by vote of the United People's Assembly 3 August 2005).

**Council of State:** Chair. RUNALDO RONALD VENETIAAN (President of the Republic); 14 mems; 10 to represent the political parties in the National Assembly, one for the Armed Forces, two for the trade unions and one for employers.

## CABINET OF MINISTERS
(July 2005)

**Vice-President:** JULES RATTANKOEMAR AJODHIA (VHP).
**Minister of Finance:** HUMPHREY HILDENBERG (NPS).
**Minister of Foreign Affairs:** MARIE E. LEVENS (NPS).
**Minster of Defence:** RONALD ASSEN (NPS).
**Minister of the Interior:** URMILA JOELLA-SEWNUNDUN (VHP).
**Minister of Justice and the Police:** SIEGFRIED F. GLIDS (SPA).
**Minister of Planning and Development Co-operation:** KEREMCHAND RAGHOEBARSINGH (VHP).
**Minister of Agriculture, Livestock and Fisheries:** GEETAPERSAD GANGARAM PANDAY (VHP).
**Minister of Transport, Communications and Tourism:** GUNO H. G. CASTELEN (SPA).
**Minister of Public Works:** DEWANAND BALESAR (VHP).
**Minister of Social Affairs and Housing:** SAMUEL PAWIRONADI (PL).
**Minister of Trade and Industry:** MICHAEL JONG TJIEN FA (PL).
**Minister of Regional Development:** ROMEO W. VAN RUSSEL (NPS).
**Minister of Education and Community Development:** WALTER T. SANDRIMAN (PL).
**Minister of Health:** MOHAMED RAKIEB KHUDABUX (VHP).
**Minister of Labour, Technological Development and the Environment:** CLIFFORD MARICA (SPA).
**Minister of Natural Resources and Energy:** FRANCO R. DEMON (NPS).

## MINISTRIES

**Ministry of Agriculture, Livestock and Fisheries:** Letitia Vriesdelaan, Paramaribo; tel. 477698; fax 470301.

**Ministry of Defence:** Kwattaweg 29, Paramaribo; tel. 474244; fax 420055.

**Ministry of Education and Community Development:** Dr Samuel Kafilludistraat 117–123, Paramaribo; tel. 498383; fax 495083.

**Ministry of Finance:** Onafhankelykheidsplein 2, Paramaribo; tel. 472610; fax 476314.

**Ministry of Foreign Affairs:** 25 Lim A Po St, POB 25, Paramaribo; tel. 477030; fax 471209.

**Ministry of Health:** Gravenstraat 64 boven, POB 201, Paramaribo; tel. 474841; fax 410702.

**Ministry of the Interior:** Wilhelminastraat 3, Paramaribo; tel. and fax 425354; e-mail gensur@sr.net.

**Ministry of Justice and the Police:** Gravenstraat 1, Paramaribo; tel. 473033; fax 412109.

**Ministry of Labour, Technological Development and the Environment:** Verlengde Gemenelandsweg 132B, POB 911, Paramaribo; tel. 432921; fax 433167; e-mail nvb@sr.net.

**Ministry of Natural Resources and Energy:** Dr J. C. de Mirandastraat 11–15, Paramaribo; tel. 473428; fax 472911.

**Ministry of Planning and Development Co-operation:** Dr S. Redmondstraat 118, Paramaribo; tel. 473628; fax 457833.

**Ministry of Public Works:** Verlengde Coppenamestraat 167, Paramaribo; tel. 462500; fax 464901.

**Ministry of Regional Development:** Van Rooseveltkade 2, Paramaribo; tel. 471574; fax 424517.

**Ministry of Social Affairs and Housing:** Waterkant 30–32, Paramaribo; tel. 472610; fax 476314.

**Ministry of Trade and Industry:** Havenlaan 3, POB 9354, Paramaribo; tel. 402886; fax 402602; e-mail dhisur@yahoo.com; internet www.sr.net/users/dirhi.

**Ministry of Transport, Communications and Tourism:** Prins Hendrikstraat 26–28, Paramaribo; tel. 411951; fax 420425; e-mail mintct@sr.net; internet www.mintct.sr.

# Legislature

## NATIONAL ASSEMBLY

**Chairman:** PAUL SALAM SOMOHARDJO.

**General Election, 25 May 2005**

| Party | Seats | % of votes cast |
|---|---|---|
| Nieuwe Front* | 23 | 39.37 |
| Nationale Democratische Partij | 15 | 22.20 |
| Volksalliantie Voor Vooruitgang† | 5 | 13.79 |
| A-Combinatie‡ | 5 | 7.21 |
| A1§ | 3 | 5.86 |
| Unie van Progressieve Surinamers/Partij voor Democratie en Ontwikkeling in Eenheid | — | 4.67 |
| Nieuw Suriname | — | 1.57 |
| Progressieve Arbeiders en Landbouwers Unie | — | 0.89 |
| Progressieve Politieke Partij | — | 0.17 |
| **Total** | **51** | **100.00‖** |

* An alliance of the Nationale Partij Suriname (NPS), the Pertajah Luhur (PL), the Surinaamse Partij van de Arbeid (SPA) and the Vooruitstrevende Hervormingspartij (VHP).
† An alliance of the Democratisch Nationaal Platform 2000 (DNP 2000), the Basispartij voor Vernieuwing en Democratie (BVD), the Kerukanan Tulodo Pranatan Ingigil (KTPI) and the Democratisch Alternatief.
‡ Including candidates of the Algemene Bevrijdings- en Ontwikkelingspartij (ABOP) and the Broederschap en Eenheid in Politiek (DOE).
§ Including candidates from the Democratisch Alternatief 1991 (DA '91), the Democraten van de 21 (D21) and the Politieke Vleugel van de FAL (PVF).
‖ Including invalid votes.

# Political Organizations

**Algemene Bevrijdings- en Ontwikkeling Partij (ABOP)** (General Liberation and Development Party): Jaguarstraat 15, Paramaribo; f. 1986; contested the 2005 election as part of the A-Combinatie electoral list; Pres. RONNIE BRUNSWIJK.

**Alternatief Forum (AF)** (Alternative Forum): Gladiolenstraat 26–28, Paramaribo; tel. 432342; Chair. GERARD BRUNINGS.

**Amazone Partij Suriname (APS)** (Suriname Amazon Party): Wilhelminastraat 91, Paramaribo; tel. 452081; Pres. KENNETH VAN GENDEREN.

**Broederschap en Eenheid in Politiek (BEP):** Ariestraat BR 34, S. O. B. Projekt, Paramaribo; tel. 494466; f. 1986; contested the 2005 election as part of the A-Combinatie electoral list; Chair. CAPRINO ALLENDY.

**Democraten Van de 21 (D21)** (Democrats of the 21st Century): Goudstraat 22, Paramaribo; f. 1986; contested the 2005 election as part of the A1 electoral list; Chair. SOEWARTO MUSTADJA.

**Democratisch Alternatief 1991 (DA '91)** (Democratic Alternative 1991): POB 91, Paramaribo; tel. 470276; fax 493121; e-mail info@da91.sr; internet www.da91.sr; f. 1991; contested the 2005 election as part of the A1 electoral list; social-democratic; Chair. DJAGENDRE RAMKHELAWAN.

**Hernieuwde Progressieve Partij (HPP)** (Renewed Progressive Party): Tourtonnelaan 51, Paramaribo; tel. 426965; e-mail hpp@cq-link.sr; f. 1986; Chair. HARRY KISOENSINGH.

**Nationale Democratische Partij (NDP)** (National Democratic Party): Benjaminstraat 17, Paramaribo; tel. 499183; fax 432174; e-mail ndpsur@sr.net; internet www.ndp.sr; f. 1987 by Standvaste (the 25 February Movement); army-supported; Chair. DESIRÉ (DESI) BOUTERSE.

**Nationale Partij Voor Leiderschap en Ontwikkeling (NPLO)** (National Party for Leaderhip and Development): Tropicaweg 1, Paramaribo; tel. 551252; f. 1986; Chair. OESMAN WANGSABESARIE.

**Naya Kadam** (New Choice): Naarstraat 5, Paramaribo; tel. 482014; fax 481012; e-mail itsvof@sr.net; Chair. INDRA DJWALAPERSAD; Sec. Ing. WALDO RAMDIHAL.

**Nieuw Front (NF)** (New Front): Paramaribo; f. 1987 as Front voor Demokratie en Ontwikkeling (FDO—Front for Democracy and Development); Pres. RONALD R. VENETIAAN; an alliance comprising:

**Nationale Partij Suriname (NPS)** (Suriname National Party): Wanicastraat 77, Paramaribo; tel. 477302; fax 475796; e-mail nps@sr.net; internet www.nps.sr; f. 1946; predominantly Creole; Sec. OTMAR ROEL RODGERS.

**Pertajah Luhur (PL)** (Full Confidence Party): Hoek Gemenlandsweg-Daniel Coutinhostraat, Paramaribo; tel. 401087; fax 420394; Pres. PAUL SOMOHARDJO.

**Surinaamse Partij van de Arbeid (SPA)** (Suriname Labour Party): Rust en vredestraat 64, Paramaribo; tel. 425912; fax 420394; f. 1987; affiliated with C-47 trade union; social democratic party; joined FDO in early 1991; Leader SIEGFRIED GILDS.

**Vooruitstrevende Hervormings Partij (VHP)** (Progressive Reformation Party): Coppenamestraat 130, Paramaribo; tel. 425912; fax 420394; internet www.parbo.com/vhp; f. 1949 as Verenigde Hindostaanse Partij (United Indian Party), name changed as above in 1973; leading left-wing party; predominantly Indian; Leader R. SARDJOE.

**Nieuw Suriname (NS)** (New Suriname): Paramaribo; contested the 2005 election.

**Partij voor Demokratie en Ontwikkeling in Eenheid (DOE)** (Party for Democracy through Unity and Development): Kamperfoeliestraat 23, Paramaribo; internet www.angelfire.com/nv/DOE; f. 1999.

**Pendawa Lima:** Bonistraat 115, Geyersvlij, Paramaribo; tel. 551802; f. 1975; predominantly Indonesian; Chair. RAYMOND SAPOEN.

**Politieke Vleugel van de FAL (PVF):** Keizerstraat 150, Paramaribo; political wing of farmers' org. Federatie van Agrariërs en Landarbeiders; contested the 2005 election as part of the A1 electoral list; Chair. JIWAN SITAL.

**Progressieve Arbeiders en Landbouwers Unie (PALU)** (Progressive Workers' and Farm Labourers' Union): Dr S. Kafiluddistraat 27, Paramaribo; tel. 400115; e-mail palu@sr.net; socialist party; Chair. JIM K. HOK; Vice-Chair. HENK RAMNANDANLAL.

**Progressieve Bosneger Partij (PBP):** f. 1968; resumed political activities 1987; represents members of the Bush Negro (Boschneger) ethnic group; associated with the Pendawa Lima (see above).

**Progressieve Politieke Partij (PPP)** (Progressive Political Party): Paramaribo; contested the 2005 election; Chair. SURINDER MUNGRA.

**Progressieve Surinaamse Volkspartij (PSV)** (Suriname Progressive People's Party): Keizerstraat 122, Paramaribo; tel. 472979; f. 1946; resumed political activities 1987; Christian democratic party; Chair. W. WONG LOI SING; Pres. HARRY KISOENSINGH.

**Unie van Progressieve Surinamers/Partij voor Democratie en Ontwikkeling in Eenheid (UPS/DOE):** Paramaribo; contested the 2005 election.

**Volksalliantie Voor Vooruitgang** (People's Alliance for Progress): Paramaribo; contested the 2005 election; Leader JULES WIJDENBOSCH; alliance comprising:

**Basispartij voor Vernieuwing en Democratie (BVD)** (Base Party for Renewal and Democracy): Hoogestraat 28–30, Paramaribo; tel. 422231; e-mail info@bvdsuriname.org; internet www.bvdsuriname.org; Chair. TJAN GOBARDHAN.

**Democratisch Alternatief** (Democratic Alternative): Jadnanansinghlaan 5, Paramaribo.

**Democratisch Nationaal Platform 2000 (DNP 2000)** (National Democratic Platform 2000): Gemenlandsweg 83, Paramaribo; f. 2000; Pres. JULES WIJDENBOSCH; alliance including:

**Democratische Partij (DP)** (Democratic Party): Paramaribo; f. 1992; Leader FRANK PLAYFAIR.

**Democraten Van de 21 (D21)** (Democrats of the 21st Century): Goudstraat 22, Paramaribo; f. 1986; contested the 2005 election as part of the A1 electoral list; Chair. SOEWARTO MUSTADJA.

**Kerukanan Tulodo Pranatan Ingigil (KTPI)** (Party for National Unity and Solidarity): Bonistraat 64, Geyersvlijt, Paramaribo; tel. 456116; f. 1947 as the Kaum-Tani Persuatan Indonesia; largely Indonesian; Leader WILLY SOEMITA.

Insurgent groups are as follows:

**Mandela Bush Negro Liberation Movement (BBM):** Upper Saramacca region; f. 1989 by mems of the Mauriër Bush Negro clan; Leader 'BIKO' (LEENDERT ADAMS).

**Tucayana Amazonica:** Bigi Poika; f. 1989 by Amerindian insurgents objecting to Kourou Accord between Govt and Bush Negroes of the SLA; Leader THOMAS SABAJO (alias 'Commander Thomas'); Chair. of Tucayana Advisory Group (Commission of Eight) ALEX JUBITANA.

**Union for Liberation and Democracy (UBD):** Moengo; f. 1989 by radical elements of SLA; Bush Negro; Leader KOFI AJONGPONG.

## Diplomatic Representation

### EMBASSIES IN SURINAME

**Brazil:** Maratakkastraat 2, POB 925, Paramaribo; tel. 400200; fax 400205; e-mail brasemb@sr.net; Ambassador CARVALHO DO NASCIMENTO BORGES.

**China, People's Republic:** Anton Dragtenweg 154, POB 3042 Paramaribo; tel. 451570; fax 452540; e-mail chinaemb_sr@mfa.gov.cn; Ambassador CHEN JINGHUA.

**France:** Gravenstraat 5–7 boven, POB 2648, Paramaribo; tel. 476455; fax 471208; e-mail ambafrance.paramaribo@diplomatie.gouv.fr; internet www.amfrance@sr.net; Ambassador JEAN-MARIE BRUNO.

**Guyana:** Henck Arronstraat 82, POB 785, Paramaribo; tel. 475209; fax 472679; e-mail guyembassy@sr.net; Ambassador KARSHANJEE ARJUN.

**India:** Rode Kruislaan 10, POB 1329, Paramaribo; tel. 498344; fax 491106; e-mail india@sr.net; Ambassador ASHOK KUMAR SHARMA.

**Indonesia:** Van Brussellaan 3, POB 157, Paramaribo; tel. 431230; fax 498234; e-mail unitkom@aksaranesia.sr; Ambassador SUPARMIN SUNJOYO.

**Japan:** Henck Arronstraat 23–25, POB 2921, Paramaribo; tel. 474860; fax 412208; e-mail eojparbo@sr.net; Ambassador YASUO MATSUI (resident in Venezuela).

**Netherlands:** Van Roseveltkade 5, POB 1877, Paramaribo; tel. 477211; fax 477792; e-mail nlgovprm@sr.net; internet www.cq-link.sr/nedamb; Ambassador HENDRIK J. W. SOETERS.

**Russia:** Anton Dragtenweg 7, POB 8127, Paramaribo; tel. 472387; fax 472387; Ambassador VLADIMIR LVOVITCH TYURDENEV.

**USA:** Dr Sophie Redmondstraat 129, POB 1821, Paramaribo; tel. 472900; fax 425-690; e-mail embuscen@sr.net; internet paramaribo.usembassy.gov; Ambassador MARSHA BARNES.

**Venezuela:** Henck Arronstraat 23–25, POB 3001, Paramaribo; tel. 475401; fax 475602; e-mail vzla@sr.net; internet www.embajadavzla.org.sr; Ambassador FRANCISCO DE JESÚS SIMANCAS.

## Judicial System

The administration of justice is entrusted to a Court of Justice, the six members of which are nominated for life, and three Cantonal Courts.

**President of the Court of Justice:** (vacant).

**Attorney-General:** SUBHAAS PUNWASI.

## Religion

Many religions are represented in Suriname. According to official sources, Christians represent approximately 40% of the population, Hindus 27% and Muslims 22%.

### CHRISTIANITY

**Committee of Christian Churches:** Paramaribo; Chair. Rev. JOHN KENT (Praeses of the Moravian Church).

#### The Roman Catholic Church

For ecclesiastical purposes, Suriname comprises the single diocese of Paramaribo, suffragan to the archdiocese of Port of Spain (Trinidad and Tobago). The Bishop participates in the Antilles Episcopal Conference (currently based in Port of Spain, Trinidad and Tobago). At 31 December 2003 there were an estimated 110,664 adherents in the diocese, representing about 23% of the population.

**Bishop of Paramaribo:** WILHELMUS ADRIANUS JOSEPHUS MARIA DE BEKKER, Bisschopshuis, Gravenstraat 12, POB 1230, Paramaribo; tel. 425918; fax 471602; e-mail azichem@sr.net.

#### The Anglican Communion

Within the Church in the Province of the West Indies, Suriname forms part of the diocese of Guyana. The Episcopal Church is also represented.

**Anglican Church:** St Bridget's, Hoogestraat 44, Paramaribo.

SURINAME — Directory

### Protestant Churches

**Evangelisch-Lutherse Kerk in Suriname:** Waterkant 102, POB 585, Paramaribo; tel. 425503; fax 481856; e-mail elks@sr.net; Pres. WIM LOOR; 4,000 mems.

**Moravian Church in Suriname** (Evangelische Broeder Gemeente): Maagdenstraat 50, POB 1811, Paramaribo; tel. 473073; fax 475797; e-mail ebgs@sr.net; f. 1735; Praeses MAARLEN MINGUERN; 40,000 mems (2004).

Adherents to the Moravian Church consitute some 15% of the population. Also represented are the Christian Reformed Church, the Dutch Reformed Church, the Baptist Church, the Evangelical Methodist Church, Pentecostal Missions, the Seventh-day Adventists and the Wesleyan Methodist Congregation.

### HINDUISM

**Sanatan Dharm:** Koningstraat 31–33, POB 760, Paramaribo; tel. 404190; f. 1930; Pres. Dr R. M. NANNAN PANDAY; over 150,000 mems.

### ISLAM

**Surinaamse Moeslim Associatie:** Kankantriestraat 55–57, Paramaribo; Javanese Islamic org.; Chair. A. ABDOELBASHIRE.

**Surinaamse Islamitische Organisatie (SIO):** Watermolenstraat 10, POB 278, Paramaribo; tel. 475220; f. 1978; Pres. Dr I. JAMALUDIN; Sec. C. HASRAT; 6 brs.

**Stichting Islamitische Gemeenten Suriname:** Verlengde Mahonielaan 39, Paramaribo; Indonesian Islamic org.; Chair. Dr T. SOWIRONO.

**Federatie Islamitische Gemeenten in Suriname:** Paramaribo; Indonesian Islamic org.; Chair. K. KAAIMAN.

### JUDAISM

The Dutch Jewish Congregation and the Dutch Portuguese-Jewish Congregation are represented in Suriname.

**Jewish Community:** The Synagogue Neve Shalom, Keizerstraat, POB 1834, Paramaribo; tel. 400236; fax 402380; e-mail rene-fernandes@cq-link.sr; f. 1854; Pres. RENÉ FERNANDES.

### OTHER RELIGIONS

**Arya Dewaker:** Dr S. Kafilludistraat 1, Paramaribo; tel. 400706; members preach the Vedic Dharma; disciples of Maha Rishi Swami Dayanand Sarswati, the founder of the Arya Samaj in India; f. 1929; Chair. Dr R. BANSRADJ; Sec. D. CHEDAMMI.

The Bahá'í faith is also represented.

## The Press

### DAILIES

**De Ware Tijd:** Malebatrumstraat 9, POB 1200, Paramaribo; tel. 472823; fax 411169; e-mail webmanager@dwt.net; internet www.dwtonline.com; f. 1957; morning; Dutch; independent/liberal; Dir STEVE JONG TJIEN FA; Editor-in-Chief DESI TRUIDMAN.

**De West:** Dr J. C. de Mirandastraat 2–6, POB 176, Paramaribo; tel. 6923338; fax 470322; e-mail dewest@cq-link.sr; internet www.dewestonline.cq-link.sr; f. 1909; midday; Dutch; liberal; Editors G. D. C. FINDLAY, L. KETTIE; circ. 15,000–18,000.

### PERIODICALS

**Advertentieblad van de Republiek Suriname:** Gravenstraat 120, POB 56, Paramaribo; tel. 473501; fax 454782; f. 1871; 2 a week; Dutch; government and official information bulletin; Editor E. D. FINDLAY; circ. 1,000.

**CLO Bulletin:** Gemenelandsweg 95, Paramaribo; f. 1973; irregular; Dutch; labour information published by civil servants' union.

**Kerkbode:** Burenstraat 17–19, POB 219, Paramaribo; tel. 473079; fax 475635; e-mail stadje@sr.net; f. 1906; weekly; religious; circ. 2,000.

**Omhoog:** Gravenstraat 21, POB 1802, Paramaribo; tel. 425992; fax 426782; e-mail rkomhoog@sr.net; weekly; f. 1952; Dutch; Catholic bulletin; Editor S. MULDER; circ. 5,000.

**Xtreme Magazine:** Uranusstraat 49, Paramaribo; tel. 456969.

### NEWS AGENCIES

**Surinaams Nieuws Agentschap (SNA)** (Suriname News Agency): Paramaribo; two daily bulletins in Dutch, one in English; Dir E. G. J. DE MEES.

### Foreign Bureau

**Inter Press Service (IPS)** (Italy): Malebatrumstraat 1–5, POB 5065, Paramaribo; tel. 471818; Correspondent ERIC KARWOFODI.

## Publishers

**Afaka International NV:** Residastraat 23, Paramaribo; tel. and fax 530640; internet www.afaka.com; f. 1996; Dir GERRIT BARRON.

**Educatieve Uitgeverij Sorava NV:** Latourweg 10, POB 8382, Paramaribo; tel. and fax 480808.

**IMWO, Universiteit van Suriname:** Universiteitscomplex, Leysweg 1, POB 9212, Paramaribo; tel. 465558; fax 462291; e-mail bmhango@yahoo.com.

**Ministerie van Onderwijs en Volksontwikkeling** (Ministry of Education and Community Development): Dr Samuel Kafilludistraat 117-123, Paramaribo; tel. 498850; fax 495083.

**Okopipi Publ.** (Publishing Services Suriname): Van Idsingastraat 133, Paramaribo; tel. 472746; e-mail pssmoniz@sr.net; fmrly I. Krishnadath.

**Papaya Media:** Plutostraat 30, POB 8304, Paramaribo; tel. and fax 454530; e-mail roy_bhikharie@sr.net; f. 2002; Dir ROY BIKHARIE.

**Stichting Wetenschappelijke Informatie** (Foundation for Information and Development): Prins Hendrikstraat 38, Paramaribo; tel. 475232; fax 422195; e-mail swin@sr.net; f. 1977.

**Tabiki Productions:** Weidestraat 34, Paramaribo; tel. 478525; fax 478526; e-mail insightsuriname@yahoo.com.

**VACO, NV:** Domineestraat 26, POB 1841, Paramaribo; tel. 472545; fax 410563; f. 1952; Dir EDUARD HOGENBOOM.

### PUBLISHERS' ASSOCIATION

**Publishers' Association Suriname:** Domineestraat 32, POB 1841, Paramaribo; tel. 472545; fax 410563.

## Broadcasting and Communications

### TELECOMMUNICATIONS

**Telecommunication Corporation Suriname (Telesur):** Heiligenweg 1, POB 1839, Paramaribo; tel. 473944; fax 404800; internet www.telesur.sr; supervisory body; Man. Dir IRIS STRUIKEN-WIJDENBOSCH.

### BROADCASTING

#### Radio

**ABC Radio** (Ampie's Broadcasting Corporation): Maystraat 57, Paramaribo; tel. 464609; fax 464680; e-mail info@abcsuriname.com; internet www.abcsuriname.com; f. 1975; re-opened in 1993; commercial; Dutch and some local languages.

**Radio Apintie:** Verlengde Gemenelandsweg 37, POB 595, Paramaribo; tel. 498855; fax 400684; e-mail apintie@sr.net; internet www.apintie.sr; f. 1958; commercial; Dutch and some local languages; Gen. Man. CHARLES VERVUURT.

**Radio Bersama:** Bonniestraat 115, Paramaribo; tel. 551804; fax 551803; f. 1997; Dir AJOEB MOENTARI.

**Radio Boskopou:** Roseveltkade 1, Paramaribo; tel. 410300; govt-owned; Sranang Tongo and Dutch; Head Mr VAN VARSEVELD.

**Radio Garuda:** Goudstraat 14–16, Paramaribo; tel. 422422; Dir TOMMY RADJI.

**Radio Nickerie (RANI):** Waterloostraat 3, Nieuw Nickerie; tel. 231462; commercial; Hindi and Dutch.

**Radio Paramaribo** (Rapar): Verlengde Coppenamestraat 34, POB 975, Paramaribo; tel. 499995; fax 493121; e-mail rapar@sr.net; f. 1957; commercial; Dutch and some local languages; Dir RASHIED PIERKHAN.

**Radio Radika:** Indira Gandhiweg 165, Paramaribo; tel. 482800; fax 482910; e-mail radika@sr.net; re-opened in 1989; Dutch, Hindi; Dir ROSHNI RADHAKISUN.

**Radio Sangeet Mala:** Indira Gandhiweg 73, Paramaribo; tel. 485893; Dutch, Hindi; Dirs RADJEN SOEKHRADJ, SOEDESH RAMSARAN.

**Radio SRS** (Stichting Radio Omroep Suriname): Jacques van Eerstraat 20, POB 271, Paramaribo; tel. 498115; fax 498116; e-mail radiosrs@sr.net; f. 1965; commercial; govt-owned; Dutch and some local languages; Dir LEOPOLD DARTHUIZEN.

**Radio Ten:** Letitia Vriesdelaan 5, Paramaribo; tel. 410881; fax 410885; e-mail radio10@cq-link.sr; internet www.radio10.cq-link.sr.

Other stations include: Radio KBC, Radio Koyeba, Radio Pertjaya, Radio Shalom, Radio Zon, Ramasha Radio, Rasonic Radio and Trishul Radio.

### Television

**ABC Televisie** (Ampie's Broadcasting Corporation): Maystraat 57, Paramaribo; tel. 464609; fax 464680; e-mail info@abcsuriname.com; internet www.abcsuriname.com; Channel 4.

**Algemene Televisie Verzorging (ATV):** Adrianusstraat 55, POB 2995, Paramaribo; tel. 404611; fax 402660; e-mail info@atv.sr; internet www.atv.sr; f. 1985; govt-owned; commercial; Dutch, English, Portuguese, Spanish and some local languages; Channel 12; Man. GUNO COOMAN.

**STVS** (Surinaamse Televisie Stichting): Letitia Vriesdelaan 5, POB 535, Paramaribo; tel. 473031; fax 477216; e-mail adm@stvs.info.sr; internet www.parbo.com/stvs; f. 1965; govt-owned; commercial; local languages, Dutch and English; Channel 8; Dir KENNETH OOSTBURG.

## Finance

(cap. = capital; res = reserves; dep. = deposits; m. = million; amounts in Suriname guilders unless otherwise stated)

### BANKING

#### Central Bank

**Centrale Bank van Suriname:** 18–20 Waterkant, POB 1801, Paramaribo; tel. 473741; fax 476444; e-mail cbvsprv@sr.net; internet www.cbvs.sr; f. 1957; cap. and res 34.5m. (Dec. 1987); Pres. ANDRE E. TELTING.

#### Commercial Banks

**Finabank NV:** Dr Sophie Redmondstraat 59–61, Paramaribo; tel. 476111; fax 410471.

**Handels-Krediet- en Industriebank (Hakrinbank NV):** Dr Sophie Redmondstraat 11–13, POB 1813, Paramaribo; tel. 477722; fax 472066; e-mail hakrindp@sr.net; internet www.hakrinbank.com; f. 1936; cap. 69,854m., res 11,901m., dep. 86,874m. (Dec. 2003); Pres. and Chair. A. K. R. SHYAMNARAIN; Man. Dirs M. TJON-A-TEN, J. D. BOUSAID; 6 brs.

**Landbouwbank NV:** FHR Lim A Postraat 34, POB 929, Paramaribo; tel. 475945; fax 411965; e-mail lbbank@sr.net; f. 1972; govt-owned; agricultural bank; cap. 5m., res 4.6m., dep. 76.0m. (Dec. 1987); Chair. D. FERRIER; Pres. R. MERHAI; 5 brs.

**RBTT Bank NV:** Kerkplein 1, Paramaribo; tel. 471555; fax 411325.

**De Surinaamsche Bank NV:** Henck Arronstraat 26–30, POB 1806, Paramaribo; tel. 471100; fax 411750; internet www.dsbbank.sr; f. 1865; cap. 17.5m., res 26,558.1m., dep. 572,666.4m. (Dec. 2003); Chair. S. SMIT; Pres. Dr E. J. MÜLLER; 7 brs.

**Surinaamse Postspaarbank:** Knuffelsgracht 10–14, POB 1879, Paramaribo; tel. 472256; fax 472952; e-mail spsbdir@sr.net; f. 1904; savings and commercial bank; cap. and res 1,044m., dep. 66,533m. (Dec. 2004); Man. ALWIN R. BAARH (acting); 2 brs.

**Surinaamse Volkscredietbank:** Waterkant 104, POB 1804, Paramaribo; tel. 472616; fax 473257; e-mail btlsvcb@sr.net; f. 1949; cap. and res 170.3m. (Dec. 1997); Man. Dir THAKOERDIEN RAMLAKHAN; 3 brs.

#### Development Bank

**Nationale Ontwikkelingsbank van Suriname NV:** Coppenamestraat 160–162, POB 677, Paramaribo; tel. 465000; fax 497192; f. 1963; govt-supported development bank; cap. and res 34m. (Dec. 1992); Man. Dir J. TSAI MEU CHONG.

### INSURANCE

**Assuria NV:** Grote Combeweg 37, POB 1501, Paramaribo; tel. 477955; fax 472390; e-mail assuria@sr.net; internet www.assuria.sr; f. 1961; life and indemnity insurance; Man. Dir Dr S. SMIT.

**Assuria Schadeverzekering NV:** Henck Arronstraat 5–7, POB 1030, Paramaribo; tel. 473400; fax 476669; e-mail assuria@sr.net; internet www.assuria.net; Dir M. R. CABENDA.

**British American Insurance Company:** Klipstenenstraat 29, POB 370, Paramaribo; tel. 476523; Man. H. W. SOESMAN.

**CLICO Life Insurance Company (SA) Ltd:** Klipstenenstraat 29, POB 3026, Paramaribo; tel. 472525; fax 476777; e-mail clicosur@sr.net; internet www.clico.com/suriname; COO GEETA SINGH.

**Fatum Levensverzekering NV:** Noorderkerkstraat 5–7, Paramaribo; tel. 471541; fax 410067; e-mail fatum@sr.net; internet www.fatum-suriname.com.

**Hennep Verzorgende Vezekering NV:** Dr Sophie Redmondstraat 246, Paramaribo.

**The Manufacturers' Life Insurance Company:** c/o Assuria NV, Grote Combeweg 37, Paramaribo; tel. 473400; fax 472390; Man. S. SMIT.

**Parsasco NV:** Gravenstraat 117 boven, Paramaribo; tel. 421212; e-mail parsasco@sr.net; internet www.parsasco.com; Dir L. KHEDOE.

**Self Reliance:** Herenstraat 22, Paramaribo; tel. 472582; fax 472475; e-mail self-reliance@sr.net; internet www.self-reliance.sr; life insurance; Dir N. J. VEIRA.

## Trade and Industry

### DEVELOPMENT ORGANIZATIONS

**Centre for Industry and Export Development:** Rust en Vredestraat 79–81, POB 1275, Paramaribo; tel. 474830; fax 476311; f. 1981; Man. R. A. LETER.

**Stichting Planbureau Suriname** (National Planning Bureau of Suriname): Dr Sophie Redmondstraat 118, POB 172, Paramaribo; tel. 447408; fax 475001; e-mail dirsps@sr.net; internet www.planbureau.net; responsible for regional and socio-economic long- and short-term planning; Man. Dir LILIAN J. M. MONSELS-THOMPSON.

### CHAMBER OF COMMERCE

**Kamer van Koophandel en Fabrieken** (Chamber of Commerce and Industry): Dr J. C. de Mirandastraat 10, POB 149, Paramaribo; tel. 474536; fax 474779; e-mail chamber@sr.net; internet www.surinamedirectory.com; f. 1910; Pres. R. L. A. AMEERALI; Sec. R. RAMDAT; 16,109 mems.

**Surinaams–Nederlandse Kamer voor Handel en Industrie** (Suriname–Netherlands Chamber of Commerce and Industry): Coppenamestraat 158, Paramaribo; tel. 463201; fax 463241.

### INDUSTRIAL AND TRADE ASSOCIATIONS

**Associatie van Surinaamse Fabrikanten (ASFA)** (Suriname Manufacturers' Asscn): Jaggernath Lachmonstraat 187, POB 3046, Paramaribo; tel. 439797; fax 439798; e-mail asfa@sr.net; Chair. KATHLEEN LIEUW KIE SONG; 317 mems.

**Vereniging Surinaams Bedrijfsleven** (Suriname Trade and Industry Association): Prins Hendrikstraat 18, POB 111, Paramaribo; tel. 475286; fax 475287; e-mail vsbtia@sr.net; internet www.vsbstia.org; Pres. MARCEL A. MEYER; 290 mems.

### MAJOR COMPANIES

The following are some of the major enterprises operating in Suriname:

**BHP Billiton Suriname:** Nickeriestraat 8, POB 1053, Paramaribo; tel. 423277; subsidiary of Royal Dutch-Shell (Netherlands); bauxite mining and refining; Man. Dir JAAP ZWAAN; Supervisor and Dir PAUL MICHAEL EVERARD; 1,000 employees.

**British American Tobacco Company Ltd Suriname:** Kan Kawastraat 7, POB 1913, Paramaribo; tel. 481444; fax 483170; internet www.batca.com; subsidiary of BAT Industries (United Kingdom); cigarette manufacturers.

**Bruynzeel Suriname Houtmaatschappij NV:** Slangenhoutstraat, POB 1831, Paramaribo; tel. 478811; fax 411304; e-mail bruynzeel@sr.net; timber merchants; producers of sawnwood, plywood and precut and prefabricated houses.

**CHM Suriname NV:** Dr Sophie Redmondstraat 2–14, POB 1819, Paramaribo; tel. 471166; fax 471534; e-mail chmbuy@sr.net; internet www.chmsuriname.com; subsidiary of Handelen Industrie Mij Ceteco NV (Netherlands); electrical appliances, televisions and radios; importer of Toyota automobiles; Gen. Man. EDMUND KASIMBEG; 160 employees.

**NV Consolidated Industries Corporation:** Industrieweg-zuid BR 34, POB 365, Paramaribo; tel. 482050; fax 481431; e-mail info@cicsur.com; internet www.cicsur.com; manufacturers of detergents and disinfectants, packaging materials and cosmetics and toiletries; Man. Dir Dr ANTOINE A. BRAHMIN; 131 employees.

**Fernandes Concern Beheer NV:** Klipstenenstraat 2–10, Paramaribo; holding co; subsidiary cos: Fernandes Bottling Co NV, Ephraimszegan, Paramaribo (fax 471154); Fernandes & Sons Bakery, Kernkampweg (fax 492177); Handelmaatschappij Fernandes & Sons NV, Klipstenenstraat 2–10, Paramaribo (fax 471154); Gen. Man. RENE FERNANDES; 800 employees.

**H. J. de Vries Beheersmaatschappij NV:** Waterkant 92–96, Paramaribo; tel. 471222; fax 475718; e-mail devries@sr.net; internet www.hj-devries.com; f. 1903; durable goods; CEO MARK GOEDE; 250 employees.

**Jong A Kiem NV:** POB 272, Steenbakkerijstraat 66, Paramaribo; tel. 471600; fax 471855; e-mail info@jongakiem.com; internet www.jongakiem.com; manufacturer of pharmaceutical products; owned by AstraZeneca PLC, UK.

**C. Kersten en Co NV:** Domineestraat 36–38, Paramaribo; tel. 471150; fax 478524; holding company for distributors of durable goods; CEO R. VAN ESSEN; CFO U. BABB; 700 employees.

**Kirplani's Kleding Industrie NV:** Domineestraat 52–56, POB 1917, Paramaribo; tel. 471400; fax 410527; e-mail kirpa@sr.net; internet www.kirpalani.com; textile and clothing producers; Gen. Man. JHAMATMAL T. KIRPALANI; 636 employees.

**Kuldipsingh Group:** Anamoestraat, POB 8089, Paramaribo; tel. 551204; fax 550669; e-mail hkm@kuldipsingh.net; internet www.kuldipsingh.net; f. 1979; manufactures and distributes metal and building products; Man. Dir SWITRANG KULDIPSINGH; 310 employees.

**Nameco NV Nationale Metaal-& Construktie Maatschappij NV:** POB 1560, Paramaribo; tel. 482014; construction and civil engineering projects.

**Shell Suriname Verkoopmaatschappij NV:** POB 849, Paramaribo; tel. 482027; fax 482569; f. 1975; acquired by Sol Group in ; producers of fuel oil and lubricants; Chair. KYFFIN SIMPSON; 190 employees.

**Staatsolie Maatschappij Suriname NV:** Dr Ir H. S. Adhinstraat 21, POB 4069, Paramaribo; tel. 499649; fax 491105; e-mail mailstaatsolie@staatsolie.com; internet www.staatsolie.com; f. 1981; state petroleum exploration and exploitation co; Pres. Dr S. E. JHARAP; 635 employees.

**Stichting Behoud Bananensector Suriname (SBBS):** Paramaribo; banana-producers' co-operative; fmrly known as Surland until closure in 2002, restructured and reopened under present name in 2004; scheduled for privitization.

**Suriname Aluminium Company (Suralco):** van't Hogerhuysstraat, POB 1810, Paramaribo; tel. 323281; internet www.alcoa.com/suriname; subsidiary of Alcoa (USA); bauxite mining and refining, alumina production; Gen. Man. GEORGE BIJNOE.

**Suriname-Amerikaanse Industrie NV (SAIL):** Cornelis Jongbawstraat 48, POB 3045, Paramaribo; tel. 474014; fax 473521; e-mail sail@sr.net; f. 1955; subsidiary of Castle & Cooke Inc (USA); processors and exporters of marine and farm-raised shrimps and fish; Dir E. K. MANNES; 137 employees.

**NV Suriname Food and Flavor Industries:** Rossignolaan 3, POB 863, Paramaribo; tel. 442090; fax 464889; f. 1982; food processors; sauces, flavourings and spices; Man. Dir HANS SINGH.

**Varossieau Suriname NV:** van't Hoogerhuysstraat 63, POB 995, Paramaribo; tel. 478308; manufacturers of industrial paints and enamels; Pres. HUERA MILANO CLEVIUS BERGEN.

### TRADE UNIONS

**Council of the Surinamese Federation of Trade Unions (RAVAKSUR):** f. 1987; comprises:

**Algemeen Verbond van Vakverenigingen in Suriname 'De Moederbond' (AVVS)** (General Confederation of Trade Unions): Verlengde Coppenamestraat 134, POB 2951, Paramaribo; tel. 463501; fax 465116; e-mail moederbonds51@hotmail.com; rightwing; Pres. IMRO GREP; Chair. A. W. KOORNAAR; 15,000 mems.

**Centrale 47 (C-47):** Wanicastraat 230, Paramaribo; includes bauxite workers; Chair. (vacant); 7,654 mems.

**Centrale Landsdienaren Organisatie (CLO)** (Central Organization for Civil Service Employees): Gemenelandsweg 743, Paramaribo; tel. 499839; Pres. HENDRIK SYLVESTER; 13,000 mems.

**Organisatte van Samenwerkende Autonome Vakbonden (OSAV):** Noorderkerkstraat 2–10, Paramaribo; Pres. SONNY CHOTKAN; Gen. Sec. RONNY HEK.

**Progressieve Werknemers Organisatie (PWO)** (Progressive Workers' Organization): Limesgracht 80, POB 406, Paramaribo; tel. 475840; f. 1948; covers the commercial, hotel and banking sectors; Pres. ANDRE KOORNAAR; Sec. EDWARD MENT; 4,000 mems.

**Progressive Trade Union Federation (C-47):** Wanicastraat 230, Paramaribo; tel. 494365; fax 401149; Pres. FRED DERBY; Gen. Sec. R NAARENDORP.

**Federation of Civil Servants Organization (CLO):** Verlengde Gemenelandsweg 74, Paramaribo; tel. 464200; fax 493918; Pres. RONALD HOOGHART; Gen. Sec. FREDDY WATERBERG.

**Federation of Farmers and Agrarians (FAL):** Keizerstraat 150, Paramaribo; tel. 464200; fax 499839; Pres. JIWAN SITAL; Gen. Sec. ANAND DWARKA.

## Transport

### RAILWAYS

There are no public railways operating in Suriname.

**Paramaribo Government Railway:** Paramaribo; single-track narrow-gauge railway from Onverwacht, via Zanderij, to Brownsweg (87 km—54 miles); operational between Onverwacht and Republiek (13 km—8 miles); Dir S. R. C. VITERLOO.

**Suriname Bauxite Railway:** POB 1893, Paramaribo; fax 475148; 70 km (44 miles), standard gauge, from the Backhuis Mountains to Apoera on the Corantijn river; owing to the abandonment of plans to mine bauxite in the Backhuis mountains, the railway transports timber and crushed stone; Pres. EMRO R. HOLDER.

### ROADS

In 2000 Suriname had an estimated 4,492 km (2,750 miles) of roads, of which 1,220 km (758 miles) were main roads. The principal east–west road, 390 km (242 miles) in length, links Albina, on the eastern border, with Nieuw Nickerie, in the west.

### SHIPPING

Suriname is served by many shipping companies and has about 1,500 km (930 miles) of navigable rivers and canals. A number of shipping companies conduct regular international services (principally for freight) to and from Paramaribo including EWL, Fyffesgroup, the Alcoa Steamship Co, Marli Marine Lines, Bic Line, Nedlloyd Line, Maersk Lines and Tecmarine Lines (in addition to those listed below). There are also two ferry services linking Suriname with Guyana, across the Corentijn river, and with French Guiana, across the Marowijne river.

**Dienst voor de Scheepvaart (Suriname Maritime Authority):** Cornelis Jongbawstraat 2, POB 888, Paramaribo; tel. 476769; fax 472940; e-mail dvsmas@sr.net; government authority supervising and controlling shipping in Surinamese waters; Man. of Nautical Affairs NAOMI EERSEL.

**Scheepvaart Maatschappij Suriname NV (SMS)** (Suriname Shipping Line Ltd): Waterkant 44, POB 1824, Paramaribo; tel. 472447; fax 474814; e-mail surinam_line@sr.net; f. 1936; regular cargo and passenger services in the interior; Chair. F. VAN DER JAGT; Dir J. AMARELLO-WILLIAMS.

**NV VSH Scheepvaartmij United Suriname Shipping Company:** Van het Hogerhuysstraat 9–11, POB 1860, Paramaribo; tel. 402558; fax 403515; e-mail united@sr.net; internet www.vshunited.com/shipping.html; shipping agents and freight carriers; Man. PATRICK HEALY.

**Staatsolie Maatschappij Suriname NV:** Dr Ir Adhinstraat, POB 4069, Paramaribo; tel. 499649; fax 491105; e-mail mailstaatsolie@staatsolie.com; internet www.staatsolie.com; Chair. H. G. COLERIDGE; Man. Dir Dr S. E. JHARAP.

**Suriname Coast Traders NV:** Flocislaan 4, Industrieterrein Flora, POB 9216, Paramaribo; tel. 463020; fax 463040; f. 1981.

### CIVIL AVIATION

The main airport is Johan Adolf Pengel International Airport (formerly Zanderij International Airport), 45 km from Paramaribo. Domestic flights operate from Zorg-en-Hoop Airport, located in a suburb of Paramaribo. There are 35 airstrips throughout the country.

**Surinaamse Luchtvaart Maatschappij NV (SLM)** (Suriname Airways): Mr. Jagernath Lachmonstraat 136, POB 2029, Paramaribo; tel. 465700; fax 491213; e-mail publicrelations@slm.firm.sr; internet www.surinamairways.net; f. 1962; services to Amsterdam (the Netherlands) and to destinations in North America, South America and the Caribbean; Vice-Pres. CLYDE CAIRO; Gen. Man. RONALD J. SAHTOE.

**Gonini Air Service Ltd:** Doekhiweg 1, Zorg-en-Hoop Airport, POB 1614, Paramaribo; tel. 499098; fax 498363; f. 1976; privately-owned; licensed for scheduled and unscheduled national and international services (charters, lease, etc.); Man. Dir GERARD BRUNINGS.

**Gum Air NV:** Rijweg naar Kwatta 254, Paramaribo; tel. 498888; fax 497670; privately owned; unscheduled domestic flights; Man. Mr GUMMELS.

## Tourism

Efforts were made to promote the previously undeveloped tourism sector in the 1990s. Attractions include the varied cultural activities, a number of historical sites and an unspoiled interior with many varieties of plants, birds and animals. There are 13 nature reserves and one nature park. There were an estimated 68,000 foreign tourist arrivals at the international airport in 2001. In 2003 tourism receipts totalled US $18m.

**Suriname Tourism Foundation:** Dr J. F. Nassylaan 2, Paramaribo; tel. 410357; fax 477786; e-mail stsur@sr.net; internet www.parbo.com/tourism; f. 1996; Exec. Dir A. LI A. YOUNG.

## Defence

The National Army numbered an estimated 1,840 men and women in August 2004. There is an army of 1,400, a navy of 240 and an air force of some 200.

**Defence Budget:** an estimated US $7.3m. in 2004.

**Commander-in-Chief:** Col ERNST MERCUUR.

## Education

Education is compulsory for children between the ages of seven and 12. Primary education lasts for six years, and is followed by a further seven years of secondary education, comprising a junior secondary cycle of four years followed by a senior cycle of three years. All education in government and denominational schools is provided free of charge. In 2001 the total enrolment in primary education was equivalent to 97.4% of children in the relevant age group (males 96.7%; females 98.1%), while the total enrolment in secondary education was equivalent to 63.4% of children in the relevant age group (males 52.4%; females 74.8%). The traditional educational system, inherited from the Dutch, was amended after the 1980 revolution to place greater emphasis on serving the needs of Suriname's population. This included a literacy campaign and programmes of adult education. Higher education was provided by four technical and vocational schools and by the University of Suriname at Paramaribo, which had faculties of law, social and technological sciences, economics and medicine. Expenditure on education by all levels of government in 1996 was an estimated Sf 3,840m. (3.2%). In March 2003 the Inter-American Development Bank approved a US $12.5m. loan to fund the reform of the basic education system into a single 10-year cycle. It was hoped that the funds would result in a 10% increase in the number of pupils who finished sixth grade and a 20% reduction in drop-out and repetition rates. A further $13m. grant for the sector was approved by the Dutch Government in January 2005.

## Bibliography

For works on the region generally, see Select Bibliography (Books)

Brown, E. *Suriname and the Netherlands Antilles*. Lanham, MD, Scarecrow Press, 1992.

Colchester, M. *Forest Politics in Suriname*. Chicago, IL, International Books, 1996.

Dew, E. M. *The Trouble in Suriname, 1975–1993*. New York, NY, Praeger Publrs, 1995.

Hoefte, R. A. L. *Suriname* (World Bibliographical Series). Oxford, ABC Clio, 1990.

*In Place of Slavery: A Social History of British Indian and Javanese Labourers in Suriname*. Gainesville, FL, University Press of Florida, 1999.

Kambel, E.-R., and MacKay, F. *The Rights of Indigenous Peoples and Maroons in Suriname*. Copenhagen, International Work Group for Indigenous Affairs, 2000.

Mackenzie, G. A. (Ed.). *Suriname*. Washington, DC, International Monetary Fund, 1999.

Meel, P. *Tussen autonomie en onafhankelijkheid: Nederlands-Surinaamse betrekkingen 1954–1961*. Leiden, KITLV Uitgeverij, 1999.

Schultz, G. *Turmoil and Triumph: Diplomacy, Power and the Victory of the American Ideal*. New York, NY, Touchstone, 1996.

Sizer, N., and Rice, R. *Backs to the Wall in Suriname: Forest Policy in a Country in Crisis*. Washington, DC, World Resources Institute, 1995.

*Suriname: Policies to Meet the Social Debt*. Santiago (Chile), PREALC, 1992.

Thoden Van Velzen, H. U. E., Van Wetering, W., Van Wetering, I., and Van Der Elst, D. *In the Shadow of the Oracle: Religion As Politics in a Suriname Maroon Society*. Prospect Heights, IL, Waveland Press Inc, 2004.

USA Ibp. *Suriname Central Bank and Financial Policy Handbook*. Milton Keynes, Lightning Source UK Ltd, 2005.

# TRINIDAD AND TOBAGO

## Geography

### PHYSICAL FEATURES

Trinidad and Tobago lies in the West Indies, its constituent parts being considered the southernmost islands of the Caribbean, although, geologically, they are extensions of the South American continent. Venezuela (specifically, the Paria peninsula) lies only 11 km (seven miles) from the island of Trinidad, the two countries embracing the Gulf of Paria. The Gulf is entered by sea channels known as the Bocas, in the north by the Dragon's Mouths and in the south by the slightly broader Serpent's Mouth, which joins the eastward-widening Columbus Channel between the southern coast of Trinidad and the delta lands of the Orinoco in Venezuela. Some 145 km (90 miles) to the north of Trinidad is Grenada, the next nearest neighbour. Tobago, which accounts for only 300 sq km (116 sq miles) of the country's total area of 5,128 sq km, lies some 32 km to the north-east of Trinidad island.

Trinidad (80 km by 60 km in extent) is an anvil-shaped island, with tapering extensions in both the south-west and the north-west that reach further westward towards the coast of Venezuela and help enclose the Gulf of Paria. This provides a vast sheltered anchorage in the lee of Trinidad island, and it is on this shore that the capital, Port of Spain, is located (in the north). West of here, near the tip of the north-western peninsula, is the only natural harbour, Chaguaramas (central): the northern coast is rocky, the eastern coast exposed to heavy seas directly off the Atlantic and the southern coast steep. Three ranges of highlands cross the island from east to west, the highest being in the north, where El Cerro del Aripo reaches 940 m (3,085 ft), amid the densely forested slopes of the Northern Range. There are also the Central Range and the Trinity Hills in the south-east. Mid-way along the northern coast of the south-western peninsula is the 47 ha (116 acre) Pitch Lake near La Brea, the largest natural asphalt reservoir in the world, which testifies to the country's hydrocarbons wealth, while another geographical oddity, mud volcanoes, indicates the volcanic origins of the island. Plains dominate the rest of the south and the land along the Caroni river between the Central and Northern Ranges. Here the landscape is predominantly agricultural, giving way to wetlands near the coast, such as the Caroni Swamp (the roosting ground for the scarlet ibis and the egret) in the west and the Nariva Swamp (the largest freshwater swamp on the island, and home to the weeping or Trinidad capuchin, the red howler monkey and 32 species of bat on the east coast. Still-extensive natural habitats such as these, as well as the dense northern rainforest, are widely protected, as both Trinidad and Tobago, once attached separately to the continental landmass, host probably the greatest ecological diversity in the insular Caribbean.

Tobago, the summit of a single mountain mass rising from the sea floor, is aligned from south-west to north-east, its northerly shores, facing the Caribbean, bearing the leeward beaches protected from the Atlantic weather. The island is about 42 km in length and up to 14 km wide, with plains of a coralline origin in the south rising towards the backbone of a central 29-km Main Ridge range, rising towards the north-east and reaching its highest point at Pigeon Peak (594 m). At this end of the island the peaks are cloaked with the oldest protected rainforest region in the Americas (since 1764). Tobago boasts an even greater variety of bird and animal life than the more developed Trinidad, claiming to be home to 210 bird species (notably the beautiful crowned blue mot mot), 123 butterfly species, 24 types of snake, 16 lizard kinds, 14 species of frog and a variety of spectacled cayman (the smallest alligator species). The country as a whole is reckoned to have 433 species of birds and 622 of butterflies, more than any other Caribbean island nation. Wildlife also thrives on the motley collection of smaller islands and islets, mostly north of Tobago, that the country also has under its jurisdiction. The marine environment is also flourishing,

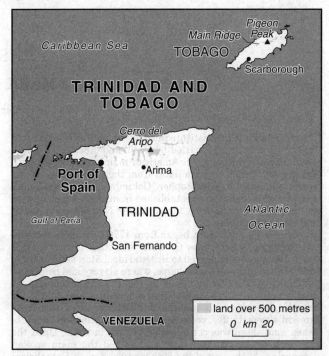

encouraged by the confluence of Atlantic currents with the warmer Caribbean and the rich discharge of the Orinoco.

### CLIMATE

The climate is tropical, with a rainy season from June to November and a slightly cooler season between December and April. The islands are not in the hurricane belt, but fairly constant breezes off the Atlantic keep the temperatures from becoming too hot. The average maximum temperature is 32°C (89°F), although Tobago is slightly cooler, owing to its more constant exposure to the trade winds. Average annual rainfall is about 2,050 mm (80 ins) over most of the country, being higher in the mountains.

### POPULATION

The population is one of the most varied in the Caribbean, a complex colonial history of competing Spanish, French and, finally, British rule augmented by the immigration of black Africans (originally as slaves), Indians (mainly from north India as indentured labour) and many others. The largest single ethnic group is now 'East' Indian, followed closely by those of African descent (each constituting about 40% of the population), with 18% of mixed heritage and the rest consisting of whites, Chinese, Lebanese and others. The East Indians tend to dominate agriculture still, but are also important in business. Blacks are more dominant on Tobago. Culturally, there are two folk traditions in the country, the Creole (black and colonial) and the East Indian, the latter including the presence of Hindus and Muslims, resulting in the official celebration of those religious holidays as well as the Christian ones. Christianity remains the main religion, with Roman Catholicism the leading denomination (30%), but Hindus represent about 23% of the population, while Anglican Christians claim the adherence of 8% and various Protestant groups 14% (Pentecostalists alone have 7%). Muslims account for 6% of the population. This diversity is reflected in the variety of languages spoken by the populace, with a French Creole patois surviving in some areas, Hindi being widely spoken among East Indians, and others using

Chinese or Spanish. However, the most widely spoken tongue and the official language of the country is English.

In September 2002 the total population was put at 1,275,700, with only some 4% living on Tobago. The capital, Port of Spain (including its suburbs), had a population of some 54,000 in 2001, while San Fernando, at the southern end of the west coast, is also a major population centre, followed by Arima, inland to the east of Port of Spain. These are the three administrative municipalities of Trinidad, which also has eight counties. Tobago, which has some autonomy within the republic, is constituted as a ward. Its chief town is Scarborough, on the southern coast at the south-western end of the island. The other important towns are Plymouth, across on the north coast from Scarborough, and Roxborough, towards the northern end of the south-facing coast.

# History

## MARK WILSON

Based on an earlier article by ROD PRINCE

The two islands that constitute the modern state of Trinidad and Tobago were both first settled by Carib and other Amerindian populations from the South American mainland. Contact with Europe and North America dates from the third voyage of the Spanish navigator Christopher Columbus, in 1498. Small Spanish settlements were established from 1592, which formed an insignificant part of the Viceroyalty of New Granada, based in Bogotá, Colombia.

Larger-scale settlement began from 1776, when Spain began to encourage settlers from Grenada and other former French possessions that had passed to British rule. Later, settlers came from Martinique and Guadeloupe, where slavery had been temporarily abolished as a result of the French Revolution. A Spanish cedula of 1783 established a system of land grants for French planters, with additional allocations for those who brought slaves. Coffee, cocoa and sugar plantations were established, and there was considerable growth in trade. Until the late 19th century French creole remained the main spoken vernacular. At the end of the 20th century Roman Catholicism remained the main Christian denomination.

Trinidad was captured by the United Kingdom in 1797 and was formally ceded to that country by Spain in 1802. Tobago was colonized by the Dutch in 1628, but then claimed by a succession of European countries until the British took possession in 1762, following 100 years of French occupation; France ceded the island to the United Kingdom in the following year. However, it was not until 1814 that the British gained the island in perpetuity. Throughout the 19th century in Trinidad there remained a latent conflict between the British administration and a mainly French planter and commercial class. For this reason, Trinidad, unlike most West Indian colonies, had no elected Assembly.

The abolition of slavery in 1834 was followed by four years in which slaves were forced to remain on the plantations under an 'apprenticeship' system. From 1838, however, many former slaves moved to Port of Spain or established peasant farms on unoccupied land; as did a large number of immigrants from other West Indian colonies. With a labour shortage developing on the plantations, from 1845 the Government encouraged the immigration of indentured labourers, mainly Indian, but also Chinese and Madeiran. On the expiry of a 10-year period, of which the first five were contracted to a single employer, the workers received either their passage home or title to a small plot of land. This form of immigration continued until 1917 and the majority of the rural population in modern Trinidad was descended from Indian indentured labourers. Tobago's sugar industry performed badly after emancipation, and was close to collapse by the 1880s. For this reason, the smaller colony was attached to Trinidad from 1889.

Agitation for increased political rights began in Trinidad from the 1880s, earlier than in most other West Indian colonies. From 1896, the Trinidad Workingmen's Association played an important role in political life. Against this background, in 1899 the colonial authorities unwisely abolished the elected Borough Council in Port of Spain. In 1914 the Council was re-established and, from 1925, there were elected members on the Legislative Council, although the electoral roll was limited by property qualifications.

### THE MOVE TO INDEPENDENCE

Disturbances resulting from the depressed economic situation of the 1930s provided suitable conditions for the foundation of the labour movements. These, in turn, evolved into political movements, particularly after the introduction of universal adult suffrage in 1946. The People's National Movement (PNM), founded in 1956 by Dr Eric Williams, won 39% of the votes cast in legislative elections held in 1956, and gained control of the Legislative Council (securing 13 of the 24 elected seats), under the provision of the new constitutional arrangements that provided for self-government. Dr Williams became Trinidad and Tobago's Chief Minister.

Along with most other British Caribbean colonies, Trinidad and Tobago joined the Federation of the West Indies in 1958. However, when Jamaica left the grouping in 1961 in order to seek independence individually, Trinidad and Tobago also withdrew, reluctant to take financial responsibility for the other, and, at that time, poorer, islands, in the north of the region, and the Federation collapsed. Internal self-government for Trinidad and Tobago in 1961 was followed by independence on 31 August 1962.

### THE PNM AND PARTY POLITICS

Elective politics in Trinidad and Tobago mostly ran along racial lines, although there was some attempt by all parties to win at least token support from other groups. The PNM was mainly urban and Afro-Trinidadian, with some backing from Muslim and Christian Indians. Its early opponent, the Democratic Labour Party (DLP), led by Rudranath Capildeo, was based in the Hindu section of the Indian-origin community. Although Indo- and Afro-Trinidadians were roughly equal numerically, support from minority and mixed-race voters, together with the pattern of constituency boundaries, was enough to give the PNM a secure parliamentary majority over the DLP and its ethnically based successor parties until 1986. Dr Williams remained Prime Minister until his death in 1981.

The PNM's strongest challenge in this period came from the 'Black Power' movement, which came to prominence on the islands in 1970. Influenced to some extent by African-American radicalism, support for the movement stemmed mainly from a well-established perception that, eight years after independence, there was significant discrimination against Afro-Trinidadians in private-sector employment, and that insufficient respect was given to black people and their culture. A series of demonstrations in 1970 was accompanied by some violence, and a state of emergency was declared. Unrest extended to junior army officers, who unsuccessfully attempted to overthrow the Government. The leaders of the Black Power movement were imprisoned without trial, and marches were banned. Nevertheless, many of their aims were achieved in the 1970s, when there were broader economic opportunities and more attention to cultural development.

Political difficulties persisted for some time, however, and there was an opposition boycott of the 1971 elections. In September 1973 Dr Williams announced his resignation, but was persuaded to rescind it. A new Constitution came into effect on 1 August 1976, making Trinidad and Tobago a republic within the Commonwealth. In elections to the House of Representatives in the following month, the PNM won 24 of the 36 seats, while an alliance of petroleum and sugar workers within the newly formed United Labour Front (ULF), a mainly Indian party with trade-union support led by Basdeo Panday, secured 10 seats. The two Tobago seats were won by the Democratic Action Congress (DAC), led by Arthur N. R. Robinson, a former PNM cabinet minister. The former Governor-General, Ellis Clarke, was sworn in as the islands' first President in December.

At this time, the PNM was presiding over a buoyant economy, following steep rises in the world price of petroleum in 1973 and 1974 that transformed Trinidad and Tobago's relatively small reserves of petroleum into a major financial and political asset. Petroleum prices remained high until 1981, by which time the PNM's dominance of national politics was such that, at his death in March of that year, Williams was without serious rivals.

## THE RISE OF THE NATIONAL ALLIANCE FOR RECONSTRUCTION

Williams was succeeded as Prime Minister by George Chambers, one of the three deputy leaders of the PNM. Chambers was seen as moderate and generally fair, but failed to command enthusiastic support. He presided over another election victory for the PNM in November 1981, winning 26 seats in the Parliament. The DAC and the ULF, as well as an intellectual pressure group, the Tapia House Movement, formed the National Alliance and took the two Tobago seats and eight, mainly Indian, rural seats, gaining 20.7% of the votes cast. The middle class, conservative, Organization for National Reconstruction (ONR), led by another former PNM minister, Karl Hudson-Phillips, won 22.3% of the ballot but took no seats, because its mainly middle-class support was evenly spread across a large number of constituencies.

In August 1983 the four opposition parties formed an alliance to contest local elections and won 66 of the 120 local council seats. A united opposition party, the National Alliance for Reconstruction (NAR), was formed a year later. The NAR aimed to offer a credible alternative to the Government, at a time when stagnant and then decreasing petroleum prices were severely reducing government revenues and highlighting the defects of the policies pursued during the increase in prices of the previous decade. In the November 1984 elections to the Tobago House of Assembly, which had been formed in 1980, the DAC (part of the NAR) reduced the PNM's share of the 12 elective seats from four to one. Prime Minister Chambers was increasingly losing political respect, and his position was not strengthened by his failure to support the US-led military intervention in Grenada in October 1983. A 33% devaluation of the currency in December 1985 was deeply unpopular, and was seen as further eroding purchasing power in what was still an import-dependent economy.

Accusations of government economic mismanagement and corruption increased in the months preceding the December 1986 general election, intensified by the Government's failure to publish an official report on drugs-trafficking. The NAR, led by Robinson, won an overwhelming victory, gaining 33 of the 36 seats in the House of Representatives, and 67% of the votes. The PNM's parliamentary representation was reduced to just three seats, owing to residual support in urban strongholds in eastern Port of Spain and in San Fernando. The party retained 33% of the popular vote, nevertheless, which formed the basis for a later recovery under the former energy minister, Patrick Manning, who was appointed PNM leader in January 1987. Robinson was appointed Prime Minister.

## THE GOVERNMENT OF A. N. R. ROBINSON

The NAR Government was an unwieldy coalition of the ULF, the ONR, the DAC and the Tapia House Movement, a grouping of university radicals. Divisions emerged almost immediately over the allocation of cabinet portfolios and positions on state boards, and extended to policy matters, as the Government was forced to implement unpopular measures in response to a deepening economic crisis. There were also accusations that Indo-Trinidadians, who formed a majority of NAR voters, were being excluded from positions of real power.

The small Tapia House Movement left the NAR in June 1987, while John Humphrey, a former ULF member, was removed from his post as Minister of Public Works and Settlement in November. The external affairs minister and former ULF leader, Basdeo Panday, was dismissed in February 1988, along with one cabinet colleague and a junior minister. All three accused the NAR leadership of racism and were expelled from the party in October. In April 1989 Panday announced the formation of a new party, the United National Congress (UNC), which became the official opposition in September 1990, by virtue of holding six seats to the PNM's three. The UNC at this time was widely perceived to be a rural, Indo-Trinidadian party with strong links to the sugar workers' trade union.

The Government's unpopularity increased as it was forced to take further austerity measures. A compensatory financing agreement with the IMF was followed by stand-by agreements in 1989 and 1990. There was a further 15% currency devaluation in August 1988, which increased retail-price levels. Government expenditure was cut, and the budget of January 1989 imposed a 10% reduction on public-sector pay. This last measure aroused the hostility of the trade unions, which successfully challenged its legality.

In July 1990 a group of insurgents from the Jamaat al Muslimeen, a sect of mainly Afro-Trinidadian Muslim converts led by Yasin Abu Bakr, a former policeman, stormed the parliament building during a session of the House of Representatives, taking 46 hostages including the Prime Minister and most of the Cabinet. At the same time, they blew up the police headquarters building and took over what was then the sole television station, using it to broadcast their demands that the Prime Minister should resign, and his deputy lead an interim Government into elections within 90 days. Widespread looting and fires began almost immediately in the capital and some suburban centres. The siege lasted for five days, during which time 23 people were killed and 500 wounded, most of them looters shot by the police. The Prime Minister was shot in the leg by the rebels after refusing to sign a letter of resignation. However, the acting President was induced to sign an amnesty for the rebels, which was delivered to them in the parliament building.

Following the surrender of the Jamaat al Muslimeen, the Government announced that the amnesty agreement had been made under duress and was therefore invalid. Abu Bakr and 113 others were charged with murder and treason. The Judicial Committee of the Privy Council (the country's final court of appeal, and based in London, United Kingdom) ruled, in November 1992, that the validity of the presidential amnesty should be determined before the case came to trial. In July 1992 the High Court ruled that the pardon was valid, and ordered the release of the accused. In October 1993 the Government lost an appeal against the ruling, but a year later the Privy Council ruled that the Jamaat al Muslimeen had invalidated the pardon by failing to surrender as soon as it was agreed, instead continuing the siege in an attempt to win further concessions. However, it was also ruled that it would be an abuse of process for the accused to be rearrested and tried. In January 2000 the Jamaat al Muslimeen was ordered to pay TT $20m. for damage incurred during the insurrection. In May 2001, however, the High Court awarded TT $625,000 in compensation to the group for damage done to its headquarters, in addition to an earlier payment of TT $1.5m. In 2005 Abu Bakr remained a prominent figure, and was reputedly linked to much of the island's violent crime, as well as to lucrative quarrying operations in north-eastern Trinidad. After a preliminary magistrate's court hearing in July 2004, he was committed to stand trial for conspiracy to murder a former member of the Jamaat al Muslimeen, who had been shot dead in June 2003. The trial jury was unable in March 2005 to agree a verdict, and a retrial was expected. In 2004 another leading figure of the Jamaat was extradited to the USA to face arms-smuggling charges.

## THE RETURN OF THE PNM

One member of parliament was killed in the siege, and the NAR lost the by-election for his successor. This highlighted the Government's loss of support, but there was some surprise at the scale of the NAR's rout in the general election that followed in December 1991, when the party gained only 24% of the valid votes cast and secured only the two seats on Robinson's home island, Tobago. The PNM won 21 legislative seats and 45% of the popular vote, mainly in urban areas with an Afro-Trinidadian majority, while the UNC took 29% of the ballot and 13, mainly rural, seats in central and southern Trinidad.

The new PNM Government, led by Patrick Manning, contained few members of previous PNM administrations. Economic policies reversed many aspects of the party's previous policies, focusing on financial and trade liberalization, divestment of state enterprises and foreign investment, particularly in heavy industries based on offshore natural gas resources.

In April 1993 the Trinidad and Tobago dollar was made convertible, in a strictly managed float against the US dollar. In contrast with Jamaica and Guyana, the currency remained stable, in spite of occasional shortages on the foreign-exchange markets. The fiscal current account was in surplus from 1993 and the economy moved back into steady growth from 1994. A protracted dispute with the trade unions over repayment of the TT $3,000m. in salary arrears incurred through the NAR's attempt in 1991 to reduce public-sector salaries was settled in 1995 for teachers, and in the following year for other public servants, who were able to claim either no-interest bonds, additional leave, or other non-cash benefits.

A state of emergency was briefly declared on 3 August 1995, in order to allow the Government to remove from office the Speaker of the House of Representatives, Occah Seapaul, who was the subject of damaging financial allegations. This, in turn, led to the defection from the Government, on the following day, of her brother, Ralph Maraj, the Minister of Public Utilities (and former Minister of Foreign Affairs). With only a narrow majority now remaining, the Prime Minister responded by announcing that a general election was to be held on 6 November 1995, one year earlier than was necessary.

## THE ELECTION OF THE UNC

In the election, although the PNM increased its share of the votes to 49%, its support was concentrated heavily in its urban strongholds. The UNC received enthusiastic backing from some important members of the business community (three of the most prominent of these supporters were, in 2002, formally charged with corruption-related offences), and raised its share of the ballot to 46%. The UNC secured control of three marginal constituencies, which left each party with 17 of the 34 Trinidad seats. The NAR's remaining support in Trinidad had collapsed, but with the support of the two Tobago members, the UNC was able to form a Government and Basdeo Panday became Prime Minister. In spite of the electoral defeat, Manning resisted continuing demands from within the PNM for his resignation as party leader, winning a further five-year term in the post in 1996. However, he lost the support of important sections of the PNM, which weakened his ability to organize an effective opposition to Panday's Government.

The new administration continued most of the economic policies of its predecessor. The new finance minister, Brian Kuei Tung, and the foreign minister, Ralph Maraj, had both been members of the previous Cabinet. In February 1997 the Government's parliamentary position was strengthened by the defection of two PNM members, who subsequently sat as independents and were appointed junior ministers. The Government presided over a period of strong economic growth, but there were persistent and widespread reports of corruption and mismanagement. On 14 February an electoral college of both houses of Parliament elected Robinson as President. He immediately relinquished his parliamentary and cabinet seats, as well as the leadership of the NAR. The PNM was opposed to a head of state who had, until his nomination, played an active role in party politics, and for the first time a presidential election was contested. However, the PNM candidate, a serving high court judge, received 18 votes to Robinson's 46. The NAR retained President Robinson's former seat at a by-election in May. However, in April the representative of the other Tobago seat, Pamela Nicholson, resigned from the Government and from the NAR to sit as an independent. As a result of these changes, from mid-1997 the Government could command the support of 20 members of the House of Representatives.

A serious dispute between the Government and the Chief Justice, Michael de la Bastide, began in September 1999 when he claimed that the independence of the judiciary was threatened by an attempt by the Attorney-General, Ramesh Lawrence Maharaj, to tighten administrative and financial controls. The Chief Justice was supported by 28 of the country's 29 judges, but not by the President of the Bar Association, Karl Hudson-Phillips. The Government responded in December by announcing that a former British Lord Chancellor, Lord Mackay of Clashfern, would lead a three-member Commission of Inquiry into the administration of justice on the islands. President Robinson, clearly unhappy with this decision, did not appoint the Commission until February 2000, signalling an unprecedented breach between the largely ceremonial head of state and the Cabinet. This was all the more surprising owing to the fact that President Robinson had himself been a senior member of Panday's Cabinet in 1995–97 and had been nominated as President by Panday's Government. The Commission announced its findings and recommendations in October 2000, clearing the Attorney-General of seeking to undermine the independence of the judiciary, but apparently failing to settle the fundamental disagreement between de la Bastide and Maharaj.

The conflict between Prime Minister Panday and President Robinson intensified in 2000 and 2001; both the President and the Chief Justice were alleged by Panday to be 'enemies' of his administration, along with Ken Gordon, Chairman of the Caribbean Communications Network media group, who, as a consequence, won damages of TT $600,000 in a libel suit against Panday in October 2000.

In January 2000 the Prime Minister decided to revoke the appointment of two Tobagonians as government senators, on the basis that they had voted against his administration; the President, himself a Tobagonian, delayed before acting on this decision, believing that their dismissal would be a breach of the understanding reached in 1995 between the UNC and the NAR. Following the mediation of the Archbishop of Port of Spain, the Most Rev. Anthony Pantin, Robinson agreed to revoke the senators' appointments, replacing them with two Tobagonians nominated by Panday.

In early January 2000 Hansraj Sumairsingh, a member of the UNC and leader of one of Trinidad's nine regional corporations, was found murdered. It was subsequently reported that Sumairsingh had written to the Prime Minister several times regarding threats made to him by the Minister of Local Government, Dhanraj Singh, following Sumairsingh's revelation of alleged corruption in the unemployment relief programme administered by Singh's ministry. Although Panday said that he could not remember having received such letters, a copy of one was later produced by a PNM member. A full investigation was initiated and in October Singh was dismissed from his government post. Singh left unexpectedly for the USA for medical treatment for stress and high blood pressure. However, he returned to Trinidad in January 2001 and was charged with corruption in the unemployment relief programme and, in February, with the murder of Sumairsingh. After preliminary hearings, in April 2002 he was committed for trial.

## THE RE-ELECTION OF THE UNC

The general election of 11 December 2000 was won by the UNC with 52% of the votes cast, compared to 46% for the PNM. The UNC gained two marginal seats on Trinidad, bringing its total in the 36-seat House of Representatives to 19, while the PNM won 16 seats and the NAR held one of the two Tobago seats. Panday was appointed Prime Minister for the second time.

At elections to the Tobago House of Assembly, held on 29 January 2001, the PNM took control of the Assembly for the first time, winning eight of the 12 seats. The NAR secured the remaining four seats, while the UNC, contesting the election for the first time, won minimal support, as did the People's Empowerment Party, formed by former NAR dissidents.

The December 2000 general election was marred by allegations of fraud. Opposition charges that the electoral list had been manipulated were highlighted by police investigations of prominent government supporters and arrests were made following attempted manipulation of the electoral register. In January 2001 Panday announced his Cabinet and his nominations to the Senate. However, President Robinson at first refused to approve the senate nominations of seven UNC members, declaring them to be unconstitutional, as they had been defeated as candidates in the congressional election. His refusal was widely felt to be outside the President's discretionary powers under the Constitution.

Despite its election victory and the continuing growth of the economy, the UNC Government continued to be dogged by allegations of corruption in early 2001, including alleged fraud within the North West Regional Health Authority and irregularities at the state-owned Petroleum Company of Trinidad and Tobago Ltd (Petrotrin). In March, owing to health problems, Panday reduced his workload, relinquishing parts of his portfolio.

Relations between Panday and UNC deputy leader (and Attorney-General) Maharaj began to deteriorate in 2001. Panday pointedly failed to appoint Maharaj to act as Prime Minister when he travelled overseas. During the September budget debate, Maharaj and two cabinet colleagues, Ralph Maraj and Trevor Sudama, were sharply critical of the Government on corruption and other issues. Although the three dissidents did vote in favour of the budget proposals, Maharaj resigned as Attorney-General on 1 October, and his two supporters also subsequently resigned their posts. They formed a temporary alliance with the PNM, proposing to form a new Government; however, President Robinson instead agreed to a request by Panday on 10 October to dissolve Parliament and hold a general election on 10 December. Maharaj and his majority on the UNC executive were unable to establish their right to the party's name and symbol, and fought the election as a new grouping, Team Unity.

## A HUNG PARLIAMENT

Following a long and bitter electoral campaign, Panday's fiercely loyal ethnic Indian support was barely diminished in the legislative elections of 10 December 2001, despite the divisions within the UNC and the allegations of corruption. The UNC retained 50% of the popular vote, with 46% for the PNM. However, the PNM narrowly gained a marginal seat in Trinidad from the UNC, as well as the NAR's remaining Tobago seat. This left each main party with 18 legislative seats each.

As a result, on 16 December 2001 the UNC and PNM leaders signed the so-called 'Crowne Plaza Accord' in a Port of Spain hotel, which, if implemented, would have allowed Parliament to function. They agreed on the choice of Speaker for the House of Representatives and to accept the choice of President Robinson for Prime Minister. Following several days' deliberation, on 24 December the President announced that Manning should be Prime Minister. Panday immediately declared the choice illegal and unconstitutional, and maintained that the candidate for Speaker previously agreed in the Crowne Plaza Accord was closely linked to the PNM. With no Speaker, Parliament could not function, nor could an electoral college be convened to choose a successor for President Robinson, whose term in office was due to expire on 19 March 2002.

In spite of deep political polarization, society continued to function more or less as normal. The budget passed in September 2001 provided a financial basis for administration to continue. Panday's threats to launch a mass campaign of 'civil disobedience' were not carried out, possibly because of pressure from his business supporters. Robinson agreed to remain in office for up to one year, or until an electoral college could be convened. In line with normal practice in Trinidad and Tobago, the PNM replaced UNC appointees with its own supporters in key positions in the state sector; but with both parties agreeing on most points of economic and social policy, there were few other changes. According to the Constitution, a six month interval is allowed between meetings of Parliament. Close to this deadline, on 5 April, an attempt was made to convene the House of Representatives. The UNC voted against the PNM's nominees for Speaker, further complicating the issue by nominating several thousand nominees of its own, while at the same time voting against them, in case any were acceptable to the Government. After two days, Parliament was prorogued.

The new Government appointed Commissions of Inquiry to investigate some of the allegations of corruption made against the UNC Government. The most prominent of these related to the construction of a new airport terminal, but was not made public on completion in August 2003, to avoid the danger of prejudicing possible legal proceedings. In some cases, formal charges were made by the Director of Public Prosecutions, an independent and non-political official. Brian Kuei Tung, Panday's Minister of Finance in 1995–2000, was in March 2002 charged, together with five associates, with corruption, conspiracy and misbehaviour in public office. The charges were brought in connection with contracts for a new airport terminal, and carried a maximum sentence of 10 years' imprisonment. Further charges were laid in May 2004 against Kuei Tung, former UNC Minister of Works Sadiq Baksh and 11 others. Panday himself was charged in September 2002 with failure to note large deposits made to his wife's bank account in the United Kingdom in his statutory declaration of financial assets. However, defence lawyers in these and other cases raised a series of procedural issues, so that progress in the cases was slow. The UNC and the PNM continued to exchange allegations of corruption and malpractice. Panday likened the operations of the Commissions of Inquiry to those of a police state, and there were allegations of criminal activity by two more UNC members of parliament. The former works minister, Carlos John, was also accused of ignoring tender procedures during his term in office.

The Commission of Inquiry into the Elections and Boundaries Commission (EBC) in early June 2002 reported countless flaws in its practices and recommended that all its members should resign. Despite Manning's pledge to implement the reforms recommended by the report, all the EBC's members refused to relinquish their posts. Government was able to function on the basis of the 2001 budget until the end of October 2002, but not beyond. Increasing pressure, both domestic and from other governments in the region, as well as the parliamentary impasse over the appointment of a Speaker, eventually forced Manning to announce, on 28 August, after the failure of a second attempt to elect a Speaker, that a further general election would be held on 7 October—the third in 22 months.

The PNM gained 50.7% of the popular vote in the election of October 2002 and 20 seats, compared with 46.6% for the UNC (and the remaining 16 seats). In February 2003 the PNM's majority enabled an electoral college of the members of both houses of Parliament to elect George Maxwell Richards, a former principal of the local campus of the University of the West Indies, as President, allowing Robinson to retire.

From the early part of 2003 political attention was focused on the rate of murders and kidnappings, which had accelerated since 2001. The number of murders rose from an average of 108 per year in 1991–99 to 260 in 2000–04, although the per-head murder rate of 20 per 100,000 remained substantially below that of Jamaica (55) and was the sixth highest in the English-speaking Caribbean. Much of the violence stemmed from feuding between rival drugs gangs in a small district of eastern Port of Spain. There was also a sharp increase in kidnappings for ransom in 2002–03, with a high proportion apparently also drugs-related. The authorities viewed the control of drugs-transhipment as an essential element of the control of crime. Installation of a new coastal radar system was completed in 2005, and the Government also planned to double the strength of the Defence Force and coast guard, and to establish a new air guard. The Government proposed the introduction of harsher penalties for abduction in an Anti-Kidnapping Bill presented early in 2005, and had, in 2004, reintroduced proposals for police reform originally formulated by the UNC Government. These measures each required a constitutional amendment, and hence opposition support. Panday stated repeatedly that the PNM was 'part of the criminal activities', and had formed an alliance with criminal elements in order to win the 2002 elections. The UNC had stated that it would oppose any constitutional measures unless there was immediate and wide-ranging constitutional reform. Manning responded positively to Panday's proposal of an executive in place of a ceremonial President, but stressed the need for wide public consultation before instituting major changes. Inter-party talks on constitutional reform began in February 2004, but were suspended in May by the UNC, which insisted on an immediate package of reform, including proportional representation. The UNC's parliamentary strength was, however, reduced to 14 in April 2005, when two members of parliament realigned themselves as independents in protest at Panday's handling of an incident in which a UNC member of parliament, Chandresh Sharma, alleged that the Minister of Housing, Keith Rowley, had thrown a teacup at him in September 2004. Sharma was suspended from the House in May 2005 for failing to apologize for his allegations. In June 2004 the EBC proposed boundary changes, which were agreed by Parliament in the following June; the reforms would increase the number of constituencies from 36 to 41, thereby preventing the recurrence of a 'hung' parliament and increasing the number of marginal seats, while reducing the influence of the two Tobago members.

In an election to the Tobago House of Assembly in January 2005, the PNM increased its legislative strength from eight to 11 of the 12 seats; the remaining seat was taken by the Democratic Action Congress, a group formed by a former NAR chairman of the island assembly, Hochoy Charles. With little local support, the UNC did not contest the island election.

Corruption allegations continued to dog both political parties in 2005. In April the UNC raised allegations originally made by a former PNM regional government councillor against the Minister of Works and Transport, Franklyn Khan, and the Minister of Energy and Energy Industries, Eric Williams (no relation to the former Prime Minister). Khan resigned in May, pending an inquiry, while Williams remained in office. In that month further corruption charges were filed against Panday, his wife, his former Minister of Works and Transport, Carlos John, and a business associate. Panday initially refused to pay bail, spending eight days in prison as a result, but later relented and was released. Earlier charges against Panday and several of his former ministers remained before the courts in 2005, but appeared likely to be resolved before the next general election.

## INTERNATIONAL RELATIONS

In April 2000 Trinidad and Tobago received widespread international condemnation following a government announcement that the country was to withdraw from the first optional protocol to the International Covenant on Civil and Political Rights, owing to its continued support of capital punishment. The Government stated that the withdrawal was intended to prevent condemned murderers from addressing lengthy appeals. Both major parties, while in office, pressed for the use of capital punishment, but, in most cases, were unsuccessful in overcoming the legal obstacles. In 2005 the Government also raised the possibility of reintroducing corporal punishment, provision for which has been on the statute books since colonial times.

Talks have been held since 1995 to demarcate the maritime boundary between Trinidad and Tobago and Barbados and to revise a 1991 agreement granting Barbados limited fishing rights. Although fishing issues were prominently reported, oil and gas-bearing geological structures close to the probable boundary line were believed to be of great interest to both sides. Differences escalated sharply from January 2004; on 14 February Barbados imposed limited licensing requirements on some imports from Trinidad and Tobago, then, on 16 February, referred the maritime dispute to the UN Law of the Sea Tribunal for resolution. With procedures slow, this process was expected to take some years to complete. However, by late 2004, constructive bilateral relations had been restored on other issues.

On 14 February 2001, at a Caribbean Community and Common Market (CARICOM) summit in Barbados, the leaders of 11 Caribbean states signed an agreement to establish the Caribbean Court of Justice. The Court, to be based in Trinidad and Tobago, would replace the Privy Council in the United Kingdom as the final court of appeal, and would also adjudicate on disputes relating to the implementation of CARICOM agreements. In July 2002 the Caribbean Development Bank announced that it would assist in the establishment of a US $100m. trust, from which the Court's costs would be funded, as soon as borrowing member states had passed the necessary legislation, a process that was completed in 2004. As the required minimum of three member states had already ratified the treaty to establish the Court, it was expected to come into existence soon afterwards. The Court was formally inaugurated in April 2005. However, by mid-2005 only Barbados and Guyana had adopted the appellate jurisdiction of the Court, with other member states initially using it only to settle CARICOM matters. Ironically, Trinidad and Tobago was likely to retain use of the Privy Council, as the opposition UNC stated that it would not support the Court, reversing the position it adopted while in office. A former President, Arthur N. R. Robinson, played a prominent role in the establishment of the International Court of Justice. Along with most of its CARICOM associates, Trinidad and Tobago remained a strong supporter of this body, and has refused to sign an 'Article 98' agreement exempting US military personnel from its proceedings. For this reason, the USA has, since July 2003, blocked certain types of military assistance.

# Economy

## MARK WILSON

At the beginning of the 21st century the economy of Trinidad and Tobago contrasted sharply with those of its Caribbean neighbours. Although no longer dependent on the extraction and refining of petroleum, it was increasingly underpinned by the energy sector, with the extraction of natural gas and crude oil providing the basis for processing and manufacturing industries. Tourism, 'offshore' finance and agriculture, by contrast, were less well developed.

A phase of rapid economic development began when petroleum prices rose from US $1.30 in 1970 to US $33.50 in 1982. This led to rapid growth in gross domestic product (GDP) in this period, at an average annual rate of 5.5%. Expansion of GDP, in turn, had strong positive effects on foreign-exchange earnings and on government revenue, which grew from TT $888m. in 1974 to TT $4,253m. in 1981. The Government attempted to spread the benefits of the increase in petroleum prices through spending on infrastructure, investment in state-owned, heavy industries and other capital projects. Imports were restricted in an effort to encourage local manufacturing, while protection also extended to services such as insurance. Real wages grew by an average of 5.2% per year from 1974 to 1984.

World oil prices fell sharply in the 1980s, adversely affecting domestic petroleum production, while new facilities for crude petroleum imports to the USA had already prompted the removal of that country's strategically motivated tax incentives for petroleum products refined in the Caribbean. In 1988–90 production averaged 150,000 barrels per day (b/d), with a further decline to 113,500 b/d in 2001. However, there were also some important discoveries, including, in January 1998, the largest newly identified reservoir for 25 years and another large oilfield in 2001; there were also substantial amounts of condensate associated with recent gas finds. Production rose to 134,200 b/d in 2003, slipped back to 122,900 b/d in 2004, but increased sharply from the start of 2005, with newly developed reserves coming on-stream. In early 2004 proven oil reserves were 990m. barrels, with 324m. barrels of discounted probable reserves and 2,000m. barrels of possible reserves.

Owing to recession, in 1986 the National Alliance for Reconstruction (NAR) Government was forced adopted a structural-adjustment programme and negotiated two successive stand-by agreements with the IMF. This was followed in 1990 by debt-rescheduling agreements with lending agencies and a US $850m. loan agreement with the Inter-American Development Bank (IDB). These measures, along with enhanced use of natural gas resources, contributed to a general economic recovery in the 1990s. In 1995–2004 GDP increased, in real terms, at an average annual rate of 7.6% (on the basis of growth data rebased in 2005). Growth reached 13.2% in 2003, but the rate fell to 6.2% in 2004. Import restrictions were gradually liberalized from the early 1990s, while foreign-exchange controls were abolished in April 1993 as the Trinidad and Tobago dollar moved from a fixed rate against the US currency to a strictly managed float. The currency moved sharply from a rate of TT $4.25 to TT $5.74, but remained stable thereafter, with the rate at or close to TT $6.29 from early 1997 to mid-2005. There was increased activity from the early 1990s in other manufacturing sectors, as well as the construction and financial sectors. Unemployment fell to 9.0% in the first quarter of 2005.

From mid-1999 to 2005 a strong recovery in international petroleum prices had a positive effect on government revenues. Chemicals and steel prices also recovered from declines in the late 1990s. Methanol, a key product, decreased from US $268 per metric ton in 1995 to a low point of US $94 per ton in mid-1999, but had recovered to an average of US $265 by 2004. The price of ammonia declined from US $199 per ton in 1995 to US $88 in the first quarter of 2002, but averaged US $252 per ton in 2004.

The population was estimated at 1.3m. in mid-2004, with a population density (per sq km) of 252. Approximately one-half of the population lived in an urbanized 'east–west corridor', stretching from Diego Martin in the west, through Port of Spain to Arima in the east. One-sixth of the population lived in other urban areas, principally the San Fernando conurbation in southern Trinidad and Chaguanas in the centre of the country, with 54,100 in the smaller island of Tobago. The average annual population growth rate was 0.3% in 1994–2004.

## AGRICULTURE

The strength of the energy-based economy led to the relative neglect of agriculture, particularly in the 1970s and early 1980s, while agricultural wage rates could not compete with other areas of employment. As a result, in 2004 the sector contributed only 0.8% of GDP (down from 6.9% in 1972), but employed 5.0% of the employed labour force.

The principal commercial crop has been sugar, which generated 2.0% of total employment in the first half of 2002 (including seasonal employees), a figure which fell to 0.5% by the first quarter of 2005. Until mid-2003 the main producer was a state-owned company, Caroni (1975) Ltd, which operated two sugar factories. However, after several years of operating at a loss, the company dissolved. The Government wrote off the company's accumulated debts of TT $2,400m. in 1994, but by 2002 continuing losses had increased the debt again, to TT $2,300m. In late July 2003 the Government closed the company down, offering its 10,000 mainly Indo-Trinidadian staff redundancy payments totalling some US $145m. and opportunities to retrain or lease land for their own use. One of its successors, the state-owned Sugar Manufacturing Company, operated the remaining sugar factory at Sainte Madeleine but did not grow sugar cane, buying the crop from independent farmers. Exports, which went mainly to Europe at the preferential price agreed under the European Union's (EU) Cotonou Agreement accounted for 60,900 metric tons in 2002, but fell to 42,900 tons in 2004; some sugar for the local market is imported. Partly because of problems with pest control, dry weather and technical difficulties at the sugar factories, production was consistently below target, while most sugar used on the local market in 2004 was imported from Guyana and Belize.

Premium quality cocoa was a traditional export, but production fell from 7,542 metric tons in 1972 to 1,321 tons in 2004. Low-grade robusta coffee was grown: production was 4,586 tons in 1968, but had fallen to 699 tons in 2004. There were some floriculture exports, mainly of orchids and anthuriums to North America. A wide variety of vegetables and fruits, were grown mainly for the local market, as well as some rice; however, citrus production fell sharply from 2003 and was 3,148 metric tons in 2004. Local producers supplied virtually the entire domestic market with eggs and broiler chicken (some 27.5m. birds in 2003).

Teak was the main forestry product, and was used extensively in the yacht-repair industry. Most lumber requirements, however, were imported. In 2002 roundwood removals totalled 86,300 cu m. The fishing sector also was small scale, employing approximately 3,140 people in 2000; most of the catch was used locally, but frozen shrimps were exported. Trinidad was also used as a base by Asian vessels for deep-sea fishing in the Caribbean and mid-Atlantic.

## PETROLEUM AND GAS

Energy-based industries were of central importance to the Trinidad and Tobago economy. Commercial petroleum production started in 1908, although the first oil well was sunk as early as 1857. Refining of local and imported petroleum was well established by the 1930s, and the sector accounted for 29.2% of GDP by 1955. The energy sector's contribution to GDP increased to 42.8% in 1980, at the height of the increase in petroleum prices. In 2004 the petroleum sector (including mining, refining and petrochemicals) accounted for an estimated 42.3% of GDP, but employed only 3.4% of the working population. In 2004 fuels and chemicals accounted for 84.1% of total exports, while the oil sector raised 37.0% of government revenue, almost 250% of the sum raised by personal income tax. Offshore petroleum production began in 1955. By the 1980s most petroleum and gas was produced on the east coast continental shelf.

In 2002 16 international companies were involved in petroleum and gas exploration and production in Trinidad and Tobago's offshore areas. The state-owned Petroleum Company of Trinidad and Tobago Ltd (Petrotrin) and the National Gas Company of Trinidad and Tobago Ltd (NGC), as well as small, local, privately owned producers, operated both onshore and offshore. Most crude petroleum output came from BP Energy Company of Trinidad and Tobago (which is owned 70% by British Petroleum, BP, and 30% by Repsol-YPF of Spain) and from Petrotrin's Trinmar. However, from 2005, the Australian company BHP Billiton became an increasingly significant producer, with its Angostura field coming on stream, while BP was planning to sell three mature oil fields to one of its minority partners, Repsol-YPF. Following the closure of two smaller plants, in the late 1990s there was just one refinery, at Pointe-à-Pierre, owned by Petrotrin, which had been upgraded in the mid-1990s, with a further US $873m. upgrade to be completed in 2004–09, and which refined imported, as well as local, crude petroleum. Most of BP's local crude was refined overseas, although, from 1999, some was processed by Petrotrin.

Consultants completed a master plan for the use of natural gas resources in 2002. Natural gas use increased at an average annual rate of 18% in 1997–2001. The discovery in the late 1990s of several significant deposits of natural gas led to the development of a major liquefied natural gas (LNG) plant. The Atlantic LNG Company of Trinidad and Tobago, came on-stream in April 1999, purchasing 450m. cu ft per day of natural gas directly from BP, 30% of total national gas production. The plant was owned by a consortium of BP, British Gas (BG) Trinidad and Tobago Ltd, Repsol-YPF, Tractebel and the NGC, which had a 10% shareholding. The capital cost of the first-phase project was US $950m. In March 2000 an agreement was reached with BP, BG and Repsol for an expansion project, completed in April 2003. The expansion increased gas consumption to approximately 1,400m. cu ft per day, with 62.5% of gas used by the LNG plant supplied by BP and 37.5% by BG and its partners, using a new pipeline that allowed Trinidad's north coast gas fields to be exploited for the first time. Sales contracts were with the USA, Spain, Puerto Rico and the Dominican Republic. A further US $1,200m. expansion was agreed in June 2003 with BP, BG, Repsol and the NGC, which was expected to make Trinidad the world's leading LNG producer by 2005, with daily gas usage of 800m. cu ft and annual output of 15m. metric tons; construction was expected to be completed by the end of 2005. After presenting a tough stance on project benefits in negotiations, the Government projected an annual revenue of TT $58,000m. over 20 years from the latest expansion, assuming a natural gas price in the USA of US $5 per million British thermal units (btu). However, the actual out-turn would depend on realized LNG prices in export markets. With US imports of LNG likely to rise from 2%–5% of gas demand in 2004 to 15%–25% by 2025, proposals for a fifth and sixth train or phase were under discussion in 2005; for these, new private-sector partners would be likely to replace the existing consortium.

Concerns centred on the likely extent of gas reserves, and the most prudent depletion rate. Large new discoveries were made in recent years, and the potential for new deep-water finds was thought to be good. In September 2000 BP announced the discovery of a natural gas deposit of 3,000,000m. cu ft, and in May 2001 a deposit of 1,000,000m. cu ft was discovered; BHP Billiton announced a further major discovery in October 2003. Proven reserves were 18,800,000m. cu ft in late 2004, an amount that would be exhausted in 15 years with Train IV of the LNG plant added to existing gas users, with a further 14,500,000m. cu ft in discounted probable and possible reserves. This represented a sharp decrease from a reserves to production ratio of 58 years in 2001, and prompted some concern that reserves were being exploited too rapidly. However, other industry sources believed that undiscovered exploration potential would bring the total to a level that would comfortably accommodate the proposed level of production. There was also some confidence that the Venezuelan Government would agree to the processing of its otherwise inaccessible gas reserves in Trinidad.

Just under one-half of the natural gas produced in Trinidad and Tobago was purchased by the NGC, which operated a 620-km pipeline system, supplying the needs of all end-users except the LNG plant; a new 77-km 56-inch pipeline to supply new energy users in south-western Trinidad was under construction in 2005. Gas sales by the NGC averaged 1,126m. cu ft per day in 2003, of which 55% was supplied by BP, 22% by BG, 16% by Enron Gas and Oil Trinidad Ltd and 8% by the NGC's own offshore compression operations. An additional 1,364 cu ft per day was supplied directly to Atlantic LNG. In 2004 some 55% of gas production was used in the manufacture of LNG, 32% for ammonia and methanol, 8% in electric power generation, with most of the remainder used in the iron and steel industry and cement manufacturing. There were no plans to establish a piped gas supply system for domestic household use. A committee was established in April 2004 to develop proposals to increase the local content of energy sector projects, a strategic aim of the Government. The first locally constructed oil-drilling platform was inaugurated in the same month. Further proposals included increased use of local service companies in engineering, fabrication or instrumentation.

## MANUFACTURING

Trinidad and Tobago became the world's largest methanol exporter in 2000, and in 2005 accounted for an estimated 14% of world production. The first methanol plant was established in 1984, under state ownership, and was sold in 1997 to Methanol Holdings (Trinidad) Ltd, a consortium of a local financial company, CL Financial, and two German companies, Ferrostaal AG and Helm AG. In mid-2005 these companies owned four methanol plants. The world's largest methanol plant, the Atlas plant owned by Methanex and BP, was completed at Point Lisas in June 2004, and will be followed by an even larger plant to be opened by Methanol Holdings in August 2005. This will take Trinidad and Tobago to seven plants with a total capacity of 6.5m. metric tons, an increase from 963,000 metric tons in 1995.

Trinidad and Tobago was also the world's principal exporter of ammonia, with the Russian Federation a close rival; production started in 1959, and in 2005 plants were owned by both local and overseas investors. From 2004 the Government no longer welcomed proposals for stand-alone methanol or ammonia plants without a further downstream component. Accordingly, there were two proposals for urea-ammonium-nitrate fertilizer plants, and one for an ammonia-urea-melamine plant.

The state-owned Iron and Steel Company of Trinidad and Tobago, established in 1981, made large losses until the commencement of a lease arrangement with the local subsidiary of an Indian company, Ispat, in 1989. Caribbean Ispat Ltd subsequently acquired the plant in 1994; it comprised three directly reduced iron units with a total capacity of 2.2m. metric tons, as well as mills for billets and wire rods. Not all metals projects have been successful. An innovative iron carbide plant opened by a US steel company Nucor was closed in 1999, while a Cliffs and Associates iron briquette plant that also used new technology was closed in 2001. However, the former Cliffs and Associates plant was reopened and sold to the parent company of Caribbean Ispat, now Mittal Steel Point Lisas Ltd, while Nucor was expected to open a new, directly reduced iron plant in 2006. In May 2004 the Government signed an outline memorandum of understanding for a US $1,000m. aluminium smelter with a US company, Alcoa, and Sural of Venezuela, with a proposed capacity of 250,000 metric tons. Employment was expected to be 2,000 in the construction phase, and 600 on completion; the Government expected this project to lead to the development of a range of downstream fabrication industries; however, proposals had not been finalized by mid-2005.

Low-cost natural gas assisted the development of other manufacturing industries. Trinidad Cement Ltd (TCL) used locally quarried limestone in gas-fired kilns; in 2004 cement production totalled 768,500 metric tons, of which 244,800 tons was exported to regional markets. A three-year expansion programme announced in 2003 was expected to double the capacity of the Trinidad plant to 1.5m. tons. The company also owned cement plants in Barbados and Jamaica. Carib Glassworks was another manufacturing concern to benefit from cheap energy, producing bottles for national and export markets, while low-cost electricity from gas-powered generating stations allowed local soft-drink manufacturers to operate competitively for export to the wider Caribbean. There was also a wide range of consumer goods, foods and other industries.

In contrast to Jamaica and some other Caribbean islands, there were very few labour-intensive export industries in Trinidad and Tobago. The clothing industry and the informatics sector were both underdeveloped. In 2004 the manufacturing sector, excluding sugar and petroleum, contributed 7.0% of GDP and employed 10.4% of the employed labour force.

## TRANSPORT AND COMMUNICATIONS

In 1999 there were 7,900 km (4,910 miles) of roads in Trinidad and Tobago. Major routes were covered by four-lane highways, which, however, suffered from heavy congestion, with an estimated 450,000 vehicles. The road system had to accommodate a huge increase in use after the growth of the petroleum sector in the 1970s. The main international ports were at Port of Spain (container, cargo and cruise ships), where a TT $500m. upgrade was in progress in 2004–06; Point Lisas (mainly specialized bulk cargo piers, but also container cargo); and, in Tobago, Scarborough (general cargo, ferry service and cruise ships). There were also specialized port facilities at Point Fortin (LNG), Pointe-à-Pierre (crude petroleum and refinery products), Claxton Bay (cement) and Tembladora (transhipment of Guyanese bauxite).

Trinidad's airport at Piarco offered direct connections to North America, Europe, most other Caribbean islands, Guyana, Suriname and Venezuela. A far reaching but controversial US $250m. improvement project was completed in May 2001; however, plans to develop Trinidad as a regional air transport hub were delayed when the US Federal Aviation Authority downgraded Trinidad and Tobago to Category II status on air safety grounds, preventing local carriers from expanding services to the USA. In 2004 some 1.54m. international passengers used the airport. There was a frequent service to Crown Point airport in Tobago, carrying some 0.7m. passengers per year; Tobago was also served by direct connections to the United Kingdom, Puerto Rico and some neighbouring Caribbean islands. These services carried 169,000 passengers in 2004. The state-owned airline, BWIA International Airways Ltd, was privatized in 1996, with the Government retaining a 33.5% share. The airline experienced severe financial problems in 2003 and requested government intervention when, in May, its creditors temporarily seized two of its aircraft. In July the Government agreed to pay BWIA's debts on the condition that the company undertook a programme of restructuring. In April 2004 a rights issue was announced, allowing the conversion to equity of US $40m. of BWIA's debt to the Government. As private-sector shareholders did take up their allotment, the airline returned to government control. With losses continuing, options under consideration in mid-2005 included a further restructuring, as well as closure of the airline.

Trinidad and Tobago was linked to the Americas I and II fibre-optic systems and also benefited from international satellite telecommunications links. The telecommunications sector was still a monopoly in mid-2005, controlled by Telecommunication

Services of Trinidad and Tobago (TSTT) Ltd, 51% of which was owned by the Government and 49% by a British company, Cable & Wireless. Charges were high and new connections were not available in many suburban and rural areas. There were independent internet providers, however, while after repeated delays since the late 1990s, two new cellular telephone operators won an auction for new licences in June 2005, and were expected to begin operations in early 2006. New licences for international services were expected to follow. There was limited growth of call centres, with some serving the US market.

## TOURISM

For many years the full tourist potential of Trinidad and Tobago was not promoted. From the early 1980s, however, successive Governments placed greater emphasis on the sector, providing new facilities, including a cruise-ship terminal in Port of Spain, and conducting more effective promotional campaigns. Stop-over arrivals increased at an annual average rate of 7.5% in 1990–2000, fell in 2001–02, and were believed to have reached a new high with 442,555 arrivals in 2004. The industry remained significantly less developed than that of other Caribbean islands; 18.1% of 2001 stop-over arrivals were business travellers, while 55.0% were visiting friends and relatives and only 21.1% were vacation tourists staying in hotels and guest houses. Net earnings from tourism were a negative balance of payments item until 1996, but were equivalent to 3.6% of merchandise imports in 2004. Hotels and guest houses accounted for 0.3% of GDP in 2004 and tourism receipts totalled US $249m. in 2003. The number of cruise passenger arrivals peaked at 82,243 in 2001, of whom 71,467 went to the main island and 10,776 to Tobago; the total for both islands was 54,254 in 2004, with cruise lines by then preferring short routes to the northern Caribbean for security reasons.

The officially reported number of hotel rooms increased from 2,314 in 1992 to 4,850 in 2001. On Trinidad the main hotels catered principally for business visitors, although the annual pre-Lenten Carnival was a major attraction. There was an important yacht- and powerboat-service industry based at marinas on the north-west peninsula of the island. For insurance purposes, Trinidad was outside the hurricane belt. This factor, as well as a combination of competitive wage rates and engineering skills, produced an attractive environment for repair services, which generated an estimated US $24m. per year. Yacht arrivals increased from 637 in 1990 to 3,249 in 2000, falling, however, to 1,788 in 2004.

## PUBLIC FINANCE

From a very strong fiscal position in the 1970s, government finances deteriorated sharply from 1983. Successive austerity and emergency budgets reduced expenditure and increased taxation, while the Government made use of IMF compensatory financing and stand-by credit. The 1989 budget announced a 10% reduction in public-sector pay, which was later ruled illegal by the courts. Value-added tax was introduced at the rate of 15% from January 1990. This was followed, in April 1990, by a further stand-by agreement with the IMF, providing US $111m. over 11 months. In the same month the IDB agreed to lend US $850m. over four years for housing, infrastructure development and major energy investments, including a full upgrade of the Pointe-à-Pierre petroleum refinery. The IDB's conditions emphasized relaxation of import and foreign-exchange controls, reduced tariffs, currency liberalization and privatization of state-owned industries.

The overall fiscal account was in deficit from 1982 to 1992. There were smaller deficits in 1993 and 1994, although the recurrent finances registered a surplus. A small, overall surplus was achieved in 1995–97. Partly as a result of a fall in energy and chemical prices, the overall account moved into a deficit equivalent to 3.1% of GDP in 1999, with the current account also in deficit. In 2000 the fiscal account returned to surplus, equivalent to 1.6% of GDP, with a recovery in energy-based revenues more than compensating for an increase in both current and capital expenditure. The overall fiscal account was close to balance in 2001–02, but there was an overall surplus of 2.4% in 2003/04, partly because of stronger energy prices. There were however some concerns over the medium-term fiscal implications of increased spending on salaries and infrastructural projects, and of debts incurred by state-owned enterprises, as well as the use of extra-budgetary expenditure to fund capital projects. An interim Revenue Stabilization Fund was established in 2000, and by October 2004 contained TT $2,835m. of energy sector revenue, equivalent to 3.9% of GDP. Detailed statutory proposals for a renamed Heritage and Stabilization Fund were to be presented in 2005.

The inflation rate remained moderate in the early 2000s, ranging from 3.4% to 5.6% in 1998–2004, but increased from late 2004 as a result of high food pirices. The annual average rate of increase in consumer prices was 3.9% in 2004.

## FINANCIAL SERVICES

Trinidad and Tobago had a strong domestic banking sector. Of the five main commercial banks, the two largest, Republic Bank Ltd and RBTT Ltd, were in local private-sector ownership; Scotiabank Trinidad and Tobago Ltd was the local subsidiary of the Canadian Bank of Nova Scotia; First Citizens Bank Ltd was state-owned; and Citibank (Trinidad and Tobago) Ltd specialized in services for international corporate clients. In the late 1990s and early 2000s there was increased activity in the commercial banking and finance sector.

Trinidad and Tobago was a regional centre for some financial services. RBTT Ltd had subsidiaries in most English- and Dutch-speaking Caribbean countries, while Republic Bank had subsidiaries in Grenada, Guyana, Barbados, Saint Lucia and the Dominican Republic. In 2003 First Citizens established a subsidiary in Saint Lucia. All major Trinidad and Tobago banks provided public- and private-sector merchant-banking services throughout the region. In January–December 2004 financial institutions based in Trinidad and Tobago led 29 bond issues with a total value of TT $3,238m. in local currency and US $690m. in other currencies for the local and regional private and public sectors. Some concerns were raised over increased exposure to country risk in other Caribbean markets. In mid-2005 the Trinidad and Tobago Stock Exchange listed 26 local and seven Barbadian or Jamaican companies.

A small number of private-sector companies played a dominant role in the local economy. The largest local conglomerate, ANSA McAL, had life- and general-insurance subsidiaries, as well as interests in brewing and glass-making, importing, distribution and media sectors, with plans well advanced for a major ammonia plant. There were a number of locally owned insurance companies, of which the largest were active in other regional markets and had important local investments in real estate and equities. CL Financial, an unlisted company that grew out of the Colonial Life Insurance Company, had a controlling 53% shareholding in the Republic Bank and interests in property, supermarkets, media, rum, foods and methanol and ammonia plants. It also had subsidiaries in several other Caribbean countries, as well as property companies in the USA, and wine and spirits companies in the USA, the United Kingdom, France and elsewhere. Guardian Holdings, also insurance-based, had important real-estate interests in Trinidad and Tobago and, in 1999, bought three insurance companies in Jamaica. Neal & Massy Holdings Ltd scaled down its activities in the mid-1990s, but retained important regional interests, and in 1998 bought a 20% shareholding in the dominant Barbados conglomerate, Barbados Shipping and Trading Company Ltd.

## FOREIGN TRADE, DEBT AND BALANCE OF PAYMENTS

From the mid-1980s the trade balance was in surplus, albeit a fluctuating one. While imports declined steadily from 1984 to 1990, reflecting the severity of the recession, exports generally increased in the rest of the decade. However, the subsequent decrease in exports and increase in imports reduced the trade surplus by 1996. In 1997 and 1998 visible trade deficits of US $528.6m. and US $740.8m., respectively, resulted from imports of capital goods for investment in heavy manufacturing. This deficit was covered on the balance-of-payments capital account by large foreign direct investment inflows. In 1999, with the completion of several projects, exports increased and imports of capital goods decreased sharply, resulting in a trade surplus of US $63.6m. This surplus increased to an estimated US $969m. in 2000, as a result of higher export prices and with some new plants in operation for the first full year. The merchandise trade surplus was US $718m. in 2001, and after falling to US $238m in 2002, recovered US $1,334m. in 2003. In 2004 the surplus was an estimated US $1,272m. in 2004, with new energy sector projects on-stream and prices strong. In that year the current-account surplus was US $919m.

With the energy sector strong and the tourism sector weak, Trinidad and Tobago's balance of payments contrasted in its structure with that of most other Caribbean islands. There was a positive balance in merchandise trade from 1987, with the exception of 1997–98, when there were exceptional imports of capital goods for energy-related investment projects. In 2004 the principal merchandise exports were: fuels (67.8%); chemicals (23.8%); manufactured goods (9.0%) foods (2.2%) and beverages, mainly rum (1.3%). The services account showed an estimated positive balance of US $511.6m. on non-factor services in 2004, sufficient to cover 10.4% of merchandise imports. The main positive items were transport, and insurance services, followed by travel and communications. Net investment income was a strong outflow item, equivalent, in 2004, to 9.2% of merchandise imports. In 2004 capital goods made up 36.8% of merchandise imports; the other major categories were raw materials and construction materials (14.8%), fuels, mainly crude petroleum for local refining (24.4%), non-food consumer goods (10.0%), and foods (6.3%).

External debt increased through the 1980s, reaching US $2,510m. in 1990. As a result of debt relief and amortization, this figure was reduced to US $1,283m. at the end of 2003. Borrowing for capital projects in 1998–2001 increased total domestic and foreign public debt to 59.3% of GDP at the end of 2002; however, with GDP expanding steadily, this figure came down to 52.7% at the end of 2004. Debt servicing peaked in 1993, with total debt service equivalent to 49.2% of current government revenue and foreign debt service to 30.6% of exports of goods and services in that year. By 2004, however, the external debt service ratio had declined to 5.7%, while interest payments made up 11.7% of current government expenditure.

The balance of payments capital account was in deficit by US $736.8m. in 2004, equivalent to 6.5% of GDP. This deficit was caused by outflows of US $690.1m. to finance Caribbean bond issues led by Trinidad and Tobago banks, and because of direct investments in other Caribbean markets, as well as US $216.1m. in net repayment of official borrowing. Strong exports and foreign direct investment inflows increased net official reserves to US $3,366.9m. by the end of May 2005.

## CONCLUSION

Trinidad and Tobago's natural gas resources provided the basis for strong economic growth from 1994, with falling unemployment, a stable exchange rate, high investment inflows and increasing foreign exchange reserves. High energy prices in 1999–2005 greatly improved the fiscal position, allowing an increase in both current and capital spending; however, there were concerns over the increasing public-sector debt, while inefficient state enterprises, including some, such as telecommunications, which were partially divested, remained a burden on the economy. The thorough restructuring of some nationalized companies would, it was hoped, go some way towards alleviating the problem, but it was clear that an acceleration in the rate of privatization and the introduction of competition for public- and private-sector monopolies were necessary. Also of concern were the standard of public services and infrastructure, including health, education, water supply and sewerage, as well the police force and the rising crime rate. However, the strong resource base would continue to attract additional foreign investment.

# Statistical Survey

Sources (unless otherwise stated): Central Statistical Office, National Statistics Building, 80 Independence Sq., POB 98, Port of Spain; tel. 623-6945; fax 625-3802; e-mail info@cso.gov.tt; internet www.cso.gov.tt; Central Bank of Trinidad and Tobago, POB 1250, Port of Spain; tel. 625-4835; fax 627-4696; e-mail info@central-bank.org.tt; internet www.central-bank.org.tt.

## Area and Population

### AREA, POPULATION AND DENSITY

| | |
|---|---:|
| Area (sq km) | 5,128* |
| Population (census results) | |
| 2 May 1990 | 1,213,733 |
| 15 May 2000 | |
|   Males | 633,051 |
|   Females | 629,315 |
|   Total | 1,262,366 |
| Population (official estimates at September) | |
| 2000 | 1,262,400 |
| 2001 | 1,266,800 |
| 2002 | 1,275,700 |
| Density (per sq km) at September 2002 | 248.8 |

* 1,980 sq miles. Of the total area, Trinidad is 4,828 sq km (1,864 sq miles) and Tobago 300 sq km (116 sq miles).

### POPULATION BY ETHNIC GROUP
(1990 census*)

| | Males | Females | Total | % |
|---|---:|---:|---:|---:|
| African | 223,561 | 221,883 | 445,444 | 39.59 |
| Chinese | 2,317 | 1,997 | 4,314 | 0.38 |
| 'East' Indian | 226,967 | 226,102 | 453,069 | 40.27 |
| Lebanese | 493 | 441 | 934 | 0.08 |
| Mixed | 100,842 | 106,716 | 207,558 | 18.45 |
| White | 3,483 | 3,771 | 7,254 | 0.64 |
| Other | 886 | 838 | 1,724 | 0.15 |
| Unknown | 2,385 | 2,446 | 4,831 | 0.43 |
| **Total** | 560,934 | 564,194 | 1,125,128 | 100.00 |

* Excludes some institutional population and members of unenumerated households, totalling 44,444.

# TRINIDAD AND TOBAGO

*Statistical Survey*

## ADMINISTRATIVE DIVISIONS
(population at 2000 census)

|  | Population | Capital |
|---|---|---|
| *Trinidad* | 1,208,282 | Port of Spain |
| Port of Spain (city, capital) | 49,031 | — |
| San Fernando (city) | 55,419 | — |
| Arima (borough) | 32,278 | Arima |
| Chaguanas (borough) | 67,433 | Chaguanas |
| Point Fortin (borough) | 19,056 | Point Fortin |
| Diego Martin | 105,720 | Petit Valley |
| San Juan/Laventille | 157,295 | Laventille |
| Tunapuna/Piarco | 203,975 | Tunapuna |
| Couva/Tabaquite/Talparo | 162,779 | Couva |
| Mayaro/Rio Claro | 33,480 | Rio Claro |
| Sangre Grande | 64,343 | Sangre Grande |
| Princes Town | 91,947 | Princes Town |
| Penal/Debe | 83,609 | Penal |
| Siparia | 81,917 | Siparia |
| *Tobago* | 54,084 | Scarborough |

## BIRTHS AND DEATHS
(official estimates, annual averages)

|  | 1997 | 1998 | 1999 |
|---|---|---|---|
| Birth rate (per 1,000) | 14.5 | 14.0 | 14.3 |
| Death rate (per 1,000) | 7.2 | 7.5 | 7.8 |

**Expectation of life** (WHO estimates, years at birth): 70 (males 67; females 73) in 2003 (Source: WHO, *World Health Report*).

## EMPLOYMENT
('000 persons aged 15 years and over, October–December)

|  | 2003 | 2004 |
|---|---|---|
| Agriculture, forestry, hunting and fishing* | 25.9 | 29.1 |
| Mining and quarrying† | 17.7 | 19.6 |
| Manufacturing‡ | 58.7 | 60.4 |
| Electricity, gas and water | 5.6 | 7.4 |
| Construction | 74.5 | 91.4 |
| Wholesale and retail trade, restaurants and hotels | 100.5 | 102.1 |
| Transport, storage and communication | 41.0 | 45.6 |
| Finance, insurance, real estate and business services | 45.7 | 45.7 |
| Community, social and personal services | 168.7 | 174.8 |
| Activities not adequately defined | 3.4 | 4.4 |
| **Total employed** | **541.8** | **580.7** |
| Males | 334.1 | 348.6 |
| Females | 207.7 | 232.1 |
| Unemployed | 61.3 | 48.8 |
| **Total labour force** | **603.1** | **629.5** |

* Includes sugar manufacture.
† Includes oil manufacture.
‡ Excludes sugar and oil manufacture.

# Health and Welfare

## KEY INDICATORS

| | |
|---|---|
| Total fertility rate (children per woman, 2003) | 1.6 |
| Under-5 mortality rate (per 1,000 live births, 2003) | 20 |
| HIV/AIDS (% of persons, aged 15–49, 2003) | 3.2 |
| Physicians (per 1,000 head, 1997) | 0.79 |
| Hospital beds (per 1,000 head, 1996) | 5.11 |
| Health expenditure (2002): US $ per head (PPP) | 428 |
| Health expenditure (2002): % of GDP | 3.7 |
| Health expenditure (2002): public (% of total) | 37.3 |
| Access to water (% of persons, 2002) | 91 |
| Access to sanitation (% of persons, 2002) | 100 |
| Human Development Index (2002): ranking | 54 |
| Human Development Index (2002): value | 0.801 |

For sources and definitions, see explanatory note on p. vi.

# Agriculture

## PRINCIPAL CROPS
('000 metric tons)

|  | 2001 | 2002 | 2003 |
|---|---|---|---|
| Rice (paddy) | 3.3 | 3.9 | 2.9 |
| Maize | 1.8 | 3.3 | 3.0 |
| Taro (Coco yam) | 2.3 | 6.9 | 4.7 |
| Sugar cane | 1,029.0 | 1,339.0 | 873.0 |
| Pigeon peas | 1.6 | 2.8 | 2.9 |
| Coconuts* | 23.5 | 21.0 | 16.5 |
| Cabbages | 2.3 | 1.8 | 1.7 |
| Lettuce | 0.7 | 1.4 | 1.4 |
| Tomatoes | 2.4 | 1.2 | 1.7 |
| Pumpkins, squash and gourds | 5.8 | 5.8 | 5.9* |
| Cucumbers and gherkins | 4.7 | 3.6 | 3.7* |
| Aubergines | 1.9 | 1.9 | 2.0* |
| Other fresh vegetables* | 5.1 | 4.9 | 5.1 |
| Watermelons | 2.5 | 0.9 | 1.0* |
| Bananas* | 6.2 | 6.5 | 6.7 |
| Plantains* | 4.2 | 4.3 | 4.4 |
| Oranges | 1.7 | 5.0 | 5.0* |
| Lemons and limes* | 1.4 | 1.4 | 1.5 |
| Grapefruit and pomelo | 2.2 | 2.5 | 2.6* |
| Pineapples* | 3.5 | 3.6 | 3.8 |
| Other fruit* | 47.0 | 39.7 | 39.7 |
| Coffee (green) | 0.4 | 0.2 | 0.6 |
| Cocoa beans | 0.6 | 1.6 | 1.0 |

* FAO estimate(s).

Source: FAO.

## LIVESTOCK
('000 head, year ending September)

|  | 2001 | 2002 | 2003 |
|---|---|---|---|
| Horses* | 1.0 | 1.1 | 1.2 |
| Mules* | 1.7 | 1.8 | 1.9 |
| Asses* | 2.0 | 2.1 | 2.2 |
| Cattle | 30.3* | 31.6* | 29.0 |
| Buffaloes* | 5.5 | 5.6 | 5.7 |
| Pigs | 63.0 | 80.0* | 75.7 |
| Poultry | 24,433 | 31,016 | 27,500* |
| Sheep* | 5.5 | 7.1 | 7.1 |
| Goats* | 16.3 | 21.7 | 23.0 |

* FAO estimate(s).
Source: FAO.

## LIVESTOCK PRODUCTS
('000 metric tons)

|  | 2001 | 2002 | 2003 |
|---|---|---|---|
| Beef and veal | 0.8 | 0.9 | 0.8 |
| Pig meat | 2.0 | 2.9 | 2.8 |
| Poultry meat | 48.4 | 60.0 | 56.5 |
| Cows' milk | 10.4 | 10.0 | 8.9 |
| Hen eggs* | 3.4 | 3.7 | 3.7 |

* Unofficial figures.
Source: FAO.

# Forestry

## ROUNDWOOD REMOVALS
('000 cubic metres, excl. bark)

|  | 2001 | 2002 | 2003 |
|---|---|---|---|
| Sawlogs, veneer logs and logs for sleepers | 56.0 | 51.0 | 51.0* |
| Fuel wood* | 36.1 | 35.7 | 35.3 |
| **Total** | **92.1** | **86.7** | **86.3** |

* FAO estimate(s).
Source: FAO.

# TRINIDAD AND TOBAGO

## SAWNWOOD PRODUCTION
('000 cubic metres, incl. railway sleepers)

|  | 2001 | 2002 | 2003 |
|---|---|---|---|
| **Total** (all broadleaved) | 41 | 43 | 43* |

\* FAO estimate.
Source: FAO.

## Fishing

('000 metric tons, live weight of capture)

|  | 2001 | 2002 | 2003 |
|---|---|---|---|
| Demersal percomorphs | 2.3 | 2.5 | 2.4 |
| King mackerel | 0.6 | 1.5 | 0.9 |
| Serra Spanish mackerel | 2.7 | 2.5 | 2.0 |
| Tuna-like fishes | 0.7 | 0.8 | 0.6 |
| Sharks, rays, skates, etc. | 0.8 | 1.0 | 0.9 |
| Other marine fishes | 2.6 | 3.0 | 2.0 |
| Penaeus shrimps | 0.9 | 0.9 | 0.7 |
| **Total catch** (incl. others) | 10.8 | 12.6 | 9.7 |

Source: FAO.

## Mining

('000 barrels, unless otherwise indicated)

|  | 2001 | 2002 | 2003 |
|---|---|---|---|
| Crude petroleum | 41,374 | 47,684 | 48,947 |
| Natural gas liquids | 7,521 | 8,505 | 10,500* |
| Natural gas (million cu m)† | 16,599 | 19,172 | 26,810 |

\* Estimate.
† Figures refer to the gross volume of output; marketed production (in million cu m) was: 15,173 in 2001; 17,777 in 2002; 26,046 in 2003.

Source: US Geological Survey.

## Industry

### SELECTED PRODUCTS
('000 metric tons, unless otherwise indicated)

|  | 2001 | 2002 | 2003 |
|---|---|---|---|
| Raw sugar | 91 | 98 | 68 |
| Fertilizers | 4,209 | 4,721 | 4,965 |
| Methanol | 2,789 | 2,829 | 2,846 |
| Cement | 697 | 744 | 766 |
| Iron (direct reduced) | 2,187 | 2,316 | 2,275 |
| Steel: |  |  |  |
|   billets | 668 | 817 | 896 |
|   wire rods | 605 | 705 | 641 |
| Electric energy (million kWh)* | 5,460 | 5,643 | n.a. |

\* Source: UN Economic Commission for Latin America and the Caribbean.

## Finance

### CURRENCY AND EXCHANGE RATES
**Monetary Units**
100 cents = 1 Trinidad and Tobago dollar (TT $).

**Sterling, US Dollar and Euro Equivalents** (31 May 2005)
£1 sterling = TT $11.3580;
US $1 = TT $6.2472;
€1 = TT $7.7034;
TT $1,000 = £88.04= US $160.07 = €129.81.

**Average Exchange Rate** (TT $ per US $)
2002   6.2487
2003   6.2951
2004   6.2990

### CENTRAL GOVERNMENT BUDGET
(TT $ million)

| Revenue | 2001 | 2002 | 2003 |
|---|---|---|---|
| Oil sector | 3,682.4 | 3,931.0 | 6,873.3 |
|   Corporation tax | 1,941.7 | 2,201.2 | 4,713.1 |
|   Withholding tax | 164.1 | 123.8 | 150.8 |
|   Royalties | 708.4 | 644.4 | 1,078.4 |
|   Unemployment levy | 139.1 | 141.5 | 335.2 |
|   Excise duties | 421.4 | 537.1 | 561.5 |
| Non-oil sector | 9,697.5 | 10,586.1 | 10,979.4 |
|   Taxes on income | 4,531.4 | 4,788.5 | 5,418.2 |
|     Taxes on companies | 1,636.0 | 1,645.5 | 2,138.2 |
|     Taxes on individuals | 2,526.9 | 2,701.9 | 2,803.3 |
|   Taxes on property | 69.5 | 84.9 | 77.1 |
|   Taxes on goods and services | 3,109.7 | 3,387.3 | 3,176.3 |
|     Excise tax | 348.7 | 399.4 | 407.0 |
|     Taxes on motor vehicles | 218.3 | 204.8 | 201.5 |
|     Value-added tax | 2,178.7 | 2,401.0 | 2,272.2 |
|   Taxes on international trade | 834.8 | 885.3 | 1,040.5 |
|   Non-tax revenue of non-oil sector | 1,152.1 | 1,440.1 | 1,267.3 |
| Capital revenue and grants | 35.6 | 38.7 | 5.8 |
| **Total** | 13,415.5 | 14,555.9 | 17,858.5 |

| Expenditure | 2001 | 2002 | 2003 |
|---|---|---|---|
| Current expenditure | 12,594.9 | 13,697.4 | 15,179.4 |
|   Wages and salaries | 4,091.3 | 4,140.8 | 4,627.8 |
|   Goods and services | 1,542.2 | 1,810.9 | 2,059.5 |
|   Interest payments | 2,222.2 | 2,469.0 | 2,459.3 |
|     Domestic | 1,453.9 | 1,644.5 | 1,732.3 |
|     External | 768.3 | 824.5 | 727.0 |
|   Transfers and subsidies | 4,739.3 | 5,276.7 | 6,032.8 |
|     Households | 1,560.9 | 1,988.6 | 2,069.3 |
|     Public sector bodies | 1,259.5 | 1,310.1 | 1,556.0 |
| Capital expenditure and net lending | 861.2 | 671.7 | 844.1 |
| **Total** | 13,456.1 | 14,369.1 | 16,023.5 |

### INTERNATIONAL RESERVES
(US $ million at 31 December)

|  | 2002 | 2003 | 2004 |
|---|---|---|---|
| Gold* | 20.9 | 25.5 | 26.8 |
| IMF special drawing rights | 0.4 | 1.1 | 2.7 |
| Reserve position in IMF | 103.8 | 192.2 | 172.5 |
| Foreign exchange | 1,923.5 | 2,257.8 | 2,993.0 |
| **Total** | 2,048.6 | 2,476.6 | 3,195.0 |

\* National valuation of gold reserves.

Source: IMF, *International Financial Statistics*.

### MONEY SUPPLY
(TT $ million at 31 December)

|  | 2002 | 2003 | 2004 |
|---|---|---|---|
| Currency outside banks | 1,501.8 | 1,708.6 | 1,957.4 |
| Demand deposits at commercial banks | 5,620.0 | 5,108.0 | 5,795.3 |
| **Total money** (incl. others) | 7,834.2 | 7,723.0 | 8,375.4 |

Source: IMF, *International Financial Statistics*.

### COST OF LIVING
(Consumer Price Index; base: 2000 = 100)

|  | 2001 | 2002 |
|---|---|---|
| Food | 114.0 | 125.6 |
| Heat and light | 100.3 | 101.2 |
| Clothing | 98.7 | 96.4 |
| Rent | 101.1 | 102.5 |
| **All items** (incl. others) | 105.6 | 109.9 |

**All items:** 114.2 in 2003; 118.3 in 2004.

Source: ILO.

# TRINIDAD AND TOBAGO

*Statistical Survey*

## NATIONAL ACCOUNTS
(TT $ million at current prices)

### National Income and Product

|  | 1998 | 1999 | 2000 |
|---|---|---|---|
| Compensation of employees | 17,864.3 | 19,087.3 | 20,004.0 |
| Operating surplus | 11,918.8 | 13,569.3 | 24,233.0 |
| **Domestic factor incomes** | 29,783.1 | 32,656.6 | 44,237.0 |
| Consumption of fixed capital | 4,562.8 | 4,787.0 | 6,063.0 |
| **Gross domestic product (GDP) at factor cost** | 34,345.9 | 37,443.6 | 50,300.0 |
| Indirect taxes | 4,070.4 | 3,989.3 | 1,379.0* |
| Less Subsidies | 379.6 | 398.9 | 807.0 |
| **GDP in purchasers' values** | 38,036.7 | 41,034.0 | 50,872.0 |
| Net factor income | −2,527.0 | −2,518.0 | −3,953.0 |
| **Gross national product (GNP)** | 35,509.7 | 38,516.0 | 46,919.0 |
| Less Consumption of fixed capital | 4,410.0 | 5,177.0 | 6,063.0 |
| **National income at market prices** | 31,099.7 | 33,339.0 | 40,856.0 |

* Figure obtained as a residual.

Source: UN, Economic Commission for Latin America and the Caribbean, *Statistical Yearbook*.

**Gross national product** (TT $ million at current prices): 51,604.0 (GDP in purchasers' values 55,009.1, Net factor income −3,402.2) in 2001; 56,456.6 (GDP in purchasers' values 59,486.9, Net factor income −3,030.3) in 2002; 65,402.8 (GDP in purchasers' values 67,692.2, Net factor income −2,289.4) in 2003 (preliminary figures).

### Expenditure on the Gross Domestic Product

|  | 2001 | 2002 | 2003* |
|---|---|---|---|
| Government final consumption expenditure | 7,925.2 | 9,525.5 | 10,954.3 |
| Private final consumption expenditure | 30,572.4 | 36,136.4 | 34,709.0 |
| Gross capital formation | 10,696.0 | 10,825.6 | 11,908.0 |
| **Total domestic expenditure** | 49,193.6 | 56,487.5 | 57,571.3 |
| Exports of goods and services | 30,731.4 | 28,467.8 | 36,702.5 |
| Less Imports of goods and services | 24,915.9 | 25,468.4 | 26,581.6 |
| **GDP in purchasers' values** | 55,009.1 | 59,486.9 | 67,692.2 |

* Preliminary figures.

### Gross Domestic Product by Economic Activity

|  | 2001 | 2002 | 2003* |
|---|---|---|---|
| Agriculture, hunting, forestry and fishing | 707.5 | 713.7 | 768.2 |
| Petroleum | 16,434.2 | 17,268.6 | 21,237.3 |
| Manufacturing | 3,920.6 | 4,278.0 | 4,634.6 |
| Electricity, gas and water | 880.6 | 802.6 | 798.5 |
| Construction | 4,447.0 | 4,535.9 | 5,352.4 |
| Transport, storage and communications | 4,528.0 | 5,294.5 | 6,101.5 |
| Distribution | 8,724.3 | 9,286.7 | 9,648.5 |
| Finance, insurance and, real estate | 8,480.1 | 9,926.9 | 10,916.2 |
| Government | 4,714.1 | 4,944.0 | 5,694.2 |
| Other services | 2,797.9 | 2,980.8 | 3,274.2 |
| **Sub-total** | 55,634.3 | 60,031.7 | 68,425.6 |
| Less Imputed bank service charges | 2,803.9 | 3,026.1 | 3,312.7 |
| Value-added tax | 2,178.7 | 2,481.3 | 2,579.3 |
| **GDP in purchaser's values** | 55,009.1 | 59,486.9 | 67,692.2 |

* Preliminary figures.

## BALANCE OF PAYMENTS
(US $ million)

|  | 2002 | 2003 | 2004* |
|---|---|---|---|
| Exports of goods f.o.b. | 3,920 | 5,256 | 6,523 |
| Imports of goods f.o.b. | −3,682 | −3,922 | −5,251 |
| **Trade balance** | 238 | 1,334 | 1,272 |
| Services (net) | 264 | 313 | 353 |
| **Balance on goods and services** | 502 | 1,647 | 1,625 |
| Other income (net) | −480 | −362 | −773 |
| **Balance on goods, services and income** | 22 | 1,285 | 852 |
| Current transfers (net) | 55 | 66 | 68 |
| **Current balance** | 76 | 1,351 | 919 |
| Direct investment abroad | −106 | −225 | −200 |
| Direct investment from abroad | 791 | 1,234 | 1,826 |
| Official, medium and long-term disbursements | 18 | 101 | 251 |
| Official, medium and long-term amortization | −79 | −130 | −241 |
| Commercial banks (net) | −79 | 92 | — |
| Other private sector capital (incl. net errors and omissions) | −572 | −2,089 | −2,072 |
| **Overall balance** | 49 | 334 | 483 |

* Estimates.

Source: IMF, *Trinidad and Tobago: 2004 Article IV Consultation—Staff Report; Staff Statement; Public Information Notice on the Executive Board Discussion; and Statement by the Executive Director for Trinidad and Tobago* (January 2005).

# External Trade

## PRINCIPAL COMMODITIES
(TT $ million, distribution by major SITC groups)

| Imports c.i.f. | 2000 | 2001 | 2002 |
|---|---|---|---|
| Food and live animals | 1,519.9 | 1,831.5 | 1,681.3 |
| Crude materials (inedible) except fuels | 504.4 | 409.3 | 775.1 |
| Mineral fuels, lubricants, etc. | 6,731.3 | 5,737.2 | 6,324.9 |
| Chemicals and related products | 1,642.7 | 1,828.5 | 1,795.5 |
| Basic manufactures | 2,759.4 | 3,041.7 | 2,875.4 |
| Machinery and transport equipment | 6,375.5 | 7,835.5 | 7,925.2 |
| Miscellaneous manufactured articles | 1,066.8 | 1,288.4 | 1,256.5 |
| **Total (incl. others)** | 20,841.9 | 22,210.8 | 22,872.9 |

| Exports f.o.b. | 2000 | 2001 | 2002 |
|---|---|---|---|
| Food and live animals | 962.4 | 907.3 | 897.8 |
| Mineral fuels, lubricants, etc. | 17,574.8 | 15,430.3 | 14,457.3 |
| Chemicals and related products | 4,665.9 | 5,102.2 | 4,019.4 |
| Basic manufactures | 2,394.0 | 2,728.3 | 2,964.4 |
| **Total (incl. others)*** | 26,923.5 | 25,748.7 | 24,062.3 |

* Excluding re-exports; including ships' stores and bunkers (TT $ million): 290.8 in 2000; 354.5 in 2001; 253.6 in 2002.

## PRINCIPAL TRADING PARTNERS
(TT $ million)

| Imports c.i.f. | 2000 | 2001 | 2002 |
|---|---|---|---|
| Barbados | 228.8 | 144.8 | 191.0 |
| Brazil | 613.7 | 1,241.4 | 1,305.8 |
| Canada | 544.7 | 568.6 | 647.0 |
| European Free Trade Association (EFTA) | 203.9 | 170.0 | 252.4 |
| European Union (EU, excl. United Kingdom) | 1,370.1 | 2,571.6 | 2,675.3 |
| Japan | 680.3 | 833.0 | 1,001.5 |
| United Kingdom | 746.9 | 989.9 | 817.1 |
| USA | 7,293.9 | 8,158.1 | 7,679.8 |
| Venezuela | 3,834.0 | 2,716.0 | 2,470.5 |
| **Total (incl. others)** | 20,841.9 | 22,210.8 | 22,872.9 |

# TRINIDAD AND TOBAGO

| Exports f.o.b. | 2000 | 2001 | 2002 |
|---|---|---|---|
| Barbados | 1,270.2 | 1,658.9 | 1,009.2 |
| Canada | 354.2 | 611.2 | 574.1 |
| European Union (EU, excl. United Kingdom) | 1,145.0 | 805.5 | 774.7 |
| Guyana | 584.3 | 548.2 | 485.2 |
| Jamaica | 2,088.6 | 2,192.7 | 1,792.9 |
| Puerto Rico and US Virgin Islands | 923.6 | 916.2 | 849.3 |
| United Kingdom | 450.9 | 395.6 | 354.3 |
| USA | 11,592.1 | 11,029.9 | 11,201.3 |
| **Total** (incl. others)* | **26,632.7** | **25,394.2** | **23,808.6** |

* Excluding ships' stores and bunkers.

## Transport

### ROAD TRAFFIC
(estimated motor vehicles in use)

| | 1994 | 1995 | 1996 |
|---|---|---|---|
| Passenger cars | 122,000 | 122,000 | 122,000 |
| Lorries and vans | 24,000 | 24,000 | 24,000 |

Source: IRF, *World Road Statistics*.

**Total number of registered vehicles:** 292,908 in 1999; 316,163 in 2000; 331,595 in 2001 (provisional figure).

### SHIPPING
**Merchant Fleet**
(registered at 31 December)

| | 2002 | 2003 | 2004 |
|---|---|---|---|
| Number of vessels | 69 | 76 | 95 |
| Total displacement ('000 grt) | 26.8 | 28.1 | 33.5 |

Source: Lloyd's Register-Fairplay, *World Fleet Statistics*.

**International Sea-borne Freight Traffic**
(estimates, '000 metric tons)

| | 1988 | 1989 | 1990 |
|---|---|---|---|
| Goods loaded | 7,736 | 7,992 | 9,622 |
| Goods unloaded | 4,076 | 4,091 | 10,961 |

Source: UN, *Monthly Bulletin of Statistics*.

**1998:** Port of Spain handled 3.3m. metric tons of cargo.

### CIVIL AVIATION
(traffic on scheduled services)

| | 1999 | 2000 | 2001 |
|---|---|---|---|
| Kilometres flown (million) | 25 | 28 | 29 |
| Passengers carried ('000) | 1,112 | 1,254 | 1,388 |
| Passenger-km (million) | 2,720 | 2,765 | 2,723 |
| Total ton-km (million) | 309 | 300 | 288 |

Source: UN, *Statistical Yearbook*.

## Tourism

**FOREIGN TOURIST ARRIVALS**

| Country of origin | 2001 | 2002 | 2003 |
|---|---|---|---|
| Barbados | 27,878 | 33,989 | 37,320 |
| Canada | 43,291 | 41,506 | 43,036 |
| Germany | 11,371 | 5,659 | 7,491 |
| Grenada | 15,130 | 16,539 | 19,220 |
| Guyana | 20,062 | 22,299 | 22,783 |
| Saint Vincent and Grenadines | 8,405 | 9,636 | 11,041 |
| United Kingdom | 48,570 | 51,688 | 57,566 |
| USA | 118,962 | 133,565 | 138,935 |
| Venezuela | 10,207 | 11,107 | 10,273 |
| **Total** (incl. others) | **383,101** | **384,212** | **409,069** |

**Tourism receipts** (US $ million, incl. passenger transport): 371 in 2000; 361 in 2001; 402 in 2002.

Sources: World Tourism Organization.

## Communications Media

| | 2000 | 2001 | 2002 |
|---|---|---|---|
| Television receivers ('000 in use) | 443 | 449 | n.a. |
| Telephones ('000 main lines in use) | 316.8 | 311.8 | 325.1 |
| Mobile cellular telephones ('000 subscribers) | 161.9 | 256.1 | 361.9 |
| Personal computers ('000 in use) | 80 | 90 | 104 |
| Internet users ('000) | 100 | 120 | 138 |

**Radio receivers** (1997, '000 in use): 680.

**Facsimile machines** (1998, number in use): 5,024.

**Daily newspapers:** 4 in 1997 (average circulation: 191,000 in 2001).

**Non-daily newspapers:** 5 in 1997 (average circulation 167,000 in 2001).

**2003** ('000 subscribers): Mobile cellular telephones 520.0.

Sources: International Telecommunication Union; UN, *Statistical Yearbook*; UNESCO, *Statistical Yearbook*.

## Education

(2001/02, unless otherwise indicated)

| | Institutions | Teachers | Students Males | Students Females | Students Total |
|---|---|---|---|---|---|
| Pre-primary | 50* | 1,790 | 13,702 | 8,398 | 22,100 |
| Primary | 480 | 7,975 | 79,023 | 75,924 | 154,947 |
| Secondary | 101† | 5,443 | 47,150 | 49,075 | 96,225 |
| University and equivalent | 3 | 550 | 3,931 | 5,935 | 9,866 |

* Government schools and assisted schools only, in 1992/93.
† 1993/94.

Source: UNESCO.

**Adult literacy rate** (UNESCO estimates): 98.5% (males 99.0%; females 97.9%) in 2002 (Source: UN Development Programme, *Human Development Report*).

# Directory

## The Constitution

Trinidad and Tobago became a republic, within the Commonwealth, under a new Constitution on 1 August 1976. The Constitution provides for a President and a bicameral Parliament comprising a Senate and a House of Representatives. The President is elected by an Electoral College of members of both the Senate and the House of Representatives. The Senate consists of 31 members appointed by the President: 16 on the advice of the Prime Minister, six on the advice of the Leader of the Opposition and nine at the President's own discretion from among outstanding persons from economic, social or community organizations. The House of Representatives consists of 36 members who are elected by universal adult suffrage. The duration of a Parliament is five years. The Cabinet, presided over by the Prime Minister, is responsible for the general direction and control of the Government. It is collectively responsible to Parliament.

## The Government

### HEAD OF STATE

**President:** Prof. GEORGE MAXWELL RICHARDS (took office 17 March 2003).

### THE CABINET
(July 2005)

**Prime Minister, Minister of Finance and Minister of Tobago Affairs:** PATRICK MANNING.
**Attorney-General:** JOHN JEREMIE.
**Minister of Legal Affairs:** CHRISTINE KANGALOO.
**Minister of Community Development, Culture and Gender Affairs:** JOAN YUILLE-WILLIAMS.
**Minister of Social Development:** ANTHONY ROBERTS.
**Minister of Education:** HAZEL ANNE MARIE MANNING.
**Minister of Energy and Energy Industries:** ERIC WILLIAMS.
**Minister of Agriculture, Land and Marine Resources:** JARRETTE NARINE.
**Minister of Foreign Affairs:** KNOWLSON GIFT.
**Minister of Health:** JOHN RAHAEL.
**Minister of Housing:** Dr KEITH ROWLEY.
**Minister of Local Government:** RENNIE DUMAS.
**Minister of Planning and Development:** CAMILLE ROBINSON REGIS.
**Minister of Public Utilties and the Environment:** PENELOPE BECKLES.
**Minister of Labour and Small and Micro Enterprise Development:** DANNY MONTANO.
**Minister of National Security:** MARTIN JOSEPH.
**Minister of Public Administration and Information:** LENNY SAITH.
**Minister of Science, Technology and Tertiary Education:** MUSTAPHA ABDUL HAMID.
**Minister of Sport and Youth Affairs:** ROGER BOYNES.
**Minister of Tourism:** HOWARD CHIN LEE.
**Minister of Trade, Industry and Consumer Affairs:** KENNETH VALLEY.
**Minister of Works and Transport:** COLM IMBERT.

### MINISTRIES

**Office of the President:** President's House, Circular Rd, St Ann's, Port of Spain; tel. 624-1261; fax 625-7950; e-mail presoftt@carib-link.net.

**Office of the Prime Minister:** Whitehall, Maraval Rd, Port of Spain; tel. 622-1625; fax 622-0055; e-mail opm@ttgov.gov.tt; internet www.opm.gov.tt.

**Ministry of Agriculture, Land and Marine Resources:** St Clair Circle, St Clair, Port of Spain; tel. 622-1221; fax 622-8202; e-mail apdmalmr@trinidad.net; internet www.agriculture.gov.tt.

**Ministry of the Attorney-General:** Cabildo Chambers, 23–27 St Vincent St, Port of Spain; tel. 623-7010; fax 625-0470; e-mail ag@ag.gov.tt; internet www.ag.gov.tt.

**Ministry of Community Development, Culture and Gender Affairs:** ALGICO Bldg, Jerningham Ave, Belmont, Port of Spain; tel. 625-3012; fax 625-3278; e-mail cdcga@tstt.net.tt.

**Ministry of Education:** Alexandra Street, St Clair; tel. 622-2181; fax 622-4892; e-mail mined@tstt.net.tt.

**Ministry of Energy and Energy Industries:** Level 9, Riverside Plaza, cnr Besson and Piccadilly Sts, Port of Spain; tel. 623-6708; fax 623-2726; e-mail admin@energy.gov.tt; internet www.energy.gov.tt.

**Ministry of Finance:** Eric Williams Finance Bldg, Independence Sq., Port of Spain; tel. 627-9700; fax 627-6108; e-mail mofcmu@tstt.net.tt; internet www.finance.gov.tt.

**Ministry of Foreign Affairs:** Knowsley Bldg, 10–14 Queen's Park West, Port of Spain; tel. 623-4116; fax 624-4220; e-mail permanentsecretary@foreign.gov.tt.

**Ministry of Health:** IDC Bldg, 10–12 Independence Sq., Port of Spain; tel. 627-0012; fax 623-9528; e-mail scdda@carib-link.net; internet www.healthsectorreform.gov.tt.

**Ministry of Housing:** 44–46 South Quay, Port of Spain; tel. 624-5058; fax 625-2793; e-mail mohas@tstt.tt.

**Ministry of Labour and Small and Micro Enterprise:** Level 11, Riverside Plaza, Cnr Besson and Piccadilly Sts, Port of Spain; tel. 623-4451; fax 624-4091; e-mail rplan@tstt.net.tt; internet www.labour.gov.tt.

**Ministry of Legal Affairs:** Registration House, Huggins Bldg, South Quay, Port of Spain; tel. 623-7163; fax 625-9805.

**Ministry of Local Government:** Kent House, Maraval Rd, Port of Spain; tel. 622-1669; fax 622-4783; e-mail molg2@carib-link.net.

**Ministry of National Security:** Temple Court, 31–33 Abercromby St, Port of Spain; tel. 623-2441; fax 627-8044; e-mail mns@tstt.net.tt.

**Ministry of Planning and Development:** Telly Paul Bldg, St Vincent St, Port of Spain; tel. 627-0403; fax 623-0341.

**Ministry of Public Administration and Information:** 45A-C St Vincent St, Port of Spain; tel. 623-8578; fax 623-6027; internet nict.gov.tt.

**Ministry of Public Utilities and the Environment:** Sacred Heart Bldg, 16–18 Sackville St, Port of Spain; tel. 625-6083; fax 625-7003; e-mail environment@tstt.net.tt.

**Ministry of Science, Technology and Tertiary Education:** Cnr Agra and Patna Sts, St James; tel. 628-9925; fax 622-0775; e-mail stte@stte.gov.tt; internet www.stte.gov.tt.

**Ministry of Social Development:** Ansa McAl Bldg, 69 Independence Sq., Port of Spain; tel. 625-8565; fax 627-4853; e-mail infor@msd.gov.tt; internet www.msd.gov.tt.

**Ministry of Sport and Youth Affairs:** ISSA Nicholas Bldg, Cnr Frederick and Duke Sts, Port of Spain; tel. 625-8874; fax 623-5006.

**Ministry of Tobago Affairs:** Whitehall, Queens Park West, Port of Spain; tel. 622-1625; fax 622-0055; e-mail opm@ttgov.gov.tt.

**Ministry of Tourism:** 51–55 Frederick St, Port of Spain; tel. 624-1403; fax 625-0437; e-mail mintourism@tourism.gov.tt.

**Ministry of Trade, Industry and Consumer Affairs:** Level 15, Riverside Plaza, Cnr Besson and Piccadilly St, Port of Spain; tel. 623-2931; fax 627-8488; e-mail info@tradeind.gov.tt; internet www.tradeind.gov.tt.

**Ministry of Works and Transport:** Cnr Richmond and London Sts, Port of Spain; tel. 625-1225; fax 625-8070.

## Legislature

### PARLIAMENT

#### Senate
**President:** Dr LINDA BABOOLAL.

#### House of Representatives
**Speaker:** BARENDRA SINANAN.

# TRINIDAD AND TOBAGO

**Election, 7 October 2002**

| Party | % of votes | Seats |
|---|---|---|
| People's National Movement | 50.7 | 20 |
| United National Congress | 46.6 | 16 |
| National Alliance for Reconstruction | 1.1 | — |
| Citizens' Alliance | 1.0 | — |
| **Total** (incl. others) | 100.0 | 36 |

### TOBAGO HOUSE OF ASSEMBLY

The House is elected for a four-year term of office and consists of 12 elected members and three members selected by the majority party.

**Chief Secretary:** ORVILLE LONDON.

**Election, 17 January 2005**

| Party | Seats |
|---|---|
| People's National Movement | 11 |
| Democratic Action Congress | 1 |
| **Total** | 12 |

## Political Organizations

**Citizens' Alliance:** 2 Gray St, St Clair, Port of Spain; f. to contest the October 2002 elections; Leader WENDELL MOTTLEY.

**Democratic Action Congress:** Scarborough; f. Jan. 2003 by faction of National Alliance for Reconstruction (q.v.); only active in Tobago; Leader HOCHOY CHARLES.

**Democratic Party of Trinidad and Tobago (DPTT):** Port of Spain; f. March 2002; Leader STEVE ALVAREZ.

**Lavantille Out-Reach for Vertical Enrichment:** L. P. 50, Juman Dr., Morvant; tel. 625-7840; f. 2000; Leader LENNOX SMITH; Gen. Sec. VAUGHN CATON.

**The Mercy Society:** L. P. 216, Mount Zion, Luango Village, Maracas; tel. 628-1753.

**National Alliance for Reconstruction (NAR):** 71 Dundonald St, Port of Spain; tel. 627-6163; f. 1983 as a coalition of moderate opposition parties; reorganized as a single party in 1986; Leader LENNOX SANKERSINGH; Chair. CHRISTO GIFT.

**National Democratic Organization (NDO):** L. P. 2, Freedom St, Enterprise, Chaguanas; tel. 750 9063; e-mail ndotrini@yahoo.com; f. 2000; Leader ENOCH JOHN; Sec. RICHARD JOODEEN.

**National Democratic Party (NDP):** Port of Spain; Leader CARSON CHARLES.

**National Joint Action Committee (NJAC):** 17 School Rd, Point Fortin; tel. 648-2749; Leader MAKANDAL DAAGA.

**People's Empowerment Party (PEP):** Miggins Chamber, Young St, Scarborough; tel. 639-3175; by independent mems of the Tobago House of Assembly; Leader DEBORAH MOORE-MIGGINS; Gen. Sec. RICHARD ALFRED.

**People's National Movement (PNM):** Balsier House, 1 Tranquillity St, Port of Spain; tel. 625-1533; e-mail pnm@carib-link.net; internet www.pnm.org.tt; f. 1956; moderate nationalist party; Leader PATRICK MANNING; Chair. Dr LINDA BABOOLAL; Gen. Sec. MARTIN JOSEPH.

**Republican Party:** Port of Spain; Leader NELLO MITCHELL.

**Team Unity National:** Port of Spain; Leader RAMESH MAHARAJ.

**United National Congress (UNC):** Rienzi Complex, 78–81 Southern Main Rd, Couva; tel. 636-8145; e-mail info@unc.org.tt; internet www.unc.org.tt; f. 1989; social democratic; Leader BASDEO PANDAY; Deputy Leader RAMESH LAWRENCE MAHARAJ; Chair. WADE MARK; CEO TIM GOPEESINGH.

## Diplomatic Representation

### EMBASSIES AND HIGH COMMISSIONS IN TRINIDAD AND TOBAGO

**Argentina:** Tatil Bldg, 4th Floor, 411 Maraval Rd, POB 162; Port of Spain; tel. 628-7557; fax 628-7544; e-mail embargen-pos@carib-link.net; Ambassador JOSÉ LUIS VIGNOLO.

**Brazil:** 18 Sweet Briar Rd, St Clair, POB 382, Port of Spain; tel. 622-5779; fax 622-4323; e-mail brastt@wow.net.tt; Ambassador GILDA MARÍA RAMOS GUIMARÃES.

**Canada:** Maple House, 3-3A Sweet Briar Rd, St Clair, POB 1246, Port of Spain; tel. 622-6232; fax 628-1830; e-mail pspan@dfait-maeci.gc.ca; internet www.portofspain.gc.ca; High Commissioner DEXTER BISHOP (acting).

**China, People's Republic:** 39 Alexandra St, St Clair, Port of Spain; tel. 622-6976; fax 622-7613; e-mail tian@wow.net; Ambassador WANG ZHIQUAN.

**Cuba:** Furness Bldg, 2nd Floor, 90 Independence Sq., Port of Spain; tel. 627-1306; fax 627-3515; e-mail cuba_tt@cablenet.tt; Ambassador FÉLIX RAÚL ROJAS CRUZ.

**Dominican Republic:** Suite 8, 1 Dere St, Queen's Park West, Port of Spain; tel. 624-7930; fax 623-7779; e-mail embadom@tstt.net.tt; Ambassador JOSÉ MANUEL CASTILLO BETANCES.

**France:** Tatil Bldg, 6th Floor, 11 Maraval Rd, Port of Spain; tel. 622-7447; fax 628-2632; e-mail francett@wow.net; internet www.ambafrance-tt.org; Ambassador CHARLEY CAUSERET.

**Germany:** 7–9 Marli St, Newtown, POB 828, Port of Spain; tel. 628-1630; fax 628-5278; e-mail germanembassy@tstt.net.tt; internet www.port-of-spain.diplo.de; Ambassador Dr HELMUT OHLRAUN.

**Holy See:** 11 Mary St, St Clair, POB 854, Port of Spain; tel. 622-5009; fax 628-5457; e-mail apnun@trinidad.net; Apostolic Nuncio Most Rev. THOMAS E. GULLICKSON (Titular Archbishop of Bomarzo).

**India:** 6 Victoria Ave, POB 530, Port of Spain; tel. 627-7480; fax 627-6985; e-mail hcipos@tstt.net.tt; internet www.hcipos.com; High Commissioner J.S. SAPRA.

**Jamaica:** 2 Newbold St, St Clair, Port of Spain; tel. 622-4995; fax 628-9043; e-mail jhctnt@tstt.net.tt; High Commissioner LORNE McDONNOUGH.

**Japan:** 5 Hayes St, St Clair, POB 1039, Port of Spain; tel. 628-5991; fax 622-0858; e-mail jpemb@wow.net; Ambassador YOSHIO YAMAGISHI.

**Mexico:** Algico Bldg, 4th Floor, 91–93 St Vincent St, Port of Spain; tel. 627-7047; fax 627-1028; e-mail embamex@carib-link.net; Ambassador BENITO ANDION SANCHO.

**Netherlands:** Life of Barbados Bldg, 3rd Floor, 69–71 Edward St, POB 870, Port of Spain; tel. 625-1210; fax 625-1704; e-mail info@holland.tt; internet www.holland.tt; Ambassador MAARTEN M. VAN DER GAAG.

**Nigeria:** 3 Maxwell-Phillip St, St Clair, Port of Spain; tel. 622-4002; fax 622-7162; High Commissioner NNE FURO KURUBO.

**Panama:** Suite 6, 1A Dere St, Port of Spain; tel. 623-3435; fax 623-3440; e-mail embapatt@wow.net; Chargé d'affaires a.i. JOSÉ RODRIGO DE LA ROSA.

**Suriname:** Tatil Bldg, 5th Floor, 11 Maraval Rd, Port of Spain; tel. 628-0704; fax 628-0086; e-mail surinameembassy@tstt.net.tt; Ambassador NEVILLE J. VEIRA.

**United Kingdom:** 19 St Clair Circle, St Clair, POB 778, Port of Spain; tel. 622-2748; fax 622-4555; e-mail csbbhc@opus.co.tt; High Commissioner RONALD NASH.

**USA:** 15 Queen's Park West, POB 752, Port of Spain; tel. 622-6371; fax 625-5462; e-mail usispos@trinidad.net; internet usembassy.state.gov/trinidad; Ambassador Dr ROY L. AUSTIN.

**Venezuela:** 16 Victoria Ave, POB 1300, Port of Spain; tel. 627-9821; fax 624-2508; e-mail embaveneztt@carib-link.net; Ambassador HECTOR AZOCAR.

## Judicial System

The Chief Justice, who has overall responsibility for the administration of justice in Trinidad and Tobago, is appointed by the President after consultation with the Prime Minister and the Leader of the Opposition. The President appoints and promotes judges on the advice of the Judicial and Legal Service Commission. The Judicial and Legal Service Commission, which comprises the Chief Justice as chairman, the chairman of the Public Service Commission, two former judges and a senior member of the bar, appoints all judicial and legal officers. The Judiciary comprises the higher judiciary (the Supreme Court) and the lower judiciary (the Magistracy).

**Chief Justice:** SATNARINE SHARMA.

**Supreme Court of Judicature:** Knox St, Port of Spain; tel. 623-2417; fax 627-5477; e-mail ttlaw@wow.net; internet www.ttlawcourts.org; the Supreme Court consists of the High Court of Justice and the Court of Appeal. The Supreme Court is housed in three locations: Port of Spain, San Fernando and Tobago. There are 23 Supreme Court Puisne Judges who sit in criminal, civil, and matrimonial divisions; Registrar EVELYN PETERSEN.

**Court of Appeal:** The Court of Appeal hears appeals against decisions of the Magistracy and the High Court. Further appeals are

# TRINIDAD AND TOBAGO

directed to the Judicial Committee of the Privy Council of the United Kingdom, sometimes as of right and sometimes with leave of the Court. The Court of Appeal consists of the Chief Justice, who is President, and six other Justices of Appeal.

### The Magistracy and High Court of Justice

The Magistracy and the High Court exercise original jurisdiction in civil and criminal matters. The High Court hears indictable criminal matters, family matters where the parties are married, and civil matters involving sums over the petty civil court limit. High Court judges are referred to as either Judges of the High Court or Puisne Judges. The Masters of the High Court, of which there are four, have the jurisdiction of judges in civil chamber courts. The Magistracy (in its petty civil division) deals with civil matters involving sums of less than TT $15,000. It exercises summary jurisdiction in criminal matters and hears preliminary inquiries in indictable matters. The Magistracy, which is divided into 13 districts, consists of a Chief Magistrate, a Deputy Chief Magistrate, 13 Senior Magistrates and 29 Magistrates.

**Chief Magistrate:** SHERMAN MCNICHOLLS, Magistrates' Court, St Vincent St, Port of Spain; tel. 625-2781.

**Director of Public Prosecutions:** GEOFFREY HENDERSON.

**Attorney-General:** JOHN JEREMIE.

## Religion

In 2000 it was estimated that 24.6% of the population was Protestant, including Anglican (7.8%), Pentecostal (6.8%), Seventh-day Adventist (4%) and Presbyterian (3.3%), while some 22.5% of the population was Hindu, 5.8% Muslim and 5.4% Shouter Baptist.

### CHRISTIANITY

**Caribbean Conference of Churches:** 4 Francis Lau St, St James, POB 876, Port of Spain; tel. 628-2028; fax 628-2031; e-mail caconftt@Trinidad.net; internet www.cariblife.com/pub/ccc.

**Christian Council of Trinidad and Tobago:** Hayes Court, 21 Maraval Rd, Port of Spain; tel. 637-9329; f. 1967; church unity organization formed by the Roman Catholic, Anglican, Presbyterian, Methodist, African Methodist, Spiritual Baptist and Moravian Churches, the Church of Scotland and the Salvation Army, with the Ethiopian Orthodox Church and the Baptist Union as observers; Pres. The Rt Rev. CALVIN BESS (Anglican Archbishop of Trinidad and Tobago); Sec. GRACE STEELE.

### The Anglican Communion

Anglicans are adherents of the Church in the Province of the West Indies, comprising eight dioceses. The Archbishop of the West Indies is the Bishop of Nassau and the Bahamas.

**Bishop of Trinidad and Tobago:** The Rt Rev. RAWLE E. DOUGLIN, Hayes Court, 21 Maraval Rd, Port of Spain; tel. 622-7387; fax 628-1319; e-mail red@trinidad.net.

### Protestant Churches

**Presbyterian Church in Trinidad and Tobago:** POB 92, Paradise Hill, San Fernando; tel. and fax 652-4829; e-mail pctt@tstt.net.tt; f. 1868; Moderator Rt Rev. ALLISON KEN NOBBEE; 45,000 mems.

**Baptist Union of Trinidad and Tobago:** 104 High St, Princes Town; tel. 655-2291; f. 1816; Pres. Rev. ALBERT EARL-ELLIS; Gen. Sec. Rev. JOHN S. C. BRAMBLE; 24 churches, 3,300 mems.

### The Roman Catholic Church

For ecclesiastical purposes, Trinidad and Tobago comprises the single archdiocese of Port of Spain. At 31 December 2003 there were some 383,302 adherents in the country, representing about 30% of the total population.

**Antilles Episcopal Conference:** 9A Gray St, Port of Spain; tel. 622-2932; fax 628-3688; e-mail aec@carib-link.net; internet www.catholiccaribbean.org; f. 1975; 21 mems from the Caribbean and Central American regions; Pres. Most Rev. LAWRENCE BURKE (Archbishop of Kingston, Jamaica); Gen. Sec. Rev. GERARD E. FARFAN.

**Archbishop of Port of Spain:** EDWARD J. GILBERT, 27 Maraval Rd, Port of Spain; tel. 622-1103; fax 622-1165; e-mail abishop@carib-link.net.

### HINDUISM

Hindu immigrants from India first arrived in Trinidad and Tobago in 1845. The vast majority of migrants, who were generally from Uttar Pradesh, were Vishnavite Hindus, who belonged to sects such as the Ramanandi, the Kabir and the Sieunaraini. The majority of Hindus currently subscribe to the doctrine of Sanathan Dharma, which evolved from Ramanandi teaching.

*Directory*

**Arya Pratinidhi Sabha of Trinidad Inc (Arya Samaj):** Seereeram Memorial Vedic School, Old Southern Main Rd, Montrose Village, Chaguanas; tel. 663-1721; e-mail sadananramnarine@hotmail.com; Pres. SADANAN RAMNARINE.

**Pandits' Parishad** (Council of Pandits): Maha Sabha Headquarters, Eastern Main Rd, St Augustine; tel. 645-3240; works towards the co-ordination of temple activities and the standardization of ritual procedure; affiliated to the Maha Sabha; 200 mems.

**Sanathan Dharma Maha Sabha of Trinidad and Tobago Inc:** Maha Sabha Headquarters, Eastern Main Rd, St Augustine; tel. 645-3240; e-mail mahasabha@ttemail.com; internet www.websitetech.com/mahasabha; f. 1952; Hindu pressure group and public organization; organizes the provision of Hindu education; Pres. Dr D. OMAH MAHARAJH; Sec. Gen. SATNARAYAN MAHARAJ.

## The Press

### DAILIES

**Newsday:** 19–21 Chacon St, Port of Spain; tel. 623-2459; fax 657-5008; internet www.newsday.co.tt; f. 1993; CEO and Editor-in-Chief THERESE MILLS; circ. 2,200,000.

**Trinidad Guardian:** 22 St Vincent St, POB 122, Port of Spain; tel. 623-8871; fax 625-5702; e-mail letters@ttol.co.tt; internet www.guardian.co.tt; f. 1917; morning; independent; Editor-in-Chief DOMINIC KALIPERSAD; circ. 52,617.

**Trinidad and Tobago Express:** 35 Independence Sq., Port of Spain; tel. 623-1711; fax 627-1451; e-mail express@trinidadexpress.com; internet www.trinidadexpress.com; f. 1967; morning; CEO KEN GORDON; Editor KEITH SMITH; circ. 55,000.

### PERIODICALS

**Blast:** 5–6 Hingoo Lane, El Socorro, San Juan; tel. 674-4414; weekly; Editor ZAID MOHAMMED; circ. 22,000.

**The Boca:** Crews Inn Marina and Boatyard, Village Sq., Chaguaramas; tel. 634-2055; fax 634-2056; e-mail boca@boatersenterprise.com; internet www.boatersenterprise.com/boca; monthly; magazine of the sailing and boating community; Man. Dir JACK DAUSEND.

**The Bomb:** Southern Main Rd, Curepe; tel. 645-2744; weekly.

**Caribbean Beat:** 6 Prospect Ave, Maraval, Port of Spain; tel. 622-3821; fax 628-0639; e-mail info@meppublishers.com; internet www.caribbean-beat.com; 6 a year; distributed by BWIA International Airways Ltd; Editor JEREMY TAYLOR; Ma. HELEN SHAIR-SINGH.

**Catholic News:** 31 Independence Sq., Port of Spain; tel. 623-6093; fax 623-9468; e-mail cathnews@trinidad.net; internet www.catholicnews-tt.net; f. 1892; weekly; Editor JUNE JOHNSTON; circ. 16,000.

**Economic Bulletin:** 35–41 Queen St, POB 98, Port of Spain; tel. 623-7069; fax 625-3802; e-mail statinfo@wow.net; f. 1950; issued 3 times a year by the Central Statistical Office.

**Energy Caribbean:** 6 Prospect Ave, Maraval, Port of Spain; tel. 622-3821; fax 628-0639; e-mail dchin@meppublishers.com; internet www.meppublishers.com; f. 2002; bimonthly; Editor DAVID RENWICK.

**Showtime:** Cnr 9th St and 9th Ave, Barataria; tel. 674-1692; fax 674-3228; circ. 30,000.

**Sunday Express:** 35 Independence Sq., Port of Spain; tel. 623-1711; fax 627-1451; e-mail express@trinidadexpress.com; internet www.trinidadexpress.com; f. 1967; Editor OMATIE LYDER; circ. 51,405.

**Sunday Guardian:** 22 St Vincent St, POB 122, Port of Spain; tel. 623-8870; fax 625-7211; e-mail esunday@ttol.co.tt; internet www.guardian.co.tt; f. 1917; independent; morning; Editor-in-Chief DOMINIC KALIPERSAD; circ. 48,324.

**Sunday Punch:** Cnr 9th St and 9th Ave, Barataria; tel. 674-1692; fax 674-3228; internet www.tntmirror.com; weekly; Editor ANTHONY ALEXIS; circ. 40,000.

**Tobago News:** Milford Rd, Scarborough; tel. 639-5565; fax 625-4480; f. 1985; weekly; Editor COMPTON DELPH.

**Trinidad and Tobago Gazette:** 2–4 Victoria Ave, Port of Spain; tel. 625-4139; weekly; official government paper; circ. 3,300.

**Trinidad and Tobago Mirror:** Cnr 9th St and 9th Ave, Barataria; tel. 674-1692; fax 674-3228; internet www.tntmirror.com; 2 a week; Editors KEN ALI, KEITH SHEPHERD; circ. 35,000.

**Tropical Agriculture:** Faculty of Agriculture and Natural Sciences, University of the West Indies, St Augustine; tel. and fax 645-3640; e-mail tropicalagri@fans.uwi.tt; f. 1924; journal of the School of Agriculture (fmrly Imperial College of Tropical Agriculture); quarterly; Editor-in-Chief Prof. FRANK A. GUMBS.

**Weekend Heat:** Southern Main Rd, Curepe; tel. 625-4583; weekly; Editor STAN MORA.

## Publishers

**Caribbean Children's Press:** 7 Coronation St, St James; tel. and fax 628-4248; e-mail caripres@tstt.net.tt; f. 1987; educational publishers for primary schools.

**Caribbean Educational Publishers:** Gulf View Link Rd, La Romaine; tel. 657-9613; fax 652-5620; e-mail mbscep@tstt.net.tt; Pres. TEDDY MOHAMMED.

**Charran Publishing House Ltd:** 60 South Quay, Port of Spain; tel. 625-9821; fax 623-6597; e-mail publishing@charran.com; Man. RICO CHARRAN.

**Daiman Publishers:** 10 Eccles Trace, Curepe; tel. 645-4143; fax 645-4143; e-mail daiman@cablenett.net; educational books; Dir DAISY SEEGOBIN.

**Lexicon Trinidad Ltd:** 48 Boundary Rd, San Juan; tel. 675-3395; fax 675-3360; e-mail lexicon@tstt.net.tt; Dir KEN JAIKARANSINGH.

**Morton Publishing:** 97 Saddle Rd, Maraval; tel. 348-37777; fax 762-9923; e-mail morton@morton-pub.com; internet www.morton-pub.com; f. 1977; educational books; Pres. DOUG MORTON; Dir JULIE MORTON.

**Prospect Press** (Media and Editorial Projects): 6 Prospect Ave, Maraval, Port of Spain; tel. 622-3821; fax 628-0639; e-mail info@meppublishers.com; internet www.meppublishers.com; f. 1991; magazine and book publishing; Man. Dir JEREMY TAYLOR.

**Trinidad Publishing Co Ltd:** 22–24 St Vincent St, Port of Spain; tel. 623-8870; fax 625-7211; e-mail business@ttol.co.tt; internet www.guardian.co.tt; f. 1917; Man. Dir GRENFELL KISSOON.

**University of the West Indies:** St Augustine; tel. 662-2002; fax 663-9684; e-mail isersta@trinidad.net; f. 1960; academic books and periodicals; Editor PATRICIA SAMPSON.

## Broadcasting and Communications

### TELECOMMUNICATIONS

**Digicel Trinidad and Tobago:** 11–13 Victoria Ave, Port of Spain; internet www.digiceltrinidadandtobago.com; owned by an Irish consortium; mobile cellular telephone licence granted in 2005; Chair. DENIS O'BRIEN.

**Telecommunication Services of Trinidad and Tobago (TSTT) Ltd:** 1 Edward St, POB 3, Port of Spain; tel. 625-4431; fax 627-0856; e-mail tsttceo@tstt.net.tt; internet www.tstt.co.tt; 51% state-owned, 49% by Cable & Wireless (United Kingdom); 51% privatization pending; CEO SAMUEL A. MARTIN.

**TSTT Cellnet:** 114 Frederick St, Port of Spain; fax 625-5807; internet www.tstt.co.tt; f. 1991; mobile cellular telephone operator.

**Caribbean Communications Network (CCN):** 35 Independence Sq., Port of Spain; tel. 623-1711; fax 627-2721; e-mail express@trinidadexpress.com; CEO CRAIG REYNALD.

**Open Telecom Ltd:** 88 Edward St, Port of Spain; tel. 622-6736.

### BROADCASTING

#### Radio

**Central Radio 90.5:** 1 Morequito Ave, Valsayn, Port of Spain; tel. 662-4309.

**Hott 93 FM:** Cumulus Broadcasting Inc, 3A Queens Park West, Port of Spain; tel. 623-4688; fax 624-3234; e-mail studio@hott93.com.

**Love 94.1 and Power 102 FM:** 88–90 Abercromby St, Port of Spain; tel. 627-6937; fax 624-8223.

**Music Radio 97 FM:** Long Circular Rd, St James; tel. 622-9797; fax 624-3234.

**National Broadcasting Network Ltd (FM 100, Yes 98.9 FM, Swar Milan 91.1, Radio 610):** 11A Maraval Rd, Port of Spain; tel. 662-4141; fax 622-0344; e-mail bdesilva@nbn.co.tt; f. 1957; AM and FM transmitters at Chaguanas, Cumberland Hill, Hospedales and French Fort, Tobago; govt-owned; CEO DOMINIC BEAUBRUN; Man. BRENDA DE SILVA; est. regular audience 105,000.

**Trinidad Broadcasting Co Ltd (Radio Trinidad, Radio Nine Five):** Broadcasting House, 11B Maraval Rd, POB 716, Port of Spain; tel. 622-1151; fax 622-2380; commercial.

**Trinidad Broadcasting Company (Radio Trinidad 730 AM, Rhythm Radio, Caribbean Tempo and WEFM 96.1 FM):** 22 St Vincent St, Port of Spain; tel. 623-9202; fax 622-2380; f. 1947; commercial; four programmes; Man. Dir GRENFELL KISSOON.

**Trinidad and Tobago Radio Network:** 35 Independence Sq., Port of Spain; tel. 624-7078.

#### Television

**CCN TV6:** 35 Express House, Independence Sq., Port of Spain; tel. 627-8806; fax 627-1451; e-mail jlumyoung@trinidadexpress.com; internet www.trinidadexpress.com; f. 1991; operates channels 6 and 18; owned by Caribbean Communications Network (CCN); CEO CRAIG REYNALD; Gen. Man. RASHIDAN BOLAI.

**National Broadcasting Network Ltd:** 11A Maraval Rd, POB 665, Port of Spain; tel. 622-4141; fax 622-0344; e-mail bdesilva@nbn.co.tt; f. 1962; state-owned commercial station; operates channels 2, 4, 13, 16, TIC and The Information Channel; CEO DOMINIC BEAUBRUN.

## Finance

(cap. = capital; dep. = deposits; res = reserves; m. = million; brs = branches; amounts in TT $ unless otherwise stated)

### BANKING

#### Central Bank

**Central Bank of Trinidad and Tobago:** Eric Williams Plaza, Brian Lara Promenade, POB 1250, Port of Spain; tel. 625-4835; fax 627-4696; e-mail info@central-bank.org.tt; internet www.central-bank.org.tt; f. 1964; cap. 100.0m., res 100.0m., dep. 10,637.2m. (Sept. 2003); Gov. EWART S. WILLIAMS.

#### Commercial Banks

**Citibank (Trinidad and Tobago) Ltd:** 12 Queen's Park East, POB 1249, Port of Spain; tel. 625-1046; fax 624-8131; internet www.citicorp.com; f. 1983; fmrly The United Bank of Trinidad and Tobago Ltd; name changed as above 1989; owned by Citicorp Merchant Bank Ltd; cap. 30.0m., res 14.7m., dep. 455.5m. (Dec. 1996); Chair. IAN E. DASENT; Man. Dir STEVE BIDESHI; 2 brs.

**Citicorp Merchant Bank Ltd:** 12 Queen's Park East, POB 1249, Port of Spain; tel. 623-3344; fax 624-8131; cap. 57.1m., res 28.5m., dep 473.6m. (Dec. 2002); owned by Citibank Overseas Investment Corpn; Chair. IAN E. DASENT; Man. Dir KAREN DARBASIE.

**First Citizens Bank Ltd:** 62 Independence Sq., Port of Spain; tel. 625-2893; fax 623-3393; e-mail itc@firstcitizenstt.com; internet www.firstcitizenstt.com; f. 1993; as merger of National Commercial Bank of Trinidad and Tobago Ltd, Trinidad Co-operative Bank Ltd and Workers' Bank of Trinidad and Tobago; state-owned; cap. 340.0m., res 907.3m., dep. 3,237.5m. (Sept. 2002); Chair. KENNETH GORDON; CEO LARRY HOWAI; 22 brs.

**RBTT Ltd:** Royal Court, 19–21 Park St, POB 287, Port of Spain; tel. 623-1322; fax 625-3764; e-mail royalinfo@rbtt.co.tt; internet www.rbtt.com; f. 1972 as Royal Bank of Trinidad and Tobago to take over local branches of Royal Bank of Canada, present name adopted April 2002; cap. 403.9m., res 219.3m., dep. 5,184.0m. (March 2003); Chair. PETER J. JULY; CEO TERRENCE A. J. MARTINS; 21 brs.

**Republic Bank Ltd:** 9–17 Park St, POB 1153, Port of Spain; tel. 623-1056; fax 624-1323; e-mail email@republictt.com; internet www.republictt.com; f. 1837 as Colonial Bank; 1972 became Barclays Bank; 1981 name changed as above; 1997 merged with Bank of Commerce Trinidad and Tobago Ltd; cap. 481.2m., res 252.5m., dep. 21,123.3m. (Sept. 2003); Chair. and Man. Dir RONALD HARFORD; Exec. Dir GREGORY I. THOMSON; 47 brs.

**Republic Finance & Merchant Bank Ltd:** 9–17 Park St, POB 1153, Port of Spain; tel. 623-1056; fax 624-1323; e-mail email@republictt.com; internet www.republictt.com; f. 1965; owned by Republic Bank Ltd (q.v.); cap. 30.0m., res 35.0m., dep. 2,066.1m. (Sept. 2002); Chair. RONALD HARFORD; Man. Dir CHERYL F. GREAVES.

**Scotiabank Trinidad and Tobago Ltd:** Cnr of Park and Richmond Sts, POB 621, Port of Spain; tel. 625-3566; fax 627-5278; e-mail scotiamain@carib.link.net; internet www.scotiabanktt.com; cap. 117.6m., res 207.9m., dep. 5,690.5m. (Oct. 2003); Chair. ROBERT H. PITFIELD; Man. Dir R. P. YOUNG; 23 brs.

#### Development Banks

**Agricultural Development Bank of Trinidad and Tobago:** 87 Henry St, POB 154, Port of Spain; tel. 623-6261; fax 624-3087; e-mail adbceo@tstt.net.tt; f. 1968; provides long-, medium- and short-term loans to farmers and the agri-business sector; Chair. HUBERT ALLEYNE; CEO JACQUELINE RAWLINS.

TRINIDAD AND TOBAGO

**DFL Caribbean:** 10 Cipriani Blvd, POB 187, Port of Spain; tel. 623-4665; fax 624-3563; e-mail dfl@dflcaribbean.com; internet www.dflcaribbean.com; provides short- and long-term finance, and equity financing for projects in manufacturing, agro-processing, tourism, industrial and commercial enterprises; total assets US $84.2m. (Dec. 1998); Man. Dir GERARD M. PEMBERTON.

### Credit Unions

**Co-operative Credit Union League of Trinidad and Tobago Ltd:** 32–34 Maraval Rd, St Clair; tel. 622-3100; fax 622-4800; e-mail culeague@tstt.net.tt; internet www.ccultt.org.

**Hindu Credit Union Co-operative Society Ltd:** Ramlals Bldg, Main Rd, Chaguanas; tel. 671-3718; e-mail e-mail@hinducreditunion.com; internet www.hinducreditunion.com/hcu.html; Pres. HARRY HARNARINE.

### STOCK EXCHANGE

**Trinidad and Tobago Stock Exchange Ltd:** 1st Floor, 1 Ajax St, Wrightson Rd, Port of Spain; tel. 625-5107; fax 623-0089; e-mail ttstockx@tstt.net.tt; internet www.stockex.co.tt; f. 1981; 38 companies listed (2005); electronic depository system came into operation in 1999; Chair. KATHLEEN DHANNYRAM; Gen. Man. HUGH EDWARDS.

### INSURANCE

**American Life and General Insurance Co (Trinidad and Tobago) Ltd:** ALGICO Plaza, 91–93 St Vincent St, POB 943, Port of Spain; tel. 625-4425; fax 623-6218; e-mail algico@wow.net.

**Bankers Insurance Co of Trinidad and Tobago Ltd:** 177 Tragarete Rd, Port of Spain; tel. 622-4613; fax 628-6808.

**Barbados Mutual Life Assurance Society:** The Mutual Centre, 16 Queen's Park West, POB 356, Port of Spain; tel. 628-1636; Gen. Man. HUGH MAZELY.

**Capital Insurance Ltd:** 38–42 Cipero St, San Fernando; tel. 657-8077; fax 652-7306; f. 1958; motor and fire insurance; total assets TT $65m.; 10 brs and 9 agencies.

**Colonial Life Insurance Co (Trinidad) Ltd:** Colonial Life Bldg, 29 St Vincent St, POB 443, Port of Spain; tel. 623-1421; fax 627-3821; e-mail info@clico.com; internet www.clico.com; f. 1936; Chair. LAWRENCE A. DUPREY; CEO CLAUDIUS DACON.

**CUNA Caribbean Insurance Society Ltd:** 37 Wrightson Rd, POB 193, Port of Spain; tel. 623-7963; fax 623-6251; e-mail cunains@trinidad.net; internet www.cunacaribbean.com; f. 1991; marine aviation and transport; motor vehicle, personal accident, property; Gen. Man. ANTHONY HALL; 3 brs.

**Furness Anchorage General Insurance Ltd:** 11–13 Milling Ave, Sea Lots, POB 283, Port of Spain; tel. 623-0868; fax 625-1243; e-mail furness@wow.net; internet www.furnessgroup.com; f. 1979; general; Chair. WILLIAM A. FERREIRA.

**GTM Fire Insurance Co Ltd:** 95–97 Queen St, Port of Spain; tel. 623-1525.

**Guardian General Insurance Ltd:** Princes Court, Keate St, Port of Spain, Port of Spain; tel. 623-4741; fax 623-4320; e-mail info@guardiangenerallimited.com; internet www.guardiangenerallimited.com; founded by merger of NEMWIL and Caribbean Home; Chair. HENRY PETER GANTEAUME; CEO RICHARD ESPINET.

**Guardian Life of the Caribbean:** 1 Guardian Dr., West Moorings, Port of Spain; tel. 625-5433; internet www.guardianlife.co.tt; Chair. ARTHUR LOK JACK; Pres. and CEO DOUGLAS CAMACHO.

**Gulf Insurance Ltd:** 1 Gray St, St Clair, Port of Spain; tel. 622-5878; fax 628-0272; e-mail gulf@wow.net; f. 1974; general.

**Maritime Financial Group:** Maritime Centre, 10th Ave, POB 710, Barataria; tel. 674-0130; fax 638-6663; f. 1978; property and casualty; CEO JOHN SMITH.

**Motor and General Insurance Co Ltd:** 1–3 Havelock St, St Clair, Port of Spain; tel. 622-2637; fax 622-5345.

**New India Assurance Co (T & T) Ltd:** 22 St Vincent St, Port of Spain; tel. 623-1326; fax 625-0670; e-mail newindia@wow.net.

**Presidential Insurance Co Ltd:** 54 Richmond St, Port of Spain; tel. 625-4788; e-mail pic101@tstt.net.tt.

**Trinidad and Tobago Export Credit Insurance Co Ltd:** 30 Queen's Park West, Port of Spain; tel. and fax 628–2762; e-mail eximbank@wow.net; internet www.eximbankt.com; Gen. Man. JOSEPHINE IBLE.

**Trinidad and Tobago Insurance Ltd (TATIL):** 11 Maraval Rd, POB 1004, Port of Spain; tel. 622-5351; fax 628-0035; e-mail info@tatil.co.tt; internet www.tatil.co.tt; Chair. JOHN JARDIM; CEO RELNA VIRE.

*Directory*

### INSURANCE ORGANIZATIONS

**Association of Trinidad and Tobago Insurance Companies:** 28 Sackville St, Port of Spain; tel. 624-2817; fax 625-5132; e-mail jsc-attic@trinidad.net; internet www.attic.org.tt; Chair. INEZ SINANAN.

**National Insurance Board:** Cipriani Pl., 2A Cipriani Blvd, POB 1195, Port of Spain; tel. 625-2171; fax 624-0276; internet www.nibtt.co.tt; f. 1971; statutory corporation; Chair. CALDER HART; Exec. Dir JEFFREY MCFARLANE.

## Trade and Industry

### GOVERNMENT AGENCIES

**Cocoa and Coffee Industry Board:** 27 Frederick St, POB 1, Port of Spain; tel. 625-0298; fax 627-4172; e-mail ccib@tstt.net.tt; f. 1962; marketing of coffee and cocoa beans, regulation of cocoa and coffee industry; Man. KENT VILLAFANA.

**Export-Import Bank of Trinidad and Tobago Ltd (EXIMBANK):** 30 Queens Park West, Port of Spain; tel. 628-2762; fax 622-3545; e-mail eximbank@wow.net; internet www.eximbankt.com; Chair. CLARRY BENN; Gen. Man. JOSEPHINE IBLE.

**Trinidad and Tobago Forest Products Ltd (TANTEAK):** Connector Rd, Carlsen Field, Chaguanas; tel. 665-0078; fax 665-6645; f. 1975; harvesting, processing and marketing of state plantation-grown teak and pine; privatization pending; Chair. RUSKIN PUNCH; Man. Dir CLARENCE BACCHUS.

### DEVELOPMENT ORGANIZATIONS

**National Energy Corporation of Trinidad and Tobago Ltd:** PLIPDECO House, Orinoco Dr., POB 191, Point Lisas, Couva; tel. 636-4662; fax 679-2384; e-mail infocent@carib-link.net; owned by the National Gas Co of Trinidad and Tobago Ltd (q.v.); f. 1979; Chair. KENNETH BIRCHWOOD.

**National Housing Authority:** 44–46 South Quay, POB 555, Port of Spain; tel. 627-1703; fax 625-3963; e-mail info@housing.gov.tt; internet www.housing.gov.tt; f. 1962; Chair. ANDRE MONYEIL; CEO NOEL GARCIA.

**Point Lisas Industrial Port Development Corporation Ltd (PLIPDECO):** PLIPDECO House, Orinoco Dr., POB 191, Point Lisas, Couva; tel. 636-2201; fax 636-4008; e-mail plipdeco@plipdeco.com; internet www.plipdeco.com; f. 1966; privatized in the late 1990s; deep-water port handling general cargo, liquid and dry bulk, to serve adjacent industrial estate, which now includes iron and steel complex, methanol, ammonia, urea and related downstream industries; Chair. Commdr KAYAM MOHAMMED; CEO Capt. RAWLE BADDALOO.

**Tourism and Industrial Development Company of Trinidad and Tobago (TIDCO):** 10–14 Philipps St, POB 222, Port of Spain; tel. 623-6022; fax 624-3848; e-mail invest-info@tidco.co.tt; internet www.tidco.co.tt; Marketing Man. NEEMAH PERSAD-CELESTINE.

### CHAMBERS OF COMMERCE

**South Trinidad Chamber of Industry and Commerce:** Suite 311, Cross Crossing Shopping Centre, Lady Hailes Ave, San Fernando; tel. 652-5613; e-mail execoffice@southchamber.org; internet www.southchamber.org; f. 1956; Pres. JIM LEE YOUNG; CEO Dr THACKWRAY DRIVER.

**Trinidad and Tobago Chamber of Industry and Commerce (Inc):** Chamber Bldg, Columbus Circle, Westmoorings, POB 499, Port of Spain; tel. 637-6966; fax 637-7425; e-mail chamber@chamber.org.tt; internet www.chamber.org.tt; f. 1891; Pres. CHRISTIAN MOUTTET; CEO JOAN FERREIRA; 600 mems.

### INDUSTRIAL AND TRADE ASSOCIATIONS

**Agricultural Society of Trinidad and Tobago:** 1st Floor, Henry St, Port of Spain; tel. 623-7797; fax 623-3087; e-mail leeyuen@tstt.net.tt.

**Coconut Growers' Association (CGA) Ltd:** Eastern Main Rd, POB 229, Laventille, Port of Spain; tel. 623-5207; fax 623-2359; e-mail cgaltd@tstt.net.tt; f. 1936; 354 mems; Chair. PHILLIP AGOSTINI.

**Co-operative Citrus Growers' Association of Trinidad and Tobago Ltd:** Eastern Main Rd, POB 174, Laventille, Port of Spain; tel. 623-2255; fax 623-2487; e-mail ccga@wow.net; internet www.ccga.co.tt; f. 1932; 437 mems; Pres. AINSLEY NICHOLS; Gen. Man. MUMTAZ ALI.

**Pan Trinbago:** Victoria Park Suites, 14–17 Park St, Port of Spain; tel. 623-4486; fax 625-6715; e-mail admin@pantrinbago.co.tt; internet www.pantrinbago.co.tt; f. 1971; official body for Trinidad

# TRINIDAD AND TOBAGO  *Directory*

and Tobago steelbands; Pres. PATRICK LOUIS ARNOLD; Sec. RICHARD FORTEAN.

**Shipping Association of Trinidad and Tobago:** 15 Scott Bushe St, Port of Spain; tel. 623-3355; fax 623-8540; e-mail agm@shipping.co.tt; internet www.shipping.co.tt; f. 1938; Pres. NOEL JENVEY; Gen. Man. JENNIFER GONZÁLEZ.

**Sugar Association of the Caribbean:** Brechin Castle, Couva; tel. 636-2449; fax 636-2847; f. 1942; promotes and protects sugar industry in the Caribbean; 6 mem. asscns; Chair. IAN MCDONALD; Sec. A. MOHAMMED.

**Trinidad and Tobago Contractors' Association:** Morequito Ave, Valsayn Park, Valsayn; tel. 637-2967; fax 637-2963; e-mail sec@ttca.com; internet www.ttca.com; f. 1968; represents contractors and manufacturers and suppliers to the sector; Pres. HUGH SCHAMBER.

**Trinidad and Tobago Manufacturers' Association:** 122–124 Frederick St, Port of Spain; tel. 623-1029; fax 623-1031; e-mail ttmagm@opus.co.tt; internet www.ttma.com; f. 1956; 260 mems; Pres. ANTHONY ABOUD.

## EMPLOYERS' ORGANIZATION

**Employers' Consultative Association of Trinidad and Tobago (ECA):** 43 Dundonald St, POB 911, Port of Spain; tel. 625-4723; fax 625-4891; e-mail ecatt@tstt.net.tt ; internet www.ecatt.org; f. 1959; Chair. CLARENCE RAMBHARAT; CEO LINDA BESSON; 156 mems.

## STATE HYDROCARBONS COMPANIES

**National Gas Company of Trinidad and Tobago Ltd (NGC):** Orinoco Dr., Point Lisas Industrial Estate, POB 1127, Port of Spain; tel. 636-4662; fax 679-2384; e-mail ngc@ngc.co.tt; internet www.ngc.co.tt; f. 1975; purchases, sells, compresses, transmits and distributes natural gas to consumers; Pres. FRANK LOOK KIN; Chair. KEITH AWONG.

**Petroleum Company of Trinidad and Tobago Ltd (Petrotrin):** Petrotrin Administration Bldg, Cnr Queen's Park West and Cipriani Blvd, Port of Spain; tel. 625-5240; fax 624-4661; e-mail kharnanan@petrotrin.com; internet www.petrotrin.com; f. 1993 following merger between Trinidad and Tobago Oil Company Ltd (Trintoc) and Trinidad and Tobago Petroleum Company Ltd (Trintopec); govt-owned; petroleum and gas exploration and production; operates refineries and a manufacturing complex, producing a variety of petroleum and petrochemical products; Gen. Man. WAYNE BERTRAND; Vice-Pres KAIN LOOK YEE, KELVIN HARNANAN.

**Petrotrin Trinmar Operations:** Petrotrin Administration Bldg, Point Fortin; tel. 648-2127; fax 648-2519; f. 1962; owned by Petrotrin; marine petroleum and natural gas co; Gen. Man. VICTOR MICHELLE; 705 employees.

**Trintomar Ltd:** Petrotrin Administration Bldg, Pointe-à-Pierre; tel. 649-5500; e-mail lisle.ramyad@petrotrin.com; 80% owned by Petrotrin, 20% owned by NGC; develops offshore petroleum sector; Sec. ADRIAN JEFFERS.

## MAJOR COMPANIES

The following is a selection of major industrial and commercial companies operating in Trinidad and Tobago:

### Food and Beverages

**Angostura Holdings Ltd:** Cnr Eastern Main Rd and Trinity Ave, POB 62, Port of Spain; tel. 623-1841; fax 623-1847; e-mail corphq@angostura.com; internet www.angostura.com; f. 1921; manufacturers of rum, Angostura aromatic bitters and other alcoholic beverages; gross sales of TT $1,351m. (2003); Chair. LAWRENCE A. DUPREY; Man. Exec. Dir PATRICK B. PATEL; 35 employees.

**Bermudez Biscuit Co Ltd:** 6 Maloney St, POB 885, Mount Lambert; tel. 638-3336; fax 638-5911; e-mail bermudez@tstt.net.tt; f. 1950; producers of biscuits and other food products; Man. Dir NOBLE PHILIP; Gen. Man. NICHOLAS MOUTTET; 367 employees.

**Kiss Baking Co Ltd:** 12–14 Gaston St, POB 776, Lange Park, Chaguanas; tel. 671-2253; fax 672-3840; e-mail kissbaking@cariblink.net; f. 1975; bakery; Chair. ROBERT BERMUDEZ; Man. DEBBIE LEE CHARLES; 2,890 employees.

**National Flour Mills Ltd:** 27–29 Wrightson Rd, POB 1154, Port of Spain; tel. 625-2416; fax 625-4389; e-mail flourmil@tstt.net.tt; f. 1972; milling of flour and grains; manufacturers of rice, edible oils and animal feed; sales of TT $738.3m. (1996); Chair WINSTON CONNELL; 275 employees.

**Nestlé Trinidad and Tobago Ltd:** Churchill Roosevelt Highway Valsayn, POB 172, Port of Spain; tel. 663-6832; fax 663-6840; manufacture and distribution of dairy products; subsidiary of Nestlé SA, Switzerland.

**Catelli Primo Ltd:** Churchill Roosevelt Highway, Valsayn; tel. 625-1414; fax 625-3915; fruit and vegetable canners; subsidiary of Nestlé Trinidad and Tobago Ltd; Gen. Man. MICHAEL HACKSHAW.

**Sugar Manufacturing Company (SMCL):** Usine Sainte Madeleine, Victoria; tel. 652-3441; fax 657-5421; e-mail smclchair@tstt.net.tt; state-owned; f. 2003, following dissolution of Caroni Ltd; Chair. PREM NANDLAL; CEO ANDRE NAYADEE.

### Metals

**Bhagwansingh's Hardware and Steel Industries Ltd:** Beetham Highway, Sea Lots, Port of Spain; tel. 627-8335; fax 623-0804; e-mail bhsil@trinidad.net; internet www.trinidad.net/bhsil; f. 1974; manufacturers of aluminium products; wholesale of steel and electrical products; sales of TT $153.0m. (1996); Chair. HELEN BHAGWANSINGH; Man. Dir HUBERT BHAGWANSINGH; 310 employees.

**Mittal Steel Point Lisas Ltd:** Mediterranean Dr., Point Lisas Industrial Estate, POB 476, Couva; tel. 636-2211; fax 636-5696; tel. www.ispat.com; f. 2005; Dutch-owned; fmrly Iron and Steel Co of Trinidad and Tobago Ltd; name changed to Caribbean Ispat Ltd in 1998; name changed as above in Jan. 2005; acquired International Steel Group (USA) in April 2005; production of iron and steel wire and rods; Chair. LAKSHMI MITTAL; CEO JOHN KURIYAN; 732 employees.

### Petroleum, Natural Gas and Asphalt
(see also State Hydrocarbons Companies above)

**Atlantic LNG Co of Trinidad and Tobago:** POB 1337 Port of Spain; tel. 624-2916; fax 624-8057; e-mail atlanticinfo@atlanticlng.com; internet www.atlanticlng.com; f. 1995; 3 natural gas trains in operation and another under construction in 2004; production of liquefied natural gas; Pres. RICHARD CAPE.

**BG (British Gas Trinidad and Tobago Ltd):** BG House, 5 St Clair Ave, Port of Spain; tel. 627-8106; fax 627-8102; e-mail jacques.robinson@cabgep.co.uk; extraction of natural gas; Pres. CRAIG MCKENZIE; 110 employees.

**BP Energy Company of Trinidad and Tobago:** 5–5A Queen's Park West Plaza, Port of Spain; tel. 623-2862; fax 628-5058; internet www.bp.com; f. 1960; exploration and extraction of natural gas and petroleum; CEO ROBERT RILEY.

**Lake Asphalt of Trinidad and Tobago (1978) Ltd:** Brighton, La Brea; tel. 648-7555; fax 648-7433; e-mail latt@trinidadlakeasphalt.com; internet www.trinidadlakeasphalt.com; state-owned; manufacturers of asphalt; Dep. Chair. IAN RAJACK.

**Mora Oil Ventures (MORAVEN):** Suite 405, Long Circular Mall, Port of Spain; tel. 622-0427; fax 628-3708; e-mail kpamora@trinidad.net; oil production; Chair. TREVOR BOOPSINGH; CEO GEORGE NICHOLAS.

### Petrochemicals

**Conoco Trinidad B.V.:** POB 225, Port of Spain; tel. 624-9205; fax 623-6025; natural gas liquid processing; Man. Dir PAUL WARWICK.

**Farmland Mississippi Chemical Corporation:** North Caspian Dr., PO Bag 38, Point Lisas, Couva; tel. 679-4045; fax 679-2452; e-mail fmcl@carib-link.net; opened ammonia plant in 1999; Pres. LARRY HOLLEY.

**Methanol Holdings (Trinidad) Ltd:** Atlantic Ave, POB 457, Point Lisas Industrial Estate, Couva; tel. 636-2803; fax 636-4501; e-mail mhtlweb@methanl.com; internet www.ttmethanol.com; f. 1999 to consolidate the overall management of the Trinidad and Tobago Methanol Co Ltd (TTMC), the Caribbean Methanol Co Ltd (CMC) and the Methanol IV Co Ltd (MIV); consortium of CL Financial, Ferrostaal AG (Germany) and Helm AG (Germany); 7 methanol production plants in operation in 2005; CEO RAMPERSAND MOTILAL; 300 emps.

**Trinidad and Tobago Methanol Co Ltd (TTMC):** Atlantic Ave, POB 457, Couva, Port of Spain; tel. 623-2101; fax 623-1847; e-mail mhtlweb@methanl.com; internet www.ttmethanol.com; f. 1984; fmrly state-owned, privatized in 1994–97; Chair. LAWRENCE A. DUPREY; CEO RAMPERSAD MOTILAL; 300 employees.

**PCS (Potash Corpn of Saskatchewan) Nitrogen Ltd:** Goodrich Bay Rd, Point Lisas, Couva; tel. 636-2205; fax 636-2052; internet www.potashcorp.com; f. 1977; formerly state-owned Fertilizers of Trinidad and Tobago (Fertrin) Ltd, bought by Arcadian Partners (USA) in 1993; Arcadian Partners bought by PCS in 1997; manufacturers of fertilizers; Chair. WILLIAM J. DOYLE; Man. Dir IAN WELCH; 395 employees.

**Titan Methanol:** Maracaibo Dr., Point Lisas, Couva; tel. 679-5052; fax 679-5065; e-mail titanmc@carib-link.net; operates Trinidad and Tobago's largest methanol plant, owned by Methanex (Canada); Pres. MALCOLM JONES.

**Trinidad and Tobago Urea Company Ltd (TTUC):** Port of Spain; fmrly state-owned, bought by Arcadian Partners (USA) in 1993.

## Miscellaneous

**A. S. Bryden and Sons (Trinidad) Ltd:** POB 607, Port of Spain; tel. 623-2312; fax 627-5655; e-mail asbryden@trinidad.net; f. 1928; importers and distributors of alcoholic beverages, food products, household goods and pharmaceuticals; sales of TT $121.0m. (1996/97); Chair. HAROLD L. BRYDEN; Man. Dir KEITH MAINGOT; 92 employees.

**Agostini's Ltd:** 4 Nelson St, POB 191, Port of Spain; tel. 623-2236; fax 624-6751; e-mail marketing@agostini-mktg.com; internet www.agostini-mktg.com; f. 1925; importers and wholesale distributors of construction materials, foodstuffs and pharmaceuticals; sales of TT $287m. (2003); Chair. JOE ESAU; Man. Dir GEOFFREY AGOSTINI; 500 employees.

**Automotive Components Ltd:** O'Meara Rd, POB 1298, Arima; tel. 642-3268; fax 646-1956; e-mail autocomp@neal-and-massy.com; internet www.neal-and-massy.com; f. 1964; manufacturers of automobile components; owned by Neal & Massy Holdings Ltd (q.v.); sales of TT $80m. (1996); Chair. BERNARD DULAL-WHITEWAY; Man. Dir GORDON RAUSEO; 255 employees.

**Berger Paints Trinidad Ltd:** 11 Concessions Rd, Sea Lots, POB 546, Port of Spain; tel. 623-2231; fax 623-1682; e-mail berger@tstt.net.tt; internet www.bergercaribbean.com; f. 1760; manufacturers of paint, wood stains, wood preservatives, auto refinishes; Man. Dir KISHORE S. ADVANI; 100 employees.

**Carib Glassworks Ltd:** Eastern Main Rd, POB 1287, Champs Fleurs, Port of Spain; tel. 662-2231; fax 663-1779; e-mail carglass@ttol.co.tt; f. 1955; glass manufacturers; Chair. SURESH DUTTA; 420 employees.

**Conglomerates ANSA McAL:** 69 Independence Sq., Port of Spain; tel. 625-3670; fax 624-8753; activities include glass making, construction, finance, media; owns Caribbean Devt Co brewery; Group Chair. ANTHONY SABGA; CEO A. NORMAN SABGA.

**Fujitsu Transaction Solutions (Trinidad) Ltd:** 19-20 Victoria Sq. West, POB 195, Port of Spain; tel. 623-2826; fax 623-4314; e-mail services@fj-icl.com; internet www.fujitsu.com/caribbean; subsidiary of Fujitsu Ltd of Japan; manufacturers of computers and telecommunications equipment; Pres. MERVYN EYRE; Vice-Pres. IAN GALT.

**Johnson and Johnson (Trinidad) Ltd:** New Trincity Industrial Estate, Trincity, St George's; tel. 640-3772; fax 640-3777; f. 1965; subsidiary of Johnson and Johnson of the USA; manufacturers of pharmaceuticals; Chair. SURT SELQUIST; Man. CLIVE ANNANDSINGH; 135 employees.

**Lever Brothers (West Indies) Ltd:** Eastern Main Rd, POB 295, Champ Fleurs, Port of Spain; tel. 663-1787; fax 662-1780; f. 1964; subsidiary of Unilever PLC of the United Kingdom; manufacturers of soaps, detergents, cosmetic products and foods; Man. Dir and Chair. PABLO GARRIDO; 600 employees.

**Neal & Massy Holdings Ltd:** 63 Park St, POB 1544, Port of Spain; tel. 625-3426; fax 627-9061; e-mail nmh@neal-and-massy.com; internet www.neal-and-massy.com; f. 1923; industrial, trading and financial group involved in metals, engineering and automobile assembly; revenue of TT $1,464m. (2004); Chair. ARTHUR LOK JACK; Man. Dir JESÚS PAZOS; 6,500 employees.

**Thomas Peake and Co Ltd:** 177 Western Main Rd, Cocorite, Port of Spain; tel. 662-8816; fax 622-7288; manufacturers and repairers of air conditioning equipment; Dir PAUL PEAKE.

**Trinidad Cement Ltd (TCL):** Southern Main Rd, Claxton Bay; tel. 659-2381; fax 659-2540; internet www.tclgroup.com; f. 1954; manufacture and sale of Portland, sulphate-resisting and oil-well cement and paper sacks and bags; Chair. DAVID DULAL-WHITEWAY; Gen. Man. Dt ROLLIN BERTRAND; 400 employees.

**Trinidad and Tobago National Petroleum Marketing Co Ltd:** National Drive, POB 666, Sea Lots, Port of Spain; tel. 627-2975; fax 627-4028; e-mail simarine@np.co.tt; marketing of petroleum products; state-owned; Chair. CAROLYN SEEPERSAD-BACHAN.

**L. J. Williams Ltd:** 122 St Vincent St, Port of Spain; tel. 623-2865; fax 625-6782; e-mail sales@ljw.co.tt; f. 1925; manufacturers of glues and sealants; distribution of groceries and beverages; sales of TT $135.0m. (1996); Chair. J. G. FURNESS-SMITH; Man. Dir P. J. WILLIAMS; 500 employees.

**West Indian Tobacco Co Ltd (WITCO):** Eastern Main Rd, POB 177, Champ Fleurs, Port of Spain; tel. 662-2271; fax 663-5451; f. 1904; subsidiary of British-American Tobacco Co Ltd of the United Kingdom; cigarette manufacturers; Man. Dir ANTHONY PHILLIP; 165 emps.

# UTILITIES

## Regulatory Authority

**Regulated Industries Commission:** 90 Independence Sq., Port of Spain; tel. 627-0821; fax 624-2027; e-mail ricoffice@ric.org.tt; internet www.ric.org.tt; Chair. DENNIS PANTIN; Exec. Dir HARJINDER S. ATWAL.

## Electricity

**Inncogen Ltd:** 10 Marine Villas, Columbus Blvd, West Moorings by the Sea; tel. 632-7339; fax 632-7341; internet www.centennialenergy.com/inncogen.html; 49.9% share bought by Centennial Energy in Feb. 2004; Chair. ROBERT PALEDINE.

**Power Generation Co of Trinidad and Tobago (PowerGen):** 6A Queen's Park West, Port of Spain; tel. 624-0383; fax 625-3759; owned by Mirant (USA); Gen. Man. GARTH CHATOOR.

**Trinidad and Tobago Electricity Commission (TTEC):** 63 Frederick St, POB 121, Port of Spain; tel. 623-2611; fax 623-3759; e-mail ttecisd@trinidad.net; internet www.ttec.co.tt; state-owned electricity transmission and distribution company; 51% owned by Power Generation Company of Trinidad and Tobago, 49% owned by Southern Electric Int./Amoco; Chair. DEVANAND RAMLAL; Gen. Man. DENIS SINGH.

## Gas

**National Gas Co of Trinidad and Tobago Ltd:** see State Hydrocarbons Companies.

## Water

**Water and Sewerage Authority (WASA):** Farm Rd, St Joseph; tel. 662-9272; fax 652-1253; internet www.wasa.gov.tt; CEO ERROL GRIMES; Gen. Man. WAYNE JOSEPH.

# TRADE UNIONS

**National Trade Union Centre (NATUC):** 16 New St, Port of Spain; tel. 625-3023; fax 627-7588; e-mail natuc@carib-link.net; f. June 1991 as umbrella organization unifying entire trade-union movement, including former Trinidad and Tobago Labour Congress and Council of Progressive Trade Unions; Pres. ROBERT GIUSEPPI; Gen. Sec. VINCENT CARBERA.

## Principal Affiliates

**Airline Superintendent's Association:** c/o Data Centre Bldg, BWIA, Piarco; tel. 664-3401; fax 664-3303; Pres. JEFFERSON JOSEPH; Gen. Sec. THEO OLIVER.

**All-Trinidad Sugar and General Workers' Trade Union (ATSGWTU):** Rienzi Complex, Exchange Village, Southern Main Rd, Couva; tel. 636-2354; fax 636-3372; e-mail atsgwtu@tstt.net.tt; f. 1937; Pres. RUDRANATH INDARSINGH; Gen. Sec. SYLVESTER MARAJH; 2,000 mems.

**Amalgamated Workers' Union:** 16 New St, Port of Spain; tel. 627-6717; fax 627-8993; f. 1953; Pres.-Gen. CYRIL LOPEZ; Sec. FLAVIUS NURSE; c. 7,000 mems.

**Association of Technical, Administrative and Supervisory Staff:** Brechin Castle, Couva; Pres. Dr WALLY DES VIGNES; Gen. Sec. ISAAC BEEPATH.

**Aviation, Communication and Allied Workers' Union:** Aero Services Bldg, Orange Grove Rd, Tacarigua; tel. and fax 640-6518; f. 1982; Pres. CHRISTOPHER ABRAHAM; Gen. Sec. SIEUNARINE BALROOP.

**Banking, Insurance and General Workers' Union:** 27 Borde St, Woodbrook, Port of Spain; tel. 627-0278; fax 627-3931; e-mail bgwu@carib-link.net; internet www.bigwu.org; f. 1974 as Bank and General Workers' Union; name changed as above following merger with Bank Employees' Union in April 2003; Pres. VINCENT CABRERA; Dep. Pres. WAYNE CORBIE.

**Communication, Transport and General Workers' Trade Union:** Aero Services Credit Union Bldg, Orange Grove Rd, Tacarigua; tel. and fax 640-8785; e-mail cattu@tstt.net.tt; Pres. JAGDEO JAGROOP; Gen. Sec. RAYMOND SMALL.

**Communication Workers' Union:** 146 Henry St, Port of Spain; tel. 623-5588; fax 625-3308; e-mail cwutdad@tstt.net.tt; f. 1953; Pres. PATRICK HALL; Gen. Sec. LYLE TOWNSEND; c. 2,100 mems.

**Contractors' and General Workers' Trade Union (CGWTU):** 37 Rushworth St, San Fernando; tel. 657-8072; fax 657-6834; Pres. OWEN HINDS; Gen. Sec. AINSLEY MATTHEWS.

**Customs and Excise Extra Guard Association:** Nicholas Court, Abercromby St, Port of Spain; tel. 625-3311; Pres. ALEXANDER BABB; Gen. Sec. NATHAN HERBERT.

# TRINIDAD AND TOBAGO

**Fire Services Association (FSA):** 52 Lewis St, Woodbrook, Port of Spain; tel. 628-1033; Pres. LENNOX LONDON; Sec. JULES MOORE.

**National Farmers' and Workers' Union:** 25 Coffee St, San Fernando; tel. 652-4348; Pres. Gen. RAFFIQUE SHAH; Gen. Sec. DOOLIN RANKISSOON.

**National General Workers' Union:** c/o 143 Charlotte St, Port of Spain; tel. 623-0694; Pres. JIMMY SINGH; Gen. Sec. CHRISTOPHER ABRAHAM.

**National Union of Domestic Employees (NUDE):** 53 Wattley Circular Rd, Mount Pleasant Rd, Arima; tel. 667-5247; fax 664-0546; e-mail domestic@tstt.net.tt; Gen. Sec. IDA LE BLANC.

**National Union of Government and Federated Workers:** 145–147 Henry St, Port of Spain; tel. 623-4591-4; fax 625-7756; e-mail headoffice@nugfw.org.tt; internet nugfw.org.tt; f. 1937; Pres. Gen. ROBERT GUISEPPI; Gen. Sec. JACQUELINE JACK; c. 20,000 mems.

**Oilfield Workers' Trade Union (OWTU):** Paramount Bldg, 99A Circular Rd, San Fernando; tel. 652-2701; fax 652-7170; e-mail owtu@owtu.org; internet www.owtu.org; f. 1937; Pres. ERROL MCLEOD; Gen. Sec. WENDY JOY WHITE; 9,000 mems.

**Public Services Association:** 89–91 Abercromby St, POB 353, Port of Spain; tel. 623-7987; fax 627-2980; e-mail psa@tstt.net.tt; f. 1938; Pres. JENNIFER BAPTISTE; Sec. PATRICK ROUSSEAU; c. 15,000 mems.

**Seamen and Waterfront Workers' Trade Union:** 1D Wrightson Rd, Port of Spain; tel. 625-1351; fax 625-1182; e-mail swwtu@tstt.net.tt; f. 1937; Pres.-Gen. MICHAEL ANNISETTE; Sec.-Gen. ROSS ALEXANDER; c. 3,000 mems.

**Steel Workers' Union of Trinidad and Tobago:** c/o ISPAT, Point Lisas, Couva; tel. 679-4666; fax 679-4175; e-mail swutt@tstt.net.tt; Pres. GRAFTON WOODLEY; Gen. Sec. WAYNE ROBERTS.

**Transport and Industrial Workers' Union:** 114 Eastern Main Rd, Laventille, Port of Spain; tel. 623-4943; fax 623-2361; f. 1962; Pres. ALDWYN BREWSTER; Gen Sec. JUDY CHARLES; c. 5,000 mems.

**Trinidad and Tobago Airline Pilots' Association (TTALPA):** 35A Brunton Rd, St James; tel. 628-6556; fax 628-2418; e-mail info@ttalpa.org; internet www.ttalpa.org; Chair. Capt. ANTHONY WIGHT; Man. CHRISTINE DAVIS.

**Trinidad and Tobago Postal Workers' Union:** c/o General Post Office, Wrightson Rd, POB 692, Port of Spain; tel. 625-2121; fax 642-4303; Pres. KENNETH SOOKOO; Gen. Sec. EVERALD SAMUEL.

**Trinidad and Tobago Unified Teachers' Association:** Cnr Fowler and Southern Main Rd, Curepe; tel. 645-2134; fax 662-1813; e-mail ttuta@trinidad.net; Pres. CLYDE PERMELL; Gen. Sec. DAVID LEWIS.

**Union of Commercial and Industrial Workers:** TIWU Bldg, 114 Eastern Main Rd, POB 460, Port of Spain; tel. and fax 626-2285; f. 1951; Pres. KELVIN GONZALES; Gen. Sec. ROSALIE FRASER; c. 1,500 mems.

## Transport

### RAILWAYS
The railway service was discontinued in 1968.

### ROADS
In 1999 there were 7,900 km (4,910 miles) of roads in Trinidad and Tobago.

**Public Transport Service Corporation:** Railway Bldgs, South Quay, POB 391, Port of Spain; tel. 623-2341; fax 625-6502; f. 1965 to operate national bus services; CEO EDISON ISAAC; operates a fleet of buses.

### SHIPPING
The chief ports are Port of Spain, Pointe-à-Pierre and Point Lisas in Trinidad and Scarborough in Tobago. Port of Spain handles 85% of all container traffic, and all international cruise arrivals. In 1998 Port of Spain handled 3.3m. metric tons of cargo. Port of Spain and Scarborough each have a deep-water wharf. Port of Spain possesses a dedicated container terminal, with two large overhead cranes. Plans were put in place in 2002 for an expansion of operations, through the purchase of an additional crane, the computerization of operations, and the deepening of the harbour (from 9.75m to 12m).

**Caribbean Drydock Ltd:** Port Chaguaramas, Western Main Rd, Chaguaramas; tel. 634-4226; fax 625-1215; e-mail info@ttdockyard.com; internet www.caridoc.org; ship repair, marine transport, barge and boat construction.

**Point Lisas Industrial Port Development Corporation Ltd (PLIPDECO):** see Development Organizations.

**Port Authority of Trinidad and Tobago:** Dock Rd, POB 549, Port of Spain; tel. 623-2901; fax 627-2666; e-mail vilmal@patnt.com; internet www.patnt.com; f. 1962; Chair. NOEL GARCIA; CEO CHRISTOPHER MENDEZ.

**Shipping Association of Trinidad and Tobago:** 15 Scott Bushe St, Port of Spain; tel. 623-3355; fax 623-8570; e-mail satt@wow.net; internet shipping.co.tt; Pres. SONJA VOISIN-TOM; Gen. Man. JENNIFER GONZÁLEZ.

### CIVIL AVIATION
Piarco International Airport is situated 25.7 km (16 miles) south-east of Port of Spain and is used by numerous airlines. The airport was expanded and a new terminal was constructed in 2001. Piarco remains the principal air transportation facility in Trinidad and Tobago. However, following extensive aerodrome development at Crown Point Airport (located 13 km from Scarborough) in 1992 the airport was opened to jet aircraft. It is now officially named Crown Point International Airport. There is a domestic service between Trinidad and Tobago.

**Airports Authority of Trinidad and Tobago (AATT):** Airport Administration Centre, Piarco International Airport; tel. 669-8047; fax 669-2319; e-mail airport@tntairports.com; internet www.airporttnt.com; administers Piarco and Crown Point International Airports; Chair. LINUS ROGERS.

**BWIA West India Airways Ltd (Trinidad and Tobago):** Administration Bldg, Golden Grove Rd, Piarco International Airport, POB 604, Port of Spain; tel. 669-3000; fax 669-1865; e-mail mail@bwee.com; internet www.bwee.com; f. 1980 by merger of BWIA International (f. 1940) and Trinidad and Tobago Air Services (f. 1974); state-owned; operates scheduled passenger and cargo services linking destinations in the Caribbean region, South America, North America and Europe; operates Tobago Express service between Trinidad and Tobago; Chair. LAWRENCE DUPREY; Gen. Man. NELSON TOM YEW.

## Tourism

The climate and coastline attract visitors to Trinidad and Tobago. The latter island is generally believed to be the more beautiful and is less developed. The annual pre-Lenten carnival is a major attraction. In 2003 there were an estimated 409,069 foreign visitors, excluding cruise-ship passengers, and tourist receipts in 2002 were estimated at US $402m. In 2000 there were an estimated 82,859 cruise-ship passenger arrivals. There were 4,532 hotel rooms in Trinidad and Tobago in that same year. The Government announced plans to increase first-class hotel room accommodation by 8,000 in 2001, marketing the islands as a sophisticated cultural destination. There were additional tourism developments in Chaguaramas and Toco Bay.

**Tourism and Industrial Development Company of Trinidad and Tobago (TIDCO):** see Development Organizations.

**Trinidad Hotels, Restaurants and Tourism Association (THRTA):** c/o Trinidad & Tobago Hospitality and Tourism Institute, Airway Rd, Chaguaramas; tel. 634-1174; fax 634-1176; e-mail info@tnthotels.com; internet www.tnthotels.com; Pres. ERNEST LITTLES; Exec. Dir GREER ASSAM.

**Tobago Chapter:** Goddard's Bldg, Auchenskeoch, Carnbee; tel. and fax 639-9543; e-mail tthtatob@tstt.net.tt; Pres. RENE SEEPERSADSINGH.

## Defence

At 1 August 2004 the Trinidad and Tobago Defence Force consisted of 2,700 active members: an army of an estimated 2,000 men and a coastguard of 700. Included in the coastguard was an air wing of 50.

**Defence Budget:** TT $198m. (US $32m.) in 2004.

**Chief of Defence Staff:** Brig.-Gen. ANCIL ANTOINE.

## Education

Primary and secondary education is free. Many schools are run jointly by the State and religious bodies. Attendance at school is officially compulsory for children between five and 12 years of age. Primary education begins at the age of five and lasts for seven years. In 2001 94% of children in this age group were enrolled at primary schools. Secondary education, beginning at 12 years of age, lasts for up to five years, comprising a first cycle of three years and a second of two years. The ratio for secondary enrolment in 2001 was 68% of those in the relevant age-group (males 67%; females 69%). Entrance to secondary schools is determined by the Common Entrance Examination. Many schools are administered jointly by the State and religious bodies. In 2000 the Government announced an education reform programme to improve access to and levels of education. A school-to-work apprenticeship programme was also to be established.

The Trinidad campus of the University of the West Indies (UWI), at St Augustine, includes an engineering faculty. The UWI Institute of Business offers postgraduate courses and develops programmes for local companies. Other institutions of higher education are the Eric Williams Medical Sciences Complex, the Polytechnic Institute and the Eastern Caribbean Farm Institute. The country has one teacher-training college and three government technical institutes and vocational centres, including the Trinidad and Tobago Hotel School. In the late 1990s the Government established the Trinidad and Tobago Institute of Technology, and the College of Science, Technology and Applied Arts of Trinidad and Tobago. Distance learning was also being expanded. Government expenditure on education in 2000 was TT $2,111.0m.

# Bibliography

For works on the Caribbean generally, see Select Bibliography (Books)

Anthony, M. *Historical Dictionary of Trinidad and Tobago.* Lanham, MD, Scarecrow Press, 1997.

Birbalsingh, F. (Ed.). *Indo-Caribbean Resistance.* Toronto, ON, TSAR, 1993.

Harrison, P. *The Impact of Macroeconomic Policies in Trinidad and Tobago.* New York, NY, Palgrave Macmillan, 2002.

Khan, A. *Callaloo Nation: Metaphors of Race and Religious Identity Among South Asians in Trinidad.* Durham, NC, Duke University Press, 2004.

Klass, M. *Singing with Sai Baba: Politics of Revitalization in Trinidad.* Prospect Heights, IL, Waveland Press, 1996.

Maharaj, P. D. *Clash of Cultures: The Indian–African Competition in Trinidad.* Arima, Indian Free Press, 2002.

Meighoo, K. P. *Politics in a Half Made Society: Trinidad and Tobago, 1925-2002.* New York, NY, Markus Wiener Publrs, 2003.

Munasinghe, V. *Callaloo or Tossed Salad? East Indians and the Cultural Politics of Identity in Trinidad.* Ithaca, NY, Cornell University Press, 2001.

Purdy, J. M. *Common Law and Colonised Peoples.* Aldershot, Ashgate Publishing Ltd, 1997.

Quamina, A. *The Silent Revolution: Book I: A Study of the Functioning of Dominant Political Parties in Emerging Societies, Using the PNM of Trinidad and Tobago as the Primary Case Study with Comparisons with Dominant Parties of India, Ghana and Tanganyika.* Frederick, MD, PublishAmerica, 2005.

Reddock, R. *Women, Labour and Politics in Trinidad and Tobago.* London, Zed Books, 1994.

Regis, L. *The Political Calypso: True Opposition in Trinidad and Tobago.* Gainesville, FL, University Press of Florida, 1999.

*Report of the Commission of Inquiry into the Functioning of the Elections and Boundaries Commission of Trinidad and Tobago.* Port of Spain, Government of Trindad and Tobago, 2002.

Seerattan, D. *Tax Reform and Financial Development in Trinidad and Tobago.* St Augustine, Caribbean Centre for Monetary Studies, University of the West Indies, 2002.

Stuempfle, S. *The Steelband Movement: The Forging of a National Art in Trinidad and Tobago.* University Park, PA, University of Pennsylvania Press, 1996.

Williams, E. *History of the People of Trinidad and Tobago.* Fineshade, A&B Books, 1996.

# THE TURKS AND CAICOS ISLANDS

## Geography

### PHYSICAL FEATURES

The Turks and Caicos Islands is an Overseas Territory of the United Kingdom, a British crown colony, once a dependency of Jamaica, until the latter's independence in 1962, and then administered from the Bahamas between 1965 and 1973. The two groups of islands, the Caicos to the west and the Turks to the east, form the south-eastern extremity of the Atlantic archipelago dominated by the Bahamas. The territory continues the main chain some 63 km (39 miles) to the south-east of Mayaguana, although it lies directly east from the two more isolated Inagua islands. There is only about 45 km between West Caicos and Little Inagua. The Turks and Caicos Islands lies 145 km north of the island of Hispaniola (above the border between the western part, Haiti, and the Dominican Republic, which occupies the greater, eastern part). Some 40 islands, islets and cays, with a total coastline of 389 km, cover 430 sq km (166 sq km) of territory.

The two groups of islands are separated by the Columbus or Turks Island Passage (35 km long and reaching a depth of some 7,000 ft—over 2,130 m) between the Atlantic and the Caribbean. The islands are low and coralline, with thin, poor soil on limestone bases, but this dry and semi-barren surface (particularly in the east) belies the submarine luxuriance of the third largest coral-reef system in the world. Moreover, the flat terrain, prone to marshiness and mangrove swamps, and dotted with salt pans and salinas, is attractive to birds and hardy wildlife. Although the rocky interiors, generally covered in scrub, cacti and thorny acacias (some of the islands were deliberately denuded of what trees they had by the original, salt-harvesting Bermudan colonists, to discourage rainfall), remain bleak, the creeks, flats, marshlands and sand flats have delicate and unique ecosystems, teeming with attractions for resident and migratory birds, for instance. As a result, the natural and historical environment of the islands and their reefs is protected by over 30 reserves (including wildlife such as the endemic rock iguana or the pygmy boa). All the islands tend to higher terrain, with limestone cliffs and sand dunes, on the northern or north-eastern weather sides. The highest point on the islands is unclear, but is about 49 m (161 ft), be it the peak of the Ridge on Grand Turk or the only recently named Blue Hills on Providenciales. Flamingo Hill (48 m), on East Caicos, is also sometimes quoted as the highest point in the territory.

The Turks Islands (named for the fez-like Turks Head cactus) define the south-eastern edge of the territory, forming an arc of mainly small islands heading from the south-west and culminating in the northward-pointing, and much larger, Grand Turk. Although only 18 sq km in area, the island is the second largest population centre and the site of the capital, Cockburn Town. There is also an argument that it is the real San Salvador, where Christopher Columbus, the Genoese (Italian) navigator working for Spain, first set foot in the Americas. The only other one of the Turks of any size or population is Salt Cay, a designated UNESCO World Heritage Site, 11 km southwards from Grand Turk.

South Caicos (sometimes known as East Harbour or the Rock) is 35 km west of Grand Turk. It is the eastern-most bulge of the Caicos Islands (named for the Lucayan—Taino—word for a chain of islands), which box in the infamous Caicos Banks, where the sea depth goes from some 2,000 m to less than 10 m in under 1 km. The main landmasses define the northern edge, while South Caicos and a line of cays (such as the Ambergris and Seal Cays) facing the Turks form the eastern edge. There are more islands and cays to the west, but only treacherous reefs mark the south, with dry land rarely breaching the sea surface. The main islands are East Caicos, Middle Caicos and North Caicos, stretching north-westwards, before another chain of islets heads south-westwards to the hooked island of Providenciales ('Provo'). The more isolated and uninhabited West

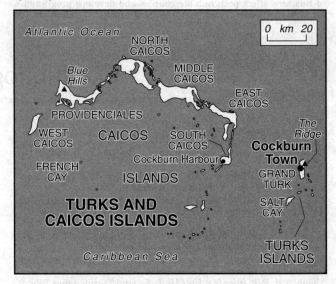

Caicos is just south-west of Providenciales. East Caicos covers about 47 sq km, and is typical with a northern ridge protecting wetlands of creeks, mudflats and mangrove swamp. Middle or Grand Caicos is the largest island in the territory (124 sq km—larger than Anguilla or Montserrat), with more dry land and dramatic coasts than most. North Caicos, at the north-western apex of the territory, has relatively good tree cover and good rainfall, making it the greenest of the islands, and it is known as a sanctuary for flamingos and the West Indian whistling duck. Providenciales, about 40 km in length and 5 km wide, has long been settled, but has grown to be the leading population centre only since tourist development began in the late 1960s, and particularly in the 1990s.

### CLIMATE

The climate is subtropical marine, relatively dry and sunny, and exposed to the trade winds off the Atlantic. There are frequently hurricanes. The average annual temperature is 27°C (81°F), with little seasonal variation. Rainfall annually averages about 530 mm (21 ins) in the Turks and eastern Caicos, but reaches almost 1,000 mm in the west.

### POPULATION

Almost all of the native population (known as 'belongers') is black and Christian, and officially speaks English, although there is also a local Creole in use. Most belongers are Protestants, with 40% of the total Baptist (1990 figures), 16% Methodist and 12% Church of God. About 18% are Anglican, and Cockburn Town now has a pro-cathedral. More recent figures, if available, would probably show a large increase in the numbers of Roman Catholics, as many of the immigrants arriving since then have come from Haiti or the Dominican Republic.

Estimates vary, but the total population in mid-2002 was around 20,900, of whom less than one-half (about 8,000) are belongers. About one-fifth of the belongers live on Grand Turk and Salt Cay. Most of the rest of the population resides in the Caicos Islands and is from Haiti, the Dominican Republic and North America. Six islands are permanently inhabited: Grand Turk and Salt Cay, and South Caicos, Middle Caicos, North Caicos and the tourist-dominated Providenciales. Parrot Cay and Pine Cay have resorts on them. The main towns are the capital, Cockburn Town (Grand Turk), which only has a population of about 5,000, and Cockburn Harbour on South Caicos. South Caicos was once the main population centre, but this

… position is now held by the flourishing tourist centre of Providenciales or Provo. In addition to the many immigrants who work legally, particularly on Provo, the islands also suffer from illegal immigration (and drugs-transhipping), as Haitians particularly use them as a transit point for reaching the Bahamas and, thence, the USA.

# History

The Turks and Caicos Islands constitute a United Kingdom Overseas Territory and the British monarch is represented locally by a Governor who presides over the Executive Council. From February 1998 the British Dependent Territories were referred to as the Overseas Territories, following the announcement of the interim findings of a British Government review of the United Kingdom's relations with its remaining imperial possessions.

The islands, which had been a Jamaican dependency from 1874 until 1959, became a separate colony in 1962, following the dissolution of the West Indian Federation and Jamaican independence. They were accorded their own resident Governor in the early 1970s. At the 1976 elections the pro-independence People's Democratic Movement (PDM) won a majority of seats in the Legislative Council, and the party leader, J. A. G. S. McCartney, became Chief Minister. In 1980 the PDM reached an agreement with the United Kingdom that, if it won the next election, the islands would be granted independence. However, the election was won by the opposition Progressive National Party (PNP), which is committed to continued dependent status. In the 1984 election the PNP, led by Norman Saunders, maintained its lead.

In March 1985 Saunders was arrested in Miami, FL, USA, on charges involving illicit drugs and accepting a bribe. He resigned as Chief Minister and was replaced by Nathaniel Francis. In July 1986 an official report on the destruction of a government building by fire in December 1985 forced the resignation of Francis and two other ministers and discredited the Government amid allegations of unconstitutional behaviour and ministerial malpractice. The Governor proceeded to dissolve the Government and an Order-in-Council authorized him to replace the Executive Council with an interim Advisory Council. In March 1987 the British Government announced that it had accepted a constitutional commission's central recommendations and a general election was held in March 1988, after which there was a return to ministerial government.

The PDM won 11 of the 13 seats on the Legislative Council in the 1988 elections and Oswald Skippings, the leader of the PDM, was appointed Chief Minister. In the April 1991 general election the PNP secured eight seats and Washington Misick, the PNP leader, replaced Skippings as Chief Minister. In January 1995 the PDM, now led by Derek Taylor, gained eight seats, while the PNP won four and the remaining seat went to Saunders, who had stood as an independent.

At legislative elections in March 1999 the PDM increased its representation in the Legislative Council to nine seats, after Saunders, who had secured nomination as a PDM candidate, retained the seat that he had held as an independent. The PNP won the other four seats. In the same month the British Government published draft legislation on its future relationship with its overseas dependencies, which had been referred to as United Kingdom Overseas Territories since February 1998. The Government announced that in future citizens of Overseas Territories would have the right to reside in the United Kingdom, provided they meet international standards in the areas of human rights and the regulation of the financial services sector.

In May 2002 the British Overseas Territories Act, having received royal assent in the United Kingdom in February, came into force and granted British citizenship rights to the people of the Overseas Territories, including the Turks and Caicos Islands. Under the new law Turks and Caicos Islanders were able to hold British passports and work in the United Kingdom and anywhere else in the European Union. In October 2002 capital punishment for treason and piracy, the only remaining capital crimes in the British Overseas Territories, was abolished in the Turks and Caicos Islands. In December James Poston replaced Mervyn Jones as Governor.

At a general election held on 24 April 2003 the PDM won seven of the 13 seats in the Legislative Council, two fewer than in 1999. The PNP took the remaining six seats, but challenged the results in two of the more closely fought constituencies. The election was largely fought over Chief Minister Taylor's claims of economic progress, which Michael Misick, leader of the PNP, argued had benefited expatriates rather than Turks and Caicos 'belongers'.

In June 2003 the Supreme Court ruled in favour of the defeated PNP candidates in the two disputed constituencies; Chief Justice Ground found evidence of bribery by supporters of the PDM in one constituency and irregularities in voter-registration lists in the other. Following the judicial ruling, by-elections, in which the PNP secured victory and therefore wrested overall control of the Legislative Council from the PDM, were held in the two constituencies on 7 August. Taylor resigned as Chief Minister on 15 August and Misick was sworn in as his replacement on the same day. Misick's first Executive Council included Floyd Hall as the Minister of Finance and National Insurance.

In April 2004 the Chief Minister spoke to his counterpart in Canada, Paul Martin, about the creation of a closer relationship between Canada and the Turks and Caicos Islands, possibly involving the island nation's full political integration into the Canadian federation. The likelihood of such a union being established was open to debate: public support on the islands was not believed to be strong; the lack of income tax in the territory would complicate a fiscal union; and the United Kingdom would need to agree the proposal. In October 2004 Galmo Williams replaced Karen Delancy in the slightly altered portfolio of Natural Resources and Social Services. In April 2005 a new parliament building was inaugurated (dedicated to N. J. S. Francis). On 14 July Richard Tauwhare was sworn into office as Governor, in succession to James Poston (Tauwhare forsook the official ceremonial dress at the Chief Minister's request).

# Economy

Salt produced by solar evaporation was the principal export until 1964; however, by the early 1990s the activity had virtually ceased. In the late 1990s the underwater extraction of aragonite was being investigated. Agriculture is not practised on any significant scale in the Turks Islands or on South Caicos (the most populous island of the Territory). The other islands of the Caicos group grow some beans, maize and a few fruits and vegetables. There is some livestock-rearing, but the islands' principal natural resource is fisheries, which account for almost all commodity exports, the principal species caught being the spiny lobster (an estimated 247 metric tons in 2003) and the conch (an estimated 5,646 tons in that year). Conches are now being developed commercially (on the largest conch farm in the world), and there is potential for larger-scale fishing. Exports of lobster and conch earned US $4.9m. in 2004, according to provisional estimates. However, agricultural possibilities were limited and most foodstuffs were imported. Industrial activity consists mainly of construction (especially for the tourist industry) and fish-processing.

The principal economic sector is the service industry. This is dominated by the expanding tourist sector, which is concentrated on the island of Providenciales. The market is for wealthier visitors, most of whom come from the USA. Tourist arrivals increased from 120,898 in 1999 to 165,341 in 2003. A decline in the number of tourist arrivals was observed in 2002, owing to the weakening of demand for air travel in the USA, although the sector recovered in 2003 when estimated tourist arrivals totalled 163,600. As a result of the growth in tourism throughout the last decade many new hotels and resorts have been developed, and in January 2004 it was announced that a new US $35m. cruise-ship terminal would be constructed in Grand Turk, enabling, for the first time, large passenger liners to stop in the territory. However, concern has been expressed that the islands are in danger of becoming overdeveloped, thereby damaging their reputation as an unspoilt tourist location. In January 2005 plans were announced for the construction of a resort on West Caicos; although popular as a day-trip destination, owing to its nature reserve, this island was undeveloped hitherto. The resort was scheduled for completion in late 2007 and was to include a village and marina.

An 'offshore' financial sector was encouraged in the 1980s, and new regulatory legislation was ratified in 1989. In early 2000 there were some 8,000 overseas companies registered in the islands. Earnings from some 2,858 new registrations totalled US $2.3m. in 1995. However, in June 2000 the Turks and Caicos Islands were included in the list of so-called 'unco-operative tax havens' compiled by the Organisation for Economic Co-operation and Development (OECD). OECD urged the jurisdictions included on the list to improve their legal and administrative transparency to prevent companies using the jurisdictions' tax systems in an attempt to launder money or avoid paying income tax. The Turks and Caicos Islands were removed from the list in March 2002, when OECD declared that the Government had made sufficient commitments to improve transparency and to achieve effective exchange of information on tax matters by the end of 2005. As part of its regulatory overhaul, the Government established a Financial Services Commission in 2002. In early 2004 the financial sector faced further regulatory disruption when the territory came under pressure to implement the European Union's Savings Tax Directive, which would involve the Government disclosing the identities and account details of Europeans holding private savings accounts on the islands.

During 1990–2001 the islands' gross domestic product increased, in real terms, by 7.0% per year; growth in 2001 and 2002 was 3.7% and 3.4%, respectively. Following a recovery in the tourist sector, growth was provisionally estimated at 6.1% in 2003. Overall, the economy, and also the population, were estimated to have doubled in size since the early 1990s, making the islands one of the region's most dynamic economies. However, a source of disquiet is the exclusion of many inhabitants from the benefits of economic growth; it is estimated that the economic situation of the majority of 'belongers' improved only marginally in that period, as newly created jobs were often taken by low-wage migrant workers or by highly skilled expatriate workers. It has also been noted that the Territory's growth is highly dependent on exterior factors, and the pressure exerted by Europe on the 'offshore' financial-services sector is therefore of particular concern. Recurrent revenue increased by 12% in 2004/05, to US $117.0m.; recurrent expenditure was some $121.99m. In 2005 it was announced that the Government had concluded arrangements for the creation of a national bank, to be known as TCI Bank Ltd.

# Statistical Survey

Source: Chief Secretary's Office, South Base, Grand Turk; tel. 946-2702; fax 946-2886.

## AREA AND POPULATION

**Area:** 430 sq km (166 sq miles).

**Population:** 7,435 at census of 12 May 1980; 12,350 (males 6,289, females 6,061) at census of 31 May 1990; *Mid-2002* (estimate): 20,900 (Source: Caribbean Development Bank, *Social and Economic Indicators*). *By Island* (1980): Grand Turk 3,098; South Caicos 1,380; Middle Caicos 396; North Caicos 1,278; Salt Cay 284; Providenciales 977. *1990:* Grand Turk 3,761; Providenciales 5,586.

**Density:** 48.6 per sq km (mid-2002).

**Principal Towns:** Cockburn Town (capital, on Grand Turk), population 2,500 (estimate, 1987); Cockburn Harbour (South Caicos), population 1,000. *Mid-2003* (UN estimate, incl. suburbs): Grand Turk 5,680 (Source: UN, *World Urbanization Prospects: The 2003 Revision*).

**Births and Deaths** (2000): Live births 290; Deaths 67 (Source: UN, *Population and Vital Statistics report*). *2004:* Birth rate 22.9 per 1,000; Death rate 4.3 per 1,000 (Source: Pan American Health Organization).

**Expectation of Life** (years at birth, 2004): 74.3 (males 72.1; females 76.6). Source: Pan American Health Organization.

**Economically Active Population** (1990 census): 4,848 (males 2,306, females 2,542).

## HEALTH AND WELFARE

**Total Fertility Rate** (children per woman, 2004): 3.1.

**Under-5 Mortality Rate** (per 1,000 live births, 1997): 22.0.

**Hospital Beds** (per 1,000 head, 2003): 94.7.

**Physicians** (per 1,000 head, 1999): 0.73.

**Access to Water** (% of persons, 2002): 100.

**Access to Sanitation** (% of persons, 2002): 96.

Source: Pan American Health Organization.
For definitions see explanatory note on p. vi.

## AGRICULTURE, ETC.

**Fishing** (metric tons, live weight, 2003): Capture 6,093* (Marine fishes 200*, Caribbean spiny lobster 247, Stromboid conchs 5,646); Aquaculture 25; *Total catch* 6,118*. Figures exclude aquatic plants and mammals.
*FAO estimate.
Source: FAO.

## INDUSTRY

**Electric Energy** (estimated production, million kWh): 5 in 1999; 5 in 2000; 5 in 2001. Source: UN, *Industrial Commodity Statistics Yearbook*.

## FINANCE

**Currency and Exchange Rate:** United States currency is used: 100 cents = 1 US dollar ($). *Sterling and Euro Equivalents* (31 May 2005): £1 sterling = US $1.8181; €1 = US $1.2331; $100 = £55.00 = €81.10.

**Budget** (US $ million, 2003): Total revenue and grants 122.5 (current revenue 104.3, capital revenue and grants 18.2); Total expenditure 114.9 (current expenditure 93.5, capital expenditure 21.4). Source: Caribbean Development Bank, *Social and Economic Indicators*.

**Gross Domestic Product** (US $ million at current prices): 222.7 in 2000; 266.1 in 2001; 280.1 in 2002. Source: Caribbean Development Bank, *Social and Economic Indicators*.

# THE TURKS AND CAICOS ISLANDS

**Expenditure on the Gross Domestic Product** (US $ million, 2002): Government final consumption expenditure 53.5; Public final consumption expenditure 339.9; Gross fixed capital formation 55.4; *Total domestic expenditure* 448.8; Exports of goods and services 8.7; *Less* Imports of goods and services 177.5; *GDP in market prices* 280.1.

**Balance of Payments** (US $ million, 2002): Exports of goods f.o.b. 8.7; Imports of goods f.o.b. –177.5; *Trade balance* –168.8; Exports of services 163.1; Imports of services –81.8; *Balance on goods and services* –87.5. Source: Caribbean Development Bank, *Social and Economic Indicators*.

### EXTERNAL TRADE

**Principal Trading Partners** (US $ million, 2003): *Imports c.i.f.*: USA 167.7; Total 170.7. *Exports f.o.b.*: USA 9.2; Total 9.8. Source: Caribbean Development Bank, *Social and Economic Indicators*.

### TRANSPORT

**Road Traffic** (1984): 1,563 registered motor vehicles.

**Shipping:** *International Freight Traffic* (estimates in '000 metric tons, 1990) Goods loaded 135; Goods unloaded 149. *Merchant Fleet* (vessels registered at 31 December 2004): 5; Total displacement 975 grt. Sources: UN, *Monthly Bulletin of Statistics*; Lloyd's Register-Fairplay, *World Fleet Statistics*.

### TOURISM

**Tourist Arrivals** ('000): 165.4 in 2001; 155.6 in 2002; 163.6 in 2003.

**Tourism Receipts** (estimates, US $ million): 314 in 2000; 341 in 2001; 319 in 2002.
Source: Caribbean Development Bank, *Social and Economic Indicators*.

### COMMUNICATIONS MEDIA

**Radio Receivers** (1997): 8,000 in use.

**Telephones** (1994): 3,000 main lines in use.

**Facsimile Machines** (1992): 200 in use.

**Non-daily Newspapers** (1996): 1 (estimated circulation 5,000).
Sources: UNESCO, *Statistical Yearbook*; UN, *Statistical Yearbook*.

### EDUCATION

**Pre-primary** (2001/02): 21 schools (1996/97); 70 teachers; 886 pupils.

**Primary** (2003): 21 schools (1996/97); 119 teachers (2001/02); 3,003 pupils.

**General Secondary** (2003): 5 schools (1990); 141 teachers (UNESCO estimate, 2001/02); 1,391 pupils.

**Vocational Education** (1993/94): 30 teachers; 89 pupils.

**Adult Literacy Rate** (UNESCO estimates): 99% (males 99%; females 98%) in 1998.
Sources: UNESCO Institute for Statistics; Caribbean Development Bank, *Social and Economic Indicators*.

# Directory

## The Constitution

The Order in Council of July 1986 enabled the Governor to suspend the ministerial form of government, for which the Constitution of 1976 made provision. Ministerial government was restored in March 1988, following amendments to the Constitution, recommended by a constitutional commission.

The revised Constitution of 1988 provides for an Executive Council and a Legislative Council. Executive authority is vested in the British monarch and is exercised by the Governor (the monarch's appointed representative), who also holds responsibility for external affairs, internal security, defence, the appointment of any person to any public office and the suspension and termination of appointment of any public officer.

The Executive Council comprises: two *ex officio* members (the Chief Secretary and the Attorney-General); a Chief Minister (appointed by the Governor) who is, in the judgement of the Governor, the leader of the political party represented in the Legislative Council that commands the support of a majority of the elected members of the Council; and four other ministers, appointed by the Governor, on the advice of the Chief Minister. The Executive Council is presided over by the Governor.

The Legislative Council consists of the Speaker, the two *ex officio* members of the Executive Council, 13 members elected by residents aged 18 and over, and three nominated members (appointed by the Governor, one on the advice of the Chief Minister, one on the advice of the Leader of the Opposition and one at the Governor's discretion).

For the purposes of elections to the Legislative Council, the islands are divided into five electoral districts. In 1988 and 1991 a multiple voting system was used, whereby three districts elected three members each, while the remaining five districts each elected two members. However, from the 1995 election a single-member constituency system was used.

## The Government

**Governor:** RICHARD TAUWHARE (sworn in 11 July 2005).

### EXECUTIVE COUNCIL
(July 2005)

**President:** RICHARD TAUWHARE (The Governor).

**Chief Minister and Minister of Development, Planning, Tourism and District Administration:** Dr MICHAEL EUGENE MISICK.

**Deputy Chief Minister and Minister of Finance, Health and National Insurance:** FLOYD BASIL HALL.

**Minister of Housing, Immigration and Labour:** JEFFREY CHRISTOVAL HALL.

**Minister of Education, Youth, Sports, Culture, Arts and Gender Affairs:** LILLIAN ELAINE ROBINSON-BEEN.

**Minister of Natural Resources and Social Services:** GALMO WILLIAMS.

**Minister of Communications, Works and Utilities:** MCALLISTER EUGENE HANCHELL.

*Ex Officio* **Members:**

**Attorney-General:** KURT DE FREITES.

**Chief Secretary:** MAHALA WYNNS.

### GOVERNMENT OFFICES

**Office of the Governor:** Government House, Grand Turk; tel. 946-2308; fax 946-2903; e-mail govhouse@tciway.tc.

**Office of the Chief Minister:** Government Sq., Grand Turk; tel. 946-2801; fax 946-2777.

**Chief Secretary's Office:** South Base, Grand Turk; tel. 946-2702; fax 946-2886; e-mail cso@tciway.tc.

**Office of the Permanent Secretary:** Finance Dept, Chief Minister's Office, Government Bldgs, Front St, Grand Turk; tel. 946-1115; fax 946-2777.

### LEGISLATIVE COUNCIL

**Speaker:** GLENNEVANS CLARKE.

**Election, 24 April 2003**

| Party | Seats* |
|---|---|
| People's Democratic Movement (PDM) | 7 |
| Progressive National Party (PNP) | 6 |
| **Total** | **13** |

*Following two by-elections on 7 August 2003, the PDM held five seats and the PNP held eight seats.

There are two *ex officio* members (the Chief Secretary and the Attorney-General), three appointed members, and a Speaker (assisted by a Deputy Speaker) chosen from outside the Council.

THE TURKS AND CAICOS ISLANDS                                                                    Directory

## Political Organizations

**People's Democratic Movement (PDM):** POB 38, Grand Turk; favours internal self-government and eventual independence; Leader DEREK H. TAYLOR.

**Progressive National Party (PNP):** Providenciales; tel. 941-4663; fax 946-3673; e-mail pnp@tciway.tc; internet www.votepnp.com; supports full internal self-government; Chair. SANDRA GARLAND; Leader MICHAEL EUGENE MISICK.

**United Democratic Party (UDP):** Grand Turk; f. 1993; Leader WENDAL SWANN.

## Judicial System

Justice is administered by the Supreme Court of the islands, presided over by the Chief Justice. There is a Chief Magistrate resident on Grand Turk, who also acts as Judge of the Supreme Court. There are also three Deputy Magistrates.

The Court of Appeal held its first sitting in February 1995. Previously the islands had shared a court of appeal in Nassau, Bahamas. In certain cases, appeals are made to the Judicial Committee of the Privy Council (based in the United Kingdom).

### Judicial Department
Grand Turk; tel. 946-2114; fax 946-2720; e-mail court@gov.tc.

**Chief Justice:** (vacant).

**Magistrate:** DEREK REDMAN.

**Attorney-General's Chambers:** South Base, Grand Turk; tel. 946-2882; fax 946-2588; e-mail attorneygeneral@tciway.tc.

**Attorney-General:** KURT DE FRIET.

## Religion

### CHRISTIANITY

#### The Anglican Communion
Within the Church in the Province of the West Indies, the territory forms part of the diocese of Nassau and the Bahamas. The Bishop is resident in Nassau. According to census results, there were 1,465 adherents in 1990.

**Anglican Church:** St Mary's Church, Front St, Grand Turk; tel. 946-2289; internet bahamas.anglican.org; Archbishop Rev. DREXEL GOMEZ.

#### The Roman Catholic Church
The Bishop of Nassau, Bahamas (suffragan to the archdiocese of Kingston in Jamaica), has jurisdiction in the Turks and Caicos Islands, as Superior of the Mission to the Territory (founded in June 1984).

**Roman Catholic Mission:** Leeward Highway, POB 340, Providenciales; tel. and fax 946-1888; e-mail rcmission@tciway.tc; internet www.catholic.tc; churches on Grand Turk, South and North Caicos, and on Providenciales; 132 adherents in 1990 (according to census results); Chancellor Fr PETER BALDACCHINO.

#### Other Christian Churches

**Baptist Union of the Turks and Caicos Islands:** South Caicos; tel. 946-3220; 3,153 adherents in 1990 (according to census results).

**Jehovah's Witnesses:** Kingdom Hall, Intersection of Turtle Cove and Bridge Rd, POB 400, Providenciales; tel. 941-5583.

**Methodist Church:** The Ridge, Grand Turk; tel. 946-2115; 1,238 adherents in 1990 (according to census results).

**New Testament Church of God:** Orea Alley, Grand Turk; tel. 946-2175.

**Seventh-day Adventists:** Grand Turk; tel. 946-2065; Pastor PETER KERR.

## The Press

**Times of the Islands Magazine:** Caribbean Pl., POB 234, Providenciales; tel. and fax 946-4788; e-mail timespub@tciway.tc; internet www.timespub.tc; f. 1988; quarterly; circ. 10,000; Editor KATHY BORSUK.

**Turks and Caicos Free Press:** Market Pl., POB 179, Providenciales; tel. 941-5615; fax 941-3402; e-mail freepress@tciway.tc; internet www.freepress.tc; f. 1991; bi-weekly; Editor CINDI ROUSE; Man. KATHI BARRINGTON; circ. 2,000.

**Turks & Caicos Islands Real Estate Association Real Estate Magazine:** Caribbean Pl., POB 234, Providenciales; tel. and fax 946-4788; e-mail timespub@tciway.tc; internet www.timespub.tc; circ. 15,000; Editor KATHY BORSUK; 3 a year.

**Turks and Caicos Weekly News:** Leeward Highway, Cheshire House, POB 52, Providenciales; tel. 946-4664; fax 946-4661; e-mail tcnews@tciway.tc.

**Where, When, How:** POB 192, Providenciales; tel. 946-4815; fax 941-3497; e-mail wwh@provo.net; internet www.wherewhenhow.com; monthly; travel magazine.

## Broadcasting and Communications

### TELECOMMUNICATIONS

**Cable and Wireless (Turks and Caicos) Ltd:** Leeward Highway, POB 78, Providenciales; tel. 946-2200; fax 946-2497; e-mail cwtci@tciway.tc; internet www.cw.tc; f. 1973.

### Radio

**Power 92.5 FM:** Providenciales; e-mail kenny@power925fm.com; internet www.power925fm.com.

**Radio Providenciales:** Leeward Highway, POB 32, Providenciales; tel. 946-4496; fax 946-4108; commercial.

**Radio Turks and Caicos (RTC):** POB 69, Grand Turk; tel. 946-2007; fax 946-1600; e-mail rtc@tciway.tc; internet www.turksandcaicos.tc/rtc/; govt-owned; commercial; broadcasts 105 hrs weekly; Man. LYNETTE SMITH.

**Radio Visión Cristiana Internacional:** North End, South Caicos; tel. 946-6601; fax 946-6600; e-mail radiovision@tciway.tc; internet www.radiovision.net; commercial; Man. WENDELL SEYMOUR.

### Television

Television programmes are available from a cable network, and broadcasts from the Bahamas can be received in the islands.

**Turks and Caicos Television:** Pond St, POB 80, Grand Turk; tel. 946-1530; fax 946-2896.

**WIV Cable TV:** Tower Raza, Leeward Highway, POB 679, Providenciales; tel. 946-4273; fax 946-4790.

## Finance

### REGULATORY AUTHORITY

**Financial Services Commission (FSC):** Harry E. Francis Bldg, Pond St, POB 173, Grand Turk; tel. 946-2791; fax 946-2821; e-mail fsc@tciway.tc; f. 2002; regulates local and 'offshore' financial services sector; Man. Dir NEVILLE CADOGAN.

### BANKING

**Belize Bank:** Providenciales; tel. 941-5028; fax 941-5029.

**Bordier International Bank and Trust Ltd:** Caribbean Pl., Leeward Highway, POB 5, Providenciales; tel. 946-4535; fax 946-4540; e-mail enquiries@bibt.com; internet www.bibt.com; Man. ELISE HARTSHORN.

**FirstCaribbean International Bank (Bahamas) Ltd:** Leeward Highway, POB 698, Providenciales; tel. 946-2831; fax 946-2695; internet www.firstcaribbeanbank.com; f. 2002 following merger of Caribbean operations of Barclays Bank PLC and CIBC; Exec. Chair. MICHAEL MANSOOR; CEO CHARLES PINK.

**Scotiabank** (Canada): Cherokee Rd, POB 15, Providenciales; tel. 946-4750; fax 946-4755; e-mail bns.turkscaicos@scotiabank.com; Man. Dir DAVID TAIT; br. on Grand Turk.

**Turks and Caicos Banking Co Ltd:** Duke St North, Cockburn Town, POB 123, Grand Turk; tel. 946-2368; fax 946-2365; e-mail tcbc@tciway.tc; internet www.turksandcaicosbanking.tc; f. 1980; cap. US $2.7m., dep. US $9.7m.; Man. Dir ANTON FAESSLER.

### TRUST COMPANIES

**Berkshire Trust Co Ltd:** POB 657, Caribbean Pl., Providenciales; tel. 946-4324; fax 946-4354; e-mail berkshire.trust@tciway.tc; internet www.berkshire.tc; Pres. GORDON WILLIAMSON.

**Chartered Trust Co:** Town Centre Bldg, Butterfield Sq., POB 125, Providenciales; tel. 946-4881; fax 946-4041; e-mail reception@

# THE TURKS AND CAICOS ISLANDS — Directory

chartered-tci.com; internet www.chartered-tci.com; Man. Dir Peter A. Savory.

**Meridian Trust Co Ltd:** Caribbean Pl., Leeward Highway, POB 599, Providenciales; tel. 941-3082; fax 941-3223; e-mail mtcl@tciway.tc; internet www.meridiantrust.tc.

**M & S Trust Co Ltd:** POB 260, Butterfield Sq., Providenciales; tel. 946-4650; fax 946-4663; e-mail mslaw@tciway.tc; internet www.mslaw.tc/trusts.htm; Man. Timothy P. O'Sullivan.

**Temple Trust Co Ltd:** 228 Leeward Highway, Providenciales; tel. 946-5740; fax 946-5739; e-mail info@templefinancialgroup.com; internet www.templefinancialgroup.com; f. 1985; CEO David C. Knipe.

### INSURANCE

**Turks and Caicos Islands National Insurance Board:** Misick's Bldg, POB 250, Grand Turk; tel. 946-1048; fax 946-1362; e-mail nib@tciway.tc; internet www.nib.tc; 4 brs.

Several foreign (mainly US and British) companies have offices in the Turks and Caicos Islands. More than 2,000 insurance companies were registered at the end of 2002.

## Trade and Industry

### GOVERNMENT AGENCIES

**Financial Services Commission (FSC):** see Finance.

**General Trading Company (Turks and Caicos) Ltd:** PMBI, Cockburn Town, Grand Turk; tel. 946-2464; fax 946-2799; shipping agents, importers, air freight handlers; wholesale distributor of petroleum products, wines and spirits.

**Turks Islands Importers Ltd (TIMCO):** Front St, POB 72, Grand Turk; tel. 946-2480; fax 946-2481; f. 1952; agents for Lloyds of London, importers and distributors of food, beer, liquor, building materials, hardware and appliances; Dir H. E. Magnus.

### DEVELOPMENT ORGANIZATION

**Turks and Caicos Islands Investment Agency (TC Invest):** Hon. Headley Durham Bldg, Church Folly, POB 105, Grand Turk; tel. 946-2058; fax 946-1464; e-mail tcinvest@tciway.tc; internet www.tcinvest.tc; f. 1974 as Development Board of the Turks and Caicos Islands; statutory body; development finance for private sector; promotion and management of internal investment; Chair. Lillian Misick; Pres. and CEO Colin R. Heartwell.

### CHAMBERS OF COMMERCE

**Grand Turk Chamber of Commerce:** POB 148, Grand Turk; tel. 946-2324; fax 946-2504; e-mail gtchamberofcommerce@tciway.tc; internet www.turksandcaicos.tc/GrandTurkChamber; f. 1974; Pres. and Exec. Dir Glennevans Clarke; Hon. Sec. Sherlin Williams.

**North Caicos Chamber of Commerce:** tel. 231-1232; Pres. Franklyn Robinson.

**Providenciales Chamber of Commerce:** POB 361, Providenciales; tel. 231-2110; fax 946-4582; internet www.provochamber.com; Pres. Doug Parnell.

### UTILITIES

#### Electricity and Gas

**Atlantic Equipment and Power (Turks and Caicos) Ltd:** New Airport Rd, Airport Area, South Caicos; tel. 946-3201; fax 946-3202.

**PPC Ltd:** Town Centre Mall, POB 132, Providenciales; tel. 946-4313; fax 946-4532.

**Turks and Caicos Utilities Ltd:** Pond St, POB 80, Grand Turk; tel. 946-2402; fax 946-2896.

#### Water

**Provo Water Co:** Grace Bay Rd, POB 124, Providenciales; tel. 946-5202; fax 946-5204.

**Turks and Caicos Water Co:** Provo Golf Clubhouse, Grace Bay Rd, POB 124, Providenciales; tel. 946-5126; fax 946-5127.

### TRADE UNION

**Turks and Caicos Workers' Labour Trade Union:** Grand Turk; all professions; f. 2000; Leader Calvin Handfield.

## Transport

### ROADS

There are 121 km (75 miles) of roads in the islands, of which 24 km (15 miles), on Grand Turk, South Caicos and Providenciales, are surfaced with tarmac.

### SHIPPING

There are regular freight services from Miami, Florida, USA. The main sea ports are Grand Turk, Providenciales, Salt Cay and Cockburn Harbour on South Caicos. In January 2004 it was announced that a new US $35m. cruise-ship terminal would be constructed in Grand Turk, enabling large passenger liners to stop in the territory. The terminal was scheduled to open in November 2005.

**Cargo Express Shipping Service Ltd:** South Dock Rd, Providenciales; tel. 941-5006; fax 941-5062.

**Seacair Ltd:** Churchill Bldg, Front St, POB 170, Grand Turk; tel. 946-2591; fax 946-2226.

**Tropical Shipping:** South Dock Rd, Providenciales; tel. 941-5006; fax 941-5062; e-mail nbeen@tropical.com; internet www.tropical.com; Pres. Rick Murrell.

### CIVIL AVIATION

There are international airfields on Grand Turk, South Caicos, North Caicos and Providenciales, the last being the most important; there are also landing strips on Middle Caicos, Pine Cay, Parrot Cay and Salt Cay.

**Department of Civil Aviation:** POB 168, Grand Turk; tel. 946-2138; fax 946-1185; e-mail cad@tciway.tc; Dir Thomas Swann.

**Air Turks and Caicos (2003) Ltd:** 1 InterIsland Plaza, Old Airport Rd, POB 191, Providenciales; tel. 941-5481; fax 946-4040; e-mail fly@airturksandcaicos.com; internet www.airturksandcaicos.com.

**Caicos Caribbean Airlines:** South Caicos; tel. 946-3283; fax 946-3377; freight to Miami (USA).

**Cairsea Services Ltd:** Old Airport Rd, POB 138, Providenciales; tel. 946-4205; fax 946-4504; e-mail info@cairsea.com; internet www.cairsea.com.

**SkyKing Ltd:** POB 398, Providenciales; tel. 941-5464; fax 941-4264; e-mail king@tciway.tc; internet www.skyking.tc; f. 1985; daily inter-island passenger services, and services to Haiti and the Dominican Republic; CEO Harold Charles; Gen. Man. Mallory McComish.

**Turks Air Ltd:** Providenciales; Head Office: 6111 North West 72nd Ave, Miami, FL 33166, USA; tel. 946-4504; fax 946-4504; tel. (305) 593-8847; fax (305) 871-1622; e-mail turksair@earthlink.net; twice-weekly cargo service to and from Miami (USA); Grand Turk Local Agent Cris Newton.

**Turks and Caicos Airways Ltd:** Providenciales International Airport, POB 114, Providenciales; tel. 946-4255; fax 946-4438; f. 1976 as Air Turks and Caicos, privatized 1983; scheduled daily inter-island service to each of the Caicos Islands, charter flights; Chair. Albray Butterfield; Dir-Gen. C. Moser.

## Tourism

The islands' main tourist attractions are the numerous unspoilt beaches, and the opportunities for diving. Salt Cay has been designated a World Heritage site by UNESCO. Hotel accommodation is available on Grand Turk, Salt Cay, South Caicos, Parrot Cay, Pine Cay and Providenciales. In 2003 there were 163,600 tourist arrivals (an increase of 5.1% compared with the previous year). In 2000 73.9% of tourists were from the USA. In 1998 there were 1,522 hotel rooms (some 80% of which were on Providenciales). Revenue from the sector in 2002 totalled an estimated US $319m.

**Turks and Caicos Hotel and Tourism Association:** Ports of Call, Providenciales; tel. 941-5787; fax 946-4001; e-mail tchta@tciway.tc; internet www.tchta.com; fmrly Turks and Caicos Hotel Asscn; Chair. Andre Niederhauser; CEO Caesar Campbell.

**Turks and Caicos Islands Tourist Board:** Front St, POB 128, Grand Turk; tel. 946-2321; fax 946-2733; e-mail tci.tourism@tciway.tc; internet www.turksandcaicostourism.com; f. 1970; br. in Providenciales.

## Defence

The United Kingdom is responsible for the defence of the Turks and Caicos Islands.

## Education

Primary education, beginning at seven years of age and lasting seven years, is compulsory, and is provided free of charge in government schools. Secondary education, from the age of 14, lasts for five years, and is also free. In 1996/97 there were 21 primary schools, and in 1990 one private secondary school and four government secondary schools. In 1997 the Caribbean Development Bank approved a loan of almost US $4m. to fund the establishment of a permanent campus for the Turks and Caicos Community College.

# THE UNITED STATES VIRGIN ISLANDS

## Geography

### PHYSICAL FEATURES

The United States Virgin Islands is an unincorporated territory of the USA, a Caribbean colony formerly known as the Danish West Indies (purchased in 1917). The Virgin Islands are the first main group of the Lesser Antilles and are divided between two sovereignties, that of the USA and of the United Kingdom. The main island of the British Virgin Islands, Tortola, lies across a narrow sea channel, to the north-east of St John. About 64 km (40 miles) to the west is the mainland of the US commonwealth territory of Puerto Rico (its offshore island of Culebra lying about mid-way between it and St Thomas). St Thomas is about 8 km to the west of St John, while the third main island of the US Virgins, St Croix, is some distance to the south (64 km from St Thomas and 56 km from St John), making it, geographically, not truly part of the Virgin Islands. St Croix is also the largest of the islands, covering 215 sq km (83 sq miles), out of a total area for the territory of 347 sq km (including about 3 sq km of inland waters).

There are 68 islands in all, but only the main islands are permanently inhabited. All the main islands are mountainous, fertile and of volcanic origin, but much territory has been added by reef action, and there are numerous coralline islands, islets and cays (keys). St Croix (Santa Cruz to the Spanish, but first seriously settled by the Knights of St John of Malta for the French—there were also Dutch and British attempts at colonization) is the largest island, with a less indented coast than the other islands and rolling green hills. St Croix is rockier and more arid in the east, but with wetter, wooded heights in the west. St Thomas (80 sq km) is dominated by an east–west central ridge running the length of the island and towards the west reaching the highest point in the territory, at Crown Mountain (474 m or 1,556 ft). Numerous smaller islands surround St Thomas, the largest being Water Island, to the south of the central coast, which is the other inhabited island of the territory and sometimes called the 'fourth Virgin Island'. East of St Thomas is St John, similarly steep and island- and reef-fringed, but covering only 52 sq km. Two-thirds of the island is national park, forming the main part of the wide-ranging protected areas in the islands. Marine ecology is generally more important in the US Virgin Islands (the area is important for leatherback turtles, for instance, and has extensive reefs), as development has long since damaged the land environment. Thus, as happened in several other islands of the West Indies, the introduction of the mongoose (supposedly to deal with rats—although rats are usually nocturnal and the mongoose is not), which is partial to eggs, severely affected many parrot and snake species, although iguanas have survived in places.

### CLIMATE

The climate is subtropical marine, as the heat and humidity is tempered by the Atlantic trade winds. There is little seasonal variation in temperature, with the cooler 'winter' months (December–March) averaging about 25°C (77°F), rising to 28°C (82°F) in summer. There is a rainy season from May to November, average annual precipitation being 1,030 mm (40 ins). The islands can suffer from hurricanes and the afflictions of both drought and flood.

### POPULATION

Ethnically, the local population is 80% black and 15% white. By place of birth, 74% are West Indian (49% born in the Virgin

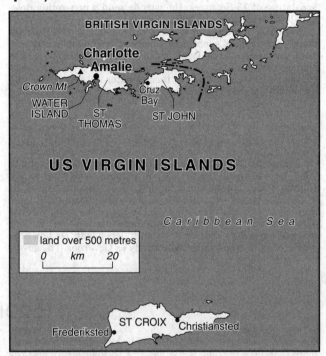

Islands, the remaining 29% from elsewhere in the West Indies), 13% are from the US mainland and 5% are from Puerto Rico. The high level of immigration is illustrated on St Croix, where the population is described as either native Crucian or North American 'Continentals'. In terms of religion, most people are Christian, with 42% claiming to be Baptist, 28% Roman Catholic and 17% Episcopalian (Anglican Communion). The longest established denomination is the Dutch Reformed Church (17th century), and there are also groups of Lutherans and other Protestants. The Jewish congregation is also venerable, having been established in the late 18th century. The cosmopolitan history and nature of the islands is visually apparent, particularly the long period of Danish rule, but, culturally, US rule has proved decisive. The official language is English, which is generally spoken, although there are some Spanish-speakers (some on St Croix and Hispanic immigrants) and a local patois—St Croix Creole or Crucian—has survived.

The total population was estimated at 110,026 in 2002, the most populous island being St Croix (53,234 at the 2000 census), closely followed by St Thomas (including Water Island—51,181). St John had only 4,197 people on it at the time of the census. However, the islands also host some 2m. visitors annually, with the territorial capital, Charlotte Amalie, the most-visited port in the Caribbean. The town, with just over 11,000 residents, is located on the central south coast of St Thomas and was originally named for the consort of the then Danish king. The main towns of St Croix are Christiansted, the island headquarters (on the north coast, where the island begins to narrow towards the east) and Frederiksted (on the west coast). Cruz Bay, at the western end of St John, is the centre of activity on that island.

# History

The Virgin Islands constitute a US external territory. Executive power is vested in the popularly elected Governor and there is a legislature of 15 Senators; the islands send a non-voting delegate to the US House of Representatives.

Originally inhabited by Carib and Arawak Indians (Amerindians), the islands eventually came under Danish control. In 1917 they were sold by Denmark to the USA for US $25m. The islands were granted a measure of self-government in 1954, but subsequent attempts to give them greater autonomy were all rejected by popular referendum. The Democratic Party held a majority in the legislature for many years, although a member of the Independent Citizens' Movement (which split from the Democratic Party), Cyril E. King, was elected Governor in 1974. On King's death in 1978, Juan Luis became Governor, being elected in his own right in 1982. The governorship returned to the Democratic Party with the election of Alexander Farrelly in 1986; he was re-elected in 1990. A further referendum on the islands' status, postponed following the disruption caused by 'Hurricane Hugo', eventually took place in October 1993; however, the result was invalidated by the low turn-out: only 27.4% of voters took part (50% participation was required for the referendum to be valid). In 1994 an Independent, Dr Roy Schneider, was elected to succeed Farrelly as Governor. The governorship was regained by a Democrat, Charles Turnbull, in the 1998 elections; he was re-elected to the post in 2000 and 2002. At legislative elections in November 2002, the Democratic Party won an overall majority of seats in the Senate. In November 2003 a teachers' union petitioned to have the Governor recalled for incompetence after the territory's public schools lost their federal accreditation; however, the union was only able to collect 600 out of a required 17,000 signatures.

Following the presentation of legislation to create a constitutional convention of the US Virgin Islands Senate, in May 2000 a committee of the US House of Representatives was considering a range of measures to enlarge the scope of local self-government in the territory. In November 2004 Governor Turnbull approved legislation allowing the creation of a further constituent assembly to redraft the Constitution.

In February 2005 a 7,000-signature petition was submitted to the US Congress by residents of St Croix, in support of making the island a separate US External Territory from St Thomas and St John. Organizers of the petition claimed that such a move would generate more federal funding for the island, which was affected by a higher unemployment rate than the other two islands, despite being the location for one of the world's largest petroleum refineries.

# Economy

The islands are heavily dependent on links to the US mainland; more than 90% of trade is conducted with Puerto Rico and the USA. St Croix has the world's second largest petroleum refinery, Hovensa LLC (a joint venture between the US oil company Amerada Hess and Petróleos de Venezuela, SA), which produced some 484,000 barrels per day (b/d) in 2004. In 2003 Hovensa exported refined petroleum products to the USA worth US $4,800m. In response to stricter environmental controls, in 2005 the company planned to build a US $400m. desulphurization unit; construction had been scheduled to begin in late 2002 but was delayed by industrial action in Venezuela, the origin of more than 50% of the refinery's crude petroleum imports. Efforts have been made to introduce labour-intensive and non-polluting manufacturing industries. An alumina-processing plant on St Croix, closed in 1994, was acquired by the US company Alcoa, in 1995 and reopened briefly before closing again in 2001; the plant site was subsequently redeveloped as an eco-industrial business and recreation park that operated on a 'zero-waste' policy, aiming to reduce, eliminate or recycle its waste emissions. Rum is an important product, earning money from refunded US excise duty at the rate of about US $75m. per year, but considered likely to suffer from increasing competition from Mexico as the North Atlantic Free Trade Agreement (NAFTA) internal market becomes established. In 2003 the industry was based around a single company, Virgin Islands Rum Industries Ltd, which, in that year alone, exported some 5,973.3m. gallons of rum to the USA. Some fruit and food crops (notably sorghum) are grown for local consumption but the land is unsuitable for large-scale cultivation.

According to the 2000 census figures, 0.7% of the economically active population was engaged in agriculture, forestry, fishing and mining. Industry (including construction and mining) engaged some 12.0% of the non-agricultural labour force in 2002, according to official figures; while services employed some 88.0% of the non-agricultural labour force, of which about 33% was employed in public administration.

The major source of income and employment is tourism, with the emphasis on the visiting cruise-ship business and the sale of duty-free products to visitors from the US mainland. Major expansion of hotel and resort facilities was under way in 2002 on St Croix and St Thomas, at a cost of more than US $1,000m. In June 2004 Governor Charles Turnbull proposed the introduction of a $3 per day 'environmental fee' for hotel guests that, it was envisaged, would generate some $3m. annually to offset infrastructural damage attributed to visitors; representatives from the tourism industry were lobbying against the proposal. The tourism sector is estimated to account for more than 60% of gross domestic product (GDP), although storm damage can cause visitor numbers to fluctuate, as in 1995 when 'Hurricane Marilyn' saw tourist arrivals decrease by 9.7% on the previous year. In 1989 'Hurricane Hugo' was estimated to have caused $1,000m. in property damage, although subsequent rebuilding revitalized employment in the construction industry, albeit funded by substantial loans from the US Federal Emergency Management Agency (FEMA). 'Hurricane Marilyn' caused considerable damage in September 1995, destroying an estimated 80% of houses on St Thomas. Serious damage was caused again in the late 1990s and in November 2003 a state of emergency was called after a week-long storm caused damage amounting to more than $5m. In 2003 tourist arrivals were an estimated 623,723 and tourism receipts amounted to $1,271m.

The budget deficit, which reached an estimated US $305m. by December 1999, was reduced to nearer $200m. in 2001 following the introduction of a Five-Year Strategic and Financial Operating Plan to cut government expenditure and enhance the effectiveness of procedures for revenue collection; total tax collection was estimated to have increased by 19% in 2001. By 2003/04 the budget deficit had been reduced to $4.6m.; however, the Government forecast that the deficit would rise to $9.6m. in 2004/05 and to $92.4m. in 2005/06. In 2003/04 the islands' debt amounted to some $1,000m. The average rate of unemployment rose from 8.8% in 2002 to 9.4% in 2003. The islands were expected in the foreseeable future to continue to receive grants and other remittances from the US Government.

Owing to the islands' heavy reliance on imported goods, local prices and inflation are higher than on the mainland, and the islands' economy, in contrast to that of the USA, remained in recession for most of the 1990s. In 2002 the Territory recorded a trade deficit of US $336.9m. In that year the USA provided 13.8% of imports and took 90.8% of exports. Venezuela was also a major source of imports. Other major sources of imports in 1995 included the United Arab Emirates, the Republic of the Congo and Nigeria. Of total exports to the USA in 2003, 91.9%

# THE UNITED STATES VIRGIN ISLANDS

was refined petroleum products. Crude petroleum accounted for 75.8% of the islands' total imports in that year.

In 1999 the islands' administration successfully negotiated the suspension of debt repayments to FEMA, and in late 2001 the US Congress repaid the balance in full, after the territory had applied for tax relief under the Federal Credit Reform Act. In March 2002 the US Virgin Islands was removed from the Organisation for Economic Co-operation and Development's list of 'unco-operative tax havens', after the Government pledged to improve the transparency of its financial services sector by the end of 2005. In October 2004 the US Congress approved legislation to reform corporate tax in the US Virgin Islands. There were concerns that the measures would deter companies from using the Territory as a tax 'haven', at a cost to the local economy of up to US $63m. in lost revenue and as many as 600 jobs.

# Statistical Survey

Sources (unless otherwise stated): Office of Public Relations, Office of the Governor, Charlotte Amalie, VI 00802; tel. (340) 774-0294; fax (340) 774-4988; Bureau of Economic Research, Department of Economic Development and Agriculture, POB 6400, Charlotte Amalie, VI 00804; tel. (340) 774-8784; internet www.usviber.org.

## AREA AND POPULATION

**Area:** 347.1 sq km (134 sq miles): St Croix 215 sq km (83 sq miles); St Thomas 80.3 sq km (31 sq miles); St John 51.8 sq km (20 sq miles).

**Population:** 101,809 at census of 1 April 1990; 108,612 (males 51,684, females 56,748) at census of 1 April 2000. *By Island* (2000 census): St Croix 53,234, St Thomas 51,181, St John 4,197. Source: US Bureau of the Census. *2002* (official estimate): 110,026.

**Density** (2002): 317.0 per sq km.

**Principal Towns** (population at census of 1 April 2000): Charlotte Amalie (capital) 11,004; Christiansted 2,637; Frederiksted 732. Source: Thomas Brinkhoff, *City Population* (internet www.citypopulation.de).

**Births and Deaths** (2001): Registered live births 1,669 (birth rate 13.7 per 1,000); Registered deaths (preliminary) 605 (death rate 5.6 per 1,000). Source: US National Center for Health Statistics.

**Expectation of Life** (estimates, years at birth): 78.3 (males 74.4; females 82.4) in 2001. Source: US Bureau of the Census.

**Economically Active Population** (persons aged 16 years and over, 2000 census): Agriculture, forestry, fishing, hunting and mining 324; Manufacturing 2,754; Construction 4,900; Wholesale trade 912; Retail trade 6,476; Transportation, warehousing and utilities 3,321; Information 931; Finance, insurance, real estate, rental and leasing 2,330; Professional, scientific, management, administrative and waste management services 3,058; Educational, health, and social services 6,742; Arts, entertainment, recreation, accommodation and food services 7,351; Public administration 4,931; Other services 2,535; *Total employed* 46,565. Source: US Bureau of the Census. *2004* (official figures): Total employed 43,640; Total unemployed 4,530; Total labour force 48,170.

## HEALTH AND WELFARE

**Under-5 Mortality Rate** (per 1,000 live births, 1997): 15.0.

**Physicians** (per 1,000 head, 1999): 1.65 (Source: Pan American Health Organization).

For definitions, see explanatory note on p. vi.

## AGRICULTURE, ETC.

**Livestock** (FAO estimates, 2003): Cattle 8,000; Sheep 3,200; Pigs 2,600; Goats 4,000; Chickens 35,000.

**Fishing** (metric tons, live weight, 2003): Total catch (all capture) 1,492 (Snappers, jobfishes, etc. 183; Parrotfishes 197; Caribbean spiny lobster 130; Stromboid conchs 493).
Source: FAO.

## INDUSTRY

**Production** (estimates, '000 metric tons, unless otherwise indicated, 2001): Jet fuels 1,620; Motor spirit (petrol) 2,370; Kerosene 70; Gas-diesel (distillate fuel) oil 2,910; Residual fuel oils 3,825; Liquefied petroleum gas 240; Electric energy 1,090 million kWh. Source: UN, *Industrial Commodity Statistics Yearbook*.

## FINANCE

**Currency and Exchange Rates:** 100 cents = 1 United States dollar (US $). *Sterling and Euro Equivalents* (31 May 2005): £1 sterling = US $1.8181; €1 = US $1.2331; $100 = £55.00 = €81.10.

**Budget** (projections US $ million, year ending 30 September 2002): Operating budget 580.2 (Net revenues from taxes, duties and other sources 521.8); Rum excise taxes (Federal remittance) 70.9; Direct Federal expenditures 573.

## EXTERNAL TRADE

**Total** (US $ million): *Imports*: 5,349.9 in 2000; 4,608.7 in 2001; 4,213.2 in 2002. *Exports*: 5,200.4 in 2000; 4,234.2 in 2001; 3,876.3 in 2002. The main import is crude petroleum ($3,192.7m. in 2002), while the principal exports are refined petroleum products ($3,233.9m. in 2002).

**Trade with the USA** (US $ million): *Imports*: 592.3 in 2000; 730.1 in 2001; 582.7 in 2002. *Exports*: 4,964.3 in 2000; 3,960.3 in 2001; 3,519.9 in 2002.

## TRANSPORT

**Road Traffic** (registered motor vehicles, 1993): 63,332.

**Shipping:** *Cruise-ship Arrivals:* 1,014 in 2000; 976 in 2001; 845 in 2002. *Passenger Arrivals:* 1,738,703 in 2002; 1,773,948 in 2003; 1,963,609 in 2004.

**Civil Aviation** (visitor arrivals): 598,019 in 2002; 620,814 in 2003; 661,800 in 2004.

## TOURISM

**Tourist Arrivals** (estimates, excl. excursionists and cruise-passengers): 597,437 in 2001; 585,684 in 2002; 623,723 in 2003.

**Tourism Receipts** (US $ million, incl. passenger transport): 1,323 in 2001; 1,195 in 2002; 1,271 in 2003.
Source: World Tourism Organization.

## COMMUNICATIONS MEDIA

**Radio Receivers** (1997): 107,100 in use.

**Television Receivers** (1999): 71,000 in use.

**Telephones** (2000): 69,000 main lines in use.

**Mobile Cellular Telephones** (2000): 35,000 subscribers.

**Internet Users** (1999): 12,000.

**Daily Newspapers** (1996): 3; average circulation 42,000 copies.

**Non-daily Newspapers** (1988 estimates): 2; average circulation 4,000 copies.
Sources: UNESCO, *Statistical Yearbook*; International Telecommunication Union.

## EDUCATION

**Pre-primary** (1992/93): 62 schools; 121 teachers; 4,714 students (2000).

**Elementary** (1992/93): 62 schools; 790 teachers (public schools only); 13,421 students (2001).

**Secondary:** 541 teachers (public schools only, 1990); 5,359 students (2001).

**Higher Education:** 266 teachers (2003/04); 2,610 students (2004). Sources: UNESCO, *Statistical Yearbook*; US Bureau of the Census.

THE UNITED STATES VIRGIN ISLANDS

# Directory

## The Constitution

The Government of the US Virgin Islands is organized under the provisions of the Organic Act of the Virgin Islands, passed by the Congress of the United States in 1936 and revised in 1954 and 1984. Subsequent amendments provided for the popular election of the Governor and Lieutenant-Governor of the Virgin Islands in 1970 and, since 1973, for representation in the US House of Representatives by a popularly elected Delegate. The Delegate has voting powers only in committees of the House. Executive power is vested in the Governor, who is elected for a term of four years by universal adult suffrage and who appoints, with the advice and consent of the legislature, the heads of the executive departments. The Governor may also appoint administrative assistants as his representatives on St John and St Croix. Legislative power is vested in the legislature of the Virgin Islands, a unicameral body comprising 15 Senators, elected for a two-year term by popular vote. Legislation is subject to the approval of the Governor, whose veto can be overridden by a two-thirds vote of the Legislature. All residents of the islands, who are citizens of the USA and at least 18 years of age, have the right to vote in local elections but not in national elections. In 1976 the Virgin Islands were granted the right to draft their own constitution, subject to the approval of the US President and Congress. A constitution permitting a degree of autonomy was drawn up in 1978 and gained the necessary approval, but was then rejected by the people of the Virgin Islands in a referendum in March 1979. A fourth draft, providing for greater autonomy than the 1978 draft, was rejected in a referendum in November 1981.

## The Government

### EXECUTIVE
(July 2005)

**Governor:** CHARLES WESLEY TURNBULL.

**Lieutenant-Governor:** VARGRAVE A. RICHARDS, II.

**Commissioner of Agriculture:** LAWRENCE LEWIS (acting).

**Commissioner of Education:** NOREEN MICHAEL.

**Commissioner of Finance:** BERNICE TURNBULL.

**Commissioner of Health:** DARLENE CARTY.

**Commissioner of Housing, Parks and Recreation:** IRA M. HOBSON.

**Commissioner of Human Services:** SEDONIE HALBERT.

**Commissioner of Labor:** CECIL R. BENJAMIN.

**Commissioner of Licensing and Consumer Affairs:** ANDREW RUTNIK.

**Commissioner of Planning and Natural Resources:** DEAN C. PLASKETT.

**Commissioner of Police:** ELTON LEWIS.

**Commissioner of Property and Procurement:** MARC BIGGS.

**Commissioner of Public Works:** WAYNE D. CALLWOOD.

**Commissioner of Tourism:** PAMELA C. RICHARDS.

**Attorney-General:** IVER A. STRIDIRON.

**US Virgin Islands Delegate to the US Congress:** DONNA M. CHRISTENSEN.

### GOVERNMENT OFFICES

**Office of the Governor:** Government House, 21–22 Kongens Gade, Charlotte Amalie, VI 00802; tel. (340) 774-0001; fax (340) 774-1361.

**Office of the Lieutenant-Governor:** Government Hill, 18 Kongens Gade, Charlotte Amalie, VI 00802; tel. (340) 773-6449; fax (340) 773-0330; tel. (340) 774-2991; fax (340) 774-6953; internet www.ltg.gov.vi.

**Department of Agriculture:** Estate Lower Love, Kingshill, St Croix, VI 00850; tel. (340) 774-0997; fax (340) 774-5182; internet www.usvi.org/agriculture.

**Department of Economic Development:** 81 AB Kronprindsens Gade, POB 6400, Charlotte Amalie, VI 00801; tel. (340) 774-8784; fax (340) 774-4390.

**Department of Education:** 44–46 Kongens Gade, Charlotte Amalie, VI 00802; tel. (340) 774-0100; fax (340) 779-7153; e-mail education@usvi.org; internet www.doe.vi.

**Department of Finance:** GERS Bldg, 2nd Floor, 76 Kronprindsens Gade, Charlotte Amalie, VI 00802; tel. (340) 774-4750; fax (340) 776-4028; internet www.usvi.org/finance.

**Department of Health:** 48 Sugar Estate, Charlotte Amalie, VI 00802; tel. (340) 774-0117; fax (340) 777-4001; internet www.usvi.org/health.

**Department of Housing, Parks and Recreation:** Property & Procurement Bldg No. 1, Sub Base, 2nd Floor, Rm 206, Charlotte Amalie, VI 00802; tel. (340) 774-0255; fax (340) 774-4600.

**Department of Human Services:** Knud Hansen Complex Bldg A 1303, Hospital Ground, Charlotte Amalie, VI 00802; tel. (340) 774-0930; fax (340) 774-3466; e-mail humanservices@usvi.org; internet www.usvi.org/humanservices.

**Department of Justice:** GERS Bldg, 2nd Floor, 48B–50C Kronprindsens Gade, Charlotte Amalie, VI 00802; tel. (340) 774-5666; fax (340) 774-9710; internet www.usvi.org/justice.

**Department of Labor:** POB 302608, St Thomas, VI 00803; tel. (340) 776-3700; fax (340) 774-5908; e-mail customersupport@vidol.gov; internet www.vidol.gov.

**Department of Licensing and Consumer Affairs:** Property & Procurement Bldg No. 1, Sub Base, Rm 205, Charlotte Amalie, VI 00802; tel. (340) 774-3130; fax (340) 776-0675; e-mail commissioner@dlca.gov.vi; internet www.dlca.gov.vi.

**Department of Planning and Natural Resources:** 396-1 Anna's Retreat–Foster Bldg, Charlotte Amalie, VI 00802; tel. (340) 774-3320; fax (340) 775-5706; e-mail stt-office@vidpnr-dep.org; internet www.dpnr.gov.vi.

**Department of Police:** 8172 Sub Base, St Thomas, VI 00802; tel. (340) 774-2211; fax (340) 778-2373; e-mail police@usvi.org; internet www.vipd.gov.vi.

**Department of Property and Procurement:** Property & Procurement Bldg No. 1, Sub Base, 3rd Floor, Charlotte Amalie, VI 00802; tel. (340) 774-0828; fax (340) 774-9704; e-mail pnp@usvi.org; internet www.usvi.org/pnp.

**Department of Public Works:** No. 8 Sub Base, Charlotte Amalie, VI 00802; tel. (340) 773-1290; fax (340) 774-5869; e-mail publicworks@usvi.org; internet www.usvi.org/publicworks.

**Department of Tourism:** POB 6400, St Thomas, VI 00804; tel. (340) 774-8784; fax (340) 774-4390; e-mail info@usvitourism.vi; internet www.usvitourism.vi.

## Legislature

### LEGISLATIVE ASSEMBLY

**Senate**
(15 members)

**President of the Senate:** LORRAINE BERRY.

**Election, 2 November 2004**

| Party | Seats |
|---|---|
| Democrats | 10* |
| Independent Citizens Movement | 3 |
| Independent | 2 |
| **Total** | **15** |

* In late November 2004 three Democrat candidates defected to form a consensus majority coalition with the Independents and Independent Citizens Movement.

## Political Organizations

**Democratic Party of the Virgin Islands:** POB 2033, St Thomas Democratic District, VI 00803; affiliated to the Democratic Party in the USA; Chair. CECIL R. BENJAMIN, Jr.

**Independent Citizens Movement:** Charlotte Amalie, VI 00801; Chair. USIE RICHON.

# THE UNITED STATES VIRGIN ISLANDS

**Republican Party of the Virgin Islands:** POB 1532, St Thomas, VI 00804; tel. (340) 776-0583; affiliated to the Republican Party in the USA; Chair. JAMES M. OLIVER.

## Judicial System

**US Federal District Court of the Virgin Islands:** Federal Bldg and US Courthouse, 5500 Veteran's Dr., Charlotte Amalie, VI 00802; tel. (340) 774-0640; internet www.vid.uscourts.gov; jurisdiction in civil, criminal and federal actions; the judges are appointed by the President of the USA with the advice and consent of the Senate.

**US Territorial District Court of the Virgin Islands:** Alexander A. Farrelly Justice Center, POB 70, Charlotte Amalie, VI 00802; tel. (340) 774-6680; internet www.vid.uscourts.gov; exclusive jurisdiction in violations of police and executive regulations; original jurisdiction over all local criminal matters and civil actions, regardless of the amount in controversy; and concurrent jurisdiction with the US Federal District Court of the Virgin Islands over federal question and diversity cases, and over local Virgin Islands criminal offences based on the same act or transaction if the act or transaction also constitutes an offence against federal law. The Court is also vested with jurisdiction to promulgate rules and regulations governing its practice and procedure, and to regulate the admission of attorneys to the Virgin Islands Bar.

**Judges:** RAYMOND L. FINCH (Chief Judge), CURTIS V. GOMEZ.

## Religion

The population is mainly Christian. The main churches with followings in the islands are Baptist, Roman Catholic, Episcopalian, Lutheran, Methodist, Moravian and Seventh-day Adventist. There is also a small Jewish community.

### CHRISTIANITY

#### The Roman Catholic Church

The US Virgin Islands comprises a single diocese, suffragan to the archdiocese of Washington, DC, USA. At 31 December 2003 there were an estimated 30,000 adherents in the territory (27.6% of the population).

**Bishop of St Thomas:** Most Rev. GEORGE V. MURRY, Bishop's Residence, 29A Princesse Gade, POB 301825, Charlotte Amalie, VI 00803-1825; tel. (340) 774-3166; fax (340) 774-5816; e-mail chancery@islands.vi.

#### The Anglican Communion

**Episcopal Church of the Virgin Islands:** Bishop: Rt Rev. THEODORE A. DANIELS, POB 10437, St Thomas, VI 00801; fax (340) 777-8485; e-mail tad@aol.com.

## The Press

**Pride Magazine:** 22A Norre Gade, POB 7908, Charlotte Amalie, VI 00801; tel. (340) 776-4106; f. 1983; monthly; Editor JUDITH WILLIAMS; circ. 4,000.

**St Croix Avis:** La Grande Princesse, Christiansted, St Croix, VI 00820; tel. (340) 773-2300; f. 1944; morning; independent; Editor RENA BROADHURST-KNIGHT; circ. 10,000.

**St John Tradewinds:** Tradewinds Bldg, Garden Level Unit 4, POB 1500, Cruz Bay, St John, VI 00831; tel. (340) 776-6496; fax (340) 693-8885; e-mail editor@tradewinds.vi; internet www.stjohntradewindsnews.com; f. 1972; fortnightly; Publr MALINDA NELSON; circ. 2,500.

**Virgin Islands Business Journal:** 69 Kronprindsens Gade, POB 1208, Charlotte Amalie, VI 00804; tel. (340) 776-2874; fax (340) 774-3636; internet www.vibj.com; f. 1986; weekly; independent; Man. Editor CARTER HAGUE; circ. 10,000.

**Virgin Islands Daily News:** 9155 Estate Thomas, VI 00802; tel. (340) 774-8772; fax (340) 776-0740; e-mail dailynews@vipowernet.net; internet www.virginislandsdailynews.com; f. 1930; bought by Innovative Communication Corpn in 1997; morning; CEO and Exec. Editor J. LOWE DAVIS; circ. 15,000.

# Broadcasting and Communications

### TELECOMMUNICATIONS

**Innovative Telephone:** POB 1141, Charlotte Amalie, VI 00801; tel. (340) 774-5555; internet www.iccvi.com; f. 1959 as Virgin Islands Telephone Corpn (Vitelco), present name adopted in 2001; bought by Innovative Communication Corpn in 1987; provides telephone services throughout the islands; Chair. JEFFREY J. PROSSER; Pres. and CEO DAVID SHARP.

**Innovative Wireless:** St Croix, 4006 Estate Diamond, Christiansted, VI 00820; fax (340) 778-6011; internet www.vitelcellular.com; f. 1989 as Vitelcellular; owned by Innovative Communication Corpn; cellular telephone services; Gen. Man. NEIL WILLIAMS.

### RADIO

**WAVI—FM:** POB 25016, Gallows Bay Station, St Croix, VI 00824; tel. (340) 773-3693; commercial; Gen. Man. DOUG HARRIS.

**WGOD:** Crown Mountain, POB 5012, Charlotte Amalie, VI 00803; tel. (340) 774-4498; fax (340) 776-0877; commercial; Gen. Man. PETER RICHARDSON.

**WJKC-WMNG-WVIQ:** 5020 Anchor Way, POB 25680, Christiansted, St Croix, VI 00824; tel. (340) 773-0995; e-mail jkc95@aol.com; internet www.wjkcisle95.com; commercial; Gen. Man. JONATHAN COHEN.

**WSTA:** Sub Base 121, POB 1340, St Thomas, VI 00804; tel. (340) 774-1340; fax (340) 776-1316; e-mail addie@wsta.com; internet www.wsta.com; commercial; Gen. Man. ATHNIEL C. OTTLEY.

**WVVI, WVJZ, WWKS** (Knight Quality Stations): 13 Crown Bay Fill, POB 305678, Charlotte Amalie, VI 00803; tel. (340) 776-1000; fax (340) 776-5357; e-mail kqs@viaccess.net; f. 1962; commercial; Pres. and Gen. Man. RANDOLPH H. KNIGHT.

### TELEVISION

**Caribbean Communications Corporation:** 1 Beltjen Pl., Charlotte Amalie, VI 00802-6735; tel. (340) 776-2150; fax (340) 774-5029; f. 1966; cable service, 72 channels; Gen. Man. ANDREA L. MARTIN.

**St Croix Cable TV:** Heron Commercial Park, POB 5968, Sunny Isle, St Croix, VI 00823; tel. (340) 778-6701; f. 1981; bought by Innovative Cable Television in 1997; 32 channels; Gen. Man. JACK WHITE.

**TV2:** 1 Beltjen Pl., St Thomas, VI 00802; tel. (340) 774-2200; fax (340) 776-1957; e-mail bbutler@tv2.vi; internet www.tv2.vi; f. 2000; owned by Innovative Communication Corpn; Gen. Man. BRENT A. BUTLER.

**WSVI—TV:** Sunny Isle, POB 8ABC, Christiansted, St Croix, VI 00823; tel. (340) 778-5008; fax (340) 778-5011; f. 1965; one channel and one translator; Gen. Man. BARAKAT SALEH.

**WTJX—TV** (Public Television Service): Barbel Plaza, POB 7879, Charlotte Amalie, VI 00801; tel. (340) 774-6255; fax (340) 774-7092; e-mail lelskoe@wtjx.org; internet www.wtjx.org; one channel; Gen. Man. OSBERT POTTER.

# Finance

### BANKING

**Banco Popular of the Virgin Islands:** 193 Altona and Welgunst, Charlotte Amalie, VI 00802; tel. (340) 693-2702; fax (340) 693-2702; Regional Man. VALENTINO I. MCBEAN; 7 brs.

**Bank of Nova Scotia** (Canada): 214C Altona and Welgunst, POB 420, Charlotte Amalie, VI 00804; tel. (340) 774-6393; fax (340) 693-5994; Man. R. HAINES; 10 brs.

**Bank of St Croix:** POB 24240, Gallows Bay, St Croix 00824; tel. (340) 773-8500; fax (340) 773-8508; Man. JAMES BRISBOIS; 1 br.

**Chase Manhattan Bank, NA** (USA): Waterfront, Charlotte Amalie, POB 6620, VI 00801; tel. (340) 776-2222; Gen. Man. WARREN BEER; brs in St Croix and St John.

**Citibank, NA** (USA): Grand Hotel Bldg, 43–46 Norrie Gade, POB 5167, Charlotte Amalie, VI 00801; tel. (340) 776-353; fax (340) 774-6609; Vice-Pres. KEVIN SZOT.

**First Bank of Puerto Rico:** POB 3126, St Thomas; tel. (340) 774-2022; fax (340) 776-1313; Pres. and CEO JAMES E. CRITES; 4 brs.

**First Federal Savings Bank:** Veterans Dr., Charlotte Amalie, VI; tel. (340) 774-2022; fax (340) 776-1313; Man. ALFRED LESLIE; br. in St Croix.

# THE UNITED STATES VIRGIN ISLANDS

**Virgin Islands Community Bank:** 12–13 King St, Christiansted, St Croix, VI 00820; tel. (340) 773-0440; fax (340) 773-4028; Pres. and CEO MICHAEL J. DOW; 2 brs.

### INSURANCE

A number of mainland US companies have agencies in the Virgin Islands.

## Trade and Industry

### GOVERNMENT AGENCY

**US Virgin Islands Economic Development Commission:** 1050 Norre Gade, Government Development Bank Bldg, 1050 Norre Gade, POB 305038, St Thomas, VI 00803; tel. (340) 774-8104; e-mail edc@usvieda.org; internet www.usvieda.org/idc; govt agency; offices in St Thomas and St Croix.

### CHAMBERS OF COMMERCE

**St Croix Chamber of Commerce:** 3009 Orange Grove, Suite 12, Christiansted, St Croix, VI 00820; tel. and fax (340) 773-1435; fax (340) 773-8172; e-mail info@stxchamber.org; f. 1924; 250 mems; Pres. DIANE BUTLER; Exec. Dir RACHEL HAINES.

**St Thomas–St John Chamber of Commerce:** POB 324, Charlotte Amalie, VI 00804; tel. (340) 776-0100; fax (340) 776-0588; e-mail chamber@islands.vi; internet www.usvichamber.com; c. 700 mems; Pres. THADDEUS BAST; Exec. Dir JOSEPH S. AUBAIN.

### UTILITIES

#### Electricity

**Virgin Islands Energy Office:** 45 Mars Hill, Frederiksted, St Croix, VI 00840; tel. (340) 773-3450; e-mail vieo0441@viaccess.net; internet www.vienergy.org; Dir VICTOR SOMME, III.

#### Water

**Virgin Islands Water and Power Authority (WAPA):** POB 1450, Charlotte Amalie, VI 00804; tel. (340) 774-3552; fax (340) 774-3422; f. 1964; public corpn; manufactures and distributes electric power and desalinated sea water; Chair. DARYL 'MICKEY' LYNCH; Exec. Dir ALBERTO BRUNO-VEGA; c. 50,000 customers.

## Transport

### ROADS

The islands' road network totals approximately 855.5 km (531.6 miles). Throughout 2004 the Department of Public Works implemented a programme, valued at US $17.5m., to repair roads damaged in storms in November 2003.

### SHIPPING

The US Virgin Islands are a popular port of call for cruise ships. The bulk of cargo traffic is handled at a container port on St Croix. A ferry service provides frequent daily connections between St Thomas and St John and between St Thomas and Tortola (British Virgin Islands). In June 2004 the Port Authority approved a US $9.3m. project to expand freight, vehicle and passenger facilities at Red Hook.

**Virgin Islands Port Authority:** POB 301707, Charlotte Amalie, VI 00803; also at POB 1134, Christiansted, St Croix, VI 00821; tel. (340) 774-1629; fax (340) 774-0025; tel. (340) 778-1012; fax (340) 778-1033; e-mail info@viport.com; internet www.viport.com; f. 1968; semi-autonomous govt agency; maintains, operates and develops marine and airport facilities; Exec. Dir DARLAN BRIN.

### CIVIL AVIATION

There are airports on St Thomas and St Croix, and an airfield on St John. Seaplane services link the three islands. The runways at Cyril E. King Airport, St Thomas, and Alexander Hamilton Airport, St Croix, can accommodate intercontinental flights.

## Tourism

The islands have a well-developed tourism infrastructure, offering excellent facilities for fishing, yachting and other aquatic sports. A National Park covers about two-thirds of St John. There were 5,059 hotel rooms in 2004. In 2002 some 585,684 tourists visited the islands, increasing to 623,723 in 2003. During the same period tourism expenditure increased from US $1,195m. to $1,271m.

**St Croix Hotel Association:** POB 24238, St Croix, VI 00824; tel. (340) 773-7117; fax (340) 773-5883.

**St Thomas–St John Hotel and Tourism Association:** POB 2300, Charlotte Amalie, VI 00803; tel. (340) 774-6835; fax (340) 774-4993; e-mail stsjhta@vipowernet.net; internet www.stsjhta.com; Pres. and Exec. Dir BEVERLY NICHOLSON.

## Education

Education is compulsory up to the age of 16 years. It generally comprises eight years at primary school and four years at secondary school. In 1990 there were 28,691 elementary and secondary students at public and private schools, and 1,762 public school teachers; in 2002 the number of elementary and secondary students totalled 24,934. The University of the Virgin Islands, with campuses on St Thomas and St Croix, had 2,610 students in 2004; the total number of students in higher education in 2000 was 3,107.

# URUGUAY

## Geography

### PHYSICAL FEATURES

The Oriental Republic of Uruguay is in south-eastern South America, on the east bank of the River Uruguay (hence described as Oriental). On the other side of the Uruguay is Argentina, which also faces the country across the great estuary of the Río de la Plata (River Plate) to the south. Originally the east-bank (Banda Oriental) territory of the Argentine possessions of Spain, it proclaimed independence in 1825 and was recognized in 1828. There remains only a dispute, unresolved but uncontested, over some river islands along the border with Brazil (which stretches for about 1,000 km—620 miles; the Argentine border is almost 60% this length), to the north-east. Uruguay is the second smallest country in South America (after Suriname), with an area of 176,215 sq km (68,037 sq miles), making it only slightly smaller than the US state of Florida. There are 2,600 sq km of inland waters, mainly from coastal lagoons and lakes and the great reservoirs along the Río Negro.

Uruguay sits on the Plate, the estuary formed where the Uruguay and the Paraná rivers enter the Atlantic, although it also has a stretch of south-east-facing coastline running up into Brazil. The country has 660 km of seashore and 813 km of river boundaries (435 km along the Uruguay alone). It is roughly triangular in shape, extending for over 560 km north–south and 480 km east–west. Although the defining feature of Uruguay is the river from which it takes its name, the central feature is its tributary, the Negro, which traverses the middle of the country from east to west, ending in the south-west. Its basin lies between two low ranges, the Cuchilla Grande, to the south, and the Cuchilla de Haedo, striking down from the north to bisect the northern part of the country. These ranges rise from plateau uplands of rolling hills, which occupy most of the country. The coastal plains, which only account for 15% of Uruguay's territory, lie along a shore of lagoons, and sandy beaches and dunes that can extend inland for up to 8 km. Beyond this sandy coast the land is very fertile, supporting the vast pasturelands of tall, bluish-tinted prairie grass (which gives way to shorter, scrubbier grasses on the ridges). Pastureland, most of it ideal for ranching, accounts for three-quarters of the country, while woodland can only be found on about 3% (the lowest proportion in South America), mainly along the river banks. In the eastern third of the country the uplands mount into a gentle arc of low mountains (Serranas) and the Cuchilla Grande forms a watershed between the narrow river valleys cutting across the rolling plains to drain into the Uruguay and the shorter, steeper valleys heading directly for the Atlantic. In this region is the country's highest point, Cerro Catedral (only 514 m or 1,687 ft). The highlands of the north and east are continuations of the basaltic plateau underlying the Brazilian highlands, while the plains begin the landscape of the Argentine Pampas. Once the plains, hills and coasts were rich with wildlife, but subsequent to European colonization they have become scarcer—pumas, rheas, tapirs and seals are now seldom seen, although there are deer, alligators, armadillos, wild pigs and many birds (swans, cranes, burrowing owls and parakeets, for instance).

### CLIMATE

Uruguay is the only country in Latin America to lie completely outside the tropics. The maritime influence also ensures warm summers and mild winters, although the flatness of the country makes flooding likely in the heavy rain to which a country without significant mountains can be exposed. Droughts are also not uncommon. In winter (June–September) there can be cold wind currents, pamperos, blowing from the south-west, but there is seldom frost anywhere. The summer features the hot zonda wind from the north. While the climate overall is temperate, average figures can obscure the experience of actual day- and night-time temperatures, for instance, and the interior has

more of these extremes. The average temperature for the mid-summer months of January and February is some 22°C (72°F), and for the coldest month, June, 10°C (50°F). It is hotter in the north-west, where it is also wetter. It is slightly less wet in the south. Generally, rainfall is well distributed and totals about 1,000 mm (about 40 ins) per year.

### POPULATION

Unlike in most Latin American countries, most of the population of Uruguay is white, with very few Mestizos (about 8%) and no indigenous Amerindians. A large number of people are foreign-born. Most originate from Spain (54%), but also from Italy (22%), Brazil, Argentina and France, with some coming from Eastern Europe after the Second World War. Apart from those of Spanish and Italian descent, perhaps 12% more originate in Europe, the rest of the population being Mestizo, mulatto, black (4%), Arab or Asian. About 61% of Uruguayans are nominally Roman Catholic, but less than 50% of the adult population attends church regularly. Only about 2% are Protestant, with Jews as the largest non-Christian minority (almost 2%), and many people are professedly non-believers. The official and most widely spoken language is Spanish, although there are some speakers of a Spanish–Portuguese patois, Brazilero, near the eastern borders. English and French are the most popular second languages.

The total population was estimated at 3.2m. in mid-2004. Despite the importance of agriculture to the economy, this population is very urbanized—only 9% live in the country (and most are very poor). However, Uruguayans are well educated (the country is one of the few territories in the Western hemisphere where all education is free). Most people live near the coast and almost one-half in Greater Montevideo alone. By far the largest city, port and economic centre, Montevideo is the national capital. It is located on the south coast near the mouth of the Plate. The next-largest cities are both on the River Uruguay, Paysandú and, further north, Salto. The main city on the Brazilian border is Rivera, mid-way along the frontier, in the north of Uruguay. The country is divided into 19 departments.

# History

## HELEN SCHOOLEY

Revised for this edition by KATHARINE MURISON

During the early 20th century Uruguay acquired a reputation as one of the most stable, prosperous and democratic countries in Latin America. That image, albeit tarnished by a period of military rule in 1973–85, was at variance with the country's early history. From the 16th century to the early 19th century the territory changed hands between the Spanish and Portuguese several times. In 1776 the territory of Uruguay became part of the newly created Spanish Viceroyalty of Río de la Plata. Uruguay's independence from Spain, proclaimed in 1825 and recognized in 1828, was promoted by the British Government to provide a buffer state between Brazil and Argentina. However, both countries continued to influence domestic affairs through their support of rival rulers (caudillos). The first 50 years of nationhood was a period of anarchy, as the rival bands of Blancos ('Whites') and Colorados ('Reds'), the respective forerunners of the modern conservative and liberal political groupings, struggled for land and power.

The transition from anarchy to stability began in the 1870s, with the creation of the apparatus of a modern state. As world demand for wool and hides increased, Uruguayan landowners sought protection from the damage to livestock caused by persistently warring factions. Trading interests, strong in the capital, Montevideo, because of the port's natural advantages in the River Plate region, also sought peace. The replacement of weak civilian government by a military regime (1876–86) and the introduction of modern armaments, railways and telegraph services secured the dominance of central authority in Montevideo over regional caudillos. Public utilities received substantial British capital investment, and foreign trade expanded. Ironically, the modernization of the rural export economy displaced much of the labour force, many of whom were to form the armies of the Blanco rebels and the Colorado Government, which fought Uruguay's most recent civil war, in 1904.

### THE RULE AND INFLUENCE OF JOSÉ BATLLE Y ORDÓÑEZ

The Colorados' victory confirmed the authority of José Batlle y Ordóñez, who was twice President (1903–07 and 1911–15) and the dominant figure in Uruguayan politics until his death in 1929. His policies were nationalist and reformist and, under his rule, Uruguay became South America's first welfare state. Women were given the vote, the Church was disestablished and the death penalty abolished. The country's social structure was transformed by immigrants from the Mediterranean countries of Europe, and its economic prosperity was founded on export growth. Within the country, Batlle y Ordóñez tried to limit the power of British capital by encouraging domestic investment in manufacturing and cultivating closer relations with the USA as a counter-weight.

The radicalism of Batlle y Ordóñez drew support from the middle and working classes, but antagonized foreign and domestic investors and alarmed his own Partido Colorado (Colorado Party). His proposals for constitutional reform, which, in modified form, were embodied in the 1919 Constitution, led to the emergence of the Batllistas and other factions in the party. The Constitution, which divided the executive branch of government between a President and eight other members of a National Council of Government, remained in force until its overthrow by a *coup d'état*, led by Gabriel Terra, in 1933. Terra claimed he was acting against the inadequacies of the Council, but it was more probable that the coup was a reflection of the anxiety of landowners to maintain access to the British market for their beef exports.

Terra attempted to follow the policies of Benito Mussolini, the Fascist leader of Italy. For most of the Second World War Uruguay was formally neutral, although its political sympathies (and commercial advantage) lay with the Allies, led by the USA and the United Kingdom. In 1942 political and constitutional change saw the restoration to power of the Batllistas, who represented the interests of the urban economy, rather than those of rural exporters. As in the First World War, Uruguay found easy markets for its exports, and commercial relations with the USA were strengthened. Following the War, the United Kingdom again became an important export market, but the special relationship between the two countries declined with the sale of the mainly British-built railways to the Uruguayan Government in 1948.

### THE DEVELOPMENT OF TWO-PARTY DOMINANCE

The manufacturing sector expanded rapidly in the decade following the War. However, export earnings did not keep pace with industrial growth, and in 1956 the country entered a period of economic stagnation, which lasted until the early 1970s. The political parties proved to be adept at staying in power by distributing rewards to political clienteles, but showed little ability to devise new policies. In 1959, however, the Partido Nacional (National Party, the Blancos) defeated the Colorados and took power for the first time in the 20th century. Although the Government adopted an IMF-sponsored austerity programme, the economic decline continued. Political and social frustration intensified in the 1960s, and the political left, small in electoral terms but strong in the trade-union movement, became more militant.

In 1966 a new Constitution reintroduced the presidential system of government, following 15 years of a collegiate executive (whereby the presidency was rotated annually among the nine members of the National Council of Government). The presidential election was won by Gen. Oscar Daniel Gestido of the Colorados, who took office in March 1967. After his death in December, his successor, Jorge Pacheco Areco, used emergency powers to confront students and organized labour and to reduce the influence of professional politicians. With political life radicalized, the Tupamaro urban guerrilla movement increased its operations and remained active until the armed forces took control of internal security in 1972.

### MILITARY DICTATORSHIP, 1973–85

In the 1971 presidential election Juan María Bordaberry Arocena, one of the Colorado candidates, was declared victorious, even though Wilson Ferreira Aldunate of the Blancos was the most popular individual candidate. He took office in March 1972, but was little more than a figurehead President: in February 1973 he accepted the armed forces' demand for military participation in political affairs. In June both chambers of the elected Congreso (Congress) were dissolved, and in December Bordaberry appointed a new legislature, the Council of State, composed of 25 (later 35) nominated members. All left-wing groups were banned and political activity was suspended. With an estimated 6,000 political prisoners detained by 1976, the military-supported regime came to be regarded as one of the most repressive in South America. In June 1976 Bordaberry was deposed by the army because of his refusal to consider any return to constitutional rule. The newly formed Council of the Nation, comprising the Council of State and the 21-member Joint Council of the Armed Forces, chose Dr Aparicio Méndez Manfredini (a former Minister of Health) as the new President, and he took office in September. Under military rule economic growth resumed, but real wages declined sharply and about 10% of the population emigrated, for economic or political reasons. Close political and commercial links developed with Brazil and Argentina, but the previously good relations with the USA were strained during US President Jimmy Carter's Administration

(1977–81), as a result of his criticism of the Uruguayan regime's flagrant violations of human rights.

In November 1980 the military leadership submitted a draft constitution to a referendum. Large sections of the Colorado and Blanco parties campaigned against the military's proposals, on the grounds that they institutionalized the army's role in security matters. When the proposed document was rejected, by 57.8% of those voting, the military leadership was forced to consult with leaders of the recognized parties on the constitutional future of the country.

## THE RETURN TO CIVILIAN RULE

The Joint Council of the Armed Forces appointed Lt-Gen. Gregorio Álvarez Armellino to the presidency in September 1981. Restrictions on political activity were eased, and in June 1982 the Partido Colorado, the Partido Nacional and the smaller Unión Cívica (Civic Union) were allowed to operate, although left-wing parties remained banned, and censorship was increased. Constitutional negotiations foundered in 1983 over attempts by the military leaders to preserve their authority in future governments, and in August the Government suspended all political activity. Distinctions between the two main parties, the liberal Colorados and the more conservative Blancos, gradually eroded, and pro- and anti-military tendencies developed within each of them. Two national days of protest were organized by the opposition in November 1983, and the Plenario Intersindical de Trabajadores—Convención Nacional de Trabajadores (PIT—CNT, Intersyndical Plenary of Workers—National Convention of Workers), the country's main trade-union federation, organized the first industrial action in the country for 10 years in September. The PIT—CNT was banned in January 1984, following a 24-hour strike protesting against a 28% increase in prices and continued political repression.

Negotiations between the Government and the Partido Colorado, the still-proscribed Frente Amplio (Broad Front, a left-wing coalition) and the Unión Cívica resumed in July 1984. The Partido Nacional refused to participate while its leader, Wilson Ferreira Aldunate, was detained (he had been arrested in June, on returning from an 11-year exile). In August the Government withdrew its suspension of political activity and began to release some political prisoners (at that time numbering 600–900), but Ferreira Aldunate and Gen. Líber Seregni Mosquera, the Frente Amplio leader, were banned from political activity. In November the presidential election was won by Dr Julio María Sanguinetti Cairolo of the Colorado's Unidad y Reforma (Unity and Reform) faction. In the legislative elections, however, the Partido Colorado did not gain an absolute majority in either congressional chamber. In preparation for the full restoration of civilian rule, most remaining political prisoners, including Ferreira Aldunate, were released in November, and the Tupamaro leader, Raúl Sendic, was freed in March 1985. In December the Tupamaros formally agreed to become a political party, the Movimiento de Liberación Nacional (MLN—National Liberation Movement), and voted to join the Frente Amplio.

## THE AMNESTY LAW

President Sanguinetti was inaugurated on 1 March 1985, and formed a Council of Ministers including two members of the Partido Nacional and one of the Unión Cívica. The Partido Comunista (Communist Party) and other previously outlawed organizations were legalized, and remaining restrictions on press freedom were lifted. Trade unions were quick to demand more rapid liberalization, and the Blancos, the Unión Cívica and the Frente Amplio criticized the Government's management of the economy.

The most controversial issue for the Sanguinetti administration was the introduction in August 1986 of a draft amnesty law, which protected all members of the security forces from prosecution for alleged violations of human rights during the period of military rule. After considerable debate a revised version of the original proposals was passed by both houses in December, despite violent outbursts in the streets of Montevideo and within the Chamber of Representatives itself. The 'Punto Final' (Full Stop) law, as it was known, ended current military trials and made President Sanguinetti responsible for any further investigations into the whereabouts of the 'disappeared'. It was claimed that under military rule more than 170 Uruguayan citizens had been seized and subsequently murdered by the military; 33 had 'disappeared' within Uruguay, 132 in Argentina and the remainder in Paraguay and Chile. Public opposition to the legislation increased, and during 1988 a sufficient number of signatures (a minimum of 25% of the electorate) was gathered to force a referendum, which was held in April 1989. For the first time in a national referendum voting was declared compulsory, and the law was confirmed by a vote of 57% in favour.

## ECONOMIC LIBERALIZATION

With the resolution of the amnesty question, economic policy again became the Government's chief concern. The trade unions criticized the Government for adhering to the dictates of international monetary agencies, at the expense of the domestic work-force, and organized general strikes in late 1987 and mid-1988. The Government's economic performance cost the Colorados the elections, held on 26 November 1989 (the first free elections since 1972), even though there was little material difference between the economic manifestos of the two leading presidential candidates. Luis Alberto Lacalle Herrera of the Blancos won 37% of the votes cast, compared with 30% for Jorge Batlle Ibáñez of the Colorados and 21% for Líber Seregni of the Frente Amplio.

The Partido Nacional could form only a minority Government, with 39 out of 99 seats in the Chamber of Representatives and 13 of the 31 Senate seats. Although Lacalle reached a power-sharing agreement with the Colorados, taking office in March 1990, the Partido Colorado publicly expressed its opposition to parts of the Government's economic policy, as did a number of Blanco deputies. The most contentious economic issue was the proposed privatization programme, which was suspended in 1992. President Lacalle remained committed to privatization, however, and the process of eroding the country's large state sector was set in motion, albeit at a far gentler pace than in many neighbouring countries.

The Government provoked further outcry in 1992 when it attempted to impose a programme of wage restraint on public-sector workers, but exempted the police and the armed forces. In March 1993 two Blanco factions withdrew their congressional support, and in May the Government was defeated in a vote on finance policy, leaving President Lacalle to rely on his own faction of the Partido Nacional and a right-wing Colorado faction for support in the Congreso. The divisions within the Partido Nacional were exacerbated by other Blanco politicians distancing themselves from the Lacalle administration ahead of the 1994 elections. Lacking reliable political support, Lacalle's Government became more sympathetic to the armed forces. The Government prepared legislation extending military powers and appeared reluctant to pursue investigations into violent incidents implicating members of the military.

## ELECTORAL REFORM

The presidential and legislative elections of November 1994 marked an end to traditional two-party politics. The vote was almost evenly split between the Partido Colorado (32.5%), the Partido Nacional (31.4%) and a third political force, the Encuentro Progresista (Progressive Congress, 30.8%), a predominantly left-wing alliance of the parties of the Frente Amplio, Blanco dissidents and other minority parties. In the Chamber of Representatives the Colorados won 32 seats, and the Blancos and Encuentro Progresista 31 seats each. Sanguinetti, the leading Colorado presidential candidate, was re-elected President, and was obliged to form a coalition administration, leaving the Encuentro Progresista as the main opposition force. However, the principal member of this alliance, the Frente Amplio, reportedly gave an undertaking to co-operate with the new administration in return for government consideration of its views when determining social policy.

The new Government immediately tackled electoral reform. Under the existing system, each party could present multiple presidential candidates, with the leading candidate in each party assuming the total number of votes for candidates in that party. This system had long been criticized for promoting factionalism and for yielding an unclear result. Under the proposed reform each party would present a single candidate and, in the event of no candidate securing an absolute majority, a second round of voting would take place. Although members of the Frente Amplio claimed that the change was a deliberate attempt to deny any left-wing group the chance to win the presidency, the amendment was passed in the Congreso in October 1996, and narrowly approved (by 50.5%) in a plebiscite held in December. It came into effect in January 1997.

In March 1995 President Sanguinetti launched a restructuring of the social-security system, his fifth attempt at such a reform (see also Economy). The fiscal burden of the pension system was growing to such an extent that on this occasion the proposed reform received support from the two larger parties. The draft legislation was approved in August, in spite of trade-union opposition and resistance from the Frente Amplio. The new system proved to have far greater appeal than expected, and within a year of its introduction nearly one-third of the work-force had subscribed.

In May 1996 the Colorados and the Blancos strengthened their 'governability pact'. The attempt to establish a common political agenda was the first such occurrence in 160 years and underlined the shift away from the traditional two-party rivalry. Ironically, this move was accompanied by the emergence of divisions within the Frente Amplio. In September 1997 the Front's leader, Tabaré Vázquez, threatened to resign after party in-fighting over the issue of privatization. Although his grouping generally opposed privatization, the Frente Amplio-controlled authorities in Montevideo had begun to offer contracts to private tender, antagonizing the more left-wing member organizations.

In April 1997 Frente Amplio senator Felipe Michelini, whose father was among the 'disappeared', called for an investigation into allegations that 32 of the victims of the dictatorship had been interred in the grounds of military property and covertly removed from the burial sites in 1985–86. A judicial ruling in the same month promised to facilitate such an inquiry, and in May 20,000 people attended a rally in Montevideo to demand that the Government and the armed forces acquiesce. However, in June a court of appeal overturned the ruling on the grounds that it contravened the Punto Final amnesty law. Under pressure, in October the Government conceded that it had not kept its undertaking in the 1986 amnesty to ascertain where the 'disappeared' had been buried. Nevertheless, the armed forces made it clear that they would obstruct any attempts by the civilian judiciary to conduct investigations involving military property or personnel, and no investigations into human rights abuses during military rule were launched during the Sanguinetti administration.

The human rights issue continued to be prominent in 1998. In May the country's principal religious and human rights groups published a National Reconciliation Declaration, which made explicit their opposition to the amnesty law. In July a former intelligence chief, Rear-Adm. (retd) Eladio Moll, caused a diplomatic crisis when he alleged that US military officers had instructed the Uruguayan security forces on the detention and torture of Tupamaro suspects, and that Uruguayan officers had resisted US demands that the suspects be killed after interrogation.

## BATLLE IBÁÑEZ ELECTED PRESIDENT

In accordance with the Sanguinetti Government's electoral reforms, a presidential primary election was held in April 1999. The Partido Colorado secured 38% of the total votes, and its nomination was won by Jorge Batlle Ibáñez. The Encuentro Progresista—Frente Amplio (EP—FA) obtained 30% of the total votes and its leader, Tabaré Vázquez, was selected as the coalition's candidate. For the Partido Nacional, the campaign for the primary election was seriously divisive, which contributed to its third place overall, with 29% of the total ballot. During the campaign the Blanco candidate, former President Luis Alberto Lacalle, had been accused of corruption, and in the months preceding the elections it became clear that the presidential election would be, in effect, a contest between Batlle and Vázquez. The Colorados and Blancos attempted to portray Vázquez (an eminent oncologist) as a Marxist, and claimed that he would destroy the country's economy. His main manifesto promises were government-led job creation, and the rather less popular introduction of an income-tax system.

In presidential and legislative elections held in October 1999 the EP—FA won a simple majority in both the Chamber of Representatives (40 of the 99 seats) and in the Senate (12 of the 30 seats). The Partido Colorado secured 33 and 10 seats, respectively, and the Partido Nacional obtained only 22 seats in the lower house and seven Senate seats (although the party performed well in local and regional elections in May 2000, taking control of 13 of the 19 departments). In the presidential ballot Vázquez won 38.5% of the vote, compared with 31.3% for Batlle and 21.3% for Lacalle. Under the system operating until 1995 Vázquez would have been elected President, but the new electoral law required a second round of voting, which was held on 28 November. Most of the Blanco support was transferred to the Colorado candidate, and Batlle won with 51.6% of the votes cast.

Batlle's inauguration speech of 1 March 2000 included undertakings to reduce public spending, address the farming crisis, investigate the fate of the 'disappeared' and institutionalize Mercosur (Mercado Común del Sur—Southern Common Market—see Economy). His new Government comprised eight Colorados and five Blancos. In April he dismissed Gen. Manuel Fernández, a senior army official who had sought to justify the military repression of the 1970s and had spoken of a renewed struggle against left-wing extremists, and in August a six-member Commission for Peace was formed to investigate the disappearance of 170 people during the military dictatorship. In May 2001 the bodies of two Uruguayan exiles were found in Argentine soil. The Commission released its first report in October 2002, which detailed instances of collusion between the Argentine and Uruguayan armed forces. Of the 39 cases examined it found that 26 of the 33 named Uruguayan citizens had been tortured and killed, and of the six named Argentine nationals, one had died in Uruguay and the remainder had been returned to Argentina. It was reported that many of Uruguay's 'disappeared' had been killed in Argentina by members of that country's armed forces.

## ECONOMIC CRISIS

Against a background of a prolonged economic downturn since 1998 and with the economic crisis of 2002, the dominant issue of the Batlle administration became its management of the economy. There was a series of protests, including two general strikes in June 2000 and July 2001 led by the PIT—CNT, over public-spending cuts and privatization proposals. There was also unrest in rural areas, with farmers appealing for aid to mitigate the disastrous effects on the rural economy of the outbreak of foot-and-mouth disease in April 2001. In January 2002, however, the problems escalated with the Argentine financial crisis. In May Argentine withdrawals from deposits held in Uruguay increased sharply, triggering a run on Uruguayan banks by Uruguayan nationals, and the Government introduced highly unpopular austerity measures. In July this wave of unpopularity resulted in the resignation of the Minister of Economy and Finance, Alberto Bensión. In the last week of July the Government closed the banks (see Economy), and there were fears that the country might be obliged to default on the national debt. Emergency international funding enabled the banks to reopen a few days later.

The economic crisis deepened the rift between the two governing parties, and an opinion poll conducted in August 2002 gave the EP—FA over 50% of public confidence. The Blancos became increasingly critical of government policy, objecting to what they termed the 'high social cost' of financing the national debt and demanding more rigorous measures on extending taxation and curbing administrative costs. All five Blanco ministers resigned from the Council of Ministers in November; President Batlle then reshuffled the cabinet and reduced the number of ministerial posts from 13 to 11. In early 2003 the opposition was more conciliatory and did not obstruct the Government's negotiations with the IMF and its efforts to restructure debt payments.

The advent of presidential and legislative elections in October 2004 brought the Batlle administration under further pressure, and broad support for the IMF programme began to wane. By mid-2003 various elements of the Government's economic policy were being criticized, and the Frente Amplio was voicing its opposition to plans to broaden the national tax base. Trade unions and opposition groups held a one-day general strike in August in protest at the Government's reform programme, in particular at plans to increase private-sector participation in the economy. In December, following a campaign by the EP—FA, a referendum was held regarding controversial legislation enacted in 2001 allowing private-sector involvement in the petroleum, cement and alcohol monopoly, the Administración Nacional de Combustible, Alcohol y Portland (ANCAP); 62% voted to repeal the law.

## SOCIAL CONCERNS

Traditionally, Uruguay enjoyed relative wealth within Latin America, with more equal distribution of income, a lower rate of poverty and well-funded education, social-security and health systems. It also had a reputation for more liberal legislation on social issues, and Batlle advocated a measured decriminalization of certain illegal drugs. An opinion poll conducted in 2001 indicated that support for the principle of democratic government ran at one of the highest levels on the continent, despite criticism for current government practice. By 2002, however, an increasing number of citizens were unable to meet the cost of health insurance and, in common with other countries in the region, there was a marked rise in street violence and organized crime. Although Uruguay retained the lowest incidence of poverty in Latin America, living standards were in decline even before the 2002 crisis. The poverty index rose from 9.4% in 1999 to 11.4% in 2001, while the extreme poverty index rose from 1.8% to 2.4%. The economic crisis also precipitated a sharp increase in the number of suicides in late 2002 and accelerated emigration. Many Uruguayan citizens were expected to take advantage of a law introduced in Spain in January 2003 allowing the children and grandchildren of Spanish citizens to emigrate to Spain.

## THE 2004 ELECTIONS

In the presidential primary election, held in June 2004, the EP—FA received the greatest share of the votes (43%), and Vázquez was confirmed as the grouping's candidate. The Blancos took 41% of the votes, with Jorge Larrañaga defeating Lacalle to secure the party's nomination, while the Colorado candidate, Guillermo Stirling, won only 15% of the votes.

Vázquez was elected to the presidency in a first round of voting on 31 October 2004, narrowly securing an absolute majority, with 50.7% of the votes cast. His nearest rival, Larrañaga, won 34.1% of the vote, while Stirling only received 10.3%, in the worst ever result for the Partido Colorado. Vázquez's victory, which brought an end to more than 170 years of rule by the Blancos and Colorados, was widely attributed to the outgoing administration's mismanagement of the economy. The EP—FA was also successful in the concurrent legislative elections, winning an absolute majority in both chambers: 52 of the 99 seats in the Chamber of Representatives and 16 of the 30 seats in the Senate. The Partido Nacional won 35 and 11 seats, respectively, and the Partido Colorado's congressional representation declined from 33 to 10 seats in the House of Representatives and from 10 to three seats in the Senate.

Also on 31 October 2004 a constitutional reform guaranteeing that the supply of water and sanitation services would remain under state control was approved by 64.5% of voters in a referendum. The proposal was supported by the leaders of the EP—FA and the Partido Nacional, but opposed by the Colorados. President-elect Vázquez gave assurances during the campaign that the reform would not be retroactive and would therefore not affect existing private-sector involvement in the water sector; a decree to that effect was issued in May 2005.

Prior to taking office Vázquez sought co-operation with other parties represented in the Congreso, and in mid-February 2005 an agreement was signed on cross-party support for the incoming Government's policies on the economy, foreign affairs and education. The opposition parties were also to accept Vázquez's offer of a number of positions at state institutions. However, in late February the Partido Colorado withdrew from this arrangement after the opposition was allocated only one seat on the five-member board of the education directorate. The Partido Nacional followed suit in mid-March when it was offered only one place on the board of a state-owned bank. Members of the opposition had traditionally occupied two seats in every five-strong governing board.

## THE VÁZQUEZ PRESIDENCY, 2005–

Vázquez took office on 1 March 2005 as the country's first left-wing President. His new Council of Ministers was largely composed of members of the various factions of the EP—FA, although two independents received posts, including Jorge Lepra, a former executive of the US energy company Chevron-Texaco Corporation, who became Minister of Industry, Energy and Mining. Reforming ANCAP, the privatization of which the EP—FA had opposed, was expected to be a priority for Lepra. Danilo Astori, the leader of the Asamblea Uruguay (the most right-wing party in the ruling coalition), was appointed Minister of Economy and Finance, inheriting an economy that, although recovering, remained burdened by a large external debt. The appointments of both Lepra and Astori, apparently aimed at strengthening investor confidence, proved controversial with some of the more leftist parties in the EP—FA. The significant portfolios of the interior, foreign affairs and national defence were all allocated to members of Vázquez's own Partido Socialista del Uruguay, while former Tupamaro guerrilla José 'Pepe' Mujica, the leader of the Movimiento de Participación Popular (MPP), which had won some 30% of the EP—FA's vote in the elections, was appointed as Minister of Livestock, Agriculture and Fishing.

Following his inauguration, Vázquez pledged to investigate 'disappearances' during the military dictatorship, announcing that excavations would be conducted at military units in an attempt to determine the fate of the missing detainees. Vázquez also announced the introduction of the Plan de Atención Nacional de Emergencia Social, a two-year social welfare programme aimed at reducing poverty, in which the Government was initially to invest some US $100m. Vázquez was, however, criticized by the MPP in late April 2005 for commitments made to the IMF on the fiscal balance and external debt. The party accused the President of prioritizing financial stability at the expense of his social welfare programme in order to satisfy the Fund. In response, Vázquez travelled to one of the country's poorest towns with his Council of Ministers and announced that further such visits were planned.

The EP—FA performed well in municipal elections held on 8 May 2005, securing control of eight of the 19 departments, comprising some 70% of the country's population, having previously only held Montevideo. The mayoralty of Montevideo was won by Ricardo Ehrlich of the MPP, which was again the dominant faction of the ruling coalition, taking 25% of the party's vote nation-wide. The Partido Nacional retained 10 departments, but lost three, while the Partido Colorado suffered a further reverse, only managing to hold onto one. At the end of June the PIT—CNT led a one-day strike in support of demands for increased wages, but emphasized that its action was targeted at employers, rather than the Government.

## FOREIGN RELATIONS

Relations with Argentina have been close, with many Argentines taking holidays and owning second homes in Uruguay. However, Uruguayan-Argentine relations were strained in 2002, especially after the run on the Uruguayan banks. In January 2004, moreover, Argentine President Néstor Kirchner announced he would prosecute former members of the Uruguayan military held responsible for the deaths of Argentine citizens during the military dictatorship; Batlle, meanwhile, ordered an investigation into the fate of 80 Uruguayan citizens who had 'disappeared' in Argentina during the same period.

The new administration of President Tabaré Vázquez, which took office in March 2005, emphasized the importance of strengthening relations with other Latin American countries, particularly within Mercosur. On the day after his inauguration, at meetings in Montevideo, Vázquez signed agreements on trade and the supply of energy with his Venezuelan counterpart, President Hugo Chávez Frías, and an accord on human rights with Argentina's President Kirchner, which was aimed at facilitating the exchange of information on abuses stemming from both countries' military dictatorships. Vázquez's first foreign visit was to Brazil in the following month, when he signed six agreements on co-operation with Brazilian President Lula da Silva.

In one of his first official acts, President Vázquez restored full diplomatic ties with Cuba in March 2005. In April 2002 Uruguay had severed diplomatic relations with Cuba (established in 1986), after Cuba had taken exception to Uruguayan support for a UN resolution calling for greater respect for human rights in Cuba.

The delay in the ratification by the Uruguayan Congreso of an investment treaty with the USA threatened to strain bilateral relations in mid-2005. The treaty, which had been signed by outgoing President Batlle in December 2004, was opposed by some members of the ruling EP—FA.

# Economy

## HELEN SCHOOLEY

Revised for this edition by the editorial staff

From the late 19th century Uruguay developed as an agricultural export economy by selling wool, beef and hides, mainly to European countries. With favourable natural conditions and a low density of population, export values per head were high. Government policy also encouraged domestic manufacturing, and protection was intensified in the 1930s and after the Second World War. However, the rate of growth of exports was low and the small size of the domestic market meant that the strategy of industrialization by import substitution was exhausted by the mid-1950s. A period of economic stagnation followed and continued into the early 1970s. The economic policies of the military regime after 1973, based on reduction in the public sector and incentives to non-traditional exports, helped stimulate growth in the gross domestic product (GDP), averaging 4.5% per year in 1974–80. However, attempts to control inflation and the general effects of the world recession halted economic expansion, and GDP registered negative growth in 1982–84. Overall, GDP expanded by an average of 0.3% per year in 1980–90.

Traditionally, the chief factor in the performance of the economy was export growth and the balance of trade. GDP growth resumed in the 1990s, but the economy then entered a decline towards the end of the decade. Real GDP fell by 2.7% in 1999, under the influence of a devaluation of the Brazilian currency, and by a further 1.4% in 2000, because of the recession in Argentina and continued low world prices for agricultural exports. In 2001 production was drastically affected by an outbreak of foot-and-mouth disease, and the economy contracted by 3.4%. In 2002 the decline became a crisis when Argentina's debt default spilled over, precipitating a run on Uruguayan banks, devaluation and a major increase in the national debt. In 2002 the economy contracted by 11.2%. The economy recovered much faster than originally anticipated in 2003, with growth of 2.4%, and dramatic growth of 12.3% was recorded in 2004.

Although the economy was always closely linked to the economies of neighbouring countries, the establishment in 1995 of the Southern Common Market (Mercado Común del Sur— Mercosur, or, in Portuguese, Mercado Comum do Sul— Mercosul) accentuated the influence of markets in Brazil and Argentina. Uruguay maintained one of the highest levels of GDP per head in Latin America, although this fell from US $6,000 in 2000 to an estimated $3,915 in 2004. Consumption rose by 62% over the 10 years to 1997 and Uruguay was the only country in Latin America where the gap between the wealthiest and the poorest did not increase during the early and mid-1990s. Consumption fell dramatically in 2001 and 2002, owing principally to the economic crisis in neighbouring Argentina, and although the poverty rate rose, it remained the lowest rate in Latin America. There was a relatively large public sector, the level of trade-union membership was substantial for the region, and the official rate of unemployment increased from 11.3% in November 1999 to almost 20% by late 2002, the highest on record. By early 2005 it stood at an estimated 12.6%.

A relatively small country, with an area of 176,215 sq km (68,037 sq miles), Uruguay had an estimated population of 3.2m. in mid-2004. The average annual population growth of 0.7% in 1990–2003 was the lowest in Latin America, whereas the proportion of people living in urban areas was over 90% of the total, the highest in the region, with 39% of the population living in Montevideo. Some 85% of the national territory was devoted to sheep- and cattle-raising, with only about 7% used for arable land. In many places the soils were thin, and in 1995 it was reported that the country was facing an erosion problem, losing more than 50m. metric tons of topsoil per year.

### AGRICULTURE

Agriculture traditionally formed the basis of the Uruguayan economy, and accounted (directly or indirectly) for almost all commodity export earnings. Some 4.5% of the working population were employed in the sector in 2003, compared with 20% in 1965, although the 2003 figure only took into account urban areas. The sector performed strongly in the early and mid-1990s, largely owing to substantial investment in infrastructure and production. In 2003 agriculture (including forestry and fishing) accounted for a provisional 11.9% of GDP.

Uruguay's two main export products were meat (especially beef) and wool. Meat was also an important item of the local diet. The main markets for Uruguayan beef were Brazil, Chile, the European Union (EU) and Israel. In 2003 exports of live animals, animal products, skins and hides contributed 43.5% of export revenue. The dairy sector, centred in the south-western departments of San José and Colonia, was comparatively small, although it grew rapidly in the 1980s and produced exportable surpluses. Milk yields more than doubled between 1981 and 1997, and wool production rose by over 50% from the early 1970s to the mid-1990s. In 1988–94 Uruguay's share of wool production in the Southern Cone region increased from 32% to 41%, although the region's total output declined by 30% in the same period.

The main crops were rice, sugar cane, wheat, barley, potatoes, sorghum and maize, with cereal production based in the west of the country, but in general the yields were low. Rice was grown in the east. Owing principally to demand from Brazil, rice output trebled between 1990 and 2000. Citrus fruit was grown mainly in the department of Salto, while other fruits in production included grapes, apples, pears, peaches and nectarines. Fruit crop levels increased in the 1990s, as a result of favourable weather conditions and the cultivation of virgin land. Flowers and vines were grown around Montevideo and Canelones to the north. There were also efforts to increase timber production, with a view to the export market: following government investment in the industry's infrastructure in the early 2000s, output increased by 25% in 2001–03.

Although small scale, the fishing industry grew rapidly from the late 1970s. In 1982 an estimated 119,201 metric tons were caught, compared with 26,000 tons in 1975, and about 90% of the total catch was exported. The total catch in 2003 was 116,900 tons. Concern over dwindling fishery reserves in the late 1990s prompted attempts at diversification, including a plan to begin farming caviar in the Baygorria reservoir, 200 km north of Montevideo, using sturgeon eggs imported from Russia.

By the end of the 1990s weak international markets for Uruguay's traditional meat and dairy products had precipitated a crisis in the farming sector. The Government responded by offering tax breaks and refinancing arrangements, but farmers continued their protests. Problems in the livestock sector were compounded by an outbreak of foot-and-mouth disease in October 2000, resulting in a temporary halt of meat exports and the loss of 6,500 jobs in the meat and dairy industry. Although exports resumed following the slaughter of some 20,000 animals, they were suspended again in April 2001, when a second outbreak of the disease spread rapidly across the country. The Government was forced to adopt a mass vaccination programme, thereby reducing the value of future meat exports. Exports to the EU and Israel did not resume until November, and during that period the meat share of export earnings fell from 20% to 12.8%. In 2002 the agricultural sector grew by 6.7% but declined by 0.4% in the following year.

## MINING AND POWER

The absence of any commercially viable deposits of petroleum, natural gas or coal made Uruguay heavily dependent on the import of crude petroleum, which accounted for about two-thirds of the country's energy requirements. Mining, quarrying and power accounted for a provisional 0.2% of GDP and accounted for 1.2% of the economically active population in 2003. Prospecting for petroleum began in the 1940s and resumed in the 1970s in the wake of the oil crisis, when attention was focused on offshore possibilities, but these were judged to be economically unviable. In 1999, however, after a further round of oil-price increases, an initiative was announced within Mercosur to investigate the deposits. Uruguayan refining capacity was 50,000 barrels per day at the end of 2003. Controversial legislation enacted in 2001 to allow private-sector involvement in the state oil concern, Administración Nacional de Combustibles, Alcohol y Portland (ANCAP), was repealed following the law's rejection in a referendum in December 2003.

In an attempt to reduce the dependence on imported petroleum, successive Governments promoted the use of hydroelectricity; in 2002 energy was derived principally from hydroelectric power (60% of total electricity production, according to the US Energy Information Administration). Exploitation of hydroelectric resources had begun in the 1930s, but substantial development came in the 1970s, as a result of major investments in the 1,800 MW Salto Grande project with Argentina, and the 300 MW Palmar installation with Brazil. The output from these two plants and two others on the Río Negro (Rincón del Bonete and Baygorria) made Uruguay a net exporter of electric energy by the 1980s, and hydroelectric power became an important factor in the country's balance of payments from 1992. Net electricity exports for 2000 totalled 135m. kWh, and in 2002 the Government called for the construction of further generation plants in order to ease dependency on Argentina. During 2004, however, the hydroelectric sector was generating well below capacity, owing to adverse climatic conditions. Uruguay also developed its gas resources for export. A US $135m. gas pipeline from Buenos Aires, Argentina, to Montevideo was completed in 2002, as part of a greater scheme to link the gas network of the Mercosur countries, but its successful commercial operation was threatened by a dispute with Argentina over the price of this gas. In June 2005, however, the Presidents of the four Mercosur member countries signed an agreement creating an 'energy ring' (*anillo energético*), which would facilitate the supply and storage of natural gas from the Camisea field in Peru. Uruguay consumed an estimated 700m. cu ft of natural gas in 2002.

Apart from quarrying for marble and other construction materials, and a limited extraction of semi-precious stones, there was little mining activity. There were small reserves of gold, manganese, lead and iron ore. According to the Central Bank, the GDP of the mining sector increased by an annual average of 9.1% in 1990–2001. Mining GDP decreased sharply in 2001 and 2002, by an estimated 8.8% and 11.2%, respectively.

## MANUFACTURING

From the 1970s manufacturing became a principal sector of the Uruguayan economy, contributing an estimated 18.0% of GDP in 2003. In the early 1990s the country's widening trade deficit prompted a programme of investment and reorganization in the sector. Industrial activity was concentrated in Montevideo and accounted for about one-quarter of the capital's total employed population. By comparison with other Latin American countries the sector was subject to high levels of taxation. The most important branches of manufacturing were food products (mainly processed meat for export), beverages and tobacco (an estimated 45.4% of the total), chemicals (22.8%), metal products, machinery and equipment, and textiles, clothing and leather products. By 2000 the number of people working in industry was only one-half the level of 1988, reflecting both greater productivity (with a consequent loss of jobs) and a major expansion of the service sector; furthermore, in the textile sector the number fell to below one-third. In 2003 industry accounted for 21.5% of the working population. Between 1973 and 1980 industrial output increased by 36%. The sector was adversely affected by the recession of the early 1980s, but grew rapidly in the later part of the decade, chiefly as a result of an increase in car manufacturing and, to a lesser degree, in textiles and parts for the petroleum industry. During the 1990s the sector fluctuated, with expansion in the food-processing and textile sectors in 1992, an increase in car production in 1994–95, and a marked upturn in the chemical sector in 1997. The sector declined after that year, and in 2001 many meat and dairy processing jobs were lost as a result of the outbreak of foot-and-mouth disease. The sector contracted sharply in 2002 and early 2003, but began to recover from mid-2003.

## TRANSPORT, INFRASTRUCTURE AND COMMUNICATIONS

Montevideo was the focal point of all Uruguay's transport systems, and the city's port was the principal gateway for foreign trade, but during the 1980s the transport systems suffered from neglect. Major improvements began in the 1990s, stimulated by the growth of tourism and investment in non-traditional exports; by the end of the decade there was a rolling programme of road, rail and port extension. The sector was a growth area even though the country's freight shipment costs were the highest in Mercosur.

The railway system was relatively extensive, with 3,002 km of track radiating from Montevideo, and links to the Argentine and Brazilian networks. After a decade of decline, passenger services were halted completely in 1988, but a passenger rail link between Montevideo, Florida and Canelones was reopened in 1993. Apart from livestock, principal freight items were stone, cement, rice, and fuel, and in 1996 a project was launched for a freight railway scheme to connect the forestry industry in the north-east of the country with the coast. Like the railways, Uruguay's roads linked interior cities with the capital, rather than with each other. Although in August 1999 the Congreso (Congress) gave final approval for the construction of a 41-km toll bridge over the Río de la Plata (River Plate), between Colonia del Sacramento in Uruguay and Punta Lara in Argentina, subsequently the project was suspended indefinitely.

The transport, storage and communications sector contributed an estimated 9.2% of GDP in 2003, and during the 1990s recorded an average annual growth rate of 8%, owing chiefly to the telecommunications sector. By 2003 Uruguay had around 507,000 mobile phone owners and a similar number of internet users. The communications sector was in the vanguard of government free market reforms. Although opposition to privatization halted the proposed sale of the state telecommunications concern, Administración Nacional de Telecomunicaciones (ANTEL, see below), cellular-telephone services, road maintenance, port and airport administration and the new passenger railway services were opened to private-sector participation in the mid-1990s.

## PRIVATIZATION

The process of privatization in Uruguay was much slower than in other South American countries, partly because the state sector had been more efficient than its counterparts and partly because of more effective political opposition to the policy. Government attempts to introduce a major privatization programme in the early 1990s met with resistance from the Frente Amplio and the trade unions, which initially were successful in reversing plans by demanding a referendum on individual sales. Consequently, many initiatives were presented on more than one occasion. In 1992 the Government was forced to suspend the sale of shares in ANTEL, and to adopt a more gradual approach to privatization, encouraging competition and the use of private capital in state concerns. The state monopolies on insurance and the financing and construction of new homes were ended in 1993 and 1996, respectively, and the national airline, Primeras Líneas Uruguayas de Navegación Aérea (PLUNA) was privatized in 1995.

The restructuring of the social-security system undertaken by the Government of Julio María Sanguinetti (1995–2000) proved less controversial. President Sanguinetti had made four previous attempts at such a reform, and his arguments were reinforced by the pressing need to alleviate the fiscal pressure exerted by pensions. In 1995 only 65% of the social-security budget was raised by contributions from those currently employed, while the number of people of pensionable age was over one-half of the economically active population. The new legislation, approved in August 1996, increased the period of contribution, raised the retirement age and established a mixed capitalization and distribution system. The reforms aimed to eliminate the system's structural deficit within 10 years. The first pension-fund administration company (fondo de ahorro previsional—Afap) was inaugurated in April 1996 and by 2000 some 85% of the work-force had joined a new private pension scheme.

In 1996 ANCAP lost its monopoly over the manufacture of alcohol and cement, and in 1997 the Government ended the state monopoly on the generation of electricity and introduced legislation allowing the use of private capital in electricity distribution. Despite a campaign on the part of the Frente Amplio and the trade unions, the energy reform law received a positive vote in a referendum held in June 1998.

When President Jorge Batlle Ibáñez came to office in March 2000 he continued a cautious liberalization policy, opting for the introduction of competition rather than the dismantling of state concerns, although this policy was still opposed by the Frente Amplio and the trade unions. Further efforts were made to deregulate the telecommunications sector and the Government made it possible for ANCAP to use private-sector partners to develop its refining capacity. The privatization programme gained impetus in 2002 when the Government sought IMF assistance to counter the effects of the Argentine default. Under the agreement reached in March, the Government undertook to end the remaining state monopolies in the energy and telecommunications sectors, to eliminate the pension-fund deficit and to allow private companies into the administration of roads and airports. In December 2003, however, the legislation to allow private-sector involvement in ANCAP was rejected in a referendum.

## INVESTMENT AND FINANCE

In the 1970s Uruguay experienced a high level of public-sector investment, largely in hydroelectricity and bridge-construction programmes, but during the 1980s and 1990s an increasing amount of investment came from the private sector. A law passed in 1987 allowed for the establishment of free trade zones, both publicly and privately owned, for various manufacturing and service industries. In the mid-1990s the opening up of the public sector to private capitalization and joint-venture schemes encouraged significant levels of foreign direct investment for the first time. A new investment law, approved in 1998, promised further opportunities for private finance, with incentives including tax and tariff exemptions and reduction of employers' social-security contributions. Its main consequence was to boost foreign investment, although indications were that investment funds were growing in response to the decline in interest rates and inflation and the 'pegging' of the exchange rate. Like internal investment, domestic savings and stock-exchange trading were relatively low for the region. In 2003 the financial and insurance sector accounted for an estimated 10.3% of GDP.

In conformity with the terms of agreements reached with the IMF in 1985, 1990, 1992 and 1996, successive Governments reduced the fiscal deficit, to 1.4% of GDP in 1998. Notably, the number of public-sector employees was reduced from 240,000 in 1990 to 150,000 in 1997, the social-security system was reformed in 1996, taxes were increased and performance-related pay was introduced for civil servants. Further fiscal measures were attached to IMF stand-by credits approved in January 1999 and May 2000, but with economic decline the fiscal deficit soared to 4.3% of GDP in 2001. According to the IMF, in 2003 the fiscal deficit stood at 3.2%. However, according to official budget figures, the fiscal deficit stood at 7.6% of GDP in the same year. It was expected to fall in 2004 to 2.2% of GDP, following the impressive economic growth recorded in the previous year. As part of an agreement with the IMF for a stand-by loan worth some US $1,130m., formally approved in July 2005, a primary fiscal surplus equivalent to 3.5% of GDP was set as a target for 2005.

During the 1990s successive Governments also made considerable efforts to counter inflation, which had risen as high as 112.5% in 1990. The rate was brought down to 44.7% in 1994, mainly through the use of exchange controls and public-sector wage restraints. In 1993 the peso uruguayo replaced the peso nuevo at the rate of one to 1,000. The currency entered a floating exchange-rate system, supervised by the Government at a monthly devaluation rate of approximately 1%. The inflation rate continued to fall, to 4.8% in 2000 and 4.4% in 2001, assisted by the recession in Brazil and Argentina. Consumer prices soared in 2002 and 2003, however, in the wake of the financial crisis, averaging 13.9% and 19.4%, respectively. In 2004 the average annual increase in consumer prices was 9.2%. In 2001 and January 2002 the Batlle administration adjusted the 'crawling peg' on the currency as a result of severe economic problems. New taxes were introduced, and value-added tax (VAT) was extended to the health sector, leaving only a few basic foodstuffs exempt. Banking reforms were adopted, and in early 2002 the Government intervened in two ailing banks (Banco Comercial, where considerable funds had been misappropriated, and Banco de Crédito). In March and June agreements were reached with the IMF for loans of US $781m. and $1,500m., respectively, while new legislation was introduced to increase taxation on salaries and pensions.

The Uruguayan economy experienced additional difficulties following the Argentine default of January 2002, as Argentine citizens began to access their assets in Uruguayan banks. In May this gentle flow turned into a flood when the Argentine Government drastically restricted monthly withdrawals from Argentine banks, triggering a run on Uruguay's banking sector. The Government's decision to float the peso on 19 July increased the pressure on the banks, and 10 days later the Government closed all banks for four days and suspended four banks (including the two already under its administration); national reserves had fallen to US $655m. from $3,000m. in January. During the bank closure, legislation was passed to prevent customers from accessing long-term foreign-exchange deposits at the two state banks for periods ranging from one to three years. The banks reopened with the assistance of an emergency loan of $1,500m. from the US Administration, which also facilitated the disbursement of IMF and other international funds. By the end of 2004 total international reserves were estimated at $2,512m., compared with $772m. at the end of 2002.

The effective devaluation of the peso had a dramatic impact on the Government's economic targets, particularly on the national debt, and for several months it was anticipated that the country might be obliged to default. Before the flotation the exchange rate had been 17 pesos to the US dollar, and the rate dropped to 35 pesos before stabilizing at around 28 pesos. In December 2002 Uruguay failed to meet IMF requirements; the Fund expressed particular concern that the future of the four suspended banks had not been resolved. The two sides held further discussions to allow for the release of US $3,000m. in IMF funding. The IMF agreed a plan in February 2003, which included a commitment from the Uruguayan Government to implement banking reforms and restructure at least part of the debt. In March, in compliance with IMF demands, a new state-owned bank (Nuevo Banco Comercial—NBC) was created from the merger of three of the suspended banks. In May 2005 the Government announced plans to privatize the NBC by the year's end. In many respects the economy began to recover faster than anticipated, and after an IMF team visited the country in June 2003 it revised its forecasts favourably. Later that month Uruguay regained its investment-grade rating.

## DEBT

Uruguay's total foreign debt rose from US $2,156m. in 1981 to $4,279m. by the end of 1987; nearly 90% of it was publicly held.

A series of agreements with creditors reduced interest payments, and by 1990 the total debt had been reduced to $3,707m. By 1997 the country's debt profile had improved sufficiently for it to be accorded an investment-grade rating. The picture changed drastically in 2002. The investment-grade rating was forfeited in February, and the flotation of the peso increased the debt from $8,500m. in July to $10,000m. in August. In 2003 the total reached $11,764m., and debt payments due during 2003 amounted to some $1,500m. In relation to GDP, the debt rose from the equivalent of 38% in 1998 to some 90% in 2002 and 2003. In May 2003 the Government reached a debt-exchange agreement on about one-half of the current debt, with a grace period of up to five years. Total debt reached $11,764m. at the end of 2003, some $7,903m. of which was long-term debt. In that year the cost of debt-servicing was equivalent to 26.3% of the value of exports of goods and services.

## FOREIGN TRADE AND THE BALANCE OF PAYMENTS

Uruguay's traditional exports were meat, wool and other unprocessed animal products, which, until the mid-1970s, provided about 80% of the country's total export earnings. By the 1990s that proportion had fallen to 20%–25%, although these primary commodities were still crucial to overall export performance. Non-traditional exports included textiles, leather goods, fish and rice, and by the 1990s hydroelectric power and tourism had also become major contributors. Despite this growth, the trade balance, which had recorded surpluses in 1990 and 1991, registered a deficit in the rest of the 1990s, chiefly because of rising imports of capital goods. In 1999 the imbalance increased to US $897m., largely as a result of the Brazilian devaluation; reduced demand in Brazil helped to cut exports by 24%, while imports declined by only 14%. In 2000 the trade deficit reached $927m., but there was a major change with the recession of 2002: a sharp reduction in imports resulted in a visible trade surplus of $48.3m. This surplus rose to $181.8m. in 2003.

The main export markets for Uruguayan meat were Brazil, the EU and Middle Eastern countries, and during the 1990s trade with the USA grew by 50%. Bilateral trade agreements were concluded with Argentina and Brazil in the 1970s, and by 1992 these two countries accounted for almost 43% of Uruguay's imports and about one-third of the country's exports. Recession in Brazil and then Argentina helped to boost other trading links, and in 2003 the leading market for exports was the EU (22.9%), followed by Brazil (21.4%) and Asia and the Middle East (14.1%).

After more than a decade in deficit, the current account of the balance of payments was in surplus every year from 1986 to 1991, with the exception of 1987. From 1992, however, the current account stayed in deficit as a result of the deteriorating trade balance, rising to US $566.2m. in 2000, before declining to $487.7m. in 2001. The balance regained a surplus (of $322.2m.) in 2002, equivalent to 2.1% of GDP. In 2003 this surplus declined to $52.1m.

## MERCOSUR

Argentina, Brazil, Paraguay and Uruguay signed the Treaty of Asunción in March 1991, forming a regional common market, Mercosur, with its headquarters in Montevideo, which came into effect on 1 January 1995. In preparation for the implementation of Mercosur, Uruguay continued to liberalize its trade, reducing its three-tier tariffs in March 1992 and again in January 1993. The country generally recorded a trade deficit within Mercosur, and closer ties increased Uruguay's vulnerability to economic events in the other member countries, notably the Argentine economic crises in 1995 and 2001, austerity measures in Brazil in 1996 and the Brazilian devaluation in January 1999. Membership of the common market did, however, bring economic advantages through the presence of the Mercosur secretariat in Uruguay, reduced tariffs and interest in investment from a number of multinational companies. By the end of the 1990s most duties between the member countries had been eliminated and a common external tariff had been established for trade outside Mercosur. By the time of the Mercosur Heads of State Summit in December 2000 there were evident strains within the organization. In June 2005 the Minister of Economy and Finance, Danilo Astori, expressed grave concerns that progress in free trade and customs union within Mercosur was being reversed and that macroeconomic co-ordination had been extremely poor.

## TOURISM

Tourism and financial services had become the major contributors to the country's balance of payments by the mid-1990s, with tourism revenues outstripping receipts from exports of wool and meat. Uruguay's principal attractions for tourists were the fashionable resort of Punta del Este and the chain of beaches between it and Montevideo. The sector expanded considerably in the 1980s, with the number of summer-season visitors rising from just over 316,000 in 1980/81, to nearly 690,000 in 1990/91. Although the 1990s saw an increase in visitors from other neighbouring countries, and also the USA and Europe, in 2003 Argentine tourists still accounted for 57% of total arrivals. Economic recession in Argentina and Brazil severely affected tourism receipts from 1999; in January 2002 the number of Argentine tourists in Punta del Este was estimated at only 10% of its usual level, and this downward trend continued. The total number of visitors declined steadily, from 2.3m. in 1998 to 1.4m. in 2002. A slight increase, to 1.5m. visitors, was recorded in 2003. Receipts from tourism also decreased, from US $700m. in 2001 to $406m. in 2003.

## THE ENVIRONMENT

By the mid-1990s certain recreational areas in Montevideo were severely damaged by industrial pollution from meat-processing and tannery plants in the Miguelete river. There was also concern over the possible consequences of a proposed five-nation (the Mercosur countries and Bolivia) scheme to dredge a waterway to allow large shipping into the interior up the River Plate Estuary as far as the Bolivian border. In December 2000 a Brazilian judge suspended licences for dredging and construction projects along the waterway out of concern for wildlife in the Pantanal, a 140,000 sq km area in western Brazil. A petroleum spill, which occurred 20 km off the coast of the resort of Punta del Este in February 1998, polluted not only the beach, but also affected local seal and sea-lion colonies. The oil leak halted the harvesting of mussels, which accounted for a major part of the country's shellfish catch. Like other Latin American countries, Uruguay suffered from abnormal weather conditions caused by El Niño in 1997–98, with severe flooding in the west of the country, and considerable damage to the rice crop. In mid-2003 Uruguay, Brazil, Argentina and Paraguay began work on an agreement covering the future usage and protection of the Guaraní aquifer, with its administrative headquarters in Montevideo.

## CONCLUSION

On taking office in March 2000 President Batlle inherited an economy with the advantages of the lowest rate of inflation for over 50 years, a substantially reduced budget deficit, and the future benefits of the pension reform for budget resources. However, the lower inflation rate was also a measure of the economic difficulties facing the country, especially the increase in the price of oil, the effects of economic difficulties in Brazil, with their consequent repercussions in Argentina, and the weak international market for Uruguay's main exports.

The vulnerability of the economy to the misfortunes of neighbouring countries was highlighted by the crisis of 2002. Obliged to turn to the IMF, Uruguay was, from mid-2003, showing signs of recovery, fuelled by an increase in trade from Argentina and Brazil. In March 2005 Dr Tabaré Ramón Vázquez Rosas assumed the presidency to form a Government of the left-wing Encuentro Progresista—Frente Amplio (EP—FA) coalition. International financial markets initially reacted cautiously to the incoming administration, although Vázquez's appointment of orthodox economist Danilo Astori as finance minister gave some reassurance that Uruguay's international obligations would continue to be met and that the process of economic liberalization would continue. The economic growth of 12.3% in 2004 aided President Vázquez, who had committed to an ambitious plan of social spending with the aim of eradicating poverty. In July 2005 the IMF formally approved a new stand-by arrangement with Uruguay worth some US $1,130m. Implementation of the arrangement was contingent on the successful continuation of reforms, including, notably, rationalization of the public sector and of government spending, liberalization of the banking sector, continuation of the privatization programme and the reduction of the national debt (to less than an equivalent of 60% of GDP) by 2009. It was thought, however, that some of the reforms, especially those regarding privatization and social spending, would meet strong opposition from within the ruling EP—FA coalition. Growth of 6% was predicted for 2005.

# Statistical Survey

Sources (unless otherwise stated): Instituto Nacional de Estadística, Río Negro 1520, 11100 Montevideo; tel. (2) 9027303; internet www.ine.gub.uy; Banco Central del Uruguay, Avda Juan P. Fabini, esq. Florida 777, Casilla 1467, 11100 Montevideo; tel. (2) 9017112; fax (2) 9021634; e-mail info@bcu.gub.uy; internet www.bcu.gub.uy; Cámara Nacional de Comercio y Servicios del Uruguay, Edif. Bolsa de Comercio, Rincón 454, 2°, Casilla 1000, 11000 Montevideo; tel. (2) 9161277; fax (2) 9161243; e-mail info@cncs.com.uy; internet www.cncs.com.uy.

## Area and Population

### AREA, POPULATION AND DENSITY

| | |
|---|---:|
| Area (sq km) | |
| Land area | 175,016 |
| Inland water | 1,199 |
| Total | 176,215* |
| Population (census results)† | |
| 22 May 1996 | 3,163,763 |
| May–July 2004 | |
| Males | 1,565,533 |
| Females | 1,675,470 |
| Total | 3,241,003 |
| Density (per sq km) at 2004 census | 18.4 |

* 68,037 sq miles.
† Excluding adjustment for underenumeration.

### DEPARTMENTS
(population at 2004 census)

| | Area (sq km) | Population | Density (per sq km) | Capital |
|---|---:|---:|---:|---|
| Artigas | 11,928 | 78,019 | 6.6 | Artigas |
| Canelones | 4,536 | 485,240 | 107.0 | Canelones |
| Cerro Largo | 13,648 | 86,564 | 6.3 | Melo |
| Colonia | 6,106 | 119,266 | 19.5 | Colonia del Sacramento |
| Durazno | 11,643 | 58,859 | 5.1 | Durazno |
| Flores | 5,144 | 25,104 | 4.9 | Trinidad |
| Florida | 10,417 | 68,181 | 6.5 | Florida |
| Lavalleja | 10,016 | 60,925 | 6.1 | Minas |
| Maldonado | 4,793 | 140,192 | 29.2 | Maldonado |
| Montevideo | 530 | 1,325,968 | 2,501.8 | Montevideo |
| Paysandú | 13,922 | 113,244 | 8.1 | Paysandú |
| Río Negro | 9,282 | 53,989 | 5.8 | Fray Bentos |
| Rivera | 9,370 | 104,921 | 11.2 | Rivera |
| Rocha | 10,551 | 69,937 | 6.6 | Rocha |
| Salto | 14,163 | 123,120 | 8.7 | Salto |
| San José | 4,992 | 103,104 | 20.7 | San José de Mayo |
| Soriano | 9,008 | 84,563 | 9.4 | Mercedes |
| Tacuarembó | 15,438 | 90,489 | 5.9 | Tacuarembó |
| Treinta y Tres | 9,529 | 49,318 | 5.2 | Treinta y Tres |
| **Total** | **175,016*** | **3,241,003** | **18.5** | |

* Land area only.

### PRINCIPAL TOWNS
(population at 22 May 1996 census)

| | | | | |
|---|---:|---|---:|
| Montevideo (capital) | 1,378,707 | Mercedes | 50,800 |
| Salto | 93,420 | Maldonado | 50,420 |
| Paysandú | 84,160 | Melo | 47,160 |
| Las Piedras | 66,100 | Tacuarembó | 42,580 |
| Rivera | 63,370 | | |

**2003** (estimate): Montevideo 1,340,560 (Source: UN, *World Urbanization Prospects: The 2003 Revision*).

### BIRTHS, MARRIAGES AND DEATHS

| | Registered live births* | | Registered deaths | |
|---|---:|---:|---:|---:|
| | Number | Rate (per 1,000) | Number | Rate (per 1,000) |
| 1996 | 56,928 | 17.8 | 31,108 | 9.5 |
| 1997 | 58,032 | 18.0 | 30,459 | 9.3 |
| 1998 | 54,760 | 16.6 | 32,082 | 9.8 |
| 1999 | 54,055 | 16.3 | 32,430 | 9.8 |
| 2000 | 52,720 | 15.9 | 30,456† | 9.2† |
| 2001 | 51,959 | 15.5 | 31,228† | 9.3† |
| 2002 | 51,997 | 15.5 | 31,628† | 9.4† |
| 2003† | 50,538 | 15.0 | 32,427 | 9.6 |

* Data are tabulated by year of registration rather than by year of occurrence (Source: mainly UN, *Demographic Yearbook*).
† Preliminary.

**Registered marriages:** 13,888 (marriage rate 4.2 per 1,000) in 2000; 13,988 in 2001; 14,073 in 2002.

**Expectation of life** (WHO estimates, years at birth): 75 (males 71; females 80) in 2003 (Source: WHO, *World Health Report*).

### ECONOMICALLY ACTIVE POPULATION
(ISIC major divisions, '000 persons aged 14 years and over, urban areas)

| | 2001 | 2002 | 2003 |
|---|---:|---:|---:|
| Agriculture, hunting, forestry and fishing | 45.4 | 43.7 | 46.9 |
| Mining and quarrying | 1.3 | 1.2 | 1.2 |
| Manufacturing (incl. electricity, gas and water) | 167.1 | 154.2 | 151.1 |
| Construction | 87.9 | 77.4 | 69.6 |
| Trade, restaurants, hotels and repair of vehicles and household goods | 240.8 | 228.9 | 225.4 |
| Transport, storage and communications | 66.8 | 62.4 | 61.1 |
| Financing, insurance, real estate and business services | 97.4 | 96.5 | 91.0 |
| Public administration and defence, compulsory social security | 85.1 | 86.9 | 91.2 |
| Education | 57.2 | 62.2 | 61.6 |
| Health and social work | 72.5 | 76.7 | 76.7 |
| Community, social and personal services | 56.2 | 52.2 | 54.8 |
| Private households with employed persons | 98.5 | 96.0 | 100.9 |
| Activities not adequately defined | — | — | 0.3 |
| **Total employed** | **1,076.2** | **1,038.3** | **1,032.0** |
| Males | 617.7 | 597.9 | 589.7 |
| Females | 458.5 | 440.4 | 442.3 |
| Unemployed | 193.2 | 211.3 | 208.5 |
| **Total labour force** | **1,269.4** | **1,250.1** | **1,240.5** |

Source: ILO.

# URUGUAY

## Health and Welfare

**KEY INDICATORS**

| | |
|---|---|
| Total fertility rate (children per woman, 2003) | 2.3 |
| Under-5 mortality rate (per 1,000 live births, 2003) | 14 |
| HIV/AIDS (% of persons aged 15–49, 2003) | 0.3 |
| Physicians (per 1,000 head, 2002) | 3.65 |
| Hospital beds (per 1,000 head, 1996) | 4.39 |
| Health expenditure (2002): US $ per head (PPP) | 805 |
| Health expenditure (2002): % of GDP | 10.0 |
| Health expenditure (2002): public (% of total) | 29.0 |
| Access to water (% of persons, 2002) | 98 |
| Access to sanitation (% of persons, 2002) | 94 |
| Human Development Index (2002): ranking | 46 |
| Human Development Index (2002): value | 0.833 |

For sources and definitions, see explanatory note on p. vi.

## Agriculture

**PRINCIPAL CROPS**
('000 metric tons)

| | 2001 | 2002 | 2003 |
|---|---|---|---|
| Wheat | 143.6 | 205.8 | 326.0 |
| Rice (paddy) | 1,030.2 | 939.5 | 1,250.0 |
| Barley | 117.7 | 217.4 | 323.7 |
| Maize | 266.8 | 163.4 | 178.5 |
| Oats | 45.0* | 50.0* | 40.0 |
| Sorghum | 142.6 | 62.0 | 60.2 |
| Potatoes | 121.1 | 147.0 | 151.0 |
| Sweet potatoes | 63.0* | 63.5* | 65.0 |
| Sugar cane | 176.5 | 187.7 | 165.0 |
| Sunflower seed | 58.2 | 150.3 | 234.0 |
| Tomatoes | 37.0* | 41.2 | 36.0 |
| Dry onions | 20.5* | 26.9 | 25.0 |
| Carrots | 20.5* | 32.5 | 37.0 |
| Oranges | 189.0 | 115.8 | 122.0 |
| Tangerines, mandarins, clementines and satsumas | 97.0 | 74.5 | 75.0 |
| Lemons and limes | 46.5 | 38.5 | 40.0 |
| Apples | 53.8 | 73.8 | 72.0 |
| Pears | 17.7 | 14.2 | 16.0 |
| Peaches and nectarines | 13.7 | 10.6 | 11.0 |
| Grapes | 113.4 | 93.8 | 103.0 |

* FAO estimate.
Source: FAO.

**LIVESTOCK**
('000 head, year ending September)

| | 2001 | 2002 | 2003 |
|---|---|---|---|
| Cattle | 10,595 | 11,274 | 11,689 |
| Sheep | 12,083 | 10,986 | 9,780 |
| Pigs | 282* | 270 | 240 |
| Horses* | 590 | 390 | 380 |
| Chickens | 13,100* | 13,200* | 13,300 |

* FAO estimate(s).
Source: FAO.

**LIVESTOCK PRODUCTS**
('000 metric tons)

| | 2001 | 2002 | 2003 |
|---|---|---|---|
| Beef and veal | 317.0* | 411.8 | 424.2 |
| Mutton and lamb | 51.2† | 48.0† | 41.5 |
| Pigmeat | 22.6 | 19.5 | 16.8 |
| Poultry meat† | 52.7 | 53.2 | 53.7 |
| Cows' milk | 1,495.0† | 1,490.0† | 1,495.0 |
| Eggs | 31.6* | 32.5† | 42.5† |
| Wool: greasy | 56.7 | 41.3 | 43.0† |
| Cattle hides† | 41.0 | 52.0 | 55.2 |
| Sheepskins† | 18.5 | 17.2 | 17.2 |

* Unofficial figure.
† FAO estimate(s).
Source: FAO.

## Forestry

**ROUNDWOOD REMOVALS**
('000 cubic metres, excl. bark)

| | 2001 | 2002 | 2003 |
|---|---|---|---|
| Sawlogs and veneer logs | 548 | 591 | 485 |
| Pulpwood | 960 | 1,151 | 1,637 |
| Other industrial wood | 90 | 90 | 10 |
| Fuel wood | 1,387 | 1,607 | 1,607 |
| **Total** | **2,985** | **3,439** | **3,739** |

Source: FAO.

**SAWNWOOD PRODUCTION**
('000 cubic metres, incl. railway sleepers)

| | 2001 | 2002 | 2003 |
|---|---|---|---|
| Coniferous (softwood) | 100 | 120 | 82 |
| Broadleaved (hardwood) | 103 | 104 | 148 |
| **Total** | **203** | **224** | **230** |

Source: FAO.

## Fishing

('000 metric tons, live weight)

| | 2001 | 2002 | 2003 |
|---|---|---|---|
| Capture | 105.1 | 108.8 | 116.9 |
| Argentine hake | 27.7 | 32.2 | 35.2 |
| Striped weakfish | 10.9 | 9.0 | 7.1 |
| Whitemouth croaker | 27.3 | 26.7 | 30.7 |
| Patagonian toothfish | 7.8 | 6.5 | 4.9 |
| Argentine shortfin squid | 7.4 | 11.8 | 6.4 |
| Red crab | 2.1 | 2.0 | 3.0 |
| Aquaculture | 0.0 | 0.0 | 0.0 |
| **Total catch** | **105.1** | **108.8** | **116.9** |

Source: FAO.

## Mining

('000 metric tons, unless otherwise indicated)

| | 2001 | 2002 | 2003* |
|---|---|---|---|
| Gold (kg) | 2,083 | 2,079 | 1,730 |
| Gypsum | 1,127 | 1,130* | 1,130 |
| Feldspar (metric tons) | 4,722 | 4,700* | 4,700 |

* Estimate(s).
Source: US Geological Survey.

## Industry

**SELECTED PRODUCTS**
('000 metric tons, unless otherwise indicated)

| | 2001 | 2002 | 2003 |
|---|---|---|---|
| Raw sugar* | 7.0 | 6.7 | 6.3 |
| Wine | 87.3 | 71.4 | 72.0† |
| Cigarettes (million) | 9,616 | n.a. | n.a. |
| Motor spirit (petrol) ('000 barrels)‡ | 2,200 | 2,200 | 2,200 |
| Kerosene ('000 barrels)‡ | 500 | 500 | 500 |
| Distillate fuel oils ('000 barrels)‡ | 4,100 | 4,100 | 4,200 |
| Residual fuel oils ('000 barrels)‡ | 3,600 | 3,600 | 3,650 |
| Cement (hydraulic) | 1,015 | 1,000‡ | 1,050‡ |
| Electric energy (million kWh) | 9,259 | n.a. | n.a. |

* Unofficial figures.
† FAO estimate.
‡ US Geological Survey estimate(s).

Sources: FAO; US Geological Survey; UN, *Industrial Commodity Statistics Yearbook*.

URUGUAY

*Statistical Survey*

# Finance

## CURRENCY AND EXCHANGE RATES

**Monetary Units**
100 centésimos = 1 peso uruguayo.

**Sterling, Dollar and Euro Equivalents** (31 May 2005)
£1 sterling = 43.6344 pesos;
US $1 = 24.0000 pesos;
€1 = 29.5944 pesos;
1,000 pesos uruguayos = £22.92 = $41.67 = €33.79.

**Average Exchange Rate** (pesos per US $)
2002   21.2570
2003   28.2087
2004   28.7037

Note: On 1 March 1993 a new currency, the peso uruguayo (equivalent to 1,000 former new pesos), was introduced.

## BUDGET*
(million pesos uruguayos)

| Revenue | 2001 | 2002 | 2003 |
|---|---|---|---|
| Tax revenue | 40,923 | 43,825 | 54,921 |
| Taxes on income, profits, etc. | 9,539 | 11,568 | 13,560 |
| Individual taxes | 4,198 | 6,307 | 8,065 |
| Corporate taxes | 5,149 | 5,056 | 5,153 |
| Taxes on property | 3,913 | 5,162 | 7,095 |
| Social security contributions | 505 | 84 | — |
| Domestic taxes on goods and services | 23,820 | 23,791 | 30,309 |
| General sales, take-over or value-added tax | 15,443 | 15,827 | 21,106 |
| Taxes on international trade and transactions | 1,795 | 1,364 | 1,665 |
| Taxes on leisure activities | 470 | 120 | 26 |
| Other taxes and rates | 881 | 1,736 | 2,266 |
| Non-tax revenue | 3,645 | 4,410 | 7,034 |
| Transfers | 3,226 | 4,504 | 2,703 |
| Other income | 307 | 358 | 899 |
| **Total** | 48,102 | 53,098 | 65,558 |

| Expenditure | 2001 | 2002 | 2003 |
|---|---|---|---|
| General public services | 16,270 | 17,724 | 18,616 |
| Government administration | 8,551 | 9,819 | 9,564 |
| Defence | 3,303 | 3,332 | 4,157 |
| Public order and safety | 4,416 | 4,573 | 4,895 |
| Special and community services | 31,903 | 33,808 | 36,369 |
| Education | 8,597 | 8,890 | 10,558 |
| Health | 5,228 | 5,336 | 6,117 |
| Social security and welfare | 16,155 | 17,837 | 17,986 |
| Housing and community amenities | 1,260 | 1,126 | 1,084 |
| Recreational, cultural and religious affairs | 664 | 620 | 625 |
| Economic affairs and services | 4,696 | 4,318 | 8,140 |
| Fuel and energy | 178 | 143 | 175 |
| Agriculture, fishing, forestry and hunting | 1,008 | 1,057 | 1,096 |
| Mining and mineral resources | 244 | 221 | 231 |
| Transport and communications | 2,737 | 2,306 | 2,690 |
| Other economic services | 528 | 591 | 3,948 |
| Debt-servicing and governmental transfers | 8,175 | 13,893 | 26,263 |
| **Total** | 61,044 | 69,744 | 89,389 |

* Figures represent the consolidated accounts of the central Government, which include social security revenue and expenditure.

## INTERNATIONAL RESERVES
(US $ million at 31 December)

| | 2002 | 2003 | 2004 |
|---|---|---|---|
| Gold | 3 | 3 | 4 |
| IMF special drawing rights | 6 | 4 | 1 |
| Reserve position in the IMF | — | — | — |
| Foreign exchange | 763 | 2,079 | 2,507 |
| **Total** | 772 | 2,086 | 2,512 |

Source: IMF, *International Financial Statistics*.

## MONEY SUPPLY
(million pesos uruguayos at 31 December)

| | 2002 | 2003 | 2004 |
|---|---|---|---|
| Currency outside banks | 7,673.4 | 9,440.5 | 10,803.7 |
| Demand deposits at commercial banks | 6,556.6 | 9,706.8 | 10,928.0 |
| **Total money** (incl. others) | 14,278.4 | 19,177.4 | 22,225.4 |

Source: IMF, *International Financial Statistics*.

## COST OF LIVING
(Consumer Price Index for Montevideo; base: 2000 = 100)

| | 2001 | 2002 | 2003 |
|---|---|---|---|
| Food | 103.1 | 117.2 | 142.5 |
| Fuel and light | 105.4 | 125.4 | 175.4 |
| Clothing and footwear | 100.1 | 108.0 | 127.5 |
| Rent | 101.9 | 103.0 | 100.8 |
| **All items** | 104.4 | 118.9 | 142.0 |

**2004:** Food 159.2; All items 155.0.

Source: ILO.

## NATIONAL ACCOUNTS
(million pesos uruguayos at current prices)

**Expenditure on the Gross Domestic Product**

| | 2002 | 2003 | 2004 |
|---|---|---|---|
| Government final consumption expenditure | 33,622 | 35,833 | 41,137 |
| Private final consumption expenditure | 192,167 | 235,348 | 281,324 |
| Increase in stocks | 3,707 | 9,947 | 7,173 |
| Gross fixed capital formation | 26,360 | 29,785 | 43,222 |
| **Total domestic expenditure** | 255,856 | 310,913 | 372,856 |
| Exports of goods and services | 57,325 | 82,301 | 112,461 |
| *Less* Imports of goods and services | 52,214 | 77,535 | 105,999 |
| **GDP in purchasers' values** | 260,967 | 315,681 | 379,317 |
| **GDP at constant 1983 prices** | 246 | 252 | 283 |

Source: IMF, *International Financial Statistics*.

**Gross Domestic Product by Economic Activity**

| | 2001 | 2002* | 2003* |
|---|---|---|---|
| Agriculture | 14,709.6 | 23,518.2 | 39,320.5 |
| Fishing | 409.5 | 724.2 | 1,062.8 |
| Mining and quarrying | 668.7 | 553.6 | 604.7 |
| Manufacturing | 40,284.0 | 45,599.8 | 59,620.1 |
| Electricity, gas and water | 10,919.9 | 12,567.5 | 14,956.8 |
| Construction | 13,467.5 | 10,994.2 | 10,924.6 |
| Trade, restaurants and hotels | 32,052.3 | 31,605.2 | 37,874.8 |
| Transport, storage and communications | 22,513.8 | 24,019.5 | 30,494.4 |
| Finance and insurance | 29,318.0 | 29,818.3 | 34,087.1 |
| Real estate and business services | 43,799.2 | 44,111.6 | 44,046.9 |
| General government services | 25,213.8 | 25,466.1 | 26,993.9 |
| Other community, social and personal services | 28,237.1 | 28,204.0 | 30,730.7 |
| **Sub-total** | 261,593.2 | 277,182.3 | 330,717.4 |
| Import duties | 10,944.3 | 9,359.2 | 13,574.4 |
| *Less* Imputed bank service charge | 25,326.1 | 25,574.8 | 28,845.9 |
| **GDP in purchasers' values** | 247,211.4 | 260,966.7 | 315,445.9 |

* Provisional figures.

# URUGUAY

*Statistical Survey*

## BALANCE OF PAYMENTS
(US $ million)

|  | 2001 | 2002 | 2003 |
|---|---|---|---|
| Exports of goods f.o.b. | 2,139.4 | 1,922.1 | 2,273.3 |
| Imports of goods f.o.b. | −2,914.7 | −1,873.8 | −2,091.5 |
| **Trade balance** | **−775.3** | **48.3** | **181.8** |
| Exports of services | 1,122.5 | 753.7 | 777.9 |
| Imports of services | −801.4 | −600.3 | −615.3 |
| **Balance on goods and services** | **−454.2** | **201.7** | **344.4** |
| Other income received | 832.0 | 453.4 | 237.8 |
| Other income paid | −895.2 | −405.1 | −601.7 |
| **Balance on goods, services and income** | **−517.4** | **250.0** | **−19.5** |
| Current transfers received | 48.0 | 83.7 | 79.8 |
| Current transfers paid | −18.3 | −11.5 | −8.2 |
| **Current balance** | **−487.7** | **322.2** | **52.1** |
| Direct investment abroad | −6.2 | −53.8 | −3.7 |
| Direct investment from abroad | 320.3 | 174.6 | 274.6 |
| Portfolio investment assets | 236.7 | 95.2 | −521.7 |
| Portfolio investment liabilities | 264.4 | 204.5 | 22.9 |
| Other investment assets | −2,275.2 | 1,825.5 | −1,252.8 |
| Other investment liabilities | 1,966.6 | −4,173.5 | 1,138.1 |
| Net errors and omissions | 285.1 | −2,291.8 | 1,248.5 |
| **Overall balance** | **304.0** | **−3,897.1** | **957.9** |

Source: IMF, *International Financial Statistics*.

# External Trade

## PRINCIPAL COMMODITIES
(US $ million)

| Imports | 2001 | 2002 | 2003 |
|---|---|---|---|
| Live animals and animal products | 30.3 | 26.2 | 29.7 |
| Plants and plant products (incl. cereals) | 92.4 | 118.6 | 114.5 |
| Food and beverages (incl. tobacco) | 225.1 | 137.4 | 126.7 |
| Mineral products | 389.9 | 306.8 | 501.8 |
| Petroleum and derivatives | 375.6 | 295.5 | 487.4 |
| Chemical products | 469.2 | 342.3 | 380.8 |
| Plastics, rubber and manufactures thereof | 227.5 | 158.8 | 198.3 |
| Hides and leather goods | 91.1 | 63.0 | 60.0 |
| Wood, plant fibres and manufactures thereof | 23.5 | 12.2 | 10.4 |
| Wood pulp, paper and cardboard | 131.6 | 81.2 | 72.9 |
| Textiles | 156.0 | 97.8 | 120.3 |
| Stone, cement, plaster, glass and manufactures thereof | 53.1 | 29.4 | 28.5 |
| Base metals and manufactures thereof | 138.7 | 77.4 | 83.4 |
| Machinery and electronic goods | 574.4 | 296.9 | 271.1 |
| Transportation materials (incl. vehicles) | 249.9 | 97.3 | 77.4 |
| **Total** (incl. others) | **3,060.8** | **1,964.3** | **2,190.4** |

| Exports | 2001 | 2002 | 2003 |
|---|---|---|---|
| Live animals and animal derivatives | 514.0 | 543.4 | 687.9 |
| Plants and plant derivatives (incl. cereals) | 292.8 | 270.1 | 385.4 |
| Mineral products | 51.4 | 16.5 | 38.4 |
| Chemicals and manufactures thereof | 101.3 | 84.2 | 93.6 |
| Plastics, rubber and manufactures thereof | 90.0 | 84.7 | 104.7 |
| Hides and leather goods | 276.1 | 251.2 | 267.9 |
| Wood, plant fibres and manufactures thereof | 49.0 | 51.9 | 71.9 |
| Wood pulp, paper and cardboard | 72.4 | 53.2 | 49.3 |
| Textiles | 253.1 | 221.5 | 229.9 |
| Transportation goods and materials (incl. vehicles) | 112.3 | 64.3 | 35.7 |
| **Total** (incl. others) | **2,057.6** | **1,861.0** | **2,198.0** |

## PRINCIPAL TRADING PARTNERS
(US $ million)

| Imports c.i.f. | 2001 | 2002 | 2003 |
|---|---|---|---|
| Argentina | 705.8 | 540.6 | 571.7 |
| Brazil | 625.9 | 389.6 | 459.8 |
| Chile | 67.1 | 40.1 | 41.5 |
| China, People's Republic | 121.5 | 75.3 | 86.0 |
| France (incl. Monaco) | 128.6 | 65.9 | 58.5 |
| Germany | 93.3 | 81.2 | 61.7 |
| Italy | 100.3 | 46.4 | 50.2 |
| Japan | 49.7 | 26.7 | 28.1 |
| Korea, Republic | 45.2 | 32.5 | 36.7 |
| Mexico | 37.1 | 23.7 | 20.5 |
| Nigeria | 36.8 | — | 27.3 |
| Russia | 25.9 | 111.2 | 257.0 |
| Spain | 100.8 | 61.9 | 42.8 |
| Switzerland | 28.5 | 20.7 | 21.2 |
| Taiwan | 22.4 | 12.8 | 20.4 |
| United Kingdom | 38.3 | 24.6 | 25.0 |
| USA | 271.4 | 164.3 | 165.6 |
| Venezuela | 174.5 | 44.2 | 1.1 |
| **Total** (incl. others) | **3,060.8** | **1,964.3** | **2,190.4** |

| Exports f.o.b. | 2001 | 2002 | 2003 |
|---|---|---|---|
| Argentina | 316.4 | 113.3 | 154.9 |
| Brazil | 440.7 | 431.8 | 470.8 |
| Canada | 64.9 | 27.1 | 86.4 |
| Chile | 44.4 | 53.0 | 71.5 |
| China, People's Republic | 102.9 | 103.6 | 95.2 |
| Germany | 96.5 | 108.5 | 145.1 |
| Hong Kong | 24.4 | 14.6 | 14.0 |
| Iran | 24.3 | 38.1 | 15.8 |
| Israel | 45.5 | 51.4 | 40.1 |
| Italy | 70.9 | 88.2 | 88.9 |
| Japan | 12.1 | 14.0 | 12.4 |
| Malaysia | 32.7 | 29.5 | — |
| Mexico | 76.2 | 71.3 | 90.8 |
| Netherlands | 32.2 | 38.6 | 37.1 |
| Paraguay | 82.8 | 61.7 | 47.8 |
| Spain | 64.2 | 67.8 | 73.5 |
| United Kingdom | 65.6 | 78.6 | 78.4 |
| USA | 176.3 | 140.8 | 233.4 |
| Venezuela | 22.4 | 11.2 | 3.9 |
| **Total** (incl. others) | **2,057.6** | **1,861.0** | **2,198.0** |

Source: UN, *International Trade Statistics Yearbook*.

# Transport

## RAILWAYS
(traffic)

|  | 1998 | 1999 | 2000 |
|---|---|---|---|
| Passenger-km (million) | 14 | 10 | 9 |
| Net ton-km (million) | 244 | 272 | 239 |

Sources: UN, *Statistical Yearbook*.

## ROAD TRAFFIC
(motor vehicles in use at 31 December)

|  | 1995 | 1996 | 1997 |
|---|---|---|---|
| Passenger cars | 464,547 | 485,109 | 516,889 |
| Buses and coaches | 4,409 | 4,752 | 4,984 |
| Lorries and vans | 41,417 | 43,656 | 45,280 |
| Road tractors | 12,511 | 14,628 | 15,514 |
| Motorcycles and mopeds | 300,850 | 328,406 | 359,824 |

Source: International Road Federation: *World Road Statistics*.

**2003:** Passenger cars and vans 526,236; Coaches and minibuses 6,681; Lorries and road tractors 53,615; Motorcycles and mopeds 427,286; Taxis and hire cars 5,447.

# URUGUAY

## SHIPPING

**Merchant Fleet**
(registered at 31 December)

|  | 2002 | 2003 | 2004 |
|---|---|---|---|
| Number of vessels | 90 | 90 | 94 |
| Total displacement ('000 grt) | 74.7 | 75.6 | 76.0 |

Source: Lloyd's Register-Fairplay, *World Fleet Statistics*.

## CIVIL AVIATION
(traffic on scheduled services)

|  | 2001 | 2002 | 2003 |
|---|---|---|---|
| Kilometres flown (million) | 7.5 | 7.1 | 8.4 |
| Passengers carried ('000) | 558.6 | 525.0 | 463.9 |
| Passenger-km (million) | 582.3 | 576.5 | 1,028.8 |
| Total ton-km (million) | 13.3 | 11.7 | 23.2 |

Source: UN Economic Commission for Latin America and the Caribbean, *Statistical Yearbook*.

## Tourism

**ARRIVALS BY NATIONALITY***
('000)

|  | 2001 | 2002 | 2003 |
|---|---|---|---|
| Argentina | 1,478.6 | 813.3 | 866.6 |
| Brazil | 121.9 | 118.4 | 151.4 |
| **Total** (incl. others) | 2,136.4 | 1,353.9 | 1,508.1 |

* Figures refer to arrivals at frontiers of visitors from abroad, including Uruguayan nationals permanently resident elsewhere.

**Tourism receipts** (US $ million, incl. passenger transport): 700 in 2001; 409 in 2002; 406 in 2003.

Source: World Tourism Organization.

## Communications Media

|  | 1999 | 2000 | 2001 |
|---|---|---|---|
| Television receivers ('000 in use) | 1,760 | 1,770 | n.a. |
| Telephones ('000 main lines in use) | 896.8 | 929.1 | 950.9 |
| Mobile cellular telephones ('000 subscribers) | 316.1 | 440.2 | 520.0 |
| Personal computers ('000 in use) | 330 | n.a. | 370 |
| Internet users ('000) | 330 | 370 | 400 |

**2003:** Telephones ('000 main lines in use) 938.2; Mobile cellular telephones ('000 subscribers) 507.5.

**Radio receivers** (1997): 1,970,000 in use.

**Book production** (1996): 934 titles.

**Daily newspapers** (1996): 36 (estimated average circulation 950,000).

**Facsimile machines** (1995): 11,000 in use.

Sources: UNESCO, *Statistical Yearbook*; UN, *Statistical Yearbook*; International Telecommunication Union.

## Education
(2003)

|  | Institutions | Teachers | Students |
|---|---|---|---|
| Pre-primary | 1,507 | 3,425 | 103,619 |
| Primary | 2,396 | 16,605 | 354,843 |
| Secondary: general | 413 | 25,168* | 229,404 |
| Secondary: vocational | 124 | n.a. | 68,779 |
| University and equivalent institutions* | 6 | 7,723 | 72,100 |

* Public education only.

**Adult literacy rate** (UNESCO estimates): 97.7% (males 97.3%; females 98.1%) in 2002 (Source: UN Development Programme, *Human Development Report*).

# Directory

## The Constitution

The Constitution of Uruguay was ratified by plebiscite, on 27 November 1966, when the country voted to return to the presidential form of government after 15 years of 'collegiate' government. The main points of the Constitution, as amended in January 1997, are as follows:

### GENERAL PROVISIONS

Uruguay shall have a democratic republican form of government, sovereignty being exercised directly by the Electoral Body in cases of election, by initiative or by referendum, and indirectly by representative powers established by the Constitution, according to the rules set out therein.

There shall be freedom of religion; there is no state religion; property shall be inviolable; there shall be freedom of thought. Anyone may enter Uruguay. There are two forms of citizenship: natural, being persons born in Uruguay or of Uruguayan parents, and legal, being people established in Uruguay with at least three years' residence in the case of those with family, and five years' for those without family. Every citizen has the right and obligation to vote.

### LEGISLATURE

Legislative power is vested in the Congreso (Congress or General Assembly), comprising two houses, which may act separately or together according to the dispositions of the Constitution. It elects in joint session the members of the Supreme Court of Justice, of the Electoral Court, Tribunals, Administrative Litigation and the Accounts Tribunal.

Elections for both houses, the President and the Vice-President shall take place every five years on the last Sunday in October; sessions of the Assembly begin on 1 March each year and last until 15 December (15 September in election years, in which case the new Congress takes office on 15 February). Extraordinary sessions can be convened only in case of extreme urgency.

### CHAMBER OF REPRESENTATIVES

The Chamber of Representatives has 99 members elected by direct suffrage according to the system of proportional representation, with at least two representatives for each Department. The number of representatives can be altered by law by a two-thirds' majority in both houses. Their term of office is five years and they must be over 25 years of age and be natural citizens or legal citizens with five years' exercise of their citizenship. Representatives have the right to bring accusations against any member of the Government or judiciary for violation of the Constitution or any other serious offence.

### SENATE

The Senate comprises 31 members, including the Vice-President, who sits as President of the Senate, and 30 members elected directly by proportional representation on the same lists as the representatives, for a term of five years. They must be natural citizens or legal citizens with seven years' exercise of their rights, and be over 30 years of age. The Senate is responsible for hearing cases brought by the representatives and can deprive a guilty person of a post by a two-thirds' majority.

### THE EXECUTIVE

Executive power is exercised by the President and the Council of Ministers. There is a Vice-President, who is also President of the Congress and of the Senate. The President and Vice-President are directly elected by absolute majority, and remain in office for five years. They must be over 35 years of age and be natural citizens.

The Council of Ministers comprises the office holders in the ministries or their deputies, and is responsible for all acts of government and administration. It is presided over by the President of the Republic, who has a vote.

### THE JUDICIARY

Judicial power is exercised by the five-member Supreme Court of Justice and by Tribunals and local courts; members of the Supreme

URUGUAY

Court must be over 40 years of age and be natural citizens, or legal citizens with 10 years' exercise and 25 years' residence, and must be lawyers of 10 years' standing, eight of them in public or fiscal ministry or judicature. Members serve for 10 years and can be re-elected after a break of five years. The Court nominates all other judges and judicial officials.

## The Government

### HEAD OF STATE

**President:** Dr Tabaré Ramón Vázquez Rosas (took office 1 March 2005).
**Vice-President:** Rodolfo Nin Novoa (AP).

### COUNCIL OF MINISTERS
(July 2005)

**Minister of the Interior:** José Díaz (PS).
**Minister of Foreign Affairs:** Reinaldo Gargano (PS).
**Minister of National Defence:** Azucena Berruti (PS).
**Minister of Economy and Finance:** Danilo Astori (AU).
**Director of Planning and the Budget Office:** Carlos Viera (VA).
**Minister of Industry, Energy and Mining:** Jorge Lepra (Ind.).
**Minister of Livestock, Agriculture and Fishing:** José 'Pepe' Mujica (MPP).
**Minister of Tourism:** Héctor Lescano (AP).
**Minister of Transport and Public Works:** Víctor Rossi (AP).
**Minister of Labour and Social Security:** Eduardo Bonomi (MPP).
**Minister of Education, Culture, Youth and Sport:** Jorge Brovetto (Ind.).
**Minister of Public Health:** Dr María Julia Muñoz (VA).
**Minister of Housing, Territorial Regulation and the Environment:** Mariano Arana (VA).

### MINISTRIES

**Office of the President:** Casa de Gobierno, Edif. Libertad, Avda Luis Alberto de Herrera 3350, Montevideo; tel. (2) 4872110; fax (2) 4809397; internet www.presidencia.gub.uy.

**Ministry of Economy and Finance:** Colonia 1089, 3°, 11100 Montevideo; tel. (2) 9021017; fax (2) 9021277; internet www.mef.gub.uy.

**Ministry of Education, Culture, Youth and Sport:** Reconquista 535, 11000 Montevideo; tel. (2) 9150103; fax (2) 9162632; e-mail webmaster@mec.gub.uy; internet chana.mec.gub.uy.

**Ministry of Foreign Affairs:** Avda 18 de Julio 1205, 11100 Montevideo; tel. (2) 9021007; fax (2) 9021327; e-mail webmaster@mrree.gub.uy; internet www.mrree.gub.uy.

**Ministry of Housing, Territorial Regulation and the Environment:** Zabala 1427, 11000 Montevideo; tel. (2) 9150211; fax (2) 9162914; e-mail computos@mvotma.gub.uy; internet www.mvotma.gub.uy.

**Ministry of Industry, Energy and Mining:** Rincón 723, 11000 Montevideo; tel. (2) 9000231; fax (2) 9021245; internet www.miem.gub.uy; internet www.mintur.gub.uy.

**Ministry of the Interior:** Mercedes 993, 11100 Montevideo; tel. (2) 9089024; fax (2) 9001626; internet www.minterior.gub.uy.

**Ministry of Labour and Social Security:** Juncal 1511, 4°, 11000 Montevideo; tel. (2) 9162681; fax (2) 9162708; e-mail webmtss@mtss.gub.uy; internet www.mtss.gub.uy.

**Ministry of Livestock, Agriculture and Fishing:** Avda Constituyente 1476, 11900 Montevideo; tel. (2) 4104155; fax (2) 4184051; e-mail informatica@mgap.gub.uy; internet www.mgap.gub.uy.

**Ministry of National Defence:** Edif. General Artigas, Avda 8 de Octubre 2628, Montevideo; tel. (2) 4809707; fax (2) 4809397.

**Ministry of Public Health:** Avda 18 de Julio 1892, 11100 Montevideo; tel. (2) 4000101; fax (2) 4088676; e-mail msp@msp.gub.uy; internet www.msp.gub.uy.

**Ministry of Tourism:** Rambla 25 de Agosto, 1825 esq. Yacaré s/n, Montevideo; tel. (2) 1885100; fax (2) 9021624; e-mail webmaster@mintur.gub.uy; internet www.turismo.gub.uy.

**Ministry of Transport and Public Works:** Rincón 575, 4° y 5°, 11000 Montevideo; tel. (2) 9150509; fax (2) 9152438; e-mail apena@mtop.gub.uy; internet www.mtop.gub.uy.

**Office of Planning and the Budget:** Edif. Libertad, Luis A. de Herrera 3350, Montevideo; tel. (2) 4872110; fax (2) 2099730; e-mail webmaster@opp.gub.uy; internet www.opp.gub.uy.

## President and Legislature

### PRESIDENT

Election, 31 October 2004

| Candidate | % of vote |
|---|---|
| Tabaré Ramón Vázquez Rosas (Encuentro Progresista—Frente Amplio) | 50.70 |
| Jorge Larrañaga (Partido Nacional) | 34.06 |
| Guillermo Stirling (Partido Colorado) | 10.32 |
| Others | 2.53 |
| Invalid votes | 2.39 |
| **Total** | **100.00** |

### CONGRESO

#### Cámara de Senadores
(Senate)

Election, 31 October 2004

| Party | Seats |
|---|---|
| Encuentro Progresista—Frente Amplio | 16 |
| Partido Nacional | 11 |
| Partido Colorado | 3 |
| **Total*** | **30** |

*An additional seat is reserved for the Vice-President, who sits as President of the Senate.

#### Cámara de Representantes
(Chamber of Representatives)

**President:** Nora Castro.

Election, 31 October 2004

| Party | Seats |
|---|---|
| Encuentro Progresista—Frente Amplio | 52 |
| Partido Nacional | 35 |
| Partido Colorado | 10 |
| Other parties | 2 |
| **Total** | **99** |

## Political Organizations

**Alianza Libertadora Nacionalista:** Montevideo; extreme right-wing; Leader Osvaldo Martínez Jaume.

**Encuentro Progresista—Frente Amplio (EP—FA):** Col. 1367, 2°, 11100 Montevideo; tel. (2) 9022176; e-mail prensaepfa@montevideo.com.uy; internet www.epfaprensa.org; f. 1971; left-wing grouping; Pres. Dr Tabaré Ramón Vázquez Rosas; Vice-Pres. Jorge Brovetto; members include:

**Alianza Progresista 738 (AP):** Col. 1831, Montevideo; tel. and fax (2) 4016365; e-mail a738@alianza738.org.uy; internet www.alianza738.org.uy; left-wing; Leader Rodolfo Nin Novoa.

**Asamblea Uruguay (AU):** Carlos Quijano 1271, Montevideo; e-mail jmahia@parlamento.gub.uy; internet www.asamblea.org.uy; centre-left; Leader Danilo Astori.

**Frente Izquierda de Liberación (FIDEL):** Montevideo; f. 1962; socialist; Leader Adolfo Aguirre González.

**Grupo Pregón:** Montevideo; left-wing liberal party; Leaders Sergio Previtali, Enrique Moras.

**Movimiento de Acción Nacionalista (MAN):** Montevideo; left-wing nationalist org.; Leader José Durán Matos.

**Movimiento Blanco Popular y Progresista (MBPP):** Montevideo; moderate left-wing; Leader A. Francisco Rodríguez Camusso.

**Movimiento de Participación Popular (MPP):** Germán Barbato 1491; tel. (2) 9086948; e-mail correos@mppuruguay.org; internet www.mppuruguay.org; f. 1989 by the MLN—Tupamaros

# URUGUAY

(see below); grouping of left-wing parties; Leader JOSÉ 'PEPE' MUJICA; members include:

**Movimiento de Liberación Nacional (MLN)—Tupamaros:** Tristán Narvaja 1578, CP 11.200, Montevideo; tel. (2) 4092298; fax (2) 4099957; e-mail mln@chasque.apc.org; internet www.chasque.net/mlnweb; f. 1962; radical socialist; during 1962–73 the MLN, operating under its popular name of the **Tupamaros**, conducted a campaign of urban guerrilla warfare until it was defeated by the Armed Forces in late 1973; following the return to civilian rule, in 1985, the MLN announced its decision to abandon its armed struggle; legally recognized in May 1989; Sec.-Gen. JOSÉ 'PEPE' MUJICA.

**Movimiento 26 de Marzo:** Durazno 1118, 11200 Montevideo; tel. (2) 9011584; f. 1971; socialist; Pres. EDUARDO RUBIO; Sec.-Gen. FERNANDO VÁZQUEZ.

**Partido Comunista:** Río Negro 1525, 11100 Montevideo; tel. (2) 9017171; fax (2) 9011050; f. 1920; Sec.-Gen. MARINA ARISMENDI; 42,000 mems (est.).

**Partido de Democracia Avanzada:** Montevideo; Communist.

**Partido Socialista del Uruguay (PS):** Casa del Pueblo, Soriano 1218, 11100 Montevideo; tel. (2) 9013344; fax (2) 9082548; e-mail info@ps.org.uy; internet www.ps.org.uy; f. 1910; Pres. REINALDO GARGANO; Sec.-Gen. ROBERTO CONDE.

**Vertiente Artiguista (VA):** San José 1191, CP 11200, Montevideo; tel. (2) 9000177; e-mail vertiente@vertiente.org.uy; internet www.vertiente.org.uy; f. 1989; left-wing; Leader MARIANO ARANA; Sec.-Gen. LAURA FERNÁNDEZ.

**Nuevo Espacio:** Eduardo Acevedo 1615, 11200 Montevideo; tel. (2) 4026989; fax (2) 4026991; e-mail internacionales@nuevoespacio.org.uy; internet www.nuevoespacio.org.uy; f. 1994; social-democratic; allied to the Frente Amplio since Dec. 2002; moderate left-wing; Leader RAFAEL MICHELINI; Sec. EDGARDO CARVALHO.

**Partido Azul (PA):** Paul Harris 1722, Montevideo; tel. and fax (2) 6016327; e-mail hablacon@partidoazul.s5.com; internet www.partidoazul.s5.com; liberal; f. 1993; Leader Dr ROBERTO CANESSA; Gen. Sec. Ing. ARMANDO VAL.

**Partido Colorado:** Andrés Martínez Trueba 1271, 11100 Montevideo; tel. (2) 4090180; f. 1836; Sec.-Gen. JORGE LUIS BATLLE IBÁÑEZ; factions include:

**Foro Batllista:** Colonia 1243, 11100 Montevideo; tel. (2) 9030154; e-mail info@forobatllista.com; internet www.forobatllista.com; Leader Dr JULIO MARÍA SANGUINETTI CAIROLO.

**Lista 15:** Leader JORGE LUIS BATLLE IBÁÑEZ.

**Unión Colorada y Batllista (Pachequista):** Buenos Aires 594, 11000 Montevideo; tel. (2) 9164648; right-wing.

**Vanguardia Batllista:** Casa de Vanguardia, Paysandú 1333, entre Ejido y Curiales, Montevideo; tel. (2) 9027779; e-mail info@scavarelli.com; internet www.scavarelli.com; Leader Dr ALBERTO SCARAVELLI.

**Partido Demócrata Cristiano (PDC):** Aquiles Lanza 1318 bis, Montevideo; tel. and fax (2) 9030704; e-mail pdc@chasque.apc.org; internet www.chasque.apc.org/pdc; fmrly Unión Cívica del Uruguay; allied to the Alianza Progresista since 1999; f. 1962; Pres. Dr HÉCTOR LESCANO; Sec.-Gen. FRANCISCO OTTONELLI.

**Partido Justiciero:** Montevideo; extreme right-wing; Leader BOLÍVAR ESPÍNDOLA.

**Partido Nacional (Blanco):** Juan Carlos Gómez 1384, Montevideo; tel. (2) 9163831; fax (2) 9163758; e-mail partidonacional@partidonacional.com.uy; internet www.partidonacional.com.uy; f. 1836; Exec. Pres. LUIS ALBERTO LACALLE; Sec.-Gen. ALBERTO ZUMARÁN; tendencies within the party include:

**Alianza Nacional:** Leader JORGE LARRAÑAGA.

**Consejo Nacional Herrerista:** Leader LUIS ALBERTO LACALLE.

**Desafío Nacional:** Leader JUAN ANDRÉS RAMÍREZ.

**Linea Nacional de Florida:** Leader ARTURO HEBER.

**Manos a la Obra:** Plaza Cagancha 1145, 11100 Montevideo; tel. (2) 9028149; Leader ALBERTO VOLONTÉ.

**Movimiento Nacional de Rocha—Corriente Popular Nacionalista:** Avda Uruguay 1324, Montevideo; tel. (2) 9027502; Leader CARLOS JULIO PEREYRA.

**Partido del Sol:** Peatonal Yi 1385, 11000 Montevideo; tel. (2) 9001616; fax (2) 9006739; e-mail partidodelsol@adinet.com.uy; internet www.partidodelsoluruguay.org; ecologist, federal, pacifist; Leader HOMERO MIERES.

**Partido de los Trabajadores:** Convención 1196, 11100 Montevideo; tel. (2) 9082624; f. 1980; extreme left-wing; Leader JUAN VITAL ANDRADE.

**Unión Cívica:** Montevideo; tel. (2) 9005535; e-mail info@unioncivica.org; internet www.dreamsmaker.com.uy/trabajos/union-civica; f. 1912; recognized Christian Democrat faction, split from the Partido Demócrata Cristiano in 1980; Leader W. GERARDO AZAMBUYA.

# Diplomatic Representation

## EMBASSIES IN URUGUAY

**Argentina:** Cuareim 1470, 11800 Montevideo; tel. (2) 9028166; fax (2) 9028172; e-mail emargrou@adinet.com.uy; internet emb-uruguay.mrecic.gov.ar; Ambassador HERNÁN MARÍA PATIÑO MAYER.

**Bolivia:** Dr Prudencio de Peña 2469, 11300 Montevideo; tel. (2) 7083573; fax (2) 7080066; e-mail embolivia-montevideo@rree.gov.bo; Ambassador ARMANDO LOAYZA MARIACA.

**Brazil:** Blvr Artigas 1328, 11300 Montevideo; tel. (2) 7072119; fax (2) 7072086; e-mail montevideo@brasemb.org.uy; internet www.brasil.org.uy; Ambassador EDUARDO DOS SANTOS.

**Canada:** Plaza Independencia 749, Of. 102, 11100 Montevideo; tel. (2) 9022030; fax (2) 9022029; e-mail mvdeo@dfait-maeci.gc.ca; internet www.dfait-maeci.gc.ca/uruguay; Ambassador PATRICIA FULLER.

**Chile:** Calle 25 de Mayo 575, Montevideo; tel. (2) 9164090; fax (2) 9153804; e-mail echileuy@netgate.com.uy; Ambassador CARLOS APPELGREN BALBONTIN.

**China, People's Republic:** Miraflores 1508, Carrasco, Casilla 18966, Montevideo; tel. (2) 6016126; fax (2) 6018508; e-mail embchina@adinet.com.uy; Ambassador WANG YONGZHAN.

**Colombia:** Edif. Tupí, Juncal 1305, 18°, 11000 Montevideo; tel. (2) 9161592; fax (2) 9161594; e-mail euruguay@minrelext.gov.co; Ambassador CLAUDIA DINORA OLGA MARIA TURBAY QUINTERO.

**Costa Rica:** José Benito Lamas 2963, Montevideo; tel. (2) 7083645; fax (2) 7089727; e-mail embarica@adinet.com.uy; Ambassador ALEXANDER SALAS ARAYA.

**Cuba:** Francisco Visal 667, Montevideo; tel. (2) 7124668; fax (2) 7124670; e-mail cancilleria@netgate.com.uy; Consul-Gen. RAÚL GORTÁZAR MARRERO; diplomatic relations severed in 2002; restored in March 2005.

**Czech Republic:** Luis B. Cavia 2996, Casilla 12262, 11300 Montevideo; tel. (2) 7087808; fax (2) 7096410; e-mail montevideo@embassy.mzv.cz; internet www.mzv.cz/montevideo; Ambassador VÍT KORSELT.

**Dominican Republic:** Tomás de Tezanos 1186, 11300 Montevideo; tel. (2) 6287766; fax (2) 6289655; e-mail embajadomuruguay@hotmail.com; Ambassador RAFAEL ANTONIO JULIÁN CEDANO.

**Ecuador:** Pedro Berro 1217, entre Guayaquí y Pereira, Montevideo; tel. (2) 7076463; fax (2) 7076465; e-mail embajadaecuador@netgate.com.uy; Ambassador LEONARDO CARRIÓN EGUIGUREN.

**Egypt:** Avda Brasil 2663, 11300 Montevideo; tel. (2) 7096412; fax (2) 7080977; e-mail boustanemontevideo@easymail.com.uy; Ambassador Dr OHEIB ANWAR ELSOKARY.

**El Salvador:** Melitón González 1157, Apto 501, Montevideo 11300; tel. (2) 6222005; fax (2) 6226842; e-mail embasauy@dedicado.net.uy; Ambassador ERNESTO FERREIRO RUSCONI.

**France:** Avda Uruguay 853, Casilla 290, 11100 Montevideo; tel. (2) 9020077; fax (2) 9023711; e-mail ambfra@adinet.com.uy; internet www.amb-montevideo.fr; Ambassador LAURENT-JOSEPH RAPIN.

**Germany:** La Cumparsita 1417/1435, Casilla 20014, 11200 Montevideo; tel. (2) 9025222; fax (2) 9023422; e-mail info@embajadaalemana-montevideo.info; internet www.embajadaalemana-montevideo.info; Ambassador Dr VOLKER ANDING.

**Greece:** Edif. Artigas, Rincón 487, 2°, 11100 Montevideo; tel. (2) 9165191; fax (2) 9150795; e-mail gremb.mvd@mfa.gr; Ambassador NICOLAOS DICTAKIS.

**Guatemala:** Edif. Plaza Marítima, 3°, Of. 342, 6a Avda 20–25, Montevideo; tel. (2) 3680810; fax (2) 3337553; e-mail uruguate@guate.net.gt; Chargé d'affaires a.i. ALEJANDRO VELA AQUINO.

**Holy See:** Blvr Artigas 1270, Casilla 1503, Montevideo (Apostolic Nunciature); tel. (2) 7072016; fax (2) 7072209; Apostolic Nuncio Most Rev. JANUSZ BOLONEK (Titular Archbishop of Madaurus).

**Iran:** Blvr Artigas 531, 11300 Montevideo; tel. (2) 7116657; fax (2) 7116659; e-mail emb.iran.secretaria@multitel.com.uy; Ambassador MOHAMMAD FARAJI.

# URUGUAY

**Israel:** Blvr Artigas 1585, 11200 Montevideo; tel. (2) 4004164; fax (2) 4095821; e-mail ambassador@montevideo.mfa.gov.il; Ambassador JOEL SALPAK.

**Italy:** José Benito Lamas 2857, Casilla 268, 11300 Montevideo; tel. (2) 7084916; fax (2) 7084148; e-mail ambasciata.montevideo@esteri.it; internet www.ambitalia.com.uy; Ambassador GIORGIO MALFATTI DI MONTE TRETTO.

**Japan:** Blvr Artigas 953, 11300 Montevideo; tel. (2) 4187645; fax (2) 4187980; e-mail embjapon@adinet.com.uy; internet www.uy.emb-japan.go.jp; Ambassador YOSHIHIRO NAKAMURA.

**Korea, Republic:** Edif. World Trade Center, Avda Luis Alberto de Herrera 1248, Torre 2, 10°, Montevideo; tel. (2) 6289374; fax (2) 6289376; e-mail ecorea@adinet.com.uy; Ambassador TAE-SHIN JAN.

**Lebanon:** Avda General Rivera 2278, Montevideo; tel. (2) 4086640; fax (2) 4086365; e-mail embliban@adinet.com.uy; Ambassador VICTOR BITAR.

**Mexico:** Andes 1365, 7°, 11100 Montevideo; tel. (2) 9020791; fax (2) 9021232; e-mail embmexur@netgate.com.uy; Ambassador PERLA MARÍA CARVALHO SOTO.

**Netherlands:** Leyenda Patria 2880, Of. 202, 2°, Casilla 1519, 11300 Montevideo; tel. (2) 7112956; fax (2) 7113301; e-mail embajada@holanda.org.uy; internet www.holanda.org.uy; Ambassador R.H. MEYS.

**Panama:** Juan Benito Blanco 3388, 11300 Montevideo; tel. (2) 6230301; fax (2) 6230300; e-mail empauru@dedicado.net.uy; Chargé d'affaires JOSÉ DE JESÚS MARTÍNEZ GONZÁLEZ.

**Paraguay:** Blvr Artigas 1256, Montevideo; tel. (2) 7072138; fax (2) 7083682; e-mail embapur@netgate.com.uy; internet www.geocities.com/embapur; Ambassador Dr CARLOS RIVEROS SALCEDO.

**Peru:** Obligado 1384, 11300 Montevideo; tel. (2) 7076862; fax (2) 7077793; e-mail emba8@embaperu.org.uy; internet www.angelfire.com/country/embaperu; Ambassador WILLIAM BELEVÁN MCBRIDE.

**Poland:** Jorge Canning 2389, Casilla 1538, 11600 Montevideo; tel. (2) 4801313; fax (2) 4873389; e-mail ambmonte@netgate.com.uy; internet www.embajadapoloniauruguay.com; Ambassador LECH KUBIAK.

**Portugal:** Avda Dr Francisco Soca 1128, Casilla 701, 11300 Montevideo; tel. (2) 7084061; fax (2) 7096456; e-mail portmont@netgate.com.uy; Ambassador Dr DOMINGO TOMAS VILA GARRIDO SERRA.

**Romania:** Echevarriarza 3452, Casilla 12040, 11000 Montevideo; tel. (2) 6220876; fax (2) 6220135; e-mail ambromvd@adinet.com.uy; Ambassador VASILE MACOVEI.

**Russia:** Blvr España 2741, 11300 Montevideo; tel. (2) 7081884; fax (2) 7086597; e-mail embaru@montevideo.com.uy; Ambassador YAN A. BURLIAY.

**South Africa:** Echevarriarza 3335, Casilla 498, 11000 Montevideo; tel. (2) 6230161; fax (2) 6230066; e-mail safem@netgate.com.uy; Ambassador Prof. MLUNGISI MAKALIMA (resident in Argentina).

**Spain:** Avda Libertad 2738, 11300 Montevideo; tel. (2) 7086010; fax (2) 7083291; e-mail embespuy@correo.mae.es; Ambassador FERNANDO VALDERRAMA PAREJA.

**Switzerland:** Ing. Federico Abadie 2936/40, 11°, Casilla 12261, 11300 Montevideo; tel. (2) 7115545; fax (2) 7115031; e-mail vertretung@mtv.rep.admin.ch; Chargé d'affaires a.i. DANIEL VON MURALT.

**United Kingdom:** Marco Bruto 1073, Casilla 16024, 11300 Montevideo; tel. (2) 6223630; fax (2) 6227815; e-mail bemonte@internet.com.uy; internet www.britishembassy.org.uy; Ambassador Dr HUGH SALVESSEN.

**USA:** Lauro Muller 1776, 11200 Montevideo; tel. (2) 4187777; fax (2) 4188611; internet uruguay.usembassy.gov; Ambassador MARTIN J. SILVERSTEIN (until Aug. 2005).

**Venezuela:** José Agustín Iturriaga 3589, Montevideo; tel. (2) 6221262; fax (2) 6282530; e-mail embaven@adinet.com.uy; Ambassador MARÍA LOURDES URBANEJA DURANT.

## Judicial System

The Supreme Court of Justice comprises five members appointed at the suggestion of the executive, for a period of five years. It has original jurisdiction in constitutional, international and admiralty cases, and hears appeals from the appellate courts, of which there are seven, each with three judges.

Cases involving the functioning of the state administration are heard in the ordinary Administrative Courts and in the Supreme Administrative Court, which consists of five members appointed in the same way as members of the Supreme Court of Justice.

In Montevideo there are 19 civil courts, 10 criminal and correctional courts, 19 courts presided over by justices of the peace, three juvenile courts, three labour courts and courts for government and other cases. Each departmental capital, and some other cities, have a departmental court; each of the 224 judicial divisions has a justice of the peace.

The administration of justice became free of charge in 1980, with the placing of attorneys-at-law in all courts to assist those unable to pay for the services of a lawyer.

### Supreme Court of Justice

H. Gutiérrez Ruiz 1310, Montevideo; tel. (2) 9001041; fax (2) 902350; e-mail secparga@poderjudicial.gub.uy; internet www.poderjudicial.gub.uy.

**President of the Supreme Court of Justice:** Dr DANIEL GUTIÉRREZ.

**Supreme Administrative Court:** Mercedes 961, 11100 Montevideo; tel. (2) 9008047; fax (2) 9080539.

## Religion

Under the Constitution, the Church and the State were declared separate and toleration for all forms of worship was proclaimed. Roman Catholicism predominates.

### CHRISTIANITY

**Federación de Iglesias Evangélicas del Uruguay:** Estero Bellaco 2676, 11600 Montevideo; tel. (2) 4873375; tel. (2) 4800984; fax (2) 4872181; internet www.chasque.net/obra/skontakt.htm; f. 1956; eight mem. churches; Pres. LUIS ROSSO; Sec. ALFREDO SERVETTI.

#### The Roman Catholic Church

Uruguay comprises one archdiocese and nine dioceses. At 31 December 2003 there were an estimated 2,316,495 adherents in the country, representing about 61% of the total population.

**Bishops' Conference**

Conferencia Episcopal Uruguaya, Avda Uruguay 1319, 11100 Montevideo; tel. (2) 9002642; fax (2) 9011802; e-mail ceusecre@adinet.com.uy; internet www.iglesiauruguaya.com.

f. 1972; Pres. Mgr PABLO JAIME GALIMBERTI DI VIETRI (Bishop of San José de Mayo).

**Archbishop of Montevideo:** Most Rev. NICOLÁS COTUGNO FANIZZI, Arzobispado, Treinta y Tres 1368, Casilla 356, 11000 Montevideo; tel. (2) 9158127; fax (2) 9158926; e-mail vicario@arquidiocesis.net.

#### The Anglican Communion

Uruguay is the newest diocese in the Province of the Southern Cone of America, having been established in 1988. The presiding Bishop of the Iglesia Anglicana del Cono Sur de América is the Bishop of Northern Argentina.

**Bishop of Uruguay:** Rt Rev. MIGUEL TAMAYO ZALDÍVAR, Centro Diocesano, Reconquista 522, Casilla 6108, 11000 Montevideo; tel. (2) 9159627; fax (2) 9162519; e-mail mtamayo@netgate.com.uy; internet www.uruguay.anglican.org.

#### Other Churches

**Baptist Evangelical Convention of Uruguay:** Mercedes 1487, 11100 Montevideo; tel. and fax (2) 2167012; e-mail suspasos@adinet.com.uy; f. 1948; 4,500 mems; Pres Dr JUAN CARLOS OTORMÍN.

**Iglesia Adventista** (Adventist Church): Castro 167, Montevideo; f. 1901; 4,000 mems; Principal officers Dr GUILLERMO DURÁN, Dr ALEXIS PIRO.

**Iglesia Evangélica Metodista en el Uruguay** (Evangelical Methodist Church in Uruguay): San José 1457, 11200 Montevideo; tel. (2) 4136552; fax (2) 4136554; e-mail iemu@adinet.com.uy; internet www.gbgm-umc.org/iemu; f. 1878; 1,193 mems (1997); Pres. Rev. OSCAR BOLIOLI.

**Iglesia Evangélica Valdense** (Waldensian Evangelical Church): Avda 8 de Octubre 3039, 11600 Montevideo; tel. and fax (2) 4879406; e-mail ievm@internet.com.uy; f. 1952; 15,000 mems; Pastor ALVARO MICHELIN SALOMÓN.

**Iglesia Pentecostal Unida Internacional en Uruguay** (United Pentecostal Church International in Uruguay): Helvecia 4032, Piedras Blancas, 12200 Montevideo; tel. (2) 5133618; e-mail lrodrigu@montevideo.com.uy; internet members.tripod.com/~lrodrigu; Pastor LUIS RODRÍGUEZ.

**Primera Iglesia Bautista** (First Baptist Church): Avda Daniel Fernández Crespo 1741, Casilla 5051, 11200 Montevideo; tel. (2)

# URUGUAY

4098744; fax (2) 4094356; e-mail piebu@adinet.com.uy; f. 1911; 314 mems; Pastor LEMUEL J. LARROSA.

Other denominations active in Uruguay include the Iglesia Evangélica del Río de la Plata and the Iglesia Evangélica Menonita (Evangelical Mennonite Church).

## BAHÁ'Í FAITH

**National Spiritual Assembly of the Bahá'ís:** Blvr Artigas 2440, 11600 Montevideo; tel. (2) 4875890; fax (2) 4802165; e-mail bahai@multi.com.uy; f. 1938; mems resident in 140 localities.

# The Press

## DAILIES

### Montevideo

**El Diario:** Rincón 712, 11000 Montevideo; tel. (2) 9030465; fax (2) 9030637; e-mail joterom@adinet.com.uy; f. 1923; evening; independent; Editor JORGE OTERO; circ. 12,000.

**El Diario Español:** Cerrito 551–555, Casilla 899, 11000 Montevideo; tel. (2) 9159481; fax (2) 9157389; f. 1905; morning (except Monday); newspaper of the Spanish community; Editor MARCELO REINANTE; circ. 20,000.

**Diario Oficial:** Avda 18 de Julio 1373, Montevideo; tel. (2) 9085042; fax (2) 9023098; e-mail impo@impo.com.uy; internet www.impo.com.uy; f. 1905; morning; publishes laws, official decrees, parliamentary debates, judicial decisions and legal transactions; Dir ZAIN NASSIF DE ZARUMBE.

**La Mañana:** Casilla 5005, Suc. 2, Montevideo 11100; tel. (2) 9029055; fax (2) 9021326; f. 1917; supports the Partido Colorado; Editor Dr SALVADOR ALABÁN DEMARE; circ. 50,000.

**Mundocolor:** Cuareim 1287, 11800 Montevideo; f. 1976; evening (except Sunday); Dir DANIEL HERRERA LUSSICH; circ. 4,500.

**El Observador:** Cuareim 2052, 11800 Montevideo; tel. (2) 9247000; fax (2) 9248698; e-mail elobservador@observador.com.uy; internet www.observador.com.uy; f. 1991; morning; Editor RICARDO PEIRANO; circ. 26,000.

**Observador Económico:** Soriano 791, 11100 Montevideo; tel. (2) 9030690; fax (2) 9030691.

**El País:** Plaza Cagancha 1356, 11100 Montevideo; tel. (2) 9020115; fax (2) 9020464; e-mail cartas@elpais.com.uy; internet www.elpais.com.uy; f. 1918; morning; supports the Partido Nacional; Editor MARTÍN AGUIRRE; circ. 106,000.

**La República:** Avda Gral Garibaldi 2579, 11600 Montevideo; tel. (2) 4873565; fax (2) 4873823; e-mail redaccion@diariolarepublica.com; internet www.diariolarepublica.com; f. 1988; morning; Editor FEDERICO FASANO MERTENS; circ. 20,000.

**Ultimas Noticias:** Paysandú 1179, 11100 Montevideo; tel. (2) 9020452; fax (2) 9024669; internet www.ultimasnoticias.com.uy; f. 1981; evening (except Saturday); owned by Impresora Polo; Publr Dr ALPHONSE EMANUILOFF-MAX; circ. 25,000.

### Florida

**El Heraldo:** Independencia 824, 94000 Florida; tel. (35) 22229; fax (35) 24546; e-mail heraldo@adinet.com.uy; internet www.elheraldo.com.uy; f. 1919; morning; independent; Dir ALVARO RIVA REY; circ. 20,000.

### Maldonado

**Correo de Punta del Este:** Zelmar Michelini 815 bis, 20000 Maldonado; tel. and fax (42) 35633; e-mail gallardo@adinet.com.uy; internet www.diariocorreo.com; f. 1993; morning; Editor MARCELO GALLARDO; circ. 2,500.

### Minas

**La Unión:** Florencio Sánchez 569, Minas; tel. (442) 2065; fax (442) 4011; e-mail union@chasque.apc.org; f. 1877; evening (except Sunday); Dir LAURA PUCHET MARTÍNEZ; Editor ALEJANDRO MAYA SOSA; circ. 2,600.

### Paysandú

**El Telégrafo:** 18 de Julio 1027, 60000 Paysandú; tel. (722) 3141; fax (722) 7999; e-mail correo@eltelegrafo.com; internet www.eltelegrafo.com; f. 1910; morning; independent; Dir FERNANDO M. BACCARO; circ. 8,500.

### Salto

**El Pueblo:** 18 de Julio 15, Salto; e-mail dipueblo@adinet.com.uy; internet www.diarioelpueblo.com.uy; morning; Dir ADRIANA MARTÍNEZ.

**Tribuna Salteña:** Joaquín Suárez 71, Salto; f. 1906; morning; Dir MODESTO LLANTADA FABINI; circ. 3,000.

## PERIODICALS

### Montevideo

**Aquí:** Zabala 1322, Of. 102, 11000 Montevideo; weekly; supports the Encuentro Progresista—Frente Amplio; Dir FRANCISCO JOSÉ O'HONELLI.

**Brecha:** Avda Uruguay 844, 11100 Montevideo; tel. (2) 9025042; fax (2) 9020388; e-mail brecha@brecha.com.uy; internet www.brecha.com.uy; f. 1985; weekly; politics, current affairs; circ. 8,500; Dir IVONNE TRÍAS; Editor-in-Chief DANIEL GATTI.

**Búsqueda:** Avda Uruguay 1146, 11100 Montevideo; tel. (2) 9021300; fax (2) 9022036; e-mail busqueda@adinet.com.uy; f. 1972; weekly; independent; politics and economics; Dir DANILO ARBILLA FRACHIA; circ. 25,000.

**Charoná:** Gutiérrez Ruiz 1276, Of. 201, Montevideo; tel. (2) 9086665; f. 1968; fortnightly; children's; Dir SERGIO BOFFANO; circ. 25,000.

**Colorín Colorado:** Dalmiro Costa 4482, Montevideo; f. 1980; monthly; children's; Dir SARA MINSTER DE MURNINKAS; circ. 3,000.

**Crónicas Económicas:** Avda Libertador Brig.-Gen. Lavalleja 1532, Montevideo; tel. (2) 9004790; fax (2) 9020759; e-mail cronicas@netgate.com.uy; internet www.cronicas.com.uy; f. 1981; weekly; independent; business and economics; Dirs JULIO ARIEL FRANCO, WALTER HUGO PAGÉS, JORGE ESTELLANO.

**La Gaceta Militar Naval:** Montevideo; monthly.

**Guambia:** Rimac 1576, 11400 Montevideo; tel. (2) 6132703; fax (2) 6132703; e-mail info@guambia.com.uy; internet www.guambia.com.uy; f. 1983; monthly; satirical; Dir and Editor ANTONIO DABEZIES.

**Indice Industrial-Anuario de la Industria Uruguaya:** Sarandí 456, 11000 Montevideo; tel. (2) 9151963; f. 1957; annual; Dir W. M. TRIAS; circ. 6,000.

**La Justicia Uruguaya:** 25 de Mayo 555, 11000 Montevideo; tel. (2) 9157587; fax (2) 9159721; e-mail lajusticiauruguaya@lju.com.uy; internet www.lajusticiauruguaya.com.uy; f. 1940; bimonthly; jurisprudence; Dirs EDUARDO ALBANELL, ADOLFO ALBANELL, SUSANA ARIAS; circ. 3,000.

**La Juventud:** 18 de Julio 1357, Of. 202, Montevideo; tel. (2) 9030305; e-mail juventud@chasque.apc.org; internet www.chasque.apc.org/juventud; weekly; supports the Movimiento 26 de Marzo; Dir GUILLERMO FERNÁNDEZ; Editor JOSÉ L. BORGES.

**Marketing Directo:** Mario Cassinoni 1157, 11200 Montevideo; tel. (2) 4012174; fax (2) 4087221; e-mail consumo@adinet.com.uy; f. 1988; monthly; Dir EDGARDO MARTÍNEZ ZIMARIOFF; circ. 9,500.

**Mate Amargo:** Tristán Narvaja 1578 bis, Montevideo; f. 1986; organ of the Movimiento de Liberación Nacional; circ. 22,500.

**Opción:** J. Barrios Amorín 1531, Casilla 102, 11100 Montevideo; f. 1981; weekly; Dir FRANCISCO JOSÉ OTTONELLI; circ. 15,000.

**Patatín y Patatán:** Montevideo; f. 1977; weekly; children's; Dir JUAN JOSÉ RAVAIOLI; circ. 3,000.

**Patria:** Montevideo; internet www.patria.com.uy; weekly; organ of the Partido Nacional; right-wing; Dir LUIS ALBERTO LACALLE HERRERA; Editor Dr JOSÉ LUIS BELLANI.

**Revista Naval:** Soriano 1117, 11100 Montevideo; tel. (2) 9087884; f. 1988; 3 a year; military; Editor GUSTAVO VANZINI; circ. 1,000.

## PRESS ASSOCIATIONS

**Asociación de Diarios del Uruguay:** Río Negro 1308, 6°, 11100 Montevideo; f. 1922; Pres. BATLLE T. BARBATO.

**Asociación de la Prensa Uruguaya:** Colonia 1086, Of. 903, Montevideo; tel. and fax (2) 9013695; e-mail apu@adinet.com.uy; f. 1944; Pres. MANUEL MÉNDEZ.

## PRESS AGENCIES

### Foreign Bureaux

**Agence France-Presse (AFP):** Plaza Independencia 831, 11100 Montevideo; tel. (2) 9005095; Chief JUPITER PUYO.

**Agencia EFE (Spain):** Wilson Ferreira Aldunate 1294, Of. 501, 11200 Montevideo; tel. (2) 9020322; fax (2) 9026726; e-mail mhurtado@efe.com.uy; Correspondent MARTA HURTADO GÓMEZ.

URUGUAY

**Agenzia Nazionale Stampa Associata (ANSA)** (Italy): Florida 1408, Montevideo; tel. (2) 9011032; fax (2) 9081950; Bureau Chief JUAN ATELLA.

**Associated Press (AP)** (USA): Avda 18 de Julio 1076, Montevideo; tel. (2) 9018291; Correspondent DANIEL GIANELLI.

**Deutsche Presse-Agentur (dpa)** (Germany): Avda 18 de Julio 994, 4°, Montevideo; tel. (2) 9028052; fax (2) 9022662; e-mail dpaurc@montevideo.com.uy; Correspondent MARÍA ISABEL RIVERO DE LOS CAMPOS.

**Inter Press Service (IPS)** (Italy): Juan Carlos Gómez 1445, Of. 102, 1°, 11000 Montevideo; tel. (2) 9164397; fax (2) 9163598; e-mail ips@tips.org.uy; internet www.ips.org; f. 1964; Dir MARIO LUBETKIN.

**Reuters** (United Kingdom): Plaza Independencia 831, Of. 907-908, 11100 Montevideo; tel. (2) 9020336; fax (2) 9027912; Correspondent ANAHÍ RAMA.

## Publishers

**Autores Uruguayos:** Paysandú 1561,11200 Montevideo; e-mail mensajes@autoresuruguayos.com.uy; internet www.autoresuruguayos.com; publishes works by Uruguayan authors; Man. ADRIANA DOS SANTOS.

**Editorial Arca:** Ana Monterroso 2231, Montevideo; tel. (2) 4099796; fax (2) 4099788; f. 1963; general literature, social science and history; Man. Dir ENRIQUE PIQUÉ.

**Ediciones de la Banda Oriental:** Gaboto 1582, 11200 Montevideo; tel. (2) 4083206; fax (2) 4098138; e-mail ebo@chasque.net; general literature; Man. Dir HEBER RAVIOLO.

**CENCI—Uruguay** (Centro de Estadísticas Nacionales y Comercio Internacional): Misiones 1361, Casilla 1510, 11000 Montevideo; tel. (2) 9152930; fax (2) 9154578; e-mail cenci@cenci.com.uy; f. 1956; economics, statistics; Dir KENNETH BRUNNER.

**Editorial y Librería Jurídica Amalio M. Fernández SRL:** 25 de Mayo 589, 11000 Montevideo; tel. and fax (2) 9151782; e-mail amflibrosjurid@movinet.com.uy; f. 1951; law and sociology; Man. Dir CARLOS W. DEAMESTOY.

**Editorial La Flor del Itapebí:** Luis Piera 1917/401, Montevideo; tel. and fax (2) 7109267; internet www.itapebi.com.uy; f. 1991; cultural, technical, educational.

**Fundación de Cultura Universitaria:** 25 de Mayo 568, Casilla 1155, 11000 Montevideo; tel. (2) 9152532; fax (2) 9152549; e-mail ventas@fcu.com.uy; internet www.fcu.com.uy; f. 1968; law and social sciences; Pres. Dr PABLO DONNÁNGELO.

**Hemisferio Sur:** Buenos Aires 335, Casilla 1755, 11000 Montevideo; tel. (2) 9164515; fax (2) 9164520; e-mail librperi@adinet.com.uy; f. 1951; agronomy and veterinary science.

**Editorial Idea:** Misiones 1424, 5°, 11000 Montevideo; tel. (2) 9165456; fax (2) 9150868; e-mail vescovi@fastlink.com.uy; law; Dir Dr GUILLERMO VESCOVI.

**Librería Linardi y Risso:** Juan C. Gómez 1435, 11000 Montevideo; tel. (2) 9157129; fax (2) 9157431; e-mail lyrbooks@linardiyrisso.com.uy; internet www.linardiyrisso.com.uy; f. 1944; general; Man. Dirs ALVARO RISSO, ANDRÉS LINARDI.

**Editorial Medina SRL:** Gaboto 1521, Montevideo; tel. (2) 4085800; f. 1933; general; Pres. MARCOS MEDINA VIDAL.

**A. Monteverde & Cía, SA:** Treinta y Tres 1475, Casilla 371, 11000 Montevideo; tel. (2) 9152939; fax (2) 9152012; f. 1879; educational; Man. Dir LILIANA MUSSINI.

**Mosca Hermanos SA:** Avda 18 de Julio 1578, 11300 Montevideo; tel. (2) 4093141; fax (2) 4088059; e-mail mosca@attmail.com.uy; f. 1888; general; Pres. Lic. ZSOLT AGARDY.

**Librería Selecta Editorial:** Guayabo 1865, 11200 Montevideo; tel. (2) 4086989; fax (2) 4086831; f. 1950; academic books; Dir FERNANDO MASA.

**Ediciones Trilce:** Durazno 1888, 11200 Montevideo; tel. (2) 4127662; fax (2) 4127722; e-mail trilce@trilce.com.uy; internet www.trilce.com.uy; f. 1985; science, politics, history.

**Vintén Editor:** Hocquart 1771, Casilla 11804, Montevideo; tel. (2) 2090223; internet vinten-uy.com; poetry, theatre, history, art, literature.

### PUBLISHERS' ASSOCIATION

**Cámara Uruguaya del Libro:** Juan D. Jackson 1118, 11200 Montevideo; tel. (2) 4015732; fax (2) 4011860; e-mail camurlib@adinet.com.uy; f. 1944; Pres. ERNESTO SANJINÉS; Man. ANA CRISTINA RODRÍGUEZ.

## Broadcasting and Communications

### TELECOMMUNICATIONS

**Administración Nacional de Telecomunicaciones (ANTEL):** Complejo Torre de las Telecomunicaciones, Guatemala 1075, Montevideo; e-mail antel@antel.com.uy; internet www.antel.com.uy; f. 1974; state-owned; Pres. Ing. MARÍA SIMON; Gen. Man. Ing. JOSÉ LUIS SALDÍAS.

  **ANCEL:** Pablo Galarza 3537, Montevideo; internet www.ancel.com.uy; f. 1994; state-owned mobile telephone co; pending partial (40%) privatization.

**Movicom:** Avda Constituyente, Edif. Torre el Gaucho, 1467 Montevideo; tel. (2) 4087502; internet www.movicom.com.uy; owned by Telefónica Móviles, SA (Spain); mobile telephone services.

**Unidad Reguladora de Servicios de Comunicaciones (URSEC):** Uruguay 988, Casilla 11100, Montevideo; tel. (2) 9028082; fax (2) 9005708; e-mail bergara@ursec.gub.uy; internet www.ursec.gub.uy; regulates telecommunications and postal sectors; Pres. Dr FERNANDO OMAR PÉREZ TABÓ.

### BROADCASTING

#### Regulatory Authority

**Asociación Nacional de Broadcasters Uruguayos (ANDEBU):** Carlos Quijano 1264, 11100 Montevideo; tel. (2) 9021525; fax (2) 9021540; e-mail andebu@internet.com.uy; internet www.andebu.com.uy; f. 1933; 101 mems; Pres. CARLOS FALCO; Vice-Pres. Dr WALTER C. ROMAY.

#### Radio

**El Espectador:** Río Branco 1481, 11100 Montevideo; tel. (2) 9027106; fax (2) 9083044; e-mail am810@espectador.com.uy; internet www.espectador.com; f. 1923; commercial; Pres. MARÍA A. MASTRÁNGELO; Dir-Gen. JAVIER MASSA.

**Radio Carve:** Mercedes 973, 11100 Montevideo; tel. (2) 9026162; fax (2) 9020126; e-mail carve@portalx.com.uy; internet www.carve.com.uy; f. 1928; commercial; Dir PABLO FONTAINA MINELLI.

**Radio Montecarlo:** Avda 18 de Julio 1224, 1°, 11100 Montevideo; tel. (2) 9030703; fax (2) 9017762; f. 1928; commercial; Dir DANIEL ROMAY.

**Radio Sarandí:** Enriqueta Compte y Riqué 1250, 11800 Montevideo; tel. (2) 2082612; fax (2) 2036906; e-mail sarandi@netgate.com.uy; f. 1931; commercial; Pres. RAMIRO RODRÍGUEZ VALLAMIL RIVIERE.

**Radio del Sol:** tel. (2) 6283314; e-mail comoestamos@fmdelsol.com; internet www.comoestamos.com.uy.

**Radio Universal:** Avda 18 de Julio 1220, 3°, 11100 Montevideo; tel. (2) 9026022; fax (2) 9026050; e-mail info@22universal.com; internet www.22universal.com; f. 1929; commercial; Pres. OSCAR IMPERIO.

**Radiodifusión Nacional SODRE:** Sarandí 430, 11000 Montevideo; tel. (2) 957865; fax (2) 9161933; f. 1929; state-owned; Pres. JULIO CÉSAR OCAMPOS.

In 2002 there were some 16 AM and six FM radio stations in the Montevideo area. In addition, there were approximately 41 AM and 56 FM radio stations outside the capital.

#### Television

The Uruguayan Government holds a 10% stake in the regional television channel Telesur (q.v.), which began operations in May 2005 and is based in Caracas, Venezuela.

**Canal 4 Monte Carlo:** Paraguay 2253, 11800 Montevideo; tel. (2) 9244444; fax (2) 9247929; e-mail secretarias@montecarlotv.com.uy; internet www.canal4.com.uy; f. 1961; Dir HUGO ROMAY SALVO.

**SAETA TV—Canal 10:** Dr Lorenzo Carnelli 1234, 11200 Montevideo; tel. (2) 4102120; fax (2) 4009771; internet www.canal10.com.uy; f. 1956; Pres. JORGE DE FEO.

**SODRE** (Servicio Oficial de Difusión Radiotelevisión y Espectáculos): Blvr Artigas 2552, 11600 Montevideo; tel. (2) 4806448; fax (2) 4808515; e-mail direccion@tveo.com.uy; internet www.sodre.gub.uy; f. 1963; Pres. Dra NELLY GOITIÑO.

**Teledoce Televisora Color—Canal 12:** Enriqueta Compte y Riqué 1276, 11800 Montevideo; tel. (2) 2083555; fax (2) 2037623; e-mail latele@teledoce.com; internet www.teledoce.com; f. 1962; Gen. Man. HORACIO SCHECK.

**TV Ciudad:** Javier Barrios Amorín 1460, Montevideo; tel. (2) 4021908; fax (2) 4001908; e-mail tvciudad@tvciudad.imm.gub.uy; internet www.montevideo.gub.uy/teveciudad; f. 1996; state-owned.

In 1999 there were 21 television stations outside the capital.

# Finance

## BANKING
(cap. = capital; res = reserves; dep. = deposits; m. = million; amounts in pesos, unless otherwise indicated)

### State Banks

**Banco Central del Uruguay:** Avda Juan P. Fabini 777, Casilla 1467, 11100 Montevideo; tel. (2) 9085629; fax (2) 9021634; e-mail info@bcu.gub.uy; internet www.bcu.gub.uy; f. 1967; note-issuing bank, also controls private banking; Pres. WALTER CANCELA; Vice-Pres. CÉSAR FAILACHE; Gen. Man. ANDRÉS PIERONI; Gen. Sec. Dr AURELIANO BERRO.

**Banco Hipotecario del Uruguay (BHU):** Avda Daniel Fernández Crespo 1508, Montevideo; tel. (2) 4090000; fax (2) 4090782; e-mail info@bhu.net; internet www.bhu.net; f. 1892; state mortgage bank; in 1977 assumed responsibility for housing projects in Uruguay; Pres. MIGUEL PIPERNO.

**Banco de la República Oriental del Uruguay (BROU):** Cerrito y Zabala 351, 11000 Montevideo; tel. (2) 9150157; fax (2) 9162064; e-mail broupte@adinet.com.uy; internet www.brounet.com.uy; f. 1896; a state institution; cap. and res 10,848.1m., dep. 114,686.5m. (Dec. 2002); Pres. DANIEL CAIRO VILA; Gen. Man. FERNANDO JORAJURÍA; 107 brs.

**Nuevo Banco Comercial (NBC):** Casilla 34, Cerrito 400, 11000 Montevideo; tel. (2) 9160541; fax (2) 9168955; internet www.nbc.com.uy; f. 2003 by merger of Banco Comercial, Banco La Caja Obrera and Banco de Montevideo; state-run, scheduled for privatization; dep. US $667m., assets US $882m. (July 2003); privatization due by end 2005; Pres. EDUARDO ARRUABARRENA; Gen. Man. JOSÉ FUENTES; 46 brs.

### Principal Commercial Banks

**Banco Bilbao Vizcaya Argentaria Uruguay SA (BBVA):** 25 de Mayo 401, esq. Zabala, 11000 Montevideo; tel. (2) 9161444; fax (2) 9162821; internet www.bbvabanco.com.uy; f. 1968; fmrly Unión de Bancos del Uruguay, and later Banesto Banco Uruguay, SA and Banco Francés Uruguay, SA; adopted current name in 2000 following merger with Banco Exterior de América, SA; cap. 1,299.8m., res 1,091.1m., dep. 13,377.4m. (Dec. 2002); Chair. TOMÁS DEANE; Gen. Man. VICENTE BOGLIOLO DEL RÍO; 14 brs.

**Banco Galicia Uruguay, SA:** World Trade Center, Luis A. Herrera 1248, 22°, Montevideo; tel. 6281230; e-mail contactenos@bancogalicia.com.uy; internet www.bancogalicia.com.uy; f. 1999.

**Banco Surinvest SA:** Rincón 530, 11000 Montevideo; tel. (2) 9160177; fax (2) 9160241; e-mail bancosurinvest@surinvest.com.uy; internet www.surinvest.com.uy; f. 1981 as Surinvest Casa Bancaria, name changed as above 1991; cap. 202.2m., res 90.6m., dep. 3,104.1m. (Dec. 2003); Gen. Man. ALBERTO A. MELLO.

### Foreign Banks

**ABN AMRO Bank Uruguay NV** (Netherlands): Julio Herrera y Obes 1365, Casilla 888, 11100 Montevideo; tel. (2) 9031073; fax (2) 9025011; internet www.abnamro.com.uy; f. 1952; Country Rep. FRANCISCO DI ROBERTO, Jr; 24 brs.

**Banco de la Nación Argentina:** Juan Carlos Gómez 1372, 11000 Montevideo; tel. (2) 9158760; fax (2) 9164582; e-mail bna@bna.com.uy; f. 1961; Gen. Man. Dr OSCAR JORGE VISSANI; 2 brs.

**Banco Santander, SA:** Cerrito 449, esq. Misiones, 11000 Montevideo; tel. (2) 9160656; fax (2) 9163685; e-mail santander@santander.com.uy; internet www.santander.com.uy; 100% owned by Banco Santander Central Hispano (Spain); cap. 161.1m., res. 477.9m., dep. 9,283.8m. (Dec. 2000); Vice-Pres. MIGUEL ESTRUGO SANTAEUGENIA; Dir JORGE JOURDÁN PEYRONEL; 28 brs.

**Banco Sudameris** (Brazil): Rincón 500, 11000 Montevideo; tel. (2) 9150095; fax (2) 9164292; e-mail suduruguay@sudameris.com.uy; internet www.sudameris.com.uy; acquired by Banco ABN AMRO Real of Brazil in 2003; Pres. Dr SAGUNTO PÉREZ FONTANA; Gen. Man. ALEJANDRO SUZACQ; 6 brs.

**BankBoston NA** (USA): Zabala 1463, Casilla 90, 11000 Montevideo; tel. (2) 9160127; fax (2) 9162209; internet www.bankboston.com.uy; f. 1976; cap. US $10.8m., dep. $179.9m. (May 1990); Gen. Man. HORACIO VILARÓ; 13 brs.

**BNP Paribas (Uruguay) SA** (France): Rincón 477, Of. 901/5, Montevideo; tel. (2) 9162768; fax (2) 9162609; e-mail uruguay@bnpparibas.com.ar; f. 1989 as BNP (Uruguay) SA; adopted present name in 2001.

**Citibank NA** (USA): Cerrito 455, esq. Misiones, Casilla 690, 11000 Montevideo; tel. (2) 91550374; fax (2) 9150374; internet www.citibank.com.uy/uruguay; Vice-Pres. PAOLA FEOLI DE MELLO; 2 brs.

**Discount Bank (Latin America), SA** (USA): Rincón 390, 11000 Montevideo; tel. (2) 9164848; fax (2) 9160890; e-mail mensajes@discbank.com.uy; internet www.discbank.com.uy; f. 1978; cap. US $12.8m., res $0.62m., dep. $179.3m. (Dec. 2002); Pres. and Chair. ARIE SHEER; Gen. Man. VALENTIN D. MALACHOWSKI; 4 brs.

**HSBC Bank (Uruguay), SA** (United Kingdom): Ituzaingó 1389, 11000 Montevideo; tel. (2) 9153395; fax (2) 9160125; f. 1995; fmrly Banco Roberts (Argentina); CEO FERNANDO GRASSI.

**Lloyds TSB Bank PLC** (United Kingdom): Zabala 1500, Casilla 204, 11000 Montevideo; tel. (2) 9161370; fax (2) 9161262; e-mail lloydsm@lloydstsb-americas.com.uy; internet www.lloydstsb.com.uy; f. 1862; fmrly Bank of London and South America; Gen. Man. EDUARDO ANGULO.

### Credit Co-operatives

There are several credit co-operatives, which permit members to secure small business loans at preferential rates.

**Cooperativa Nacional de Ahorro y Crédito (COFAC):** Zabala 1338, 11000 Montevideo; tel. (2) 9160100; fax (2) 9160826; e-mail mensajes@cofac.com.uy; internet www.cofac.com.uy; 200,000 mems; f. 1986; operations suspended in March 2005; Gen. Man. GUSTAVO JAVIER MARTON AMEAL; 37 brs.

**Federación Uruguaya de Cooperativas de Ahorro y Credito (FUCAC):** Blvr Artigas 1472, Montevideo; tel. (2) 7088888; fax (2) 7088888; e-mail empresas@fucac.com.uy; internet www.fucac.com.uy; f. 1972; Pres. CARLOS ALBERTO ICASURIAGA SAMANO; Gen. Man. JAVIER HUMBERTO PI LEÓN.

### Bankers' Association

**Asociación de Bancos del Uruguay** (Bank Association of Uruguay): Rincón 602, 5°, 11000 Montevideo; tel. (2) 9162342; fax (2) 9162329; e-mail uy34042@adinet.com.uy; internet www.abu.org.uy; f. 1945; 7 mem. banks; Dir OSCAR JORGE VISSANI.

## STOCK EXCHANGE

**Bolsa de Valores de Montevideo:** Edif. Bolsa de Comercio, Misiones 1400, 11000 Montevideo; tel. (2) 9165051; fax (2) 9161900; e-mail info@bolsademontevideo.com.uy; internet www.bolsademontevideo.com.uy; f. 1867; 75 mems; Pres. IGNACIO ROSPIDE.

## INSURANCE

From mid-1994, following the introduction of legislation ending the state monopoly of most types of insurance, the Banco de Seguros del Estado lost its monopoly on all insurance except life, sea transport and fire risks, which have been traditionally open to private underwriters.

**AIG Uruguay Compañía de Seguros, SA** (USA): Colonia 993, 1°, Montevideo; tel. (2) 9000330; fax (2) 9084552; e-mail aig.uruguay@aig.com; internet www.aig.com; f. 1996; all classes; Gen. Man. JORGE FERRANTE.

**Alico Compañía de Seguros de Vida, SA** (USA): 18 de Julio 1738, Montevideo; tel. (2) 4033939; fax (2) 4033938; e-mail alico@alico.com.uy; internet www.alico.com; f. 1996; life; Gen. Man. JUAN ETCHEVERRY.

**Axa Seguros Uruguay, SA** (France): Misiones 1549, Montevideo; tel. (2) 9160850; fax (2) 9160847; e-mail gabriel.penna@axa-seguros.com.uy; internet www.axa-seguros.com.uy; f. 1998; general; fmrly UAP Seguros, SA; Gen. Man. GABRIEL PENNA.

**Banco de Seguros del Estado:** Avda Libertador 1465, Montevideo; tel. (2) 9089303; fax (2) 9017030; e-mail directorio@bse.com.uy; internet www.bse.com.uy; f. 1912; state insurance org.; all risks; Pres. ENRIQUE ROIG CURBELO; Gen. Man. CARLOS VALDÉS.

**Compañía de Seguros Aliança da Bahia Uruguay, SA** (Brazil): Río Negro 1394, 7°, Montevideo; tel. (2) 9021086; fax (2) 9021087; e-mail avivo@netgate.com.uy; f. 1995; transport; Gen. Man. BERNARDO VIVO.

**Mapfre Compañía de Seguros, SA** (Spain): Blvr Artigas 459, Montevideo; tel. (2) 7116595; fax (2) 7116595; e-mail info@mapfre.com.uy; internet www.mapfre.com.uy; f. 1994; general; Gen. Man. DIEGO SOBRINI.

**Porto Seguro, Seguros del Uruguay SA** (Brazil): Blvr Artigas 2025; tel. (2) 4028000; fax (2) 4030097; e-mail admin@portoseguro.com.uy; internet www.portoseguro.com.uy; f. 1995; property; Pres. LEANDRO SUÁREZ.

**Real Uruguaya de Seguros SA** (Netherlands): Avda 18 de Julio 988, Montevideo; tel. (2) 9025858; fax (2) 9024515; e-mail realseguros@abnamro.com; internet www.realseguros.com.uy; f. 1900; life and property; part of the ABN AMRO Group; Gen. Man. JOSÉ LUIZ TOMAZINI.

# URUGUAY

**Royal & SunAlliance Seguros, SA** (United Kingdom): Peatonal Sarandí 620, Montevideo; tel. (2) 9170505; fax (2) 9170490; internet www.royalsunalliance.com.uy; f. 1997; life and property; Dir Dr JUAN QUARTINO.

**Surco, Compañía Cooperativa de Seguros:** Blvr Artigas 1320, Montevideo; tel. (2) 7090089; fax (2) 7077313; e-mail surco@surco.com.uy; internet www.surco.com.uy; f. 1995; insurance co-operative; all classes; Gen. Man. ANDRÉS ELOLA.

### INSURANCE ASSOCIATION

**Asociación Uruguaya de Empresas Aseguradoras (AUDEA):** Juncal 1305, Of. 1901, 11000 Montevideo; tel. (2) 9161465; fax (2) 9165991; e-mail audea@adinet.com.uy; Pres. MANUEL RODRÍGUEZ; Gen. Man. MAURICIO CASTELLANOS.

## Trade and Industry

### GOVERNMENT AGENCIES

**Oficina de Planeamiento y Presupuesto de la Presidencia de la República:** Edif. Libertad, Luis A. de Herrera 3350, Montevideo; tel. (2) 4872110; fax (2) 2099730; e-mail diropp@presidencia.gub.uy; internet www.opp.gub.uy; f. 1976; responsible for the implementation of devt plans; co-ordinates the policies of the various ministries; advises on the preparation of the budget of public enterprises; Dir CARLOS VIERA; Sub-Dir DANIEL MESA PELUFFO.

**Uruguay XXI** (Instituto de Promoción de Inversiones y Exportaciones de Bienes y Servicios): Yaguarón 1407, Of. 1103, 11100 Montevideo; tel. (2) 9002912; fax (2) 9008298; e-mail info@uruguayxxi.gub.uy; internet www.uruguayxxi.gub.uy; govt agency to promote economic investment; f. 1996; Exec. Dir VICTOR ANGENSCHEIDT.

### DEVELOPMENT ORGANIZATIONS

**Corporación Nacional para el Desarrollo (CND):** Rincón 528, 7°, Casilla 977, 11000 Montevideo; tel. (2) 9162680; fax (2) 9159662; e-mail cnd01@adinet.com.uy; internet www.cnd.org.uy; f. 1985; national devt corpn; mixed-capital org.; obtains 60% of funding from state; Pres. ALVARO GARCÍA; Vice-Pres. RICARDO PUGLIA; Gen. Man. MARTÍN J. DIBARBOURE ROSSINI.

**Asociación Nacional de Micro y Pequeños Empresarios (ANMYPE):** Miguelete 1584, Montevideo; tel. (2) 9241010; e-mail anmype@anmype.net.uy; internet www.anmype.net.uy; promotes small businesses; f. 1988; Pres. RICARDO POSADA.

**Asociación Nacional de Organizaciones No Gubernmentales Orientadas al Desarrollo:** Avda del Libertador 1985 escalera 202, Montevideo; tel. and fax (2) 9240812; e-mail anong@anong.com.uy; internet www.anong.org.uy; f. 1992; umbrella grouping of devt NGOs; Pres. MARÍA ELENA MARTÍNEZ.

**Centro Interdisciplinario de Estudios sobre el Desarrollo, Uruguay (CIEDUR):** 18 de Julio 1645-7, Casilla 11200, Montevideo; tel. and fax (2) 408 4520; e-mail ciedur@chasque.net; internet www.chasque.net/ciedur; devt studies and training; Exec. Sec. ALFREDO BLUM.

**Fundación Uruguaya de Cooperación y Desarrollo Solidario (FUNDASOL)** (Uruguayan Foundation for Supportive Co-operation and Development): Blvr Artigas 1119, esq. Maldonado, Montevideo 11200; tel. (2) 4002020; fax (2) 4081485; e-mail consultas@fundasol.org.uy; internet www.fundasol.org.uy; f. 1979; Gen. Man. JORGE NAYA.

**Programa Alianzas para el Desarrollo Local en América Latina (ALOP):** tel. and fax (2) 9007194; e-mail info@desarrollolocal.org; internet www.desarrollolocal.org; umbrella grouping of local devt orgs; Dir ENRIQUE GALLICCHIO.

### CHAMBERS OF COMMERCE

**Cámara de Industrias del Uruguay** (Chamber of Industries): Avda Italia 6101, Montevideo 11500; tel. (2) 6040464; fax (2) 6040501; e-mail ciu@ciu.com.uy; internet www.ciu.com.uy; f. 1898; Pres. WASHINGTON BURGHI.

**Cámara Nacional de Comercio y Servicios del Uruguay** (National Chamber of Commerce): Edif. Bolsa de Comercio, Rincón 454, 2°, Casilla 1000, 11000 Montevideo; tel. (2) 9161277; fax (2) 9161243; e-mail info@cncs.com.uy; internet www.cncs.com.uy; f. 1867; 1,500 mems; Pres. JOSÉ LUIS PUIG; Sec. and Man. Dr CLAUDIO PIACENZA.

**Cámara Mercantil de Productos del País** (Chamber of Commerce for Local Products): Avda General Rondeau 1908, 1°, 11800 Montevideo; tel. (2) 9240644; fax (2) 9240673; e-mail info@camaramercantil.com.uy; internet www.camaramercantil.com.uy; f. 1891; 180 mems; Pres. RICARDO SEIZER; Gen. Man. GONZALO GONZÁLEZ PIEDRAS.

### EMPLOYERS' ORGANIZATIONS

**Asociación de Importadores y Mayoristas de Almacén** (Importers' and Wholesalers' Asscn): Edif. Bolsa de Comercio, Of. 317/319, Rincón 454, 11000 Montevideo; tel. (2) 9156103; fax (2) 9160796; e-mail fmelissori@nidera.com.uy; f. 1926; 52 mems; Pres. FERNANDO MELISSORI.

**Asociación Rural del Uruguay (ARU):** Avda Uruguay 864, 11100 Montevideo; tel. (2) 9020484; fax (2) 9020489; e-mail consultas@aru.com.uy; internet www.aru.com.uy; f. 1871; 1,800 mems; Pres. FERNANDO MATTOS COSTA; Vice-Pres. ROBERTO SYMONDS.

**Comisión Patronal del Uruguay de Asuntos Relacionados con la OIT** (Commission of Uruguayan Employers for Affairs of the ILO): Edif. Bolsa de Comercio, Rincón 454, 2°, Casilla 1000, 11000 Montevideo; tel. (2) 9161277; fax (2) 9161243; f. 1954; mem. of Cámara Nacional de Comercio y Servicios del Uruguay; 8,000 mems; Sec. and Man. Dr CLAUDIO PIACENZA.

**Federación Rural del Uruguay:** Avda 18 de Julio 965, 1°, Montevideo; tel. (2) 9005583; fax (2) 9004791; e-mail fedrural@adinet.com.uy; f. 1915; 2,000 mems; Pres. ALEJANDRO TEDESCO.

**Unión de Exportadores del Uruguay** (Uruguayan Exporters' Asscn): Edif. Nacional de Aduanas, Yacaré s/n, 11000 Montevideo; tel. (2) 9170105; fax (2) 9165967; e-mail info@uruguayexporta.com; internet www.uruguayexporta.com; Pres. DANIEL SOLODUCHO; Exec. Sec. TERESA AISHEMBERG.

### MAJOR COMPANIES

**Acindar Uruguay:** World Trade Center Montevideo, Avda Luis A. de Herrera 1248, Of. 321, 11300 Montevideo; tel. (2) 6286316; e-mail acindar@multi.com.uy; internet www.acindar.com.ar; production of iron and steel.

**Azucarera del Litoral, SA:** Meriggi y Libertad, Paysandú 60; tel. (2) 9160868; fax (2) 9161192; f. 1943; processors of raw cane sugar; Gen. Man. RAÚL CONCELO; 495 employees.

**Compañía Industrial de Tabacos Monte Paz, SA:** San Ramón 716, Montevideo; tel. (2) 2008821; fax (2) 2037890; e-mail info@montepaz.com.uy; internet www.montepaz.com.uy; f. 1930; tobacco and cigarette manufacturers; Pres. JORGE LUIS MAILHOS; 425 employees.

**Compañía Sudamericana de Empresas Eléctricas, Mecánicas y Obras Públicas (SACEEM):** Treinta y Tres 1468, 11800 Montevideo; tel. (2) 9160208; fax (2) 9163939; e-mail saceem@uyweb.com.uy; f. 1951; construction of industrial buildings and warehouses; Chair. MARTÍN CARRIQUIRY; 800 employees.

**Compañía Uruguaya de Transportes Colectivos, SA (CUTCSA):** Sarandí 528, 11000 Montevideo; tel. (2) 9157422; fax (2) 9162807; e-mail cac@cutcsa.com; internet www.cutcsa.com; f. 1937; passenger transport services; Pres. JUAN SALGADO; 5,000 employees.

**Cooperativa Nacional de Productores de Leche, SA (CONAPROLE):** Magallanes 1871, 11200 Montevideo; tel. (2) 9247171; fax (2) 9246672; e-mail jfernandez@conaprole.com.uy; internet www.conaprole.com.uy; f. 1936; manufacturers and wholesalers of milk and dairy products; Pres. JORGE PANIZZA TORRENS; Gen. Man. RUBEN NÚÑEZ HERNÁNDEZ; 2,200 employees.

**Cybaran, SA** (Frigorífico la Caballada, SA): Lima 1200, 11800 Montevideo; tel. (2) 9240687; fax (2) 9240721; f. 1962; processors and wholesalers of meat; Pres. MARCO ANDRÉS DUTRA; 700 employees.

**Dymac, SA:** Thompson 3077, 11600 Montevideo; tel. (2) 4870812; fax (2) 4872786; f. 1962; clothing manufacturers; Pres. DANIEL DYMENSTEIN; 520 employees.

**Fábrica Nacional de Papel, SA (FANAPEL):** Rincón 477, 6°, 11000 Montevideo; tel. (2) 9150917; fax (2) 9163096; e-mail fanapel@fanapel.com.uy; internet www.fanapel.com.uy; f. 1898; pulp and paper mill; Pres. RICARDO ZERBINO CAVAJANI; 680 employees.

**Fábrica Uruguaya de Neumáticos, SA (FUNSA):** Corrales 3076, Casilla 15175, Montevideo; tel. (2) 5083141; fax (2) 5070611; e-mail funsa@ciu.com.uy; f. 1935; manufacturers of rubber tyres, shoes and insulated electrical cables; 81% owned by Titan Corpn of the USA; Gen. Man. OSCAR VÁZQUEZ; 600 employees.

**Fábricas Nacionales de Cerveza, SA (FNC):** Entre Ríos 1060, Montevideo; tel. (2) 2001683; fax (2) 2034525; e-mail fnc@multi.com.uy; internet www.fnc.com.uy; f. 1932; brewery; Pres. JUAN D. GANG; 400 employees.

**FRIPUR, SA:** Avda General Rondeau 2260, Montevideo; tel. (2) 9245821; fax (2) 9243149; e-mail informes@fripur.com.uy; internet www.fripur.com.uy; f. 1976; foodstuffs and fish processing; Pres. MAXIMO FERNÁNDEZ ALONSO; 1,185 employees.

# URUGUAY

**Industria Lanera del Uruguay, SA (ILDUSA):** José de Bejar 2600, Montevideo; tel. (2) 5083161; fax (2) 5071068; f. 1932; manufacturers of woollen textiles; Pres. ALBERTO PUIG TERRA; 980 employees.

**Industrias Philips del Uruguay, SA:** Avda Uruguay 1248, 10°, Montevideo; tel. (2) 6281111; fax (2) 6287777; f. 1957; subsidiary of NV Philips (Netherlands); manufacturers of lighting and other electrical goods; Pres. ANTONIO MOLEMAR; 300 employees.

**Montevideo Refrescos, SA:** Camino Carrasco 6173, Montevideo; tel. (2) 6008401; fax (2) 6008186; f. 1946; owned by Coca-Cola Corpn of the USA; producers of carbonated beverages; Pres. GREGORIO AZNARES; 560 employees.

**Motociclo, SA:** Avda Uruguay 1171, Montevideo 11100; tel. (2) 9020070; fax (2) 902 1702; e-mail ventas@motociclo.com.uy; internet www.motociclo.com.uy; f. 1931; bicycle and motorcycle manufacturers; annual capacity of 370,000 units; Pres. LEONARDO ROZENBLUM; 450 employees.

**Paysandú Industrias del Cuero, SA (PAYCUEROS):** Rincón 487, 9°, Montevideo; tel. (2) 9160023; fax (2) 9162919; e-mail infomvd@sadesa.com; f. 1946; tannery, manufacturing handbags and other leather products; owned by SADESA of Argentina; Pres. ISAAC TOLODUCHO; 700 employees.

**Shell Uruguay:** San Fructuoso 921, Casilla 11800, Montevideo; tel. (2) 2009920; fax (2) 2088468; e-mail elizabeth.mountford@shell.com; internet www.shell.com.uy; f. 1919 in Uruguay; Gen. Man. JUAN CARLOS ESPONDA.

**Sociedad Anónima Arroceros Nacionales (SAMAN):** Rambla Baltasar Brum 2772, Montevideo; tel. (2) 2081421; fax (2) 2037007; e-mail info@saman.com.uy; internet www.saman.com.uy; f. 1942; rice mills; Pres. RICARDO FERRÉS BLANCO; 538 employees.

**Supermercados Disco del Uruguay:** Jaime Zudáñez 2627, Montevideo; tel. (2) 7107421; fax (2) 7117903; internet www.disco.com.uy; f. 1960; Uruguayan subsidiary of Disco SA, Argentina; Pres. LUIS EDUARDO CORDUZO.

## UTILITIES

### Electricity

**Administración Nacional de Usinas y Transmisiones Eléctricas (UTE):** Paraguay 2431, 10°, 11100 Montevideo; tel. (2) 2003424; fax (2) 2037082; e-mail ute@ute.com.uy; internet www.ute.com.uy; f. 1912; autonomous state body; sole purveyor of electricity until 1997; Pres. Ing. BENO RUCHANSKY; Gen. Man. CARLOS POMBO.

### Gas

**Conecta:** Sanlúcar 1631, esq. Avda Rivera, Montevideo; tel. (2) 6008400; fax (2) 6006732; internet www.conecta.com.uy; gas distribution.

**Gaseba Uruguay (Gaz de France):** 25 de Mayo 702, 11000 Montevideo; tel. (2) 9017454; internet www.gaseba.com.uy; gas producers and service providers.

### Petroleum

**Administración Nacional de Combustibles, Alcohol y Portland (ANCAP):** Paraguay 1598, 11100 Montevideo; tel. (2) 9020608; fax (2) 9021136; e-mail webmaster@ancap.com.uy; internet www.ancap.com.uy; f. 1931; deals with transport, refining and sale of petroleum products, and the manufacture of alcohol, spirit and cement; tanker services, also river transport; Pres. DANIEL MARTÍNEZ; Gen. Man. SERGIO LATTANZIO; Sec.-Gen. JORGE URRUTIA.

### Water

**Aguas de la Costa:** Maldonado; subsidiary of Aguas de Barcelona (Spain); operating in Uruguay since 1994, contract due to expire in 2019; management of water supply in Maldonado dept.

**Obras Sanitarias del Estado (OSE):** Carlos Roxlo 1275, 11200 Montevideo; tel. (2) 4001151; fax (2) 4088069; e-mail oserou@adinet.com.uy; internet www.ose.com.uy; f. 1962; processing and distribution of drinking water, sinking wells, supplying industrial zones of the country; Pres. Ing. JORGE CARLOS COLACCE MOLINARI; Vice-Pres. FERNANDO DANIEL NOPITSCH D'ANDREA.

## TRADE UNION

**Plenario Intersindical de Trabajadores—Convención Nacional de Trabajadores (PIT—CNT):** Avda 18 de Julio 2190, Montevideo; tel. (2) 4096680; fax (2) 4004160; e-mail pitcnt@adinet.com.uy; internet www.chasque.net/icudu; f. 1966; org. comprising 83 trade unions, 17 labour federations; 320,000 mems; Pres. JORGE CASTRO; Exec. Sec. JUAN CASTILLO.

## Transport

**Dirección Nacional de Transporte:** Rincón 575, 5°, 11000 Montevideo; tel. (2) 9162940; fax (2) 9163122; e-mail dntdinac@adinet.com.uy; internet www.dnt.gub.uy; co-ordinates national and international transport services.

### RAILWAYS

**Administración de los Ferrocarriles del Estado (AFE):** La Paz 1095, Casilla 419, Montevideo; tel. (2) 9240805; fax (2) 9240847; e-mail affegg@adinet.com.uy; f. 1952; state org.; 3,002 km of track connecting all parts of the country; there are connections with the Argentine and Brazilian networks; passenger services ceased in 1988; passenger services linking Montevideo with Florida and Canelones were resumed in mid-1993; Pres. NILO OJEDA.

### ROADS

In 2003 Uruguay had an estimated 8,733 km of motorways (forming the most dense motorway network in South America), connecting Montevideo with the main towns of the interior and the Argentine and Brazilian frontiers. There was also a network of approximately 40,000 km of paved roads under departmental control.

**Corporación Vial del Uruguay, SA:** Rincón 528, 7°, 11000 Montevideo; tel. (2) 9170114; e-mail cvu@cnd.org.uy; state road construction agency; 100% owned by the Corporación Nacional para el Desarrollo; Gen. Man. CRISTINA MONTES; Pres. ALDO BONSIGNORE.

### INLAND WATERWAYS

There are about 1,250 km of navigable waterways, which provide an important means of transport.

**Nobleza Naviera, SA:** Avda General Rondeau 2257, Montevideo; tel. (2) 9243222; fax (2) 9243218; e-mail nobleza@netgate.com.uy; operates cargo services on the River Plate, and the Uruguay and Paraná rivers; Chair. AMÉRICO DEAMBROSI; Man. Dir DORIS FERRARI.

### SHIPPING

**Administración Nacional de Combustibles, Alcohol y Portland (ANCAP):** Paraguay 1598, 11100 Montevideo; tel. (2) 9020608; fax (2) 9021136; e-mail webmaster@ancap.com.uy; internet www.ancap.com.uy; f. 1931; deals with transport, refining and sale of petroleum products, and the manufacture of alcohol, spirit and cement; tanker services, also river transport; Pres. DANIEL MARTÍNEZ; Gen. Man. SERGIO LATTANZIO; Sec.-Gen. JORGE URRUTIA.

**Administración Nacional de Puertos (ANP):** Rambla 25 de Agosto de 1825 160, Montevideo; tel. (2) 9151441; fax (2) 9161704; e-mail presidencia@anp.com.uy; www.anp.com.uy; f. 1916; national ports admin.; Pres. LUIS E. LOUREIRO.

**Prefectura Nacional Naval:** Edif. Comando General de la Armada, 5°, Rambla 25 de Agosto de 1825 s/n, esq. Maciel, Montevideo; tel. (2) 9160741; fax (2) 9163969; internet www.armada.gub.uy/prena; f. 1829; maritime supervisory body, responsible for rescue services, protection of sea against pollution, etc.; Commdr Rear-Adm. OSCAR OTERO IZZI.

**Navegación Atlántida, SA:** Río Branco 1373, 11100 Montevideo; tel. (2) 9084449; f. 1967; ferry services for passengers and vehicles between Argentina and Uruguay; Pres. H. C. PIETRANERA.

**Transportadora Marítima de Combustibles, SA (TRAMACO, SA):** Rincón 540, Puerta Baja, 11000 Montevideo; tel. (2) 9165754; fax (2) 9165755; e-mail tramaco@tramaco.com.uy; Pres. J. FERNÁNDEZ BAUBETA.

### CIVIL AVIATION

Civil aviation is controlled by the Dirección General de Aviación Civil and the Dirección General de Infraestructura Aeronáutica. The main airport is at Carrasco, 21 km from Montevideo, and there are also airports at Paysandú, Rivera, Salto, Melo, Artigas, Punta del Este and Durazno.

**Primeras Líneas Uruguayas de Navegación Aérea (PLUNA):** Colonia 1013, 11000 Montevideo; tel. (2) 9030273; fax (2) 9023916; e-mail info@pluna.com.uy; internet www.pluna.aero; f. 1936; nationalized 1951; partially privatized in 1994; 49% stake acquired by Aerolíneas Argentinas in 2004; operates international services to Argentina, Brazil, Chile, El Salvador, Paraguay, Spain and the USA; Dir VÍCTOR MESA; Man. JORGE NEVES.

**Aeromás, SA:** Aeropuerto Internacional de Carrasco s/n, Of. 101, Montevideo; tel. (2) 6040294; fax (2) 6040013; e-mail aeromas@aeromas.com; internet www.aeromas.com; private hire, cargo, and air ambulance flights; internal mass transit services to Salto, Paysandú, Rivera, Tacuarembó and Artigas; f. 1983; Dir DANIEL DALMÁS.

## Tourism

The sandy beaches and woodlands on the coast and the grasslands of the interior, with their variety of fauna and flora, provide the main tourist attractions. About 57% of tourists come from Argentina, and a further 10% from Brazil. Uruguay received 1.5m. visitors in 2003. In that year tourism revenues totalled US $406m.; however, these figures represented a significant decrease from the the 2.1m. visitors received in 2001, when revenues totalled $700m. The decline was owing to economic difficulties in neighbouring Argentina,which accounted for over one-half of all tourist arrivals. Brazil accounted for a further 10% of tourists.

**Asociación Uruguaya de Agencias de Viajes (AUDAVI):** Río Branco 1407, Of. 205, 11100 Montevideo; tel. (2) 9012326; fax (2) 9021972; e-mail audavi@netgate.com.uy; f. 1951; 100 mems; Pres. FEDERICO GAMBARDELLA; Man. MÓNICA W. DE RAIJ.

**Cámara Uruguaya de Turismo:** La Paz 3052, 11800 Montevideo; tel. (2) 4016013; fax (2) 4016013.

**Uruguay Natural:** Rambla 25 de Agosto de 1825, esq. Yacaré s/n, Montevideo; tel. (2) 1885100; e-mail webmaster@mintur.gub.uy; internet www.uruguaynatural.com; f. 2003; state-run tourism promotion agency; Dir-Gen. MARTHA CASAL.

**Uruguayan Hotel Association:** Gutiérrez Ruiz 1213, Montevideo; tel. (2) 9080141; fax (2) 9082317; e-mail ahru@montevideo.com.uy; internet www.ahru.org; Pres. EDGARDO BENZO.

## Defence

In August 2004 Uruguay's Armed Forces consisted of 24,000 volunteers between the ages of 18 and 45 who contract for one or two years of service. There was an Army of 15,200, a Navy of 5,700 (including a coastguard service of 1,600) and an Air Force of 3,100. There were also paramilitary forces numbering 920.

**Defence Budget:** an estimated 3,000m. pesos uruguayos in 2004.

**Commander-in-Chief of the Army:** Lt.-Gen. SANTIAGO H. POMOLI.

**Commander-in-Chief of the Navy:** Vice-Adm. TABARÉ YAMANDÚ DANERS EYRAS.

**Commander-in-Chief of the Air Force:** Lt-Gen. ENRIQUE ATILIO BONELLI.

## Education

All education in Uruguay, including university tuition, is provided free of charge. Education is officially compulsory for six years between six and 14 years of age. Primary education begins at the age of six and lasts for six years. Secondary education, beginning at 12 years of age, lasts for a further six years, comprising two cycles of three years each. In 1996 the total enrolment at both primary and secondary schools was equivalent to 97% of the school-age population. Primary enrolment in that year included an estimated 93% of children in the relevant age-group (males 92%; females 93%), while secondary enrolment was equivalent to 85% of the population in the appropriate age-group (males 77%; females 92%). The programmes of instruction are the same in both public and private schools and private schools are subject to certain state controls. There are two universities. Expenditure on education by the central Government in 2003 was 10,558m. pesos uruguayos (11.8% of total government spending).

## Bibliography

For works on South America generally, see Select Bibliography (Books)

Achard, D. *La Transición en Uruguay*. Montevideo, Ingenio de Servicios de Comunicación y Marketing, 1992.

Alexander, R. J. *A History of Organized Labor in Uruguay and Paraguay*. Westport, CT, Praeger Publrs , 2005.

Barahona de Brito, A. *Human Rights and Democratization in Latin America: Uruguay and Chile*. Oxford, Oxford University Press, 1997.

de Castro Gomes, A. *Estado, Corporativismo y Acción Social en Brasil, Argentina y Uruguay*. Buenos Aires, Editorial Biblos, 1992.

Gillespie, C. G. 'Negotiating Democracy: Politicians and Generals in Uruguay', in *Cambridge Latin American Studies*, No. 72. Cambridge, Cambridge University Press, 1992.

González, L. E. *Political Structures and Democracy in Uruguay*. Notre Dame, University of Notre Dame Press, 1991.

Heinz, W., and Fruhling, H. *Determinants of Gross Human Rights Violations by State and State-sponsored Actors in Brazil, Uruguay, Chile and Argentina*. Zoetermeer, Martinus Nijhoff Publrs, 1999.

Lavin, A. *Women, Feminism and Social Change in Argentina, Chile and Uruguay, 1840–1940*. Lincoln, NE, University of Nebraska, 1998.

Minsburg, N. *El Mercosur, un Problema Complejo*. Buenos Aires, Centro Editor de America Latina, 1993.

Roniger, L., and Sznajder, M. *The Legacy of Human-Rights Violations in the Southern Cone: Argentina, Chile and Uruguay*. Oxford, Oxford University Press, 1999.

*Uruguay, Trade Reform and Economic Efficiency*. Washington, DC, World Bank, 1993.

Villareal, N. *La Izquierda en Uruguay, Impactos y Reformulaciones, 1989–1992*. Montevideo, Observatorio del Sur, 1992.

Weschler, L. *A Miracle, A Universe: Settling Accounts with Torturers*. Chicago, IL, University of Chicago Press, 1998.

# VENEZUELA

## Geography

### PHYSICAL FEATURES

The Bolivarian Republic of Venezuela is on the northern coast of South America. Colombia lies to the west of the country, pushing into it in the south-west. This border (2,050 km or 1,73 miles) is only a little shorter than that with Brazil (2,200 km), which lies to the south. To the east is Guyana—beyond a 743-km frontier, which Venezuela claims should be further east still, along the Essequibo river. There is also a dispute over maritime boundaries in the Gulf of Venezuela with Colombia, and several Caribbean nations object to Venezuelan possession of the isolated Isla des Aves (Island of Birds), 565 km north of the mainland, on a similar latitude to northern Dominica (over 200 km to the east). There are islands that are not Venezuelan territory much closer to the mainland: northern Trinidad (Trinidad and Tobago) is 11 km off shore, to the east of the Paria peninsula and dropping south above the Orinoco delta; at the other end of the country, in the west, the Dutch island of Aruba is 25 km north of the Paraguaná peninsula (at the mouth of the Gulf of Venezuela); while a little further east is Curaçao (55 km off shore), and then Bonaire (80 km), both part of the Netherlands Antilles, although Bonaire is further from the mainland than it is from the Venezuelan dependencies in the Lesser Antilles (specifically, the Islas Las Aves—not to be confused with the single, northerly Aves island mentioned above). Aves is the most northerly of the 72 Caribbean islands, islets or cays included within the territory of Venezuela, which totals 916,445 sq km (353,841 sq miles).

Venezuela stretches along a mainly north-facing coast, the west penetrating less deeply inland than the east, but with a southern extension in the centre of the country thrusting towards the Amazon basin to include the headwaters of the Orinoco. More than 2,700 km of coast is mainly along the Caribbean; it is narrow, steep and deeply indented, owing to the mountains coming so close to the sea. Only in the far east is it low and marshy, around the delta through which the Orinoco debouches into the Atlantic. Just north of this delta is the Gulf of Paria, between Venezuela and the island of Trinidad, and most of the shore runs west from here, to the Gulf of Venezuela, near the Colombian border. Parallel to this coast are most of Venezuela's Caribbean islands, the largest being Isla de Margarita. The Gulf of Venezuela is between two peninsulas, that of Paraguaná and that of Guajira, the latter on the west, its head being Colombian territory. Leading south from the Gulf is a channel of 8–15 km in width (dredged to a depth of 11 m—36 ft—in 1956, to make it navigable for larger vessels) into Lake Maracaibo, the largest lake in South America (210 km by 120 km). In all, Venezuela has about 30,000 sq km of inland waters. The country can be divided into four topographical regions: the Maracaibo lowlands; the Andean highlands of the north-west and the Caribbean coast; the vast plains (Llanos) of the centre-north; and the Guianan highlands, which dominate the east and south.

The lowlands around Lake Maracaibo, surrounded by heights and giving on to coastal lowlands along the Gulf of Venezuela, constitute the smallest natural region of the country (about 15% of its territory). The fairly brackish lake is in the far north-west, and the original stilt-supported villages of the local Amerindians along its shores inspired the Spanish name for the country ('little Venice'). Now Maracaibo is famous for its petroleum wealth, although it also has a diverse natural environment—semi-arid brush cloaks the dry north, with wooded savannah intervening before the tropical forest and swampy lagoons of the south. Forests continue up the flanks of the mountains, as Maracaibo is cupped between two Andean ranges. The Andes of Colombia's Cordillera Oriental split just before they enter western Venezuela, the crest of the more westerly Sierra de Perijá heading northwards to form much of the border between the two countries. Some of the Perijá peaks reach over 3,400 m, but the higher range runs in a north-easterly direction, towards the Caribbean, and is known as the Cordillera de Mérida. Here is the highest point in Venezuela, the Pico Bolívar, or La Columna, at 5,007 m (16,433 ft). Before hitting the Caribbean coast, the general alignment of the heavily forested mountains turns east, and the main range runs between the sea and the Orinoco plains, to peter out and then briefly rear up again before descending into the flat, marshy Orinoco delta. The south-eastern half of western Venezuela is occupied by the uplands of the broad Llanos, tropical grasslands that lap the foothills of the Cordillera de Mérida and the coastal range. The Llanos, which covers about one-third of the country, seldom itself exceeds 215 m above sea level and slopes steadily towards the Orinoco delta. All this area, indeed four-fifths of the country, is drained by the Orinoco, which forms some of the western border of Venezuela's southern extension, before heading east and a little north, between the coastal and Guianan highlands, towards the sea. The plains stretch a considerable distance to the north and west of the main river, but vast tracts are flood prone in May–November, yet parched in summer. The lower reaches remain wetter and there are permanent wetlands and mangroves in the delta. A further one-third of the country comprises the Guianan highlands of the south-east. The centre of the country is dominated by the rugged, hilly plateau that marks the start of the highlands that stretch eastwards to divide the Guiana coast from the Amazon basin. In Venezuela the central heights push south to cup the westward-opening basin of the Orinoco headwaters and east, to fall steeply into the Orinoco valley to the north. The landscape alternates between open grasslands and dense forests. Some of the mountains top 2,700 m, and the precipitous terrain allows for some dramatic scenery, such as the Angel Falls, the highest waterfall in the world. Deforestation and irresponsible mining are particular threats to the environment of this region, with industrial pollution more of an issue in the west, but over one-half of the country is still wooded, and one-fifth is classed as pastureland. The type of vegetation is determined more by elevation than latitude, with plants common to a temperate zone established above 900 m, tropical forests in the lower country, mangroves in the Orinoco delta and long prairie grass on the Llanos.

## CLIMATE

The country lies entirely in the Tropic of Cancer, and its climate is tropical on the Llanos and on the coast, but temperate in the mountains. Altitude is very important, with hot, temperate and cold climates distinguished locally. There is a wet season in May–November, with more rain falling on the southern slopes than the northern, annual averages ranging from about 1,400 mm (55 ins) in the Andes to 280 mm on the coast. Caracas, the national capital in the mountains immediately above the Caribbean coast, has an average daily temperature range of 59°–78°F (15°–26°C) in January and 63°–80°F (17°–26°C) in July. Lowland Maracaibo has ranges of 73°–90°F (23°–32°C) and 76°–94°F (24°–34°C), respectively.

## POPULATION

The population is predominantly Mestizo (67%), with 21% as largely unmixed European (predominantly of Spanish descent, but also Italian, Portuguese and German, for instance). Most of the rest of the population are black or mulatto, although there are small communities from other ethnic groups, like Arabs, and 2% are Amerindian. There are at least 40 Amerindian groups, mainly in the Amazon basin, the more isolated ones retaining principal use of their native languages, rather than the more widely used official language, Spanish. The country is overwhelmingly Roman Catholic (84%), with about 2% of Protestants, and some Jews and adherents of other faiths (including native, animist faiths). The cultural dominance of the Roman Catholic, Spanish-speaking majority is not unduly disturbed by the presence of a large expatriate community, as it is mainly constituted of Colombians.

The total population was estimated at 26.1m. in mid-2004. Most people live in the coastal highlands, and 87% are urbanized. Society is also divided between extremes of rich and poor. Caracas, the federal capital (with a population of some 2.0m. at mid-2000), is in the central coastal highlands, its port being La Guaira. The second city of the country is Maracaibo (1.8m.), in the west, on the north-western shore of the lake that shares its name. Another important city is the manufacturing centre of Valencia (1.3m.), mid-way between Caracas and Barquisimeto (0.9m.), a transport hub situated at the northern end of the main Cordillera de Mérida and the western end of the coastal range. The next city in size is the largest in eastern Venezuela, Ciudad Guayana (0.7m.). The country is a federal republic constituted of 23 states, a Federal District (Caracas) and 11 federally controlled Caribbean island groups (totalling 72 islands) described as federal dependencies.

# History

## Dr JULIA BUXTON

Europeans first discovered Venezuela in 1498, during the navigator Christopher Columbus's third Spanish expedition to the New World. In 1499 a Spanish conquistador, Alonso de Ojeda, reached Lake Maracaibo. The Amerindian villages constructed on poles over the lake reminded him of a little Venice—hence the name Venezuela. The country was an economic disappointment to the Spanish conquistadors, as it lacked mining potential and a settled indigenous population, but the mythical land of El Dorado was thought to lie in the Orinoco delta, and this sustained interest in the territory.

## INDEPENDENCE

Until the foundation of Nueva Granada as a Viceroyalty in 1739, Venezuela was administered from Lima, Peru. In 1777 Venezuela became a Spanish Captaincy-General, with an enhanced degree of administrative autonomy from Bogotá, the capital of Nueva Granada. In 1724 a company of Basque merchants, the Caracas Company, obtained a monopoly of the territory's foreign trade and developed new markets in Europe and the Caribbean for local produce, including cocoa and coffee. The export market fostered a small élite of European planters, the so-called *Marqueses de Chocolate*, and it was a member of this class, Simón Bolívar, who emerged as the leader of the movement for independence in the Andean region. Venezuela gained independence from Spain in 1819 and joined with Colombia, Ecuador and Panama to form the 'Gran Colombia' federation. Bolívar perceived regional unification as an essential counter to the emerging power of the USA, but this objective was undermined by in-fighting between Venezuelan and Colombian élites, and in 1830 the federation was dissolved and Venezuela became a separate republic.

Independence did not bring democracy to Venezuela and for the following 100 years military oligarchs fought each other for control of the country. Economic growth following from the discovery and exploitation of oil reserves at the end of the 19th century catalysed rapid social change. A fledgling democratic opposition emerged from the middle-class student movement during the dictatorship of Juan Vicente Gómez (1908–35). Repression failed to stem the reformist impetus, and influential student leaders Rómulo Betancourt and Dr Rafael Caldera Rodríguez formed what were to become the country's leading political parties: the social-democrat Acción Democrática (AD— Democratic Action) party and the Christian-democrat Comité de Organización Política Electoral Independiente (COPEI— Committee of Independent Electoral Political Organization) in 1936 and 1946, respectively. The student movement and the political parties after them found it difficult to organize in the repressive political conditions and developed hierarchical and centralized internal structures controlled by the party élite, or *cogollo*, to counter infiltration. Collaboration between AD and progressive elements of the military led to a coup in 1945 that brought democratic elections and propelled AD to power for three years. The Government was, however, weakened by intense partisan conflict and, after a military coup in 1948, Gen. Marcos Pérez Jiménez seized power. Pérez Jiménez occupied the presidency for 10 years. His regime was notable for its nation-building projects but also its corruption, economic mismanagement and oppression of the opposition. A military rebellion and general strike forced the dictator to flee the country in January 1958. Prior to the fall of Pérez Jiménez, AD, COPEI, and representatives from the private sector, the union movement and the Roman Catholic Church had signed the Pact of Punto Fijo in 1957. This established a political alliance and centrist policy consensus, the so-called *coincidencia* between the two parties. This institutional engineering aimed to create political stability and avoid further military dictatorships in Venezuela. The Pact committed AD and COPEI to share appointments to the state administration (including the election administration and senior positions in the military and judiciary), to respect the outcome of democratic elections and to control the demands of their respective constituencies. Petroleum export revenues played a fundamental role in the consolidation of the new democratic system. The distribution of the petrodollars through the network of clientelist interests affiliated to AD and COPEI created strong support for the maintenance of this model of democracy and the two parties that had created it. The revenues ensured that all groups' demands could be met, from the business élite to the working classes, enabling Venezuela to avoid class-based conflicts and political mobilization.

## THE DEVELOPMENT OF PETROLEUM RESOURCES

Venezuela had been the world's third largest producer of coffee in the 19th century, after Brazil and Java (the latter now part of Indonesia). Following the discovery of petroleum resources at the end of that century, oil overtook coffee as the country's primary export commodity. Venezuela's importance as a petroleum exporter was enhanced by the nationalization of Mexico's oil industry in 1938 and the outbreak of the Second World War in Europe in 1939. Venezuela subsequently became increasingly skilled in international negotiation, and, with the transition to democracy in 1958, adopted the so-called 'fifty-fifty' profit-sharing agreement with those foreign oil companies that had received concessions to exploit the country's reserves during the Governments of Gómez and Pérez Jiménez. This 'oil nationalism' that characterized the Punto Fijo period also led Venezuela to undertake much of the preliminary planning that culminated in the creation of the Organization of the Petroleum Exporting Countries (OPEC), of which Venezuela was one of five founder member nations. OPEC was formally constituted at a conference in Venezuela in January 1961. Venezuela's petroleum industry was nationalized in 1976 following the expiry of private-sector concessions.

## THE ADMINISTRATION OF CARLOS ANDRÉS PÉREZ RODRÍGUEZ

Venezuela's democracy was seen to be rapidly consolidated after the democractic elections of 1958 that brought the AD leader, Rómulo Betancourt to the presidency. For the following two decades, major social progress was made as successive AD and COPEI Governments channelled petroleum export earnings into state subsidies and social investment. A comprehensive welfare-state system was introduced and a programme of land distribution implemented. Following the national elections of 1973, which brought the AD candidate, Carlos Andrés Pérez, to power, Venezuela experienced a dramatic change in economic fortunes as the Arab oil embargo led to a sharp rise in world oil prices. This coincided with the nationalization of the Venezuelan petroleum industry, leading to an increase in central government revenues of an estimated 170%. However, the oil 'boom' laid the foundations of a subsequent economic crisis that in the longer term weakened the legitimacy of the Punto Fijo model. The extraordinary levels of revenue accruing to the state in the 1970s exacerbated corruption and led the national administration to become excessively bureaucratic, interventionist and inefficient. In addition, the Government began to borrow from international creditors in order to sustain state investment projects when the price of petroleum began to fall, towards the end of Pérez's term in office. Despite the steady deterioration in the oil price and increasingly negative international borrowing conditions, successive COPEI and AD Governments were reluctant to reduce fiscal spending, owing to concerns that economic adjustment would undermine support for the parties, and no progress was made in diversifying the economic base away from dependence on oil. During the administration of Luis Herrera Campins (1979–1984), the Government seized the profits of the state petroleum company, Petróleos de Venezuela, SA (PDVSA), in an attempt to stabilize the economy after a devaluation of the currency in 1983. The move was fiercely but unsuccessfully resisted by PDVSA managers, who subsequently began to move PDVSA investment funds and profits into 'offshore' ventures in the USA and Germany in order to prevent the Government using PDVSA as a 'cash cow'.

In the 1980s previously high levels of electoral participation declined and there was mounting popular disaffection with AD and COPEI as the economy went into recession and the network of welfare coverage and oil-revenue distribution began to shrink. Despite the increasingly negative evaluation of their performance, AD and COPEI remained the dominant political forces as a result of their institutionalized control of the state and election administration and their ability to privilege a progressively narrowing circle of élite clientelist interests.

In 1988 Pérez was re-elected to the presidency. Hopes that his return to power would stabilize the political situation were disappointed and the political crisis deepened during his administration. Expectations that his second term would herald a renewed era of prosperity were immediately frustrated, as an orthodox programme of liberal economic policies was introduced within a month of him taking office. The strict austerity measures, and in particular increased fuel and transport prices, provoked serious civil disturbances. The security forces killed at least 400 demonstrators during riots in the capital in February 1989. President Pérez rapidly lost the support of his own party, AD, and the pro-AD Confederación de Trabajadores de Venezuela (CTV—Venezuelan Workers' Confederation), which organized the first general strike for 31 years in May. In an attempt to enhance the legitimacy of the political system, in that year the Pérez Government introduced a series of major reforms, including administrative and political decentralization and changes to the electoral system. The political opening allowed minor parties of the centre-left such as La Causa R (LCR—Radical Cause), which had made inroads into the CTV's control of organized labour, and the Movimiento al Socialismo (MAS—Movement to Socialism) to win control of a number of state and municipal governments. Decentralization also enabled independent 'populist' figures running on anti-AD and COPEI platforms to win elective office and gain national prominence, such as Irene Saez, the mayor of Chacao municipality in Caracas, and Henrique Salas Römer, who was twice elected Governor of Carabobo state. The reforms failed, however, to increase support for the discredited political parties, AD and COPEI, and the opening proved to be a double-edged sword for new parties and independent politicians, as it led them to be associated with the discredited Punto Fijo model of consensus and dominant party control.

### Failed Military Coup

The depth of the developing political crisis in Venezuela was revealed by an attempted *coup d'état* in February 1992. The rebels belonged to a nationalist faction of junior officers known as the Movimiento Bolivariano Revolucionario-200 (MBR-200—Bolivarian Revolutionary Movement 200), formed by Lt-Col Hugo Rafael Chávez Frías and three colleagues 10 years earlier. MBR-200 was highly critical of the Punto Fijo system created by the 1958 Pact and proposed a 'Bolivarian' alternative that addressed the needs of the poor and marginalized. The officers organized the coup attempt in response to increased social unrest, and received the support of a number of minor left-wing parties including LCR. Although the coup failed, its size and level of support were significant and, despite his subsequent imprisonment, the rebellion transformed Chávez into a popular hero.

The coup attempt and a second one organized by forces indirectly linked to MBR-200 in November of that year seriously weakened President Pérez, who came under pressure to resign. He was suspended from office by the AD-controlled Congreso (Congress) in May 1993 following allegations of misappropriation of public funds. In May 1996 he was found guilty and sentenced to two years and four months under house arrest.

## THE ADMINISTRATION OF RAFAEL CALDERA RODRÍGUEZ

Following the suspension of Pérez, in June 1993 the Congreso elected Ramón José Velásquez, an independent senator, as interim President until fresh elections were held in December. Rafael Caldera Rodríguez, who contested the election as an independent candidate following his expulsion from COPEI in June, won the presidency on an anti-party and anti-economic-liberalization platform. He was supported by a 17-party alliance, Convergencia Nacional (CN—National Convergence). The presidential contest was marred by allegations of electoral fraud against the LCR candidate, Andrés Velásquez. The election was also significant in that it was the first time since democratization that neither AD nor COPEI controlled the executive. However, they remained the dominant force in the bicameral legislature. Caldera's immediate priority was to placate the military, particularly junior officers disaffected by corruption at senior levels of the service: on taking office he replaced the entire military high command and released the coup plotters of 1992. Chávez was pardoned for his role in the February coup, but was not allowed to re-enter the army. Chávez subsequently focused on promoting abstention and non-participation in politics as a means of forcing the collapse of the political system.

Although Caldera had promised to overhaul the 1961 Constitution and address growing levels of poverty and employment in the informal sector, little was achieved, adding to growing popular frustration with the political status quo. The Caldera administration was initially distracted by a severe crisis in the banking sector, which followed from the collapse of the country's second largest commercial bank, Banco Latino, one month before Caldera assumed office. The Government intervened in Banco Latino and 15 other poorly regulated financial institutions, but this did not contain a major economic crisis that the Government sought to avert through the introduction of a fixed exchange rate and price controls. Ultimately, Caldera's administration was forced to renege on electoral promises and it signed a US $1,400m. stand-by agreement with the IMF in July 1996. The series of economic and structural reforms, the so-called *Agenda Venezuela*, exacerbated popular discontent and there were numerous anti-Government demonstrations. Caldera became increasingly reliant on the AD and COPEI parties in the Congreso in order to pass legislation as his Convergencia coalition began to fragment. Caldera also looked to the AD-affiliated CTV to support a modification of labour legislation. This ended the system of retroactive severance payments, a crucial form of welfare provision. Social-security reforms and minimum wage increases were negotiated with the CTV but these did not extend to workers in the informal labour sector, which, by 1995, constituted one-half of the country's total work-force.

A rise in the oil price afforded the Government fiscal leeway in 1997 and structural adjustment measures were postponed. However, in 1998, the year of national elections, the oil price fell, forcing the Government into spending cuts of US $6,000m. As an indication of the profound popular hostility to the established political parties, all the leading presidential candidates contested the election as independents. The two front-running candidates, Henrique Salas Römer and Irene Saez who both ran on neo-liberal platforms, saw their support levels collapse after they accepted the endorsement of AD and COPEI and as the popular sentiment turned against orthodox economic proposals. This benefited Chávez, a little-noticed candidate until the closing stages of the election campaign, who was supported by a multiparty alliance, the Polo Patriótico (PP—Patriotic Pole). This comprised the Movimiento V República (MVR—Fifth Republic Movement), which Chávez had established in 1997, a dissident faction of La Causa R, the Patria Para Todos (PPT—Homeland For All), the MAS and the Partido Comunista de Venezuela (PCV—Venezuelan Communist Party). The Chávez manifesto promised a complete break from the Punto Fijo model of government and a 'Bolivarian revolution' that would rewrite the Constitution, restructure state institutions and terminate the internationalization strategy of PDVSA, thereby ensuring that the oil-export revenues flowed directly to the state, for the benefit of all Venezuelans. Chávez pledged to create a 'just' economy and rejected economic liberalization, which he condemned as 'savage'. Chávez enjoyed enormous popular support, owing to his nationalist, populist rhetoric. He had an enormous appeal among those who were not part of the clientelist framework, particularly the socially marginalized and informal-sector workers. Chávez was of mixed race and had no historic ties to AD or COPEI, and this further enhanced his credentials as an anti-party candidate from the grassroots of Venezuelan society.

COPEI and AD manoeuvred to prevent a victory by Chávez, using their congressional majority to approve a revision of the electoral schedule. None the less, in December 1998 Chávez won a decisive victory in a run-off contest against the AD- and COPEI-endorsed candidate Henrique Salas Römer; Chávez won 56% of the votes cast, while Salas Römer received some 40%. AD secured the most congressional seats (55 of the 189 seats in the Cámara de Diputados and 19 of the 48 seats in the Senado) in congressional elections held in November. The PP also performed well, winning 18 upper-house seats and 70 seats in the Cámara de Diputados.

## THE ADMINISTRATIONS OF LT-COL (RETD) HUGO RAFAEL CHÁVEZ FRÍAS

President Chávez introduced his Bolivarian revolution on the day that he assumed office in February 1999. He used his inauguration to decree a national referendum on the convening of a constituent assembly to draft a new constitution. His preference for an 'organic' assembly to replace the existing state institutions, in contrast to a 'derivative' assembly, which would work alongside existing institutions, generated opposition from AD, COPEI and Salas Römer's Proyecto Venezuela, which appealed unsuccessfully against the referendum in the Supreme Court. In the plebiscite, held in April 1999, 81.5% of voters endorsed Chávez's proposals.

In an election to the 131-member Constituent Assembly, held in July 1999, supporters of Chávez won 121 of the seats, an accomplishment facilitated by the decision of AD and COPEI not to participate. The Constituent Assembly was convened in August and, in a symbolic act intended to demonstrate the sovereignty of the Assembly, Chávez resigned. The Assembly immediately reappointed him to the presidency. The Congreso declared itself in recess, thereby avoiding an impasse between the executive and the legislature. The Constituent Assembly completed its work in November and the new Constitution was approved by 71% of voters in a second popular referendum held in December. The Bolivarian Constitution introduced radical changes to the institutional framework of the Venezuelan state. These included: the introduction of a renewable six-year term for the President (replacing the traditional non-renewable five-year term); the replacement of the bicameral legislative arrangements with a 165-seat unicameral chamber (the Asamblea Nacional, or National Assembly); and the abolition of the Supreme Court, which was to be superseded by the Tribunal Suprema de Justicia (TSJ—Supreme Tribunal of Justice). Two new state powers were created: the Consejo Moral Republicano (Moral Republican Council), the principal duty of which was to uphold the Constitution; and the Consejo Nacional Electoral (CNE—National Electoral Council), which was awarded constitutional status. The post of Vice-President was created, and serving military officers were given the right to vote, with military promotions removed from the hands of the legislature and given to the President. The Army, Navy and Air Force were merged into a single unified command. The Constitution changed the official name of the country to the Bolivarian Republic of Venezuela and there was a strong emphasis throughout the document on citizen participation, through a variety of mechanisms that included recall referendums and participation by civil society groups in state appointments.

Following the promulgation of the new Constitution, the Constituent Assembly was replaced by an unelected 21-person *congresillo* (mini-congress), pending fresh elections, scheduled for May 2000, to re-legitimize all elective authorities. In a move strongly criticized by the opposition and civil society groups, the *congresillo* began filling the new posts created by the Constitution with individuals closely identified with the PP. This was justified on the grounds that the country was in a 'transitional phase' and that AD and COPEI loyalists were impeding the introduction of policy and institutional change. The appointments fuelled allegations of authoritarianism from the increasing number of anti-Government groups. These concerns were deepened following the appointment of a number of military officers to senior cabinet and administrative positions.

Chávez viewed the military as an integral player in his project for national reconstruction and this was reflected in the Social Emergency and Internal Defence and Development Plan, known as the 'Plan Bolívar 2000', which was launched in February 1999. The Plan was intended to rehabilitate public property and land, and led to the deployment of the Armed Forces in local communities to build schools and hospitals. Intended as a means of bypassing the ineffective state administration, the Plan was, nevertheless, repudiated by senior and retired military officers, who began organizing their own anti-Chávez groups.

Throughout 1999 opposition to President Chávez was uncoordinated, discredited and inchoate. In February 2000, on the eighth anniversary of the 1992 military coup attempt, Francisco Arias Cárdenas, a former close colleague of Chávez, and five other prominent figures within MBR-200 issued the Declaration of Maracay. This criticized Chávez for betraying the democratic ideals of MBR-200. Arias subsequently announced his intention to contest the presidency.

### The 2000 Elections

The elections scheduled for May 2000 were popularly known as the 'mega-elections', since they were the largest and most complex in the country's history, with more than 6,000 posts to be decided. An unruly coalition of left-wing groups supported Arias, who was the only significant challenger to Chávez following the decision by AD and COPEI not to present candidates. Three days before the elections were to be held, the TSJ ruled in favour of an injunction introduced by civil society organizations against the elections, on the grounds that the CNE was not technically competent to administer the process. Following the postponement, the CNE executive was replaced by new authorities, selected by a committee comprising members of the opposition, civil society groups and the MVR. The CNE decided to separate the election of the president, legislative assembly, state governors and mayors from those for regional and municipal legislatures.

The presidential election proceeded on 30 July 2000. Chávez received 59.7% of the votes cast, compared with the 37.5% secured by Arias. A severe economic contraction in 1999 did not diminish support for President Chávez, who consolidated his support among the marginalized sectors. In an election to the new 165-seat Asamblea Nacional the MVR-led PP alliance won the three-fifths' majority required to make appointments to the positions of Fiscal- and Comptroller-General and to the judiciary. AD emerged as the largest opposition party, winning 32 seats, while COPEI was eliminated as a significant political force, securing just five assembly seats. In early August the *congresillo* was dissolved and the Asamblea Nacional convened for the first time. The primary task of the legislature was to modify existing legislation in accordance with the new Constitution. In November the Asamblea Nacional granted 'enabling powers' to the executive for one year, allowing President Chávez to legislate by decree in a range of areas.

There was a chronic deterioration of political stability at the beginning of President Chávez's second term, as the administration pressed ahead with its project of radical reform and moved to introduce legislation in line with the new Bolivarian Constitution. In October 2000 President Chávez decreed a referendum to approve measures to reform the trade-union movement through the introduction of direct internal elections and the creation of a single, unified confederation. The International Labour Organization and the CTV criticized the proposed referendum as a violation of the constitutional provision of freedom from state interference in private organizations. The changes were interpreted as an attempt to replace the AD-affiliated CTV with a new trade-union movement loyal to President Chávez called the Frente Bolivariana de Trabajadores (FBT—Bolivarian Workers Front), fuelling claims that the Government was replacing the Punto Fijo network of clientelist ties with a Bolivarian model that benefited those excluded from the old Punto Fijo arrangements. Opposition to the referendum led to a resurrection in the political fortunes of the discredited CTV, which convoked a series of stoppages by affiliated trade unions. This included a costly four-day strike by petroleum workers in October 2000, which culminated in Chávez dismissing the President of PDVSA for his handling of the dispute. The referendum went ahead in December and the proposals were supported by 63% of voters. However, with an abstention rate of 78%, the legitimacy of the Government's reforms was questioned. Elections to a unified confederation proceeded in October 2001 and, in a major humiliation for the Government, the opposition candidate, Carlos Ortega, won the union's presidency amid claims of violence and intimidation by conflicting union organizations. The Government refused to recognize Ortega's election, leading to a legal stand-off between the union and the Government.

Anti-Government demonstrations organized by the CTV throughout 2000 and 2001 were supported by a diverse array of groups, whose interests were affected by government policy. This included teachers, parents and the Roman Catholic Church, protesting against changes to the Education Act introduced in December 2000. The introduction of a package of 49 legislative changes decreed in November 2001 as President Chávez's enabling powers were scheduled to expire, served to weld diverse anti-Government interests into an increasingly unified and coherent opposition movement. The CTV assumed leadership of the umbrella movement with the Federación Venezolana de Cámaras y Asociaciones de Comercio y Producción (Fedecámaras), the leading private-sector lobby group. Lobby groups and sectoral interests substituted for the weak and disorganized opposition political parties that remained discredited and incapable of mounting a viable challenge to the Government. The legislation covered a range of areas, including tourism, fishing, banking and hydrocarbons. Particularly contentious was the Land Reform Bill, which Chávez claimed would allow for the expropriation of idle land in the national interest. As with the changes to the Education Act, debate on the underlying rationale of the reform, which, in this instance, was to reduce the highest concentration of landownership in Latin America, was undermined by the conflict between an opposition alleging authoritarianism and the Government, which attacked the resistance of the 'old oligarchy'.

During the administration of President Caldera, there had been a selective opening to foreign capital of the national oil industry, previously the monopoly of the state petroleum company, PDVSA. The *apertura*, or opening, aroused intense nationalist opposition and created hostility towards the executives managing PDVSA. Chávez had condemned PDVSA as a 'state within a state', and he introduced a revision of oil policy and a review of PDVSA's structures and operations with the aim of bringing the company under closer control of the Ministry of Energy and Petroleum. The 1999 Constitution reasserted state ownership of hydrocarbons and, under the 2001 Hydrocarbons Law, PDVSA was required to have a 51% stake in new production and exploration activities. Strengthening relations with oil-exporting countries became a central aspect of the Government's energy and foreign policy, and the then Minister of Energy and Mines, Alí Rodríguez Araque, who served as OPEC Secretary-General from March 2000 until June 2002, was successful in introducing a petroleum price-band system within the Organization. This helped to lift the oil price to the band's US $28 'ceiling' in 2001.

Changes to oil policy were resisted fiercely from within PDVSA. Senior managers and white-collar workers looked to the CTV and other anti-Chávez groups for solidarity in their campaign to prevent implementation of government policy. A general strike was launched on 10 December 2001 in protest at all aspects of government policy. The economic élite used their control of the private-sector media to attack the Government and mobilize support for the union-led demonstrations. Claims that the Government was seeking to install a Cuban-style communist regime in Venezuela increased support for the protests, which were justified as a defence of democracy. There was a second general strike in January 2002 and persistent appeals for the military to intervene and remove Chávez from power.

President Chávez rejected calls for negotiations over the contents of the November 2001 legislation and this repudiation of a more consensual course led to divisions in the ruling coalition. A section of the MAS defected to the opposition in the legislative assembly, reducing the Government's majority to single figures, and the leader of the MVR, Chávez's ideological mentor, Luis Miquilena, left the Government and created his own party, Solidaridad (Solidarity). Inflammatory attacks by Chávez on his opponents increased hostility toward the Government, but the fiery speeches did have the effect of deepening support for Chávez among the poor, who actively supported pro-Government demonstrations. This mass mobilization by both supporters and opponents of the administration was acutely destabilizing and was characterized by escalating levels of violence. The cabinet became increasingly focused on directing day-to-day political activities rather than policy elaboration and governmental matters. This added to a sense of drift, which was aggravated by frequent changes in the Council of Ministers.

## The April 2002 Coup

Bolstered by broad popular support and positive signals from the US authorities that appeared to endorse its action, at the beginning of 2002 the opposition revised its demands for legislative reform and pressed for President Chávez to resign. On 9 April the CTV and Fedecámaras held a third strike in support of striking workers at PDVSA, who were protesting at changes made by Chávez to the executive and presidency of the state oil company.

In a highly disputed series of events, 17 people were killed on 11 April 2002 as Chávez loyalists clashed with demonstrators near the presidential palace. Amid claims that the President had ordered the security services to fire on the demonstrators, senior military officers, including the Commander-in-Chief of the Armed Forces, Gen. Lucas Rincón Romero, announced that President Chávez had resigned and been placed under arrest.

Chávez was subsequently transferred to a military base on a nearby Caribbean island. Senior cabinet figures were arrested and an interim junta was installed, headed by the President of Fedecámaras, Pedro Carmona. The break with the constitutional order in Venezuela was widely condemned by the international community, with the notable exception of the US, British and Spanish Governments. The military and trade unions withdrew their support from the junta, however, when Carmona appointed a cabinet composed of right-wing economists, dissolved the democratically elected legislature and regional assemblies, and suspended the Constitution. The junta collapsed on 13 April, following a revolt by the presidential guard and mass popular demonstrations calling for the return of the democratically elected President. Chávez returned to office the following day and he emphasized his commitment to pursuing a more consensual approach to government.

### The 2004 Recall Referendum

In the immediate aftermath of the coup, the Government convened a series of committees to promote national dialogue, and the well-respected Rodríguez was appointed President of PDVSA. Managers who had been dismissed from the oil company during the protests were reinstated and the economic team in the cabinet was replaced. However, the measures failed to reduce polarization and political tension. The opposition, which created the Coordinadora Democrática (CD—Democratic Coordinator) as an umbrella organization for anti-Chávez groups, filed legal suits against President Chávez in relation to the deaths on 11 April. This focused attention on the judiciary, which the opposition claimed was pro-Chávez in orientation. This argument was undermined by a number of key rulings issued by the TSJ that went against the Government, the most important being the ruling in August 2002 that the military officers involved in the April putsch were not guilty of military rebellion. As a result of the continuing conflict, in September the international community was invited to Venezuela to mediate between pro- and anti-Government forces. The 'tripartite group', comprising the Carter Center (based in the USA), the Organization of American States (OAS) and the UN Development Programme, created dialogue tables chaired by the OAS President, César Gaviria. External mediation was complicated as both the opposition and pro-Chávez groups were divided between moderates and radicals resisting any concessions to their opponents. Within the anti-Government alliance, radicals maintained that Chávez had to be removed by any means possible, while moderates supported a 'constitutional' approach that focused on a recall referendum, as provided for under the Bolivarian Constitution, as a means for resolving the political impasse.

In November 2002 the opposition convened an indefinite general strike in support of their demand that President Chávez stand down. The strike was organized by the CTV and Fedecámaras and it was observed by managers and technicians at PDVSA. The stoppage lasted until February 2003 and came at a grave fiscal cost to Venezuela. Oil production fell from an average of 2.65m. barrels per day (b/d) to fewer than 250,000 b/d. The Government was deprived of US $6,000m. in petroleum export revenue, and PDVSA was forced to import petroleum at a loss. The strategy exacerbated splits in the opposition and reduced popular support for the anti-Chávez lobby among the wider population.

At the end of January 2003 the international community increased mediation efforts and a six-country 'Friends of Venezuela' delegation was convened, comprising senior officials from the USA, Brazil, Mexico, Chile, Spain and Portugal. As the strike fragmented, the Friends and the tripartite group stepped up pressure for an accord between the Government and the opposition. This was signed in May. The opposition hailed the agreement as a major breakthrough, but it merely restated the constitutional provision for a recall referendum after the midway point of the President's term.

The disintegration of the strike strengthened the Government. The administration dismissed 18,000 of the 33,000 full-time workers at PDVSA for participating in the stoppage. Oil production was restarted under PDVSA's contingency plan, and three months after the strike the Government claimed production levels had reached pre-stoppage levels. The President of Fedecámaras was later arrested and charged with a number of crimes, including treason, although he left for medical treatment in the USA, while the CTV President, Carlos Ortega, sought political asylum in Costa Rica.

In order for a recall referendum on the executive to be convened, 20% of the electorate (over 2.4m. voters) were constitutionally required to sign a petition. A civil society group, Súmate (Join In!), financed by the US National Endowment for Democracy, took a lead role in distributing information to voters ahead of the referendum and preparing databases of the electorate. In June 2004 the CNE ruled that the opposition had succeeded in gathering the requisite number of signatures after a petition in April and a 'repair' of signatures that had technical irregularities at the end of May. Despite allegations that President Chávez would seek to prevent the referendum, the Government accepted the validity of the results and a recall referendum was scheduled for 15 August 2004. If a greater proportion of the electorate voted against Chávez than had supported him in the 2000 elections (59.7% of voters), then fresh elections were to be convened within 30 days of the recall result being issued. The opposition went into the recall campaign deeply divided, with the various party factions vying for the CD candidacy in the event that fresh presidential elections were held. By contrast, the Government approached the referendum with a solid organizational base, comprising grassroots organizations grouped in the Círculos Bolivarianos (CBs—Bolivarian Circles) and the Comando Ayacucho. Higher than expected oil prices in 2004 further strengthened the position of the Government, which used the petroleum revenue to fund a number of socially popular literacy, health and house-building projects, called 'missions'. The popularity of Chávez, combined with the failure of the CD to draw support from the so-called *ni ni* voters—those neither pro-Chávez nor in favour of the opposition—enabled the President to retain executive control, with 59% voting in favour of his presidency and 42% against. Participation in the referendum was high, at 70%.

In the aftermath of his referendum victory, Chávez announced a deepening of the Bolivarian revolution. Land-distribution programmes were accelerated and the petroleum revenue-financed 'missions' were extended into new areas of welfare provision. The opposition was weak after the recall defeat and was unable to sustain the protest movement that had previously prevented the full application of the Bolivarian agenda. The CD went into the regional elections of October 2004 fragmented and without a coherent policy platform, a situation that facilitated a overwhelming victory for pro-Chávez candidates, who won control of all but one of the country's 23 state governments. The capacity of the CD alliance to convoke anti-Government mobilizations was further restricted in December, when the Government introduced a 'media responsibility law' and other revisions to the Penal Code that made defamation of a public official, incitement to violence and the distribution of inaccurate reports offences punishable by imprisonment. The introduction of the legislation followed the assassination of public prosecutor Danilo Anderson in November. Anderson had been investigating the events surrounding the coup attempt of April 2002 and the suspected links between extremist anti-Chávez elements in Venezuela and anti-communist groups in Miami, USA. In what was interpreted as a further curtailment of opposition activities, in July 2005 the TSJ announced its intention to prosecute the executive of Súmate for violating domestic laws relating to foreign financing of political activity.

Having assumed a position of uncontested dominance over the opposition, divisions began to emerge within the pro-Chávez movement at the beginning of 2005. While anti-US radical elements pushed the Government to adopt a more overtly left-wing identity and apply Bolivarianism at a more rapid pace, moderates supported consensus-building to centrist elements in the opposition. Divisions also emerged between the grassroots pro-Chávez groups and the MVR party, with the wide network of community-based organizations critical of the MVR's control of candidacies for the municipal and legislative elections scheduled to be held in August and December, respectively. Despite the internal wranglings, it was expected that Chávez would face no significant opposition challenge in the presidential election scheduled for 2006, despite the lack of definition around his Bolivarian revolution. Barring a catastrophic decline in the oil price, the administration was expected to retain the loyalty of its core support base, the socio-economically marginalized, despite concerns regarding the democratic credentials of President Chávez and the capacity of his administration to institutionalize its revolutionary project.

## INTERNATIONAL RELATIONS

President Chávez instituted a dramatic change in Venezuela's international relations. Three central ideas underpinned the Government's approach: Bolivarianism; state sovereignty; and the need for close ties with other oil-producing nations. Bolivarianism is named after South America's independence leader, Simón Bolívar. Drawing on Bolívar's vision of an integrated Latin America as an essential counter to the emerging USA, Chávez pursued regional integration based on principles of social justice. Chávez has repeatedly stated his admiration for both the Cuban leader, Dr Fidel Castro Ruz, and for the Cuban revolution. Several reciprocal agreements in areas such as investment, trade and social provision have been signed by Chávez and Castro, providing, most notably, for the supply of Venezuelan oil to Cuba (averaging some 100,000 b/d) at preferential rates, and for the assistance of Cuban doctors and teachers in Venezuelan social 'missions'. Chávez emerged as an outspoken opponent of the planned Free Trade Area of the Americas (FTAA) promoted by the USA, and posited a Bolivarian Alternative for the Americas as a counter model that excluded the USA. Initially, the proposal was viewed as utopian and unhelpful by other regional Governments, but by 2005 it had gained some interest from other left of centre administrations in the region. The Venezuelan Government sought to forge strong links with these left-of-centre Governments in Brazil, Argentina and Uruguay, and a number of economic and cultural integration projects were launched, including the broadcasting initiative, Telesur, and the regional petroleum company, Petrosur.

President Chávez also sought to reduce Venezuela's trade dependence on the USA and to revise the close cultural and political relationship that existed between the two countries during the Cold War. A critical stance was adopted toward the US embargo of Cuba, which Chávez condemned as a violation of Cuban sovereignty and international law. The Chávez Government was also critical of the USA's 'war on terror' and condemned the US-led invasions of Afghanistan and Iraq. During the Administration of President George W. Bush, relations between the two countries became increasingly strained amid allegations by Chávez that the Bush Administration was working with the CD opposition alliance to destabilize his Government and remove him from power. Leading opposition figures were received by the US President after 2001 and extensive financial resources channelled to the anti-Government movement, through the USA's National Endowment for Democracy. The USA maintained a hostile stance toward Chávez and his Bolivarian revolution, which were condemned as destabilizing for the region, authoritarian in approach and damaging to the security interests of the USA. Suspicion of President Chávez was exacerbated by the decision by the Venezuelan Government to deny the US Air Force permission to enter Venezuelan airspace to undertake anti-narcotics and aerial-surveillance exercises. Venezuela initially assumed a hostile position towards US-sponsored 'Plan Colombia', intended to combat the Colombian narcotics trade, on the grounds that the Plan would lead to an escalation, rather than a resolution, of the Colombian civil conflict and that it would lead to the displacement of the narcotics industry into Venezuela. However, critics charged that Chávez was motivated primarily by his ideological sympathy for Colombia's leftist rebels, who were also a target of the Plan. However, in May 2001 Venezuela revised its opposition to Plan Colombia in exchange for financial assistance for border areas. Nevertheless, concerns that Venezuela would be affected by an overspill of the Colombian conflict were resurrected in May 2004, when more than 70 members of a right-wing Colombian paramilitary group were discovered in Caracas, in an apparent plot to assassinate Chávez. Despite persistent allegations by US government officials that the Chávez administration was financing 'terrorist' and other violent groups in the region (including the radical Bolivian opposition movement), no evidence was presented to substantiate these claims, and attempts to isolate Venezuela through the OAS failed. Venezuelan weaponry purchases from Spain and Russia in 2005 initially provoked concerns of an arms race in the Andean region, but this was dismissed by the Venezuelan administration, which claimed the purchases were intended to modernize outdated weaponry, thereby contributing to enhanced border security.

Antagonisms between the USA and Venezuela were particularly pronounced in relation to oil. The Chávez Government pursued close relations with other oil-producing nations, including Iraq, Libya and Iran. In August 2000 President Chávez undertook a controversial tour of the Middle East, during which he visited Iraq and Libya. Chávez became the first elected head of state to meet the Iraqi leader since sanctions were imposed on that country after the war in the region of the Persian (Arabian) Gulf in 1991. The USA condemned the meeting and was also critical of the Caracas Energy Accord of 2000, which extended to Cuba the preferential terms for oil offered by Venezuela and Mexico to other countries in the region. In 2005 Chávez publicly stated his commitment to co-operating with the Iranian Government in the field of nuclear energy. In improving the traditionally poor relations between Venezuela and other OPEC member countries and seeking close co-operation within the cartel to achieve high and stable oil prices, Venezuela also went against the interests and energy security of the USA. These were seen to be most conspicuously threatened by the forging of amicable links with the People's Republic of China, despite the Venezuelan Government's stated commitment to remaining a reliable supplier of crude to the US market.

President Chávez was strongly supportive of regional integration. As a member of the Comunidad Andina de Naciones (CAN—Andean Community of Nations), Venezuela took a leading role in negotiations between fellow members (Bolivia, Colombia, Ecuador and Peru) and the countries of the Southern Common Market (Mercado Común del Sur—Mercosur) to form a single free trade area. In April 1998 a trade liberalization agreement, committing the two regional trading blocs to the creation of a free trade area by 1 January 2000, was signed. Integration, however, was delayed, prompting the Venezuelan Government to pursue unilateral entry into Mercosur, of which it was an associate member. In July 2005 Chávez assumed the presidency of the CAN and stated his intention vigorously to pursue projects of regional integration.

Relations between Venezuela and neighbouring Colombia repeatedly swung from strained to amicable during President Chávez's terms in office. The Venezuelan President supported a peaceful resolution of Colombia's civil conflict and sought to facilitate dialogue between the Colombian Government of President Andrés Pastrana Arango (1998–2002) and the left-wing guerrilla group, the Fuerzas Armadas Revolucionarias de Colombia—Ejército del Pueblo (FARC—EP, or the Revolutionary Armed Forces of Colombia—People's Army). Colombian President Alvaro Uribe Vélez, who was elected in May 2002, adopted a military solution to Colombia's conflict, which was supported by the USA and thereafter conceived as a 'war on terror'. This definition was rejected by the Venezuelan Government, which accused right-wing paramilitary 'terrorists' of colluding with the Venezuelan opposition. Relations between the two countries deteriorated in December 2004 when a senior figure in the Colombian FARC, Rodrigo Granda Escobar, was abducted in Venezuela by Venezuelan security officials paid by Colombian authorities. After initially denying any wrongdoing in the affair, the Colombian Government subsequently expressed its regret and relations between the two countries were normalized in February 2005 at a meeting between the two countries' Presidents in Caracas.

Relations with Venezuela's other neighbour, Guyana, were strained by territorial and border disputes. Venezuela had historical claims to all territory west of the Essequibo river in Guyana, and in November 1989 a UN mediator was appointed to resolve this territorial dispute. Limited progress was made. Relations deteriorated in July 2000 when Venezuela strongly protested against the decision by the Government of Guyana to lease land to a US company in the disputed Essequibo delta for the purpose of developing a satellite-launch facility. In March 2004 President Chávez met his Guyanese counterpart, Bharrat Jagdeo; no progress was made on resolving the dispute, but the Venezuelan Government indicated its willingness to 'authorize' Guyanese mineral exploration in the Essequibo region.

# Economy

## Dr JULIA BUXTON

Venezuela has the sixth largest proven petroleum reserves in the world, and petroleum has been the mainstay of the economy since the discovery of deposits at the end of the 19th century. Successive Governments failed to devise an economic development plan appropriate for the country's status as a leading petroleum exporter and, as a result, Venezuela's dependence on petroleum revenues has become entrenched. This rendered the country vulnerable to variations in the international economy, with the performance of the domestic economy characterized by 'boom and bust' cycles. High petroleum prices led to expansionary spending policies, which, in turn, led to severe fiscal problems when the petroleum price fell. The channelling of petroleum revenue through central government spending resulted in high rates of economic growth in the 1960s and 1970s. Real gross domestic product (GDP) grew by an annual average of 3.7% between 1965 and 1980. Between 1985 and 2003 GDP growth averaged only 1% per year. Compared with annual population growth of 2.1%, this led to a marked fall in average real incomes, which in 2003 were estimated to be on a par with income levels in 1956. There has been a number of attempts to diversify the economic base. These included a programme of state-led industrialization focusing on heavy industrial development in the mid-1970s, policies to encourage non-oil export growth, led by the private sector in the early 1990s, and the promotion of small and medium industries and diversification into 'downstream' and agricultural activities by the Government of Hugo Rafael Chávez Frías from 1999. All of these programmes were implemented ineffectively and they failed to break dependence on oil export revenue. In 2003 petroleum accounted for one-quarter of GDP and 85% of export revenue.

The 10-fold increase in world petroleum prices in the early 1970s, which coincided with the nationalization of the oil industry in Venezuela, led to a reversal in the trend of steady economic growth of the previous two decades. The temporary abundance of petroleum revenue led to heavy external borrowing to finance state-led expansion and political patronage. This increased debt and exacerbated import dependence. Difficulties developed as petroleum prices declined and interest repayments on the debt increased in the 1980s; however, successive Governments were unwilling to pay the political price of reducing public spending. As a result, austerity policies were deferred repeatedly, and foreign debt accumulated.

The adoption of an IMF austerity programme in 1989 contributed to a 7.8% fall in real GDP. This was reversed between 1990 and 1993, when an increase in world petroleum prices (following the Iraqi invasion of Kuwait in August 1990), coupled with a strong performance by the non-oil sector, led to an average annual economic growth of 7.4%. In 1996 there was a contraction of 0.2% in real GDP, corresponding with the adoption of a one-year stand-by agreement with the IMF in July. Nevertheless, the petroleum sector experienced an expansion of 8.8% after its opening to private investors, leading to real GDP growth of 6.4% in 1997. Following the decline of world petroleum prices and a weak performance by the non-petroleum sector, growth contracted by 0.2% in 1998. Despite a sharp rise in oil prices in 1999, GDP contracted by a further 6.1% in that year following a 7.4% decline in the oil sector, owing to production cuts, an 8% decline in the non-oil export sector and recession in the domestic economy. This negative performance was reversed in 2000, with growth in the non-oil sector, led by telecommunications, combining with a 3.4% expansion of the oil sector to produce GDP growth of 3.2%. There was a slowdown in the economy in 2001 and growth of 2.7% was below the Government's target of 4.5%. The contraction was related to cuts in oil production, which led to a 0.9% fall in oil GDP and reduced export demand. Political tension, heightened perception of investor 'risk' and protracted strike activity by opponents of the Government had a catastrophic effect on the economy in 2002 and 2003, when GDP contracted by 8.9% and a provisional 7.7%, respectively. However, the economy rebounded in 2004, with estimated growth of 17.3% as the oil price strengthened and consumer demand recovered. The oil sector grew by 8.7% over that year and the non-oil sector by 17.8%. This strong performance was expected to continue, although at a slower pace, into 2005, with GDP growth of 6% projected for the year.

As a result of a high birth rate, a comparatively low death rate and significant immigration, the Venezuelan population increased five-fold between 1958 and 2004, to 26.1m., of which some 15.5m. were under the age of 30. The total population was expected to reach 28.8m. in 2010. Immigration slowed down significantly after 1998, while emigration to the USA and Western Europe increased considerably. In 2004 the labour force totalled an estimated 12.2m., with 10.4m. in employment. At an average of 4.4% growth per year since 2001, the increase in the labour force has outpaced the rate of job creation. Creating employment remained a serious challenge as a result of limited economic growth and the capital- rather than labour-intensive nature of the petroleum economy. Political instability and the economic recession of 2002 and 2003 negatively affected employment creation in the public and private sector. Unemployment increased from 13.2% in 2001 to 18.2% in 2003. In 2004 the rate fell to 15.1% following strong economic growth and an increase in state and private-sector investment. Limited progress was made in reducing informal-sector employment, despite a government programme to formalize conditions in the sector. Venezuela's informal sector remained one of the largest in Latin America in 2004, accounting for an estimated 51% of total employment in that year.

The structure of employment and the distribution of the population in the early 2000s were markedly different from the 1960s. Agricultural employment declined from 35% of total employment in 1960 to about 10% in 2004. Service-sector employment showed the largest growth and accounted for an estimated 70% of the labour force, with the remaining 20% employed in industry. Trade union-led resistance to the rationalization of public-sector employment ensured that the figure remained high, at 17% of the total work-force. The major industry in economic terms, petroleum production and processing, employed only around 40,000 workers in late 2002, less than 1% of the working population. Some 18,000 state petroleum workers were dismissed by the Government in early 2003 for participating in a general strike.

Poverty in Venezuela has increased more rapidly than in any other country in the region. In 1978 10% of the population lived in general poverty, of which 2% lived in conditions of extreme poverty. By 2004 54% of the population lived in poverty, while those living in extreme poverty (less than US $2 per day) was 42%. Poverty levels increased during the first years of the Chávez Government, despite the launch of a number of initiatives intended to increase employment and reduce rural and urban poverty. The inadequacy of the social security system, economic recession in 2002 and 2003 and the growth of informal-sector employment explained the increase. The monthly salaries of formal sector workers are 62% higher than those in the informal sector. The salaries of female workers in the informal sector are also 70% less than those earned by men. Rural poverty is a major problem, with 82% of residents in rural areas living in impoverished conditions. New social security legislation, which was intended to incorporate provisions for informal-sector workers and new socio-economic rights enshrined in the 1999 Constitution, was under consideration after the Chávez Government opposed changes to the social security system formulated by the previous administration. However, by 2005 no progress had been made in developing the new social security framework.

Tripartite committees representing the Government, business and the labour sector were traditionally responsible for setting minimum and public wage levels, which, of course, did not cover workers in the informal sector. The Chávez Government disbanded the tripartite committees, arguing that the main trade union confederation, the opposition-dominated Confederación de Trabajadores de Venezuela (CTV) was unrepresentative of the labour sector. Minimum wage increases below the level of inflation were unilaterally decreed by the Government. They rose by 10% in 1999, 20% in 2000, 10% in 2001 and 20% in 2002. In 2003 a phased 30% increase was introduced that lifted minimum monthly earnings from US $119 to $155. In April 2005 the minimum wage was increased by 27%, to $202 per month. Industrial relations remained volatile after a referendum on the reform of the union movement, decreed by the administration and supported by voters, was held in December 2000. Opponents of the Government retained control of the CTV after disputed elections in September 2001, despite the administration's refusal to recognize the legitimacy of the results. The CTV convoked four general strikes in 2002 and 2003 as part of wider anti-Government protests. A strike in April 2002 culminated in the temporary overthrow of President Chávez (see History), while a follow-up strike in December lasted three months and caused grave fiscal damage to the country. Following Chávez's success in a recall referendum held in August 2004, the opposition alliance disintegrated and the focus of industrial conflict shifted to disputes between moderate and radical *Chavista* union organizations.

## AGRICULTURE

The agriculture sector went into steep and seemingly irreversible decline after the discovery and exploitation of the petroleum sector at the beginning of the 20th century. It was characterized by inefficiency, limited investment in modern farm technology and inequalities in the distribution of land, despite a distribution programme initiated in the 1960s. Venezuela has the second highest level of land concentration in Latin America and the Caribbean, with 70% of agricultural land in the hands of 3% of agricultural proprietors; it also failed to achieve self-sufficiency in agricultural production. The country imported an estimated 71% of food consumed in 2000, mainly from the USA and Colombia. Of the 35m. hectares (ha) of land suitable for agriculture, 58% was in the hands of large farms. Rural unrest has been pronounced since the end of the 1990s, with reports of increased levels of squatting on privately owned land. This has in turn been met by violence from landowners, with some 100 peasant leaders reported to have been murdered since the introduction of the Agricultural Development and Land Reform Law in 2001. The powerful agro-industrial lobby traditionally benefited from state protection. This, in addition to the dependence of small farmers on inferior land, meant that an estimated 30% of the 7.3m. ha of land used for arable production failed to achieve maximum potential output, according to FAO. This situation was paralleled in the 18.4m. ha under livestock grazing, which was underutilized by 40%. Limited opportunities in the agricultural sector had led to a decline in the rural population and by 2004 only 10% of the total labour force were employed in agriculture, while the sector accounted for less than 5% of GDP.

The Government of President Chávez committed itself to a programme of agricultural reform and development. This was intended to achieve self-sufficiency in food production, reverse the trend of rural to urban migration, reduce rural poverty and contribute to a diversification of the country's economic base. Over 5m. ha of state-held land controlled by the Instituto Agrario Nacional (IAN—National Agricultural Institute) was distributed to 75,000 peasant families between 2001 and 2004 and five areas—fishing, forestry, maize, palm oil and sugar—received special funding. The 1999 Constitution mandated protectionist tariffs on imports of these products, which led to protests from the World Trade Organization and the Comunidad Andina de Naciones (CAN—Andean Community of Nations). Article 305 of the Constitution also looked to the development of a national programme to increase agricultural and livestock production. In 2001 the Government introduced a plan to accelerate agricultural development. The administration was charged with finding new markets for Venezuelan agriculture, particularly in non-traditional products such as coffee and cocoa. The Government additionally developed a Programme for Food Security and Rural Development in collaboration with FAO under the organization's Special Programme for Food Security. This led to the modernization of 13 irrigation systems. Despite its ambitions for the sector, however, the Government presided over a sharp deterioration in production during the political and economic turmoil of 2002 and 2003. In 2002 production of rice, corn, coffee and sorghum all declined significantly. The animal population failed to return to the level of the mid-1990s and consumption of domestically produced goods, including chicken and beef, fell sharply. The area under cultivation contracted and during 1998–2002 an estimated 240,000 jobs in the sector were lost. Aside from the negative political conditions, a number of factors accounted for this set-back, the most important of which was the 2001 Agricultural Development and Land Reform Law. The controversial legislation required landowners to register with the Instituto Nacional de Tierras (National Land Institute), which replaced the IAN, for ownership certificates and to declare land as productive, to be improved, or idle. Many refused to comply with the new legislation, which generated ownership uncertainty and reduced the incentive to invest in agriculture. Moves by the Government in December 2004 to expropriate privately held land in those instances where ownership could not be proven served to exacerbate insecurity in the sector.

The Government imported agricultural produce to overcome problems in the domestic supply chain and establish popular food markets, administered by the military, to distribute food in urban areas. The introduction of fixed prices was a further deterrent to productivity in the sector, as was the introduction of a fixed exchange rate in February 2003. This spurred imports of cheaper products from the USA and Colombia. There was a notable recovery in the agricultural sector in 2004: under the Plan Nacional de Siembra y Producción (National Sowing and Production Plan), the amount of utilized farmland rose by 10.4%, to 1,982,114 ha and production increases were registered for black beans (48.2%), corn (13.5%), cotton (73.6%), rice (45.8%), beans (55.9%) and potatoes (4.8%).

## MINING AND POWER

### Metallic Minerals

Venezuela possesses vast metallic mineral wealth, the majority of which is controlled by the state-owned Corporación Venezolana de Guayana (CVG). Recent Governments have sought to exploit the country's minerals in order to reduce dependence on petroleum. In 2004 estimated reserves of iron ore and bauxite totalled 14.6m. metric tons and 5.2m. tons, respectively, and there are deposits of zinc, copper, lead, phosphorus, nickel, diamond, silver, uranium and gold reserves estimated at 10,000 tons. In 1997 and 1998 a series of measures to encourage foreign investment was introduced. Mining activites were taxed at a new rate of 34%, but foreign companies were exempt from paying 16.5% in value-added tax (VAT) during the first five years of pre-production activities. In 2000 a reform of the 1941 mining law enhanced legal security and streamlined the concession-granting process. This did not lead to a significant expansion of the sector, however, and in 2001 mining grew by just 1.1%, down from growth of 10.4% and 8.2% recorded in 1999 and 2000, respectively. The Government pursued a number of public-private partnerships with foreign-owned state and private companies in order to increase aluminium and iron-ore production. Growth in 2002 and 2003 was disrupted by industrial unrest, particularly in the iron ore sector. However, there was a robust recovery in 2004 and record production levels of liquefied aluminium (438,373 tons) were recorded.

Venezuela has 12% of the world's gold reserves, although exploitation of the sector has suffered from prohibitive mining legislation, environmental impact disputes and the activities of an estimated 50,000 illegal miners. A complex and bitter legal conflict over concessions for the development of Las Crisitinas mine, which involved two Canadian companies, set back investment and production and exacerbated private-sector concerns over the security of contracts. These problems led to a contraction in exploration spending and a consequent decline in production, which fell from 22.3m. kg in 1997 to 5.9m. kg in 1999. Production rose sharply in 2004 to an estimated 20m. kg, of which 6m. kg was attributed to unofficial mining activities. CVG Minerven extracted 3.2 metric tons of gold per year and reached its highest production level in seven years in 2004, with gold production rising by 29%.

### Coal

Venezuela is the second largest coal producer in Latin America after Colombia with estimated reserves of 10,200m. metric tons. Poor transport links in the east of the country, where 80% of reserves are located, and low investment ratios, limited the development of the sector. Until 2004 Petróleos de Venezuela, SA (PDVSA) produced and marketed coal through its subsidiary Carbozulia. Following a reorganization of PDVSA, the company's shareholding in Carbozulia was transferred to the development agency of Zulia state, Corpozulia, enabling PDVSA to concentrate on oil and gas operations. Carbozulia has outsourced mining activities to German and South African commercial partners in a bid to meet a coal production target of 21m. tons by 2008. Coal production increased from 6.6m. tons in 1999 to 7.0m. tons in 2003. The largest purchasers are the USA and Canada, which account for around 50% of sales, and Western Europe, which accounts for around 40%.

### Electricity

Electricity consumption in Venezuela is among the highest in Latin America (more than 90% of households have electricity, of which an estimated one-quarter is obtained through illegal tapping), a trend that has been encouraged by state subsidization of energy prices. Supply has grown at an average of 3.3% per year over the past 10 years but it has not kept up with demand, which in 2001 grew by 6.2%. According to the US Energy Information Administration (EIA), 62% of electricity in 2002 was generated by the state-run company Electrificación de Caroní (Edelca)—which increased its energy generation by 16% in 2004—and 30% by other private and public companies. In 2002 an estimated 62% of electricity was generated by hydropower projects, including the Guri Dam in Guayana, dams on the Caroní River and the Macagua plant. The remainder is generated with oil or gas. Work began in 2002 on two new hydroelectric facilities, the Yacambu-Quibor project in the west of Venezuela and a US $2,000m. dam next to the Guri site. It was expected that these projects would be completed by 2006 and 2010, respectively. There are three cross-border transmission lines between Venezuela and Colombia and one with Brazil. Declining government revenues and a reluctance to increase electricity tariffs made it difficult to maintain investment in the sector. Investment requirements were estimated at $12,000m. in 2005. There were major problems within the transmission and distribution infrastructure and shortages and blackouts were relatively frequent in some areas. Unusually low rainfall levels in 2003 compounded electricity-generation problems. In 1998 the Government introduced the National Electrical Service Law. This formed the regulatory basis for the privatization of the sector and in October the sale of electricity assets began with the privatization of Sistema Eléctrico de Nueva Esparta (Seneca), at a cost of $63m. (US-owned CMS Electric and Gas acquired an 86% stake). This was followed in 1999 by the privatization of a 70% stake in several regional electricity distributors and in June 2000 an 86% stake in the electricity distributor for the capital, La Electricidad de Caracas (EDC), was acquired by US energy group AES Corporation. The privatization programme slowed thereafter as profit and investment incentives were reduced by the gas regulator's refusal to adjust tariffs. The Government hoped that the development of Venezuela's offshore natural gasfields would allow the country to generate electricity relatively cheaply and reduce dependence on its ageing hydroelectric plants.

### Petroleum and Natural Gas

Production of petroleum declined following the political turmoil of 2001–03. The country has total crude reserves estimated at 221,000m. barrels, of which 77,200m. were proven at the end of 2004. Of these proven reserves, some 69% are heavy and extra-heavy crudes, and 31% are condensates, light and medium crudes, which are in greater demand internationally. The petroleum industry is the mainstay of the economy, providing 80% of government revenue and 90% of export revenues. The oil industry was nationalized in 1975 and the state oil company, PDVSA, and its subsidiaries were exclusively responsible for extracting and refining Venezuelan petroleum. The Government extracted revenues from PDVSA through taxes, royalties and special dividend payments. Successive Governments and management teams of PDVSA have differed over the question of how the country should best exploit and manage its oil wealth. In the 1990s a strategy was pursued of maximizing output, breaking production quotas set by the Organization of the Petroleum Exporting Countries (OPEC) and incorporating the private sector into production, exploration and refining projects. At the same time, PDVSA executives pursued a strategy of internationalizing the assets of PDVSA in order to prevent the Government from drawing on the company's profits to fund fiscal spending and debt payments, as it had done in the first half of the 1980s. The Government of President Chávez (1999– ) reversed this approach and sought to reduce the operating autonomy of PDVSA, which was criticized by Chávez for acting like 'a state within a state' and for not responding to the national interest. This led to conflict between PDVSA and the Government and the near-paralysis of production following a general strike in the industry in December 2002.

Production of crude petroleum, which derived mostly from the Maracaibo, Apure-Barinas and Eastern Venezuela basins, declined from a peak annual average of 3.7m. barrels per day (b/d) in 1970, to 1.6m. b/d in 1985. By 1988 Venezuelan crude output remained at 1.57m. b/d, in line with increases in OPEC production quotas. Venezuela's quota increased to just below 2m. b/d in 1990 following the Iraqi invasion of Kuwait, which resulted in a suspension of supplies from those two countries. Under President Rafael Caldera Rodríguez (1994–99) a strategy of maximizing output beyond OPEC quotas was pursued. The steep decline in petroleum prices in 1998 forced Venezuela to embrace OPEC-negotiated production cuts in an attempt to rescue prices. Under President Chávez the Venezuelan Government has pursued high and stable prices through reduction of production levels and co-operation with OPEC and non-OPEC countries. In 1998 and 1999 OPEC and non-OPEC countries agreed to reductions in petroleum production which were effective in bolstering prices to their highest level since 1991. In March 2000 OPEC adopted a Venezuelan proposal to regulate output in order to maintain a price band of US $22–$28 per barrel, and member countries agreed to a total production increase of 1.7m. b/d. This was followed by a further 710,000 b/d increase in June. Venezuela's share of the increase in production amounted to 81,000 b/d, raising its overall quota to 2.93m. b/d. In 2000 Venezuelan petroleum production averaged 3.1m. b/d. OPEC production cuts in July 2001 reduced Venezuelan output to 2.67m. b/d. The country exceeded its OPEC quota of 2.45m. b/d during 2002 in order to improve the Government's disastrous fiscal position, although production collapsed during the national general strike, which began in December. Production temporarily fell from 2.8m. b/d as a result of the stoppage, leading to oil export revenue losses in the region of $6,700m. Venezuela was forced to import oil to cover its domestic needs and PDVSA declared *force majeure*. PDVSA claimed to have restored production to pre-strike levels within five weeks of the January 2003 general strike. A lack of clarity over production figures led to some doubt over PDVSA's claim that production in 2004 stood at 3.2m. b/d. Many industry experts believed the real production figure in 2004 was significantly lower and on a par with the production level recorded in 2003 of 2.6m. b/d.

The largest market for petroleum exports was the USA, which absorbed 1.4m. b/d (62%) of Venezuelan petroleum exports in 2003, according to the EIA. However, the oil strike damaged Venezuela's reputation as a reliable supplier. Exports to the Caribbean and Central America were also significant, owing to the San José Agreement, under which Venezuela and Mexico sell 160,000 b/d on preferential financial terms to 11 Caribbean and Central American nations, the 2000 Caracas Energy Accord, under which subsidized petroleum is supplied to Caribbean nations, including Cuba, and the Caribbean regional oil initiative, Petrocaribe, launched by Venezuela in 2005 with the aim of enhancing regional energy security by offering discounted oil to less developed countries in the region. The agreement to supply Cuba proved politically controversial and supplies were halted temporarily in the wake of the April 2002 coup against the Chávez Government, although they were resumed later in the year and consolidated through a series of bilateral trade agreements signed between the two countries at the beginning of 2005. Energy agreements with the People's Republic of China, Brazil, Russia and Argentina were also signed in 2005 as the Government expanded its network of partners in domestic and overseas energy ventures.

An increase in Venezuela's production output, together with rising petroleum prices, generated a substantial increase in petroleum export revenue in the 1980s and early 1990s, and by 1996 revenue reached US $23,400m., while the average export price of Venezuelan crude petroleum was $18.4 per barrel. In 1997 and 1998 this fell, respectively, to $16.3 and $10.6 per barrel. Consensus on production cuts between OPEC and non-OPEC countries increased the price of Venezuelan crude petroleum to an average of $16.1 per barrel in 1999, producing export earnings of $16,343m. for the year. Compliance with production quotas and higher global petroleum demand led to a steep increase in the price of Venezuelan petroleum in 2000, to $25.91 per barrel, raising export earnings to $26,643m. This favourable scenario was reversed when a decline in the price of Venezuelan petroleum, combined with production cuts and political instability, reduced oil-export earnings to $20,758m. in 2001, $21,530m. in 2002 and $20,831m. in 2003. Although there was a marginal fall in export revenues between 2002 and 2003, full year profits increased from $3,540m. to $4,230m. as a result of a 16% rise in the value of the Venezuelan export basket and cost-cutting measures implemented by PDVSA. In 2004 the average export price of Venezuelan oil increased by 29% to $33.22 per barrel and this, rather than production increases, led to a 48% increase in oil-export revenue. PDVSA contributed some $3,700m. to government finances in 2004, of which $2,000m. was channelled through the Fondo para el Desarrollo Económico y Social del País (Fondespa—Economic and Social Development Fund) for infrastructural development. A further $600m. was assigned to the Government's social welfare programmes, or 'missions', as part of the Government's plans to create a national oil industry at the service of national development.

In 1991 a policy of opening (*apertura*) the petroleum sector to private participation was initiated. Under legislation introduced in 1992 and 1995, foreign companies were permitted to hold a majority stake in future joint ventures. However, PDVSA was required to maintain a shareholding in any private-sector venture. The opening enhanced PDVSA's operational autonomy, which managers extended by adopting a system of transfer pricing that enabled the company to relocate additional profits to its network of overseas assets. This further reduced the Treasury's share of oil revenues.

The Chávez Government reversed the approach followed in the 1990s. The 1999 Constitution enshrined state ownership of PDVSA, dashing hopes among supporters of the *apertura* that the company would eventually be privatized. In 2001 the new Hydrocarbons Law was introduced, requiring PDVSA to have a 51% stake in new production and exploration agreements. The legislation also increased royalty rates on oil production from 16.6% to 30%, but reduced the level of income tax from 67.7% to 50% in a move designed to stabilize tax revenues. Private sector involvement in the oil sector falls within three frameworks: operating service agreements in marginal and low-yielding fields that were signed in the early 1990s; strategic associations for longer-term projects; and profit-sharing arrangements under which the private sector would bear exploration costs. In early 2005 the profit-sharing arrangements came under scrutiny and the Ministry of Energy and Petroleum announced that 32 operating-service agreements signed with foreign companies would be reviewed in order to bring them into line with the new hydrocarbons legislation and to ensure that all taxation owed to the Venezuelan Government had been paid. The companies affected were informed in April that they could sign new contracts drawn up in line with the Constitution of 1999 and the 2001 Hydrocarbons Law which gave PDVSA a 51% share in the ventures. The new contracts increased the income-tax rate designated for these special projects from 34% to 50%, to be applied retroactively from the time of introduction of the 2001 hydrocarbons legislation. The Government reasserted control over oil policy by reducing PDVSA's operational and decision-making autonomy, by restructuring operations into regional divisions and by transferring control to the Ministry of Energy and Petroleum, a move reflected in the appointment of the Minister of Energy and Petroleum, Rafael Darío Ramírez Carreño, to the presidency of PDVSA in early 2005. The increase in government control from 2001, combined with planned redundancies and the appointment of supporters of the Government's oil policy to the company's executive, led to sustained protests by both white- and blue-collar workers at PDVSA, culminating in senior PDVSA managers participating in strike activity in late 2002. The Government moved decisively against opponents within PDVSA after the disintegration of the strike in 2003, dismissing some 18,000 workers who had participated in the protest actions. The Ministry of Energy and Petroleum progressed with plans to restructure PDVSA and its operations, and in early 2005 President Chávez gave public consideration to selling off eight PDVSA refineries in the USA and a 50% stake held by PDVSA in four German refineries, a move that would represent a reversal of the internationalization strategy pursued by PDVSA for the previous decade. As the Government consolidated its control over PDVSA, there were concerns that the company had become subject to corruption and political patronage, while it was considered that the PDVSA executive had made slow progress in training new staff and inculcating both them and existing staff with the new ethos of nationalization. The transfer of PDVSA profits to the Government also raised concerns that the company would not have sufficient resources for investment and upgrades. In 2005 PDVSA's projected capital expenditure totalled some US $5,600m., of which 70% was to be reinvested in exploration and production.

The Government of President Chávez came to office committed to implementing a series of policies intended to reduce Venezuelan dependence on petroleum revenue through diversification into petrochemicals and the production of coal and gas. In its 10-year business plan, the PDVSA petrochemical subsidiary Pequiven envisaged raising production levels to 19.4m. metric tons, with investments from PDVSA and the private sector totalling US $8,700m. Limited progress was made, however, in raising petrochemical production. At the beginning of 2004 ExxonMobil announced a $3,000m. project to construct a petrochemical complex in Anzoátegui state. In 2005 Pequiven was renamed Corporación Petroquímica de Venezuela (CPV) and was to operate independently from PDVSA. In its strategic plan for 2005–12, CPV looked to increase investment in petrochemicals by $10,000m., with the aim of raising annual sales to $12,000m. per year. The plan included the modernization of the El Tablazo and Morón plants, the construction of a new site in Guiria and a new $700m. fertilizer plant near to the Morón site. New legislation intended to increase the domestic petrochemicals sector was being drawn up by the National Assembly. This followed Chávez's criticism of the high domestic costs of petrochemical products manufactured in joint ventures with foreign companies.

As part of its diversification plans, the Government sought to capitalize on Venezuela's estimated 42,000 metric tons of bitumen reserves, through increased production of Orimulsion (an alternative boiler fuel derived from a mixture of bitumen and water) to 20m. tons by 2006. Following the signing of supply contracts worth some US $200m. per year with Canada, the People's Republic of China, Denmark, Germany, Italy, Japan, the Republic of Korea (South Korea) and Lithuania, production was at full installed capacity of 6.23m. tons in 2001, of which 6.17m. tons was exported. A 30-year joint venture with the Chinese National Petroleum Corporation was agreed in 2002 for the construction of a 6.5m.-ton plant in the Orinoco belt and the supply to China of 1.8m. tons of Orimulsion per year. At the end of 2003 PDVSA announced that the low level of profit margins in the production of Orimulsion had forced the company to reduce operations with a view to ending production entirely. However, existing supply contracts would be honoured.

The exploitation of the country's natural gas reserves was a central element of the Government's diversification programme.

At 149,027,892m. cu ft (4,220,000m. cu m), Venezuela had the ninth largest proven gas reserves in the world at the end of 2003. These resources were underdeveloped, and in 2003 Venezuela produced only 61,657m. cu m. From 1971 the exploitation and production of natural gas was undertaken by Corpovén, a subsidiary of PDVSA (known from 1998 as PDV Servicios). The price of natural gas was fixed at artificially low levels, making it an attractive energy source: output and domestic sales increased significantly from the mid-1980s. The petroleum industry remained the largest consumer, absorbing some 60% of domestically produced gas for reinjection into oilfields for flaring. PDVSA introduced a series of investment and production plans in the 1990s to increase capacity. In 1990 the Nurgas pipeline was completed. This carried an estimated 22m. cu m of gas per day between extraction terminals in the Paria peninsula and industrial towns in the central and western parts of the country. A US $3,000m. public-private initiative, the Cristóbal Colón project, was also inaugurated to develop gas reserves in the Gulf of Paria. The project was relaunched in early 2002 as the Mariscal Sucre project and concessions were awarded to a consortium of Shell and Mitsubishi. Other concessions in the Deltana Platform offshore gasfield near the maritime border with Trinidad and Tobago were awarded to ChevronTexaco of the USA and Statoil of Norway in early 2003, and a further round of auctions for offshore licences was announced in July of that year. Legislation introduced in 1999 permitted foreign investment in exploration and production, distribution, transmission and gasification. Under the 2001 Hydrocarbons Law there are no restrictions on foreign ownership of projects to exploit non-associated natural gas. This formed the regulatory framework for a series of ambitious plans. These included expanding the 4,830 km of domestic pipeline and the construction of export pipelines, a project to allow the export of liquid natural gas by 2008 and the expansion of the natural gas grid. This involved transporting natural gas and crude oil from the Anaco area and interconnecting the Anaco–Barquisimeto gas-transportation system with the Ule–Amuya grid in the west of the country.

## FINANCE

Venezuela's financial sector underwent major restructuring following a banking crisis in 1994, which forced the Government to intervene in one-third of the country's financial institutions, which were typically family-run. After the introduction of legislation in 1996 to improve supervision of the sector and create universal banks, many of the largest banks were privatized. This increased the foreign presence in the sector and led to a series of mergers and acquisitions that resulted in an estimated 40,000 job losses. In 2003 there were 17 universal banks, accounting for 70% of the total assets in the financial sector. There were an additional 43 financial institutions, including a small number of state-run banks created by the Chávez Government. These state-run institutions included: the Banco del Pueblo Soberano, the Fondo de Desarrollo Microfinanciero and BANMUJER. They were intended to develop the 'social economy' by boosting access to credit for specifically targeted sectors, including small and medium-sized industries, indigenous groups and women. BANMUJER was recognized as one of the most socially progressive of these financial facilities. In 2004 it granted to women over 12,000 microcredits, totalling some US $3.5m. These, in addition to credits distributed through joint initiatives with the Banco de Desarrollo Económico y Social de Venezuela (BANDES), led to the creation of an estimated 15,832 direct jobs and 31,664 indirect jobs.

Relations between the financial sector and the Chávez administration were strained after the Government demanded a reduction in interest rates in order to expand credit access, and ordered at the beginning of 2001 an inquiry by the Competition Superintendency into 'cartelistic' practices in the industry; the findings did not support the Government's allegations. Relations were further soured following the introduction of a new banking law in November 2001. This reduced the operating autonomy of the Central Bank, required banks to increase agricultural loans from 8% to 15% of loan portfolios and from 1% to 3% in the case of small businesses. It also increased minimum capital requirements by US $8m. and empowered the President of Venezuela to decree a financial emergency. At the beginning of 2002 banks were prevented from issuing variable interest rate loans following a Supreme Court ruling, which ordered the Central Bank to set a fixed-interest rate for mortgages acquired after 1996. The sector was adversely affected by political instability in 2001 and 2002, which led to an increase in deposit withdrawals and a reduction in savings accounts. After the imposition of exchange controls in 2003 deposits grew strongly, while credit demand remained low. As a result, the banking system increased its purchase of government securities, such as central bank *certificados de depósito*, which accounted for 40% of total assets in 2003, overtaking total loans.

Lending and deposit rates continued to fall from 1998 to 2004. After reaching a high of 77% in 1998, lending rates fell back progressively, dropping to 22.5% in 2001. In 2002 the deposit rate increased to 29% (from 15.5% in 2001) and the lending rate rose to 36.6% amid escalating capital flight and mounting perceptions of political and economic risk. In April 2003 the Central Bank, the Government and representatives of the banking system signed an accord that partially regulated interest rates. Under the agreement private banks were to lend 6% of their outstanding loan portfolio (an estimated US $300m.) to the private sector at a preferential rate set at 85% of the average lending rate of the six largest banks. In 2003 both the average lending rates and deposit rates declined, to 26.2% and 16.7%, respectively. The trend of state intervention continued into 2005 when, at the end of April, the Central Bank established a maximum lending rate and minimum deposit rates. As a result, the maximum that could be charged for loans was 28%, while savings accounts and time deposits had to pay a minimum of 6.5% and 10%, respectively. The average lending rate at the end of December 2004 was 15.4%, compared with 20.5% at the end of 2003.

Foreign participation in the insurance sector increased following the introduction of the Insurance and Reinsurance Company Law of 1994, which removed restrictions on foreign shareholdings. As with the banking system, this led to an increase in mergers and acquisitions by foreign companies. None the less, the sector remained fragmented and 53 companies were operating in Venezuela in 2003. The five largest companies held one-half of all premiums. In November 2001 a revised Insurance and Reinsurance Law was introduced. This greatly extended regulation of the sector by the Superintendency of Insurance, and raised reserves and minimum capital requirements. Take-up of private life and other insurance remained low, owing to a lack of public trust in private provision. Take-up was highest in the non-life sector, which represented nearly three-quarters of premiums, specifically car and hospital insurance.

The development of the equity market was fostered by the structural-reform programme introduced in 1989, which led to an increase in trading activity. In 1992 an automated trading system was introduced and changes in legislation allowed the shares of privatized companies to be traded on the Caracas stock exchange (Bolsa de Valores de Caracas—BVC). By 1996 the BVC recorded an average trading volume of US $20m. per day. Political and economic uncertainty reduced activity on the severely undercapitalized stock market thereafter and the equity market dwindled, with the BVC recording one of the worst performances in the world in 1998, with a dollar loss of 51%. Two of the country's three stock markets closed in 2000, rendering the BVC the only forum to trade equities and fixed-income instruments. After a strong performance in 2000, growth was reversed in 2001 and the general index declined by 11.3% in US dollar terms, with an average daily trading volume of $1m. Venezuela was removed by the International Finance Corporation from its Latin America Investible Index for failing to meet minimum liquidity requirements. The market fell by a further 30% in 2002. However, interest in the stock exchange grew rapidly in 2003, when investors realized that equities could be exchanged for US currency, circumventing the Government's exchange-rate restrictions. The decision by many of the largest Venezuelan companies to withdraw from the local market in favour of issuing US or global depository receipts has further weakened the Venezuelan equity market and despite the absence of restrictions on foreign firms, none have opted to issue equity over the past five years.

## TELECOMMUNICATIONS

Following the introduction of what was recognized as a model telecommunicators law in 2000, the sector has been one of the most dynamic areas of the economy, although growth slowed in 2002 and 2003 as a result of currency devaluation and recession. Dissatisfaction with the fixed-line services monopolized by the state-owned telecoms company, Compañía Anónima Nacional Teléfonos de Venezuela (CANTV), which was privatized in 1991, led to a surge in mobile cellular telephone use in the 1990s. As a result, Venezuela had one of the highest mobile phone penetration rates in South America, with 25 in every 100 people subscribing to a mobile-phone service. Over 90% of mobile subscribers used pre-paid services. By 2003 three service providers shared the cellular market: CANTV's subsidiary, Movilnet (40%), TelCel CA (47%), a subsidiary of US corporation Bellsouth, and Digitel (12%), owned by Italy's TIM. In June 2000 legislation was introduced that ended CANTV's monopoly on basic telephone services in accord with the 1991 privatization legislation. TelCel CA and Digicel CA, owned by Santander Central Hispano of Spain, entered the fixed-line market, although they failed to acquire a significant share and cellular phones remained the preferred option of consumers. In May 2005 the telecommunications regulator, the Comisión Nacional de Telecomunicaciones opposed a proposal by CANTV to purchase Digitel on competition grounds, as the fusion of the two operators would have provided CANTV with control of over 50% of the mobile telephone market.

Internet access experienced strong growth in the late 1990s and early 2000s. However, levels of computer ownership were low, with 1.5m. in use in 2002, and this has been an impediment to the expansion of internet services. More recently, 'cybercafes' and some 250 government-funded internet centres have flourished: more than 1,000 had been opened by 2002, the majority of which were run by CANTV. Growth of cable subscription television was also pronounced, with an estimated 600,000 of the 4.5m. televisions in Venezuela connected to a cable network in 2002. In early 2005 the Government launched Televisora del Sur (Telesur—Television of the South), a media venture in partnership with the Governments of Cuba, Argentina and Uruguay. The Venezuelan Government held a 51% stake in Telesur. The initiative was intended to promote regional integration and a news agenda ready to contest US foreign policy in the region and economic policies of the so-called 'Washington Consensus'. In late July the US Congress approved an amendment allowing the USA to commence broadcasts to the region intended to act as a check on the influence of Telesur.

## MANUFACTURING

Development of the country's manufacturing sector was initially fostered by the use of protectionist and interventionist measures. Small and medium-sized industries in the private sector concentrated on the production of consumer goods for the domestic market, while the major capital-intensive industries were state-owned and located in the Ciudad Guayana development region, in the east of the country. Structural reform in 1989 and free trade agreements within the CAN generated strong manufacturing growth at the beginning of the 1990s. This was not consolidated, owing to the volatility of the exchange rate and economic recession. Manufacturing GDP increased by 2.1% in 2000, but declined by 0.2% in 2001. Political uncertainty and instability in 2002 and 2003 had a deleterious effect on manufacturing and the main groups representing private-sector manufacturers supported and participated in industrial action and anti-Government protests launched by the opposition. In 2003 it was estimated that 5,000 businesses had been closed over the previous five years, contributing to job losses in the region of 200,000. Underscoring the severity of the contraction in the sector, real manufacturing output in 2003 was more than 25% below output a decade earlier. However, the economic growth of 2004 led to a strong improvement in manufacturing performance and growth of 25.4% was recorded. This expansion continued into the first quarter of 2005, although further growth of the sector continued to be constrained by a lack of private investment. In 2004 the Government invested US $217m. in the small and medium-sized industrial sector to stimulate manufacturing growth and capacity.

## CONSTRUCTION

Growth in Venezuela's construction industry has followed trends in the petroleum economy. Periods of high oil prices have led to increased levels of public spending and growth of the sector, with subsequent drops in the oil price leading to pronounced contractions. The industry expanded in the 1970s as a result of public-works projects, only to decline in the 1980s as the petroleum price fell and the economy went into recession. Despite major government building projects and official encouragement of low-cost housing, the industry remained depressed in the 1980s, especially in the private sector. Conditions improved in the 1990s as the liberalization of the petroleum sector in 1997 led to a surge in capital investment. In 1998 the sector contributed 6.1% of GDP, despite a steep rise in interest rates. In 1999 the sector contracted by 16.5%, and unemployment among construction workers rose to 30%. A further contraction of 5.1% was experienced in the sector in 2000. After three consecutive years of recession, construction expanded by 13% in 2001; however, this performance was reversed in 2002 and 2003, when economic recession, capital flight and political instability led to a 20% contraction in the sector. Robust economic growth in 2004 and increased government spending into the first quarter of 2005 led to a strong performance by the construction sector, which, with estimated growth of 32.1% in 2004, was one of the fastest growing non-oil sectors. The sector has, however, struggled to meet the demand for new housing and infrastructure, and delays led the Government to create a new state-owned construction company in 2005, in a joint venture with the Cuban Government as part of a wider series of trade and commercial agreements signed at the beginning of that year.

## TRANSPORT

Historically high levels of public investment in the transport infrastructure allowed for the development of an extensive highway network that totalled 96,200 km in 2004, of which 32,300 km were paved. A decline in public spending has led to a deterioration of the road network, which was the principal means of transport for freight and people. This situation has not been reversed despite the decentralization of responsibility for public highways to state governors in 1998. Schemes to boost revenue, such as road tolls, lacked popular support and attempts to increase private-sector participation were constrained by the weak regulatory environment and political uncertainty. Serious deficiencies in the country's infrastructure, particularly the main highway between Maiquetía and Caracas, remained a concern and the Government was under increased pressure to deliver promised major infrastructural repairs before the main transport arteries reached a state of absolute disrepair.

The railway network was underdeveloped and small at 584 km, one-half of which was privately owned. In line with the objectives set out in the 2001–07 National Development Plan, a railroad project linking industrial towns in the east to the northern Caribbean coast was announced in December 2000, and in 2002 construction of a 9.5-km railroad between Caracas and the commuter belt area of Los Teques began, at a cost of US $384m. In 1982 an underground system opened in Caracas, which was efficiently managed, and later added new lines in the 1990s. Work on further extensions to the system began in 2002. Long-term plans to construct an underground system in the second city of Maracaibo were halted by a lack of finance, and had still not restarted by mid-2005.

Deregulation of the airline industry in 1989 and decentralization of responsibility for airports in 1992 has led to growth in the industry and related support services and infrastructure. The country has 366 airports, of which 122 have paved runways. There are 11 international airports, with 90% of international flights handled by Maiquetía in Caracas, which is undergoing a major upgrade. In 2003 12 airlines provided domestic services and the Government announced plans for the creation of a national state airline, Conviasa, with an investment of US $60m. The country has 13 major ports and harbours. La Guaira, Puerto Cabello and Maracaibo handle 80% of bulk trade. The Orinoco river was navigable for about 1,120 km, and in 2000 a $10m. study began to investigate suitable areas for development in order to increase barge transport along the Orinoco and Apure rivers. The Chávez administration had ambitious plans for the development of the country's transport infrastructure, but there were delays in realizing these, owing to the economic contraction in 2002 and 2003 and power facility constraints.

## TOURISM

Despite a vast array of potential tourist attractions, including the world's highest waterfall (Angel Falls, with an overall drop of 979 m), a 2,718-km Caribbean coastline and an Andean mountain range running from the south-west to the north-east of the country, the Venezuelan tourism industry remained underdeveloped, with the exception of Margarita island, which is visited by 90% of tourists to Venezuela. Facilities for tourists were limited, and the strong currency increased the cost of travel to Venezuela. In the late 1990s the sector accounted for an estimated 6% of GDP, with the USA the principal market. Political instability, civil unrest and problems of personal security caused tourist numbers to fall, despite vigorous marketing campaigns and the introduction of the Tourism Law in 1998, which provided tax incentives for the development of the industry. In 1997 the number of tourists that visited Venezuela peaked at 811,285. By 2003 this had dropped to 336,974, as political conflict deepened and President Chávez's anti-US rhetoric contributed to a fall in tourists from the USA. The decision by a number of foreign embassies to issue advice against travel to the country in 2002 and 2003 also proved a major reverse for the industry. Nevertheless, the Government's vigorous promotion of the tourism sector and specifically domestic tourism, contributed to growth of 32.2% in the sector from January to November 2004, with foreign tourist arrivals increasing by 46.4% over the same period, to 430,000. Tourism accounted for an estimated 4.9% of GDP in 2004. Investments worth over US $200m. were planned for 2005. After an upturn in hotel development in the mid-1990s, particularly in the provision of five-star facilities (of which there are 29) by foreign operators, investment decreased and there has been no subsequent expansion of hotel capacity. In 2002 there were 25,000 rooms available.

## PUBLIC FINANCE

In the 1970s increased income from the petroleum sector allowed for high levels of public spending. Budget deficit financing formed a major part of the Government's expansionary policies and provided an important stimulus to economic growth. Reduced petroleum revenues and a growing public-sector debt resulted in a deficit in 1982. Budget surpluses were recorded in 1983 and 1984, but falling government revenues, increasing outflows of capital and an expanding public debt became a major problem. By 1988 the fiscal deficit was the equivalent of 8.0% of GDP.

Following the adoption of an IMF-sponsored austerity programme in 1989, the deficit was reduced to 1.1% of GDP. This was transformed into modest surpluses of 0.2% and 0.7% of GDP in 1990 and 1991, respectively. An increase in public spending and a decline in petroleum exports generated a deficit equivalent to 5.8% of GDP in 1992, which was reduced to 2.9% of GDP in 1993. A slump in petroleum prices in 1994 and the imposition of a fixed exchange-rate regime contributed to a fiscal deficit of 6% of GDP in 1995. An increase in world petroleum prices improved revenue in 1996 and a budget surplus was maintained in 1997. A collapse in petroleum prices in 1998 forced the Government into a drastic revision of its budget and attempts were made to increase ordinary fiscal revenues. This failed to overcome the shortfall created by the decline in petroleum income, and a budget deficit of 4.2% was recorded. Petroleum production cuts and economic recession offset an increase in the petroleum price in 1999, leading to a deficit of 3.1%. In 2000 a dramatic recovery in the petroleum price led to a 106.5% increase in petroleum revenue. The fiscal accounts, however, registered a deficit of 1.6% of GDP, owing to a 46% increase in public spending as the Government attempted to accelerate economic recovery. In 2001 current revenue increased by 13.3% following an increase in taxation of PDVSA, but the deficit doubled to 4.3% of GDP as current government spending remained high, increasing by 32%, to 20.6% of GDP. The deficit narrowed to 2.7% of GDP in 2002, despite the sustained strength of the oil price and a reduction in government spending. The transfer of foreign-exchange profits from the Central Bank and a dividend payment from PDVSA helped the Government to bridge a shortfall in oil revenues caused by the paralysis of petroleum production during the January 2003 general strike. At the end of 2003 the fiscal deficit had widened to an estimated 4.2% of GDP. In 2004, despite a dramatic increase in the international oil price, financial transfers from the Government and PDVSA, and a strong improvement in non-oil taxation collection, an estimated deficit of 3.0% of GDP was recorded as the Government accelerated its public-spending plans. It was expected that the deficit would increase in 2005 as the Government maintained its expansionary fiscal policy in the months preceding the legislative elections scheduled for December.

Corruption, tax evasion and high levels of informal-sector employment rendered the non-petroleum tax base in Venezuela a source of fiscal weakness that the Chávez Government had pledged to address. Non-oil-related tax revenues accounted for the equivalent of just 8.7% of GDP in 2001, despite major efforts to improve collection. Legislation introduced in that year imposed custodial sentences on tax evaders and improved auditing techniques. Plans were also announced to collect VAT from informal street vendors, and automated procedures were introduced in the customs service to increase efficiency. These measures contributed to a 23% year-on-year increase in tax collection in 2001, with revenue from VAT increasing by 20%, income taxes by 40% and customs duties by 14%. Non-oil tax revenues increased to the equivalent of 9.0% of GDP in 2002, despite a non-payment of taxes campaign organized by the opposition. Tax revenue in the first six months of 2003 was 11% higher than in the same period of 2002, although this was 8.2% lower in real terms, given the massive devaluation of the bolívar during the intervening period. In 2004 non-oil taxation revenues increased by 85%.

## COST OF LIVING

Inflation in Venezuela was historically low. The strategy adopted in the 1980s of devaluing the currency in order to boost the bolívar equivalent of dollar oil revenue generated inflationary cycles, which subsequent administrations sought to contain through the periodic imposition of price controls. This in turn fed inflation, which remained in double figures and one of the highest on the continent. The adjustment programme and liberalization of the exchange rate in 1989 led to a surge in inflation, which reached 84% at the end of the year. The devaluation of the bolívar in 1994, coupled with wage rises, contributed to inflation increasing to 60.8%. In 1995 the average annual rate of inflation decreased slightly, to 59.9%, but in the following year it reached a record 103.2%, as a result of two further devaluations and the liberalization of the exchange rate under the Agenda Venezuela adjustment programme.

In 1996 the Central Bank adopted a new anti-inflationary strategy. A 'crawling peg' banded exchange rate was introduced. This succeeded in reducing the rate of inflation, but at the cost of an increasingly overvalued domestic currency. Depressed domestic demand after 1999 allowed for a progressive fall in the inflation rate, which fell to an annual average of 12.5% in 2001. In February 2002 the crawling peg was abandoned with the devaluation of the currency leading to an acceleration of price increases. Annual inflation, however, increased to only 22.4%. This was lower than expected and reflected continued weak demand. Following the general strike in December 2002 and the disruption to oil production, the Government imposed price controls and a new exchange-rate regime in February 2003 to stabilize foreign reserves. The regime restricted the availability of dollars to private-sector opponents of the Government and fixed the exchange rate at 1,600 bolívares to US $1. The measures were successful in containing escalating prices and allaying fears of a hyperinflationary cycle, although they created a 'black' market for currency where US $1 was valued at around 3,000 bolívares. In 2003 inflation averaged 31.1%, largely as a of result producers paying premium exchange rates on the 'black' market to facilitate import purchases. In 2004 further progress was made in reducing inflation, which averaged 21.8%. The reduction in price pressures was linked to the growth of state food-distribution programmes and to a strong foreign reserve position, which stabilized the exchange rate.

In 2004 the Central Bank came under intense pressure from the Government to transfer its profits from the foreign-exchange regime in order to fund social projects and development programmes. Following intense political dispute, the Central Bank transferred US $1,700m. to the Government in February 2005. In mid-July the Asamblea Nacional approved legislation allowing the Government to access up to $5,000m. in international reserves. Opposition groups vowed to challenge the new law, owing to fears (also held by many international investors and financial commentators) that it would lead to increased inflation and devaluation of the bolívar, not least because the Government planned to allocate about one-third of the additional revenue to current expenditure, rather than to development programmes.

## FOREIGN DEBT

Declining petroleum revenues, an overvalued bolívar and the accumulation of debt maturities made it increasingly difficult for Venezuela to service its foreign debt during the 1980s, 85% of which was owed to foreign creditors. In 1986 and in 1990 the debt was restructured under a Brady debt-relief programme and the debt-service was lightened. As a result of the crisis in the financial sector in 1994, Venezuela fell behind on repayment of its 'Paris Club' obligations. In April 1996 a structural-adjustment programme was introduced, which secured a US $1,400m. stand-by agreement from the IMF and a strong rise in petroleum export revenues enabled the Government to pay back the bulk of the debt that had been acquired during the previous three years. The Chávez Government pledged to reschedule the debt burden. However, little progress was made, and Venezuela maintained a heavy foreign debt-repayment schedule in the early 2000s. Heightened perceptions of political and economic risk surrounding the country restricted access to foreign borrowing and, as a result, the stock of external debt remained broadly stable at $32,000m.–$36,000m. in 2002. Sentiment towards Venezuela changed in 2003 and 2004 as international reserves rose and the international oil price increased. In September 2003 and January 2004 Venezuela issued, respectively, $1,600m. and $1,000m. in 10-year bonds and 30-year papers. In March 2005 this was followed by a highly successful sale of €1,000m. in 10-year bonds, Venezuela's first issue in the euro domination, with the revenues raised contributing to interest and principal payments falling due in 2005. A further $170m. of floating-rate notes due in 2006 were exchanged for new securities maturing in 2011. The external debt to GDP ratio was an estimated 38% of GDP in 2003. The Government relied increasingly on domestic debt, which rose from $4,000m. in 1997 to $7,900m. in 2002.

In November 1998 the Congreso approved the capitalization of a macroeconomic stabilization fund (Fondo de Inversión para la Estabilidad Macroecónomica—FIEM). This set aside 'windfall' oil revenues proceeding from periods when the average oil price exceeded the budgeted level. The surplus was divided between two separate accounts: the debt-redemption fund, used to pay outstanding external liabilities, and a macroeconomic stabilization fund to cover government revenue shortfalls in periods of depressed oil prices. The credibility of the mechanism was diminished by changes in the rules governing the allocation of funding and the Government's decision to draw repeatedly on FIEM funds to cover the fiscal deficit. At the end of 2001 FIEM funds totalled US $6,227m. However, in December of that year the Government withdrew $894m. and, as a result, total reserve assets, including the FIEM, fell from $20,471m. at the end of 2000 to $18,522m. in 2001. The deterioration in the reserve position followed repeated interventions by the Central Bank to support the bolívar amid accelerating levels of capital flight. Further withdrawals in 2002–03 reduced the balance of FIEM funds to $1,000m. in July 2003. By 2004 the FIEM fund totalled just $700m., despite the strong rise in the oil price. The Government's handling of the FIEM attracted significant criticism. This was increased in May 2002 when it was revealed that the Government had failed to make deposits to the FIEM that had been mandated by the Asamblea Nacional and had instead spent the money on paying public-sector wages. It was anticipated that transfers to the fund would recommence in 2006 following the introduction of new legislation that proposed to reallocate management of the FIEM from the Central Bank to the state development bank, the Banco de Desarrollo Económico y Social de Venezuela (BANDES). At the end of 2004 international reserves, excluding gold, totalled $24,200m., as a result of the increase in the oil price and restrictions on capital flight and dollar-exchange availability. As reserves continued to increase in the first quarter of 2005, President Chávez reignited an earlier controversy by proposing a ceiling for international reserves of between $18,000m. and $20,000m., with any excess finances to be channelled into development projects.

## BALANCE OF PAYMENTS AND TRADE

Venezuela's external trade is dominated by petroleum. Petroleum traditionally ensured that the country maintained a large trade and current-account surplus. The performance of the non-oil sector was weak owing to the small and uncompetitive profile of the private sector and the overvaluation of the bolívar. This increased reliance on petroleum revenues, which remained volatile and generated a high level of import dependence. The Venezuelan import bill expanded at an average annual rate of 22% during the 1970s. This was offset by rising petroleum revenues, but recurrent falls in world petroleum prices in the 1980s, combined with rising external debt, narrowed the balance of trade, culminating in balance-of-payments deficits in 1987 and 1988. The trade and current accounts moved into surplus. In 1994 exchange controls were introduced to stem foreign reserve losses after the collapse of the banking sector. This generated a surplus on the trade and current accounts as imports fell. In 1995–98 imports grew strongly as the exchange controls were abolished and the currency became increasingly overvalued. The petroleum price was halved in 1998 as imports reached an historic high. As a result, the current account registered a deficit of $3,253m. In 1999 an increase in the price of petroleum, in conjuction with a deep recession (which reduced import demand), moved the current account back into a surplus. Non-traditional exports registered strong performance in 2000 following a recovery in demand in the Andean market. Imports grew by 21.7% as a result of improved domestic demand. This was offset by a 72% increase in petroleum-export revenues as petroleum prices rose sharply and export volumes increased. The current account registered US $12,106m., while the trade surplus was $16,664m. The surplus on the current account declined to $1,987m. in 2001 as international petroleum prices fell, while the trade surplus similarly declined, to $7,460m. Economic recession led to a 30% fall in imports in 2002. With exports totalling $26,781m., this generated a healthy current-account surplus of $7,599m. and a trade surplus of $13,421m. In 2003 the imposition of exchange controls led to a fall in imports, while export earnings rose slightly, to $27,170m., generating an improvement on the current account, which was $11,448m. in surplus. The current account position improved further in 2004, when it widened to $14,575m. as a result of a 48% increase in oil export revenue and a 33% increase in non-oil exports. Imports rose strongly to $17,318m. as the Government made more US dollars available and consumer demand showed robust recovery from the deterioration of 2002 and 2003. The full year trade surplus for 2004 was $22,053m.

The USA was Venezuela's main trading partner in 2003, absorbing 70% of Venezuela's exports and accounting for 33.7% of the country's imports. Progressive reductions in petroleum production led Venezuela to lose its market share in the USA and by 2003 Venezuela had fallen from second to fourth largest supplier, behind Canada, Saudi Arabia and Mexico. In an attempt to diversify its trading partners, the Venezuelan Government signed bilateral accords with several countries, including Russia, China, Iran and Cuba, although these did not significantly alter the country's trading profile. In 2000 Venezuela became Cuba's main trading partner. Trade with Colombia increased from the late 1980s, and was augmented by the signing of a free trade agreement in 1991 and a series of trade accords among members of the CAN. In 1998 the balance of trade tipped in favour of Colombia and imports from the country continued to increase, from 5% of Venezuela's total imports in 1998 to 10.4% by 2002. In contrast, the ratio of Venezuelan exports to Colombia declined from 8.2% of total exports in 1998 to 3% by 2002. Cross-border trade was negatively affected by political instability in Venezuela and the imposition of protectionist measures by the Venezuelan Government in 1999 and 2002. Imports from Colombia and the USA grew strongly as the economy recovered in 2004 and the official supply of US dollars expanded.

As well as being a member of the CAN, Venezuela was, with Mexico and Colombia, a member of the 'Group of Three', which implemented a free trade agreement in 1994. Venezuela participated in the Latin American Economic System, which promoted intra-regional economic and social co-operation. In April 1998 the CAN and the Southern Common Market (Mercado Común del Sur—Mercosur) signed an accord to establish a free trade area by 1 January 2000. The agreement, which would lead to the creation of the largest market in Latin America, was delayed subsequently by technical problems and Venezuela received associate membership of the organization. Venezuela maintained a critical stance towards the proposed Free Trade Area of the Americas and posited a 'Bolivarian' alternative model of integration that excluded the USA.

## CONCLUSION

In 1999 the new Government of Hugo Chávez inherited a dire economic legacy. Growing poverty, informality and unemployment were the main trends after two decades of political and economic turmoil. The objectives of reducing dependence on petroleum export revenue and improving distribution to the most deprived sectors of society shaped government economic policy. Strategies to overcome the profound inequalities that characterized Venezuelan society, including land reform and improved taxation collection, were negatively received by the wealthiest sectors. The perceived attack on historically privileged groups, compounded by the reluctance of the Government to negotiate with affected sectoral interests, led to a spiral of conflict between the Government and the opposition, that culminated in the temporary displacement of the Government in April 2002 and a three-month general strike from December 2002. Perceptions of political risk, the introduction of protectionist measures and economic uncertainty reduced private-sector confidence in the country and accelerated capital flight. The discretionary dispersion of US dollars under the exchange-rate regime introduced in February 2003 exacerbated hostility towards the Government while generating serious concerns as to the coherence of economic policy. In 2004 the dire economic performance recorded in 2002 and 2003 was reversed, as the oil price increased and political conflict diminished following the President's successful defence of the recall referendum held in August of that year. Despite making considerable progress in identifying and addressing the needs of the poor and excluded, by 2005 the Government had failed to reduce the country's dependence on revenue from petroleum exports, thereby throwing into doubt the sustainability of the administration's projects to create food self-sufficiency and a 'social' economy.

# Statistical Survey

Sources (unless otherwise stated): Instituto Nacional de Estadística (formerly Oficina Central de Estadística e Informática), Edif. Fundación La Salle, Avda Boyacá, Caracas 1050; tel. (212) 782-1133; fax (212) 782-2243; e-mail ocei@platino.gov.ve; internet www.ine.gov.ve; Banco Central de Venezuela, Avda Urdaneta esq. de las Carmelitas, Caracas 1010; tel. (212) 801-5111; fax (212) 861-0048; e-mail mbatista@bcv.org.ve; internet www.bcv.org.ve.

## Area and Population

### AREA, POPULATION AND DENSITY

| | |
|---|---:|
| Area (sq km) | 916,445* |
| Population (census results) | |
| 20 October 1990† | 18,105,265 |
| 30 October 2001‡ | |
| Males | 11,402,869 |
| Females | 11,651,341 |
| Total | 23,054,210 |
| Population (official estimates at mid-year)§ | |
| 2002 | 25,219,910 |
| 2003 | 25,673,550 |
| 2004 | 26,127,351 |
| Density (per sq km) at mid-2004 | 28.5 |

* 353,841 sq miles.
† Excluding Indian jungle population and adjustment for underenumeration, estimated at 6.7%.
‡ Excluding Indian jungle population, enumerated at 183,143 in a separate census of indigenous communities in 2001. Also excluding adjustment for underenumeration, estimated at 6.7%.
§ Based on results of 2001 census, including Indian jungle population and adjustment for underenumeration.

### ADMINISTRATIVE DIVISIONS
(30 October 2001)

| | Area (sq km) | Population (provisional) | Density (per sq km) | Capital |
|---|---:|---:|---:|---|
| Federal District | 433 | 1,836,286 | 4,240.8 | Caracas |
| Amazonas | 177,617 | 70,464 | 0.4 | Puerto Ayacucho |
| Anzoátegui | 43,300 | 1,222,225 | 28.2 | Barcelona |
| Apure | 76,500 | 377,756 | 4.9 | San Fernando |
| Aragua | 7,014 | 1,449,616 | 206.7 | Maracay |
| Barinas | 35,200 | 624,508 | 17.7 | Barinas |
| Bolívar | 240,528 | 1,214,846 | 5.1 | Ciudad Bolívar |
| Carabobo | 4,650 | 1,932,168 | 415.5 | Valencia |
| Cojedes | 14,800 | 253,105 | 17.1 | San Carlos |
| Delta Amacuro | 40,200 | 97,987 | 2.4 | Tucupita |
| Falcón | 24,800 | 763,188 | 30.8 | Coro |
| Guárico | 64,986 | 627,086 | 9.6 | San Juan de los Morros |
| Lara | 19,800 | 1,556,415 | 78.6 | Barquisimeto |
| Mérida | 11,300 | 715,268 | 63.2 | Mérida |
| Miranda | 7,950 | 2,330,872 | 293.2 | Los Teques |
| Monagas | 28,900 | 712,626 | 24.6 | Maturín |
| Nueva Esparta | 1,150 | 373,851 | 325.1 | La Asunción |
| Portuguesa | 15,200 | 725,740 | 47.7 | Guanare |
| Sucre | 11,800 | 786,483 | 66.7 | Cumaná |
| Táchira | 11,100 | 992,669 | 89.4 | San Cristóbal |
| Trujillo | 7,400 | 608,563 | 82.2 | Trujillo |
| Vargas | 1,497 | 298,109 | 199.1 | La Guaira |
| Yaracuy | 7,100 | 499,049 | 70.3 | San Felipe |
| Zulia | 63,100 | 2,983,679 | 47.3 | Maracaibo |
| Federal Dependencies | 120 | 1,651 | 13.8 | — |
| **Total** | 916,445 | 23,054,210 | 25.2 | |

# VENEZUELA

## PRINCIPAL TOWNS
(city proper, estimated population at 1 July 2000)

| | | | |
|---|---|---|---|
| Caracas (capital) | 1,975,787 | Mérida | 230,101 |
| Maracaibo | 1,764,038 | Barinas | 228,598 |
| Valencia | 1,338,833 | Turmero | 226,084 |
| Barquisimeto | 875,790 | Cabimas | 214,000 |
| Ciudad Guayana | 704,168 | Baruta | 213,373 |
| Petare | 520,982 | Puerto la Cruz | 205,635 |
| Maracay | 459,007 | Los Teques | 183,142 |
| Ciudad Bolívar | 312,691 | Guarenas | 170,204 |
| Barcelona | 311,475 | Puerto Cabello | 169,959 |
| San Cristóbal | 307,184 | Acarigua | 166,720 |
| Maturín | 283,318 | Coro | 158,763 |
| Cumaná | 269,428 | | |

## BIRTHS, MARRIAGES AND DEATHS*

| | Registered live births | | Registered marriages | | Registered deaths | |
|---|---|---|---|---|---|---|
| | Number | Rate (per 1,000) | Number | Rate (per 1,000) | Number | Rate (per 1,000) |
| 1995 | 571,670 | 26.1 | n.a. | n.a. | 102,451 | 4.7 |
| 1996 | 572,503 | 25.7 | n.a. | 3.7 | 104,416 | 4.7 |
| 1997 | 573,503 | 25.1 | n.a. | 3.8 | 106,141 | 4.7 |
| 1998 | 574,553 | 24.7 | n.a. | 3.7 | 108,077 | 4.7 |
| 1999 | 576,073 | 24.3 | 90,220 | 3.8 | 109,999 | 4.6 |
| 2000 | 577,346 | 23.9 | 91,088 | 3.8 | 111,907 | 4.6 |
| 2001 | 579,544 | 23.5 | 81,516 | 3.3 | 113,889 | 4.6 |
| 2002 | 581,325 | 23.2 | 73,163 | 2.9 | 115,884 | 4.6 |

* Figures for numbers of births and deaths include adjustment for under-enumeration. Rates are calculated using adjusted population data.

**2003:** Marriages 74,562 (2.9 per 1,000).

**Expectation of life** (WHO estimates, years at birth): 74 (males 71; females 77) in 2003 (Source: WHO, *World Health Report*).

## ECONOMICALLY ACTIVE POPULATION
(household surveys, '000 persons aged 15 years and over)*

| | 2000 | 2001 | 2002 |
|---|---|---|---|
| Agriculture, hunting, forestry and fishing | 899.5 | 892.1 | 949.0 |
| Mining and quarrying | 54.6 | 49.7 | 47.3 |
| Manufacturing | 1,147.0 | 1,207.1 | 1,150.3 |
| Electricity, gas and water | 59.5 | 57.7 | 51.4 |
| Construction | 707.5 | 774.3 | 775.8 |
| Wholesale and retail trade, restaurants and hotels | 2,292.6 | 2,455.3 | 2,585.3 |
| Transport, storage and communications | 612.1 | 654.3 | 703.6 |
| Financing, insurance, real estate business services | 455.6 | 482.2 | 481.8 |
| Community, social and personal services | 2,583.5 | 2,820.7 | 2,932.7 |
| Activities not adequately defined | 9.7 | 11.2 | 21.8 |
| **Total employed** | **8,821.8** | **9,404.6** | **9,698.9** |
| Unemployed | 1,423.5 | 1,435.8 | 1,822.6 |
| **Total labour force** | **10,245.3** | **10,840.4** | **11,521.5** |

* Figures exclude members of the armed forces.

Source: ILO.

# Health and Welfare

## KEY INDICATORS

| | |
|---|---|
| Total fertility rate (children per woman, 2003) | 2.7 |
| Under-5 mortality rate (per 1,000 live births, 2003) | 21 |
| HIV/AIDS (% of persons aged 15–49, 2003) | 0.7 |
| Physicians (per 1,000 head, 2001) | 1.94 |
| Hospital beds (per 1,000 head, 1996) | 1.47 |
| Health expenditure (2002): US $ per head (PPP) | 272 |
| Health expenditure (2002): % of GDP | 4.9 |
| Health expenditure (2002): public (% of total) | 46.9 |
| Access to water (% of persons, 2002) | 83 |
| Access to sanitation (% of persons, 2002) | 68 |
| Human Development Index (2002): ranking | 68 |
| Human Development Index (2002): value | 0.778 |

For sources and definitions, see explanatory note on p. vi.

# Agriculture

## PRINCIPAL CROPS
('000 metric tons)

| | 2001 | 2002 | 2003 |
|---|---|---|---|
| Rice (paddy) | 787 | 668 | 701 |
| Maize | 1,801 | 1,392 | 1,505 |
| Sorghum | 554 | 509 | 600 |
| Potatoes | 329 | 351 | 298 |
| Cassava (Manioc) | 606 | 521 | 490 |
| Yautia (Cocoyam) | 64 | 65 | 42 |
| Yams | 88 | 93 | 74 |
| Sugar cane | 7,900 | 7,320 | 8,865 |
| Coconuts | 106 | 175 | 190 |
| Oil palm fruit | 348 | 320 | 314 |
| Cabbages | 54 | 61 | 60 |
| Tomatoes | 182 | 197 | 168 |
| Chillies and green peppers | 62 | 69 | 62 |
| Dry onions | 236 | 277 | 237 |
| Carrots | 178 | 185 | 194 |
| Other vegetables | 228* | 218 | 214* |
| Watermelons | 168 | 205 | 173 |
| Cantaloupes and other melons | 131 | 182 | 192 |
| Bananas | 735 | 591 | 639 |
| Plantains | 754 | 461 | 497 |
| Oranges | 456 | 342 | 316 |
| Tangerines, mandarins, clementines and satsumas | 177 | 195 | 188* |
| Lemons and limes | 77 | 85 | 81* |
| Mangoes | 75 | 74 | 75 |
| Avocados | 44 | 50 | 52 |
| Pineapples | 300 | 347 | 384 |
| Papayas | 130 | 153 | 175 |
| Other fruit* | 214 | 222 | 222 |
| Coffee (green) | 92 | 84 | 82 |

* FAO estimate(s).

Source: FAO.

## LIVESTOCK
('000 head, year ending September)

| | 2001 | 2002 | 2003 |
|---|---|---|---|
| Horses* | 500 | 500 | 500 |
| Asses* | 440 | 440 | 440 |
| Cattle | 15,474 | 15,791 | 15,989 |
| Pigs | 2,780 | 2,825 | 2,916 |
| Sheep | 816 | 650* | 820* |
| Goats | 4,013 | 2,900* | 2,700* |
| Chickens* | 145,000 | 147,000 | 110,000 |

* FAO estimate(s).

Source: FAO.

# VENEZUELA

## LIVESTOCK PRODUCTS
('000 metric tons)

|  | 2001 | 2002 | 2003 |
|---|---|---|---|
| Beef and veal | 418 | 429 | 435 |
| Pig meat | 119 | 119 | 120 |
| Poultry meat | 860 | 865 | 644 |
| Cows' milk | 1,400 | 1,389 | 1,238 |
| Cheese | 97 | 98 | 104* |
| Hen eggs | 161 | 161 | 167 |
| Cattle hides (fresh)* | 48 | 48 | 50 |

* FAO estimate(s).
Source: FAO.

## Forestry

### ROUNDWOOD REMOVALS
('000 cubic metres, excl. bark)

|  | 2001 | 2002 | 2003 |
|---|---|---|---|
| Sawlogs, veneer logs and logs for sleepers | 737 | 1,227 | 1,058 |
| Pulpwood | 233 | 184 | 231 |
| Fuel wood* | 3,650 | 3,697 | 3,745 |
| **Total** | 4,620 | 5,108 | 5,034 |

* FAO estimates.
Source: FAO.

### SAWNWOOD PRODUCTION
('000 cubic metres, incl. railway sleepers)

|  | 2001 | 2002 | 2003 |
|---|---|---|---|
| Coniferous (softwood) | 90 | 101 | 260 |
| Broadleaved (hardwood) | 211 | 263 | 241 |
| **Total** | 301 | 364 | 501 |

Source: FAO.

## Fishing

('000 metric tons, live weight)

|  | 2001 | 2002 | 2003 |
|---|---|---|---|
| Capture | 411.6 | 513.8 | 524.4 |
| Round sardinella | 71.2 | 158.1 | 141.9 |
| Yellowfin tuna | 122.4 | 127.2 | 98.1 |
| Ark clams | 43.7 | 44.7 | 45.9 |
| Aquaculture | 16.6 | 17.9 | 15.7 |
| **Total catch** | 428.3 | 531.7 | 540.2 |

Note: Figures exclude crocodiles, recorded by number rather than by weight. The number of spectacled caimans caught was: 19,215 in 2001; 20,349 in 2002; 31,636 in 2003.
Source: FAO.

## Mining
('000 metric tons, unless otherwise indicated)

|  | 2001 | 2002 | 2003 |
|---|---|---|---|
| Hard coal | 7,685 | 8,097 | 7,034 |
| Crude petroleum ('000 barrels) | 1,155,075 | 1,105,793 | 964,695 |
| Natural gas (million cu metres)* | 62,941 | 61,982 | 61,657 |
| Iron ore: gross weight | 16,902 | 16,684 | 17,954 |
| Iron ore: metal content | 10,817 | 11,092 | 11,936 |
| Nickel ore (metric tons)† | 13,600 | 18,600 | 20,700 |
| Bauxite | 4,585 | 5,191 | 5,446 |
| Gold (kilograms)† | 9,076 | 9,465 | 8,190 |
| Phosphate rock | 399 | 390 | 260 |
| Salt (evaporated)‡ | 350 | 350 | 350 |
| Amphilbolite | 14,230 | 18,610 | 3,520 |
| Diamonds (carats): Gem | 14,321 | 45,707 | 11,080 |
| Diamonds (carats): Industrial | 27,826 | 61,060 | 23,710 |

* Figures refer to the gross volume of output. Marketed production (in million cu metres) was: 35,347 in 2001; 33,124 in 2002; 26,060 in 2003.
† Figures refer to the metal content of ores and concentrates.
‡ Estimated production.
Source: US Geological Survey.

## Industry

### PETROLEUM PRODUCTS
('000 barrels)

|  | 2001 | 2002 | 2003* |
|---|---|---|---|
| Motor spirit (petrol) | 74,128 | 68,565 | 55,000 |
| Kerosene | 157 | — | 125 |
| Jet fuel | 32,233 | 32,113 | 26,000 |
| Distillate fuel oils | 110,642 | 114,584 | 92,000 |
| Residual fuel oils | 92,914 | 81,475 | 65,000 |

* Estimated production.
Source: US Geological Survey.

### SELECTED OTHER PRODUCTS
('000 metric tons, unless otherwise indicated)

|  | 2001 | 2002 | 2003 |
|---|---|---|---|
| Raw sugar* | 612 | 637 | 736‖ |
| Fertilizers†‡ | 525 | 576 | n.a. |
| Cement§ | 8,700‖ | 7,000 | 7,000 |
| Crude steel§ | 3,814 | 4,164 | 3,930 |
| Aluminium§ | 571 | 605 | 601 |
| Electric energy (million kWh)† | 85,211 | 87,406 | n.a. |

* FAO figures.
† Data from UN Economic Commission for Latin America and the Caribbean.
‡ Including phosphatic, nitrogenous and potassic fertilizers, year beginning 1 July.
§ Data from US Geological Survey.
‖ Estimate.

## Finance

### CURRENCY AND EXCHANGE RATES
**Monetary Units**
 100 céntimos = 1 bolívar.
**Sterling, Dollar and Euro Equivalents** (31 May 2005)
 £1 sterling = 3,903.46 bolívares;
 US $1 = 2,147.00 bolívares;
 €1 = 2,647.47 bolívares;
 10,000 bolívares = £2.56 = $4.66 = €3.78.
**Average Exchange Rate** (bolívares per US dollar)
 2002  1,160.95
 2003  1,606.96
 2004  1,891.33

# VENEZUELA

*Statistical Survey*

## BUDGET
('000 million bolívares, preliminary figures)

| Revenue | 2002 | 2003 | 2004 |
|---|---|---|---|
| Tax revenue | 11,447.7 | 15,145.4 | 26,931.4 |
|   Petroleum | 990.8 | 1,967.9 | 3,802.4 |
|   Other tax revenue | 10,456.9 | 13,177.5 | 23,129.0 |
| Non-tax revenue | 12,441.5 | 16,239.6 | 22,755.4 |
|   Petroleum | 10,331.9 | 13,586.4 | 19,915.9 |
|   Other non-tax revenue | 2,109.6 | 2,653.2 | 2,839.5 |
| **Total revenue** | 23,889.2 | 31,384.9 | 49,686.8 |

| Expenditure | 2002 | 2003 | 2004 |
|---|---|---|---|
| Current expenditure | 20,704.2 | 27,888.0 | 40,536.1 |
|   Wages and salaries | 4,507.1 | 5,709.1 | 9,496.5 |
|   Interest payments | 4,950.7 | 6,300.1 | 7,061.6 |
|   Goods and services | 1,740.9 | 2,064.4 | 1,906.5 |
|   Transfers | 9,467.8 | 13,511.2 | 22,071.5 |
|   Other current expenditure | 37.7 | 303.1 | — |
| Capital expenditure | 4,798.7 | 7,364.8 | 10,628.7 |
|   Acquisition of fixed capital | 931.1 | 2,302.1 | 560.1 |
|   Capital transfers | 3,867.6 | 5,062.7 | 10,068.5 |
| Extrabudgetary expenditure | 1,385.8 | 1,718.6 | 155.7 |
| **Total expenditure*** | 26,888.7 | 36,971.4 | 51,320.5 |

* Excluding net lending (preliminary figures): 846.6 in 2002; 315.3 in 2003; 245.4 in 2004.

Source: Central Office of Budget, Caracas.

## CENTRAL BANK RESERVES
(US $ million at 31 December)

| | 2002 | 2003 | 2004 |
|---|---|---|---|
| Gold* | 3,515 | 4,632 | 5,122 |
| IMF special drawing rights | 11 | 10 | 9 |
| Reserve position in IMF | 438 | 478 | 500 |
| Foreign exchange | 8,038 | 15,546 | 17,867 |
| **Total** | 12,002 | 20,667 | 23,497 |

* National valuation.

Source: IMF, *International Financial Statistics*.

## MONEY SUPPLY
('000 million bolívares at 31 December)

| | 2002 | 2003 | 2004 |
|---|---|---|---|
| Currency outside banks | 3,795.9 | 4,777.5 | 6,506.3 |
| Demand deposits at commercial banks | 7,029.9 | 13,651.2 | 20,036.5 |
| **Total** (incl. others) | 10,973.5 | 19,055.8 | 27,927.3 |

Source: IMF, *International Financial Statistics*.

## COST OF LIVING
(Consumer Price Index for Caracas; Base: 2000 = 100)

| | 2002 | 2003 | 2004 |
|---|---|---|---|
| Food | 149.0 | 205.2 | 274.6 |
| **All items** (incl. others) | 137.8 | 180.6 | 219.9 |

Source: ILO.

## NATIONAL ACCOUNTS
**National Income and Product**
(million bolívares at current prices)

| | 2001 | 2002 | 2003* |
|---|---|---|---|
| Compensation of employees | 31,261.0 | 35,636.6 | 41,187.3 |
| Net operating surplus | 33,024.0 | 41,474.9 | 57,035.0 |
| Net mixed income | 12,196.8 | 14,368.3 | 17,064.9 |
| **Domestic primary incomes** | 76,481.8 | 91,479.8 | 115,287.2 |
| Consumption of fixed capital | 5,367.0 | 7,113.0 | 8,284.6 |
| **Gross domestic product (GDP) at factor cost** | 81,848.8 | 98,592.8 | 123,571.9 |
| Taxes on production and imports | 7,386.0 | 9,669.2 | 12,484.7 |
| *Less* Subsidies | 289.3 | 421.8 | 1,839.3 |
| **GDP in market prices** | 88,945.6 | 107,840.2 | 134,217.3 |
| Primary incomes received from abroad | 1,882.0 | 1,781.3 | 2,771.8 |
| *Less* Primary incomes paid abroad | 3,351.0 | 4,973.0 | 6,659.2 |
| **Gross national income (GNI)** | 87,476.6 | 104,648.5 | 130,329.9 |
| *Less* Consumption of fixed capital | 5,367.0 | 7,113.0 | 8,284.6 |
| **Net national income** | 82,109.6 | 97,535.4 | 122,045.2 |
| Current transfers from abroad | 257.9 | 335.0 | 416.9 |
| *Less* Current transfers paid abroad | 365.8 | 525.6 | 388.7 |
| **Net national disposable income** | 82,001.7 | 97,344.9 | 122,073.4 |

* Preliminary figures.

**Expenditure on the Gross Domestic Product**
('000 million bolívares at current prices)

| | 2001 | 2002 | 2003* |
|---|---|---|---|
| Final consumption expenditure | 61,502.2 | 71,767.3 | 90,772.0 |
|   Households | 47,993.2 | 56,775.7 | 72,275.5 |
|   Non-profit institutions serving households | 845.6 | 964.5 | 1,208.2 |
|   General government | 12,663.4 | 14,027.2 | 17,288.3 |
| Gross capital formation | 24,481.5 | 22,817.8 | 20,885.0 |
|   Gross fixed capital formation | 21,389.0 | 23,643.3 | 20,943.0 |
|   Changes in inventories | 3,089.9 | −826.6 | −58.6 |
|   Acquisitions, less disposals, of valuables | 2.6 | 1.0 | 0.6 |
| **Total domestic expenditure** | 85,983.7 | 94,585.1 | 111,657.0 |
| Exports of goods and services | 20,222.3 | 32,819.6 | 45,344.7 |
| *Less* Imports of goods and services | 17,260.5 | 19,564.5 | 22,784.4 |
| **GDP in market prices** | 88,945.6 | 107,840.2 | 134,217.3 |
| **GDP at constant 1997 prices** | 42,405.4 | 38,650.1 | 35,667.5 |

* Preliminary figures.

# VENEZUELA

## Gross Domestic Product by Economic Activity
('000 million bolívares at current prices)

|  | 2001 | 2002 | 2003* |
|---|---|---|---|
| Agriculture, hunting and forestry | 3,507.2 | 3,813.7 | 5,202.2 |
| Fishing | 256.5 | 335.0 | 556.0 |
| Mining and quarrying† | 12,482.0 | 20,243.1 | 31,443.9 |
| Manufacturing‡ | 15,111.3 | 17,727.4 | 22,961.4 |
| Electricity, gas and water | 2,075.0 | 2,508.1 | 2,757.0 |
| Construction | 8,498.6 | 9,834.1 | 8,307.3 |
| Wholesale and retail trade; repair of motor vehicles, motorcycles and personal and household goods | 7,019.9 | 8,030.7 | 10,833.2 |
| Hotels and restaurants | 1,581.2 | 1,788.2 | 2,303.1 |
| Transport, storage and communications | 5,813.9 | 6,766.9 | 8,369.6 |
| Financial intermediation | 2,213.4 | 3,019.8 | 3,646.4 |
| Real estate, renting and business activities | 9,739.4 | 11,216.6 | 12,039.5 |
| Public administration and defence; compulsory social security | 5,065.0 | 5,456.3 | 6,451.0 |
| Education | 6,682.9 | 7,626.7 | 9,016.4 |
| Health and social work | 2,837.9 | 3,209.8 | 3,849.6 |
| Other community, social and personal services | 1,596.3 | 1,817.1 | 2,058.8 |
| Private households with employed persons | 517.3 | 594.2 | 703.0 |
| **Sub-total** | 84,997.8 | 103,987.7 | 130,498.4 |
| *Less* Financial intermediation services indirectly measured | 2,172.7 | 2,883.1 | 3,539.4 |
| **Gross value added in basic prices** | 82,824.9 | 101,104.6 | 126,959.0 |
| Taxes on products | 6,402.3 | 7,078.5 | 9,066.3 |
| *Less* Subsidies on products | 281.6 | 342.9 | 1,808.0 |
| **GDP in market prices** | 88,945.6 | 107,840.2 | 134,217.3 |

* Preliminary figures.
† Includes crude petroleum and natural gas production.
‡ Includes petroleum refining.

## BALANCE OF PAYMENTS
(US $ million)

|  | 2002 | 2003 | 2004 |
|---|---|---|---|
| Exports of goods f.o.b. | 26,781 | 27,170 | 39,371 |
| Imports of goods f.o.b. | −13,360 | −10,687 | −17,318 |
| **Trade balance** | 13,421 | 16,483 | 22,053 |
| Exports of services | 1,013 | 878 | 1,098 |
| Imports of services | −3,922 | −3,522 | −4,724 |
| **Balance on goods and services** | 10,512 | 13,839 | 18,427 |
| Other income received | 1,474 | 1,729 | 1,554 |
| Other income paid | −4,230 | −4,140 | −5,216 |
| **Balance on goods, services and income** | 7,756 | 11,428 | 14,765 |
| Current transfers received | 288 | 257 | 180 |
| Current transfers paid | −445 | −237 | −370 |
| **Current balance** | 7,599 | 11,448 | 14,575 |
| Direct investment abroad | −1,026 | −1,318 | 158 |
| Direct investment from abroad | 782 | 2,659 | 1,144 |
| Portfolio investment assets | −1,354 | −812 | −1,090 |
| Portfolio investment liabilities | −956 | −143 | −1,158 |
| Other investment assets | −7,169 | −4,328 | −7,656 |
| Other investment liabilities | 477 | −1,000 | −1,552 |
| Net errors and omissions | −2,781 | −1,052 | −2,266 |
| **Overall balance** | −4,428 | 5,454 | 2,155 |

Source: IMF, *International Financial Statistics*.

# External Trade

## PRINCIPAL COMMODITIES
(US $ million)

| Imports f.o.b. | 1999 | 2000 | 2001 |
|---|---|---|---|
| **Food and live animals** | 1,387.6 | 1,327.7 | 1,506.7 |
| Cereals and cereal preparations | 476.1 | 446.4 | 467.0 |
| **Mineral fuels** | 345.8 | 538.6 | 703.6 |
| Petroleum and products | 339.3 | 526.3 | 687.7 |
| **Chemicals and related products** | 1,595.9 | 2,010.4 | 2,315.2 |
| Organic chemicals | 390.8 | 449.5 | 459.4 |
| Medicinal and pharmaceutical products | 326.7 | 461.7 | 610.4 |
| **Basic manufactures** | 2,160.0 | 2,276.4 | 2,475.1 |
| Iron and steel | 545.7 | 516.7 | 607.6 |
| Other metals and metal manufactures | 624.0 | 648.0 | 636.2 |
| **Machinery and transport equipment** | 5,847.2 | 6,059.7 | 6,847.4 |
| Machinery specialized for particular industries | 399.1 | 418.3 | 546.5 |
| General industrial machinery equipment and parts | 1,883.2 | 1,450.3 | 1,299.2 |
| Heating and cooling equipment and parts | 719.6 | 488.2 | 271.7 |
| Machinery, plant and laboratory equipment for heating and cooling | 566.2 | 346.9 | 109.2 |
| Pumps, compressors, centrifuges and filtering apparatus | 443.4 | 297.1 | 292.8 |
| Office machines and automatic data-processing equipment | 408.2 | 387.8 | 386.9 |
| Telecommunications and sound recording and reproducing equipment | 684.6 | 760.3 | 766.7 |
| Telecommunications equipment parts and accessories | 551.6 | 617.0 | 559.0 |
| Other electrical machinery, apparatus, etc. | 737.2 | 852.9 | 849.5 |
| Road vehicles | 1,116.9 | 1,453.5 | 2,293.7 |
| Passenger motor vehicles (except buses) | 573.8 | 893.0 | 1,535.6 |
| **Miscellaneous manufactured articles** | 1,567.5 | 623.5 | 1,913.3 |
| **Total** (incl. others) | 13,554.0 | 14,584.2 | 16,345.6 |

| Exports f.o.b. | 1999 | 2000 | 2001 |
|---|---|---|---|
| **Mineral fuels, lubricants and related materials** | 16,343.1 | 26,642.5 | 21,014.2 |
| Petroleum, petroleum products and related materials | 16,230.7 | 26,487.8 | 20,758.0 |
| Crude petroleum | 10,775.3 | 18,238.0 | 14,755.9 |
| Petroleum products, refined | 5,451.2 | 8,237.1 | 5,980.3 |
| **Chemicals and related products** | 712.4 | 861.5 | 944.2 |
| **Basic manufactures** | 1,845.1 | 2,177.3 | 2,065.8 |
| Iron and steel | 649.5 | 822.8 | 789.7 |
| Non-ferrous metals | 663.1 | 786.2 | 765.7 |
| Aluminium | 653.1 | 770.4 | 748.6 |
| **Total** (incl. others) | 20,076.2 | 30,948.1 | 25,304.3 |

Source: UN, *International Trade Statistics Yearbook*.

# VENEZUELA

*Statistical Survey*

## PRINCIPAL TRADING PARTNERS
(US $ million)

| Imports f.o.b. | 1999 | 2000 | 2001 |
|---|---|---|---|
| Argentina | 220.4 | 226.6 | 226.8 |
| Belgium | 125.7 | 157.5 | 114.4 |
| Bolivia | 37.0 | 40.6 | 178.2 |
| Brazil | 458.2 | 727.4 | 975.0 |
| Canada | 376.6 | 403.4 | 473.2 |
| Chile | 192.2 | 242.8 | 281.9 |
| China, People's Republic | 68.7 | 184.9 | 335.8 |
| Colombia | 738.1 | 1,083.7 | 1,432.3 |
| France (incl. Monaco) | 245.1 | 354.3 | 302.7 |
| Germany | 614.6 | 519.6 | 579.0 |
| Hong Kong | 193.2 | 169.5 | 151.0 |
| Italy | 1,523.2 | 640.8 | 549.5 |
| Japan | 460.8 | 499.8 | 749.2 |
| Korea, Republic | 265.5 | 339.0 | 352.4 |
| Mexico | 497.0 | 627.4 | 773.8 |
| Netherlands | 148.3 | 222.1 | 168.7 |
| Netherlands Antilles | 161.2 | 320.5 | 368.2 |
| Panama | 239.6 | 281.8 | 375.8 |
| Spain | 432.5 | 365.3 | 448.2 |
| United Kingdom | 215.5 | 315.1 | 344.9 |
| USA | 5,221.9 | 5,508.5 | 5,572.2 |
| **Total** (incl. others) | 13,554.0 | 14,584.2 | 16,435.6 |

| Exports f.o.b. | 1999 | 2000 | 2001 |
|---|---|---|---|
| Brazil | 818.2 | 1,129.1 | 674.9 |
| Canada | 573.6 | 452.4 | 474.8 |
| Colombia | 788.8 | 853.5 | 730.7 |
| Costa Rica | 227.5 | 332.7 | 258.5 |
| Dominican Republic | 562.3 | 859.4 | 715.4 |
| Germany | 202.2 | 279.6 | 215.7 |
| India | 3.4 | 177.4 | 461.7 |
| Japan | 233.1 | 234.9 | 139.5 |
| Mexico | 172.2 | 275.3 | 360.9 |
| Netherlands | 332.8 | 270.4 | 291.7 |
| Netherlands Antilles | 1,418.4 | 1,735.8 | 1,543.9 |
| Peru | 285.6 | 532.2 | 295.7 |
| Spain | 194.6 | 431.4 | 426.3 |
| Trinidad and Tobago | 284.6 | 540.1 | 420.2 |
| USA | 11,234.7 | 18,442.0 | 14,280.3 |
| **Total** (incl. others) | 20,076.2 | 30,948.1 | 25,304.3 |

Source: UN, *International Trade Statistics Yearbook*.

## Transport

### RAILWAYS
(traffic)

| | 1994 | 1995 | 1996 |
|---|---|---|---|
| Passenger-kilometres (million) | 31.4 | 12.5 | 0.1 |
| Freight ton-kilometres (million) | 46.8 | 53.3 | 45.5 |

**Freight ton-kilometres** (million): 54.5 in 1997; 79.5 in 1998.

Source: UN Economic Commission for Latin America and the Caribbean.

### ROAD TRAFFIC
('000 motor vehicles in use)

| | 1999 | 2000 | 2001 |
|---|---|---|---|
| Passenger cars | 1,420.0 | 1,326.2 | 1,372.0 |
| Commercial vehicles | 846.0 | 1,078.6 | 1,107.9 |

Source: UN, *Statistical Yearbook*.

### SHIPPING
**Merchant Fleet**
(registered at 31 December)

| | 2002 | 2003 | 2004 |
|---|---|---|---|
| Number of vessels | 274 | 275 | 286 |
| Total displacement ('000 grt) | 865.4 | 847.0 | 1,010.9 |

Source: Lloyd's Register-Fairplay, *World Fleet Statistics*.

### CIVIL AVIATION
(traffic on scheduled services)

| | 2001 | 2002 | 2003 |
|---|---|---|---|
| Kilometres flown (million) | 79.7 | 77.5 | 50.4 |
| Passengers carried ('000) | 4,051.7 | 5,446.8 | 3,823.7 |
| Passenger-km (million) | 3,680.5 | 3,301.5 | 2,042.8 |
| Freight ton-km (million) | 30.5 | 6.6 | 2.0 |

Source: UN Economic Commission for Latin America and the Caribbean.

## Tourism

### ARRIVALS BY NATIONALITY

| | 2001 | 2002 | 2003 |
|---|---|---|---|
| Argentina | 27,396 | 15,133 | 14,108 |
| Belgium | 11,798 | 11,300 | 7,660 |
| Brazil | 18,909 | 11,022 | 9,929 |
| Canada | 32,982 | 29,106 | 20,588 |
| Chile | 12,340 | 6,786 | 6,345 |
| Colombia | 20,029 | 11,855 | 10,576 |
| Denmark | 10,434 | 8,636 | 6,325 |
| France | 24,792 | 18,526 | 14,362 |
| Germany | 67,168 | 60,426 | 42,320 |
| Italy | 37,421 | 23,396 | 20,166 |
| Mexico | 13,730 | 8,722 | 7,447 |
| Netherlands | 56,341 | 52,310 | 36,039 |
| Spain | 20,060 | 14,288 | 11,389 |
| United Kingdom | 32,299 | 26,880 | 19,624 |
| USA | 121,135 | 80,007 | 66,711 |
| **Total** (incl. others) | 584,399 | 431,677 | 336,974 |

**Tourism receipts** (US $ million, incl. passenger transport): 677 in 2001; 484 in 2002; 368 in 2003.

Source: World Tourism Organization.

# VENEZUELA

## Communications Media

|  | 2001 | 2002 | 2003 |
|---|---|---|---|
| Television receivers ('000 in use) | 4,600 | n.a. | n.a. |
| Telephones ('000 main lines in use) | 2,704.9 | 2,841.8 | 2,842.6 |
| Mobile cellular telephones ('000 subscribers) | 6,472.6 | 6,463.6 | 7,015.7 |
| Personal computers ('000 in use) | 1,300 | 1,536 | n.a. |
| Internet users ('000) | 1,152.5 | 1,274.4 | 1,549.5 |

**Facsimile machines** ('000 in use, 1997): 70.

**Radio receivers** ('000 in use, 1997, estimate): 10,750.

**Book production** (titles, 1997): 3,851*.

**Daily newspapers** (1996): 86 (estimated average circulation 4,600,000).

* First editions only.

Sources: UNESCO Institute for Statistics; UN, *Statistical Yearbook*; International Telecommunication Union.

## Education

(2001/02, unless otherwise indicated)

|  | Institutions* | Teachers | Students |
|---|---|---|---|
| Pre-primary | 14,366 | 49,977† | 863,364 |
| Primary | 18,992 | 186,658† | 4,818,201 |
| Secondary | 3,174 | 61,761† | 499,706 |
| Higher | 144‡ | 53,590§ | 650,000‡‖ |

* Data may be duplicated for institutions where education is offered at more than one level.
† 1997/98 figure, refers to teaching posts.
‡ 2001/02 figure.
§ 1999/2000 figure.
‖ UNESCO estimate.

Source: partly UNESCO Institute for Statistics.

**Adult literacy rate** (UNESCO estimates): 93.1% (males 93.5%; females 92.7%) in 2002 (Source: UN Development Programme, *Human Development Report*).

# Directory

## The Constitution

The Bolivarian Constitution of Venezuela was promulgated on 30 December 1999.

The Bolivarian Republic of Venezuela is divided into 22 States, one Federal District and 72 Federal Dependencies. The States are autonomous but must comply with the laws and Constitution of the Republic.

### LEGISLATURE

Legislative power is exercised by the unicameral National Assembly (Asamblea Nacional). This replaced the bicameral Congreso Nacional (National Congress) following the introduction of the 1999 Constitution.

Deputies are elected by direct universal and secret suffrage, the number representing each State being determined by population size on a proportional basis. A deputy must be of Venezuelan nationality and be over 21 years of age. Indigenous minorities have the right to select three representatives. Ordinary sessions of the Asamblea Nacional begin on the fifth day of January of each year and continue until the fifteenth day of the following August; thereafter, sessions are renewed from the fifteenth day of September to the fifteenth day of December, both dates inclusive. The Asamblea is empowered to initiate legislation. The Asamblea also elects a Comptroller-General to preside over the Audit Office (Contraloría General de la República), which investigates Treasury income and expenditure, and the finances of the autonomous institutes.

### GOVERNMENT

Executive power is vested in a President of the Republic elected by universal suffrage every six years, who may serve one additional term. The President is empowered to discharge the Constitution and the laws, to nominate or remove Ministers, to take supreme command of the Armed Forces, to direct foreign relations of the State, to declare a state of emergency and withdraw the civil guarantees laid down in the Constitution, to convene extraordinary sessions of the Asamblea Nacional and to administer national finance.

### JUDICIARY

Judicial power is exercised by the Supreme Tribunal of Justice (Tribunal Suprema de Justicia) and by the other tribunals. The Supreme Tribunal forms the highest court of the Republic and the Magistrates of the Supreme Tribunal are appointed by the Asamblea Nacional following recommendations from the Committee for Judicial Postulations, which consults with civil society groups. Magistrates serve a maximum of 12 years.

The 1999 Constitution created two new elements of power. The Moral Republican Council (Consejo Moral Republicano) is comprised of the Comptroller-General, the Attorney-General and the Peoples' Defender (or ombudsman). Its principal duty is to uphold the Constitution. The National Electoral Council (Consejo Nacional Electoral) administers and supervises elections.

## The Government

### HEAD OF STATE

**President of the Republic:** Lt-Col (retd) HUGO RAFAEL CHÁVEZ FRÍAS (took office 2 February 1999; re-elected 30 July 2000).

### COUNCIL OF MINISTERS
(August 2005)

**Vice-President:** JOSÉ VICENTE RANGEL.

**Minister of Finance:** NELSON MERENTES DÍAZ.

**Minister of the Interior and Justice:** Lt (retd) JESSE CHACÓN ESCAMILLO.

**Minister of National Defence:** Adm. RAMÓN ORLANDO MANGLIA FERREIRA.

**Minister of Basic Industry and Mining:** VÍCTOR ALVAREZ RODRÍGUEZ.

**Minister of Light Industry and Trade:** EDMEÉ BETANCOURT DE GARCÍA.

**Minister of Energy and Petroleum:** RAFAEL DARÍO RAMÍREZ CARREÑO.

**Minister of Foreign Affairs:** ALÍ RODRÍGUEZ ARAQUE.

**Minister of Labour:** MARÍA CRISTINA IGLESIAS.

**Minister of the Mass Economy:** ELÍAS JAUA.

**Minister of Food:** Gen. JOSÉ RAFAEL OROPEZA.

**Minister of Education and Sport:** ARISTÓBULO ISTÚRIZ.

**Minister of Tourism:** WILMAR CASTRO SOTELDO.

**Minister of Health:** Dr FRANCISCO ARMADA.

**Minister of Higher Education:** SAMUEL MONCADA.

**Minister of Infrastructure:** RAMÓN ALFONSO CARRIZALES RENGIFO.

**Minister of the Environment and Natural Resources:** JACQUELINE COROMOTO FARÍA PINEDA.

**Minister of Agriculture and Lands:** ANTONIO ALBARRÁN MORENO.

**Minister of Planning and Development:** JORGE GIORDANI.

**Minister of Science and Technology:** YADIRA CÓRDOVA.

**Minister of Social Development and Popular Participation:** Gen. JORGE LUIS GARCÍA CARNEIRO.

**Minister of Communications and Information:** YURI PIMENTEL.

**Secretary of the Presidency:** RODRIGO CHÁVEZ.

### MINISTRIES

**Ministry of Agriculture and Lands:** Avda Lecuna, Torre Este, 7°, Parque Central, San Agustín, Caracas; tel. (212) 509-0405; internet www.mat.gov.ve.

**Ministry of Basic Industry and Mining:** Caracas.

# VENEZUELA

**Ministry of Communications and Information:** Edif. Administrativo, 1°, Palacio de Miraflores, Caracas; tel. (212) 806-3702; fax (212) 806-3244; e-mail miraflores@hotmail.com; internet www.minci.gov.ve.

**Ministry of Education and Sport:** Edif. Ministerio de Educación, Mezzanina, esq. de Salas, Parroquia Altagracia, Caracas 1010; tel. (212) 564-0025; fax (212) 562-1096; e-mail prensa@gov.me.ve; internet www.me.gov.ve.

**Ministry of Energy and Petroleum:** Edif. Petróleos de Venezuela, Torre Oeste, Avda Libertador, La Campiña, Apdo 169, Caracas 1010-A; tel. (212) 708-1299; fax (212) 708-7014; e-mail dazaroy@hotmail.com; internet www.mem.gov.ve.

**Ministry of the Environment and Natural Resources:** Torre Sur, 25°, Centro Simón Bolívar, Caracas 1010; tel. (212) 408-1002; fax (212) 408-1009; e-mail jfaria@marn.gov.ve; internet www.marn.gov.ve.

**Ministry of Finance:** Edif. Ministerio de Finanzas, esq. de Carmelitas, Avda Urdaneta, Caracas; tel. (212) 802-1404; fax (212) 802-1413.

**Ministry of Food:** Antiguo Edif. Seguros Orinoco, 11°, Avda Fuerzas Armadas, esq. Socarras, Caracas 1010; tel. (212) 564-8303; internet www.minal.gob.ve.

**Ministry of Foreign Affairs:** Torre MRE, esq. Carmelitas, Avda Urdaneta, Caracas 1010; tel. (212) 862-1085; fax (212) 864-3633; e-mail criptogr@mre.gov.ve; internet www.mre.gov.ve.

**Ministry of Health:** Torre Sur, 9°, Centro Simón Bolívar, Caracas 1010; tel. (212) 484-3725; fax (212) 484-3725; internet www.msds.gov.ve.

**Ministry of Higher Education:** Torre MCT, 6°, Avda Universidad, esq. el Chorro, Caracas 1010; tel. (212) 596-5293; internet www.mes.gov.ve.

**Ministry of Infrastructure:** Torre Este, 50°, Parque Central, Caracas 1010; tel. (212) 509-1076; fax (212) 509-3682.

**Ministry of the Interior and Justice:** Edif. Ministerio del Interior y Justicia, esq. de Platanal, Avda Urdaneta, Caracas 1010; tel. (212) 506-1101; fax (212) 506-1559; internet www.mij.gov.ve.

**Ministry of Labour:** Torre Sur, 5°, Centro Simón Bolívar, Caracas 1010; tel. (212) 481-1368; fax (212) 483-8914.

**Ministry of Light Industry and Trade:** Torre Este, 18°, Avda Lecuna, Parque Central, Caracas 1010; tel. (212) 509-0445; fax (212) 574-2432; internet www.mpc.gov.ve.

**Ministry of National Defence:** Edif. 17 de Diciembre, planta baja, Base Aérea Francisco de Miranda, La Carlota, Caracas; tel. (212) 908-1264; fax (212) 237-4974; e-mail prensa@ven.net; internet www.mindefensa.gov.ve.

**Ministry of Planning and Development:** Torre Oeste, 22°, Avda Lecuna, Parque Central, Caracas 1010; tel. (212) 507-7645; fax (212) 573-3076; internet www.mpd.gov.ve.

**Ministry of Science and Technology:** Edif. Maploca 1, planta baja, Final Avda Principal de los Cortijos de Lourdes, Caracas; tel. (212) 239-6475; fax (212) 239-6056; e-mail mct@mct.gov.ve; internet www.mct.gov.ve.

**Ministry of Social Development and Popular Participation:** Caracas.

**Ministry of Tourism:** Caracas.

### State Agencies

**Consejo de Defensa de la Nación (Codena):** Palacio Blanco, 3°, Avda Urdaneta frente del Palacio Miraflores, Caracas; tel. (212) 806-3104; fax (212) 806-3151; national defence council.

**Consejo Nacional Electoral (CNE):** Centro Simón Bolívar, Nivel Avenida, El Silencio, Caracas; tel. (212) 484-4145; fax (212) 484-5269; internet www.cne.gov.ve; national electoral board; Pres. FERNANDO CARRASQUERO.

**Contraloría General de la República (CGR):** Edif. Contraloría, Avda Andrés Bello, Guaicaipuro, Caracas; tel. (212) 481-3437; fax (212) 571-8131; national audit office for Treasury income and expenditure, and for the finances of the autonomous institutes.

**Defensoría del Pueblo:** Edif. Defensoría del Pueblo, Plaza Morelos, Los Caobos, Caracas; tel. (212) 578-3862; fax (212) 578-3862; e-mail prensadefensoria@hotmail.com; acts as an ombudsman and investigates complaints between citizens and the authorities.

**Procuraduría General de la República:** Paseo Los Ilustres con Avda Lazo Martí, Santa Mónica, Caracas; tel. (212) 693-0911; fax (212) 693-4657; e-mail procuraduria@platino.gov.ve.

## President

**Presidential Election, 30 July 2000**

| Candidates | % of valid votes |
|---|---|
| Lt-Col (retd) Hugo Rafael Chávez Frías (MVR) | 59.7 |
| Francisco Arias Cárdenas (La Causa R) | 37.5 |
| Claudio Fermín (Encuentro Nacional) | 2.7 |
| Others | 0.1 |
| **Total** | **100.0** |

## Legislature

**ASAMBLEA NACIONAL**
(National Assembly)

**President:** NICOLÁS MADURO MOROS.
**First Vice-President:** RICARDO ANTONIO GUTIÉRREZ BRICEÑO.
**Second Vice-President:** PEDRO MIGUEL CARREÑO ESCOBAR.

**Election, 30 July 2000**

| Party | Seats |
|---|---|
| Movimiento V República (MVR) | 93 |
| Acción Democrática (AD) | 32 |
| Proyecto Venezuela (PRVZL) | 8 |
| Movimiento al Socialismo (MAS) | 6 |
| Partido Social-Cristiano (COPEI) | 5 |
| Primero Justicia (PJ) | 5 |
| Causa Radical (La Causa R) | 3 |
| Indigenous groups | 3 |
| Others | 10 |
| **Total** | **165** |

## Political Organizations

**Acción Democrática (AD):** Casa Nacional Acción Democrática, Calle Los Cedros, La Florida, Caracas 1050; internet www.acciondemocratica.org.ve; f. 1936 as Partido Democrático Nacional; adopted present name and obtained legal recognition in 1941; social democratic; Sec-Gen. HENRY RAMOS ALLUP; Pres. JESÚS MENDEZ QUIJADA.

**Alianza Bravo Pueblo:** Caracas; oppositionist; Pres. ANTONIO LEDEZMA.

**Asamblea del Pueblo** (Asamblea Popular): Caracas; internet www.aporrea.org; f. 2005; opposition group formed by dissident left-wing mems of AD, Bandera Roja and CD; Leader CLAUDIO FERMÍN.

**Bandera Roja:** Caracas; f. 1968; Marxist-Leninist grouping.

**Bloque Democrático:** Caracas; e-mail contacto@bloquedemocratico.org; internet www.bloquedemocratico.org; hardline opposition faction; split from Coordinadora Democrática in 2004; Dir ROBERT ALONSO.

**La Causa Radical (La Causa R):** Santa Teresa a Cipreses, Residencias Santa Teresa, 2°, Ofs 21 y 22, Caracas; tel. (212) 545-7002; internet www.lacausar.org.ve; f. 1971; radical democratic; Leader ANDRÉS VELÁSQUEZ.

**Convergencia Nacional (CN):** Edif. Tajamar, 2°, Of. 215, Parque Central, Avda Lecuna, El Conde, Caracas 1010; tel. (212) 578-1177; fax (212) 578-0363; e-mail convergencia@convergencia.org.ve; internet www.convergencia.org.ve; f. 1993; Leader Dr RAFAEL CALDERA RODRÍGUEZ; Gen. Co-ordinator JUAN JOSÉ CALDERA.

**Coordinadora Democrática (CD):** Caracas; internet www.coordinadora-democratica.org; f. 2002; umbrella organization for 27 anti-Govt political parties and opposition groups; Leader ENRIQUE MENDOZA.

**Movimiento Electoral del Pueblo (MEP):** Caracas; f. 1967 by left-wing AD dissidents; 100,000 mems; Pres. Dr LUIS BELTRÁN PRIETO FIGUEROA; Sec-Gen. Dr JESÚS ÁNGEL PAZ GALARRAGA.

**Movimiento Al Socialismo (MAS):** Quinta Alemar, Avda Valencia, Las Palmas, Caracas 1050; tel. (212) 555-5555; fax (212) 761-9297; e-mail asamblea07@cantv.net; internet www.mas.org.ve; f. 1971 by PCV dissidents; democratic-socialist party; split in 1997 over issue of support for presidential campaign of Lt-Gen. (retd) Hugo Rafael Chávez Frías; Pres. FELIPE MÚJICA; Sec.-Gen. LEOPOLDO PUCHI.

**Movimiento de Integración Nacional (MIN):** Edif. José María Vargas, 1°, esq. Pajarito, Caracas; tel. (212) 563-7504; fax (212) 563-7553; f. 1977; Sec.-Gen. GONZALO PÉREZ HERNÁNDEZ.

**Movimiento de Izquierda Revolucionaria (MIR):** c/o Fracción Parlamentaria MIR, Edif. Tribunales, esq. Pajaritos, Caracas; f. 1960 by splinter group from AD; left-wing; Sec.-Gen. MOISÉS MOLEIRO.

**Movimiento V República (MVR):** Calle Lima, cruce con Avda Libertador, Los Caobos, Caracas; tel. (212) 782-3808; fax (212) 782-9720; f. 1998; promotes Bolivarian revolution; mem. of the Polo Patriótico (q.v.); Leader Lt-Col (retd) HUGO RAFAEL CHÁVEZ FRÍAS.

**Partido Comunista de Venezuela (PCV):** Edif. Cantaclaro, esq. San Pedro, San Juan, Apdo 20428, Caracas; tel. (212) 484-0061; fax (212) 481-9737; internet www.pcv-venezuela.org; f. 1931; Sec.-Gen. OSCAR FIGUERA.

**Partido Social-Cristiano (Comité de Organización Política Electoral Independiente) (COPEI):** esq. San Miguel, Avda Panteón cruce con Fuerzas Armadas, San José, Caracas 1010; f. 1946; Christian Democratic; more than 1,500,000 mems; Leader ENRIQUE MENDOZA; Sec.-Gen. CÉSAR PÉREZ.

**Partido Unión:** Caracas; f. 2001; Leader Lt-Col (retd) FRANCISCO ARIAS CÁRDENAS; Sec-Gen. LUIS MANUEL ESCULPÍ.

**Patria Para Todos (PPT):** Caracas; tel. (212) 577-4545; e-mail ppt@cantv.net; f. 1997; breakaway faction of La Causa Radical; revolutionary humanist party; Leaders ARISTÓBULO ISTÚRIZ, PABLO MEDINA.

**Polo Patriótico (PP):** f. 1998; grouping of small, mainly left-wing and nationalist political parties, including the MVR (q.v.), in support of presidential election campaign of Lt-Col (retd) Hugo Rafael Chávez Frías.

**Primero Justicia:** Planta de Oficinas, Centro Comercial Chacaito, Chacaito, Caracas; tel. (212) 952-9733; internet www.primerojusticia.org.ve; f. 2000; no fixed ideological stance; Sec.-Gen. ARMANDO BRIQUET MARMOL.

**Proyecto Venezuela (PRVZL):** internet www.proyectovenezuela.org.ve; f. 1998; humanist party; ended alliance with Govt in Jan. 2002; Leader HENRIQUE SALAS RÖMER.

**Solidaridad:** Caracas; f. 2001; Leader LUIS MIQUILENA.

### OTHER ORGANIZATIONS

**Asociación Civil Queremos Elegir:** Edif. Industrial, 4°, Avda Sucre, Los Dos Caminos, Municipio Sucre. Caracas; tel. (212) 286-9785; e-mail info@queremoselegir.org; internet www.queremoselegir.org; f. 1991; opposition grouping promoting citizens' rights; mem of Alianza Cívica de la Sociedad Venezolana; Principal Co-ordinator ELÍAS SANTANA.

**COFAVIC:** Edif. El Candil, 1°, Of. 1-A, Avda Urdaneta, esq. El Candilito, Apdo 16150, La Candelaria, Caracas 1011-A; tel. (212) 572-9631; fax (212) 572-9908; e-mail lortega@cofavic.org.ve; internet www.cofavic.org.ve; promotes human rights; Exec. Dir L. ORTEGA.

**Movimiento 1011:** tel. (414) 304-0432; e-mail contacto@movimiento1011.com; internet www.movimiento1011.com; civil asscn promoting educational reform; mem of Alianza Cívica de la Sociedad Venezolana.

**Súmate:** Caracas; e-mail info@sumate.org; internet www.sumate.org; f. 2002; opposition grouping promoting citizens' rights; Leader MARÍA CORINA MACHADO.

## Diplomatic Representation

### EMBASSIES IN VENEZUELA

**Algeria:** 8va Transversal con 3ra Avda, Quinta Azahar, Urb. Altamira, Caracas 1060; tel. (212) 263-2092; fax (212) 261-4254; e-mail ambalgcar@cantv.net; Ambassador MOHAMMED KHELLADI.

**Argentina:** Edif. Fedecámaras, 3°, Avda El Empalme, El Bosque, Apdo 569, Caracas; tel. (212) 731-3311; fax (212) 731-2659; e-mail argentina@impsat.net.ve; e-mail embargentina@cantv.net; internet www.embargentina.org.ve; Ambassador EDUARDO ALBERTO SADOUS.

**Australia:** Avda Francisco de Miranda, cruce con Avda Sur Altamira, 1°, Caracas 1060-A; tel. (212) 263-4033; fax (212) 261-3448; e-mail caracas@dfat.gov.au; Ambassador JOHN MAGNUS L. WOODS.

**Austria:** Edif. Torre Las Mercedes, 4°, Of. 408, Avda La Estancia, Chuao, Apdo 61381, Caracas 1060-A; tel. (212) 991-3863; fax (212) 993-2275; e-mail e.austria@cantv.net; Ambassador Dr ERICA LIEBENWEIN.

**Barbados:** Edif. Los Frailes, 5°, Of. 501, Avda Principal con Calle La Guairita, Chuao, Caracas; tel. (212) 992-0545; fax (212) 991-0333; e-mail caracas@foreign.gov.bb; Ambassador KEITH MACPHERSON FRANKLIN.

**Belgium:** Quinta la Azulita, Avda 11, entre 6 y 7 transversales, Apdo del Este 61550, Altamira, Caracas 1060; tel. (212) 263-3334; fax (212) 261-0309; e-mail caracas@diplobel.org; internet www.diplobel.org/venezuela; Ambassador BADOUIN VANDERHULST.

**Bolivia:** Avda Luis Roche con 6 transversal, Altamira, Caracas; tel. (212) 263-3015; fax (212) 261-3386; e-mail embaboliviaven@hotmail.com; Ambassador RÉNE RECACOCHEA SALINAS.

**Brazil:** Avda Mohedano con Calle Los Chaguaramos, Centro Gerencial Mohedano, 6°, La Castellana, Caracas; tel. (212) 261-5505; fax (212) 261-9601; e-mail brasembcaracas@cantv.net; internet www.embajadabrasil.org.ve; Ambassador JOÃO CARLOS DE SOUZA-GOMES.

**Bulgaria:** Quinta Sofia, Calle Las Lomas, Urb. Las Mercedes, Apdo 68389, Caracas; tel. (212) 993-2714; fax (212) 993-4839; e-mail embulven@cantv.net; Ambassador LAZAR KOPRINAROV.

**Canada:** Edif. Embajada de Canadá, Avda Francisco de Miranda con Avda Sur, Altamira, Caracas 1060-A; Apdo 62302, Caracas 1060; tel. (212) 600-3000; fax (212) 261-8741; e-mail crcas@dfait-maeci.gc.ca; internet www.caracas.gc.ca; Ambassador ALLAN CULHAM.

**Chile:** Edif. Torre La Noria, 10°, Calle Paseo Enrique Eraso, Las Mercedes, Caracas; tel. (212) 992-3378; fax (212) 992-0614; e-mail echileve@cantv.net; internet www.embachileve.org; Ambassador FABIO VÍO UGARTE.

**China, People's Republic:** Avda El Paseo, Quinta El Oriente, Prados del Este, Caracas; tel. (212) 977-4949; fax (212) 978-0876; e-mail embcnven@att.com.ven; Ambassador YU YIJIE.

**Colombia:** Torre Credival, 11°, 2A Calle de Campo Alegre con Avda Francisco de Miranda, Apdo 60887, Caracas; tel. (212) 261-6596; fax (212) 261-1358; e-mail ecaracas@minrelext.gov.co; Ambassador MARÍA ANGELA HOLGUÍN CUELLAR.

**Costa Rica:** Edif. For You P.H., Avda San Juan Bosco, entre 1 y 2 transversal, Altamira, Apdo 62239, Caracas; tel. (212) 267-1104; fax (212) 265-4660; e-mail embcostar@cantv.net; Ambassador WALTER RUBÉN HERNÁNDEZ JUÁREZ.

**Croatia:** Calle Río Ticoporo, Res. Patricia, Urb. La Ciudadela, Redoma de Prados del Este, Caracas 1080; tel. (212) 977-3967; fax (212) 979-0064; Ambassador ZDRAVKO SANCEVIC.

**Cuba:** Calle Roraima e Rio de Janeiro y Choroni, Chuao, Caracas 1060; tel. (212) 991-6611; fax (212) 993-5695; e-mail embacubavzla@cantv.net; Ambassador GERMÁN SÁNCHEZ OTERO.

**Czech Republic:** Calle Los Cedros, Quinta Isabel, Urb. Country Club, Altamira, Caracas 1060; tel. (212) 261-8528; fax (212) 266-3987; e-mail caracas@embassy.mzv.cz; internet www.mfa.cz/caracas; Ambassador Dr JIŘÍ JIRÁNEK.

**Dominican Republic:** Edif. Humboldt, 6°, Of. 26, Avda Francisco de Miranda, Altamira, Caracas; tel. (212) 284-2443; fax (212) 283-3965; e-mail embaredo@telcel.net.ve; Ambassador RICARDO A. DE MOYA DESPRADEL.

**Ecuador:** Centro Andrés Bello, Torre Oeste, 13°, Avda Andrés Bello, Maripérez, Apdo 62124, Caracas 1060; tel. (212) 265-0801; fax (212) 264-6917; e-mail embajadaecuador@cantv.net; Ambassador PAULINA GARCÍA DONOSO DE LARREA.

**Egypt:** Calle Caucagua con Calle Guaicaipuro, Quinta Maribel, Urb. San Román, Caracas; tel. (212) 992-6259; fax (212) 993-1555; e-mail egyptianembassy@cantv.net; Ambassador MOHAMED FADEL AL-KADI.

**El Salvador:** Centro Comercial Ciudad Tamanaco (CCCT), Torre C, 4°, Of. 406, Chuao, Caracas; tel. (212) 959-0817; fax (212) 959-3920; e-mail embsalv@viptel.com; Ambassador JOSÉ ARRIETA PERALTA.

**Finland:** Calle Sorocaima, entre Avda Venezuela y Avda Tamanaco, Edif. Atrium, 1°, El Rosal, Caracas 1060; tel. (212) 952-4111; fax (212) 952-7536; e-mail sanomat.car@formin.fi; Ambassador ORA HERMAN MERES-WUORI.

**France:** Calle Madrid con Avda Trinidad, Las Mercedes, Apdo 60385, Caracas 1060; tel. (212) 909-6500; fax (212) 909-6630; e-mail consulat@telul.net.ve; internet www.francia.org.ve; Ambassador PIERRE-JEAN VANDOORNE.

**Germany:** Avda Eugenio Mendoza y Avda José Angel Lamas, Torre La Castellana, 10°, La Castellana, Caracas 1010-A, Caracas; tel. (212) 261-0181; fax (212) 261-0641; e-mail diplogermacara@cantv.net; internet www.embajada-alemana.org.ve; Ambassador HERMANN ERATH.

**Greece:** Quinta Maryland, Avda Principal del Avila, Alta Florida, Caracas 1050; tel. (212) 730-3833; fax (212) 731-0429; e-mail embgrccs@cantv.net; Ambassador ATHANASSIOS M. VALASSIDIS.

**Grenada:** Avda Norte 2, Quinta 330, Los Naranjos del Cafetal, Caracas; tel. (212) 985-5461; fax (212) 985-6391; e-mail egrenada@cantv.net; Chargé d'affaires a.i. DUNCAN McPHAIL.

# VENEZUELA

**Guatemala:** Avda de Francisco de Miranda, Torre Dozsa, 1°, Urb. El Rosal, Caracas; tel. (212) 952-1166; fax (212) 954-0051; e-mail embvenezuela@minex.gob.gt; Ambassador LUIS PEDRO QUEZADA CÓRDOVA.

**Guyana:** Quinta 'Roraima', Avda El Paseo, Prados del Este, Apdo 51054, Caracas 1050; tel. (212) 977-1158; fax (212) 977-1158; fax (212) 976-3765; e-mail embaguy@caracas.org.ve; Ambassador BAYNEY RAM KARRAN.

**Haiti:** Quinta Flor 59, Avda Las Rosas, La Florida, Caracas; tel. (212) 730-7220; fax (212) 730-4605; Chargé d'affaires a.i. JEAN-ROBERT HERARD.

**Holy See:** Avda La Salle, Los Caobos, Apdo 29, Caracas 1010-A (Apostolic Nunciature); tel. (212) 781-8939; fax (212) 793-2403; e-mail nunapos@cantv.net; Apostolic Nuncio Most Rev. ANDRÉ DUPUY (Titular Archbishop of Selsea).

**Honduras:** Edif. Excélsior, Avda San Juan Bosco, 5°, Altamira, Apdo 68259, Caracas; tel. (212) 263-3184; fax (212) 263-4379; e-mail honduven@cantv.net; Ambassador CARLOS ALBERTO TURCIOS OREAMUNO.

**India:** Quinta Tagore, No. 12, Avda San Carlos, La Floresta, Caracas, Venezuela; tel. (212) 285-7887; fax (212) 286-5131; e-mail hoc.caracas@mea.gov.ve; internet www.embindia.org; Ambassador N.N. DESAÍ.

**Indonesia:** Quinta La Trinidad, Avda El Paseo, Prados del Este, Apdo 80807, Caracas 1080; tel. (212) 976-2725; fax (212) 976-0550; e-mail kbri@telcel.net.ve; Ambassador CORNELIS MANOPPO.

**Iran:** Quinta Ommat, Calle Kemal Ataturk, Valle Arriba, Caracas; tel. (212) 992-3575; fax (212) 992-9989; e-mail embairanve@cantv.net; Ambassador AHMAD SOBHANI.

**Iraq:** Quinta Babilonia, Avda Nicolás Cópernico con Calle Los Malabares, Valle Arriba, Caracas; tel. (212) 991-1627; fax (212) 992-0268.

**Israel:** Centro Empresarial Miranda, 4°, Avda Principal de los Ruices cruce con Francisco de Miranda, Apdo 70081, Los Ruices, Caracas; tel. (212) 239-4511; fax (212) 239-4320; e-mail caracas@israel.org; Ambassador ARIE TENNE.

**Italy:** Edif. Atrium, Calle Sorocaima, entre Avdas Tamanaco y Venezuela, El Rosal, Apdo 3995, Caracas; tel. (212) 952-7311; fax (212) 952-7120; e-mail ambcara@italamb.org.ve; internet www.italamb.org.ve; Ambassador GERARDO CARANTE.

**Jamaica:** Edif. Los Frailes, 5°, Calle La Guairita, Urb. Chuao, Caracas 1062; tel. (212) 991-6741; fax (212) 991-5708; e-mail embjaven@cantv.net; Ambassador AUDREY RODRÍQUEZ.

**Japan:** Edif. Bancaracas, 10°, Avda San Felipe con 2da Transveral, La Castellana, Caracas; tel. (212) 261-8333; fax (212) 261-6780; e-mail ajapon@genesisbci.net; internet www.ve.emb-japan.go.jp/esp/index.htm; Ambassador YASUO MATSUI.

**Korea, Democratic People's Republic:** Ambassador PAK TONG CHUN (resident in Cuba).

**Korea, Republic:** Avda Francisco de Miranda, Centro Lido, Torre B, 9°, Ofs 91-B y 92-B, El Rosal, Caracas; tel. (212) 954-1270; fax (212) 954-0619; e-mail venadmi@2net-uno.net; Ambassador SUNG-CHULL SHIN.

**Kuwait:** Quinta El-Kuwait, Avda Las Magnolias con Calle Los Olivos, Los Chorros, Caracas; tel. (212) 239-4234; fax (212) 238-3878; Ambassador YOUSEF HUSSAIN AL-GABANDI.

**Lebanon:** Prolongación Avda Parima, Edif. Embajada del Líbano, Colinas de Bello Monte, Calle Motatán, Caracas 1050; tel. (212) 751-5943; fax (212) 753-0726; e-mail emblibano@telcel.net.ve; Ambassador NICOLAS BECHARA KHAWAJA.

**Malaysia:** Centro Profesional Eurobuilding, 6°, Ofs 6F–G, Calle La Guairita, Apdo 65107, Chuao, Caracas 1060; tel. (212) 992-1011; fax (212) 992-1277; e-mail mwcaracas@malaysia.org.ve; Ambassador Datuk JOHN TENEWI NUEK.

**Mexico:** Edif. Forum, Calle Guaicaipuro con Principal de las Mercedes, 5°, El Rosal, Apdo 61371, Caracas; tel. (212) 952-5777; fax (212) 952-3003; e-mail mexico@embamex.com.ve; internet www.embamex.com.ve; Ambassador ENRIQUE M. LOAEZA Y TOVAR.

**Morocco:** Torre Multinvest, Plaza Isabel La Católica, Avda Eugenio Mendoza, 2°, La Castellana, Caracas; tel. (212) 266-7543; fax (212) 266-4681; e-mail embamaroccaracas@cantv.net; Ambassador Dr IBRAHIM HOUSSEIN MOUSSA.

**Netherlands:** Edif. San Juan Bosco, 9°, San Juan Bosco con 2 transversal de Altamira, Caracas; tel. (212) 263-3622; fax (212) 263-0462; e-mail nlgovcar@internet.ve; Ambassador DIRK CORNELIS DEN HAAS.

**Nicaragua:** Avda Altamira Sur 1, Edif. Terepaima, 2°, Of. 207, Chacao, Caracas; tel. (212) 263-0904; fax (212) 263-8875; e-mail embanic@cantv.net; internet www.ibw.com.net; Ambassador MANUEL SALVADOR ABAUZA.

**Nigeria:** Calle Chivacoa cruce con Calle Taría, Quinta Leticia, Urb. San Román, Caracas; tel. (212) 993-1520; fax (212) 993-7648; e-mail embnig@cantv.net; Chargé d'affaires a.i. VICTOR UCHEM HALLIDAY.

**Norway:** Centro Lido, Torre A-92A, Avda Francisco de Miranda, El Rosal, Apdo 60532, Caracas 1060-A; tel. (212) 953-0269; fax (212) 953-6877; e-mail emb.caracas@mfa.no; internet www.noruega.org.ve; Ambassador DAG MORK ULNES.

**Panama:** Edif. Los Frailes, 6°, Calle La Guairita, Chuao, Apdo 1989, Caracas; tel. (212) 992-9093; fax (212) 992-8107; Ambassador CARMEN GABRIELA MENÉNDEZ GONZÁLEZ.

**Paraguay:** Quinta Paraguay, Avda Principal Macaracuay 1960, entre Avda Cuicas y Carretera del Este, Caracas; tel. (212) 257-2747; fax (212) 257-7256; e-mail embaparven@cantv.net; Ambassador ANA MARÍA FIGUEREDO.

**Peru:** Andres Bello, 7°, Ofs 71-72 (Torre Oeste) y 73-74 (Torre Este), Mariperez, Caracas; tel. (212) 264-1483; fax (212) 265-7592; e-mail leprucaracas@cantv.net; Ambassador CARLOS URRUTIA BOLOÑA.

**Philippines:** 5ta Transversal de Altamira, Quinta Filipinas, Altamira, Caracas 1060; tel. (212) 266-4725; fax (212) 266-6443; e-mail caracas@embassyph.com; Ambassador RONALD B. ALLAREY.

**Poland:** Quinta Ambar, Final Avda Nicolás Copérnico, Sector Los Naranjos, Las Mercedes, Apdo 62293, Caracas; tel. (212) 991-1461; fax (212) 992-2164; e-mail ambcarac@ambasada.org.ve; Ambassador ADAM SKRYBANT.

**Portugal:** Edif. Fedecámaras, 1°, Avda El Empalme, El Bosque, Caracas 1062; tel. (212) 731-0320; fax (212) 731-0543; e-mail embajadaportugal@cantv.net; Ambassador VASCO LUIS PEREIRA BRAMÃO RAMOS.

**Qatar:** Hotel Tamanaco Internacional, 8°, Suite 800, Final Avda Principal Las Mercedes, Caracas; tel. (212) 909-7800; fax (212) 993-7925; Ambassador NASER RASHID MUHAMMAD A. AN-NUAMI.

**Romania:** 4a Avda entre 8a y 9a Transversales, Quinta Guardatinajas 94-14, Altamira, Caracas; tel. (212) 261-9480; fax (212) 263-7161; e-mail ambasadaccs@cantv.net; Ambassador PETRU PEPELEA.

**Russia:** Quinta Soyuz, Calle Las Lomas, Las Mercedes, Caracas; tel. (212) 993-4395; fax (212) 993-6526; e-mail rusemb95@infoline.wtfe.com; Ambassador ALEKSEI ERMAKOV.

**Saudi Arabia:** Calle Andrés Pietri, Quinta Makkah, Los Chorros, Caracas 1071; tel. (212) 239-0290; fax (212) 239-6494; e-mail saudiembassycaracas@cantv.net; Ambassador SALEH A. AL-HEGELAN.

**Serbia and Montenegro:** 4a Avda de Campo Alegre 13, Urb. Campo Alegre, Caracas; tel. (212) 266-7995; fax (212) 266-9957; Chargé d'affaires VERA MAVRIĆ.

**South Africa:** Centro Profesional Eurobuilding, 4°, Of. 4B-C, Calle La Guairita, Chuao, Apdo 2613, Caracas 1064; tel. (212) 991-6822; fax (212) 991-5555; e-mail rsaven@ifxnw.com.ve; Chargé d'affaires a.i. J. SWANEPOEL.

**Spain:** Avda Mohedano entre 1a y 2a transversal, La Castellana, Caracas; tel. (212) 263-2855; fax (212) 261-0892; Ambassador RAÚL MORODO LEONCIO.

**Suriname:** 4a Avda entre 7a y 8a transversal, Quinta 41, Altamira, Caracas; Apdo 61140, Chacao, Caracas; tel. (212) 261-2724; fax (212) 263-9006; e-mail emsurl@cantv.net; Ambassador GLENN ANTONIUS ALVARES.

**Sweden:** Torre Phelps, 19°, Plaza Venezuela, Caracas; tel. (212) 781-6976; fax (212) 781-5932; Ambassador OLOF SKOOG.

**Switzerland:** Centro Letonia, Torre Ing-Bank, 15°, La Castellana, Apdo 62555, Chacao, Caracas 1060-A; tel. (212) 267-9585; fax (212) 267-7745; e-mail vertretung@car.rep.admin.ch; Ambassador WALTER SUTER.

**Syria:** Avda Casiquiare, Quinta Damasco, Colinas de Bello Monte, Caracas; tel. (212) 753-5375; fax (212) 751-6146; Ambassador MOHAMMAD SALEH KHAFIF.

**Trinidad and Tobago:** Quinta Serrana, 4a Avda entre 7 y 8 transversales, Altamira, Caracas; tel. (212) 261-5796; fax (212) 261-9801; e-mail embtt@caracas.c-com.net; Ambassador SHEELAGH MARILYN DE OSUNA.

**Turkey:** Calle Kemal Atatürk, Quinta Turquesa 6, Valle Arriba, Apdo 62078; Caracas 1060-A; tel. (212) 991-0075; fax (212) 992-0442; e-mail turkishemb@cantv.net; Ambassador DOĞAN AKDUR.

**United Kingdom:** Torre La Castellana, 11°, Avda Principal La Castellana, Caracas 1061; tel. (212) 263-8411; fax (212) 267-1275; e-mail britishembassy@internet.ve; internet www.britain.org.ve; Ambassador DONALD ALEXANDER LAMONT.

**USA:** Calle Suapure con Calle F, Colinas de Valle Arriba, Caracas; tel. (212) 975-6411; fax (212) 975-6710; e-mail embajada@state.gov;

# VENEZUELA

*Directory*

internet www.embajadausa.org.ve; Ambassador WILLIAM R. BROWNFIELD.

**Uruguay:** Torre Delta, 8°, Of. A y B, Avda Francisco de Miranda, Altamira Sur, Apdo 60366, Caracas 1060-A; tel. (212) 261-7603; fax (212) 266-9233; Ambassador JUAN JOSÉ ARTEAGA SAENZ DE ZUMARÁN.

## Judicial System

The judicature is headed by the Supreme Tribunal of Justice, which replaced the Supreme Court of Justice after the promulgation of the December 1999 Constitution. The judges are divided into penal and civil and mercantile judges; there are military, juvenile, labour, administrative litigation, finance and agrarian tribunals. In each state there is a superior court and several secondary courts which act on civil and criminal cases. A number of reforms to the judicial system were introduced under the Organic Criminal Trial Code of March 1998. The Code replaced the inquisitorial system, based on the Napoleonic code, with an adversarial system in July 1999. In addition, citizen participation as lay judges and trial by jury was introduced, with training financed by the World Bank.

### SUPREME TRIBUNAL OF JUSTICE

The Supreme Tribunal comprises 20 judges appointed by the Asamblea Nacional for 12 years. In May 2004 the number of judges was increased to 32. It is divided into six courts, each with three judges: political-administrative, civil, constitutional, electoral, social and criminal. When these act together the court is in full session. It has the power to abrogate any laws, regulations or other acts of the executive or legislative branches conflicting with the Constitution. It hears accusations against members of the Government and high public officials, cases involving diplomatic representatives and certain civil actions arising between the State and individuals.

**Tribunal Supremo de Justicia**
Final Avda Baralt, esq. Dos Pilitas, Foro Libertador, Caracas 1010; tel. (212) 801-9178; fax (212) 564-8596; e-mail cperez@tsj.gov.ve; internet www.tsj.gov.ve.

**President:** OMAR ALFREDO MORA DÍAZ.

**President of the Constitutional Court:** LUISA ESTELA MORALES.

**President of the Court of Administrative Policy:** EVELYN MARRERO ORTÍZ.

**President of the Court of Civil Cassation:** CARLOS OBERTO VÉLEZ.

**President of the Court of Penal Cassation:** ELADIO APONTE APONTE.

**President of the Court of Social Cassation:** OMAR ALFREDO MORA DÍAZ.

**President of the Electoral Court:** JUAN JOSÉ NÚÑEZ CALDERÓN.

**Attorney-General:** ISAÍAS RODRÍGUEZ DÍAZ.

## Religion

Roman Catholicism is the religion of the majority of the population, but there is complete freedom of worship.

### CHRISTIANITY

#### The Roman Catholic Church

For ecclesiastical purposes, Venezuela comprises nine archdioceses, 23 dioceses and four Apostolic Vicariates. There are also apostolic exarchates for the Melkite and Syrian Rites. At 31 December 2003 there were an estimated 23.8m. adherents, of whom 25,000 were of the Melkite Rite, accounting for 84% of the total population.

*Latin Rite*

**Bishops' Conference**
Conferencia Episcopal de Venezuela, Prolongación Avda Páez, Montalbán, Apdo 4897, Caracas 1010; tel. (212) 471-6284; fax (212) 472-7029; e-mail prensa@cev.org.ve; internet www.cev.org.ve.
f. 1985; statutes approved 2000; Pres. Most Rev. BALTAZAR ENRIQUE PORRAS CARDOZO (Archbishop of Mérida).

**Archbishop of Barquisimeto:** Most Rev. TULIO MANUEL CHIRIVELLA VARELA, Arzobispado, Venezuela con Calle 29 y 30 Santa Iglesia Catedral, Nivel Sotano, Barquisimeto 3001; tel. (251) 231-3446; fax (251) 231-3724; e-mail arquidiocesisdebarquisimeto@hotmail.com.

**Archbishop of Calabozo:** Most Rev. ANTONIO JOSÉ LÓPEZ CASTILLO, Arzobispado, Calle 4, No 11–82, Apdo 954, Calabozo 2312; tel. (246) 871-0483; fax (246) 871-2097; e-mail el.real@telcel.net.ve.

**Archbishop of Caracas (Santiago de Venezuela):** (vacant), Arzobispado, Plaza Bolívar, Apdo 954, Caracas 1010-A; tel. (212) 542-1611; fax (212) 542-0297; e-mail arzobispado@cantv.net.

**Archbishop of Ciudad Bolívar:** Most Rev. MEDARDO LUIS LUZARDO ROMERO, Arzobispado, Avda Andrés Eloy Blanco con Calle Naiguatá, Apdo 43, Ciudad Bolívar 8001; tel. (285) 654-4960; fax (285) 654-0821; e-mail arzcb@cantv.net.ve.

**Archbishop of Coro:** Most Rev. ROBERTO LÜCKERT LEÓN, Arzobispado, Calle Federación esq. Palmasola, Apdo 7342, Coro; tel. (268) 251-7024; fax (268) 251-1636; e-mail dioceco@reaccium.ve.

**Archbishop of Cumaná:** Most Rev. DIEGO RAFAEL PADRÓN SÁNCHEZ, Arzobispado, Calle Bolívar 34 con Catedral, Apdo 134, Cumaná 6101-A; tel. (293) 431-4131; fax (293) 433-3413; e-mail dipa@cantv.net.

**Archbishop of Maracaibo:** Most Rev. UBALDO RAMÓN SANTANA SEQUERA, Arzobispado, Calle 95, entre Avdas 2 y 3, Apdo 439, Maracaibo; tel. (261) 722-5351; fax (261) 721-0805; e-mail ubrasan@hotmail.com.

**Archbishop of Mérida:** Most Rev. BALTAZAR ENRIQUE PORRAS CARDOZO, Arzobispado, Avda 4, Plaza Bolívar, Apdo 26, Mérida 5101-A; tel. (274) 252-5786; fax (274) 252-1238; e-mail arquimer@latinmail.com.

**Archbishop of Valencia:** Most Rev. JORGE LIBERATO UROSA SAVINO, Arzobispado, Avda Urdaneta 100-54, Apdo 32, Valencia 2001-A; tel. (241) 858-5865; fax (241) 857-8061; e-mail arqui_valencia@cantv.net.

*Melkite Rite*

**Apostolic Exarch:** Rt Rev. GEORGES KAHHALÉ ZOUHAÏRATY, Iglesia San Jorge, Final 3a Urb. Montalbán II, Apdo 20120, Caracas; tel. (212) 443-3019; fax (212) 443-0131; e-mail georges@cev.org.ve.

*Syrian Rite*

**Apostolic Exarch:** LOUIS AWAD, Parroquia Nuestra Señora de la Asunción, 1A Calle San Jacinto, Apdo 11, Maracay; tel. (243) 235-0821; fax (243) 235-7213.

#### The Anglican Communion

Anglicans in Venezuela are adherents of the Episcopal Church in the USA, in which the country forms a single, extra-provincial diocese attached to Province IX.

**Bishop of Venezuela:** Rt Rev. ORLANDO DE JESÚS GUERRERO, Avda Caroní 100, Apdo 49-143, Colinas de Bello Monte, Caracas 1042-A; tel. (212) 753-0723; fax (212) 751-3180; e-mail iglanglicanavzla@cantv.net.

#### Protestant Churches

**Iglesia Evangélica Luterana en Venezuela:** Apdo 68738, Caracas 1062-A; tel. and fax (212) 264-1868; e-mail ielv@telcel.net.ve; Pres. AKOS V. PUKY; 4,000 mems.

**National Baptist Convention of Venezuela:** Avda Santiago de Chile 12–14, Urb. Los Caobos, Caracas 1050; Apdo 61152, Chacao, Caracas 1060-A; tel. (212) 782-2308; fax (212) 781-9043; e-mail cnbv@telcel.net.ve; f. 1951; Pres. Rev. ENRIQUE DÁMASO; Gen. Man. Rev. DANIEL RODRÍGUEZ.

### ISLAM

**Mezquita Sheikh Ibrahim bin-Abdulaziz bin-Ibrahim:** Calle Real de Quebrada Honda, Los Caobos, Caracas; tel. (212) 577-7382; internet www.mezquitaibrahim.org; f. 1994; Leader OMAR KADWA.

### BAHÁ'Í FAITH

**National Spiritual Assembly of the Bahá'ís:** Colinas de Bello Monte, Apdo 49133, Caracas; tel. and fax (212) 751-7669; f. 1961; mems resident in 954 localities.

## The Press

### PRINCIPAL DAILIES

#### Caracas

**Abril:** Edif. Bloque DeArmas, final Avda San Martín cruce con Avda La Paz, Caracas; tel. (212) 406-4376; fax (212) 443-1575; e-mail ldelosreyes@dearmas.com; internet www.abril.com.ve; f. 1997; independent; morning; Mon. to Sat.; Pres. ANDRÉS DE ARMAS S.; Vice-Pres. MARTÍN DE ARMAS S.

# VENEZUELA

**Así es la Noticia:** Maderero a Puente Nuevo, Caracas; tel. (212) 408-3444; fax (212) 408-3911; e-mail ipacheco@el-nacional.com; f. 1996; morning; Editor ERNESTINA HERRERA.

**The Daily Journal:** Avda Principal de Boleíta Norte, Apdo 76478, Caracas 1070-A; tel. (212) 237-9644; fax (212) 232-6831; e-mail redaccion@dj.com.ve; f. 1945; morning; in English; Chief Editor RUSSELL M. DALLEN, Jr.

**Diario 2001:** Edif. Bloque DeArmas, 2°, final Avda San Martín cruce con Avda La Paz, Caracas; tel. (212) 406-4111; fax 443-4961; e-mail jtirado@dearmas.com; www.2001.com.ve; f. 1973; Pres. ANDRÉS DE ARMAS S.; Dir ISRAEL MÁRQUEZ.

**El Diario de Caracas:** Calle Los Laboratorios, Torre B, 1°, Of. 101, Los Ruices, Caracas 1070; tel. (212) 238-0386; e-mail editor@eldiariodecaracas.net; internet www.eldiariodecaracas.net; f. 2003; Editor JULIO AUGUSTO LÓPEZ.

**El Globo:** Avda Principal de Maripérez, transversal Colón con Avda Libertador, Apdo 16415, Caracas 1010-A; tel. (2) 576-4111; fax (212) 576-1730; f. 1990; Dir ANÍBAL J. LATUFF.

**Meridiano:** Edif. Bloque DeArmas, final Avda San Martín cruce con Avda La Paz, Caracas 1010; tel. (212) 406-4040; fax (212) 442-5836; e-mail meridian@dearmas.com; internet www.meridiano.com.ve; f. 1969; morning; sport; Dir ANDRÉS DE ARMAS S.; Vice-Pres. MARTÍN DE ARMAS S.

**El Mundo:** Torre de la Prensa, 4°, Plaza del Panteón, Apdo 1192, Caracas; tel. (212) 596-1911; fax (212) 596-1478; e-mail kico@la-cadena.com; internet www.elmundo.com.ve; f. 1958; evening; independent; Pres. MIGUEL ANGEL CAPRILES LÓPEZ; Vice-Pres. ARMANDO CAPRILES LÓPEZ.

**El Nacional:** Edif. El Nacional, Puente Nuevo a Puerto Escondido, El Silencio, Apdo 209, Caracas; tel. (212) 408-3111; fax (212) 408-3169; e-mail contactenos@el-nacional.com; internet www.el-nacional.com; f. 1943; morning; right-wing; independent; Pres. and Editor MIGUEL HENRIQUE OTERO; Asst Editor YELITZA LINARES; circ. 12,000.

**El Nuevo País:** Pinto a Santa Rosalía 44, Caracas; tel. (212) 541-5211; fax (212) 545-9675; e-mail enpais1@telcel.net.ve; f. 1988; Exec. Pres. ZORAIDA GARCÍA VARA; Dir and Editor FRANCISCO ORTA.

**Puerto:** Avda Soublette, Maiquetia, Caracas; tel. (212) 331-2275; fax (212) 331-0886; e-mail diariopuerto@telcel.net.ve; f. 1975; Pres. CARLOS PARMIGIANI.

**La Religión:** Edif. Juan XXIII, Torre a Madrices, Caracas; tel. (212) 563-0600; fax (212) 563-5583; e-mail religion@cantv.ve; internet www.iglesia.org.ve; f. 1890; morning; independent; Dir ENNIO TORRES.

**Reporte (Diario de la Economía):** Torre Británica, 12°, Of. B, Avda Luis Roche, Altamira Sur, Caracas; tel. (212) 264-0591; fax (212) 264-6023; e-mail info@reporte.com.ve; internet www.reporte.com.ve; f. 1988; Pres. TANNOUS GERGES; Editor WILLIAM BECERRA.

**Tal Cual:** Avda Francisco de Miranda, Edif. Menegrande, 5°, Of. 51, Caracas; tel. (212) 286-7446; fax (212) 232-7446; e-mail tpekoff@talcualdigital.com; internet www.talcualdigital.com; f. 2000; morning; right-wing; Pres. TEODORO PETKOFF; Editor-in-Chief EDMUNDO BRACHO.

**Ultimas Noticias:** Torre de la Prensa, 3°, Plaza del Panteón, Apdo 1192, Caracas; tel. (212) 596-1911; fax (212) 596-1433; e-mail edrangel@la-cadena.com; internet www.ultimasnoticias.com.ve; f. 1941; morning; independent; Pres. MIGUEL ANGEL CAPRILES LÓPEZ.

**El Universal:** Edif. El Universal, Avda Urdaneta esq. de Animas, Apdo 1909, Caracas; tel. (212) 505-2314; fax (212) 505-3710; e-mail diario@eluniversal.com; internet www.eluniversal.com; f. 1909; morning; Dir ANDRÉS MATA OSORIO; Chief Editor ELIDES ROJAS.

**Vea:** Sótano Uno, Edif. San Martín, Parque Central, Caracas 1010; tel. (212) 516-1004; fax (212) 578-3031; e-mail redaccion@diariovea.com; internet www.diariovea.com; f. 2003; morning; left-wing; Editor and Dir GUILLERMO GARCÍA PONCE; Asst Editor MANUEL PÉREZ RODRÍGUEZ.

**La Voz:** C. C. Nueva Guarenas, Mezzanina, Urb. Trapichito, Sector 2, Caracas; tel. (212) 362-9702; fax (212) 362-0851; e-mail diariolavoz@cantv.net; internet www.diariolavoz.net; morning; right-wing; Exec. Dir FREDDY BLANCO; Editor ALEXIS CASTRO BLANDÍN.

### Barcelona

**El Norte:** Avda Intercomunal Andrés Bello, Sector Colinas del Neverí, Barcelona; tel. (281) 286-1653; fax (281) 286-2884; e-mail info@elnorte.com.ve; internet www.elnorte.com.ve; f. 1989; morning; Dir OMAR GONZALES MORENO; circ. 133,000.

### Barquisimeto

**El Impulso:** Avda Los Comuneros, entre Avda República y Calle 1a, Urb. El Parque, Apdo 602, Barquisimeto; tel. (251) 250-2222; fax (251) 250-2129; e-mail reaccion@elimpulso.com; internet www.elimpulso.com; f. 1904; morning; independent; Dir and Editor JUAN MANUEL CARMONA PERERA.

**El Informador:** Carrera 21, esq. Calle 23, Barquisimeto; tel. (251) 231-1811; fax (251) 231-0624; e-mail informad@telcel.net.ve; f. 1968; morning; Dir ALEJANDRO GÓMEZ SIGALA.

### Ciudad Bolívar

**El Bolivarense:** Calle Igualdad 26, Apdo 91, Ciudad Bolívar; tel. (285) 632-2378; fax (285) 632-4878; e-mail abi28@unete.com.ve; f. 1957; morning; independent; Dir ALVARO NATERA.

**El Expreso:** Paseo Gáspari con Calle Democracia, Ciudad Bolívar; tel. (285) 632-0334; fax (285) 632-0334; e-mail webmaster@diarioelexpreso.com.ve; internet www.diarioelexpreso.com.ve; f. 1969; morning; independent; Dir LUIS ALBERTO GUZMÁN.

### Maracaibo

**Panorama:** Avda 15 No 95–60, Apdo 425, Maracaibo; tel. (261) 725-6888; fax (261) 725-6911; e-mail editor@panodi.com; internet www.panodi.com; f. 1914; morning; independent; Pres. ESTEBAN PINEDA BELLOSO; Dir ROBERTO BAITTINER; circ. 16,000.

### Maracay

**El Aragueño:** Calle 3a Oeste con Avda 1 Oeste, Urb. Ind. San Jacinto, Maracay; tel. (243) 235-9018; fax (243) 235-7866; f. 1972; morning; Editor EVERT GARCÍA.

**El Siglo:** Edif. 'El Siglo', Avda Bolívar Oeste 244, La Romana, Maracay; tel. (243) 554-9265; fax (243) 554-5910; e-mail elsiglo@telcel.net.ve; f. 1973; morning; independent; Editor TULIO CAPRILES.

**El Periodiquito:** Calle Páez Este 178, Maracay; tel. (243) 321-422; fax (243) 336-987; e-mail periodiquito@cantv.ve; internet www.elperiodiquito.com; f. 1986; Dir and Editor GUSTAVO URBINA; circ. 45,000.

### Puerto la Cruz

**El Tiempo:** Avda Constitución, Paseo Miranda 39, Apdo 4733, Puerto la Cruz; tel. (281) 265-4344; fax (281) 269-9224; e-mail buzon@eltiempo.com.ve; f. 1958; independent; Pres. MARÍA A. MÁRQUEZ; Editor Dra GIOCONDA DE MÁRQUEZ.

### San Cristóbal

**Diario Católico:** Carrera 4a No 3–41, San Cristóbal; tel. (276) 343-2819; fax (276) 343-4683; e-mail catolico@truevision.net; f. 1924; morning; Catholic; Man. Dir Mgr NELSON ARELLANO.

**Diario La Nación:** Edif. La Nación, Calle 4 con Carrera 6 bis, La Concordia, Apdo 651, San Cristóbal; tel. (276) 346-4263; fax (276) 346-5051; e-mail lanacion@lanacion.com.ve; f. 1968; morning; independent; Editor JOSÉ RAFAEL CORTEZ.

### El Tigre

**Antorcha:** Torre Alcabala, 5°, Of. 53-A, Alcabala a Peligro, El Tigre; tel. (283) 235-2383; fax (283) 235-3923; f. 1954; morning; independent; Pres. and Editor ANTONIO BRICEÑO AMPARÁN.

### Valencia

**El Carabobeño:** Edif. El Carabobeño, Avda Universidad, Urb. La Granja, Naguanagua, Valencia; tel. (241) 867-2918; fax (241) 867-3450; e-mail general@el-carabobeno.com; internet www.el-carabobeno.com; f. 1933; morning; Dir EDUARDO ALEMÁN PÉREZ.

**Notitarde:** Avda Boyacá entre Navas Spínola y Flores, Valencia; tel. (241) 850-1666; fax (241) 850-1534; e-mail lauodr@notitarde.com; internet www.notitarde.com; evening; Dir LAURENTZI ODRIOZOLA ECHEGARAY.

## PERIODICALS

**Ambiente:** Edif. Sur, Nivel Plaza Caracas, Local 9, Centro Simón Bolívar, Caracas; tel. (212) 408-1549; fax (212) 408-1546; e-mail fundamb@cantv.net; environmental issues; Gen. Man. AVRA MARINA SÁNCHEZ.

**Artesanía y Folklore de Venezuela:** C. C. Vista Mar, Local 20, Urbaneja, Lecherías, Estado Anzoátegui; tel. and fax (212) 286-2857; e-mail ismandacorrea@cantv.net; handicrafts and folklore; Dir ISMANDA CORREA.

**Automóvil de Venezuela:** Avda Caurimare, Quinta Expo, Colinas de Bello Monte, Caracas 1041; tel. (212) 751-1355; fax (212) 751-1122; e-mail ortizauto@cantv.net; internet www.automovildevenezuela.com; f. 1961; monthly; automotive trade; circ. 6,000; Editor MARÍA A. ORTIZ T.

**Barriles:** Centro Parque Carabobo, Torre B, 20°, Of. 2003, Avda Universidad, La Candelaria, Caracas; e-mail informaciones@

# VENEZUELA

camarapetrolera.org; publ. of the Cámara Petrolera de Venezuela; Editor Haydée Reyes.

**Bohemia:** Edif. Bloque DeArmas, Final Avda San Martín cruce con Avda La Paz, Apdo 575, Caracas; tel. (212) 406-4040; fax (212) 451-0762; e-mail bohemia@dearmas.com; f. 1966; weekly; general interest; Dir Pedro Ramón Romera.

**Business Venezuela:** Torre Credival, Avda de Campo Alegre, Apdo 5181, Caracas 1010-A; tel. (212) 263-0833; fax (212) 263-2060; e-mail bramirez@venancham.org; every 2 months; business and economics journal in English published by the Venezuelan-American Chamber of Commerce and Industry; Gen. Man. Antonio Herrera.

**Computer World:** Edif. MaryStella, Avda Carabobo, El Rosal, Caracas; tel. (212) 952-7427; fax (212) 953-3950; e-mail cernic@cwv.com.ve; Editor Clelia Santambrogio.

**Contrapunto:** Edif. Unión, 1°, Of. 2, Avda El Parque, Las Acacias Sur, Caracas; tel. (212) 690-0431; Dir Víctor Vera Morales.

**Convenciones:** Torre Nonza, 4°, Plaza Venezuela, Caracas; tel. (212) 793-1962; e-mail redacta@tutopia.com; Dir Mario Ernesto Arbeláez.

**El Corresponsal:** Conj. Residencial El Naranjal, Torre D, 8°, No 82-D, Urb. Los Samanes, Caracas; tel. and fax (212) 941-0409; Editor Domingo García Pérez.

**Dinero:** Edif. Aco, Entrada A, 7°, Avda Principal Las Mercedes, Caracas; tel. (212) 993-5011; fax (212) 991-3132; e-mail lomonaco@gep.com.ve; business and finance; Dir Salvatore Lomonaco.

**Exceso:** Edif. Karam, Avda Urdaneta, 5°, Caracas; tel. (212) 564-1702; fax (212) 564-6760; e-mail baf-exceso@cantv.net; internet www.exceso.net; lifestyle; Dir Ben Amí Fihman; Editor Armando Coll.

**Gerente Venezuela:** Avda Orinoco 3819, entre Muchuchies y Monterrey, Las Mercedes, Caracas 1060; tel. (212) 267-3733; fax (212) 267-6583; business and management; Editor Luis Rodán; circ. 15,000.

**El Mirador:** Edif. Pascal, Torre A, 1°, Of. 12-A, Avda Rómulo Gallegos, Santa Eduvigis, Caracas; tel. (212) 286-1661; fax (212) 283-5823; e-mail diarioelmirador@cantv.net; Dir Jesús Couto.

**Mujer-Mujer:** Centro Banaven, Nivel Sótano, No 13-A, Avda La Estancia, Chuao, Caracas; tel. (212) 959-5393; fax (212) 959-7864; e-mail ceciliapicon@hotmail.com.ve; women's interest; Dir and Editor Cecilia Picón de Torres.

**Nueva Sociedad:** Edif. IASA, 6°, Of. 606, Plaza La Castellana, Apdo 61712, Caracas; tel. (212) 265-9975; fax (212) 267-3397; e-mail nusoven@nuevasoc.org.ve; internet www.nuevasoc.org.ve; f. 1972; Latin American affairs; Editor Dietmar Dirmoser; Editor Sergio Chejfec.

**Primicia:** Edif. El Nacional, 4°, Puente Nuevo a Puerto Escondido, Caracas; tel. (212) 408-3434; fax (212) 408-3485; e-mail primicia@el-nacional.com; current affairs and business; Editor Franchesca Cordido.

**La Razón:** Edif. Valores, Sótano 'A', Avda Urdaneta, esq. de Urapal, Apdo 16362, La Candelaria, Caracas; tel. (212) 578-3143; fax (212) 578-2397; e-mail larazon@internet.ve; internet www.razon.com; weekly on Sundays; independent; Dir Pablo López Ulacio.

**La Red:** Urb. Vista Alegre, Calle 7, Quinta Luisa Amelia, Caracas; tel. (212) 472-0703; fax (212) 471-7749; e-mail ldavila@lared.com.ve; internet www.lared.com.ve; f. 1996; information technology; Editor Luis Manuel Dávila.

**Reporte Petrolero:** Torre Británica, 12°, Of. B, Altamira Sur, Caracas; tel. (212) 264-6023; fax (212) 266-9991; e-mail reporte2002@hotmail.com; journal of the petroleum industry; Dir Enrique Romai; Editor Miguel López Trocelt.

**Sic:** Edif. Centro de Valores, esq. de Luneta, Centro Gumilla, Caracas; tel. (212) 564-9803; fax (212) 564-7557; e-mail sic@gumilla.org.ve; internet www.gumilla.org.ve/sic.htm; f. 1938; monthly; liberal Jesuit publication; Dir Jesús María Aguirre.

**Variedades:** Edif. Bloque DeArmas, final Avda San Martín cruce con Avda La Paz, Caracas 1020; tel. (212) 406-43-90; fax (212) 451-0762; e-mail mgonzalez@dearmas.com; women's weekly; Dir Gloria Fuentes de Valladares.

**VenEconomía:** Edif. Gran Sabana, 1°, Avda Abraham Lincoln 174, Blvr de Sabana Grande, Caracas; tel. (212) 761-8121; fax (212) 762-8160; e-mail editor@veneconomia.com; internet www.veneconomia.com; f. 1982; monthly; Spanish and English; business and economic issues; Editor Toby Bottome.

**Venezuela Gráfica:** Torre de la Prensa, Plaza del Panteón, Apdo 2976, Caracas 101; tel. (212) 81-4931; f. 1951; weekly; illustrated news magazine, especially entertainment; Dir Diego Fortunato; Editor Miguel Angel Capriles.

**Zeta:** Pinto a Santa Rosalía 44, Apdo 14067, Santa Rosalía, Caracas; tel. (212) 541-5211; fax (212) 545-9675; e-mail enpaiscolumna@hotmail.com; f. 1974; weekly; politics and current affairs; Dir Jurate Rosales; Editor Rafael Poleo.

**Zulia Deportivo:** Torre Luali, 5°, Of. 3, Piñango a Muñoz, Avda Baralt, Maracaibo; tel. (261) 787-2809; sports; Dir Manuel Colina Hidalgo.

## PRESS ASSOCIATIONS

**Asociación de Prensa Extranjera en Venezuela (APEX):** Hotel Caracas Hilton, Avda México, Torre Sur, 3°, Of. 301, Caracas; tel. (212) 503-5301; fax (212) 576-9284; e-mail caracashilton@hotmail.com; Pres. Phil Gunson.

**Bloque de Prensa Venezolano (BEV):** Edif. El Universal, Avda Urdaneta, 5°, Of. C, Caracas; tel. (212) 561-7704; fax (212) 561-9409; e-mail luichi@telcel.net.ve; asscn of newspaper owners; Pres. Miguel Angel Martínez.

**Colegio Nacional de Periodistas (CNP):** Casa Nacional del Periodista, Avda Andrés Bello, 2°, Caracas; tel. and fax (212) 781-7601; e-mail cnpjdn@cantv.net; journalists' asscn; Pres. Levy Benshimol; Sec.-Gen. Noel Molina.

## STATE PRESS AGENCY

**Venpres:** Torre Oeste, 16°, Parque Central, Caracas; tel. (212) 572-7175; fax (212) 571-0563; internet www.venpres.gov.ve; Dir Orlando Utrera Reyes.

## FOREIGN PRESS AGENCIES

**Agence France-Presse (AFP):** Torre Provincial, Torre A, 14°, Of. 14-1, Avda Francisco de Miranda, Chacao, Caracas; tel. (212) 264-2945; fax (212) 267-7797; e-mail jorge.calmet@afp.com; Bureau Chief Jacques Thomet.

**Agencia EFE** (Spain): Calle San Cristóbal, Quinta Altas Cumbres, Urb. Las Palmas, Caracas 1050; tel. (212) 793-7118; fax (212) 793-4920; e-mail efered1@cantv.net.

**Agenzia Nazionale Stampa Associata (ANSA)** (Italy): Centro Financiero Latino, 7°, Of. 2, Animas a Plaza España, Avda Urdaneta, Caracas; tel. (212) 564-2059; fax (212) 564-2516; e-mail ansaven@infoline.wtfe.com; Dir Natacha Salazar.

**Associated Press (AP)** (USA): Edif. El Universal, Avda Urdaneta, esq. Animas, 2°, Of. D, Caracas; tel. (212) 564-1834; fax (212) 564-7124; e-mail janderson@ap.org; Chief James Anderson.

**BBC Latin Service:** Res. Los Tulipanes, Torre A, 6°, No 6-B, Urb. Guaicai, Los Samanes; tel. (212) 941-7564; e-mail mariusareyes@hotmail.com; internet www.bbcmundo.com; Correspondent Mariusa Reyes.

**Bloomberg:** Torre Edicampo, 5°, Of. 1-2, Avda Francisco de Miranda, Caracas; tel. (212) 263-3355; fax (212) 264-2171; e-mail pewilson@bloomberg.net; Chief Correspondent Peter Wilson.

**Deutsche Presse-Agentur (dpa)** (Germany): Edif. El Universal, 4°, Of. 4-E, Avda Urdaneta, Caracas; tel. (212) 561-9776; fax (212) 562-9017; e-mail dpacaracas@tutopia.com; Correspondent Néstor Rojas.

**Inter Press Service (IPS)** (Italy): Edif. El Universal, 8°, Of. 3, Avda Urdaneta Caracas; tel. (212) 564-6386; fax (212) 564-6374; e-mail caracas@ipsenespanol.org; Dir Andrés Cañizález.

**Prensa Latina** (Cuba): Edif. Fondo Común, Torre Sur, Of. 20-D, Avda de las Fuerzas Armadas y Urdaneta, Apdo 4400, Carmelitas, Caracas; tel. (212) 561-9733; fax (212) 564-7960; e-mail prelaccs@telcel.net.ve; Correspondent Javier Rodríguez.

**Reuters** (United Kingdom): Torre la Castellana, 4°, Avda Principal La Castellana, Caracas; tel. (212) 277-2700; fax (212) 277-2664; e-mail caracas.newsroom@reuters.com; Chief Correspondent Pascal Fletcher.

**Xinhua (New China) News Agency** (People's Republic of China): Quinta Xinjua, Avda Maracaibo, Prados del Este, Apdo 80564, Caracas; tel. (212) 8977-2489; fax (212) 978-1664; e-mail xinhua@c-com.ve; Bureau Chief Quanfu Wuanc.

# Publishers

**Alfadil Ediciones:** Calle Las Flores con Calle Paraíso, Sábana Grande, Apdo 50304, Caracas 1020-A; tel. (212) 762-3036; fax (212) 762-0210; f. 1980; general; Pres. Leonardo Milla A.

**Armitano Editores, CA:** Centro Industrial Boleita Sur, 4a Transversal de Boleita, Apdo 50853, Caracas 1070; tel. (212) 234-2565; fax (212) 234-1647; e-mail armiedit@telcel.net.ve; internet www.alfagrupo.com; art, architecture, ecology, botany, anthropology, history, geography; Pres. Ernesto Armitano.

# VENEZUELA

*Directory*

**Ediciones La Casa Bello:** Mercedes a Luneta, Apdo 134, Caracas 1010; tel. (212) 562-7100; f. 1973; literature, history; Pres. Oscar Sambrano Urdaneta.

**Editorial El Ateneo, CA:** Complejo Cultural, Plaza Morelos, Los Caobos, Apdo 662, Caracas; tel. (212) 573-4622; f. 1931; school-books and reference; Pres. María Teresa Castillo; Dir Antonio Polo.

**Editorial Cincel Kapelusz Venezolana, SA:** Avda Cajigal, Quinta K No 29, entre Avdas Panteón y Roraima, San Bernardino, Apdo 14234, Caracas 1011-A; f. 1963; school-books; Pres. Dante Toni; Man. Mayela Morgado.

**Colegial Bolivariana, CA:** Edif. COBO, 1°, Avda Diego Cisneros (Principal), Los Ruices, Apdo 70324, Caracas 1071-A; tel. (212) 239-1433; f. 1961; Dir Antonio Juzgado Arias.

**Ediciones Ekaré:** Edif. Banco del Libro, Final Avda Luis Roche, Altamira Sur, Apdo 68284, Caracas 1062; tel. (212) 264-7615; fax (212) 263-3291; e-mail editorial@ekare.com.ve; internet www.ekare.com; f. 1978; children's; Pres. Carmen Diana Dearden; Exec. Dir María Francisca Mayobre.

**Editora Ferga, CA:** Torre Bazar Bolívar, 5°, Of. 501, Avda Francisco de Miranda, El Marqués, Apdo 16044, Caracas 1011-A; tel. (212) 239-1564; fax (212) 234-1008; e-mail ddex1@ibm.net; internet www.ddex.com; f. 1971; Venezuelan Exporters' Directory; Dir Nelson Sánchez Martínez.

**Fundación Biblioteca Ayacucho:** Centro Financiero Latino, 12°, Of. 1, 2 y 3, Avda Urdaneta, Animas a Plaza España, Apdo 14413, Caracas 1010; tel. (212) 561-6691; fax (212) 564-5643; e-mail biblioayacucho@cantv.net; f. 1974; literature; Pres. Stefania Mosca.

**Fundación Bigott:** Casa 10-11, Calle El Vigia, Plaza Sucre, Centro Histórico de Petare, Caracas 1010-A; tel. (212) 272-2020; fax (212) 272-5942; e-mail contacto@fundacionbigott.com; internet www.fundacionbigott.com; f. 1936; Venezuelan traditions, environment, agriculture; Admin. Co-ordinator Nelson Reyes.

**Fundarte:** Edif. Tajamar, P. H., Avda Lecuna, Parque Central, Apdo 17559, Caracas 1015-A; tel. (212) 573-1719; fax (212) 574-2794; f. 1975; literature, history; Pres. Alfredo Gosen; Dir Roberto Lovera de Sola.

**Editorial González Porto:** Sociedad a Traposos 8, Avda Universidad, Caracas; Pres. Dr Pablo Perales.

**Ediciones IESA:** Edif. IESA, 3°, Final Avda IESA, San Bernardino, Apdo 1640, Caracas 1010-A; f. 1984; economics, business; Pres. Ramón Piñango.

**Ediciones María Di Mase:** Caracas; f. 1979; children's books; Pres. María di Mase; Gen. Man. Ana Rodríguez.

**Monte Avila Editores Latinoamericana, CA:** Avda Principal La Castellana, Quinta Cristina, Apdo 70712, Caracas 1070; tel. (212) 265-6020; fax (212) 263-8783; e-mail maelca@telcel.net.ve; f. 1968; general; Pres. Mariela Sánchez Urdaneta.

**Nueva Sociedad:** Edif. IASA, 6°, Of. 606, Plaza La Castellana, Apdo 61712, Chacao, Caracas 1060-A; Apdo 61712, Caracas 1060-A; tel. (212) 265-0593; fax (212) 267-3397; e-mail nuso@nuevasoc.org.ve; internet www.nuevasoc.org.ve; f. 1972; social sciences; Dir Dietmar Dirmoser.

**Ediciones Panamericanas EP, SRL:** Edif. Freites, 2°, Avda Libertador cruce con Santiago de Chile, Apdo 14054, Caracas; tel. (212) 782-9891; Man. Jaime Salgado Palacio.

**Editorial Salesiana, SA:** Paradero a Salesianos 6, Apdo 369, Caracas; tel. (212) 571-6109; fax (212) 574-9451; e-mail administracion@salesiana.com.ve; internet www.salesiana.com.ve; f. 1960; education; Gen. Man. Clarencio García.

**Oscar Todtmann Editores:** Avda Libertador, Centro Comercial El Bosque, Local 4, Caracas 1050; tel. (212) 763-0881; fax (212) 762-5244; science, literature, photography; Dir Carsten Todtmann.

**Vadell Hermanos Editores, CA:** Edif. Golden, Avda Sur 15, esq. Peligro a Pele el Ojo, Caracas; tel. (212) 572-3108; f. 1973; science, social science; Gen. Man. Manuel Vadell Graterol.

**Ediciones Vega S.R.L.:** Edif. Odeon, Plaza Las Tres Gracias, Los Chaguaramos, Caracas 1050-A; tel. (212) 662-2092; fax (212) 662-1397; f. 1965; educational; Man. Dir Fernando Vega Alonso.

## PUBLISHERS' ASSOCIATION

**Cámara Venezolana del Libro:** Centro Andrés Bello, Torre Oeste, 11°, Of. 112-0, Avda Andrés Bello, Caracas 1050-A; tel. (212) 793-1347; fax (212) 793-1368; f. 1969; Pres. Hans Schnell; Sec. Isidoro Duarte.

# Broadcasting and Communications

## TELECOMMUNICATIONS

### Regulatory Authority

**Comisión Nacional de Telecomunicaciones (CONATEL):** Ministerio de Transportes y Comunicaciones, Torre Este, 35°, Parque Central, Caracas; tel. (212) 993-5389; e-mail conatel@conatel.gov.ve; internet www.conatel.gov.ve; regulatory body for telecommunications; Dir-Gen. Alvin Reinaldo Lezama Pereira.

### Service Providers

**Compañía Anónima Nacional Teléfonos de Venezuela (CANTV):** Edif. NEA, 20, Avda Libertador, Caracas 1010-A; tel. (212) 500-3016; fax (212) 500-3512; e-mail amora@cantv.com.ve; internet www.cantv.net; privatized in 1991; 49% state-owned, 40% owned by Verizon (USA); 11% owned by employees; Pres. Gustavo Roosen; Gen. Man. Eduardo Menascé.

**Movilnet:** Edif. NEA, 20, Avda Libertador, Caracas 1010-A; tel. (202) 705-7901; e-mail info@movilnet.com.ve; internet www.movilnet.com.ve; f. 1992; mobile cellular telephone operator; owned by CANTV; Pres. Guillermo Olaizola.

**CVG Telecomunicaciones, CA** (CVG Telecom): Puerto Ordaz; f. 2004; state-owned telecommunications co.

**Digicel CA:** Caracas; internet www.digicel.com.ve; owned by Banco Santander Central Hispano of Spain; fixed-line telecommunications.

**Digitel TIM:** Caracas; e-mail 0412empres@digitel.com.ve; internet www.digitel.com.ve; f. 2000; mobile cellular telephone operator; owned by Telecom Italia Mobile (TIM) of Italy.

**Intercable:** Avda La Pedregosa Sur, cruce con Avda Los Próceres, Tapias; e-mail jguerrero@multimedios.net; internet www.intercable.net; cable, internet and telecommunications services; Dir Juan Gerardo Guerrero.

**NetUno:** Caracas; f. 1995; voice, data and video transmission services; Pres. Gilbert Minionis.

**TelCel CA:** Edif. Parque Cristal, Torre Este, 14°, Avda Francisco de Miranda, Caracas; tel. (212) 200-8201; internet www.telcel.net.ve; f. 1991; acquired by Telefónica Móviles, SA (Spain) in 2004; mobile cellular telephone operator; Pres. Enrique García Viamonte.

## BROADCASTING

### Regulatory Authorities

**Cámara Venezolana de la Industria de Radiodifusión:** Avda Antonio José Istúriz entre Mohedano y Country Club, La Castellana, Caracas; tel. (212) 263-2228; fax (212) 261-4783; e-mail camradio@camradio.org.ve; internet www.camradio.org.ve; Pres. Miguel Angel Martínez.

**Cámara Venezolana de Televisión:** Edif. Venevisión, 4°, Colinas de Los Caobos, Caracas; tel. (212) 708-9223; fax (212) 708-9146; e-mail esalinas@cisneros.com; regulatory body for private stations; Pres. Eduardo Salinas.

### Radio

**Radio Nacional de Venezuela (RNV):** Final Calle Las Marías, entre Chapellín y Country Club, La Florida, Caracas 1050; tel. (212) 730-6022; fax (212) 731-1457; e-mail ondacortavenezuela@hotmail.com; internet www.rnv.gov.ve; f. 1936; state broadcasting org.; 15 stations; Gen. Man. Helena Salcedo.

There are also 20 cultural and some 500 commercial stations.

### Television

#### Government Stations

**Telesur** (Televisora del Sur): Edif. Anexo VTV, 4°, Avda Principal Los Ruices, Caracas; tel. (212) 716-5605; e-mail contactenos@telesurtv.net; internet www.telesurtv.net; f. 2005; jtly owned by govts of Venezuela (51%), Argentina (20%), Cuba (19%) and Uruguay (10%); regional current affairs and general interest; Pres. Lt (retd) Andrés Izarra; Vice-Pres. Aram Aharonian.

**Venezolana de Televisión (VTV)—Canal 8:** Edif. VTV, Avda Principal Los Ruices, Caracas; tel. (212) 239-4870; fax (212) 239-8102; internet www.venezuela.gov.ve/vtv; 26 relay stations; Pres. Vladimir Villegas Poljak.

#### Private Stations

**Corporación Televén—Canal 10:** Edif. Televén, 4ta Transversal con Avda Rómulo Gallegos, Urb. Horizonte, Apdo 1070, Caracas; tel. (212) 280-0011; fax (212) 280-0204; e-mail aferro@televen.com; internet www.televen.com; f. 1988; Pres. Omar Camero Zamora.

VENEZUELA — Directory

**Corporación Venezolana de Televisión (Venevisión)—Canal 4:** Edif. Venevisión, Final Avda La Salle, Colinas de los Caobos, Apdo 6674, Caracas; tel. (212) 708-9224; fax (212) 708-9535; e-mail mponce@venevision.com.ve; internet www.venevision.net; f. 1961; commercial; Pres. GUSTAVO CISNEROS.

**Globovisión—Canal 33:** Quinta Globovisión, Avda Los Pinos, Urb. Alta Florida, Caracas; tel. (212) 730-2290; fax (212) 731-4380; e-mail info@globovision.com; internet www.globovision.com; f. 1994; 24-hour news and current affairs channel; Pres. GUILLERMO ZULOAGA; Gen. Man. ALBERTO FEDERICO RAVELL.

**Puma TV:** Edif. Puma, Calle Sanatorio del Avila, Boleita Norte, Caracas 1070; tel. (212) 232-5656; fax (212) 237-8655; e-mail recheverria@pumatv.net; subscription TV channel; Pres. JOSÉ RODRÍGUEZ.

**Radio Caracas Televisión (RCTV)—Canal 2:** Edif. RCTV, Dolores a Puente Soublette, Quinta Crespo, Caracas; tel. (212) 401-2222; fax (212) 401-2647; e-mail marriaga@rctv.net; internet www.rctv.net; f. 1953; commercial station; station in Caracas and 13 relay stations throughout the country; Pres. MARCEL GRANIER.

**Televisora Andina de Mérida (TAM)—Canal 6:** Edif. Imperador, Entrada Independiente, Avda 6 y 7, Calle 23, Mérida 5101; tel. and fax (274) 251-0660; fax (274) 251-0660; f. 1982; Pres. Most Rev. BALTAZAR ENRIQUE PORRAS CARDOZO.

**VALE TV (Valores Educativos)—Canal 5:** Quinta VALE TV, Final Avda La Salle, Colinas de los Caobos, Caracas 1050; tel. (212) 793-9215; fax (212) 708-9743; e-mail webmaster@valetv.com; internet www.valetv.com; f. 1998; Pres. Mgr NICOLÁS BERMÚDEZ; Man. MARÍA ISABEL ROJAS.

**Zuliana de Televisión—Canal 30:** Edif. 95.5 América, Avda 11 (Veritas), Maracaibo; tel. (265) 641-0355; fax (265) 641-0565; e-mail elregionalredac@iamnet.com; Pres. GILBERTO URDANETA FIDOL.

# Finance

(cap. = capital; res = reserves; dep. = deposits; m. = million; brs = branches; amounts in bolívares unless otherwise indicated)

## BANKING

### Regulatory Authority

**Superintendencia de Bancos (SUDEBAN):** Edif. SUDEBAN, Avda Universidad, Apdo 6761, Caracas 1010; tel. (212) 505-0933; e-mail sudeban@sudeban.gov.ve; internet www.sudeban.gov.ve; regulates banking sector; Supt TRINO A. DÍAZ.

### Central Bank

**Banco Central de Venezuela:** Avda Urdaneta esq. de Carmelitas, Caracas 1010; tel. (212) 801-5111; fax (212) 861-0048; e-mail info@bcv.org.ve; internet www.bcv.org.ve; f. 1940; bank of issue and clearing house for commercial banks; granted autonomy 1992; controls international reserves, interest rates and exchange rates; cap. 10.0m., res 9,662,937m., dep. 4,374,707m. (Dec. 2002); Pres. DIEGO LUIS CASTELLANOS ESCALONA.

### Commercial Banks, Caracas

**Banco del Caribe, CA:** Edif. Banco del Caribe, 1°, Dr Paúl a esq. Salvador de León, Apdo 6704, Carmelitas, Caracas 1010; tel. (212) 505-5103; fax (212) 562-0460; e-mail producto@bancaribe.com.ve; internet www.bancaribe.com.ve; f. 1954; cap. 10,666.7m., res 12,831.3m., dep. 425,340.7m. (Dec. 1999); Pres. JOSÉ ANTONIO ELOSEGUI; Gen. Man. Dr LUIS E. DE LLANO; 70 brs and agencies.

**Banco de Comercio Exterior (Bancoex):** Central Gerencial Mohedano, 1°, Calle Los Chaguaramos, La Castellana, Caracas 1060; tel. (212) 265-1433; fax (212) 265-6722; e-mail exports@bancoex.com; internet www.bancoex.com; f. 1997; principally to promote non-traditional exports; state-owned; cap. US $200m.; Pres. VÍCTOR ALVAREZ.

**Banco de los Trabajadores de Venezuela (BTV) CA:** Edif. BTV, Avda Universidad, esq. Colón a esq. Dr Díaz, Caracas; tel. (212) 541-7322; f. 1968 to channel workers' savings for the financing of artisans and small industrial firms; came under state control in 1982; Pres. JOSÉ SÁNCHEZ PIÑA; Man. SILVERIO ANTONIO NARVÁEZ; 11 agencies.

**Banco Exterior, CA—Banco Universal:** Edif. Banco Exterior, 1°, Avda Urdaneta esq. Urapal a Río, Candelaria, Apdo 14278, Caracas 1011-A; tel. (212) 501-0441; fax (212) 575-3798; e-mail presidencia@bancoexterior.com; internet www.bancoexterior.com; f. 1958; cap. 45,360m., res 80,492.9m., dep. 533,973.6m. (Dec. 2002); Chair. FRANCISCO LÓPEZ HERRERA; Pres. VÍCTOR ALVAREZ; 72 brs.

**Banco Industrial de Venezuela, CA:** Torre Financiera BIV, Avda Las Delicias de Sabana Grande, cruce con Avda Francisco Solano López, Caracas 1010; tel. (212) 952-4051; fax (212) 952-6282; e-mail webmaster@biv.com.ve; internet www.biv.com.ve; f. 1937; 98% state-owned; cap. 100,000m. (Dec. 2001); Pres. LEONARDO GONZÁLEZ DELLÁN; 60 brs.

**Banco Mercantil, CA:** Edif. Mercantil, 35°, Avda Andrés Bello 1, San Bernardino, Apdo 789, Caracas 1010-A; tel. (212) 503-1111; fax (212) 503-1075; e-mail mercan24@bancomercantil.com; internet www.bancomercantil.com; f. 1925; cap. 134,172.0m., res 216,594.0m., dep. 4,703,246.0m. (Dec. 2003); Chair. and CEO Dr GUSTAVO A. MARTURET; 293 brs.

**Banco Nacional de Ahorro y Préstamo (BANAP):** Torre Banco Nacional de Ahorro y Préstamo, Avda Venezuela, El Rosal, Caracas 1060; tel. (212) 951-6222; fax (212) 951-1308; e-mail banap@banap.com; internet www.banap.com; f. 1966; Pres. FRANCISCO ZÚNIGA ROBLES; Gen. Man. GAMAL CHACÓN MOLINA.

**Banco Standard Chartered:** Edif. Banaven, Torre D, 5°, Avda la Estancia A, Chuao, Caracas 1060; tel. (212) 993-3293; fax (212) 993-3130; internet www.standardchartered.com/ve/; f. 1980 as Banco Exterior de los Andes y de España, current name adopted in 1998 following acquisition by Standard Chartered Bank (UK); representative office only; Chair. DAVID LORETTA.

**Banco de Venezuela (Grupo Santander):** Torre Banco de Venezuela, 16°, Avda Universidad, esq. Sociedad a Traposos, Apdo 6268, Caracas 1010-A; tel. (212) 501-2556; fax (212) 501-2546; internet www.bancodevenezuela.com; f. 1890; fmrly Banco de Venezuela CA, 93.38% share purchased by Banco Santander (Spain) in Dec. 1996; changed name to above in 1998; acquired Banco Caracas, CA in Dec. 2000; cap. 40,523.7m., res 319,642.9m., dep. 4,208,032.6m. (Dec. 2003); Pres. MICHEL J. GOGUIKIAN; 202 brs and agencies.

**Banesco:** Torre Banesco, Avda Guaicaipuro con Avda Principal de Las Mercedes, Caracas; tel. (212) 952-4972; fax (212) 952-7124; e-mail atcliente@banesco.com; internet www.banesco.com; Chair. JUAN CARLOS ESCOTET RODRÍGUEZ; Exec. Pres. LUIS XAVIER LUJÁN PUIGBÓ.

**BBVA Banco Provincial, SA:** Centro Financiero Provincial, 27°, Avda Vollmer con Avda Este O, San Bernadino, Apdo 1269, Caracas 1011; tel. (212) 504-5098; fax (212) 574-9408; e-mail calidad@provincial.com; internet www.provincial.com; f. 1952; 55.14% owned by Banco Bilbao Vizcaya Argentaria, 26.27% owned by Grupo Polar; cap. 91,945.4m., res 621,387.0m., dep. 4,586,027.8m. (Dec. 2003); Pres. HERNÁN ANZOLA GIMÉNEZ; Exec. Pres. JOSÉ CARLOS PLA ROYO.

**Corp Banca, CA Banco Universal:** Torre Corp Banca, Plaza la Castellana, Chacao, Caracas 1060; tel. (212) 206-3333; fax (212) 206-4950; e-mail calidad@corpbanca.com.ve; internet www.corpbanca.com.ve; f. 1969; fmrly Banco Consolidado, current name adopted in 1997; cap. 40,000m., res 36,842.2m., dep. 709,751.5m. (Dec. 2003); Chair. JORGE SELUME ZAROR; CEO MARIO CHAMORRO; 116 brs.

**Unibanca Banco Universal, CA:** Torre Grupo Unión, Avda Universidad, esq. El Chorro, Apdo 2044, Caracas; tel. (212) 501-7031; fax (212) 563-0986; internet www.unibanca.com.ve; f. 2001 by merger of Banco Unión (f. 1943) and Caja Familia; Pres. Dr IGNACIO SALVATIERRA; Vice-Pres. JOSÉ Q. SALVATIERRA; 174 brs.

**Venezolano de Crédito SA—Banco Universal:** Edif. Banco Venezolano de Crédito, Avda Alameda, San Bernadino, Caracas 1011; tel. (212) 806-6111; fax (212) 541-2757; e-mail jurbano@venezolano.com; internet www.venezolano.com; f. 1925, as Banco Venezolano de Crédito, SACA; name changed as above in 2001; cap. 42,000.0m., res 71,052.4m., dep. 974.008.3m. (Dec. 2003); Pres. Dr OSCAR GARCÍA MENDOZA; 59 brs.

### Barquisimeto, Lara

**Banco Capital, CA:** Carrera 17 Cruce con Calle 26, Frente a La Plaza Bolívar, Barquisimeto, Lara; tel. (251) 31-4979; fax (251) 31-1831; e-mail eximport@bancocapital.net; internet www.bancocapital.net; f. 1980; cap. 1,500.0m., res 1,854.7m., dep. 60,428.3m. (Dec. 1997); Chair. and Pres. JOSÉ REINALDO FURIATI; Gen. Man. VICENTE M. FURIATI; 13 brs.

### Ciudad Guayana, Bolívar

**Banco Caroní:** Edif. Multicentro Banco Caroní, Vía Venezuela, Puerto Ordaz, Estado Bolívar; tel. (286) 23-2230; fax (286) 22-0995; e-mail carupsis@telcel.net.ve; Pres. ARÍSTIDES MAZA TIRADO.

**Banco Guayana, CA:** Edif. Los Bancos, Avda Guayana con Calle Caura, Puerto Ordaz, Bolívar; f. 1955; state-owned; Pres. OSCAR EUSEBIO JIMÉNEZ AYESA.

### Coro, Falcón

**Banco Federal, CA:** Avda Manaure, cruce con Avda Ruiz Pineda, Coro, Falcón; tel. (268) 51-4011; e-mail masterbf@bancofederal.com;

# VENEZUELA

internet www.bancofederal.com; f. 1982; cap. 50m., dep. 357m. (Dec. 1986); Pres. NELSON MEZERHANE; Exec. Pres. ROGELIO TRUJILLO.

**Banco de Fomento Regional Coro, CA:** Avda Manaure, entre Calles Falcón y Zamora, Coro, Falcón; tel. (268) 51-4421; f. 1950; transferred to private ownership in 1994; Pres. ABRAHAM NAÍN SENIOR URBINA.

### Maracaibo, Zulia

**Banco Occidental de Descuento, SACA:** Avda 5 de Julio esq. Avda 17, Apdo 695, Maracaibo 4001-A, Zulia; tel. (261) 759-3011; fax (261) 750-2274; e-mail oficina.virtual@bodinternet.com; internet www.bodinternet.com; f. 1957; transferred to private ownership in 1991; Pres. VÍCTOR VARGAS IRAUSQUIN; Exec. Dir CÁNDIDO RODRÍGUEZ LOSADA; 17 brs.

### Nuevo Esparta

**Banco Confederado:** Edif. Centro Financiero Confederado, Blvd Gómez cruce con Calle Marcano, Porlamar, Estado Nuevo Esparta; tel. (95) 65-4230; fax (95) 63-7033; e-mail dbanel002@bancoconfederado.com; internet www.bancoconfederado.com; Pres. HASSAN SALEH SALEH.

### San Cristóbal, Táchira

**Banco de Fomento Regional Los Andes, CA:** Avda 8, La Concordia con Calle 4, San Cristóbal, Táchira; tel. (276) 43-1269; f. 1951; Pres. EDGAR A. HERNÁNDEZ; Exec. Vice-Pres. PEDRO ROA SÁNCHEZ.

### Mortgage and Credit Institutions, Caracas

**Banco Hipotecario de la Vivienda Popular, SA:** Intersección Avda Roosevelt y Avda Los Ilustres, frente a la Plaza Los Símbolos, Caracas; tel. (212) 62-9971; f. 1961; cap. 100m., res 68.2m., dep. 259.3m. (Dec. 1987); Pres. HELY MALARET M.; First Vice-Pres. ALFREDO ESQUIVAR.

**Banco Hipotecario Unido, SA:** Edif. Banco Hipotecario Unido, Avda Este 2, No 201, Los Caobos, Apdo 1896, Caracas 1010; tel. (212) 575-1111; fax (212) 571-1075; f. 1961; cap. 230m., res 143m., dep. 8,075m. (May 1990); Pres. ARTURO J. BRILLEMBOURG; Gen. Man. ALFONSO ESPINOSA M.

### Maracaibo, Zulia

**Banco Hipotecario del Zulia, CA:** Avda 2, El Milagro con Calle 84, Maracaibo, Zulia; tel. (261) 91-6055; f. 1963; cap. 120m., res 133.5m., dep. 671.5m. (Nov. 1986); Pres. ALBERTO LÓPEZ BRACHO.

### Development Banks

**Banco de Desarrollo Económico y Social de Venezuela (BANDES):** Torre Bandes, Avda Universidad, Traposos a Colón, Caracas 1010; tel. (212) 505-8010; fax (212) 505-8030; e-mail apublicos@bandes.gov.ve; internet www.bandes.gov.ve; state-owned development bank; Pres. NELSON MERENTES; Gen. Man. MARITZA BALZA.

**Banco del Pueblo Soberano CA:** Edif. El Gallo de Oro, Gradillas a San Jacinto Parroquia Catedral, Caracas; tel. (212) 505-2800; fax (212) 505-2995; e-mail abarrera@bancodelpueblo.com.ve; internet www.bancodelpueblo.com.ve; f. 1999; microfinance; Pres. HUMBERTO ORTEGA DÍAZ; Dir JORGE GONZÁLEZ.

**BANMUJER:** Edif. Sudameris, planta baja, Avda Urdaneta con Avda Fuerzas Armadas, esq. Plaza España, Caracas 1010; tel. (212) 564-3015; e-mail banmujer@banmujer.gov.ve; internet www.banmujer.gov.ve; f. 2001; state-owned bank offering loans to women; Pres. NORA CASTAÑEDA.

**Fondo de Desarrollo Microfinanciero (FONDEMI):** Edif. Sudameras, 2°, Avda Urdaneta, esq. Fuerzas Armadas, Caracas 1030; tel. (212) 564-4327; fax 564-0170; e-mail promocion@fondemi.gov.ve; internet www.fondemi.gov.ve; f. 2001; Micro-financing development fund; Pres. ISA MERCEDES SIERRA FLORES.

### Foreign Banks

**ABN AMRO Bank NV** (Netherlands): Edif. Centro Seguros Sud América, 1°, Avda Francisco de Miranda, El Rosal, Apdo 69179, Caracas 1060; tel. (212) 957-0300; fax (212) 953-5758.

**Banco do Brasil SA** (Brazil): Edif. Centro Lido, 9°, Of. 93A, Avda Francisco de Miranda, El Rosal, 1067-A Caracas; tel. (212) 952-2674; fax (212) 952-5251; e-mail caracas@bb.com.br; internet www.bb.com.br.

**Citibank NA** (USA): Edif. Citibank, esq. de Carmelitas a esq. de Altagracia, Apdo 1289, Caracas; tel. (212) 81-9501; fax (212) 81-6493; Pres. VICTOR J. MENEZES; 4 brs.

## Directory

### Banking Association

**Asociación Bancaria de Venezuela:** Torre Asociación Bancaria de Venezuela, Avda Venezuela, 1°, El Rosal, Caracas; tel. (212) 951-4711; fax (212) 951-3696; e-mail abvinfo@asobanca.com.ve; internet www.asobanca.com.ve; f. 1959; 49 mems; Pres. ARÍSTIDES MAZA TIRADO.

## STOCK EXCHANGE

**Bolsa de Valores de Caracas, CA:** Edif. Atrium, Nivel C-1, Calle Sorocaima entre Avdas Tamanaco y Venezuela, Urb. El Rosal, Apdo 62724-A, Caracas 1060-A; tel. (212) 905-5511; fax (212) 952-2640; e-mail bvc@caracasstock.com; internet www.caracasstock.com; f. 1947; 65 mems; Pres. NELSON ORTIZ.

## INSURANCE

### Supervisory Board

**Superintendencia de Seguros:** Edif. Torre del Desarrollo, PH, Avda Venezuela, El Rosal, Chacao, Caracas 1060; tel. (212) 905-1611; fax (212) 953-8615; e-mail sudeseg@sudeseg.gov.ve; internet www.sudeseg.gov.ve; Supt Dr LUCIANO OMAR ARIAS.

### Principal Insurance Companies

**Adriática, CA de Seguros:** Edif. Adriática de Seguros, Avda Andrés Bello, esq. de Salesianos, Caracas; tel. (212) 571-5702; fax (212) 571-0812; e-mail adriatica@adriatica.com.ve; internet www.adriatica.com.ve; f. 1952; Pres. FRANÇOIS THOMAZEAU; Exec. Vice-Pres. GHISLAIN FABRE.

**Avila, CA de Seguros:** Edif. Centro Seguros La Paz, 7°, Avda Francisco de Miranda, Caracas; tel. (212) 239-7911; fax (212) 238-2470; f. 1936; Pres. RAMÓN RODRÍGUEZ; Vice-Pres. JUAN LUIS CASAÑAS.

**Carabobo, CA de Seguros:** Edif. Centro Empresarial Sábana Grande, 12°, Ofs 1 y 2, Calle Negrín, entre Avda Francisco Solano y Blvd de Sábana Grande, El Recreo; tel. (212) 761-8514; fax (212) 761-5727; f. 1955; Pres. PAUL FRAYND; Gen. Man. ENRIQUE ABREU.

**La Occidental, CA de Seguros:** Edif. Seguros Occidental, Avda 4 (Bella Vista) esq. con Calle 71, No 10126, Maracaibo, Zulia; tel. (261) 798-4780; fax (261) 797-5422; e-mail relaciones@laoccidental.com; internet www.laoccidental.com; f. 1956; Pres. TOBÍAS CARRERO NÁCAR; Dir CARLOS MONÍZ ROCHA.

**La Oriental, CA de Seguros:** Torre Sede Gerencial La Castellana, Avda Francisco de Miranda, cruce con Avda Principal de la Castellana, Chacao, Caracas; tel. (212) 277-5000; fax (212) 263-1501; internet www.laoriental.com; f. 1975; Pres. GONZALO LAURÍA ALCALÁ.

**Seguros Los Andes, CA:** Edif. Seguros Los Andes, Avda Las Pilas, Santa Inés, San Cristóbal, Táchira; tel. (276) 340-2611; fax (276) 340-2596; Pres. RAMÓN RODRÍGUEZ.

**Seguros Caracas de Liberty Mutual, CAV:** Torre Seguros Caracas C-4, Centro Comercial El Parque, Avda Francisco de Miranda, Los Palos Grandes, Caracas; tel. (212) 209-9111; fax (212) 209-9556; f. 1943; Pres. ROBERTO SALAS.

**Seguros Catatumbo, CA:** Edif. Seguros Catatumbo, Avda 4 (Bella Vista), No 77–55, Apdo 1083, Maracaibo; tel. (261) 700-5555; fax (261) 216-0037; e-mail mercado@seguroscatatumbo.com; internet www.seguroscatatumbo.com; f. 1957; cap. 9,300m. (2003); Pres. ATENÁGORAS VERGEL RIVERA.

**Seguros Mapfre La Seguridad, CA:** Calle 3A, Frente a La Torre Express, La Urbina Sur, Apdo 473, Caracas 1010; tel. (212) 204-8000; fax (212) 204-8751; f. 1943; owned by Seguros Mapfre (Spain); Pres. ARISTÓBULO BAUSELA.

**Seguros Mercantil, CA:** Edif. Seguros Mercantil, Avda Libertador con calle Andrés Galarraga, Chacao, Caracas; tel. (212) 276-2000; fax (212) 276-2596; internet www.segurosmercantil.com; f. 1988; acquired Seguros Orinoco in 2002; Pres. ALBERTO BENSHIMOL; Gen. Man. RAFAEL CUBILLÁN.

**Seguros Nuevo Mundo, SA:** Edif. Seguros Nuevo Mundo, Avda Luis Roche con 3 transversal, Altamira, Apdo 2062, Caracas; tel. (212) 201-1111; fax (212) 263-1435; internet www.nmbc.com.ve; f. 1856; cap. 100m. (2003); Pres. RAFAEL PEÑA ALVAREZ; Exec. Vice-Pres. RAFAEL VALENTINO.

**Seguros La Previsora, CNA:** Torre La Previsora, Avda Abraham Lincoln, Sábana Grande, Caracas; tel. (212) 709-1555; fax (212) 709-1976; internet www.previsora.com; f. 1914; Pres. ALBERTO QUINTANA; Exec. Vice-Pres. JUAN CARLOS MALDONADO.

**Seguros Venezuela, CA:** Edif. Seguros Venezuela, 8° y 9°, Avda Francisco de Miranda, Urb. Campo Alegre, Caracas; tel. (212) 901-7111; fax (212) 901-7218; e-mail carmen.guillen@segurosvenezuela.com; internet www.segurosvenezuela.com; part of American International group; Exec. Pres. ENRIQUE BANCHIERI ORTIZ.

# VENEZUELA

**Universitas de Seguros, CA:** Edif. Impres Médico, 2°, Avda Tamanaco, El Rosal, Caracas; tel. (212) 951-6711; fax (212) 901-7506; e-mail tbarrera@universitasdeseguros.com; internet www.universitasdeseguros.com; cap. 6,500m.; Pres. ANA TERESA FERRINI.

### Insurance Association

**Cámara de Aseguradores de Venezuela:** Torre Taeca, 2°, Avda Guaicaipuro, El Rosal, Apdo 3460, Caracas 1010-A; tel. (212) 952-4411; fax (212) 951-3268; e-mail rrpp@camaraseg.org; internet www.camaraseg.org; f. 1951; 42 mems; Pres. PEDRO LUIS GARMENDIA; Exec. Pres. JUAN B. BLANCO-URIBE.

## Trade and Industry

### GOVERNMENT AGENCIES

**Comisión de Administración de Divisas (CADIVI):** Antiguo Edif. PDVSA Servicios, 6°, Avda Leonardo Da Vinci, Los Chaguaramos, Caracas; tel. (212) 606-3904; fax (212) 606-3026; e-mail info@cadivi.gov.ve; internet www.cadivi.gov.ve; f. 2003; regulates access to foreign currency; Pres. MARÍA ESPINOZA DE ROBLES.

**Corporación Venezolana de Guayana (CVG):** Edif. General, 2°, Avda La Estancia, Apdo 7000, Chuao, Caracas; tel. (212) 992-9764; fax (212) 993-0554; e-mail presidenciaccs@cvg.com; internet www.cvg.com; f. 1960 to organize development of Guayana area, particularly its metal ore and hydroelectric resources; Pres. VÍCTOR ALVAREZ RODRÍGUEZ (Minister of Basic Industry and Mining).

**Dirección General Sectorial de Hidrocarburos:** Avda Lecuna, Torre Oeste, 12°, Parque Central, Caracas 1010; tel. (212) 507-5212; division of Ministry of Energy and Mines responsible for determining and implementing national policy for the exploration and exploitation of petroleum reserves and for the marketing of petroleum and its products; Vice-Minister LUIS VIERMA.

**Dirección General Sectorial de Minas y Geología:** Avda Lecuna, Torre Oeste, 4°, Parque Central, Caracas; tel. (212) 708-7108; fax (212) 575-2497; division of Ministry of Energy and Mines responsible for formulating and implementing national policy on non-petroleum mineral reserves; Vice-Minister ORLANDO ORTEGANO.

**Fondo Intergubermental para la Decentralización (FIDES):** Avda Las Acacias, cruce con Avda Casanova, Torre Banhorient, Sábana Grande, Caracas; tel. (212) 708-0000; fax (212) 708-3642; e-mail info@fides.gov.ve; internet www.fides.gov.ve; f. 1993; part of the Ministry of Planning and Development; co-ordinates investment; Pres. RICHARD CANÁN.

**Instituto Nacional de Nutrición (INN):** Edif. INN, Avda Baralt, esq. El Carmen. Qta, Crespo, Caracas; tel. (212) 483-5142; Dir RHAITZA MENDOZA.

**Instituto Nacional de Tierras (INTI):** Avda Lecuna, Torre Oeste, 37°, Parque Central, Caracas; tel. (212) 574-8554; fax (212) 576-2201; internet www.inti.gov.ve; f. 1945, as Instituto Agrario Nacional (IAN); present name adopted in 2001; under Agrarian Law to assure ownership of the land to those who worked on it; now authorized to expropriate and redistribute idle or unproductive lands; Pres. RICAURTE LEONETT.

**Instituto Nacional de la Vivienda:** Torre Inavi, Avda Francisco de Miranda entre Guaicaipuro y San Ignacio de Loyola, Chacao, Caracas; tel. (212) 206-9279; e-mail sugerencias@inavi.gov.ve; internet www.inavi.gov.ve; f. 1975; administers government housing projects; Pres. JESÚS HERNÁNDEZ GONZÁLEZ.

**Mercal, CA:** Edif. Torres Seguros Orinoco, Avda Fuerzas Armadas, esq. Socarras, Caracas; tel. (212) 564-3856; e-mail gestioncomunicacional@mercal.gov.ve; internet www.mercal.gov.ve; responsible for marketing agricultural products; frmly Corporación de Mercadeo Agrícola; Pres. MARÍA MILAGROS TORO LANDAETA.

**Palmaven:** Edif. PDVSA La Floresta, Torre Palmaven, Avda Principal, Urb. La Floresta, Caracas; tel. (212) 208-0309; fax (212) 208-0212; promotes agricultural development, provides agricultural and environmental services and technical assistance to farmers; Man. Dir EDDIE RAMÍREZ.

**Superintendencia de Inversiones Extranjeras (SIEX):** Edif. La Perla, Bolsa a Mercaderes, 3°, Apdo 213, Caracas 1010; tel. (212) 483-6666; fax (212) 484-4368; e-mail siexdespacho@cantv.net; internet www.siex.gov.ve; f. 1974; supervises foreign investment in Venezuela; Supt MIRIAM BEATRIZ AGUILERA DE BLANCO.

### DEVELOPMENT ORGANIZATIONS

**Corporación de Desarrollo de la Pequeña y Mediana Industria (Corpoindustria):** Aragua; tel. (243) 23459; internet www.sain.org.ve/corpoind/cedinco.htm; promotes the development of small and medium-sized industries; Pres. Dr CARLOS GONZÁLEZ-LÓPEZ.

*Directory*

**CVG Bauxita Venezolana (Bauxivén):** Caracas; f. 1978 to develop the bauxite deposits at Los Pijiguaos; financed by the FIV and the CVG which has a majority holding; Pres. JOSÉ TOMÁS MILANO.

**Fondo de Desarrollo Agropecuario, Pesquero, Forestal y Afines (FONDAFA):** Edif. FONDAFA, esq. Salvador de León a Socarras, La Hoyada, Caracas; tel. (212) 542-3570; fax (212) 542-5887; e-mail fondafa@fondafa.gov.ve; internet www.fondafa.gov.ve; f. 1974; devt of agriculture, fishing and forestry; Pres. ALIRIO RONDÓN.

**Fondo de Desarrollo Microfinanciero (FONDEMI):** Edif. Sudameras, 2°, Avda Urdaneta, esq. Fuerzas Armadas, Caracas 1030; tel. (212) 564-4327; fax 564-0170; e-mail promocion@fondemi.gov.ve; internet www.fondemi.gov.ve; f. 2001; Micro-financing development fund; Pres. ISA MERCEDES SIERRA FLORES.

### CHAMBERS OF COMMERCE AND INDUSTRY

**Federación Venezolana de Cámaras y Asociaciones de Comercio y Producción (Fedecámaras):** Edif. Fedecámaras, Avda El Empalme, Urb. El Bosque, Apdo 2568, Caracas; tel. (212) 731-1711; fax (212) 730-2097; e-mail direje@fedecameras.org.ve; internet www.fedecamaras.org.ve; f. 1944; 307 mems; Pres. ALBIS MUÑOZ; Exec. Dir MARIO TEPEDINO RAVEN.

**Cámara de Comercio de Caracas:** Edif. Cámara de Comercio de Caracas, Avda Andrés Eloy Blanco 215, 8°, Los Caobos, Caracas; tel. (212) 571-3222; fax (212) 571-0050; e-mail comercio@ccc.com.ve; internet www.ccc.com.ve; f. 1893; 650 mems; Exec. Dir VLADIMIR CHELMINSKI.

**Cámara de Industriales de Caracas:** Edif. Cámara de Industriales, 3°, Avda Las Industrias esq. Pte Anauco, La Candelaria, Apdo 14255, Caracas 1011; tel. (212) 571-4224; fax (212) 571-2009; e-mail ciccs@telcel.net.ve; internet www.cic.org.ve; f. 1939; Pres. ROBERTO J. BALL; 550 mems.

**Cámara Venezolano-Americana de Industria y Comercio (Venamcham):** Torre Credival, 10°, Of. A, 2a Avda Campo Alegre, Apdo 5181, Caracas 1010-A; tel. (212) 263-0833; fax (212) 263-2060; e-mail venam@venamcham.org; internet www.venamcham.org; f. 1950; Pres. IMELDA CISNEROS; Gen. Man. ANTONIO HERRERA VAILLANT.

There are chambers of commerce and industry in all major provincial centres.

### EMPLOYERS' ORGANIZATIONS

#### Caracas

**Asociación Nacional de Comerciantes e Industriales:** Plaza Panteón Norte 1, Apdo 33, Caracas; f. 1936; traders and industrialists; Pres. Dr HORACIO GUILLERMO VILLALOBOS; Sec. R. H. OJEDA MAZZARELLI; 500 mems.

**Asociación Nacional de Industriales Metalúrgicos y de Minería de Venezuela:** Edif. Cámara de Industriales, 9°, Puente Anauco a Puente República, Apdo 14139, Caracas; metallurgy and mining; Pres. JOSÉ LUIS GÓMEZ; Exec. Dir LUIS CÓRDOVA BRITO.

**Asociación Textil Venezolana:** Edif. Textilera Gran Colombia, Calle el Club 8, Los Cortijos de Lourdes, Caracas; tel. (212) 238-1744; fax (212) 239-4089; f. 1957; textiles; Pres. DAVID FIHMAN; 68 mems.

**Asociación Venezolana de Exportadores (AVEX):** Centro Comercial Coneresa, Redoma de Prados del Este 435, 2°, Prados del Este, Caracas; tel. (212) 979-5042; fax (212) 979-4542; e-mail directorejecutivo@avex.com.ve; internet www.avex.com.ve; Pres. FRANCISCO MENDOZA.

**Cámara Petrolera:** Torre Domus, 3°, Of. 3-A, Avda Abraham Lincoln con Calle Olimpo, Sábana Grande, Caracas; tel. (212) 794-1222; fax (212) 793-8529; e-mail informacion@camarapetrolera.org; internet www.camarapetrolera.org; f. 1978; asscn of petroleum-sector cos; Pres. ANTONIO VINCENTELLI.

**Confederación Venezolana de Industriales (CONINDUSTRIA):** Edif. CIEMI, Avda Principal de Chuao, Caracas 1061; tel. (212) 991-2116; fax (212) 991-7737; e-mail comunicaciones@conindustria.org; internet www.conindustria.org; asscn of industrialists; Pres. EDUARDO GÓMEZ SIGALA; Exec. Pres. JUAN FRANCISCO MEJÍA B.

**Confederación Nacional de Asociaciones de Productores Agropecuarios (FEDEAGRO):** Edif. Casa de Italia, planta baja, Avda La Industria, San Bernardino, Caracas 1010; tel. (212) 571-4035; fax (212) 573-4423; e-mail fedeagro@fedeagro.org; internet www.fedeagro.org; f. 1960; agricultural producers; 133 affiliated asscns; Pres. GUSTAVO MORENO LLERAS.

**Federación Nacional de Ganaderos de Venezuela (FEDENAGA):** Avda Urdaneta, Centro Financiero Latino P-18, Ofc. 18-2 y 18-4, La Candelaria, Caracas; tel. (212) 563-2153; fax (212) 564-7273; e-mail fedenagat@cantv.net; cattle owners; Pres. JOSÉ LUIS BETANCOURT.

# VENEZUELA

**Unión Patronal Venezolana del Comercio:** Edif. General Urdaneta, 2°, Marrón a Pelota, Apdo 6578, Caracas; tel. (582) 561-7025; fax (582) 561-4321; trade; Sec. H. Espinoza Banders.

## Other Towns

**Asociación de Comerciantes e Industriales del Zulia (ACIZ):** Edif. Los Cerros, 3°, Calle 77 con Avda 3c, Apdo 91, Maracaibo, Zulia; tel. (261) 91-7174; fax (261) 91-2570; f. 1941; traders and industrialists; Pres. Jorge Avila.

**Asociación Nacional de Cultivadores de Algodón (ANCA)** (National Cotton Growers' Association): Edif. Portuguesa, Avda Los Pioneros, Sector Aspiga-Acarigua; tel. (255) 621-5111; fax (255) 621-4368; Sec. Concepción Quijada G.

**Asociación Nacional de Empresarios y Trabajadores de la Pesca:** Cumaná; fishermen.

**Unión Nacional de Cultivadores de Tabaco:** Urb. Industrial La Hamaca, Avda Hustaf Dalen, Maracay; tobacco growers.

## STATE HYDROCARBONS COMPANIES

**Corporación Petroquímica de Venezuela (CPV):** Torre Pequiven, Avda Francisco de Miranda, Chacao, Apdo 2066, Caracas 1010-A; tel. (212) 201-4011; fax (212) 201-3189; e-mail webmaster@pdvsa.com; internet www.pequiven.com; f. 1956 as Instituto Venezolano de Petroquímica; became Pequiven in 1977; name changed as above in 2005 following independence from PDVSA; involved in many joint ventures with foreign and private Venezuelan interests for expanding petrochemical industry; active in regional economic integration; an affiliate of PDVSA from 1978 until 2005; Pres. Saúl Ameliach.

**Petróleos de Venezuela, SA (PDVSA):** Edif. Petróleos de Venezuela, Torre Este, Avda Libertador, La Campiña, Apdo 169, Caracas 1010-A; tel. (212) 708-4743; fax (212) 708-4661; e-mail saladeprensa@pdvsa.com; internet www.pdvsa.com; f. 1975; holding company for national petroleum industry; responsible for petrochemical sector since 1978 and for development of coal resources in western Venezuela since 1985; Pres. Rafael Ramírez Carreño; Vice-Pres. of Exploration and Production Luis Vierma; Vice-Pres. of Refining Alejandro Granado; The following are subsidiaries of PDVSA:

**Bariven, SA:** Edif. PDVSA Los Chaguaramos, Avda Leonardo Da Vinci, 6°, Urb. Los Chaguaramos, Apdo 1889, Caracas 1010-A; tel. (212) 606-4060; fax (212) 606-2741; handles the petroleum, petrochemical and hydrocarbons industries' overseas purchases of equipment and materials.

**Bitúmenes Orinoco, SA (BITOR):** Edif. PDVSA Exploración, Producción y Mejoramiento, 9°, Avda Ernesto Blohm, La Estancia, Chuao, Apdo 3470, Caracas 1010-A; tel. (212) 908-2811; fax (212) 908-3982; e-mail abreuew@pdvsa.com; plans, develops and markets the bitumen resources of the Orinoco belt; produces Venezuela's trademark boiler fuel Orimulsion.

**Carbozulia, SA:** Edif. PDVSA Exploración, Producción y Mejoramiento, Avda 5 de Julio, esq. Avda 11, Apdo 1200, Maracaibo 4001; tel. (261) 806-8600; fax (261) 806-8790; f. 1978; responsible for the commercial exploitation of the Guasare coalfields in Zulia.

**Deltaven:** Edif. PDVSA Deltaven, Avda Principal de La Floresta, La Floresta, Caracas 1060; tel. (212) 208-1111; markets PDVSA products and services within Venezuela.

**Interven, SA:** Edif. Petróleos de Venezuela, Torre Oeste, 11°, Avda Libertador, La Campiña, Apdo 2103, Caracas 1010-A; tel. (212) 708-1111; fax (212) 706-6621; f. 1986; to manage PDVSA's joint ventures overseas.

**Intevep:** Edif. Sede Central, Urb. Santa Rosa, Sector El Tambor, Los Teques, Apdo 76343, Caracas 1070-A; tel. (212) 908-6111; fax (212) 908-6447; e-mail webmaster@intevep.com; internet www.intevep.pdv.com; f. 1979; research and development branch of PDVSA; undertakes applied research and development in new products and processes and provides specialized technical services for the petroleum and petrochemical industries.

**PDV Marina:** Edif. Petróleos de Venezuela Refinación, Suminstro y Comercio, Torre Oeste, 9°, Avda Libertador, La Campiña, Apdo 2103, Caracas 1010-A; tel. (212) 708-1111; fax (212) 708-2200; Pres. Fernando Camejo Arenas.

**PDV Servicios:** Edif. PDV Servicios, Avda La Estancia, Chuao, Caracas; tel. (212) 606-3637; previously known as Corpoven; present name adopted in 1998; extraction and exploitation of natural gas.

**PDVSA Gas:** Edif. Sucre, Avda Francisco de Miranda, La Floresta, Caracas; tel. (212) 208-6212; fax (212) 208-6288; e-mail messina@pdvsa.com; Pres. Nelson Martínez; Dir-Gen. Juan José García.

**Productos Especiales CA (Proesca):** Torre Pequiven, 13°, Avda Francisco de Miranda, Chacao, Apdo 2066, Caracas 1060-A; tel. (212) 201-3152; fax (212) 201-3008.

## MAJOR COMPANIES

The following is a selection of major industrial and commercial enterprises, in terms of sales and employment, operating in Venezuela:

### Metals and Mining

**Conduven, CA:** Edif. Torre Financiera, 9°, Avda Beethoven, Urb. Colinas de Bello Monte, Caracas 1060-A; tel. (212) 752-4111; fax (212) 751-1542; f. 1959; manufacturers of welded pipe for use in petroleum industry, fluid conduction, electrical installations; Pres. Dezider Weisz W.; 1,200 employees.

**CVG Aluminio del Caroní, SA (ALCASA):** Avda Fuerzas Armadas, Zona Industrial Matanzas, Apdo 115, Ciudad Guayana, Bolívar; tel. (286) 980-1567; fax (286) 980-1891; internet www.alcasa.com.ve; state-owned manufacturer of aluminium products; Pres. Carlos Lanz Rodríguez; 1,700 employees.

**CVG Bauxilum, CA (Industria Integrado de Aluminio):** Avda Fuerzas Armadas, Zona Industrial Matanzas, Ciudad Guayana, Bolívar; tel. (286) 950-6271; fax (286) 950-6270; internet www.bauxilum.com; f. 1994; state-owned manufacturer of aluminium products; Pres. Dr Jesús Alberto Imery Buiza.

**CVG Ferrominera Orinoco, CA:** Apdo 399, Vía Caracas, Edif. Administrativo I, Puerto Ordaz, Bolívar 8015; tel. (286) 930-3111; fax (286) 930-3775; e-mail contacto@ferrominera.com; internet www.ferrominera.com; f. 1975; subsidiary of the state-owned Corporación Venezolana de Guayana; see section on Trade and Industry—Government Agencies; iron ore mining; Pres. César Bertani; 4,100 employees.

**Siderúrgica del Orinoco, CA (SIDOR):** Edif. General de Seguros, 7°, Avda La Estancia, Chuao, Caracas; tel. (212) 902-3700; fax (212) 993-9906; internet www.sidor.com; f. 1964; formerly state-owned, privatized in 1997; steel processing; CEO Martín Berardi; Exec. Pres. Daniel Novegil; 11,406 employees.

**Siderúrgica del Turbio, SA (SIDETUR):** Edif. Torre América, 11°, Avda Venezuela, Urb. Bello Monte, Caracas 1060; tel. (212) 407-0300; fax (212) 407-0372; internet www.sidetur.com.ve; f. 1972; manufacturers of steel products including galvanized wire and steel rods; owned by SIVENSA (see below); Gen. Man. Nicolás Izquierdo; 600 employees.

**Siderúrgica Venezolana SACA (SIVENSA):** Edif. Torre América, 12°, Avda Venezuela, Urb. Bello Monte, Caracas 1060; tel. (212) 707-6280; fax (212) 707-6352; e-mail antonio.osorio@sivensa.com; internet www.sivensa.com.ve; f. 1948; manufacturers of briquetted iron, steel products, wire and wire products; Corp. Man. Antonio Osorio; 2,560 employees.

**Unión Industrial Venezolana, SA (UNIVENSA):** Carrera 3, No 30–30 Zona Industrial Comdibar 1, Barquisimeto, Lara; tel. (251) 454-033; fax (251) 451-860; f. 1965; manufacturers of stainless steel tubing; Pres. Miguel González; 400 employees.

### Rubber and Tobacco

**Bridgestone Firestone Venezolana, CA:** Carrera Valencia-Los Guayos, cruce con San Diego, Zona Industrial, Valencia, Carabobo; tel. (241) 874-7758; fax (241) 832-8254; internet www.bfvz.com; f. 1954; subsidiary of Bridgestone Corpn (USA); rubber tyre producers; Pres. Rosendo Torrades; 1,111 employees.

**Tabacalera Nacional, CA:** Edif. Seguros Venezuela, 2°, 3° y 4°, Avda Francisco de Miranda, Campo Alegre, Caracas 1060; tel. (212) 901-7824; fax (212) 901-7766; f. 1953; subsidiary of Philip Morris Int. (USA); cigarette manufacturers; Pres. José Antonio Cordibo-Freytés; 1,200 employees.

### Food and Drink

**Cervecería Polar:** Centro Empresarial Polar, 4a Transversal, Urb. Los Cortijos de Lourdes, Caracas; tel. (212) 202-3111; fax (212) 203-3037; e-mail webmaster@empresas-polar.com; internet www.empresas-polar.com; brewing and foods co; Pres. Gustavo Jiménez.

**CA Cervecería Regional:** Avda 17 (Los Haticos) 112-13, Apdo 255, Maracaibo; tel. (261) 65-1411; fax (261) 65-1477; brewing co; Pres. José R. Odon; 135 employees.

**Mavesa, SA:** Edif. Mavesa, Avda Principal, Urb. Industrial Los Cortijos de Lourdes, Caracas 1071; tel. (212) 238-1633; fax (212) 239-2506; f. 1949; acquired by Empresas Polar in 2001; manufacturers and distributors of consumer processed food and cleaning products; Chair. Jonathan Coles; 2,827 employees.

**Parmalat Venezuela (INDULAC):** Edif. Parmalat, Avda San Francisco y Palmarito, Apdo 1546, Urb. Colinas de la California,

# VENEZUELA

Caracas; tel. (212) 257-1422; fax (212) 257-7195; internet jperezor@parmalat.com.ve; internet www.parmalat.com.ve; f. 1966; manufacturers and distributors of dairy products; owned by Parmalat of Italy; 2,198 employees.

### Textiles

**Rori Internacional, SA:** Edif. Rori, Avda Principal, Urb. Los Cortijos de Lourdes, 4a Transversal, Caracas; tel. (212) 239-4037; fax (212) 239-3480; f. 1964; manufacturers of men's clothing; Man. ROBERTO RIMERIS; 1,200 employees.

**Sudamtex de Venezuela, CA:** Edif. Karam, 1° y 2°, Ibarras a Pelota, Avda Urdaneta, Caracas 1010; tel. (212) 562-9222; fax (212) 564-1987; e-mail slugo@sudamtex.com; f. 1946; textile manufacturers; Pres. ALEXANDER FURTH; 2,210 employees.

**Telares Los Andes, SA:** Cuji a Punceres 7, Caracas 1010; tel. (212) 574-9522; fax (212) 576-2025; e-mail info@telareslosandes.com; internet www.telareslosandes.com; f. 1944; manufacturers of synthetic fabrics; Pres. JAMES VICTOR LEVY; 1,800 employees.

### Chemicals

**Clariant (Venezuela), SA:** Zona Industrial San Vicente, Avda Anton Philips, Apdo 34, Maracay 2101; tel. (212) 550-3111; fax (212) 550-3127; internet www.clariant.com; f. 1952; subsidiary of Clariant Int. of Switzerland; manufacturers and distributors of chemicals, textiles, leather, paper, paint, adhesives, plastics; Gen. Man. DARIO GAETA; 207 employees.

**Corimón, CA:** Edif. Corimón, Urb. Los Cortijos de Lourdes, Calle Hans Neumann, Caracas; tel. (212) 400-5530; e-mail info@corimon.com; internet www.corimon.com; f. 1949; holding co. for subsidiaries producing paint, packaging and processed food; Pres. and CEO CARLOS GILL; 2,655 employees.

**Du Pont de Venezuela, SA:** Avda Eugencio Mendoza, Zona Industrial Carabobo, Valencia; tel. (41) 407-200; fax (41) 333-425; e-mail houtmajl@rennerdupont.com; internet www.dupont.com; f. 1956; manufacturers of industrial chemicals, plastics, pesticides, resins and films; Chair. and CEO CHARLES HOLLIDAY, Jr; Pres. for Latin America EDUARDO W. WANICK; Gen. Man. GERRARD BARRETO; 260 employees.

**Pfizer, SA:** Avda Principal La Castellana, Torre Banco Lara, 12°, Chacao, Caracas; tel. (212) 201-2611; fax (212) 201-2776; f. 1953; subsidiary of Pfizer Inc. (USA); manufacturers of pharmaceutical products; Pres. JOSÉ CLAVIER; 250 employees.

**Plastiflex, CA:** Torre Phelps, 4°, Avda La Salle, Plaza Venezuela, Caracas 1050; tel. (212) 793-3133; fax (212) 793-3636; f. 1957; manufacturers of plastic sheeting and film; Man. JONA MISHAAN; 415 employees.

**Procter and Gamble de Venezuela, SA:** Edif. P&G, Sorokaima, Trinidad, Caracas 1080; tel. (212) 919-777; fax (212) 206-6364; internet www.pg.com; subsidiary of Procter and Gamble Co (USA); manufacturers of soaps, detergents and pharmaceuticals; Man. JOSÉ R. RIVAS; 380 employees.

### Electrical Equipment

**Asea Brown Boveri, SA:** Edif. ABB, Avda Diego Cisneros, Los Ruices, Caracas; tel. (212) 203-1920; fax (212) 237-9164; f. 1957; manufacturers of power transmission machinery and equipment; Pres. ARMANDO BASAVE; 300 employees.

**General Electric de Venezuela, SA:** Edif. Centro Banaven, Torre A, 6°, Avda La Estancia, Caracas; tel. (212) 902-5122; fax (212) 902-5300; internet www.ge.com/venezuela; f. 1927; subsidiary of General Electric Corpn (USA); manufacturers of television sets, radio receivers and household electrical appliances; Man. JOSÉ SERRA; 2,600 employees.

### Miscellaneous

**Cerámica Carabobo, SA:** Avda Venezuela, El Rosal, Torre Asociación Bancaria Mezanina, Caracas; tel. (241) 825-6166; fax (241) 825-6731; e-mail infocc@ceramica-carabobo.com; internet www.ceramica-carabobo.com; f. 1956; manufacturers of ceramic floor and wall tiles; also owns Cerámica Industrial del Caribe and Pan-American Ceramics; Pres. ALBERTO SOSA SCHLAGETER; 3,000 employees.

**Ford Motors de Venezuela, SA:** Avda Henry Ford, Zona Industrial Sur, Valencia, Carabobo 1041; tel. (241) 406-111; fax (241) 406-483; internet www.ford.com.ve; f. 1962; subsidiary of Ford Motor Co (USA); assembly and production of motor vehicles, trucks and farm machinery; Pres. JOSÉ BISOGNO; 1,500 employees.

**Nardi de Venezuela, CA:** Zona Industrial Comdivar, Avda 3 con Calle 30, Barquisimeto, Lara; tel. (251) 454-460; fax (251) 454-942; f. 1971; manufacturers of agricultural sub-soilers, skimmer scoops and trench diggers; Man. GIANCARLO VITALI; 155 employees.

## UTILITIES

### Electricity

**La Electricidad de Caracas (EDC):** Edif. La Electricidad de Caracas, Avda Vollmer, San Bernadino, Caracas; tel. (212) 502-2111; e-mail info@edc-ven.com; internet www.edc-ven.com; supplies electricity to capital; owned by AES Corpn; Pres. ANDRÉS GLUSKI.

**Electrificación del Caroní, CA (Edelca):** Edif. General, planta baja, Avda La Estancia, Chuao, Caracas; tel. (212) 950-2111; fax (212) 950-2808; e-mail wriera@edelca.com.ve; internet www.edelca.com.ve; affiliate of Corporación Venezolana de Guayana; supplies some 70% of the country's electricity; Pres. DANIEL MACHADO GÓMEZ.

**Sistema Eléctrico de Nueva Esparta (Seneca):** electricity co; privatized in Oct. 1998.

### Water

**Compañía Anónima Hidrológica de la Región Capital (Hidrocapital):** Edif. Hidroven, Avda Augusto César Sandino con 9a Transversal, Maripérez, Caracas; tel. (212) 793-1638; fax (212) 793-6794; e-mail 73070.2174@compuserve.com; internet www.hidrocapital.com.ve; f. 1992; owned by Hidroven; operates water supply in Federal District and states of Miranda and Vargas; Pres. ALEJANDRO HITCHER.

**Hidroven:** Edif. Hidroven, Avda Augusto César Sandino con 9a Transversal, Maripérez, Caracas; tel. (212) 781-4778; fax (212) 781-6424; e-mail hvenpres@cantv.net; internet www.hidroven.gov.ve; national water co.; owns Hidroandes, Hidrocapital, Hidrocaribe, Hidrofalcon, Hidrolago, Hidrollanos, Hidropaez, Hidrosuroeste, Aguas de Monagas, Aguas de Ejido, Hidrolara, Aguas de Anaco, Aguas de Cojedes, Aguas de Mérida, Aguas de Apure, Aguas de Yaracuy, Aguas de Portuguesa; Pres. CRISTÓBAL FRANCISCO ORTIZ; Vice-Pres. FRANCISCO DURÁN.

## TRADE UNIONS

About one-quarter of the labour force in Venezuela belongs to unions, more than one-half of which are legally recognized. Venezuela's union movement is strongest in the public sector.

**Confederación de Trabajadores de Venezuela (CTV)** (Confederation of Venezuelan Workers): Edif. José Vargas, 17°, Avda Este 2, Los Caobos, Caracas; tel. (212) 575-0005; e-mail mcova@ctv.org.ve; internet www.ctv.org.ve; f. 1936; largest trade union confederation; principally active in public sector; Chávez Govt disputes legitimacy of election of CTV leadership; Pres. CARLOS ALFONSO ORTEGA CARVAJAL; Sec.-Gen. MANUEL JOSÉ COVA FERMÍN; 1,000,000 mems from 24 regional and 16 industrial feds.

**Fedepetrol:** union of petroleum workers; Pres. RAFAEL ROSALES.

**Federación Campesina (FC):** peasant union; CTV affiliate; Leader RUBÉN LANZ.

**Fetrametal:** union of metal workers; CTV affiliate; Leader JOSÉ MOLLEGAS.

**Fuerza Bolivariana de Trabajadores (FBT):** f. 2000; pro-Govt union.

**Movimiento Nacional de Trabajadores para la Liberación (MONTRAL):** Edif. Don Miguel, 6°, esq. Cipreses, Caracas; f. 1974; affiliated to CLAT and WFTU; Pres. LAUREANO ORTIZ BRAEAMONTE; Sec.-Gen. DAGOBERTO GONZÁLEZ; co-ordinating body for the following trade unions:

**Central Nacional Campesina (CNC):** Pres. REINALDO VÁSQUEZ.

**Cooperativa Nacional de Trabajadores de Servicios Múltiples (CNTSM).**

**Federación Nacional de Sindicatos Autónomos de Trabajadores de la Educación de Venezuela (FENASATREV):** Pres. LUIS EFRAÍN ORTA.

**Federación de los Trabajadores de Hidrocarburos de Venezuela (FETRAHIDROCARBUROS).**

**Frente de Trabajadores Copeyanos (FTC):** Sec.-Gen. DAGOBERTO GONZÁLEZ.

**Movimiento Agrario Social-Cristiano (MASC):** Sec.-Gen. GUSTAVO MENDOZA.

**Movimiento Magisterial Social-Cristiano (MMSC):** Sec.-Gen. FELIPE MONTILLA.

**Movimiento Nacional de Trabajadores de Comunicaciones (MONTRAC).**

**Movimiento Nacional de Trabajadores Estatales de Venezuela (MONTREV).**

# VENEZUELA

## Transport

### RAILWAYS

In 1999 work began on lines linking Acarigua and Turén (45 km) and Morón and Riecito (100 km). Construction of the first section (Caracas–Cúa) of a line linking the capital to the existing network at Puerto Cabello (219 km in total) was also begun. Services on the underground system began in 1983 on a two-line system: east to west from Palos Verdes to Propatria; north to south from Capitolio/El Silencio to Zoológico. A southern extension from Plaza Venezuela to El Valle opened in 1995. Further extensions to the lines were under way in 2005.

**CVG Ferrominera Orinoco, CA:** Vía Caracas, Puerto Ordaz, Apdo 399, Bolívar; tel. (286) 30-3451; fax (286) 30-3333; e-mail 104721.2354@compuserve.com; f. 1976; state company; operates two lines San Isidro mine–Puerto Ordaz (316 km) and El Pao–Palua (55 km) for transporting iron ore; Pres. Ing. LEOPOLDO SUCRE FIGARELLA; Man. M. ARO G.

**Ferrocarril de CVG Bauxilum—Operadora de Bauxita:** Caracas; tel. (212) 40-1716; fax (212) 40-1707; f. 1989; state company; operates line linking Los Pijiguaos with river Orinoco port of Gumilla (52 km) for transporting bauxite; Pres. P. MORALES.

**Instituto Autónomo de Ferrocarriles del Estado (IAFE):** Edif. Torre Británica de Seguros, 7° y 8°, Avda José Féliz Sosa, Urb. Altamira, Chacao, Caracas 1062-A; tel. (212) 201-8911; e-mail relacp@cantv.net; state co; 336 km; Pres. RAFAEL ALVAREZ; Vice-Pres. INOVA CASTRO PÉREZ.

**CA Metro de Caracas:** Multicentro Empresarial del Este, Edif. Miranda, Torre B, 7°, Avda Francisco de Miranda, Apdo 61036, Caracas; tel. (212) 206-7111; fax (212) 266-3346; internet www.metrodecaracas.com.ve; f. 1976 to supervise the construction and use of the underground railway system; state-owned; Pres. DANIEL DAVIS.

### ROADS

In 2004 there were an estimated 96,200 km of roads, of which 32,300 km were paved.

Of the three great highways, the first (960 km) runs from Caracas to Ciudad Bolívar. The second, the Pan-American Highway (1,290 km), runs from Caracas to the Colombian frontier and is continued as far as Cúcuta. A branch runs from Valencia to Puerto Cabello. The third highway runs southwards from Coro to La Ceiba, on Lake Maracaibo.

A new 'marginal highway' was under construction along the western fringe of the Amazon Basin in Venezuela, Colombia, Ecuador, Peru, Bolivia and Paraguay. The Venezuelan section now runs for over 440 km and is fully paved.

### INLAND WATERWAYS

**Instituto Nacional de Canalizaciones:** Edif. INC, Calle Caracas, al lado de la Torre Diamen, Chuao, Caracas; tel. (212) 908-5106; fax (212) 959-6906; internet www.incanal.gov.ve; f. 1952; semi-autonomous institution connected with the Ministry of Infrastructure; Pres. Cmmdr WOLFGANG LÓPEZ CARRASQUEL; Vice-Pres. LUZKARIM CORNETT PABÓN.

**Compañía Anónima La Translacustre:** Maracaibo; freight and passenger service serving Lake Maracaibo, principally from Maracaibo to the road terminal from Caracas at Palmarejo.

### SHIPPING

There are 13 major ports, 34 petroleum and mineral ports and five fishing ports. Formerly the main port for imports, La Guaira, the port for Caracas, was affected by mudslides caused by heavy rains in November 1999. Venezuela's main port is now Puerto Cabello, which handles raw materials for the industrial region around Valencia. Maracaibo is the chief port for the petroleum industry. Puerto Ordaz, on the Orinoco River, was also developed to deal with the shipments of iron from Cerro Bolívar.

**Instituto Nacional de Puertos:** Caracas; tel. (212) 92-2811; f. 1976 as the sole port authority; Pres. Vice-Adm. FREDDY J. MOTA CARPIO.

**Consolidada de Ferrys, CA (CONFERRY):** Torre Banhorient, 3°, Avda Las Acacias y Avda Casanova, Apdo 87, Sabana Grande, Caracas 1010-A; tel. (212) 781-9711; fax (212) 781-8739; f. 1970; Dir RAFAEL MATA.

**Consorcio Naviero Venezolano (Conavén):** Torre Uno, 4°, Avda Orinoco, Las Mercedes, Caracas; tel. (212) 993-2922; fax (212) 993-1636; e-mail conavent@conaven.com.

**Corpoven, SA:** Edif. Petróleos de Venezuela, Avda Libertador, La Campiña, Apdo 61373, Caracas 1060-A; tel. (212) 708-1111; fax (212) 708-1833; Pres. Dr ROBERTO MANDINI; Vice-Pres. JUAN CARLOS GÓMEZ; 2 oil tankers.

**Lagoven, SA:** Edif. Lagovén, Avda Leonardo da Vinci, Los Chaguaramos, Apdo 889, Caracas; tel. (212) 606-3311; fax (212) 606-3637; f. 1978 as a result of the nationalization of the petroleum industry; fmrly Creole Petroleum Group; transports crude petroleum and by-products between Maracaibo, Aniba and other ports in the area; Pres. B. R. NATERA; Marine Man. P. D. CAREZIS; 10 tankers.

**Tacarigua Marina, CA:** Torre Lincoln 7A-B, Avda Lincoln, Apdo 51107, Sabana Grande, Caracas 1050-A; tel. (212) 781-1315; Pres. R. BELLIZZI.

**Transpapel, CA:** Edif. Centro, 11°, Of. 111, Centro Parque Boyaca, Avda Sucre, Los Dos Caminos, Apdo 61316, Caracas 1071; tel. (212) 283-8366; fax (212) 285-7749; e-mail nmaldonado@cantv.net; Chair. GUILLERMO ESPINOSA F.; Man. Dir Capt. NELSON MALDONADO A.

**Transporte Industrial, SA:** Carretera Guanta, Km 5, Planta Vencemos, Pertigalete-Edif. Anzoátegui, Apdo 4356, Puerto la Cruz; tel. (281) 68-5607; fax (281) 68-5683; f. 1955; bulk handling and cement bulk carrier; Chair. VÍCTOR ROMO; Man. Dir RAFAEL ANEE.

### CIVIL AVIATION

There are two adjacent airports 13 km from Caracas; Maiquetía for domestic and Simón Bolívar for international services. There are 61 commercial airports, 11 of which are international airports.

#### National Airlines

**Aeropostal (Alas de Venezuela):** Torre Polar Oeste, 22°, Plaza Venezuela, Los Caobos, Caracas 1051; tel. (212) 708-6211; fax (212) 782-6323; internet www.aeropostal.com; f. 1933; transferred to private ownership in Sept. 1996, acquired by Venezuela/US consortium Corporación Alas de Venezuela; domestic services and flights to destinations in the Caribbean, South America and the USA; Pres. and CEO NELSON RAMIZ.

**Aerovías Venezolanas, SA (AVENSA):** Torre Humboldt, 1°, Avda Rio Caura, Prados del Este, Caracas; tel. (212) 976-5240; fax (212) 563-0225; e-mail info@avensa.com.ve; internet www.avensa.com.ve; f. 1943; provides domestic services from Caracas and services to Europe and the USA; govt-owned; Pres. WILMAR CASTRO SOTELDO.

**Aserca Airlines:** Torre Exterior, 8°, Avda Bolívar Norte, Valencia, Carabobo 2002; tel. (241) 237-111; fax (241) 220-210; e-mail rsv@asercaairlines.com; internet www.asercaairlines.com; f. 1968; domestic services and flights to Caribbean destinations; Pres. SIMEÓN GARCÍA.

**LASER** (Línea Aérea de Servicio Ejecutivo Regional): Torre Bazar Bolívar, 8°, Avda Francisco de Miranda, El Marqués, Caracas; tel. (212) 202-0100; fax (212) 235-8359; internet www.laser.com.ve; f. 1994; scheduled and charter services to domestic and international destinations, passenger and cargo; Pres. INOCENCIO ALVAREZ; Gen. Man. JORGE ANDRADE HIDALGO.

**Línea Turística Aereotuy, CA:** Edif. Gran Sábana, 5°, Blvd de Sabana Grande, Apdo 2923, Carmelitas, Caracas 1050; tel. (212) 761-6231; fax (212) 762-5254; e-mail tuysales@etheron.net; internet www.tuy.com; f. 1982; operates on domestic and international routes; Pres. PETER BOTTOME; Gen. Man. JUAN C. MÁRQUEZ.

**Santa Barbara Airlines:** Avda 3H, No 78-51, Res. República, Local 01, Maracaibo; tel. (261) 922-090; fax (261) 927-977; internet www.sbairlines.com; f. 1996; domestic and international services; Pres. FRANCISCO GONZÁLEZ YANES.

## Tourism

In 2003 Venezuela received 366,974 tourists. Receipts from tourism in that year amounted to US $368m. Venezuela's high crime levels, political instability and high prices relative to regional competitors have deterred tour operators. An estimated 90% of tourists visit the island of Margarita, while only 20% of tourists visit the mainland.

**Corporación Nacional de Hoteles y Turismo (CONAHOTU):** Caracas; f. 1969; govt agency; Pres. ERASTO FERNÁNDEZ.

**INATUR:** Hotel Caracas Hilton, Torre Sur, 4°, Of. 424, Caracas; tel. (212) 503-5423; fax (212) 503-5424; e-mail inatur@inatur.gov.ve; internet www.inatur.gov.ve; govt tourist devt agency; Exec. Dir SILVIA ARTEAGA.

**Viceministerio de Turismo:** c/o Central Information Office of the Presidency, Torre Oeste, 18°, Parque Central, Caracas; tel. (212) 509-0959; fax (212) 509-0941; e-mail vtur@mpc.gov.ve; Vice-Minister DALILA MONSERRATT.

## Defence

In August 2004 the Armed Forces numbered 82,300 (including the 23,000-strong National Guard and an estimated 31,000 conscripts): an Army of 34,000 (including 27,000 conscripts), a Navy of 18,300 men (including 500 naval aviation, 7,800 marines, 1,000 coast guard and an estimated 4,000 conscripts), and an Air Force of 7,000 men. There were an estimated 8,000 army reserves. In April 2005 President Chávez swore in the General Command of a new Army Reserve, projected to number some 1.5m. reserves. Military service is selective and the length of service varies by region for all services.

**Defence Budget:** 2,400,000m. bolívares in 2004.

**Commander-in-Chief of the Armed Forces:** Gen. (retd) JORGE LUIS GARCÍA CARNEIRO.

**Inspector-General of the National Armed Forces:** Gen. MELVIN LÓPEZ HIDALGO.

**Commander-General of the Navy:** Vice-Adm. ARMANDO LAGUNA LAGUNA.

**Commander of the Army:** Gen. RAÚL ISAÍAS BADUEL.

**Commander-General of the Air Force:** Gen. ROGER RAFAEL CORDERO LARA.

**Commander-General of the Army Reserves:** Gen. JULIO QUINTERO VILORIA.

## Education

Primary education in Venezuela is free and compulsory between the ages of six and 15 years. Secondary education begins at the age of 15 years and lasts for a further two years. In 1993 the total enrolment of children at primary and secondary schools was equivalent to 83% of the school-age population (males 80%; females 85%). In 1996 primary enrolment was equivalent to 91% of children in the relevant age-group (males 90%; females 93%), while secondary enrolment was 40% (males 33%; females 41%). Only 50% of pupils complete their basic education. In 2001 there were 20 universities. Expenditure by the central Government on education and sport was an estimated 29,601.6m. bolívares in 2004.

## Bibliography

For works on the region generally, see Select Bibliography (Books)

Baena, C. E. *The Policy Process in a Petro-State: An Analysis of PDVSA's (Petróleos de Venezuela, SA's) Internationalisation Strategy*. Aldershot, Ashgate Publishing Ltd, 1999.

Boué, J. C. *Venezuela: The Political Economy of Oil*. Oxford, Oxford University Press (for Oxford Unit for Energy Studies), 1993.

Buxton, J. 'Venezuela', in Buxton, J., and Phillips, N. *Case Studies in Latin American Political Economy*. Manchester, Manchester University Press, 1999.

*The Failure of Political Reform in Venezuela*. Aldershot, Ashgate Publishing Ltd, 2001.

Canache, D. *Venezuela: Public Opinion and Protest in a Fragile Democracy*. Boulder, CO, Lynne Rienner Publrs, 2001.

Chávez, H., Deutschmann, D., and Salado, J. *Chávez: Venezuela and the New Latin America*. New York, NY, Consortium, 2004.

Corrales, J. *Presidents Without Parties: The Politics of Economic Reform in Argentina and Venezuela in the 1990s*. University Park, PA, Penn State University Press, 2002.

Crisp, B. F. *Democratic Institutional Design: The Powers and Incentives of Venezuelan Politicians and Interest Groups*. Stanford, CA, Stanford University Press, 2000.

Ellner, S., and Hellinger, D. (Eds). *Venezuelan Politics in the Chávez Era*. Boulder, CO, Lynne Rienner Publrs, 2003.

Ewell, J. *Venezuela and the United States*. Athens, GA, University of Georgia Press, 1996.

*Venezuela's Movimiento al Socialismo: From Guerrilla Defeat to Innovative Politics*. Durham, NC, Duke University Press, 1988.

Friedman, E. J. *Unfinished Transitions: Women and the Gendered Development of Democracy in Venezuela, 1936–1996*. University Park, PA, Penn State University Press, 2000.

Guevara, A. *Chávez: Venezuela and the New Latin America—Hugo Chávez Interviewed by Aleida Guevara*. New York, NY, Ocean Press, 2005.

Gott, R. *In the Shadow of the Liberator: The Impact of Hugo Chávez on Venezuela and Latin America*. London, Verso Books, 2000.

*Hugo Chávez and the Bolivarian Revolution*. London, Verso Books, 2005.

Harnecker, M. *Understanding the Venezuelan Revolution: Hugo Chávez Talks to Marta Harnecker*. New York, NY, Fordham University Press, 2005.

Kelly, J., and Romero, C. A. *United States and Venezuela: Rethinking a Relationship*. London, Routledge, 2001.

*United States and Venezuela: Relations Between Friends*. London, Routledge, 2002.

McBeth, B. S. *Juan Vicente Gómez and the Oil Companies in Venezuela, 1908–1935*. Cambridge, Cambridge University Press, 2002.

McCaughan, M. *The Battle of Venezuela*. London, Latin American Bureau, 2004.

McCoy, J., Myers, D. J. (Eds). *The Unraveling of Representative Democracy in Venezuela*. Baltimore, MD, Johns Hopkins University Press, 2005.

Salazar, J. *Oil and Development in Venezuela during the Twentieth Century*. London, Praeger Publrs, 1994.

Tarver, H. M. *The Rise And Fall Of Venezuelan President Carlos Andrés Pérez: The Early Years, 1936–1973*. Lewiston, NY, Edwin Mellen Press, 2001.

# PART THREE
# Regional Information

# REGIONAL ORGANIZATIONS

## The United Nations

**Address:** United Nations, New York, NY 10017, USA.

**Telephone:** (212) 963-1234; **fax:** (212) 963-4879; **internet:** www.un.org.

The United Nations (UN) was founded on 24 October 1945. The organization, which has 191 member states, aims to maintain international peace and security and to develop international co-operation in addressing economic, social, cultural and humanitarian problems. The principal organs of the UN are the General Assembly, the Security Council, the Economic and Social Council (ECOSOC), the International Court of Justice and the Secretariat. The General Assembly, which meets for three months each year, comprises representatives of all UN member states. The Security Council investigates disputes between member countries, and may recommend ways and means of peaceful settlement: it comprises five permanent members (the People's Republic of China, France, Russia, the United Kingdom and the USA) and 10 other members elected by the General Assembly for a two-year period. The Economic and Social Council comprises representatives of 54 member states, elected by the General Assembly for a three-year period: it promotes co-operation on economic, social, cultural and humanitarian matters, acting as a central policy-making body and co-ordinating the activities of the UN's specialized agencies. The International Court of Justice comprises 15 judges of different nationalities, elected for nine-year terms by the General Assembly and the Security Council: it adjudicates in legal disputes between UN member states.

In March 2005 the Secretary-General announced a series of proposals for extensive reforms of the UN.

**Secretary-General:** KOFI ANNAN (Ghana) (1997–2006).

### MEMBER STATES IN SOUTH AMERICA, CENTRAL AMERICA AND THE CARIBBEAN
(with assessments for percentage contributions to UN budget for 2004–06, and year of admission)

| Country | Assessment | Year |
|---|---|---|
| Antigua and Barbuda | 0.003 | 1981 |
| Argentina | 0.956 | 1945 |
| Bahamas | 0.013 | 1973 |
| Barbados | 0.010 | 1966 |
| Belize | 0.001 | 1981 |
| Bolivia | 0.009 | 1945 |
| Brazil | 1.523 | 1945 |
| Chile | 0.223 | 1945 |
| Colombia | 0.155 | 1945 |
| Costa Rica | 0.030 | 1945 |
| Cuba | 0.043 | 1945 |
| Dominica | 0.001 | 1978 |
| Dominican Republic | 0.035 | 1945 |
| Ecuador | 0.019 | 1945 |
| El Salvador | 0.022 | 1945 |
| Grenada | 0.001 | 1974 |
| Guatemala | 0.030 | 1945 |
| Guyana | 0.001 | 1966 |
| Haiti | 0.003 | 1945 |
| Honduras | 0.005 | 1945 |
| Jamaica | 0.008 | 1962 |
| Mexico | 1.883 | 1945 |
| Nicaragua | 0.001 | 1945 |
| Panama | 0.019 | 1945 |
| Paraguay | 0.012 | 1945 |
| Peru | 0.092 | 1945 |
| Saint Christopher and Nevis | 0.001 | 1983 |
| Saint Lucia | 0.002 | 1979 |
| Saint Vincent and the Grenadines | 0.001 | 1980 |
| Suriname | 0.001 | 1975 |
| Trinidad and Tobago | 0.022 | 1962 |
| Uruguay | 0.048 | 1945 |
| Venezuela | 0.171 | 1945 |

## Diplomatic Representation

### PERMANENT MISSIONS TO THE UNITED NATIONS
(August 2005)

**Antigua and Barbuda:** 610 Fifth Ave, Suite 311, New York, NY 10020; tel. (212) 541-4117; fax (212) 757-1607; e-mail antigua@un.int; internet www.un.int/antigua; Permanent Representative JOHN W. ASHE.

**Argentina:** 1 United Nations Plaza, 25th Floor, New York, NY 10017; tel. (212) 688-6300; fax (212) 980-8395; e-mail argentina@un.int; internet www.un.int/argentina; Permanent Representative CÉSAR MAYORAL.

**Bahamas:** 231 East 46th St, New York, NY 10017; tel. (212) 421-6925; fax (212) 759-2135; e-mail bahamas@un.int; Permanent Representative PAULETTE A. BETHEL.

**Barbados:** 800 Second Ave, 2nd Floor, New York, NY 10017; tel. (212) 867-8431; fax (212) 986-1030; e-mail barbados@un.int; Permanent Representative CHRISTOPHER F. HACKETT.

**Belize:** 800 Second Ave, Suite 400G, New York, NY 10017; tel. (212) 593-0999; fax (212) 593-0932; e-mail blzun@undp.org; Permanent Representative STUART LESLIE.

**Bolivia:** 211 East 43rd St, 8th Floor, Room 802, New York, NY 10017; tel. (212) 682-8132; fax (212) 682-8133; e-mail bolivia@un.int; Permanent Representative ERNESTO ARANÍBAR QUIROGA.

**Brazil:** 747 Third Ave, 9th Floor, New York, NY 10017; tel. (212) 372-2600; fax (212) 371-5716; e-mail braun@delbrasonu.org; internet www.un.int/brazil; Permanent Representative RONALDO MOTA SARDENBERG.

**Chile:** 3 Dag Hammarskjöld Plaza, 305 East 47th St, 10th/11th Floor, New York, NY 10017; tel. (212) 832-3323; fax (212) 832-0236; e-mail chile@un.int; internet www.un.int/chile; Permanent Representative JUAN GABRIEL VALDES.

**Colombia:** 140 East 57th St, 5th Floor, New York, NY 10022; tel. (212) 355-7776; fax (212) 371-2813; e-mail colombia@colombiaun.org; internet www.colombiaun.org; Permanent Representative MARIA ANGELA HOLGUÍN.

**Costa Rica:** 211 East 43rd St, Room 903, New York, NY 10017; tel. (212) 986-6373; fax (212) 986-6842; e-mail pmnu@rree.go.cr; internet www.un.int/costarica; Permanent Representative BERND H. NIEHAUS.

**Cuba:** 315 Lexington Ave and 38th St, New York, NY 10016; tel. (212) 689-7215; fax (212) 779-1697; e-mail cuba@un.int; internet www.un.int/cuba; Permanent Representative ORLANDO REQUEIJO GUAL.

**Dominica:** 800 Second Ave, Suite 400H, New York, NY 10017; tel. (212) 949-0853; fax (212) 808-4975; e-mail dominica@un.int; Permanent Representative SIMON PAUL RICHARDS.

**Dominican Republic:** 144 East 44th St, 4th Floor, New York, NY 10017; tel. (212) 867-0833; fax (212) 986-4694; e-mail drun@un.int; internet www.un.int/dr; Permanent Representative ERASMO LARA.

**Ecuador:** 866 United Nations Plaza, Room 516, New York, NY 10017; tel. (212) 935-1680; fax (212) 935-1835; e-mail ecuador@un.int; internet www.un.int/ecuador; Permanent Representative JAIME MONCAYO.

**El Salvador:** 46 Park Ave, New York, NY 10016; tel. (212) 679-1616; fax (212) 725-7831; e-mail elsalvador@un.int; Permanent Representative CARMEN MARÍA GALLARDO HERNÁNDEZ.

**Grenada:** 800 Second Ave, Suite 400K, New York, NY 10017; tel. (212) 599-0301; fax (212) 599-1540; e-mail grenada@un.int; Permanent Representative Dr RUTH ELIZABETH ROUSE.

**Guatemala:** 57 Park Ave, New York, NY 10016; tel. (212) 679-4760; fax (212) 685-8741; e-mail guatemala@un.int; internet www.un.int/guatemala; Permanent Representative JORGE SKINNER-KLÉE ARENALES.

**Guyana:** 866 United Nations Plaza, Suite 555, New York, NY 10017; tel. (212) 527-3232; fax (212) 935-7548; e-mail guyana@un.int; Permanent Representative SAMUEL R. INSANALLY.

**Haiti:** 801 Second Ave, Room 600, New York, NY 10017; tel. (212) 370-4840; fax (212) 661-8698; e-mail haiti@un.int; Permanent Representative Léo Mérorès.

**Honduras:** 866 United Nations Plaza, Suite 417, New York, NY 10017; tel. (212) 752-3370; fax (212) 223-0498; e-mail m.suazo@worldnet.att.net; internet www.un.int/honduras; Permanent Representative Manuel Acosta Bonilla.

**Jamaica:** 767 Third Ave, 9th Floor, New York, NY 10017; tel. (212) 935-7509; fax (212) 935-7607; e-mail jamaica@un.int; internet www.un.int/jamaica; Permanent Representative Stafford Oliver Neil.

**Mexico:** 2 United Nations Plaza, 28th Floor, New York, NY 10017; tel. (212) 752-0220; fax (212) 688-8862; e-mail mexico@un.int; internet www.un.int/mexico; Permanent Representative Enrique Berruga Filloy.

**Nicaragua:** 820 Second Ave, Suite 801, New York, NY 10017; tel. (212) 490-7997; fax (212) 286-0815; e-mail nicaragua@un.int; internet www.un.int/nicaragua; Permanent Representative Eduardo J. Sevilla Somoza.

**Panama:** 866 United Nations Plaza, Suite 4030, New York, NY 10017; tel. (212) 421-5420; fax (212) 421-2694; e-mail emb@panama_msun.org; Permanent Representative Ricardo Alberto Arias.

**Paraguay:** 211 East 43rd St, Suite 400, New York, NY 10017 ; tel. (212) 687-3490; fax (212) 818-1282; e-mail paraguay@un.int; Permanent Representative Eladio Loizaga.

**Peru:** 820 Second Ave, Suite 1600, New York, NY 10017; tel. (212) 687-3336; fax (212) 972-6975; e-mail peru@un.int; Permanent Representative Oswaldo de Rivero.

**Saint Christopher and Nevis:** 414 East 75th St, 5th Floor, New York, NY 10021; tel. (212) 535-1235; fax (212) 535-6858; e-mail sknmission@aol.com; Permanent Representative Dr Joseph R. Christmas.

**Saint Lucia:** 800 Second Ave, 9th Floor, New York, NY 10017; tel. (212) 697-9360; fax (212) 697-4993; e-mail stlucia@un.int; internet www.un.int/stlucia; Permanent Representative Julian Robert Hunte.

**Saint Vincent and the Grenadines:** 801 Second Ave, 21st Floor, New York, NY 10017; tel. (212) 687-4490; fax (212) 949-5946; e-mail stvg@un.int; Permanent Representative Margaret Hughes Ferrari.

**Suriname:** 866 United Nations Plaza, Suite 320, New York, NY 10017; tel. (212) 826-0660; fax (212) 980-7029; e-mail suriname@un.int; internet www.un.int/suriname; Permanent Representative Ewald Wensley Limon.

**Trinidad and Tobago:** 820 Second Ave, 5th Floor, New York, NY 10017; tel. (212) 697-7620; fax (212) 682-3580; e-mail tto@un.int; Permanent Representative Philip R.A. Sealy.

**Uruguay:** 866 United Nations Plaza, Suite 322, New York, NY 10017; tel. (212) 752-8240; fax (212) 593-0935; e-mail uruguay@un.int; internet www.un.int/uruguay; Permanent Representative Alejandro Artucio Rodriguez.

**Venezuela:** 335 East 46th St, New York, NY 10017; tel. (212) 557-2055; fax (212) 557-3528; e-mail venezuela@un.int; internet www.un.int/venezuela; Permanent Representative Fermin Toro Jimenez.

### OBSERVERS

Non-member states, inter-governmental organizations, etc., which have received an invitation to participate in the sessions and the work of the General Assembly as Observers, maintaining permanent offices at the UN.

**Caribbean Community:** 97-40 62nd Drive, 15k, Rego Park, NY 11374-1336; tel. and fax (718) 896-1179; Permanent Representative Hamid Mohammed.

**Commonwealth Secretariat:** 800 Second Ave, 4th Floor, New York, NY 10017; tel. (212) 599-6190; fax (212) 808-4975; e-mail comsec@thecommonwealth.org.

**World Conservation Union—IUCN:** 406 West 66th St, New York, NY 10021; tel. and fax (212) 734-7608.

African, Caribbean and Pacific Group of States;
African Development Bank;
Agency for the Prohibition of Nuclear Weapons in Latin America and the Caribbean;
Andean Community;
Association of Caribbean States;
Central American Integration System;
Commonwealth of Independent States;
Council of Europe;
East African Community;
Economic Co-operation Organization;
Eurasian Economic Community;
GUAAM;
International Criminal Police Organization (Interpol);
International Institute for Democracy and Electoral Assistance;
Inter-Parliamentary Union;
Latin American Economic System;
Latin American Parliament;
Organisation for Economic Co-operation and Development;
Organization for Security and Co-operation in Europe;
Organization of American States;
Organization of the Black Sea Economic Co-operation;
Permanent Court of Arbitration;
Pacific Islands Forum.

## United Nations Information Centres/Services

**Argentina:** Junín 1940, 1°, 1113 Buenos Aires; tel. (1) 4803-7671; fax (1) 4804-7545; e-mail buenosaires@unic.org.ar; internet www.unic.org.ar; also covers Uruguay.

**Bolivia:** Apdo 9072, Calle 14 esq. S. Bustamante, Ed. Metrobol II, Calacoto, La Paz; tel. (2) 2795544; fax (2) 2795963; e-mail unicbol@nu.org.bo; internet www.no.org.bo/cinu.

**Brazil:** Palacio Itamaraty, Avda Marechal Floriano 196, 20080-002 Rio de Janeiro; tel. (21) 253-2211; fax (21) 233-5753; e-mail infounic@unicrio.org.br; internet www.unicrio.org.br.

**Chile:** Edif. Naciones Unidas, Avda Dag Hammarskjöld, Casilla 179-D, Santiago; tel. (2) 210-2000; fax (2) 208-1946; e-mail dpisantiago@eclac.cl.

**Colombia:** Apdo Aéreo 058964 Calle 100, No. 8A-55, 10°, Santafé de Bogotá 2; tel. (1) 257-6044; fax (1) 257-6244; e-mail uniccol@mbox.unicc.org; internet www.onucolombia.org; also covers Ecuador and Venezuela.

**Mexico:** Presidente Masaryk 29, 6°, México 11 570, DF; tel. (55) 5263-9724; fax (55) 5203-8238; e-mail unicmex@un.org.mx; internet www.cinu.org.mx; also covers Cuba and the Dominican Republic.

**Nicaragua:** Apdo 3260, Palacio de la Cultura, Managua; tel. (2) 664253; fax (2) 222362; e-mail cedoc@sdnnic.org.ni.

**Panama:** POB 6-9083, El Dorado; Banco Central Hispano Edif., 1°, Calle Gerardo Ortega y Av. Samuel Lewis, Panama City; tel. (7) 233-0557; fax (7) 223-2198; e-mail cinup@cciglobal.net.pa; internet www.cinup.org.

**Paraguay:** Casilla de Correo 1107, Edif. City, 3°, Asunción; tel. (21) 614443; fax (21) 449611; e-mail unic.py@undp.org.

**Peru:** POB 14-0199, Lord Cochrane 130, San Isidro, Lima 27; tel. (1) 441-8745; fax (1) 441-8735; e-mail informes@uniclima.org.pe; internet www.uniclima.org.pe.

**Trinidad and Tobago:** POB 130, Bretton Hall, 16 Victoria Ave, Port of Spain; tel. 623-4813; fax 623-4332; e-mail unicpos@unicpos.org.tt; also covers Antigua and Barbuda, the Bahamas, Barbados, Belize, Dominica, Grenada, Guyana, Jamaica, the Netherlands Antilles, Saint Christopher and Nevis, Saint Lucia, Saint Vincent and the Grenadines and Suriname.

# Economic Commission for Latin America and the Caribbean—ECLAC

**Address:** Edif. Naciones Unidas, Avda Dag Hammarskjöld, Casilla 179D, Santiago, Chile.
**Telephone:** (2) 2102000; **fax:** (2) 2080252; **e-mail:** dpisantiago@eclac.cl; **internet:** www.eclac.org.

The UN Economic Commission for Latin America was founded in 1948 to co-ordinate policies for the promotion of economic development in the Latin American region. The current name of the Commission was adopted in 1984.

### MEMBERS

Antigua and Barbuda
Argentina
Bahamas
Barbados
Belize
Bolivia
Brazil
Canada
Chile
Colombia
Costa Rica
Cuba
Dominica
Dominican Republic
Ecuador
El Salvador
France
Germany
Grenada
Guatemala
Guyana
Haiti
Honduras
Italy
Jamaica
Mexico
Netherlands
Nicaragua
Panama
Paraguay
Peru
Portugal
Saint Christopher and Nevis
Saint Lucia
Saint Vincent and the Grenadines
Spain
Trinidad and Tobago
United Kingdom
USA
Uruguay
Venezuela

### ASSOCIATE MEMBERS

Anguilla
Aruba
British Virgin Islands
Montserrat
Netherlands Antilles
Puerto Rico
United States Virgin Islands

## Organization

(August 2005)

### COMMISSION

The Commission normally meets every two years. The 30th session was held in San Juan, Puerto Rico, in June–July 2004 and the 31st session was to be held in Montevideo, Uruguay, in 2006. The Commission has established the following permanent bodies:

**Caribbean Development and Co-operation Committee.**

**Central American Development and Co-operation Committee.**

**Committee of High-Level Government Experts.**

**Committee of the Whole.**

**Regional Conference on the Integration of Women into the Economic and Social Development of Latin America and the Caribbean.**

**Regional Council for Planning.**

### SECRETARIAT

The Secretariat employs more than 500 staff and is headed by the Offices of the Executive Secretary and of the Secretary of the Commission. ECLAC's work programme is carried out by the following divisions: Economic Development; Social Development; International Trade; Production, Productivity and Management; Statistics and Economic Projections; Sustainable Development and Human Settlements; Natural Resources and Infrastructure; Documents and Publications; and Population. There are also units for information and conference services, women and development and special studies, an electronic information section, and a support division of administration.

**Executive Secretary:** José Luis Machinea.

**Deputy Executive Secretary (and Head of the Office of the Secretary of the Commission):** Alicia Bárcena.

### SUB-REGIONAL OFFICES

**Caribbean:** 1 Chancery Lane, POB 1113, Port of Spain, Trinidad and Tobago; tel. 623-5595; fax 623-8485; e-mail registry@eclacpos.org; internet www.eclacpos.org; f. 1956; covers non-Spanish-speaking Caribbean countries; functions as the Caribbean Development and Co-operation Committee and as the secretariat for the Programme of Action for the Sustainable Development of Small Island Developing States; Dir Neil Pierre.

**Central America and Spanish-speaking Caribbean:** Avda Presidente Masaryk 29, 11570 México, DF; tel. (55) 5263-9600; fax (55) 5531-1151; e-mail cepal@un.org.mx; internet www.eclac.org.mx; f. 1951; covers Central America and Spanish-speaking Caribbean countries; Dir Rebeca Grynspan.

There are also national offices, in Santafé de Bogotá, Brasília, Buenos Aires and Montevideo and a liaison office in Washington, DC.

## Activities

ECLAC collaborates with regional governments in the investigation and analysis of regional and national economic problems, and provides guidance in the formulation of development plans. Its activities include research; analysis; publication of information; provision of technical assistance; participation in seminars and conferences; training courses; and co-operation with national, regional and international organizations.

The 26th session of the Commission, which was held in San José, Costa Rica, in April 1996, considered means of strengthening the economic and social development of the region, within the framework of a document prepared by ECLAC's Secretariat, and adopted a resolution which defined ECLAC as a centre of excellence, charged with undertaking an analysis of specific aspects of the development process, in collaboration with member governments. In May 1998 the 27th session of the Commission, held in Oranjestad, Aruba, approved the ongoing reform programme, and in particular efforts to enhance the effectiveness and transparency of ECLAC's activities. The main topics of debate at the meeting were public finances, fiscal management and social and economic development. The Commission adopted a Fiscal Covenant, incorporating measures to consolidate fiscal adjustment and to strengthen public management, democracy and social equity, which was to be implemented throughout the region and provide the framework for further debate at national and regional level. ECLAC's 28th session, convened in Mexico City in April 2000, debated a document prepared by the Secretariat which proposed that the pursuit of social equity, sustainable development and 'active citizenship' (with emphasis on the roles of education and employment) should form the basis of future policy-making in the region.

ECLAC's 29th session, which was held in Brasília, Brazil, in May 2002, focused on the process of globalization. The meeting adopted the Brasilia Resolution, which outlined a strategic agenda to meet the challenges of globalization. Proposed action included the consolidation of democracy, strengthening social protection, the formulation of policies to reduce macroeconomic and financial vulnerability, and the development of sustainable and systemic competitiveness. The objectives of the agenda were to achieve a guaranteed supply of general public goods, to overcome, steadily, the imbalances in the world order, and to build, gradually, an international social agenda based on rights. The Resolution requested that ECLAC strengthen its work in the relevant areas. The 30th session was held in June–July 2004 in San Juan, Puerto Rico. Among the issues discussed were proposals in a new ECLAC study entitled *Productive Development in Open Economies*, activities to be undertaken, with UNESCO, to promote education in the region, and other areas of activity in 2006–07.

ECLAC works closely with other agencies within the UN system and with other regional and multinational organizations. ECLAC is co-operating with the OAS and the Inter-American Development Bank in the servicing of intergovernmental groups undertaking preparatory work for the establishment of a Free Trade Area of the Americas (FTAA). In May 2001 ECLAC hosted the first meeting of the Americas Statistics Conference. In January 2002 ECLAC hosted an Interregional Conference on Financing for Development, held in Mexico City, which it had organized as part of the negotiating process prior to the World Summit on Financing for Development, held in March. In June senior representatives of ECLAC, UNDP, the World Bank and the Inter-American Development Bank signed a Protocol of Intentions with a commitment to co-ordinate activities in pursuit of the development goals proclaimed by the so-called Millennium Summit meeting of the General Assembly in September 2000. ECLAC was to adapt the objectives of the goals to the reality of countries in the region. In July 2004 the 30th session of the Commission approved the establishment of an intergovernmental forum to monitor the implementation of decisions emerging from the World Summit on Sustainable Development, held in Johannesburg, South Africa, in September 2002. ECLAC provides regional support to the UN Information and Communication Technologies Task Force, which was established in November 2001. A working meeting on the establishment of a regional network under the Task Force, in order to plan and monitor the development of digital technology in the region, was held in June 2002. In January 2003 a regional conference was convened, in the Dominican Republic, in preparation for the World Summit on the Information Society, the first phase of which was held in December, in Geneva, Switzerland. In July 2004 delegates to the 30th session of the Commission requested that ECLAC co-ordinate a regional preparatory meeting to define objectives and proposals for the second phase of the Summit, scheduled to be convened in November 2005 in Tunis, Tunisia. In November 2003 ECLAC launched 'Redesa', a web-based network of institutions and experts in social and environmental statistics.

**Latin American and Caribbean Institute for Economic and Social Planning (ILPES):** Edif. Naciones Unidas, Avda Dag Hammarskjöld 3477, Vitacura, Santiago, Chile; tel. (2) 2102506; fax (2) 2066104; e-mail paul.dekock@cepal.org; internet www.ilpes.org; f. 1962; supports regional governments through the provision of training, advisory services and research in the field of public planning policy and co-ordination; Dir JOSÉ LUIS MACHINEA.

**Latin American Demographic Centre (CELADE):** Edif. Naciones Unidas, Avda Dag Hammarskjöld, Casilla 179D, Santiago, Chile; tel. (2) 2102002; fax (2) 2080252; e-mail mvilla@eclac.cl; internet www.eclac.org/celade; f. 1957, became an integral part of ECLAC in 1975; provides technical assistance to governments, universities and research centres in demographic analysis, population policies, integration of population factors in development planning, and data processing; conducts three-month courses on demographic analysis for development and various national and regional seminars; provides demographic estimates and projections, documentation, data processing, computer packages and training; Dir DIRK JASPERS-FAIJER.

## Finance

For the two-year period 2004–05 ECLAC's proposed regular budget, an appropriation from the UN, amounted to US $81.2m. In addition, extra-budgetary activities are financed by governments, other organizations, and UN agencies, including UNDP, UNFPA and UNICEF.

## Publications

*Boletín del Banco de Datos del CELADE* (annually).
*Boletín demográfico* (2 a year).
*Boletín de Facilitación del Comercio y el Transporte* (monthly).
*CEPAL News* (digital, monthly).
*CEPAL Review* (Spanish and English, 3 a year).
*CEPALINDEX* (annually).
*Co-operation and Development* (Spanish and English, quarterly).
*DOCPAL Resúmenes* (population studies, 2 a year).
*ECLAC Notes / Notas de la CEPAL* (every 2 months).
*Economic Survey of Latin America and the Caribbean* (Spanish and English, annually).
*Foreign Investment in Latin America and the Caribbean* (annually).
*Latin America and the Caribbean in the World Economy* (annually).
*Latin American Projections 2001–02.*
*Notas de Población* (2 a year).
*PLANINDEX* (2 a year).
*Preliminary Overview of the Economies of Latin America and the Caribbean* (annually).
*Social Panorama of Latin America* (annually).
*Statistical Yearbook for Latin America and the Caribbean* (Spanish and English).
Studies, reports, bibliographical bulletins.

# United Nations Development Programme—UNDP

**Address:** One United Nations Plaza, New York, NY 10017, USA.
**Telephone:** (212) 906-5295; **fax:** (212) 906-5364; **e-mail:** hq@undp.org; **internet:** www.undp.org.

The Programme was established in 1965 by the UN General Assembly. Its central mission is to help countries to eradicate poverty and achieve a sustainable level of human development, an approach to economic growth that encompasses individual well-being and choice, equitable distribution of the benefits of development, and conservation of the environment. UNDP advocates for a more inclusive global economy. UNDP is the focus of UN efforts to achieve the Millennium Development Goals (see below).

## Organization
(August 2005)

UNDP is responsible to the UN General Assembly, to which it reports through ECOSOC.

### EXECUTIVE BOARD

The Executive Board is responsible for providing intergovernmental support to, and supervision of, the activities of UNDP and the UN Population Fund (UNFPA). It comprises 36 members: eight from Africa, seven from Asia and the Pacific, four from eastern Europe, five from Latin America and the Caribbean and 12 from western Europe and other countries. Members serve a three-year term.

### SECRETARIAT

Offices and divisions at the Secretariat include: an Operations Support Group; Offices of the United Nations Development Group, the Human Development Report, Development Studies, Audit and Performance Review, Evaluation, and Communications; and Bureaux for Crisis Prevention and Recovery, Resources and Strategic Partnerships, Development Policy, and Management. Five regional bureaux, all headed by an assistant administrator, cover: Africa; Asia and the Pacific; the Arab states; Latin America and the Caribbean; and Europe and the Commonwealth of Independent States.

**Administrator:** KEMAL DERVIŞ (Turkey).
**Associate Administrator:** Dr ZÉPHIRIN DIABRÉ (Burkina Faso).
**Assistant Administrator and Director, Regional Bureau for Latin America and the Caribbean:** ELENA MARTÍNEZ (Cuba).

### COUNTRY OFFICES

In almost every country receiving UNDP assistance there is an office, headed by the UNDP Resident Representative, who usually also serves as UN Resident Co-ordinator, responsible for the co-ordination of all UN technical assistance and operational development activities, advising the Government on formulating the country programme, ensuring that field activities are undertaken, and acting as the leader of the UN team of experts working in the country. The offices function as the primary presence of the UN in most developing countries.

#### OFFICES OF UNDP REPRESENTATIVES IN SOUTH AMERICA, CENTRAL AMERICA AND THE CARIBBEAN

**Argentina:** Casilla 2257, 1000 Capital Federal, Buenos Aires; tel. (1) 4320-8700; fax (1) 4320-8754; e-mail fo.arg@undp.org; internet www.undp.org.ar.

**Barbados:** POB 625C, Bridgetown; tel. 429-2521; fax 429-2448; e-mail fo.brb@undp.org; internet www.bb.undp.org; also covers Anguilla, Antigua and Barbuda, British Virgin Islands, Dominca, Grenada, Montserrat, Saint Christopher and Nevis, Saint Lucia, Saint Vincent and the Grenadines.

**Bolivia:** Avda Sánchez Bustamante, Casilla 9072, La Paz, Bolivia; tel. (2) 279-5544; fax (2) 279-5820; e-mail registry.bo@undp.org; internet www.pnud.bo.

**Brazil:** CP 0285, 70359 Brasília, DF; tel. (61) 329-2000; fax (61) 329-0099; e-mail fo.bra@undp.org; internet www.br.undp.org.

**Chile:** Avda Dag Hammarskjöld, 3241 Vitacura, Santiago; tel. (2) 337-2400; e-mail fo.chl@undp.org; internet www.pnud.cl.

**Colombia:** Apdo Aéreo 091369, Bogotá; tel. (1) 214-2200; fax (1) 214-0110; e-mail informacion@pnud.org.co; internet www.pnud.org.co.

**Costa Rica:** Apdo Postal 4540-1000, San José; tel. 2961544; fax 2961545; e-mail fo.cr@undp.org; internet www.nu.or.cr/pnud.

**Cuba:** Calle 18 No. 110 (entre 1ra y 3ra), Miramar, Playa, Havana; tel. (7) 33-1512; fax (7) 33-1516; e-mail fo.cub@undp.org; internet www.undp.org/cuba.

**Dominican Republic:** Apdo 1424, Santo Domingo; tel. 531-3403; fax 531-3507; e-mail fo.dom@undp.org; internet www.undp.org/fodom.

**Ecuador:** Avda Amazonas 2889 y la Granja, Quito; tel. (2) 2460-330; fax (2) 2461-960; e-mail registry.ec@undp.org; internet www.pnud.org.ec.

**El Salvador:** POB 1114, San Salvador; tel. 2790366; fax 2791929; e-mail fo.slv@undp.org; also covers Belize.

**Guatemala:** Apdo Postal 23A, 01909 Guatemala City; tel. (2) 3335416; fax (2) 3370304; e-mail fo.gtm@undp.org; internet www.pnud.org.gt.

**Guyana:** 42 Brickdam and Boyle Place, POB 10960, Georgetown, Guyana; tel. (2) 64040; fax (2) 62942; e-mail fo.guy@undp.org; internet www.sdnp.org.gy.

**Haiti:** BP 557, Port-au-Prince; tel. 231400; fax 239340; e-mail fo.hti@undp.org; internet www.ht.undp.org.

**Honduras:** Apdo Postal 976, Tegucigalpa DC; tel. 220-1100; fax 239-8010; e-mail fo.hnd@undp.org; internet www.undp.un.hn.

**Jamaica:** 1-3 Lady Musgrave Rd, POB 280, Kingston; tel. 978-2390; fax 946-2163; e-mail registry.jm@undp.org; internet www.undp.org/fojam; also covers Bahamas, Cayman Islands, Turks and Caicos Islands).

**Mexico:** Apdo Postal 105-39, 11581 México, DF; tel. (55) 5250-1555; fax (55) 5255-0095; e-mail fo.mex@undp.org; internet www.pnud.org.mx.

**Nicaragua:** Apdo Postal 3260, Managua, JR; tel. (2) 661701; fax (2) 666909; e-mail registry.ni@undp.org; internet www.undp.org.ni.

**Panama:** Apdo 6314, Panama 5; tel. 265-0838; fax 263-1444; e-mail registry@fopan.undp.org; internet aleph.onu.org.pa/pnud.

**Paraguay:** Estrella 345, Edificio Citibank, Asunción; tel. (21) 493025; fax (21) 444325; e-mail registry@undp.org.py; internet www.undp.org.py.

**Peru:** Apdo 27-0047, Lima 27; tel. (1) 2213636; fax (1) 4404166; e-mail fo.per@undp.org; internet www.onu.org.pe.

**Trinidad and Tobago:** POB 812, Port of Spain; tel. 623-7056; fax 623-1658; e-mail registry@undp.org.tt; internet www.undp.org.tt; also covers Aruba, Netherlands Antilles and Suriname.

**Uruguay:** J. Barrias Amorín 870, 3°, Montevideo; tel. (2) 4123357; fax (2) 4123360; e-mail onunet@undp.org.uy; internet www.undp.org.uy.

**Venezuela:** Avda Francisco de Miranda, Torre Hewlett-Packard, 6°, oficina 6A, Urb. Los Palos Grandes, Caracas; tel. (212) 208-4444; fax (212) 263-8179; e-mail pnud.ven@undp.org; internet www.pnud.org.ve.

## Activities

UNDP provides advisory and support services to governments and UN teams with the aim of advancing sustainable human development and building national development capabilities. Assistance is mostly non-monetary, comprising the provision of experts' services, consultancies, equipment and training for local workers. Developing countries themselves contribute significantly to the total project costs in terms of personnel, facilities, equipment and supplies. UNDP also supports programme countries in attracting aid and utilizing it efficiently. A network of nine Sub-regional Resource Facilities (SURFs) has been established to strengthen and co-ordinate UNDP's role as a global knowledge provider and channel for sharing knowledge and experience.

During the late 1990s UNDP undertook an extensive internal process of reform, 'UNDP 2001', which placed increased emphasis on its activities in the field and on performance and accountability. In 2001 UNDP established a series of Thematic Trust Funds to enable increased support of priority programme activities. In accordance with the more results-oriented approach developed under the 'UNDP 2001' process UNDP introduced a new Multi-Year Funding Framework (MYFF), which outlined the country-driven goals around which funding was to be mobilized, integrating programme objectives, resources, budget and outcomes. The MYFF was to provide the basis for the Administrator's Business Plans for the same duration and enables policy coherence in the implementation of programmes at country, regional and global levels. A Results-Oriented Annual Report (ROAR) was produced for the first time in 2000 from data compiled by country offices and regional programmes. In September 2000 the first ever Ministerial Meeting of ministers of development co-operation and foreign affairs and other senior officials from donor and programme countries, convened in New York, USA, endorsed UNDP's shift to a results-based orientation.

In accordance with the second phase of the MYFF, covering 2004–07, UNDP was to focus on the following five practice areas: democratic governance; poverty reduction; energy and the environment; crisis prevention and recovery; and combating HIV/AIDS. Other important 'cross-cutting' themes, to be incorporated throughout the programme areas, included gender equality, information and communication technologies, and human rights.

From the mid-1990s UNDP also assumed a more active and integrative role within the UN system-wide development framework. UNDP Resident Representatives—usually also serving as UN Resident Co-ordinators, with responsibility for managing interagency co-operation on sustainable human development initiatives at country level—were to play a focal role in implementing this approach. In order to promote its co-ordinating function UNDP allocated increased resources to training and skill-sharing programmes. In 1997 the UNDP Administrator was appointed to chair the UN Development Group (UNDG), which was established as part of a series of structural reform measures initiated by the UN Secretary-General, with the aim of strengthening collaboration between all UN funds, programmes and bodies concerned with development. The UNDG promotes coherent policy at country level through the system of UN Resident Co-ordinators (see above), the Common Country Assessment mechanism (CCA, a country-based process for evaluating national development situations), and the UN Development Assistance Framework (UNDAF, the foundation for planning and co-ordinating development operations at country level, based on the CCA). Within the framework of the Administrator's Business Plans for 2000–03 a new Bureau for Resources and Strategic Partnerships was established to build and strengthen working partnerships with other UN bodies, donor and programme countries, international financial institutions and development banks, civil society organizations and the private sector. The Bureau was also to serve UNDP's regional bureaux and country offices through the exchange of information and promotion of partnership strategies.

### MILLENNIUM DEVELOPMENT GOALS

UNDP, through its leadership of the UNDG and management of the Resident Co-ordinating system, has a co-ordinating function as the focus of UN system-wide efforts to achieve the so-called Millennium Development Goals (MDGs), pledged by 189 governments attending a summit meeting of the UN General Assembly in September 2000. The objectives were to establish a defined agenda to reduce poverty and improve the quality of lives of millions of people and to serve as a framework for measuring development. There are eight MDGs, as follows, for which one or more specific targets have been identified:

i) to eradicate extreme poverty and hunger, with the aim of reducing by 50% the number of people with an income of less than US $1 a day and those suffering from hunger by 2015;

ii) to achieve universal primary education by 2015;

iii) to promote gender equality and empower women, in particular to eliminate gender disparities in primary and secondary education by 2005 and at all levels by 2015;

iv) to reduce child mortality, with a target reduction of two-thirds in the mortality rate among children under five by 2015;

v) to improve maternal health, and specifically to reduce by 75% the numbers of women dying in childbirth;

vi) to combat HIV/AIDS, malaria and other diseases;

vii) to ensure environmental sustainability, including targets to integrate the principles of sustainable development into country policies and programmes, to reduce by 50% the number of people without access to safe drinking water by 2015, to achieve significant improvement in the lives of at least 100m. slum dwellers by 2020;

viii) to develop a global partnership for development, including efforts to deal with international debt, to address the needs of least developed countries and landlocked and small island developing states, to develop decent and productive youth employment, to provide access to affordable, essential drugs in developing countries, and to make available the benefits of new technologies.

UNDP plays a leading role in efforts to integrate the MDGs into all aspects of UN activities at country level and to ensure the MDGs are incorporated into national development strategies. The Programme supports reporting by countries, as well as regions and sub-regions, on progress towards achievement of the goals, and on specific social, economic and environmental indicators, through the formulation of MDG reports. These form the basis of a global report, issued annually by the UN Secretary-General since mid-2002. UNDP provides administrative and technical support to the Millennium Project, an independent advisory body established by the UN Secretary-General in 2002 to develop a practical action plan to achieve the MDGs. Financial support of the Project is channelled through a Millennium Trust Fund, administered by UNDP. In January 2005 the Millennium Project presented its report, based on extensive research conducted by teams of experts, which included recommendations for the international system to support country level development efforts and identified a series of Quick Wins to bring conclusive benefit to millions of people in the short-term. UNDP also works to raise awareness of the MDGs and to support advocacy efforts at all levels, for example through regional publicity campaigns, target-specific publications and support for the Millennium Campaign to generate support for the goals in developing and developed countries.

## DEMOCRATIC GOVERNANCE

UNDP supports national efforts to ensure efficient and accountable governance, improve the quality of democratic processes, and to build effective relations between the state, the private sector and civil society, which are essential to achieving sustainable development. As in other practice areas, UNDP assistance includes policy advice and technical support, capacity-building of institutions and individuals, advocacy and public information and communication, the promotion and brokering of dialogue, and knowledge networking and sharing of good practices. In March 2002 a UNDP Governance Centre was inaugurated in Oslo, Norway, to enhance the role of UNDP in support of democratic governance and to assist countries to implement democratic reforms in order to achieve the MDGs.

UNDP works to strengthen parliaments and other legislative bodies as institutions of democratic participation. It assists with constitutional reviews and reform, training of parliamentary staff, and capacity-building of political parties and civil organizations as part of this objective. UNDP undertakes missions to help prepare for and ensure the conduct of free and fair elections. Increasingly, UNDP is also focused on building the long-term capacity of electoral institutions and practices within a country, for example voter registration, election observation, the establishment of electoral commissions, and voter and civic education projects. Similarly, UNDP aims to ensure an efficient, independent and fair judicial system is available to all, in particular the poor and disadvantaged. Within its justice sector programme UNDP undertakes a variety of projects to improve access to justice. UNDP also works to promote access to information, the integration of human rights issues into activities concerned with sustainable human development, as well as support for the international human rights system.

Since 1997 UNDP has been mandated to assist developing countries to fight corruption and improve accountability, transparency and integrity (ATI). It has worked to establish national and international partnerships in support of its anti-corruption efforts and used its role as a broker of knowledge and experience to uphold ATI principles at all levels of public financial management and governance. UNDP publishes case studies of its anti-corruption efforts and assists governments to conduct self-assessments of their public financial management systems.

In February 2004 a UNDP-sponsored report entitled *Democracy in Latin America: Towards a Citizens' Democracy* was published, with the aim of promoting regional debate on means of strengthening democratic institutions.

Within the democratic governance practice area UNDP supports more than 300 projects at international, country and city levels designed to improve conditions for the urban poor, in particular through improvement in urban governance. The Local Initiative Facility for Urban Environment (LIFE) undertakes small-scale projects in low-income communities, in collaboration with local authorities, the private sector and community-based groups, and promotes a participatory approach to local governance. UNDP also works closely with the UN Capital Development Fund to implement projects in support of decentralized governance, which it has recognized as a key element to achieving sustainable development goals.

UNDP aims to ensure that, rather than creating an ever-widening 'digital divide', ongoing rapid advancements in information technology are harnessed by poorer countries to accelerate progress in achieving sustainable human development. UNDP advises governments on technology policy, promotes digital entrepreneurship in programme countries and works with private-sector partners to provide reliable and affordable communications networks. The Bureau for Development Policy operates the Information and Communication Technologies for Development Programme, which aims to promote sustainable human development through increased utilization of information and communications technologies globally. The Programme aims to establish technology access centres in developing countries. A Sustainable Development Networking Programme focuses on expanding internet connectivity in poorer countries through building national capacities and supporting local internet sites. UNDP has used mobile internet units to train people even in isolated rural areas. In 1999 UNDP, in collaboration with an international communications company, Cisco Systems, and other partners, launched NetAid, an internet-based forum (accessible at www.netaid.org) for mobilizing and co-ordinating fundraising and other activities aimed at alleviating poverty and promoting sustainable human development in the developing world. With Cisco Systems and other partners, UNDP has worked to establish academies of information technology to support training and capacity-building in developing countries. By September 2003 88 academies had been established. UNDP and the World Bank jointly host the secretariat of the Digital Opportunity Task Force, a partnership between industrialized and developing countries, business and non-governmental organizations that was established in 2000. UNDP is a partner in the Global Digital Technology Initiative, launched in 2002 to strengthen the role of information and communications technologies in achieving the development goals of developing countries. In January 2004 UNDP and Microsoft Corporation announced an agreement to develop jointly information and communication technology (ICT) projects aimed at assisting developing countries to achieve the MDGs.

## POVERTY REDUCTION

UNDP's activities to facilitate poverty eradication include support for capacity-building programmes and initiatives to generate sustainable livelihoods, for example by improving access to credit, land and technologies, and the promotion of strategies to improve education and health provision for the poorest elements of populations (with a focus on women and girls). UNDP aims to help governments to reassess their development priorities and to design initiatives for sustainable human development. In 1996, following the World Summit for Social Development, which was held in Copenhagen, Denmark, in March 1995, UNDP launched the Poverty Strategies Initiative (PSI) to strengthen national capacities to assess and monitor the extent of poverty and to combat the problem. All PSI projects were to involve representatives of governments, the private sector, social organizations and research institutions in policy debate and formulation. Following the introduction, in 1999, by the World Bank and IMF of Poverty Reduction Strategy Papers (PRSPs), UNDP has tended to direct its efforts to helping governments draft these documents, and, since 2001, has focused on linking the papers to efforts to achieve and monitoring progress towards the MDGs. In early 2004 UNDP inaugurated the International Poverty Centre, in Brasília, Brazil, which aimed to foster the capacity of countries to formulate and implement poverty reduction strategies and to encourage South-South co-operation in all relevant areas of research and decision-making. In particular, the Centre aimed to assist countries to meet Millennium goals and targets through the research and implementation of pro-poor growth policies and social protection and human development strategies, and the monitoring of poverty and inequality.

UNDP country offices support the formulation of national human development reports (NHDRs), which aim to facilitate activities such as policy-making, the allocation of resources and monitoring progress towards poverty eradication and sustainable development. In addition, the preparation of Advisory Notes and Country Co-operation Frameworks by UNDP officials helps to highlight country-specific aspects of poverty eradication and national strategic priorities. In January 1998 the Executive Board adopted eight guiding principles relating to sustainable human development that were to be implemented by all country offices, in order to ensure a focus to UNDP activities. Since 1990 UNDP has published an annual *Human Development Report*, incorporating a Human Development Index, which ranks countries in terms of human development, using three key indicators: life expectancy, adult literacy and basic income required for a decent standard of living. In 1997 a Human Poverty Index and a Gender-related Development Index, which assesses gender equality on the basis of life expectancy, education and income, were introduced into the Report for the first time. Also in 1997 a UNDP scheme to support private-sector and community-based initiatives to generate employment opportunities, MicroStart, became operational.

In 1996 UNDP initiated a process of collaboration between city authorities world-wide to promote implementation of the commitments made at the 1995 Copenhagen summit for social development and to help to combat aspects of poverty and other urban problems, such as poor housing, transport, the management of waste disposal, water supply and sanitation. The so-called World Alliance of Cities Against Poverty was formally launched in October 1997, in the context of the International Decade for the Eradication of Poverty. The first Forum of the Alliance was convened in October 1998, in Lyon, France; it has subsequently been held every two years.

UNDP sponsors the International Day for the Eradication of Poverty, held annually on 17 October.

## ENERGY AND THE ENVIRONMENT

UNDP plays a role in developing the agenda for international co-operation on environmental and energy issues, focusing on the relationship between energy policies, environmental protection, poverty and development. UNDP promotes the development of national capacities and other strategies that support sustainable development practices, for example through the formulation and implementation of Poverty Reduction Strategies and National Strategies for Sustainable Development. UNDP works to ensure the effective governance of freshwater and aquatic resources, and promotes co-operation in transboundary water management. It works closely with other agencies to promote safe sanitation, ocean and coastal management, and community water supplies. Another priority area of UNDP's Energy and Environment Practice is to promote clean energy technologies (through the Clean Development Mechanism) and to extend access to sustainable energy services, including the introduction of renewable alternatives to conventional fuels, as well as access to investment financing for sustainable energy.

UNDP recognizes that desertification and land degradation is a major cause of rural poverty and promotes sustainable land management, drought preparedness and reform of land tenure as means of addressing the problem. It also aims to reduce poverty caused by land degradation through implementation of environmental conventions at a national and international level. In 2002 UNDP inaugurated an Integrated Drylands Development Programme which aimed to ensure that the needs of people living in drylands are met and considered at a local and national level. By 2005 the programme was being implemented, by the Drylands Development Centre, in 19 African, Arab and West Asian countries. UNDP is also concerned with sustainable management of forestries, fisheries and agriculture. Its Biodiversity Global Programme assists developing countries and communities to integrate issues relating to sustainable practices and biodiversity into national and global practices. Since 1992 UNDP has administered a Small Grants Programme, funded by the Global Environment Facility, to support community-based initiatives concerned with biodiversity conservation, prevention of land degradation and the elimination of persistent organic pollutants. The Equator Initiative was inaugurated in 2002 as a partnership between UNDP, representatives of governments, civil society and businesses, with the aim of reducing poverty in communities along the equatorial belt by fostering local partnerships, harnessing local knowledge and promoting conservation and sustainable practices.

### CRISIS PREVENTION AND RECOVERY

UNDP collaborates with other UN agencies in countries in crisis and with special circumstances to promote relief and development efforts, in order to secure the foundations for sustainable human development and thereby increase national capabilities to prevent or pre-empt future crises. In particular, UNDP is concerned to achieve reconciliation, reintegration and reconstruction in affected countries, as well as to support emergency interventions and management and delivery of programme aid. Special development initiatives include the demobilization of former combatants and destruction of illicit small armaments, rehabilitation of communities for the sustainable reintegration of returning populations and the restoration and strengthening of democratic institutions. UNDP is seeking to incorporate conflict prevention into its development strategies. The Conflict-related Development Analysis (CDA), building upon conflict assessment activities undertaken during 2001–02 in countries including Guatemala, Guinea-Bissau, Nepal, Nigeria and Tajikistan, is being developed as a tool for country offices to use in formulating and analyzing programmes in conflict zones. UNDP has established a mine action unit within its Bureau for Crisis Prevention and Recovery in order to strengthen national de-mining capabilities including surveying, mapping and clearance of anti-personnel landmines. UNDP also works closely with UNICEF to raise mine awareness and implement risk reduction education programmes.

UNDP is the focal point within the UN system for strengthening national capacities for natural disaster reduction (prevention, preparedness and mitigation relating to natural, environmental and technological hazards). UNDP's Bureau of Crisis Prevention and Recovery oversees the system-wide Disaster Management Training Programme, which was established in 1991. By the end of 2004 the Programme had conducted more than 70 workshops benefiting some 6,000 participants. In February 2004 UNDP introduced a Disaster Risk Index that enabled vulnerability and risk to be measured and compared between countries and demonstrated the correspondence between human development and death rates following natural disasters. UNDP was actively involved in preparations for the second World Conference on Disaster Reduction which was held in Kobe, Japan, in January 2005.

In September 2002 a meeting on strengthening regional and national capacities for natural disaster reduction over the period 2003–05, convened jointly by UNDP and the Co-ordination Centre for the Prevention of Natural Disasters in Central America (CEPREDENAC), identified 81 projects requiring donor funding. UNDP also co-operates with the Caribbean Disaster Emergency Response Agency (CDERA). In 2005 UNDP was supporting national recovery activities in Colombia.

### HIV/AIDS

UNDP regards the HIV/AIDS pandemic as a major challenge to development, and advocates for making HIV/AIDS a focus of national planning; supports decentralized action against HIV/AIDS at community level; helps to strengthen national capacities at all levels to combat the disease; and aims to link support for prevention activities, education and treatment with broader development planning and responses. UNDP places a particular focus on combating the spread of HIV/AIDS through the promotion of women's rights. UNDP is a co-sponsor, jointly with WHO, the World Bank, UNICEF, UNESCO, UNODC, ILO, UNFPA, WFP and UNHCR, of the Joint UN Programme on HIV/AIDS (UNAIDS), which became operational on 1 January 1996. UNAIDS co-ordinates UNDP's HIV and Development Programme.

## Finance

UNDP and its various funds and programmes are financed by the voluntary contributions of members of the United Nations and the Programme's participating agencies, as well as through cost-sharing by recipient governments and third-party donors. In 2004–05 total voluntary contributions were projected at US $3,500m., of which a projected $1,700m. constituted regular (core) resources and $1,807m. third-party co-financing and thematic trust fund income. Cost-sharing by programme country governments was projected at $2,100m., bringing total resources (both donor and local) to a projected $5,600m.

## Publications

*Annual Report of the Administrator.*
*Choices* (quarterly).
*Global Public Goods: International Co-operation in the 21st Century.*
*Human Development Report* (annually, also available on CD-ROM).
*Poverty Report* (annually).
*Results-Oriented Annual Report.*

## Associated Funds and Programmes

UNDP is the central funding, planning and co-ordinating body for technical co-operation within the UN system. A number of associated funds and programmes, financed separately by means of voluntary contributions, provide specific services through the UNDP network. UNDP manages a trust fund to promote economic and technical co-operation among developing countries.

### CAPACITY 2015

UNDP initiated Capacity 2015 at the World Summit for Sustainable Development, which was held in August–September 2002. Capacity 2015 aims to support developing countries in expanding their capabilities to meet the Millennium Development Goals pledged by governments at a summit meeting of the UN General Assembly in September 2000.

### GLOBAL ENVIRONMENT FACILITY (GEF)

The GEF, which is managed jointly by UNDP, the World Bank and UNEP, began operations in 1991 and was restructured in 1994. Its aim is to support projects concerning climate change, the conservation of biological diversity, the protection of international waters, reducing the depletion of the ozone layer in the atmosphere, and (since October 2002) arresting land degradation and addressing the issue of persistent organic pollutants. The GEF acts as the financial mechanism for the Convention on Biological Diversity and the UN Framework Convention on Climate Change. UNDP is responsible for capacity-building, targeted research, pre-investment activities and technical assistance. UNDP also administers the Small Grants Programme of the GEF, which supports community-based activities by local non-governmental organizations, and the Country Dialogue Workshop Programme, which promotes dialogue on national priorities with regard to the GEF. Some 32 donor countries pledged US $2,920m. for the third periodic replenishment of GEF funds (GEF-3), covering the period 2002–06. During 1991–2003 the GEF allocated $4,500m. in grants and raised $14,474m. in co-financing from other sources in support of more than 1,400 projects.

**Chair. and CEO:** Dr LEONARD GOOD (Canada).

## MONTREAL PROTOCOL

Through its Montreal Protocol Unit UNDP collaborates with public and private partners in developing countries to assist them in eliminating the use of ozone-depleting substances (ODS), in accordance with the Montreal Protocol to the Vienna Convention for the Protection of the Ozone Layer, through the design, monitoring and evaluation of ODS phase-out projects and programmes. In particular, UNDP provides technical assistance and training, national capacity-building and demonstration projects and technology transfer investment projects. By mid-2003, through the Executive Committee of the Montreal Protocol, UNDP had implemented projects and activities resulting in the elimination of 33,529 metric tons of ODS.

## UNDP DRYLANDS DEVELOPMENT CENTRE (DDC)

The Centre, based in Nairobi, Kenya, was established in February 2002, superseding the former UN Office to Combat Desertification and Drought (UNSO). (UNSO had been established following the conclusion, in October 1994, of the UN Convention to Combat Desertification in Those Countries Experiencing Serious Drought and/or Desertification, Particularly in Africa; in turn, UNSO had replaced the former UN Sudano-Sahelian Office.) The DDC was to focus on the following areas: ensuring that national development planning takes account of the needs of dryland communities, particularly in poverty reduction strategies; helping countries to cope with the effects of climate variability, especially drought, and to prepare for future climate change; and addressing local issues affecting the utilization of resources.

**Director:** PHILIP DOBIE (United Kingdom).

## UNITED NATIONS CAPITAL DEVELOPMENT FUND (UNCDF)

The Fund was established in 1966 and became fully operational in 1974. It invests in poor communities in least-developed countries through local governance projects and microfinance operations, with the aim of increasing such communities' access to essential local infrastructure and services and thereby improving their productive capacities and self-reliance. UNDCF encourages participation by local people and local governments in the planning, implementation and monitoring of projects. The Fund aims to promote the interests of women in community projects and to enhance their earning capacities. In 1998 the Fund nominated 15 less-developed countries in which to concentrate subsequent programmes. A Special Unit for Microfinance (SUM), established in 1997 as a joint UNDP/UNCDF operation, was fully integrated into UNCDF in 1999. UNDCF/SUM helps to develop financial services for poor communities and supports UNDP's MicroStart initiative. In 2003 the UN Secretary-General designated UNCDF, with the UN Department of Economic and Social Affairs, as a co-ordinator of the International Year of Microcredit in 2005. UNCDF's annual programming budget amounts to some US $40m.

**Officer-in-Charge:** HENRIETTE KEIJZERS.

## UNITED NATIONS DEVELOPMENT FUND FOR WOMEN (UNIFEM)

UNIFEM is the UN's lead agency in addressing the issues relating to women in development and promoting the rights of women worldwide. The Fund provides direct financial and technical support to enable low-income women in developing countries to increase earnings, gain access to labour-saving technologies and otherwise improve the quality of their lives. It also funds activities that include women in decision-making related to mainstream development projects. In 2001 UNIFEM's Trust Fund in Support of Actions to Eliminate Violence Against Women (established in 1996) provided grants to 21 national and regional programmes. During 1996–2004 the Trust Fund awarded grants totalling US $8.3m. in support of 175 initiatives in 96 countries. UNIFEM has supported the preparation of national reports in 30 countries and used the priorities identified in these reports and in other regional initiatives to formulate a Women's Development Agenda for the 21st century. Through these efforts, UNIFEM played an active role in the preparation for the UN Fourth World Conference on Women, which was held in Beijing, People's Republic of China, in September 1995. UNIFEM participated at a special session of the General Assembly convened in June 2000 to review the conference, entitled Women 2000: Gender Equality, Development and Peace for the 21st Century (Beijing + 5). In March 2001 UNIFEM, in collaboration with International Alert, launched a Millennium Peace Prize for Women. UNIFEM maintains that the empowerment of women is a key to combating the HIV/AIDS pandemic, in view of the fact that women and adolescent girls are often culturally, biologically and economically more vulnerable to infection and more likely to bear responsibility for caring for the sick. In March 2002 UNIFEM launched a three-year programme aimed at making the gender and human rights dimensions of the pandemic central to policy-making in ten countries. A new online resource (www.genderandaids.org) on the gender dimensions of HIV/AIDS was launched in February 2003. UNIFEM was a co-founder of WomenWatch (accessible online at www.un.org/womenwatch), a UN system-wide resource for the advancement of gender equality. Following the massive earthquake and tsunamis that struck parts of the Indian Ocean in late December 2004, UNIFEM undertook to promote the needs and rights of women and girls in all emergency relief and reconstruction efforts, in particular in Indonesia, Sri Lanka and Somalia, and supported capacity-building of grass-roots organizations. Programme expenditure in 2003 totalled $27.0m.

**Director:** NOELEEN HEYZER (Singapore), Headquarters: 304 East 45th St, 15th Floor, New York, NY 10017, USA; tel. (212) 906-6400; fax (212) 906-6705; e-mail unifem@undp.org; internet www.unifem.org.

## UNITED NATIONS VOLUNTEERS (UNV)

The United Nations Volunteers is an important source of middle-level skills for the UN development system supplied at modest cost, particularly in the least-developed countries. Volunteers expand the scope of UNDP project activities by supplementing the work of international and host-country experts and by extending the influence of projects to local community levels. UNV also supports technical co-operation within and among the developing countries by encouraging volunteers from the countries themselves and by forming regional exchange teams comprising such volunteers. UNV is involved in areas such as peace-building, elections, human rights, humanitarian relief and community-based environmental programmes, in addition to development activities.

The UN International Short-term Advisory (UNISTAR) Programme, which is the private-sector development arm of UNV, has increasingly focused its attention on countries in the process of economic transition. Since 1994 UNV has administered UNDP's Transfer of Knowledge Through Expatriate Nationals (TOKTEN) programme, which was initiated in 1977 to enable specialists and professionals from developing countries to contribute to development efforts in their countries of origin through short-term technical assignments.

At the end of March 2005 4,884 UNVs were serving in 135 countries. At that time the total number of people who had served under the initiative amounted to more than 30,000 in some 140 countries.

**Executive Co-ordinator:** AD DE RAAD (Netherlands), Headquarters: POB 260111, 53153 Bonn, Germany; tel. (228) 8152000; fax (228) 8152001; e-mail information@unvolunteers.org; internet www.unv.org.

# United Nations Environment Programme—UNEP

**Address:** POB 30552, Nairobi, Kenya.
**Telephone:** (20) 621234; **fax:** (20) 624489; **e-mail:** cpiinfo@unep.org; **internet:** www.unep.org.

The United Nations Environment Programme was established in 1972 by the UN General Assembly, following recommendations of the 1972 UN Conference on the Human Environment, in Stockholm, Sweden, to encourage international co-operation in matters relating to the human environment.

## Organization
(July 2005)

### GOVERNING COUNCIL

The main functions of the Governing Council, which meets every two years, are to promote international co-operation in the field of the environment and to provide general policy guidance for the direction and co-ordination of environmental programmes within the UN system. It comprises representatives of 58 states, elected by the UN General Assembly, for four-year terms, on a regional basis. The Council is assisted in its work by a Committee of Permanent Representatives.

### HIGH-LEVEL COMMITTEE OF MINISTERS AND OFFICIALS IN CHARGE OF THE ENVIRONMENT

The Committee was established by the Governing Council in 1997, with a mandate to consider the international environmental agenda and to make recommendations to the Council on reform and policy issues. In addition, the Committee, comprising 36 elected members, was to provide guidance and advice to the Executive Director, to enhance UNEP's collaboration and co-operation with other multilateral bodies and to help to mobilize financial resources for UNEP.

### SECRETARIAT

Offices and divisions at UNEP headquarters include the Office of the Executive Director; the Secretariat for Governing Bodies: Offices for Evaluation and Oversight, Programme Co-ordination and Management, and Resource Mobilization; and divisions of communications and public information, early warning and assessment, policy development and law, policy implementation, technology and industry and economics, regional co-operation and representation, environmental conventions, and Global Environment Facility co-ordination.

**Executive Director:** Dr KLAUS TÖPFER (Germany).

### REGIONAL OFFICES

**Latin America and the Caribbean:** Blvd de los Virreyes 155, Lomas Virreyes, 11000 México, DF, Mexico; tel. (55) 52024841; fax (55) 52020950; e-mail registro@pnuma.org; internet www.pnuma.org.

**UNEP Brasília Office:** SCN 0.2 Bloco A-11 andar, Brasília, Brazil; tel. (61) 329-2113; e-mail unep.brazil@undp.org.br.

### OTHER OFFICES

**Convention on International Trade in Endangered Species of Wild Fauna and Flora—CITES:** 15 chemin des Anémones, 1219 Châtelaine, Geneva, Switzerland; tel. (22) 9178139; fax (22) 7973417; e-mail cites@unep.ch; internet www.cites.org; Sec.-Gen. WILLEM WOUTER WIJNSTEKERS (Netherlands).

**Global Programme of Action for the Protection of the Marine Environment from Land-based Activities:** POB 16227, 2500 BE The Hague, Netherlands; tel. (70) 3114460; fax (70) 3456648; e-mail gpa@unep.nl; internet www.gpa.unep.org; Co-ordinator Dr VEERLE VANDEWEERD.

**Regional Co-ordinating Unit for the Caribbean Environment Programme:** 14–20 Port Royal St, Kingston, Jamaica; tel. 9229267; fax 9229292; e-mail uneprcuja@cwjamaica.com; internet www.cep.unep.org; Co-ordinator NELSON ANDRADE COLMENARES.

**Secretariat of the Basel Convention:** CP 356, 13–15 chemin des Anémones, 1219 Châtelaine, Geneva, Switzerland; tel. (22) 9178218; fax (22) 7973454; e-mail sbc@unep.ch; internet www.basel.int; Exec. Sec. SACHIKO KUWABARA-YAMAMOTO.

**Secretariat of the Multilateral Fund for the Implementation of the Montreal Protocol:** 1800 McGill College Ave, 27th Floor, Montréal, QC, Canada H3A 3J6; tel. (514) 282-1122; fax (514) 282-0068; e-mail secretariat@unmfs.org; internet www.multilateralfund.org; Chief MARIA NOLAN (United Kingdom).

**Secretariat of the UN Framework Convention on Climate Change:** Haus Carstanjen, Martin-Luther-King-Str. 8, 53175 Bonn, Germany; tel. (228) 815-1000; fax (228) 815-1999; e-mail secretariat@unfccc.de; internet www.unfccc.de; Exec. Sec. JOKE WALLER-HUNTER (Netherlands).

**UNEP/CMS (Convention on the Conservation of Migratory Species of Wild Animals) Secretariat:** Martin-Luther-King-Str. 8, 53175 Bonn, Germany; tel. (228) 8152402; fax (228) 8152449; e-mail secretariat@cms.int; internet www.cms.int; Exec. Sec. ROBERT HEPWORTH.

**UNEP Chemicals:** International Environment House, 11–13 chemin des Anémones, 1219 Châtelaine, Geneva, Switzerland; tel. (22) 9178192; fax (22) 7973460; e-mail chemicals@unep.ch; internet www.chem.unep.ch; Dir JAMES B. WILLIS.

**UNEP Division of Technology, Industry and Economics:** Tour Mirabeau, 39–43, Quai André Citroën, 75739 Paris Cédex 15, France; tel. 1-44-37-14-41; fax 1-44-37-14-74; e-mail unep.tie@unep.fr; internet www.uneptie.org/; Dir MONIQUE BARBUT (France).

**UNEP International Environmental Technology Centre—IETC:** 2–110 Ryokuchi koen, Tsurumi-ku, Osaka 538-0036, Japan; tel. (6) 6915-4581; fax (6) 6915-0304; e-mail ietc@unep.or.jp; internet www.unep.or.jp.

**UNEP Ozone Secretariat:** POB 30552, Nairobi, Kenya; tel. (20) 624691; fax (20) 623913; e-mail ozoneinfo@unep.org; internet www.unep.org/ozone/; Exec. Sec. MARCO GONZÁLEZ (Costa Rica).

**UNEP-SCBD (Convention on Biological Diversity—Secretariat):** 413 St Jacques St, Office 800, Montréal, QC, Canada H2Y 1N9; tel. (514) 288-2220; fax (514) 288-6588; e-mail secretariat@biodiv.org; internet www.biodiv.org; Exec. Sec. HAMDALLAH ZEDAN (until 31 Dec. 2005), AHMED DJOGHLAF (Algeria) (designate).

**UNEP Secretariat for the UN Scientific Committee on the Effects of Atomic Radiation:** Vienna International Centre, Wagramerstrasse 5, POB 500, 1400 Vienna, Austria; tel. (1) 26060-4330; fax (1) 26060-5902; e-mail norman.gentner@unvienna.org; internet www.unscear.org; Sec. Dr NORMAN GENTNER.

## Activities

UNEP serves as a focal point for environmental action within the UN system. It aims to maintain a constant watch on the changing state of the environment; to analyse the trends; to assess the problems using a wide range of data and techniques; and to promote projects leading to environmentally sound development. It plays a catalytic and co-ordinating role within and beyond the UN system. Many UNEP projects are implemented in co-operation with other UN agencies, particularly UNDP, the World Bank group, FAO, UNESCO and WHO. About 45 intergovernmental organizations outside the UN system and 60 international non-governmental organizations have official observer status on UNEP's Governing Council, and, through the Environment Liaison Centre in Nairobi, UNEP is linked to more than 6,000 non-governmental bodies concerned with the environment. UNEP also sponsors international conferences, programmes, plans and agreements regarding all aspects of the environment.

In February 1997 the Governing Council, at its 19th session, adopted a ministerial declaration (the Nairobi Declaration) on UNEP's future role and mandate, which recognized the organization as the principal UN body working in the field of the environment and as the leading global environmental authority, setting and overseeing the international environmental agenda. In June a special session of the UN General Assembly, referred to as 'Rio + 5', was convened to review the state of the environment and progress achieved in implementing the objectives of the UN Conference on Environment and Development (UNCED), held in Rio de Janeiro, Brazil, in June 1992. The meeting adopted a Programme for Further Implementation of Agenda 21 (a programme of activities to promote sustainable development, adopted by UNCED) in order to intensify efforts in areas such as energy, freshwater resources and technology transfer. The meeting confirmed UNEP's essential role in advancing the Programme and as a global authority promoting a coherent legal and political approach to the environmental challenges of sustainable development. An extensive process of restructuring and realignment of functions was subsequently initiated by UNEP, and a new organizational structure reflecting the decisions of the Nairobi Declaration was implemented during 1999. UNEP played a leading role in preparing for the World Summit on Sustainable Development (WSSD), held in August–September 2002 in Johannesburg, South

Africa, to assess strategies for strengthening the implementation of Agenda 21. Governments participating in the conference adopted the Johannesburg Declaration and WSSD Plan of Implementation, in which they strongly reaffirmed commitment to the principles underlying Agenda 21 and also pledged support to all internationally-agreed development goals, including the UN Millennium Development Goals adopted by governments attending a summit meeting of the UN General Assembly in September 2000. Participating governments made concrete commitments to attaining several specific objectives in the areas of water, energy, health, agriculture and fisheries, and biodiversity. These included a reduction by one-half in the proportion of people world-wide lacking access to clean water or good sanitation by 2015, the restocking of depleted fisheries by 2015, a reduction in the ongoing loss in biodiversity by 2010, and the production and utilization of chemicals without causing harm to human beings and the environment by 2020. Participants determined to increase usage of renewable energy sources and to develop by 2005 integrated water resources management and water efficiency plans. A large number of partnerships between governments, private-sector interests and civil society groups were announced at the conference.

In May 2000 UNEP sponsored the first annual Global Ministerial Environment Forum (GMEF), held in Malmö, Sweden, and attended by environment ministers and other government delegates from more than 130 countries. Participants reviewed policy issues in the field of the environment and addressed issues such as the impact on the environment of population growth, the depletion of earth's natural resources, climate change and the need for fresh water supplies. The Forum issued the Malmö Declaration, which identified the effective implementation of international agreements on environmental matters at national level as the most pressing challenge for policy-makers. The Declaration emphasized the importance of mobilizing domestic and international resources and urged increased co-operation from civil society and the private sector in achieving sustainable development. The second GMEF, held in Nairobi in February 2001, addressed means of strengthening international environmental governance, establishing an Open-Ended Intergovernmental Group of Ministers or Their Representatives (IGM) to prepare a report on possible reforms. GMEF-6, scheduled to be held in February 2006 in Dubai, United Arab Emirates, was to consider energy and the environment and chemicals management.

## ENVIRONMENTAL ASSESSMENT AND EARLY WARNING

The Nairobi Declaration resolved that the strengthening of UNEP's information, monitoring and assessment capabilities was a crucial element of the organization's restructuring, in order to help establish priorities for international, national and regional action, and to ensure the efficient and accurate dissemination of emerging environmental trends and emergencies.

In 1995 UNEP launched the Global Environment Outlook (GEO) process of environmental assessment. UNEP is assisted in its analysis of the state of the global environment by an extensive network of collaborating centres. The *GEO Year Book* is published annually. The following regional and national *GEO* reports have been produced in recent years: *Africa Environment Outlook* (2002), *Brazil Environment Outlook* (2002), *Caucasus Environment Outlook* (2002), *North America's Environment* (2002), *Latin America and the Caribbean Environment Outlook* (2003), *Andean Environment Outlook* (2003), *Pacific Environment Outlook* (2005), *Caribbean Environment Outlook* (2005), and *Atlantic and Indian Oceans Environment Outlook* (2005). UNEP is leading a major Global International Waters Assessment (GIWA) to consider all aspects of the world's water-related issues, in particular problems of shared transboundary waters, and of future sustainable management of water resources. UNEP is also a sponsoring agency of the Joint Group of Experts on the Scientific Aspects of Marine Environmental Pollution and contributes to the preparation of reports on the state of the marine environment and on the impact of land-based activities on that environment. In November 1995 UNEP published a Global Biodiversity Assessment, which was the first comprehensive study of biological resources throughout the world. The UNEP—World Conservation Monitoring Centre (UNEP—WCMC), established in June 2000, provides biodiversity-related assessment. UNEP is a partner in the International Coral Reef Action Network—ICRAN, which was established in 2000 to manage and protect coral reefs world-wide. In June 2001 UNEP launched the Millennium Ecosystems Assessment, which was completed in March 2005. Other major assessments undertaken included GIWA (see above); the Assessment of Impact and Adaptation to Climate Change; the Solar and Wind Energy Resource Assessment; the Regionally-Based Assessment of Persistent Toxic Substances; the Land Degradation Assessment in Drylands; and the Global Methodology for Mapping Human Impacts on the Biosphere (GLOBIO) project.

UNEP's environmental information network includes the Global Resource Information Database (GRID), which converts collected data into information usable by decision-makers. The UNEP-INFOTERRA programme facilitates the exchange of environmental information through an extensive network of national 'focal points'. By March 2005 177 countries were participating in the network. Through UNEP-INFOTERRA UNEP promotes public access to environmental information, as well as participation in environmental concerns. UNEP aims to establish in every developing region an Environment and Natural Resource Information Network (ENRIN) in order to make available technical advice and manage environmental information and data for improved decision-making and action-planning in countries most in need of assistance. UNEP aims to integrate its information resources in order to improve access to information and to promote its international exchange. This has been pursued through UNEPnet, an internet-based interactive environmental information- and data-sharing facility, and Mercure, a telecommunications service using satellite technology to link a network of 16 earth stations throughout the world.

UNEP's information, monitoring and assessment structures also serve to enhance early-warning capabilities and to provide accurate information during an environmental emergency.

## POLICY DEVELOPMENT AND LAW

UNEP aims to promote the development of policy tools and guide-lines in order to achieve the sustainable management of the world environment. At a national level it assists governments to develop and implement appropriate environmental instruments and aims to co-ordinate policy initiatives. Training workshops in various aspects of environmental law and its applications are conducted. UNEP supports the development of new legal, economic and other policy instruments to improve the effectiveness of existing environmental agreements.

UNEP was instrumental in the drafting of a Convention on Biological Diversity (CBD) to preserve the immense variety of plant and animal species, in particular those threatened with extinction. The Convention entered into force at the end of 1993; by April 2005 187 countries and the European Community were parties to the CBD. The CBD's Cartagena Protocol on Biosafety (so called as it had been addressed at an extraordinary session of parties to the CBD convened in Cartagena, Colombia, in February 1999) was adopted at a meeting of parties to the CBD held in Montréal, Canada, in January 2000, and entered into force in September 2003; by April 2005 the Protocol had been ratified by 118 countries and the European Community. The Protocol regulates the transboundary movement and use of living modified organisms resulting from biotechnology in order to reduce any potential adverse effects on biodiversity and human health. It establishes an Advanced Informed Agreement procedure to govern the import of such organisms. In January 2002 UNEP launched a major project aimed at supporting developing countries with assessing the potential health and environmental risks and benefits of genetically modified (GM) crops, in preparation for the Protocol's entry into force. In February the parties to the CBD and other partners convened a conference, in Montréal, to address ways in which the traditional knowledge and practices of local communities could be preserved and used to conserve highly threatened species and ecosystems. The sixth conference of parties to the CBD, held in April 2002, adopted detailed voluntary guide-lines concerning access to genetic resources and sharing the benefits attained from such resources with the countries and local communities where they originate; a global work programme on forests; and a set of guiding principles for combating alien invasive species. UNEP supports co-operation for biodiversity assessment and management in selected developing regions and for the development of strategies for the conservation and sustainable exploitation of individual threatened species (e.g. the Global Tiger Action Plan). It also provides assistance for the preparation of individual country studies and strategies to strengthen national biodiversity management and research. UNEP administers the Convention on International Trade in Endangered Species of Wild Flora and Fauna (CITES), which entered into force in 1975.

UNEP is the lead UN agency for promoting environmentally sustainable water management. It regards the unsustainable use of water as the most urgent environmental and sustainable development issue, and estimates that two-thirds of the world's population will suffer chronic water shortages by 2025, owing to rising demand for drinking water as a result of growing populations, decreasing quality of water because of pollution, and increasing requirements of industries and agriculture. In 2000 UNEP adopted a new water policy and strategy, comprising assessment, management and co-ordination components. The Global International Waters Assessment (see above) is the primary framework for the assessment component. The management component includes the Global Programme of Action (GPA) for the Protection of the Marine Environment from Land-based Activities (adopted in November 1995), and UNEP's freshwater programme and regional seas programme. The GPA for the Protection of the Marine Environment for Land-based Activities focuses on the effects of activities such as pollution on freshwater resources, marine biodiversity and the coastal ecosystems of small-island developing states. UNEP aims to develop a similar global instrument to ensure the integrated management of freshwater resources. It promotes international co-operation in the management of river basins and coastal areas and for the development of tools and guide-lines to achieve the sustainable management of freshwater and coastal resources. UNEP provides scientific, technical and administrative support to facilitate the implementation and co-ordination of 14 regional seas conventions and 13 regional plans of action, and is developing a strategy to strengthen collaboration in their implementation. The new water policy and strategy emphasizes the need for improved co-ordination of existing activities. UNEP aims to play an enhanced role within relevant co-ordination mechanisms, such as the UN open-ended informal consultation process on oceans and the law of the sea.

In 1996 UNEP, in collaboration with FAO, began to work towards promoting and formulating a legally binding international convention on prior informed consent (PIC) for hazardous chemicals and pesticides in international trade, extending a voluntary PIC procedure of information exchange undertaken by more than 100 governments since 1991. The Convention was adopted at a conference held in Rotterdam, Netherlands, in September 1998, and entered into force in February 2004. It aims to reduce risks to human health and the environment by restricting the production, export and use of hazardous substances and enhancing information exchange procedures.

In conjunction with UN-Habitat, UNDP, the World Bank and other organizations and institutions, UNEP promotes environmental concerns in urban planning and management through the Sustainable Cities Programme, as well as regional workshops concerned with urban pollution and the impact of transportation systems. In 1994 UNEP inaugurated an International Environmental Technology Centre (IETC), with offices in Osaka and Shiga, Japan, in order to strengthen the capabilities of developing countries and countries with economies in transition to promote environmentally sound management of cities and freshwater reservoirs through technology co-operation and partnerships.

UNEP has played a key role in global efforts to combat risks to the ozone layer, resultant climatic changes and atmospheric pollution. UNEP worked in collaboration with the World Meteorological Organization to formulate the UN Framework Convention on Climate Change (UNFCCC), with the aim of reducing the emission of gases that have a warming effect on the atmosphere, and has remained an active participant in the ongoing process to review and enforce the implementation of the Convention and of its Kyoto Protocol. UNEP was the lead agency in formulating the 1987 Montreal Protocol to the Vienna Convention for the Protection of the Ozone Layer (1985), which provided for a 50% reduction in the production of chlorofluorocarbons (CFCs) by 2000. An amendment to the Protocol was adopted in 1990, which required complete cessation of the production of CFCs by 2000 in industrialized countries and by 2010 in developing countries; these deadlines were advanced to 1996 and 2006, respectively, in November 1992. In 1997 the ninth Conference of the Parties (COP) to the Vienna Convention adopted a further amendment which aimed to introduce a licensing system for all controlled substances. The eleventh COP, meeting in Beijing, People's Republic of China, in November–December 1999, adopted the Beijing Amendment, which imposed tighter controls on the import and export of hydrochlorofluorocarbons, and on the production and consumption of bromochloromethane (Halon-1011, an industrial solvent and fire extinguisher). The Beijing Amendment entered into force in December 2001. A Multilateral Fund for the Implementation of the Montreal Protocol was established in June 1990 to promote the use of suitable technologies and the transfer of technologies to developing countries. UNEP, UNDP, the World Bank and UNIDO are the sponsors of the Fund, which by 30 June 2004 had approved financing for more than 4,300 projects in 134 developing countries at a cost of US $1,631m. Commitments of $474m. were made to the fifth replenishment of the Fund, covering the three-year period 2003–05.

## POLICY IMPLEMENTATION

UNEP's Division of Environmental Policy Implementation incorporates two main functions: technical co-operation and response to environmental emergencies.

With the UN Office for the Co-ordination of Humanitarian Assistance (OCHA), UNEP has established a joint Environment Unit to mobilize and co-ordinate international assistance and expertise for countries facing environmental emergencies and natural disasters. In mid-1999 UNEP and UN-Habitat jointly established a Balkan Task Force (subsequently renamed UNEP Balkans Unit) to assess the environmental impact of NATO's aerial offensive against the Federal Republic of Yugoslavia (now Serbia and Montenegro). In November 2000 the Unit led a field assessment to evaluate reports of environmental contamination by debris from NATO ammunition containing depleted uranium. A final report, issued by UNEP in March 2001, concluded that there was no evidence of widespread contamination of the ground surface by depleted uranium and that the radiological and toxicological risk to the local population was negligible. It stated, however, that considerable scientific uncertainties remained, for example as to the safety of groundwater and the longer-term behaviour of depleted uranium in the environment, and recommended precautionary action. In December 2001 UNEP established a new Post-conflict Assessment Unit, which replaced, and extended the scope of, the Balkans Unit. In 2005 the Post-conflict Assessment Unit was undertaking activities in Afghanistan, and was compiling desk assessments of the state of the environment in Iraq, Liberia and the Palestinian territories.

UNEP, together with UNDP and the World Bank, is an implementing agency of the Global Environment Facility (GEF), which was established in 1991 as a mechanism for international co-operation in projects concerned with biological diversity, climate change, international waters and depletion of the ozone layer. UNEP services the Scientific and Technical Advisory Panel, which provides expert advice on GEF programmes and operational strategies.

## TECHNOLOGY, INDUSTRY AND ECONOMICS

The use of inappropriate industrial technologies and the widespread adoption of unsustainable production and consumption patterns have been identified as being inefficient in the use of renewable resources and wasteful, in particular in the use of energy and water. UNEP aims to encourage governments and the private sector to develop and adopt policies and practices that are cleaner and safer, make efficient use of natural resources, incorporate environmental costs, ensure the environmentally sound management of chemicals, and reduce pollution and risks to human health and the environment. In collaboration with other organizations and agencies UNEP works to define and formulate international guide-lines and agreements to address these issues. UNEP also promotes the transfer of appropriate technologies and organizes conferences and training workshops to provide sustainable production practices. Relevant information is disseminated through the International Cleaner Production Information Clearing House. UNEP, together with UNIDO, has established 27 National Cleaner Production Centres to promote a preventive approach to industrial pollution control. In October 1998 UNEP adopted an International Declaration on Cleaner Production, with a commitment to implement cleaner and more sustainable production methods and to monitor results; the Declaration had 529 signatories at January 2005, including representatives of 54 national governments. In 1997 UNEP and the Coalition for Environmentally Responsible Economies initiated the Global Reporting Initiative, which, with participation by corporations, business associations and other organizations and stakeholders, develops guide-lines for voluntary reporting by companies on their economic, environmental and social performance. In April 2002 UNEP launched the 'Life-Cycle Initiative', which aims to assist governments, businesses and other consumers with adopting environmentally sound policies and practice, in view the upward trend in global consumption patterns.

UNEP provides institutional servicing to the Basel Convention on the Control of Transboundary Movements of Hazardous Wastes and their Disposal, which was adopted in 1989 with the aim of preventing the disposal of wastes from industrialized countries in countries that have no processing facilities. In March 1994 the second meeting of parties to the Convention determined to ban the exportation of hazardous wastes between industrialized and developing countries. The third meeting of parties to the Convention, held in 1995, proposed that the ban should be incorporated into the Convention as an amendment. The resulting so-called Ban Amendment (prohibiting exports of hazardous wastes for final disposal and recycling from states and/or parties also belonging to OECD and, or, the European Union, and from Liechtenstein, to any other state party to the Convention) required ratification by three-quarters of the 62 signatory states present at the time of adoption before it could enter into effect; by April 2005 the Ban Amendment had been ratified by 56 parties. In 1998 the technical working group of the Convention agreed a new procedure for clarifying the classification and characterization of specific hazardous wastes. The fifth full meeting of parties to the Convention, held in December 1999, adopted the Basel Declaration outlining an agenda for the period 2000–10, with a particular focus on minimizing the production of hazardous wastes. At April 2005 the number of parties to the Convention totalled 165. In December 1999 132 states adopted a Protocol to the Convention to address issues relating to liability and compensation for damages from waste exports. The governments also agreed to establish a multilateral fund to finance immediate clean-up operations following any environmental accident.

The UNEP Chemicals office was established to promote the sound management of hazardous substances, central to which has been the International Register of Potentially Toxic Chemicals (IRPTC). UNEP aims to facilitate access to data on chemicals and hazardous wastes, in order to assess and control health and environmental risks, by using the IRPTC as a clearing house facility of relevant information and by publishing information and technical reports on the impact of the use of chemicals.

In 2005 a Pollutant Release and Transfer Register (PRTR), for collecting and disseminating data on toxic emissions, was in effect in Mexico. Regulations for the establishment of a new mandatory system for publishing data on emissions in that country were finalized in that year; a new *Registro de Emisiones y Transferencia de Contaminantes* was to be made public in 2006.

UNEP's OzonAction Programme works to promote information exchange, training and technological awareness. Its objective is to strengthen the capacity of governments and industry in developing countries to undertake measures towards the cost-effective phasing-out of ozone-depleting substances. UNEP also encourages the development of alternative and renewable sources of energy. To achieve this, UNEP is supporting the establishment of a network of centres to research and exchange information of environmentally sound energy technology resources.

### REGIONAL CO-OPERATION AND REPRESENTATION

UNEP maintains six regional offices. These work to initiate and promote UNEP objectives and to ensure that all programme formulation and delivery meets the specific needs of countries and regions. They also provide a focal point for building national, subregional and regional partnership and enhancing local participation in UNEP initiatives. Following UNEP's reorganization a co-ordination office was established at headquarters to promote regional policy integration, to co-ordinate programme planning, and to provide necessary services to the regional offices.

UNEP provides administrative support to several regional conventions, for example the Lusaka Agreement on Co-operative Enforcement Operations Directed at Illegal Trade in Wild Flora and Fauna, which entered into force in December 1996 having been concluded under UNEP auspices in order to strengthen the implementation of the CBD and CITES in Eastern and Central Africa. UNEP also organizes conferences, workshops and seminars at national and regional levels, and may extend advisory services or technical assistance to individual governments.

### CONVENTIONS

UNEP aims to develop and promote international environmental legislation in order to pursue an integrated response to global environmental issues, to enhance collaboration among existing convention secetariats, and to co-ordinate support to implement the work programmes of international instruments.

UNEP has been an active participant in the formulation of several major conventions (see above). The Division of Environmental Conventions is mandated to assist the Division of Policy Development and Law in the formulation of new agreements or protocols to existing conventions. Following the successful adoption of the Rotterdam Convention in September 1998, UNEP played a leading role in formulating a multilateral agreement to reduce and ultimately eliminate the manufacture and use of Persistent Organic Pollutants (POPs), which are considered to be a major global environmental hazard. The agreement on POPs, concluded in December 2000 at a conference sponsored by UNEP in Johannesburg, South Africa, was adopted by 127 countries in May 2001; it entered into force in May 2004, three months after its ratification by the requisite 50 states.

UNEP has been designated to provide secretariat functions to a number of global and regional environmental conventions (see above for list of offices).

### COMMUNICATIONS AND PUBLIC INFORMATION

UNEP's public education campaigns and outreach programmes promote community involvement in environmental issues. Further communication of environmental concerns is undertaken through the media, an information centre service and special promotional events, including World Environment Day, photography competitions, and the awarding of the Sasakawa Prize (to recognize distinguished service to the environment by individuals and groups) and of the Global 500 Award for Environmental Achievement. In 1996 UNEP initiated a Global Environment Citizenship Programme to promote acknowledgment of the environmental responsibilities of all sectors of society.

## Finance

UNEP derives its finances from the regular budget of the United Nations and from voluntary contributions to the Environment Fund. A budget of US $144m. was proposed for the two-year period 2006–07, of which $122m. was for programme activities, $5.8m. for programme support, $10.2m. for management and administration, and $6m. for fund programme reserves.

## Publications

*Annual Report.*
*APELL Newsletter* (2 a year).
*Cleaner Production Newsletter* (2 a year).
*Climate Change Bulletin* (quarterly).
*Connect* (UNESCO-UNEP newsletter on environmental degradation, quarterly).
*Earth Views* (quarterly).
*Environment Forum* (quarterly).
*Environmental Law Bulletin* (2 a year).
*Financial Services Initiative* (2 a year).
*GEF News* (quarterly).
*GEO Year Book* (annually).
*Global Water Review.*
*GPA Newsletter.*
*IETC Insight* (3 a year).
*Industry and Environment Review* (quarterly).
*Leave it to Us* (children's magazine, 2 a year).
*Managing Hazardous Waste* (2 a year).
*Our Planet* (quarterly).
*OzonAction Newsletter* (quarterly).
*Tierramerica* (weekly).
*Tourism Focus* (2 a year).
*UNEP Chemicals Newsletter* (2 a year).
*UNEP Update* (monthly).
*World Atlas of Biodiversity.*
*World Atlas of Coral Reefs.*
*World Atlas of Desertification.*

Studies, reports (including *Andean Environment Outlook*, *Latin America and the Caribbean Environment Outlook*, and *Caribbean Environment Outlook*), legal texts, technical guide-lines, etc.

# United Nations High Commissioner for Refugees—UNHCR

**Address:** CP 2500, 1211 Geneva 2 dépôt, Switzerland.
**Telephone:** (22) 7398111; **fax:** (22) 7397312; **e-mail:** unhcr@unhcr.ch; **internet:** www.unhcr.ch.

The Office of the High Commissioner was established in 1951 to provide international protection for refugees and to seek durable solutions to their problems.

## Organization
(August 2005)

### HIGH COMMISSIONER

The High Commissioner is elected by the United Nations General Assembly on the nomination of the Secretary-General, and is responsible to the General Assembly and to the UN Economic and Social Council (ECOSOC).

**High Commissioner:** ANTÓNIO MANUEL DE OLIVEIRA GUTERRES (Portugal).
**Deputy High Commissioner:** WENDY CHAMBERLIN (USA).

### EXECUTIVE COMMITTEE

The Executive Committee of the High Commissioner's Programme (ExCom), established by ECOSOC, gives the High Commissioner policy directives in respect of material assistance programmes and advice in the field of international protection. In addition, it oversees UNHCR's general policies and use of funds. ExCom, which comprises representatives of 66 states, both members and non-members of the UN, meets once a year.

### ADMINISTRATION

Headquarters include the Executive Office, comprising the offices of the High Commissioner, the Deputy High Commissioner and the Assistant High Commissioner. The Inspector General, the Director of the UNHCR liaison office in New York, and the Director of the Department of International Protection report directly to the High Commissioner. The other principal administrative units are the Division of Financial and Supply Management, the Division of Human Resources Management, the Division of External Relations, the Division of Information Systems and Telecommunications, the Emergency and Security Service, and the Department of Operations, which is responsible for the five regional bureaux covering Africa; Asia and the Pacific; Europe; the Americas and the Caribbean; and Central Asia, South-West Asia, North Africa and the Middle East. At July 2004 there were 252 UNHCR offices in 116 countries worldwide. At that time UNHCR employed 6,143 people, including short-term staff, of whom 5,109 (or 83%) were working in the field.

### OFFICES IN SOUTH AND CENTRAL AMERICA AND THE CARIBBEAN

**Regional Office for the USA and the Caribbean:** 1775 K St, NW, Suite 300, Washington, DC 20006, USA; e-mail usawa@unhcr.ch.

**Regional Office for Northern South America:** Apdo 69045, Caracas 1062-A, Venezuela; e-mail venca@unhcr.ch.

**Regional Office for South America:** Cerrito 836, 10°, Buenos Aires 1010, Argentina; e-mail argbu@unhcr.ch.

## Activities

The competence of the High Commissioner extends to any person who, owing to well-founded fear of being persecuted for reasons of race, religion, nationality or political opinion, is outside the country of his or her nationality and is unable or, owing to such fear or for reasons other than personal convenience, remains unwilling to accept the protection of that country; or who, not having a nationality and being outside the country of his or her former habitual residence, is unable or, owing to such fear or for reasons other than personal convenience, is unwilling to return to it. This competence may be extended, by resolutions of the UN General Assembly and decisions of ExCom, to cover certain other 'persons of concern', in addition to refugees meeting these criteria. Refugees who are assisted by other UN agencies, or who have the same rights or obligations as nationals of their country of residence, are outside the mandate of UNHCR.

In recent years there has been a significant shift in UNHCR's focus of activities. Increasingly UNHCR has been called upon to support people who have been displaced within their own country (i.e. with similar needs to those of refugees but who have not crossed an international border) or those threatened with displacement as a result of armed conflict. In addition, greater support has been given to refugees who have returned to their country of origin, to assist their reintegration, and UNHCR is working to enable local communities to support the returnees, frequently through the implementation of Quick Impact Projects (QIPs). In 2004 UNHCR led the formulation of a UN system-wide Strategic Plan for internally displaced persons (IDPs).

UNHCR has been increasingly concerned with the problem of statelessness, where people have no legal nationality, and promotes new accessions to the 1954 Convention Relating to the Status of Stateless Persons and the 1964 Convention on the Reduction of Statelessness.

At December 2004 the total population of concern to UNHCR, based on provisional figures, amounted to 19.2m., compared with 17.0m. in the previous year. At the end of 2004 the refugee population world-wide totalled 9.24m. UNHCR was also concerned with some 1.49m. recently returned refugees, 5.43m. IDPs, 839,100 asylum seekers, 148,000 returned IDPs and 2.05m. others (mostly stateless persons).

World Refugee Day, sponsored by UNHCR, is held annually on 20 June.

### INTERNATIONAL PROTECTION

As laid down in the Statute of the Office, UNHCR's primary function is to extend international protection to refugees and its second function is to seek durable solutions to their problems. In the exercise of its mandate UNHCR seeks to ensure that refugees and asylum-seekers are protected against *refoulement* (forcible return), that they receive asylum, and that they are treated according to internationally recognized standards. UNHCR pursues these objectives by a variety of means that include promoting the conclusion and ratification by states of international conventions for the protection of refugees. UNHCR promotes the adoption of liberal practices of asylum by states, so that refugees and asylum-seekers are granted admission, at least on a temporary basis.

The most comprehensive instrument concerning refugees that has been elaborated at the international level is the 1951 United Nations Convention relating to the Status of Refugees. This Convention, the scope of which was extended by a Protocol adopted in 1967, defines the rights and duties of refugees and contains provisions dealing with a variety of matters which affect the day-to-day lives of refugees. The application of the Convention and its Protocol is supervised by UNHCR. Important provisions for the treatment of refugees are also contained in a number of instruments adopted at the regional level. These include the 1969 Convention Governing the Specific Aspects of Refugee Problems adopted by OAU (now AU) member states in 1969, the European Agreement on the Abolition of Visas for Refugees, and the 1969 American Convention on Human Rights.

UNHCR has actively encouraged states to accede to the 1951 United Nations Refugee Convention and the 1967 Protocol: 145 states had acceded to either or both of these basic refugee instruments by February 2005. An increasing number of states have also adopted domestic legislation and/or administrative measures to implement the international instruments, particularly in the field of procedures for the determination of refugee status. UNHCR has sought to address the specific needs of refugee women and children, and has also attempted to deal with the problem of military attacks on refugee camps, by adopting and encouraging the acceptance of a set of principles to ensure the safety of refugees. In recent years it has formulated a strategy designed to address the fundamental causes of refugee flows. In 2001, in response to widespread concern about perceived high numbers of asylum-seekers and large-scale international economic migration and human trafficking, UNHCR initiated a series of Global Consultations on International Protection with the signatories to the 1951 Convention and 1967 Protocol, and other interested parties, with a view to strengthening both the application and scope of international refugee legislation. A consultation of 156 Governments, convened in Geneva, in December, reaffirmed commitment to the central role played by the Convention and Protocol. The final consultation, held in May 2002, focused on durable solutions and the protection of refugee women and children. Subsequently, based on the findings of the Global Consultations process, UNHCR developed an Agenda on Protection with six main objectives: strengthening the implementation of the 1951 Convention and 1967 Protocol; the protection of refugees within broader migration movements; more equitable sharing of burdens and responsibilities and building of capacities to receive and protect refugees; addressing more effectively security-related concerns; increasing efforts to find durable solutions; and meeting the protection needs of refugee women and children. The Agenda was endorsed by the Executive Council in October 2002. In September of that year the High Commissioner for Refugees launched the *Convention Plus* initiative, which aims to address contemporary global asylum issues by developing, on the basis of the Agenda on Protection, international agreements and measures to supplement the 1951 Convention and 1967 Protocol.

In June 2004 UNHCR became the 10th co-sponsor of UNAIDS.

## ASSISTANCE ACTIVITIES

The first phase of an assistance operation uses UNHCR's capacity of emergency response. This enables UNHCR to address the immediate needs of refugees at short notice, for example, by employing specially trained emergency teams and maintaining stockpiles of basic equipment, medical aid and materials. A significant proportion of UNHCR expenditure is allocated to the next phase of an operation, providing 'care and maintenance' in stable refugee circumstances. This assistance can take various forms, including the provision of food, shelter, medical care and essential supplies. Also covered in many instances are basic services, including education and counselling.

As far as possible, assistance is geared towards the identification and implementation of durable solutions to refugee problems—this being the second statutory responsibility of UNHCR. Such solutions generally take one of three forms: voluntary repatriation, local integration or resettlement in another country. Where voluntary repatriation, increasingly the preferred solution, is feasible, the Office assists refugees to overcome obstacles preventing their return to their country of origin. This may be done through negotiations with governments involved, or by providing funds either for the physical movement of refugees or for the rehabilitation of returnees once back in their own country. In 2005 UNHCR was supporting the implementation of the Guidance Note on Durable Solutions for Displaced Persons, adopted in 2004 by the UN Development Group.

When voluntary repatriation is not an option, efforts are made to assist refugees to integrate locally and to become self-supporting in their countries of asylum. This may be done either by granting loans to refugees, or by assisting them, through vocational training or in other ways, to learn a skill and to establish themselves in gainful occupations. One major form of assistance to help refugees re-establish themselves outside camps is the provision of housing. In cases where resettlement through emigration is the only viable solution to a refugee problem, UNHCR negotiates with governments in an endeavour to obtain suitable resettlement opportunities, to encourage liberalization of admission criteria and to draw up special immigration schemes. During 2004 an estimated 30,000 refugees were resettled under UNHCR auspices.

In the early 1990s UNHCR aimed to consolidate efforts to integrate certain priorities into its programme planning and implementation, as a standard discipline in all phases of assistance. The considerations include awareness of specific problems confronting refugee women, the needs of refugee children, the environmental impact of refugee programmes and long-term development objectives. In an effort to improve the effectiveness of its programmes, UNHCR has initiated a process of delegating authority, as well as responsibility for operational budgets, to its regional and field representatives, increasing flexibility and accountability. An Evaluation and Policy Analysis Unit reviews systematically UNHCR's operational effectiveness.

All UNHCR personnel are required to sign, and all interns, contracted staff and staff from partner organizations are required to acknowledge, a Code of Conduct, to which is appended the UN Secretary-General's bulletin on special measures for protection from sexual exploitation and sexual abuse (issued in October 2003). The post of Senior Adviser to the High Commissioner on Gender Issues, within the Executive Office, was established in 2004.

## THE AMERICAS AND THE CARIBBEAN

In May 1989, when an International Conference on Central American Refugees (CIREFCA) was held in Guatemala, there were some 146,400 refugees receiving UNHCR assistance (both for emergency relief and for longer-term self-sufficiency programmes) in the region, as well as an estimated 1.8m. other refugees and displaced persons. UNHCR and UNDP were designated as the principal UN organizations to implement the CIREFCA plan of action for the repatriation or resettlement of refugees, alongside national co-ordinating committees. UNHCR QIPs were implemented in the transport, health, agricultural production and other sectors to support returnee reintegration (of both refugees and internally displaced persons) into local communities, and to promote the self-sufficiency of the returning populations. Implementation of UNHCR's programme for the repatriation of some 45,000 Guatemalan refugees in Mexico began in January 1993 with a convoy of 2,400 people. UNHCR initiated projects to support the reintegration of Guatemalan returnees, and in 1994–95 undertook a campaign to clear undetonated explosives in forest areas where they had resettled. The CIREFCA process was formally concluded in June 1994, by which time some 118,000 refugees had voluntarily returned to their countries of origin under the auspices of the programme, while thousands of others had integrated into their host countries. In December a meeting was held, in San José, Costa Rica, to commemorate the 10th anniversary of the Cartagena Declaration, which had provided a comprehensive framework for refugee protection in the region. The meeting adopted the San José Declaration on Refugees and Displaced Persons, which aimed to harmonize legal criteria and procedures to consolidate actions for durable solutions of voluntary repatriation and local integration in the region. UNHCR's efforts in the region subsequently emphasized legal issues and refugee protection, while assisting governments to formulate national legislation on asylum and refugees. Since 1996 several thousand Guatemalan refugees have received Mexican citizenship under a fast-track naturalization programme. At December 2004 4,343 refugees remained in Mexico. During 2005 UNHCR was to assist the Mexican authorities with completing the naturalization of some 2,000 Guatemalan refugees, and also with the naturalization of 2,800 refugees of other nationalities sheltering in urban areas. In December 2004 regional leaders met in Mexico City to commemorate the 20th anniversary of the Cartagena Declaration, and launched a new plan of action aimed at addressing current refugee problems in Central America, with a particular focus on the humanitarian crisis in Colombia and the border areas of its neighbouring countries (see below), and the increasing numbers of refugees concentrated in urban centres in the region.

In 1999 the Colombian Government approved an operational plan proposed by UNHCR to address a massive population displacement that has arisen in that country in recent years (escalating from 1997), as a consequence of ongoing internal conflict and alleged human rights abuses committed by armed groups. Significant cross-border movements of Colombian refugees into neighbouring countries prompted UNHCR to intensify its border-monitoring activities during the early 2000s to enhance its capacity to forecast and react to new population movements. UNHCR has assisted with the implementation of an IDP registration plan, provided training in emergency response to displacements, and supported ongoing changes in Colombia's legislative framework for IDPs. UNHCR has also co-operated with UNICEF to improve the provision of education to displaced children. During 2005 UNHCR aimed to support compliance with national legislation and policies on IDPs and to promote durable solutions for IDPs such as return, relocation and integration into local communities. The Office has focused on building up stockpiles of relief items in neighbouring countries and has developed contingency plans with other partners to enable the rapid deployment of personnel to border areas should the exodus of refugees from Colombia intensify further. At the end of 2004 Costa Rica was hosting 8,750 Colombian refugees (from a total refugee population of 10,413) and Ecuador was accommodating 8,270 Colombian refugees and 1,660 asylum-seekers (also mainly from Colombia). At that time there were some 2m. IDPs within Colombia of concern to UNHCR.

Canada and the USA are major countries of resettlement for refugees. UNHCR provides counselling and legal services for asylum-seekers in these countries. At 31 December 2004 the estimated refugee populations totalled 141,398 in Canada and 420,854 in the USA, while asylum-seekers numbered 27,290 and 263,710 respectively.

## CO-OPERATION WITH OTHER ORGANIZATIONS

UNHCR works closely with other UN agencies, intergovernmental organizations and non-governmental organizations (NGOs) to increase the scope and effectiveness of its operations. Within the UN system UNHCR co-operates, principally, with the World Food Programme in the distribution of food aid, UNICEF and the World Health Organization in the provision of family welfare and child immunization programmes, OCHA in the delivery of emergency humanitarian relief, UNDP in development-related activities and the preparation of guide-lines for the continuum of emergency assistance to development programmes, and the Office of the UN High Commissioner for Human Rights. UNHCR also has close working relationships with the International Committee of the Red Cross and the International Organization for Migration. In 2004 UNHCR worked with 565 NGOs as 'implementing partners', enabling UNHCR to broaden the use of its resources while maintaining a co-ordinating role in the provision of assistance.

## TRAINING

UNHCR organizes training programmes and workshops to enhance the capabilities of field workers and non-UNHCR staff, in the following areas: the identification and registration of refugees; people-orientated planning; resettlement procedures and policies; emergency response and management; security awareness; stress management; and the dissemination of information through the electronic media.

## Finance

The United Nations' regular budget finances a proportion of UNHCR's administrative expenditure. The majority of UNHCR's programme expenditure (about 98%) is funded by voluntary contributions, mainly from governments. The Private Sector and Public Affairs Service aims to increase funding from non-governmental donor sources, for example by developing partnerships with foundations and corporations. Following approval of the Unified Annual Programme Budget any subsequently identified requirements are managed in the form of Supplementary Programmes, financed by separate appeals. The total Unified Annual Programme Budget for 2005 was projected at US $982m.

## Publications

*Refugees* (quarterly, in English, French, German, Italian, Japanese and Spanish).
*Refugee Resettlement: An International Handbook to Guide Reception and Integration.*
*Refugee Survey Quarterly.*
*Sexual and Gender-based Violence Against Refugees, Returnees and Displaced Persons: Guide-lines for Prevention and Response.*
*The State of the World's Refugees* (every 2 years).
*UNHCR Handbook for Emergencies.*
Press releases, reports.

## Statistics

**PERSONS OF CONCERN TO UNHCR IN LATIN AMERICA AND THE CARIBBEAN\***
(at 31 December 2004, provisional figures)

| Country | Refugees | Asylum-seekers | Returnees | Others† |
|---|---|---|---|---|
| Argentina | 2,916 | 990 | 4 | — |
| Brazil | 3,345 | 446 | — | — |
| Colombia | 141 | 36 | 67 | 2,000,000 |
| Costa Rica | 10,413 | 223 | — | — |
| Ecuador | 8,450 | 1,660 | 3 | — |
| Mexico | 4,343 | 161 | — | — |
| Panama | 1,608 | 271 | — | — |
| Peru | 766 | 232 | 2 | — |
| Venezuela | 244 | 3,904 | — | 26,350 |

\* Countries with fewer than 1,000 persons of concern to UNHCR are not listed.
† Mainly internally displaced persons.

# United Nations Peace-keeping

**Address:** Department of Peace-keeping Operations, Room S-3727-B, United Nations, New York, NY 10017, USA.
**Telephone:** (212) 963-8077; **fax:** (212) 963-9222; **internet:** www.un.org/Depts/dpko/.

United Nations peace-keeping operations have been conceived as instruments of conflict control. The UN has used these operations in various conflicts, with the consent of the parties involved, to maintain international peace and security, without prejudice to the positions or claims of parties, in order to facilitate the search for political settlements through peaceful means such as mediation and the good offices of the UN Secretary-General. Each operation is established with a specific mandate, which requires periodic review by the UN Security Council. United Nations peace-keeping operations fall into two categories: peace-keeping forces and observer missions.

Peace-keeping forces are composed of contingents of military and civilian personnel, made available by member states. These forces assist in preventing the recurrence of fighting, restoring and maintaining peace, and promoting a return to normal conditions. To this end, peace-keeping forces are authorized as necessary to undertake negotiations, persuasion, observation and fact-finding. They conduct patrols and interpose physically between the opposing parties. Peace-keeping forces are permitted to use their weapons only in self-defence.

Military observer missions are composed of officers (usually unarmed), who are made available, on the Secretary-General's request, by member states. A mission's function is to observe and report to the Secretary-General (who, in turn, informs the Security Council) on the maintenance of a cease-fire, to investigate violations and to do what it can to improve the situation.

The UN's peace-keeping forces and observer missions are financed in most cases by assessed contributions from member states of the organization. In recent years a significant expansion in the UN's peace-keeping activities has been accompanied by a perpetual financial crisis within the organization, as a result of the increased financial burden and some member states' delaying payment. At 30 April 2005 outstanding assessed contributions to the peace-keeping budget amounted to some US $2,220m.

The report of the UN Secretary-General entitled 'In Larger Freedom: Towards Development, Security and Human Rights for All', issued in March 2005, proposed the creation of an inter-governmental Peace-building Commission, as well as a Peace-building Support Office within the Secretariat, with a view to ensuring the 'sustained and sustainable' implementation of peace agreements.

### UNITED NATIONS STABILIZATION MISSION IN HAITI—MINUSTAH

**Address:** Port-au-Prince, Haiti.

**Special Representative of the UN Secretary-General:** Juan Gabriel Valdés (Chile).
**Force Commander:** Lt-Gen. Augusto Heleno Ribeiro Pereira (Brazil).

In early 2004 political tensions within Haiti escalated as opposition groups demanded political reforms and the resignation of President Jean-Bertrand Aristide. Increasingly violent public demonstrations took place throughout the country, in spite of diplomatic efforts by regional organizations to resolve the crisis, and in February armed opposition forces seized control of several northern cities. At the end of that month, with opposition troops poised to march on the capital and growing pressure from the international community, President Aristide tendered his resignation and fled the country. On that same day the UN Security Council, acting upon a request by the interim President, authorized the establishment of a Multinational Interim Force (MIF) to help to secure law and order in Haiti. The Council also declared its readiness to establish a follow-on UN mission. In late April the Security Council agreed to establish MINUSTAH, which was to assume authority from the MIF with effect from 1 June. MINUSTAH was mandated to create a stable and secure environment, to support the transitional government in institutional development and organizing and monitoring elections, and to monitor the human rights situation. Among its declared objectives was the improvement of living conditions of the population through security measures, humanitarian actions and economic development. In September MINUSTAH worked closely with other UN agencies and non-governmental organizations to distribute food and other essential services to thousands of people affected by a severe tropical storm. By the end of 2004 MINUSTAH's priority continued to be the security situation in the country. However, the following civil units were fully operational: electoral assistance; child protection; gender; civil affairs; human rights; and HIV/AIDS. In January 2005 MINUSTAH, with the UN Development Programme, the Haitian Government and the Provisional Electoral Council, signed an agreement on the organization of a general election, to be held later in that year. In May the UN Secretary-General expressed concern at the security environment with respect to achieving political transition. In the following month the Security Council approved a temporary reinforcement of MINUSTAH to provide increased security in advance of local, parliamentary and presidential elections, scheduled to be held in October and November. The military component was to comprise up to 7,500 troops and the civilian police force up to 1,897 officers. The Council requested that the Secretary-General devise a strategy for the progressive reduction of MINUSTAH force levels in the post-election period.

At 31 May 2005 MINUSTAH comprised 6,207 troops, 1,437 civilian police, 1,222 international and local civilian staff and 139 UN Volunteers. The mission is financed by assessments in respect of a Special Account. The approved budget for the period 1 July 2005–30 June 2006 amounted to US $494.89m.

# World Food Programme—WFP

**Address:** Via Cesare Giulio Viola 68, Parco dei Medici, 00148 Rome, Italy.
**Telephone:** (06) 6513-1; **fax:** (06) 6513-2840; **e-mail:** wfpinfo@wfp.org; **internet:** www.wfp.org.

WFP, the principal food aid organization of the United Nations, became operational in 1963. It aims to alleviate acute hunger by providing emergency relief following natural or man-made humanitarian disasters, and supplies food aid to people in developing countries to eradicate chronic undernourishment, to support social development and to promote self-reliant communities.

## Organization

(August 2005)

### EXECUTIVE BOARD

The governing body of WFP is the Executive Board, comprising 36 members, 18 of whom are elected by the UN Economic and Social Council (ECOSOC) and 18 by the Council of the Food and Agriculture Organization (FAO). The Board meets four times each year at WFP headquarters.

### SECRETARIAT

WFP's Executive Director is appointed jointly by the UN Secretary-General and the Director-General of FAO and is responsible for the management and administration of the Programme. At December 2003 there were 8,770 staff members, more than 90% of whom were working in the field. WFP administers some 87 country offices, in order to provide operational, financial and management support at a more local level, and has established seven regional bureaux, located in Bangkok, Thailand (for Asia), Cairo, Egypt (for the Middle East, Central Asia and the Mediterranean), Rome, Italy (for Eastern Europe), Managua, Nicaragua (for Latin America and the Caribbean), Yaoundé, Cameroon (for Central Africa), Kampala, Uganda (for Eastern and Southern Africa), and Dakar, Senegal (for West Africa).

**Executive Director:** JAMES T. MORRIS (USA).

## Activities

WFP is the only multilateral organization with a mandate to use food aid as a resource. It is the second largest source of assistance in the UN, after the World Bank group, in terms of actual transfers of resources, and the largest source of grant aid in the UN system. WFP handles more than one-third of the world's food aid. WFP is also the largest contributor to South–South trade within the UN system, through the purchase of food and services from developing countries. WFP's mission is to provide food aid to save lives in refugee and other emergency situations, to improve the nutrition and quality of life of vulnerable groups and to help to develop assets and promote the self-reliance of poor families and communities. WFP aims to focus its efforts on the world's poorest countries and to provide at least 90% of its total assistance to those designated as 'low-income food-deficit'. At the World Food Summit, held in November 1996, WFP endorsed the commitment to reduce by 50% the number of undernourished people, no later than 2015. During 2004 WFP food assistance benefited some 113m. people world-wide, of whom 24m. received aid through development projects, 38m. through emergency operations, 25m. through Protracted Relief and Recovery Operations, and 26m. through the Iraqi bilateral operation. Total food deliveries in 2004 amounted to 5.1m. metric tons, compared with 4.6m. metric tons in 2003.

WFP aims to address the causes of chronic malnourishment, which it identifies as poverty and lack of opportunity. It emphasizes the role played by women in combating hunger, and endeavours to address the specific nutritional needs of women, to increase their access to food and development resources, and to promote girls' education. It also focuses resources on supporting the food security of households and communities affected by HIV/AIDS and on promoting food security as a means of mitigating extreme poverty and vulnerability and thereby combating the spread and impact of HIV/AIDS. In February 2003 WFP and the Joint UN Programme on HIV/AIDS (UNAIDS) concluded an agreement to address jointly the relationship between HIV/AIDS, regional food shortages and chronic hunger, with a particular focus on Africa, South-East Asia and the Caribbean. In October of that year WFP became a co-sponsor of UNAIDS. WFP urges the development of new food aid strategies as a means of redressing global inequalities and thereby combating the threat of conflict and international terrorism.

WFP food donations must meet internationally-agreed standards applicable to trade in food products. In May 2003 WFP's Executive Board approved a new policy on donations of genetically-modified (GM) foods and other foods derived from biotechnology, determining that the Programme would continue to accept donations of GM/biotech food and that, when distributing it, relevant national standards would be respected.

In the early 1990s there was a substantial shift in the balance between emergency relief ('food-for-life') and development assistance ('food-for-growth') provided by WFP, owing to the growing needs of victims of drought and other natural disasters, refugees and displaced persons. By 1994 two-thirds of all food aid was for relief assistance and one-third for development, representing a direct reversal of the allocations five years previously. In addition, there was a noticeable increase in aid given to those in need as a result of civil war, compared with commitments for victims of natural disasters. Accordingly, WFP has developed a range of mechanisms to enhance its preparedness for emergency situations and to improve its capacity for responding effectively to situations as they arise. A new programme of emergency response training was inaugurated in 2000, while security concerns for personnel was incorporated as a new element into all general planning and training activities. Through its Vulnerability Analysis and Mapping (VAM) project, WFP aims to identify potentially vulnerable groups by providing information on food security and the capacity of different groups for coping with shortages, and to enhance emergency contingency-planning and long-term assistance objectives. In 2003 VAM field units were operational in more than 50 countries. WFP also co-operates with other UN agencies including FAO (collaborating on 77 projects in 41 countries in 2003), IFAD (collaborating on 21 projects in that year), UNHCR and UNICEF. The key elements of WFP's emergency response capacity are its strategic stores of food and logistics equipment, stand-by arrangements to enable the rapid deployment of personnel, communications and other essential equipment, and the Augmented Logistics Intervention Team for Emergencies (ALITE), which undertakes capacity assessments and contingency-planning. During 2000 WFP led efforts, undertaken with other UN humanitarian agencies, for the design and application of local UN Joint Logistics Centre facilities, which aimed to co-ordinate resources in an emergency situation. In 2001 a new UN Humanitarian Response Depot was opened in Brindisi, Italy, under the direction of WFP experts, for the storage of essential rapid response equipment. In that year the Programme published a set of guidelines on contingency planning.

Through its development activities, WFP aims to alleviate poverty in developing countries by promoting self-reliant families and communities. Food is supplied, for example, as an incentive in development self-help schemes and as part-wages in labour-intensive projects of many kinds. In all its projects WFP aims to assist the most vulnerable groups and to ensure that beneficiaries have an adequate and balanced diet. Activities supported by the Programme include the settlement and resettlement of groups and communities; land reclamation and improvement; irrigation; the development of forestry and dairy farming; road construction; training of hospital staff; community development; and human resources development such as feeding expectant or nursing mothers and schoolchildren, and support for education, training and health programmes. No individual country is permitted to receive more than 10% of the Programme's available development resources. During 2001 WFP initiated a new Global School Feeding Campaign to strengthen international co-operation to expand educational opportunities for poor children and to improve the quality of the teaching environment. In

December 2003 WFP launched a *19-Cents-a-day* campaign to encourage donors to support its school feeding activities (19 US cents being the estimated cost of one school lunch). During 2004 school feeding projects benefited 16.6m. children.

Following a comprehensive evaluation of its activities, WFP is increasingly focused on linking its relief and development activities to provide a continuum between short-term relief and longer-term rehabilitation and development. In order to achieve this objective, WFP aims to integrate elements that strengthen disaster mitigation into development projects, including soil conservation, reafforestation, irrigation infrastructure, and transport construction and rehabilitation; and to promote capacity-building elements within relief operations, e.g. training, income-generating activities and environmental protection measures. In 1999 WFP adopted a new Food Aid and Development policy, which aims to use food assistance both to cover immediate requirements and to create conditions conducive to enhancing the long-term food security of vulnerable populations. During that year WFP began implementing Protracted Relief and Recovery Operations (PRROs), where the emphasis is on fostering stability, rehabilitation and long-term development for victims of natural disasters, displaced persons and refugees. PRROs are introduced no later than 18 months after the initial emergency operation and last no more than three years. When undertaken in collaboration with UNHCR and other international agencies, WFP has responsibility for mobilizing basic food commodities and for related transport, handling and storage costs. The 20 PRROs undertaken in 2004 involved the provision of 3.53m. metric tons of food, at a cost of some US $1,870m.

In 2004 WFP operational expenditure in Latin America and the Caribbean amounted to US $59.3m., compared with $49.2m. in 2003. Of the total expenditure in 2004 $26.2m. was for emergency relief operations, $30.2m. or agricultural, rural and human resource development projects, and $2.9m. for special operations. WFP estimates that some 46% of Haiti's population is malnourished and, in early 2005, was supporting vulnerable people in that country, who had been affected by the political and civil unrest as well as natural disasters in 2004, through education, nutrition and health, and disaster mitigation activities. A new two-year PRRO, to assist some 550,000 Haitians, was initiated in May 2005. In November 2004 WFP initiated a six-month operation in Peru to assist 12,500 vulnerable families affected by severe weather conditions in high altitude areas.

## Finance

The Programme is funded by voluntary contributions from donor countries, intergovernmental bodies such as the European Commission, and the private sector. Contributions are made in the form of commodities, finance and services (particularly shipping). Commitments to the International Emergency Food Reserve (IEFR), from which WFP provides the majority of its food supplies, and to the Immediate Response Account of the IEFR (IRA), are also made on a voluntary basis by donors. WFP's operational expenditures in 2004 amounted to some US $3,100m. Contributions by donors in that year totalled $2,206m.

## Publications

*Annual Report.*
*Food and Nutrition Handbook.*
*School Feeding Handbook.*

# Food and Agriculture Organization of the United Nations—FAO

**Address:** Viale delle Terme di Caracalla, 00100 Rome, Italy.
**Telephone:** (06) 5705-1; **fax:** (06) 5705-3152; **e-mail:** fao-hq@fao.org; **internet:** www.fao.org.

FAO, the first specialized agency of the UN to be founded after the Second World War, aims to alleviate malnutrition and hunger, and serves as a co-ordinating agency for development programmes in the whole range of food and agriculture, including forestry and fisheries. It helps developing countries to promote educational and training facilities and the creation of appropriate institutions.

## Organization

(August 2005)

### CONFERENCE

The governing body is the FAO Conference of member nations. It meets every two years, formulates policy, determines the Organization's programme and budget on a biennial basis, and elects new members. It also elects the Director-General of the Secretariat and the Independent Chairman of the Council. Every other year, FAO also holds conferences in each of its five regions (Africa, Asia and the Pacific, Europe, Latin America and the Caribbean, and the Near East).

### COUNCIL

The FAO Council is composed of representatives of 49 member nations, elected by the Conference for staggered three-year terms. It is the interim governing body of FAO between sessions of the Conference. The most important standing Committees of the Council are: the Finance and Programme Committees, the Committee on Commodity Problems, the Committee on Fisheries, the Committee on Agriculture and the Committee on Forestry.

### SECRETARIAT

The number of FAO staff in 2005 was around 3,450, of whom 1,450 were professional staff and 2,000 general service staff. About one-half of the Organization's staff were based at headquarters. Work is supervised by the following Departments: Administration and Finance; General Affairs and Information; Economic and Social Policy; Agriculture; Forestry; Fisheries; Sustainable Development; and Technical Co-operation.

**Director-General:** JACQUES DIOUF (Senegal).

### REGIONAL AND SUB-REGIONAL OFFICES

**Regional Office for Latin America and the Caribbean:** Avda Dag Hammarskjöld 3241, Casilla 10095, Vitacura, Santiago, Chile; tel. (2) 337-2102; fax (2) 337-2101; e-mail fao-rlc@field.fao.org; internet www.fao.org/regional/lamerica/default.htm; Regional Rep. GUSTAVO GORDILLO DE ANDA.

**Sub-regional Office for the Caribbean:** POB 631-C, Bridgetown, Barbados; tel. 426-7110; fax 427-6075; e-mail fao-slac@field.fao.org; Sub-regional Rep. WINSTON RUDDER.

### JOINT DIVISION AND LIAISON OFFICE

**Joint FAO/IAEA Division of Nuclear Techniques in Food and Agriculture:** Wagramerstrasse 5, 1400 Vienna, Austria; tel. (1) 2600-0; fax (1) 2600-7.

**United Nations:** Suite DC1-1125, 1 United Nations Plaza, New York, NY 10017, USA; tel. (212) 963-6036; fax (212) 963-5425; e-mail fao-lony@field.fao.org; Dir HOWARD W. HJORT.

## Activities

FAO aims to raise levels of nutrition and standards of living by improving the production and distribution of food and other commodities derived from farms, fisheries and forests. FAO's ultimate objective is the achievement of world food security, 'Food for All'. The organization provides technical information, advice and assistance by disseminating information; acting as a neutral forum for discussion of food and agricultural issues; advising governments on policy and planning; and developing capacity directly in the field.

In November 1996 FAO hosted the World Food Summit, which was held in Rome and was attended by heads of state and senior government representatives of 186 countries. Participants approved the Rome Declaration on World Food Security and the World Food Summit Plan of Action, with the aim of halving the number of people afflicted by undernutrition, at that time estimated to total 828m. world-wide, by no later than 2015. A review conference to assess progress in achieving the goals of the summit, entitled World Food Summit: Five Years Later, held in June 2002, reaffirmed commitment to this objective, which is also incorporated into the UN Millennium Development Goal of eradicating extreme poverty and hunger. During that month FAO announced the formulation of a global 'Anti-Hunger Programme', which aimed to promote investment in the agricultural sector and rural development, with a particular focus on small farmers, and to enhance food access for those most in need, for example through the provision of school meals, schemes to feed pregnant and nursing mothers and food-for-work programmes. In late 2003 FAO reported that an estimated 842m. people world-wide were undernourished; of these 798m. resided in developing countries.

In November 1999 the FAO Conference approved a long-term Strategic Framework for the period 2000–15, which emphasized national and international co-operation in pursuing the goals of the 1996 World Food Summit. The Framework promoted interdisciplinarity and partnership, and defined three main global objectives: constant access by all people to sufficient nutritionally adequate and safe food to ensure that levels of undernourishment were reduced by 50% by 2015 (see above); the continued contribution of sustainable agriculture and rural development to economic and social progress and well-being; and the conservation, improvement and sustainable use of natural resources. It identified five corporate strategies (each supported by several strategic objectives), covering the following areas: reducing food insecurity and rural poverty; ensuring enabling policy and regulatory frameworks for food, agriculture, fisheries and forestry; creating sustainable increases in the supply and availability of agricultural, fisheries and forestry products; conserving and enhancing sustainable use of the natural resource base; and generating knowledge. In November 2001 the FAO Conference adopted a medium-term plan covering 2002–07, based on the Strategic Framework. In November 2004 the FAO Council adopted a set of voluntary Right to Food Guide-lines that aimed to 'support the progressive realization of the right to adequate food in the context of national food security' by providing practical guidance to countries in support of their efforts to achieve the 1996 World Food Summit commitment and UN Millennium Development Goal relating to hunger reduction.

FAO organizes an annual series of fund-raising events, 'TeleFood', some of which are broadcast on television and the internet, in order to raise public awareness of the problems of hunger and malnutrition. Since its inception in 1997 public donations to TeleFood have reached almost US $14m., financing more than 1,750 'grass-roots' projects in 124 countries. The projects have provided tools, seeds and other essential supplies directly to small-scale farmers, and have been especially aimed at helping women.

In 1999 FAO signed a memorandum of understanding with UNAIDS on strengthening co-operation. In December 2001 FAO, IFAD and WFP determined to strengthen inter-agency collaboration in developing strategies to combat the threat posed by the HIV/AIDS epidemic to food security, nutrition and rural livelihoods. During that month experts from those organizations and UNAIDS held a technical consultation on means of mitigating the impact of HIV/AIDS on agriculture and rural communities in affected areas.

In September 2004 FAO published *Recommendations for the Prevention, Control and Eradication of Highly Pathogenic Avian Influenza (HPAI) in Asia*. In the same month, following new outbreaks of the disease in the People's Republic of China, Cambodia, Viet Nam, Malaysia and Thailand, FAO and the World Health Organization declared the avian influenza epidemic to be a 'crisis of global importance'. FAO was working closely with the World Organisation for Animal Health to study and help to contain the disease.

The Technical Co-operation Department has responsibility for FAO's operational activities, including policy development assistance to member countries; investment support; and the management of activities associated with the development and implementation of country, sub-regional and regional programmes. The Department manages the technical co-operation programme (TCP, which funds 13% of FAO's field programme expenditures), and mobilizes resources.

## AGRICULTURE

FAO's most important area of activity is crop production, accounting annually for about one-quarter of total field programme expenditure. FAO assists developing countries in increasing agricultural production, by means of a number of methods, including improved seeds and fertilizer use, soil conservation and reforestation, better water resource management techniques, upgrading storage facilities, and improvements in processing and marketing. FAO places special emphasis on the cultivation of under-exploited traditional food crops, such as cassava, sweet potato and plantains.

In 1985 the FAO Conference approved an International Code of Conduct on the Distribution and Use of Pesticides, and in 1989 the Conference adopted an additional clause concerning 'Prior Informed Consent' (PIC), whereby international shipments of newly banned or restricted pesticides should not proceed without the agreement of importing countries. Under the clause, FAO aims to inform governments about the hazards of toxic chemicals and to urge them to take proper measures to curb trade in highly toxic agrochemicals while keeping the pesticides industry informed of control actions. In 1996 FAO, in collaboration with UNEP, publicized a new initiative which aimed to increase awareness of, and to promote international action on, obsolete and hazardous stocks of pesticides remaining throughout the world (estimated in 2001 to total some 500,000 metric tons). In September 1998 a new legally-binding treaty on trade in hazardous chemicals and pesticides was adopted at an international conference held in Rotterdam, Netherlands. The so-called Rotterdam Convention required that hazardous chemicals and pesticides banned or severely restricted in at least two countries should not be exported unless explicitly agreed by the importing country. It also identified certain pesticide formulations as too dangerous to be used by farmers in developing countries, and incorporated an obligation that countries halt national production of those hazardous compounds. The treaty entered into force in February 2004. FAO was co-operating with UNEP to provide an interim secretariat for the Convention. In July 1999 a conference on the Rotterdam Convention, held in Rome, established an Interim Chemical Review Committee with responsibility for recommending the inclusion of chemicals or pesticide formulations in the PIC procedure. As part of its continued efforts to reduce the environmental risks posed by over-reliance on pesticides, FAO has extended to other regions its Integrated Pest Management (IPM) programme in Asia and the Pacific on the use of safer and more effective methods of pest control, such as biological control methods and natural predators (including spiders and wasps), to avert pests. In February 2001 FAO warned that some 30% of pesticides sold in developing countries did not meet internationally accepted quality standards. A revised International Code of Conduct on the Distribution and Use of Pesticides, adopted in November 2002, aimed to reduce the inappropriate distribution and use of pesticides and other toxic compounds, particularly in developing countries.

FAO's Joint Division with the International Atomic Energy Agency (IAEA) tests controlled-release formulas of pesticides and herbicides that gradually free their substances and can limit the amount of agrochemicals needed to protect crops. The Joint FAO/IAEA Division is engaged in exploring biotechnologies and in developing non-toxic fertilizers (especially those that are locally available) and improved strains of food crops (especially from indigenous varieties). In the area of animal production and health, the Joint Division has developed progesterone-measuring and disease diagnostic kits, of which thousands have been delivered to developing countries. FAO's plant nutrition activities aim to promote nutrient management, such as the Integrated Plant Nutritions Systems (IPNS), which are based on the recycling of nutrients through crop production and the efficient use of mineral fertilizers.

The conservation and sustainable use of plant and animal genetic resources are promoted by FAO's Global System for Plant Genetic Resources, which includes five databases, and the Global Strategy on the Management of Farm Animal Genetic Resources. An FAO programme supports the establishment of gene banks, designed to maintain the world's biological diversity by preserving animal and plant species threatened with extinction. FAO, jointly with UNEP, has published a document listing the current state of global livestock genetic diversity. In June 1996 representatives of more than 150 governments convened in Leipzig, Germany, at a meeting organized by FAO (and hosted by the German Government) to consider the use and conservation of plant genetic resources as an essential means of enhancing food security. The meeting adopted a Global Plan of Action, which included measures to strengthen the development of plant varieties and to promote the use and availability of local varieties and locally-adapted crops to farmers, in particular following a natural disaster, war or civil conflict. In November 2001 the FAO Conference adopted the International Treaty on Plant Genetic Resources for Food and Agriculture, which was to provide a framework to ensure access to plant genetic resources and to related knowledge, technologies and funding. The Treaty entered into force on 29 June 2004, having received the required number of ratifications (40) by signatory states.

The Emergency Prevention System for Transboundary Animal and Plant Pests and Diseases (EMPRES) was established in 1994 to strengthen FAO's activities in the prevention, early warning of, control and, where possible, eradication of pests and highly contagious livestock diseases (which the system categorizes as epidemic diseases of strategic importance, such as rinderpest or foot-and-mouth; diseases requiring tactical attention at international or regional level, e.g. Rift Valley fever; and emerging diseases, e.g. bovine spongiform encephalopathy—BSE). EMPRES has a desert locust component, and has published guide-lines on all aspects of desert locust monitoring. FAO has assumed responsibility for technical leadership and co-ordination of the Global Rinderpest Eradication Programme (GREP), which has the objective of eliminating the disease by 2010. Following technical consultations in late 1998, an Intensified GREP was launched. In November 1997 FAO initiated a Programme Against African Trypanosomiasis, which aimed to counter the disease affecting cattle in almost one-third of Africa. EMPRES promotes Good Emergency Management Practices (GEMP) in animal health. The system is guided by the annual meeting of the EMPRES Expert Consultation. In May 2004 FAO and the World Organisation for Animal Health signed an agreement in which they clarified their respective areas of competence and paved the way for improved co-operation, in response to an increase in contageous transboundary animal diseases (such as foot-and-mouth disease and avian influenza, see below). The two bodies agreed to establish a global framework on the control of transboundary animal diseases, entailing improved international collaboration and circulation of information. FAO advises countries on good agricultural practices, disease control and eradication methods and co-operates with the World Organisation for Animal Health in building national surveillance and early warning systems. In September FAO issued a set of recommendations on the prevention, control and eradication of highly pathogenic avian influenza, which had spread to humans in south-east Asia in late 2003. During 2004 FAO provided technical co-operation assistance to Cambodia, Indonesia, Laos, Thailand and Viet Nam aimed at preparing a post-avian influenza rehabilitation programme.

FAO's organic agriculture programme provides technical assistance and policy advice on the production, certification and trade of organic produce. In July 2001 the FAO/WHO Codex Alimentarius Commission adopted guide-lines on organic livestock production, covering organic breeding methods, the elimination of growth hormones and certain chemicals in veterinary medicines, and the use of good quality organic feed with no meat or bone meal content.

### ENVIRONMENT

At the UN Conference on Environment and Development (UNCED), held in Rio de Janeiro, Brazil, in June 1992, FAO participated in several working parties and supported the adoption of Agenda 21, a programme of activities to promote sustainable development. FAO is responsible for the chapters of Agenda 21 concerning water resources, forests, fragile mountain ecosystems and sustainable agriculture and rural development. FAO was designated by the UN General Assembly as the lead agency for co-ordinating the International Year of Mountains (2002), which aimed to raise awareness of mountain ecosystems and to promote the conservation and sustainable development of mountainous regions.

### FISHERIES

FAO's Fisheries Department consists of a multi-disciplinary body of experts who are involved in every aspect of fisheries development from coastal surveys, conservation management and use of aquatic genetic resources, improvement of production, processing and storage, to the compilation and analysis of statistics, development of computer databases, improvement of fishing gear, institution-building and training. In March 1995 a ministerial meeting of fisheries adopted the Rome Consensus on World Fisheries, which identified a need for immediate action to eliminate overfishing and to rebuild and enhance depleting fish stocks. In November the FAO Conference adopted a Code of Conduct for Responsible Fishing, which incorporated many global fisheries and aquaculture issues (including fisheries resource conservation and development, fish catches, seafood and fish processing, commercialization, trade and research) to promote the sustainable development of the sector. In February 1999 the FAO Committee on Fisheries adopted new international measures, within the framework of the Code of Conduct, in order to reduce over-exploitation of the world's fish resources, as well as plans of action for the conservation and management of sharks and the reduction in the incidental catch of seabirds in longline fisheries. The voluntary measures were endorsed at a ministerial meeting, held in March and attended by representatives of some 126 countries, which issued a declaration to promote the implementation of the Code of Conduct and to achieve sustainable management of fisheries and aquaculture. In March 2001 FAO adopted an international plan of action to address the continuing problem of so-called illegal, unreported and unregulated fishing (IUU). In that year FAO estimated that about one-half of major marine fish stocks were fully exploited, one-quarter under-exploited, at least 15% over-exploited, and 10% depleted or recovering from depletion. IUU was estimated to account for up to 30% of total catches in certain fisheries. In October FAO and the Icelandic Government jointly organized the Reykjavik Conference on Responsible Fisheries in the Marine Ecosystem, which adopted a declaration on pursuing responsible and sustainable fishing activities in the context of ecosystem-based fisheries management (EBFM). EBFM involves determining the boundaries of individual marine ecosystems, and maintaining or rebuilding the habitats and biodiversity of each of these so that all species will be supported at levels of maximum production. In March 2005 FAO's Committee of Fisheries adopted voluntary guide-lines for the so-called eco-labelling and certification of fish and fish products, i.e. based on information regarding capture management and the sustainable use of resources. FAO promotes aquaculture (which contributes almost one-third of annual global fish landings) as a valuable source of animal protein and income-generating activity for rural communities. In February 2000 FAO and the Network of Aquaculture Centres in Asia and the Pacific (NACA) jointly convened a Conference on Aquaculture in the Third Millennium, which was held in Bangkok, Thailand, and attended by participants representing more than 200 governmental and non-governmental organizations. The Conference debated global trends in aquaculture and future policy measures to ensure the sustainable development of the sector. It adopted the Bangkok Declaration and Strategy for Aquaculture Beyond 2000.

### FORESTRY

FAO focuses on the contribution of forestry to food security, on effective and responsible forest management and on maintaining a balance between the economic, ecological and social benefits of forest resources. The Organization has helped to develop national forestry programmes and to promote the sustainable development of all types of forest. FAO administers the global Forests, Trees and People Programme, which promotes the sustainable management of tree and forest resources, based on local knowledge and management practices, in order to improve the livelihoods of rural people in developing countries. FAO's Strategic Plan for Forestry was approved in March 1999; its main objectives were to maintain the environmental diversity of forests, to realize the economic potential of forests and trees within a sustainable framework, and to expand access to information on forestry.

### NUTRITION

The International Conference on Nutrition, sponsored by FAO and WHO, took place in Rome in December 1992. It approved a World Declaration on Nutrition and a Plan of Action, aimed at promoting efforts to combat malnutrition as a development priority. Since the conference, more than 100 countries have formulated national plans of action for nutrition, many of which were based on existing development plans such as comprehensive food security initiatives, national poverty alleviation programmes and action plans to attain the targets set by the World Summit for Children in September 1990. In October 1996 FAO, WHO and other partners jointly organized the first World Congress on Calcium and Vitamin D in Human Life, held in Rome. In January 2001 a joint team of FAO and WHO experts issued a report concerning the allergenicity of foods derived from biotechnology (i.e. genetically modified—GM—foods). In July the Codex Alimentarius Commission agreed the first global principles for assessing the safety of GM foods, and approved a series of maximum levels of environmental contaminants in food. FAO and WHO jointly convened a Global Forum of Food Safety Regulators in Marrakesh, Morocco, in January 2002. In April the two organizations announced a joint review of their food standards operations, including the activities of the Codex Alimentarius Commission. In July 2004 the Codex Alimentarius Commission adopted a definition of product tracing, increasingly regarded as an important component of national and international food regulatory systems. In October FAO and WHO jointly launched the International Food Safety Authorities Network (INFOSAN), which aimed to promote the exchange of food safety information and to advance co-operation among food safety authorities.

### PROCESSING AND MARKETING

An estimated 20% of all food harvested is lost before it can be consumed, and in some developing countries the proportion is much higher. FAO helps reduce immediate post-harvest losses, with the introduction of improved processing methods and storage systems. It also advises on the distribution and marketing of agricultural produce and on the selection and preparation of foods for optimum nutrition. Many of these activities form part of wider rural development projects. Many developing countries rely on agricultural products as their main source of foreign earnings, but the terms under which they are traded are usually more favourable to the industrialized countries. FAO continues to favour the elimination of export subsidies and related discriminatory practices, such as protectionist measures that hamper international trade in agricultural commodities. FAO has organized regional workshops and national projects in order to help member states to implement World Trade Organization regulations, in particular with regard to agricultural policy, intellectual property rights, sanitary and phytosanitary measures, technical barriers to trade and the international standards of the Codex Alimentarius. FAO evaluates new market trends and helps to develop improved plant and animal quarantine procedures. In November 1997 the FAO Conference adopted new guide-lines on surveillance and on export certification systems in order to harmonize plant quarantine standards. FAO participates in PhAction, a forum of 12 agencies that was established in 1999 to promote post-

## FOOD SECURITY

FAO's policy on food security aims to encourage the production of adequate food supplies, to maximize stability in the flow of supplies, and to ensure access on the part of those who need them. In 1994 FAO initiated the Special Programme for Food Security (SPFS), designed to assist low-income countries with a food deficit to increase food production and productivity as rapidly as possible, primarily through the widespread adoption by farmers of improved production technologies, with emphasis on areas of high potential. FAO was actively involved in the formulation of the Plan of Action on food security that was adopted at the World Food Summit in November 1996, and was to be responsible for monitoring and promoting its implementation. In March 1999 FAO signed agreements with IFAD and WFP that aimed to increase co-operation within the framework of the SPFS. A budget of US $10.5m. was allocated to the SPFS for the two-year period 2004–05. In 2004 the SPFS was operational in 100 countries, of which 42 were in Africa. About 70 of these countries were categorized as 'low-income food-deficit'. The Programme promotes South-South co-operation to improve food security and the exchange of knowledge and experience. By September 2003 28 bilateral co-operation agreements were in force, for example, between Egypt and Cameroon, and Viet Nam and Benin.

In 2004 five countries in Central America and the Caribbean were categorized as 'low-income food-deficit': Belize, Cuba, Haiti, Honduras and Nicaragua. At that time food insecurity was estimated to affect some 54m. people in Latin America and the Caribbean.

FAO's Global Information and Early Warning System (GIEWS), which become operational in 1975, maintains a database on and monitors the crop and food outlook at global, regional, national and sub-national levels in order to detect emerging food supply difficulties and disasters and to ensure rapid intervention in countries experiencing food supply shortages. It publishes regular reports on the weather conditions and crop prospects in sub-Saharan Africa and in the Sahel region, issues special alerts which describe the situation in countries or sub-regions experiencing food difficulties, and recommends an appropriate international response. FAO's annual publication *State of Food Insecurity in the World* is based on data compiled by the Organization's Food Insecurity and Vulnerability Information and Mapping Systems programme.

In January 2005 GIEWS published a report on food shortages in Haiti caused by poor infrastructure and civil unrest as well as by damage inflicted by tropical storm 'Jeanne' in September 2004.

## FAO INVESTMENT CENTRE

The Investment Centre was established in 1964 to help countries to prepare viable investment projects that will attract external financing. The Centre focuses its evaluation of projects on two fundamental concerns: the promotion of sustainable activities for land management, forestry development and environmental protection, and the alleviation of rural poverty. By December 2004 the Centre had approved a total of 1,400 projects for 140 developing countries, representing a total investment of around US $76,000m.

## EMERGENCY RELIEF

FAO works to rehabilitate agricultural production following natural and man-made disasters by providing emergency seed, tools, and technical and other assistance. Jointly with the United Nations, FAO is responsible for WFP, which provides emergency food supplies and food aid in support of development projects. FAO's Division for Emergency Operations and Rehabilitation was responsible for preparing the emergency agricultural relief component of the 2005 UN inter-agency appeals for 14 countries and regions.

In January 2005, following a massive earthquake in the Indian Ocean in December 2004, which caused a series of tidal waves, or tsunamis, that devastated coastal regions in 11 countries in South and South-East Asia and East Africa, FAO requested emergency funding of US $26m. to support an initial six-month rehabilitation operation to restore the livelihoods of fishermen and farmers affected by the natural disaster.

## INFORMATION

FAO collects, analyses, interprets and disseminates information through various media, including an extensive internet site. It issues regular statistical reports, commodity studies, and technical manuals in local languages (see list of publications below). Other materials produced by the FAO include information booklets, reference papers, reports of meetings, training manuals and audio-visuals.

FAO's internet-based interactive World Agricultural Information Centre (WAICENT) offers access to agricultural publications, technical documentation, codes of conduct, data, statistics and multimedia resources. FAO compiles and co-ordinates an extensive range of international databases on agriculture, fisheries, forestry, food and statistics, the most important of these being AGRIS (the International Information System for the Agricultural Sciences and Technology) and CARIS (the Current Agricultural Research Information System). Statistical databases include the GLOBEFISH databank and electronic library, FISHDAB (the Fisheries Statistical Database), FORIS (Forest Resources Information System), and GIS (the Geographic Information System). In addition, FAOSTAT provides access to updated figures in 10 agriculture-related topics. The AGORA (Access to Global Online Research in Agriculture) initiative, launched in November 2003 by FAO and other partners, aims to provide free or low-cost access to more than 400 scientific journals in agriculture, nutrition and related fields for researchers from developing countries.

In June 2000 FAO organized a high-level Consultation on Agricultural Information Management (COAIM), which aimed to increase access to and use of agricultural information by policy-makers and others. The second COAIM was held in September 2002; a third meeting, scheduled to be held in June 2004, was postponed.

World Food Day, commemorating the foundation of FAO, is held annually on 16 October.

## FAO Councils and Commissions

(Based at the Rome headquarters unless otherwise indicated)

**Caribbean Plant Protection Commission:** f. 1967 to preserve the existing plant resources of the area; 13 member states.

**Commission for Inland Fisheries of Latin America:** f. 1976 to promote, co-ordinate and assist national and regional fishery and limnological surveys and programmes of research and development leading to the rational utilization of inland fishery resources; 21 member states.

**FAO/WHO Codex Alimentarius Commission:** internet www.codexalimentarius.net; f. 1962 to make proposals for the co-ordination of all international food standards work and to publish a code of international food standards; established Intergovernmental Task Force on Foods Derived from Biotechnology in 1999; Trust Fund to support participation by least-developed countries was inaugurated in February 2003; 165 member states.

**Latin American and Caribbean Forestry Commission:** f. 1948 to advise on formulation of forest policy and review and co-ordinate its implementation throughout the region to exchange information and advise on technical problems; meets every two years; 31 member states.

## Finance

FAO's Regular Programme, which is financed by contributions from member governments, covers the cost of FAO's Secretariat, its Technical Co-operation Programme (TCP) and part of the cost of several special action programmes. The proposed budget for the two-year period 2004–05 totalled US $749m. Much of FAO's technical assistance programme is funded from extra-budgetary sources, predominantly by trust funds that come mainly from donor countries and international financing institutions. The single largest contributor is the United Nations Development Programme (UNDP).

## Publications

*Animal Health Yearbook.*
*Commodity Review and Outlook* (annually).
*Environment and Energy Bulletin.*
*Ethical Issues in Food and Agriculture.*
*Fertilizer Yearbook.*
*Food Crops and Shortages* (6 a year).
*Food Outlook* (5 a year).
*Food Safety and Quality Update* (monthly; electronic bulletin).
*Forest Resources Assessment.*
*Plant Protection Bulletin* (quarterly).
*Production Yearbook.*
*Quarterly Bulletin of Statistics.*
*The State of Food and Agriculture* (annually).
*The State of Food Insecurity in the World* (annually).
*The State of World Fisheries and Aquaculture* (every two years).
*The State of the World's Forests* (every 2 years).
*Trade Yearbook.*

*Unasylva* (quarterly).
*Yearbook of Fishery Statistics*.
*Yearbook of Forest Products*.

*World Animal Review* (quarterly).
*World Watch List for Domestic Animal Diversity*.
Commodity reviews; studies, manuals.

# International Bank for Reconstruction and Development—IBRD (World Bank)

**Address:** 1818 H St, NW, Washington, DC 20433, USA.
**Telephone:** (202) 473-1000; **fax:** (202) 477-6391; **e-mail:** pic@worldbank.org; **internet:** www.worldbank.org.

The IBRD was established in December 1945. Initially it was concerned with post-war reconstruction in Europe; since then its aim has been to assist the economic development of member nations by making loans where private capital is not available on reasonable terms to finance productive investments. Loans are made either directly to governments, or to private enterprises with the guarantee of their governments. The World Bank, as it is commonly known, comprises the IBRD and the International Development Association (IDA). The affiliated group of institutions, comprising the IBRD, the IDA, the International Finance Corporation (IFC), the Multilateral Investment Guarantee Agency (MIGA) and the International Centre for Settlement of Investment Disputes (ICSID, see below), is now referred to as the World Bank Group.

## Organization

(August 2005)

Officers and staff of the IBRD serve concurrently as officers and staff in the IDA. The World Bank has offices in New York, Brussels, Paris (for Europe), Frankfurt, London, Geneva and Tokyo, as well as in more than 100 countries of operation. Country Directors are located in some 30 country offices.

### BOARD OF GOVERNORS

The Board of Governors consists of one Governor appointed by each member nation. Typically, a Governor is the country's finance minister, central bank governor, or a minister or an official of comparable rank. The Board normally meets once a year.

### EXECUTIVE DIRECTORS

The general operations of the Bank are conducted by a Board of 24 Executive Directors. Five Directors are appointed by the five members having the largest number of shares of capital stock, and the rest are elected by the Governors representing the other members. The President of the Bank is Chairman of the Board.

### PRINCIPAL OFFICERS

The principal officers of the Bank are the President of the Bank, two Managing Directors, three Senior Vice-Presidents and 24 Vice-Presidents.

**President and Chairman of Executive Directors:** PAUL WOLFOWITZ (USA).

**Vice-President, Latin America and the Caribbean Regional Office:** DAVID DE FERRANTI (USA).

## Activities

### FINANCIAL OPERATIONS

IBRD capital is derived from members' subscriptions to capital shares, the calculation of which is based on their quotas in the International Monetary Fund. At 30 June 2004 the total subscribed capital of the IBRD was US $189,718m., of which the paid-in portion was $11,483m. (6.1%); the remainder is subject to call if required. Most of the IBRD's lendable funds come from its borrowing, on commercial terms, in world capital markets, and also from its retained earnings and the flow of repayments on its loans. IBRD loans carry a variable interest rate, rather than a rate fixed at the time of borrowing.

IBRD loans usually have a 'grace period' of five years and are repayable over 15 years or fewer. Loans are made to governments, or must be guaranteed by the government concerned, and are normally made for projects likely to offer a commercially viable rate of return. In 1980 the World Bank introduced structural adjustment lending, which (instead of financing specific projects) supports programmes and changes necessary to modify the structure of an economy so that it can restore or maintain its growth and viability in its balance-of-payments over the medium-term.

The IBRD and IDA together made 245 new lending and investment commitments totalling US $20,080.1m. during the year ending 30 June 2004, compared with 240 (amounting to $18,513.2m.) in the previous year. During 2003/04 the IBRD alone approved commitments totalling $11,045.4m. (compared with $11,230.7m. in the previous year), of which $4,981.6m. (45%) was allocated to Latin America and the Caribbean. Disbursements by the IBRD in the year ending 30 June 2004 amounted to $10,109m.

IBRD operations are supported by medium- and long-term borrowings in international capital markets. During the year ending 30 June 2004 the IBRD's net income amounted to US $4,328m.

The World Bank's primary objectives are the achievement of sustainable economic growth and the reduction of poverty in developing countries. In the context of stimulating economic growth the Bank promotes both private-sector development and human resource development and has attempted to respond to the growing demands by developing countries for assistance in these areas. In March 1997 the Board of Executive Directors endorsed a 'Strategic Compact' to increase the effectiveness of the Bank in achieving its central objective of poverty reduction. The reforms included greater decentralization of decision-making, and investment in front-line operations, enhancing the administration of loans, and improving access to information and co-ordination of Bank activities through a knowledge management system comprising four thematic networks: the Human Development Network; the Environmentally and Socially Sustainable Development Network; the Finance, Private Sector and Infrastructure Development Network; and the Poverty Reduction and Economic Management Network. In 2000/01 the Bank adopted a new Strategic Framework which emphasized two essential approaches for Bank support: strengthening the investment climate and prospects for sustainable development in a country, and supporting investment in the poor. In September 2001 the Bank announced that it was to join the UN as a full partner in implementing the so-called Millennium Development Goals (MDGs), and was to make them central to its development agenda. The objectives, which were approved by governments attending a special session of the UN General Assembly in September 2000, represented a new international consensus to achieve determined poverty reduction targets. These included reducing by 50% the number of people with an income of less than US $1 a day and those suffering from hunger and lack of safe drinking water by 2015, achieving education for all, reducing maternal mortality, and combating HIV/AIDS, malaria and other major diseases. The Bank was closely involved in preparations for the International Conference on Financing for Development, which was held in Monterrey, Mexico, in March 2002. The meeting adopted the Monterrey Consensus, which outlined measures to support national development efforts and to achieve the MDGs. During 2002/03 the Bank, with the IMF, undertook to develop a monitoring framework to review progress in the MDG agenda. The first *Global Monitoring Report* was issued by the Bank and IMF in April 2004. Other efforts to support a greater emphasis on development results were also undertaken by the Bank during 2003/04 as part of a new strategic action plan.

The Bank's efforts to reduce poverty include the compilation of country-specific assessments and the formulation of country assistance strategies (CASs) to review and guide the Bank's country programmes. Since August 1998 the Bank has published CASs, with the approval of the government concerned. A new results-based CAS initiative was piloted in 2003/04. In 1998/99 the Bank's Executive Directors endorsed a Comprehensive Development Framework (CDF) to effect a new approach to development assistance based on partnerships and country responsibility, with an emphasis on the interdependence of the social, structural, human, governmental, economic and environmental elements of development. The Framework, which aimed to enhance the overall effectiveness of development assistance, was formulated after a series of consultative meetings organized by the Bank and attended by representatives of governments, donor agencies, financial institutions, non-governmental organizations, the private sector and academics.

In December 1999 the Bank introduced a new approach to implement the principles of the CDF, as part of its strategy to enhance the debt relief scheme for heavily indebted poor countries (see below). Applicant countries were requested to formulate, in consultation with external partners and other stakeholders, a results-oriented national strategy to reduce poverty, to be presented in the form of a Poverty Reduction Strategy Paper (PRSP). In cases where there might be some delay in issuing a full PRSP, it was permissible for a country to submit a less detailed 'interim' PRSP (I-PRSP) in order to secure the preliminary qualification for debt relief. By June 2004 42 countries had finalized full PRSPs. The approach also requires the publication of annual progress reports. In 2000/01 the Bank introduced a new Poverty Reduction Support Credit to help low-income countries to implement the policy and institutional reforms outlined in their PRSP. The first credits were approved for Uganda and Viet Nam in May and June respectively. In January 2002 a PRSP public review conference, attended by more than 200 representatives of donor agencies, civil society groups, and developing country organizations was held as part of an ongoing review of the scheme by the Bank and the IMF. Increasingly, PRSPs have been considered by the international community to be the appropriate country-level framework to assess progress towards achieving the MDGs.

In September 1996 the World Bank/IMF Development Committee endorsed a joint initiative to assist heavily indebted poor countries (HIPCs) to reduce their debt burden to a sustainable level, in order to make more resources available for poverty reduction and economic growth. A new Trust Fund was established by the World Bank in November to finance the initiative. The Fund, consisting of an initial allocation of US $500m. from the IBRD surplus and other contributions from multilateral creditors, was to be administered by IDA. Of the 41 HIPCs identified by the Bank, 33 were in sub-Saharan Africa. In April 1997 the World Bank and the IMF announced that Uganda was to be the first beneficiary of the initiative, enabling the Ugandan Government to reduce its external debt by some 20%, or an estimated $338m. In early 1999 the World Bank and IMF initiated a comprehensive review of the HIPC initiative. By April meetings of the Group of Seven industrialized nations (G-7) and of the governing bodies of the Bank and IMF indicated a consensus that the scheme needed to be amended and strengthened, in order to allow more countries to benefit from the initiative, to accelerate the process by which a country may qualify for assistance, and to enhance the effectiveness of debt relief. In June the G-7 and Russia, meeting in Cologne, Germany, agreed to increase contributions to the HIPC Trust Fund and to cancel substantial amounts of outstanding debt, and proposed more flexible terms for eligibility. In September the Bank and IMF reached an agreement on an enhanced HIPC scheme, with further revenue to be generated through the revaluation of a percentage of IMF gold reserves. Under the enhanced initiative it was agreed that, during the initial phase of the process to ensure suitability for debt relief, each applicant country should formulate a PRSP, and should demonstrate prudent financial management in the implementation of the strategy for at least one year, with support from the IDA and IMF. At the pivotal 'decision point' of the process, having thus developed and successfully applied the poverty reduction strategy, applicant countries still deemed to have an unsustainable level of debt were to qualify for interim debt relief from the IMF and IDA, as well as relief on highly concessional terms from other official bilateral creditors and multilateral institutions. During the ensuing 'interim period' countries were required successfully to implement further economic and social development reforms, as a final demonstration of suitability for securing full debt relief at the 'completion point' of the scheme. Data produced at the decision point was to form the base for calculating the final debt relief (in contrast to the original initiative, which based its calculations on projections of a country's debt stock at the completion point). In the majority of cases a sustainable level of debt was targeted at 150% of the net present value (NPV) of the debt in relation to total annual exports (compared with 200%–250% under the original initiative). Other countries with a lower debt-to-export ratio were to be eligible for assistance under the scheme, providing that their export earnings were at least 30% of GDP (lowered from 40% under the original initiative) and government revenue at least 15% of GDP (reduced from 20%). By May 2005 18 countries had reached completion point under the enhanced HIPC initiative, including Bolivia ($1,302m. in NPV terms approved in June 2001), Guyana ($1,553m. in April 2003) and Nicaragua ($3,308m. in January 2004). Honduras reached completion point in April 2005. At that time a further nine countries had reached their decision point. At mid-March total assistance committed under the HIPC initiative amounted to US $32,003m., or $54,102m. in total estimated nominal debt service relief, of which the Bank had committed $12,930m. In June finance ministers of the G-8 announced their intention to provide debt relief amounting to some $40,000m., over a three-year period. The 18 countries that had reached their completion point were to qualify for immediate assistance. In July the heads of state and government of G-8 countries requested the Bank to ensure the effective delivery of the additional resources and to provide a framework for performance measurement.

During 2000/01 the World Bank strengthened its efforts to counter the problem of HIV and AIDS in developing countries. In November 2001 the Bank appointed its first Global HIV/AIDS Adviser. In 2001 a Multi-Country HIV/AIDS Prevention and Control Programme for the Caribbean was launched, with an allocated budget of US $155m. By mid-2004 loans and grants had been made available under the initiative to Barbados, Dominican Republic, Jamaica, Grenada, Guyana, Saint Christopher and Nevis, Trinidad and Tobago and a Pan-Caribbean Partnership against HIV/AIDS. In March 2003 a Post-Conflict Grant for Haiti, totalling $2.9m, was approved to prevent and control infectious diseases, including HIV. In addition, the Bank has extended some $425m. to an AIDS/sexually transmitted disease control project in Brazil, financing for the third phase of which was approved in June 2003.

In addition to providing financial services, the Bank also undertakes analytical and advisory services, and supports learning and capacity-building, in particular through the World Bank Institute (see below), the Staff Exchange Programme and knowledge-sharing initiatives. The Bank has supported efforts, such as the Global Development Gateway, to disseminate information on development issues and programmes, and, since 1988, has organized the Annual Bank Conference on Development Economics (ABCDE) to provide a forum for the exchange and discussion of development-related ideas and research. In September 1995 the Bank initiated the Information for Development Programme (InfoDev) with the aim of fostering partnerships between governments, multilateral institutions and private-sector experts in order to promote reform and investment in developing countries through improved access to information technology.

### TECHNICAL ASSISTANCE

The provision of technical assistance to member countries has become a major component of World Bank activities. The economic and sector work (ESW) undertaken by the Bank is the vehicle for considerable technical assistance and often forms the basis of CASs and other strategic or advisory reports. In addition, project loans and credits may include funds earmarked specifically for feasibility studies, resource surveys, management or planning advice, and training. The Economic Development Institute has become one of the most important of the Bank's activities in technical assistance. It provides training in national economic management and project analysis for government officials at the middle and upper levels of responsibility. It also runs overseas courses aiming to build up local training capability, and administers a graduate scholarship programme.

The Bank serves as an executing agency for projects financed by the UN Development Programme. It also administers projects financed by various trust funds.

Technical assistance (usually reimbursable) is also extended to countries that do not need Bank financial support, e.g. for training and transfer of technology. The Bank encourages the use of local consultants to assist with projects and stimulate institutional capability.

The Project Preparation Facility (PPF) was established in 1975 to provide cash advances to countries to prepare projects that may be financed by the Bank. In December 1994 the PPF's commitment authority was increased from US $220m. to $250m. In 1992 the Bank established an Institutional Development Fund (IDF), which became operational on 1 July; the purpose of the Fund was to provide rapid, small-scale financial assistance, to a maximum value of $500,000, for capacity-building proposals. In 2002 the IDF was reoriented to focus on good governance, in particular financial accountability and system reforms.

### ECONOMIC RESEARCH AND STUDIES

In the 1990s the World Bank's research, conducted by its own research staff, was increasingly concerned with providing information to reinforce the Bank's expanding advisory role to developing countries and to improve policy in the Bank's borrowing countries. The principal areas of current research focus on issues such as maintaining sustainable growth while protecting the environment and the poorest sectors of society, encouraging the development of the private sector, and reducing and decentralizing government activities.

The Bank chairs the Consultative Group on International Agricultural Research (CGIAR), which was founded in 1971 to raise financial support for international agricultural research work for improving crops and animal production in developing countries; it supports 16 research centres.

### CO-OPERATION WITH OTHER ORGANIZATIONS

The World Bank co-operates with other international partners with the aim of improving the impact of development efforts. It collaborates with the IMF in implementing the HIPC scheme and the two agencies work closely to achieve a common approach to development initiatives. The Bank has established strong working relationships with many other UN bodies, in particular through a mutual commitment to poverty reduction objectives. In May 2000 the Bank signed a joint statement of co-operation with the OECD. The Bank holds regular consultations with other multilateral development banks and with the European Union with respect to development issues. The Bank-NGO Committee provides an annual forum for discussion with non-governmental organizations (NGOs). Strengthening co-operation with external partners was a fundamental element of the Comprehensive Development Framework, which was adopted in 1998/99 (see above). In 2001/02 a Partnership Approval and Tracking System was implemented to provide information on the Bank's regional and global partnerships.

In 1997 a Partnerships Group was established to strengthen the Bank's work with development institutions, representatives of civil society and the private sector. The Group established a new Development Grant Facility, which became operational in October, to support partnership initiatives and to co-ordinate all of the Bank's grant-making activities. In 2003/04 the Facility had funds of US $178.2m. Also in 1997 the Bank, in partnership with the IMF, UNCTAD, UNDP, the World Trade Organization (WTO) and International Trade Commission, established an Integrated Framework for Trade-related Assistance to Least Developed Countries, at the request of the WTO, to assist those countries to integrate into the global trading system and improve basic trading capabilities.

In June 1995 the World Bank joined other international donors (including regional development banks, other UN bodies, Canada, France, the Netherlands and the USA) in establishing a Consultative Group to Assist the Poorest (CGAP), which was to channel funds to the most needy through grass-roots agencies. An initial credit of approximately US $200m. was committed by the donors. The Bank manages the CGAP Secretariat, which is responsible for the administration of external funding and for the evaluation and approval of project financing. The CGAP provides technical assistance, training and strategic advice to microfinance institutions and other relevant bodies. As an implementing agency of the Global Environment Facility (GEF) the Bank assists countries to prepare and supervise GEF projects relating to biological diversity, climate change and other environmental protection measures. It is an example of a partnership in action which addresses a global agenda, complementing Bank country assistance activities. Other Trust Funds administered by the Bank include the Global Program to Eradicate Poliomyelitis, launched during the financial year 2002/03, the Least Developed Countries Fund for Climate Change, established in September 2002, and an Education for All Fast-Track Initiative Catalytic Trust Fund, established in 2003/04.

The Bank is a lead organization in providing reconstruction assistance following natural disasters or conflicts, usually in collaboration with other UN agencies or international organizations, and through special trust funds. In May–June 2004 the Bank, jointly with the Inter-American Development Bank, the European Commission and the UN, assisted the Haitian Government to prepare an Interim Co-operation Framework (ICF) as an assessment of the country's technical and financial needs in the next two years. In July the ICF was presented to an International Donor Conference on Haiti, held in Washington, DC Participants to the conference, which was hosted by the four lead institutions, pledged some US $1,085m. to support Haiti's economic, social and political recovery.

The Bank conducts co-financing and aid co-ordination projects with official aid agencies, export credit institutions, and commercial banks to leverage additional concessional funds for recipient countries. During the year ending 30 June 2004 the Bank's main partners were the Inter-American Development Bank (which provided US $3,700m.) and the European Commission ($640m.).

### EVALUATION

The Operations Evaluation Department is an independent unit within the World Bank. It conducts Country Assistance Evaluations to assess the development effectiveness of a Bank country programme, and studies and publishes the results of projects after a loan has been fully disbursed, so as to identify problems and possible improvements in future activities. In addition, the department reviews the Bank's global programmes and produces the *Annual Review of Development Effectiveness*. In 1996 a Quality Assurance Group was established to monitor the effectiveness of the Bank's operations and performance.

In September 1993 the Bank established an independent Inspection Panel, consistent with the Bank's objective of improving project implementation and accountability. The Panel, which became operational in September 1994, was to conduct independent investigations and report on complaints from local people concerning the design, appraisal and implementation of development projects supported by the Bank. By mid-2004 the Panel had received 33 formal requests for inspection and had recommended investigations in 14 of those cases.

### IBRD INSTITUTIONS

**World Bank Institute (WBI):** founded in March 1999 by merger of the Bank's Learning and Leadership Centre, previously responsible for internal staff training, and the Economic Development Institute (EDI), which had been established in 1955 to train government officials concerned with development programmes and policies. The new Institute aimed to emphasize the Bank's priority areas through the provision of training courses and seminars relating to poverty, crisis response, good governance and anti-corruption strategies. From 2004 the Institute was to place greater emphasis on individual country needs and on long-term institutional capacity-building. During 2003/04 WBI activities reached some 78,000 participants in 124 countries. The Institute has continued to support a Global Knowledge Partnership, which was established in 1997 to promote alliances between governments, companies, other agencies and organizations committed to applying information and communication technologies for development purposes. Under the EDI a World Links for Development programme was also initiated to connect schools in developing countries with partner establishments in industrialized nations via the internet. In 1999 the WBI expanded its programmes through distance learning, a Global Development Network, and use of new technologies. A new initiative, Global Development Learning Network (GDLN), aimed to expand access to information and learning opportunities through the internet, video-conferences and organized exchanges. At late-2004 there were 70 GDLN centres, or affiliates. At mid-2004 formal partnership arrangements were in place between WBI and some 125 learning centres and public, private and non-governmental organizations; Vice-Pres. FRANNIE LÉAUTIER (Tanzania/France).

**International Centre for Settlement of Investment Disputes (ICSID):** founded in 1966 under the Convention of the Settlement of Investment Disputes between States and Nationals of Other States. The Convention was designed to encourage the growth of private foreign investment for economic development, by creating the possibility, always subject to the consent of both parties, for a Contracting State and a foreign investor who is a national of another Contracting State to settle any legal dispute that might arise out of such an investment by conciliation and/or arbitration before an impartial, international forum. The governing body of the Centre is its Administrative Council, composed of one representative of each Contracting State, all of whom have equal voting power. The President of the World Bank is (*ex officio*) the non-voting Chairman of the Administrative Council. At the end of 2004 142 countries had signed and ratified the Convention to become ICSID Contracting States. By mid-2004 the Centre had concluded 83 cases, while 78 were pending; Sec.-Gen. ROBERTO DAÑINO (Peru).

## Publications

*Abstracts of Current Studies: The World Bank Research Program* (annually).
*Annual Report on Operations Evaluation.*
*Annual Report on Portfolio Performance.*
*Annual Review of Development Effectiveness.*
*Doing Business* (annually).
*EDI Annual Report.*
*Global Commodity Markets* (quarterly).
*Global Development Finance* (annually, also on CD-Rom and online).
*Global Economic Prospects* (annually).
*ICSID Annual Report.*
*ICSID Review—Foreign Investment Law Journal* (2 a year).
*Joint BIS-IMF-OECD-World Bank Statistics on External Debt* (quarterly, also available online at www.worldbank.org/data/jointdebt.html).
*New Products and Outreach* (EDI, annually).
*News from ICSID* (2 a year).
*Poverty Reduction and the World Bank* (annually).
*Poverty Reduction Strategies Newsletter* (quarterly).
*Research News* (quarterly).
*Staff Working Papers.*
*Transition* (every 2 months).
*World Bank Annual Report.*
*World Bank Atlas* (annually).
*World Bank Economic Review* (3 a year).
*The World Bank and the Environment* (annually).
*World Bank Research Observer.*

*World Development Indicators* (annually, also on CD-Rom and online).
*World Development Report* (annually, also on CD-Rom).

## Statistics

**IBRD LOANS APPROVED IN LATIN AMERICA AND THE CARIBBEAN, 1 JULY 2003–30 JUNE 2004**
(US $ million)

| Country | Purpose | Amount |
|---|---|---|
| Argentina | Economic recovery support structural adjustment loan | 500.0 |
| | National highway asset management | 200.0 |
| | Provincial maternal-child health programme | 135.8 |
| | Maternal and child health sector adjustment loan | 750.0 |
| Bolivia | First programmatic bank and corporate sector restructuring programme | 15.0* |
| Brazil | Bolsa Família (family grants programme) | 572.2 |
| | Second disease surveillance and control adaptable programme loan | 100.0 |
| | Integrated Maranhão poverty reduction specific investment loan | 30.0 |
| | Programmatic loan for sustainable and equitable growth | 505.1 |
| | Tocantins sustainable regional development sector investment and maintenance loan | 60.0 |
| Chile | Social protection technical assistance loan | 10.7 |
| | Social protection sector adjustment loan | 200.0 |
| Colombia | First peace and development adaptable programme loan | 30.0 |
| | Integrated mass transit systems | 250.0 |
| | Second programmatic fiscal and institutional structural adjustment loan | 150.0 |
| | First programmatic labour reform and social sector adjustment loan | 200.0 |
| | Cundinamarca education quality investment loan | 15.0 |
| Dominica | Economic recovery support | 3.0* |
| Dominican Republic | Financial sector technical assistance | 12.5 |
| | Power sector technical assistance | 7.3 |
| | Social crisis response adjustment loan | 100.0 |
| Ecuador | Institutional reform | 20.0 |
| | Second indigenous and Afri-Ecuadorian peoples development investment loan | 34.0 |
| Mexico | Second phase savings and rural finance investment loan | 75.5 |
| | Decentralized infrastructure reform and development | 108.0 |
| | Affordable housing and urban poverty sector adjustment loan | 100.0 |
| | Second community forestry investment loan | 21.3 |
| | Integrated irrigation modernization | 303.0 |
| | E-business for small business development sector investment and maintenance loan | 58.4 |
| Paraguay | Economic recovery structural adjustment loan | 30.0 |
| | Education reform | 24.0 |
| Peru | Justice services improvement technical assistance loan | 12.0 |
| | Lima urban transport | 45.0 |
| | Programmatic decentralization and competitiveness structural adjustment loan | 150.0 |
| | Third programmatic social reform structural adjustment loan | 150.0 |
| Saint Lucia | Second St Lucia disaster management investment loan | 3.7* |
| Saint Vincent and the Grenadines | OECS education development | 3.1* |

* IBRD/IDA joint operation.

Source: *World Bank Annual Report 2004.*

# International Development Association—IDA

**Address:** 1818 H Street, NW, Washington, DC 20433, USA.
**Telephone:** (202) 473-1000; **fax:** (202) 477-6391; **internet:** www.worldbank.org/ida.

The International Development Association began operations in November 1960. Affiliated to the IBRD, IDA advances capital to the poorer developing member countries on more flexible terms than those offered by the IBRD.

## Organization

(August 2005)

Officers and staff of the IBRD serve concurrently as officers and staff of IDA.

**President and Chairman of Executive Directors:** PAUL WOLFOWITZ (*ex officio*).

## Activities

IDA assistance is aimed at the poorer developing countries (i.e. those with an annual GNP per capita of less than US $865 in 2002 dollars were to qualify for assistance in 2003/04) and support their poverty reduction strategies. Under IDA lending conditions, credits can be extended to countries whose balance of payments could not sustain the burden of repayment required for IBRD loans. Terms are more favourable than those provided by the IBRD; credits are for a period of 35 or 40 years, with a 'grace period' of 10 years, and carry no interest charges. At mid-2004 81 countries were eligible for IDA assistance, including several small-island economies with a GNP per head greater than $865, but which would otherwise have little or no access to Bank funds, and 15 so-called 'blend borrowers' which are entitled to borrow from both the IDA and IBRD. IDA administers a Trust Fund, which was established in November 1996 as part of a World Bank/IMF initiative to assist heavily indebted poor countries (HIPCs).

IDA's total development resources, consisting of members' subscriptions and supplementary resources (additional subscriptions and contributions), are replenished periodically by contributions from the more affluent member countries. Discussions on the 13th replenishment of IDA funds commenced in February 2001, and for the first time involved representatives of borrowing countries, civil society and other public groups. A final commitment, providing for some US $23,000m. in resources for the period 1 July 2002–30 June 2005, was concluded in July 2002 by 38 donor countries. The IDA-13 lending framework emphasized the following objectives: promoting sound policies for growth and poverty reduction; ensuring effective assistance and measurable results; improving co-ordination, transparency, and consultation; and providing for substantial replenishment of resources. The replenishment programme also provided for greater use of grants to address the problems of the poorest recipient countries, for example those most vulnerable to debt, those in post-conflict situations, as well as reconstruction projects after a natural disaster and HIV/AIDS programmes. Negotiations on the 14th replenishment of IDA funds commenced in February 2004. An agreement to provide a substantial replenishment of funds was concluded in February 2005, when 40 donor countries committed $34,000m. for the period 1 July 2005–30 June 2008.

During the year ending 30 June 2004 IDA credits totalling US $9,034.6 were approved, compared with $7,282.5m. in the previous year. Of the total new lending in 2003/04 some $1,712m. (or 19%) was in the form of grants for the poorest or most vulnerable countries.

## Publication

Annual Report.

## Statistics

**IDA CREDITS APPROVED IN SOUTH AMERICA AND THE CARIBBEAN, 1 JULY 2003–30 JUNE 2004**
(US $ million)

| Country | Purpose | Amount |
|---|---|---|
| Bolivia | Social sectors programmatic structural adjustment credit | 25.0 |
| | First programmatic bank and corporate sector restructuring programme | 15.0* |
| | Emergency economic recovery credit | 14.0 |
| Caribbean | HIV/AIDS prevention and control | 9.0 |
| Dominica | Economic recovery support | 3.0* |
| Guyana | HIV/AIDS prevention and control | 10.0 |
| Honduras | Poverty reduction | 58.8 |
| | Poverty reduction support technical assistance credit | 8.0 |
| | Forests and rural productivity | 20.0 |
| | Nuestras Raíces community development programme | 15.0 |
| | Trade facilitation and productivity improvement | 28.1 |
| Nicaragua | Broad-based access to financial services technical assistance credit | 7.0 |
| | Public sector technical assistance credit | 23.5 |
| | Poverty reduction support | 70.0 |
| Saint Lucia | Second St Lucia disaster management investment loan | 3.8* |
| Saint Vincent and the Grenadines | OECS education development | 3.1* |

* Joint IBRD/IDA funded project.

Source: *World Bank Annual Report 2004*.

# International Finance Corporation—IFC

**Address:** 2121 Pennsylvania Ave, NW, Washington, DC 20433, USA.

**Telephone:** (202) 473-3800; **fax:** (202) 974-4384; **e-mail:** information@ifc.org; **internet:** www.ifc.org.

IFC was founded in 1956 as a member of the World Bank Group to stimulate economic growth in developing countries by financing private-sector investments, mobilizing capital in international financial markets, and providing technical assistance and advice to governments and businesses.

## Organization

(August 2005)

IFC is a separate legal entity in the World Bank Group. Executive Directors of the World Bank also serve as Directors of IFC. The President of the World Bank is *ex officio* Chairman of the IFC Board of Directors, which has appointed him President of IFC. Subject to his overall supervision, the day-to-day operations of IFC are conducted by its staff under the direction of the Executive Vice-President.

**PRINCIPAL OFFICERS**

**President:** Paul Wolfowitz (USA).

**Executive Vice-President:** Assaad Jabre.

**Director, Latin America and the Caribbean Department:** Atul Mehta.

**MISSIONS AND OFFICES IN SOUTH AMERICA, CENTRAL AMERICA AND THE CARIBBEAN**

**Argentina:** Bouchard 680, Torre Fortabat, 11°, 1106 Buenos Aires; tel. (11) 4114-7200; fax (11) 4312-7184; Senior Man. Yolande Duhem.

**Bolivia:** Edif. Victor, 9°, Calle Fernando Guachalla 342, La Paz; tel. (2) 244-3133; fax (2) 212-5065; Country Man. Serge Devieux.

**Brazil:** Rua Redentor 14, Ipanema, 22421-030, Rio de Janeiro; tel. (21) 5185-6888; fax (21) 5181-8252; Resident Dir Atul Mehta.

**Colombia:** Carrera 7, 71-21, Torre A, 16°, Edif. Fiduagraria, Santafé de Bogotá, DC; tel. (1) 319-2330; fax (1) 319-2359; Country Man. Serge Devieux.

**Dominican Republic:** Calle Virgilio Diaz Ordoñez 36, esq. Gustavo Mejía Ricart, Edif. Mezzo Tempo, Suite 401, Santo Domingo; tel. 566-6815; fax 566-7746; Resident Rep. Salem Rohana.

**Mexico:** Prado Sur 240, Col. Lomas de Chapultepec, 11000 México, DF; tel. (55) 5520-6191; fax (55) 5520-5629; Senior Man. Paolo Martelli.

**Peru:** Avda Alvarez Calderón 185, 7°, San Isidro, Lima; tel. (11) 6150660; fax (11) 4217241; Resident Rep. Per Kjellerhaug; Gen. Man. LAC Facility Anita Bhatia.

**Trinidad and Tobago:** SW Penthouse, SAGICOR Bldg, 3rd Floor, 16 Queen's Park West, POB 751, Port of Spain; tel. 628-5074; fax 622-1003; Resident Rep. Kirk B. Ifill.

## Activities

IFC aims to promote economic development in developing member countries by assisting the growth of private enterprise and effective capital markets. It finances private sector projects, through loans, the purchase of equity, quasi-equity products, and risk management services, and assists governments to create conditions that stimulate the flow of domestic and foreign private savings and investment. IFC may provide finance for a project that is partly state-owned, provided that there is participation by the private sector and that the project is operated on a commercial basis. IFC also mobilizes additional resources from other financial institutions, in particular through syndicated loans, thus providing access to international capital markets. IFC provides a range of advisory services to help to improve the investment climate in developing countries and offers technical assistance to private enterprises and governments.

To be eligible for financing, projects must be profitable for investors, as well as financially and economically viable; must benefit the economy of the country concerned; and must comply with IFC's environmental and social guide-lines. IFC aims to promote best corporate governance and management methods and sustainable business practices, and encourages partnerships between governments, non-governmental organizations and community groups. In 2001/02 IFC developed a Sustainability Framework to help to assess the longer-term economic, environmental and social impact of projects. The first Sustainability Review was published in mid-2002. In 2002/03 IFC assisted 10 international banks to draft a voluntary set of guide-lines (the Equator Principles), based on IFC's environmental, social and safeguard monitoring policies, to be applied to their global project finance activities. By 30 June 2004 24 financial institutions had signed up to the Equator Principles.

IFC's authorized capital is US $2,450m. At 30 June 2004 paid-in capital was $2,362m. The World Bank was originally the principal source of borrowed funds, but IFC also borrows from private capital markets. IFC's net income amounted to $993m. in 2003/04, compared with $487m. in the previous year.

In the year ending 30 June 2004 project financing approved by IFC amounted to US $5,633m. for 217 projects (compared with $5,033m. for 204 projects in the previous year). Of the total approved, $4,753m. was for IFC's own account, while $880m. was in the form of loan syndications and underwriting of securities issues and investment funds by more than 100 participant banks and institutional investors. Generally, the IFC limits its financing to less than 25% of the total cost of a project, but may take up to a 35% stake in a venture (although never as a majority shareholder). Disbursements for IFC's account amounted to $3,152m. in 2003/04 (compared with $2,959m. in the previous year).

The largest proportion of investment commitments in 2003/04 was allocated to Europe and Central Asia (36%). Latin America and the Caribbean received 28%, East Asia and the Pacific 14%, South Asia 9%, sub-Saharan Africa 7% and Middle East and North Africa 4%. In that year almost one-third of total financing committed (30%) was for financial services. Other commitments included warehousing and utilities (13%), oil, gas, mining (11%) and information (6%).

IFC has identified the following as priority strategic areas for future activity in Latin America and the Caribbean: expansion of private-sector participation in infrastructure and social sector activities; development of domestic capital markets; provision of resources to companies lacking access to international capital flows; and diversification of activities to less-developed national or local economies. In September 2000 IFC, with the Bank of Nova Scotia, inaugurated a new loan facility for the Caribbean to provide long-term financing, with loans in the range of $500,000 to $5m., to small and medium-sized enterprises undertaking expansion or restructuring, particularly for export-oriented projects. In May 2003 IFC established a Latin America and Caribbean Small and Medium Enterprise Facility (LAC-SME), initially serving Bolivia, Honduras, Nicaragua and Peru. The Facility was to focus on the following areas: strengthening SME competitiveness; simplifying business regulations; broadening access to finance; and fostering indigenous and social enterprises. In 2003/04 IFC signed financing commitments for 45 projects in 16 Latin American and Caribbean countries amounting to $1,593m., compared with $2,180m. for 54 projects in the previous financial year.

Since 1990 IFC has undertaken risk-management services, in order to assist institutions to avoid financial risks that arise from changes in interest rates, in exchange rates or in commodity prices. In 2003/04 IFC committed 10 risk-management transactions for companies and banks.

IFC's Private Sector Advisory Services (PSAS), jointly managed with the World Bank, advises governments and private enterprises on policy, transaction implementation and foreign direct investment. The Foreign Investment Advisory Service (FIAS), also jointly operated and financed with the World Bank, provides advice on promoting foreign investment and strengthening the country's investment framework at the request of governments. During 2003/04 FIAS completed 60 advisory projects. At the end of that year the service had assisted more than 130 countries since it commenced operations in 1986. Under the Technical Assistance Trust Funds Program (TATF), established in 1988, IFC manages resources contributed by various governments and agencies to provide finance for feasibility studies, project identification studies and other types of technical assistance relating to project preparation. By mid-2004 contributions to the TATF programme totalled US $188m. and more than 1,380 technical assistance projects had been approved.

IFC support for private sector development includes technical assistance and advisory services for small-scale entrepreneurs to assist the development of business proposals and efforts to generate funding for their projects. IFC has established, with the support of external donors, a network of Small and Medium Enterprise (SME) facilities and programmes which aim to provide support services at enterprise level; to assist the development of local private sector support institutions; and to advocate ways to improve the business-enabling environment. The Africa Project Development Facility (APDF) was established in 1986 by IFC, UNDP and the African Development Bank, and has headquarters in Johannesburg, South Africa, with other offices in Cape Town (South Africa), Cameroon, Ghana, Kenya, and Nigeria. The Facility also promotes capacity-building for SMEs, local business associations and financial institutions. It works closely with the African Management Services Company (AMSCo, established in 1989, with headquarters in Amsterdam, the Netherlands) which helps to find qualified senior executives and technical personnel to work with African companies, assist in the training of local managers, and provide management support. The South Pacific Project Facility, based in Sydney, Australia, was established in 1991, mainly to assist local businesses in the IFC Pacific Island member countries. A separate office in Port Moresby, Papua New Guinea, was opened in 1997. The Facility was renamed as the Pacific Enterprise Development Facility from August 2003. The Mekong Private Sector Development Facility (MPDF) became operational in 1997 (initially as a five-year technical assistance programme, the Mekong Project Development Facility), specifically to support the development of SMEs in Cambodia, Laos and Viet Nam. In September 2000 the Southeast Europe Enterprise Development (SEED) initiative was formally launched at its headquarters in Sarajevo, Bosnia and Herzegovina, as a five-year scheme to support the development of the private-sector in Albania, Bosnia and Herzegovina, Kosovo, the former Yugoslav republic of Macedonia, and the Federal Republic of Yugoslavia (now Serbia and Montenegro). A successor facility, the Private Enterprise Partnership Southeast Europe, covering, additionally, Bulgaria, Moldova and Romania, became operational on 1 July 2005. In May 2002 a China Project Development Facility, with headquarters in Chengdu, Sichuan Province, became operational. In October 2002 the SouthAsia Enterprise Development Facility, based in Dhaka, Bangladesh, became operational. A Program for Eastern Indonesian SME Development, based in Bali, Indonesia, was established in September 2003. In October 2004 the Private Enterprise Partnership for the Middle East and North Africa (PEP-MENA) was launched to provide technical assistance and other SME and investment support to 18 countries across the region, incorporating the functions of the previously-established North Africa Enterprise Development facility and the Private Enterprise Partnership for the Middle East.

## Publications

Annual Report.
Emerging Stock Markets Factbook (annually).
Impact (quarterly).
Lessons of Experience (series).
Results on the Ground (series).
Review of Small Businesses (annually).
Discussion papers and technical documents.

# Multilateral Investment Guarantee Agency—MIGA

**Address:** 1818 H Street, NW, Washington, DC 20433, USA.
**Telephone:** (202) 473-6163; **fax:** (202) 522-2630; **internet:** www.miga.org.

MIGA was founded in 1988 as an affiliate of the World Bank. Its mandate is to encourage the flow of foreign direct investment to, and among, developing member countries, through the provision of political risk insurance and investment marketing services to foreign investors and host governments, respectively.

## Organization

(August 2005)

MIGA is legally and financially separate from the World Bank. It is supervised by a Council of Governors (comprising one Governor and one Alternate of each member country) and an elected Board of Directors (of no less than 12 members).

**President:** PAUL WOLFOWITZ (USA).
**Executive Vice-President:** YUKIKO OMURA (Japan).

## Activities

The convention establishing MIGA took effect in April 1988. Authorized capital was US $1,082m. In April 1998 the Board of Directors approved an increase in MIGA's capital base. A grant of $150m. was transferred from the IBRD as part of the package, while the capital increase (totalling $700m. callable capital and $150m. paid-in capital) was approved by MIGA's Council of Governors in April 1999. A three-year subscription period then commenced, covering the period April 1999–March 2002 (later extended to March 2003). At 30 June 2004 102 countries had subscribed $682.6m. of the new capital increase. At that time total subscriptions to the capital stock amounted to $1,818.0m., of which $347.6m. was paid-in.

MIGA guarantees eligible investments against losses resulting from non-commercial risks, under four main categories:

(i) transfer risk resulting from host government restrictions on currency conversion and transfer.

(ii) risk of loss resulting from legislative or administrative actions of the host government.

(iii) repudiation by the host government of contracts with investors in cases in which the investor has no access to a competent forum.

(iv) the risk of armed conflict and civil unrest.

Before guaranteeing any investment, MIGA must ensure that it is commercially viable, contributes to the development process and is not harmful to the environment. During the fiscal year 1998/99 MIGA and IFC appointed the first Compliance Advisor and Ombudsman to consider the concerns of local communities directly affected by MIGA- or IFC-sponsored projects. In February 1999 the Board of Directors approved an increase in the amount of political risk insurance available for each project, from US $75m. to $200m.

During the year ending 30 June 2004 MIGA issued 55 investment insurance contracts for 35 projects with a value of US $1,076m., compared with 59 contracts valued at $1,372m. in the previous financial year. The amount of direct investment associated with the contracts in 2003/04 totalled approximately $5,200m. (compared with $3,900m. in 2002/03). Since 1988 the total investment facilitated amounted to some $49,700m. in 85 countries, through 656 contracts.

MIGA works with local insurers, export credit agencies, development finance institutions and other organizations to promote insurance in a country, to ensure a level of consistency among insurers and to support capacity-building within the insurance industry. By mid-2004 MIGA had signed memoranda of understanding with 37 partners.

MIGA also offers technical assistance and investment marketing services to help to promote foreign investment in developing countries and in transitional economies, and to disseminate information on investment opportunities. In October 1995 MIGA established a new network on investment opportunities, which connected investment promotion agencies (IPAs) throughout the world on an electronic information network. The so-called IPAnet aimed to encourage further investments among developing countries, to provide access to comprehensive information on investment laws and conditions and to strengthen links between governmental, business and financial associations and investors. A new version of IPAnet was launched in 1997 (and can be accessed at www.ipanet.net). In June 1998 MIGA initiated a new internet-based facility, 'PrivatizationLink', to provide information on investment opportunities resulting from the privatization of industries in developing economies. In October 2000 a specialized facility within the service was established to facilitate investment in Russia (russia.privatizationlink.com). During 2000/01 an office was established in Paris, France, to promote and co-ordinate European investment in developing countries, in particular in Africa and Eastern Europe. In March 2002 MIGA opened a regional office, based in Johannesburg, South Africa. In September a new regional office was inaugurated in Singapore, in order to facilitate foreign investment in Asia.

In April 2002 MIGA launched a new service, 'FDIXchange', to provide potential investors, advisors and financial institutions with up-to-date market analysis and information on foreign direct investment opportunities in emerging economies (accessible at www.fdixchange.com). An FDIXchange Investor Information Development Programme was launched in January 2003. In January 2004 a new FDI Promotion Centre became available on the internet (www.fdipromotion.com) to facilitate information exchange and knowledge-sharing among investment promotion professionals, in particular in developing countries. (A Serbian language version was launched in June 2005.) During 2003/04 MIGA established a new fund, the Invest-in-Development Facility, to enhance the role of foreign investment in attaining the Millennium Development Goals. In July 2004 an Afghanistan Investment Guarantee Facility, to be administered by MIGA, became operational to provide political risk guarantees for foreign investors in that country.

## Publications

*Annual Report.*
*Investment Promotion Quarterly.*
*MIGA News* (quarterly).

# International Fund for Agricultural Development—IFAD

**Address:** Via del Serafico 107, 00142 Rome, Italy.
**Telephone:** (06) 54591; **fax:** (06) 5043463; **e-mail:** ifad@ifad.org; **internet:** www.ifad.org.

IFAD was established in 1977, following a decision by the 1974 UN World Food Conference, with a mandate to combat hunger and eradicate poverty on a sustainable basis in the low-income, food-deficit regions of the world. Funding operations began in January 1978.

## Organization

(August 2005)

### GOVERNING COUNCIL

Each member state is represented in the Governing Council (the Fund's highest authority) by a Governor and an Alternate. Sessions are held annually with special sessions as required. The Governing Council elects the President of the Fund (who also chairs the Executive Board) by a two-thirds majority for a four-year term. The President is eligible for re-election.

### EXECUTIVE BOARD

Consists of 18 members and 18 alternates, elected by the Governing Council, who serve for three years. The Executive Board is responsible for the conduct and general operation of IFAD and approves loans and grants for projects; it holds three regular sessions each year. An independent Office of Evaluation reports directly to the Board.

The governance structure of the Fund is based on the classification of members. Membership of the Executive Board is distributed as follows: eight List A countries (i.e. industrialized donor countries), four List B (petroleum-exporting developing donor countries), and six List C (recipient developing countries), divided equally among the three Sub-List C categories (i.e. for Africa, Europe, Asia and the Pacific, and Latin America and the Caribbean).

**President and Chairman of Executive Board:** LENNART BÅGE (Sweden).

**Vice-President:** CYRIL ENWEZE (Nigeria).

## Activities

IFAD provides financing primarily for projects designed to improve food production systems in developing member states and to strengthen related policies, services and institutions. In allocating resources IFAD is guided by: the need to increase food production in the poorest food-deficit countries; the potential for increasing food production in other developing countries; and the importance of improving the nutrition, health and education of the poorest people in developing countries, i.e. small-scale farmers, artisanal fishermen, nomadic pastoralists, indigenous populations, rural women, and the rural landless. All projects emphasize the participation of beneficiaries in development initiatives, both at the local and national level. Issues relating to gender and household food security are incorporated into all aspects of its activities. IFAD is committed to achieving the so-called Millennium Development Goals, pledged by governments attending a special session of the UN General Assembly in September 2000, and, in particular, the objective to reduce by 50% the proportion of people living in extreme poverty by 2015. In 2001 the Fund introduced new measures to improve monitoring and impact evaluation, in particular to assess its contribution to achieving the Millennium Goals. IFAD's Strategic Framework for 2002–06 reiterates its commitment to enabling the rural poor to overcome their poverty. Accordingly, the Fund's efforts were to focus on the following objectives: strengthening the capacity of the rural poor and their organizations; improving equitable access to productive natural resources and technology; and increasing access to financial services and markets. Within this Framework the Fund has also formulated regional strategies for rural poverty reduction, based on a series of regional poverty assessments. In 2003 a new Policy Division was established under the External Affairs Department. The new Division was to co-ordinate policy work at the corporate level and aimed to launch a Policy Forum in 2004, comprising IFAD senior management and staff.

IFAD is a leading repository in the field of knowledge, resources and expertise in the field of rural hunger and poverty alleviation. In 2001 it renewed its commitment to becoming a global knowledge institution for rural poverty-related issues. Through its technical assistance grants, IFAD aims to promote research and capacity-building in the agricultural sector, as well as the development of technologies to increase production and alleviate rural poverty. In recent years IFAD has been increasingly involved in promoting the use of communication technology to facilitate the exchange of information and experience among rural communities, specialized institutions and organizations, and IFAD-sponsored projects. Within the strategic context of knowledge management, IFAD has supported initiatives to support regional electronic networks, such as ENRAP (see below) in Asia and the Pacific and FIDAMERICA in Latin America and the Caribbean, as well as to develop other lines of communication between organizations, local agents and the rural poor.

In September 1998 the Executive Board endorsed a second phase of FIDAMERICA (established in 1995, see above), which envisaged linking all IFAD projects in the region, specialist and technical staff, and more than 100 local organizations, working in the field.

IFAD is empowered to make both grants and loans. Grants are limited to 7.5% of the resources committed in any one financial year. Loans are available on highly concessionary, intermediate and ordinary terms. Highly concessionary loans carry no interest but have an annual service charge of 0.75% and a repayment period of 40 years, including a 10-year grace period. Intermediate term loans are subject to a variable interest charge, equivalent to 50% of the interest rate charged on World Bank loans, and are repaid over 20 years. Ordinary loans carry a variable interest charge equal to that charged by the World Bank, and are repaid over 15–18 years. In 2003 highly concessionary loans represented some 76% of total lending in that year. In order to increase the impact of its lending resources on food production, the Fund seeks as much as possible to attract other external donors and beneficiary governments as co-financiers of its projects. In 2003 external cofinancing accounted for some 17.5% of all project funding, while domestic contributions, i.e. from recipient governments and other local sources, accounted for almost 25.8%.

IFAD's development projects usually include a number of components, such as infrastructure (e.g. improvement of water supplies, small-scale irrigation and road construction); input supply (e.g. improved seeds, fertilizers and pesticides); institutional support (e.g. research, training and extension services); and producer incentives (e.g. pricing and marketing improvements). IFAD also attempts to enable the landless to acquire income-generating assets: by increasing the provision of credit for the rural poor, it seeks to free them from dependence on the capital market and to generate productive activities.

In addition to its regular efforts to identify projects and programmes, IFAD organizes special programming missions to certain selected countries to undertake a comprehensive review of the constraints affecting the rural poor, and to help countries to design strategies for the removal of these constraints. In general, projects based on the recommendations of these missions tend to focus on institutional improvements at the national and local level to direct inputs and services to small farmers and the landless rural poor. Monitoring and evaluation missions are also sent to check the progress of projects and to assess the impact of poverty reduction efforts.

The Fund supports projects that are concerned with environmental conservation, in an effort to alleviate poverty that results from the deterioration of natural resources. In addition, it extends environmental assessment grants to review the environmental consequences of projects under preparation. In October 1997 IFAD was appointed to administer the Global Mechanism of the Convention to Combat Desertification in those Countries Experiencing Drought and Desertification, particularly in Africa, which entered into force in December 1996. The Mechanism was envisaged as a means of mobilizing and channelling resources for implementation of the Convention. A series of collaborative institutional arrangements were to be concluded between IFAD, UNDP and the World Bank in order to facilitate the effective functioning of the Mechanism. In May 2001 the Global Environmental Facility approved IFAD as an executing agency.

In Latin America and the Caribbean IFAD has aimed to formulate and implement projects that integrate the rural poor into the mainstream economy as well as local and centralized decision-making processes, enhance the productivity and market competitiveness of small-scale farmers, promote sustainable production and utilization of natural resources in environmentally fragile areas, and encourage the participation of women in rural development programmes. During 2003 IFAD approved three loans for projects in the Latin America and Caribbean region amounting to US $74.0m. (or 18.3% of total IFAD lending in that year).

In February 1998 IFAD inaugurated a new Trust Fund to complement the multilateral debt initiative for Heavily Indebted Poor Countries (HIPCs). The Fund was intended to assist IFAD's poorest

REGIONAL ORGANIZATIONS — The United Nations in South America, Central America and the Caribbean

members deemed to be eligible under the initiative to channel resources from debt repayments to communities in need. In February 2000 the Governing Council approved full participation by IFAD in the enhanced HIPC debt initiative agreed by the World Bank and IMF in September 1999.

During 1998 the Executive Board endorsed a policy framework for the Fund's provision of assistance in post-conflict situations, with the aim of achieving a continuum from emergency relief to a secure basis from which to pursue sustainable development. In July 2001 IFAD and UNAIDS signed a memorandum of understanding on developing a co-operation agreement. A meeting of technical experts from IFAD, FAO, WFP and UNAIDS, held in December, addressed means of mitigating the impact of HIV/AIDS on food security and rural livelihoods in affected regions. In January 2005 IFAD announced that it aimed to mobilize some US $100m. to assist countries affected by the devastating tsunamis which had struck coastal regions in 11 countries in the Indian Ocean in late December 2004. IFAD was also participating in needs assessments in Indonesia, Sri Lanka and the Maldives, which were to provide the basis for longer-term resource allocation and mobilization.

During the late 1990s IFAD established several partnerships within the agribusiness sector, with a view to improving performance at project level, broadening access to capital markets, and encouraging the advancement of new technologies. Since 1996 it has chaired the Support Group of the Global Forum on Agricultural Research (GFAR), which facilitates dialogue between research centres and institutions, farmers' organizations, non-governmental bodies, the private sector and donors. In October 2001 IFAD became a co-sponsor of the Consultative Group on International Agricultural Research (CGIAR).

## Finance

In accordance with the Articles of Agreement establishing IFAD, the Governing Council periodically undertakes a review of the adequacy of resources available to the Fund and may request members to make additional contributions. The sixth replenishment of IFAD funds, covering the period 2003–04 and amounting to US $560m., was approved in February 2003 and became effective in December. The provisional budget for administrative expenses for 2004 amounted to $57m.

## Publications

*Annual Report.*
*IFAD Update* (2 a year).
*Rural Poverty Report 2001.*
*Staff Working Papers* (series).

# International Monetary Fund—IMF

**Address:** 700 19th St, NW, Washington, DC 20431, USA.
**Telephone:** (202) 623-7300; **fax:** (202) 623-6278; **e-mail:** publicaffairs@imf.org; **internet:** www.imf.org.

The IMF was established at the same time as the World Bank in December 1945, to promote international monetary co-operation, to facilitate the expansion and balanced growth of international trade and to promote stability in foreign exchange.

## Organization

(August 2005)

**Managing Director:** RODRIGO DE RATO Y FIGAREDO (Spain).
**First Deputy Managing Director:** ANNE KRUEGER (USA).
**Deputy Managing Directors:** TAKATOSHI KATO (Japan), AGUSTÍN CARSTENS (Mexico).
**Director, Western Hemisphere Department:** ANOOP SINGH (India).

### BOARD OF GOVERNORS

The highest authority of the Fund is exercised by the Board of Governors, on which each member country is represented by a Governor and an Alternate Governor. The Board normally meets annually. The voting power of each country is related to its quota in the Fund. An International Monetary and Financial Committee (IMFC, formerly the Interim Committee) advises and reports to the Board on matters relating to the management and adaptation of the international monetary and financial system, sudden disturbances that might threaten the system and proposals to amend the Articles of Agreement.

### BOARD OF EXECUTIVE DIRECTORS

The 24-member Board of Executive Directors is responsible for the day-to-day operations of the Fund. The USA, the United Kingdom, Germany, France and Japan each appoint one Executive Director. There is also one Executive Director from the People's Republic of China, Russia and Saudi Arabia, while the remainder are elected by groups of the remaining countries.

## Activities

The purposes of the IMF, as defined in the Articles of Agreement, are:

(i) To promote international monetary co-operation through a permanent institution which provides the machinery for consultation and collaboration on monetary problems.

(ii) To facilitate the expansion and balanced growth of international trade, and to contribute thereby to the promotion and maintenance of high levels of employment and real income and to the development of members' productive resources.

(iii) To promote exchange stability, to maintain orderly exchange arrangements among members, and to avoid competitive exchange depreciation.

(iv) To assist in the establishment of a multilateral system of payments in respect of current transactions between members and in the elimination of foreign exchange restrictions which hamper the growth of trade.

(v) To give confidence to members by making the general resources of the Fund temporarily available to them, under adequate safeguards, thus providing them with the opportunity to correct maladjustments in their balance of payments, without resorting to measures destructive of national or international prosperity.

(vi) In accordance with the above, to shorten the duration of and lessen the degree of disequilibrium in the international balances of payments of members.

In joining the Fund, each country agrees to co-operate with the above objectives. In accordance with its objective of facilitating the expansion of international trade, the IMF encourages its members to accept the obligations of Article VIII, Sections two, three and four, of the Articles of Agreement. Members that accept Article VIII undertake to refrain from imposing restrictions on the making of payments and transfers for current international transactions and from engaging in discriminatory currency arrangements or multiple currency practices without IMF approval. By March 2005 164 members had accepted Article VIII status.

The financial crises of the late 1990s, notably in several Asian countries, Brazil and Russia, contributed to widespread discussions concerning the strengthening of the international monetary system. In April 1998 the Executive Board identified the following fundamental aspects of the debate: reinforcing international and domestic financial systems; strengthening IMF surveillance; promoting greater availability and transparency of information regarding member countries' economic data and policies; emphasizing the central role of the IMF in crisis management; and establishing effective procedures to involve the private sector in forestalling or resolving financial crises. During 1999/2000 the Fund implemented several measures in connection with its ongoing efforts to appraise and reinforce the global financial architecture, including, in March 2000, the adoption by the Executive Board of a strengthened framework to safeguard the use of IMF resources. During 2000 the Fund established the IMF Center, in Washington, DC, which aimed to promote awareness and understanding of its activities. In September the Fund's new Managing Director announced his intention to focus and streamline the principles of conditionality (which links Fund financing with the implementation of specific economic policies by the recipient countries) as part of the wider reform of the international financial system. A comprehensive review was undertaken, during which the issue was considered by public forums and representatives of civil society. New guide-lines on conditionality, which, *inter alia*, aimed to promote national ownership of policy reforms and to introduce specific criteria for the implementation of conditions given different states' circumstances, were approved by the Executive Board in September 2002. In 2000/01 the Fund established an International Capital Markets Department to improve its understanding of financial markets and a separate Consultative Group on capital markets to serve as a forum for regular dialogue between the Fund and representatives of the private sector.

In 2002 a position of Director for Special Operations was created to enhance the Fund's ability to respond to critical situations affecting member countries. In February the newly appointed Director immediately assumed leadership of the staff team working with the authorities in Argentina to help that country to overcome its extreme economic and social difficulties. In September the IMFC approved further detailed consideration of a sovereign debt restructuring mechanism (SDRM), which aimed to establish a procedure to enable countries with an unsustainable level of debt to renegotiate loans more effectively. In January 2003 the IMF hosted a conference for representatives from the financial sector and civil society and other public officials and academics to discuss aspects of the SDRM. In April, after further discussion of the issue by the Board of Directors, the IMFC stated that the SDRM would not be implemented, although other means of orderly resolution of financial crises were to remain under consideration. In their meeting the Directors determined that the Fund promote more actively the use of Collective Action Clauses in international bond contracts, as a voluntary measure to facilitate debt restructuring should the need arise.

## SURVEILLANCE

Under its Articles of Agreement, the Fund is mandated to oversee the effective functioning of the international monetary system. Accordingly, the Fund aims to exercise firm surveillance over the exchange rate policies of member states and to assess whether a country's economic situation and policies are consistent with the objectives of sustainable development and domestic and external stability. The Fund's main tools of surveillance are regular, bilateral consultations with member countries conducted in accordance with Article IV of the Articles of Agreement, which cover fiscal and monetary policies, balance-of-payments and external debt developments, as well as policies that affect the economic performance of a country, such as the labour market, social and environmental issues and good governance, and aspects of the country's capital accounts, and finance and banking sectors. In April 1997, in an effort to improve the value of surveillance by means of increased transparency, the Executive Board agreed to the voluntary issue of Press Information Notices (PINs) (on the internet and in *IMF Economic Reviews*), following each member's Article IV consultation with the Board, to those member countries wishing to make public the Fund's views. Other background papers providing information on and analysis of economic developments in individual countries continued to be made available. In addition, World Economic Outlook discussions are held, normally twice a year, by the Executive Board to assess policy implications from a multilateral perspective and to monitor global developments.

The rapid decline in the value of the Mexican peso in late 1994 and the financial crisis in Asia, which became apparent in mid-1997, focused attention on the importance of IMF surveillance of the economies and financial policies of member states and prompted the Fund to enhance the effectiveness of its surveillance and to encourage the full and timely provision of data by member countries in order to maintain fiscal transparency. In April 1996 the IMF established the Special Data Dissemination Standard (SDDS), which was intended to improve access to reliable economic statistical information for member countries that have, or are seeking, access to international capital markets. In March 1999 the IMF undertook to strengthen the Standard by the introduction of a new reserves data template. By mid-2005 61 countries had subscribed to the Standard. In December 1997 the Executive Board approved a new General Data Dissemination System (GDDS), to encourage all member countries to improve the production and dissemination of core economic data. The operational phase of the GDDS commenced in May 2000. By mid-2005 83 countries had participated in the GDDS. The Fund maintains a Dissemination Standards Bulletin Board (accessible at dsbb.imf.org), which aims to ensure that information on SDDS subscribing countries is widely available.

In April 1998 the then Interim Committee adopted a voluntary Code of Good Practices on Fiscal Transparency: Declaration of Principles, which aimed to increase the quality and promptness of official reports on economic indicators, and in September 1999 it adopted a Code of Good Practices on Transparency in Monetary and Financial Policies: Declaration of Principles. The IMF and World Bank jointly established a Financial Sector Assessment Programme (FSAP) in May 1999, initially as a pilot project, which aimed to promote greater global financial security through the preparation of confidential detailed evaluations of the financial sectors of individual countries. It remained under regular review by the Boards of Governors of the Fund and World Bank. As part of the FSAP, Fund staff may conclude a Financial System Stability Assessment (FSSA), addressing issues relating to macroeconomic stability and the strength of a country's financial system. A separate component of the FSAP are Reports on the Observance of Standards and Codes (ROSCs), which are compiled after an assessment of a country's implementation and observance of internationally recognized financial standards.

In March 2000 the IMF Executive Board adopted a strengthened framework to safeguard the use of IMF resources. All member countries making use of Fund resources were to be required to publish annual central bank statements audited in accordance with internationally accepted standards. It was also agreed that any instance of intentional misreporting of information by a member country should be publicized. In the following month the Executive Board approved the establishment of an Independent Evaluation Office (IEO) to conduct objective evaluations of IMF policy and operations. The Office commenced activities in July 2001. In 2003/04 the Office issued reports on the role of the IMF in the capital account crises in Brazil, Indonesia and the Republic of Korea, and on fiscal adjustment in IMF-supported programmes.

In April 2001 the Executive Board agreed on measures to enhance international efforts to counter money-laundering, in particular through the Fund's ongoing financial supervision activities and its programme of assessment of offshore financial centres (OFCs). In November the IMFC, in response to the terrorist attacks against targets in the USA, which had occurred in September, resolved, *inter alia*, to strengthen the Fund's focus on surveillance, and, in particular, to extend measures to counter money-laundering to include the funds of terrorist organizations. It determined to accelerate efforts to assess offshore centres and to provide technical support to enable poorer countries to meet international financial standards. In March 2004 the Board of Directors resolved that an anti-money laundering and countering the financing of terrorism (AML/CFT) component be introduced into regular OFC and FSAP assessments conducted by the Fund and the World Bank, following a pilot programme undertaken from November 2002 with the World Bank, the Financial Action Task Force and other regional supervisory bodies. By early 2005 41 initial OFC assessments had been concluded (of 44 contacted jurisdictions).

## QUOTAS

**IMF Membership and Quotas in Latin America and the Caribbean**
(million SDR*)

| Country | August 2005 |
|---|---|
| Antigua and Barbuda | 13.5 |
| Argentina | 2,117.1 |
| Bahamas | 130.3 |
| Barbados | 67.5 |
| Belize | 18.8 |
| Bolivia | 171.5 |
| Brazil | 3,036.1 |
| Chile | 856.1 |
| Colombia | 774.0 |
| Costa Rica | 164.1 |
| Dominica | 8.2 |
| Dominican Republic | 218.9 |
| Ecuador | 302.3 |
| El Salvador | 171.3 |
| Grenada | 11.7 |
| Guatemala | 210.2 |
| Guyana | 90.9 |
| Haiti | 81.9 |
| Honduras | 129.5 |
| Jamaica | 273.5 |
| Mexico | 2,585.8 |
| Nicaragua | 130.0 |
| Panama | 206.6 |
| Paraguay | 99.9 |
| Peru | 638.4 |
| Saint Christopher and Nevis | 8.9 |
| Saint Lucia | 15.3 |
| Saint Vincent and the Grenadines | 8.3 |
| Suriname | 92.1 |
| Trinidad and Tobago | 335.6 |
| Uruguay | 306.5 |
| Venezuela | 2,659.1 |

*The Special Drawing Right (SDR) was introduced in 1970 as a substitute for gold in international payments, and was intended eventually to become the principal reserve asset in the international monetary system. Its value (which was US $1.44999 at 25

July 2005, and averaged $1.48201 in 2004) is based on the currencies of the five largest exporting countries. Each member is assigned a quota related to its national income, monetary reserves, trade balance and other economic indicators; the quota approximately determines a member's voting power and the amount of foreign exchange it may purchase from the Fund. A member's subscription is equal to its quota. Quotas are reviewed at intervals of not more than five years, to take into account the state of the world economy and members' different rates of development. In January 1998 the Board of Governors approved an increase of some 45% of total IMF resources, bringing the total value of quotas to approximately SDR 212,000m. By January 1999 member states having at least 85% of total quotas (as at December 1997) had consented to the new subscriptions enabling the increase to enter into effect. The Twelfth General Review was concluded at the end of January 2003 without an increase in quotas. At August 2005 total quotas in the Fund amounted to SDR 213,478.4m.

## RESOURCES

Members' subscriptions form the basic resource of the IMF. They are supplemented by borrowing. Under the General Arrangements to Borrow (GAB), established in 1962, the 'Group of Ten' industrialized nations (G-10—Belgium, Canada, France, Germany, Italy, Japan, the Netherlands, Sweden, the United Kingdom and the USA) and Switzerland (which became a member of the IMF in May 1992 but which had been a full participant in the GAB from April 1984) undertake to lend the Fund as much as SDR 17,000m. in their own currencies to assist in fulfilling the balance-of-payments requirements of any member of the group, or in response to requests to the Fund from countries with balance-of-payments problems that could threaten the stability of the international monetary system. In 1983 the Fund entered into an agreement with Saudi Arabia, in association with the GAB, making available SDR 1,500m., and other borrowing arrangements were completed in 1984 with the Bank for International Settlements, the Saudi Arabian Monetary Agency, Belgium and Japan, making available a further SDR 6,000m. In 1986 another borrowing arrangement with Japan made available SDR 3,000m. In May 1996 GAB participants concluded an agreement in principle to expand the resources available for borrowing to SDR 34,000m., by securing the support of 25 countries with the financial capacity to support the international monetary system. The so-called New Arrangements to Borrow (NAB) was approved by the Executive Board in January 1997. It was to enter into force, for an initial five-year period, as soon as the five largest potential creditors participating in NAB had approved the initiative and the total credit arrangement of participants endorsing the scheme had reached at least SDR 28,900m. While the GAB credit arrangement was to remain in effect, the NAB was expected to be the first facility to be activated in the event of the Fund's requiring supplementary resources. In July 1998 the GAB was activated for the first time in more than 20 years in order to provide funds of up to US $6,300m. in support of an IMF emergency assistance package for Russia (the first time the GAB had been used for a non-participant). The NAB became effective in November, and was used for the first time as part of an extensive programme of support for Brazil, which was adopted by the IMF in early December. (In March 1999, however, the activation was cancelled.) In November 2002 NAB participants agreed to renew the arrangement for a further five-year period from November 2003, and approved Chile's Central Bank as the 26th participant.

## DRAWING ARRANGEMENTS

Exchange transactions within the Fund take the form of members' purchases (i.e. drawings) from the Fund of the currencies of other members for the equivalent amounts of their own currencies. Fund resources are available to eligible members on an essentially short-term and revolving basis to provide members with temporary assistance to contribute to the solution of their payments problems. Before making a purchase, a member must show that its balance of payments or reserve position makes the purchase necessary. Apart from this requirement, reserve tranche purchases (i.e. purchases that do not bring the Fund's holdings of the member's currency to a level above its quota) are permitted unconditionally.

With further purchases, however, the Fund's policy of 'conditionality' means that a member requesting assistance must agree to adjust its economic policies, as stipulated by the IMF. All requests other than for use of the reserve tranche are examined by the Executive Board to determine whether the proposed use would be consistent with the Fund's policies, and a member must discuss its proposed adjustment programme (including fiscal, monetary, exchange and trade policies) with IMF staff. Purchases outside the reserve tranche are made in four credit tranches, each equivalent to 25% of the member's quota; a member must reverse the transaction by repurchasing its own currency (with SDRs or currencies specified by the Fund) within a specified time. A credit tranche purchase is usually made under a 'Stand-by Arrangement' with the Fund, or under the Extended Fund Facility. A Stand-by Arrangement is normally of one or two years' duration, and the amount is made available in instalments, subject to the member's observance of 'performance criteria'; repurchases must be made within three-and-a-quarter to five years. An Extended Arrangement is normally of three years' duration, and the member must submit detailed economic programmes and progress reports for each year; repurchases must be made within four-and-a-half to 10 years. A member whose payments imbalance is large in relation to its quota may make use of temporary facilities established by the Fund using borrowed resources, namely the 'enlarged access policy' established in 1981, which helps to finance Stand-by and Extended Arrangements for such a member, up to a limit of between 90% and 110% of the member's quota annually. Repurchases are made within three-and-a-half to seven years. In October 1994 the Executive Board approved a temporary increase in members' access to IMF resources, on the basis of a recommendation by the then Interim Committee. The annual access limit under IMF regular tranche drawings, Stand-by Arrangements and Extended Fund Facility credits was increased from 68% to 100% of a member's quota, with the cumulative access limit remaining at 300% of quota. The arrangements were extended, on a temporary basis, in November 1997.

In addition, special-purpose arrangements have been introduced, all of which are subject to the member's co-operation with the Fund to find an appropriate solution to its difficulties. The Compensatory Financing Facility (CCF) provides compensation to members whose export earnings are reduced as a result of circumstances beyond their control, or which are affected by excess costs of cereal imports. In December 1997 the Executive Board established a new Supplemental Reserve Facility (SRF) to provide short-term assistance to members experiencing exceptional balance-of-payments difficulties resulting from a sudden loss of market confidence. The SRF was activated immediately to provide SDR 9,950m. to the Republic of Korea, as part of a Stand-by Arrangement amounting to SDR 15,550m. (at that time the largest amount ever committed by the Fund). In July 1998 SDR 4,000m. was made available to Russia under the SRF and, in December, some SDR 9,100m. was extended to Brazil under the SRF as part of a new Stand-by Arrangement. In January 2001 some SDR 2,100m. in SRF resources were approved for Argentina as part of an SDR 5,187m. Stand-by Arrangement augmentation. (In January 2002 the Executive Board approved an extension of one year for Argentina's SRF repayments.) The SDR 22,821m. Stand-by credit approved for Brazil in September 2002 included some SDR 7,600m. committed under the SRF. In April 1999 an additional facility, the Contingent Credit Lines (CCL), was established to provide short-term financing on similar terms to the SRF in order to prevent more stable economies being affected by adverse international financial developments and to maintain investor confidence. No funds were ever committed under the CCL, however, and in November 2003 the Executive Board resolved to allow the facility to terminate, as scheduled, at the end of that month. The Board requested further consideration of other precautionary arrangements to limit the risk of financial crises. In April 2004 the Board approved a new financing policy, the Trade Integration Mechanism, to assist countries experiencing short-term balance-of-payments shortfalls as a result of trade liberalization by other countries.

In October 1995 the Interim Committee of the Board of Governors endorsed recent decisions of the Executive Board to strengthen IMF financial support to members requiring exceptional assistance. An Emergency Financing Mechanism was established to enable the IMF to respond swiftly to potential or actual financial crises, while additional funds were made available for short-term currency stabilization. (The Mechanism was activated for the first time in July 1997, in response to a request by the Philippines Government to reinforce the country's international reserves, and was subsequently used during that year to assist Thailand, Indonesia and the Republic of Korea, and, in July 1998, Russia.) Emergency assistance was also to be available to countries in a post-conflict situation, extending the existing arrangements for countries having been affected by natural disasters, to facilitate the rehabilitation of their economies and to improve their eligibility for further IMF concessionary arrangements. Assistance, typically, was to be limited to 25% of a member's quota, although up to 50% would be permitted in certain circumstances. In May 2001 a post-conflict emergency assistance account was established to administer contributions from bilateral donors. During 2003/04 one country, Burundi, drew SDR 9.6m. from the account.

In November 1999 the Fund's existing facility to provide balance-of-payments assistance on concessional terms to low-income member countries, the Enhanced Structural Adjustment Facility, was reformulated as the Poverty Reduction and Growth Facility (PRGF), with greater emphasis on poverty reduction and sustainable development as key elements of growth-orientated economic strategies. Assistance under the PRGF (for which 77 countries were deemed eligible) was to be carefully matched to specific national requirements. Prior to drawing on the facility each recipient country was, in collaboration with representatives of civil society, non-governmental organizations and bilateral and multilateral institutions, to develop a national poverty reduction strategy, which was to be presented in a Poverty Reduction Strategy Paper (PRSP). PRGF loans carry an interest rate of 0.5% per year and are repayable over 10 years, with a five-and-a-half-year grace period; each eligible country is normally permitted to borrow up to 140% of its quota (in exceptional circumstances the maximum access can be raised to 185%). A PRGF Trust replaced the former ESAF Trust.

During 2003/04 new PRGF arrangements were approved for Dominica (amounting to SDR 7.7m.) and Honduras (SDR 71.2m.).

The PRGF supports, through long-maturity loans and grants, IMF participation in a joint initiative, with the World Bank, to provide exceptional assistance to heavily indebted poor countries (HIPCs), in order to help them to achieve a sustainable level of debt management. The initiative was formally approved at the September 1996 meeting of the Interim Committee, having received the support of the 'Paris Club' of official creditors, which agreed to increase the relief on official debt from 67% to 80%. In all, 41 HIPCs were identified, of which 33 were in sub-Saharan Africa. In April 1997 Uganda was approved as the first beneficiary of the initiative (see World Bank). Resources for the HIPC initiative are channelled through the PRGF Trust. In early 1999 the IMF and World Bank initiated a comprehensive review of the HIPC scheme, in order to consider modifications of the initiative and to strengthen the link between debt relief and poverty reduction. A consensus emerged among the financial institutions and leading industrialized nations to enhance the scheme, in order to make it available to more countries, and to accelerate the process of providing debt relief. In September the IMF Board of Governors expressed its commitment to undertaking an off-market transaction of a percentage of the Fund's gold reserves (i.e. a sale, at market prices, to central banks of member countries with repayment obligations to the Fund, which were then to be made in gold), as part of the funding arrangements of the enhanced HIPC scheme; this was undertaken during the period December 1999–April 2000. Under the enhanced initiative it was agreed that countries seeking debt relief should first formulate, and successfully implement for at least one year, a national poverty reduction strategy (see above). In May 2000 Uganda became the first country to qualify for full debt relief under the enhanced scheme. By June 2005 18 countries had reached completion point under the enhanced HIPC initiative, while a further nine eligible countries had reached their decision point. At mid-March the Fund had committed an estimated SDR 1,825.5m. to the initiative.

During 2003/04 the IMF approved funding commitments for new arrangements amounting to SDR 15,486m., compared with SDR 30,571m. in the previous year. Of the total amount, SDR 14,518.4m. was committed under five new Stand-by Arrangements and the augmentation of one already in place for Brazil. The arrangement approved in September 2002, in support of the Brazilian Government's efforts to secure economic and financial stability, was the largest ever stand-by credit agreed by the Fund, amounting to SDR 22,821m. Ten new PRGF arrangements were approved in 2003/04, and an existing commitment (for Madagascar) was augmented, amounting to SDR 966.9m. During 2003/04 members' purchases from the general resources account amounted to SDR 17,830m., compared with SDR 21,784m. in the previous year, with the main users of IMF resources being Brazil (SDR 9,594m.) and Argentina (SDR 5,372m.). Outstanding IMF credit at 30 April 2004 totalled SDR 69,031m., compared with SDR 72,879m. the previous year.

During the financial year 2003/04 new Stand-by Arrangements were agreed for Argentina (SDR 8,981.0m.), Dominican Republic (SDR 437.8m.), Guatemala (SDR 84.0m.), and Paraguay (SDR 50.0m.). In addition, an augmentation of an existing arrangement was approved for Brazil, amounting to SDR 4,554.0m.

### TECHNICAL ASSISTANCE

Technical assistance is provided by special missions or resident representatives who advise members on every aspect of economic management, while more specialized assistance is provided by the IMF's various departments. In 2000/01 the IMFC determined that technical assistance should be central to the IMF's work in crisis prevention and management, in capacity-building for low-income countries, and in restoring macroeconomic stability in countries following a financial crisis. Technical assistance activities subsequently underwent a process of review and reorganization to align them more closely with IMF policy priorities and other initiatives, for example the Financial Stability Assessment Programme. During 2003/04 the largest area of technical assistance involvement was sub-Saharan Africa, with other post-conflict assistance extended to Cambodia and Timor-Leste. Several countries in central and eastern Europe received technical assistance related to preparations for joining the European Union on 1 May 2004. In January 2005 the IMF Managing Director announced the Fund's commitment to providing expert advice and assistance to countries affected by tsunamis that had devastated coastal regions throughout the Indian Ocean in the previous month, in particular to assess the macroeconomic impact of the disaster and immediate budgetary and balance-of-payments needs.

The majority of technical assistance is provided by the Departments of Monetary and Exchange Affairs, of Fiscal Affairs and of Statistics, and by the IMF Institute. The Institute, founded in 1964, trains officials from member countries in financial analysis and policy, balance-of-payments methodology and public finance; it also gives assistance to national and regional training centres.

The IMF Institute also co-operates with other established regional training centres and institutes in order to refine its delivery of technical assistance and training services. During 2000/01 the Institute agreed to establish a regional training centre for Latin America in Brazil. A Caribbean Regional Technical Assistance Centre (CARTAC), located in Barbados, began operations in November 2001.

## Publications

*Annual Report.*
*Balance of Payments Statistics Yearbook.*
*Direction of Trade Statistics* (quarterly and annually).
*Emerging Markets Financing* (quarterly).
*Finance and Development* (quarterly).
*Financial Statements of the IMF* (quarterly).
*Global Financial Stability Report* (2 a year).
*Government Finance Statistics Yearbook.*
*IMF Commodity Prices* (monthly).
*IMF Research Bulletin* (quarterly).
*IMF Survey* (2 a month).
*International Financial Statistics* (monthly and annually, also on CD-ROM).
*Joint BIS-IMF-OECD-World Bank Statistics on External Debt* (quarterly).
*Quarterly Report on the Assessments of Standards and Codes.*
*Staff Papers* (3 a year).
*World Economic Outlook* (2 a year).
Other country reports, economic and financial surveys, occasional papers, pamphlets, books.

# United Nations Educational, Scientific and Cultural Organization—UNESCO

**Address:** 7 place de Fontenoy, 75352 Paris 07 SP, France.
**Telephone:** 1-45-68-10-00; **fax:** 1-45-67-16-90; **e-mail:** scg@unesco.org; **internet:** www.unesco.org.

UNESCO was established in 1946 'for the purpose of advancing, through the educational, scientific and cultural relations of the peoples of the world, the objectives of international peace and the common welfare of mankind'.

## Organization
(August 2005)

### GENERAL CONFERENCE
The supreme governing body of the Organization, the Conference meets in ordinary session once in two years and is composed of representatives of the member states.

### EXECUTIVE BOARD
The Board, comprising 58 members, prepares the programme to be submitted to the Conference and supervises its execution; it meets twice or sometimes three times a year.

### SECRETARIAT
**Director-General:** KOÏCHIRO MATSUURA (Japan).

### CO-OPERATING BODIES
In accordance with UNESCO's constitution, national Commissions have been set up in most member states. These help to integrate work within the member states and the work of UNESCO.

### REGIONAL OFFICES

**Caribbean Network of Educational Innovation for Development:** The Towers, 3rd Floor, 25 Dominica Drive, Kingston, Jamaica; tel. 929-7087; fax 929-8468; e-mail kingston@unesco.org; internet www.unesco.org/ext/field/carneid; Dir of Caribbean Office HÉLÈNE-MARIE GOSSELIN.

**International Institute for Higher Education in Latin America and the Caribbean (IESALC):** Ave Los Chorros, c/c Calle Acueducto, Edif. Asovincar, Altos de Sebucan, Apdo 68394, Caracas 1062 A, Venezuela; tel. (2) 283-1411; fax (2) 283-1454; e-mail caracas@unesco.org; internet www.iesalc.unesco.org.ve; Dir CLAUDIO RAMA VITALE.

**Regional Office for Culture in Latin America and the Caribbean (ORCALC):** Apdo 4158, Havana 4, Cuba; tel. (7) 832-7741; fax (7) 833-3144; e-mail havana@unesco.org; internet www.unesco.org.cu; f. 1950; activities include research and programmes of cultural development and cultural tourism; maintains a documentation centre and a library of 14,500 vols; Dir FRANCISCO JOSÉ LACAYO PARAJÓN; publs *Oralidad* (annually), *Boletín Electrónico* (quarterly).

**Regional Office for Education in Latin America and the Caribbean (OREALC):** Calle Enrique Delpiano 2058, Providencia, Santiago, Chile; Casilla 127, Correo 29, Providencia, Santiago, Chile; tel. (2) 472-4600; fax (2) 655-1046; e-mail unesco@unesco.cl; internet www.unesco.cl; Dir ANA LUIZA MACHADO PINHEIRO.

**Regional Office for Science for Latin America and the Caribbean:** Calle Dr Luis Piera 1992, 2°, Casilla 859, 11000 Montevideo, Uruguay; tel. (2) 413-2075; fax (2) 413-2094; e-mail orcyt@unesco.org.uy; internet www.unesco.org.uy; also cluster office for Argentina, Brazil, Chile, Paraguay, Uruguay; Dir JORGE GRANDI.

**UNESCO Office Quito:** Juan León Mera 130 y ave Patria, Edif. CFN, 6°, Quito, Ecuador; tel. (2) 2529085; fax (2) 2504435; e-mail lopez@unesco.org.ec; cluster office for Bolivia, Colombia, Ecuador, Peru, Venzuela; Dir GUSTAVO LOPEZ OSPINA.

## Activities

In November 2001 the General Conference approved a medium-term strategy to guide UNESCO during the period 2002–07. The Conference adopted a new unifying theme for the organization: 'UNESCO contributing to peace and human development in an era of globalization through education, the sciences, culture and communication'. UNESCO's central mission as defined under the strategy was to contribute to peace and human development in the globalized world through its four programme domains (Education, Natural and Social and Human Sciences, Culture, and Communication and Information), incorporating the following three principal dimensions: developing universal principles and norms to meet emerging challenges and protect the 'common public good'; promoting pluralism and diversity; and promoting empowerment and participation in the emerging knowledge society through equitable access, capacity-building and knowledge-sharing. Programme activities were to be focused particularly on supporting disadvantaged and excluded groups or geographic regions. The organization aimed to decentralize its operations in order to ensure more country-driven programming. The UN General Assembly designated UNESCO as the lead agency for co-ordinating the International Decade for a Culture of Peace and Non-Violence for the Children of the World (2001–10), with a focus on education, and the UN Literacy Decade (2003–12). In 2004 UNESCO was the lead agency in promoting the International Year to Commemorate the Struggle Against Slavery and its Abolition. In the implementation of all its activities UNESCO aims to contribute to achieving the UN Millennium Development Goal (MDG) of halving levels of extreme poverty by 2015, as well as other MDGs concerned with education and sustainable development (see below).

### EDUCATION
Since its establishment UNESCO has devoted itself to promoting education in accordance with principles based on democracy and respect for human rights. The Associated Schools Project (ASPnet—comprising some 7,509 institutions in 175 countries in early 2005) has, since 1953, promoted the principles of peace, human rights, democracy and international co-operation through education.

In March 1990 UNESCO, with other UN agencies, sponsored the World Conference on Education for All. 'Education for All' was subsequently adopted as a guiding principle of UNESCO's contribution to development. In April 2000 several UN agencies, including UNESCO and UNICEF, and other partners sponsored the World Education Forum, held in Dakar, Senegal, to assess international progress in achieving the goal of 'Education for All' and to adopt a strategy for further action (the 'Dakar Framework'), with the aim of ensuring universal basic education by 2015. The Forum launched the Global Initiative for Education for All. The Dakar Framework emphasized the role of improved access to education in the reduction of poverty and in diminishing inequalities within and between societies. UNESCO was appointed as the lead agency in the implementation of the Framework. UNESCO's role in pursuing the goals of the Dakar Forum was to focus on co-ordination, advocacy, mobilization of resources, and information-sharing at international, regional and national levels. It was to oversee national policy reforms, with a particular focus on the integration of 'Education for All' objectives into national education plans. UNESCO's work programme on Education aims to promote an effective follow-up to the Forum and comprises the following two main components: Basic education for all: meeting the commitments of the Dakar World Education Forum; and Building knowledge societies through quality education and a renewal of education systems. 'Basic Education for All', signifying the promotion of access to learning opportunities throughout the lives of all individuals, including the most disadvantaged, is the principal theme of the programme, and deemed to require urgent action. The second part of the strategy aims to improve the quality of educational provision and renew and diversify education systems, with a view to ensuring that educational needs at all levels were met. This component includes updating curricular programmes in secondary education, strengthening science and technology activities and ensuring equal access to education for girls and women. (UNESCO supports the UN Girls' Education Initiative, established following the Dakar Forum.) A particular focus is placed on the importance of knowledge, information and communication in the increasingly globalized world, and the significance of education as a means of empowerment for the poor and of enhancing basic quality of life.

UNESCO advocates 'Literacy for All' as a key component of 'Education for All', regarding literacy as essential to basic education and to social and human development. In December 2001 the UN General Assembly appointed UNESCO to be the co-ordinating agency of the UN Literacy Decade (2003–12), which aimed to formulate an international plan of action to raise literacy standards throughout the world. UNESCO was also designated as the lead agency for the UN Decade of Education for Sustainable Development (2005–14), which aimed to promote learning and awareness in support of environmental protection, economical development and social and cultural development.

In December 1993 the heads of government of nine highly-populated developing countries (Bangladesh, Brazil, the People's Republic of China, Egypt, India, Indonesia, Mexico, Nigeria and Pakistan), meeting in New Delhi, India, agreed to co-operate, with the objective of achieving comprehensive primary education for all children and of expanding further learning opportunities for children and adults. By September 1999 all of the so-called 'E-9' (or Education-9) countries had officially signed the 'Delhi Declaration' issued by the meeting. UNESCO is working towards the UN MDGs of eliminating gender disparity in primary and secondary education by 2005 and attaining universal primary education in all countries by 2015.

Within the UN system, UNESCO is responsible for providing technical assistance and educational services in the context of emergency situations. This includes providing education to refugees and displaced persons, as well as assistance for the rehabilitation of national education systems.

UNESCO is concerned with improving the quality, relevance and efficiency of higher education. It assists member states in reforming their national systems, organizes high-level conferences for Ministers of Education and other decision-makers, and disseminates research papers. A World Conference on Higher Education was convened in October 1998 in Paris, France. The Conference adopted a World Declaration on Higher Education for the 21st Century, incorporating proposals to reform higher education, with emphasis on access to education, and educating for individual development and active participation in society. The Conference also approved a framework for Priority Action for Change and Development of Higher Education, which comprised guide-lines for governments and institutions to meet the objectives of greater accessibility, as well as improved standards and relevancy of higher education.

The April 2000 World Education Forum recognized the global HIV/AIDS pandemic to be a significant challenge to the attainment of 'Education for All'. UNESCO, as a co-sponsor of UNAIDS, takes an active role in promoting formal and non-formal preventive health education.

## NATURAL SCIENCES

In November 1999 the General Conference endorsed a Declaration on Science and the Use of Scientific Knowledge and an agenda for action, which had been adopted at the World Conference on Science, held in June–July 1999, in Budapest, Hungary. UNESCO was to coordinate the follow-up to the conference and, in conjunction with the International Council for Science, to promote initiatives in international scientific partnership. The following are priority areas of UNESCO's work programme on Natural Sciences: Science and technology: capacity-building and management; and Sciences, environment and sustainable development. Water Security in the 21st Century is the principal theme, involving addressing threats to water resources and their associated ecosystems. UNESCO was the lead UN agency involved in the preparation of the first *World Water Development Report*, issued in March 2003. In that year the UNESCO Institute for Water Education was inaugurated in Delft, the Netherlands. UNESCO was a joint co-ordinator of the International Year of Freshwater (2003), which aimed to raise global awareness of the importance of improving the protection and management of freshwater resources. The Science and technology component of the programme focuses on the follow-up of the World Conference on Science, involving the elaboration of national policies on science and technology; strengthening science education; improving university teaching and enhancing national research capacities; and reinforcing international co-operation in mathematics, physics, chemistry, biology, biotechnology and the engineering sciences. UNESCO aims to contribute to bridging the divide between community-held traditional knowledge and scientific knowledge. UNESCO supports the UN MDG concerning the implementation, by 2005, of national strategies for sustainable development with a view to achieving by 2015 the reversal of current trends in the loss of environmental resources.

UNESCO aims to improve the level of university teaching of the basic sciences through training courses, establishing national and regional networks and centres of excellence, and fostering co-operative research. In carrying out its mission, UNESCO relies on partnerships with non-governmental organizations and the world scientific communities. With the International Council of Scientific Unions and the Third World Academy of Sciences, UNESCO operates a short-term fellowship programme in the basic sciences and an exchange programme of visiting lecturers. In September 1996 UNESCO initiated a 10-year World Solar Programme, which aimed to promote the application of solar energy and to increase research, development and public awareness of all forms of ecologically sustainable energy use.

UNESCO has over the years established various forms of intergovernmental co-operation concerned with the environmental sciences and research on natural resources, in order to support the recommendations of the June 1992 UN Conference on Environment and Development and, in particular, the implementation of 'Agenda 21' to promote sustainable development. The International Geological Correlation Programme, undertaken jointly with the International Union of Geological Sciences, aims to improve and facilitate global research of geological processes. In the context of the International Decade for Natural Disaster Reduction (declared in 1990), UNESCO conducted scientific studies of natural hazards and means of mitigating their effects and organized several disaster-related workshops. The International Hydrological Programme considers scientific aspects of water resources assessment and management; and the Intergovernmental Oceanographic Commission (IOC) focuses on issues relating to oceans, shorelines and marine resources, in particular the role of the ocean in climate and global systems. The IOC has been actively involved in the establishment of a Global Coral Reef Monitoring Network and is developing a Global Ocean Observing System. In June 2005 the IOC established an Intergovernmental Co-ordination Group to support and manage the development of an Indian Ocean Tsunami Warning and Mitigation System. Similar arrangements were also approved for the Caribbean region, the north-east Atlantic, and the Mediterranean and connected seas. A long-standing International Co-operation Group for the Tsunami Warning System in the Pacific was first convened, under IOC auspices, in 1968. An initiative on Environment and Development in Coastal Regions and in Small Islands is concerned with ensuring environmentally-sound and sustainable development by strengthening management of the following key areas: freshwater resources; the mitigation of coastline instability; biological diversity; and coastal ecosystem productivity. A working group on coral bleaching and local ecological responses was initiated in September 2000. UNESCO hosts the secretariat of the World Water Assessment Programme on freshwater resources.

UNESCO's Man and the Biosphere Programme supports a worldwide network of biosphere reserves (comprising 482 sites in 102 countries in June 2005), which aim to promote environmental conservation and research, education and training in biodiversity and problems of land use (including the fertility of tropical soils and the cultivation of sacred sites). In October 2002 UNESCO announced that the 138 biospheres in mountainous areas would play a leading role in a new Global Change Monitoring Programme aimed at assessing the impact of global climate changes. Following the signing of the Convention to Combat Desertification in October 1994, UNESCO initiated an International Programme for Arid Land Crops, based on a network of existing institutions, to assist implementation of the Convention.

## SOCIAL AND HUMAN SCIENCES

UNESCO is mandated to contribute to the world-wide development of the social and human sciences and philosophy, which it regards as of great importance in policy-making and maintaining ethical vigilance. The structure of UNESCO's Social and Human Sciences programme takes into account both an ethical and standard-setting dimension, and research, policy-making, action in the field and future-oriented activities. UNESCO's work programme on Social and Human Sciences comprises three main components: Ethics of science and technology; Promotion of human rights, peace and democratic principles; and Improvement of policies relating to social transformations and promotion of anticipation and prospective studies. The priority Ethics of science and technology element aims to reinforce UNESCO's role as an intellectual forum for ethical reflection on challenges related to the advance of science and technology; oversee the follow-up of the Universal Declaration on the Human Genome and Human Rights (see below); promote education in science and technology; ensure UNESCO's role in promoting good practices through encouraging the inclusion of ethical guiding principles in policy formulation and reinforcing international networks; and to promote international co-operation in human sciences and philosophy. The Social and Human Sciences programme has the main intellectual and conceptual responsibility for the transdisciplinary theme 'eradication of poverty, especially extreme poverty'.

UNESCO aims to promote and protect human rights and acts as an interdisciplinary, multicultural and pluralistic forum for reflection on issues relating to the ethical dimension of scientific advances, for example in biogenetics, new technology, and medicine. In May 1997 the International Bioethics Committee, a group of 36 specialists who meet under UNESCO auspices, approved a draft version of a Universal Declaration on the Human Genome and Human Rights, in an attempt to provide ethical guide-lines for developments in human genetics. The Declaration, which identified some 100,000 hereditary genes as 'common heritage', was adopted by the UNESCO General Conference in November and committed states to promoting the dissemination of relevant scientific knowledge and co-operating in genome research. In October 2003 the General Conference adopted an International Declaration on Human Genetic Data, establishing standards for scientists working in that field. UNESCO hosts the secretariat of the 18-member World Commission on the Ethics of Scientific Knowledge and Technology (COMEST), which aims to serve as a forum for the exchange of information and ideas and to promote dialogue between scientific communities, decision-makers and the public. COMEST met for the first time in April 1999 in Oslo, Norway. Its second meeting, which took place in December 2001 in Berlin, Germany, focused on the ethics of energy, fresh water and outer space. The third meeting of COMEST, held in Rio de Janeiro, Brazil, in December 2003, inaugurated a new regional-focused approach. An extraordinary meeting of COMEST was held in May 2004, and the fourth regular session was convened in Bangkok, Thailand, in March 2005.

In 1994 UNESCO initiated an international social science research programme, the Management of Social Transformations (MOST), to promote capacity-building in social planning at all levels of decision-making. UNESCO sponsors several research fellowships in the social sciences. In other activities UNESCO promotes the rehabilitation of underprivileged urban areas, the research of sociocultural factors affecting demographic change, and the study of family issues.

UNESCO aims to assist the building and consolidation of peaceful and democratic societies. An international network of institutions and centres involved in research on conflict resolution is being established to support the promotion of peace. Other training, workshop and research activities have been undertaken in countries that have suffered conflict. An International Youth Clearing House and Information Service (INFOYOUTH) aims to increase and consolidate the information available on the situation of young people in society, and to heighten awareness of their needs, aspirations and potential among public and private decision-makers. UNESCO also focuses on the educational and cultural dimensions of physical education and sport and their capacity to preserve and improve health. Fundamental to UNESCO's mission is the rejection of all forms of discrimination. It disseminates scientific information aimed at combating racial prejudice, works to improve the status of women and their access to education, and promotes equality between men and women.

## CULTURE

In undertaking efforts to preserve the world's cultural and natural heritage UNESCO has attempted to emphasize the link between culture and development. In November 2001 the General Conference adopted the UNESCO Universal Declaration on Cultural Diversity, which affirmed the importance of intercultural dialogue in establishing a climate of peace. The work programme on Culture includes the following interrelated components: Reinforcing normative action in the field of culture; Protecting cultural diversity and promoting cultural pluralism and intercultural dialogue; and Strengthening links between culture and development. A particular focus is placed on all aspects of cultural heritage, and on the encouragement of cultural diversity and dialogue between cultures and civilizations. In January 2002 UNESCO inaugurated a six-year initiative, the Global Alliance on Cultural Diversity, to promote partnerships between governments, non-governmental bodies and the private sector with a view to supporting cultural diversity through the strengthening of cultural industries and the prevention of cultural piracy. UNESCO is formulating a draft International Convention on the Protection of the Diversity of Cultural Contents and Artistic Expressions.

UNESCO's World Heritage Programme, inaugurated in 1978, aims to protect landmarks of outstanding universal value, in accordance with the 1972 UNESCO Convention Concerning the Protection of the World Cultural and Natural Heritage, by providing financial aid for restoration, technical assistance, training and management planning. At July 2005 the 'World Heritage List' comprised 812 sites in 137 countries, of which 628 had cultural significance, 160 were natural landmarks, and 24 were of 'mixed' importance. Examples include the city of Potosí (Bolivia), the Galapagos Islands (Ecuador), the Inca city of Machu Picchu in Peru, the Pitons Management Area in Saint Lucia, and numerous other historic sites and nature reserves in the region. UNESCO also maintains a list of World Heritage in Danger; at July 2005 this numbered 33 sites including the Humberstone and Santa Laura saltpeter works in Chile (newly inscribed), the Rio Platano Biosphere Reserve in Honduras and the Chan Chan Archaeological Zone in Peru.

The formulation of a Declaration against the Intentional Destruction of Cultural Heritage was authorized by the General Conference in November 2001. In addition, the November General Conference adopted the Convention on the Protection of the Underwater Cultural Heritage, covering the protection from commercial exploitation of shipwrecks, submerged historical sites, etc., situated in the territorial waters of signatory states. By mid-2005 the Convention had been ratified by five states (requiring 20 ratifications to enter into force). UNESCO also administers the 1954 Hague Convention on the Protection of Cultural Property in the Event of Armed Conflict and the 1970 Convention on the Means of Prohibiting and Preventing the Illicit Import, Export and Transfer of Ownership of Cultural Property. In 1992 a World Heritage Centre was established to enable rapid mobilization of international technical assistance for the preservation of cultural sites. Through the World Heritage Information Network (WHIN), a world-wide network of more than 800 information providers, UNESCO promotes global awareness and information exchange.

UNESCO supports efforts for the collection and safeguarding of humanity's non-material 'intangible' heritage, including oral traditions, music, dance and medicine. In May 2001 UNESCO awarded the title of 'Masterpieces of the Oral and Intangible Heritage of Humanity' to 19 cultural spaces (i.e. physical or temporal spaces hosting recurrent cultural events) and popular forms of expression deemed to be of outstanding value. UNESCO produces an *Atlas of the World's Languages in Danger of Disappearing*. The most recent edition, issued in February 2002, reported that of some 6,000 languages spoken world-wide, about one-half were endangered. In October 2003 the UNESCO General Conference adopted a Convention for the Safeguarding of Intangible Cultural Heritage, which provided for the establishment of an intergovernmental committee and for participating states to formulate national inventories of intangible heritage.

UNESCO encourages the translation and publication of literary works, publishes albums of art, and produces records, audiovisual programmes and travelling art exhibitions. It supports the development of book publishing and distribution, including the free flow of books and educational material across borders, and the training of editors and managers in publishing. UNESCO is active in preparing and encouraging the enforcement of international legislation on copyright.

In December 1992 UNESCO established the World Commission on Culture and Development, to strengthen links between culture and development and to prepare a report on the issue. The first World Conference on Culture and Development was held in June 1999, in Havana, Cuba. Within the context of the UN's World Decade for Cultural Development (1988–97) UNESCO launched the Silk Roads Project, as a multi-disciplinary study of the interactions among cultures and civilizations along the routes linking Asia and Europe, and established an International Fund for the Promotion of Culture, awarding two annual prizes for music and the promotion of arts. In April 1999 UNESCO celebrated the completion of a major international project, the *General History of Africa*.

## COMMUNICATION AND INFORMATION

In 2001 UNESCO introduced a major programme, 'Information for All', as the principal policy-guiding framework for the Communication and Information sector. The organization works towards establishing an open, non-exclusive knowledge society based on information-sharing and incorporating the socio-cultural and ethical dimensions of sustainable development. It promotes the free flow of, and universal access to, information, knowledge, data and best practices, through the development of communications infrastructures, the elimination of impediments to freedom of expression, and the promotion of the right to information; through encouraging international co-operation in maintaining libraries and archives; and through efforts to harness informatics for development purposes and strengthen member states' capacities in this field. Activities include assistance with the development of legislation and training programmes in countries where independent and pluralistic media are emerging; assistance in the monitoring of media independence, pluralism and diversity; promotion of exchange programmes and study tours; and improving access and opportunities for women in the media. UNESCO recognizes that the so-called global 'digital divide', in addition to other developmental differences between countries, generates exclusion and marginalization, and that increased participation in the democratic process can be attained through strengthening national communication and information capacities. UNESCO promotes the upholding of human rights in the use of cyberspace. The organization participated in the first phase of the World Summit on the Information Society, held in Geneva, Switzerland, in December 2003; the second phase was scheduled to take place in Tunis, Tunisia, in November 2005. The work programme on Communication and Information comprises the following components: Promoting equitable access to information and knowledge, especially in the public domain, and Promoting freedom of expression and strengthening communication capacities. UNESCO's Memory of the World project, established in 1992, aims to preserve in digital form, and thereby to promote wide access to, the world's documentary heritage. By June 2005 120 inscriptions had been included on the project's Register.

In regions affected by conflict UNESCO supports efforts to establish and maintain an independent media service. This strategy is largely implemented through an International Programme for the Development of Communication (IPDC). In Cambodia, Haiti and Mozambique UNESCO participated in the restructuring of the media in the context of national reconciliation, and in Bosnia and

Herzegovina it assisted in the development of independent media. In December 1998 the Israeli-Palestinian Media Forum was established, to foster professional co-operation between Israeli and Palestinian journalists. IPDC provides support to communication and media development projects in the developing world, including the establishment of news agencies and newspapers and training editorial and technical staff. Since its establishment in 1982, IPDC has financed some 1,000 projects in more than 130 countries.

In March 1997 the first International Congress on Ethical, Legal and Societal Aspects of Digital Information ('INFOethics') was held in Monte Carlo, Monaco. At the second INFOethics Congress, held in October 1998, experts discussed issues concerning privacy, confidentiality and security in the electronic transfer of information. In November 2000 a third INFOethics conference was held, on the theme of the 'Right to universal access to information in the 21st century'. UNESCO maintains an Observatory on the Information Society, which provides up-to-date information on the development of new information and communications technologies, analyses major trends, and aims to raise awareness of related ethical, legal and societal issues. A UNESCO Institute for Information Technologies in Education was established in Moscow, Russia in 1998. In 2001 the UNESCO Institute for Statistics was established in Montréal, Canada.

## Finance

UNESCO's activities are funded through a regular budget provided by contributions from member states and extrabudgetary funds from other sources, particularly UNDP, the World Bank, regional banks and other bilateral Funds-in-Trust arrangements. UNESCO co-operates with many other UN agencies and international non-governmental organizations.

UNESCO's Regular Programme budget for the two years 2004–05 was US $610m. Extrabudgetary funds for 2004–05 were estimated at $347m.

## Publications

(mostly in English, French and Spanish editions; Arabic, Chinese and Russian versions are also available in many cases)

*Atlas of the World's Languages in Danger of Disappearing.*
*Copyright Bulletin* (quarterly).
*Encyclopedia of Life Support Systems* (internet-based).
*International Review of Education* (quarterly).
*International Social Science Journal* (quarterly).
*Museum International* (quarterly).
*Nature and Resources* (quarterly).
*The New Courier* (quarterly).
*Prospects* (quarterly review on education).
*UNESCO Sources* (monthly).
*UNESCO Statistical Yearbook.*
*World Communication Report.*
*World Educational Report* (every 2 years).
*World Heritage Review* (quarterly).
*World Information Report.*
*World Science Report* (every 2 years).

Books, databases, video and radio documentaries, statistics, scientific maps and atlases.

# World Health Organization—WHO

**Address:** Ave Appia 20, 1211 Geneva 27, Switzerland.
**Telephone:** (22) 7912111; **fax:** (22) 7913111; **e-mail:** info@who.int; **internet:** www.who.int.

WHO, established in 1948, is the lead agency within the UN system concerned with the protection and improvement of public health.

## Organization

(August 2005)

### WORLD HEALTH ASSEMBLY

The Assembly meets in Geneva, once a year. It is responsible for policy making and the biennial programme and budget; appoints the Director-General; admits new members; and reviews budget contributions.

### EXECUTIVE BOARD

The Board is composed of 32 health experts designated by, but not representing, their governments; they serve for three years, and the World Health Assembly elects 10–12 member states each year to the Board. It meets at least twice a year to review the Director-General's programme, which it forwards to the Assembly with any recommendations that seem necessary. It advises on questions referred to it by the Assembly and is responsible for putting into effect the decisions and policies of the Assembly. It is also empowered to take emergency measures in case of epidemics or disasters.
**Chairman:** D. Á. Gunnarson (Iceland).

### SECRETARIAT

**Director-General:** Dr Jong-Wook Lee (Republic of Korea).
**Assistant Directors-General:** Denis Aitken (United Kingdom) (Director of the Office of the Director-General), Liu Peilong (People's Republic of China) (Adviser to the Director-General), Anarfi Asamoa-Baah (Ghana) (Communicable Diseases), Kazem Behbehani (Kuwait) (External Relations and Governing Bodies), Jack C. Chow (USA) (HIV/AIDS, TB and Malaria), Timothy G. Evans (Canada) (Evidence and Information for Policy), Catherine le Galès-Camus (France) (Non-Communicable Diseases and Mental Health), Kerstin Leitner (Germany) (Sustainable Development & Healthy Environments), Vladimir Lepakhin (Russia) (Health Technology and Pharmaceuticals), Anders Nordström (Sweden) (General Management), Joy Phumaphi (Botswana) (Family & Community Health).

### REGIONAL OFFICE

**Americas:** Pan-American Health Organization, 525 23rd St, NW, Washington, DC 20037, USA; tel. (202) 974-3000; fax (202) 974-3663; e-mail director@paho.org; internet www.paho.org; Dir Dr Mirta Roses Periago (Argentina).

## Activities

WHO's objective is stated in its constitution as 'the attainment by all peoples of the highest possible level of health'. 'Health' is defined as 'a state of complete physical, mental and social well-being and not merely the absence of disease and infirmity'. In November 2001 WHO issued the International Classification of Functioning, Disability and Health (ICF) to act as an international standard and guide-lines for determining health and disability.

WHO acts as the central authority directing international health work, and establishes relations with professional groups and government health authorities on that basis.

It provides, on request from member states, technical and policy assistance in support of programmes to promote health, prevent and control health problems, control or eradicate disease, train health workers best suited to local needs and strengthen national health systems. Aid is provided in emergencies and natural disasters.

A global programme of collaborative research and exchange of scientific information is carried out in co-operation with about 1,200 national institutions. Particular stress is laid on the widespread communicable diseases of the tropics, and the countries directly concerned are assisted in developing their research capabilities.

It keeps diseases and other health problems under constant surveillance, promotes the exchange of prompt and accurate information and of notification of outbreaks of diseases, and administers the International Health Regulations. It sets standards for the quality control of drugs, vaccines and other substances affecting health. It formulates health regulations for international travel.

It collects and disseminates health data and carries out statistical analyses and comparative studies in such diseases as cancer, heart disease and mental illness.

It receives reports on drugs observed to have shown adverse reactions in any country, and transmits the information to other member states.

It promotes improved environmental conditions, including housing, sanitation and working conditions. All available information on effects on human health of the pollutants in the environment is critically reviewed and published.

Co-operation among scientists and professional groups is encouraged. The organization negotiates and sustains national and global partnerships. It may propose international conventions and agreements, and develops and promotes international norms and standards. The organization promotes the development and testing of new technologies, tools and guide-lines. It assists in developing an informed public opinion on matters of health.

## HEALTH FOR ALL

WHO's first global strategy for pursuing 'Health for all' was adopted in May 1981 by the 34th World Health Assembly. The objective of 'Health for all' was identified as the attainment by all citizens of the world of a level of health that would permit them to lead a socially and economically productive life, requiring fair distribution of available resources, universal access to essential health care, and the promotion of preventive health care. In May 1998 the 51st World Health Assembly renewed the initiative, adopting a global strategy in support of 'Health for all in the 21st century', to be effected through regional and national health policies. The new approach was to build on the primary health care approach of the initial strategy, but was to strengthen the emphasis on quality of life, equity in health and access to health services. The following have been identified as minimum requirements of 'Health for all':

Safe water in the home or within 15 minutes' walking distance, and adequate sanitary facilities in the home or immediate vicinity;

Immunization against diphtheria, pertussis (whooping cough), tetanus, poliomyelitis, measles and tuberculosis;

Local health care, including availability of essential drugs, within one hour's travel;

Trained personnel to attend childbirth, and to care for pregnant mothers and children up to at least one year old.

WHO's technical programmes are divided into the following groups, or 'clusters': Communicable Diseases; Non-communicable Diseases and Mental Health; Family and Community Health; Sustainable Development and Healthy Environments; Health Technology and Pharmaceuticals; and Evidence and Information for Policy. In 2004–05 the following areas of work were designated as organization-wide priorities: HIV/AIDS; TB; malaria; cancer, cardiovascular diseases and diabetes; tobacco; mental health; making pregnancy safer and children's health; health and environment; food safety; health systems, including essential medicines; and blood safety. In 2000 WHO adopted a new corporate strategy, entailing a stronger focus on performance and programme delivery through standardized plans of action, and increased consistency and efficiency throughout the organization.

The Tenth General Programme of Work, for the period 2002–05, defined a policy framework for pursuing the principal objectives of building healthy populations and combating ill health. The Programme took into account: increasing understanding of the social, economic, political and cultural factors involved in achieving better health and the role played by better health in poverty reduction; the increasing complexity of health systems; the importance of safeguarding health as a component of humanitarian action; and the need for greater co-ordination among development organizations. It incorporated four interrelated strategic directions: lessening excess mortality, morbidity and disability, especially in poor and marginalized populations; promoting healthy lifestyles and reducing risk factors to human health arising from environmental, economic, social and behavioural causes; developing equitable and financially fair health systems; and establishing an enabling policy and an institutional environment for the health sector and promoting an effective health dimension to social, economic, environmental and development policy.

## COMMUNICABLE DISEASES

WHO identifies infectious and parasitic communicable diseases as a major obstacle to social and economic progress, particularly in developing countries, where, in addition to disabilities and loss of productivity and household earnings, they cause nearly one-half of all deaths. Emerging and re-emerging diseases; those likely to cause epidemics; increasing incidence of zoonoses (diseases passed from animals to humans either directly or by insects), attributable to environmental changes; outbreaks of unknown etiology; and the undermining of some drug therapies by the spread of antimicrobial resistance are main areas of concern. In recent years WHO has noted the global spread of communicable diseases through international travel, voluntary human migration and involuntary population displacement.

WHO's Communicable Diseases group works to reduce the impact of infectious diseases world-wide through surveillance and response; prevention, control and eradication strategies; and research and product development. The group seeks to identify new technologies and tools, and to foster national development through strengthening health services and the better use of existing tools. It aims to strengthen global monitoring of important communicable disease problems. The group advocates a functional approach to disease control. It aims to create consensus and consolidate partnerships around targeted diseases and collaborates with other groups at all stages to provide an integrated response. In April 2000 WHO and several partner institutions in epidemic surveillance established a Global Outbreak Alert and Response Network. Through the Network WHO aims to maintain constant vigilance regarding outbreaks of disease and to link world-wide expertise to provide an immediate response capability. From March 2003 WHO, through the Network, was co-ordinating the international investigation into the global spread of Severe Acute Respiratory Syndrome (SARS), a previously unknown atypical pneumonia. From the end of that year WHO was monitoring the spread through several Asian countries of the virus A/H5N1 (a rapidly mutating strain of zoonotic highly pathogenic avian influenza). In February 2005 WHO issued guide-lines for the global surveillance of the spread of A/H5N1 infection in human and animal populations. WHO has urged all countries to develop influenza pandemic preparedness plans.

One of WHO's major achievements was the eradication of smallpox. Following a massive international campaign of vaccination and surveillance (begun in 1958 and intensified in 1967), the last case was detected in 1977 and the eradication of the disease was declared in 1980. In May 1996 the World Health Assembly resolved that, pending a final endorsement, all remaining stocks of the smallpox virus were to be destroyed on 30 June 1999, although 500,000 doses of smallpox vaccine were to remain, along with a supply of the smallpox vaccine seed virus, in order to ensure that a further supply of the vaccine could be made available if required. In May 1999, however, the Assembly authorized a temporary retention of stocks of the virus until 2002. In late 2001, in response to fears that illegally-held virus stocks could be used in acts of biological terrorism (see below), WHO reassembled a team of technical experts on smallpox. In January 2002 the Executive Board determined that stocks of the virus should continue to be retained, to enable research into more effective treatments and vaccines.

In 1988 the World Health Assembly declared its commitment to the eradication of poliomyelitis by the end of 2000 and launched the Global Polio Eradication Initiative. In August 1996 WHO, UNICEF and Rotary International, together with other national and international partners, initiated a campaign to 'Kick Polio out of Africa', with the aim of immunizing more than 100m. children in 46 countries against the disease over a three-year period. In 2000 WHO adopted a strategic plan for the eradication of polio covering the period 2001–05, which envisaged the effective use of National Immunization Days (NIDs) to secure global interruption of polio transmission by the end of 2002, with a view to achieving certification of the global eradication of polio by the end of 2005. (In conflict zones so-called 'days of tranquility' have been negotiated to facilitate the implementation of NIDs. Vitamin A has also been administered during NIDS in order to combat nutritional deficiencies in children and thereby boost their immunity.) Meanwhile, routine immunization services were to be strengthened. A post-certification immunization policy for polio was to be formulated. In January 2004 the ministers of health of six countries (Afghanistan, Egypt, India, Niger, Nigeria and Pakistan) that had been designated as still 'polio-endemic' at the end of 2003, and global partners, meeting under the auspices of WHO and UNICEF, adopted the Geneva Declaration on the Eradication of Poliomyelitis, in which they made a commitment to accelerate the drive towards eradication of the disease, by improving the scope of vaccination programmes. Significant progress in eradication of the virus was reported in Asia during that year. In sub-Saharan Africa, however, an outbreak originating in northern Nigeria had spread, by mid-2004, to 12 previously polio-free countries. These included Côte d'Ivoire and Sudan, where ongoing civil unrest and population displacements were impeding control efforts. By that time 257 new cases had been reported in the northern Nigerian outbreak, which had been caused by a temporary cessation of vaccination activities in response to local opposition to the vaccination programme. During September and October the governments of affected West and Central African countries organized a number of synchronized immunization days, which resulted in the vaccination of more than 80m. children. A renewed co-ordinated mass vaccination drive across West and Central Africa was launched in late February 2005; this was expected to reach 100,000 children in 22 countries, with further vaccination rounds scheduled for April and May. At December 2004 the number of confirmed polio cases world-wide stood at 1,252, of which 786 were in Nigeria and 135 were in India. (In 1988 35,000 cases had been confirmed in 125 countries, with the actual number of cases estimated at around

350,000.) WHO has declared the following regions 'polio-free': the Americas (1994); Western Pacific (2000); and Europe (2002).

The Onchocerciasis Elimination Programme in the Americas (OEPA), launched in 1992, co-ordinates work to control the disease in six endemic countries of Latin America. In January 1998 a new 20-year programme to eliminate lymphatic filariasis was initiated, with substantial funding and support from two major pharmaceutical companies, and in collaboration with the World Bank, the Arab Fund for Economic and Social Development and the governments of Japan, the United Kingdom and the USA. South American trypanosomiasis ('Chagas disease') is endemic in Central and South America, causing the deaths of some 45,000 people each year and infecting a further 16m.–18m. A regional intergovernmental commission is implementing a programme to eliminate Chagas from the Southern Cone region of Latin America; it is hoped that this goal will be achieved by 2010. The countries of the Andean region of Latin America initiated a plan for the elimination of transmission of Chagas disease in February 1997, and a similar plan was launched by Central American governments in October.

WHO is committed to the elimination of leprosy (the reduction of the prevalence of leprosy to less than one case per 10,000 population). The use of a highly effective combination of three drugs (known as multi-drug therapy—MDT) resulted in a reduction in the number of leprosy cases world-wide from 10m.–12m. in 1988 to 458,000 in 2004. The number of countries having more than one case of leprosy per 10,000 had declined to 10 by 2003, compared with 122 in 1985. The Global Alliance for the Elimination of Leprosy, launched in November 1999 by WHO, in collaboration with governments of affected countries and several private partners, including a major pharmaceutical company, aimed to bring about the eradication of the disease by the end of 2005, through the continued use of MDT treatment. In March 2005 WHO transferred responsibility for its global leprosy programme from the Organization's Geneva headquarters to the South-East Asia Regional Office, based in New Delhi, India. Responsibility for supplying MDT remained with WHO headquarters. At that time leprosy control activities were being intensified in the most endemic countries: Angola, Brazil, India, Madagascar, Mozambique, Nepal and Tanzania. (India had more than one-half of all active leprosy cases, while Brazil was the second most endemic country.) In 1998 WHO launched the Global Buruli Ulcer Initiative, which aimed to co-ordinate control of and research into Buruli ulcer, another mycobacterial disease. In July of that year the Director-General of WHO and representatives of more than 20 countries, meeting in Yamoussoukro, Côte d'Ivoire, signed a declaration on the control of Buruli ulcer. In May 2004 the World Health Assembly adopted a resolution urging improved research into, and detection and treatment of, Buruli ulcer.

The Special Programme for Research and Training in Tropical Diseases, established in 1975 and sponsored jointly by WHO, UNDP and the World Bank, as well as by contributions from donor countries, involves a world-wide network of some 5,000 scientists working on the development and application of vaccines, new drugs, diagnostic kits and preventive measures, and an applied field research on practical community issues affecting the target diseases.

The objective of providing immunization for all children by 1990 was adopted by the World Health Assembly in 1977. Six diseases (measles, whooping cough, tetanus, poliomyelitis, tuberculosis and diphtheria) became the target of the Expanded Programme on Immunization (EPI), in which WHO, UNICEF and many other organizations collaborated. As a result of massive international and national efforts, the global immunization coverage increased from 20% in the early 1980s to the targeted rate of 80% by the end of 1990. This coverage signified that more than 100m. children in the developing world under the age of one had been successfully vaccinated against the targeted diseases, the lives of about 3m. children had been saved every year, and 500,000 annual cases of paralysis as a result of polio had been prevented. In 1992 the Assembly resolved to reach a new target of 90% immunization coverage with the six EPI vaccines; to introduce hepatitis B as a seventh vaccine (with the aim of an 80% reduction in the incidence of the disease in children by 2001); and to introduce the yellow fever vaccine in areas where it occurs endemically.

In June 2000 WHO released a report entitled 'Overcoming Antimicrobial Resistance', in which it warned that the misuse of antibiotics could render some common infectious illnesses unresponsive to treatment. At that time WHO issued guide-lines which aimed to mitigate the risks associated with the use of antimicrobials in livestock reared for human consumption.

## HIV/AIDS, TB AND MALARIA

Combating HIV/AIDS, tuberculosis (TB) and malaria are organization-wide priorities and, as such, are supported not only by their own areas of work but also by activities undertaken in other areas. In July 2000 a meeting of the Group of Seven industrialized nations and Russia (G-8), convened in Genoa, Italy, announced the formation of a new Global Fund to Fight AIDS, TB and Malaria (as previously proposed by the UN Secretary-General and recommended by the World Health Assembly). By early 2005 the Fund, a partnership between governments, UN bodies (including WHO) and other agencies, and private-sector interests, had committed around US $3,000m. in grants to prevention and treatment programmes in 128 countries.

The HIV/AIDS epidemic represents a major threat to human well-being and socio-economic progress. Some 95% of those known to be infected with HIV/AIDS live in developing countries, and AIDS-related illnesses are the leading cause of death in sub-Saharan Africa. At December 2004 an estimated 39.4m. people world-wide were living with HIV/AIDS (including some 2.2m. children under 15 years); 4.9m. people were newly infected during that year. WHO's Global Programme on AIDS, initiated in 1987, was concluded in December 1995. A Joint UN Programme on HIV/AIDS (UNAIDS) became operational on 1 January 1996, sponsored by WHO and other UN agencies. The UNAIDS secretariat is based at WHO headquarters. WHO established an Office of HIV/AIDS and Sexually-Transmitted Diseases in order to ensure the continuity of its global response to the problem, which included support for national control and education plans, improving the safety of blood supplies and the care and support of AIDS patients. In addition the Office was to liaise with UNAIDS and to make available WHO's research and technical expertise. HIV/AIDS are an organization-wide priority. Sufferers of HIV/AIDS in developing countries have often failed to receive advanced antiretroviral (ARV) treatments that are widely available in industrialized countries, owing to their high cost. In May 2000 the World Health Assembly adopted a resolution urging WHO member states to improve access to the prevention and treatment of HIV-related illnesses and to increase the availability and affordability of drugs. A WHO-UNAIDS HIV Vaccine Initiative was launched in that year. In June 2001 governments participating in a special session of the UN General Assembly on HIV/AIDS adopted a Declaration of Commitment on HIV/AIDS. WHO, with UNAIDS, UNICEF, UNFPA, the World Bank, and major pharmaceutical companies, participates in the 'Accelerating Access' initiative, which aims to expand access to care, support and ARVs for people with HIV/AIDS. In March 2002, under its 'Access to Quality HIV/AIDS Drugs and Diagnostics' programme, WHO published a comprehensive list of HIV-related medicines deemed to meet standards recommended by the Organization. In April WHO issued the first treatment guide-lines for HIV/AIDS cases in poor communities, and endorsed the inclusion of HIV/AIDS drugs in its *Model List of Essential Medicines* (see below) in order to encourage their wider availability. The secretariat of the International HIV Treatment Access Coalition, founded in December of that year by governments, non-governmental organizations, donors and others to facilitate access to ARVs for people in low- and middle-income countries, is based at WHO headquarters. WHO, jointly with UNAIDS and the Global Fund to Fight AIDS, TB and Malaria, supports the so-called '3-by-5' target of providing 3m. people in developing countries with ARVs by the end of 2005. WHO supports governments in developing effective health-sector responses to the HIV/AIDS epidemic through enhancing the planning and managerial capabilities, implementation capacity, and resources of health systems. In February 2003 WHO and FAO jointly published a manual on nutritional care for people living with HIV/AIDS.

At December 2003 some 430,000 people in the Caribbean region and 1.6m. in Latin America were reported to have HIV/AIDS. The Caribbean has the second highest rate of HIV prevalence in the world, after sub-Saharan Africa. At the end of 2003 national prevalence rates exceeded 1% in 12 Caribbean countries and had reached 5.6% in Haiti.

**Joint UN Programme on HIV/AIDS (UNAIDS):** 20 ave Appia, 1211 Geneva 27, Switzerland; tel. (22) 7913666; fax (22) 7914187; e-mail unaids@unaids.org; internet www.unaids.org; established in 1996 to lead, strengthen and support an expanded response to the global HIV/AIDS pandemic; activities focus on prevention, care and support, reducing vulnerability to infection, and alleviating the socioeconomic and human effects of HIV/AIDS; launched the Global Coalition on Women and AIDS in Feb. 2004; co-sponsors: WHO, UNICEF, UNDP, UNFPA, UNODC, ILO, UNESCO, the World Bank, WFP, UNHCR; Exec. Dir PETER PIOT (Belgium).

In 1995 WHO established a Global Tuberculosis Programme to address the challenges of the TB epidemic, which had been declared a global emergency by the Organization in 1993. According to WHO estimates, one-third of the world's population carries the TB bacillus, generating around 9m. active cases and killing around 2m. people each year. The largest concentration of TB cases is in south-east Asia. WHO provides technical support to all member countries, with special attention given to those with high TB prevalence, to establish effective national tuberculosis control programmes. WHO's strategy for TB control includes the use of DOTS (direct observation treatment, short-course), standardized treatment guide-lines, and result accountability through routine evaluation of treatment outcomes. Simultaneously, WHO is encouraging research with the aim of further disseminating DOTS, adapting DOTS for wider use, developing new tools for prevention, diagnosis and treatment, and containing new threats such as the HIV/TB co-epidemic. 'Stop TB', launched by WHO in 1999, in partnership with the World Bank, the US Government and a coalition of non-governmental organizations, aims to promote the use of DOTS to ensure a one-half reduction in global TB cases by 2010 (compared with levels at 2000). However, inadequate control of DOTS in some areas, leading to partial and inconsistent treatments, has resulted in the development of drug-resistant and, often, incurable strains of the disease. The incidence of so-called multidrug-resistant TB (MDR-TB) strains, that are unresponsive to at least two of the four most commonly used anti-TB drugs, has risen in recent years; in 2004 WHO estimated that, annually, 300,000 new MDR-TB cases were arising world-wide, of which some 79% were 'super strains', resistant to at least three of the main anti-TB drugs. WHO has developed DOTS-Plus, a strategy for controlling the spread of MDR-TB in areas of high prevalence. TB is the principal cause of death for people infected with the HIV virus and an estimated one-third of people living with HIV/AIDS globally are co-infected with TB. In March 2001 the Global TB Drug Facility was launched under the 'Stop TB' initiative; this aims to increase access to high-quality anti-TB drugs for sufferers in developing countries. The 'Stop TB' partnership has co-ordinated the development of the Global Plan to Stop TB, which represents a 'roadmap' for TB control, envisaging the expansion of access to DOTS; the advancement of MDR-TB prevention measures; the development of anti-TB drugs entailing a shorter treatment period; and the implementation of new strategies for treating people with HIV and TB. The first phase of the Global Plan covered the period 2000–05, while the second stage was to cover 2006–15.

In 2004 WHO reported that the People's Republic of China, Ecuador, Estonia, Israel, Kazakhstan, Latvia, Lithuania, parts of the Russian Federation, South Africa and Uzbekistan had the highest rates of MDR-TB infection in the world.

In October 1998 WHO, jointly with UNICEF, the World Bank and UNDP, formally launched the Roll Back Malaria (RBM) programme, which aimed to halve the prevalence of malaria by 2010. The disease acutely affects at least 350m.–500m. people, and kills an estimated 1m. people, every year. Some 90% of all malaria cases occur in sub-Saharan Africa. It is estimated that the disease directly causes 18% of all child deaths in that region. The global RBM Partnership, linking governments, development agencies, and other parties, aims to mobilize resources and support for controlling malaria. An RBM Partnership Global Strategic Plan for the period 2005–15 was expected to be adopted in November 2005. In January 2004 WHO's RBM department recommended a number of guide-lines for malaria control, focusing on the need for prompt, effective antimalarial treatment, and the issue of drug resistance; vector control, including the use of insecticide-treated bednets; malaria in pregnancy; malaria epidemics; and monitoring and evaluation activities. WHO, with several private- and public-sector partners, supports the development of more effective anti-malaria drugs and vaccines through the 'Medicines for Malaria' venture.

## NON-COMMUNICABLE DISEASES AND MENTAL HEALTH

The Non-communicable Diseases and Mental Health group comprises departments for the surveillance, prevention and management of uninfectious diseases, such as those arising from an unhealthy diet, and departments for health promotion, disability, injury prevention and rehabilitation, mental health and substance abuse. Surveillance, prevention and management of non-communicable diseases, tobacco, and mental health are organization-wide priorities.

Tobacco use, unhealthy diet and physical inactivity are regarded as common, preventable risk factors for the four most prominent non-communicable diseases: cardiovascular diseases, cancer, chronic respiratory disease and diabetes. WHO aims to monitor the global epidemiological situation of non-communicable diseases, to co-ordinate multinational research activities concerned with prevention and care, and to analyse determining factors such as gender and poverty. In 1998 the organization adopted a resolution on measures to be taken to combat non-communicable diseases; their prevalence was anticipated to increase, particularly in developing countries, owing to rising life expectancy and changes in lifestyles. For example, between 1995 and 2025 the number of adults affected by diabetes world-wide was projected to increase from 135m. to 300m. In 2001 chronic diseases reportedly accounted for about 59% of the estimated 56.5m. total deaths globally and for 46% of the global burden of disease. In February 1999 WHO initiated a new programme, 'Vision 2020: the Right to Sight', which aimed to eliminate avoidable blindness (estimated to be as much as 80% of all cases) by 2020. Blindness was otherwise predicted to increase by as much as twofold, owing to the increased longevity of the global population. In May 2004 the World Health Assembly endorsed a Global Strategy on Diet, Physical Activity and Health; it was estimated at that time that more than 1,000m. adults world-wide were overweight, and that, of these, some 300,000 were clinically obese. WHO has studied obesity-related issues in co-operation with the International Association for the Study of Obesity (IASO). The International Task Force on Obesity, affiliated to the IASO, aims to encourage the development of new policies for managing obesity. WHO and FAO jointly commissioned an expert report on the relationship of diet, nutrition and physical activity to chronic diseases, which was published in March 2003.

WHO's programmes for diabetes mellitus, chronic rheumatic diseases and asthma assist with the development of national initiatives, based upon goals and targets for the improvement of early detection, care and reduction of long-term complications. WHO's cardiovascular diseases programme aims to prevent and control the major cardiovascular diseases, which are responsible for more than 14m. deaths each year. It is estimated that one-third of these deaths could have been prevented with existing scientific knowledge. The programme on cancer control is concerned with the prevention of cancer, improving its detection and cure, and ensuring care of all cancer patients in need. In May 2004 the World Health Assembly adopted a resolution on cancer prevention and control, recognizing an increase in global cancer cases, particularly in developing countries, and stressing that many cases and related deaths could be prevented. The resolution included a number of recommendations for the improvement of national cancer control programmes. WHO is a co-sponsor of the Global Day Against Pain, which was held for the first time in 2004 and was to take place thereafter annually on 11 October. The Global Day highlights the need for improved pain management and palliative care for sufferers of diseases such as cancer and AIDS, with a particular focus on patients living in low-income countries with minimal access to opioid analgesics, and urges recognition of access to pain relief as a basic human right.

The WHO Human Genetics Programme manages genetic approaches for the prevention and control of common hereditary diseases and of those with a genetic predisposition representing a major health importance. The Programme also concentrates on the further development of genetic approaches suitable for incorporation into health care systems, as well as developing a network of international collaborating programmes.

WHO works to assess the impact of injuries, violence and sensory impairments on health, and formulates guide-lines and protocols for the prevention and management of mental problems. The health promotion division promotes decentralized and community-based health programmes and is concerned with developing new approaches to population ageing and encouraging healthy life-styles and self-care. It also seeks to relieve the negative impact of social changes such as urbanization, migration and changes in family structure upon health. WHO advocates a multi-sectoral approach—involving public health, legal and educational systems—to the prevention of injuries, which represent 16% of the global burden of disease. It aims to support governments in developing suitable strategies to prevent and mitigate the consequences of violence, unintentional injury and disability. Several health promotion projects have been undertaken, in collaboration between WHO regional and country offices and other relevant organizations, including: the Global School Health Initiative, to bridge the sectors of health and education and to promote the health of school-age children; the Global Strategy for Occupational Health, to promote the health of the working population and the control of occupational health risks; Community-based Rehabilitation, aimed at providing a more enabling environment for people with disabilities; and a communication strategy to provide training and support for health communications personnel and initiatives. In 2000 WHO, UNESCO, the World Bank and UNICEF adopted the joint Focusing Resources for Effective School Health (FRESH Start) approach to promoting life skills among adolescents.

In July 1997 the fourth International Conference on Health Promotion (ICHP) was held in Jakarta, Indonesia, where a declaration on 'Health Promotion into the 21st Century' was agreed. The fifth ICHP was convened in June 2000, in Mexico City, Mexico.

Mental health problems, which include unipolar and bipolar affective disorders, psychosis, epilepsy, dementia, Parkinson's disease, multiple sclerosis, drug and alcohol dependency, and neuropsychiatric disorders such as post-traumatic stress disorder, obsessive compulsive disorder and panic disorder, have been identified by WHO as significant global health problems. Although, overall, physical health has improved, mental, behavioural and social health problems are increasing, owing to extended life expectancy and improved child mortality rates, and factors such as war and poverty. WHO aims to address mental problems by increasing awareness of mental health issues and promoting improved mental health services and primary care.

The Substance Abuse department is concerned with problems of alcohol, drugs and other substance abuse. Within its Programme on Substance Abuse (PSA), which was established in 1990 in response to the global increase in substance abuse, WHO provides technical support to assist countries in formulating policies with regard to the prevention and reduction of the health and social effects of psychoactive substance abuse. PSA's sphere of activity includes epidemiological surveillance and risk assessment, advocacy and the dissemination of information, strengthening national and regional prevention and health promotion techniques and strategies, the development of cost-effective treatment and rehabilitation approaches, and also encompasses regulatory activities as required under the international drugs-control treaties in force.

The Tobacco or Health Programme aims to reduce the use of tobacco, by educating tobacco-users and preventing young people from adopting the habit. In 1996 WHO published its first report on the tobacco situation world-wide. According to WHO, about one-third of the world's population aged over 15 years smoke tobacco, which causes approximately 3.5m. deaths each year (through lung cancer, heart disease, chronic bronchitis and other effects). In 1998 the 'Tobacco Free Initiative', a major global anti-smoking campaign, was established. In May 1999 the World Health Assembly endorsed the formulation of a Framework Convention on Tobacco Control (FCTC) to help to combat the increase in tobacco use (although a number of tobacco growers expressed concerns about the effect of the convention on their livelihoods). The FCTC entered into force in February 2005. The greatest increase in tobacco use is forecast to occur in developing countries.

## FAMILY AND COMMUNITY HEALTH

WHO's Family and Community Health group addresses the following areas of work: child and adolescent health, research and programme development in reproductive health, making pregnancy safer and men and women's health. Making pregnancy safer is an organization-wide priority. The group's aim is to improve access to sustainable health care for all by strengthening health systems and fostering individual, family and community development. Activities include newborn care; child health, including promoting and protecting the health and development of the child through such approaches as promotion of breast-feeding and use of the mother-baby package, as well as care of the sick child, including diarrhoeal and acute respiratory disease control, and support to women and children in difficult circumstances; the promotion of safe motherhood and maternal health; adolescent health, including the promotion and development of young people and the prevention of specific health problems; women, health and development, including addressing issues of gender, sexual violence, and harmful traditional practices; and human reproduction, including research related to contraceptive technologies and effective methods. In addition, WHO aims to provide technical leadership and co-ordination on reproductive health and to support countries in their efforts to ensure that people: experience healthy sexual development and maturation; have the capacity for healthy, equitable and responsible relationships; can achieve their reproductive intentions safely and healthily; avoid illnesses, diseases and injury related to sexuality and reproduction; and receive appropriate counselling, care and rehabilitation for diseases and conditions related to sexuality and reproduction.

In September 1997 WHO, in collaboration with UNICEF, formally launched a programme advocating the Integrated Management of Childhood Illness (IMCI), following successful regional trials in more than 20 developing countries during 1996–97. IMCI recognizes that pneumonia, diarrhoea, measles, malaria and malnutrition cause some 70% of the approximately 11m. childhood deaths each year, and recommends screening sick children for all five conditions, to obtain a more accurate diagnosis than may be achieved from the results of a single assessment. WHO's Division of Diarrhoeal and Acute Respiratory Disease Control encourages national programmes aimed at reducing childhood deaths as a result of diarrhoea, particularly through the use of oral rehydration therapy and preventive measures. The Division is also seeking to reduce deaths from pneumonia in infants through the use of a simple case-management strategy involving the recognition of danger signs and treatment with an appropriate antibiotic.

## SUSTAINABLE DEVELOPMENT AND HEALTHY ENVIRONMENTS

The Sustainable Development and Healthy Environments group focuses on the following areas of work: health in sustainable development; nutrition; health and environment; food safety; and emergency preparedness and response. Food safety is an organization-wide priority.

WHO promotes recognition of good health status as one of the most important assets of the poor. The Sustainable Development and Healthy Environment group seeks to monitor the advantages and disadvantages for health, nutrition, environment and development arising from the process of globalization (i.e. increased global flows of capital, goods and services, people, and knowledge); to integrate the issue of health into poverty reduction programmes; and to promote human rights and equality. Adequate and safe food and nutrition is a priority programme area. WHO collaborates with FAO, the World Food Programme, UNICEF and other UN agencies in pursuing its objectives relating to nutrition and food safety. An estimated 780m. people world-wide cannot meet basic needs for energy and protein, more than 2,000m. people lack essential vitamins and minerals, and 170m. children are estimated to be malnourished. In December 1992 WHO and FAO hosted an international conference on nutrition, at which a World Declaration and Plan of Action on Nutrition was adopted to make the fight against malnutrition a development priority. Following the conference, WHO promoted the elaboration and implementation of national plans of action on nutrition. WHO aims to support the enhancement of member states' capabilities in dealing with their nutrition situations, and addressing scientific issues related to preventing, managing and monitoring protein-energy malnutrition; micronutrient malnutrition, including iodine deficiency disorders, vitamin A deficiency, and nutritional anaemia; and diet-related conditions and non-communicable diseases such as obesity (increasingly affecting children, adolescents and adults, mainly in industrialized countries), cancer and heart disease. In 1990 the World Health Assembly resolved to eliminate iodine deficiency (believed to cause mental retardation); a strategy of universal salt iodization was launched in 1993. In collaboration with other international agencies, WHO is implementing a comprehensive strategy for promoting appropriate infant, young child and maternal nutrition, and for dealing effectively with nutritional emergencies in large populations. Areas of emphasis include promoting healthcare practices that enhance successful breast-feeding; appropriate complementary feeding; refining the use and interpretation of body measurements for assessing nutritional status; relevant information, education and training; and action to give effect to the International Code of Marketing of Breast-milk Substitutes. The food safety programme aims to protect human health against risks associated with biological and chemical contaminants and additives in food. With FAO, WHO establishes food standards (through the work of the Codex Alimentarius Commission and its subsidiary committees) and evaluates food additives, pesticide residues and other contaminants and their implications for health. The programme provides expert advice on such issues as food-borne pathogens (e.g. listeria), production methods (e.g. aquaculture) and food biotechnology (e.g. genetic modification). In July 2001 the Codex Alimentarius Commission adopted the first global principles for assessing the safety of genetically modified (GM) foods. In March 2002 an intergovernmental task force established by the Commission finalized 'principles for the risk analysis of foods derived from biotechnology', which were to provide a framework for assessing the safety of GM foods and plants. In the following month WHO and FAO announced a joint review of their food standards operations. In February 2003 the FAO/WHO Project and Fund for Enhanced Participation in Codex was launched to support the participation of poorer countries in the Commission's activities.

WHO's programme area on environment and health undertakes a wide range of initiatives to tackle the increasing threats to health and well-being from a changing environment, especially in relation to air pollution, water quality, sanitation, protection against radiation, management of hazardous waste, chemical safety and housing hygiene. Some 1,100m. people world-wide have no access to clean drinking water, while a further 2,400m. people are denied suitable sanitation systems. WHO helped launch the Water Supply and Sanitation Council in 1990 and regularly updates its *Guidelines for Drinking Water Quality*. In rural areas, the emphasis continues to be on the provision and maintenance of safe and sufficient water supplies and adequate sanitation, the health aspects of rural housing, vector control in water resource management, and the safe use of agrochemicals. In urban areas, assistance is provided to identify local environmental health priorities and to improve municipal governments' ability to deal with environmental conditions and health problems in an integrated manner; promotion of the 'Healthy City' approach is a major component of the Programme. Other Programme activities include environmental health information development and management, human resources development, environmental health planning methods, research and work on problems relating to global environment change, such as UV-radiation. A report considering the implications of climate change on human health, prepared jointly by WHO, WMO and UNEP, was published in July 1996. The WHO Global Strategy for Health and Environment, developed in response to the WHO Commission on Health and Environment which reported to the UN Conference on Environment and Development in June 1992, provides the framework for programme activities. In December 2001 WHO published a report on the relationship between macroeconomics and health.

Through its International EMF Project WHO is compiling a comprehensive assessment of the potential adverse effects on human health deriving from exposure to electromagnetic fields (EMF). In June 2004 WHO organized a workshop on childhood sensitivity to EMF.

WHO's work in the promotion of chemical safety is undertaken in collaboration with ILO and UNEP through the International Programme on Chemical Safety (IPCS), the Central Unit for which is located in WHO. The Programme provides internationally evaluated scientific information on chemicals, promotes the use of such information in national programmes, assists member states in establishment of their own chemical safety measures and programmes, and helps them strengthen their capabilities in chemical emergency preparedness and response and in chemical risk reduction. In 1995 an Inter-organization Programme for the Social Management of Chemicals was established by UNEP, ILO, FAO, WHO, UNIDO and OECD, in order to strengthen international co-operation in the field of chemical safety. In 1998 WHO led an international assessment of the health risk from bendocine disruptors (chemicals which disrupt hormonal activities).

Since the major terrorist attacks perpetrated against targets in the USA in September 2001, WHO has focused renewed attention on the potential malevolent use of bacteria (such as bacillus anthracis, which causes anthrax), viruses (for example, the variola virus, causing smallpox) or toxins, or of chemical agents, in acts of biological or chemical terrorism. In September 2001 WHO issued draft guide-lines entitled 'Health Aspects of Biological and Chemical Weapons'.

Within the UN system, WHO's Department of Emergency and Humanitarian Action co-ordinates the international response to emergencies and natural disasters in the health field, in close co-operation with other agencies and within the framework set out by the UN's Office for the Co-ordination of Humanitarian Affairs. In this context, WHO provides expert advice on epidemiological surveillance, control of communicable diseases, public health information and health emergency training. Its emergency preparedness activities include co-ordination, policy-making and planning, awareness-building, technical advice, training, publication of standards and guide-lines, and research. Its emergency relief activities include organizational support, the provision of emergency drugs and supplies and conducting technical emergency assessment missions. The Division's objective is to strengthen the national capacity of member states to reduce the adverse health consequences of disasters. In responding to emergency situations, WHO always tries to develop projects and activities that will assist the national authorities concerned in rebuilding or strengthening their own capacity to handle the impact of such situations. Under the UN's Consolidated Appeals Process (CAP) for 2005, launched in November 2004, WHO appealed for US $51.5m. to fund emergency activities in 14 countries and regions.

### HEALTH TECHNOLOGY AND PHARMACEUTICALS

WHO's Health Technology and Pharmaceuticals group, made up of the departments of essential drugs and other medicines, vaccines and other biologicals, and blood safety and clinical technology, covers the following areas of work: essential medicines—access, quality and rational use; immunization and vaccine development; and world-wide co-operation on blood safety and clinical technology. Blood safety and clinical technology are an organization-wide priority.

In January 1999 the Executive Board adopted a resolution on WHO's Revised Drug Strategy which placed emphasis on the inequalities of access to pharmaceuticals, and also covered specific aspects of drugs policy, quality assurance, drug promotion, drug donation, independent drug information and rational drug use. Plans of action involving co-operation with member states and other international organizations were to be developed to monitor and analyse the pharmaceutical and public health implications of international agreements, including trade agreements. In April 2001 experts from WHO and the World Trade Organization participated in a workshop to address ways of lowering the cost of medicines in less developed countries. In the following month the World Health Assembly adopted a resolution urging member states to promote equitable access to essential drugs, noting that this was denied to about one-third of the world's population. WHO participates with other partners in the 'Accelerating Access' initiative, which aims to expand access to antiretroviral drugs for people with HIV/AIDS (see above).

WHO reports that 2m. children die each year of diseases for which common vaccines exist. In September 1991 the Children's Vaccine Initiative (CVI) was launched, jointly sponsored by the Rockefeller Foundation, UNDP, UNICEF, the World Bank and WHO, to facilitate the development and provision of children's vaccines. The CVI has as its ultimate goal the development of a single oral immunization shortly after birth that will protect against all major childhood diseases. An International Vaccine Institute was established in Seoul, Republic of Korea, as part of the CVI, to provide scientific and technical services for the production of vaccines for developing countries. In September 1996 WHO, jointly with UNICEF, published a comprehensive survey, entitled *State of the World's Vaccines and Immunization*. In 1999 WHO, UNICEF, the World Bank and a number of public- and private-sector partners formed the Global Alliance for Vaccines and Immunization (GAVI), which aimed to expand the provision of existing vaccines and to accelerate the development and introduction of new vaccines and technologies, with the ultimate goal of protecting children of all nations and from all socio-economic backgrounds against vaccine-preventable diseases.

WHO supports states in ensuring access to safe blood, blood products, transfusions, injections, and healthcare technologies.

### EVIDENCE AND INFORMATION FOR HEALTH POLICY

The Evidence and Information for Health Policy group addresses the following areas of work: evidence for health policy; health information management and dissemination; and research policy and promotion and organization of health systems. Through the generation and dissemination of evidence the Evidence and Information for Health Policy group aims to assist policy-makers assess health needs, choose intervention strategies, design policy and monitor performance, and thereby improve the performance of national health systems. The group also supports international and national dialogue on health policy.

WHO co-ordinates the Health InterNetwork Access to Research Initiative (HINARI), which was launched in July 2001 to enable relevant authorities in developing countries to access more than 2,000 biomedical journals through the internet at no or greatly reduced cost, in order to improve the world-wide circulation of scientific information; some 28 medical publishers participate in the initiative.

## Finance

WHO's regular budget is provided by assessment of member states and associate members. An additional fund for specific projects is provided by voluntary contributions from members and other sources, including UNDP and UNFPA.

A regular budget of US $901.5m. was proposed for 2004–05, of which some 8.7%, or $75.4m., was provisionally allocated to the Americas.

## Publications

*Bulletin of the World Health Organization* (monthly).
*Eastern Mediterranean Health Journal* (annually).
*International Classification of Functioning, Disability and Health—ICF.*
*Model List of Essential Medicines* (biennially).
*Pan-American Journal of Public Health* (annually).
*3 By 5 Progress Report.*
*Toxicological Evaluation of Certain Veterinary Drug Residues in Food* (annually).
*Weekly Epidemiological Record* (in English and French, paper and electronic versions available).
*WHO Drug Information* (quarterly).
*WHO Global Atlas of Traditional, Complementary and Alternative Medicine.*
*WHO Model Formulary.*
*World Health Report* (annually, in English, French and Spanish).
*World Malaria Report (with UNICEF).*
*Zoonoses and Communicable Diseases Common to Man and Animals.*
Technical report series; catalogues of specific scientific, technical and medical fields available.

# Other UN Organizations Active in the Region

### OFFICE FOR THE CO-ORDINATION OF HUMANITARIAN AFFAIRS—OCHA

**Address:** United Nations Plaza, New York, NY 10017, USA.
**Telephone:** (212) 963-1234; **fax:** (212) 963-1312; **e-mail:** ochany@un.org; **internet:** ochaonline.un.org.

The Office was established in January 1998 as part of the UN Secretariat, with a mandate to co-ordinate international humanitarian assistance and to provide policy and other advice on humanitarian issues. It administers the Humanitarian Early Warning System, as well as Integrated Regional Information Networks (IRIN) to monitor the situation in different countries and a Disaster Response System. A complementary service, Reliefweb, which was launched in 1996, monitors crises and publishes information on the internet.

**Under-Secretary-General for Humanitarian Affairs and Emergency Relief Co-ordinator:** Jan Egeland (Norway).

### UNITED NATIONS OFFICE ON DRUGS AND CRIME—UNODC

**Address:** Vienna International Centre, POB 500, 1400 Vienna, Austria.
**Telephone:** (1) 26060-0; **fax:** (1) 26060-5866; **e-mail:** unodc@unodc.org; **internet:** www.unodc.org.

The Office was established in November 1997 (as the UN Office of Drug Control and Crime Prevention) to strengthen the UN's integrated approach to issues relating to drug control, crime prevention and international terrorism. It comprises two principal components: the United Nations Drug Programme and the Crime Programme.

**Executive Director:** Antonio Maria Costa (Italy).

### OFFICE OF THE UNITED NATIONS HIGH COMMISSIONER FOR HUMAN RIGHTS—OHCHR

**Address:** Palais Wilson, 52 rue de Paquis, 1201 Geneva, Switzerland.
**Telephone:** (22) 9179290; **fax:** (22) 9179022; **e-mail:** infodesk@ohchr.org; **internet:** www.ohchr.org.

The Office is a body of the UN Secretariat and is the focal point for UN human-rights activities. Since September 1997 it has incorporated the Centre for Human Rights. The High Commissioner is the UN official with principal responsibility for UN human rights activities.

**High Commissioner:** Louise Arbour (Canada).

### UNITED NATIONS HUMAN SETTLEMENTS PROGRAMME—UN-HABITAT

**Address:** POB 30030, Nairobi, Kenya.
**Telephone:** (20) 621234; **fax:** (20) 624266; **e-mail:** infohabitat@unhabitat.org; **internet:** www.unhabitat.org.

UN-Habitat was established, as the United Nations Centre for Human Settlements, in October 1978 to service the intergovernmental Commission on Human Settlements. It became a full UN programme on 1 January 2002, serving as the focus for human settlements activities in the UN system.

**Executive Director:** Anna Kajumulo Tibaijuka (Tanzania).

### UNITED NATIONS CHILDREN'S FUND—UNICEF

**Address:** 3 United Nations Plaza, New York, NY 10017, USA.
**Telephone:** (212) 326-7000; **fax:** (212) 888-7465; **e-mail:** info@unicef.org; **internet:** www.unicef.org.

UNICEF was established in 1946 by the UN General Assembly as the UN International Children's Emergency Fund, to meet the emergency needs of children in post-war Europe and China. In 1950 its mandate was changed to emphasize programmes giving long-term benefits to children everywhere, particularly those in developing countries who are in the greatest need.

**Executive Director:** Ann Veneman (USA).

### UNITED NATIONS CONFERENCE ON TRADE AND DEVELOPMENT—UNCTAD

**Address:** Palais des Nations, 1211 Geneva 10, Switzerland.
**Telephone:** (22) 9171234; **fax:** (22) 9070043; **e-mail:** info@unctad.org; **internet:** www.unctad.org.

UNCTAD was established in 1964. It is the principal organ of the UN General Assembly concerned with trade and development, and is the focal point within the UN system for integrated activities relating to trade, finance, technology, investment and sustainable development. It aims to maximize the trade and development opportunities of developing countries, in particular least-developed countries, and to assist them to adapt to the increasing globalization and liberalization of the world economy. UNCTAD undertakes consensus-building activities, research and policy analysis and technical co-operation.

**Secretary-General:** Rubens Ricúpero (Brazil).

### UNITED NATIONS POPULATION FUND—UNFPA

**Address:** 220 East 42nd St, New York, NY 10017, USA.
**Telephone:** (212) 297-5020; **fax:** (212) 297-4911; **internet:** www.unfpa.org.

Created in 1967 as the Trust Fund for Population Activities, the UN Fund for Population Activities (UNFPA) was established as a Fund of the UN General Assembly in 1972 and was made a subsidiary organ of the UN General Assembly in 1979, with the UNDP Governing Council (now the Executive Board) designated as its governing body. In 1987 UNFPA's name was changed to the United Nations Population Fund (retaining the same acronym).

**Executive Director:** Thoraya A. Obaid (Saudi Arabia).

## UN Specialized Agencies

### INTERNATIONAL CIVIL AVIATION ORGANIZATION—ICAO

**Address:** 999 University St, Montréal, QC H3C 5H7, Canada.
**Telephone:** (514) 954-8219; **fax:** (514) 954-6077; **e-mail:** icaohq@icao.org; **internet:** www.icao.int.

ICAO was founded in 1947, on the basis of the Convention on International Civil Aviation, signed in Chicago, in 1944, to develop the techniques of international air navigation and to help in the planning and improvement of international air transport.

**Secretary-General:** Taïeb Chérif (Algeria).

### INTERNATIONAL LABOUR ORGANIZATION—ILO

**Address:** 4 route des Morillons, 1211 Geneva 22, Switzerland.
**Telephone:** (22) 7996111; **fax:** (22) 7988685; **e-mail:** ilo@ilo.org; **internet:** www.ilo.org.

ILO was founded in 1919 to work for social justice as a basis for lasting peace. It carries out this mandate by promoting decent living standards, satisfactory conditions of work and pay and adequate employment opportunities. Methods of action include the creation of international labour standards; the provision of technical co-operation services; and training, education, research and publishing activities to advance ILO objectives.

**Director-General:** Juan O. Somavía (Chile).

**Regional Office for Latin America and the Caribbean:** Las Flores 275 San Isidro, Apdo 14-124 Lima, Peru; tel. (1) 6150300; fax (1) 6150400; e-mail oit@oit.org.pe; Regional Dir Daniel Martinez (acting).

### INTERNATIONAL MARITIME ORGANIZATION—IMO

**Address:** 4 Albert Embankment, London, SE1 7SR, United Kingdom.
**Telephone:** (20) 7735-7611; **fax:** (20) 7587-3210; **e-mail:** info@imo.org; **internet:** www.imo.org.

The Inter-Governmental Maritime Consultative Organization (IMCO) began operations in 1959, as a specialized agency of the UN to facilitate co-operation among governments on technical matters affecting international shipping. Its main aims are to improve the safety of international shipping, and to prevent pollution caused by ships. IMCO became IMO in 1982.

**Secretary-General:** Efthimios Mitropoulos (Greece).

### INTERNATIONAL TELECOMMUNICATION UNION—ITU

**Address:** Place des Nations, 1211 Geneva 20, Switzerland.
**Telephone:** (22) 7305111; **fax:** (22) 7337256; **e-mail:** itumail@itu.int; **internet:** www.itu.int.

Founded in 1865, ITU became a specialized agency of the UN in 1947. It acts to encourage world co-operation for the improvement and use of telecommunications, to promote technical development,

# REGIONAL ORGANIZATIONS

to harmonize national policies in the field, and to promote the extension of telecommunications throughout the world.

**Secretary-General:** YOSHIO UTSUMI (Japan).

## UNITED NATIONS INDUSTRIAL DEVELOPMENT ORGANIZATION—UNIDO

**Address:** Vienna International Centre, POB 300, 1400 Vienna, Austria.

**Telephone:** (1) 260260; **fax:** (1) 2692669; **e-mail:** unido@unido.org; **internet:** www.unido.org.

UNIDO began operations in 1967 and became a specialized agency in 1985. Its objectives are to promote sustainable and socially equitable industrial development in developing countries and in countries with economies in transition. It aims to assist such countries to integrate fully into global economic system by mobilizing knowledge, skills, information and technology to promote productive employment, competitive economies and sound environment.

**Director-General:** CARLOS ALFREDO MAGARIÑOS (Argentina).

## UNIVERSAL POSTAL UNION—UPU

**Address:** Weltpoststr., 3000 Bern 15, Switzerland.

**Telephone:** (31) 3503111; **fax:** (31) 3503110; **e-mail:** info@upu.int; **internet:** www.upu.int.

The General Postal Union was founded by the Treaty of Berne (1874), beginning operations in July 1875. Three years later its name was changed to the Universal Postal Union. In 1948 UPU became a specialized agency of the UN. It aims to develop and unify the international postal service, to study problems and to provide training.

**Director-General:** EDOUARD DAYAN (France).

## WORLD INTELLECTUAL PROPERTY ORGANIZATION—WIPO

**Address:** 34 chemin des Colombettes, 1211 Geneva 20, Switzerland.

**Telephone:** (22) 3389111; **fax:** (22) 7335428; **e-mail:** wipo.mail@wipo.int; **internet:** www.wipo.int.

WIPO was established in 1970. It became a specialized agency of the UN in 1974 concerned with the protection of intellectual property (e.g. industrial and technical patents and literary copyrights) throughout the world. WIPO formulates and administers treaties embodying international norms and standards of intellectual property, establishes model laws, and facilitates applications for the protection of inventions, trademarks etc. WIPO provides legal and technical assistance to developing countries and countries with economies in transition and advises countries on obligations under the World Trade Organization's agreement on Trade-Related Aspects of Intellectual Property Rights (TRIPS).

**Director-General:** Dr KAMIL IDRIS (Sudan).

## WORLD METEOROLOGICAL ORGANIZATION—WMO

**Address:** 7 bis, ave de la Paix, 1211 Geneva 2, Switzerland.

**Telephone:** (22) 7308111; **fax:** (22) 7308181; **e-mail:** ipa@wmo.int; **internet:** www.wmo.int.

WMO was established in 1950 and was recognized as a Specialized Agency of the UN in 1951, aiming to improve the exchange of information in the fields of meteorology, climatology, operational hydrology and related fields, as well as their applications. WMO jointly implements, with UNEP, the UN Framework Convention on Climate Change.

**Secretary-General:** MICHEL JARRAUD (France).

## WORLD TOURISM ORGANIZATION

**Address:** Capitán Haya 42, 28020 Madrid, Spain.

**Telephone:** (91) 5678100; **fax:** (91) 5713733; **e-mail:** omt@world-tourism.org; **internet:** www.world-tourism.org.

The World Tourism Organization was established in 1975 and was recognized as a Specialized Agency of the UN in December 2003. It works to promote and develop sustainable tourism, in particular in support of socio-economic growth in developing countries.

**Secretary-General:** FRANCESCO FRANGIALLI (France).

# ANDEAN COMMUNITY OF NATIONS
## (COMUNIDAD ANDINA DE NACIONES—CAN)

**Address:** Paseo de la República 3895, San Isidro, Lima 27; Apdo 18-1177, Lima 18, Peru.
**Telephone:** (1) 4111400; **fax:** (1) 2213329; **e-mail:** contacto@comunidadandina.org; **internet:** www.comunidadandina.org.

The organization was established in 1969 as the Acuerdo de Cartagena (the Cartagena Agreement), also referred to as the Grupo Andino (Andean Group) or the Pacto Andino (Andean Pact). In March 1996 member countries signed a Reform Protocol of the Cartagena Agreement, in accordance with which the Andean Group was superseded in August 1997 by the Andean Community of Nations (CAN, generally referred to as the Andean Community). The Andean Community was to promote greater economic, commercial and political integration under a new Andean Integration System (Sistema Andino de Integración), comprising the organization's bodies and institutions. The Community covers an area of 4,710,000 sq km, with some 115m. inhabitants.

### MEMBERS
Bolivia　Colombia　Ecuador　Peru　Venezuela

Note: Chile, which withdrew from the Andean Group in 1976, was granted observer status with the Andean Community in December 2004. Panama also has observer status with the Community.

## Organization
(August 2005)

### ANDEAN PRESIDENTIAL COUNCIL
The presidential summits, which had been held annually since 1989, were formalized under the 1996 Reform Protocol of the Cartagena Agreement as the Andean Presidential Council. The Council is the highest-level body of the Andean Integration System, and provides the political leadership of the Community.

### COMMISSION
The Commission consists of a plenipotentiary representative from each member country, with each country holding the presidency in turn. The Commission is the main policy-making organ of the Andean Community, and is responsible for co-ordinating Andean trade policy.

### COUNCIL OF FOREIGN MINISTERS
The Council of Foreign Ministers meets annually or whenever it is considered necessary, to formulate common external policy and to co-ordinate the process of integration.

### GENERAL SECRETARIAT
The General Secretariat (formerly the Junta) is the body charged with implementation of all guide-lines and decisions issued by the bodies listed above. It submits proposals to the Commission for facilitating the fulfilment of the Community's objectives. Members are appointed for a three-year term. They supervise technical officials assigned to the following Departments: External Relations, Agricultural Development, Press Office, Economic Policy, Physical Integration, Programme of Assistance to Bolivia, Industrial Development, Programme Planning, Legal Affairs, Technology. Under the reforms agreed in March 1996 the Secretary-General is elected by the Council of Foreign Ministers for a five-year term, and has enhanced powers to adjudicate in disputes arising between member states. In August 1997 the General Secretariat assumed the functions of the Board of the Cartagena Agreement.
**Secretary-General:** ALLAN WAGNER TIZÓN (Peru).

### PARLIAMENT
**Parlamento Andino:** Avda 13, No. 70–61, Bogotá, Colombia; tel. (1) 249-3400; fax (1) 348-2805; e-mail pandino@cable.net.co; internet www.parlamentoandino.org; f. 1979; comprises five members from each country, and meets in each capital city in turn; makes recommendations on regional policy; in April 1997 a new protocol was adopted which provided for the election of members by direct and universal voting; in November 1998 Venezuela put the new voting mechanism into practice; the remaining Community countries were expected to complete the process by 2007; Pres. VICTOR ENRIQUE URQUIDI HODGKINSON.

### COURT OF JUSTICE
**Tribunal de Justicia de la Comunidad Andina:** Calle Roca 450 y Seis de Diciembre, Quito, Ecuador; tel. (2) 529998; fax (2) 565007; e-mail tjca@tribunalandino.org.ec; internet www.tribunalandino.org.ec; f. 1979, began operating in 1984; a protocol approved in May 1996 (which came into force in August 1999) modified the Court's functions; its main responsibilities are to resolve disputes among member countries and interpret community legislation; comprises five judges, one from each member country, appointed for a renewable period of six years; the Presidency is assumed annually by each judge in turn.

## Activities

In May 1979, at Cartagena, Colombia, the Presidents of the five member countries signed the 'Mandate of Cartagena', which envisaged greater economic and political co-operation, including the establishment of more sub-regional development programmes (especially in industry). In May 1989 the Group undertook to revitalize the process of Andean integration, by withdrawing measures that obstructed the programme of trade liberalization, and by complying with tariff reductions that had already been agreed upon. In May 1991, in Caracas, Venezuela, a summit meeting of the Andean Group agreed the framework for the establishment of a free-trade area on 1 January 1992 (achieved in February 1993) and for an eventual Andean common market (see below, under Trade).

In March 1996 heads of state, meeting in Trujillo, Peru, agreed to a substantial restructuring of the Andean Group. The heads of state signed the Reform Protocol of the Cartagena Agreement, providing for the establishment of the Andean Community of Nations, which was to have more ambitious economic and political objectives than the Group. Consequently, in August 1997 the Andean Community was inaugurated, and the Group's Junta was replaced by a new General Secretariat, headed by a Secretary-General with enhanced executive and decision-making powers. The initiation of these reforms was designed to accelerate harmonization in economic matters, particularly the achievement of a common external tariff. In April 1997 the Peruvian Government announced its intention to withdraw from the Cartagena Agreement, owing to disagreements about the terms of Peru's full integration into the Community's trading system. Later in that month the heads of state of the four other members attended a summit meeting, in Sucre, Bolivia, and reiterated their commitment to strengthening regional integration. A high-level group of representatives was established to pursue negotiations with Peru regarding its future relationship with the Community (agreement was reached in June—see below).

At the 13th presidential summit, held in Valencia, Venezuela, in June 2001, heads of state adopted an Andean Co-operation Plan for the Control of Illegal Drugs and Related Offences, which was to promote a united approach to combating these problems. An executive committee was to be established under the accord to oversee implementation of an action plan. It was also agreed that an Andean passport system should enter into effect no later than December 2005. In January 2002 a special Andean presidential summit, held in Santa Cruz, Bolivia, reiterated the objective of creating a common market and renewing efforts to strengthen sub-regional integration, including the adoption of a common agricultural policy and the standardization of macroeconomic policies.

In June 2002 ministers of defence and of foreign affairs of the Andean Community approved an Andean Charter for Peace and Security, establishing principles and commitments for the formulation of a policy on sub-regional security, the establishment of a zone of peace, joint action in efforts to counter terrorism, and the limitation of external defence spending. Other provisions of the Charter included commitments to eradicate illegal trafficking in firearms, ammunition and explosives, to expand and reinforce confidence-building measures, and to establish verification mechanisms to strengthen dialogue and efforts in those areas. In January 2003 the Community concluded a co-operation agreement with Interpol providing for collaboration in combating national and transnational crime, and in June the presidential summit adopted an Andean Plan for the Prevention, Combating and Eradication of Small, Light Weapons. The heads of state, convened in Quirama, Colombia, also endorsed a new strategic direction for the Andean integration process based on the following core themes: developing the Andean common market, common foreign policy and social agenda, the physical integration of South America, and sustainable development. In July 2004 the 15th presidential summit, held in Quito, Ecuador, formulated priority objectives for a New Strategic Scheme. A sub-regional workshop to formulate an Andean Plan to Fight Corruption was held in April 2005, organized by the General Secretariat and the European Commission.

## TRADE

A council for customs affairs met for the first time in January 1982, aiming to harmonize national legislation within the group. In December 1984 the member states launched a common currency, the Andean peso, aiming to reduce dependence on the US dollar and to increase regional trade. The new currency was to be supported by special contributions to the Fondo Andino de Reservas (now the Fondo Latinoamericano de Reservas) amounting to US $80m., and was to be 'pegged' to the US dollar, taking the form of financial drafts rather than notes and coins.

The 'Caracas Declaration' of May 1991 provided for the establishment of an Andean free-trade area, which entered into effect (excluding Peru—see below) in February 1993. Heads of state also agreed in May 1991 to create a common external tariff (CET), to standardize member countries' trade barriers in their dealings with the rest of the world, and envisaged the eventual creation of an Andean common market. In December heads of state defined four main levels of external tariffs (between 5% and 20%). In August 1992 the Group approved a request by Peru for the suspension of its rights and obligations under the Pact, thereby enabling the other members to proceed with hitherto stalled negotiations on the CET. Peru was readmitted as a full member of the Group in 1994, but participated only as an observer in the ongoing negotiations.

In November 1994 ministers of trade and integration, meeting in Quito, Ecuador, concluded a final agreement on a four-tier structure of external tariffs (although Bolivia was to retain a two-level system). The CET agreement came into effect on 1 February 1995. The agreement covered 90% of the region's imports which were to be subject to the following tariff bands: 5% for raw materials; 10%–15% for semi-manufactured goods; and 20% for finished products. In order to reach an agreement, special treatment and exemptions were granted, while Peru, initially, was to remain a 'non-active' member of the accord: Bolivia was to maintain external tariffs of 5% and 10%, Ecuador was permitted to apply the lowest rate of 5% to 990 items and was granted an initial exemption from tariffs on 400 items, while Colombia and Venezuela were granted 230 items to be subject to special treatment for four years. In June 1997 an agreement was concluded to ensure Peru's continued membership of the Community, which provided for that country's integration into the free-trade area. The Peruvian Government determined to eliminate customs duties on some 2,500 products with immediate effect, and it was agreed that the process be completed by 2005. However, negotiations were to continue with regard to the replacement of Peru's single tariff on products from outside the region with the Community's scale of external duties.

In May 1999 the 11th presidential summit agreed to establish the Andean Common Market by 2005; the Community adopted a policy on border integration and development to prepare the border regions of member countries for the envisaged free circulation of people, goods, capital and services, while consolidating sub-regional security. In June 2001 the Community agreed to recognize national identification documents issued by member states as sufficient for tourist travel in the sub-region. Community heads of state, meeting in January 2002 at a special Andean presidential summit, agreed to consolidate and improve the free-trade zone by mid-2002 and apply a new CET (with four levels, i.e. 0%, 5%, 10% and 20%). To facilitate this process a common agricultural policy was to be adopted and macro-economic policies were to be harmonized. In June 2002 ministers of foreign affairs approved a schedule of activities relating to the new CET. In October member governments determined the new tariff levels applicable to 62% of products and agreed the criteria for negotiating levels for the remaining 38%. The new CET was to become effective on 1 January 2004. This date was subsequently postponed until 1 March, and then until 10 May. In early May ministers of foreign trade agreed to postpone the effective date of the CET until 10 May 2005. The value of intra-Community trade totalled US $7,765.6m. in 2004, more than 50% than in 2003.

## EXTERNAL RELATIONS

In September 1995 heads of state of member countries identified the formulation of common positions on foreign relations as an important part of the process of relaunching the integration initiative. A Protocol Amending the Cartagena Agreement was signed in June 1997 to confirm the formulation of a common foreign policy. During 1998 the General Secretariat held consultations with government experts, academics, representatives of the private sector and other interested parties to help formulate a document on guide-lines for a common foreign policy. The guide-lines, establishing the principles, objectives and mechanisms of a common foreign policy, were approved by the Council of Foreign Ministers in 1999. In July 2004 Andean ministers of foreign affairs approved new guide-lines for an Andean common policy on external security, which aimed to prevent and counter new security threats more effectively through co-operation and co-ordination. The ministers, meeting in Quito, Ecuador, also adopted a Declaration on the Establishment of an Andean Peace Zone, free from nuclear, chemical or biological weapons. In April 2005 the Community Secretariat signed a memorandum of understanding with the Organization for the Prohibition of Chemical Weapons, which aimed to consolidate the Andean Peace Zone, assist countries to implement the Chemical Arms Convention and promote further collaboration between the two groupings.

The Group has sought to strengthen relations with the European Union, and a co-operation agreement was signed between the two blocs in April 1993. A Euro-Andean Forum is held periodically to promote mutual co-operation, trade and investment. In February 1998 the Community signed a co-operation and technical assistance agreement with the EU in order to combat drugs trafficking. At the first summit meeting of Latin American, Caribbean and EU leaders held in Rio de Janeiro, Brazil, in June 1999, Community-EU discussions were held on strengthening economic, trade and political co-operation and on the possibility of eventually concluding an Association Agreement. In May 2002 the European Union adopted a Regional Strategy for the Andean Community covering the period 2002–06. The second Latin America, Caribbean and EU summit meeting, held in May 2002 in Madrid, Spain, welcomed a new initiative to negotiate an accord on political dialogue and co-operation, envisaging that this would strengthen the basis for subsequent bilateral negotiations. Consequently, the Political Dialogue and Co-operation Agreement was negotiated during May–October 2003, and signed in December. In May 2004 a meeting of the two sides held during the third Latin American, Caribbean and EU summit, in Guadalajara, Mexico, confirmed that an EU-CAN Association Agreement was a common strategic objective. It was agreed that a Joint Committee would meet in late 2004 to define a timetable of activities and objectives which would lead to the start of substantive negotiations on the Association Agreement. In January 2005 an *ad hoc* working group was established in order to undertake a joint assessment exercise on regional economic integration.

In March 2000 the Andean Community concluded an agreement to establish a political consultation and co-operation mechanism with the People's Republic of China. At the first ministerial meeting within this framework, which took place in October 2002, it was agreed that consultations would be held thereafter on a biennial basis. The first meeting of the Council of Foreign Ministers with the Chinese Vice-President took place in January 2005. A high-level meeting between senior officials from Community member states and Japan was organized in December 2002; further consultations were to be convened, aimed at cultivating closer relations.

In April 1998, at the 10th Andean presidential summit, an agreement was signed with Panama establishing a framework for negotiations providing for the conclusion of a free-trade accord by the end of 1998 and for Panama's eventual associate membership of the Community. Also in April 1998 the Community signed a framework agreement with the Mercado Común del Sur (Mercosur) on the establishment of a free-trade accord. Although negotiations between the Community and Mercosur were subsequently delayed, bilateral agreements between the countries of the two groupings were extended. A preferential tariff agreement was concluded between Brazil and the Community in July 1999; the accord entered into effect, for a period of two years, in August. In August 2000 a preferential tariff agreement concluded with Argentina entered into force. The Community commenced negotiations on drafting a preferential tariff agreement with (jointly) El Salvador, Guatemala and Honduras in March of that year. In September leaders of the Community and Mercosur, meeting at a summit of Latin American heads of state, determined to relaunch negotiations, with a view to establishing a free-trade area. In July 2001 ministers of foreign affairs of the two groupings approved the establishment of a formal mechanism for political dialogue and co-ordination in order to facilitate negotiations and to enhance economic and social integration. In December 2003 Mercosur and the Andean Community signed an Economic Complementary Agreement providing for free-trade provisions, according to which tariffs on 80% of trade between the two groupings to be phased out by 2014 and tariffs to be removed from the remaining 20% of, initially protected, products by 2019. The entry into force of the accord, scheduled for 1 July 2004, was postponed owing to delays in drafting the tariff reduction schedule. Members of the Latin American Integration Association remaining outside Mercosur and the Andean Community—Cuba, Chile and Mexico—were to be permitted to apply to join the envisaged larger free-trade zone. In December 2004 the Andean Community agreed to grant observer status to Chile. In 2005 the Community and Mercosur were working towards a reciprocal association agreement, allowing members of one grouping automatically to be associate members of the other. In December 2004 leaders from 12 Latin American countries (excluding Argentina, Ecuador, Paraguay and Uruguay) attending a pan-South American summit, convened in

Cuzco, Peru, approved in principle the creation of a new South American Community of Nations. It was envisaged that negotiations on the formation of the new Community, which was to entail the merger of the Andean Community, Mercosur and the Rio Group, would be completed within 15 years. A South American summit meeting was scheduled to be held in August 2005.

In March 1998 ministers of trade from 34 countries, meeting in San José, Costa Rica, concluded an agreement on the structure of negotiations for the establishment of a Free Trade Area of the Americas (FTAA). The process was formally initiated by heads of state, meeting in Santiago, Chile, in the following month. The Community negotiated as a bloc to obtain chairmanship of three of the nine negotiating groups: on market access (Colombia), on competition policy (Peru), and on intellectual property (Venezuela). The Community insisted that the final declaration issued by the meeting include recognition that the varying levels of development of the participating countries should be taken into consideration throughout the negotiating process. In April 2001, convened in Québec, Canada, leaders of the participating countries determined to conclude negotiations on the FTAA by January 2005. At a special summit of the Americas, held in January 2004 in Monterrey, Mexico, the leaders adopted a declaration committing themselves to its eventual establishment although failed to specify a completion date for the process. Negotiations were suspended in March and remained stalled at mid-2005.

In August 1999 the Secretary-General of the Community visited Guyana in order to promote bilateral trading opportunities and to strengthen relations with the Caribbean Community. The Community held a meeting on trade relations with the Caribbean Community during 2000.

## INDUSTRY

Negotiations began in 1970 for the formulation of joint industrial programmes, particularly in the petrochemicals, metal-working and motor vehicle sectors, but disagreements over the allocation of different plants, and the choice of foreign manufacturers for co-operation, prevented progress and by 1984 the more ambitious schemes had been abandoned. Instead, emphasis was to be placed on assisting small and medium-sized industries, particularly in the agro-industrial and electronics sectors, in co-operation with national industrial organizations.

An Andean Agricultural Development Programme was formulated in 1976 within which 22 resolutions aimed at integrating the Andean agricultural sector were approved. In 1984 the Andean Food Security System was created to develop the agrarian sector, replace imports progressively with local produce, and improve rural living conditions. In April 1998 the Presidential Council instructed the Commission, together with ministers of agriculture, to formulate an Andean Common Agricultural Policy, including measures to harmonize trade policy instruments and legislation on animal and plant health. The 12th Andean presidential summit, held in June 2000, authorized the adoption of the concluded Policy and the enforcement of a plan of action for its implementation. In January 2002, at the special Andean presidential summit, it was agreed that all countries in the bloc would adopt price stabilization mechanisms for agricultural products.

In May 1987 member countries signed the Quito Protocol, modifying the Cartagena Agreement, to amend the strict rules that had formerly been imposed on foreign investors in the region. The Protocol entered into force in May 1988. Accordingly, each government was to decide which sectors were to be closed to foreign participation, and the period within which foreign investors must transfer a majority shareholding to local investors was extended to 30 years (37 years in Bolivia and Ecuador). In March 1991 the Protocol was amended, with the aim of further liberalizing foreign investment and stimulating an inflow of foreign capital and technology. External and regional investors were to be permitted to repatriate their profits (in accordance with the laws of the country concerned) and there was no stipulation that a majority shareholding must eventually be transferred to local investors. A further directive, adopted in March, covered the formation of 'Empresas Multinacionales Andinas' (multinational enterprises) in order to ensure that at least two member countries have a shareholding of 15% or more of the capital, including the country where the enterprise was to be based. These enterprises were entitled to participate in sectors otherwise reserved for national enterprises, subject to the same conditions as national enterprises in terms of taxation and export regulations, and to gain access to the markets of all member countries.

In November 1988 member states established a bank, the Banco Intermunicipal Andino, which was to finance public works.

In May 1995 the Group initiated a programme to promote the use of cheap and efficient energy sources and greater co-operation in the energy sector. The programme planned to develop a regional electricity grid. During 2003 efforts were undertaken to establish an Andean Energy Alliance, with the aim of fostering the development of integrated electricity and gas markets.

In September 1999 Colombia, Ecuador and Venezuela signed an accord to facilitate the production and sale of vehicles within the region. The agreement became effective in January 2000, with a duration of 10 years.

## TRANSPORT AND COMMUNICATIONS

The Andean Community has pursued efforts to improve infrastructure throughout the region. In 1983 the Commission formulated a plan to assist land-locked Bolivia, particularly through improving roads connecting it with neighbouring countries and the Pacific Ocean. An 'open skies' agreement, giving airlines of member states equal rights to airspace and airport facilities within the grouping, was signed in May 1991. In June 1998 the Commission approved the establishment of an Andean Commission of Land Transportation Authorities, which was to oversee the operation and development of land transportation services. Similarly, an Andean Committee of Water Transportation Authorities has been established to ensure compliance with Community regulations regarding ocean transportation activities. The Community aims to facilitate the movement of goods throughout the region by the use of different modes of transport ('multimodal transport') and to guarantee operational standards. It also intends to harmonize Community transport regulations and standards with those of Mercosur countries.

In August 1996 a regulatory framework was approved for the development of a commercial Andean satellite system. In December 1997 the General Secretariat approved regulations for granting authorization for the use of the system; the Commission subsequently granted the first Community authorization to an Andean multinational enterprise (Andesat), comprising 48 companies from all five member states. In 1994 the Community initiated efforts to establish digital technology infrastructure throughout the Community: the resulting Andean Digital Corridor comprises ground, underwater and satellite routes providing a series of cross-border interconnections between the member countries. The Andean Internet System, which aims to provide internet protocol-based services throughout the Community, was operational in Colombia, Ecuador and Venezuela in 2000, and was due to be extended to all five member countries. In May 1999 the Andean Committee of Telecommunications Authorities agreed to remove all restrictions to free trade in telecommunications services (excluding sound broadcasting and television) by 1 January 2002. The Committee also determined to formulate provisions on interconnection and the safeguarding of free competition and principles of transparency within the sector.

**Asociación de Empresas de Telecomunicaciones de la Comunidad Andina (ASETA):** Calle La Pradera 510 y San Salvador, Casilla 17-1106042, Quito, Ecuador; tel. (2) 256-3812; fax (2) 256-2499; e-mail info@aseta.org; internet www.aseta.org; f. 1974; co-ordinates improvements in national telecommunications services, in order to contribute to the further integration of the countries of the Andean Community; Sec.-Gen. MARCELO LÓPEZ ARJONA.

## SOCIAL INTEGRATION

Several formal agreements and institutions have been established within the framework of the grouping to enhance social development and welfare (see below). The Community aims to incorporate these bodies into the process of enhanced integration and to promote greater involvement of representatives of civil society. In May 1999 the 11th Andean presidential summit adopted a 'multidimensional social agenda' focusing on job creation and on improvements in the fields of education, health and housing throughout the Community. In June 2000 the 12th presidential summit instructed the Andean institutions to prepare individual programmes aimed at consolidating implementation of the Community's integration programme and advancing the development of the social agenda. At a special presidential summit in January 2002, corresponding ministers were directed to meet during the first half of the year to develop a Community strategy to complement national efforts in this area. In June 2003 ministers of foreign affairs and foreign trade adopted 16 legal provisions aimed at giving maximum priority to the social dimension of integration within the Community, including a measure providing for mobility of workers between member countries (see above). In July 2004 Community heads of state declared support for a new Andean Council of Social Development Ministers and a draft Comprehensive Social Development Plan. Other bodies established in 2003/04 included Councils of Ministers of Education and of Ministers responsible for Cultural Policies, and a Consultative Council of Municipal Authorities. In April 2005 the first meeting of an Andean Community Council of Ministers of the Environment and Sustainable Development was convened, in Paracas, Peru.

REGIONAL ORGANIZATIONS

## INSTITUTIONS

**Consejo Consultivo Empresarial Andino** (Andean Business Advisory Council): Paseo de la República 3895, Lima, Peru; tel. (1) 4111400; fax (1) 2213329; e-mail rsuarez@comunidadandina.org; first meeting held in November 1998; an advisory institution within the framework of the Sistema Andino de Integración; comprises elected representatives of business organizations; advises Community ministers and officials on integration activities affecting the business sector.

**Consejo Consultivo Laboral Andino** (Andean Labour Advisory Council): c/o CGTP, Plaza 2 de Mayo, 4 Lima, Peru; tel. (1) 4242357; fax (1) 4234180; e-mail cgtp1@terra.com.pe; internet www.ccla.org.pe; an advisory institution within the framework of the Sistema Andino de Integración; comprises elected representatives of labour organizations; advises Community ministers and officers on related labour issues; Chair. JUAN JOSÉ GORRITTI VALLE (Peru).

**Convenio Andrés Bello** (Andrés Bello Agreement): Paralela Autopista Norte, Avda 13 85–60, Bogotá, Colombia; tel. (1) 618-1701; fax (1) 610-0139; e-mail ecobello@col1.telecom.com.co; internet www.cab.int.co; f. 1970, modified in 1990; aims to promote integration in the educational, technical and cultural sectors; a new Interinstitutional Co-operation Agreement was signed with the Secretariat of the CAN in August 2003; mems: Bolivia, Chile, Colombia, Cuba, Ecuador, Panama, Paraguay, Peru, Spain, Venezuela.

**Convenio Hipólito Unanue** (Hipólito Unanue Agreement): Edif. Cartagena, Paseo de la República 3832, 3°, San Isidro, Lima, Peru; tel. (1) 2210074; fax (1) 4409285; e-mail postmaster@conhu.org.pe; internet www.conhu.org.pe; f. 1971 on the occasion of the first meeting of Andean ministers of health; became part of the institutional structure of the Community in 1998; aims to enhance the development of health services, and to promote regional co-ordination in areas such as environmental health, disaster preparedness and the prevention and control of drug abuse.

**Convenio Simón Rodríguez** (Simón Rodríguez Agreement): Paseo de la República 3895, esq. Aramburú, San Isidro, Lima 27, Peru; tel. (1) 4111400; fax (1) 2213329; promotes a convergence of social and labour conditions throughout the Community, for example, working hours and conditions, employment and social security policies, and to promote the participation of workers and employers in the sub-regional integration process.

**Corporación Andina de Fomento (CAF)** (Andean Development Corporation): Torre CAF, Avda Luis Roche, Altamira, Apdo 5086, Caracas, Venezuela; tel. (2) 2092111; fax (2) 2092394; e-mail infocaf@caf.com; internet www.caf.com; f. 1968, began operations in 1970; aims to encourage the integration of the Andean countries by specialization and an equitable distribution of investments; conducts research to identify investment opportunities, and prepares the resulting investment projects; gives technical and financial assistance; and attracts internal and external credit; auth. cap. US $5,000m.; subscribed or underwritten by the governments of member countries, or by public, semi-public and private-sector institutions authorized by those governments; the Board of Directors comprises representatives of each country at ministerial level; mems: the Andean Community, Argentina, Brazil, Chile, Jamaica, Mexico, Panama, Paraguay, Spain, Trinidad and Tobago, and 22 private banks in the Andean region; Exec. Pres. ENRIQUE GARCÍA RODRÍGUEZ (Bolivia).

**Fondo Latinoamericano de Reservas (FLAR)** (Latin American Reserve Fund): Avda 82, 12–18, 7°, Bogotá, Colombia; tel. (1) 634-4360; fax (1) 634-4384; e-mail flar@flar.net; internet www.flar.net; f. 1978 as the Fondo Andino de Reservas to support the balance of payments of member countries, provide credit, guarantee loans, and contribute to the harmonization of monetary and financial policies; adopted present name in 1991, in order to allow the admission of other Latin American countries; in 1992 the Fund began extending credit lines to commercial for export financing; it is administered by an Assembly of the ministers of finance and economy of the member countries, and a Board of Directors comprising the Presidents of the central banks of the member states; subscribed cap. US $2,109.4m. (30 June 2004); Exec. Pres. JULIO VELARDE; Gen. Sec. ANA MARÍA CARRASQUILLA.

**Universidad Andina Simón Bolívar** (Simón Bolívar Andean University): Calle Real Audiencia 73, Casilla 608-33, Sucre, Bolivia; tel. (64) 60265; fax (64) 60833; e-mail uasb@uasb.edu.bo; internet www.uasb.edu.bo; f. 1985; institution for postgraduate study and research; promotes co-operation between other universities in the Andean region; branches in Quito (Ecuador), La Paz (Bolivia), Caracas (Venezuela) and Cali (Colombia); Pres. Dr JULIO GARRET AILLÓN.

## Publications

*Andean Apparel for the Third Millennium.*
*Trade and Investment Guide.*
Reports, working papers.

# CARIBBEAN COMMUNITY AND COMMON MARKET—CARICOM

**Address:** POB 10827, Avenue of the Republic, Georgetown, Guyana.
**Telephone:** (2) 226-9280; **fax:** (2) 226-7816; **e-mail:** carisec3@caricom.org; **internet:** www.caricom.org.

CARICOM was formed in 1973 by the Treaty of Chaguaramas, signed in Trinidad, as a movement towards unity in the Caribbean; it replaced the Caribbean Free Trade Association (CARIFTA), founded in 1965. A revision of the Treaty of Chaguaramas (by means of nine separate Protocols), in order to institute greater regional integration and to establish a CARICOM single market and economy (CSME), was instigated in the 1990s and completed in July 2001.

### MEMBERS

| | |
|---|---|
| Antigua and Barbuda | Jamaica |
| Bahamas* | Montserrat |
| Barbados | Saint Christopher and Nevis |
| Belize | Saint Lucia |
| Dominica | Saint Vincent and the Grenadines |
| Grenada | Suriname |
| Guyana | Trinidad and Tobago |
| Haiti | |

*The Bahamas is a member of the Community but not the Common Market.

### ASSOCIATE MEMBERS

| | |
|---|---|
| Anguilla | Cayman Islands |
| Bermuda | Turks and Caicos Islands |
| British Virgin Islands | |

Note: Aruba, Colombia, the Dominican Republic, Mexico, the Netherlands Antilles, Puerto Rico and Venezuela have observer status with the Community.

## Organization
(August 2005)

### HEADS OF GOVERNMENT CONFERENCE AND BUREAU

The Conference is the final authority of the Community and determines policy. It is responsible for the conclusion of treaties on behalf of the Community and for entering into relationships between the Community and international organizations and states. Decisions of the Conference are generally taken unanimously. Heads of government meet annually, although inter-sessional meetings may be convened.

At a special meeting of the Conference, held in Trinidad and Tobago in October 1992, participants decided to establish a Heads of Government Bureau, with the capacity to initiate proposals, to update consensus and to secure the implementation of CARICOM decisions. The Bureau became operational in December, comprising the Chairman of the Conference, as Chairman, as well as the incoming and outgoing Chairmen of the Conference, and the Secretary-General of the Conference, in the capacity of Chief Executive Officer.

## COMMUNITY COUNCIL OF MINISTERS

In October 1992 CARICOM heads of government agreed that a Caribbean Community Council of Ministers should be established to replace the existing Common Market Council of Ministers as the second highest organ of the Community. Protocol I amending the Treaty of Chaguaramas, to restructure the organs and institutions of the Community, was formally adopted at a meeting of CARICOM heads of government in February 1997 and was signed by all member states in July. The inaugural meeting of the Community Council of Ministers was held in Nassau, the Bahamas, in February 1998. The Council consists of ministers responsible for community affairs, as well as other government ministers designated by member states, and is responsible for the development of the Community's strategic planning and co-ordination in the areas of economic integration, functional co-operation and external relations.

## COURT OF JUSTICE

**Caribbean Court of Justice (CCJ):** Unit Trust Building, 5th Floor, Independence Sq., Port-of-Spain, Trinidad and Tobago; tel. 623-2225; e-mail info@caribbeancourtofjustice.org; internet www.caribbeancourtofjustice.org; inaugurated in April 2005; an agreement establishing the Court was formally signed by 10 member countries in February 2001; in January 2004 a revised agreement on the establishment of the CCJ, which incorporated provision for a Trust Fund, entered into force; serves as a tribunal to enforce rights and to consider disputes relating to the CARICOM Single Market and Economy; intended to replace the Judicial Committee of the Privy Council as the Court of Final Appeal (effective for Barbados and Guyana at April 2005); Pres. MICHAEL DE LA BASTIDE.

## MINISTERIAL COUNCILS

The principal organs of the Community are assisted in their functions by the following bodies, established under Protocol I amending the Treaty of Chaguaramas: the Council for Trade and Economic Development (COTED); the Council for Foreign and Community Relations (COFCOR); the Council for Human and Social Development (COHSOD); and the Council for Finance and Planning (COFAP). The Councils are responsible for formulating policies, promoting their implementation and supervising co-operation in the relevant areas.

## SECRETARIAT

The Secretariat is the main administrative body of the Caribbean Community. The functions of the Secretariat are: to service meetings of the Community and of its Committees; to take appropriate follow-up action on decisions made at such meetings; to carry out studies on questions of economic and functional co-operation relating to the region as a whole; to provide services to member states at their request in respect of matters relating to the achievement of the objectives of the Community.

**Secretary-General:** EDWIN W. CARRINGTON (Trinidad and Tobago).

**Deputy Secretary-General:** LOLITA APPLEWHAITE (Barbados).

# Activities

## REGIONAL INTEGRATION

In 1989 CARICOM heads of government established the 15-member West Indian Commission to study regional political and economic integration. The Commission's final report, submitted in July 1992, recommended that CARICOM should remain a community of sovereign states (rather than a federation), but should strengthen the integration process and expand to include the wider Caribbean region. It recommended the formation of an Association of Caribbean States (ACS), to include all the countries within and surrounding the Caribbean Basin. In November 1997 the Secretaries-General of CARICOM and the ACS signed a Co-operation Agreement to formalize the reciprocal procedures through which the organizations work to enhance and facilitate regional integration. The Heads of Government Conference that was held in October 1992 established an Inter-Governmental Task Force, which was to undertake preparations for a reorientation of CARICOM. In February 1993 it presented a draft Charter of Civil Society for the Community, which set out principles in the areas of democracy, government, parliament, freedom of the press and human rights. The Charter was signed by Community heads of government in February 1997. Suriname was admitted to the organization in July 1995. In July 1997 the Heads of Government Conference agreed to admit Haiti as a member, although the terms and conditions of its accession to the organization had yet to be negotiated. These were finalized in July 1999. In July 2001 the CARICOM Secretary-General formally inaugurated a CARICOM Office in Haiti, which aimed to provide technical assistance in preparation of Haiti's accession to the Community. In January 2002 a CARICOM special mission visited Haiti, following an escalation of the political violence which had started in the previous month. Ministers of foreign affairs emphasized the need for international aid for Haiti when they met their US counterpart in February. Haiti was admitted as the 15th member of CARICOM at the Heads of Government Conference, held in July.

In August 1998 CARICOM and the Dominican Republic signed a free-trade accord, covering trade in goods and services, technical barriers to trade, government procurement, and sanitary and phytosanitary measures and standards. A protocol to the agreement was signed in April 2000, following the resolution of differences concerning exempted items. The accord was ratified by the Dominican Republic in February 2001 and entered partially into force on 1 December (except in Guyana, Suriname and the Bahamas).

In July 1999 CARICOM heads of government endorsed proposals to establish a Caribbean Court of Justice (CCJ), which, it was provisionally agreed, would be located in Port of Spain, Trinidad and Tobago. The Court was intended to replace the Judicial Committee of the Privy Council as the Court of Final Appeal for those countries recognizing its jurisdiction, and was also to adjudicate on trade disputes and on the interpretation of the CARICOM Treaty. It was finally inaugurated in April 2005 (see above).

In November 2001 the CARICOM Secretary-General formally inaugurated a Caribbean Regional Technical Assistance Centre (CARTAC), in Barbados. The Centre was intended to provide technical advice and training to officials from member countries and the Dominican Republic in support of the region's development, with particular focus on fiscal management, financial sector supervision and regulation, and the compilation of statistics. The IMF was to manage the Centre's operations, while UNDP was to provide administrative and logistical support.

In July 2002 a conference was held, in Liliendaal, Guyana, attended by representatives of civil society and the CARICOM heads of government. The meeting issued a statement of principles on 'Forward Together', recognizing the role of civil society in meeting the challenges to the region. It was agreed to hold regular meetings and to establish a task force to develop a regional strategic framework for pursuing the main recommendations of the conference.

## CO-ORDINATION OF FOREIGN POLICY

The co-ordination of foreign policies of member states is listed as one of the main objectives of the Community in its founding treaty. Activities include: strengthening of member states' position in international organizations; joint diplomatic action on issues of particular interest to the Caribbean; joint co-operation arrangements with third countries and organizations; and the negotiation of free-trade agreements with third countries and other regional groupings. This last area of activity has assumed increasing importance since the agreement in 1994 by almost all the governments of countries in the Americas to establish a 'Free Trade Area of the Americas' (FTAA). In April 1997 CARICOM inaugurated a Regional Negotiating Machinery body to co-ordinate and strengthen the region's presence at external economic negotiations. The main focus of activities has been the establishment of the FTAA, ACP relations with the European Union (EU), and multilateral trade negotiations under the World Trade Organization (WTO).

In July 1991 Venezuela applied for membership of CARICOM, and offered a non-reciprocal free-trade agreement for CARICOM exports to Venezuela, over an initial five-year period. In October 1993 the newly-established Group of Three (Colombia, Mexico and Venezuela) signed joint agreements with CARICOM and Suriname on combating drugs-trafficking and environmental protection. In June 1994 CARICOM and Colombia concluded an agreement on trade, economic and technical co-operation, which, *inter alia*, gives special treatment to the least-developed CARICOM countries. CARICOM has observer status in the Latin American Rio Group.

In 1992 Cuba applied for observer status within CARICOM, and in July 1993 a joint commission was inaugurated to establish closer ties between CARICOM and Cuba and to provide a mechanism for regular dialogue. In July 1997 the heads of government agreed to pursue consideration of a free-trade accord between the Community and Cuba. A Trade and Economic Agreement was signed by the two sides in July 2000, and a CARICOM office was established in Cuba, in February 2001. In July 2004 a meeting of the two sides was held in Havana, Cuba, at ministerial level. A CARICOM-Cuba meeting of heads of state and government was scheduled to be convened in December 2005, in Barbados. In February 1992 ministers of foreign affairs from CARICOM and Central American states met to discuss future co-operation, in view of the imminent conclusion of the North American Free Trade Agreement (NAFTA) between the USA, Canada and Mexico. It was agreed that a consultative forum would be established to discuss the possible formation of a Caribbean and Central American free-trade zone. In October 1993 CARICOM declared its support for NAFTA, but requested a 'grace period', during which the region's exports would have parity with Mexican products, and in March 1994 requested that it should be considered for early entry into NAFTA. In July 1996 the heads of government expressed strong concern over the complaint lodged with the WTO by the USA, Ecuador, Guatemala and Honduras regarding the European Union's import regime on bananas, which gives preferential access to bananas from the ACP countries (see the European Union). CARICOM requested the US Government to withdraw its complaint and to negotiate a settlement. Nevertheless, WTO panel hearings on the complaint were initiated in September. Banana producers from the ACP countries were granted third-party status, at the insistence of the Eastern Caribbean ambassador to the EU, Edwin Laurent. In December a special meeting of the Heads of Government Conference was convened, in Barbados, in order to formulate a common position on relations with the USA, in particular with respect to measures to combat illegal drugs-trafficking, following reports that the US Government was planning to impose punitive measures against certain regional authorities, owing to their perceived failure to implement effective controls on illicit drugs.

In May 1997 a meeting of CARICOM heads of government and the US President established a partnership for prosperity and security, and arrangements were instituted for annual consultations between the ministers of foreign affairs of CARICOM countries and the US Secretary of State. However, the Community failed to secure a commitment by the USA to grant the region's exports 'NAFTA-parity' status, or to guarantee concessions to the region's banana industry, following a temporary ruling of the WTO, issued in March, upholding the US trade complaint. The WTO ruling was confirmed in May and endorsed by the WTO dispute settlement body in September. The USA's opposition to a new EU banana policy (which was to terminate the import licensing system, extending import quotas to 'dollar' producers, while maintaining a limited duty-free quota for Caribbean producers) was strongly criticized by CARICOM leaders, meeting in July 1998. In March 1999 the Inter-Sessional meeting of the Conference of Heads of Government issued a statement condemning the imposition by the USA of sanctions against a number of EU imports, in protest at the revised EU banana regime, and the consequences of this action on Caribbean economies, and agreed to review its co-operation with the USA under the partnership for prosperity and security.

During 1998 CARICOM was particularly concerned by the movement within Nevis to secede from its federation with Saint Christopher. In July heads of government agreed to dispatch a mediation team to the country (postponed until September). The Heads of Government Conference held in March 1999 welcomed the establishment of a Constitutional Task Force by the local authorities to prepare a draft constitution, on the basis of recommendations of a previous constitutional commission and the outcome of a series of public meetings. In July 1998 heads of government expressed concern at the hostility between the Government and opposition groupings in Guyana. The two sides signed an agreement, under CARICOM auspices, and in September a CARICOM mediation mission visited Guyana to promote further dialogue. CARICOM has declared its support for Guyana in its territorial disputes with Venezuela and Suriname. In June 2000 CARICOM initiated negotiations following Suriname's removal of petroleum drilling equipment from Guyanan territorial waters. In March 2000 heads of government issued a statement supporting the territorial integrity and security of Belize in that country's ongoing border dispute with Guatemala. CARICOM subsequently urged both countries to implement the provisions of an agreement signed in November. In December 2001 a CARICOM mission observed a general election in Trinidad and Tobago. Following an inconclusive outcome to the election, a delegation from CARICOM visited that country in late January 2002.

In July 2000 the Heads of Government meeting issued a statement strongly opposing the OECD Harmful Tax Initiative, under which punitive measures had been threatened against 35 countries, including CARICOM member states, if they failed to tighten taxation legislation. The meeting also condemned a separate list, issued by the OECD's Financial Task Force on Money Laundering (FATF), which identified 15 countries, including five Caribbean states, of failing to counter effectively international money-laundering. The statement reaffirmed CARICOM's commitment to fighting financial crimes and support for any necessary reform of supervisory practices or legislation, but insisted that national taxation jurisdictions, and specifically competitive regimes designed to attract offshore business, was not a matter for OECD concern. CARICOM remained actively involved in efforts to counter the scheme, and in April 2001 presented its case to the US President. In September the FATF issued a revised list of 19 'unco-operative jurisdictions', including Dominica, Grenada, St Christopher and Nevis and St Vincent and the Grenadines. In early 2002 most Caribbean states concluded a provisional agreement with the OECD to work to improve the transparency and supervision of offshore sectors.

In February 2002 the first meeting of heads of state and of government of CARICOM and the Central American Integration System convened in Belize City. The meeting aimed to strengthen co-operation between the groupings, in particular in international negotiations, efforts to counter transnational organized crime, and support for the regions' economies. In March 2004 CARICOM signed a free-trade agreement with Costa Rica.

In January 2004 CARICOM heads of government resolved to address the escalating political crisis in Haiti. Following a visit by a high-level delegation to that country early in the month discussions were held with representatives of opposition political parties and civil society groups. A few days later the Prime Minister of the Bahamas met with Haiti's President Aristide. At the end of January several CARICOM leaders met with Aristide and members of his government and announced a Prior Action Plan, incorporating opposition demands for political reform. The Plan, however, was rejected by opposition parties since it permitted President Aristide to complete his term-in-office. CARICOM, together with the OAS, continued to pursue diplomatic efforts to secure a peaceful solution to the crisis. On 29 February Aristide resigned and left the country and a provisional president was appointed. In July CARICOM heads of government resolved to send a five-member ministerial team to Haiti to discuss developments in that country with the new interim administration, and to discuss the conditions of participation by that authority in the councils of the Community. In July 2005 CARICOM heads of government expressed concern at the deterioration of the situation in Haiti, but reiterated their readiness to provide technical assistance for the electoral process, under the auspices of the UN mission.

In July 2005 CARICOM heads of government issued a statement protesting against proposals by the European Commission, issued in the previous month, to reform the EU sugar regime. Particular concern was expressed at a proposed price reduction in the cost of refined sugar of 39% over a four-year period. The heads of government insisted that, in accordance with the ACP-EU Cotonou Agreement, any review of the Sugar Protocol was required to be undertaken with the agreement of both parties and with regard to safeguarding benefits.

## ECONOMIC CO-OPERATION

The Caribbean Community's main field of activity is economic integration, by means of a Caribbean Common Market which replaced the former Caribbean Free Trade Association. The Secretariat and the Caribbean Development Bank undertake research on the best means of facing economic difficulties, and meetings of the Chief Executives of commercial banks and of central bank officials are also held with the aim of strengthening regional co-operation.

In July 1984 heads of government agreed to establish a common external tariff (CET) on certain products, in order to protect domestic industries. They also urged the necessity of structural adjustment in the economies of the region, including measures to expand production and reduce imports. In 1989 the Conference of Heads of Government agreed to implement, by July 1993, a series of measures to encourage the creation of a single Caribbean market. These included the establishment of a CARICOM Industrial Programming Scheme; the inauguration of the CARICOM Enterprise Regime; abolition of passport requirements for CARICOM nationals travelling within the region; full implementation of the rules of origin and the revised scheme for the harmonization of fiscal incentives; free movement of skilled workers; removal of all remaining regional barriers to trade; establishment of a regional system of air and sea transport; and the introduction of a scheme for regional capital movement. A CARICOM Export Development Council, established in November 1989, undertook a three-year export development project to stimulate trade within CARICOM and to promote exports outside the region.

In August 1990 CARICOM heads of government mandated the governors of CARICOM members' central banks to begin a study of the means to achieve a monetary union within CARICOM; they also institutionalized meetings of CARICOM ministers of finance and senior finance officials, to take place twice a year.

The initial deadline of 1 January 1991 for the establishment of a CET was not achieved, and in July a new deadline of 1 October was set for those members which had not complied—Antigua and Barbuda, Belize, Montserrat, Saint Christopher and Nevis and Saint Lucia, whose governments feared that the tariff would cause an increase in the rate of inflation and damage domestic industries. This deadline was later (again unsuccessfully) extended to February 1992. The tariff, which imposed a maximum level of duty of 45% on imports, was also criticized by the World Bank, the IMF and the US Government as being likely to reduce the region's competitiveness. At a special meeting, held in October 1992, CARICOM heads of government agreed to reduce the maximum level of tariffs to between 30% and 35%, to be in effect by 30 June 1993 (the level was to be further lowered, to 25%–30% by 1995). The Bahamas, however, was not party to these trading arrangements (since it is a member of the Community but not of the Common Market), and Belize was granted an extension for the implementation of the new tariff levels. At the Heads of Government Conference, held in July 1995 in Guyana, Suriname was admitted as a full member of CARICOM and acceded to the treaty establishing the Common Market. It was granted until 1 January 1996 for implementation of the tariff reductions.

The 1995 Heads of Government Conference approved additional measures to promote the single market. The free movement of skilled workers (mainly graduates from recognized regional institutions) was to be permitted from 1 January 1996. At the same time an agreement on the mutual protection and provision of social security benefits was to enter into force. In July 1996 the heads of government decided that CARICOM ministers of finance, central bank governors and planning agencies should meet more frequently to address single market issues and agreed to extend the provisions of free movement to sports people, musicians and others working in the arts and media.

In July 1997 the heads of government, meeting in Montego Bay, Jamaica, agreed to accelerate economic integration, with the aim of completing a single market by 1999. At the meeting 11 member states signed Protocol II amending the Treaty of Chaguaramas, which constituted a central element of a CARICOM Single Market and Economy (CSME), providing for the right to establish enterprises, the provision of services and the free movement of capital and labour throughout participating countries. A regional collaborative network was established to promote the CSME. In July 1998, at the meeting of heads of government, held in Saint Lucia, an agreement was signed with the Insurance Company of the West Indies to accelerate the establishment of a Caribbean Investment Fund, which was to mobilize foreign currency from extra-regional capital markets for investment in new or existing enterprises in the region. Some 60% of all funds generated were to be used by CARICOM countries and the remainder by non-CARICOM members of the ACS.

In November 2000 a special consultation on the single market and economy was held in Barbados, involving CARICOM and government officials, academics, and representatives of the private sector, labour organizations, the media, and other regional groupings. In February 2001 heads of government agreed to establish a new high-level sub-committee to accelerate the establishment of the CSME and to promote its objectives. The sub-committee was to be supported by a Technical Advisory Council, comprising representatives of the public and private sectors. By June all member states had signed and declared the provisional application of Protocol II, which had received two ratifications. By March 2005 12 countries had completed the fourth phase of the CET. The CSME was scheduled to be in effect by the end of that year. However, issues remaining under consideration at mid-2005 included the request of the Government of the Bahamas to opt-out of the agreement, and demands for differential treatment by certain member countries of the Organisation of Eastern Caribbean States. Work was progressing on the establishment of a Regional Development Fund, also scheduled to be operational by December, which aimed to remove economic inequalities within the single market

In October 2001 CARICOM heads of government, meeting in a special emergency meeting, considered the impact on the region's economy of the terrorist attacks perpetrated against targets in the USA in the previous month. The meeting resolved to enhance aviation security, to implement promotion and marketing campaigns in support of the tourist industry, and to approach international institutions to assist with emergency financing. The economic situation, which had been further adversely affected by the reduced access to the EU banana market, the economic downturn in the USA, and the effects on the investment climate of the OECD Harmful Taxation Initiative, was considered at the Heads of Government Conference, held in Guyana, in July 2002. Heads of government agreed to meet in August in special session to elaborate a programme to revive the economy, on the basis of the work of a newly-appointed technical team. A technical committee was also established in July to develop proposals for a regional stabilization programme, and a Stabilization Fund, with initial capital of US $60m. An inter-sessional Heads of Government Conference that was held in March 2004, however, agreed that there was insufficient support for the Fund to be established at that time.

### CRIME AND SECURITY

In December 1996 CARICOM heads of government determined to strengthen comprehensive co-operation and technical assistance to combat illegal drugs-trafficking. The Conference decided to establish a Caribbean Security Task Force to help formulate a single regional agreement on maritime interdiction, incorporating agreements already concluded by individual members. A Regional Drugs Control Programme at the CARICOM Secretariat aims to co-ordinate regional initiatives with the overall objective of reducing the demand and supply of illegal substances. In July 2001 the Prime Minister of Antigua and Barbuda, Lester Bird, proposed the establishment of a rapid response unit to deal with drugs-related and other serious crimes. Heads of government agreed, instead, to establish a task force to be responsible for producing recommendations for a forthcoming meeting of national security advisers. In October heads of government convened an emergency meeting in Nassau, the Bahamas, to consider the impact of the terrorist attacks against the USA which had occurred in September. The meeting determined to convene immediately the so-called Task Force on Crime and Security in order to implement new policy directives. It was agreed to enhance co-ordination and collaboration of security services throughout the region, in particular in intelligence gathering, analysis and sharing in relation to crime, illicit drugs and terrorism, and to strengthen security at airports, seaports and borders. In July 2002 heads of government agreed on a series of initiatives recommended by the Task Force to counter the escalation in crime and violence. These included strengthening border controls, preparing national anti-crime master plans, establishing broad-based National Commissions on law and order and strengthening the exchange of information and intelligence. In July 2005 CARICOM heads of government endorsed a new Management Framework for Crime and Security, which provided for regular meetings of a Council of Ministers responsible for national security and law enforcement, a Security Policy Advisory Committee, and an Implementation Agency for Crime and Security.

### INDUSTRY AND ENERGY

A protocol relating to the CARICOM Industrial Programming Scheme (CIPS), approved in 1988, is the Community's instrument for promoting the co-operative development of industry in the region. Protocol III amending the Treaty of Chaguaramas, with respect to industrial policy, was opened for signature in July 1998.

The Secretariat has established a national standards bureau in each member country to harmonize technical standards, and supervises the metrication of weights and measures. In 1999 members agreed to establish a new CARICOM Regional Organization of Standards and Quality (CROSQ) to develop common regional standards and resolve disputes. The agreement to establish CROSQ, to be located in Barbados, was signed in February 2002. By March 2005 the agreement was being provisionally applied in 12 member countries, and was to enter into force upon the signature of Montserrat.

The CARICOM Alternative Energy Systems Project provides training, assesses energy needs and conducts energy audits. Efforts in regional energy development are directed at the collection and analysis of data for national energy policy documents. Implementation of a Caribbean Renewable Energy Development Programme, a project initiated in 1998, commenced in 2004. The Programme aimed to remove barriers to renewable energy development, establish a foundation for a sustainable renewable energy industry, and to create a framework for co-operation among regional and national renewable energy projects. A Caribbean Renewable Energy Fund was to be established to provide equity and development financing for renewable energy projects.

### TRANSPORT, COMMUNICATIONS AND TOURISM

A Caribbean Confederation of Shippers' Councils represents the interests of regional exporters and importers. In July 1990 the Caribbean Telecommunications Union was established to oversee developments in regional telecommunications.

In 1988 a Consultative Committee on Caribbean Regional Information Systems (CCCRIS) was established to evaluate and monitor the functioning of existing information systems and to seek to co-ordinate and advise on the establishment of new systems.

A Summit of Heads of Government on Tourism, Trade and Transportation was held in Trinidad and Tobago, in August 1995, to which all members of the ACS and regional tourism organizations were invited. In 1997 CARICOM heads of government considered a number of proposals relating to air transportation, tourism, human resource development and capital investment, which had been identified by Community ministers of tourism as critical issues in the sustainable development of the tourist industry. The heads of government requested ministers to meet regularly to develop tourism policies, and in particular to undertake an in-depth study of human resource development issues in early 1998. A new fund to help train young people from the region in aspects of the tourist industry was inaugurated in July 1997, in memory of the former Prime Minister of Jamaica, Michael Manley. A regional summit on tourism, in recognition of the importance of the industry to the economic development of the region, was held in the Bahamas, in December 2001.

A Multilateral Agreement Concerning the Operations of Air Services within the Caribbean Community entered into force in November 1998, providing a formal framework for the regulation of the air transport industry and enabling CARICOM-owned and -controlled airlines to operate freely within the region. In July 1999 heads of government signed Protocol VI amending the Treaty of Chaguaramas providing for a common transportation policy, with harmonized standards and practices, which was to be an integral component of the development of a single market and economy. In November 2001 representatives of national civil aviation authorities signed a memorandum of understanding, providing for the estab-

## AGRICULTURE AND FISHERIES

In July 1996 the CARICOM summit meeting agreed to undertake wide-ranging measures in order to modernize the agricultural sector and to increase the international competitiveness of Caribbean agricultural produce. The CARICOM Secretariat was to support national programmes with assistance in policy formulation, human resource development and the promotion of research and technology development in the areas of productivity, marketing, agri-business and water resources management. During 1997 CARICOM Governments continued to lobby against a complaint lodged at the WTO with regard to the EU's banana import regime (offering favourable conditions to ACP producers—see above) and to generate awareness of the economic and social importance of the banana industry to the region. Protocol V amending the Treaty of Chaguaramas, which was concerned with agricultural policy, was opened for signature by heads of government in July 1998. In July 2002 heads of government approved an initiative to develop a CARIFORUM Special Programme for Food Security.

In February 2003 the CARICOM Secretariat was mandated to draft a proposal for a common fisheries policy.

## HEALTH AND EDUCATION

In 1986 CARICOM and the Pan-American Health Organization launched 'Caribbean Co-operation in Health' with projects to be undertaken in six main areas: environmental protection, including the control of disease-bearing pests; development of human resources; chronic non-communicable diseases and accidents; strengthening health systems; food and nutrition; maternal and child health care; and population activities. In 2001 CARICOM co-ordinated a new regional partnership to reduce the spread and impact of HIV and AIDS in member countries. All countries were to prepare national strategic plans to facilitate access to funding to combat the problem. A meeting of the so-called Pan-Caribbean Partnership against HIV/AIDS (PANCAP) was convened in November. In February 2002 PANCAP initiated regional negotiations with pharmaceutical companies to secure reductions in the cost of anti-retroviral drugs. A Caribbean Environmental Health Institute (see below) aims to promote collaboration among member states in all areas of environmental management and human health. In July 2001 heads of government, meeting in the Bahamas, issued the Nassau Declaration on Health, advocating greater regional strategic co-ordination and planning in the health sector, institutional reform, and increased resources.

CARICOM educational programmes have included the improvement of reading in schools through assistance for teacher-training; and ensuring the availability of low-cost educational material throughout the region. In July 1997 CARICOM heads of government adopted the recommendations of a ministerial committee, which identified priority measures for implementation in the education sector. These included the objective of achieving universal, quality secondary education and the enrolment of 15% of post-secondary students in tertiary education by 2005, as well as improved training in foreign languages and science and technology. From the late 1990s youth activities have been increasingly emphasized by the Community. These have included new programmes for disadvantaged youths, a mechanism for youth exchange and the convening of a Caribbean Youth Parliament.

## EMERGENCY ASSISTANCE

A Caribbean Disaster Emergency Response Agency (CDERA) was established in 1991 to co-ordinate immediate disaster relief, primarily in the event of hurricanes. In May 2004 the Community determined to contribute to relief and rehabilitation efforts following the devastation caused by flooding and mudslides in the Dominican Republic and Haiti, which had killed more than 1,000 people. In January 2005, meeting on the sidelines of the fifth Summit of the Alliance of Small Island States, in Port Louis, Mauritius, the Secretaries-General of CARICOM, the Commonwealth, the Pacific Islands Forum and the Indian Ocean Commission determined to take collective action to strengthen the disaster preparedness and response capabilities of their member countries in the Caribbean, Pacific and Indian Ocean areas.

## INSTITUTIONS

The following are among the institutions formally established within the framework of CARICOM.

**Assembly of Caribbean Community Parliamentarians:** c/o CARICOM Secretariat; an intergovernmental agreement on the establishment of a regional parliament entered into force in August 1994; inaugural meeting held in Barbados, in May 1996. Comprises up to four representatives of the parliaments of each member country, and up to two of each associate member. It aims to provide a forum for wider community involvement in the process of integration and for enhanced deliberation on CARICOM affairs; authorized to issue recommendations for the Conference of Heads of Government and to adopt resolutions on any matter arising under the Treaty of Chaguaramas.

**Caribbean Agricultural Research and Development Institute (CARDI):** UWI Campus, St Augustine, Trinidad and Tobago; tel. 645-1205; fax 645-1208; e-mail info@cardi.org; internet www.cardi.org; f. 1975; aims to contribute to the competitiveness and sustainability of Caribbean agriculture by generating and transferring new and appropriate technologies and by developing effective partnerships with regional and international entities; Exec. Dir FRANK B. LAUCKNER; publs CARDI Weekly, CARDI Review, technical bulletin series.

**Caribbean Centre for Development Administration (CARICAD):** 1st Floor, Weymouth Corporate Centre, Roebuck St, St Michael, Barbados; tel. 427-8535; fax 436-1709; e-mail caricad@caribsurf.com; internet www.caricad.org; f. 1980; aims to assist governments in the reform of the public sector and to strengthen their managerial capacities for public administration; promotes the involvement of the private sector, non-governmental organizations and other bodies in all decision-making processes; Exec. Dir Dr P. I. GOMES.

**Caribbean Disaster Emergency Response Agency (CDERA):** Bldg 1, Manor Lodge, Lodge Hill, St Michael, Barbados; tel. 425-0386; fax 425-8854; e-mail cdera@caribsurf.com; internet www.cdera.org; f. 1991 for activities, see Emergency Assistance above; Regional Co-ordinator JEREMY COLLYMORE.

**Caribbean Environmental Health Institute (CEHI):** POB 1111, The Morne, Castries, St Lucia; tel. 4522501; fax 4532721; e-mail cehi@candw.lc; internet www.cehi.org.lc; f. 1980 (began operations in 1982); provides technical and advisory services to member states in formulating environmental health policy legislation and in all areas of environmental management (for example, solid waste management, water supplies, beach and air pollution, and pesticides control); promotes, collates and disseminates relevant research; conducts courses, seminars and workshops throughout the region; Exec. Dir VINCENT SWEENEY.

**Caribbean Food and Nutrition Institute (CFNI):** UWI Campus, POB 140, St Augustine, Trinidad and Tobago; tel. 663-1544; e-mail cfni@cablenett.net; internet www.cfni.paho.org; f. 1967 to serve the governments and people of the region and to act as a catalyst among persons and organizations concerned with food and nutrition through research and field investigations, training in nutrition, dissemination of information, advisory services and production of educational material; mems: all English-speaking Caribbean territories, including the mainland countries of Belize and Guyana; Dir Dr FITZROY HENRY; publs Cajanus (quarterly), Nyam News (monthly), Nutrient-Cost Tables (quarterly), educational material.

**Caribbean Meteorological Organization (CMO):** POB 461, Port of Spain, Trinidad and Tobago; tel. 624-4481; fax 623-3634; e-mail hqcmo@tstt.net.tt; f. 1951 to co-ordinate regional activities in meteorology, operational hydrology and allied sciences; became a specialized institution of CARICOM in 1973; comprises a Council of Government Ministers, a Headquarters Unit, the Caribbean Meteorological Foundation and the Caribbean Institute for Meteorology and Hydrology, located in Barbados; mems: govts of 16 countries and territories represented by the National Meteorological and Hydrometeorological Services; Co-ordinating Dir TYRONE W. SUTHERLAND.

## ASSOCIATE INSTITUTIONS

**Caribbean Development Bank:** POB 408, Wildey, St Michael, Barbados; tel. 431-1600; fax 426-7269; e-mail info@caribank.org; internet www.caribank.org; f. 1969 to stimulate regional economic growth through support for agriculture, industry, transport and other infrastructure, tourism, housing and education; subscribed cap. US $705.0m. (April 2004); in 1999 net approvals totalled $146.6m. for 16 projects; at the end of 1999 cumulative grant and loan disbursements totalled $1,275.0m.; the Special Development Fund was replenished in 1996; mems: CARICOM states, and Canada, the People's Republic of China, Colombia, Germany, Italy, Mexico, United Kingdom, Venezuela; Pres. Dr. COMPTON BOURNE.

**Caribbean Law Institute:** University of the West Indies, Cave Hill Campus, POB 64, Bridgetown, Barbados; tel. 417-4560; fax 417-4138; f. 1988 to harmonize and modernize commercial laws in the region; Exec. Dir Dr WINSTON C. ANDERSON.

ns of CARICOM, in accordance with its constitution, are the University of Guyana, the University of the West Indies and the Secretariat of the Organisation of Eastern Caribbean States.

## Publications

*CARICOM Perspective* (annually).
*CARICOM View* (6 a year).

# CENTRAL AMERICAN INTEGRATION SYSTEM
## (SISTEMA DE LA INTEGRACIÓN CENTROAMERICANA—SICA)

**Address:** Blv. Orden de Malta 470, Urb. Santa Elena, Antiguo Cuscatlán, San Salvador, El Salvador.
**Telephone:** 2248-8800; **fax:** 2248-8899; **e-mail:** info@sgsica.org; **internet:** www.sgsica.org.

Founded in December 1991, when the heads of state of six Central American countries signed the Protocol of Tegucigalpa to the agreement establishing the Organization of Central American States (f. 1951), creating a new framework for regional integration. A General Secretariat of the Sistema de la Integración Centroamericana (SICA) was inaugurated in February 1993 to co-ordinate the process of political, economic, social cultural and environmental integration and to promote democracy and respect for human rights throughout the region.

### MEMBERS

Belize
Costa Rica
El Salvador
Guatemala
Honduras
Nicaragua
Panama

### OBSERVERS

Dominican Republic
Taiwan

## Organization
(August 2005)

### SUMMIT MEETINGS
The meetings of heads of state of member countries serve as the supreme decision-making organ of SICA.

### COUNCIL OF MINISTERS
Ministers of Foreign Affairs of member states meet regularly to provide policy direction for the process of integration.

### CONSULTATIVE COMMITTEE
The Committee comprises representatives of business organizations, trade unions, academic institutions and other federations concerned with the process of integration in the region. It is an integral element of the integration system and assists the Secretary-General in determining the policies of the organization.

**President:** RICARDO SOL.

### GENERAL SECRETARIAT
The General Secretariat of SICA was established in February 1993 to co-ordinate the process of enhanced regional integration. It comprises the following divisions: inter-institutional relations; research and co-operation; legal and political affairs; economic affairs; and communications and information.

In September 1997 Central American Common Market (CACM) heads of state, meeting in the Nicaraguan capital, signed the Managua Declaration in support of further regional integration and the establishment of a political union. A commission was to be established to consider all aspects of the policy and to formulate a timetable for the integration process. In February 1998 SICA heads of state resolved to establish a Unified General Secretariat to integrate the institutional aspects of SICA (see below) in a single office, to be located in San Salvador. The process was ongoing in 2005.

**Secretary-General:** ANÍBAL ENRIQUE QUIÑÓNEZ ABARCA.

### SPECIALIZED TECHNICAL SECRETARIATS

**Secretaría Ejecutiva de la Comisión Centroamericana de Ambiente y Desarrollo (SE-CCAD):** Blv. Orden de Malta 470, Santa Elena, Antiguo Cuscatlán, San Salvador, El Salvador; tel. 2289-6131; fax 2289-6126; e-mail mcastro@sgsica.org; internet www .ccad.ws; f. 1989 to enhance collaboration in the promotion of sustainable development and environmental protection; Exec. Sec. MARCO GONZÁLEZ PASTORA.

**Secretaría General de la Coordinación Educativa y Cultural Centroamericana (SG-CECC):** 175m norte de la esquina oeste del ICE, Sabana Norte, San José, Costa Rica; tel. 232-3783; fax 231-2366; e-mail sgcecc@sol.racsa.co.cr; f. 1982; promotes development of regional programmes in the fields of education and culture; Sec.-Gen. MARVIN HERRERA ARAYA.

**Secretaría de Integración Económica Centroamericana (SIECA):** 4A Avda 10–25, Zona 14, Apdo 1237, 01901 Guatemala City, Guatemala; tel. 2368-2151; fax 2368-1071; e-mail info@sieca .org.gt; internet www.sieca.org.gt; f. 1960 to assist the process of economic integration and the creation of a Central American Common Market (CACM—established by the organization of Central American States under the General Treaty of Central American Economic Integration, signed in December 1960 and ratified by Costa Rica, Guatemala, El Salvador, Honduras and Nicaragua in September 1963); supervises the correct implementation of the legal instruments of economic integration, carries out relevant studies at the request of the CACM, and arranges meetings; comprises departments covering the working of the CACM: negotiations and external trade policy; external co-operation; systems and statistics; finance and administration; also includes a unit for co-operation with the private sector and finance institutions, and a legal consultative committee; Sec.-Gen. HAROLDO RODAS MELGAR; pubs *Anuario Estadístico Centroamericano de Comercio Exterior*, *Carta Informativa* (monthly), *Cuadernos de la SIECA* (2 a year), *Estadísticas Macroeconómicas de Centroamérica* (annually), *Series Estadísticas Seleccionadas de Centroamérica* (annually), *Boletín Informativo* (fortnightly).

**Secretaría de la Integración Social Centroamericana (SISCA):** Blv. Orden de Malta 470, Santa Elena, Antiguo Cuscatlán, San Salvador, El Salvador; tel. 2289-6131; fax 2289-6124; e-mail hmorgado@sgsica.org.sv; f. 1995; Dir-Gen. Dr HUGO MORGADO.

### OTHER SPECIALIZED SECRETARIATS

**Secretaría de Integración Turística Centroamericana (SITCA):** Blv. Orden de Malta 470, Santa Elena, Antiguo Cuscatlán, San Salvador, El Salvador; tel. 2248-8837; fax 2248-8897; e-mail csilva@sgsica.org; f. 1965 to develop regional tourism activities; Dir. MERCEDES DE MENA.

**Secretaría del Consejo Agropecuario Centroamericano (SCAC):** Sacretarion Isidro de Coronado, Apdo Postal 55-2200, San José, Costa Rica; tel. 216-0303; fax 216-0285; e-mail coreca@iica.ac .cr; internet www.coreca.org; f. 1991 to determine and co-ordinate regional policies and programmes relating to agriculture and agro-industry; Sec-Gen. RÓGER GUILLÉN BUSTOS.

**Secretaría Ejecutiva del Consejo Monetario Centroamericano (SECMCA)** (Central American Monetary Council): Ofiplaza del Este, Edif. C, 75m. oeste de la Rotonda la Bandera, San Pedro Montes de Oca, Apdo Postal 5438-1000, San José, Costa Rica; tel. 280-9522; fax 280-2511; e-mail secma@secma.org; internet www .secma.org; f. 1964 by the presidents of Central American central banks, to co-ordinate monetary policies; Exec. Sec. MIGUEL CHORRO; publs *Boletín Estadístico* (annually), *Informe Económico* (annually).

**Comisión Centroamericana de Transporte Marítimo (COCATRAM):** Frente al Costado Oeste del Hotel Mansión Teodolinda, Barrio Bolonia, Apdo Postal 2423, Managua, Nicaragua; tel. 222-3482; fax 222-2759; e-mail cocatram@ibw.com.ni; internet www .cocatram.org.ni; f. 1981; Exec. Dir ALFONSO BREUILLET GALINDO; publ. *Boletín Informativo*.

### PARLIAMENT

**Address:** 12A Avda 33-04, Zona 5, Guatemala City, Guatemala 01005.
**Telephone:** 2339-0466; **fax:** 2334-6670; **e-mail:** guatemala@ parlacen.org.gt; **internet:** www.parlacen.org.gt.

# REGIONAL ORGANIZATIONS

*Central American Integration System*

Officially inaugurated in 1991. Comprises representatives of El Salvador, Guatemala, Honduras, Nicaragua and Panama. In 2004 Honduras announced its intention temporarily to withdraw from the Parliament.

**President:** FABIO GADEA MANTILLA.

**Secretary-General:** WERNER VARGAS.

## COURT OF JUSTICE

**Address:** Apdo Postal 907 Managua, Nicaragua.

**Telephone:** 266-6273; **fax:** 266-8486; **e-mail:** cortecen@tmx.com.ni; **internet:** www.ccj.org.ni.

Officially inaugurated in 1994. Tribunal authorized to consider disputes relating to treaties agreed within the regional integration system. In February 1998 Central American heads of state agreed to limit the number of magistrates in the Court to one per country.

**President:** ADOLFO LEÓN GÓMEZ.

**Secretary-General:** ORLANDO GUERRERO MAYORGA.

## AD HOC INTERGOVERNMENTAL SECRETARIATS

**Comisión de Ciencia y Tecnología de Centroamérica y Panamá (CTCAP)** (Committee for Science and Technology of Central America and Panama): Antigua base de Clayton, Edif. 213, Panamá; tel. 317-0014; fax 317-0026; e-mail espinoza@ns.hondunet .net; internet www.senacyt.gob.pa/ctcap; f. 1976.

**Consejo Centroamericano de Instituciones de Seguridad Social (COCISS)** (Central American Council of Social Security Institutions): Apdo 10105, San José, Costa Rica; tel. 257-0122; fax 233-1850; e-mail cociss@ccss.sa.cr; internet www.ccss.sa.cr/cociss/ idcociss.htm; f. 1992.

**Consejo del Istmo Centroamericano de Deportes y Recreación (CODICADER)** (Committee of the Central American Isthmus for Sport and Recreation): Palacio Nacional de los Deportes, Edif. Administrativo, 3°, San Salvador, El Salvador; tel. 2231-9993; fax 2231-9990; e-mail jorgehernandez@indes.gob.sv; f. 1992.

**Unidad Técnica del Consejo Centroamericano de Vivienda y Asentamientos Humanos (CCVAH)** (Central American Council on Housing and Human Settlements): Avda la Paz 244, Tegucigalpa, Honduras; tel. 236-5804; fax 236-6560; f. 1992.

**Secretaría Ejecutiva del Consejo de Electrificación de América Central (CEAC)** (Central American Electrification Council): 9A Calle Pte 950, Edif. CEL, Centro de Gobierno, San Salvador, El Salvador; tel. 2211-6175; fax 2211-6239; e-mail jmontesi@cel.gob.sv; internet www.ceac-ca.org.sv; f. 1985.

**Organización Centroamericana de Entidades Fiscalizadores Superiores:** e-mail cdcr@es.com.sv; internet www.sgsica.org.ir/ ocefs.htm.

## OTHER REGIONAL INSTITUTIONS

### Finance

**Banco Centroamericano de Integración Económica (BCIE)** (Central American Bank for Economic Integration): Blv. Suyapa, Contigua a Banco de Honduras, Apdo 772, Tegucigalpa, Honduras; tel. 228-2243; fax 228-2185; e-mail jarevalo@bcie.org; internet www .bcie.hn; f. 1961 to promote the economic integration and balanced economic development of member countries; finances public and private development projects, particularly those related to industrialization and infrastructure; loan approvals in 2002/03 totalled US $728.5m.; auth. cap. $2,000m; regional mems: Costa Rica, Dominican Republic, El Salvador, Guatemala, Honduras, Nicaragua; non-regional mems: Argentina, the People's Republic of China, Colombia, Mexico; Pres. HARRY BRAUTIGAM; publs *Annual Report*, *Revista de la Integración y el Desarrollo de Centroamérica*.

### Public Administration

**Centro de Coordinación para la Prevención de Desastres Naturales en América Central (CEPREDENAC):** Antigua base de Howard, Ave Rencher, Edif. 707, Panamá, Panama; tel. 316-0065; fax 316-0074; e-mail secretaria@cepredenac.org; internet www .cepredenac.org; f. 1988, integrated into SICA in 1995; aims to strengthen the capacity of the region to reduce its vulnerability to natural disasters; Exec. Sec. GERONIMO GIUSTO-ROBELO.

**Instituto Centroamericano de Administración Pública (ICAP)** (Central American Institute of Public Administration): Apdo Postal 10025-1000 San José, Costa Rica; tel. 234-1011; fax 225-2049; e-mail icapcr@racsa.co.cr; internet www.icap.ac.cr; f. 1954 by the five Central American Republics and the United Nations, with later participation by Panama; the Institute aims to train the region's public servants, provide technical assistance and carry out research leading to reforms in public administration; Dir Dr HUGO ZELAYA CÁLIX.

**Secretaría Ejecutiva de la Comisión Regional de Recursos Hidráulicos (SE-CRRH):** Apdo Postal 21–2300, Curridabat, San José, Costa Rica; tel. 231-5791; fax 296-0047; e-mail crrhcr@sol .racsa.co.cr; internet aguayclima.com; f. 1966; mems: Belize, Costa Rica, El Salvador, Guatemala, Honduras, Nicaragua, Panama; Exec. Sec. MAX CAMPOS ORTIZ.

### Agriculture and Fisheries

**Organismo Internacional Regional de Sanidad Agropecuaria (OIRSA)** (International Regional Organization of Plant Protection and Animal Health): Avda Las Camelias 14, Col. San Francisco, Apdo 61, San Salvador, El Salvador; tel. 2223-9545; fax 2279-0189; e-mail orgoirsa@gbm.net; internet ns1.oirsa.org.sv; f. 1953 for the prevention of the introduction of animal and plant pests and diseases unknown in the region; research, control and eradication programmes of the principal pests present in agriculture; technical assistance and advice to the ministries of agriculture and livestock of member countries; education and qualification of personnel; mems: Belize, Costa Rica, Dominican Republic, El Salvador, Guatemala, Honduras, Mexico, Nicaragua, Panama; Exec. Dir Dr CELIO HUMBERTO BARRETO ORTEGA.

**Unidad Coordinadora de la Organización del Sector Pesquero y Acuícola del Istmo Centroamericano:** Blv. Orden de Malta 470, Santa Elena, Antiguo Cuscatlán, San Salvador, El Salvador; tel. 2248-8800; fax 2248-8899; e-mail mgonzalez@sgsica.org; f. 1995, incorporated into SICA in 1999; Regional Co-ordinator MARIO GONZÁLEZ RECINOS.

### Education and Health

**Comité Coordinador Regional de Instituciones de Agua Potable y Saneamiento de Centroamérica, Panamá y República Dominicana (CAPRE):** Avda Bo. 1A, El Obelisco Comayagüela, Tegucigalpa, Honduras; tel. 237-8552; fax 237-2575; e-mail capregtz@sol.racsa.co.cr; f. 1979; Dir LILIANA ARCE UMAÑA.

**Consejo Superior Universitario Centroamericano (CSUCA)** (Central American University Council): Avda Las Américas 1–03, Zona 14, International Club Los Arcos, Guatemala City, Guatemala; tel. 2367-1833; fax 2367-4517; e-mail sp@csuca.edu.gt; internet www .csuca.edu.gt; f. 1948 to guarantee academic, administrative and economic autonomy for universities and to encourage regional integration of higher education; maintains libraries and documentation centres; Council of 32 mems; mems: 16 universities, in Belize, Costa Rica (four), El Salvador, Guatemala, Honduras (two), Nicaragua (four) and Panama (three); Sec.-Gen. EFRAIN MEDINA GUERRA; publs *Estudios Sociales Centroamericanos* (quarterly), *Cuadernos de Investigación* (monthly), *Carta Informativa de la Secretaría General* (monthly).

**Instituto de Nutrición de Centroamérica y Panamá (INCAP)** (Institute of Nutrition of Central America and Panama): Calzada Roosevelt, Zona 11, Apdo Postal 1188, Guatemala City, Guatemala; tel. 2472-3762 ; fax 2473-6529; e-mail hdelgado@incap.ops-oms.org; internet www.incap.org.gt; f. 1949 to promote the development of nutritional sciences and their application and to strengthen the technical capacity of member countries to reach food and nutrition security; provides training and technical assistance for nutrition education and planning; conducts applied research; disseminates information; maintains library (including about 600 periodicals); administered by the Pan American Health Organization (PAHO) and the World Health Organization; mems: CACM mems and Belize and Panama; Dir Dr HERNÁN L. DELGADO; publ. *Annual Report*.

### Transport and Communications

**Comisión Técnica de Telecomunicaciones (COMTELCA)** (Technical Commission for Telecommunications): Col. Palmira, Edif. Alfa, 608 Avda Brasil, Apdo 1793, Tegucigalpa, Honduras; tel. 220-6666; fax 220-1197; e-mail sec@comtelca.hn; internet www.comtelca .hn; f. 1966 to co-ordinate and improve the regional telecommunications network; Dir-Gen. HÉCTOR LEONEL RODRÍGUEZ MILLA.

**Corporación Centroamericana de Servicios de Navegación Aérea (COCESNA)** (Central American Air Navigation Service Corporation): Apdo 660, Aeropuerto de Toncontín, Tegucigalpa, Honduras; tel. 234-3360; fax 234-2550; e-mail sec-interna@cocesna.org; internet www.cocesna.org; f. 1960; offers radar air traffic control services, aeronautical telecommunications services, flight inspections and radio assistance services for air navigation; administers the Central American Aeronautical School; Exec. Pres. EDUARDO JOSÉ MARÍN.

## Activities

In June 1990 the presidents of the Central American Common Market (CACM) countries (Costa Rica, El Salvador, Guatemala, Honduras and Nicaragua) signed a declaration welcoming peace initiatives in El Salvador, Guatemala and Nicaragua, and appealing for a revitalization of CACM, as a means of promoting lasting peace in the region. In December the presidents committed themselves to the creation of an effective common market, proposing the opening of negotiations on a comprehensive regional customs and tariffs policy by March 1991, and the introduction of a regional 'anti-dumping' code by December 1991. They requested the support of multilateral lending institutions through investment in regional development, and the cancellation or rescheduling of member countries' debts. In December 1991 the heads of state of the five CACM countries and Panama signed the Protocol of Tegucigalpa, and in February 1993 the General Secretariat of SICA was inaugurated to co-ordinate the integration process in the region.

In February 1993 the European Community (EC, now European Union—EU) signed a new framework co-operation agreement with the CACM member states extending the programme of economic assistance and political dialogue initiated in 1984; a further co-operation agreement with the EU (EC, now European Union—EU) was signed in early 1996.

In October 1993 the presidents of the CACM countries and Panama signed a protocol to the 1960 General Treaty, committing themselves to full economic integration in the region (with a common external tariff of 20% for finished products and 5% for raw materials and capital goods) and creating conditions for increased free trade. The countries agreed to accelerate the removal of internal non-tariff barriers, but no deadline was set. Full implementation of the protocol was to be 'voluntary and gradual', owing to objections on the part of Costa Rica and Panama. In May 1994, however, Costa Rica committed itself to full participation in the protocol. In March 1995 a meeting of the Central American Monetary Council discussed and endorsed a reduction in the tariff levels from 20% to 15% and from 5% to 1%. However, efforts to adopt this as a common policy were hindered by the implementation of these tariff levels by El Salvador on a unilateral basis, from 1 April, and the subsequent modifications by Guatemala and Costa Rica of their external tariffs. In March 2002 Central American leaders adopted the San Salvador Plan of Action for Central American Economic Integration, establishing several objectives as the basis for the creation, by 1 January 2004, of a regional customs union, with a single tariff.

In December 1994 the SICA member states and the USA signed the Central American-United States of America Joint Declaration (CONCAUSA), covering co-operation in the following areas: conservation of biodiversity, sound management of energy, environmental legislation, and sustainable economic development. In June 2001 the SICA states and the USA signed a renewed and expanded CONCAUSA, now also covering co-operation in addressing climate change, and in disaster preparedness. In May 1997 the heads of state of CACM member countries, together with the Prime Minister of Belize, conferred with the then US President, Bill Clinton, in San José, Costa Rica. The leaders resolved to establish a Trade and Investment Council to promote trade relations; however, Clinton failed to endorse a request from CACM members that their products receive preferential access to US markets, on similar terms to those from Mexico agreed under the NAFTA accord. During the 1990s the Central American Governments pursued negotiations to conclude free-trade agreements with Mexico, Panama and the members of the Caribbean Community and Common Market (CARICOM). Nicaragua signed a bilateral accord with Mexico in December (Costa Rica already having done so in 1994). El Salvador, Guatemala and Honduras jointly concluded a free-trade arrangement with Mexico in May 2000. In November 1997, at a special summit meeting of CACM heads of state, an agreement was reached with the President of the Dominican Republic to initiate a gradual process of incorporating that country into the process of Central American integration, with the aim of promoting sustainable development throughout the region. The first sectors for increased co-operation between the two sides were to be tourism, health, investment promotion and air transport. A free-trade accord with the Dominican Republic was concluded in April 1998, and formally signed in November.

In November 1998 Central American heads of state held an emergency summit meeting to consider the devastation in the region caused by 'Hurricane Mitch'. The Presidents urged international creditors to write off the region's estimated debts of US $16,000m. to assist in the economic recovery of the countries worst-affected. They also reiterated requests for preferential treatment for the region's exports within the NAFTA framework. In October 1999 the heads of state adopted a strategic framework for the period 2000–04 to strengthen the capacity for the physical, social, economic and environmental infrastructure of Central American countries to withstand the impact of natural disasters. In particular, programmes for the integrated management and conservation of water resources, and for the prevention of forest fires were to be implemented.

In June 2001 the heads of state and representatives of Belize, Costa, Rica, El Salvador, Guatemala, Honduras, Mexico, Nicaragua and Panama agreed to activate the Puebla-Panama Plan (PPP) to promote sustainable social and economic development in the region and to reinforce integration efforts among Central America and the southern states of Mexico (referred to as Mesoamerica). The heads of state identified the principal areas for PPP initiatives, including tourism, road integration, telecommunications, energy interconnection, and the prevention and mitigation of disasters. In June 2002 the heads of state of seven countries, and the Vice-President of Panama, convened in Mérida, Mexico during an investment fair to promote the Plan and reiterated their support for the regional initiatives. The meeting was also held within the framework of the Tuxtla dialogue mechanism, so-called after an agreement signed in 1991 between Mexico and Central American countries, to discuss co-ordination between the parties, in particular in social matters, health, education and the environment. Regular 'Tuxtla' summit meetings have subsequently been convened. A Centre for the Promotion of Small and Medium-sized Enterprises was established in June 2001.

In April 2001 Costa Rica concluded a free-trade accord with Canada; the other four CACM countries commenced negotiations with Canada in November with the aim of reaching a similar agreement. In February 2002 heads of state of SICA countries convened an extraordinary summit meeting in Managua, Nicaragua, at which they resolved to implement measures to further the political and economic integration of the region. The leaders determined to pursue initial proposals for a free-trade pact with the USA during the visit to the region of US President George W. Bush in the following month, and, more generally, to strengthen trading relations with the EU. They also pledged to resolve all regional conflicts by peaceful means. Earlier in February the first meeting of heads of state or government of SICA and CARICOM countries took place in Belize, with the aim of strengthening political and economic relations between the two groupings. The meeting agreed to work towards concluding common negotiating positions, for example in respect of the FTAA and World Trade Organization.

In May 2002 ministers of foreign affairs of Central America and the EU agreed upon a new agenda for a formalized dialogue and on priority areas of action, including environmental protection, democracy and governance, and poverty reduction. The meeting determined to work towards the eventual conclusion of an Association Agreement. It was agreed that meetings between the two sides, at ministerial level, were to be held each year. In December 2003 a new EU-Central America Political Dialogue and Co-operation Agreement was signed to replace an existing (1993) framework accord. The EU has allocated €74.5m. to finance co-operation programmes under its Regional Strategy for Central America covering the period 2002–06; these were to focus on strengthening the role of civil society in the process of regional integration, reducing vulnerability to natural disasters, and environmental improvement.

The summit meeting of CACM heads of state, convened in December 2002, in San José, Costa Rica, adopted the 'Declaration of San José', supporting the planned establishment of the Central American customs union (see above), and endorsing the initiation of negotiations with the USA on the creation of a new Central American Free-Trade Area (CAFTA). The establishment of a new Central American Tourism Agency was also announced at the summit.

Negotiations on CAFTA between the CACM countries and the USA were initiated in January 2003. An agreement was concluded between the USA and El Salvador, Guatemala, Honduras and Nicaragua in December, and with Costa Rica in January 2004. Under the resulting US-Central America Free-Trade Agreement some 80% of US exports of consumer and industrial goods and more than 50% of US agricultural exports to CAFTA countries were to become duty-free immediately upon its entry into force, with remaining tariffs to be eliminated over a 10-year period for consumer and industrial goods and over a 15-year period for agricultural exports. Almost all CAFTA exports of consumer and industrial products to the USA were to be duty-free on the Agreement's entry into force. The Agreement was signed by the US Trade Representative and CACM ministers of trade and economy, convened in Washington, DC, in May 2004. It required ratification by all national legislatures before entering into effect. Negotiations on a US-Dominican Republic free-trade agreement, to integrate the Dominican Republic into CAFTA, were concluded in March and the agreement was signed in August. The so-called DR-CAFTA accord was ratified by El Salvador in December, and by the USA in July 2005.

A meeting of SICA heads of state, held in December 2004, was concerned with economic and regional security issues. In March 2005 SICA ministers responsible for security, defence and the interior resolved to establish a special regional force to combat crime, drugs and arms trafficking and terrorism. A new regional agricultural and fisheries policy was inaugurated in July.

# THE COMMONWEALTH

**Address:** Commonwealth Secretariat, Marlborough House, Pall Mall, London, SW1Y 5HX, United Kingdom.
**Telephone:** (20) 7747-6500; **fax:** (20) 7930-0827; **e-mail:** info@commonwealth.int; **internet:** www.thecommonwealth.org.

The Commonwealth is a voluntary association of 53 independent states (at August 2005), comprising more than one-quarter of the world's population. It includes the United Kingdom and most of its former dependencies, and former dependencies of Australia and New Zealand (themselves Commonwealth countries). All Commonwealth countries accept Queen Elizabeth II as the symbol of the free association of the independent member nations and as such the Head of the Commonwealth.

### MEMBERS IN SOUTH AMERICA AND THE CARIBBEAN

Antigua and Barbuda
Bahamas
Barbados
Belize
Dominica
Grenada
Guyana
Jamaica
Saint Christopher and Nevis
Saint Lucia
Saint Vincent and the Grenadines
Trinidad and Tobago
United Kingdom Dependencies:
Anguilla
Bermuda
British Virgin Islands
Cayman Islands
Falkland Islands
Montserrat
Turks and Caicos Islands

## Organization

(August 2005)

The Commonwealth is not a federation: there is no central government nor are there any rigid contractual obligations such as bind members of the United Nations.

The Commonwealth has no written constitution but its members subscribe to the ideals of the Declaration of Commonwealth Principles unanimously approved by a meeting of heads of government in Singapore in 1971. Members also approved the Gleneagles Agreement concerning apartheid in sport (1977); the Lusaka Declaration on Racism and Racial Prejudice (1979); the Melbourne Declaration on relations between developed and developing countries (1981); the New Delhi Statement on Economic Action (1983); the Goa Declaration on International Security (1983); the Nassau Declaration on World Order (1985); the Commonwealth Accord on Southern Africa (1985); the Vancouver Declaration on World Trade (1987); the Okanagan Statement and Programme of Action on Southern Africa (1987); the Langkawi Declaration on the Environment (1989); the Kuala Lumpur Statement on Southern Africa (1989); the Harare Commonwealth Declaration (1991); the Ottawa Declaration on Women and Structural Adjustment (1991); the Limassol Statement on the Uruguay Round of multilateral trade negotiations (1993); the Millbrook Commonwealth Action Programme on the Harare Declaration (1995); the Edinburgh Commonwealth Economic Declaration (1997); the Fancourt Commonwealth Declaration on Globalization and People-centred Development (1999); the Coolum Declaration on the Commonwealth in the 21st Century: Continuity and Renewal (2002); the Aso Rock Commonwealth Declaration and Statement on Multilateral Trade (2003); and the Declaration of the Nairobi Meeting of Commonwealth National Committees on International Humanitarian Law (2005).

### MEETINGS OF HEADS OF GOVERNMENT

Commonwealth Heads of Government Meetings (CHOGMs) are private and informal and operate not by voting but by consensus. The emphasis is on consultation and exchange of views for co-operation. A communiqué is issued at the end of every meeting. Meetings are normally held every two years in different capitals in the Commonwealth. The last meeting was held in Abuja, Nigeria, in December 2003. The next meeting was scheduled to be held in Malta, in November 2005.

### OTHER CONSULTATIONS

Meetings at ministerial and official level are also held regularly. Since 1959 finance ministers have met in a Commonwealth country in the week prior to the annual meetings of the IMF and the World Bank. Meetings on education, legal, women's and youth affairs are held at ministerial level every three years. Ministers of health hold annual meetings, with major meetings every three years, and ministers of agriculture meet every two years. Ministers of finance, trade, labour and employment, industry, science, and the environment also hold periodic meetings. The first meeting of Commonwealth ministers of tourism was convened in March 2004.

Senior officials—cabinet secretaries, permanent secretaries to heads of government and others—meet regularly in the year between meetings of heads of government to provide continuity and to exchange views on various developments.

### COMMONWEALTH SECRETARIAT

The Secretariat, established by Commonwealth heads of government in 1965, operates as an international organization at the service of all Commonwealth countries. It organizes consultations between governments and runs programmes of co-operation. Meetings of heads of government, ministers and senior officials decide these programmes and provide overall direction. A Board of Governors, on which all eligible member governments are represented, meets annually to review the Secretariat's work and approve its budget. The Board is supported by an Executive Committee which convenes four times a year to monitor implementation of the Secretariat's work programme. The Secretariat is headed by a secretary-general, elected by heads of government.

In 2002 the Secretariat was restructured, with a view to strengthening the effectiveness of the organization to meet the priorities determined by the meeting of heads of government held in Coolum, Australia, in March 2002. Under the reorganization the number of deputy secretaries-general was reduced from three to two. Certain work divisions were amalgamated, while new units or sections, concerned with youth affairs, human rights and good offices, were created to strengthen further activities in those fields. Accordingly, the new divisional structure was as follows: Legal and constitutional affairs; Political affairs; Corporate services; Communications and public affairs; Strategic planning and evaluation; Economic affairs; Governance and institutional development; Social transformation programmes; and Special advisory services. In addition there were units responsible for human rights, youth affairs, and project management and referrals, and an Office of the Secretary-General. In 2004 the youth affairs unit acquired divisional status.

The Secretariat's strategic plan for 2004/05–2007/08 set out two main, long-term objectives for the Commonwealth: to support member countries in preventing or resolving conflicts, to strengthen democracy and the rule of law, and to achieve greater respect for human rights; and to support pro-poor policies for economic growth and sustainable development in member countries. Four programmes were to facilitate the pursuit of the first objective, 'Peace and Democracy': Good Offices for Peace; Democracy and Consensus Building; Rule of Law; and Human Rights. The second objective—'Pro-Poor Growth and Sustainable Development'—was to be achieved through the following nine programmes: International Trade; Investment; Finance and Debt; Public Sector Development; Environmentally Sustainable Development; Small States; Education; Health; and Young People.

**Secretary-General:** Rt Hon. DONALD (DON) C. MCKINNON (New Zealand).

**Deputy Secretaries-General:** FLORENCE MUGASHA (Uganda), WINSTON A. COX (Barbados).

## Activities

### INTERNATIONAL AFFAIRS

In October 1991 heads of government, meeting in Harare, Zimbabwe, issued the Harare Commonwealth Declaration, in which they reaffirmed their commitment to the Commonwealth Principles declared in 1971, and stressed the need to promote sustainable development and the alleviation of poverty. The Declaration placed emphasis on the promotion of democracy and respect for human rights and resolved to strengthen the Commonwealth's capacity to assist countries in entrenching democratic practices. In November 1995 Commonwealth heads of government, convened in New Zealand, formulated and adopted the Millbrook Commonwealth Action Programme on the Harare Declaration, to promote adherence by member countries to the fundamental principles of democracy and human rights (as proclaimed in the 1991 Declaration). The Programme incorporated a framework of measures to be pursued in support of democratic processes and institutions, and actions to be taken in response to violations of the Harare Declaration principles, in particular the unlawful removal of a democratically-elected government. A Commonwealth Ministerial Action Group on the Harare Declaration (CMAG) was to be established to implement this process and to assist the member country involved to comply with the

## REGIONAL ORGANIZATIONS
## The Commonwealth

Harare principles. On the basis of this Programme, the leaders suspended Nigeria from the Commonwealth with immediate effect, following the execution by that country's military Government of nine environmental and human rights protesters and a series of other violations of human rights. The meeting determined to expel Nigeria from the Commonwealth if no 'demonstrable progress' had been made towards the establishment of a democratic authority by the time of the next summit meeting. In addition, the Programme formulated measures to promote sustainable development in member countries, which was considered to be an important element in sustaining democracy, and to facilitate consensus-building within the international community.

In December 1995 CMAG convened for its inaugural meeting in London, United Kingdom. The Group, comprising the ministers of foreign affairs of Canada, Ghana, Jamaica, Malaysia, New Zealand, South Africa, the United Kingdom and Zimbabwe (with membership to be reconstituted periodically), commenced by considering efforts to restore democratic government in the three Commonwealth countries then under military regimes, i.e. The Gambia, Nigeria and Sierra Leone. At the second meeting of the Group, in April 1996, ministers commended the conduct of presidential and parliamentary elections in Sierra Leone and the announcement by The Gambia's military leaders that there would be a transition to civilian rule. In June a three-member CMAG delegation visited The Gambia to reaffirm Commonwealth support of the transition process in that country and to identify possible areas of further Commonwealth assistance. In August the Gambian authorities issued a decree removing the ban on political activities and parties, although shortly afterwards they prohibited certain parties and candidates involved in political life prior to the military take-over from contesting the elections. CMAG recommended that in such circumstances no Commonwealth observers should be sent to either the presidential or parliamentary elections, which were held in September 1996 and January 1997 respectively. Following the restoration of a civilian Government in early 1997, CMAG requested the Commonwealth Secretary-General to extend technical assistance to The Gambia in order to consolidate the democratic transition process. In April 1996 it was noted that the human rights situation in Nigeria had continued to deteriorate. CMAG, having pursued unsuccessful efforts to initiate dialogue with the Nigerian authorities, outlined a series of punitive and restrictive measures (including visa restrictions on members of the administration, a cessation of sporting contacts and an embargo on the export of armaments) that it would recommend for collective Commonwealth action in order to exert further pressure for reform in Nigeria. Following a meeting of a high-level delegation of the Nigerian Government and CMAG in June, the Group agreed to postpone the implementation of the sanctions, pending progress on the dialogue. (Canada, however, determined, unilaterally, to impose the measures with immediate effect; the United Kingdom did so in accordance with a decision of the European Union to implement limited sanctions against Nigeria.) A proposed CMAG mission to Nigeria was postponed in August, owing to restrictions imposed by the military authorities on access to political detainees and other civilian activists in that country. In September the Group agreed to proceed with the visit and to delay further a decision on the implementation of sanction measures. CMAG, without the participation of the representative of the Canadian Government, undertook its ministerial mission in November. In July 1997 the Group reiterated the Commonwealth Secretary-General's condemnation of a military coup in Sierra Leone in May, and decided that the country's participation in meetings of the Commonwealth should be suspended pending the restoration of a democratic government.

In October 1997 Commonwealth heads of government, meeting in Edinburgh, United Kingdom, endorsed CMAG's recommendation that the imposition of sanctions against Nigeria be held in abeyance pending the scheduled completion of a transition programme towards democracy by October 1998. It was also agreed that CMAG be formally constituted as a permanent organ to investigate abuses of human rights throughout the Commonwealth. Jamaica and South Africa were to be replaced as members of CMAG by Barbados and Botswana, respectively.

In March 1998 CMAG commended the efforts of the Economic Community of West African States (ECOWAS) in restoring the democratically-elected Government of President Ahmed Tejan Kabbah in Sierra Leone, and agreed to remove all restrictions on Sierra Leone's participation in Commonwealth activities. Later in that month, a representative mission of CMAG visited Sierra Leone to express its support for Kabbah's administration and to consider the country's needs in its process of reconstruction. At the CMAG meeting held in October members agreed that Sierra Leone should no longer be considered under the Group's mandate; however, they urged the Secretary-General to continue to assist that country in the process of national reconciliation and to facilitate negotiations with opposition forces to ensure a lasting cease-fire. A Special Envoy of the Secretary-General was appointed to co-operate with the UN, ECOWAS and the Organization of African Union (OAU, now African Union—AU) in monitoring the implementation of the Sierra Leone peace process, and the Commonwealth has supported the rebuilding of the Sierra Leone police force. In September 2001 CMAG recommended that Sierra Leone be removed from its remit, but that the Secretary-General should continue to monitor developments there.

In April 1998 the Nigerian military leader, Gen. Sani Abacha, confirmed his intention to conduct a presidential election in August, but indicated that, following an agreement with other political organizations, he was to be the sole candidate. In June, however, Abacha died suddenly. His successor, Gen. Abdulsalam Abubakar, immediately released several prominent political prisoners, and in early July agreed to meet with the Secretaries-General of the UN and the Commonwealth to discuss the release of the imprisoned opposition leader, Chief Moshood Abiola. Abubakar also confirmed his intention to abide by the programme for transition to civilian rule by October. In mid-July, however, shortly before he was to have been liberated, Abiola died. The Commonwealth Secretary-General subsequently endorsed a new transition programme, which provided for the election of a civilian leader in May 1999. In October 1998 CMAG, convened for its 10th formal meeting, acknowledged Abubakar's efforts towards restoring a democratic government and recommended that member states begin to remove sanctions against Nigeria and that it resume participation in certain Commonwealth activities. The Commonwealth Secretary-General subsequently announced a programme of technical assistance to support Nigeria in the planning and conduct of democratic elections. Staff teams from the Commonwealth Secretariat observed local government, and state and governorship elections, held in December and in January 1999, respectively. A Commonwealth Observer Group was also dispatched to Nigeria to monitor preparations and conduct of legislative and presidential elections, held in February. While the Group reported several irregularities in the conduct of the polling, it confirmed that, in general, the conditions had existed for free and fair elections and that the elections were a legitimate basis for the transition of power to a democratic, civilian government. In April CMAG voted to readmit Nigeria to full membership on 29 May, upon the installation of the new civilian administration.

In 1999 the Commonwealth Secretary-General appointed a Special Envoy to broker an agreement in order to end a civil dispute in Honiara, Solomon Islands. An accord was signed in late June, and it was envisaged that the Commonwealth would monitor its implementation. In October a Commonwealth Multinational Police Peace Monitoring Group was stationed in Solomon Islands; this was renamed the Commonwealth Multinational Police Assistance Group in February 2000. Following further internal unrest, however, the Group was disbanded. In June CMAG determined to send a new mission to Solomon Islands in order to facilitate negotiations between the opposing parties, to convey the Commonwealth's concern and to offer assistance. The Commonwealth welcomed the peace accord concluded in Solomon Islands in October, and extended its support to the International Peace Monitoring Team that was established to oversee implementation of the peace accords. CMAG welcomed the conduct of parliamentary elections held in Solomon Islands in December 2001. CMAG removed Solomon Islands from its agenda in December 2003 but was to continue to receive reports from the Secretary-General on future developments.

In June 1999 an agreement was concluded between opposing political groups in Zanzibar, having been facilitated by the good offices of the Secretary-General; however, this was only partially implemented. In mid-October a special meeting of CMAG was convened to consider the overthrow of the democratically-elected Government in Pakistan in a military coup. The meeting condemned the action as a violation of Commonwealth principles and urged the new authorities to declare a timetable for the return to democratic rule. CMAG also resolved to send a four-member delegation, comprising the ministers of foreign affairs of Barbados, Canada, Ghana and Malaysia, to discuss this future course of action with the military regime. Pakistan was suspended from participation in meetings of the Commonwealth with immediate effect. The suspension, pending the restoration of a democratic government, was endorsed by heads of government, meeting in November, who requested that CMAG keep the situation in Pakistan under review. At the meeting, held in Durban, South Africa, CMAG was reconstituted to comprise the ministers of foreign affairs of Australia, Bangladesh, Barbados, Botswana, Canada, Malaysia, Nigeria and the United Kingdom. It was agreed that no country would serve for more than two consecutive two-year terms. CMAG was requested to remain actively involved in the post-conflict development and rehabilitation of Sierra Leone and the process of consolidating peace. In addition, it was urged to monitor persistent violations of the Harare Declaration principles in all countries. Heads of government also agreed to establish a new ministerial group on Guyana and to reconvene a ministerial committee on Belize, in order to facilitate dialogue in ongoing territorial disputes with neighbouring countries. The meeting established a 10-member Commonwealth High Level Review Group to evaluate the role and activities of the Commonwealth. In 2000 the Group initiated a programme of consultations to proceed with its mandate and established a working group of experts to consider the Commonwealth's role in supporting information technology capabilities in member countries.

In June 2000, following the overthrow in May of the Fijian Government by a group of armed civilians, and the subsequent illegal detention of members of the elected administration, CMAG suspended Fiji's participation in meetings of the Commonwealth pending the restoration of democratic rule. In September, upon the request of CMAG, the Secretary-General appointed a Special Envoy to support efforts towards political dialogue and a return to democratic rule in Fiji. The Special Envoy undertook his first visit in December. In December 2001, following the staging of democratic legislative elections in August–September, Fiji was readmitted to Commonwealth meetings on the recommendation of CMAG. Fiji was removed from CMAG's agenda in May 2004, although the Group determined to continue to note developments there, as judgments were still pending in the Fiji Supreme Court on unresolved matters concerning the democratic process.

In March 2001 CMAG resolved to send a ministerial mission to Zimbabwe, in order to relay to the government the Commonwealth's concerns at the ongoing violence and abuses of human rights in that country, as well as to discuss the conduct of parliamentary elections and extend technical assistance. The mission was rejected by the Zimbabwe Government, which queried the basis for CMAG's intervention in the affairs of an elected administration. In September, under the auspices of a group of Commonwealth foreign ministers partly derived from CMAG, the Zimbabwe Government signed the Abuja Agreement, which provided for the cessation of illegal occupations of white-owned farms and the resumption of the rule of law, in return for financial assistance to support the ongoing process of land reform in that country. In January 2002 CMAG expressed strong concern at the continuing violence and political intimidation in Zimbabwe. The summit of Commonwealth heads of government convened in early March (see below) also expressed concern at the situation in Zimbabwe, and, having decided on the principle that CMAG should be permitted to engage with any member Government deemed to be in breach of the organization's core values, mandated a Commonwealth Chairperson's Committee on Zimbabwe to determine appropriate action should an impending presidential election (scheduled to be held during that month) be found not to have been conducted freely and fairly. Following the publication by a Commonwealth observer team of an unfavourable report on the conduct of the election, the Committee decided to suspend Zimbabwe from meetings of the Commonwealth for one year. In March 2003 the Committee concluded that the suspension should remain in force pending consideration by the next summit of heads of government.

In March 2002, meeting in Coolum, near Brisbane, Australia, Commonwealth heads of government adopted the Coolum Declaration on the Commonwealth in the 21st Century: Continuity and Renewal, which reiterated commitment to the organization's principles and values. Leaders at the meeting condemned all forms of terrorism; welcomed the Millennium Development Goals (MDGs) adopted by the UN General Assembly; called on the Secretary-General to constitute a high-level expert group on implementing the objectives of the Fancourt Declaration; pledged continued support for small states; and urged renewed efforts to combat the spread of HIV/AIDS. The meeting adopted a report on the future of the Commonwealth drafted by the High Level Review Group. The document recommended strengthening the Commonwealth's role in conflict prevention and resolution and support of democratic practices; enhancing the good offices role of the Secretary-General; better promoting member states' economic and development needs; strengthening the organization's role in facilitating member states' access to international assistance; and promoting increased access to modern information and communications technologies. The meeting expanded CMAG's mandate to enable the Group to consider action against serious violations of the Commonwealth's core values perpetrated by elected administrations (such as that in Zimbabwe, see above) as well as by military regimes. At the summit CMAG was reconstituted to comprise the ministers of foreign affairs of Australia, the Bahamas, Bangladesh, Botswana, India, Malta, Nigeria and Samoa.

A Commonwealth team of observers dispatched to monitor legislative and provincial elections that were held in Pakistan, in October 2002, found them to have been well-organized and conducted in a largely transparent manner. The team made several recommendations on institutional and procedural issues. CMAG subsequently expressed concern over the promulgation of new legislation in Pakistan following the imposition earlier in the year of a number of extra-constitutional measures. CMAG determined that Pakistan should continue to be suspended from meetings of the Commonwealth, pending a review of the role and functioning of its democratic institutions. Pakistan's progress in establishing democratic institutions was welcomed by a meeting of CMAG in May 2003. In November 2002 a Commonwealth Expert Group on Papua New Guinea, established in the previous month to review the electoral process in that country (in view of unsatisfactory legislative elections that were conducted there in July), made several recommendations aimed at enhancing the future management of the electoral process.

In December 2003 the meeting of heads of government, held in Abuja, Nigeria, resolved to maintain the suspension of Pakistan and Zimbabwe from participation in Commonwealth meetings. President Mugabe of Zimbabwe responded by announcing his country's immediate withdrawal from the Commonwealth and alleging a pro-Western bias within the grouping. Support for Zimbabwe's position was declared by a number of members, including South Africa, Mozambique, Namibia and Zambia. A Commonwealth committee, consisting of six heads of government, was established to monitor the situation in Zimbabwe and only when the committee believed sufficient progress had been made towards consolidating democracy and promoting development within Zimbabwe would the Commonwealth be consulted on readmitting the country. At the meeting CMAG was reconstituted to comprise the ministers of foreign affairs of the Bahamas, Canada, India, Lesotho, Malta, Samoa, Sri Lanka and Tanzania.

In concluding the 2003 meeting heads of government issued the Aso Rock Commonwealth Declaration, which emphasized their commitment to strengthening development and democracy, and incorporated clear objectives in support of these goals. Priority areas identified included efforts to eradicate poverty and attain the MDGs, strengthening democratic institutions, empowering women, promoting the involvement of civil society, combating corruption and recovering assets (for which a working group was to be established), facilitating finance for development, efforts to address the spread of HIV/AIDS and other diseases, combating the illicit trafficking in human beings, and promoting education. The leaders also adopted a separate statement on multilateral trade, in particular in support of the stalled Doha round of World Trade Organization negotiations.

In response to the earthquake and tsunami that devastated coastal areas of several Indian Ocean countries (including Bangladesh, India, Malaysia, the Maldives and Sri Lanka) in late December 2004, the Commonwealth Secretary-General appealed for assistance from Commonwealth Governments for the mobilization of emergency humanitarian relief. In early January 2005 the Secretariat dispatched a Disaster Relief Co-ordinator to the Maldives to assess the needs of that country and to co-ordinate ongoing relief and rehabilitation activities, and later in that month the Secretariat sent emergency medical doctors from other member states to the Maldives. In mid-January, meeting during the fifth Summit of the Alliance of Small Island States, in Port Louis, Mauritius, the Secretaries-General of the Commonwealth, the Caribbean Community and Common Market (CARICOM), the Pacific Islands Forum and the Indian Ocean Commission determined to take collective action to strengthen the disaster-preparedness and response capacities of their member countries in the Caribbean, Pacific and Indian Ocean areas.

In February 2005 CMAG expressed serious concern that President Musharraf of Pakistan had failed to relinquish the role of chief of army staff (at meetings held in May and September 2004 CMAG had urged the separation of the military and civilian offices held by the President, deeming this arrangement to be undemocratic). Noting President Musharraf's own undertaking not to continue as chief of army staff beyond 2007, the Group stated its view that the two offices should not be combined in one person beyond the end (in that year) of the current presidential term. CMAG recommended that the Secretary-General should maintain high-level contacts with Pakistan.

**Political Affairs Division:** assists consultation among member governments on international and Commonwealth matters of common interest. In association with host governments, it organizes the meetings of heads of government and senior officials. The Division services committees and special groups set up by heads of government dealing with political matters. The Secretariat has observer status at the United Nations, and the Division manages a joint office in New York to enable small states, which would otherwise be unable to afford facilities there, to maintain a presence at the United Nations. The Division monitors political developments in the Commonwealth and international progress in such matters as disarmament and the Law of the Sea. It also undertakes research on matters of common interest to member governments, and reports back to them. The Division is involved in diplomatic training and consular co-operation.

In 1990 Commonwealth heads of government mandated the Division to support the promotion of democracy by monitoring the preparations for and conduct of parliamentary, presidential or other elections in member countries at the request of national governments. During 2004 Commonwealth observer groups attended legislative elections in Antigua and Barbuda (in March); legislative elections in Sri Lanka (April); legislative and presidential elections in Malawi (May); local government elections in Sierra Leone (also May); a presidential election in Cameroon (October); legislative elections in Saint Christopher and Nevis (also October); and legislative and presidential elections in Mozambique (December). In 2005 an observer group was dispatched to observe legislative elections in the Maldives (January); local government elections in Lesotho (April); and, jointly with members of a Pacific Islands Forum as part of a wider international team, the first general election for the Autonomous Government of Bougainville (May–June).

Under the reorganization of the Secretariat in 2002 a Good Offices Section was established within the Division to strengthen and support the activities of the Secretary-General in addressing political conflict in member states and in assisting countries to adhere to the principles of the Harare Declaration. The Secretary-General's good offices may be directed to preventing or resolving conflict and assisting other international efforts to promote political stability.

**Human Rights Unit:** undertakes activities in support of the Commonwealth's commitment to the promotion and protection of fundamental human rights. It develops programmes, publishes human rights materials, co-operates with other organizations working in the field of human rights, in particular within the UN system, advises the Secretary-General, and organizes seminars and meetings of experts. The Unit aims to integrate human rights standards within all divisions of the Secretariat.

## LAW

**Legal and Constitutional Affairs Division:** promotes and facilitates co-operation and the exchange of information among member governments on legal matters and assists in combating financial and organized crime, in particular transborder criminal activities. It administers, jointly with the Commonwealth of Learning, a distance training programme for legislative draftsmen and assists governments to reform national laws to meet the obligations of international conventions. The Division organizes the triennial meeting of ministers, Attorneys General and senior ministry officials concerned with the legal systems in Commonwealth countries. It has also initiated four Commonwealth schemes for co-operation on extradition, the protection of material cultural heritage, mutual assistance in criminal matters and the transfer of convicted offenders within the Commonwealth. It liaises with the Commonwealth Magistrates' and Judges' Association, the Commonwealth Legal Education Association, the Commonwealth Lawyers' Association (with which it helps to prepare the triennial Commonwealth Law Conference for the practising profession), the Commonwealth Association of Legislative Counsel, and with other international non-governmental organizations. The Division provides in-house legal advice for the Secretariat. The *Commonwealth Law Bulletin*, published four times a year, reports on legal developments in and beyond the Commonwealth. The Division promotes the exchange of information regarding national and international efforts to combat serious commercial crime through its other publications, *Commonwealth Legal Assistance News* and *Crimewatch*.

The heads of government meeting held in Coolum, Australia, in March 2002 endorsed a Plan of Action for combating international terrorism. A Commonwealth Committee on Terrorism, convened at ministerial level, was subsequently established to oversee its implementation.

A new expert group on good governance and the elimination of corruption in economic management convened for its first meeting in May 1998. In November 1999 Commonwealth heads of government endorsed a Framework for Principles for Promoting Good Governance and Combating Corruption, which had been drafted by the group. The conference of heads of government that met in Coolum in March 2002 endorsed a Commonwealth Local Government Good Practice Scheme, to be managed by the Commonwealth Local Government Forum (established in 1995).

## ECONOMIC CO-OPERATION

In October 1997 Commonwealth heads of government, meeting in Edinburgh, United Kingdom, signed an Economic Declaration that focused on issues relating to global trade, investment and development and committed all member countries to free-market economic principles. The Declaration also incorporated a provision for the establishment of a Trade and Investment Access Facility within the Secretariat in order to assist developing member states in the process of international trade liberalization and promote intra-Commonwealth trade.

In May 1998 the Commonwealth Secretary-General appealed to the Group of Eight industrialized nations (G-8) to accelerate and expand the initiative to ease the debt burden of the most heavily indebted poor countries (HIPCs—see World Bank and IMF). In October Commonwealth finance ministers, convened in Ottawa, Canada, reiterated their appeal to international financial institutions to accelerate the HIPC initiative. The meeting also issued a Commonwealth Statement on the global economic crisis and endorsed proposals to help to counter the difficulties experienced by several countries. These measures included a mechanism to enable countries to suspend payments on all short-term financial obligations at a time of emergency without defaulting, assistance to governments to attract private capital and to manage capital market volatility, and the development of international codes of conduct regarding financial and monetary policies and corporate governance. In March 1999 the Commonwealth Secretariat hosted a joint IMF-World Bank conference to review the HIPC scheme and initiate a process of reform. In November Commonwealth heads of government, meeting in South Africa, declared their support for measures undertaken by the World Bank and IMF to enhance the HIPC initiative. At the end of an informal retreat the leaders adopted the Fancourt Commonwealth Declaration on Globalization and People-Centred Development, which emphasized the need for a more equitable spread of wealth generated by the process of globalization, and expressed a renewed commitment to the elimination of all forms of discrimination, the promotion of people-centred development and capacity-building, and efforts to ensure that developing countries benefit from future multilateral trade liberalization measures. In June 2002 the Commonwealth Secretary-General urged more generous funding of the HIPC initiative. Meetings of ministers of finance from African Commonwealth member countries participating in the HIPC initiative are convened periodically (most recently in March 2005). The Secretariat aims to assist HIPCs and other small economies through its Debt Recording and Management System, which was first used in 1985 and updated in 2002. In July 2005 the Commonwealth Secretary-General welcomed an initiative of the G-8 to eliminate the debt of those HIPCs which had reached their completion point in the process, in addition to a commitment substantially to increase aid to Africa.

In February 1998 the Commonwealth Secretariat hosted the first Inter-Governmental Organizations Meeting to promote co-operation between small island states and the formulation of a unified policy approach to international fora. A second meeting was convened in March 2001, where discussions focused on the forthcoming WTO ministerial meeting and OECD's Harmful Tax Competition Initiative. In September 2000 Commonwealth finance ministers, meeting in Malta, reviewed the OECD initiative and agreed that the measures, affecting many member countries with offshore financial centres, should not be imposed on governments. The ministers mandated the involvement of the Commonwealth Secretariat in efforts to resolve the dispute; a joint working group was subsequently established by the Secretariat with the OECD. In April 2002 a meeting on international co-operation in the financial services sector, attended by representatives of international and regional organizations, donors and senior officials from Commonwealth countries, was held under Commonwealth auspices in Saint Lucia.

The first meeting of governors of central banks from Commonwealth countries was held in June 2001 in London, United Kingdom.

The Commonwealth Secretariat was to contribute funding of €4m. towards the €20m. 'Hub and Spokes' project, launched in October 2004 by the European Commission as a capacity-building initiative in the areas of trade policy formulation, mainstreaming trade in poverty reduction strategies, and participation in international trade negotiations for the African, Caribbean and Pacific (ACP) group of countries. The Secretariat was to manage the project in 55 of the 78 ACP member states.

**Economic Affairs Division:** organizes and services the annual meetings of Commonwealth ministers of finance and the ministerial group on small states and assists in servicing the biennial meetings of heads of government and periodic meetings of environment ministers. It engages in research and analysis on economic issues of interest to member governments and organizes seminars and conferences of government officials and experts. The Division undertook a major programme of technical assistance to enable developing Commonwealth countries to participate in the Uruguay Round of multilateral trade negotiations and has assisted the African, Caribbean and Pacific (ACP) group of countries in their trade negotiations with the European Union. It continues to help developing countries to strengthen their links with international capital markets and foreign investors. The Division also services groups of experts on economic affairs that have been commissioned by governments to report on, among other things, protectionism; obstacles to the North-South negotiating process; reform of the international financial and trading system; the debt crisis; management of technological change; the impact of change on the development process; environmental issues; women and structural adjustment; and youth unemployment. A separate section within the Division addresses the specific needs of small states and provides technical assistance. The work of the section covers a range of issues including trade, vulnerability, environment, politics and economics. A Secretariat Task Force services a Commonwealth Ministerial Group of Small States which was established in 1993 to provide strategic direction in addressing the concerns of small states and to mobilize support for action and assistance within the international community. The Economic Affairs Division also co-ordinates the Secretariat's environmental work and manages the Iwokrama International Centre for Rainforest Conservation and Development.

The Division played a catalytic role in the establishment of a Commonwealth Equity Fund, initiated in September 1990, to allow developing member countries to improve their access to private institutional investment, and promoted a Caribbean Investment Fund. The Division supported the establishment of a Commonwealth Private Investment Initiative (CPII) to mobilize capital, on a regional basis, for investment in newly-privatized companies and in small and medium-sized businesses in the private sector. The first regional fund under the CPII was launched in July 1996. The Commonwealth Africa Investment Fund (Comafin), was to be managed by the United Kingdom's official development institution, the Commonwealth Development Corporation, to assist businesses in 19 countries in sub-Saharan Africa, with initial resources of US $63.5m. In August 1997 a fund for the Pacific Islands was launched, with an initial capital of $15.0m. A $200m. South Asia Regional Fund was established at the heads of government meeting in October. In October 1998 a fund for the Caribbean states was inaugurated, at a meeting of Commonwealth finance ministers. The 2001 summit of Commonwealth heads of government authorized the establishment of a new fund for Africa (Comafin II): this was inaugurated in March 2002, and attracted initial capital in excess of $200m.

### SOCIAL WELFARE

**Social Transformation Programmes Division:** consists of three sections concerned with education, gender and health.

The **Education Section** arranges specialist seminars, workshops and co-operative projects, and commissions studies in areas identified by ministers of education, whose three-yearly meetings it also services. Its present areas of emphasis include improving the quality of and access to basic education; strengthening the culture of science, technology and mathematics education in formal and non-formal areas of education; improving the quality of management in institutions of higher learning and basic education; improving the performance of teachers; strengthening examination assessment systems; and promoting the movement of students between Commonwealth countries. The Section also promotes multi-sectoral strategies to be incorporated in the development of human resources. Emphasis is placed on ensuring a gender balance, the appropriate use of technology, promoting good governance, addressing the problems of scale particular to smaller member countries, and encouraging collaboration between governments, the private sector and other non-governmental organizations.

The **Gender Affairs Section** is responsible for the implementation of the Commonwealth Plan of Action for Gender Equality, covering the period 2005–15, which succeeded the Commonwealth Plan of Action on Gender and Development (adopted in 1995 and updated in 2000). The Plan of Action supports efforts towards achieving the MDGs, and the objectives of gender equality adopted by the 1995 Beijing Declaration and Platform for Action and the follow-up Beijing + 5 review conference held in 2000. Gender equality, poverty eradication, promotion of human rights, and strengthening democracy are recognized as intrinsically inter-related, and the Plan has a particular focus on the advancement of gender mainstreaming in the following areas: democracy, peace and conflict; human rights and law; poverty eradication and economic empowerment; and HIV/AIDS. In February–March 2005 Commonwealth ministers responsible for gender affairs attended the Beijing + 10 review conference.

The **Health Section** organizes ministerial, technical and expert group meetings and workshops, to promote co-operation on health matters, and the exchange of health information and expertise. The Section commissions relevant studies and provides professional and technical advice to member countries and to the Secretariat. It also supports the work of regional health organizations and promotes health for all people in Commonwealth countries.

**Youth Affairs:** A Youth Affairs unit, reporting directly to a Deputy Secretary-General, was established within the Secretariat in 2002. The unit acquired divisional status in 2004.

The Division administers the **Commonwealth Youth Programme (CYP)**, which was initiated in 1973 to promote the involvement of young people in the economic and social development of their countries. The CYP, funded through separate voluntary contributions from governments, was awarded a budget of £2.5m. for 2005/06. The Programme's activities are centred on four key programmes: Youth Enterprise Development; Youth Networks and Governance; Youth Participation; and Youth Work, Education and Training. Regional centres are located in Zambia (for Africa), India (for Asia), Guyana (for the Caribbean), and Solomon Islands (for the South Pacific). The Programme administers a Youth Study Fellowship scheme, a Youth Project Fund, a Youth Exchange Programme (in the Caribbean), and a Youth Service Awards Scheme. It also holds conferences and seminars, carries out research and disseminates information. The Commonwealth Youth Credit Initiative, launched in 1995, provides funds, training and advice to young entrepreneurs. A Plan of Action on Youth Empowerment to the Year 2005 was approved by a Commonwealth ministerial meeting held in Kuala Lumpur, Malaysia, in May 1998.

In March 2002 Commonwealth heads of government approved the Youth for the Future initiative to encourage and use the skills of young people throughout the Commonwealth. It was to comprise four main components: Youth enterprise development; Youth volunteers; Youth mentors; and Youth leadership awards.

### TECHNICAL ASSISTANCE

**Commonwealth Fund for Technical Co-operation (CFTC):** f. 1971 to facilitate the exchange of skills between member countries and to promote economic and social development; it is administered by the Commonwealth Secretariat and financed by voluntary subscriptions from member governments. The CFTC responds to requests from member governments for technical assistance, such as the provision of experts for short- or medium-term projects, advice on economic or legal matters, in particular in the areas of natural resources management and public-sector reform, and training programmes. The CFTC also administers the Langkawi awards for the study of environmental issues, which is funded by the Canadian Government; the CFTC budget for 2005/06 amounted to £24.1m.

CFTC activities are mainly implemented by the following divisions:

**Governance and Institutional Development Division:** strengthens good governance in member countries, through advice, training and other expertise in order to build capacity in national public institutions. The Division administers the Commonwealth Service Abroad Programme (CSAP), which is funded by the CFTC. The Programme extends short-term technical assistance through highly qualified volunteers. The main objectives of the scheme are to provide expertise, training and exposure to new technologies and practices, to promote technology transfers and sharing of experiences and knowledge, and to support community workshops and other grassroots activities.

**Special Advisory Services Division:** advises on economic and legal issues, such as debt and financial management, natural resource development, multilateral trade issues, export marketing, trade facilitation, competitiveness and the development of enterprises.

## Finance

The Secretariat's budget for 2005/06 was £12.4m. Member governments meet the cost of the Secretariat through subscriptions on a scale related to income and population.

## Publications

*Commonwealth Currents* (quarterly).

*International Development Policies* (quarterly).

*Report of the Commonwealth Secretary-General* (every 2 years).

Numerous reports, studies and papers (catalogue available).

## Commonwealth Organizations

(In the United Kingdom, unless otherwise stated)

### PRINCIPAL BODIES

**Commonwealth Business Council:** 18 Pall Mall, London, SW1Y 5LU; tel. (20) 7024-8200; fax (20) 7930-3944; e-mail info@cbcglobelink.org; internet www.cbcglobelink.org; f. 1997 by the Commonwealth Heads of Government Meeting to promote co-operation between governments and the private sector in support of trade, investment and development; the Council aims to identify and promote investment opportunities, in particular in Commonwealth developing countries, to support countries and local businesses to work within the context of globalization, to promote capacity-building and the exchange of skills and knowledge (in particular through its Information Communication Technologies for Development programme), and to encourage co-operation among Commonwealth members; promotes good governance; supports the process of multilateral trade negotiations and other liberalization of trade and services; represents the private sector at government level; Dir-Gen. and CEO Dr MOHAN KAUL.

# REGIONAL ORGANIZATIONS

**Commonwealth Foundation:** Marlborough House, Pall Mall, London, SW1Y 5HY; tel. (20) 7930-3783; fax (20) 7839-8157; e-mail geninfo@commonwealth.int; internet www.commonwealthfoundation.com; f. 1966; intergovernmental body promoting people-to-people interaction, and collaboration within the non-governmental sector of the Commonwealth; supports non-governmental organizations, professional associations and Commonwealth arts and culture; awards an annual Commonwealth Writers' Prize; funds are provided by Commonwealth govts; Chair. Prof. GUIDO DE MARCO (Mozambique); Dir Dr MARK COLLINS (United Kingdom); publ. *Commonwealth People* (quarterly).

**Commonwealth of Learning (COL):** 1055 West Hastings St, Suite 1200, Vancouver, BC V6E 2E9, Canada; tel. (604) 775-8200; fax (604) 775-8210; e-mail info@col.org; internet www.col.org; f. 1987 by Commonwealth Heads of Government to promote the devt and sharing of distance education and open learning resources, including materials, expertise and technologies, throughout the Commonwealth and in other countries; implements and assists with national and regional educational programmes; acts as consultant to international agencies and national governments; conducts seminars and studies on specific educational needs; COL is financed by Commonwealth governments on a voluntary basis; in 1999 heads of government endorsed an annual core budget for COL of US $9m; Pres. and CEO Sir JOHN DANIEL (Canada/UK); publs *Connections, EdTech News*.

The following represents a selection of other Commonwealth organizations:

## AGRICULTURE AND FORESTRY

**Commonwealth Forestry Association:** 2 Webbs Barn Cottage, Witney Rd, Kingston Bagpuize, Abingdon, Oxon, OX13 5AN; tel. (1865) 820935; fax (870) 0116645; e-mail cfa@cfa-international.org; internet www.cfa-international.org; f. 1921; produces, collects and circulates information relating to world forestry and promotes good management, use and conservation of forests and forest lands throughout the world; mems: 1,200; Pres. DAVID BILLS (Australia/UK); publs *International Forestry Review* (quarterly), *Commonwealth Forestry News* (quarterly), *Commonwealth Forestry Handbook* (irregular).

**Standing Committee on Commonwealth Forestry:** Forestry Commission, 231 Corstorphine Rd, Edinburgh, EH12 7AT; tel. (131) 314-6137; fax (131) 316-4344; e-mail libby.jones@forestry.gsi.gov.uk; f. 1923 to provide continuity between Confs, and to provide a forum for discussion on any forestry matters of common interest to mem. govts which may be brought to the Cttee's notice by any mem. country or organization; 54 mems; 2005 Conference: Sri Lanka; Sec. LIBBY JONES; publ. *Newsletter* (quarterly).

## COMMONWEALTH STUDIES

**Institute of Commonwealth Studies:** 28 Russell Sq., London, WC1B 5DS; tel. (20) 7862-8844; fax (20) 7862-8820; e-mail ics@sas.ac.uk; internet www.sas.ac.uk/commonwealthstudies; f. 1949 to promote advanced study of the Commonwealth; provides a library and meeting place for postgraduate students and academic staff engaged in research in this field; offers postgraduate teaching; Dir Prof. TIMOTHY SHAW; publs *Annual Report, Collected Seminar Papers, Newsletter, Theses in Progress in Commonwealth Studies*.

## COMMUNICATIONS

**Commonwealth Telecommunications Organization:** Clareville House, 26–27 Oxendon St, London, SW1Y 4EL; tel. (20) 7930-5511; fax (20) 7930-4248; e-mail info@cto.int; internet www.cto.int; f. 1967 as an international development partnership between Commonwealth and non-Commonwealth governments, business and civil society organizations; aims to help to bridge the digital divide and to achieve social and economic development by delivering to developing countries knowledge-sharing programmes in the use of information and communication technologies (ICT) in the specific areas of telecommunications, IT, broadcasting and the internet; CEO Dr EKWOW SPIO-GARBRAH; publs *CTO Update* (quarterly), *Annual Report, Research Reports*.

## EDUCATION AND CULTURE

**Association of Commonwealth Universities (ACU):** John Foster House, 36 Gordon Sq., London, WC1H 0PF; tel. (20) 7380-6700; fax (20) 7387-2655; e-mail info@acu.ac.uk; internet www.acu.ac.uk; f. 1913; promotes international co-operation and understanding; provides assistance with staff and student mobility and development programmes; researches and disseminates information about universities and relevant policy issues; organizes major meetings of Commonwealth universities and their representatives; acts as a liaison office and information centre; administers scholarship and fellowship schemes; operates a policy research unit; mems: c. 500 universities in 36 Commonwealth countries or regions; Sec.-Gen. Dr JOHN ROWETT; publs include *Yearly Review, Commonwealth Universities Yearbook, ACU Bulletin* (quarterly), *Report of the Council of the ACU* (annually), *Who's Who of Executive Heads: Vice-Chancellors, Presidents, Principals and Rectors, International Awards*, student information papers (study abroad series).

**Commonwealth Association for Education in Journalism and Communication (CAEJAC):** c/o Faculty of Law, University of Western Ontario, London, ON N6A 3K7, Canada; tel. (519) 661-3348; fax (519) 661-3790; e-mail caejc@julian.uwo.ca; f. 1985; aims to foster high standards of journalism and communication education and research in Commonwealth countries and to promote co-operation among institutions and professions; c. 700 mems in 32 Commonwealth countries; Pres. Prof. SYED ARABI IDID (Malaysia); Sec. Prof. ROBERT MARTIN (Canada); publ. *CAEJAC Journal* (annually).

**Commonwealth Association of Science, Technology and Mathematics Educators (CASTME):** 7 Lion Yard, Tremadoc Rd, London, SW4 7NQ; tel. (20) 7819-3932; fax (20) 7720-5403; e-mail ann.powell@lect.org.uk; internet www.castme.org; f. 1974; special emphasis is given to the social significance of education in these subjects; organizes an Awards Scheme to promote effective teaching and learning in these subjects, and biennial regional seminars; Hon. Sec. Dr LYN HAINES; publ. *CASTME Journal* (quarterly).

**Commonwealth Council for Educational Administration and Management:** Department of Education, University of Cyprus, POB 20537, 1678 Lefkosia, Cyprus; tel. 22753739; fax 22377950; e-mail edpetros@ucy.ac.cy; internet www.cceam.org; f. 1970; aims to foster quality in professional development and links among educational administrators; holds nat. and regional confs, as well as visits and seminars; mems: 24 affiliated groups representing 3,000 persons; Pres. Dr PETROS PASHIARDIS; publ. *International Studies in Educational Administration* (2 a year).

**Commonwealth Institute:** New Zealand House, 80 Haymarket, London, SW1Y 4TQ; tel. (20) 7024-9822; fax (20) 7024-9833; e-mail information@commonwealth-institute.org; internet www.commonwealth.org.uk; f. 1893 as the Imperial Institute; restructured as an independent pan-Commonwealth agency Jan. 2000; governed by a Bd of Trustees elected by the Bd of Governors; Commonwealth High Commissioners to the United Kingdom act as *ex-officio* Governors; the Inst. houses a Commonwealth Resource and Literature Library and a Conference and Events Centre; supplies educational resource materials and training throughout the United Kingdom; provides internet services to the Commonwealth; operates as an arts and conference centre, running a Commonwealth-based cultural programme; a five-year strategic plan, entitled 'Commonwealth 21', was inaugurated in 1998; in 2004 the Institute, in collaboration with Cambridge University, established a Centre of Commonwealth Education; Chair. JUDITH HANRATTY; Chief Exec. DAVID FRENCH; publ. *Annual Review*.

**League for the Exchange of Commonwealth Teachers:** 7 Lion Yard, Tremadoc Rd, London, SW4 7NQ; tel. (0870) 770-2636; fax (0870) 770-2637; e-mail info@lect.org.uk; internet www.lect.org.uk; f. 1901; promotes educational exchanges between teachers throughout the Commonwealth; Dir ANNA TOMLINSON; publs *Annual Review* (annually).

## HEALTH

**Commonwealth Medical Trust (COMMAT):** BMA House, Tavistock Sq., London, WC1H 9JP; tel. (20) 7272-8492; fax (20) 7272-8569; e-mail office@commat.org; internet www.commat.org; f. 1962 (as the Commonwealth Medical Association) for the exchange of information; provision of tech. co-operation and advice; formulation and maintenance of a code of ethics; promotes the Right to Health; liaison with WHO and other UN agencies on health issues; meetings of its Council are held every three years; mems: medical asscns in Commonwealth countries; Dir MARIANNE HASLEGRAVE.

**Commonwealth Pharmaceutical Association:** 1 Lambeth High St, London, SE1 7JN; tel. (20) 7572-2364; fax (20) 7572-2508; e-mail admin@commonwealthpharmacy.org; internet www.commonwealthpharmacy.org; f. 1970 to promote the interests of pharmaceutical sciences and the profession of pharmacy in the Commonwealth to maintain high professional standards, encourage links between members and the creation of nat. asscns; and to facilitate the dissemination of information; holds confs (every four years) and regional meetings; mems: pharmaceutical asscns from over 40 Commonwealth countries; Pres. GRACE ALLEN YOUNG; publ. *Quarterly Newsletter*.

**Commonwealth Society for the Deaf:** 34 Buckingham Palace Rd, London, SW1W 0RE; tel. (20) 7233-5700; fax (20) 7233-5800; e-mail sound.seekers@btinternet.com; internet www.sound-seekers.org.uk; undertakes initiatives to establish audiology services in developing Commonwealth countries, including mobile clinics to provide outreach services; aims to educate local communities in aural hygiene and the prevention of ear infection and deafness; provides audiological equipment and organizes the training of audiological maintenance technicians; conducts research into the causes and prevention of deafness; Chief Exec. GARY WILLIAMS; publ. *Annual Report.*

**Sight Savers International:** Grosvenor Hall, Bolnore Rd, Haywards Heath, West Sussex, RH16 4BX; tel. (1444) 446600; fax (1444) 446688; e-mail generalinformation@sightsavers.org.uk; internet www.sightsavers.org; f. 1950 to prevent blindness and restore sight in developing countries, and to provide education and community-based training for incurably blind people; operates in collaboration with local partners, with high priority given to training local staff; Chair. Sir JOHN COLES; Chief Exec. Dr CAROLINE HARPER; publ. *Sight Savers News.*

## INFORMATION AND THE MEDIA

**Commonwealth Broadcasting Association:** 17 Fleet St, London, EC4Y 1AA; tel. (20) 7583-5550; fax (20) 7583-5549; e-mail cba@cba.org.uk; internet www.cba.org.uk; f. 1945; gen. confs are held every two years (2006: India); mems: c. 100 in more than 50 countries; Pres. GEORGE VALARINO; Sec.-Gen. ELIZABETH SMITH; publs *Commonwealth Broadcaster* (quarterly), *Commonwealth Broadcaster Directory* (annually).

**Commonwealth Institute:** see under Education and Culture.

**Commonwealth Journalists Association:** 305 Goodwood Heights, Diego Martin, Trinidad and Tobago; tel. (868) 633-3397; fax (868) 633-0152; f. 1978 to promote co-operation between journalists in Commonwealth countries, organize training facilities and confs, and foster understanding among Commonwealth peoples; Exec. Dir JOSANNE LEONARD; publ. *Newsletter* (3 a year).

**Commonwealth Press Union** (Association of Commonwealth Newspapers, News Agencies and Periodicals): 17 Fleet St, London, EC4Y 1AA; tel. (20) 7583-7733; fax (20) 7583-6868; e-mail lindsay@cpu.org.uk; internet www.cpu.org.uk; f. 1950; promotes the welfare of the Commonwealth press; provides training for journalists and organizes biennial confs; mems: c. 750 newspapers, news agencies, periodicals in 49 Commonwealth countries; Exec. Dir LINDSAY ROSS; publ. *Annual Report.*

## LAW

**Commonwealth Lawyers' Association:** c/o Institute of Commonwealth Studies, 28 Russell Sq., London, WC1B 5DS; tel. (20) 7862-8824; fax (20) 7862-8816; e-mail cla@sas.ac.uk; internet www.commonwealthlawyers.com; f. 1983 (fmrly the Commonwealth Legal Bureau); seeks to maintain and promote the rule of law throughout the Commonwealth, by ensuring that the people of the Commonwealth are served by an independent and efficient legal profession; upholds professional standards and promotes the availability of legal services; assists in organizing the triennial Commonwealth law confs; Exec. Sec. CLAIRE MARTIN; publs *The Commonwealth Lawyer*, *Clarion.*

**Commonwealth Legal Advisory Service:** c/o British Institute of International and Comparative Law, Charles Clore House, 17 Russell Sq., London, WC1B 5DR; tel. (20) 7636-5802; fax (20) 7323-2016; e-mail bicl@dial.pipex.com; f. 1962; financed by the British Institute and by contributions from Commonwealth govts; provides research facilities for Commonwealth govts and law reform commissions; Chair. Rt Hon. Lord BROWNE-WILKINSON; publ. *New Memoranda* series.

**Commonwealth Legal Education Association:** c/o Legal and Constitutional Affairs Division, Commonwealth Secretariat, Marlborough House, Pall Mall, London, SW1Y 5HX; tel. (20) 7747-6415; fax (20) 7747-6406; e-mail clea@commonwealth.int; internet www.clea.org.uk; f. 1971 to promote contacts and exchanges and to provide information regarding legal education; Gen. Sec. JOHN HATCHARD; publs *Commonwealth Legal Education Association Newsletter* (3 a year), *Directory of Commonwealth Law Schools* (every 2 years).

**Commonwealth Magistrates' and Judges' Association:** Uganda House, 58–59 Trafalgar Sq., London, WC2N 5DX; tel. (20) 7976-1007; fax (20) 7976-2394; e-mail info@cmja.org; internet www.cmja.org; f. 1970 to advance the administration of the law by promoting the independence of the judiciary, to further education in law and crime prevention and to disseminate information; confs and study tours; corporate membership for asscns of the judiciary or courts of limited jurisdiction; assoc. membership for individuals; Pres. Rt Hon. Lord HOPE OF CRAIGHEAD; Exec. Vice-Pres. MICHAEL A. LAMBERT; publs *Commonwealth Judicial Journal* (2 a year), *CMJA News.*

## PARLIAMENTARY AFFAIRS

**Commonwealth Parliamentary Association:** Westminster House, Suite 700, 7 Millbank, London, SW1P 3JA; tel. (20) 7799-1460; fax (20) 7222-6073; e-mail pirc@cpahq.org; internet www.cpahq.org; f. 1911 to promote understanding and co-operation between Commonwealth parliamentarians; organization: Exec. Cttee of 35 MPs responsible to annual Gen. Assembly; 170 brs in national, state, provincial and territorial parliaments and legislatures throughout the Commonwealth; holds annual Commonwealth Parliamentary Confs and seminars; also regional confs and seminars; Pres. Hon. Ratu EPELI NAILATIKAU; publ. *The Parliamentarian* (quarterly).

## PROFESSIONAL AND INDUSTRIAL RELATIONS

**Commonwealth Association of Architects:** POB 508, Edgware, HA8 9XZ; tel. (20) 8951-0550; fax (20) 8951-0550; e-mail info@comarchitect.org; internet comarchitect.org; f. 1964; an asscn of 39 socs of architects in various Commonwealth countries; objectives: to facilitate the reciprocal recognition of professional qualifications; to provide a clearing house for information on architectural practice; and to encourage collaboration. Plenary confs every three years; regional confs are also held; Exec. Dir TONY GODWIN; publs *Handbook*, *Objectives and Procedures: CAA Schools Visiting Boards*, *Architectural Education in the Commonwealth* (annotated bibliography of research), *CAA Newsnet* (2 a year), a survey and list of schools of architecture.

**Commonwealth Association for Public Administration and Management (CAPAM):** 1075 Bay St, Suite 402, Toronto, ON M5S 2B1, Canada; tel. (416) 920-3337; fax (416) 920-6574; e-mail capam@capam.ca; internet www.capam.org; f. 1994; aims to promote sound management of the public sector in Commonwealth countries and to assist those countries undergoing political or financial reforms; an international awards programme to reward innovation within the public sector was introduced in 1997, and is awarded every 2 years; more than 1,200 individual mems and 80 institutional memberships in some 80 countries; Pres. Hon. JOCELYNE BOURGON (Canada); Exec. Dir GILLIAN MASON (Canada).

## SCIENCE AND TECHNOLOGY

**Commonwealth Engineers' Council:** c/o Institution of Civil Engineers, One Great George St, London, SW1P 3AA; tel. (20) 7222-7722; fax (20) 7223-1806; e-mail international@ice.org.uk; internet www.ice.org.uk/cec; f. 1946; the Council meets every two years to provide an opportunity for engineering institutions of Commonwealth countries to exchange views on collaboration; Pres. Prof. TONY RIDLEY; Sec.-Gen. TOM FOULKES.

**Commonwealth Geological Surveys Forum:** c/o Commonwealth Science Council, CSC Earth Sciences Programme, Marlborough House, Pall Mall, London, SW1Y 5HX; tel. (20) 7839-3411; fax (20) 7839-6174; e-mail comsci@gn.apc.org; f. 1948 to promote collaboration in geological, geochemical, geophysical and remote sensing techniques and the exchange of information; Geological Programme Officer Dr SIYAN MALOMO.

## SPORT

**Commonwealth Games Federation:** 2nd Floor, 138 Piccadilly, London, W1J 7NR; tel. (20) 7491-8801; fax (20) 7409-7803; e-mail info@thecgf.com; internet www.thecgf.com; the Games were first held in 1930 and are now held every four years; participation is limited to competitors representing the mem. countries of the Commonwealth; 2006 games: Melbourne, Australia, in March; mems: 72 affiliated bodies; Pres. MICHAEL FENNELL; CEO MICHAEL HOOPER.

## YOUTH

**Commonwealth Youth Exchange Council:** 7 Lion Yard, Tremadoc Rd, London, SW4 7NQ; tel. (20) 7498-6151; fax (20) 7622-4365; e-mail mail@cyec.demon.co.uk; internet www.cyec.org.uk; f. 1970; promotes contact between groups of young people of the United Kingdom and other Commonwealth countries by means of educational exchange visits, provides information for organizers and allocates grants; provides host governments with technical assistance for delivery of the Commonwealth Youth Forum, held every two years (2005: Malta); 222 mem. orgs; Chief Exec. V. S. G. CRAGGS; publs *Contact* (handbook), *Exchange* (newsletter), *Final Communiqués* (of the Commonwealth Youth Forums), *Safety and Welfare* (guide-lines for Commonwealth Youth Exchange groups).

**Duke of Edinburgh's Award International Association:** Award House, 7–11 St Matthew St, London, SW1P 2JT; tel. (20) 7222-4242; fax (20) 7222-4141; e-mail sect@intaward.org; internet www.intaward.org; f. 1956; offers a programme of leisure activities for young people, comprising Service, Expeditions, Physical Recreation, and Skills; operates in more than 60 countries (not confined to the

Commonwealth); International Sec.-Gen. DAVID MANSON; publs *Award World* (2 a year), *Annual Report*, handbooks and guides.

### MISCELLANEOUS

**Commonwealth Countries League:** 7 The Park, London, NW11 7SS; tel. (20) 8451-6711; e-mail info@ccl-int.org.uk; internet www.ccl-int.org.uk; f. 1925 to secure equal opportunities and status between men and women in the Commonwealth, to act as a link between Commonwealth women's orgs, and to promote and finance secondary education of disadvantaged girls of high ability in their own countries, through the CCL Educational Fund; holds meetings with speakers and an annual conf., organizes the annual Commonwealth Fair for fund-raising; individual mems and affiliated socs in the Commonwealth; Hon. Sec. RUTH E. WHITEHOUSE; publs *CCL Newsletter* (3 a year), *Annual Report*.

**Commonwealth War Graves Commission:** 2 Marlow Rd, Maidenhead, Berks, SL6 7DX; tel. (1628) 634221; fax (1628) 771208; internet www.cwgc.org; casualty and cemetery enquiries e-mail casualty.enq@cwgc.org; f. 1917 (as Imperial War Graves Commission); responsible for the commemoration in perpetuity of the 1.7m. members of the Commonwealth Forces who died during the wars of 1914–18 and 1939–45; provides for the marking and maintenance of war graves and memorials at some 23,000 locations in 150 countries; mems: Australia, Canada, India, New Zealand, South Africa, United Kingdom; Pres. HRH The Duke of KENT; Dir-Gen. RICHARD KELLAWAY.

**Joint Commonwealth Societies' Council:** c/o Royal Commonwealth Society, 18 Northumberland Ave, London, WC2N 5BJ; tel. (20) 7930-6733; fax (20) 7930-9705; e-mail jcsc@rcsint.org; internet www.commonwealthday.com; f. 1947; provides a forum for the exchange of information regarding activities of mem. orgs which promote understanding among countries of the Commonwealth; co-ordinates the distribution of the Commonwealth Day message by Queen Elizabeth, organizes the observance of the Commonwealth Day and produces educational materials relating to the occasion; mems: 13 unofficial Commonwealth orgs and four official bodies; Chair. Sir PETER MARSHALL; Sec. Sir DAVID THORNE.

**Royal Commonwealth Ex-Services League:** 48 Pall Mall, London, SW1Y 5JG; tel. (20) 7973-7263; fax (20) 7973-7308; e-mail mgordon-roe@commonwealthveterans.org.uk; internet www.commonwealthveterans.org.uk; links the ex-service orgs in the Commonwealth, assists ex-servicemen of the Crown who are resident abroad; holds triennial confs; 57 mem. orgs in 48 countries; Grand Pres. HRH The Duke of EDINBURGH; publ. *Annual Report*.

**Royal Commonwealth Society:** 18 Northumberland Ave, London, WC2N 5BJ; tel. (20) 7930-6733; fax (20) 7930-9705; e-mail info@rcsint.org; internet www.rcsint.org; f. 1868 to promote international understanding of the Commonwealth and its people; organizes meetings and seminars on topical issues, and cultural and social events; library housed by Cambridge University Library; more than 10,000 mems; Chair. Baroness PRASHAR; Dir STUART MOLE; publs *Annual Report, Newsletter* (3 a year), conference reports.

**Royal Over-Seas League:** Over-Seas House, Park Place, St James's St, London, SW1A 1LR; tel. (20) 7408-0214; fax (20) 7499-6738; e-mail info@rosl.org.uk; internet www.rosl.org.uk; f. 1910 to promote friendship and understanding in the Commonwealth; clubhouses in London and Edinburgh; membership is open to all British subjects and Commonwealth citizens; Chair. Sir COLIN IMRAY; Dir-Gen. ROBERT F. NEWELL; publ. *Overseas* (quarterly).

**Victoria League for Commonwealth Friendship:** 55 Leinster Sq., London, W2 4PU; tel. (20) 7243-2633; fax (20) 7229-2994; e-mail victorialeaguehq@btconnect.com; internet www.victorialeague.co.uk; f. 1901; aims to further personal friendship among Commonwealth peoples and to provide hospitality for visitors; maintains Student House, providing accommodation for students from Commonwealth countries; has brs elsewhere in the UK and abroad; Chair. JOHN KELLY; Gen. Sec. JOHN M. W. ALLAN; publ. *Annual Report*.

# EUROPEAN UNION—REGIONAL RELATIONS

The European Community (EC) signed a non-preferential trade agreement with Uruguay in 1974 and economic and commercial co-operation agreements with Mexico in 1975 and Brazil in 1980. A five-year co-operation agreement with the members of the Central American Common Market (and with Panama) entered into force in 1987, as did a similar agreement with the member countries of the Andean Group (now the Andean Community). Priority was given to technology transfer, rural development, training, promotion of trade and investment, and co-operation in the energy sector. Co-operation agreements were signed with Argentina and Chile in 1990, and in that year tariff preferences were approved for Bolivia, Colombia, Ecuador and Peru in support of those countries' efforts to combat drugs-trafficking. The first ministerial conference between the EC and the then 11 Latin American states of the Rio Group took place in April 1991; thereafter high-level joint ministerial meetings have been held on a regular basis.

In December 1991 tariff preferences were extended to Costa Rica, El Salvador, Guatemala, Honduras, Nicaragua and Panama for a three-year period. In February 1993 the EC signed a new framework co-operation agreement with the countries of the Central American Common Market; a further co-operation agreement with the countries of Central America was signed in early 1996. In May 2002 Central American ministers of foreign affairs and the EU agreed to start negotiations on an association agreement by the end of 2004 and, in December 2003, a mutual dialogue and co-operation agreement was signed, widening the scope of the existing co-operation. In April 1993 a new co-operation agreement was signed with the countries of the Andean Group. Further trade benefits were extended to those countries in April 1997 to encourage the production of agricultural food crops in substitution of illegal drugs. An EU-Andean Community Political Dialogue and Co-operation Agreement was signed in December 2003. At the third EU-LAC summit meeting held in Guadalajara, Mexico, in May 2004 it was agreed by the two parties that an association agreement, which included a free-trade area, was a common objective. A joint assessment exercise on regional economic integration was initiated by the two groupings in January 2005.

In October 1995 the EU (as the EC became in 1993) announced its intention to forge closer links with Latin America, by means of strengthened political ties, an increase in economic integration and free trade, and co-operation in other areas. The first summit meeting of all EU and Latin American heads of state or government was held in Rio de Janeiro, Brazil, in June 1999. A second summit meeting was convened in Madrid, Spain, in May 2002, and covered co-operation in political, economic, social and cultural fields. Leaders attending the third such summit meeting, held in Guadalajara, Mexico, in May 2004, made commitments in the areas of social cohesion, multilateralism and regional integration.

In December 1995 a framework agreement for commercial and economic co-operation was signed with the Southern Common Market (Mercosur). The agreement also provided for formal political dialogue between the two groupings, in particular to promote regional integration and respect for democratic principles and human rights. In July 1998 the European Commission voted to commence negotiations towards an inter-regional association agreement with Mercosur and an association agreement with Chile. (In June 1996 a framework agreement on political and economic co-operation had been concluded with Chile, providing for a process of bilateral trade liberalization, as well as co-operation in other industrial and financial areas. An EU-Chile Joint Council was established.) Association negotiations with Mercosur commenced in April 2000. They were ongoing at mid-2005. In November 2002 an EU-Chile association agreement and an EU-Chile free-trade agreement were signed. The free-trade agreement provided for the liberalization of trade within seven years for industrial products and 10 years for agricultural products.

In early 1995 the EU registered a formal complaint against proposed US legislation to sanction countries and businesses trading with or investing in Cuba. The so-called 'Helms-Burton' legislation was approved by US President Clinton in March 1996, although he postponed implementation of a part of the law. In February 1997 the World Trade Organization (WTO) appointed a disputes panel, at the request of the EU, to determine whether the Helms-Burton Act contravened multilateral trade provisions; however, in April the EU agreed temporarily to withdraw its petition, and commence negotiations on a new multilateral accord on investment, while the USA was to continue to defer implementation of the law. In December 1996 the EU adopted a common position on its relations with Cuba, making any further economic co-operation contingent on that country's progress towards democracy and respect for human rights. This position has been reiterated at subsequent meetings of the European Council. In April 2000 Cuba rejected the Cotonou Agreement (see below), following criticism by some European governments of its human rights record. Subsequent improvements in bilateral relations between Cuba and the EU led to the opening of an EU office in Havana in March 2003 and support for Cuba's renewed application to join the Cotonou Agreement. However, the imprisonment of a large number of dissidents by the Cuban regime in the following month led to the downgrading of diplomatic relations with Cuba by the EU, the instigation of an EU policy of inviting dissidents to embassy receptions in Havana (the so-called 'cocktail wars') and the indefinite postponement of Cuba's application to join the Cotonou Agreement. In May Cuba withdrew its application for membership and in July the Cuban President announced that the Government would not accept aid from the EU and would terminate all political contact with the organization. In January 2005 Cuba announced that it was restoring diplomatic ties with all EU states. In February the EU temporarily suspended the diplomatic sanctions imposed on Cuba in 2003 and announced its intention to resume a constructive dialogue with the Cuban authorities.

In July 1997 an economic partnership, political co-ordination and co-operation agreement was concluded by the EU and Mexico, together with an interim agreement on trade. The accords were signed in December and negotiations on a free-trade agreement were launched in July 1998. The agreement was to provide for the elimination of tariffs on industrial products by January 2007. After nine rounds, negotiations were completed in early 2000, and the main part of the interim agreement entered into effect in July. The co-operation agreement entered into force on 1 October.

In 2003 the EU pledged €637m. to finance co-operation with Latin America, including €249m. for financial and technical co-operation and €13m. for rehabilitation and construction measures in developing countries. A four-year programme providing aid for refugees and displaced persons was launched in 2000, with funds of €15.9m. The programme focused on Mexico, Nicaragua and Guatemala. The EU's natural disaster prevention and preparedness programme (Dipecho) has targeted earthquake, flood, hurricane, and volcanic eruption preparedness throughout Latin America and the Caribbean. In February 2001 the European Parliament adopted a resolution opposing Plan Colombia, a US-backed initiative which aimed to combat the Colombian illegal narcotics trade through measures that included military-supported aerial crop spraying to destroy coca production; the Plan had reportedly resulted in the forced displacement of some farming communities. The Parliament urged support for the ongoing peace process in Colombia and for the adoption of structural reforms in that country as the preferred means of addressing the drugs problem. In April the EU announced a €30m. aid package designed to support the peace process. (In July the USA proposed the inauguration of a new Andean Regional Initiative, which shifted emphasis from the military-backed focus of Plan Colombia to the region-wide promotion of democracy and development.) Humanitarian aid granted to countries in the region in 2004 included €8m. to assist displaced persons in Colombia. The EU has adopted the following decentralized programmes to provide economic assistance to Latin America: AL-INVEST (supporting European investment in Latin America-based small and medium-sized enterprises that seek to operate internationally); ALFA (promoting bilateral co-operation in higher education); URB-AL (promoting links between European and Latin American cities); ALURE (promoting the best use of energy); @LIS (supporting the use of information technologies); OREAL (aimed at establishing a network of non profit-making institutions from both regions); and EURO-SociAL (inaugurated in May 2004 to assist Latin American countries with developing and implementing social policies aimed at strengthening social cohesion).

## ACP-EU Partnership

From 1976–February 2000 the principal means of channeling Community aid to developing countries were the successive Lomé conventions, concluded by the Community and the group of African, Caribbean and Pacific (ACP) states. The first Lomé Convention (Lomé I), which came into force on 1 April 1976, replaced the Yaoundé Conventions and the Arusha Agreement, and was designed to provide a new framework of co-operation, taking into account the varying needs of developing countries. Under Lomé I the Community committed ECU 3,052.4m. for aid and investment in developing countries, through the European Development Fund (EDF) and the European Investment Bank (EIB). Provision was made for over 99% of ACP (mainly agricultural) exports to enter the EC market duty free, while certain products that compete directly with Community agriculture were given preferential treatment but not free access: for some commodities, such as sugar, imports of fixed quantities at internal Community prices were guaranteed. The Stabex (stabilization of export earnings) scheme was designed to help developing countries withstand fluctuations in the price of their agricultural products, by paying compensation for reduced export earnings.

The second Lomé Convention (January 1981–February 1985) envisaged Community expenditure of ECU 5,530m.; it extended some of the provisions of Lomé I, and introduced new fields of co-operation. One of the most important innovations was a scheme (Sysmin), similar to Stabex, to safeguard exports of mineral products. Lomé III provided a total of ECU 8,500m. (about US $6,000m. at 30 January 1985) in assistance to the ACP states over the five years from March 1985.

The fourth Lomé Convention entered partially into force (trade provisions) on 1 March 1990, and fully into force on 1 November 1991. It covered the period 1990–February 2000. The financial protocol for 1990–95 made commitments of ECU 12,000m. (US $13,700m.), of which ECU 10,800m. was from the EDF (including ECU 1,500m. for Stabex and ECU 480m. for Sysmin) and ECU 1,200m. from the EIB. Under Lomé IV the obligation of most of the ACP states to contribute to the replenishment of Stabex resources, including the repayment of transfers made under the first three Conventions, was removed. In addition, special loans made to ACP member countries were to be cancelled, except in the case of profit-orientated businesses. Other innovations included the provision of assistance for structural adjustment programmes, measures to avoid increasing the recipient countries' indebtedness (e.g. by providing Stabex and Sysmin assistance in the form of grants, rather than loans), and increased support for the private sector, environmental protection, and control of population growth.

In September 1993 the Community announced plans to revise the Lomé Convention, with a view to establishing 'more open, equitable and transparent relations among the signatories'. In May 1994 a mid-term review of Lomé IV was initiated; however, in February 1995 a joint EU–ACP ministerial council, which was scheduled to conclude the negotiations, was adjourned, owing to significant discord among EU member states concerning reimbursement of the EDF for the period 1995–2000. In June 1995 EU heads of government reached an agreement, subsequently endorsed by an EU–ACP ministerial group, whereby ECU 14,625m. was to be allocated for the second phase of Lomé IV (ECU 12,967m. from the EDF and ECU 1,658m. from the EIB). Agreement was also reached on revision of the country-of-origin rules for manufactured goods; expansion of the preferential system of trade for ACP products; and a new protocol on the sustainable management of forest resources. At the same time, a joint declaration on support for the banana industry was issued, amid growing tension relating to the EU's banana import regime. This followed the introduction by the EC, in 1993, of an import regime designed to protect the banana industries of ACP countries (mostly in the Caribbean), which were threatened by the availability of cheaper bananas produced by countries in Latin America. Latin American countries, and later, the USA, lodged complaints against the new regime with international bodies. In April 1999, despite efforts by the EU to amend the regime, an arbitration panel of the WTO confirmed that the EU had failed to conform with WTO rules and formally authorized the USA to impose trade sanctions. In May 2000 the WTO further authorized Ecuador to impose punitive tariffs on copyrighted material from the EU, to offset losses incurred from the regime. An accord, involving the adoption by the EU of a new system of licences and quotas to cover 2001–06, and the introduction of a tariff-only system from 2006, was eventually reached by the USA and the EU in April 2001, subject to approval by EU member states and the European Parliament; the USA consequently agreed to suspend its punitive trade sanctions. The accord reduced by 100,000 metric tons the quota reserved primarily for ACP producers, adding this to the quota to be filled primarily by Latin American bananas. The Cotonou Agreement of June 2000 (see below) stated that the EU would try to ensure 'the continued viability of (ACP) banana export industries and the continuing outlet for (ACP) bananas on the Community market'. In 2003 the ACP Council expressed concerns about the continuing decline in prices for ACP bananas on the European market, which threatened banana production in the Windward Islands and Belize. The ACP Council requested further funds to ensure that the increase in quotas pursuant to EU enlargement in May 2004 did not further jeopardize the future of ACP banana producers. In August 2005 the WTO rejected the proposed EU banana tariff.

In November 1997 the first summit of heads of state of ACP countries was held in Libreville, Gabon. The principal issues under consideration at the meeting were the strategic challenges confronting the ACP group of countries and, in particular, relations with the EU beyond 2000, when Lomé IV was scheduled to expire. The summit mandated ACP ministers of finance and of trade and industry to organize a series of regular meetings in order to strengthen co-ordination within the grouping.

Formal negotiations on the conclusion of a successor agreement to the Lomé Convention were initiated in September 1998 and concluded in February 2000; the new partnership accord was signed by ACP and EU Heads of State and Government in June, in Cotonou, Benin. The so-called Cotonou Agreement covered the period 2000–20 and was subject to revision every five years. (The Agreement entered into force on 1 April 2003 following ratification by the then 15 EU member states and more than the requisite two-thirds of the ACP countries; however, many of its provisions had been applicable for a transitional period since August 2000.) It comprised the following

# REGIONAL ORGANIZATIONS

main elements: increased political co-operation; the enhanced participation of civil society in ACP–EC partnership affairs; a strong focus on the reduction of poverty (addressing the economic and technical marginalization of developing nations as a primary concern); a reform of the existing structures for financial co-operation; and a new framework for economic and trade co-operation. Under the provisions of the new accord, the EU was to negotiate free-trade arrangements (replacing the previous non-reciprocal trade preferences) with the most developed ACP countries during 2000–08; these would be structured around a system of regional free-trade zones, and would be designed to ensure full compatibility with WTO provisions. Once in force, the agreements would be subject to revision every five years. An assessment to be conducted in 2004 would identify those mid-ranking ACP nations also capable of entering into such free-trade deals. Meanwhile, the least-developed ACP nations were to benefit from an EU initiative to allow free access for most of their products by 2005. The preferential agreements currently in force would be retained initially (phase I), in view of a waiver granted by the WTO; thereafter ACP-EU trade was to be gradually liberalized over a period of 12–15 years (phase II). It was envisaged that Stabex and Sysmin would be eliminated gradually. In February 2001 the EU agreed to phase out trade barriers on imports of everything but military weapons from the world's 48 least-developed countries, 39 of which were in the ACP group. Duties on sugar, rice, bananas and some other products were to remain until 2009. The review process was initiated by the ACP-EU Council of Ministers, meeting in Gaborone, Botswana, in May 2004. In June 2005 representatives of ACP sugar-supplying countries expressed concern at EU proposals to reform the sugar regime, which envisaged a reduction in prices by 39% over a four year period.

A financial protocol was attached to the Cotonou Agreement which indicated the funds available to the ACP through the EDF. The first financial protocol, covering the initial five-year period from March 2000, provided a total budget of €13,500m., of which €1,300m. was allocated to regional co-operation and €2,200m. was for the new investment facility for the development of the private sector. In addition, uncommitted balances from previous EDFs amounted to a further €2,500m. The first meeting of the ACP–EU Joint Parliamentary Assembly following the signing of the Cotonou Agreement was held in Brussels in October 2000. Resolutions were adopted on subjects including the banana dispute and AIDS. The meeting also called for increased funding for decentralized co-operation to be made available in the EU budget. In total, the EU provided €2,543m. in financing for ACP countries in 2002. Humanitarian aid to the ACP was projected at €165m. in 2004.

At the fourth plenary session of the joint ACP-EU Joint Parliamentary Assembly, held in Cape Town, South Africa, in March 2002, the ninth EDF was discussed. The plan was to be organized along sub-regional lines. One major programme set up on behalf of the ACP countries and financed by the EDF was the new ProInvest programme, which was launched in 2002, with funding of €110m. over a seven-year period. The first general stage of negotiations for Economic Partnership Agreements (EPAs), involving discussions with all ACP countries regarding common procedures, began in September 2002; the Cotonou Agreement provided for a system of EPAs to replace all non-reciprocal trade preferences with the most developed ACP countries. The regional phase of EPA negotiations to establish a new framework for trade and investment commenced in October 2003, including discussions with the Caribbean region from mid-April 2004.

In January 2001 Article 96 of the Cotonou Agreement, which provides for suspension of the Agreement in specific countries in the event of violation of one of its essential elements (respect for human rights, democratic principles and the rule of law), was invoked against Haiti. The measures were partially removed in September 2004.

## SIGNATORY STATES TO THE COTONOU AGREEMENT

### European Union
Austria, Belgium, Cyprus, Czech Republic, Denmark, Estonia, Finland, France, Germany, Greece, Hungary, Ireland, Italy, Latvia, Lithuania, Luxembourg, Malta, Netherlands, Poland, Portugal, Slovakia, Slovenia, Spain, Sweden, United Kingdom

### Caribbean ACP states
Antigua and Barbuda, Bahamas, Barbados, Belize, Cuba*, Dominica, Dominican Republic, Grenada, Guyana, Haiti, Jamaica, Saint Christopher and Nevis, Saint Lucia, Saint Vincent and the Grenadines, Suriname, Trinidad and Tobago

The ACP states also comprise 48 African and 15 Pacific countries.
*Cuba is not a signatory of the Cotonou Agreement.

## ACP-EU INSTITUTIONS

**Council of Ministers:** one minister from each signatory state; one co-chairman from each of the two groups; meets annually.

**Committee of Ambassadors:** one ambassador from each signatory state; chairmanship alternates between the two groups; meets at least every six months.

**Joint Parliamentary Assembly:** EU and ACP are equally represented; attended by parliamentary delegates from each of the ACP countries and members of the European Parliament; one co-chairman from each group; meets twice a year.

**Secretariat of the ACP–EU Council of Ministers:** 175 rue de la Loi, 1048 Brussels, Belgium; tel. (2) 285-61-11; fax (2) 285-74-58.

**Centre for the Development of Enterprise (CDE):** 52 ave Hermann Debroux, 1160 Brussels, Belgium; tel. (2) 679-18-11; fax (2) 675-19-03; e-mail info@cde.int; internet www.cde.int; f. 1977 to encourage and support the creation, expansion and restructuring of industrial companies (mainly in the fields of manufacturing and agro-industry) in the ACP states by promoting co-operation between ACP and European companies, in the form of financial, technical or commercial partnership, management contracts, licensing or franchising agreements, sub-contracts, etc.; manages the ProInvest programme; Dir HAMED SOW (Mali).

**Technical Centre for Agricultural and Rural Co-operation:** Postbus 380, 6700 AJ Wageningen, Netherlands; tel. (317) 467100; fax (317) 460067; e-mail cta@cta.int; internet www.cta.int; f. 1983 to provide ACP states with better access to information, research, training and innovations in agricultural development and extension; Dir CARL B. GREENIDGE.

## ACP INSTITUTIONS

**ACP Council of Ministers.**

**ACP Committee of Ambassadors.**

**ACP Secretariat:** ACP House, 451 ave Georges Henri, 1200 Brussels, Belgium; tel. (2) 743-06-00; fax (2) 735-55-73; e-mail info@acp.int; internet www.acpsec.org; Sec.-Gen. Sir JOHN KAPUTIN (Papua New Guinea).

# INTER-AMERICAN DEVELOPMENT BANK—IDB

**Address:** 1300 New York Ave, NW, Washington, DC 20577, USA.
**Telephone:** (202) 623-1000; **fax:** (202) 623-3096; **e-mail:** pic@iadb.org; **internet:** www.iadb.org.

The Bank was founded in 1959 to promote the individual and collective development of Latin American and Caribbean countries through the financing of economic and social development projects and the provision of technical assistance. From 1976 membership was extended to include countries outside the region.

## MEMBERS

Argentina, Austria, Bahamas, Barbados, Belgium, Belize, Bolivia, Brazil, Canada, Chile, Colombia, Costa Rica, Croatia, Denmark, Dominican Republic, Ecuador, El Salvador, Finland, France, Germany, Guatemala, Guyana, Haiti, Honduras, Israel, Italy, Jamaica, Japan, Republic of Korea, Mexico, Netherlands, Nicaragua, Norway, Panama, Paraguay, Peru, Portugal, Slovenia, Spain, Suriname, Sweden, Switzerland, Trinidad and Tobago, United Kingdom, USA, Uruguay, Venezuela

## Organization
(August 2005)

### BOARD OF GOVERNORS

All the powers of the Bank are vested in a Board of Governors, consisting of one Governor and one alternate appointed by each member country (usually ministers of finance or presidents of central banks). The Board meets annually, with special meetings when necessary. The 46th annual meeting of the Board of Governors took place in Okinawa, Japan, in April 2005.

### BOARD OF EXECUTIVE DIRECTORS

The Board of Executive Directors is responsible for the operations of the Bank. It establishes the Bank's policies, approves loan and technical co-operation proposals that are submitted by the President of the Bank, and authorizes the Bank's borrowings on capital markets.

There are 14 executive directors and 14 alternates. Each Director is elected by a group of two or more countries, except the Directors representing Canada and the USA. The USA holds 30% of votes on the Board, in respect of its contribution to the Bank's capital. The Board has five permanent committees, relating to: Policy and evaluation; Organization, human resources and board matters; Budget, financial policies and audit; Programming; and a Steering Committee.

### ADMINISTRATION

The Bank comprises three Regional Operations Departments, as well as the following principal departments: Finance; Legal; Research; Integration and Regional Operations; Private Sector; External Relations; Sustainable Development; Information Technology and General Services; Strategic Planning and Budget; and Human Resources. There are also Offices of the Auditor-General, the Vice-President for Planning and Administration, and of Evaluation and Oversight. In October 2003 an Office of Institutional Integrity was established. In March 2004 the Board of Governors approved the establishment of a new Office of Development Effectiveness. The Bank has country offices in each of its borrowing member states, and special offices in Paris, France and in Tokyo, Japan. At the end of 2004 there were 1,884 Bank staff (excluding the Board of Executive Directors and the Evaluation Office), of whom 540 were based in country offices. The administrative budget for 2004 amounted to US $430m.

**President:** Luis Alberto Moreno (Colombia) (from 1 October 2005).
**Executive Vice-President:** Dennis E. Flannery (USA).

## Activities

Loans are made to governments, and to public and private entities, for specific economic and social development projects and for sectoral reforms. These loans are repayable in the currencies lent and their terms range from 12 to 40 years. Total lending authorized by the Bank by the end of 2004 amounted to US $135,937m. During 2004 the Bank approved loans totalling $6,020m., compared with $6,810m. in 2003. Disbursements on authorized loans amounted to $4,232m. in 2004, compared with $8,902m. in the previous year. In April 2005 the Board of Governors approved a new lending framework for the period 2005–08. The framework, which amended the limits on investment and policy-based loans, aimed to strengthen the country focus in lending programmes, to assess and meet more closely the needs of borrowing countries, and to harmonize procedures with other multilateral lending operations.

The subscribed ordinary capital stock, including inter-regional capital, which was merged into it in 1987, totalled US $100,951.4m. at the end of 2004, of which $4,340.2m. was paid-in and $96,611.2m. was callable. The callable capital constitutes, in effect, a guarantee of the securities which the Bank issues in the capital markets in order to increase its resources available for lending. In July 1995 the eighth general increase of the Bank's authorized capital was ratified by member countries: the Bank's resources were to be increased by $41,000m. to $102,000m.

In 2004 the Bank borrowed the equivalent of US $4,710m. on the international capital markets, bringing total borrowings outstanding to more than $48,886m. at the end of the year. During 2004 net earnings amounted to $862m. in ordinary capital resources and $104m. from the Fund for Special Operations (see below), and at the end of that year the Bank's total reserves were $13,563m.

The Fund for Special Operations enables the Bank to make concessional loans for economic and social projects where circumstances call for special treatment, such as lower interest rates and longer repayment terms than those applied to loans from the ordinary resources. The Board of Governors approved US $200m. in new contributions to the Fund in 1990, and in 1995 authorized $1,000m. in extra resources for the Fund. During 2004 the Fund made 27 loans totalling $552m., bringing the cumulative total of Fund lending to $17,392m. for 1,148 loans. Assistance may be provided to countries adversely affected by economic crises or natural disasters through an Emergency Lending Program. In September 2004 the Bank announced emergency relief to assist five Caribbean countries affected by Hurricane Ivan, including $200,000 for construction materials and medical supplies in Grenada, and $200,000 to assist the rehabilitation of housing in Jamaica. A Disaster Prevention Sector Facility assists countries to reduce their vulnerability to natural disasters and provides grants for identification, mitigation and preparedness projects.

In 1998 the Bank agreed to participate in an initiative of the International Monetary Fund and the World Bank to assist heavily indebted poor countries (HIPCs) to maintain a sustainable level of debt. Four member countries were eligible for assistance under the initiative (Bolivia, Guyana, Honduras and Nicaragua). Also in 1998, following projections of reduced resources for the Fund for Special Operations, borrowing member countries agreed to convert about US $2,400m. in local currencies held by the Bank, in order to maintain a convertible concessional Fund for poorer countries, and to help to reduce the debt-servicing payments under the HIPC initiative. In mid-2000 a committee of the Board of Governors endorsed a financial framework for the Bank's participation in an enhanced HIPC initiative, which aimed to broaden the eligibility criteria and accelerate the process of debt reduction. The Bank was to provide $896m. (in net present value), in addition to $204m. committed under the original scheme, of which $307m. was for Bolivia, $65m. for Guyana, $391m. for Nicaragua and $133m. for Honduras. The Bank assisted the preparation of national Poverty Reduction Strategy Papers, a condition of reaching the 'completion point' of the process.

The Bank supports a range of consultative groups, in order to strengthen donor co-operation with countries in the Latin America and Caribbean region. In December 1998 the Bank established an emergency Consultative Group for the Reconstruction and Transformation of Central America to co-ordinate assistance to countries that had suffered extensive damage as a result of Hurricane Mitch. The Bank hosted the first meeting of the group in the same month, which was attended by government officials, representatives of donor agencies and non-governmental organizations and academics. A total of US $6,200m. was pledged in the form of emergency aid, longer-term financing and debt relief. A second meeting of the group was held in May 1999, in Stockholm, Sweden, at which the assistance package was increased to some $9,000m., of which the Bank and World Bank committed $5,300m. In March 2001 the group convened, in Madrid, Spain, to promote integration and foreign investment in Central America. The meeting, organized by the Bank, was also used to generate $1,300m. in commitments from international donors to assist emergency relief and reconstruction efforts in El Salvador following an earthquake earlier in the year. In October the Bank organized a consultative group meeting in support of a Social Welfare and Alternative Preventive Development Programme for Ecuador, to which donor countries committed $266m. Other consultative efforts co-ordinated by the Bank include groups to support the peace process in Colombia, and in Guatemala. In November 2001 the Bank hosted the first meeting of a Network for the Prevention and Mitigation of Natural Disasters in Latin America and the Caribbean, which was part of a regional policy dialogue, sponsored by the Bank to promote broad debate on strategic issues. In July 2004 the Bank co-hosted an international donor conference, together with the World Bank, the European Union and the United Nations, to consider the immediate and medium-term needs for Haiti following a period of political unrest. Some $1,080m. was pledged at the conference, of which the Bank's contribution was $260m.

An increasing number of donor countries have placed funds under the Bank's administration for assistance to Latin America, outside the framework of the Ordinary Resources and the Bank's Special Operations. These trust funds, totalling 55 in 2004, include the Social Progress Trust Fund (set up by the USA in 1961); the Venezuelan Trust Fund (set up in 1975); the Japan Special Fund (1988); and other funds administered on behalf of Austria, Belgium, Canada, Denmark, Finland, France, Israel, Italy, Japan, the Netherlands, Norway, Portugal, Spain, Sweden, Switzerland, the United Kingdom and the EU. A Program for the Development of Technical Co-operation was established in 1991, which is financed by European countries and the EU. Total cumulative lending from all these trust funds was US $1,748m. for loans approved by the end of 2004. During 2004 cofinancing by bilateral and multilateral sources amounted to $3,101.1m., which helped to finance some 30 projects in 11 countries.

The Bank provides technical co-operation to help member countries to identify and prepare new projects, to improve loan execution, to strengthen the institutional capacity of public and private agencies, to address extreme conditions of poverty and to promote small- and micro-enterprise development. The Bank has established a special co-operation programme to facilitate the transfer of experience and technology among regional programmes. In 2004 the Bank approved 340 technical co-operation operations, totalling US $56.7m., mainly financed by income from the Fund for Special Operations and donor trust funds. The Bank supports the efforts of the countries of the region to achieve economic integration and has provided extensive technical support for the formulation of integration strategies in the Andean, Central American and Southern Cone regions. The Bank is supporting the initiative to establish a Free Trade Area of the Americas (FTAA) and has provided technical assistance, developed programming strategies and produced a number of studies on relevant integration and trade issues. In 2003/04 the Bank granted some $600m. in loans and other technical assistance to Central American countries and the Dominican Republic for the preparation of national action plans for trade-related capacity building, in support of negotiations with the USA to conclude a Central America Free Trade Agreement (DR-CAFTA). From March 2003 the Bank supported the transfer of the Administrative Secretariat of the FTAA from Panama City, Panama, to Puebla, Mexico. In 2001 the Bank took a lead role in a Central American regional initiative, the Puebla-Panama Plan, which aimed to consolidate integration and support for social and economic development. The Bank is also a member of the technical co-ordinating committee of the Integration of Regional Infrastructure in South America initiative, which aimed to promote multinational development projects, capacity-building and integration in that region. In March 2004 the Bank's Board of Executive Directors approved an Initiative for the Promotion of Regional Public Goods, to support the development of regional public goods and to increase competitiveness in the global economy.

### AFFILIATES

**Inter-American Investment Corporation (IIC):** 1350 New York Ave, NW, Washington, DC 20577, USA; tel. (202) 623-3900; fax (202) 623-2360; e-mail iicmail@iadb.org; internet www.iadb.org/iic; f. 1986 as a legally autonomous affiliate of the Inter-American Development Bank, to promote the economic development of the region; commenced operations in 1989; initial capital stock was US $200m., of which 55% was contributed by developing member nations, 25.3% by the USA, and the remainder by non-regional members; in total, the IIC has 42 shareholders (26 Latin American and Caribbean countries, 13 European countries, Israel, Japan and the USA); in 2001 the Board of Governors of the IADB agreed to increase the IIC's capital to $500m.; places emphasis on investment in small and medium-sized enterprises without access to other suitable sources of equity or long-term loans; in 2004 the IIC approved 31 direct loan and equity transactions amounting to $163.6m.; Gen. Man. JACQUES ROGOZINSKI; publ. *Annual Report* (in English, French, Portuguese and Spanish).

**Multilateral Investment Fund (MIF):** 1300 New York Ave, NW, Washington, DC 20577, USA; tel. (202) 942-8211; fax (202) 942-8100; e-mail mifcontact@iadb.org; internet www.iadb.org/mif; f. 1993 as an autonomous fund administered by the Bank, to promote private sector development in the region; the 21 Bank members who signed the initial draft agreement in 1992 to establish the Fund pledged to contribute US $1,200m.; the Fund's activities are undertaken through three separate facilities concerned with technical co-operation, human resources development and small enterprise development; in 2000 a specialist working group, established to consider MIF operations, recommended that it target its resources on the following core areas of activity: small business development; market functioning; and financial and capital markets; in June 2002 the Fund approved $1.2m. to assist the financial intelligence units in eight South American countries in their efforts to counter money laundering; during 2004 the Fund approved some $116m. for 82 projects; the Bank's Social Entrepreneurship Program makes available credit to individuals or groups without access to commercial or development loans; some $10m. is awarded under the programme to fund projects in 26 countries; in April 2005 38 donor countries agreed to establish MIF II, and replenish the Fund's resources with commitments totalling $502m.; Man. DONALD F. TERRY.

### INSTITUTIONS

**Instituto para la Integración de América Latina y el Caribe** (Institute for the Integration of Latin America and the Caribbean): Esmeralda 130, 17°, 1035 Buenos Aires, Argentina; tel. (11) 4320-1850; fax (11) 4320-1865; e-mail int/inl@iadb.org; internet www.iadb.org/intal; f. 1965 under the auspices of the Inter-American Development Bank; forms part of the Bank's Integration and Regional Programmes Department; undertakes research on all aspects of regional integration and co-operation and issues related to international trade, hemispheric integration and relations with other regions and countries of the world; activities come under four main headings: regional and national technical co-operation projects on integration; policy fora; integration fora; and journals and information; hosts the secretariat of the Integration of Regional Infrastructure in South America (IIRSA) initiative; maintains an extensive Documentation Center and various statistical databases; Dir RICARDO CARCIOFI; publs *Integración y Comercio/Integration and Trade* (2 a year), *Intal Monthly Newsletter*, *Informe Andino/Andean Report*, *CARICOM Report*, *Informe Centroamericano/Central American Report*, *Informe Mercosur/Mercosur Report* (2 a year).

**Inter-American Institute for Social Development (INDES):** 1350 New York Ave, NW, Washington, DC 20057, USA; fax (202) 623-2008; e-mail indes@iadb.org; internet www.iadb.org/indes; commenced operations in 1995; aims to support the training of senior officials from public sector institutions and organizations involved with social policies and social services; organizes specialized subregional courses and seminars and national training programmes; produces teaching materials and also serves as a forum for the exchange of ideas on social reform; during 2004 almost 2,000 people participated in regional and national training courses; Dir NOHRA REY DE MARULANDA.

## Publications

*Annual Report* (in English, French, Portuguese and Spanish).

*Annual Report on Oversight and Evaluation*.

*Annual Report on the Environment and Natural Resources* (in English and Spanish).

*Equidad* (2 a year).

*Ethics and Development* (weekly).

*IDBamérica* (monthly, English and Spanish).

*IDB Projects* (10 a year, in English).

*IFM (Infrastructure and Financial Markets) Review* (quarterly).

*Latin American Economic Policies* (quarterly).

*Micro Enterprise Américas* (annually).

*Proceedings of the Annual Meeting of the Boards of Governors of the IDB and IIC* (in English, French, Portuguese and Spanish).

*Social Development Newsletter* (2 a year).

Brochure series, occasional papers, working papers, reports.

## Statistics

**Distribution of loans**
(US $ million)

| Sector | 2004 | % | 1961–2004 | % |
|---|---|---|---|---|
| **Competitiveness** | | | | |
| Energy | 146.0 | 2.4 | 18,792.2 | 13.8 |
| Transportation and communication | 582.7 | 9.7 | 15,168.3 | 11.2 |
| Agriculture and fisheries | 90.8 | 1.5 | 13,723.4 | 10.1 |
| Industry, mining and tourism | 0.8 | 0.0 | 13,063.5 | 9.6 |
| Multisector credit and preinvestment | 1,000.0 | 16.6 | 3,663.1 | 2.7 |
| Science and technology | 0.0 | 0.0 | 1,642.1 | 1.2 |
| Trade financing | 37.5 | 0.6 | 1,650.8 | 1.2 |
| Multisector infrastructure | 100.0 | 1.7 | 614.8 | 0.5 |
| Capital markets | 145.0 | 2.4 | 160.3 | 0.1 |
| **Social Sector Reform** | | | | |
| Social investment | 2,545.3 | 42.3 | 16,490.6 | 12.1 |
| Water and sanitation | 61.8 | 1.0 | 9,063.4 | 6.7 |
| Urban development | 164.7 | 2.7 | 7,857.2 | 5.8 |
| Education | 91.0 | 1.5 | 5,566.2 | 4.1 |
| Health | 136.5 | 2.3 | 2,789.4 | 2.1 |
| Environment | 34.0 | 0.6 | 2,809.2 | 2.1 |
| Microenterprise | 0.0 | 0.0 | 432.8 | 0.3 |
| **Reform and Modernization of the State** | | | | |
| Public sector planning and reform | 78.7 | 1.1 | 10,063.3 | 7.8 |
| Financial sector reform | 125.0 | 2.1 | 5,905.4 | 4.3 |
| Fiscal reform | 43.0 | 0.7 | 3,759.7 | 2.8 |
| E-government | 13.2 | 0.2 | 13.9 | 0.0 |
| Decentralization policies | 30.4 | 0.5 | 671.2 | 0.5 |
| Modernization and administration of justice | 3.0 | 0.0 | 313.5 | 0.2 |
| Planning and state reform | 0.0 | 0.0 | 119.4 | 0.1 |
| Parliamentary modernization | 4.8 | 0.1 | 76.2 | 0.1 |
| Civil society | 0.0 | 0.0 | 23.1 | 0.0 |
| Trade policy support | 0.0 | 0.0 | 13.6 | 0.0 |
| **Total** | 6,019.9 | 100.0 | 135,937.2 | 100.0 |

**Yearly and cumulative loans and guarantees, 1961–2004**
(US $ million; after cancellations and exchange adjustments)

| Country | Total Amount 2004 | Total Amount 1961–2004 | Ordinary Capital* 2004 | Ordinary Capital* 1961–2004 | Fund for Special Operations 2004 | Fund for Special Operations 1961–2004 | Funds in Administration 2004 | Funds in Administration 1961–2004 |
|---|---|---|---|---|---|---|---|---|
| Argentina | 528.0 | 20,399.0 | 528.0 | 19,705.0 | — | 644.9 | — | 49.1 |
| Bahamas | 3.5 | 358.5 | 3.5 | 356.4 | — | — | — | 2.0 |
| Barbados | — | 426.9 | — | 367.4 | — | 40.7 | — | 19.0 |
| Belize | — | 92.0 | — | 92.0 | — | — | — | — |
| Bolivia | 92.6 | 3,624.6 | 31.0 | 1,288.3 | 61.6 | 2,263.1 | — | 73.2 |
| Brazil | 2,609.7 | 28,869.0 | 2,609.7 | 27,178.0 | — | 1,558.2 | — | 132.8 |
| Chile | 23.2 | 5,381.4 | 23.2 | 5,133.0 | — | 204.7 | — | 43.7 |
| Colombia | 737.3 | 11,905.6 | 737.3 | 11,074.2 | — | 767.6 | — | 63.8 |
| Costa Rica | 11.0 | 2,379.0 | 11.0 | 1,879.7 | — | 361.3 | — | 138.0 |
| Dominican Republic | 337.0 | 2,978.4 | 337.0 | 2,147.1 | — | 743.9 | — | 87.4 |
| Ecuador | 17.4 | 4,262.7 | 17.4 | 3,207.9 | — | 965.0 | — | 89.8 |
| El Salvador | 20.0 | 3,050.4 | 20.0 | 2,119.4 | — | 785.9 | — | 145.1 |
| Guatemala | 100.6 | 2,776.5 | 100.6 | 2,046.5 | — | 659.9 | — | 70.1 |
| Guyana | 117.3 | 978.9 | — | 121.2 | 117.3 | 850.7 | — | 6.9 |
| Haiti | — | 984.3 | — | — | — | 978.0 | — | 6.3 |
| Honduras | 228.8 | 2,742.9 | — | 582.4 | 228.8 | 2,095.1 | — | 65.4 |
| Jamaica | 56.8 | 1,872.6 | 56.8 | 1,503.6 | — | 170.1 | — | 198.9 |
| Mexico | 485.4 | 17,644.3 | 485.4 | 17,026.7 | — | 559.0 | — | 58.6 |
| Nicaragua | 143.8 | 2,292.6 | — | 285.4 | 143.8 | 1,941.0 | — | 66.2 |
| Panama | — | 2,103.3 | — | 1,767.9 | — | 292.8 | — | 42.8 |
| Paraguay | — | 1,946.6 | — | 1,325.5 | — | 608.2 | — | 12.6 |
| Peru | 351.1 | 6,966.5 | 351.1 | 6,310.3 | — | 435.1 | — | 221.2 |
| Suriname | 10.8 | 101.6 | 10.8 | 99.6 | — | — | — | 2.0 |
| Trinidad and Tobago | — | 1,069.1 | — | 1,013.4 | — | 30.6 | — | 25.2 |
| Uruguay | 99.5 | 3,814.0 | 99.5 | 3,667.8 | — | 104.3 | — | 41.8 |
| Venezuela | 6.0 | 4,035.1 | 6.0 | 3,860.9 | — | 101.4 | — | 72.9 |
| Regional | 40.0 | 2,881.4 | 40.0 | 2,639.9 | — | 227.9 | — | 13.6 |
| **Total** | 6,019.9 | 135,937.2 | 5,468.3 | 116,799.2 | 551.6 | 17,391.5 | — | 1,746.7 |

*Includes private sector loans, net of participations.

Source: IADB *Annual Report*, 2004.

# LATIN AMERICAN INTEGRATION ASSOCIATION—LAIA
## (ASOCIACIÓN LATINOAMERICANA DE INTEGRACIÓN—ALADI)

**Address:** Cebollatí 1461, Casilla 577, 11200 Montevideo, Uruguay. **Telephone:** (2) 410-1121; **fax:** (2) 419-0649; **e-mail:** sgaladi@aladi.org; **internet:** www.aladi.org.

The Latin American Integration Association was established in August 1980 to replace the Latin American Free Trade Association, founded in February 1960.

### MEMBERS

| | | |
|---|---|---|
| Argentina | Colombia | Paraguay |
| Bolivia | Cuba | Peru |
| Brazil | Ecuador | Uruguay |
| Chile | Mexico | Venezuela |

**Observers:** People's Republic of China, Costa Rica, Dominican Republic, El Salvador, Guatemala, Honduras, Italy, Japan, Republic of Korea, Nicaragua, Panama, Portugal, Romania, Russia, Spain and Switzerland; also the UN Economic Commission for Latin America and the Caribbean (ECLAC), the UN Development Programme (UNDP), the European Union, the Inter-American Development Bank, the Organization of American States, the Andean Development Corporation, the Inter-American Institute for Co-operation on Agriculture, the Latin American Economic System, and the Pan American Health Organization.

## Organization
(August 2005)

### COUNCIL OF MINISTERS

The Council of Ministers of Foreign Affairs is responsible for the adoption of the Association's policies. It meets when convened by the Committee of Representatives.

### CONFERENCE OF EVALUATION AND CONVERGENCE

The Conference, comprising plenipotentiaries of the member governments, assesses the integration process and encourages negotiations between members. It also promotes the convergence of agreements and other actions on economic integration. The Conference meets when convened by the Committee of Representatives.

### COMMITTEE OF REPRESENTATIVES

The Committee, the permanent political body of the Association, comprises a permanent and a deputy representative from each member country. Permanent observers have been accredited by 15 countries and eight international organizations (see above). The Committee is the main forum for the negotiation of ALADI's initiatives and is responsible for the correct implementation of the Treaty and its supplementary regulations. There are the following auxiliary bodies:

**Advisory Commission for Financial and Monetary Affairs.**

**Advisory Commission on Customs Valuation.**

**Advisory Council for Enterprises.**

**Advisory Council for Export Financing.**

**Advisory Council for Customs Matters.**

**Budget Commission.**

**Commission for Technical Support and Co-operation.**

**Council for Financial and Monetary Affairs:** comprises the Presidents of member states' central banks, who examine all aspects of financial, monetary and exchange co-operation.

**Council of National Customs Directors.**

**Council on Transport for Trade Facilitation.**

**Labour Advisory Council.**

**Nomenclature Advisory Commission.**

**Sectoral Councils.**

**Tourism Council.**

### GENERAL SECRETARIAT

The General Secretariat is the technical body of the Association; it submits proposals for action, carries out research and evaluates activities. The Secretary-General is elected for a three-year term, which is renewable. There are two Assistant Secretaries-General.

**Secretary-General:** Dr Didier Opertti Badán (Uruguay).

## Activities

The Latin American Free Trade Association (LAFTA) was an intergovernmental organization, created by the Treaty of Montevideo in February 1960 with the object of increasing trade between the Contracting Parties and of promoting regional integration, thus contributing to the economic and social development of the member countries. The Treaty provided for the gradual establishment of a free-trade area, which would form the basis for a Latin American Common Market. Reduction of tariff and other trade barriers was to be carried out gradually until 1980.

By 1980, however, only 14% of annual trade among members could be attributed to LAFTA agreements. In June it was decided that LAFTA should be replaced by a less ambitious and more flexible organization, the Latin American Integration Association (Asociación Latinoamericana de Integración—ALADI), established by the 1980 Montevideo Treaty, which came into force in March 1981, and was fully ratified in March 1982. The Treaty envisaged an area of economic preferences, comprising a regional tariff preference for goods originating in member states (in effect from 1 July 1984) and regional and partial scope agreements (on economic complementation, trade promotion, trade in agricultural goods, scientific and technical co-operation, the environment, tourism, and other matters), taking into account the different stages of development of the members, and with no definite timetable for the establishment of a full common market.

The members of ALADI are divided into three categories: most developed (Argentina, Brazil and Mexico); intermediate (Chile, Colombia, Peru, Uruguay and Venezuela); and least developed (Bolivia, Cuba—which joined the Association in August 1999–Ecuador and Paraguay), enjoying a special preferential system. By 2004 intra-ALADI exports were estimated to total US $56,871m., an increase of almost 30% compared with the previous year.

Certain LAFTA institutions were retained and adapted by ALADI, e.g. the Reciprocal Payments and Credits Agreement (1965, modified in 1982) and the Multilateral Credit Agreement to Alleviate Temporary Shortages of Liquidity, known as the Santo Domingo Agreement (1969, extended in 1981 to include mechanisms for counteracting global balance-of-payments difficulties and for assisting in times of natural disaster).

By August 1998 98 agreements had entered into force. Seven were 'regional agreements' (in which all member countries participate). These agreements included a regional tariff preference agreement, whereby members allow imports from other member states to enter with tariffs 20% lower than those imposed on imports from other countries, and a Market Opening Lists agreement in favour of the three least developed member states, which provides for the total elimination of duties and other restrictions on imports of certain products. The remaining 91 agreements were 'partial scope agreements' (in which two or more member states participate), including: renegotiation agreements (pertaining to tariff cuts under LAFTA); trade agreements covering particular industrial sectors; the agreements establishing (Mercosur) and the Group of Three (G3); and agreements covering agriculture, gas supply, tourism, environmental protection, books, transport, sanitation and trade facilitation. A new system of tariff nomenclature, based on the 'harmonized system', was adopted from 1 January 1990 as a basis for common trade negotiations and statistics. General regimes on safeguards and rules of origin entered into force in 1987.

The Secretariat convenes meetings of entrepreneurs in various private industrial sectors, to encourage regional trade and co-operation. In early 2001 ALADI conducted a survey on small and medium-sized enterprises in order to advise the Secretary-General in formulating a programme to assist those businesses and enhance their competitiveness.

ALADI has worked to establish multilateral links or agreements with Latin American non-member countries or integration organizations, and with other developing countries or economic groups outside the continent. In February 1994 the Council of Ministers of Foreign Affairs urged that ALADI should become the co-ordinating body for the various bilateral, multilateral and regional accords (with the Andean Community, Mercosur and G3, etc.), with the aim of eventually forming a region-wide common market. The General

Secretariat initiated studies in preparation for a programme to undertake this co-ordinating work. At the same meeting in February there was a serious disagreement regarding the proposed adoption of a protocol to the Montevideo Treaty to enable Mexico to participate in the North American Free Trade Agreement (NAFTA), while remaining a member of ALADI. Brazil, in particular, opposed such a solution. However, in June the first Interpretative Protocol to the Montevideo Treaty was signed by the Ministers of Foreign Affairs: the Protocol allows member states to establish preferential trade agreements with developed nations, with a temporary waiver of the most-favoured nation clause, subject to the negotiation of unilateral compensation.

Mercosur (comprising Argentina, Brazil, Paraguay and Uruguay) aims to conclude free-trade agreements with the other members of ALADI. In March 2001 ALADI signed a co-operation agreement with the Andean Community to facilitate the exchange of information and consolidate regional and subregional integration. In December 2003 Mercosur and the Andean Community signed an Economic Complementary Agreement, and in April 2004 concluded a free-trade agreement, to come into effect on 1 July 2004 (although later postponed). Those ALADI member states remaining outside Mercosur and the Andean Community—Cuba, Chile and Mexico—would be permitted to apply to join the envisaged larger free-trade zone. In December 2004 leaders from 12 Latin American countries (excluding Argentina, Ecuador, Paraguay and Uruguay) attending a pan-South American summit, convened in Cuzco, Peru, approved in principle the creation of a new South American Community of Nations (SACN). In April 2005 the ALADI Secretary-General attended the first joint meeting of ministers of foreign affairs, held within the framework of establishing the SACN.

## Publications

*Empresarios en la Integración* (monthly, in Spanish).
*Noticias ALADI* (monthly, in Spanish).
*Estadísticas y Comercio* (quarterly, in Spanish).
Reports, studies, brochures, texts of agreements.

# NORTH AMERICAN FREE TRADE AGREEMENT—NAFTA

**Address: Canadian section:** Royal Bank Centre, 90 Sparks St, Suite 705, Ottawa, ON K1P 5B4.
**Telephone:** (613) 992-9388; **fax:** (613) 992-9392; **e-mail:** canada@nafta-sec-alena.org; **internet:** www.nafta-sec-alena.org/canada.
**Address: Mexican section:** Blvd Adolfo López Mateos 3025, 2°, Col Héroes de Padierna, 10700 México, DF.
**Telephone:** (5) 629-9630; **fax:** (5) 629-9637; **e-mail:** mexico@nafta-sec-alena.org.
**Address: US section:** 14th St and Constitution Ave, NW, Room 2061, Washington, DC 20230.
**Telephone:** (202) 482-5438; **fax:** (202) 482-0148; **e-mail:** usa@nafta-sec-alena.org; **internet:** www.nafta-sec-alena.org.

The North American Free Trade Agreement (NAFTA) grew out of the free-trade agreement between the USA and Canada that was signed in January 1988 and came into effect on 1 January 1989. Negotiations on the terms of NAFTA, which includes Mexico in the free-trade area, were concluded in October 1992 and the Agreement was signed in December. The accord was ratified in November 1993 and entered into force on 1 January 1994. The NAFTA Secretariat is composed of national sections in each member country.

### MEMBERS

Canada         Mexico         USA

### MAIN PROVISIONS OF THE AGREEMENT

Under NAFTA almost all restrictions on trade and investment between Canada, Mexico and the USA were to be gradually removed over a 15-year period. Most tariffs were eliminated immediately on agricultural trade between the USA and Mexico, with tariffs on 6% of agricultural products (including corn, sugar, and some fruits and vegetables) to be abolished over the 15 years. Tariffs on automobiles and textiles were to be phased out over 10 years in all three countries. Mexico was to open its financial sector to US and Canadian investment, with all restrictions to be removed by 2007. Barriers to investment were removed in most sectors, with exemptions for petroleum in Mexico, culture in Canada and airlines and radio communications in the USA. Mexico was to liberalize government procurement, removing preferential treatment for domestic companies over a 10-year period. In transport, heavy goods vehicles were to have complete freedom of movement between the three countries by 2000. An interim measure, whereby transport companies could apply for special licences to travel further within the borders of each country than the existing limit of 20 miles (32 km), was postponed in December 1995, shortly before it was scheduled to come into effect. The postponement was due to concerns, on the part of the US Government, relating to the implementation of adequate safety standards by Mexican truck-drivers. The 2000 deadline for the free circulation of heavy goods vehicles was not met, owing to the persistence of these concerns. In February 2001 a five-member NAFTA panel of experts appointed to adjudicate on the dispute ruled that the USA was violating the Agreement. In December the US Senate approved legislation entitling Mexican long-haul trucks to operate anywhere in the USA following compliance with rigorous safety checks to be enforced by US inspectors. In April 1998 the fifth meeting of the three-member ministerial Free Trade Commission (see below), held in Paris, France, agreed to remove tariffs on some 600 goods, including certain chemicals, pharmaceuticals, steel and wire products, textiles, toys, and watches, from 1 August. As a result of the agreement, a number of tariffs were eliminated as much as 10 years earlier than had been originally planned.

In April 2003 the Mexican Government announced that, in order to protect the livelihoods of local producers, it was planning to renegotiate provisions of the Agreement that would permit, with effect from January 2008, the tariff-free importation of maize and beans from Canada and the USA.

In the case of a sudden influx of goods from one country to another that adversely affects a domestic industry, the Agreement makes provision for the imposition of short-term 'snap-back' tariffs.

Disputes are to be settled in the first instance by intergovernmental consultation. If a dispute is not resolved within 30 to 40 days, a government may call a meeting of the Free Trade Commission. In October 1994 the Commission established an Advisory Committee on Private Commercial Disputes to recommend procedures for the resolution of such disputes. If the Commission is unable to settle the issue a panel of experts in the relevant field is appointed to adjudicate. In June 1996 Canada and Mexico announced their decision to refer the newly-enacted US 'Helms-Burton' legislation on trade with Cuba to the Commission. They claimed that the legislation, which provides for punitive measures against foreign companies that engage in trade with Cuba, imposed undue restrictions on Canadian and Mexican companies and was, therefore, in contravention of NAFTA. However, at the beginning of 1997 certain controversial provisions of the Helms-Burton legislation were suspended for a period of six months by the US administration. In April these were again suspended, as part of a compromise agreement with the European Union. The relevant provisions continued to be suspended at six-monthly intervals, and remained suspended in 2005. An Advisory Committee on Private Commercial Disputes Regarding Agricultural Goods was formed in 1998.

In December 1994 NAFTA members issued a formal invitation to Chile to seek membership of the Agreement. Formal discussions on Chile's entry began in June 1995, but were stalled in December when the US Congress failed to approve 'fast-track' negotiating authority for the US Government, which was to have allowed the latter to negotiate a trade agreement with Chile, without risk of incurring a line-by-line veto from the US Congress. In February 1996 Chile began high-level negotiations with Canada on a wide-ranging bilateral free-trade agreement. Chile, which already had extensive bilateral trade agreements with Mexico, was regarded as advancing its position with regard to NAFTA membership by means of the proposed accord with Canada. The bilateral agreement, which provided for the extensive elimination of customs duties by 2002, was signed in November 1996 and ratified by Chile in July 1997. However, in November 1997 the US Government was obliged to request the removal of the 'fast-track' proposal from the legislative agenda, owing to insufficient support within Congress.

In April 1998 heads of state of 34 countries, meeting in Santiago, Chile, agreed formally to initiate the negotiating process to establish a Free Trade Area of the Americas (FTAA). The US Government had originally proposed creating the FTAA through the gradual extension of NAFTA trading privileges on a bilateral basis. However, the framework agreed upon by ministers of trade of the 34 countries, meeting in March, provided for countries to negotiate and accept FTAA provisions on an individual basis and as part of a sub-regional economic bloc. It was envisaged that the FTAA would exist alongside the sub-regional associations, including NAFTA. In April 2001, meeting in Québec, QC, Canada, leaders of the participating countries agreed to conclude the negotiations on the FTAA by January 2005 and implement it by the end of that year. At a special summit of the Americas, held in January 2004 in Monterrey, Mexico, however, the leaders did not specify a completion date for the negotiations, although they adopted a declaration committing themselves to its eventual establishment. Negotiations were suspended in March and remained stalled at mid-2005.

### ADDITIONAL AGREEMENTS

During 1993, as a result of domestic pressure, the new US Government negotiated two 'side agreements' with its NAFTA partners, which were to provide safeguards for workers' rights and the environment. A Commission for Labour Co-operation was established under the North American Agreement on Labour Co-operation (NAALC) to monitor implementation of labour accords and to foster co-operation in that area. The North American Commission for Environmental Co-operation (NACEC) was initiated to combat pollution, to ensure that economic development was not environmentally damaging and to monitor compliance with national and NAFTA environmental regulations. Panels of experts, with representatives from each country, were established to adjudicate in cases of alleged infringement of workers' rights or environmental damage. The panels were given the power to impose fines and trade sanctions, but only with regard to the USA and Mexico; Canada, which was opposed to such measures, was to enforce compliance with NAFTA by means of its own legal system. In 1995 the North American Fund for Environmental Co-operation (NAFEC) was established. NAFEC, which is financed by the NACEC, supports community environmental projects.

In February 1996 the NACEC consented for the first time to investigate a complaint brought by environmentalists regarding non-compliance with domestic legislation on the environment. Mexican environmentalists claimed that a company that was planning to build a pier for tourist ships (a project that was to involve damage to a coral reef) had not been required to supply adequate environmental impact studies. The NACEC was limited to presenting its findings in such a case, as it could only make a ruling in the case of complaints brought by one NAFTA government against another. The NACEC allocates the bulk of its resources to research undertaken to support compliance with legislation and agreements on the environment. However, in October 1997 the council of NAFTA ministers of the environment, meeting in Montréal, Canada, approved a new structure for the NACEC's activities. The NACEC's main objective was to be the provision of advice concerning the environmental impact of trade issues. It was also agreed that the Commission was further to promote trade in environmentally-sound products and to encourage private-sector investment in environmental trade issues.

With regard to the NAALC, National Administration Offices have been established in each of the three NAFTA countries in order to monitor labour issues and to address complaints about non-compliance with domestic labour legislation. However, punitive measures in the form of trade sanctions or fines (up to US $20m.) may only be imposed in the specific instances of contravention of national legislation regarding child labour, a minimum wage or health and safety standards. A Commission for Labour Co-operation has been established (see below) and incorporates a council of ministers of labour of the three countries.

In August 1993 the USA and Mexico agreed to establish a Border Environmental Co-operation Commission (BECC) to assist with the co-ordination of projects for the improvement of infrastructure and to monitor the environmental impact of the Agreement on the US–Mexican border area, where industrial activity was expected to intensify. The Commission is located in Ciudad Juárez, Mexico. By April 2005 the BECC had certified 105 projects, at a cost of US $2180.6m. In October 1993 the USA and Mexico concluded an agreement to establish the North American Development Bank (NADB or NADBank), which was mandated to finance environmental and infrastructure projects along the US–Mexican border.

**Commission for Labour Co-operation:** 1211 Connecticut Ave, NW Suite 200, Washington, DC 20036, USA; tel. (202) 464-1100; fax (202) 464-9487; e-mail info@naalc.org; internet www.naalc.org; f. 1994; Exec. Dir MARK S. KNOUSE (USA); publ. *Annual Report*.

**North American Commission for Environmental Co-operation (NACEC):** 393 rue St Jacques Ouest, Bureau 200, Montréal, QC H2Y IN9, Canada; tel. (514) 350-4300; fax (514) 350-4314; e-mail info@ccemtl.org; internet www.cec.org; f. 1994; Exec. Dir WILLIAM V. KENNEDY; publs *Annual Report*, *Taking Stock* (annually), industry reports, policy studies.

**North American Development Bank (NADB/NADBank):** 203 St Mary's, Suite 300, San Antonio, TX 78205, USA; tel. (210) 231-8000; fax (210) 231-6232; internet www.nadbank.org; at 31 March 2004 the NADB had authorized capital of US $3,000m., subscribed equally by Mexico and the USA, of which $349m. was paid-up; Man. Dir RAUL RODRIGUEZ (Mexico); publs *Annual Report*, *NADBank News*.

# ORGANIZATION OF AMERICAN STATES—OAS
## (ORGANIZACIÓN DE LOS ESTADOS AMERICANOS—OEA)

**Address:** 17th St and Constitution Ave, NW, Washington, DC 20006, USA.

**Telephone:** (202) 458-3000; **fax:** (202) 458-6319; **e-mail:** pi@oas.org; **internet:** www.oas.org.

The ninth International Conference of American States (held in Bogotá, Colombia, in 1948) established the Organization of American States by adopting the Charter (succeeding the International Union of American Republics, founded in 1890). The Charter was subsequently amended by the Protocol of Buenos Aires (creating the annual General Assembly), signed in 1967, enacted in 1970, and by the Protocol of Cartagena de Indias, which was signed in 1985 and enacted in 1988. The purpose of the Organization is to strengthen the peace and security of the continent; to promote and consolidate representative democracy, with due respect for the principle of non-intervention; to prevent possible causes of difficulties and to ensure the peaceful settlement of disputes that may arise among the member states; to provide for common action in the event of aggression; to seek the solution of political, juridical and economic problems that may arise among them; to promote, by co-operative action, their economic, social and cultural development; to achieve an effective limitation of conventional weapons; and to devote the largest amount of resources to the economic and social development of the member states.

### MEMBERS

| | |
|---|---|
| Antigua and Barbuda | Guyana |
| Argentina | Haiti |
| Bahamas | Honduras |
| Barbados | Jamaica |
| Belize | Mexico |
| Bolivia | Nicaragua |
| Brazil | Panama |
| Canada | Paraguay |
| Chile | Peru |
| Colombia | Saint Christopher and Nevis |
| Costa Rica | Saint Lucia |
| Cuba* | Saint Vincent and the Grenadines |
| Dominica | Suriname |
| Dominican Republic | Trinidad and Tobago |
| Ecuador | USA |
| El Salvador | Uruguay |
| Grenada | Venezuela |
| Guatemala | |

*The Cuban Government was suspended from OAS activities in 1962.

**Permanent Observers:** Algeria, Angola, Armenia, Austria, Azerbaijan, Belgium, Bosnia and Herzegovina, Bulgaria, People's Republic of China, Croatia, Cyprus, Czech Republic, Denmark, Egypt, Equatorial Guinea, Estonia, Finland, France, Georgia, Germany, Ghana, Greece, Holy See, Hungary, India, Ireland, Israel,

Italy, Japan, Kazakhstan, Republic of Korea, Latvia, Lebanon, Luxembourg, Morocco, Netherlands, Nigeria, Norway, Pakistan, Philippines, Poland, Portugal, Qatar, Romania, Russia, Saudi Arabia, Serbia and Montenegro, Slovakia, Slovenia, Spain, Sri Lanka, Sweden, Switzerland, Thailand, Tunisia, Turkey, Ukraine, United Kingdom, Yemen and the European Union.

## Organization
(August 2005)

### GENERAL ASSEMBLY

The Assembly meets annually and may also hold special sessions when convoked by the Permanent Council. As the supreme organ of the OAS, it decides general action and policy. The 35th regular session of the General Assembly was convened in Fort Lauderdale, USA, in June 2005.

### MEETINGS OF CONSULTATION OF MINISTERS OF FOREIGN AFFAIRS

Meetings are convened, at the request of any member state, to consider problems of an urgent nature and of common interest to member states, or to serve as an organ of consultation in cases of armed attack or other threats to international peace and security. The Permanent Council determines whether a meeting should be convened and acts as a provisional organ of consultation until ministers are able to assemble.

### PERMANENT COUNCIL

The Council meets regularly throughout the year at OAS headquarters. It is composed of one representative of each member state with the rank of ambassador; each government may accredit alternate representatives and advisers and when necessary appoint an interim representative. The office of Chairman is held in turn by each of the representatives, following alphabetical order according to the names of the countries in Spanish. The Vice-Chairman is determined in the same way, following reverse alphabetical order. Their terms of office are three months.

The Council acts as an organ of consultation and oversees the maintenance of friendly relations between members. It supervises the work of the OAS and promotes co-operation with a variety of other international bodies including the United Nations. It comprises a General Committee and Committees on Juridical and Political Affairs, Hemispheric Security, Inter-American Summits Management and Civil Society Participation in OAS Activities, and Administrative and Budgetary Affairs. There are also *ad hoc* working groups. The official languages are English, French, Portuguese and Spanish.

### INTER-AMERICAN COUNCIL FOR INTEGRAL DEVELOPMENT (CIDI)

The Council was established in 1996, replacing the Inter-American Economic and Social Council and the Inter-American Council for Education, Science and Culture. Its aim is to promote co-operation among the countries of the region, in order to accelerate economic and social development. CIDI's work focuses on eight areas: social development and education; cultural development; the generation of productive employment; economic diversification, integration and trade liberalization; strengthening democratic institutions; the exchange of scientific and technological information; the development of tourism; and sustainable environmental development. An Executive Secretariat for Integral Development provides CIDI with technical and secretarial services.

**Executive Secretary:** Dr Brian J. R. Stevenson.

**Inter-American Agency for Co-operation and Development:** f. November 1999 as a subsidiary body of CIDI to accelerate the development of Latin America and the Caribbean through technical co-operation and training programmes. In particular, the Agency aimed to formulate strategies for mobilizing external funds for OAS co-operation initiatives; establish criteria for the promotion and exchange of co-operation activities; prepare co-operation accords and evaluate project requests and results; and review mechanisms for promoting scholarships and professional exchange programmes. The Agency has developed an Educational Portal of the Americas providing comprehensive electronic access to information regarding distance learning; a Virtual Classroom was launched in May 2003. A Trust for the Americas works within the Agency to enable the private sector to participate in OAS development programmes. The Executive Secretary for Integral Development serves as the Agency's Director-General.

### GENERAL SECRETARIAT

The Secretariat, the central and permanent organ of the Organization, performs the duties entrusted to it by the General Assembly, Meetings of Consultation of Ministers of Foreign Affairs and the Councils.

**Secretary-General:** José Miguel Insulza (Chile).

**Assistant Secretary-General:** Albert R. Ramdin (Suriname).

### INTER-AMERICAN COMMITTEES AND COMMISSIONS

**Inter-American Commission on Human Rights** (Comisión Interamericana de Derechos Humanos): 1889 F St, NW, Washington, DC 20006, USA; tel. (202) 458-6002; fax (202) 458-3992; e-mail cidhoea@oas.org; internet www.cidh.oas.org; f. 1960; comprises seven members; promotes the observance and protection of human rights in the member states of the OAS; examines and reports on the human rights situation in member countries; considers individual petitions relating to alleged human rights violations by member states; in Feb. 2005 established a Special Rapporteurship on the Rights of People of Afro-Descendants, and against Racial Discrimination; Pres. Clare Kamau Roberts (Antigua and Barbuda); Exec. Sec. Santiago Canton.

**Inter-American Committee on Ports** (Comisión Interamericana de Puertos—CIP): 1889 F St, NW, Washington, DC 20006, USA; tel. (202) 458-3871; fax (202) 458-3517; e-mail cip@oas.org; internet www.oas.org/cip; f. 1998 to further OAS activities in the sector (previously undertaken by Inter-American Port Conferences); aims to develop and co-ordinate member state policies in port administration and management; the first meeting of the Committee took place in October 1999; three technical advisory groups were established to advise on port operations, port security, and navigation control and environmental protection; an Executive Board meets annually; Sec. Carlos M. Gallegos.

**Inter-American Court of Human Rights (IACHR)** (Corte Interamericana de Derechos Humanos): Apdo Postal 6906-1000, San José, Costa Rica; tel. (506) 234-0581; fax (506) 234-0584; e-mail corteidh@corteidh.or.cr; internet www.corteidh.or.cr; f. 1978, as an autonomous judicial institution whose purpose is to apply and interpret the American Convention on Human Rights (which entered into force in 1978: at late 2004 the Convention had been ratified by 25 OAS member states, of which 21 had accepted the competence of the Court); comprises seven jurists from OAS member states; Pres. Sergio García-Ramírez (Mexico); Sec. Pablo Saavedra-Alessandri (Chile); publ. *Annual Report*.

**Inter-American Drug Abuse Control Commission** (Comisión Interamericana para el Control del Abuso de Drogas—CICAD): 1889 F St, NW, Washington, DC 20006, USA; tel. (202) 458-3178; fax (202) 458-3658; e-mail oidcicad@oas.org; internet www.cicad.oas.org; f. 1986 by the OAS to promote and facilitate multilateral co-operation in the control and prevention of the trafficking, production and use of illegal drugs, and related crimes; mems: 34 countries; Exec. Sec. James F. Mack; publs *Statistical Survey* (annually), *Directory of Governmental Institutions Charged with the Fight Against the Illicit Production, Trafficking, Use and Abuse of Narcotic Drugs and Psychotropic Substances Evaluation of Progress in Drug Control Progress Report in Drug Control—Implementation in Recommendations* (twice a year).

**Inter-American Juridical Committee (IAJC)** (Comissão Jurídica Interamericana/Comité Jurídico Interamericano): Av. Marechal Floriano 196, 3° andar, Palácio Itamaraty, Centro, 20080-002, Rio de Janeiro, RJ, Brazil; tel. (21) 2206-9903; fax (21) 2203-2090; e-mail cjioea.trp@terra.com.br; composed of 11 jurists, nationals of different member states, elected for a period of four years, with the possibility of re-election. The Committee's purposes are: to serve as an advisory body to the Organization on juridical matters; to promote the progressive development and codification of international law; and to study juridical problems relating to the integration of the developing countries in the hemisphere, and, in so far as may appear desirable, the possibility of attaining uniformity in legislation; Chair. Mauricio Herdocia Sacasa (Nicaragua).

**Inter-American Telecommunication Commission** (Comisión Interamericana de Telecomunicaciones—CITEL): 1889 F St, NW, Washington, DC 20006, USA; tel. (202) 458-3004; fax (202) 458-6854; e-mail citel@oas.org; internet www.citel.oas.org; f. 1993 to promote the development of telecommunications in the region; mems: 35 countries; Exec. Sec. Clovis José Baptista Neto.

## Activities

In December 1994 the first Summit of the Americas was convened in Miami, USA. The meeting endorsed the concept of a Free Trade Area of the Americas (FTAA), and also approved a Plan of Action to strengthen democracy, eradicate poverty and promote sustainable development throughout the region. The OAS subsequently embarked on an extensive process of reform and modernization to strengthen its capacity to undertake a lead role in implementing the Plan. The Organization realigned its priorities in order to respond to the mandates emerging from the Summit and developed a new institutional framework for technical assistance and co-operation, although many activities continued to be undertaken by the specialized or associated organizations of the OAS (see below). In 1998, following the second Summit of the Americas, held in April, in Santiago, Chile, the OAS established an Office of Summit Follow-Up, in order to strengthen its servicing of the meetings, and to co-ordinate tasks assigned to the Organization. The third Summit, convened in Québec, QC, Canada, in April 2001, reaffirmed the central role of the OAS in implementing decisions of the summit meetings and instructed the Organization to pursue the process of reform in order to enhance its operational capabilities, in particular in the areas of human rights, combating trade in illegal drugs, and enforcement of democratic values. The Summit declaration stated that commitment to democracy was a requirement for a country's participation in the summit process. The meeting also determined that the OAS was to be the technical secretariat for the process, assuming many of the responsibilities previously incumbent on the host country. Further to its mandate, the OAS established a Summits of the Americas Secretariat, which assists countries in planning and follow-up and provides technical, logistical and administrative support for the Summit Implementation Review Group and the summit process. A Special Summit of the Americas was held in January 2004, in Monterrey, Mexico, to reaffirm commitment to the process and to advance the implementation of measures, in particular to combat poverty to promote social development and equitable economic growth and to strengthen democratic governance. The fourth Summit was scheduled to be convened in Argentina, in November 2005.

## TRADE AND ECONOMIC INTEGRATION

A trade unit was established in 1995 in order to strengthen the Organization's involvement in trade issues and the process of economic integration, which became a priority area following the first Summit of the Americas. The unit was to provide technical assistance in support of the establishment of the FTAA, and to co-ordinate activities between regional and sub-regional integration organizations. Subsequently the unit has provided technical support to the nine FTAA negotiating groups: market access; investment; services; government procurement; dispute settlement; agriculture; intellectual property rights; subsidies, anti-dumping and countervailing duties; and competition policy. In April 2001 the third Summit of the Americas requested the OAS to initiate an analysis of corporate social responsibility. At the Summit it was agreed that negotiations to establish the FTAA should be concluded by January 2005. At a special summit of the Americas, held in January 2004 in Monterrey, Mexico, however, the leaders failed to specify a completion date for the negotiations, although they adopted a declaration committing themselves to its eventual establishment. Negotiations on the FTAA stalled during 2004. The trade unit supports a Hemispheric Co-operation Programme, which was established by ministers of trade of the Americas, meeting in November 2002, to assist smaller economies to gain greater access to resources and technical assistance.

The unit operates in consultation with a Special Committee on Trade, which was established in 1993, comprising high-level officials representing each member state. The Committee studies trade issues, provides technical analyses of the economic situation in member countries and the region, and prepares reports for ministerial meetings of the FTAA. The OAS also administers an Inter-American Foreign Trade Information System which facilitates the exchange of information.

## DEMOCRACY AND CIVIL SOCIETY

Two principal organs of the OAS, the Inter-American Commission on Human Rights and the Inter-American Court of Human Rights, work to secure respect for human rights in all member countries. The OAS aims to encourage more member governments to accept the jurisdiction of the Court. The OAS also collaborates with member states in the strengthening of representative institutions within government and as part of a democratic civil society. The third Summit of the Americas, convened in April 2001, mandated the OAS to formulate an Inter-American Democratic Charter. The Charter was adopted in September at a special session of the Assembly. Central to the Charter's five chapters was democracy and its relationship to human rights, integral development and combating poverty. It aimed to establish procedures to strengthen and promote democracy and to respond to violations of or threats to democracy in member countries.

Through its Unit for the Promotion of Democracy, established in 1990, the OAS provides electoral technical assistance to member states and undertakes missions to observe the conduct of elections. By mid-2005 the OAS had conducted more than 70 electoral missions in 19 member countries. In the first half of 2005 missions included observing political party primary elections in Honduras and a general election in Suriname. The OAS also supports societies in post-conflict situations and recently-installed governments to promote democratic practices. In November 2003 the OAS sponsored the third Inter-American Forum on Political Parties, held in Cartagena, Colombia, during which leaders of political parties considered means of enhancing campaign transparency, implementing structural party reforms and strengthening links to grass-roots membership.

In June 1991 the OAS General Assembly approved a resolution on representative democracy, which authorized the Secretary-General to summon a session of the Permanent Council in cases where a democratically-elected government had been overthrown or democratic procedures abandoned in member states. The Council could then convene an *ad hoc* meeting of ministers of foreign affairs to consider the situation. The procedure was invoked following political developments in Haiti, in September 1991, and Peru, in April 1992. Ministers determined to impose trade and diplomatic sanctions against Haiti and sent missions to both countries. The resolution was incorporated into the Protocol of Washington, amending the OAS charter, which was adopted in December 1992 and entered into force in September 1997. A high-level OAS mission was dispatched to Peru in June 2000 to assist with the process of 'strengthening its institutional democratic system', following allegations that the Peruvian authorities had manipulated the re-election of that country's President in May. The mission subsequently co-ordinated negotiations between the Peruvian Government and opposition organizations.

In August 2000 the OAS Secretary-General undertook the first of several high-level missions to negotiate with the authorities in Haiti in order to resolve the political crisis resulting from a disputed general election in May. (An OAS electoral monitoring team was withdrawn from Haiti in July prior to the second round of voting owing to concern at procedural irregularities.) In January 2001, following a meeting with the Haitian Prime Minister, the Assistant Secretary-General recommended that the OAS renew its efforts to establish a dialogue between the government, opposition parties and representatives of civil society in that country. In May the OAS and CARICOM undertook a joint mission to Haiti in order to assess prospects for a democratic resolution to the political uncertainties, and in June the OAS General Assembly issued a resolution urging all parties in Haiti to respect democratic order. At the end of that month the OAS Secretary-General led a visit of the joint mission to Haiti, during which further progress was achieved on the establishment of a new electoral council. Following political and social unrest in Haiti in December 2001, the OAS and CARICOM pledged to conduct an independent investigation into the violence, and in March 2002 an agreement to establish a Special OAS Mission for Strengthening Democracy in Haiti was signed in the capital, Port-au-Prince. The independent commission of inquiry reported to the OAS at the beginning of July, and listed a set of recommendations relating to law reform, security and other confidence-building measures to help to secure democracy in Haiti. In September, in acknowledgement of the perceived efforts of the Government to resolve the political *impasse* (including a commitment by the President to staging legislative and municipal elections by the end of 2003), the OAS issued a resolution releasing aid to Haiti that had been blocked since the controversial May 2000 elections. Release of the funds, however, was contingent upon certain political and social reforms (such as reform of the police force and the disarming of partisans), and by mid-2003 it appeared that few of the conditions set by the OAS had been met. In January 2004 the OAS Special Mission condemned the escalation of political violence in Haiti and in February took a lead in drafting a plan of action to implement a CARICOM-brokered action plan to resolve the crisis. In late February the Permanent Council met in special session, and urged the United Nations to take necessary and appropriate action to address the deteriorating situation in Haiti. On 29 February President Aristide resigned and left the country; amidst ongoing civil unrest, a provisional president was sworn in. The OAS Mission continued to attempt to maintain law and order, in co-operation with a UN-authorized Multinational Interim Force, and facilitated political discussions on the establishment of a transitional government. From March the Special Mission participated in the process to develop an Interim Co-operation Framework, identifying the urgent and medium-term needs of Haiti, which was presented to a meeting of international donors held in July. In June the OAS General Assembly adopted a resolution instructing the Permanent Council to undertake all necessary diplomatic initiatives to foster the restoration of democracy in Haiti, and called upon the Special Mission to work with the new UN Stabilization Mission in Haiti in preparing, organizing and monitoring future elections. In the first half of 2005 OAS technical experts, together with UN counterparts, assisted Haiti's Provisional Electoral Council (PEC) with the process of voter registration for legislative and presidential elections scheduled for later in that year. In May the then acting OAS Secretary-General proposed the creation of a tripartite Haiti International Commission in order to achieve progress in the case of the former Haitian Prime Minister, Yvon Neptune, the circumstances of whose detention by the Haitian authorities had prompted strong criticism by the Inter-American Commission on Human Rights. In July the newly-elected OAS Secretary-General visited Haiti and expressed his support for the PEC and the electoral process there.

In April 2002 a special session of the General Assembly was convened to discuss the ongoing political instability in Venezuela. The Assembly applied its authority granted under the Inter-American Democratic Charter to condemn the alteration of the constitutional order in Venezuela which forced the temporary eviction of President Hugo Chávez from office. In January 2003 the OAS announced the establishment of a Group of Friends, composed of representatives from Brazil, Chile, Mexico, Spain, Portugal and the USA, to support its efforts to resolve the ongoing crisis in Venezuela. In March the OAS Secretary-General was invited by Venezuelan opposition groupings to mediate negotiations with the Government. The talks culminated in May with the signing of an OAS-brokered agreement which, it was hoped, would lead to mid-term referendums on elected officials, including the presidency. The OAS, with the Carter Center, subsequently oversaw and verified the collection of signatures to determine whether referendums should be held. Following the staging of a recall referendum on the Venezuelan presidency in August 2004, OAS member states urged that there should be a process of reconciliation in that country. An OAS mission was to observe legislative elections scheduled to be held there in December 2005.

In April 2005 a special session of the Permanent Council expressed concern at recent political developments in Ecuador and determined to send a high-level mission to work with officials and political and civil organizations in that country in order to strengthen democratic processes. In May, in a report on its visit to Ecuador, the high-level mission urged all Ecuadorians to engage in an urgent national dialogue, facilitated by the OAS. In the same month, the Permanent Council of the OAS approved a resolution supporting the country and encouraging the Government and all political, social and economic sectors to maintain a wide-ranging dialogue.

An OAS Assistance Programme for Demining in Central America was established in 1992, based in Managua, Nicaragua, to contribute to the social and economic rehabilitation of the region. The programme was to provide training, oversee clearance of anti-personnel devices, undertake risk education, and assist victims of mines, including assistance with medical treatment and retraining activities. The mine clearance operation in Costa Rica was concluded in December 2002 and in Honduras by the end of 2004. In 2002 OAS began to support demining operations in Ecuador and Peru, and in 2003 opened an office in Bogotá, Colombia, to help support mine action programmes, in particular victim rehabilitation and the development of a mine action database. Since mid-2002 the programme, together with the Inter-American Defense Board (see below), had assisted the destruction of an estimated 1m. landmines from military stock-piles in Argentina, Colombia, Chile, Ecuador, Honduras, Nicaragua, Peru and Suriname (which, in April 2005, became the third member state to conclude its demining programme). In November 1997 an Inter-American Convention against the Illicit Manufacture of and Trafficking in Firearms, Ammunition, Explosives, and Other related materials was adopted; it entered into force in July 1998. By June 2005 the Convention had been ratified by 26 countries. An Inter-American Convention on Transparency in Conventional Weapons was adopted in 1999.

The OAS formulated an Inter-American Programme of Co-operation to Combat Corruption in order to address the problem at national level and, in 1996, adopted a Convention against Corruption. The first conference of the parties to the Convention was held in Buenos Aires, Argentina, in May 2001. In June the General Assembly approved the proposed establishment of a verification mechanism, including a policy-making annual conference and an intergovernmental committee of experts. At June 2005 the Convention had been ratified by 30 member states. A working group on transparency aims to promote accountability throughout the public sector and supports national institutions responsible for combating corruption. In 1997 the OAS organized a meeting of experts to consider measures for the prevention of crime. The meeting adopted a work programme which included commitments to undertake police training in criminology and crime investigation, to exchange legal and practical knowledge, and to measure crime and violence in member countries.

### REGIONAL SECURITY

In 1991 the General Assembly established a working group to consider issues relating to the promotion of co-operation in regional security. A Special Commission on Hemispheric Security was subsequently established, while two regional conferences have been held on security and confidence-building measures. Voluntary practices agreed by member states include the holding of border meetings to co-ordinate activities, co-operation in natural disaster management, and the exchange of information on military exercises and expenditure. From 1995 meetings of ministers of defence have been convened regularly, which provide a forum for discussion of security matters. The OAS aims to address the specific security concerns of small-island states, in particular those in the Caribbean, by adopting a multidimensional approach to counter their vulnerability, for example through efforts to strengthen democracy, to combat organized crime, to mitigate the effects of natural disasters and other environmental hazards, and to address the problem of HIV/AIDS.

In June 2000 the OAS General Assembly, convened in Windsor, Canada, established a Fund for Peace in support of the peaceful settlement of territorial disputes. In the early 2000s the Fund supported efforts to resolve disputes between Belize and Guatemala and between Honduras and Nicaragua. In June 2001 an agreement was concluded to enable an OAS Civilian Verification Mission to visit Honduras and Nicaragua in order to monitor compliance with previously agreed confidence-building measures. The Mission was to be financed by the Fund for Peace.

The OAS is actively involved in efforts to combat the abuse and trafficking of illegal drugs, and in 1996 members approved a Hemispheric Anti-drug Strategy, reiterating their commitment to addressing the problem. In 1998 the Inter-American Drug Abuse Control Commission (CICAD) established a Multilateral Evaluation Mechanism (MEM) to measure aspects of the drug problem and to co-ordinate an evaluation process under which national plans of action to combat drugs trafficking were to be formulated. The first hemispheric drugs report was published by MEM in January 2001 and in February 34 national reports produced under MEM were issued together with a series of recommendations for action. In January 2002 MEM published a progress report on the implementation of the recommendations, in which it stated that the countries of the Americas had made a 'significant effort' to adopt the recommendations, although in some instances they had encountered difficulties owing to a lack of technical or financial resources. It was reported that advances had been made in developing national anti-drugs plans, measuring land under illicit cultivation, and adopting procedures against money-laundering. Since 1996 an OAS group of experts has undertaken efforts to assist countries in reducing the demand for illegal substances. Activities include the implementation of prevention programmes for street children; the development of communication strategies; and education and community projects relating to the prevention of drug dependence.

The first Specialized Inter-American Conference on Terrorism was held in Lima, Peru, in April 1996. A Declaration and Plan of Action were adopted, according to which member states agreed to co-operate and implement measures to combat terrorism and organized crime. A second conference was held in Mar del Plata, Argentina, in 1998, which culminated in the adoption of the Commitment of Mar del Plata. Member states recommended the establishment of an Inter-American Committee against Terrorism (CICTE) to implement decisions relating to judicial, police and intelligence co-operation. The Committee held its first session in Miami, USA, in October 1999; a second meeting was convened in January 2002. Two special sessions of the CICTE were held in October and November 2001, following the terrorist attacks on targets in the USA in September. A CICTE On-Line Anti-terrorism Database came into operation during 2002. In June 2002, at the OAS General Assembly, held in Bridgetown, Barbados, 30 heads of state and of government signed an Inter-American Convention against Terrorism. The Assembly also adopted a Declaration of Bridgetown on the multidimensional approach to hemispheric security. The Convention entered into force in July 2003, following its ratification by the required six member states. By February 2005 the Convention had been signed by all member states and ratified by 12 states. A Special Conference on Security was convened in October 2003 to review the security structure throughout the Americas.

### SOCIAL DEVELOPMENT AND EDUCATION

In June 1996 the OAS established a specialized unit for social development and education to assist governments and social institutions of member states to formulate public policies and implement initiatives relating to employment and labour issues, education development, social integration and poverty elimination. It was also to provide technical and operational support for the implementation of inter-American programmes in those sectors, and to promote the exchange of information among experts and professionals working in those fields. In June 1997 the OAS approved an Inter-American Programme to Combat Poverty and Discrimination. The unit serves as the technical secretariat for annual meetings on social development that were to be convened within the framework of the Programme. The unit also administers the Social Networks of Latin America and the Caribbean project, and its co-ordinating committee, which promotes sub-regional co-operation to combat poverty and discrimination. From 1999 the unit implemented a project funded by the Inter-American Development Bank to place interns and trainees within the Social Network institutions and to promote exchanges between the institutions.

The first meeting of ministers of education of the Americas was held in Brasília, Brazil, in July 1998, based on the mandate of the second Summit of the Americas. The meeting approved an Inter-American Education Programme, formulated by the unit for social development and education, which incorporated the following six priority initiatives: education for priority social sectors; strengthening educational management and institutional development; education for work and youth development; education for citizenship and sustainability in multicultural societies; training in the use of new technologies in teaching the official languages of the OAS; and training of teachers and education administrators. Other programmes in the education sector are undertaken with international agencies and non-governmental organizations (NGOs). In April 2004 OAS convened a special meeting on the promotion of a democratic culture through education, which was attended by ministers of education and other officials, education experts and representatives of NGOs. The meeting agreed upon guide-lines for the development of an Inter-American Values Programme, which was to carry out goals identified by OAS ministers of education and principles expressed in the Inter-American Democratic Charter.

The OAS supports member states to formulate and implement programmes to promote productive employment and vocational training, to develop small businesses and other employment generation initiatives, and to regulate labour migration. In 1998 the OAS initiated the Labour Market Information System project, which aimed to provide reliable and up-to-date indicators of the labour situation in member countries, to determine the impact of economic policy on the labour situation, and to promote the exchange of information among relevant national and regional institutions. Labour issues were addressed by the second Summit of the Americas, and, following an Inter-American Conference of Labour Ministers, held in Viña del Mar, Chile, in October 1998, two working groups were established to consider the globalization of the economy and its social and labour dimension and the modernization of the state and labour administration. In June 2002 the Pan-American Health Organization (PAHO) and Inter-American Institute for Co-operation on Agriculture (IICA) signed an agreement to increase co-operation in matters relating to health and agriculture, in order to combat poverty in rural communities.

### SUSTAINABLE DEVELOPMENT AND THE ENVIRONMENT

In 1996 a summit meeting on social development adopted a plan of action, based on the objectives of the UN Conference on the Environment and Development, which was held in Rio de Janeiro, Brazil, in June 1992. The OAS was to participate in an inter-agency group to monitor implementation of the action plan. The OAS has subsequently established new financing mechanisms and networks of experts relating to aspects of sustainable development. Technical co-operation activities include multinational basin management; a strategic plan for the Amazon; natural disaster management; and the sustainable development of border areas in Central America and South America. In December 1999 the Inter-American Council for Integral Development approved a policy framework and recommendations for action of a new Inter-American Strategy for Public Participation in Decision-making for Sustainable Development.

The following initiatives have also been undertaken by the OAS Unit for Sustainable Development and Environment: a Caribbean Disaster Mitigation Project, to help those countries to counter and manage the effects of natural disasters; a Post-Georges Disaster Mitigation initiative specifically to assist countries affected by 'Hurricane Georges'; a Natural Hazards Project to provide a general programme of support to assess member states' vulnerability, to provide technical assistance and training to mitigate the effects of a disaster, and to assist in the planning and formulation of development and preparedness policies; the Renewable Energy in the Americas initiative to promote co-operation and strengthen renewable energy and energy efficiency; an Inter-American Water Resources Network, which aims to promote collaboration, training and the exchange of information within the sector; and a Water Level Observation Network for Central America to provide support for coastal resources management, navigation and disaster mitigation in the countries affected by 'Hurricane Mitch'. The OAS has also initiated projects to assist Caribbean countries with formulating policies to reduce their vulnerability to natural disasters. In mid-2005 the OAS and other international agencies launched a multi-year environmental project, aimed at assisting the countries of the Amazon Basin with the conservation and management of the region's water resources, forests and wildlife. The project was to focus in particular on assisting countries to confront the challenge of acute climate change.

### SCIENCE AND TECHNOLOGY

The OAS supports and develops activities to contribute to the advancement of science and technology throughout the Americas, and to promote its contribution to social and sustainable development. In particular, it promotes collaboration, dissemination of information and improved communication between experts and institutions working in the sector. Specialized bodies and projects have been established to promote activities in different fields, for example metrology; co-operation between institutions of accreditation, certification and inspection; the development of instruments of measurements and analysis of science and technology; chemistry; the development of technical standardization and related activities; and collaboration between experts and institutions involved in biotechnology and food technology. The OAS also maintains an information system to facilitate access to databases on science and technology throughout the region.

### TOURISM AND CULTURE

A specialized unit for tourism was established in 1996 in order to strengthen and co-ordinate activities for the sustainable development of the tourism industry in the Americas. The unit supports regional and subregional conferences and workshops, as well as the Inter-American Travel Congress, which was convened for the first time in 1993 to serve as a forum to formulate region-wide tourism policies. The unit also undertakes research and analysis of the industry.

In 1998 the OAS approved an Inter-American Programme of Culture to support efforts being undertaken by member states and to promote co-operation in areas such as cultural diversity; protection of cultural heritage; training and dissemination of information; and the promotion of cultural tourism. The OAS also assists with the preparation of national and multilateral cultural projects, and co-operates with the private sector to protect and promote cultural assets and events in the region.

### COMMUNICATIONS

In June 1993 the OAS General Assembly approved the establishment of an Inter-American Telecommunication Commission. The body has technical autonomy, within the statute and mandate agreed by the Assembly. It aims to facilitate and promote the development of communications in all member countries, in collaboration with the private sector and other organizations, and serves as the principal advisory body of the OAS on matters related to telecommunications.

## Finance

The OAS budget for 2006, approved by the General Assembly in mid-2005, amounted to US $84.1m., of which $76.3m. was to come from the regular fund, and $7.9m. from the Special Fund for Integral Development (FEMCIDI).

## Publications

(in English and Spanish)

*Américas* (6 a year).
*Annual Report.*
*Catalog of Publications* (annually).
*Newsletter* (online, 6 a year).
Numerous cultural, legal and scientific reports and studies.

## Specialized Organizations and Associated Agencies

**Inter-American Children's Institute** (Instituto Americano del Niño—IIN): Avda 8 de Octubre 2904, POB 16212, Montevideo 11600, Uruguay; tel. (2) 487-2150; fax (2) 487-3242; e-mail iin@oas.org; internet www.iin.oea.org; f. 1927; promotes the regional implementation of the Convention on the Rights of the Child, assists in the development of child-oriented public policies; promotes co-operation between states; and aims to develop awareness of problems affecting children and young people in the region. The Institute organizes workshops, seminars, courses, training programmes and conferences on issues relating to children, including, for example, the rights of children, children with disabilities, and the child welfare system. It also provides advisory services, statistical data and other relevant information to authorities and experts throughout the region; Pres. of the Directing Council ANA TERESA ARANDA OROZCO; publ. *iinfancia* (annually).

**Inter-American Commission of Women** (Comisión Interamericana de Mujeres—CIM): 1889 F St, NW, Suite 350 Washington, DC 20006, USA; tel. (202) 458-6084; fax (202) 458-6094; e-mail spcim@oas.org; f. 1928 for the extension of civil, political, economic, social and cultural rights for women; in 1991 a Seed Fund was established to provide financing for grass-roots projects consistent with the Commission's objectives; Pres. NILCÉA FREIRE (Brazil); Exec. Sec. CARMEN LOMELLIN (USA).

**Inter-American Committee Against Terrorism** (Comité Interamericano Contra el Terrorismo—CICTE): 17th St and Constitution Ave, NW, Washington, DC 20006, USA; tel. (202) 458-3000; fax (202) 458-3857; internet www.cicte.oas.org; f. 1999 to enhance the exchange of information via national authorities (including the establishment of an Inter-American database on terrorism issues), formulate proposals to assist member states in drafting counter-terrorism legislation in all states, compile bilateral, sub-regional, regional and multilateral treaties and agreements signed by member states and promote universal adherence to international counter-terrorism conventions, strengthen border co-operation and travel documentation security measures, and develop activities for training and crisis management; Exec. Sec. STEVEN MONBLATT (USA).

**Inter-American Defense Board** (Junta Interamericana de Defensa—JID): 2600 16th St, NW, Washington, DC 20441, USA; tel. (202) 939-6041; fax (202) 387-2880; e-mail iadc-registrar@jid.org; internet www.jid.org; works in liaison with member governments to plan and train for the common security interests of the western hemisphere; operates the Inter-American Defense College; Dir Maj.-Gen. KEITH M. HUBER (USA).

**Inter-American Indigenous Institute** (Instituto Indigenista Interamericano—III): Avda de las Fuentes 106, Col. Jardines del Pedregal, Delegación Álvaro Obregón, 01900 México, DF, Mexico; tel. (55) 5595-8410; fax (55) 5595-4324; e-mail ininin@data.net.mx; internet www.indigenista.org; f. 1940; conducts research on the situation of the indigenous peoples of America; assists the exchange of information; promotes indigenous policies in member states aimed at the elimination of poverty and development within Indian communities, and to secure their position as ethnic groups within a democratic society; Hon. Dir Dr GUILLERMO ESPINOSA VELASCO (Mexico); publs *América Indígena* (quarterly), *Anuario Indigenista*.

**Inter-American Institute for Co-operation on Agriculture (IICA)** (Instituto Interamericano de Cooperación para la Agricultura): Apdo Postal 55–2200 San Isidro de Coronado, San José, Costa Rica; tel. (506) 216-0222; fax (506) 216-0233; e-mail iicahq@iica.ac.cr; internet www.iica.int; f. 1942 (as the Inter-American Institute of Agricultural Sciences, present name adopted 1980); supports the efforts of member states to improve agricultural development and rural well-being; encourages co-operation between regional organizations, and provides a forum for the exchange of experience; Dir-Gen. Dr CHELSTON W. D. BRATHWAITE (Barbados).

**Pan American Development Foundation (PADF)** (Fundación Panamericana para el Desarrollo): 1889 F St, NW, Washington, DC 20006, USA; tel. (202) 458-3969; fax (202) 458-6316; e-mail padf-dc@padf.org; internet www.padf.org; f. 1962 to promote and facilitate economic and social development in Latin America and the Caribbean by means of innovative partnerships and integrated involvement of the public and private sectors; provides low-interest credit for small-scale entrepreneurs, vocational training, improved health care, agricultural development and reafforestation, and strengthening local non-governmental organizations; provides emergency disaster relief and reconstruction assistance; Pres. RUTH ESPEY-ROMERO; Sec. YOLANDA MELLON SUÁREZ.

**Pan American Health Organization (PAHO)** (Organización Panamericana de la Salud): 525 23rd St, NW, Washington, DC 20037, USA; tel. (202) 974-3000; fax (202) 974-3663; e-mail webmaster@paho.org; internet www.paho.org; f. 1902; co-ordinates regional efforts to improve health; maintains close relations with national health organizations and serves as the Regional Office for the Americas of the World Health Organization; Dir Dr MIRTA ROSES PERIAGO (Argentina).

**Pan-American Institute of Geography and History (PAIGH)** (Instituto Panamericano de Geografía e Historia–IPGH): Ex-Arzobispado 29, 11860 México, DF, Mexico; tel. (55) 5277-5888; fax (55) 5271-6172; e-mail secretariageneral@ipgh.org.mx; internet www.ipgh.org.mx; f. 1928; co-ordinates and promotes the study of cartography, geophysics, geography and history; provides technical assistance, conducts training at research centres, distributes publications, and organizes technical meetings; Sec.-Gen. SANTIAGO BORRERO MUTIS (Colombia); Publs *Revista Cartográfica* (2 a year), *Revista Geográfica* (2 a year), *Revista de Historia de América* (2 a year), *Revista Geofísica* (2 a year), *Revista de Arqueología Americana* (annually), *Folklore Americano* (annually), *Boletín de Antropología Americana* (annually).

# SOUTHERN COMMON MARKET— MERCOSUR/MERCOSUL
## (MERCADO COMÚN DEL SUR/MERCADO COMUM DO SUL)

**Address:** Edif. Mercosur, Luis Piera 1992, 1°, 11200 Montevideo, Uruguay.

**Telephone:** (2) 412-9024; **fax:** (2) 418-0557; **e-mail:** sam@mercosur.org.uy; **internet:** www.mercosur.org.uy.

Mercosur (known as Mercosul in Portuguese) was established in March 1991 by the heads of state of Argentina, Brazil, Paraguay and Uruguay with the signature of the Treaty of Asunción. The primary objective of the Treaty is to achieve the economic integration of member states by means of a free flow of goods and services between member states, the establishment of a common external tariff, the adoption of common commercial policy, and the co-ordination of macroeconomic and sectoral policies. The Ouro Preto Protocol, which was signed in December 1994, conferred on Mercosur the status of an international legal entity with the authority to sign agreements with third countries, group of countries and international organizations.

### MEMBERS
Argentina   Brazil   Paraguay   Uruguay

Bolivia, Chile, Colombia, Ecuador, Peru and Venezuela are associate members.

## Organization
(August 2005)

### COMMON MARKET COUNCIL
The Common Market Council (Consejo del Mercado Común) is the highest organ of Mercosur and is responsible for leading the integration process and for taking decisions in order to achieve the objectives of the Asunción Treaty.

### COMMON MARKET GROUP
The Common Market Group (Grupo Mercado Común) is the executive body of Mercosur and is responsible for implementing concrete measures to further the integration process.

### TRADE COMMISSION
The Trade Commission (Comisión de Comercio del Mercosur) has competence for the area of joint commercial policy and, in particular, is responsible for monitoring the operation of the common external tariff (see below). The Brasília Protocol may be referred to for the resolution of trade disputes between member states.

### JOINT PARLIAMENTARY COMMISSION
The Joint Parliamentary Commission (Comisión Parlamentaria Conjunta) is made up of parliamentarians from the member states and is charged with accelerating internal national procedures to implement Mercosur decisions, including the harmonization of country legislation.

## CONSULTATIVE ECONOMIC AND SOCIAL FORUM

The Consultative Economic and Social Forum (Foro Consultivo Económico-Social) comprises representatives from the business community and trade unions in the member countries and has a consultative role in relation to Mercosur.

## ADMINISTRATIVE SECRETARIAT

**Director:** REGINALDO BRAGA ARCURI (Brazil) (until 31 Dec. 2005).

# Activities

Mercosur's free-trade zone entered into effect on 1 January 1995, with tariffs removed from 85% of intra-regional trade. A regime of gradual removal of duties on a list of special products was agreed, with Argentina and Brazil given four years to complete this process while Paraguay and Uruguay were allowed five years. Regimes governing intra-zonal trade in the automobile and sugar sectors remained to be negotiated. Mercosur's customs union also came into force at the start of 1995, comprising a common external tariff of 0%–20%. A list of exceptions from the common external tariff was also agreed; these products were to lose their special status and be subject to the general tariff system concerning foreign goods by 2006. The value of intra-Mercosur trade was estimated to have tripled during the period 1991–95 and was reported to have amounted to US $20,400m. in 1998. However, intra-subregional trade was reported to have declined to $15,200m. in 1999, stabilizing at about 15,000m. in 2000. The financial crisis that escalated in Argentina during late 2001 had a detrimental effect on intra-Mercosur trade.

In December 1995 Mercosur presidents affirmed the consolidation of free trade as Mercosur's 'permanent and most urgent goal'. To this end they agreed to prepare norms of application for Mercosur's customs code, accelerate paper procedures and increase the connections between national computerized systems. It was also agreed to increase co-operation in the areas of agriculture, industry, mining, energy, communications, transport and tourism, and finance. At this meeting Argentina and Brazil reached an accord aimed at overcoming their dispute regarding the trade in automobiles between the two countries. They agreed that cars should have a minimum of 60% domestic components and that Argentina should be allowed to complete its balance of exports of cars to Brazil, which had earlier imposed a unilateral quota on the import of Argentine cars. In June 1995 Mercosur ministers responsible for the environment agreed to harmonize environmental legislation and to form a permanent sub-group of Mercosur

In May 1996 Mercosur parliamentarians met with the aim of harmonizing legislation on patents in member countries. In December Mercosur heads of state, meeting in Fortaleza, Brazil, approved agreements on harmonizing competition practices (by 2001), integrating educational opportunities for post-graduates and human resources training, standardizing trading safeguards applied against third-country products (by 2001) and providing for intra-regional cultural exchanges. An Accord on Subregional Air Services was signed at the meeting (including by the heads of state of Bolivia and Chile) to liberalize civil transport throughout the region. In addition, the heads of state endorsed texts on consumer rights that were to be incorporated into a Mercosur Consumers' Defence Code, and agreed to consider the establishment of a bank to finance the integration and development of the region.

In June 1996 the Joint Parliamentary Commission agreed that Mercosur should endorse a 'Democratic Guarantee Clause', whereby a country would be prevented from participation in Mercosur unless democratic, accountable institutions were in place. The clause was adopted by Mercosur heads of state at the summit meeting held in San Luis de Mendoza, Argentina, later in the month. The presidents approved the entry into Mercosur of Bolivia and Chile as associate members. An Economic Complementation Accord with Bolivia, which includes Bolivia in Mercosur's free-trade zone, but not in the customs union, was signed in December 1995 and was to come into force on 1 January 1997. In December 1996 the Accord was extended until 30 April 1997, when a free-trade zone between Bolivia and Mercosur was to become operational. Measures of the free-trade agreement, which was signed in October 1996, were to be implemented over a transitional period commencing on 28 February 1997 (revised from 1 January). Chile's Economic Complementation Accord with Mercosur entered into effect on 1 October 1996, with duties on most products to be removed over a 10-year period (Chile's most sensitive products were given 18 years for complete tariff elimination). Chile was also to remain outside the customs union, but was to be involved in other integration projects, in particular infrastructure projects designed to give Mercosur countries access to both the Atlantic and Pacific Oceans (Chile's Pacific coast was regarded as Mercosur's potential link to the economies of the Far East).

In June 1997 the first meeting of tax administrators and customs officials of Mercosur member countries was held, with the aim of enhancing information exchange and promoting joint customs inspections. During 1997 Mercosur's efforts towards regional economic integration were threatened by Brazil's adverse external trade balance and its Government's measures to counter the deficit, which included the imposition of import duties on certain products. In November the Brazilian Government announced that it was to increase its import tariff by 3%, in a further effort to improve its external balance. The measure was endorsed by Argentina as a means of maintaining regional fiscal stability. The new external tariff, which was to remain in effect until 31 December 2000, was formally adopted by Mercosur heads of state at a meeting held in Montevideo, Uruguay, in December 1997. At the summit meeting a separate Protocol was signed providing for the liberalization of trade in services and government purchases over a 10-year period. In order to strengthen economic integration throughout the region, Mercosur leaders agreed that Chile, while still not a full member of the organization, be integrated into the Mercosur political structure, with equal voting rights. In December 1998 Mercosur heads of state agreed on the establishment of an arbitration mechanism for disputes between members, and on measures to standardize human, animal and plant health and safety regulations throughout the grouping. In March 1998 the ministers of the interior of Mercosur countries, together with representatives of the Governments of Chile and Bolivia, agreed to implement a joint security arrangement for the border region linking Argentina, Paraguay and Brazil. In particular, the initiative aimed to counter drugs-trafficking, money-laundering and other illegal activities in the area.

Tensions within Mercosur were compounded in January 1999 owing to economic instability in Brazil and its Government's decision effectively to devalue the national currency, the real. In March the grouping's efforts at integration were further undermined by political instability in Paraguay. As a consequence of the devaluation of its currency, Brazil's important automotive industry became increasingly competitive, to the detriment of that of Argentina. In April Argentina imposed tariffs on imports of Brazilian steel and, in July, the Argentine authorities approved a decree permitting restrictions on all imports from neighbouring countries, in order to protect local industries, prompting Brazil to suspend negotiations to resolve the trading differences between the two countries. Argentina withdrew the decree a few days later, but reiterated its demand for some form of temporary safeguards on certain products as compensation for their perceived loss of competitiveness resulting from the devalued real. An extraordinary meeting of the Common Market Council was convened, at Brazil's request, in August, in order to discuss the dispute, as well as measures to mitigate the effects of economic recession throughout the subregion. However, little progress was made and the bilateral trade dispute continued to undermine Mercosur. Argentina imposed new restrictions on textiles and footwear, while, in September, Brazil withdrew all automatic import licences for Argentine products, which were consequently to be subject to the same quality control, sanitary measures and accounting checks applied to imports from non-Mercosur countries. In January 2000, however, the Argentine and Brazilian Governments agreed to refrain from adopting potentially divisive unilateral measures and resolved to accelerate negotiations on the resolution of ongoing differences. In March Mercosur determined to promote and monitor private accords to cover the various areas of contention, and also established a timetable for executing a convergence of regional macroeconomic policies. In June Argentina and Brazil signed a bilateral automobile agreement. The motor vehicle agreement, incorporating new tariffs and a nationalization index, was endorsed by all Mercosur leaders at a meeting convened in Florianopolis, Brazil, in December. The significant outcome of that meeting was the approval of criteria, formulated by Mercosur finance ministers and central bank governors, determining monetary and fiscal targets to achieve economic convergence. Annual inflation rates were to be no higher than 5% in 2002–05, and reduced to 4% in 2006 and 3% from 2007 (with an exception for Paraguay). Public debt was to be reduced to 40% of GDP by 2010, and fiscal deficits were to be reduced to no more than 3% of GDP by 2002. The targets aimed to promote economic stability throughout the region, as well as to reduce competitive disparities affecting the unity of the grouping. The Florianopolis summit meeting also recommended the formulation of social indicators to facilitate achieving targets in the reduction of poverty and the elimination of child labour. However, political debate surrounding the meeting was dominated by the Chilean Government's announcement that it had initiated bilateral free-trade discussions with the USA, which was considered, in particular by the Brazilian authorities, to undermine Mercosur's unified position at multilateral free-trade negotiations. Procedures to incorporate Chile as a full member of Mercosur were suspended. (Chile and the USA concluded negotiations on a bilateral free-trade agreement in December 2002.)

In early 2001 Argentina imposed several emergency measures to strengthen its domestic economy, in contradiction of Mercosur's external tariffs. In March Brazil was reported to have accepted the measures, which included an elimination of tariffs on capital goods and an increase in import duties on consumer goods, as an exceptional temporary trade regime; this position was reversed by mid-2001 following Argentina's decision to exempt certain countries from import tariffs. In February 2002, at a third extraordinary meeting of the Common Market Council, held in Buenos Aires, Argentina, Mercosur heads of state expressed their support for Argentina's application to receive international financial assistance, in the wake of that country's economic crisis. Although there were fears that the crisis might curb trade and stall economic growth across the region, Argentina's adoption of a floating currency made the prospect of currency harmonization between Mercosur member countries appear more viable. During the meeting it was also agreed that a permanent panel to consider trade disputes would be established in Asunción, Paraguay, comprising one legal representative from each of Mercosur's four member countries, plus one 'consensus' member. The summit meeting held in July in Buenos Aires, Argentina, adopted an agreement providing for reduced tariffs and increased quotas in the grouping's automotive sector, with a view to establishing a fully liberalized automotive market by 2006. In December 2002 Mercosur ministers of justice signed an agreement permitting citizens of Mercosur member and associate member states to reside in any other Mercosur state, initially for a two-year period. At a summit convened in in June 2003, in Asunción, heads of state of the four member countries agreed to strengthen integration of the bloc and to harmonize all import tariffs by 2006, thus creating the basis for a single market. They also agreed to establish a directly-elected Mercosur legislature by 2006. The July 2004 summit of Mercosur heads of state announced that an Asunción-based five-member tribunal responsible for ruling on appeals in cases of disputes between member countries was to become operational in the following month. In June 2005 Mercosur heads of state announced a US $100m. structural convergence fund to support education, job creation and infrastructure projects in the poorest regions, in particular in Paraguay and Uruguay, in order to remove some economic disparities within the grouping. The meeting also endorsed a multilateral energy project to link gasfields in Camisea, Peru to existing supply pipelines in Argentina, Brazil and Uruguay, via Tocopilla, Chile.

**EXTERNAL RELATIONS**

In December 1995 Mercosur and the EU signed a framework agreement for commercial and economic co-operation, which provided for co-operation in the economic, trade, industrial, scientific, institutional and cultural fields and the promotion of wider political dialogue on issues of mutual interest. In June 1997 Mercosur heads of state, convened in Asunción, reaffirmed the group's intention to pursue trade negotiations with the EU, Mexico and the Andean Community, as well as to negotiate as a single economic bloc in discussions with regard to the establishment of a Free Trade Area of the Americas (FTAA). Chile and Bolivia were to be incorporated into these negotiations. Negotiations between Mercosur and the EU on the conclusion of an Interregional Association Agreement commenced in 1999 and were ongoing in 2005. During 1997 negotiations to establish a free-trade accord with the Andean Community were hindered by differences regarding schedules for tariff elimination and Mercosur's insistence on a local content of 60% to qualify for rules of origin preferences. However, in April 1998 the two groupings signed an accord that committed them to the establishment of a free-trade area by January 2000. Negotiations in early 1999 failed to conclude an agreement on preferential tariffs between the two blocs, and the existing arrangements were extended on a bilateral basis. In March the Andean Community agreed to initiate free-trade negotiations with Brazil; a preferential tariff agreement was concluded in July. In August 2000 a similar agreement between the Community and Argentina entered into force. In September leaders of Mercosur and the Andean Community, meeting at a summit of Latin American heads of state, determined to relaunch negotiations. The establishment of a mechanism to support political dialogue and co-ordination between the two groupings, which aimed to enhance the integration process, was approved at the first joint meeting of ministers of foreign affairs in July 2001. In April 2004 Mercosur and the Andean Community signed a free-trade accord, providing for tariffs on 80% of trade between the two groupings to be phased out by 2014 and for tariffs to be removed from the remaining 20% of, initially protected, products by 2019. The entry into force of the accord, scheduled for 1 July 2004, was postponed owing to delays in drafting the tariff reduction schedule. Peru and Venezuela became Mercosur's third and fourth associate members, respectively, in December 2003 and July 2004. Colombia and Ecuador were granted associate membership in December 2004. In July of that year Mexico was invited to attend all meetings of the organization with a view to future accession to associate membership. Bilateral negotiations on a free-trade agreement between Mexico and Mercosur were initiated in 2001. In 2005 Mercosur and the Andean Community were formulating a reciprocal association agreement, to extend associate membership to all member states of both groupings.

In March 2003 Argentina and Brazil, with the backing of other Mercosur member states, formed the Southern Agricultural Council (CAS), which was to represent the interests of the grouping as a whole in negotiations with third countries. In December 2004 leaders from 12 Latin American countries (excluding Argentina, Ecuador, Paraguay and Uruguay) attending a pan-South American summit, convened in Cuzco, Peru, approved in principle the creation of a new South American Community of Nations (SACN). It was envisaged that negotiations on the formation of the new Community, which was to entail the merger of Mercosur, the Andean Community and the Rio Group, would be completed within 15 years. In April 2005 a region-wide meeting of ministers of foreign affairs was convened, within the framework of establishing the SACN. A joint SACN communiqué was released, expressing concern at the deterioration of constitutional rule and democratic institutions in Ecuador and announcing its intention to send a ministerial mission to that country. An SACN summit meeting was scheduled to be convened in August.

In March 1998 ministers of trade of 34 countries agreed a detailed framework for negotiations on the establishment of the FTAA. Mercosur secured support for its request that a separate negotiating group be established to consider issues relating to agriculture, as one of nine key sectors to be discussed. The FTAA negotiating process was formally initiated by heads of state of the 34 countries meeting in Santiago, Chile, in April 1998. In June Mercosur and Canada signed a Trade and Investment Co-operation Arrangement, which aimed to remove obstacles to trade and to increase economic co-operation between the two signatories. In July the European Commission proposed obtaining a mandate to commence negotiations with Mercosur and Chile towards a free-trade agreement, which, it was envisaged, would provide for the elimination of tariffs over a period of 10 years. However, Mercosur requested that the EU abolish agricultural subsidies as part of any accord. Negotiations between Mercosur, Chile and the EU were initiated in April 2000. Specific discussion of tariff reductions and market access commenced at the fifth round of negotiations, held in July 2001, at which the EU proposed a gradual elimination of tariffs on industrial imports over a 10-year period and an extension of access quotas for agricultural products. The summit meeting held in December 2000 was attended by the President of South Africa, and it was agreed that Mercosur would initiate free-trade negotiations with that country. (These commenced in October 2001.) In June 2001 Mercosur leaders agreed to pursue efforts to conclude a bilateral trade agreement with the USA, an objective previously opposed by the Brazilian authorities, while reaffirming their commitment to the FTAA process. Leaders attending a special summit of the Americas, convened in January 2004 in Monterrey, Mexico, failed to specify a completion date for the FTAA process, although they adopted a declaration committing themselves to its eventual establishment. However, negotiations were suspended in March, and remained stalled at mid-2005.

# Finance

The annual budget for the secretariat is contributed by the four full member states.

# Publication

*Boletín Oficial del Mercosur* (quarterly).

# OTHER INTERNATIONAL ORGANIZATIONS

## Agriculture, Food, Forestry and Fisheries

(For organizations concerned with agricultural commodities, see Commodities)

**CAB International (CABI):** Wallingford, Oxon, OX10 8DE, United Kingdom; tel. (1491) 832111; fax (1491) 833508; e-mail cabi@cabi.org; internet www.cabi.org; f. 1929 as the Imperial Agricultural Bureaux (later Commonwealth Agricultural Bureaux), current name adopted in 1985; aims to improve human welfare world-wide through the generation, dissemination and application of scientific knowledge in support of sustainable development; places particular emphasis on sustainable agriculture, forestry, human health and the management of natural resources, with priority given to the needs of developing countries; compiles and publishes extensive information (in a variety of print and electronic forms) on aspects of agriculture, forestry, veterinary medicine, the environment and natural resources, Third World rural development and others; maintains regional centres in the People's Republic of China, India, Kenya, Malaysia, Pakistan, Switzerland, Trinidad and Tobago, and the United Kingdom; mems: 41 countries; Dir-Gen. Dr DENIS BLIGHT.

**CABI Bioscience:** Bakeham Lane, Egham, Surrey, TW20 9TY, United Kingdom; tel. (1491) 829080; fax (1491) 829100; e-mail bioscience.egham@cabi.org; internet www.cabi-bioscience.org; f. 1998 by integration of the following four CABI scientific institutions: International Institute of Biological Control; International Institute of Entomology; International Institute of Parasitology; International Mycological Institute; undertakes research, consultancy, training, capacity-building and institutional development measures in sustainable pest management, biosystematics and molecular biology, ecological applications and environmental and industrial microbiology; maintains centres in Kenya, Malaysia, Pakistan, Switzerland, Trinidad and Tobago, and the United Kingdom; Dir DAVID DENT.

**Inter-American Association of Agricultural Librarians, Documentalists and Information Specialists** (Asociación Interamericana de Bibliotecarios, Documentalistas y Especialistas en Información Agrícolas—AIBDA): c/o IICA-CIDIA, Apdo 55-2200 Coronado, Costa Rica; tel. 216-0290; fax 216-0291; e-mail aibda@iica.ac.cr; internet www.iica.int/servicios/aibda; f. 1953 to promote professional improvement through technical publications and meetings, and to promote improvement of library services in agricultural sciences; mems: c. 400 in 29 countries and territories; Pres. GERARDO SÁNCHEZ AMBRIZ (Mexico); Exec. Sec. AURA MATA (Costa Rica); publs *Boletín Informativo* (3 a year), *Boletín Especial* (irregular), *Revista AIBDA* (2 a year), *AIBDA Actualidades* (4 or 5 a year).

**Inter-American Tropical Tuna Commission (IATTC):** 8604 La Jolla Shores Drive, La Jolla, CA 92037-1508, USA; tel. (858) 546-7100; fax (858) 546-7133; e-mail rallen@iattc.org; internet www.iattc.org; f. 1950; administers two programmes, the Tuna-Billfish Programme and the Tuna-Dolphin Programme. The principal responsibilities of the Tuna-Billfish Programme are: to study the biology of the tunas and related species of the eastern Pacific Ocean to estimate the effects of fishing and natural factors on their abundance; to recommend appropriate conservation measures in order to maintain stocks at levels which will afford maximum sustainable catches; and to collect information on compliance with Commission resolutions. The principal functions of the Tuna-Dolphin Programme are: to monitor the abundance of dolphins and their mortality incidental to purse-seine fishing in the eastern Pacific Ocean; to study the causes of mortality of dolphins during fishing operations and promote the use of fishing techniques and equipment that minimize these mortalities; to study the effects of different fishing methods on the various fish and other animals of the pelagic ecosystem; and to provide a secretariat for the International Dolphin Conservation Programme; mems: Costa Rica, Ecuador, El Salvador, France, Guatemala, Japan, Mexico, Nicaragua, Panama, Peru, Spain, USA, Vanuatu, Venezuela; Dir ROBIN ALLEN; publs *Bulletin* (irregular), *Annual Report*, *Fishery Status Report*, *Stock Assessment Report* (annually), *Special Report* (irregular).

**International Centre for Tropical Agriculture** (Centro Internacional de Agricultura Tropical—CIAT): Apdo Aéreo 6713, Cali, Colombia; tel. (2) 445-0000; fax (2) 445-0073; e-mail ciat@cgiar.org; internet www.ciat.cgiar.org; f. 1967 to contribute to the alleviation of hunger and poverty in tropical developing countries by using new techniques in agriculture research and training; focuses on production problems in field beans, cassava, rice and tropical pastures in the tropics; Dir-Gen. Dr JOACHIM VOSS; publs *Annual Report*, *Growing Affinities* (2 a year), *Pasturas Tropicales* (3 a year), catalogue of publications.

**World Organisation of Animal Health:** 12 rue de Prony, 75017 Paris, France; tel. 1-44-15-18-88; fax 1-42-67-09-87; e-mail oie@oie.int; internet www.oie.int; f. 1924 as Office International des Epizooties (OIE); objectives include promoting international transparency of animal diseases; collecting, analysing and disseminating scientific veterinary information; providing expertise and promoting international co-operation in the control of animal diseases; promoting veterinary services; providing new scientific guide-lines on animal production, food safety and animal welfare; experts in a network of 156 collaborating centres and reference laboratories; 167 mems; Dir-Gen. BERNARD VALLAT; publs *Disease Information* (weekly), *World Animal Health* (annually), *Scientific and Technical Review* (3 a year), other manuals, codes etc.

## Arts and Culture

**Inter-American Music Council** (Consejo Interamericano de Música—CIDEM): 2511 P St NW, Washington, DC 20007, USA; f. 1956 to promote the exchange of works, performances and information in all fields of music, to study problems relative to music education, to encourage activity in the field of musicology, to promote folklore research and music creation, and to establish distribution centres for music material of the composers of the Americas; mems: national music societies of 33 American countries; Sec.-Gen. EFRAÍN PAESKY.

**Organization of World Heritage Cities:** 15 rue Saint-Nicolas, Québec, QC G1K 1MB, Canada; tel. (418) 692-0000; fax (418) 692-5558; e-mail secretariat@ovpm.org; internet www.ovpm.org; f. 1993 to assist cities inscribed on the UNESCO World Heritage List to implement the Convention concerning the Protection of the World Cultural and Natural Heritage (1972); promotes co-operation between city authorities, in particular in the management and sustainable development of historic sites; holds a General Assembly, comprising the mayors of member cities, at least every two years; mems: 208 cities world-wide; Sec.-Gen. DENIS RICARD.

## Commodities

**Cocoa Producers' Alliance (CPA):** National Assembly Complex, Tafawa Balewa Sq., POB 1718, Lagos, Nigeria; tel. (1) 2635574; fax (1) 2635684; e-mail info@copal-cpa.com; internet www.copal-cpa.org; f. 1962 to exchange technical and scientific information, to discuss problems of mutual concern to producers, to ensure adequate supplies at remunerative prices and to promote consumption; mems: Brazil, Cameroon, Côte d'Ivoire, Dominican Republic, Gabon, Ghana, Malaysia, Nigeria, São Tomé and Príncipe, Togo; Sec.-Gen. HOPE SONA EBAI.

**International Cocoa Organization (ICCO):** 22 Berners St, London, W1P 3DB, United Kingdom; tel. (20) 7637-3211; fax (20) 7631-0114; e-mail exec.dir@icco.org; internet www.icco.org; f. 1973 under the first International Cocoa Agreement, 1972; the ICCO supervises the implementation of the agreements, and provides member governments with up-to-date information on the world cocoa economy; the sixth International Cocoa Agreement (2001) entered into force in October 2003; mems: 12 exporting countries and 18 importing countries; and the European Union; Council Chair. (2004–05) S. P. ESSOMBA ABANDA (Cameroon); Exec. Dir Dr J. VINGERHOETS (Netherlands); publs *Quarterly Bulletin of Cocoa Statistics*, *Annual Report*, *World Cocoa Directory*, *Cocoa Newsletter*, studies on the world cocoa economy.

**International Coffee Organization (ICO):** 22 Berners St, London, W1T 4DD, United Kingdom; tel. (20) 7580-8591; fax (20) 7580-6129; e-mail info@ico.org; internet www.ico.org; f. 1963 under the International Coffee Agreement, 1962, which was renegotiated in 1968, 1976, 1983, 1994 (extended in 1999) and 2001; aims to improve international co-operation and provide a forum for intergovernmental consultations on coffee matters; to facilitate international trade in coffee by the collection, analysis and dissemination of statistics; to act as a centre for the collection, exchange and publication of coffee information; to promote studies in the field of coffee; and to encourage an increase in coffee consumption; mems: 40 exporting and 15 importing countries; Chair. of Council JACQUES THINSY (Belgium); Exec. Dir NÉSTOR OSORIO (Colombia).

**International Sugar Organization:** 1 Canada Sq., Canary Wharf, London, E14 5AA, United Kingdom; tel. (20) 7513-1144; fax (20) 7513-1146; e-mail exdir@isosugar.org; internet www.isosugar.org; administers the International Sugar Agreement (1992), with the objectives of stimulating co-operation, facilitating trade and encouraging demand; aims to improve conditions in the sugar market through debate, analysis and studies; serves as a forum for discussion; holds annual seminars and workshops; sponsors projects from developing countries; mems: 72 countries producing some 83% of total world sugar; Exec. Dir Dr PETER BARON; publs *Sugar Year Book*, *Monthly Statistical Bulletin*, *Market Report and Press Summary*, *Quarterly Market Outlook*, seminar proceedings.

**International Tropical Timber Organization (ITTO):** International Organizations Center, 5th Floor, Pacifico-Yokohama, 1-1-1, Minato-Mirai, Nishi-ku, Yokohama 220-0012, Japan; tel. (45) 223-1110; fax (45) 223-1111; e-mail itto@itto.or.jp; internet www.itto.or.jp; f. 1985 under the International Tropical Timber Agreement (1983); a new treaty, ITTA 1994, came into force in 1997; provides a forum for consultation and co-operation between countries that produce and consume tropical timber, and is dedicated to the sustainable development and conservation of tropical forests; facilitates progress towards 'Objective 2000', which aims to move as rapidly as possible towards achieving exports of tropical timber and timber products from sustainably managed resources; encourages, through policy and project work, forest management, conservation and restoration, the further processing of tropical timber in producing countries, and the gathering and analysis of market intelligence and economic information; mems: 33 producing and 26 consuming countries and the EU; Exec. Dir Dr MANOEL SOBRAL FILHO; publs *Annual Review and Assessment of the World Timber Situation*, *Tropical Timber Market Information Service* (every 2 weeks), *Tropical Forest Update* (quarterly).

**Organization of the Petroleum Exporting Countries—OPEC:** 1020 Vienna, Obere Donaustrasse 93, Austria; tel. (1) 211-12-279; fax (1) 214-98-27; e-mail info@opec.org; internet www.opec.org; f. 1960 to unify and co-ordinate members' petroleum policies and to safeguard their interests generally; holds regular conferences of member countries to set reference prices and production levels; conducts research in energy studies, economics and finance; provides data services and news services covering petroleum and energy issues; mems: Algeria, Indonesia, Iran, Iraq, Kuwait, Libya, Nigeria, Qatar, Saudi Arabia, United Arab Emirates, Venezuela; Sec.-Gen. a.i. Dr ADNAN SHIBAB-ELDIN; publs *Annual Report*, *Annual Statistical Bulletin*, *OPEC Bulletin* (monthly), *OPEC Review* (quarterly), *Monthly Oil Market Report*.

**Regional Association of Oil and Natural Gas Companies in Latin America and the Caribbean** (Asociación Regional de Empresas de Petróleo y Gas Natural en Latinoamérica y el Caribe—ARPEL): Javier de Viana 2345, Casilla de correo 1006, 11200 Montevideo, Uruguay; tel. (2) 4106993; fax (2) 4109207; e-mail bsettembri@arpel.org.uy; internet www.arpel.org; f. 1965 as the Mutual Assistance of the Latin American Oil Companies; aims to initiate and implement activities for the development of the oil and natural gas industry in Latin America and the Caribbean; promotes the expansion of business opportunities and the improvement of the competitive advantages of its members; promotes guide-lines in support of competition in the sector; and supports the efficient and sustainable exploitation of hydrocarbon resources and the supply of products and services. Works in co-operation with international organizations, governments, regulatory agencies, technical institutions, universities and non-governmental organizations; mems: state enterprises, representing more than 90% of regional operations, in Argentina, Bolivia, Brazil, Canada, Chile, Colombia, Costa Rica, Cuba, Ecuador, Jamaica, Mexico, Nicaragua, Paraguay, Peru, Suriname, Trinidad and Tobago, Uruguay, Venezuela; Exec. Sec. JOSÉ FÉLIX GARCÍA GARCÍA; publ. *Boletín Técnico*.

**Sugar Association of the Caribbean (Inc.):** c/o Caroni (1975) Ltd, Brechin Castle, Conva, Trinidad and Tobago; tel. 636-2449; fax 636-2847; f. 1942; mems: national sugar cos of Barbados, Belize, Guyana, Jamaica and Trinidad and Tobago, and Sugar Asscn of St Kitts–Nevis–Anguilla; Chair. T. FALLOON (Jamaica); CEO Dr IAN MCDONALD; publs *SAC Handbook*, *SAC Annual Report*, *Proceedings of Meetings of WI Sugar Technologists*.

**Union of Banana-Exporting Countries** (Unión de Paises Exportadores de Banano—UPEB): Apdo 4273, Bank of America, 7°, Panamá 5, Panama; tel. 263-6266; fax 264-8355; e-mail iicapan@pan.gbm.net; f. 1974 as an intergovernmental agency to assist in the cultivation and marketing of bananas and to secure prices; collects statistics; mems: Colombia, Costa Rica, Guatemala, Honduras, Nicaragua, Panama, Venezuela; publs *Informe UPEB*, *Fax UPEB*, *Anuario de Estadísticas*, biblio-graphies.

**West Indian Sea Island Cotton Association (Inc.):** c/o Barbados Agricultural Development Corporation, Fairy Valley, Christ Church, Barbados; mems: organizations in Antigua-Barbuda, Barbados, Jamaica, Montserrat and St Christopher and Nevis; Pres. LEROY ROACH; Sec. MICHAEL I. EDGHILL.

# Development and Economic Co-operation

**Association of Caribbean States (ACS):** 5–7 Sweet Briar Rd, St Clair, POB 660, Port of Spain, Trinidad and Tobago; tel. 622-9575; fax 622-1653; e-mail communications@acs-aec.org; internet www.acs-aec.org; f. 1994 by the Governments of the 13 CARICOM countries and Colombia, Costa Rica, Cuba, Dominican Republic, El Salvador, Guatemala, Haiti, Honduras, Mexico, Nicaragua, Suriname and Venezuela; aims to promote economic integration, sustainable development and co-operation in the region; to co-ordinate participation in multilateral forums; to undertake concerted action to protect the environment, particularly the Caribbean Sea; and to co-operate in the areas of science and technology, health, trade, transport, tourism, education and culture. Policy is determined by a Ministerial Council and implemented by a Secretariat based in Port of Spain. In December 2001 a third Summit of Heads of State and Government was convened in Venezuela, where a Plan of Action focusing on issues of sustainable tourism, trade, transport and natural disasters was agreed. The fourth ACS Summit was held in Panama, in July 2005; a final Declaration included resolutions to strengthen co-operation mechanisms with the EU and to promote a strategy for the Caribbean Sea Zone to be recognized as a special area for the purposes of sustainable development programmes, support for a strengthened social agenda and efforts to achieve the Millennium Development Goals, and calls for member states to sign or ratify the following accords: an ACS Agreement for Regional Co-operation in the area of Natural Disasters; a Convention Establishing the Sustainable Tourism Zone of the Caribbean; and an ACS Air Transport Agreement; mems: 25 signatory states, three associate members, 18 observer countries; Sec.-Gen. Dr RUBÉN ARTURO SILIÉ VALDEZ.

**Caribbean-Britain Business Council:** Westminster Palace Gdns, Suite 18, 1–7 Artillery Row, London, SW1P 1RR, United Kingdom; tel. (20) 7799-1521; e-mail admin@caribbean-council.org; f. 2001; promotes trade and investment development between the United Kingdom, the Caribbean and the European Union; Chair. JEFFRIES BRIGINSHAW; Exec. Dir S. MONTEATH; publ. *Caribbean Briefing* (weekly).

**Council of American Development Foundations** (Consejo de Fundaciones Americanas de Desarrollo—SOLIDARIOS): Calle 6 No. 10 Paraíso, Apdo Postal 620, Santo Domingo, Dominican Republic; tel. 549-5111; fax 544-0550; e-mail solidarios@codetel.net.do; f. 1972; exchanges information and experience, arranges technical assistance, raises funds to organize training programmes and scholarships; administers development fund to finance programmes carried out by members through a loan guarantee programme; provides consultancy services. Mem. foundations provide technical and financial assistance to low-income groups for rural, housing and microenterprise development projects; mems: 18 institutional mems in 14 Latin American and Caribbean countries; Pres. MERCEDES P. DE CANALDA; Sec.-Gen. ISABEL C. ARANGO; publs *Solidarios* (quarterly), *Annual Report*.

**Group of Three (G3):** c/o Secretaría de Relaciones Exteriores, 1 Tlatelolco, Del. Cuauhtémoc 06995 México, DF; e-mail gtres@sre.gob.mx; internet g3.sre.gob.mx; f. 1990 by Colombia, Mexico and Venezuela to remove restrictions on trade between the three countries; the trade agreement covers market access, rules of origin, intellectual property, trade in services, and government purchases, and entered into force in early 1994. Tariffs on trade between member states were to be removed on a phased basis. Co-operation was also envisaged in employment creation, the energy sector and the fight against cholera. The secretariat function rotates between the three countries on a two-yearly basis.

**Inter-American Planning Society** (Sociedad Interamericana de Planificación—SIAP): c/o Revista Interamericana de Planificación, Casilla 01-05-1978, Cuenca, Ecuador; tel. (7) 823-860; fax (7) 823-949; e-mail siap1@siap.org.ec; f. 1956 to promote development of comprehensive planning; mems: institutions and individuals in 46 countries; Pres. Prof. PATRICIA A. WILSON (USA); Exec. Sec. LUIS E. CAMACHO (Colombia); publs *Correo Informativo* (quarterly), *Inter-American Journal of Planning* (quarterly).

**Latin American Association of Development Financing Institutions** (Asociación Latinoamericana de Instituciones Financieras para el Desarrollo—ALIDE): Apdo Postal 3988, Paseo de la República 3211, Lima 100, Peru; tel. (1) 4422400; fax (1) 4428105; e-mail sg@alide.org.pe; internet www.alide.org.pe; f. 1968 to promote co-operation among regional development financing bodies; programmes: technical assistance; training; studies and research; technical meetings; information; projects and investment promotion; mems: 62 active, 8 assoc. and 8 collaborating (banks and financing institutions and development organizations in 22 Latin American countries, Slovenia and Spain); Sec.-Gen. ROMMEL ACEVEDO; publs *ALIDE Bulletin* (6 a year), *ALIDENOTICIAS Newsletter* (monthly), *Annual Report*, *Latin American Directory of Development Financing Institutions*.

**Latin American Economic System** (Sistema Económico Latinoamericano—SELA): Torre Europa, 4°, Urb. Campo Alegre, Avda Francisco de Miranda, Caracas 1061, Venezuela; Apdo 17035, Caracas 1010-A, Venezuela; tel. (212) 955-7111; fax (212) 951-5292; e-mail difusion@sela.org; internet www.sela.org; f. 1975 in accordance with the Panama Convention; aims to foster co-operation and integration among the countries of Latin America and the Caribbean, and to provide a permanent system of consultation and co-ordination in economic and social matters; conducts studies and other analysis and research; extends technical assistance to sub-regional and regional co-ordination bodies; provides library, information service and data bases on regional co-operation. The Latin American Council, the principal decision-making body of the System, meets annually at ministerial level and high-level regional consultation and co-ordination meetings are held; there is also a Permanent Secretariat; mems: 28 countries; Perm. Sec. ROBERTO GUARNIERI (Venezuela); publs *Capítulos del SELA* (3 a year), *Bulletin on Latin America and Caribbean Integration* (monthly), *SELA Antenna in the United States* (quarterly).

**Organization of the Cooperatives of America** (Organización de las Cooperativas de América): Apdo Postal 241263, Carrera 11, No 86-32, Of. 101, Bogotá, Colombia; tel. (1) 6103296; fax (1) 6101912; f. 1963 for improving socio-economic, cultural and moral conditions through the use of the co-operatives system; works in every country of the continent; regional offices sponsor plans and activities based on the most pressing needs and special conditions of individual countries; mems: national or local orgs in 23 countries and territories; Exec. Sec. Dr CARLOS JULIO PINEDA SUÁREZ; publs *América Cooperativa* (monthly), *OCA News* (monthly).

**Pacific Basin Economic Council (PBEC):** 900 Fort St, Suite 1080, Honolulu, HI 96813, USA; tel. (808) 521-9044; fax (808) 521-8530; e-mail info@pbec.org; internet www.pbec.org; f. 1967; an asscn of business representatives aiming to promote business opportunities in the region, in order to enhance overall economic development; advises governments and serves as a liaison between business leaders and government officials; encourages business relationships and co-operation among members; holds business symposia; mems: 20 economies (Australia, Canada, Chile, People's Republic of China, Colombia, Ecuador, Hong Kong, Indonesia, Japan, Republic of Korea, Malaysia, Mexico, New Zealand, Peru, Philippines, Russia, Singapore, Taiwan, Thailand, USA); Chair. DAVID ELDON; Pres. STEPHEN OLSEN; publs *Pacific Journal* (quarterly), *Executive Summary* (annual conference report).

## Economics and Finance

**Centre for Latin American Monetary Studies** (Centro de Estudios Monetarios Latinoamericanos—CEMLA): Durango 54, Col. Roma, Del. Cuauhtémoc, 06700 México, DF, Mexico; tel. (55) 5533-0300; fax (55) 5525-4432; e-mail estudios@cemla.org; internet www.cemla.org; f. 1952; organizes technical training programmes on monetary policy, development finance, etc; runs applied research programmes on monetary and central banking policies and procedures; holds regional meetings of banking officials; mems: 31 associated members (Central Banks of Latin America and the Caribbean), 28 co-operating members (supervisory institutions of the region and non-Latin American Central Banks); Dir-Gen. KENNETH GILMORE COATES SPRY; publs *Bulletin* (every 2 months), *Monetaria* (quarterly), *Money Affairs* (2 a year).

**Eastern Caribbean Central Bank (ECCB):** POB 89, Basseterre, St Christopher and Nevis; tel. 465-2537; fax 465-9562; e-mail eccbinfo@caribsurf.com; internet www.eccb-centralbank.org; f. 1983 by OECS governments; maintains regional currency (Eastern Caribbean dollar) and advises on the economic development of member states; mems: Anguilla, Antigua and Barbuda, Dominica, Grenada, Montserrat, Saint Christopher and Nevis, Saint Lucia, Saint Vincent and the Grenadines; Gov. Sir K. DWIGHT VENNER.

**Financial Action Task Force on Money Laundering (FATF)** (Groupe d'action financière sur le blanchiment de capitaux—GAFI): 2 rue André-Pascal, 75775 Paris Cédex 16, France; tel. 1-45-24-79-45; fax 1-45-24-17-60; e-mail contact@fatf-gafi.org; internet wwwfatf-gafi.org; f. 1989, on the recommendation of the Group of Seven industrialized nations (G-7), to develop and promote policies to combat money laundering and the financing of terrorism; formulated a set of recommendations (40+9) for countries world-wide to implement; established partnerships with regional task forces in the Caribbean, Asia-Pacific, Central Asia, Europe, East and South Africa, the Middle East and North Africa and South America; mems: 31 state jurisdictions, the European Commission, and the Co-operation Council for the Arab States of the Gulf; observer: People's Republic of China; Pres. Prof. KADER ASMAL (South Africa); Exec. Sec. ALAIN DAMAIS; publ. *Annual Report*.

**Latin American Banking Federation** (Federación Latinoamericana de Bancos—FELABAN): Bogotá, DC, Colombia Cra 11A No. 93-67 Of. 202 A.A 091959; tel. (1) 6218617; fax (1) 6217659; internet www.latinbanking.com; f. 1965 to co-ordinate efforts towards wide and accelerated economic development in Latin American countries; mems: 19 Latin American national banking asscns; Pres. of Board IGNACIO SALVATIERRA (Venezuela); Sec.-Gen. MARICIELO GLEN DE TOBÓN (Colombia).

## Education

**Association of Caribbean University and Research Institutional Libraries (ACURIL):** Apdo postal 23317, San Juan 00931-3317, Puerto Rico; tel. 790-8054; fax 764-2311; e-mail acuril@rrpac.upr.clu.edu; internet acuril.rrp.upr.edu; f. 1968 to foster contact and collaboration between mem. universities and institutes; holds conferences, meetings and seminars; circulates information through newsletters and bulletins; facilitates co-operation and the pooling of resources in research; encourages exchange of staff and students; mems: 250; Pres. LUISA VIGO-CEPEDA (Puerto Rico); Exec.-Sec. ONEIDA R. ORTIZ (Puerto Rico); publ. *Newsletter* (2 a year).

**Caribbean Council of Legal Education:** POB 323, Tunapuna, Trinidad and Tobago; tel. 662-5860; fax 662-0927; f. 1971; responsible for the training of members of the legal profession; mems: govts of 12 countries and territories.

**Caribbean Examinations Council:** The Garrison, St Michael 20, Barbados; tel. 436-6261; fax 429-5421; e-mail ezo@cxc.org; internet www.cxc.org; f. 1972; develops syllabuses and conducts examinations; mems: govts of 16 English-speaking countries and territories.

**Inter-American Centre for Research and Documentation on Vocational Training** (Centro Interamericano de Investigación y Documentación sobre Formación Profesional—CINTER-FOR): Avda Uruguay 1238, Casilla de correo 1761, Montevideo, Uruguay; tel. (2) 9020557; fax (2) 9021305; e-mail biblio@cinterfor.org.uy; internet www.ilo.org/public/english/region/ampro/cinterfor; f. 1964 by the International Labour Organization for mutual help among the Latin American and Caribbean countries in planning vocational training; services are provided in documentation, research, exchange of experience; holds seminars and courses; Dir PEDRO DANIEL WEINBERG; publs *Bulletin CINTERFOR/OIT Heramientas para la transformación*, *Trazos de la formación*, studies, monographs and technical papers.

**Inter-American Confederation for Catholic Education** (Confederación Interamericana de Educación Católica—CIEC): Calle 78 No 12–16 (of. 101), Apdo Aéreo 90036, Bogotá 8 DE, Colombia; tel. (1) 255-3676; fax (1) 255-0513; e-mail ciec@cable.net.co; internet www.ciec.to; f. 1945 to defend and extend the principles and rules of Catholic education, freedom of education, and human rights; organizes congress every three years; Pres. RAMÓN E. RIVAS TOMASI (Venezuela); Sec.-Gen. MARIA CONSTANZA ARANGO; publ. *Educación Hoy*.

**Inter-American Organization for Higher Education (IOHE):** Édifce Vieux Séminaire, local 1244, 1 Côte de la Fabrique, Québec, QC, Canada G1R 3V6; tel. (418) 650-1515; fax (418) 650-1519; e-mail secretariat@oui-iohe.qc.ca; internet www.oui-iohe.qc.ca; f. 1980 to promote co-operation among universities of the Americas and the development of higher education; mems: 390 in 24 countries; Exec. Dir a.i. MARCEL HAMELIN.

**International Institute of Iberoamerican Literature** (Instituto Internacional de Literatura Iberoamericana): 1312 CL, University of Pittsburgh, PA 15260, USA; tel. (412) 624-0829; e-mail iilit@apvtt-edn; f. 1938 to advance the study of Iberoamerican literature, and intensify cultural relations among the peoples of the Americas; mems: scholars and artists in 37 countries; publs *Revista Iberoamericana Memorias*.

**Organization of Ibero-American States for Education, Science and Culture** (Organización de Estados Iberoamericanos para la Educación, la Ciencia y la Cultura—OEI): Centro de Recursos Documentales e Informáticos, Calle Bravo Murillo 38, 28015 Madrid, Spain; tel. (91) 594-43-82; fax (91) 594-32-86; e-mail oeimad@oei.es; internet www.oei.es; f. 1949 (as the Ibero-American Bureau of Education); promotes peace and solidarity between member countries, through education, science, technology and culture; provides information, encourages exchanges and organizes training courses; the General Assembly (at ministerial level) meets every four years; mems: govts of 20 countries; Sec.-Gen. Francisco José Piñón; publ. *Revista Iberoamericana de Educación* (quarterly).

**Organization of the Catholic Universities of Latin America** (Organización de Universidades Católicas de América Latina—ODUCAL): c/o Dr J. A. Tobías, Universidad del Salvador, Viamonte 1856, CP 1056, Buenos Aires, Argentina; tel. (11) 4813-1408; fax (11) 4812-4625; e-mail udes-rect@salvador.edu.ar; f. 1953 to assist the social, economic and cultural development of Latin America through the promotion of Catholic higher education in the continent; mems: 43 Catholic universities in 15 Latin American countries; Pres. Dr Juan Alejandro Tobías (Argentina); publs *Anuario*, *Sapientia*, *Universitas*.

**Union of Latin American Universities** (Unión de Universidades de América Latina—UDUAL): Edificio UDUAL, Apdo postal 70-232, Ciudad Universitaria, Del. Coyoacán, 04510 México, DF, Mexico; tel. (55) 5622-0991; fax (55) 5616-1414; e-mail udual@servidor.unam.mx; internet www.unam.mx.udual; f. 1949 to organize exchanges between professors, students, research fellows and graduates and generally encourage good relations between the Latin American universities; arranges conferences; conducts statistical research; maintains centre for university documentation; mems: 165 universities; Pres. Dr Salomón Lerner Febres (Peru); Sec.-Gen. Dr Juan José Sánchez Sosa (Mexico); publs *Universidades* (2 a year), *Gaceta UDUAL* (quarterly), *Censo* (every 2 years).

## Environmental Conservation

**Caribbean Conservation Association:** Savannah Lodge, The Garrison, St Michael, Barbados; tel. 426-5373; fax 429-8483; e-mail cca@caribsurf.com; f. 1967; aims to conserve the environment and cultural heritage of the region through education, legislation, and management of museums and sites; mems: 20 governments, 92 organizations and individuals in 33 countries and territories; Exec. Dir Calvin Howell; publ. *Caribbean Conservation News* (quarterly).

**IUCN—The World Conservation Union:** 28 rue Mauverney, 1196 Gland, Switzerland; tel. (22) 9990000; fax (22) 9990002; e-mail mail@hq.iucn.org; internet www.iucn.org; f. 1948, as the International Union for Conservation of Nature and Natural Resources; supports partnerships and practical field activities to promote the conservation of natural resources, to secure the conservation of biological diversity as an essential foundation for the future; to ensure wise use of the earth's natural resources in an equitable and sustainable way; and to guide the development of human communities towards ways of life in enduring harmony with other components of the biosphere, developing programmes to protect and sustain the most important and threatened species and eco-systems and assisting governments to devise and carry out national conservation strategies; maintains a conservation library and documentation centre and units for monitoring traffic in wildlife; mems: more than 1,000 states, government agencies, non-governmental organizations and affiliates in some 140 countries; Pres. Yolanda Kakabadse Navarro (Ecuador); Dir-Gen. Achim Steiner; publs *World Conservation Strategy*, *Caring for the Earth*, *Red List of Threatened Plants*, *Red List of Threatened Species*, *United Nations List of National Parks and Protected Areas*, *World Conservation* (quarterly), *IUCN Today*.

**International Seabed Authority:** 14–20 Port Royal St, Kingston, Jamaica; tel. 922-9105; fax 922-0195; e-mail postmaster@isa.org.jm; internet www.isa.org.jm; f. Nov. 1994 upon the entry into force of the 1982 United Nations Convention on the Law of the Sea; the Authority is the institute through which states party to the Convention organize and control activities in the international seabed area beyond the limits of national jurisdiction, particularly with a view to administering the resources of that area; Sec.-Gen. Satya N. Nandan; publs *Annual Report*, *WWF News* (quarterly).

**WWF International:** ave du Mont-Blanc, 1196 Gland, Switzerland; tel. (22) 3649111; fax (22) 3645358; e-mail info@wwfint.org; internet www.panda.org; f. 1961 (as World Wildlife Fund), name changed to World Wildlife Fund for Nature 1986, current nomenclature adopted 2001; aims to stop the degradation of the natural environment, conserve bio-diversity, ensure the sustainable use of renewable resources, and promote the reduction of both pollution and wasteful consumption; addresses six priority issues: forests, fresh water programmes, endangered seas, species, climate change, and toxins; has identified, and focuses its activities in, 200 'ecoregions' (the 'Global 200'), believed to contain the best part of the world's remaining biological diversity; actively supports and operates conservation programmes in 90 countries; mems: 28 national organizations, four associates, c. 5m. individual mems world-wide; Pres. Chief Emeka Anyaoku (Nigeria); Dir-Gen. Claude Martin; publs *Annual Report*, *Living Planet Report*.

## Government and Politics

**Agency for the Prohibition of Nuclear Weapons in Latin America and the Caribbean** (Organismo para la Proscripción de las Armas Nucleares en la América Latina y el Caribe—OPANAL): Schiller 326, 5°, Col. Chapultepec Morales, 11570 México, DF, Mexico; tel. (55) 5255-2914; fax (55) 5255-3748; e-mail info@opanal.org; internet www.opanal.org; f. 1969 to ensure compliance with the Treaty for the Prohibition of Nuclear Weapons in Latin America (Treaty of Tlatelolco), 1967 to ensure the absence of all nuclear weapons in the application zone of the Treaty; to contribute to the movement against proliferation of nuclear weapons; to promote general and complete disarmament; to prohibit all testing, use, manufacture, acquisition, storage, installation and any form of possession, by any means, of nuclear weapons; the organs of the Agency comprise the General Conference, meeting every two years, the Council, meeting every two months, and the secretariat; a General Conference is held every two years; mems: 33 states that have fully ratified the Treaty; the Treaty has two additional Protocols: the first signed and ratified by France, the Netherlands, the United Kingdom and the USA, the second signed and ratified by China, the USA, France, the United Kingdom and Russia; Sec.-Gen. Edmundo Vargas Carreño (Chile).

**Alliance of Small Island States (AOSIS):** c/o 211 East 43rd St, 15th Floor, New York, NY 10017, USA; tel. (212) 949-0190; fax (212) 697-3829; e-mail mauritius@un.int; internet www.sidsne.org/aosis; f. 1990 as an *ad hoc* intergovernmental grouping to focus on the special problems of small islands and low-lying coastal developing states; mems: 43 island nations; Chair. Jagdish Koonjul (Mauritius); publ. *Small Islands, Big Issues*.

**Comunidade dos Países de Língua Portuguesa (CPLP)** (Community of Portuguese-Speaking Countries): rua S. Caetano 32, 1200-829 Lisbon, Portugal; tel. (21) 392-8560; fax (21) 392-8588; e-mail comunicacao@cplp.org; internet www.cplp.org; f. 1996; aims to produce close political, economic, diplomatic and cultural links between Portuguese-speaking countries and to strengthen the influence of the Lusophone commonwealth within the international community; mems: Angola, Brazil, Cape Verde, Guinea-Bissau, Mozambique, Portugal, São Tomé e Príncipe, Timor-Leste; Exec. Sec. Luis Fonseca (Cape Verde).

**International Institute for Democracy and Electoral Assistance (IDEA):** Strömsborg, S-103 34 Stockholm, Sweden; tel. (8) 698-3700; fax (8) 20-2422; e-mail info@idea.int; internet www.idea.int; f. 1995; aims to promote sustainable democracy in new and established democracies; provides world-wide electoral assistance and focuses on broader democratic issues in Africa, the Caucasus and Latin America; 21 mem. states; Sec.-Gen. a.i. Massimo Tommasoli (Italy).

**Latin American Parliament** (Parlamento Latinoamericano): Avda Auro Soares de Moura Andrade 564, São Paulo, Brazil; tel. (11) 3824-6325; fax (11) 3824-0621; internet www.parlatino.org.br; f. 1965; permanent democratic institution, representative of all existing political trends within the national legislative bodies of Latin America; aims to promote the movement towards economic, political and cultural integration of the Latin American republics, and to uphold human rights, peace and security; Sec.-Gen. Rafael Correa Flores; publs *Acuerdos, Resoluciones de las Asambleas Ordinarias* (annually), *Parlamento Latinoamericano–Actividades de los Órganos*, *Revista Patria Grande* (annually), statements and agreements.

**Non-aligned Movement (NAM):** c/o Permanent Representative of Malaysia to the UN, 313 East 43rd St, New York, NY 10016, USA (no permanent secretariat); tel. (212) 986-6310; fax (212) 490-8576; e-mail malaysia@un.int; internet www.namkl.org.my; f. 1961 by a meeting of 25 Heads of State, with the aim of linking countries that had refused to adhere to the main East/West military and political blocs; co-ordination bureau established in 1973; works for the establishment of a new international economic order, and especially for better terms for countries producing raw materials; maintains special funds for agricultural development, improvement of food production and the financing of buffer stocks; South Commission promotes co-operation between developing countries; seeks changes in the United Nations to give developing countries greater decision-making power; holds summit conference every three years; 13th conference (February 2003): Kuala Lumpur, Malaysia; mems: 116 countries.

**Organisation of Eastern Caribbean States (OECS):** POB 179, The Morne, Castries, Saint Lucia; tel. 452-2537; fax 453-1628; e-mail oesec@oecs.org; internet www.oecs.org; f. 1981 by the seven states which formerly belonged to the West Indies Associated States (f. 1966); aims to promote the harmonized development of trade and industry in member states; single market created on 1 January 1988; principal institutions are: the Authority of Heads of Government (the supreme policy-making body), the Foreign Affairs Com-

# REGIONAL ORGANIZATIONS

*Other Regional Organizations*

mittee, the Defence and Security Committee, and the Economic Affairs Committee; there is also an Export Development and Agricultural Diversification Unit—EDADU (based in Dominica); an OECS Technical Mission to the World Trade Organization in Geneva, Switzerland, was inaugurated in June 2005; mems: Antigua and Barbuda, Dominica, Grenada, Montserrat, Saint Christopher and Nevis, Saint Lucia, Saint Vincent and the Grenadines; assoc. mems: Anguilla, British Virgin Islands; Dir-Gen. Dr LEN ISHMAEL.

**Organization of Solidarity of the Peoples of Africa, Asia and Latin America (OSPAAAL)** (Organización de Solidaridad de los Pueblos de Africa, Asia y América Latina): Apdo 4224, Calle C No 670 esq. 29, Vedado, Havana 10400, Cuba; tel. (7) 830-5136; fax (7) 833-3985; e-mail ospaal1966@enet.cu; internet www.tricontinental.cubaweb.cu; f. 1966 at the first Conference of Solidarity of the Peoples of Africa, Asia and Latin America, to unite, co-ordinate and encourage national liberation movements in the three continents, to oppose foreign intervention in the affairs of sovereign states, colonial and neo-colonial practices, and to fight against racialism and all forms of racial discrimination; favours the establishment of a new international economic order; mems: 56 organizations in 46 countries; Sec.-Gen. HUMBERTO HERNÁNDEZ; publ. *Tricontinental* (quarterly).

**Rio Group:** f. 1987 at a meeting in Acapulco, Mexico, of eight Latin American government leaders, who agreed to establish a 'permanent mechanism for joint political action'; additional countries subsequently joined the Group (see below); holds annual summit meetings at presidential level. At the ninth presidential summit (Quito, Ecuador, September 1995) a 'Declaration of Quito' was adopted, which set out joint political objectives, including the strengthening of democracy; combating corruption, drugs-production and -trafficking and 'money-laundering'; and the creation of a Latin American and Caribbean free trade area by 2005 (supporting the efforts of the various regional groupings). Opposes US legislation (the 'Helms-Burton' Act), which provides for sanctions against foreign companies that trade with Cuba; a pan-South American summit convened in December 2004, in Cuzco, Peru, approved in principle the creation within 15 years of a new South American Community of Nations, entailing the eventual merger of the Rio Group, the Andean Community and Mercosur; also concerned with promoting sustainable development in the region, the elimination of poverty, and economic and financial stability. The Rio Group holds annual ministerial conferences with the European Union (11th meeting held in Greece in March 2003; third summit of heads of state and govt held in Guadalajara, Mexico in May 2004); mems: Argentina, Bolivia, Brazil, Chile, Colombia, Costa Rica, Dominican Republic, Ecuador, El Salvador, Guatemala, Guyana, Honduras, Mexico, Nicaragua, Panama, Paraguay, Peru, Uruguay, Venezuela.

## Industrial and Professional Relations

**Caribbean Congress of Labour:** NUPW Bldg, Dalkeith Rd, St Michael, Barbados; tel. 429-5517; fax 427-2496; e-mail cclres@caribsurf.com; f. 1960 to fight for the recognition of trade union organizations to build and strengthen the ties between the free trade unions of the Caribbean and the rest of the world; to support the work of the International Confederation of Free Trade Unions; to encourage the formation of national groupings and centres; mems: 29 unions in 17 countries; Pres. LLOYD GOODLEIGH (Jamaica); Sec.-Treas. KERTIST AUGUSTUS (Dominica).

**Central American Confederation of Workers** (Confederación Centroamericana de Trabajadores): Apdo 226, 2200 Coronado, San José, Costa Rica; tel. 229-0152; fax 229-3893; e-mail icaesca@sol.racsa.co.cr; f. 1963; mems: national confederations in seven countries; Sec.-Gen. ALSIMIRO HERRERA TORRES.

**Inter-American Regional Organization of Workers (ORIT)** (Organización Regional Interamericana de Trabajadores): Edif. José Vargas, Avda Andrés Eloy Blanco No 2, 15°, Los Caobos, Caracas, Venezuela; tel. (212) 578-3538; fax (212) 578-1702; e-mail sedeorit@cioslorit; internet www.cioslorit.org/; f. 1951 by the International Confederation of Free Trade Unions, to link and represent democratic labour organizations in the region; sponsors training; mems: trade unions in 28 countries (including Canada and the USA) with over 40m. individuals; Gen. Sec. VICTOR BÁEZ.

**International Confederation of Free Trade Unions (ICFTU):** 5 blvd Roi Albert II, 1210 Brussels, Belgium; tel. (2) 224-02-11; fax (2) 201-58-15; e-mail internetpo@icftu.org; internet www.icftu.org; f. 1949 by trade union federations which had withdrawn from the World Federation of Trade Unions; aims to promote the interests of working people and to secure recognition of workers' organizations as free bargaining agents; mems: 225 organizations in 148 countries; Gen. Sec. GUY RYDER (UK); publs *Survey of Violations of Trade Union Rights* (annually, in English, French and Spanish), *Trade Union World* (monthly, in English, French and Spanish).

**Latin American Confederation of Trade Unions** (Central Latinoamericano de Trabajadores—CLAT): Apdo 6681, Caracas 1010A, Venezuela; tel. (212) 372-0794; fax (212) 372-0463; e-mail clat@telcel.net.ve; internet www.clat.org; f. 1954; affiliated to the World Confederation of Labour; mems: over 50 national and regional organizations in Latin America and the Caribbean; Sec.-Gen. EDUARDO GARCÍA MOURE.

**Latin American Federation of Agricultural Workers** (Federación Latinoamericana de Trabajadores Agrícolas, Pecuarios y Afines—FELTRA): Antiguo Local Conadi, B° La Granja, Comayaguela, Tegucigalpa, Honduras; tel. 2252526; fax 2252525; internet www.acmoti.org; f. 1999 by reorganization of FELTACA (f.1961) to represent the interests of workers in agricultural and related industries in Latin America; mems: national unions in 28 countries and territories; Sec.-Gen. MARCIAL REYES CABALLERO; publ. *Boletín Luchemos* (quarterly).

**World Confederation of Labour (WCL):** 33 rue de Trèves, 1040 Brussels, Belgium; tel. (2) 285-47-00; fax (2) 230-87-22; e-mail info@cmt-wcl.org; internet www.cmt-wcl.org; f. 1920 as the International Federation of Christian Trade Unions (IFCTU); reconstituted under present title in 1968; mems: about 26m. in 116 countries, incl. Czech Republic, Hungary, Lithuania, Poland, Romania and Slovakia; Sec.-Gen. WILLY THYS (Belgium); publ. *Tele-flash* (every 2 weeks).

## Law

**Inter-American Bar Association (IABA):** 1211 Connecticut Ave, NW, Suite 202, Washington, DC 20036, USA; tel. (202) 466-5944; fax (202) 466-5946; e-mail iaba@iaba.org; internet www.iaba.org; f. 1940 to promote the rule of law and to establish and maintain relations between asscns and organizations of lawyers in the Americas; mems: 90 asscns and 3,500 individuals in 27 countries; Sec.-Gen. HARRY A. INMAN (USA); publs *Newsletter* (quarterly), *Conference Proceedings*.

**International Union of Latin Notaries** (Union Internationale du Notariat Latin—UINL): Alsina 2280, 2°, 1090 Buenos Aires, Argentina; tel. (11) 4952-8848; fax (11) 4952-7094; internet www.iunl.org; f. 1948 to study and standardize notarial legislation and promote the progress, stability and advancement of the Latin notarial system; mems: organizations and individuals in 70 countries; Sec. BERNARDO PÉREZ FERNÁNDEZ DEL CASTILLO (Mexico); publs *Revista Internacional del Notariado* (quarterly), *Notarius International*.

## Medicine and Health

**Inter-American Association of Sanitary and Environmental Engineering** (Asociación Interamericana de Ingeniería Sanitaria y Ambiental—AIDIS): Rua Nicolau Gagliardi 354, 05429-010 São Paulo, SP, Brazil; tel. (11) 3812-4080; fax (11) 3814-2441; e-mail aidis@aidis.org.br; internet www.aidis.org.br; f. 1948 to assist in the development of water supply and sanitation; mems: 32 countries; Pres. ALEX CHECHILNITZKY (Chile); Exec. Dir LUIZ AUGUSTO DE LIMA PONTES (Brazil); publs *Revista Ingeniería Sanitaria* (quarterly), *Desafío* (quarterly).

**Latin American Association of National Academies of Medicine:** Carrera 7, Bogotá, Colombia; tel. and fax (1) 2493122; e-mail alanam_colombia@hotmail.com; f. 1967; mems: 11 national Academies; Exec. Sec. Dr ZOÍLO CUÉLLAR-MONTOYA (Colombia).

**Latin American Odontological Federation** (Federación Odontológica Latinoamericana): c/o Federación Odontológica Colombiana, Calle 71 No 11-10, Of. 1101, Apdo Aéreo 52925, Bogotá, Colombia; internet www.folaoral.org; f. 1917; linked to FDI World Dental Federation; mems: national organizations in 12 countries; Sec. Dr M. E. VILLEGAS MEJIA.

**Pan-American Association of Ophthalmology (PAAO):** 1301 South Bowen Rd, Suite 365, Arlington, TX 76013, USA; tel. (817) 275-7553; fax (817) 275-3961; e-mail info@paao.org; internet www.paao.org; f. 1939 to promote friendship within the profession and the dissemination of scientific information; holds biennial Congress (2005: Chile); mems: national ophthalmological societies and other bodies in 39 countries; Pres. Dr BRONWYN BATEMAN; Exec. Dir Dr CRISTIÁN LUCO (Chile); publ. *Vision Panamerica* (quarterly).

**Pan-American Medical Association** (Asociación Médica Panamericana): 745 Fifth Ave, New York, NY 10151, USA; tel. (212) 753-6033; f. 1925; holds inter-American congresses, conducts seminars and grants post graduate scholarships to Latin American physicians; mems: 6,000 in 30 countries; Sec. FREDERIC C. FENIG.

**World Self-Medication Industry (WSMI):** Centre International de Bureaux, 13 chemin du Levant, 01210 Ferney-Voltaire, France; tel. 4-50-28-47-28; fax 4-50-28-40-24; e-mail dwebber@wsmi.org; internet www.wsmi.org; Dir-Gen. Dr DAVID E. WEBBER.

## Posts and Telecommunications

**International Telecommunications Satellite Organization (INTELSAT):** 3400 International Drive, NW, Washington, DC 20008-3098, USA; tel. (202) 944-6800; fax (202) 944-7860; internet www.intelsat.com; f. 1964 to establish a global commercial satellite communications system; Assembly of Parties attended by representatives of member governments, meets every two years to consider policy and long-term aims and matters of interest to members as sovereign states; meeting of Signatories to the Operating Agreement held annually; 24 INTELSAT satellites in geosynchronous orbit provide a global communications service; provides most of the world's overseas traffic; in 1998 INTELSAT agreed to establish a private enterprise, incorporated in the Netherlands, to administer six satellite services; mems: 143 governments; Dir-Gen. and Chief Exec. CONNY L. KULLMAN (Sweden).

**Internet Corporation for Assigned Names and Numbers (ICANN):** 4676 Admiralty Way, Suite 330, Marina del Rey, CA 90292-6601, USA; tel. (310) 823-9358; fax (310) 823-8649; e-mail icann@icann.org; internet www.icann.org; f. 1998; non-profit, private-sector body; aims to co-ordinate the technical management and policy development of the internet; comprises three Supporting Organizations to assist, review and develop recommendations on internet policy and structure relating to addresses, domain names, and protocol; Pres. and CEO PAUL TWOMEY (Australia).

**Postal Union of the Americas, Spain and Portugal** (Unión Postal de las Américas, España y Portugal): Cebollatí 1468/70, 1°, Casilla de Correos 20.042, Montevideo, Uruguay; tel. (2) 4100070; fax (2) 4105046; e-mail secretaría@upaep.com.uy; internet www.upaep.com.uy; f. 1911 to extend, facilitate and study the postal relationships of member countries; mems: 27 countries; Sec.-Gen. MARIO FELMER KLENNER (Chile).

## Press, Radio and Television

**Inter-American Press Association (IAPA)** (Sociedad Interamericana de Prensa): Jules Dubois Bldg, 1801 SW 3rd Ave, Miami, FL 33129, USA; tel. (305) 634-2465; fax (305) 635-2272; e-mail info@sipiapa.org; internet www.sipiapa.org; f. 1942 to guard the freedom of the press in the Americas to promote and maintain the dignity, rights and responsibilities of the profession of journalism; to foster a wider knowledge and greater interchange among the peoples of the Americas; mems: 1,400; Exec. Dir JULIO E. MUÑOZ; publ. *IAPA News* (monthly).

**International Association of Broadcasting** (Asociación Internacional de Radiodifusión—AIR): Edif. Torre Uruguay, Cnel Brandzen 1961, Of. 402, 11200 Montevideo, Uruguay; tel. (2) 4088129; fax (2) 4088121; e-mail airiab@distrinet.com.uy; internet www.airiab.com; f. 1946 to preserve free and private broadcasting to promote co-operation between the corporations and public authorities; to defend freedom of expression; mems: national asscns of broadcasters; Pres. Dr LUIS H. TARSITANO; Dir-Gen. Dr HÉCTOR OSCAR AMENGUAL; publ. *La Gaceta de AIR* (every 2 months).

**Latin American Catholic Organization for the Cinema and Audiovisuals** (Organización Católica Internacional del Cine y del Audiovisual): Apdo Postal 17-21-178, Quito, Ecuador; tel. (2) 548-046; fax (2) 501-658; e-mail ocic-al@seccom.ec; f. 1961 to promote and develop the work of Roman Catholic video- and film-makers; mems: national associations of producers, video- and film-makers, cinema critics; Exec. Sec. CARLOS CORTÉS.

**Latin-American Catholic Press Union:** Apdo Postal 17-21-178, Quito, Ecuador; tel. (2) 548-046; fax (2) 501-658; f. 1959 to co-ordinate, promote and improve the Catholic press in Latin America; mems: national asscns and local groups in most Latin American countries; Pres. ISMAR DE OLIVEIRA SOARES (Brazil); Sec. CARLOS EDUARDO CORTÉS (Colombia).

**Latin American Catholic Radio and Television Association** (Asociación Católica Latinoamericana para la Radio, la TV y los Medios Afines): Apdo Postal 17-21-178, Quito, Ecuador; tel. (2) 548-046; fax (2) 501-658; e-mail scc@seccom.ecx.ec; f. 1957 to promote activities in radio, TV and similar media in Latin America; mems: national associations of Roman Catholic radio and TV stations, and individuals engaged in radio and TV work in 19 countries; Exec. Sec. CARLOS CORTÉS.

## Religion

**Caribbean Conference of Churches:** 8 Gallus St, Woodbrook, POB 876, Port of Spain, Trinidad and Tobago; tel. 623-0588; fax 624-9002; e-mail cchq@tstt.net.tt; internet www.cariblife.com/pub/ccc; f. 1973; holds Assembly every five years; the recognized Regional Ecumenical Organization of the Caribbean and a major development agency in the region; initiatives include human development programmes (designated as Priority Regional Initiatives), sustainable socio-economic development programmes, an advocacy and communications programme and the international relations programme; maintains the Caribbean Ecumenical Institute; mems: 34 churches; Gen. Sec. GERARD GRANADO; publ. *Ecuscope Caribbean*.

**Latin American Council of Churches** (Consejo Latinoamericano de Iglesias—CLAI): Casilla 17-08-8522, Calle Inglaterra N.32–113 y Mariana de Jesús, Quito, Ecuador; tel. (2) 255-3996; fax (2) 252-9933; e-mail israel@clai.org.ec; internet www.clai.org.ec; f. 1982; mems: 147 churches in 19 countries; Gen. Sec. Rev. ISRAEL BATISTA.

**Latin American Episcopal Council** (Consejo Episcopal Latinoamericano—CELAM): Apartado Aéreo 51086, Bogotá, Colombia; tel. (1) 6121620; fax (1) 6121929; e-mail celam@celam.org; internet www.celam.org; f. 1955 to co-ordinate Church activities in and with the Latin American and the Caribbean Catholic Bishops' Conferences; mems: 22 Episcopal Conferences of Central and South America and the Caribbean; Pres. Cardinal FRANCISCO JAVIER ERRÁZURIZ O.; publ. *Boletín* (6 a year).

## Science

**International Council for Science (ICSU):** 51 blvd de Montmorency, 75016 Paris, France; tel. 1-45-25-03-29; fax 1-42-88-94-31; e-mail secretariat@icsu.org; internet www.icsu.org; f. 1919 as International Research Council; present name adopted 1931; new statutes adopted 1996; to co-ordinate international co-operation in theoretical and applied sciences and to promote national scientific research through the intermediary of affiliated national organizations; General Assembly of representatives of national and scientific members meets every three years to formulate policy. The following committees have been established: Cttee on Science for Food Security, Scientific Cttee on Antarctic Research, Scientific Cttee on Oceanic Research, Cttee on Space Research, Scientific Cttee on Water Research, Scientific Cttee on Solar-Terrestrial Physics, Cttee on Science and Technology in Developing Countries, Cttee on Data for Science and Technology, Programme on Capacity Building in Science, Scientific Cttee on Problems of the Environment, Steering Cttee on Genetics and Biotechnology and Scientific Cttee on International Geosphere-Biosphere Programme. The following services and Inter-Union Committees and Commissions have been established: Federation of Astronomical and Geophysical Data Analysis Services, Inter-Union Commission on Frequency Allocations for Radio Astronomy and Space Science, Inter-Union Commission on Radio Meteorology, Inter-Union Commission on Spectroscopy, Inter-Union Commission on Lithosphere; national mems: academies or research councils in 98 countries; Scientific mems and assocs: 26 international unions and 28 scientific associates; Pres. W. ARBER; Sec.-Gen. H. A. MOONEY; publs *ICSU Yearbook*, *Science International* (quarterly), *Annual Report*.

## Social Sciences

**International Peace Academy (IPA):** 777 United Nations Plaza, New York, NY 10017, USA; tel. (212) 687-4300; fax (212) 983-8246; e-mail ipa@ipacademy.org; internet www.ipacademy.org; f. 1970 to promote the prevention and settlement of armed conflicts between and within states through policy research and development; educates government officials in the procedures needed for conflict resolution, peace-keeping, mediation and negotiation, through international training seminars and publications; off-the-record meetings are also conducted to gain complete understanding of a specific conflict; Chair. RITA E. HAUSER; Pres. TERJE ROED-LARSEN.

## Social Welfare and Human Rights

**Co-ordinator of the Indigenous Organizations of the Amazon Basin (COICA):** e-mail com@coica.org; internet www.coica.org; f. 1984; aims to co-ordinate the activities of national organizations concerned with the indigenous people and environment of the Amazon basin, and promotes respect for human rights and the self-determination of the indigenous populations; nine member orgs; Co-ordinator-General SEBASTIÃO MANCHINERI; publ. *Nuestra Amazonia* (quarterly, in English, Spanish, French and Portuguese).

**Global Commission on Migration:** rue Richard-Wagner 1, 1202 Geneva, Switzerland; tel. (22) 748-48-50; fax (22) 748-48-48-51; internet www.test.gcim.org; f. 2003 to place international migration issues on the global agenda, to analyse migration policy, to examine links with other fields, and to present recommendations for consideration by the UN Secretary-General, governments and other parties; 19 commissioners; Exec. Dir Dr ROLF K. JENNY (Switzerland).

**Inter-American Conference on Social Security** (Conferencia Interamericano de Seguridad Social—CISS): Calle San Ramon s/n esq. Avda San Jerónimo, UCol. San Jerónimo Lídice, Del. Magdalena Contreras, CP 10100 México, DF, Mexico; tel. (55) 5595-0011; fax (55) 5683-8524; e-mail ciss@ciss.org.mx; internet www.ciss.org.mx; f. 1942 to contribute to the development of social security in the countries of the Americas and to co-operate with social security institutions; CISS bodies are: the General Assembly, the Permanent Inter-American Committee on Social Security, the Secretariat General, six American Commissions of Social Security and the Inter-American Center for Social Security Studies; mems: 66 social security institutions in 36 countries; Pres. Santiago Levy Algazi (Mexico); Sec.-Gen. Dr Gabriel Martínez González (Mexico); publs *Social Security Journal/Seguridad Social* (every 2 months), *The Americas Social Security Report* (annually), *Social Security Bulleting* (monthly, online), monographs, study series.

## Sport and Recreations

**International Federation of Association Football** (Fédération internationale de football association—FIFA): FIFA House, Hitzigweg 11, POB 85, 8030 Zürich, Switzerland; tel. (1) 3849595; fax (1) 3849696; e-mail media@fifa.org; internet www.fifa.com; f. 1904 to promote the game of association football and foster friendly relations among players and national asscns; to control football and uphold the laws of the game as laid down by the International Football Association Board to prevent discrimination of any kind between players; and to provide arbitration in disputes between national asscns; organizes World Cup competition every four years; mems: 204 national asscns, six continental confederations; Pres. Joseph S. Blatter (Switzerland); Gen. Sec. Urs Linsi; publs *FIFA News* (monthly), *FIFA Magazine* (every 2 months) (both in English, French, German and Spanish ), *FIFA Directory* (annually), *Laws of the Game* (annually), *Competitions' Regulations* and *Technical Reports* (before and after FIFA competitions).

## Technology

**Latin American Association of Pharmaceutical Industries** (Asociación Latinoamericana de Industrias Farmaceuticas—ALIFAR): Av. Libertador 602, 6°, 1001 ABT Buenos Aires, Argentina; tel. and fax (11) 4812-4532; e-mail mlevis@cilfa.org.ar; f. 1980; mems: about 400 enterprises in 15 countries; Sec.-Gen. Rubén Abete; Exec. Sec. Mirta Levis.

**Latin-American Energy Organization** (Organización Latinoamericana de Energía—OLADE): Avda Mariscal Antonio José de Sucre, No N58–63 y Fernándes Salvador, Edif. OLADE, Sector San Carlos, POB 17-11-6413 CCI, Quito, Ecuador; tel. (2) 2598-122; fax (2) 2531-691; e-mail oladel@olade.org.ec; internet www.olade.org.ec; f. 1973 to act as an instrument of co-operation in using and conserving the energy resources of the region; mems: 26 Latin-American and Caribbean countries; Exec. Sec. Dr Diego Pérez Pallares; publ. *Energy Magazine*.

**Latin-American Iron and Steel Institute:** Benjamín 2944, 5°, Las Condes, Santiago, Chile; tel. (2) 233-0545; fax (2) 233-0768; e-mail ilafa@ilafa.org; internet www.ilafa.org; f. 1959 to help achieve the harmonious development of iron and steel production, manufacture and marketing in Latin America; conducts economic surveys on the steel sector; organizes technical conventions and meetings; disseminates industrial processes suited to regional conditions; prepares and maintains statistics on production, end uses, prices, etc., of raw materials and steel products within this area; mems: 18 hon. mems; 63 mems; 68 assoc. mems; Chair. Daniel Novegíl; Sec.-Gen. Guillermo Moreno; publs *Acero Latinoamericano* (every 2 months), *Statistical Year Book*, *Directory of Latin American Iron and Steel Companies* (every 2 years).

## Tourism

**Caribbean Tourism Organization:** One Financial Pl., Collymore Rock, St Michael, Barbados; tel. 427-5242; fax 429-3065; e-mail ctobar@caribsurf.com; internet www.onecaribbean.org; f. 1989, by merger of the Caribbean Tourism Association (f. 1951) and the Caribbean Tourism Research and Development Centre (f. 1974); aims to encourage tourism in the Caribbean region; organizes annual Caribbean Tourism Conference, Sustainable Tourism Development Conference and Tourism Investment Conference; conducts training and other workshops on request; maintains offices in New York, Canada and London; mems: 33 Caribbean governments, 400 allied mems; Sec.-Gen. Vincent Vanderpool-Wallace (Bahamas); publs *Caribbean Tourism Statistical News* (quarterly), *Caribbean Tourism Statistical Report* (annually).

**Latin-American Confederation of Tourist Organizations** (Confederación de Organizaciones Turísticas de la América Latino—COTAL): Viamonte 640, 8°, 1053 Buenos Aires, Argentina; tel. (11) 4322-4003; fax (11) 4393-5696; e-mail cotal@cscom.com.ar; internet www.cotal.org.ar; f. 1957 to link Latin American national asscns of travel agents and their members with other tourist bodies around the world; mems: in 21 countries; Pres. Enzo U. Furnari; publ. *Revista COTAL* (every 2 months).

## Trade and Industry

**Cairns Group:** (no permanent secretariat); internet www.cairnsgroup.org; f. 1986 by major agricultural exporting countries; aims to bring about reforms in international agricultural trade, including reductions in export subsidies, in barriers to access and in internal support measures; represents members' interests in WTO negotiations; mems: Argentina, Australia, Bolivia, Brazil, Canada, Chile, Colombia, Costa Rica, Fiji, Guatemala, Indonesia, Malaysia, New Zealand, Paraguay, Philippines, South Africa, Thailand, Uruguay; Chair. Mark Vaile (Australia).

**Caribbean Association of Industry and Commerce (CAIC):** POB 442, Trinidad Hilton and Conference Centre, Rooms 1238–1241, Lady Young Rd, St Ann's, Trinidad and Tobago; tel. (868) 623-4830; fax (868) 623-6116; e-mail sifill@wow.net; internet www.caic.wow.net; f. 1955; aims to encourage economic development through the private sector; undertakes research and training and gives assistance to small enterprises; encourages export promotion; mems: chambers of commerce and enterprises in 20 countries and territories; Exec. Dir Sean Ifill; publ. *Caribbean Investor* (quarterly).

**CropLife International:** ave Louise 143, 1050 Brussels, Belgium; tel. (2) 542-04-10; fax (2) 542-04-19; e-mail croplife@croplife.org; internet www.croplife.org; f. 1960 as European Group of National Asscns of Pesticide Manufacturers, international body since 1967, present name adopted in 2001, evolving from Global Crop Protection Federation; represents the plant science industry, with the aim of promoting sustainable agricultural methods; aims to harmonize national and international regulations concerning crop protection products and agricultural biotechnology; promotes observation of the FAO Code of Conduct on the Distribution and Use of Pesticides; holds an annual General Assembly; mems: 6 regional bodies and national asscns in 85 countries; Dir-Gen. Christian Verschueren.

**Federación de Cámaras de Comercio del Istmo Centroamericano** (Federation of Central American Chambers of Commerce): 10A Calle 3-80, Zona 1, 01001 Guatemala City, Guatemala; internet www.fecamco.com; f. 1961; plans and co-ordinates industrial and commercial exchanges and exhibitions; Pres. Emilio Bruce Jiménez (Costa Rica); Sec. Jorge E. Briz Abularach (Guatemala).

**Instituto Centroamericano de Administración de Empresas (INCAE)** (Central American Institute for Business Administration): Apdo 960, 4050 Alajuela, Costa Rica; tel. 443-0506; fax 433-9101; e-mail artaviar@mail.incae.ac.cr; internet www.incae.ac.cr; f. 1964; provides a postgraduate programme in business administration; runs executive training programmes; carries out management research and consulting; maintains a second campus in Nicaragua; libraries of 85,000 vols; Rector Dr Roberto Artavia; publs *Alumni Journal* (in Spanish), *Bulletin* (quarterly), books and case studies.

**Inter-American Commercial Arbitration Commission:** OAS Administration Bldg, Rm 211, 19th and Constitution Ave, NW, Washington, DC 20006, USA; tel. (202) 458-3249; fax (202) 458-3293; f. 1934 to establish an inter-American system of arbitration for the settlement of commercial disputes by means of tribunals; mems: national committees, commercial firms and individuals in 22 countries; Dir-Gen. Dr Adriana Polania.

## Transport

**Pan American Railway Congress Association** (Asociación del Congreso Panamericano de Ferrocarriles): Av. 9 de Julio 1925, 13°, 1332 Buenos Aires, Argentina; tel. (11) 4381-4625; fax (11) 4814-1823; e-mail acpf@nat.com.ar; f. 1907, present title adopted 1941; aims to promote the development and progress of railways in the American continent; holds Congresses every three years; mems: government representatives, railway enterprises and individuals in 21 countries; Pres. Juan Carlos de Marchi (Argentina); Gen. Sec. Luis V. Donzelli (Argentina); publ. *Boletín ACPF* (5 a year).

## Youth and Students

**Latin American and Caribbean Confederation of Young Men's Christian Associations** (Confederación Latinoamericana y del Caribe de Asociaciones Cristianas de Jóvenes): Culpina 272, 1406 Buenos Aires, Argentina; tel. (11) 4373-4156; fax (11) 4374-4408; e-mail clacj@wamani.apc.org; f. 1914; aims to encourage the moral, spiritual, intellectual, social and physical development of young men; to strengthen the work of national Asscns and to sponsor the establishment of new Asscns; mems: affiliated YMCAs in 25 countries (comprising 350,000 individuals); Pres. GERARDO VITUREIRA (Uruguay); Gen. Sec. MARCO ANTONIO HOCHSCHEIT (Brazil); publs *Diecisiete/21* (bulletin), *Carta Abierta*, *Brief*, technical articles and other studies.

**WFUNA Youth:** c/o Palais des Nations, 16 ave Jean-Tremblay, 1211 Geneva 10, Switzerland; tel. (22) 7985850; fax (22) 7334838; internet www.wfuna-youth.org; f. 1948 by the World Federation of United Nations Associations (WFUNA) as the International Youth and Student Movement for the United Nations (ISMUN), independent since 1949; an international non-governmental organization of students and young people dedicated especially to supporting the principles embodied in the United Nations Charter and Universal Declaration of Human Rights; encourages constructive action in building economic, social and cultural equality and in working for national independence, social justice and human rights on a world-wide scale; maintains regional offices in Austria, France, Ghana, Panama and the USA; mems: asscns in 53 countries world-wide; Head Co-ordinator TIM JARMAN.

# MAJOR COMMODITIES OF LATIN AMERICA

Note: For each of the commodities in this section, there are generally two statistical tables: one relating to recent levels of production, and one indicating recent trends in prices. Each production table shows estimates of output for the world and for Latin America. In addition, the table lists the main Latin American producing countries and, for comparison, the leading producers from outside the region. In most cases, the table referring to prices provides indexes of export prices, calculated in US dollars. The index for each commodity is based on specific price quotations for representative grades of that commodity in countries that are major traders (excluding countries of Eastern Europe and the former USSR).

## Aluminium and Bauxite

Aluminium is the most abundant metallic element in the earth's crust, comprising about 8% of the total. However, it is much less widely used than steel, despite having about the same strength and only half of the weight. Aluminium has important applications as a metal because of its lightness, ease of fabrication and other desirable properties. Other products of alumina (aluminium oxide trihydrate, into which bauxite, the commonest aluminium ore, is refined) are materials used in glass manufacture, refractories, abrasives, ceramic products, catalysts and absorbers. Alumina hydrates are used for the production of aluminium chemicals, as fire retardants in carpet backing, and as industrial fillers in plastics and related products.

The major markets for aluminium are in transportation, building and construction, electrical machinery and equipment, consumer durables and the packaging industry, which in 2000 accounted for more than 20% of all aluminium use. Although the production of aluminium is energy-intensive, its light weight results in a net saving, particularly in the transportation industry. About one-quarter of aluminium output is consumed in the manufacture of transport equipment, particularly road motor vehicles and components, where the metal is increasingly being used as a substitute for steel. In the early 1990s steel substitution accounted for about 16% of world aluminium consumption, and it has been forecast that aluminium demand by the motor vehicle industry alone could more than double, to exceed 5.7m. metric tons in 2010, from around 2.4m. tons in 1990. Aluminium is of great value to the aerospace industry for its weight-saving characteristics and its low cost relative to alternative materials. Aluminium-lithium alloys command considerable potential for use in this sector, although the traditional dominance of aluminium in the aerospace sector has been challenged since the 1990s by 'composites' such as carbonepoxy, a fusion of carbon fibres and hardened resins, whose lightness and durability can exceed that of many aluminium alloys.

Until recently world markets for finished and semi-finished aluminium products were dominated by six Western producers—Alcan (Canada), Alcoa Inc., Reynolds Metals Co, Kaiser (all USA), Pechiney (France) and algroup (Switzerland). Proposals for a merger between Alcan, algroup and Pechiney, and between Alcoa and Reynolds, were announced in August 1999. However, the proposed terms of the Pechiney-Alcan-algroup merger encountered opposition from the European Commission, on the grounds that the combined grouping could restrict market competition and adversely affect the interests of consumers. The tripartite merger plan was abandoned in April 2000, although Alcan and algroup were permitted to merge in October. In 2003, having agreed to meet conditions imposed by the European Commission and the US Department of Justice in respect of safeguarding free competition, Alcan was permitted to purchase Pechiney. One of the most significant of the conditions imposed in respect of Alcan's purchase of Pechiney was a requirement that it divest some of its rolled aluminium products assets. In late 2004, as a consequence of this divestment, a new rolled aluminium products group, Novelis, was emerging. In the USA Alcoa Inc. and Reynolds Metals Co merged in mid-2000. In 2002, after its purchase of Germany's VAW, Norway's Norsk Hydro became the world's third largest integrated aluminium concern. Prior to the mergers detailed above the level of dominance of the six major Western producers had been reduced by a significant shift in the location of alumina and aluminium production to areas where cheap power is available, such as Australia, Brazil, Norway, Canada and Venezuela. The Gulf states of Bahrain and Dubai (United Arab Emirates), with the advantage of low energy costs, also produce primary aluminium. Since the mid-1990s Russia has become a significant force in the world aluminium market (see below), and in 2000 the country's principal producers, together with a number of plants located in the Commonwealth of Independent States, merged to form the Russian Aluminium Co (Rusal), whose facilities in 2005 were the source of 75% of Russian and 10% of global primary aluminium output. Sual is Russia's other major producer.

Bauxite is the principal aluminium ore, but nepheline syenite, kaolin, shale, anorthosite and alunite are all potential alternative sources of alumina, although not currently economic to process. Of all bauxite mined, approximately 85% is converted to alumina ($Al_2O_3$) for the production of aluminium metal. Developing countries, in which at least 70% of known bauxite reserves are located, supply some 50% of the ore required. The industry is structured in three stages: bauxite mining, alumina refining and smelting. While the high degree of 'vertical integration' (i.e. the control of successive stages of production) in the industry means that a significant proportion of trade in bauxite and alumina is in the form of intra-company transfers, and the increasing tendency to site alumina refineries near to bauxite deposits has resulted in a shrinking bauxite trade, there is a growing free market in alumina, catering for the needs of the increasing number of independent (i.e. non-integrated) smelters.

The alumina is separated from the ore by the Bayer process. After mining, bauxite is fed to process directly if mine-run material is adequate (as in Jamaica) or is crushed and beneficiated. Where the ore 'as mined' presents handling problems, or weight reduction is desirable, it may be dried prior to shipment.

**Production of Bauxite**
(crude ore, '000 metric tons)

|  | 2003 | 2004* |
|---|---|---|
| **World total** (excl. USA) | 146,000 | 156,000 |
| Latin America | 37,420 | 43,400 |
| **Latin American producers** | | |
| Brazil | 13,100 | 18,500 |
| Guyana† | 1,500 | 1,700 |
| Jamaica† | 13,400 | 13,500 |
| Suriname | 4,220 | 4,200 |
| Venezuela | 5,200 | 5,500 |
| **Other leading producers** | | |
| Australia | 55,600 | 56,000 |
| China, People's Repub. | 12,500* | 15,000 |
| Guinea† | 15,500 | 15,500 |
| India | 10,000 | 10,000 |
| Kazakhstan | 4,377‡ | n.a. |
| Russia | 4,000* | 5,000 |

* Estimated production.
† Dried equivalent of crude ore.
‡ 2002 figure.

Source: US Geological Survey.

At the alumina plant the ore is slurried with spent-liquor directly, if the soft Caribbean type is used, or, in the case of other types, it is ball-milled to reduce it to a size which will facilitate the extraction of the alumina. The bauxite slurry is then digested with caustic soda to extract the alumina from the ore while leaving the impurities as an insoluble residue. The digest conditions depend on the aluminium minerals in the ore and the impurities. The liquor, with the dissolved alumina, is then separated from the insoluble impurities by combinations of sedimentation, decantation and filtration, and the residue is washed to minimize the soda losses. The clarified liquor is concentrated and the alumina precipitated by seeding with hydrate. The precipitated alumina is then filtered, washed and calcined to produce alumina. The ratio of bauxite to alumina is approximately 1.95:1.

The smelting of the aluminium is generally by electrolysis in molten cryolite. Because of the high consumption of electricity by the smelting process, alumina is usually smelted in areas where low-cost electricity is available. However, most of the electricity now used in primary smelting in the Western world is generated by hydroelectricity—a renewable energy source.

The recycling of aluminium is economically, as well as environmentally, desirable, as the process uses only 5% of the electricity required to produce a similar quantity of primary aluminium. Aluminium recycled from scrap currently accounts for almost 30% of the total annual world output of primary aluminium. With the added impetus of environmental concerns, considerable growth occurred world-wide in the recycling of used beverage cans (UBC) during the 1990s. In the middle of that decade, according to aluminium industry estimates, the recycling rate of UBC amounted to at least 55% world-wide. In the USA in 2003 a UBC recycling rate of 50% was reported.

In 2004, according to the International Aluminium Institute (IAI), world output of primary aluminium totalled an estimated 22.6m. metric tons, of which Latin American producers (Argentina, Brazil, Mexico, Suriname and Venezuela) accounted for about 2.4m. tons. The USA normally accounts for more than one-quarter of total aluminium consumption (excluding communist and former communist countries). The USA was for long the world's principal producing country, but in 2001 US output of primary aluminium was surpassed by that of Russia and the People's Republic of China. In 2002 and, it was estimated, in 2003 Canadian production, in addition to that of Russia and China, exceeded that of the USA. In 2003 production of primary aluminium by China was estimated at more than double that of the USA.

In 2004 Brazil was the world's second largest producer of bauxite (after Australia). The country possesses extensive bauxite reserves in Minas Gerais state and in the Amazon region, where deposits of some 4,600m. metric tons have been identified. Brazil's huge Albrás aluminium plant, located in the state of Pará, was founded in 1978 and constructed in two phases, each with a nominal annual capacity of 160,000 tons. The first phase commenced operations in 1985, while the second became fully operational in 1991. By 2002 successive technological improvements had raised total annual metal production capacity to 406,000 tons. The plant's initial construction was largely financed by Japan. In 1997 Albrás was privatized: now, via its subsidiary Vale do Rio Doce Alumínio, Companhia Vale do Rio Doce (CVRD) disposes of 51% of the capital, the remaining 49% being held by Nippon Amazon Aluminium Co Ltd. During the first 10 years after it became operational, Albrás was dependent on imported alumina. Since 1995, however, alumina has been supplied by the nearby Alunorte plant, where bauxite mined in the Trombetas river region in north-west Pará is refined. Albrás receives some 800,000 tons of alumina annually from Alunorte, which supplies additional quantities to other Brazilian and overseas aluminium industries. Most of the metal produced by Albrás is exported through the port of Vila do Conde to the USA, Japan and Europe. In the north-eastern state of Maranhão Alcoa Alumínio manages the Alumar complex, comprising the largest alumina refinery in South America and the second largest South American smelter. Annual output of alumina from Alumar currently amounts to some 1.3m. tons, and that of metal to about 365,000 tons. In 2004 CVRD concluded a framework agreement with the Aluminium Corporation of China Ltd to construct an additional alumina refinery in north-eastern Brazil. Brazil's annual output of primary aluminium was maintained at about 1.2m. tons during 1992–2000, but declined slightly, to 1.1m. tons, in 2001. In 2002 Brazilian production reached 1.3m. tons, and it was estimated to have risen again, to 1.4m. tons, in 2003.

Jamaica was the world's fifth largest producer of bauxite in 2004. A production levy on the Jamaican bauxite industry, which has operated since 1974, provides an important source of government revenue. Production capacity was increased in the second half of the 1980s in order to supply the strategic mineral stockpiles of both the USA and the USSR. However, as a result of depressed conditions in the international market (see below), sales of aluminium ores and concentrates, including alumina, accounted for only 56.4% of total export earnings in 2000, compared with 63% in 1990. Measures to revitalize Jamaica's bauxite and alumina industry were announced in July 1998 by representatives of the Government, industry and the trade unions. Output of bauxite in 1998 was 12.6m. metric tons, its highest level for 25 years, but production declined to 11.7m. tons in 1999, owing to the closure of a US refinery that used Jamaican ore. Plans to expand bauxite production and improve efficiency were under discussion in 2000. In that year output declined again, to 11.1m. tons, but it recovered in 2001, to 12.4m. tons. In 2002 Jamaican production was estimated to have risen once more, to some 13.1m. tons. Output totalled 13.4m. tons in 2003, and was estimated at 13.5m. tons in 2004.

The economy of Suriname relies heavily on the mining and export of bauxite. In 1993 bauxite, alumina and aluminium accounted for 91.5% of Suriname's annual export earnings, whereas in 1978 the corresponding proportion had been 83.8%. In 2003 exports of alumina represented about 60% of receipts accruing from major exports. In early 2005 the expansion of the alumina refinery operated by Suriname Aluminium Co, LLC (Suralco)—owned by Alcoa World Alumina and Chemicals—at Paranam was completed, raising the facility's annual capacity to some 2.2m. metric tons.

Venezuela's aluminium industry achieved rapid growth in the 1980s, as a result of the availability of raw materials and cheap hydroelectric power. Aluminium production, based on imported alumina, subsequently overtook iron ore to become Venezuela's main export industry after petroleum. The exploitation of bauxite reserves of 500m. metric tons was expected to enable Venezuela to produce 3.4m.–4.0m. tons of high-grade ore annually from mines in Bolívar State when output reached full capacity in 1992, by which time Venezuelan aluminium plants were to be supplied solely with local bauxite. In 1992 Venezuela's production of bauxite exceeded 1m. tons. The subsequent expansion of the bauxite sector industry occurred rapidly, output advancing to 2.5m. tons in 1993, and rising to 5.1m. tons per year in 1997 and 1998. Production averaged about 4.3m. tons annually in 1999–2001. In 2002 output was estimated to have risen substantially, to 5m. tons, and increased again, to 5.2m. tons, in 2003. In 2004 Venezuela mined an estimated 5.5m. tons of bauxite. In 2000 the country's state-owned aluminium complex was reported to be undergoing further preparation for transfer to the private sector, following unsuccessful attempts in 1998 and 1999 to dispose of the Government's 70% holding by auction. In 2001, meanwhile, Pechiney was reported to have undertaken to modernize and extend state-owned facilities centred on the Pijiguaos mine in Bolívar state, the first substantial private investment in the sector since the Government's failure to privatize it. According to the US Geological Survey (USGS), Venezuela's production of primary aluminium increased from 561,000 tons in 1992 to a record 634,000 tons in 1997. It subsequently decreased, to about 570,000 tons annually, before rising significantly, to 605,000 tons, in 2002. In 2003 output was estimated to have fallen slightly, to 601,000 tons. Aluminium's contribution to the country's export revenue was about 3% in 2001. Venezuela reportedly exported 76% of all the aluminium it produced in 2003.

Guyana's nationalized bauxite production industry is a major source of the country's export revenue (excluding revenue from re-exports), of which bauxite provided 8.8% in 2003. Most of the country's production is by Aroaima Bauxite Co, in which Reynolds formerly held a 50% stake. Following Alcoa Inc.'s merger with Reynolds Metals Co in 2000, this stake was sold to the Guyana Government, which already held the remaining 50%, in 2001. The Government has announced its intention to prepare Aroaima for privatization by reducing its production costs. In early 2004 the Government signed a memorandum of understanding (MOU) with Russia's Rusal in which Guyana's potential to supply Rusal with some 500,000–600,000 metric tons of bauxite annually from 2005 was acknowledged. The two parties also agreed to examine the possibility of reactivating alumina production at disused Guyanese facilities. Later in the year the Government was reported to be studying the feasibility of constructing new alumina production facilities, with the aim, possibly, of supplying alumina to Trinidad, which has signed an MOU with Alcoa to construct an aluminium smelter. However, although Trinidad has reportedly expressed its willingness to supply new Guyanese bauxite refining facilities with cheap power, their economic viability remains uncertain.

In November 1998 a preliminary agreement was signed for the construction of a primary aluminium smelter, with an eventual capacity of 474,000 metric tons per year, at Point Lisas, Trinidad. It was envisaged that this development, with a projected cost of US $1,500m., would proceed in two stages. The first phase, with an annual capacity of 237,000 tons, was to be completed in 2002. Its metal output would, according to the proposals, be exported to Europe and North America. However, in early 1999, with international prices of aluminium at their lowest for about five years (see below), plans for the Trinidad smelter project were postponed. In 2004 the Government of Trinidad and Tobago and Alcoa signed an MOU that envisaged the construction of a primary aluminium smelter at La Brea, in the south-west of Trinidad island.

Although world demand for aluminium advanced by an average of 3% annually from the late 1980s until 1994 (see below), industrial recession began, in 1990, to create conditions of over-supply. Despite the implementation of capacity reductions at an annual rate of 10% by the major Western producers, stock levels began to accumulate. The supply problem was exacerbated by a rapid rise, beginning in 1991, of exports by the USSR and its successor states, which had begun to accumulate substantial stocks of aluminium as a consequence of the collapse of the Soviet arms industry. The requirements of these countries for foreign exchange to restructure their economies led to a rapid acceleration in low-cost exports of high-grade aluminium to Western markets. These sales caused considerable dislocation of the market and involved the major Western producers in heavy financial losses. Producing members of the European Community (EC, now the European Union—EU) were particularly severely affected, and in August 1993 the EC imposed quota arrangements, under which aluminium imports from the former USSR were to be cut by 50% for an initial three-month period, while efforts were made to negotiate an agreement that would reduce the flow of low-price imports and achieve a reduction in aluminium stocks (by then estimated to total 4.5m. metric tons world-wide).

These negotiations, which involved the EC, the USA, Canada, Norway, Australia and Russia (but in which the minor producers, Brazil, the Gulf states, Venezuela and Ukraine were not invited to take part), began in October 1993. Initially, the negotiations made little progress, and in November the market price of high-grade aluminium ingots fell to an 11-year 'low'. Following further meetings in January 1994, however, an MOU was finalized on a plan whereby Russia was to 'restructure' its aluminium industry and reduce its output by 500,000 metric tons annually. By March the major Western producers had agreed to reduce annual production by about 1.2m. tons over a maximum period of two years. Additionally, Russia was to receive US $2,000m. in loan guarantees. The MOU provided for participants to monitor world aluminium supplies and prices on a regular basis. In March the EU quota was terminated. By July the world price had recovered by about 50% on the November 1993 level.

The successful operation of the MOU, combined with a strong recovery in world aluminium demand, led to the progressive reduction of stock levels and to a concurrent recovery in market prices during 1994 and 1995. Consumption in the Western industrialized countries and Japan rose by an estimated 10.3% to 17.3m. metric tons, representing the highest rate of annual growth since 1983. This recovery was attributed mainly to a revival in demand from the motor vehicle sector in EU countries and the USA, and to an intensified programme of public works construction in Japan. Increased levels of demand were also reported in China, and in the less industrialized countries of the South and East Asia region. Demand in the industrialized countries advanced by 11% in 1994, and by 2.2% in 1995.

Levels of world aluminium stocks were progressively reduced during 1994, and by late 1995 it was expected that the continuing fall in stock levels could enable Western smelters to resume full capacity operation during 1996. In May stock levels were reported to have fallen to their lowest level since March 1993, and exports of aluminium from Russia, totalling 2m. metric tons annually, were viewed as essential to the maintenance of Western supplies. Meanwhile, progress continued to be made in arrangements under the MOU for the modernization of Russian smelters and their eventual integration into the world aluminium industry. International demand for aluminium rose by 0.2% in 1995 and by 0.8% in 1996. In 1997 aluminium consumption by industrialized countries advanced by an estimated 5.4%. This growth in demand was satisfied by increased primary aluminium production, combined with sustained reductions in world levels of primary aluminium stocks. Demand in 1998, however, was adversely affected by the economic crisis in East Asia, and consumption of aluminium in established market economy countries (EMEC) rose by only 0.1%: the lowest growth in aluminium demand since 1982. However, consumption in the EMEC area increased by an estimated 3.9% in 1999, with demand for aluminium rising strongly in the USA and in much of Asia. Compared with 1998, growth in consumption was, however, reduced in Europe and Latin America. World-wide, the fastest growing sector of aluminium demand in 1999 was the transport industry (the largest market for the metal), in which consumption increased by about 9%. In 2000, according to USGS data, production of primary aluminium grew by 3.4%. At the end of the year, according to the IAI, total world inventories (comprising unwrought aluminium, unprocessed scrap, metal in process and finished, semi-fabricated metal) had declined slightly, compared with the previous year. Stocks of primary aluminium held by the London Metal Exchange (LME), meanwhile, had fallen heavily. Demand from the USA and Asia was characterized by the USGS as weak during 2000, especially during the second half of the year, while European demand remained firm. In 2001 IAI data indicated a slight fall in total world inventories of aluminium, while, conversely, those of primary aluminium held by the LME rose substantially. Production of primary aluminium in 2001 was slightly lower than in 2000. In 2002, however, output rose by about 3.2%. In that year, according to analysis by Alcan, consumption of primary aluminium in the West rose by more than 3.5%, to almost 20m. tons. IAI data indicated a decline in total aluminium stocks of about 2.6% in 2002. In 2003, when prices fell to their lowest ever level in real (i.e. constant US dollar) terms, it was evident that reductions in costs had enabled larger, integrated producers to safeguard the viability of their enterprises. (Older and, generally, smaller producers, meanwhile, had been forced into closure.) Analysts cited as evidence of this plans to create substantial new primary metal capacity world-wide up to 2010. According to IAI data, production of primary metal increased by 3.5% in 2003, while stocks held world-wide increased by 1.1%. In 2004 IAI data indicated a substantial increase, of 7%, in global stocks of aluminium, while output of primary metal increased by 3%.

**Export Price Index for Aluminium**
(base: 1980 = 100)

|  | Average | Highest month(s) | Lowest month(s) |
|---|---|---|---|
| 1990 | 93 | — | — |
| 1995 | 104 | — | — |
| 2000 | 88 | 95 (Jan., Feb.) | 83 (March, April) |
| 2001 | 82 | 92 (Jan.) | 73 (Oct.) |
| 2002 | 77 | 80 (March) | 74 (Aug., Sept.) |

In November 1993 the price of high-grade aluminium (minimum purity 99.7%) on the LME was quoted at US $1,023.5 (£691) per metric ton, its lowest level for about eight years. In July 1994 the London price of aluminium advanced to $1,529.5 (£981) per ton, despite a steady accumulation in LME stocks of aluminium, which rose to a series of record levels, increasing from 1.9m. tons at mid-1993 to 2.7m tons in June 1994. In

November 1994, when these holdings had declined to less than 1.9m. tons, aluminium was traded at $1,987.5 (£1,269) per ton.

In January 1995 the LME aluminium price rose to US $2,149.5 (£1,346) per metric ton, its highest level since 1990. The price of aluminium was reduced to $1,715.5 (£1,085) per ton in May 1995, but recovered to $1,945 (£1,219) in July. It retreated to $1,609.5 (£1,021) per ton in October. Meanwhile, on 1 May the LME's holdings of aluminium were below 1m. tons for the first time since January 1992. In October 1995 these stocks stood at 523,175 tons, their lowest level for more than four years and only 19.7% of the June 1994 peak. Thereafter, stock levels moved generally higher, reaching 970,275 tons in October 1996. During that month the London price of aluminium fell to $1,287 (£823) per ton, but later in the year it exceeded $1,500.

In March 1997 the London price of aluminium reached US $1,665.5 (£1,030) per metric ton. This was the highest aluminium price recorded in the first half of the year, despite a steady decline in LME stocks of the metal. After falling to less than $1,550 per ton, the London price of aluminium rose in August to $1,787.5 (£1,126). In that month the LME's aluminium holdings were reduced to 620,475 tons, but in October they reached 744,250 tons. Stocks subsequently decreased, but at the end of December the metal's price was $1,503.5 (£914) per ton, close to its lowest for the year. The average price of aluminium on the LME in 1997 was 72.5 US cents per lb, compared with 68.3 cents per lb in 1996 and 81.9 cents per lb in 1995.

The decline in aluminium stocks continued during the early months of 1998, but this had no major impact on prices, owing partly to forecasts of long-term over-supply. In early May the LME's holdings stood at 511,225 metric tons, but later that month the price of aluminium fell to less than US $1,350 per ton. In June LME stocks increased to more than 550,000 tons, and in early July the price of the metal was reduced to $1,263.5 (£768) per ton. In early September the LME's holdings decreased to about 452,000 tons (their lowest level since July 1991), and the aluminium price recovered to $1,409.5 (£840) per ton. However, the market remained depressed by a reduction in demand from some consuming countries in Asia, affected by the economic downturn, and in December 1998 the London price of aluminium declined to $1,222 (£725) per ton. For the year as a whole, the average price was 61.6 US cents per lb.

During the first quarter of 1999 the aluminium market continued to be over-supplied, and in March the London price of the metal fell to US $1,139 (£708) per metric ton: its lowest level, in terms of US currency, since early 1994. Later that month the LME's stocks of aluminium rose to 821,650 tons. By the end of July 1999 these holdings had been reduced to 736,950 tons, and the price of aluminium had meanwhile recovered to $1,433.5 (£902) per ton. Stock levels subsequently rose, but, following the announcement in August of proposed cost-cutting mergers among major producers (see above), the aluminium price continued to increase, reaching $1,626.5 (£1,009) per ton at the end of the year. The steady rise was also attributable to a sharp increase in the price of alumina in the second half of the year. Following an explosion at (and subsequent closure of) a US alumina refinery in July, the price of the material advanced from $160 per ton in that month to $400 per ton (its highest level for 10 years) in December. However, the average LME price of aluminium for the year (61.7 US cents per lb) was almost unchanged from the 1998 level.

With alumina remaining in short supply, prices of aluminium continued to rise during the opening weeks of 2000, and in early February the London quotation reached US $1,743.5 (£1,079) per metric ton: its highest level for more than two years. However, the LME's stocks of the metal also increased, reaching 868,625 tons later that month. The London price of aluminium declined to $1,413 (£891) per ton in April, but recovered to $1,599 (£1,070) in July. Throughout this period there was a steady decrease in LME holdings, which were reduced to less than 700,000 tons in April, under 600,000 tons in May and below 500,000 tons in July. At the end of July aluminium stocks were 461,975 tons: only 53% of the level reached in February; and by the end of August they had fallen to 399,925 tons. In September prices continued to rise, reaching $1,644 per ton on 13 September, even though stocks fell steadily throughout the month, which they ended at 361,050 tons. Thereafter, in October and November, the London quotation for aluminium was somewhat weaker, falling to respective 'lows' in those months of $1,446 and $1,443. Stocks declined simultaneously, standing at 331,250 tons on 31 October, and 320,725 tons on 30 November. Prices strengthened in December, reaching a 'high' for that month of $1,632.5 per ton. The firmer quotation was accompanied by the beginning of a sustained rise in stocks of aluminium held by the LME. Having fallen to 298,925 tons on 18 December, stocks ended the month slightly higher, at 321,850 tons, than at the end of the previous month.

At the end of January 2001 the London quotation rose to US $1,737 per metric ton, thus approaching the highest level recorded in 2000. By the end of January stocks had recovered to 394,075 tons. By the end of February LME holdings stood at 483,200 tons, and the London price had declined to $1,553 per ton. In both March and April, on a month-on-month basis, stocks fell, while prices were generally weaker, the London quotation reaching $1,540 per ton on 27 April. May 2001 marked the beginning of a very substantial accumulation of stocks. Although the London quotation rose as high as $1,586 per ton on 4 May 2001, prices fell precipitously in June–November, reaching a 'low' of $1,243 per ton on 7 November, but recovering to $1,430 at the end of that month. The sustained decline was attributed to slow economic growth world-wide, which had caused the market to be over-supplied, aggravated by the attacks on mainland USA on 11 September. For the whole of 2001 the LME average monthly 'spot' price for high-grade aluminium was US cents 65.5 per pound.

On 28 March 2002 LME stocks rose to 1,029,400 metric tons. The London quotation weakened in both April and May, falling to US $1,318 per ton on 23 May. In early June the price recovered to $1,398, but had fallen to $1,364.5 per ton by the end of the month, when stocks of aluminium held by the LME totalled more than 1.2m. tons. By the end of July the price of aluminium traded on the LME had fallen to $1,310 per ton, by which time stocks held by the Exchange had risen to 1,291,000 tons. On 14 August the London quotation fell to $1,279 per ton, but had recovered to $1,293.5 per ton by the end of the month. On 13 August, meanwhile, stocks rose to 1,300,125 tons, and remained above 1,290,000 until the first day of October, when they fell to 1,288,200. On 11 September the London quotation recovered to $1,340.5 per ton, but subsequently weakened, ending the month at $1,280.5 per ton. At the end of the first week of October the price of aluminium fell to $1,275.5 per ton, but had risen to $1,337.5 by 31 October. From early November, the London quotation began to recover somewhat, reaching $1,370.5 per ton on 4 November and ending the month at $1,378 per ton. This upward movement continued into December: on 13 December the London quotation was just short of $1,400 per ton. Stocks declined further in the final month of 2002, falling to 1,238,000 tons on 12 December and ending the year at 1,241,350 tons. The price of aluminium, meanwhile, declined to $1,344.5 per ton on 31 December.

On 22 January 2003 the London quotation for primary aluminium closed at more than US $1,400 per metric ton for the first time since 22 March 2002, and it had risen further, to $1,427 per ton, by the end of the month. By the end of January 2003 stocks held by the LME had fallen to 1,199,550 tons and they continued to fall thereafter until about the middle of February. With the exception of 19 February, the London quotation closed at more than $1,400 per ton on each day of that month. The price weakened during March, however, at the same time as stocks were rising. On 31 March the London quotation closed at $1,350 per ton, while stocks were recorded at 1,252,775 tons. The London price continued to fall until around mid-April, subsequently strengthening to finish the month at $1,356.5 per ton. From around mid-May this stronger trend became more pronounced, the London quotation closing at more than $1,400 per ton on each day of the month after 12 May. By the end of May stocks of metal held by the LME had declined to 1,130,625 tons. On 16 June stocks fell to 1,115,150 tons. Prices remained stable throughout that month, at the end of which the London quotation was $1,389 per ton.

On 14 July 2003 the London quotation for primary aluminium closed at US $1,463 per metric ton. Prices were thereafter generally weaker until towards the end of the month. On 28 July the closing price for the metal was $1,484.5 per ton. On 1 August the London quotation closed at $1,505 per ton, the first time a closing price in excess of $1,500 per ton had been recorded since May 2001. On 29 August, the last trading day of that month, the London price closed at $1,432 per ton, having fallen as low as $1,427.5 in the interim. On 11 September the price declined to $1,378 per ton, but it had recovered by the end of the month to $1,407.5 per ton. Generally, from October until the end of 2003, the London price strengthened, reaching $1,552 per ton on 5 December and ending the year at $1,592.5 per ton. During the second half of 2003 stocks of aluminium held by the LME rose steadily. At the end of July they totalled 1,304,450 tons; by the end of December they had increased to 1,423,275 tons.

On 2 January 2004 the London quotation for primary aluminium closed at US $1,600 per metric ton, the first time the quotation had reached that level since February 2001. On 30 January a closing price of $1,636.5 per ton was recorded. The London quotation strengthened further in February, closing at $1,754 per ton on 18 February, but thereafter declined somewhat to end the month at $1,702 per ton. The London price remained above $1,625 per ton throughout March, ending the month at $1,688.5 per ton. Sharp increases occurred from early April, and on 16 April a closing price greater than $1,800 per ton ($1,802) was recorded. Prices in April rose to their highest levels for more than eight years. By 10 May, however, the quotation had fallen to $1,575 per ton. On 2 June a closing price of $1,703.5 was recorded, rising to $1,721 per ton on 21 June. Stocks of aluminium held by the LME rose as high as 1,453,125 tons in January. From February until the end of June, however, they declined steadily. At the end of February they totalled 1,393,675 tons, but had fallen to 940,200 tons by the end of June. For the whole of 2004 the average quotation for primary aluminium traded on the LME was $1,717 per ton, 19.9% higher than the average price recorded in 2003. The higher price in 2004 was attributed to a substantial increase in global demand for aluminium, in particular from China, that had outstripped, and led to a heavy fall in, world inventories—the global market was in deficit for the first time since 2000. Growth in demand worldwide from the aerospace and automotive sectors was especially strong in 2004. Aluminium also benefited, in the early part of the year, from even sharper increases in the prices of those metals, such as copper and steel, for which it can be substituted. At the end of 2004 stocks of aluminium held by the LME, at 692,775 tons, were more than 50% lower than at the end of 2003.

During the first six months of 2005 the price of primary aluminium traded on the LME ranged between US $1,694 per metric ton (13 and 14 June) and $2,031.5 per ton (15 March). Stocks of metal held by the Exchange declined steadily during the first half of 2005, from 654,025 tons at the end of January to 528,650 tons at 30 June. The decline in the price of aluminium in the second quarter of the year was attributed to weaker demand while, according to the IAI, primary production was increasing, especially in the USA and Canada. In mid-2005 increases in European production were reported to be under threat from increased energy prices. Analysts forecast that prices would be supported in the second half of 2005 by a continued overall deficit in aluminium supply.

The IAI, based in London, United Kingdom, is a global forum of producers of aluminium dedicated to the development and wider use of the metal. In 2005 the IAI had 26 member companies, representing every part of the world, including Russia and China, and responsible for about 80% of global primary aluminium production and a significant proportion of the world's secondary output.

## Banana (*Musa sapientum*)

Although it is often erroneously termed a 'tree', the banana plant is, in fact, a giant herb. It grows to a height of 3 m–9 m (10 ft–30 ft) and bears leaves which are very long and broad. The stem of the plant is formed by the overlapping bases of the leaves above. Bananas belong to the genus *Musa* but the cultivated varieties are barren hybrid forms which cannot, therefore, be assigned specific botanical names. These banana hybrids, producing edible seedless fruits, are now grown throughout the tropics, but originally diversified naturally or were developed by humans in prehistoric times from wild bananas which grow in parts of South-East Asia. The plantain hybrid has grown in Central Africa for thousands of years, and traders and explorers gradually spread this and other varieties to Asia Minor and East Africa; the Spanish and Portuguese introduced them to West Africa and took them across the Atlantic to the Caribbean islands and the American continent. However, the varieties which are now most commonly traded internationally were not introduced to the New World until the 19th century.

The banana is propagated by the planting of suckers or shoots growing from the rhizome, which is left in the ground after the flowering stem, having produced its fruit, has died and been cut down. Less than one year after planting, a flowering stem begins to emerge from the tip of the plant. As it grows, the stem bends and hangs downwards. The barren male flowers which grow at the end of the stem eventually wither and fall off. The seedless banana fruits develop, without fertilization, from the clusters of female flowers further up the stem. Each stem usually bears between nine and 12 'hands' of fruit, each hand comprising 12–16 fruits. Before it is ripe, the skin of the banana fruit is green, turning yellow as it ripens. To obtain edible white flesh, the skin is peeled back. The process of fruiting and propagation can repeat itself indefinitely. In commercial cultivation the productive life of a banana field is usually limited to between five and 20 years before it is replanted, although small producers frequently allow their plants to continue fruiting for up to 60 years. Banana plantations are vulnerable to disease and to severe weather (particularly tropical storms), but the banana plant is fast-growing, and a replanted field can be ready to produce again within a year, though at a high cost.

**Production of Bananas**
('000 metric tons, excluding plantains)

|  | 2003 | 2004* |
|---|---|---|
| **World total** | 70,424 | 71,343 |
| Latin America | 24,380 | 24,280 |
| **Leading Latin American producers** | | |
| Brazil | 6,775 | 6,603 |
| Colombia | 1,511 | 1,450† |
| Costa Rica† | 2,220 | 2,230 |
| Ecuador | 5,883 | 5,900† |
| Guatemala† | 960 | 1,000 |
| Honduras† | 965 | 965 |
| Mexico | 2,027 | 2,027† |
| Panama | 509 | 525† |
| Venezuela | 560 | 550 |
| **Other leading producers** | | |
| Burundi† | 1,600 | 1,600 |
| China, People's Repub. | 5,784 | 5,827‡ |
| India† | 16,820 | 16,820 |
| Indonesia | 4,177 | 4,394 |
| Philippines | 5,369 | 5,638 |
| Thailand† | 1,900 | 1,900 |

* Provisional figures.
† FAO estimate(s).
‡ Unofficial figure.

The volume and value of exported bananas are greater than those of any other fresh fruit export. The varieties of dessert banana on which international trade is predominantly based, and which are most usually eaten raw owing to their high sugar content (17%–19%), comprise only a few of the many varieties which are grown—on the basis of gross value of production, bananas are the world's fourth most important food crop after rice, wheat and maize. Many types of sweet banana, unsuitable for export, are consumed locally. Apart from the trade in fresh, sweet bananas, the fruit has few commercial uses, although there is a small industry in dried bananas and banana flour. The numerous high-starch varieties with a lower sugar content, which are not eaten in their raw state, are used in cooking, mostly in the producing areas. Such varieties are picked when their flesh is unripe, although they are occasionally of a type which would become sweet if left to ripen. Cooking bananas, sometimes called 'plantains' (though this term is also applied to types of dessert banana in some countries), form the staple diet of millions of East Africans. Bananas can also be used for making beer, and in East Africa special varieties are cultivated for that purpose. Advances in production methods, packaging, storage and transport (containerization) have made bananas available world-wide. Although international trade is principally in the sweet dessert fruit, this type comprises less than one-fifth of total annual world banana production.

The banana was introduced into Latin America and the Caribbean by the Spanish and the Portuguese during the 16th century. The expansion of the banana industry in the small Latin American (so-called 'green' or 'banana') republics between 1880 and 1910 had a decisive effect in establishing this region as the centre of the world banana trade. Favourable soil and climatic conditions, combined with the ease of access around the Caribbean and to a major market in the USA, were important factors in the initial commercial success of the Latin American banana industry. Although advances in storage and transportation made the US market more accessible to producers in other areas, they also made available an equally important market in Europe (see below). Owing to considerably lower production costs than in other producing areas, bananas from Central and South America command the major portion of the world export market. It is there that the large multinational companies (notably United Brands, Standard Fruit and Del Monte) which dominate the world banana market are established.

Bananas are grown in most Latin American countries. Brazil has traditionally been the region's leading producer, although they contribute only a small proportion of the country's total export earnings. The banana industry is of far greater economic importance in Ecuador, where production in 1996–2000 surpassed that of Brazil and where bananas are the leading cash crop (as they are in Honduras—in most years—and Costa Rica), and in Colombia where, after coffee, they are the second most important source of export revenue. Ecuador is the world's leading exporter of bananas, with shipments valued at an estimated US $1,023m. in 2004. In that year bananas accounted for 13.4% of Ecuador's export earnings. Banana shipments contribute significantly to the export receipts of Honduras (providing 13.4% of the total in 2004) and Costa Rica (9% in 2000). In October 1998, however, about 70% of the Honduran crop was destroyed by 'Hurricane Mitch', necessitating extensive replanting during 1999. In that year bananas contributed only 3.3% of Honduran export revenue. Among other Latin American countries where bananas are a major export are Panama (13% in 2003) and Belize (12.9% in 2004). Bananas are also a useful source of foreign exchange in Colombia (where they provided 3.7% of total export earnings in 2000), Guatemala (10.8% in 2002) and Nicaragua (an estimated 1.4% in 2004). The territories most dependent on bananas for export receipts are Caribbean islands, including Dominica (where bananas represented 21.8% of total exports, by value, in 2001), Guadeloupe (21.9% in 1997), Martinique (36.6% in 1997), Saint Lucia (35.8% of domestic exports in 2002) and Saint Vincent and the Grenadines (30.6% of total exports, including re-exports, in 2002).

There is no international agreement governing trade in bananas, but there are various associations—such as the Union of Banana-Exporting Countries (UPEB), comprising Colombia, Costa Rica, Guatemala, Honduras, Nicaragua, Panama and Venezuela; the Caribbean Banana Exporters' Association, comprising Jamaica, Belize, the Windward Islands and Suriname; and WIBDECO, the Windward Islands Banana Development Co—which have been formed by producer countries to protect their commercial interests. Prices vary greatly, depending on relative wages and yields in the producing countries, freight charges, and various trade agreements negotiated under the Lomé Convention between the European Community (EC, now the European Union—EU) and 71 African, Caribbean and Pacific (ACP) countries. The Lomé arrangements include a banana protocol, which ensures producers an export market, a fixed quota and certain customs duty concessions. In this connection, the prospect of a withdrawal of internal barriers to intra-EC trade, as a result of the completion of a single market with effect from January 1993, was the cause of considerable anxiety to most major banana exporters (see below).

The USA and the EU are the world's largest markets for bananas. In 2003 imports by EU member states (excluding intra-trade) totalled some 4.9m. metric tons, representing about 34% of world banana imports. Under the market arrangements prior to the end of 1992, just over one-half of banana imports were subject to quota controls under the Lomé Convention. Approximately 20% of market supplies were imported, duty free, from former British, French and Italian colonies. The Windward Islands of Saint Lucia, Dominica and Grenada together accounted for some 90% of banana exports to the EC under the Lomé Convention's banana protocol, Jamaica and Belize also being parties to the Convention. Around 70% of bananas from the Windward Islands were exported to the United Kingdom. France maintained similar arrangements with its traditional suppliers (overseas territories such as Guadeloupe and Martinique), as did Italy (Somalia). Another 25% of EC demand was satisifed internally, mainly from Spain (the Canary Islands), with smaller quantities from Portugal (Madeira and the Azores) and Greece (Crete). The remainder of quota imports (representing 10% of overall market demand) were imported, duty free, into Germany, mainly from Latin and Central American countries. Banana consumption in Germany, at some 18 kg per head annually, has been approximately twice that of other EU countries. Banana imports into the residual free market originated mainly from the same countries, but attracted a 20% import tariff.

This complex market structure, which strongly favoured the Caribbean islands, was maintained by barriers to internal trade. These prevented the re-export of imports into Germany from, for instance, Honduras or Ecuador to the United Kingdom and France. Without these restrictions, Central and Latin American exporters, enjoying the cost advantages of modern technology and large-scale production, and subject to a 20% import tariff on their EC exports, would have substantially expanded their market share, and possibly displaced Caribbean producers altogether in the longer term.

The intention of the EC Commission that the tariff-free quota system would cease after 1992 held serious implications for Caribbean producers, who, with the ACP secretariat, sought to persuade the EC to devise a new preference system to protect their existing market shares, possibly taking the form of quota allocations, which would accord with the obligations of the EC countries in relation to trade with the ACP under the terms of the Lomé Convention. These guarantee that 'no ACP state shall be placed, as regards access to its traditional markets...in a less favourable position than in the past or present'.

In December 1992 it was announced by the EC Commission that ACP banana producers were to retain their preferential status under the EC's single-market arrangements that were to enter into force in July 1993. These would guarantee 30% of the European banana market to ACP producers, by way of an annual duty-free quota of approximately 750,000 metric tons. Imports of Latin American bananas were to be limited to 2m. tons per year at a tariff of 20%, with any additional shipments to be subject to a tariff rate of 170%, equivalent to 850 European Currency Units (ECU) per ton. It was asserted that, as the proposals linked tariffs with quotas, the EC was not in contravention of the General Agreement on Tariffs and Trade (GATT) regulations on the restriction of market access.

These arrangements were opposed by Germany, Denmark, Belgium and Luxembourg, as well as by the Latin American banana producers, who forecast that their shipments to the EC, totalling approximately 2.6m. metric tons annually, could decline by as much as 20%. In early 1993 the German Government unsuccessfully sought a declaration from the Court of Justice of the European Communities that the EC Commission was in violation of GATT free-trade regulations. In June the governments of Ecuador, Guatemala, Honduras, Mexico and Panama obtained agreement by GATT to examine the validity of the EC proposals. In the mean time, the new arrangements covering EC banana imports duly took effect on 1 July.

In February 1994 a GATT panel ruled in favour of the five Latin American producers, declaring that the EU policy on bananas unfairly favoured the ACP banana exporters and was in contravention of free-trade principles. The Latin American producers accordingly demanded that the EU increase their annual quota to 3.2m. metric tons and reduce the tariff level on excess shipments. The EU responded to the ruling by offering to increase the Latin American banana quota to 2.1m. tons in 1994 (with effect from 1 October) and to 2.2m. tons in 1995, and to reduce the tariff rate by one-quarter. This compromise was accepted by Colombia, Costa Rica, Nicaragua and Venezuela, but rejected by Ecuador and the other Latin American producers. The Latin American exporters assenting to the plan were each to receive specific quotas, based on their past share of the market, within the overall 2.1m.-ton quota. In October 1994 the Court of Justice of the European Communities rejected a petition by Germany seeking a quota of 2.5m. tons in that year, with subsequent annual increases of 5%.

With the accession to the EU in January 1995 of Austria, Finland and Sweden, it was anticipated that the quota for duty-free bananas would be increased by up to 350,000 metric tons annually to accommodate the community's enlarged membership. (This proposed level of increase, however, was viewed by Germany as inadequate.) In the same month, the US Government indicated that it was contemplating the imposition of retaliatory trade measures against both the members of the EU and those Latin American banana exporters which had agreed to the 1994 quota compromise arrangements. (Although the USA is only a marginal producer of bananas, US business interests hold substantial investments in the multinational companies operating in Central and South America.) In February 1995 the EU came under further pressure from a number of African banana-producing countries, led by Cameroon and Côte d'Ivoire, which sought further improvements in their access to EU markets under ACP arrangements. The USA, meanwhile, with the support of Germany, Belgium, the Netherlands and Finland, declared in July that renewed efforts should be made to formulate a new banana import regime that would reconcile the interests of the Latin American and ACP producers. France and the United Kingdom, however, reiterated their intention to maintain their perceived obligations towards their former Caribbean colonies, whose economies would be severely affected by the operation of a free market in bananas.

In September 1995 the USA formally instituted a complaint against the EU with the World Trade Organization (WTO), the successor organization to GATT. Further representations to the WTO by the US Government, supported by the governments of Ecuador, Guatemala, Honduras and Mexico, followed in February 1996. In the following month the EU, which had raised its annual duty-free banana quota to 2.35m. metric tons (see above), proposed a compromise reallocation of quotas, under which the Latin American producers would obtain 70.5% of the new quota (as against their existing proportion of 66.5%) while the ACP growers would receive 26% (down from 30%). An additional allowance of 90,000 tons annually was also to be extended to certain categories of bananas from Belize, Cameroon and Côte d'Ivoire. In May, however, the USA, Ecuador, Guatemala, Honduras and Mexico made a concerted approach to the WTO to disallow the EU banana import regime in its entirety. Efforts by the contending groups of producers to achieve a compromise followed in July, when it was announced that ministerial meetings between the ACP and Latin American producing countries were to take place in October. In late July the EU announced that its annual quota for Latin American bananas was to be raised to 2.2m. tons to reflect increased demand from the enlarged EU membership.

An interim report issued by the WTO in March 1997 upheld the complaint instituted in 1996 by the USA and four Latin American producers, and declared the EU to be in violation of its international trade treaty obligations. Representations by the EU against the ruling were rejected by the WTO appeals committee in September 1997, and the EU was informed that it must formulate a new system for banana imports by early 1998. This ruling was accepted in principle by the EU, which in January 1998 proposed new arrangements under which it would apply a system of quotas and tariffs to both groups of producers, while retaining an import-licensing system. A quota increase to 2.5m. metric tons annually was to be offered to the Latin American exporters. It was also proposed to offer aid totalling ECU 450m. to the Caribbean producers to assist their banana industries in achieving the ability to compete efficiently in a proposed 'free' market in bananas by 2008.

In June 1998 the USA expressed dissatisfaction with the terms of the EU plan. The EU, however, refused to agree to further consultations with the WTO appeals committee. The US position was reinforced in September when the 'Group of Seven' major industrial countries sought, without success, to persuade the EU to accept the terms of the WTO ruling. However, the EU offered a new compromise plan, whereby the Latin American banana producers would, with effect from early 1999, receive an annual quota of 2.53m. metric tons, at an import duty rate of US $90 per ton, with the ACP growers retaining free access for annual shipments of 857,000 tons, which the EU stated to be in compliance with the WTO decision. This was rejected in October 1998 by the USA and the Latin American petitioners (who had been joined in February by Panama), and it was announced that the US Government was to seek authorization by the WTO for a programme of trade retaliation against the EU. Following further unsuccessful negotiations, and a challenge by the EU to the legality of the proposed sanctions, the US Government published in early December a list of 16 EU exports, including cashmere clothing, biscuits, candles and electric coffee-makers, on which it intended to impose tariffs of 100% with effect from early 1999.

These measures, which were directed principally against the United Kingdom, France and Germany, and from which Denmark and the Netherlands (which had advocated acceptance of the WTO ruling) were exempted, provoked considerable anxiety internationally, on the grounds that they could lead to a wider disruption of world trade. Negotiations between representatives of the USA and the EU were resumed in January 1999, while the implementation of the proposed tariffs was suspended pending consideration by the WTO of their validity in international law. The WTO also agreed to re-examine the terms of the compromise plan put forward by the EU in the previous September.

Following sustained but unsuccessful attempts by the USA and the EU to reach a negotiated settlement, the USA, supported by the five Latin American petitioners, announced in early March 1999 that it was to invoke the punitive tariffs. Concurrently, a number of countries in the Caribbean Community and Common Market (CARICOM) grouping announced that they were considering the abrogation of a narcotics-control agreement with the USA, which permitted US enforcement agencies to pursue suspected drugs-traffickers into their territorial waters and air space.

In April 1999 the WTO ruled that the revised EU banana import proposals represented an attempt to avoid compliance with the original WTO ruling, and formally authorized the USA to impose trade sanctions against EU goods, valued at US $191.4m. annually (compared with damages of $520m. originally claimed), to compensate for losses incurred by US companies as a consequence of the EU banana regime. (This authorization was only the second in the 51-year history of the WTO and its predecessor, GATT.) The EU announced its acceptance of the decision, although indicating that the formulation of reforms could span the period to January 2000. The USA, meanwhile, accepted the need for this period of consultation and further extended the suspension of its import tariff proposals. However, in March 2000, following a complaint by the EU, a WTO panel ruled that the sanctions against EU goods imposed by the USA were a violation of international trade rules.

In mid-2000 the European Commission proposed the possible introduction of a tariff quota system involving the distribution of licences on a 'first come, first served' basis. However, it also suggested that, if a consensus were not reached, a tariff-only solution, whereby no restrictions would be imposed on quantities but which would give ACP countries preferable rates, would be the only viable solution; the latter option was rejected by a number of EU governments as potentially damaging to EU banana-growers. Moreover, the granting of licences on a 'first come, first served' basis was subsequently rejected by representatives of Latin American banana-producing states, while the USA stated that the proposed regime did not adequately address its concerns. In July a WTO panel ruled that the sanctions against EU goods had been imposed prematurely by the USA, some six weeks before it had received the authorization of the WTO to apply them. The European Commission indicated at this time that it would take no further action itself, but that companies affected by sanctions before the WTO authorization were free to pursue claims for compensation in the US courts. The USA, meanwhile, was reported to be planning to 'rotate' the sanctions onto a new list of goods, the US Congress having ordered a so-called 'carousel' sanctions regime under which the list would change every six months. In addition, in May, the WTO had authorized Ecuador, the world's largest banana exporter, to impose sanctions of up to US $201.6m. in the form of tariffs applied on EU service providers, industrial designs and copyrights.

In October 2000 the European Commission formally approved the introduction of a tariff quota system for bananas on a 'first come, first served' basis. Such a system would retain the use of three quotas until 2006, the Commission favouring the introduction of a tariff-only regime thereafter. The third quota would permit annual imports of 850,000 metric tons and would be available to all exporters. However, ACP exporters would be granted a tariff preference of US $264 per ton, rather than $275 per ton as had previously been proposed. The proposed new tariff quota system was immediately rejected by Colombia, Costa Rica, Guatemala, Honduras, Nicaragua and Panama, which threatened renewed action with the WTO. The USA also questioned whether the new regime was 'WTO-compatible'. Shortly after the announcement of the details of the proposed new regime. EU governments declared that it provided 'a basis for settling the banana dispute'.

In March 2001 the EU announced that it would delay the introduction of the new tariff quota system for bananas in order to consider an alternative proposal, by the US fruit trading company, Chiquita Brands International, which would effectively grant quotas based on historical market share. Chiquita Brands International had, in January, initiated legal action against the European Commission in respect of losses allegedly incurred as a result of EU restrictions on banana imports. The USA had also threatened to implement new trade sanctions if the EU proceeded with the introduction of the new regime. In April the dispute over bananas appeared finally to have been resolved after the EU agreed temporarily to reallocate to Chiquita Brands International a larger share of the European banana market from 1 July 2001, the date set as a target for the introduction of the new tariff quota system. In return, the USA was reported to have agreed to suspend, from that date, the sanctions it had imposed on EU goods; and to support a waiver in the WTO that would allow a specific quota for bananas from traditional ACP exporters. Under the settlement negotiated, the EU agreed to transfer 100,000 metric tons of bananas from the quota reserved mainly for ACP producers to that set for Latin American producers. From 1 July most import licences to the EU would be allocated on the basis of traditional shipment levels. Some 17% of licences would be reserved for new exporters or for companies which had significantly increased their imports since a 1994–96 reference period. The arrangements detailed above were supplemented in 2004 by what was termed an 'additional quantity' quota in order to allow imports of bananas into the 10 new member states of the enlarged EU.

In November 2004 negotiations between the EU and Latin American banana-exporting countries commenced in respect of the transfer to a tariff-only EU banana import regime from 1 January 2006, the European Commission stating that it would 'seek to maintain a level of protection equivalent to that currently existing in order to ensure that Community production is maintained and that these producers are not put in a less favourable situation [than] before the entering into force of the import quota regime in 1993'. With regard to ACP banana suppliers, the Commission stated its commitment to 'examine appropriate ways to address their specific situation, including preferential access for ACP products, and seek to maintain a level of preference to the ACP countries equivalent to that afforded by the enlarged Community of 25'. In January 2005 Colombia, Costa Rica, Ecuador, Guatemala, Honduras, Nicaragua and Panama rejected the tariff—€230 per metric ton, on a most-favoured nation basis—proposed by the EU for the new banana import regime, arguing that it was not consistent with the obligations that the EC had assumed at a ministerial meeting of the WTO held in Doha, Qatar, in 2001. (At the Doha meeting the EU obtained a waiver from the WTO that allows it to grant ACP countries a quota for duty-free exports of bananas to its market.) According to the European Commission, the tariff had been set by computing the gap between internal and external EC prices. Many Latin American banana-exporting countries have reportedly urged that the tariff be set no higher than €75 per ton, while ACP banana-exporting countries have been reported to favour a new single tariff of €275 per ton. Generally, analysts have indicated that a high tariff would place Latin American exporters at a competitive disadvantage vis-à-vis their ACP counterparts and EU producers; that a low tariff would favour Latin American exporters relative to ACP exporters and EU producers; and that an intermediate tariff might result in increased exports from Latin American countries and some ACP countries to the EU and a fall in EU domestic prices.

Under the terms of the Cotonou Agreement, the successor convention to the fourth Lomé Convention that covers the period 2000–20, the system of trade preferences hitherto pertaining was to be gradually replaced by new economic partnerships based on the progressive and reciprocal removal of trade barriers. In a protocol to the Cotonou Agreement (the Second Banana Protocol), the 'overwhelming importance to the ACP banana suppliers of their exports to the Community (EU) market' was acknowledged, and the EU agreed to seek to ensure the continued viability of their banana export industries, and the continued access of their bananas to the EU market.

**Export Price Index for Bananas**
(base: 1980 = 100)

|  | Average | Highest month(s) | Lowest month(s) |
| --- | --- | --- | --- |
| 1990 | 150 | — | — |
| 1995 | 111 | — | — |
| 2000 | 122 | 166 (Feb.) | 90 (July) |
| 2001 | 162 | 192 (Feb.) | 136 (Oct., Nov.) |
| 2002 | 147 | 174 (June) | 112 (Nov.) |

Bananas were included by the UN Conference on Trade and Development (UNCTAD) on a list of 18 commodities in its proposed integrated programme for the regulation of international trade in primary products, based on commodity agreements backed by a common fund for financial support. The UN's Food and Agriculture Organization (FAO) and UNCTAD are seeking, through an inter-governmental group, to formulate such an international agreement on bananas, involving producers and consumers, which would ensure a proper balance of supply and demand, regulating trade in order to provide regular supplies for consumers at a price remunerative to producers. The main impediments to a trade pact are disagreements on the means by which the market would be stabilized, in particular on the definition and use of export quotas, which producers consider to be restrictive, and on the lowering of trade barriers to allow reciprocal trade, which is opposed by countries with high production costs.

Since 1999 world banana prices have, on the whole, declined as a result of over-supply and also of adverse economic conditions in East Asia and Russia in 1997–2000, resulting in reduced imports by those markets. Within the EU prices have likewise tended to decline, but have remained comparatively high—the EU banana market is one of the world's most profitable—owing to the operation of a tariff quota import regime. The average import price (f.o.r.) at US East Coast ports of main-brand bananas from Central America in 1999 was US $6.90 per 18.14 kg box. In 2000 the price rose to $7.58 per box and in 2001, when supplies of bananas were restricted as a result of adverse weather conditions in Latin America, an average price of $10.42 per box was recorded. The average price declined slightly in 2002, to $9.41 per box. In 2003 it fell to only $6.81 per box, but recovered to $9.92 per box in 2004. According to FAO, preliminary data indicated that total Latin American exports of bananas had increased slightly in 2004, compared with 2003, but prices had none the less risen owing to higher demand and to a depreciation of the US dollar. The average import price (f.o.r.) at US West Coast ports of main-brand bananas from Central America was $7.27 per box in 1999. An average import price of $7.64 per box was recorded in 2000, and in 2001 the price rose to $10.58 per box. In 2002 the average price was $9.40 per box. The average import price declined to only $6.72 per box in 2003, but recovered to $9.85 per box in 2004.

### Cocoa (Theobroma cacao)

This tree, up to 14 m (46 ft) tall, originated in the tropical forests of Central and South America. The first known cocoa plantations were in southern Mexico around AD 600, although the crop may have been cultivated for some centuries earlier. Cocoa first came to Europe in the 16th century, when Spanish explorers found the beans being used in Mexico as a form of primitive currency as well as the basis of a beverage. The Spanish and Portuguese introduced cocoa into West Africa at the beginning of the 19th century.

# REGIONAL INFORMATION

*Major Commodities of Latin America*

Cocoa is now widely grown in the tropics, usually at altitudes less than 300 m above sea-level, where it needs a fairly high rainfall and good soil. Cocoa trees can take up to four years from planting before producing enough fruit to merit harvesting. They may live for 80 years or more, although the fully-productive period is usually about 20 years. The tree is very vulnerable to pests and diseases, and it is highly sensitive to climatic changes. Its fruit is a large pod, about 15 cm–25 cm (6 in–10 in) long, which at maturity is yellow in some varieties and red in others. The ripe pods are cut from the tree, on which they grow directly out of the trunk and branches. When opened, cocoa pods disclose a mass of seeds (beans) surrounded by white mucilage. After harvesting, the beans and mucilage are scooped out and fermented. Fermentation lasts several days, allowing the flavour to develop. The mature fermented beans, dull red in colour, are then dried, ready to be bagged as raw cocoa which may be further processed or exported.

Cultivated cocoa trees may be broadly divided into three groups. Most cocoas belong to the Amazonian Forastero group, which now accounts for more than 80% of world cocoa production. It includes the Amelonado variety, suitable for chocolate-manufacturing. Criollo cocoa is not widely grown and is used only for luxury confectionery. The third group is Trinitario, which comprises about 15% of world output and is cultivated mainly in Central America and the northern regions of South America.

Cocoa-processing takes place mainly in importing countries. The processes include shelling, roasting and grinding the beans. Almost half of each bean after shelling consists of a fat called cocoa butter. In the manufacture of cocoa powder for use as a beverage, this fat is largely removed. Cocoa is a mildly stimulating drink, because of its caffeine content, and, unlike coffee and tea, is highly nutritional.

The most important use of cocoa is in the manufacture of chocolate, of which it is the main ingredient. About 90% of all cocoa produced is used in chocolate-making, for which extra cocoa butter is added, as well as other substances such as sugar—and milk in the case of milk chocolate. Proposals that were initially announced in December 1993 (and subsequently amended in November 1997) by the consumer countries of the European Union (EU), permitting chocolate manufacturers in member states to add as much as 5% vegetable fats to cocoa solids and cocoa fats in the manufacture of chocolate products, have been perceived by producers as potentially damaging to the world cocoa trade. In 1998 it was estimated that the implementation of this plan could reduce world demand for cocoa beans by 130,000–200,000 metric tons annually. In July 1999, despite protests from Belgium, which, with France, Germany, Greece, Italy, Luxembourg, the Netherlands and Spain, prohibits the manufacture or import of chocolate containing non-cocoa-butter vegetable fats, the European Commission cleared the way to the abolition of this restriction throughout the EU countries. The implementation of the new regulations took effect in May 2000.

Latin America (and the Caribbean) is the third most important producing area for cocoa beans after West Africa and Asia and Oceania, providing 12.9% of the estimated total world crop in 2004. In that year Brazil was the world's fifth largest producer of cocoa beans, after Côte d'Ivoire, Ghana, Indonesia and Nigeria. In 2003 Brazil's exports of cocoa beans totalled 1,851 metric tons, equivalent to less than 0.1% of total world exports of 2,425,337 tons in that year. Brazil's earnings from cocoa exports, which totalled about US $2m. in 2000, are relatively insignificant to the country's economy compared with those from coffee. In 2000 Ecuador obtained 1.5% of its export revenue from cocoa, compared with 2.3% in 1999 and 1% in 1998. The total volume of exports declined from 71,100 tons in 1996 to 12,135 tons in 1998, owing to the devastation caused to the cocoa crop as a result of El Niño (an aberrant current which periodically causes the warming of the Pacific coast of South America, disrupting usual weather patterns) in 1997–98. Cocoa bean exports increased to 63,600 tons in 1999, but declined again, to 49,047 tons, in 2000. In 2001 Ecuador's exports of cocoa beans totalled 55,420 tons. They rose slightly in 2002, to 55,598 tons, and again, more substantially, to 64,756 tons, in 2003. In 2004 the Dominican Republic derived 4.5% of its export earnings from cocoa and cocoa products (excluding exports from free-trade zones), compared with 7.2% in 2003. The country's cocoa crop was severely damaged by 'Hurricane Georges' in September 1998.

The principal importers of cocoa are developed countries with market economies, which generally account for more than 80% and sometimes for as much as 90% of cocoa imports from developing countries. Recorded world imports of cocoa beans in 2003 totalled 2,691,723 metric tons. The principal importing countries in that year were the Netherlands (with 558,710 tons, representing 21% of the total), the USA (384,217 tons) and Germany (209,894 tons).

World prices for cocoa are highly sensitive to changes in supply and demand, making its market position volatile. Negotiations to secure international agreement on stabilizing the cocoa industry began in 1956. Full-scale cocoa conferences, under UN auspices, were held in 1963, 1966 and 1967, but all proved abortive. A major difficulty was the failure to agree on a fixed minimum price. In 1972 the fourth UN Cocoa Conference took place in Geneva, Switzerland, and resulted in the first International Cocoa Agreement (ICCA), adopted by 52 countries, although the USA, the world's principal cocoa importer at that time, did not sign. The ICCA took formal effect in October 1973. It operated for three quota years and provided for an export quota system for producing countries, a fixed price range for cocoa beans and a buffer stock to support the agreed prices. In accordance with the ICCA, the International Cocoa Organization (ICCO), based in London, United Kingdom, was established in 1973. In mid-2005 the membership of the 2001 ICCA (see below) comprised 40 countries (13 exporting members, 27 importing members), representing about 80% of world cocoa production and some 60% of world cocoa consumption. The European Union is also an inter-governmental party to the 2001 Agreement. However, the USA, a leading importer of cocoa, is not a member. Nor is Indonesia, whose production and exports of cocoa have expanded rapidly in recent years. The governing body of the ICCO is the International Cocoa Council (ICC), established to supervise implementation of the ICCA. It is planned to relocate the ICCO to Abidjan, Côte d'Ivoire.

A second ICCA operated during 1979–81. It was followed by an extended agreement, which was in force in 1981–87. A fourth ICCA took effect in 1987. (For detailed information on these agreements, see *South America, Central America and the Caribbean 1991*.) During the period of these ICCAs, the effective operation of cocoa price stabilization mechanisms was frequently impeded by a number of factors, principally by crop and stock surpluses, which continued to overshadow the cocoa market in the early 1990s. In addition, the achievement of ICCA objectives was affected by the divergent views of producers and consumers, led by Côte d'Ivoire, on one side, and by the USA, on the other, as to appropriate minimum price levels. Disagreements also developed over the allocation of members' export quotas and the conduct of price support measures by means of the buffer stock (which ceased to operate during 1983–88), and subsequently over the disposal of unspent buffer stock funds.

**Production of Cocoa Beans**
('000 metric tons)

|  | 2003 | 2004* |
|---|---|---|
| **World total** | 3,415 | 3,607 |
| Latin America | 470 | 465 |
| **Leading Latin American producers** | | |
| Brazil | 170 | 169 |
| Colombia | 54 | 49† |
| Dominican Republic | 50 | 45† |
| Ecuador | 88 | 88 |
| Mexico | 48 | 48 |
| **Other leading producers** | | |
| Cameroon | 130‡ | 130† |
| Côte d'Ivoire | 1,352 | 1,331 |
| Ghana‡ | 497 | 736 |
| Indonesia | 453 | 430‡ |
| Malaysia | 36 | 33 |
| Nigeria | 361 | 366 |

* Provisional figures.
† FAO estimate.
‡ Unofficial figure(s).

The effectiveness of financial operations under the fourth ICCA was severely limited by the accumulation of arrears of individual members' levy payments, notably by Côte d'Ivoire and Brazil. The fourth ICCA was extended for a two-year period from October 1990, although the suspension of the economic clauses relating to cocoa price support operations rendered the agreement ineffective in terms of exerting any influence over cocoa market prices.

Preliminary discussions on a fifth ICCA, again held under UN auspices, ended without agreement in May 1992, when consumer members, while agreeing to extend the fourth ICCA for a further year (until October 1993), refused to accept producers' proposals for the creation of an export quota system as a means of stabilizing prices, on the grounds that such arrangements would not impose sufficient limits on total production to restore equilibrium between demand and supply. Additionally, no agreement was reached on the disposition of cocoa buffer stocks, then totalling 240,000 metric tons. In March 1993 ICCO delegates abandoned efforts to formulate arrangements whereby prices would be stabilized by means of a stock-withholding scheme. At a further negotiating conference in July, however, terms were finally agreed for a new ICCA, to take effect from October, subject to its ratification by at least five exporting countries (accounting for at least 80% of total world exports) and by importing countries (representing at least 60% of total imports). Unlike previous commodity agreements sponsored by the UN, the fifth ICCA aimed to achieve stable prices by regulating supplies and promoting consumption, rather than through the operation of buffer stocks and export quotas.

The fifth ICCA, operating until September 1998, entered into effect in February 1994. Under the new agreement, buffer stocks totalling 233,000 metric tons that had accrued from the previous ICCA were to be released on the market at the rate of 51,000 tons annually over a maximum period of four-and-a-half years, beginning in the 1993/94 crop season. At a meeting of the ICCO, held in October 1994, it was agreed that, following the completion of the stocks reduction programme, the extent of stocks held should be limited to the equivalent of three months' consumption. ICCO members also assented to a voluntary reduction in output of 75,000 tons annually, beginning in 1993/94 and terminating in 1998/99. Further measures to achieve a closer balance of production and consumption, under which the levels of cocoa stocks would be maintained at 34% of world grindings during the 1996/97 crop year, were introduced by the ICCO in September 1996. The ICCA was subsequently extended until September 2001. In April 2000 the ICCO agreed to implement measures to remedy low levels of world prices (see below), which were to centre on the elimination of sub-grade cocoa in world trade: these cocoas were viewed by the ICCO as partly responsible for the downward trend in prices. In mid-July Côte d'Ivoire, Ghana, Nigeria and Cameroon disclosed that they had agreed to destroy a minimum of 250,000 tons of cocoa at the beginning of the 2000/01 crop season, with a view to assisting prices to recover and to 'improving the quality of cocoa' entering world markets.

**Export Price Index for Cocoa**
(base: 1980 = 100)

|  | Average | Highest month(s) | Lowest month(s) |
|---|---|---|---|
| 1990 | 49 | — | — |
| 1995 | 55 | — | — |
| 2000 | 35 | 38 (Jan.) | 32 (Nov.) |
| 2001 | 40 | 48 (Dec.) | 36 (Feb., June, July) |
| 2002 | 66 | 82 (Oct.) | 49 (Jan.) |

A sixth ICCA was negotiated, under the auspices of the UN, in February 2001. Like its predecessor, the sixth ICCA aimed to achieve stable prices through the regulation of supplies and the promotion of consumption. The Agreement took effect on 1 October 2003. In December, in accordance with its provisions, the ICC established a Consultative Board on the World Cocoa Economy, a private-sector board comprising seven exporting and seven importing members, with a mandate to 'contribute to the development of a sustainable cocoa economy; identify threats to supply and demand and propose actions to meet the challenges; facilitate the exchange of information of production, consumption and stocks; and advise on other cocoa-related matters within the scope of the Agreement'.

As the above table indicates, international prices for cocoa have generally been very low in recent years. In 1992 the average of the ICCO's daily prices (based on selected quotations from the London and New York markets) was US $1,099.5 per metric ton (49.9 US cents per lb), its lowest level since 1972. The annual average price per ton subsequently rose steadily, reaching $1,456 in 1996 and $1,619 in 1997. The average rose in 1998 to $1,676 per ton, its highest level since 1987, but slumped in 1999 to $1,140 (a fall of 32.0%). In 2000 the average of the ICCO's daily prices again declined steeply, falling well below that recorded in 1992, to only $888 per ton, a reduction of some 22%. Prices recovered somewhat in 2001, but, even so, the ICCO's daily quotation averaged only $1,089 per ton in that year, the second lowest average price recorded since 1972. In 2002, however, the average price rebounded by almost 63%, compared with the previous year, to reach $1,778 per ton. In 2003 the ICCO's average daily price fell slightly, by 1.3%, to $1,755 per ton. A more substantial decline, of 11.8%, was recorded in 2004, when the ICCO's daily quotation averaged $1,548 per ton. In 1996 the monthly average ranged from $1,339 per ton (in March) to $1,538 (June). In 1997 it varied from $1,373 per ton (February) to $1,770 (September). The average increased in May 1998 to $1,794 per ton (its highest monthly level since February 1988), but fell in December to $1,515. In 1999 the highest monthly average was $1,455 per ton in January, and the lowest was $919 in December. In 2000 the average ICCO quotation ranged between $942 per ton in June, and $801 per ton in November, the lowest monthly average since March 1973. The corresponding average prices for 2001 were $1,337 per ton, recorded in December of that year, and $967 per ton, recorded in January. In 2002 the average monthly ICCO quotation ranged between $2,205 per ton, recorded in October, and $1,384 in January of that year. In 2003 the highest monthly average price, $2,239 per ton, was recorded in February, and the lowest, $1,482 per ton, in October. The ICCO average quotation ranged between $1,408 per ton (June) and $1,729 per ton (August) in 2004. In March 2005 an average quotation of $1,748 per ton was recorded.

On the London Commodity Exchange (LCE, merged with the London International Financial Futures and Options Exchange—LIFFE—under the latter name in September 1996) the price of cocoa for short-term delivery increased from £637 (US $983) per metric ton in May 1993 to £1,003.5 in November, but it later retreated. In July 1994, following forecasts that the global production decifit would rise, the price reached £1,093.5 ($1,694) per ton.

In late February 1995 the London cocoa quotation for March delivery stood at £1,056.5 per metric ton, but in March the price was reduced to £938 (US $1,498). The downward trend continued, and in late July the LCE cocoa price was £827.5 ($1,321) per ton. Prices under short-term contracts remained below £1,000 per ton until the end of December, when the 'spot' quotation (for cocoa for immediate delivery) stood at £847.5 ($1,319) per ton.

During the first quarter of 1996 London cocoa prices continued to be depressed, but in April the short-term quotation rose to more than £1,000 per metric ton. In May the LCE 'spot' price reached £1,104.5 (US $1,672) per ton. Cocoa prices had increased in spite of the ICCO's forecast that supply would exceed demand in 1995/96, following four consecutive years of deficits. In July 1996, however, the 'spot' quotation in London declined from £1,049 ($1,630) per ton to £924 ($1,438). In December the 'spot' price was reduced to £848.5 ($1,419) per ton.

In January and February 1997 short-term quotations for cocoa were at similarly low levels, but in March the 'spot' price on the London market rose from £894.5 (US $1,449) per metric ton to £1,012.5 ($1,621) in less than two weeks. On 1 July the London 'spot' price stood at £1,143 ($1,895) per ton: its highest level, in terms of sterling, for more than nine years. Three weeks later, however, the price declined to £963 ($1,615) per ton. By late August international cocoa prices had recovered strongly, in response to fears that crops would suffer storm damage, and in early September the LCE 'spot' quotation reached £1,133.5 ($1,798) per ton, while prices for longer-term contracts were at their highest for almost a decade. Thereafter, the trend in cocoa prices was generally downward. In the first half of December, however, the London price advanced from £987 ($1,663) per ton to £1,117 ($1,824).

In February 1998 the LCE price of cocoa for short-term delivery was reduced to less than £1,000 per metric ton. Following political unrest in Indonesia and forecasts of an increased global supply deficit, the cocoa market rallied in the first half of May, with the London 'spot' price rising from £1,072.5 (US $1,787) per ton to £1,140 ($1,857). Meanwhile, cocoa under long-term contracts was being traded at more than £1,200 per ton. However, in late June the London price of cocoa for July delivery declined to £1,002.5 per ton. During July the 'spot' price reached £1,070 ($1,752) per ton, before easing to £1,026.5 ($1,684). London cocoa prices remained above £1,000 per ton until late September, when the 'spot' quotation fell to £970 ($1,651). Later in the year there was a steady downward trend, and in late December the price of cocoa was about £860 ($1,440) per ton.

During the early weeks of 1999 the London cocoa market remained relatively stable, but in March the 'spot' price declined to £803 (US $1,307) per metric ton. The slump later intensified, following forecasts of plentiful crops and a weakening in consumption trends, and by late May the London price of cocoa had fallen to only £602.5 ($962) per ton. Prices subsequently rallied, and in June, with the EU failing to resolve an impasse over common rules on chocolate products (see above), the quotation for July delivery reached £819 per ton. In July, after the EU agreed to allow chocolate manufacturers to include vegetable fats, the 'spot' price of cocoa eased to £694 ($1,089) per ton, although it later recovered to £754 ($1,194). A further decline in cocoa prices ensued, and in September the 'spot' quotation fell to £601.5 ($975) per ton. After a slight recovery, the downward trend continued. In November the London cocoa price for short-term delivery was reduced to £527.5 per ton. In December the 'spot' quotation reached £570.5 ($926) per ton, but later in the month the price retreated to £530.5 ($854): its lowest level, in terms of sterling, since 1992.

Despite the coup in Côte d'Ivoire in December 1999, the cocoa market weakened further during the opening weeks of 2000, and in late February the London price for short-term delivery stood at only £590 per metric ton. Meanwhile, the equivalent New York price of cocoa fell in February to only US $734 per ton: its lowest level for more than 25 years. In March the London 'spot' quotation advanced to £598.5 ($940) per ton, before easing to £549 ($874). Comparable prices in May ranged from £575.5 ($880) to £606.5 ($911) per ton, and those in July were between £582 ($881) and £599 ($907). In August the London price of cocoa for short-term delivery fell to £564 per ton. In September, however, the 'spot' quotation rallied, ranging between £586 ($855) and £593 ($860) in that month. In December a further downward movement occurred, the 'spot' quotation in that month ranging between £556 ($803) per ton and £534 ($774). Early in the same month the New York second position 'futures' price declined to $707 per ton, its lowest level for 27 years.

In January 2001 the London price of cocoa for short-term delivery rose to £833 per metric ton, this upward movement being attributed to a steeper decline in deliveries from Côte d'Ivoire than had earlier been forecast. This recovery was sustained in February, when speculative buying and estimates of potential shortages in supply boosted the short-term price to £945. In March, however, as a result of improved forecasts of production by Côte d'Ivoire, the London 'spot' quotation declined, ranging between £799 (US $1,153) and £917 ($1,346) per ton. Comparable prices in May ranged from £745 ($1,068) to £799 ($1,136). In early May fund and speculative buying, together with renewed pessimism about the level of production in Côte d'Ivoire, boosted the price of the London July 'futures' contract to a three-week high of £815 per ton at one point. In June, however, the London price of cocoa for short-term delivery declined steadily, falling to £676 per ton late in the month, while the price of the second position 'futures' contract fell to a 'low' of £666 per ton. During the first two weeks of July the London 'spot' market quotation ranged between £728 ($1,024) per ton and £674 ($947) per ton. In August reports of a poor conclusion to the main crop in Côte d'Ivoire caused the price of the second 'futures' position on the London market to rise above £750 per ton, but this recovery was cancelled out in September by substantial selling of new crop West African cocoa. Late in the month forecasts of reduced production by Côte d'Ivoire in 2001/02 caused the London December 'futures' contract to rise as high as £758 per ton at one point, the rally attributed to short-covering by commodity funds. In early October the price of the London second position 'futures' contract rose to £838 per ton, the highest price since mid-May. This strengthening of the second position contract was, once again, attributed to fund buying. London 'futures' prices increased rapidly from mid-November, in response to renewed pessimism with regard to production by Côte d'Ivoire, and forecasts of a deficit in the season's cocoa crop of some 200,000 tons. In the third week of November London 'futures' prices reached their highest levels for three years. The closing price of the London March contract on 16 November was £952 per ton, representing an increase of some £200 per ton over the preceding 14 days. The upward trend in prices continued during 2002, amid further reports of poor weather in West Africa and political unrest in the principal cocoa-growing areas of Côte d'Ivoire. The maintenance by one trading company of an unusually 'long' position in cocoa, thereby causing scarcity elsewhere in the market, exacerbated this trend throughout the first half of 2002. In March the price of the London second position contract reached £1,261 per ton, the highest price for a second position contract since September 1987. By the end of July 2002 the London 'futures' price had risen to £1,317. At the end of 2002 the price of the London March contract was £1,287 per ton. From September political crisis in Côte d'Ivoire propelled London cocoa prices to very high levels—in October the March contract price rose to more than £1,600 per ton. Prices subsequently fell back, however, as a result of confidence that any disruptions to supply arising from the situation in Côte d'Ivoire would be manageable.

The declines in the average daily quotation of the ICCO in 2002/03 and 2003/04 detailed above have occurred as a consequence of surpluses in cocoa production relative to grindings. According to the ICCO, world production of cocoa beans reached a record level of 3.5m. metric tons in 2003/04—some 10% higher than in 2002/03, in which year output had likewise risen to a record level—when production exceeded that of the previous year for a fourth successive year. The increase in world cocoa production in 2003/04 arose to a large extent from growth in the output of Côte d'Ivoire and from a huge increase, of 48%, in that of Ghana. Production by Ecuador was also reported to have risen, by some 16%. At the same time, according to the ICCO, in 2003/04 grindings of cocoa beans world-wide rose to a record level of 3.2m. tons, compared with 3.1m. tons in 2002/03. Surplus production nevertheless caused world stocks of cocoa to increase to 1.4m. tons, and the stocks-to-grindings ratio to 45%. In its assessement of the cocoa year 2003/04, the ICCO noted that Europe and North America remained the main centres for cocoa-processing, and the Netherlands and the USA the two principal cocoa-processing countries. However, grindings in countries of origin continued to increase and accounted for 36% of total world grindings in 2003/04. This trend—exemplified by Côte d'Ivoire's displacement, since the late 1990s, of Germany as the world's third largest cocoa-processing country—was a result of government policies that had sought to promote exports of value-added, semi-finished products, prompting multinational companies to make substantial investments in cocoa-processing facilities in countries of origin.

The Cocoa Producers' Alliance (COPAL), with headquarters in Lagos, Nigeria, had 10 members as of 2005, including Brazil and the Dominican Republic. The alliance was formed in 1962 with the aim of preventing excessive price fluctuations by regulating the supply of cocoa. Members of COPAL currently account for about 78% of world cocoa production. COPAL has acted in concert with successive ICCAs.

The principal centres for cocoa-trading in the industrialized countries are the London Cocoa Terminal Market, in the United Kingdom, and the New York Coffee, Sugar and Cocoa Exchange, in the USA.

**Coffee** (*Coffea*)

The coffee plant is an evergreen shrub or small tree, generally 5 m–10 m tall, indigenous to Asia and tropical Africa. Wild trees grow to 10 m, but cultivated shrubs are usually pruned to a maximum of 3 m. The dried seeds (beans) are roasted, ground and brewed in hot water to provide the most popular of the world's non-alcoholic beverages. Coffee is drunk in every country in the world, and its consumers comprise an estimated one-third of the world's adult population. Although it has little nutrient value, coffee acts as a mild stimulant, owing to the presence of caffeine, an alkaloid also present in tea and cocoa.

There are about 40 species of *Coffea*, most of which grow wild in the eastern hemisphere. The two species of chief economic importance are *C. arabica* (native to Ethiopia), which, in the early 2000s, accounted for about 60%–65% of world production, and *C. canephora* (the source of robusta coffee), which accounted for almost all of the remainder. Arabica coffee is more aromatic, but robusta, as the name implies, is a stronger plant. Coffee grows in the tropical belt, between 20°N and 20°S, and from sea-level to as high as 2,000 m above. The optimum growing conditions are found at 1,250 m–1,500 m above sea-level, with an average temperature of around 17°C and an average annual rainfall of 1,000 mm–1,750 mm. Trees begin bearing fruit three to five years after planting, depending upon the variety, and give their maximum yield (up to 5 kg of fruit per year) from the sixth to the 15th year. Few shrubs remain profitable beyond 30 years.

Arabica coffee trees are grown mostly in the American tropics and supply the largest quantity and the best quality of coffee beans. In Africa and Asia arabica coffee is vulnerable in lowland areas to a serious leaf disease and consequently cultivation has been concentrated on highland areas. Some highland arabicas, such as those grown in Kenya, are renowned for their high quality.

In recent years Latin American arabica production has been inhibited by the coffee berry borer beetle, which has been described as the most damaging pest to coffee world-wide. The beetle has been estimated to cost Latin American producers US $500m. annually in lost production. Infestation has been particularly acute in Colombia, where, at one time, the beetle, which cannot be effectively eliminated by pesticides, depredated two-thirds of the total area under coffee. During the late 1990s experimental research, aimed at eliminating the beetle through biotechnological methods, was proceeding in Ecuador, Guatemala, Honduras, Jamaica, Mexico and India.

The robusta coffee tree, grown mainly in East and West Africa, and in the Far East, has larger leaves than arabica, but the beans are generally smaller and of lower quality and, consequently, fetch a lower price. However, robusta coffee has a higher yield than arabica as the trees are more resistant to disease. Robusta is also more suitable for the production of soluble ('instant') coffee. About 60% of African coffee is of the robusta variety. Soluble coffee accounts for more than one-fifth of world coffee consumption.

Each coffee berry, green at first but red when ripe, usually contains two beans (white in arabica, light brown in robusta) which are the commercial product of the plant. To produce the best quality arabica beans—known in the trade as 'mild' coffee—the berries are opened by a pulping machine and the beans fermented briefly in water before being dried and hulled into green coffee. Much of the crop is exported in green form. Robusta beans are generally prepared by dry-hulling. Roasting and grinding are usually undertaken in the importing countries, for economic reasons, and because roasted beans rapidly lose their freshness when exposed to air.

Apart from beans, coffee produces a few minor by-products. When the coffee beans have been removed from the fruit, what remains is a wet mass of pulp and, at a later stage, the dry material of the 'hull' or fibrous sleeve that protects the beans. Coffee pulp is used as cattle feed, the fermented pulp makes a good fertilizer and coffee bean oil is an ingredient in soaps, paints and polishes.

More than one-half of the world's coffee is produced on small-holdings of less than 5 ha. In most producing countries, coffee is almost entirely an export crop, for which (with the exception of Brazil, after the USA the world's second largest coffee consumer) there is little domestic demand. Green coffee accounts for some 96% of all the coffee that is exported, with soluble and roasted coffee comprising the balance. Tariffs on green/raw coffee are usually low or non-existent, but those applied to soluble coffee may be as high as 30%. The USA is the largest single importer, although its volume of coffee purchases was overtaken in 1975 by the combined imports of the (then) nine countries of the European Community (EC, now the European Union—EU).

After petroleum, coffee is the major raw material of world trade, and the single most valuable agricultural export of the tropics. Coffee is the most important cash crop of Latin America, with a number of countries heavily dependent on it as a source of foreign exchange. Of the estimated total world crop of coffee beans in 2004/05, Latin American and Caribbean countries accounted for 62.2%. Africa, which formerly ranked second, was overtaken in 1992/93 by Asian producers. In 2004/05 African producers accounted for 13.5% of the estimated world coffee crop, compared with 24.3% for Asian countries.

Brazil and Colombia are among the world's main producers of coffee beans, together consistently accounting for more than 32% of world trade in green (unroasted) coffee during the 1990s. In 1999 their share of the world coffee trade rose to 39%, reflecting the bumper crop from Brazil, and in 2002 more than 43% of all coffee traded world-wide emanated from Brazil and Colombia. In the 2004/05 crop year exports by Brazil and Colombia represented about 42% of all coffee exported world-wide. Coffee was formerly Brazil's most important agricultural export, but its relative importance has declined in recent years. In 1991 coffee (including extracts and essences) provided 4.8% of the country's total exports, but in 1992, when prices were lower (see below), the proportion was only 3.1%. Coffee and coffee substitutes provided 5.1% of Brazil's total exports in 1999, 3.2% in 2000 and 2.4% in 2001. In terms of volume, Brazilian coffee exports reached a record level of 29.9m. bags (each of 60 kg) in 2002/03. After a devastating frost in 1975, Brazil initiated a major programme to replant coffee trees. This concentrated on the state of Minas Gerais, north of another important coffee-growing area, Paraná state, which is more prone to frost. Replanting on a smaller scale took place after the less severe frost damage which coffee plantations suffered in 1981, although efforts have continued to move the centre of production further northwards. In the late 1990s Minas Gerais usually accounted for more than 50% of total Brazilian coffee production. In 2001/02, according to official estimates, Minas Gerais accounted for 45.3% of total Brazilian output.

The Colombian economy continues to depend heavily on coffee. 'Colombian Mild' arabica coffee beans, to which the country has given its name, are regarded as being of a superior quality to other coffee types and are grown primarily in Colombia, Kenya and Tanzania. (Colombian Milds are one of four internationally designated coffee groups, the others being Other Milds, which are produced primarily in Central America, Brazilian Natural Arabicas, which are produced primarily in Brazil and Ethiopia, and Robustas, of which the main producers are Viet Nam, Indonesia and Brazil.) Shipments of coffee from Colombia accounted for 6.2% of the country's total export earnings in 2001, 6.4% in 2002 and 6.2% in 2003. An earthquake in Colombia in January 1999 caused heavy damage to coffee-processing facilities, although plantations were largely unaffected. In 2001/02, according to International Coffee Organization (ICO) data, Colombian production of coffee totalled almost 12m. bags. Output fell slightly in 2002/03, but still totalled some 11.9m. bags. In the 2003/04 crop year, however, Colombian output declined steeply, by about 7.6%, to only 11m. bags. Provisional ICO data indicated that production fell further, to 10.5m. bags, in the 2004/05 crop year. According to the same source, Colombian exports of coffee totalled 10.3m. bags in 2001/02, 10.3m. bags in 2002/03 and 10.6m. bags in 2003/04. Exports were estimated at about 10.8m. bags in 2004/05. The USA is consistently the most important American market for Colombian coffee, accounting for an estimated 31.2% of total shipments in 2000/01. Overall, however, Europe, in particular the countries of the EU, is the most important market for Colombian coffee. In 2000/01 Germany was the most important European destination for shipments, accounting for an estimated 19.0% of Colombia's total exports of coffee.

In the early 1990s coffee was the principal export crop in El Salvador, Guatemala, Haiti, Mexico, Nicaragua and Peru, and the second most important crop in Costa Rica, the Dominican Republic, Ecuador, Honduras and Puerto Rico. Sales of coffee contributed 33.2% of El Salvador's total export earnings in 1996, and 38.2% in 1997. Unfavourable weather, disease (rust), civil war and falling prices, however, have reduced production in recent years. Sales of coffee contributed 25.8% of El Salvador's export revenue in 1998. The contribution of coffee (including roasted and decaffeinated coffee) to El Salvador's total export revenue (excluding exports from *maquila* zones) was 20.8% in 1999 and 22.5% in 2000. In 2001, however, coffee contributed only 9.8% of total export revenue, and the proportion fell again, to 8.9%, in 2002. Sales of coffee and coffee substitutes accounted for 21.3% of Guatemala's export earnings in 2000, 12.7% in 2001 and 11.7% in 2002. Sales of Nicaraguan coffee accounted for 13.1% of the country's export revenue in 2002, an estimated 14.1% in 2003 and an estimated 16.8% in 2004. In Costa Rica coffee ranks second to bananas as the principal export commodity, supplying 10.8% of total export revenue in 1997. In 1998, however, coffee's contribution to Costa Rica's total export revenue declined to 7.4%, and in 1999 to only 4.3%. In 2000 coffee accounted for 4.7% of Costa Rica's total export earnings. In Honduras coffee has displaced the banana as the country's principal agricultural export commodity, contributing 13.8% of export revenue in 2002, 13.6% in 2003 and 16.2% in 2004. Coffee provided 23.1% of Haiti's export earnings in 1989/90, but its contribution has declined since then, amounting to less than 1% in 2002 and 2003 and to 1.2% in 2004. The neighbouring Dominican Republic earned only 7.3% of its total export revenue from coffee (including exports from free-trade zones) in 1998, 1.8% in 1999 and 2.2% in 2000. Coffee's share in Ecuador's export revenue, which was 10.8% in 1994, declined to 3.3% in 1996 and 2.2% in 1997, before rising to 2.5% in 1998. In 1999, however, the contribution of coffee to Ecuador's total export earnings declined to only 1.8%, and it fell again in 2000, to 0.9%. Coffee accounted for 3.3% of Peru's total export earnings in 2000, 2.6% in 2001 and 2.5% in 2002.

According to the ICO, in a situation assessment for coffee world-wide in May 2005, output of green coffee by ICO exporting members totalled some 113.4m. bags in the 2004/05 crop year, an increase of 9% compared with the 2003/04 crop year, when exporting members' output amounted to some 103.2m. bags. Production was forecast to decline to about 106m. bags in 2005/06, on the basis of preliminary data from exporting countries. The rise in global output in 2004/05 was propelled by increased production in South America, in particular in Brazil where output rose by more than 34% to some 38.7m. bags. Peruvian production also rose markedly, by about 21%, to some 3.1m. bags, while that of Colombia increased slightly, by about 3.6%, to 11.5m. bags. Ecuador's output declined slightly, by about 2.2%, to 750,000 bags. In all of the exporting countries of North and Central America, however, with the exception of Costa Rica, production either declined or stagnated in 2004/05. In Nicaragua, notably, output fell by more than 34%, to 920,000 bags, compared with the previous crop year. As for the other major producing regions, the output of Africa increased while that of Asia/Oceania declined. Production of Arabicas amounted to about 75m. bags in 2004/05, an increase of about 10.9% compared with 2003/04. Arabicas thus accounted for some 66% of world coffee production in 2004/05, a marginally higher proportion than in 2003/04. While output of Colombian Milds rose by about 2.2% in 2004/05, their share of world production declined to 11.36%, compared with 11.87% in the previous crop year. Output of Other Milds declined by 1.61% in 2004/05, when they accounted for 22.94% of world coffee production. Output of Brazilian Naturals rose by 23.75% to 36.01m. bags, representing 31.76% of world production. Production of Robustas increased by 8.08%, to 38.47m. bags, in 2004/05, when they accounted for 33.93% of global coffee output. Opening stocks of coffee in exporting countries at the beginning of the 2004/05 crop year totalled 17.43m. bags, some 21% lower than at the start of the preceding crop year. With regard to stocks in importing countries, the ICO noted that these had stood at 10.6m. bags at the end of December 1999, but were estimated at 21.5m. bags at the end of December 2004. Exports of coffee in the calendar year 2004 totalled some 89.62m. bags, compared with 85.87m. bags in 2003. According to the ICO, the total value of coffee exports in 2004 was US $7,050m., compared with $5,560m. in 2003. Domestic consumption in exporting countries in 2004/05 was estimated at 28.52m. bags. Brazil's consumption, which accounts for about one-half of total consumption by exporting countries, increased by 3.64% in 2004/05, compared with the previous crop year, to 14.25m. bags. Consumption by importing countries totalled 86.18m. bags in 2004.

Effective international attempts to stabilize coffee prices began in 1954, when a number of producing countries made a short-term agreement to fix export quotas. After three such agreements, a five-year International Coffee Agreement (ICA), covering both producers and consumers and introducing a quota system, was signed in 1962. This led to the establishment in 1963 of the ICO, with its headquarters in London, United Kingdom. In May 2005 the International Coffee Council, the highest authority of the ICO, comprised 74 members (i.e. participants in the 2001 ICA—44 exporting countries and 30 importing countries; the USA withdrew from the ICO in 1993 and did not rejoin until 2005). Successive ICAs took effect in 1968, 1976, 1983, 1994 and 2001 (see below), but the system of export quotas to stabilize prices was eventually abandoned in July 1989 (for detailed information on these agreements, see *South America, Central America and the Caribbean 1991*). During each ICA up to and including the one implemented in 1994, contention arose over the allocation of members' export quotas, the operation of price support mechanisms, and, most importantly, illicit sales by some members of surplus stocks to non-members of the ICO (notably to the USSR and to countries in Eastern Europe and the Middle East). These 'leaks' of low-price coffee, often at less than one-half of the official ICA rate, also found their way to consumer members of the ICO through free ports, depressing the general market price and making it more difficult for exporters to fulfil their quotas.

**Production of Green Coffee Beans**
('000 bags, each of 60 kg, coffee years, ICO members only)

| | 2003/04 | 2004/05 |
|---|---|---|
| **World total** | 103,110 | 112,673 |
| Latin America | 61,243 | 70,075 |
| **Leading Latin American producers** | | |
| Brazil | 28,820 | 38,264* |
| Colombia | 11,000 | 10,500 |
| Costa Rica | 1,802 | 1,924 |
| Dominican Repub. | 454 | 555† |
| Ecuador | 767 | 750* |
| El Salvador | 1,488 | 1,430 |
| Guatemala | 3,610 | 3,450† |
| Honduras | 2,968 | 2,750† |
| Mexico | 4,550 | 4,500 |
| Nicaragua | 1,442 | 900 |
| Peru | 2,525 | 3,067* |
| Venezuela | 839 | 920† |
| **Other leading producers** | | |
| Cameroon | 900 | 1,100† |
| Côte d'Ivoire | 2,674 | 1,475 |
| Ethiopia | 3,874 | 5,000 |
| India | 4,445 | 4,850† |
| Indonesia | 6,464 | 5,750* |
| Kenya | 673 | 917† |
| Philippines | 433 | 443† |
| Thailand | 846 | 1,056† |
| Uganda | 2,510 | 2,750† |
| Viet Nam | 14,830 | 14,000 |

*Figure derived on the basis of gross opening stocks at 31 March 2005.
†Estimated figure.

Source: International Coffee Organization.

The issue of coffee export quotas became further complicated in the 1980s, as consumers in the main importing market, the USA, and, to a lesser extent, in the EC came to prefer the milder arabica coffees grown in Central America at the expense of the robustas exported by Brazil and the main African producers. Disagreements over a new system of quota allocations, taking account of coffee by variety, had the effect of undermining efforts in 1989 to preserve the economic provisions of the ICA, pending the negotiation of a new agreement. The ensuing deadlock between consumers and producers, as well as among the producers themselves, led in July to the collapse of the quota system and the suspension of the economic provisions of the ICA. The administrative clauses of the agreement, however, continued to operate and were subsequently extended until October 1993, pending an eventual settlement of the quota issue and the entry into force of a successor ICA.

With the abandonment of the ICA quotas, coffee prices fell sharply in world markets, and were further depressed by a substantial accumulation of coffee stocks held by consumers. The response by some Latin American producers was to seek to revive prices by imposing temporary suspensions of exports; this strategy, however, merely increased losses of coffee revenue. By early 1992 there had been general agreement among the ICO exporting members that the export quota mechanism should be revived. However, disagreements persisted over the allocation of quotas, and in April 1993 it was announced that efforts to achieve a new ICA with economic provisions had collapsed. In the following month Brazil and Colombia, the two largest coffee producers at that time, were joined by some Central American producers in a scheme to limit their annual coffee production and exports in the 1993/94 coffee year. Although world consumption of coffee exceeded the level of shipments, prices were severely depressed by surpluses of coffee stocks totalling 62m. bags, with an additional 21m. bags held in reserve by consumer countries. Prices, in real terms, stood at historic 'lows'.

In September 1993 the Latin American producers announced the formation of an Association of Coffee Producing Countries (ACPC) to implement an export-withholding, or coffee retention, plan. In the following month the 25-member Inter-African Coffee Organization (IACO) agreed to join the Latin American producers in a new plan to withhold 20% of output whenever market prices fell below an agreed level. With the participation of Asian producers, a 28-member ACPC, with headquarters in London, was formally established in August. Its signatory member countries numbered 28 in 2001, 14 of which were ratified. Production by the 14 ratified members in 1999/2000 accounted for 61.4% of coffee output world-wide.

The ACPC coffee retention plan came into operation in October 1993 and gradually generated improved prices; by April 1994 market quotations for all grades and origins of coffee had achieved their highest levels since 1989. In June and July 1994 coffee prices escalated sharply, following reports that as much as 50% of the 1995/96 Brazilian crop had been damaged by frosts. In July 1994 both Brazil and Colombia announced a temporary suspension of coffee exports. The occurrence of drought following the Brazilian frosts further affected prospects for its 1994/95 harvest, and ensured the maintenance of a firm tone in world coffee prices during the remaining months of 1994.

The intervention of speculative activity in the coffee futures market during early 1995 led to a series of price falls, despite expectations that coffee consumption in 1995/96, at a forecast 93.4m. bags, would exceed production by about 1m. bags. In an attempt to restore prices, the ACPC announced in March 1995 that it was to modify the price ranges of the export-withholding scheme. In May the Brazilian authorities, holding coffee stocks of about 14.7m. bags, introduced new arrangements under which these stocks would be released for export only when the 20-day moving average of the ICO arabica coffee indicator rose to about US $1.90 per lb. Prices, however, continued to decline, and in July Brazil joined Colombia, Costa Rica, El Salvador and Honduras in imposing a reduction of 16% in coffee exports for a one-year period. Later in the same month the ACPC collectively agreed to limit coffee shipments to 60.4m. bags from July 1995 to June 1996. This withholding measure provided for a decrease of about 6m. bags in international coffee exports during this period. In July 1997 the ACPC announced that the withholding scheme was to be replaced by arrangements for the restriction of exports of green coffee. Total exports for 1997/98 were to be limited to 52.75m. bags. Following the withdrawal in September 1998 of Ecuador from the export restriction scheme (and subsequently from the ACPC) and the accession of India to membership in September 1999, there were 14 ratified member countries participating in the withholding arrangements. The continuing decline in world coffee prices (see below) prompted the ACPC to announce in February 2000 that it was considering the implementation of a further scheme involving the withholding of export supplies. In the following month the members indicated their intention to withdraw supplies of low-grade beans (representing about 10% of annual world exports), and on 19 May announced arrangements under which 20% of world exports would be withheld until the ICO 15-day composite price reached 95 US cents per lb (at that time the composite price stood at 69 cents per lb). Retained stocks would only be released when the same indicator price reached 105 cents per lb. Five non-member countries, Guatemala, Honduras, Mexico, Nicaragua and Viet Nam, also signed a so-called London Agreement pledging to support the retention plan. Implementation of the plan, which had a duration of up to two years, was initiated by Brazil in June, with Colombia following in September. In December 2000 the ACPC identified a delay in the full implementation of the retention plan as one of the factors that had caused the average ICO composite indicator price in November to fall to its lowest level since April 1993, and the ICO robusta indicator price to its lowest level since August 1969. In May 2001 the ACPC reported that exchange prices continued to trade at historical 'lows'. Their failure to recover, despite the implementation of the retention plan, was partly attributed to the hedging of a proportion of the 7m. bags of green coffee retained by that time. On the physical market, meanwhile, crop problems and the implementation of the retention plan had significantly increased differentials for good quality coffees, in particular those of Central America. In April 2001 the ICO daily composite indicator price averaged 47.13 cents per lb (compared with an average of 64.24 cents per lb for the whole of 2000, the lowest annual average since 1973), the lowest monthly average since September 1992. In October 2001 the ACPC announced that it would dissolve itself in January 2002. The Association's relevance had been increasingly compromised by the failure of some of its members to comply with the retention plan in operation at that time, and by some members' inability to pay operating contributions to the group owing to the depressed state of the world market for coffee.

In June 1993 the members of the ICO agreed to a further extension of the ICA, to September 1994. However, the influence of the ICO, from which the USA withdrew in October 1993, was increasingly perceived as having been eclipsed by the ACPC. In 1994 the ICO agreed provisions for a new ICA, again with primarily consultative and administrative functions, to operate for a five-year period, until September 1999. In November of that year it was agreed to extend this limited ICA until September 2001. A successor ICA took effect, provisionally, in October 2001, and definitively in May 2005. By May 2005 the new ICA had been endorsed by 74 members of the International Coffee Council (44 exporting members and 30 importing members). Among the principal objectives of the ICA of 2001 were the promotion of international co-operation with regard to coffee, and the provision of a forum for consultations, both intergovernmental and with the private sector, with the aim of achieving a reasonable balance between world supply and demand in order to guarantee adequate supplies of coffee at fair prices for consumers, and markets for coffee at remunerative prices for producers.

**Export Price Index for Coffee**
(base: 1980 = 100)

|      | Average | Highest month(s) | Lowest month(s) |
| ---- | ------- | ---------------- | --------------- |
| 1990 | 46      | —                | —               |
| 1995 | 83      | —                | —               |
| 2000 | 44      | 55 (Jan.)        | 33 (Dec.)       |
| 2001 | 30      | 34 (Jan.)        | 25 (Dec.)       |
| 2002 | 28      | 32 (Nov.)        | 25 (Aug.)       |

International prices for coffee beans in the early 1990s were generally at very low levels, even in nominal terms (i.e. without taking inflation into account). On the London Commodity Exchange (LCE) the price of raw robusta coffee for short-term delivery fell in May 1992 to US $652.5 (£365) per metric ton, its lowest level, in terms of dollars, for more than 22 years. By December the London coffee price had recovered to $1,057.5 per ton (for delivery in January 1993). The LCE quotation eased to $837 (£542) per ton in January 1993, and remained within this range until August, when a sharp increase began. The coffee price advanced in September to $1,371 (£885) per ton, its highest level for the year. In April 1994 a further surge in prices began, and in May coffee was traded in London at more than $2,000 per ton for the first time since 1989. In late June 1994 there were reports from Brazil that frost had damaged the potential coffee harvest for future seasons, and the LCE quotation exceeded $3,000 per ton. In July, after further reports of frost damage to Brazilian coffee plantations, the London price reached $3,975 (£2,538) per ton. Market conditions then eased, but in September, as drought persisted in Brazil, the LCE price of coffee increased to $4,262.5 (£2,708) per ton: its highest level since January 1986. In December 1994, following forecasts of a rise in coffee production and a fall in consumption, the London quotation for January 1995 delivery stood at $2,481.5 per ton.

The coffee market subsequently revived, and in March 1995 the LCE price reached US $3,340 (£2,112) per metric ton. However, in early July coffee traded in London at $2,400 (£1,501) per ton, although later in the month, after producing countries had announced plans to limit exports, the price rose to $2,932.5 (£1,837). During September the LCE 'spot' quotation (for coffee for immediate delivery) was reduced from $2,749 (£1,770) per ton to $2,227.5 (£1,441), but in November it advanced from $2,370 (£1,501) to $2,739.5 (£1,786). Coffee for short-term delivery was traded in December at less than $2,000 per ton, while longer-term quotations were considerably lower.

In early January 1996 the 'spot' price of coffee in London stood at US $1,798 (£1,159) per metric ton, but later in the month it reached $2,050 (£1,360). The corresponding quotation rose to $2,146.5 (£1,401) per ton in March, but declined to $1,844.5 (£1,220) in May. The 'spot' contract in July opened at $1,730.5 (£1,112) per ton, but within four weeks the price fell to $1,487 (£956), with the easing of concern about a threat of frost damage to Brazilian coffee plantations. In November the 'spot' quotation rose to $1,571 (£934) per ton, but slumped to $1,375.5 (£819) within a week. By the end of the year the London price of coffee (for delivery in January 1997) had been reduced to $1,259 per ton.

In early January 1997 the 'spot' price for robusta coffee stood at only US $1,237 (£734) per metric ton, but later in the month it reached $1,597.5 (£981). The advance in the coffee market continued in February, but in March the price per ton was reduced from $1,780 (£1,109) to $1,547.5 (£960) within two weeks. In May coffee prices rose spectacularly, in response to concerns about the scarcity of supplies and fears of frost in Brazil. The London 'spot' quotation increased from $1,595 (£986) per ton to $2,502.5 (£1,526) by the end of the month. Meanwhile, on the New York market the price of arabica coffee for short-term delivery exceeded $3 per lb for the first time since 1977. However, the rally was short-lived, and in July 1997 the London price for robusta coffee declined to $1,490 (£889) per ton. In the first half of November the coffee price rose from $1,445 (£862) per ton to $1,658 (£972). During December the price for January 1998 delivery reached $1,841 per ton, but a week later it decreased to $1,657.

The coffee market rallied in January 1998, with the London 'spot' quotation rising from US $1,746.5 (£1,066) per metric ton to $1,841 (£1,124). Coffee prices for the corresponding contract in March ranged from $1,609 (£977) per ton to $1,787 (£1,065). Following reports of declines in the volume of coffee exports by producing countries (owing to inadequate rainfall), the upward trend in prices continued in April, with the price of robusta for short-term delivery reaching $1,992 per ton. In the first half of May there was another surge in prices (partly as a result of political unrest in Indonesia, the main coffee-producing country in Asia at that time), with the London quotation rising from $1,881.5 (£1,129) per ton to $2,202.5 (£1,351). Later in the month, however, the price was reduced to $1,882.5 (£1,155) per ton. Coffee prices subsequently fell further, and in late July the London 'spot' contract stood at only $1,505.5 (£909) per ton, before recovering to $1,580 (£963). The quotation per ton for September delivery reached $1,699.5 at the beginning of August, having risen by $162 in a week. In September the 'spot' price advanced from $1,640 (£974) per ton to $1,765 (£1,036) a week later. In late October a further sharp rise in coffee prices began, following storm damage in Central America, and in November the London 'spot' quotation for robusta increased from $1,872.5 (£1,123) per ton to $2,142.5 (£1,278). Meanwhile, trading in other contracts continued at less than $1,800 per ton until December, when the London price of coffee (for delivery in January 1999) rose to $1,977.

Coffee prices retreated in January 1999, with the London 'spot' quotation falling from US $1,872.5 (£1,131) per metric ton to $1,639 (£995). During March the price was reduced from $1,795.5 (£1,111) per ton to $1,692.5 (£1,030), but recovered to $1,795 (£1,112) within a week. As before, the market for longer-term deliveries was considerably more subdued, with coffee trading mainly within a range of $1,490–$1,590 per ton. Thereafter, a generally downward trend was evident, and in May the 'spot' price declined to $1,376.5 (£850) per ton, although it had reached $1,536.5 (£962) by the end of the month. The advance was short-lived, with prices for most coffee contracts standing at less than $1,400 per ton in late June. The 'spot' price in July fell to only $1,255 (£805) per ton. In August the London quotation for September 'futures', which had been only $1,282.5 per ton in July, rose to $1,407. However, the 'spot' price in September retreated from $1,323 (£825) per ton to $1,212.5 (£754). In October the price for short-term delivery was reduced to less than $1,200 per ton. The 'spot' quotation in November advanced from $1,212 (£736) per ton to $1,399.5 (£866). Prices strengthened further in December, with the London quotation for short-term delivery reaching $1,557 per ton. Meanwhile, the market for longer-term contracts was more stable, with prices remaining below $1,400 per ton. In that month the Brazilian Government's forecast for the country's coffee output in the year beginning April 2000 was higher than some earlier predictions, despite fears that the crop would have been damaged by the unusually dry weather there since September 1999. For 1999 as a whole, average prices of robusta coffee declined by 18.3% from the previous year's level, while arabica prices fell by 23.2%.

In January 2000 the 'spot' price of coffee in London rose strongly towards the end of the month, increasing from US $1,401.5 (£848) per metric ton to $1,727.5 (£1,067) within a week. However, prices of coffee 'futures' continued to be much lower: at the end of January the quotation for March delivery was $1,073.5 per ton. In February prices of robusta coffee 'futures' were below $1,000 per ton for the first time for nearly seven years. In March the 'spot' quotation eased from $993 (£628) per ton to $944 (£593). Prices continued to weaken in April, with the quotation for short-term delivery falling to less than $900 per ton. The 'spot' price in May declined to $891.5 (£602) per ton, but subsequently recovered to $941 (£639). Another downward movement ensued, and by early July the London 'spot' quotation stood at only $807 (£532) per ton. Later that month, prices briefly recovered, owing to concerns about the possible danger of frost damage to coffee crops in Brazil. The 'spot' quotation rose to $886.5 (£585) per ton, while prices of coffee 'futures' advanced to more than $1,000. However, the fear of frost was allayed, and on the next trading day the 'spot' price of coffee in London slumped to $795 (£525) per ton: its lowest level, in terms of US currency, since September 1992.

The weakness in the market was partly attributed to the abundance of supplies, particularly from Viet Nam, which has substantially increased its production and export of coffee in recent years. By mid-2000 Viet Nam had overtaken Indonesia to become the world's leading supplier of robusta coffee and was rivalling Colombia as the second largest coffee-producing country overall. Viet Nam and Mexico were the most significant producers outside the ACPC, but their representatives supported the Association's plan for a coffee retention scheme to limit exports and thus attempt to raise international prices. The plan was also endorsed by the Organisation africaine et malgache du café (OAMCAF), a Paris-based grouping of nine African coffee-producing countries.

In the first week of September 2000 the 'spot' market quotation rallied to $829 (£577) per metric ton, remaining at this level until 21 September, when another downward movement occurred. Towards the end of the month the 'spot' quotation declined to $776 (£530) per ton. At the beginning of November the London 'spot' quotation stood at only $709 (£490) per ton and was to decline steadily throughout the month, reaching $612 (£432) on 30 November. High consumer stocks and uncertainty about the size of the Brazilian crop were cited as factors responsible for the substantial decline in November, when the average ICO robusta indicator price fell to its lowest level since August 1969.

In early January 2001 the 'spot' quotation on the London market rallied, rising as high as $677 (£451) per metric ton. This recovery, which was attributed to concern about the lack of availability of new-crop Central American coffees and reports that producers in some countries were refusing to sell coffee for such low prices, was sustained, broadly, until March, when the downward trend resumed. On 23 March the London 'spot' quotation declined to only $570 (£399) per ton. On 17 April the London price of robusta coffee 'futures' for July delivery declined to a life-of-contract low of $560 per ton, the lowest second-month contract price ever recorded.

By May 2001 the collapse in the price of coffee had been described as the most serious crisis in a global commodity market since the 1930s, with prices at their lowest level ever in real terms. The collapse of the market was regarded, fundamentally, as the result of an ongoing increase in world production at twice the rate of growth in consumption, this oversupply having led to an overwhelming accumulation of stocks. During May the London 'spot' quotation fell from $584 (£407) per metric ton to $539 (£378) per ton.

In June 2001 producers in Colombia, Mexico and Central America were reported to have agreed to destroy more than 1m. bags of low-grade coffee in a further attempt to boost prices. The ACPC hoped that this voluntary initiative would eventually be adopted by all of its members. By this time the ACPC's retention plan was widely regarded as having failed, with only Brazil, Colombia, Costa Rica and Viet Nam having fully implemented it.

In early July 2001 the price of the robusta coffee contract for September delivery fell below $540 per metric ton, marking a record 30-year 'low'. At about the same time the ICO recorded its lowest composite price ever, at 43.80 US cents per lb. Despite a recovery beginning in October, the average composite price recorded by the ICO for 2001 was 45.60 cents per lb, 29% lower than the average composite price (64.25 cents per lb) recorded in 2000. In 2001 coffee prices were at their lowest level since 1973 in nominal terms, and at a record low level in real terms. The decline in the price of robusta coffees was especially marked in 2001, the ICO recording an average composite price of only 27.54 cents per lb, compared with 41.41 cents per lb in 2000, and 67.53 cents per lb in 1999. In 1996–98 the ICO composite price for robusta varieties had averaged 81.11 cents per lb. In 2002 the ICO recorded an average composite price for robustas of 30.02 cents per lb, 9% higher than the average price recorded in 2001. The average composite price for all coffees recorded in 2002 was 47.74 cents per lb. In 2003 the average composite price rose by 8.7%, to 51.91 cents per lb.

In 2004 the average composite price recorded by the ICO, at 62.15 US cents per lb, was 19.7% higher than in 2003. In 2004 the average composite price recorded for robustas declined by 2.6% while all of the average prices recorded for arabicas increased. During the first six months of 2005 the ICO recorded an average composite price for all coffees of 94.1 cents per lb, 51.4% higher than the average composite price recorded for the whole of 2004. In the first half of 2005 the average composite price ranged between 79.35 cents per lb (January) and 101.44 cents per lb (March). Assessing the situation for coffee in May 2005, the ICO noted that since the final quarter of 2004 coffee prices had risen to their highest levels since the last quarter of 1998 and that prices for the second and third positions on the London and New York coffee futures markets had displayed the same trend. The Organization concluded that fundamental factors, in particular a deficit between production and consumption of some 8m. bags in 2005, were likely to support current price levels. In the longer term, however, the maintenance of prices at satisfactorily remunerative levels for producers would depend on their ability to match their output to demand. In June the ICO noted that while fundamental factors continued to support prices, a correction had occurred in respect of arabicas. Prices on the New York futures market, from which the situation for arabicas may be gauged, had fallen to 102.5 cents per lb at the end of June, compared to 130.5 cents per lb at the end of May. Robustas, meanwhile, the situation for which is reflected on the London coffee futures market, had resisted this correction, the average price of the market's second and third positions having risen by 8% in June compared with May. According to the ICO, weaker prices for arabicas in June were a consequence of profit-taking by speculative market participants. Robusta prices were supported by a lower estimate of total production in 2005/06 owing to the effects of drought in Viet Nam.

## Copper

The ores containing copper are mainly copper sulphide or copper oxide. They are mined both underground and by open-cast or surface mining. After break-up of the ore body by explosives, the lumps of ore are crushed, ground and mixed with reagents and water, in the case of sulphide ores, and then subjected to a flotation process by which copper-rich minerals are extracted. The resulting concentrate, which contains about 30% copper, is then dried, smelted and cast into anode copper, which is further refined to about 99.98% purity by electrolysis (chemical decomposition by electrical action). The cathodes are then cast into convenient shapes for working or are sold as such. Oxide ores, less important than sulphides, are treated in ways rather similar to the solvent extraction process described below.

Two alternative copper extraction processes have been developed in recent years. The first of these techniques, and as yet of minor importance in the industry, is known as 'Torco' (treatment of refractory copper ores) and is used for extracting copper from silicate ores which were previously not treatable.

The second, and relatively low-cost, technique is the solvent extraction process. This is suited to the treatment of very low-grade oxidized ores and is currently being used on both new ores and waste dumps that have accumulated over previous years from conventional copper working. The copper in the ore or waste material is dissolved in acid and the copper-bearing leach solution is then mixed with a special organic-containing chemical reagent which selectively extracts the copper. After allowing the two layers to separate, the layer containing the copper is separated from the acid leach solution. The copper is extracted from the concentrated leach solution by means of electrolysis to produce refined cathodes.

Copper is ductile, resists corrosion and is an excellent conductor of heat and electricity. Its industrial uses are mainly in the electrical industry (about 60% of copper is made into wire for use in power cables, telecommunications, domestic and industrial wiring) and the building, engineering and chemical industries. Bronzes and brasses are typical copper alloys used for both industrial and decorative purposes. There are, however, substitutes for copper in almost all of its industrial uses, and in recent years aluminium has presented a challenge in the electrical and transport industries.

Since 1982, when it overtook the USA, Chile (in which, according to the US Geological Survey (USGS), about 30% of world copper reserves are located) has been the world's leading producer of copper. It is also the biggest exporter, and within the country are located the world's three largest copper mines (Chuquicamata, El Teniente and La Escondida). In 1996 La Escondida overtook Chuquicamata as the world's largest copper mine, with an annual production capacity exceeding 800,000 metric tons. With the opening of new, large-scale mines in the north of the country, Chilean copper production rose to 3.7m. tons in 1998 and to 4.4m. tons in 1999. In 2000, according to the USGS, Chile's output of copper amounted to 4.6m. tons, rising again, in 2001, to 4.7m. tons. In 2002 production totalled 4.6m. tons, and rose to 4.9m. tons in 2003. In 2004 Chilean copper production was estimated to have risen again, to some 5.4m. tons. The Chilean economy relies heavily on the copper industry, which accounted for an annual average of 44.1% of the country's exports in the period 1984–88, and in 2001 was the source of 37.2% of Chile's export revenue. Although the industry has been vulnerable to labour unrest in recent years, foreign investment in Chilean mining development has been rising: during 1989–95 more than one-half of all foreign investment was directed towards this sector. In 1999 the share of private mines in total copper output was 63.1%, while the proportion represented by the state-owned copper corporation, CODELCO, was 36.9%, compared with about 84% in 1980. Among the principal externally financed ventures is the Zaldivar deposit in northern Chile, believed to be one of the richest undeveloped copper ore bodies in Latin America. Chile has refinery capacity for processing more than 1m. tons of copper annually.

# REGIONAL INFORMATION

*Major Commodities of Latin America*

**Production of Copper Ore**
(copper content, '000 metric tons)

|  | 2003 | 2004* |
|---|---|---|
| **World total** | 13,600 | 14,500 |
| Latin America | 6,092 | 6,780 |
| **Leading Latin American producers†** | | |
| Chile | 4,900 | 5,380 |
| Mexico | 361 | 400 |
| Peru | 831 | 1,000 |
| **Other leading producers** | | |
| Australia | 830 | 850 |
| Canada | 558 | 560 |
| China, People's Repub. | 610 | 620 |
| Indonesia | 979 | 860 |
| Kazakhstan | 485 | 485 |
| Poland | 495 | 500 |
| Russia | 675 | 675 |
| USA | 1,120 | 1,160 |
| Zambia | 330 | 400 |

\* Estimated production.
† Figures represent the sum of output in listed countries. Smaller quantities of copper-bearing ores are also produced in Argentina, Bolivia, Brazil, Colombia, Cuba, Ecuador and Honduras.

Source: US Geological Survey.

After Chile, the Latin American country to which the copper industry is most important is Peru. Over a decade until the mid-1990s, however, the industry was adversely affected by the country's economic instability and guerrilla attacks on the mines. The restoration of relative internal stability, together with the phased privatization of the state-owned mining corporation, Centromín, has stimulated foreign investment in mining exploration ventures. In June 1999 it was announced that financing had been secured for the proposed development of the Antamina copper-zinc mine in Peru. The Antamina scheme, involving Canadian interests, is the country's largest investment project, with an estimated cost of US $2,260m. The open-pit mine, in the Andes mountains, began operating in late 2001, and is expected eventually to yield about 270,000 metric tons of copper annually for 20 years. By April 2002 the Antamina mine was already the most productive in Peru. Earnings from sales of copper manufactures, ores and concentrates represented 16.8% of Peru's total export revenue in 2002. The development of deposits of copper in Argentina, estimated to total 700m. tons, commenced in 1997; production totalled 30,000 tons in that year. In 2000 output amounted to 145,197 tons, some 27% lower than in 1999, but it recovered to more than 190,000 tons in 2001. In 2002 Argentina's production was about 204,000 tons, but was estimated to have declined slightly, to about 199,000 tons, in 2003. Brazil, a relatively minor producer of copper in the mid-1990s, was preparing in 2003 to bring into production a new project at Marabá in the state of Pará, with potential annual output of 250,000 tons of refined copper, with by-products of gold, silver and molybdenum. The project at Marabá is a joint venture between Companhia Vale do Rio Doce (99%) and the Banco do Desenvolvimento Econômico e Social. In Panama, as of 2003, two of the world's largest copper deposits, one in Cerro Colorado district, Chiriqui province, the other in Cerro Petaquilla district, province of Colon, remained undeveloped.

The major copper-importing countries are the member states of the European Union (EU), Japan and the USA. At the close of the 1980s, demand for copper was not being satisfied in full by current production levels, which were being affected by industrial and political unrest in some producing countries, with the consequence that levels of copper stocks were declining. Production surpluses, reflecting the lower levels of industrial activity in the main importing countries, occurred in the early 1990s, but were followed by supply deficits, exacerbated by low levels of copper stocks. In 2004, according to provisional figures from the International Copper Study Group (ICSG), world-wide usage of refined copper increased by 5.6% from its 2003 level, rising to 16,523,000 metric tons, while production, including secondary output (recovery from scrap), increased by 3.5%, to 15,234,000 tons. There was consequently a copper deficit for the year of 1,289,000 tons, compared with a deficit of 407,000 tons in 2003. Identified stocks of refined copper throughout the world fell by 857,000 tons in 2004, to total 923,000 tons at the end of the year. Provisional data from the ICSG indicated that world-wide usage of refined copper during January–March 2005 reached 4,032,000 tons, a decline of 5.6% compared with the corresponding period of 2004. Over the same period, total production (primary and secondary) was 3,973,000 tons: 4.1% higher than in January–March 2003. As a result, there was a deficit of 59,000 tons in world copper supplies for the first three months of 2005. Identified stocks of refined copper also declined, and at the end of March they stood at 887,000 tons.

There is no international agreement between producers and consumers governing the stabilization of supplies and prices. Although most of the world's supply of primary and secondary copper is traded direct between producers and consumers, prices quoted on the London Metal Exchange (LME) and the New York Commodity Exchange (COMEX) provide the principal price-setting mechanism for world copper trading.

**Export Price Index for Copper**
(base: 1980 = 100)

|  | Average | Highest month(s) | Lowest month(s) |
|---|---|---|---|
| 1990 | 123 | — | — |
| 1995 | 135 | — | — |
| 2000 | 86 | 94 (Sept.) | 80 (April) |
| 2001 | 75 | 86 (Jan.) | 66 (Oct.) |
| 2002 | 74 | 79 (June) | 70 (Aug., Sept.) |

On the LME the price of Grade 'A' copper (minimum purity 99.95%) per metric ton declined from £1,563.5 (US $2,219) in February 1993 to £1,108.5 ($1,746) in May. From 1 July 1993 the LME replaced sterling by US dollars as the basis for pricing its copper contract. In September the London copper quotation increased to $2,011.5 (£1,304) per ton, but in October, with LME stocks of copper at a 15-year 'high', the price slumped to $1,596 (£1,079). The copper price subsequently revived, with the LME quotation exceeding $1,800 per ton by the end of the year. The market remained buoyant in January 1994, although, during that month, copper stocks in LME warehouses reached 617,800 tons, their highest level since February 1978. However, stocks were quickly reduced, and the London copper price moved above $2,000 per ton in May 1994. It continued to rise, reaching $2,533.5 (£1,635) per ton in July. In December, with LME stocks of copper below 300,000 tons, the London price of the metal exceeded $3,000 per ton for the first time in more than five years).

In January 1995 the London copper price reached US $3,055.5 (£1,939) per metric ton, but in May it fell to $2,721.5 (£1,728), although LME stocks were then less than 200,000 tons. However, in July, with copper stocks reduced to about 141,000 tons, the LME price advanced to $3,216 (£2,009) per ton, its highest level, in terms of US currency, since early 1989.

The LME's holdings of copper rose to 356,800 metric tons in February 1996, when the price of the metal eased to US $2,492.5 (£1,609) per ton. After increasing again, the copper price fell in April to $2,479.5 (£1,624) per ton, although in early May it recovered to $2,847.5 (£1,872). In late May and June the copper market was gravely perturbed by reports that the world's largest copper-trading company, Sumitomo Corporation of Japan, had transferred, and later dismissed, its principal trader, following revelations that he had incurred estimated losses of $1,800m. in unauthorized dealings (allegedly to maintain copper prices at artificially high levels) on international markets over a 10-year period. This news led to widespread selling of copper: in late June the LME price was reduced to $1,837.5 (£1,192) per ton, although it quickly moved above $2,000 again. In July, after LME stocks had declined to 224,100 tons, the copper price reached $2,102.5 (£1,352) per ton. The London price of copper advanced in November to $2,547.5 (£1,522) per ton, following a decline in LME stocks of the metal to 90,050 tons, their lowest level since July 1990. The copper price at the end of 1996 was $2,217 (£1,304) per ton. Meanwhile, the extent of the losses incurred in the Sumitomo scandal was revised to $2,600m.

In January 1997 the London quotation for high-grade copper reached US $2,575.5 (£1,594) per metric ton, despite a steady increase in LME stocks, which rose to 222,500 tons in February. In June, with copper stocks reduced to 121,550 tons, the price of the metal reached $2,709.5 (£1,644) per ton. However, LME stocks were soon replenished, rising by 82% (to about 235,000 tons) in July and exceeding 300,000 tons in September. The LME's holdings of copper increased to more than 340,000 tons in October. By the end of that month the London price of copper had fallen below $2,000 per ton, and in late December it stood at $1,696.5 (£1,015). At the end of the year the copper price, in US currency, was 23.4% lower than it had been at the beginning. For 1997 as a whole, however, the average price per ton ($2,276) was only slightly less than in the previous year ($2,294).

The copper market remained depressed in the early weeks of 1998. In February the LME's copper stocks reached 379,325 metric tons (their highest level since June 1994), while the London price of the metal was reduced to US $1,601.5 (£973) per ton. The copper price recovered to $1,878 (£1,122) per ton in April. LME stocks of copper decreased to less than 247,000 tons in July, and in that month the price advanced from $1,571 (£963) per ton to $1,756.5 (£1,058). London copper prices remained within this range until late October, when LME stocks exceeded 450,000 tons. At the end of November the COMEX price of high-grade copper for short-term delivery fell below 70 US cents per lb for the first time since 1987. In December 1998, with LME stocks amounting to about 550,000 tons, the London price of copper was reduced to $1,437.5 (£851) per ton. The average London copper price for the year was $1,653 per ton: 27% lower than in 1997 and, in real terms, the lowest annual price level since 1935. The decline in copper prices was attributed to a reduction in imports by some Asian countries, affected by severe financial and economic problems.

In late January 1999 the LME's stocks of copper surpassed the previous record of 645,300 metric tons, established in January 1978. In March 1999, with these holdings standing at more than 700,000 tons, the London price of copper fell to US $1,351.5 (£832) per ton: its lowest level, in terms of dollars, since 1987. In the same month the COMEX price slumped to only 61 cents per lb. Despite a continuing rise in stocks, the London copper quotation increased in May 1999 to $1,581.5 (£964) per ton, although later in the month, when the LME's holdings reached 776,375 tons, the price retreated to $1,354.5 (£845). Copper prices subsequently recovered, and in July, following reports of proposed reductions in refinery output, the London quotation reached $1,689.5 (£1,085) per ton. In August the LME's copper stocks increased to 795,375 tons, but in September the price exceeded $1,700 per ton. The advance in copper prices continued, and at the end of the year the London quotation was $1,854 (£1,150) per ton. For 1999 as a whole, however, the average London price of copper was $1,572 per ton: 4.9% lower than in 1998 and the lowest annual level, in real terms, for more than 60 years.

On 21 January 2000 the LME's stocks of copper exceeded 800,000 metric tons for the first time. Nevertheless, on the same day, the London price of the metal rose to US $1,893.5 (£1,147) per ton: its highest level for more than two years. The accumulation of stocks continued, and in early March the LME's holdings reached a record 842,975 tons. The copper price was reduced to $1,619 (£1,021) per ton in April, but recovered to $1,859 (£1,228) in July. Throughout this period the LME's stocks of copper steadily declined, falling to less than 700,000 tons in April, under 600,000 tons in June and below 500,000 tons in July. At the end of July copper stocks were 487,750 tons: less than 58% of the level reached in March. By the end of August the LME's holdings had fallen further, to 449,050 tons, while the price of copper had continued to recover, reaching $1,900.0 per ton on 31 August. On 13 September the price rose above $2,000 per ton, and closed, on the following day, at $2,009.0 per ton. The LME's holdings of copper fell steadily throughout September, and on 2 October stood at 399,300 tons. On 3 October the London price of the metal closed at $1,919.0 per ton. During the remainder of October the cash price for copper weakened somewhat, falling to $1,810.0 per ton on 27 October, and closing on the final day of the month at $1,839.0 per ton. During November the LME's holdings of copper and the London quotation declined in tandem, the cash price for the metal closing at a 'low' of $1,759.0 per ton on 23 November, but recovering to $1,820.5 per ton on 28 November. On the final day of the month the LME's stocks of copper stood at 349,300 tons. By the end of December there had been little change in either the London cash price or the level of stocks held by the LME. The London quotation for the metal closed at a 'high' of $1,903.5 per ton on 11 December, but had fallen back to $1,808.5 per ton on the final trading day of 2000, by which time the LME's copper holdings had risen slightly, to 357,225 tons.

During January 2001 the London quotation for copper moved within a range of US $1,720.5 per metric ton, recorded on 3 January, and $1,837.0 per ton, the closing price on 23 January. Stocks of the metal, meanwhile, rose as high as 370,225 tons on 10 January, but had declined to 349,825 tons by the end of the month. The LME's holdings of the metal declined steadily during February, and stood at 327,900 tons on the final day of the month. At the same time, the price of copper weakened, falling to $1,736.0 per ton on 26 February. On 6 March, when the LME's holdings of copper declined to 322,775 tons, the London quotation for the metal rose to $1,822.5 per ton. From 7 March, however, stocks recovered steadily, reaching 400,325 tons on 30 March, while the price of copper fell to $1,664.5 per ton on 29 March. By 29 June the price of the metal had fallen to $1,550.5 per ton, while the LME's stocks of copper had recovered to 464,550 tons. The price of copper fell steadily during July, closing at $1,465.0 per ton on 25 July. The LME's holdings of the metal increased rapidly during July, reaching 654,325 tons on 25 July, and 673,225 tons on 10 August, before declining during the remainder of the month, to 662,825 tons on 30 August. The London quotation for copper closed at a 'low' for the month of $1,426.0 per ton on 14 August, but had recovered to $1,507.0 per ton on the final trading day of the month. During September, however, the price declined sharply, descending to $1,403.0 per ton on 20 September, as stocks continued to rise, reaching 707,100 tons on 24 September. By the end of October the London price of copper stood at only $1,360 per ton, and closed at only $1,319.0 per ton on 7 November. The price of copper 'futures' was boosted in late October by the announcement by US producer Phelps Dodge of its intention to reduce output of the metal by some 220,000 tons per year, in an attempt to stem losses incurred as a result of the price of the metal falling below the price of production. Stocks rose as high as 780,225 tons on 30 November, and, on the final trading day of 2001, stood at 799,225 tons, by which time the London price of copper had recovered to $1,462.0 per ton.

During the second week of January 2002 the London quotation for high-grade copper recovered to US $1,543 per metric ton, and remained above $1,500 per ton for most of the remainder of the month. The LME's holdings of copper continued to rise during January, reaching 857,675 tons by the end of the month. On 1 February the price of copper rose as high as $1,588 per ton and two weeks later closed at $1,609.5 per ton, the first time it had surpassed $1,600 per ton since June 2001. For the remainder of February the price of copper remained comfortably above $1,500 per ton, reaching $1,610.5 per ton at the end of the first week of March. By mid-February stocks of copper had risen as high as 882,425 tons, and on 27 February rose above 900,000 tons. On 19 March the London quotation for copper closed at $1,650.5 per ton, but from the second week of April the price began to weaken, falling as low as $1,551 on 15 April. Stocks rose during March and continued to rise during April, reaching 973,550 tons by the end of the month. During the final week of April, nevertheless, the London quotation for copper was either above or a little below $1,600 per ton. During May the quotation ranged between $1,562 (7 May) and $1,645 per ton, the higher price recorded on the penultimate day of the month. The rate of increase in the LME's holdings of copper slackened slightly during May, but stocks nevertheless remained substantially above 900,000 tons for the whole of the month. Throughout the whole of June the price of copper remained above $1,600 per ton, closing at $1,689.5 on 6 June, falling to $1,614.5 on 25 June and ending the month at $1,654 per ton. By 21 June the LME's stocks of copper stood at 898,375 tons. On 15 July 2002 the London price of copper fell to US $1,606 per ton, declining to $1,597 per ton on 18 July and to $1,491 per ton on 24 July. Stocks of copper fell to 874,075 tons on 16 July, but by the end of the month were once again approaching 900,000 tons. For almost the whole of August the London quotation remained below $1,500 per ton, ending the month at $1,500 per ton exactly. Stocks reached 896,675 tons on 21 August and ended the month at 896,425 tons. By the end of

September the London quotation had fallen to only $1,434.5 per ton. Stocks fell somewhat during the first three weeks of September, declining to 870,375 tons by the end of the month. On 18 October the price of copper rose above $1,500 per ton and remained above that level for the rest of the month, closing at $1,536 per ton on 31 October. Stocks continued to decline throughout October, falling to 860,000 tons on 15 October. The LME's holdings of copper were stable during most of November, ending the month at 862,550 tons. On 20 November the London price of copper closed at $1,600.5 per ton, rising further, to $1,626.5 per ton, on 29 November. During December the quotation ranged between $1,536 per ton, recorded on the final day of 2002, and $1,649.5, recorded on 2 December. Stocks were little changed at the end of the year, compared with one month previously, totalling 855,625 tons.

The London quotation for high-grade copper rose substantially during January 2003, reaching US $1,713.5 per metric ton on the final day of that month, by which time the LME's holdings of copper had declined to 833,225 tons. During February the price of copper ranged between $1,627.5 (19 February) and $1,728 (3 and 27 February). Stocks had fallen to 825,650 tons by the end of February and fell a little further during March, ending that month at 812,950 tons. By the end of March the London quotation for copper had weakened to $1,587.5 per ton. The price of copper rose above $1,600 per ton again on 15 April, ending the month at $1,604 per ton. On 10 April the LME's holdings of copper fell below 800,000 tons and stood at 768,075 tons by the end of the month. By late May the London quotation was once again approaching $1,700 per ton, and on 2 June closed at $1,705. A sudden increase in the LME's copper stocks in early May subsequently gave way to a decline, and by the end of the month holdings totalled 740,600 tons. On 11 June stocks fell below 700,000 tons and had descended to 665,650 tons by the end of the month. For the whole of June the London price of copper ranged between $1,657.5 (26 June) and $1,711.5 (3 June).

During the second half of July 2003, generally, the London quotation for high-grade copper increased, closing at US $1,781 per metric ton on the final trading day of that month. On 1 August the London cash price rose to $1,824.5 per ton, but thereafter, during the remainder of the month, declined somewhat. In September the price of copper ranged between $1,815 per ton (24 September) and $1,759 per ton (1 September). During October the London quotation rose consistently, ending the month at $2,057 per ton, and remaining above $2,000 per ton until 20 November, when it declined to $1,996 per ton. On the final trading day of November a price of $2,073.5 was recorded. During December the price continued to rise, reaching $2,321 per ton on the final trading day of 2003. Stocks of copper held by the LME declined steadily in the second half of 2003, from 612,425 tons at the end of July to 430,525 at the end of the year.

For most of January and February 2004 the London quotation for high-grade copper remained above US $2,400 per metric ton, and on 25 February it rose above $3,000 per ton (to $3,001) for the first time in the 2000s. The London price remained above $3,000 per ton during most of March, declining marginally in April, when it ranged between $3,170 per ton (19 April) and $2,929.5 per ton (13 April). For the whole of May the London quotation remained below $3,000 per ton. In June the London quotation ranged between $2,878 per ton (2 June) and $2,554 per ton (15 June). During the first half of 2004 stocks of copper held by the LME declined steadily, from slightly less than 400,000 on 16 January to 101,475 tons on 30 June. For the whole of 2004 the average price of copper traded on the London exchange was $2,868 per ton, more than 60% higher than the average price recorded in 2003. Stocks of copper held by the LME declined steadily and substantially during the second half of the year, to total 48,875 tons at 31 December. Demand for copper, both for physical metal and as an investment asset, increased markedly in 2004. In the physical sphere, as with other base metals in 2004, strong demand from the People's Republic of China was a key market factor.

During the first half of 2005 the price of copper traded on the LME ranged between US $3,072 per metric ton (5 January) and $3,670 per ton (17 and 24 June). Stocks of the metal held by the Exchange had risen to 61,000 tons by the end of April, but declined to less than 30,000 tons by the end of June, their lowest level for more than 30 years. Analysts forecast a decline in the price of copper in the second half of 2005 as increased output world-wide began to replenish inventories and return markets to surplus.

The ICSG, initially comprising 18 producing and importing countries, was formed in 1992 to compile and publish statistical information and to provide an intergovernmental forum on copper. In 2005 ICSG members and observers totalled 22 countries, plus the EU, accounting for more than 80% of world trade in copper. The ICSG, which is based in Lisbon, Portugal, does not participate in trade or exercise any form of intervention in the market.

## Iron Ore

The main economic iron-ore minerals are magnetite and haematite, which are used almost exclusively to produce pig-iron and direct-reduced iron (DRI). These comprise the principal raw materials for the production of crude steel. Most iron ore is processed after mining to improve its chemical and physical characteristics and is often agglomerated by pelletizing or sintering. The transformation of the ore into pig-iron is achieved through reduction by coke in blast furnaces; the proportion of ore to pig-iron yielded is usually about 1.5:1 or 1.6:1. Pig-iron is used to make cast iron and wrought iron products, but most of it is converted into steel by removing most of the carbon content. Particular grades of steel (e.g. stainless) are made by the addition of ferro-alloys such as chromium, nickel and manganese. In the late 1990s processing technology was being developed in the use of high-grade ore to produce DRI, which, unlike the iron used for traditional blast furnace operations, requires no melting or refining. The DRI process, which is based on the use of natural gas, has expanded rapidly in Venezuela, but, owing to technological limitations, is not expected within the foreseeable future to replace more than a small proportion of the world's traditional blast-furnace output. In 1998 Mexico overtook Venezuela as the leading producer of DRI.

Iron is, after aluminium, the second most abundant metallic element in the earth's crust, and its ore volume production is far greater than that of any other metal. Some ores contain 70% iron, while a grade of only 25% is commercially exploitable in certain areas. As the basic feedstock for the production of steel, iron ore is a major raw material in the world economy and in international trade. Because mining the ore usually involves substantial long-term investment, about 60% of trade is conducted under long-term contracts, and the mine investments are financed with some financial participation from consumers.

Iron ore is widely distributed throughout Latin America. Brazil is by far the dominant producer in the region, and its ore is of a high quality (68% iron). In terms of iron content, Brazil is the largest producer in the world, accounting for some 22% of estimated global output in 2003. Brazil's giant open-cast mine in the Serra do Carajás began production in 1986. The project was developed by the Companhia Vale do Rio Doce (CVRD), which also operates several other mines in the state of Minas Gerais. The Carajás deposit has been stated to have 18,000m. metric tons of high-grade reserves (66% iron). Improved rail links have facilitated the transportation of iron ore for export from Carajás, in the country's interior. Following the construction of the Igarapava dam, in north-east Minas Gerais, the Carajás complex was expected to become partially self-sufficient in energy. In early 2004 CVRD planned to begin to increase annual production capacity at its Carajás facility by 14m. tons, to 70m. tons, at a cost of US $144m. At the same time, as the result of strong demand for iron ore, CVRD planned to develop new mines at Fabrica Nova—in a joint venture with Japan's JFE Steel Corporation—and, from 2006, at Brucutu. The new mines were to be complemented by an expansion of export capacity at the ports of Ponta da Madeira and Tubarão. In early 2005 the Government was reported to be preparing to authorize Rio Tinto, an Anglo-Australian mining group, to proceed with the expansion of its iron ore mine at Corumbá, in south-western Brazil, near the border with Bolivia. The cost of the expansion, whose objective would be to raise production from 1.3m. tons in 2004 to 15m. tons by 2009, was estimated at US $1,000m. Rio Tinto's participation in the proposed construction of a smelter and steel plant close to the mine was also reportedly under consideration. Iron ore is the main mineral export commodity in Brazil, which is the major supplier to the world market after

## REGIONAL INFORMATION

*Major Commodities of Latin America*

Australia. Sales of iron ore and concentrates provided 5% of Brazil's total export earnings in 2001. Brazil also has a major iron and steel industry, accounting for a further 5.5% of export revenue in 2001.

**Production of Iron Ore**
(iron content, '000 metric tons)

|  | 2002 | 2003* |
|---|---|---|
| World total | 614,774 | 640,754 |
| Latin America | 164,934 | 166,963 |
| **Leading Latin American producers** | | |
| Brazil | 140,000 | 140,000 |
| Chile | 4,398 | 4,420 |
| Mexico | 5,943 | 6,747 |
| Peru | 3,105 | 3,542 |
| Venezuela | 11,100 | 11,900 |
| **Other leading producers** | | |
| Australia | 113,548 | 116,355 |
| Canada† | 19,820 | 19,800 |
| China, People's Repub. | 76,200* | 83,200 |
| India | 60,300* | 67,500 |
| Russia | 49,000 | 53,000 |
| Ukraine | 32,300 | 34,300 |
| USA | 32,499 | 29,286 |

* Estimated production.
† Including the metal content of by-product ore.

Source: US Geological Survey.

Venezuela has proven reserves of more than 4,000m. metric tons. About one-half of total proven reserves is classified as high-grade ore. Iron ore has been overtaken by aluminium, however, as the second most important export industry, after petroleum. In the early 1990s there was installed capacity for the production of as much as 20m. tons annually, but, generally, output has been declining since 1997. In 2003, however, production of iron ore by Ferrominera Orinoco, the country's sole producer, reached its highest level since 1997. Ferrominera reportedly planned to increase output by more than 50% from its 2004 level by 2009, and to invest some US $460m. in a concentration plant that was scheduled for completion in 2007. Sales of iron ore and concentrates provided only 0.3% of Venezuela's total export earnings in 2000. Sales of iron and steel, however, provided 2.7% of total export revenue in that year, and 3.1% in 2001.

In Mexico construction of the Tehuantepec steel project, in the state of Oaxaca, began in 1999. The scheme, which was originally scheduled for completion in April 2001, at an estimated total cost of US $1,850m., involves the development of the country's largest iron-ore mine, with proven reserves of about 180m. metric tons (averaging 55% iron). At the planning stage it was envisaged that the mine would have an annual output of about 6m. tons of iron ore, while the associated steel plant, using DRI technology, would produce about 4m. tons of steel per year.

In 2003, according to data from the International Iron and Steel Institute (IISI), world exports of iron ore totalled about 586m. metric tons, compared with some 531m. tons in 2002. Australia and Brazil were by far the two dominant exporting countries in 2003, jointly accounting for more than 63% of global iron ore shipments. India and Canada, which jointly accounted for a further 14% of world iron ore exports in 2003, were also significant exporters. There has been tremendous growth in imports of iron ore by the People's Republic of China in recent years. In 2003, as the destination for shipments totalling some 148m. tons—compared with about 70m. tons in 1999—China was the world's leading importer, followed by Japan (132m. tons), the Republic of Korea (41m. tons) and Germany (34m. tons). In 2003 imports of iron ore world-wide totalled about 583m. tons, compared with some 531m. tons in 2002. World iron-ore reference prices are decided annually at a series of meetings between producers and purchasers (the steel industry accounts for about 95% of all iron ore consumption). The USA and the republics of the former USSR, although major steel producers, rely on domestic ore production and take little part in the price negotiations. It is generally accepted that, because of its diversity in form and quality, iron ore is ill-suited to price stabilization through an international buffer stock arrangement.

**Export Price Index for Iron Ore**
(base: 1980 = 100)

|  | Average | Highest month(s) | Lowest month(s) |
|---|---|---|---|
| 1990 | 124 | — | — |
| 1995 | 140 | — | — |
| 2000 | 115 | 126 (Jan.) | 107 (Oct., Nov.) |
| 2001 | 111 | 117 (Jan.) | 106 (June, July) |
| 2002 | 117 | 126 (Dec.) | 108 (Feb.) |

In early 2005 Rio Tinto announced that its subsidiary, Hamersley Iron, had negotiated an increase of 71.5% in the f.o.b. price paid for all iron ores with Nippon Steel of Japan in the (Japanese) financial year commencing 1 April 2005. The reference price of fine ore, which accounts for some 60% of the traded iron ore market, increased from US $35.99 per dry metric ton (dmt) to a record nominal price of $61.72 per dmt. Brazil's CVRD was reported to have negotiated a comparable increase with its Chinese, European and Japanese customers, as was BHP Billiton. The increase in the price of iron ore has been attributed to growth in demand for steel arising from China's ongoing industrialization—the world indicator price of hot rolled coil steel increased from $247 per ton in December 2001 to $648 per ton in December 2004. According to the IMF, the average price of Brazilian ores (Carajás fines, 67.55% Fe content, contract price to Europe—f.o.b. Ponta da Madeira, CVRD, Rio de Janeiro) in 2004 was 38 US cents per dmt unit, compared with 32 cents per dmt unit in 2003 and 29 cents in 2002. The corresponding prices in the first two quarters of 2005 were, respectively, 47 cents per dmt unit and 65 cents per dmt unit.

The Association of Iron Ore Exporting Countries (Association des Pays Exportateurs de Minerai de Fer—APEF) was established in 1975 to promote close co-operation among members, to safeguard their interests as iron ore exporters, to ensure the orderly growth of international trade in iron ore and to secure 'fair and remunerative' returns from its exploitation, processing and marketing. In 1995 APEF, which also collects and disseminates information on iron ore from its secretariat in Geneva, Switzerland, had nine members, including Peru and Venezuela but not Brazil. The UN Conference on Trade and Development (UNCTAD) compiles statistics on iron ore production and trade, and in recent years has sought to establish a permanent international forum for discussion of issues affecting the industry.

**Maize** (Indian Corn, Mealies) (*Zea mays*)

Maize is one of the world's three principal cereal crops, with wheat and rice. The main varieties are dent maize (which has large, soft, flat grains) and flint maize (which has round, hard grains). Dent maize is the predominant type world-wide, but flint maize is widely grown in southern Africa and parts of South America. Maize may be white or yellow (there is little nutritional difference) but the former is preferred for human consumption. Native to the Americas, maize was brought to Europe by Columbus and has since been dispersed to many parts of the world. It is an annual crop, planted from seed, and matures within three to five months. It requires a warm climate and ample water supplies during the growing season. Genetically modified varieties of maize, with improved resistance to pests, are now being cultivated, particularly in the USA and also in Argentina and the People's Republic of China. However, further development of genetically modified maize may be slowed by consumer resistance in importing countries and doubts about its possible environmental impact.

Maize is an important food source in regions such as sub-Saharan Africa and the tropical zones of Latin America, where the climate precludes the extensive cultivation of other cereals. It is, however, inferior in nutritive value to wheat, being especially deficient in lysine, and tends to be replaced by wheat in diets when the opportunity arises. As food for human consumption, the grain is ground into meal, or it can be made into (unleavened) corn bread and breakfast cereals. In Latin America maize meal is made into cakes, called 'tortillas'. Maize is also the source of an oil used in cooking.

The high starch content of maize makes it highly suitable as a compound feed ingredient, especially for pigs and poultry. Animal feeding is the main use of maize in the USA, Europe and Japan, and large amounts are also used for feed in developing countries in Far East Asia, Latin America and, to some extent, in North Africa. Maize has a large variety of industrial uses, including the preparation of ethyl alcohol (ethanol), which may be added to petrol to produce a blended motor fuel. Maize is also a source of dextrose and fructose, which can be used as artificial sweeteners, many times sweeter than sugar. The amounts of maize used for these purposes depend, critically, on its price to the users relative to those of petroleum, sugar and other potential raw materials. Maize cobs, previously regarded as a waste product, may be used as feedstock to produce various chemicals (e.g. acetic acid and formic acid).

In recent years world production has averaged about 630m. metric tons annually. The USA is by far the largest producer, with annual harvests of, on average, about 256m. tons in recent years. In years of drought or excessive heat, however, US output can fall significantly: in 1995, for example, the maize crop totalled only 188m. tons. China, whose maize output has been expanding rapidly, is the second largest producer; its harvest has averaged about 118m. tons in recent years.

Maize production in Latin America increased rapidly from the early 1990s, stabilized in the second half of that decade and began to rise again in the early 2000s. In 2004 production was estimated to have totalled some 89m. metric tons, compared with about 94m. tons in 2003. In most years, three countries—Argentina, Brazil and Mexico—between them account for a clear majority of Latin America's annual maize crop (about 87% in 2004). Brazil is the largest producer, with output totalling an estimated 41.9m. tons in 2004. In the north and north-east of the country, which account for about 10% of Brazil's production, maize is an important food for human consumption and is largely grown by subsistence farmers. The droughts to which these regions are prone can occasion considerable hardship. Most of Brazil's maize is grown commercially for use as animal feed in the centre and south of the country. Production there varies according to the amount of government support (mostly in the form of subsidized credit) and the prices of alternative crops, especially oilseeds. Although Brazil's animal feed requirements are growing rapidly, many of the feed mills are located in the far south of the country, where maize supplies may be obtained more cheaply from Argentina than from domestic sources. Brazil's maize imports totalled 797,670 tons in 2003, compared with 345,256 tons in 2002.

Mexico is usually Latin America's second largest maize producer, with harvests averaging about 19m. metric tons annually in recent years. Much of the Mexican crop consists of white corn rather than the yellow variety. The establishment of CIMMYT (Centro Internacional del Mejoramiento de Maíz y Trigo—the International Wheat and Maize Improvement Centre) at Sonora, in northern Mexico, has made the country the testing-ground for many of the technical advances in the development of different maize varieties since the 1960s. Local production, however, has been hampered by the small size of most agricultural holdings, competition for irrigated land from other crops, and the inability of small producers to afford enough fertilizers. Shortages of irrigation water commonly result in poor crops. Maize in Mexico is mainly used for human consumption, particularly in the form of tortillas. Domestic maize is mostly reserved for this purpose, animal feed manufacturers traditionally preferring to use sorghum and, increasingly, yellow corn imported from the USA.

Recent policy changes will affect maize production and consumption trends in Mexico. The Compañía Nacional de Subsistencias Populares (CONASUPO), the parastatal food distribution company, no longer directly supports the maize market, while controls on tortilla prices have been lifted. Competition between food manufacturers for local maize will benefit the larger commercial farmers and help them finance productivity improvements, but many smaller producers will probably turn to other crops. Under the North American Free Trade Agreement (NAFTA), Mexico's grain market is increasingly being opened to imports from the USA. In 2003 Mexico's imports of maize totalled about 5.8m. metric tons.

Argentina is Latin America's only substantial maize exporting country, and, in world terms, ranks second only to the USA in amounts sold. Market liberalization in the early 1990s, particularly the abolition of export taxes, encouraged maize production. Farmers were able to plan their activities more rationally, and make longer-term investments in land improvement and up-to-date equipment. At the same time, privatization and decontrol of the ports and transport systems resulted in much greater efficiency in grain movement. However, farmers are no longer shielded from international price trends, and the low prices of the late 1990s caused many to switch from maize to oilseeds or other more profitable crops. Argentina's maize production was estimated at 15m. metric tons in 2004, equal to output in 2003 and 2002. Commercial plantings of genetically modified maize started in 1999. In 2003 Argentina exported an estimated 12m. tons of maize.

Maize is grown widely as a subsistence crop in Central America. It is one of the most important foods in El Salvador and Guatemala, where consumption per head is around 100 kg per year. In South America food use is generally declining, although it remains important in some countries, notably Bolivia, Colombia, Paraguay and Venezuela, in each of which consumption averages 40 kg–50 kg per year.

**Production of Maize**
('000 metric tons)

|  | 2003 | 2004* |
|---|---|---|
| **World total** | 641,269 | 721,379 |
| Latin America | 94,382 | 88,983 |
| **Leading Latin American producers** | | |
| Argentina | 15,040 | 15,000 |
| Brazil | 47,988 | 41,864 |
| Chile | 1,190 | 1,321 |
| Colombia | 1,209 | 1,458† |
| Guatemala† | 1,054 | 1,072 |
| Mexico | 19,652 | 20,000† |
| Paraguay | 1,056 | 1,120 |
| Peru | 1,357 | 1,181 |
| Venezuela | 1,823 | 2,068 |
| **Other leading producers** | | |
| Canada | 9,587 | 8,836 |
| China, People's Repub. | 115,998 | 132,160 |
| France | 11,991 | 16,391 |
| India | 14,720 | 14,000† |
| Italy | 8,702 | 10,983 |
| Romania | 9,577 | 14,542 |
| USA | 256,905 | 299,917 |

\* Provisional figures.
† Unofficial figure(s).

World trade in maize totalled about 88m. metric tons in 2003, compared with some 85m. tons in 2002, and around 79m. tons in 2001. In the mid-1990s growth in trade was curtailed by adverse economic conditions in eastern Asia, which caused a decline in the region's meat consumption and, consequently, in its demand for animal feed. Recovery in the Asian economies, together with growing demand from Latin America and North Africa, subsequently restored that demand.

The pre-eminent world maize exporter is the USA, which, with exports estimated at about 43m. metric tons, accounted for some 49% of total world exports of maize in 2003. In recent years, however, the USA's dominance of world trade in maize has begun to be eroded by increased competition from Argentina, China and very low-priced supplies from a number of countries in Central and Eastern Europe. US exports in 2000–03 averaged about 47m. tons annually.

Argentina's exports fell to only around 3m. metric tons per year in the late 1980s, but grew rapidly in the 1990s. In 2003 they amounted to 11.9m. tons, compared with some 9.5m. tons in 2002, and 10.9m. tons in 2001. Argentina's main markets are in Latin America, particularly Brazil, Chile and Peru, but it also regularly makes large sales to Japan and other importers in Asia. The world's principal maize importer is Japan. Its trade has shown little growth in recent years, however, as the domestic livestock industry has been rationalized to compete with imported meat. Japanese imports of maize totalled about 17.1m. tons in 2003, compared with about 16.4m. tons in 2002 and 16.2m. tons in 2001. Rapidly growing livestock industries elsewhere in eastern Asia have made the region the major world market for maize. Feed users in the Republic of Korea (South Korea) are willing to substitute other grains for maize, particularly feed wheat, when prices are attractive, so that maize imports are variable, averaging about 8.8m. tons per year in

2000–03. Taiwan regularly imports about 5m. tons of maize annually, but imports by other countries in the region, notably Indonesia, Malaysia and the Philippines, have fluctuated considerably in recent years, as a result of economic instability. In the 1980s the USSR was a major market, but the livestock industries in its successor republics declined very sharply during the 1990s, greatly reducing feed needs. Maize imports by Latin America grew considerably during the 1990s. In 2003 the region accounted for some 17% of world maize imports, compared with less than 10% in the late 1980s. Latin America's biggest importer is Mexico, whose annual purchases averaged 5.7m. tons in 2000–03. Brazil, Chile, Colombia, Peru and Venezuela also regularly import substantial quantities.

Massive levels of carry-over stocks of maize were accumulated in the USA during the mid-1980s, reaching a high point of 124m. metric tons in 1987. Government support programmes were successful in discouraging surplus production, but several poor harvests also contributed to the depletion of these stocks, which were reduced to only 11m. tons at the close of the 1995/96 marketing year. A succession of good crops in the late 1990s, and increased competition from Argentina and China, led to a substantial rebuilding of US maize stocks, despite growth in domestic feed requirements associated with the strong economy. Carry-overs reached 48.2m. tons at the end of the 2000/01 marketing year.

Export prices of maize are mainly influenced by the level of supplies in the USA, and the intensity of competition between the exporting countries. Record quotations were achieved in April 1996, when the price of US No. 2 Yellow Corn (f.o.b. Gulf ports) reached US $210 per metric ton. For the whole of 1996 the price averaged $165.1 per ton. In 1997 an average price of $117.2 per ton was recorded, but this declined in 1998 to only $102.0 per ton, and again, in 1999, to only $91.7 per ton. In 2000 an average price of $88.4 per ton was recorded, rising to $89.6 per ton in 2001, $99.2 per ton in 2002 and $105.2 per ton in 2003. The upward trend continued in 2004, when an average price of $111.7 per ton was recorded. During the first six months of 2005 the average monthly price of US No. 2 Yellow Corn (f.o.b. Gulf ports) ranged between $95.2 per ton (January) and $100.2 (March).

**Export Price Index for Maize**
(base: 1980 = 100)

|  | Average | Highest month(s) | Lowest month(s) |
|---|---|---|---|
| 1990 | 92 | — | — |
| 1995 | 80 | — | — |
| 2000 | 54 | 59 (Jan.–April) | 47 (Sept.) |
| 2001 | 59 | 63 (Jan.) | 56 (June) |
| 2002 | 67 | 80 (Sept.) | 56 (April) |

## Petroleum

Crude oils, from which petroleum is derived, consist essentially of a wide range of hydrocarbon molecules which are separated by distillation in the refining process. Refined oil is treated in different ways to make the different varieties of fuel. More than four-fifths of total world oil supplies are used as fuel for the production of energy in the form of power or heating.

Petroleum, together with its associated mineral fuel, natural gas, is extracted both from onshore and offshore wells in many areas of the world. It is the leading raw material in international trade. World-wide, demand for this commodity totalled 80.8m. barrels per day (b/d) in 2004, an increase of 3.1% compared with the previous year. The world's 'published proven' reserves of petroleum and natural gas liquids at 31 December 2004 were estimated to total 161,900m. metric tons, equivalent to about 1,188,600m. barrels (1 metric ton is equivalent to approximately 7.3 barrels, each of 42 US gallons or 34.97 imperial gallons, i.e. 159 litres). The dominant producing region is the Middle East, whose proven reserves in December 2004 accounted for 61.8% of known world deposits of crude petroleum and natural gas liquids. The Middle East accounted for 30.7% of world output in 2004. Latin America contained 16,400m. tons of proven reserves (10.1% of the world total) at the end of 2004, and accounted for 13.8% of world production in that year.

From storage tanks at the oilfield wellhead, crude petroleum is conveyed, frequently by pumping for long distances through large pipelines, to coastal depots where it is either treated in a refinery or delivered into bulk storage tanks for subsequent shipment for refining overseas. In addition to pipeline transportation of crude petroleum and refined products, natural (petroleum) gas is, in some areas, also transported through networks of pipelines. Crude petroleum varies considerably in colour and viscosity, and these variations are a determinant both of price and of end-use after refining.

In the refining process, crude petroleum is heated until vaporized. The vapours are then separately condensed, according to their molecular properties, passed through airless steel tubes and pumped into the lower section of a high, cylindrical tower, as a hot mixture of vapours and liquid. The heavy unvaporized liquid flows out at the base of the tower as a 'residue' from which is obtained heavy fuel and bitumen. The vapours passing upwards then undergo a series of condensation processes that produce 'distillates', which form the basis of the various petroleum products.

The most important of these products is fuel oil, composed of heavy distillates and residues, which is used to produce heating and power for industrial purposes. Products in the kerosene group have a wide number of applications, ranging from heating fuels to the powering of aviation gas turbine engines. Gasoline (petrol) products fuel internal combustion engines (principally in road motor vehicles), and naphtha, a gasoline distillate, is a commercial solvent that can also be processed as a feedstock. Propane and butane, the main liquefied petroleum gases, have a wide range of industrial applications and are also used for domestic heating and cooking.

Mexico was the world's leading petroleum producer in 1921 but by 1938, when the industry was nationalized (it remains in the control of a government agency, Petróleos Mexicanos—PEMEX), production had fallen dramatically. The discovery of extensive deposits of petroleum in the states of Tabasco and Chiapas, and off shore in the Bay of Campeche, enabled output to increase significantly in the 1970s. Mexico's proven reserves of petroleum stood at 2,000m. metric tons at the end of 2004. Mexico was the world's fifth largest producer of petroleum in 2004 and had the second largest refinery capacity in Latin America, after Brazil. Since the mid-1980s Mexico's petroleum revenues have been affected by lower levels of exports, together with a series of price reductions which were authorized by the Government, and the commodity's contribution to the country's earnings of foreign exchange has, accordingly, reflected the level of prices in international petroleum markets. Mexico's sales of petroleum and its derivatives, which provided 17.2% of foreign earnings in 1992, accounted for 7.0% in 1999, 9.6% in 2000 and 7.9% in 2001. In 2004 the total value of exports of crude oil by PEMEX was US $26,200m. In that year Mexico's heavy crude petroleum, Maya-22, accounted for about 87% of all petroleum exports, for which the USA was the destination of some 80%.

Brazil's proven reserves of petroleum—the second largest in the region after those of Venezuela—were assessed at 1,500m. metric tons at the end of 2004. It is anticipated, too, that Brazilian reserves will rise substantially once Petróleo Brasileiro (PETROBRÁS), the state-owned petroleum company, has fully evaluated recent discoveries in the Campos and Santos Basins. PETROBRÁS operates almost all of the country's petroleum refineries, whose total capacity as of January 2004 was some 2m. b/d, the largest in the region. The company has plans to construct a new refinery and to upgrade existing ones, with the aim of increasing their ability to refine the predominantly heavy crudes that Brazil produces. Although Brazil's production of these crudes has increased steadily in recent years, it remains insufficient to meet domestic demand and has to be supplemented by imports of lighter crudes. Brazil is also an exporter of heavy crudes to, for example, the People's Republic of China. PETROBRÁS's monopoly of the Brazilian petroleum sector was ended in 1997 by the adoption of legislation allowing private-sector investment in all parts of the industry. At the same time, a National Petroleum Agency (Agência National do Petróleo) was established. These and other reforms of the sector were undertaken with the aim of achieving self-sufficiency in petroleum supply by 2006. In mid-2004, however, PETROBRÁS remained by far the country's dominant producer, many potential foreign producers having reportedly been deterred from entering the sector by disappointing results of exploration, a perceived lack of transparency in the sector's regulation and unfavourable licensing awards. Since 2003 Royal Dutch/Shell has been in a partnership with PETROBRÁS to produce oil (and natural gas) in the Campos Basin. In 2004 ChevronTexaco, likewise in partnership with PETROBRÁS, was involved in the development of Brazil's Frade oilfield.

The production of petroleum is the dominant economic activity in Venezuela, although the contribution of petroleum and its derivatives to the country's total export earnings, which had been more than 90% until 1986, declined, to 77% in 1989, before recovering to 80.3% in 1991. The proportion fell to 75.2% in 1994, but rose to 81.0% in 1996. Petroleum's contribution to total export earnings was 80.8% in 1999, 85.6% in 2000 and 82.0% in 2001. In 2003 Venezuela was the fourth largest supplier of petroleum to the USA—more than 90% of the country's petroleum exports were to the USA in that year. In 2004, when Venezuela accounted for 28.8% of Latin American production, it ranked as the world's seventh largest producer. As of mid-2004, however, it still remained unclear to what degree production had been restored, and at what cost, following industrial action by employees of Petróleos de Venezuela, SA (PDVSA), the national oil company, in late 2002 and early 2003. The country's proven reserves were estimated at 11,100m. metric tons at the end of 2004. In early 2004 PDVSA announced plans to invest US $26,000m. in oil (and natural gas) exploration and production activities in 2004–09. PDVSA was nationalized in 1976. In 1992, for the first time since nationalization, exploration and production were opened up to foreign participation. Since 1998, however, the Government of President Hugo Chávez Frías has apparently reversed this policy, while emphasizing that agreements already concluded with international oil companies will be honoured. New hydrocarbons legislation adopted in late 2001 stipulated that PDVSA must take a 51% share in all new production and exploration agreements. Since early 2004 the restructuring of PDVSA, with the aim of making the company more efficient, has been under way. In mid-2004 Venezuela's domestic petroleum refining capacity was assessed at some 1.3m. b/d. This is supplemented by additional capacity overseas, including in the USA and Europe.

**Production of Crude Petroleum**
('000 metric tons, including natural gas liquids)

|  | 2003 | 2004 |
|---|---|---|
| **World total** | 3,702,900 | 3,867,900 |
| Latin America | 510,800 | 532,700 |
| **Leading Latin American producers** | | |
| Argentina | 40,200 | 37,900 |
| Brazil | 77,000 | 76,500 |
| Colombia | 27,900 | 27,300 |
| Ecuador | 21,700 | 27,300 |
| Mexico | 188,800 | 190,700 |
| Peru | 4,500 | 4,400 |
| Trinidad and Tobago | 7,900 | 7,400 |
| Venezuela | 134,900 | 153,500 |
| **Other leading producers** | | |
| Canada | 142,700 | 147,600 |
| China, People's Repub. | 169,600 | 174,500 |
| Iran | 197,900 | 202,600 |
| Iraq | 66,100 | 99,700 |
| Kuwait | 110,200 | 119,800 |
| Nigeria | 110,300 | 122,200 |
| Norway | 153,000 | 149,900 |
| Russia | 421,400 | 458,700 |
| Saudi Arabia | 487,900 | 505,900 |
| United Arab Emirates | 119,600 | 125,800 |
| United Kingdom | 106,100 | 95,400 |
| USA | 338,400 | 329,800 |

Source: BP, *Statistical Review of World Energy 2005*.

Ecuador's petroleum industry has been a significant contributor to the economy since 1972, when petroleum was exported for the first time, after the completion of the 480-km (300-mile) trans-Andean pipeline (capacity 390,000 b/d in 2000), linking the oilfields of Oriente Province with the tanker-loading port of Esmeraldas. Government-owned Petroecuador is the country's most important producer of crude, accounting for more than one-third of all output in 2004. However, private producers, such as Occidental, are beginning to play a greater part in the sector and Petroecuador's production has declined in recent years. As of early 2005 Ecuador's three refineries had a total capacity of some 176,000 b/d. The Government reportedly plans to construct a new refinery and to upgrade the capacity of the three existing ones. In 2001 the construction of a new pipeline, the Oleoducto de Crudos Pesados (OCP), to transport heavy crudes commenced. Costing some US $1,100m. and with a maximum capacity of 450,000 b/d, OCP links oilfields in the Amazon region with port facilities on the Pacific Ocean. The entry into operation of OCP, the country's second pipeline, in September 2003 raised Ecuador's potential oil transportation capacity to 850,000 b/d and had already, by the end of 2003, permitted a significant increase in production. More than 50% of Ecuador's exports of petroleum are destined for the USA, to which it is the second largest South American supplier of crude after Venezuela. In 1993 Ecuador's proven reserves of petroleum were almost tripled, following discoveries in the Amazon region, and in 1994 and 1995 the Government signed contracts with numerous companies for further exploration and drilling (including rights to explore areas in the Eastern Amazon, which had previously been withheld, owing to indigenous and environmental protests). Ecuador's proven published reserves at the end of 2004 amounted to 700m. metric tons. Petroleum and petroleum products accounted for 36.3% of export revenue in 1996 but by 1998 the proportion had declined to 23.6%, reflecting lower prices. Since 2000, however, their contribution has risen steadily, totalling 40.8% in 2002, 43.2% in 2003 and 55.4% in 2004.

The economy of Trinidad and Tobago has relied heavily on the petroleum industry since the 1940s. The industry received a fillip from the discovery of offshore fields in 1955, and by the 1970s petroleum accounted for about 50% of the country's gross domestic product and about 90% of its export income. Until 1998 no significant onshore discoveries had been made since the early 1970s, although there are substantial reserves of natural gas. A series of offshore discoveries in 1998–2000 was reported to constitute the largest discovery of gas and petroleum in the history of the nation, equivalent to 630m. barrels. The petroleum industry remains a vital sector of the economy, accounting, with asphalt, for an estimated 27.1% of gross domestic product in 1997. Mineral fuels and lubricants accounted for 65.3% of the country's export revenue (excluding re-exports) in 2000, 59.9% in 2001, and 60.1% in 2002.

Colombia's importance as a producer of petroleum increased dramatically in 1984, when vast reserves of petroleum were discovered at Caño-Limón, near the Venezuelan border. This deposit, the largest petroleum discovery ever made in Colombia (with initial proven reserves of 192m. metric tons, of which 80m.–140m. tons were assessed as recoverable), doubled the country's recoverable petroleum reserves and transformed its prospects as a producer. The Caño-Limón discovery was followed in 1991 by the discovery of the Cusiana field, in the Andes foothills, with reserves estimated at 178m. tons. A further discovery, north of Cusiana at Cupiagua, with reserves estimated at 70m. tons, followed in 1992. Although the reserves at Cusiana and Cupiagua maintained Colombia's production and export capacity until the end of the 1990s, production declined from its highest ever recorded level of 41.6m. tons in 1999 to 35.3m. tons in 2000, and to 31.0m. tons in 2001. Output has since continued to fall, to 29.7m. tons in 2002, 27.9m. tons in 2003 and 27.3m. tons in 2004. Naturally declining production levels (by mid-2002 reserves at Caño-Limón were reported to be close to exhaustion), technical difficulties and political instability have raised fears that, without new discoveries, for which there is substantial, as yet unexplored, potential, Colombia could soon become a net importer of petroleum. Colombia currently exports about 50% of all of the petroleum it produces, mainly to the USA. Export earnings from petroleum and its derivatives accounted for 26.6% of total foreign revenue in 2001, 27.3% in 2002 and 25.8% in 2003. In December 2004 Colombia's proven reserves were estimated at 200m. tons. The state-owned Empresa Colombiana de Petróleos (ECOPETROL) accounts for about one-half of Colombia's crude output. In recent years the Government has attempted to make investment in the sector more attractive to foreign companies by, for example, permitting 100% participation in oil projects. In 2004 the Government reportedly concluded a record number of production agreements with foreign operators. Colombia's refining capacity was assessed at about 285,000 b/d in 2005, all of it controlled by ECOPETROL. The country is obliged to import petroleum products as domestic refining capacity is insufficient to meet local demand.

Argentina had estimated proven reserves of crude petroleum totalling 400m. metric tons in December 2004. Production increased substantially during the 1990s, and the country, the fourth largest consumer of petroleum in Latin America, is now a significant regional exporter, shipping mainly to Brazil and Chile. In 1993 the Government relinquished its monopoly in the petroleum sector and transferred the state corporation, Yacimientos Petrolíferos Fiscales (YPF), to private-sector ownership. In 1999 Repsol of Spain merged with YPF and Repsol-YPF is now the principal producer, explorer and refiner in the petroleum sector. In 2000 Chile imported more than 90% of its petroleum requirements. The national oil company, ENAP, was reported in 1999 to have begun to search for partners in proposed joint ventures, with the objective of increasing domestic production. Deposits in the Strait of Magellan constitute Chile's richest reserves, but access to new technology is required in order to maximize their and other fields' potential. In 2001 proven reserves amounted to 21m. tons. Peru had proven petroleum reserves of 100m. tons in December 2004. Production has steadily declined in recent years: in 1994 output amounted to 6.3m. tons, but fell to only 4.4m. tons in 2004. Formerly an exporter of petroleum that was surplus to domestic demand, Peru has become a net importer, purchasing from Colombia, Ecuador and Venezuela. Foreign participation in the development of new deposits had, by 2002, failed to check Peru's increasing reliance on imported petroleum, prompting the Government, in January, to introduce new legislation with the aim of encouraging exploration by improving the terms for investment. Exploration has since burgeoned and in 2004 Occidental announced a substantial discovery in the Amazon Basin. Guatemala, the only petroleum producer in Central America, commenced petroleum exports in 1980. The level of published proven reserves was 26m. tons in December 1997. Domestic consumption exceeds both domestic production and the country's refining capacity, and the country is, accordingly, a net importer of petroleum. Most domestic crude is exported to the USA. Petroleum accounted for only 3.3% of Guatemala's export earnings in 1999, this proportion rising to 5.9% in 2000. Petroleum accounted for 4.2% of export earnings in 2001 and for 6.7% in 2002. Although a minor regional producer, Cuba's output of 1.7m. tons in 1998 satisfied about 22% of its domestic petroleum requirements in that year.

International petroleum prices are strongly influenced by the Organization of the Petroleum Exporting Countries (OPEC), founded in 1960 to co-ordinate the production and marketing policies of those countries whose main source of export earnings is petroleum. In 2005 OPEC had 11 members. Venezuela is the only Latin American participant, although Mexico, together with other non-members, has collaborated closely with OPEC in recent years.

**Export Price Index for Crude Petroleum**
(base: 1980 = 100)

|  | Average | Highest month(s) | Lowest month(s) |
| --- | --- | --- | --- |
| 1990 | 69 | — | — |
| 1995 | 54 | — | — |
| 2000 | 86 | 96 (Sept.) | 73 (April) |
| 2001 | 72 | 82 (May) | 55 (Nov.) |
| 2002 | 76 | 88 (Dec.) | 58 (Jan.) |

The average price of crude petroleum was about US $20 per barrel in 1992, declined to only about $14 in late 1993, and recovered to more than $18 in mid-1994. Thereafter, international petroleum prices remained relatively stable until the early months of 1995. The price per barrel reached about $20 again in April and May 1995, but eased to $17 later in the year. In April 1996 the London price of the standard grade of North Sea petroleum for short-term delivery rose to more than $23 per barrel, following reports that stocks of petroleum in Western industrialized countries were at their lowest levels for 19 years. After another fall, the price of North Sea petroleum rose in October to more than $25 per barrel, its highest level for more than five years. The price per barrel was generally in the range of $22–$24 for the remainder of the year.

Petroleum traded at US $24–$25 per barrel in January 1997, but the short-term quotation for the standard North Sea grade fell to less than $17 in June. The price increased to more than $21 per barrel in October, in response to increased tension in the Persian (Arabian) Gulf region. However, the threat of an immediate conflict in the region subsided, and petroleum prices eased. The market was also weakened by an OPEC decision, in November, to raise the upper limit on members' production quotas, and by the severe financial and economic problems affecting many countries in eastern Asia. By the end of the year the price for North Sea petroleum had again been reduced to less than $17 per barrel.

Petroleum prices declined further in January 1998, with the standard North Sea grade trading at less than US $15 per barrel. Later in the month the price recovered to about $16.5 per barrel, but in March some grades of petroleum were trading at less than $13. Later that month three of the leading exporting countries—Saudi Arabia, Venezuela and Mexico—agreed to reduce petroleum production, in an attempt to revive prices. In response, the price of North Sea petroleum advanced to about $15.5 per barrel. Following endorsement of the three countries' initiative by OPEC, however, there was widespread doubt that the proposals would be sufficient to have a sustained impact on prices, in view of the existence of large stocks of petroleum. Under the plan, OPEC members and five other countries (including Mexico and Norway) agreed to reduce their petroleum output between 1 April and the end of the year. The proposed reductions totalled about 1.5m. b/d (2% of world production), with Saudi Arabia making the greatest contribution (300,000 b/d). In early June, having failed to make a significant impact on international prices, the three countries that had agreed in March to restrict their production of petroleum announced a further reduction in output, effective from 1 July. However, petroleum prices continued to be depressed, and in mid-June some grades sold for less than $12 per barrel. A new agreement between OPEC and other producers, concluded later that month, envisaged further reductions in output, totalling more than 1m. b/d, but this attempt to stimulate a rise in petroleum prices had little effect. The price on world markets was generally in the range of $12–$14 per barrel over the period July–October. Subsequently there was further downward pressure on petroleum prices, and in December the London quotation for the standard North Sea grade was below $10 per barrel for the first time since the introduction of the contract in 1986. For 1998 as a whole, the average price of North Sea petroleum was $13.37 per barrel: more than 30% less than in 1997 and the lowest annual level since 1976. In real terms (i.e. taking inflation into account), international prices for crude petroleum in late 1998 were at their lowest level since the 1920s.

During the early months of 1999 there was a steady recovery in the petroleum market. In March five leading producers (including Mexico) announced plans to reduce further their combined output by about 2m. b/d. Later that month an OPEC meeting agreed reductions in members' quotas totalling 1.7m. b/d (including 585,000 b/d for Saudi Arabia), to operate for 12 months from 1 April. The new quotas represented a 7% decrease from the previous levels (applicable from July 1998). At the same time, four non-members agreed voluntary cuts in production, bringing total proposed reductions in output to about 2.1m. b/d. In May 1999 the London price of North Sea petroleum rose to about US $17 per barrel. After easing somewhat, the price advanced again, reaching more than $20 per barrel in August. The upward trend continued, and in late September the price of North Sea petroleum (for November delivery) was just above $24 per barrel. After easing somewhat, prices rose again in November, when North Sea petroleum (for delivery in January 2000) was traded at more than $25 per barrel. The surge in prices followed indications that, in contrast to 1998, the previously agreed limits on output were, to a large extent, being implemented by producers and thus having an effect on stock levels. Surveys found that the rate of compliance among the 10 OPEC countries participating in the scheme to restrict production was 87% in June and July 1999, although it fell to 83% in October.

International prices for crude petroleum rose steadily during the opening weeks of 2000, with OPEC restrictions continuing to operate and stocks declining in industrial countries. In early March the London price for North Sea petroleum exceeded US $31.5 per barrel, but later in the month nine OPEC members agreed to restore production quotas to pre-March 1999 levels from 1 April 2000, representing a combined increase of about 1.7m. b/d. The London petroleum price fell in April to less than $22 per barrel, but the rise in OPEC production was insufficient to increase significantly the stocks held by major consuming countries. In June the price of North Sea petroleum rose to about $31.5 per barrel again, but later that month OPEC ministers agreed to a further rise in quotas (totalling about 700,000 b/d) from 1 July. By the end of July the North Sea petroleum price was below $27 per barrel, but in mid-August it rose to more than $32 for the first time since 1990 (when prices had surged in response to the Iraqi invasion of Kuwait). In New York in the same month, meanwhile, the September contract for light crude traded at a new record level of $33 per barrel at one point. The surge in oil prices in August was attributed to continued fears regarding supply levels in coming months, especially in view of data showing US inventories to be at their lowest level for 24 years, and of indications by both Saudi Arabia and Venezuela that OPEC would not act to raise production before September. Deliberate attempts to raise the price of the expiring London September contract were an additional factor.

In early September 2000 the London price of North Sea petroleum for October delivery climbed to a new 10-year 'high' of US $34.55, reflecting the view that any production increase that OPEC might decide to implement would be insufficient to prevent tight supplies later in the year. In New York, meanwhile, the price of light crude for October delivery rose beyond the $35 per barrel mark. This latest bout of price volatility reflected the imminence of an OPEC meeting, at which Saudi Arabia was expected to seek an agreement to raise the Organization's production by at least 700,000 b/d in order to stabilize the market. In the event, OPEC decided to increase production by 800,000 b/d, with effect from 1 October, causing prices in both London and New York to ease. This relaxation was short-lived, however. Just over a week after OPEC's decision was announced the price of the New York October contract for light crude closed at $36.88 per barrel, in response to concerns over tension in the Persian (Arabian) Gulf area between Iraq and Kuwait. The same contract had at one point risen above $37 per barrel, its highest level for 10 years. These latest increases prompted OPEC representatives to deliver assurances that production would be raised further in order to curb price levels regarded as economically damaging in the USA and other consumer countries. Towards the end of September the London price of North Sea petroleum for November delivery fell below $30 per barrel for the first time in a month, in response to the decision of the USA to release petroleum from its strategic reserve in order to depress prices.

In the first week of October 2000, however, the price of the November contract for both North Sea petroleum traded in London and New York light crude had stabilized at around US $30 per barrel, anxiety over political tension in the Middle East preventing the more marked decline that had been anticipated. This factor exerted stronger upward pressure during the following week, when the London price of North Sea petroleum for November delivery rose above $35 per barrel for the first time since 1990. In early November 2000 crude oil continued to trade at more than $30 per barrel in both London and New York, despite the announcement by OPEC of a further increase in production, this time of 500,000 b/d, and a lessening of political tension in the Middle East. Prices were volatile throughout November, with the price of both London and New York contracts for January delivery remaining in excess of $30 per barrel at the end of the month.

During the first week of December 2000 the price of the January 'futures' contract in both London and New York declined substantially, in response, mainly, to an unresolved dispute between the UN and Iraq over the pricing of Iraqi oil. On 8 December the closing London price of North Sea petroleum for January delivery was US $26.56 per barrel, while the equivalent New York price for light crude was $28.44. Trading during the second week of December was characterized by further declines, the London price of North Sea petroleum for January delivery falling below $25 per barrel at one point. Analysts noted that prices had fallen by some 20% since mid-November, and OPEC representatives indicated that the Organization might decide to cut production in January 2001 if prices fell below its preferred trading range of $22–$28 per barrel. At $23.51 per barrel on 14 December, the price of the OPEC 'basket' of crudes was at its lowest level since May. During the third week of December the price of the OPEC 'basket' of crudes declined further, to $21.64 per barrel. Overall, during December, the price of crudes traded in both London and New York declined by some $10 per barrel, and remained subject to pressure at the end of the month.

Continued expectations that OPEC would decide to reduce production later in the month caused 'futures' prices to strengthen in the first week of January 2001. The London price of North Sea petroleum for February delivery closed at US $25.18 on 5 January, while the corresponding price for light crude traded in New York was $27.95. Prices remained firm in the second week of January, again in anticipation of a decision by OPEC to reduce production. Paradoxically, prices fell immediately after OPEC's decision to reduce production by 1.5m. b/d was announced on 17 January. However, it was widely recognized that the reduction had been factored into markets by that time.

Oil prices rose significantly at the beginning of February 2001, although the gains were attributed mainly to speculative purchases rather than to any fundamental changes in market conditions. On 2 February the London price of North Sea petroleum for March delivery closed at US $29.19 per barrel, while the corresponding price for light crude traded in New York was $31.19. On 8 February prices rose to their highest levels for two months, the London price for North Sea petroleum for March delivery exceeding $30 per barrel at one point. The upward movement came in response to a forecast, issued by the US Energy Information Administration (EIA), that the 'spot' price for West Texas Intermediate (WTI—the US 'marker' crude) would average close to $30 per barrel throughout 2001. During the remainder of the month prices in both London and New York remained largely without direction.

During the early part of March 2001 oil prices in both London and New York drifted downwards while it remained unclear whether OPEC would decide to implement a further cut in production, and what the effect of such a reduction might be. On 9 March the London price of North Sea petroleum for April delivery was US $26.33 per barrel, while the corresponding price for light crude traded in New York was $28.01 per barrel. By late March prices in both London and New York had declined further, the London price of North Sea petroleum for May delivery closing at almost $24.82 per barrel on 30 March, while the corresponding price for New York light crude was $26.35.

During the first week of April 2001 crude oil prices on both sides of the Atlantic strengthened in response to fears of a gasoline shortage in the USA later in the year. On 6 April the London price of North Sea petroleum for May delivery was US $25.17 per barrel, while New York light crude for delivery in May closed at $27.06. Strong demand for crude oil by US gasoline refiners was the strongest influence on markets throughout April as gasoline 'futures' rose markedly. Towards the end of April the price of the New York contract for gasoline for May delivery rose to the equivalent of $1.115 per gallon, higher than the previous record price recorded in August 1990.

Fears of a shortage of gasoline supplies in the USA remained the key influence on oil markets in early May 2001, with 'futures' prices rising in response to successive record prices for gasoline 'futures'. On 4 May the London price of North Sea petroleum for June delivery was US $28.19 per barrel, while that of the corresponding contract for light crude traded in New York closed at $28.36. Prices were prevented from rising further during the week ending 4 May by the report of an unexpected increase in US inventories of crude oil. A further check came in the following week, when the International Energy Agency (IEA) reduced its forecast of world growth in demand for crude oil by 300,000 b/d to 1.02m. b/d. On 18 May, however, demand for crude oil by gasoline refiners raised the price of New York crude for June delivery to its highest level for three months, while the price of the July contract rose above $30 per barrel. On 5 June, at an extraordinary conference held in Vienna, Austria, OPEC voted to defer a possible adjustment of its production level for one month, noting that stocks of both crude oil and products were at a satisfactory level, the market in balance and that the year-to-date average of the OPEC reference 'basket' of crudes had been $24.8 per barrel (i.e. within the trading range of $22–$28 per barrel targeted by the Organization). OPEC nevertheless decided to hold a further extraordinary conference in early July in order to take account of future developments. At that meeting OPEC oil ministers once again opted to maintain production at the prevailing level, emphasizing that they would continue to monitor the market and take further measures, if deemed necessary, to maintain prices within the Organization's preferred trading range. The conference appealed to other oil exporters to continue to collaborate with OPEC in order to minimize price volatility and safeguard stability. Towards the end of the month, as prices declined steadily towards (and briefly below) $23 per barrel, the Secretary-General of OPEC, Dr Ali Rodríguez Araque, indicated that he was consulting OPEC ministers regarding the possibility of holding a further extraordinary conference early in August—ahead of the next ordinary OPEC meeting scheduled for September. Two days later, on 25 July, OPEC agreed to reduce production by a further 1m. b/d, to 23.2m. b/d, with effect from 1 September; the Organization reiterated that it was retaining the option to convene an extraordinary meeting if the market warranted it (this latest reduction, which had been agreed without a full meeting of OPEC, had been ratified by oil ministers by telephone). The Organization again expressed confidence that its action in reducing output would be matched by non-OPEC producing/exporting countries, and recognized in particular Mexico's support for its efforts. While there was general agreement that the production cut would reduce inventories, the consensus remained that demand would also decline in view of the prevailing world economic outlook.

A decline of 1.1% in US inventories of crude oil, reported by the American Petroleum Institute (API) in late July 2001, apparently indicated that US demand was resisting, for the time being, a deceleration in economic growth, and was cited as the main reason for a recovery in the price of Brent blend North Sea petroleum, to US $24.97 per barrel, on 1 August. A further decline of the same order, reported on 7 August, raised the price of Brent to $27.94 per barrel, and that of the OPEC 'basket' of crudes to $24.99 per barrel. Throughout most of the remainder of August declining US inventories of both crude and refined products appeared to suggest a strength of demand that belied pessimistic assessments of US (and global) economic prospects in the near term, combining with anticipation of the reduction of OPEC production by 1m. b/d from 1 September to support the price of the Brent reference blend and the OPEC 'basket' of crudes.

Immediately following the terrorist attacks carried out against targets in New York and Washington, DC, on 11 September 2001, as the price of Brent blend North Sea petroleum rose above US $30 per barrel, OPEC's Secretary-General was swift to emphasize the Organization's commitment to 'strengthening market stability and ensuring that sufficient supplies are available to satisfy market needs', by utilizing its spare capacity, if necessary. However, there was virtual unanimity among commentators, following the attacks, that their effect would be to worsen the prospects of the global economy, if not plunge it into recession, causing a considerable decline in demand for oil. By mid-October 2001 the price of the OPEC 'basket' of crudes had remained below $22 per barrel, the minimum price the Organization's market management strategy was designed to sustain, since late September, and it was clear that, at the risk of adding to recessionary pressures, OPEC would have to implement a further cut in production if it was to bring the price back into its preferred trading range. In late October Venezuela, Iran, Saudi Arabia, the United Arab Emirates (UAE) and non-OPEC Oman all declared themselves in favour of a further cut in production. However, diplomacy undertaken by President Chávez of Venezuela and Saudi Arabia's Minister of Petroleum and Mineral Resources, Ali ibn Ibrahim an-Nuaimi, had apparently made no progress in achieving its objective of persuading key non-OPEC producers Mexico, Norway and Russia to support the Organization's management strategy.

In the first week of November 2001 the price of Brent blend North Sea petroleum fell to fractionally above US $19 per barrel, while that of OPEC's 'basket' of crudes declined to $17.56 per barrel. In Vienna, on 14 November, an extraordinary meeting of the OPEC conference observed that 'as a result of the global economic slowdown and the aftermath of the tragic events of 11 September 2001, in order to achieve a balance in the oil market, it will be necessary to reduce the supply from all oil producers by a further 2m. b/d, bringing the total reduction in oil supply to 5.5m. b/d from the levels of January 2001, including the 3.5m. b/d reduction already effected by OPEC this year. In this connection, and reiterating its call on other oil exporters to co-operate so as to minimize price volatility and ensure market stability, the Conference decided to reduce production by an additional 1.5m. b/d, with effect from 1 January 2002, subject to a firm commitment from non-OPEC oil producers to cut their production by a volume of 500,000 b/d simultaneously'. The meeting acknowledged the positive response of Mexico and Oman to OPEC's efforts to balance the market. However, it was widely recognized that the success of a collaboration of the kind envisaged depended on the co-operation of Russia. Prior to the extraordinary conference held in November, Russia had indicated that it would be willing to reduce its production, estimated at more than 7m. b/d, by no more than 30,000 b/d, far less than would be necessary for OPEC's strategy to be effective. It was not until early December that Russia's Prime Minister announced the country's commitment to reducing its exports of crude by up to 150,000 b/d, and it was uncertain, in any case, whether a reduction of that magnitude could be enforced, owing to the Russian Government's lack of control over Russia's oil industry.

On 18 December 2001 the price of the OPEC 'basket' of crudes fell to US $16.62 per barrel. Thereafter, in December, however, the price recovered, in response to commitments by major non-OPEC producers to collaborate with the Organization by reducing either output or exports. On 28 December, at a consultative meeting of the OPEC conference, convened in Cairo, Egypt, OPEC confirmed its decision to implement a reduction of 1.5m. b/d in its overall production from 1 January 2002, having received assurances that Angola, Mexico, Norway, Oman and Russia would reduce their output, or, in the case of Russia, exports, of crude petroleum by a total of 462,500 b/d. BP's *Statistical Review of World Energy 2002* noted that, for the whole of 2001, the price of Brent blend North Sea petroleum had averaged $24.77 per barrel. The average price of the reference blend was substantially less than $20 per barrel in October–December 2001, however. In 2001, for the first time since 1993, consumption of oil world-wide declined, albeit marginally.

In January 2002 the decline in the price of the OPEC 'basket' of crudes was halted for the first time since the terrorist attacks in September 2001. According to the Organization's own data, the 'basket' price rose by 4.6% in January 2002, compared with December 2001, but was 24% lower when considered on a year-on-year basis. The price of Brent blend North Sea petroleum, meanwhile, averaged US $19.48 per barrel in January. Reviewing the state of the market in January 2002, OPEC noted that it was still too early to assess the effectiveness of the reduction in output and exports that had begun on 1 January. The recovery in the average price of the OPEC 'basket' had been uneven throughout the month. Low demand, as indicated by data published by the US Department of Energy's EIA and the API, recording rises in US crude inventories, and either rises or lower-than-expected declines in inventories of distillate products, combined in the second week of January with uncertainty regarding Russia's commitment to reducing its exports by 150,000 b/d (as it had pledged to do) to exert pressure on prices. During the third week of January, however, prices were supported, according to OPEC, by the strength of product prices, and by the USA's decision to add 22m. barrels of crude to its Strategic Petroleum Reserve. Among other factors, apparently good adherence by OPEC members to the revised quotas announced in December 2001 helped to sustain the upward trend in the final week of January 2002.

In February 2002 the average price of the OPEC 'basket' of crudes rose for the second consecutive month, recording an increase of 3.1% compared with January. As OPEC noted in its review of crude price movements in February, however, the average price of the 'basket' was 25.7% lower when considered on a year-on-year basis. The price of the OPEC 'basket' rose steadily during the first two weeks of February, supported, among other factors, by reports of an explosion at oil-gathering facilities in Kuwait that was initially expected to remove some 600,000 b/d from the market; and by increased political tension between the USA and Iraq. In the third week of February the OPEC 'basket' price moved up and down, but, overall, was weaker compared with the previous week. Continued doubts over Russia's commitment to reducing its exports was one factor that exerted downward pressure on prices. In the final week of the month prices strengthened considerably as a result of the interplay between reportedly higher product inventories, renewed political tension between the West and Iraq, and a dispute between the Venezuelan Government and employees of PDVSA.

In early March 2002 a meeting took place between a delegation of senior OPEC representatives and Russian Government and energy officials. OPEC's objective at the meeting was to persuade Russia to continue to limit its exports of crude to 150,000 b/d during the second quarter of 2002. OPEC regarded the continued limitation of Russia exports as imperative if market stability was to be maintained at a time of seasonally weak markets for oil. Russia's initial commitment to reduce its exports, however, applied only to the first quarter of 2002, with any continuation of the restriction subject to a review of market conditions. Norway had, by this time, already agreed to continue to restrict its production of crude to 150,000 b/d during April–June 2002. Moreover, it was clear in March that the market stability measures undertaken since the beginning of the year had been successful. The price of the OPEC 'basket' of crudes rose by 20% in March, compared with the previous month, although it was 5% lower when considered on a year-on-year basis. As OPEC noted, in its review of markets for crude in March, prices rose consistently throughout the month. Evidence of economic recovery in the USA was a positive factor in early March—data published by the API indicated declining US inventories of gasoline and distillate products—as was the apparent likelihood of Russia agreeing to carry over into the second quarter the restriction applied to its exports of crude. In the second week of March an intensification of the conflict between Israel and the Palestinians, and OPEC's announcement that it would maintain production at the prevailing level until the end of June at least, were additional factors that supported prices. In the third week of March prices rose above US $25 per barrel, owing, according to OPEC, to technical factors that were subsequently cancelled out by profit-taking. The upward movement continued towards the end of March, when positive inventory data combined with optimism about the sustainability of economic recovery in the USA to support prices.

From late March 2002 the escalating conflict between Israel and the Palestinians was cited as a key factor supporting crude prices. On 1 April, for the first time since 11 September 2001, the price of the OPEC 'basket' of crudes rose above US $25 per barrel. On 8 April Iraq added to the increased political tension in the Middle East by suspending its exports of crude for a period of 30 days in response to attacks by Israeli armed forces against Palestinian targets in the West Bank. Although Iraq's action was of little real consequence for crude markets, since other producers could easily compensate for the loss of its exports if necessary, it came at a time when a number of countries, including Iraq itself, Iran and Libya, had expressed their support for an embargo to be placed on the supply of oil by OPEC, in support of the Palestinian struggle against Israeli occupation. It was generally acknowledged in April that prices were inflated by a so-called war premium of some $4–$6 per barrel, without which they would decline towards the lower end of OPEC's preferred trading range. Political upheaval in Venezuela also lent a degree of volatility to prices in mid-April, when the brief removal of President Chávez from power caused them to decline sharply. Following Chávez's reinstatement as President, the key market influence for the remainder of the month was the perception, supported by a reported decline in US inventories of crude, that demand for oil was growing in response to improved economic conditions in the USA.

A combination of apparently stronger US demand and tighter supplies was regarded as the most significant determinant of the direction of markets for crude in early May 2002. On 7 May the API announced that US inventories of crude had declined by some 4.5m. barrels in the week to 3 May, and on the day of the announcement the price of Brent blend North Sea petroleum reached US $27.14 per barrel, an increase of more than $1 per barrel compared with the previous week. Among other factors contributing to the tightening of supplies at this time was OPEC's apparent decision not to compensate for the 30-day suspension of Iraqi exports. In the second week of May prices declined somewhat in response to data published by the US Department of Energy that indicated an increase in US inventories of crude. Towards the end of the month prices weakened again, in response to the publication of data that appeared to cast doubt on the strength of the US economic recovery.

Prices remained under pressure at the beginning of June 2002, owing to the publication of data indicating a further, unexpected increase in US inventories of crude and of distillate products in late May. Iraqi exports had also risen substantially in late May and early June. Most commentators appeared to agree with OPEC, representatives of whose members referred to a balanced market for crude in statements released early in the month, and indicated that OPEC would not alter its production quotas at its forthcoming extraordinary ministerial conference. In its assessment of the oil market in June, the IEA noted that geopolitical factors (i.e. violence in the Middle East) were now perceived as less of a risk to supplies of crude and predicted that prices would continue to weaken. US inventories of crude and gasoline were reported to have declined slightly in the first week of June. Crude stocks fell again during the second week of June, but this decline was balanced by substantial increases in inventories of gasoline and distillate products, indicating the ongoing weakness of economic recovery in the USA. At an extraordinary ministerial conference, held in Vienna on 26 June, OPEC, as expected, agreed to maintain production at the prevailing level until the end of September 2002. The Organization noted that its 'reduction measures during 2001 and 2002, supported by similar measures from some non-OPEC producers over the first half of the year, had restored relative market balance'. At the same time, OPEC observed that 'the relative strength in current market prices is partially a reflection of the prevailing political situation rather than solely the consequence of market fundamentals', and undertook to continue carefully to monitor market conditions and to take further action, if necessary, to maintain market stability.

The price of OPEC's 'basket' of crudes eased in the final week of July and the first week of August 2002, but rose to about US $26 per barrel in the second week of August owing to increased political tension in the Middle East. The threat of US military action against Iraq continued to support prices during the remainder of August and the first two weeks of September. Prior to the OPEC ministerial conference held in Osaka, Japan, on 19 September, Iraq's expressed willingness to allow the return of UN weapons inspectors caused prices to weaken, but they were subsequently supported by the Organization's decision to maintain production at the prevailing level until 12 December 2002, when the conference would meet again to review the market. The average price of the OPEC 'basket' of crudes in September, at $27.38 per barrel, was the third highest recorded for that month since 1984. At one point during the final week of the month the closing price of the 'basket', at $28.11 per barrel, exceeded the upper limit of the Organization's targeted trading range for the first time since November 2000. The average 'spot' quotation for Brent increased by $1.6 to $28.28 per barrel. Meanwhile, WTI for immediate delivery traded at an average of $29.52 per barrel, compared with $28.41 per barrel in August. In its review of the markets for crude petroleum in September, OPEC noted that the price of the 'basket' had averaged $23.48 per barrel during the first nine months of the year: $1.23 per barrel lower than the average price recorded during the corresponding period of 2001. In the final week of September 2002 the front-month (October) WTI 'futures' contract traded on the New York Mercantile Exchange (NYMEX) rose to its highest level for 19 months, closing above $30 per barrel at one point, having displayed considerable volatility during the month in accordance with the perceived likelihood of US military action against Iraq.

In October 2002 the average 'spot' quotation for the crudes comprising OPEC's reference 'basket' declined slightly, by US $0.06, to $27.32 per barrel, for the first time since July. The average weekly price of the 'basket' reached its highest level, $28.24 per barrel, in the first week of the month, but declined quite steeply, by $1 per barrel and $1.45 per barrel, in, respectively, the third and final weeks. The average 'spot' quotation for Brent declined by $0.59 to $27.69 per barrel, while that of WTI fell by $0.52 to $29.0 per barrel. Accordingly, during the first 10 months of 2002 the cumulative average price of the OPEC reference 'basket' was $23.91 per barrel, $0.02 below the average price recorded during the corresponding period of 2001. In its monthly review, OPEC again identified political developments pertaining to the Middle East, in particular a statement by US President George W. Bush early in the month which appeared to lessen the likelihood of US military action against Iraq, as a key influence—at the expense of fundamental factors, such as a sharp and greater-than-anticipated increase in both OPEC and non-OPEC supplies—on markets for crude petroleum. The price of 'futures' contracts for crude petroleum responded similarly to the perceived reduction in political tension, in combination with a reported increase in OPEC production and rising US stocks. On NYMEX the front-month WTI 'futures' contract declined from $30.83 per barrel (1 October) to only $27.22 per barrel (31 October).

In November 2002 the average price of the OPEC reference 'basket' declined steeply, by more than US $3, to $24.29 per barrel. In the same month, nevertheless, the year-to-date average price of the 'basket' rose above the corresponding price for 2001 for the first time, reaching $23.94 per barrel. The price of the 'basket' was at its weakest in the second week of the month. During the second half of November quotations recovered to the extent that the average price of the 'basket' re-entered OPEC's targeted price range, by a narrow margin, in the final week of the month. In its monthly assessment of markets for crude oil, OPEC attributed the steep decline in prices in the first half of November to the dissipation of the so-called 'political/war premium' after the UN Security Council's approval, and Iraq's subsequent unconditional acceptance of, Resolution 1441. Their recovery, in the second half of the month, was attributed to the perception that OPEC would take action to curb over-production, and to colder weather conditions in northern America and northern Asia. The average 'spot' quotation for Brent fell by $3.7 to $23.99 per barrel in November, while that of WTI fell by $2.69 to $26.31 per barrel. The front-month NYMEX WTI 'futures' contract, meanwhile, fell to a 'low' of $25.16 per barrel on 13 November, subsequently recovering by some $2 per barrel before the end of the month.

During December 2002 the average price of the OPEC reference 'basket' of crudes rose by more than US $4 to $28.39 per barrel, its highest level for two years. Average 'spot' quotations rose consistently during the month, in particular during the third and final weeks. By the end of the month the average price exceeded $30 per barrel, one of the highest levels ever recorded in December. The average 'spot' quotation for Brent rose by $4.84 to $28.83 per barrel, while that of WTI increased by $3.35 to $29.66 per barrel. In its monthly review of markets for crude petroleum, in addition to the continued threat of military conflict in Iraq, OPEC identified declining crude oil inventories (especially in the USA) and a sharp fall in Venezuelan production and exports as a consequence of strike action as the principal market influences. Before an extraordinary meeting of the OPEC conference took place in Vienna on 12 December, there was speculation that the Organization would seek to reassert its credibility by increasing its formal quotas (or overall 'ceiling') while, at the same time, making clear its intent to bring actual production into line with the new (raised) production level in order to restore discipline. In its assessment of market conditions prior to the extraordinary conference, the IEA concluded that an increase of some 1.5m. b/d in OPEC production that had occurred over the previous three months had probably been necessary in order to avert astronomical prices and to prevent a perilous depletion of industrial stocks at a time of geopolitical tension and as winter approached. In the event, as some commentators had predicted, the decision was taken at the conference to raise the production 'ceiling' for member states (excluding Iraq) from 21.7m. b/d to 23m. b/d, with effect from 1 January 2003, and to take steps to ensure that *de facto* production was reduced to within the new 'ceiling'. In its *Statistical Review of World Energy 2003*, BP noted that, as a result of production restraint and various unforeseen disruptions, OPEC's output had fallen by some 1.8m. b/d, or 6.4%, in the course of 2002. Oil demand in 2002 had been exceptionally weak for a third consecutive year, with consumption expanding by only 290,000 b/d.

At US $30.34 per barrel, the average price of the OPEC reference 'basket' in January 2003 was the highest recorded for that month since 1983. Declines in the average 'spot' quotation during the first two weeks of the month were offset by an increase in the third week and a further increase, followed by a correction, in the final week. By the end of January the average price of the 'basket' had been above $28 per barrel—the upper limit of OPEC's targeted trading range—for more than 33 consecutive days. The average 'spot' quotation for Brent increased by $2.48 to $31.31 per barrel, while that of WTI rose by $3.42 to $33.08 per barrel. The continued rise in the price of crude petroleum was attributed by OPEC to the combination of preparations for military action against Iraq in the Middle East, ongoing strike action by Venezuelan oil workers and a consequent steep decline in US inventories of crude petroleum, and cold weather conditions in the northern hemisphere. In response to these key market characteristics, at an extraordinary meeting of the OPEC conference convened in Vienna on 12 January, the Organization agreed to raise its production 'ceiling' by 1.5m. b/d to 24.5m. b/d, with effect from 1 February 2003. Production under the new 'ceiling' was to be distributed as follows (b/d, former production level in parentheses): Algeria 782,000 (735,000); Indonesia 1,270,000 (1,192,000); Iran 3,597,000 (3,377,000); Kuwait 1,966,000 (1,845,000); Libya 1,312,000 (1,232,000); Nigeria 2,018,000 (1,894,000); Qatar 635,000 (596,000); Saudi Arabia 7,963,000 (7,476,000); UAE 2,138,000 (2,007,000); Venezuela 2,819,000 (2,647,000).

During February 2003 the average price of the OPEC reference 'basket' rose by a further US $1.20 per barrel to $31.45: the third highest average price recorded in February since 1982 and $12.65 per barrel higher than in February 2002. 'Spot' quotations rose on a weekly basis throughout the month. The average 'spot' quotation for Brent increased by $1.24 to $32.54 per barrel, while that for WTI rose by $2.55 to $35.63 per barrel. As in the previous month, prices were boosted by the continued likelihood of war in Iraq and by very low inventories of crude petroleum and products in the USA. In its overview of market conditions in February, the IEA noted that 'The issues of high oil prices, stocks and spare capacity have assumed a greater urgency in advance of a potential military invasion of Iraq'. Despite an increase in production of some 2m. b/d in February (of which OPEC had contributed 1.5m. b/d), producers had been unable to restrain prices and their capacity to take further action was now limited by the consequent significant reduction in surplus production capacity.

Markets for crude petroleum were subject to a correction in March 2003. As the US-led military operation against the regime of Saddam Hussain in Iraq commenced, the so-called 'war premium' which had been a key characteristic of markets for many months evaporated. During March the average price of the OPEC reference 'basket' fell to US $29.78 per barrel, $1.76 per barrel lower than in February. Even so, this was highest average price recorded in March for 20 years, and the cumulative average price for the first quarter of 2003 was, at more than $30 per barrel, the highest ever recorded. The average 'spot' quotation for Brent fell by $1.56 to $30.98 per barrel, while that of WTI declined by $1.75 to $33.88 per barrel. The correction to prices occurred in spite of ongoing or recent disruptions to supplies from Iraq, Venezuela and Nigeria and apparently reflected consumers' confidence that measures taken by producers (such as the strategic locating of crude in major consuming areas) to offset these disruptions would be effective. At a meeting of the OPEC ministerial conference held in Vienna on 11 March it was agreed to maintain the Organization's production at its existing level, which was deemed adequate, in view of the restoration of Venezuelan production to normal levels, to meet demand.

The price of crude petroleum declined even more sharply in April 2003, the average price of the OPEC reference 'basket' falling by almost 15%, compared with the previous month, to US $25.34 per barrel. However, as OPEC noted in its monthly market review, despite the steep, consecutive monthly declines in March and April, the average price remained solidly within the Organization's targeted trading range and, indeed, the cumulative average price for the first four months of 2003 exceeded that of the corresponding period of 2002 by some $7.79 per barrel, almost 37%. In April the greatest decline occurred in the final week of the month, when the 'basket' lost some 7% of its value, the average price having moved both up and down during the preceding three weeks. The average 'spot' quotation of Brent declined by $5.91 to $25.07 per barrel, while that of WTI fell by $5.48 to $28.40 per barrel. OPEC noted that the US-led military campaign in Iraq remained the key influence on markets for crude, with prices weakening as the perceived likelihood of protracted hostilities diminished. Other factors that exerted downward pressure on markets for crude during the second half of April were the gradual return of Nigerian light-sweet crude to the market and the collapse of European refiners' margins. At a consultative meeting of the OPEC conference held in Vienna on 24 April, it was decided to reduce the Organization's actual production by 2m. b/d and to set a new 'ceiling' for output at 25.4m. b/d, effective from 1 June 2003. Quotas within the new 'ceiling' were as follows (b/d): Algeria 811,000; Indonesia 1,317,000; Iran 3,729,000; Kuwait 2,038,000; Libya 1,360,000; Nigeria 2,092,000; Qatar 658,000; Saudi Arabia 8,256,000; UAE 2,217,000; Venezuela 2,923,000.

In May 2003 the average price of OPEC's reference 'basket' of crude oils rose by US $0.26 to $25.60 per barrel. In its monthly review of markets for crude, the Organization noted that the cumulative average for 2003, at $28.37 per barrel, was almost 30% higher than the average for the corresponding period of 2002. The value of the 'basket' increased consistently throughout the month, in particular in the second week when it rose by 6.4%. The average 'spot' quotation of Brent increased by $0.72 to $25.79 per barrel, while that of WTI declined by $0.17 to $28.23 per barrel. In its review of market developments in May, OPEC noted the declining influence of events in Iraq and the re-establishment of fundamental factors as the key market drivers. The most important of these was the low level of US stocks of crude, reformulated gasoline (RFG) and distillates.

On 11 June 2003, at an extraordinary meeting of the OPEC conference convened in Doha, Qatar, it was agreed to maintain production at the prevailing level of 25.4m. b/d, with strict compliance. The conference noted that, while markets had been stable since OPEC had reduced its actual production to 25.4m. b/d and remained well supplied, prices had recently displayed an upward trend as a consequence of the slower-than-anticipated recovery in Iraqi output and unusually low inventory levels. The average price of the OPEC reference 'basket' did, in fact, rise substantially in June: by 4.5% to US $26.74 per barrel. At the end of June the cumulative average price of the 'basket' stood at $28.11, 27% higher than the average recorded for the first half of 2002. The greatest increase in the value of OPEC crudes occurred in the second week of the month, when the price of the 'basket' rose by 2.5%. This, together with smaller increases in the first and final weeks of June, was sufficient to compensate for a 5% decline in the price that occurred in the third week of the month. The average price of Brent rose by $1.65 to $27.44 per barrel, while that of WTI increased by $2.48 to $30.71 per barrel. Prices were supported in June by a further decline in US stocks of crude petroleum, especially in the early part of the month. OPEC's decision, on 11 June, to maintain production at the prevailing level had been anticipated to a large extent and its influence on prices was regarded as relatively insignificant. Unanticipated delays in the recovery of Iraqi production were another factor that supported prices in June.

In July 2003 the average price of the OPEC reference 'basket' was US $27.43 per barrel, 2.5% higher than the average price recorded in June. During the first seven months of 2003, accordingly, the cumulative average price of the 'basket' was $27.99 per barrel, some 24% higher than the average price recorded in the corresponding period of 2002 and just below the upper limit of OPEC's targeted trading range. In August the price of the 'basket' increased further, averaging $28.63 per barrel. The cumulative average price for the first eight months of 2003, at $28.07 per barrel, was accordingly 22% higher than that recorded in the corresponding period of 2002.

At a meeting of the OPEC conference held in Vienna on 24 September 2003, and attended by an Iraqi delegation headed by the newly appointed Minister of Oil of that country, Ibrahim Bahr al-Ulum, the decision was taken to reduce the Organization's production ceiling to 24.5m. b/d, with effect from 1 November. This decision was made in the light of OPEC's assessment of markets for crude as well supplied, and its observation that 'only normal, seasonal growth in demand [was] expected for the fourth quarter…'. As a result of continued increases in non-OPEC output and an ongoing recovery in Iraqi supplies, stocks were reported to be rapidly approaching normal seasonal levels. Furthermore, the supply/demand balance in the final quarter of 2003 and first quarter of 2004 indicated a 'contra-seasonal stock build-up' which, it was feared, could destabilize markets. OPEC's decision aimed to avert that threat, and the Organization appealed to non-OPEC producers to support it by likewise restraining increases in output.

In September 2003 the average price of the OPEC reference 'basket' of crude oils declined sharply, by US $2.31, to $26.32 per barrel. In the same month the average 'spot' quotation of Brent fell by $2.46, to $27.32 per barrel, while that of WTI declined by as much as $3.05, to $27.32 per barrel. The decline in the average price of the OPEC 'basket' was most marked in the first two weeks of the month, when it averaged, respectively, $27.61 per barrel and $26.42 per barrel. By 25 September the cumulative decline for the month amounted to more than 12%, but in the final week of September an increase of $1.45 per barrel was recorded. In its review of market developments in September, OPEC noted that the price of crude petroleum had hitherto drawn support from firm speculative US gasoline prices. From the beginning of the month, however, these had declined to a surprising extent, thus removing that support. At mid-September prices were regarded as 'steady', the major fundamental influence being concern over the level of US and EU heating oil stocks. The Organization defended its decision to reduce the production ceiling to 24.5m. b/d—which had taken speculators by surprise—as a 'reasoned response to market fundamentals and as a proactive effort…to accommodate the return of Iraqi production'.

In October 2003 the average price of the OPEC 'basket' increased by US $2.22 per barrel, to $28.54. Taking this increase into account, the cumulative average price of the OPEC 'basket' in 2003 stood at $27.91 and was thus approaching the upper limit—$28 per barrel—of OPEC's targeted trading range. The average 'spot' quotation of Brent, meanwhile, rose by $2.53 per barrel, to $29.85, and that of WTI by $1.88, to $30.43 per barrel. Analysts attributed the surge in prices in October to the continued psychological effect of the Organization's unexpected decision in September to reduce production. Another factor in the first three weeks of October was the perception that US stocks of heating oil and distillates, though rising, would be inadequate to meet demand during the long, severe winter that was forecast. In the final week, however, prices fell in response to more realistic formulations of the supply/demand equation likely to pertain in coming months.

The average price of the OPEC 'basket' declined marginally, by US $0.09, to $28.45 per barrel, in November 2003. The average 'spot' quotation for Brent declined by $1.17 per barrel, to $28.68, while that of WTI rose by $0.51, to $30.94 per barrel. Quotations, which had been relatively weak in the second half of October, strengthened during most of November, before weakening again at the end of the month. The most influential fundamental factor remained continued concern regarding the level of US stocks of heating oil as the winter approached. In mid-November speculation was identified as the factor behind surges in the prices of WTI and Brent. In its review of market developments in November, OPEC noted that the speculative rally was fuelled by 'fears of inadequate crude oil and product inventories in the USA and Europe, preliminary figures showing OPEC-10 [i.e. all OPEC members except Iraq] was implementing the September 24 Agreement calling for production cuts, and the dramatic increase in speculators' long positions at the NYMEX, which indicates that the market expects prices to rise in the future'. The speculative rally ended at the close of the month with profit-taking by market participants and reduction of exposure in advance of a forthcoming extraordinary meeting of OPEC.

On 4 December 2003, at the extraordinary conference held in Vienna, OPEC decided to maintain production at its current level until further notice. During the month the average price of the OPEC 'basket' of crude oils rose by US $0.99 per barrel, to $29.44, its highest level since March. At the same time, the average quotation of Brent rose by $1.14, to $29.82 per barrel, and that of WTI by $1.21, to $32.15. In its review of market developments in December, OPEC noted that the cumulative average 'spot' quotation of its reference 'basket' in 2003 was, at $28.10, the highest nominal annual average since 1984. During December 2003 prices were initially supported by very strong Asian, in particular Chinese, demand for petroleum products, by declines in US commercial inventories of crude and by indications that economic recovery was well established in the USA. As the month progressed these factors were reinforced by cold weather conditions. BP noted, in its *Statistical Review of World Energy 2004*, that oil prices in 2003 had been at their highest level in nominal terms (i.e. without taking inflation into account) for 20 years. Consumption world-wide had also risen strongly, by 2.1%. Despite interruptions to the output of Iraq and Venezuela, OPEC production had increased substantially in 2003, by some 1.9m. b/d.

In January 2004 the average price of the OPEC reference 'basket' of crudes rose by US $0.89 per barrel, to $30.33. The average 'spot' quotation of Brent rose by $1.51, to $31.33, in that month, while the average quotation of WTI increased by as much as $2.18, to $34.33. The surge in the price of WTI was attributed to a combination of very low US inventories of crude oil, which in mid-January reportedly fell below the minimum operational level that had been established in 1998, and very cold weather in eastern areas of the USA early in the month. In spite of the substantial discount in the price of North Sea and West African crudes relative to WTI, deliveries of these crudes to US markets were restricted throughout most of January 2004 by very high freight rates.

On 10 February 2004, at an extraordinary conference held in Algiers, Algeria, OPEC decided to reduce its production 'ceiling' from 24.5m. b/d to 23.5m. b/d, with effect from 1 April 2004. This decision was taken in response to projections of a 'significant supply surplus in the seasonally low demand second quarter [of 2004]', in order to avert downward pressure on prices. Production under the new 'ceiling' was to be distributed as follows (b/d, former production level in parentheses): Algeria 750,000 (782,000); Indonesia 1,218,000 (1,270,000); Iran 3,450,000 (3,597,000); Kuwait 1,886,000 (1,966,000); Libya 1,258,000 (1,312,000); Nigeria 1,936,000 (2,018,000); Qatar 609,000 (635,000); Saudi Arabia 7,638,000 (7,963,000); UAE 2,051,000 (2,138,000); Venezuela 2,704,000 (2,819,000). During February the average price of the OPEC 'basket' of crudes declined by US $0.77, to $29.56 per barrel. The average 'spot' quotation of Brent declined by $0.68 per barrel, to $30.65, in February, while that of WTI rose by $0.29, to $34.62 per barrel. In its review of market developments in February, OPEC indicated the increasing importance as an influence on markets for crude of the US market for gasoline, which faced a potential supply shortage owing to a combination of steady and rising demand, the inability of Asia-Pacific refiners to supply it, and low domestic (US) inventories.

In March 2004 the average price of the OPEC reference 'basket' increased by US $2.49, to $32.05 per barrel. It was the first time that an average price in excess of $32 per barrel had been recorded since October 1990. The average 'spot' quotation of Brent rose by $3.05, to $33.70 per barrel in March 2004, while that of WTI increased by $1.97, to $36.59. According to OPEC, the US gasoline market remained the key influence on markets for crude, while strong global demand for crude petroleum and economic growth were other important factors. At a conference held in Vienna on 31 March, OPEC confirmed that it would adjust its production 'ceiling' to 23.5m. b/d from the beginning of April, in accordance with the decision announced in February. In the view of the Organization, prevailing high prices for crude petroleum did not reflect supply/demand fundamental factors, but rather were 'predominantly a consequence of long positions of market speculators in the futures markets coupled with a tightening in the US gasoline market in some regions, and exacerbated by uncertainties arising from prevailing geopolitical concerns…'.

The average price of the OPEC reference 'basket' in April 2004, at US $32.35 per barrel, was the second highest ever recorded—the highest was $34.32, registered in October 1990. The average 'spot' quotation of Brent fell by $0.47 in April, to $33.23 per barrel, while that of WTI rose by $0.21, to $36.80. OPEC indicated that the cumulative average price of its reference 'basket' up to 30 April 2004 was $31.13 per barrel, compared with $29.02 per barrel for the corresponding period of 2003. The continued strength of markets for crude was attributed to, among other things, low US inventories of gasoline, in particular RFG, stocks of which were reportedly some 32% lower at 30 April 2004 than at 30 April 2003, and some 41% lower than the five-year average, according to the EIA. Continued strong US demand for gasoline was likely to face further pressure, as it rose during the summer months, from new specifications, introduced from January 2004, banning the use of methyl-tertiary-butane ether (MTBE) for the production of RFG in the states of California, Connecticut and New York. OPEC also identified very strong demand on the part of China for gasoline and its consequent reduced availability for export as an additional factor supporting gasoline markets. In addition to these fundamental factors, greater concern over unrest in petroleum-producing countries had reportedly led to increased speculative activity on crude markets, pushing prices further upward.

In the week ending 6 May 2004 the average price of the OPEC 'basket' of crudes rose once again, to US $34.91 per barrel, and a further increase, to $36.16 per barrel, was recorded in the week ending 13 May. In early June, following attacks by militant Islamists who were suspected of having links with the al-Qa'ida (Base) organization on foreign workers in the Saudi Arabian city of al-Khobar, an important centre for the Saudi oil industry, the price of crude petroleum traded in the USA rose to the record level of $42.45 per barrel, while that of Brent approached $40 per barrel. On 3 June, at an extraordinary conference held in Beirut, Lebanon, OPEC decided to raise its production 'ceiling' to 25.5m. b/d with effect from 1 July, and to 26m. b/d with effect from 1 August. The Organization noted that prices had continued to escalate, in spite of its efforts to ensure that markets were well supplied, as a result of continued growth in demand in the USA and China, geopolitical tensions, problems in respect of refining and distribution in some consuming regions and more stringent product specifications. Production under the new 'ceiling' was to be distributed as follows from 1 July (b/d): Algeria 814,000; Indonesia 1,322,000; Iran 3,744,000; Kuwait 2,046,000; Libya 1,365,000; Nigeria 2,101,000; Qatar 661,000; Saudi Arabia, 8,288,000; UAE 2,225,000; Venezuela 2,934,000. From 1 August the distribution of OPEC production was to be: Algeria, 830,000; Indonesia 1,347,000; Iran 3,817,000; Kuwait 2,087,000; Libya 1,392,000; Nigeria 2,142,000; Qatar 674,000; Saudi Arabia 8,450,000; UAE 2,269,000; Venezuela 2,992,000.

In the immediate aftermath of OPEC's announcement of its new production 'ceilings' the price of both US-traded crudes and of Brent declined. However, some analysts expressed doubts over whether OPEC's action was sufficient to exercise a sustained calming effect on markets, noting that most of its members were already producing at close to capacity, in some cases in breach of prevailing quotas. The new 'ceilings' would thus, in the view of those observers, simply legitimize over-production. Representatives of some OPEC member states, meanwhile, conceded that the new production limits would not necessarily bring prices back within the Organization's preferred trading range, but would counter any perception of shortages.

In July 2004 the average price of the OPEC reference 'basket' rose to a new record level of US $36.29 per barrel—in June it had declined slightly in comparison with May, to $34.61 per barrel; this was, none the less, the highest average price recorded in the month of June for 22 years. In July the average price of Brent increased by $3.21 per barrel, to $38.33, while that of WTI rose by $2.51 per barrel, to $40.69. The cumulative average price of the OPEC reference basket for January–July 2004 rose accordingly, to $33.45 per barrel, some 18% higher than the average price recorded in the corresponding period of 2003. In its assessment of markets for crude in July, OPEC attempted to clarify the market's perception of tightness in supplies, indicating that while there had been a scarcity of light, sweet crudes, 'it is also true that sour crudes are inundating the market…'. OPEC also noted that the global refining system was operating at close to full capacity and concluded its assessment by indicating that apparently 'the market has entered a new reality, one where tightness in upstream spare capacity due to lack of capacity expansion and surprisingly robust oil demand growth promises to set the scene for a new market dynamic'.

In a statement issued in early August 2004, OPEC noted that production by its members (excluding Iraq) had continued to increase in order to meet greater-than-expected growth in demand. Output was estimated to have been substantially greater than 29m. b/d in July 2004, and it was forecast that total OPEC production, including that of Iraq, would approach 30m. b/d in August. With reference to possible future market-stabilization measures, the Organization indicated that it retained spare production capacity of some 1m. b/d–1.5m. b/d, which would permit an immediate further increase in output, and that OPEC members intended to raise production capacity by some 1m. b/d in late 2004 and during 2005. Oil continued to trade at record levels in August 2004: the average price of the OPEC reference 'basket' of crude oils was, for the first time since the Organization adopted the reference price system, in excess of US $40 per barrel, having risen by 11% compared with the previous month. In the same month the average 'spot' quotation of Brent rose by 11.8%, to $42.87 per barrel, and that of WTI by 10%, to $44.77 per barrel. The cumulative average price of the OPEC 'basket' in January–August 2004 was 21% higher than in the corresponding period in 2003.

As oil prices reached record levels in the third week of August 2004, some analysts warned of a future energy crisis—in which oil prices would continue to rise, perhaps above US $65 per barrel—that might give rise to economic recession world-wide. It was noted at this time that some traders did not appear to view the current 'spike' in prices as a cyclical high: since May 2003 futures prices for oil, for delivery in two years' time, had risen by almost as much as those of oil for immediate or imminent delivery. Factors cited as responsible for the record prices recorded in August 2004 included the perceived insecurity of supplies from important producing countries such as Iraq, Saudi Arabia, Russia and Venezuela, and the economic boom that was occurring in China. Prices, according to some observers, were likely to remain at unprecedented levels until they had triggered a recession which would bring growth in demand for oil back into line with existing capacity for production, refining and distribution.

In the view of OPEC, the dramatic increases in the price of crudes that took place in August 2004 were attributable to a combination of both fundamental and non-fundamental factors. The estimated extent of the total increase in demand for oil world-wide in 2004 continued to be revised upwards, and was assessed in August at some 2.5m. b/d, to be followed by a further increase in 2005 of some 1.7m. b/d. The forecast increase in demand from China for the whole of 2004, at more than 0.8m. b/d, was already characterized by OPEC as 'spectacular' and had the potential to rise further before the end of the year. There were indications, according to OPEC, that growth in demand in the second quarter of 2004, at some 3.80m. b/d (4.95% higher than in the corresponding three-month period of 2003), had been higher than at any time since 1985. With regard to supply, intermittent interruptions to Iraqi output had delayed and continued to delay the return of that country's production to 2.8m. b/d. Other factors that had combined to strengthen the prices of crude oil and petroleum products in August included the maintenance of stocks at comparatively low levels and fears of increased unrest in Iraq and in the Middle East generally. These concerns had, according to OPEC, been exacerbated by speculative activity on futures markets for crude oil: with regard to fundamental factors, OPEC concluded that supply was more than sufficient to meet the increases in demand forecast to occur during the remainder of 2004 and in early 2005.

In late August and early September 2004 the price of the OPEC reference 'basket' of crudes eased somewhat, as concerns regarding supplies (in particular those from Venezuela and Iraq) subsided. As September progressed, however, adequate supplies and OPEC's express commitment to raising its production 'ceiling' to 27m. b/d from 1 November 2004 (see below) were insufficient to calm continued fears of disruption to Russian supplies (arising from the so-called Yukos affair—from late 2003 senior officials of Russia's prominent petroleum company, Yukos, and its subsidiaries were accused of tax evasion and fraud, resulting in the company being dismantled by the Russian authorities in late 2004), and these, in combination with the actual disruption of production and refining in the Gulf of Mexico by hurricane weather conditions, propelled the average price of the 'basket' to a new record level of US $40.36 per barrel for the whole of September, 0.2% higher than the average price recorded in August. Average quotations for Brent and WTI rose, respectively, by 1.3%, to $43.43 per barrel, and by 2.7%, to $45.98 per barrel. Compared with January–September 2003, the cumulative average price of the OPEC 'basket' in the first nine months of 2004 was almost 25% higher, at $34.98 per barrel.

In mid-September 2004, at a meeting of the OPEC conference in Vienna, it was agreed to increase the production 'ceiling' for the Organization (excluding Iraq) by 1m. b/d, to 27m. b/d, effective from 1 November 2004, with the aim of reducing the high prices recently recorded to 'a more sustainable level'. Commenting on market conditions at the time of the conference, OPEC summarized the factors responsible for recent price increases as: the rise in demand that had occurred earlier in the year, especially in North America, China and other Asian countries—preliminary figures published by OPEC in September 2004 indicated that demand for oil world-wide had increased by 1.8m. b/d, or 2.3%, in the first quarter of 2004, and by 4m. b/d, more than 5%, in the second quarter; geopolitical considerations; doubts over the adequacy of existing spare capacity to meet possible supply disruptions; speculative investment activity; and constraints within the downstream industry.

In October 2004 the average price of the OPEC reference 'basket' of crude oils rose by 12.4%, compared with the previous month, to US $45.37 per barrel, its highest ever level. During the third week of the month, moreover, the highest ever weekly price, $46.04 per barrel, was recorded. The average 'spot' quotation of Brent increased by 14.5% in October, to $49.74 per barrel, while that of WTI rose by 16%, to $53.32 per barrel. (During the second and third weeks of October the price of WTI rose to successive record levels of $55.08 per barrel and $56.42 per barrel, respectively, boosted by, among other factors, low US stocks of heating oil and industrial action in Norway.) In Europe record prices for Brent were recorded owing to concerns that there would be shortages of winter fuels. Hurricane weather conditions on the US Gulf coast and concern that supplies from Nigeria would be restricted were identified as the main factors driving prices upwards in the early part of the month. In the third week of October fears that winter fuels would be in short supply combined with industrial action in Nigeria to support prices.

As concerns over possible shortages of winter fuels, especially in the USA, subsided in response to mild weather conditions world-wide and the accumulation of stocks of middle distillates, and crude production by OPEC members reached record levels, the average price of the OPEC reference 'basket' of crudes in November 2004 was, at US $38.96 per barrel, some 14% lower than in October. At the same time, the average 'spot' quotations of Brent and WTI declined, respectively, by some 14%, to $42.80 per barrel, and by 9.6%, to $48.22 per barrel. The average price of the OPEC reference 'basket' in November was the lowest recorded since July 2003, and the fall in the price in November 2004, compared with October, was the steepest month-on-month decline ever recorded. In January–November 2004 the cumulative average price of the OPEC reference 'basket', at $36.07 per barrel, was 29% higher than in the first 11 months of 2003. Heavy selling of crude oil 'futures' contracts by speculative investors, in order to realize profits, was reported to have occurred in November 2004, as the level of OPEC production, in combination with slower growth in demand, allowed stocks to accumulate in advance of winter in the northern hemisphere.

The ready availability of Middle Eastern crudes remained an important influence on the market in December 2004. At an extraordinary meeting held in Cairo on 10 December, the OPEC conference observed that its recent initiative to raise supplies had caused 'current crude oil prices [to] reflect convergence towards market fundamentals'. The conference agreed to maintain OPEC's production 'ceiling' and individual production levels unchanged, but reported that member countries 'which have responded to the market need for additional supply over the course of this year by producing above their allocations, have agreed to collectively reduce the over-production by 1m. b/d from their current actual output, effective 1 January 2005'. At the beginning of December adequate Middle Eastern supplies had applied downward pressure to prices. As the month progressed, other depressive factors, including a resumption of Nigerian output and forecast mild winter weather conditions, prevailed over factors, such as concerns about the security of Middle Eastern oil infrastructure and the decision of the OPEC conference to maintain its production 'ceiling' unchanged, that tended to strengthen prices: overall in December the average price of the OPEC reference 'basket' of crude oils declined by 8.4%, compared with November, to US $35.70 per barrel. At the same time, the average 'spot' quotation of Brent declined by 7.9%, to $39.43 per barrel, while that of WTI fell by 10.6%, to $43.12 per barrel. The average price of the OPEC reference 'basket' for the whole of 2004, at $36.05, was 28% higher than in 2003, in spite of the consecutive monthly declines recorded in November and December 2004. According to data published by OPEC at the end of the year, demand for oil world-wide in 2004 averaged 81.99m. b/d.

At the beginning of 2005, assessing the prospects for oil in the next 12 months, some analysts noted that since mid-2003 economic growth, in particular in North America, China and other emerging Asian economies, such as India, had caused the price of oil and other commodities to increase sharply. If the world economy were to conform to typical, cyclical patterns, however, it was anticipated that growth would decline overall by about 20% in 2005, curbing demand for oil. It was noted, too, that during the course of 2004 there had been a substantial rise in the supply of oil world-wide, from both OPEC members and other countries. The perceived vulnerability of supplies to, *inter alia*, developments in the Yukos affair in Russia and hurricane weather conditions in the Gulf of Mexico, in combination with the insecurity of Iraq's oil infrastructure, social and political instability in Nigeria and Mexico, and the continued, gradual decline in North Sea output, had applied upward pressure to prices; and these were the factors to which OPEC was attempting to respond by raising its members' output, to the point that actual production, at 29.1m. b/d, was at its highest level for 25 years. Globally, in 2004, supply grew by about 1m. b/d more than demand, thus meeting demand and allowing the recovery of commercial stocks and strategic reserves.

In January 2005 the average price of the OPEC reference 'basket' of crude oils, at US $40.24 per barrel, was 12.7% higher than in December 2004. At the same time, the average 'spot' quotation of Brent increased by 11.6%, to $44.01 per barrel, and that of WTI by 8.2%, to $44.39 per barrel. The average price of the OPEC reference 'basket' in January 2005 was 33% higher than in the corresponding month of 2004. Among the factors that were cited as having contributed to the increase in prices at the beginning of 2005 were cold weather conditions in the northern hemisphere, instability in Iraq, expectations of industrial action in Venezuela and declines in US stocks of crude. At an extraordinary meeting of the OPEC conference held in Vienna on 30 January, OPEC noted that supplies of oil worldwide exceeded demand, and that commercial stocks had been replenished to above their five-year average levels. An element of volatility remained in the market, however, owing to concerns over interruption of supply and the predicted persistence of strong demand. With reference to forecasts indicating that markets would remain in balance during the first quarter of 2005, OPEC agreed to maintain output of crude at its prevailing level. The conference noted, furthermore, that prices had remained outside of the Organization's preferred trading range of $22–$28 per barrel for more than one year, owing to changes in the market that had made the range 'unrealistic'. The decision was taken therefore to suspend the price range 'pending completion of further studies on the subject'.

OPEC's decision to suspend the preferred trading range of US $22–$28 per barrel was interpreted by some analysts as a further sign that the recent unprecedentedly high price of oil should be viewed as the result of a fundamental change that was occurring in the supply/demand equation rather than as a temporary phenomenon. It was argued that spare production capacity in the oil industry world-wide had been almost exhausted. (This argument had been used earlier to explain the apparently disproportionate effect of perceived threats to supplies from Norway and Nigeria on the price of oil in October 2004: analysts had suggested that oil producers in general and OPEC member countries in particular had neglected to expand and upgrade their production capacity to enable it to accommodate rapid rises in global demand for oil. OPEC's production capacity was estimated by some observers to have risen to close to its upper limit, where other members would be unable to compensate for, for example, disruption to Nigerian output.) It was reported at the end of 2004 that production was declining in some 70% of the world's largest oil-producing countries, and that in 2004 only Russia and the Gulf states had been able to raise output. In July 2004 Saudi Arabian officials had announced that they regarded the price of oil at that time—about $35 per barrel—as fair, suggesting that, like other members of OPEC, Saudi Arabia no longer favoured the trading range of $22–$28 per barrel. Some analysts regarded this as a tacit acknowledgement by Saudi Arabia of its inability to increase production from its ageing oilfields to any great extent beyond current levels, and thus as a step closer towards so-called 'peak' production and, in turn, the end of 'cheap' oil. Other observers, however, remained confident that the issue of spare capacity would decline in significance. As in the past, high prices for oil would encourage new exploration initiatives and world oil production would continue to rise for the indefinite future.

In February 2005 the average price of the OPEC reference 'basket' of crude oils, at US $41.68 per barrel, was 3.6% higher than the average price recorded in the previous month. The average 'spot' quotation of Brent increased by almost 2% in February, to $44.87 per barrel, while that of WTI rose by 2.3%, to $47.69 per barrel. During the first two weeks of February prices generally declined in response to factors such as the predicted warmer weather conditions. Gains occurred in the latter part of the month, in response to industry concerns over a possible reduction of OPEC output, colder weather conditions, increased demand for winter fuels and a decline in the value of the US dollar.

At US $49.07 per barrel, 17.7% higher than in February 2005, the average price of the OPEC reference 'basket' of crude oils in March was the highest ever recorded. During March the average 'spot' quotations of Brent and WTI rose, respectively, by 17.2%, to $52.60 per barrel, and by 13.4%, to $54.09 per barrel. Projected high demand was a key determinant of market movements in the early part of March, in combination with continued cold weather conditions. Fears that demand would exceed supply propelled prices to a level close to $50 per barrel in the second week of March. In the second half of the month prices rose above $50 per barrel in response, among other things, to a substantial decline in US stocks of gasoline, distillates and heating oil. Towards the end of March the price of the OPEC 'basket' of crudes rose to a record high level of $50.72 per barrel after an accident at a US refinery prompted fears of a shortage of refined products. The price declined in the final days of the month, however. On 16 March, meeting in Esfahan, Iran, the OPEC conference agreed to increase the Organization's production 'ceiling' to 27.5m. b/d, with immediate effect. Moreover, the conference authorized OPEC's President to announce a further increase of 500,000 b/d in the 'ceiling' before the next meeting of the conference, should 'oil prices remain at current levels or continue to rise further'. In its assessment of market conditions prior to this decision, OPEC noted that world crude oil prices had begun to rise again even though all indicators pointed to a market that was fundamentally well supplied. In explanation, OPEC's concluding statement referred to late cold weather conditions in the northern hemisphere, the anticipated continued strength of demand, speculative activity in oil futures markets, geopolitical tensions and downstream bottlenecks. Another factor cited by analysts in explanation of the so-called spring 'spike' in oil prices was the continued decline in the value of the US dollar in relation to the euro.

In April 2005 the average price of the OPEC reference 'basket' of crude oils increased by 1.1%—the lowest monthly increase for four months—to US $49.63 per barrel. The average 'spot' quotation of both Brent and WTI also declined in April, by 1.4%, to $51.87 per barrel, and by 1.8%, to $53.09 per barrel, respectively. Early in the month prices rose in response to a report by the investment bank Goldman Sachs in which the possibility that oil prices might double from their current levels was discussed. Indeed, in the first week of April the average price of OPEC 'basket' crudes, at $52.07 per barrel, was the highest ever recorded. In the second week of April, however, continued increases in US stocks of crude, high OPEC output and a revised, lower forecast of world demand for oil in 2005 by the IEA combined to bring down prices. During the third week of April markets for oil lacked clear direction, responding to developments in the Yukos affair by rising, for instance, and declining in response to the ready availability of OPEC crudes.

In May 2005 the average price of the OPEC reference 'basket' of crude oils fell by more than 5%, to US $46.96 per barrel. At the same time, the average 'spot' quotation of Brent declined by 5.7%, compared with the previous month, to $48.90 per barrel, while that of WTI declined by 5.3%, to $50.25 per barrel. During the first three weeks of May OPEC's reference price fell steadily in response to such factors as the ready availability of OPEC crudes and, in particular, the high level of US crude inventories. In the final week of the month, however, prices were supported by fears that supplies of refined products would be disrupted in advance of the US driving season. May marked the first time for six months when an average price lower than that of the preceding month had been recorded.

At an extraordinary meeting of the OPEC conference held in Vienna on 15 June 2005, the Organization took the decision to raise its production 'ceiling' by 500,000 b/d, with effect from 1 July. Commenting on prevailing market conditions, OPEC noted that although markets were well supplied, world crude prices had remained high and volatile owing to concerns over a lack of effective global oil refining capacity. Refiners' difficulties in meeting strong growth in demand for distillates, moreover, were being exacerbated by geopolitical developments and speculative investment activity on futures markets. OPEC's decision to raise the production 'ceiling' was taken in the light of this analysis and in response to anticipated continued strong growth in demand for crude oil in 2005 and renewed price increases. The conference also authorized its President, after having consulted with heads of delegations, to announce an additional increase of 500,000 b/d in the production 'ceiling' if crude prices remained at their current levels or continued to rise. Separately, the conference announced its decision to change the composition of the OPEC reference 'basket' from 16 June 2005. The new 'basket' would comprise the main export crudes (Saharan Blend—Algeria; Minas—Indonesia; Iran Heavy—Iran; Basra Light—Iraq; Kuwait Export—Kuwait; Es Sider—Libya; Bonny Light—Nigeria; Qatar Marine—Qatar; Arabian Light—Saudi Arabia; Murban—UAE; BCF 17—Venezuela) of all OPEC member states and would be weighted according to production and exports to the principal markets. The new 'basket' would also better reflect the average quality of member countries' crudes.

In June 2005 the average price of the OPEC reference 'basket' of crudes (calculated on the basis of both the old and the new components of the 'basket'—see above) rose by 11%, to US $52.04 per barrel. The average price of the 'basket' as composed until 16 June increased by 12%, to $52.72 per barrel, while that of its newly defined counterpart rose by 13%, to $52.92 per barrel. In June the average 'spot' quotation of Brent increased by 11.9%, to $54.73 per barrel, while that of WTI rose by 12.6%, to $56.60 per barrel. Concern over possible shortages of refined products was identified as the key determinant of price movements in June. Assessing the prospects for world petroleum markets in 2006 in mid-2005, OPEC noted that growth in demand in 2005 was estimated at, on average, 1.6m. b/d. Forecasts of growth in non-OPEC supplies had been reduced owing, chiefly, to a fall in Russian output. OPEC production, meanwhile, was estimated to have averaged 30m. b/d in the first half of 2005, an increase of 1m. b/d compared with the first half of 2004, and this had led to counter-seasonal growth in inventories in the first quarter and the first two months of the second quarter of 2005. Record price levels had thus been recorded in spite of abundant supplies of crude and increased stocks.

In the first half of July 2005 stormy weather conditions in the Gulf of Mexico boosted the price of US light crudes to more than US $61 per barrel, while in Europe Brent blend traded at close to $60 per barrel.

The Regional Association of Oil and Natural Gas Companies in Latin America and the Caribbean (Asociación Regional de Empresas de Petróleo y Gas Natural en Latinoamérica y el Caribe—ARPEL) exists to promote co-operation in matters of technical and economic development. ARPEL, with headquarters in Montevideo, Uruguay, has 26 members, including companies based outside the region.

## Silver

Known since prehistoric times, silver is a white metal which is extremely malleable and ductile. It is one of the best metallic conductors of heat and electricity, hence its use in electrical contacts and in electroplating. Silver's most important compounds are the chloride and bromide which darken on exposure to light, the basis of photographic emulsions.

World-wide, about 70% of silver production in 2004 was generated as a by-product, or co-product, of gold, copper, lead, zinc and other mining operations. Methods of recovery depend upon the composition of the silver-bearing ore. The exploitation of primary sources of silver ore accounted for about 30% of silver output in 2004.

In 2004 the world-wide use of silver in fabricated products totalled 836.7m. troy ounces (a decline of 2% from the 2003 level). The manufacture of coins and medals absorbed 41.1m. oz (4.9% of total fabrication demand) in 2004. Of the remainder, 367.1m. oz were consumed for industrial purposes (compared with 350.5m. oz in 2003). The production of photographic materials (including X-ray film) absorbed 181m. oz in 2004 (compared with 192.9m. oz in 2003), while 247.5m. oz (29.6%) went into jewellery and silverware. The manufacture of electronic equipment and batteries in the major industrialized countries absorbed 166.5m. oz of silver (19.9% of world demand) in 2004. Other industrial uses for silver include the production of brazing alloys, mirrors and catalysts. In 2004 the principal user of silver in fabricated products (including coinage) was the USA, which consumed 180.3m. oz (21.5% of the world total), followed by Japan, with 125.1m. oz (15%), and India, with 79.2m. oz (9.5%). Other major users of silver in 2004 were Italy (55m. oz) and the People's Republic of China (52m. oz).

World mine production of silver was estimated to have increased by 25% during the period 1979–85, as a result of generally higher (although widely fluctuating) prices. However, while output advanced, consumption declined, resulting in a world surplus of silver in each year during 1980–89. In 1990 this trend was reversed, and, according to the Silver Institute (an international association of miners, refiners, fabricators and manufacturers, with its headquarters in Washington, DC, USA), world demand exceeded mine and secondary (recycled and scrap silver) production, creating the first silver market supply deficit since 1978. In each of the years 1991–97 world consumption of silver outpaced mine production, which declined by 13.8% over the period 1990–94. Output recovered by 6.2% in 1995 and by 1.7% in 1996, partly as a result of new mines entering production, together with the reopening of a number of primary silver mines as a result of improved prices. A third consecutive year of increased production was achieved in 1997, when mine output rose by 6%, and in 1998 production advanced by a further 4.5% to a record 544m. oz. In 1999 the output of silver from the world's mines rose slightly, by 0.8%, to 548.5m. oz. In 2000 world output of mined silver rose by 7.1% to a new record level of 587.3m. oz. Record mined production was again achieved in 2001, when output increased by 4.2% to 611.8m. oz. In 2002, however, world mined production fell slightly, by 0.7%, to 607.4m. oz. Sharp falls in the mined output of the USA and Chile were among the factors contributing to the world-wide decline in 2002. In 2003 world mined production increased by 0.6%, to 595.6m. oz. World output of silver rose by 3.8% to a new record level of 634.4m. oz in 2004. Increased output by the world's four major producers—Mexico, Peru, Australia and China—was reported to have accounted for most of the increase in 2004.

Mexico is the world's leading producer of silver. Output has fluctuated somewhat in recent years, declining from 91.6m. oz in

1998 to 75.2m. oz in 1999. Production recovered to 88.3m. oz in 2000. In 2001 output rose again, to 97.4m. oz, and again, in 2002, to 101.2m. oz, but declined in 2003, by 6.4%, to 94.7m. oz. In 2004 Mexican production of silver increased by 4.8%, to 99.2m. oz. Mexico's Proaño mine, near Fresnillo (in the state of Zacatecas), was the second largest primary silver mine in the world (in terms of production) in 2003 and 2004. An expansion programme at the Proaño mine was completed in the latter year. In 2004 Mexico's reserves of silver ore were estimated to represent about 13.7% of total reserves world-wide. Peru is second to Mexico in terms of silver production world-wide, accounting for about 16% of output in 2004. In the same year Chile, the third most important regional producer, accounted for about 7% of world production. Although a relatively minor producer, Argentina has significantly increased its silver output since 1998 as the result of two new mines entering into production in that year. In 2004 output amounted to 5m. oz, compared with 4.8m. oz in 2003. Following a positive feasibility study, development of one of the world's largest open-pit silver projects commenced in 2000 at the San Cristóbal deposits in the Potosí department of southern Bolivia. Proven and probable reserves at San Cristóbal, where production was expected to begin by the end of 2004, total 240m. metric tons of ore, containing 2.0 oz of silver per ton. The San Bartolomé silver project, also located in the Potosí department, was expected to commence operations in mid-2004.

As an investment metal (an estimated 80% of the world's silver bullion stocks are speculative holdings), silver is highly sensitive to factors other than the comparative levels of supply and demand. Silver, like gold and platinum, is customarily measured in troy weight. The now obsolete troy pound contains 12 ounces (oz), each of 480 grains. One troy ounce is equal to 31.1 grams (1 kg = 32.15 troy oz), compared with the avoirdupois ounce of 28.3 grams.

Fluctuations in the price of silver bullion traditionally tended to follow trends in prices for gold and other precious metals. However, silver has come to be viewed increasingly as an industrial raw material and hence likely to decrease in price at times of economic recession. Two of the main centres for trading in silver are the London Bullion Market (LBM) and the New York Commodity Exchange (COMEX). Dealings in silver on the LBM are only on the basis of 'spot' contracts (for prompt delivery), while COMEX contracts are also for silver 'futures' (options to take delivery at specified future dates). Over a 15-day period in January 1992 the LBM price of silver per troy ounce increased from £2.06 (US $3.87) to £2.41 ($4.335), and until the middle of the year it remained within this range. Thereafter, the silver market slumped, and in August the LBM quotation was reduced to £1.83 ($3.65) per oz. In terms of sterling, the London silver price later moved sharply higher, but the trend was largely a reflection of the British currency's depreciation in relation to the US dollar. In February 1993 the LBM price reached £2.65 (then $3.77) per oz, but later in the month it stood at £2.46 (only $3.58). In March the London silver price in US currency was only $3.56 per oz: just above the 17-year 'low' of $3.55 recorded in February 1991. However, a vigorous recovery ensued, with the LBM quotation advancing to £3.66 ($5.42) per oz in August 1993, before declining to £2.55 ($3.92) in September. With demand continuing to exceed supply, another advance took the silver price to £3.45 ($5.115) per oz at the end of the year.

**Silver Prices on the New York Commodity Exchange**
(average 'spot' quotations, US $ per troy ounce)

|  | Average | Highest month(s) | Lowest month(s) |
|---|---|---|---|
| 1990 | 4.817 | — | — |
| 1995 | 5.185 | — | — |
| 2002 | 4.601 | 5.125 (June) | 4.223 (Jan.) |
| 2003 | 4.896 | 5.993 (Dec.) | 4.346 (Mar.) |
| 2004 | 6.658 | 8.290 (April) | 5.496 (May) |

Sources: The Silver Institute; Gold Fields Mineral Services Ltd.

The recovery in the silver market continued in the early months of 1994, partly as a result of speculative activity, and by the end of March the London price had reached £3.87 (US $5.735) per oz. A week earlier, the price in US currency was $5.75 per oz. After another decline, the LBM quotation stood at £3.62 ($5.71) per oz in September. Thereafter, the price of silver moved generally downwards, declining to £2.965 ($4.64) per oz in December.

Silver prices fell further in the early months of 1995, with the metal trading in London at £2.68 (US $4.42) per oz in March. The price later rose sharply, and in May it reached £3.75 ($6.04) per oz. In August the LBM price of silver per oz stood at £3.76, then equivalent to $5.80. As on previous occasions, the surge in prices was attributed mainly to speculative investment.

Another advance took the LBM silver quotation to £3.84 (US $5.83) per oz in February 1996, but the price moved generally downward during the remainder of the year, reaching £2.81 ($4.73) in December. For the year as a whole, the average London price of silver was $5.199 per oz.

There were extreme fluctuations in the price of silver in 1997. The London price declined to £2.75 (US $4.66) per oz in January, but recovered to £3.30 ($5.26) in March. However, in early July the London quotation was reduced to £2.51 ($4.25) per oz, and a week later the price in US currency was only $4.22: its lowest level since November 1993. Prices subsequently rallied, and in December 1997, with silver in short supply after a steady decline in stocks (see below), the London quotation rose to £3.76 per oz, equivalent to $6.27: a nine-year 'high' in the US currency price. Despite the strong advance in the second half of the year, the average London silver price in 1997 was $4.897 per oz, nearly 6% lower than in the previous year. The average price of gold in 1997 was 67.6 times that of silver, but by the end of the year the ratio was only 47.

During the early weeks of 1998 there was intense speculative activity, amid allegations of manipulation, in the silver market. In early February it was disclosed that one major US investor had purchased 129.7m. oz (more than 4,000 metric tons) of the metal, equivalent to about 15% of annual world demand, during the preceding six months. Three days after this report, the London silver price per oz increased to £4.76, equivalent to US $7.81: its highest level, in terms of US currency, since July 1988. Meanwhile, the price of gold (which had fallen to its lowest level for more than 18 years) was only 38.3 times that of silver: the smallest ratio since 1983. The surge in silver prices was short-lived, despite the continuing scarcity of supplies, and in May 1998, following a sharp fall in demand for the metal from India (the main importing country), the London quotation was reduced to £3.04 ($4.96) per oz. Silver prices remained fairly stable for the next few months, but they declined further in August. At the beginning of September the LBM price per oz was £2.82 ($4.73), although by the end of the month it had recovered to £3.175 ($5.385). After another fall, the price in early December stood at £2.81 ($4.69) per oz, the lowest for the year. The average London price of silver in 1998 was $5.544 per oz: 13.2% above the 1997 average and the highest annual level since 1988. The difference between the maximum and minimum prices per oz in 1998 was $3.12, equivalent to 56.3% of the annual average: the widest trading range since 1987. For 1998 as a whole, the average price of gold was 53.0 times that of silver: the lowest annual ratio since 1985.

The London silver price advanced to US $5.75 (£3.525) per oz in February 1999, but declined to $4.88 (£3.044) in June. The price recovered to $5.71 (£3.470) per oz in September and ended the year at $5.33 (£3.298). For the whole of 1999, the London price of silver averaged $5.22 per oz. In late March 2000 the price of silver stood at $4.935 (£3.094) per oz. In terms of US currency, it was lower in early June, at $4.895 (£3.271) per oz. For the whole of 2000, the London price of silver averaged $4.95 per oz, a decline of 5.1% compared with the average price recorded in 1999. In 2001, the decline was even greater, the London price of silver averaging only $4.37 per oz, 12% lower than the average recorded in 2000. In 2002 the London price of silver averaged $4.60 per oz, an increase of 5.3% compared with the average price recorded in 2001. In 2003 the London price of silver averaged $4.85 per oz, an increase of 5.4% compared with 2002. In 2004 the average price of silver traded on the LBM, at $6.656 per oz, was some 37% higher than that recorded in 2003. According to the Silver Institute, investment activity was the key determinant of the increase in the price of silver in 2004. Additional factors were a high level of industrial fabrication and

# REGIONAL INFORMATION

## Major Commodities of Latin America

a decline in government sales of silver, notably those of the Chinese Government.

**Production of Silver Ore**
(provisional figures, silver content, metric tons)

|  | 2003 | 2004 |
|---|---:|---:|
| **World total** | 19,011 | 19,731 |
| Latin America | 7,866 | 8,110 |
| **Leading Latin American producers** | | |
| Argentina | 150 | 156 |
| Bolivia | 466 | 407 |
| Chile | 1,309 | 1,330 |
| Honduras | 54 | 50 |
| Mexico | 2,946 | 3,085 |
| Peru | 2,921 | 3,060 |
| **Other leading producers** | | |
| Australia | 1,872 | 2,237 |
| Canada | 1,282 | 1,262 |
| China, People's Repub. | 1,828 | 1,985 |
| Poland | 1,376 | 1,362 |
| USSR (former)* | 1,890 | 1,927 |
| USA | 1,240 | 1,250 |

*Estimated production. The main producers of silver among the successor states of the USSR are Russia (1,180 metric tons in 2004) and Kazakhstan (642 metric tons in 2004).

Sources: The Silver Institute; Gold Fields Mineral Services Ltd.

## Sugar

Sugar is a sweet crystalline substance which may be derived from the juices of various plants. Chemically, the basis of sugar is sucrose, one of a group of soluble carbohydrates which are important sources of energy in the human diet. It can be obtained from trees, including the maple and certain palms, but virtually all manufactured sugar is derived from two plants, sugar beet (*Beta vulgaris*) and sugar cane, a giant perennial grass of the genus *Saccharum*.

Sugar cane, found in tropical areas, grows to a height of up to 5 m (16 ft). The plant is native to Polynesia but its distribution is now widespread. It is not necessary to plant cane every season as, if the root of the plant is left in the ground, it will grow again in the following year. This practice, known as 'ratooning', may be continued for as long as three years, when yields begin to decline. Cane is ready for cutting 12–24 months after planting, depending on local conditions. Much of the world's sugar cane is still cut by hand, but rising costs are hastening the change-over to mechanical harvesting. The cane is cut as close as possible to the ground, and the top leaves, which may be used as cattle fodder, are removed.

After cutting, the cane is loaded by hand or by machine into trucks or trailers and transported directly to a factory for processing. Sugar cane rapidly deteriorates after it has been cut and should be processed as soon as possible. At the factory the cane passes first through shredding knives or crushing rollers, which break up the hard rind and expose the inner fibre, and then to squeezing rollers, where the crushed cane is subjected to high pressure and sprayed with water. The resulting juice is heated, and lime is added for clarification and the removal of impurities. The clean juice is then concentrated in evaporators. This thickened juice is next boiled in steam-heated vacuum pans until a mixture or 'massecuite' of sugar crystals and 'mother syrup' is produced. The massecuite is then spun in centrifugal machines to separate the sugar crystals (raw cane sugar) from the residual syrup (cane molasses).

The production of beet sugar follows the same process, except that the juice is extracted by osmotic diffusion. Its manufacture produces white sugar crystals which do not require further refining. In most producing countries, it is consumed domestically, although the European Union (EU), which generally accounts for about 13% of total world sugar production, is a net exporter of white refined sugar. Beet sugar accounts for more than one-third of world production. Production data for sugar cane generally cover all crops harvested, except crops grown explicitly for feed. The second table covers the production of raw sugar by the centrifugal process. In the late 1990s global output of non-centrifugal sugar (i.e. produced from sugar cane which has not undergone centrifugation) was about 14m. tons per year.

Most of the raw cane sugar which is produced in the world is sent to refineries outside the country of origin, unless the sugar is for local consumption. Cuba, Thailand, Brazil and India are among the few cane-producers that export part of their output as refined sugar. The refining process further purifies the sugar crystals and eventually results in finished products of various grades, such as granulated, icing or castor sugar. The ratio of refined to raw sugar is usually about 0.9:1.

**Production of Sugar Cane**
('000 metric tons)

|  | 2003 | 2004* |
|---|---:|---:|
| **World total** | 1,353,518 | 1,323,952 |
| Latin America | 598,230 | 621,662 |
| **Leading Latin American producers** | | |
| Argentina‡ | 19,250 | 19,300 |
| Brazil | 389,849 | 410,983 |
| Colombia‡ | 37,000 | 37,100 |
| Cuba | 22,902 | 24,000‡ |
| Dominican Repub. | 5,019 | 5,256 |
| Ecuador | 4,623 | 5,400‡ |
| Guatemala | 17,400‡ | 18,000 |
| Mexico | 45,127 | 45,127‡ |
| Peru‡ | 9,650 | 7,950 |
| Venezuela | 9,950 | 9,832 |
| **Other leading producers** | | |
| Australia | 37,968 | 36,892 |
| China, People's Repub. | 92,039 | 90,635 |
| India | 281,600 | 244,800 |
| Indonesia† | 24,500 | 24,600 |
| Pakistan | 52,056 | 53,419 |
| Philippines‡ | 30,000 | 28,000 |
| Thailand | 74,258 | 67,900 |
| USA | 30,715 | 9,832 |

*Provisional figures.
†Unofficial figures.
‡FAO estimate(s).

**Production of Centrifugal Sugar**
(raw value, '000 metric tons)

|  | 2003 | 2004* |
|---|---:|---:|
| **World total** | 149,819 | 146,383 |
| Latin America | 46,586 | 48,876 |
| **Leading Latin American producers** | | |
| Argentina† | 1,925 | 1,740 |
| Brazil† | 26,400 | 28,150 |
| Colombia | 2,646 | 2,664 |
| Cuba | 2,205 | 2,530† |
| Guatemala† | 1,801 | 1,900 |
| Mexico† | 5,442 | 5,540 |
| **Other leading producers** | | |
| Australia | 5,461 | 5,022 |
| China† | 10,944 | 11,391 |
| France† | 4,340 | 4,613 |
| Germany† | 4,120 | 4,456 |
| India | 22,140 | 15,450 |
| Pakistan† | 4,004 | 4,064 |
| South Africa | 2,412† | 2,247‡ |
| Thailand† | 7,670 | 7,460 |
| Turkey† | 1,989 | 1,960 |
| USA | 8,118 | 7,100‡ |

*Provisional figures.
†Unofficial figure(s).
‡FAO estimate(s).

As well as providing sugar, quantities of cane are grown in some countries for seed, feed, fresh consumption, the manufacture of alcohol and other uses. Molasses may be used as cattle feed or fermented to produce alcoholic beverages for human consumption, such as rum, a distilled spirit manufactured in Caribbean countries. Sugar cane juice may be used to produce ethyl alcohol (ethanol). This chemical can be utilized, either exclusively or mixed with petroleum derivatives, as a fuel for motor vehicles. The steep rise in the price of petroleum after 1973 made the large-scale conversion of sugar cane into ethanol economically attractive (particularly to developing nations), especially as sugar, unlike petroleum, is a renewable source of energy. Several countries developed ethanol production by this means in order to reduce petroleum imports and to support cane growers. Ethanol-based fuel, which generates fewer harmful exhaust hydrocarbons than petroleum-based fuel, may be known as 'gasohol', 'alcogas', 'green petrol' or, as in Brazil, simply as 'alcohol'. Brazil was the pioneer in this field, establishing the largest ethanol-based fuel production programme—'PROALCOOL'—in the world. Public subsidies and tax concessions encouraged farmers to plant more sugar cane, investors to construct more distilleries and designers to blueprint cars fuelled exclusively by ethanol. At the same time, the Government created an extensive distribution network to transport ethanol-based fuel to filling stations, where its price was maintained low through subsidization. By the mid-1980s almost every new car sold in Brazil was fuelled exclusively by ethanol. In the 1990s, however, a shortage of ethanol, in conjunction with lower world petroleum prices and the Government's withdrawal of ethanol subsidies resulted in a sharp fall in Brazil's output of these vehicles. In mid-2005, however, as the Brazilian Government strove to achieve self-sufficiency in petroleum supply, the popularity of ethanol-fuelled motor vehicles was reported to be rising once more. At that time, about 40% of all vehicle fuel utilized in Brazil was ethanol-based and retailed for about one-half of the price of petroleum-based fuel. Current legislation requires all gasoline sold in Brazil to contain 25% ethanol and a new generation of vehicles has been designed to run on either gasoline, ethanol or combinations of both. The consequence of the renewed demand for ethanol has been a huge increase in investment in sugar production and in facilities for its conversion into ethanol.

After the milling of sugar, the cane has dry fibrous remnants known as bagasse, which is usually burned as fuel in sugar mills but can also be pulped and used for making fibreboard, particle board and many grades of paper. As the costs of imported wood pulp have risen, cane-growing regions have turned increasingly to the manufacture of paper from bagasse. In view of rising energy costs, some countries (such as Cuba) have encouraged the use of bagasse as fuel for electricity production in order to conserve foreign exchange expended on imports of petroleum. Another by-product, cachaza, has also been utilized as an animal feed.

In recent years sugar has encountered increased competition from other sweeteners, including maize-based products, such as isoglucose (a form of high-fructose corn syrup or HFCS), and chemical additives, such as saccharine, aspartame and xylitol. Aspartame (APM) was the most widely used high-intensity artificial sweetener in the early 1990s, although its market dominance was under challenge from sucralose, which is about 600 times as sweet as sugar (compared with 200–300 times for other intense sweeteners) and is more resistant to chemical deterioration than APM. From the late 1980s research was conducted in the USA to formulate means of synthesizing thaumatin, a substance derived from the fruit of the West African katemfe plant, *Thaumatococcus daniellii*, which is about 2,500 times as sweet as sugar. As of 2005, the use of thaumatin had been approved in the EU, Israel and Japan, while in the USA its use as a flavouring agent had been endorsed. In 1998 the US Government approved the domestic marketing of sucralose, the only artificial sweetener made from sugar. Sucralose was stated to avoid many of the taste problems associated with other artificial sweeteners.

The area under sugar cultivation in the whole of Latin America has increased in an attempt to satisfy greater domestic consumption, to diversify from predominant industries (e.g. coffee, cocoa) and to reduce dependence on imported sugar. By its sheer geographical extent, the region offers a wide range of conditions for the cultivation of sugar cane and, to a limited degree, of sugar beet (in Chile and Uruguay). The sugar industries of the region operate a diversity of modern mechanized and labour-intensive cultivation and processing techniques. Producer costs vary from country to country, according to production methods, the availability of labour and the level of wages. Latin American countries have exported, on average, more than 18m. metric tons (in terms of raw sugar equivalence) annually of raw and refined sugar in recent years, making theirs the largest exporting region in the world. The highest yields in the region (and among the highest in the world) have been obtained in Peru, where intensive use of fertilizers and irrigation has compensated for the natural disadvantages of light soil and aridity. Peruvian output of sugar cane has averaged about 9m. tons in recent years. The sugar industry in Guatemala has been expanded rapidly and annual production of sugar cane has averaged about 17m. tons in recent years. Guatemala is now generally the third largest regional exporter, after Brazil and Cuba. About 70%–75% of Guatemala's total output is exported, mainly, in recent years, to the Republic of Korea and Russia. Guatemala has also supplied a higher percentage than any other country of 'above quota' sugar to the USA, under that country's re-export programme. International labour organizations have criticized some aspects of working conditions in the Guatemalan sugar industry. Guatemala is a party to the Central American Free Trade Agreement (CAFTA, see below) and one of the principal objections to it of opponents of that Agreement, both in Guatemala and the USA, has been its scant provisions for safeguarding and enforcing national labour legal rights. Venezuela has become a substantial importer of sugar (about 40% of its requirements were imported in 1997), owing partly to the demands of its petroleum industry, which have reduced the availability of labour to other industries and necessitated the introduction of costly mechanized techniques. South America is also a significant producer of non-centrifugal sugar, mainly from Colombia and Brazil, all of which is consumed locally.

Brazil is firmly established as the world's largest producer of sugar cane, and as a leading producer of sugar and ethanol (see above). Output of sugar cane has risen rapidly in recent years: in 2004 it totalled an estimated 411m. metric tons, compared with about 328m. tons in 2000. FAO forecast that Brazil's output of centrifugal sugar would total 27.5m. tons in 2004/05. Brazil is one of the world's most efficient sugar producers and, because it possesses the infrastructure to process cane into either sugar or ethanol, is able to respond swiftly to market conditions. As the world's largest producer of sugar cane, Brazil exerts considerable influence on the world market price of sugar, of which the balance in the country between ethanol and raw sugar production is a major determinant. Although as much as one-half of the country's total production of sugar cane may be processed into ethanol, Brazil is regularly among the world's leading three exporters of raw sugar. The central-southern region is the most important producing area—the state of São Paulo alone may account for up to 60% of national output—followed by the northeast, in particular the states of Pernambuco and Alagoas. Owing to the relatively greater economic significance of sugar production to the north-eastern region, the Government has tended to grant it the entire annual US sugar import quota allocation, for which premium prices are obtained. Domestic demand for ethanol and for sugar itself determine export availability—Brazil is a leading consumer as well as exporter of the commodity. In recent years the competitiveness of Brazilian sugar on the world market has been boosted by such factors as the steady decline in the value of the currency, the real, relative to the US dollar, and improved land transport and port facilities. Brazil's foreign exchange earnings from sales of sugar and sugar preparations amounted to US $2,403.8m. (about 4.1% of total export revenue) in 2001, compared with $1,295.2m. (about 2.3% of total export revenue) in 2000.

The Cuban economy has traditionally relied heavily on sales of sugar and sugar products. The share of these exports in Cuba's total earnings from trade has been as high as 90%, but low world prices in the 1980s, occasionally not even covering producers' costs, affected all sugar exporters. Sugar and sugar products accounted for 73.2% of Cuban export revenue in 1989, compared with about 43% in the late 1990s. In 2000 exports of raw sugar accounted for 26.7% of Cuba's export revenue. Prior to the revolution of 1959–60, Cuba exported more than 50% of its sugar to the USA. Following the trade embargo imposed by the USA, approximately 60% of annual exports were taken by the USSR and other Eastern bloc countries, mostly under long-term trade and barter agreements at preferential prices. (In 1990 the USSR supplied Cuba with 95% of its total petroleum requirements.) Since 1991, however, these arrangements have ceased, and the Cuban sugar trade with the successor republics of the USSR, as well as with Eastern European countries and the People's Republic of China, is now conducted on the basis of full market prices, and Cuba has sought new markets in Canada,

North Africa, and in the Middle East and Far East. However, Cuban sugar production has declined sharply since the early 1990s, owing both to adverse weather conditions, especially drought, and to disruptions in the procurement of fuel, fertilizers, mill equipment and other essential production inputs, as a result of the US embargoes. These factors, which resulted in the 1994/95 harvest declining to the lowest level in 50 years, necessitated over-extended harvests and the use of reserves of cane intended for future crops. Consequently, in 1993 sugar exports were temporarily suspended. Difficulties have arisen in meeting subsequent export commitments, which included an agreement to supply Russia with 1m. metric tons of sugar in the year to March 1996 in exchange for 3m. tons of petroleum products. Russia has remained the principal market for Cuban sugar, absorbing approximately 28% of the total exported during 1994–96. Cuba was, however, unable to fulfil a contract to supply China with 400,000 tons of sugar during 1995. In that year the Cuban Government negotiated loans totalling US $100m. from private European sources for the purchase of fertilizers and sugar factory equipment. Additional foreign finance was obtained to assist a further recovery in the 1996/97 season, when exports of sugar were estimated to have earned $950m. in foreign exchange. A disappointing crop in 1997/98, however, was expected to generate only $432m. in export revenue. In 2002 a restructuring programme was initiated that involved the closure of 71 of Cuba's 156 sugar mills and, reportedly, the reassignment of as much as 60% of sugar plantation land to other crops. In 2005, however, the failure of the programme to revitalize the sector was apparent and the closure of more mills was forecast. Output in 2005, at an estimated 1.3m. tons, was at its lowest level since 1908, necessitating imports in order to satisfy contractual obligations and domestic consumption.

Mexico was a sugar exporter until 1979. The low domestic price of sugar, which the Government traditionally set below the cost of production, discouraged the purchase of modern equipment and the renovation of sugar plants. The result was stagnation in sugar production, a steady rise in domestic sugar consumption to a level among the highest in the world (45 kg per head per year), and the need to import sugar. Since 1980, as part of a plan to increase production, to slow the rate of increase in demand and to allow the Government to reduce the subsidy on sugar, domestic sugar prices have been set above the level of production costs. During the period 1980–85, as much as an additional 170,000 ha were planted to sugar cane. Under the North American Free Trade Agreement (NAFTA), which entered into operation in January 1994, Mexico was permitted to substantially increase its sugar exports to the USA and Canada after NAFTA arrangements for the duty-free access of Mexican sugar to the US market took effect in 2001. Since 2002, however, Mexico has been involved in a dispute with the USA over a tax of 20% that it has applied since that year to soft drinks that are sweetened with anything other than Mexican sugar cane syrup. The Mexican Government ostensibly imposed the tax in order to protect the Mexican sugar industry from competition from US HFCS, but it has also been viewed by some observers as a retaliatory response to the USA's interpretation of the sugar-related provisions of the NAFTA, which has prevented Mexico from obtaining a significant sugar quota for the US market. In mid-2005 the dispute remained under consideration by the World Trade Organization (WTO). Mexican output of sugar cane has averaged about 45m. metric tons in recent years. FAO forecast that Mexican production of centrifugal sugar would total 5.5m. tons in 2004/05.

The sugar sector in Colombia has undergone a major expansion since the 1960s, in an attempt to diversify the economy away from coffee. Production of sugar cane has averaged about 35m. metric tons in recent years. About one-half of Colombian production is exported, and in 2002 Colombia ranked as Latin America's fourth largest exporter of sugar. In the main production zone, the Cauca river valley in south-western Colombia, climatic conditions permit the cultivation of sugar uninterruptedly throughout the year, resulting in exceptionally low fixed-investment costs per ton of sugar produced.

In some of the smaller Latin American countries and the Caribbean islands, sugar cultivation is the main bulwark of the economy. Sugar is the most important agricultural product of the Caribbean Community and Common Market (CARICOM) economic and political grouping, for instance. In Belize, sugar cane plantations occupy about 50% of the total cultivated area, while the sugar industry employs about one-quarter of the labour force. In 2004 sales of sugar contributed 19.9% of total export revenue. Sugar cane is the principal commercial crop in the Dominican Republic, but inefficient production techniques, lack of investment and falling international sugar prices depressed the industry throughout the 1980s. The country's output of raw sugar increased from 580,000 metric tons in 1995 to 741,000 tons in 1996 and to 813,000 tons in 1997. However, 'Hurricane Georges' caused severe damage to cane plantations and sugar mills in September 1998. Exports of sugar and sugar cane derivatives accounted for an estimated 7% of the Dominican Republic's total export revenue (excluding exports from free-trade zones) in 2004. Sugar has traditionally been a major export commodity in Barbados, where it and molasses contributed an estimated 9.2% of total export revenue (including re-exports) in 2003; in Guyana, where it contributed 25.6% of total export revenue (including re-exports) in 2003; and in Saint Christopher and Nevis, where 21% of total export revenue was derived from sales of raw sugar in 2001. In 2005, however, it was reported that Barbadian production had declined to the point that the country was unable to meet its EU export quota. Saint Christopher and Nevis has announced that the 2005 sugar harvest will be the country's last as rising production costs and a fall in revenues have left the state-owned sugar company there heavily in debt.

Representatives of US sugar cane and sugar beet producers have strongly opposed the CAFTA concluded between the USA and Costa Rica, El Salvador, Guatemala, Honduras, Nicaragua and the Dominican Republic in 2004. Under the terms of the CAFTA, which was approved by the US Congress in July 2005, most tariffs on goods traded between the USA and the Central American and Caribbean participants would be eliminated. The US Administration of George W. Bush has undertaken to maintain a 'ceiling' on sugar imports from its six CAFTA partners for the duration of current US farm legislation, and to investigate the feasibility of instituting a sugar-based ethanol programme—US ethanol production is currently maize-based.

The first International Sugar Agreement (ISA) was negotiated in 1958, and its economic provisions operated until 1961. A second ISA did not come into operation until 1969. It included quota arrangements and associated provisions for regulating the price of sugar traded on the open market, and established the International Sugar Organization (ISO) to administer the agreement. However, the USA and the six original members of the European Community (EC, now the EU) did not participate in the ISA, and, following its expiry in 1974, it was replaced by a purely administrative interim agreement, which remained operational until the finalization of a third ISA, which took effect in 1978. The new agreement's implementation was supervised by an International Sugar Council (ISC), which was empowered to establish price ranges for sugar-trading and to operate a system of quotas and special sugar stocks. Owing to the reluctance of the USA and EC countries (which were not a party to the agreement) to accept export controls, the ISO ultimately lost most of its power to regulate the market, and since 1984 the activities of the organization have been restricted to recording statistics and providing a forum for discussion between producers and consumers. Subsequent ISAs, without effective regulatory powers, have been in operation since 1985. (For detailed information on the successive agreements, see *South America, Central America and the Caribbean 1991*.)

Special arrangements for exports of sugar from 15 Caribbean states (including Belize and Guyana) existed in the successive Lomé Conventions, in operation since 1975, between the EU and a group of African, Caribbean and Pacific (ACP) countries, whereby a special Protocol on sugar, forming part of each Convention, required the EU to import specified quantities of raw sugar annually from ACP countries. In June 1998 it was announced that preferential sugar prices paid by the EU to the ACP countries would fall to open-market levels within three years. Under the terms of the Cotonou Agreement (covering the period 2000–20), the successor convention to the fourth Lomé Convention, the EU undertook to purchase and import, at guaranteed prices, specific quantities of sugar originating in ACP states for an indefinite period.

In tandem with world output of cane and beet sugars, stock levels (of centrifugal sugar) are an important factor in determining the prices at which sugar is traded internationally. These stocks, which were at relatively low levels in the late 1980s, increased significantly in the 1990s, although not, according to USDA data, in each successive trading year (September–August). In the early 1990s rises in stocks were due partly to the disruptive effects of the Gulf War on demand in the Middle East (normally a major sugar-consuming area), and were also a result of considerably increased production in Mexico and the Far East. Another factor was the increased area under sugar cane and beet in the EU and in Australia. In 2002/03, according to data released by USDA, when world sugar production totalled about 149m. metric tons and consumption some 138m. tons, stocks of sugar held world-wide rose to about 41m. tons, compared with about 37m. tons in 2001/02. This increase was succeeded, in 2003/04, when world sugar production was estimated at about 142m. tons and consumption at about 139m. tons, by a decline in stocks held to 39m. tons. In 2004/05, when the output of sugar world-wide again totalled about 142m. tons and consumption rose to about 141m. tons, stocks declined again, to some 36m. tons. In 2005/06 USDA forecast, on the basis of estimated world production of 146m. tons and estimated world consumption of some 143m. tons, that world sugar stocks would fall for a third consecutive year, to about 34m. tons.

The world market for sugar has been described as one of the most distorted of all commodity markets, and reference has frequently been made to the sugar regime of the EU to explain how this distortion has arisen. EU sugar producers have been protected, through the Union's Common Agricultural Policy, by means of import duties, production quotas and export subsidies. First, international competition has been excluded from EU markets by the application of very high duties on imported sugar. Second, the EU has guaranteed member states' producers high production quotas for which it has paid prices substantially in excess of the prevailing world price of sugar, leading to overproduction. Third, export subsidies have then been paid to allow surplus EU sugar to be sold on world markets at prices that are lower than the cost of production—a practice known as 'dumping'. As a consequence, the EU has become the world's largest exporter of refined sugar. The disposal of large quantities of sugar in this way has depressed the world price of sugar, on which depend smaller producers world-wide, who have, for the most part, no access to government subsidies. Sugar prices fluctuate constantly since it is traded on commodity markets. It has been calculated that in 1980–2000, in real terms, the world price of sugar declined by 76%, and this situation has been largely attributed to the increase in supply caused by regimes like that of the EU while demand, owing to such factors as the greater popularity of sugar substitutes, has been falling. In mid-2005 the European Commission announced proposals radically to reform the EU's sugar regime. In April the WTO, in response to a complaint by Brazil, Australia and Thailand, had ruled that some of the EU's subsidized exports of sugar originating in ACP countries were illegal. ACP sugar producers, however, have criticized the EU's intention to remove the trade preferences their sugars enjoy (see above) as part of the proposed reform of the sugar regime.

**Export Price Index for Sugar**
(base: 1980 = 100)

|  | Average | Highest month(s) | Lowest month(s) |
|---|---|---|---|
| 1990 | 45 | — | — |
| 1995 | 48 | — | — |
| 2000 | 28 | 34 (Oct.) | 20 (March) |
| 2001 | 30 | 34 (Jan., May) | 24 (Oct.) |
| 2002 | 25 | 29 (Dec.) | 21 (June) |

In 2001 world sugar prices rose in response to, among other factors, reduced output in Cuba as a consequence of hurricane damage to crops there and reduced sugar beet production in the EU. For the whole of the year the ISA daily price averaged 8.64 US cents per lb. In 2002, however, the average ISA daily price declined to only 6.9 cents per lb. The decline in 2002 was particularly pronounced in the early part of the year as it became clear that there would be a substantial increase in Brazil's output of sugar cane and, consequently, in that country's export potential. The ISA recorded a marginally higher average daily price, of 7.10 cents per lb, in 2003. In 2004, again, a marginally higher daily price, of 7.16 cents per lb, was recorded. On a monthly basis, however, the ISA daily price rose markedly in the final quarter of 2004, averaging 8.41 cents per lb in October, 8.14 cents per lb in November and 8.26 cents per lb in December. In February 2005 the average ISA daily price rose to 9.12 cents per lb. An average daily price of 8.81 US cents per lb was recorded in the first six months of 2005. The average price of the No. 5 sugar contract (refined sugar, f.o.b. Europe, for immediate delivery) traded on the London International Financial Futures Exchange (LIFFE) was 11.29 cents per lb in 2001. Successive declines in the average price, to, respectively, 10.35 cents per lb and 9.74 cents per lb, were recorded in 2002 and 2003. In 2004 the average price recovered to 10.87 cents per lb. During the first four months of 2005 the average price of the contract was 11.63 cents per lb in January, 12.09 cents per lb in February, 12.02 cents per lb in March and 11.76 cents per lb in April.

The Group of Latin American and Caribbean Sugar Exporting Countries (GEPLACEA), with a membership of 23 Latin American and Caribbean countries, together with the Philippines, and representing about 50% of world cane production and some 45% of sugar exports, complements the activities of the ISO as a forum for co-operation and research. At the end of 1992 the USA withdrew from the ISO, following a disagreement over the formulation of members' financial contributions. The USA had previously provided about 9% of the organization's annual budget.

## Tin

The world's tin deposits, estimated by the US Geological Survey (USGS) to total 6.1m. metric tons in 2004, are located mainly in the equatorial zones of Asia, in central South America and in Australia. Cassiterite is the only economically important tin-bearing mineral, and it is generally associated with tungsten, silver and tantalum minerals. There is a clear association of cassiterite with igneous rocks of granitic composition, and 'primary' cassiterite deposits occur as disseminations, or in veins and fissures in or around granites. If the primary deposits are eroded, as by rivers, cassiterite may be concentrated and deposited in 'secondary', sedimentary deposits. These secondary deposits form the bulk of the world's tin reserves. The ore is treated, generally by gravity method or flotation, to produce concentrates prior to smelting.

Tin owes its special place in industry to its unique combination of properties: low melting point, the ability to form alloys with most other metals, resistance to corrosion, non-toxicity and good appearance. Its main uses are in tinplate (about 40% of world tin consumption), in alloys (tin-lead solder, bronze, brass, pewter, bearing and type metal), and in chemical compounds (in paints, plastics, medicines, coatings and as fungicides and insecticides). In the late 1990s a number of possible new applications for tin were under study, including its use in fire-retardant chemicals and as an environmentally preferable substitute for cadmium in zinc alloy anti-corrosion coatings on steel. The possible development of a lead-free tin solder was also receiving consideration.

In 2004, according to the USGS, Latin America accounted for 28.4% of estimated world output of tin concentrates, compared with 32.5% of world output in 2003. Bolivia's state-owned mining company, Corporación Minera de Bolivia (COMIBOL), ceased production of tin in 2000, and in 2003 small mines and co-operatives were reported to have accounted for more than 50% of mined output. In 2003 Bolivia was the world's seventh largest producer of primary tin, and this commodity accounted for 3.8% of the country's export revenue in 2001. Since the 1970s output has been in decline, mainly as a consequence of internal economic conditions and low world prices. In 1978 tin sales accounted for almost 60% of Bolivia's export revenues; by 1996 the comparable proportion was only 9.3%. Brazil possesses the world's largest tin mine, located at Pitinga, in Amazonas state, with identified reserves of 420,000 metric tons, although it was suggested in the late 1990s that the mine's total reserves could exceed 800,000 tons. Since the early 1990s, however, as a result of adverse market conditions, Brazilian tin output has been declining. In 2003 Brazil's revenue from exports of tin (mainly semi-manufactured, non-alloyed metal) was reported to have totalled US $22.3m., compared with $62.5m. in 1997. In 1994 Peru emerged as the region's leading producer of tin and the world's third largest source. Until recently, almost all of Peru's tin exports were in the form of ores and concentrates, which provided 1.5% of the country's total export earnings in 1996. However, Peru's first tin smelter, at Pisco, began operating in 1997, with an initial production capacity of 15,000 tons of metal per year. In 2003 Peru was the third largest producer of primary tin in the world, with estimated output of 39,181 tons, 9.4% higher than in 2002, and equivalent to about 14% of total world production. Peru exported 36,500 tons of tin, worth $175.2m., in 2003. Plans announced in 1998 for the resumption of tin-mining in Argentina had not been realized by the end of 2003.

Over the period 1956–85, much of the world's tin production and trade was covered by successive international agreements, administered by the International Tin Council (ITC), based in London, United Kingdom. The aim of each successive International Tin Agreement (ITA), of which there were six, was to stabilize prices within an agreed range by using a buffer stock to regulate the supply of tin. (For detailed information on the last of these agreements, see *South America, Central America and the Caribbean 1991*.) The buffer stock was financed by producing countries, with voluntary contributions by some consuming countries. 'Floor' and 'ceiling' prices were fixed, and market operations conducted by a buffer stock manager who intervened, as necessary, to maintain prices within these agreed limits. For added protection, the ITA provided for the imposition of export controls if the 'floor' price was being threatened. The ITA was effectively terminated in October 1985, when the ITC's buffer stock manager informed the London Metal Exchange (LME) that he no longer had the funds with which to support the tin market. The factors underlying the collapse of the ITA included its limited membership (Bolivia and the USA, leading producing and consuming countries, were not signatories) and the accumulation of tin stocks which resulted from the widespread circumvention of producers' quota limits. The LME responded by suspending trading in tin, leaving the ITC owing more than £500m. to some 36 banks, tin smelters and metals traders. The crisis was eventually resolved in March 1990, when a financial settlement of £182.5m. was agreed between the ITC and its creditors. The ITC was itself dissolved in July. Transactions in tin contracts were resumed on the LME in 1989.

These events lent new significance to the activities of the Association of Tin Producing Countries (ATPC), founded in 1983 by Malaysia, Indonesia and Thailand and later joined by Bolivia, Nigeria, Australia and Zaire (now the Democratic Republic of the Congo, DRC). Prior to the withdrawal of Australia and Thailand in 1996 (see below), members of the ATPC accounted for almost two-thirds of world production. The ATPC, which was intended to operate as a complement to the ITC and not in competition with it, introduced export quotas for tin for the year from 1 March 1987. Brazil and the People's Republic of China agreed to co-operate with the ATPC in implementing these supply restrictions, which, until their suspension in 1996 (see below), were renegotiated to cover succeeding years, with the aim of raising prices and reducing the level of surplus stocks. The ATPC membership also took stringent measures to control smuggling. Brazil and China (jointly accounting for more than one-third of world tin production) both initially held observer status at the ATPC. China became a full member in 1994, but Brazil remained as an observer, together with Peru and Viet Nam. China and Brazil agreed to participate in the export quota arrangements, for which the ATPC had no power of enforcement.

The ATPC members' combined export quota was fixed at 95,849 metric tons for 1991, and was reduced to 87,091 tons for 1992. However, the substantial level of world tin stocks, combined with depressed demand, led to mine closures and reductions in output, with the result that members' exports in 1991 were below quota entitlements. The progressive depletion of stock levels prompted a forecast by the ATPC, in May 1992, that export quotas would be removed in 1994 if these disposals continued at their current rate. The ATPC had previously set a target level for stocks of 20,000 tons, representing five weeks of world tin consumption. Projections that world demand for tin would remain at about 160,000 tons annually, together with continued optimism about the rate of stock disposals, led the ATPC to increase its members' 1993 export quota to 89,700 tons. The persistence, however, of high levels of annual tin exports by China (estimated to have exceeded 30,000 tons in 1993 and 1994, compared with its ATPC quota of 20,000 tons), together with sales of surplus defence stocks of tin by the US Government, necessitated a reduction of the quota to 78,000 tons for 1994. In late 1993 prices had fallen to a 20-year 'low' and world tin stocks were estimated at 38,000–40,000 tons, owing partly to the non-observance of quota limits by Brazil and China, as well as to increased production by non-ATPC members. World tin stocks resumed their rise in early 1994, reaching 48,000 tons in June. However, the effects of reduced output, from both ATPC and non-ATPC producing countries, helped to reduce stock levels to 41,000 tons at the end of December. In 1995 exports by ATPC members exceeded the agreed voluntary quotas by 10%, and in May 1996, when world tin stocks were estimated to have been reduced to 31,000 tons, the ATPC suspended its quota arrangements. Shortly before the annual meeting of the ATPC was convened in September, Australia and Thailand announced their withdrawal from the organization. Although China and Indonesia indicated that they would continue to support the ATPC, together with Bolivia, Malaysia and Nigeria (Zaire had ceased to be an active producer of tin), the termination of its quota arrangements in 1996, together with the continuing recovery of the tin market, indicated that its future role would be that of a forum for tin-producers and consumers. Malaysia, Australia and Indonesia left the ATPC in 1997, and Brazil became a full member in 1998. In June 1999, when the organization's headquarters were moved from Kuala Lumpur to Rio de Janeiro, the membership comprised Brazil, Bolivia, China, the DRC and Nigeria. It was hoped that Peru would join the ATPC following its relocation to South America.

**Production of Tin Concentrates**
(tin content, metric tons)

|  | 2003 | 2004* |
|---|---|---|
| **World total** (excl. USA) | 207,000 | 250,000 |
| Latin America† | 67,200 | 71,100 |
| **Leading Latin American producers** | | |
| Bolivia | 15,000 | 16,900 |
| Brazil | 14,200 | 14,000 |
| Peru | 38,000 | 40,200 |
| **Other leading producers** | | |
| Australia | 6,500 | 1,200 |
| China, People's Repub. | 50,000 | 100,000 |
| Indonesia | 70,000 | 70,000 |
| Malaysia | 3,400 | 3,500 |
| Portugal | 1,000 | 500 |
| Russia | 2,000 | 2,500 |
| Viet Nam | 4,600 | n.a. |

* Estimated production.
† Figures represent the sum of output in listed countries. Mexico is also a minor producer of tin-bearing ores and concentrates.

Source: US Geological Survey.

**Export Price Index for Tin**
(base: 1980 = 100)

|  | Average | Highest month(s) | Lowest month(s) |
|---|---|---|---|
| 1990 | 42 | — | — |
| 1995 | 37 | — | — |
| 2000 | 29 | 32 (Jan.) | 27 (Oct.–Dec.) |
| 2001 | 23 | 27 (Jan., Feb.) | 19 (Sept., Oct.) |
| 2002 | 22 | 23 (June, July, Oct., Dec.) | 19 (Feb.) |

In 2000 the average quotation for tin for immediate delivery traded on the LME was US $5,435 per metric ton. At the end of that year stocks of the metal held by the Exchange totalled 12,885 tons. In 2001 the average price of tin fell to $4,483 per ton, while at 31 December the LME's tin stocks had more than doubled to total 30,550 tons. The average 'spot' quotation of London tin declined to $4,062 per ton in 2002, while stocks had fallen to 25,610 by the end of the year. The price of tin recovered in 2003, the metal trading at an average price of $4,896 per ton. Stocks of tin held by the LME fell heavily during the course of 2003, totalling 14,475 tons at the end of the year. Demand for tin world-wide was estimated to have increased by about 7% in 2003, and provisional figures indicated a similar level of growth in 2004, when output surged in response to the growth that had occurred in the previous year. The price of tin in London increased by almost 75% in 2004—a larger increase than that recorded for any other metal traded on the LME—to an average of $8,513 per ton. The increase in the average price of the metal was especially marked in the second quarter of 2004, when it rose to $9,202 per ton, compared with only $6,951 in January–March. Stocks of tin held by the Exchange declined precipitously in 2004, totalling 8,160 tons at the end of the year. On the LME, during the first six months of 2005, tin traded within a range of $7,075 per ton (6 January) and $8,700 per ton (14 March). Inventories of the metal held by the LME continued to decline in the early part of the year, and at the end of February totalled only 3,155 tons. In mid-2005 the market for tin remained characterized by uncommonly low stocks—3,855 tons at the end of June. However, the price of tin was reportedly under pressure at that time in spite of strong global demand.

The success, after 1985, of the ATPC in restoring orderly conditions in tin trading (partly by the voluntary quotas and partly by working towards the reduction of tin stockpiles) unofficially established it as the effective successor to the ITC as the international co-ordinating body for tin interests. The International Tin Study Group (ITSG), comprising 36 producing and consuming countries, was established by the ATPC in 1989 to assume the informational functions of the ITC. In 1991 the secretariat of the United Nations Conference on Trade and Development (UNCTAD) assumed responsibility for the publication of statistical information on the international tin market. The International Tin Research Institute (ITRI), founded in 1932 and based in London, United Kingdom, promotes scientific research and technical development in the production and use of tin.

## Acknowledgements

We gratefully acknowledge the assistance of the following organizations in the preparation of this section:

Food and Agriculture Organization of the UN;
International Cocoa Organization;
International Coffee Organization;
International Copper Study Group;
International Iron and Steel Institute;
International Monetary Fund;
International Aluminium Institute;
International Sugar Organization;
The Silver Institute;
US Department of Energy;
US Geological Survey, US Department of the Interior.

Sources for Agricultural Production Tables (unless otherwise indicated): FAO. Source for Export Price Indexes: UN, *Monthly Bulletin of Statistics*.

# RESEARCH INSTITUTES

## ASSOCIATIONS AND INSTITUTIONS STUDYING LATIN AMERICA AND THE CARIBBEAN

### ARGENTINA

**Centro Argentino de Datos Oceanográficos (CEADO)** (Argentine Centre of Oceanographic Data): Avda Montes de Oca 2124, 1271 Buenos Aires; tel. and fax (11) 4303-2240; e-mail ceado@hidro.gov.ar; internet www.hidro.gov.ar; f. 1974; stores oceanographic data of national area, provides information to the scientific community, private and public enterprises and other marine users; Dir ARIEL HERNÁN TROISI.

**Centro Argentino de Información Científica y Tecnológica (CAICYT)** (Argentine Centre for Scientific and Technological Information): Saavedra 15, 1°, 1083 Buenos Aires; tel. (11) 4951-6975; fax (11) 4951-7310; e-mail caicyt@caicyt.gov.ar; internet www.caicyt-conicet.gov.ar; f. 1958; attached to Consejo Nacional de Investigaciones y Técnicas; Dir CECILIA MABRAGAÑA.

**Centro de Investigaciones Económicas (CIE)** (Economic Research Centre): Instituto Torcuato de Tella, Miñones 2177, 1428 Buenos Aires; tel. (11) 4783-8680; fax (11) 4783-3061; e-mail postmaster@itdtar.edu.ar; internet www.itdt.edu; library of 60,000 vols and 1,400 domestic and foreign periodicals; Dir ADOLFO CANITROT; publ. *Documentos de Trabajo*.

**Consejo Argentino para las Relaciones Internacionales (CARI)** (Argentine Council for International Relations): Uruguay 1037, 1°, 1016 Buenos Aires; tel. (11) 4811-0071; fax (11) 4815-4742; e-mail cari@cari1.org.ar; internet www.cari1.org.ar; f. 1978; Pres. CARLOS M. MUÑIZ.

**Instituto Interamericano de Estadística** (Inter-American Statistical Institute): Balcarce 184, 2°, Of. 211, 1327 Buenos Aires; tel. (11) 4349-5772; fax (11) 4349-5776; e-mail efabb@indec.mecon.gov.ar; internet www.indec.mecon.gov.ar/proyectos/index.htm; f. 1940; research, seminars, technical meetings; co-ordination and co-operation with the Organization of American States; consultative status with the Economic and Social Council; affiliated to the International Statistical Institute; Pres.-elect PEDRO LUIS DO NASCIMENTO SILVA; publ. *Estadística* (2 a year).

**Instituto para la Integración de América Latina y el Caribe** (Institute for the Integration of Latin America and the Caribbean): Esmeralda 130, 17°, 1035 Buenos Aires; tel. (11) 4320-1850; fax (11) 4320-1865; e-mail int/inl@iadb.org; internet www.iadb.org/intal; f. 1965 under auspices of IDB's Integration and Regional Program Dept; research on all aspects of regional integration and co-operation; activities are channelled through four lines of action: regional and national technical projects on integration; policy forums; integration forums; journals and information; documentation centre includes 100,000 documents, 12,000 books, 400 periodicals; Dir RICARDO CARICROFT; publs *Integración y Comercio* (2 a year), *Carta Mensual* (monthly newsletter), *Serie Informes Subregionales de Integración*.

### AUSTRALIA

**Australian Institute of International Affairs:** 32 Thesiger Court, Deakin, ACT 2600; tel. (2) 6282-2133; fax (2) 6285-2334; e-mail ceo@aiia.asn.au; internet www.aiia.asn.au; f. 1933; 1,800 mems; brs in all states; Pres. NEAL BLEWETT; Exec. Dir CHARLES HAMILTON STUART; publ. *The Australian Journal of International Affairs* (4 a year).

### AUSTRIA

**Österreichische Forschungsstiftung für Entwicklungshilfe** (Austrian Foundation for Development Research): 1090 Vienna, Berggasse 7; tel. (1) 317-40-10; fax (1) 317-40-15; e-mail office@oefse.at; internet www.oefse.at; f. 1967; documentation and information on development aid, developing countries and international development, particularly relating to Austria; library of 40,000 vols, 250 periodicals; Pres. Prof. K. ZAPOTOCZKY; publs *Ausgewählte Neue Literatur zur Entwicklungspolitik* (3 a year), *Österreichische Entwicklungspolitik* (annually), Läanderprofile.

**Österreichische Gesellschaft für Aussenpolitik und Internationale Beziehungen** (Austrian Association for Foreign Policy and International Relations): 1010 Vienna, Hofburg/Schweizerhof/Brunnenstiege; tel. (1) 535-46-27; fax (1) 532-26-05; e-mail oega@afa.at; internet afa.at/oega; f. 1958; lectures, discussions; approx. 400 mems; Pres. Dr WOLFGANG SCHALLENBERG; publ. *Österreichisches Jahrbuch für Internationale Politik*.

**Österreichisches Institut für Entwicklungshilfe und technische Zusammenarbeit mit den Entwicklungsländern** (Austrian Institute for Development Aid and Technical Co-operation with the Developing Countries): 1010 Vienna, Wipplingerstr. 35; tel. (1) 42-65-04; f. 1963; projects for training of industrial staff; Pres Dr HANS IGLER, ERICH HOFSTETTER.

**Österreichisches Lateinamerika Institut** (Austrian Latin American Institute): 1090 Vienna, Schlickgasse 1; tel. (1) 310-74-65; fax (1) 310-74-65-21; e-mail office@lai.at; internet www.lai.at; f. 1965; Dir Dr SIEGFRIED HITTMAIR; publs *Atención: Jahrbuch des Österreichischen Lateinamerika Instituts* (annually), *Diálogo Austria-América Latina* (annually).

### BARBADOS

**Sir Arthur Lewis Institute of Social and Economic Studies:** University of the West Indies, Cave Hill Campus, POB 64, Bridgetown; tel. 417-4478; fax 424-7291; e-mail salises@uwichill.edu.bb; internet www.uwichill.edu.bb/ISER; f. 1948; applied research and graduate teaching programme relating to the Caribbean; Dir Prof. ANDREW DOWNES; publ. *Journal of Eastern Caribbean Studies* (quarterly).

### BELGIUM

**Académie Royale des Sciences d'Outre-Mer/Koninklijke Academie voor Overzeese Wetenschappen** (Royal Academy of Overseas Sciences): rue Defacqz 1/3, 1000 Brussels; tel. (2) 538-02-11; fax (2) 539-23-53; e-mail kaowarsom@skynet.be; internet www.kaowarsom.be; f. 1928; the promotion of scientific knowledge of overseas areas, especially those with particular development problems; 154 hon. mems, 140 mems; Perm. Sec. Prof. DANIELLE SWINNE; publs *Bulletin des Séances/Mededelingen der Zittingen*, *Mémoires/Verhandelingen*, *Recueils d'Etudes Historiques/Historische bijdragen*, *Biographie belge d'Outre-Mer/Belgische Overzeese Biographie*, *Actes Symposiums/Acta Symposia*.

**Institut d'Etudes du Développement** (Institute for Development Studies): Université Catholique de Louvain, 1 place des Doyens, 1348 Louvain-La-Neuve; tel. (10) 47-39-35; fax (10) 47-28-05; e-mail vandenbossche@dvlp.ucl.ac.be; internet www.dvlp.ucl.ac.be; f. 1961; incorporates the Groupe de Recherches Interdisciplinaires sur l'Amerique Latine (Interdisciplinary Latin America Research Group); Pres. J. M. WAUTELET.

**Institut Royal des Relations Internationales (IRRI-KIIB):** 69 rue de Namur, 1000 Brussels; tel. (2) 223-41-14; fax (2) 223-41-16; e-mail info@irri-kiib.be; internet www.irri-kiib.be; f. 1947; research in foreign policy, international relations, law, economics, European issues, environment and defence; specialized library containing 700 vols and 1,200 periodicals; archives; lectures and conferences are held; Dir-Gen. CLAUDE MISSON; publs *Studia Diplomatica* (6 a year).

**Institute of Development Policy and Management:** Venusstraat 35, 2000 Antwerp; tel. (3) 220-49-98; fax (3) 220-44-81;

e-mail dev@ua.ac.be; internet www.ma.ac.be/dev; f. 1965; autonomous institution of the University of Antwerp; courses in development studies; library of 11,000 vols, 300 periodicals, 700 theses; Pres. Prof. Dr F. REYNTJENS.

## BRAZIL

**Centro de Estatística e Informações** (Statistics and Information Centre): Centro Administrativo da Bahia, 4a Avda 435, 41750-300 Salvador, BA; tel. (71) 371-9665; fax (71) 371-9664; f. 1938; statistics, natural resources, economic indicators; library of 15,448 vols; Dir RENATA PROSERPIO; publ. *Bahia Análise e Dados* (every 4 months).

**Instituto Brasileiro de Economia** (Brazilian Institute of Economics): Getúlio Vargas Foundation, Praia de Botafogo 190, Botafogo, 22253-900 Rio de Janeiro, RJ; tel. (21) 2559-6087; fax (21) 2553-6372; e-mail abrandao@fgvrj.br; internet www.fgv.br/ibre; f. 1951; Dir A. SALAZAR BRANDÃO; publs *National Accounts* (annually), *Conjuntura Econômica* (monthly), *Agroanalysis* (monthly).

**Instituto Brasileiro de Relações Internacionais** (Brazilian Institute of International Relations): CP 4400, 70919-970 Brasília, DF; tel. (21) 3114-1557; fax (21) 3114-1560; e-mail info@ibri-rbpi.org.br; internet www.puc-rio.br/iri; f. 1954; 1,000 vols; Dir Gen. JOSÉ FLÁVIO SOMBRA SARAIVA; Sec. Gen. ANTÔNIO CARLOS LESSA; publ. *Revista Brasileira de Política Internacional* (quarterly).

**Instituto Nacional de Pesquisas Da Amazônia (INPA)** (National Institute for Amazonian Research): Avda André Araújo 2936, Petrópolis, CEP 69083-000, Manaus, AZ; tel. (92) 643-3398; fax (92) 642-1840; e-mail gab@inpa.gov.br; internet www.inpa.gov.br; f. 1952; basic and applied research on Amazonian biodiversity, including botany, entomology, aquatic biology, ecology, earth sciences, human health, agriculture, aquaculture, forestry, forest products, natural products and food technology; postgraduate programme in tropical biology and natural resources; Dir JOSÉ ANTÔNIO ALVES GOMES.

**Instituto de Pesquisa Econômica Aplicada** (Institute of Applied Economic Research): Avda Presidente Antônio Carlos 51, 13° andar, CP 2672, 20020-010 Rio de Janeiro, RJ, Brazil; tel. (21) 3804-8000; fax (21) 2220-1920; internet www.ipea.gov.br; library of 60,000 vols; Pres. GLAUCO ARBIX; publs *Jornal* (monthly).

**Instituto de Relações Internacionais** (Institute of International Relations): Pontifícia Universidade Católica do Rio do Janeiro (PUC-Rio), Rua Marquês de São Vicente 225, Vila dos Diretórios, Casa 20, Gávea, 22453-900 Rio de Janeiro, RJ; tel. (21) 3114-1558; fax (21) 3114-1560; e-mail iripuc@rdc.puc-rio.br; internet www.puc-rio.br/iri; f. 1979; education, research and publications; Dir MÔNICA HERZ; publs *Contexto Internacional* (2 a year).

## CANADA

**Canadian Association for Latin American and Caribbean Studies/Association Canadienne des Etudes Latino-Américaines et des Caraïbes:** Université de Montréal, CP 6128, Succ. Centre-Ville, Montréal, QC H3C 3J7; tel. (514) 343-6569; fax (514) 343-7716; e-mail calacs@cetase.@umontreal.ca; internet www.calacs.umontreal.ca; f. 1969; Pres. ALBERT BERRY; publ. *Canadian Journal of Latin American and Caribbean Studies* (2 a year), *Newsletter* (2 a year).

**Canadian Association of Latin American Studies/Association Canadienne des Etudes Latino-Américaines:** c/o Prof. ANTONIO URRELLO, Dept of French, Hispanic and Italian Studies, University of British Columbia, Vancouver, BC V6T 1W5; f. 1969; 200 mems; Pres. J. C. M. OGELSBY; Vice-Pres. CLAUDE MORIN; publs *North South/Nord Sud/Norte Sur*, *Canadian Journal of Latin American Studies* (2 a year), *Newsletter* (quarterly), *Directory of Canadian Scholars and Universities interested in Latin American Studies*.

**Canadian Council for International Co-operation/Conseil canadien pour la coopération internationale:** 1 Nicholas St, Suite 300, Ottawa, ON K1N 7B7; tel. (613) 241-7007; fax (613) 241-5302; e-mail info@ccic.ca; internet www.ccic.ca; f. 1968 (formerly Overseas Institute of Canada, f. 1961); co-ordination centre for voluntary agencies working in international development; 100 mems; Chair. JEAN-PIERRE MASSÉ; publs *Newsletter* (2 a year), *Directory of Canadian NGOs*.

**Canadian Institute of International Affairs/Institut Canadien des Affaires Internationales:** Glendon Manor, 2275 Bayview Ave, Toronto, ON M4N 3M6; tel. (416) 487-6830; fax (416) 487-6831; e-mail mailbox@ciia.org; internet www.ciia.org; f. 1928; 2,100 mems in 23 brs; library of 8,000 vols; Pres. and CEO BARBARA MCDOUGALL; Chair. ROY MACLAREN; publs *Behind the Headlines* (quarterly), *International Journal* (quarterly), Annual Report.

**Canadian Institute of Island Studies:** University of Prince Edward Island, 550 University Ave, Charlottetown, PE C1A 4P3; tel. (902) 566-0386; fax (902) 566-0756; e-mail iis@upei.ca; internet www.upei.ca/iis; f. 1985; public policy research and facilitation of public debate; comparative island studies; Dir IRENE NOVACZEK.

**Centre for Research on Latin America and the Caribbean (CERLAC):** York University, 240 York Lanes, 4700 Keele St, North York, ON M3J 1P3; tel. (416) 736-5237; fax (416) 736-5737; e-mail cerlac@yorku.ca; internet www.yorku.ca/cerlac/; f. 1978; interdisciplinary research organization; seeks to build academic and cultural links with the region; research findings made available through publs, lectures, seminars, etc.; Dir VIVIANA PATRONI.

## CHILE

**Centro de Estudios Públicos:** Monseñor Sotero Sanz 175, Providencia, Santiago; tel. (2) 2315324; fax (2) 2335253; e-mail biblioteca@cepchile.cl; Librarian CARMEN LUZ SALVESTRINI.

**Consorcio Latinoamericano sobre Agroecologia y Desarrollo (CLADES)** (Latin American Consortium for Agroecology and Sustainable Development): Casilla 97, Correo 9, Santiago; tel. (2) 2341141; fax (2) 2338918; e-mail adm@clades.mic.cl; f. 1989; grouping of Latin American NGOs; aims to prevent the collapse of rural agriculture by the adoption of efficient and sustainable practices; undertakes research, oversees training programmes and facilitates information exchange; Exec. Sec. ANDRÉS YURJEVIC.

**Corporación de Investigaciones Económicas para Latinoamérica:** MacIver 125, 5°, Casilla 16496, Correo 9, Santiago; tel. (2) 6333836; fax (2) 6334411; e-mail cieplan@lascar.puc.cl; f. 1976; economic research; Dir JOAQUIN VIAL.

**Facultad Latinoamericana de Ciencias Sociales (FLACSO)—Programa Santiago:** Avda Dag Hammarskjöld 3269, Vitacura, Santiago; tel. (2) 2900200; fax (2) 2900270; e-mail flacso@flacso.cl; internet www.flacso.cl; f. 1957; research in sociology, education, political science, international affairs; library of 18,000 vols, 592 periodicals; Dir FRANCISCO ROJAS ARAVENA; publs *Nueva Serie*, *Serie Libros FLASCO*, *Fuerzas Armadas y Sociedad* (quarterly).

**Instituto Antártico Chileno (INACH):** Luis Thayer Ojeda 814, Casilla 16521, Correo 9, Providencia, Santiago; tel. (2) 2318177; fax (2) 2320440; e-mail inach@inach.cl; internet www.inach.cl; f. 1963; a centre for technological and scientific development on matters relating to the Antarctic; 43 mems; library of 2,550 vols and 400 periodicals; Dir OSCAR PINOCHET DE LA BARRA; publs *Serie Científica* (annually), *Boletítin Antártico Chileno* (2 a year).

**Instituto de Estudios Internacionales:** Universidad de Chile, Avda Condell 249, Suc. 21, Providencia, Casilla 14187, Santiago; tel. (2) 2745377; fax (2) 2740155; e-mail inesint@uchile.cl; internet www.uchile.cl/facultades/estinter; f. 1966; research and teaching institute for international relations, political science, international law, economics and studies on Pacific Basin; Dir Prof. JEANETTE IRIGOIN B.; publ. *Revista de Estudios Internacionales* (quarterly).

**Instituto Latinoamericano y del Caribe de Planificación Económica y Social (ILPES)** (Latin American and Caribbean Institute for Economic and Social Planning): Edif. Naciones Unidas, Avda Dag Hammarskjöld s/n, Vitacura, Casilla 1567, Santiago; tel. (2) 2102506; fax (2) 2066104; e-mail pdekock@eclac.cl; internet www.eclac.cl/ilpes; f. 1962 by UN Economic

Commission for Latin America; provides technical assistance, training for govt officials and research on planning techniques; acts as technical Secretariat of the Regional Council for Planning (CRP), the meetings of the Presiding Officers of the CRP and the System of Co-operation among Planning Bodies of Latin America and the Caribbean; Dir FERNANDO SÁNCHEZ-ALBAVERA; publs *Cuadernos del ILPES*, *Boletín* (2 a year).

**Latin American Demographic Centre (CELADE):** UN Economic Commission for Latin American and the Caribbean (ECLAC) Population Division, Avda Dag Hammarskjöld 3477, Casilla 179-D, Santiago; tel. (2) 2102002; fax (2) 2080196; e-mail celade@eclac.cl; internet www.eclac.cl; f. 1957; analysis of demographic trends, population and development research, teaching and training, and diverse information on population; Dir DIRK JASPARS-FAIJER.

## PEOPLE'S REPUBLIC OF CHINA

**Institute of Latin American Studies, Chinese Academy of Social Sciences:** 3 Zhang Zi Zeng Rd, Dong Cheng District, POB 1113, Beijing; tel. (10) 64014011; e-mail latinlat@public.bta.net.cn; f. 1961; 60 mems; library of 40,000 vols; Dir LI MINGDE; publ. *Latin-American Studies*.

**Institute of World Economics and Politics, Chinese Academy of Social Sciences:** 5 Jianguomen Nei Da Jie, Beijing 100732; tel. (10) 65137744; fax (10) 65126105; e-mail webmaster@iwep.org.co; internet www.iwep.org.cn; Dir PU SHAN.

## COLOMBIA

**Centro de Estudios sobre Desarrollo Económico (CEDE)** (Centre for Economic Development Studies): Carrera 1E, No 18A-10, Apdo Aéro 4976, Bogotá, DC; tel. (1) 341-2240; fax (1) 281-5771; e-mail infcede@uniandes.edu.co; internet www.prof.uniandes.edu.co/dependencias/centros-investigacion/cede; f. 1948; research in all aspects of economic development; 40 research staff; library of 40,000 vols; Dir JOSÉ LEIBOVICH; publs *Desarrollo y Sociedad* (quarterly), *Cuadernos CEDE*, documents series.

**Centro de Información y Documentación Biblioteca José Fernández de Madrid:** Universidad de Cartagena, Centro Carrera 6, No 36-100, Cartagena; tel. (95) 660-0682; fax (95) 664-5756; e-mail biblioteca@unicartagena.edu.co; f. 1827; Librarian RAQUEL MIRANDA AVENDAÑO.

**Centro Regional para el Fomento del Libro en América Latina y el Caribe (CERLALC)** (Regional Centre for the Promotion of Books in Latin America and the Caribbean): Calle 70, No 9-52, Apdo Aereo 57348, Bogotá, DC; tel. (1) 321-7501; fax (1) 321-7503; e-mail libro@cerlalc.com; internet www.cerlalc.org; f. 1972 by UNESCO and Colombian Govt; promotes production and circulation of books and development of libraries; provides training; promotes protection of copyright; 21 mem. countries; Dir ALMA BYINGTON DE ARBOLEDA; publs *El Libro en America Latina y el Caribe* (quarterly), *Boletín Informativo CERLALC* (quarterly).

**Latin American Centre:** Pontifíca Universidad Javeriana, Carrera 10, No 65-48, Of. 424, Bogotá, DC; tel. (1) 542-1660; fax (1) 640-1759; e-mail dl-clam@javercol.javeriana.edu.co; Dir E. CRISTINA MONTAÑA.

## COSTA RICA

**Centro Agronómico Tropical de Investigación y Enseñanza (CATIE)** (Tropical Agricultural Research and Higher Education Center): Apdo 65, 7170 Turrialba; tel. 558-2000; fax 556-1533; e-mail catie@catie.ac.cr; internet www.catie.ac.cr; f. 1973; applied research, graduate and short-term training; mems: Inter-American Institute for Co-operation on Agriculture—IICA, Belize, Brazil, Colombia, Costa Rica, Dominican Republic, Ecuador, El Salvador, Guatemala, Honduras, Mexico, Nicaragua, Panama, Venezuela; library of 80,000 vols; Dir-Gen. Dr PEDRO FERREIRA ROSSI; publs *Boletín de Semillas Forestales*, *Revista MIP*, *Revista Forestal Centroamericana*, *Revista Agroforestería—las Américas*, *Noticias de Turrialba* (quarterly), *Informe Anual*.

**Instituto Centroamericano de Administración Pública (ICAP)** (Central American Institute of Public Administration): Apdo 10025-1000, San José; tel. 234-1011; fax 225-2049; e-mail icapcr@racsa.co.cr; internet www.icap.ac.cr; f. 1954; technical assistance from UNDP; public administration, economic development and integration; library of 30,800 vols; Dir Dr HUGO ZELAYA CALIX; publ. *Revista* (2 a year).

**Instituto de Estudios Centroamericanos** (Institute of Central American Studies): Apdo 1524, 2050 San Pedro; tel. 253-3195; fax 234-7682; e-mail mesoamerica@ice.co.cr; internet www.mesoamericaonline.net; f. 1981; Exec. Dir LINDA J. HOLLAND; publ. *Mesoamérica* (monthly).

**Instituto de Estudios de Desarrollo Centroamericanos (ICADS):** Apdo 03-20-70, Sabanilla de Montes de Oca, San José; tel. 234-9410; fax 234-1337; e-mail icads@netbox.com; internet www.icadscr.com; Dir SANDRA NIEL KINGHOM.

**Inter-American Institute for Co-operation on Agriculture:** Apdo 55, 2200 San Isidro de Coronado, San José; tel. 216-0222; fax 216-0233; e-mail iicahq@iica.ac.cr; internet www.iica.int; f. 1942; agricultural development and rural well-being; mems: 32 countries of the Americas and Caribbean; library of 75,000 vols; Dir-Gen. Dr CHELSTON W. D. BRAITHWAITE; publ. *Turrialba* (quarterly).

## CUBA

**Centro de Estudios Sobre América (CEA)** (Centre for Studies on the Americas): Calle 18, No 316E, entre 3ra y 5ta, Playa, Havana 13; tel. (7) 33-2716; fax (7) 33-1490; e-mail cea@tinored.cu; f. 1978; Dir LUIS SUÁREZ SALAZAR.

**Centro de Información Bancaria y Económica, Banco Central de Cuba** (Banking and Economic Information Centre, Central Bank of Cuba): Cuba 410, Havana 10100; tel. (7) 62-8318; fax (7) 66-6661; f. 1950; library of 32,000 vols; Man. NUADIS PLANAS GARCIA; Library Dept Chief JORGE FERNÁNDEZ PÉREZ; publs *Cuba: Half Yearly Economic Report* (annually), *Journal of the Central Bank of Cuba*.

**Instituto de Política Internacional** (Institute of International Politics): Ministerio de Relaciones Exteriores, Calzada 360, Vedado, Havana; tel. (7) 32-3279; f. 1962; 11 mems; Dir RENÉ ALVÁREZ RÍOS.

## CZECH REPUBLIC

**Ústav mezinárodních vztahů** (Institute of International Relations): 118 50 Prague 1, Nerudova 3; tel. (2) 51108111; fax (2) 51108222; e-mail umv@iir.cz; internet www.iir.cz; f. 1957; research on international relations and foreign and security policy of the Czech Republic, publishing, training, education; Dir PETR DRULÁK; Deputy Dir Ing. Mgr PETR KRATOCHVÍL; publs include *International Relations* (quarterly) in Czech, *International Politics* (monthly) in Czech, and *Perspectives* (2 a year) in English.

## DENMARK

**Institute for International Studies (IIS):** Gammel Kongevej 5, 1610 Copenhagen V; tel. 33-85-46-00; fax 33-25-81-10; e-mail cdr@cdr.dk; internet www.cdr.dk; f. 2002 following merger of Centre for Development Research, Danish Institute of International Affairs, Copenhagen Peace Research Institute and Danish Centre for Holocaust and Genocide Studies; forms part of Danish Centre for International Studies and Human Rights; Dir PER CARLSEN; publs *Newsletter* (6 a year).

**Udenrigspolitiske Selskab** (Foreign Policy Society): Amaliegade 40A, 1256 Copenhagen K; tel. 33-14-88-86; fax 33-14-85-20; e-mail udenrigs@udenrigs.dk; internet www.udenrigs.dk; f. 1946; studies, debates, courses and conferences on international affairs; library of 150 periodicals and publs from UN, OECD, WTO, EU; Dir KLAUS CARSTEN PEDERSEN; Chair. UFFE ELLEMANN-JENSEN; publs *Udenrigs*, *Udenrigspolitiske SkrifterLandeLommoformal*.

## DOMINICAN REPUBLIC

**Centro de Investigación Económica para el Caribe (CIECA)** (Economic Research Centre for the Caribbean): Calle

Juan Parada Bonilla 8A, Plaza Winnie, La Arboleda Ans. Naco, Apdo Postal 3117, Santo Domingo, DN; tel. 565-6362; fax 227-2533; e-mail ciecard@codetel.net.do; internet www.cieca.org; f. 1987; Pres. PAVEL ISA CONTRERAS.

## ECUADOR

**Centro Internacional de Estudios Superiores de Comunicación para América Latina (CIESPAL):** Diego de Almagro 2155 y Andrade Marín, Apdo 484, Quito; tel. (2) 2254-8011; fax (2) 2502-487; e-mail info@ciespal.net; internet www.ciespal.net/chasqui79.htm; f. 1959; research in communications and training of communicators; library of 16,500 documents, 2,000 vols; Dir-Gen. Dr EDGAR JARAMILLO; publ. *Revista Chasqui* (quarterly).

**Instituto Latinoamericano de Investigaciones Sociales (ILDIS)** (Latin American Social Sciences Research Institute): Calle Calama 354 y Juan León Mera, Casilla 17-03-367, Quito; tel. (2) 2562-103; fax (2) 2504-337; e-mail ildis1@ildis.org.ec; internet www.ildis.org.ec; f. 1974; research in economics, sociology, political science and education; library of 15,000 vols; Dir HANS-ULRICH BÜNGER.

## FRANCE

**Académie du Monde Latin:** 217 blvd Saint-Germain, 75007 Paris; aims to encourage contact between leading personalities of countries whose language, culture and civilization are of Latin origin; 100 co-opted mems; Pres. PAULO DE BERREDO CARNEIRO (Brazil).

**Association d'Etudes et d'Informations Politiques Internationales:** 86 blvd Haussmann, 75008 Paris; f. 1949; Dir G. ALBERTINI; publs *Est & Ouest* (Paris, 2 a month), *Documenti sul Comunismo* (Rome), *Este y Oeste* (Caracas).

**Centre de Coopération Internationale en Recherche Agronomique pour le Developpement (CIRAD):** 42 rue Scheffer, 75116 Paris; tel. 1-53-70-20-00; fax 1-47-55-15-30; internet www.cirad.fr; f. 1992; scientific and technical research; experimental stations, industrial plantations; researchers based in over 50 countries; Dir-Gen. BENOÎT LESAFFRE DESPRÉAUX; publ. *Plantations Recherche Développement* (every 2 months).

**Centre d'Etudes Prospectives et d'Informations Internationales:** 9 rue Georges Pitard, 75740 Paris Cédex 15; tel. 1-53-68-55-00; fax 1-53-68-55-01; e-mail hurion@cepii.fr; internet www.cepii.fr; f. 1978; study of international economics; 50 mems; library of 20,000 vols, 400 periodicals; Dir LIONEL FONTAGNÉ; Sec-Gen. FRÉDÉRIQUE ABIVEN; publs *Economie Internationale* (quarterly), *La Lettre du CEPII*, *Electronic News-letter for foreign correspondents* (quarterly), books, working papers.

**Ecole des Hautes Etudes Internationales:** 4 place Saint-Germain des Prés, 77553 Paris Cédex II; tel. 1-42-22-68-06; f. 1904; Pres. M. SCHUMANN; Dir P. CHAIGNEAU.

**Ibero-American Centre for Study and Research:** Institut Catholique de Paris, 21 rue d'Assas, Paris Cédex 06; Dirs J. DESCOLA, J. PINGLE.

**Institut d'Etudes Ibéroaméricaines:** Université de Bordeaux III, Domaine Universitaire, 33607 Pessac Cédex; tel. 5-57-12-46-26; fax 5-57-12-45-74; e-mail berton@montaigne.u_bordeaux.fr; f. 1943; teaching and research; 30 staff; library of 51,000 vols; Dir F. BRAVO.

**Institut Européen des Hautes Etudes Internationales (IEHEI):** 10 ave des Fleurs, 06000 Nice; tel. 4-93-97-93-70; fax 4-93-97-93-71; e-mail iehei@wanadoo.fr; internet www.iehei.org; f. 1964; library of 4,000 vols; Dir CLAUDE NIGOUL.

**Institut de Recherche pour le Développement (IRD):** 213 rue La Fayette, 75480 Paris Cédex; tel. 1-48-03-77-77; fax 1-48-03-08-29; e-mail editions@paris.ird.fr; internet www.ird.fr; f. 1944; a public corporation charged to aid developing countries by means of research, with special application to human environment problems, tropical climate and diseases, water resources, biodiversity and food production; library and documentation centre; Pres. JEAN-FRANÇOIS GIRARD; Dir-Gen. SERGE CALABRE.

**Institut de Recherches Agronomiques Tropicales et des Cultures Vivrières (IRAT):** 110 rue de l'Université, 75340 Paris Cédex; tel. 1-45-50-32-10; fax 1-93-37-79-39; f. 1960; works in numerous stations in Africa, the Antilles, French Guiana, Brazil; research into general agronomy, the cultivation of food crops, sugar cane, forages, spices, etc.; 220 research workers and technicians; library of 28,000 vols; Dir F. BOUR; publ. *L'Agronomie Tropicale* (quarterly).

**Institut des Hautes Etudes de l'Amérique Latine:** 28 rue Saint-Guillaume, 75007 Paris; tel. 1-44-39-86-00; fax 1-45-48-79-58; internet www.iheal.univ-paris3.fr; teaching and research unit of Université de Paris III Sorbonne Nouvelle; shares its library, publications service and website with Centre de recherche et de documentation sur l'Amerique latine—CREDAL; Dir JEAN-MICHEL BLANQUER; publs *Cahiers des Amériques latines* (3 a year), *Travaux et Mémoires*.

**Institut Français des Relations Internationales:** 27 rue de la Procession, 75740 Paris Cédex 15; tel. 1-40-61-60-00; fax 1-40-61-60-60; e-mail accueil@ifri.org; internet www.ifri.org; f. 1979; international politics and economy, security issues and regional studies; library of 30,000 vols and 200 periodicals; Pres. THIERRY DE MONTBRIAL; publs *Politique Etrangère* (quarterly), *Notes*, *Travaux et recherches*, *Cahiers et Conférences*, *Rapport Annuel sur le Système Economique et les Stratégies—RAMSES* (annually).

**Institut Pluridisciplinaire d'Etudes sur l'Amérique Latine de Toulouse:** Maison de la Recherche, Université de Toulouse Le Mirail, 5 allée Antonio Machado, 31058 Toulouse Cédex; tel. 5-61-50-43-93; fax 5-61-50-36-25; e-mail marin@univ-tlse2.fr; internet www.univ-tlse2.fr/ipealt; f. 1985; specialized research; economic documentation centre; 31 staff; Dir R. MARIN; publs *Caravelle*, *L'Ordinaire latino américain*, *Les Ateliers de Caravelle*.

**Laboratoire Interdisciplinaire de Recherche sur Les Amériques (LIRA):** Université de Rennes II, 6 ave Gaston Berger, 35043 Rennes Cédex; tel. 2-99-14-16-06; e-mail jean-pierre.sanchez@uhb.fr; f. 1966; general and musical studies on region; Dir JEAN-PIERRE SANCHEZ.

**Musée de l'Homme:** Palais de Chaillot, place du Trocadéro, 75116 Paris; tel. 1-44-05-72-72; fax 1-44-05-72-91; f. 1878; library of 250,000 vols, 5,000 periodicals; ethnography, anthropology, prehistory; attached to the Muséum National d'Histoire Naturelle; also a research and education centre; Profs BERNARD DUPZIGNE, ANDRÉ LANGANEY, HENRY DE LUMLEY; publ. *Objets et mondes* (quarterly).

**Societé des Américanistes:** Musée de l'Homme, 17 place du Trocadéro, 75116 Paris; tel. 1-47-04-63-11; e-mail jsa@mae.u-paris10.fr; f. 1896; 500 mems; Pres. P. DESCOLA; Gen. Sec. D. MICHELET; publ. *Journal*.

## GERMANY

**Deutsche Gesellschaft für Auswärtige Politik eV** (German Council on Foreign Relations): 10787 Berlin, Rauchstr. 17–18; tel. and fax (30) 254231; e-mail info@dgap.org; internet www.dgap.org; f. 1955; 1,600 mems; discusses and promotes research on problems of international politics; research library of 62,000 vols; Pres. Dr AREUD OETHER; Exec. Vice-Pres. FRITJOF VON NORDENSKJOLD; Dir, Research Institute Prof. Dr EBERHARD SANDSCHNEIDER; publs *Die Internationale Politik* (annually), *Internationale Politik: Transatlantic Edition* (quarterly), *Internationale Politik* (monthly).

**Deutsches Übersee-Institut (DÜI)** (German Overseas Institute): 20354 Hamburg 36, Neuer Jungfernstieg 21; tel. (40) 42825593; fax (40) 42825547; e-mail duei@duei.de; internet www.duei.de; f. 1966; incl. the Institute for Ibero-American Studies; Head Dr ANDREAS MEHLER.

**Ibero-Amerikanisches Institut Preussischer Kulturbesitz:** 10785 Berlin, Potsdamer Str. 37, Postfach 1247; tel. (30) 2662500; fax (30) 2662503; e-mail iai@iai.spk-berlin.de; internet www.iai.spk-berlin.de; f. 1930; library and research institute; 1.2m. vols (830,000 monographs); Dir Dr BARBARA GOEBEL; publs *Ibero-Amerikanisches Archiv*, *Indiana* (reviews), *Iberoamericana*, *Monumenta Americana*, *Quellenwerke*, *Bibliotheca Ibero-Americana*, *Miscellanea* (monograph series).

**Lateinamerika-Institut der Freien Universität Berlin:** 14197 Berlin, Rüdesheimer Str. 54–56; tel. (30) 83853072; fax (30) 83855464; e-mail lai@zedat.fu-berlin.de; internet www.fu-berlin.de/lai; f. 1970; teaching and research; Dir Prof. Dr MARIANNE BRAIG.

**Stiftung Wissenschaft und Politik (SWP):** Deutsches Institut für Internationale Politik und Sicherheit (German Institute for International and Security Affairs), 10719 Berlin, Ludwigkirchpl. 3–4; tel. (30) 880070; fax (30) 88007100; internet www.swp-berlin.org; f. 1962; Dir Dr CHRISTOPH BERTRAM.

## GUADELOUPE

**Caribbean Research Centre:** Université des Antilles et de la Guyane, 97159 Pointe-à-Pitre Cédex; tel. 938640; fax 938639; Dir ALAIN YACOU.

## GUATEMALA

**Instituto Centroamericano de Investigación y Tecnología Industrial (ICAITI)** (Central American Research Institute for Industry): Guatemala City; tel. 2331-0631; fax 2331-7470; e-mail general@icaiti.org.gt; f. 1956; research on marketing, development of new industries and manufacturing techniques, microbiology, geology and energy research projects, establishment of Central American standards, information services to industry, and professional advice; library of 36,000 vols; Dir Ing. LUIS FIDEL CIFUENTES E. (acting).

**Instituto de Nutrición de Centroamérica y Panamá (INCAP)** (Institute of Nutrition of Central America and Panama): Calzada Roosevelt, Zona 11, Apdo 1188, 01001 Guatemala City; tel. 2472-6518; fax 2473-6529; e-mail email@incap.ops-oms.org; internet www.incap.org.gt; f. 1949; represents the following countries: Belize, Costa Rica, El Salvador, Guatemala, Honduras, Nicaragua, Panama; administered by Pan American Health Organization (PAHO)/World Health Organization (WHO); programmes to promote food and nutrition security among Central American countries through: technical co-operation; human resources; development research; dissemination of information and resources mobilization; main areas of interest are: Food Protection; Nutritionally Improved Foods; Food Nutrition and Security in Disaster Areas; Health and Nutrition of Vulnerable Groups; library of 70,500 vols; Dir Dr HERNÁN L. DELGADO; publs various documents.

**Instituto de Relaciones Internacionales y de Investigaciones para la Paz** (International Relations and Peace Research Institute): 1A Calle 9-52, Zona 1, 01001 Guatemala City; tel. 2232-8260; fax 2253-6635; e-mail iripazguatemala@hotmail.com; f. 1989; research, training and lobbying in international relations and peace studies; Dir Dr LEONEL A. PADILLA; Project Co-ordinator JACOBO VARGAS-FORONDA.

## GUYANA

**Guyana Institute of International Affairs:** POB 101176, Georgetown; tel. (2) 77768; fax (2) 29542; f. 1965; 175 mems; library of 5,000 vols; Pres. DONALD A. B. TROTMAN; publs *Annual Journal of International Affairs*, occasional papers.

## INDIA

**Indian Council of World Affairs:** Sapru House, Barakhamba Rd, New Delhi 110 001; tel. (211) 3317246; fax (211) 331208; f. 1943; non-governmental institution for the study of Indian and international questions; 2,625 mems; library of 124,122 vols; Pres. HARCHARAN SINGH JOSH; Hon. Sec.-Gen. S. C. PARASHER; publs *India Quarterly*, *Foreign Affairs Reports* (monthly).

## ISRAEL

**International Institute for Development, Labour and Co-operative Studies (Afro-Asian Institute of the Histradut):** 7 Nehardea St, POB 16201, Tel-Aviv 64235; tel. and fax 3-5229195; f. 1958; seminars and courses for leadership of trade unions, co-operative and development institutions, community organizations, women's and youth groups, etc., in African, Asian, Caribbean and Pacific regions; library of 15,000 vols; Dir and Principal Dr Y. PAZ.

## ITALY

**Istituto Affari Internazionali (IAI):** Palazzo Rondinini, Via Angelo Brunetti 9, 00186 Roma; tel. (06) 3224360; fax (06) 3224363; e-mail iai@iai.it; internet www.iai.it; f. 1965; library of 13,000 vols; Pres. STEFANO SILVESTRI; Dir GIANNI BONVICINI; publs *The International Spectator* (quarterly, in English), *L'Italia e la politica internazionale* (annually).

**Istituto Italo-Latino Americano:** Palazzo Santacroce, Pasolini Piazza Benedetto Cairoli 3, 00186 Roma; tel. (06) 684921; fax (06) 6872834; e-mail info@iila.org; internet www.iila.org; f. 1966 by 20 Latin American states and Italy; cultural activities, commercial exchanges, economic and sociological, scientific and technical research, etc.; awards student grants; library of 90,000 vols; Sec.-Gen. LUDOVICO INCISA DI CAMERANA.

**Istituto per gli Studi di Politica Internazionale (ISPI):** Palazzo Clerici, Via Clerici 5, 20121 Milano; tel. (02) 8633131; fax (02) 8692055; e-mail ispi.eventi@ispionline.it; internet www.ispionline.it; f. 1933 for the promotion of the study and knowledge of all problems concerning international relations; seminars at postgraduate level; library of 100,000 vols; Pres. BORIS BIANCHERI; Man. Dir Dr GIOVANNI ROGGERO FOSSATI; publs *Relazioni Internazionali* (quarterly), *Working Papers*.

**Istituto per le relazioni tra l'Italia e i paesi dell'Africa, America Latina e Medio Oriente (IPALMO)** (Institute for Relations between Italy and the Countries of Africa, Latin America and the Middle East): Via degli Scipioni 147, 00192 Roma; tel. (06) 32699701; fax (06) 32699750; e-mail ipalmo@ipalmo.com; internet www.ipalmo.com; f. 1971; Pres. GILBERTO BONALUMI; publ. *Politica Internazionale* (every 2 months).

## JAMAICA

**Asociación de Universidades e Institutos de Investigación del Caribe (UNICA)** (Association of Caribbean Universities and Research Institutes): c/o Office of Administration and Special Initiatives, University of the West Indies, Kingston 7; tel. 977-6065; fax 977-7525; e-mail unica@uwimona.edu.jm; f. 1968 to foster contact and collaboration between member universities and institutes; conferences, meetings, seminars, etc.; circulation of information; facilitates co-operation and the pooling of resources in research; encourages exchanges of staff and students; mems: 50 institutions; Pres. Dr ORVILLE KEAN; Sec.-Gen. GERARD LATORTUE; publ. *Caribbean Educational Bulletin* (quarterly).

**Caribbean Food and Nutrition Institute:** University of the West Indies, POB 140, Mona, Kingston 7; tel. 927-1540; fax 927-2657; internet www.cfni.paho.org; f. 1967; research and field investigations, training in nutrition, dissemination of information, advisory services, production of educational material; mems: all English-speaking Caribbean territories, Belize, Guyana and Suriname; library of 4,500 vols; Dir Dr ADELINE WYNANTE PATTERSON; publs *Cajanus* (quarterly), *Nyam News Nutrient-Cost Tables* (quarterly).

## JAPAN

**Ajia Keizai Kenkyusho** (Institute of Developing Economies): Wakaba 3-2-2, Mihamaku, Chiba 261-8545; tel. (4) 3299-9500; fax (4) 3299-9724; e-mail info@ide.go.jp; internet www.ide.go.jp; f. 1960; merged with Japan External Trade Organization in 1998; researches industrial devt and political change in Latin America; library of 545,000 vols; Pres. MASAHISA FUJITA; publ. *Developing Economies* (quarterly).

**Centre for Latin American Studies:** Nanzan University, 18 Yamazato-cho, Showa-ku, Nagoya 466; e-mail cfls@ic.nanzan-u.ac.jp; internet www.nanzan-u.ac.jp/English/centers_latin.htm; f. 1983; an institute specializing in the study of contemporary Latin America (social sciences).

**Tokyo University of Foreign Studies:** 3-11-1, Asahi-cho, Fuchu-shi, Tokyo 183; tel. (3) 3917-6111; e-mail ml-zenhp@tufs.ac.jp; internet www.tufs.ac.jp; f. 1899; programmes of study into world languages, cultures and international relations; Pres. SETSUHO IKEHATA.

## REPUBLIC OF KOREA

**Institute of Brazilian Studies:** Kyung Hee University, 1 Hoiki Dong, Dongdaemun-ku, Seoul 131; tel. (2) 965-8000; f. 1978.

**Institute of Latin American Studies:** Hankuk University of Foreign Studies, 270 Imun-dong, Dongdaemun-ku, Seoul; tel. (2) 961-4114; fax (2) 959-7898; Dir Prof. CHUNG KYU HO.

## MEXICO

**Centro Coordinador y Difusor Estudios Latinoamericanos:** Torre II de Humanidades 8°, Ciudad Universitaria, 04510 México, DF; tel. (55) 5623-0211; fax (55) 5623-0219; f. 1978; attached to Universidad Nacional de México; study of Latin America and the Caribbean in all disciplines (history, literature, philosophy, etc.); library of 11,898 monographs, 8,700 magazines, 3,000 pamphlets, 160 theses, 150 records; Dir Dra ESTELA MORALES CAMPOS; publs *Anuario Estudios Latinoamericanos*, *Serie Nuestra América* (3 a year).

**Centro de Cooperación Regional para la Educación de Adultos en América Latina y el Caribe (CREFAL)** (Regional Co-operation Centre for Adult Education in Latin America and the Caribbean): Avda Lázaro Cardenas s/n, Col. Revolución, 61609 Pátzcuaro, Mich.; tel. (434) 342-8112; fax (434) 342-8151; e-mail crefal@crefal.edu.mx; internet www.crefal.edu.mx; f. 1951 by UNESCO and OAS; admin. by Board of Directors from mem. countries; regional technical assistance, specialist training in literary and adult education, research; library of 42,799 vols; Dir Lic. HUMBERTO SALAZAR HERRERA; publs *Revista Interamericana de Educación de Adultos* (quarterly), *Circular Informativa* (quarterly), *Boletín de Resúmenes Analíticos* (2 a year).

**Centro de Estudios Educativos, AC** (Education Studies Centre): Avda Revolución 1291, Col. Tlacopoa, San Angel, 01040 México, DF; tel. (55) 5593-5719; fax (55) 5651-6374; e-mail ceemexico@compuserve.com.mx; f. 1963; scientific research into the problems of education in Mexico and Latin America; 20 researchers; library of 35,000 vols and 600 periodicals; Dir-Gen. LUIS MORFÍN LÓPEZ; publ. *Revista Latinoamericana de Estudios Educativos* (quarterly).

**Centro de Estudios Históricos** (Historical Studies Centre): Colegio de México, AC, Camino al Ajusco 20, Col. Pedregal de Santa Teresa, 10740 México, DF; tel. (55) 5449-3066; fax (55) 5645-0464; e-mail direccion.ceh@colmex.mx; internet www.colmex.mx; Dir GUILLERMO PALACIOS Y OLIVARES.

**Centro de Estudios Internacionales** (Centre for International Studies): Colegio de México, Camino al Ajusco 20, Col. Pedregal de Santa Teresa, 10740 México, DF; tel. (55) 5449-3000; fax (55) 5645-0464; e-mail ancova@colmex.mx; internet www.colmex.mx; f. 1960; research and teaching in international relations and public administration; Dir MARÍA CELIA TORO.

**Centro de Estudios Monetarios Latinoamericanos** (Centre for Latin American Monetary Studies): Durango 54, Col. Roma, Del. Cuauhtémoc, 06700 México, DF; tel. (55) 5533-0300; fax (55) 5525-4432; e-mail estudios@cemla.org; internet www.cemla.org; f. 1952; organizes technical training programmes on monetary policy, development finance, etc.; applied research programmes, regional meetings of banking officials; 59 mems; Dir Gen. Dr KENNETH GILMORE COATES SPRY; Deputy Dir JOSÉ LINALDO GOMES DE AGUIAR; publs *Boletín* (every 2 months), *Monetaria* (quarterly), *Money Affairs* (2 a year).

**Centro de Relaciones Internacionales (CRI)** (Centre for International Relations): Ciudad Universitaria, FCPM, 04510 México, DF; tel. (55) 5655-1344; attached to the Faculty of Political and Social Sciences of the Universidad Nacional Autónoma de México; f. 1970; co-ordinates and promotes research in all aspects of international relations and Mexico's foreign policy, as well as the training of researchers in different fields: disciplinary construction problems, co-operation and international law, developing nations, actual problems in world society, Africa, Asia, peace research; 30 full mems; library of 16,000 vols; Dir Lic. ILEANA CID CAPETILLO; publs *Relaciones Internacionales* (quarterly), *Cuadernos*, *Boletín Informativo del CRI*.

**Consejo Latinamericano de Investigación para la Paz (CLAIP)** (Latin American Peace Research Council): Calle Magnolia 39, Col. San Jeronimo Lidice, 10200 México, DF; internet www.coprí.dk/ipra/claip.html; f. 1978; holds conferences; Sec.-Gen. PAULA PIRES DE NIELSON; publ. *Boletín Informativo CLAIP*.

**Pan American Institute of Geography and History:** Ex-Arzobispado 29, Col. Observatorio, 11860 México, DF; tel. (55) 5277-5888; fax (55) 5271-6172; e-mail secretariageneral@ipgh.org.mx; internet www.ipgh.org; f. 1928; promotion of professional improvement and application of modern methodology in cartography, geography, geophysics, history, anthropology and archaeology in the Western Hemisphere; mems: nations of the Organization of American States; library of 300,000 vols; Sec.-Gen. CARLOS SANTIAGO BORRERO MUTIS (Colombia); publs *Revista Cartográfica*, *Revista Geográfica*, *Revista de Historia de América*, *Revista Geofísica*, *Boletín de Antropología Americana*, *Folklore Americano*, *Revista de Arqueología Americana*, 400 books and monographs.

## THE NETHERLANDS

**Institute of Social Studies:** POB 29776, 2502 LT The Hague; tel. (70) 4260460; fax (70) 4260799; e-mail promotions@iss.nl; internet www.iss.nl; f. 1952; postgraduate courses, research and consultancy in development studies; Rector Prof. L. DE LA RIVE BOX; publs *Development and Change* (5 a year), *Development Issues* (3 a year), *Working Papers*.

## NICARAGUA

**Coordinadora Regional de Investigaciones Económicas y Sociales (CRIES)** (Regional Co-ordinating Committee of Economic and Social Research): De Iglesia El Carmen 1c. al Largo, Apdo 3516, Managua; tel. (2) 22-5137; fax (2) 22-6180; research into economic development and other socio-economic and socio-political issues in Central America and the Caribbean; Pres. XAVIER GOROSTIAGA; publs *Cuadernos de Pensamiento Propio*, *Revista Pensamiento Propio* (monthly), *Servicios Especiales* (2 a month).

**Instituto Histórico Centroamericano (IHCA)** (Central American Historical Institute): Universidad Centroamericana, Apdo A-194, Managua; tel. (2) 78-2557; fax (2) 78-1402; e-mail envio@ns.uca.rain.ni; Dir P. ANDREU OLIVA; publ. *Envío*, (monthly).

## PAKISTAN

**Area Study Centre for Africa, North and South America:** Quaidi-i-Azam University, Islamabad; tel. (51) 230834; fax (51) 230833; f. 1978; teaching and research; library of 10,000 vols, microfilm/microfiche collection; Dir Dr RASUL BAKHSH RAIS; publ. *Pakistan Journal of American Studies* (2 a year).

## PANAMA

**Centro de Estudios Latinamericanos 'Justo Arosamena'** (Justo Arosamena Centre for Latin American Studies): Apdo 63093, El Dorado, Panamá; tel. 223-0028; fax 269-2032; f. 1977 for the analysis and dissemination of international agreements and intervention and other foreign affairs issues; Exec. Sec. M. A. GANDÁSEGUI.

## PARAGUAY

**Servicio Técnico Interamericano de Cooperación Agrícola** (Inter-American Technical Service for Agricultural Co-operation): Casilla de Correo 819, Asunción; f. 1943; 10,000 vols; Librarian LUCILA M. I. CARDUS.

## PERU

**Centro de Investigaciones Económicas y Sociales (CIESUL)** (Economic and Social Research Centre): Universidad de Lima, Avda Javier Prado Este s/n, Monterrico, Lima 33; tel. (1) 4376767; fax (1) 4378066; internet www.ulima.edu.pe.

**Centro Peruano de Estudios Internacionales (CEPEI)** (Peruvian Centre for International Studies): San Ignacio de Loyola 554, Miraflores 8, Lima 18; tel. (1) 4457225; fax (1) 4451094; f. 1983; external relations, incl. Peru's border relations; Exec. Pres. Dr EDUARDO FERRERO COSTA; publ. *Cronología de Las Relaciones Internacionales del Peru* (quarterly).

**Instituto de Economía de Libre Mercado (IELM)** (Institute of Free Market Economics): Avda Santa Cruz 398, San Isidro, Lima 27; fax (1) 4216242; e-mail mail@ielm.org; internet www.ielm.org; f. 1993; studies economic, political and social history of the area; Exec. Pres. CARLOS BOLOÑA BEHR.

### POLAND

**Centre for Latin American Studies (CESLA):** University of Warsaw, ul. Smyczkowa 14, 02-678 Warsaw; tel. (22) 5534209; fax (22) 5534210; e-mail cesla@uw.edu.pl; internet www.cesla.uw.edu.pl; f. 1988; documentation, publications and library service; Dir Prof. ANDRZEJ DEMBICZ; publs *CESLA 'Estudios y Memorias' Series, Documentos de Trabajo, Revista de CESLA*.

**Polski Instytut Spraw Międzynarodowych** (Polish Institute for International Affairs): 00-950 Warsaw, Warecka 1A; tel. (22) 5568000; fax (22) 5568099; e-mail pism@pism.pl; internet www.pism.pl; f. 1999; international relations; library of 155,000 vols; Dir RYSZARD STEMPLOWSKI; publs *Polski Przeglad Dyplomatyczny* (6 a year), *Polish Foreign Affairs Digest* (quarterly), *Europa* (quarterly, in Russian).

### PUERTO RICO

**Institute of Caribbean Studies:** POB 23361, University Station, Río Piedras, PR 00931; tel. (787) 764-0000; fax (787) 764-3099; e-mail iec@rrpac.upr.clu.edu; f. 1959; research and publishing; 10 mems; library of 150 vols; Dir Dr PEDRO J. RIVERA GUZMÁN; publ. *Caribbean Studies*.

### RUSSIA

**Institute of Latin America of the Russian Academy of Sciences:** 113035 Moscow, B. Ordynka 21; tel. (095) 951-53-23; fax (095) 953-40-70; e-mail ilaran@pol.ru; concerned with the economic, social, political and cultural development of Latin American countries; Dir Dr V. M. DAVYDOV.

**Institute of World Economics and International Relations:** 117859 Moscow, Profsoyuznaya 23; tel. (095) 120-43-32; fax (095) 310-70-27; e-mail imemoran@glasnet.ru; f. 1956; Dir V. A. MARTYNOV (acting).

### SENEGAL

**Centre des Hautes Etudes Afro-Ibéro-Américaines:** Université Cheikh Anta Diop de Dakar, Dakar-Fann, Dakar; tel. 22-05-30; concerned with all matters relating to Africa and Latin America in the fields of law, science and the arts.

### SERBIA AND MONTENEGRO

**Institute of International Politics and Economics:** 11000 Belgrade, POB 750, Makedonska 25; tel. (11) 3373633; fax (11) 3373835; e-mail iipe@diplomacy.bg.ac.yu; internet www.diplomacy.bg.ac.yu; f. 1947; international relations, world economy, international law, social, economic and political development in all countries; library of 250,000 vols; Dir Prof. VATROSLAV VEKARIĆ; publs *International Problems* (annually), *Review of International Affairs* (quarterly).

### SPAIN

**Agencia Española de Cooperación Internacional (AECI)** (Spanish Agency for International Co-operation): Avda de los Reyes Católicos 4, Ciudad Universitaria, 28040 Madrid; tel. (91) 5838100; fax (91) 5838310; e-mail infoaeci@aeci.es; internet www.aeci.es; f. 1988; promotes cultural understanding and promotes international co-operation by organizing conferences, exhibitions and exchanges, scholarships; finances programmes of cultural, scientific, economic and technical co-operation; information department; library of 510,000 vols and 7,000 periodicals; Pres. MIGUEL ANGEL CORTES; Sec.-Gen. RAFAEL RODRÍGUEZ-PONGA; Dir-Gen. for Co-operation with Latin America JUAN LÓPEZ-DÓRIGA; numerous publs on international development and co-operation.

**Escuela de Estudios Hispanoamericanos:** Alfonso XII 16, 41002 Seville; tel. (95) 4500972; fax (95) 4224331; e-mail bibescu@cica.es; f. 1943; studies history of the Americas; Library Dir ISABEL REAL DÌAZ.

**Instituto de Cuestiones Internacionales y Política Exterior (INCIPE)** (Institute of International Affairs and Foreign Policy): Alberto Aguilera 7, 6°, 28015 Madrid; tel. (91) 4455847; fax (91) 4457489; e-mail info@incipe.org; internet www.incipe.org; f. 1988; Dir VICENTE GARRIDO; publ. *Ensayos, Estudios, Informe Política Exterior Española* (occasional).

**Instituto de Relaciones Europeo-Latinoamericanas (IRELA)** (Institute for European-Latin American Relations): Calle Pedro de Valdivia 10, Apdo 2600, 28002 Madrid; tel. (91) 5617200; fax (91) 5626499; e-mail info@ivcla.org; f. 1984; research, conferences, etc.; Pres. ROLF LINKOHR; Dir WOLF GRABENDORFF; publs *Dossiers* (8 a year), *Working Papers* (6 a year).

**Real Academia Hispano-Americana** (Royal Spanish-American Academy): Calle Almirante Vienna 14, Apdo 16, 11009 Cádiz; f. 1910; 29 mems; Dir ANTONIO OROZCO ACUAVIVA; publs *Anuario, Boletín*.

### SWEDEN

**Iberoamerican Institute:** University of Göteborg, POB 200 405 30 Göteborg; tel. (31) 773-18-03; fax (31) 773-18-04; internet www.hum.gu.se/ibero; f. 1939; information, research, courses; library of 50,000 vols; Dir Dr MAJ-LIS FOLLÉR; publs *Anales, Haina*, regular series.

**Latinamerika-Institutet i Stockholm** (Institute of Latin American Studies, Stockholm University): Universitetsvägen 10B, 106 91 Stockholm; tel. (8) 16-28-82; fax (8) 15-65-82; e-mail lai@lai.su.se; internet www.lai.su.se; f. 1951; research on economic, political and social development in the region; library of 50,000 vols; Dir MONA ROSENDAHL; publ. *Iberoamericana: Nordic Journal of Latin American and Caribbean Studies* (2 a year).

### SWITZERLAND

**Institut Universitaire d'Etudes du Développement** (Graduate Institute of Development Studies): 24 rue Rothschild, CP 136, 1211 Geneva 21; tel. 229065940; fax 229065947; e-mail iued@unige.ch; internet www.unige.ch/iued/; f. 1961; African history, Middle Eastern and Latin American studies, international relations, Switzerland-Third World economic relations; Dir MICHAEL GARTON.

**Institut Universitaire des Hautes Etudes Internationales** (Graduate Institute of International Studies): 132 rue de Lausanne, 1211 Geneva 21; tel. 229085700; fax 229085710; e-mail info@hei.unige.ch; internet heiwww.unige.ch; f. 1927; a research and teaching institution studying international questions from the juridical, political and economic viewpoints; Dir (vacant).

**Schweizerisches Institut für Auslandforschung** (Swiss Institute of International Studies): Seilergraben 49, 8001 Zürich; tel. 16326362; fax 16321947; e-mail siafucd@pw.unizh.ch; internet www.siaf.ch; f. 1943; Man. Dir Prof. Dr DIETER RULOFF.

**Zentrum für Vergleichende und Internationale Studien** (Centre for Comparative and International Studies): Seilergraben 45-53, 8001 Zürich; tel. 6327968; fax 6321942; e-mail cispostmaster@sipo.gess.ethz.ch; internet www.cis.ethz.ch; f. 1997; international relations, comparative politics, security studies and conflict research; Dir Prof. THOMAS BERNAUER.

### TRINIDAD AND TOBAGO

**Caribbean Agricultural Research and Development Institute (CARDI):** University of the West Indies, St Augustine Campus, St Augustine; tel. 645-1205; fax 645-1208; e-mail info@cardi.org; internet www.cardi.org; f. 1975; mems: CARICOM countries (see Regional Organizations); provides technical assistance, technology devt and transfer in agriculture and animal sciences; library of 3,000 vols; Exec. Dir FRANK B. LAUCKNER; publs *CARDI Weekly, CARDI Review*, technical bulletins and papers.

**Caribbean Association of Industry and Commerce (CAIC):** Hilton Trinidad and Conference Centre, POB 442, Lady Young Rd, St Ann's, Port of Spain; tel. 623-4830; fax 623-6116; e-mail caic@wow.net; internet caic.wow.net; policy advocacy to improve trading conditions for regional private sector; CEO SEAN IFILL; publ. *Caribbean Investor and the CAIC Times* (quarterly).

**Institute of International Relations:** University of the West Indies, St Augustine Campus, St Augustine; tel. 645-3232; fax 663-9685; e-mail iirt@fss.uwi.tt; internet hostings.diplomacy.edu/iirt/; f. 1966; diplomatic training and postgraduate teaching and research; Dir Dr ANTHONY P. GONZALES (acting).

## UNITED KINGDOM

**Centre for Caribbean Studies:** University of Warwick, Coventry, CV4 7AL; tel. (24) 7652-3443; fax (24) 7652-3473; e-mail m.davies@warwick.ac.uk; internet www.warwick.ac.uk/fac/arts/CCS; f. 1984; MA and PhD programme, conferences and symposia, lectures, publishing; Dir Prof. GAD HEUMAN; publs *Warwick/Macmillan Caribbean Series*.

**Centre of Latin American Studies:** 17 Mill Lane, Cambridge, CB2 1RX; tel. (1223) 335390; fax (1223) 335397; e-mail general@latin-american.cam.ac.uk; internet www.latin-american.cam.ac.uk; f. 1969; research and graduate teaching, mainly in comparative history and anthropology, and in Latin American culture, sociology, politics and economics; Dir Dr CHARLES JONES (until Sept. 2005), Dr GEOFFREY KANTARIS (from Sept. 2005); publs *Cambridge Latin American Miniatures, Working Paper Series*.

**Commonwealth Institute:** Kensington High St, London, W8 6NQ; tel. (20) 7603-4535; fax (20) 7602-4525; e-mail information@commonwealth.org.uk; internet www.commonwealth.org.uk; aims to advance primary and secondary education in the Commonwealth through the Centre for Commonwealth Education, est. in partnership with Cambridge University; permanent exhibitions of Commonwealth countries including the Caribbean, open to the public with special provision for visiting parties; CEO DAVID FRENCH.

**David Livingstone Institute of Overseas Development Studies:** University of Strathclyde, Livingstone Tower, 26 Richmond St, Glasgow, G1 1LH; tel. (141) 552-4400; fax (141) 522-0775; e-mail g.zawdle@strath.ac.uk; f. 1973; economic, scientific and technological research; Dir Dr JAMES PICKETT.

**Hispanic and Luso-Brazilian Council:** Canning House, 2 Belgrave Sq., London, SW1X 8PJ; tel. (20) 7235-2303; fax (20) 7235-3587; e-mail enquiries@canninghouse.com; internet www.canninghouse.com; f. 1943; cultural, educational and economic links with Latin America, Spain and Portugal; 130 corporate mems; library of 60,000 vols; Dir-Gen. BARRY HAMILTON; publ. *British Bulletin of Publications on Latin America, the Caribbean, Portugal and Spain* (2 a year).

**Institute of Commonwealth Studies:** 28 Russell Sq., London, WC1B 5DS; tel. (20) 7862-8844; fax (20) 7862-8820; e-mail ics@sas.ac.uk; internet www.sas.ac.uk/commonwealthstudies; f. 1949; attached to University of London; for postgraduate research in social sciences and recent history relating to the Commonwealth; lead institution of CASBAH project (Caribbean Studies Black and Asian History); library of 160,000 vols, includes library of West India Committee; Dir Prof. TIM SHAW.

**Institute of Development Studies:** University of Sussex, Falmer, Brighton, BN1 9RE; tel. (1273) 606261; fax (1273) 621202; e-mail ids@ids.ac.uk; internet www.ids.ac.uk; f. 1966; research, training, post-graduate teaching, advisory work, information services; Dir Prof. LAWRENCE HADDAD; publs *IDS Bulletin* (quarterly), research reports, working papers development bibliographies, discussion papers, policy briefings, annual report, publications catalogue.

**Institute of Latin American Studies:** University of Liverpool, 86 Bedford St South, Liverpool, L69 7WW; tel. (151) 794-3079; fax (151) 794-3080; e-mail smurph@liverpool.ac.uk; internet www.liv.ac.uk/ILAS/; f. 1966; university centre for the development of teaching and research on Latin America; 11 faculty mems; library of 50,000 vols; Dir Dr NIKKI CRASKE; publs monographs, research papers.

**Institute for the Study of the Americas:** 31 Tavistock Sq., London, WC1H 9HA; tel. (20) 7862-8870; fax (20) 7862-8886; e-mail americas@sas.ac.uk; internet americas.sas.ac.uk; f. 2004; graduate study centre within the School of Advanced Study of the University of London; co-ordinates national information on the Americas in the United Kingdom; post-graduate courses on politics, economics, history, sociology, globalization and development of Latin America and the Caribbean; library of bibliographies, guides and research aids; wide range of seminars, workshops and conferences on Latin America and the Caribbean; Dir Prof. JAMES DUNKERLEY; publs monographs, research papers and miscellaneous documents.

**International Development Centre:** Queen Elizabeth House, 21 St Giles, Oxford, OX1 3LA; tel. (1865) 273600; fax (1865) 273607; e-mail qeh@qeh.ox.ac.uk; internet www.qeh.ox.ac.uk; f. 1986; attached to the University of Oxford; specializes in development studies; Dir Prof. FRANCES STEWART.

**Latin American Centre:** St Antony's College, Oxford, OX2 6JF; tel. (1865) 274486; fax (1865) 274489; e-mail enquiries@lac.ox.ac.uk; internet www.lac.ox.ac.uk; f. 1964; promotes research on Latin America, particularly with regard to the post-Independence period and in the fields of history, the social sciences, literature and geography; organizes seminars; Dir ROSEMARY THORPE.

**Overseas Development Institute:** Overseas Development Institute, 111 Westminster Bridge Rd, London, SE1 7JD; tel. (20) 7922-0300; fax (20) 7922-0399; e-mail odi@odi.org.uk; internet www.odi.org.uk; f. 1960 to act as a research and information centre on overseas development issues and problems; library of 16,000 vols; Dir SIMON MAXWELL; publs *Development Policy Review* (quarterly), *Disasters: the Journal of Disaster Studies, Policy and Management* (quarterly), books, pamphlets, briefing papers.

**Royal Commonwealth Society:** 18 Northumberland Ave, London, WC2N 5BJ; tel. (20) 7930-6733; fax (20) 7930-7905; e-mail info@rcsint.org; internet www.rcsint.org; f. 1868; organizes public affairs programme; Chair. Baroness PRASHAR; Dir Gen. STUART MOLE; publs *Annual Report, Newsletter* (3 a year), conference reports.

**Royal Institute of International Affairs:** Chatham House, 10 St James's Sq., London, SW1Y 4LE; tel. (20) 7957-5700; fax (20) 7957-5710; e-mail contact@riia.org; internet www.riia.org; f. 1920; an independent body which aims to promote the study and understanding of international affairs; over 300 corporate mems and many individual mems; library of 160,000 vols, 650 periodicals; Chair. Dr DEANNE JULIUS; Dir Prof. VICTOR BULMER-THOMAS; publs *The World Today* (monthly), *International Affairs* (6 a year).

## USA

**Brookings Institution:** 1775 Massachusetts Ave, NW, Washington, DC 20036; tel. (202) 797-6000; fax (202) 797-6004; e-mail brookinfo@brookings.edu; internet www.brookings.edu; f. 1916; research, education and publishing in the fields of economics, government and foreign policy; library of 80,000 vols; Pres. STROBE TALBOTT.

**Center for International Policy (CIP):** 1717 Massachusetts Ave NW, Suite 801, Washington, DC 20036; tel. (202) 232-3317; fax (202) 232-3440; e-mail cip@ciponline.org; internet www.ciponline.org; f. 1975; promotes international co-operation and demilitarization; Pres. ROBERT E. WHITE; publ. *International Policy Reports*.

**Center for International Studies:** Massachusetts Institute of Technology, Bldg E38, 292 Main St, Cambridge, MA 02139; tel. (617) 253-8093; internet web.mit.edu/cis/; f. 1952; development, migration, defence and arms control studies, environment, trade, political economy; Dir RICHARD SAMUELS.

**Center for Latin American Studies:** University of Florida, 319 Grinter Hall, POB 115530, Gainesville, FL 32611-5530; tel. (352) 392-0375; fax (352) 392-7682; e-mail www@latam.ufl.edu; internet www.latam.ufl.edu; f. 1931; graduate teaching and research; tropical resources and development, business environment and social change, religion and the state, Haitian Creole language programmes; extensive Latin American collection in library; Dir CARMEN DIANA DEERE; publ. *The Latinamericanist* (2 a year).

## REGIONAL INFORMATION

### Research Institutes

**Center of Economic Research:** Florida International University, University Park, DM 319-B, Miami, FL 33199; tel. (305) 348-3283; fax (305) 348-1524; e-mail salazar@fin.edu; f. 1982 to conduct economic research, with emphasis on the Caribbean basin and Cuba; Dir Prof. SALAZAR-CARILLO; publ. *Caribbean Basin Country Projections*.

**Council on Foreign Relations, Inc:** The Harold Pratt House, 58 East 68th St, New York, NY 10021; tel. (212) 434-9400; fax (212) 434-9800; e-mail communications@cfr.org; internet www.cfr.org; f. 1921; 4,000 mems; Foreign Relations Library of 5,000 vols, 300 periodicals; Pres. RICHARD N. HAASS; Chair. PETER G. PETERSON; publs *Foreign Affairs* (6 a year) and books on major issues of US foreign policy.

**Council on Hemispheric Affairs:** Suite 1010, 1730 M St, NW, Washington, DC 20036; tel. (202) 216-9261; fax (202) 223-6035; e-mail coha@coha.org; internet www.coha.org; f. 1975; conducts research into relations between North and South America; Dir LAURENCE BIRNS; publ. *News and Analysis* (2 a week).

**Helen Kellogg Institute for International Studies:** University of Notre Dame, 130 Hesburgh Center for International Studies, Notre Dame, IN 46556-5677; tel. (574) 631-6580; fax (574) 631-6717; e-mail kellogg@nd.edu; internet www.nd.edu/~kellogg; f. 1982; international research, particularly focused upon Latin America; Dir SCOTT MAINWARING; publs *Working Papers*, *Newsletter* (2 a year), monograph series.

**Hispanic Society of America:** 613 West 155th St, New York, NY 10032; tel. (212) 926-2234; e-mail info@hispanicsociety.org; internet www.hispanicsociety.org; f. 1904; maintains a public museum, rare book room, research staff, publishing section; 400 hon. mems; library of 250,000 vols and 15,000 rare books; Dir MITCHELL A. CODDING, Jr.

**Institute for Latin American Studies:** Sid Richardson Hall 1.310, University of Texas at Austin, Austin, TX 78712; tel. (512) 471-5551; fax (512) 471-3090; e-mail ilas@uts.cc.utexas.edu; internet www.utexas.edu/cola/llilas/index.html; Dir Dr NICOLAS SHUMWAY.

**Institute of Latin American and Iberian Studies:** Columbia University, 420 West 118th St, Rm 830, New York, NY 10027; tel. (212) 854-4643; fax (212) 854-4607; internet www.columbia.edu/cu/ilas; f. 1961; co-ordinates events, lectures and seminars on subjects relating to Latin America and Spain; Dir Prof. DOUGLAS A. CHALMERS; publs *Newsletter* (3 a year), working paper series.

**Inter-American Dialogue:** 1211 Connecticut Ave, Suite 510, Washington, DC 20036; tel. (202) 822-9002; fax (202) 822-9553; e-mail iad@thedialogue.org; internet www.iadialog.org; f. 1982; centre for policy analysis, communication and exchange on Western affairs; 100 mems; Pres. PETER HAKIM.

**Latin American and Caribbean Center:** Florida International University, University Park, Miami, FL 33199; tel. (305) 348-2000; e-mail lacc@fiu.edu; internet lacc.fiu.edu; f. 1979; university research institute; Dir EDUARDO A. GAMARRA; publ. *Hemisphere* (3 a year).

**Latin American Center:** University of California, Los Angeles (UCLA), 10343 Bunche Hall, Hilgard Ave, POB 951447, Los Angeles, CA 90095-1447; tel. (310) 825-4571; e-mail latinamctr@isop.ucla.edu; internet www.isop.ucla.edu/lac; Dir CARLOS ALBERTO TORRES.

**Latin American and Iberian Institute:** University of New Mexico, 801 Yale NE, MSC02 1690, Albuquerque, NM 87131-0001; tel. (505) 277-2961; fax (505) 277-5989; e-mail laii@unm.edu; internet www.laii.unm.edu; Dir CYNTHIA RADDING; publ. research papers.

**Library of International Relations:** 565 West Adams, Chicago, IL 60661; tel. (312) 906-5600; fax (312) 906-5685; f. 1932; aims to stimulate interest and research in international problems; conducts seminars, etc.; library of over 200,000 books, documents and periodicals; Pres. HOKEN SEKI.

**Middle American Research Institute:** Tulane University, New Orleans, LA 70118; tel. (504) 865-5110; fax (504) 862–8778; e-mail mari@tulane.edu; internet tulane.edu/~mari; f. 1924; publs on archaeology in Mesoamerica and related subjects; Dir E. WYLLYS ANDREWS V.; publs books, miscellaneous papers.

**North-South Center:** University of Miami, POB 248205, Coral Gables, FL 33124; tel. (305) 284-6868; fax (305) 284-6370; e-mail nscenter@miami.edu; internet www.miami.edu/nsc; f. 1984; research on contemporary intra-American relations; Dir AMBLER J. MOSS, Jr; publs working papers, policy reports.

**Paul H. Nitze School of Advanced International Studies:** Johns Hopkins University, 1740 Massachusetts Ave, N.W. Washington, DC 20036; tel. (202) 663-5734; fax (202) 663-5737; e-mail rroett@jhu.edu; internet www.sais-jhu.edu; Dean JESSICA EINHORN.

**Pre-Columbian Studies, Dumbarton Oaks:** 1703 32nd St, NW, Washington, DC 20007; tel. (202) 339-6440; fax (202) 625-0284; e-mail pre-Columbian@doaks.org; internet www.doaks.org/Pre-Columbian.html; f. 1962; residential fellowships, annual symposia, seminars, etc.; Pre-Columbian art collection; library of 26,000 vols on Pre-Columbian history; Dir of Studies JOANNE PILLSBURY; publs annual symposia vols and occasional monographs.

**Princeton Institute for International and Regional Studies (PIIRS):** 116 Bendheim Hall, Princeton University, Princeton, NJ 08544; tel. (609) 258-4851; fax (609) 258-3988; internet www.princeton.edu/~piirs; f. 2003; an academic institute of Princeton University; international relations; 65 faculty associates; Dir MIGUEL CENTENO; publs *World Politics* (quarterly), monographs, occasional papers.

**School of International and Public Affairs:** Columbia University, 420 West 118th St, Rm 1414, New York, NY 10027; tel. (212) 854-5406; fax (212) 864-4847; internet www.sipa.columbia.edu; Dean LISA ANDERSON; publ. *Journal of International Affairs* (2 a year).

**Woodrow Wilson Center—Latin American Program:** 1 Woodrow Wilson Plaza, 1300 Pennsylvania Ave, NW, Washington, DC 20004; tel. (202) 691-4170; fax (202) 691-4001; e-mail fellowships@wwic.si.edu; internet wwics.si.edu; f. 1977; residential fellowship programme: inter-American dialogue, inter-American economic issues, conferences, history and culture of Latin America, administration of social policy and governance, resolution of civil conflict; Dir JOSEPH S. TULCHIN; publ. *Working Paper Series*.

### URUGUAY

**Asociación Sudamericana de Estudios Geopolíticos e Internacionales** (South American Association of Geopolitical and International Studies): Quiebrayugos 4814, Casilla Correo 18.112, 11400 Montevideo; tel. (2) 6192953; fax (2) 9161923; f. 1979; research into inter-American issues, including that of economic integration in the Southern Cone; Sec.-Gen. Prof. BERNARDO QUAGLIOTTI DE BELLIS; publ. *Geosur* (6 a year).

**Centro de Estadísticas Nacionales y Comercio Internacional del Uruguay (CENCI Uruguay)** (Centre for National Statistics and International Trade): Misiones 1361, 2°, Montevideo; tel. (2) 9154578; e-mail cenci@cenci.com.uy; internet www.cenci.com.uy; f. 1957; economic and statistical information on all Latin American countries; mem. of ALADI and CEPAL; library of 1,500 vols; Dir KENNETH BRUNNER; publs *Manual Práctico del Importador*, *Manual Práctico del Exportador*, *Manual Práctico Aduanero*, *Manual Práctico del Contribuyente*, *Valor en Aduana*, *Análisis Estadístico de Importación-Exportación del Uruguay*, *Indice Alfabético del Sistema Armonizado*, *Dictámenes de Clasificacíon Arancelaria*, etc.

**Centro Latinoamericano de Economía Humana (CLAEH)** (Latin American Centre for Human Economy): Zelmar Michelini 1220, POB 5021, 11100 Montevideo; tel. (2) 9007194; e-mail info@claeh.org.uy; internet www.claeh.org.uy; f. 1958 to conduct research into economics and other social sciences; Dir ALVARO ARROYO; publ. *Cuadernos del CLAEH* (3 a year).

### VENEZUELA

**Centro de Estudios del Desarrollo (CENDES)** (Centre for Development Studies): Universidad Central de Venezuela, Edif. FUNDAVAC, Avda Neverí, Colinas de Bello Monte, Apdo 47604, Caracas 1041-A; tel. (212) 753-3475; fax (212) 751-2691; e-mail cendes@reacciun.ve; internet www.cendes.ucv.edu.ve; f. 1961; centre for research and graduate studies on all aspects of

## REGIONAL INFORMATION — Research Institutes

development in Venezuela and Latin America; library of 30,000 vols; Dir Carmen García-Guadilla; publs *Anuario de Estudios del Desarrollo*, *CENDES Newsletter* (3 a year), *Cuadernos del CENDES* (3 a year).

**Centro Experimental de Estudios Latinoamericanos (CEELA)** (Experimental Centre for Latin American Studies): Universidad del Zulia, Apdo de Correos 526, Maracaibo 4011, Zulia; tel. (261) 596703; research in socio-economic development especially the Andean Pact model, inflation and crises in Latin America; conferences and seminars; Dir Dr Gaston Parra Luzardo; publ. *Cuadernos Latinoamericanos*.

**Centro de Historia del Estado Carabobo:** Valencia, Carabobo; f. 1979 to conduct research in national and regional history, preserve and improve regional archives, conserve monuments, encourage and publicize celebrations of national historic events, and establish cultural relations with similar Venezuelan and foreign organizations; 24 mems; Pres. Luis Cubillán; Sec. Dr Marco Tulio Mérida; publ. *Boletín*.

**Instituto de Altos Estudios de América Latina** (Institute for Advanced Latin American Studies): Apdo 17271, El Conde, Caracas 1010; tel. (212) 573-8824; f. 1975; research, seminars, publs, on Latin America; attached to the Universidad Simón Bolívar; library of 3,000 vols; Dir Miguel Angel Burelli Rivas; publs *Mundo Nuevo: Revista de Estudios Latinoamericanos* (quarterly), working papers, books.

**Instituto de Investigaciones Económicas y Sociales (IIES)** (Institute of Economic and Social Research): Universidad Católica Andrés Bello, Urb. Montalbán, La Vega, Caracas 1020; tel. (212) 407-4174; fax (212) 407-4349; e-mail lespana@ucab.edu.ve; internet www.ucab.edu.ve/investigacion/iies/; studies labour and demographic economics; Dir Luis Pedro España.

**Instituto Latinoamericano de Investigaciones Sociales (ILDIS)** (Latin American Institute for Social Research): Edif. Parsa, 1°, Plaza la Castellana, 61712 CLACO, Caracas 1060; tel. (212) 33-3741; f. 1974; conducts research in the social sciences, particularly economics, education, politics, and sociology.

**Instituto Venezolano de Estudios Sociales y Politicos (INVESP)** (Venezuelan Institute of Social and Political Studies): Edif. Centro Parque Caraboo, Torre B, 21°, Of. 2115, Avda Universidad, Caracas; tel. and fax (212) 574-6549; f. 1986; affiliated to the South American Commission on Peace, Regional Security and Democracy in Chile; research on foreign policy, including relations with the Caribbean and Latin America; Dir Dr Andrés Serbín.

# SELECT BIBLIOGRAPHY (BOOKS)

## South America

Adler, E. *The Power of Ideology: The Quest for Technological Autonomy in Argentina and Brazil*. Berkeley and Los Angeles, CA, University of California Press, 1987.

Albert, B. *South America and the First World War*. Cambridge, Cambridge University Press, 2002.

Allison, G. T. *Essence of Decision: Explaining the Cuban Missile Crisis*. Boston, MA, Little Brown, 1971.

Almond, G. A., and Verba, S. *The Civic Culture: Political Attitudes and Democracy in Five Nations*, 2nd edn. Newbury Park, CA, Sage Publications, 1989.

Angell, A. et al. *Decentralizing Development: The Political Economy of Institutional Change in Colombia and Chile*. Oxford, Oxford University Press, 2001.

Anglade, C., and Fortín, C. *The State and Capital Accumulation in Latin America*. London, Macmillan, 1985.

Arceneaux, C. L. *Bounded Missions: Military Regimes and Democratization in the Southern Cone and Brazil*. Pennsylvania, PA, Penn State University Press, 2001.

Archetti, E. P., Cammack, P., and Roberts, B. (Eds). *Sociology of 'Developing Societies': Latin America*. Basingstoke, Macmillan, 1987.

Aviel, J. F. 'Political Participation of Women in Latin America', in *Western Political Quarterly*, Vol. 34. 1981.

Baloyra, E. A. (Ed.). *Comparing New Democracies, Transition and Consolidation in Mediterranean Europe and the Southern Cone*. Boulder, CO, Westview Press, 1987.

Bethell, L., and Roxborough, I. (Eds). *Latin America between the Second World War and the Cold War, 1944-1948*. Cambridge, Cambridge University Press, 1993.

Black, J. K. 'Elections and Other Trivial Pursuits: Latin America and the New World Order', in *Third World Quarterly*, Vol. 14, No. 3. 1993.

*Latin America: Its Problems and Its Promise*, 3rd edn. Boulder, CO, Westview Press, 1998.

Bouvier, V. *Alliance or Compliance, Implications of the Chilean Experience for the Catholic Church in Latin America*. Syracuse, NY, Syracuse University Press, 1983.

Boville, B. *The Cocaine War: Drugs, Politics, and the Environment*. New York, NY, Algora Publishing, 2004.

Bowman, K. S. *Militarization, Democracy and Development: The Perils of Praetorianism in Latin America*. Pennsylvania, PA, Penn State University Press, 2004.

Brannstrom, C. (Ed.). *Territories, Commodities and Knowledges: Latin American Environmental Histories in the Nineteenth and Twentieth Centuries*. London, Institute for the Study of the Americas, 2005.

Brass, T. *Latin American Peasants*. London, Frank Cass, 2003.

Bruneau, T. C. *The Political Transformation of the Brazilian Catholic Church*. Cambridge, Cambridge University Press, 1974.

Bulmer-Thomas, V. *The Economic History of Latin America Since Independence*. Cambridge, Cambridge University Press, 1994.

*The New Economic Model in Latin America and its Impact on Income Distribution and Poverty*. Basingstoke, Macmillan, 1996.

Calvert, P. *A Political and Economic Dictionary of Latin America*. London, Europa, 2004.

*A Study of Revolution*. Oxford, Clarendon Press, 1970.

'Latin America: Laboratory of Revolution', in *Revolutionary Theory and Political Reality*, edited by Noel O'Sullivan. Brighton, Harvester Press, 1983.

'Demilitarisation in Latin America', in *Third World Quarterly*, Vol. 7. 1985.

(Ed.). *Political and Economic Encyclopedia of South America and the Caribbean*. Harlow, Essex, Longman, 1991.

*The International Politics of Latin America*. Manchester, Manchester University Press, 1994.

Calvert, P., and Calvert, S. *Latin America in the Twentieth Century*, 2nd edn. Basingstoke, Macmillan, 1993.

Calvert, P., and Milbank, S. *The Ebb and Flow of Military Government in South America*, (Conflict Studies No. 198). London, Institute for the Study of Conflict, 1987.

Camp, R. A. *Democracy in Latin America: Patterns and Cycles*. Wilmington, DE, SR Books, 1996.

Castañeda, J. G. *Utopia Unarmed: The Latin American Left after the Cold War*. New York, NY, Vintage Books, 1994.

Castro, D. (Ed.). *Revolution and Revolutionaries, Guerrilla Movements in Latin America*. Scholarly Review Books, 1999.

Clapham, C., and Philip, G. (Eds). *The Political Dilemmas of Military Regimes*. London, Croom Helm, 1985.

Clawson, P., and Lee, R. *The Andean Cocaine Industry*. Hampshire, Palgrave Macmillan, 1999.

Clissold, S. *Soviet Relations with Latin America, 1918-1968: A Documentary Survey*. London, Oxford University Press for Royal Institute of International Affairs, 1969.

Collier, D. (Ed.). *The New Authoritarianism in Latin America*. Princeton, NJ, Princeton University Press, 1979.

Collinson, H. (Ed.). *Green Guerrillas: Environmental Conflicts and Initiatives in Latin America*. London, Latin American Bureau, 1996.

Conniff, M. L. *Latin American Populism in Comparative Perspective*. Albuquerque, NM, University of New Mexico Press, 1982.

Cubitt, T. *Latin American Society*, 2nd edn. Harlow, Longman, 1995.

De Janvry, A. *The Agrarian Question and Reformism in Latin America*. Baltimore, MD, Johns Hopkins University Press, 1981.

Desch, M. C. *When the Third World Matters: Latin America and United States Grand Strategy*. Baltimore, MD, Johns Hopkins University Press, 1993.

De Soto, H. *The Other Path: The Invisible Revolution in the Third World*. New York, NY, Harper Row, 1989.

Deutsch, S. M. *Las Derechas: The Extreme Right in Argentina, Brazil and Chile, 1890-1939*. Stanford, CA, Stanford University Press, 1999.

Di Tella, T. S. *Latin American Politics: A Theoretical Framework*. Austin, TX, University of Texas Press, 1990.

Dix, R. H. 'The Breakdown of Authoritarian Regimes', in *Western Political Quarterly*, Vol. 35. 1982.

'Populism: Authoritarian and Democratic', in *Latin American Research Review*, Vol. 20. 1985.

Domínguez, F. (Ed.). *Identity and Discursive Practices: Spain and Latin America*. Bern, Peter Lang AG, 2000.

Domínguez, F., and Guedes de Oliveira, M. (Eds). *Mercosur: Between Integration and Democracy*. New York, NY, Peter Lang Publrs, Inc, 2004.

Domínguez, J., and Shifter, M. (Eds). *Constructing Democratic Governance in Latin America*. Baltimore, MD, Johns Hopkins University Press, 2003.

Duran, E. *European Interests in Latin America*. London, Royal Institute of International Affairs, 1985.

Eckstein, S. E. *Back from the Future: Cuba under Castro*. Princeton, NJ, Princeton University Press, 1994.

(Ed.). *Power and Popular Protest: Latin American Social Movements*. Berkeley, CA, University of California Press, 1989.

Farcau, B. W. *The Ten Cents War: Chile, Peru, and Bolivia in the War of the Pacific, 1879–1884*. Westport, CT, Praeger Publrs, 2000.

Feinberg, R. E. *The Intemperate Zone: The Third World Challenge to US Foreign Policy*. New York, NY, W. W. Norton, 1983.

Ferrell, R. H. *Latin American Diplomacy: The Twentieth Century*. New York, NY, W. W. Norton, 1988.

Figueira, D. *Cocaine and Heroin Trafficking in the Caribbean: The Case of Trinidad and Tobago, Jamaica and Guyana*. Lincoln, NE, iUniverse, Inc, 2004.

Finer, S. E. *The Man on Horseback: The Role of the Military in Politics*, 2nd revised edn. Harmondsworth, Penguin, 1976.

Fisher, J. *Out of the Shadows: Women, Resistance and Politics in South America*. London, Latin American Bureau, 1993.

Foders, F., and Feldsieper, M. *The Transformation of Latin America: Economic Development in the Early 1990s*. Northampton, MA, Edward Elgar Publishing, 2000.

Foweraker, J. *Theorizing Social Movements*. London, Pluto Press, 1995.

Fowler, W. *Ideologues and Ideologies in Latin America*. Westport, CT, Greenwood Press, 1997.

Frieden, J. A. *Debt, Development and Democracy: Modern Political Economy and Latin America, 1965-1985*. Princeton, NJ, Princeton University Press, 1992.

Frieden, J. A., Pastor, M., and Tomz, M. *Modern Political Economy and Latin America: Theory and Policy*. Boulder, CO, Westview Press, 2000.

Gilbert, A. *Latin America*. London, Routledge, 1990.

Gill, L. *School of the Americas: Military Training and Political Violence in the Americas (American Encounters/Global Interactions)*. Durham, NC, Duke University Press, 2004.

*The Latin American City*, revised edn. London, Latin American Bureau, 1998.

Gillespie, R. *Soldiers of Perón: Argentina's Montoneros*. Oxford, Clarendon Press, 1982.

Gleijeses, P. *The Dominican Crisis: The 1965 Constitutional Revolt and American Intervention*. Baltimore, MD, Johns Hopkins University Press, 1979.

Graham-Yooll, A. *A State of Fear: Memories of Argentina's Nightmare*. London, Eland, 1986.

Green, D. *Silent Revolution: The Rise of Market Economics in Latin America*. London, Cassell/Latin American Bureau, 1995.

*Faces of Latin America*. London, Latin American Bureau, 1997.

Grosse, R. *Government Responses to the Latin American Debt Problem*. Boulder, CO, Lynne Rienner Publrs, 1996.

Guillermoprieto, A. *Looking for History: Dispatches from Latin America*. New York, NY, Pantheon Books, 2001.

Gutierrez, G. *A Theology of Liberation*. London, SLM Press, 1973.

Gwynne, R. N., and Kay, C. (Eds). *Latin America Transformed: Globalization and Modernity*. London, Arnold, 1999.

Hall, A. (Ed.). *Global Impact, Local Action: New Environmental Policy in Latin America*. London, Institute for the Study of the Americas, 2005.

Hall, A., and Patrinos, H. A. (Eds). *Indigenous Peoples, Poverty and Human Development in Latin America: 1994-2004*. Basingstoke, Palgrave Macmillan, 2005.

Harrison, L. E. *Underdevelopment is a State of Mind: The Latin American Case*. Lanham, MD, University Press of America, 1985.

Heinz, W. S., and Fruhling, H. *Determinants of Gross Human Rights Violations by State and State-sponsored Actors in Brazil, Uruguay, Chile and Argentina*. Martinus Nijhoff Publrs, 1999.

Hinds, H. E., Jr, and Tatum, C. M. (Eds). *Handbook of Latin American Popular Culture*. Westport, CT, Greenwood Press, 1985.

Jones, G. A., and Varley, A. 'The Contest for the City Centre: Street Traders versus Buildings', in *Bulletin of Latin American Research*, Vol. 13, No. 1 (Jan.). 1994.

Keegan, J. (Ed.). *World Armies*. London, Macmillan, 1983.

Kennedy, J. J. *Catholicism, Nationalism and Democracy in Argentina*. South Bend, IN, University of Notre Dame Press, 1958.

Kilty, K. M., Segal, E. (Eds). *Poverty and Inequality in the Latin American-U.S. Borderlands: Implications of U.S. Interventions*. Binghamton, NY, Haworth Press, Inc, 2005.

Kirk, J. 'John Paul II and the Exorcism of Liberation Theology...', in *Bulletin of Latin American Research*, Vol. 4, No. 1. 1985.

Larson, B. *Trials of Nation Making: Liberalism, Race and Ethnicity in the Andes, 1810–1910*. Cambridge, Cambridge University Press, 2004.

Lehmann, D. *Democracy and Development in Latin America*. Cambridge, Polity Press, 1990.

LeoGrande, W. 'Enemies Evermore: US Policy Towards Cuba After Helms-Burton', in *Journal of Latin American Studies*, Vol. 29, 1997.

Levine, D. H. (Ed.). *Churches and Politics in Latin America*. Beverley Hills, CA, Sage Publications, 1979.

Linz, J. J., and Stepan, A. (Eds). *The Breakdown of Democratic Regimes: Latin America*. Baltimore, MD, Johns Hopkins University Press, 1978.

Lipset, S. M., and Solari, A. (Eds). *Elites in Latin America*. New York, NY, Oxford University Press, 1967.

Lockhart, J., and Schwartz, S. B. *Early Latin America: A History of Colonial Spanish America and Brazil*. New York, NY, Cambridge University Press, 1983.

Loveman, B., and Davies, T. M. (Eds). *The Politics of Antipolitics: The Military in Latin America*. Lincoln, NE, University of Nebraska Press, 1978.

Loveman, B. *The Constitution of Tyranny: Regimes of Exception in Latin America*. Pittsburgh, PA, University of Pittsburgh Press, 1994.

Lowenthal, A. F. (Ed.). *Armies and Politics in Latin America*. New York, NY, Holmes and Meier, 1976.

Lynch, J. *Argentine Dictator: Juan Manuel de Rosas, 1829-1852*. Oxford, Clarendon Press, 1981.

MacDonald, S. B., and Fauriol, G. A. *Fast Forward: Latin America on the Edge of the 21st Century*. New Brunswick, NJ, and London, Transaction Publrs, 1997.

Maier, J., and Weatherhead, R. W. (Eds). *The Latin American University*. Albuquerque, NM, University of New Mexico Press, 1979.

Mainwaring, S., O'Donnell, G., and Valenzuela, J. S. (Eds). *Issues in Democratic Consolidation: The New South American Democracies in Comparative Perspective*. South Bend, IN, University of Notre Dame Press, 1992.

Malloy, J. M. (Ed.). *Authoritarianism and Corporatism in Latin America*. Pittsburgh, PA, University of Pittsburgh Press, 1977.

Malloy, J. M., and Seligson, M. A. (Eds). *Authoritarians and Democrats: Regime Transition in Latin America*. Pittsburgh, PA, University of Pittsburgh Press, 1987.

Martz, J. D. (Ed.). *United States Policy in Latin America: A Quarter Century of Crisis and Challenge, 1961-1986*. Lincoln, NE, University of Nebraska Press, 1988.

Martz, J. D., and Schoultz, L. (Eds). *Latin America, the United States and the Inter-American System*. Boulder, CO, Westview Press, 1980.

Meso-Lago, C. *Market, Socialist, and Mixed Economies: Comparative Policy and Performance in Chile, Cuba, and Costa Rica*. Baltimore, MD, Johns Hopkins University Press, 2000.

Middlebrook, K. J. *Conservative Parties, the Right and Democracy in Latin America*. Baltimore, MD, Johns Hopkins University Press, 2000.

Morgenstern, S., and Nacif, B. *Legislative Politics in Latin America*. Cambridge, Cambridge University Press, 2002.

Morris, M. A., and Millán, V. *Controlling Latin American Conflicts: Ten Approaches*. Boulder, CO, Westview Press, 1983.

Mouzelis, N. P. *Politics in the Semi-Periphery: Early Parliamentarism and Late Industrialization in The Balkans and Latin America*. Basingstoke, Macmillan, 1986.

Munck, R. *Politics and Dependency in the Third World: The Case of Latin America*. London, Zed Books, 1984.

Muñoz, H., and Tulchin, J. S. (Eds). *Latin American Nations in World Politics*. Boulder, CO, Westview Press, 1984.

Murillo, M. V. *Labour Unions, Partisan Coalitions and Market Reforms in Latin America*. Cambridge, Cambridge University Press, 2001.

Nunn, F. M. *The Time of the Generals: Latin American Professional Militarism in World Perspective*. Lincoln, NE, University of Nebraska Press, 1992.

*Yesterday's Soldiers: European Military Professionalism in South America, 1890–1940*. Lincoln, NE, University of Nebraska Press, 1983.

O'Brien, P., and Cammack, P. (Eds). *Generals in Retreat: The Crisis of Military Rule in Latin America*. Manchester, Manchester University Press, 1985.

O'Donnell, G. *Delegative Democracy*. South Bend, IN, University of Notre Dame Press, 1992.

*Counterpoints: Selected Essays on Authoritarianism and Democratization*. Notre Dame, IN, University of Notre Dame, 2000.

O'Donnell, G., Schmitter, P., and Whitehead, L. (Eds). *Transitions from Authoritarian Rule*. Baltimore, MD, Johns Hopkins University Press, 1986.

Oxhorn, P., and Starr, P. *Markets and Democracy in Latin America: Conflict or Convergence?* Boulder, CO, Lynne Rienner Publrs, 1998.

Painter, M., and Durham, W. H. *The Social Causes of Environmental Destruction in Latin America*. Ann Arbor, MI, University of Michigan Press, 1995.

Pang, E. *The International Political Economy of Transformation in Argentina, Brazil and Chile since 1960*. Hampshire, Palgrave Macmillan, 2002.

Parkinson, F. *Latin America, the Cold War and the World Powers, 1945-1973*. Beverley Hills, CA, Sage Publications, 1974.

Pastor, R. A. *Condemned to Repetition: The United States and Nicaragua*. Princeton, NJ, Princeton University Press, 1987.

Pearce, J. (Ed.). *The European Challenge: Europe's New Role in Latin America*. London, Latin American Bureau, 1982.

Perkins, D. *A History of the Monroe Doctrine*. London, Longman, 1960.

Petras, J., and Morley, M. *Latin America in the Time of Cholera: Electoral Politics, Market Economy, and Permanent Crisis*. New York, NY, Routledge, 1992.

Philip, G. *Oil and Politics in Latin America: Nationalist Movements and State Companies*. Cambridge, Cambridge University Press, 1982.

*The Military and South American Politics*. London, Croom Helm, 1985.

'The New Economic Liberalism in Latin America: Friends or Enemies?', in *Third World Quarterly*, Vol. 14, No. 3. 1993.

Phillips, N. *The Southern Cone Model: The Political Economy of Regional Capitalist Development in Latin America*. London, Routledge, 2004.

Pike, F. B. 'The Catholic Church and Modernization in Peru and Chile', in *Journal of Inter-American Affairs*, Vol. 20, No. 272. 1966.

*The United States and Latin America: Myths and Stereotypes of Civilization and Nature*. Austin, TX, University of Texas Press, 1992.

Poppino, R. *International Communism in Latin America: A History of the Movement, 1917–1963*. New York, NY, The Free Press, 1964.

Portes, A. 'Latin American Urbanization During the Years of the Crisis', in *Latin American Research Review*, Vol. 24, No. 3. 1989.

Przeworski, A., and Wallerstein, M. 'Structural Dependence on the State for Capital', in *American Political Science Review*, Vol. 82, (March). 1988.

Posada-Carbó, E., and Malamud, C. (Eds). *The Financing of Politics: Latin American and European Perspectives*. London, Institute for the Study of the Americas, 2005.

Rakowski, C. A. (Ed.). *Contrapunto: The Informal Sector Debate in Latin America*. Albany, NY, State University of New York Press, 1994.

Randall, L. 'Lies, Damn Lies and Argentine GDP', in *Latin American Research Review*, Vol. 11. 1974.

Ritter, A. R. M., and Kirk, J. M. *Cuba in the International System: Normalization and Integration*. London, Macmillan, 1995.

Roberts, B. R. *The Making of Citizens: Cities of Peasants Revisited*. London, Arnold, 1995.

Roberts, K. M. *Deepening Democracy? The Modern Left and Social Movements in Chile and Peru*. Stanford, CA, Stanford University Press, 2000.

Rouquié, A. *The Military and the State in Latin America*. Berkeley, CA, University of California Press, 1987.

Ruhl, M. 'Social Mobilisation, Military Tradition and Current Patterns of Civil-Military Relations in Latin America', in *Western Political Quarterly*, Vol. 35. 1982.

Sarmiento, D. F. *Life in the Argentine Republic in the Days of the Tyrants (or Civilization and Barbarism)*. New York, NY, Collier Books, 1961.

Schmitter, P. C. (Ed.). *Military Rule in Latin America: Function, Consequences and Perspectives*. Beverly Hills, CA, Sage Publications, 1973.

Scranton, M. E. *The Noriega Years: US-Panamanian Relations, 1981-1990*. Boulder, CO, Lynne Rienner Publrs, 1991.

Seckinger, R. 'The Central American Militaries: A Survey of the Literature', in *Latin American Research Review*, Vol. 16. 1981.

Shafer, D. M. *Deadly Paradigms: The Failure of US Counterinsurgency Policy*. Princeton, NJ, Princeton University Press, 1988.

Sheahan, J. *Patterns of Development in Latin America: Poverty, Repression and Economic Strategy*. Princeton, NJ, Princeton University Press, 1987.

Sherman, J. W. *Latin America in Crisis*. Boulder, CO, Westview Press, 2000.

Sigmund, P. E. *The United States and Democracy in Chile*. Baltimore, MD, Johns Hopkins for Twentieth Century Fund, 1993.

Silvert, K. H. *The Conflict Society: Reaction and Revolution in Latin America*. New York, NY, American Universities Field Staff Inc., 1966.

Skidmore, T. E., and Smith, P. H. *Modern Latin America*. Oxford, Oxford University Press, 2000.

Smith, B. *The Church and Politics in Chile, Challenges to Modern Catholicism*. Princeton, NJ, Princeton University Press, 1982.

Spalding, H. A. *Organised Labor in Latin America: Historical Case Studies of Urban Workers in Dependent Societies*. New York, NY, Harper and Row, 1977.

Sunkel, O. (Ed.). *Development from Within: Towards a Neostructuralist Approach for Latin America*. Boulder, CO, and London, Lynne Rienner Publrs, 1993.

Tamarin, D. *The Argentine Labor Movement, 1930-1945: A Study in the Origins of Peronism*. Albuquerque, NM, University of New Mexico Press, 1985.

Teichman, J. A. *The Politics of Freeing Markets in Latin America: Chile, Argentina and Mexico*. Chapel Hill, NC, University of North Carolina Press, 2001.

Thomas, J. R. *Bibliographical Dictionary of Latin American Historians and Historiography*. Westport, CT, Greenwood Press, 1984.

Thorp, R. (Ed.). *Latin America in the 1930s: The Role of the Periphery in World Crisis*. Basingstoke, Macmillan, 1984.

Thorp, R., and Whitehead, L. (Eds). *Latin American Debt and the Adjustment Crisis*. Basingstoke, Macmillan, 1987.

Tiano, S. 'Authoritarianism and Political Culture in Argentina and Chile in the mid 1960s', in *Latin American Research Review*, Vol. 31. 1986.

Timerman, J. *Prisoner without a Name, Cell without a Number*. Harmondsworth, Penguin, 1982.

Tokman, V. E., and Klein, E. *Regulation and the Informal Economy: Microenterprises in Chile, Ecuador and Jamaica*. Boulder, CO, Lynne Rienner Publrs, 1995.

Trubowitz, P. *Defining the National Interest: Conflict and Change in American Foreign Policy*. Chicago, IL, University of Chicago Press, 1998.

Tulchin, J. S., and Garland, A. M. (Eds). *Social Development in Latin America*. Boulder, CO, Lynne Rienner Publrs, 2000.

Tulchin, J. S., and Espach, R. H. *Latin America in the New International System*. Boulder, CO, Lynne Rienner Publrs, 2000.

Turner, B. (Ed.). *Latin America Profiled: Essential Facts on Society, Business and Politics in Latin America (Syb Factbook)*. New York, NY, St Martin's Press, 2000.

Van Cott, D. L. *The Friendly Liquidation of the Past: The Politics of Diversity in Latin America (Pitt Latin American Series)*. Pittsburgh, PA, University of Pittsburgh Press, 2000.

Wesson, R. (Ed.). *The Latin American Military Institution*. New York, NY, Praeger Publrs, 1986.

Weyland, K. G. *The Politics of Market Reform in Fragile Democracies: Argentina, Brazil, Peru and Venezuela*. Princeton, NJ, Princeton University Press, 2002.

Wilgus, A. C. (Ed.). *South American Dictators During the First Century of Independence*. New York, NY, Russell and Russell, 1963.

Wilkie, J. W., and Perkal, A. (Eds). *Statistical Abstract of Latin America*. Los Angeles, CA, University of California (Los Angeles) Latin American Center, annual.

Wood, B. *The Dismantling of the Good Neighbor Policy*. Austin, TX, University of Texas Press, 1985.

Wright, T. C. *Latin America in the Era of the Cuban Revolution*. New York, NY, Praeger Publrs, 2000.

Youngers, C., and Rosin, E. *Drugs and Democracy in Latin America: The Impact of US Policy*. London, Lynne Rienner Publrs, 2004.

## Central America

Adelman, A., and Reading, R. (Eds). *Confrontation in the Caribbean Basin: International Perspectives on Security, Sovereignty and Survival*. Pittsburgh, PA, Centre for Latin American Studies, 1984.

Aguilera, G. *El fusil y el olivo: la cuestión militar en Centroamérica*. USA, FLACSO/DEI, 1988.

Anderson, T. P. *Politics in Central America: Guatemala, El Salvador, Honduras and Nicaragua*. New York, NY, Praeger Publrs, 1988.

Barry, T. *Roots of Rebellion: Land and Hunger in Central America*. USA, South End Press, 1987.

Bendaña, A. *Demobilization and Reintegration in Central America: Peace Building Challenges and Responses*. Managua, Centro de Estudios Internacionales, 1999.

*Binational Study: The State of Migration Flows Between Costa Rica and Nicaragua—An Analysis of Economic and Social Implications for Both Countries*. Geneva, Intergovernmental Committee for Migration, 2003.

Blachman, M., et al. *Confronting Revolution: Security through Diplomacy in Central America*. New York, NY, Pantheon, 1986.

Blakemore, H. *Central American Crisis: Challenge to US Diplomacy*. London, Institute for the Study of Conflict, 1984.

Booth, J. A., and Walker, T. W. *Understanding Central America*, 3rd edn. Boulder, CO, Westview Press, 1999.

Booth, J., and Seligson, M. *Elections and Democracy in Central America*. Chapel Hill, NC, University of North Carolina Press, 1989.

Brockett, C. D., et al. (Eds). *Political Movements and Violence in Central America*. Cambridge, Cambridge University Press, 2005.

Bulmer-Thomas, V. *Studies in the Economics of Central America*. London, Macmillan, 1989.

Chomsky, A., and Lauria-Santiago, A. (Eds). *Identity and Struggle at the Margins of the Nation-State: The Laboring Peoples of Central America and the Hispanic Caribbean*. Durham, NC, Duke University Press, 1998.

Dunkerley, J. *Power in the Isthmus: A Political History of Modern Central America*. London: Verso, 1988.

*The Pacification of Central America*. London and New York, NY, Verso, 1994.

Flora, J., and Torres-Rivas, E. (Eds). *Central America*. Austin, TX, Central America Resource Center, 1989.

Goodman, L. W., Leogrande, W. M., and Mendelson Forman, J. (Eds). *Political Parties and Democracy in Central America*. Boulder, CO, Westview Press, 1992.

Holden, R. H. *Armies Without Nations: Public Violence and State Formation in Central America, 1821–1960*. Oxford, Oxford University Press, 2004.

Holland, S., and Anderson, D. *Kissinger's Kingdom? A Counter-report on Central America*. Nottingham, Spokesman, 1984.

Karnes, T. L. *The Failure of Union in Central America, 1824-1960*. Chapel Hill, NC, University of North Carolina Press.

Keeley, J. *Containing the Communists: America's Foreign Policy Entanglements*. San Diego, CA, Lucent Books, 2003.

Kirk, J. M., and Schuyler, G. V. (Eds). *Central America: Democracy, Development and Change*. New York, NY, Praeger Publrs, 1989.

Krenn, M. L. *The Chains of Interdependence: US Policy toward Central America, 1945–1954*. M. E. Sharpe, 1996.

LaFeber, W. *Inevitable Revolutions: The United States in Central America*. London, W. W. Norton, 1983.

Landau, S. *The Guerrilla Wars of Central America*. London, Weidenfeld & Nicolson, 1993.

Leiken, R. S. (Ed.). *Central America: Anatomy of Conflict*. New York, NY, Pergamon Press, 1984.

Mahoney, J. *The Legacies of Liberalism: Path Dependence and Political Regimes in Central America*. Baltimore, MD, Johns Hopkins University Press, 2002.

Paige, J. M. *Coffee and Power: Revolution and the Rise of Democracy in Central America*. Cambridge, MA, Harvard University Press, 1998.

Parker, F. D. *Central American Republics*. New York, NY, and London, Greenwood Press, 1981.

Pérez-Brignoli, H. *A Brief History of Central America*. Berkeley, CA, University of California Press, 1989.

Pearce, J. *Under the Eagle: US Intervention in Central America and the Caribbean*, 2nd edn. London, Latin America Bureau, 1982.

*The Report of the President's National Bipartisan Commission on Central America*. London, Collier Macmillan, 1984.

Putnam, L. *The Company They Kept: Migrants and the Politics of Gender in Caribbean Costa Rica, 1870–1960*. Chapel Hill, NC, University of North Carolina Press, 2002.

Rockwell, R. J., and Janus, N. *Media Power in Central America*. Champaign, IL, University of Illinois Press, 2003.

Sandoval-García, C. *Threatening Others: Nicaraguans and the Formation of National Identities in Costa Rica*. Columbus, OH, Ohio University Press, 2004.

Schooley, H. *Conflict in Central America*. Harlow, Keesing's International—Longman, 1987.

Schulz, D. E., and Graham, D. H. (Eds). *Revolution and Counterrevolution in Central America and the Caribbean*. Boulder, CO, Westview Press, 1984.

Sieder, R. (Ed.). *Central America: Fragile Transition*. London and Basingstoke, Macmillan with Institute of Latin American Studies Series, 1996.

Torres-Rivas, E. *Repression and Resistance: The Struggle for Democracy in Central America*. Boulder, CO, Westview Press, 1989.

Vilas, C. *Between Earthquakes and Volcanoes, Market, State, and the Revolutions in Central America*. New York, NY, Monthly Review Press, 1995.

Wearne, P., and R. Menchu. *Return of the Indian: Conquest and Revival in the Americas*. Philadelphia, PA, Temple University Press, 1996.

Wiarda, H. J. (Ed.). *US Policy in Central America: Consultant Papers for the Kissinger Commission*. Washington, DC, American Enterprise Institute for Public Policy Research, 1984.

Woodward, R. L. *Central America: Historical Perspectives on the Contemporary Crises*. London, Greenwood, 1988.

## The Caribbean

Adelman, A., and Reading, R. (Eds). *Confrontation in the Caribbean Basin: International Perspectives on Security, Sovereignty and Survival*. Pittsburgh, PA, Center for Latin American Studies, 1984.

Ahmed, B., and Afroz, S. *The Political Economy of Food and Agriculture in the Caribbean*. Kingston, Ian Randle Publrs, 1996.

Anderson, T. D. *Geopolitics of the Caribbean: Ministates in a Wider World*. New York, NY, Praeger Publrs, 1985.

*Annual Report of the Secretary-General of the Caribbean Community*. Georgetown, Caribbean Community Secretariat, 1981–.

Ayala, C. J. *American Sugar Kingdom: The Plantation Economy of the Spanish Caribbean 1898-1934*. Chapel Hill, NC, University of North Carolina Press, 1999.

Birbalsingh, F. (Ed.). *Indo-Caribbean Resistance*. Toronto, TSAR, 1993.

Buxton, J., and Phillips, N. *Case Studies in Latin American Political Economy*. Manchester, Manchester University Press, 1999.

Chamberlain, M. (Ed.). *Caribbean Migration: Globalised Identities*. London, Routledge, 1998.

Desch, M. C., Domínguez, J. I., Serbin, A. (Eds). *From Pirates to Druglords, the Post-Cold War Caribbean Security Environment*. Oxford, Heinemann, 1998.

Domínguez, J. I. *Democratic Politics in Latin America and the Caribbean*. Baltimore, MD, Johns Hopkins University Press, 1998.

Dubois, L. *A Colony of Citizens: Revolution and Slave Emancipation in the French Caribbean, 1787–1804*. Chapel Hill, NC, University of North Carolina Press, 2004.

Dunn, H. S. (Ed.). *Globalization, Communications and Caribbean Identity*. Kingston, Ian Randle Publrs, 1995.

Ferguson, J. *Far From Paradise: An Introduction to Caribbean Development*. London, Latin America Bureau, 1990.

Frazier, E. F., and Williams, E. *The Economic Future of the Caribbean*. Majority Press, 2004.

Gaspar, D. B., and Geggus, D. P. (Eds). *A Turbulent Time: The French Revolution and the Greater Caribbean*. Bloomington, IN, Indiana University Press, 1997.

Graham, N. A., and Edwards, K. L. *The Caribbean Basin to the Year 2000: Demographic, Economic and Resource-use Trends in Seventeen Countries: A Compendium of Statistics and Projections*. Boulder, CO, Westview Press, 1984.

Griffith, I. L., and Sedoc-Dahlberg, B. N. (Eds). *Democracy and Human Rights in the Caribbean*. Boulder, CO, Westview Press, 1997.

Grossman, L. S. *The Political Ecology of Bananas, Contract Farming, Peasants and Agrarian Change in the Eastern Caribbean*. Chapel Hill, NC, University of North Carolina Press, 1998.

Hall, K., and Benn, D. (Eds.). *Contending with Destiny: The Caribbean in the 21st Century*. Kingston, Ian Randle Publrs, 2000.

Harrison, M. *King Sugar: Jamaica, the Caribbean and the World Sugar Industry*. New York, NY, New York University Press, 2001.

Hennessy, A. (Ed.). *Intellectuals in the Twentieth Century Caribbean—Unity in Variety*, Vol. II: *The Hispanic and Francophone Caribbean*. Basingstoke, Macmillan, 1992.

Heron, T. *The New Political Economy of United States-Caribbean Relations: The Apparel Industry and the Politics of Nafta Parity*. Aldershot, Ashgate Publishing, 2004.

Hodge, A. *The Caribbean*. Hove, East Sussex, Macdonald Young, 1998.

Holme, P. *Colonial Encounters: Europe and the Native Caribbean 1492–1797*. London, Methuen, 1986.

Hope, K. *Urbanization in the Commonwealth Caribbean*. Boulder, CO, Westview Press, 1986.

Hope, K. R. *Economic Development in the Caribbean*. New York, NY, Praeger Publrs, 1986.

*Development Finance and the Development Process: A Case Study of Selected Caribbean Countries*. London, Greenwood, 1987.

Klak, T. *Globalization and Neoliberalism: The Caribbean Context*. Lanham, MD, Rowman & Littlefield Publrs, 1998.

Klein, A., Harriott, A., and Day, M. (Eds). *Caribbean Drugs: From Criminalization to Harm Reduction*. London, Zed Books, 2004.

Klein, H. S. *African Slavery in Latin America and the Caribbean*. New York, NY, Oxford University Press, 1986.

Lewis, G. K., and Maingot, A. P. *Main Currents in Caribbean Thought: The Historical Evolution of Caribbean Society in its Ideological Aspects, 1492–1900*. Lincoln, NE, University of Nebraska Press, 2004.

Maingot, A. P. *US Power and Caribbean Security: Geopolitics in a Sphere of Influence*. London, Lynne Rienner Publrs, 1989.

Mandle, J. R. *Persistent Underdevelopment: Change and Economic Modernization in the West Indies*. Newark, NJ, Gordon & Breach, 1996.

Mars, P., and Young, A. H. *Caribbean Labor and Politics: Legacies of Cheddi Jagan and Michael Manley*. Detroit, MI, Wayne State University Press, 2004.

Marshall, D. D. *Caribbean Political Economy at the Crossroads: NAFTA and Regional Developmentalism*. Basingstoke, Macmillan, 1998.

Martínez-Fernández, L. *Protestantism and Political Conflict in the Nineteenth-Century Hispanic Caribbean*. Piscataway, NJ, Rutgers University Press, 2002.

Mintz, S. *Caribbean Transformations*. Baltimore, MD, Johns Hopkins University Press, 1974.

Mora, F. O., and Hey, J. A. K. (Eds). *Latin America and Caribbean Foreign Policy*. Lanham, MD, Rowman and Littlefield Publrs, 2003.

Munro, D. G. *Intervention and Dollar Diplomacy in the Caribbean, 1900–1921*. Princeton, NJ, Princeton University Press, 1964.

*The United States and the Caribbean Republics*. Princeton, NJ, Princeton University Press, 1974.

Olwig, K. F. (Ed.). *Small Islands, Large Questions: Society, Culture and Resistance in the Post-Emancipation Caribbean*. London, Frank Cass, 1995.

Palmer, R. W. (Ed.). *US-Caribbean Relations, Their Impact on Peoples and Culture*. Westport, CT, Greenwood Press, 1998.

Pattullo, P. *Last Resorts*. London, Cassell, 1996.

*Fire from the Mountain: The Story of the Montserrat Volcano*. London, Constable and Co Ltd, 2000.

Payne, A., and Sutton, P. (Eds). *Modern Caribbean Politics*. Baltimore, MD, Johns Hopkins University Press, 1993.

Potter, R. B. *The Contemporary Caribbean*. Harlow, Prentice Hall, 2004.

Randall, S. J., and Mount, G. S. *The Caribbean Basin: An International History*. London, Routledge, 1998.

Richardson, B. C. *The Caribbean in the Wider World, 1492–1992: A Regional Geography*. Cambridge, Cambridge University Press, 1992.

*Economy and Environment in the Caribbean*. Jamaica, University of the West Indies Press, 1997.

Sutton, P. *Dual Legacies in the Contemporary Caribbean: Continuing Aspects of British and French Domination*. London, Frank Cass, 1986.

Thomas, C. Y. *The Poor and the Powerless: Economic Policy and Change in the Caribbean*. New York, NY, Monthly Review Press, 1989.

Thomas-Hope, E. M. (Ed.). *Perspectives on Caribbean Regional Identity*. Liverpool, Centre for Latin American Studies, 1984.

*Explanation in Caribbean Migration, Perception and the Image: Jamaica, Barbados, Saint Vincent*. London, Macmillan, 1992.

Thompson, A. O. *The Haunting Past: Politics, Economics and Race in Caribbean Life*. Oxford, James Curry Publrs, 1997.

Williams, E. E. *From Columbus to Castro: The History of the Caribbean 1492-1969*. London, André Deutsch, 1970.

Worrell, D. *Small Island Economies; Structure and Performance in the English-Speaking Caribbean since 1970*. New York, NY, Praeger Publrs, 1987.

Wucker, M. *Why the Cocks Fight: Dominicans, Haitians, and the Struggle for Hispaniola*. New York, NY, Hill & Wang Publishing, 2000.

Young, A. H., and Phillips, D. E. *Militarization in the Non-Hispanic Caribbean*. London, Lynne Rienner Publrs, 1986.

# SELECT BIBLIOGRAPHY (PERIODICALS)

*Amazonía Peruana.* Gonzales Prada 626, Magdalena, Apdo 14-0166, Lima 14, Peru; tel. (1) 4615223; f. 1976; Amazon anthropology, ethnology and linguistics, community development, bilingual education; Spanish, but with abstracts in English, French and German; Dir PIERRE GUÉRIG; 2 a year.

*AméricaEconomía.* Santiago, Chile; tel. (2) 290-9442; internet www.americaeconomia.com; f. 1986; Latin American business, economics and finance; Spanish; monthly; Editorial Dir RAÚL FERRO.

*América Indígena.* Nubes 232, Col. Pedregal de San Angel, México, DF, Mexico; tel. (55) 5568-0819; fax (55) 5652-1274; f. 1940; anthropology, rural development, Indians of the Americas, ethnology; Spanish and Portuguese, with abstracts in English; Editorial Dir Dr JOSÉ MATOS MAR (Instituto Indigenista Interamericano); quarterly.

*América Latina Internacional.* Latin American Faculty of Social Science, Argentina—FLACSO, Ayacucho 551, 1026 Buenos Aires, Argentina; tel. (11) 4375-2438; fax (11) 4375-1373; e-mail postmaster@flacso.cci.org.ar; f. 1993; Latin American international relations, international trade and finances, regional integration and Mercosur; Spanish; Editor S. R. L. MIÑO Y DAVILA; 2 a year.

*Américas Magazine.* Organization of American States, Suite 300, 19th and Constitution Ave, NW, Washington, DC 20006, USA; tel. (202) 458-6218; fax (202) 458-6217; f. 1949; culture, history, literature, travel, art, music, book reviews, Inter-American System; English and Spanish editions; Editor JAMES PATRICK KIERNAN; 6 a year.

*Americas Review.* CEB Ltd, 2 Market St, Saffron Walden, CB10 1HZ, United Kingdom; tel. (1799) 521150; fax (1799) 524805; e-mail queries@worldinformation.com; internet www.worldinformation.com; f. 1979; business, economic and political issues concerning North, South and Central America and the Caribbean; English; Editor TONY AXON; annual.

*Americas Update.* 427 Bloor St West, Toronto ON MSS 1X7, Canada; tel. (416) 967-5562; fax (416) 922-8587; e-mail amupdate@web.net; f. 1979; politics and socio-economic development of Latin America and the Caribbean, with emphasis on Canada's relations with the region; English; quarterly.

*Análisis Semanal.* Elizalde 119, 10°, Apdo 4925, Guayaquil, Ecuador; tel. (4) 232-6590; fax (4) 232-6842; analysis of economy and politics of Ecuador; Spanish; Editor WALTER SPURRIER BAQUIERZO; weekly.

*Andean Report.* Pasaje Los Pinos 156, Of. B6, Casilla 531, Miraflores, Lima 100, Peru; tel. (1) 4472552; fax (1) 4467888; f. 1975; commerce, development, economics and politics of Peru, including mining and petroleum; English; Editor NICHOLAS ASHESHOV; monthly.

*Andes: Revista Interamerican.* Casilla 4171, La Paz, Bolivia; f. 1967; politics and government in Latin America; Spanish.

*Anuario de Estudios Americanos.* Consejo Superior de Investigaciones Científicas, Departamento de Publicaciones, Vitruvio 8, 28006 Madrid; tel. (91) 5612833; fax (91) 5629634; e-mail publ@orgc.csic.es; internet www.csic.es/publica/revistas/anuario_estudios_americanos; f. 1944; humanities and social sciences of the Americas; Spanish, Portuguese, English and French; Dir ROSARIO SEVILLA SOLER; 2 a year.

*Anuario de Estudios Centroamericanos.* Instituto de Investigaciones Sociales, 2060 Universidad de Costa Rica, Costa Rica; tel. 2073505; internet cariari.ucr.ac.cr/~anuario/index.html; f. 1974; history, society, politics and economics relating to Central America; Spanish; Dir OSCAR FERNÁNDEZ; Editor RONALDO SOLANO; 2 a year.

*Anuario Indigenista.* Mexico; tel. (55) 5568-0819; fax (55) 5652-1274; f. 1962; Indians of the Americas, policies, anthropology, government, minorities; Spanish, Portuguese; Dir JOSÉ MATOS MAR (Instituto Indigenista Interamericano); annual.

*Apuntes del CENES.* Centro de Estudios Económicos—CENES, Universidad Pedagógica y Tecnológica de Colombia, Faculty of Economics, Apdo Aéreo 1234, Carretera Central del Norte, Tunja, Boyacá, Colombia; tel. and fax (87) 42-5237; f. 1983; national economic and development studies, politics and culture; Spanish; Editor LUIS E. VALLEJO ZAMUDIO; 2 a year.

*Archivo Ibero Americano.* Joaquín Costa 36, 28002 Madrid, Spain; tel. (91) 5619900; f. 1914; history of Spain and Hispanic America, mainly relating to the Franciscan Order; quarterly.

*¡Basta!.* Chicago Religious Task Force on Central America, Suite 1400, 59 East Van Buren, Chicago, IL 60605, USA; tel. (312) 663-4398; fax (312) 427-4171; f. 1984; analysis of social and political events in Central America: theological debate, news and information; English; 3 a year.

*Boletín de la Academia Nacional de la Historia.* Balcarce 139, 1064 Buenos Aires, Argentina; tel. (11) 4343-4416; fax (11) 4331-4633; e-mail admite@an-historia.org.ar; f. 1924; history of Argentina and America, principally as review of activities of the Academia Nacional de la Historia; Spanish; annual.

*Boletín Americanista.* Universidad de Barcelona, Departamento de Historia de America, Gran Vía 585, 08007 Barcelona, Spain; tel. (93) 3333466; fax (93) 4498510; f. 1959; anthropology, economics, geography, history of America; Spanish, English and French; Editor Dr MIQUEL IZARD; annual.

*Boletín del Instituto Latinoamericano y del Caribe de Planificación Económica y Social.* Edif. Naciones Unidas, Av. Dag Hammarskjöld s/n, Vitacura, Casilla 1567, Santiago, Chile; tel. (2) 2102506; fax (2) 2066104; e-mail pdekock@eclac.cl; internet www.eclac.cl/ilpes; f. 1976, publication suspended 1982, refounded 1996; economics; Spanish and English; 2 a year.

*Brazil.* Brasília, DF, Brazil; tel. (61) 223-5180; deals with trade and industry of Brazil and is published by Fundação Visconde de Cabo Frio under the auspices of the Trade Promotion Department of the Ministry of Foreign Affairs and of the Vice-Presidency of Resources and Operations of the Banco do Brasil; Editor FERNANDO LUZ; Portuguese, German and French editions (quarterly), English and Spanish editions (monthly).

*Brazil Service.* International Reports, 11300 Rockville Pike, Rockville, MD 20852-3030, USA; tel. (301) 816-8950; Brazilian business and investment; English; Editor-in-Chief ROBERT G. TAYLOR; 2 a month.

*Brennpunkt Lateinamerika.* Institut für Iberoamerika-Kunde, 20354 Hamburg, Alsterglacis 8, Germany; tel. (40) 41478201; fax (40) 41478241; e-mail sekretariat@iik.duei.de; internet www.duei.de/iik; f. 1999; political, economic and social development; German; fortnightly.

*Bulletin of Hispanic Studies.* University of Liverpool, 18 Oxford St, Liverpool, L69 7ZN, United Kingdom; tel. and fax (151) 794-2774; fax (151) 794-1459; e-mail bhs@liverpool.ac.uk; internet www.bulletinofhispanicstudies.org; f. 1923; language, literature and civilization of Spain, Portugal and Latin America; mainly English and Spanish, occasionally Portuguese, Catalan and French; Editor Prof. DOROTHY SHERMAN SEVERIN; 6 a year.

*Bulletin de l'Institut Français d'Etudes Andines.* IFEA, Ave Arequipa 4595, Lima 18, Peru; tel. (1) 4476070; fax (1) 4457650; e-mail postmaster@ifea.org.pe; internet www.ifeanet.org; f. 1972; geology and human and social sciences in the Andes; French, Spanish and English; Editor ANNE-MARIE BROUGÈRE; 3 a year.

*Bulletin of Latin American Research.* Blackwell Publishing, 9600 Garsington Rd, Oxford, OX4 2DQ, United Kingdom; tel. (1865) 776868; fax (1865) 791347; internet www.blackwellpublishing.com; on behalf of The Society for Latin American Studies; current interest research in social sciences and humanities; English; Editor NICOLA MILLER; quarterly.

*Business Latin America.* Business International Corpn, 1 Dag Hammarskjöld Plaza, New York, NY 10017, USA; tel. (212) 750-6300; English; Editor JAN ROVIRA BURCH; weekly.

*Cahiers des Amériques Latines.* Institut des Hautes Etudes de l'Amérique Latine, 28 rue Saint-Guillaume, 75007 Paris, France; tel. 1-44-39-86-00; fax 1-45-48-79-58; e-mail iheal.edition@univ-paris3.fr; f. 1968; political science, economy, urbanism, geography, history, sociology, ethnology, etc.; mainly French, but also Spanish and English; 3 a year.

*Canadian Journal of Latin American and Caribbean Studies / Revue canadienne des études latino-américaines et caraïbes.* Canadian Association of Latin American and Caribbean Studies, Université de Montréal, CP 6128, Succ. Centre-Ville, Montréal QC H3C 3J7, Canada; tel. (514) 343-6569; fax (514) 343-7716; internet www.calacs.umontreal.ca; f. 1969; political, economic, cultural, etc.; English, French, Portuguese and Spanish; Editor W. E. HEWITT; 2 a year.

*Caribbean Affairs.* 93 Frederick St, Port of Spain, Trinidad and Tobago; tel. 624-2477; fax 627-3013; f. 1988; business, political and social affairs of the Caribbean; English; Editor OWEN BAPTISTE; quarterly.

*Caribbean Business.* Casiano Communications, 1700 Fernández Juncos Ave, San Juan, PR 00909, Puerto Rico; tel. 728-3000; fax 268-1001; f. 1975; business and finance; English; Man. Editor MANUEL A. CASIANO, Jr; weekly.

*Caribbean Handbook.* FT Caribbean (BVI) Ltd, 19 Mercers Rd, London, N19 4PH, United Kingdom; tel. (20) 7281-5746; fax (20) 7281-7157; e-mail ftcaribbean@btinternet.com; f. 1983; business, economic and political information on the Caribbean region, including country profiles; English; Editor LINDSAY MAXWELL; annual.

*Caribbean Insight.* c/o Caribbean Council, Suite 18, Westminster Palace Gardens, 1–7 Artillery Row, London, SW1P 1RR, United Kingdom; tel. (20) 7799-1521; fax (20) 7340-1050; e-mail insight@caribbean-council.org; business, political and social; English; Editor Dr CHRIS BROGAN; weekly.

*Caribbean Investor and the CAIC Times.* Caribbean Association of Industry and Commerce, Hilton Trinidad and Conference Centre, POB 442, Lady Young Rd, St Ann's, Trinidad and Tobago; tel. 623-4830; fax 623-6116; e-mail caic@wow.net; internet caic.wow.net; finance and trading; quarterly.

*Caribbean Newsletter.* Friends for Jamaica Collective, Park West Station, POB 20392, New York, NY 10025, USA; f. 1981; analysis of events in the Caribbean; Caribbean-American issues; English; quarterly.

*Caribbean Quarterly.* POB 130, Kingston 7, Jamaica; tel. 977-1689; fax 970-3261; e-mail cq@uwimena.edu.jm; f. 1949; general; English; Editor REX NETTLEFORD; quarterly.

*Caribbean Review.* 9700 SW 67th Ave, Miami, FL 33156-3272, USA; tel. (305) 284-8466; fax (305) 284-1019; f. 1969; all subjects relating to the Caribbean, Latin America and their emigrant groups; quarterly.

*Caribbean Studies.* Institute of Caribbean Studies, POB 23361, University Station, Río Piedras, PR 00931, Puerto Rico; tel. (787) 764-0000; fax (787) 764-3099; e-mail iec@rrpac.upr.clu.edu; Caribbean affairs; English; Editor OSCAR MENDOZA; 2 a year.

*Caribbean Update.* 116 Myrtle Ave, Millburn, NJ 07041, USA; tel. (973) 376-2314; e-mail mexcarib@cs.com; internet www.caribbeanupdate.org; f. 1985; business and economic news and opportunities in the Caribbean and Central America; English; Editor and Publr KAL WAGENHEIM; monthly.

*El Caribe Contemporáneo.* Centro de Estudios Latinoamericanos, Facultad de Ciencias Políticas y Sociales, Universidad Nacional Autónoma de México, 04510 México, DF, Mexico; tel. (55) 5655-1344; f. 1980; political organization and government; Spanish; Dir Dr PABLO A. MARÍÑEZ A.; 2 a year.

*Carta Internacional.* Programa de Pesquisa de Relações Internacionais, Universidade de São Paulo, Rua do Anfiteatro 181, Colméia, Cidade Universitária, 05508-900 São Paulo, SP, Brazil; tel. (11) 818-3061; fax (11) 210-4154; e-mail guilhon@usp.br; internet www.usp.br/melint; Brazilian foreign and economic policy, regional integration, NAFTA and Mercosur; English, Portuguese and Spanish; Dir J. A. GUILHON ALBUQUERQUE; monthly.

*Central America Report.* Inforpress Centroamericana, Guatemala City, Guatemala; tel. 2221-0301; fax 2232-0306; e-mail inforpre@inforpressca.com; internet www.inforpressca.com/CAR/; f. 1972; review of economics and politics of Central America; English; Dir RICHARD WILSON-GRAU; weekly.

*Centroamerica Internacional.* Latin American Faculty of Social Sciences, Costa Rica—FLACSO, General Secretariat, Apdo 5429, 1000 San José, Costa Rica; tel. 257-0533; fax 221-5671; f. 1989; Central American, South American and Caribbean international relations, regional integration; Spanish; 2 a month.

*CEPAL Review / Revista de la CEPAL.* Casilla 179-D, Santiago, Chile; tel. (2) 2102000; fax (2) 2080252; f. 1976; a publication of the UN Economic Commission for Latin America and the Caribbean dealing with socio-economic topics; English and Spanish; Editor OSCAR ALTIMIR; 3 a year.

*Chile Ahora.* Ministry of Foreign Affairs, Palacio de la Moneda, Santiago, Chile; all aspects of life in Chile; Spanish; Editor M. ANGELICA HUIDOBRO G. H.; monthly.

*Colombia Internacional.* Centro de Estudios Internacionales, Universidad de los Andes, Calle 19, No 1–46, Apdo 4976, Bogotá, Colombia; tel. (1) 286-7504; fax (1) 284-1890; international cooperation, Latin American integration, drugs-trafficking controls; Spanish; quarterly.

*Colonial Latin American Review.* Routledge, Taylor & Francis, 4 Park Sq., Milton Park, Abingdon, Oxon, OX14 4RN, United Kingdom; tel. (20) 7017-6000; fax (20) 7017-6336; internet www.tandf.co.uk/journals; f. 1992; colonial period in Latin America; English, with articles in Portuguese and Spanish; Gen. Editor Prof. FREDERICK LUCIANI; 2 a year.

*Comercio Exterior.* Banco Nacional de Comercio Exterior, SNC, Camino a Santa Teresa 1679, Col. Jardines del Pedregal, 01900 México, DF, Mexico; tel. (55) 5481-6220; e-mail revcomer@bancomext.gob.mx; internet www.bancomext.com; f. 1951; international trade, analysis of Latin America's economics, general economics; Spanish with English abstracts; monthly.

*Contexto Internacional.* Instituto de Relações Internacionais, Pontifica Universidade Católica de Rio de Janeiro, Rua Marquês de São Vicente 225, Gávea, 22453-900 Rio de Janeiro, RJ, Brazil; tel. (21) 3114-1559; fax (21) 3114-1560; e-mail iripuc@rdc.puc-rio.br; internet www.puc-rio.br/iri; international relations, Brazilian foreign policy, Latin American and European integration, US-Latin American relations; Portuguese; Editor SONIA DE CAMARGO; 2 a year.

*The Courier ACP-European Union.* European Commission, 200 rue de la Loi, 1049 Brussels, Belgium; tel. (2) 299-11-11; fax (2) 299-30-02; internet www.europa.eu.int/comm/development/body/publications_courier_en.htm; affairs of the African, Caribbean and Pacific countries and the European Union; English and French editions; Editor SIMON HORNER; 6 a year.

*Cronología de las Relaciones Internacionales del Peru.* Peruvian Centre for International Studies, San Ignacio de Loyola 554, Miraflores 8, Lima 18, Peru; tel. (1) 4457225; fax (1) 4451094; economic and political international relations of Peru; Spanish; quarterly.

*CrossCurrents.* Washington Office on Latin America, Suite 200, 1630 Connecticut Ave, NW, Washington, DC, USA; tel. (202) 797-2171; fax (202) 797-2171; e-mail wola@wola.org; internet www.wola.org; human rights, Latin American politics and economics; English; quarterly.

*Cuadernos de Economía* (American Journal of Economics). Instituto de Economía, Pontíficia Universidad Católica de Chile, Casilla 76, Correo 17, Santiago, Chile; tel. (2) 3544314; fax (2) 3544312; e-mail cuadecon@faceapuc.cl; internet www.cuadernoseconomia.cl; f. 1963; applied economies as contribution to economic policy, with special emphasis in Latin America; Spanish, with English abstracts; Editors JOSÉ MIGUEL SÁNCHEZ, JUAN PABLO MONTERO, RAIMUNDO SOTO; 2 a year.

*Cuadernos Hispanoamericanos.* Avda de los Reyes Católicos 4, 28040 Madrid, Spain; tel. (91) 5838399; e-mail cuadernos@hispanoamericanosaeci.es; f. 1948; humanities, particularly

## REGIONAL INFORMATION

relating to Hispanic America; Spanish; Editor BLAS MATAMORO; monthly.

*Cuba Internacional.* Calle 21, No. 406, Vedado, Havana 4; Apdo B603, Havana 3, Cuba; tel. (7) 329-3531; f. 1959; politics and foreign affairs; Spanish and Russian; monthly.

*Cuba Update.* Center for Cuban Studies, 124 West 23rd St, New York, NY 10011, USA; tel. (212) 242-0559; fax fax (212) 242-1937; e-mail cubanctr@igc.org; internet www.cubaupdate.org; f. 1977; Cuban foreign affairs, culture and development; English; Editor SANDRA LEVINSON; quarterly.

*The Developing Economies.* Nihon Boeki Shinkokiko Ajia Keizai Kenkyusho (Institute of Developing Economies, Japan External Trade Organization), 3-2-2 Wakaba, Mihama-ku, China-shi, Chiba 261-8545, Japan; tel. (43) 299-9500; fax (43) 299-9726; e-mail journal@ide.go.jp; internet www.ide.go.jp; f. 1960; English; quarterly.

*Development Policy Review.* Overseas Development Institute, 111 Westminster Bridge Rd, London, SE1 7JD, United Kingdom; tel. (20) 7922-0399; fax (20) 7922-0399; internet www.odi.org.uk; f. 1982; Editor DAVID BOOTH; quarterly.

*Economia Brasileira e Suas Perspectivas.* Associação Promotora de Estudos de Economia, Rua Sorocaya 295, Botafogo, Rio de Janeiro, RJ, Brazil; fax (21) 266-3597; f. 1962; Brazilian economic issues, published by the Association for the Promotion of Economic Studies; Portuguese and English; Editor BASÍLIO MARTINS; annual.

*Economía Mexicana Nueva Época.* Centro de Investigación y Docencia Económicas—CIDE, Carretera México-Toluca 3655 (km 16.5), Col. Lomas de Santa Fe, Del. Alvaro Obregón, 01210 México, DF, Mexico; tel. (55) 5727-9800; fax (55) 5727-9878; e-mail ecomex@cide.edu; internet www.cide.edu; f. 1992; economic problems in Mexico and Latin America; Spanish and English; Editor DAVID MAYER FOULKES; 2 a year.

*Economic Development and Cultural Change.* University of Chicago, Judd Hall, Suite 318, 5835 South Kimbark Ave, Chicago, IL 60637, USA; tel. (773) 702-7951; fax (773) 834-3832; e-mail edcc@uchicago.edu; internet www.journals.uchicago.edu/EDCC; f. 1952; various aspects of economic development and cultural change; English; Editor (vacant); quarterly.

*Economic and Social Progress in Latin America.* Inter-American Development Bank, 1300 New York Ave, NW, Washington, DC 20577, USA; tel. (202) 623-1403; internet www.iadb.org; f. 1961; socio-economic conditions and development; topical reports; English and Spanish; annual.

*Enlace: Política y Derechos Humanos en las Americas.* Washington Office on Latin America, Suite 200, 1630 Connecticut Ave, NW, Washington, DC, USA; tel. (202) 797-2171; fax (202) 797-2172; e-mail wola@wola.org; internet www.wola.org; f. 1991; US foreign-policy implications for human rights and democratization in Latin America; Spanish; quarterly.

*Environment Watch: Latin America.* Cutter Information Corporation, Suite 1, 37 Broadway, Arlington, MA 02174, USA; tel. (617) 641-5125; fax (617) 648-1950; e-mail jstoub@igc.apc.org; f. 1991; environmental policy, law and corporate initiatives in Latin America; English; Editor JEFFREY STOUB; monthly.

*Estadística y Economía.* Instituto Nacional de Estadísticas, Avda Presidente Bulnes 418, Casilla 498, Correo 3, Santiago, Chile; tel. (2) 366-7777; fax (2) 671-2169; e-mail ine@ine.cl; internet www.ine.cl; f. 1843; commerce and economics; Spanish; 2 a year.

*Estudios de Coyuntura.* Universidad del Zulia, Facultad de Ciencias Ecónomicas y Sociales, Apdo 526, Maracaibo 4011, Estado Zulia, Venezuela; tel. (261) 51-7697; fax (261) 51-2525; f. 1989; business and economics; Editor HERNAN PARDO; Spanish; 2 a year.

*Estudios de Cultura Maya.* Centro de Estudios Mayas, Instituto de Investigaciones Filológicas, Circuito Mario de la Cueva, Ciudad Universitaria, 04510 México, DF, Mexico; tel. and fax (55) 5622-7490; e-mail cem@servidor.uma.mx; f. 1961; anthropology, archaeology, history, epigraphy and linguistics of the Mayan groups; Spanish, English and French; Editor MARIO HUMBERTO RUZ; 2 a year.

*Estudios Económicos.* Universidad Nacional del Sur, Departamento de Economía, 12 de Octubre y San Juan, 8000 Bahía Blanca, Argentina; tel. and fax (91) 25432; f. 1962; Spanish; Editor RICARDO BARA; annual.

*Estudios de Historia Moderna y Contemporánea de México.* Instituto de Investigaciones Históricas, Circuito Metropolitano Mario de la Cueva, Del. Coyoacán, Ciudad Universitaria, 04510 México, DF, Mexico; tel. (55) 5622-7520; f. 1967; history of Mexico from independence war (1810) to present; Spanish; Editor MARCELA TERRAZAS Y BASANTE; 2 a year.

*Estudios Internacionales.* International Relations and Peace Research Institute, 1a Calle 9-52, Zona 1, 01001 Guatemala City, Guatemala; international relations; Spanish; 2 a year.

*Estudios Internacionales.* Institute of International Studies, University of Chile, Avda Condell 249, Suc. 21, Casilla 14187, Santiago 9, Chile; tel. (2) 4961200; fax (2) 2740155; e-mail inesint@uchile.cl; internet www.uchile.cl/facultades/estinter; f. 1966; contemporary international relations, particularly concerning Latin America; Spanish; Editor ROSE CAVE S.; quarterly.

*Estudios Paraguayos.* Universidad Católica, Casilla 1718, Asunción, Paraguay; tel. (21) 446251; fax (21) 445245; f. 1973; philosophy, politics, law, history, linguistics, economics, literature; Spanish; 2 a year.

*Estudios Políticos.* Facultad de Ciencias Políticas y Sociales, Universidad Nacional Autónoma de México, Ciudad Universitaria, Apdo 70-266, 04510 México, DF, Mexico; tel. (55) 5665-1233; fax (55) 5666-8334; e-mail igarza@sociolan.politicas.unam.mx; f. 1975; politics of government, political science; Spanish; quarterly.

*Estudios Públicos.* Centro de Estudios Públicos, Monseñor Sótero Sanz 175, Santiago, Chile; tel. (2) 2315324; fax (2) 2335253; e-mail dparra@cepchile.cl; f. 1980; forum of ideas and commentary on economic development and public policy in Latin America; English, Spanish; 4 a year.

*Estudios Sociales.* Miguel Claro 1460, Santiago, Chile; tel. (2) 2043418; fax (2) 2741828; e-mail cpu@cpu.cl; internet www.cpu.cl; f. 1973; sociology, history, anthropology, economics, political science, education, philosophy, social psychology, law; Spanish; Editor RAÚL ATRIA; 2 a year.

*EURE Revista Latinoamericana de Estudios Urbanos y Regionales.* Instituto de Estudios Urbanos, Universidad Católica de Chile, Casilla 16002, Correo 9, Santiago, Chile; tel. (2) 6865511; fax (2) 2328805; e-mail eure@puc.cl; internet www.scielo.cl/eure.htm; f. 1970; urban and regional development in Latin America; Spanish, with English abstracts; Editor DIEGO CAMPOS ALVAREZ; quarterly.

*FIDE, Coyuntura y Desarrollo.* Development Research Foundation—FIDE, Sánchez de Loria 1338, 1241 Buenos Aires, Argentina; tel. and fax (11) 4931-9257; national and international socio-economic analysis and economic theory; Spanish; monthly.

*Foro Internacional.* Colegio de México, Camino al Ajusco 20, Col. Pedregal de Santa Teresa, 10740 México, DF, Mexico; tel. (5) 5449-3011; fax (5) 5645-0464; internet www.colmex.mx/centros/cei/foro_index.htm; Latin American politics; Mexican domestic affairs, public administration and policy; Spanish; Editor CARLOS VEGA ALBA; quarterly.

*Fuerzas Armadas y Sociedad.* Latin American Faculty of Social Sciences—FLACSO Chile, Avda Dag Hammarskjöld 3269, Vitacura, Santiago; tel. 2900200; fax 2900270; e-mail flacso@flacso.cl; internet www.flacso.cl; military affairs and international relations; Spanish; quarterly.

*Global Studies: Latin America.* Dushkin Publishing, McGraw Hill 2460, Kerper Blvd, Dubuque, IA 52001, USA; tel. (203) 453-4351; fax (203) 453-6000; e-mail nichole_altman@mcgraw-hill.com; internet www.dushkin.com; f. 1991; articles on Mexico, Central America, South America and the Caribbean; English; Editor PAUL GOODWIN; every 2 years.

*Hemisphere.* Latin American and Caribbean Center, Florida International University, University Park, Miami, FL 33199, USA; tel. (305) 348-2894; fax (305) 349-3593; e-mail bottap@fiu.edu; internet lacc.fiu.edu; f. 1988; Latin American and Caribbean Affairs; English; Editors EDUARDO A. GAMARRA, PEDRO D. BOTTA; 2 a year.

*Hispanic American Historical Review (HAHR).* Duke University Press, 905 W Main St, Suite 18B, Durham, NC 27701, USA; tel. (919) 687-3600; fax (919) 688-4574; e-mail subscriptions@dukeupress.edu; internet www.dukeupress.edu; f. 1918; Editors GILBERT M. JOSEPH, STUART B. SCHWARTZ; quarterly.

*Historia.* Casilla 6277, Santiago 22, Chile; tel. (2) 3547831; e-mail revhist@puc.cl; internet www.hist.puc.cl; f. 1961; history of Chile and related subjects; mainly Spanish; Editor NICOLÁS CRUZ; 2 a year.

*Historia Mexicana.* Colegio de México, Camino al Ajusco 20, Pedregal de Santa Teresa 10740, México, DF, Mexico; tel. (55) 5449-3067; fax (55) 5645-0464; f. 1951; history of Mexico; Spanish; Editor OSCAR MAZIN; quarterly.

*Iberoamericana.* 60318 Frankfurt am Main, Wielandstr. 40, Verlag Klaus Dieter Vervuert, Germany; tel. (69) 5974617; fax (69) 5978743; e-mail info@iberoamericanalibros.com; internet www.ibero-americana.net; f. 2001; comprises formerly separate journals *Iberoamericana, Ibero-Amerikanisches Archiv* and *Notas*; Latin American literature, history and social sciences; Spanish, Portuguese, English and German; Editors W. BERNECKER, K. BODEMER, M. BRAIG, O. ETTE, F. GEWECKE, R. LIEHR, J.-M. LÓPEZ DE ABIADA, G. MAIHOLD, D. MESSNER, A. PAGNI, H.-J. PUHLE, C. STROSETSKI, M. TIETZ; quarterly.

*Iberoamericana: Nordic Journal of Latin American and Caribbean Studies.* Institute of Latin American Studies, Stockholm University, Universitetsvägen 10B, 106 91 Stockholm, Sweden; tel. (8) 16-28-82; fax (8) 15-65-82; e-mail lai@lai.su.se; internet www.lai.su.se; articles on economic, political and local developments in Latin America and the Caribbean; English, Portuguese, Spanish; 2 a year; Prof. JAIME BEHAR.

*Indicadores Económicos.* Contraloría-General de la República, Dirección de Estadística y Censo, Apdo 5213, Panamá 5, Panama; tel. 210-4800; fax 210-4801; e-mail cgrdec@contraloria.gob.pa; internet www.contraloria.gob.pa; f. 1996; economic indicators for Panama; Spanish; annual.

*Industri-Noticias.* Publi-News Latinoamericano, SACV, Colina 436, México, DF, Mexico; f. 1966; business, economics and industry; Spanish; Editor ROBERTO J. MÁRQUEZ; monthly.

*Industria.* Sindicato de Industriales de Panamá, Apdo 6-4798, El Dorado, Panamá, Panama; tel. 230-0169; fax 230-0805; e-mail sip@cableonda.net; internet www.industriales.org; f. 1953; economics and industry in Panama; Spanish; Editor FLOR ORTEGA; quarterly.

*Industria Venezolana.* Editorial Guía Industrial, Apdo 60772, Chacao, Caracas 101, Venezuela; f. 1971; economics and industry in Venezuela; Spanish; Editor JOSÉ PRECEDO; 6 a year.

*Informe Latinoamericano.* Latin American Newsletters, 61 Old St, London, EC1V 9HW, United Kingdom; tel. (20) 7251-0012; fax (20) 7253-8193; internet www.latinnews.com; f. 1982; Latin American News; book reviews; Spanish; Editor EDUARDO CRAWLEY; weekly.

*Información Sistemática.* Valencia 84, Insurgentes Mixcoac, Del. Benito Juárez, 03920 México, DF, Mexico; tel. (55) 5598-6043; fax (55) 5598-6325; f. 1976; clippings archive since 1976; electronic database since 1988; daily summary of 11 newspapers; Spanish; Dirs BERNARDA AVALOS, LUPITA FLORES; customized frequency.

*Information Services Latin America.* POB 6103, Albany, CA 94706, USA; tel. (510) 996-23181; e-mail isla@lmi.net; internet www.igc.org/isla; f. 1970; reprinted articles from major daily news sources; available in English and Spanish; English; Dir KAREN CRUMP; monthly.

*Integración Financiera—pasado, presente y futuro de las finanzas en Colombia y el mundo.* Medios and Medios Publicidad Cía Ltda, No. 11–45, Of. 802, Calle 63, Apdo 036943, Bogotá, DC, Colombia; tel. (1) 255-0992; fax (1) 249-4696; f. 1984; financial-sector development in Colombia and the rest of Latin America; Spanish; Editor RAÚL RODRÍGUEZ PUERTO; 6 a year.

*Integración y Comercio.* Institute for the Integration of Latin America and the Caribbean, Esmeralda 130, 17°, 1035 Buenos Aires, Argentina; tel. (11) 4320-1850; fax (11) 4320-1865; e-mail int/inl@iadb.org; internet www.iabd.org/intal; f. 1965; Latin American integration; Spanish and English; 2 a year.

*Investigaciones y Ensayos.* Balcarce 139, 1064 Buenos Aires, Argentina; tel. (11) 4343-4416; e-mail adminte@an-historia.org.ar; internet www.an-historia.org.ar; f. 1966; history of Argentina and America; Spanish; published by the Academia Nacional de la Historia; 2 a year.

*Jahrbuch für Geschichte Lateinamerikas.* 20146 Hamburg, Von-Melle-Park 6, Universität Hamburg Historisches Seminar, Germany; e-mail horst.pietschmann@uni-hamburg.de; internet www-gewi.uni-graz.at/jbla; published by Böhlau-Verlag, Cologne and Vienna; f. 1964; political, economic, social and cultural history of Latin America from colonial period to present; articles in Spanish, English, French, German and Portuguese; Dir Prof. Dr HORST PIETSCHMANN; annual.

*Journal of Development Studies.* Routledge, Taylor & Francis, 4 Park Sq., Milton Park, Abingdon, Oxon, OX14 4RN, United Kingdom; tel. (20) 7017-6000; fax (20) 7017-6336; internet www.tandf.co.uk/journals; f. 1964; Editors CHRISTOPHER COLCLOUGH, JOHN HARRISS, CHRIS MILNER; 6 a year.

*Journal of Iberian and Latin American Studies.* Routledge, Taylor & Francis, 4 Park Sq., Milton Park, Abingdon, Oxon, OX14 4RN, United Kingdom; tel. (20) 7017-6000; fax (20) 7017-6336; internet www.tandf.co.uk/journals; fmrly known as *Tesserae*; language, literature, history and culture of Latin America and Iberian peninsula; Editor JORDI LARIOS; English, with articles in Catalan and Spanish; 3 a year.

*Journal of Interamerican Studies and World Affairs.* University of Miami, POB 248123, Coral Gables, FL 33124C, USA; tel. (1305) 284-5554; fax (305) 284-4406; f. 1959; Latin American-US relations and Latin America's international relations; English; Editor WILLIAM SMITH; quarterly.

*Journal of Latin American Cultural Studies.* Routledge, Taylor & Francis, 4 Park Sq., Milton Park, Abingdon, Oxon, OX14 4RN, United Kingdom; tel. (20) 7017-6000; fax (20) 7017-6336; internet www.tandf.co.uk/journals; history and analysis of Latin American culture; English; Editors JENS ANDERMANN, BEN BOLLIG, CATHERINE BOYLE, JOHN KRANIAUSKAS, PHILLIP DERBYSHIRE, HERMANN HERLINGHAUS, LORRAINE LEU.

*Journal of Latin American Studies.* Institute for the Study of the Americas, 31 Tavistock Sq., London, WC1H 9HA, United Kingdom; tel. (20) 7862-8877; fax (20) 7862-8886; e-mail jlas@sas.ac.uk; internet www.sas.ac.uk/ilas/jlas.htm; sponsored by Centres or Insts of Latin American Studies at Univs of Cambridge, Essex, Liverpool, London and Oxford; Editors Prof. JAMES DUNKERLEY, Prof. PAUL CAMMACK, RACHEL SIEDER; 4 a year.

*Journal de la Société des Américanistes.* Musée de l'Homme, 17 place du Trocadéro, 75116 Paris, France; tel. 1-46-69-26-34; fax 1-46-69-25-08; e-mail jsa@mae.u-paris10.fr; f. 1896; archaeology, ethnology, ethnohistory and linguistics of the American continent; French, Spanish, English and Portuguese; Editor DOMINIQUE MICHELET; annual.

*Kañina.* Faculdad de Letras, Ciudad Universitaria Rodrigo Facio, Universidad de Costa Rica, San Pedro, San José, Costa Rica; tel. 207-5107; fax 207-5089; e-mail kanina@cariari.ucr.ac.cr; f. 1976; arts and literature; mainly Spanish, but also French, English and Italian; Editor VICTOR SÁNCHEZ CORRALES; 3 a year.

*Lagniappe Letter / Quarterly Monitor.* 159 W 53rd St, 28th Floor, New York, NY 10019, USA; tel. (212) 765-5520; fax (212) 765-2927; e-mail rwerret@pipeline.com; internet www.lais.com; Latin American business; English; Editor ROSEMARY WERRETT; fortnightly/quarterly.

*Lagniappe Monthly on Latin American Projects and Finance.* 159 W 53rd St, 28th Floor, New York, NY 10019, USA; tel. (212) 765-5520; fax (212) 765-2927; e-mail rwerret@pipeline.com; internet www.lais.com; Latin American business; Editor ROSEMARY WERRETT; monthly.

*LARF Report.* Latin America Reserve Fund—LARF, Bogotá, DC, Colombia; tel. (1) 285-8511; fax (1) 288-1117; f. 1979; economic summary of Bolivia, Colombia, Ecuador, Peru and Venezuela; Spanish and English; annual.

*Lateinamerika Analysen.* Institut für Iberoamerika-Kunde, 20354 Hamburg, Alsterglacis 8, Germany; tel. (40) 41478201; fax (40) 41478241; e-mail sekretariat@iik.duei.de; internet www.duei.de/iik; f. 2002; formerly *Lateinamerika: Analysen-Daten-*

*Dokumentation*; political, economic and social development, regional integration and international relations of Latin America; German, Spanish and Portuguese; 3 a year.

*Lateinamerika Anders Panorama.* Informationsgruppe Lateinamerika, Postfach 557, 1060 Vienna, Austria; e-mail igla2@compuserve.com; f. 1976; news and analysis of Latin American affairs; German; Editors Werner Hörtner, Hermann Klosius; 6 a year.

*Lateinamerika Jahrbuch.* Institut für Iberoamerika-Kunde, 20354 Hamburg, Alsterglacis 8, Germany; tel. (40) 41478201; fax (40) 41478241; e-mail sekretariat@iik.duei.de; internet www.duei.de/iik; f. 1992; political, business and social review; annual.

*Latin America and Caribbean Contemporary Record.* Holmes and Meier Publishers Inc, 160 Broadway East Wing, New York, NY 10038, USA; tel. (212) 374-0100; fax (212) 374-1313; e-mail info@holmesandmeier.com; internet www.holmesandmeier.com; f. 1981; analysis of events and trends in Latin America and the Caribbean; English; Editors James Malloy, Eduardo Gamarra; annual.

*Latin American Economy and Business.* Latin American Newsletters, 61 Old St, London, EC1V 9HW, United Kingdom; tel. (20) 7251-0012; fax (20) 7253-8193; e-mail subs@latinnews.com; internet www.latinnews.com; economic data and indicators; English; Editor Will Ollard; monthly.

*Latin American Index.* Welt Publishing LLC, Washington, DC, USA; tel. (202) 371-0555; fax (202) 408-9369; political, economic and social affairs of Latin America, and US-Latin American relations; English; Editor William Knepper; 23 a year.

*Latin American Monitor.* Business Monitor International, Mermaid House, 2 Puddle Dock, Blackfriars, London, EC4V 3DS, United Kingdom; tel. (20) 7248-0468; fax (20) 7248-0467; e-mail subs@businessmonitor.com; internet www.businessmonitor.com; publishes 6 regional reports; (i) Mexico; (ii) Brazil; (iii) Central America; (iv) Andean Group; (v) Southern Cone; (vi) Caribbean; English; monthly.

*Latin American Perspectives.* University of California, POB 5703, Riverside, CA 92517-5703, USA; tel. (951) 827-1571; fax (951) 827-5685; e-mail laps@ucr.edu; internet www.latinamericanperspectives.com; economics, political science, international relations, philosophy, history, sociology, geography, anthropology and literature; English; Editor Ronald H. Chilcote; 6 a year.

*Latin American Regional Reports.* Latin American Newsletters, 61 Old St, London, EC1V 9HW, United Kingdom; tel. (20) 7251-0012; fax (20) 7253-8193; e-mail subs@latinnews.com; internet www.latinnews.com; regional reports on: Andean Group; Mexico and NAFTA; Southern Cone and Brazil; Caribbean and Central America; English; Editors Jon Farmer, Will Ollard; monthly.

*Latin American Research Review (LARR).* University of Texas at Austin, 1 University Station D0800, Austin, TX 78712, USA; tel. (512) 471-8770; fax (512) 471-3090; e-mail larr@uts.cc.utexas.edu; internet larr.lanic.utexas.edu; f. 1965; articles, research notes and essays dealing with contemporary issues; Editor Peter M. Ward; 3 a year.

*Latin American Special Reports.* Latin American Newsletters, 61 Old St, London, EC1V 9HW, United Kingdom; tel. (20) 7251-0012; fax (20) 7253-8193; e-mail subs@latinnews.com; internet www.latinnews.com; each edition provides detailed information on and analysis of one specific subject; English and Spanish; Editor Eduardo Crawley; 6 a year.

*Latin American Studies in Asia.* K. K. Roy (Pvt) Ltd, 55 Gariahat Rd, POB 10210, Calcutta 700 019, India; tel. (33) 4754872; f. 1991; Latin American and Caribbean affairs; English; Editor Dr K. K. Roy; quarterly.

*Latin American Weekly Report.* Latin American Newsletters, 61 Old St, London, EC1V 9HW, United Kingdom; tel. (20) 7251-0012; fax (20) 7253-8193; e-mail subs@latinnews.com; internet www.latinnews.com; political, economic and general news; English and Spanish; Editor Eduardo Crawley; weekly.

*Latinskaya Amerika.* Institute of Latin American Studies, Russian Academy of Sciences, 119034 Moscow, per. Kropotkinskii 24, Russia; tel. (095) 201-56-64; fax (095) 200-42-14; e-mail revistala@mtu-net.ru; f. 1969; politics, economic development, history and culture of Latin American countries; Russian; Editor Vladimir E. Travkin; monthly.

*Latin Studies Journal.* Center for Latino Research, Northeastern University, Dept of Sociology and Anthropology, 521 Holmes Hall, Boston, MA 02115, USA; tel. (617) 373-2000; f. 1990; Latin American affairs; Latin American-North American relations; Editor Felix Padilla; 3 a year.

*Lecturas de Economía.* Centro de Investigaciones Económicas, Universidad de Antioquia, Apdo Aéreo 1226, Ciudad Universitaria, Medellín, Antioquia, Colombia; tel. (4) 210-5841; fax (4) 263-8282; f. 1980; economic issues; Spanish; Editor Jorge Lotero Contreras; 2 a year.

*Luso-Brazilian Review.* University of Wisconsin Press, Journals Division, 1930 Monroe St, Madison, WI 53711, USA; tel. (608) 263-0668; e-mail journals@uwpress.wisc.edu; internet www.wisc.edu/wisconsinpress/journals/Luso-Brazilian_Review.html; f. 1964; history, social sciences and literature; English and Portuguese; Editors Peter M Beattie, Ellen W. Sapega, Severino J. Albuquerque; 2 a year.

*Mesoamérica.* Centro de Investigaciones Regionales de Mesoamérica (CIRMA)/Plumsock Mesoamerican Studies, POB 38, South Woodstock, VT 05071, USA; tel. (802) 457-1199; fax (802) 457-2212; e-mail pmsvt@aol.com; f. 1980; anthropology, history, archaeology, linguistics and social sciences of southern Mexico and Central America; Spanish; Editors Armando J. Alfonzo U., W. George Lovell; 2 a year.

*Mexico Service.* International Reports, 11300 Rockville Pike, Rockville, MD 20852-3030, USA; tel. (301) 816-8950; f. 1980; business and finance in Brazil; English; Editor-in-Chief Robert G. Taylor; Editor Andrew Vogel; every 2 weeks.

*Mexico Watch.* Orbis Publications Ltd, 1924 47th St, NW, Washington, DC 20007, USA; tel. (202) 298-7936; fax (202) 298-7938; e-mail orbis@orbispub.com; politics, economics and business of Mexico; English; Publr Stephen M. Foster; monthly.

*Mundo Nuevo. Revista de Estudios Latinoamericanos.* Apdo 17271, Caracas 1015-A, Venezuela; tel. (212) 573-8824; internet www.iaeal.usb.ve; f. 1975; international relations, politics, economy of Latin America; Spanish, but some articles are published in their original languages, English, French and Portuguese; published by the Instituto de Altos Estudios de América Latina, Universidad Simón Bolívar; quarterly.

*Mundus.* Wissenschaftliche Verlagsgesellschaft GmbH, Birkenwaldstr. 44, 70191 Stuttgart, Germany; tel. (711) 25820; fax (711) 2582290; f. 1965; review of German research on Latin America, Asia and Africa; English; Editor Jürgen Hohnholz (Institute for Scientific Co-operation); quarterly.

*NACLA Report on the Americas.* North American Congress on Latin America, 38 Greene St, 4th Floor, New York, NY 10013, USA; tel. (646) 613-1440; fax (646) 613-1443; e-mail nacla@igv.apc.org; internet www.nacla.org; f. 1966; publication of North American Congress on Latin America covering US foreign policy towards Latin America and the Caribbean, and domestic developments in the region; English; Editor Jo Marie Burt; 6 a year.

*Negobancos-Negocios y Bancos.* Bolívar 8-103, Apdo 1907, 06000 México, DF, Mexico; tel. (55) 5510-1884; fax (55) 5512-9411; f. 1951; business and economics; Spanish; Dir Alfredo Farrugia Reed; every 2 weeks.

*Nordeste: Análise Conjuntural.* Banco do Nordeste do Brasil, Escritório Técnico de Estudos Econômicos do Nordeste, Praça Murilo Borges, No 1, CP 628, Fortaleza, CE, Brazil; e-mail nego_bancos@mexico.com; f. 1972 as Análise Conjuntural de Economia Nordestina; adopted existing name in 1974; Brazilian economy, including statistics; Portuguese; Editor João de Aquino Limaverde.

*NotiCen.* University of New Mexico, Latin American and Iberian Institute, Latin American Data Base, 801 Yale NE, Albuquerque, NM 87131, USA; tel. (505) 277-6839; fax (505) 277-6837; e-mail info@ladb.unm.edu; internet ladb.unm.edu/noticen; online news digest; sustainable development, economic and political affairs in Central America and the Caribbean; Editor Michael Leffert; weekly.

# REGIONAL INFORMATION

*Noticias Indigenistas de América / Indian News of the Americas.* Instituto Indigenista Interamericano, Avda Insurgentes Sur 1690, Col. Florida, 01030 México, DF, Mexico; tel. (55) 5660-0007; fax (55) 5534-8090; internet www.ini.gob.mx/iii/noticiasindi.html; f. 1978; news of the Indians of the Americas, government policies, anthropology; Spanish and English editions; Editor Dr José Matos Mar; 3 a year.

*Noticias de Latino América Documentos.* 41 rue de Suède, 1060 Brussels, Belgium; tel. (2) 538-78-81; extracts from Latin American press; Spanish; Editor Kristin Minne; every two months.

*NotiSur.* University of New Mexico, Latin American and Iberian Institute, Latin America Data Base, 801 Yale, NE, Albuquerque, NM 87131, USA; tel. (505) 277-6389; fax (505) 277-6387; e-mail info@ladb.unm.edu; internet ladb.unm.edu/notisur; online news digest; provides alternative viewpoints on political and economic affairs in South America, particularly on human rights and peace issues and sustainable development; English; Editor Joe Gardner Wessely; weekly.

*Opciones.* Territorial, esq. Gen Suárez, Plaza de la Revolución, Havana, Cuba; economics and politics; Spanish; weekly.

*Panorama Económico.* Bancomer, SA, Grupo Investigaciones Económicas, Centro Bancomer, Avda Universidad 1200, 03339 México, DF, Mexico; tel. (55) 5534-0034; fax (55) 5621-3230; f. 1966; Mexican economy, in particular the automobile and textiles industries; Spanish and English; Editor Eduardo Millan Lozano; 6 a year.

*Panorama Económico Latinoamericano (PEL).* Ediciones Cubanas, Obispo 527, Apdo 605, Havana, Cuba; tel. (7) 63-1981; fax (7) 33-8943; f. 1960; book reviews and statistics; available on micro-film; Spanish; Editor José Bodes Gómez; 2 a month.

*Panorama Latinoamericano.* Moscow, Zubovskii bul. 4, Russia; tel. (095) 201-55-79; review of economic, political, cultural, and social life of Latin America drawn from Russian Press and published by Novosti Press Agency; Spanish; fortnightly.

*Paz.* South American Commission for Peace, Regional Security and Democracy, Santiago, Chile; tel. (2) 232-8329; fax (2) 233-3502; Spanish; 2 a year.

*Pesquisa e Planejamento Econômico.* Institute of Applied Economic Research, Avda Antônio Carlos 51, 16° andar, CP 2672, 20.020-010 Rio de Janeiro, RJ, Brazil; tel. (21) 3804-8118; fax (21) 2220-5533; e-mail editrj@ipea.gov.br; internet www.ipea.gov.br; f. 1970; economics and planning; Portuguese and English; Editors Octávio Augusto Fontes Tourinho, Marco Antonio Cavalcanti; quarterly.

*Política Externa.* Programa de Pesquisa de Relações Internacionais, Universidade de São Paulo, Rua do Anfiteatro 181, Colméia, favo 14, Cidade Universitária, 05508-900 São Paulo, SP, Brazil; tel. (11) 818-3061; fax (11) 814-7342; e-mail guilhon@usp.br; Brazilian foreign and economic policy, regional integration, NAFTA and Mercosur; Portuguese.

*Problemas del Desarrollo—revista latinoamericana de economía.* Instituto de Investigaciones Económicas de la Universidad Nacional Autónoma de México, Ciudad Universitaria, Del. Coyoacán, 04510, México, DF, Mexico; tel. (55) 5623-0105; fax (55) 5623-0097; e-mail revprode@servidor.unam.mx; f. 1969; economic, political and social affairs of Mexico, Latin America and the developing world; Spanish, with English and French abstracts; quarterly.

*Puerto Rico Business Review / Puerto Rico Economic Indicators.* Government Development Bank for Puerto Rico, POB 42001, San Juan, PR 00940-2001, Puerto Rico; tel. 722-2525; fax 268-5496; e-mail gdbcomm.bgf.gobierno.pr; internet www.gdbpur.com; f. 1976; English; Editor Anabel Hernández; quarterly.

*Quarterly Economic Review.* Central Bank of the Bahamas, Research Dept, Frederick St, POB N-4868, Nassau, Bahamas; tel. 322-2193; fax 356-4321; e-mail cbob@centralbankbahamas.com; internet www.bahamascentralbank.com; review of Bahamian economy; English; quarterly.

*Quehacer.* Centro de Estudios y Promoción del Desarrollo—DESCO, León de la Fuente 110, Magdalena del Mar, Lima 17, Peru; tel. (1) 6168300; e-mail postmaster@desco.org.pe; internet www.desco.org.pe; f. 1979; business development and research;

## Select Bibliography (Periodicals)

Spanish; Dir Abelardo Sánchez-Léon; Editor Juan Larco Guichard; 6 a year.

*Relaciones Internacionales.* Facultad de Ciencias Políticas y Sociales, Universidad Nacional Autónoma de México, México, DF, Mexico; tel. (55) 5622-9412; f. 1973; international relations; Spanish; Editor Roberto Domínguez; Dir Consuelo Dávila; 3 a year.

*Review: Literature and Arts of the Americas.* Routledge, Taylor & Francis, 4 Park Sq., Milton Park, Abingdon, Oxon, OX14 4RN, United Kingdom; tel. (20) 7017-6000; fax (20) 7017-6336; internet www.tandf.co.uk/journals; f. 1967; Latin American literature in translation, articles on visual arts, theatre, music, cinema, book reviews; English, but poetry is, in addition, published in the vernacular; Editor Alfred Mac Adam; 2 a year.

*Revista Análisis.* Revistas Interamericanos SA, Apdo 8038, Panama 7, Panama; tel. 26-0073; fax 26-3758; f. 1979; economic and political analysis; academic; Spanish; Editor Mario A. Rognoni; monthly.

*Revista Argentina de Estudios Estratégicos.* Argentine Centre of Strategic Studies, Viamonte 494, 3°, Of. 11, 1053 Buenos Aires, Argentina; tel. (11) 4312-1605; fax (11) 4312-5802; the armed forces; Spanish; quarterly.

*Revista de Biología Tropical.* Universidad de Costa Rica, Ciudad Universitaria Rodrigo Facio, San José, Costa Rica; tel. and fax 2075550; e-mail rbt@cariari.ucr.ac.cr; f. 1953; biology, ecology, taxonomy, etc. of Neotropics and African tropics; Spanish and English; Editor Bernal Morera Brenes; quarterly, with supplements, and online edition.

*Revista Brasileira de Economia.* Escola de Pós-Graduação em Economia, Praia de Botafogo 190, Of. 1100, 22.253-900 Rio de Janeiro, RJ, Brazil; tel. (21) 2559-5860; fax (21) 2552-4898; e-mail rbe@fgv.br; internet www.fgv.br/epge/home/publi/RBE; f. 1947; economic theory, economic policy and econometrics; Portuguese and English; Editor Ricardo Cavalcanti; quarterly.

*Revista Brasileira de Estatística.* Fundação Instituto Brasileiro de Geografia e Estatística, Avda Chile 500, 10° andar, 20.031-170 Rio de Janeiro, RJ, Brazil; tel. (21) 21420501; fax (21) 21424548; e-mail pedrosilva@ibge.gov.br; f. 1940; statistical subjects by means of articles, analysis, etc.; Portuguese; 2 a year.

*Revista Brasileira de Estudos Políticos.* Av. Alvares Cabral 211, salá 1206, CP 1301, 30170-000 Belo Horizonte, MG, Brazil; tel. (31) 224-8507; f. 1956; public law, political science, economics, history; Portuguese; published by Universidade Federal de Minas Gerais; Dirs Prof. Orlando M. Carvalho, Prof. Raul Machado Horta; 2 a year.

*Revista Brasileira de Geografia.* Fundação Instituto Brasileiro de Geografia e Estatistica, Directoria de Geociências, Av. Brasil 15.671, bloco 3-B, térreo, Parada de Lucas, 21.241-051 Rio de Janeiro, RJ, Brazil; tel. (21) 391-1420; fax (21) 391-770; f. 1939; advanced geographic, socio-economic and scientific articles, also news and translations; Portuguese, with English summaries; 2 a year.

*Revista Chilena de Historia y Geografía.* Sociedad Chilena de Historia y Geografía, Londres 65, Casilla 1386, Santiago, Chile; tel. (2) 6382489; f. 1911; history, geography, anthropology, archaeology, genealogy, numismatics; Spanish; Dir Ysidoro Vásquez de Amiña; annual.

*Revista de Ciencias Sociales de la Universidad de Costa Rica.* Vicerectoría de Investigación, Apdo 49-2060, Montes de Oca 2050, Costa Rica; tel. 2073450; fax 2075569; e-mail ceciliaa@cariari.ucr.ac.cr; internet cariari.ucr.ac.cr/~revicsoc/pagina.htm; f. 1956; sociology, anthropology, geography, history, etc., with special reference to Costa Rica and Central America; Spanish; Editor Cecilia Arguedas; Dir Daniel Camacho; quarterly.

*Revista Ecuador Debate.* Centro Andino de Acción Popular—CAAP, Utreras 733 y Selva Alegre, Quito, Ecuador; tel. (2) 252-2763; fax (2) 256-8452; e-mail caapl@caap.org.ec; f. 1983; economic conditions and agriculture; Spanish; Editor Freddy Rivera Vélez; 3 a year.

*Revista Educación.* Instituto de Investigaciones en Educación, Ciudad Universitaria Rodrigo Facio, Universidad de Costa Rica,

2060 San Pedro, Montes de Oca, San José, Costa Rica; tel. 207-4790; fax 207-4679; e-mail reduca@cariari.ucr.ac.cr; f. 1977; education; Spanish; Dir MARTA ROJAS PORRAS; Editor MARÍA ELENA CAMACHO; 2 a year.

*Revista Estudios Sociales Centroamericanos.* CSUCA, Apdo 37 Ciudad Universitaria Rodrigo Facio, San José, Costa Rica; tel. 252744; f. 1972; socio-economic and political aspects of Central America; Spanish; Editor MARIO LUNGO UCLÉS; 3 a year.

*Revista Europea de Estudios Latinoamericanos y del Caribe / European Review of Latin American and Caribbean Studies.* Centre for Latin American Research and Documentation (CEDLA), Keizersgracht 395-397, 1016 EK Amsterdam, Netherlands; tel. (20) 525-34-98; fax (20) 625-51-27; e-mail secretariat@cedla.uva.nl; internet www.cedla.uva.nl; f. 1989; social scientific research on Latin America and the Caribbean (anthropology, economics, geography, history, politics, sociology, etc.); English and Spanish; Editor KATHLEEN WILLINGHAM; 2 a year.

*Revista Geológica de América Central.* Escuela Centroamericana de Geología, Apdo 214-2060 Universidad de Costa Rica, San José, Costa Rica; tel. 207-3310; fax 234-2347; e-mail skussmau@geologia.ucr.ac.cr; internet www.geologia.ucr.ac.cr; f. 1983; geology and geophysics of Central America; Spanish, with abstracts in English, and English, with abstracts in Spanish; 2 a year.

*Revista Homines.* Dept of Social Sciences, Universidad Interamericana de Puerto Rico, Apdo 191293, Hato Rey, PR 00919, Puerto Rico; internet coqui/metro.inter.edu/homines/homines.htm; f. 1957; sociology, anthropology, history, economics, geography, political sciences, psychology; Spanish; Editorial Dir ALINE FRAMBES BUXEDA DE ALZÉRRECA; 2 a year.

*Revista do Instituto Histórico e Geográfico Brasileiro.* Av. Augusto Severo 8, 10° andar, 20.021-040 Rio de Janeiro, RJ, Brazil; tel. (21) 232-1312; fax (21) 252-4430; f. 1838; history and geography; Portuguese; quarterly.

*Revista do Mercado Comum do Sul—Mercosul / Revista del Mercado Común del Sur—Mercosur.* Editora Terceiro Mundo, Rua da Gloria 122, 105-106, 20241 Rio de Janeiro, RJ, Brazil; tel. (21) 242-0763; fax (21) 252-8455; e-mail etm@etm.com.br; internet www.uol.com.br/revistadomercosul; f. 1992; Latin American integration; Portuguese and Spanish; monthly.

*Revista Mexicana de Ciencias Políticas y Sociales.* Facultad de Ciencias Políticas y Sociales, Universidad Autónoma de México, 04510 México, DF, Mexico; tel. (55) 5622-9433; fax (55) 5665-1786; f. 1955; political and social sciences; Spanish; Assoc. Dir JUDIT BOKSER MISSES; quarterly.

*Revista Mexicana de Política Exterior.* Matias Romero Institute for Diplomatic Studies, Reforma Norte 707, esq. Avda Peralvillo, Col. Morales, 06200 México, DF, Mexico; tel. (55) 5529-9514; fax (55) 5327-3031; Spanish; Editor DULCE MARÍA MÉNDEZ; quarterly.

*Revista Repertorio Americano.* Instituto de Estudios Latinoamericanos, Universidad Nacional, Apdo 86-3000, Heredia, Costa Rica; tel. and fax 277-3430; e-mail idela@una.ac.cr; f. 1919; Latin American and Spanish culture; Spanish; Editor M. L. JULIÁN GONZÁLEZ ZÚÑIGA; 2 a year.

*Revista Venezolana de Análisis de Coyuntura.* Research Institute of the Faculty of Economic and Social Sciences, Universidad Central de Venezuela, Of. de Publicaciones, Apdo 54057, Caracas 1051-A, Venezuela; tel. and fax (212) 605-2523; e-mail coyuntura@cantv.net; f. 1980 as Boletín de Indicadores Socioeconómicos; adopted existing name in 1995; socio-economic issues; Spanish; 2 a year.

*Revista Venezolana de Economía y Ciencias Sociales.* Research Institute of the Faculty of Economic and Social Sciences, Universidad Central de Venezuela, Of. de Publicaciones, Apdo 54057, Caracas 1051-A, Venezuela; tel. and fax (212) 605-2523; f. 1958 as Economía y Ciencias Sociales; changed name as above in 1995; Spanish; Editor DICK PARKER; 3 a year.

*Semana Económica.* Apdo 671, Lima, Peru; tel. (1) 4455237; fax (1) 4455946; f. 1985; economic affairs; Spanish; Editor AUGUSTO ALVARE-RODRICH; weekly.

*Síntesis Económica.* Bogotá, DC, Colombia; tel. (1) 212-5121; fax (1) 212-8365; f. 1975; Spanish; Dir FÉLIX LAFAURIE RIVERA; weekly.

*Sourcemex.* University of New Mexico, Latin American and Iberian Institute, Latin America Data Base, 801 Yale, NE, Albuquerque, NM 87131, USA; tel. (505) 277-6389; fax (505) 277-6387; e-mail info@ladb.unm.net; internet ladb.unm.edu/sourcemex; online news digest; political and economic news and analysis in Mexico; weekly.

*Suplemento Antropológico.* Centro de Estudios Antropológicos de la Universidad Católica, Casilla 1718, Asunción, Paraguay; tel. (21) 446251; fax (21) 445245; f. 1965; practical and theoretical problems of the indigenous peoples of the River Plate basin (Bolivia, Brazil, Uruguay, Argentina and Paraguay); Spanish; 2 a year.

*Terra Ameriga.* Salita Santa Maria della Sanità 43, 16122 Genova, Italy; tel. (010) 814737; f. 1964; archaeology, anthropology, ethnology, history, linguistics, literature, etc. of ancient America, published by Associazione Italiana Studi Americanistici; annual.

*Third World Quarterly.* Routledge, Taylor & Francis, 4 Park Sq., Milton Park, Abingdon, Oxon, OX14 4RN, United Kingdom; tel. (20) 7017-6000; fax (20) 7017-6336; internet www.tandf.co.uk/journals; Editor SHAHID QADIR; 8 a year.

*Tricontinental.* Organization of Solidarity of the Peoples of Asia, Africa and Latin America, Calle C, No 668, esq. 29, Vedado, Apdo 4224, Havana, Cuba; tel. (7) 30-4941; fax (7) 33-3985; f. 1967; analysis of cultural, political and social developments in Cuba and other developing countries; Spanish, French and English; Editor ANA MARÍA PELLÓN SÁEZ; quarterly.

*El Trimestre Económico.* Fondo de Cultura Económica, Carretera Picacho Ajusco 227, Col. Bosques del Pedregal, Del. Tlaplan, 14200 México, DF, Mexico; tel. (55) 5227-4671; fax (55) 5227-4649; f. 1934; economic development and economic theory, employment and investment policy; Spanish; Editor FAUSTO HERNÁNDEZ TRILLO; quarterly.

*Uruguay Económico.* Ministerio de Economía y Finanzas, Asesoría Economico-Financiera, Col. 1013, 8°, Montevideo, Uruguay; f. 1978; economy of Uruguay; published by the Economic and Financial Advisory Office of the Uruguayan Ministry of Economy and Finance; Spanish; 2 a year.

*Uruguay Síntesis Económico.* Banco Central del Uruguay, Departamento de Estadísticas Económicas, Paysando y Florida, Casilla 1467, Montevideo, Uruguay; tel. (2) 917117; fax (2) 921634; economic conditions in Uruguay; publication of the Economic Statistics Department of the Central Bank of Uruguay; Spanish; 2 a year.

*Visión* (La Revista Latinoamericana Visión). Arguímedes 199, 6° y 7°, Col. Polanco, 11570 México, DF, Mexico; tel. (5) 203-6091; f. 1950; news and analysis of Latin America; Spanish Gen. Man. ROBERTO BELLO; 2 a month.

*Washington Letter on Latin America.* 1117 North 19th St, Arlington, VA 22209, USA; tel. (703) 247-3433; political and economic; English; Editor BARBARA ANNIS; 24 a year.

# INDEX OF REGIONAL ORGANIZATIONS

(Main reference only)

## A

ACP Committee of Ambassadors, 976
ACP Council of Ministers, 976
ACP Secretariat, 976
ACP States, 975
Agency for the Prohibition of Nuclear Weapons in Latin America and the Caribbean, 993
ALADI, 980
Alliance of Small Island States, 993
Andean Business Advisory Council, 959
Andean Community of Nations, 956
Andean Development Corporation, 959
Andean Labour Advisory Council, 959
Andean Parliament, 956
Andrés Bello Agreement, 959
AOSIS, 993
Asociación Católica Latinoamericana para la Radio, la TV y los Medios Afines, 995
Asociación de Empresas de Telecomunicaciones de la Comunidad Andina (ASETA), 958
Asociación del Congreso Panamericano de Ferrocarriles, 996
Asociación Interamericana de Bibliotecarios, Documentalistas y Especialistas en Información Agrícolas, 990
Asociación Interamericana de Ingeniería Sanitaria y Ambiental, 994
Asociación Internacional de Radiodifusión, 995
Asociación Latinoamericana de Industrias Farmaceuticas, 996
Asociación Latinoamericana de Integración, 980
Asociación Médica Panamericana, 994
Asociación Regional de Empresas de Petróleo y Gas Natural en Latinoamérica y el Caribe (ARPEL), 991
Assembly of Caribbean Community Parliamentarians, 963
Association of Caribbean States (ACS), 991
Association of Caribbean University and Research Institutional Libraries, 992
Association of Commonwealth Newspapers, News Agencies and Periodicals, 973
Association of Commonwealth Universities (ACU), 972

## B

Banco Centroamericano de Integración Económica (BCIE), 965

## C

CAB International (CABI), 990
CABI Bioscience, 990
CACM, 966
CAFTA, 966
Cairns Group, 996
CAN, 956
Capacity 2015, 919
Caribbean Agricultural Research and Development Institute (CARDI), 963
Caribbean Association of Industry and Commerce (CAIC), 996
Caribbean-Britain Business Council, 991
Caribbean Centre for Development Administration (CARICAD), 963
Caribbean Community and Common Market, 959
Caribbean Conference of Churches, 995
Caribbean Congress of Labour, 994
Caribbean Conservation Association, 993
Caribbean Council of Legal Education, 992
Caribbean Court of Justice (CCJ), 960
Caribbean Development Bank, 963
Caribbean Disaster Emergency Response Agency (CDERA), 963
Caribbean Environmental Health Institute (CEHI), 963
Caribbean Examinations Council, 992
Caribbean Food and Nutrition Institute (CFNI), 963
Caribbean Law Institute, 963
Caribbean Meteorological Organization (CMO), 963
Caribbean Network of Educational Innovation for Development, 945
Caribbean Plant Protection Commission, 932
Caribbean Tourism Organization, 996
CARICOM, 959
CELADE, 916
Central American Air Navigation Service Corporation, 965
Central American Bank for Economic Integration, 965
Central American Common Market (CACM), 966
Central American Confederation of Workers, 994
Central American Council of Social Security Institutions, 965
Central American Council on Housing and Human Settlements, 965
Central American Court of Justice, 965
Central American Electrification Council, 965
Central American Free Trade Area (CAFTA), 966
Central American Institute for Business Administration, 996
Central American Institute of Public Administration, 965
Central American Integration System, 964
Central American Monetary Council, 964
Central American Parliament, 964
Central American University Council, 965
Central Latinoamericano de Trabajadores (CLAT), 994
Centre for Latin American Monetary Studies, 992
Centre for the Development of Enterprise, 976
Centro de Coordinación para la Prevención de Desastres Naturales en América Central, 965
Centro de Estudios Monetarios Latinoamericanos, 992
Centro Interamericano de Investigación y Documentación sobre Formación Profesional (CINTER-FOR), 992
Centro Internacional de Agricultura Tropical, 990
CMAG, 967
Co-ordinator of the Indigenous Organizations of the Amazon Basin (COICA), 995
Cocoa Producers' Alliance (CPA), 990
Comisión Centroamericana de Transporte Marítimo (COCATRAM), 964
Comisión de Ciencia y Tecnología de Centroamérica y Panamá (CTCAP), 965
Comisión Interamericana de Derechos Humanos, 983
Comisión Interamericana de Mujeres, 987
Comisión Interamericana de Puertos, 983
Comisión Interamericana de Telecomunicaciones (CITEL), 983
Comisión Interamericana para el Control del Abuso de Drogas (CICAD), 983
Comisión Técnica de Telecomunicaciones (COMTELCA), 965
Comissão Jurídica Interamericana/Comité Jurídico Interamericano, 983
Comité Coordinador Regional de Instituciones de Agua Potable y Saneamiento de Centroamérica, Panamá y República Dominicana (CAPRE), 965
Comité Interamericano Contra el Terrorismo (CICTE), 987
Commission for Inland Fisheries of Latin America, 932
Commission for Labour Co-operation, 982
Commission for Technical Support and Co-operation, 980
Committee for Science and Technology of Central America and Panama, 965
Committee of the Central American Isthmus for Sport and Recreation, 965
Commonwealth, 967
Commonwealth Association for Education in Journalism and Communication (CAEJAC), 972
Commonwealth Association for Public Administration and Management (CAPAM), 973
Commonwealth Association of Architects, 973
Commonwealth Association of Science, Technology and Mathematics Educators (CASTME), 972
Commonwealth Broadcasting Association, 973
Commonwealth Business Council, 971
Commonwealth Council for Educational Administration and Management, 972
Commonwealth Countries League, 974
Commonwealth Engineers' Council, 973
Commonwealth Forestry Association, 972
Commonwealth Foundation, 972
Commonwealth Fund for Technical Co-operation (CFTC), 971
Commonwealth Games Federation, 973
Commonwealth Geological Surveys Forum, 973
Commonwealth Institute, 972
Commonwealth Journalists Association, 973
Commonwealth Lawyers' Association, 973
Commonwealth Legal Advisory Service, 973
Commonwealth Legal Education Association, 973
Commonwealth Magistrates' and Judges' Association, 973
Commonwealth Medical Trust (COMMAT), 972
Commonwealth Ministerial Action Group on the Harare Declaration (CMAG), 967

# REGIONAL INFORMATION — Index of Regional Organizations

Commonwealth of Learning (COL), 972
Commonwealth Parliamentary Association, 973
Commonwealth Pharmaceutical Association, 972
Commonwealth Press Union, 973
Commonwealth Secretariat, 967
Commonwealth Society for the Deaf, 972
Commonwealth Telecommunications Organization, 972
Commonwealth War Graves Commission, 974
Commonwealth Youth Exchange Council, 973
Community of Portuguese-Speaking Countries, 993
Comunidad Andina de Naciones, 956
Comunidade dos Países de Língua Portuguesa, 993
Confederación Centroamericana de Trabajadores, 994
Confederación de Organizaciones Turísticas de la América Latino (COTAL), 996
Confederación Interamericana de Educación Católica, 992
Confederación Latinoamericana y del Caribe de Asociaciones Cristianas de Jóvenes, 996
Conferencia Interamericano de Seguridad Social, 995
Consejo Centroamericano de Instituciones de Seguridad Social, 965
Consejo Consultivo Empresarial Andino, 959
Consejo Consultivo Laboral Andino, 959
Consejo de Fundaciones Americanas de Desarrollo, 991
Consejo del Istmo Centroamericano de Deportes y Recreación (CODICADER), 965
Consejo Episcopal Latinoamericano, 995
Consejo Interamericano de Música, 990
Consejo Latinoamericano de Iglesias, 995
Consejo Superior Universitario Centroamericano, 965
Consultative Group on International Agricultural Research (CGIAR), 934
Consultative Group to Assist the Poorest (CGAP), 935
Convenio Andrés Bello, 959
Convenio Hipólito Unanue, 959
Convenio Simón Rodríguez, 959
Convention on International Trade in Endangered Species of Wild Fauna and Flora (CITES), 921
Corporación Andina de Fomento (CAF), 959
Corporación Centroamericana de Servicios de Navegación Aérea (COCESNA), 965
Corte Interamericana de Derechos Humanos, 983
Cotonou Agreement, 975
Council of American Development Foundations, 991
CPLP, 993
CropLife International, 996

## D

Duke of Edinburgh's Award International Association, 973

## E

Eastern Caribbean Central Bank (ECCB), 992
Economic Commission for Latin America and the Caribbean (ECLAC), 915
European Union, 974

## F

FAO, 929
FAO/WHO Codex Alimentarius Commission, 932
FATF, 992
Federación de Cámaras de Comercio del Istmo Centroamericano, 996
Federación Latinoamericana de Bancos, 992
Federación Latinoamericana de Trabajadores Agrícolas, Pecuarios y Afines, 994
Federación Odontológica Latinoamericana, 994
Fédération internationale de football association (FIFA), 996
Federation of Central American Chambers of Commerce, 996
Financial Action Task Force on Money Laundering, 992
Fondo Latinoamericano de Reservas (FLAR), 959
Food and Agriculture Organization of the United Nations, 929
Free Trade Agreement of the Americas (FTAA), 983
Fundación Panamericana para el Desarrollo, 987

## G

Global Commission on Migration, 995
Global Environment Facility (GEF), 919
Global Programme of Action for the Protection of the Marine Environment from Land-based Activities, 921
Group of Three (G3), 991
Groupe d'action financière sur le blanchiment de capitaux (GAFI), 992

## H

HIPC, 934
Hipólito Unanue Agreement, 959

## I

IBRD, 933
ICAO, 954
ICSID, 935
IDA, 936
IDB, 977
IFAD, 940
IFC, 937
ILO, 954
IMF, 941
IMO, 954
Institute for the Integration of Latin America and the Caribbean, 978
Institute of Commonwealth Studies, 972
Institute of Nutrition of Central America and Panama, 965
Instituto Americano del Niño, 986
Instituto Centroamericano de Administración de Empresas (INCAE), 996
Instituto Centroamericano de Administración Pública (ICAP), 965
Instituto de Nutrición de Centroamérica y Panamá (INCAP), 965
Instituto Indigenista Interamericano, 987
Instituto Interamericano de Cooperación para la Agricultura, 987
Instituto Internacional de Literatura Iberoamericana, 992
Instituto Panamericano de Geografía e Historia, 987
Instituto para la Integración de América Latina y el Caribe, 978
Inter-American Agency for Co-operation and Development, 983
Inter-American Association of Agricultural Librarians, Documentalists and Information Specialists, 990
Inter-American Association of Sanitary and Environmental Engineering, 994
Inter-American Bar Association (IABA), 994
Inter-American Centre for Research and Documentation on Vocational Training, 992
Inter-American Children's Institute, 986
Inter-American Commercial Arbitration Commission, 996
Inter-American Commission of Women, 987
Inter-American Commission on Human Rights, 983
Inter-American Committee Against Terrorism, 987
Inter-American Committee on Ports, 983
Inter-American Confederation for Catholic Education, 992
Inter-American Conference on Social Security, 995
Inter-American Court of Human Rights (IACHR), 983
Inter-American Defense Board, 987
Inter-American Development Bank, 977
Inter-American Drug Abuse Control Commission, 983
Inter-American Indigenous Institute, 987
Inter-American Institute for Co-operation on Agriculture (IICA), 987
Inter-American Institute for Social Development (INDES), 978
Inter-American Investment Corporation (IIC), 978
Inter-American Juridical Committee (IAJC), 983
Inter-American Music Council, 990
Inter-American Organization for Higher Education, 992
Inter-American Planning Society, 991
Inter-American Press Association (IAPA), 995
Inter-American Regional Organization of Workers (ORIT), 994
Inter-American Telecommunication Commission, 983
Inter-American Tropical Tuna Commission (IATTC), 990
International Association of Broadcasting, 995
International Bank for Reconstruction and Development—IBRD (World Bank), 933
International Centre for Settlement of Investment Disputes, 935
International Centre for Tropical Agriculture, 990
International Civil Aviation Organization, 954
International Cocoa Organization (ICCO), 990
International Coffee Organization (ICO), 990
International Confederation of Free Trade Unions (ICFTU), 994
International Council for Science (ICSU), 995
International Development Association, 936
International Federation of Association Football, 996
International Finance Corporation, 937
International Fund for Agricultural Development, 940
International Institute for Democracy and Electoral Assistance (IDEA), 993
International Institute for Higher Education in Latin America and the Caribbean, 945
International Institute of Iberoamerican Literature, 992
International Labour Organization, 954
International Maritime Organization, 954
International Monetary Fund, 941
International Peace Academy, 995

International Regional Organization of Plant Protection and Animal Health, 965
International Seabed Authority, 993
International Sugar Organization, 991
International Telecommunication Union, 954
International Telecommunications Satellite Organization (INTELSTAT), 995
International Tropical Timber Organization (ITTO), 991
International Union of Latin Notaries, 994
Internet Corporation for Assigned Names and Numbers (ICANN), 995
ITU, 954
IUCN—The World Conservation Union, 993

## J

Joint Commonwealth Societies' Council, 974
Joint UN Programme on HIV/AIDS, 950
Junta Interamericana de Defensa, 987

## L

LAFTA, 980
LAIA, 980
Latin American and Caribbean Confederation of Young Men's Christian Associations, 996
Latin American and Caribbean Forestry Commission, 932
Latin American and Caribbean Institute for Economic and Social Planning, 916
Latin American Association of Development Financing Institutions, 991
Latin American Association of National Academies of Medicine, 994
Latin American Association of Pharmaceutical Industries, 996
Latin American Banking Federation, 992
Latin American Catholic Organization for the Cinema and Audiovisuals, 995
Latin-American Catholic Press Union, 995
Latin American Catholic Radio and Television Association, 995
Latin-American Confederation of Tourist Organizations, 996
Latin American Confederation of Trade Unions, 994
Latin American Council of Churches, 995
Latin American Demographic Centre, 916
Latin American Economic System, 992
Latin-American Energy Organization, 996
Latin American Episcopal Council, 995
Latin American Federation of Agricultural Workers, 994
Latin American Free Trade Association (LAFTA), 980
Latin American Integration Association, 980
Latin-American Iron and Steel Institute, 996
Latin American Odontological Federation, 994
Latin American Parliament, 993
Latin American Reserve Fund, 959
League for the Exchange of Commonwealth Teachers, 972
Lomé Convention, 975

## M

Mercado Común del Sur/Mercado Comum do Sul, 987
MERCOSUR/MERCOSUL, 987
MIGA, 939
Millennium Development Goals, 917
MINUSTAH, 927
Montreal Protocol, 919
Multilateral Investment Fund (MIF), 978
Multilateral Investment Guarantee Agency, 939

## N

NAALC, 982
NAFTA, 981
NAM, 993
Non-aligned Movement, 993
North American Agreement on Labour Co-operation, 982
North American Commission for Environmental Co-operation, 982
North American Development Bank (NADB/NADBank), 982
North American Free Trade Agreement, 981

## O

OAS, 982
OCHA, 954
OEA, 982
OECS, 993
Office for the Co-ordination of Humanitarian Affairs, 954

Office of the United Nations High Commissioner for Human Rights, 954
OHCHR, 954
Organisation of Eastern Caribbean States, 993
Organismo Internacional Regional de Sanidad Agropecuaria (OIRSA), 965
Organismo para la Proscripción de las Armas Nucleares en la América Latina y el Caribe, 993
Organización Católica Internacional del Cine y del Audiovisual, 995
Organización Centroamericana de Entidades Fiscalizadores Superiores, 965
Organización de Estados Iberoamericanos para la Educación, la Ciencia y la Cultura (OEI), 992
Organización de las Cooperativas de América, 992
Organización de los Estados Americanos, 982
Organización de Solidaridad de los Pueblos de Africa, Asia y América Latina, 994
Organización de Universidades Católicas de América Latina, 993
Organización Latino-americana de Energía, 996
Organización Panamericana de la Salud, 987
Organización Regional Interamericana de Trabajadores, 994
Organization of American States, 982
Organization of Ibero-American States for Education, Science and Culture, 992
Organization of Solidarity of the Peoples of Africa, Asia and Latin America, 994
Organization of the Catholic Universities of Latin America, 993
Organization of the Cooperatives of America, 992
Organization of the Petroleum Exporting Countries (OPEC), 991
Organization of World Heritage Cities, 990

## P

Pacific Basin Economic Council (PBEC), 992
Pan-American Association of Ophthalmology, 994
Pan American Development Foundation (PADF), 987
Pan American Health Organization (PAHO), 987
Pan-American Institute of Geography and History, 987
Pan-American Medical Association, 994
Pan American Railway Congress Association, 996
Parlamento Andino, 956
Parlamento Latinoamericano, 993
Postal Union of the Americas, Spain and Portugal, 995
Puebla-Panama Plan (PPP), 966

## R

Regional Association of Oil and Natural Gas Companies in Latin America and the Caribbean, 991
Regional Co-ordinating Unit for the Caribbean Environment Programme, 921
Regional Office for Culture in Latin America and the Caribbean, 945
Regional Office for Education in Latin America and the Caribbean, 945
Regional Office for Science for Latin America and the Caribbean, 945
Rio Group, 994
Royal Commonwealth Ex-Services League, 974
Royal Commonwealth Society, 974
Royal Over-Seas League, 974

## S

SACN, 989
Secretaría de Integración Económica Centroamericana, 964
Secretaría de Integración Turística Centroamericana (SITCA), 964
Secretaría de la Integración Social Centroamericana (SISCA), 964
Secretaría del Consejo Agropecuario Centroamericano (SCAC), 964
Secretaría Ejecutiva de la Comisión Centroamericana de Ambiente y Desarrollo (SE-CCAD), 964
Secretaría Ejecutiva de la Comisión Regional de Recursos Hidráulicos (SE-CRRH), 965
Secretaría Ejecutiva del Consejo de Electrificación de América Central (CEAC), 965
Secretaría Ejecutiva del Consejo Monetario Centroamericano (SECMCA), 964
Secretaría General de la Coordinación Educativa y Cultural Centroamericana (SG-CECC), 964
Secretariat of the ACP–EU Council of Ministers, 976
Secretariat of the Basel Convention, 921
Secretariat of the Multilateral Fund for the Implementation of the Montreal Protocol, 921

Secretariat of the UN Framework Convention on Climate Change, 921
SELA, 992
SICA, 964
SIECA, 964
Sight Savers International, 973
Simón Bolívar Andean University, 959
Simón Rodríguez Agreement, 959
Sistema de la Integración Centroamericana, 964
Sistema Económico Latinoamericano, 992
Sociedad Interamericana de Planificación, 991
Sociedad Interamericana de Prensa, 995
SOLIDARIOS, 991
South American Community of Nations, 989
Southern Common Market, 987
Standing Committee on Commonwealth Forestry, 972
Sugar Association of the Caribbean (Inc.), 991
Summits of the Americas, 984

T

Technical Centre for Agricultural and Rural Co-operation, 976
Technical Commission for Telecommunications, 965
Tribunal de Justicia de la Comunidad Andina, 956

U

UN-Habitat, 954
UNAIDS, 950
UNCTAD, 954
UNDP, 916
UNDP Drylands Development Centre (DDC), 920
UNEP, 921
UNEP International Environmental Technology Centre, 921
UNEP Ozone Secretariat, 921
UNEP-SCBD (Convention on Biological Diversity—Secretariat), 921
UNEP Secretariat for the UN Scientific Committee on the Effects of Atomic Radiation, 921
UNEP/CMS (Convention on the Conservation of Migratory Species of Wild Animals) Secretariat, 921
UNESCO, 945
UNFPA, 954
UNHCR, 925
UNICEF, 954
Unidad Coordinadora de la Organización del Sector Pesquero y Acuícola del Istmo Centroamericano, 965
Unidad Técnica del Consejo Centroamericano de Vivienda y Asentamientos Humanos, 965
UNIDO, 955
UNIFEM, 920
Unión de Paises Exportadores de Banano, 991
Unión de Universidades de América Latina, 993
Union Internationale du Notariat Latin, 994
Union of Banana-Exporting Countries, 991
Union of Latin American Universities, 993

Unión Postal de las Américas, España y Portugal, 995
United Nations, 913
United Nations Capital Development Fund (UNCDF), 920
United Nations Children's Fund, 954
United Nations Conference on Trade and Development, 954
United Nations Development Fund for Women, 920
United Nations Development Programme, 916
United Nations Diplomatic Representation, 913
United Nations Economic Commission for Latin America and the Caribbean, 915
United Nations Educational, Scientific and Cultural Organization, 945
United Nations Environment Programme, 921
United Nations High Commissioner for Refugees, 925
United Nations Human Settlements Programme, 954
United Nations Industrial Development Organization, 955
United Nations Information Centres/Services, 914
United Nations Millennium Development Goals, 917
United Nations Observers, 914
United Nations Office for the Co-ordination of Humanitarian Affairs, 954
United Nations Office on Drugs and Crime, 954
United Nations Peace-keeping, 927
United Nations Population Fund, 954
United Nations Stabilization Mission in Haiti, 927
United Nations Volunteers, 920
Universal Postal Union, 955
Universidad Andina Simón Bolívar, 959
UNODC, 954
UNV, 920
UPU, 955

V

Victoria League for Commonwealth Friendship, 974

W

WBI, 935
West Indian Sea Island Cotton Association (Inc.), 991
WFP, 928
WFUNA Youth, 997
WHO, 948
WIPO, 955
WMO, 955
World Bank Institute, 935
World Confederation of Labour, 994
World Food Programme, 928
World Health Organization, 948
World Heritage Programme, 947
World Intellectual Property Organization, 955
World Meteorological Organization, 955
World Organisation of Animal Health, 990
World Self-Medication Industry, 994
World Tourism Organization, 955
WWF International, 993

# The Europa Regional Surveys of the World 2006

'Europa's Regional Surveys of the World are justly renowned for their exceptionally high levels of production, content and accuracy.' - *Reference Reviews*

A nine-volume library of historical, political, geographical and economic data, providing an accurate and impartial overview of the major regions of the world.

These exhaustive surveys bring together a unique collection of information, from description and analysis of the principal issues affecting each region, to statistical data and directory information.

Meticulously researched and updated every year, these Surveys are an indispensable information source on which you can rely.

Available as a set or individually, the nine titles that make up the series are as follows: *Africa South of the Sahara; Central and South-Eastern Europe; Eastern Europe, Russia and Central Asia; The Far East and Australasia; The Middle East and North Africa; South America, Central America and the Caribbean; South Asia; The USA and Canada; Western Europe.*

**Bibliographic Details**
November 2005: 285x220mm: 9 volumes
Set: 1-85743-320-3: **£2,700.00**

To order your set or for further information about individual volumes please contact:
Tel: + 44 (0) 20 7017 6629
Fax: + 44 (0) 20 7017 6720
E-mail: sales.europa@tandf.co.uk

## Also available online!

# Europa World *Plus*
*Europa World and the Europa Regional Surveys of the World online*
**www.europaworld.com**

Europa World Plus enables you to subscribe to Europa World together with as many of the nine Europa Regional Surveys of the World online as you choose, in one simple annual subscription.

The Europa Regional Surveys of the World complement and expand upon the information in Europa World with in-depth, expert analysis at regional, sub-regional and country level.

For further information and to register for a free trial please contact us at:
Tel: + 44 (0) 20 7017 6629; Fax: + 44 (0) 20 7017 6720
E-mail: reference.online@tandf.co.uk

**Routledge**
Taylor & Francis Group

# The International Who's Who
## Part of the Europa Biographical Reference Series

**69TH EDITION**
**THE INTERNATIONAL WHO'S WHO 2006**

'Remains quite indispensable as the one single source of up-to-date biographical information about important people in the world.'
- *The Times Literary Supplement*

Published annually since 1935, *The International Who's Who* is now in its 69th edition. Providing hard-to-find biographical details for more than 20,000 of the world's most prominent and influential personalities, this outstanding reference work records the lives and achievements of men and women from almost every profession and activity - from heads of state to sporting greats.

**Bibliographic Details**
June 2005: 285x220mm; 2,250pp
Hb: 1-85743-307-6: **£325.00**

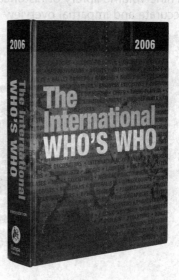

**To order your copy and for further information:**
Tel: + 44 (0) 20 7017 6629; Fax: + 44 (0) 20 7017 6720; E-mail: sales.europa@tandf.co.uk

**Free trial available!**

## The International Who's Who online
### www.worldwhoswho.com

All of the content of this exceptional reference source is available online.

* Clickable e-mail and web addresses
* Core content now updated quarterly!
* Full text search to quickly locate information
* Tailor your search by name, nationality, place and date of birth, and profession
* Browse a full A-Z listing of entrants
* A section detailing royal families
* Backdated entries and obituaries from the previous year

For further information and to register for a free trial please contact us at:
tel: + 44 (0) 20 7017 6608 /6131
fax: + 44 (0) 20 7017 6720
e-mail: reference.online@tandf.co.uk

# The Europa World of Learning ONLINE

## www.worldoflearning.com
## Available Now

 Routledge
Taylor & Francis Group

Locate academic institutions of every type, world-wide

- Universities and Colleges
- Schools of Art, Music and Architecture
- Learned Societies
- Research Institutes
- Libraries and Archives
- Museums and Art Galleries

Instant access to educational contacts around the globe

- Professors
- University Chancellors and Rectors
- Deans
- Librarians
- Curators
- Directors

Additional features of The Europa World of Learning online

- Multiple-user facility available
- A full range of sophisticated search and browse functions
- Regularly updated

Free 30-day trial available
For further information e-mail: info.europa@tandf.co.uk

# The European Union Information Series

**6th Edition**
**European Union Encyclopedia and Directory 2006**
Thoroughly updated, this extensive reference source provides in-depth information on all matters relating to the European Union: the expansion of the EU under the Nice Treaty is covered, and the future of the union is addressed.
November 2005

**6th Edition**
**The EU Institutions Register**
This fully revised and updated directory provides accurate and reliable information on the institutions involved in the running of the EU.
October 2005

**3rd Edition**
**The Practical Guide to Foreign Direct Investment in the European Union**
This title provides detailed coverage of national and EU financial incentives, and draws a comparison between each Member State's corporate and personal taxation, labour costs, social security charges and employment regulations.
October 2005

**4th Edition**
**Lobbying in the European Union**
An essential directory of the process of lobbying within the EU.
October 2005

**3rd Edition**
**The Guide to EU Information Sources on the Internet**

Divided thematically, this guide covers both institutional and non-institutional websites. Each entry includes: site name; web address; publisher's details; description of contents; languages; cost; and useful notes.
July 2005

**14th Edition**
**The Directory of EU Information Sources**
This major directory contains in-depth information on each of the constituent institutions of the EU, as well as diplomats in Brussels, Press Agencies and many other information sources.
February 2005

**2nd Edition**
**The EU Capital Guide**
Lists over 12,000 key decision makers in Brussels.
2004

**2nd Edition**
**A Dictionary of the European Union**
David Phinnemore and Lee McGowan
Provides concise definitions and explanations on all aspects of the EU.
2004

**6th Edition**
**The Directory of Trade and Professional Associations in the European Union**
Contact details, including e-mail and web addresses, of some 750 EU-level associations, and 11,700 national associations.
2004

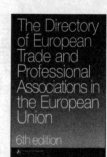

**The Rome, Maastricht, Amsterdam and Nice Treaties: Comparative Texts**
Highlights amendments and new Articles in the Nice Treaty.
2003

**Influence and Interests in the European Union: the New Politics of Persuasion and Advocacy**
Clearly discusses the impact and uses of interest representation in the EU.
2002

For further details please contact our marketing department:
Tel: +44 (0)20 7017 6629  Fax: +44 (0)20 7017 6720
E-mail: info.europa@tandf.co.uk
www.europapublications.com

# THE EUROPA POLITICAL AND ECONOMIC DICTIONARIES SERIES

**This new series provides extensive, up-to-date regional information, examining some of the major political and economic issues of modern international affairs.**

Each book in the series reflects the unique perspective of each of the world's regions, providing invaluable information specific to that particular area. They provide, in one easy-to-use format, information that might otherwise take time and many resources to locate.

**Each volume offers:**
- Concise and cross-referenced entries, providing contact details where appropriate
- Entries on the history and economy of each constituent country of the region
- Entries detailing distinct territories, ethnic groups, religions, political parties, prime ministers, presidents, politicians, businesses, international organizations, multinationals and major NGOs.

Each of these titles is compiled by acknowledged experts in the political economy of the region, and is produced to Europa's high standards.

International Affairs journal called the first edition of *A Political and Economic Dictionary of Eastern Europe* "*a fascinating and useful volume which is much more than a dictionary*".

It was also named one of **Reference Reviews Top Ten Reference Sources.**

**The set includes:**
A Political and Economic Dictionary of Africa
A Political and Economic Dictionary of East Asia
A Political and Economic Dictionary of Eastern Europe
A Political and Economic Dictionary of Latin America
A Political and Economic Dictionary of the Middle East
A Political and Economic Dictionary of South Asia
A Political and Economic Dictionary of South-East Asia
A Political and Economic Dictionary of Western Europe

For further information on any of the above titles contact our marketing department on:
tel: + 44 (0) 20 7017 6649; fax: + 44 (0) 20 7017 6720
e-mail: info.europa@tandf.co.uk
web: www.europapublications.com

# The Europa Biographical Reference Series

### WHO'S WHO IN INTERNATIONAL AFFAIRS 2005

Invaluable biographical details of the thousands of major figures involved in all aspects of international affairs.

- Over 6,000 entries
- Includes an A-Z guide to international organizations and the officials working in them
- Covers diplomats, politicians, government ministers, heads of state, academics, journalists and writers who are prominent in the world of international affairs.

### INTERNATIONAL WHO'S WHO OF WOMEN 2006

The most influential and distinguished women throughout the world are brought together in this unique publication.

- Lists recognized personalities as well as those women who are rising to prominence
- Some 6,000 entries are listed
- Easy to use, this one-stop reference source for information on the world's leading women is also indexed by profession.

### INTERNATIONAL WHO'S WHO IN CLASSICAL MUSIC 2005

An invaluable and practical source of biographical information on classical musicians, composers and conductors and the organizations behind them.

- Over 8,000 detailed biographical entries cover the Classical and Light Classical fields
- Entries include details of career, repertoire, recordings and compositions.

### INTERNATIONAL WHO'S WHO IN POETRY 2005

Profiles the careers of leading and emerging poets.

- Over 4,000 biographical entries
- Each entry provides full career history and publication details
- Contact details are provided for poetry organizations, poetry publishers and for organizations that chair poetry awards and prizes.

### INTERNATIONAL WHO'S WHO IN POPULAR MUSIC 2005

Comprehensive biographical information covering the leading names in all aspects of popular music.

- Over 5,000 entries
- Profiles pop, rock, folk, jazz, dance, world and country artists
- Provides full biographical information: major career details, concerts, recordings and compositions, honours and contact address
- Includes full contact details for companies and organizations throughout the popular music industry.

### INTERNATIONAL WHO'S WHO OF AUTHORS AND WRITERS 2006

An invaluable source of information on the personalities and organizations of the literary world.

- Over 8,000 entries
- Provides concise biographical information on novelists, authors, playwrights, columnists, journalists, editors and critics
- Each entry details career, works published, literary awards and prizes, membership and contact addresses where available.

For further information on any of the above titles contact our marketing department on:
tel: + 44 (0) 20 7017 6649
fax: + 44 (0) 20 7017 6720
e-mail: info.europa@tandf.co.uk
web: www.europapublications.com